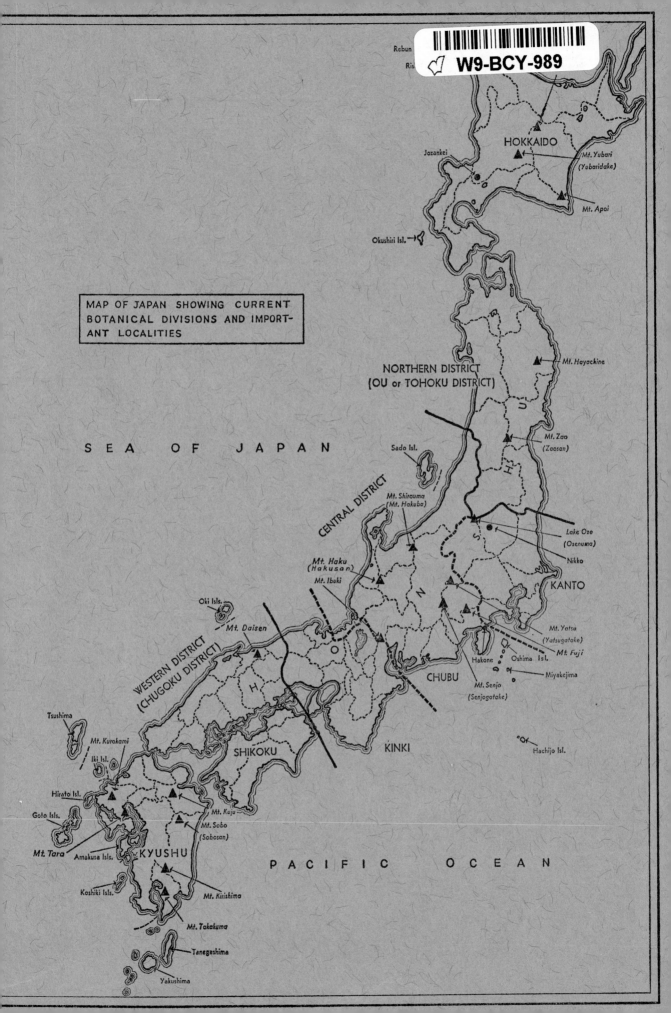

Rebun
Ris...

HOKKAIDO

Jozankei

Mt. Yubari
(Yubaridake)

Mt. Apoi

Okushiri Isl.

MAP OF JAPAN SHOWING CURRENT
BOTANICAL DIVISIONS AND IMPORT-
ANT LOCALITIES

NORTHERN DISTRICT
(OU or TOHOKU DISTRICT)

Mt. Hayachine

O U

Mt. Zao
(Zaosan)

SEA   OF   JAPAN

Sado Isl.

CENTRAL DISTRICT

Mt. Shirouma
(Mt. Hakuba)

Lake Oze
(Ozenuma)

Nikko

KANTO

Mt. Haku
(Hakusan)

Mt. Ibuki

H O N

Mt. Yatsu
(Yatsugatake)

Mt. Fuji

Oki Isls.

Mt. Daisen

WESTERN DISTRICT
(CHUGOKU DISTRICT)

H

O

Hakone   Oshima Isl.

Miyakejima

CHUBU

Tsushima

Mt. Senjo
(Senjogatake)

KINKI

Mt. Kurokami

Iki Isl.

SHIKOKU

Hachijo Isl.

Hirato Isl.

Goto Isls.

Mt. Kuju

Mt. Sobo
(Sobosan)

Mt. Tara   Amakusa Isls.   KYUSHU

PACIFIC   OCEAN

Koshiki Isls.

Mt. Kirishima

Mt. Takakuma

Tanegeshima

Yakushima

Lysichiton camtschatcense Schott (Araceae). Edge of pond near Shiobara, Shimotsuke Prov., Honshu. (Photo J. Ohwi, April 30, 1961.)

FLORA OF JAPAN

(in English)

by

JISABURO OHWI

National Science Museum, Tokyo, Japan

A combined, much revised, and extended translation by the author of his
日 本 植 物 誌 FLORA OF JAPAN (1953)
and
日 本 植 物 誌 シ ダ 篇 FLORA OF JAPAN—PTERIDOPHYTA (1957)

Edited by

Frederick G. Meyer
Research Botanist, U.S. National Arboretum

and

Egbert H. Walker
Research Associate, Smithsonian Institution

SMITHSONIAN INSTITUTION

WASHINGTON, D.C.

1965

CONTENTS

LIST OF ILLUSTRATIONS

PLATES

FIGURES

AUTHOR'S PREFACE TO THE ENGLISH EDITION

The original Japanese language edition of my Flora of Japan was published in Tokyo in 1953. This work included the indigenous and adventive spermatophytes but not the pteridophytes found in present-day Japan. Various authors have published floras of our realm, the first were by Europeans who visited Japan under Dutch auspices in the 17th, 18th, and 19th centuries. Under the influence of Europeans our own people began to collect plants in earnest after the middle of the 19th century. Soon thereafter the first flora listing all the plants known at that time was published by Jinzō Matsumura in 1884. Since that time other floras have been produced.

My Flora is a culmination of more than 30 years of study. Based largely upon my own field studies, this work is designed as a manual for students and for others less technically trained, who, from time to time, require a reference work to the flora of our islands. The present translation, the only flora of Japan in the English language and the first in a European tongue since Franchet and Savatier's flora of 1875–79, is an emended and in several respects a revised version of my original work. Inclusion of the pteridophytes (ferns and fern allies), the photographs, and the maps are features of the English version not found in my original Japanese edition. Also included in this English edition are some species recently recorded and not included in my Japanese edition. The nomenclature in the present work is in accordance with the International Code (International Rules Bot. Nom., 1961).

I shall be pleased if this English edition of my Flora is found to be useful to others outside my country. The translation might never have been undertaken except for my acquaintance with Leopold A. Charette of Burlington, Vermont, who, as a member of the U.S. Air Force, collected plants in parts of western Honshu and came to me for assistance shortly after publication of my flora in 1953. He urged me to prepare translations of certain genera for him, which I did. Very soon thereafter, I accepted the proposal to translate the entire work for publication in English.

I am deeply grateful to Drs. Tetsuo Koyama and Siro Kitamura for contributing full treatments included here. The former wrote up the Araceae, Eriocaulaceae, and Juncaceae; the latter the Compositae. Much assistance has been generously given by various other Japanese and American botanists. In addition, much of the translation of this English version from my Japanese volume of 1953 was made by Dr. Koyama, a labor for which I am very grateful.

I wish to thank my American sponsors for making this translation of my Flora possible and for their continued and devoted interest over the long period since this translation project began in 1954. The Missouri Botanical Garden, St. Louis, Missouri, sponsored the first two grants-in-aid received from the National Science Foundation, Washington, D.C., for translation in Japan of the Japanese text. A third and final grant, again from the National Science Foundation, together with funds provided by the Smithsonian Institution, made publication possible. Final editing and all arrangements for publication were entrusted to Frederick G. Meyer, Takoma Park, Maryland (formerly of the Missouri Botanical Garden), and Egbert H. Walker of the U.S. National Museum, Smithsonian Institution, Washington, D.C. (retired). To friends and colleagues in my country I wish to extend thanks for providing photographs which appear in the present work and for assisting me in various other ways.

JISABURŌ OHWI
National Science Museum, Tokyo

January 1965

EDITORS' PREFACE TO THE ENGLISH EDITION

The flora of Japan first became known to westerners through the agency of the Dutch East India Company, established in Nagasaki in 1609. The first list of Japanese plants published in Europe was the "Amoenitatum Exoticarum" of Engelbert Kaempfer, published in 1712. Europeans under Dutch auspices continued to study the Japanese flora until Japan was opened to world commerce in 1859. Relatively soon thereafter, and continuing up to the present day, Japanese botanists have been busily engaged in the study of their flora.

To agriculturists, foresters, and horticulturists in America and Europe, and in other warm-temperate areas of the world, the Japanese flora long has been an important source of plant materials of economic importance. To botanists, the Japanese flora gained lasting prominence among students of the North Temperate boreal flora with Asa Gray's now classic paper published in 1859,[1] which emphasized the relationships of the Japanese flora to parts of eastern United States.

The flora of Japan is perhaps the best known of any country in eastern Asia. Since about 1868, each period of activity has seen new floristic works published. Floras published in Japanese by Jinzō Matsumura, Tomitarō Makino, and Takenoshin Nakai, for example, are well known in Japan, although they have been of relatively little use to western botanists, principally because of language barriers.

This English language edition of a Flora of Japan, by Dr. Jisaburō Ohwi, is an attempt to bridge the language barrier. The last floristic work to cover the Japanese archipelago in a western language was Franchet and Savatier's two-volume "Enumeratio Plantarum in Japonia Sponte Crescentium Hucusque Rite Cognitarum," published 1875–79. Unlike other floras of Japan of the past, Dr. Owhi's work is the first to include synoptical keys of all taxa through the level of the species. With this English translation, botanists, horticulturists, agriculturists, and others not fluent in the Japanese language have available a modern floristic work of Japan which covers the ferns through the phanerograms. We should point out here that the English language edition of this Flora of Japan is an emended account and not merely a verbatim translation of Dr. Ohwi's original work in Japanese.

The Romaji or Japanese vernacular names are included for all taxa listed in this English edition, although to westerners the supplying of a vernacular name for every taxon might seem superfluous. This is due to the relative similarity of western languages to Latin, the basis of the scientific names. But to the nonbotanist in Japan there is a far greater need for a vernacular name in Japanese because the great difference between that language and Latin precludes the use of the latter by the uninitiated. The vernacular names supplied by the author have been altered by the junior editor, with the aid of Mr. Hisao Matsumoto of the Library of Congress in Washington, D.C., by the insertion of additional hyphens, more than are commonly used by Japanese scholars. In this way the attempt has been made to aid westerners who may be interested in these names to comprehend more readily their structure and meaning. Japanese plants in cultivation often bear the original Romaji name and for this reason inclusion of the Japanese names is justified and useful.

In preparing the manuscript for publication, the editors have at all times endeavored to render faithfully and accurately the full intentions of the author, realizing fully the pitfalls of editing another author's work. All changes, corrections, and additions have been carefully checked by the author himself. Various specialists in the United States have offered editorial assistance: Agnes Chase (Gramineae); F. A. McClure (Bambuseae); C. V. Morton (ferns); Lincoln Constance (Umbelliferae); Robert E. Woodson, Jr. (Apocynaceae and Asclepiadaceae); Rogers McVaugh (Campanulaceae); F. Raymond Fosberg (Rubiaceae); and S. F. Blake (Compositae). Without the full cooperation and cordial relationships between the editors and the author, this project would not have been possible.

FREDERICK G. MEYER
EGBERT H. WALKER, *Editors*

January 1965

---

[1] Gray, A. Diagnostic characters of new species of phaenogamous plants, collected in Japan by Charles Wright, Botanist of the U.S. North Pacific Exploring Expedition, with observations upon the relations of the Japanese flora to that of North America, and of other parts of the northern temperate zone. Mem. Amer. Acad., n. ser. 6: 377–452, 1859.

FLORA OF JAPAN

INTRODUCTION

In Europe and North America, many exhaustive floristic works are available, but in our country the lack of such works frequently has been keenly felt by us. Most of the early floristic investigations on the Japanese flora, beginning in the 18th century, were made by Europeans. Serious study and collecting by Japanese botanists began first in the 1860's. The early published works by our own botanists were mostly regional floristic studies, although several floras of the entire Japanese archipelago from time to time have been published. One of the earliest was that of Jinzō Matsumura, "Nippon shokubutsu mei-i," published first in 1884, which went through nine editions. The "Nippon shokubutsu-dzukan" ["Illustrated Flora of Japan"], by Tomitarō Makino, first published in 1925, went through several editions and reprints, the latest in 1963. The "Nippon-shokubutsu-sōran" ["Flora of Japan"], by Tomitarō Makino and Kwanji Nemoto, first published in 1925, was issued the same year as Makino's Illustrated Flora. The "Nova Flora Japonica," by Takenoshin Nakai and M. Honda, 1935–51, unfortunately was never completed.

The writing of my original Japanese edition began in 1947 after more than 30 years' study on our flora. The aim was to produce a manual for botanists, dendrologists, foresters, and agriculturists, and a guide book for students who require a ready source of taxonomic information about the plants of Japan.

The present work enumerates all spontaneous plants inclusive of the ferns and fern allies, gymnosperms, and phanerogams. Synoptical keys are included for all taxa to the level of the species. A conservative interpretation of the taxa has been attempted. In complex groups, such as *Sasa, Aconitum, Hosta,* and others, where innumerable microspecies have been recorded by specialists, it has not been possible to include these in my Flora. Trivial variations in flower color, horticultural variants, local aberrations in vegetative morphology, such as dwarfs and monstrosities, are generally excluded. Exceptions to this are in instances where garden plants, long known to us, are enumerated as having originated from elements of our indigenous flora. Wherever such garden plants appear, the nomenclature is in accordance with modern usage in the naming of horticultural plants (International Code for Cultivated Plants, 1961). A few plants of Chinese origin, long established in our country, such as *Mahonia japonica, Prunus japonica, Ginkgo biloba, Magnolia liliflora, Clematis florida,* and some others, are also included.

The Engler and Prantl system of classification, as outlined in "Syllabus der Pflanzenfamilien," has been adopted for the phanerogams, and Copeland's "Genera Filicum" as the guide in the treatment of pteridophytes. The aim was to construct the analytical keys on the basis of phylogeny, but this has not always been practical or possible. Purely artificial keys often have been constructed for the convenience of the user. The diagnoses of the taxa and the keys are based almost wholly on available herbarium specimens.

The geographical areas covered in this flora include all of present-day Japan, excluding the Tokara Islands. The Japanese archipelago is divided into eight segments: Hokkaido (including Rishiri and Rebun islands), northern Tōhoku, Kantō, central, Kinki, and western Chūgoku districts of Honshu, Shikoku, and Kyushu, including the adjacent islands of Tanegashima and Yakushima as the southern boundary.

I am deeply grateful to the late Dr. G. Koidzumi, my mentor, for his kind instructions over a long time and to many others for their many valuable suggestions. Dr. Y. Satake, S. Okuyama, and K. Hisauchi have rendered kind assistance to the author in various ways ever since he became a staff member of the National Science Museum. Dr. T. Koyama helped in preparing the drawings, reading proof, and making indexes, as well as in various other ways. In the preparation of accounts of complex plant groups various monographic studies by specialists were very helpful. To all authors of these works the writer is greatly obligated.

I wish to mention that the original Japanese edition of this work was partially sponsored by a Publications Subsidy of the Ministry of Education, for which support I am deeply grateful.

PHYTOGEOGRAPHICAL RÉSUMÉ

Japan supports a very rich flora in proportion to its size, a circumstance of historical importance in relation to Japan's early closer relationship with mainland Asia and to its subsequent development as an insular province with highly distinctive geographical characteristics. Historically, the Japanese home islands were a part of the continental landmass of Asia, at least down to the Quaternary Period. For this reason, our flora is most closely related to the Chinese flora, especially to plants of the mountains of China, where many species exist as relics or as remnants of the much older Tertiary floras. That Japan was not greatly affected by Pleistocene glaciation is a factor which favored the preservation of older floras that might otherwise have vanished. Moreover, the approximately 850-mile length of the land area of Japan, extending over nearly 15 degrees of latitude from about 30° to 45° N.——from the subtropical belt of the southern areas to the alpine summits of numerous mountain peaks——together with a complex mountain system that covers nearly 70 percent of the total land mass, are factors that determine the component elements of the flora.

The close proximity of the sea produces an insular climate over much of the country. The warm Japan Current or Black Stream (*Kuroshio*), as it flows from southwest to northeast along the Pacific side of the archipelago, influences all southern and southeastern areas, giving them a relatively high precipitation and little or no frost in areas near the coast. In these southern areas are found many plants which occur principally in areas farther south. In the north, the cool Kurile (*Oyashio*) Current has a pronounced cooling effect on the climate of northern Honshu. Likewise, in our northern areas and on the higher mountains occur many boreal species extending into this area from more northerly latitudes.

WARM-TEMPERATE REGION

This area embraces the coastal areas of southern Honshu, Shikoku, and the Pacific side of Kyushu, with a warm-temperate to subtropical climate influenced by the warm Japan Current (*Kuroshio*) from the southwest Pacific Ocean. This area is dominated largely by evergreen woody species, especially broad-leaved angiosperms. The heavy well-distributed precipitation of usually more than 60 inches per annum in this region results in a lush vegetation containing many plants with more southern affinities. Some common plants characteristic of this warm-temperate region are:

BROAD-LEAVED EVERGREEN TREES AND SHRUBS

Camellia japonica	Itea japonica
Cinnamomum spp.	Machilus spp.
Cleyera japonica	Myrica rubra
Daphniphyllum macropodum	Prunus spinulosa
Distylium racemosum	Prunus zippeliana
Elaeocarpus japonicus	Quercus spp.
Elaeocarpus sylvestris	Symplocos spp.
Eurya japonica	Trochodendron aralioides

CONIFERS

Cryptomeria japonica	Podocarpus macrophyllus
Pinus densiflora	Podocarpus nagi
(in the secondary forest belt)	

LIANAS

Actinidia rufa	Piper kadsura
Hedera rhombea	Trachelospermum asiaticum
Hosiea japonica	Uncaria rhynchophylla
Kadsura japonica	

DECIDUOUS TREES

Clethra barbinervis	Rhus javanica
Euscaphis japonica	Rhus succedanea
Idesia polycarpa	Rhus trichocarpa
Mallotus japonicus	Styrax japonica
Prunus jamasakura	Zanthoxylum ailanthoides
Prunus lannesiana	

SHRUBBY AND HERBACEOUS SPECIES

Alocasia spp.	Debregeasia edulis
Ardisia crenata	Dicliptera japonica
Ardisia sieboldii	Fatsia japonica
Arundinaria	Nandina domestica
(in forest understory)	Pseudopyxis sp.
Arundo donax	Rhododendron spp.
Broussonetia kaempferi	Rubus trifidus
Broussonetia kazinoki	Villebrunea frutescens

AQUATICS

Euryale ferox	Nymphaea tetragona
Nelumbo nucifera	Potamogeton spp.
Nuphar japonicum	

MARSH PLANTS

Epilobium pyrricholophum	Polygonum spp.
Eriocaulon spp.	Rhynchospora spp.
Lycopus lucidus	Typha spp.
Miscanthus sacchariflorus	Zizania latifolia
Phragmites spp.	

PLANTS OF ROCKY PLACES IN MOUNTAINS

Calamagrostis hakonensis	Potentilla spp.
Carex spp.	Rhododendron spp.

STAGES OF SUCCESSION IN THE WARM-TEMPERATE REGION

The pioneer herbaceous plants which invade cut-over lands consist chiefly of *Miscanthus sinensis, M. floridulus, Imperata, Themeda, Smilax china, Lespedeza, Rubus,* and *Rhododendron.* The commonest pioneer woody species in many areas are *Pinus densiflora, Mallotus japonicus, Clethra barbinervis, Callicarpa japonica,* and deciduous species of *Symplocos.* In the final stages of forest succession, evergreen species of *Quercus* and *Cinnamomum* reappear as climax dominants.

COASTAL DUNES

Arundo donax	Limonium tetragonum
Calystegia soldanella	Messerschmidia sibirica
Canavalia lineata	Wedelia prostrata
Cnidium japonicum	Zoysia macrostachys
Ischaemum aristatum	

ZONE BETWEEN THE DUNES AND CONIFEROUS FOREST

Euonymus japonicus	Pittosporum tobira
Eurya emarginata	Quercus phillyraeoides
Hibiscus hamabo	Rosa wichuraiana
Juniperus chinensis var.	Ternstroemia gymnanthera
procumbens	Vitex rotundifolia
Litsea japonica	

SUBTROPICAL ELEMENTS OF SOUTHERN KYUSHU, TANEGASHIMA, YAKUSHIMA, AND AOSHIMA

Ardisia sieboldii	Livistona chinensis
Caesalpinia nuga	Melastoma candidum
Cassytha filiformis	Messerschmidia argentea
Cycas revoluta	Microstegium ciliatum
Entada phaseoloides	Myoporum bontioides
Ficus microcarpa	Osbeckia chinensis
Glochidion hongkongense	Schefflera octophylla
Ipomoea pes-caprae	Spinifex littoreus
Kandelia candel	Tree-ferns

On rocks in fast-running streams in the low mountains of southern Kyushu and Yakushima are found *Cladopus* and *Hydrobryum,* both of the Podostemaceae. These are quite remarkable members of the Japanese flora, since these genera are primarily tropical with a distribution centering in India and Malaysia.

The Laurisylvae or broad-leaved evergreen forest extends northward along both the Pacific side and the Japan Sea coast of Honshu to about 38° N. latitude, where the deciduous forest species become the dominant element of the woody vegetation. Deciduous forest species, such as *Quercus acutissima, Q. serrata, Castanea crenata, Acer* spp., *Carpinus laxiflora, C. tschonoskii, Alnus japonica,* and *A. sieboldiana,* are mixed with evergreen species. Evergreen trees native in the vicinity of Tokyo are represented only by a few species of evergreen *Quercus, Castanopsis, Machilus,* and a few others.

Areas of continental eastern Asia that correspond most closely with the warm-temperate parts of Japan are the lowland and hilly areas from southern Korea to the Yangtze Valley and the mountainous regions of Chekiang, Fukien, Hunan, Szechuan, Kwangsi, Kweichow, and Yunnan, and from Formosa to the western part of Sikang to the highlands of the Himalayas, Burma, and Indochina.

Significant genera found in both China and in Japan include *Trochodendron, Loropetalum, Chikusichloa, Cryptomeria, Elli-*

siophyllum, Chionographis, Nandina, Hosiaea, Skimmia, Stauntonia, Hovenia, Liriope, Ophiopogon, Shibataea, Heterosmilax, and Phaenosperma. Genera such as Buxus, Camellia, Broussonetia, Aulacolepis are distributed even more widely over parts of southeastern Asia and India. While many of the species in the warm-temperate parts of Japan are endemic, very few of the genera are confined to this region. Alectorurus and Neofinetia are believed to be the only genera endemic of the warm-temperate region of Japan.

In the warm-temperate region, endemism at the species level is perhaps best developed in the Fuji Volcanic Range and the Fossa Magna region. These areas, especially the Fuji Volcanic Range, as explained by Dr. F. Maekawa, were subject to a long period of volcanic activity during the Tertiary when they became pioneer areas for the development of new taxa. Campanula punctata var. microdonta, Carex hachijoensis, Rhododendron tsusiophyllum, and Astilbe simplicifolia are examples of endemic taxa found there.

Endemic Plants of the Fuji Volcanic Range with Their Vicarious Ancestors in the Mainland (Honshu, Shikoku, and Kyushu)

Calamagrostis autumnalis
Campanula punctata var.
  microdonta
Carex doenitzii var. okuboi
Carex oshimensis
Hydrangea macrophylla var.
  normalis
Lilium auratum var.
  platyphyllum
Meliosma hachijoensis

Polygonum cuspidatum var.
  terminale
Prunus lannesiana var. speciosa
Rhododendron tsusiophyllum
Saxifraga fortunei var.
  crassifolia
Styrax japonica var.
  jippei-kawamurae
Weigela coraeënsis var.
  fragrans

TEMPERATE REGION

The temperate region begins at about 1,000–1,500 m. above sea level in Kyushu and southern Shikoku, thence gradually decreasing in altitude northeastward to the low mountains of the Kantō District. The temperate region descends to sea level at about 38° N. latitude and continues northward along the coastal areas to the southwestern part of Hokkaido, including the southern province of Tokachi. The intermediate forest zone between the warm-temperate and temperate regions are usually dominated by Quercus acutissima, Q. serrata, Castanea crenata, and Pinus densiflora. In the temperate region Cryptomeria japonica, Chamaecyparis obtusa, and C. pisifera are widely cultivated, but the larger bamboos are no longer very evident. Plants common in the temperate region are:

Deciduous Trees

Acer japonicum
Acer mono
Acer palmatum
Aesculus turbinata
Alnus hirsuta
Betula grossa
Betula maximowicziana
Betula schmidtii
Carpinus cordata
Carpinus japonica
Fagus crenata
Fagus japonica
Fraxinus mandschurica var.
  japonica

Fraxinus sieboldiana
Hamamelis japonica
Kalopanax septemlobus
Magnolia obovata
Prunus sargentii
Prunus verecunda
Quercus mongolica
Sorbus alnifolia
Sorbus commixta
Sorbus japonica
Styrax shiraiana
Tilia japonica
Ulmus laciniata
Viburnum furcatum

Conifers

Abies firma
Abies homolepis
Larix leptolepis
Pinus koraiensis
Pinus pentaphylla
Pseudotsuga japonica

Sciadopitys verticillata
Taxus cuspidata
Thuja standishii
Thujopsis dolabrata
Tsuga diversifolia
Tsuga sieboldii

Broad-leaved Evergreen Shrubs

Aucuba japonica
Daphniphyllum humile
Euonymus fortunei

Ilex leucoclada
Ilex sugeroki
Skimmia japonica

Deciduous Shrubs

Hydrangea macrophylla var.
  acuminata
Ilex serrata
Kerria japonica
Lespedeza spp.

Lonicera spp.
Sorbus gracilis
Spiraea japonica
Tripetaleia paniculata
Vaccinium spp.

Deciduous Lianas

Actinidia polygama
Akebia quinata
Akebia trifoliata
Hydrangea petiolaris

Rhus ambigua
Schizophragma hydrangeoides
Tripterygium regelii
Vitis coignetiae

Herbaceous Plants

Cacalia spp.
Chrysosplenium spp.
Circaea spp.
Cirsium spp.
Gentiana scabra var. buergeri

Melampyrum laxum
Plectranthus trichocarpus
Saussurea spp.
Trillium spp.
Viola spp.

Endemic Genera Found in the Temperate Region

Anemonopsis (Ranunculaceae)
Deinanthe (Saxifragaceae)
Glaucidium (Ranunculaceae)
Hakonechloa (Gramineae)
Kirengeshoma (Saxifragaceae)

Peltoboykinia (Saxifragaceae)
Ranzania (Berberidaceae)
Sciadopitys (Coniferae)
Thujopsis (Coniferae)
Tripetaleia (Ericaceae)

Elements of the Warm-temperate Region Found in the Littoral Zone of the Temperate Region

Calystegia soldanella
Carex kobomugi
Carex pumila

Imperata cylindrica
Ixeris repens
Zoysia macrostachya

Plants Found in the Temperate Region More Widespread Farther South

Arabis stelleri
Lathyrus maritimus
Linaria japonica

Messerschmidia sibirica
Rosa rugosa
Thermopsis lupinoides

Plants Commonly Found in Rocky Places Along the Coast

Chrysanthemum yezoense
Lysimachia mauritiana

Sedum kamtschaticum

Dominant Components of the Coastal Forests

Pinus thunbergii

Quercus dentata

Marsh Plants of Coastal Areas

Aster tripolium
Carex rugulosa
Carex scabrifolia

Fimbristylis subbispicata
Triglochin maritimum

Aquatic Plants

Aldrovanda vesiculosa
Brasenia purpurea
Myriophyllum spicatum
Najas marina
Najas minor
Nuphar subintegerrimum

Potamogeton spp.
Ranunculus nipponicus
Sparganium spp.
Utricularia spp.
Vallisneria asiatica

MARSH PLANTS AND OTHERS COMMON ALONG STREAMS

Caltha palustris	*Scirpus juncoides*
Carex dickinsii	*Scirpus lacustris* var. *creber*
Carex dispalata	*Scirpus mitsukurianus*
Cyperus glomeratus	*Scirpus preslii*
Eleocharis spp.	*Scirpus triqueter*
Eriocaulon spp.	*Scirpus wichurae*
Lycopus uniflorus	*Sparganium stoloniferum*
Rhynchospora spp.	*Typha latifolia*

The temperate flora of Japan shows close relationship to that of certain areas of mainland eastern Asia, especially with the mountainous areas of southern Korea and the lowlands of central Korea and the Huan River valley, also with the mountainous regions of central China and the high mountains of the Himalaya and Malaysia.

GENERA COMMON TO THE TEMPERATE FLORAS OF JAPAN AND THE MAINLAND OF EASTERN ASIA

Actinidia	*Euptelea*
Ainsliaea	*Helwingia*
Akebia	*Hosta*
Aucuba	*Hovenia*
Cercidiphyllum	*Peracarpa*
Cryptomeria	*Tricyrtis*
Damnacanthus	*Weigela*
Deutzia	

GENERA FOUND IN THE TEMPERATE FLORA OF JAPAN WITH OUTLYING TAXA EXTENDING ACROSS ASIA, SOME INTO EUROPE

Adenophora	*Hedera*
Adonis	*Ilex*
Bothriospermum	*Pseudostellaria*
Kengia	*Syringa*
Eranthis	*Thelygonum*
Forsythia	

The relationship of the flora of Japan to that of North America, especially eastern North America, was first elucidated by Asa Gray in his now classic paper published in the Memoirs of the American Academy of Arts and Sciences (see footnote p. ix). Gray first elcuidated the close affinities that exist between the two areas. This floristic relationship is now explained on a historical basis of a former land-bridge connection between Asia and North America across the Bering Sea in preglacial times. Most of the American representatives of genera common to both areas are considered by most authors to be distinct from those in Japan, but in some instances the American and Japanese taxa may be distinguished only with difficulty.

GENERA COMMON TO JAPAN AND EASTERN NORTH AMERICA

Apios	*Meehania*
Boykinia	*Menispermum*
Buckleya	*Menziesia*
Caulophyllum	*Muhlenbergia*
Clethra	*Osmorhiza*
Croomia	*Pachysandra*
Cryptotaenia	*Phryma*
Diarrhena	*Shortia*
Diphylleia	*Stewartia*
Epigaea	*Tipularia*
Hamamelis	*Torreya*
Itea	*Trautvetteria*
Leucothoë	*Tsuga*
Lespedeza	*Wisteria*
Magnolia	*Zizania*

RELICT SPECIES OF TEMPERATE JAPAN, WHOSE PRINCIPAL DISTRIBUTION IS IN NORTHERN KOREA, MANCHURIA, AND AMUR

Adenophora palustris	*Carex onoei*
Astragalus adsurgens	*Lilium callosum*
Campanula glomerata	*Polygonatum inflatum*
Carex cinerascens	*Senecio flammeus*
Carex leiorhyncha	*Trigonotis nakaii*
Carex lithophila	*Triosteum sinuatum*
Carex meyeriana	*Viburnum carlesii*
Carex neurocarpa	

The coastal areas of the Japan Sea side of Honshu, centering around Hokuriku from San'in District as far north as the west coast of Ugo Province, contrast rather sharply in climate with areas of the Pacific Coast at the same latitude. The winters on the Japan Sea side are considerably more humid than those on the Pacific side with much more snow in the mountains and along coastal areas. Coniferous forests of *Abies*, *Picea*, and *Tsuga* are charasteristic of the Pacific side of the country.

ENDEMIC OR HOKURIKU ELEMENTS IN THE FLORA OF JAPAN SEA SIDE

Agrostis hideoi	*Hamamelis japonica* var.
Alnus fauriei	*obtusata*
Berchemia longeracemosa	*Ilex leucoclada*
Calamagrostis fauriei	*Iris gracilipes*
Calamagrostis gigas	*Pedicularis nipponica*
Camellia rusticana	*Poa fauriei*
Carex aphyllopus	*Ranzania japonica*
Chrysosplenium fauriei	*Tripterygium regelii*
Corydalis capillipes	*Viola faurieana*
Epimedium sempervirens	

Several gigantic herbaceous plants are found in the wet coastal areas of the Japan Sea side, the most common being *Petasites japonicus* var. *giganteus*, *Polygonum sachalinense*, *Cacalia hastata* var. *orientalis*, *Urtica platyphylla*, *Angelica matsumurae*, *A. edulis*, *A. ursina*, and *Filipendula kamtschatica*.

In the mountains near the Japan Sea coast where winter snows are deep, the alpine zone descends to a relatively low elevation and the development of moors is a prominent aspect of the high mountainous districts of this region. Among the common species in wet alpine meadows are *Fauria crista-galli*, *Tofieldia japonica*, *Narthecium asiaticum*, *Phyllodoce aleutica*, *Boykinia lycoctonifolia*, *Scirpus caespitosus*, *Scirpus hondoensis*, *Juncus beringensis*, *Geum pentapetalum*, *Plantago hakusanensis*, and *Primula cuneifolia* var. *hakusanensis*. This mountain flora appears to be most closely related to that of more northern areas in Kamchatka and Alaska where snowfall is heavy and the climate is moist.

In the Hokuriku region of the Japanese alps where heavy winter snow weighs down all shrubby vegetation, most of the understory shrubs are nearly prostrate or at least decumbent. The occurrence of broad-leaved evergreen shrubs in this region, such as *Ilex leucoclada*, *Camellia rusticana*, *Daphniphyllum humile*, *Cephalotaxus harringtonia* var. *nana*, *Ilex crenata* var. *paludosa*, is possible because of a protective covering of deep snow in winter. Also found here are *Chikusichloa aquatica* and *Diplaziopsis cavaleriana*, both represented more widely in the warm-temperate areas of western Japan.

BOREAL REGION

The boreal region of Japan is characterized by a coniferous forest belt composed of *Abies mariesii*, *A. homolepis*, *A. mayri-*

ana, A. veitchii, Picea jezoensis, Larix kaempferi, Tsuga diversifolia, and *Taxus cuspidata,* which occurs at altitudes up to 2,000 m, in the western parts of Honshu and Shikoku. In the central district of Honshu this zone occurs from 1,500 to 2,000 m, altitude. In the northern districts of Honshu the coniferous belt is found from 1,000 to 1,500 m. altitude and to near sea level in the eastern and northern parts of Hokkaido. Common plants of this boreal region are:

Deciduous Trees

Acer tschonoskii	*Betula playtyphylla*
Alnus matsumurae	*Prunus ssiori*
Alnus maximowiczii	*Sorbus commixta*
(upper zone)	*Sorbus matsumurana*
Betula ermanii	

Deciduous Shrubs

Euonymus tricarpus	*Salix reinii*
Oplopanax Japonicus	*Vaccinium yatabei*

Broad-leaved Evergreen Shrubs

Ilex rugosa	*Rhododendron brachycarpum*
Ilex sugeroki	*Rhododendron degronianum*

Herbaceous Plants

Circaea alpina var. *caulescens*	*Maianthemum dilatatum*
Cornus canadensis	*Microstylis monophyllos*
Epilobium spp.	*Platanthera ophrydioides*
Glyceria alnasteretum	*Trientalis europaea*

Lianas

Actinidia kolomikta	*Rhus ambigua*

Plants of Coastal Dunes

Arabis stelleri var.	*Linaria japonica*
Artemisia stelleriana	*Matricaria matricarioides*
Carex gmelinii	*Matricaria tetragonosperma*
Elymus mollis	*Mertensia asiatica*
Glehnia littoralis	*Rosa rugosa*
Honkenya peploides	*Scutellaria strigillosa*
Lathyrus maritimus	

Plants of Sea Cliffs

Chrysanthemum arcticum	*Trifolium lupinaster*
Potentilla megalantha	

Plants of Littoral Swamps in Coastal Areas

Carex lyngbyei	*Scirpus planiculmis*
Carex mackenziei	*Triglochin maritimum*
Glaux maritima	*Triglochin palustre*
Salicornia europaea	

Aquatics

Nuphar pumilum	*Scirpus tabernaemontani*
Polygonum amphibium	*Sparganium gramineum*
Potamogeton heterophyllus	

Plants of Sphagnum Bogs

Andromeda spp.	*Ledum* spp.
Carex curta	*Narthecium asiaticum*
Carex limosa	*Rhychospora alba*
Carex middendorffii	*Scheuchzeria*
Drosera angelica	*Tofieldia japonica*
Drosera rotundifolia	*Vaccinium oxycoccus*
Eriophorum spp.	

Plants of Fresh-water Marshes and Stream-margins

Alnus hirsuta	*Polygonum sachalinense*
Calamagrostis langsdorffii	*Salix* spp.
Epilobium angustifolium	*Scirpus wichurae*
Phragmites communis	

The genera found in the boreal region mostly are those with a wide circumboreal distribution. Only *Pteridophyllum, Dactylostalix,* and *Tripetaleia* are endemic of this floristic province.

Coniferous forests are widely scattered in the high mountains of the Japan Sea side, consisting, when they do occur, mainly of *Thuja standishii* and *Larix leptolepis.* Deciduous woody species predominate in the vegetation of this region, include *Acer tschonoskii, Alnus maximowiczii, Betula ermanii, Hamamelis japonica* var., *Magnolia salicifolia,* and *Sorbus matsumurana.*

In lowlands, especially near the seacoast of eastern Hokkaido centering in the provinces of Nemuro and Tokachi, fogs in summer are of frequent occurrence as a result of the cold Kurile Current which flows southwestward along the south coast. Moors in this area are well developed. *Carex subspathacea* and *C. mackenziei* occur in the littoral swamps, the only areas where these plants appear in Japan. In the coastal area are *Saxifraga bracteata, Potentilla megalantha, Cochlearia oblongifolia, Rhododendron parvifolium,* and *Fritillaria camtschatcensis,* these having reached our area from farther north.

ARCTIC-ALPINE REGION

Extensive arctic-alpine areas are not represented in Japan, although treeless tundralike areas of limited extent occur on several mountain peaks scattered over various parts of the country. In the central district of Honshu, the alpine zone occurs at elevations of about 2,500 m. In the northern district of Honshu it begins at about 2,000 m. and at 1,500 m. in Hokkaido. Characteristically, the alpine zone is represented by a shrubby pine, *Pinus pumila,* except on the more recent volcanoes, such as Mount Fuji and Mount Asama. Immediately below the *Pinus pumila* zone deciduous shrubs are prominently represented, mainly *Alnus maximowiczii, Betula ermanii, Prunus nipponica, Vaccinium uliginosum, V. axillare, Sorbus sambucifolia,* and *S. matsumurana,* mixed with many high-altitude herbaceous plants. In the alpine zone, the plants consist mainly of Ericaceae, Primulaceae, Caryophyllaceae, Gentianaceae, Scrophulariaceae, Ranunculaceae, Rosaceae, and Cruciferae. The genera of the alpine zone mostly are those widely distributed throughout the northern parts of the Northern Hemisphere. The only endemic genus of the alpine area of Japan is *Japonolirion,* found on serpentine rocks. Most of the species of alpine plants found in Japan occur also in eastern Siberia, Alaska, the Aleutians, and Kamchatka, or represent widely distributed circumpolar species. Very little relationship is shown between the alpine plants of Japan and those of the Sino-Himalayan area, although *Polystichum lachenense* (Filices) and the genus *Androcorys* (Orchidaceae) occur in both areas.

Recent vulcanism in Mount Fuji, Mount Asama, Mount Iwate, Mount Chokai and in some other areas may be responsible for the occurrence of alpine species at lower elevations than on nonvolcanic mountain peaks found elsewhere.

Local endemism on Mount Hayachine, Mount Shibutsu, and Mount Apoi is associated with serpentine rocks found on these mountain peaks. The incidence of alpine plants is relatively high on Mount Apoi as a result of a lowered summer temperature brought on by heavy fog during the growing months of summer. The flora of Rebun Island, adjacent to Hokkaido, is known for the occurrence of floristic elements from Sakhalin as a result of a lowered summer temperature brought on by the cold current that flows southward through the Mamiya Channel.

HISTORICAL RÉSUMÉ OF FLORISTIC WORK IN JAPAN

When Linnaeus wrote the first edition of his "Species Plantarum" in 1753, he knew only a few of the plants of Japan, all taken from the "Amoenitatum Exoticarum" (1712) of **Engelbert Kaempfer** (1651–1716). Kaempfer was a German naturalist who lived in Japan from 1690 to 1692 as a medical officer of the Dutch East India Company. The plants listed by Kaempfer were: *Chenopodium scoparia, Rhus vernix, Laurus camphora, Thea sinensis, Uvaria japonica, Camellia japonica, Morus papyrifera, Xanthium strumarium, Ficus pumila, Smilax china, Taxus nucifera, Epidendrum moniliforme, Azalea indica,* etc., all well known among the indigenous plants of Japan.

In volume I of Linnaeus's "Mantissa Plantarum" (1767), *Sophora japonica, Prenanthes japonica,* and *Tussilago japonica* were described on the basis of actual specimens collected in 1759 by **Christiaan Kleynhoff,** a Hollander of German birth. The next year Kleynhoff's collection was reported upon by N. L. Burman (1734–1793), also of Holland, in his "Flora Indica" (1768). In Burman's work *Azalea rosmarinifolia, Basella japonica, Arnica tussilaginea,* and *Ficus pyrifolia* are described from Japan as new taxa based upon Kleynhoff's original specimens.

C. P. Thunberg (1743–1828), a Swedish naturalist, physician, and disciple of Linnaeus, came to Japan in August of 1775 at the age of 32 years, as a medical officer of the Dutch East India Company. Thunberg traveled from Nagasaki to Yedo (Tokyo) and back before his return home via Java in December of 1776. Thunberg's Japanese collections amounted to approximately 1,000 species. The best of his collection was from Hakone. The major work of Thunberg culminated in his now classic work "Flora Japonica," published in 1784, the cornerstone of taxonomic botany in Japan. Thunberg published separately on parts of his Japanese collections in "Kaempferus Illustratus," 1–2 (1780, 1783); "Nova Genera Plantarum," 1 and 3 (1781, 1783), and several other works. Several novelties from his collection were published in "Supplementum Plantarum Systematis Vegetabilium" (1781) by the son of Linnaeus and in part 2, volumes 8–14, of the "Natuurlyke Historie" (1773–83) of Martinus Houttuyn. Nearly all of Thunberg's collections are now preserved in the University of Uppsala in Sweden.

Through the Dutch East India Company, established in Nagasaki in 1609, Japanese plants found their way to Europe via Dejima Island, the only Japanese port opened to foreigners at that time. Only ships of Dutch nationality were permitted entry until restrictions were removed in 1859. Floristic studies on the Japanese flora during this period, although of a relatively limited scope, were carried on under Dutch sponsorship. Botanical collecting by the Dutch was greatly accelerated after the beginning of the 19th century by the opening of a botanical garden at Buitenzorg (now Bogor) in Java and by the exploits of Siebold and others. Japanese plants from various collectors were studied by K. L. von Blume (1796–1862) at Buitenzorg. We find that publication on some of Siebold's early collections began to appear in volume 2 of Blume's "Bijdrangen tot de Flora van Nederlandsch Indie" (1825–26).

P. F. von Siebold (1796–1866), of German birth, came to Japan in 1823 when he was 27 years of age as a medical officer of the Dutch East India Company. He remained in Japan until 1829. During his residence of six years he collected assiduously at Nagasaki and vicinity, and in 1826 he traveled to Yedo (Tokyo) and back. Siebold was assisted at various times by the able Japanese naturalists Keiske Itō (1803–1901), Yōan Udagawa (1798–1846), and Hōbun Mizutani (1779–1833). Upon his return to Europe in 1830, Siebold lived at Leiden and continued to receive plants from Japan sent through his Japanese acquaintances, from H. Buerger who had earlier collected plants with Siebold, and from Jacques Pierot (1812–41) who visited Japan in 1840. Plants also came from Otto Gottlieb Johan Mohnike (1814–87). For a time Siebold operated the commercial nursery of Siebold & Co. at Leiden and sold many Japanese plants widely over Europe. This undertaking was the most important effort up to this time to introduce Japanese plants into Europe. Ch. F. A. Mooren (1807–58), Joseph Decaisne (1807–82), and Justus Karl Hasskarl (1811–94) made studies on the living plants brought from Japan by Siebold.

The principal works published by Siebold on the Japanese flora are his "Plantarum, quas in Japonia Collegit Dr. Ph. Fr. de Siebold, Genera Nova, Notis Characteristicis Delineationibusque Illustrata Proponunt" (1843) and his "Florae Japonicae Familiae Naturales" (1845–46) both published jointly with J. G. Zuccarini, and his "Flora Japonica" (1826–70) published also under joint authorship with Zuccarini, except for the second volume, of which F. A. W. Miquel was joint author. Most of the herbarium specimens collected by Siebold are preserved in the Rijksherbarium at Leiden. A somewhat smaller set, purchased by Maximowicz, is in Leningrad along with Siebold's unpublished drawings prepared for the "Flora Japonica."

Siebold's Japanese collections were cited in several works by other authors, especially by K. L. Blume in the "Museum Botanicum Lugduno-Batavum," 1–2 (1849–56) and in his "Flora Javae et Insularum Adjacentium Nova Series" (Orchidaceae), 1858, and by E. G. Steudel in his "Synopsis Plantarum Glumacearum," (1854–55). F. A. W. Miquel published the most amplified account of Siebold's collection in his "Prolusio Florae Japonicae" (1865–67) issued in his serial "Annales Musei Botanici Lugduno-Batavi."

Heinrich Zollinger, a Swiss botanist who worked in the herbarium at Buitenzorg from 1841 to 1848, published on the Japanese plants he found deposited in the Buitenzorg herbarium. This work, entitled "Systematisches Verzeichnis der im Indischen Archipel in den Jahren 1842–48 gesammelten, sowie der aus Japan empfangenen Pflanzen" published in Zurich in 1854–55, included an account of the Japanese collections of P. F. W. Göring.

Philip Friedrich Wilhelm Göring (1809–79) was a botanist of German birth who collected under Dutch auspices at Nagasaki. Göring's collections were studied by various specialists, but principally by Zuccarini, who published on them in the periodical "Flora," volume 29 (1846). Steudel reported on the grasses and sedges, also in volume 29 of "Flora." The German botanist H. G. Reichenbach *f.* published on the orchids in the "Botanische Zeitung," volume 3 (1845). Some new species based on Göring's collections were described by N. S.

Turczaninov, a Russian botanist, in various issues of the "Bulletin de la Société des Naturalistes de Moscou" for 1846 and 1848. The exact dates of Göring's Japanese itinerary have not been definitely established. It is known, however, that he was in Java from 1844 to 1856.

The middle of the century saw the beginning of a new era of botanical activity in Japan, when foreign plant collectors, other than Dutch, began to visit the country. Among the first can be mentioned **S. Wells Williams** and **James Morrow** from the the United States of America who were attached to the Perry Expedition of 1852–54. Collections of plants were made chiefly at Shimoda, a harbor in the southern part of the Idzu Peninsula, and at Hakodate in Hokkaido. In the wake of these collectors came **Charles Wright** and **J. Small,** who visited Japan as members of the U.S. North Pacific Exploring (Ringold and Rodgers) Expedition of 1853–56. The collections from these two American expeditions were reported on by Asa Gray under the title "List of Dried Plants Collected in Japan by S. Wells Williams, Esq., and Dr. James Morrow" (1856) and "Diagnostic Characters of New Species of Phaenogamous Plants, Collected in Japan by Charles Wright" (1859). In the latter paper, Gray presented his views on the relationships between the floras of Japan and of eastern North America (see footnote p. ix).

C. P. Hodgson, the English consul at Hakodate, 1859–60, sent to W. J. Hooker at Kew his collection from Hakodate and vicinity. Hooker in 1861 published a "Catalogue of Japan Plants, Systematically Arranged" based upon Hodgson's collection.

C. J. Maximowicz (1827–91), a Russian botanist, first came to Hakodate in 1860 soon after publication of his "Primitiae Florae Amurensis" (1859). Maximowicz's itinerary in Japan covered at least the following localities as cited on his herbarium specimens: the southwestern part of Hokkaido, Yokohama, Hakone, Mount Fuji, Mount Kuju, Nagasaki, Mount Aso, and Kumamoto. He left Japan in 1864. In 1866 Maximowicz purchased from Siebold's widow part of his Japanese collections. At this period Maximowicz, through his many acquaintances in Japan, brought together the largest collection of Japanese plants made up to this time. Tschonoski Sugawa (1841–1925), who was Maximowicz's attendant during his stay in Japan, continued to send many plants to Maximowicz after his departure from Japan. Michael Albrecht, a medical emissary to the Russian consul at Hakodate, offered to Maximowicz his collection of plants gathered at Hakodate. In addition to these, much material was sent to Maximowicz by the leading Japanese botanists of the period, including Yasusada Tashiro (1856–1928), Yoshio Tanaka (1838–1916), Ryōkichi Yatabe (1851–99), Jinzō Matsumura (1856–1928), Tokutarō Itō (1868–1941), Kingo Miyabe (1860–1951), and Tomitarō Makino (1862–1957).

Maximowicz published his first paper on the Japanese flora in 1866. He continued to publish intermittently from then on until his death in 1891. His most elaborate publications are "Diagnoses breves plantarum novarum Japoniae et Mandshuriae" (1866–77) and "Diagnoses plantarum novarum Asiaticarum" (1877–93). These were scholarly works, each a great contribution to Japanese botany. It should be mentioned that Maximowicz made extensive use of the classical work "Sōmoku-dzusetsu" by Yokusai Iinuma (1783–1865), the first illustrated flora of Japan (ed. 1, 1856), in which Maximowicz assigned Latin names to many of the figures. The principal set of Maximowicz's Japanese collections is in

Leningrad, with nearly complete sets at Kew and in Paris. Partial sets exist in the U.S. National Museum, Washington, D.C., and in various other herbaria.

From 1857 to 1859 **C. Wilford,** an Englishman, traveled in eastern Asia and visited Japan as well as the north China coastal area, Formosa, and Korea. Wilford's collection was sent to Kew and was studied later by Maximowicz.

Two British horticulturists, **James Gould Veitch,** who visited Japan from 1859 to 1862, and **Robert Fortune,** who came in 1861, sent many living plants and seeds to England.

Richard Oldham (1837–1864), an Englishman, botanized in the neighborhood of Nagasaki in 1862 and 1863. His collection was studied by Daniel Oliver at Kew, by Miquel in Leiden, and by Maximowicz in Leningrad.

Otto Schottmüller (?–1864) and **Max Ernst Wichura** (1817–66), as members of a German scientific exploring expedition dispatched to Japan and China between 1859 and 1861, botanized in the Ryukyus, Hokkaido, and Nagasaki in 1860.

P. A. L. Savatier, a Frenchman, collected plants in Japan from 1866 to 1871 and again from 1873 to 1876, during his tenure as a medical officer of the Iron Works at Yokosuka. The collections of Savatier were gathered chiefly in Yokosuka and its suburbs, Mount Fuji, Hakone, Atami, Nikko, and Hakodate. Savatier, like his predecessors, obtained assistance from several Japanese botanists of the period, as well as from several visiting foreigners. Japanese botanists assisting Savatier included Keiske Itō, Motoyoshi Ono, Ichirō Saba, and Yoshio Tanaka. The foreigners who collected for Savatier included F. L. Verny (Mount Asama, Tomioka in Kōzuke Province, and Niigata), de Brandt, the Prussian minister to Japan (Hakodate and Kyushu), a Mr. Robert, an army surgeon (Hakodate), and F. Hilgendorf (chiefly Hakodate). Other names among foreign contributors include Hogg, Kramer, Vidal (Tomioka, Niigata), and Dickins (Atami). The two-volume work by Franchet and Savatier, "Enumeratio plantarum in Japonia sponte crescentium," published in 1875–79, was the most extensive enumeration up to this time of the plants of Japan. Franchet and Savatier's flora included a useful bibliographical compilation of classical studies on Japanese plants, such as Kwa-wi, Honzō-dzufu, and Sōmoku-dzusetsu. Savatier prepared a translation of Kwa-wi under the French title "Livres Kwa-wi traduits du japonais avec l'aide de M. Saba" (1873). The first set of Savatier's Japanese collections is kept in the Muséum National d'Histoire Naturelle, in Paris. Partial sets are at Kew, in the U.S. National Museum, Washington, D.C., and elsewhere.

Several small collections by Europeans were made during the latter part of the 19th century as follows:

Emanuel Weiss (1835–70) and **J. von Xanthus** were members of a Hungarian expedition in 1869–70, led by H. Wawra von Fernsee. Their collections are mainly from Yokohama. A publication on their collections was made by A. Kanitz, entitled "Expeditio Austro-Hungarica ad oras Asiae orientalis. Anthophyta quae in Japonia legit beat" (1878).

James Bisset (1843–1911), an Englishman, collected in Japan at various times from 1866 to 1886, chiefly in Hakone, on Mount Ōyama, and at Yokohama. Spencer le Marchant Moore published on these in parts I and II of his "Alabastra Diversa" in the Journal of Botany British and Foreign (1877–78).

Doenitz, a German, visited Japan from 1876 to 1880. He climbed Mount Nantaisan in Nikko in 1875, Mount Fuji in

1875, and Mount Kanosan in 1876, and went to Yumoto near Nikko in 1876.

J. J. Rein (1835–1918), a German geographer, came to Japan in 1874 and made tours of Mount Ontake in Shinano province, Hakone, Nikko, Shikoku, and Amami-Ōshima. His specimens are cited by Franchet and Savatier in their "Enumeratio."

F. Hilgendorff (1839–1904), a German teacher of zoology, botanized in Japan in connection with his zoological investigations.

L. H. Doederlein (1855–1936), a German, in company with Yasusada Tashiro visited Amami-Ōshima and some other areas of Japan in 1880. His publication "Botanische Mitteilungen aus Japan" appeared in Botanisches Centralblatt in 1881.

Otto H. Warburg (1859–1938) came to Tokyo in 1887 after collecting previously in Korea, the Tsushima Islands, and the Gotō Islands.

Charles S. Sargent, an American botanist, collected in various parts of Japan in 1892. The results of his investigations on the Japanese forest flora were summarized in his "Forest Flora of Japan," published in 1894.

Urbain Faurie (1847–1914), came to Japan in 1874 at the age of 27 years, as a French missionary. Faurie must be extolled as perhaps the most energetic of 19th-century collectors of Japanese plants. His collections of flowering plants and cryptogams, amounting to several hundred thousand herbarium sheets, remain as a monumental legacy of his many energetic years spent in the country.

Faurie first took up residence at Niigata in 1874. In 1883 he moved to Hakodate in Hokkaido and began a tour of that part of Japan. In 1897 he lived in Aomori on Honshu. By this time Faurie had collected extensively over much of the country as well as the adjoining areas of the South Kuriles, southern Korea, and Amami-oshima. He died from illness in Formosa in 1914. His collections were sent mostly to the Muséum National d'Histoire Naturelle in Paris. A nearly complete set of Faurie's collections is also kept in the herbarium of the University of Kyoto. Duplicates were widely dispersed to the leading herbaria of Europe, and large sets exist also in the older herbaria of America.

Faurie's collections of his Niigata period were cited in Franchet and Savatier's "Enumeratio." After Franchet's death the Faurie collections came into the hands of H. Léveillé (1863–1918), who worked on the Cyperaceae. This latter set of Faurie's collections was purchased by the British Museum (Natural History) in London after Léveillé's death. Faurie's collections were studied by many specialists, including B. Hayata, T. Nakai, G. Koidzumi, H. Christ (ferns), E. Rosenstock (ferns), C. Christensen (ferns), E. Hackel (grasses), E. Koehne (Rosaceae), M. Petitmengin, G. Kükenthal (Cyperaceae), W. Becker (Violaceae), H. de Boissieu (Saxifragaceae and Cruciferae), C. B. Clarke (Cyperaceae), A. Bennett (Potamogetonaceae), G. Bonati (Scrophulariaceae), A. Finet (Orchidaceae), O. von Seemen (*Salix*), C. K. Schneider (trees), R. Keller, F. N. Williams, and M. T. Masters (Coniferae).

Early in the 20th century many European and American botanists visited Japan, including Hans Hallier (1903), Netherlands; E. B. Copeland (1907), U.S.A.; E. D. Merrill (1907), U.S.A.; H. Lecomte (1911), France; G. Finet (1911), France;

Adolf Engler (1912), Germany; E. H. Wilson (1914), U.S.A.; and W. T. Swingle (1915), U.S.A.

In the latter period of the Tokugawa government the introduction into Japan of European and American techniques of taxonomy changed the traditional herbalist approach to botany. Leading this movement were Keiske Itō, Yoshio Tanaka, and Motoyoshi Ōno. A school founded in 1873 as the Kaisei-Gakko, together with the Medical School of Tokyo, were united in 1877 as the new Imperial University of Tokyo, established as a center of Western learning. Here Ryōkichi Yatabe (1851–99), who had studied at Cornell University in the United States, lectured on botany with the aid of Jinzō Matsumura and Saburō Okubo as assistant professors. At this time Kingo Miyabe and Tomitarō Makino were engaged in botanical research at the Botanical Institute of the University.

The Tokyo Botanical Society, presently the Botanical Society of Japan, was founded in 1882, with the Botanical Magazine of Tokyo as its organ for publication.

Outstanding among the taxonomic botanists of the new era in Japan are such names as M. Miyoshi (1861–1939); T. Makino (1862–1957); M. Shirai (1863–1932); K. Shibata (1877–1949); Y. Yabe (1866–1931); B. Hayata (1874–1934); T. Nakai (1882–1952); R. Kanehira (1882–1948); G. Koidzumi (1883–1953); H. Takeda (1882–); and Y. Kudō (1887–1932). Botanical publications at the end of the 19th century increased greatly to include floristic studies on the Japanese flora as well as on outlying areas, such as Korea, Formosa, the Bonin Islands, Sakhalin, Manchuria, and China. Collecting expeditions, such as Kudo's Sakhalin Expedition and Hayata's Indo-Chinese Expedition, were sent to various parts. Dr. Hayata's, for instance, during his three trips to Indochina, Yunnan, and Siam, sent back several thousand sheets of vascular plants.

Among the publications we should mention by the botanists just cited are Miyabe's "The Flora of the Kurile Islands" (1890); Miyabe and Kudo's "Flora of Hokkaido and Saghalien" (1930–34); Yabe's "Icones Florae Manchuriae" (1914–22); B. Hayata's "Icones Plantarum Formosanarum" (1911–21); Nakai's various floristic notes on the Korean flora and his critical monographic studies on *Aconitum, Viola, Lespedeza, Arisaema, Euonymus, Camellia,* Myrsinaceae, Polygonaceae, Caprifoliaceae, Bambuseae, and Pteridophyta; Takeda's many editions of an "Alpine Flora"; and Koidzumi's monographs on Rosaceae, Aceraceae, and *Morus,* and his phytogeographical works.

Nakai's prolific researches resulted in numerous important floristic works not only on the Japanese flora but also of Korea. His Korean Flora, "Chosen Shokubutsu" ["Flora of Chosen"] (1914), "Flora Sylvatica Koreana" (1915–39), and his incomplete "Iconographia Plantarum Asiae-Orientalis" (1935–52) are especially important. The "Nova Flora Japonica" (1938–51) is an unfinished work, jointly authored with M. Honda, an authority on Japanese grasses. T. Makino edited the "Journal of Japanese Botany" started by him in 1916, and G. Koidzumi edited the "Acta Phytotaxonomica et Geobotanica" started by him in Kyoto in 1932. These two periodicals remain today the leading Japanese journals for phytotaxonomy. The Botanical Magazine of Tokyo in recent years has greatly reduced its publication of taxonomic papers. The leading centers for botanical research in Japan from about 1930 onward have been the University of Tokyo and the University of Kyoto. The younger taxonomists of present-day Japan are products largely of these two institutions.

GENERAL KEY TO THE FAMILIES

I. Phylum PTERIDOPHYTA   Shida-Shokubutsu   Vascular Cryptogams

Plants without true flowers and seeds, reproducing from spores borne in a sporangium; archegonia and antheridia produced on the same or different prothallia.——About 26 families with more than 10,000 species, especially abundant in the Tropics. The sequence of the genera of true ferns follows the classification of Copeland, "Genera Filicum," 1947, with few exceptions.

II. Phylum SPERMATOPHYTA Shushi-Shokubutsu   Seed Plants

Plants with true flowers and seeds; female gametophyte developed within the macrosporangium (ovule), at maturity forming, with the enclosed embryonic sporophyte, the seed.

1A. Ovules naked, borne on the surface of a scale (sporophyll), sometimes arillate; flowers usually unisexual (plants monoecious or dioecious); trees or shrubs, mostly evergreen with needlelike, scalelike to subulate or rarely flabellate or pinnately compound leaves; stigmas absent; plants cone-bearing. .... Class 1. Gymnospermae,   109

1B. Ovules borne within an inclosed cavity (ovary); flowers perfect or unisexual; trees, shrubs, or herbs; leaves various; plants with true flowers consisting of a stigma, style, ovary and stamens plus a floral envelope, or modifications of this condition.
Class 2. Angiospermae,   118

Class 1. GYMNOSPERMAE

1A. Plants with palmlike leaves, dioecious; leaves pinnately compound, leathery, shiny, evergreen; ovules borne on pinnately compound naked sporophylls; microsporangia borne in a large terminal cone. .... Fam. 24. Cycadaceae,   109

1B. Plants with leaves otherwise, not palmlike; leaves simple.

2A. Leaves flabellate, often deeply bifid at apex, long-petiolate, deciduous. .... Fam. 25. Ginkgoaceae,   109

2B. Leaves acicular, linear, or scalelike, rarely flat and oblong, sessile or nearly so, evergreen.

3A. Ovules 1 to few, enveloped by a relatively soft pulpy drupe or berrylike fruit; cotyledons 2.

4A. Anthers 2-locular; leaves narrowly lanceolate to oblong, 5–30 mm. wide; ovules solitary. ..... Fam. 27. Podocarpaceae,   110

4B. Anthers 3- to 9-locular; leaves linear, 2.0–3.5 mm. wide.

5A. Anthers 3-locular; leaves 2–5 cm. long; ovules several, paired. .... Fam. 28. Cephalotaxaceae,   111

5B. Anthers 4- to 9-locular; leaves 1.5–2.5 cm. long; ovules solitary. .... Fam. 26. Taxaceae,   109

3B. Ovules numerous, borne on scales enclosed in a woody cone; cotyledons 2 or more.

6A. Cone-scales spirally arranged; leaves alternate or in fascicles.

7A. Cone-scales 2-seeded, flattened, bracteate; leaves acicular or linear. .... Fam. 29. Pinaceae,   111

7B. Cone-scales 2- to 9-seeded, flattened or peltate, ebracteate; leaves subulate or linear by cohesion of a pair.
Fam. 30. Taxodiaceae,   115

6B. Cone-scales and leaves opposite or verticillate, the latter scalelike or short-subulate. .... Fam. 31. Cupressaceae,   116

Class 2. ANGIOSPERMAE

Subclass 1. MONOCOTYLEDONEAE

Leaves usually parallel-veined; flowers typically 3-merous, sometimes 6-merous, rarely 4-merous (Stemonaceae).

1A. Plants with chlorophyll and characteristically with green leaves.

2A. Flowers in axils of chaffy or scaly bracts; perianth obsolete or bristle-shaped; ovary superior, 1-locular, 1-ovuled.

3A. Leaves 2-ranked; internodes hollow; culms round; leaf sheath split. .... Fam. 41. Gramineae,   131

3B. Leaves 3-ranked; internodes mostly solid; culms predominantly 3-angled; leaf sheath not split. ...... Fam. 42. Cyperaceae,   195

2B. Flowers not in axils of scaly or chaffy bracts; perianth either present or absent.

4A. Ovary superior or partly inferior.

5A. Carpels 2 to many, free, each containing 1 ovule, rarely 6 ovules.

6A. Perianth-segments 6.

7A. Outer 3 perianth-segments sepallike, the inner 3 petallike. .... Fam. 38. Alismataceae,   126

7B. Perianth-segments all alike. .... Fam. 37. Scheuchzeriaceae,   125

Subclass 2.   DICOTYLEDONEAE

Leaves mostly net-veined; flowers typically 5-merous but variously modified, rarely 3- or 6-merous.

100B.   Flowers variable, not as above.

  101A.   Stamens perigynous.

    102A.   Stipules usually prominent; carpels one to many. .... Fam. 104. Rosaceae,   517

    102B.   Stipules usually absent; carpels frequently fewer than the sepals. .... Fam. 101. Saxifragaceae,   498

  101B.   Stamens hypogynous.

    103A.   Trees with pinnate leaves. .... Fam. 111. Simaroubaceae,   585

    103B.   Shrubs or woody climbers; leaves never pinnate.

      104A.   Leaves palmately or ternately compound. .... Fam. 91. Lardizabalaceae,   461

      104B.   Leaves simple.

        105A.   Leaves alternate.

          106A.   Stipules none; flowers usually greenish, small, rarely solitary, unisexual.

                Fam. 93. Menispermaceae,   465

          106B.   Stipules large, deciduous leaving distinctive scars; flowers usually white or colored, large, solitary, perfect. .... Fam. 94. Magnoliaceae,   467

        105B.   Leaves opposite. .... Fam. 118. Coriariaceae,   596

98B.   Carpel 1, or if more than 1, these united into a compound ovary.

  107A.   Styles 2–5.

    108A.   Leaves opposite or verticillate.

      109A.   Sepals 2; plants rather fleshy. .... Fam. 83. Portulacaceae,   422

      109B.   Sepals 3–5.

        110A.   Herbs with entire leaves.

          111A.   Flowers 3-merous; plants small, flaccid, with stipules, growing in mud or in water.

                Fam. 138. Elatinaceae,   633

          111B.   Flowers 4- or 5-merous; stipules commonly absent.

            112A.   Leaves mostly with pellucid resinous glands; stamens frequently numerous and clustered into fascicles of 3 to 5. .... Fam. 137. Guttiferae,   630

            112B.   Leaves without resinous glands; stamens twice as many as the petals or fewer.

                Fam. 84. Caryophyllaceae,   422

        110B.   Herbs or woody plants with serrate leaves.

          113A.   Leaves simple. .... Fam. 101. Saxifragaceae,   498

          113B.   Leaves pinnate or ternate-pinnate. .... Fam. 122. Staphyleaceae,   605

    108B.   Leaves cauline and alternate or basal.

      114A.   Sepals 2; leaves pinnately parted. .... Fam. 96. Papaveraceae,   474

      114B.   Sepals 3–5.

        115A.   Sepals 3; much-branched procumbent alpine shrubs; leaves evergreen, linear, small and densely arranged on branchlets. .... Fam. 117. Empetraceae,   595

        115B.   Sepals 4–5.

          116A.   Trees or shrubs.

            117A.   Leaves compound. .... Fam. 119. Anacardiaceae,   596

            117B.   Leaves simple. .... Fam. 103. Hamamelidaceae,   516

          116B.   Herbs.

            118A.   Leaves trifoliolate. .... Fam. 107. Oxalidaceae,   580

            118B.   Leaves simple.

              119A.   Leaves entire, linear to lanceolate. .... Fam. 108. Linaceae,   580

              119B.   Leaves ovate to cordate-orbicular.

                120A.   Leaves chiefly cauline, with stellate hairs. .... Fam. 134. Sterculiaceae,   625

                120B.   Leaves chiefly basal, without stellate hairs. .... Fam. 101. Saxifragaceae,   498

  107B.   Style 1.

    121A.   Ovary 2- to 5-locular.

      122A.   Flowers distinctly zygomorphic.

        123A.   Trees; leaves opposite, palmately compound. .... Fam. 125. Hippocastanaceae,   611

        123B.   Herbs; leaves alternate, simple.

          124A.   Flowers spurred; ovary 5-locular, the ovules numerous; valves of capsule coiling and ejecting the seeds elastically in dehiscence. .... Fam. 128. Balsaminaceae,   613

          124B.   Flowers not spurred; ovary 2-locular, each containing 1 ovule; capsules much flattened, the seeds arillate. .... Fam. 113. Polygalaceae,   585

      122B.   Flowers actinomorphic.

        125A.   Leaves with pellucid oil-glands. .... Fam. 110. Rutaceae,   581

        125B.   Leaves without pellucid oil-glands.

          126A.   Climbing shrubs with tendrils. .... Fam. 130. Vitaceae,   618

          126B.   Plants without tendrils.

            127A.   Plants herbaceous.

              128A.   Petals 4; stamens 6, didynamous; fruit a silicle or silique. .... Fam. 98. Cruciferae,   479

              128B.   Petals more than 4; stamens as many as the petals or more numerous; fruit various, not as above.

                129A.   Plants saprophytic, without chlorophyll. .... Fam. 158. Pyrolaceae,   690

                129B.   Plants not saprophytic.

                  130A.   Fruits spiny. .... Fam. 109. Zygophyllaceae (*Tribulus*),   581

                  130B.   Fruits not spiny.

Phylum PTERIDOPHYTA
Ferns and Fern Allies

Fam. 1. **PSILOTACEAE**   Matsuba-ran Ka   Psilotum Family

Frequently epiphytic, herbaceous, erect or pendulous perennials with dichotomously branched creeping rhizomes; leaves (branch-tips) flattened and leaflike or scalelike; sporangia homosporous, solitary and terminal on the branch-tips, rather large and prominent, 2- or 3-locular and 2- or 3-lobed; spores many.——Two genera with about 3 species, in the Tropics and sub-tropics.

1. **PSILOTUM** Sw.   Matsuba-ran Zoku

Rhizomes wiry, creeping, branched; stems simple in the lower half, much forked toward the top into slender flattened or triquetrous branchlets; leaves (branch tips) scalelike, minute, 3- or 2-ranked; sporangia coriaceous, sessile, depressed at apex, 3-locular, 3-lobed, loculicidally split; spores oblong, 1-ribbed.——Two species, of wide distribution, mostly in warmer regions.

1. Psilotum nudum (L.) Beauv. *Lycopodium nudum* L.; *P. triquetrum* Sw.——Matsuba-ran.   Rhizomes 1–2 mm. across, brown-villous; stems 10–35 cm. long, 1–1.5 mm. thick, erect, slender, green, glabrous, loosely scaly, dichotomously forking in upper half; sterile scales entire, the fertile bifid; spor-angia depressed-globose, about 2 mm. across, 3-grooved and obtusely 3-angled.——Epiphytic on tree trunks or on rocky cliffs; Honshu (Kantō Distr. and westw.), Shikoku, Kyushu; rare.——s. Korea, Ryukyus, Formosa, China, and generally in the Tropics.

Fam. 2. **EQUISETACEAE**   Tokusa Ka   Horsetail Family

Perennial herbs with creeping rhizomes; stems green, all alike or of two kinds (the sterile stems green, the fertile without chlorophyll), simple or branched, grooved, with a ring of smaller cavities around the central cylinder; leaves very small, connate and forming a sheath above the nodes, the sheaths with as many teeth as the grooves on stems; spikes terminal, the stipitate-peltate sporangiophores spirally arranged on a common axis; spores all alike, green, provided with 4 elastic, hygroscopic, clavate bands.——A single genus.

1. **EQUISETUM** L.   Tokusa Zoku

Characters of the family.——About 23 species, nearly cosmopolitan, absent in Australia.

1A. Spikes obtuse to subacute, not apiculate; stems annual.
  2A. Stems dimorphic, the fertile whitish or flesh-colored, soft and succulent, the sterile green and branched.
    3A. Internodes nearly smooth or with minute scabrid tubercles; teeth of primary sheaths brown throughout, those on the branches subulate-triangular. .... 1. *E. arvense*
    3B. Internodes minutely spiny on the ridges; teeth of primary sheaths with broad, scarious, white margins, those on the branches broadly deltoid. .... 2. *E. pratense*
  2B. Stems all alike, green.
    4A. Stems spinulose on the ridges; teeth of primary sheaths with very broad membranous margins; branches spreading or drooping.
      5A. Sheaths of branches with erect deltoid teeth; branches simple. .... 2. *E. pratense*
      5B. Sheaths of branches with spreading subulate-linear teeth; branches ramulose, drooping above. .... 3. *E. sylvaticum*
    4B. Stems smooth, with minute tubercules; teeth of primary sheaths with scarious margins; branches ascending, usually simple.
      6A. Primary sheaths loose, with 10 or fewer broadly lanceolate teeth rather prominently scarious on the margin; stems with slender central cavity. .... 4. *E. palustre*
      6B. Primary sheaths tight, with 15–20 subulate-lanceolate teeth scarcely or very narrowly scarious on margin; stems with a large central cavity. .... 5. *E. fluviatile*
1B. Spikes apiculate to obtuse; stems evergreen.
  7A. Teeth caducous.
    8A. Sheaths nearly as wide as long, becoming blackish at both ends, with 12–30 caducous teeth; stems mostly simple. .. 6. *E. hyemale*
    8B. Sheaths longer than wide, pale green, with 5–15 caducous teeth; stems branched at least near base. .... 7. *E. ramosissimum*
  7B. Teeth persistent.
    9A. Stems hollow; teeth elongate, much longer than broad. .... 8. *E. variegatum*
    9B. Stems solid; teeth as broad as long. .... 9. *E. scirpoides*

1. Equisetum arvense L. Sugina.   Sterile stems 20–80 cm. long, erect or decumbent at base, green, 3–4 mm. across, the internodes nearly smooth or with minute scabrid tubercles, 6- to 15-grooved, the central cavity less than half the diameter of the stem; primary sheaths 3–6 mm. long, green, the teeth as many as the grooves, subulate, acuminate, brown throughout; branches spreading, simple, verticillate, (3-)4-grooved, the sheaths pale green, 3- or 4-toothed, the teeth subulate-triangular, acuminate, black-tinged at the tip; fertile stems 10–30 cm. long, smooth, terete, simple, flesh-colored to pale brown, the sheaths 10–25 mm. long, loose, pale green to brownish, 6- to 12-toothed; spikes 2–4 cm. long, pedunculate.——Fruiting Mar.–June. Sunny banks and waste grounds in lowlands to mountains; Hokkaido, Honshu, Shikoku, Kyushu; very common.——Circumboreal.

2. Equisetum pratense Ehrh. Yachi-sugina.   Sterile stems 20–60 cm. long, green, 2–3 mm. across, 8- to 20-grooved, the internodes minutely spiny on the ridges; primary sheaths 3–8 mm. long, pale green, the teeth subulate-deltoid, pale brown, scarious, with a darker colored midrib and white margins; branches horizontally spreading, simple, verticillate, the sheaths pale, with 3 or 4 short, deltoid, acute teeth; fertile stems 10–25 cm. long, simple or with short branches at maturity of the spike, the sheaths 5–10 mm. long, pale yellow-green, with 10–20 lanceolate white-scarious teeth, the midrib brown; spikes 1.5–4 cm. long, pedunculate.——Hokkaido (n. and e. distr.); rare.——Sakhalin, n. Kuriles, Korea, Siberia to Europe, and N. America.

3. Equisetum sylvaticum L. Fusa-sugina.   Sterile stems 30–70 cm. long, erect, 2–4 mm. across, green, rather soft, 10- to 18-grooved, the ridges with 2 rows of minute spinules, the central cavity about half the diameter of the stem; sheaths green at base, the teeth irregularly connate into 3–6 subacute lobes, rufous-brown; branches slender, verticillate, drooping above, loosely spinulose on the angles, ramulose, 3- or 4-grooved, the sheaths with 3 or 4 long, spreading, subulate-linear teeth; fertile stems with short branches; spikes 1–2 cm. long, pedunculate.——Reported to occur in Hokkaido (Shiribeshi Prov.).——Sakhalin, n. Kuriles, n. Korea, Siberia to Europe, and N. America.

4. Equisetum palustre L. Inu-sugina.   Stems 20–50 cm. long, 1.5–3 mm. across, green, usually branched, densely tuberculate, 4- to 10-grooved, the central cavity small; sheaths 5–12 mm. long, loose, usually green, the teeth 10 or fewer, acuminate, broadly lanceolate, blackish, the margins narrowly scarious; branches ascending, simple, 4- or 5-grooved, the sheaths with lanceolate-deltoid appressed brown-tipped acuminate teeth; spikes 1–3 cm. long, pedunculate.——Fruiting May–June. Boggy places; Hokkaido, Honshu, Shikoku, Kyushu; more common northward.——Sakhalin, Kuriles, Korea, Formosa, China, Siberia to Europe, and N. America.

5. Equisetum fluviatile L. *E. limosum* L.; *E. heleo-charis* Ehrh.——Mizu-sugina, Mizu-dokusa.   Stems 50–100 cm. long, erect, 3–10 mm. across, green, simple or with irregular whorls of branches in the middle, smooth, slenderly 10- to 30-grooved, the central cavity 4/5 the diameter of the stem; sheaths 5–12 mm. long, green, tight, the teeth 15–20, small, subulate-lanceolate, acuminate, blackish, scarcely or narrowly scarious on margin; branches ascending, slender, simple, 5-angled, the sheaths green, the teeth 4 or 5, subulate, green, ascending and acuminate; spikes 1–2 cm. long, pedunculate or subsessile.——Fruiting May–July. Boggy places; Hokkaido, Honshu (centr. and n. distr.).——s. Kuriles, Sakhalin, Korea, Siberia to Europe, and N. America.

6. Equisetum hyemale L.   Tokusa.   Stems 20–75 cm. long, erect, firm, simple, evergreen, 3–8 mm. across, shallowly 14- to 26-grooved, scaberulous, the ridges with 2 rows of tubercles, the central cavity about 2/3 the diameter of the stem; sheaths 5.5–14 mm. long, tight, a little longer than broad (3–8 mm. wide), with a black band below the middle and a black border, the teeth 14–26, the apical portion 3–6 mm. long, membranous, dark brown, caducous, the basal portion persistent on the sheath; spikes 6–13 mm. long, sessile.——Fruiting Aug.–Dec. Sandy shaded places in valleys and ravines; Hokkaido, Honshu (n. and centr. distr.).——Sakhalin, Kuriles, Korea, Himalayas, Siberia to Europe, and N. America.

7. Equisetum ramosissimum Desf. *E. ripense* Nakai & F. Maekawa.——Inu-dokusa.   Stems 40–120 cm. long, usually evergreen, minutely tuberculate, gray-green, 8- to 24-grooved, branched at base or near the middle, the central cavity about half the diameter of the stem; sheaths 5.5–21 mm. long, rather loose, pale green, much longer than broad (2.5–8.5 mm. wide), the teeth membranous, 8–24, dark brown with white margins, linear, 3–5 mm. long, caducous, the basal portion persistent on the sheath; branches 6- to 10-grooved; spikes 8–17 mm. long, sessile or nearly so.——Fruiting May–Sept. Honshu, Shikoku, Kyushu; rather common.——Eurasia and Africa.

8. Equisetum variegatum Schleich.   Chishima-hime-dokusa.   Stems slender, ascending, branched at base, simple above, 3- to 12-grooved, the central cavity about 1/3 the diameter of the stem or much reduced or absent in small stems; sheaths green with a black border, the teeth 4–10, lanceolate to lanceolate-deltoid, persistent; spikes 5–10 mm. long.——Hokkaido; rare.——Sakhalin, Kuriles to N. America, and Europe.

9. Equisetum scirpoides Michx. Hime-dokusa.   Stems 5–28 cm. long, tufted, simple or slightly branched at base, green, decumbent or ascending, slender, firm, wiry, 0.5–1 mm. thick, solid, slightly flexuous, 6-ribbed, tuberculate on the ribs; sheaths short, rather loose, green at base, black upwardly, the teeth pale brown, ovate, broadly scarious on margin, with an awnlike excurrent midrib; spikes 2–3 mm. long, sessile or short-pedunculate.——Along valleys in mountains; Hokkaido; rare.——Sakhalin, Siberia to Europe, and N. America.

Fam. 3. **LYCOPODIACEAE**   Hikage-no-kazura Ka   Clubmoss Family

Herbs with small simple 1-nerved evergreen leaves spirally arranged or 4-ranked; sporangia solitary in the axils of the leaves or in terminal spikes in the axils of modified leaves (bracts), 2-valved, homosporous, the spores globose, granular, usually with 3 lines radiating from the apex.——Two genera, *Phylloglossum,* a monotypic genus of Australia and New Zealand, and the following.

1. LYCOPODIUM L.   HIKAGE-NO-KAZURA ZOKU

Stems erect or creeping, much branched; leaves small, imbricate or subverticillate, 4- to 16-ranked, usually all alike, sometimes dimorphic and distichous; sporangia coriaceous, 1-locular, 2-valved, reniform to subglobose.——About 180 species, widely distributed.

1A. Stems erect or drooping, not creeping; sporangia in axils of leaflike or scalelike bracts of main stem.
2A. Plants terrestrial, 3–25 cm. high, usually erect from the decumbent or ascending base, often bearing axillary bulbils in the upper part.
3A. Leaves lanceolate, serrate, costate, petiolelike at base. . . . . .1. L. serratum
3B. Leaves linear-lanceolate, entire, ecostate, not narrowed at base. . . . . 2. L. selago
2B. Plants epiphytic, usually pendulous, sometimes erect, to 70 cm. long, often tufted.
4A. Leaves less than 4 mm. long, nearly all alike.
5A. Leaves broadly subulate, 1.5–3 mm. long, strongly incurved above the middle, 1.5–2 times as long as the sporangia.
. . . . 3. L. fargesii
5B. Leaves triangular-ovate, about 1.5 mm. long, closely appressed to the stem, about as long as the sporangia. . . . . . . 4. L. sieboldii
4B. Leaves 8–20 mm. long, nearly all alike or dimorphic.
6A. Sterile leaves linear to lanceolate; fertile leaves similar but usually slightly smaller, much longer than the sporangia.
7A. Leaves spirally arranged, broadly linear to linear-lanceolate, the sterile leaves about 1.5–2 mm. wide. . . . 5. L. cryptomerinum
7B. Leaves subdistichous, lanceolate, the sterile leaves 2.5–3 mm. wide. . . . . 6. L. fordii
6B. Sterile leaves deltoid-ovate to broadly deltoid-lanceolate, spreading, 6–13 mm. long, 4–6 mm. wide, lustrous, broadest at the base; fertile leaves scalelike, deltoid or deltoid-ovate, as long as or slightly longer than the sporangia. . . . . 7. L. phlegmaria
1B. Stems creeping or long-climbing; sporangia in special fertile terminal spikes on the lateral branches.
8A. Fertile stems unbranched, arising directly from short-creeping, simple or sparsely branched primary stems; sterile branches creeping, not erect.
9A. Leaves (bracts) of fertile stems (peduncles) dense, simulating those of the creeping stems; sporangia subglobose. . 8. L. inundatum
9B. Leaves (bracts) of fertile stems (peduncles) scattered, scalelike, much smaller than those of the creeping stems; sporangia reniform. . . . . .9. L. carolinianum
8B. Fertile stems branched, arising from much elongated primary stems; sterile branches often erect.
10A. Spikes nodding to deflexed.
11A. Stems much elongate, climbing on trees; spikes in groups of 2–10, pedunculate; branchlets much flattened; leaves of sterile branches linear-lanceolate, 1.5–2 mm. long, appressed, with a long hairlike tip; fertile leaves (bracts) obsoletely toothed, with a hyaline hairlike tip. . . . . 10. L. casuarinoides
11B. Stems erect, not twining or climbing, arcuate above; spikes solitary, sessile; branchlets not flattened; leaves of sterile stems linear-subulate, 3–4 mm. long, with an awnlike tip; fertile leaves (bracts) fimbriate, with a long awnlike tip. 11. L. cernuum
10B. Spikes erect.
12A. Leaves spirally to somewhat distichously arranged, all alike or nearly so.
13A. Stems long-creeping below ground; branches erect, unbranched at base, much branched above, of treelike form.
12. L. obscurum
13B. Stems creeping on surface of ground; branches sparsely ramulose near the ascending base, not of treelike form.
14A. Free portion of sterile leaves linear to linear-lanceolate, 3.5–6(–7) mm. long, toothed or the tip filiform; spikes 2–6 cm. long.
15A. Spikes 1–6, long-pedunculate; leaves of sterile branches entire, with a filiform hairlike tip.
13. L. clavatum var. nipponicum
15B. Spikes solitary, sessile; leaves of sterile branches toothed, with a pungent awnlike tip. . . . . 14. L. annotinum
14B. Free portion of sterile leaves subulate, 2–3 mm. long, entire, acuminate; spikes 1–2 cm. long. 15. L. sitchense var. nikoense
12B. Leaves in 4 rows, somewhat dimorphic.
16A. Spikes solitary, 1–2 cm. long, sessile; sterile branches slightly flattened, the ventral leaves trowel-shaped or subulate, the lateral leaves strongly incurved on outer margin, with the free portion about as long as the decurrent base.
16. L. alpinum
16B. Spikes 2–5, 2–3 cm. long, long-pedunculate; sterile branchlets much flattened, the ventral leaves reduced to a minute appressed subulate tip, the lateral leaves not or only slightly incurved on outer margin, with the free portion much shorter than the decurrent base. . . . . 17. L. complanatum

1. Lycopodium serratum Thunb. var. **serratum.** *L. lucidulum* sensu auct. Japon., non Michx.——HOSOBA-TŌGE-HIBA. Stems erect, ascending to shortly decumbent at base, simple or sparsely forked, 7–25 cm. long, striate, bulbils in upper part deep green, consisting of 3 oblong, obtuse, carnose, erect leaves about 5 mm. long; leaves rather thin and loose, spreading to subdeflexed, narrowly to broadly lanceolate, 6–20 mm. long, 1–5 mm. wide, acuminate to short-cuspidate, green, flat, sometimes unequal in size, costate on both sides especially beneath, serrate and sometimes crisped on margin, nearly sessile; sporangia in axils of ordinary leaves, reniform, sessile. ——Woods in mountains; Hokkaido, Honshu, Shikoku, Kyushu; rather common and variable.——Sakhalin, s. Kuriles, Korea, Ryukyus, Formosa, China to India, and Malaysia.

Var. **longepetiolatum** Spring. *L. javanicum* Sw.; *L. serratum* var. *javanicum* (Sw.) Makino; *L. serratum* forma *intermedium* Nakai; *L. serratum* var. *intermedium* (Nakai) Miyabe & Kudō——Ō-TŌGESHIBA, HIROHA-NO-TŌGESHIBA. Leaves broader, distinctly narrowed to a petiolelike base.——Honshu (s. distr.), Shikoku, Kyushu.

2. Lycopodium selago L. var. **appressum** Desv. KO-SUGI-RAN. Stems nearly simple or sparsely forked, erect from a short ascending base, 3–10 cm. long, 4–12 mm. wide including the leaves, with bulbils in upper portion; leaves rather thick and dense, linear-lanceolate, 3–5 mm. long, 0.8–1 mm. wide, ascending, incurved to appressed toward apex, ecostate, acuminate, entire, lustrous, scarcely broadened at base; sporangia reniform, in axils of ordinary leaves.——Mossy sunny slopes in alpine regions; Hokkaido, Honshu (centr. and n. distr.), Kyushu (Yakushima); rather rare.——Kuriles, Kamchatka to Europe and N. America.

Var. **chinense** (Christ) Ohwi. *L. chinense* Christ; *L. selago* forma *chinense* (Christ) E. Pritz.; *L. miyoshianum* Makino; *L. selago* var. *miyoshianum* (Makino) Makino——HIME-SUGI-RAN. Stems 8–15 cm. long, rather slender; leaves broadly linear, 3–7 mm. long, 0.5–0.8 mm. wide, spreading to obliquely ascending, the lower ones sometimes slightly deflexed, slightly incurved above, acuminate.——Coniferous woods in mountains; Hokkaido, Honshu, Shikoku, Kyushu; rather rare.——s. Kuriles, Korea, and China.

Var. **somae** (Hayata) Ohwi. *L. somae* Hayata; *L. chinense* var. *somae* (Hayata) Masam.——KO-SUGI-TŌGESHIBA. Stems slender, often rubescent at base, 3–10 cm. long, 3–5 mm. across including the leaves; leaves smaller, narrowly oblong to narrowly lanceolate, 2–3 mm. long, 0.5–0.8 mm. wide, short-acuminate, deflexed.——Woods in mountains; Kyushu (Yakushima); rare.——Formosa.

3. **Lycopodium fargesii** Hert. *L. tereticaule* Hayata; *L. fauriei* Rosenst.; *L. quasiprimaevum* Koidz.——HIMO-SUGI-RAN, HOSOHIMO-YŌRAKU. Stems slender, tufted, pendulous, 30–40 cm. long, sparingly forked, 2–3.5 mm. across including the leaves, prominently short-striate; leaves rather densely imbricate, broadly subulate, 1.5–3 mm. long, 0.2–0.5 mm. wide, acute, entire, strongly incurved above the middle, obtusely carinate on back at base, slightly concave on inside, more or less adnate to the stem at base; sporangia reniform-orbicular, more than half as long as the fertile leaves.——Kyushu (Yakushima); rare.——Formosa and China.

4. **Lycopodium sieboldii** Miq. HIMO-RAN. Stems tufted, filiform, pendulous, about 1.5 mm. across including the leaves, slightly angular, sparsely forked, 20–40 cm. long, leafy; leaves scalelike, closely appressed and adnate to the stem in lower two-thirds, the free portion triangular-ovate to ovate, about 1.5 mm. long, very acute, about as long as the sporangia, convex and scarcely keeled on back; fertile leaves simulating the scale leaves but more dilated and barely acute to short-mucronate; sporangia reniform-orbicular, as long as to slightly longer than the leaves.——Honshu (Sagami, Izu, and Kii Prov.), Kyushu (s. distr.); rare.——Korea (Quelpaert Isl.), Ryukyus (var.).

5. **Lycopodium cryptomerinum** Maxim. SUGI-RAN. Epiphytic; stems solitary or few together, rather stout, 15–30 cm. long, erect, short-ascending at base, 3–5 mm. across near base, or 2–3 cm. across including the leaves, sparsely forked; sterile leaves coriaceous, rather dense, spirally arranged, broadly linear to linear-lanceolate, 12–18 mm. long, 1.5–2 mm. wide, gradually narrowed and obtusish at tip, shortly adnate to the stem at base, entire, slightly recurved on margin, flat, lustrous above, the midrib broad, obsoletely raised at tip; fertile leaves simulating the sterile, very slightly smaller; sporangia reniform.——Hokkaido (Oshima and Hidaka Prov.), Honshu, Shikoku, Kyushu; rare.

6. **Lycopodium fordii** Bak. *L. hamiltonii* var. *petiolatum* C. B. Clarke; *L. petiolatum* (C. B. Clarke) Hert.; *L. henryi* Bak.; *L. subdistichum* Makino——NANKAKŪ-RAN. Rather soft pendulous epiphyte; stems usually tufted, branched once or twice or simple, rather slender, 20–40 cm. long, 1–1.7 cm. wide (including the leaves) in the sterile part; sterile leaves lanceolate, 8–12 mm. long, 2.5–3 mm. wide, obliquely spreading, subacute, slightly narrowed at the base, entire,

1-nerved, the basal ones small and suberect; fertile leaves nearly like the sterile ones, often smaller and narrower, frequently ascending, 3–10 mm. long, 1–2.5 mm. wide; sporangia cordate, broadly rounded on upper margin.——Honshu (Hachijo Isl. and Kii Prov.), Shikoku, Kyushu; rare.——Ryukyus, Formosa, and China.

7. **Lycopodium phlegmaria** L. YŌRAKU-HIBA. Pendulous epiphyte; stems loosely forked, rather stout, deeply grooved, 40–70 cm. long, to 8 mm. across at base excluding the leaves, or 2–2.5 cm. wide including the leaves; fertile branches terminal, several times forked, 10–25 cm. long, 1.5–2 mm. across; sterile leaves coriaceous, narrowly deltoid-ovate to broadly deltoid-lanceolate, 6–13 mm. long, 4–6 mm. wide, pungent at apex, rounded-truncate at base, very short-petiolate, spreading, lustrous, entire, silghtly recurved on margin, midrib not prominent; fertile leaves (bracts) dense, scalelike, about 1 mm. long and as wide, deltoid or deltoid-ovate, obtuse to rounded at apex, slightly shorter than to slightly longer than the reniform-cordate sporangia.——Kyushu (Yakushima and Tanegashima); rather rare.——Ryukyus and Formosa, and generally throughout the Old World Tropics.

8. **Lycopodium inundatum** L. YACHI-SUGI-RAN. Rather soft, pale green, creeping herb; stems slender, simple or sparingly branched, 5–12 cm. long, about 1.5 mm. across, or 5–10 mm. across including the leaves; leaves linear, soft, spreading somewhat ascending, 4–6 mm. long, 0.5–0.8 mm. wide, acuminate, entire or with few minute teeth, midnerve obsolete; fertile branches arising directly from the creeping stems, simple usually solitary, erect, 4–10 cm. long, leafy; spikes usually solitary, erect, cylindric, 2–4 cm. long, 6–15 mm. wide inclusive of the fertile leaves (bracts), the bracts many, similar to the sterile leaves, spreading to recurved, dilated at base; sporangia subglobose.——Boggy places; Hokkaido, Honshu (Kinki Distr. and northw.); rather rare.——China, Europe, and N. America.

9. **Lycopodium carolinianum** L. *L. subinundatum* Tagawa——INU-YACHI-SUGI-RAN. Similar to the preceding in habit, but firmer on all parts; stems creeping, 5–15 cm. long, simple or sparingly forked, rather densely leaved; leaves subulate-lanceolate or linear, 4–7 mm. long, about 1 mm. wide, acuminate, entire, the lateral ones ascending; scapes (peduncles) arising directly from creeping stems, solitary, simple, erect, 4–7 cm. long, about 1.5 mm. across; scalelike leaves of scape loosely disposed, erect to suberect, subulate, about 3 mm. long; spike terminal, solitary, erect, cylindric, 2–5 cm. long, about 3 mm. wide, densely many-leaved; bracts imbricate, scalelike, deltoid-ovate, about 3 mm. long, scarcely 2 mm. wide, acuminate erose on membranous lower margin, pale yellowish, obliquely spreading, with a broad entire cusp at apex; sporangia reniform.——Wet boggy places; Honshu (Mount Tenganzan in Oomi Prov.); very rare.——N. and S. America, S. Africa, Mauritius, Ceylon, and Malaysia.

10. **Lycopodium casuarinoides** Spring. *L. casuarinoides* var. *japonicum* Nakai——HIMO-ZURU. Climbing on trees; stems much elongate, terete, wiry; leaves of sterile stems scalelike, appressed, linear-lanceolate, convex on back not keeled, 1.5–2 mm. long, with a long hairlike tip 2–4 mm. long; branches rather loosely disposed, 20–40 cm. long, ramified, pendulous, the branchlets much flattened, 1.5–4 mm. wide inclusive of the loosely arranged leaves; leaves of branchlets prominently decurrent, firm, the free portion deltoid and appressed to lanceolate and ascending, 0.5–2 mm. long, ob-

soletely toothed, with a hyaline tardily deciduous apical hair; spikes in groups of 2–10, pedunculate, erect, pedicelled, cylindric, 2–5 cm. long, the bracts ovate-deltoid, 2.5–3 mm. long, undulate-toothed, with a hyaline apical hair.——Pine forests; Honshu (Kii Prov.), Kyushu; very rare.——Formosa, China, Philippines, and Malaysia to n. India.

11. Lycopodium cernuum L. MIZU-SUGI. Stems rather slender, 20–50 cm. long, erect at base, arcuate above, often rooting and bearing new shoots at apex, densely leafy, pale green; branches much ramified, relatively short, slender, somewhat arcuate at the tip; leaves of sterile branches all alike, linear-subulate, entire, spreading to ascending, 3–4 mm. long, 0.2–0.3 mm. wide, attenuate to an awnlike tip; spikes solitary on the branch tips, sessile, nodding, ovoid to ovoid-cylindric, densely bracteate, 5–10 mm. long, about 3 mm. across; bracts deltoid, fimbriate, awnlike at the tip.——Wet sunny slopes in hills and mountains; Hokkaido (Iburi Prov.), Honshu, Shikoku, Kyushu.——Ryukyus, Formosa, and the tropics of both hemispheres.

12. Lycopodium obscurum L. MANNEN-SUGI. Stems long-creeping, subterranean, slender, about 2.5 mm. wide, loosely scaly; branches erect, 15–30 cm. long, simple at base, much branched above, rather loosely to densely leafy; leaves all alike, linear to broadly so, 3–4 mm. long, 0.5–0.7 mm. wide, green, spreading, often incurved at apex, minutely spine-tipped; spikes 1 to several on a branch, solitary and terminal on the branchlets, erect, 2–5 cm. long, the bracts cordate-deltoid, undulate, more or less scarious-margined, with a short cusp at the tip.——Damp coniferous woods in mountains; Hokkaido, Honshu, Shikoku, Kyushu; rather common.——s. Kuriles, Sakhalin, Korea, e. Siberia, and N. America.—— Forma **strictum** (Milde) D. C. Eaton. *L. dendroideum* var. *strictum* Milde; *L. juniperoideum* Sw.; *L. obscurum* var. *dendroideum* (Michx.) D. C. Eaton; *L. dendroideum* Michx.—— TACHI-MANNEN-SUGI. Branchlets short, erect; plant with a treelike aspect.—— Forma **flabellatum** (Milde) Takeda. *L. dendroideum* vor. *flabellatum* Milde——UCHIWA-MANNEN-SUGI Branchlets obliquely spreading, subflabellate.

13. Lycopodium clavatum L. var. **nipponicum** Nakai. *L. japonicum* Thunb.——HIKAGE-NO-KAZURA. Stems extensively creeping, leafy, 3–4 mm. across excluding the leaves, alternately branched; branches forked, ascending, densely leafy, 6–8 mm. across inclusive of the leaves; leaves all alike, spreading, linear to broadly so, 4–5 mm. long, 0.5–1 mm. wide, more or less incurved and filiform at the tip, entire, green, slightly lustrous; spikes erect 1–6, cylindric, 2–6 cm. long; peduncles 7–20 cm. long, erect, loosely leafy, the leaves ascending or suberect, linear, with a long hairlike tip; bracts ovate-deltoid, with a hairlike tip, spreading, the margins narrowly scarious, undulate-erose.——Woods in hills to high mountains; Hokkaido, Honshu, Shikoku, Kyushu; common and very variable. Our plant, sometimes differentiated into 2 or 3 varieties, usually has leaves with a shorter hairlike tip than European plants. ——The species occurs in Sakhalin, Kuriles, Korea, Formosa, China, India, Malaysia to Polynesia and Hawaii, N. America, Europe, and Africa.

14. Lycopodium annotinum L. SUGI-KAZURA. Stems long-creeping, leafy; branches ascending, somewhat forked at base, erect, to 20 cm. long, densely leafy, 8–13 mm. across; leaves of sterile branches spreading to subreflexed, 4–7 mm. long, 0.6–1.3 mm. wide, with an incurved tip, toothed, obsoletely costate on both sides, with a pungent awnlike tip; spikes

solitary, terminal on the branchlets, 2.5–4 cm. long, 4–5 mm. across, sessile, erect; bracts dense, broadly deltoid-ovate, abruptly acuminate with a short deciduous filiform tip, the margins erose-dentate, more or less hyaline.——Coniferous woods: Hokkaido, Honshu, Shikoku, Kyushu; rather common. ——Kuriles, Sakhalin, Kamchatka, Korea to China and the Himalayas, Siberia to Europe, and N. America. Our plants may be separated into the following:

Var. **pungens** Desv. TAKANE-SUGI-KAZURA. Leaves ascending, about 0.5–0.7 mm. wide, incurved at the tip, nearly entire.

Var. **latifolium** Takeda. HIROHA-NO-SUGI-KAZURA. Leaves horizontally to obliquely spreading, lanceolate, usually more than 1 mm. wide, slightly incurved at tip.

Var. **angustatum** Takeda. SHINNO-SUGI-KAZURA. Leaves reflexed, linear-lanceolate, about 1 mm. wide or sometimes narrower.

15. Lycopodium sitchense Rupr. var. **nikoense** (Fr. & Sav.) Takeda. *L. nikoense* Fr. & Sav.; *L. alpinum* var. *nikoense* Fr. & Sav.; *L. sabinaefolium* var. *sitchense* subvar. *nikoense* (Fr. & Sav.) Koidz.;? *L. alpinum* var. *planiramulosum* Takeda——TAKANE-HIKAGE-NO-KAZURA. Stems long-creeping, loosely leafy, about 2 mm. across exclusive of the leaves; branches forked nearly to the base, 3–15 cm. long (inclusive of the fertile branchlets which are usually longer than the sterile), 3–5 mm. across inclusive of the leaves, somewhat glaucescent, ascending at base; leaves of sterile branches ascending to suberect, incurved at tip, subulate, 2–3 mm. long, 0.5–1 mm. wide, acuminate, entire, convex on back, concave inside, not keeled; spikes solitary on the branchlets, yellowish, cylindric, erect, sessile; branchlets 1–2 cm. long; bracts deltoid-orbicular, cuspidate, erose to undulate on margin.——Sunny slopes among mosses and lichens in alpine regions; Hokkaido, Honshu (centr. and n. distr.), Kyushu (Yakushima); rather rare. ——Kuriles. The typical variety occurs in Kamchatka, the Aleutians, and N. America.

16. Lycopodium alpinum L. CHISHIMA-HIKAGE-NO-KAZURA. Stems long-creeping, loosely leafy; branches erect from an ascending base, forked below and becoming flabellate or bushlike, 4–10 cm. long, the branchlets slightly flattened, 2.5–3.5 mm. wide; leaves rather densely disposed in 4 rows, lanceolate-subulate, acutish with a short hyaline tip, the lateral ones more or less flattened, strongly incurved on margin, ascending, the dorsal leaves shorter, convex on back, appressed, the ventral ones trowel-shaped or subulate; spikes 1–2 cm. long, few on each branch, nearly equal in height, sessile; bracts ascending, narrowly deltoid-ovate, hyaline and subentire on margin, acutish.——Alpine slopes; Hokkaido, Honshu (centr. mountains); rare.——Sakhalin, Kuriles, Kamchatka to Siberia and Europe and N. America.

17. Lycopodium complanatum L. *L. complanatum* var. *anceps* Asch.——ASUHI-KAZURA. Stems elongate and long-creeping, loosely leafy; branches ascending, 10–30 cm. long, usually flabellately ramulose; branchlets much flattened, green, 2-grooved above, 3–4 mm. wide; leaves of sterile branches rather loosely disposed in 4 rows, small, long-decurrent and adnate to the axis, spinelike at tip, the lateral ones flattened, the free part 1–1.5 mm. long, ascending to suberect, the dorsal leaves subulate, appressed, adnate most of the length on inner side to the axis, convex on back, the ventral leaves almost wholly adnate, reduced to a minute appressed subulate tip; peduncles erect, 8–13 cm. long, the leaves loose, linear, ascending, often hyaline at the tip; spikes 2–5 on each peduncle, 2–3 cm. long, pedicelled; bracts deltoid-cordate, with a linear hyaline short tip, erose and membranous on margin.——

Mountains; Hokkaido, Honshu (centr. and n. distr.)——Kuriles, Sakhalin, Korea, China, Himalayas, Siberia to Europe, and N. America.

Fam. 4. SELAGINELLACEAE   Iwa-hiba Ka   Selaginella Family

Herbs with long, usually creeping stems; rhizophores leafless, giving rise to adventitious roots; leaves small, 1-nerved, ligulate at base, all alike and spirally arranged or of two sizes arranged in 4 rows; spikes terminal, usually 4-angled, the bracts all alike or dimorphic; sporangia axillary, of two kinds, the larger megasporangia usually borne in lower axils, fewer than the smaller microsporangia borne above.——One genus.

1. SELAGINELLA Beauv.   Iwa-hiba Zoku

Characters of the family.——A cosmopolitan genus comprising about 700 species.

1A. Leaves all alike, spirally arranged.
2A. Leaves spreading, minutely spine-tipped; spikes pedunculate, solitary, the bracts spreading, spirally arranged, 3.5–5 mm. long. .... 1. S. selaginoides
2B. Leaves imbricate, appressed to ascending, hair-tipped; spikes sessile on short ascending branchlets, the bracts in 4 rows.
3A. Branchlets about 1.5 mm. across including the leaves; megaspores with prominent primary reticulations. .... 2. S. sibirica
3B. Branchlets about 1 mm. across including the leaves; megaspores with prominent primary and secondary reticulations. .... 3. S. shakotanensis
1B. Leaves of two sizes, in 4 rows, those of the 2 dorsal rows smaller, appressed, those of the 2 ventral rows larger, spreading to ascending.
4A. Stems densely tufted, 5–12 cm. long; roots confined to the base, forming a short false stem. .... 4. S. tamariscina
4B. Stems not tufted, creeping or ascending.
5A. Main stems without green leaves, shallowly subterranean, wiry, with small scarious scales, unbranched at base, branched in upper part. .... 5. S. pachystachys
5B. Main stems with green leaves, ascending or creeping at ground level, branched from the base.
6A. Stems erect to ascending, with aerial roots in the lower half, deep green above, 6–9 mm. wide. .... 6. S. doederleinii
6B. Stems prostrate, creeping on the ground.
7A. Bracts all alike, loosely arranged, or if two forms present then the larger ones in the same plane as the large leaves, entire, ciliolate, or obsoletely toothed.
8A. Stems long-creeping, loosely leaved, scantily forked. .... 7. S. remotifolia var. japonica
8B. Stems short-creeping, rather densely leaved, prominently forked.
9A. Fertile branches sparsely branched, the leaves not smaller than the bracts; bracts abruptly acuminate; stems and sterile branches 4–6 mm. wide; sterile leaves of upper plane cuspidate. .... 8. S. nipponica
9B. Fertile branches simple or once forked at tip, scapiform, the leaves smaller than the bracts; bracts acute; stems and sterile branches 2.5–3.5 mm. wide; sterile leaves of upper plane obtuse to acutish, not cuspidate. .... 9. S. helvetica
7B. Bracts of two forms, densely arranged, the smaller ones in the same plane as the larger leaves, spine-toothed.
10A. Stems mat-forming, densely leaved, 3–5 mm. wide; spikes 1.5–3 mm. wide. .... 10. S. heterostachys
10B. Stems long-creeping, loosely leaved, 7–8 mm. wide; spikes 2.5–3 mm. wide. .... 11. S. boninensis

1. **Selaginella selaginoides** (L.) Link. *Lycopodium selaginoides* L.; *S. spinosa* Beauv.——Koke-sugi-ran. Stems creeping and appressed to the ground, slender, 3–5 mm. wide, with short ascending branches; leaves multifarious, all alike, spreading, rather loose, more dense in upper part, lanceolate, 2–3 mm. long, acute, minutely spine-tipped, remotely ciliate, the midrib obsolete; fertile stems erect from a creeping base, scapiform, 5–8 cm. long, inclusive of the spike, simple; leaves of fertile stems similar to the sterile ones; spike solitary, erect, 1–2 cm. long, pedunculate, rather densely bracteate; bracts multifarious, similar to the sterile leaves but slightly larger, spreading to ascending, 3.5–5 mm. long, 1–1.5 mm. wide, prominently serrate-ciliate; sporangia globose.——June–July. Alpine slopes; Hokkaido, Honshu (centr. to n. distr.); rather rare.——Kuriles, Kamchatka, Siberia to Europe, and N. America.

2. **Selaginella sibirica** (Milde) Hieron. *S. rupestris* forma *sibirica* Milde; *S. schmidtii* Hieron.——Ezo-no-himo-kazura. Stems rather short, much branched, forming a dense mat, the branchlets ascending; leaves all alike, multifarious, persistent, subulate-linear, about 2 mm. long, 0.5 mm. wide, abruptly acuminate, filiform-tipped, ciliate, gray-green, narrowly and deeply grooved on back, appressed or ascending

and incurved at tip when dried; spikes 4-angled, sessile, the bracts all alike, in 4 rows, narrowly deltoid, 1.5–2 mm. long, short awn-tipped, ciliate; megaspores with prominent primary reticulations.——Rocks; Hokkaido (Hidaka and Kitami Prov. including Rebun and Riishiri Isls.); rare.——Sakhalin, n. Korea, e. Siberia to Alaska.

3. **Selaginella shakotanensis** (Franch.) Miyabe & Kudō. *S. rupestris* sensu auct. Japon., non Spring; *S. rupestris* var. *shakotanensis* Franch.——Himo-kazura. Stems creeping, much branched, forming small dense mats, the branchlets short, ascending, dark gray-green; leaves persistent, all alike, multifarious, subulate-linear, 1–1.3 mm. long, about 0.3 mm. wide, acute, ciliate, appressed or ascending, incurved and white-filiform at tip, narrowly and deeply grooved on back, slightly concave on inner side; spikes as thick as the sterile ones, 4-angled, the bracts narrowly deltoid-ovate, about 1.5 mm. long, ciliate, short awn-tipped; megaspores with prominent primary and secondary reticulations.——Rocks in mountains; Hokkaido, Honshu (Tango, Yamato and Iwashiro Prov. northeastw.); rare.——Sakhalin and s. Kuriles.

4. **Selaginella tamariscina** (Beauv.) Spring. *Lycopodium tamariscinum* Beauv.; *S. involvens* sensu auct. Japon., non Spring (?); *S. veitchii* MacNab; *S. involvens* var. *veitchii*

(MacNab) Bak.——IWA-HIBA. Evergreen; stems densely tufted, densely leaved, ascending to spreading, 5–12 cm. long, 2.5–3 mm. wide, branched, gray-green on upper side, pale green beneath, flat, the branches sparingly forked in one plane; roots all at the base, interwoven to form a simple or rarely few-branched false stem to 15 cm. long; leaves of two forms, in 4 rows, ascending, the larger or lower ones ovate, 1.5–2 mm. long, short-acuminate, filiform-tipped, serrulate, the margins cartilaginous, the smaller or upper leaves slightly smaller, sub-erect, scarcely oblique; spikes slightly narrower than the sterile branchlets, regularly 4-angled, the bracts ovate-deltoid, about 1.5 mm. long, acuminate, awn-tipped, serrulate, green, white-cartilaginous on margin.——Shaded rocks; Hokkaido (Oshima Prov. including Okushiri Isl.), Honshu, Shikoku, Kyushu; rather common.——Korea, Ryukyus, Formosa, Philippines and China to n. India.

5. Selaginella pachystachys Koidz. *S. caulescens* sensu auct. Japon., non Spring; *S. japonica* Moore ex MacNab, non Miq.; *S. caulescens* var. *japonica* (Moore) Bak.——KATA-HIBA. Plant pale green, sometimes rubescent; basal stems slender, shallowly creeping, unbranched at base, branched on upper part, 0.5–0.7 mm. across, whitish, with ovate scarious scales about 1 mm. long; aerial stems 8–20 cm. long, erect from an ascending base, 15–30 cm. long, unbranched in lower half, pinnately much branched above in one plane, flat, frondlike, the branches 3–10 cm. long, 2–2.5 mm. wide, the branchlets 1.5–2 mm. wide, flat, densely leaved; leaves on aerial stems rather loose, all alike in the petiolelike lower part, dimorphic on upper part; leaves of lower plane ovate to broadly so, about 1.5 mm. long, acute, oblique, serrulate-ciliolate at least on the upper side near base, those of the upper plane 1/2–1/3 as long as the lower ones, appressed, ovate, cuspidate, the midrib prominent on upper side; bracts uniform, dense, in 4 rows, deltoid-ovate, about 1 mm. long, cuspidate, minutely serrulate toward base.——Mountains; Honshu (Kantō Distr. and westw.), Shikoku, Kyushu; common.——s. Korea, Ryukyus, Formosa, and (?) China.

6. Selaginella doederleinii Hieron. *S. atroviridis* sensu auct. Japon., non Spring——ONI-KURAMAGOKE. Stems erect to ascending, flexuous, much branched, to 30 cm. long or more, deep green above, pale green beneath, 1.5–2.5 mm. across, or 6–9 mm. wide including the leaves, 3-grooved on upper side, the rhizophores to 10 cm. long on the lower half; branches forked; leaves of lower plane larger, narrowly oblong-ovate, denticulate, slightly oblique, spreading, rounded at base on upper side, those of upper plane appressed, ovate, about 1.5 mm. long excluding the short-awned tip, denticulate; fertile branchlets 4-angled, the bracts all alike, ascending, deltoid-ovate, 1.5–2 mm. long, denticulate, green.——Woods in lowlands and hills; Honshu (Izu Isls.), Kyushu (Koshiki, Yaku-shima, Tanegashima).——Ryukyus, Formosa, China, and Indochina.

7. Selaginella remotifolia Spring var. **japonica** (Miq.) Koidz. *S. japonica* Miq.; *S. kraussiana* sensu auct. Japon., non A. Br.——KURAMAGOKE. Stems long-creeping, loosely leafy, 0.5–0.7 mm. wide, or 5–6 mm. wide inclusive of the leaves; branches short, 1–3 times forked, 3–5 cm. long, prostrate, with ascending tips, densely leafy; leaves of the lower plane horizontally spreading, flat, ovate, about 3 mm. long, about 2 mm. wide, acute or acutish, rounded at base, entire, those of the upper plane appressed, 2–2.5 mm. long, short-acuminate, strongly auriculate at base; fertile branchlets slender, scarcely more than

1 mm. across, 4-angled, the bracts deltoid-lanceolate, suberect, about 1.3 mm. long, gradually acuminate, entire or obsoletely denticulate.——July–Oct. Damp woods in hills and lowlands; Honshu, Shikoku, Kyushu; rather common.——The typical variety occurs in Formosa, China, and Malaysia.

Selaginella uncinata (Desv.) Spring. *Lycopodium uncinatum* Desv.——KONTERI-KURAMAGOKE. Leaves iridescent, those of the upper plane cuspidate. Frequently cultivated in greenhouses.——China.

8. Selaginella nipponica Fr. & Sav. *S. savatieri* Bak.; *S. hachijoensis* Nakai——TACHI-KURAMAGOKE. Delicate herb; stems weak, appressed-creeping, densely leafy, 4–6 mm. wide; leaves of lower plane broadly ovate, 2–2.5 mm. long, spreading or slightly recurved, acute, slightly unequal, obsoletely ciliate, rounded on upper margin near base, those of the upper plane narrowly ovate, about 1 mm. long, cuspidate, denticulate; fertile branches 5–10 cm. long, erect, scapiform, simple to sparsely branched, the bracts abruptly acuminate; leaves on fertile branches all alike or slightly dimorphic, more densely disposed than on sterile branches, ovate; sporangia in axils.——May–July. Damp shaded places in lowlands and hills; Honshu (Kantō Distr. and westw.), Shikoku, Kyushu; common.——Korea and China.

9. Selaginella helvetica (L.) Link. *Lycopodium helveticum* L.; *S. mariesii* Bak.——EZO-NO-HIME-KURAMAGOKE. Delicate herb; stems and branches short, 2.5–3.5 mm. wide, appressed to the ground, forming a loose mat; leaves of lower plane spreading to slightly deflexed, oblong to ovate-oblong, 1.2–1.7 mm. long, about 1 mm. wide, obtuse to subacute, loosely ciliolate, pale green, slightly dilated on upper margin at base, those of upper plane narrowly ovate, obtuse to sub-acute, about 1 mm. long, ascending; fertile branches erect, simple or once forked, scapiform, 4–8 cm. long, the leaves all alike, ovate, 1.2–1.5 mm. long, acute, ciliolate, pale yellow-green, ascending, rather loosely disposed; spikes slender, solitary or in pairs, 1–3 cm. long, the bracts membranous, similar to the leaves of the scape, obliquely spreading.——July–Aug. Alpine slopes; Hokkaido, Honshu, Kyushu; rare.——s. Kuriles, Korea, China, Siberia to Asia Minor, Europe.

10. Selaginella heterostachys Bak. *S. recurvifolia* Warb.; *S. integerrima* sensu auct. Japon., non Spring——HIME-TACHI-KURAMAGOKE. Closely resembling No. 8 in habit, but differing in the more numerous sterile leaves of the upper plane, more rounded at base, the keeled bracts dimorphic, strongly differing from the leaves on the fertile branches, and the more numerous somewhat flattened spikes; leaves of lower plane on branches and stems ovate, 2–3 mm. long, obtusish to acute, oblique at base, white-margined, those of upper plane narrowly deltoid-ovate, rounded at base, nearly equilateral; spikes 7–25 mm. long, 1.5–3 mm. wide, dorsiventral, the bracts of two forms, many, dense, the longer ones deltoid-lanceolate, ciliolate, acuminate, keeled, the smaller ones slightly keeled.——July–Oct. Honshu (Izu Isls., Kinki Distr. and westw.), Shikoku, Kyushu.——Ryukyus, Formosa, China, and Indochina.

11. Selaginella boninensis Bak. *S. longicauda* Warb.; *S. somae* Hayata——HIBAGOKE. Stems long-creeping, prostrate, slender, loosely leafy, 7–8 mm. wide inclusive of the leaves, the branches creeping, simple to thrice forked; leaves of the lower plane membranous, narrowly oblong-ovate, 3–4 mm. long, acute, cordate at base, horizontally spreading, green, toothed, those of the upper plane ovate, 2–2.5 mm. long, cuspidate, rounded at base, appressed; spikes dorsiventral, 2.5–3 mm. wide, the bracts of upper plane larger, obliquely spreading, broadly deltoid-lanceolate, about 2 mm. long, acuminate.

——Reported to occur in Honshu (Hachijo Isl. in Izu Prov.); rare.——Bonins, Formosa, and Philippines.

Fam. 5. ISOËTACEAE   Mizu-nira Ka   Quillwort Family

Small grasslike or sedgelike perennial herbs, aquatic or of wet places; stem a short thick cormlike axis becoming grooved or lobed with age; leaves tufted, linear, with septate air-canals, with 1 central and often 4 or more peripheral vascular bundles, the leaf-base spoon-shaped with a small persistent ligule inside above the cavity; sporangia in the cavity at base of the leaves, covered with a velum, of two kinds, the megasporangia usually on the outer leaves, the microsporangia on inner ones; megaspores tetrahedral, variously sculptured or spined.——One genus, of about 60 widely distributed mostly temperate species, absent or rare in the Tropics.

1. ISOËTES L.   Mizu-nira Zoku

Characters of the family.

1A. Megaspores (under a lens) deeply and subequally alveolate; corms 3-lobed at base; leaves 20–100 cm. long; velum absent; slowly running water of streams and ditches. . . . . 1. *I. japonica*
1B. Megaspores densely spiny all over; corms 2-lobed at base; leaves 5–20 cm. long; velum orbicular to elliptic; shallow water of lakes and ponds. . . . . 2. *I. asiatica*

1. Isoëtes japonica A. Br.   *I. edulis* Sieb. ex Miq.—— Mizu-nira.   Aquatic perennial herb; corms 3-lobed at base, 1–5 cm. across, densely rooting on inner side of the lobes; leaves tufted, many, slender, usually 20–100 cm. long, 2–5 mm. wide at base, gradually narrowed to an obtusish tip, bright green, subterete in transverse section, the phyllopodia white, nearly square to ovate, 4–15 mm. long, 5–12 mm. wide, thinly membranous on margin, closely imbricate; velum absent; ligule thinly membranous, cordate-deltoid, 3–8 mm. long, attenuate-acuminate; megaspores white, globose, with 3 commissures in one hemisphere, deeply and subregularly alveolate.—— Slowly running water of streams and ditches; Honshu, Shikoku, Kyushu; rather rare.——Korea.

2. Isoëtes asiatica (Makino) Makino.   *I. echinospora* var. *asiatica* Makino——Hime-mizu-nira.   Small dark green aquatic herb; corms 5–15 mm. across, 2-lobed at base, densely rooting on inner side of the lobes; leaves 10–30, tufted, erect or ascending, subulate, 5–20 cm. long, 1.5–2.5 mm. wide near the middle, gradually narrowed to an obtusish tip, subterete, with a narrow flat face inside, the phyllopodia deltoid, 4–7 mm. long, 5–10 mm. wide, closely imbricate, membranous on margin, white; velum orbicular to elliptic with a suborbicular to elliptic opening at base extending to the middle; ligule cordate-suborbicular or broadly cordate; megaspores white, globose, marked in one hemisphere with 3 commissures, densely spiny. ——Shallow water of lakes and ponds; Hokkaido, Honshu (centr. and n. distr.); rare.——s. Sakhalin.

Fam. 6. OPHIOGLOSSACEAE   Hana-yasuri Ka   Grapefern Family

Mostly terrestrial (rarely epiphytic) scaleless herbs with short fleshy stems; leaves solitary or few, straight or bent, with a sterile foliar segment at base and an apical spicate or paniculate nonfoliar fertile part; sporangia bivalvate, relatively large, without an annulus.——Four genera, with about 90 species, of wide distribution.

1A. Sporangia connate, forming a simple spike; foliar (sterile) leaf simple, entire or forked, with reticulate veins. . . . . . . 1. *Ophioglossum*
1B. Sporangia free, in panicles; foliar (sterile) leaf often much dissected and toothed, with free veins. . . . . 2. *Botrychium*

1. OPHIOGLOSSUM L.   Hana-yasuri Zoku

Rhizomes erect, fleshy, sometimes tuberous; sterile and fertile leaves arising from a common stipe, the sterile foliar leaf simple or sometimes forked at the tip, entire, reticulately veined, the fertile spike terminal, simple, pedunculate; sporangia marginal, coalescent in 2 ranks, transversely dehiscent.——About 30 species, of wide distribution in both hemispheres.

1A. Epiphytic on tree-trunks, evergreen, with solitary or paired fertile spikes on upper side of an elongate entire or forked ribbonlike sterile frond. . . . . 1. *O. pendulum*
1B. Terrestrial, with a solitary terminal fertile spike and a lower simple, entire sterile blade, the latter sometimes much reduced or absent.
  2A. Sterile blade present; stipes 1–3 together.
    3A. Sterile blade broadly linear to oblong or navicular, narrowly cuneate at base; exospore with rough reticulation. . . . . 2. *O. thermale*
    3B. Sterile blade ovate, elliptic, or nearly orbicular.
      4A. Sterile blade usually distinctly petiolate; exospore with rough reticulation. . . . . 3. *O. petiolatum*
      4B. Sterile blade sessile or short-petiolate; exospore with fine reticulations and the spores apparently smooth under magnification.
                                             4. *O. vulgatum*
  2B. Sterile blade absent; stipes 5–7 together. . . . . 5. *O. kawamurae*

1. Ophioglossum pendulum L.   *Ophioderma pendulum* (L.) Presl——Kobu-ran.   Evergreen pendulous rhizomatous epiphyte; sterile blades tufted, green, rather fleshy, linear to

ribbonlike, 30–80 cm. long, 1.5–3 cm. wide, obtuse to rounded at apex, slightly narrowed toward the base, flat, unlobed or more often 1 or 2 times forked at tip, entire or subentire, margins somewhat whitish, the nerves slender, parallel in lower half, the veinlets anastomosing; peduncle of fertile blade on the upper side of frond near the middle, arcuate, rather stout, thickened above, simple or once forked, 1–4 cm. long; fertile spikes linear, straight, slightly flattened, with a deep groove on each side, 5–10 cm. long, 5–6 mm. wide, obtuse with a small blackish mucro at apex.——Tree trunks; Kyushu (Yaku-shima); rather rare.——Ryukyus, Formosa, Philippines, Malaysia to India, Polynesia, and Australia.

2. **Ophioglossum thermale** Komar. *O. vulgatum* var. *thermale* (Komar.) C. Chr.; *O. littorale* Makino; *O. nipponicum* var. *littorale* (Makino) Nishida ex Ohwi——HAMA-HANA-YASURI. Stipes 7–20 cm. long; sterile blade broadly linear to oblong or navicular, 1–3 (–5) cm. long, 0.3–1 cm. wide, obtuse to acute at apex, narrowly cuneate at base, sessile or indistinctly short-petiolate; venation fine, with small oblong areoles and a few secondary free veinlets.——Apr.–Dec. Wet sandy seashores; Hokkaido, Honshu, Shikoku, Kyushu.——Formosa and Kamchatka.

Var. **nipponicum** (Miyabe & Kudo) Nishida. *O. nipponicum* Miyabe & Kudō; *O. japonicum* Prantl, non Thunb.; *O. angustatum* Maxon; *O. savatieri* Nakai——KO-HANA-YASURI. Sterile blade thinner, usually oblong, sometimes elliptic, more often lanceolate, 2.5–6 cm. long, 1–2 cm. wide, distinctly short-petiolate.——Sunny ground in inland areas; Hokkaido, Honshu, Shikoku, Kyushu.——Korea and China.

3. **Ophioglossum petiolatum** Hook. FUJI-HANA-YASURI. Stipes 8–20(–30) cm. long; sterile blade ovate to broadly so, or oblong, 1.5–6 cm. long, 1–3 cm. wide, usually distinctly petiolate; areoles rather large, without intermediate veinlets.—— Apr.–Nov. Hokkaido, Honshu, Shikoku, Kyushu.——Tropical Asia, Korea, Ryukyus, Formosa, Bonins, Hawaii, Micronesia, and tropical America.

4. **Ophioglossum vulgatum** L. HIROHA-HANA-YASURI. Stipes 15–30 cm. long; sterile blade fleshy and softly herbaceous, ovate to deltoid-ovate, rarely oblong, 6–12 cm. long, 3–7 cm. wide, usually sessile, with a cordate to sometimes truncate base embracing the base of peduncle of the sporophyll, the venation with oblong areoles and sometimes with secondary areoles and numerous secondary connecting veinlets.——Apr.–July. Hokkaido, Honshu, Shikoku, Kyushu.——Widely distributed in the temperate areas of the N. Hemisphere.

5. **Ophioglossum kawamurae** Tagawa. SAKURAJIMA-HANA-YASURI. Small somewhat tufted herb with short erect rhizomes; sterile blade wholly suppressed; fertile spike broadly linear, 3–10 mm. long, cuspidate; peduncle slightly thickened at apex, green.——July–Aug. Kyushu (Sakurajima); rare.

2. BOTRYCHIUM Sw. HANA-WARABI ZOKU

Rhizomes short, erect or ascending, with rather fleshy roots; sterile blade sessile or petiolate, pinnately or ternately divided, the segments toothed, with free veins; fertile blades 1–3 times pinnately divided; ultimate branchlets with 2 rows of naked sporangia; sporangia free, rather coriaceous, globose, mostly sessile, without an annulus, opening transversely into 2 valves.—— A cosmopolitan genus with about 30 or more species.

1A. Small glabrous herbs, usually not more than 20 cm. high, of alpine regions; sterile blade simply pinnate or ternate, not more than 8 cm. long.
  2A. Sterile blades oblong to ovate, simply pinnate to ternate, the pinnae oblong, obovate, or flabellate, obtuse.
    3A. Sterile blade sessile or nearly so, pinnate, inserted at or above the middle of the stipe, the pinnae 7–11. . . . . 1. *B. lunaria*
    3B. Sterile blade petiolate, inserted near the top of the stipe; pinnae fewer. . . . . 2. *B. simplex* var. *tenebrosum*
  2B. Sterile blades deltoid, sessile, the lowest pair of pinnae largest, the blade apparently ternately divided, the pinnae lanceolate, acute.
                                                            3. *B. lanceolatum*
1B. Larger herbs more or less hairy on the buds, usually 15–60 cm. high or more, of mountains and lowlands; sterile blade ternate, the segments 2-pinnate or ternately 2-pinnate, 3–30 cm. long.
  4A. Sterile blade petiolate, usually evergreen.
    5A. Segments of the sterile blade obtuse, the ultimate segments mucronate- to crenate-toothed.
      6A. Fertile stalk without a sterile frond or scale of the previous year at base; sterile segments scarcely to very slightly white-margined. . . . . 4. *B. ternatum*
      6B. Fertile stalk with a sterile frond or white membranous scale of previous year at base; sterile segments more or less white-margined.
        7A. Plant 15–30 cm. high; sterile blade 3- or 4-ternate, 5–10 cm. long, 6–12 cm. wide. . . . . 5. *B. robustum*
        7B. Plant 8–13 cm. high; sterile blade 2- or rarely 3-ternate, 2.5–4 cm. long, 3–5 cm. wide. . . . . 6. *B. multifidum*
    5B. Segments of the sterile blade acute, the ultimate segments mucronate- to aristulate-toothed.
      8A. Sterile blade aristulate-toothed; lowest posterior lobe of the pinnules of sterile blade flabellate-orbicular, decurrent on the rachis of the main segment beyond the base of the pinnules. . . . . 7. *B. japonicum*
      8B. Sterile blade mucronate-toothed; lowest posterior lobe of the pinnules of sterile blade obliquely cuneate and sometimes only slightly decurrent, not entirely so on the rachis of the main segment. . . . . 8. *B. nipponicum*
  4B. Sterile blade sessile, deciduous.
    9A. Fertile panicle once pinnate, linear or cylindric, scarcely broadened at base, scarcely or not overtopping the sterile blade; pinnae with an accessory lobe on the posterior side below the insertion of the pinnae. . . . . 9. *B. strictum*
    9B. Fertile panicle compound, pyramidal, broadened at base, longer than and overtopping the sterile blade; pinnae without an accessory lobe on the posterior side below the insertion of the pinnae. . . . . 10. *B. virginianum*

1. **Botrychium lunaria** (L.) Sw. *Osmunda lunaria* L. ——HIME-HANA-WARABI. Glabrous rather fleshy herb 5–25 cm. high; stipe solitary, covered with brown membranous slightly fibrous sheaths at base; sterile blade at or above the middle of the stipe, simply pinnate, 3–6 cm. long, 1.2–2.2 cm. wide, sessile or nearly so; pinnae 7–11, flabellate, reniform, or broadly cuneate, 7–13 mm. long, 5–15 mm. wide, truncate to cuneate and entire on lower margin, crenulate or rarely incised on upper margin, palmately veined, short-petiolulate; fertile spike 1.5–6 cm. long, on a peduncle twice as long.——July–

Sept. Sunny alpine slopes; Hokkaido, Honshu (centr. and n. distr.); rare.——Korea, Kuriles, Sakhalin, Kamchatka, Siberia, Himalayas, Europe, N. and S. America.

2. **Botrychium simplex** E. Hitchc. var. **tenebrosum** (A. A. Eaton) Clausen. *B. tenebrosum* A. A. Eaton——KOKE-HANA-WARABI. Delicate glabrous herb, 4–6 cm. high (in ours); stipes very slender; sterile blade obovate, sometimes ovate, 3–5 mm. long, 2–3 mm. wide, usually ternate (in ours), the petiole about 1–2 mm. long, the segments obovate or cuneate, obscurely undulate on upper margin, few-nerved; fertile part consisting of a short slender peduncle with few pairs of sporangia.——Aug. Mossy forests in high mountains; Honshu (Mount Yatsugatake in Shinano); very rare.——e. N. America and Europe (Austria).

3. **Botrychium lanceolatum** (Gmel.) Ångstr. *Osmunda lanceolata* Gmel.; *B. palmatum* Presl; *B. rutaceum* var. *tripartitum* Ledeb.——MIYAMA-HANA-WARABI. Rather fleshy glabrous herb 6–20 cm. high, with brown membranous withering sheaths at base; sterile blade below the fertile panicle, sessile, deltoid, 2–3 cm. long and as wide, pinnately divided, the segments linear to lanceolate, pinnately lobed to entire, sessile, obtuse, the lowest lateral segments pinnately veined, 1.5–2 cm. long; fertile panicle rather dense, pyramidal, subsessile or very shortly pedunculate, usually not longer than the sterile blade. ——Aug. Alpine regions; Honshu (centr. distr.); very rare. ——Sakhalin, n. Kuriles, Kamchatka, Siberia to Europe, and N. America.

4. **Botrychium ternatum** (Thunb.) Sw. *Osmunda ternata* Thunb.——FUYU-NO-HANA-WARABI. Rather fleshy evergreen herb 15–40 cm. high, very sparsely long-hairy or glabrate; stems naked at base, the bud hidden, hairy; sterile blade thick-herbaceous, 5-angled in outline, 3- or 4-ternate, 4–15 cm. long, 5–20 cm. wide, glabrous or nearly so, the primary and secondary segments petiolulate, ovate or ovate-deltoid, obtuse, the ultimate segments ovate, lobed, crenate- to mucronate-toothed, obtuse, the petiole rather stout, arising immediately above the base, 3–15 cm. long; fertile panicle long-pedunculate, erect, pyramidal, 4–10 cm. long, dense.——Sept.–Nov. Thickets and grassy places in lowlands and low mountains; Honshu, Shikoku, Kyushu; rather common.——Ryukyus, Formosa, China to the Himalayas.

5. **Botrychium robustum** (Rupr.) Underw. *B. rutaefolium* var. *robustum* Rupr.; *B. multifidum* (Gmel.) Rupr. var. *robustum* (Rupr.) C. Chr.; *B. matricariae* sensu auct. Japon., non Spring——EZO-FUYU-NO-HANA-WARABI. Rather fleshy herb 15–30 cm. high, thinly long-hairy or rarely glabrate; stems with a whitish basal sheath or scale, or with last year's sterile blade persistent; sterile blade thick-herbaceous, 5-angled in outline, 3- or 4-ternate, 5–10 cm. long, 6–12 cm. wide, petiolate, the segments obtuse, the ultimate ones ovate, lobed, crenulate- to mucronate-toothed, the petiole arising immediately above the base; fertile panicle long-pedunculate, pyramidal, dense, 4–8 cm. long.——Aug.–Oct. Hokkaido, Honshu (mountains of centr. and n. distr.).——Korea, Manchuria, Ussuri, Sakhalin, Kuriles, Kamchatka to Alaska.

6. **Botrychium multifidum** (Gmel.) Rupr. *Osmunda multifida* Gmel.; *B. matricariae* (Schrank) Spreng.; *O. matricariae* Schrank; *B. rutaceum* Sw.; *B. rutaefolium* A. Br.—— AZUSA-HANA-WARABI. Similar to the preceding but smaller, 8–13 cm. high, very thinly white-hairy to glabrescent; stems usually with a whitish membranous basal scale, sometimes with last year's sterile blade persistent; bud white-hairy; sterile blade 3- or 5-angled in outline, 2 or 3 times ternate, 2.5–4 cm. long,

3–5 cm. wide, the segments deltoid-ovate, obtuse, the ultimate segments ovate, obtuse, pinnately lobed, crenate-toothed, the petiole 3–6 cm. long, arising immediately above the base; fertile panicle rather dense, pyramidal, 2–3 cm. long, the peduncle 7–12 cm. long.——Aug. Mountains; Honshu (Shinano Prov.); very rare.——Europe to Siberia, and N. America.

7. **Botrychium japonicum** (Prantl) Underw. *B. daucifolium* var. *japonicum* Prantl——Ō-HANA-WARABI. Stout slightly hairy herb 25–40 cm. high; stems naked at base (rarely with an old persistent frond attached); sterile blade herbaceous, 5-angled in outline, petiolate, 8–15 cm. long, 10–20 cm. wide, 3 or 4 times ternate, the pinnae deltoid or deltoid-ovate, acute to subacuminate, petiolulate, the pinnules oblong-ovate to broadly lanceolate, pinnately lobed to parted, acuminate, aristulate-toothed, the lowest basal lobe flabellate-orbicular, the petioles 5–15(–20) cm. long, stout, green to brownish or yellowish, arising 3–10 cm. above the base of the stem; fertile panicle 7–15 cm. long, yellow-brown to brown, the peduncle erect, rather stout, slightly longer to twice as long as the sterile blade.—— Sept.–Nov. Honshu (Kantō Distr. and westw.), Shikoku, Kyushu.

8. **Botrychium nipponicum** Makino. AKA-HANA-WARABI. Similar to the preceding but more fleshy and red-brown tinged, very thinly hairy or glabrous; stems naked at base; sterile blade 5-angled in outline, 8–15 cm. long, 10–20 cm. wide, 3 or 4 times ternate, the pinnae and pinnules petiolulate, acute to acuminate, obtuse at the tip, the ultimate segments deltoid-lanceolate, pinnately lobed to parted, mucronate-toothed, the lowest basal lobes decurrent beyond the base of the segment but not inserted on the main rachis, the petioles 7–12 cm. long, arising 2–5 cm. above the base of the stem; fertile panicle 3–15 cm. long, narrowly pyramidal, twice as long as the sterile blade.——Oct.–Nov. Honshu (Kantō Distr.); rare.

9. **Botrychium strictum** Underw. *Osmundopteris stricta* (Underw.) Nishida——NAGABO-NO-HANA-WARABI, NAGABO-NO-NATSU-NO-HANA-WARABI. Deciduous, thinly hairy to nearly glabrous herb, 30–80 cm. high; stems terete, rather stout, slightly thickened, commonly with a membranous ovate entire brown scale 2–3 cm. long at base; sterile blade sessile, ternately divided, the main lateral branches ovate, 15–30 cm. long, 10–20 cm. wide, bipinnate, short-acuminate, short-petiolate, the rachis narrowly winged; pinnae acuminate, sessile, the pinnules oblong-ovate, toothed, pinnatiparted, the terminal primary segment deltoid; fertile panicle linear-cylindric, 10–30 cm. long, scarcely or not overtopping the sterile blade.——Aug.–Oct. Hokkaido, Honshu, Shikoku, Kyushu; rare.——Korea and China.

10. **Botrychium virginianum** (L.) Sw. *Osmunda virginiana* L.; *Osmundopteris virginiana* (L.) Small; *B. gracile* Pursh——NATSU-NO-HANA-WARABI. Deciduous, thinly hairy rather soft herb, 30–60 cm. high; stems with a brown membranous scale 2–3 cm. long at base; sterile blade sessile, ternately divided, green, the main lateral branches ovate, 10–20 cm. long, 5–15 cm. wide, short-acuminate, short-petiolulate, bipinnate; pinnae lanceolate to oblong-lanceolate, the pinnules oblong to narrowly ovate, pinnately lobed and toothed, sessile or somewhat adnate on the prominently winged rachis, the primary terminal segment deltoid; fertile panicle ovate to broadly lanceolate, 7–13 cm. long, compound, longer than and overtopping the sterile blade.——May–June. Woods; Hokkaido, Honshu, Shikoku, Kyushu; rather rare.——Korea, Manchuria, China, Himalayas, Siberia to Europe, and N. America.

Fam. 7. **MARATTIACEAE**   Ryūbintai Ka   Marattia Family

Terrestrial; fronds with circinnate vernation, compound, joined on the rhizome or short stem by the enlarged stipulelike base of the stipe; pinnae jointed at base; sporangia in elongate or round dorsal sori, the sporangia separate, derived from several cells, opening by a ventral longitudinal split.——Four genera, with about 150 species, widely distributed in the Tropics of both hemispheres.

1. **ANGIOPTERIS** Hoffm.   Ryūbintai Zoku

Large ferns with short stout stems; fronds bipinnate, the veins free; sori dorsal, near the margin, the sporangia usually 7 to 13, in 2 rows, contiguous but not coherent.——Madagascar to Polynesia and Japan. A complex genus represented by more than 100 species according to some authors or reduced to a single species by others.

1. Angiopteris lygodiifolia Rosenst. *A. fauriei* Hieron.; *A. oschimensis* Hieron.; *A. sakuraii* Hieron.; *A. evecta* sensu auct. Japon., non Hoffm.; *A. suboppositifolia* sensu auct. Japon., non De Vriese——RYŪBINTAI. Nearly glabrous, with a short stout stem or rhizome; fronds tufted, to 1.5 m. long or sometimes longer; stipes stout, fleshy, terete, green, about 2.5 cm. thick above the base, the basal portion dilated, obovate-orbicular, rounded on back, with a semirounded auricle on each side at top; blades oblong-ovate to ovate, bipinnate, to 1 m. long, 50–60 cm. wide; pinnae 3–6 pairs, alternate, the rachis swollen and jointed at base; pinnules 15–25 pairs, alternate, narrowly to broadly lanceolate, 5–10 cm. long, 1–2.5 cm. wide, acuminate, broadly cuneate to subtruncate at base, subsessile, minutely toothed, green above, paler beneath, the costa prominent on both sides, the veinlets many, parallel, spreading, distinct, usually once forked; sori oblong, about 1.5 mm. long, with 7–13 sessile ellipsoidal sporangia, each with an opening on the inner side, solitary on the veinlets near the margin of the pinnae.——Woods in low mountains; Honshu (Izu Isls., Izu Peninsula, Kii and Ise Prov.), Shikoku, Kyushu; rarer eastw.——Ryukyus and Formosa.

Fam. 8. **OSMUNDACEAE**   Zenmai Ka   Osmunda Family

Terrestrial with erect or ascending large scaleless rhizomes; fronds with circinnate vernation, pinnately divided, the veins free; sporangia large, maturing simultaneously, the annulus incomplete, with thick cells on one side near the tip, dehiscent by a vertical slit across the apex.——Three genera and about 20 species, 2 of the genera confined to S. Africa, Australia, and New Guinea; *Osmunda* of wide distribution.

1. **OSMUNDA** L.   Zenmai Zoku

Fronds dimorphic; fertile pinnules terminal, medial, or occupying the whole blade; sporangia marginal on much-reduced pinnules.——Species about 15, nearly cosmopolitan, absent in Australia.

1A. Sterile blade ovate, bipinnate; pinnae and pinnules few, the pinnules lanceolate to narrowly ovate, minutely toothed.
  2A. Pinnules of sterile blade obliquely truncate to rounded at base, sessile, chartaceous to membranous. . . . . 1. *O. japonica*
  2B. Pinnules of sterile blade acuminate to cuneate at base, sometimes indistinctly petiolulate, coriaceous to chartaceous. . . . . 2. *O. lancea*
1B. Sterile blade lanceolate, pinnate; pinnae many, coarsely acute-toothed or pinnatiparted.
  3A. Fronds deciduous, chartaceous to herbaceous; pinnae pinnatiparted, nearly truncate and sessile or subsessile at base.
    4A. Fertile pinnae on separate fronds, red-brown; sterile pinnae green, paler beneath, acuminate; woolly hairs red-brown. 3.*O. asiatica*
    4B. Fertile and sterile pinnae on the same blade, the fertile medial; sterile pinnae slightly glaucescent especially beneath, acute; woolly hairs of young plants light brown. . . . . 4. *O. claytoniana*
  3B. Fronds evergreen, coriaceous, lustrous; pinnae coarsely acute-toothed, long-acuminate, narrowed to a petiolelike base; fertile pinnae medial. . . . . 5. *O. banksiifolia*

1. Osmunda japonica Thunb. *O. regalis* sensu auct. Japon., non L.; *O. regalis* var. *japonica* (Thunb.) Milde; *O. regalis* var. *biformis* Benth.; *O. biformis* (Benth.) Makino; *O. nipponica* Makino——ZENMAI. Rhizomes short, ascending, stout, covered with bases of old withered leaves, the woolly hairs cinnamon-brown, mixed with blackish ones, soon deciduous; fronds dimorphic, sterile fronds to 1 m. long, tufted; stipes stramineous, smooth, with a chartaceous wing on each side near base; blades ovate or triangular-ovate, bipinnate, glabrous or nearly so; pinnae 20–30 cm. long, oblong-ovate, the lower ones short-petiolulate; pinnules oblong to broadly lanceolate, 4–10 cm. long, 1–2.5 cm. wide, obtuse to acute, obliquely truncate and rounded at base, minutely toothed, green above, glaucescent beneath, sessile, the costa distinct, penninerved, the veinlets spreading, parallel, 1–3 times forked; fertile fronds 20–50 cm. long, rising among the sterile fronds, the sporophylls paniculate, erect, rather loose, cinnamon-brown.——Apr.–May

(–Oct.). Hokkaido, Honshu, Shikoku, Kyushu; common.——s. Sakhalin, Korea, Ryukyus, Formosa, China to the Himalayas.

2. Osmunda lancea Thunb.   *O. japonica* Houtt., non Thunb.; *Osmundastrum lanceum* (Thunb.) Presl——YASHA-ZENMAI. Closely resembling the preceding but somewhat smaller and less densely woolly; sterile fronds to 1 m. long; pinnules slightly thicker, linear-lanceolate to broadly lanceolate, acuminate to cuneate at base, petiolulate (sessile in var. **lati-pinnula**. Tagawa).——Apr.–May. Hokkaido (s. distr.), Honshu, Shikoku, Kyushu; rather rare.

3. Osmunda asiatica (Fern.) Ohwi. *O. cinnamomea* sensu auct. Japon., non L.; *O. cinnamomea* var. *fokiensis* Copel.; *O. cinnamomea* var. *asiatica* Fern.; *Osmundastrum cinnamomeum* var. *fokiense* (Copel.) Tagawa——YAMADORI-ZENMAI. Rhizomes stout; woolly hairs red-brown, mixed with some blackish ones in the fertile areas, soon deciduous except at base of pinnae; sterile fronds tufted, erect, green; stipe shorter than the blade, flattened and dilated near base; blades oblong-lanceolate, 30–60 cm. long, 12–20 cm. wide, abruptly contracted and short-acuminate at apex, not or very shortly narrowed at base; pinnae many, horizontally spreading, linear-lanceolate, 16–23 mm. wide, acuminate, pinnatiparted, sessile on the posterior side, truncate and subsessile on the anterior side; segments rather numerous, oblong-ovate, 4–7 mm. wide, sub-entire, obtuse to rounded at the tip, 1-nerved, the veinlets slender, usually once forked, pinnately arranged; fertile blades several, central, erect, strict, narrow, the pinnae ascending, red-brown, soon withering after spore dissemination.——May–July. Wet places; Hokkaido, Honshu, Shikoku, Kyushu; locally abundant.——s. Kuriles, Sakhalin, Amur, Ussuri, Manchuria, Korea, China, and Formosa.

4. Osmunda claytoniana L. *Osmundastrum claytonianum* (L.) Tagawa——ONI-ZENMAI. Closely allied to the preceding, the woolly hairs light brown; sterile blades acute; pinnae somewhat glaucescent beneath; fertile pinnae olive-brown to blackish in interrupted pairs in the middle portion of the blade; sterile blades 15–25 cm. wide; pinnae 2–3 cm.

wide, the segments 6–8 mm. wide.——May–July. Wet places; Honshu (centr. distr.); locally abundant.——Korea, Manchuria, Ussuri, and e. N. America; a variety occurs in Formosa and sw. China to the Himalayas.

5. Osmunda banksiifolia (Presl) Kuhn. *Nephrodium banksiaefolium* Presl; *Plenasium banksiaefolium* (Presl) Presl; *O. javanica* sensu auct. Japon., non Bl.——SHIROYAMA-ZENMAI. Rhizomes stout, erect or ascending; woolly hairs scanty, red-brown, early deciduous; sterile fronds evergreen, 1–1.5 m. long, tufted; stipes lustrous, brownish, with a wing on each side near base; blades oblong-lanceolate, simply pinnate; pinnae 10–20 pairs, coriaceous, linear-lanceolate, 15–25 cm. long, 1–2 cm. wide, gradually long-acuminate, narrowed to a petiolelike base, coarsely acute-toothed, obliquely spreading, jointed at base, except the terminal one, lustrous, glabrous, slightly thickened and whitish on the margin, the costa and veinlets raised on both sides, the veinlets 1–4 times forked, parallel, obliquely spreading; fertile pinnae medial, linear-cylindric, 7–10 cm. long, dark brown.——June–Dec. Wet places; Honshu (Izu Prov.), Shikoku, Kyushu.——Ryukyus, Formosa, s. China, Indochina, and Malaysia.

Fam. 9. **SCHIZAEACEAE**   KANI-KUSA KA   Climbing Fern Family

Terrestrial; rhizomes creeping or ascending; fronds often scandent and much elongate; sporangia marginal, sometimes apparently superficial, with a complete distal annulus, opening by a longitudinal slit.——Four genera, with about 150 species, chiefly in the S. Hemisphere, 1 in our area.

1. **LYGODIUM** Sw.   KANI-KUSA ZOKU

Rhizomes creeping, hairy; rachis of fronds much elongate, twining; pinnae alternate, pinnately or palmately divided, short-petiolulate; veins usually free; sporangia subtended by an outgrowth simulating an indusium, in 2 rows on margin of the ultimate segments; spores tetrahedral.——About 40 species, in New Zealand, S. Africa, Asia, Malaysia, Polynesia, and 1 species in e. N. America.

1. Lygodium japonicum (Thunb.) Sw. *Ophioglossum japonicum* Thunb.; *Hydroglossum japonicum* (Thunb.) Willd.——KANI-KUSA. Scandent; rhizomes creeping, 3–4 mm. across, densely covered with soft blackish hairs about 1 mm. long; rachis much elongate, twining, 2-striate, pale brown, lustrous; pinnae alternate, with a pair of pinnate segments and a hairy bud at the tip of a short petiolule, the segments deltoid to ovate, ternately 1- or 2-pinnate, with the rachis puberulent in upper part; sterile pinnae thinly chartaceous, often 3- to

5-lobed to -parted, minutely toothed, the terminal lobe or segments elongate deltoid- to linear-lanceolate, to 60 mm. long, 4–8 mm. wide, acute to obtuse; veins obliquely ascending, once to thrice forked, the veinlets ending in the marginal teeth; sporangia in 2 closely adjacent rows on margin beneath the ultimate segments.——Aug.–Jan. Thickets and hedges in lowlands and low mountains; Honshu (Kantō Distr., Shinano, Etchu Prov. and westw.), Shikoku, Kyushu; common.——Ryukyus, Formosa, Korea and China.

Fam. 10. **GLEICHENIACEAE**   URAJIRO KA   Gleichenia Family

Terrestrial ferns with long-creeping rhizomes; fronds often falsely forked by the abortion of the terminal bud, the lateral branches usually again dichotomous, the ultimate branches pinnatifid or pinnate-pinnatifid; veins free, pinnate, simple or forked; sori exindusiate, dorsal, the sporangia 4–15, with a complete oblique annulus, opening by a longitudinal slit.——Six genera, with about 130 species, of wide distribution in the Tropics and subtropics.

1A. Plant with jointed hairs on rhizomes and buds; pinnae pinnatiparted. .... 1. *Dicranopteris*
1B. Plant with flattened paleae on rhizomes and buds; pinnae bipinnatiparted. .... 2. *Gleichenia*

1. **DICRANOPTERIS** Bernh.   KO-SHIDA ZOKU

Rhizomes with jointed hairs; fronds evergreen, the primary rachis elongate, the lateral branches several times dichotomous, usually with a pair of foliaceous stipulelike outgrowths at the base of each fork, the secondary rachises naked, only the ultimate branches (and the stipulelike branches) leafy; veins 2–4 times forked; sori superficial, usually with 6 or more sporangia.——A pantropic genus with about 10 species, also in New Zealand.

1. Dicranopteris linearis (Burm. f.) Underw. *Polypodium lineare* Burm. f.; *Gleichenia linearis* (Burm. f.) C. B. Clarke; *P. pedatum* Houtt.; *P. dichotomum* Thunb.; *D. dicho-*

toma (Thunb.) Bernh.; *G. dichotoma* (Thunb.) Hook.; *D. linearis* var. *dichotoma* (Thunb.) Holttum——KO-SHIDA. Rhizomes firm, terete, wiry, 3–4 mm. across, with sparse brown jointed hairs while young; stipes pale brown, terete; fronds sometimes more than 1 m. long, firm; ultimate branches (pinnae) chartaceous, linear- to oblong-lanceolate, pinnatiparted, 15–30 cm. long, 3–7 cm. wide, gradually acute, obtuse at the tip, sessile, green above, glaucous beneath; segments many, nearly horizontally spreading, linear, 3–4 mm. wide, obtuse to retuse, entire, slightly hairy beneath, the hairs jointed, brown, soon deciduous, pinnately veined, the veins 2–4 times forked, slightly raised on both sides; sori small, rounded, solitary near the middle portion of the uppermost vein.——Oct.–Jan. Dry thin woods and thickets; Honshu (Iwaki and Echigo Prov. westw.), Shikoku, Kyushu; common. ——Korea, Ryukyus, Formosa, China, and widely distributed in the Tropics and subtropics, also in Australia.

2. GLEICHENIA J. E. Smith   Urajiro Zoku

Rhizomes long-creeping, firm, scaly while young; fronds large, stipitate, pinnate or falsely forked; pinnae bipinnatiparted; veins once forked; sori superficial, with about 4 sporangia.——About 10 species, Malaysia to India and Japan, 1 in tropical America.

1A. Scales ciliate; pinnules nearly horizontally spreading, glaucous and glabrous beneath; ultimate segments narrowly oblong, obtuse; stipes terete. .... 1. *G. glauca*
1B. Scales entire; pinnules obliquely spreading, slightly glaucous with scattered glandular hairs beneath; ultimate segments broadly linear, cuspidate; stipes flattened and with 2 longitudinal ridges to the base. .... 2. *G. laevissima*

1. Gleichenia japonica Spreng. *Polypodium glaucum* Houtt.; *G. glauca* (Houtt.) Hook., non Sw.; *Dicranopteris glauca* (Houtt.) Underw.; *Diplopterygium glaucum* (Houtt.) Nakai; *Hicriopteris glauca* (Houtt.) St. John——Urajiro. Rhizomes long-creeping, terete, 5–6 mm. across, wiry, densely scaly while young; stipes terete; scales lanceolate-deltoid, 6–7 mm. long, caudate, ciliate, lustrous, dark brown; fronds scaly while very young, the dormant terminal bud densely scaly, rufous to dark brown, ciliate; pinnae chartaceous, opposite, 50–100 cm. long, 20–30 cm. wide, bipinnatiparted, green, slightly lustrous above, glaucous beneath, glabrous; pinnules many, horizontally spreading, linear-lanceolate to broadly linear, 15–30 mm. wide, sessile, pinnatiparted, the ultimate segments spreading, narrowly oblong, obtuse; costa impressed above, the costules slender; veins once forked; sori median; on the anterior branches of the veins.——Nov.–Jan. Dry, thin woods; Honshu (Iwaki and Echigo Prov. westw.), Shikoku, Kyushu; common.——Korea, Ryukyus, Formosa, China, the Tropics of Asia, Australia, and Polynesia.

2. Gleichenia laevissima Christ. *G. kiusiana* Makino; *Dicranopteris laevissima* (Christ) Nakai; *Diplopterygium laevissimum* (Christ) Nakai; *Hicriopteris laevissima* (Christ) St. John——Kaneko-shida. Closely allied to the preceding; scales brown to yellow-brown, membranous, entire, glabrous; stipes with a narrow flat face and 2 longitudinal ridges to the base; pinnules obliquely spreading, the ultimate segments broadly linear, acute or cuspidate, green above, slightly glaucous and with short subclavate minute glandular hairs beneath.——Oct.–Nov. Kyushu; rare.——China, Indochina, and Philippines.

Fam. 11. HYMENOPHYLLACEAE   Koke-shinobu Ka   Filmyfern Family

Epiphytic or terrestrial ferns; rhizomes slender, long-creeping, nearly naked or with jointed hairs; fronds usually small, thin, without stomata, usually with circinnate vernation, very rarely straight in extremely reduced fronds; veins nearly always free; sporangia in marginal sori elevated with a slender receptacle surrounded by a cup-shaped sometimes 2-valved indusium; annulus complete; spores tetrahedral to globose.——About 34 genera, with about 650 species, chiefly in the Tropics and subtropics, few in colder regions.

1A. Indusium 2-valved; rhizomes very scantily hairy; fronds fulvous to brownish green.
  2A. Ultimate segments of the blades entire. .... 1. *Mecodium*
  2B. Ultimate segments of the blades toothed. .... 2. *Hymenophyllum*
1B. Indusium cup-shaped to tubular, entire or shallowly 2-lobed at mouth; rhizomes prominently hairy; fronds green.
  3A. Rhizomes long-creeping, remotely leafy.
    4A. False veinlets absent.
      5A. Fronds pinnately compound. .... 3. *Vandenboschia*
      5B. Fronds palmately lobed to parted. .... 4. *Gonocormus*
    4B. False veinlets present. .... 5. *Crepidomanes*
  3B. Rhizomes short, erect to ascending, densely leafy. .... 6. *Selenodesmium*

1. MECODIUM Presl   Koke-shinobu Zoku

Epiphytic or growing among mosses on rocks; rhizomes very slender, long-creeping, thinly hairy to subglabrate; fronds usually thin, pinnately compound, the ultimate segments entire, 1-veined; indusium 2-valved, usually nearly to the base, the receptacle included.——A pantropic genus comprising about 100 species.

1A. Stipes prominently winged.
  2A. Plants usually 10–30 cm. high; ultimate segments of frond 1–2 mm. wide; wings of stipes entire to slightly crispate-undulate.
    3A. Ultimate segments of blades 1–1.5 mm. wide; main rachis 1.5–2 mm. wide inclusive of the wings; fronds narrowly to broadly lanceolate. .... 1. *M. flexile*
    3B. Ultimate segments of blades about 2 mm. wide; main rachis 2–3 mm. wide inclusive of the wings; fronds oblong to ovate-oblong.
                                                  2. *M. badium*
  2B. Plants 5–10 cm. high; ultimate segments of frond 0.5–1 mm. wide; wings of stipes strongly crispate-undulate. .... 3. *M. riukiuense*

1B. Stipes not or scarcely winged.
    4A.   Stipes and nerves on underside of blades and rachis with yellow-brown flexuous hairs; plants about 5 cm. high. .. 4. *M. oligosorum*
    4B.   Stipes and blades glabrous or nearly so; plants taller.
        5A.   Pinnae disposed at an acute angle on the rachis, simple or the segments 2–5, elongate, 1.2–1.7 mm. wide. ...... 5. *M. wrightii*
        5B.   Pinnae disposed at an obtuse angle on the rachis, the segments narrower, 0.7–1 mm. wide.
            6A.   Fronds 10–25 cm. long; stipes narrowly winged. .... 6. *M. polyanthos*
            6B.   Fronds 2–5 cm. long; stipes not winged. .... 7. *M. paniculiflorum*

1. Mecodium flexile (Makino) Copel.   *Hymenophyllum flexile* Makino——Ō-KOKE-SHINOBU, NACHI-KOKE-SHINOBU. Rhizomes long-creeping, filiform, remotely leafy; fronds 10–30 cm. long; stipes winged except at base; blades narrowly to broadly lanceolate, 7–20 cm. long, 1.5–3(–4) cm. wide, usually narrowed at tip, 2- or 3-pinnatiparted, the rachis 1.5–2 mm. wide inclusive of the entire or slightly crispate wing; pinnae rather numerous, narrowly ovate, somewhat oblique, pinnately to bipinnately parted, the pinnules 1–1.5 mm. wide; sori few on each pinna, the valves orbicular to slightly depressed, entire, 1.5–2 mm. across.——Among mosses; Honshu (Izu Prov. through Tōkaidō to Kinki Distr.), Shikoku, Kyushu.——Formosa.

2. Mecodium badium (Hook. & Grev.) Copel.   *Hymenophyllum badium* Hook. & Grev.; *H. crispatum* sensu auct. Japon., non Wall.——ONI-KOKE-SHINOBU.   Plant very sparingly pilose, mostly while young; rhizomes long-creeping, 0.5–0.7 mm. across; fronds oblong to ovate-oblong, 15–30 cm. long, 4–7 cm. wide; stipes 5–10 cm. long, 2–3 mm. wide inclusive of the entire sometimes slightly crispate wings; blades ovate to narrowly so, usually narrowed at tip, 3-pinnatiparted, the rachis winged, 2–3 mm. wide; pinnae obliquely spreading, usually somewhat elongate; ultimate pinnules or segments entire, broadly linear, about 2 mm. wide, obtuse; indusia depressed-orbicular, the valves obsoletely toothed on upper margin or entire, 2–3 mm. wide.——Among mosses on rocks and on tree trunks; Honshu (Kii Prov.), Shikoku, Kyushu; rare.——Formosa, China to Malaysia, and India.

3. Mecodium riukiuense (H. Chr.) Copel.   *Hymenophyllum riukiuense* H. Chr.——RYUKYU-KOKE-SHINOBU, OKINAWA-KOKE-SHINOBU.   Rhizomes creeping, nearly filiform; fronds 5–10 cm. long, 17–30 mm. wide; stipes at first with an entire, soon crispate-undulate wing on both sides; blades narrowly ovate, sometimes somewhat elongate and narrowed at tip, 3-pinnatiparted, the rachis winged; ultimate pinnules or segments narrow, linear, 0.5–1 mm. wide, obtuse; valves of indusia ovate to elliptic, few-toothed on upper margin, about 1 mm. wide.——Shikoku, Kyushu (s. distr.); rare.——Ryukyus.

4. Mecodium oligosorum (Makino) H. Itō.   *Hymenophyllum oligosorum* Makino——KIYOSUMI-KOKE-SHINOBU. Rhizomes, rachis, and costa of the fronds rufous-pubescent; rhizomes capillary, long-creeping, sparsely branched; fronds small, about 5 cm. long; stipes filiform, 1–3 cm. long, not winged; blades broadly to narrowly ovate, 2–4 cm. long, about 1.5 cm. wide, obtuse, bipinnately parted, the rachis winged; pinnae obliquely spreading, palmately to pinnately parted, the ultimate segments obtuse, entire; indusia orbicular, about 1.5 mm. wide, the valves obsoletely few-toothed on upper margin.——Among mosses on rocks in damp woods; Honshu (Kantō Distr. and Mino Prov. westw.), Shikoku, Kyushu.

5. Mecodium wrightii (v. d. Bosch) Copel.   *Hymenophyllum wrightii* v. d. Bosch——KOKE-SHINOBU.   Delicate, very thinly hairy, especially while young; rhizomes capillary, branched, forming a loose mat; fronds small; stipes 7–20 mm. long, glabrous, narrowly winged in upper part; blades deltoid-ovate to narrowly so, 1.5–3 cm. long, 1–1.5 cm. wide, bipinnatiparted, the rachis winged; pinnae rather few, ascending, divided or simple in the upper ones; ultimate segments more or less elongate, 1.2–1.7 mm. wide, entire, retuse or sometimes rounded at apex; indusia orbicular to broadly ovate, the valves 1.2–1.5 mm. wide, entire or obsoletely bilobed on upper margin.——Damp woods in mountains; Hokkaido, Honshu, Shikoku, Kyushu.——Sakhalin and Korea.

6. Mecodium polyanthos (Sw.) Copel. *Trichomanes polyanthos* Sw.; *Hymenophyllum polyanthos* Sw.; *H. blumeanum* Spreng.; *H. integrum* v. d. Bosch; *H. fujisanense* Nakai——HOSOBA-KOKE-SHINOBU,   HOSOBA-HIME-KOKE-SHINOBU. Fronds 10–25 cm. long; rhizomes filiform to capillary, hairy, more especially while young; stipes 2–10 cm. long, narrowly winged, with a tuft of dark brown straight hairs at base; blades broadly lanceolate to ovate, 5–17 cm. long, 2–4 cm. wide, 2- or 3-pinnatiparted, the rachis narrowly winged; pinnae usually many; pinnules short, entire or few-parted; ultimate segments narrow, usually elongate, about 0.7–1 mm. wide, obtuse; indusia broadly ovate to suborbicular, about 1 mm. long, the valves entire, obtuse to rounded at apex.——Honshu (Iwashiro and westw.), Shikoku, Kyushu; more common southward.——Ryukyus, Formosa, widely distributed in the tropics and subtropics.

7. Mecodium paniculiflorum (Presl) Copel. *Hymenophyllum paniculiflorum* Presl——HIME-KOKE-SHINOBU, FUJI-KOKE-SHINOBU.   Closely allied to the preceding; rhizomes capillary, branched, thinly brown-hairy at the tip; fronds 2–5 cm. long, nearly glabrous; stipes capillary, 1–2 cm. long; blades ovate to oblong-ovate, 1.5–3.5 cm. long, 1.2–1.5 cm. wide, obtuse, the rachis winged; pinnae ovate to ovate-deltoid, 2- to 9-parted, obliquely spreading, the segments emarginate, about 1 mm. wide; indusia orbicular, the valves entire, about 1 mm. wide.——Mountains; Honshu (centr. and Kantō Distr.); rare.——sw. China, Malay Peninsula, and Malaysia.

2. HYMENOPHYLLUM J. E. Smith   KŌYA-KOKE-SHINOBU ZOKU

Epiphytic or terrestrial, with slender creeping scantily hairy rhizomes; fronds small, thin, pinnately compound, the ultimate segments toothed; false veinlets absent; indusia 2-valved nearly or to the base; receptacle not or only slightly exserted.——About 25 species, of wide distribution, except in cold regions.

1A.   Fronds 3-pinnatiparted, ovate to broadly lanceolate, sparingly brownish hairy on veins and costas on underside and on stipes; ultimate segments about 1.2–1.7 mm. wide. .... 1. *H. barbatum*
1B.   Fronds 2-pinnatiparted, lanceolate, nearly glabrous; ultimate segments 2–3 mm. wide. .... 2. *H. simonsianum*

1. Hymenophyllum barbatum (v. d. Bosch) Bak. *H. japonicum* Miq.; *Leptocionium barbatum* v. d. Bosch.—— Kōya-koke-shinobu.   Costas beneath, stipes, and rhizomes with red-brown crisped rather coarse hairs especially while young; rhizomes filiform, loosely branched, long-creeping; fronds 3-pinnatiparted, 6–12 cm. long, ovate to broadly lanceolate; stipes scarcely thicker than the rhizomes, 2–5 cm. long, wingless or with a very narrow deciduous wing on each side; blades broadly to narrowly ovate, the sterile obtuse, the fertile usually more or less elongate, 3–8 cm. long, 2–3 cm. wide; pinnae ovate, once or sometimes twice pinnatiparted; ultimate segments 1.2–1.7 mm. wide, rounded to obtuse at apex, spinulose; sori usually in upper portion of blade; indusia about 1.2 mm. wide, broadly ovate, the valves distinctly toothed on upper margin.——Honshu (Ugo Prov. and southw.), Shikoku, Kyushu.——Korea (Quelpaert Isl.) and China.

2. Hymenophyllum simonsianum Hook. Shimon-koke-shinobu.   Plant slender, nearly glabrous; rhizomes filiform, creeping; stipes remote, slender, 1.5–4 cm. long; fronds 2-pinnatiparted; blades broadly lanceolate to linear-oblong, 4–10 cm. long, 15–25 mm. wide, often narrowed at tip, the rachis winged; pinnae flabellate to rhombic-cuneate, shallowly parted, the lobes elliptic, obtuse, spinulose-toothed; ultimate segments 2–3 mm. wide; sori in upper part of blade; indusia oblong, the valves prominently toothed on margin.—— Reported to occur in Kyushu (Yakushima); rare.——Formosa to India.

3. VANDENBOSCHIA Copel.   Hai-horagoke Zoku

Usually epiphytic, sometimes terrestrial with hairy elongate rhizomes; fronds pinnately compound, thin; ultimate segments 1-veined, without false veinlets; indusia cylindric to cup-shaped, with an entire mouth, the receptacle slender, elongate.——More than 25 species, mostly tropical, but also in N. America, southernmost S. America, S. Africa, and Europe.

1A. Blades lanceolate to narrowly so, subsessile or very short-stiped, simply pinnate. . . . . 1. *V. auriculata*
1B. Blades ovate to broadly lanceolate, long-stiped, bipinnatiparted.
  2A. Stipes distinctly winged at top; fronds 10–20 cm. long, the ultimate segments 0.5–0.7 mm. wide. . . . . 2. *V. radicans*
  2B. Stipes scarcely winged; fronds 1.5–4 cm. long, the ultimate segments 0.8–1 mm. wide. . . . . 3. *V. titibuensis*

1. Vandenboschia auriculata (Bl.) Copel. *Trichomanes auriculatum* Bl.——Tsuru-horagoke.   Rhizomes long-creeping, wiry, 1.5–2 mm. across, while young with dark brown, lustrous, jointed, straight hairs 1.5–2 mm. long; fronds rather remote; blades lanceolate to narrowly so, 15–35 cm. long, 3–5 cm. wide, very short-stiped or subsessile, simply pinnate, the rachis with dark brown spreading hairs while young, often bisulcate by virtue of the very narrow upturned wings; pinnae many, rhombic-ovate, 15–30 mm. long, 6–13 mm. wide, obtuse, broadly cuneate at base, auriculate on anterior side at base, the pinnules parallel, entire, irregularly toothed at the truncate apex; veinlets ascending to suberect; indusia exserted, about 1.7 mm. long, tubular, narrowed at base, truncate at apex.——Tree trunks; Honshu (Izu Isls. and Kii Prov.), Shikoku, Kyushu.——Ryukyus, Formosa, China to India, and Malaysia.

2. Vandenboschia radicans (Sw.) Copel. var. **orientalis** (C. Chr.) H. Itô.   *Trichomanes orientale* C. Chr.; *T. japonicum* Fr. & Sav., non Thunb.; *T. amabile* Nakai——Hai-horagoke.   Plant nearly glabrous; rhizomes slender, 0.5–1 mm. across, with dark brown lustrous jointed hairs, remotely leaved; fronds 10–20 cm. long; stipes 4–7 cm. long, winged; blades more or less 3-angled, narrowly ovate to broadly lanceolate, 8–15 cm. long, 3–6 cm. wide, the rachis narrowly winged; pinnae bipinnatiparted; pinnules linear, usually 0.5–0.7 mm.

wide, obtuse; indusia 1.2–1.5 mm. long, exserted, tubular, narrowed at base, the mouth somewhat dilated and truncate.—— Honshu (Awa, Izu Isls. and westw.), Shikoku, Kyushu; rather common.——Ryukyus and Formosa.

Var. **nipponica** (Nakai) H. Itô. *Trichomanes nipponicum* Nakai——Hime-hai-horagoke.   Rhizomes 0.2–0.5 mm. across; fronds 5–10 cm. long; blades 3–8 cm. long.——Hokkaido, Honshu, Kyushu.——Korea (Quelpaert Isl).

Var. **naseana** (Christ) H. Itô. *Trichomanes naseanum* H. Chr.——Ryukyu-kogane, Ō-hai-horagoke.   Rhizomes 1–1.5 mm. across; fronds 20–40 cm. long; blades 15–30 cm. long; indusia tubular, not distinctly constricted or dilated above, 1.5–2 mm. long.——Honshu (Izu and Kii Prov.), Kyushu.——Ryukyus and Formosa. The typical variety is pantropic.

3. Vandenboschia titibuensis H. Itô. Chichibu-horagoke.   Rhizomes long-creeping, 0.2–0.3 mm. across, with dark brown hairs, remotely leaved; fronds 1.5–4 cm. long; stipes 2–10 mm. long, scarcely winged; blades bi- or partially subtripinnatiparted, oblong-ovate, the rachis winged; pinnules rather close or somewhat imbricate, linear, entire, subacute, 0.8–1 mm. wide; indusia short, tubular-campanulate, entire, dilated above, slightly recurved on margin, 2-winged nearly to the top, about 2 mm. long.——Honshu (Kantō to Kinki Distr.), Shikoku, Kyushu; rare.

4. GONOCORMUS v. d. Bosch   Uchiwagoke Zoku

Usually epiphytic, with freely branched filiform or capillary rhizomes sometimes proliferous; fronds very small, palmately lobed to parted, glabrous, with flabellate venation; false veinlets absent; indusia immersed, elongate with an entire dilated mouth; receptacle extruded.——Seven species or perhaps more, Africa to Japan, Queensland to Hawaii.

1. Gonocormus minutus (Bl.) v. d. Bosch. *Trichomanes minutum* Bl.; *T. parvulum* sensu auct. Japon., non Poir. ——Uchiwagoke.   Plant very small; rhizomes capillary, branched and intricate, dark brown villous; fronds interrupted; stipes 1–2 cm. long, about as thick as the rhizomes, pale green, nearly glabrous on the upper half, densely dark brown hairy at base; blades flabellate, 5–12 mm. long, 7–15 mm. wide, cordate to truncate, rarely broadly cuneate at base, subglabrous, palmately and irregularly parted to lobed, the segments linear, 0.5–1 mm. wide, entire, very obtuse; indusia few, tubular, dilated at apex, entire, obscurely papillose on margin.——Hokkaido, Honshu, Shikoku, Kyushu.——Korea, Manchuria, Formosa, China to Malaysia.

5. CREPIDOMANES Presl　　Ao-horagoke Zoku

Usually epiphytic; rhizomes filiform; fronds small, usually pinnately compound, rarely digitately divided by reduction, glabrous, the ultimate segments entire, with intramarginal or irregular false veinlets (striae); indusia obconical to trumpet-shaped, winged, with an extruded receptacle from a bilobulate mouth.——About 12 species, Madagascar to Japan and Polynesia.

1A. Ultimate segments of frond obtuse. . . . . 1. *C. latealatum*
1B. Ultimate segments of frond very acute to subacuminate. . . . . 2. *C. makinoi*

1. Crepidomanes latealatum (v. d. Bosch) Copel. *Trichomanes latealatum* v. d. Bosch; *T. bipunctatum* sensu auct. Japon., non Poir.——Ao-horagoke. Rhizomes filiform, 0.2–0.3 mm. across, loosely branched, densely blackish hairy; fronds 3–10 cm. long; stipes 7–25 mm. long, winged, with blackish hairs at base; blades narrowly ovate, 2–7 cm. long, 1.5–2 mm. wide, nearly glabrous on both sides, the rachis winged; pinnae sometimes subbipinnately parted, the ultimate segments 0.7–1 mm. wide, entire, obtuse; indusia obovoid, about 1.5 mm. long, winged on lower half of both margins, bilobed to the middle, the lobes semirounded-deltoid, very obtuse, entire.——Honshu (Uzen and Rikuzen Prov. southw.), Shikoku, Kyushu; rather rare.——Ryukyus, Formosa to Malaysia.

2. Crepidomanes makinoi (C. Chr.) Copel. *Trichomanes makinoi* C. Chr.; *T. acutum* sensu auct. Japon., non Poir.——Koke-horagoke. Rhizomes capillary, branched, densely dark brown short-hairy; fronds 2–4 cm. long; stipes 3–15 mm. long, winged at the top; blades broadly ovate to broadly lanceolate, 15–25 mm. long, prominently cuneate at base, bipinnately or sometimes pinnately parted, the rachis winged; pinnae obliquely ascending, 2- to 5-parted, the segments linear-lanceolate, 1–1.5 mm. wide, acute to subacuminate, entire; indusia few, winged only at base, obovoid, nearly 2 mm. long, 2-lobed to the middle, the lobes deltoid, acutish to obtuse, entire or undulate.——Honshu (Kantō Distr. and westw.), Shikoku, Kyushu; rare.

6. SELENODESMIUM Copel.　　Oni-horagoke Zoku

Terrestrial, with stout short-creeping to erect rhizomes; stipes elongate with short deciduous bristlelike hairs; blades pinnately compound; pinnules pinnatifid, the cell walls typically thick and coarsely pitted; indusia cylindric, with an entire mouth, the receptacle extruded.——About 19 species, in the Tropics and New Zealand.

1. Selenodesmium obscurum (Bl.) Copel. *Trichomanes obscurum* Bl.; *T. cupressoides* sensu auct. Japon., non Sw.; *T. rigidum* sensu auct. Japon., non Sw.——Oni-horagoke. Rhizomes short-creeping or ascending, 2–3 mm. across, rather densely leafy, with blackish lustrous straight hairs about 2 mm. long; fronds tufted, erect, 15–30 cm. long; stipes wiry, terete, not winged, with blackish hairs especially toward the base, dark brown, 1–1.2 mm. across; blades dark green, narrowly ovate, 7–17 cm. long, 3.5–7 cm. wide, 3-pin-natiparted, the rachis wingless, with short scurfy hairs and scattered multicellular jointed hairs; pinnae rather numerous, broadly lanceolate, narrowed at tip; pinnules oblong, pinnately parted, the ultimate segments linear, 0.2–0.4 mm. wide, obtuse to subacute, suberect; indusia narrowly tubular, about 2 mm. long, shallowly bilobed at apex, the lobes deltoid-rounded, ciliolate.——Kyushu (Yakushima); rare.——Ryukyus, Formosa to India and Malaysia.

Fam. 12. PTERIDACEAE　　I-no-moto-so Ka　　Pteris Family

Usually terrestrial; rhizomes creeping, ascending, or sometimes nearly erect; fronds usually pinnate, not jointed with the rhizome; sori typically marginal and indusiate or exindusiate, the reflexed margin of the pinnules often enclosing the sori, or the sori rarely along the veins and exindusiate; sporangia sometimes covering the entire undersurface of the pinnules, opening by a transverse slit through a definite stomium; annulus longitudinal and interrupted in most genera.——A cosmopolitan family, with about 63 genera and about 1,500 species.

1A. Indusia present.
　2A. Indusium opening exteriorly.
　　3A. Ultimate pinnules not dimidiate.
　　　4A. Ultimate pinnules not cuneiform.
　　　　5A. Sori marginal. . . . . 1. *Dennstaedtia*
　　　　5B. Sori intramarginal. . . . . 2. *Microlepia*
　　　4B. Ultimate pinnules usually cuneate at base. . . . . 5. *Sphenomeris*
　　3B. Ultimate pinnules dimidiate. . . . . 4. *Lindsaea*
　2B. Indusium consisting of the reflexed margin of the segment.
　　6A. Rhizomes hairy, not scaly.
　　　7A. Sori not continuous along the margin. . . . . 6. *Hypolepis*
　　　7B. Sori continuous along the margin. . . . . 7. *Pteridium*

6B. Rhizomes scaly, sometimes also hairy.
  8A. Sori protected by but not borne on the indusium.
    9A. Veinlets joined in the sori.
      10A. Ultimate pinnules 3–20 mm. wide; marginal indusia not in contact with the adjacent side by the strongly reflexed margins.
        11A. Rhizomes long-creeping, solenostelic; fronds remote. . . . . 8. *Histiopteris*
        11B. Rhizomes short, dictyostelic; fronds tufted. . . . . 9. *Pteris*
      10B. Ultimate pinnules 1–1.5 mm. wide; marginal indusia strongly reflexed, the adjacent indusia often in contact along the midrib beneath, especially when young. . . . . 14. *Onychium*
    9B. Veinlets free.
      12A. Fronds often dimorphic; alpine. . . . . 13. *Cryptogramma*
      12B. Fronds all alike; hills and lowlands.
        13A. Fronds farinose beneath; rachis nearly smooth. . . . . 12. *Aleuritopteris*
        13B. Fronds not farinose beneath; rachis with 2 raised prominently scaly ridges on upper side. . . . . 11. *Cheilanthes*
  8B. Sori borne on the reflexed margin of the pinnules. . . . . 16. *Adiantum*
1B. Indusia absent.
  14A. Sorus rounded, not elongate.
    15A. Blades membranous, glabrous, often with bulbils in axils of the pinnae or proliferous at the tip. . . . . 3. *Monachosorum*
    15B. Blades herbaceous, pilose, without bulbils, not proliferous at the tip. . . . . 6. *Hypolepis*
  14B. Sorus elongate along the vein, superficial.
    16A. Fronds 3–7 cm. long, the blades hairy. . . . . 15. *Pleurosoriopsis*
    16B. Fronds 70–130 cm. long, the blades glabrous or thinly hairy. . . . . 10. *Coniogramme*

1. DENNSTAEDTIA Bernh.   Koba-no-ishi-kaguma Zoku

Terrestrial; rhizomes creeping, hairy, solenostelic; fronds large, pinnately compound, hairy or naked, the pinnules oblique; veins free; sori marginal, usually in the sinuses at the ends of the veinlets; indusia fused with a minute tooth of the frond to form an entire or slightly bivalved sometimes deflexed cup; sporangia slender-stalked.——About 70 species, of wide distribution, mainly in the Tropics.

1A. Fronds deciduous, somewhat dimorphic, broadly lanceolate, bipinnatiparted or bipinnate; lowest pinnae not or only slightly larger than the others, less than 10 cm. long.
  2A. Fronds glabrous. . . . . 1. *D. wilfordii*
  2B. Fronds pubescent. . . . . 2. *D. hirsuta*
1B. Fronds evergreen, all alike, narrowly deltoid, 3- or 4-pinnate; lowest pinnae largest, 10–30 cm. long. . . . . 3. *D. scabra*

1. Dennstaedtia wilfordii (Moore) Koidz. *Microlepia wilfordii* Moore; *Davallia wilfordii* (Moore) Bak.; *Coptidipteris wilfordii* (Moore) Nakai & Momose; *Davallia nipponica* Miq.; *M. nipponica* (Miq.) C. Chr.——Ō-REN-SHIDA. Rhizomes slender, creeping, sometimes stolonlike, 1–1.5 mm. across, with brownish multicellular jointed hairs while young; fronds slightly dimorphic, glabrous, the sterile ones usually smaller than the fertile; stipes 3–20 cm. long, pale green in upper part, purple-brown or dark brown and lustrous toward base, grooved on upper side; blades membranous, broadly lanceolate, 7–25 cm. long, 2–6 cm. wide, acuminate to acute; fertile pinnae deltoid-ovate, acute or obtuse in the sterile pinnae, short-petiolulate; ultimate pinnules usually cuneate at base, broadly lanceolate to ovate in the fertile ones, ovate to broadly ovate in the sterile, obtusely toothed; sori marginal; indusia cup-shaped, membranous, glabrous, entire, about 1 mm. long and as wide.——June–Oct. Hokkaido, Honshu, Shikoku, Kyushu.——Korea, Manchuria, Ussuri, and China.

2. Dennstaedtia hirsuta (Sw.) Mett. *Davallia hirsuta* Sw.; *Humata hirsuta* (Sw.) Desv.; *Microlepia hirsuta* (Sw.) Diels, non Presl; *Trichomanes japonicum* Poir., non Thunb.; *Davallia pilosella* Hook.; *M. pilosella* (Hook.) Moore; *Fujiifilix pilosella* (Hook.) Nakai & Momose——INU-SHIDA. Plants soft-pubescent throughout; rhizomes 1–2 mm. across, short-creeping; stolons elongate, remotely leaved; fronds 7–35 cm. long; stipes 3–15 cm. long, stramineous, grooved on the upper side; fertile blades herbaceous, 15–30 cm. long, 4–8 cm. wide, acuminate; pinnae broadly deltoid-lanceolate to narrowly deltoid-ovate, obtuse to subacute, nearly sessile, pinnatiparted to

nearly pinnate, slightly unequal at base; ultimate pinnules oblong-ovate, 3–10 mm. long, obtuse, acute- to mucronate-toothed; sori marginal; indusia cup-shaped, pale brown, prominently hispid on the underside, green above, about 0.7 mm. across.——May–Oct. Hokkaido, Honshu, Shikoku, Kyushu; common.——Korea, Manchuria, and China.

3. Dennstaedtia scabra (Wall.) Moore. *Dicksonia scabra* Wall. ex Hook.; *Dicksonia deltoidea* Hook.; *Dennstaedtia deltoidea* (Hook.) Moore——KOBA-NO-ISHI-KAGUMA. Evergreen; rhizomes long-creeping, 2–3 mm. across, hairy at least while young; fronds more or less scabrous; stipes reddish brown to dark brown-purple, prominently scabrous at base, 15–40 cm. long, somewhat shiny, grooved on upper side; blades narrowly deltoid, 20–45 cm. long, 15–35 cm. wide, acuminate, green above, slightly paler beneath, 3- or 4-pinnate, the rachis brownish, sparsely hispid on both sides; pinnae deltoid-lanceolate to broadly lanceolate, 10–30 cm. long, 4–10 cm. wide, acuminate, short-petiolate; ultimate pinnules oblong-ovate, 7–15 mm. long, 3–8 mm. wide, obtuse, oblique at base, sessile, pinnately lobed to toothed, thinly hispid on nerves of both sides; sori marginal; indusia cup-shaped, about 0.7 mm. across, the lower side membranous, glabrous, entire, the upper margin more or less incurved.——Mar.–Oct. Honshu (Kantō Distr. and Echigo Prov. westw.), Shikoku, Kyushu.——Formosa, China to Malaysia and India.

Var. **glabrescens** (Ching) Tagawa. *D. glabrescens* Ching ——USUGE-KOBA-NO-ISHI-KAGUMA. Nearly glabrous.——Distributed with the species.

2. MICROLEPIA Presl   FUMOTO-SHIDA ZOKU

Terrestrial, usually hairy, with creeping solenostelic rhizomes; fronds moderate to large, pinnate to decompound, the ultimate pinnules obliquely incised; veins free; sori intramarginal, terminating the veinlets; indusia half cup-shaped, glabrous to hispid, the receptacle short; spores tetrahedral.——More than 45 species, chiefly in the Tropics and subtropics of the Old World.

1A. Indusia prominently setose, often ciliate.
  2A. Blades pinnate; rachis uniformly pubescent on both sides. . . . . 1. *M. marginata*
  2B. Blades bipinnate; rachis glabrous on upper side except upper part. . . . . 2. *M. pseudostrigosa*
1B. Indusia glabrous or with 1 or 2 long hispid hairs near base.
  3A. Blades bipinnate; pinnae 1.5–3 cm. wide. . . . . 3. *M. strigosa*
  3B. Blades tripinnate; pinnae 5–10 cm. wide.
    4A. Pinnae lanceolate, the lower about 5 cm. wide; pinnules obtuse to acute. . . . . 4. *M. substrigosa*
    4B. Pinnae ovate-deltoid, the lower 5–10 cm. wide; pinnules long-acuminate. . . . . 5. *M. yakusimensis*

1. **Microlepia marginata** (Houtt.) C. Chr. *Polypodium marginatum* Houtt.; *P. marginale* Thunb., non L.; *Dicksonia marginalis* (Thunb.) Sw.; *Davallia marginalis* (Thunb.) Baker; *M. marginalis* (Thunb.) Bedd.——FUMOTO-SHIDA. Fronds evergreen, pilose; rhizomes creeping, 4–7 mm. across, thinly hairy; stipes 30–50 cm. long, more or less scabrous, dull purple-brown; blades broadly lanceolate, 30–60 cm. long, 15–30 cm. wide, gradually acuminate, pinnate, the rachis uniformly pubescent on both sides, dull purpie-brown; pinnae chartaceous-herbaceous, linear-lanceolate, 1.5–3 cm. wide, pinnatilobed, acuminate, the pinnae of smaller fronds obtuse, subsessile, obliquely truncate to broadly cuneate, auriculate or with a larger lobe on anterior side at base, spreading setosehirsute beneath on costas and veinlets, strigose on costas above; ultimate pinnules ovate-deltoid to ovate-oblong, toothed; sori 1–7 on a lobe, submarginal; indusia depressed-orbicular, whitish, prominently setose, often ciliate, about 1.5 mm. wide.—— Honshu (Hitachi and Etchu Prov. westw.), Shikoku, Kyushu; common.——Ryukyus, Formosa, China to India.

2. **Microlepia pseudostrigosa** Makino. *M. marginata* var. *bipinnata* Makino——FUMOTO-KAGUMA, KUJAKU-FUMOTO-SHIDA.   Closely allied to the preceding, evergreen, but stouter; blades bipinnate, the rachis usually glabrous and smooth on upper side except near the top; pinnules sessile, oblique, ovate to oblong-ovate, 1–2 cm. long, 6–12 mm. wide, obtuse to subacute, toothed, often lobulate.——Honshu (Awa and Sagami Prov. and westw.), Shikoku, Kyushu; rare.

3. **Microlepia strigosa** (Thunb.) Presl. *Trichomanes strigosum* Thunb.; *Dicksonia strigosa* (Thunb.) Thunb.; *Davallia strigosa* (Thunb.) Kunze; *Dennstaedtia strigosa* (Thunb.) J. Smith; *Dicksonia japonica* Sw.; *M. japonica* (Sw.) Presl——ISHI-KAGUMA.   Resembling the preceding in habit; rhizomes short-creeping, firm, 3–5 mm. across, while young with brownish multicellular jointed hairs; fronds evergreen, 50–150 cm. long, rather coarse, strigose; stipes 30–60 cm. long, dull purple on the lower side, stramineous on the upper side, densely brownish hairy at base, sparsely so above, the callose base of the hairs persistent; blades chartaceous-herbaceous, dark green, paler beneath, broadly lanceolate, 40–100 cm. long,

15–40 cm. wide, abruptly acuminate, scarcely narrowed at base, bipinnate, the rachis glabrous and grooved on the stramineous smooth upperside, densely brownish pilose on the dull purplish underside; pinnae linear-lanceolate, 1.5–3 cm. wide, long-acuminate, thinly hispid on veins especially beneath; pinnules oblong-ovate, 10–20 mm. long, 5–10 mm. wide, acute to obtuse, unequally cuneate at base, sessile, toothed, sometimes lobulate; sori nearly marginal, in the sinus; indusia membranous, semirounded, hyaline, entire, glabrous or sometimes with 1 or 2 hispid hairs near base, about 0.7 mm. wide.——Honshu (Izu Isls., Awa Prov. and westw.), Shikoku, Kyushu.——Korea (Quelpaert Isl.), Ryukyus, Formosa, China to India, Malaysia, and Polynesia.

4. **Microlepia substrigosa** Tagawa. USUBA-ISHI-KAGUMA.   Closely allied to the preceding but the rhizomes with dark brown hairs; fronds thinner, the lower pinnae broader, lanceolate, about 5 cm. wide; pinnules narrowly ovate to broadly lanceolate-deltoid, 2–3 cm. long, 6–12 mm. wide, obtuse to acute, oblique, short-petiolulate, pinnate to pinnatiparted; ultimate pinnules oblong, toothed; sori nearly marginal, suborbicular; indusia hyaline, glabrous or with few setose hairs near base, undulate to entire.——Shikoku, Kyushu; rare.——Formosa.

5. **Microlepia yakusimensis** Tagawa. *M. obtusiloba* sensu auct. Japon., non Hayata; *M. majuscula* sensu auct. Japon., non Moore——YAKUSHIMA-KAGUMA.   Rhizomes creeping, about 5 mm. across, chestnut-brown, hairy; stipes 30–50 cm. long, scattered short-pubescent, glabrescent, fuscous-brown; blades thinly herbaceous, 30–50 cm. long, 25–30 cm. wide, subtripinnate, the rachis subglabrous, grooved above, fuscous-brown, densely short-pubescent beneath; pinnae ovate-deltoid, acuminate, petiolulate, the lower pinnae 15–20 cm. long, 5–10 cm. wide, pinnules pinnatiparted, long-acuminate in the lower, acute to obtuse in the upper ones, hirsute beneath on costas and veins, short-pubescent above on the costas; ultimate pinnules oblong, obtuse, oblique, crenately toothed, sometimes lobulate; sori small; indusia pale, with a few hispid hairs on back.——Kyushu; rare.

3. MONACHOSORUM Kunze   FUJI-SHIDA ZOKU

Terrestrial, with ascending short rhizomes covered at the tip by a mucous excretion; fronds thinly brown-scurfy while young; stipes tufted; blades simply or 2- to 4-pinnate, often with bulbils in axils of pinnae or proliferous at the tip; ultimate pinnules toothed or lobed, membranous, blackish when dried, glabrous except for sparse glandular short hairs on the axis, the lobes with a single veinlet ending short of the tip; sori terminal or nearly so on the veinlets, intramarginal, rounded; indusia absent, the receptacle scarcely raised; sporangia mixed with glandular hairs.——About 5 species, Malaysia to Japan.

1A. Blades simply pinnate, not broadened at base; pinnae obtuse, oblique at base. . . . . 1. *M. maximowiczii*
1B. Blades 2- to 4-pinnate, broadest at base; pinnae acute to acuminate, equilateral.
   2A. Blades 10–25 cm. wide, 2- or 3-pinnatiparted to 2- or 3-pinnate, the rachis often extended and gemmate at the tip; bulbils absent
      in the axils at base of pinnae. . . . . 2. *M. flagellare*
   2B. Blades larger, 25–40 cm. wide, acuminate, 3-pinnate, not gemmate at the tip; bulbils present in axils at base of median and upper
      pinnae. . . . . 3. *M. arakii*

1. Monachosorum maximowiczii (Bak.) Hayata. *Polypodium maximowiczii* Bak.; *Ptilopteris maximowiczii* (Bak.) Hance; *Polystichum maximowiczii* (Bak.) Diels; *Phegopteris maximowiczii* (Bak.) H. Chr.; *Monachosorella maximowiczii* (Bak.) Hayata——Fuji-shida. Fronds slender, evergreen, sparsely brownish scurfy beneath while young; rhizomes short, ascending, covered with the persistent bases of old stipes; stipes brown, lustrous, 6–20 cm. long, about 2 mm. across; blades herbaceous-membranous, linear-lanceolate, 20–50 cm. long, 2–4(–5) cm. wide, simply pinnate, contracted at base, gradually long-attenuate and gemmate at the tip on the naked extension of the rachis; pinnae many, horizontally spreading, broadly lanceolate, 1–2.5 cm. long, 3–7 mm. wide, obtuse, obliquely cuneate and auriculate at base, obtusely toothed, sessile, the lower pinnae smaller and deflexed, the upper ones gradually reduced; veins pinnate, short, oblique, veinlets simple; sori intramarginal, solitary on the teeth, sometimes slightly incurved on margin; indusia absent.——Shaded rocky places in woods in mountains; Honshu (Kantō Distr. and westw.), Shikoku, Kyushu; rather rare.——Formosa and China.

2. Monachosorum flagellare (H. Chr.) Hayata. *Polystichum flagellare* H. Chr.; *Monachosorella flagellaris* (H. Chr.) Hayata; *Monachosorum nipponicum* Makino; *Monachosorella nipponica* (Makino) Hayata; *Monachosorella flagellaris* var. *nipponica* (Makino) Tagawa——Ō-fuji-shida, Kishū-shida. Fronds evergreen, thinly brown scurfy beneath; rhizomes short, ascending, densely clothed with the basal remains of old stipes; stipes 10–30 cm. long, 2–3 mm. across, stramineous, brownish at base; blades narrowly 3-angled, 20–60 cm. long, 10–25 cm. wide, 2- or 3-pinnatiparted to nearly 2- or 3-pinnate, narrowed at tip, broadest at base, the rachis often extended and gemmate at the tip; pinnae many, membranous, linear to lanceolate, 7–17 cm. long in the lower ones, 2–4 cm. wide, narrowed at tip, sessile or subsessile, green on both sides, bulbils present in axils at base of median and upper pinnae; ultimate pinnules oblong, obtuse, toothed, oblique; veinlets few on each side of the ultimate pinnules, not prominent; sori intramarginal, solitary on the teeth; indusia absent.——Rocky places in damp woods in mountains; Honshu (Kantō Distr. and westw.), Shikoku, Kyushu; rare.

3. Monachosorum arakii Tagawa. Hime-mukago-shida. Brown-scurfy throughout except for the upper side of frond; rhizomes short, covered with the persistent remains of old stipes; stipes 55–60 cm. long, about 5 mm. across at base, stramineous to rusty-brown toward base; blades thinly herbaceous, narrowly triangular-ovate, 50–70 cm. long, 25–40 cm. wide, tripinnate, the rachis grooved on upper side, bulbils present in axils of median and upper pinnae; pinnae broadly lanceolate, to 30 cm. long and 10 cm. wide in the lower ones, long-acuminate, spreading, short-petiolate; pinnules lanceolate, to 7 cm. long and 1.5 cm. wide, acute to acuminate, sessile; ultimate pinnules and segments oblong to ovate, to 1 cm. long and 5 mm. wide, obtuse, obliquely cuneate at base, crenate-incised to pinnatifid; sori submarginal, naked.——Honshu (Tanba, Kii, and Suwō Prov.); very rare.

4. LINDSAEA Dryand.   Hongū-shida Zoku

Terrestrial or epiphytic; rhizomes creeping, elongate or short, with narrow small scalelike hairs; fronds stipitate, pinnate or pinnately compound, rarely simple; pinnules usually thin, usually dimidiate, rarely equilateral, glabrous; veins free or sometimes loosely anastomosing; sori on the upper and outer margins, intramarginal or apparently marginal, terminal, solitary or more often united into transverse coenosori; indusia rounded, more often transversely elongate by the union of adjacent ones.——About 200 species, of wide distribution in the Tropics.

1A. Stipes sulcate in upper part, not 4-angled; blades simply pinnate; rhizomes slender and long-creeping, remotely leaved.
   2A. Stipes 3–15 cm. long, usually red-brown at base; blades 7–25 cm. long; median pinnae truncate on the anterior side at base; sori
      usually interrupted by sinuses of lobes; rhizomes 1.5–2 mm. across, the scales about 2 mm. long. . . . . 1. *L. cultrata*
   2B. Stipes 1–8 cm. long, dark purple-brown at base; blades 3–10 cm. long; median pinnae broadly cuneate on anterior side at base; sori
      usually uninterrupted, nearly straight, sometimes interrupted by a shallow sinus between the teeth; rhizomes 1–1.5 mm. across,
      the scales about 1 mm. long. . . . . 2. *L. japonica*
1B. Stipes 4-angled, not sulcate; blades simply pinnate or bipinnate; rhizomes short-creeping, closely leaved.
   3A. Blades pinnate or bipinnate, the ultimate divisions oblique, elliptic or ovate, rounded to broadly cuneate at base.
                                                              3. *L. orbiculata* var. *chienii*
   3B. Blades pinnate, the pinnae flabellate, nearly equilateral, truncate at base. . . . . 4. *L. simulans*

1. Lindsaea cultrata (Willd.) Sw. *Adiantum cultratum* Willd.——Hongū-shida. Fronds slender, evergreen, glabrous; rhizomes slender, creeping, 1.5–2 mm. across, with linear-lanceolate, dark brown, lustrous, entire scales about 2 mm. long; stipes 3–15 cm. long, green and sulcate on the upper portion, reddish brown in lower half, glabrous or with few scales toward the base; blades linear-lanceolate, 7–25 cm. long, 1.5–3.5 cm. wide, gradually acute to subacuminate, pinnate, green; rachis usually pale green; pinnae thinly herbaceous, ovate, 8–18 mm. long, 4–7 mm. wide, subacute to obtuse, nearly truncate on anterior margin at base, minutely impressed-punctate beneath, short-petiolulate, the posterior margin nearly straight and only slightly incurved toward the apex, entire, the anterior margin toothed or few-lobulate, gently curved to nearly straight; sori on the anterior margin of the pinnae, interrupted, usually with an obscure mucro on both sides; indusia undulate to erose.——Wet shaded rocks along streams in mountains; Honshu (Izu Isls., Ise and Kii

Prov.), Shikoku, Kyushu; rare.——Formosa and widespread in tropical Asia.

2. Lindsaea japonica (Bak.) Bak. ex Diels. *L. cultrata* var. *japonica* Bak.——Saigoku-hongū-shida.　　Fronds evergreen, small; rhizomes creeping, 1–1.5 mm. across, with sparse, dark brown, linear-lanceolate scales about 1 mm. long; stipes 1–8 cm. long, dark purple-brown, few-scaled at base, not or scarcely lustrous, sulcate on upper side; blades lanceolate, 3–10 cm. long, 1.5–2 cm. wide, acute, glabrous, pinnate, the rachis pale green, sulcate on upper side, rounded on back; pinnae 7–12 pairs, the lower ones spreading, the upper ones ascending, broadly subovate or broadly subelliptic, 6–12 mm. long, 3–6 mm. wide, rounded to obtuse, broadly cuneate or rarely subtruncate at base on anterior side, nearly truncate (crenate-toothed in the sterile pinnae) on anterior margin, the posterior margin entire, straight in lower half, incurved on upper half; sori on the anterior margin of pinnae, straight, usually continuous; indusia nearly entire, reaching the edge of the pinnae. ——Wet shaded rocks along streams in mountains; Honshu (Hachijō Isl. and Kii Prov.), Kyushu; rare.——Ryukyus, Formosa, Korea (Quelpaert Isl.).

3. Lindsaea orbiculata (Lam.) Mett. var. **chienii** (Ching) Ohwi. *L. chienii* Ching; *L. tenera* var. *chienii* (Ching) C. Chr. & Tard.-Bl.; *L. orbiculata* sensu auct. Japon., non Mett.; *L. commixta* Tagawa——Edauchi-hongū-shida. Fronds glabrous, evergreen; rhizomes short, creeping or ascending, firm, densely leafy, while young with red-brown lustrous linear-lanceolate scales about 1.5 mm. long; stipes 2–20 cm. long, dark purple-brown, lustrous, 4-angled, smooth, sparsely scaly at the base; blades broadly lanceolate, simply pinnate, 8–20 cm. long, 2–3 cm. wide, or bipinnate and narrowly triangular, 5–12 cm. wide, gradually acuminate in larger fronds, the rachis 4-angled, lustrous, pale green or often purple-brown beneath on lower half; pinnules chartaceous-herbaceous, elliptic, squarrose or ovate, 8–20(–25) mm. long, 4–10(–13) mm. wide, oblique, rounded to very obtuse, broadly cuneate at base, short-petiolulate, entire and nearly at right angles to the axis on the lower margin, acutely toothed and/or lobulate toward the tip, palmately veined; sori usually interrupted, short; indusia narrow, marginal.——Honshu (Izu Pen. and Izu Isls. through Tōkaidō to s. Kinki Distr.), Shikoku, Kyushu; rather rare.——Ryukyus, Formosa, China, and Indochina. Polymorphic.

Var. **deltoidea** Wu.　*L. chienii* var. *deltoidea* (Wu) Tagawa——Edauchi-hongu-shida-modoki.　　Terminal pinna elongate-rhombic, lobulate, large.——Honshu (Kii Prov.), Kyushu.——Ryukyus, Formosa, and China.

4. Lindsaea simulans Ching. *L. orbiculata* var. *orbiculata* sensu Ohwi——Uchiwa-hongū-shida.　　Blades simply pinnate, gradually narrowed at the tip to a single obtuse small segment; pinnae scarcely lobulate, paler beneath, the lower pinnae flabellate, nearly equilateral, truncate at base; sori continuous.——Shikoku (Tosa Prov.), Kyushu (Yakushima); rere.——China. Closely allied to the preceding.

5. SPHENOMERIS Maxon　　Hora-shinobu Zoku

Terrestrial; rhizomes creeping, with dark brown narrow hairlike scales; fronds erect, glabrous, pinnately compound; pinnules usually cuneate at base; sori submarginal, terminating the veinlets, solitary or in pairs, sometimes in threes, usually united; veinlets free; indusia fixed by the base and sides.——About 20 species, in the Tropics.

1. Sphenomeris chinensis (L.) Maxon. *Trichomanes chinense* L.; *Davallia chinensis* (L.) Smith; *Stenoloma chinense* (L.) Bedd.; *Odontosoria chinensis* (L.) J. Smith; *Lindsaea chinensis* (L.) Mett.; *D. tenuifolia* var. *chinensis* (L.) Moore; *Adiantum chusanum* L.; *Sphenomeris chusana* (L.) Copel.; *Stenoloma chusanum* (L.) Ching; *O. chinensis* var. *tenuifolia* Matsum., excl. syn.; *O. chusana* var. *tenuifolia* Masam., excl. syn.; *D. tenuifolia* sensu auct. Japon., non Sw.; *Stenoloma gracile* Tagawa——Hora-shinobu.　　Fronds evergreen, glabrous, often somewhat brownish red; rhizomes short-creeping, densely leafy; fronds 10–80 cm. long, tufted; stipes 4–30 cm. long, pale brown, smooth, scaly at base; blades smooth, ovate to broadly lanceolate, 5–50 cm. long, 3–20 cm. wide, acute to acuminate, slightly or not contracted at base, 3- or 4-pinnate, the rachis sulcate on upper side; pinnae narrowly ovate to broadly lanceolate, more or less 3-angled, usually narrowed at tip, petiolulate; ultimate segments coriaceous to thick-herbaceous, 3–4 mm. long, 2–3 mm. wide at the truncate and few-toothed broad tip, cuneate at base; veins not prominent, once or twice forked; sori marginal, 1 to 3 on the ultimate pinnules; indusia cup-shaped, depressed, often partially united.——Sunny exposed rocks and roadsides in low mountains and hills; Honshu (Kantō Distr. and westw.), Shikoku, Kyushu; common.——Ryukyus, Formosa, China to India, Polynesia, and Madagascar.

Var. **littorale** (Tagawa) Ohwi. *Stenoloma biflorum* (Kaulf.) Ching; *Davallia biflora* Kaulf.; *Sphenomeris biflora* (Kaulf.) Tagawa; *S. littorale* Tagawa; *S. chusanum* var. *littorale* (Tagawa) H. Itō——Hama-hora-shinobu.　　Rhizomes with longer and broader scales; blades thickly coriaceous, ovate to broadly so.——Along the sea; Honshu (Shimosa Prov. and westw.), Shikoku, Kyushu.——Ryukyus to the Philippines.

6. HYPOLEPIS Bernh.　　Iwa-hime-warabi Zoku

Terrestrial; fronds moderately large; rhizomes creeping, solenostelic, usually with reddish hairs; fronds herbaceous, compound or decompound, hairy or glabrous; veins free; sori usually nearly marginal, protected by a reflexed tooth, rarely inframarginal and naked.——Pantropic, with about 45 species, north to Japan, south to New Zealand and South Africa.

1. Hypolepis punctata (Thunb.) Mett. *Polypodium punctatum* Thunb.; *Phegopteris punctata* (Thunb.) Mett.; *Nephrodium punctatum* (Thunb.) Diels; *Dryopteris punctata* (Thunb.) C. Chr., non O. Kuntze——Iwa-hime-warabi. Fronds deciduous, large, more or less pilose; rhizomes long-creeping, remotely leaved, with red-brown jointed callose flexuous short hairs; stipes 10–60 cm. long, sulcate on upper side, yellow-brown, dull, 2.5–4 mm. across, scabrous; blades deltoid-ovate to oblong, 20–120 cm. long, 16–60 cm. wide, acuminate, 3- or 4-pinnate, the pilose indument jointed, callose at the base; pinnae broadly lanceolate to oblong-ovate, acuminate, short-petiolate, spreading, the lower 2 or 3 pairs larger, 13–35 cm.

long, 5–15 cm. wide; pinnules thinly herbaceous to membranous, oblong-ovate, 1–3 cm. long, 4–12 mm. wide, obtuse, sessile, pinnately lobed to parted, often toothed, green above, paler beneath, the lobes spreading, ovate, obtuse to rounded; veinlets reaching the tip of the teeth; sori intramarginal, naked, rounded, 1–1.5 mm. across, yellow-brown, often covered by the slightly recurved tooth of the pinnules.——Hills and low mountains; Honshu (Ugo and Rikuzen Prov. southw.), Shi-

koku, Kyushu.——Pantropic, also in Korea, Ryukyus, Formosa, and China.

Hypolepis bamleriana Rosenst.   SHIMA-IWA-HIME-WARABI. A larger plant, 1.5–2 m. high with stout stipes often 1.2 cm. in diameter; lowest pinna much larger and broader than the others.——Recorded from Kyushu (Yakushima).——New Guinea and Bonin Isls.

7. PTERIDIUM Scop.   WARABI ZOKU

Terrestrial; rhizomes long-creeping, solenostelic, with jointed hairs; fronds pinnately compound, coriaceous to thick-herbaceous, more or less hairy; veins free except for a marginal strand; sori continuous along the margin, on a connecting vein; indusium double; the outer one false, formed by the reflexed margin of the pinnules, the inner one distinct or obsolete. ——A single cosmopolitan species with several geographically distinct variants.

1. Pteridium aquilinum (L.) Kuhn var. **latiusculum** (Desv.) Underw.   *Pteris lanuginosa* Spreng.; *Pteris sprengelii* Steud.; *Pteris latiuscula* Desv.; *Pteridium latiusculum* (Desv.) Hieron.; *Pteridium aquilinum* var. *japonicum* Nakai; *P. aquilinum* sensu auct. Japon., non Kuhn——WARABI.   Fronds coarse, deciduous; rhizomes long-creeping, terete, 4–5 mm. across, remotely leafy, usually with pale brown jointed hairs at the tip; stipes 20–80 cm. long, rather stout, pale green, glabrous above, with soft hairs at the dark brownish base; blades broadly ovate-triangular, 20–100 cm. long, 17–70 cm. wide, short-acuminate, green above, slightly paler beneath, usually ternate-pinnate, the rachis sulcate and pubescent on upper side especially in the axils at base of pinnae, pale green; pinnae of

the lowest pair manifestly larger than the others, reaching 2/3 of the entire length of the blade, spreading, 2- to 4-pinnatiparted, triangular-ovate, acuminate; pinnules subcoriaceous to thick-chartaceous, horizontally spreading, straight, oblong to linear-lanceolate, 3–6 mm. wide in the fertile, 5–8 mm. wide in the sterile ones, obtuse, usually glabrous above, thinly white-pilose at least along the costas beneath, slightly recurved on margins; sori marginal, continuous; indusia scarious, glabrous or nearly so, narrow.——Aug.–Oct.   Sunny slopes in hills and mountains; Hokkaido, Honshu, Shikoku, Kyushu; very common.——Ryukyus, China, Korea, Sakhalin, s. Kuriles, Kamchatka to Europe, and e. N. America.

8. HISTIOPTERIS J. Smith   YUNOMINE-SHIDA ZOKU

Terrestrial; rhizomes long-creeping, solenostelic, with narrow dark brown hairlike scales; fronds usually large, glabrous; pinnae chartaceous to coriaceous, opposite, usually sessile, often glaucous, with stipulelike basal pinnules; veins anastomosing without included veinlets; sori continuous along a marginal connecting vein, covered by a scarious reflexed false indusium, freely paraphysate.——Pantropic, with about 7 species.

1. Histiopteris incisa (Thunb.) J. Smith.   *Pteris incisa* Thunb.; *P. aurita* Bl.——YUNOMINE-SHIDA.   Fronds large, glabrous, evergreen; rhizomes creeping, remotely leaved, 4–7 mm. across, with dark brown lustrous jointed crisped hairs; stipes shorter than the blades, lustrous, dark purple-brown at least on the lower half, 5–15 mm. across; blades large, to 1 m. long, triangular-ovate, obtuse to short-acute, glabrous, green above, glaucous beneath, bipinnate or bipinnatiparted, the rachis terete, brownish or stramineous; pinnae opposite, 5–10

pairs, nearly horizontally spreading, broadly lanceolate to oblong-ovate, acute with an obtuse tip, sessile, pinnatiparted to pinnate, in large fronds the lowest pair of pinnules much reduced and stipulelike; ultimate pinnules opposite, narrowly ovate to elliptic, 5–20 mm. wide, obtuse to rounded, horizontally spreading, entire or sinuately lobed; veins anastomosing slender; sori marginal, elongate, covered by the reflexed marginal false indusium.——Honshu (Izu and Kii Prov.), Kyushu; rare.——Pantropic.

9. PTERIS L.   I-NO-MOTO-SO ZOKU

Terrestrial; rhizomes short, dictyostelic, scaly or sometimes hairy; fronds herbaceous to coriaceous or chartaceous, tufted, pinnate to decompound, glabrous or rarely hairy; veins free except in the sori, or anastomosing without included veinlets; sori continuous along the margin except at the apex and in the sinuses of the pinnules, covered by the scarious reflexed margin of the segments, with paraphyses among the sporangia.——About 280 species, nearly all tropical, some in New Zealand, Tasmania, S. Africa, Japan, Europe, and the United States.

1A. Ultimate divisions of blades linear, unlobed; rachis and costas without spines on upper side.
  2A. Blades narrowed at base; stipes prominently brown-scaly. . . . . 1. *P. vittata*
  2B. Blades not narrowed at base; stipes smooth or nearly so.
    3A. Upper pinnae decurrent on the rachis; sterile pinnae 6–12 mm. wide. . . . . 2. *P. multifida*
    3B. Upper pinnae not decurrent on the rachis.
      4A. Sterile pinnae 8–25 mm. wide.
        5A. Fronds nearly all alike; sterile pinnae linear-lanceolate, 10–25 cm. long, 1–2.5 cm. wide, acuminate. . . . . 3. *P. cretica*
        5B. Fronds dimorphic; sterile pinnae lanceolate to oblong-ovate, 1.5–10 cm. long, 7–12 mm. wide, obtuse. . . . . 4. *P. ryukyuensis*
      4B. Sterile pinnae linear, 3–20 cm. long, 3–6 mm. wide. . . . . 5. *P. yamatensis*

1B.   Ultimate divisions of blades pectinately pinnatiparted, at least the terminal.
   6A.   Costas and costules without spines at base of segments on upper side; veinlets reaching the awn-toothed sterile margin of the blade.
      7A.   Blades 40–60 cm. long, 20–25 cm. wide; stipes 4-angled, 3–4 mm. across near base; rachis rounded on back; stipes and rachis dark purple-brown, nearly concolorous; ultimate segments 5–10 mm. wide. . . . . 6. *P. semipinnata*
      7B.   Blades 25–40 cm. long, 8–15 cm. wide; stipes subtrapezoid or 3-angled, 2–2.5 mm. across near base; rachis 3-angled; stipes and rachis yellow-brown, with paler angles; ultimate segments 4–5 mm. wide. . . . . 7. *P. dispar*
   6B.   Costas and costules with a spine at base of segments on upper side.
      8A.   Veinlets reaching the margin of sterile part of blade; sterile segments entire. . . . . 8. *P. quadriaurita*
      8B.   Veinlets not reaching the margin; sterile segments toothed.
         9A.   Blades 20–30 cm. long, bipinnatiparted, the lowest pinnae sometimes with 1 or 2 sessile branches on posterior side near base.
            10A.   Pinnae gray- or brownish green; veinlets free. . . . . 9. *P. inaequalis*
            10B.   Pinnae vivid green; lowest veinlets of the segments anastomosing, forming elongate areoles along the costas of the pinnules.
                                                                           10. *P. naḳasimae*
         9B.   Blades often more than 1 m. long, with 5 bipinnatiparted, short-stalked, large pinnae; lowest veinlets of the segments anastomosing and forming areoles along the costas of the pinnules. . . . . 11. *P. wallichiana*

1.   Pteris vittata L.   *P. longifolia* sensu auct. Asiat., non L.——MOEJIMA-SHIDA.   Rhizomes short, rather stout, with dense, pale brown, membranous, broadly linear scales about 3 mm. long; stipes 3–15 cm. long, pale brown or stramineous, sulcate on upper side, prominently scaly, the callose base of scales persistent; blades obovate-oblong to broadly oblanceolate, 20–50 cm. long inclusive of the elongate terminal pinna, narrowed at base, simply pinnate, the rachis stramineous, sulcate on upper side, rounded and more or less scaly or scabrous on underside; pinnae 12–26 pairs, spreading, simple, the lower ones 1–2 cm. long, the median 4–8 cm. long, truncate and sometimes slightly auriculate at base, sessile, glabrous and gray-green above, the costa thinly pubescent beneath, the sterile pinnae linear-lanceolate to oblong-ovate, 7–8 mm. wide, usually obtuse, minutely toothed, the fertile pinnae elongate, 4–7 mm. wide.——Honshu (Kii Prov.), Shikoku, Kyushu (s. distr.); rare.——Ryukyus, Formosa, s. China, and generally in the Tropics and subtropics of the Old World.

2.   Pteris multifida Poir.   *P. serrulata* L. f., non Forsk.——I-NO-MOTO-SŌ.   Fronds slender, glabrous; rhizomes short, with black-brown lustrous broadly linear entire scales about 2.5 mm. long; fronds tufted, the sterile ones shorter; stipes slender, pale green, 3-angled, glabrous in upper part, thinly scaly at base; blades broadly ovate in outline, the rachis winged; sterile pinnae thinly chartaceous, 2 or 3 on each side, 3- to 5-parted, the sterile divisions linear to lanceolate, 5–15 cm. long, 6–12 mm. wide, long-acuminate, sessile or decurrent on the rachis, irregularly aristulate-toothed; fertile pinnae linear, entire, 10–20 cm. long, 3–7 mm. wide.——Lowlands; Honshu (Kantō Distr. and westw.), Shikoku, Kyushu; common. ——China and Indochina.

3.   Pteris cretica L.   *P. nervosa* Thunb.——ŌBA-NO-I-NO-MOTO-SŌ.   Fronds evergreen, glabrous, nearly all alike; rhizomes short, with dark purple-brown lustrous broadly linear scales about 3 mm. long; stipes 15–60 cm. long, pale stramineous, sometimes somewhat brownish toward the base, obtusely 4-angled, sulcate on the upper side, glabrous, 2–2.5 mm. across; blades thinly coriaceous, 20–40 cm. long, simply pinnate, with 1–5 pairs of pinnae, the rachis not winged; lower pinnae usually 2- or 3-parted to -divided; fertile lobes broadly linear, 5–12 mm. wide, gradually long-acuminate, sterile at the tip, the sterile lobes linear-lanceolate, 10–25 cm. long, 1–2.5 cm. wide, acuminate, awn-toothed, green; veinlets reaching the cartilaginous margin of the sterile pinnae.——Honshu (Kantō Distr. and Echigo Prov. westw.), Shikoku, Kyushu; common. Widely distributed in the tropics and warm-temperate regions of the N. Hemisphere.——Cv. **Albolineata**. *P. cretica* var. *albolineata* Hook. MATSUZAKA-SHIDA.   Sterile lobes white-variegated along the costas.——Cultivated.

4.   Pteris ryukyuensis Tagawa.   RYUKYU-I-NO-MOTO-SŌ. Fronds dimorphic, glabrous, evergreen; rhizomes small, short, with dark purple-brown, lustrous, broadly linear scales, about 1 mm. long; stipes firm, slender, obtusely angled, lustrous, brown, paler on the angles, 0.7–1 mm. across, 3–12 cm. long in the sterile fronds, 5–25 cm. long in the fertile; blades ternate or sometimes with a pair of remote pinnae at base, the rachis not winged; sterile pinnae chartaceous to thinly coriaceous, lanceolate to oblong-ovate, 2–7(–10) cm. long, 7–12 mm. wide, obtuse, the lower ones sometimes 2-parted, the fertile pinnae linear, 4–15 cm. long, 3–4 mm. wide.——Stone walls; Kyushu (s. distr.); rare.——Ryukyus.

5.   Pteris yamatensis (Tagawa) Tagawa.   *P. angustipinna* var. *yamatensis* Tagawa——HIME-I-NO-MOTO-SŌ.   Somewhat resembling the preceding in habit; rhizomes short, with linear-lanceolate black-brown pale margined scales 2–2.5 mm. long; fronds tufted, the sterile shorter than the fertile; stipes slender, trigonous, 0.5–1 mm. across, chestnut-brown, 2–8 cm. long in the sterile fronds, about 12 cm. long in fertile; blades ternately or pedately 3- to 5-foliolate; sterile pinnae thinly chartaceous, linear, 3–13 cm. long, 3–6 mm. wide, the terminal pinna 10–20 cm. long, acutely toothed, gradually narrowed at both ends, subsessile, glabrous but with scattered hairlike scales along costas on both sides while young; veins simple or once forked, not reaching the margin; fertile pinnae 2–3 mm. wide, entire.——Calcareous rocks in mountains; Honshu (Mount Ōmine in Yamato Prov.); rare.

6.   Pteris semipinnata L.   Ō-AMAKUSA-SHIDA.   Stipes 30–60 cm. long, 3–4 mm. across near base, lustrous, 4-angled, dark purple-brown or deep brown, smooth on upper part, scaly at base; blades 40–60 cm. long, 20–25 cm. wide, bipinnatiparted, with a large terminal pinna, the rachis deep brown, lustrous, rounded on back, smooth; lateral pinnae 5–8 pairs, 10–17 cm. long, subsessile, the long-acuminate slightly falcate tip 5–10 cm. long, unlobed and entire on anterior margin or sometimes with a small auricle at base, the terminal pinna regularly pinnatiparted, the costas with a sparsely scabrous line above on each side; pinnules linear-lanceolate, about 8 cm long, 5–10 mm. wide, acuminate, incurved-toothed on sterile margins; costules sparsely scabrous on the upper side.——Kyushu (Yakushima); rare.——Ryukyus, Formosa, China, and tropical Asia.

7.   Pteris dispar Kunze.   *P. semipinnata* var. *dispar* (Kunze) Bak.——AMAKUSA-SHIDA.   Fronds evergreen; rhizomes short-creeping, with dark brown linear scales 3–3.5 mm. long; stipes tufted, 20–40 cm. long, yellow-brown, subtrapezoid or 3-angled, lustrous, 2–2.5 mm. across near base, smooth, obtuse on the dorsal angle; blades chartaceous, lanceolate to narrowly oblong-ovate, 25–40 cm. long, 8–15 cm. wide, bipinnatiparted, with a large terminal pinna, the rachis yellow-brown, 3-angled, lustrous, smooth; lateral pinnae 3–7 pairs, narrowly deltoid, 5–10 cm. long, sessile or subsessile, pinnatiparted or unlobed on anterior side, the acutely toothed tip 1–3 cm. long, the terminal pinna triangular-lanceolate, 10–20 cm. long, 3–5 cm. wide, regularly pinnatiparted with a linear acutely toothed tip, the costas on each side above with a sparsely scabrous elevated white line; pinnules linear- to oblong-lanceolate, 1–3 cm. long, 4–5 mm. wide, obtuse to acute, slightly falcate.——Honshu (s. Kantō Distr. and westw.), Shikoku, Kyushu.—— Ryukyus and Formosa.

8. Pteris quadriaurita Retz. *P. biaurita* var. *quadriaurita* (Retz.) Luerss.——HACHIJŌ-SHIDA. Rhizomes short, with deep brown, linear, lustrous scales about 12 mm. long; stipes 30–70 cm. long, tufted, stramineous, often brownish on lower surface, usually dark brown toward the base, smooth or with scattered linear scales at base, 3–6 mm. across, 2-grooved on upper side, rounded on underside; blades broadly ovate to oblong-ovate, 25–60 cm. long, 20–35 cm. wide, bipinnatiparted; lowest pinnae usually with 1 or 2 branches on posterior side near base; veinlets reaching the margin of sterile part of blade; pinnae 3–8 pairs, chartaceous to subcoriaceous, linear-lanceolate to broadly lanceolate, 12–30 cm. long, 2–4 cm. wide, caudate, regularly pinnatiparted; costas and sometimes the costules on upper side with a prominent subulate spine at base of the segment or of the veins; sterile pinnules narrowly oblong, 5–7 mm. wide, obtuse to rounded, entire, usually slightly falcate, 3–5 mm. wide in the fertile pinnules.——Honshu (s. Kantō Distr. and westw.), Shikoku, Kyushu.——Ryukyus, Formosa, China, and generally in the Tropics and subtropics. Sometimes segregated as microspecies are: *P. kiuschiuensis* Hieron.; *P. oshimensis* Hieron.; *P. fauriei* Hieron.; *P. hachijoensis* Nakai; and *P. natiensis* Tagawa, probably all synonymous with *P. quadriaurita.*

9. Pteris inaequalis Bak. var. **inaequalis.** *P. longipinnula* Wall. forma *inaequalis* (Bak.) Makino; *P. inaequalis* var. *simplicior* Tagawa——ŌBA-NO-AMAKUSA-SHIDA. Rhizomes stout, short, ascending, with brown, membranous, linear scales 1–1.3 cm. long; stipes stramineous, 20–50 cm. long, 3–6 mm. across near the dark purplish brown base, sulcate on the upper side, glabrous; blades oblong-ovate, 30–70 cm. long, 15–30 cm. wide, bipinnatiparted, the rachis stramineous, smooth; pinnae 3–7 pairs, more or less falcate, with a linear elongate tip 7–15 cm. long, pinnatiparted on the posterior side, unlobed or remotely lobed on the anterior side, opaque, the terminal pinna distinct, narrowly elongate, 15–30 cm. long, 8–15 cm. wide, regularly pinnatiparted; costas with a spine at the base of the

pinnules; pinnules lanceolate, 2–8 cm. long, 8–13 mm. wide, sometimes linear-lanceolate or oblong, 2–8 cm. long, 8–13 mm. wide, acute to obtuse, toothed on the sterile margins; veinlets free, indistinct, not reaching the margin.——Honshu (s. Kantō Distr. and westw.), Shikoku, Kyushu.——China.

Var. **aequata** (Miq.) Tagawa. *P. semipinnata* var. *aequata* Miq.; *P. longipinnula* sensu auct. Japon., non Wall.——ŌBA-NO-HACHIJŌ-SHIDA. Usually larger, often 1 m. high or more; stipes stouter, to 1 cm. across at base; pinnae pinnatiparted on both sides, the lowest sometimes with a sessile branch on posterior side near base, the pinnules 6–10 mm. wide, usually acute.——More common than the typical phase; Honshu (Ugo Prov. and Kantō Distr. westw.), Shikoku, Kyushu.——Korea and Formosa.

10. Pteris nakasimae Tagawa. *P. mcclurei* var. *nakasimae* Tagawa——HINOTANI-SHIDA. Rhizomes short-creeping, densely scaly; stipes 35–40 cm. long, purple-brown to purplish, lustrous, very sparingly scabrous; blades thinly herbaceous, ovate-oblong, 50–60 cm. long, 20–30 cm. wide, bipinnatiparted, glabrous, the rachis purple-brown or purplish, lustrous, sparsely scabrous; pinnae 5–7 pairs, spreading, lanceolate, 15–20 cm. long, 3–6 cm. wide, acuminate, slightly oblique, short-cuneate at base, sessile, the lowest pinna 2-parted; costas with a soft spine on upper side at base of the pinnule; pinnules 2–4 cm. long, about 7 mm. wide, obtuse or rounded, the sterile portion incurved-toothed; lowest veinlets anastomose, forming elongate areoles along the costas of the pinnules.——Kyushu (Satsuma Prov.); rare.

11. Pteris wallichiana Agardh. *P. wallichiana* var. *magna* (Christ) Tagawa; *P. tripartita* var. *magna* Christ—— NACHI-SHIDA. Fronds large, deciduous; stipes stout, nearly 2 cm. across, yellow- or purple-brown, lustrous, trisulcate on upper side when dried, rounded on back, smooth, with brown membranous scales at base; blades large, 5-angled, often more than 1 m. long and nearly as wide, with pinnatifid pinnae, the rachis yellow-brown, smooth, sulcate on upper side, the terminal pinna longest, 50–90 cm. long, 25–40 cm. wide; pinnules many, herbaceous, linear-lanceolate, 15–25 cm. long, 2.5–4 cm. wide, acuminate, sessile, regularly pinnatiparted; costas with a soft spine at base of the pinnules; pinnules spreading, 1–2.5 cm. long, 3–5 mm. wide, acute to obtuse, the sterile portion toothed; veinlets ending short of the margin, the lowest ones anastomosing and forming areoles along the costas of the pinnules.——Honshu (Awa, Izu and Mikawa Prov. to s. Kinki Distr.), Shikoku, Kyushu.——Ryukyus and Formosa.

10. CONIOGRAMME Fée IWAGANE-SŌ ZOKU

Terrestrial; rhizomes creeping, scaly, dictyostelic; fronds large, herbaceous, pinnate to tripinnate, with entire or finely toothed divisions, mostly glabrous; veins free, or sometimes anastomosing, without included veinlets, ending in hydathodes; sori elongate along the veins except near the margin; indusia absent.——About 20 species, in Africa, Polynesia to Japan, 1 in Mexico.

1A. Veins free. .... 1. *C. fraxinea* var. *intermedia*
1B. Veins anastomosing. .... 2. *C. japonica*

1. Coniogramme fraxinea (Don) Diels var. **intermedia** (Hieron.) C. Chr. *C. intermedia* Hieron.; *C. fraxinea* sensu auct. Japon., non Diels; *Gymnogramme javanica* sensu auct. Japon., non Bl.; *C. fraxinea* var. *serrulata* Nakai, non Hieron. ——IWAGANE-ZENMAI. Rhizomes creeping, about 5 mm. across, with membranous brown lanceolate entire acuminate

scales 3–4 mm. long; fronds 70–130 cm. long; stipes 30–60 cm. long, glabrous, thinly scaly at base, sulcate on upper part, rounded and often purplish on lower side; blades ovate, 40–70 cm. long, 25–30 cm. wide, simply pinnate; pinnae 5–8 on each side, chartaceous or thinly herbaceous, oblong- to linear-lanceolate, 12–20 cm. long, 2–4.5 cm. wide, abruptly contracted with a linear tail at the tip, rounded to abruptly acute at base, minutely toothed, glabrous on both sides or very thinly pubescent along the costas and veins beneath, the lower pinnae often with 1 or 2 smaller divisions on each side near base, petiolate; veins closely parallel, once or twice forked, free, reaching the base of the marginal tooth; sori along the lower 2/3–4/5 of the dorsal part of the veins, yellow.——Woods in low mountains; Hokkaido, Honshu, Shikoku, Kyushu; common.—— Sakhalin, s. Kuriles, Korea, Manchuria, China to India. The typical phase occurs in India to Malaysia and Hawaii.

2. Coniogramme japonica (Thunb.) Diels. *Hemionitis japonica* Thunb.; *Gymnogramme japonica* (Thunb.) Desv.; *Dictyogramme japonica* (Thunb.) Fée; *Notogramme japonica* (Thunb.) Presl——IWAGANE-SŌ. Closely allied to the preceding; pinnae usually more or less gradually acuminate; veins obliquely spreading, anastomosing in lower half, not reaching the teeth of the pinnae.——Woods in low mountains; Honshu, Shikoku, Kyushu; rather common.——Ryukyus (Amamioshima), Formosa, China, and Korea.

Var. **fauriei** (Hieron.) Tagawa. *C. fauriei* Hieron.——INU-IWAGANE-SŌ. The veins barely anastomosing along the costas in lower half of the pinnae.——Honshu, Kyushu; rare.—— Korea.

11. CHEILANTHES Sw.   EBIGARA-SHIDA ZOKU

Terrestrial; rhizomes short-creeping or erect, scaly; fronds small, usually tufted, pinnate to decompound, hairy or scaly, rarely smooth; veins free; sori marginal, on the tips of the veins, often in contact but not laterally confluent; indusia formed from the more or less modified reflexed margin of the segments, often obsolete or wanting.——About 180 species, in tropical and warm temperate regions.

1. Cheilanthes chusana Hook. *C. mysurensis* sensu auct. pro parte, non Wall.——EBIGARA-SHIDA. Rhizomes short, ascending, densely covered with the persistent bases of old stipes and with lustrous dark brown linear scales about 3 mm. long; fronds tufted; stipes 1.5–10 cm. long, purple-brown, lustrous, terete, 2-striate on the upper side, prominently scaly; scales on stipes and rachis rather unequal, broadly linear, 1–2 mm. long, spreading, deep brown, entire; blades oblong-ovate in the smaller ones, nearly linear in the larger ones, 7–30 cm. long, 18–30 mm. wide, acute, scarcely narrowed at base, bipinnatiparted, the rachis purple-brown, with a prominently scaly raised line on the margins above; pinnae 7–15 on each side, herbaceous, opposite or nearly so, triangular- to oblong-ovate, 1–2.5 cm. long, obtuse, sessile, glabrous, spreading to ascending, the costa impressed on upper side, raised and purple-brown beneath; pinnules oblong or oblong-ovate, 2–7 mm. long, obtuse to rounded, crenately few-toothed; veins and veinlets indistinct on both sides; sori marginal, on the ends of the veinlets; indusia apparently soon confluent, rusty-brown.—— Honshu (Chūgoku Distr.), Shikoku, Kyushu.——s. Korea and China.

12. ALEURITOPTERIS Fée   HIME-URAJIRO ZOKU

Terrestrial small ferns with short, ascending, scaly rhizomes; fronds tufted; stipes purple- to black-brown, polished, naked or scaly; blades deltoid to ovate, bipinnatiparted with large bipinnatifid basal pinnae, more or less densely white- or yellow-farinose beneath; veins free; sori marginal, on the tips of the veins, covered by the reflexed margins of the segments.——About 15 species, Africa to Malaysia, China, Japan, and Mexico.

1A. Blades ternately parted, shorter than the stipes, 2–8 cm. long, as wide or slightly wider; pinnae not or only slightly shorter than the blade, decurrent on the rachis on anterior side. . . . . 1. *A. argentea*
1B. Blades bipinnate to 2- or 3-pinnately parted, manifestly longer than wide; lowest pinnae much shorter than the blade, sessile, not decurrent.
2A. Stipes 7–20 cm. long, longer than the blades; blades triangular-ovate; scales on stipe obsoletely striate. . . . . 2. *A. krameri*
2B. Stipes 3–15 cm. long, usually shorter than the blades; blades ovate to oblong-ovate; scales very finely striate.
                                                                                                    3. *A. kuhnii* var. *brandtii*

1. Aleuritopteris argentea (Gmel.) Fée *Pteris argentea* Gmel.; *Cheilanthes argentea* (Gmel.) Kunze——HIME-URA-JIRO. Rhizomes short, clothed with old stipes and dark brown lustrous scales about 2 mm. long; fronds evergreen, tufted; stipes lustrous, fragile, terete, 10–25 cm. long, 1–1.5 mm. across, sparsely scaly toward the base; blades 5-angled, firmly herbaceous, 2–8 cm. long, shorter than the stipes, as wide to slightly wider than long, ternately parted, glabrous and gray-green above, white- to pale yellow-farinose beneath; pinnae slightly shorter than the blade, obtuse, decurrent on the rachis on the anterior side, pinnatiparted, the 2 lowest posterior pinnae elongate, sometimes pinnatilobed, the terminal segment decurrent on the rachis; sori marginal; false indusia (reflexed margin of the pinnae) hyaline, entire or crenate-undulate.—— Dry rocks; Honshu (Rikuchu, Kantō Distr., Kii), Shikoku, Kyushu.——Korea, Manchuria, Siberia, China, Formosa to Malaysia, and India.

2. Aleuritopteris krameri (Fr. & Sav.) Ching. *Cheilanthes krameri* Fr. & Sav.——IWA-URAJIRO. Rhizomes short, densely covered with stubs of old stipes and dark brown lustrous linear scales 3–4 mm. long; fronds glabrous, fragile, tufted; stipes 7–20 cm. long, longer than the blades, sparsely scaly at base; scales obsoletely striate, terete, lustrous, dark purple-brown; blades thickly herbaceous, triangular-ovate, 4–15 cm. long, acuminate, bipinnatiparted in the lower portion, pinnatiparted in the upper portion, the rachis sulcate on upper side; pinnae spreading, obtuse, sessile, gray-green and glabrous on upper side, white-farinose beneath, the lowest ones longer on the posterior side, the upper pinnae decurrent on the rachis; pinnules lobed to parted, crenate, rounded at apex; veins obsolete; sori marginal; false indusia confluent and lobed.—— Rocks; Honshu (Kantō Distr.); rare.

3. Aleuritopteris kuhnii (Milde) Ching var. **brandtii** (Fr. & Sav.) Tagawa. *Cheilanthes brandtii* Fr. & Sav.; *C.*

kuhnii var. *brandtii* (Fr. & Sav.) Tagawa——Miyama-urajiro. Rhizomes short, covered with stubs of old stipes and dark brown lanceolate scales; fronds fragile, tufted; stipes 3–15 cm. long, terete, purple- to red-brown, lustrous, the scales at base membranous, lanceolate, 4–6 mm. long, finely striate; blades usually longer than the stipes, ovate to oblong-ovate, 8–30 cm. long, 4–12 cm. wide, glabrous on upper side, more or less white-farinose beneath, 2- or 3-pinnatiparted, acuminate, acute at the tip, the rachis purplish, sulcate on upper side; pinnae sessile; pinnules oblong-ovate to oblong, obtuse to subacute, lobed to parted, the lower ones usually sessile, the upper ones and sometimes the lower pinnules decurrent on the rachis, crenate; veins very slender, usually invisible on underside; sori marginal; false indusia lobulate, scarious.——Mountains; Honshu (centr. and Kantō Distr.); rather rare.——Forma **efarinosa** (Makino) Tagawa. *Cheilanthes brandtii* var. *efarinosa* Makino. Blades green beneath.——The typical phase occurs in Korea, China, Manchuria, and Amur.

13. CRYPTOGRAMMA R. BR. Rishiri-shinobu Zoku

Terrestrial with solenostelic or dictyostelic rhizomes; fronds tufted or remote, small, herbaceous, glabrous, pinnately dissected, often dimorphic, the fertile fronds with narrower and longer pinnules than the sterile; veins free; sori submarginal, on the branches of the forked veins, covered by a mostly continuous reflexed margin (false indusium); paraphyses absent.——Four species, widely distributed in cold regions of the N. Hemisphere.

1A. Rhizomes rather short, densely leafy and thickly clothed with stubs of old stipes; fronds tufted; blades 2- to 4-pinnate, herbaceous.
1. *C. crispa*
1B. Rhizomes very slender, sparsely leafy, without old stipes; fronds not tufted; blades 1- or 2-pinnate, membranous. ...... 2. *C. stelleri*

1. **Cryptogramma crispa** (L.) R. Br. *Osmunda crispa* L.; *Pteris crispa* (L.) All.; *Allosorus crispus* (L.) Bernh.—— Rishiri-shinobu. Rhizomes rather short, ascending, densely clothed with stubs of old stipes; fronds glabrous, tufted; stipes 4–20 cm. long, light brown, 1.5–2 mm. across at base, smooth, sulcate on upper side, sparsely scaly at base; scales on stipes and rhizomes membranous, broadly lanceolate to narrowly deltoid-ovate, 4–6 mm. long, pale brown, deciduous; blades herbaceous, ovate to narrowly so, 5–15 cm. long, 2.5–6 cm. wide, acute, yellowish green, 2- to 4-pinnate; pinnae 4–7 on each side, oblong-ovate, short-petiolate; ultimate sterile pinnules ovate to oblong, 5–8 mm. long, 2–6 mm. wide, few-toothed or lobed to parted; veinlets not reaching the margin of the tooth; ultimate fertile pinnules lanceolate, 4–10 mm. long, 2–4 mm. wide, subobtuse, crenate; sori terminal on the veinlets; false indusium hyaline, undulate.——July–Oct. Rocky alpine slopes; Hokkaido, Honshu (n. distr.); rare.——s. Kuriles, s. Sakhalin, Siberia to Europe.

2. **Cryptogramma stelleri** (Gmel.) Prantl. *Pteris stelleri* Gmel.; *Allosorus stelleri* (Gmel.) R. Br.; *Pellaea stelleri* (Gmel.) Bak.; *Pteris gracilis* Michx.——Yatsuga-take-shinobu. Rhizomes creeping, very slender, about 1.5 mm. across, sparsely leafy and scaly; scales broadly lanceolate, 1–2 mm. long, fugacious, pale brown; fronds delicate, glabrous; stipes 3–10 cm. long, 0.5–0.7 mm. across, light brown, slightly lustrous, smooth, sparsely scaly at base; blades membranous, oblong-ovate or ovate, 3–7 cm. long, 2–3 cm. wide, 1- or 2-pinnate; pinnae simple or ternate, very short-petiolulate to subsessile; sterile pinnules ovate to obovate or flabellate, 4–7 mm. long, rounded to crenate-toothed on upper half, broadly cuneate and entire on lower portion; fertile lobes broadly lanceolate, 5–12 mm. long, 3–4 mm. wide, crenate to subentire.——July–Sept. Wet shaded rocks, alpine regions; Honshu (centr. distr.); very rare.——Siberia, the Himalayas, and N. America.

14. ONYCHIUM Kaulf. Tachi-shinobu Zoku

Terrestrial; rhizomes creeping, or short and compact, solenostelic, scaly; fronds tripinnate to decompound; pinnules small, herbaceous to subcoriaceous, narrow, glabrous; veins free except the fertile commissure connecting the tips; sori continuous along both margins, covered by the scarious reflexed marginal or submarginal false indusium; paraphyses absent.——Few species, in the Old World.

1. **Onychium japonicum** (Thunb.) Kunze. *Trichomanes japonicum* Thunb.; *Caenopteris japonica* (Thunb.) Thunb.; *Cryptogramma japonica* (Thunb.) Prantl——Tachi-shinobu, Kan-shinobu. Rhizomes creeping, sparsely leafy, 2.5–4 mm. across, with brown membranous broadly lanceolate scales 2.5–3 mm. long; fronds evergreen, glabrous; stipes slender, 10–50 cm. long, 1.5–3 mm. across near base, pale green to straw-colored, smooth, striate on the upper side, purple-brown toward the base; blades triangular-ovate, 15–40 cm. long, 8–20 cm. wide, acuminate, deep green on upper side, paler beneath, 3- or 4-pinnate; pinnae obliquely ascending, gradually smaller in the upper ones, alternate, the lowest ones 1/2–2/3 as long as the blade, triangular-ovate, long-acuminate, prominently petiolulate; pinnules narrowly lanceolate, 5–8 mm. long, 1–1.5 mm. wide, acute, entire, the sterile ones often few-toothed; veins unbranched in the sterile pinnules, pinnately branched in fertile ones; sori marginal; false indusia hyaline, paired, often overlapping while young, entire, 3–8 mm. long.——Low mountains; Honshu (Kantō Distr. and westw.), Shikoku, Kyushu; rather common.——s. Korea, China, Formosa, Malaysia to India.

15. PLEUROSORIOPSIS Fomin Karakusa-shida Zoku

Terrestrial rhizomatous fern; fronds remote, small; blades ovate to ovate-oblong, herbaceous, 1(-2)-pinnate, with numerous spreading brown multicellular hairs; veins free, not prominent on both sides, not reaching the margin of the pinnules; sporangia along the lower and middle portions of the veins, naked.——A single species.

1. Pleurosoriopsis makinoi (Maxim.) Fomin. *Gymnogramme makinoi* Maxim.; *Anogramme makinoi* (Maxim.) Christ——KARAKUSA-SHIDA.   Terrestrial, evergreen fern; rhizomes slender, scarcely 1 mm. across, long-creeping, densely red-brown woolly, loosely leaved, with few fugacious jointed short hairs at the tip; fronds 3–7 cm. long; stipes slightly shorter to longer than the blade, straw-colored, with red-brown somewhat woolly hairs 1–2 mm. long on lower half, with spreading pale brown jointed flexuous hairs 0.2–0.3 mm. long on upper half; blades herbaceous, ovate to ovate-oblong, 2–5 cm. long, 1–1.5 cm. wide, obtuse to subacute, 1(–2)-pinnate, with numerous spreading brown jointed hairs about 0.3 mm.

long on the rachis and on the segments of both sides, especially on margin; pinnae 3–5 pairs, obliquely spreading, ovate, obtuse, cut into a few segments, usually short-petiolulate; pinnules oblong, 1–1.5 mm. wide, obtuse, entire or 2- or 3-lobed; veins obsolete on both sides, solitary in the lobes, not reaching the margin; sori along the veins on underside of the pinnules, margin; sori along the veins on underside of the pinnules, often confluent with the adjacent ones, rusty-brown, exindusiate.——Among mosses on rocks in mountains; Hokkaido, Honshu, Shikoku, Kyushu; rare.——Korea, China, Manchuria, and e. Siberia.

16. ADIANTUM L.   HAKONE-SHIDA ZOKU

Terrestrial; rhizomes long-creeping or short, scaly, solenostelic; stipes scaly only at base, lustrous; blades membranous to coriaceous, 1- to 3-pinnate, or decompound, or sometimes simple, glabrous or rarely hairy; veins free or rarely anastomosing; sori marginal, the sporangia along the veins, on underside of the reflexed margins (false indusium) of pinnules.——More than 200 species, widely distributed, most abundant in S. America.

1A.   Ultimate pinnules cuneate to orbicular, equilateral; sori on outer margin.
    2A.   Blades 2- to 4-pinnate, rarely simply pinnate, not proliferous at the tip; pinnules cuneate at base; rhizomes creeping.
        3A.   Ultimate pinnules not lobed, subcoriaceous; sori mostly solitary on the center of upper margin. . . . . 1. *A. monochlamys*
        3B.   Ultimate pinnules mostly 3- to 5-palmately lobed, membranous; sori few on a pinnule, usually solitary on a lobe.
                                             2. *A. capillus-veneris*
    2B.   Blades simply pinnate, often elongate and proliferous at the tip; pinnae orbicular, thinly chartaceous, usually shallowly 2- to 4-lobulate; rhizomes short. . . . . 3. *A. capillus-junonis*
1B.   Ultimate pinnules broadly falcate-lanceolate to obovate, dimidiate; sori on anterior margin.
    4A.   Blades simply pinnate, glabrous, often elongate and proliferous at the tip; false indusium glabrous. . . . . 4. *A. edgeworthii*
    4B.   Blades with ternately to pedately divided pinnae, if simply pinnate, the pinnae and false indusium more or less setose.
        5A.   Blades, at least in part, sometimes setose.
            6A.   Rachis glabrous; pinnules membranous, thinly setose at least along the posterior margin; false indusium cordate, distant, prominently setose on back. . . . . 5. *A. diaphanum*
            6B.   Rachis densely setulose on upper side; pinnules somewhat coriaceous, glabrous; false indusium transversely linear-oblong, contiguous, glabrous. . . . . 6. *A. flabellulatum*
        5B.   Blades glabrous throughout. . . . . 7. *A. pedatum*

1. Adiantum monochlamys Eaton. *A. venustum* var. *monochlamys* (Eaton) Luerss.; *A. veitchii* Hance——HAKONE-SHIDA, HAKONE-SŌ.   Rhizomes short-creeping, 3–4 mm. across, with purple-brown to dark brown broadly linear scales 3–4 mm. long; fronds evergreen, glabrous; stipes 10–20 cm. long, purple-brown to dark brown, lustrous, terete; blades narrowly triangular-ovate, 10–25 cm. long, 4–8 cm. wide, smooth, 3- or 4-pinnate, the rachis lustrous, purple-brown, flexuous; pinnae 4–6 on each side, the upper ones gradually reduced; pinnules subcoriaceous, obtriangular, 5–12(–15) mm. long, 4–8(–12) mm. wide at the truncate-rounded toothed apex, cuneate in lower half, short-petiolate, slightly lustrous on upper side, often slightly glaucous beneath; veins slender, obsolete, palmate; sori mostly solitary in the center of upper margin; false indusia reniform, 2–3 mm. long, 2.5–4 mm. wide, glabrous, entire.——Mountains; Honshu (Uzen and Rikuzen Prov. southw.), Shikoku, Kyushu; common.——s. Korea, Formosa, and China.

2. Adiantum capillus-veneris L.   HŌRAI-SHIDA.   Rhizomes creeping, 2–3 mm. across, with lustrous, brown, broadly linear scales 2–2.5 mm. long; fronds glabrous; stipes 2–15 cm. long, slender, dark purple-brown, lustrous, scaly at base; blades broadly lanceolate to triangular-ovate, 4–20 cm. long, 2–10 cm. wide, simply pinnate, bi- to partially tripinnate, the rachis purple-brown, lustrous, slightly flexuous, slender; pinnae usually 7–9 pairs, 3–4 pairs in depauperate blades, petiolulate; pinnules membranous, flabellate to broadly cuneate, 7–15 mm. long, nearly as wide, palmately 3- to 5-lobed, green above,

usually glaucous beneath, slenderly veined on both sides, minutely toothed along the upper margin; sori few on a pinnule, usually solitary on each of the ultimate pinnules; false indusia transversely oblong or slightly reniform, glabrous, about 1.5 mm. long, 2–3.5 mm. wide.——Rocky cliffs; Honshu (Sagami Prov., Izu Isls.), Shikoku, Kyushu; rare.——Widely distributed in the tropics and subtropics of both hemispheres.

3. Adiantum capillus-junonis Rupr.   *A. cantoniense* Hance——HŌRAI-KUJAKU.   Rhizomes small, short, clothed with broadly linear brown scales about 3 mm. long; fronds tufted, slender, glabrous; stipes 1–7 cm. long, dark purple-brown, lustrous, 0.3–0.5 mm. across; blades oblong-lanceolate, 6–20 cm. long, 2–3 cm. wide, simply pinnate, the rachis purple-brown, very slender, sometimes much elongate and proliferous at the tip; pinnae thinly chartaceous, orbicular, 1–2 cm. long and as wide or slightly wider than long, minutely undulate-crenate, shallowly 2- to 4-lobulate on upper margin, the petiolules 2–3 mm. long; veins numerous, slender, parallel; sori 2–4 on a pinnule, transversely oblong, 1.2–1.5 mm. long, 1.5–2.5 mm. wide, usually free, rarely confluent.——Kyushu (Buzen Prov.); very rare.——China and Manchuria.

4. Adiantum edgeworthii Hook.   *A. caudatum* var. *rhizophorum* Wall. ex C. B. Clarke; *A. caudatum* var. *edgeworthii* (Hook.) Bedd.——OTOME-KUJAKU.   Rhizomes small, short, the scales about 2 mm. long, linear, lustrous, black-brown with paler margins; fronds glabrous, tufted; stipes 4–8 cm. long, slender, purple-brown, sparsely scaly, about 0.5 mm. across; blades simply pinnate, lanceolate, 5–12 cm. long, 15–20 mm. wide, glabrous, the rachis often much extended beyond the upper pinnae and proliferous at the tip; pinnae 15–20 on each side, membranous, obliquely flabellate, 7–13 mm. long, 4–6 mm. wide near base, obtuse, the upper ones gradually reduced, spreading, subentire to lobulate on anterior margin, entire on the posterior side, glaucescent beneath, very short-petiolulate; veins evident, slender; sori on the lobules of the pinnae; false indusia reniform, brown, about 1 mm. long, 1–2 mm. wide, glabrous.——Kyushu (Buzen Prov.); rare.—— Manchuria, China, Formosa to Indochina, India, and Philippines.

5. Adiantum diaphanum Bl. *A. setulosum* J. Smith—— SUKIYA-KUJAKU. Rhizomes short, the scales broadly linear, membranous, about 1 mm. long; fronds slender; stipes purple-brown, slender, lustrous, 10–20 cm. long, 0.5–1 mm. across, with scattered spreading scales toward base; blades simply pinnate, bipinnate, or with 1 or 2 pairs of elongate basal pinnae, the terminal pinna lanceolate, 7–12 cm. long, 2–2.5 cm. wide, acute, the rachis slender, glabrous; pinnules membranous, 7–18 pairs, close together or overlapping, oblong-squarrose, oblique, obtuse to rounded, cuneate at base, obtusely and irregularly toothed except the lower two-thirds on posterior side, with scattered dark brown bristles on both sides and on posterior margin, the petioles about 1 mm. long; sori in deep marginal sinuses near tip; false indusia cordate, gray-brown, entire, hyaline on margin, about 1 mm. long and as wide, prominently setose on back.——Kyushu (Iki, Hirato, Yakushima); very rare. ——Bonins, Formosa, China, Malaysia, Polynesia to Australia, and New Zealand.

6. Adiantum flabellulatum L. *A. fuscum* Retz.; *A. amoenum* Wall. ex Hook. & Grev.——OKINAWA-KUJAKU. Rhizomes rather stout, short, densely clothed with yellow-brown, membranous, linear to broadly linear lustrous scales 4–5 mm. long; fronds firm, tufted; stipes 10–30 cm. long, black-purple to dark brown, lustrous, with 2 slender ridges on upper side, 1–2 mm. across, scaly near base; blades pedately divided, bipinnate; pinnae 3–10, lanceolate, 5–10 cm. long, 1.5–2.5 cm. wide, obtuse to subacute, short-petiolulate, the rachis densely dark brown setulose on upper side; pinnules 5–18 pairs, somewhat coriaceous, elliptic, obovate to ovate, 7–15 mm. long, 4–7 mm. wide, oblique, rounded to obtuse, broadly cuneate at base, minutely toothed on upper margins, glabrous, obsoletely veined on both sides; sori few on each pinnule, contiguous; false indusia transversely linear-oblong, 1–1.2 mm. long, 1–2 mm. wide, contiguous, glabrous, dark brown, entire. ——Kyushu (Yakushima); rare.——Ryukyus, Formosa, China to India, and Malaysia.

7. Adiantum pedatum L. *A. pedatum* var. *kamtschaticum* Rupr. & var. *aleuticum* Rupr.——KUJAKU-SHIDA. Rhizomes short-creeping, with dark brown, lustrous, narrowly lanceolate scales 4–5 mm. long; fronds glabrous throughout, deciduous; stipes dark purple to purple-brown, lustrous, smooth, 20–50 cm. long, 2–3.5 mm. across at base, bifid at apex; blades flabellate, pedately divided, bipinnate; pinnae 8–23, short-petioluled, linear-lanceolate, 18–30 cm. long in the median, 2–4 cm. wide, obtuse, the rachis purple- to red-brown, lustrous, nearly terete; pinnules many, membranous, spreading, subovate, obtuse, straight to slightly recurved and entire on posterior margin, truncate to convex and lobulate on anterior margin, somewhat glaucous beneath, very short-petiolulate, slenderly veined; sori few on the anterior margin of the pinnules, solitary on a lobule; false indusia transversely oblong to lanceolate, about 1 mm. long, 1.5–3 mm. wide, hyaline on margin.——June-Oct. Woods in mountains; Hokkaido, Honshu, Shikoku; rather common.——China, Korea, Manchuria, e. Siberia, Sakhalin, s. Kuriles, Kamchatka, and N. America.

Fam. 13. **PARKERIACEAE** MIZU-WARABI KA Waterfern Family

Aquatic or subaquatic; rhizomes short, erect, sparsely brown-scaly; fronds pinnate, or pinnately decompound, glabrous, dimorphic, the fertile ones larger, more finely divided, with longer and narrower pinnules than the sterile, often proliferous in the axils; veins anastomosing; sporangia sessile on the veins, large, enclosed by the reflexed margins of the pinnules.——One genus, comprising a few species in the warmer regions of the world.

1. **CERATOPTERIS** Brongn. MIZU-WARABI ZOKU

Characters of the family.

1. Ceratopteris thalictroides (L.) Brongn. *Acrostichum thalictroides* L.; *A. siliquosum* L.; *Pteris thalictroides* (L.) Sw.; *C. gaudichaudii* Brongn.——MIZU-WARABI. Fronds tufted, usually dimorphic; sterile blades ovate to ovate-deltoid, 3–20 cm. long, 2.5–17 cm. wide, 1- to 4-pinnatiparted; pinnules lanceolate, with few coarse teeth or lobes; veins anastomosing; fertile blades ovate, 1- to 3-pinnate, erect, the stipes usually shorter than the blades, the ultimate pinnules linear-subulate, 2–7 cm. long, 2–3 mm. wide, obtuse, entire, sessile.——Aug.-Oct. Ponds and paddy fields; Honshu (Kantō Distr. and westw.), Shikoku, Kyushu; common.——s. Korea, Ryukyus, Formosa, China to India, Malaysia, and Polynesia.

Fam. 14. **DAVALLIACEAE** SHINOBU KA Davallia Family

Mostly epiphytic ferns; rhizomes creeping, rarely suberect, dictyostelic, scaly; stipes mostly interrupted and jointed with the rhizomes, rarely (*Nephrolepis*) approximate and not jointed; fronds pinnate, simple to decompound; veins free; sori submarginal or dorsal, terminal or very rarely dorsal on the vein, usually indusiate.——About 12 genera with about 300 species in the warmer regions of the world.

1A. Stipes jointed at base with the rhizomes; pinnae not jointed at base.
  2A. Indusium fixed by base and sides. . . . . 1. *Davallia*
  2B. Indusium attached only at base. . . . . 2. *Humata*
1B. Stipes not jointed at base; pinnae jointed at base. . . . . 3. *Nephrolepis*

1. DAVALLIA J. E. Smith   SHINOBU ZOKU

Epiphytic; rhizomes elongate, with peltate ciliate scales; stipes interrupted, jointed at the base; blades firm to rigidly coriaceous, deltoid to narrowly ovate, uniform or subdimorphic, mostly decompound and rather finely dissected, mostly glabrous, the ultimate rachises winged; veins free, sometimes terminating in the cartilaginous margin; sori terminal; indusia fixed by the base and sides, somewhat elongate.——About 40 species, Atlantic Islands, S. Africa, Madagascar, abundant in s. Asia and Polynesia.

1. Davallia mariesii Moore. *D. bullata* sensu auct. Japon., non Wall.——SHINOBU. Rhizomes long-creeping, rather stout, densely scaly, 4–5 mm. across, the scales brown, membranous, linear-lanceolate, 6–7 mm. long, about 1 mm. wide near base, long-acuminate, ciliate, scarious-margined; fronds deciduous, glabrous; stipes 8–13 cm. long, slender, jointed at base, smooth, light-green, sometimes reddish, with 2 lines on upper surface; blades broadly ovate-deltoid, 15–20 cm. long, 10–15 cm. wide, short-acuminate, 3- or 4-pinnatiparted, the rachis with pale brown peltate scales 2–3 mm. long while young; pinnae ovate-oblong, subacute, short–petiolulate, the lowest ones largest, ovate-deltoid, 7–12 cm. long, dilated on posterior side; pinnules oblong to oblanceolate, 1–2.5 mm. wide, subobtuse, entire or 2- or 3-lobulate at the tip; veins usually once, sometimes twice forked, or simple; indusia obovate, hyaline, about 1.5 mm. long, solitary on the pinnules.—— Aug.–Oct. Rocks and tree trunks in mountains; Hokkaido (Oshima Prov.), Honshu, Shikoku, Kyushu.——Korea, Formosa, and China.

2. HUMATA Cav.   KIKU-SHINOBU ZOKU

Epiphytic; rhizomes long-creeping, covered with appressed peltate scales; fronds coriaceous, interrupted, jointed at base with the rhizomes, simple and lanceolate or pinnately dissected and usually deltoid, glabrous; veins free, often very broad; sori terminal on the veins, usually submarginal; indusia rounded to broadly reniform, attached at base and sometimes laterally.—— About 50 species, in Malaysia and Polynesia to the Himalayas and Madagascar.

1. Humata repens (L. f.) Diels. *Adiantum repens* L. f.; *Davallia repens* (L. f.) Kuhn; *D. pedata* Smith; *D. chrysanthemifolia* Hayata——KIKU-SHINOBU. Evergreen; rhizomes long-creeping, wiry, 2–2.5 mm. across inclusive of the scales; scales brown, chartaceous, slightly lustrous, gradually narrowed at the tip, 4–5 mm. long, about 1 mm. wide, with a ciliate, hyaline, membranous margin; fronds 3–20 cm. long; stipes 1–12 cm. long, firm, pale green to pale brown, sparsely scaly; blades coriaceous, deltoid-ovate to narrowly so, 2–10 cm. long, 1.5–5 cm. wide, obtuse to acuminate, pinnate, glabrous, the veins on lower side flat, translucent, the rachis flat, with scattered ovate-peltate ciliate scales while young; pinnae few, oblong to lanceolate, obtuse, sessile or adnate at base, the sterile ones crenately toothed to subentire, the fertile ones dentate, the lowest usually dilated and pinnately lobed to parted on the posterior side; veins ascending; sori inserted in the sinus of the teeth; indusia orbicular or broadly elliptic, attached at base, about 0.7 mm. wide.——On rocks; Honshu (s. Kinki Distr.), Shikoku, Kyushu.——Ryukyus, Formosa, China, India, Mascarene Isls. to Polynesia, and Australia.

3. NEPHROLEPIS Schott   TAMA-SHIDA ZOKU

Terrestrial, usually stoloniferous ferns; rhizomes erect, short, scaly; fronds tufted, scaly, often hairy to glabrescent, simply pinnate; pinnae jointed on the rachis; veins free; sori terminal on the veins, dorsal or marginal; indusia fixed by a point or along the base.——Pantropic, with about 30 species.

1. Nephrolepis auriculata (L.) Trimen. *Polypodium auriculatum* L.; *N. cordifolia* sensu auct. Asiatic., non Presl—— TAMA-SHIDA. Evergreen; rhizomes short, ascending, covered with the persistent stubs of old stipes; stolons wiry, with a few globose densely scaly tubers, the scales thinly membranous, pale brown, linear-lanceolate, 7–10 mm. long, about 1 mm. wide, filiform at the tip, sparsely ciliate; fronds glabrous, tufted; stipes short, 5–10 cm. long, 2–3 mm. wide, nearly terete, the scales prominent, rounded, long-ciliate; blades linear-lanceolate to broadly linear, 25–50 cm. long, 3–6 cm. wide, narrowed at both ends, simply pinnate; pinnae many, spreading, narrowly oblong, 5–7 mm. wide, obtuse, appressed-toothed, sessile, rounded to subtruncate on posterior side at base, triangularly auriculate and subcordate on anterior side at base, the lower pinnae broader and shorter, oblong to elliptic, rounded at the tip; sori submarginal or nearly median between the costa and the margin; indusia obliquely reniform, attached at base, entire, about 1 mm. wide.——Thin woods and roadsides in hills near the sea; Honshu (Izu, Kii, and Nagato Prov.), Shikoku, Kyushu; locally common.——Ryukyus, Formosa, China, and generally throughout the Tropics and subtropics of the Old World.

Fam. 15. PLAGIOGYRIACEAE   KIJI-NO-O KA   Plagiogyria Family

Terrestrial, glabrous, scaleless; rhizomes short, erect, symmetrical, dictyostelic; stipes crowded, enlarged at base, triangular in cross section, with a double row of glandlike pneumatodes; blades herbaceous or coriaceous, pinnatifid to pinnate, glabrous, somewhat dimorphic; pinnae of the fertile blades narrow; veins forked, free except in the sorus.——A single genus with more than 20 species, New Guinea to the Himalayas and e. Asia, also in Central and S. America.

1A. Sterile and fertile pinnae short-petiolulate except for the few upper ones. .... 1. *P. euphlebia*
1B. Sterile pinnae sessile, the fertile sessile or short-petiolulate.

2A. Sterile pinnae contracted at base, at least the lower ones; fertile pinnae short-petioluled, at least the lower ones.
  3A. Pinnae in upper third to fourth of the sterile blade decurrent on the winged rachis; terminal pinna pinnately lobulate at the base; lateral pinnae falcate. .... 2. *P. adnata*
  3B. Pinnae only of the last few pairs toward tip decurrent on the winged rachis; terminal pinna usually unlobed; lateral pinnae straight. .... 3. *P. japonica*
2B. Sterile pinnae broadly decurrent on the rachis, not contracted at base; fertile pinnae sessile or short-petiolulate.
  4A. Fertile pinnae short-petiolulate, apiculate; rachis 2-angled beneath; blades usually with a few pairs of much reduced deflexed flabellate or deltoid sterile pinnae at base. .... 4. *P. stenoptera*
  4B. Fertile pinnae slightly decurrent on the rachis; rachis of sterile blades flat; blades without reduced pinnae at base.

                                                                                        5. *P. matsumureana*

1. PLAGIOGYRIA (Kunze) Mett.   KIJI-NO-O-SHIDA ZOKU

Characters of the family.

1. Plagiogyria euphlebia (Kunze) Mett. *Lomaria euphlebia* Kunze; *Stenochlaena triquetra* J. Smith; *P. triquetra* (J. Smith) Mett.——Ō-KIJI-NO-O.   Evergreen; rhizomes stout, short; fronds crowded, glabrous, simply pinnate, to 1 m. high; stipes 30–60 cm. long; sterile blades 30–60 cm. long, 20–40 cm. wide, the rachis 2-grooved on upper side, rounded on back; pinnae chartaceous, spreading, 10–15 pairs, linear-lanceolate, 10–20 cm. long, 12–20 mm. wide, long-acuminate, acute at the tip, cuneate-rounded at base, short-petiolulate, minutely toothed toward base, more prominently so toward the tip, the terminal pinna usually distinct, sometimes few-lobed at the base, the costa slightly raised, the veins distinct on both sides, usually once-forked, horizontally spreading, closely parallel; fertile pinnae short-petiolulate, 7–29 cm. long.——Honshu (Echigo and Hitachi Prov. and westw.), Shikoku, Kyushu. ——s. Korea (Quelpaert Isl.), Ryukyus, Formosa, China to India.

2. Plagiogyria adnata (Bl.) Bedd. *Lomaria adnata* Bl.; *P. rankanensis* Hayata; *P. yakushimensis* K. Satō; *P. adnata* var. *yakushimensis* (K. Satō) Tagawa——TAKASAGO-KIJI-NO-O. Rhizomes stout, ascending to erect; fronds crowded; stipes 5–25 cm. long in the sterile fronds, 20–50 cm. long in the fertile ones, 3-angled at the base, somewhat 4-angled in upper part; sterile blades 15–30 cm. long, 6–15 cm. wide, the rachis winged at least in the upper portion; pinnae subcoriaceous, spreading, narrowly lanceolate, acute or sometimes the tip elongate, somewhat falcate, serrulate near the tip, decurrent, the basal not decurrent, the terminal pinna pinnately lobulate at the base; veins slender, spreading, usually once-forked, slightly raised on both sides; fertile pinnae narrowly linear, 4–6 cm. long, the lower ones short-petiolulate sessile but the upper ones scarcely decurrent.——Honshu (Izu Prov. and westw.), Shikoku, Kyushu.——Ryukyus, Formosa, China to Malaysia, and India.

3. Plagiogyria japonica Nakai. KIJI-NO-O-SHIDA.   Rhizomes stout, short, densely clothed with the stubs of old stipes; fronds crowded, the sterile ones ascending, the fertile ones longer, erect; stipes 15–40 cm. long in sterile fronds, 30–60 cm. long in the fertile, 3-angled at base; sterile blades 20–40 cm. long, 10–15 cm. wide, the rachis 2-grooved on upper side, rounded on lower side, not winged except near top; sterile pinnae 12–16 pairs, spreading, chartaceous-coriaceous, narrowly lanceolate, 5–8 cm. long, 8–12 mm. wide, gradually acuminate, subentire or crenate-toothed, sessile, the upper ones not or very slightly decurrent on the rachis; veins spreading, once forked or simple; fertile pinnae 5–10 cm. long, short-petiolulate.—— Honshu, Shikoku, Kyushu.

4. Plagiogyria stenoptera (Hance) Diels. *Blechnum stenopterum* Hance; *Lomaria stenoptera* (Hance) Bak.; *L. concinna* Bak.——SHIMA-YAMA-SOTETSU.   Rhizomes short, stout, erect or ascending, densely covered with the stubs of old stipes; stipes 5–15 cm. long in sterile fronds, 10–30 cm. long in the fertile, acutely trigonous, narrowly winged on the angles while young; sterile blades lanceolate, 25–45 cm. long, 5–12 cm. wide, acuminate, abruptly contracted at base, with few pairs of reduced flabellate or deltoid often deflexed pinnae, the rachis flat and scarcely raised above, prominently elevated and 2-angled beneath; pinnae of sterile blades 25–32 pairs, herbaceous, horizontally spreading, linear-lanceolate, acute to short acuminate, widely decurrent, crenate-toothed in lower portion, incurved-toothed toward the tip, the upper pinnae smaller, forming a pinnatilobed terminal pinna with a short tail; fertile fronds erect, with few deflexed pairs of flabellate to deltoid much reduced sterile pinnae at base; fertile pinnae linear, 4–8 cm. long, apiculate, short-petiolulate.——Kyushu (Yaku-shima).——Formosa, China, Indochina, and Philippines.

5. Plagiogyria matsumureana Makino. *Lomaria fauriei* Christ; *P. fauriei* (Christ) Matsumura——YAMA-SOTETSU. Rhizomes stout, short, erect or ascending, densely covered with the stubs of old stipes; fronds crowded; stipes of sterile fronds 7–15 cm. long, compressed-trigonous at base; sterile blades 30–60 cm. long, 10–20 cm. wide, abruptly acuminate, slightly narrowed at base, the rachis slightly raised, with flat faces on both sides; pinnae horizontally spreading, linear-lanceolate or broadly linear, the median ones 5–10 cm. long, 8–15 mm. wide, widely decurrent, acutely incurved-toothed in lower portion, irregularly and more coarsely so toward the long-acuminate tip; fertile pinnae linear, 3–8 cm. long, obtuse, widely decurrent.——Woods in mountains; Hokkaido, Honshu, Shikoku, Kyushu; rare southw.——s. Kuriles.

Fam. 16. CYATHEACEAE   HEGO KA   Treefern Family

Usually arborescent, with stout erect scaly dictyostelic trunks; fronds mostly large and pinnately decompound; sori dorsal, rounded; indusia globose, opening at the top, partial or wanting; pedicel of sporangium short, of more than three rows of cells. ——Seven genera, with about 1,000 species, of wide distribution in the tropics and subtropics of both hemispheres.

1A. Indusia present; axes not dark and polished; arborescent, with a distinct trunk. .... 1. *Cyathea*
1B. Indusia wanting; axes dark and polished; scarcely arborescent, without a distinct trunk. .... 2. *Gymnosphaera*

1. CYATHEA J. E. Smith   HEGO ZOKU

Arborescent, with erect trunks, scaly at apex; fronds large, crowded at the summit of the trunk, bipinnate or decompound, usually coriaceous, variously scaly and sometimes hairy, or nearly glabrous; veins free; sori dorsal on the veins or in the vein axils, usually with filamentose paraphyses among the sporangia; receptacle elevated, hemispheric, globose, or columnar; indusia complete and globose to partial or wanting.——About 800 species, mostly in humid tropical regions.

1. Cyathea fauriei (Christ) Copel. *Alsophila fauriei* Christ; *C. boninsimensis* auct. Japon., non Copel.; *C. spinulosa* auct. Japon., non Wall.——HEGO.   Trunk simple, 3 m. high or more, 10–20 cm. across near the top, densely covered with coarse wiry aerial roots; fronds crowded at the top of trunk, widely spreading, large, 1.5–2 m. long; stipes and rachis dark brown, slightly lustrous, spiny and asperous, with pale brown scurfy hairs while young, the stipes often 4 cm. across near base; pinnae many, chartaceous-herbaceous, 40–60 cm. long, 20–25 cm. wide, abruptly caudate-acuminate, green and gla-brous above, paler beneath, equilateral, bipinnatiparted, the rachis densely short-pilose on upper side; pinnules many, horizontally spreading, linear-lanceolate, 7–10 cm. long, about 20 mm. wide, gradually acuminate, approximate, closely pinnati-parted, with deciduous ovate thin scales 1–2 mm. long along the costae and costules beneath; ultimate segments oblong, sub-acute, slightly falcate, serrulate; indusia thinly membranous, globose, hyaline, easily broken when dried.——Honshu (Ha-chijo Isl.), Shikoku, Kyushu.——Ryukyus and Formosa.

2. GYMNOSPHAERA Bl.   MARUHACHI ZOKU

Mostly arborescent (ours without a distinct trunk), brown-scaly in upper part; fronds mostly bipinnate, rarely simply pinnate or tripinnate, the rachis black or sometimes brown, commonly lustrous, naked, scaly or rarely hairy; pinnae and pinnules sub-coriaceous; veinlets typically simple, free; sori dorsal on the veins, variously paraphysate; indusia wanting.——More than 30 species, in Madagascar to e. Asia and Fiji.

1. Gymnosphaera denticulata (Bak.) Copel. *Alsophila denticulata* Bak.; *Alsophila acaulis* Makino——KUSA-MARUHA-CHI.   Stems or rhizomes about 10 cm. long, 2–3 cm. across, creeping, densely covered with lustrous yellow-brown scales about 10 mm. long; fronds about 1.5 m. long; stipes 30–50 cm. long, dark brown, lustrous, scurfy-puberulent while young, densely scaly near base, the scales linear-lanceolate, 10–12 mm. long, membranous, yellow-brown, lustrous, ciliate near base; blades narrowly ovate, gradually narrowed toward tip, 2- or 3-pinnatiparted; pinnae broadly lanceolate, 20–35 cm. long, 4–10 cm. wide, long-acuminate, short-petiolulate or sessile, the rachis brown, lustrous, with dense incurved setulose hairs on upper side; pinnules broadly lanceolate, acute, sessile or sub-sessile, sometimes decurrent, toothed, usually pinnately lobed to parted, minutely scurfy on the costa and costules beneath, minutely setulose-hairy on the costas above; sori between the costule and margin of the pinnules or segments, rounded, naked.——Honshu (s. Kinki Distr.), Shikoku, Kyushu, rare.——Ryukyus, Formosa, and China.

Fam. 17. ASPIDIACEAE   Ō-SHIDA KA   Aspidium Family

Mostly terrestrial or rarely epiphytic; rhizomes creeping to erect, dictyostelic, scaly; fronds usually pinnate or simple to de-compound, sometimes dimorphic; sori mostly dorsal, very rarely marginal or extramarginal, rounded, sometimes elongate, rarely diffuse along veins or over the surface; indusia basal and opening around the margin, mostly rounded to reniform, or elongate, or peltate, rarely absent; annulus longitudinal, interrupted by the pedicel.——About 66 genera, with about 3,000 species, cosmopolitan.

1A. Fronds strongly dimorphic.
  2A. Blades pinnate.
    3A. Sori covering the entire lower surface of pinnae. .... 9. *Bolbitis*
    3B. Sori enclosed by the revolute margin of the pinnules.
      4A. Veins free. .... 1. *Matteuccia*
      4B. Veins anastomosing. .... 2. *Onoclea*
  2B. Blades simple; sori naked, occupying the entire undersurface of blade. .... 10. *Elaphoglossum*
1B. Fronds alike or nearly so; sori on veins.
  5A. Veins free or nearly so.
    6A. Sori elongate along the veins.
      7A. Pinnae not jointed at base.
        8A. Fronds glabrous or sparsely soft pilose. .... 18. *Athyrium*
        8B. Fronds densely hispid. .... 13. *Lastrea*
      7B. Pinnae jointed at base. .... 14. *Gymnocarpium*
    6B. Sori round.
      9A. Indusia inferior below and surrounding sporangia, globose, opening above, symmetrical. .... 3. *Woodsia*
      9B. Indusia superior (i.e. lateral to sporangia) or absent.
        10A. Indusia reniform, attached on the sinus, or orbicular to ovate and basal.
          11A. Indusia ovate to orbicular, basal. .... 17. *Cystopteris*
          11B. Indusia reniform, attached at a sinus, rarely absent.
            12A. Nodes of rachis enlarged. .... 4. *Acrophorus*
            12B. Nodes of rachis not enlarged.

13A. Rachis with colored articulate hairs. . . . . 12. *Ctenitis*
13B. Rachis glabrous or the hairs unicellular.
   14A. Base of stipe inflated, with numerous elongate scales; fronds with spreading hairs and sometimes with minute capitate glands; calcareous rocks. . . . . 5. *Hypodematium*
   14B. Base of stipe not inflated.
      15A. Blades and stipes with straight, spreading, translucent, hispid hairs. . . . . 13. *Lastrea*
      15B. Blades and stipes without straight hispid hairs.
         16A. Blades anadromous, i.e., the lowest pinnules always on the acroscopic (anterior) side. . . . . . . 8. *Arachniodes*
         16B. Blades catadromous, i.e., the lowest pinnules always on the basiscopic (posterior) side. . . . . . . 11. *Dryopteris*
  10B. Indusia peltate. . . . . 6. *Polystichum*
5B. Veins anastomosing.
  17A. Sori numerous, orbicular to linear; indusia usually present.
    18A. Indusia peltate; fronds glabrous, coriaceous, simply pinnate. . . . . 7. *Cyrtomium*
    18B. Indusia reniform to linear, not peltate.
      19A. Sori orbicular, rarely elongate; indusia not as below; fronds usually hispid. . . . . 15. *Cyclosorus*
      19B. Sori linear; indusia bursting longitudinally along the back of sorus; fronds glabrous. . . . . 19. *Diplaziopsis*
  17B. Sori diffuse on veins and veinlets; indusia absent; fronds hairy; blades pinnately lobed to parted. . . . . 16. *Dictyocline*

1. MATTEUCCIA Todaro   KUSA-SOTETSU ZOKU

Terrestrial; rhizomes stout, ascending to erect, dictyostelic, scaly; fronds crowded, dimorphic; blades of sterile fronds herbaceous to chartaceous-membranous, pinnate, the pinnae pinnatifid, the veins free; blades of fertile fronds smaller, on a longer stipe, less divided; sori nearly terminal on the veinlets, longitudinally contiguous in a single or double row on the pinnae, enveloped by the revolute margin and by a thin scalelike indusium, the receptacle somewhat raised; paraphyses absent. ——Few species, Europe, Himalayas, e. Asia, and e. N. America.

1A. Pinnae 8–20 pairs, the lower ones scarcely reduced; blades of sterile fronds ovate-elliptic to broadly deltoid-lanceolate, the pinnae pinnately lobed; fertile pinnae broadly linear, not torulose. . . . . 1. *M. orientalis*
1B. Pinnae 30–50 pairs, the lower ones gradually reduced; blades of sterile fronds broadly oblanceolate, the pinnae pinnatiparted; fertile pinnae narrowly linear, slightly torulose. . . . . 2. *M. struthiopteris*

1. Matteuccia orientalis (Hook.) Trevir. *Struthiopteris orientalis* Hook.; *Onoclea orientalis* (Hook.) Hook.; *Pentarhizidium orientale* (Hook.) Hayata; *Pteretis orientalis* (Hook.) Ching; *Pentarhizidium japonicum* Hayata; *M. japonica* (Hayata) C. Chr.; *Pteretis japonica* (Hayata) Ching ——INU-GANSOKU. Rhizomes stout; fronds deciduous, crowded, the sterile longer than the fertile; stipes of sterile fronds 20–60 cm. long, prominently scaly, straw-colored, the scales membranous, pale brown, linear-lanceolate to narrowly ovate, 7–20 mm. long, 1.5–4 mm. wide, long-acuminate, lustrous, entire; blades ovate-elliptic to broadly lanceolate-deltoid, 30–50 cm. long, 20–30 cm. wide, pinnate, abruptly short-acuminate, the rachis scaly; pinnae 8–20 pairs, narrowly lanceolate, 10–15 cm. long, 1.5–4(–5) cm. wide, nearly truncate at base, sessile, closely pinnatilobed, glabrous on both sides or brown paleate along the costa beneath; segments ovate, 4–7 mm. wide, obtuse to acute, obliquely spreading, toothed, the veins simple; fertile fronds 30–70 cm. long, the stipes terete, usually longer than the 1-sided blade, the pinnae close, ascending, dark brown, broadly linear, 5–10 cm. long, 5–6 mm. wide, sessile. ——Sept.–Nov. Hokkaido, Honshu, Shikoku, Kyushu.——s. Kuriles, Korea, China to the Himalayas.

2. Matteuccia struthiopteris (L.) Todaro. *Osmunda struthiopteris* L.; *Onoclea struthiopteris* (L.) Hoffm.; *Struthiopteris germanica* Willd.——KUSA-SOTETSU. Rhizomes short, erect, covered with the basal stubs of old stipes; sterile fronds tufted; stipes 10–20 cm. long, 3-angled, pale green, scaly toward the base, the scales brown, membranous, lanceolate, 1–1.5 cm. long, entire, often somewhat blackish striolate; blades of sterile fronds thinly herbaceous, broadly oblanceolate, 40–80 cm. long, 15–30 cm. wide, abruptly short-acuminate, gradually narrowed at base, vivid green; pinnae 30–50 pairs, broadly linear, 12–20 mm. wide, long-acuminate, sessile, closely pinnatiparted; segments oblong to narrowly ovate, 2.5–5 mm. wide, obliquely spreading, obtuse to acute, crenulate, the lowest pair sometimes slightly deflexed and stipulelike, the veins simple; fertile fronds 30–60 cm. long, erect, the stipes shorter than the blades, broadly sulcate on the upper side; pinnae narrowly linear, 4–6 cm. long, slightly torulose.——Mountains; Hokkaido, Honshu, Shikoku, Kyushu; locally abundant.——s. Kuriles, Kamchatka, Sakhalin, Korea, China, Manchuria, Siberia to Europe.

2. ONOCLEA L.   KŌYA-WARABI ZOKU

Terrestrial, with creeping dictyostelic rhizomes; fronds deciduous, dimorphic, the sterile pinnae with a mostly winged rachis; pinnae sinuate to lobed, thinly herbaceous, glabrous; veins freely anastomosing; fertile fronds bipinnate; pinnules sessile, interrupted, lobed, strongly involute and forming globular berrylike divisions; sorus solitary in each lobe, enclosed also by an interior indusium, the receptacle prominent, paraphyses absent.——A single widely distributed species.

1. Onoclea sensibilis L. var. **interrupta** Maxim. *O. sensibilis* auct. Asiat. non L.——KŌYA-WARABI. Glabrous fern, scantily paleate on young stipes and rhizomes with thinly membranous, fugacious, ovate, acute scales 2–3 mm. long; rhizomes elongate, creeping, 4–5 mm. across, remotely leaved; stipes 20–50 cm. long, straw-colored, brownish toward base, lustrous; sterile fronds broadly ovate-deltoid, 15–30 cm. long and nearly as wide, simply pinnate, contracted at tip to a short subacute point, the rachis prominently winged except near base, the pinnae 5–10 pairs, lanceolate, 10–20 cm. long, 1.5–3 cm. wide, obliquely spreading, shallowly and obtusely lobed, crenate or subentire, minutely scabrous on margin under magnification, lower 1 or 2 pairs gradually narrowed and sessile at base, broadly decurrent on the adjacent ones, green above, glaucous beneath; fertile fronds 10–20 cm. long, the pinnae erect, 3–7 cm. long, with sessile, berrylike, globular pinnules about 2 mm. across in 2 loose series, the pinnules with 3 or 4 thin-indusiate sori surrounded by the involute lobes of the pinnules.——Sept.–Oct. Wet places; Hokkaido, Honshu, Shikoku, Kyushu; rather common.——Korea, Manchuria, s. Kuriles, s. Sakhalin to e. Siberia. The typical variety occurs in N. America.

3. WOODSIA R. Br.   Iwa-denda Zoku

Terrestrial; rhizomes erect, dictyostelic, covered with broad thin scales; fronds herbaceous, pinnate or bipinnate, hairy and scaly, or glabrescent; veins free; sori rounded, dorsal on the blade, subterminal or dorsal on the veins; receptacle slightly raised; indusia inferior, fragile, globose, enclosing the sorus, irregularly parted at the top, or small, basal, and parted into hairlike segments; paraphyses absent.——About 23 species, in cold regions and high mountains of the N. Hemisphere, also in S. America, 1 species in S. Africa.

1A. Stipes not jointed; indusia shallowly and irregularly lobed, not ciliate; blades thinly herbaceous, green when dried, slightly whitish to
      glaucescent beneath. . . . . 1. *W. manchuriensis*
1B. Stipes jointed; indusia lobed to parted, ciliate.
   2A. Blades uniformly scaly or hairy.
      3A. Pinnae prominently oblique, auricled on anterior side at base.
         4A. Pinnae sessile, not decurrent. . . . . .2. *W. polystichoides*
         4B. Upper pinnae distinctly decurrent. . . . . 3. *W. intermedia*
      3B. Pinnae nearly equilateral, not auricled at base.
         5A. Rachis and pinnae hairy, but not scaly; indusia rather large, subglobose; stipe jointed at apex. . . . . 4. *W. macrochlaena*
         5B. Rachis and pinnae hairy and scaly; indusia small, salverform; stipe jointed below the apex.
            6A. Stipes jointed below the middle nearly at the base. . . . . 5. *W. ilvensis*
            6B. Stipes jointed above the middle. . . . . 6. *W. subcordata*
   2B. Blades glabrous.
      7A. Scales ovate to ovate-lanceolate, minutely toothed; indusia with 5 or 6 irregularly fimbriate lobes. . . . . 7. *W. hancockii*
      7B. Scales elliptic to ovate, entire; indusia finely dissected into hairlike segments. . . . . 8. *W. glabella*

1. **Woodsia manchuriensis** Hook. *Physematium manchuriense* (Hook.) Nakai; *Protowoodsia manchuriensis* (Hook.) Ching; *W. insularis* Hance——Fukuro-shida. Rhizomes short, erect, densely scaly; fronds tufted, green to somewhat whitish or glaucescent; stipes 1–5 cm. long, lustrous, red-brown, minutely hairy to nearly glabrous, densely scaly near base, the scales pale brown, membranous, lanceolate to linear, 2–4 mm. long; blades thinly herbaceous, lanceolate, 10–30 cm. long, 2–5 cm. wide, acuminate, narrowed at base, slightly whitish to glaucescent beneath, bipinnatiparted, nearly glabrous; pinnae broadly triangular-lanceolate, obtuse to subacute, broadly cuneate to nearly truncate at base, sessile; pinnules oblong, obtuse, usually obliquely spreading, usually obtusely toothed; sori 1–6 on each pinnule, nearer the margin than the costule; indusia large, thinly membranous, glabrous, globose, shallowly and irregularly lobed.——Shaded rocks in mountains; Hokkaido, Honshu, Shikoku, Kyushu.——Korea, Manchuria, China, Amur, and Ussuri.

2. **Woodsia polystichoides** Eaton. *W. polystichoides* var. *nudiuscula* Hook. and var. *veitchii* Hance——Iwa-denda. Rhizomes short, erect or ascending, densely scaly; fronds tufted; stipes 5–10 cm. long, red-brown, obliquely jointed at the tip, hairy and scaly throughout, the scales pale brown, lanceolate, acuminate; blades firmly herbaceous, narrowly lanceolate, acuminate with an obtuse tip, slightly narrowed at base, simply pinnate, hairy on both sides, thinly scaly beneath; pinnae spreading, oblong-lanceolate, 1–2.5 cm. long, 3–7 mm. wide, obtuse to acute, often falcate, obliquely cuneate at base, sessile, with an auricle on the anterior side, entire to obtusely dentate; sori in 1 series along margin on both sides; indusia subglobose, shallowly lobed, ciliate.——Exposed rocks in mountains; Hokkaido, Honshu, Shikoku, Kyushu; rather common.——s. Kuriles, Sakhalin, Amur, Ussuri, Manchuria, Korea, China, and Formosa.

3. **Woodsia intermedia** Tagawa. *W. brantii* Fr. & Sav. (?)——Inu-iwa-denda. Rhizomes short, densely scaly; fronds tufted; stipes 5–12 cm. long, straw-colored to brown, obliquely jointed at apex, hairy and scaly throughout, the basal scales pale brown, membranous, lanceolate, 4–5 mm. wide, ciliate; blades herbaceous, broadly lanceolate, 5–13 cm. long, 2–4 cm. wide, acute to subobtuse, very slightly narrowed at base; pinnae 4–12 pairs, spreading, oblong-ovate, 1.5–2.5 cm. long, 7–10 mm. wide, obtuse to subacute, obliquely cuneate at base, subauriculate at base on anterior side, undulate to shallowly lobulate, thinly hairy on both sides, with narrow hairlike scales beneath, the upper pinnae decurrent on the rachis; sori marginal; indusia subglobose, irregularly lobed, ciliate on the lobes.——Honshu (Bitchu Prov.); rare.——Korea and Manchuria.

4. **Woodsia macrochlaena** Mett. *W. sinuata* Makino; *W. japonica* Makino; *W. frondosa* Christ——Kogane-shida. Rhizomes short, densely scaly; fronds tufted; stipes 3–10 cm. long, straw-colored to red-brown, obliquely jointed at apex, with hairs and scales, the basal scales red-brown, linear-lanceolate, acuminate, scattered ciliate; blades narrowly deltoid-ovate to broadly lanceolate, 5–15 cm. long, 2–5 cm. wide, acuminate with an obtuse tip, scarcely narrowed and simply pinnate at base, pinnatiparted toward the tip, the rachis hairy; pinnae obliquely spreading, ovate or ovate-oblong, 1–3 cm. long, 5–15 mm. wide, obtuse, sessile, pinnatilobed to crenately toothed, hairy, the upper ones decurrent on the winged rachis; sori near the margin, rusty-brown; indusia subglobose, irregularly lobed and ciliate.——Honshu, Shikoku, Kyushu.——Korea, Manchuria and China.

Var. **glabrata** (Nakai) Nemoto. *W. sinuata* var. *glabrata*

Nakai——JŌSHŪ-KOGANE-SHIDA. Pinnae glabrescent, horizontally spreading.——Honshu (centr. distr.).

5. Woodsia ilvensis (L.) R. Br. *Acrostichum ilvense* L.; *W. hyperborea* var. *rufidula* W. Koch——MIYAMA-IWA-DENDA, RISHIRI-DENDA. Rhizomes erect or ascending, densely scaly; fronds tufted; stipes red-brown, slightly lustrous, 3–10 cm. long, jointed below the middle nearly to the base, loosely hairy and thiny scaly, the basal scales membranous, ovate-lanceolate, acuminate, loosely ciliate; blades herbaceous, lanceolate to oblong-lanceolate, 5–15 cm. long, 1.5–5 cm. wide, acute with an obtuse tip, bipinnate to bipinnatiparted, hairy on both sides, with narrow scales beneath; pinnae 5–15 pairs, obliquely spreading, narrowly triangular-ovate, 1–3 cm. long, 5–10 mm. wide, sessile; pinnules oblong to elliptic, obtuse, obtusely toothed to subentire, decurrent; sori rusty-brown; indusia small, salverform, irregularly parted into 5 or 6 narrow segments.——Rocky places; Hokkaido.——Sakhalin, Korea, Manchuria, Kamchatka, Siberia to Europe, and N. America.

6. Woodsia subcordata Turcz. *W. eriosora* Christ; *W. kitadakensis* Ohwi——KITADAKE-DENDA. Rhizomes short, erect; fronds tufted; stipes 3–6 cm. long, red-brown, with an oblique joint above the middle, loosely hairy and scaly, the scales pale red-brown, membranous, loosely ciliate, the lower ones ovate, the upper ones linear-lanceolate; blades lanceolate to broadly linear, 5–12 cm. long, 1–2 cm. wide, acute to subobtuse, gradually narrowed at base, pinnate; pinnae 15–20 pairs, spreading, herbaceous, narrowly deltoid-ovate to broadly lanceolate, 5–10 mm. long, 3–5 mm. wide, obtuse, truncate to broadly cuneate at base, sessile, pinnately lobed to parted, loosely hairy on both sides, loosely scaly beneath, the median ones larger; pinnules elliptic to oblong, entire or crenately dentate; sori submarginal; indusia salverform, with 4 or 5 finely laciniate lobes.——Alpine slopes; Honshu (centr. distr.); rare.——Korea, Manchuria, Amur, and Ussuri.

7. Woodsia hancockii Bak. *W. tsurugisanensis* Makino; *W. gracillima* C. Chr.——KENZAN-DENDA. Rhizomes short, erect or ascending, scaly; fronds glabrous, the stipes tufted, slender, 1.5–3 cm. long, straw-colored, brownish at base, jointed below the middle, scaly near the base, the scales brown, membranous, ovate to ovate-lanceolate, minutely toothed; blades 5–12 cm. long, 1–1.5 cm. wide, acuminate, pinnate to bipinnatiparted, slightly narrowed at base, the rachis greenish, smooth, glabrous; lower pinnae ovate-deltoid, sessile, the median larger, ovate-deltoid to narrowly ovate, 5–12 mm. long, 5–6 mm. wide, subacute, obliquely cuneate to subtruncate at base, short-petiolulate; pinnules few, obtuse to acute, sparsely crenate to lobulate; sori 1–3 on a pinnule; indusia small, salverform, with 5 or 6 irregularly fimbriate lobes.——Shikoku (Mount Tsurugi in Awa Prov.); rare.——n. Korea, Manchuria, and China.

8. Woodsia glabella R. Br. *W. yazawae* Makino——TOGAKUSHI-DENDA, KARAFUTO-IWA-DENDA. Rhizomes short, erect to ascending; fronds glabrous; stipes tufted, slender, jointed slightly above the base, 1–2.5 cm. long, sparsely scaly near base, the scales pale brown, membranous, elliptic to ovate, acuminate, entire; blades thinly herbaceous, narrowly lanceolate, 3–7 cm. long, 6–12 mm. wide, acute to acuminate, slightly narrowed at base, smooth; pinnae sessile, spreading, ovate-orbicular in the lower ones, ovate-triangular to rhombic-ovate in median and upper ones, 3–6 mm. long, 3–5 mm. wide, obtuse to acute, broadly cuneate to nearly truncate at base, sessile, pinnatiparted, rarely simply dentate; pinnules elliptic to obovate, obtuse, with few obtuse teeth; sori small, 1–3 on a segment, yellow-brown; indusia small, salverform, finely dissected into hairlike segments.——Hokkaido, Honshu (centr. mountains); rare.——Sakhalin, Kamchatka, Korea, China, Manchuria, Siberia to northern N. America.

4. ACROPHORUS Presl   TAIWAN-HIME-WARABI ZOKU

Terrestrial; rhizomes short-ascending or erect, dictyostelic, with broad dark brown scales; fronds clustered; stipes elongate, scabrous; blades deltoid-ovate, 4-pinnate, the nodes of rachis enlarged; ultimate pinnules small, herbaceous; veins free, with jointed hairs or linear scales on upper surface; sori dorsal on the pinnules, terminal or dorsal on the veins, the receptacle rounded, slightly raised; indusia reniform, entire, erose, or short-fimbriate, attached at the sinus on the basal side of the sorus; paraphyses absent.——Few species, in Malaysia, India, e. Asia to Polynesia.

1. Acrophorus stipellatus Moore. *A. nodosus* sensu auct. Japon., non Presl——TAIWAN-HIME-WARABI. Rhizomes short-creeping; fronds large, 1–2 m. long; stipes straw-colored, stout, lustrous, sparsely scaly in upper part, densely so in lower part, the scales brown, ovate to lanceolate, 8–15 mm. long, tardily deciduous and leaving a small scar on the stipe; blades thinly herbaceous, large, deltoid-ovate, acuminate, 4-pinnate, the rachis and its divisions smooth, often somewhat chestnut-brown, brown-pilose on nodes above; pinnae and pinnules sessile, spreading, acuminate, the lowest pinnae large, the posterior side broader than the anterior; ultimate pinnules narrowly ovate, 6–10 mm. long, obtuse, pinnately parted to lobed, with fleshy red-brown hairs on costas and costules, glabrous beneath; lobes and pinnules obtuse, entire, 1-veined, cuneate and decurrent at base; sori subterminal on the veins, dorsal, rounded; indusia small, suborbicular, erose, 0.2–0.3 mm. long.——Kyushu (Yakushima); rare.——Formosa, China to India.

5. HYPODEMATIUM Kunze   KINMŌ-WARABI ZOKU

Rhizomes short-creeping, densely clothed with large membranous scales; fronds with spreading hairs and sometimes with minute capitate glands; stipes inflated and with many elongate scales at base; blades deltoid-ovate, 3- or 4-pinnatiparted, white-setose; veins free; sori dorsal on the veins, the receptacle prominent; indusia vaulted, setose, reniform, usually asymmetrical, sometimes athyrioid or obsolete.——Few species, Africa to Malaysia and e. Asia. Often on calcareous rocks.

1A. Stipes glabrous at base, the scales 2.5–4 cm. long; fronds without glandular hairs; ultimate pinnules not overlapping; veins scarcely impressed on upper side. .... 1. *H. fauriei*
1B. Stipes pilose at the base, the scales 1–1.5 cm. long; fronds with short glandular hairs; ultimate pinnules usually more or less overlapping; veins strongly impressed on upper side. .... 2. *H. glandulosopilosum*

1. Hypodematium fauriei (Kodama) Tagawa. *Dryopteris fauriei* Kodama; *D. crenata* sensu auct. Japon., non O. Kuntze; *H. crenatum* sensu auct. Japon., non Kuhn——KINMŌ-WARABI. Rhizomes stout, short-creeping, the scales ascending, membranous, linear- to broadly lanceolate, 2.5–4 cm. long, 3–4 mm. wide, long-acuminate, lustrous, entire; fronds deciduous; stipes 7–30 cm. long, straw-colored, thickened and densely scaly but not hairy at base, with very spare spreading eglandular hairs toward the top; blades herbaceous, ovate-deltoid, 10–40 cm. long, 8–30 cm. wide, usually acuminate and subobtuse at the tip, 3- or 4-pinnate, vivid green, thinly pilose; pinnae to 30 cm. long, about 15 cm. wide, petiolate, the lowest largest, dilated on posterior side at base; ultimate pinnules oblong-ovate, 6–20 mm. long, obtuse, crenate or lobed to parted; veins scarcely impressed on upper side; sori 1–10 on each pinnule; indusia cordate, sometimes oblique, about 1 mm. wide, setulose on back.——Aug.–Oct. Calcareous rocks; Honshu (Kantō Distr. and Shinano Prov.), Shikoku, Kyushu; rare.

2. Hypodematium glandulosopilosum (Tagawa) Ohwi. *H. fauriei* forma *glandulosopilosum* Tagawa——KEKINMŌ-WARABI. Closely related to the preceding; rhizomes densely clothed with brown membranous linear-lanceolate long-acuminate scales 1–1.5 cm. long; stipes straw-colored, angled, slender, 10–15 cm. long, pilose and prominently scaly toward base; blades gray-green, 7–20 cm. long, 6–13 cm. wide, acute to obtuse, with white, spreading eglandular and short-glandular hairs on rachis and underside of blades; pinnae narrowly ovate, the lower ones prominently petiolulate, usually obtuse; ultimate pinnules obtuse, numerous, slightly imbricate.——July–Oct. Honshu (Chūgoku Distr.), Shikoku.——Korea.

6. POLYSTICHUM Roth   INODE ZOKU

Mostly terrestrial; rhizomes usually short, ascending to erect, usually with lacerate scales; stipes densely clustered, scaly; blades anadromous, not dilated at base, pinnately compound; pinnules mostly mucronate- to spine-toothed; scales usually filiform, stellate-peltate at base; veins mostly free; sori dorsal, sometimes apparently subterminal on the veins, orbicular; indusia peltate or rarely absent.——With more than 180 species, cosmopolitan.

1A. Blades pinnate; pinnae strongly oblique, the lowest pair again pinnate in No. 4.
  2A. Fronds firmly coriaceous; sori dorsal on veins.
    3A. Blades widest at base, often gemmate and producing a plantlet at tip of the much-prolonged rachis; pinnae 3–10 cm. long, subentire or undulate-toothed, with ovate to lanceolate ciliate scales beneath. .... 1. *P. lepidocaulon*
    3B. Blades narrowed at base, widest near the middle, not gemmate; pinnae 1.2–3 cm. long, often spine-toothed, with linear to linear-lanceolate entire scales beneath. .... 2. *P. lonchitis*
  2B. Fronds thinly coriaceous to herbaceous; sori terminal on veins.
    4A. Fronds suberect, 20–80 cm. long, not gemmate at tip; scales on rachis broadly lanceolate to ovate, erose-dentate; indusia about 1 mm. across, fugacious or soon shriveling.
      5A. Blades uniformly once-pinnate, much longer than the stipes; pinnae thinly coriaceous, narrowly oblong, 1–3 cm. long, obtuse to abruptly acute, mucronate- to spine-toothed. .... 3. *P. deltodon*
      5B. Blades bipinnate only at the base, as long as to slightly longer than the stipes; pinnae herbaceous, lanceolate, 2.5–5 cm. long, acuminate, coarsely awn-toothed or lobulate. .... 4. *P. tripteron*
    4B. Fronds spreading, 5–23 cm. long, often gemmate at the tip of the much prolonged rachis; scales on rachis linear to filiform, entire; pinnae 1–1.5 cm. long, obtuse; indusia about 2 mm. across, persistent, inflated. .... 5. *P. craspedosorum*
1B. Blades pinnate to bipinnate; pinnae equilateral or nearly so.
  6A. Blades not more than 2.5 cm. wide, usually simply pinnate.
    7A. Fronds scarcely tufted; blades 3–7 cm. long, 8–12 mm. wide. .... 6. *P. inaense*
    7B. Fronds tufted; blades 8–20 cm. long, 1.5–2.5 cm. wide. .... 7. *P. lachenense*
  6B. Blades more than 3 cm. wide.
    8A. Blades firmly coriaceous, lustrous above.
      9A. Scales on rachis black-brown, chartaceous, lanceolate to broadly linear; blades as long as to slightly longer than the stipes, not narrowed at base.
        10A. Scales on rachis and upper half of stipes lanceolate, 1–1.5 mm. wide, prominently ciliate; indusia impressed in the center. .... 8. *P. rigens*
        10B. Scales on rachis and upper half of stipes linear-lanceolate to nearly filiform, less than 1 mm. wide, sparsely short-ciliate; indusia flat.
          11A. Scales on rachis and upper half of stipes linear-lanceolate; fronds coarse; pinnules 8–15 mm. long, 5–8 mm. wide. .... 9. *P. mayebarae*
          11B. Scales on rachis and upper half of stipes linear to nearly filiform; fronds slender; pinnules 7–15 mm. long, 3–7 mm. wide. .... 10. *P. tsussimense*
      9B. Scales on rachis brown, membranous, linear-filiform, often contorted; blades much longer than the stipes, slightly narrowed at base. .... 11. *P. neolobatum*
    8B. Blades herbaceous to herbaceous-coriaceous.
      12A. Fronds all alike, not gemmate; pinnules spine-toothed.
        13A. Pinnules mostly not decurrent on the rachis, except those in very depauperate fronds, usually with finely appressed teeth.
          14A. Blades not or only slightly narrowed at base.
            15A. Some of the scales on lower part of stipes black or with a black center.
              16A. Teeth of pinnules ending in a spreading spine. .... 12. *P. doianum*
              16B. Teeth of pinnules ending in an appressed spine.
                17A. Scales on lower portion of rachis ovate-oblong. .... 13. *P. kurokawae*
              17B. Scales on lower portion of rachis lanceolate to linear-lanceolate.

1. Polystichum lepidocaulon (Hook.) J. Smith. *As-pidium lepidocaulon* Hook.; *Dryopteris lepidocaulon* (Hook.) O. Kuntze; *Cyrtomidictyum lepidocaulon* (Hook.) Ching; *Nephrodium faberi* Bak.——ORIZURU-SHIDA. Rhizomes short, ascending; fronds gray-green, dull; stipes few, 10–40 cm. long; scales dull brown, thinly chartaceous, scarious on margin, the larger ones tardily deciduous, acuminate, spreading, elliptic to narrowly deltoid-ovate, subentire, the median scales on stipes and rachis appressed, broadly ovate, 3–4 mm. long, acute, ciliate, the smaller ones membranous, densely ciliate, orbicular, 1–2 mm. long, appressed, rarely prolonged and taillike at the tip; blades narrowly ovate, 15–40 cm. long, 6–15 (–18) cm. wide, widest at base, often rooting and proliferous at the tip of the prolonged rachis, simply pinnate; pinnae 10–20 on each side, spreading, lanceolate, 3–10 cm. long, 6–20 mm. wide, acuminate, subentire to undulate-toothed, sessile, glabrate above, with ovate to lanceolate ciliate scales beneath, auricled on anterior side at base with a deltoid acute lobe or rarely a free segment; veinlets rarely somewhat anastomosing; sori in 2 series on the pinnae, nearer the costa than the margin, or in 3 or 4 somewhat irregular series; indusia small, orbicular, fugacious.——Thickets on hillsides; Honshu (s. Kantō Distr. and Tōkaidō), Shikoku, Kyushu; locally common.——s. Korea, China, and Formosa.

2. Polystichum lonchitis (L.) Roth. *Polypodium lonchitis* L.; *Aspidium lonchitis* (L.) Sw.; *Polystichum lonchitis* var. *japonicum* Nakai & H. Ito.——HIIRAGI-SHIDA, KUMOI-KA-GUMA. Rhizomes stout, erect, 2–4 cm. across; fronds tufted; stipes 4–6 cm. long, 2–2.5 mm. wide, erect, lustrous, dark brown and densely scaly especially at base, chestnut-brown above, the scales membranous, lanceolate to ovate, 1–1.3 cm. long, erose-ciliolate; blades linear-oblanceolate, 15–30 cm. long, 2.5–5 cm. wide, acuminate, gradually narrowed at base, simply pinnate; pinnae 30–50 pairs, firmly coriaceous, spreading, often spine-toothed, firmly spine-tipped, glabrous on upper side, with linear-lanceolate to linear, hyaline, entire scales beneath, sessile, auriculate-truncate on anterior side, cuneate on posterior side at base, the lower pinnae deltoid, 5–10 mm. long, the median oblong-ovate to broadly lanceolate, 12–27 mm. long, 5–8 mm. wide; sori in 2 series on the pinnae, between the costule and margin; indusia concealed among the sporangia. (Description based upon European specimens.)——Alpine slopes; Honshu (Mount Kitadake in Kai Prov.); very rare.——Sakhalin, n. Kuriles, Kamchatka, Himalayas to Europe, and N. America.

3. Polystichum deltodon (Bak.) Diels. *Aspidium deltodon* Bak.; *A. auriculatum* var. *submarginale* Bak.; *P. deltodon* var. *submarginale* (Bak.) C. Chr.; *A. tosaense* Makino; *P. tosaense* (Makino) Makino——TACHI-DENDA. Rhizomes

stout, erect to subascending, densely clothed with the basal stubs of old stipes; fronds tufted; stipes 5–12 cm. long, straw-colored, sparsely scaly, the scales membranous, pale brown, lanceolate, 3–5 mm. long, often attenuate with a long taillike tip, denticulate-ciliate; blades linear-lanceolate, 15–40 cm. long, 2.5–5 cm. wide, acuminate to elongate-acuminate, simply pinnate, the rachis straw-colored, thinly scaly; pinnae 15–30 pairs, firmly chartaceous, spreading, narrowly oblong to narrowly ovate-oblong, 1–3 cm. long, 5–10 mm. wide, glabrous above, thinly scaly beneath, sessile, obtuse to abruptly acute, mucronate- to spine-toothed, awnlike at the tip, the anterior side nearly straight, truncate at base and auricled, narrowly cuneate on posterior side at base; sori on apical and anterior side of pinnae, much nearer the margin than the costa; indusia thin, orbicular, about 1 mm. across, erose.——Calcareous rocks; Honshu (Chūgoku Distr.), Shikoku, Kyushu; rare.——China and Formosa (var.).

Polystichum formosanum Rosenst. *P. obtusoauriculatum* Hayata——SHIMA-NOKOGIRI-SHIDA, TAIWAN-NOKOGIRI-SHIDA. Differs from the preceding in the loosely disposed acute pinnae.——Reported to occur in Kyushu (Yakushima).——Formosa and Ryukyus.

4. Polystichum tripteron (Kunze) Presl. *Aspidium tripteron* Kunze; *Dryopteris triptera* (Kunze) O. Kuntze; *Ptilopteris triptera* (Kuntze) Hayata——JŪMONJI-SHIDA. Rhizomes erect to ascending, clothed with the basal stubs of old stipes; fronds tufted; stipes erect, 10–40 cm. long, pale green, brownish at base, loosely scaly, the scales membranous, brown, entire, long-filiform at the tip, the larger ones broadly lanceolate to oblong, 7–15 mm. long, evanescent on upper part of stipe, persistent at base, smaller ones ovate, 3–5 mm. long; blades simply pinnate in upper part, lanceolate, 20–45 cm. long, 5–10 cm. wide, acuminate, the rachis pale green, sparsely scaly, bipinnate in the lowest pinnae; pinnae 20–35 on each side, herbaceous, lanceolate, 2.5–5 cm. long, 5–12 mm. wide, often falcate toward the acuminate tip or straight, strongly oblique on anterior side at base, ascending-toothed, ending in a soft slender awnlike tip, sometimes lobulate, glabrous above, with scattered thin scales beneath; pinnules 8–15 pairs, smaller than the pinnae; sori in 2(–4) series on the pinnules, between the costule and the margin or slightly nearer the costule; indusia membranous, orbicular, erose.——Woods in mountains; Hokkaido, Honshu, Shikoku, Kyushu; common.——s. Kuriles, Korea, China to e. Siberia.

5. Polystichum craspedosorum (Maxim.) Diels. *As-pidium craspedosorum* Maxim.; *A. craspedosorum* var. *japonicum* Maxim. and var. *mandschuricum* Maxim.; *Ptilopteris craspedosora* (Maxim.) Hayata——TSURU-DENDA. Rhizomes short; stipes few, spreading, short, 0.5–5 cm. long, densely scaly, the scales firmly membranous, spreading, rusty-brown, linear-lanceolate to linear, 2.5–4 mm. long, long-acuminate, with woollike hairs on margin toward base; blades lanceolate, 5–18 cm. long, 1.5–3.5 cm. wide, slightly narrowed at base, the rachis often much-prolonged, proliferous and rooting at the tip; pinnae 15–35 pairs, firmly herbaceous, spreading, dark green and glabrate above, paler and with hairlike scales beneath, sessile, narrowly oblong to narrowly ovate-oblong, 1–1.5 cm. long, obtuse, with ascending, short, awn-tipped teeth, the anterior side oblique, broadly cuneate at base, auricled; sori in 1 series on anterior side near margin of the pinnae; indusia large, brown, orbicular, about 2 mm. across, depressed, incurved on margin, entire, persistent, inflated.——Shaded rocks in mountains; Hokkaido, Honshu, Shikoku, Kyushu. ——Korea, China, Manchuria, to e. Siberia.

6. Polystichum inaense (Tagawa) Tagawa. *Dryopteris inaensis* Tagawa——INA-DENDA. Rhizomes short; stipes spreading, slender, 1–4 cm. long, pale green, the scales brown, narrowly lanceolate to linear, about 2 mm. long, entire; blades thinly herbaceous, lanceolate to linear-lanceolate, 3–7 cm. long, 8–12 mm. wide, acute, slightly narrowed at base, simply pinnate, the rachis slender, green, with brown linear scales; pinnae 12–17 pairs, spreading, ovate, acute with a mucro at the tip, broadly cuneate at base, with scattered short linear scales on both sides, coarsely mucronate-toothed, the lower ones usually pinnately parted to lobed; sori 1 to few on a pinna, between the costa and the margin; indusia membranous, orbicular-reniform, about 1 mm. across, deeply erose-dentate.——Rocks in high mountains; Honshu (s. Shinano Prov. and w. Musashi Prov.); very rare.

7. Polystichum lachenense (Hook.) Bedd. *Aspidium lachenense* Hook.; *Dryopteris lachenensis* (Hook.) O. Kuntze ——TAKANE-SHIDA. Rhizomes rather stout; fronds tufted; stipes erect, short, 3–10 cm. long, sparsely scaly in upper part, densely so near base; scales on upper part of stipe and rachis membranous, brown, spreading, linear to linear-lanceolate, 2–4 mm. long, entire, ciliate at the more or less dilated base, deciduous; scales on the lower part of stipe ovate, deep brown, those on the underside of pinnae linear, about 1 mm. long; blades linear-lanceolate, 8–20 cm. long, 1.5–2.5 cm. wide, acute, slightly narrowed at base; pinnae 15–25 pairs, herbaceous, spreading, green and glabrate on upper side, thinly scaly beneath, ovate, obtuse to acute, sessile, with rather coarse awn-tipped teeth, sometimes lobed; costas and veins not prominent; sori to 12, in 2 series on the pinnules, between the costa and the margin or slightly nearer the costa; indusia about 1 mm. across, erose, slightly impressed.——Rocky slopes in alpine regions; Honshu (centr. mountains); very rare.——China to the Himalayas.

8. Polystichum rigens Tagawa. ONI-INODE. Stipes 20–40 cm. long, 2.5–3 mm. wide at base, straw-colored in upper part, darker at base, rather densely scaly, the scales brown, the lower ones often blackish, 5–12 mm. long, long-acuminate, ciliate-toothed, the basal ones narrowly ovate, to 15 mm. long, about 5 mm. wide; blades firmly coriaceous, narrowly ovate-oblong, 30–45 cm. long, 10–20 cm. wide, acuminate, bipinnate, the scales of the rachis gradually long-acuminate, slightly falcate, lanceolate, ciliate, 4–6 mm. long; pinnae lanceolate, 15–25 mm. wide, the lower ones very short-petiolulate; pinnules obliquely ovate, 10–12 mm. long, 4–7 mm. wide, acute with a firm spine at the tip, spine-toothed to subentire, glabrous on upper side, with scattered filiform-tipped scales beneath,

decurrent on the narrowly winged rachis; sori in 2 series on the pinnules, between the costule and the margin; indusia caducous, membranous, about 1.5 mm. across, impressed in the center.——Honshu (Kantō Distr. and westw.).

9. Polystichum mayebarae Tagawa. *P. tsussimense* var. *mayebarae* (Tagawa) Kurata——Ō-KIYOZUMI-SHIDA. Closely allied to the preceding; rhizomes stout, short, densely clothed with the basal stubs of old stipes; stipes straw-colored, rather densely scaly, the scales black-brown, lustrous, spreading, linear-lanceolate, 7–10 mm. long, sparsely long-attenuate, spinose-ciliate on lower margin, the basal scales dark brown, narrowly oblong, abruptly acuminate; blades ovate-lanceolate, 30–40 cm. long, 10–18 cm. wide, bipinnate, the rachis prominently scaly, the scales chartaceous, spreading, black-brown, linear-lanceolate to broadly linear, thinly spinulose on lower half; pinnae lanceolate, 1.5–3 cm. wide; pinnules rhombic-ovate, 8–15 mm. long, 5–8 mm. wide, acute with a short spine at the tip, spinulose-toothed on upper margin; sori 2–10, in 2 series on the pinnules, between the costule and the margin or slightly nearer the costule; indusia about 1.5 mm. across.—— Honshu (Kantō through Tōkaidō to Kinki and Chūgoku Distr.), Shikoku, Kyushu; rare.

10. Polystichum tsussimense (Hook.) J. Smith. *Aspidium tsussimense* Hook.; *Polystichum monotis* Christ——HIME-KANA-WARABI. Rhizomes rather stout, short, covered with the basal stubs of old stipes; fronds tufted; stipes pale green to straw-colored, brownish at base, rather sparsely scaly in upper part, densely so at base; scales on stipes and rachis black-brown, linear to nearly filiform, 3–5 mm. long, the basal scales oblong-lanceolate, 1–1.2 cm. long, abruptly contracted at the tip, acuminate; blades broadly lanceolate to oblong-ovate, 25–40 cm. long, 10–20 cm. wide, acuminate, the rachis prominently scaly; pinnules oblique, ovate to oblong-ovate, 7–15 mm. long, 3–7 mm. wide, acute and spine-tipped, sessile, gray-green and glabrous above, with scattered hairlike scales beneath, spine-toothed; sori in 2 series on the pinnules, inserted in a shallow pit; indusia deciduous, flat, about 1 mm. across.——Honshu (Kantō Distr. and westw.), Shikoku, Kyushu.——Korea, China, and Formosa.

11. Polystichum neolobatum Nakai. *P. lobatum* var. *chinense* Christ——INA-INODE. Stipes 5–30 cm. long, rather stout, straw-colored, brown at base, densely scaly, the scales spreading, membranous, brown, lustrous, ciliate, the larger ones broadly ovate, 1–2.2 cm. long, 3–10 mm. wide, abruptly acuminate, the smaller ones lanceolate to linear-lanceolate, 3–10 mm. long, 0.5–2 mm. wide; blades broadly lanceolate, 20–80 cm. long, gradually acuminate, slightly narrowed at base, bipinnate or the smaller ones bipinnatiparted, the scales on rachis dense, linear-filiform, 2.5–4 mm. long, contorted, mixed with broader scales in lower portion; pinnae 15–40 pairs, firmly coriaceous, spreading, lanceolate to ovate, acuminate, subsessile, sparsely scaly on upper side, more densely so beneath, the scales pale brown, hairlike, 1–2 mm. long; pinnules obliquely ovate, 7–15 mm. long, 3–7 mm. wide, acute with a pungent spine at the tip, with few spine-tipped teeth on margin, lustrous on both sides; veinlets slender, slightly impressed beneath; sori in 2 series on the pinnules of upper portion of blade; indusia flat, brownish, entire, about 1.2 mm. across.——Honshu (s. Shinano and Sagami Prov.); very rare.——China, Formosa, and India.

12. Polystichum doianum Tagawa. SAKURAJIMA-INODE. Stipes 25–30 cm. long, straw-colored, with 3 kinds of scales, the largest rather sparse, chestnut-brown, spreading, narrowly ovate, 10–12 mm. long, abruptly acuminate, lustrous, sparsely ciliolate, subentire, often very narrowly hyaline-margined, scales of intermediate size very sparse, linear-lanceolate, 12–15 mm. long, long-filiform at the tip, the smallest ones very numerous and dense, membranous, hairlike, pale rufous-brown, 2–4 mm. long, ciliate at base, deflexed to spreading; blades thick-herbaceous, narrowly oblong-ovate, 40–45 cm. long, 12–15 cm. wide, acuminate; scales on rachis dense, hairlike, mixed with larger, linear-lanceolate darker ones on lower half; pinnae sessile, 15–17 mm. wide; pinnules spreading, oblong-ovate, 5–10 mm. long, 3–5 mm. wide, acute with a short spine at the tip, auricled at base on anterior side, with hairlike scales beneath, spine-toothed; sori in 2 series on the pinnules, erose-fimbriate, 0.2–0.5 mm. across.——Kyushu (Sakurajima in Satsuma Prov.); rare.

13. Polystichum kurokawae Tagawa.   Aкame-inode. Stipes 25–30 cm. long, rather stout, with 3 kinds of scales, the lower and basal broadly lanceolate to ovate-oblong, 1.5–2 cm. long, 4–7 mm. wide, acuminate, ciliate, the upper ones membranous, narrowly ovate, 8–15 mm. long, abruptly acuminate, ciliate, the smallest scales uniformly distributed over the stipe, broadly lanceolate, 3–6 mm. long, long-acuminate, ciliate; blades broadly lanceolate to oblong-ovate, 40–80 cm. long, 20–25 cm. wide, slightly narrowed at base, the scales on rachis spreading, lustrous, pale red-brown to dull brown, the larger ones narrowly ovate, 5–8 mm. long, ciliate, the smaller ones linear to linear-lanceolate, 3–5 mm. long; pinnae 18–28 mm. wide; pinnules firmly herbaceous, oblique, oblong-ovate to ovate, auricled on anterior side, obtuse to subacute, spine-tipped, spine-toothed on margin, filiform-scaly beneath, sparsely so above; sori nearly median, in 2 series on the pinnules; indusia flat, about 0.7 mm. across.——Honshu (Kantō Distr. through Tōkaidō to Kinki Distr.).

14. Polystichum pseudomakinoi Tagawa.   Saikoku-inode.   Fronds tufted; stipes 20–40 cm. long, pale straw-colored, densely scaly; scales membranous, the largest pale brown or pale red-brown, lustrous, lanceolate, 6–8 mm. long, 1.5–3 mm. wide, gradually acuminate, long-ciliate, the smallest ones filiform, 2–5 mm. long, entire at the tip, broadened and long-ciliate at base, the lower and basal scales deep chestnut-brown in the center, pale brown on margin, oblong-ovate to broadly lanceolate, 7–12 mm. long, 2–4 mm. wide, abruptly acuminate, sparsely ciliate to subentire; blades oblong-ovate to broadly lanceolate, 30–50 cm. long, 10–20 cm. wide, the scales on rachis dense, spreading to slightly deflexed, lanceolate to broadly linear, 2–3 mm. long, long-attenuate, long-ciliate; pinnules firmly herbaceous, oblique, oblong- to elliptic-ovate, 8–15 mm. long, 4–6 mm. wide, spine-toothed, with hairlike scales scattered on upper side, prominent on under side; sori submarginal; indusia flat, 0.7–1 mm. across, with a small dark center. ——Woods in mountains; Honshu (Kantō Distr. through Tōkaidō to Kinki and Chūgoku Distr.), Shikoku.——China.

Var. **ambiguum** Tagawa.   Inode-modoki.   Scales concolorous, those on the rachis densely fimbriate-dentate.——Honshu (Kantō Distr. and westw.), Shikoku, Kyushu.

15. Polystichum makinoi (Tagawa) Tagawa.   P. aculeatum var. nigropaleaceum Makino, non Christ; P. aculeatum var. makinoi Tagawa——Kata-inode.   Fronds tufted; stipes 20–40 cm. long, straw-colored, with 3 kinds of scales, the lower and basal rather firm and lustrous, chestnut-brown, with a pale brown membranous margin, lanceolate to broadly so, 7–12

mm. long, 2–3.5 mm. wide, sparsely ciliate to subentire, the upper ones membranous, brown, concolorous, ciliate, the smallest ones dense over the whole stipe, brown, spreading to deflexed, 2–7 mm. long; blades 30–60 cm. long, 10–20 cm. wide, the scales on rachis dense, spreading to deflexed, lanceolate to filiform, ciliate in lower half or at base; pinnae acute to acuminate; pinnules coriaceous-herbaceous, obliquely ovate, obtuse to subacute, 7–15 mm. long, 3–6 mm. wide, short spine-tipped, appressed spine-toothed, with fibrouslike scales beneath; sori in 2 series on the pinnules; indusia about 0.7 mm. across.—— Honshu (Kantō Distr. and westw.), Shikoku, Kyushu.

16. Polystichum retrorsopaleaceum (Kodama) Tagawa.   P. aculeatum var. retrorsopaleaceum Kodama——Sakage-inode.   Rhizomes stout, short; fronds tufted; stipes 20–40 cm. long, densely scaly, the scales membranous, pale brown, lustrous, ciliate, the largest spreading, ovate, 8–20 mm. long, 5–10 mm. wide, abruptly contracted and short-tailed at tip, more dense toward the base, the median scales numerous, appressed to spreading, orbicular-ovate to ovate, 2–6 mm. long, rounded to obtuse, short-tailed at the tip, the smallest ones orbicular to ovate, 1–2 mm. long, appressed, without or with a filiform apical tail much longer than the lower broadened portion of scale; blades 40–100 cm. long, 15–30 cm. wide, abruptly acuminate, slightly narrowed at base, the scales on rachis beneath orbicular-ovate to ovate, 2–4 mm. long, retrorsely appressed, scarious on margin, with hairlike spreading to deflexed scales 2–3 mm. long especially on upper side; pinnules herbaceous, oblique, ovate to ovate-oblong, obtuse, short spine-tipped, with an obtuse to subacute auricle on anterior side, the teeth ascending to appressed, spine-tipped, the scales hairlike, very sparse and tardily deciduous on upper side, rather prominent on under side; sori in 2 series on the pinnules of upper half of blade; indusia impressed.——Woods in mountains; Hokkaido, Honshu, Shikoku.

Var. **ovatopaleaceum** (Kodama) Tagawa.   P. aculeatum var. ovatopaleaceum Kodama——Tsuyanashi-inode.   Smallest scales on the stipe usually lanceolate, 2–4 mm. long, spreading to deflexed, gradually narrowed to a rather short tail at tip, scales on the under side of rachis antrorsely spreading to ascending; pinnae more densely scaly, appressed-toothed.—— Woods; Honshu (Kantō Distr. and westw.), Shikoku, Kyushu.

Var. **coraiense** (Christ) Tagawa.   P. aculeatum var. coraiense Christ——Iwashiro-inode.   Differs from the preceding variety in the narrower and gradually acuminate scales on the underside of rachis.——Honshu; rare.——Korea.

17. Polystichum igaense Tagawa.   Chabo-inode. Rhizomes short, erect; stipes crowded, 5–15 cm. long, very densely scaly, the scales on upper part firmly membranous, brown, concolorous, spreading, broadly lanceolate to nearly linear, 4–8 mm. long, gradually filiform at tip, contorted, short-ciliate, unequal, the lower scales ovate to narrowly so, 6–10 mm. long, short-acuminate; blades lanceolate, 25–40 cm. long, 6–10 cm. wide, gradually long-acuminate, slightly narrowed at base, bipinnate, the rachis densely scaly; pinnae approximate, oblong-lanceolate, acute, often abruptly so, sessile; pinnules firmly herbaceous, obliquely ovate or rhombic-ovate, 5–10 mm. long, 3–5 mm. wide, obtuse, short spine-tipped, sessile, spine-toothed, with a short depressed-deltoid auricle on anterior side, deep green and nearly glabrous on upper side, paler and with hairlike scales beneath; sori submarginal, on upper 2/3 of the blade; indusia orbicular, about 0.7 mm. across, entire.—— Honshu (Kantō to Kinki Distr.); rare.

18. Polystichum polyblepharum (Roem.) Presl. *Aspidium polyblepharum* Roem. ex Kunze; *Aspidium aculeatum* var. *japonicum* Fr. & Sav.; *P. aculeatum* var. *japonicum* (Fr. & Sav.) Christ; *P. japonicum* (Fr. & Sav.) Diels——INODE. Rhizomes ascending to erect, densely clothed with basal stubs of old stipes; fronds tufted; stipes 20–30 cm. long, stout, straw-colored, densely scaly, the scales membranous, brown, concolorous, lustrous, dentate-ciliate, the largest narrowly to broadly lanceolate, 10–25 mm. long, gradually acuminate, spreading to deflexed, the basal ones broader, rather abruptly contracted at the tip, the smallest ones linear-lanceolate, 7–15 mm. long, ciliate in lower half, gradually attenuate forming a contorted filiform tip; blades narrowly oblong-ovate, 30–80 cm. long, 15–25 cm. wide, short-acuminate, slightly narrowed at base, deep green and slightly lustrous on upper side, the scales on rachis dense, deflexed, narrowly lanceolate to broadly linear, filiform at the tip; pinnae 15–25 mm. wide; pinnules firmly herbaceous, oblique, oblong-ovate, 8–15 mm. long, 4–6 mm. wide, obtuse to acute, spine-tipped, appressed spine-toothed, glabrate on upper side, with fibrouslike scales beneath; sori median between the costule and the margin.——Woods; Honshu, Shikoku, Kyushu; common.——s. Korea.

Var. **fibrillosopaleaceum** (Kodama) Tagawa. *P. aculeatum* var. *fibrillosopaleaceum* Kodama; *P. fibrillosopaleaceum* (Kodama) Tagawa——ASUKA-INODE. Larger scales on stipes linear-lanceolate, entire to sparsely spinulose-ciliolate, gradually narrowed at tip, usually contorted, the scales on rachis linear to filiform.——Honshu, Shikoku.

19. Polystichum braunii (Spenner) Fée. *Aspidium braunii* Spenner; *Dryopteris braunii* (Spenner) Underw.——Hoso-INODE. Rhizomes short, erect, fronds tufted; stipes 5–25 cm. long, densely scaly, the scales membranous, pale brown, lustrous, spreading, the larger ones ovate to lanceolate, 8–15 mm. long, 2–7 mm. wide, abruptly to gradually acuminate, subentire, the smaller ones broadly linear to filiform, 2–4 mm. long, sparsely ciliate; blades broadly lanceolate to narrowly ovate, 25–60 cm. long, 7–22 cm. wide, narrowed at base, the scales on rachis spreading to reflexed, lanceolate to linear, often ciliate at base; pinnae lanceolate, acute to short-acuminate, the lower ones shorter, often slightly deflexed, acute to obtuse; pinnules thickly herbaceous, approximate, obliquely deltoid-ovate to elliptic, 6–15 mm. long, 3–7 mm. wide, with hairlike scales on upper side, with hairlike dilated scales on under side, coarsely spine-toothed; sori in 2 series on pinnules on upper half of blade, nearer the costule than the margin; indusia flat, 0.3–0.5 mm. wide.——Hokkaido, Honshu (n. and centr.

distr.).——Korea, Manchuria, China, Kuriles, Sakhalin, Kamchatka and Siberia to Europe, and N. America.

20. Polystichum microchlamys (Christ) Kodama. *Aspidium microchlamys* Christ; *P. braunii* var. *kamtschaticum* C. Chr. & Hult.——KARAKUSA-INODE. Fronds large, often 1 m. long or more; stipes stout, straw-colored, 20–30 cm. long, densely scaly; scales membranous, spreading, pale- to red-brown, lustrous, entire, the largest narrowly ovate, 18–25 mm. long, 4–8 mm. wide, abruptly acuminate, the intermediate ones lanceolate, 10–20 mm. long, 2–4 mm. wide, gradually acuminate, the smallest ones linear to filiform, 3–8 mm. long, slightly dilated at base, contorted; blades 50–100 cm. long, 12–25 cm. wide, scarcely distinctly narrowed at base, the scales on rachis pale brown, filiform to broadly linear, 3–6 mm. long, sometimes mixed with lanceolate scales on lower portion; pinnae 1.5–3.5 cm. wide; pinnules narrowly ovate, acute, short awn-tipped, somewhat decurrent on the narrowly winged axis, coarsely awn-toothed, with pale brown hairlike scales on both sides; sori usually in 2 series on pinnules on upper half of blade, slightly nearer the costule than the margin; indusia small, membranous, about 0.7 mm. across, erose.——Coniferous woods in mountains; Hokkaido, Honshu (n. and centr. distr.).——Kamchatka.

21. Polystichum eximium (Mett.) C. Chr. var. **minus** Tagawa. *P. gemmiferum* Tagawa——KOMOCHI-INODE. Rhizomes densely clothed with basal stubs of old stipes; stipes 22–25 cm. long, straw-colored, densely scaly, the larger scales ovate-oblong, lustrous, dark chestnut-brown, ovate-oblong, 8–10 mm. long, 3–4 mm. wide, acuminate, irregularly ciliate-fimbriate on a brown scarious margin, the smaller ones linear-subulate or filiform, ferrugineous, ciliate; blades dimorphic, lanceolate, acuminate, bipinnate in the lower portion, slightly narrowed at base, the sterile ones longer and wider, often 40 cm. long or more, about 25 cm. wide, the fertile about 15 cm. wide, gemmate in upper axils, the scales on rachis ferrugineous, filiform, ciliate, contorted; pinnae subsessile, the sterile ones broadly lanceolate, to 13 cm. long, 3–3.5 cm. wide, acuminate, the fertile narrowly lanceolate, 1.5–1.8 cm. wide, acute; sterile pinnules herbaceous-coriaceous, ovate-oblong, to 2 cm. long, 8 mm. wide, obtuse, oblique, glabrous on upper side, with scattered filiform ciliate scales on lower side, depressed-toothed; fertile pinnules smaller, entire to crenate; sori in 2 series on the pinnules; indusia entire, about 1 mm. across.——Kyushu (Yakushima); rare.——The typical phase occurs in Formosa, China to India.

Polystichum kiusiuense Tagawa, KYŪSHŪ-INODE, may belong here. Reported from Kyushu (s. distr.).

7. CYRTOMIUM Presl   YABU-SOTETSU ZOKU

Terrestrial; rhizomes short, ascending to erect, densely scaly; fronds simply pinnate with a terminal pinna or pinnatifid at tip, the pinnae mostly acuminate, usually falcate, commonly auriculate on the anterior side, entire to toothed, the veins anastomosing to form areoles, or nearly free; sori dorsal, or sometimes terminal on the veins; indusia peltate, persistent or caducous.——About 20 species, Hawaii, e. Asia to S. Africa, Central and S. America.

1A. Terminal pinna distinct, entire to 3-cleft; areoles in several rows, with 1–3 free veinlets in each areole.
  2A. Pinnae coriaceous, entire near apex. . . . . 1. *C. falcatum*
  2B. Pinnae chartaceous, toothed near apex.
    3A. Pinnae 12–26 pairs, the terminal smaller than the lateral; stipes densely scaly throughout. 2. *C. fortunei*
    3B. Pinnae usually 2–8, rarely to 10 pairs, the terminal larger than to as large as the lateral; stipes densely scaly on lower half.
      4A. Pinnae finely and regularly spine-toothed, with an acute auricle on anterior side, or sometimes on both sides; indusium laciniate-fimbriate. . . . . 3. *C. caryotideum*

4B. Pinnae subentire or irregularly mucronate-toothed, slightly unequal, without an auricle; indusium nearly entire.

　　　　　　　　　　　　　　　　　　　　　　　　　　　　　　　4. *C. macrophyllum*

1B. Terminal pinna pinnately lobed to cleft; areoles of veins in 1 to 2 rows, with a single, rarely paired, free veinlet in each areole.

5A. Pinnae broadly falcate-lanceolate, auriculate on the anterior side. .... 5. *C. vittatum*

5B. Pinnae linear-lanceolate, not auricled. .... 6. *C. tachiroanum*

1. Cyrtomium falcatum (L. f.) Presl. *Polypodium falcatum* L. f.; *Aspidium falcatum* (L. f.) Sw.; *Phanerophlebia falcata* (L. f.) Copel.; *Polystichum falcatum* (L. f.) Diels; *Polypodium japonicum* Houtt.——ONI-YABU-SOTETSU. Rhizomes stout, short; stipes tufted, 15–40 cm. long, the scales very dense on lower part, ovate to broadly lanceolate or linear, 1–1.5(–2) cm. long, abruptly acuminate, dark brown; blades firmly coriaceous, 20–60 cm. long, 10–25 cm. wide, lustrous, dark green; pinnae 3–11 pairs, ovate to narrowly oblong-ovate, 7–13 cm. long, 2.5–5 cm. wide, abruptly narrowed to a broad tail at tip, rounded to broadly cuneate at base, oblique, falcate, short-petiolulate, nearly entire near apex, sinuate-toothed at base, with rufous woolly-scales beneath while very young; sori uniformly covering undersurface; indusia orbicular, nearly entire.——Rocks near seashores, rarely on inland hills; Hokkaido (s. distr.), Honshu, Shikoku, Kyushu; common.——s. Korea, Ryukyus, Formosa, China, Malaysia, India to e. and S. Africa, and Hawaii.

Var. **devexiscapulae** (Koidz.) Tagawa. *Polystichum devexiscapulae* Koidz.; *Cyrtomium devexiscapulae* Koidz.; *Phanerophlebia falcata* var. *devexiscapulae* (Koidz.) Ohwi——NAGABA-YABU-SOTETSU. Pinnae narrower, broadly lanceolate, 8–13 cm. long, 2–4 cm. wide, nearly equilateral and cuneate at base.——Hills and low mountains; Honshu (Echigo Prov. to Kantō Distr. and westw.).——Korea, China, and Indochina.

2. Cyrtomium fortunei J. Smith. *Phanerophlebia fortunei* (J. Smith) Copel.; *Aspidium falcatum* var. *fortunei* (J. Smith) Makino; *Polystichum falcatum* var. *fortunei* (J. Smith) Matsum.; *Polystichum fortunei* (J. Smith) Nakai——YABU-SOTETSU. Rhizomes short, stout; stipes tufted, 15–30 cm. long, densely scaly throughout, the scales dark brown, rather firm, lustrous, lanceolate to ovate, sometimes hairlike, 7–12 mm. long, acuminate, pubescent on margin at tip; blades broadly lanceolate, 30–60 cm. long, 10–15 cm. wide; pinnae chartaceous, 12–26 pairs, broadly lanceolate to narrowly ovate-oblong, 5–7 cm. long, 1–3 cm. wide, gradually attenuate, subsessile, usually auriculate on anterior side, subentire to minutely toothed, with jointed short hairlike scales on both sides while young; indusia orbicular, subentire.——Thickets in hills and low mountains; Hokkaido (Okushiri Isl.), Honshu, Shikoku, Kyushu; common.——s. Korea and China.

Var. **clivicola** (Makino) Tagawa. *Polystichum caryotideum* var. *clivicola* Makino; *Polystichum clivicola* (Makino) Makino; *Cyrtomium clivicola* (Makino) Tagawa——YAMA-YABU-SOTETSU. Pinnae (5–)10–15 pairs, the lower ones largest, about 15 cm. long, 6 cm. wide, coarsely crenate, often minutely toothed; indusia erose.——Woods in mountains; Honshu, Shikoku, Kyushu.

Var. **intermedium** Tagawa. *Phanerophlebia fortunei* var. *intermedia* (Tagawa) Ohwi; *C. fortunei* forma *intermedium* (Tagawa) Ching; *C. yamamotoi* Tagawa——MIYAKO-YABU-SOTETSU. Pinnae 10–12 pairs, usually broadly lanceolate, 10–15 cm. long, 2.5–4 cm. wide, gradually narrowed to tip; indusia slightly undulate on margin.——Honshu (s. Kantō Distr. and westw.), Kyushu.

3. Cyrtomium caryotideum (Wall.) Presl. *Phanerophlebia caryotidea* (Wall.) Copel.; *Aspidium caryotideum* Wall. ex Hook. & Grev.; *Aspidium anomophyllum* Zenker; *Polystichum anomophyllum* (Zenker) Nakai; *Aspidium falcatum* var. *caryotideum* (Wall.) Hook. & Bak.; *Polystichum caryotideum* (Wall.) Diels; *Polystichum falcatum* var. *caryotideum* (Wall.) Matsum.——ME-YABU-SOTETSU. Rhizomes short, erect or ascending; stipes tufted, 20–40 cm. long, densely scaly near base, sparsely so above, the scales lanceolate to oblong, 1–1.5 cm. long, dark brown, smaller in the upper ones; blades 20–50 cm. long, 15–22 cm. wide; pinnae 3–6 pairs, oblong-ovate to ovate, 8–15 cm. long, 3–7 cm. wide, long-acuminate, short-petiolulate, finely and regularly spine-toothed, usually with a prominent auricle on anterior side, or sometimes on both sides in the lower ones, the terminal pinna large, 3- or 2-cleft; indusia flat, laciniate-fimbriate.——Calcareous areas; Honshu (Kantō Distr. and westw.), Shikoku, Kyushu.——China to India and Hawaii.

4. Cyrtomium macrophyllum (Makino) Tagawa. *Phanerophlebia macrophylla* (Makino) Okuyama; *Aspidium falcatum* var. *macrophyllum* Makino; *Polystichum falcatum* var. *macrophyllum* (Makino) Matsum.; *Polystichum macrophyllum* (Makino) Tagawa; *C. falcatum* var. *muticum* Christ; *C. caryotideum* var. *intermedium* C. Chr.; *C. muticum* (Christ) C. Chr.——HIROHA-YABU-SOTETSU. Fronds tufted; stipes 20–30 cm. long, prominently scaly in lower half, sparsely so above, the scales spreading, dark brown, lustrous, narrowly ovate to linear-lanceolate, 7–10 mm. long, abruptly acuminate, the upper ones gradually narrowed; blades 20–50 cm. long, 15–25 cm. wide; pinnae 2–8 pairs, thinly chartaceous, the upper ones ovate to oblong-ovate, abruptly acuminate, usually rounded at base, short-petiolulate, slightly unequal, subentire or irregularly mucronate-toothed, the lower larger, 10–20 cm. long, 4–7 (–10) cm. wide, the terminal entire or 3-cleft; indusia nearly entire.——Honshu (Kantō Distr. and Echigo Prov. westw.), Shikoku, Kyushu.——Formosa (var.), China to the Himalayas.

Cyrtomium tukusicola Tagawa. TSUKUSHI-YABU-SOTETSU. Intermediate between Nos. 2 and 4, differs from var. *intermedium* of No. 2 in the broader, fewer, 8–10(–13) not auricled pinnae, and the larger terminal pinna, and from No. 4 in the smaller pinnae and erose indusia.——Honshu (s. Kantō Distr. and westw.), and Kyushu.

5. Cyrtomium vittatum Christ. *Phanerophlebia vittata* (Christ) Copel.; *Polystichum vittatum* (Christ) C. Chr.; *Polystichum miyasimense* Kodama; *Polystichum anomophyllum* var. *miyasimense* (Kodama) Nakai; *Polystichum balansae* Christ; *Cyrtomium balansae* (Christ) C. Chr.——MIYAJIMA-SHIDA. Fronds tufted; stipes 15–40 cm. long, rather densely scaly, the scales membranous, lanceolate, acuminate, scarcely lustrous, dull-brown, obsoletely denticulate, the basal ones linear-lanceolate, about 1 cm. long; blades lanceolate, 30–50 cm. long, 10–15 cm. wide, the terminal segment lanceolate-deltoid, acuminate, pinnatilobed; rachis with small ovate, dentate scales 1–2 mm. long; pinnae 10–20 pairs, firmly chartaceous, deltoid-lanceolate, 5–10 cm. long, the lower ones 12–20 mm. wide, long-acuminate, falcate, obliquely broad-cuneate at base, with short awn- or callose-tipped teeth on margin except near base, auriculate on anterior side, with small ovate scales beneath while young; sori in 2 nearly regular series or irregularly dispersed; indusia small.——Honshu (Miyajima in Aki Prov.), Kyushu; rare.——China to Indochina.

6. Cyrtomium tachiroanum (Luerss.) C. Chr. *Phanerophlebia tachiroana* (Luerss.) Copel.; *Polypodium tachiroanum* Luerss.; *Polystichum tachiroanum* (Luerss.) Tagawa; *Polystichum integripinnum* Hayata; *C. integripinnum* (Hayata) Copel.——HOSOBA-YABU-SOTETSU.   Stipes tufted, 20–40 cm. long, straw-colored, scaly throughout, the scales thinly membranous, linear- to ovate-lanceolate, pale brown, fugacious; blades oblong-lanceolate, abruptly short-acuminate, the termi-

nal segment pinnately lobed to parted; rachis pale green, the scales sparse, lanceolate, 3–5 mm. long, lobulate-dentate, pale brown; pinnae 13–17 pairs, firmly coriaceous, dull, linear-lanceolate, 8–13 cm. long, 12–20 mm. wide, subequilateral, acuminate, falcate or nearly straight, cuneate at base, very short-petiolulate, entire or crenate toward base, glabrous above, the scales beneath ovate, appressed, less than 1 mm. long, scales of the costas lanceolate to ovate, 1–2 mm. long, ciliate near base; sori 2-seriate or in a few irregular series; indusia small, subentire, flat.——Kyushu (s. distr.); rare.——Formosa, China, and Indochina.

8. ARACHNIODES Bl.   KANA-WARABI ZOKU

Terrestrial; rhizomes mostly long-creeping or sometimes short and ascending, dictyostelic, the scales entire or subentire; fronds deltoid or ovate, with a broad base, bipinnate or more profusely divided, anadromous in branching, the ultimate pinnules rhomboid, awned or spinose on the margin and apex; veins free; sori dorsal or subterminal on the veins; indusia mostly orbicular-reniform.——About 40 species, abundant in e. Asia, few in Malaysia, and New Zealand.

1A. Blades herbaceous to coriaceous, ovate to ovate-oblong.
  2A. Blades simply pinnate to 4-pinnate, subcoriaceous; sori borne from tip to the base of pinnae.
    3A. Scales of stipe black-brown, narrowly oblong to broadly lanceolate, 3–5 mm. wide, abruptly filiform at tip; blades mostly longer
        than the stipe; rachis very scaly; pinnules mucronate- to obtuse-toothed. . . . . 1. *A. mutica*
    3B. Scales of stipe lanceolate to linear, scarcely more than 2.5 mm. wide; blades as long as to shorter than the stipe; rachis sparsely
        scaly to nearly glabrous.
      4A. Blades 2- to 4-pinnate; pinnae with posterior side mostly broader than the anterior.
        5A. Rhizomes long-creeping, with few stipes remotely placed. . . . . 2. *A. aristata*
        5B. Rhizomes short-creeping, with many stipes closely placed.
          6A. Pinnules spine-tipped; indusia 1–1.5 mm. across.
            7A. Pinnules firmly chartaceous to coriaceous, dark green when dried.
              8A. Sori nearer the costule than the margin; pinnules firmly coriaceous, pungent-toothed.
                9A. Blades 3-pinnate. . . . . 3. *A. pseudoaristata*
                9B. Blades 2-pinnate. . . . . 4. *A. simplicior*
              8B. Sori nearer the margin than the costule; pinnules firmly chartaceous to subcoriaceous, the teeth not pungent.
                                                                                                        5. *A. amabilis*
            7B. Pinnules herbaceous to firmly so, vivid to pale green when dried. . . . . 6. *A. nipponica*
          6B. Pinnules crenate-toothed; indusia 1.5–2.5 mm. across. . . . . 7. *A. cavaleriei*
      4B. Blades simply pinnate, sometimes bipinnate to pinnatiparted at base; pinnae not broadened on the posterior side. . 8. *A. assamica*
  2B. Blades 4-pinnatiparted to 4-pinnate, thinly herbaceous; sori borne mostly at base of pinnae. . . . . 9. *A. standishii*
1B. Blades membranous, 5-angled or deltoid.
  10A. Blades glabrous. . . . . 10. *A. maximowiczii*
  10B. Blades hairy on both sides. . . . . 11. *A. miqueliana*

1. Arachniodes mutica (Fr. & Sav.) Ohwi. *Rumohra mutica* (Fr. & Sav.) Ching; *Aspidium muticum* Fr. & Sav.; *Dryopteris mutica* (Fr. & Sav.) C. Chr.; *Polystichopsis mutica* (Fr. & Sav.) Tagawa——SHINOBU-KAGUMA.   Rhizomes stout, rather short, erect to ascending; stipes tufted, 15–35 cm. long, straw-colored; scales dense, narrowly oblong to broadly lanceolate, 1.5–2.5 cm. long, 3–5(–7) mm. wide, abruptly filiform at the tip, sparsely ciliolate, lustrous, black-brown, those on upper part and on rachis linear-lanceolate to broadly linear, 3–6 mm. long, 0.2–1 mm. wide, spreading, contorted; blades ovate, 25–50 cm. long, 13–25 cm. wide, short-acuminate, tripinnate; pinnae obliquely spreading, narrowly ovate, acuminate, the lowermost 5–10 cm. long, the posterior side broader than the anterior; pinnules narrowly ovate, 1–2.5 cm. long, obtuse to subacute, mucronate- to obtuse-toothed, pinnately divided to parted, deep green and lustrous above, paler and minutely scaly beneath; sori on the upper part of blade, usually descending to the lower penultimate pair of pinnae, median; indusia orbicular-reniform, erose.——Coniferous woods in mountains; Hokkaido, Honshu (Kinki Distr. and northw.), Shikoku, Kyushu (Yakushima). Sakhalin and Korea (Dagelet Isl.).

2. Arachniodes aristata (Forst.) Tindale. *Rumohra aristata* (Forst.) Ching; *Polypodium aristatum* Forst.; *Aspidium exile* Hance; *Polystichum aristatum* (Forst.) Presl; *Arachniodes exilis* (Hance) Ching; *Polystichopsis aristata* (Forst.) Holttum; *Byrsopteris aristata* (Forst.) Morton——HOSOBA-KANA-WARABI.   Rhizomes densely scaly, long-creeping; scales on rhizomes and basal portion of stipe membranous, linear-lanceolate to lanceolate, 4–7 mm. long, about 1 mm. wide, gradually narrowed to a long filiform point, entire, yellowish brown; scales on stipes and rachis deep brown, linear, 3–4 mm. long, subentire, filiform at tip; stipes few, remote, 30–60 cm. long, longer than the blade, pale green to straw-colored, rather sparsely scaly; blades ovate to broadly so, 20–40 cm. long, 12–25 cm. wide, distinctly tailed at apex, bipinnate or subtripinnate, with 1 (to few) pinnate branch on the lowest pinnae; pinnules coriaceous, oblong-ovate to narrowly ovate, 7–20 mm. long, abruptly spine-tipped, spine-toothed, glabrate, lustrous above; sori median, in 2 series on the pinnules; indusia orbicular-reniform, about 1 mm. across.——Woods; Honshu (Kantō Distr. and westw.), Shikoku, Kyushu.——Korea, Ryukyus, Formosa, China, Malaysia to India, and Polynesia.

3. Arachniodes pseudoaristata (Tagawa) Ohwi. *Ru-*

mohra pseudoaristata (Tagawa) H. Itō; *Polystichum pseudo-aristatum* Tagawa; *Aspidium aristatum* var. *subdimorphum* Christ and var. *davalliaeforme* Christ; *Polystichopsis pseudo-aristata* (Tagawa) Tagawa——KOBANO-KANA-WARABI. Differs from the preceding in the short-creeping rhizomes and less remotely placed stipes; blades subdimorphic, tripinnate, gradually short-acuminate.——Honshu (Kantō Distr. and westw.), Shikoku, Kyushu.—— s. Korea, Ryukyus, and Formosa.

4. Arachniodes simplicior (Makino) Ohwi. *Rumohra simplicior* (Makino) Ching; *Aspidium aristatum* var. *simplicius* Makino; *Polystichum aristatum* var. *simplicius* (Makino) Matsum.; *Polystichum simplicius* (Makino) Tagawa; *Polystichopsis simplicior* (Makino) Tagawa——HAKATA-SHIDA. Rhizomes short-creeping, densely scaly, knotty, covered with the basal stubs of old stipes; stipes 30–60 cm. long, straw-colored or pale green, rather densely scaly; scales on stipe and rachis tardily deciduous, broadly linear, 2–8 mm. long, less than 1 mm. wide, gradually narrowed to a filiform tip, ciliate at base, black-brown; blades broadly ovate, 25–40 cm. long, 15–25 cm. wide, pinnate, the lowest pinnae with elongate pinnate branches longer on posterior than on anterior side, the terminal pinna coriaceous, linear-lanceolate, 15–25 cm. long, abruptly contracted at tip, acuminate, short-petiolulate, dark green and lustrous above, paler beneath; pinnae 2–5 pairs, gradually acuminate, short-petiolulate; pinnules oblong-ovate, 1–2.5 cm. long, oblique, obscurely auricled on anterior side, obtuse, spine-tipped and -toothed, sometimes lobulate, glabrate on both sides; sori median, in 2 rows on the pinnules; indusia orbicular-reniform, crenulate, about 1 mm. across.——Honshu (Kantō Distr. and westw.), Shikoku, Kyushu.——China.

Var. **major** (Tagawa) Ohwi. *Rumohra simplicior* var. *major* (Tagawa) H. Itō; *Polystichum simplicius* var. *majus* Tagawa; *Polystichopsis simplicior* var. *major* (Tagawa) Tagawa ——ONI-KANA-WARABI. Terminal pinna lanceolate-deltoid, broadened at base, gradually merging into the upper lateral pinnae.——Honshu (Kantō Distr. and westw.), Shikoku, Kyushu.

5. Arachniodes amabilis (Bl.) Tindale. *Rumohra amabilis* (Bl.) Ching; *Aspidium amabile* Bl.; *Polystichum amabile* (Bl.) J. Smith; *Dryopteris amabilis* (Bl.) O. Kuntze; *Polystichopsis amabilis* (Bl.) Tagawa; *Polystichum rhomboideum* Schott; *A. rhomboideum* Wall. ex Mett.——Ō-KANA-WARABI. Rhizomes short-creeping; stipes approximate, 20–40 cm. long, often slightly rubescent; scales on rhizomes and at base of stipes dense, membranous, reddish brown, lanceolate, 8–10 mm. long, to 2 mm. wide, gradually narrowed to a filiform tip, entire, scales on stipes and rachis filiform, deciduous, 2–5 mm. long; blades ovate to broadly so, 25–45 cm. long, 15–25 cm. wide, simply pinnate or more often with a pinnate branch on posterior side in the lowest (or lower 2) pairs, the pinnae 3–7 pairs, lanceolate, 12–20 cm. long, 2–3.5 cm. wide, acuminate, petiolulate; terminal pinnae acuminate, short-petiolulate; pinnules firmly chartaceous to subcoriaceous, deltoid-ovate to oblong-ovate, 1.5–3 cm. long, 8–12 mm. wide, oblique, short-petiolulate, subacute, spine-tipped and -toothed, slightly lustrous above, glabrate on both sides; sori submarginal, in 2 series on the pinnules, nearer the margin than the costule; indusia orbicular-reniform, 1–1.5 mm. across, ciliate, sparsely puberulent on back.——Honshu (Kantō Distr. and westw.), Shikoku, Kyushu.——Ryukyus, Formosa, China to Malaysia, and India.

Var. **yakusimensis** (H. Itō) Ohwi. *Rumohra amabilis* var. *yakusimensis* H. Itō; *Polystichum aristatum* var. *yakusimense* (H. Itō) Masam.; *Polystichopsis amabilis* var. *yakusimensis* (H. Itō) Tagawa——YAKUSHIMA-KANA-WARABI. Fronds narrower, 45–60 cm. long; pinnae 7–9 pairs, the pinnules smaller, coriaceous, often lobed to parted, 7–15 mm. long; indusia not ciliate.——Honshu (Kii Prov.), Shikoku, Kyushu.——Ryukyus, Formosa, and China.

6. Arachniodes nipponica (Rosenst.) Ohwi. *Rumohra nipponica* (Rosenst.) Ching; *Polystichum nipponicum* Rosenst.; *Polystichopsis nipponica* (Rosenst.) Tagawa——MIDORI-KANA-WARABI. Rhizomes creeping, densely scaly; stipes approximate, 30–45 cm. long, straw-colored to reddish; scales membranous, dense toward base, sparse above, lanceolate, the lower 5–10 mm. long, the upper 3–5 mm. long, entire, deciduous, the base persistent, callose, red-brown; blades vivid green above, paler beneath, firmly herbaceous to subcoriaceous, narrowly ovate, 40–60 cm. long, 25–35 cm. wide, long-acuminate; scales on rachis sparse, 1–2 mm. long, linear-lanceolate, deciduous; pinnae 6–8 pairs, 15–30 cm. long, pinnate or partially bipinnate, petiolulate, long-acuminate, the lateral gradually grading into the terminal pinna; pinnules herbaceous to firmly so, ovate, sometimes oblong, oblique, 12–30 mm. long, 6–15 mm. wide, acute, short-petiolulate, pinnately lobed or parted, mucronate-toothed, nearly glabrous above, with hairlike scales beneath while young; sori nearly median; indusia orbicular–reniform, sparsely ciliolate, about 1 mm. across.——Honshu (from Izu Prov., Tōkaidō to Kinki and Chūgoku Distr.), Shikoku, Kyushu.——China.

7. Arachniodes cavaleriei (Christ) Ohwi. *Rumohra cavaleriei* (Christ) Ching; *Aspidium cavaleriei* Christ; *Dryopteris cavaleriei* (Christ) C. Chr.; *Polystichopsis cavaleriei* (Christ) Tagawa; *Polystichum globisorum* Hayata; *Rumohra globisora* (Hayata) H. Itō; *Dryopteris sphaerosora* Tagawa ——YAKUSHIMA-KANA-WARABI. Rhizomes short-creeping to ascending; stipes tufted, purple-brown, 30–60 cm. long, sparsely scaly above and on rachis, densely scaly at base; scales lanceolate to linear-lanceolate, 10–15 mm. long, 1.5–2.5 mm. wide, much elongate at tip, contorted; blades broadly ovate, 30–55 cm. long, 25–40 cm. wide, acuminate; pinnae about 5 pairs, deltoid-ovate, 20–30 cm. long, 15–20 cm. wide, acuminate, petiolulate, bipinnate at base, pinnate in upper part; pinnules herbaceous to coriaceous, narrowly deltoid-ovate to broadly deltoid-lanceolate, usually acuminate, crenate-toothed, petiolulate, pinnately lobed to parted, obtusely toothed; sori in 2 series along the costule; indusia orbicular-reniform, 1.5–2.5 mm. across, ciliolate.——Kyushu (Yakushima).——Formosa, China, and Indochina.

8. Arachniodes assamica (Kuhn) Ohwi. *Rumohra assamica* (Kuhn) Ching; *Aspidium assamicum* Kuhn; *Polystichum assamicum* (Kuhn) Ching ex C. Chr.; *Polystichopsis assamica* (Kuhn) Tagawa; *Aspidium yoshinagae* Makino; *Polystichum yoshinagae* (Makino) Makino——OTOKO-SHIDA. Rhizomes short-creeping; stipes approximate, 20–50 cm. long, straw-colored, slightly lustrous, sparsely scaly; scales dense toward the base, broadly linear, 3–6 mm. long, long-attenuate, entire, purple-brown; blades narrowly oblong to broadly deltoid-lanceolate, 30–60 cm. long, 10–20 cm. wide, short-acuminate, pinnate, sometimes bipinnate to bipinnati-parted; pinnae 10–13 pairs, lanceolate, acuminate, the petioles 5–20 mm. long, the simple pinnae 6–10 cm. long, 1.5–2 cm. wide, cuneate at base, slightly oblique, coarsely short **awn-**

toothed, the compound pinnae 15–25 cm. long, 4–6 cm. wide, with a narrowly winged axis, the pinnules 2.5–5 cm. long, 1–1.5 cm. wide, acute, cuneate at base, slightly oblique and somewhat adnate to the axis, toothed, nearly glabrous on both sides; sori median, remote, in 2 series on the ultimate pinnule or in 2–4 series on the simple pinnae; indusia orbicular-reniform, subentire, 1–1.5 mm. across.——Honshu (Izu, Ise, Kii, and Suwo Prov.), Shikoku, Kyushu; rare.——China, Burma, and India.

9. **Arachniodes standishii** (Moore) Ohwi. *Rumohra standishii* (Moore) Ching; *Lastrea standishii* Moore; *Polystichum standishii* (Moore) C. Chr.; *Dryopteris standishii* (Moore) C. Chr.; *Polystichopsis standishii* (Moore) Tagawa; *Aspidium laserpitiifolium* Mett.; *D. viridescens* sensu auct. Japon., non O. Kuntze——Ryōmen-shida. Rhizomes short-creeping; stipes closely approximate, pale green to straw-colored, often brownish, 30–50 cm. long, densely scaly at base, sparsely so above, the lower scales pale brown, membranous, lanceolate, 1–1.5 cm. long, 1.5–2.5(–3.5) mm. wide, gradually acuminate, filiform at tip, entire, the upper scales 3–10 mm. long, linear-lanceolate, broadly cordate; fronds vivid green, the blades thinly herbaceous, oblong-ovate, 40–60 cm. long, 20–30 cm. wide, short-acuminate, 4-pinnatiparted to 4-pinnate, the rachis sparsely scaly; pinnae 12–15 pairs, approximate, obliquely spreading, short-petiolulate, the lowest with a broad posterior side; ultimate pinnules thinly herbaceous, oblong-ovate, 5–10 mm. long, 3–6 mm. wide, sometimes lobed or parted, acute to obtuse, acutely toothed, vivid green above, slightly paler and with hairlike short scales beneath, very sparsely so above; sori median in 2 series on the ultimate pinnules; indusia orbicular-reniform, often reddish when young, entire, 1–1.5 mm. across.——Woods in mountains; Hokkaido, Honshu, Shikoku, Kyushu; rather common.——Korea.

10. **Arachniodes maximowiczii** (Bak.) Ohwi. *Rumohra maximowiczii* (Bak.) Ching; *Nephrodium maximowiczii* Bak.; *Dryopteris maximowiczii* (Bak.) O. Kuntze; *Lastrea maximowiczii* (Bak.) Moore; *Polystichopsis maximowiczii* (Bak.) Tagawa; *Aspidium commutatum* Fr. & Sav.; *A. callopsis* Fr. & Sav.; *N. callopsis* (Fr. & Sav.) Palib.; *D.*

callopsis (Fr. & Sav.) C. Chr.——Nantai-shida. Rhizomes short-creeping, densely scaly; stipes 15–35 cm. long, pale green, usually chestnut-brown toward base and often on under side, lustrous; scales brown, thinly herbaceous to submembranous, ovate, (2–)4–8 mm. long, short-acuminate, entire, spreading; blades 5-angled, 15–25 cm. long and as wide, tripinnate to tripinnatiparted, glabrous, uniformly green, the rachis pale green, the scales scanty, about 2 mm. long; pinnae petiolulate, spreading, narrowly deltoid-ovate, 10–18 cm. long, short-attenuate, with a broader posterior side, the upper ones abruptly smaller; ultimate pinnules oblong-ovate, 1–2.5 cm. long, obtuse, oblique, short-petiolulate, acutely denticulate; sori sparse, few, at base of the teeth especially on anterior margin of the segments, orbicular-reniform, about 1 mm. across, entire, brown, incurved on margin.——Woods; Hokkaido, Honshu (Kinki Distr. and eastw.).——Korea.

11. **Arachniodes miqueliana** (Maxim.) Ohwi. *Aspidium miquelianum* Maxim.; *Dryopteris miqueliana* (Maxim.) C. Chr.; *Leptorumohra miqueliana* (Maxim.) H. Itō; *Polystichopsis miqueliana* (Moore) Tagawa——Narai-shida. Rhizomes elongate, long-creeping, fugacious; stipes 20–50 cm. long, straw-colored to brown, lustrous, dark brown near base, rather densely scaly; scales tardily deciduous, thinly membranous, pale brown, spreading, the larger ones lanceolate, entire, abruptly long-acuminate, 4–10 mm. long, the smaller ones narrowly ovate, 1–2 mm. long; blades deltoid, remotely 5-angled, 3- or 4-pinnate, 20–50 cm. long, slightly narrower to nearly as broad as long, acuminate, cordate, hairy on both sides, the rachis pale green to yellow-brown, thinly pilose especially in the axils, the lowest pinnae narrowly deltoid-ovate, 15–35 cm. long, oblique, with a dilated base on posterior side, the petioles 1.5–3 cm. long; ultimate pinnules membranous, elliptic-ovate to oblong or ovate, sometimes lobed to parted, obtuse or sometimes acute to rounded, cuneate at base, oblique, sessile, vivid green above, slightly paler beneath; sori 1–4 on each lobe or segment, orbicular-reniform, reddish in the center when young, 0.8–1 mm. across.——Hokkaido, Honshu, Shikoku, Kyushu.——Korea and China.

9. BOLBITIS Schott   Hetsuka-shida Zoku

Terrestrial; rhizomes creeping, dictyostelic, scaly; stipes sparsely scaly; fronds dimorphic, usually pinnate, rarely simple or bipinnatifid, glabrous, the segments crenate to lobed or incised, with regular areoles along the costas and costules and regular or irregular areoles beyond, free veinlets usually present within the areoles; fertile fronds long-stipitate; sori uniformly covering underside, exindusiate, with paraphyses.——About 85 species, in the Tropics, abundant in the Indo-Malayan region.

1. **Bolbitis subcordata** (Copel.) Ching. *Campium subcordatum* Copel.; *B. nakaii* H.Itō; *Leptochilus virens* sensu auct. Japon., non C. Chr.; *B. quoyana* sensu auct. Japon., non Ching——Hetsuka-shida. Rhizomes short-creeping, 4–6 mm. across; fronds dimorphic; stipes 30–50 cm. long in the sterile, slightly longer and more slender in the fertile, pale straw-colored; scales appressed, purple-brown, ovate to broadly lanceolate, 2–3 mm. long, most abundant toward base of stipe; sterile blades lanceolate, 40–80 cm. long, 15–30 cm. wide, pinnate, the terminal pinna abruptly narrowed, lanceolate, pinnaticleft, caudately long-acuminate, sometimes proliferous at

tip, the rachis while young with purple-brown dentate scales about 1 mm. long; lateral pinnae 8–12 pairs, thinly chartaceous, linear-oblong, 10–15 cm. long, 1.5–3 cm. wide, caudately long-acuminate, rounded-lobulate to crenate-toothed, the lobules usually with an awnlike mucro at base of the sinus, truncate to cuneate and very short-petiolulate at base; costa slender, often slightly rubescent; fertile pinnae 5–9 cm. long, 5–10 mm. wide, obtuse, very short-petiolulate, entire to undulate, entirely covered with sporangia.——Woods in hills and low mountains; Kyushu (s. distr.).——Ryukyus, Formosa, s. China to Indochina.

10. ELAPHOGLOSSUM Schott   Atsuita Zoku

Epiphytic or terrestrial; rhizomes short-creeping, scaly, dictyostelic; stipes often tufted, articulate to phyllopodous or not jointed; fronds coriaceous, simple, entire, sometimes with a cartilaginous margin, scaly or glabrate, the veins usually immersed, mostly forked, then straight and parallel, normally free; fertile fronds usually smaller and narrower than the sterile, often longer-stiped; sori entirely covering underneath surface; paraphyses absent.——More than 400 species, in all warm regions, abundant in the Andes.

1A. Scales on rhizomes and fertile stipes ovate, obtuse to acute; blades oblanceolate to lanceolate, obtuse, gradually acuminate at base; stipes of sterile fronds short, about half as long as the blade or shorter. . . . . 1. *E. yoshinagae*
1B. Scales on rhizomes and fertile stipes lanceolate, acuminate; blades oblong-lanceolate or narrowly oblong, subacute to very obtuse, acute to abruptly acuminate at base; stipes of fertile fronds as long as to half as long as the blade. . . . . 2. *E. tosaense*

1. Elaphoglossum yoshinagae (Yatabe) Makino. *Acrostichum yoshinagae* Yatabe——ATSUITA.   Fronds somewhat dimorphic, evergreen; stipes rather stout, prominently scaly, slightly flattened, narrowly winged above, 2–10 cm. long in the sterile, to about 15 cm. long in the fertile; scales brown, membranous, ovate, 3–6 mm. long, obtuse to acute, sparsely ciliate; sterile blades thick-coriaceous, lanceolate to oblanceolate, 10–30 cm. long, 2–4 cm. wide, obtuse at tip, gradually narrowed at base, decurrent on the wing of the stipe, glabrous above, with sparse deciduous scales beneath, entire, prominently white-cartilaginous on margin, the costa broad but not prominently raised on both sides; veins invisible; fertile blades smaller, usually 3.5–17 cm. long, 1–2.5 cm. wide.——Rocks and tree trunks; Honshu (Izu Isls. and s. Kinki Distr.), Shikoku, Kyushu; rare.——Ryukyus, Formosa, China, and Indochina.

2. Elaphoglossum tosaense (Yatabe) Makino. *Acrostichum tosaense* Yatabe——HIROHA-ATSUITA.   Closely allied to the preceding; scales on rhizomes and stipes lanceolate to linear-lanceolate, 4–6 mm. long, about 1 mm. wide, brown, gradually attenuate, very sparsely long-ciliate; stipes 3–14 cm. long in the sterile blades or slightly longer in the fertile, sparsely scaly in upper part, rather densely so at base; sterile blades 5–15 cm. long, 2–4 cm. wide, prominently white-cartilagineous on margin, obtuse or subacute, acute to short-acuminate at base, slightly decurrent on the stipe; fertile fronds usually shorter than the sterile but the stipe longer.——Rocks and trees; Honshu (Izu Isls. and s. Kinki Distr.), Shikoku, Kyushu (Yakushima); rare.——Ryukyus.

11. DRYOPTERIS Adans.   Ō-SHIDA ZOKU

Terrestrial; rhizomes short or elongate, ascending to erect, sometimes creeping, scaly; scales entire or toothed; stipes elongate, usually scaly; blades bipinnatifid to decompound, catadromous, usually broad at base, naked or scaly on both sides; veins free, forked; sori mostly dorsal on veins, rounded; indusia orbicular-reniform, attached by the inner end of the sinus, very rarely absent.——With about 150 species, cosmopolitan.

1A. Pinnules provided with multicellular spinelike hairs on upper side.
  2A. Indusium absent; ultimate pinnules entire or undulate. . . . . 1. *sikokiana*
  2B. Indusium present; ultimate pinnules dentate. . . . . 2. *D. hendersonii*
1B. Pinnules without multicellular spinelike hairs on upper side.
  3A. Scales on blades beneath and on rachis not saccate, nearly absent in Nos. 17 to 19.
    4A. Scales on blades usually abundant, at least on rachis.
      5A. Scales on stipes and rachis ovate; blades herbaceous, (bi-) tripinnate.
        6A. Fronds few, ovate-oblong to deltoid, annual, deciduous; lowest pinnae largest; sori distant; indusium flat, not more than 1 mm. across.
          7A. Fronds herbaceous, broadly ovate to ovate-oblong, usually longer than broad; lowest pinnae ⅓–⅔ as long as the blade; scales on stipe abundant, a few with a chestnut-colored band down the center. . . . . 3. *D. austriaca*
          7B. Blades membranous, deltoid, usually as wide as to slightly wider than long; lowest pinnae three-fourths to nearly as long as the frond; scales on stipe sparse, concolorous. . . . . 4. *D. amurensis*
      6B. Fronds many, crowded, broadly oblanceolate or broadly lanceolate, persistent over one winter, marcescent, smaller than Nos. 3 and 4; lowermost pinnae smaller than the median; sori close; indusia convex, with incurved margin, 1.5–2 mm. across, often overlapping. . . . . 5. *D. fragrans*
      5B. Scales on stipes and rachis linear to lanceolate; blades coriaceous to firmly herbaceous, pinnate to bipinnate.
        8A. Blades bipinnatiparted.
          9A. Blades oblong-lanceolate to oblanceolate, usually much narrowed at base; rachis densely scaly; pinnae sessile; lower segments of lower pinnae broadly adnate, not contracted at base.
            10A. Sori submarginal; fronds to 65 cm. long; scales black or black-brown; veinlets simple. . . . . 6. *D. polylepis*
            10B. Sori medial to subcostal; fronds 50–130 cm. long; scales brown; veinlets, at least some of them forked.
              11A. Blades 40–100 cm. long, narrowed at base; pinnae herbaceous, acuminate; indusium not split. . . 7. *D. crassirhizoma*
              11B. Blades 100–150 cm. long, scarcely narrowed at base; pinnae subcoriaceous, caudately long-acuminate; indusium breaking into 2 pieces when mature. . . . . 8. *D. wallichiana*
        9B. Blades broadly lanceolate to oblong-ovate, not or scarcely narrowed at base; rachis usually sparsely scaly; pinnae very short-petiolulate to subsessile; lower segments of lower pinnae usually contracted at base.
           12A. Scales black-brown. . . . . 9. *D. uniformis*
           12B. Scales light brown to chestnut-brown.
              13A. Rhizomes erect; sori on upper ¼–⅓ of blade, the fertile area deciduous from the sterile portion in winter; fertile pinnules much smaller than the sterile pinnules. . . . . 10. *D. lacera*
              13B. Rhizomes short-creeping; sori on the upper ½–⅔ of blade, the fertile area not deciduous from the sterile portion in winter; fertile pinnules scarcely smaller than the sterile pinnules. . . . . 11. *D. monticola*
    8B. Blades simply pinnate.

    14A. Pinnae more than 10 pairs.
       15A. Pinnae slightly dilated at base; fronds gradually narrowed at base; sori along the costa, usually in 2 series; indusium 1.2–1.5 mm. across. .... 12. *D. tokyoensis*
       15B. Pinnae not dilated at base; fronds slightly or not narrowed at base; sori between the costa and margin, in irregular series; indusium about 0.5 mm. across.
          16A. Sori nearer the margin than the costa; scales on stipes and rachis brown. .... 13. *D. dickinsii*
          16B. Sori nearer the costa than the margin; scales on stipes and rachis black-brown to purplish brown. ..... 14. *D. atrata*
    14B. Pinnae 2–7 pairs.
       17A. Pinnae 2–5 pairs, unlobed. .... 15. *D. sieboldii*
       17B. Pinnae 5–7 pairs, pinnately parted in the lower portion, lobed in the upper portion; terminal pinnae pinnatiparted.
                                        16. *D. toyamae*
4B. Scales on blades nearly absent.
    18A. Stipes densely scaly on lower half, the basal scales ovate, abruptly acuminate; axis of pinnae winged in upper portion.
                                          17. *D. sabae*
    18B. Stipes sparsely scaly throughout, the basal scales lanceolate, gradually acuminate; axis of pinnae more or less winged throughout.
       19A. Basal scales chartaceous, 6–10 mm. long; indusium strongly incurved on margin, at maturity more or less 2- or 3-lobed.
                                          18. *D. hayatae*
       19B. Basal scales membranous, 10–15 mm. long; indusium slightly incurved on margin, not breaking apart at maturity.
                                          19. *D. sparsa*
3B. Scales on blades beneath and on upper portion of stipes prominently dilated or saccate, nearly absent in Nos. 30 and 31.
    20A. Blades simply pinnate .... 20. *D. decipiens*
    20B. Blades bi- to tri (-4)-pinnate.
       21A. Lowermost posterior pinnules of lowest pinnae not larger than the adjacent ones.
          22A. Scales of stipes and rachis linear- to lanceolate-deltoid, prominently toothed, brown. .... 21. *D. championii*
          22B. Scales of stipes and rachis scattered, linear to linear-lanceolate, entire or rarely obsoletely toothed, brown to black-brown.
             23A. Pinnules entire or crenately mucronate-toothed, broadly lanceolate, elliptic to oblong, acute to obtuse or truncate at apex.
                                          22. *D. fuscipes*
          23B. Pinnules coarsely toothed to pinnatiparted, lanceolate, linear-lanceolate, or oblong.
             24A. Blades deltoid to deltoid-ovate; pinnae and pinnules horizontally spreading; pinnules deltoid-ovate to narrowly oblong, at least the lower ones pinnatiparted. .... 23. *D. indusiata*
             24B. Blades oblong to ovate-oblong; pinnae and pinnules oblique to nearly horizontally spreading; pinnules linear-lanceolate, oblong, or oblong-ovate.
                25A. Pinnae sessile or nearly so. .... 24. *D. erythrosora*
                25B. Pinnae distinctly petiolulate .... 25. *D. hondoensis*
       21B. Lowermost posterior pinnules of lowest pinnae larger than the adjacent ones.
          26A. Indusium present.
             27A. Blades ovate to oblong-ovate, much longer than wide; lowest pinnae mostly less than half as long as the entire blade; scales on stipes and rachis linear to linear-lanceolate.
                28A. Scales on rachis and stipes appressed to ascending.
                   29A. Scales on stipes and rachis ascending, mostly black-brown at base, rarely brown and concolorous. ...... 26. *D. varia*
                   29B. Scales on stipes and rachis appressed, nearly uniformly black, concolorous. .... 27. *D. sordidipes*
                28B. Scales on rachis and stipes spreading. .... 28. *D. saxifraga*
             27B. Blades 5-angled, or deltoid with an angle at base on each side, nearly as wide as long; lowest pinnae mostly more than half as long as the entire blade, rarely less, if so, then the scales on stipes broadly lanceolate.
                30A. Scales on rachis and on underside of secondary axis rather dense, subulate-linear, black-brown, saccate at base; lowest pinnae subsessile to very short-petiolate; pinnules with incurved short awn-tipped teeth. .... 29. *D. formosana*
                30B. Scales on rachis and secondary axis very scanty or nearly absent; lowest pinnae distinctly petiolulate; pinnules with ascending acute teeth or mucronate-toothed.
                   31A. Scales thinly scattered beneath on costules of pinnules, stipes, and rachis; pinnae thinly herbaceous, the lowest with petiolules 1–2.5 cm. long. .... 30. *D. chinensis*
                   31B. Plant glabrate; pinnae thickly herbaceous, the lowest with petiolules 2–5 cm. long. .... 31. *D. gymnophylla*
          26B. Indusium absent. .... 32. *D. gymnosora*

1. Dryopteris sikokiana (Makino) C. Chr. *Ctenitis sikokiana* (Makino) H. Itō; *Nephrodium sikokianum* Makino——HŌNOKAWA-SHIDA. Rhizomes short-creeping; stipes 20–50 cm. long, brownish, densely scaly; scales on lower half of the stipes broadly lanceolate to narrowly ovate, 7–10 mm. long, entire, dark chestnut-brown, lustrous, spreading to slightly deflexed, those on upper half of stipe and those of the rachis linear to linear-lanceolate, 3–8 mm. long; blades ovate, 40–75 cm. long, acuminate, 3-pinnatiparted; pinnae 3–10 pairs, narrowly oblong-ovate to broadly lanceolate, equilateral, obliquely spreading, caudate-acuminate, short-petiolulate; pinnules narrowly oblong, obtuse in the smaller ones, narrowly ovate-oblong, 2–6 cm. long, 1–2.5 cm. wide, short-acuminate in larger ones, pinnatilobed, sometimes slightly whitish beneath, nearly glabrous on both sides, sometimes with a few scattered fleshy hairs on upper side, the costules above with an elevated ridge on each side, the lowest pinnule largest, sometimes bipinnatiparted; ultimate pinnules oblong to oblong-lanceolate, 5–20 mm. long, 2–7 mm. wide, obtuse, crenate; sori nearer the costa (or to the veins) of the pinnules; indusia obsolete or absent. ——Honshu (Izu, Echizen, and Kii Prov.), Shikoku, Kyushu.

2. Dryopteris hendersonii (Bedd.) C. Chr. *Ctenitis hendersonii* (Bedd.) H. Itō; *Lastrea hendersonii* Bedd.; *Nephrodium spectabile* C. B. Clarke; *L. spectabilis* (C. B. Clarke) Bedd.; *D. leptorhachia* Hayata——HŌRAI-HIME-WARABI. Rhizomes short-creeping; stipes 40–60(–100) cm. long, deep brown, 3–6 mm. across, scaly; scales spreading to reflexed, brown, lanceolate to linear-lanceolate, 5–15 mm. long, entire;

scales on rachis 5–6 mm. long, narrowly linear-lanceolate, long-acuminate, spreading, rather dense; blades chartaceous-herbaceous, ovate, 40–50(–80) cm. long, 30–40 cm. wide, 3-pinnatiparted; pinnae 5–8 pairs, linear-oblong, 6–10 cm. wide, long-acuminate, the petiolules 5–10 mm. long, the lowest pinnae largest, nearly deltoid, the petiolules 2–3 cm. long; pinnules linear-oblong, obtuse to acute, cuneate at base, pinnatiparted, lowest posterior pinnules largest, 8–10 cm. long, the costas beneath with narrow, linear-lanceolate scales; ultimate pinnules obliquely spreading, oblong-linear, rounded to subtruncate at apex, serrulate; sori in 1–3 pairs near base of the segments; indusia erose.——Kyushu (Yakushima); rare.——Formosa and n. India.

3. Dryopteris austriaca (Jacq.) Woynar. *Polypodium austriacum* Jacq.; *P. dilatatum* Hoffm.: *Polystichum dilatatum* (Hoffm.) Schumach.; *Aspidium dilatatum* (Hoffm.) Smith; *D. dilatata* (Hoffm.) A. Gray; *A. spinulosum* Sw.; *D. siranensis* Nakai; *A. spinulosum* var. *dilatatum* (Hoffm.) Hook.; *D. spinulosa* (Sw.) Watt; *D. spinulosa* var. *dilatata* (Hoffm.) Underw.; *A. spinulosum* var. *deltoideum* Milde and var. *oblongum* Milde; *D. dilatata* var. *deltoidea* (Milde) Takeda and var. *oblonga* (Milde) Takeda——SHIRANE-WARABI. Rhizomes short-creeping, densely scaly; fronds few, deciduous; stipes 20–40 cm. long, pale green to straw-colored, densely scaly toward base; scales membranous, ovate to broadly lanceolate, 8–15 mm. long, acuminate, cordate, brown, with a broad lustrous chestnut-brown band in the center of those immediately above the base of stipe; blades herbaceous, broadly ovate to oblong-ovate, 30–60 cm. long, 25–35 cm. wide, short-acuminate, 3-pinnatiparted or sometimes nearly 3-pinnate, the rachis especially in axils above with linear to broadly linear membranous scales; pinnae 6–11 pairs, spreading, acuminate, short-petiolulate, subdeltoid, the posterior side much broader than the anterior; pinnules oblong-ovate to broadly lanceolate, 1.5–4 cm. long, 8–20 mm. wide, acute to obtuse, pinnatiparted, toothed, the lowest pinnule of the lowest pinnae largest, often again pinnate, 8–13 cm. long, acuminate; sori on upper half of the lower veinlets; indusia orbicular-reniform, entire, about 1 mm. across.——Coniferous woods in mountains; Hokkaido, Honshu (n. and centr. distr.), Shikoku (high mts.), Kyushu (high mts.).——Korea, Manchuria to the Himalayas, Sakhalin, Kuriles, Kamchatka, Siberia to Europe, and N. America.

4. Dryopteris amurensis Christ. *Aspidium spinulosum* subsp. *genuinum* var. *amurense* Milde——OKUYAMA-SHIDA. Rhizomes short-creeping; stipes 20–50 cm. long, straw-colored, often dark brown toward base, thinly scaly on lower half; scales on stipes and rhizomes pale brown, concolorous, thinly membranous, ovate to broadly so, 3–7 mm. long, acute to abruptly acuminate; blades membranous, deltoid, more or less 5-angled, 15–20 cm. long, 15–22 cm. wide, 3-pinnate or -pinnatiparted, the rachis and secondary axes thinly scaly; pinnae 5–8 pairs, narrowly deltoid, acuminate, distinctly petiolulate, much broader on the posterior than on the anterior side; pinnules oblong-ovate to narrowly oblong, 7–20 mm. long, 4–10 mm. wide, pinnatiparted, obtuse or acute, softly awn-toothed, glabrous above, with small ovate scales on veins beneath; sori on upper half of lower veinlets; indusia subreniform, entire, about 0.8 mm. across.——Coniferous woods; Hokkaido, Honshu (n. distr. and n. Kantō Distr.); rare.——Sakhalin and Amur.

5. Dryopteris fragrans (L.) Schott. *Polypodium fragrans* L.; *Aspidium fragrans* (L.) Sw.; *Nephrodium fragrans* (L.) J. Rich.; *D. fragrans* var. *remotiuscula* Komar. and var. *lepidota* Komar.——NIOI-SHIDA. Rhizomes short, densely covered with brownish marcescent fronds; fronds small, uniformly capitate-glandular; stipes crowded, many, short, 2–8 cm. long, spreading to ascending, straw-colored, brownish toward base; scales membranous, brown, rather many, ovate to broadly so, 3–7 mm. long, short-acuminate, minutely toothed; blades broadly lanceolate to broadly oblanceolate, 8–20 cm. long, 2.5–4 cm. wide, gradually acute to acuminate at both ends, 2- or 3-pinnatiparted, nearly scaleless above, with lanceolate scales on costas beneath; pinnae firmly herbaceous, horizontally spreading, lanceolate, 4–8 mm. wide, obtuse, sessile; pinnules elliptic to oblong, 1.5–2 mm. wide, crenately or obtusely toothed; indusia large, brown, lustrous, persistent, very close and overlapping, orbicular-reniform, 1–1.5 mm. across, convex, the margins incurved.——Rocks in mountains; Hokkaido, Honshu (centr. and n. distr.); rare.——Sakhalin, Kuriles, Kamchatka, Korea, Manchuria, Siberia to Europe, and N. America.

6. Dryopteris polylepis (Fr. & Sav.) C. Chr. *Aspidium polylepis* Fr. & Sav.; *Nephrodium polylepis* (Fr. & Sav.) Bak.; *N. filix-mas* var. *polylepis* (Fr. & Sav.) Makino——MIYAMA-KUMA-WARABI, MIYAMA-INODE. Rhizomes stout, short; fronds deciduous; stipes short, rather stout, 15–20 cm. long, tufted, sulcate above, densely scaly; scales on stipes of 3 forms, the largest spreading, black-brown or black, lustrous, thinly chartaceous, ovate to lanceolate, 5–10 mm. long, acuminate, finely ciliate, the smallest 1–3 mm. long, finely laciniate, the basal scales and those on rhizomes membranous, dense, lanceolate, 1–2 cm. long, gradually acuminate, entire; blades oblanceolate, 40–65 cm. long, 15–20 cm. wide, short-acuminate, gradually narrowed at base, bipinnatiparted, the scales of rachis broadly lanceolate, black-brown, ciliate; pinnae many, horizontally spreading or the lower (smaller) ones slightly deflexed, linear lanceolate, 1–2 cm. wide, acuminate, sessile, glabrous above, with appressed long-ciliate small scales on costas and costules beneath; pinnules many, subfalcate, linear-oblong to lanceolate, 1.5–3 mm. wide, obtuse, crenate; sori on upper 1/3 of frond, subterminal on the veinlets, submarginal; indusia orbicular-reniform, about 0.5 mm. across, entire.——Woods in mountains; Honshu, Shikoku, Kyushu.

7. Dryopteris crassirhizoma Nakai. *Aspidium filix-mas* sensu auct. Japon., non Schott; *D. filix-mas* var. *setosa* Christ; *D. crassirhizoma* var. *setosa* (Christ) Miyabe & Kudō; *D. setosa* (Christ) Kudō——Ō-SHIDA. Rhizomes short, stout; fronds deciduous; stipes stout, membranous, crowded, densely scaly, 10–25 cm. long, straw-colored; scales on stipes, rhizomes, and rachis narrowly lanceolate to linear, the lower 10–30(–45) mm. long, those of the rachis 5–15 mm. long, gradually narrowed at the tip, spreading to deflexed, rather unequal, brown, lustrous; blades oblanceolate to broadly so, 40–100 cm. long, 15–25 cm. wide, short-acuminate, narrowed at base, bipinnatiparted, with hairlike scales; pinnae many, herbaceous, spreading, linear-lanceolate, 15–30 mm. wide, acuminate, sessile, the costas beneath with small appressed ciliate brownish scales; pinnules narrowly oblong, 2.5–5 mm. wide, rounded to very obtuse, crenate; sori on upper half of blade between the costule and margin, disposed in 2 series on lower half of the pinnules on lower half of the forked veinlets; indusia orbicular-reniform, about 1.2 mm. across.——Wooded slopes; Hokkaido, Honshu, Shikoku; locally common.——Sakhalin, s. Kuriles, Korea, and Manchuria.

8. Dryopteris wallichiana (Spreng.) Hylander *Aspidium wallichianum* Spreng.; *D. paleacea* sensu auct. Japon., non C. Chr.; *D. doiana* Tagawa; *D. cyrtolepis* var. *doiana* (Tagawa) H. Itō——Ō-YAGURUMA-SHIDA. Rhizomes erect, stout, to 20 cm. long; fronds large; stipes 30–40 cm. long, about 1.5 cm. across near base, brown, densely scaly; scales chartaceous, lanceolate to linear, 1–2 cm. long, 1–3 mm. wide, gradually long-attenuate, entire or sparsely and minutely toothed, lustrous, dark brown with darker striations; blades lanceolate, 1–2 m. long, 30–40 cm. wide, acuminate, gradually narrowed at base, bipinnatiparted; rachis straw-colored, with dense, linear scales; pinnae subcoriaceous, linear-lanceolate, 2–3 cm. wide, caudately long-acuminate, sessile, horizontally spreading, the costa with filiform scales above, with linear to broadly linear scales beneath; pinnules narrowly oblong, 6–8 mm. wide, rounded to subtruncate at apex, cartilaginous-toothed; sori median, or slightly nearer the costules than the margin of the segments; indusia orbicular-reniform, about 1 mm. across, entire, glabrous.——Kyushu (Sakurajima); rare. ——Formosa and China to the Himalayas.

9. Dryopteris uniformis (Makino) Makino. *Nephrodium lacerum* var. *uniforme* Makino; *D. lacera* var. *subtripartita* Miyabe & Kudō; *D. dentipalea* Nakai——Ō-KUMA-WARABI. Rhizomes stout, short, erect; fronds evergreen; stipes 15–30 cm. long, stout, densely scaly; scales black-brown, linear to lanceolate, 1–2 cm. long, spinulose on margin or nearly entire, lustrous, spreading to deflexed, unequal, the larger ones sometimes paler on upper half, the basal ones often brown and entire; blades herbaceous, broadly lanceolate to oblong-ovate, 40–60 cm. long, 15–22 cm. wide, acuminate, not or scarcely narrowed at the truncate base, bipinnatiparted to somewhat bipinnate, the scales on rachis linear to linear-lanceolate, spreading to deflexed, spinulose; pinnae rather numerous, spreading, lanceolate, the sterile broadly so, 2–3.5 cm. wide, often linear-lanceolate, the fertile about 1 cm. wide, with small appressed scales beneath especially on costas, glabrous above; pinnules oblong to broadly oblong-lanceolate, 4–7 mm. wide, rounded to obtuse, the lowest pinnule of the lower pinnae often much contracted and subsessile at base, the veinlets forked or simple; sori covering ¼–¾ of blade; indusia orbicular-reniform, entire, 0.7–1 mm. across.——Woods in mountains; Hokkaido (Okushiri Isl.), Honshu, Shikoku, Kyushu.——s. Korea and China.

Dryopteris × kominatoensis Tagawa. TANI-HEGO-MODOKI. An alleged hybrid of *D. tokyoensis* and *D. uniformis*. Differs from *D. uniformis* in the brown, thinner scales with entire paler thinly membranous margins.——Honshu (centr. and n. distr.).

10. Dryopteris lacera (Thunb.) O. Kuntze. *Polypodium lacerum* Thunb.; *Aspidium lacerum* (Thunb.) Sw.; *Nephrodium lacerum* (Thunb.) Bak.; *A. filix-mas* var. *lacerum* (Thunb.) Christ——KUMA-WARABI. Rhizomes stout, short, erect; fronds evergreen; stipes short, tufted, 7–25 cm. long, pale brown, prominently scaly; scales rusty- to dark-brown, membranous, spreading, lustrous, linear-lanceolate to narrowly ovate, subentire, the basal scales broadly ovate, 6–15 mm. long, dense; blades thickly herbaceous to subcoriaceous, oblong to narrowly so, 30–60 cm. long, 15–25 cm. wide, short-acuminate, bipinnatiparted to somewhat bipinnate, the rachis with linear to lanceolate scales; sterile pinnae pinnate at base, pinnatiparted at tip, oblong-lanceolate, 3–5 cm. wide, acuminate, short-petiolulate, oblique, whitish beneath, nearly glabrous on both sides; sterile pinnules broadly lanceolate, 2–3 cm. long, 5–8

mm. wide, acute to subacuminate, more or less falcate, crenately toothed, mucro of the teeth incurved; fertile pinnules much smaller than the sterile pinnules; sori on the upper 1/4–1/3 of the frond, the fertile area deciduous apart from the sterile portion in winter; indusia orbicular-reniform, about 1.2 mm. across, entire.——Honshu, Shikoku, Kyushu.——Korea and Manchuria.

11. Dryopteris monticola (Makino) C. Chr. *Nephrodium monticola* Makino; *Aspidium monticola* (Makino) Christ; *A. filix-mas* var. *deorsolobatum* Christ; *N. erythrosorum* var. *manshuricum* Komar.; *D. submonticola* Nakai——MIYAMA-BENI-SHIDA. Rhizomes stout, short-creeping; fronds deciduous; stipes 20–50 cm. long, straw-colored, lustrous, densely scaly toward base; scales broadly linear to narrowly oblong, 1–2 cm. long, deep brown, spreading, lustrous, entire; blades oblong-ovate to oblong, 40–80 cm. long, 20–30 cm. wide, bipinnatiparted to bipinnatisected, abruptly short-acuminate, pale green and somewhat whitish beneath, the rachis with linear, thinly membranous, pale brown scales; pinnae 12–20 pairs, herbaceous, linear-lanceolate, 2–4 cm. wide, acuminate, very short-petiolulate, glabrous on both sides or the costas beneath with linear to lanceolate-deltoid, pale brown scales; pinnules oblong-lanceolate, 1–2 cm. long, 4–6 mm. wide, acute to obtuse, adnate at base, the lowermost ones contracted and sessile, coarsely crenate, softly awn-tipped; sori in 2 series on the upper 1/2–2/3 of frond, slightly nearer the costules than the margin; indusia orbicular-reniform, 1–1.2 mm. across.——Woods in mountains; Hokkaido, Honshu (Kinki Distr. and eastw.), Shikoku.——Korea and s. Manchuria.

12. Dryopteris tokyoensis (Makino) C. Chr. *Nephrodium tokyoense* Makino; *Aspidium transitorium* Christ——TANI-HEGO. Rhizomes short, erect; fronds deciduous; stipes crowded, stout, pale straw-colored, 15–30 cm. long, densely scaly near base; scales membranous, lanceolate to elliptic, 1–1.5 cm. long, abruptly acuminate, lustrous, pale brown, often slightly darker toward base, entire, deciduous; blades oblanceolate, 70–90 cm. long, 17–25 cm. wide, gradually narrowed at both ends, the rachis with small, thin, linear to lanceolate scales; pinnae 30–40 pairs, pinnately lobed, spreading, slightly dilated at base, the median broadly linear to linear-lanceolate, 1.5–2(–2.5) cm. wide, acuminate, rounded-cordate, very short-petiolulate, slightly paler beneath, glabrate on both sides; pinnules elliptic, 3–7 mm. long, 4–6 mm. wide, to 12 mm. long and 10 mm. wide in the lowermost, rounded at apex, acutely toothed at tip, the lower pinnae much-reduced, pinnately lobed, deltoid-ovate, obtuse, 2–7 cm. long, 15–40 mm. wide; sori usually in 2 series, remote to rather close, nearer the costas than the margin; indusia orbicular-reniform, 1.2–1.5 mm. across, entire.——Woods; Hokkaido, Honshu, Shikoku, Kyushu.——Korea.

13. Dryopteris dickinsii (Fr. & Sav.) C. Chr. *Aspidium dickinsii* Fr. & Sav.; *Nephrodium dickinsii* (Fr. & Sav.) Bak.; *D. okushirensis* Miyabe & Kudō; *D. hirtipes* var. *japonica* Nakai——Ō-KUJAKU-SHIDA. Rhizomes ascending to erect; stipes 10–25 cm. long, rather slender, straw-colored, moderately scaly; scales membranous, linear to narrowly oblong, 4–10 mm. long, abruptly acuminate, obsoletely toothed, sometimes with a dark brown band in the center; blades thick-chartaceous, lanceolate or oblanceolate, 40–70 cm. long, 12–20 cm. wide, acuminate, slightly narrowed at base, the rachis with narrow, spreading, ciliate scales; pinnae 20–30 pairs, the median linear-lanceolate to broadly linear, 1–2 cm. wide, acuminate, truncate or very broadly cuneate at base, horizontally spreading, sessile or very short-petiolulate, toothed and lobulate, with linear to filiform scales on costas (and costules) beneath, base of veins on upper side with a fleshy short spinelike hair; veins single in each lobule, subpinnate, the veinlets simple; lowest pinnae slightly deflexed, deltoid-lanceolate, 3–5 cm. long, 1–1.5 cm. wide, acute; sori on both sides nearer the margin than to the costa; indusia orbicular-reniform, about 0.5 mm. across, entire.——Honshu, Shikoku, Kyushu.——China.

Dryopteris tasiroi Tagawa. TSUKUSHI-Ō-KUJAKU. Said to differ from the preceding in the smaller fronds, oblong-lanceolate pinnae with parallel sides on the lower two-thirds, and very short sometimes double, incumbent, long-acuminate teeth, with 2 or 3 pairs of veinlets closer together, and smaller intramarginal sori disposed in 2 irregular series on the pinnae. Poorly understood.——Kyushu.

14. Dryopteris atrata (Wall.) Ching. *Aspidium atratum* Wall. ex Kunze; *Dryopteris cycadina* (Fr. & Sav.) C. Chr.; *Aspidium cycadinum* Fr. & Sav.; *D. hirtipes* sensu auct. Japon., non O. Kuntze——IWA-HEGO. Rhizomes stout, short; fronds evergreen; stipes crowded, 20–45 cm. long, pale brown, prominently scaly; scales spreading, chartaceous, black-brown to purplish brown, lustrous, linear-lanceolate, the largest sometimes lanceolate, 7–20 mm. long, with a long filiform tip, spinulose-ciliate, the basal scales ascending, brown, subentire; blades chartaceous, oblanceolate to oblong-oblanceolate, 40–60 cm. long, 15–25 cm. wide, abruptly contracted and long-acuminate, slightly narrowed at base, the rachis with smaller, broadly linear, sparsely ciliate scales; pinnae 20–30 pairs, horizontally spreading, broadly linear, 1–2 cm. wide, long-acuminate, truncate to subcordate at base, sessile or very short-petiolulate, sparsely scaly beneath, more densely so on the costas, nearly glabrous above, coarsely toothed, with a broad mucro on anterior margin near tip; veins single to a tooth, subpinnate, the veinlets simple; lowest pinna sometimes smaller than the others, deltoid-lanceolate, 5–13 cm. long, slightly deflexed; sori nearer the costa than the margin; indusia orbicular-reniform, about 0.5 mm. across, nearly entire.——Honshu (s. Kantō through Tōkaidō to Kinki Distr. and westw.), Shikoku, Kyushu.——Quelpaert Isl., Formosa, China to n. India.

15. Dryopteris sieboldii (Moore) C. Chr. *Pycnopteris sieboldii* Moore; *Aspidium sieboldii* Van Houtte: *Lastrea sieboldii* (Moore) Moore; *Nephrodium sieboldii* (Moore) Hook. ——Ō-MITSUDE. Rhizomes stout, short-creeping; stipes straw-colored or pale brown, rather stout, 30–60 cm. long; scales deciduous, dark-brown, linear, 5–12 mm. long, tapering to a long filiform tip, sparsely spinulose-ciliate, the basal scales membranous, many, dense, brown, linear-lanceolate, 15–20 mm. long, subentire; blades broadly ovate, 20–50 cm. long, 20–35 cm. wide, simply pinnate, the rachis straw-colored, glabrate, rather stout, 5–20 cm. long; pinnae 2–5 pairs, coriaceous-chartaceous, unlobed, linear-lanceolate to oblong-lanceolate, oblique, abruptly or gradually acuminate, usually rounded at base, with short, linear, hairlike scales beneath, glabrous above, very short-petiolulate or sessile and the upper ones often narrowly adnate at base, the costa raised beneath; fertile pinnae 10–25 cm. long, 15–35 mm. wide, subentire or crenate, the sterile broader, 10–30 cm. long, 3–7 cm. wide, crenate to mucronate-toothed; sori mostly toward the margin; indusia orbicular-reniform, entire, about 1.5 mm. across.——Woods; Honshu (s. Kantō to Tōkaidō and Chūgoku Distr.), Shikoku, Kyushu.——Formosa.

16. Dryopteris toyamae Tagawa. *D. sieboldii* var. *toyamae* (Tagawa) Kurata——NAGASAKI-SHIDA-MODOKI. Very

closely allied to the preceding; rhizomes stout, ascending; stipes straw-colored, 20–30 cm. long, densely scaly toward the base; basal scales brown, membranous, lustrous, lanceolate to linear-lanceolate, 1–1.5 cm. long, long-acuminate, entire, those on the upper part of stipes and lower part of rachis very sparse, dark-brown, linear, 3–7 mm. long, sparsely spinulose-ciliate; fertile blades broadly ovate, 30–40 cm. long, 25–40 cm. wide, the rachis 20–25 cm. long; pinnae 5–7 pairs, linear-lanceolate, 2–3 cm. wide, gradually acuminate, the lower ones pinnately parted, the upper lobed, mucronate-toothed, the lobes oblique, ovate-deltoid to oblong, abruptly acute to obtuse, the basal pinnae often smaller and cut nearly to the base; veins subpinnate, single to a lobe, the veinlets elongate, simple; terminal pinna pinnatiparted; sori scattered, sometimes more so toward margin; indusia orbicular-reniform, entire, about 1.5 mm. across.——Honshu (reported from Mount Kiyosumi in Awa Prov.), Kyushu (Hizen and Higo Prov.); rare.——Formosa.

17. Dryopteris sabae (Fr. & Sav.) C. Chr. *Aspidium sabae* Fr. & Sav.; *Nephrodium sabae* (Fr. & Sav.) Hand.-Mazz.; *A. filix-mas* var. *sabae* (Fr. & Sav.) Christ——MIYAMA-ITACHI-SHIDA. Rhizomes short, erect; stipes many, tufted, 15–30 cm. long, deep-brown, lustrous on lower half, prominently scaly especially toward base; scales brown, narrowly to broadly ovate, 5–10 mm. long, abruptly acuminate, lustrous, spreading, entire; blades ovate, 20–40 cm. long, 15–25 cm. wide, abruptly acuminate, bipinnate, in the lower part often tripinnate, the rachis straw-colored, sparsely scaly while young; pinnae 5–7 pairs, narrowly oblong-ovate, the median equilateral, the lowest deltoid-ovate, oblique, acuminate, short-petiolulate; pinnules thick-chartaceous, deltoid-oblong to lanceolate, 1.5–4 cm. long, 8–25 mm. wide, obtuse to acute, acute- or mucronate-toothed, glabrous on both sides, pinnately parted to lobed, the lowest posterior pinnule of the lowest pinnae largest, acuminate, 4–8 cm. long; sori on upper half of the blade, mostly nearly median, in 2 series on the ultimate pinnule; indusia orbicular-reniform, 1–1.2 mm. wide, entire, slightly incurved and thinner on margin.——Hokkaido, Honshu, Shikoku, Kyushu.

18. Dryopteris hayatae Tagawa. *D. subexaltata* sensu auct. Japon., non C. Chr.——INU-TAMA-SHIDA. Rhizomes short; stipes tufted, slender, lustrous, 10–30 cm. long, chestnut-brown, sparsely scaly toward base, nearly naked in upper part; scales linear-lanceolate, gradually acuminate, brown, often with few dark striations, spreading, 2–3 mm. long, the basal ones chartaceous, 6–10 mm. long; blades narrowly ovate, 15–30 cm. long, 8–15 cm. wide, acuminate to caudate with an obtuse to acute tip, bipinnate; pinnae 5–8 pairs, petiolulate, oblong-ovate, rarely linear-oblong, 2–4 cm. long, often caudate at the tip, the lowest pinnae ovate-deltoid, to 5 cm. wide, oblique; pinnules chartaceous, narrowly oblong to deltoid-lanceolate, 1–2 cm. long, 5–10 mm. wide, rounded to obtuse, cuneate at base, acute- to mucronate-toothed, often lobulate or parted, the lowest pinnule of the basal pinnae largest, usually deltoid-ovate, about 5 cm. long, 2 cm. wide; sori median in 2 series on the ultimate pinnules; indusia subreniform, entire, strongly incurved on margin, 2- or 3-lobed at maturity, about 1 mm. across.——Honshu (Hachijo Isl.), Kyushu (Osumi Prov. incl. Yakushima).——Ryukyus and Formosa.

19. Dryopteris sparsa (Hamilt. ex D. Don) O. Kuntze. *Nephrodium sparsum* Hamilt. ex D. Don; *Aspidium sparsum* (Hamilt.) Spreng.; *Lastrea sparsa* (Hamilt.) Moore; *N. viridescens* Bak.; *D. viridescens* (Bak.) O. Kuntze——NAGABA-NO-

ITACHI-SHIDA.   Rhizomes short; stipes 20–60 cm. long, brown, sparsely scaly throughout while young, rather densely so at base; scales pale brown, deciduous, spreading, broadly lanceolate, 3–10 mm. long, acuminate, entire; basal scales membranous, linear-lanceolate, 10–15 mm. long; blades narrowly ovate or oblong-ovate, 30–50 cm. long, 15–25 cm. wide, bipinnate, or the larger often tripinnate at base, acuminate, the rachis straw-colored, usually suffused red-purple, often pale brown on lower half beneath, smooth, lustrous; pinnae 7–10 pairs, deltoid-ovate to -lanceolate, obliquely spreading, short-caudate at apex, petiolulate, the lowest pinnae oblique; pinnules broadly lanceolate to broadly oblong-ovate, 1–3 cm. long, 4–20 mm. wide, obtuse to acute, mucronate-toothed, nearly glabrous on both sides, usually pinnately parted to lobed, the lowest pinnule of the basal pinnae largest, 5–12 cm. long, often pinnate; divisions of the pinnules narrowly oblong, 2.5–5 mm. wide, obtuse; sori in 2 distinct or in more or less irregular series, median on the ultimate pinnules; indusia orbicular-reniform, about 1 mm. across, entire, slightly incurved on margin. ——Honshu (s. Kantō through Tōkaidō to Kinki Distr.), Shikoku, Kyushu.——Ryukyus, Formosa, China to Malaysia, and India.

20. Dryopteris decipiens (Hook.) O. Kuntze. *Aspidium decipiens* (Hook.) Luerss.; *Nephrodium decipiens* Hook. ——NACHI-KUJAKU, IWA-URABOSHI.   Rhizomes ascending to short-creeping, densely covered with the basal part of old stipes; fronds evergreen; stipes 10–30 cm. long; basal scales rather dense, black-brown, broadly linear to linear-lanceolate, 8–10 mm. long, the upper ones 5–8 mm. long, entire; blades broadly lanceolate, 25–40 cm. long, 8–18 cm. wide, acuminate, simply pinnate; rachis slender, the scales linear, black-brown, spreading, saccate at base; pinnae deltoid-lanceolate, acuminate, crenate, somewhat falcate, glabrous above, with short hairlike scales beneath, short-petiolulate, pinnately lobed to lobulate, the costas with small saccate scales beneath; veins pinnate, oblique, the veinlets simple or forked; sori in 2 series, nearer the costa than the margin; indusia orbicular-reniform, 1–1.2 mm. across, brown, entire.——Honshu (Tōkaidō to s. Kinki and Chūgoku Distr.), Shikoku, Kyushu.——Ryukyus and China.

21. Dryopteris championii (Benth.) Ching. *Aspidium championii* Benth.; *D. erythrosora* var. *cavaleriei* Rosenst.; *D. pseudoerythrosora* Kodama; *D. kinkiensis* Koidz.——SAIKOKU-BENI-SHIDA.   Rhizomes short; fronds evergreen; stipes 20–50 cm. long, densely scaly; scales spreading, membranous, brown, slightly lustrous, linear- to lanceolate-deltoid or sometimes ovate, 8–15 mm. long, narrowed to a filiform tip, prominently spine-toothed, except at the tip; blades oblong-ovate to ovate, 30–50 cm. long, 20–30 cm. wide, gradually or abruptly caudate-acuminate, broadest near base, bipinnate, the rachis densely scaly; pinnae 10–13 pairs, deltoid-lanceolate, 10–20 cm. long, 2.5–5 cm. wide, oblique, acuminate, usually broadest at base, short-petiolulate; pinnules thick-chartaceous, oblong-ovate to ovate, obtuse to rounded or subacute, broadly cuneate to rounded at base, usually with a small auricle on each side at base, sessile or very short-petiolulate, toothed, glabrous above, sometimes lobed, with short hairlike scales beneath while young; sori in 2 series, nearly median between the costule and margin; indusia orbicular-reniform, 1–1.2 mm. across, subentire.——Honshu (s. Kantō through Tōkaidō and s. Shinano to Kinki Distr. and westw.), Shikoku, Kyushu.——s. Korea and China.

22. Dryopteris fuscipes C. Chr. *Nephrodium erythrosorum* var. *obtusum* Makino; *D. erythrosora* var. *obtusa* (Makino) Makino; *D. obtusissima* Makino; *D. bipinnata* C. Chr., non Copel.; *D. medioxima* Koidz.; *D. makinoi* Koidz.; *D. inuyamensis* H. Itō——MARUBA-BENI-SHIDA.   Rhizomes short, ascending; fronds evergreen, stipes 20–40 cm. long, straw-colored or brownish, rather densely scaly; scales spreading, linear-lanceolate, 5–10(–15) mm. long, brown, gradually long-acuminate, entire; blades oblong- to deltoid-ovate, 25–50 cm. long, 15–30 cm. wide, bipinnate, the rachis rather densely scaly; pinnae 10–13 pairs, lanceolate to broadly so, 2–4 cm. wide, long-acuminate, the rachis with small brown ovate-saccate scales beneath; pinnules broadly lanceolate, elliptic to oblong, acute to obtuse or truncate at apex, sessile, entire or crenately mucronate-toothed, sometimes auricled on anterior side at base, nearly glabrous above, with hairlike appressed scales beneath; sori mostly median, few, in 2 rows on the lower portion of the pinnules, 1–1.5 mm. across, entire, membranous on margin. ——Honshu (Kantō Distr. and westw.), Shikoku, Kyushu.——Formosa and China.

23. Dryopteris indusiata (Makino) Makino & Yamamoto. *Nephrodium gymnosorum* var. *indusiatum* Makino; *D. gymnosora* var. *indusiata* (Makino) Makino——NUKA-ITACHI-SHIDA-MODOKI, NUKA-ITACHI-SHIDA-MAGAI.   Rhizomes short, stout; stipes 20–45 cm. long, scaly; scales near base brown to black-brown, linear-lanceolate, 6–10 mm. long, entire, the upper ones membranous, deflexed, lanceolate-deltoid, 3–5 mm. long, entire; blades narrowly deltoid or deltoid-ovate, 20–50 cm. long, 15–30 cm. wide, acuminate, bipinnate, the rachis with small lanceolate-deltoid scales cordate or saccate at base; pinnae deltoid-lanceolate, 15–20 cm. long, 3–7 cm. wide, acuminate, sessile or nearly so, usually horizontally spreading, the rachis and costules beneath with appressed ovate-saccate brown scales about 1 mm. long; pinnules deltoid-ovate to narrowly oblong, 1.5–4.5 cm. long, 5–15 mm. wide, obtuse to rounded at apex, truncate, sometimes rounded at base, somewhat whitish beneath, sessile, usually pinnately parted to lobed, at least the lower, toothed; veins indistinct; sori median, usually in 2 series on the pinnules; indusia orbicular-reniform, entire, about 1 mm. across.——Honshu (Izu, Mikawa Prov. and westw.), Shikoku, Kyushu.——Formosa.

24. Dryopteris erythrosora (Eaton) O. Kuntze. *Aspidium erythrosorum* Eaton; *Nephrodium erythrosorum* (Eaton) Hook.; *A. filix-mas* var. *erythrosorum* (Eaton) Christ; *A. prolificum* Maxim.; *D. erythrosora* var. *prolifica* (Maxim.) Makino——BENI-SHIDA.   Rhizomes ascending, stout; fronds evergreen; stipes 30–60 cm. long, often reddish or brownish; scales linear to linear-lanceolate, nearly entire, brown to black-brown, 10–15 mm. long in the basal, the upper ones smaller; blades broadly ovate to oblong, 30–70 cm. long, 15–35 cm. wide, acuminate, bipinnate, the rachis with linear-lanceolate to linear scales 2–3.5 mm. long; pinnae 8–12 pairs, sessile or nearly so, acuminate, the rachis with appressed ovate-saccate small scales with or without a linear tail; pinnules narrowly oblong to linear-lanceolate, 2–4 cm. long, 4–7 mm. wide, acute to rounded at apex, toothed, sometimes pinnately lobed, nearly glabrous above, with appressed hairlike short scales on veins beneath; sori mostly median, in 2 series on the pinnules; indusia often reddish while young, orbicular-reniform, entire, about 1.5 mm. across.——Woods in low mountains and hills; Honshu, Shikoku, Kyushu; very common and variable.——Korea, Ryukyus, Formosa, China to Philippines.

Var. **caudipinna** (Nakai) H. Itō. *D. caudipinna* Nakai ——HACHIJŌ-BENI-SHIDA. Blades longer, with ascending pinnatilobed pinnules to 6 cm. long; scales more dense.—— Warmer parts of our area.

Var. **purpurascens** H. Itō. *D. purpurella* Tagawa—— MURASAKI-BENI-SHIDA. Stipes and rachis suffused purple.—— Kyushu (Yakushima).——Ryukyus.

Var. **cystolepidota** (Miq.) Nakai. *Aspidium cystolepidotum* Miq.; *D. cystolepidota* (Miq.) C. Chr.; *D. nipponensis* Koidz.——Ō-TŌGOKU-SHIDA. Blade usually abruptly acuminate, the lowest pinnae wider with a wider posterior side; pinnules broader, often incurved-toothed to parted; sori smaller. ——Honshu (Kantō Distr. and westw.), Shikoku, Kyushu.

Var. **ambigens** (Koidz.) Nakai. *D. cystolepidota* var. *ambigens* (Koidz.) H. Itō; *D. nipponensis* var. *ambigens* Koidz. ——Ō-TŌGOKU-SHIDA. Similar to the preceding variety, the blades elliptic, the pinnules obtuse, the fertile blades linear-oblong, often subentire and slightly crisped on margin.—— Honshu (Kantō Distr. and westw.), Shikoku, Kyushu; rare.

25. Dryopteris hondoensis Koidz. *D. rhombeo-ovata* H. Itō——Ō-BENI-SHIDA. Rhizomes ascending, stout; stipes 20–50 cm. long, brownish, moderately scaly; scales more numerous near the base, lanceolate or narrowly so, the lower ones 8–12 mm. long, upper ones 3–5 mm. long, long-acuminate, lustrous, entire, spreading; blades ovate to narrowly so, 30–50 cm. long, 20–30 cm. wide, acuminate, the rachis usually brownish, with sparse, broadly linear to filiform, brown scales 2–4 mm. long; pinnae obliquely spreading, deltoid-lanceolate, 3–5 cm. wide, acuminate, distinctly petiolulate, the lowest pinnae slightly broader, oblique; pinnules broadly lanceolate to narrowly oblong-ovate, 15–40 mm. long, 6–15 mm. wide, acute to obtuse, broadly cuneate at base, with scattered, short-appressed, hairlike scales beneath, glabrous above, the upper ones coarsely incurved-toothed, the lower pinnately parted, the costules beneath with small appressed ovate-saccate scales; sori more or less median in 2 rows on the pinnules; indusia orbicular-reniform, entire, about 1 mm. across.——Honshu (s. Kantō through Tōkaidō to Kinki and Chūgoku Distr.), Shikoku, Kyushu.

26. Dryopteris varia (L.) O. Kuntze. *Polypodium varium* L.; *Nephrodium varium* (L.) Desv.; *Polystichum varium* (L.) Presl; *Lastrea opaca* Hook.; *D. yabei* Hayata; *D. matsuzoana* Koidz.; *D. ogawae* H. Itō——ITACHI-SHIDA-MODO-KI. Rhizomes stout; fronds evergreen; stipes few, 20–60 cm. long; scales linear to subulate, 3–10 mm. long, filiform at the tip, entire, black-brown, dilated, lobed, ciliate, closely appressed to the stipe at base, the basal scales dense, 1–2.5 cm. long, 1–2 mm. wide, lustrous, brown to black-brown, scarcely dilated at base; blades broadly ovate to elliptic, 25–50 cm. long, 20–30 cm. wide, abruptly contracted and often subtruncate with a deltoid-lanceolate acuminate tail at the tip, bipinnate; pinnae 6–9 pairs, sometimes more, broadly linear, 10–25 cm. long, 2–3.5 cm. wide, gradually acuminate, short-petiolulate; pinnules oblong to linear-oblong, more or less falcate, 1–2.5 (–3.5) cm. long, 5–8 mm. wide, acute to obtuse, usually auricled, subcordate to cuneate at base, few-toothed, sometimes lobulate or pinnatiparted, sessile; sori in 2 rows, nearer the margin than the costule; indusia orbicular-reniform, nearly entire, about 1.2 mm. across.——Honshu (Kantō and Tōkaidō to Kinki and Chūgoku Distr.), Shikoku, Kyushu; rather rare.——Ryukyus, Formosa, China to Indochina, and Philippines; very variable.

Var. **setosa** (Thunb.) Ohwi. *Polypodium setosum* Thunb.; *Polystichum setosum* (Thunb.) Presl, non Schott; *P. thunbergii* Koidz.——ITACHI-SHIDA. Blades gradually narrowed from below the middle to the tip.——Woods in hills and low mountains; Honshu (Kantō Distr. and westw.), Shikoku, Kyushu; very common.——Korea and China.

Var. **subtripartita** (Fr. & Sav.) H. Itō. *Aspidium lacerum* var. *subtripartitum* Fr. & Sav.; *A. lacerum* var. *bipinnatum* Fr. & Sav.; *Polystichum varium* var. *subtripartitum* (Fr. & Sav.) Nakai; *Nephrodium bissetianum* Bak.; *D. bissetiana* (Bak.) C. Chr.; *P. bissetianum* (Bak.) Nakai; *P. pacificum* Nakai——Ō-ITACHI-SHIDA. Blades often larger, often tripinnately parted to tripinnate, the pinnules often broadly lanceolate, acute, toothed; sori median or somewhat closer to the costules than to the margin.——Honshu (Kantō Distr. and westw.), Shikoku, Kyushu.

Var. **sacrosancta** (Koidz.) Ohwi. *D. sacrosancta* Koidz.; *Polystichum sacrosanctum* (Koidz.) Koidz.; *D. kobayashii* Kitag.——HIME-ITACHI-SHIDA. Similar to the preceding variety but the blades thinner, firmly herbaceous to subcoriaceous, dull; pinnules more deeply and finely parted.——Honshu (Kantō Distr. and westw.), Shikoku, Kyushu.——Korea and Manchuria.

27. Dryopteris sordidipes Tagawa. YOGORE-ITACHI-SHIDA. Closely allied to the preceding species, especially var. *setosa*; stipes 20–50 cm. long, with dense blackish appressed scales; blades firmly herbaceous to thinly coriaceous, ovate to oblong-ovate, 25–45 cm. long, 20–30 cm. wide, acuminate, bipinnate, the scales of rachis dense, black, appressed, linear to nearly filiform, dilated; pinnae petiolulate, the lowest on a rather prominent petiolule 1.5–2 cm. long, the rachis beneath with appressed rusty-brown to blackish nearly filiform scales dilated at base; pinnules broadly lanceolate, deltoid-lanceolate, or narrowly oblong, 1–3 cm. long, 3–20 mm. wide, obtuse to acute, often auricled at base, toothed, pinnately lobed to parted and the lower ones very short-petiolulate, the lowest pinnules of the lowest pinnae largest, to nearly 10 cm. long, pinnate in lower half; sori median; indusia orbicular-reniform, entire, usually less than 1 mm. across.——Kyushu (s. distr. incl. Yakushima).——Ryukyus and Formosa.

28. Dryopteris saxifraga H. Itō. IWA-ITACHI-SHIDA. Rhizomes stout, covered with the basal stubs of old stipes; stipes several, 10–20 cm. long, rather stout, densely scaly, straw-colored, brown and thickened at base; scales spreading, linear to broadly so, black-brown, the upper ones and those of the rachis 3–7 mm. long, with a pale-brown dilated or saccate base, the lower ones longer, 7–12 mm. long, not prominently dilated at base; blades oblong-ovate to ovate, 15–30 cm. long, 10–15 cm. wide, acuminate; pinnae short-petiolulate; pinnules linear-oblong to lanceolate, acute, cuneate at base or decurrent on the axis, the lower ones pinatifid to pinnatiparted; scales on costas beneath and on rachis membranous, ovate, inflated, 0.5–1 mm. long, rusty-brown, often with a hairlike tail at the tip; sori in 2 rows on the pinnules, median or nearer to the costules; indusia orbicular-reniform, entire, about 1 mm. across.—— Rocks; Hokkaido, Honshu, Shikoku, Kyushu (high mts.).—— Korea and Manchuria.

29. Dryopteris formosana (Christ) C. Chr. *Aspidium formosanum* Christ; *Polystichum constantissimum* Hayata; *D. phaeolepis* Hayata; *D. kodamae* Hayata; *D. takeuchiana* Koidz.; *Polystichum varium* var. *eurylepidotum* Rosenst.—— TAKASAGO-SHIDA. Rhizomes short, ascending, covered with the basal stubs of old stipes; stipes few, 20–60 cm. long, brown, scaly; scales spreading, the lower lanceolate, 10–15 mm. long, entire, black-brown, lustrous, the upper ones often deflexed, broadly linear, 2–7 mm. long, dilated at base; blades bipinnate, broadly ovate, 25–40 cm. long, 25–30 cm. wide, abruptly long-acuminate, mostly 5-angled; scales on rachis rather numerous, subulate-linear, black-brown, saccate at base; lowest pinnae lanceolate to broadly so, broadest at base, caudate-acuminate, short-petiolulate to subsessile, the lowest subdeltoid, with a very broad posterior side, the petiolule 1–1.5 cm. long; pinnules and ultimate segments narrowly ovate-oblong or lanceolate, 1.5–4 cm. long, 7–15 mm. wide, obtuse to subacute, auriculate on anterior side and obliquely cuneate at base, with short awn-tipped teeth, subsessile or adnate and slightly decurrent at base, the lower ones pinnately lobed to parted, the lowest posterior pinnules of the lowest pinnae caudate-acuminate, pinnate, 10–15 cm. long, 2.5–3 cm. wide; sori in 2 series; indusia orbicular-reniform, about 1 mm. across, entire.——Honshu (Tōtōmi Prov. and Kinki Distr. westw.), Kyushu. ——Formosa.

30. Dryopteris chinensis (Bak.) Koidz. *Nephrodium chinense* Bak.; *N. subtripinnatum* Bak., excl. syn.; *Aspidium subtripinnatum* sensu auct. Japon., non Miq.; *D. subtripinnata* sensu auct. Japon., non O. Kuntze——Misaki-kaguma. Rhizomes creeping, short, usually densely covered with stubs of old stipes; fronds deciduous; stipes 15–30 cm. long, slender, pale green to brownish, scaly; scales brown, broadly lanceolate, 3–10 mm. long, short-acuminate, remotely toothed to entire; blades 5-angled, broadly ovate, 15–25 cm. long, 10–22 cm. wide, bipinnate, the rachis slender, with spreading scales 1.5–2.5 mm. long; pinnae 5–6 pairs, thinly herbaceous, deltoid-ovate to -lanceolate, 7–15 cm. long, 5–10 cm. wide, acuminate, spreading, the petiolules 1–2.5 cm. long; pinnules ovate-oblong, 1–2 cm. long, 4–12 mm. wide, obtuse, very short-petiolulate, pinnately lobed to parted, or the lowest one pinnatifid; ultimate pinnules oblong, toothed, 2–3 mm. wide, with minute hairlike scales beneath; sori submarginal, near sinus of the teeth; indusia orbicular-reniform, about 0.7 mm. across, entire.—— Hokkaido, Honshu, Shikoku, Kyushu.——Korea and China.

31. Dryopteris gymnophylla (Bak.) C. Chr. *Nephrodium gymnophyllum* Bak.; *D. subtripinnata* var. *sakuraii* Rosenst.; *D. koraiensis* Tagawa; *D. chinensis* var. *sakuraii* (Rosenst.) H. Itō; *D. sakuraii* (Rosenst.) Tagawa——Sakurai-kaguma. Rhizomes creeping, short to somewhat elongate; stipes slender, pale green, lustrous, 15–30 cm. long, nearly naked except for few scales near base; scales brown, lanceolate, 4–8 mm. long; blades 5-angled, broadly ovate, 20–40 cm. long, 15–30 cm. wide, bipinnate, abruptly long-acuminate, nearly scaleless and smooth, the rachis pale green; pinnae 5–8 pairs, thick-herbaceous, spreading, the lower ones distinctly petiolulate, broadly deltoid-lanceolate, the lowest subdeltoid with a dilated posterior side, the petiolules 2–5 cm. long; pinnules oblong-lanceolate, 1.5–4 cm. long, 5–15 mm. wide, acute to obtuse, cuneate to truncate at base, pinnately lobed to parted, crenate, the lowest one 5–15 cm. long, 2–5 cm. wide, pinnatiparted to bipinnatiparted, petiolulate; sori near the sinus of the teeth, median or sometimes subcostal; indusia orbicular-reniform or reniform, hyaline, about 0.7 mm. across, erose-dentate.—— Honshu (Kantō Distr. to Kai, s. Shinano, and Mikawa Prov.). ——Korea, China, and Manchuria.

32. Dryopteris gymnosora (Makino) C. Chr. *Nephrodium gymnosorum* Makino——Nuka-itachi-shida. Rhizomes short; stipes 30–50 cm. long, slender, sparsely scaly, straw-colored, often brownish purple on under side, the scales at base linear, 6–8 mm. long, nearly entire, brown to black-brown, the upper ones very sparse, linear to filiform, 3–4 mm. long, brown; blades narrowly to broadly ovate, sometimes nearly deltoid, 25–45 cm. long, 18–30 cm. wide, somewhat contracted in upper part, acuminate; pinnae lanceolate-deltoid to broadly lanceolate, 3–6 cm. wide, caudate-acuminate, sessile, spreading, the lowest ones with a broader posterior side, the rachis and underside of costules with small, brown, globose-saccate scales with or without a filiform tip; pinnules herbaceous, broadly lanceolate to narrowly deltoid-ovate, 1.5–5 cm. long, 5–22 mm. wide, obtuse to subacute, sessile, pinnately lobed to parted, ascending-toothed; indusia absent.——Honshu (Izu Prov. through Tōkaidō to Kinki Distr.), Shikoku, Kyushu.

12. CTENITIS (C. Chr.) C. Chr.   Katsumō-inode Zoku

Terrestrial fern of moderate to large size; rhizomes short, ascending, erect, or rarely creeping, scaly; fronds bipinnatifid to decompound, broadest at base, the ultimate divisions usually rounded to obtuse, the rachis usually with toothed scales, the upper side mostly with jointed multicellular hairs; veins free, simple or branched; sori on the veins; indusia orbicular-reniform, sometimes absent.——About 150 species in the warmer regions of the world.

1A. Lowest posterior pinnule of the lowest pinnae smaller than the others. .... 1. *C. maximowicziana*
1B. Lowest posterior pinnule of the lowest pinnae the largest.
  2A. Indusium present; fronds 2- to 4-pinnate.
    3A. Fronds 3- to 4-pinnate; scales on rachis and upper half of the stipes appressed. .... 2. *C. subglandulosa*
    3B. Fronds 2-pinnate; scales on rachis and upper half of the stipes spreading. .... 3. *C. eatonii*
  2B. Indusium absent; fronds once-pinnate, but the lowest pinnae often again pinnate in large specimens. .... 4. *sinii*

1. Ctenitis maximowicziana (Miq.) Ching. *Aspidium maximowiczianum* Miq.; *Dryopteris maximowicziana* (Miq.) C. Chr.; *A. matsumurae* Makino; *Nephrodium matsumurae* (Makino) Makino; *D. matsumurae* (Makino) C. Chr.——Shiraga-shida, Kiyosumi-hime-warabi. Rhizomes short; stipes 20–40 cm. long, densely scaly; scales membranous, spreading to deflexed, brown, broadly lanceolate to linear-lanceolate, 7–15 mm. long, those on the rachis 4–10 mm. long, entire; blades ovate to

broadly so, 40–80 cm. long, 25–50 cm. wide, acuminate, 3-pinnatiparted to subtripinnate; pinnae broadly lanceolate to narrowly oblong-ovate, 5–8 cm. wide, acuminate, short-petiolulate; pinnules narrowly oblong-ovate, 2.5–4 cm. long, 1–1.2 cm. wide, obtuse, pinnately parted, entire to crenate or lobulate, sparingly pilose on both sides, more densely so on the costule above, the lowest posterior pinnule smaller than the adjacent ones; ultimate pinnules oblong, obtuse, crenate to lobulate; sori in 2 rows along the costule, sometimes with 1 or 2 irregular rows nearer the margin and a single sorus (sometimes 2-3) near the base of the segment; indusia membranous, reniform, lacinulate-ciliate, about 0.7 mm. across.——Woods; Honshu (s. Kantō through Tōkaidō to s. Kinki Distr. and westw.), Shikoku, Kyushu.——Formosa.

2. **Ctenitis subglandulosa** (Hance) Ching. *Alsophila subglandulosa* Hance; *Aspidium subtripinnatum* Miq.; *Nephrodium subtripinnatum* (Miq.) Bak.; *Dryopteris subtripinnata* (Miq.) O. Kuntze; *C. subtripinnata* (Miq.) H. Itō; *Polypodium oldhamii* Bak.; *Dryopteris subglandulosa* (Hance) Hayata; *D. lepigera* sensu auct. Japon., saltem pro parte, non O. Kuntze——KATSUMŌ-INODE. Rhizomes short-creeping, densely scaly; stipes 30-70 cm. long, brown; scales at base of stipes and on rhizomes very numerous, membranous, lustrous, linear, 2-3 cm. long, about 2 mm. wide, nearly entire, glabrous; scales on upper part of stipes and rachis ovate to broadly lanceolate, 2-4 mm. long, narrowed at tip, appressed, lustrous, ciliate; blades ovate, 40-70 cm. long, 20-40 cm. wide, 3-pinnate, short-acuminate, the rachis densely scaly; pinnae obliquely spreading, equilateral, short-petiolulate or sessile, the lowest pinnae largest, longer petiolulate, the posterior side broader than the anterior; pinnules oblong-ovate to lanceolate, usually 2-5 cm. long, obtuse to subacute, crenate to lobed, sparsely pilose on upper surface, with brownish short hairs on the costas, the costules beneath with small appressed ovate scales; sori costal or near base of veinlets; indusia membranous, ovate-orbicular or orbicular-reniform, about 0.2 mm. across, erose-ciliate.——Honshu (s. Kantō and s. Kinki Distr.), Shikoku, Kyushu.——Ryukyus, Formosa, China to India.

3. **Ctenitis eatonii** (Bak.) Ching. *Nephrodium eatonii* Bak.; *Dryopteris eatonii* (Bak.) O. Kuntze; *N. leucostipes* Bak.——HORA-KAGUMA. Rhizomes short, ascending; stipes 15-40 cm. long; scales on stipes and rachis dark brown, chartaceous, linear-subulate, 5-10 mm. long, subentire, auriculate at base; blades ovate, 20-45 cm. long, 13-30 cm. wide, bipinnate to tripinnatiparted; pinnae 9-10 pairs, the upper ones sessile to subsessile, the lower spreading, narrowly ovate-deltoid, 5-14 cm. wide, acuminate, short-petiolulate; pinnules broadly lanceolate to oblong-ovate, obtuse to acute, usually prominently short-pubescent on both sides especially on the costas, often lobulate to parted; ultimate pinnules oblong to elliptic, crenate; sori median or slightly nearer the costule than the margin of the ultimate pinnules; indusia orbicular-reniform, entire, glandular-ciliolate, about 0.3 mm. across.——Kyushu (Yakushima).——Ryukyus, Formosa, and s. China.

4. **Ctenitis sinii** (Ching) Ohwi. *Tectaria sinii* Ching; *Ctenitopsis sinii* (Ching) Ching——SATSUMA-SHIDA. Stipes 30-45 cm. long, brown; scales at base of stipes numerous, thinly membranous, linear or nearly filiform, 10-20 mm. long, about 1 mm. wide, brown, entire; scales on lower half of stipes lanceolate, 2-7 mm. long, long-acuminate, entire, purplish brown, deflexed, with elliptic to round cells, the scales on upper half of stipe and on rachis similar to the lower ones but smaller, ascending to appressed, few-toothed, narrowly deltoid-ovate to broadly lanceolate, 2-3 mm. long; blades herbaceous-chartaceous, ovate, 30-50 cm. long, 20-25 cm. wide, acuminate, bipinnatiparted at base, simply pinnate in upper half; pinnae several pairs, the uppermost linear-lanceolate, forming a large pinnatiparted terminal pinna, the median linear-oblong, 2-5 cm. wide, acuminate, broadly cuneate at base, sessile to short-petiolulate, pinnately lobed to parted, sparingly pilose above, with a few appressed small scales on costas beneath; pinnules obliquely spreading, rounded to oblong, 6-12 mm. wide, subentire to obsoletely crenate; lowest pinnae ovate-deltoid, to 25 cm. long, 15 cm. wide, distinctly petiolulate, oblique, the segments on posterior side linear-lanceolate, 2-4 times as long as those of the anterior side, usually pinnately lobed to coarsely crenate; sori scattered or in 2 rows slightly nearer the costule than the margin; indusia obsolete.——Kyushu (Satsuma Prov.); rare.——China.

13. LASTREA Bory ŌBA-SHORIMA ZOKU

Small to moderate terrestrial ferns; rhizomes creeping to erect, dictyostelic, sparsely scaly; fronds often hairy; blades usually bipinnatifid, sometimes more profusely divided, often narrowed at base, with simple, straight, 1-celled hairs; veins free, usually simple, reaching to or short of the margin; sori on the veins, dorsal, rarely terminal, small, rounded or rarely elongate; indusia orbicular-reniform, or absent. Cosmopolitan with about 500 species.

1A. Indusium absent, or very minute.
  2A. Sori linear. . . . . 1. *L. pozoi*
  2B. Sori orbicular to elliptic.
    3A. Pinnae adnate at base on the winged rachis.
      4A. Blades deltoid, shorter than the stipe. . . . . 2. *L. phegopteris*
      4B. Blades lanceolate, much longer than the stipe. . . . . 3. *L. decursivepinnata*
    3B. Pinnae sessile, not adnate at base, except the uppermost.
      5A. Pinnules coarsely crenate to lobulate; pneumatophores absent at the base of pinnae; hairs or some of them branched at base.
                      4. *L. bukoensis*
      5B. Pinnules subentire to crenate; pneumatophores subglobose at the base of pinnae beneath; hairs simple. . . . . 5. *L. omeiensis*
1B. Indusium present.
  6A. Scales on stipes and rachis copious; stipes tufted, on a short rhizome. . . . . 6. *L. quelpaertensis*
  6B. Scales very scanty except at base of stipes and on rhizomes; stipes mostly interrupted, on creeping rhizomes.
    7A. Pinnae beneath with a subglobose pneumatophore at base. . . . . 7. *L. subochthodes*
    7B. Pinnae without a globose pneumatophore at base.
      8A. Blades bipinnatiparted; pinnae sessile or nearly so.
        9A. Veinlets reaching the margin of the segments.
          10A. Veinlets forked; blades subglandular; indusium caducous. . . . . 8. *L. thelypteris*
          10B. Veinlets simple; blades usually glandular-dotted; indusium persistent.

1. Lastrea pozoi (Lag.) Ohwi. *Hemionitis pozoi* Lag.; *Leptogramma pozoi* (Lag.) Heywood; *Thelypteris pozoi* (Lag.) Morton; *Stenogramma pozoi* (Lag.) Iwatsuki; *Polypodium africanum* Desv.; *P. tottum* Willd., non Thunb.; *Gymnogramma totta* Schltdl.; *Leptogramma totta* (Schltdl.) J. Smith; *Dryopteris africana* (Desv.) C. Chr.; *Lastrea africana* (Desv.) Ching; *Lastrea mollissima* (Fisch.) Okuyama; *Gymnogramma mollissima* Fisch.; *Leptogramma mollissima* (Fisch.) Ching; *Leptogramma amabilis* Tagawa——MIZO-SHIDA. Rhizomes elongate; fronds short-pilose, deciduous, sparsely scaly; stipes 20–40 cm. long, straw-colored, brownish at base, densely pilose; scales brown, ovate to broadly lanceolate, 6–8 mm. long, 1–3 mm. wide, ciliolate, puberulous outside; blades membranous, oblong-lanceolate, 30–45 cm. long, 12–22 cm. wide, acuminate, not or slightly narrowed at base, densely white spreading-pilose; pinnae 10–15 pairs, linear-lanceolate, 15–30 cm. wide, acuminate, sessile, pinnately parted to cleft, the lobes oblong, obtuse to rounded, subentire to crenate, 4–6 mm. wide, the veins pinnate, the veinlets usually simple; sori linear; indusia absent.——Lowlands and hills; Honshu, Shikoku, Kyushu; common.——Korea, Ryukyus, Formosa, China, Malaysia, India to Europe and Africa.

2. Lastrea phegopteris (L.) Bory. *Polypodium phegopteris* L.; *Gymnocarpium phegopteris* (L.) Newm.; *Phegopteris polypodioides* Fée; *P. vulgaris* Mett.; *Dryopteris phegopteris* (L.) C. Chr.; *Thelypteris phegopteris* (L.) Sloss.——MIYAMA-WARABI. Rhizomes slender, long-creeping, short-pilose, sparsely scaly; fronds slender, deciduous; stipes slender, 10–30 cm. long, straw-colored, sparsely pilose to nearly glabrous, sparsely scaly; scales thinly membranous, lanceolate to linear-lanceolate, 2–4 mm. long, 0.5–1.5 mm. wide, long-attenuate; blades thinly herbaceous, deltoid, 10–18 cm. long, nearly as wide, acuminate, vivid green, sparsely pilose on both sides, prominently so on the costas and upper side of rachis, the rachis sparsely scaly beneath; pinnae lanceolate, 1–2.5 cm. wide, acuminate, sessile, pinnatiparted, adnate to the rachis and auriculate, except the two lowest pinnae which are sessile and not adnate on the rachis; pinnules oblong, 3–6 mm. wide, rounded at apex, coarsely crenate to subentire, the veins pinnate, the veinlets simple or forked; sori submarginal, in 2 series on the segments, orbicular; indusia absent.——Coniferous woods; Hokkaido, Honshu (Kinki Distr. and eastw.), Shikoku, Kyushu; common northw.——Korea, China, Formosa, Himalayas, Siberia to Asia Minor and Europe, and N. America.

3. Lastrea decursivepinnata (Van Hall) J. Smith. *Polypodium decursivepinnatum* Van Hall; *Aspidium decursivepinnatum* (Van Hall) Kunze; *Phegopteris decursivepinnata* (Van Hall) Fée; *Dryopteris decursivepinnata* (Van Hall) O. Kuntze; *Thelypteris decursivepinnata* (Van Hall) Ching ——GEJIGEJI-SHIDA. Rhizomes short, erect to ascending; fronds pilose, deciduous; stipes tufted, 5–25 cm. long, sparsely scaly, thinly whitish spreading-pilose, pale straw-colored; scales spreading, pale brown, linear, 2–4 mm. long, slightly broadened at base, long-ciliate; blades lanceolate, 15–50 cm. long, 2.5–15 cm. wide, long-acuminate, gradually narrowed at base, more or less whitish pilose on both sides and on margin, densely so above on rachis and on costas, the rachis beneath with spreading ciliate scales; pinnae broadly linear to deltoid-lanceolate, 5–13 mm. wide, acute to acuminate, coarsely obtuse-toothed, more often pinnately lobed, sessile, broadly adnate to the rachis, auriculate; sori orbicular; indusia minute, long-pilose.——Walls and rocks in lowlands and low mountains; Honshu (Kantō Distr. and westw.), Shikoku, Kyushu; common.——Korea, Ryukyus, Formosa, China to Indochina, and India.

4. Lastrea bukoensis (Tagawa) H. Itō. *Dryopteris bukoensis* Tagawa; *Thelypteris bukoensis* (Tagawa) Ching; *Phegopteris bukoensis* (Tagawa) Tagawa——TACHI-HIME-WARABI. Rhizomes long-creeping, very sparsely scaly, glabrous or thinly pilose; fronds sparsely stellate-pilose; stipes straw-colored, 20–30 cm. long, 3–3.5 mm. wide, sparsely scaly, sometimes glabrate; scales membranous, broadly lanceolate, 3–5 mm. long, 1–2.5 mm. wide, acuminate, scarcely lustrous, glabrous, brown; blades membranous, broadly lanceolate, 40–60 cm. long, 12–20 cm. wide, acuminate, slightly narrowed at base, bipinnatiparted; pinnae horizontally spreading, lanceolate, 2–4 cm. wide, subopposite, acuminate, sessile, the lowest smallest, ovate-deltoid, 2–3 cm. wide; pinnules broadly lanceolate to oblong-ovate, 4–6 mm. wide, obtuse to acute, lobulate to coarsely crenate, broadly adnate at base, more or less auriculate and winged on the rachis, pilose especially on the costas of both sides; sori in 2 rows on the pinnules, nearer the costules than the margin, orbicular-elliptic; indusia absent.——July-Sept. Honshu (Hida, Etchu Prov. and northeastw.); rare.

5. Lastrea omeiensis (Bak.) Copel. *Polypodium omeiense* Bak.; *Nephrodium omeiense* (Bak.) Diels; *Dryopteris omeiensis* (Bak.) C. Chr.; *Leptogramma omeiensis* (Bak.) Tagawa; *Thelypteris omeiensis* (Bak.) Ching; *Cyclogramma omeiensis* (Bak.) Tagawa; *Glaphyropteris omeiensis* (Bak.) H. Itō; *Dryopteris izuensis* Kodama; *D. pseudoafricana* Makino & Ogata; *Leptogramma izuensis* (Kodama) H. Itō——MIZO-SHIDA-MODOKI. Rhizomes creeping, scaly, thinly pilose; fronds thinly pilose; stipes slender, wiry, 20–45 cm. long, straw-brown, sparsely scaly toward base; scales broadly lanceolate, 1–2 mm. wide, dull brown, often thinly pilosulous on one side; blades broadly lanceolate to oblong-ovate, 20–50 cm. long,

10–25 cm. wide, acuminate, abruptly contracted at base, the rachis short spreading-pilose and longer hispid; pinnae thinly herbaceous, oblong-lanceolate, 4–12 cm. long, 1.5–2.5 cm. wide, gradually contracted at the tip, caudate, sessile, pinnatiparted, short-hairy on margin and costas on upper side, hispid beneath especially on the costas; pneumatophores at the base of the pinnae beneath subglobose; pinnules narrowly oblong, 4–6 mm. wide, rounded at apex, crenate to subentire, the veins pinnate, the veinlets simple; sori orbicular to elliptic, between the costule and the margin; indusia absent.——Wet slopes; Honshu (Izu, Suruga, and Kii Prov.), Kyushu; rare.——Formosa and China.

6. Lastrea quelpaertensis (Christ) Copel. *Dryopteris quelpaertensis* Christ; *Athyrium quelpaertense* (Christ) Ching; *Thelypteris quelpaertensis* (Christ) Ching; *Ctenitis quelpaertensis* (Christ) H. Itō; *Nephrodium montanum* var. *fauriei* Christ; *D. oreopteris* var. *fauriei* (Christ) Miyabe & Kudō; *D. kamtschatica* Komar.; *D. christiana* Kodama; *D. oreopteris* sensu auct. Japon., non Maxon——ŌBA-SHORIMA. Rhizomes short, erect to short-creeping; fronds often more than 1 m. long, deciduous, prominently scaly; stipes tufted, 10–30 cm. long, light green; scales thinly membranous, brown, lanceolate, linear, or narrowly ovate, 3–8 mm. long, entire; blades oblanceolate, 50–80 cm. long, 12–25 cm. wide, abruptly acuminate, gradually narrowed toward the base, bipinnatiparted, the rachis sparsely short-pilose on upper side; scales linear to linear-lanceolate, 2–5 mm. long; pinnae rather numerous, spreading, broadly linear to linear-lanceolate, 1–2(–2.5) cm. wide, acuminate, nearly truncate at base, sessile, sparsely scaly beneath glabrous above except on the costas, the lowermost pinnae ovate-deltoid, obtuse, 1–2 cm. long; pinnules spreading, oblong to elliptic, 3–5 mm. wide, obtuse to rounded, the lowest slightly longer than the others, the veinlets simple or forked; sori submarginal; indusia usually reniform-orbicular, sometimes quadrate to deltoid, 0.2–0.5 mm. across, frequently oblique, erose.——Mountains; Hokkaido, Honshu, Shikoku, Kyushu.——Korea (Dagelet Isl.), Kuriles, and Kamchatka.

Var. **yakumontana** (Masam.) Tagawa. *Dryopteris yakumontana* Masam.; *Thelypteris quelpaertensis* var. *yakumontana* (Masam.) Tagawa; *Ctenitis quelpaertensis* var. *yakumontana* (Masam.) H. Itō——YAKUSHIMA-SHORIMA. Fronds smaller, 6–25 cm. long; pinnules ovate to ovate-oblong.——Kyushu (Yakushima).

7. Lastrea subochthodes (Ching) Tagawa. *Thelypteris subochthodes* Ching; *Dryopteris ochthodes* auct. Japon., non C. Chr.; *Dryopteris ligulata* auct. Japon., non O. Kuntze; *L. falciloba* auct. Japon., non Hook.——IBUKI-SHIDA. Rhizomes creeping, glabrous, thinly scaly while young; fronds deciduous; stipes straw-colored, 20–30 cm. long, usually short-pilose, thinly scaly toward the base or nearly naked; scales ovate, 3–4 mm. long, entire, puberulous; blades oblong-lanceolate, 30–80 cm. long, 10–30 cm. wide, abruptly acuminate, abruptly contracted at base, the rachis usually short-pilose; pinnae rather numerous, broadly linear, 1–2 cm. wide, acuminate, acute to obtuse at the tip, sessile, regularly pinnatiparted, sparsely short-pilose on both sides and margin, densely so on costas of both sides; pneumatophores subglobose, at the base of pinnae beneath; pinnules narrowly oblong, 2–3 mm. wide, acute, subfalcate, crenate to subentire, the veins pinnate, the veinlets simple; sori submarginal, in 2 rows on the pinnules; indusia orbicular-reniform, entire, slightly incurved on margin, about 0.5 mm. across.——July–Nov. Honshu (s. Kantō Distr.

and westw.), Shikoku, Kyushu.——s. Korea, Ryukyus, Formosa, China.

8. Lastrea thelypteris (L.) Bory. *Acrostichum thelypteris* L.; *Polypodium palustre* Salisb.; *Polystichum thelypteris* (L.) Roth; *Aspidium thelypteris* (L.) Sw.; *Dryopteris thelypteris* (L.) A. Gray; *Thelypteris palustris* Schott——HIMESHIDA. Rhizomes slender, wiry, long-creeping; fronds somewhat dimorphic, slender, scattered-pilose, deciduous; stipes erect, slender, 10–50 cm. long, sometimes black-green or purplish red; scales thin, deciduous, sparse, broadly lanceolate to ovate; blades chartaceous-membranous, lanceolate, 30–50 cm. long, 5–15 cm. wide, scarcely narrowed at base, vivid green, often somewhat glaucous, the rachis slender, lustrous, usually glabrous; pinnae 10–25 pairs, linear to lanceolate, the fertile 3–10 mm. wide, the sterile 10–15 mm. wide, acute, subsessile, pinnatiparted, the veins pinnate, the veinlets mostly forked; indusia small, orbicular-reniform, about 0.2 mm. across, ciliate, caducous.——Aug.–Oct. Wet meadows; Honshu, Shikoku,——Korea, China to n. India and Siberia, Europe, Africa, New Zealand, and N. America.

9. Lastrea glanduligera (Kunze) Moore. *Aspidium glanduligerum* Kunze; *Dryopteris glanduligera* (Kunze) Christ; *Thelypteris glanduligera* (Kunze) Ching; *Nephrodium gracilescens* var. *glanduligerum* (Kunze) Hook. & Bak.; *Dryopteris gracilescens* sensu auct. Japon., non O. Kuntze; *D. gracilescens* var. *glanduligera* (Kunze) Makino——HASHIGO-SHIDA. Rhizomes slender, long-creeping, thinly scaly; fronds evergreen; stipes slender, lustrous, pale green, thinly short-pilose, brownish, sparsely scaly near base; scales pale brown, lanceolate, 2–3 mm. long; blades narrowly deltoid to broadly lanceolate, 15–45 cm. long, 4–17 cm. wide, acuminate, not narrowed at base; pinnae spreading, linear-lanceolate, 2–8 cm. long, 7–15 mm. wide, acuminate, sessile, regularly pinnatiparted, thinly pilose on both sides, densely so on costas above, with yellow sessile glands beneath; pinnules oblong-linear to narrowly oblong, 1.5–2.5 mm. wide, obtuse, subentire, the veinlets simple, reaching the margin; sori submarginal, in 2 rows; indusia orbicular-reniform or elliptic to flabellate, about 0.5 mm. across, ciliate.——Woods; Honshu (Kantō Distr. and westw.), Shikoku, Kyushu.——Ryukyus, Formosa, Korea, China to India, and Indochina.

Var. **hyalostegia** (Copel.) Ohwi. *Athyrium hyalostegia* Copel.; *Thelypteris glanduligera* var. *hyalostegia* (Copel.) H. Itō; *Aspidium angustifrons* Miq.; *T. angustifrons* (Miq.) Ching; *L. miqueliana* Tagawa——KO-HASHIGO-SHIDA. Blades smaller, narrowed at base; pinnae smaller, oblong; indusia often long-ciliate and long-pilose, the hairs nearly as long as the width of the indusium.——Woods; Honshu (Kantō Distr. and westw.), Shikoku, Kyushu.——Ryukyus, Formosa, China, and Philippines.

10. Lastrea cystopteroides (Eaton) Copel. *Athyrium cystopteroides* Eaton; *Asplenium cystopteroides* (Eaton) Hook.; *Dryopteris cystopteroides* (Eaton) Kodama; *Thelypteris cystopteroides* (Eaton) Ching; *D. gracilescens* subsp. *glanduligera* var. *abbreviata* Kodama; *D. abbreviatipinna* Makino & Ogata——HIME-HASHIGO-SHIDA. Rhizomes long-creeping, slender, usually less than 1 mm. across, thinly scaly; fronds small, delicate, sparsely short-pilose on upper side of stipe, rachis, and costas of the pinnae; stipes nearly filiform, 2–10 cm. long, pale green, brownish and thinly scaly at base; scales on stipes and rhizomes lanceolate, about 1 mm. long, brown; blades lanceolate to oblong-ovate, 2–8 cm. long, 12–30 mm. wide, acuminate to acute with an obtuse tip, scarcely narrowed at base; pinnae 4–8 pairs, membranous, alternate, ovate to ovate-deltoid, 5–15 mm. long, 5–10 mm. wide, obtuse to rounded at the tip, sessile, thinly pilose on both sides; pinnules elliptic to rounded; veinlets reaching the margin, simple to once forked; sori 1–5 on each pinnule, nearer the margin than the costule; indusia membranous, orbicular-reniform, about 0.5 mm. across, sometimes oblique, ciliate, pilose on back.——Shaded walls and moist rocks; Honshu (s. Kinki Distr. and westw.), Shikoku, Kyushu.——Ryukyus, Formosa, and Korea.

11. Lastrea japonica (Bak.) Copel. *Nephrodium japonicum* Bak.; *Dryopteris japonica* (Bak.) C. Chr.; *D. formosa* Nakai, non Maxon; *D. castanea* Tagawa; *D. japonica* var. *formosa* (Nakai) C. Chr.——HARIGANE-WARABI. Rhizomes creeping, black-brown, sparsely scaly, 3–5 mm. across; fronds deciduous; stipes rather slender, 20–60 cm. long, pale green to dark purple-brown, lustrous, sparsely scaly; scales membranous, lanceolate, 4–6 mm. long, remotely dentate, pubescent to nearly glabrous, brown, deciduous; blades deltoid-ovate to oblong, 20–50 cm. long, 15–25 cm. wide, acuminate, the rachis and costas prominently short-hairy on upper side; pinnae 10–18 pairs, thinly herbaceous, broadly linear, 8–20 mm. wide, acuminate, mostly reflexed, slightly narrowed at base, sessile, pinnatiparted, with spreading short hairs especially on the costas beneath, thinly ascending-pilose above; pinnules narrowly oblong, 2.5–3.5 mm. wide, crenate to subentire; veinlets simple; sori in 2 series between the costule and margin; indusia membranous, rounded-reniform, about 1 mm. across, pale, short-pilose on back.——Hokkaido, Honshu, Shikoku, Kyushu; common.——Korea, China, and Formosa.

Var. **musashiensis** (Hiyama) Honda. IWA-HARIGANE-WARABI. Scales darker brown; indusia nearly glabrous.—— Honshu (w. Kantō to s. Shinano Prov.).

Var. **glabrata** (Ching) Ohwi. *Thelypteris japonica* var. *glabrata* Ching——KŌRAI-YAWARA-SHIDA. Plant Glabrescent. Occurs rarely in our area.

Lastrea simozawae (Tagawa) Tagawa. *Thelypteris simozawae* Tagawa——CHŪREI-HASHIGO-SHIDA. Rhizomes short-creeping; blade nearly destitute of sessile discoid glands on undersurface.——A Formosan species reported to occur in Kyushu (Yakushima).

12. Lastrea nipponica (Fr. & Sav.) Copel. *Aspidium nipponicum* Fr. & Sav.; *Dryopteris nipponica* (Fr. & Sav.) C. Chr.; *Thelypteris nipponica* (Fr. & Sav.) Ching; *D. nipponica* var. *borealis* Hara——NIKKŌ-SHIDA. Rhizomes creeping, rather densely covered with the basal stubs of old stipes; stipes slender, 15–30 cm. long, the fertile up to 40 cm. long, straw-colored, dark brown at base, very sparsely scaly; scales ovate, 2–4 mm. long, abruptly acuminate, dark brown, with a few spinelike teeth; blades thinly herbaceous, broadly oblanceolate to oblong, 20–40 cm. long, 8–15 cm. wide, fertile ones often narrower, acuminate, somewhat narrowed at base; sterile pinnae contiguous, the fertile distant, broadly linear to lanceolate, 8–15 mm. wide, acuminate with an acute to subobtuse tip, sessile, with a stipulelike lobe at base, thinly pilose, pinnatiparted, the fertile pinnae often with recurved margins, the costas pilose above, white-villous beneath at least on lower half; veinlets usually simple, reaching the margin; indusia membranous, orbicular-reniform, glandular and sometimes pilose, 0.5–0.7 mm. across.——Hokkaido, Honshu (centr. and n. distr.).——Korea and China.

13. Lastrea beddomei (Bak.) Bedd. *Nephrodium beddomei* Bak.; *Dryopteris beddomei* (Bak.) O. Kuntze——HO-SOBA-SHORIMA. Rhizomes long-creeping, about 2 mm. across, sparsely scaly while young; sterile fronds 20–50 cm. long, the fertile longer, 40–60 cm. long; stipes slender, straw-colored, dark purple-brown toward base, lustrous, 5–15 cm. long, very thinly scaly at base; scales dull brown, ovate, subentire; blades broadly lanceolate, 6–10 cm. wide, long-acuminate, abruptly narrowed at base, with 5–10 pairs of much reduced auriclelike pinnae at base, bipinnatiparted, the rachis pale green, slender, short-pilose on upper side; pinnae spreading, thinly herbaceous to membranous, linear-lanceolate, 5–7 mm. wide, long-acuminate, sessile, thinly appressed-pilose above and on margin, pubescent beneath and on costas, the lower reduced pinnae 5–15 mm. long; pinnules narrowly oblong, nearly entire, 2–3 mm. wide; veinlets simple, reaching the margin; sori in 2 series, nearer the margin than the costule; indusia small, orbicular-reniform to oblong-cordate, 0.2–0.4 mm. across, usually long-pilose.——Honshu (Suruga Prov.), Shikoku, Kyushu.——s. Korea, Formosa to Malaysia, and India.

14. Lastrea laxa (Fr. & Sav.) Copel. *Aspidium laxum* Fr. & Sav.; *Dryopteris laxa* (Fr. & Sav.) C. Chr.; *Thelypteris laxa* (Fr. & Sav.) Ching——YAWARA-SHIDA. Rhizomes long-creeping, sparsely pilose, scaly, 2–3 mm. across; fronds deciduous, softly short-pilose; stipes 15–35 cm. long, slender, brownish toward base, thinly scaly while young; scales lanceolate, 1.5–2 mm. long, minutely pilose; blades broadly lanceolate to deltoid-ovate, 25–50 cm. long, 10–20(–25) cm. wide, acuminate, not narrowed at base, the rachis and costas short-pilose above; pinnae membranous, lanceolate or linear-lanceolate, 1–2 cm. wide, long-acuminate, the lower pinnae usually narrowed at base, sessile, short-pilose on lower side; pinnules oblong to oblong-lanceolate, 2.5–5 mm. wide, obtuse to acute, coarsely and regularly toothed; veinlets simple to forked, not reaching the margin; sori median, in 2 rows on the pinnules; indusia membranous, orbicular-reniform, pilose.——Honshu, Shikoku, Kyushu; common.——Formosa, Korea, and China.

15. Lastrea gracilescens (Bl.) Moore. *Aspidium gracilescens* Bl.; *Thelypteris gracilescens* (Bl.) Ching; *Dryopteris gracilescens* (Bl.) O. Kuntze; *D. sublaxa* Hayata; *D. arisanensis* Rosenst.——SHIMA-YAWARA-SHIDA. Rhizomes short-creeping; stipes 15–40 cm. long, the sterile shorter than the fertile, slender, puberulous; scales lanceolate, 1–2 mm. long, pale brown, entire; blades lanceolate, 25–35 cm. long, 7–12 cm. wide, long-acuminate, not narrowed at base, the rachis with short spreading-hairs on upper side; pinnae firmly membranous, linear-lanceolate, 1–1.5 cm. wide, falcate, caudate, sessile, nearly glabrous on upper side, with prominent short ascending-hairs on the costas, very thinly and minutely pilose beneath and on margin; pinnules narrowly oblong, subentire; veins pinnate, the veinlets simple, not reaching the margin; sori in 2 series, median; indusia orbicular-reniform, about 0.5 mm. across, often unequal, erose to subentire.——Kyushu (Yakushima); rare.——Formosa to Malaysia and Polynesia.

16. Lastrea hattorii (H. Itō) Tagawa. *Dryopteris hattorii* H. Itō; *Thelypteris hattorii* (H. Itō) Tagawa——YOKO-GURA-HIME-WARABI. Closely related to No. 14, differing chiefly in the deltoid tripinnate blades, the pinnae broadly lanceolate, 4–5 cm. wide, the lowest ones petiolate, the lower pinnules sessile, contracted at base; stipes pale brown, smooth, lustrous, very short-pilose; blades deltoid to deltoid-ovate, 20–35 cm. long, 15–28 cm. wide, long-acuminate; pinnae about

10 pairs, nearly horizontally spreading, lanceolate to oblong, caudate-acuminate; pinnules lanceolate, acute or obtuse, pinnately parted to obtusely toothed, decurrent in the upper ones, short-pilose beneath; sori in 2 series, between the costule and margin of the pinnules; indusia orbicular-reniform, pilose.—— Honshu (Mikawa and Harima Prov.), Shikoku, Kyushu; rare. ——China.

17. Lastrea oligophlebia (Bak.) Copel. *Nephrodium oligophlebium* Bak.; *Dryopteris oligophlebia* (Bak.) C. Chr.; *D. elegans* Koidz.; *D. oligophlebia* var. *elegans* (Koidz.) H. Itō; *Thelypteris oligophlebia* var. *elegans* (Koidz.) Ching—— HIME-WARABI. Rhizomes very short, creeping or ascending; fronds deciduous; stipes 40–60 cm. long, green to straw-colored, sparsely scaly at base; scales linear-lanceolate, 6–10 mm. long, entire, short-pilose; blades yellowish to bright green, ovate to elliptic-ovate, 30–80 cm. long, 20–50 cm. wide, 3- or 4-pinnatiparted; pinnae broadly lanceolate, 5–8 cm. wide in the upper ones, about 10 cm. wide in the lowest, acuminate, short-petiolate in the lower; pinnules narrowly to broadly lanceolate, acuminate, short-pilose on the costas and costules above, sparingly long-pilose beneath, the rachis more or less narrowly winged except near the base; pinnules obliquely spreading, oblong to lanceolate, obtuse to acuminate, subentire or lobulate to pinna-tiparted, sessile, frequently more or less adnate to the axis; veinlets simple to once forked; sori between the veins and margin; indusia orbicular-reniform, glandular, often pilose.——Thickets in lowlands and hills; Honshu (Kantō Distr. and westw.), Shikoku, Kyushu; common.——Korea, Ryukyus, Formosa, China.

Var. **lasiocarpa** (Hayata) H. Itō. *Aspidium uliginosum* Kunze; *Dryopteris uliginosa* (Kunze) C. Chr.; *Thelypteris uliginosa* (Kunze) Ching; *Dryopteris lasiocarpa* Hayata; *D. oligophlebia* var. *lasiocarpa* (Hayata) Nakai; *Thelypteris oligophlebia* var. *lasiocarpa* (Hayata) H. Itō——ARAGE-HIME-WARABI. Differs from the typical phase in the longer 1- to 5-celled hairs of the indusia, rachis, and costas of blades.—— Honshu (Hachijo Isl.), Kyushu.——Ryukyus and Formosa.

Var. **subtripinnata** (Tagawa) Ohwi. *Dryopteris elegans* var. *subtripinnata* Tagawa; *D. oligophlebia* var. *subtripinnata* (Tagawa) H. Itō; *Thelypteris oligophlebia* var. *subtripinnata* (Tagawa) H. Itō; *T. viridifrons* Tagawa; *D. viridifrons* (Tagawa) Tagawa; *L. viridifrons* (Tagawa) Tagawa——MIDORI-HIME-WARABI. Blades mostly larger, bright green, membranous; pinnules spaced, broader, with a distinct, short petiolule, at least in the lower ones.——Honshu (Kantō Distr. and westw.), Kyushu.——Korea.

14. GYMNOCARPIUM Newm. USAGI-SHIDA ZOKU

Slender, glabrous ferns; rhizomes wiry, slender, long-creeping; stipes slender, often minutely glandular; blades thinly herbaceous, deltoid to ovate-deltoid, pinnatifid to 2- or 3-pinnatiparted, jointed at the base or at base of the pinnae; veins pinnate, free; sori dorsal on the veins, rounded to oblong; indusia absent.——Few species, in the temperate regions of the N. Hemisphere.

1A. Blades 2- or 3-pinnatiparted.
  2A. Blades 5-angled, membranous; lowest lateral pinnae nearly or as large as the remainder (terminal pinna) of the blade.
                                                     1. *G. dryopteris*
  2B. Blades deltoid-ovate, chartaceous-membranous; lowest lateral pinnae distinctly smaller than the remainder (terminal pinna) of the blade. . . . . 2. *G. robertianum* var. *longum*
1B. Blades simply pinnatiparted to -cleft. . . . . 3. *G. oyamense*

1. Gymnocarpium dryopteris (L.) Newm. *Polypodium dryopteris* L.; *Lastrea dryopteris* (L.) Bory; *Phegopteris dryopteris* (L.) Fée; *Polypodium disjunctum* Rupr.; *Dryopteris disjuncta* (Rupr.) Mort.; *G. robertianum* var. *disjunctum* (Rupr.) Ching; *D. linnaeana* C. Chr.——USAGI-SHIDA. Rhizomes long-creeping, wiry, about 1 mm. across, scaly when young; fronds deciduous, slender, glabrous, often slightly glaucescent on the blades beneath; stipes slender, pale green, lustrous, sparsely scaly, 10–25 cm. long, dark brown at base; scales pale brown, thinly membranous, narrowly ovate, 2–3 mm. long, glabrous, those at base of stipes and on rhizomes broadly ovate, 3–4 mm. long; blades membranous, 5-angled, deltoid, 8–15 cm. long, slightly broader than long, glabrous, smooth; pinnae 3, the terminal as large as to only slightly larger than the lateral ones, deltoid, 5–10 cm. wide, gradually acute, bipinnatiparted or in lower part bipinnate, the petioles 1.5–4 cm. long; pinnules deltoid-lanceolate, subacute to obtuse, sessile except for the lowest pair on the terminal pinna, sometimes petiolate, the ultimate pinnules narrowly oblong, obtuse, crenate to lobulate; sori in 2 series on the pinnules, nearer the margin than the costule.——Coniferous woods; Hokkaido, Honshu (centr. and n. distr.).——Sakhalin, Korea, Himalayas, Siberia to Europe and N. America.

2. Gymnocarpium robertianum (Hoffm.) Newm. var. **longulum** (Christ) H. Itō. *Aspidium dryopteris* var. *longulum* Christ; *Dryopteris jessoensis* Koidz.; *G. jessoense* (Koidz.) Koidz.; *G. longulum* (Christ) Kitag.; *Lastrea robertiana* var. *longula* (Christ) Ohwi——IWA-USAGI-SHIDA. Closely allied to the preceding but frequently sparsely grandular on the stipes and blades, especially so on the nodes of the rachis; blades chartaceous-membranous, firmer, deltoid-ovate, 18–20 cm. long, 14–18 cm. wide; lateral pinnae narrowly ovate-deltoid, 9–10 cm. long, 6–8 cm. wide, the petioles 15–20 mm. long, the lowest pinnae distinctly smaller than the remainder of the blade, the terminal pinna 13–15 cm. long, 10–11 cm. wide; pinnules linear-oblong, rounded to obtuse, pinnately cleft to toothed.——Coniferous woods; Hokkaido, Honshu (Kantō and centr. distr.), Shikoku (high mts.).——Korea, n. China, and e. Siberia. The typical phase occurs from Siberia to Afghanistan and Europe, also in N. America.

3. Gymnocarpium oyamense (Bak.) Ching. *Polypodium oyamense* Bak.; *Dryopteris oyamensis* (Bak.) C. Chr.; *Currania oyamensis* (Bak.) Copel.; *Phegopteris oyamensis* (Bak.) Rosenb.; *Polypodium krameri* Fr. & Sav., incl. var. *incisum* Fr. & Sav.; *Phegopteris krameri* (Fr. & Sav.) Makino ——EBIRA-SHIDA. Rhizomes wiry, long-creeping, 1.5–2 mm. across, the scales scattered, pale brown, broadly lanceolate, 3–4 mm. long, long-acuminate; fronds deciduous, slender, glabrous, slightly glaucous; stipes interrupted, 12–20 cm. long, slender, pale green, dark brown at base; scales sparse, pale brown, thinly membranous, deciduous; blades chartaceous-membranous to membranous, ovate-deltoid, 10–15 cm. long,

8–12 cm. wide, acute, cordate, jointed with the stipe, pinnati-parted to -cleft, smooth; pinnules 7–12 pairs, oblong-lanceo-late, 1–1.5 cm. wide, horizontally spreading, obtuse to sub-acute, pinnately lobed, crenate-toothed, the lowest pinnules broadly lanceolate, to 2 cm. wide, more or less deflexed, slightly broadened on the posterior side; veins pinnately branched, the veinlets simple or once forked; sori elliptic, 1–2 mm. long, sometimes oblong or rounded, in 2 series on the nearer pinnules, the costa in several somewhat irregular series.——Honshu (Kantō to s. Kinki distr.). Shikoku.——China (Philippines to New Guinea?).

15. CYCLOSORUS Link   HO-SHIDA ZOKU

Rhizomes erect to creeping; blades bipinnatiparted; veins pinnately branched, the single or sometimes few lower veinlets united with those of the adjacent veins at or below the sinus; hairs short, acicular and glandless or glandular, often mixed to-gether; sori on the veinlets rounded; indusia orbicular-reniform, usually large, the sporangia often with few, simple or glandular hairs.——About 300 species, in the Tropics and subtropics.

1A. Lowest veinlet only fused to the adjacent vein; blades regularly bipinnatiparted.
  2A. Pinnae coriaceous to firmly chartaceous, with small ovate scales on costas beneath; pinnules broadly deltoid-ovate, mucronate; vein-lets very close; sori marginal; seaside plant. . . . . 1. *C. goggilodus*
  2B. Pinnae chartaceous, not scaly; pinnules ovate to oblong, acute to very obtuse, sometimes mucronate; veinlets not very close; sori median, or nearer the margin than the costa.
    3A. Stipes tufted on short ascending rhizomes; blades narrowed at base; lowest anterior segment not obviously larger than the others; scales puberulous. . . . . 2. *C. dentatus*
    3B. Stipes distant on long-creeping rhizomes; blades not at all narrowed at base; lowest anterior segment larger than the others; scales glabrous.
      4A. Pinnae sparsely hairy, often glabrescent above except for the costas. . . . . 3. *C. acuminatus*
      4B. Pinnae prominently hairy on both sides. . . . . 4. *C. parasiticus*
1B. Nearly all of the veinlets fused to those of the adjacent veins; blades simple or with a few unlobed pinnae. . . . . 5. *C. triphyllus*

1. Cyclosorus goggilodus (Schk.) Link. *Aspidium goggilodus* Schk.; *Dryopteris goggiloda* (Schk.) O. Kuntze; *Dryopteris unita* sensu auct. Japon., non O. Kuntze——TETSU-HO-SHIDA.   Plant nearly glabrous; rhizomes long-creeping, with scattered brown ovate scales 2–3 mm. long; stipes 30–90 cm. long, 3–4 mm. across, lustrous, brownish, nearly naked; blades lanceolate to narrowly oblong, 30–70 cm. long, 10–25 cm. wide, abruptly contracted at tip; pinnae 20–28 pairs, coriaceous to firmly chartaceous, broadly linear, 8–15 mm. wide, acuminate, cuneate at base, subsessile, subglabrous to short-pilose beneath, nearly glabrous above, with small ovate scales 0.5–1 mm. long on costas beneath, the terminal pinna similar to the lateral ones; segments broadly deltoid-ovate, mucronate; veins 6–8 pairs per segment, very close; sori sub-marginal; indusia orbicular-reniform, 0.2–0.5 mm. across, nearly glabrous to short-pilose.——Wet places near the sea; Honshu (Izu Prov. and westw. in warmer parts), Shikoku, Kyushu.——Ryukyus, Formosa, and China; widely distributed in the tropics and subtropics.

2. Cyclosorus dentatus (Forsk.) Ching. *Polypodium dentatum* Forsk.; *C. oblancifolius* (Tagawa) Tagawa; *Dryopteris oblancifolia* Tagawa——INU-KEHO-SHIDA.   Rhizomes short, ascending; stipes tufted, short-pilose, sparsely scaly; scales dark brown, linear, about 10 mm. long, puberulous; blades herbaceous, oblong-lanceolate to oblanceolate-oblong, 30–60 cm. long, 10–16 cm. wide, gradually narrowed at base, rather abruptly narrowed toward tip, the terminal pinna deltoid-lanceolate, 10–12 cm. long, 2–3 cm. wide; rachis densely pubescent on upper side, less so beneath; pinnae 13–15 pairs, spreading, broadly linear-lanceolate, 6–8 cm. long, 10–17 mm. wide, abruptly acuminate, truncate at base, sessile, pinnately cleft, short-pilose on both sides, densely long-pilose on the costas, sparsely so on the veinlets, the lower pinnae reduced, ovate, 1.5–2 cm. long, about 1 cm. wide, obtuse; pinnules oblong, about 3 mm. wide, obtuse or truncate-rounded at apex, entire; veinlets simple, not reaching the margin; sori median; indusia orbicular-reniform, densely short-pilose.——

Honshu (Kii Prov.), Shikoku, Kyushu (s. distr.); rare.——Formosa and widespread in the Tropics.

3. Cyclosorus acuminatus (Houtt.) Nakai. *Polypodium acuminatum* Houtt.; *Dryopteris acuminata* (Houtt.) Nakai; *Polypodium sophoroides* Thunb.; *Aspidium sophoroides* (Thunb.) Sw.; *Nephrodium sophoroides* (Thunb.) Desv.; *D. sophoroides* (Thunb.) O. Kuntze; *C. sophoroides* (Thunb.) Tard.-Bl.; *A. oshimense* Christ; *D. oshimensis* (Christ) C. Chr.; *D. ogatana* Koidz.; *D. ensipinna* Tagawa——HO-SHIDA.   Evergreen; rhizomes long-creeping, 3–4 mm. across, sparsely pilose and thinly scaly; stipes slender, 25–50 cm. long, straw-colored, sometimes dull purplish, sparsely short-pilose, sparsely scaly at base; scales lanceolate, 4–6 mm. long, brown; blades chartaceous, lanceolate to nar-rowly oblong-ovate, 30–60 cm. long, pinnate, deep green above, slightly paler beneath, abruptly and caudately long-acuminate, the terminal pinna linear- to deltoid-lanceolate, the rachis prominently short-pilose, somewhat hispid on underside; pin-nae linear- to broad-lanceolate, 7–20 mm. wide, gradually acuminate, spreading, truncate at base, sessile, pinnately cleft, costas and veins strigose on upper side, spreading-pilose be-neath; pinnules ovate to oblong-ovate, 3–4 mm. wide, obtuse to subacute with a short mucro at tip, the lowest anterior pinnule longer than the others; sori in 2 rows, nearer the margin than the veins; indusia small, orbicular-reniform, short-pilose on back, about 0.5 mm. across, entire, brownish.——Thickets and sunny slopes; Honshu (Kantō Distr. through Tōkaidō to Kinki and westw.), Shikoku, Kyushu.——Ryukyus, Formosa, s. Korea, and China.

4. Cyclosorus parasiticus (L.) Farw. *Polypodium parasiticum* L.; *Aspidium parasiticum* L.; *Nephrodium para-siticum* (L.) Desv.; *Dryopteris parasitica* (L.) O. Kuntze——KE-HO-SHIDA.   Rhizomes long-creeping; stipes 20–80 cm. long, pilose, scaly toward the base; scales dark brown, linear to linear-lanceolate, 8–15 mm. long, entire; blades chartaceous, oblong-lanceolate, 40–70 cm. long, 15–30 cm. wide, abruptly contracted and acuminate at the tip, prominently pubescent on both sides; pinnae 18–24 pairs, broadly linear to linear-lanceolate, 1–1.5 cm. wide, gradually long-acuminate, truncate at base, sessile, pinnately cleft; pinnules broadly oblong to narrowly ovate-oblong, 2–4 mm. wide, rounded at apex, crenate; sori median or slightly nearer the margin than the vein; indusia orbicular-reniform, 0.5–0.7 mm. wide, pilose.—— Honshu (Izu Isls.), Shikoku, Kyushu.——Ryukyus, Formosa, China, Tropics of Asia, Africa, and Australia.

5. Cyclosorus triphyllus (Sw.) Tard.-Bl. *Meniscium triphyllum* Sw.; *Phegopteris triphylla* (Sw.) Mett.; *Dryopteris triphylla* (Sw.) C. Chr.; *Abacopteris triphylla* (Sw.) Ching; *M. simplex* Hook.; *Polypodium simplex* (Hook.) Lowe; *Phegopteris simplex* (Hook.) Mett.; *D. simplex* (Hook.) C. Chr.——Kōmori-shida.   Rhizomes slender, long-creeping, 2–3 mm. across, short-pilose, scaly toward tip, black-brown; scales brown, linear-lanceolate, 2–5 mm. long, entire, short-puberulent; stipes slender, straw-colored to pale brown, scaly toward base, short-pilose especially on upper portion, 7–20 cm. long in sterile fronds, to 40 cm. long in the fertile; blades with 1–3, rarely 5 simple pinnae; terminal pinna of sterile blades thinly chartaceous, broadly lanceolate to broadly oblong-lanceolate, 10–22 cm. long, 2.5–4 cm. wide, long-acuminate, sometimes subcordate or rounded at base, undulate or entire, pinnately veined, short-pilose on the costas of both sides, the petiolule about 1 cm. long; terminal pinna of fertile blades usually lanceolate; lateral pinnae of sterile and fertile blades usually smaller, 1/4 to 1/3 as long as the terminal pinna, sometimes greatly reduced and adnate to the base of the terminal pinna; veins obliquely spreading, parallel, slightly arcuate, raised especially beneath, the venation meniscioid, i.e. the opposite veinlets uniting and sending out an excurrent veinlet, this often meeting the adjacent veinlets, thus forming two rows of areoles between the main veins; sori on the veinlets, linear to oblong, the sporangia with 1–3 hispid hairs (under magnification); indusia absent.——Kyushu (Yakushima).——Ryukyus, Formosa, s. China to India, Malaysia to Australia.

16. DICTYOCLINE Moore   Ami-shida Zoku

Terrestrial, pilose ferns; rhizomes short, ascending, dictyostelic; scales on rhizomes and lower part of stipes narrow, dark brown, ciliate; blades deltoid-ovate, pinnatifid to pinnate; pinnae entire, chartaceous-herbaceous, with somewhat irregular goniopteroid or meniscioid venation; veins sometimes branched and anastomosing to form 3 or 4 rows of areoles between the costas; sori elongate, on the veins; indusia absent.——One species, Japan to India.

1. Dictyocline griffithii Moore var. **wilfordii** (Hook.) Moore. *Hemionitis wilfordii* Hook.; *D. wilfordii* (Hook.) J. Smith; *D. griffithii* var. *pinnatifida* (Hook.) Bedd.; *Hemionitis griffithii* var. *pinnatifida* Hook.——Ami-shida.   Rhizomes short-creeping; fronds dull with spreading hairs; stipes 15–30 cm. long, slender, straw-colored, dark brown and thinly scaly toward base; scales dark brown, linear-lanceolate, 4–6 mm. long, long-acuminate, loosely ciliate; blades thinly chartaceous, narrowly deltoid, simply pinnaticleft to lobulate, 15–30 cm. long, 10–18 cm. wide, acuminate to acute, with a pair of distinct sessile pinnae at base, these narrowly oblong-ovate, 1.5–2.5 cm. wide, obtuse to subacute, subentire, sometimes slightly wider than the others, rounded at base on the posterior side; rachis, costas, veins and veinlets prominent on both sides; sporangia on the veinlets.——Woods; Honshu (s. Kinki Distr.), Shikoku, Kyushu.——Ryukyus, Formosa. The typical variety occurs in Formosa, China to India.

17. CYSTOPTERIS Bernh.   Nayo-shida Zoku

Terrestrial; rhizomes somewhat creeping, dictyostelic, with brown rather broad scales; fronds bipinnate or more compound, stipitate, glabrous or sparsely polise; veins free; sori round; indusia attached to the base, basiscopic, thin; paraphyses absent.—— Cosmopolitan, with about 18 species.

1A. Pinnules obliquely cuneate, short-petiolulate; fronds nearly glabrous.
  2A. Rhizomes short-creeping, the fronds numerous, clustered; basal scales broadly lanceolate to linear, acuminate; blades oblong to oblong-ovate, much longer than broad, not broadest at base. . . . . 1. *C. fragilis*
  2B. Rhizomes slender, long-creeping, the fronds remote; basal scales ovate, acute; blades deltoid to deltoid-ovate, broadest at base.
                                                  2. *C. sudetica*
1B. Pinnules truncate at base, sessile; fronds sparsely pubescent.
  3A. Pinnae obliquely spreading, sparsely pubescent on both sides. . . . . 3. *C. japonica*
  3B. Pinnae horizontally spreading, densely pubescent on both sides. . . . . 4. *C. tenuisecta*

1. Cystopteris fragilis (L.) Bernh. *Polypodium fragile* L.; *Athyrium fragile* (L.) Spreng.——Nayo-shida.   Rhizomes short-creeping, densely covered with the basal stubs of old stipes; fronds deciduous; stipes 3–10(–18) cm. long, tufted, straw-colored, brown and rather densely scaly at base; scales broadly lanceolate to linear, acuminate, pale brown, entire; blades thinly membranous, oblong to oblong-ovate, 7–18 cm. long, 2–6 cm. wide, acuminate to acute, bipinnatiparted to bipinnate, not broadened at base, nearly glabrous; pinnae ovate to broadly lanceolate, sometimes the lowest deltoid-ovate, 6–12 mm. wide, acute to obtuse, short-petiolulate; pinnules oblong to obovate, 3–8 mm. long, 2–4 mm. wide, obtuse, sessile, usually more or less decurrent on posterior side at base, few-toothed to lobulate; veinlets not reaching the margin; sori on upper portion of the simple veinlets; indusia thinly membranous, whitish, nearly orbicular to ovate-lanceolate.——Hokkaido, Honshu (centr. and n. distr.), Shikoku (Mount Tsurugi in Awa Prov.).——Sakhalin, Kuriles, Korea, Formosa (alpine), and the colder regions and mountains of the N. and S. Hemispheres.

2. Cystopteris sudetica A. Br. & Milde. *C. leucosoria* Schur——Yama-hime-warabi.   Rhizomes slender, long-creeping, 1.5–2 mm. across, sparsely scaly, remote; stipes 10–20 cm. long, slender, straw-colored, smooth, lustrous, with few scales toward the dark brown base, very sparsely so above; scales ovate, acute; blades deltoid to deltoid-ovate, 10–15 cm. long, 7–12 cm. wide, acuminate, glabrous or sometimes with scattered hairlike scales on the rachises; pinnae thinly membranous, ovate-lanceolate, 5–10 cm. long, 2–3 cm. wide, acuminate, short-petiolulate; pinnules obliquely ovate, 5–10 mm. long, 3–6 mm. wide, obtuse, obliquely cuneate at base, very short-petiolulate, pinnately parted, obtusely toothed; veinlets ending in the sinus between the teeth; sori round; indusia orbicular to elliptic, thinly membranous, white.——Coniferous woods in high mountains; Honshu (Shinano and Suruga Prov.); rare.——n. Korea to Siberia and n. Europe.

3. Cystopteris japonica Luerss. *Acystopteris japonica* (Luerss.) Nakai——Usu-hime-warabi.   Rhizomes creeping, 3–4 mm. across, scaly, the fronds close to remote; fronds deciduous; stipes 20–40 cm. long, rather slender, lustrous, 2–3 mm. across, rather densely scaly and with minute scurf-like hairs; scales spreading, ovate to broadly lanceolate, 1–3.5 mm. long, acute, rounded at base, with oblong to elliptic cells; blades thinly herbaceous, deltoid-ovate, 30–50 cm. long, 25–40 cm. wide, acuminate, 3-pinnate, the rachises with jointed curved hairs; pinnae usually opposite, obliquely to nearly horizontally spreading, broadly lanceolate, 2.5–6 cm. wide, acuminate, nearly sessile, the lowest pinnae narrowly deltoid-ovate, 4–10 cm. wide; pinnules narrowly oblong-ovate, 1.5–4 cm. long, 7–15 mm. wide, obtuse to acute, sessile, horizontally spreading, pinnatifid to pinnate, sparsely pubescent; ultimate pinnules oblong, obtuse to rounded, obtusely toothed to lobulate; veinlets reaching the margin; sori near the base of the sinus; indusia minute, ovate-orbicular to ovate.——Woods in mountains; Honshu (Iwashiro Prov. and westw.), Shikoku, Kyushu.——Formosa (var.).

4. Cystopteris tenuisecta (Bl.) Mett. *Aspidium tenuisectum* Bl.; *Athyrium tenuisectum* (Bl.) Moore; *Acystopteris tenuisecta* (Bl.) Tagawa; *Lastrea setosa* Bedd.; *C. setosa* (Bedd.) Bedd.; *C. formosana* Hayata; *Acystopteris formosana* (Hayata) Tagawa——Hōrai-hime-warabi.   Resembles the preceding in habit; rhizomes short-creeping to ascending; stipes 20–40 cm. long, with scattered scales; scales thinly membranous, spreading, broadly lanceolate, 2–3 mm. long, short-acuminate, dull; blades thinly herbaceous, ovate- to deltoid-lanceolate, to 40 cm. long and 30 cm. wide, acuminate, 3-pinnate, the rachises straw-colored, short-pubescent, also with a few hairlike scales; pinnae usually opposite, horizontally spreading, lanceolate, the lowest to 17 cm. long, 6 cm. wide, acuminate, sessile; pinnules lanceolate to narrowly oblong, acute, sessile, pinnately parted to pinnate; ultimate pinnules oblong, obtuse, obtusely toothed to lobed, short-pubescent on both sides.——Kyushu (Yakushima).——Formosa, China to India, and Malaysia.

18. ATHYRIUM Roth   Inu-warabi Zoku

Rhizomes erect and short, or creeping and elongate, the scales membranous to somewhat firm, pale brown to nearly black, sometimes spinulose; blades membranous to coriaceous, usually pinnately compound, rarely simple; veins and veinlets mostly free, rarely anastomose; sori dorsal, elongate, rarely short and roundish along the veinlets; indusia elongate and straight on one side (asplenioid), curved across the veinlets (athyrioid), or interrupted at the distal end and equally disposed on both sides of the veinlets (diplazioid), rarely rudimentary or absent.——Nearly cosmopolitan, with about 600 species.

1A. Indusia absent or fugacious, soon concealed under the sorus.
    2A. Ultimate pinnules awn-toothed; blades deltoid; indusia minute, reniform. .... 22. *A. spinulosum*
    2B. Ultimate pinnules crenate to acute-toothed.
       3A. Rhizomes short, not obviously creeping; fronds tufted; blades bright green; indusia usually absent. .... 23. *A. alpestre*
       3B. Rhizomes creeping, elongate; fronds not tufted, remote, toward end of the rhizomes; blades glaucous to deep green.
          4A. Indusia absent.
             5A. Sori linear to oblong, elongate. .... 1. *A. decurrentialatum*
             5B. Sori round to elliptic, short.
                6A. Sori elliptic. .... 2. *A. hakonense*
                6B. Sori round.
                   7A. Stipe brownish; pinnae green, glabrous to pubescent beneath; sori submarginal. .... 3. *A. crenulatoserrulatum*
                   7B. Stipe yellowish green; pinnae more or less whitish beneath; sori subcostal. .... 4. *A. fluviale*
          4B. Indusia present, reniform to broadly lanceolate.
             8A. Pinnae simply lobed or 2- or 3-pinnatiparted; rachises at least above the middle and rachillas prominently winged.
                9A. Pinnae pinnatiparted to nearly bipinnate, the pinnules crenate-toothed; blades thinly membranous. .... 5. *A. viridifrons*
                9B. Pinnae pinnatilobed to pinnatiparted, the pinnules oblong, very obtuse to rounded, crenate-toothed; blades thinly charta-ceous-membranous. .... 6. *A. unifurcatum*
             8B. Pinnae 3- or 4-pinnatiparted; rachises not winged. .... 7. *A. atkinsonii*
1B. Indusia persistent.
    10A. Blades simple, pinnatiparted, or with 3–7 simple pinnae.
       11A. Blades pinnate, pinnae 3–7.
          12A. Pinnae uniformly awn-toothed; scales small, entire. .... 8. *A. pinfaense*
          12B. Pinnae obsoletely toothed only toward apex; scales spinulose. .... 9. *A. aphanoneuron*
       11B. Blades simple and entire, or pinnatiparted.
          13A. Blades subcoriaceous to thick-herbaceous, glabrous on upper surface. .... 10. *A. dubium*
          13B. Blades herbaceous, with few fleshy bristlelike hairs on upper surface. .... 17. *A. grammitoides*
    10B. Blades with more than 11 pinnae.
       14A. Pinnae simple, toothed to lobulate, strongly oblique at the base.
          15A. Indusia reniform to orbicular.
             16A. Scales at base of stipe ovate-triangular to broadly lanceolate, 2–3 mm. long; pinnae obtuse to rounded. .... 11. *A. nakanoi*
             16B. Scales at base of stipe linear-lanceolate, gradually acuminate; pinnae acuminate.
                17A. Pinnae deflexed, except the upper ones. .... 26. *A. reflexipinnum*
                17B. Pinnae not deflexed, except sometimes the lowest ones.

18A.   Indusia erose-ciliate; sori with paraphyses. .... 27. *A. rupestre*
18B.   Indusia entire; sori without paraphyses. .... 28. *A. yokoscense*
15B.   Indusia linear, usually 4–8 mm. long.
19A.   Pinnae membranous, the median and upper ones slightly adnate to the rachis; scales on stipes lanceolate to linear-lanceolate.
12. *A. okudairae*
19B.   Pinnae herbaceous to somewhat chartaceous, mostly not adnate to the rachis; scales on stipes narrowly ovate. 13. *A. wichurae*
14B.   Pinnae simple to much dissected, not at all or only slightly oblique at the base, usually equilateral.
20A.   Pinnae sessile, lanceolate to linear-lanceolate, equilateral, regularly lobulate, sometimes dissected, the pinnules on a prominently winged rachis.
21A.   Pinnae deeply pinnatiparted, the sinus between the segments broad; indusia solitary, simple, sometimes athyrioid.
22A.   Blades 35–45 cm. wide; pinnae 15–22 cm. long, 3.5–4 cm. wide. .... 14. *A. pterorachis*
22B.   Blades 17–22 cm. wide; pinnae less than 15 cm. long, 2–3.5 cm. wide. .... 15. *A. henryi*
21B.   Pinnae lobulate to pinnatiparted, the sinus between the segments narrow; indusia numerous and parallel on the segments or lobes, usually asplenioid, sometimes diplazioid on the lowest veinlet, very rarely athyrioid.
23A.   Rhizomes short and thick; fronds tufted. .... 16. *A. pycnosorum*
23B.   Rhizomes long-creeping, slender; fronds remote.
24A.   Pinnae obtuse to acute, 1–3 cm. long. .... 17. *A. grammitoides*
24B.   Pinnae acuminate, 5–15 cm. long. .... 18. *A. japonicum*
20B.   Pinnae at least the lower petiolulate, if sessile the pinnules oblique or pinnately compound on a wingless rachis.
25A.   Pinnules while young with few fleshy deciduous bristles along cost on the upper side.
26A.   Pinnae long-acuminate, deeply bipinnaticleft to bipinnate.
27A.   Stipes and rachis greenish; pinnules oblique; indusia asplenioid to athyrioid. .... 19. *A. iseanum*
27B.   Stipes and rachis purplish; pinnules equilateral or nearly so; indusia sometimes diplazioid. .... 20. *A. frangulum*
26B.   Pinnae obtuse to acute, pinnately lobulate to parted. .... 21. *A. tozanense*
25B.   Pinnules without deciduous fleshy bristles on upper side.
28A.   Pinnae sessile or nearly so.
29A.   Scales on stipes broadly lanceolate to narrowly ovate, to 5 mm. wide or more.
30A.   Scales at base of stipes broadly lanceolate, fuscous. .... 24. *A. filix-femina* var. *longipes*
30B.   Scales at base of stipes narrowly ovate, nearly black. .... 25. *A. melanolepis*
29B.   Scales on stipes subulate to narrowly lanceolate, gradually narrowed to a long filiform tip, to 2(–3) mm. wide.
31A.   Pinnae deflexed except the upper ones, usually not more than 2.5 cm. long. .... 26. *A. reflexipinnum*
31B.   Pinnae spreading, not deflexed except sometimes the lowest ones.
32A.   Scales at base of stipes fuscous to pale brown; indusia asplenioid and athyrioid.
33A.   Blades oblong-ovate to broadly lanceolate; pinnae sessile.
34A.   Indusia erose-ciliate; sori with paraphyses. .... 27. *A. rupestre*
34B.   Indusia entire; sori without paraphyses.
35A.   Scales at base of stipes 7–12 mm. long, 1–1.5 mm. wide. .... 28. *A. yokoscense*
35B.   Scales at base of stipes about 15 mm. long, 0.5–2 mm. wide. .... 29. *A. tashiroi*
33B.   Blades deltoid to deltoid-ovate; pinnae short-petiolulate to subsessile.
36A.   Blades 40–70 cm. long, 30–50 cm. wide; pinnae short-petiolulate; pinnules acute. .... 30. *A. multifidum*
36B.   Blades 17–30 cm. long, 15–25 cm. wide; pinnae nearly sessile; pinnules subobtuse to acute. .. 31. *A. pinetorum*
32B.   Scales at base of stipes nearly black; indusia mostly asplenioid.
37A.   Pinnae lanceolate to linear-lanceolate, 20–40 mm. wide; pinnules 12–20 mm. long, usually acute; lowest anterior indusia sometimes athyrioid. .... 32. *A. otophorum*
37B.   Pinnae broadly linear, about 15 mm. wide; pinnules less than 12 mm. long, obtuse to rounded; lowest anterior indusia diplazioid. .... 33. *A. subrigescens*
28B.   Lower pinnae more or less petiolulate.
38A.   Ultimate pinnules prominently oblique.
39A.   Stipes greenish to straw-colored, herbaceous, dull; pinnae equilateral.
40A.   Rhizomes creeping. .... 34. *A. niponicum*
40B.   Rhizomes short, erect or ascending; fronds tufted.
41A.   Indusia asplenioid or sometimes athyrioid; blades ovate to deltoid-ovate. .... 35. *A. vidalii*
41B.   Indusia mostly asplenioid, rarely diplazioid; blades narrowly to broadly ovate. .... 36. *A. wardii*
39B.   Stipes brownish, coriaceous, lustrous; pinnae oblique; sori costal; indusia asplenioid. .... 37. *A. mesosorum*
38B.   Ultimate pinnules nearly equilateral.
42A.   Rachis and stipes prominently scaly throughout; sori costal.
43A.   Pinnae nearly opposite, thinly herbaceous; stipes and blades smooth.
44A.   Rhizomes long-creeping, slender, the fronds somewhat remote; pinnae more or less oblique; lower pinnules pinnatiparted; indusia 1–3 mm. long, asplenioid. .... 38. *A. crenatum* var. *glabrum*
44B.   Rhizomes thick, short, erect or ascending, the fronds tufted; pinnae equilateral; pinnules sometimes pinnatiparted to -lobed, the lobes orbicular to elliptic; indusia 3–4 mm. long, frequently diplazioid and asplenioid.
39. *A. squamigerum*
43B.   Pinnae alternate, chartaceous-herbaceous; stipes and blades scabrous. .... 40. *A. procerum*
42B.   Rachis and stipes densely scaly only at base, sparsely so or glabrous above.
45A.   Sori mostly elongate; indusia linear, solitary or sometimes diplazioid; pinnae crenate- or mucronate-toothed, the upper ones only slightly decurrent.
46A.   Lower veinlets united to those of the adjacent veins; blades thinly chartaceous, brownish when dried; pinnules lobulate to shallowly lobed. .... 41. *A. esculentum*
46B.   Veinlets free.
47A.   Blades 2- to 4-pinnate.

48A. Most of the ultimate pinnules pinnately cleft to parted.
  49A. Sori more or less median to subcostal, near middle of the veinlets; blades usually brownish green to green when dried.
    50A. Stipes prominently scaly at base; ultimate pinnules broadly lanceolate to narrowly oblong, 2–3 mm. wide. . . . . 42. *A. naganumanum*
    50B. Stipes sparsely scaly at base; ultimate pinnules ellipitic to oblong, 3.5–5 mm. wide.
      51A. Ultimate pinnules remote. . . . . 43. *A. bittyuense*
      51B. Ultimate pinnules close. . . . . 44. *A. nipponicola*
  49B. Sori along the costas on lower half of veinlets; blades brownish green when dried. . . 45. *A. doederleinii*
48B. Most of the ultimate pinnules coarsely toothed to pinnately lobed.
  52A. Sori near middle of the veinlets, short to elongate; blades herbaceous-chartaceous; ultimate pinnules rarely to 10 cm. long.
    53A. Scales at base of stipes withering, fugacious, brown, nearly smooth on margin. . . . . 46. *A. hachijoense*
    53B. Scales at base of stipes persistent, nearly black, prominently spinulose. . . . . 47. *A. virescens*
  52B. Sori on lower half of the veinlets; blades thinly chartaceous; pinnules (3–)5–20 cm. long; scales at base of stipes more or less persistent, fuscous, linear, about 1 cm. long, about 1 mm. wide, spinulose-ciliolate. . . . . 48. *A. maximum*
47B. Blades simply pinnate.
  54A. Veins nearly at right angles to the costa; sori subequal in length, asplenioid. . . . . 49. *A. petri*
  54B. Veins forming an obtuse angle to the costa; sori unequal in length, the sorus of lowest veinlet longest and nearly always diplazioid. . . . . 50. *A. mettenianum*
45B. Sori orbicular or reniform, sometimes obovate; indusia reniform or obovate; pinnae aristate-toothed, the upper ones strongly decurrent; blades simply pinnate. . . . . 51. *A. sheareri*

1. Athyrium decurrentialatum (Hook.) Copel. *Diplazium decurrentialatum* (Hook.) C. Chr.; *Gymnogramme decurrentialata* Hook.; *Dryopteris decurrentialata* (Hook.) C. Chr.; *Cornopteris decurrentialata* (Hook.) Nakai; *D. hookerianum* Koidz.; *C. opaca* Tagawa, sensu auct. Japon., excl. basion.——SHIKECHI-SHIDA. Rhizomes creeping, very thinly scaly; fronds glabrous, 30–70 cm. long; stipes usually shorter than the blades, pale green to straw-green, purplish, sparsely scaly; scales thinly membranous, narrowly to broadly lanceolate, sometimes narrowly ovate, 4–5 mm. long, pale brown, entire, deciduous; blades membranous, deltoid-ovate to oblong-ovate, 25–40 cm. long, 17–30 cm. wide, bipinnatiparted, or sometimes in lower part bipinnate, grayish green, slightly brownish when dried; pinnae 6–9 pairs, nearly opposite, spreading, broadly lanceolate, 2.5–5 cm. wide, sessile or subsessile, acuminate; pinnules oblong to ovate-oblong, 2–3.5 cm. long, 5–10 mm. wide, obtuse to rounded or subacute, usually decurrent on the broadly winged rachis, obsoletely toothed, entire or sometimes lobulate to parted; sori linear, naked, 1–3 mm. long, often forked.——Honshu (Ugo Prov., Kantō Distr. and westw.), Shikoku, Kyushu; rather common.——s. Korea, Formosa, and China.

Var. **pilosellum** (H. Itō) Ohwi. *Cornopteris decurrentialata* Nakai var. *pilosella* H. Itō; *Diplazium christensenianum* Koidz.; *C. christenseniana* (Koidz.) Tagawa; *C. musashiensis* Nakai; *A. musashiense* (Nakai) C. Chr.——TAKAO-SHIKECHI-SHIDA. With multicellular hairs on stipes, rachis, and on the underside of blade.——Honshu, Shikoku, Kyushu.

2. Athyrium hakonense (Makino) C. Chr. *Cornopteris hakonensis* (Makino) Nakai; *A. crenulatoserrulatum* forma *hakonense* Makino——HAKONE-SHIKECHI-SHIDA. Resembles the preceding species, differing chiefly in shorter sori; fronds glabrous, the costas often short-hairy on the underside of the ultimate pinnules; blades ovate-deltoid, bipinnate; pinnules oblong-ovate, 2–4 cm. long, 7–12 mm. wide, acute to obtuse, truncate at base, sessile or very slightly adnate to the scarcely winged rachis, lobulate to parted, obsoletely toothed; sori elliptic, median, about 1 mm. long.——Honshu (Kantō to s. Kinki Distr.); rather rare.

3. Athyrium crenulatoserrulatum Makino. *Dryopteris crenulatoserrulata* (Makino) C. Chr.; *Cornopteris crenulatoserrulata* (Makino) Nakai——IPPON-WARABI, Ō-MIYAMA-INU-WARABI. Rhizomes creeping; stipes usually as long as the blades, distinctly scaly especially toward the base, straw-colored, greenish; scales pale brown, membranous, broadly to narrowly lanceolate, 5–10 mm. long; blades nearly membranous, deltoid, 25–40 cm. long, nearly as wide, bipinnatiparted, thinly short-hairy on the costas of segments beneath, sometimes also on the rachis of pinnae beneath; pinnae 6–8 pairs, nearly opposite, narrowly oblong, 10–25 cm. long, acuminate, the upper ones smaller, the lowest pinnae broader, ovate-oblong, with a much-reduced pinnule near base, subsessile or the petiolules 1–2 cm. long; pinnules oblong to broadly lanceolate, 1–2 cm. wide, acuminate to acute, very short-petiolulate or sessile, pinnatiparted, the ultimate pinnules oblong, 4–8 mm. long, 2–4 mm. wide, obtuse, toothed; sori rounded, 2–5 on each side of the pinnules, fuscous-brown, about 1 mm. across, naked.——Hokkaido, Honshu (centr. and n. distr.).——Korea and Manchuria.

4. Athyrium fluviale (Hayata) C. Chr. *Dryopteris fluvialis* Hayata; *D. athyriiformis* Rosenst.; *Cornopteris fluvialis* (Hayata) Tagawa; *C. tashiroi* Tagawa; *A. tagawae* C. Chr.——HOSOBA-SHIKECHI-SHIDA, ŌBA-MIYAMA-INU-WARABI. Rhizomes ascending, with the basal stubs of old stipes toward the tip; stipes rather slender, brownish green, glabrous, thinly scaly toward base; scales membranous, broadly lanceolate to narrowly deltoid-ovate, 2–4 mm. long, brown, crisped, deciduous; blades narrowly deltoid, 20–30 cm. long, 15–25 cm. wide, short-acuminate, 3-pinnatiparted, blackened when dry, slightly glaucescent beneath, thinly puberulous while very young; pinnae 6–8 pairs, the lower opposite, 7–13 cm. long, 3–6 cm. wide, acuminate, spreading, subsessile, the median ones not contracted at base, the lowest pinnae contracted at base, short-petiolulate; pinnules oblong, 15–40 mm. long, 7–15 mm. wide, acute to obtuse, sessile, pinnatiparted, the ultimate pinnules oblong, rounded at apex, obsoletely toothed to lobulate; sori 3–8 in 2 series on each pinnule, rounded, naked, less than 1 mm. across, brown.——Mountains; Kyushu (Yakushima).——Formosa and Bonins.

5. Athyrium viridifrons Makino var. **okuboanum** (Makino) Ohwi. *A. okuboanum* Makino; *Dryopteris okuboana* (Makino) Koidz.; *A. viridifrons* forma *okuboanum* (Makino) Makino; *Lunathyrium unifurcatum* var. *okuboanum* (Makino) Kurata; *Athyrium henryi* auct. Japon., non Diels——Ō-HIME-WARABI. Rhizomes creeping; stipes light green, 30–50 cm. long, with deciduous scales; scales membranous, linear to broadly lanceolate, 3–10 mm. long, about 2 mm. wide, entire, unequal, fuscous-brown; blades thinly herbaceous to membranous, deltoid-ovate, or sometimes oblong-ovate, 40–80 cm. long, 25–50 cm. wide, acuminate, to bipinnatiparted to bipinnate, green, the rachis sometimes with minute dark brown scales; pinnae 7–10 pairs, opposite to alternate, oblong-ovate, 20–40 cm. long, subsessile, the lower pinnae often short-petiolulate with a pair of smaller pinnules at base, the costas on both sides with some hairlike scales, the rachis winged; pinnules broadly lanceolate to oblong, 3–6 cm. long, 1–1.5 cm. wide, acuminate to obtuse, usually broadly adnate at base to the rachis, crenate-toothed or pinnately lobed to cleft; sori 10–20, rounded; indusia reniform or sometimes hamate-oblong, about 0.5 mm. wide, erose-dentate.——Woods; Honshu, Shikoku, Kyushu.

Var. **viridifrons**. *Dryoathyrium viridifrons* (Makino) Ching; *Lunathyrium viridifrons* (Makino) Kurata——MIDORI-WARABI. Blades more finely cut, bipinnate-pinnatifid, the pinnules with a broad costular wing. Honshu (Kantō to Kinki Distr.), Shikoku, Kyushu; rare.——Korea (Quelpaert Isl.) and centr. China.

6. **Athyrium unifurcatum** (Bak.) C. Chr. *Nephrodium unifurcatum* Bak.; *Dryopteris unifurcata* (Bak.) C. Chr.; *D. tosensis* Kodama; *Lunathyrium unifurcatum* (Bak.) Kurata; *Dryoathyrium unifurcatum* (Bak.) Ching——Ō-HIME-WARABI-MODOKI. Rhizomes creeping, naked; stipes light brown to pale green, 20–40 cm. long, rather sparsely scaly in upper part while young, naked at base; scales narrowly lanceolate to linear, 2–5 mm. long, black-brown, slightly lustrous, subentire, the scales on rachis often smaller; blades herbaceous, deltoid- to oblong-ovate, 40–60 cm. long, 25–35 cm. wide, acuminate, pale- to yellow-green when dried, pinnate; pinnae 7–9 pairs, alternate to subopposite, lanceolate, 3–5 cm. wide, gradually acuminate, sessile, sometimes with sparse hairlike scales on the costas on both sides, pinnately parted to cleft, the upper ones gradually smaller; pinnules oblong, subacute to rounded, ascending to spreading, obsoletely crenate-toothed, sometimes lobulate; sori rounded, 5–12 on each pinnule; indusia reniform to arcuate-oblong, about 0.5 mm. across, gray-brown, erose-denticulate.——Honshu, Shikoku, Kyushu.——Formosa, s. China, and Indochina.

7. **Athyrium atkinsonii** Bedd. *Asplenium atkinsonii* (Bedd.) C. B. Clarke; *Aspidium senanense* Fr. & Sav.; *Dryopteris senanensis* (Fr. & Sav.) C. Chr.; *Athyrium senanense* (Fr. & Sav.) Koidz. & Tagawa; *Athyrium microsorum* Makino——TEBAKO-WARABI, MIYAMA-Ō-INU-WARABI. Fronds large, glabrous, vivid green; stipes pale brown, rather stout, glabrous, sparsely scaly at base; scales thinly membranous, unequal, lanceolate to linear-lanceolate, 5–10 mm. long, entire, pale brown, deciduous; blades thinly herbaceous, broadly ovate or deltoid-ovate, 70–100 cm. long, tripinnate to tripinnate-pinnatifid, glabrous; pinnae ovate, 20–40 cm. long, 10–20 cm. wide, acuminate, petiolulate; ultimate pinnules rather unequal, oblong, 8–25 mm. long, 5–15 mm. wide, obtuse, broadly cuneate at base, often short-petiolulate, pinnatiparted to lobed;

sori rounded, 2–4 on each side along the costules; indusia minute, membranous, reniform to short-lanceolate, pale gray-brown, erose, 0.5–0.7 mm. long.——Woods; Honshu (centr. distr. and Tango Prov.), Shikoku, Kyushu; rather rare.——Formosa, China, and the Himalayas.

8. **Athyrium pinfaense** (Ching) Ohwi. *Diplazium pinfaense* Ching——FUKUREGI-SHIDA. Rhizomes short, erect; fronds 4 or 5 in a tuft; stipes slender, to 17 cm. long, 2 mm. across, stramineous, sparsely scaly at base, smooth above; scales membranous, ovate-oblong, about 2 mm. long, acuminate, entire, brown; blades ovate, 20–26 cm. long, to 16 cm. wide, simply pinnate, the rachis slightly flexuous; pinnae thinly herbaceous, 2 or 3 pairs, rarely 1 pair, alike, oblong-lanceolate, 10–14 cm. long, 3–3.5 cm. wide, acuminate, rounded to rounded-cuneate at base, regularly aristate-toothed, glabrous on both sides, the petiolules to 6 mm. long; veinlets forked, slender, extending to the teeth, the anterior branch soriferous; sori 1–1.6 cm. long, arising near the base of the branch, not reaching the margin; indusia membranous, grayish while young, brownish with age.——Kyushu (Fukuregi in Amakusa Isl. in Higo Prov.); very rare.——s. China.

9. **Athyrium aphanoneuron** (Ohwi) Ohwi. *Diplazium aphanoneuron* Ohwi; *D. donianum* sensu auct. Japon., non Tard.-Bl.; *D. bantamense* sensu auct. Japon., non Bl.——ATSUBA-KINOBORI-SHIDA. Rhizomes short-creeping, rather stout; fronds about 60 cm. long, glabrous; stipes 30–40 cm. long, about 3 mm. across, terete, with two deep grooves in upper part, very sparsely scaly in upper part, those at base broadly linear, 4–5 mm. long, acuminate, spinulose, black-brown; blades ovate, pinnate, the rachis 10–15 cm. long; pinnae 5–7, alternate, except for the lowest pair, firmly herbaceous, broadly lanceolate to narrowly oblong, 15–18 cm. long, 3–4.5 cm. wide, acuminate, acute, slightly unequal at base, nearly entire except for a few teeth near tip, slightly recurved on margin, the petiolules of the lateral pinnae 2–10 mm. long, the terminal one 15–30 mm. long; costa prominent especially beneath; sori usually double (diplazioid), linear; indusia unequal, usually about 20 mm. long, erose and ciliolate.——Kyushu (Yakushima and Tanegashima); rare.——Ryukyus, China, and Tonkin.

10. **Athyrium dubium** (G. Don) Ohwi. *Asplenium lanceum* Thunb.; *Diplazium lanceum* (Thunb.) Presl, non Bory; *Athyrium lanceum* (Thunb.) Milde, non Moore; *Scolopendrium dubium* G. Don; *Asplenium subsinuatum* Wall. ex Hook & Grev.; *Diplazium subsinuatum* (Wall.) Tagawa——HERA-SHIDA. Rhizomes slender, long-creeping, about 2 mm. across, remotely leafy, while young with black-brown linear entire spreading scales 2–3 mm. long; stipes slender, pale green, sparingly scaly, 3–25 cm. long; blades coriaceous-herbaceous, lanceolate to linear-lanceolate, 7–20 (–30) cm. long, 1–2.5 (–3) cm. wide, acuminate, obtuse to acute at base, entire to obsoletely undulate, margin slightly recurved, glabrous, dark green above, paler beneath, flat except for the raised midrib beneath; veins once to thrice forked near base, the veinlets parallel, nearly at right angles to the costa; sori linear, parallel; indusia linear, asplenioid, sometimes also diplazioid, 5–12 mm. long, entire.——Shaded places; Honshu, Shikoku, Kyushu; common.——Ryukyus, Formosa, China, India to Ceylon.

Var. **crenatum** (Makino) Ohwi. *Diplazium lanceum* var. *crenatum* Makino; *D. tomitaroanum* Masam.; *D. lanceum* var.

grandicrenatum Nakai ex H. Itō——NOKOGIRI-HERA-SHIDA. Blades pinnately lobed to parted, the lobes rounded at apex, entire; sori few pairs on the lobes, short-linear.——Honshu (Izu Isls.), Shikoku, Kyushu.——Ryukyus and Formosa.

11. Athyrium nakanoi Makino. HIME-HŌBI-SHIDA. Rhizomes short, with the remains of old fronds; fronds 15–30 cm. long; stipes 2–5 cm. long, slender, slightly lustrous, brownish in lower part, with membranous, pale brown, entire, ovate-triangular to broadly lanceolate scales 2–3 mm. long; blades membranous-herbaceous, lanceolate, 13–25 cm. long, about 3 cm. wide, gradually acuminate, slightly contracted at base, deep green, thinly puberulent while very young on underside and on rachis; pinnae 14–18 pairs, oblong to ovate, 1–1.7 cm. long, 5–8 mm. wide, obtuse, auriculate and nearly truncate on upper side at base, cuneate on lower side, crenately toothed, spreading, penninerved, the lower 1 or 2 pairs smaller and subreflexed; veinlets usually simple except the one on the auricle and this pinnately branched; sori 5–10 on a pinnule, median; indusia membranous, reniform or obovate, erose-dentate.——Mountains; Kyushu (Yakushima); rare.——Formosa.

12. Athyrium okudairae (Makino) Ohwi. *Diplazium okudairae* Makino——IYO-KUJAKU. Rhizomes creeping, nearly naked, 3–4 mm. across, rather loosely leaved; fronds glabrous, flaccid; stipes 20–30 cm. long, pale green, brownish at base, sparingly scaly; scales linear-lanceolate, dark brown, 3–5 mm. long, about 1 mm. wide, entire or obsoletely toothed on margin; blades deltoid-lanceolate, 25–35 cm. long, 15–20 cm. wide, simply pinnate, caudate-acuminate, broad at base; pinnae 10–12 pairs, thinly herbaceous, lanceolate, 7–11 cm. long, 2–2.5 cm. wide, caudate-acuminate, falcate, the upper ones adnate and decurrent on the winged rachis, the lower ones short-petiolulate, auriculate and nearly truncate on outer side at base, cuneate on lower side, pinnately lobulate and acutely denticulate or doubly toothed; costa raised beneath, the veins pinnate, the veinlets very slender, distinct on both sides; sori subcostal, linear, to 1 cm. long; indusia linear, entire, straight, solitary, rarely diplazioid.——Woods; Honshu (Izu Prov. through Tōkaidō and s. Kinki to Chūgoku Distr.), Shikoku, Kyushu; rare.

13. Athyrium wichurae (Mett.) Ohwi. *Asplenium wichurae* Mett.; *Diplazium wichurae* (Mett.) Diels; *D. wichurae* var. *amabile* Tagawa——NOKOGIRI-SHIDA. Rhizomes creeping, rather loosely leafy, 2–2.5 mm. across, wiry, nearly naked; fronds glabrous; stipes firm, 10–25 cm. long, pale green to brownish, very sparsely scaly; scales subulate-lanceolate, 4–6 mm. long, 1–1.5 mm. wide, entire, brown; blades broadly lanceolate, 20–35 cm. long, 10–20 cm. wide, short-caudate, not narrowed at base, the rachis not winged; pinnae 10–15 pairs, 6–10 cm. long, 1–2 cm. wide, caudate-acuminate, short-petiolulate or a few of the upper ones sessile and adnate at base, slightly falcate, denticulate, often pinnately lobulate, strongly auriculate and truncate at base on upper side, cuneate on lower side; costa and veins often impressed above, distinct but scarcely raised beneath; sori on the lowest veinlets, linear; indusia linear, solitary, very rarely diplazioid, slightly arcuate, obliquely ascending, entire, 5–8 mm. long, about 1 mm. wide.——Woods; Honshu (s. Kantō Distr. and westw.), Shikoku, Kyushu; rather rare.——Ryukyus, Formosa, and China.

14. Athyrium pterorachis Christ. *Dryoathyrium pterorachis* (Christ) Ching; *Lunathyrium pterorachis* (Christ) Kurata; *Parathyrium pterorachis* (Christ) Holttum——Ō-ME-SHIDA. Stipes stout, light brown, tufted, about 60 cm. long, dilated near base, to 15 mm. wide above the base, very sparsely scaly in the upper part, rather densely so toward the base; scales pale brown, membranous, broadly lanceolate to ovate, 1.5–2.5 cm. long, 4–8 mm. wide, entire; blades herbaceous, narrowly oblong, often more than 1 m. long, 35–45 cm. wide, short-acuminate, slightly narrowed at base, bipinnatiparted; pinnae 15–20 pairs, subopposite to alternate, spreading, linear-lanceolate, about 20 cm. long, 3.5–4 cm. wide, acuminate, scarcely narrowed at base, sessile, sparingly pubescent while young along costas, the rachis prominently winged; pinnules numerous, horizontally spreading, narrowly oblong, 4–8 mm. wide, obtuse, pinnately lobed; sori 12–20, subcostular; indusia lanceolate, 1–2 mm. long, slightly arcuate, sometimes somewhat athyrioid, minutely ciliolate.——Wet meadows in mountains; Hokkaido, Honshu (centr. and n. distr.).——Sakhalin, s. Kuriles, and Kamchatka.

15. Athyrium henryi (Bak.) Diels. *Aspidium henryi* Bak.; *Dryoathyrium henryi* (Bak.) Ching; *Lunathyrium henryi* (Bak.) Kurata; *Athyrium coreanum* Christ; *A. decursivum* Yabe; *Dryoathyrium coreanum* (Christ) Tagawa; *A. heterophyllum* Nakai——KŌRAI-INU-WARABI. Rhizomes short-creeping; fronds tufted; stipes 20–40 cm. long, thinly scaly in lower part, 4–5 mm. across near base; scales pale brown, membranous, broadly lanceolate, entire; blades thinly herbaceous, broadly lanceolate to narrowly ovate, 35–50 cm. long, 17–22 cm. wide, slightly narrowed at base, bipinnatiparted, vivid green, paler beneath; pinnae about 10 pairs, subopposite or subalternate, lanceolate, 2–3.5 cm. wide, long-acuminate, sessile, short-hairy on costas above and on underside while young or glabrous, the rachis prominently winged; pinnules rather many, oblong to oblong-ovate, 1–1.5 cm. long, 5–7 mm. wide, obtuse, lobed to coarsely toothed; sori 4–12, nearly median; indusia 1–2 mm. long, slightly arcuate, often athyrioid, minutely ciliolate.——Honshu (centr. distr.), Kyushu (n. distr.); rare.——Korea, Manchuria, and e. Siberia.

Var. **kiyozumianum** (Kurata) Tagawa. *A. coreanum* var. *kiyozumianum* (Kurata) Ohwi; *Lunathyrium henryi* var. *kiyozumianum* (Kurata) Kurata; *Dryoathyrium coreanum* var. *kiyozumianum* Kurata——KIYOZUMI-ME-SHIDA. Stipes shorter, 9–16 cm. long; scales more abundant on stipes and rachis.——Honshu (Mount Kiyozumi in Awa Prov.).

16. Athyrium pycnosorum Christ. *A. acrostichoides* sensu auct. Japon., non Diels; *Lunathyrium pycnosorum* (Christ) Koidz.; *A. acrostichoides* Diels var. *pycnosorum* (Christ) C. Chr.——MIYAMA-SHIKE-SHIDA, HAKUMŌ-INODE. Rhizomes short; fronds tufted, to 1 m. long, nearly glabrous or prominently pubescent; stipes 15–30 cm. long, pale green, often purplish; scales membranous, linear to narrowly lanceolate, the upper ones 2–5 mm. long, the basal ones 2–3 mm. wide, 8–12 mm. long; blades herbaceous, broadly oblanceolate to oblong, usually broadest above the middle, 40–70 cm. long, 10–17 cm. wide, bipinnatiparted, abruptly long-acuminate, gradually narrowed at base; pinnae 15–22 pairs, alternate, spreading, linear-lanceolate, 6–10 cm. long, 1–1.7 cm. wide, long-acuminate, truncate at base, sessile, the lower few distant, ovate-deltoid, slightly reflexed; pinnules elliptic to oblong, 3–5 mm. wide, rounded at apex, crenately toothed; sori 3–6 pairs on each pinnule, obliquely ascending; indusia oblong to lanceolate, 1–3 mm. long, solitary, rarely athyrioid and diplazioid.——Wet slopes in mountains; Hokkaido, Honshu, Shikoku, Kyushu; rather common.——s. Kuriles, Sakhalin, Korea, Manchuria, China, and Formosa.

17. Athyrium grammitoides (Presl) Milde. *Diplazium grammitoides* Presl; *Asplenium grammitoides* (Presl) Hook.; *A. conilii* Fr. & Sav.; *D. conilii* (Fr. & Sav.) Makino; *D. japonicum* var. *conilii* (Fr. & Sav.) Makino; *A. conilii* var. *coreanum* Hook. & Bak.; *D. japonicum* var. *latipes* Rosenst.; *Athyrium conilii* (Fr. & Sav.) Tagawa——HOSOBA-SHIKE-SHIDA. Rhizomes creeping, branched, slender, 1–1.5 mm. across, sparsely scaly; fronds nearly glabrous to sparsely pubescent; stipes slender, 7–20(–25) cm. long, sparsely scaly, pale green, brownish toward base; scales membranous, pale brown, broadly lanceolate to linear-lanceolate, 2–5 mm. long, to 2 mm. wide, entire; blades broadly lanceolate, 10–20 cm. long, 2.5–6 cm. wide, gradually acute to acuminate, slightly to scarcely narrowed toward base, simply pinnate, the rachis often sparsely scaly; pinnae 10–15 pairs, herbaceous, oblong-ovate to elliptic, 1–3 cm. long, 5–8 mm. wide, obtuse to rounded, sessile, spreading, crenate, the upper ones adnate to the rachis, the lower ones undivided, lobed, cleft, or rarely parted; sori often diplazioid.——Shaded places in lowlands and low mountains; Hokkaido, Honshu, Shikoku, Kyushu.——Korea, Ryukyus, Formosa, Philippines, and Malaysia. Common and variable, closely related to *A. japonicum*.

Var. **oldhamii** (Hook. & Bak.) Ohwi. *Asplenium japonicum* var. *oldhamii* Hook. & Bak.; *Diplazium oldhamii* (Hook. & Bak.) H. Chr.; *Diplazium conilii* var. *oldhamii* (Hook. & Bak.) Nakai; *Athyrium conilii* var. *oldhamii* (Hook. & Bak.) Tagawa; *D. japonicum* var. *oldhamii* (Hook. & Bak.) C. Chr. ——YABU-SHIDA. Plants smaller, with narrower fronds and fewer scales.

Var. **simplicifolium** (Makino) Ohwi. *Diplazium conilii* var. *simplicifolium* Makino; *D. lobatocrenatum* Tagawa; *D. lanceum* var. *subtripinnatum* Nakai; *Lunathyrium lobatocrenatum* (Tagawa) Kurata; *Athyrium lobatocrenatum* (Tagawa) Tagawa——HITOTSUBA-SHIKE-SHIDA. Blades simply pinnate.——Honshu (Izu Prov.), Shikoku, Kyushu; rare.

18. Athyrium japonicum (Thunb.) Copel. *Asplenium japonicum* Thunb.; *Diplazium japonicum* Bedd.; *Lunathyrium japonicum* (Thunb.) Kurata; *Athyrium oshimense* Christ; *D. oshimense* (Christ) H. Itô; *D. thunbergii* Nakai——SHIKE-SHIDA. Rhizomes creeping, branched, about 2 mm. across; stipes 15–35 cm. long, slender, the scales on lower half lanceolate to broadly so, 5–10 mm. long, 1.5–3 mm. wide, brown, entire, those on the upper half linear-lanceolate to linear, 2–3 mm. long, 0.2–1 mm. wide; blades herbaceous, broadly lanceolate to narrowly oblong-ovate, 25–40 cm. long, 8–15 cm. wide, acuminate, usually scarcely narrowed at base, pinnate, nearly glabrous to thinly pubescent on both sides; pinnae pinnately-cleft to sometimes parted, lanceolate, 1–2 cm. wide, acuminate, spreading, or the lowest pair slightly reflexed, sessile, truncate and nearly equilateral at base, sessile, the upper pinnae small and adnate to rachis at base; pinnules elliptic, 4–7 mm. wide, rounded to obtuse, crenate; sori 6–10 on each pinnule, median; indusia lanceolate to broadly linear, 2–3.5 mm. long, closely parallel, ascending, solitary or sometimes diplazioid in the lowest acroscopic one, minutely erose to ciliolate.——Shaded places in lowlands and foothills; Honshu, Shikoku, Kyushu; common.——Korea, Ryukyus, Formosa, China, and India.

Var. **dimorphophyllum** (Koidz.) Ohwi. *A. dimorphophyllum* (Koidz.) Tagawa; *Lunathyrium dimorphophyllum* (Koidz.) Kurata; *Diplazium dimorphophyllum* Koidz.——

SEITAKA-SHIKE-SHIDA. Plant stouter; fronds slightly dimorphic, usually thinly pubescent on both sides at least while young; rhizomes 3–4 mm. across; stipes 20–50 cm. long, sparingly scaly; scales linear-lanceolate to lanceolate, pale brown, 5–12 mm. long; blades ovate-oblong to ovate, 20–50 cm. long, 20–30 cm. wide, not narrowed at base; pinnae 8–10 pairs, broadly lanceolate, 1.5–3 cm. wide, the pinnules narrowly ovate-oblong.——Shaded places; Honshu, Shikoku, Kyushu.

Var. **kiusianum** (Koidz.) Ohwi. *A. kiusianum* (Koidz.) Tagawa; *Diplazium kiusianum* Koidz.——MUKUGE-SHIKE-SHIDA. Larger and denser pubescent; scales abundant on stipes and rachis.——Kyushu (centr. distr.).

19. Athyrium iseanum Rosenst. *A. goeringianum* sensu auct. Japon., non Moore; *A. iseanum* var. *angustisectum* Tagawa, and var. *obtusum* Tagawa——HOSOBA-INU-WARABI. Rhizomes short; fronds tufted; stipes 10–25 cm. long, straw-colored to pale green, sparsely scaly toward the base; scales pale brown, thinly membranous, linear-lanceolate, entire, 2–5 mm. long; blades broadly to narrowly ovate, 15–35 cm. long, 10–20 cm. wide, caudately acuminate, not at all or only slightly narrowed at base, bipinnate; pinnae 5–10 pairs, broadly lanceolate or the smaller ones oblong-ovate, 1.7–3 cm. wide, caudately acuminate, short-petiolulate; pinnules oblong-ovate to deltoid-ovate, 1–1.7 cm. long, obtuse to acute, broadly cuneate at base, oblique, rather irregularly lobulate to parted, rarely dissected to the costules, coarsely mucronate-toothed, with a few fleshy deciduous bristles on upper side along the costas; indusia usually asplenioid, 1–2 mm. long, sometimes also athyrioid, entire.——Shaded places in low mountains; Honshu (Awa, Izu, and Echigo Prov. westw.), Shikoku, Kyushu; rather common. ——Formosa.

20. Athyrium frangulum Tagawa. *A. iseanum* var. *fragile* Tagawa——MIYAKO-INU-WARABI. Rhizomes erect, short; fronds soft, thin, tufted; stipes 15–35 cm. long, purplish, glabrous, sparsely scaly at base; scales pale brown, membranous, lanceolate to linear-lanceolate, 3–4 mm. long, entire; blades oblong-ovate to ovate, 20–40 cm. long, 15–25 cm. wide, acuminate, not narrowed at base, bipinnate; pinnae broadly lanceolate to oblong-ovate, 2–5 cm. wide, long-acuminate, obliquely spreading, short-petiolulate; pinnules narrowly ovate to oblong-ovate, 1.5–2.5 cm. long, 5–10 mm. wide, acute to obtuse, slightly oblique and cuneate at base, very short-petiolulate to sessile, lobulate to parted, acutely mucronate-toothed, with a few fleshy deciduous bristles along the costas on upper side; sori in 2 rows along the costas; indusia lanceolate to oblong, straight (asplenioid), sometimes also athyrioid, very rarely diplazioid, entire.——Low mountains; Honshu (Kantō Distr. and westw.), Shikoku, Kyushu.

21. Athyrium tozanense (Hayata) Hayata. *Asplenium tozanense* Hayata——SHIMA-INU-WARABI. Rhizomes erect or ascending, the basal stubs of old stipes persistent; fronds tufted, small, slender, glabrous, sometimes sparsely puberulent on the rachis of the pinnae beneath; stipes 5–15 cm. long, slender, scaly while young, pale green, brownish toward the base, dull; scales thinly membranous, pale brown, broadly lanceolate, about 3 mm. long, scarcely 1 mm. wide, entire; blades membranous, lanceolate-deltoid, 12–20 cm. long, 5–10 cm. wide, acuminate to caudately acuminate, scarcely narrowed at base, bipinnatisect or sometimes bipinnate in the lower portion; pinnae 8 or 9 pairs, narrowly ovate, 2.5–5 cm. long, 1–1.8 cm. wide, acute to short-acuminate, spreading, oblique, very short-petiolulate; pinnules several pairs, oblong to elliptic or ovate to narrowly so, 5–8 mm. long, 2–4 mm. wide, oblique, usually slightly adnate to rachis at base, crenate or mucronate-toothed, often lobulate, with scattered fleshy deciduous bristles on upper side; sori 3–10, costular; indusia broadly lanceolate to oblong, straight, 1–1.5 mm. long, often athyrioid and rarely also diplazioid, brownish, erose-denticulate.—— Mountains; Kyushu (Yakushima); rare.——Formosa.

22. Athyrium spinulosum (Maxim.) Milde. *Cystopteris spinulosa* Maxim.; *Asplenium spinulosum* (Maxim.) Bak.——MIYAMA-INU-WARABI. Rhizomes slender, creeping, remotely leaved; stipes straw-colored, 40–50 cm. long, sparsely scaly, usually longer than the blades; scales membranous, 6–10 mm. long, 1–4 mm. wide, pale brown, entire; blades deltoid, 20–30 cm. long, slightly wider than long, abruptly acuminate, glabrous, 3-pinnate, the rachis glabrous but puberulent on upper side at insertion of the pinnae; pinnae 5–6 pairs, acuminate, the lowest ones largest, narrowly ovate, 7–12 cm. wide, short-petiolulate, with a pair of much-reduced pinnules above the base, the upper pinnae rather abruptly reduced; pinnules oblong-ovate, 7–10 mm. long, 3–5 mm. wide, obtuse, sessile, pinnatiparted, with few, rather prominent, spine-tipped teeth on upper margin; sori rounded, about 1 mm. across, subcostular; indusia thinly membranous, minute, whitish, sparsely fimbriate.——Coniferous woods in mountains; Honshu (centr. distr.); rare.——Sakhalin, Korea, Manchuria, China to e. Siberia, and the Himalayas.

23. Athyrium alpestre (Hoppe) Rylands. *Aspidium alpestre* Hoppe; *Pseudathyrium alpestre* (Hoppe) Newm.; *Phegopteris alpestris* (Hoppe) Mett.——OKUYAMA-WARABI. Rhizomes short, rather stout; fronds erect, tufted, 50–70 cm. long, glabrous, sparsely scaly while very young; stipes 20–30 cm. long, pale green to slightly reddish, more densely scaly toward base; scales membranous, brown, unequal, linear to broadly lanceolate, 7–12(–15) mm. long, 1–4 mm. wide, entire; blades herbaceous, oblong or narrowly oblong-ovate, sometimes narrowly ovate, 30–40 cm. long, 12–18 cm. wide, short-acuminate, slightly narrowed at base, bipinnate, the rachis and pinnae beneath scaly while young; pinnae 8–10 pairs, the lower ones somewhat interrupted and subopposite, broadly lanceolate, 6–10 cm. long, 2–3 cm. wide, acuminate, sessile, ascending to spreading; pinnules narrowly ovate, 1–2 cm. long, 5–8 mm. wide, acute, usually broadly cuneate at base, sessile, pinnately cleft to parted, the segments elliptic, obtuse, mucronate-toothed; sori solitary on the lower veinlets of the pinnules.——Alpine slopes; Hokkaido, Honshu (centr. distr.); locally abundant.——Kamchatka, Siberia to Europe, and N. America.

24. Athyrium filix-femina (L.) Roth var. **longipes** Hara. *A. brevifrons* Nakai; *A. melanolepis* sensu auct. pro parte, non Christ; *A. filix-femina* var. *melanolepis* sensu auct. pro parte, non Makino——EZO-ME-SHIDA. Rhizomes short; fronds tufted; stipes 25–40 cm. long, suberect, as long as to slightly shorter than the blades, sparsely scaly in upper part, rather densely so toward the base, straw-colored; scales linear-lanceolate to narrowly oblong-ovate, 10–12 mm. long, 1–5 mm. wide, fuscous, with a lighter brown entire margin, lustrous; blades herbaceous, oblong-ovate to ovate, 30–60 cm. long, 15–30 cm. wide, short-acuminate, bipinnate, glabrous, the rachis very thinly scaly while young; pinnae 10–15 pairs, lanceolate, 10–15 cm. long, 2–4 cm. wide, long-acuminate, sessile, the lower ones shorter, subopposite; pinnules spreading,

equilateral, broadly lanceolate to narrowly ovate, 8–15(–20) mm. long, 3–7 mm. wide, obtuse to short-acuminate, toothed, often slightly adnate to the rachis, pinnately lobed to cleft; sori 8–16 to a pinnule, subcostal; indusia oblong, straight to curved, athyrioid, 0.7–1 mm. long, erose, long-ciliate.—— Mountains; Hokkaido, Honshu.——s. Kuriles and Sakhalin. The typical phase and many variants occur in the N. Hemisphere.

25. Athyrium melanolepis (Fr. & Sav.) Christ. *Asplenium melanolepis* Fr. & Sav.; *Athyrium filix-femina* var. *melanolepis* (Fr. & Sav.) Makino; *A. filix-femina* var. *nigropaleaceum* Makino; *A. nigropaleaceum* (Makino) Makino—— ME-SHIDA, MIYAMA-ME-SHIDA. Closely related to the preceding; rhizomes short and thick; fronds tufted; stipes straw-colored, 20–30 cm. long, scaly, more densely so near the base; scales lanceolate, 8–12 mm. long, often crispate, nearly black to somewhat fuscous, entire, lustrous; blades herbaceous, oblong-ovate, 30–50 cm. long, 15–30 cm. wide, abruptly acuminate, contracted at base, thinly scaly beneath while young, glabrescent; pinnae 12–15 pairs, lanceolate, 2–4 cm. wide, long-acuminate, equilateral, sessile, the lower 1 or 2 pairs shorter; pinnules rather numerous, broadly lanceolate to narrowly ovate, 4–8 mm. wide, acute, often slightly adnate to the rachis, sessile, nearly equilateral, toothed and lobulate to parted; sori 6–16 to a pinnule, subcostal; indusia thin, oblong, straight or athyrioid, about 0.7 mm. long, erose, sparsely long-ciliate.—— Mountains; Honshu (n. and centr. distr.); locally common.

26. Athyrium reflexipinnum Hayata. SAKABA-INU-WA-RABI. Rhizomes ascending, short, densely covered with the stubs of old stipes; fronds slender, glabrous, tufted; stipes few, slender, straw-colored, 5–12 cm. long, sparsely scaly toward the base; scales membranous, lanceolate to linear-lanceolate, 3–4 mm. long, 0.3–0.7 mm. wide, entire, pale brown, deciduous; blades thinly herbaceous to membranous, broadly lanceolate or sometimes lanceolate-deltoid, 10–15 cm. long, 3–4.5 cm. wide, caudately acuminate, somewhat narrowed toward the base, green, paler beneath, the rachis slender; pinnae 8–11 pairs, sessile, more or less reflexed except the upper ones, narrowly deltoid-ovate to broadly lanceolate, 2–2.5 cm. long, 7–9 mm. wide, acute to subacute, acutely toothed, truncate on upper side at base, cuneate on lower side, pinnately parted to cleft; pinnules elliptic to oblong, larger on upper (anterior) side of pinnae; sori one to few on each pinnule, costal and costular; indusia ovate and straight or athyrioid, erose-dentate, about 0.5 mm. across.——Mountains; Kyushu (Yakushima); rare.——Formosa.

27. Athyrium rupestre Kodama. MIYAMA-HEBI-NO-NE-GOZA. Closely resembles *A. yokoscense* var. *fauriei*; rhizomes erect, covered with the stubs of old stipes; fronds tufted; stipes 3–15 cm. long, straw-colored in upper part, brownish and densely scaly near base; scales linear to broadly lanceolate, about 1 cm. long, 1.5–3 mm. wide; blades herbaceous, broadly lanceolate, 15–25 cm. long, 5–9 cm. wide, acuminate, narrowed at the base; pinnae 10–15 pairs, lanceolate, acute to acuminate, sessile, pinnately lobed to parted; pinnules oblong, obtuse to acute, toothed; sori median, with paraphyses among the sporangia; indusia oblong, straight, sometimes athyrioid, erose-ciliate.——Rocky cliffs in mountains; Hokkaido, Honshu (n. distr.); rare.——s. Kuriles.

28. Athyrium yokoscense (Fr. & Sav.) Christ. *Asplenium yokoscense* Fr. & Sav.; *Athyrium demissum* Christ, pro parte; *A. flaccidum* Christ; *A. yokoscense* var. *dilatatum* Tagawa; *A. yokoscense* var. *alpicola* Hiyama——HEBI-NO-NEGOZA. Rhizomes short, erect, covered with the stubs of old stipes; fronds green, glabrous, somewhat tufted; stipes 15–25 cm. long, straw-colored, scaly toward base; scales membranous, linear to linear-lanceolate, 8–15 mm. long, about 2 mm. wide, uniformly brown or with a darker somewhat lustrous thicker band in the center, entire; blades herbaceous, oblong-ovate, 25–40 cm. long, 8–20 cm. wide, acuminate, scarcely narrowed at base, sessile, 2-pinnate; pinnae 10–15 on each side, equilateral, lanceolate, 1–2.5 cm. wide, long-acuminate, the rachis scarcely winged; pinnules oblong-ovate, 5–12 mm. long, 3–5 mm. wide, obtuse to acute, toothed to lobulate, very slightly adnate to the rachis, scarcely decurrent; sori 4–10 to a pinnule, median; indusia about 1 mm. long, oblong, straight, the lower ones often hamate (athyrioid), thin, entire.——Woods in lowlands and mountains; Hokkaido, Honshu, Shikoku, Kyushu; common and variable.——Sakhalin, Korea, Manchuria, and e. Siberia.

Var. **fauriei** (Christ) Tagawa. *Nephrodium fauriei* Christ; *A. fauriei* (Christ) Makino; *A. nikkoense* Makino; *Aspidium fauriei* (Christ) Christ; *Athyrium demissum* Christ, pro parte ——IWA-INU-WARABI. Depauperate, with smaller, narrower bipinnatifid blades narrowed at base, the pinnae acute to obtuse.——Rocky places in mountains; occurs with the typical phase.——s. Kuriles and Korea.

Poorly known are: **Athyrium kirishimense** Tagawa. KIRISHIMA-HEBI-NO-NEGOZA. Reported to have larger scales, to 17 mm. long at base of stipes, and chartaceous-herbaceous blades.——Kyushu (centr. distr.).——**Athyrium satowii** H. Itô——TORA-NO-O-INU-WARABI. An aberrant phase of *A. yokoscense* with the lower pinnae broadly linear, rounded, pinnatiparted, the upper pinnae short, simple, flabellate, the pinnules rounded-obovate, the sori few near the margin.

29. Athyrium tashiroi Tagawa. USUBA-HEBI-NO-NEGOZA. Rhizomes erect or ascending, covered with the basal stubs of old fronds; stipes tufted, 15–25 cm. long, slender, straw-colored, dark brown and densely scaly toward base; scales membranous, narrowly lanceolate to linear, 1–1.5 cm. long, to 2 mm. wide, entire, uniformly brown or sometimes with a chestnut-brown longitudinal band in the center; blades herbaceous, ovate, 15–30 cm. long, 10–18 cm. wide, acuminate, bipinnate, glabrous; pinnae 10–15 pairs, spreading, broadly ovate-lanceolate, 8–20 cm. long, 2–3.4 cm. wide, caudately long-acuminate, sessile; pinnules several to rather many on a pinna, rather remote, broadly lanceolate to narrowly ovate, 1.5–2 cm. long, 4–8 mm. wide, acute to short-acuminate, oblique, cuneate at base, sessile, irregularly few-toothed, pinnately cleft to parted, pinnules ascending, acuminate; sori median; indusia membranous, athyrioid, sometimes straight, entire, about 1 mm. long.——Mountains; Kyushu; rare. Possibly only a variety of *A. yokoscense*.

30. Athyrium multifidum Rosenst. *A. filix-femina* var. *deltoideum* Makino; *A. solutum* Rosenst., non Christ; *A. multifidum* var. *latisectum* Rosenst.; *A. deltoidofrons* Makino; *A. multifidum* var. *deltoideum* (Makino) Nakai; *A. deltoidofrons* var. *multifidum* (Rosenst.) Koidz. and var. *latisectum* (Rosenst.) Koidz.——SATO-ME-SHIDA, Ō-SATO-ME-SHIDA. Fronds large, glabrous, tufted; stipes rather stout, 40–60 cm. long, straw-colored, densely scaly near base; scales membranous, pale brown, linear-lanceolate to lanceolate, 10–15 mm. long, 1–2 mm. wide, long-acuminate, entire; blades thinly herbaceous, deltoid-ovate to broadly ovate, 40–70 cm. long,

30–50 cm. wide, acuminate, 3-pinnatiparted; pinnae 9–12 pairs, equilateral, spreading, 15–30 cm. long, 4–7 cm. wide, gradually acuminate, short-petiolulate; primary pinnules narrowly deltoid-ovate to deltoid-lanceolate, 2.5–4 cm. long, 8–12 mm. wide, acute to acuminate, very short-petiolulate to subsessile, pinnatiparted nearly to the costas; ultimate pinnules oblong to ovate, 4–7 mm. long, obtuse to subacute, toothed; sori 5–10 to a pinnule, subcostal; indusia thin, oblong and straight, 0.7–1 mm. long, sometimes athyrioid, erose-ciliate.——Hokkaido, Honshu, Shikoku.

Var. **acutissimum** (Kodama) Ohwi. *A. acutissimum* Kodama; *A. deltoidofrons* var. *acutissimum* (Koidz.) Tagawa; *A. solutum* var. *acutissimum* (Kodama) Tagawa——TOGARIBA-ME-SHIDA. Plant smaller, with narrower blades, and the pinnae often horizontally spreading, usually 10-20 cm. long, and with less deeply parted pinnules.——Honshu and Shikoku.

31. Athyrium pinetorum Tagawa. TAKANE-SATO-ME-SHIDA. Rhizomes erect or nearly so, covered with the stubs of old stipes; stipes 15–30 cm. long, slender, straw-colored, dark brown toward base, thinly scaly in upper part, more densely so near base; scales near base of stipe narrowly lanceolate, 10–15 mm. long, 1.5–2 mm. wide, entire, slightly lustrous, with a dark brown longitudinal band in the center, the scales on upper part of the stipes shorter and thinner, lighter brown, often curled, lanceolate, 7–10 mm. long, 2–3 mm. wide; blades thinly herbaceous, ovate-deltoid to nearly deltoid, 17–30 cm. long, 15–25 cm. wide, acuminate, not narrowed at base, bipinnate or tripinnatiparted, sparsely scaly while young, glabrescent; pinnae 7–14 cm. long, 3–4.5 cm. wide, acuminate, equilateral, slightly narrowed at base, sessile; pinnules narrowly oblong-ovate, 15–20 mm. long, 5–9 mm. wide, obtuse to subacute, slightly oblique, sessile, pinnately cleft to parted; pinnules elliptic, 2–4 mm. long, few-toothed; sori 7–12 to a pinnule, subcostal; indusia thin, oblong, straight, often athyrioid, about 1 mm. long, erose-ciliate.——Mountains; Hokkaido, Honshu (centr. distr. to Kii Prov.), Shikoku; rather rare.

32. Athyrium otophorum (Miq.) Koidz. *Asplenium otophorum* Miq.; *Diplazium otophorum* (Miq.) C. Chr.; *Athyrium rigescens* Makino——TANI-INU-WARABI. Rhizomes erect, short, covered with the basal stubs of old fronds; fronds tufted; stipes 20–30 cm. long, straw-colored to pale brown, usually slightly purplish, densely scaly at base; scales broadly linear, 8–10 mm. long, about 1 mm. wide, nearly black, slightly lustrous, entire; blades firmly herbaceous, broadly ovate to oblong-ovate, 25–45 cm. long, 15–30 cm. wide, abruptly caudate-acuminate, not narrowed at base, glabrous, bipinnate; pinnae 8–10 pairs, lanceolate, 2–4 cm. wide, long-acuminate, spreading, sessile or nearly so; pinnules rather numerous, the basal one auriculate on anterior side, narrowly ovate, 1.2–2.5 cm. long, 5–12 mm. wide, acute to obtuse, broadly cuneate at base, sessile, toothed to lobulate; sori 8–14 to a pinnule, subcostal, parallel, ascending; indusia broadly lanceolate, subentire, often brownish, straight to gently curved, 1.5–4 mm. long, the lowest anterior one often athyrioid.——Honshu (Tōkaidō to Kinki Distr. and westw.), Shikoku, Kyushu.——China.

Athyrium yakusimense Tagawa, of Kyushu (Yakushima), with paler scales, broader blades, and scarcely developed fleshy bristles on the rachis of pinnae above at the insertion of pinnules, may be a form of *A. otophorum*.

33. Athyrium subrigescens (Hayata) Hayata ex H. Itō. *Diplazium subrigescens* Hayata; *A. elegans* Tagawa; *A. elegans* var. *purpurascens* Tagawa——OTOME-INU-WARABI, HŌRAI-INU-WARABI. Rhizomes short, erect; fronds deep green, glabrous, tufted; stipes straw-colored, 15–30 cm. long, densely scaly toward the base; scales at base of stipes broadly linear, 8–12 mm. long, 1–2 mm. wide, black- to chestnut-brown, usually with a paler margin, entire, slightly lustrous; scales on upper part of stipe very sparse, membranous, early deciduous, linear, 3–5 mm. long, pale brown; blades oblong and narrowly long-acuminate, or ovate-oblong, 20–45 cm. long, 12–20 cm. wide, not or scarcely narrowed at base, bipinnate; pinnae 10–15 pairs, equilateral, broadly linear to linear-lanceolate, 7–15 cm. long, 15–25 mm. wide, gradually acuminate, falcate, sessile or nearly so; pinnules ovate, 8–12 mm. long, rounded to obtuse, obliquely cuneate at base, subsessile, toothed to lobulate; sori 3–10 to a pinnule, costal, ascending; indusia broadly lanceolate, 1–2 mm. long, straight, rarely diplazioid. ——Mountains; Kyushu (Yakushima); rare.——Formosa.

34. Athyrium niponicum (Mett.) Hance. *Asplenium niponicum* Mett.; *Asplenium uropteron* Miq.; *Athyrium uropteron* (Miq.) C. Chr.; *A. matsumurae* Christ——INU-WARABI. Plant glabrous; rhizomes slender, creeping, rather firm, about 3 mm. across, scaly; stipes 20–40 cm. long, straw-colored, very sparingly scaly; scales pale brown, membranous, broadly linear, 4–6 mm. long, about 1 mm. wide, deciduous; blades herbaceous, ovate to narrowly so, 25–35 cm. long, 12–25 cm. wide, acuminate-caudate, usually bipinnate; pinnae 6–10 pairs, alternate except the lowest 2, broadly lanceolate, 3–6 cm. wide, long-acuminate, somewhat oblique, petiolulate; pinnules broadly lanceolate to ovate, 1–3 cm. long, 3–12 mm. wide, obtuse to acuminate, pinnately parted to lobulate, acutely toothed, the lower ones often very short-petiolulate, the upper pinnules decurrent on the rachis; sori on the ultimate pinnules, linear, closely parallel; indusia lunate-lanceolate, 1–1.5 mm. long, frequently hamate (athyrioid), irregularly dentate.——Shaded places in lowlands; Hokkaido (sw. distr.), Honshu, Shikoku, Kyushu; common.——Korea, Manchuria, China, and Formosa.

35. Athyrium vidalii (Fr. & Sav.) Nakai. *Asplenium vidalii* Fr. & Sav.; *Athyrium vidalii* var. *confusum* Miyabe & Kudō; *A. commixtum* Koidz.——YAMA-INU-WARABI. Rhizomes short, erect or ascending, densely covered with the basal stubs of old stipes; stipes 20–50 cm. long, straw-colored, densely scaly near base; basal scales broadly linear, 8–12 mm. long, about 1 mm. wide, entire, brown; blades ovate to deltoid-ovate, 25–50 cm. long, 20–40 cm. wide, acuminate, not or scarcely narrowed at base, bipinnate; pinnae broadly lanceolate, 3–5 cm. wide, long-acuminate, equilateral, short-petiolulate or sometimes subsessile; pinnules sessile or very short-petiolulate, oblique, deltoid-ovate to -lanceolate, 15–30 mm. long, 6–13 mm. wide, acute to acuminate, toothed, cuneate to subtruncate at base, pinnately lobed to parted, the ultimate pinnules elliptic to oblong, rounded; sori subcostal, 8–15 to a pinnule, lanceolate to linear; indusia lanceolate, 1–2 mm. long, ascending, subentire, straight or hamate.——Woods in mountains; Hokkaido, Honshu, Shikoku, Kyushu; rather common and variable.——Korea and Formosa.

Var. **yamadae** (Miyabe & Kudō) Miyabe & Tatew. EZO-INU-WARABI. Pinnules pinnately parted; indusia strongly erose-toothed.——Hokkaido.

Athyrium arisanense (Hayata) Tagawa. *Diplazium arisa-nense* Hayata. A Formosan species with smaller fronds, obtuse pinnules and median sori.——Reported from Kyushu (Yakushima).

36. Athyrium wardii (Hook.) Makino. *Asplenium wardii* Hook.; *Athyrium tsusimense* Koidz.——HIROHA-NO-INU-WARABI. Rhizomes short, erect, densely covered with the basal stubs of old stipes; stipes 10–20 cm. long, straw-colored, densely scaly near base; scales broadly linear, 6–10 mm. long, about 1 mm. wide, long-acuminate, with a dark chestnut-brown longitudinal band down the center; blades firmly herbaceous, bipinnate, narrowly to broadly ovate, 20–35 cm. long, 20–30 cm. wide, abruptly long-acuminate with the tip more than one-third as long as the blade; pinnae usually 4–6 pairs, broadly lanceolate, 2.5–4 cm. wide, long-acuminate, the lower ones short-petiolulate, not contracted at base except the lowest one; pinnules elliptic to narrowly ovate, 1.5–3.5 cm. long, 6–10 mm. wide, oblique, the lower ones broadly cuneate at base, sessile, the upper ones more or less adnate and slightly decurrent on the rachis, toothed to pinnately lobulate; sori 6–14 on each pinnule, nearer the costules than the margin; indusia linear to lanceolate, 2–3.5 mm. long, straight, subentire, the lowest anterior one rarely diplazioid.——Woods in mountains; Hokkaido, Honshu, Shikoku, Kyushu.——Korea and China.

Var. **majus** Makino. *A. majus* (Makino) Makino; *A. clivicolum* Tagawa; *A. wardii* var. *clivicolum* (Tagawa) Kurata——Ō-HIROHA-NO-INU-WARABI. Fronds larger; stipes to 50 cm. long; scales 10–15 mm. long, 1–1.5 mm. wide; blades 30–60 cm. long, rather abruptly narrowed at the tip but not long-tailed; pinnae 8–10 pairs; pinnules obtuse to subacute, cuneate or narrowly so at base, usually pinnately lobed on the anterior side.——Honshu (Kantō Distr. and westw.), Shikoku, Kyushu.

37. Athyrium mesosorum (Makino) Makino. *Asplenium mesosorum* Makino; *Diplazium mesosorum* (Makino) Koidz.——NURI-WARABI. Rhizomes creeping, to 15 cm. long, densely covered with the persistent stubs of old stipes, densely scaly; stipes 25–40 cm. long, brown, scaly at base; scales thinly membranous, pale brown, lanceolate to broadly so, 1.5–2 mm. wide, filiform-acuminate, entire or sparsely fimbriate toward the tip; blades herbaceous, deltoid to ovate-deltoid, 30–60 cm. long and nearly as wide, short-acuminate, glabrous, subtripinnate, the rachis fulvous, lustrous; pinnae nearly opposite or alternate, the upper ones narrowly oblong-ovate, 10–15 cm. wide, obliquely ascending, gradually acuminate, not or slightly contracted at base, distinctly petiolulate, equilateral except the lowest one, the lowest pinnae largest; pinnules narrowly deltoid-ovate, to 10 cm. long, acuminate to acute, petiolulate; ultimate pinnules oblong to narrowly ovate, 5–25 mm. long, obtuse to rounded, short-petiolulate, oblique, toothed, often lobed to parted, often slightly decurrent on the rachis; sori few on the ultimate pinnules, nearly parallel to the costule; indusia lanceolate, 2.5–3 mm. long, entire, nearly straight, solitary, whitish.——Mountains; Honshu (Uzen Prov. and southw.), Shikoku, Kyushu.——s. Korea.

38. Athyrium crenatum (Summerf.) Rupr. var. **glabrum** Tagawa. *Diplazium sibiricum* var. *glabrum* (Tagawa) Kurata.——MIYAMA-SHIDA. Rhizomes long-creeping, 2–3 mm. across; stipes 20–30 cm. long, prominently scaly especially toward the base; scales uniformly black-brown, lustrous, broadly lanceolate, 7–10 mm. long, 1–3 mm. wide, sparsely spinulose toward the tip; blades herbaceous, glabrous, deltoid, 15–30 cm. long and as wide or slightly wider than long, acuminate, bipinnatiparted, the rachis sparsely scaly; pinnae 4 to 5 pairs, the lowest ones largest, opposite, spreading, 10–20 cm. long, 5–10 cm. wide, distinctly petiolulate, slightly broader on the posterior side than on the anterior, acuminate, the uppermost pinnae smaller; pinnules equilateral, ovate to broadly lanceolate, 1.5–3.5(–5) cm. long, 7–15 mm. wide, acute to obtuse, sessile or very short-petiolulate, regularly pinnatiparted, the ultimate pinnules oblong, 5–8 mm. long, 4–5 mm. wide, crenately toothed, decurrent to the costule; indusia broadly linear to lanceolate, 1–3 mm. long, solitary, rarely forked above or diplazioid, erose-dentate.——Woods in mountains; Hokkaido, Honshu (centr. and n. distr.).——Korea. The typical phase occurs from Europe to Siberia, China, Manchuria, Korea, and Sakhalin.

39. Athyrium squamigerum (Mett.) Ohwi. *Asplenium squamigerum* Mett.; *Diplazium squamigerum* (Mett.) Matsumura——KIYOTAKI-SHIDA. Rhizomes short, ascending, densely covered with the persistent stubs of old stipes; stipes 15–35 cm. long, straw-colored to pale green, often slightly brownish and prominently scaly especially toward the base; scales rather firm, lustrous, black-brown, broadly linear to lanceolate, 8–12 mm. long, 1–1.5 mm. wide, loosely spinulose toward the tip; blades thinly herbaceous, deltoid, 30–50 cm. long, and as wide, bipinnate, the rachis usually scaly; pinnae 5–7 pairs, broadly lanceolate, equilateral, 20–30 cm. long, acuminate, not or scarcely contracted at base, petiolulate in the lower ones; pinnules equilateral, narrowly deltoid-ovate to oblong, 2–4 cm. long, 1–1.5 cm. wide, rounded to obtuse, the larger ones acute to acuminate, short-petiolulate to sessile, pinnatilobed to shallowly parted; ultimate pinnules rounded to elliptic, rarely oblong, 4–6 mm. wide, rounded, obsoletely crenate; sori costal and subcostal, 1 to few on a pinnule, rather unequal in length, ascending; indusia linear to broadly so, (2–)3–4(–6) mm. long, subentire, solitary, nearly straight to slightly curved, rarely diplazioid or forked.——Woods in mountains; Honshu, Shikoku, Kyushu.——Formosa, China, and n. India.

40. Athyrium procerum Milde. *Asplenium procerum* Wall. ex C. B. Clarke; *Asplenium umbrosum* var. *procerum* (Milde) Hook. & Bak.; *Diplazium umbrosum* var. *procerum* (Milde) Bedd.; *D. kawakamii* Hayata; *Athyrium kawakamii* (Hayata) C. Chr.; *A. allanticarpum* Rosenst.——AO-IGA-WARABI. Rhizomes creeping; stipes rather stout, 40–60 cm. long, scabrous, brown; scales black-brown, dense, 3–4 mm. long, deflexed; blades herbaceous, broadly ovate, 40–80 cm. long, 30–60 cm. wide, acuminate, 3-pinnate to 3-pinnatiparted; pinnae 7–10 pairs, equilateral, the upper ones gradually reduced, alternate, narrowly oblong-ovate, the lower ones 40–45 cm. long, acuminate, not contracted at base, distinctly petiolulate; pinnules 4–10 cm. long, 1.5–2.5 cm. wide; ultimate pinnules oblique, oblong-ovate to oblong, 6–20 mm. long, 4–8 mm. wide, acute r rounded, mucronate, entire or toothed to pinnatilobed, sessile, often slightly decurrent on the rachis; sori costal, 5–12 on each pinnule, ascending, linear; indusia oblong, solitary, straight, about 1.5 mm. long.——Kyushu (Yakushima); rare.——Formosa to India.

Var. **subglabratum** (Tagawa) Tagawa. *Diplazium kawakamii* var. *subglabratum* Tagawa——USUGE-AO-IGA-WARABI. Stipes and rachis sparsely scaly.——Kyushu (Yakushima).

41. Athyrium esculentum (Retz.) Copel. *Hemionitis esculenta* Retz.; *Diplazium esculentum* (Retz.) Sw.; *Asplenium esculentum* (Retz.) Presl; *Anisogonium esculentum* (Retz.) Presl; *D. malabaricum* Spreng.——KUWARE-SHIDA. Rhizomes creeping; stipes stout, pale brown, 25–40 cm. long, scaly at base; scales membranous, linear-lanceolate, 6–10 mm. long, about 1 mm. wide, spinulose; blades broadly ovate, 40–100 cm. long, 30–70 cm. wide, bipinnate; pinnae 10–15 pairs, chartaceous, spreading, brownish, the larger ones oblong-lanceolate, 20–40 cm. long, 8–15 cm. wide, not contracted at base, the upper pinnae simple, linear-lanceolate, pinnately lobulate, the terminal pinna narrowly deltoid, acuminate, pinnatilobed; pinnules lanceolate, 5–10 cm. long, 1–1.8 mm. wide, acuminate, truncate at base, obsoletely toothed, lobulate, sessile to very short-petiolulate; veins 1 to each lobule, pinnately branched, the lower veinlets fused at apex to those of the adjacent vein; sori linear, on the lower part of the veinlets, ascending, 2–4 mm. long, straight or slightly curved; indusia early withering, very thin, solitary or diplazioid on some of the lowest veinlets.——Kyushu (s. distr.).——Ryukyus, Formosa, s. China, Indochina, Polynesia to Malaysia, India and Ceylon.

42. Athyrium naganumanum (Makino) Ohwi. *Asplenium chinense* Bak.; *Diplazium chinense* (Bak.) C. Chr.; *D. naganumanum* Makino; *D. kodamae* Nakai——HIKAGE-WARABI. Rhizomes short-creeping; stipes 30–40 cm. long, rather stout, straw-colored or pale green to pale brown, sparsely scaly on upper part, more densely so toward base; scales lanceolate, 5–7 mm. long, 1–1.5(–2) mm. wide, black-brown, lustrous, subentire; blades membranous, deltoid, 40–60 cm. long and nearly as wide, acuminate, glabrous, 3-pinnate to 3-pinnatisect; pinnae 5–7 pairs, the lower usually opposite, spreading, acuminate, petiolulate, the upper ones rather abruptly smaller, the lowest pinnae largest, slightly oblique, narrowly ovate, 20–35 cm. long, 10–15 cm. wide, usually contracted at the base, the others narrower, smaller, equilateral, not or scarcely contracted at base; pinnules broadly lanceolate, 3–7 cm. long, 12–25 mm. wide, acuminate, short-petiolulate; ultimate pinnules broadly lanceolate to narrowly oblong, 6–8 mm. long, 2–3(–4) mm. wide, obtuse, toothed, sometimes lobulate, often slightly decurrent on the rachis; sori 4–10 on the ultimate pinnules, ascending, linear, in 2 rows nearer the costule than the margin; indusia 1–2 mm. long, linear, straight, the lower ones frequently diplazioid.——Honshu (Kantō Distr. and Wakasa Prov. westw.), Shikoku, Kyushu.——Korea (Quelpaert Isl.), China, and Indochina.

43. Athyrium bittyuense (Tagawa) Ohwi. *Diplazium bittyuense* Tagawa——BITCHŪ-HIKAGE-WARABI. Rhizomes creeping; stipes 35–40 cm. long, sparsely scaly and blackish at base; scales membranous, lanceolate, 5–7 mm. long, 1–1.5 mm. wide, entire, fuscous; blades membranous, deltoid, 55–65 cm. long, 40–45 cm. wide, acuminate, glabrous, 3-pinnatiparted; pinnae 7–10 pairs, equilateral, obliquely spreading, broadly lanceolate, acuminate, to 40 cm. long; pinnules 4–6 cm. long, 15–20 mm. wide, long caudate-acuminate, pinnatiparted; pinnules elliptic to oblong, 3.5–5 mm. wide, obtuse to acute, rather remote, toothed; sori narrowly oblong, about 2 mm. long, subcostular; indusia thin, entire, brownish.——Honshu (Bitchu and Totomi Prov.).

44. Athyrium nipponicola Ohwi. *Diplazium nipponicum* Tagawa——ONI-HIKAGE-WARABI. Rhizomes short-creeping, stout; stipes 40–70 cm. long, brown, stout, scarcely lustrous; scales ovate-lanceolate, 1–1.3 cm. long, 1.5–3 mm. wide, fugacious, dark brown, spinulose; blades glabrous, deltoid-ovate, 50–80 cm. long and nearly as wide, acuminate, 3-pinnati-parted in lower portion, bipinnate in upper portion, the rachis smooth; pinnae 7–10 pairs, spreading, equilateral, 30–40 cm. long, the lowest one 15–18 cm. wide, the middle ones 7–12 cm. wide, petiolulate; pinnules equilateral, oblong- to deltoid-lanceolate, 5–10 cm. long, 12–20 mm. wide, truncate at base, acuminate, crenate-toothed, pinnately lobed or more often -parted, petiolulate or sessile; pinnules ovate-oblong to ovate, 4–6 mm. wide, very obtuse to acutish; sori 1–12 on each pinnule, slightly nearer the costule than the margin, obliquely ascending, linear, unequal; indusia linear, 2–3(–4) mm. long, straight, single, frequently diplazioid and longer on the lowest anterior veinlets, subentire.——Honshu, Shikoku, Kyushu.

45. Athyrium doederleinii (Luerss.) Ohwi. *Asplenium doederleinii* Luerss.; *Diplazium doederleinii* (Luerss.) Makino; *D. aridum* Christ; *Athyrium nudicaule* Copel.——Shi-MA-Shiroyama-shida. Rhizomes long-creeping, stout, with the short persistent basal stubs of old stipes; stipes 40–50 cm. long, pale green, dark brown toward base, scarcely scaly; blades herbaceous, deltoid-ovate, 50–60 cm. long and as wide, bipinnate; pinnae 7–8 pairs, equilateral, alternate, ovate-oblong, 10–15 cm. wide, acuminate, petiolulate; pinnules nearly horizontally spreading, narrowly ovate-deltoid to broadly deltoid-lanceolate, 3.5–7 cm. long, 1.8–3 cm. wide, acuminate, sessile or short-petiolulate, truncate at base, pinnately cleft to shallowly parted; pinnules elliptic to obliquely ovate, rounded or with a short acutish tip; veinlets sometimes forked; sori costular, on basal portion of the veinlets; indusia unequal in length, 3.5–4.5 mm. long, sometimes diplazioid on the lowest veinlets, solitary, 1.5–3.5 mm. long in those of the other veinlets.——Honshu (Ise Prov. and westw.), Shikoku, Kyushu.——Ryukyus, Formosa, s. China, Indochina.

46. Athyrium hachijoense (Nakai) Ohwi. *Diplazium hachijoense* Nakai; *D. siroyamense* Tagawa——Shiroyama-SHIDA. Rhizomes creeping, stout, blackish; stipes 45–60 cm. long, stout, sparsely scaly and about 1 cm. across at base; scales fugacious, fuscous, membranous, lanceolate to linear-lanceolate, about 5 mm. long, 1–2 mm. wide, entire; blades deltoid-ovate, 40–80 cm. long, 30–60 cm. wide, bipinnate, the rachis greenish, glabrous or very thinly puberulous; pinnae 8–13 pairs, spreading, oblong-lanceolate, the lowest 18–40 cm. long, 8–13 cm. wide, acuminate, petiolulate; pinnules deltoid-lanceolate to linear-lanceolate, 4–7 cm. long, 1.3–2 cm. wide, acuminate, truncate at base, short-petiolulate or sessile, pinnately cleft or sometimes parted; ultimate pinnules elliptic, obsoletely toothed; veinlets simple, or lower ones forked; sori median, linear, 2–3 mm. long; indusia thin, erose, solitary, frequently diplazioid in the lowermost lobe.——Woods in low mountains; Honshu (Izu, Etchu Prov. and westw.), Shikoku, Kyushu.——Korea (Quelpaert Isl.), and Ryukyus.

47. Athyrium virescens (Kunze) Ohwi. *Diplazium virescens* Kunze; *Asplenium virescens* (Kunze) Mett.; *Asplenium wheeleri* Bak.; *D. wheeleri* (Bak.) Diels; *D. lutchuense* Koidz.; *D. allantodioides* Ching——Kokumō-kujaku. Rhizomes creeping, rather stout, black-scaly toward the tip; stipes 40–60 cm. long, straw-colored, prominently scaly and blackish toward the base; scales broadly linear, 5–10 mm. long, 1–1.5(–2) mm. wide, nearly black, lustrous, prominently spinulose; blades chartaceous to thinly coriaceous, broadly deltoid-ovate, long-acuminate, glabrous, bipinnate in the lower

portion; pinnae 5–8 pairs, alternate, the lower ones spreading, equilateral, broadly lanceolate to narrowly deltoid-ovate, 17–30 cm. long, 6–15 cm. wide, gradually acuminate, distinctly petiolulate, the lowest ones largest, the petiolules 2–5 cm. long; upper pinnae simple, lanceolate-deltoid, lobulate to parted, the terminal pinna decurrent on the rachis, pinnatilobed; pinnules of the lower pinnae horizontally spreading, lanceolate-deltoid, 3–7 cm. long, 1–1.5 cm. wide, acuminate to acute, sometimes obtuse, truncate at the base, crenately toothed, lobulate to lobed, slightly recurved on margin; veins pinnately branched, solitary in a lobule or lobe; sori linear; indusia thin, narrow, linear, 1–3 mm. long.——Honshu (Izu and Kii Prov. westw.), Shikoku, Kyushu.——Ryukyus, Formosa, s. China, and Indochina.

Athyrium okinawaense (Tagawa) Ohwi. *Diplazium okinawaense* Tagawa——Okinawa-kujaku. Stipes densely black-scaly at base; pinnae alternate, ovate-lanceolate, long-petiolulate; pinnules ovate-lanceolate, acuminate, 3-angled, pinnately parted, the ultimate pinnules oblong, obtuse to rounded; sori linear, about 2 mm. long, subcostal.——Reported to occur in Honshu (Kii Prov.), Shikoku, Kyushu (Tanegashima).——Ryukyus.

48. Athyrium maximum (G. Don) Copel. *Asplenium maximum* G. Don; *Diplazium dilatatum* Bl.; *D. latifolium* Moore; *Athyrium latifolium* (Moore) Milde; *D. uraiense* Rosenst.; *D. latifolium* var. *cyclolobum* Christ; *D. crinipes* Ching; *D. uraiense* var. *cyclolobum* (Christ) Tagawa——Hiroha-nokogiri-shida. Rhizomes short, ascending, stout; stipes 30–60 cm. long, stout, rather densely scaly toward base, green; scales lanceolate, 8–13 mm. long, about 1 mm. wide, linear to filiform at the tip, dark brown, remotely spinulose; blades chartaceous, large, deltoid, 50–80 cm. long, narrower or nearly as wide, acuminate, bipinnate in the lower portion, abruptly pinnate in the upper part, dark green or brownish when dried, paler beneath; pinnae 7–10 pairs, alternate, equilateral, the lower ones broadly deltoid-lanceolate, 30–40 cm. long, 10–20 cm. wide, gradually acuminate, pinnate, petiolulate, the upper pinnae deltoid-lanceolate, 10–20 cm. long, 2–4 cm. wide, gradually long-acuminate, truncate at base, short-petiolulate, doubly toothed, lobulate, the terminal pinna decurrent on the rachis; sori few to each lobe, linear, subcostal; indusia 3–5(–7) mm. long, the lowest anterior ones longest and often diplazioid.——Honshu (Hachijo Isl., and Kii Prov.), Kyushu (s. distr.).——Ryukyus, Formosa, China, Australia to Polynesia.

Athyrium yakumontanum (Tagawa) Ohwi. *Diplazium yakumontanum* Tagawa——Yakushima-warabi. Differs from *A. maximum* in the narrower, linear-lanceolate pinnules about 1 cm. wide near base.——Honshu (Suwo Prov.), Kyushu (Yakushima). Poorly understood.

49. Athyrium petri (Tard.-Bl.) Ohwi. *Diplazium petri* Tard.-Bl.; *D. triangulare* Tagawa——Hiroha-miyama-noko-giri-shida. Rhizomes creeping, firm; stipes light brown, 40–60 cm. long, rather prominently scaly, dark brown near base; scales linear to broadly so, 5–10 mm. long, about 1–1.5 mm. wide, spinulose; blades chartaceous, broadly deltoid-lanceolate to narrowly ovate-deltoid, 40–50 cm. long, 20–25 cm. wide, acuminate, not narrowed at base, bipinnatiparted (in ours); pinnae 12–14 pairs, spreading, acuminate, equilateral, 10–20 cm. long, 2.5–3 cm. wide, petiolulate, truncate to slightly cordate at base, pinnately parted to deeply lobed, the upper ones smaller and decurrent on the rachis; pinnules oblong to narrowly ovate-oblong, 5–8 mm. wide, rounded to obtuse or acute, crenate-toothed; sori 8–14, median or slightly costal, linear, obliquely ascending, straight; indusia 2–4 mm. long, nearly equal in length, narrow.——Kyushu (Yakushima); rare.——Ryukyus, Formosa, and Indochina.

50. Athyrium mettenianum (Miq.) Ohwi. *Asplenium mettenianum* Miq.; *Diplazium mettenianum* (Miq.) C. Chr.; *Asplenium textori* Miq.; *D. textori* (Miq.) Makino——MI-YAMA-NOKOGIRI-SHIDA. Rhizomes long-creeping, firm, 3–4 mm. across, black-brown, scaly; stipes 30–40 cm. long, pale green to pale brown, thinly scaly and dark brown toward base; scales firmly membranous, the upper ones brown, the basal nearly black, linear, 3–4 mm. long, about 0.5 mm. wide, sparsely spine-toothed toward tip; blades narrowly ovate to broadly lanceolate, 20–40 cm. long, 8–20(–30) cm. wide, long-acuminate, not contracted at base; pinnae 10–13 pairs, horizontally spreading, equilateral except sometimes the lowest one, lanceolate, 4–10(–15) cm. long, 1–3(–3.5) cm. wide, short-petiolulate, pinnatilobed, the upper pinnae smaller and decurrent on the rachis; sori linear, 3–7 mm. long, slightly arcuate, median, or those on the lowest veinlets often diplazioid and longer.——Honshu (Izu Prov., Tōkaidō, Kinki Distr. and westw.), Shikoku, Kyushu.——Ryukyus, Formosa, China, and Philippines.

Var. **fauriei** (Christ) Ohwi. *Diplazium fauriei* Christ; *D. mettenianum* var. *fauriei* (Christ) Tagawa——HOSOBA-NOKO-GIRI-SHIDA. Pinnae irregularly toothed. Occurs with the species.

Var. **isobasis** (Christ) Ohwi. *Diplazium isobasis* Christ;

D. mettenianum var. *isobasis* (Christ) Tagawa——KIREBA-NOKOGIRI-SHIDA. Pinnae deeply parted. Occurs with the species.

51. Athyrium sheareri (Bak.) Ching. *Nephrodium sheareri* Bak.; *Dryopteris sheareri* (Bak.) C. Chr.; *N. polypodiforme* Makino; *D. polypodiformis* (Makino) C. Chr.; *A. polypodiforme* (Makino) Tagawa——URABOSHI-NOKOGIRI-SHIDA. Rhizomes creeping, 3–4 mm. across, densely scaly toward the tip; stipes slender, 20–40 cm. long, glabrous, straw-colored or pale green, thinly scaly and dark brown toward base; scales brown, broadly linear, 3–5 mm. long, about 0.5 mm. wide, long-acuminate, entire; blades chartaceous, simply pinnate, broadly lanceolate to narrowly ovate, 20–30 cm. long, 10–20 cm. wide, caudately long acuminate, glabrous, the rachis and costas usually thinly white-puberulent on both sides; pinnae 5–7 pairs, alternate or subopposite, nearly equilateral, spreading, lanceolate to broadly so, caudately long-acuminate, falcate, truncate at base, sessile, the lower pinnae pinnately lobed to parted, 15–25(–30) mm. wide, short-petiolulate, the upper pinnae decurrent on the rachis, forming a pinnatilobed tail at the tip; pinnules oblong-ovate to lanceolate, 4–6 mm. wide, obtuse to acute, spine-toothed; veinlets simple or forked, ending in a tooth; sori 6–14, median or slightly nearer the costule than the margin; indusia short, obovate, straight, or more often reniform, 0.5–0.7 mm. across, dentate.——Honshu (Izu and Echigo Prov. westw.), Shikoku, Kyushu.——Korea (Quelpaert Isl.) and China.

19. DIPLAZIOPSIS C. Chr.   IWAYA-SHIDA ZOKU

Terrestrial; rhizomes ascending, brown-scaly; fronds membranous, simply pinnate; pinnae large, all alike, entire or undulate, glabrous; veins remote, free at base, then divaricately branched and anastomosing; sori on the acroscopic side of the veins, extending from the costa to the lowest areoles; indusia very thin, opening usually along the back of sorus, sometimes along the distal side.——Few species, India to Japan, and Samoa.

1. Diplaziopsis cavaleriana (Christ) C. Chr. *Allantodia cavaleriana* Christ; *Diplaziopsis javanica* var. *cavaleriana* (Christ) Tagawa; *Diplazium javanicum* sensu auct. Japon., non Makino; *Diplaziopsis javanica* sensu auct. Japon., non C. Chr.——IWAYA-SHIDA. Rhizomes creeping; stipes 30–50 cm. long, pale green to straw-colored, smooth in upper part, sparsely scaly toward the base; scales membranous, lanceolate, 5–8 mm. long, 1.5–3 mm. wide, acuminate, entire; blades broadly lanceolate to narrowly ovate-oblong, 50–70 cm. long, 17–30 cm. wide, sometimes slightly narrowed at base, simply pinnate, glabrous, green above, slightly glaucous beneath; pinnae 8–13 pairs, thinly membranous, spreading, equilateral,

oblong- to linear-lanceolate, 10–15 cm. long, 15–20 mm. wide, or the sterile ones to 3 cm. wide, gradually acuminate-tailed, truncate to very broadly cuneate at base, sessile, sometimes obsoletely auriculate at base, undulate to crenate; veins spreading, slender, forked near base and forming 4 or 5 rows of oblong to narrowly ovate somewhat 4-angled areolae on each side of the pinnae; sori linear-oblong to -lanceolate, on the lower half of the veins; indusia thinly membranous, linear-lanceolate, obtuse at both ends, nearly straight to very slightly arcuate.——Dark, damp woods; Honshu (Ugo Prov., s. Kantō Distr. and westw.), Shikoku, Kyushu.——China.

Fam. 18. BLECHNACEAE   SHISHIGASHIRA KA   Blechnum Family

Terrestrial, rarely scandent; rhizomes creeping or erect, dictyostelic, with nonclathrate scales; fronds usually pinnatifid to pinnate or decompound, rarely simple; veinlets branched and anastomosing, forming a secondary vein or veins enclosing a row of areoles on each side of the costa, or free in the sterile pinnae; sori on the secondary veins, separate or united; indusia opening on the costal side, very rarely absent; sporangia rather large.——Cosmopolitan, with about 8 genera and about 250 species.

1A. Fronds usually dimorphic, with sterile and fertile fronds (except *B. orientale*); veins free in the sterile pinnae; sori fused. 1. *Blechnum*
1B. Fronds monomorphic or nearly so; veins anastomosing; sori separate, not fused. . . . . 2. *Woodwardia*

1. BLECHNUM L.   SHISHIGASHIRA ZOKU

Terrestrial; rhizomes usually stout, erect or sometimes creeping, dictyostelic; fronds usually dimorphic, coriaceous, pinnate or sometimes pinnatifid, rarely simple or bipinnate, glabrous, entire or toothed; veins of sterile pinnae mostly free; sori on the vascular commissures parallel to the costa, fused in a single uninterrupted line on each side; indusia attached to the fertile commissure and opening on the side toward the costa.——About 200 species, abundant in the S. Hemisphere.

1A. Blades monomorphic, about 1 m. long; veins simple to once forked, closely parallel. .... 1. *B. orientale*
1B. Blades dimorphic, not over 40 cm. long; veins nearly always once forked.
  2A. Fertile fronds barely taller than the sterile; stipes and rachis pale green to pale brown; sterile pinnae without a raised costa beneath; veins ascending.
    3A. Basal scales ovate, membranous; sterile blades 2–4 cm. wide. .... 2. *B. amabile*
    3B. Basal scales linear, chartaceous; sterile blades 5–8(–10) cm. wide. .... 3. *B. niponicum*
  2B. Fertile fronds much taller than the sterile, the pinnae abruptly dilated and decurrent on the rachis; stipes and underside of rachis chestnut-brown; sterile pinnae with slightly raised costa beneath; veins obliquely spreading. .... 4. *B. castaneum*

1. **Blechnum orientale** L. *Asplenium orientale* (L.) Bernh.; *Blechnopsis orientalis* (L.) Presl——HIRYŪ-SHIDA. Evergreen, uniformly clothed with rusty-brown crispate woollike scales while very young; rhizomes stout, erect, densely scaly; stipes 30–50 cm. long inclusive of the lower part of rachis, deeply sulcate on upper side, stout, reddish brown, scaly at base, with several remote pairs of much-reduced pinnae 2–3 mm. long; scales at base of stipes and on rhizomes slightly lustrous, linear, 1–1.5 cm. long, about 1 mm. wide, long-attenuate, minutely toothed at tip; blades monomorphic, chartaceous-coriaceous, linear-lanceolate to lanceolate, about 1 m. long, 30–40 cm. wide, simply pinnate, glabrate; pinnules obliquely spreading, linear to broadly so, 15–25 cm. long, 1–1.5 cm. wide, gradually long caudate-acuminate, entire, glabrous, sessile, rounded at base on the anterior side, subcordate to rounded at base and decurrent on the rachis on posterior side, the costa raised beneath, sulcate and raised on upper side; veins simple or once forked near base, parallel; sori in longitudinal rows on the pinnae close to the costas; indusia entire, continuous, about 0.7 mm. wide.——Kyushu (Yakushima); rare.——Ryukyus, Formosa, China, Himalayas, Australia, and Polynesia.

2. **Blechnum amabile** Makino. *B. crenulatum* Makino; *Spicantopsis amabilis* (Makino) Nakai; *Lomaria amabilis* (Makino) Ching; *Struthiopteris amabilis* (Makino) Ching——OSA-SHIDA. Rhizomes creeping; fronds dimorphic, evergreen, glabrous; stipes of the sterile fronds 2–10 cm. long, the fertile 5–12 cm. long, straw-colored, often somewhat reddish, sparsely scaly in upper part, densely so toward base, shallowly sulcate on upper side; scales membranous, broadly ovate, 2–7 mm. long, abruptly acuminate, spreading, entire, brown; sterile blades lanceolate, 12–30 cm. long, 2–4(–5) cm. wide, acute, with a short tail 7–20 mm. long at the tip, narrowed at the base, simply pinnate, the rachis pale green to pale brownish; pinnae herbaceous-coriaceous, nearly horizontally spreading, broadly linear, 2.5–5 mm. wide, subacute, decurrent on the rachis, entire, pale green beneath, the lowest pinnae reduced, deltoid to ovate-deltoid, 2–5 mm. long, 4–6 mm. wide, more or less deflexed, obtuse, sterile, interrupted; fertile pinnae 8–15 mm. long, 2–5–3.5 mm. wide, subapiculate.——Woods in low mountains; Honshu, Shikoku, Kyushu.

3. **Blechnum niponicum** (Kunze) Makino. *Lomaria niponica* Kunze; *Lomaria spicanta* var. *japonica* Hook.; *B. japanense* Moore; *Struthiopteris niponica* (Kunze) Nakai; *Spicanta niponica* (Kunze) Hayata; *Spicantopsis niponica* (Kunze) Nakai; *Spicantopsis niponica* var. *minima* Tagawa——SHISHIGASHIRA. Rhizomes short, rather stout, erect, densely scaly; fronds dimorphic, evergreen, glabrous; stipes of the sterile fronds short, 1–5 cm. long, scaly, sulcate on upper side, the scales at base chartaceous, linear, 1–1.5 cm. long, about 1 mm. wide, lustrous, dark brown with a narrow pale margin, entire; sterile blades oblanceolate to oblong-oblanceolate, 20–30 cm. long, (3.5–)5–8(–10) cm. wide, pectinate, abruptly contracted at the tip to a linear subacute tail 1–2 cm. long, gradually narrowed at base, with scattered brown membranous linear-lanceolate scales while young; pinnae nearly horizontally spreading, broadly linear, 3–7 mm. wide, obtuse to subobtuse, entire, the lower ones often somewhat deflexed and slightly broader on anterior side, the lowermost ones auriclelike, 3–10 mm. long, as wide or wider at base, the lower fertile pinnae distant, linear, 1.5–3 cm. long, 2–3 mm. wide, obtuse, not or slightly contracted at base, the lowest few pairs sterile, 1–3 mm. long, distant and much-reduced.——Woods in mountains; Hokkaido, Honshu, Shikoku, Kyushu.

4. **Blechnum castaneum** Makino. *Struthiopteris castanea* (Makino) Nakai——MIYAMA-SHISHIGASHIRA. Rhizomes short-creeping to suberect, densely clothed with the stubs of old stipes; fronds dimorphic; stipes of sterile fronds 3–10 cm. long, puncticulate, the scales rather sparse, brown, broadly linear, 3–8 mm. long, entire; blades lanceolate, 15–35 cm. long, 3.5–6 cm. wide, narrowed at both ends, obtuse at the tip, pinnately divided nearly to the rachis, glabrous, the rachis deeply sulcate on upper side, puncticulate on both sides; pinnae spreading, linear-oblong, 4–7 mm. wide, rounded with an obtuse mucro at the tip, subentire, the costa more or less distinct beneath, the lower pinnae semirounded, 3–6 mm. long, 5–8 mm. wide; veins mostly once-forked, scarcely visible; fertile fronds much longer than the sterile, the blades with few nearly obsolete distant sterile pinnae at base; fertile pinnae linear, 15–30 mm. long, 1.5–2 mm. wide, obliquely spreading, spaced out, abruptly dilated and decurrent on the rachis.——Woods in mountains; Honshu (centr. and n. distr.).

2. WOODWARDIA J. E. Smith   KOMOCHI-SHIDA ZOKU

Rather large, terrestrial ferns; rhizomes ascending to erect, short, densely scaly; fronds all alike, mostly bipinnatifid, entire or minutely toothed; veinlets anastomosing, forming costal and costular areolae; sori opposite the costal or costular areolae, often deeply impressed; indusia opening along the costule.——About 12 species, in the warmer parts of Asia and Europe, also in N. America.

1A. Rhizomes rather slender, creeping, with remote stipes; blades usually deltoid in outline; sori linear, straight, unequal, 2–15 mm. long, along the costas and the rachis; lowest pinnae 10–20 cm. long, 1–3 cm. wide. .... 1. *W. harlandii* var. *takeoi*
1B. Rhizomes stout and short, with close stipes; blades ovate to oblong-ovate in outline, bipinnatiparted; sori oblong to short-linear, straight to slightly curved, 2–7 mm. long, along the costule of the pinnules and on upper part of costas of the pinnae.

2A. Stipes usually with prominent brown scales; pinnately lobed to cleft, 2–4.5 cm. wide; sori straight. . . . . 2. *W. japonica*
2B. Stipes soon becoming naked, the scales at the base persistent; pinnae pinnately cleft to parted, 4–10 cm. wide, at least in the lower ones; sori slightly curved outward in upper part.
3A. Blades with a large densely scaly gemma in axils of upper pinnae. . . . . 3. *W. unigemmata*
3B. Blades without gemmae in axils of pinnae, though sometimes with numerous adventitious plantlets on the upper side.

4. *W. orientalis*

1. **Woodwardia harlandii** Hook. var. **takeoi** (Hayata) Masam. *W. kempii* Copel.; *W. takeoi* Hayata——HOSOBA-ŌKAGUMA. Rhizomes slender, creeping, with remote stipes, with narrowly ovate, membranous scales about 3 mm. long; stipes rather slender, 15–35 cm. long, pale brown to straw-colored, naked except for a few scales toward the base; fertile blades usually deltoid in outline, 10–17 cm. long, 8–15 cm. wide, short-acuminate, pinnate, without a distinct terminal pinna, naked on both sides, the rachis broadly winged; pinnae 5–7 pairs, opposite, obliquely spreading, lanceolate, 7–15 mm. wide, broadly adnate at base, acuminate to very acute, minutely toothed, with a rounded to very obtuse lobe 5–7 mm. wide at the base, the lowest pinnae 5–10 cm. long, 2–5 cm. wide, pinnately parted on the posterior side or sometimes on both sides, the pinnules narrowly lanceolate, to 5 cm. long, 4–8 mm. wide; sterile blades similar to the fertile, the stipes usually shorter, the pinnae wider and less prominently parted; sori rusty-brown, linear, straight, unequal, 2–15 mm. long, along the costas and rachis, those along the costules of the pinnules 1–6 mm. long.——Kyushu (Yakushima).——Ryu-kyus, Formosa, and s. China. The typical variety occurs in the Ryukyus, Formosa, s. China to Indochina.

2. **Woodwardia japonica** (L. f.) J. E. Smith. *Blechnum japonicum* L. f.; *W. radicans* var. *japonica* (L. f.) Luerss.; *W. intermedia* Christ——ŌKAGUMA. Rhizomes stout, short; stipes stout, straw-colored or somewhat rubescent, 30–50 cm. long, prominently scaly; scales on upper portion of stipes deep-brown, membranous, deltoid-lanceolate to broadly linear, deflexed, or those on lower portion of stipes ascending, acuminate, 7–15 mm. long, the scales at base of stipes broadly linear, 15–20 mm. long; blades oblong-ovate, 40–80 cm. long, 20–35 cm. wide, somewhat contracted toward the tip, the terminal segment deltoid-ovate to broadly lanceolate, acuminate; pinnae 10–15 pairs, obliquely spreading, lanceolate to linear-lanceolate, 15–22 cm. long, 2–4.5 cm. wide, long-acuminate, minutely toothed, pinnately lobed to cleft, the pinnules oblong-ovate, 7–25 mm. long, 6–12 mm. wide, acute or subobtuse, obliquely spreading, sessile, with a rounded lobe on the posterior side at base, cuneate on the anterior side; sori 3–5 mm. long, along the costules of the pinnules, to 7 mm. long on the upper part of costas of the pinnae.——Honshu (Kii Prov. and Chūgoku Distr.), Shikoku, Kyushu.——Korea (Quelpaert Isl.), Formosa, China, and Indochina.

3. **Woodwardia unigemmata** (Makino) Nakai. *W. radicans* var. *unigemmata* Makino; *W. radicans* sensu auct.

Asiat., non Smith——HAI-KOMOCHI-SHIDA. Rhizomes stout; stipes stout, 30–50 cm. long, glabrate except at base, straw-colored to pale brown, the basal scales membranous, linear to linear-lanceolate, 2–3 cm. long, acuminate, entire; blades broadly ovate-lanceolate, 30–100 cm. long, 20–50 cm. wide, obtuse to acuminate, bipinnaticleft to parted, with large densely scaly gemmae in axils of upper pinnae, the rachis sparsely clothed while very young with small deciduous scales; pinnae equilateral, alternate, thinly coriaceous, ovate-lanceolate or deltoid-lanceolate, 20–30 cm. long, 5–9 cm. wide, caudate-acuminate, abruptly cuneate at base, subsessile, with or without an auricle on posterior side; pinnules obliquely spreading, ovate-deltoid to deltoid-lanceolate, 1.5–3(–5) cm. long, 6–10 mm. wide, acute or sometimes acuminate, minutely spine-toothed; sori along the costules of the pinnules, oblanceolate to linear-oblanceolate, 2–3 mm. long.——Honshu (Izu Prov.), Kyushu (Higo Prov.).——Formosa, Philippines, China, se. Asia, to the Himalayas.

4. **Woodwardia orientalis** Sw. *Blechnum japonicum* Houtt.; *W. radicans* var. *orientalis* (Sw.) Luerss.; *W. radicans* sensu auct. Japon., pro parte, non Smith——KOMOCHI-SHIDA. Rhizomes short-creeping, stout; fronds large; stipes straw-colored, 30–80 cm. long, the basal scales brown, membranous, broadly linear to linear-lanceolate, 3–4 cm. long, 3–5 mm. wide, long-acuminate, entire; blades narrowly ovate to broadly ovate-lanceolate, 30–100 cm. long, 20–35 cm. wide, acuminate, bipinnitiparted, sometimes with numerous adventitious plantlets on the upper side; pinnae obliquely spreading, ovate-lanceolate to deltoid-lanceolate, 12–30 cm. long, 4–10 cm. wide, acuminate, somewhat cuneate at base, sessile or short-petiolulate, pinnately parted, slightly wider on the anterior than on the posterior side; pinnules lanceolate, 7–10 cm. long, acute to acuminate, minutely toothed; sori along the costule of the pinnules linear-oblanceolate to narrowly oblong, 2–4 mm. long.——Honshu, (Etchu, Rikuzen Prov. and southw.), Shikoku, Kyushu.——China.

Var. **formosana** Rosenst. *W. prolifera* Hook. & Arn.; *W. orientalis* var. *prolifera* (Hook. & Arn.) Ching; *W. angustiloba* Hance; *W. exaltata* Nakai; *W. orientalis* sensu auct. Japon., pro parte, non Sw.——HACHIJŌ-KAGUMA. Fronds larger, to about 2 m. long, 50 cm. wide or more; pinnae larger, more deeply pinnatiparted; pinnules broadly linear to linear-lanceolate, to 15 cm. long, about 1 cm. wide, often caudately long-acuminate, minutely toothed.——Honshu (Izu Isls. and Awa Prov.), Shikoku, Kyushu.——Ryukyus and Formosa.

Fam. 19. **ASPLENIACEAE** TORANO-O-SHIDA KA Spleenwort Family

Usually terrestrial, sometimes epiphytic ferns; rhizomes creeping to suberect, dictyostelic, with clathrate scales; stipes not jointed, usually with two fibrovascular bundles united toward the top; blades simple to decompound, usually firm and evergreen; veins forked, free or anastomosing; sori elongate along the veinlets; indusia on the veinlets, rarely absent.——Cosmopolitan, with about 9 genera and about 720 species.

1A. Veins free or united by a marginal vein. . . . . 1. *Asplenium*
1B. Veins reticulate. . . . . 2. *Camptosorus*

1. ASPLENIUM L.   Tora-no-o-shida Zoku

Terrestrial or epiphytic; rhizomes usually short-creeping, with clathrate scales; stipes not jointed; blades simple and entire to decompound, glabrous or minutely scaly; veins usually forked, mostly free; sori lengthwise along the veinlets; indusia on one side of the sorus.——Cosmopolitan, with about 700 species.

1A. Blades compound.
  2A. Stipes, at least in the lower portion, black-brown, lustrous.
    3A. Rhizomes long-creeping; pinnae membranous.
      4A. Pinnae somewhat dimidiate, pinnately veined only on anterior side; sori on the teeth; stipes rather close. .... 1. *A. cheilosorum*
      4B. Pinnae not dimidiate, with veins mostly once forked on both sides except at the posterior base; sori dorsal; stipes remote.
                                                        2. *A. unilaterale*
    3B. Rhizomes short; pinnae usually herbaceous.
      5A. Blades broadly linear to lanceolate, not at all broadened at base, simply pinnate (often 2- or 3-pinnate in No. 8).
        6A. Pinnae very oblique, with an auricle on the anterior side at base.
          7A. Stipes slender, 0.6–1 mm. across; pinnae with few irregular coarse obtuse teeth; indusia 1–2 mm. long.
                                                       3. *A. oligophlebium*
          7B. Stipes 1–2 mm. across; pinnae crenately toothed; indusia 1–4 mm. long. .... 4. *A. normale*
        6B. Pinnae nearly equilateral.
          8A. Rachis dark brown throughout.
            9A. Rachis obtuse on back, without gemmae. .... 5. *A. trichomanes*
            9B. Rachis with 3 brown narrow wings, sometimes gemmate in upper part. .... 6. *A. tripteropus*
          8B. Rachis brown only on lower part.
            10A. Teeth of the pinnae obtuse; fronds uniform, the pinnae 3–6 mm. long. .... 7. *A. viride*
            10B. Teeth of the pinnae mucronate; fronds often more or less dimorphic, the pinnae 6–35 mm. long. ...... 8. *A. incisum*
      5B. Blades narrowly deltoid, bipinnate in lower portion, the lowest pinnae largest. .... 9. *A. coenobiale*
  2B. Stipes green throughout, rarely brown at base, dull.
    11A. Sori solitary on the 1-veined ultimate pinnules.
      12A. Fronds at most 30 cm. long.
        13A. Blades lanceolate, the rachis prolonged, rooting and forming a plantlet at the tip; pinnae usually simply pinnate.
                                                  10. *A. prolongatum*
        13B. Blades ovate to broadly so, the tail at tip 1–4 cm. long, not gemmate; lower pinnae bipinnate to bipinnatiparted.
                                                 11. *A. ritoense*
      12B. Fronds large, 50–100 cm. long. .... 12. *A. trigonopterum*
    11B. Sori few to many on the few- to many-veined ultimate pinnules.
      14A. Pinnae less than 7 cm. long.
        15A. Pinnules oblong to oblanceolate or obovate, sparsely toothed or lobed; fronds relatively large, (10–) 20–60 cm. long.
          16A. Blades herbaceous, 3–30 cm. long, 1.5–13 cm. wide, with acute to obtuse teeth.
            17A. Blades not gemmate, 2- to 4-pinnate.
              18A. Blades oblong-ovate, 3- or 4-pinnate.
                19A. Blades 5–15 cm. long. .... 13. *A. sarelii*
                19B. Blades 15–35 cm. long. .... 14. *A. wilfordii*
              18B. Blades broadly lanceolate, 2-pinnate. .... 15. *A. pseudowilfordii*
            17B. Blades with a gemma on the rachis, cut into unequally incised and toothed pinnae. .... 16. *A. yoshinagae*
          16B. Blades membranous, 2–10(–15) cm. long, 1–3(–4) cm. wide, with a few obtuse often mucronate-tipped teeth.
                                                  17. *A. varians*
        15B. Pinnules coriaceous, flabellate to rhombic-orbicular, minutely toothed toward the tip, entire at base; fronds small, 3–10 cm.
           long. .... 18. *A. ruta-muraria*
      14B. Pinnae 10–17 cm. long.
        20A. Pinnae unlobed to pinnately parted; sori mostly in 2 series on the pinnae. .... 19. *A. wrightii*
        20B. Pinnae 2- or 3-pinnate; sori solitary on the pinnules or teeth. .... 12. *A. trigonopterum*
1B. Blades simple.
  21A. Blades without a submarginal connecting vein.
    22A. Blades narrowly oblanceolate; stipes less than 10 cm. long; sori mostly facing the anterior side of the blade.
      23A. Blades appressed crenate-toothed in upper half; veins obliquely spreading. .... 20. *A. griffithianum*
      23B. Blades subentire to remotely undulate-toothed; veins suberect to ascending. .... 21. *A. ensiforme*
    22B. Blades lanceolate, cordate; stipes 10–20 cm. long; sori paired, face to face. ....22. *A. scolopendrium*
  21B. Blades with a submarginal connecting vein, large, 70–100 cm. long; sori closely parallel, always facing anteriorly. 23. *A. antiquum*

1. Asplenium cheilosorum Kunze ex Mett. *Hymen-asplenium cheilosorum* (Kunze) Tagawa; *A. heterocarpum* Wall. ex Hook.——Usuba-kujaku. Rhizomes creeping, scaly, 2–2.5 mm. across; fronds slender; stipes 10–15 cm. long, 1–1.5 mm. across, purple-brown to dark brown, lustrous, smooth except at base; scales on rhizomes and at base of stipes lanceolate, 3–4 mm. long, long-acuminate, spreading, black-brown, minutely spinulose near tip; blades broadly linear to linear-lanceolate, 20–40 cm. long, 3–4 cm. wide, gradually long-acuminate, slightly narrowed at base, smooth, simply pinnate, the rachis lustrous, sulcate on upper side; pinnae rather many, thinly herbaceous to membranous, semiovate, 5–8 mm. wide, obtuse, somewhat dimidiate, spreading, subsessile, nearly straight from base to the top, entire on posterior margin except for 1 or 2 teeth near tip, the anterior side toothed, nearly straight to slightly arcuate, broadly cuneate at base; teeth on sterile pinnae mostly regular, ovate, subobtuse; fertile pinnae slightly larger, often bidentate at the tip, the costa very slender, close to the posterior margin, pinnately veined on anterior side, the median ones once-forked, ending in a tooth; sori on the teeth, few to a pinna; indusia 1.5–2.5 mm. long.——Kyushu (Yakushima and Tanegashima).—— Formosa, China, Indochina to India, and Malaysia.

2. **Asplenium unilaterale** Lam. *Hymenasplenium unilaterale* (Lam.) Hayata; *A. resectum* Smith——HŌBI-SHIDA. Rhizomes long-creeping, scaly, 1.5–3 mm. across; stipes slender, remote, 12–20 cm. long, 1–1.5 mm. across, lustrous, purple- to chestnut-brown, glabrous except for few scales at base; scales on rhizomes and at base of stipes lanceolate, 2–3 mm. long, long-acuminate, tardily deciduous, dark brown, with few obsolete teeth on upper margin; blades broadly linear to lanceolate, 18–30 cm. long, 3.5–6(–8) cm. wide, abruptly to gradually long-acuminate, not or very slightly narrowed at base, glabrous, the rachis with deciduous hairlike dark brown scales while young, lustrous, deeply sulcate on upper side; pinnae rather many, thinly herbaceous to membranous, spreading, lanceolate to broadly so, 4–10 mm. wide, subacute to obtuse, toothed toward the tip, entire toward the base, short-petiolulate, the costa slender, sometimes dark purple-brown beneath, close to the posterior margin near base, the veins mostly once-forked; sori dorsal, sparse, linear; indusia 2–6 mm. long, nearly straight, very rarely diplazioid.—— Honshu (s. Kantō through Tōkaidō to s. Kinki Distr. and westw.), Shikoku, Kyushu.——Ryukyus, Bonins, Formosa, Malaysia, China to India, Africa, Polynesia, and Hawaii.

3. **Asplenium oligophlebium** Bak. *A. fauriei* Christ ——KAMIGAMO-SHIDA, HIME-CHASEN-SHIDA. Rhizomes short, erect to ascending, often clothed in the remains of old stipes, the scales rather scanty, lanceolate, about 1 mm. long, purple-brown, acuminate; fronds smooth, rather small; stipes crowded, 2–8 cm. long, 0.6–1 mm. across, glabrous, lustrous, deep purple-brown; blades narrowly lanceolate, 7–15 cm. long, 1.5–3.5 cm. wide, the rachis dark purple-brown, lustrous, slender, often prolonged and gemmate at the tip; pinnae thinly herbaceous, squarrose-lanceolate to semideltoid, 7–15 mm. long, 3–5 mm. wide, obtuse, obliquely cuneate at base, sessile, with scattered coarse obtuse teeth, entire on posterior margin of lower portion, the costa slender, the veins few, simple to once forked; sori few on a pinna; indusia 1–2 mm. long.——Honshu (Hokuriku to Kinki and Chūgoku Distr.), Shikoku, Kyushu.——Ryukyus.

4. **Asplenium normale** Don. *A. opacum* Kunze—— NURI-TORA-NO-O. Rhizomes short, with chartaceous, lanceolate, dark brown scales about 1.5 mm. long; fronds tufted; stipes 1–2 mm. across, dark purple- to chestnut-brown, lustrous, 3–15 cm. long, naked; blades broadly linear to linear-lanceolate, 10–40 cm. long, 15–40 mm. wide, the rachis dark purple-brown, lustrous, sulcate on upper side, often prolonged and gemmate at tip; pinnae spreading, obliquely oblong to narrowly oblong-ovate, 5–10 mm. wide near base, the lower ones deflexed, obtuse to rounded, sessile, crenately toothed at the tip, entire toward the base, often obsoletely auricled on anterior side; sori few to a pinna, straight; indusia 1–4 mm. long, very rarely diplazioid.——Honshu (Kantō Distr. and westw.), Shikoku, Kyushu.——Ryukyus, Formosa, s. China to India.

5. **Asplenium trichomanes** L. *A. trichomanoides* Houtt.——CHASEN-SHIDA. Rhizomes short, erect, with the remains of old stipes, and with linear-lanceolate, acuminate scales about 3 mm. long; fronds deep green, nearly naked, tufted; stipes 1–5 cm. long, chestnut-brown, lustrous, very narrowly winged or angled, flat on upper side, obtuse on back; blades linear to broadly so, 10–25 cm. long, 12–20 mm. wide, gradually narrowed to an obtuse tip, slightly narrowed at base, the rachis lustrous, narrowly winged, obtuse to subacute, not winged on back; pinnae nearly all alike, firmly herbaceous, elliptic, ovate or oblong, 4–8(–10) mm. long, 3–6 mm. wide, spreading, rounded at apex, obliquely cuneate at base, obtusely toothed; sori usually less than 10 to a pinna; indusia about 1.5 mm. long.——On rocks; Hokkaido, Honshu, Shikoku, Kyushu.——Formosa to the Himalayas, Caucasus, Europe, S. Africa, Australia, N. and S. America.

6. **Asplenium tripteropus** Nakai. *A. anceps* var. *proliferum* Nakai——INU-CHASEN-SHIDA. Closely allied to the preceding, the stipes and rachis somewhat stouter, narrowly winged on back and margin, sometimes gemmate on upper part of rachis; rhizomes short, the scales lanceolate, about 3 mm. long, short-acuminate; fronds tufted; stipes 1–8 cm. long, chestnut-brown, lustrous, with 3 narrow brown wings; blades linear to broadly so, 7–25 cm. long, 15–25 mm. wide, gradually narrowed to an obtuse tip, slightly narrowed at base; pinnae oblong or oblong-ovate, sometimes elliptic, 5–13 mm. long, 3–6 mm. wide near base, rounded at the tip, obliquely cuneate at base, obtusely toothed, spreading, sessile; sori 2–10 on a pinna; indusia 1–2 mm. long.——Honshu (Sagami and Etchu Prov., Kinki Distr. and westw.), Shikoku, Kyushu.——Formosa and China.

7. **Asplenium viride** Huds. *A. trichomanes* var. *ramosum* L.——AO-CHASEN-SHIDA. Rhizomes short, ascending, the scales linear-lanceolate to broadly linear, about 3 mm. long, entire, gray-brown; fronds tufted, glabrous; stipes erect, 2–5 cm. long, 0.5–1 mm. across, chestnut-brown, lustrous at least on lower half, green at tip, shallowly sulcate on upper side; blades suberect, linear to linear-lanceolate, 5–12 cm. long, 8–12 mm. wide, obtuse, not or slightly narrowed at base; pinnae 10–17 pairs, herbaceous, rhombic-ovate to rhombic-orbicular, 3–6 mm. long, 3–5 mm. wide, obtuse to very obtuse, oblique and broadly cuneate at base, with few obtuse teeth near tip, entire near base, sessile or very short-petiolulate, the costas and veins very slender, inconspicuous on both sides, the veins usually once forked; sori 2–6 on a pinna, on the lower portion of veins, subcostal; indusia 1–1.5(–2) mm. long.——On rocks in alpine regions; Hokkaido, Honshu (centr. and Kantō Distr.), Shikoku (Mount Tsurugi in Awa Prov.); rare.—— Sakhalin, Formosa to the Himalayas, Europe, and N. America.

8. **Asplenium incisum** Thunb. *A. elegantulum* Hook. ——TORA-NO-O-SHIDA. Rhizomes short, ascending, the scales membranous, broadly linear, 3–4 mm. long, gray-brown, entire; fronds smooth, somewhat dimorphic, the sterile ones smaller, 5–15 cm. long, with a short stipe 1–3 cm. long, the fertile erect, larger, 30 cm. long or more, with stipes 2–5 cm. long; blades oblanceolate, 7–30 cm. long, 2–7 cm. wide, acumibase, the sterile fronds pinnate to bipinnatiparted, the fertile ones 2- or 3-pinnatiparted, the rachis green on upper side, mostly chestnut-brown beneath near base; pinnae herbaceous, very short-petiolulate or sessile, spreading, acutely to obtusely mucronate-toothed, the median fertile pinnae lanceolate-deltoid, 6–35 mm. long, 7–15 mm. wide near base, obtuse to acute, pinnately parted to pinnate, the lower pinnae reniform to orbicular, 5–10 mm. long and as wide, cordate at base, unlobed or shallowly 3-lobed; sori few, nearer the costa or costule than the margin; indusia 1–2 mm. long.——Lowlands and

low mountains; Hokkaido, Honshu, Shikoku, Kyushu; common.——Sakhalin, s. Kuriles, Kamchatka, Ryukyus, Korea, Manchuria, and China.

9. **Asplenium coenobiale** Hance. *A. fuscipes* Bak.; *A. toramanum* Makino——KUROGANE-SHIDA. Rhizomes short, erect, with a tuft of old stipes, the scales chartaceous, broadly linear, 4–5 mm. long, black-brown, entire; fronds tufted, with minute capitate hairlike scales while young; stipes erect, terete, nearly black, lustrous, 3–10 cm. long, 0.2–0.5 mm. across; blades narrowly deltoid, 4–8 cm. long, 1.5–4 cm. wide, long-attenuate, subacute to obtuse at the tip, bipinnate in lower portion, simply pinnate in upper portion, the rachis sulcate on upper side; lower pinnae spreading, firmly herbaceous, narrowly oblong-ovate, 8–20 mm. long, 5–8 mm. wide, obtuse, sessile, the upper pinnae obliquely spreading, the lowest anterior pinnules slightly larger than the others; pinnules obliquely elliptic to oblong, 3–6 mm. long, 2–4 mm. wide, obtuse, often slightly imbricate, obtusely few-toothed, slightly decurrent; veins and costules slender; sori solitary to few on the upper portion of the pinnules, subcostular; indusia 1–2 mm. long.——Calcareous rocks; Shikoku; rare.——China and Indochina.

10. **Asplenium prolongatum** Hook. *A. achilleifolium* sensu auct. Japon., non Liebm. nec C. Chr.; *A. rutaefolium* sensu auct. Japon., non Kunze——HINOKI-SHIDA. Rhizomes short, ascending, covered with basal remains of old stipes, the scales narrowly lanceolate, 4–5 mm. long, about 1 mm. wide, acuminate, dark brown, with a narrow brown sparsely spined margin; fronds tufted; stipes flattened, sparsely scaly while young, 10–25 cm. long, the scales about 1 mm. long, spreading; blades 10–20 cm. long inclusive of the prolonged rachis gemmate at the tip, broadly lanceolate, bipinnate, the rachis green, flattened, with a slender raised midrib on upper side; pinnae 10–18 pairs, ascending, oblique, oblong to ovate, 15–40 mm. long, 6–12 mm. wide, obtuse, simply pinnate, petiolulate; pinnules about 5 pairs, herbaceous, ascending, linear, 5–10 mm. long, 1–1.5 mm. wide, obtuse, entire, the lower ones rarely 2 parted in vigorous fronds, 1-veined; sori solitary on the pinnules, linear; indusia 3–7 mm. long.——Wet shaded rocks in mountains; Honshu (Izu and Kii Prov.), Shikoku, Kyushu.——s. Korea, Ryukyus, Formosa, China to India.

11. **Asplenium ritoense** Hayata. *A. davallioides* Hook., non Tausch; *Humata dareoidea* Mett.; *A. dareoideum* (Mett.) Makino, non Desv.——KŌZAKI-SHIDA. Rhizomes short, rather stout, ascending; stipes few together, green, flattened, sparsely scaly on the lower half, 7–20 cm. long, the scales linear-lanceolate, 3–6 mm. long, acuminate, spreading, dark brown, remotely few-spined; blades 2- or 3-pinnate, ovate to broadly so, 10–18 cm. long, 4–10 cm. wide, prolonged at the tip to a simply pinnate tail 1–4 cm. long, the rachis green, flat, with a rather broad slightly raised midrib on upper side; pinnae ascending, petiolulate the lower ones bipinnate to bipinnatiparted, larger, ovate-deltoid, 2.5–9 cm. long, 1–4 cm. wide, prolonged and simply pinnate at the tip, the petiolules 3–8 mm. long; pinnules thickly herbaceous, linear-oblong to lanceolate, 3–5 mm. long, 1–1.5 mm. wide, subacute to obtuse, the apical mostly arcuately spreading, entire or rarely bifid, green on both sides, slenderly 1-veined; sori solitary on a pinnule; indusia 2–3(–4) mm. long.——Moist shaded places; Honshu (s. Kantō through Tōkaidō to s. Kinki Distr.), Shikoku, Kyushu.——s. Korea (Quelpaert Isl.), Ryukyus, Formosa, and s. China.

12. **Asplenium trigonopterum** Kunze. *A. mertensi-*

anum Kunze——ŌBANO-HINOKI-SHIDA. Rhizomes stout, ascending; fronds 50–100 cm. long; stipes stout, light green, grooved on upper side toward tip, naked on upper portion, with pale brown, lanceolate scales 8–10 mm. long, 1.5–2 mm. wide at base; blades narrowly ovate-oblong, 3- or 4-pinnatiparted, glabrous; pinnae 10–15 pairs, obliquely spreading, 12–17 cm. long, 3–6 cm. wide, 2- or 3-pinnate; pinnules firmly herbaceous, ascending, green, slightly paler beneath, the sterile ones narrowly rhombic-ovate, 15–30 mm. long, 6–10 mm. wide, subobtuse, cuneate at base, more or less irregularly obtuse-toothed, often parted in lower half, the fertile pinnules finely and pinnately cut, the segments ascending, broadly linear, 5–10 mm. long, 1.5–2 mm. wide, 1-veined, entire, the rachis flattened, green, with a slender raised midrib on upper side; sori solitary on the pinnules or teeth; indusia linear, 3–5 mm. long.——Honshu (Hachijo Isl.); rare.——Bonins and Formosa.

13. **Asplenium sarelii** Hook. *A. saulii* Hook. ex Bak.; *A. blakistonii* Bak.——KOBA-NO-HINOKI-SHIDA. Rhizomes short, erect to ascending, scaly; stipes pale green, 1.5–10 cm. long, slender, sparsely scaly on lower half while young, the basal scales lanceolate, 2–4 mm. long, long-attenuate, filiform at the tip, those on upper part of stipe and on blade very sparse, fugacious, hairlike; blades ovate to narrowly so, 5–15 cm. long, 2.5–6.5 cm. wide, bipinnate, acute to short-acuminate, not narrowed at base; pinnae spreading, deltoid-lanceolate, 1–1.5 cm. wide, acute to short-acuminate, short-petiolulate; pinnules spreading to ascending, ovate to rhombic-ovate, obtuse to subacute, cuneate at base, short-petiolulate, cut nearly to the base into a few segments; ultimate pinnules herbaceous, broadly oblanceolate, 3–4 mm. long, 1.5–2.5 mm. wide, cuneate at base, with a few acute teeth and sometimes lobes on upper margin, entire at base; sori few on a pinnule, ascending; indusia 1.5–3 mm. long.——Honshu (Kantō through Tōkaidō to Kinki Distr. and westw.), Shikoku, Kyushu.——Ryukyus, Korea, China to the Himalayas.

Var. **pekinense** (Hance) C. Chr. *A. pekinense* Hance; *A. sepulchrale* Hook.; *A. abbreviatum* Makino——TOKIWA-TORA-NO-O. Stipes usually about 5 cm. long; blades broadly lanceolate, nearly 15–30 mm. wide, acute; pinnae deltoid to broadly ovate, acute to obtuse, subsessile, the pinnules slightly thicker, slightly broader, more closely arranged, and less cuneately narrowed at base.——Occurs with the typical variety.

14. **Asplenium wilfordii** Mett. *Tarachia wilfordii* (Mett.) H. Itō——AOGANE-SHIDA. Rhizomes ascending to short-creeping, clothed with basal remains of old stipes; fronds tufted; stipes firm, pale green to purplish, dull, 2-grooved on upper side, 10–25 cm. long, about 1.5 mm. across, very sparsely scaly in upper part, densely so at base, the scales linear-lanceolate, 3–6 mm. long, deep brown, gradually narrowed to a long filiform tip, entire; blades oblong-ovate to ovate, 15–35 cm. long, 4–10 cm. wide, acute to short-acuminate, nearly glabrous, 3- or 4-pinnatiparted; pinnae 8–10 pairs, obliquely spreading, deltoid-ovate, acuminate, petiolulate; lower pinnules petiolulate; ultimate pinnules firmly herbaceous, oblanceolate-cuneate, 4–7 mm. long, 1.5–2.5 mm. wide, ascending, unlobed or deeply 3-fid, usually bidentate at the tip, deep green above, whitish green beneath; veins 1–3 in the segments, nearly parallel, slender, often once forked at the tip; sori 1–3 to a segment; indusia linear, 1–3 mm. long, closely parallel.——Honshu (Izu Prov. through Tōkaidō to Kinki Distr. and westw.), Shikoku, Kyushu.——Ryukyus, Formosa, and s. Korea.

15. **Asplenium pseudowilfordii** Tagawa. *A. calcicola*

H. Itō, non Tagawa——OKUTAMA-SHIDA. Intermediate between the preceding and No. 16 but differing in the bipinnate blades; rhizomes short-creeping to ascending, with many stipes, the scales linear-lanceolate, 4–6 mm. long, long-acuminate, entire, dark brown; stipes 5–15 cm. long, 1.5–2 mm. across, green, slightly flattened, sulcate on upper side, often dull purplish beneath, sparsely scaly, the scales spreading, deciduous, broadly linear, 2–3 mm. long; blades broadly lanceolate, 10–20 cm. long, 3.5–7 cm. wide, acuminate, bipinnate; pinnae 10–15 pairs, deltoid-ovate, 2–5 cm. long, 1.5–2.5 cm. wide, acute to acuminate with an obtuse tip, obliquely spreading, the petiolules 2–7 mm. long; pinnules firmly herbaceous, 5–15 mm. long, 2–5 mm. wide, cuneate at base, sessile, ascending, unlobed or slightly incised, entire except for a few obtuse teeth at the subtruncate tip, the costules and veins slender, nearly parallel; sori few on a pinnule; indusia 2–6 mm. long, straight.—— Honshu (Musashi, Suruga, Hida, Tanba Prov.); rare.

16. Asplenium yoshinagae Makino. *Tarachia yoshinagae* (Makino) H. Itō; *A. laciniatum* forma *viviparum* Wu; *A. planicaule* var. *yoshinagae* (Makino) Tagawa——TOKIWA-SHIDA. Rhizomes short-ascending to suberect, rather stout, densely scaly, the scales dark brown, broadly linear, 4–6 mm. long, about 0.5 mm. wide, with a long-filiform tip, entire; stipes more or less flattened, green, often dull purplish on lower side, sulcate on upper side, 5–15 cm. long, 1–1.5 mm. wide, very sparsely scaly, the scales linear, 1–2.5 mm. long, spreading; blades linear to broadly lanceolate, 10–30 cm. long, 3–5 cm. wide, acuminate, the rachis nearly glabrous, gemmate; pinnae 10–20 pairs, spreading, oblique, oblong-ovate, 15–40 mm. long, 7–10 mm. wide, obtuse or somewhat elongate and acute, obliquely cuneate at base, the posterior side entire on lower half, irregularly 2- to few-lobed, truncate and obtusely toothed at the tip, the petiolules 3–5 mm. long; sori along the costas, few to a pinna, linear; indusia 3–7 mm. long, straight, narrow.——Honshu (Kantō through Tōkaidō to Kinki Distr.), Shikoku, Kyushu.——s. China to the Himalayas.

17. Asplenium varians Hook. & Grev. *A. capillipes* Makino; *A. varians* var. *sakuraii* Rosenst.; *Gymnogramma fauriei* Christ; *Anogramme fauriei* (Christ) C. Chr.; *Asplenium subvarians* Ching; *A. siobarense* Koidz.; *A. varians* forma *obtusilobum* Miyabe & Kudō——IWA-TORA-NO-O. Rhizomes short, the scales membranous, deltoid-lanceolate, 1 mm. long, long-acuminate, entire, dark brown; stipes pale green, subcapillary to very slender, 1–10 cm. long, brownish at base, with minute hairlike scales while young; blades membranous, oblong-ovate to narrowly oblong, 2–10(–15) cm. long, 1–3(–4) cm. wide, bipinnate, the smaller ones bipinnatiparted, obtuse to acuminate; pinnae 3–8 pairs, obliquely spreading, ovate to rhombic-ovate, the smaller ones rhombic-orbicular, usually short-petiolulate, pinnate to subternate, the ultimate ternate, 4–20(–30) mm. long, obtuse; pinnules membranous, few, obovate or rhombic-ovate, 3–5 mm. long, 3–4 mm. wide, with coarse 1-veined teeth or lobes; sori ascending, few on a pinnule; indusia thinly membranous, oblong to linear, 1–3 mm. long, white.——Hokkaido (Kamuikotan, *fide* Miyabe & Kudō), Honshu, Shikoku, Kyushu.——China to India.

18. Asplenium ruta-muraria L. *Acrostichum ruta-muraria* (L.) Lam.; *Amesium ruta-muraria* (L.) Newm.; *Tarachia ruta-muraria* (L.) Presl——ICHŌ-SHIDA. Rhizomes short, densely clothed with the basal remains of old stipes, the attenuate; fronds 3–10 cm. long, tufted; stipes 2–6 cm. long,

pale green on upper part, brown at base, with capitate glands and deciduous linear scales; blades ovate to deltoid-ovate, obtuse, 1.5–4 cm. long, 1.3–2.5 cm. wide, bipinnatifid to bipinnate, the rachis sulcate on upper side; pinnae 1–3 pairs, alternate, the lowest largest, petiolulate, 2- or 3-parted or pinnate; pinnules coriaceous, 3–5, flabellate to rhombic-orbicular, or cuneate-obovate, 3–6 mm. long, nearly as wide, rounded to very obtuse, broadly cuneate and entire on lower half, minutely toothed on upper half, flat on both sides; veins palmate, concealed, parallel; sori few on the center near base of pinnules, parallel, linear, confluent; indusia 2–3 mm. long.—— Calcareous rocks; Honshu, Shikoku, Kyushu.——Formosa to the Himalayas, n. Asia to Europe, and N. America.

19. Asplenium wrightii Eaton. *A. wrightii* var. *fauriei* Christ; *A. centrochinense* Christ; *A. wrightioides* Christ—— KURUMA-SHIDA. Rhizomes stout, ascending, clothed with the basal remains of old stipes; stipes 20–40 cm. long, pale green to slightly purplish, sulcate on upper side, scaly, about 4 mm. across near base, the scales narrowly lanceolate, 5–7 mm. long, 1–1.2 mm. wide near base, brown, sparsely ciliate; blades broadly lanceolate, 30–50 cm. long, 15–25 cm. wide, simply pinnate, nearly naked, contracted in upper portion, short-acuminate; pinnae 13–20 pairs, herbaceous, obliquely spreading, short-petiolulate, linear-lanceolate, 10–17 cm. long, 1–2(–2.5) cm. wide, falcate, long-caudate, obliquely cuneate and entire at base, simply to doubly acute- or mucronate-toothed, unlobed or sometimes pinnately parted, glabrous; costas rather slender, raised on upper side, the veins ascending, rather remote, usually once to thrice forked; sori mostly in 2 series on the pinnae, ascending, very slightly curved, linear, nearer the costa than the margin, or nearly median; indusia (3–)6–12(–15) mm. long, 1–1.2 mm. wide.——Honshu (Izu Prov. through Tōkaidō to Kinki and Chūgoku Distr.), Shikoku, Kyushu.——Ryukyus, Formosa, and China.

Var. **shikokianum** (Makino) Makino. *A. shikokianum* Makino——HAYAMA-SHIDA. Pinnae deltoid-lanceolate to linear-lanceolate, petiolulate, pinnately dissected, with a rather prominently winged rachis; pinnules ovate, cuneate at base; veins pinnately branched.——Honshu (Totomi, Kii, and Ise Prov.), Shikoku (Tosa Prov.), Kyushu (Hyuga Prov.).—— China and Formosa.

20. Asplenium griffithianum Hook. *A. nakanoanum* Makino——FUSA-SAJI-RAN. Rhizomes short-creeping, 3–5 mm. across, the scales rather dense, narrowly lanceolate, about 5 mm. long, acuminate, dark brown; blades thickly herbaceous, simple, narrowly oblanceolate, 15–23 cm. long, 1.5–2 cm. wide, acuminate with a short obtuse entire tip, gradually narrowed at base to the short winged stipe, crenate-toothed on upper half, deep green and glabrous on upper side, paler beneath, with minute appressed lobed scales sometimes with a black-brown short apical tail; costas slightly raised on both sides, whitish and slender on upper side, broader beneath, the scales broadly linear, dark brown, about 3 mm. long, long-veins obliquely spreading, slender, scarcely visible, simple or forked, ending short of the slightly cartilaginous hyaline margin; sori linear, obliquely spreading, straight, parallel, between the margin and the costa; indusia 4–8 mm. long, entire.—— Kyushu (Yakushima).——Formosa, China to India.

21. Asplenium ensiforme Wall. ex Hook. & Grev. *A. bicuspe* Hayata, in syn.; *Diplazium bicuspe* Hayata——HOKO-GATA-SHIDA. Fronds tufted, 20–30 cm. long; scales on rhizomes and basal part of stipes lanceolate, 4–5 mm. long, dark brown; blades herbaceous-coriaceous, linear-oblanceolate,

1–1.5(–2) cm. wide, gradually narrowed at base to the short winged stipe, subentire to remotely undulate-toothed, scarcely cartilaginous on margin, sometimes bifid, nearly glabrous on both sides; costa slightly raised on both sides; veins parallel, once-forked; sori elongate, suberect to ascending, mostly on the anterior branchlet of veins, distant, straight; indusia 7–20 mm. long.——Kyushu (s. distr.); rare.——Formosa and sw. China to India.

22. Asplenium scolopendrium L. *Phyllitis scolopendrium* (L.) Newm.; *Scolopendrium vulgare* Smith; *S. officinarum* Sw.——Ko-tani-watari. Rhizomes ascending to short-creeping, clothed with basal remains of old stipes; stipes 10–20 cm. long, shorter than the blades, densely scaly while young, the scales membranous, deltoid-lanceolate, 5–8 mm. long, 1–2 mm. wide near base, long-filiform at the tip, brown, spreading, flexuous, very sparingly ciliate; blades chartaceous-herbaceous, simple, narrowly to broadly lanceolate, 15–40 cm. long, 3.5–6 cm. wide, acute, cordate-auriculate at base, entire to undulate, green above, slightly paler and with minute brown scales beneath while young, indistinctly hyaline on margin; costa with brown linear scales while young, especially beneath, the veins very slender, once or twice forked, the veinlets closely parallel, spreading, ending short of the margin; sori on upper two-thirds of blade, linear, paired, straight, spreading, parallel; indusia membranous, 7–25 mm. long, the pairs overlapping while young.——Shaded slopes and rocks in mountains; Hokkaido, Honshu, Shikoku, Kyushu.——s. Kuriles, Sakhalin, Caucasus and Asia Minor to Europe, and N. America.

23. Asplenium antiquum Makino. *Thamnopteris antiqua* (Makino) Makino; *Neottopteris antiqua* (Makino) Masam.; *A. nidus* L. forma *intermedia* Mett., pro parte; *A. nidus* sensu auct. Japon., non L.; *Neottopteris rigida* Fée var. *erubescens* Nakai——Tani-watari, Ō-tani-watari. Rhizomes short, erect; fronds tufted, obliquely spreading; blades coriaceous, lustrous, vivid-green, scarcely paler beneath, simple, linear-oblanceolate, 70–100 cm. long, 7–10(–12) cm. wide, entire or nearly so, abruptly acute to short-acuminate, or gradually narrowed at both ends, somewhat decurrent on the short stipe, scaly beneath while very young; stipes very short to nearly obsolete, winged in upper portion, the scales at base firmly membranous, gray-brown, linear-lanceolate, 2–2.5 cm. long, 3–5 mm. wide, acuminate, loosely toothed on margin toward tip; costa rather stout, flat on upper side, raised and obtuse beneath, often brownish toward base, the veins often once forked near base, spreading, straight, close, connected with a marginal vein; sori occupying nearly the whole length of the veinlets on upper half of the fronds; indusia entire, about 0.7 mm. wide.——Tree-trunks; Honshu (Hachijo Isl. and Kii Prov.), Shikoku, Kyushu.——Ryukyus to Formosa.

2. CAMPTOSORUS Link   Kumo-no-su-shida Zoku

Terrestrial; rhizomes short; scales linear, attenuate, clathrate; blades herbaceous, simple, lanceolate to linear, gradually prolonged and gemmate at the tip, cordate to cuneate or acute at base, glabrous; veins forking and freely anastomosing; sori rather irregularly disposed.——Two species, 1 in N. America, the other in e. Asia.

1. Camptosorus sibiricus Rupr. *Scolopendrium sibiricum* (Rupr.) Hook.; *Phyllitis sibirica* (Rupr.) O. Kuntze; *Asplenium ruprechtii* Kurata——Kumo-no-su-shida. Rhizomes small, short-ascending to erect, the scales brown, narrowly lanceolate, about 3 mm. long, acuminate, usually entire; fronds evergreen, glabrous; stipes slender, 1–10 cm. long, pale green, the larger ones usually purple-brown and lustrous on lower half, naked except for a few small basal scales; blades linear to linear-lanceolate, 5–15 cm. long, 5–10 mm. wide, long-tapering and usually rooting at the tip, usually cuneate at base, undulate to obsoletely crenate, the small sterile blades usually elliptic, 7–30 mm. long, rounded to acuminate, usually long-stipitate, costas raised beneath, the veins slender, scarcely visible on either side, once to thrice forked, ascending, the veinlets not reaching the margin, partially anastomosing and forming areolae in a few irregular rows along the costa; sori linear to oblong, ascending to spreading, 1–5 mm. long, the longer ones mostly along the costa; indusia entire, membranous, straight or slightly recurved.——Calcareous rocks; Hokkaido, Honshu, Shikoku, Kyushu.——Korea, Manchuria, n. China to s. Siberia.

Fam. 20. POLYPODIACEAE   Uraboshi Ka   Polypody Family

Mostly epiphytic, rarely terrestrial; rhizomes creeping or sometimes ascending, dictyostelic, scaly, the scales usually broad and peltate, very rarely bristlelike or hairlike; fronds usually jointed at base of stipe, simple to pinnate, with free or reticulate venation; sori exindusiate, typically round, sometimes elongate along the veins.——About 65 genera, with about 1,000 species, abundant in the Tropics.

1A. Rhizomes hairy; fronds dimorphic; fertile blades much narrower than the sterile; paraphyses capitate. . . . . 1. *Cheiropleuria*
1B. Rhizomes scaly.
  2A. Veins anastomosing, or if not the blades more than 2 cm. wide.
    3A. Fronds usually dimorphic, with stellate scales. . . . . 7. *Pyrrosia*
    3B. Fronds usually all alike, the scales not stellate.
      4A. Veins free, or if anastomosing the areolae with a simple ascending veinlet, or the veinlet absent; paraphyses none or filamentous, rarely clathrate or stellate.
        5A. Blades pinnatifid or compound, scarcely fleshy; sori round. . . . . 2. *Polypodium*
        5B. Blades simple, entire, fleshy; sori elongate. . . . . 13. *Loxogramme*
      4B. Veins anastomosing, with variously directed or branched veinlets in the areolae.
        6A. Paraphyses peltate.
          7A. Fronds strongly dimorphic, the sterile elliptic to ovate or obovate, the fertile linear to oblanceolate. . . . . 5. *Lemmaphyllum*
          7B. Fronds all alike (somewhat dimorphic in *Crypsinus*).

1. CHEIROPLEURIA Presl   SUJI-HITOTSUBA ZOKU

Terrestrial; rhizomes creeping, with soft brownish hairs; fronds dimorphic; stipes approximate, erect, slender, not jointed with the rhizomes; sterile blades ovate to rounded, often 2-lobed, glabrous, entire; veinlets forming small areolae; fertile blades simple, entire, much narrower than the sterile; sporangia covering the entire blade; paraphyses capitate.——One species.

1. **Cheiropleuria bicuspis** (Bl.) Presl. *Polypodium bicuspe* Bl.; *Acrostichum bicuspe* (Bl.) Hook.; *A. bicuspe* Hook. var. *integrifolium* Eaton ex Hook.; *C. bicuspis* var. *integrifolia* Eaton ex Matsum. & Hayata——SUJI-HITOTSUBA. Firm evergreen fern; rhizomes creeping, with close stipes, 4–8 mm. across, densely covered with pale brown hairs 4–6 mm. long; stipes 15–40 cm. long, lustrous, straw-colored, glabrous in upper part, hairy at base, those of the fertile fronds slightly longer than the sterile; sterile blades coriaceous-herbaceous, nearly concolorous, broadly lanceolate to ovate, 8–20 cm. long, 2–8 cm. wide, obtuse, entire or bifid at apex, rounded at base and then abruptly acute, 3- to 5-nerved; fertile blades broadly linear, 10–18 cm. long, about 1 cm. wide, 3-nerved, the lateral nerves marginal; sporangia densely covering the surface except on the costa.——Hills and low mountains; Honshu (Izu Isls., Totomi Prov., and s. Kinki Distr.), Shikoku, Kyushu; rare. ——Ryukyus, Formosa, and China to Malaysia.

2. POLYPODIUM L.   EZO-DENDA ZOKU

Epiphytic, rarely terrestrial; rhizomes creeping, dictyostelic, scaly; stipes jointed at base; blades usually uniform, pinnatifid or compound, glabrous or pubescent, sometimes scaly, the veins forked or branching, free or anastomosing and forming areolae each with one simple excurrent included veinlet; sori dorsal, terminal or nearly so on the lowest anterior veinlet, typically round, superficial, exindusiate; paraphyses absent or filamentous, rarely clathrate or stellate.——About 75 species, mostly in the N. Hemisphere, abundant in the American Tropics.

1. **Polypodium vulgare** L. Ō-EZO-DENDA. Rhizomes creeping, rather stout to somewhat slender, densely scaly, the scales ascending, membranous, brown, broadly lanceolate, 2.5–4.5 mm. long, 1–1.2 mm. wide, gradually narrowed, short-filiform at the tip; fronds glabrous; stipes straw-colored; blades firmly herbaceous, oblong-ovate to broadly lanceolate, 6–10 cm. long, 4–5 cm. wide, pinnately parted nearly to the rachis, glabrous, the rachis stout and raised on both sides; segments 10–15 pairs, horizontally spreading, linear-oblong, 5–7 mm. wide, obsoletely crenate, very obtuse; costa slender, raised beneath, the veins not prominent on either side; sori orbicular, medial or slightly nearer the costa.——Tree-trunks and rocks; Hokkaido, Honshu (mountains n. distr., very rare and local westw.).——Kuriles, Korea (Dagelet Isl.), China, Tibet, Siberia to Europe, and N. America.

2. **Polypodium virginianum** L. *P. vulgare* var. *virginianum* (L.) Eaton——EZO-DENDA. Rhizomes creeping, rather slender, 1.5–3 mm. across, the scales lanceolate, 3–4 mm. long, gradually narrowed to a filiform tip, slightly dilated at base, with a dark brown median band, sparsely ciliolate; fronds glabrous; stipes straw-colored, rather slender, 5–12 cm. long; blades firmly herbaceous, linear- to oblong-lanceolate, 7–20 cm. long, 2.5–4(–5) cm. wide, pinnatiparted; segments 12–25 pairs, oblong-lanceolate, 4–7 mm. wide, very obtuse, crenately toothed toward the tip; costas slender, the veins not visible on either side; sori orbicular, nearer the margin than the costa.——Tree-trunks and rocks; Hokkaido, Honshu (centr. and n. distr.); rare.——Sakhalin, Korea, China, Mongolia, Manchuria, e. Siberia to N. America.

3. **Polypodium fauriei** Christ. *P. vulgare* var. *japonicum* Fr. & Sav.; *P. japonicum* (Fr. & Sav.) Maxon, non Houtt. ——OSHAGUJI-DENDA. Rhizomes rather slender, 2–3 mm. across, long-creeping, densely scaly, the scales brown, membranous, spreading, ovate, 2–3 mm. long, abruptly contracted to a short filiform point; stipes rather short, 3–6 cm. long, straw-colored, glabrous on upper portion, slightly scaly at base; blades chartaceous to herbaceous, narrowly ovate to broadly lanceolate, 5–20 cm. long, 2.5–8 cm. wide, usually slightly narrowed at base, glabrous on upper side, sparsely crisped-pubescent beneath, pinnately parted, the rachis raised on both sides; segments 15–25 pairs, horizontally spreading, linear-lanceolate to broadly linear, 3–5 mm. wide, obtuse to subacute, usually obsoletely crenate toward the apex, the costas slender, slightly raised on both sides, the veins not visible; sori orbicular, medial.——Tree-trunks and rocks in mountains; Hokkaido, Honshu, Shikoku, Kyushu, rather rare.——s. Kuriles and Korea (Quelpaert Isl.).

4. **Polypodium niponicum** Mett. *Marginaria niponica* (Mett.) Nakai——AONE-KAZURA. Rhizomes long-creeping, 4–5 mm. across, green, somewhat glaucous, fleshy, sparsely scaly, the scales appressed, deltoid-ovate, 1.5–2 mm. long, acuminate, rounded to subcordate at base, peltately attached, clathrate; stipes straw-colored, 7–15 cm. long, nearly glabrous; blades broadly lanceolate to oblong-ovate, 15–30 cm. long, 5–10(–12) cm. wide, pinnately parted, puberulent on upper side, densely so and often velvety beneath; segments 15–25 pairs, horizontally spreading, the lower 1 or 2 pairs sometimes slightly deflexed, oblong-lanceolate to narrowly lanceolate 6–10 mm. wide, obtuse to acute, subentire, the costas very slender, the veins scarcely visible; sori orbicular, nearer the costa than the margin.——Mossy rocks and tree-trunks in hills and low mountains; Honshu (Musashi and Sagami Prov. and westw.), Shikoku, Kyushu.——s. China.

5. **Polypodium formosanum** Bak. *Marginaria formosana* (Bak.) Nakai; *P. liukiuense* Christ——TAIWAN-AONE-KAZURA. Rhizomes long-creeping, stout, 4–6 mm. across, terete, somewhat glaucous, nearly naked except for scattered minute scales toward the tip; stipes terete, pale brown to straw-colored, slightly lustrous, 15–30 cm. long, 2–3 mm. across, glabrous or nearly so; blades herbaceous, narrowly oblong-ovate, 30–50 cm. long, 10–15 cm. wide, pinnately parted, thinly puberulent on both sides, the rachis short-pilose above; segments 20–30 pairs, horizontally spreading, linear-lanceolate, 1–1.5 cm. wide, acute, nearly entire; costa slender, thinly short-pilose on upper side; sori slightly nearer the costa than the margin.——Kyushu (Yakushima); rare.——Formosa.

6. **Polypodium someyae** Yatabe. *Marginaria someyae* (Yatabe) Nakai——MYŌGI-SHIDA. Rhizomes long-creeping, 3–4 mm. across, prominently scaly, especially at the base of stipes, the scales linear-lanceolate, 4–5 mm. long, long-acuminate, ciliolate, minutely clathrate; stipes straw-colored to pale green, somewhat lustrous, slender, glabrous on upper portion with a fascicle of scales at base; blades thinly herbaceous, ovate to narrowly so, 12–25 cm. long, 7–12 cm. wide, abruptly contracted above and caudate at the tip, pinnately parted, nearly glabrous beneath, puberulent on upper side, especially on the rachis and costas; segments 7–15 pairs, spreading, linear-lanceolate, 6–15 mm. wide, obtuse, usually slightly falcate, crenately toothed, the costa slender, raised on both sides, slightly flexuous, decurrent on the rachis, the veins more or less visible on both sides; sori medial.——Mossy rocks in mountains; Honshu (Musashi and Kōtsuke Prov.), Shikoku; very rare.

3. PLEOPELTIS Humb. & Bonpl. NOKI-SHINOBU ZOKU

Epiphytic; rhizomes elongate, dictyostelic, scaly; fronds coriaceous, jointed at base to the rhizome, simple or rarely pinnatifid, entire, with peltate scales on one or both sides, or nearly naked; veins and veinlets freely anastomosing; sori borne at the union of several veinlets, typically round, rarely elongate or fused and parallel to the costa, protected at least at first by peltate paraphyses with a flat expanded apex.——About 40 species, Hawaii, Japan, Philippines, Sumatra, Africa, and tropical America.

1A. Scales of rhizomes lanceolate, gradually long-attenuate, 2.5–4 mm. long.
  2A. Fronds coriaceous, the veins not distinct.
    3A. Stipes often very short, 2–30 mm. long.
      4A. Rhizomes 2–3 mm. across; fronds 12–30 cm. long, 5–15 mm. wide; stipes 1–3 cm. long. .... 1. *P. thunbergiana*
      4B. Rhizomes about 1 mm. across; fronds 3–7 cm. long, 2–5 mm. wide; stipes 2–7 mm. long. .... 2. *P. onoei*
    3B. Stipes slender, 3–6 cm. long; blades 5–10 cm. long, 3–5 mm. wide, rather abruptly narrowed at both ends. .... 3. *P. uchiyamae*
  2B. Fronds chartaceous when dry, the veins distinct on both surfaces. .... 4. *P. clathrata*
1B. Scales of rhizomes ovate, abruptly acuminate, 1–2 mm. long.
  5A. Rhizomes long-creeping, slender; fronds remote; blades linear-lanceolate, 0.5–1.5 cm. wide. .... 5. *P. ussuriensis* var. *distans*
  5B. Rhizomes short-creeping; fronds closer and more numerous; blades usually lanceolate, wider.
    6A. Sori nearer the costa than the margin; blades often acuminate to caudate, sometimes with an isolated pair of sori at the tip.
                                                    6. *P. tosaensis*
    6B. Sori between the costa and margin; blades acuminate, sometimes with an obtuse tip, without isolated sori at the tip.
                                                    7. *P. annuifrons*

1. **Pleopeltis thunbergiana** Kaulf. *Lepisorus thunbergianus* (Kaulf.) Ching; *Polypodium lineare* Thunb., non Burm.; *Pleopeltis linearis* (Thunb.) Moore, non Kaulf.; *Polypodium lineare* var. *thunbergianum* (Kaulf.) Takeda; *Pleopeltis elongata* sensu Kunze, non Kaulf.——NOKI-SHINOBU. Evergreen glabrous fern; rhizomes long-creeping, 2–3 mm. across, densely scaly, the scales ascending, linear-subulate, 3–4 mm. long, dilated and ovate-cordate at base, dark brown, minutely dentate; fronds evergreen, glabrous, 12–30 cm. long; stipes 1–3 cm. long; blades coriaceous, linear to broadly so, 5–15 mm. wide, gradually acuminate, gradually narrowed at base, entire, somewhat recurved on margin, entire, dark green above, paler beneath, the costa prominent on both sides, usually with minute ovate appressed scales on lower half beneath; sori round, on upper half of blades, close to nearly contiguous at maturity, medial.——Tree-trunks and rocks in

lowlands and low mountains; Hokkaido, Honshu, Shikoku, Kyushu; common.——Korea, Formosa, China, Indochina, and Philippines.

2. **Pleopeltis onoei** (Fr. & Sav. Okuyama. *Polypodium onoei* Fr. & Sav.; *Lepisorus onoei* (Fr. & Sav.) Ching——HIME-NOKI-SHINOBU. Rhizomes long-creeping, slender, about 1 mm. across, rather densely scaly, the scales appressed, lanceolate, 2.5–3 mm. long, acuminate, dark brown, slightly paler toward the margin, rounded to obscurely lobed at base; fronds evergreen, remote, 3–7 cm. long, 2–5 mm. wide, glabrous; stipes 2–7 mm. long; blades coriaceous, linear to linear-spathulate, 2–5 mm. wide, mostly broadest below the rounded or very slightly obtuse apex, the margins very slightly recurved, entire, deep green above, paler beneath; costa slender, the veins invisible; sori few, on upper portion of the blade, median, round.——Mossy rocks and tree-trunks in mountains; Hokkaido, Honshu, Shikoku, Kyushu; rather common.——Korea.

3. **Pleopeltis uchiyamae** (Makino) Ohwi. *Polypodium uchiyamae* Makino; *Lepisorus uchiyamae* (Makino) H. Itō——KO-URABOSHI. Rhizomes slender, creeping, about 2 mm. across, rather densely scaly at least toward tip, the scales rather thin, clathrate, ascending to obliquely spreading, lanceolate, (2–)2.5–3 mm. long, acuminate-attenuate, rounded to cordate at base, toothed; fronds evergreen, remote, 5–15 cm. long, glabrous; stipes 3–6 cm. long; blades thinly coriaceous, narrowly lanceolate to broadly linear, 5–10 cm. long, 3–5 mm. wide, usually rather abruptly narrowed at both ends, the margins very slightly recurved, entire, paler beneath; costa raised on both sides or often scarcely so above, the veins invisible; sori round, on upper part of blade.——Rocks near seacoast; Honshu (Tōkaidō and s. Kinki Distr.), Shikoku, Kyushu; rare.——Ryukyus.

4. **Pleopeltis clathrata** (C. B. Clarke) Bedd. *Polypodium clathratum* C. B. Clarke; *Lepisorus clathratus* (C. B. Clark) Ching; *Polypodium papakense* Masam.; *L. clathratus* var. *namegatae* Kurata; *Pleopeltis clathrata* var. *namegatae* (Kurata) Ohwi——TOYOGUCHI-URABOSHI. Rhizomes creeping, densely scaly, 1–1.5 mm. across, the scales deep brown, membranous, lanceolate or ovate-lanceolate, 3–5 mm. long, 1–1.5 mm. wide, acuminate, clathrate with large oblong thin cells, rounded and peltate at base, with few slender teeth; stipes 1–4 cm. long, straw-colored, glabrous; blades lanceolate or linear-lanceolate, 5–15 cm. long, 6–15 mm. wide, gradually narrowed at both ends, entire or slightly undulate, pale green, glabrous, thinly dispersed with small scales while young, the costa distinct, raised below; sori slightly nearer the costa than the margin.——Honshu (Shinano Prov.); rare.——Formosa, China, and n. India to Siberia.

5. **Pleopeltis ussuriensis** Regel & Maack var. **distans** (Makino) Okuyama. *Polypodium lineare* var. *distans* Makino; *Polypodium distans* (Makino) Makino; *Polypodium annuifrons* var. *distans* (Makino) Nakai; *Polypodium ussuriense* sensu auct. Japon., non Regel——MIYAMA-NOKI-SHINOBU. Rhizomes slender, long-creeping, about 1.5 mm. across, densely scaly toward the tip, the scales appressed, deltoid-ovate, 0.5–0.7 mm. long, abruptly acuminate, black-brown; fronds evergreen, remote, 10–20 cm. long, glabrous; stipes 2–5 cm. long; blades thinly coriaceous, linear-lanceolate, 0.5–1.5 cm. wide, acuminate at both ends, entire, the margins sometimes slightly recurved, green, with minute appressed ovate scales along the costa beneath while young; costa raised on both sides, the veins invisible; sori on upper half of the blade, medial.——Mossy rocks and tree-trunks in mountains; Hokkaido, Honshu, Shikoku, Kyushu.——s. Kuriles. The typical variety occurs in e. Siberia, Manchuria, China, and Korea.

6. **Pleopeltis tosaensis** (Makino) Ohwi. *Polypodium tosaense* Makino; *Lepisorus tosaensis* (Makino) H. Itō; *Polypodium lineare* var. *caudatum* Makino; *Polypodium lineare* var. *contortum* sensu auct. Japon., non Christ——TSUKUSHI-NOKI-SHINOBU, ONAGA-URABOSHI. Rhizomes short-creeping, densely scaly, 2–3 mm. across, the scales brown, broadly ovate, 2(–2.5) mm. long, abruptly contracted at the tip, linear-attenuate; fronds evergreen, approximate, 15–30 cm. long; stipes 0.5–3 cm. long; blades thinly coriaceous, lanceolate to narrowly so, broadest below the middle, acuminate to caudate, often gradually attenuate at the base, deep green above, paler beneath, glabrous or thinly scaly while young along the lower half of the costa beneath; costa slender; sori mostly in 2 series, much nearer the costa than the margin, sometimes with an isolated pair on the apical tail of blade.——Honshu (s. Kinki Distr.), Shikoku, Kyushu.——Formosa and China.

7. **Pleopeltis annuifrons** (Makino) Nakai. *Polypodium annuifrons* Makino; *Lepisorus annuifrons* (Makino) Ching——HOTEI-SHIDA. Rhizomes creeping, 2–3 mm. across, densely scaly, the scales dark brown, ovate-deltoid, 1.5–2 mm. long, abruptly contracted at the tip to a linear tail; fronds deciduous, 10–25 cm. long; stipes 1–4 cm. long; blades firmly chartaceous to thinly coriaceous, lanceolate, 15–30 mm. wide, acuminate, sometimes obtuse at the tip, abruptly narrowed at base; costa slender, the veins not visible; sori on upper half of blade, medial.——Hokkaido, Honshu, Shikoku, Kyushu.

4. NEOCHEIROPTERIS Christ   KURIHA-RAN ZOKU

Terrestrial; rhizomes long-creeping, dictyostelic, with clathrate scales; fronds herbaceous, remote; stipes inconspicuously jointed; blades thin, pinnatifid or sometimes pedatisect, more commonly simple and entire; sparsely scaly; venation reticulate; sori sometimes elongate, parallel to the costas near the base, more commonly round; paraphyses peltate or laterally affixed, clathrate.——Five or six species, India to e. Asia.

1A. Blades lanceolate, 25–40 cm. long, 4–7 cm. wide, broadest near middle, unlobed or rarely irregularly pinnatilobed; sori in a subcostal row on each side at least in lower part of blade. .... 1. *N. ensata*
1B. Blades deltoid-lanceolate, 10–20 cm. long, 1.5–5 cm. wide, broadest near base, usually sinuately lobed in lower part; sori scattered.
 2. *N. subhastata*

1. **Neocheiropteris ensata** (Thunb.) Ching. *Polypodium ensatum* Thunb.; *Pleopeltis ensata* (Thunb.) Moore; *Neolepisorus ensatus* (Thunb.) Ching; *Microsorium ensatum* (Thunb.) H. Itō——KURIHA-RAN. Rhizomes long-creeping, 3–4 mm. across, the scales membranous, ascending to obliquely spreading, deltoid-lanceolate, 4–6 mm. long, acuminate, ciliate, with oblong cell walls and translucent areolae; stipes 10–30 cm. long, usually shorter than the blades, scaly especially

toward base; blades firmly chartaceous, lanceolate to broadly so, 25–40 cm. long, 4–7 cm. wide, usually broadest below the middle, acuminate, narrowed at base, decurrent on the stipe, subentire, undulate or rarely irregularly lobed, with appressed small scales beneath and on costas above; costas raised on both sides, the veins 20–30 pairs, obliquely spreading, slender, slightly flexuous, the veinlets scarcely distinct; sori in a single series on each side of the costas, often also scattered, round, sometimes elliptic, 3–5 mm. across, covered while young with numerous small peltate paraphyses.——Shaded places and on moist rocks in mountains; Honshu, Shikoku, Kyushu.——Korea, Ryukyus, and Formosa.

Var. **platyphylla** Tagawa. *N. ensata* var. *phyllomanes* Tagawa, excl. basionym——HIROHA-KURIHA-RAN.   Blades larger, 6–9 cm. wide; sori close to the lateral veins.——Honshu (Suwo Prov.).——Ryukyus.

2.   **Neocheiropteris subhastata** (Bak.) Tagawa. *Poly-*

podium subhastatum Bak.; *Microsorium subhastatum* (Bak.) Ching; *P. subhastatum* var. *longifrons* Takeda; *M. subhastatum* var. *longifrons* (Takeda) Ching——YANONE-SHIDA.   Rhizomes slender, long-creeping, wiry, 2–3 mm. across, the scales membranous, spreading, more or less lustrous, lanceolate, 2.5–4 mm. long, acuminate, sparsely denticulate, clathrate; fronds remote; stipes 2–6 cm. long; blades thinly chartaceous, broadly linear to deltoid-lanceolate, 10–15(–20) cm. long, 1.5–5 cm. wide, acuminate, usually with an obtuse angle on both sides at base, abruptly decurrent on the short stipe, subentire to undulate, often sinuately lobule on lower margin, nearly glabrous on both sides; costa raised on both sides, the veins and veinlets not distinct; sori round, or sometimes oblong, about 2 mm. across, with peltate paraphyses while very young.——Honshu (Kazusa and Izu Prov. and westw.), Shikoku, Kyushu.——China.

5.   LEMMAPHYLLUM Presl   MAMEZUTA ZOKU

Epiphytic; rhizomes long-creeping, the scales ovate-lanceolate, entire or short-ciliate; fronds dimorphic, jointed on the rhizomes; sterile blades obovate to ovate or elliptic, entire, somewhat fleshy, usually glabrous or nearly so; veins reticulate; fertile blades linear or oblanceolate; sporangia mostly in continuous coenosori, these not confluent around the apex; paraphyses peltate, clathrate.——About 4 species, Japan to the Himalayas.

1A.   Fronds 1–3 cm. long; sori elongate along the costas. . . . . 1. *L. microphyllum*
1B.   Fronds 3–5 cm. long; sori round to ovate, remote, the upper ones often nearly confluent. . . . . 2. *L. pyriforme*

1.   **Lemmaphyllum microphyllum** Presl. *Taenitis microphylla* (Presl) Mett.; *Drymoglossum microphyllum* (Presl) C. Chr.; *D. carnosum* var. *minor* Hook.; *D. nobukoanum* Makino; *D. carnosum* sensu auct. Japon., non J. Sm.—— MAMEZUTA.   Evergreen; rhizomes long-creeping, filiform, 0.7–1 mm. across, fronds remote, the scales rather sparse, dark brown, spreading, linear-filiform, 1–1.5 mm. long; blades coriaceous, orbicular to elliptic or broadly ovate to obovate, 1–2 cm. long, 6–15 mm. wide, rounded to very obtuse, rounded to broadly cuneate at base, glabrous, entire, the stipes 2–8 mm. long; costa slightly raised beneath in lower half, the veins and veinlets invisible; fertile blades broadly linear to narrowly oblanceolate, 1–3 cm. long, 3–4 mm. wide, obtuse to rounded, gradually narrowed at base, the stipes 1–3 cm. long; costa raised beneath and in lower half of upper side; sori elongate along the costas and covering the entire blade beneath, with numerous minute peltate paraphyses while young.

——Rocks and tree trunks in lowlands and low mountains; Honshu, Shikoku, Kyushu; common.——Korea, Ryukyus, and Formosa.

2.   **Lemmaphyllum pyriforme** (Ching) Ching. *Polypodium pyriforme* Ching——ONI-MAMEZUTA.   Rhizomes long-creeping, very slender, scarcely 1 mm. across, loosely scaly to nearly naked, the scales brown, broadly ovate, entire, abruptly caudate-acuminate; fronds dimorphic; sterile blades ovate to pyriform in outline, 2–4 cm. long, 1.5–3 cm. wide, glabrous, the stipe 5–10 mm. long; fertile blades firmly coriaceous, lingulate-lanceolate, 3–5 cm. long, 5–10 mm. wide, sometimes similar to the sterile, subacuminate, gradually attenuate at base, very slightly recurved on margin, the stipes 2–3 cm. long; sori in 2 rows on the blade, round to ovate, remote, the upper ones often nearly confluent, medial.——Epiphytic on tree-trunks; Kyushu (Yakushima).——China.

6.   DRYMOTAENIUM Makino   KURAGARI-SHIDA ZOKU

Evergreen, epiphytic on tree trunks; rhizomes short-creeping, dictyostelic, with black lanceolate acuminate scales; fronds approximate, all alike, jointed at base, narrowly linear, glabrous, coriaceous; veins concealed, anastomosing to form one or two rows of areolae, these with a few included veinlets; sori continuous in a groove on each side of costa, uninterrupted, paraphyses peltate, clathrate.——A single species.

1.   **Drymotaenium miyoshianum** (Makino) Makino. *Taenitis miyoshiana* Makino; *Pleurogramma robusta* Christ; *Monogramma robusta* (Christ) C. Chr.——KURAGARI-SHIDA. Rhizomes creeping, about 2.5 mm. across, densely scaly, the scales membranous, deltoid-lanceolate to narrowly deltoid-ovate, 2.5–4 mm. long, about 1 mm. wide, acuminate, slightly cordate at base, minutely toothed, clathrate, with translucent, elliptic cells and dark brown cell walls; fronds tufted, simple, fleshy, elongate, narrowly linear, 25–50 cm. long, 3–4 mm. wide, nar-

rowed to a short obtuse apex, gradually narrowed to a short stipelike base about 2 mm. across, glabrous, entire, with a rounded margin, 1-grooved above, with a raised obtuse prominent costa beneath, the veins hidden, anastomosing; sori continuous, in an elongate line along the costa, covered with small peltate paraphyses while very young.——Tree trunks in mountains; Honshu (Mikawa and Hida Prov. and westw.), Shikoku; rare.——Formosa, China.

7. PYRROSIA Mirbel   HITOTSUBA ZOKU

Epiphytic; rhizomes creeping, dictyostelic, scaly; fronds coriaceous, uniform or slightly dimorphic, jointed on the rhizome, usually simple and entire, sometimes pedately lobed, with more or less persistent stellate scales; veins reticulate; sori usually apical on the veinlets, round, or sometimes elongate and confluent; paraphyses stellate.——About 100 species, Amur and Ussuri, Africa to Polynesia and New Zealand, most abundant in se. Asia.

1A. Blades linear, 3–8(–13) cm. long, 2–5 mm. wide, sessile; sori in 2 parallel series on upper part of blade, distinct or subconfluent.
1. *P. linearifolia*
1B. Blades broader, more than 15 mm. wide, distinctly stipitate; sori scattered, occupying the entire fertile blade or the upper half, sometimes confluent.
2A. Blades simple, linear to lanceolate, acute or gradually narrowed to the base.
3A. Rhizomes with close stipes; blades 5–15 mm. wide. .... 2. *P. pekinensis*
3B. Rhizomes with remote stipes; blades 2–7 cm. wide. .... 3. *P. lingua*
2B. Blades pedately 3- to 5-lobed, hastate, truncate to subcordate at base. .... 4. *P. hastata*

1. Pyrrosia linearifolia (Hook.) Ching. *Niphobolus linearifolius* Hook.; *Polypodium linearifolium* (Hook.) Hook.; *Cyclophorus linearifolius* (Hook.) C. Chr.; *Neoniphopsis linearifolia* (Hook.) Nakai——BIRODO-SHIDA. Rhizomes long-creeping, slender, about 2 mm. across, the scales dense, ascending, linear, 4–5 mm. long, filiform at the tip, sparsely toothed; blades simple, linear, 3–8(–13) cm. long, 2–5 mm. wide, very obtuse, gradually narrowed toward base, entire, sessile, with red-brown to pale brown stellate hairs beneath, sparsely so above; costa glabrous beneath; sori in 2 parallel series along the costa on upper part of the blade, subconfluent, elliptic to round.——Rocks and tree trunks in mountains; Hokkaido, Honshu, Shikoku, Kyushu.——Korea, Manchuria, China, and Formosa (var.).

2. Pyrrosia pekinensis (Christ) Ching. *Cyclophorus pekinensis* Christ——IWADARE-HITOTSUBA. Rhizomes creeping, 1.5–2 mm. across, with close stipes, densely scaly, the scales appressed, about 2 mm. long, deltoid-lanceolate, caudately long-attenuate, dark brown at the center, gray-brown and thinner toward margin, denticulate; fronds 15–25 cm. long, firm, the stipes 5–8 cm. long, sparsely stellate-pilose, scaly at base; scales linear, membranous, pale brown, gradually caudate-acuminate, 3–4 mm. long; blades linear-lanceolate or broadly linear, 8–15 cm. long, 5–15 mm. wide, simple, entire, narrowed to the obtuse apex, gradually attenuate to the stipe, sparingly stellate-pilose but soon glabrate above, densely cinereous-brown stellate hairy, the costa slender, slightly elevated beneath; sori somewhat scattered, not confluent, round. ——Rocks; Honshu (Mikawa, Totomi, and Sagami Prov.); very rare.——n. China.

3. Pyrrosia lingua (Thunb.) Farwell. *Acrostichum lingua* Thunb.; *Polypodium lingua* (Thunb.) Sw.; *Niphobolus lingua* (Thunb.) Spreng.; *Polycampium lingua* (Thunb.) Presl——HITOTSUBA. Evergreen; rhizomes long-creeping, wiry, about 3 mm. across, the scales dense, appressed, rather firm, red- to yellow-brown, slightly lustrous, linear-lanceolate,

5–8 mm. long, ciliate; stipes 5–25 cm. long, shorter to slightly longer than the blade, straw-colored, densely brown stellate-hairy while young, glabrescent, scaly only at base; blades firmly coriaceous, mostly simple, lanceolate to broadly so, acute or subobtuse, acute to broadly cuneate at base, entire or undulate, densely brown stellate-hairy on both sides while young, soon glabrous on upper side, the sterile 10–25 cm. long, 2–7 cm. wide, the fertile 10–20 cm. long, 1.5–3 cm. wide; costa raised beneath, the veins and veinlets concealed; sori rounded, confluent, densely covering the underside of blade except the costa.——Rocks in hills and low mountains; Honshu (s. Kantō through Tōkaidō to Kinki and Chūgoku Distr.), Shikoku, Kyushu; common.——Ryukyus, Formosa, China to Indochina.——Cv. **Corymbiferus** with crested fronds is cultivated.

4. Pyrrosia hastata (Thunb.) Ching. *Acrostichum hastatum* Thunb.; *Niphobolus hastatus* (Thunb.) Kunze; *Polycampium hastatum* (Thunb.) Presl; *Polypodium tricuspe* Sw.; *Pyrrosia tricuspis* (Sw.) Tagawa——IWA-OMADAKA. Rhizomes short-creeping, densely scaly, 5–7 mm. across, the scales appressed, lustrous, black, firm, deltoid-lanceolate, slightly concave, acute, with short brown hairs toward margin; stipes 7–20(–30) cm. long, longer than the blades, with tardily deciduous brown appressed stellate hairs; blades firmly coriaceous, hastate, 5–15 cm. long, 3–10 cm. wide near base inclusive of the lateral lobes, pedately 3- to 5-lobed, truncate to subcordate at base, nearly glabrate and green on upper side, with gray- to red-brown densely appressed stellate hairs beneath; terminal lobe lanceolate, 4–12 cm. long, 1.5–4 cm. wide, acute to subobtuse, the lateral lobes obliquely spreading, narrowly deltoid-ovate to ovate, obtuse, sometimes with a small accessory lobule near base on the posterior side; costa slightly raised, the veins and veinlets concealed; sori round, densely covering the entire undersurface.——Rocks in mountains; Hokkaido, Honshu, Shikoku, Kyushu.——Korea and Manchuria.

8. MICROSORIUM Link   HOKOZAKI-URABOSHI ZOKU

Mostly epiphytic; rhizomes creeping, dictyostelic, usually with broad clathrate scales; stipes remote, jointed with the rhizomes; blades herbaceous to coriaceous, simple or pinnatifid, rarely pinnate, glabrous or rarely pubescent, not scaly, entire; venation irregularly reticulate; sori at the union of the veinlets, usually round, scattered, or in a single series on each side of the costa, without paraphyses.——About 40 species, Japan, Polynesia to Malaysia, India, and Africa.

1A. Sori large, depressed, 3–5 mm. across, usually in a single row on each side of the costa, or sometimes in a double row; rhizomes rather sparsely scaly. .... 1. *M. scolopendria*
1B. Sori smaller, not depressed, 1–2.5 mm. across, irregularly dispersed; rhizomes rather densely scaly.

2A. Rhizomes long-creeping; blades chartaceous, simple, brownish when dried, cuneately narrowed at base; sori mostly round, 2–2.5 mm. across. .... 2. *M. buergerianum*

2B. Rhizomes short-creeping; blades herbaceous, usually pinnately parted, rarely simple, green when dried, long-decurrent on the stipe; sori round to oblong, sometimes short-linear, 1–1.5 mm. across. .... 3. *M. dilatatum*

1. **Microsorium scolopendria** (Burm.) Copel. *Polypodium scolopendria* Burm.; *Phymatodes scolopendria* (Burm.) Ching; *Polypodium phymatodes* L.; *Drynaria phymatodes* (L.) Fée; *Phymatodes vulgaris* Presl——OKINAWA-URA-BOSHI, OKINAWA-KURIHA-RAN. Rhizomes long-creeping, rather stout, grayish, sparsely scaly, the fronds remote, the scales ascending, narrowly deltoid-lanceolate, 4–5 mm. long, acuminate, ciliolate, with oblong to elliptic dark brown cell walls, peltately attached; stipes 10–20 cm. long, straw-colored, naked, rather stout, lustrous, nearly terete; blades chartaceous-coriaceous, usually ovate, 10–30 cm. long, pinnately parted or the small ones sometimes simple and lanceolate, nearly truncate and abruptly narrowed at base, glabrous on both sides, somewhat lustrous on upper side; costa and costules on lobes prominently raised beneath; the veins and veinlets invisible on both sides; segments linear-lanceolate to lanceolate, 8–15 cm. long, 2–3 cm. wide, acuminate to acute, entire, slightly recurved on margin, the terminal one largest (in ours), the lateral ones obliquely spreading; sori round to elliptic, 3–5 mm. across, depressed, in a single row on each side of the costa, or sometimes in a double row.——Kyushu (Tsushima, according to H. Itō).——Ryukyus, Formosa, s. China to Polynesia, Malaysia, Australia, India, and tropical Africa.

2. **Microsorium buergerianum** (Miq.) Ching. *Polypodium buergerianum* Miq.——NUKABOSHI-KURIHA-RAN. Rhizomes long-creeping, 3–4 mm. across, densely scaly, the fronds remote, the scales spreading to ascending, membranous, rusty-brown, lanceolate, 4–5 mm. long, acuminate, ciliate, clathrate; stipes 3–12 cm. long; blades simple, chartaceous, lanceolate to narrowly so, 12–30 cm. long, 1.5–5 cm. wide, acute to subobtuse, entire or undulate, glabrous on both sides, cuneately narrowed at base, short-decurrent on the stipe; costa rather slender, raised beneath; sori round, 2–2.5 mm. across.——Tree trunks and shaded rocks; Honshu (Awa and Izu Prov. through Tōkaidō to s. Kinki Distr.), Shikoku, Kyushu.——Ryukyus, Formosa, China to Indochina.

3. **Microsorium dilatatum** (Bedd.) Sledge. *Polypodium dilatatum* Wall. ex Hook., non Hoffm.; *Pleopeltis dilatata* Bedd.; *M. hancockii* (Bak.) Ching; *Polypodium hancockii* Bak.——HOKOZAKI-URABOSHI. Rhizomes short-creeping, 5–7 mm. across, densely scaly, the scales membranous, spreading, deltoid-lanceolate, 5–7 mm. long, filiform at the tip, clathrate; fronds 30–60 cm. long; blades herbaceous, pinnately parted or rarely simple, narrowly oblanceolate, 3–5 cm. wide, abruptly acuminate, gradually narrowed from about the middle, long-decurrent on the stipe, glabrous except for a few scattered scales toward the base while young, entire or undulate; segments 1–3 pairs, spreading, lanceolate to ovate, abruptly acute to obtuse; costules rather slender, straw-colored; sori round to oblong, or sometimes short-linear, 1–1.5 mm. across, scattered on underside of blade.——Moist places along streams; Kyushu (Yakushima and Tanegashima in Osumi, and Koshiki in Satsuma Prov.).——Ryukyus, Formosa, China to Indochina.

9. COLYSIS Presl   IWA-HITODE ZOKU

Terrestrial; rhizomes creeping, dictyostelic, the fronds remote, the scales small, thin, entire or subentire; stipes jointed with the rhizome; blades herbaceous, simple or digitate to pinnate, entire, glabrous; veins anastomosing with simple or hamate veinlets; sori in a single series between the veins, solitary and elongate or few and round; paraphyses absent.——About 30 species, Africa to New Guinea, Queensland, and Japan.

1A. Sori linear, mostly solitary between the veins.
  2A. Blades regularly pinnate, scarcely decurrent on the stipe.
    3A. Blades broadly ovate, 10–25 cm. long; pinnae 2–6 pairs; veins and veinlets obscure. .... 1. *C. elliptica*
    3B. Blades narrowly ovate, 40–60 cm. long; pinnae 7–11 pairs; veins and veinlets usually slightly raised on both sides when dried.
                                                                2. *C. pothifolia*
  2B. Blades simple, irregularly pinnatiparted in lower half, prominently decurrent on the stipe.
    4A. Blades mostly pectinately parted in lower half, or simple, decurrent on the stipe, the winged part shorter than the blade.
                                                                  3. *C. shintenensis*
    4B. Blades simple, long-decurrent on the stipe, the winged part as long as to longer than the blade, at least in larger fronds.
                                                                      4. *C. wrightii*
1B. Sori round to oblong, in one series between the lateral veins, oblique to the rachis; blades simple. .... 5. *C. hemionitidea*

1. **Colysis elliptica** (Thunb.) Ching. *Polypodium ellipticum* Thunb.——IWA-HITODE. Rhizomes long-creeping, 4–6 mm. across, greenish, densely scaly, the fronds remote, the scales membranous, about 4 mm. long, acuminate, denticulate, lustrous, clathrate; stipes slender, 20–50 cm. long, straw-colored, naked except for a few deciduous scales toward base; blades herbaceous, broadly ovate, 10–25 cm. long, 10–20 cm. wide, pinnate, the rachis narrowly winged at least in upper part; pinnae 2–6 pairs, nearly alike, linear-lanceolate or sometimes the sterile lanceolate, 7–15(–20) cm. long, 8–20 mm. wide, long-acuminate, acute to obtuse at the tip, narrowed at base and decurrent on the narrowly winged rachis, entire, rarely crenate-undulate; costa slender, slightly raised, the veins and veinlets obscure; sori ascending to obliquely spreading, linear, 5–15 mm. long, rather remote.——Moist woods along streams; Honshu (Izu Prov. through Tōkaidō to s. Kinki Distr.), Shikoku, Kyushu.——Korea (Quelpaert Isl.), Ryukyus, Formosa, China to n. India, and Philippines.

2. **Colysis pothifolia** (Hamilt. ex D. Don) Presl. *Hemionitis pothifolia* Hamilt. ex D. Don; *Gymnogramme pothifolia* (Hamilt.) Spreng.; *Polypodium pothifolium* (Hamilt.) Mett.; *P. ellipticum* var. *pothifolium* (Hamilt.) Makino; *C. elliptica* var. *pothifolia* (Hamilt.) Ching——Ō-IWA-HITODE. Rhizomes long-creeping, densely scaly toward the tip, 5–10 mm.

across, the scales membranous, linear-lanceolate, 5–6 mm. long, gradually acuminate, sparsely toothed, clathrate; stipes 40–70 cm. long, 4–7 mm. across, naked except for a few withered scales toward base, straw-colored; blades narrowly ovate, 40–60 cm. long, simply pinnate, the rachis winged in upper part; pinnae thinly herbaceous, 7–11 pairs, obliquely spreading to ascending, linear-lanceolate to broadly linear, 12–25 cm. long, the fertile 1–2 cm. wide, the sterile to 3 cm. wide, gradually narrowed to the subcaudate tip, narrowed at base, more or less decurrent on the rachis, entire or undulate; costa more prominently raised beneath, the veins and veinlets very slender, usually slightly raised on both sides; sori linear, 1–1.8 cm. long, obliquely ascending to spreading.——Moist woods along streams; Shikoku, Kyushu.——Ryukyus, Formosa, China to India, and Philippines.

3. Colysis shintenensis (Hayata) H. Itō. *Polypodium shintenense* Hayata; *P. wrightii* var. *lobata* Rosenst.; *P. ellipticum* var. *simplicifrons* Christ; *C. simplicifrons* (Christ) Tagawa——HITOTSUBA-IWA-HITODE, SHINTEN-URABOSHI, WA-KAME-SHIDA. Rhizomes long-creeping, the fronds remote, the scales dense, brown, ascending, linear-lanceolate, 4–5 mm. long, clathrate; stipes slender, pale straw-colored, naked, 20–40 cm. long; blades herbaceous, mostly pectinately parted in lower half, or simple, lanceolate, 15–35 cm. long, 2–5 cm. wide in the unlobed portion, cuneate at base, decurrent on the stipe, the winged part shorter than the blade, glabrous, the segments nearly horizontally spreading, narrowly lanceolate, to 12 cm. long, 1–1.5 cm. wide, undulate to slightly crispate; sori linear, to 2.5 cm. long, spreading or obliquely ascending.——Honshu (Izu and Kii Prov.), Kyushu.——Formosa.

4. Colysis wrightii (Hook.) Ching. *Gymnogramme wrightii* Hook.; *Polypodium wrightii* (Hook.) Mett. ex Diels,

non Bak.——YARI-NO-HO-KURIHA-RAN. Rhizomes slender, long-creeping, 2–2.5 mm. across, densely scaly toward the tip, the scales linear-lanceolate, gradually long-acuminate, 3–4 mm. long; fronds 10–40 cm. long, remote; stipes slender, straw-colored, broadly winged on upper half in longer ones, winged to the base in shorter ones; blades chartaceous-herbaceous, simple, broadly lanceolate to deltoid-lanceolate, 7–25 cm. long, 2.5–6 cm. wide, gradually acute to acuminate, cuneate at base, long-decurrent on the broadly winged stipe, entire or crispate; costa straw-colored to pale brown, the veins and veinlets very slender, slightly visible on both sides; sori spreading, parallel, rather close, 1–2.5 cm. long.——Kyushu (Osumi Prov. incl. Yakushima and Tanegashima).——Ryukyus, Formosa, and China.

5. Colysis hemionitidea Presl. *Polypodium hemioniti-deum* (Presl) Mett.; *Pleopeltis hemionitidea* (Presl) Moore; *Polypodium ensatosessilifrons* Hayata; *C. hemionitidea* var. *ensatosessilifrons* (Hayata) Tagawa; *Microsorium ensato-sessilifrons* (Hayata) H. Itō——TAIWAN-KURIHA-RAN. Rhizomes long-creeping, scaly, 3–4 mm. across, the fronds remote, the scales membranous, clathrate; stipes 5–25 cm. long, straw-colored, sparsely scaly while young, usually winged; blades chartaceous-herbaceous, simple, lanceolate to broadly so, 15–40 cm. long, 4–6 cm. wide, acuminate, gradually narrowed or cuneate at base, decurrent on the stipe, entire or obsoletely undulate; costa raised, stouter beneath, the veins numerous, spreading, raised beneath, parallel, slightly flexuous, the veinlets raised beneath; sori interrupted, in a series between the lateral veins, round, sometimes oblong or elongate by fusion of several sori, about 2 mm. across.——Moist places in woods; Kyushu (Yakushima and Tanegashima); rare.——Ryukyus, Formosa, China to India, and Philippines.

10. CRYPSINUS Presl MITSUDE-URABOSHI ZOKU

Epiphytic; rhizomes creeping, dictyostelic, the scales lanceolate, attenuate or setaceous; stipes remote, jointed at base; blades firm to coriaceous, usually dimorphic, simple or pinnate, serrulate or remotely notched to toothed; main veins usually evident, connected by less evident branched cross veinlets to form areolae; sori on the cross veinlets, usually in a single row on each side of the costa and one between each pair of veins, impressed or superficial; paraphyses absent or filamentous.——More than 40 species, Japan to India and Malaysia, best developed in New Guinea.

1A. Fronds deciduous; blades membranous, pinnately parted, minutely and closely toothed. .... 1. *C. veitchii*
1B. Fronds evergreen; blades chartaceous, simple, or ternately to pinnately parted, subentire or remotely toothed or notched between the main lateral veins.
  2A. Blades pinnatiparted; segments 3–4 pairs, obtuse. .... 2. *C. yakuinsularis*
  2B. Blades simple or ternately parted.
    3A. Blades often ternately parted, sometimes simple, broadest near base, gradually narrowed to the tip, usually shorter than the stipe. .... 3. *C. hastatus*
    3B. Blades mostly simple, broadest near middle, narrowed at both ends, as long as to longer than the stipe.
      4A. Sori superficial; stipes frequently brownish purple at base. .... 4. *C. engleri*
      4B. Sori depressed; stipes straw-colored. .... 5. *C. yakushimensis*

1. Crypsinus veitchii (Bak.) Copel. *Polypodium veit-chii* Bak.; *Phymatodes veitchii* (Bak.) Ching; *Phymatopsis veitchii* (Bak.) H. Itō; *Polypodium shensiense* var. *filipes* Christ; *Phymatopsis veitchii* var. *filipes* (Christ) H. Itō——MIYAMA-URABOSHI. Rhizomes long-creeping, 2–2.5 mm. across, densely scaly, the scales ascending, brown, membranous, deltoid-lanceolate, 2–2.5 mm. long, ciliate; fronds deciduous; stipes slender, remote, 2–10 cm. long, naked or with a few scales at base; blades membranous, ovate-deltoid, 4–10 cm. long, 3–6 cm. wide, truncate at base, pinnately parted, minutely and closely toothed, glabrous on upper side, somewhat glaucous beneath, thinly scaly beneath while young;

segments 1–4 pairs, spreading, narrowly oblong to lanceolate, 6–12 mm. wide, obtuse to acute, decurrent on the winged rachis, the lowest slightly narrowed at base on posterior side, the terminal the longest, linear-lanceolate to lanceolate-ob-long, 2.5–6 cm. long, usually acuminate, often acute or obtuse in smaller fronds; sori round, about 2 mm. across, in 2 series on the terminal or upper pinnae, rarely over the whole blade, slightly nearer the costa than the margin.——Rocks in mountains; Hokkaido, Honshu, Shikoku.——Korea (Quel-paert Isl.), China to Tibet.

2. Crypsinus yakuinsularis (Masam.) Tagawa. *Poly-podium yakuinsulare* Masam.; *Phymatopsis yakuinsularis*

(Masam.) H. Itō; *Phymatodes yakuinsularis* (Masam.) Tagawa——YAKUSHIMA-URABOSHI. Evergreen; rhizomes creeping, 4–6 mm. across, glaucous, densely scaly, the scales subulate-lanceolate, about 3 mm. long, long-acuminate, nearly entire, ferrugineous at base; stipes slender, 4–6 cm. long, glabrous, lustrous; blades thinly chartaceous, broadly ovate-deltoid, 5–10 cm. long, nearly as wide, pinnatiparted, glabrous; segments 3–4 pairs, linear-lanceolate, 8–12 mm. wide, obtuse, the lowest horizontally spreading and short-decurrent, nearly entire, with remote minute incisions on the lateral veins; sori round, about 2 mm. across, in a row on each side of the costa.——Shikoku, Kyushu (Yakushima); rare.

3. Crypsinus hastatus (Thunb.) Copel. *Polypodium hastatum* Thunb.; *Phymatodes hastata* (Thunb.) Ching; *Phymatopsis hastata* (Thunb.) Kitag. ex H. Itō; *Polypodium hastatum* var. *yoshinagae* Makino——MITSUDE-URABOSHI. Evergreen; rhizomes creeping, 3–4 mm. across, densely scaly, the scales membranous, spreading, brown, linear-subulate, 3–4 mm. long, long-filiform at the tip, subentire; stipes to 22 cm. long, straw-colored, often purplish brown at base, lustrous; blades simple or frequently ternately parted, lanceolate, 15–35 cm. long, 10–15(–20) cm. wide, acuminate, the smaller blades oblong to ovate, 3–5 cm. long, obtuse; pinnae subentire or rarely lobulate, more or less glaucescent beneath, rounded to broadly cuneate at base; veins slender, spreading, the veinlets not visible; sori round, 2–3 mm. across, in 2 series on the pinnae, nearer the costa than the margin.——Rocks and walls; Hokkaido, Honshu, Shikoku, Kyushu; common.——Korea, Manchuria, Ryukyus, Formosa, and China.

4. Crypsinus engleri (Luerss.) Copel. *Polypodium engleri* Luerss.; *Phymatopsis engleri* (Luerss.) H. Itō; *Phymatodes engleri* (Luerss.) Ching——TAKA-NO-HA-URABOSHI. Evergreen; rhizomes long-creeping, about 3 mm. across, densely scaly, the scales membranous, ascending, red-brown, linear-lanceolate, 3–6 mm. long, filiform at the tip, slightly broadened and appressed to the rhizome at base, entire; stipes 5–15 cm. long, remote, usually shorter than the blades, lustrous, often brownish or purplish brown at base; blades firmly chartaceous, simple, linear-lanceolate, 10–30 cm. long, 12–25 mm. wide, obtuse, or rather abruptly acuminate, often cuneately narrowed at base, subentire to undulate, glabrous, often glaucescent beneath; costa rather stout, raised beneath, sulcate above, the veins spreading, slender, the veinlets invisible; sori superficial, round, about 2.5 mm. across, in 2 series on upper part of blade, slightly nearer the costa than the margin.——Tree trunks; Honshu (Izu Prov. and s. Kinki Distr.), Shikoku, Kyushu; rare.——Korea (Quelpaert Isl.) and Formosa.

5. Crypsinus yakushimensis (Makino) Tagawa. *Polypodium engleri* var. *yakushimense* Makino; *Polypodium yakushimense* (Makino) Makino; *Phymatopsis yakushimensis* (Makino) H. Itō; *Phymatodes yakushimensis* (Makino) Tagawa——HIME-TAKA-NO-HA-URABOSHI. Evergreen; rhizomes slender, long-creeping, 1.5–2 mm. across, rather densely scaly, the scales thinly membranous, ascending to obliquely spreading, pale- to rusty-brown, linear-subulate, 2–2.5 mm. long, contorted at the tip; stipes rather firm, straw-colored, slender, 3–10 cm. long, as long as to shorter than the blade; blades chartaceous, broadly linear to narrowly lanceolate, 5–20 cm. long, 5–10 mm. wide, gradually long-acuminate, sometimes glaucescent beneath, subobtuse at the tip, often abruptly acuminate at base, remotely crenate-undulate to subentire; costa slender, the veins ascending, very slender, scarcely visible, the veinlets obsolete; sori depressed, about 2.5 mm. across, in 2 series on upper part of blade, slightly nearer the costa than the margin.——Kyushu (Yakushima); rare.——Ryukyus and Formosa.

11. GRAMMITIS Sw.   HIME-URABOSHI ZOKU

Small epiphytes; rhizomes erect or short-creeping, rarely elongate, dictyostelic; fronds approximate or tufted, the stipes not jointed at base; blades membranous to fleshy or coriaceous, simple, lanceolate or linear, entire or rarely crenate to shallowly lobed, hairy or glabrescent, not scaly; costa usually prominent, the veins typically free, forked, sometimes somewhat anastomosing; sori typically on the lowest anterior veinlet of the forked veins, thus forming a single row on each side of the costa, rarely more of the veinlets with sori and forming several rows or the sori somewhat scattered, superficial or sometimes impressed or immersed; paraphyses filamentous or more often absent.——About 150 species, subantarctic regions north to the W. Indies, Africa, and Japan.

1. Grammitis dorsipila (Christ) C. Chr. & Tard.-Bl. *Polypodium dorsipilum* Christ; *P. asahinae* Ogata; *G. asahinae* (Ogata) H. Itō——HIME-URABOSHI. Rhizomes very short, ascending, rather densely scaly, the scales membranous, lanceolate to oblong-lanceolate, about 2 mm. long, short-acuminate, entire; fronds tufted, erect, thinly clothed with reddish brown spreading hairs about 0.5–0.7 mm. long; blades linear-oblanceolate to broadly linear, 2–8 cm. long, 2–4 mm. wide, obtuse to rounded, entire or obsoletely undulate on margin, gradually attenuate at base, long-decurrent on the very short, indistinct stipe, loosely involute on margin when dry, with a row of hydathodes on each side on upper surface; costa and veins obscure, the veins obliquely spreading, simple or once forked, the anterior branch (veinlet) slightly longer than to as long as the posterior; sori round or elliptic, dorsal or subterminal on the anterior veinlet, forming a row on each side along the costa.——Mossy tree trunks in mountains; Honshu (Izu Isls.), Kyushu (Satsuma Prov.); rare.——Ryukyus, China, and Indochina.

12. XIPHOPTERIS Kaulf.   ŌKUBO-SHIDA ZOKU

Small epiphytes; rhizomes creeping, dictyostelic, with ovate, acuminate, entire scales; fronds not jointed with the rhizome, blades mostly herbaceous, linear, pinnatifid to pinnate, hairy to glabrescent; veins solitary on the pinnules or pinnae, simple or once forked; sori dostal or terminal on the veins, on the anterior branch (veinlet), if this present; sori usually round, superficial, without paraphyses.——About 50 species, in the Tropics of both hemispheres.

1. Xiphopteris okuboi (Yatabe) Copel. *Polypodium okuboi* Yatabe; *Micropolypodium okuboi* (Yatabe) Hayata; *Grammitis okuboi* (Yatabe) Ching; *P. pseudotrichomanoides* Hayata——Ōkubo-shida. Rhizomes short, ascending, the scales membranous, brown, lanceolate, about 2 mm. long, acute, entire; fronds tufted, erect to spreading, with reddish brown spreading hairs about 1 mm. long; stipes short, 3–12(–15) mm. long; blades coriaceous, linear to linear-lanceolate, 2–12(–25) cm. long, 3–6 mm. wide, acute to obtuse, gradually narrowed at base, pectinately pinnatiparted, the segments 8–30 pairs, obliquely spreading, oblong-ovate to narrowly deltoid-ovate, 1–1.6 mm. wide, obtuse to subacute, entire; costa and veins obscure; sori round to elliptic, about 1 mm. across, solitary at the base of each pinna.——Mossy rocks and tree trunks in mountains; Honshu (Kantō Distr. through Tōkaidō to s. Kinki Distr.), Shikoku, Kyushu; rare.——Formosa.

13. LOXOGRAMME Presl Saji-ran Zoku

Small epiphytes; rhizomes creeping, dictyostelic, with ovate acuminate entire scales; fronds not jointed with the rhizome, sessile or short-stiped, simple and entire, usually lanceolate to oblanceolate or linear, fleshy, glabrous, the main veins barely evident, immersed, freely anastomosing; sori elongate, oblique in one row on each side of costa, commonly overlapping, superficial or slightly impressed, without paraphyses.——About 40 species, Polynesia, Malaysia, Japan, Madagascar, and Africa, one species in Mexico and Central America.

1A. Rhizomes very slender, 0.5–1 mm. across when dried; blades mostly under 10 cm. long, spathulate to linear-spathulate, very obtuse to subacute. . . . . 1. *L. grammitoides*
1B. Rhizomes 1–2 mm. across when dried; blades 10–30 cm. long, oblanceolate to broadly linear, acuminate to acute.
    2A. Stipes slender, pale green to straw-colored. . . . . 2. *L. salicifolia*
    2B. Stipes rather stout, purple-brown or black-purple toward base. . . . . 3. *L. saziran*

1. Loxogramme grammitoides (Bak.) C. Chr. *Gymnogramme grammitoides* Bak.; *Polypodium grammitoides* (Bak.) Diels; *Polypodium yakushimae* Christ; *L. yakushimae* (Christ) C. Chr.; *L. minor* Makino——Hime-saji-ran. Small glabrous fern; rhizomes very slender, long-creeping, 0.5–1 mm. across when dried, the scales numerous, dense, membranous, linear-lanceolate, long-acuminate, dark brown, entire, clathrate, about 1.5 mm. long; fronds remote, 4–10(–12) cm. long; blades spathulate to linear-spathulate, 5–10(–15) mm. wide, somewhat fleshy, very obtuse to subacute, broadest below the apex, entire or nearly so, the stipe very short and indistinct; veins hidden; sori 1–6 on each side of the broadest part of the blade, near the costa, oblong to linear-oblong, 1.5–2 mm. wide, very oblique.——Mossy rocks in mountains; Honshu, Shikoku, Kyushu.——Formosa and China.

2. Loxogramme salicifolia (Makino) Makino. *Gymnogramme salicifolia* Makino; *Polypodium makinoi* C. Chr.; *L. makinoi* (C. Chr.) C. Chr.; *L. fauriei* Copel.——Iwa-yanagi-shida. Glabrous evergreen fern; rhizomes long-creeping, densely scaly, 1–1.5 mm. across, the scales membranous, ovate to ovate-lanceolate, about 2 mm. long, acuminate, those at base of the stipe oblong-lanceolate, about 3 mm. long, reddish brown to brown; fronds distant, somewhat dimorphic, coriaceous, 10–30 cm. long; blades narrowly oblanceolate to broadly linear, the sterile sometimes oblanceolate, 1–2.5 cm. wide, the fertile slightly narrower, entire, slightly recurved on margin when dried, gradually narrowed to the slender pale green to straw-colored and narrowly winged stipe 2–10 cm. long, the costa stout, slightly raised beneath, prominent above, the veins hidden; sori usually on the upper half of blade, very oblique, midway between the costa and margin, usually slightly overlapping or sometimes arranged nearly end to end in a single row on each side of the costa, linear, 1–2(–2.5) cm. long, straight.——Mossy rocks and tree trunks; Honshu (Awa Prov. through Tōkaidō to Kinki Distr.), Shikoku, Kyushu.——Korea (Quelpaert Isl.), Ryukyus, Formosa, China, Indochina, and India.

3. Loxogramme saziran Tagawa. *Gymnogramme involuta* sensu Makino, non Hook.; *L. fauriei* sensu Ogata, non Copel.——Saji-ran. Stouter than the preceding; rhizomes long-creeping, scaly, 1.5–2 mm. across when dried, the scales rather firmly membranous, appressed to ascending, dark brown, oblong-ovate to ovate-lanceolate, about 2 mm. long, acuminate, entire, obsoletely clathrate, slightly longer and more distinctly clathrate at base of stipe; fronds 15–40 cm. long; blades coriaceous, rather fleshy, narrowly oblanceolate, broadest above the middle, acuminate or shortly caudate-acuminate, entire, gradually narrowed to the short and slightly flattened purple-brown to dark-purple stipe; costa prominent on upper side, the veins hidden; sori on upper half of blade, usually close together and overlapping, oblique, 1–3 cm. long, nearer the costa than the margin.——Tree trunks in mountains; Honshu (Kantō Distr. and westw.), Shikoku, Kyushu.——China and Indochina.

14. CTENOPTERIS Bl. Kireba-ōkubo-shida Zoku

Epiphytes; rhizomes short-creeping or erect, dictyostelic, densely scaly, the scales entire or ciliate; stipes fascicled, usually not jointed with the rhizome; blades herbaceous to coriaceous, usually lanceolate, contracted at both ends, usually pectinately pinnatisect to pinnate, setose or glabrescent; veins many in each pinnule, nearly always simple; sori dorsal or terminal on the veins, round or elliptic, superficial to deeply immersed, without paraphyses.——More than 200 species, in the tropics.

1. Ctenopteris sakaguchiana (Koidz.) H. Itō. *Polypodium sakaguchianum* Koidz.——Kireba-ōkubo-shida. Rhizomes very short, ascending, the scales minute, dark brown, ovate, about 0.5 mm. long, obtuse, sparsely ciliate; fronds with scattered, slender, spreading pale hairs; stipes slender, 1–1.5 cm. long; blades firmly herbaceous, linear, 5–18 cm. long, 5–8 mm. wide, obtuse to acute, pinnately divided; pinnae obliquely spreading, ovate to ovate-oblong, obtuse, decurrent on the rachis, with 1–3 teeth or short lobules on the anterior and sometimes also on the posterior margin; costa and veins more or less distinct; sori round, 2–5 on a pinna, solitary on each tooth or lobe.——Mossy rocks in mountains; Honshu (Musashi and Kii to Yamato Prov.), Kyushu; very rare.

Fam. 21. VITTARIACEAE   SHISHI-RAN KA   Vittaria Family

Mostly epiphytic ferns; rhizomes creeping to suberect, protostelic or siphonostelic; scales clathrate; fronds simple, entire or rarely cleft at the tip, glabrous, with long spicular cells in the epidermis; veins reticulate, free veinlets absent; sori mostly elongate along the veins; indusia absent; paraphyses usually present.——Eight genera with about 140 species, chiefly tropical.

1A. Costa absent or partial; sori few to several, elongate along the veins and sometimes similarly reticulate. . . . . 1. *Antrophyum*
1B. Costa extending to apex of frond; sori linear, solitary, uninterrupted, sometimes marginal or submarginal. . . . . 2. *Vittaria*

1. ANTROPHYUM Kaulf.   TAKIMI-SHIDA ZOKU

Mostly epiphytic; rhizomes creeping, short, with narrow clathrate scales and hairy roots; fronds densely tufted, sessile or stipitate, not jointed with the rhizome; blades simple, entire, glabrous, firm; costa absent or partial, the veins repeatedly dichotomous and typically anastomosing to form large elongate areolae; sori elongate along the veins, superficial, or more often immersed; paraphyses present.——Nearly 40 species in the tropics from Polynesia to Africa.

1. Antrophyum obovatum Bak. *A. japonicum* Makino ——TAKIMI-SHIDA. Evergreen, glabrous; rhizomes short, densely scaly, the scales membranous, dark brown, linear, 4–5 mm. long, with 4–6 rows of linear cells, sparsely spinulose-toothed; fronds green, smooth except near base of the stipe; stipes more or less tufted, 3–12 cm. long, 2.5–3 mm. wide, flattened, pale green, smooth; blades rhombic-obovate, 5–12 cm. long, 1.5–3(–6) cm. wide, abruptly acuminate or caudately acuminate, sometimes shallowly 2-or 3-lobed toward the tip or entire, and often slightly recurved on margin, narrowed to the stipe at base; costa absent, the veins immersed, anastomosing, the areolae linear-lanceolate to lanceolate; sori linear, impressed, sunken in a groove along the veins, often partially anastomosing; paraphyses simple or rarely branched, capitate. ——Moist rocks in mountains; Honshu (Sagami and Echizen Prov. and westw.), Shikoku, Kyushu; rare.——Formosa, China, Indochina, and n. India.

2. VITTARIA J. E. Smith   SHISHI-RAN ZOKU

Epiphytic, with short-creeping rhizomes; scales dark, clathrate; fronds crowded, narrowly linear, entire, stipitate or sessile, firm, glabrous, costate, the veins forming a single row of areolae between the costa and the submarginal fertile vein; sori continuous at maturity along the fertile vein, immersed or almost superficial; paraphyses present.——About 80 species in the Tropics.

1A. Sori in a marginal groove; costa not raised on underside. . . . . 1. *V. zosterifolia*
1B. Sori dorsal, naked or partially enclosed by the recurved margin of the blade; costa more or less raised on underside.
  2A. Sori sunken in a groove.
    3A. Sori midway between the costa and margin; blades deeply 2-grooved on upper side. 2. *V. fudzinoi*
    3B. Sori submarginal; blades 1-grooved or nearly flat on upper side. . . . . 3. *V. flexuosa*
  2B. Sori superficial, intramarginal, not sunken in a groove or protected by the margin of the blade. . . . . 4. *V. forrestiana*

1. Vittaria zosterifolia Willd. *V. elongata* var. *zosterifolia* (Willd.) Tard.-Bl. & C. Chr.; *V. formosana* Nakai—— AMAMO-SHISHI-RAN, SHIMA-SHISHI-RAN. Evergreen, glabrous; rhizomes creeping, 3–4 mm. across, densely scaly, the scales black-brown, ascending, linear-lanceolate, 5–7 mm. long, about 0.5 mm. wide, sparsely toothed, clathrate, the filiform tip about 1/3–1/2 as long as the body; fronds thinly coriaceous, pendulous, linear, flat, 50–80 cm. long, 6–12 mm. wide, obtuse, gradually narrowed at base, glabrous except for a few scales at base of stipe; costa not distinct; sori continuous in marginal grooves.——Tree trunks; Shikoku, Kyushu (Yakushima and Tanegashima); rare.——Ryukyus, Formosa, Indochina, Polynesia, Malaysia, and Mascarene Isls.

2. Vittaria fudzinoi Makino. *V. japonica* var. *sessilis* Eaton ex Yoshinaga; *V. sessilis* (Eaton ex Yoshinaga) Makino——NAKAMI-SHISHI-RAN. Rhizomes short-creeping, densely scaly, the scales membranous, brown, slightly lustrous, linear-lanceolate, 4–5 mm. long, filiform at the tip, sparsely toothed, clathrate; fronds coriaceous, fleshy, lustrous, closely approximate, 20–40 cm. long including the stipe, 3–4(–5) mm. wide, acute to subobtuse, gradually attenuate toward the base; costa stout, raised beneath, with a deep groove on each side above; sori on a groove midway between the costa and margin. ——Honshu (Izu and Kii Prov. westw.), Shikoku, Kyushu; rare.——China.

3. Vittaria flexuosa Fée. *V. japonica* Miq.; *V. lanceola* Christ; *V. flexuosa* var. *japonica* C. Chr.——SHISHI-RAN. Evergreen, glabrous; rhizomes short-creeping, densely scaly; scales dark-brown, membranous, broadly linear, 2.5–4 mm. long, sparsely toothed, filiform above; fronds approximate, deep green, lustrous, the blades linear, 25–50 cm. long, 5–8 mm. wide, gradually narrowed at tip, attenuate toward base to the short stipe, grooved above, the costa raised beneath; sori submarginal, on a groove partially covered by the recurved margin of blade.——Shaded rocky cliffs and tree trunks; Honshu (Kantō Distr. and westw.), Shikoku, Kyushu; rather common westw.——Ryukyus, Formosa, China, Indochina, and n. India.

4. Vittaria forrestiana Ching. Ōba-shishi-ran. Rhizomes creeping, scaly, the scales rusty-brown, very thin, lanceolate, about 1 cm. long, long-acuminate, ascending; fronds crowded, softly herbaceous, lanceolate, 15–25(–40) cm. long, 1.5(–3) cm. wide, gradually attenuate at both ends, short-stipitate, entire, hyaline on margin, flat; costa prominent on both surfaces, the veins somewhat evident, very oblique, slender; sori superficial, intramarginal.——Kyushu (Yakushima, according to M. Tagawa).——China and Indochina.

Fam. 22. MARSILEACEAE   Denji-sō Ka   Marsilea Family

Mostly aquatic, growing in mud, rarely floating or aerial; rhizomes creeping, solenostelic, hairy; leaves simple, linear, or with one or two pairs of leaflets at the end of a long petiole; veins freely forked and anastomosing at the apices; megasporangia and microsporangia enclosed in different sporocarps; sporocarps bony, borne on stipes or basal.——Three genera with about 70 species.

1. MARSILEA L.   Denji-sō Zoku

Leaves 4-foliolate; sori numerous on a gelatinous receptacle attached to the wall of the sporocarp.——Nearly 70 species, cosmopolitan, especially abundant in Australia and S. Africa.

1. Marsilea quadrifolia L. *Lemna quadrifolia* (L.) Desr.; *Zalusianskya quadrifolia* (L.) O. Kuntze; *Pteris quadrifoliata* L.; *M. quadrifoliata* (L.) L.——Denji-sō. Aquatic or hygrophytic summergreen perennial with brownish yellow appressed to ascending soft hairs while very young; rhizomes slender, long-creeping, 1.5–2 mm. in diameter; leaves radical, few, the petioles 5–15 cm. long; leaflets or pinnules 4, thinly chartaceous to membranous, sessile, deltoid-flabellate, 1–2 cm. long and as wide, broadly rounded at apex, broadly cuneate at base, entire, glabrous on upper surface; veins slender, nearly parallel, forked, somewhat anastomosing; pedicels short, 2–3-nate, connate, adnate to the base of stipe, erect; sporocarps solitary on a pedicel, ellipsoidal, 4–5 mm. long, densely pubescent at first, becoming glabrous, the basal tooth minute.——Sept.–Oct. Paddy fields and ponds; Hokkaido, Honshu, Shikoku, Kyushu; common.——Europe to n. India and e. Asia, introduced in N. America.

Fam. 23. SALVINIACEAE   Sanshō-mo Ka   Salvinia Family

Floating plants with a somewhat elongate and branched axis; leaves apparently distichous, straight in vernation; sporocarps very soft and thin-walled, two or more on a common peduncle, 1-locular, the central often branched receptacle with either a solitary macrospore or with numerous microspores.——A small family of 2 genera.

1A. Roots absent (some of the lower leaves finely dissected and rootlike); leaves unlobed, flat, short-petioled; stems simple. .... 1. *Salvinia*
1B. Roots present; leaves bilobed, sessile; stems pinnately branched, sessile. .... 2. *Azolla*

1. SALVINIA Adans.   Sanshō-mo Zoku

Small floating plants; leaves ternate with two lateral, foliar, green, entire, flat blades and a third finely dissected and submerged, the foliar structure simulating roots; sporocarps short-peduncled, on the floating leaves, globose or nearly so; microsporangia numerous; megasporangia few, each maturing only one megaspore.——About 10 species, mostly in tropical America and Africa, few in Eurasia.

1. Salvinia natans (L.) All. *Marsilea natans* L.; *S. europaea* Desv.——Sanshō-mo. Stems 5–10 cm. long, loosely branched, densely leaved, the two lateral foliar blades herbaceous, pale green, spreading, slightly imbricate, soft, elliptic to oblong, 8–15 mm. long, 6–10 mm. wide, flat or very slightly folded, pinnately arranged, rounded at apex, slightly cordate to rounded at base, short-petiolate, entire, with tufts of minute bristlelike hairs on upper surface, soft-puberulent beneath, the submerged leaf finely dissected and rootlike; sporangia of two kinds, fascicled, globose, pubescent.——Sept.–Nov. Ponds and paddy fields in lowlands; Honshu, Shikoku, Kyushu; common.——Europe to India and e. Asia.

2. AZOLLA Lam.   Aka-ukikusa Zoku

Small floating mosslike plants; roots simple, unbranched; stems pinnately branched, densely leafy; leaves 2-lobed, imbricate; sporocarps in pairs on the first leaf of a lateral branch, covered by the hoodlike upper lobe of the leaf, the smaller ones at base with a single macrospore, the larger ones globose, with a basal placenta and many pedicellate microsporangia bearing masses of microspores.——About 6 species.

1A. Leaves about 1.5 mm. long; plant prominently papillose, especially on the axis and lower part of leaves. .... 1. *A. imbricata*
1B. Leaves about 2 mm. long; plant nearly smooth or obsoletely papillose on leaves. .... 2. *A. japonica*

1. Azolla imbricata (Roxb.) Nakai. *Salvinia imbricata* Roxb. ex Griff.——Aka-ukikusa. Small reddish herb about 1 cm. long, deltoid or rounded in outline; stems or main axis closely branched, papillose, densely leafy, few-rooted; leaves sessile, 2-lobed, papillose, especially toward the base, deltoid-orbicular, about 1–1.5 mm. long, very obtuse to rounded at apex, hyaline on margin.——Ponds and paddy fields; Honshu (Kinki Distr. and westw.), Shikoku, Kyushu.——China to India.

2. Azolla japonica Fr. & Sav. ex Nakai. *A. pinnata* var. *japonica* Fr. & Sav.; *A. pinnata* sensu auct. Japon., non R. Br. ——Ō-AKA-UKIKUSA. Reddish herb about 2 cm. long, deltoid to orbicular-deltoid in outline; stems or main axis closely branched, with rather numerous slender roots; leaves closely imbricate, the lobes orbicular to orbicular-deltoid, about 2 mm. long, rounded at apex, broadly hyaline on margin.——Ponds and paddy fields; Honshu, Shikoku, Kyushu; common.

Plate 1

Osmunda asiatica Ohwi (Osmundaceae). Kawayu in Kushiro, Hokkaido. (Photo M. Tatewaki.)

Asplenium antiquum Makino (Aspleniaceae). Tanegashima Island, Kyushu. (Photo S. Ouchiyama.)

Plate 2

Dicranopteris linearis Underw. (Gleicheniaceae). Near Ujina in Bingo Prov., sw. Honshu. (Photo M. Tatewaki.)

Sphenomeris chinensis (L.) Maxon. (Pteridaceae). Forest near Cape Ashizuri, Tosa Prov., s. Shikoku. (Photo J. Ohwi.)

Plate 3

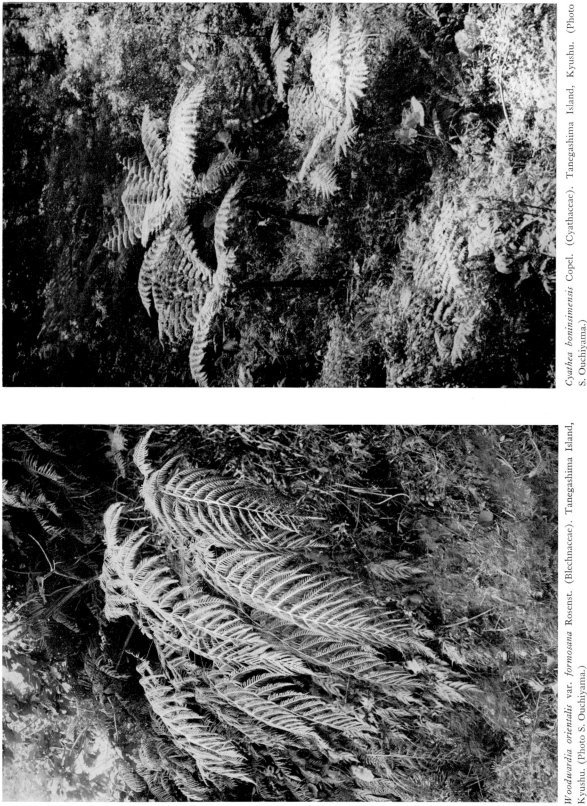

Cyathea boninsimensis Copel. (Cyatheaceae). Tanegashima Island, Kyushu. (Photo S. Ouchiyama.)

Woodwardia orientalis var. *formosana* Rosenst. (Blechnaceae). Tanegashima Island, Kyushu. (Photo S. Ouchiyama.)

Plate 4

Pinus pumila Reg. (Pinaceae). Daisetsu in Kurotake, ca. 1,900 m. alt., Hokkaido. (Photo Lindquist and Nitzelius.)

Cycas revoluta Thunb. (Cycadaceae). Cape Toi-saki, Osumi Prov., Kyushu. (Photo M. Tatewaki.)

Plate 5

Pinus thunbergii Parl. (Pinaceae). Shore of Osezaki, Idzu Prov., Honshu. (Photo J. Ohwi.)

Picea jezoensis var. *hondoensis* Rehd. (Pinaceae). Sugorokudani Valley, Hida Prov., Kyushu. (Photo Karizumi.)

Plate 6

Cryptomeria japonica (L. f.) D. Don. (Taxodiaceae). Nikko, Honshu. (Photo Lindquist and Nitzelius.)

Sciadopitys verticillata Sieb. & Zucc. (Taxodiaceae). Kii Prov., s. Honshu. (Photo M. Tatewaki.)

Phylum SPERMATOPHYTA

Seed Plants, Phanerogams or Flowering Plants

Class 1. GYMNOSPERMAE

Fam. 24. **CYCADACEAE** Sotetsu Ka Cycad Family

Palmlike, dioecious woody plants with a usually thickened and columnar rarely branched trunk; leaves closely and spirally arranged, pinnate or bipinnate with circinate vernation; male sporophylls conelike, composed of scales each with many anthers on the lower side, the female (in ours) of leaflike or scalelike segments, the ovules marginal, becoming large and drupelike, with one integument.——About 9 genera with somewhat more than 80 species chiefly in tropical and subtropical regions of both hemispheres.

1. CYCAS L. Sotetsu Zoku

Dioecious; trunk columnar, simple or sparingly branched, densely covered with bases of old petioles; leaves rosulate and terminal on the trunk, large, petiolate, pinnate, the linear entire pinnae gradually reduced to a spine toward base of rachis; male sporophylls forming an erect many-scaled cone, the scales with numerous globose anthers beneath; female sporophylls leaflike bearing ovules on the marginal notch.——About 15 species, in the warmer parts of Asia, Africa, Malaysia, and Australia.

1. Cycas revoluta Thunb. Sotetsu. Trunk 1–4 m. high; leaves densely lanate when young, becoming rigid, lustrous, rosulate at apex of stem; seeds broadly ovoid, slightly flattened, reddish, about 4 cm. long.——Usually near seashores; Kyushu.——Ryukyus.

Fam. 25. **GINKGOACEAE** Ichō Ka Ginkgo Family

Lofty deciduous dioecious tree with grayish brown branches; leaves alternate or fasciculate on short spurs, flabellate, about 7 cm. wide, parallel-veined, irregularly undulate on upper margin, often deeply notched or bifid at apex, entire at base and long-petiolate, glaucous; stamens in loose catkinlike spikes; ovules generally in pairs, long-pedunculate; seeds drupelike, about 2.5 cm. long, broadly ellipsoidal, the outer coat yellowish, fleshy at maturity and emitting a putrifying fetid odor at time of falling, the inner coat horny and whitish, peach-like.——Monogeneric. A relic family known only in cultivation.

1. GINKGO L. Ichō Zoku

One species with the characters of the family.

1. Ginkgo biloba L. Ichō. Much planted in this country, especially associated with Shinto shrines. Cultivars grown in gardens include: Cv. **Pendula** with pendulous branches; Cv. **Aureovariegata** with yellow-variegated leaves; and Cv. **Epiphylla**——Ohatsuki-ichō, with peduncles of the fruit becoming broad and winglike and adnate to the leaf petiole.

Fam. 26. **TAXACEAE** Ichii Ka Yew Family

Shrubs or trees, much branched, dioecious, rarely monoecious; leaves linear, subulate, or sometimes linear-lanceolate, alternate, often disposed in 2 rows; staminate cones strobilate, usually solitary in leaf axils, the stamens peltate with 4- to 9-locular anthers, the pollen grains without wings; ovule solitary, terminal on a short axillary branch, several bracteate at base; seed partly or entirely surrounded by a fleshy aril; cotyledons 2.——Three genera with about 15 species, in the temperate regions of the N. Hemisphere, and in New Caledonia.

1A. Fruit drupelike, the single seed completely enveloped by an aril; anthers 4-locular; branches subopposite; leaves with 2 stomatal bands beneath. .... 1. *Torreya*
1B. Fruit berrylike, the seed only partly surrounded by a fleshy red arial; anthers 5- to 9-locular; branches alternate; leaves pale green beneath. .... 2. *Taxus*

1. TORREYA Arn.　KAYA ZOKU

Evergreen mostly dioecious trees with longitudinally torn bark; branches on main axis verticillate, the lateral branches subopposite; leaves appearing 2-ranked, thick, rigid, linear, lustrous and deep green above, with 2 narrow stomatal bands beneath; stamen clusters solitary, axillary, pedunculate, consisting of 4–8 whorls of stamens, the anthers 4-locular; ovule single, sessile, subtended by several scales, surrounded by a fleshy aril; seeds ovoid, maturing the second year.——Seven or eight species in N. America and e. Asia.

1. Torreya nucifera (L.) Sieb. & Zucc. *Taxus nucifera* L.——KAYA.　Glabrous tree with spreading brownish branches; leaves linear, 15–25 mm. long, 2.2–3 mm. wide, rather rigid, gradually tapering to a short spinescent apex, abruptly narrowed to subsessile base, deep green and lustrous on more or less convex upper side, pale green with 2 narrow pale yellowish impressed stomatal bands beneath; seed narrowly ellipsoidal, about 2.5 cm. long, greenish, tinged with purple.——Honshu (Kantō Distr. and westw.), Shikoku, Kyushu.

Var. **macrosperma** (Miyoshi) Koidz.　*T. macrosperma* Miyoshi——HIDARI-MAKIGAYA.　Seed oblong, 3.5–4 cm. long.——Honshu (Omi and Iga Prov.).

Var. **radicans** Nakai.　*T. fruticosa* Nakai——CHABOGAYA. Stems less than 3 m. long, ascending, branching from base, radicant.——Mountains; Honshu.

Var. **igaensis** (Doi & Morikawa) Ohwi.　*T. igaensis* Doi & Morikawa——KOTSUBUGAYA.　Leaves 1–1.8 cm. long, abruptly spinescent; seed about 2 cm. long.——Honshu (Iga Prov.); rare.

2. TAXUS L.　ICHII ZOKU

Evergreen, dioecious, rarely monoecious trees and shrubs with red-brown scaly bark, the branches irregularly alternate; leaves spirally arranged but often appearing 2-ranked, linear, sometimes falcate, with 2 rather broad greenish stomatal bands beneath; stamens 6–14, in heads, the anthers 5- to 9-locular; ovule surrounded at base by an aril and several scales; seeds ovoid, the aril cup-shaped, fleshy, usually scarlet.——About 8 species in the temperate regions of the N. Hemisphere.

1. Taxus cuspidata Sieb. & Zucc.　*T. baccata* sensu Thunb., non L.; *T. baccata* var. *microcarpa* Trautv.; *T. baccata* subsp. *cuspidata* (Sieb. & Zucc.) Pilger——ICHII.　Erect tree with spreading or ascending branches; leaves spirally arranged or appearing 2-ranked, linear, 1.5–2.5 cm. long, 2–3 mm. wide, rather abruptly narrowed to a short spinescent apex, abruptly short-petiolate at base, dark green with the midrib conspicuous above when dry, the lower side with 2 tawny stomatal bands occupying nearly the entire area except the margins and midrib; aril scarlet when mature.——Mountains;

Hokkaido, Honshu (centr. distr. northw. and Yamato), Shikoku, Kyushu.

Var. **luteobaccata** Miyabe & Tatew. KIMI-NO-ONKO.　Aril yellowish around a greenish seed.

Var. **nana** Rehd. KYARABOKU.　Depressed low shrub with ascending, densely arranged branches and branchlets; leaves usually only spirally disposed and not appearing 2-ranked.——Mountains; Honshu (Japan Sea side); much cultivated.

Fam. 27. PODOCARPACEAE　MAKI KA　Podocarp Family

Trees and shrubs; leaves usually acicular to linear, sometimes broadly oblong to ovate; stamens usually many, terminal or axillary, on elongate peduncles, the anthers 2-locular; ovules 1 to several, with several basal scales which become a fleshy receptacle; seeds globose to ovoid, drupelike.——Seven genera, with about 100 species chiefly in the Tropics.

1. PODOCARPUS Pers.　MAKI ZOKU

Evergreen shrubs and trees, dioecious, rarely monoecious; leaves alternate, sometimes opposite, linear to broadly lanceolate or ovate; staminate cones amentlike, usually axillary, solitary or in fascicles; ovules usually solitary, axillary, enveloped by a scale which becomes a fleshy receptacle subtended by several bracts; seeds globose to ovoid, drupelike or nutlike.——More than 60 species, chiefly in mountains of tropical or subtropical regions.

1A. Leaves alternate, broadly linear, the midrib distinct. .... 1. *P. macrophyllus*
1B. Leaves opposite, lanceolate to ovate, or oblong, the midrib wanting. .... 2. *P. nagi*

1. Podocarpus macrophyllus (Thunb.) Lamb. *Taxus macrophylla* Thunb.; *P. longifolius* Hort.——INU-MAKI. Trees with rather stout branches; leaves alternate, broadly linear, 10–18 cm. long, attenuate at both ends, deep green and lustrous above, slightly yellowish below, the midrib distinct; staminate cones cylindric, in fascicles, in groups about 3 cm. long; ovules axillary, short-pedicelled; fruit furnished with an obovoid dark red fleshy receptable at base and a single broadly ovoid, greenish seed about 1 cm. long.——May. Honshu (s. Kantō to s. Kinki Distr.), Shikoku, Kyushu.——Ryukyus.

Var. **maki** Sieb. *Taxus chinensis* Roxb.; *P. chinensis*

(Roxb.) Sweet; *P. japonicus* Sieb.; *P. macrophyllus* subsp. *maki* (Sieb.) Pilger; *P. macrophyllus* var. *chinensis* (Roxb.) Maxim.——RAKAN-MAKI, MAKI.　Shrubs with ascending branches; leaves densely arranged, ascending, 4–8 cm. long, 5–9 mm. wide, obtuse.——Native of China, widely cultivated in Japan; several cultivars are grown.

2. Podocarpus nagi (Thunb.) Zoll. & Moritzi ex Makino. *Myrica nagi* Thunb.; *Nageia japonica* Gaertn.; *N. nagi* (Thunb.) O. Kuntze——NAGI.　Branched, upright tree with glabrous branches; leaves opposite or subopposite, lanceolate to ovate or oblong, 3–8 cm. long, 12–30 mm. wide, obtuse, narrowed at base, entire, deep-green and lustrous above, paler or whitish beneath with many parallel slender veinlets; stamens fasciculate; seeds globose, 10–15 mm. wide, bluish green, slightly glaucous; receptacle small and not fleshy.——May–June. Mountains; Honshu (w. Chūgoku), Shikoku, Kyushu; occasionally planted.——Ryukyus and Formosa.

Fam. 28. CEPHALOTAXACEAE   INUGAYA KA   Plum-yew Family

Evergreen shrubs and trees, dioecious, rarely monoecious; leaves linear, appearing 2-ranked; stamens 7–12, axillary, globose, in heads or spikes, with short filaments, the anthers usually 3-locular; ovulate organ axillary on lower part of branchlets, consisting of several pairs of 2-ovuled megasporophylls; seed solitary or geminate, ellipsoidal, drupelike, with fleshy pericarp and thin woody inner coat.——Two genera and about 6 species in e. Asia and the Himalayas.

1. CEPHALOTAXUS Sieb. & Zucc.   INUGAYA ZOKU

Evergreen, dioecious shrubs or trees with opposite branches; leaves appearing 2-ranked, linear, with a raised midrib above and 2 broad stomatal bands beneath; seeds large, ellipsoidal, greenish or purplish.——About 5 species in e. Asia and the Himalayas.

1. Cephalotaxus harringtonia (Knight) K. Koch. *Taxus harringtonia* Knight ex Forbes; *C. pedunculata* Sieb. & Zucc.; *C. drupacea* Sieb. & Zucc.; *C. drupacea* var. *harringtonia* (Knight) Pilger——INUGAYA.   Small tree with spreading branches; leaves appearing 2-ranked, linear, 2–5 cm. long, 2.5–3.5 mm. wide, abruptly mucronate, dark green above, with 2 rather broad grayish stomatal bands beneath; stamens on short, simple or rarely branched peduncles; seeds ovoid or obovoid, rarely globose, about 2.5 cm. long, greenish, becoming purplish when mature.——Hondo (Kantō Distr. and westw.), Shikoku, Kyushu.——Korea and n. China.

Cv. **Fastigiata**. *C. harringtonia* var. *fastigiata* (Carr.) Rehd.; *Podocarpus koraiana* Endl.; *C. koraiana* Hort. ex Gordon; *C. pedunculata fastigiata* Carr.; *C. buergeri* Miq.; *C. harringtonia* var. *koraiana* (Endl.) Koidz.——CHŌSEN-MAKI.   Shrub with erect or ascending branches, and usually spirally arranged leaves not appearing 2-ranked.——Known only in cultivation.

Var. **nana** (Nakai) Rehd.   *C. nana* Nakai——HAI-INUGAYA. Spreading shrub with ascending branches; seeds pale purplish when mature.——Mountains on the Japan Sea side; Hokkaido, Honshu.

Fam. 29. PINACEAE   MATSU KA   Pine Family

Evergreen, rarely deciduous shrubs and trees, usually monoecious, with long shoots only or with both long shoots and short spurs; leaves acicular or linear, spirally arranged; staminate inflorescence amentlike, consisting of many imbricate scales with several 2-loculed anthers on the lower side, the pollen-grains often winged; ovulate aments or cones consisting of many imbricate woody scales, with 2 usually winged seeds on the upper side, the ovules anatropous.——About 9 genera, with more than 200 species widely distributed in the N. Hemisphere.

1A. Leaves not in fascicles, disposed in 2 ranks or spiral on elongate branches; long shoots prevailingly well-developed, the short shoots obsolete or poorly developed.
  2A. Cones erect, the scales deciduous leaving a persistent axis; branches not pulvinate; pollen-grains winged. .... 1. *Abies*
  2B. Cones reflexed or pendulous, the scales persistent.
    3A. Bracts of the cone scales much-elongate, exserted; branches scarcely pulvinate; pollen-grains wingless. .... 2. *Pseudotsuga*
    3B. Bracts shorter than the scales; branches pulvinate.
      4A. Cones small, not over 2.5 cm. long, the seeds enclosed by the wing-base; leaves with solitary resin ducts; pollen-grains wingless. 3. *Tsuga*
      4B. Cones larger, never less than 3 cm. long, the seeds enclosed by the wing-base on one side only; leaves with paired resin ducts; pollen-grains winged. .... 4. *Picea*
1B. Leaves fasciculate on short (these sometimes obscure) shoots or spurs or remotely alternate on long shoots; branches of 2 kinds.
  5A. Leaves many in each fascicle; cones erect, the scales not umbonate.
    6A. Leaves deciduous; cones 2–3 cm. long, maturing within 1 year. .... 5. *Larix*
    6B. Leaves evergreen; cones 5–12 cm. long, maturing after 2–3 years. .... 6. *Cedrus*
  5B. Leaves 2–5 (–8) in each fascicle, rarely solitary, alternate, reduced to scales on long shoots; cones erect to reflexed, the scales usually umbonate. .... 7. *Pinus*

1. ABIES Mill.   MOMI ZOKU

Evergreen monoecious trees; leaves spirally arranged, often appearing 2-ranked, linear or linear-lanceolate, usually flat, with 2 stomatal bands beneath, usually retuse; staminate aments axillary, nodding or pendulous, oblong, mostly on under side of branches; pollen grains winged; ovulate aments erect, near summit of tree; cones erect, the scales larger than the bracts, deciduous from the persistent axis at maturity; seeds winged, the cotyledons 4–10.——About 40 species, in the cooler regions of the N. Hemisphere.

1A. Young branches glabrous and lustrous, with deep grooves especially on those of the second year; winter buds large, obtuse; bracts of the cones not exserted beyond the scales; leaves slightly emarginate. .... 1. *A. homolepis*
1B. Young branches more or less pubescent.

2A. Leaves acutely bifid in young trees, emarginate when mature; cones 10–12 cm. long. .... 2. *A. firma*
2B. Leaves always obtuse to emarginate; cones less than 10 cm. long.
  3A. Leaves about 1.5 mm. wide; resin ducts internal; young branches with short rusty to grayish hairs; winter buds bluish.
  3. A. sachalinensis
  3B. Leaves about 2 mm. wide; resin ducts marginal.
    4A. Young branches pubescent with rusty-brown hairs; leaves thickish, slightly broadened at tip. .... 4. *A. mariesii*
    4B. Young branches with grayish or gray-brown hairs, glabrescent; leaves rather thin, narrow, not broadened at tip.
    5. A. veitchii

1. Abies homolepis Sieb. & Zucc. *A. brachyphylla* Maxim.; *A. finhonnoskiana* Neumann ex Parl.; *Pinus harryana* MacNab——URAJIRO-MOMI, DAKE-MOMI. Bark scaly; buds ovoid, resinous; branchlets glabrous, yellowish gray; leaves linear, 15–25 mm. long, 1.7–2.2 mm. wide, rounded to emarginate, with 2 broad stomatal bands beneath; cones narrowly oblong, 7–10 cm. long, purplish when young, the scales 2–2.5 cm. wide, the bracts small, subcuneate, cuspidate, retuse, appressed on the scale, not exserted.——Mountains; Honshu (Fukushima pref. westw. to centr. distr. and Yamato), Shikoku, Kyushu.

Var. **umbellata** (Mayr) Wils. *A. umbellata* Mayr; *A. umbilicata* Mayr——MITSUMINE-MOMI. Cones greenish when young.——Honshu.

2. Abies firma Sieb. & Zucc. *A. bifida* Sieb. & Zucc.; *A. firma* var. *momi* MAST.——MOMI. Tree with scaly bark; winter buds ovoid, acute, lustrous, grayish brown, only slightly resinous; branches grayish brown, short-pubescent in the shallow grooves; leaves pectinate, 2–3.5 cm. long, obtuse and emarginate, acutely bifid in young plants, with 2 grayish stomatal bands beneath; cones short-cylindric, 10–12 cm. long, yellowish green when young, the scales about 2.5 cm wide, the bracts exserted, not reflexed, linear-lanceolate, acuminate.——Hills and mountains; Honshu, Shikoku, Kyushu.

3. Abies sachalinensis (F. Schmidt) Mast. *A. homolepis* var. *tokunaii* Carr.; *A. veitchii* var. *sachalinensis* F. Schmidt——AKA-TODO-MATSU. Tree with nearly smooth, grayish bark; buds small, subglobose; branchlets pubescent in the shallow grooves; leaves linear, 2–3.5 cm. long, about 1.5 mm. wide, rounded to emarginate, with 2 rather narrow stomatal bands beneath; cones cylindric, about 7 cm. long, the scales externally densely puberulent, entire, the bracts obcordate-quadrangular, cuspidate, exserted, straight.——Hokkaido. ——s. Kuriles and Sakhalin.

Var. **nemorensis** Mayr. *A. nemorensis* (Mayr) Miyabe & Miyake; *A. wilsonii* Miyabe & Kudo——EZO-SHIRABISO. Bracts smaller, not exserted.——Hokkaido.

Var. **mayriana** Miyabe & Kudo. *A. mayriana* (Miyabe & Kudo) Miyabe & Kudo——AO-TODO-MATSU. Leaves with 4–5 stomatal bands; cone scales conspicuously reflexed.——Hokkaido.

4. Abies mariesii Mast. AOMORI-TODO-MATSU, Ō-SHIRABISO. Bark roughened in old trees; buds small; branchlets densely pubescent with short, patent, rusty-brown hairs; leaves linear, 1–2 cm. long, about 2.5 mm. wide, slightly broadened at tip, rounded or emarginate, lustrous above, with 2 broad whitish stomatal bands beneath; cones narrowly ovoid, 4–9 cm. long, purplish when young, the scales about 2.5 cm. wide, the bracts not exserted.——Mountains; Hondo (centr. distr. northw.).

5. Abies veitchii Lindl. *Pinus solenolepis* Parl.; *A. veitchii* var. *reflexa* Koidz.; *A. sikokiana* Nakai——SHIRABE, SHIRABISO. Tree with smooth grayish bark; buds purplish; branchlets grayish or reddish brown, rather densely puberulent with patent grayish brown hairs; leaves linear, 1–2.5 cm. long, about 2 mm. wide, slightly emarginate, not broadened at tip, with 2 whitish stomatal bands beneath; cones cylindric, about 5 cm. long, bluish purple while young, the scales 10–13 mm. wide, entire, the bracts slightly exserted, reflexed; seeds with a short wing nearly as long as the seed.——Mountains; Honshu (Fukushima pref. and westw. to centr. distr. and Yamato), Shikoku.——In Shikoku, this species has smaller, more rounded cones, 4–4.5 cm. long, and leaves slightly broadened at tip. This phase sometimes is called *A. sikokiana* Nakai. SHIKOKU-SHIRABE.

Var. **nikkoensis** Mayr. Cones smaller, and the bracts nearly hidden.

Var. **olivacea** Shiras. AO-SHIRABE. Cones greenish, becoming grayish brown when mature.——Honshu (centr. distr.).

2. PSEUDOTSUGA Carr. TOGA-SAWARA ZOKU

Evergreen, monoecious trees; branches relatively smooth, with ovate leaf-scars; leaves spirally arranged, but appearing 2-ranked, linear, compressed, with a groove above and 2 stomatal bands beneath; staminate aments axillary, cylindric; ovulate cones terminal on short spurs, composed of numerous spirally arranged scales, pendulous, ovoid, the scales rounded, concave, each subtended by an exserted tricuspidate bract; seeds 2, winged; cotyledons 6–12.——Four species in e. Asia and w. N. America.

1. Pseudotsuga japonica (Shiras.) Beissn. *Tsuga japonica* Shiras.——TOGA-SAWARA, GOYŌ-TOGA, SAWARA-TOGA. Tree with longitudinally fissured bark; branches glabrous, yellowish gray but dark colored around the transversely oblong short leaf-scars; leaves rather loosely arranged, slightly emarginate, 2–2.5 cm. long, 1.5–1.7 mm. wide, with a rather deep groove above and 2 white stomatal bands beneath; cones 3.5–4.5 cm. long, the several scales entire, concave, 2–2.5 cm. wide, the bracts exserted, reflexed, 3-fid with an acicular entire midlobe and denticulate rounded-triangular lateral lobes.——Mountains; Honshu (Kii and Yamato), Shikoku; rare.

3. TSUGA Carr. TSUGA ZOKU

Evergreen monoecious trees; branches spreading, with distinct pulvini; leaves spirally arranged, appearing 2-ranked, linear, compressed, usually emarginate, grooved above, with 2 whitish stomatal bands beneath; staminate aments axillary in globose pedunculate clusters; ovulate cones pendulous, terminal; cone scales woody, persistent, concave, the subtending bracts rarely exserted; seeds small, winged; cotyledons 3–6.——About 10 species in e. Asia, the Himalayas, and N. America.

1A. Bark brown, deeply fissured; branchlets glabrous; buds acutish. . . . . 1. *T. sieboldii*
1B. Bark grayish, slightly fissured; branchlets short-pubescent; buds obtuse. . . . . 2. *T. diversifolia*

1. Tsuga sieboldii Carr. *Abies tsuga* Sieb. & Zucc.; *T. thuja* A. Murr.; *Pinus tsuga* (Sieb. & Zucc.) Ant.——TSUGA, TOGA. Tree with ciliate bud-scales; branchlets glabrous, lustrous, yellowish brown; leaves rather loosely arranged, linear, 8–20 mm. long, emarginate, lustrous and 1-grooved above, with 2 rather narrow whitish stomatal bands beneath; cones broadly ovoid, 2–2.5 cm. long, short-stalked, the scales orbicular, 1 cm. long, ciliolate, glabrous on back.——Mountains and hills; Honshu (Kantō Distr. and westw.), Shikoku, Kyushu; common.

2. Tsuga diversifolia (Maxim.) Mast. *Abies diversi-*folia Maxim.; *Pinus tsuga* var. *nana* Endl.; *T. sieboldii* var. *nana* (Endl.) Carr.——KOME-TSUGA. Tree; buds obovoid-globose to globose, with minutely pubescent and ciliate scales; branchlets short-pubescent; leaves rather densely arranged, linear to broadly so, 5–15 mm. long, emarginate, deep green, 1-grooved above, with 2 rather narrow whitish stomatal bands beneath; cones broadly ovoid, about 2 cm. long, subsessile, the scales cuneate-orbicular, puberulent.——High mountains; Honshu (centr. distr., northw., and Yamato), Shikoku, Kyushu; rather common.

4. PICEA A. Dietr.   TŌ-HI ZOKU

Evergreen, monoecious trees; branches with prominent pulvini separated by grooves; leaves spirally arranged, linear to acicular, 4-angled with stomatal bands on all faces or compressed and with stomata only on the inner (upper) side; staminate aments axillary, with numerous spirally arranged anthers; ovulate aments terminal, of numerous spirally arranged 2-ovuled scales; cones usually pendulous, ovoid to short cylindric, the scales persistent, the subtending bracts small; seeds small, winged. ——About 40 species in the N. Hemiphere.

1A. Leaves dorsally flattened, linear-rhomboid in cross section, with stomatal bands on under side; branchlets glabrous and lustrous; winter buds broadened at base, broadly conical, resinous and lustrous. . . . . 1. *P. jezoensis*
1B. Leaves 4-angled or only slightly flattened dorsally, usually with stomatal bands on all sides.
  2A. Leaves slightly compressed dorsally, broader than thick and rhomboid in cross section.
    3A. Branchlets glabrous, or with short grayish pubescence on main axis; winter buds dull to chestnut-brown, not broadened at base; cones ovoid-cylindric, 6–12 cm. long. . . . . 2. *P. bicolor*
    3B. Branchelets densely pubescent with short rusty-brown hairs; winter buds conical to ovoid, broadened at base, lustrous, chestnut-brown, resinous; cones cylindric, 5–8 cm. long. . . . . 3. *P. glehnii*
  2B. Leaves 4-angled, as broad as thick or a little narrower.
    4A. Branchlets glabrous.
      5A. Winter buds ovoid, chestnut-brown, not or slightly resinous; leaves 15–20 mm. long; cones 8–10 cm. long. . . . . 4. *P. polita*
      5B. Winter buds conical, reddish brown, conspicuously resinous; leaves 8–15 mm. long; cones 3–6 cm. long. . . . 5. *P. maximowiczii*
    4B. Branchlets usually glandular-pubescent, glabrescent or nearly so; winter buds conical, broadened at base, brown, lustrous, resinous; leaves 8–12 mm. long; cones 4–10 cm. long. . . . . 6. *P. koyamae*

1. Picea jezoensis (Sieb. & Zucc.) Carr. *Abies jezoensis* Sieb. & Zucc.; *P. ajanensis* auct. Japon., non Fisch.; *A. alcoquiana* Veitch ex Lindl. pro parte; *Pinus jezoensis* (Sieb. & Zucc.) Ant.; *A. microsperma* Lindl.; *A. microcarpa* Lindl. apud Miq. err. typogr.; *Picea microsperma* (Lindl.) Carr.——EZO-MATSU, KURO-EZO. Tree; branchlets glabrous and lustrous, yellowish brown or yellowish, with slightly swollen pulvini; leaves compressed, 1–2 cm. long, 1.5–2 mm. wide, acute, mucronate, somewhat keeled on both sides, slightly recurved, lustrous on external (upper) side, with 2 stomatal bands on inner (under) side; cones short-cylindric, 4–7.5 cm. long, the scales oblong, denticulate on upper margin; seeds about 3 mm. long, the wing 2–2.5 times as long as the seed.——Hokkaido——s. Kuriles and Sakhalin.

Var. **hondoensis** (Mayr) Rehd. *P. hondoensis* Mayr——TŌ-HI. Branchlets usually reddish brown, the pulvini much swollen; leaves a little shorter, acutish to obtuse, the terminal mucro almost entirely wanting.——High mountains; Honshu (centr. distr. and Yamato).

2. Picea bicolor (Maxim.) Mayr. *Abies bicolor* Maxim.; *A. alcoquiana* Veitch ex Lindl. pro parte; *P. alcockiana* Carr. ——IRA-MOMI, MATSU-HADA. Tree; branchlets lustrous, reddish brown, glabrous but short-pubescent on main axis; leaves slightly compressed, 1–2 cm. long, 1–1.3 mm. wide, acuminate;

cones narrowly ovoid, 6–12 cm. long, the scales obovate, rounded or slightly narrowed at apex, obscurely toothed.——Mountains; Honshu (centr. distr.).

Var. **acicularis** Shiras. & Koyama——HIME-MATSU-HADA. Branchlets minutely pubescent; leaves recurved; cone-scales subentire.——Honshu (Yatsugatake in Shinano); rare.

Var. **reflexa** Shiras. & Koyama——SHIRANE-MATSU-HADA. Branchlets minutely pubescent; leaves shorter; cone-scales smaller, entire with slightly elongate reflexed tip.——Honshu (Akaishi mountain range); rare.

3. Picea glehnii (F. Schmidt) Mast. *Abies glehnii* F. Schmidt——AKA-EZO-MATSU. Tree; branchlets densely pubescent with reddish brown short hairs; leaves compressed, 4-angled, acicular, 6–12 mm. long, 1.2–1.5 mm. wide, obtuse or slightly pointed, dark green, slightly recurved; cones cylindric, 5–8 cm. long, dark purplish, the scales suborbicular, entire or slightly denticulate.——Bogs; Hokkaido, Honshu (Mt. Hayachine in Rikuchu).——Sakhalin.——Forma **chlorocarpa** Miyabe & Kudo. AOMI-NO-AKA-EZO-MATSU. Cones green to greenish yellow.

4. Picea polita (Sieb. & Zucc.) Carr. *Abies polita* Sieb. & Zucc.; *P. torano* (Sieb.) Koehne; *P. thunbergii* Asch. & Graebn.——HARI-MOMI, BARA-MOMI. Tree with thickish, glabrous, pale yellow or pale reddish brown branchlets; leaves rigid, slightly recurved, 15–20 mm. long, about 1.8 mm. wide, short spine-tipped, 4-angled and slightly narrower than thick, the pulvini elongate, about 1.5 mm. long; cones oblong, 8–10 cm. long, the scales elliptic, irregularly denticulate.——Mountains; Honshu (centr. distr. and Yamato), Shikoku, Kyushu.

5. Picea maximowiczii Regel. *Abies obovata* var. *japonica* Maxim.; *P. obovata* var. *japonica* (Maxim.) Beissn.; *P. tschonoskii* Mayr——HIME-BARA-MOMI. Tree; branchlets rather slender, glabrous, yellowish or reddish brown, with short pulvini; leaves rigid, spreading, 8–15 mm. long, dark green, acute or rather obtuse, 4-angled, with stomatal bands on all sides; cones oblong, 3–6 cm. long, brown when mature,

lustrous, the scales rounded, entire.——Mountains; Honshu (Yatsugatake and Senjogatake in Shinano); rare.

6. Picea koyamae Shiras. YATSUGATAKE-TŌ-HI. Tree; branchlets reddish brown and slightly glaucous, glabrous or nearly so on main axis, usually glandular-pubescent in the lateral, the pulvini prominent; leaves 8–12 mm. long, thickish, acute or obtuse, sometimes slightly curved, with 2 whitish prominent stomatal bands above and 2 faint bands beneath; cones short-cylindric, 4–10 cm. long, pale brown, the scales rounded, denticulate.——Mountains; Honshu (Yatsugatake); rare.

5. LARIX Mill.   KARA-MATSU ZOKU

Deciduous, monoecious trees; leaves linear, compressed, rarely 4-angled, loosely and spirally arranged on elongate shoots, fasciculate on short spurs; staminate aments sometimes pedunculate, globose to oblong, with numerous spirally arranged anthers; ovulate aments globose, of few to rather numerous 2-ovuled scales each subtended by a large bract; cones subglobose to oblong, maturing in one season, the scales rounded to elliptic, persistent; seeds geminate, subtriangular, winged; cotyledons usually 6.——About 10 species in the cooler regions of the N. Hemisphere and the Himalayas.

1. Larix leptolepis (Sieb. & Zucc.) Gord. *Pinus kaempferi* Lamb.; *Abies leptolepis* Sieb. & Zucc.; *L. japonica* Carr.; *L. leptolepis* var. *minor* A. Murr.; *L. japonica* var. *macrocarpa* Carr.; *L. leptolepis* var. *murrayana* Maxim.; *L. kaempferi* (Lamb.) Sarg., non Carr.——KARA-MATSU. Tree; branchlets yellowish or reddish brown, glabrous or slightly short-pubescent; leaves compressed, 15–35 mm. long, 1–1.5 mm. wide, obtuse, green, slightly glaucous while young, slightly

narrowed toward base, with 2 stomatal bands beneath; cones broadly ovoid, 2–3 cm. long, the scales several, truncate and recurved at apex, minutely gland-dotted on back.——Mountains; Honshu (centr. distr., Oze, and Zaosan in Rikuzen), often planted for timber.

Larix dahurica Turcz. var. **japonica** Maxim. *L. kurilensis* Mayr——GUI-MATSU, SHIKOTAN-MATSU. Occasionally planted.——s. Kuriles and Sakhalin.

6. CEDRUS Link   HIMARAYA-SUGI ZOKU

Evergreen monoecious tree; leaves spirally arranged, remotely alternate on long shoots, fasciculate on short spurs, acicular and rather rigid, trigonous, acuminate and pointed; staminate aments cylindric, erect; ovulate aments ovoid, with numerous spirally arranged 2-ovuled scales; cones ovoid, maturing the second or third year, the scales densely imbricate, broad, finely puberulent outside; seeds with a broad membranous wing; cotyledons 9–10.——About 4 closely related species, n. Africa, Crete, Turkey, Lebanon, and Himalayas.

1. Cedrus deodara (Roxb.) Loud. *Pinus deodara* Roxb.; *C. libani* var. *deodara* (Roxb.) Hook. f.——HIMARAYA-SUGI, HIMARAYA-SHIDA. Tree; branches slender, elongate, pendulous at tip, densely pubescent; leaves dark bluish green,

acicular, usually 3–4 cm. long, subrigid, straight, acuminate; staminate flowers about 3 cm. long, appearing in late summer; cones ellipsoidal, 7–10 cm. long.——Often cultivated; introduced from the Himalayas.

7. PINUS L.   MATSU ZOKU

Evergreen monoecious trees, rarely shrubs; leaves dimorphic, the primary ones spirally arranged and reduced to small scarious bracts, the secondary leaves borne on short spurs in the axils of bracts, acicular, triangular or subterete in cross section, borne in fascicles of 2–5 (–8), rarely 1, surrounded at base by sheaths of bud-scales; stamens borne in catkinlike aments, axillary, in fascicles at base of young shoots, with numerous spirally arranged 2-locular anthers; ovulate aments lateral or subterminal, with numerous spirally arranged 2-ovuled scales, the subtending bracts small; cones globose to subcylindric, woody, the scales at the tip usually rhombic and umbonate; seeds winged or wingless; cotyledons 4–15.——About 100 species widely distributed in the N. Hemisphere, mainly extratropical, a few extending to the mountains of n. Africa, Malaysia, Central America, and the West Indies.

1A. Leaves 5 in a fascicle; vascular bundle solitary; sheaths of leaf-fascicles deciduous; seeds with or without wings.
  2A. Seeds without wings.
    3A. Procumbent shrub with creeping stems and branches; cones 3–5 cm. long, persistent for several years. . . . . 1. *P. pumila*
    3B. Erect trees; cones 5–20 cm. long.
      4A. Cones subsessile, 9–15 cm. long; young shoots densely pubescent. . . . . 2. *P. koraiensis*
      4B. Cones pedunculate, 5–10 cm. long; young shoots glabrous or nearly so. . . . . 3. *P. armandii* var. *amamiana*
  2B. Seeds winged; cones 5–7 cm. long, dehiscent, subsessile; young shoots glabrous or nearly so. 4. *P. parviflora*
1B. Leaves 2 in a fascicle, serrulate; vascular bundles 2; sheaths of leaf-fascicles persistent; seeds winged.
  5A. Bark reddish brown; terminal buds reddish brown; leaves slender. . . . . 5. *P. densiflora*
  5B. Bark blackish gray; terminal buds whitish; leaves coarser and darker green. . . . . 6. *P. thunbergii*

1. **Pinus pumila** (Pall.) Regel. *P. cembra* var. *pumila* Pall.——HAI-MATSU.   Shrub, 0.5–2 m. high; lateral and main branches long-creeping, young shoots brown-pubescent; leaves 5 in a fascicle, slightly incurved, 3–5(–7) cm. long, acute, triangular and obscurely serrulate on margin, deep green on back, glaucescent with 2 stomatal bands on inside; staminate aments dark purplish; pistillate 2 or 3 on upper part of branches, purplish; cones subsessile, ovoid or ovoid-globose, 3–5 cm. long, the scales rather numerous, broadly ovate; seeds wingless.——Alpine regions; Hokkaido, Honshu (centr. distr. northw.).——e. Siberia and Kamchatka.

2. **Pinus koraiensis** Sieb. & Zucc.   *P. cembra* var. *excelsa* Maxim.; *P. mandshurica* Rupr.; *P. cembra* var. *manchurica* Mast.——CHŌSEN-GOYŌ, CHŌSEN-MATSU.   Tree; young shoots densely brown-pubescent; leaves 5 in a fascicle, deep green, straight, 7–12 cm. long, 1–1.2 mm. wide, triangular, serrulate, acute, with 2 stomatal bands on inside; staminate aments reddish yellow; cones 9–15 cm. long, ovoid-conical or oblong-conical, 5–7 cm. across, the scales obtuse and slightly recurved at apex; seeds about 12 mm. long, wingless.——Mountains; Honshu (centr. distr.), Shikoku.——Korea, Manchuria, and Ussuri.

3. **Pinus armandii** Franch. var. **amamiana** (Koidz.) Hatus.   *P. armandii* sensu auct. Japon., non Franch.; *P. amamiana* Koidz.——AMAMI-GOYŌ, YAKUTANE-GOYŌ.   Tree; young shoots glabrous or nearly so, brownish; leaves 5 in a fascicle, rather rigid, 5–8 cm. long, about 1 mm. wide, triangular, serrulate with 2 stomatal bands on lateral face; cones short-pedunculate, ovoid or ovoid-ellipsoid, 5–10 cm. long, the scales rounded and slightly recurved at apex; seeds 12–13 mm. long, wingless.——Mountains and hills; Kyushu (Yakushima and Tanegashima).——The typical phase occurs in Formosa and China.

4. **Pinus parviflora** Sieb. & Zucc.   *P. himekomatsu* Miyabe & Kudo; *P. pentaphylla* var. *himekomatsu* (Miyabe & Kudo) Makino——HIME-KO-MATSU, GOYŌ-MATSU.   Tree with bark fissured into thin flaky scales; branches glabrous and grayish brown, or puberulent and yellowish brown while young; leaves 5 in a fascicle, rather slender, slightly curved, 2–6 cm. long, 0.7–1 mm. wide, triangular, scattered-serrulate, with 2 lateral white stomatal bands, deep green on outer side; cones ovoid-oblong, 5–7 cm. long; seeds about 1 cm. long, the wing much shorter than the seed.——Mountains; Honshu, Shikoku, Kyushu; sometimes planted.

Var. **laevis** Hara.   TODO-HADA-GOYŌ.   Bark smooth, not flaky.——Hokkaido, Honshu (centr. and n. distr.).

Var. **pentaphylla** (Mayr) Henry.   *P. pentaphylla* Mayr——KITA-GOYŌ.   Wing of seed a little shorter than to as long as the seed.——Hokkaido, Honshu (centr. and n. distr.).

Pinus × hakkodensis Makino. HAKKŌDA-GOYŌ.   Hybrid of *P. parviflora × P. pumila*.——Honshu (n. distr.).

5. **Pinus densiflora** Sieb. & Zucc.   *P. scopifera* Miq.—— AKA-MATSU, ME-MATSU.   Tree with reddish brown variously scaly bark; branchlets glabrous; leaves 2 in a fascicle, 8–12 cm. long, 0.7–1.2 mm. wide, acute, deep green, semiterete, scabrous with minute marginal teeth; staminate aments ellipsoidal, pale yellow; cones very short-pedunculate, conic-ovoid, 3–5 cm. long, the scales numerous, cuneate, the exposed part flattened and transversely rhomboidal, the umbo short-mucronate.—— Hills and low mountains; Hokkaido (s. distr.), Honshu, Shikoku, Kyushu; very common, variable. Hybridizes with the following species.——Korea.

6. **Pinus thunbergii** Parl.   *P. massoniana* sensu Sieb. & Zucc., non Lamb.——KURO-MATSU, O-MATSU.   Tree with dark gray, scaly fissured bark; young branches glabrous; leaves 2 in a fascicle, straight, rigid, dark green, subterete, sharp-pointed, 7–15 cm. long, 1.5–2 mm. wide; cones short-pedunculate, conic-ovoid, 4–6 cm. long, the scales cuneate, the exposed part flattish, with a small depressed mucro at the center.—— Seaside and lowlands; Honshu, Shikoku, Kyushu; common. ——s. Korea.

Fam. 30. **TAXODIACEAE**   SUGI KA   Taxodium Family

Trees with scaly or acicular leaves; staminate inflorescence small, terminal or axillary, the stamens with short filaments, the anthers 2- to 9-locular, the ovulate solitary and terminal, with numerous spirally arranged 2- to 9-ovulate scales; mature cones woody, with spreading, rounded scales, the bract indistinct; seeds usually with winglike margins.——About 8 genera, with about 15 species in s. Asia, N. America, and Tasmania.

1A. Leaves dimorphic, needlelike ones consisting of 2 completely connate true leaves arising from the axil of scale leaves, verticillate at apex of branchlets, narrowly linear, elongate, furrowed on both surfaces, falling individually; cone scales thick, woody, without a mucro, 7- to 9-seeded. . . . . 1. *Sciadopitys*
1B. Leaves all alike, linear-subulate, spirally arranged, rather short, incurved, decurrent, persistent; cone-scales enlarged above into a disc, mucronate, toothed, 2- to 5-seeded. . . . . 2. *Cryptomeria*

1. **SCIADOPITYS** Sieb. & Zucc.   KŌYA-MAKI ZOKU

Evergreen monoecious tree; leaves dimorphic, the scale leaves on the branchlets, the needlelike leaves 2, completely fused, narrowly linear, arising from the axils of the scale leaves, compressed, thickish, with a deeper furrow especially on under side; staminate aments densely agglomerate at tips of branchlets, the 2-locular anthers spirally arranged; ovulate aments solitary at apex of branchlets, consisting of numerous spirally arranged 7- to 9-ovulate scales each subtended by a small bract; cones narrowly ovoid, the scales broadly cuneate, thick, woody; seeds ovoid with a narrow wing; cotyledons 2.——One species in Japan.

1. **Sciadopitys verticillata** (Thunb.) Sieb. & Zucc. *Taxus verticillata* Thunb.; *Pinus verticillata* (Thunb.) Sieb. ——KŌYA-MAKI.   Branches glabrous, the scale leaves scattered on long shoots and crowded at the top, the ordinary leaves narrowly linear, 8–12 cm. long, about 3 mm. wide, lustrous, deep green, thickish, obtuse and slightly emarginate; cones 8–12 cm. long, with numerous scales, about 2.5 cm. wide, the exposed part transversely narrow-rhombic with reflexed upper margin and rather thick incurved lower margin; seeds about 12 mm. long.——Mountains; Honshu (Iwashiro and centr. distr. westw.), Shikoku, Kyushu; often planted.

2. CRYPTOMERIA D. Don   SUGI ZOKU

Evergreen monoecious tree; leaves spirally arranged in 5 rows, decurrent, linear-subulate, ascending and slightly incurved, slightly flattened laterally, anterio-posteriorly keeled, acuminate, stomatal bands on both sides; staminate aments oblong, sessile, axillary, aggregated in spikelike clusters on branchlets; stamens numerous, with 4- to 5-locular anthers; ovulate aments solitary, terminal; cones globose, woody, the scales 20–30, 2- to 5-seeded, cuneate, thickened distally, the umbo with a central spine and a few flattened toothlike spines on margin; seeds narrowly winged; cotyledons 2 or 3.——One species in Japan and China.

1. Cryptomeria japonica (L. f.) D. Don. *Cupressus japonica* L. f.; *Taxodium japonicum* (L. f.) Brongn.——SUGI. Glabrous tree with reddish brown shredding bark; leaves deep green, 3–20 mm. long, acuminate, broadened at base and de-current; cones 15–25 mm. across; seeds 5–6 mm. long.——Mountains and hills; Honshu, Shikoku, Kyushu; rarely spontaneous; extensively planted in timber plantations, parks and gardens; several cultivars are grown.——China (variety).

Fam. 31. CUPRESSACEAE   HINOKI KA   Cypress Family

Evergreen branched shrubs and trees; leaves decussate or verticillate in 3's, scaly or sometimes subulate; stamens and cones axillary or terminal on branchlets, the scales of both sexes opposite or verticillate in 3's; staminate aments with 3- to 5-locular anthers; ovulate aments with 1 to many-ovulate scales; cones woody and dehiscent or berrylike and indehiscent; seeds sometimes wing-margined; cotyledons 2, rarely 5 or 6.——About 15 genera and 150 species of wide distribution.

1A. Leaves opposite; cones woody and dehiscent.
  2A. Cones ovoid to narrowly so, the scales flattened and imbricate.
    3A. Branchlets less than 5 mm. wide, deep green, sometimes white-spotted on lower side; cone-scales 4–6 pairs, the two fertile pairs each with 1 pair of seeds. .... 1. *Thuja*
    3B. Branchlets 5–6 mm. wide, conspicuously white-spotted on lower side; cone-scales 3 or 4 pairs, 3- to 5-seeded. ...... 2. *Thujopsis*
  2B. Cones globose, the scales peltate. .... 3. *Chamaecyparis*
1B. Leaves opposite or ternate; cones berrylike, indehiscent, 1–12 seeded, composed of 2–6 fleshy connate scales. .... 4. *Juniperus*

1. THUJA L.   KUROBE ZOKU

Evergreen monoecious trees; branchlets horizontally or vertically flattened; leaves scalelike, opposite and decussate, the lateral almost covering the facial ones; staminate aments ovate, with 6–12 decussately arranged stamens; cones ovoid or oblong, the 4–6 pairs of scales ridged or mucronate at apex, only 2 (3) pairs fertile, each 2-(3-) seeded; seeds sometimes broadly winged; cotyledons 2.——About 6 species in e. Asia and N. America.

1A. Branches and branchlets horizontally flattened; cone-scales thin and woody; seeds winged. .... 1. *T. standishii*
1B. Branches and branchlets vertically flattened; cone-scales thick, rather fleshy while young; seeds wingless. .... 2. *T. orientalis*

1. Thuja standishii (Gord.) Carr. *T. japonica* Maxim.; *T. gigantea* var. *japonica* Fr. & Sav.; *Thujopsis standishii* Gord.——KUROBE, NEZU-KO. Glabrous evergreen tree; branchlets horizontally flattened, thickish, 1.5–2.5 mm. wide; leaves deep green, scalelike, deltoid, obtuse, eglandular; cones obovoid, 8–10 mm. long, the scales 6–8, broadly elliptic to cuneate; seeds 3 on each scale, oblanceolate, 5–6 mm. long, the oblong wing 6–7 mm. long, 2–2.5 mm. wide.——Mountains; Honshu, Shikoku; rarely cultivated.

2. Thuja orientalis L. *Biota orientalis* (L.) Endl.—— KO-NO-TEGASHIWA. Evergreen shrub to small tree often branched nearly to base; branchlets slender, vertically flattened, 1.5–2 mm. wide; leaves deep green, deltoid-ovate, small, with a sessile gland on back, subacute on the branches, acute on main axis; cones ovoid, 15–25 mm. long, the scales fleshy, ovate, obtuse, with a recurved mucro on back below apex; seeds 2 on each scale, ellipsoidal, wingless.——Frequently cultivated in Japan.——Korea and China.

2. THUJOPSIS Sieb. & Zucc.   ASUNARO ZOKU

Evergreen monoecious tree; branches horizontally flattened; leaves opposite and decussate, scalelike, small; staminate aments cylindric, with 6–10 pairs of stamens; ovulate cones woody, globose-ovoid, the scales 6–8, imbricate, usually mucronate below apex, with 3–5 seeds on each scale; seeds winged; cotyledons 2.——One species in Japan.

1. Thujopsis dolabrata (L. f.) Sieb. & Zucc. *Thuja dolobrata* L. f.; *Platycladus dolabrata* (L. f.) Spach; *Thujopsis dolabrata* var. *australis* Henry——ASUNARO. Tree with pendulous lateral branches, the branchlets 4–6 mm. wide; leaves lustrous, deep green above, with a conspicuous white area beneath, the lateral ones obtuse, ascending, flat and semiovate, 4–6 mm. long, the median appressed, broadly obovate; cones 1–1.6 cm. long, dehiscent, the scales 6–10 mm. long, broadly cuneate, with a deltoid, obtuse, spreading mucro below the apex; seeds elliptic, 4–5 mm. long, 3–3.5 mm. wide, with a thickish wing.——Mountains; Honshu (centr. distr. northw.), Shikoku, Kyushu; often planted.

Var. **hondae** Makino. *T. hondae* (Makino) Henry—— HINOKI-ASUNARO. Branchlets more dense; cones larger and globose; scales without a prominent mucro.——Hokkaido, Honshu (n. distr.).

Cv. **Nana.** *Thujopsis dolabrata* var. *nana* Carr.; *T. laetevirens* Lindl.; *T. dolabrata* var. *laetevirens* Lindl. Mast.—— HIME-ASUNARO. With slender branches and smaller leaves. A dwarf cultivar of gardens.

3. CHAMAECYPARIS Spach   Hinoki Zoku

Evergreen monoecious trees; branchlets horizontally flattened (except in seedlings and in certain cultivars); leaves at maturity opposite, scaly (short subulate in juvenile condition), ovate or rhomboid, entire; staminate aments small, ovoid to oblong with numerous decussate stamens, the anthers 2- to 4-locular; ovulate cones globose, the scales 6–12, peltate, mucronate; seeds with thin broad wings; cotyledons 2.——Six or seven species in N. America, Japan, and Formosa.

1A. Leaves rather obtuse, glandless; cones 10–12 mm. wide, the scales 1- to 5-seeded. .... 1. *C. obtusa*
1B. Leaves acuminate, with an obscure gland on outside; cones about 7 mm. wide, the scales 1- to 2-seeded. .... 2. *C. pisifera*

1. **Chamaecyparis obtusa** (Sieb. & Zucc.) Endl. *Retinospora obtusa* Sieb. & Zucc.; *Cupressus obtusa* (Sieb. & Zucc.) K. Koch; *Thuja obtusa* (Sieb. & Zucc.) Mast.; *Chamaecyparis breviramea* Maxim.; *C. pendula* Maxim.——Hinoki. Evergreen tree; branchlets slender, about 2 mm. wide, horizontally flattened; leaves scaly, appressed at maturity, obtuse, deep green above with a white spot beneath; cones globose, 10–12 mm. across, 8- to 10-scaled, each scale with a small mucro on the depressed back; seeds about 3 mm. long, narrowly winged. ——Mountains; Honshu (Iwaki and Kantō Distr. westw.), Shikoku, Kyushu; much planted for timber and grown extensively as an ornamental.

2. **Chamaecyparis pisifera** (Sieb. & Zucc.) Endl. *Retinospora pisifera* Sieb. & Zucc.; *Thuja pisifera* (Sieb. & Zucc.) Mast.——Sawara. Tree, closely related to the preceding; branchlets slender, flattened horizontally; leaves ascending, acuminate, deep green, with an obscure gland above, pale green and with a small white spot beneath, the lateral leaves broadly lanceolate, compressed laterally, the facial ones ovate-deltoid; cones globose, 6–7 mm. wide, 10- or 12-scaled, the scales with a triangular mucro on the depressed center; seeds narrowly obovoid to transversely elliptic, about 2 mm. long, the wing 4 mm. long.——Mountains; Honshu (Iwaki and Iwashiro westw. to Kii and Yamato), Kyushu; much planted for timber and grown extensively as an ornamental.

Cv. **Squarrosa.** *C. pisifera* var. *squarrosa* (Sieb. & Zucc.) Beissn. & Hochst.——Hi-muro. A cultivated plant with juvenile foliage, the leaves linear, soft, about 6 mm. long.

Cv. **Plumosa.** *C. pisifera* var. *plumosa* (Carr.) Beissn.; *Retinospora plumosa* Carr.——Shinobu-hiba. Leaves obliquely spreading, short-linear, 3–4 mm. long. Intermediate between the typical phase and var. *squarrosa*.

4. JUNIPERUS L.   Byakushin Zoku

Evergreen monoecious or usually dioecious shrubs and trees with congested branchlets; leaves opposite or ternate, scalelike or short acicular, sometimes both occurring together; staminate aments obovoid or oblong, with numerous opposite or ternate anthers; ovulate aments of 3–8 coalescent scales, each with 1 or 2 ovules; cones globose, berrylike, 1–12 seeded, the scales connate and more or less fleshy; cotyledons 2 or 4–5.——About 50 species widely distributed in the N. Hemisphere, extending southward to the mountains of the Tropics.

A. Leaves all acicular and ternate, spreading, articulate at base; winter buds distinct; flowers axillary.
 2A. Leaves incurved, concave above, lunate in cross section; alpine regions. .... 1. *J. communis*
 2B. Leaves straight or only slightly curved, sulcate above; seashores or in mountains.
  3A. Low procumbent shrub; cones 8–12 mm. across; leaves 8–15 mm. long. .... 2. *J. conferta*
  3B. Upright shrub or tree; cones 6–9 mm. across; leaves 12–25 mm. long. .... 3. *J. rigida*
B. Leaves all or partly scalelike or rarely all acicular, decurrent, not articulate at base; winter buds indistinct; flowers terminal.
                                                                                                     4. *J. chinensis*

1. **Juniperus communis** L. var. **montana** Ait. *J. sibirica* Burgsd.; *J. communis* var. *saxatilis* Pall. ex Willd., nom. imperf.; *J. communis* var. *sibirica* (Burgsd.) Rydb.; *J. rebunensis* Kudō & Susaki; *J. communis* var. *nana* Loud.——Rishiri-yakushin. Procumbent shrub; branchlets yellowish-brown, glabrous; leaves crowded, spreading, short, acicular, linear-oblong, curved, short-pointed, lustrous, deep green, deeply concave with a white band above; cones globose, 7–9 mm. across, bluish-black, slightly glaucous.——Alpine zone; Hokkaido; rare.——Sakhalin, Kuriles, Kamchatka, N. America to Siberia and Europe.

Var. **nipponica** (Maxim.) Wils. *J. nipponica* Maxim.; *J. sibirica* var. *nipponica* (Maxim.) Kusaka——Miyama-nezu. Similar to the preceding variety, but the leaves less strongly curved, 1–1.2 mm. wide, deeply sulcate above, keeled below; cones globose, rounded at apex.——Alpine slopes; Hokkaido, Honshu (n. distr.).

Var. **hondoensis** Satake. *J. communis* var. *nipponica* sensu auct. Japon., pro parte, non Wils.——Hondo-miyama-nezu. Very close to variety *nipponica* but differs in the broader leaves, 1.7–2 mm. wide, and flatter leaves with a broad, 0.7–0.9 mm. shallowly depressed white band above; cones globose with a rather flat apex.——Alpine slopes; Honshu (centr. distr.).

2. **Juniperus conferta** Parl. *J. litoralis* Maxim.; *J. rigida* var. *conferta* (Parl.) Patschke——Hai-nezu. Procumbent dioecious shrubs; leaves short, acicular, 8–15 mm. long, 1–2 mm. across, usually with a short point, straight or slightly incurved, rigid, with a white groove above; cones globose, 8–12 mm. across, purplish black and slightly glaucous.——Sandy seashores; Hokkaido, Honshu, Kyushu.——Sakhalin.

3. **Juniperus rigida** Sieb. & Zucc. *J. utilis* Koidz.——Nezu, Muro, Nezumi-sashi. Upright shrub or small tree; branchlets slender, pendulous, triangular; leaves spreading, acicular, 12–25 mm. long, about 1 mm. wide, straight, slender, gradually narrowed into a short spine, with a white deep furrow above; cones globose, 6–9 mm. in diameter, purplish black, slightly glaucous when young.——Hills and mountains; Honshu (Kantō Distr. westw.), Shikoku, Kyushu.——Korea and n. China.

4. **Juniperus chinensis** L. *J. cernua* Roxb.; *J. thunbergii* Hook. & Arn.; *Sabina chinensis* (L.) Ant.——Ibuki, Ibuki-byakushin, Kamakura-byakushin. Much-branched dioecious shrub to small tree; leaves dimorphic, the scalelike leaves

deep green, opposite, rhombic-ovate, obtuse, closely appressed, about 1.5 mm. long, decussate, the branchlets appearing 4-angled, with an obscure gland at back, the acicular leaves ternate, decussate, obliquely spreading or ascending, 5–10 mm. long, gradually narrowed above into a short point, with 2 stomatal bands above; cones subglobose, 6–8 mm. in diameter, brownish, slightly glaucous; seeds 2 or 3.——Near seashores; Honshu, Shikoku, Kyushu.——China and Mongolia. Much cultivated and very variable.

Cv. **Parsonsii.** *J. chinensis* var. *parsonsii* Hornibr.——PAR-SONSU-BYAKUSHIN.   Shrubby creeping habit, the branchlets elongate, scaly-leaved.——Cultivated.

Cv. **Globosa.** *J. chinensis* var. *globosa* Hornibr.——TAMA-BYAKUSHIN.   Shrubby globose habit, branching from near the base, the leaves mostly scalelike.——Cultivated.

Var. **sargentii** Henry. *J. sargentii* (Henry) Takeda ex Nakai; *J. chinensis* var. *tsukushiensis* (Masam.) Masam.; *J. tsukushiensis* Masam.; *Sabina sargentii* (Henry) Miyabe & Tatew.——MIYAMA-BYAKUSHIN.   Shrub with short creeping flexuose stems and ascending or suberect branches; leaves scalelike, dark bluish green, often acicular in young plants.——Rocky cliffs in mountains and along seashores; Hokkaido, Honshu, Shikoku, Kyushu.——s. Kuriles, and Sakhalin.

Var. **procumbens** (Sieb.) Endl. *J. procumbens* Sieb.; *J. recurva* var. *squamata* Mast., non Parl.; *Sabina pacifica* Nakai ——HAI-BYAKUSHIN, SONARE.   Low shrub with creeping, much elongate stems; leaves usually acicular, rarely partly scalelike, the acicular ones 6–8 mm. long, rigid, sharp pointed, with 2 white stomatal bands above; cones globose, 8–9 mm. ——Seashores; Kyushu; often cultivated as a ground cover.

Class 2.   ANGIOSPERMAE

Subclass 1.   MONOCOTYLEDONEAE   Monocotyledons

Fam. 32.   TYPHACEAE   GAMA KA   Cattail Family

Glabrous monoecious perennials, mostly marsh herbs with thick, elongate, creeping rhizomes; leaves radical, distichous, elongate, linear; flowers minute, very numerous, the pistillate forming a subterminal dense cylindrical spike, the staminate borne above on a prolongation of the stalk, the bracts absent or early caducous; flowers without a true perianth, surrounded by numerous long hairs at base, the filaments filiform; ovary 1-locular, 1-ovuled; style short or elongate, the stigma linear-ligulate; fruit small, nutlike, usually long-stalked, with a tuft of hairs, the seeds antropous, with a straight embryo and abundant mealy endosperm.——One genus with about 10 species widely distributed in wet places of the temperate and tropical regions of the world.

1.   TYPHA L.   GAMA ZOKU

Characters of the family.

1A.   Pistillate part of spike closely contiguous to the staminate part; pistillate flowers without bracteoles among the bristles.
2A.   Leaves 1–2 cm. wide; pistillate portion of spike 10–20 cm. long, the staminate portion 7–15 cm. long; pollen-grains in tetrads.
1. *T. latifolia*
2B.   Leaves 5–10 mm. wide; pistillate portion of spike 6–10 cm. long, the staminate portion 3–5 cm. long; pollen-grains single.
2. *T. orientalis*
1B.   Pistillate portion widely separated from the staminate by a naked axis; pistillate flowers with a bracteole among the bristles; leaves 5–10 mm. wide; pollen-grains single. . . . . 3. *T. angustata*

1.   **Typha latifolia** L. GAMA.   Rhizomes long-creeping; stems stout, solid, 1–2 m. long; leaves glaucous, flat, smooth, 0.5–1.3 m. long, 1–2 cm. wide, narrowed into an obtuse apex, thickish and rather soft, the sheaths without ligules, their margins overlapping, not closed, rounded and scarious on both sides at mouth; spike terminal, erect, the bracts leafy and caducous; staminate portion of the spike shorter than the pistillate portion, the anthers linear, 2–3 mm. long, with a globose appendage at apex, the pistillate portion contiguous on the staminate, 10–20 cm. long; fruit lanceolate on a long bristly stipe; style elongate.——July–Aug. Ponds and riversides in lowlands; Hokkaido, Honshu, Shikoku, Kyushu.——Widely distributed in the temperate regions of the N. Hemisphere.

2.   **Typha orientalis** Presl.   *T. japonica* Miq.; *T. latifolia*

var. *orientalis* (Presl) Rohrb.; *T. shuttleworthii* subsp. *orientalis* (Presl) Graebn.; *T. latifolia* var. *japonica* (Miq.) Haller f.——KO-GAMA.   Similar to but smaller than the preceding; stems 1–1.5 m. long; leaves 0.5–1 cm. wide; staminate portion of spikes 3–5 cm. long, about half as long as the pistillate portion.——July–Aug. Ponds and riversides in lowlands; Honshu, Shikoku, Kyushu.——Ussuri, China, and Philippines.

3.   **Typha angustata** Bory & Chaub. HIME-GAMA.   Similar to the preceding in habit; leaves 5–10 mm. wide, gradually narrowed to tip; staminate portion of spikes 10–30 cm. long, the pistillate 6–20 cm. long, the naked axis between them 2–6 cm. long.——June–July. Shallow ponds and riversides in lowlands; Hokkaido, Honshu, Kyushu.——Warmer parts of Eurasia and the Mediterranean region.

Fam. 33.   SPARGANIACEAE   MIKURI KA   Bur-reed Family

Glabrous monoecious herbs; stem simple, leafy at base; leaves alternate, distichous, linear, erect or floating, entire, soft and somewhat spongy, sheathed below; inflorescence globose, sessile, the heads arranged in spikes or panicles, the lower heads pistillate and bracteate; flowers sessile, unisexual, the perianth-segments 3–6, membranous, cuneate; stamens 3 or more, the filaments filiform; ovary solitary, sessile, 1- to 2-locular, the ovules solitary in each locule, basal or pendulous, anatropous; style simple or forked; fruit nutlike, indehiscent, obovoid or wedge-shaped, with a spongy exocarp; embryo cylindric, the endosperm mealy.——A single cosmopolitan genus, with about 20 species.

1. SPARGANIUM L. Mikuri Zoku

Perennial swamp or marsh herbs with the characters of the family.

1A. Stigma filiform, 3–4 mm. long; inflorescence a panicle; stout plants with leaves 7–12 mm. wide. .... 1. *S. stoloniferum*
1B. Stigma linear to narrowly ovate, 2 mm. long or less; inflorescence a spike (if the plant bearing spikes branches below, then the leaves not more than 5 mm. wide).
  2A. Styles distinct.
    3A. Lower head supra-axillary; spikes always simple.
      4A. Staminate heads usually 1–2, contiguous with the congested pistillate heads. .... 2. *S. glomeratum*
      4B. Staminate heads 2–9, remote, the pistillate heads not congested.
        5A. Leaves slender, 1–4 mm. wide, keeled not at all or only near base, usually floating, the stems often floating.
            3. *S. angustifolium*
        5B. Leaves 4–10 mm. wide, with a keel at back; stems and majority of leaves aerial.
          6A. Styles with the stigma 3–4 mm. long; fruit fusiform. .... 4. *S. simplex*
          6B. Styles with the stigma about 2 mm. long, fruit rhombic-ovoid. .... 5. *S. fallax*
    3B. Heads all axillary; spikes sometimes branched.
      7A. Fruit obovoid or broadly ovoid, subsessile; leaves 2–5 mm. wide; spikes often branched; pistillate heads sessile.
        8A. Leaves all floating, without a keel, 2–2.5 mm. wide, spikes usually branched. .... 6. *S. gramineum*
        8B. Leaves at least partly aerial, keeled beneath, 3–5 mm. wide; spikes sometimes partly branched. .... 7. *S. stenophyllum*
      7B. Fruit oblong-fusiform, slightly narrowed at the middle, stipe 1.5–2 mm. long; leaves 4–10 mm. wide, slightly keeled beneath; spikes simple; lowest head usually short-peduncled. .... 8. *S. japonicum*
  2B. Style very short; slender herb in alpine ponds; leaves slender, floating; lower pistillate heads supra-axillary, the staminate usually solitary. .... 9. *S. hyperboreum*

1. **Sparganium stoloniferum** Hamilt. *S. longifolium* sensu auct. Japon., non Turcz.; *S. ramosum* sensu auct. Japon., non Huds.; *S. ramosum* subsp. *stoloniferum* (Hamilt.) Graebn.; *S. coreanum* Lév.; *S. macrocarpum* Makino; *S. stoloniferum* var. *macrocarpum* (Makino) Hara and var. *coreanum* (Lév.) Hara——Mikuri, Ō-mikuri, Kado-hari-mikuri. Soft spongy greenish herb with slender rhizomes; stems erect, 0.7–1 m. long, rather stout; leaves flattened, keeled on back, 7–12 mm. wide, slightly narrowed upward to an obtuse apex; panicle 30–50 cm. long, the bracts leafy, the lower ones sometimes exceeding the panicle, the branches few, each with 1–3 pistillate heads in lower part and several staminate heads loosely disposed above; pistillate flowers sessile; style elongate, the stigma filiform; fruit rounded-obovoid, angular, 6–10 mm. long, 4–8 mm. in diameter, sessile.——June–Aug. Shallow water in lowlands; Hokkaido, Honshu, Shikoku, Kyushu.——Korea, China to Turkestan, and Afghanistan.

2. **Sparganium glomeratum** Laest. *S. glehnii* Meinsh. ——Tama-mikuri. Soft greenish herb 30–60 cm. high; stems usually erect, shorter than the leaves; leaves sometimes floating, the aerial ones compressed, 5–12 mm. wide, slightly keeled beneath; spikes 10–20 cm. long, simple, with leafy bracts much longer than the spike; pistillate heads approximate, the lower ones often on short supra-axillary peduncles; staminate heads 1 or 2, adjacent to the pistillate, inconspicuous; fruit numerous, on pedicels about 1.5 mm. long, narrowly oblong-fusiform, about 4 mm. long, 1.5 mm. wide, scarcely angled; style inclusive of stigma about 1.5 mm. long.——July–Aug. Shallow ponds and lake margins; Hokkaido, Honshu (mountains of centr. distr. northw.).——Eurasia and nw. America.

Var. **angustifolium** Graebn. Hosoba-tama-mikuri. Plant much more slender; leaves 2–4 mm. wide; fruit about 3 mm.

long.——Shallow ponds in high mountains; Honshu (centr. distr. northw.).——Eurasia.

3. **Sparganium angustifolium** Michx. *S. affine* Schnizl.; *S. diversifolium* sensu Miyabe & Kudo, pro parte, non Graebn.——Hosoba-uki-mikuri. Stems elongate, floating or aerial, 30–50 cm. long, the floating leaves mostly narrow, about 4 mm. wide, compressed, keeled beneath; spikes simple; pistillate heads 2–4, 7–10 mm. in diameter, the lower 1 or 2 with a supra-axillary peduncle; staminate heads commonly approximate; fruit short-pedicelled, about 4 mm. long, fusiform; stigma linear, about 1 mm. long.——Reported from Hokkaido (Mount Yubari).——Kamchatka, Europe, and N. America.

4. **Sparganium simplex** Huds. *S. affine* sensu Miyabe & Miyake, non Schnizl.——Ezo-mikuri. Stems usually erect, 20–60 cm. long; leaves elongate, 3–12 mm. wide, erect, keeled beneath; spikes with 2–5(–6) pistillate, and 3–8 staminate heads, the lowest head pedunclate; fruit 4–5 mm. long, 2–2.5 mm. in diameter, gradually narrowed to a beak; stigma linear, 1.5–2 mm. long.——Reported from Hokkaido.——Eurasia and N. America.

5. **Sparganium fallax** Graebn. *S. yamatense* Makino. ——Yamato-mikuri. Stems erect, 40–80 cm. long; leaves erect, longer than the stem, 4–8 mm. wide, keeled beneath; spikes erect, the lower bracts about as long as the spike, the pistillate heads 4–6, relatively wide apart, usually sessile, 2–2.5 cm. in diameter in fruit, the staminate 4–7, interrupted; fruit rhombic-ovoid or fusiform, 6–7 mm. long, about 2 mm. wide, the pedicels 2–3 mm. long.——Aug. Shallow ponds in lowlands; Honshu (Kantō Distr. and westw.), Kyushu.——Burma and India.

6. **Sparganium gramineum** Georgi. *S. friesii* Beurl. ——Uki-mikuri. Submerged stems slender, soft, 30–40 cm. long; leaves floating, flat, flaccid, 40–50 cm. long, 2–2.5 mm.

wide, not keeled, with a slightly dilated sheath at base; spikes aerial, 10–15 cm. long, with 1 or 2 branches; staminate heads few, contiguous; pistillate heads 1–3 on each branch, globose, 7–8 mm. across in fruit, the lower ones peduncled and provided with a leafy bract; flowers with linear-spathulate perianth segments about 1.5 mm. long; fruit broadly ovoid, conical, subsessile, not angled, about 3 mm. long, 2 mm. in diameter; stigma ovate-peltate.——July–Aug. Ponds in high mountains; rare. Hondo (Echigo).——Europe, Siberia, and Kamchatka.

7. Sparganium stenophyllum Maxim. *S. nipponicum* Makino——Hime-mikuri.	Stems erect, usually rather slender, 40–70 cm. long; leaves keeled beneath, suberect, usually somewhat longer than the stem, 3–5 mm. wide; spikes simple or with a single branch at base, the bracts nearly as long as the spike; pistillate heads 2 or 3(–4), sessile, axillary, distant in fruit, about 1.5 cm. in diameter, the staminate few, distant; fruit obovoid, subsessile, about 4 mm. long, abruptly narrowed to the style.——June–Sept. Shallow water in lowlands; Hokkaido, Honshu, Kyushu.——n. China and Ussuri.

8. Sparganium japonicum Rothert. Nagae-mikuri. Stems erect, 40–70 cm. long; leaves flat, keeled beneath, sub-

erect, as long as or slightly longer than the spikes, 4–10 mm. wide; spikes simple, more or less distant, the bracts usually longer than the spike; pistillate heads 3 or 4, 1.5–2 cm. in diameter in fruit, the lower 1 or 2 heads usually peduncled, the staminate 5–10; fruit oblong-fusiform, about 5 mm. long, 1.5–2 mm. across, not angled, with a slight constriction in the middle, narrowed toward both ends, the pedicels about 2 mm. long; style including the linear stigma, about 2 mm. long.——July–Aug. Shallow water in lowlands; Honshu, Kyushu.——Korea.

9. Sparganium hyperboreum Laest. *S. natans* var. *submuticum* Hartm.; *S. submuticum* (Hartm.) Neumann; *S. minimum* sensu auct. Japon., non Fries——Chishima-mikuri, Takane-mikuri.	Stems elongate, submersed; leaves slender, compressed, not keeled, 2–4 mm. wide, obtuse; spikes simple, commonly 2–3 cm. long, the bracts longer than the spikes; pistillate heads 2 or 3, 6–8 mm. in diameter in fruit, the lowest one usually with a supra-axillary peduncle, the staminate heads 1 or 2, close together and closely adjacent to the pistillate heads; fruit ovoid, with a narrowly ovate stigma.——Ponds; Hokkaido (Mount Taisetsu).——Subarctic regions.

## Fam. 34. **POTAMOGETONACEAE**	Hiru-mushiro Ka	Pondweed Family

Aquatic herbs growing in fresh or rarely in saline water, mostly submerged but also with floating leaves; leaves sessile or petioled, filiform or flattened, usually entire; flowers perfect or monoecious, without a perianth; inflorescence spicate or fasciculate, axillary, sometimes enveloped by a scarious sheath; stamens 1–4, the anthers extrorse, 1- to 2-locular, the connective sometimes (*Potamogeton*) bearing sepallike appendages; pistil of 1–4 separate 1-ovulate carpels each developing into a nutlet or drupelike fruit; endosperm absent.——About 4 genera with more than 100 species, of wide distribution.

1A. Flowers perfect; stamens 2 or 4.
  2A. Stamens 4; fruit sessile; peduncles straight or slightly curved; connective with a sepallike appendage. .... 1. *Potamogeton*
  2B. Stamens 2; fruit long-pedicelled; peduncles spirally curved after flowering; connective without a perianthlike appendage. .. 2. *Ruppia*
1B. Flowers monoecious; stamen 1. .... 3. *Zannichellia*

## 1. **POTAMOGETON** L.	Hiru-mushiro Zoku

Aquatic herbs, attached to the bottom by roots and rhizomes; stems elongate, sometimes branched; leaves often dimorphic, alternate, those of the inflorescence sometimes approximate, the submerged leaves usually linear, the floating lanceolate to elliptic, the stipules membranous, free or adnate to the basal portion of the blade or to the petiole, sometimes connate to form a tubular sheath; peduncles axillary; flowers small, in axillary spikes, perianth wanting; stamens 4, the anthers sessile, the connective dilated at tip into a perianthlike appendage, the anther-locules distinct or connate; ovary of (1–) 4 distinct, sessile, 1-locular and 1-ovuled carpels; stigma simple; nutlets 1–4, 1-seeded, the seeds filling the nutlet, the embryo curved.——About 100 species, in temperate regions of the world.

1A. Plants with floating leaves.
  2A. Floating leaves 5–12 cm. long, 2.5–5 cm. wide.
    3A. Submersed leaves semiterete, slenderly linear. .... 1. *P. natans*
    3B. Submersed leaves with lanceolate blades.
      4A. Submersed leaves petiolate. .... 3. *P. distinctus*
      4B. Submersed leaves sessile, the floating leaves sometimes not developed.
        5A. Petioles of the floating leaves with margin undulate near top. .... 2. *P. fryeri*
        5B. Petioles of the floating leaves without undulate margin.
          6A. Floating leaves narrowly ovate to broadly-lanceolate; stems branched; winter buds with several nodes; peduncles thickened above; submersed leaves 3–5 cm. long. .... 4. *P. heterophyllus*
          6B. Floating leaves spathulate; stems usually not branched; winter buds with 1 node; peduncles hardly thickened above; submersed leaves 8–20 cm. long. .... 5. *P. alpinus*
  2B. Floating leaves 1.5–3.5 cm. long, 5–12 mm. wide; submersed leaves all linear.
    7A. Nutlets entire or obtusely dentate on the dorsal margin; style short. .... 6. *P. octandrus*
    7B. Nutlets conspicuously cristate on the dorsal margin; style long and slender. .... 7. *P. cristatus*
1B. Plants without floating leaves; leaves all submersed.
  8A. Leaves not adnate to the stipules at base.
    9A. Leaves 7–30 mm. wide, lanceolate or narrowly so.
      10A. Leaves sessile or nearly so, obtuse or acute, not mucronate.
        11A. Leaves perfoliate. .... 8. *P. perfoliatus*

11B. Leaves narrowed at base, not perfoliate.
12A. Leaves flattened at apex.
13A. Stems branched; winter buds with several nodes; submersed leaves 4–8 cm. long. .... 4. *P. heterophyllus*
13B. Stems usually not branched; winter buds with 1 node; submersed leaves 8–20 cm. long. .... 5. *P. alpinus*
12B. Leaves cucullate at apex. .... 9. *P. praelongus*
10B. Leaves more or less petiolate, abruptly acuminate or taper-pointed.
14A. Petioles 3–12 mm. long. .... 10. *P. dentatus*
14B. Petioles 1.5–3 cm. long. .... 11. *P. malaianus*
9B. Leaves 1–6 mm. wide, linear.
15A. Leaves serrulate, 4–6 mm. wide, sometimes clasping at base. .... 12. *P. crispus*
15B. Leaves entire, 1–3 mm. wide.
16A. Margins of the stipules free, overlapping.
17A. Stems distinctly compressed, thicker than the peduncles. .... 13. *P. compressus*
17B. Stems only slightly compressed, as thick as the peduncles.
18A. Leaves 2–3 mm. wide, acuminate; nutlets 3 mm. long; rhizomes slender. .... 14. *P. oxyphyllus*
18B. Leaves 1–1.5 mm. wide, acute; nutlets about 1.7 mm. long; rhizomes absent. .... 15. *P. berchtoldii*
16B. Margins of the stipules united into a tubular sheath. .... 16. *P. pusillus*
8B. Leaves adnate to the stipules at base.
19A. Leaves 2–3 mm. wide, serrate, with a broad obtuse projection at apex. .... 17. *P. maackianus*
19B. Leaves about 1 mm. wide, entire, acute. .... 18. *P. pectinatus*

1. **Potamogeton natans** L. *P. morongii* A. Benn.— O-HIRU-MUSHIRO, ME-HIRU-MUSHIRO. Stems often elongate; floating leaves thickish, long-petiolate, ovate-oblong to elliptic, 5–10 cm. long, 2.5–5 cm. wide, many-nerved, entire, apiculate, rounded at base, the submersed leaves semiterete, slenderly linear, the stipules 5–8 cm. long, membranous and many-nerved; peduncles 5–10 cm. long, as thick as the stem throughout, bearing densely flowered spikes 4–6 cm. long; nutlets broadly ovate, 3.5–4 mm. long, nearly entire on dorsal ridge. —Apr.–Aug. Ponds and shallow rivers; Hokkaido, Honshu, Shikoku, Kyushu.—Widely distributed in temperate regions.

2. **Potamogeton fryeri** A. Benn. *P. subsessilifolius* A. Camus; *P. torquatus* Koidz.—FUTO-HIRU-MUSHIRO. Similar to the preceding, but the submersed leaves with blades linear-lanceolate, elongate, the floating leaves long-petiolate, oblanceolate to elliptic, 10–15 cm. long, rounded and decurrent at base, the stipules 5–10 cm. long; peduncles 10–15 cm. long, thickened above, bearing densely flowered spikes; nutlets 4, about 5 mm. long, narrowly winged on dorsal ridge.—Apr.–Aug. Ponds; Hokkaido, Honshu, Kyushu.—Korea.

3. **Potamogeton distinctus** A. Benn. *P. longipetiolatus* A. Camus; *P. alatus* Koidz.; *P. tepperi* auct. Japon., non A. Benn.; *P. franchetii* A. Benn.; *P. polygonifolius* auct. Japon., non Pourr.—HIRU-MUSHIRO. Stems elongate; floating leaves broadly lanceolate to ovate-elliptic, 5–10 cm. long, (1.5–)2–4 cm. wide, petioles 6–10 cm. long, the submersed leaves lanceolate, long-petiolate, acute at both ends, minutely serrulate with projecting marginal cells; stipules 3–4.5 cm. long, thinly membranous, soon decaying; peduncles not thickened above; spikes 2–5 cm. long; nutlets 3–3.5 mm. long, ridged on the dorsal margin, the style short.—June–Sept. Ponds and paddy fields; Hokkaido, Honshu, Shikoku, Kyushu. —Korea and China.

Potamogeton × malainoides Miki. AINOKO-HIRU-MU-SHIRO. Hybrid of *P. distinctus* × *P. malaianus*. Submersed leaves similar to those of No. 11.—Reported from Honshu (Kinki Distr.), Shikoku.—Formosa.

4. **Potamogeton heterophyllus** Schreb. *P. gramineus* var. *heterophyllus* (Schreb.) Fries; *P. heterophyllus* var. *foliosus* Mert. & Koch; *P. gramineus* auct., non L.—EZO-NO-HIRU-MUSHIRO. Stems elongate and often branched; floating leaves entire, long-petiolate, rather abruptly obtuse, rounded at the base, 2.5–10 cm. long, 1–3 cm. wide, the submersed leaves sessile, thin, oblanceolate, 4–8 cm. long, 6–10 mm. wide, acute, entire; stipules thinly membranous; peduncles thickened above; spikes 2–4 cm. long.—July–Sept. Ponds; Hokkaido, Honshu (centr. mts.).—Temperate regions of the N. Hemisphere.

Potamogeton nipponicus Makino. SASA-EBI-MO. Alleged hybrid similar to No. 4; stems not flexuose, very rarely with floating leaves and with slightly serrate, submersed leaves, the upper ones very short-petiolate.—Reported from Honshu (Nikko and Hakone).

5. **Potamogeton alpinus** Balbis. *P. rufescens* Schrad. —HOSOBA-HIRU-MUSHIRO. Stems simple or slightly branched; floating leaves spathulate or oblanceolate, sometimes not developed, obtuse, gradually narrowed to the base, many-nerved, the lower submersed leaves sessile, the upper short-petiolate, narrowly lanceolate, narrowed toward the base, 7-nerved, obtuse to acute; peduncles 5–20 cm. long; spikes 2.5–4 cm. long; nutlets obovate, with 3 dorsal ridges; style short.—Reported from Hokkaido.—Cooler regions of the N. Hemisphere.

6. **Potamogeton octandrus** Poir. *P. javanicus* Hassk.; *P. numasakianus* A. Benn.—HOSOBA-MIZUHIKI-MO. Plant small, slender, with filiform sometimes branched stems; floating leaves broad-lanceolate to ovate-oblong, 1.5–2.5 cm. long, 3–8 mm. wide, acute or abruptly obtuse, rather acute at base, 7-nerved, entire, the petioles shorter than the blades, submersed leaves filiform, 3–6 cm. long, 0.3–1 mm. wide, very acute; stipules thinly membranous, 6–10 mm. long, their edges overlapping; peduncles 1–2 cm. long, slightly thickened above; spikes 1–1.5 cm. long, more or less dense; nutlets sessile, broadly obovate, about 2 mm. long, 3-ridged on the back, the median ridge acute, with a few obtuse teeth; style rather short. —June–Oct. Ponds and paddy fields; Hokkaido, Honshu, Shikoku, Kyushu.—Korea, China, Ussuri, Formosa, India, and Malaysia.

Var. **miduhikimo** (Makino) Hara. *P. miduhikimo* Makino; *P. asiaticus* A. Benn.; ? *P. limosellifolius* Maxim.; ? *P. subfuscus* A. Benn.; *P. vaseyi* Miki, non Robins.—MIZU-HIKI-MO. Leaves broadly lanceolate to narrowly ovate-elliptic; median ridge of the nutlet destitute of teeth; style short. —June–Oct. Ponds and paddy fields; Honshu, Shikoku, Kyushu.

7. **Potamogeton cristatus** Regel & Maack. *P. hybridus* Makino, non Michx.——Koba-no-hiru-mushiro. Similar to the preceding, differs in the rather short and densely flowered spikes; nutlets with a short stipe, conspicuously cristate on the back; style slender, 1–1.2 mm. long.——June–Sept. Ponds; Honshu.——Ussuri, China, Korea, and Formosa.

Potamogeton kamogawaensis Miki. Ō-mizuhiki-mo. Kamogawa-mo. An alleged hybrid occurring in Honshu, Shikoku, and Kyushu, very similar to No. 7 and to No. 6, but larger in all respects and becoming slightly darker when dry; submersed leaves 1–2 mm. wide; flowers sterile.

8. **Potamogeton perfoliatus** L. Hiroha-no-ebi-mo. Stems elongate, branched, rather densely leafy; floating leaves not developed, the submersed broadly ovate to broadly lanceolate, 2–6 cm. long, 1–2.5 cm. wide, obtuse to acute, deeply cordate and perfoliate at base, entire and somewhat undulate, 3-nerved and with 2 veinlets between the nerves; stipules thinly membranous, soon withering or caducuos; peduncles slightly thickened above; spikes densely flowered, 1.5–2.5 cm. long; nutlets about 3 mm. long, broadly obovate, with an entire ridge at the back; style short.——June–Sept. Ponds and shallow rivers; Hokkaido, Honshu, Shikoku, Kyushu.——Eurasia, North America, Australia, and Argentina.

Potamogeton × leptocephalus Koidz. Hiroha-no-sennin-mo. Hybrid of *P. maackianus* × *P. perfoliatus*. Leaves narrower than in the parents, 3–3.5 cm. long, 7–8 mm. wide, rounded to very obtuse, obscurely serrulate, 5- to 7-nerved.——Reported from Kyushu.

9. **Potamogeton praelongus** Wulfen. Nagaba-ebi-mo. Stems branched; leaves all submersed, narrowly lanceolate, 10–20 cm. long, 1–2.5 cm. wide, narrowed toward the obtuse ends, cucullate and recurved at apex, rather many-nerved; stipules whitish, obtuse at apex; peduncles elongate, not thickened above; spikes 3–5 cm. long; nutlets about 4 mm. long, the keel entire.——Hokkaido.——Europe, Siberia, and N. America.

10. **Potamogeton dentatus** Hagstr. *P. teganumensis* Makino; *P. lucens* var. *teganumensis* (Makino) Makino——Gasha-moku. Stems branched; leaves all submersed, broadly lanceolate or narrowly oblong, 5–15 cm. long, 15–30 mm. wide, mucro at tip 1–4 mm. long, slightly undulate, serrulate toward tip, 7- to 13-nerved, obtuse or acute at base, the petioles 3–12 mm. long; stipules 2.5–5 cm. long, obtuse or rounded at apex; peduncles 5–20 cm. long; spikes densely flowered, 3–5 cm. long; nutlets about 3 mm. long, the keel entire.——July–Oct. Ponds; Honshu (Kantō and Kinki Distr.).

11. **Potamogeton malaianus** Miq. *P. japonicus* Fr. & Sav.; *P. wrightii* Morong; *P. tretocarpus* Maxim.; *P. miyakejimensis* Honda; *? P. gaudichaudii* Cham. & Schltdl.——Sasaba-mo, Sajiba-mo. Stems simple or slightly branched; leaves all submersed, linear-oblong to lanceolate, 8–12 cm. long, 1–2.5 cm. wide, 7- to 13-nerved, obtuse at base, with a mucro 2–3 mm. long at apex, obscurely serrulate, conspicuously undulate on margin, the petioles 1.5–3 cm. long; the stipules 2–4 cm. long; spikes rather densely flowered, 3–5 cm. long; nutlets about 3 mm. long, 3-ridged on back, the median ridge more prominent and entire.——June–Oct. Ponds and shallow rivers; Hokkaido, Honshu, Shikoku, Kyushu.——Ryukyus, China, Formosa, Malaysia, and India.

12. **Potamogeton crispus** L. *P. serratus* Huds.——Ebi-mo. Stems slender and branched; submersed leaves linear to broadly so, 4–7 cm. long, 4–6(–8) mm. wide, rounded to obtuse at apex, undulate and serrulate, sessile and obtuse or subrounded at base and sometimes slightly clasping, 3-nerved; stipules thinly membranous, about 10 mm. long; peduncles 2–5 cm. long; spikes lax and few-flowered, 1–1.5 cm. long; nutlets broadly ovate, about 3 mm. long, obscurely toothed on the median ridge of back; style elongate, about 2 mm. long. ——May–July. Ponds and shallow rivers; Hokkaido, Honshu, Shikoku, Kyushu.——Worldwide except South America.

13. **Potamogeton compressus** L. *P. zosteraefolius* K. Schum.; *P. complanatus* Willd.; *P. sibiricus* auct. Japon., non A. Benn.——Ezo-yanagi-mo. Stems prominently compressed, sometimes 2-winged, much branched; submersed leaves linear, 4–20 cm. long, 2–4 mm. wide, obtuse and mucronate or abruptly acute, laterally attached on the compressed stem, entire, 3-nerved and with veinlets between the nerves; stipules 2–4 cm. long, withering or caducous; peduncles 2–4 cm. long; spikes about 15 mm. long; nutlets broadly obovate, 2–3 mm. long, keeled.——Reported from Hokkaido.——Eurasia and N. America.

14. **Potamogeton oxyphyllus** Miq. Yanagi-mo. Stems slender and branched; leaves all submersed, linear, 5–10(–12) cm. long, 2–3 mm. wide, acuminate, entire, sessile, with very faint nerves on both sides, the stipules about 2 cm. long, obtuse, thinly membranous and their margins overlapping; peduncles 2–4 cm. long; spikes rather densely flowered, 1–1.5 cm. long; nutlets broadly elliptic-ovate, about 3.5 mm. long, subentire; style short.——May–Oct. Ponds and shallow rivers; Hokkaido, Honshu, Shikoku, Kyushu.——Korea and Manchuria.

Potamogeton × fauriei (A. Benn.) Miki. *P. oxyphyllus* var. *fauriei* A. Benn.——Ainoko-yanagi-mo. An alleged hybrid of No. 14 and No. 13. A coarser plant with leaves 17- to 19-nerved; spikes loosely flowered and sterile.

15. **Potamogeton berchtoldii** Fieber. *P. pusillus* auct., non L.——Ito-mo. Stems very slender and scarcely compressed; leaves all submersed, linear, dark green, 3–5 cm. long, 1–1.5 mm. wide, acute to subobtuse, 1- to 3-nerved or sometimes nerveless, with a somewhat reflexed gland on each side at base; stipules about 10 mm. long, margins overlapping; peduncles 2–4 cm. long; spikes rather few-flowered; nutlets broadly elliptic-ovate, about 1.5 mm. long, with 3 low ridges on back.——May–Aug. Ponds and shallow rivers; Hokkaido, Honshu, Shikoku, Kyushu.——Widely distributed in the N. Hemisphere.

16. **Potamogeton pusillus** L. *P. panormitanus* Biv.——Tsutsu-ito-mo. Very similar to the preceding; leaves vivid green; stipules tubular with an obliquely truncate mouth.——Reported from Honshu, Shikoku.——Widely distributed in the N. Hemisphere, inclusive of the West Indies and Azores.

17. **Potamogeton maackianus** A. Benn. *P. serrulatus* Regel & Maack, non Schrad. nec Opiz; *P. robbinsii* var. *japonicus* A. Benn.; *P. tenuifolius* A. Camus——Sennin-mo. Stems slender and much branched; leaves all submersed, linear, 2–6 cm. long, 2–3 mm. wide, finely serrulate, 3-nerved and faintly nervulate, becoming broadly obtuse to rounded at apex, briefly adnate to the stipules at base; stipules about 10 mm. long, membranous; peduncles 1–3 cm. long; spikes few-flowered; nutlets broadly ovate-elliptic, about 3 mm. long, with an acute keel; style short.——May–Sept. Ponds and shallow rivers; Hokkaido, Honshu, Kyushu.——e. Siberia, Korea, and Manchuria.

18. **Potamogeton pectinatus** L. Ryū-no-hige-mo.

Stems filiform, yellowish, much branched; leaves all submersed, filiform, 3–10 cm. long, 0.5–1 mm. wide, entire and acute at apex; stipules 1–2 cm. long, whitish, membranous, adnate to the base of the leaves from three-fourths to six-sevenths of their length; peduncles slender, 2–10 cm. long; spikes remotely few-flowered, 2–4 cm. long; nutlets broadly obovate, 3–3.5 mm. long, usually rounded and entire on back.——June–Oct. Ponds and shallow rivers, especially near seashores; Hokkaido, Honshu, Shikoku, Kyushu.——Eurasia, N. America, Africa, and Australia.

2. RUPPIA L. Kawa-tsuru-mo Zoku

Much branched delicate aquatic herbs with filiform branches; leaves alternate, narrow, 1-nerved, subtended by a membranous sheath; peduncles solitary, axillary, elongate after flowering; flowers without a perianth; stamens 2; anthers attached to the peduncle at the back, with 2 distinct anther-locules; stigma sessile and peltate; nutlets small, long-pedicelled, obliquely ovate.——Four or five widely distributed species.

1. Ruppia rostellata Koch. *R. maritima* L., pro parte; *R. maritima* var. *rostrata* Agardh——Kawa-tsuru-mo. Delicate aquatic herb; stems much branched, elongate; leaves filiform, slightly compressed in cross section, 0.3–0.5 mm. wide, 5–10 cm. long, acuminate, remotely serrulate toward the tip; sheath 8–15 mm. long, with a rounded mouth; nutlets 2–4, obliquely ovate, biconvex, about 2.5 mm. long, short-beaked, on a slightly curved peduncle 1.5–4 cm. long.——June–Oct. Ponds and shallow rivers; Honshu (Kantō Distr. and westw.), Shikoku, Kyushu.——Eurasia, Africa, and N. America.

Ruppia maritima L. *R. spiralis* L.; *R. maritima* var. *spiralis* Mori——Nejiri-kawa-tsuru-mo of Eurasia, N. America. and Africa, with spirally coiled, elongate peduncles in fruit, is reported from Japan and figured by Miki, in Yamashiro Suisō-shi, *fig. 29.* 1937.

Ruppia truncatifolia Miki. Known only in the sterile state; leaves 0.3–0.7 mm. wide, truncate-retuse at apex.——Reported from Hokkaido and Korea.

3. ZANNICHELLIA L. Ito-kuzu-mo Zoku

Aquatic monoecious herbs, with slender creeping rhizomes; stems delicate, elongate; leaves submersed, narrowly linear, stipulate; flowers axillary, the staminate solitary and sessile, consisting of only 1 stamen with a 2-locular anther, the pistillate consisting of 2–5 free, sessile or stipitate carpels; stigma peltate; nutlets lunate, usually with obtuse teeth at back, the embryo coiled.——Two or three widely distributed species.

1. Zannichellia palustris L. var. **indica** (Cham.) Graebn. *Z. indica* Cham.; *Z. palustris* var. *japonica* Makino ——Ito-kuzu-mo, Mikazuki-ito-mo. Stems slender; leaves less than 1 mm. wide, narrowly linear; peduncles 1–2.5 mm. long; nutlets 2–5, stipitate, about 3 mm. long, lunate, narrowly oblong, remotely cristate-dentate on back, the stipe 1.5–3.3 mm. long; style 2.5–3 mm. long, slender.——Honshu, Kyushu.——Ryukyus and India. The species is cosmopolitan.

Fam. 35. NAJADACEAE Ibara-mo Ka Naiad Family

Submersed dioecious or monoecious aquatic herbs growing in fresh or saline waters; stems much branched; leaves opposite, or verticillate, linear, spiny-margined, sheathed at base; flowers solitary in the leaf-axils, the staminate pedicellate, the perianth hyaline, bifid at apex; stamen 1, the anther 1- to 4-locular, pistillate flowers terminated by 2–4 linear stigmas; fruit achenelike, 1-seeded, the embryo straight.——With a single genus of wide distribution, nearly cosmopolitan.

1. NAJAS L. Ibara-mo Zoku

Characters of the family. About 10 species widely distributed.

1A. Margin of leaf-sheath entire; leaves 1–3 mm. wide, remotely serrate, the teeth spinescent; flowers dioecious; seeds ellipsoidal to oblong, 1.5–2.5 mm. wide.
  2A. Leaves 2–3 mm. wide, with 9–11 teeth on each side. . . . . 1. *N. marina*
  2B. Leaves about 1 mm. wide, with 2–4 teeth on each side. . . . . 2. *N. tenuicaulis*
1B. Margin of leaf-sheath denticulate; leaves usually narrow, less than 1 mm. wide, finely toothed; flowers monoecious; seeds lanceolate, less than 1 mm. wide.
  3A. Seeds prominently curved; pistillate flowers with a bracteate sheath. . . . . 3. *N. ancistrocarpa*
  3B. Seeds straight; pistillate flowers without a bracteate sheath.
    4A. Staminate flowers bracteate; leaf-sheath not distinctly extended, obliquely rounded at apex.
      5A. Leaves scarcely compressed; seeds with oblong to linear superficial cells.
        6A. Superficial cells of seeds transversely elongate.
          7A. Leaves about 3 cm. long, 0.8 mm. wide; seeds 3.5–4 mm. long. . . . . 4. *N. oguraensis*
          7B. Leaves 1.5–2 cm. long, 0.3–0.5 mm. wide; seeds 3 mm. long. . . . . 5. *N. minor*
        6B. Superficial cells of seeds longitudinally elongate.
          8A. Superficial cells of seeds narrowly oblong; seeds 2.5–3 mm. long, oblong. . . . . 6. *N. yezoensis*
          8B. Superficial cells of seeds linear; seeds about 2 mm. long, lanceolate. . . . . 7. *N. japonica*
      5B. Leaves flat; seeds with hexagonal superficial cells. . . . . 8. *N. indica*
    4B. Staminate flowers without a bracteate sheath; leaf sheath auriculate at apex. . . . . 9. *N. graminea*

1. Najas marina L. *N. major* All.; *N. fluviatilis* Poir.
——IBARA-MO. Stems much branched, smooth or sparsely spinulose; leaves rigid, linear, 2–4 cm. long, 2–3 mm. wide, flat, remotely spinescent-toothed; sheath broadly rounded, entire; fruit ellipsoidal-oblong, 4–6 mm. long; seeds minutely apiculate.——July–Sept. Ponds; Hokkaido, Honshu, Shikoku, Kyushu.——Widely distributed.

2. Najas tenuicaulis Miki. HIME-IBARA-MO. Stems much branched, sometimes sparingly spinulose; leaves 2–2.5 cm. long, about 1 mm. wide, with few marginal teeth; sheath semirounded, entire or with few minute teeth; stigmas 3; fruit 4–5 mm. long, 1.2–2 mm. wide, ellipsoidal.——July–Oct. Honshu; rare.

3. Najas ancistrocarpa A. Br. MUSASHI-MO. Stems branched; leaves 15–20 mm. long, slightly recurved, about 0.5 mm. wide, minutely toothed on margins; sheath semirounded, with few minute spines on margins; seeds prominently curved, about 2.5 mm. long, fusiform-lanceolate, longitudinally striate.——Honshu; rare.

4. Najas oguraensis Miki. Ō-TORIGE-MO. Relatively rigid annual with smooth branched stems; leaves subulate-linear, rather rigid, 3–4 cm. long, about 0.8 mm. wide, the superficial cells of seeds transversely narrow-oblong.——July–Sept. Honshu, Shikoku.

5. Najas minor All. *N. marina* L. pro parte——TO-RIGE-MO. Plants much branched; leaves subulate-linear, more or less recurved, 1–2 cm. long, 0.5 mm. wide, denticulate; sheaths semirounded, minutely spinescent-toothed; seeds about 3 mm. long, 0.8 mm. wide, lanceolate, the superficial cells transversely narrow-oblong.——July–Oct. Honshu.——Eurasia and Africa.

6. Najas yezoensis Miyabe. ITO-IBARA-MO. Stems slender, much branched; leaves rather rigid, linear-acicular, 2.3–3 cm. long, about 0.3 mm. wide, denticulate; sheaths broad, rounded, minutely spinescent on margin; seeds oblong, 2.5–3 mm. long, the superficial cells longitudinally narrow-oblong.——Sept. Hokkaido.

7. Najas japonica Nakai. *N. gracillima* Miki, non Magnus——ITO-TORIGE-MO. Stems slender and much branched; leaves narrowly linear, about 1.5 cm. long, slightly compressed, about 0.2 mm. wide, with 7–10 teeth on each side; sheath rounded, sparingly and minutely spinescent on margin; seeds lanceolate, about 2 mm. long, about 0.5 mm. wide, the superficial cells longitudinally linear-elongate.——July–Oct. Honshu, Shikoku, Kyushu.

8. Najas indica (Willd.) Cham. *Caulina indica* Willd. ——SAGAMI-TORIGE-MO. Stems elongate, diffuse; leaves linear, 1.5–2.5 cm. long, minutely spinulose on margin; sheaths semirounded, with few minute spines on margin; seeds about 2.5 mm. long, 0.6–0.7 mm. wide, superficial cells hexagonal, as broad as long.——July–Oct. Honshu.——India, Malaysia, and Philippines.

9. Najas graminea Del. *N. serristipula* Maxim.; *N. graminea* var. *serristipula* (Maxim.) Nakai——HOSSU-MO. Stems slender and much branched; leaves acicular, 1.5–2.5 cm. long, 0.3–0.5 mm. wide, minutely and rather densely denticulate; sheaths auriculate, narrowly deltoid, with minutely toothed margin; seeds broadly lanceolate, about 2.5 mm. long, 0.6 mm. wide, often striate, the superficial cells minute, angled.——July–Sept. Honshu, Shikoku, Kyushu.——Africa, Eurasia and Australia.

Fam. 36. ZOSTERACEAE   AMA-MO KA   Eelgrass Family

Dioecious or monoecious marine rhizomatous herbs; stems elongate or short, branched, often compressed; leaves distichous, flat, ribbonlike, sheathed at base; flowers on a spadix subtended by a spathe, sometimes provided with a hyaline bract; staminate flower consisting of a subsessile 1-locular anther; pistillate flower consisting of a 2-carpellate, 1-locular ovary; style terminated by 2 stigmas.——Two genera with more than 10 species.

1A. Plants monoecious, sometimes stoloniferous; ovary and fruit ovate; stems elongate. .... 1. *Zostera*
1B. Plants dioecious, rhizomatous, caespitose; ovary and fruit cordate; stems short. .... 2. *Phyllospadix*

1. ZOSTERA L.   AMA-MO ZOKU

Submersed dioecious herbs of saline water, usually with slender stolons; stems compressed, often branched; leaves flat and ribbonlike, distichous, sheathed at base; spadix surrounded by a spathe; flowers alternately arranged on one side of the spadix, the staminate consisting of one sessile, 1-locular anther, the pollen filiform; pistillate flowers consisting of 1 pistil, the ovary attached on back near the middle; style elongate, the stigmas 2, filiform; fruit flask-shaped, membranous, irregularly dehiscent, beaked; seed solitary, longitudinally ribbed, narrowly oblong.——Cosmopolitan, with about 10 species.

1A. Leaves 3–15 mm. wide.
  2A. Leaves 3–7 mm. wide.
    3A. Plants with slender, elongate stolons; leaves 3–5 mm. wide. .... 1. *Z. marina*
    3B. Plants caespitose, without stolons; leaves about 7 mm. wide. .... 2. *Z. caespitosa*
  2B. Leaves 10–15 mm. wide.
    4A. Leaves abruptly acute; seeds with indistinct longitudinal striations. .... 3. *Z. caulescens*
    4B. Leaves retuse; seeds without striations. .... 4. *Z. asiatica*
1B. Leaves 1.5–2 mm. wide; rhizomes slender and creeping. .... 5. *Z. nana*

1. Zostera marina L. AMA-MO. Rhizomes elongate; leaves elongate, 3–5 mm. wide, rounded at apex, 5–7 nerved, the sheaths without auricles; inflorescence without appendage on margin; seeds oblong, about 4 mm. long, longitudinally striolate.——June–Aug. Shallow muddy sea water; Hokkaido, Honshu, Shikoku, Kyushu; common.——Eurasia and N. America.

2. Zostera caespitosa Miki. SUGE-AMA-MO. Plant caespitose, the rhizomes not elongate, estoloniferous; leaves about 7 mm. wide, 5–7 nerved, retuse, with minutely roughened margins, the ligule obsolete; inflorescence without appendage on margin; seeds oblong, longitudinally striate.——Hokkaido, Honshu (centr. distr. and northw.).——Korea and Manchuria.

3. Zostera caulescens Miki. TACHI-AMA-MO. Rhizomes elongate and branched; stems slender; leaves about 14 mm. wide, 9(–11)-nerved, abruptly acute to rounded and mucronate at apex, smooth, the ligule obsolete; inflorescence without an appendage on margin; seeds ellipsoidal, about 5 mm. long, 3 mm. wide, with about 20 longitudinal indistinct striations.——Honshu (Rikuchu, Noto, and elsewhere).——Korea.

4. Zostera asiatica Miki. *Z. pacifica* auct. Japon., non S. Wats.——Ō-AMA-MO. Rhizomes elongate; leaves 10–15 mm. wide, elongate, 9(–11)-nerved, to 150 cm. long, entire and smooth, subretuse, the ligules not conspicuous; inflorescence not appendaged on margin; seeds oblong, about 5 mm. long, 3 mm. wide, obscurely striate.——July–Oct. Hokkaido.——Sakhalin, Kuriles, and Korea.

5. Zostera nana Roth. *Z. japonica* Asch. & Graebn.——KO-AMA-MO. Rhizomes slender, elongate; stems slender; leaves 10–40 cm. long, 1.5–2 mm. wide, 3-nerved, smooth and entire, rounded to retuse at apex; seeds oblong, about 2 mm. long, lustrous and smooth.——Sept.–Nov. Muddy places in shallow sea water; Hokkaido, Honshu, Shikoku, Kyushu.——Eurasia and Africa.

2. PHYLLOSPADIX Hook. SUGA-MO ZOKU

Perennial dioecious herbs growing in saline water, with thick rhizomes; stems short; leaves flat, ribbonlike, sheathed at base; spadix linear, naked at back, enveloped by a leaflike bract; flowers in 2 rows on one side of spadix, alternate, the perianth not developed; staminate flowers consisting of a sessile 1-locular anther; pistillate flowers consisting of one sessile ovary with a short style and two stigmas, the ovule solitary, pendulous, orthotropous; fruit broadly cordate-lobate at base.——Several species around the Pacific basin.

1A. Leaves 2–4.5 mm. wide; rhizomes with pale fulvous fibers. .... 1. *P. iwatensis*
1B. Leaves 1.5–2.5 mm. wide; rhizomes with dense blackish brown fibers. .... 2. *P. japonica*

1. Phyllospadix iwatensis Makino. *P. scouleri* auct. Japon., non Hook.——SUGA-MO. Rhizomes short, caespitose, estoloniferous, densely leafy; leaves 20–100 cm. long, 2–4.5 mm. wide, 3-nerved, rounded at apex, smooth except the upper margins minutely toothed; peduncles (stems) short, solitary, simple, compressed, to 8 cm. tall; pistillate inflorescence about 4 cm. long, the bracts 2 ranked, narrowly lanceolate, obtuse at apex; fruit rather numerous, sagittate, about 5 mm. wide.——July–Sept. Rocky places in shallow sea water; Hokkaido, Honshu (northern distr.).——Kuriles and Sakhalin.

2. Phyllospadix japonica Makino. *? P. serrulatus* auct. Japon., non Rupr.——EBI-AMA-MO. Similar to the preceding; leaves narrower, 1.5–2.5 mm. wide, 3-nerved, the basal part becoming blackish brown fibers.——July–Oct. Honshu, Shikoku, Kyushu.——? China.

Fam. 37. SCHEUCHZERIACEAE HOROMUI-SŌ KA Arrowgrass Family

Marsh herbs; leaves usually radical, linear or cylindric, sheathed at base; flowers small, bisexual, in terminal spikes or racemes; perianth-segments 4–6, biseriate, deciduous or persistent, small, greenish; stamens 3–6, filaments free, short or elongate, the anthers usually 2-locular, extrorse; carpels 3–6, each with 1–2 anatropous ovules, the embryo straight.——Three genera with more than 10 species widely distributed.

1A. Flowers numerous, ebracteate, in racemes or spikes, on naked peduncles; carpels 3–6, connate. .... 1. *Triglochin*
1B. Flowers few, loosely arranged, bracteate, on stems with few bracts; carpels 3, nearly free when mature. .... 2. *Scheuchzeria*

1. TRIGLOCHIN L. SHIBA-NA ZOKU

Marsh herbs; leaves radical, linear, semiterete, with a membranous sheath at base; flowers short-pedicelled, racemose or in spikes on leafless scapes, numerous, bractless; perianth-segments 3 or 6, deciduous, the inner whorl inserted a little above the outer; stamens 3 or 6, the anthers sessile, inserted at base of perianth, dorsifixed; carpels 3–6, 1-locular, connate, each with 1 ovule, the stigma sessile; fruit cylindric to obovoid, the mature carpels deciduous from the persistent axis; seeds erect, cylindric to ovoid.——About 10 species widely dispersed in temperate and subtropical regions.

1A. Carpels 3; fruit cylindric, narrowed to the base, on an erect pedicel. .... 1. *T. palustre*
1B. Carpels 6; fruit erect, oblong, obtusely rounded at base, on an ascending to curved pedicel. .... 2. *T. maritimum*

1. Triglochin palustre L. HOSOBA-NO-SHIBA-NA, MISA-KI-SŌ. Plant stoloniferous; rhizomes short; leaves linear, shorter than the scape, semiterete, soft, 10–25 cm. long, about 1 mm. wide, gradually narrowed to an obtuse apex, with a membranous sheath at base, the ligule ovate-rounded, about 1 mm. long; scapes 20–40 cm. long, erect, the racemes elongate after flowering, to 25 cm. long; flowers loosely arranged, small, on an erect pedicel 4–8 mm. long, the perianth-segments elliptic, concave, 1.5–2 mm. long; fruit 8–10 mm. long, about 1.2 mm. wide.——June–Aug. Bogs; Hokkaido, Honshu, (centr. distr. and northw.).——Widely distributed in the cooler regions of Eurasia, N. America, and southern S. America.

2. Triglochin maritimum L. SHIBA-NA. Perennial herb; rhizomes thick and short, caespitose; leaves 10–30 cm. long, linear and rather flat above, obtuse, 1.5–4 mm. wide, mouths of the sheaths protruding into a ligule 3–5 mm. long; scapes 20–40 cm. long, racemes rather densely many-flowered, usually 10–15 cm. sometimes to 25 cm. long, the pedicels 2–4 mm. long, commonly ascending; fruit 3–5 mm. long, oblong or ovoid-oblong.——June–Oct. Saline marshes; Hokkaido, Honshu, Shikoku, Kyushu.——Widely distributed in the N. Hemisphere and Patagonia.

Var. **asiaticum** (Kitag.) Ohwi. *T. maritimum* subsp. *asiaticum* Kitag. Leaves often broader.——Western part of Japan.

2. SCHEUCHZERIA L.   Horomui-sō Zoku

Perennial herbs with short creeping rhizomes; leaves linear, sheathed at base, semiterete, somewhat flattened above, striate, the sheaths becoming ligulate at apex; flowers small, bisexual, few, loosely arranged on a few-bracteate scape, pedicelled, bracteate; perianth-segments 6, biseriate, persistent; stamens 6, inserted at the base of the perianth-segments, the filaments elongate, the anthers linear, extrorse, basifixed; carpels 3, nearly free, 1-locular, becoming patent and dehiscing ventrally; stigma sessile; seeds without endosperm.——One species.

1. **Scheuchzeria palustris** L. Horomui-sō. Glabrous, somewhat fleshy herb; leaves mostly basal, 8–30 cm. long, 1–2 mm. wide, smooth, obtuse, the ligule 3–4 mm. long, obtuse, the cauline leaves or bracts few, 7–10 cm. long, with somewhat inflated sheaths at base; scapes 15–30 cm. long; racemes 4- to 8-flowered, simple, the pedicels suberect, 5–20 mm. long; flowers with membranous perianth-segments 2–3 mm. long; stigma oblong, sessile, inserted at the back just below the apex; carpels broadly ellipsoidal, inflated at maturity, 6–7 mm. long, horizontally spreading; seeds oblong, smooth, about 4 mm. long.——June–July. Sphagnum bogs; Hokkaido, Honshu (Nikko, Oze, and northw.).——Widely distributed in the cooler regions of the N. Hemisphere.

Fam. 38. ALISMATACEAE   Omodaka Ka   Waterplantain Family

Marsh or aquatic herbs usually glabrous throughout; leaves radical, petiolate, linear to sagittate, sometimes floating, the nerves converging above and connected by transverse veinlets; flowers in panicles or verticillate-racemose, actinomorphic, bracteate, bisexual or unisexual (plants dioecious or monoecious); sepals 3, persistent; petals 3, deciduous or somewhat persistent, imbricate in bud; stamens 6 or many; carpels usually many, separate, 1-locular and 1-ovuled; achenes usually laterally compressed, the embryo U-shaped, endosperm absent.——Cosmopolitan with about 13 genera and nearly 100 species.

1A. Carpels many, arranged in 1 ringlike series on a small receptacle; leaves lanceolate to elliptic, narrowed to or rounded at base (in ours); flowers bisexual; stamens 6. .... 1. *Alisma*
1B. Carpels in dense globose to oblong heads.
  2A. Flowers bisexual; stamens 6; carpels few, on a small receptacle; leaves cordate-rounded. .... 2. *Caldesia*
  2B. Flowers unisexual; stamens 9 or more; carpels numerous on a globose or oblong receptacle; leaves linear to sagittate.
                                                             3. *Sagittaria*

1. ALISMA L.   Saji-omodaka Zoku

Perennials, rarely annuals; leaves long-petiolate, the blades lanceolate to elliptic, entire, with several longitudinal nerves transversely connected by veinlets; scapes short or elongate; flowers small, pedicelled, in panicles; petals soon wrinkling and decaying after flowering; stamens 6; carpels many, in one ringlike series on small flattened receptacles, in fruit much flattened laterally, with 2 or 3 ribs on back.——About 10 species in temperate and tropical regions of the world.

1A. Leaves ovate-elliptic or ovate-oblong, rather thick, subrounded at base, not decurrent on the petiole; achenes with 2 shallow grooves on the back. .... 1. *A. plantago-aquatica* var. *orientale*
1B. Leaves lanceolate to narrowly oblong, rather thin, gradually narrowed to the petiole; achenes with a deep groove on the back.
                                                         2. *A. canaliculatum*

1. **Alisma plantago-aquatica** L. var. **orientale** G. Sam. *A. plantago* var. *parviflorum* sensu auct. Japon., non Torr.; *A. orientale* (G. Sam.) Juzep.——Saji-omodaka. Glabrous perennial herb with short rhizomes; leaves long-petiolate, ovate-elliptic to ovate-oblong, 5–10 cm. long, 2–6 cm. wide, 5–7 nerved, acute or abruptly so with an obtuse tip, subrounded at base and not decurrent into the petiole; scapes 50–70 cm. long; panicles large, with 3 long-acuminate bracts at each node, the pedicels verticillate, 10–15 mm. long; flowers verticillate, the pedicels 10–15 mm. long; sepals broadly ovate, obtuse, about 7-nerved, 2–2.5 mm. long; corollas white; achenes flattened, narrowly obovate, about 2 mm. long, with 2 shallow grooves on back.——Aug.–Sept. Wet places; Hokkaido, Honshu (n. distr.).——Sakhalin, Korea, Manchuria, Mongolia, and e. Siberia. The species is widely distributed in the N. Hemisphere.

2. **Alisma canaliculatum** A. Br. & Bouché; *A. plantago* var. *canaliculatum* (A. Br. & Bouché) Miyabe & Kudo: *A. plantago* var. *angustifolium* sensu auct., non Kunth——Hera-omodaka. Similar to the preceding; leaves rather thin, lanceolate to narrowly oblong, 4–15 cm. long, 1–3 cm. wide, 5- to 7-nerved, acuminate or acute with an obtuse tip, gradually narrowed to the elongate petioles; bracts acuminate; panicle-branches usually in 3's; achenes with a deep furrow on back.——Aug.–Sept. Wet places; Hokkaido, Honshu, Shikoku, Kyushu.

Var. **harimense** Makino. *A. harimense* Makino——Hosoba-hera-omodaka. Leaves narrower; anthers brownish purple, not pale green.——Honshu.

Alisma rariflorum G. Sam. is possibly a stunted form of No. 2.

2. CALDESIA Parl. MARUBA-OMODAKA ZOKU

Marsh herbs; leaves petiolate, ovate-cordate to reniform-cordate; flowers white, bisexual, pedicelled, verticillate, in panicles or racemes; sepals 3, herbaceous; petals 3, fugacious; stamens 6; carpels 6–9, nutlike, in a hemispherical head on a receptacle, the ovules erect, campylotropous, solitary in each carpel.——Few species.

1. Caldesia reniformis (D. Don) Makino. *Alisma reniforme* D. Don; *C. parnassifolia* var. *major* (M. Micheli) Buchen.; *A. parnassifolium* var. *majus* M. Micheli——MARUBA-OMODAKA. Glabrous annual; leaves radical, long-petioled, ovate to rounded, sometimes subreniform, 3–8 cm. long, 2–8 cm. wide, rounded at apex, deeply cordate at base, with 9 to 13 closely parallel nerves and transverse veinlets, patent; scapes erect, 30–100 cm. long; panicle large, 40–50 cm. long, the bracts lanceolate, acuminate, the pedicels in 3's or 4's, 2–3 cm. long; flowers white; sepals broadly elliptic, 3–4 mm. long, rounded at apex, many-nerved; achenes obovate, 3–3.5 mm. long, biconvex, longitudinally ribbed when dry, the style persistent, slender.——Aug.–Sept. Ponds; Honshu, Shikoku, Kyushu.——China, India, Australia, and Madagascar.

3. SAGITTARIA L. OMODAKA ZOKU

Marsh or aquatic herbs, monoecious or dioecious; leaves long-petiolate, linear to sagittate, sometimes floating; inflorescence racemose, terminal and solitary, on naked scapes, the pedicels verticillate, bracteate at base; flowers white, unisexual, the staminate usually above with 3 persistent sepals, 3 deciduous petals and usually many stamens on a conspicuous receptacle; pistillate flowers with many free carpels each with a solitary ovule; stigma small; achenes many, in heads, flattened, the embryo U-shaped. ——About 50 species in the temperate and tropical regions of the world.

1A. Racemes few-flowered; pistillate flowers sessile, only one in an inflorescence; staminate flowers 2–6, pedicellate; achenes cristate on the back; leaves linear to linear-oblanceolate. .... 1. *S. pygmaea*
1B. Racemes many-flowered; pistillate flowers several, pedicelled; achenes entire, rostrate.
  2A. Leaves all floating, oblong or ovate-oblong, obtuse at base or with 2 obtuse auricles; pistillate flowers short-pedicellate; achenes 2–3 mm. long. .... 2. *S. natans*
  2B. Leaves raised above water, usually sagittate, sometimes oblong; pistillate flowers with pedicels about 1–2 cm. long; achenes 3–4 mm. long.
    3A. Leaves rather soft, the lateral lobes ending in an almost filiform point; stolons with a tuber at the end, without bulbils in the leaf-axils. .... 3. *S. trifolia*
    3B. Leaves rather rigid and thick, the lateral lobes ending in an obtuse point; stolons absent; rhizomes with many bulbils in leaf axils. .... 4. *S. aginashi*

1. Sagittaria pygmaea Miq. *S. sagittifolia* var. *oligocarpa* M. Micheli; *S. sagittifolia* var. *pygmaea* (Miq.) Makino——URI-KAWA. Perennial herb; rhizomes short; stolons absent; leaves rosulate, linear or linear-oblanceolate, 10–15 cm. long, 5–8 mm. wide, acute to acuminate with an obtuse tip, gradually narrowed to a sheath at base, petioles indistinct; scapes 10–25 cm. long, few-flowered; flowers in racemes; pistillate flower solitary, sessile on the lowest node, the staminate flowers few, on pedicels 10–30 mm. long; petals rounded-obovate, 8–10 mm. long; achenes narrowly obovate, about 3 mm. long, compressed, with a broad wing.——June–Oct. Muddy places, paddy fields, and ponds; Honshu (exclusive of northernmost distr.), Shikoku, Kyushu.——Korea, Formosa, and China.

2. Sagittaria natans Pall. *S. alpina* Willd.; *S. sagittifolia* var. *tenuior* Wahlenb.——KARAFUTO-GUWAI. Perennial; leaves floating, slenderly petiolate, linear to oblong, 5–10 cm. long, obtuse at base or with obtuse auricles, 3-nerved; scapes elongate, usually to 60 cm. long; racemes simple, 1- or 2-noded; pistillate flowers 1–3, below the staminate, short-pedicelled, the staminate flowers 2–6, pedicellate; petals white; anthers yellow.——Ponds; Hokkaido; rare.——Korea, Sakhalin, Siberia, and Europe.

3. Sagittaria trifolia L. *S. sagittifolia* sensu auct. Asiat., non L.; *S. sagittifolia* var. *subaequiloba* Regel; *S. sagittifolia* var. *leucopetala* Miq.——OMODAKA. Perennial stoloniferous herb, the stolons slender, ending in a tuber; leaves long-petiolate, the blades sagittate, the basal lobes ovate to linear, the terminal lobe 5–15 cm. long, 3- to 7-nerved, acuminate but with an obtuse tip, the lateral lobes ascending, usually longer than the midlobe, acuminate with a filiform tip; scapes 20–80 cm. long; racemes sometimes branched at base; flowers rather numerous, in 3's at the nodes, the pistillate borne on the lower part, the sepals ovate, obtuse at apex, 6–8 mm. long, reflexed, the pedicels 1–2 cm. long, the staminate longer pedicelled; petals white, twice as large as the sepals; achenes broadly oblanceolate, 3–3.5 mm. long, obliquely obtriangular inclusive of the broad wing.——June–Oct. Wet places, paddy fields, riverbanks, and ponds; Hokkaido, Honshu, Shikoku, Kyushu; common.——Korea, China, and India.

Var. **edulis** (Sieb.) Ohwi. *S. sinensis* Sims; *S. sagittifolia* var. *edulis* Sieb. ex Miq.; *S. sagittifolia* forma *sinensis* (Sims) Makino; *S. trifolia* var. *sinensis* (Sims) Makino; *S. macrophylla* Bunge——KUWAI. Tubers at apex of stolons larger; leaves with broadly ovate midlobe.——Introduced from China and frequently cultivated in Japan for the edible tubers.

Var. **alismaefolia** (Makino) Makino. *S. sagittifolia* var. *alismaefolia* Makino——HITOTSUBA-OMODAKA. Leaves entire, linear to narrowly oblong, 4–6 cm. long, 4–18 mm. wide.——Nov.——s. Kyushu.

4. Sagittaria aginashi Makino. *S. sagittifolia* var. *aginashi* Makino——AGI-NASHI. Perennial herb, with short rhizomes and numerous small bulbils in leaf-axils; leaves long-petiolate, blades sagittate, rarely subsagittate and rounded at base, the terminal lobe linear or broadly lanceolate, 7–20 cm. long, 5- to 7-nerved, acuminate with an obtuse tip, the basal lobes usually a little shorter than the terminal; scapes 30–80 cm. long; inflorescence racemose, the pistillate flowers several, on the lower part, pedicelled, the staminate flowers rather long-pedicelled; sepals ovate to narrowly so, obtuse, 4–7 mm. long; anthers yellow; achenes obovate, inclusive of the broad wing, rostrate, 3–3.5 mm. long.——July–Oct. Wet places, ponds, riverbanks; Hokkaido, Honshu, Shikoku, Kyushu.

Fam. 39. HYDROCHARITACEAE Tochi-kagami Ka Frogbit Family

Fresh- or salt-water herbs; flowers on scapes or sometimes on leafy stems; leaves linear to rounded, sometimes opposite or verticillate; flowers solitary or subumbellate, unisexual or bisexual, at first enclosed in 1 or 2 sheathed bracts; outer perianth-segments 3, thin, the inner 3, or sometimes 2, petallike, or absent; stamens 3, 6, 9, 12, or rarely many, the anthers 2-locular, longitudinally dehiscent; ovary inferior, 1-locular or incompletely many-locular, the placentae parietal; styles often connate, the stigmas entire or bifid; ovules many, anatropous; seeds numerous, oblong, endosperm absent.——About 15 genera, with about 100 species in tropical and temperate regions.

1A. Marine herbs with creeping stems; leaves opposite, petiolate, ovate to oblong, penninerved; flowers solitary, unisexual, the inner
perianth-segments absent; stamens 3; ovary 1-locular; styles 3. . . . . 1. *Halophila*
1B. Fresh water herbs.
  2A. Ovary 1-locular, with scarcely protruded, parietal placentae; leaves narrow.
    3A. Leaves radical or alternate.
      4A. Flowers usually bisexual, sometimes unisexual, the spathe cylindric, staminate flowers not detached from the pedicels nor
floating at anthesis; inner perianth-segments present; stamens 3–9. . . . . 2. *Blyxa*
      4B. Flowers unisexual (plants usually dioecious), with an ovoid spathe in the staminate, cylindric in the pistillate; staminate flowers
detached from the pedicels before anthesis and floating; inner perianth-segments absent; stamens 1–3. . . . . 3. *Vallisneria*
    3B. Leaves verticillate; staminate flowers solitary, enclosed in a spathe; inner perianth-segments present; stamens 3. . . . . . . 4. *Hydrilla*
  2B. Ovary 6- to 15-locular, with protruding placentae; leaves broad.
    5A. Flowers bisexual; leaves immersed. . . . . 5. *Ottelia*
    5B. Flowers unisexual; leaves floating. . . . . 6. *Hydrocharis*

1. HALOPHILA Thou. Umi-hiru-mo Zoku

Immersed marine monoecious or dioecious herbs with long-creeping stems; leaves from the axils of hyaline scales, binate, petiolate, ovate or narrowly oblong, pinnately veined; spathe small, sessile, solitary among the binate leaves, consisting of 2 bracts enclosing a single flower; flowers unisexual, the staminate pedicelled, the outer perianth-segments 3, the inner absent; stamens 3, alternate with the outer perianth-segments, the anthers subsessile and extrorse; pistillate flowers sessile, with 3 minute outer perianth-segments, the ovary long-beaked, 1-locular, the styles 3, filiform, papillose; ovules in 2 series, on 2 parietal placentae; fruit subglobose, beaked; seeds numerous, globose, each with a coiled cotyledon.——Few species in tropical regions of the Pacific basin.

1. **Halophila ovalis** (R. Br.) Hook. f. *Caulinia ovalis* R. Br.; *H. ovata* Gaudich.; *Barkania punctata* Ehrenb.; *H. euphlebia* Makino——Umi-hiru-mo, Ō-umi-hiru-mo. Perennial herb; stems creeping, sparingly branched, rooted at the nodes; leaves petiolate, narrowly oblong to elliptic-ovate, 15–25 mm. long, 6–15 mm. wide, entire, rounded at apex, smooth, with 3 principal veins and 10–20 pairs of connecting veinlets, the petioles purplish, with a hyaline membranous scale at base; pistillate flowers axillary, sessile, 2-bracteate.——On sandy sea bottoms; Honshu (Sagami, Noto and westw.), Shikoku, Kyushu.——India, Malaysia, and Australia.

2. BLYXA Thou. Subuta Zoku

Aquatic annuals; stems rarely elongate; leaves radical or alternate, linear, sessile, acute to acuminate; staminate flowers several within a cylindrical spathe, the outer perianth-segments 3, linear, the inner 3 longer than the outer, linear, the stamens 3–9, sometimes partly reduced, with 3 reduced pistils; pistillate flowers solitary within a spathe, sessile, the perianth similar to the staminate, the stamens rudimentary or absent, the ovary cylindrical, beaked, 1-locular, the style short, the stigmas elongate; fruit cylindric, inclosed in a spathe, the seeds numerous, minute, oblong.——More than 10 species in tropical and subtropical regions of Asia and Africa.

1A. Stems distinctly elongate, sometimes branched; peduncles absent; leaves 3–5 cm. long, 2–3 mm. wide; fruit about 2 cm. long,
enclosed in a short spathe; seeds smooth, without tails. . . . . 1. *B. japonica*
1B. Stems absent or nearly so; peduncles present; leaves 5–30 cm. long, 4–8 mm. wide; fruit 2–5 cm. long.
  2A. Seeds without a taillike appendage at both ends.
    3A. Seeds smooth. . . . . 2. *B. leiosperma*
    3B. Seeds sparsely tuberculate. . . . . 3. *B. aubertii*
  2B. Seeds with a taillike appendage at both ends.
    4A. Taillike appendages shorter than or as long as the seed. . . . . 4. *B. ceratosperma*
    4B. Taillike appendages much longer than the seed. . . . . 5. *B. bicaudata*

1. **Blyxa japonica** (Miq.) Maxim. *? Hydrilla japonica* Miq.; *B. caulescens* Maxim.; *B. leiocarpa* Maxim.——Yanagi-subuta. Stems 5–20 cm. long, simple or sparingly branched, densely leafy; leaves linear, flat, gradually narrowed to the tip, minutely serrulate; flowers bisexual, solitary, sessile, the cylindrical spathe 15–20 mm. long, sessile; sepals 3, lanceolate, obtuse, about 3 mm. long; petals linear, twice as long as the sepals; stamens 3; fruit linear-cylindric, 1.5–2 cm. long; seeds lanceolate, smooth, obtuse at both ends, 2 mm. long, 0.5 mm. wide.——Aug.–Oct. Common in paddy fields; Honshu, Shikoku, Kyushu.——Korea and Formosa.

2. **Blyxa leiosperma** Koidz. Mikawa-subuta. An-

nual; stems absent or nearly so; leaves radical, rosulate, linear, 7–15 cm. long, 4–5 mm. wide, acuminate, minutely serrulate; peduncles axillary, 1–3 cm. long, terminated by a cylindrical spathe about 3.5 cm. long; flowers solitary within the spathe, sessile; sepals 3, linear, 6–10 mm. long, obtuse; petals 3, narrowly linear, about 14 mm. long; seeds numerous, oblong, smooth, obtuse at both ends.——Oct. Honshu; rare.

3. Blyxa aubertii Rich. *B. coreana* Nakai (excl. basionym ? *Hydrolirion coreanum* Lév.); *B. ecaudata* Hayata; *B. muricata* Koidz.——MARUMI-SUBUTA, Ō-SUBUTA. Stemless annual; leaves rosulate, linear, 7–20 cm. long, 4–7 mm. wide, gradually acuminate, minutely serrulate, more or less 5-veined with a number of cross veinlets; peduncles 5–7 cm. long, the spathe cylindrical, 4–5 cm. long, slightly longer than the fruit;

sepals 3, lanceolate, obtuse; fruit linear; seeds ellipsoidal, about 1.5 mm. long, 0.8 mm. wide, obscurely striate with scattered tubercles.——Shallow ponds and paddy fields; Honshu, Kyushu.——Korea, Formosa, China, India, and Australia.

4. Blyxa ceratosperma Maxim. NAGABA-SUBUTA, SUBUTA. Very similar to the preceding but the seeds provided with a taillike appendage at each end, these nearly as long as or shorter than the seed.——Aug.–Oct. Shallow ponds and paddy fields; Honshu, Kyushu; common.——Formosa.

5. Blyxa bicaudata Nakai. *B. shimadae* sensu Miki, non Hayata (?).——KO-SUBUTA, NAGA-HIGE-MI-SUBUTA. Very similar to No. 3 and No. 4, but the seeds with the taillike appendages 2–4 times as long as the seed.——Honshu; rare.

3. VALLISNERIA L.   SEKISHŌ-MO ZOKU

Stemless, usually stoloniferous dioecious herbs wholly immersed; leaves rosulate, ribbonlike, compressed, serrulate at least in upper part, short-sheathed at base; peduncles radical, with a terminal spathe; flowers apetalous, the staminate small, numerous in a short peduncled ovoid spathe, each detached from the peduncle and floating freely before anthesis, the outer perianth-segments 3, the stamens 1–3; pistillate flowers solitary in a cylindrical spathe on a long peduncle, coiling after the flower is pollinated on the surface; ovary linear, inferior, 1-locular, with 3 parietal placentas; stigmas 3, alternate with the outer perianth-segments; fruit linear, with numerous fusiform seeds.——Few species in tropical and subtropical regions, except S. America.

1A. Winter buds fusiform; leaves prominently serrulate, abruptly acute; stolons rough with small spiny protuberances; stamens 2; seeds 1.5 mm. long. . . . . 1. *V. denseserrulata*
1B. Winter buds absent; leaves slightly serrulate; stolons smooth; stamen 1.
  2A. Ovary minutely tubercled; leaves 6–13 mm. wide, 5- to 7-veined, rather obtuse, slightly serrulate only toward tip; seeds 3.5 mm. long. . . . . 3. *V. higoensis*
  2B. Ovary smooth; leaves 5–10 mm. wide, 3- to 5-veined; seeds 3 mm. long.
    3A. Ovary tightly enclosed in the spathe; leaves scarcely twisted, slightly serrulate only toward tip, acute at apex. . . . . . . 2. *V. asiatica*
    3B. Ovary loosely enclosed in the spathe; leaves much twisted, serrulate on margin, obtuse to abruptly acute at apex. . . 4. *V. biwaensis*

1. Vallisneria denseserrulata (Makino) Makino. *V. spiralis* var. *denseserrulata* Makino——KŌGAI-MO. Perennial with roughened stolons; leaves linear, 4–10 mm. wide, abruptly acute, flattened, elongate, densely and prominently serrulate, 5-veined; spathe about 1.3 cm. long, loosely enveloping the ovary at anthesis; ovary smooth, linear, about 12 cm. long in fruit; seeds narrowly obovoid, 1.5 mm. long.——Aug.–Oct. Shallow rivers and ponds; Honshu.

2. Vallisneria asiatica Miki. *V. spiralis* sensu auct. Japon., non L.; *V. spiralis* var. *subulispatha* Makino, including forma *minor* Makino; *V. subulispatha* (Makino) Koidz.——SEKISHŌ-MO. Perennial with smooth stolons; leaves elongate, linear, flattened, 5–7 mm. wide, 5-veined, loosely and obscurely serrulate toward tip; pistillate flowers enveloped by a spathe about 2 cm. long; staminate peduncles 2–3 cm. long; fruit linear, smooth, 15–20 cm. long; seeds narrowly fusiform, 3

mm. long.——Aug.–Oct. Ponds and shallow rivers; Hokkaido, Honshu, Shikoku, Kyushu; common.——China, Indochina, Korea, and Formosa.

3. Vallisneria higoensis (Miki) Ohwi. *V. asiatica* var. *higoensis* Miki——HIRA-MO, HIROHA-SEKISHŌ-MO. Similar to the preceding; leaves 6–13 mm. wide, 5- to 7-veined, obtuse to abruptly acute; ovary tubercled; fruit about 10 cm. long, 5 mm. wide; seeds about 3.5 mm. long, 0.8 mm. wide.—— Kyushu; rare.

4. Vallisneria biwaensis (Miki) Ohwi. *V. asiatica* var. *biwaensis* Miki——NEJIRE-MO. Similar to No. 2, but the leaves prominently twisted, 5–10 mm. wide, 5-veined, obtuse to abruptly acute, serrulate throughout; pistillate spathe about 15 mm. long, loosely enveloping the ovary, the staminate peduncles 3–5 cm. long.——Aug.–Oct. Shallow rivers and ponds; Honshu (Kinki Distr.).

4. HYDRILLA L. C. Rich.   KURO-MO ZOKU

Dioecious aquatic herbs with elongate stems; leaves verticillate, sometimes opposite, rather short; staminate flowers solitary, axillary, short-peduncled, enveloped by a globose spathe, detached from the spathe and freely floating above the water at anthesis, the perianth-segments 6, in 2 rows, the stamens 3; pistillate flowers enveloped by a cylindrical spathe, sessile, the ovary inferior, 1-locular, beaked, the stigmas 3; fruit linear, smooth or tubercled; seeds 1–3, fusiform with slightly elongate ends.——Two species widely distributed in tropical and temperate regions of the Old World.

1. Hydrilla verticillata (L. f.) Casp. *Serpicula verticillata* L. f.——KURO-MO. Stems elongate, loosely branching; leaves verticillate in 4's to 8's, narrowly lanceolate, spreading, 1-nerved, 1–2 cm. long, 1.5–2 mm. wide, serrate, acuminate and ending in a minute spine; pistillate flowers axillary; seeds

1–3, acute at both ends, 5–6 mm. long, short-cylindrical.—— Aug.–Oct. Ponds; Hokkaido (rare), Honshu, Shikoku, Kyushu; common.——Eurasia, Malaysia, Australia, and Madagascar.

5. OTTELIA Pers.   Mizu-ōbako Zoku

Submersed herbs; leaves radical, petiolate, lanceolate to cordate; scapes radical, elongate; flowers solitary in a spathe terminal on the scape, sessile, bisexual; outer perianth-segments 3, linear to oblong, the inner larger, petaloid, obovate or rounded, short-appendaged at base; fertile stamens 3, the anthers erect; staminodes 6–15; ovary oblong, beaked; styles 6–9, linear, bifid; ovules numerous; fruit oblong, enveloped by a spathe; seeds numerous, small, hairy.——About 10 species in tropical and subtropical regions.

1. **Ottelia alismoides** (L.) Pers. *Stratiotes alismoides* L.; *O. japonica* Miq.; *O. alismoides* var. *japonica* (Miq.) Komar.——Mizu-ōbako.   Submersed green annual without rhizomes; leaves radical, rosulate, thin, long-petioled, oblanceolate in the vernal, broad-ovate or ovate-cordate in the summer ones, 10–25 cm. long, 2–15 cm. wide, obtuse, 5- to 9-nerved, entire or obscurely dentate; scape elongate, the spathe to 4 cm. long, with or without prominent crispate wings; sepals linear-lanceolate, obtuse, green; petals thin, ovate-orbicular, much longer than the sepals, white to reddish; ovary sessile within the spathe.——Aug.–Oct. Ponds and paddy fields; Honshu, Shikoku, Kyushu.——India and Australia.

6. HYDROCHARIS L.   Tochi-kagami Zoku

Aquatic monoecious herbs; stems elongate, leafy; leaves floating or raised above the water, rounded to reniform, entire, long-petiolate; scapes axillary; staminate flowers 2 or 3 in a spathe, the outer perianth-segments 3, sepallike, the inner 3 petaloid, white, delicate, the stamens 6–18 with 3–6 staminodes; pistillate flowers solitary in a spathe, long-pedicelled, with perianth-segments similar to those of the staminate, with 6 staminodes, the ovary ovoid, 6-locular, with 6 linear bifid stigmas; fruit ovoid to oblong, fleshy; seeds numerous.——Two species in Eurasia and Australia.

1. **Hydrocharis dubia** (Bl.) Backer. *? Pontederia dubia* Bl.; *H. asiatica* Miq.——Tochi-kagami.   Aquatic perennial; stems branched, leafy at the nodes; leaves rounded-cordate, 4–8 cm. long and as wide, long-petiolate, rounded at apex; flowers raised above the water; sepals narrowly oblong, green; petals rounded-cuneate, thinly membranous, white, yellow at base, 12–15 mm. long and as wide; stamens 6–9, yellow, with short-pilose filaments, the anthers introrse; stigmas 6, yellow, deeply bifid; fruit globose.——Aug.–Oct. Ponds; Honshu, Shikoku, Kyushu; common.——se. Asia and Australia.

Fam. 40. **TRIURIDACEAE**   Hongō-sō Ka   Triuris Family

Delicate, white, pink, or reddish, saprophytic herbs with leaves reduced to scales; stems simple, loosely scaly; flowers bisexual (or plants monoecious), small, pedicelled, bracteate, in a terminal raceme or umbellate; perianth-segments 4–8, lanceolate, valvate in bud; staminate flowers with 2–6 stamens, the anthers free or impressed in a thick disc, 2- to 4-locular, the reduced pistil absent or 3; pistillate flowers with or without staminodes, with numerous carpels, sessile on a receptacle, 1-locular, 1-ovuled, the style terminal or ventral, persistent, the stigma acute or clavate; ovules erect, anatropous; fruit an achene.——Four genera, with about 40 species, mostly in the Tropics, 1 genus in Japan.

1. SCIAPHILA Bl.   Hongō-sō Zoku

Saprophytic herbs without chlorophyll, monoecious or dioecious; stems simple, filiform or very slender, loosely scaly; flowers small, unisexual or polygamous, in racemes, the pedicels bracteate at base; perianth-segments 3–8, narrow, sometimes hairy at apex, inflexed in bud; anthers sessile or with very short filaments attached on a small receptacle; carpels numerous, free, anterio-posteriorly dehiscent, the style on ventral side of the carpel; seeds erect, ellipsoidal, small.——Many species, in tropical forests of Asia and northern S. America.

1A. Perianth-segments 4, broadly lanceolate, with few hairs at tip; lower flowers bisexual. .... 1. *S. takakumensis*
1B. Perianth-segments 6, broadly lanceolate to linear, glabrous; flowers unisexual.
  2A. Very delicate herbs, with loose 4- to 15-flowered racemes; pedicels filiform, about 3.5 mm. long, longer than the flowers; staminate
      flowers with reduced pistils, the perianth-segments broadly lanceolate, acuminate. .... 2. *S. japonica*
  2B. Delicate herbs with 3- to 8-flowered racemes; pedicels shorter than the flower, 2–4 mm. long; staminate flowers without reduced
      pistils, the perianth-segments linear. .... 3. *S. tosaensis*

1. **Sciaphila takakumensis** Ohwi.   Takakuma-sō. Very delicate herb; stems 5–10 cm. long, erect, loosely few-scaled; racemes loosely 4- to 10-flowered, with minute, ovate, obtuse bracts, the pedicels filiform, 2–10 mm. long, slightly curved; perianth-segments of bisexual flowers 4, about 1 mm. long, acute and with few long hairs at apex; stamens 4; carpels about 20, oblong; style clavate, attached slightly above the base on ventral side of the carpel, about half as long as the carpel; carpels 0.8 mm. long at maturity, the seeds dark brown, glossy.——Kyushu (Mount Takakuma); very rare.

2. **Sciaphila japonica** Makino. *Seychellaria japonica* (Makino) T. Itō; *Parexuris japonica* (Makino) Nakai & F. Maekawa; *Andruris japonica* (Makino) Giesenh.——Hongō-sō. Stems 3–8 cm. long, with few appressed scales; racemes loosely 4- to 15-flowered, with minute, linear-lanceolate, acuminate bracts; staminate flowers about 2 mm. across, with 3 reduced carpels, the perianth glabrous, the segments broadly lanceolate, acuminate; stamens 3, the anthers sessile, 2-locular; pistillate flowers about 1.5 mm. across, with 6 ovate, acute, glabrous perianth-segments, the carpels numerous, the style attached above the middle on the ventral side, filiform, 0.8 mm. long, longer than the carpel; carpels about 0.7 mm. long, in a glo-

bose head 1.5–2 mm. across.——July–Sept. Honshu (Kantō Distr. and westw.), Shikoku, Kyushu; rare.

3. Sciaphila tosaensis Makino. *Seychellaria tosaensis* (Makino) T. Itō; *Parexuris tosaensis* (Makino) Nakai & F. Maekawa——Uematsu-sō. Stems 6–10 cm. long, with 1 or 2 scales; racemes 3- to 8-flowered, the flowers short-pedicellate; staminate flowers on the upper part, 5–7 mm. across, the perianth-segments 6, linear, 2.5–3.5 mm. long, caudate, glabrous; stamens 3, with very short filaments; pistillate flowers 4–5 mm. across with 6 lanceolate, acuminate perianth-segments 2–2.5 mm. long; carpels numerous, rugose above, obovoid when mature, 1.3–1.5 mm. long, collected in a globose head 5–6 mm. across.——Aug.–Sept. Honshu (centr. distr. and westw.), Shikoku, Kyushu; rare.

Fam. 41. **GRAMINEAE** Ine Ka (Kahon Ka) Grass Family

(English revision assisted by Agnes Chase; Bambuseae assisted by F. A. McClure.)

Stems (culms) jointed, with solid nodes and hollow internodes (solid in Andropogoneae and in some bamboos); leaves 2-ranked, 1 at each node, the lower part of the leaf a tubelike sheath enclosing the culm, the upper part a flat or folded, parallel-veined, usually linear blade, at the junction of sheath and blade a ligule, consisting of a small membrane or a row of hairs; inflorescence a terminal panicle, raceme or spike, bearing few to many spikelets, these consisting of a jointed axis (rachilla) with 2-ranked alternate bracts, the lower pair (glumes) empty, the succeeding bracts (lemmas) 1 to many, each subtending a 2-nerved bract (palea) borne in the axil of the lemma, and, in the axil of the palea, a flower consisting of a 1-locular, 1-ovuled ovary with 2 styles with feathery stigmas; fruit a caryopsis, a 1-seeded structure enclosed by the pericarp, the whole structure dry and indehiscent and commonly referred to as the "grain" or "seed"; embryo minute; endosperm starchy.

This is the fundamental structure of all grasses, but there are many variations. In some bamboos, in the tribe Andropogoneae, and in occasional grasses of desert areas, the culms are solid; sheaths are mostly like a cylinder split down one side, but in *Bromus* and in a few other genera the margins are closed; blades in *Arthraxon* and in a few other genera are ovate; inflorescence and spikelets are modified in various ways, but the fundamental structure is recognizable by the position of the parts. Two-ranked leaves and 2-ranked glumes and lemmas of the spikelets are constant.——About 550 genera, with about 10,000 species of worldwide distribution, one of the largest families of flowering plants.

1A. Culms woody, perennial, usually freely branched; leaf-blades jointed with the sheath; spikelets (in ours) of more than 1 flower (except in *Phyllostachys*).
  2A. Sheaths of the main culm (culm-sheath) deciduous (except in some species of *Chimonobambusa*); culms usually thick; stamens 3.
    3A. Leaf-sheaths well developed.
      4A. Buds 3 at each node of the culm; spikelets elongate, many-flowered.
        5A. Lemmas membranous; stigmas 2; culms bearing air-roots near base. .... 1. *Chimonobambusa*
        5B. Lemmas chartaceous or coriaceous; stigmas 3; culms without air-roots.
          6A. Lemmas chartaceous. .... 2. *Semiarundinaria*
          6B. Lemmas rather coriaceous below. .... 3. *Sinobambusa*
      4B. Buds 2 at each node of the culm; culms thick; spikelets 1- or 2-flowered. .... 4. *Phyllostachys*
    3B. Leaf-sheaths not developed in culms 1 year old; culms low, slender. .... 5. *Shibataea*
  2B. Sheaths of the main culm persistent; culms small or medium sized.
    7A. Buds 3–10 at each node of the culm; oral bristles flexuous, smooth; lemmas rather large; stamens 3; inflorescence racemosely fasciculate, without distinct peduncles. .... 6. *Arundinaria*
    7B. Buds solitary at each node of the culm; oral bristles typically rigid and scabrous; lemmas usually smaller; stamens 3–6; inflorescence a long-peduncled panicle. .... 7. *Sasa*
1B. Culms herbaceous, annual, rarely perennial but not woody; leaf-blades usually not jointed with the sheath.
  8A. Spikelets 1- to many-flowered (if 2-flowered the lower floret fertile, terete or laterally flattened).
    9A. Glumes not much shorter than the lowest lemma; awn of the lemma, when elongate, commonly geniculate and twisted.
      10A. Spikelet with at least the lowest floret bisexual.
        11A. Palea usually with 2 sharp keels, rarely reduced to a minute hyaline scale; lemmas membranous to coriaceous; awn, when elongate, borne on back of the lemma; lodicules 2.
          12A. Spikelets 1-flowered.
            13A. Panicle cylindric, the branches very short; florets conspicuously flattened; style and stigma much elongate, protruding from apex of floret.
              14A. Spikelets articulate below the glumes, falling entire; glumes awnless, more or less connate at base. ... 8. *Alopecurus*
              14B. Spikelets articulate above the glumes; glumes persistent, free, the keels exserted as short awns. .... 9. *Phleum*
            13B. Panicle usually open, the branches normally elongate; style very short, the stigma feathery, protruding from the side of floret.
              15A. Callus at base of floret glabrous or nearly so; rachilla not prolonged beyond the floret.
                16A. Spikelets articulate above the glumes; glumes awnless. .... 10. *Agrostis*
                16B. Spikelets articulate below the glumes.
                  17A. Glumes membranous.
                    18A. Stamens 3; spikelets only slightly flattened. .... 11. *Polypogon*
                    18B. Stamen 1; spikelets strongly flattened. .... 12. *Cinna*
                  17B. Glumes coriaceous, inflated, broadened at tip, pointed, transversely wrinkled; spikelets sessile, congested in 1-sided panicle-branches. .... 13. *Beckmannia*
              15B. Callus at base of the floret distinctly hairy; rachilla commonly prolonged beyond the floret as a hairy bristle.
                        14. *Calamagrostis*

12B. Spikelets 2- to many-flowered.
  19A. Spikelets articulate above the glumes.
    20A. Spikelets 2-flowered; lower floret staminate, the lemma long-awned from the back; upper floret perfect, the lemma awnless or nearly so. .... 15. *Arrhenatherum*
    20B. Spikelets 2- to many-flowered; florets all perfect except the uppermost.
      21A. Lemmas awnless or awned from above the middle, acute or bifid at apex.
        22A. Rachilla-joints long-hairy; lemmas awned.
          23A. Caryopsis sulcate, adhering to the lemma and palea; ovary hairy; spikelets usually more than 1 cm. long.
            24A. Annuals; spikelets pendulous; glumes 7- to 11-nerved, rounded at back, green. .... 16. *Avena*
            24B. Perennials; spikelets erect; glumes slightly keeled, 1- to 3(–5)-nerved. .... 17. *Helictotrichon*
          23B. Caryopsis not sulcate, free from the lemma and palea and enclosed by them; ovary glabrous (in ours); spikelets less than 1 cm. long; glumes 1- to 3-nerved. .... 18. *Trisetum*
        22B. Rachilla-joints glabrous or short-pubescent; spikelets less than 1 cm. long, not awned (in ours). .... 19. *Koeleria*
      21B. Lemmas awned from the middle or above the base, truncate and erose at apex; spikelets less than 1 cm. long.
                                                               20. *Deschampsia*
  19B. Spikelets articulate below the glumes; plant wholly velvety-hairy. .... 21. *Holcus*
11B. Palea rounded or only obtusely angled at back; lemmas awned from apex or from between 2 minute teeth, or awnless; lodicules 2, sometimes 3.
  25A. Rather rigid grasses; spikelets lanceolate; lemmas awned between minute teeth.
    26A. Leaf-blades narrow; lemmas herbaceous; awns slender, slightly geniculate. .... 22. *Achnatherum*
    26B. Leaf-blades flat, broader; lemmas coriaceous, with rather thick, almost straight, long awns. .... 23. *Orthoraphium*
  25B. Soft grasses; spikelets ellipsoid, awnless. .... 24. *Milium*
10B. Spikelet with the lower two florets staminate or neuter, sometimes reduced to minute scales; uppermost floret fertile.
  27A. Sterile florets scalelike, attached to the base of the bisexual floret; stamens 3. .... 25. *Phalaris*
  27B. Sterile florets larger than the bisexual floret, brownish, hairy; stamens of the bisexual florets 2.
    28A. Sterile florets, at least the upper, neuter and epaleate. .... 26. *Anthoxanthum*
    28B. Sterile florets all staminate, paleate. .... 27. *Hierochloë*
9B. Glumes equal or unequal, at least the lower shorter than the lowest lemma; awn of the lemma apical or nearly so, not geniculate nor twisted, except in *Tripogon*.
  29A. Lemmas 5- to many-nerved.
    30A. Inflorescence a spike; spikelets alternate on a flattened axis, sessile or nearly so; ligules very short, truncate, sometimes auriculate.
      31A. Spikelets all alike, 1- to many-flowered.
        32A. Spikelets solitary, rarely geminate at each node, not twisted.
          33A. Spikelets very short-pedicelled. .... 28. *Brachypodium*
          33B. Spikelets sessile.
            34A. Spikelets placed flatwise to the rachis; both glumes developed. .... 29. *Agropyron*
            34B. Spikelets placed edgewise to the rachis; first glume wanting except in the terminal spikelet. .... 30. *Lolium*
        32B. Spikelets 2 or 3, rarely solitary at each node, somewhat twisted.
          35A. Glumes well developed. .... 31. *Elymus*
          35B. Glumes wanting or reduced to short bristles. .... 32. *Asperella*
      31B. Spikelets unlike, 3 at each node, the central one perfect, the lateral reduced, sterile. .... 33. *Hordeum*
    30B. Inflorescence paniculate or digitate, rarely a spike with triangular axis bearing spikelets on 2 sides, or in a loose raceme.
      36A. Leaf-blades linear, sessile.
        37A. Glumes distinct; palea with 2 keels; spikelets 1- to many-flowered.
          38A. Caryopsis with a hairy beak at apex; style lateral; leaf-sheaths cylindric, with fused margins; spikelets usually rather large. .... 34. *Bromus*
          38B. Caryopsis not beaked; style terminal.
            39A. Lemmas 5-nerved.
              40A. Spikelets 1-flowered.
                41A. Rachilla produced behind the palea as a naked bristle. .... 35. *Brachyelytrum*
                41B. Rachilla obsolete. .... 36. *Aulacolepis*
              40B. Spikelets 2- to many-flowered.
                42A. Spikelets prominently flattened, densely arranged on one side of the upper part of panicle-branches.
                                                              37. *Dactylis*
                42B. Spikelets flattened or somewhat so, arranged not only on one side of panicle-branches.
                  43A. Lemmas glabrous or pubescent on back, often awned from apex. .... 38. *Festuca*
                  43B. Lemmas pubescent on midrib, awnless.
                    44A. Lemmas rounded and not keeled at back, not webbed at base, with parallel nerves. ...... 39. *Puccinellia*
                  44B. Lemmas keeled at back, with usually a tuft of long hairs at base, the nerves converging above. .. 40. *Poa*
            39B. Lemmas 7- to many-nerved.
              45A. Lower florets fertile, the upper few reduced; inflorescence commonly a panicle.
                46A. Callus of the florets glabrous; lemmas awnless (in ours).
                  47A. Glumes 1- to 3-nerved, usually much shorter than the lemma; lemmas all alike; grasses of wet places.
                    47AA. Leaf-sheaths closed. .... 41. *Glyceria*
                    47BB. Leaf-sheaths open. .... 41a. *Torreyochloa*
                47B. Glumes 3- to 7-nerved, nearly as long as the lemma; lemmas of few upper florets neutral and usually dissimilar to the fertile ones; grasses of mesophytic places. .... 42. *Melica*
              46B. Callus bearded; lemmas awned; grasses of forests. .... 43. *Schizachne*

45B. Lower 2 florets neutral and epaleate, the uppermost fertile and bisexual; spikelets 3-flowered; inflorescence a raceme with alternately and loosely arranged spikelets deflexed in maturity. .... 44. *Brylkinia*

37B. Glumes wanting or minute; palea of the bisexual florets 1- or 3-nerved; each spikelet consisting of one fertile floret; grasses of wet places.

48A. Spikelets all alike and bisexual.

49A. Spikelets laterally compressed; florets not stiped; stamens 3 or 6. .... 45. *Leersia*

49B. Spikelets nearly terete; florets stiped, deciduous with the stipe; stamen 1. .... 46. *Chikusichloa*

48B. Spikelets dimorphic, unisexual. .... 47. *Zizania*

36B. Leaf-blades broadly lanceolate, often short-petiolate; spikelets few-flowered, sessile, the lowest floret bisexual and fertile, the others reduced to retrorsely scabrous bristles. .... 48. *Lophatherum*

29B. Lemmas 3-nerved, sometimes nearly nerveless.

50A. Rather coarse tall grasses; spikelets prominently long-hairy; lemmas membranous, rather fragile.

51A. Callus of the floret elongate; lemma glabrous; the lowest floret much longer than the glumes, usually staminate.
49. *Phragmites*

51B. Callus of the floret very short; lemmas long-hairy; lower florets bisexual, the lowest as long as or a little longer than the glumes. .... 50. *Arundo*

50B. Grasses of small or medium size; spikelets without long hairs.

52A. Caryopsis large, visible between the lemma and palea when mature; inflorescence a panicle.

53A. Rather large grasses; spikelets 1-flowered; glumes connate at base. .... 51. *Phaenosperma*

53B. Slender grasses; spikelets 2- or 3-flowered; glumes not connate. .... 52. *Diarrhena*

52B. Caryopsis small, usually not visible.

54A. First glume developed, both glumes shorter than the lowest lemma, as rigid as the lemma or thinner.

55A. Spikelets with 2 or more bisexual florets.

56A. Lemmas entire, acute, obtuse, or short-awned at apex.

57A. Inflorescence a panicle.

58A. Callus of the floret bearded.

59A. Callus very short; culms thickened at base, with short internodes; lemmas awnless. ...... 53. *Moliniopsis*

59B. Callus elongate; culms not thickened at base; lemmas short-awned. .... 54. *Hakonechloa*

58B. Callus glabrous, short. .... 55. *Eragrostis*

57B. Inflorescence digitate; spikelets sessile, densely arranged in 2 rows on the outer side of the branches. . 56. *Eleusine*

56B. Lemmas truncate or bifid at apex.

60A. Inflorescence a panicle, branches few to many; spikelets sessile or short-pedicelled, spicately or racemosely arranged on the branches.

61A. Plants with cleistogamous flowers in the leaf axils; panicle-branches 1–7, with loosely arranged few spikelets; spikelets usually short-awned, rarely awnless. .... 57. *Kengia*

61B. Plants without cleistogamous flowers; panicle-branches with many rather densely arranged spikelets.

62A. Lemmas toothed or awned, linear-oblong, terete or subangular. .... 58. *Diplachne*

62B. Lemmas entire, awnless, laterally flattened. .... 59. *Leptochloa*

60B. Inflorescence a single terminal spike; awn of the lemmas elongate and geniculate (in ours). ...... 60. *Tripogon*

55B. Spikelets with only 1 bisexual floret.

63A. Inflorescence digitate; spikelet sessile or nearly so.

64A. Spikelets 1-flowered, awnless or nearly so, without a reduced floret. .... 61. *Cynodon*

64B. Spikelets with few reduced florets above a single bisexual one, usually awned. .... 62. *Chloris*

63B. Inflorescence a panicle; spikelets pedicelled.

65A. Lemmas awnless, nerves wanting or obsolete.

66A. Low grasses with broadly lanceolate leaf-blades; spikelets 1- or 2-flowered, the upper floret, if any, pistillate.
63. *Coelachne*

66B. Low or rather tall grasses with linear or narrowly lanceolate leaf-blades; spikelets always 1-flowered.
64. *Sporobolus*

65B. Lemmas awned (in ours); nerves distinct. .... 65. *Muhlenbergia*

54B. First glume usually wanting, rarely minute and membranous, the second one coriaceous, larger than the lemma.

67A. Spikelets sunken in hollows on opposite sides of the axis of a solitary spike. .... 66. *Lepturus*

67B. Spikelets falling singly from the pedicel of the spikelike racemes, laterally flattened. .... 67. *Zoysia*

8B. Spikelets always 2-flowered, not laterally nor dorsiventrally flattened, articulate below the glumes; upper floret bisexual, the lower staminate or neuter, bisexual only in *Isachne*, very rarely both florets unisexual (*Spinifex*).

68A. Lemma of the upper floret coriaceous, rarely herbaceous or crustaceous, more rigid than the glumes, awnless or rarely awned; lemma of lower floret usually simulating the glumes in texture, rarely in *Isachne* simulating the lemma of the upper floret in texture.

69A. Callus of the upper floret with a tuft of hairs. .... 68. *Arundinella*

69B. Callus of the upper floret glabrous.

70A. Spikelets all alike and bisexual.

71A. Spikelets subtended or surrounded by 1 to many distinct or more or less connate bristles, forming an involucre and falling with them; lemma of the fertile floret herbaceous.

72A. Bristle solitary; plant of wet places. .... 69. *Pseudoraphis*

72B. Bristles few to many; plant of mesophytic or drier places. .... 70. *Pennisetum*

71B. Spikelets, if subtended or surrounded by bristles, falling detached from the persistent bristles; lemma of the fertile floret coriaceous or crustaceous.

73A. Lower florets staminate or neuter.

74A. Inflorescence an open or cylindrical panicle; spikelets usually distinctly pedicelled, not racemose.

75A.   Branchlets of the panicle, or at least some of them, ending in a bristle. . . . . 71. *Setaria*
75B.   Branchlets not ending in a bristle.
   76A.   Panicle cylindric, dense; fertile floret pedicelled, crustaceous; upper glume swollen at base. . . . . . 72. *Sacciolepis*
   76B.   Panicle effuse; fertile floret sessile, coriaceous or nearly so; upper glume not swollen at base. . . . . . 73. *Panicum*
74B.   Inflorescence digitate or racemose; spikelets sessile or very short-pedicelled, alternate in 2 rows on one side of a winged or wingless rachis.
   77A.   Lemma of the fertile floret with hyaline membranous margins; lower glume minute or wanting. . . . . 74. *Digitaria*
   77B.   Lemma of the fertile floret without hyaline margins.
      78A.   Lower glume wanting.
         79A.   Spikelets without an annular appendage at base; convex side of the floret and the upper glume facing the rachis. . . . . 75. *Paspalum*
         79B.   Spikelets with an annular appendage at base; flat side of the floret and the lemma of sterile floret facing the rachis. . . . . 76. *Eriochloa*
      78B.   Lower glume present.
         80A.   Glumes and (or) sterile lemma awned; floret smooth.
            81A.   Leaf-blades lanceolate, membranous. . . . . 77. *Oplismenus*
            81B.   Leaf-blades linear, elongate. . . . . 78. *Echinochloa*
         80B.   Glumes and florets awnless; floret usually rugose. . . . . 79. *Brachiaria*
73B.   Both florets bisexual; inflorescence a panicle; spikelets pedicellate. . . . . 80. *Isachne*
70B.   Spikelets all unisexual; rigid littoral grasses with rigid pungent leaves; pistillate inflorescence a spiny head with long radiating branchlets each bearing a single spikelet near the base; staminate inflorescence umbellate, with several spikelets on each branch. . . . . 81. *Spinifex*
68B.   Lemma of the upper floret thinly membranous, often with a long, geniculate, twisted awn, simulating in texture the lemma of the lower floret; glumes chartaceous or rather coriaceous; spikelets usually geminate.
82A.   Spikelets bisexual, often paired with a staminate or neuter spikelet.
83A.   Spikelets solitary at each node; inflorescence of 1 to many digitate continuous racemes. . . . . 82. *Dimeria*
83B.   Spikelets geminate; racemes often articulate.
   84A.   Spikelets all alike, bisexual; rachis of racemes slender, the internodes sometimes slightly inflated above.
      85A.   Rachis of the racemes not articulate, persistent, the spikelets falling.
         86A.   Lemma of the bisexual florets bifid; lower glume coriaceous, rounded on back; racemes peduncled. . . 83. *Eccoilopus*
         86B.   Lemma of the bisexual floret not at all or scarcely bifid; lower glume membranous to coriaceous, with 2 delicate keels; racemes almost sessile.
            87A.   Inflorescence a narrow silky panicle; spikelets awnless; lower glume membranous, without distinct keels; lodicules wanting; stamens 1–2. . . . . 84. *Imperata*
            87B.   Inflorescence a panicle or corymbose panicle, sometimes with a short axis; spikelets usually awned; lower glume herbaceous or rather coriaceous, with 2 keels; lodicules present; stamens usually 3. . . . . 85. *Miscanthus*
      85B.   Rachis of racemes articulate, disarticulating with the spikelets attached.
         88A.   Racemes in panicles; spikelets 1-flowered.
            89A.   Lower glume scarcely keeled at back; fertile lemma bifid. . . . . 86. *Spodiopogon*
            89B.   Lower glume with 2 keels at back; fertile lemma nearly entire. . . . . 87. *Saccharum*
         88B.   Racemes solitary or in corymbs.
            90A.   Lower glume 2-keeled, flat or sulcate between the keels; upper glume awnless or short-awned.
               91A.   Culms decumbent and much branched at base; leaf-blades lanceolate; lower glume glabrous, impressed or sulcate. . . . . 88. *Microstegium*
               91B.   Culms erect from the base, simple, usually not branched; leaf-blades elongate, linear; lower glume mostly long-hairy, flat. . . . . 89. *Eulalia*
            90B.   Lower glume convex; racemes solitary on culms; upper glume long-awned. . . . . 90. *Pogonatherum*
   84B.   Spikelets of two forms, the sessile bisexual, the pedicellate sterile, rarely bisexual.
      92A.   Racemes slender.
         93A.   Inflorescence a panicle without bracts.
            94A.   Internodes of the rachis without a hyaline center; glumes coriaceous; fertile lemma bifid. . . . . 91. *Sorghum*
            94B.   Internodes of the rachis with a hyaline line in the center; glumes chartaceous or slightly coriaceous; fertile lemma not bifid. . . . . 92. *Bothriochloa*
        93B.   Inflorescence of 1 to many racemes arranged in a digitate corymb or of bracteate false panicles; internodes of rachis without a hyaline line in the center.
          95A.   Leaf-blades ovate, embracing the culm at base; fertile lemma entire or bidentulate at apex, awned from the back below the tip; racemes few to many, in fascicles. . . . . 93. *Arthraxon*
          95B.   Leaf-blades linear, narrow; fertile lemma awned from the tip or between the teeth.
            96A.   Lower glume of the sessile spikelets 2-keeled or sulcate in front; callus of the spikelets obtuse, not elongate.
               97A.   Lowest pair of spikelets of one of the racemes alike, staminate, or neuter. . . . . 94. *Cymbopogon*
               97B.   Lowest pair of spikelets like the upper, one fertile and sessile, the other pedicellate, staminate, or reduced.
                                           95. *Andropogon*
            96B.   Lower glume of sessile spikelets cylindric, terete, not keeled or sulcate; callus of sessile spikelets more or less elongate and acute at base; fertile lemma entire, with a thick awn at apex; 2 pairs of the spikelets at base of racemes alike, sterile, surrounding the upper spikelets as an involucre. . . . . 96. *Themeda*
      92B.   Racemes with a thick rachis.
         98A.   Fertile lemma of sessile spikelets usually awned; racemes 2(–3) (in ours), closely appressed to each other, usually pubescent. . . . . 97. *Ischaemum*
        98B.   Fertile lemmas awnless; racemes glabrous.

1. CHIMONOBAMBUSA Makino   KAN-CHIKU ZOKU

Rhizomes elongate; culms medium-sized, sparse, cylindric or obtusely squarrose, usually with air-roots from the lower nodes; buds 3 at each node; spikelets solitary or few, terminal on branchlets, many-flowered, subtended by bracts; glumes 0; rachilla slender, glabrous; florets rather distant; lemma membranous, smooth, glabrous, weakly about 7-nerved, lanceolate, acuminate; palea nearly as long as the lemma, 2-keeled, glabrous, nearly smooth, subentire; stamens 3; stigmas 2.——More than 10 species, in s. Asia.

1A. Culms smooth, terete, the nodes without prominent air-roots; culm-sheaths membranous, rather persistent, but soon withering.
   1. *C. marmorea*

1B. Culms roughened, obtusely quadrate, the lower nodes with prominent air-roots; culm sheaths rather thick, deciduous.
   2. *C. quadrangularis*

1. **Chimonobambusa marmorea** (Mitf.) Makino. *Bambusa marmorea* Mitf.; *Arundinaria matsumurae* Hack.—— KAN-CHIKU.   Culms 2–3 m. long, to 15 mm. in diameter, with rather prominent nodes; culm-sheaths membranous, brownish purple spotted, nearly bladeless, yellowish brown, bristly while young; branchlets slender, with 3–4 leaves; leaf-sheaths with spreading hairs on margin, the auricles wanting, the oral bristles soft, flexuous, very promptly deciduous, smooth; leaf-blades lanceolate, paler beneath, thin, glabrous, 6–15 cm. long, 8–12 mm. wide; spikelets 4–8 cm. long, linear, purplish; florets 5–8 mm. long, acuminate; rachilla green, 4–7 mm. long.——May–July.   Said to be spontaneous in Kyushu, but extensively planted elsewhere in Honshu (centr. distr. and westward).——Cv. **variegata**. *C. marmorea* f. *variegata* (Makino) Ohwi.   *C. marmorea* var. *variegata* Makino——CHIGO-KAN-CHIKU.   Leaves white-variegated.

2. **Chimonobambusa quadrangularis** (Fenzi) Makino. *Bambusa quadrangularis* Fenzi; *Tetragonocalamus quadrangularis* (Fenzi) Nakai——SHIHŌ-CHIKU, SHIKAKUDAKE. Culms 3–5 m. long, to 2.5 cm. in diameter, obtusely quadrate in cross section, with rather long internodes, green, becoming brownish green when dry, rough; nodes prominent, densely fringed with yellowish brown bristles; branchlets rather slender; leaves 3–5, the blades narrowly lanceolate, 15–20 cm. long, 1–2.5 cm. wide, glabrous; leaf-sheaths with spreading hairs on outer margin, the auricles wanting, the oral bristles erect, smooth.——Of Chinese origin, sometimes cultivated in Japan.

2. SEMIARUNDINARIA Makino   NARIHIRADAKE ZOKU

Rhizomes elongate; culms medium-sized, branches few at each node; culm sheaths deciduous, coriaceous, with a small blade at apex, the oral bristles rather rigid, erect; spikelets 1–3, spicately fasciculate on branchlets, 3- to 4-flowered, narrowly lanceolate, sessile or nearly so, bracteate at base; glumes wanting; rachilla short, appressed-pubescent; lemmas broadly lanceolate, acuminate, 9- to 10-nerved, rather thick, scaberulous above, with obscure transverse veinlets; palea about as long as the lemma, lanceolate, 2-keeled, shallowly bifid, scabrous especially on the keels; stamens 3; style single; stigmas 3; ovary glabrous.—— Several species in s. Asia.

1A. Prophylla (first leaf) of branchlets less than 1 cm. long, ovate-deltoid; branches densely arranged; leaf-blades narrowly lanceolate, acute or obtuse at base. .... 1. *S. fastuosa*

1B. Prophylla of branchlets 2–4 cm. long, linear, with spreading prominent hairs on the 2 keels; branches loosely arranged; leaf-blades broadly lanceolate, with rounded base. .... 2. *S. yashadake*

1. **Semiarundinaria fastuosa** (Mitf.) Makino. *Bambusa fastuosa* Mitf.; *Arundinaria narihira* Makino——NARI-HIRADAKE.   Culms rather distant on elongate rhizomes, somewhat purple tinged, 5–10 m. tall, 3–4 cm. in diameter, culm-sheaths incompletely deciduous; leaves 4–6, the blades narrowly lanceolate, gradually tapering at apex, acute to obtuse at base, glabrous; leaf-sheaths glabrous or sparingly ciliate, the auricles wanting, the oral bristle smooth, finally deciduous; bracts of inflorescence 2–5 cm. long, often with a small blade; spikelets 3–4 cm. long, 3- to 4-flowered; rachilla not visible, 6–7 mm. long; lemma broad-lanceolate, 15–20 mm. long, awn-pointed; palea ciliate on keels; anthers about 10 mm. long.——May. Honshu (centr. distr. and westw.), Shikoku, Kyushu; often cultivated.

Var. **kagamiana** (Makino) Ohwi. *S. kagamiana* Makino ——RIKUCHŪDAKE.   A smaller plant with rachilla 7–12 mm. long, 2/5 as long as the floret, less densely pubescent.——Cultivated in Honshu (Rikuchu).

2. **Semiarundinaria yashadake** (Makino) Makino. *S. fastuosa* var. *yashadake* Makino——YASHADAKE.   Culms slender, the branchlets sparse, the culm-nodes and lower part of culm-sheaths short-hairy; leaves 3–5, the blades lanceolate to broad-lanceolate, rather abruptly acuminate, rounded at base; leaf-sheaths ciliate, the auricles wanting, the oral bristles slender, smooth, at length deciduous.——Honshu (centr. distr. and westw.), Shikoku, Kyushu; spontaneous and cultivated.

3. SINOBAMBUSA Makino   TŌ-CHIKU ZOKU

Rhizomes elongate; culms medium-sized, glabrous; nodes fringed with purplish hairs when young; internodes very elongate; culm-sheaths soon deciduous, the oral bristles rigid, glabrous; buds few; spikelets pedicelled, in bracteate panicles, narrow, elongate; rachilla pilose; glumes 2 (–3), rather large, veiny; lemmas rather rigid, ovate, acuminate, 11- to 15-nerved, with transverse veinlets between the nerves; palea nearly as long as the lemma, 2-keeled, ciliate on the keel; stamens 3; ovary glabrous; stigmas 3.——Several species in s. Asia.

1. Sinobambusa tootsik Makino. *Bambos tootsik* Sieb. ——Tō-CHIKU.   Rhizome elongate; culms to 5 m. long, 3.5 cm. in diameter, with prominent nodes; internodes 40–60 cm. long; branchlets slender; leaves 3–9, approximate, the blades lanceolate, 5–20 cm. long, 1.5–3 cm. wide, puberulent below, acuminate, obtuse to acute at base; leaf-sheath ciliate, the oral bristles rigid, erect; bracts ciliate, 1.5–4 cm. long; spikelets 8–20 cm. long, many-flowered; rachilla 5–7 mm. long; glumes ovate, 7–10 mm. long, acute; lemmas about 10 mm. long, ovate, acuminate, mucronate, palea a little shorter; anthers 4–5 mm. long.——Of Chinese origin, cultivated in Japan.

4. PHYLLOSTACHYS Sieb. & Zucc.   MADAKE ZOKU

Culms tall and thick, sparse; rhizomes elongate; culm-sheaths deciduous; branches geminate with sulcate internodes; leaf-blades lanceolate, the oral bristles rigid and persistent or rather soft and soon deciduous; spikelets sessile in axils of bracts, forming a bracteate spike, few-flowered, the lateral ones prophyllate at base; glume 0 or 1, the lowest floret sometimes reduced to an empty lemma, the upper florets sterile; lemmas lanceolate, chartaceous, acuminate, the transverse veinlets obscure; palea linear-lanceolate, nearly as long as the lemma; rachilla short, often pubescent; stamens 3; ovary glabrous; style long; stigmas 3, filiform.——About 30 species in India and China; introduced in Japan, only cultivated.

1A. Culms nodding in upper part, the nodes with 2 rings, the upper very short and obscure; culm-sheath not ciliate; branches densely arranged; prophylls linear, shallowly bifid, early withering; leaf-blades approximate, to 10 cm. long, 1 cm. wide, thin; oral bristles erect, thin, early deciduous, often wanting. .... 1. *P. heterocycla*
1B. Culms erect, the nodes with 2 rings, the upper more prominent than the lower; culm-sheath ciliate or not; branches loosely arranged; leaf-blades larger, 10–12 cm. long, 1–1.5 cm. wide; oral bristles persistent.
    2A. Culm-sheath ciliate, commonly not spotted; prophylls of branchlets membranous, soon withering, deeply bifid, but soon splitting into 2 linear wrinkled filaments, 1–1.5 cm. long; oral bristles erect; auricles wanting. .... 2. *P. nigra*
    2B. Culm-sheath not ciliate, commonly spotted; prophylls of branchlets 2–3 cm. long, entire, linear, somewhat persistent; oral bristles spreading, rigid; auricles prominent. .... 3. *P. bambusoides*

1. Phyllostachys heterocycla (Carr.) Mitf. *Bambusa heterocycla* Carr.; *P. pubescens* Mazel ex Houz. de Lehaie; *P. edulis* Houz. de Lehaie (not *Bambusa edulis* Carr.); *P. mitis* ex auct.——Mōsō-CHIKU.   Culms 10–12 m. long, to 20 cm. in diameter, short-pubescent while young, later much branched, nodding at apex; nodes nearly single ringed, the upper ridge indistinct; culm sheaths clothed with purple-brown hairs, bristly at apex, the blade small, narrow; branchlets slender; leaves 2–8, the blades thin, lanceolate to narrowly so, 4–8(–10) cm. long, 4–8(–10) mm. wide, pubescent below near base, the ligule rather prominent; leaf-sheaths sometimes short-pilose above; bracts narrowly oblong to oblanceolate, 16–22 mm. long, usually with a short, linear blade at apex, glabrous; spikelets glabrous.——Aug.–Sept. Of Chinese origin, widely cultivated in Japan.

Cv. Kikko-chiku (Mitf.).   A monstrous unstable sport called KIKKŌ-CHIKU in Japan, LOHAN CHU in China or the so-called tortoise-shell bamboo. It is a curiosity without botanical standing, although the species was first known to botanists through this monstrous phase. As noted, this phase is not stable and does not reproduce itself. The lower culm-nodes are approximate, oblique, giving them a zigzag effect. This phase occurs isolated only among normal culms. Often grown in gardens but not long-lived. The names *P. mitis* var. *heterocycla* (Carr.) Makino, *P. pubescens* var. *heterocycla* (Carr.) Houz. de Lehaie, and *P. edulis* var. *heterocycla* (Carr.) Houz. de Lehaie, based on this monstrous form, must be rejected (cf. International Rules Bot. Nom. 1961, Art. 67).

2. Phyllostachys nigra (Lodd.) Munro var. **nigra.** *Bambusa nigra* Lodd.; *P. puberula* var. *nigra* (Lodd.) Houz. de Lehaie; *P. nigra* (Lodd.) Munro——KURO-CHIKU.   Culms green at first, becoming blackish, smooth. The typical phase of Chinese origin, is extensively cultivated in Japan, as well as the following.

Var. **henonis** (Bean) Stapf. *Bambusa puberula* Miq.; *P. puberula* (Miq.) Makino; *P. henonis* Bean; *P. fauriei* Hack. ——HA-CHIKU.   Culms to 10 m. long, 3–10 cm. in diameter, erect, slightly scaberulous, soon glabrescent; nodes with 2 prominent rings; culm-sheaths not spotted, ciliate, with a short blade, the auricles with scabrous bristles; leaves 3–5, the blades chartaceous, lanceolate, 5–10 cm. long, 8–12 mm. wide, acuminate, glaucescent and pubescent near base beneath, the auricles wanting, the oral bristles erect; inflorescence a fasciculate spike, 2.5–3 cm. long, the bracts shorter than the spike, lanceolate, short-pubescent on the back near tip, with a short subulate blade at apex; spikelets linear, 1- or 2-flowered, the upper floret neuter; lemmas linear-lanceolate, about 15 mm. long, pubescent, gradually tapering to the tip; palea a little shorter than the lemma, pubescent; anthers 7–8 mm. long.——May–July. Long cultivated in Japan, introduced from China.

3. Phyllostachys bambusoides Sieb. & Zucc. var. **bambusoides.** *P. megastachya* Steud.; *P. reticulata* (misapplied by K. Koch); *P. quilioi* Riv.; *P. mazelii* A. & C. Riv.——MADAKE. Culms 10–20 m. long, 5–13 cm. in diameter, smooth, erect; nodes with 2 prominent rings; culm sheath with blackish spots, glabrescent, not ciliate, with a short blade at apex; leaves 3–5, the blades chartaceous, lanceolate, 8–12 cm. long, 1–1.5(–2) cm. wide, glaucescent beneath, pubescent near base beneath; leaf-sheaths more or less ciliate, the auricles semi-rounded, spreading to slightly reflexed, with rigid spreading bristles on margin; spikes lanceolate, 4–6 cm. long, the bracts broadly oblanceolate, glabrous, 2–2.5 cm. long, with an ovate to lanceolate, acuminate blade; spikelets slightly longer than or nearly as long as the bract, glabrous, narrowly lanceolate, 1- or 2-flowered; rachilla glabrous; lemmas about 2 cm. long, aristately acuminate; palea as long as the lemma, glabrous. ——June. Of Chinese origin, widely cultivated in Japan.

Var. **aurea** (A. & C. Riv.) Makino. *P. aurea* A. & C. Riv. ——HOTEI-CHIKU.   Lower nodes of the culm approximate. Commonly cultivated.——China.

5. SHIBATAEA Makino   Okamezasa Zoku

Small erect bamboo with elongate rhizomes; culm-nodes prominent; internodes with a groove on one side; culm-sheaths thinly chartaceous, glabrous, deciduous, with a short aristate blade, without bristles; branches few, short, usually with 2 nodes, with 1 leaf (rarely 2) at apex and membranaous linear scales near the base; leaf-blades broadly lanceolate, without sheaths, not bristly at base; inflorescence a short spike, axillary on lower nodes of branches, with membranous bracts; spikelets 1–3 in the axil of each bract, sessile, 2-flowered, glabrous, broadly lanceolate or oblong, with a prophyllum at base; glumes 2, membranous, broadly lanceolate, unequal, 9- to 13-nerved, acute; rachilla glabrous; upper internodes elongate; upper florets reduced; lemmas membranous, broadly lanceolate, about 11-veined, with obscure transverse veinlets, abruptly acuminate; palea nearly as long as the lemma, with 2 keels and few slender nerves; stamens 3; style 1; stigmas 3.——Two species, one in Japan, the other in China.

1. Shibataea kumasaca (Zoll.) Makino. *Bambusa kumasaca* Zoll.; *Phyllostachys kumasaca* (Zoll.) Munro; *P. ruscifolia* Nichols.; *S. ruscifolia* (Sieb.) Makino——Okamezasa. Culms erect, 1–1.5 m. long, branches short; culm-sheaths glabrous; leaf-blades 6–10 cm. long, 15–25 mm. wide, loosely pilose beneath, abruptly acuminate, acute at base; sheaths at base of branches membranous, linear, 3–4 cm. long; prophylla of branches linear, thinly membranous, bifid, 2-keeled, pubescent on the keels; spikelets 15–18 mm. long; glumes about 4 and 8 mm. long respectively; lemmas 10–12 mm. long; palea glabrous on the keel; anthers about 8 mm. long.——Widely cultivated, reportedly spontaneous in western Japan.

6. ARUNDINARIA Michx.   Medake Zoku

Rhizomes short or elongate; culms small to medium-sized; culm-sheaths persistent; branches few (rarely 1 or 2) at each node, profusely branched; leaf-blades narrowly lanceolate, without auricles, the oral bristles slender, smooth, flexuous; inflorescence racemose-fasciculate at nodes of branches and culms; spikelets often sessile, bracteate or ebracteate, without prophylla, linear, slightly flattened, rather many-flowered; glumes 2, sometimes 1, small, chartaceous, several-veined; lemmas chartaceous, glabrous, rather large, several-nerved, with transverse veinlets between the nerves; palea as long as the lemma, 2-keeled; rachilla appressed-pubescent; stamens 3; ovary glabrous; style single; stigmas 3.——Scores of species have been described from Japan and China. The following list represents a conservative evaluation of the most important species.

1A. Rhizomes often very short; culms in scattered tufts; ligules always prominent; leaf-blades narrow, caudately long-acuminate.
  2A. Leaf-blades 4–6, 10–30 cm. long; oral bristles usually wanting; inflorescence usually axillary, not leafy at base.
    3A. Leaf-blades narrowly lanceolate, 15–30 cm. long, 1.5–2.5 cm. wide, rather coriaceous, with elevated veinlets. ...... 1. *A. hindsii*
    3B. Leaf-blades linear, 10–30 cm. long, 8–20 mm. wide, chartaceous, with only slightly raised veinlets. .... 2. *A. graminea*
  2B. Leaf-blades 5–10, 4–15 cm. long, 4–10 mm. wide; oral bristles always well developed; inflorescence terminal on leafy branchlets.
                                                            3. *A. linearis*
1B. Rhizomes typically elongate, creeping; culms solitary; ligules sometimes prominent; leaf-blades usually broader.
  4A. Culm-nodes with prominent hairs; leaf-sheath glabrous or nearly so.
    5A. Leaf-blades pubescent beneath. .... 4. *A. pygmaea*
    5B. Leaf-blades glabrous or nearly so beneath.
      6A. Leaf-blades narrowly lanceolate, gradually tapering above; culms thick, the prominent nodes densely fringed with retrorse hairs.
                                                5. *A. kiusiana*
      6B. Leaf-blades lanceolate to narrowly so, acuminate.
        7A. Culms nearly simple. .... 6. *A. pumila*
        7B. Culms much branched. .... 7. *A. argenteostriata*
  4B. Culm-nodes glabrous or nearly so.
    8A. Leaf-sheath densely short-pilose. .... 8. *A. nagashima*
    8B. Leaf-sheath glabrous to only sparsely short-pilose.
      9A. Leaf-sheath retrorsely pilose between the nerves.
        10A. Leaf- and culm-sheaths purplish, retrorsely pilose only near the margins while young; inflorescence purplish. 9. *A. vaginata*
        10B. Leaf- and culm-sheaths green, retrorsely pilose while young. .... 10. *A. virens*
      9B. Leaf-sheath not retrorsely pilose.
        11A. Culms larger, 3–5 m. tall, 1–3 cm. in diameter; branches numerous at each node; leaves rigid, long-acuminate, drooping at
            apex. .... 11. *A. simonii*
        11B. Culms smaller, 1–2 m. tall, 2–10 mm. in diameter; branches 2–3(–5) or rarely solitary; leaves thinner, not drooping at
            apex.
          12A. Leaf-blades long-acuminate. .... 12. *A. chino*
          12B. Leaf-blades abruptly acuminate. .... 4. *A. pygmaea*

1. Arundinaria hindsii Munro. *Pleioblastus hindsii* (Munro) Nakai; *Thamnocalamus hindsii* (Munro) E. G. Camus——Kanzan-chiku. Rhizomes short; culms in tufts, 3–5 m. long, 1–3 cm. in diameter, deep green; nodes not prominent, glabrous; internodes elongate; culm-sheaths glabrous; branches 3–5, erect, with densely disposed branchlets; leaves 4–5, the blades coriaceous, narrowly lanceolate, glabrous, 15–30 cm. long, 1.5–2.5 cm. wide, long-acuminate, cuneate at base; leaf-sheaths glabrous, usually without oral bristles, the ligules rather long; spikelets few together, in lateral fascicled racemes, 5–8 cm. long, more than 10-flowered; rachilla appressed-pilose; lemma 10–15 mm. long; palea with spreading hairs on the keels.——May. Spontaneous in s. China, cultivated in Japan.

2. Arundinaria graminea (Bean) Makino. *Pleioblastus gramineus* (Bean) Nakai; *A. hindsii* var. *graminea* Bean——

TAI-MIN-CHIKU.　Rhizomes short; culms in tufts, erect, 3–5 m. long, 5–20 mm. in diameter; culm-sheaths at first sparsely setose, soon glabrate; branches with densely arranged branchlets; prophylla short, ascending-hirsute; leaves 4–6, the blades narrowly lanceolate to linear, 10–30 cm. long, 8–20 mm. wide, caudately acuminate; leaf-sheaths glabrous, usually without oral bristles; spikelets 1–3, 3–5 cm. long, less than 10-flowered; lemma about 10 mm. long; palea hirsute with spreading hairs on the keels.——June. Spontaneous in the Ryukyu Islands, cultivated in Japan.

3. Arundinaria linearis Hack. *Pleioblastus linearis* (Hack.) Nakai——RYŪKYŪ-CHIKU, GYŌYŌ-CHIKU.　Culms 2–3 m. long, glabrous, smooth; culm-sheaths glabrous; branches 1–5, densely ramified; leaves 6–10, the blades broadly linear, glabrous, rather small, caudately acuminate, the oral bristles 5–6 mm. long, erect, smooth; inflorescence terminal on leafy branchlets; spikelets few, few-flowered; lemma 10–12 mm. long.——Spontaneous in the Ryukyu Islands, cultivated in Kyushu (s. distr.).

4. Arundinaria pygmaea (Miq.) Mitf. *Bambusa pygmaea* Miq.; *A. variabilis* var. *pygmaea* (Miq.) Makino; *Pleioblastus pubescens* Nakai; *A. variegata* var. *viridis* forma *pubescens* Makino——KE-NEZASA, KE-OROSHIMA-CHIKU. Rhizome elongate; culms distant, erect, 1–2 m. long, 2–5 mm. in diameter; nodes densely fringed with brownish bristles; branches 1–3 at each node; leaf-blades lanceolate to narrowly so, 3–20 cm. long, 4–30 mm. wide, abruptly acuminate, obtuse to rounded at base, pubescent beneath; leaf-sheaths glabrous or retrorsely setulose between the nerves when young, the oral bristles smooth.——May–June. Honshu (Suruga and westw.), Shikoku, Kyushu; common in hills.

Var. **glabra** (Makino) Ohwi. *A. variegata* var. *viridis* forma *glabra* Makino——NEZASA.　Leaves and nodes glabrous; occurs within the range of the typical phase.

5. Arundinaria kiusiana (Makino) Ohwi. *Pleioblastus kiusianus* Makino——FUSHIDAKA-SHINO.　Culms 2–3 m. long; nodes with dense retrorse brown bristles; leaves 3–8, the blades narrowly lanceolate, 7–25 cm. long, 8–28 mm. wide, long-acuminate, rounded to abruptly acute at base, glabrous; leaf-sheaths glabrous on the back, ciliate on the margin, the oral bristles erect, smooth.——Kyushu (Higo).

6. Arundinaria pumila Mitf. *Pleioblastus pumilus* (Mitf.) Nakai; *Nipponocalamus pumilus* (Mitf.) Nakai——SUDARE-YOSHI.　Culms 1–1.2 m. long, 2–2.5 mm. in diameter, nearly simple; nodes densely brown-hirsute; leaves 5–6, the blades narrowly lanceolate, 12–20 cm. long, 7–23 mm. wide, acuminate, abruptly acute at base, sometimes pilose beneath; leaf-sheaths glabrous, the oral bristles smooth, whitish.——Honshu (Chūgoku Distr.), Shikoku, Kyushu.

7. Arundinaria argenteostriata (Regel) Vilm. var. **argenteostriata**. *Pleioblastus argenteostriatus* (Regel) Nakai; *Bambusa argenteostriata* Regel; *A. chino* var. *argenteostriata* (Regel) Makino; *Nipponocalamus argenteostriatus* (Regel) Nakai——OKINADAKE.　A cultivated phase with variegated leaves.

Var. **communis** (Makino) Ohwi. *Pleioblastus communis* (Makino) Nakai; *A. communis* Makino——GOKIDAKE. Culms 1–3 m. long, 2–15 mm. in diameter, often branched; nodes densely hairy, later glabrate, the hairs at first erect, soon spreading; culm-sheath ciliate; leaves 3–13, the blades lanceolate, 10–30 cm. long, 10–35 mm. wide, acuminate, rounded to abruptly acute at base, glabrous; spikelets 5- to 9-flowered, 3.5–7 cm. long; lemmas 11–19 mm. long; palea 10–12 mm.

long.——Honshu (Kinki Distr. and westw.), Shikoku, Kyushu.

Cv. **disticha**. *A. argenteostriata* var. *disticha* (Mitf.) Ohwi; *Nipponocalamus argenteostriatus* var. *distichus* (Mitf.) Nakai; *Bambusa disticha* Mitf.　Plant smaller in all its parts.——Cultivated.

8. Arundinaria nagashima Mitf. *Nipponocalamus nagashima* (Mitf.) Nakai; *Pleioblastus nagashima* (Mitf.) Nakai——HIROUZASA.　Culms 0.5–1.5 m. long, 2–10 mm. in diameter, densely retrorse-hairy only while young; nodes pilose while young; leaf-blades lanceolate to narrowly so, 5–25 cm. long, 7–30 mm. wide, long-acuminate, rounded to abruptly acute at base, glabrous; leaf-sheaths densely retrorse-hairy and rather long spreading-ciliate; the oral bristles smooth, white.——Honshu (Tōkaidō, se. Kinki), Kyushu.

9. Arundinaria vaginata Hack. *Pleioblastus vaginatus* (Hack.) Nakai; *Nipponocalamus vaginatus* (Hack.) Nakai——HAKONEDAKE.　Culms 2–4 m. long, 2–12 mm. in diameter, glabrous; nodes and culm-sheaths glabrous; branches 1–3 at each node; leaf-blades narrowly lanceolate, 5–20 cm. long, 5–15 mm. wide, acuminate, rounded to acute at base, glabrous or nearly so; spikelets 2–4 cm. long, 4- to 7-flowered; lemma about 15 mm. long, purplish; palea 7–9 mm. long.——Honshu (Sagami, Izu, Suruga).

10. Arundinaria virens (Makino) Ohwi. *Pleioblastus virens* Makino; *Nipponocalamus virens* (Makino) Nakai——AO-NEZASA.　Culms erect, 2–4 m. long, 2–17 mm. in diameter, glabrous; culm-sheaths glabrous; leaf-blades lanceolate to narrowly so, 3–25 cm. long, 5–25 mm. wide, long-acuminate, rounded at base, minutely pilose along the midrib beneath; leaf-sheaths green, retrorsely short-hairy between the internerves while young, the oral bristles erect, 3–4 mm. long, smooth, flexuous.——Honshu (Tōhoku Distr.).

11. Arundinaria simonii (Carr.) A. & C. Riv. *Bambusa simonii* Carr.; *Pleioblastus simonii* (Carr.) Nakai; *Nipponocalamus simonii* (Carr.) Nakai——MEDAKE, KAWA-TAKE. Culms erect, 3–5 m. long, 1–3 cm. in diameter, glabrous, green; branches 5–10 at each node, spreading, curving above; leaves 3–6, puberulent only on the upper surface of the petiole, the blades narrowly lanceolate to broadly linear, 5–30 cm. long, 1–3 cm. wide; leaf-sheaths glabrous, the oral bristles erect, smooth; spikelets rather numerous, fasciculate in the leaf-axils, 4- to 10-flowered; lemma 10–18 mm. long, sometimes purplish.——Sometimes cultivated, spontaneous in Honshu (s. Kantō Distr. and westw.), Shikoku, Kyushu.

12. Arundinaria chino (Fr. & Sav.) Makino. *Bambusa chino* Fr. & Sav.; *Pleioblastus maximowiczii* (A. & C. Riv.) Nakai; *Nipponocalamus chino* (Fr. & Sav.) Nakai; *P. chino* (Fr. & Sav.) Nakai; *B. maximowiczii* A. & C. Riv., non Munro, nec Nichols.; *A. simonii* var. *chino* (Fr. & Sav.) Makino——AZUMA-NEZASA, SHINAGAWADAKE.　Culms 0.5–2.5 m. long, 2–7 mm. in diameter, smooth; branches 1–5 at each node; culm- and leaf-sheaths glabrous excepting the ciliate margins; leaf-blades lanceolate to narrowly so, 5–25 cm. long, 5–20 mm. wide, gradually acuminate, rounded to abruptly acute at base, glabrous or slightly puberulent on one side beneath; spikelets few, fasciculate in leaf-axils, sometimes racemose, 3–7 cm. long; lemmas green to purplish, 10–19 mm. long; palea about 10 mm. long.——Honshu (Kantō, and s. Tōhoku Distr.); very common.——Cv. **laydekeri**. *A. chino* f. *laydekeri* (Bean) Ohwi; *Pleioblastus chino* var. *laydekeri* (Bean) Nakai; *A. laydekeri* Bean——KINJŌ-CHIKU.　Leaves yellow-variegated.

7. SASA Makino & Shibata SASA ZOKU

Rhizomes much ramified and long-creeping; culms woody, small to medium-sized, with slightly to prominently thickened nodes; culm-sheaths persistent; branches solitary at each node; leaf-blades broadly lanceolate to narrowly oblong, usually lustrous above, rather large, the auricles rounded, sometimes wanting, with rigid, scabrous, straight setae on the margin; panicles pedunculate, bractless, on lower to upper part of the culms, the branches and pedicels pubescent; spikelets linear to narrowly oblong, few- to rather many-flowered, the rachilla hairy; glumes wanting to 1 or 2, membranous; lemmas rather small, ovate, 5–18 mm. long, acute or sometimes awn-pointed, few-nerved; palea about equal in length to the lemma, 2-keeled; stamens 6, sometimes 3 or 5; style simple with 3 stigmas.——Japan, Korea, Sakhalin, and Kuriles. Multitudes of species have been described from Japan, of which only the best marked are treated here.

1A. Lemmas 12–18 mm. long; oral setae smooth above; stamens usually 4 or 5; rachilla-joints much shorter than the lemma.
　　1. *S. ramosa*
1B. Lemmas 4–10(–11) mm. long; oral setae wanting, or when present, scabrous to the apex.
　2A. Auricles and oral setae prominent, but deciduous in the 2d year; stamens 6; glumes 0–1(–2), not prominent; rachilla visible.
　　3A. Culms slender, low, sparsely branched, with greatly swollen nodes and elongate internodes; panicle-branches bearing 1–3 spikelets.
　　　2. *S. nipponica*
　　3B. Culms thicker, more elongate, much branched, nodes less swollen; lower panicle-branches bearing 3–10 spikelets.
　　　4A. Leaf-blades and culm-sheaths prominently pilose; leaves narrowly oblong, those of the 2d year conspicuously white-margined; culms 80–120 cm. long, branched. . . . . 3. *S. veitchii*
　　　4B. Leaf-blades and culm-sheaths not prominently pilose.
　　　　5A. Leaves 5–8 cm. wide. . . . . 4. *S. palmata*
　　　　5B. Leaves usually less than 5 cm. wide. . . . . 5. *S. senanensis*
　2B. Auricles wanting, the oral setae usually not developed; glumes 2, prominent; spikelets densely flowered, with usually shorter, hidden rachilla.
　　6A. Stamens 6.
　　　7A. Culms ascending from base, with prominent nodes.
　　　　8A. Leaf-blades pubescent below; culm-nodes hairy while young. . . . . 6. *S. cernua*
　　　　8B. Leaf-blades glabrous; culm-nodes glabrous from the first. . . . . 7. *S. kurilensis*
　　　7B. Culms erect from base, the nodes not prominent.
　　　　9A. Culms 2–5 m. long; leaf-blades 8–30 cm. long; lemmas 10–12 mm. long; spikelets 6- to 10-flowered, 2–6 cm. long.
　　　8. *S. japonica*
　　　　9B. Culms about 1 m. long; leaf-blades 5–12 cm. long; lemmas 5–7 mm. long; spikelets 3- to 5-flowered, 1–1.5 cm. long.
　　　9. *S. owatarii*
　　6B. Stamens 3. . . . . 10. *S. borealis*

1. Sasa ramosa (Makino) Makino. *Bambusa ramosa* Makino; *Arundinaria ramosa* (Makino) Nakai; *Sasaella ramosa* (Makino) Makino——AZUMAZASA. Culms 1–2 m. long, to 9 mm. in diameter, purplish, glabrous, branched in upper part; branches solitary at the nodes; leaves 3–5, thinly coriaceous, the blades broadly lanceolate, to 15 cm. long, 2 cm. wide, acuminate, rounded at base, usually puberulent beneath, more or less white-margined during the winter; leaf-sheaths glabrous, the oral setae rigid below; spikelets rather few, in a pedunced racemelike panicle, linear, 3–6 cm. long, 5- to 10-flowered; glumes membranous, small, narrow, the first often wanting; lemmas chartaceous, 12–17 mm. long; rachilla-joints usually ⅓–⅔ as long as the floret.——May–June. Honshu.

2. Sasa nipponica (Makino) Makino. *Bambusa nipponica* Makino——MIYAKOZASA. Rhizomes elongate; culms 30–100 cm. long, solitary, sparsely branched at base, internodes long and slender, with globosely swollen nodes, usually glabrous; leaf-blades linear-oblong to narrowly lanceolate, 10–20(–25) cm. long, 2–3.5(–5) cm. wide, abruptly acuminate, cuneate to obtuse at base, white-margined during the winter, pilose beneath; leaf-sheaths usually glabrous, the auricles rounded, with spreading oral setae, deciduous in the 2d year; spikelets 3- to 6-flowered, the florets loosely arranged, 6–10 mm. long.——Honshu, Shikoku, Kyushu.

3. Sasa veitchii (Carr.) Rehd. *Bambusa veitchii* Carr.; *Phyllostachys bambusoides* var. *albomarginata* Miq.; *S. albomarginata* (Miq.) Makino & Shibata——KUMAZASA. Culms 50–120 cm. long, loosely branched in upper part, sheaths hairy while young; leaves 4–7, the blades 10–25 cm. long, 3–7 cm. wide, narrowly oblong, abruptly acuminate, rounded at base, prominently short-pilose beneath, broadly white-margined during the winter; oral setae distinct, but finally deciduous; peduncles usually elongate; spikelets many, linear.——Honshu (Chūgoku Distr.), Shikoku, Kyushu; also widely cultivated in gardens.

4. Sasa palmata (Bean) Nakai. *Arundinaria palmata* Bean——CHIMAKIZASA, KUMAIZASA. Culms 1–1.5 m. long, slightly ascending from base, 6–8 mm. in diameter; leaves 5–9, the blades 10–35 cm. long, 5–8 cm. wide, narrowly oblong, thinly coriaceous, usually glabrous on both surfaces, sometimes puberulent beneath, the auricles usually developed, bristly ciliate; panicles long-peduncled, spikelets rather many, loosely arranged; florets 7–9 mm. long, narrowly ovate.——Hokkaido, Honshu, Shikoku, Kyushu.——Sakhalin.

5. Sasa senanensis (Fr. & Sav.) Rehd. *Bambusa senanensis* Fr. & Sav.; *Arundinaria kurilensis* var. *paniculata* F. Schmidt; *S. paniculata* (F. Schmidt) Makino——NEMAGARIDAKE. Culms 2 m. long or more, ascending at base; leaf-blades 15–30 cm. long, usually 3–5 cm. wide, more or less pilose; leaf-sheaths glabrous, the oral setae sometimes wanting; panicles rather large, often purplish.——Mountains; Hokkaido, Honshu, Shikoku.

6. Sasa cernua Makino. *S. kurilensis* var. *cernua* (Makino) Nakai——OKUYAMAZASA. Culms to 1.7 m. long, 1 cm. in diameter, glabrous, ascending at base, with puberulent nodes; leaf-blades thinly coriaceous, 20–23 cm. long, 2.5–6.5 cm. wide, acuminate, acute to obtuse at base, glabrous above, pilose below; leaf-sheaths glabrous, without oral setae; peduncles elongate, panicles often nodding, 7–12 cm. long; spikelets few, pedicelled, broadly linear, purplish, densely flowered; florets 8–10 mm. long.——Hokkaido, Honshu (centr. distr. and northw.).

7. Sasa kurilensis (Rupr.) Makino & Shibata. *Arundinaria kurilensis* Rupr.; *Bambusa kurilensis* (Rupr.) Miyabe; *Pseudosasa kurilensis* (Rupr.) Makino——CHISHIMAZASA. Culms 1–2 m. long, sometimes smaller, glabrous, branched; culm-sheaths glabrous; leaves 2–4, the blades narrowly oblong, 5–20 cm. long, 1–4 cm. wide, lustrous above, glabrous; culm-sheaths glabrous, usually without oral setae; panicles ovate; peduncles usually only slightly longer than the reduced leaf-blades; spikelets few, purplish, 15–25 mm. long, densely 3- to 5-flowered, with 2 small glumes (sometimes 1) at base; lemmas 7–11 mm. long, minutely ciliate.——High mountains; Hokkaido, Honshu (centr. distr. and northw.).——Kuriles, Sakhalin, and Korea (Dagelet Isl.).

Var. **uchidae** (Makino) Makino. *S. uchidae* Makino; *Pseudosasa uchidae* (Makino) Makino——NAGABA-NEMAGARIDAKE. Blades 5–7, narrowly oblong to lanceolate.——Hokkaido, Honshu (centr. distr. and northw.).

8. Sasa japonica (Sieb. & Zucc.) Makino. *Arundinaria japonica* Sieb. & Zucc.; *A. metake* Nichols.; *Pseudosasa japonica* (Sieb. & Zucc.) Makino; *Yadakeya japonica* (Sieb. & Zucc.) Makino——YADAKE. Culms 2–5 m. long, 5–15 mm. in diameter, glabrous, erect; nodes not elevated; internodes slender; culm-sheaths persistent, elongate; branches solitary on upper part of the culm; leaf-blades approximate, narrowly lanceolate, 8–30 cm. long, 1–4.5 cm. wide, glabrous, long-acuminate, acute at base, deep green above; leaf-sheaths glabrous, sometimes purplish, the oral setae often wanting; panicles 8–15 cm. long, spikelets about 10, linear, purplish; glumes acuminate, the first 5–8 mm. long, the second 7–10 mm. long; lemmas narrowly ovate, 10–12 mm. long, acute,

scaberulous above; palea 8–12 mm. long, with ciliate keels.——May–Nov. Honshu, Shikoku, Kyushu.——s. Korea.

9. Sasa owatarii (Makino) Makino. *Arundinaria owatarii* Makino; *Yadakeya owatarii* (Makino) Makino; *Pseudosasa owatarii* (Makino) Makino——YAKUSHIMADAKE. Culms 50–100 cm. long, 2–5 mm. in diameter, branched in upper part, nodes not elevated; culm-sheaths glabrous; leaf-blades rather thick, narrowly lanceolate, 5–12 cm. long, 5–12 mm. wide, yellowish green, with well-elevated veinlets, long-acuminate, acute to obtuse at base; leaf-sheaths glabrous, the oral setae wanting; panicles terminal on leafless branchlets, 3–5 cm. long; peduncles glabrous; spikelets oblong, 3- to 5-flowered, 1–1.5 cm. long, on ascending pedicels; glumes obtuse, the first 2–3 mm., the second 4–5 mm. long; lemmas 5–7 mm. long, rather obtuse, short-mucronate; palea obtuse, with 2 ciliate keels.——High mountains; Kyushu (Yakushima).

10. Sasa borealis (Hack.) Makino. *Bambusa borealis* Hack.; *Arundinaria purpurascens* Hack.; *S. spiculosa* Makino, excl. syn.; *Pseudosasa spiculosa* Makino, excl. syn.; *P. purpurascens* (Hack.) Makino; *Sasamorpha purpurascens* (Hack.) Nakai; *Sasamorpha purpurascens* var. *borealis* (Hack.) Nakai; *Sasa purpurascens* (Hack.) E. G. Camus; *Sasa purpurascens* var. *borealis* (Hack.) Ohwi——SUZU, JIDAKE. Culms 1–2 m. long, 3–6 mm. in diameter; nodes not prominently elevated; leaf-blades oblong-lanceolate, 10–30 cm. long, 1–6 cm. wide, gradually acuminate, abruptly acute at base, rather lustrous above, usually glabrous; panicles long-peduncled, 8–15 cm. long, often purplish, usually with a whitish bloom; pedicels yellowish pilose, spikelets ascending; glumes unequal, acuminate, several-nerved, often ciliate, simulating the lemmas in texture, the first 5–10 mm., the second 8–11 mm. long; lemmas 7–10 mm. long, coriaceous, puberulent, often ciliate; palea slightly shorter than the lemma, short-ciliate on the keels, rachilla-joints short, short-pubescent, about 1/5–1/4 as long as the lemma.——Hokkaido, Honshu, Shikoku, Kyushu.

8. ALOPECURUS L.   SUZUME-NO-TEPPŌ ZOKU

Annual or perennial herbs; leaf-blades linear, flat; upper leaf-sheath often slightly inflated; panicles cylindric, dense, with very short branches; spikelets laterally flattened, 1-flowered, falling entire; glumes equal, strongly folded, usually pubescent on the keel, more or less connate at base; lemmas shorter than to as long as the glumes, 5-nerved, obtuse, the margins connate at base, the midrib exserted as an awn; rachilla not produced beyond the floret, palea wanting; lodicules reduced; ovary glabrous, caryopsis not adherent to the lemma and palea.——About 60 species, widely dispersed in the temperate and cooler regions of the N. Hemisphere.

1A. Spikelets 4–6 mm. long.
  2A. Plants annual; spikelets 5–6 mm. long, rigidly membranous; panicles dense, yellowish or pale green; anthers pale yellow, about 1 mm. long. . . . . 1. *A. japonicus*
  2B. Plants perennial with short rhizomes; spikelets 4–5 mm. long, more or less purplish, soft; panicles very dense; anthers deeper yellow, 2.5–3 mm. long. . . . . 2. *A. pratensis*
1B. Spikelets 3–3.5 mm. long; anthers orange-yellow when dried, 0.6–1 mm. long; plants annual; panicles slender, about 3–5 mm. thick.
. . . . 3. *A. aequalis* var. *amurensis*

1. Alopecurus japonicus Steud. *A. malacostachyus* A. Gray——SETOGAYA. Smooth, glabrous annual; culms erect from ascending base; usually gregarious, 20–60 cm. long; leaf-blades pale green, 4–15 cm. long, 2–5 mm. wide, ligules thinly membranous, ovate, obtuse, 2–4 mm. long; panicles cylindric, 3–6 cm. long, 5–8 mm. wide, the pedicels 0.3–0.8 mm. long; spikelets narrowly ovate, slightly lustrous, flat; glumes membranous, 3-nerved, nearly free, obtuse, long-ciliate on the keel; lemmas rather thick, rigidly membranous, narrowly ovate, 5–6 mm. long, obtuse, the awn 10–12 mm. long, arising near base, slightly geniculate; anthers pale yellow.——May. Paddy fields

and wet river banks; Honshu (Kantō Distr. and westw.), Shikoku, Kyushu; common.——China.

2. Alopecurus pratensis L. Ō-SUZUME-NO-TEPPŌ. Glabrous perennial with short rhizomes and short ascending stolons; culms tufted, 50–100 cm. long; leaf-blades flat, 3–6 mm. wide, glaucous green; ligules membranous, truncate, 1–2 mm. long; panicles cylindric, 5–8 cm. long, 7–10 mm. wide, very dense with very short branches; spikelets slightly lustrous; glumes membranous, narrowly ovate, 3-nerved, obtuse; lemmas shorter than the glumes, narrowly ovate, glabrous, smooth, the keel scabrous above, the awn from near base on back, 6–10 mm. long, slender, only very slightly geniculate. ——May–June. Cultivated for forage and widely naturalized in Japan.——Eurasia, N. Africa, now widely naturalized in temperate regions.

3. Alopecurus aequalis Sobol. var. **amurensis** (Komar.) Ohwi. *A. fulvus* var. *amurensis* Komar.; *A. amurensis* (Komar.) Komar.; *A. fulvus* sensu auct. Japon., non J. E. Smith; *A. geniculatus* sensu auct. Japon., non L.——SUZUME-NO-TEP-PŌ. Soft, glabrous, tufted annual; culms 20–40 cm. long; leaf-blades flat, 5–15 cm. long, 1.5–5 mm. wide, pale glaucous green; ligules hyaline, pale, entire, semirounded to ovate, 2–5 mm. long; panicles cylindric, very dense, 3–8 cm. long, 3–5 mm. wide, pale green; branchlets very short, slightly scabrous; spikelets broadly ovate; glumes narrowly obovate, strongly folded, obtuse, 3-nerved, very slightly connate, white-ciliate on the keel, usually appressed-pubescent on the lateral nerve beneath; lemmas nearly as long as the glumes, ovate, smooth, glabrous, obtuse, obscurely 5-nerved, the margins connate below to the middle, the awn slightly exserted, 2.5–3.5 mm. long, very slender; anthers pale yellow, changing to orange-yellow when dry, 0.6 mm. long.——Apr.–June. Cultivated fields and riverbanks; Hokkaido, Honshu, Shikoku, Kyushu; a very common weed.——Korea, China, and e. Siberia. The typical phase occurs in temperate parts of the N. Hemisphere.

9. PHLEUM L. AWAGAERI ZOKU

Annuals or perennials; leaf-blades linear; panicles cylindric, dense, with very short branches sometimes adnate to the axis; spikelets 1-flowered, laterally compressed, articulate above the glumes; glumes equal, strongly plicate, truncate with a projecting keel at apex; lemmas shorter than the glumes, 3- to 5-nerved, laterally compressed, truncate, the callus minute, glabrous; rachilla-joint not developed; palea as long as the lemma; lodicules 2; stamens 3; ovary glabrous; styles slender; stigma slender, feathery, protruding above the floret.——Cosmopolitan with more than a dozen species in temperate and cooler regions.

1A. Perennials with swollen bases; glumes membranous, the keel straight, ciliate, ending in an awn 1/3–4/5 as long as the glume.
  2A. Panicles cylindric, 3–15 cm. long. .... 1. *P. pratense*
  2B. Panicles ellipsoid to oblong, 1.5–3 cm. long. .... 2. *P. alpinum*
1B. Annual; glumes coriaceous, obovate, slightly inflated above, glabrous or short-ciliate only near the middle on back, short-cuspidate.
                                   3. *P. paniculatum*

1. Phleum pratense L. Ō-AWAGAERI. Perennial, with swollen or bulblike base; culms 50–100 cm. long, tufted, erect from ascending base; leaf-blades green, 20–60 cm. long, 5–10 mm. wide, flat, scaberulous; ligules membranous, semirounded, 1–3 mm. long; panicles cylindric, 3–15 cm. long, pale green; spikelets very densely arranged, 3–3.5 mm. long, obovate, flat; glumes narrowly obovate, strongly plicate, membranous, truncate, the keel green, long-ciliate, excurrent at the tip as a short, rigid awn 1/4–1/3 as long as the body; florets about 1.5 mm. long, smooth, glabrous; anthers yellowish.——Pastures and fields; naturalized in Japan, especially common in Hokkaido.——Europe and Siberia; widely naturalized as a pasture grass throughout all temperate regions.

2. Phleum alpinum L. MIYAMA-AWAGAERI. Perennial, from a decumbent, densely tufted base; culms 30–40 cm. long, solitary to several, smooth; leaf-blades flat, glabrous, linear or shortly so; ligules membranous, slightly yellowish, truncate to semirounded, 1–2.5 mm. long; uppermost leaf-sheath slightly inflated at tip, with a short blade; panicle long-exserted, ellipsoid or short-cylindric, 1.5–3 cm. long, 8–10 mm. wide, pale green, sometimes slightly purple-tinged; spikelets flattened, about 3 mm. long, 1.5 mm. wide, rounded at base; glumes 3-nerved, rounded-truncate, the keel with hairs about 1 mm. long, excurrent as a rigid awn about 2 mm. long; lemmas about 2 mm. long, glabrous; anthers yellow, sometimes slightly purple-tinged, oblong, about 1 mm. long.——July–Aug. Alpine regions; Hokkaido (Mount Taisetsu), Honshu (centr. distr.); rare.——Formosa, Korea, Kuriles, also in alpine and boreal regions of the N. Hemisphere and arctic America.

3. Phleum paniculatum Huds. *Phalaris aspera* Retz.; *Phleum asperum* (Retz.) Jacq.; *P. paniculatum* var. *annuum* (M. Bieb.) Griseb.; *P. annuum* Bieb.; *P. asperum* var. *annuum* (Bieb.) Griseb.; *P. japonicum* Fr. & Sav.; *P. asperum* var. *japonicum* (Fr. & Sav.) Hack. ex Matsum.; *P. paniculatum* var. *annuum* forma *japonicum* (Fr. & Sav.) Makino——AWAGAERI. Rather tufted glabrous annual; culms 15–50 cm. long; leaf-blades flat, rather soft, 2–10 cm. long, 2–5 mm. wide; ligules membranous, broadly ovate to semi-rounded, 2–4 mm. long; panicles cylindric, 2–8 cm. long, 5–6 mm. wide, very dense, yellowish green, lusterless; spikelets 2–2.5 mm. long, narrowly cuneate, rather flattened; glumes coriaceous, scaberulous, obtusely keeled, obliquely truncate at apex, rigidly mucronate; lemmas about 1 mm. long, glabrous; anthers ellipsoidal, yellowish, 0.3–0.5 mm. long.——May–June. In lowland grassy places; Honshu, Shikoku, Kyushu.——China, Siberia, and the Mediterranean region.

10. AGROSTIS L. KONUKAGUSA ZOKU

Slender, tall or short perennials; leaf-blades linear; panicles narrow or open; spikelets small, 1-flowered, articulate above the glumes; glumes usually equal, sharply to loosely plicate, acute, entire, 1- to 3-nerved; lemmas usually shorter than the glumes, membranous, 3- to 5-nerved, acute or obtuse, entire or rarely aristulate with excurrent nerves, dorsal awn when elongate, geniculate and twisted, or awnless; callus glabrous or nearly so; rachilla usually not produced; palea smaller than the lemma, often greatly reduced; stamens 3; ovary glabrous.——About 200 species, widely distributed in cooler regions, especially abundant in the N. Hemisphere.

1A. Palea more than half as long as the lemma, 2-keeled.
  2A. Branches of panicle spreading, effuse, naked at base; ligules 3–5 mm. long, truncate. .... 1. *A. gigantea*
  2B. Branches of panicle spreading and effuse only at anthesis, erect or appressed to the axis at maturity; ligules quadrangular to broadly ovate. .... 2. *A. stolonifera*

1B. Palea less than half as long as the lemma, without keels, or wholly suppressed.
  3A. Lemma distinctly shorter than the glumes, awned or not.
    4A. Anthers 0.8–1.5 mm. long, ⅔–¾ as long as the lemma.
      5A. Spikelets 2.5–3 mm. long; awn inserted near the base to ⅓ of the lemma, distinctly twisted and geniculate.
        6A. Lemmas aristulate at apex with excurrent nerves. .... 3. *A. hideoi*
        6B. Lemmas entire at apex. .... 4. *A. flaccida*
      5B. Spikelets 1.5–2.2 mm. long; awn from about the middle of the lemma or wanting. .... 5. *A. canina*
    4B. Spikelets 1.5–2.2 mm. long; awn from about the middle of the lemma or wanting.
      7A. Lemma awned; culms 10–30 cm. long; branches of panicles smooth or nearly so. .... 6. *A. borealis*
      7B. Lemma awnless; culms 20–50 cm. long.
        8A. Panicles large, to 30 cm. long or more, with capillary densely scabrous branches; spikelets mostly terminal or nearly so; radical leaves short, filiform. .... 7. *A. scabra*
        8B. Panicles usually not more than 20 cm. long, effuse to rather dense, branches filiform, less densely scabrous, with longer prickles, the spikelets not confined to the upper part of branches; radical leaves more or less flat. .... 8. *A. clavata*
  3B. Lemma slightly longer than the glumes. .... 9. *A. nipponensis*

1. Agrostis gigantea Roth. *A. nigra* With.——KURO-KONUKAGUSA. Perennial, with short rhizomes; culms erect from a often decumbent, branching base, 50–100 cm. long, smooth or slightly scabrous above; leaf-blades flat, green, slightly glaucescent, thin, 10–20 cm. long, 4–7 mm. wide, slightly scabrous; ligules thinly membranous, 3–5 mm. long, denticulate, longer than wide, truncate; panicles 10–20 cm. long, rather dense, lustrous, the branches effuse, semiverticillate, prominently scabrous, verticils naked at base; spikelets 2–2.2 mm. long, reddish brown, acute; glumes broadly lanceolate, equal, acute, 1-nerved, obscurely punctulate, the keel slightly scabrous; lemma about 1.5 mm. long, obscurely 5-nerved, obtuse; callus glabrous; anthers 1–1.5 mm. long.——May–June. Hokkaido, Honshu, Kyushu; rather rare; possibly not indigenous.——Cooler regions of the N. Hemisphere.

2. Agrostis stolonifera L. *A. maritima* Lam.; *A. coarctata* Ehrh.; *A. alba* var. *stolonifera* (L.) Smith——HAI-KONUKAGUSA. Perennial; culms long-decumbent and creeping at base, ascending above, 10–20(–30) cm. long; smooth; leaf-blades 3–10 cm. long, 1–3 mm. wide, flat, scabrous, green, slightly glaucescent, thin; ligules quadrangular to broadly ovate, obtuse; panicles narrow, lanceolate to cylindric-ovoid, 5–10 cm. long, 5–20 mm. wide, dense, the branches short, erect, unequal, 1–3 cm. long, scabrous, semiverticillate, the verticils spikelet-bearing from the base; pedicels short; spikelets 1.5–2 mm. long, sometimes purplish, rather acute, slightly lustrous; glumes broadly lanceolate, 1-nerved, punctulate, with scabrous keel; lemma 1–1.5 mm. long, obscurely 5-nerved, awnless; anthers 0.8–1 mm. long.——May–June. In wet places; naturalized in Hokkaido, Honshu; rather rare.——Cooler regions of the N. Hemisphere.

Var. **palustris** (Huds.) Farwell. *A. palustris* Huds.; *A. sylvatica* Huds.; *A. alba* sensu auct. Japon., non L.; *A. grandis* Honda; *A. exarata* sensu auct. Japon., non Trin.——KONUKA-GUSA. Culms 50–100 cm. long, short-decumbent at base; leaf-blades 10–20 cm. long, 4–7 mm. wide; ligules 3–5 mm. long; panicles ovate; branches spreading at anthesis; spikelets 2–2.5 mm. long; anthers 1–1.5 mm. long.——May–June. Wet places; Hokkaido, Honshu, Shikoku, Kyushu; naturalized; very common.——N. Hemisphere.

3. Agrostis hideoi Ohwi. *Senisetum hideoi* (Ohwi) Honda——YUKIKURA-NUKABO, OKUYAMA-NUKABO. Perennial, with short rhizomes; culms tufted, slender, smooth, 20–40 cm. long; leaf-blades flat or loosely involute, 3–10 cm. long, 1–2 mm. wide, soft, smooth beneath; ligules thin, truncate to obtuse, about 1 mm. long; panicles narrowly ovoid, 5–8 cm. long, loose, the branches 4- to 7-nate, ascending, slightly scabrous; spikelets pedicellate, broadly lanceolate, usually purplish, 3–3.5 mm. long, acuminate; glumes slightly unequal,

1-nerved, nearly smooth, the keel scaberulous above, the first long-acuminate, 3–3.5 mm. long, the second short-acuminate, 2.7–3 mm. long; lemma 1.5–1.7 mm. long, oblong, 5-nerved, the midnerve excurrent just above the base into a delicate, geniculate awn about 3–3.5 mm. long, apex with 4 lateral nerves and 1 or 2 marginal teeth excurrent as awns about 0.5–1 mm. long; callus sparsely and minutely hairy; palea wanting; anthers a little less than 1 mm. long.——June–July. Wet grassy places near mountain rivulets; Honshu (Echigo and adjacent prov.); rare.

4. Agrostis flaccida Hack. *A. canina* sensu auct. Japon., non L.——MIYAMA-NUKABO. Perennial; culms tufted, with innovation shoots at base, 15–30 cm. long, smooth; leaf-blades loosely involute or flat, 3–10 cm. long, 0.5–2 mm. wide, smooth, glabrous, soft; ligules semirounded, 0.5–2 mm. long; panicles usually 4–8 cm. long, diffuse, ovoid, the branches usually geminate, smooth or nearly so; spikelets 2.5–3 mm. long, broadly lanceolate, lustrous, purplish, rarely green; glumes slightly unequal, broadly lanceolate, smooth except the keel, the first long-acuminate, 1-nerved, 2.5–3 mm. long, the second 3-nerved, short-acuminate; lemma 1.5–2 mm. long, minutely 5-toothed, the awn exserted just above the base, delicate, slightly geniculate, 3–5 mm. long; callus minutely hairy; palea nearly obsolete; anthers 0.8–1.5 mm. long.——June–Aug. High mountains; Hokkaido, Honshu, Shikoku, Kyushu; rather common.——s. Korea, s. Kuriles (Shikotan Isl.).

5. Agrostis canina L. HIME-NUKABO. Perennial, usually tufted; culms 20–60 cm. long; leaf-blades thin, usually flat, 2–10 cm. long, 1–3 mm. wide, slightly scabrous, the radical leaves not prominent; ligules 2–3 mm. long, oblong, obtuse; panicles 5–12 cm. long, 1–5 cm. wide, diffuse, the branches 3- to 6-nate, slender, scabrous; spikelets broadly lanceolate, 1.5–2, rarely to 2.2 mm. long, acute, lustrous, reddish brown, rarely green; glumes nearly equal, short-acuminate; lemma obscurely 3-nerved, rather obtuse, the midrib excurrent from about the middle into a delicate, slightly exserted and geniculate awn about 1–2 mm. long; callus minutely hairy; palea nearly obsolete; anthers 0.8–1.2 mm. long.——Naturalized in Honshu (Yamashiro), Kyushu (Chikuzen); rare.——Temperate and cooler regions of the N. Hemisphere.

6. Agrostis borealis Hartm. *A. viridissima* Komar.——KOMIYAMA-NUKABO. Perennial, usually tufted; culms slender, 10–30 cm. long, smooth; leaf-blades nearly smooth, flat, 2–7 cm. long, 1–2 mm. wide, the radical leaves usually filiform, 0.3–1 mm. wide; ligules obtuse, 0.5–2 mm. long; panicles open, 5–10 cm. long, 1–4 cm. wide, branches ascending, nearly smooth, 2- to 3-nate; spikelets pale green to purplish, lustrous, broadly lanceolate, 2.5–3 mm. long, acute; glumes equal, acute, nearly smooth, 1-nerved, the keel scaberulous above; lemma

1.5–2 mm. long, obscurely 3-nerved, the awn slightly exserted and geniculate, 2–4 mm. long; callus minutely hairy; palea nearly obsolete; anthers ellipsoidal, 0.4–0.7 mm. long.——July. High mountains; Hokkaido, Honshu, (centr. distr. and northw.); rather rare.——Alpine and cold regions of the N. Hemisphere.

7. **Agrostis scabra** Willd. *A. hiemalis* auct.——Ezo-nukabo. Perennial, with short innovation shoots at base; culms slender, smooth, 50–80 cm. long; leaf-blades 3–10 cm. long, 1–2 mm. wide, loosely involute, the radical filiform, 2–5 cm. long, about 0.5 mm. wide; ligules obtuse, 2–3 mm. long; panicles 20–40 cm. long, broadly ovoid, strongly effuse, more than half as long as the entire culm, very loose, the branches capillary, spreading, semiverticillate, finely and closely scabrous, to 20 cm. long, with spikelets only near the ends, the verticils naked at base; spikelets pale green, lanceolate, very acute, 2 mm. long; glumes unequal, lanceolate, 1-nerved, lustrous, with scabrous keel, the first acuminate, the second slightly shorter, very acute; lemma 1–1.2 mm. long, about ⅔ as long as the spikelet, obtuse, awnless, obscurely 3-nerved; callus nearly glabrous; anthers about 0.5 mm. long.——May–July. Grassy mountain slopes; Hokkaido, Honshu (centr. distr. and northw.); rather rare.——N. America and e. Siberia.

8. **Agrostis clavata** Trin. *A. perennans* sensu auct. Japon., non Tuckerm.; *A. valvata* Steud.; *A. macrothyrsa* Hack.; *A. osakae* Honda——Yama-nukabo. Tufted perennial, with short innovation shoots at base; culms 30–70 cm. long, smooth; leaf-blades 7–15 cm. long, 1.5–5 mm. wide; ligules somewhat lacerate at apex, 1.5–3 mm. long; panicles loose, diffuse, 12–20 cm. long, the branches geminate to semiverticillate, spreading, scabrous, the verticils usually naked at base; spikelets about 2 mm. long, green; glumes unequal, lanceolate, the keel scabrous above, the first 1-nerved, acuminate, the second slightly shorter, very acute, 3-nerved beneath; lemma about 1.5 mm. long, 3-nerved, obtuse; callus nearly glabrous; anthers about 0.3 mm. long, ellipsoidal.——June–Aug. Wet grassy places in mountains; Hokkaido, Honshu, Shikoku, Kyushu; rather common.——ne. Europe to Siberia, Kamchatka and Sakhalin.

Var. **nukabo** Ohwi. *A. exarata* var. *nukabo* (Ohwi) T. Koyama; *A. matsumurae* Hack.——Nukabo. Panicle branches verticillate, scarcely naked at base, unequal; spikelets acute; glumes 1.5–1.7 mm. long, nearly equal to rather unequal, very acute to acute, slightly longer than the lemma.——May–June. Cultivated fields and waste ground; Hokkaido, Honshu, Shikoku, Kyushu; very common.——Korea, Ryukyus, Formosa, China, and possibly also in the Philippines.

9. **Agrostis nipponensis** Honda. Hime-konukagusa. Perennial, without rhizomes; culms 40–70 cm. long, smooth, rather soft; leaf-blades pale green, thin, 7–15 cm. long, 3–5 mm. wide, flat, slightly scabrous; ligules 1–2 mm. long, truncate; panicles 10–15 cm. long, loose, broadly ovoid to lanceolate, the branches 3- to 7-nate, ascending to spreading, scabrous, the verticils widely naked at base; spikelets long-pedicelled, 2.5–3 mm. long, pale green and slightly purplish; glumes equal, broadly lanceolate, obtusely folded, 1-nerved, acute, slightly scaberulous above; lemma whitish, slightly longer than to as long as the glumes, awnless, 3- to 5-nerved; callus glabrous; palea minute; anthers 0.7–1 mm. long, ellipsoidal.——June. Wet places; Honshu (Kantō Distr. and westw.), Shikoku, Kyushu; rather rare.

Agrostis dimorpholemma Ohwi. Bake-nukabo. Allied to *A. stolonifera* var. *palustris* Farwell, but the florets of two forms on the same panicle; lemmas of the awnless phase 1.7 mm. long, 3-nerved, entire or bidentulate, the callus nearly glabrous, the palea ½–⅔ as long as the lemma; anthers 1.5 mm. long; lemmas of the awned phase 3- to 5-nerved, appressed short-pubescent at base on back, minutely toothed at apex; midrib excurrent slightly above base into a scaberulous, slightly exserted and geniculate awn about 4 mm. long; callus with a tuft of hairs 1/5–1/4 as long as the lemma, the palea 1/2–3/5 as long as the lemma; rachilla pilose, shorter than the ovary.——Collected once at Hakone in Honshu; possibly a hybrid of *A. avenacea* × *A. stolonifera* var. *palustris*, but the first parent has not been reported from our area.

11. POLYPOGON Desf.　Hiegaeri Zoku

Annuals or perennials; leaf-blades flat, linear, membranous; panicles very dense, often spikelike; spikelets rather small, with rather thick pedicels, 1-flowered, articulate below the pedicel; glumes equal, membranous, sometimes bifid, awned from the apex or from the sinus; lemma thinly membranous, small, broadly obovate, truncate, often with a delicate, short, fragile awn; callus glabrous; rachilla not produced; palea as long as or slightly shorter than the lemma, 2-nerved; stamens 3; caryopsis free from the lemma and palea, scarcely flattened.——More than a dozen species; widespread in temperate regions.

1A. Awn of the glume rising from the sinus, much longer than the glume; pedicels slightly longer than broad. ...... 1. *P. monspeliensis*
1B. Awn of the glume as long as or shorter than the glume; pedicels much longer than broad.
　2A. Glumes bifid, the awn about as long as the glume. .... 2. *P. fugax*
　2B. Glumes entire, the awn much shorter than the glume. .... 3. *P. hondoensis*

1. **Polypogon monspeliensis** (L.) Desf. *Alopecurus monspeliensis* L.——Hama-hiegaeri. Tufted annual; culms 30–60 cm. long, smooth; leaf-blades green, slightly glaucous, flattened, 10–15 cm. long, 4–8 mm. wide; ligules hyaline, deltoid, 3–8 mm. long; panicles nearly cylindric, very densely flowered, not interrupted, 5–10 cm. long, 1–2.5 cm. wide; branches semiverticillate, scabrous; pedicels obconical, 0.2–0.3 mm. long; spikelets broadly oblanceolate, pale green, 2–2.2 mm. long including the pedicel; glumes oblanceolate, 1-nerved, scabrous, rounded on back, shortly bifid with an erect delicate awn between the sinus; lemma broadly elliptic, obscurely 5-nerved, about 0.8 mm. long, nearly truncate at apex, obscurely denticulate; awn deciduous, to 1 mm. long; anthers 0.5 mm. long, ellipsoidal.——June–Aug. Wet sandy and grassy places; Honshu, Shikoku, Kyushu; common near seashores.——Temperate and warmer regions of Europe, Asia, and N. Africa, now widely naturalized in the New World.

2. **Polypogon fugax** Nees ex Steud. *P. hiegaeri* Steud.; *P. demissus* Steud.; *P. littoralis,* sensu auct. Japon., non Smith; *P. miser* (Thunb.) Makino, excl. syn.; *Nowodworskya fugax*

(Nees) Nevski——HIEGAERI. Tufted annual; culms 20–50 cm. long, smooth; leaf-blades pale green, flat, scabrous, 5–15 cm. long, 3–7 mm. wide; ligules 3–8 mm. long, weakly nerved beneath; panicles nearly cylindric, more or less interrupted, 3–8 cm. long, 1–2 cm. wide; pedicels about 0.3 mm. long; spikelets about 2 mm. long, sometimes slightly purplish; glumes rounded on back, shallowly bifid, the awn erect from the sinus, about as long as the glume; lemma and palea equal, about 0.8 mm. long; anthers about 0.5 mm. long; caryopsis 0.6 mm. long, ellipsoidal, scarcely compressed.——Honshu, Shikoku, Kyushu; rather common.——Korea, China, Ryukyus, Formosa, India, s. Siberia, Asia Minor, and Africa.

3. Polypogon hondoensis Ohwi. *Agropogon hondoense* (Ohwi) Hiyama——NUKABOGAERI. Culms 20–40 cm. long; leaves glaucescent, 5–10 cm. long, 3–5 mm. wide, more or less scabrous; ligules 3–4 mm. long, glabrous; panicles narrowly ovoid, dense, 5–8 cm. long, 2–2.5 cm. wide, branches short, ascending, the pedicels clavate; spikelets about 2 mm. long, somewhat purple-tinged; glumes broadly lanceolate, scabrous, acute, entire; awn erect, ⅛–¼ as long as the glume; lemma ovate, obscurely 5-nerved, 1.5 mm. long, truncate, 4-toothed; anthers 0.6 mm. long.——Honshu (Kantō Distr. and northw.); rare.——Possibly a hybrid of *Agrostis stolonifera* var. *palustris* × *Polypogon fugax*.

12. CINNA L.   FUSAGAYA ZOKU

Perennials with rather tall culms; leaf-blades linear, flat, membranous; inflorescence a panicle; spikelets flattened, 1-flowered, articulate above the pedicel, falling entire; glumes nearly equal, strongly folded, 1- to 3-nerved, acute; lemma simulating the glumes in length and texture, folded, obscurely 3-nerved, somewhat obtuse, awnless or with a short erect awn just below the apex; rachilla minute; palea apparently 1-keeled; stamen 1; caryopsis free within the lemma and palea.——Two species in the cooler regions of the N. Hemisphere.

1. Cinna latifolia (Trevir.) Griseb. *Agrostis latifolia* Trevir.; *C. pendula* Trin.; *Muhlenbergia baicalensis* Trin.——FUSAGAYA. Slender perennial without stolons; culms 80–120 cm. long, smooth; leaf-blades flat, linear, scabrous, 20–25 cm. long, 8–12 mm. wide; ligules ovate, obtuse, hyaline, 1.3–3 mm. long; panicles exserted, nodding, narrowly ovate, 20–30 cm. long, the branches semiverticillate, slender; spikelets flattened, 3–3.5 mm. long, on slightly thickened pedicels; glumes nearly equal, lanceolate, 1-nerved, scabrous, green, acute; lemma as long as the glumes, broadly lanceolate, 2–2.5 mm. long, short-awned from just below the apex, the awn not exserted; stamens 1, rarely 2, the anthers oblong, about 0.5 mm. long; ovary glabrous.——July–Aug. Mountain woods; Hokkaido, Honshu (centr. distr. and northw.).——ne. Europe, Siberia, Manchuria, Kamchatka, Korea, and N. America.

13. BECKMANNIA Host   MINOGOME ZOKU

Annuals or perennials; leaf-blades linear; ligules hyaline; panicles narrow, often interrupted, erect, with numerous short appressed or ascending racemes; spikelets sessile in 2 rows on one side of the rachis, disarticulating below the spikelets, laterally flattened, 1-(–2) flowered; glumes equal, chartaceous, navicular, rather inflated and bowed out on the back, transversely wrinkled, mucronate, slightly connate at base, 3-nerved; lemma narrowly ovate, whitish, 5-nerved, acuminate or mucronate, slightly keeled on back; callus glabrous; palea about as long as the lemma, 2-keeled; stamens 3; caryopsis fusiform, obtusely angled.——Two species, in the cooler and temperate regions of the N. Hemisphere.

1. Beckmannia syzigachne (Steud.) Fern. *Panicum syzigachne* Steud.; *B. erucaeformis* sensu auct. Japon., non Host; *B. erucaeformis* var. *uniflora* Scribn.; *B. erucaeformis* var. *baicalensis* V. Kusn.; *B. baicalensis* (V. Kusn.) Hult.——MINOGOME, KAZUNOKO-GUSA. Rather stout, soft, glabrous annual; culms tufted, smooth, 30–90 cm. long; leaf-blades pale green, flat, scabrous, 7–20 cm. long, 5–10 mm. wide, glabrous; ligules hyaline, ovate or deltoid, 3–6 mm. long; panicles erect, 15–35 cm. long, laterally compressed and interrupted, green, the branches less than 5 cm. long, triangular; spikelets sessile, 3–3.5 mm. long, and as wide, dull, pale green, 1(2–)-flowered; lemma as long as the glumes; anthers ellipsoidal, 0.5–0.7 mm. long, pale yellow.——June–July. Paddy fields and along rivers; Hokkaido, Honshu, Shikoku, Kyushu; very common.——Korea, China, Sakhalin, Kuriles, e. Siberia, and N. America.

14. CALAMAGROSTIS Adans.   NOGARI-YASU ZOKU

Perennial, commonly tall and robust, often with stolons or rhizomes; culms slender to rather stout; leaf-blades linear; panicles erect, open to contracted; spikelets 1-flowered, articulate below the floret; glumes 2, nearly equal, keeled, acuminate to acute, 1- to 3-nerved; lemma ovate, membranous, punctulate, 3- to 5-nerved, entire to denticulate, rarely aristulate, the midrib usually excurrent on the back into a slender, geniculate awn; callus short, with a tuft of hairs; rachilla prolonged beyond the floret, rarely wanting; stamens 3; ovary glabrous; caryopsis oblong to fusiform, free from the lemma and palea.——Cooler regions and higher mountains in the Tropics.

1A. Glumes linear-lanceolate to linear; lemma 3-nerved 1/3–1/2 as long as the glumes, hyaline; callus-hairs much longer than the lemma; rachilla not prolonged.
  2A. Panicles erect; glumes equal; awn from the back of lemma; leaf-blades sparsely scabrous. . . . . 1. *C. epigeios*
  2B. Panicles nodding; glumes unequal; awn from near the tip of lemma; leaf-blades densely scabrous. . . . . 2. *C. pseudophragmites*
1B. Glumes lanceolate to narrowly ovate; lemma 3- to 5-nerved, as long as or slightly shorter than the glumes; callus-hairs much shorter than to nearly as long as the lemma; rachilla usually prolonged.
  3A. Awn included or wanting, erect or nearly so; callus-hairs usually slightly shorter than the lemma.

4A. Glumes densely scabrous.
  5A. Spikelets not closed after anthesis; ligules pale rusty-brown, scaberulous on the back; leaf-blades flat, very scabrous on both sides, slightly glaucescent. .... 3. *C. langsdorffii*
  5B. Spikelets closed after anthesis; ligules whitish, nearly smooth; leaf-blades loosely involute, scabrous above, only slightly so beneath, pale green. .... 4. *C. neglecta* var. *aculeolata*
4B. Glumes nearly glabrous except the keel.
  6A. Radical leaves not prominent; culm-leaves about equally distant.
    7A. Culms with few, coriaceous, lustrous, bladeless sheaths at base; sheaths glabrous.
      8A. Culms thickened at base; lemma awnless, 3-nerved. .... 5. *C. matsumurae*
      8B. Culms not thickened at base; lemma awned from the back, 3- to 5-nerved. .... 6. *C. sachalinensis*
    7B. Culms without scaly leaves or with few, membranous, withering scales at base; sheaths usually with a ring of short hairs at the summit.
      9A. Callus-hairs less than half as long as the lemma; leaf-blades narrowly lanceolate. .... 14. *C. tashiroi*
      9B. Callus-hairs about ¾ as long as the lemma; leaf-blades flat or loosely involute. .... 7. *C. hakonensis*
  6B. Radical leaves prominent; culm-leaves mostly on the lower portion of the culm.
    10A. Plant rigid, rather stout; leaf-blades 3–10 mm. wide; panicles dense, the branches strict, rigid, rather stout.
                                                                                                                          8. *C. autumnalis*
    10B. Plant slender; leaf-blades 1–2 mm. wide; panicles rather open, the branches slender. .... 9. *C. deschampsioides*
3B. Awn long-exserted, geniculate, twisted; callus-hairs usually much shorter than the lemma.
  11A. Awn from above the middle of the lemma; callus-hairs unequal in length, about ¼–¾ as long as the lemma.
    12A. Lemma nearly entire or minutely toothed at apex; callus-hairs ¼–⅓ as long as the lemma; awn 5–15 mm. long, usually exserted near the middle of the lemma.
      13A. Culms with 4–7 nodes, commonly rather stout and tall; leaf-sheaths pilose around the summit. .... 10. *C. gigas*
      13B. Culms with 1–3 nodes, rather slender; leaf-sheaths glabrous.
        14A. Leaf-sheath usually shorter than the internode; callus-hairs ⅔–¾ as long as the lemma. .... 11. *C. longiseta*
        14B. Leaf-sheath usually longer than the internode; callus-hairs ⅓–⅔ as long as the lemma. .... 12. *C. masamunei*
    12B. Lemma with the nerves and apical teeth excurrent into short erect awns; callus-hairs ¼–⅓ as long as the lemma; awn 15–20 mm. long, exserted slightly below the apex, rather stout, prominently geniculate; rachilla prolonged at apex into an erect glabrous awn. .... 13. *C. fauriei*
  11B. Awn from near the base of the lemma; callus-hairs shorter and fewer 1/5–1/2 as long as the lemma on the sides.
    15A. Leaf-blades narrowly lanceolate, flat, thin. .... 14. *C. tashiroi*
    15B. Leaf-blades linear, loosely involute or with loosely involute margins.
      16A. Plant 60–150 cm. high; panicles dense to loose; spikelets 3–5(–6) mm. long, slightly flattened; glumes slightly longer than the lemma; hills and mountains. .... 15. *C. arundinacea*
      16B. Plant 10–40 cm. high, panicles dense, often spikelike; spikelets 5–10 mm. long, strongly flattened; glumes narrow, conspicuously longer than the lemma; alpine slopes. .... 16. *C. purpurascens*

1. **Calamagrostis epigeios** (L.) Roth. *Arundo epigeios* L.——YAMA-AWA. Rhizomatous; culms 60–150 cm. long, rather rigid, erect; leaf-blades 5–13 mm. wide, rather rigid, scabrous; ligules 3–6 mm. long; panicles erect, 10–30 cm. long, dense, ovate to narrowly lanceolate, branches semiverticillate, prominently scabrous; spikelets narrowly lanceolate, 5–8 mm. long, sometimes purple-tinged; glumes equal, barely lustrous, acuminate, usually 1-nerved, the keel very scabrous; lemma about half as long as the glumes, hyaline, 3-nerved, the midnerve exserted near the middle into a slender included awn about 3 mm. long; callus-hairs numerous, slightly shorter than the glumes.——July–Sept. Grassy mountain slopes; Hokkaido, Honshu, Shikoku, Kyushu; common.——Temperate regions of Eurasia.

2. **Calamagrostis pseudophragmites** (Haller f.) Koeler. *Arundo pseudophragmites* Haller f.; *A. littorea* Schrad.; *C. littorea* (Schrad.) Beauv.; *C. onoei* Fr. & Sav.——HOSSUGAYA, TOSHIMA-GAYA. Tall, stoloniferous perennial; culms 100–150 cm. long, rather stout, rigid, usually scabrous in upper part; leaf-blades grayish green, 10–30 cm. long, 3–7 mm. wide, scabrous; ligules 3–8 mm. long; panicles nodding, 20–30 cm. long, dense, ovate to lanceolate, the branches semiverticillate, scabrous; spikelets linear-lanceolate, 7–8 mm. long, usually somewhat purplish; glumes prominently unequal, with conspicuously scabrous keels, the first 1-nerved, the second shorter, 4–5 mm. long, 3-nerved; lemma 2/5–1/3 as long as the first glume, about 2.5 mm. long, narrowly ovate, hyaline, 3-nerved, with an erect, included awn 1–1.5 mm. long from between 2 minute teeth; callus-hairs slightly shorter than the first glume.

——July–Aug. Wet sandy places along rivers; Hokkaido, Honshu.——Temperate regions of the Old World.

3. **Calamagrostis langsdorffii** (Link) Trin. *Arundo langsdorffii* Link; *C. halleriana* sensu Fr. & Sav., non DC.; *C. villosa* var. *langsdorffii* (Link) Hack.; *C. villosa* sensu auct. Japon., non J. F. Gmel.; *C. canadensis* var. *langsdorffii* (Link) Inman——IWA-NOGARI-YASU. Rather tall gregarious perennial; culms 80–150 cm. long; leaf-blades glaucous-green, 20–35 cm. long, 3–8 mm. wide, flat, scabrous; ligules ovate, pale rusty-brown, 4–10 mm. long, obtuse; panicles nodding, ovate to lanceolate, 10–25 cm. long, the branches semiverticillate, prominently scabrous; spikelets usually pale rusty-brown, rarely greenish, dull, 3–5(–7) mm. long; glumes equal, strongly scabrous; lemma slightly shorter than the glumes, 5-nerved, usually awned from the middle of the back, the awn included, slender, erect or nearly so, not exceeding the lemma; callus-hairs about as long as the lemma; palea 1/2–4/5 as long as the lemma.——July–Sept. Wet places in forests and burnt forest lands; Hokkaido, Honshu, and alpine slopes of Shikoku ——Siberia, Korea, Sakhalin, and Kuriles.

4. **Calamagrostis neglecta** (Ehrh.) Gaertn., Meyer, & Scherb. var. **aculeolata** (Hack.) Miyabe & Kudo. *C. stricta* var. *aculeolata* Hack.; *C. aculeolata* (Hack.) Ohwi; *C. neglecta* sensu auct. Japon., non Gaertn., Meyer, & Scherb.——CHISHI-MA-GARI-YASU. Culms 40–100 cm. long; leaf-blades 10–20 cm. long, 2–5 mm. wide, loosely involute, scabrous above, slightly so beneath; ligules ovate, obtuse, 2–4 mm. long; panicles narrowly ovate to lanceolate, erect, dense, 7–17 cm. long, the branches semiverticillate, scabrous; spikelets pale green,

usually tinged with dull purple, 3–4 mm. long, acute; glumes equal, narrowly ovate, scaberulous; lemma nearly as long as the glumes, 5-nerved, awned below the middle, the awn erect, somewhat shorter to as long as the lemma.——Aug.–Sept. Sphagnum bogs; Hokkaido, Honshu (Mount Zao, Ozenuma in Kotsuke, Mount Kirigamine in Shinano and elsewhere); rare.——Kuriles and Sakhalin. The typical phase is widely distributed in cooler regions of the N. Hemisphere.

5. Calamagrostis matsumurae Maxim. MUTSU-NOGARI-YASU.   Tufted perennial; culms rather robust, 50–100 cm. long, thickened at base, with few whitish, coriaceous, lustrous, bladeless sheaths; leaf-blades green, nearly flat, 15–30 cm. long, 3–8 mm. wide, scabrous; ligules ovate, 2–6 mm. long; panicles 7–15 cm. long, erect, lustrous, the branches semiverticillate, scabrous; spikelets broadly lanceolate, acute, 3–5 mm. long, pale green, sometimes purple-tinged; glumes nearly equal, broadly lanceolate, nearly smooth, the keel slightly scabrous; lemma slightly shorter than the glumes, 3-nerved, awnless; callus-hairs 1/3–1/2 as long as the lemma.——July–Sept. Grassy alpine slopes; Honshu (n. distr.).

6. Calamagrostis sachalinensis F. Schmidt. *C. inaequiglumis* Hack.; *C. variiglumis* Takeda; *Deyeuxia sachalinensis* (F. Schmidt) Rendle——TAKANE-NOGARI-YASU, ONOE-GARI-YASU.   Tufted perennial; culms slender, usually scabrous in upper part, 30–60 cm. long; leaf-blades rather flat, glabrous, 10–30 cm. long, 3–6 mm. wide, nearly smooth and rather lustrous beneath; ligules about 1.5 mm. long, semi-rounded; panicles 5–10 cm. long, branches 2- to 6-nate, scabrous; spikelets broadly lanceolate, acuminate, 3.5–5.5 mm. long, pale green, sometimes slightly purple-tinged; glumes unequal, nearly smooth, the first often caudately elongate and slightly recurved above, the second slightly shorter, with 1 main nerve and short lateral nerves, acute, the keel sometimes scabrid above; lemma 3- to 5-nerved, awned below the middle or just above the base, the awn short, erect, included; callus-hairs 1/2–2/3 as long as the lemma.——Aug.–Sept. Coniferous woods in high mountains; Hokkaido, Honshu (centr. distr. and northw.), Shikoku.——s. Kuriles, Sakhalin, and Kamchatka.

7. Calamagrostis hakonensis Fr. & Sav. *C. orthophylla* Hayata & Honda; *C. exaristata* Honda; *C. koidzumiana* Ohwi; *C. scaberrima* Honda; *C. hakonensis* var. *aristata* (Honda) Ohwi; *C. matsumurae* var. *aristata* Honda; *C. hiyamana* Honda——HIME-NOGARI-YASU.   Slender perennial; culms 30–60 cm. high; leaf-blades rather flat or loosely involute, 7–30 cm. long, 3–8 mm. wide, somewhat scabrous; ligules 1–1.5 mm. long; panicles 5–15 cm. long, ovoid to lanceolate, the branches 2- to 5-nate, slender, scabrous; spikelets sometimes tinged with dull purple, 3.5–6 mm. long, acute, somewhat lustrous; glumes equal; lemma nearly as long as the glumes, 5-nerved, the awn erect from the back near base, usually shorter than the lemma, sometimes obsolete; callus-hairs shorter than the lemma.——July.–Oct. Woods in mountains and hills; Hokkaido, Honshu, Shikoku, Kyushu; rather common and variable.——s. Kuriles and China.

8. Calamagrostis autumnalis Koidz. *C. kirishimensis* Honda; *C. insularis* Honda——KIRISHIMA-NOGARI-YASU. Rather rigid perennial; culms strict, 20–60 cm. long, smooth; leaf-blades rather flat, glaucous-green, 10–30 cm. long, 3–10 mm. wide, smooth, glabrous, auriculately dilated at base; sheaths pilose on the collar; ligules truncate, 1–2 mm. long; panicles strict, broadly lanceolate, 5–15 cm. long, the branches

semiverticillate, smooth, rather rigid, ascending; spikelets 3.5–5 mm. long, pale green, often dull purple-tinged, lanceolate, acuminate; glumes equal, scaberulous; lemmas as long as the glumes, 5-nerved, awned below the middle, the awn erect, straight, as long as the lemma, included; callus-hairs half as long as the lemma.——Aug.–Oct. Mountains; Honshu (Izushichito), Kyushu.

Var. **microtis** Ohwi. *C. microtis* (Ohwi) Ohwi——KUJU-GARI-YASU.   Blades not distinctly auriculate at base, with short ligules; panicle branches scabrous.——Mount Kuju in Kyushu. Thought by some authors to be a hybrid of *C. arundinacea* × *C. autumnalis*.

9. Calamagrostis deschampsioides Trin. *C. nana* Takeda; *C. levis* Takeda; *C. miyabei* Honda; *C. sekimotoi* Honda; *C. deschampsioides* var. *nana* (Takeda) Takeda——HINAGARI-YASU.   Slender perennials; culms 20–40 cm. long, ascending to short-creeping at base, with few, membranous, withering, short leaves at base; radical leaves elongate, involute, nearly filiform, the cauline few, the blades 2–8 cm. long, 1–2.5 mm. wide, loosely involute, smooth, the sheaths usually longer than the internodes; ligules 1–2 mm. long; panicles ovate to narrowly so, 5–12 cm. long, the branches 2- to 6-nate, smooth, slender, ascending; spikelets usually purple-tinged, lustrous, 4–5 mm. long, acute, broadly lanceolate; glumes equal, smooth or slightly scabrous only on the keel above; lemma slightly shorter than the glumes, 5-nerved, awned below the middle, the awn straight, slender, as long as the glumes, included; callus-hairs 1/3–1/2 as long as the lemma.——July–Aug. Wet places in alpine regions; Honshu (centr. distr. and northw.); rare.——Sakhalin, n. Kuriles, Kamchatka, Siberia, ne. Europe, and N. America.

Var. **hayachinensis** Ohwi. *C. kaialpina* Honda——ZARA-TSUKI-HINAGARI-YASU.   Panicle-branches scabrous.——Alpine regions of Honshu (Mount Hayachine and Mount Shirane in Kai); rare.

10. Calamagrostis gigas Takeda. *C. robusta* sensu Hack., non Fr. & Sav.; *C. subbiflora* Takeda; *C. alpicola* Ohwi; *C. yezoensis* Honda——ONI-NOGARI-YASU.   Stout, tall, perennial; culms 80–200 cm. long, with 5 to 7 nodes, scabrous in upper part; leaf-blades elongate, 4–12 mm. wide, rather flat; sheaths smooth to scaberulous, usually short-pilose at the mouth; ligules 3–5 mm. long; panicles large, 12–30 cm. long, the branches semiverticillate, scabrous; spikelets narrowly lanceolate, acuminate, lustrous, pale green, sometimes partly purplish; glumes usually equal, rather smooth, narrowly lanceolate, thin, 4–6 mm. long; lemma slightly shorter than the glumes, 3- to 5-nerved, the awn below the middle, slender, slightly to scarcely exserted, callus-hairs 1/2–2/3 as long as the lemma.——July–Sept. Mountains; Hokkaido, Honshu (Hokuriku Distr., from Tamba Prov. northw.).

Var. **aspera** (Honda) Ohwi. *C. aspera* Honda——IWAKI-NOGARI-YASU.   Culms, leaves, and sheaths prominently scabrous.——Honshu (Mutsu Prov.).

11. Calamagrostis longiseta Hack. *C. grandiseta* var. *breviaristata* Honda; *C. longiseta* var. *contracta* Ohwi—— HIGE-NOGARI-YASU.   Tufted perennial; culms 20–80 cm. long; leaf-blades loosely involute, 15–30 cm. long, 2–5 mm. wide, sometimes pilose above; ligules 1–3 mm. long; panicles 8–15 cm. long, branches semiverticillate, scabrous; spikelets lanceolate, acuminate, 4–5 mm. long, purplish, lustrous; glumes usually equal, nearly smooth, the keel scabrous above; lemma slightly shorter than the glumes, 5-nerved, minutely

4-toothed at apex, awned near the middle, the awn geniculate and twisted, usually twice as long as the lemma, long-exserted; callus-hairs 2/3–3/4 as long as the lemma.——July–Sept. Mountains; Honshu (Yamato, Tajima, Hoki Prov., to centr. and n. distr.); rather rare and variable.

Var. **longearistata** (Takeda) Ohwi. *C. grandiseta* Takeda, including var. *longearistata* Takeda——Ō-HIGEGARI-YASU, NAGA-HIGEGARI-YASU. Culms 20–40 cm. long, with a few rusty-brown, membranous, leafless sheaths at base; panicle-branches shorter; spikelets 5–6 mm. long, the awn thicker and longer, usually below the middle of the lemma.——High mountains; Honshu (centr. distr. and north.).

12. **Calamagrostis masamunei** Honda. YAKUSHIMA-NOGARI-YASU. Similar to the preceding, but the culms 15–40 cm. long; leaf-blades loosely involute, 5–15 cm. long, about 2 mm. wide; sheaths usually longer than the internodes, ligules about 1 mm. long; panicle about 3 cm. long, lax, the 2- to 3-nate branches and branchlets scabrous; spikelets 4–5 mm. long; lemma awned below the middle; callus-hairs 1/2–2/3 as long as the lemma.——High mountains in Kyushu (Yakushima).

13. **Calamagrostis fauriei** Hack. *Ancistrochloa fauriei* (Hack.) Honda——KANITSURI-NOGARI-YASU. Closely allied to *C. longiseta* var. *Longearistata;* culms with few membranous rusty-brown, withering, bladeless sheaths at base; leaf-blades 2–4 mm. wide, rather thin, slightly scabrous above, smooth beneath; ligules 2–4 mm. long; panicles 5–10 cm. long, broadly lanceolate to narrowly ovate; spikelets 5–7 mm. long, reddish brown, rarely pale green; lemma 5-nerved, the lateral nerves and 1 or 2 apical teeth excurrent into erect awns about 1–4 mm. long, the dorsal awn from below the apex of the lemma, rather stout, prominently geniculate and twisted, 2 or 3 times as long as the lemma; callus-hairs 1/4–1/3 as long as the lemma; rachilla prolonged into an exserted, erect, scaberulous awn.——July–Aug. Grassy alpine slopes; Honshu (centr. distr. and northw. on Japan Sea side); rare.

14. **Calamagrostis tashiroi** Ohwi. TASHIROGARI-YASU. Slender perennial, without stolons; culms 20–40 cm. long; leaf-blades flat, membranous, narrowly lanceolate, 8–13 cm. long, 4–10 mm. wide, the sheaths pilose around the mouth; ligules 1–2(–3) mm. long; panicles narrowly ovate, 5–8 cm. long, the branches 2- to 3-nate, scabrous; spikelets pale green, narrowly lanceolate, 5–6 mm. long, acuminate; glumes equal, nearly smooth; lemma slightly shorter than the glumes, broadly lanceolate, 3-nerved; awn just below the tip of lemma, included or slightly exserted, geniculate, twisted; callus-hairs very short at the back, 1/4–1/3 as long as the lemma on the sides.——Aug.–Sept. Mountains; Shikoku, Kyushu; rare.

15. **Calamagrostis arundinacea** (L.) Roth. var. **brachytricha** (Steud.) Hack. *C. nipponica* Fr. & Sav.; *C. sciuroides*

Fr. & Sav.; *C. robusta* Fr. & Sav.; *C. arundinacea* var. *nipponica* (Fr. & Sav.) Hack.; *C. arundinacea* var. *robusta* (Fr. & Sav.) Nakai ex Honda——NOGARI-YASU, SAIDOGAYA. Slender to stout usually tufted perennial; culms 60–150 cm. long, rather rigid; leaf-blades rather flat to loosely involute, 30–60 cm. long, 6–12 mm. wide, sometimes pilose; ligules 2–5 mm. long; panicles 10–50 cm. long, the branches semiverticillate, scabrous; spikelets broadly lanceolate, very acute, 4–5(–6) mm. long; glumes equal; lemma nearly as long as the glumes, 5-nerved, the awn arising just above base, exserted, geniculate, twisted; callus-hairs unequal, very short at back of the lemma, 1/4–1/2 as long as the lemma at the sides.——Aug.–Oct. Mountains and hills; Hokkaido, Honshu, Shikoku, Kyushu; very common and variable.——e. Asia.

Var. **inaequata** Hack. SAISHŪ-NOGARI-YASU. Culms slender; panicles pale green; first glume longer, short-caudate.——Kyushu.——Korea.

Var. **adpressiramea** (Ohwi) Ohwi. *C. adpressiramea* Ohwi——KOBANA-NOGARI-YASU. Culms stout, smooth; panicles more strict, densely spiculose, the branches erect, appressed to the axis; spikelets 3–3.5 mm. long, dull.——Mountains; Kyushu; rare.——The typical phase is widespread in Eurasia.

16. **Calamagrostis purpurascens** R. Br. *C. arundinacea* f. *purpurascens* (R. Br.) Gelert ex Ostenf.; *C. urelytra* Hack.——MIYAMA-NOGARI-YASU. Glabrous, low, rather stout perennial, with erect or ascending innovation shoots; culms loosely tufted, ascending, covered at base with the sheaths of radical leaves; radical leaves to 30 cm. long, the blades involute or loosely so, 3–8 cm. long, 2–4 mm. wide, smooth beneath; ligules ovate, 1.5–3 mm. long; panicles narrowly ovate to lanceolate-cylindric, 2–8 cm. long, densely spiculose, the branches short, semiverticillate, usually scabrous; spikelets short-pedicelled, 5–10 mm. long, acuminate, sometimes purplish, rather strongly flattened; glumes equal or unequal, narrowly lanceolate, scaberulous, attenuately acuminate; lemma much shorter than the glumes, 5-nerved, minutely 4-toothed, the awn arising just above the base, rather stout, geniculate, twisted, exserted; callus-hairs on the sides 1/5–1/3 as long as the lemma; rachilla rarely terminating in a small, usually rudimentary second floret.——July–Aug. Alpine regions; Hokkaido, Honshu (centr. distr. and northw.).——Kuriles, Kamchatka, e. Siberia, and N. America.

The following hybrids or reputed hybrids occur in our area:
——Calamagrostis hakonensis × C. langsdorffii. *C. aristata* Ohwi——SHIKOKU-GARI-YASU. Calamagrostis epigeios × C. langsdorffii; Calamagrostis langsdorffii × C. matsumurae; Calamagrostis langsdorffii × C. neglecta var. aculeolata.

15. ARRHENATHERUM Beauv. Ō-KANI-TSURI ZOKU

Perennial; leaf-blades flat, linear; spikelets 2-flowered, disarticulating above the glumes, slightly compressed laterally, the lower floret staminate, long-awned, the upper bisexual, awnless or nearly so; glumes 2, unequal, membranous, 1- to 3-nerved, the first shorter than the spikelet, the second as long as the spikelet; lemmas 5- to 7-nerved, acute, not keeled, the lower long-awned from the base; callus hairy; rachilla prolonged beyond the upper floret, glabrous; palea nearly as long as the lemma; stamens 3; ovary hairy.——Species few, in temperate regions of the Old World.

1. **Arrhenatherum elatius** (L.) Presl. *Avena elatior* L.; *Holcus avenaceus* Scop.; *Arrhenatherum avenaceum* (Scop.) Beauv.——Ō-KANI-TSURI. Tall perennial; culms 70–125 cm. long; leaf-blades membranous, 15–25 cm. long, 7–10 mm. wide, scabrous; ligules hyaline, 1–2 mm. long; panicles 10–25 cm. long, erect, the branches semiverticillate; spikelets 8–9 mm. long, lustrous, pale green to fulvous green; glumes keeled, the lower 4–5 mm. long, the lemma sparingly long-pilose on back,

the lemma of the staminate floret awned from the base, the awn geniculate, about twice as long as the lemma; callus-hairs short; rachilla glabrous.——May–June. Naturalized; Hokkaido, Honshu, Shikoku, Kyushu; rather common. Native of Europe, now naturalized in the temperate regions.

Var. **bulbosum** (Willd.) Spenner. *Avena tuberosa* Gilib.;

Avena bulbosa Willd.; *Holcus avenaceus* var. *bulbosus* (Willd.) Gaudin; *Avena elatior* var. *tuberosa* (Gilib.) Aschers.——CHOROGIGAYA. Basal 1–3 internodes of the culms much swollen, as much as 10 mm. thick.——Rarely cultivated in Japan.

16. AVENA L. KARASU-MUGI ZOKU

Annual; leaves linear; panicles rather large, with nodding, 2- to 6-flowered, scarcely compressed spikelets; glumes 2, nearly equal, membranous, 7- to 11-nerved, not prominently keeled; lemmas rather coriaceous, rounded on back, 2-toothed, the awn geniculate and twisted, elongate, sometimes reduced; rachilla-joints hairy on one side; callus short; palea 2-keeled; stamens 3; ovary hairy; style short; caryopsis appressed-pubescent, grooved on one side.——About 50 species, mostly in cooler regions.

1A. Spikelets usually 3-flowered; florets readily deciduous; rachilla and lemmas often brown-bearded, sometimes glabrous; awn stout, geniculate and twisted. .... 1. *A. fatua*
1B. Spikelets usually 2-flowered; florets hardly deciduous; lemmas glabrous; awn wanting or only 1 to a spikelet, erect. ...... 2. *A. sativa*

1. **Avena fatua** L. KARASU-MUGI, CHA-HIKI. Annual; culms 60–100 cm. long, stout; leaf-blades soft, flat, 10–25 cm. long, 7–15 mm. wide; panicles 15–30 cm. long, open, loose, the branches scabrous, semiverticillate; spikelets green, about 2.5 cm. long, 2- to 3-flowered; glumes broadly lanceolate, acuminate, green, not keeled; lemma brown-bearded to glabrous, the awn elongate, strongly geniculate and twisted, dark brown on the lower half; callus-hairs short.——June–July. Naturalized throughout Japan; common.——Europe, w. Asia,

and N. Africa. Now widely naturalized in all temperate regions.

2. **Avena sativa** L. MA-KARASU-MUGI, ŌTO, ŌTO-MUGI. Annual; culms 60–100 cm. long, stout; leaf-blades linear, flat, 15–30 cm. long, 6–15 mm. wide; panicles open, loose, 20–30 cm. long; spikelets green; lemma glabrous, the awn wanting or reduced.——June–July. Widely cultivated in Japan. Supposed to be derived at least in part from the preceding.

17. HELICTOTRICHON Bess. MISAYAMA-CHA-HIKI ZOKU

Perennial, tufted; leaf-blades linear; panicles open or contracted; spikelets rather large, cylindric or slightly compressed, 2- to 8-flowered; rachilla bearded, disarticulating below the florets; glumes as long as the lemma or slightly longer, membranous, with a broad hyaline margin, more or less keeled, 1- to 3 (–5)-nerved; lemma coriaceous with hyaline margins, 5-nerved, rounded on the back, bifid, awned from about the middle, the awn long, geniculate and twisted; callus obtuse, with a tuft of short hairs; rachilla short-hairy on one side; palea 2-keeled; stamens 3; ovary pubescent at apex.——Many species in Eurasia and Africa, few in N. America and in high mountains of the Tropics.

1. **Helictotrichon hideoi** (Honda) Ohwi. *Avena hideoi* Honda——MISAYAMA-CHA-HIKI. Rhizomes short; culms slender, 70–90 cm. long, smooth; leaf-blades of innovation shoots elongate, to 30 cm. long or more, loosely involute, 1–2 mm. wide, smooth, glabrous or sparingly pilose beneath; culm-blades rather flat, 5–10 cm. long, 2–3 mm. wide; sheaths pilose with spreading white hairs; ligules truncate, ciliolate, 0.5–1 mm. long; panicles loose, nodding, rather few-flowered 6–8 cm. long, the branches spreading, geminate, scabrous; spike-

lets moderately compressed, 7–9 mm. long, 2- to 3-flowered, pale yellowish green, often somewhat purple-tinged; glumes unequal, broadly lanceolate, acuminate, the first 4–5 mm. long, 1- to 3-nerved, the second 7–8 mm. long, 3-nerved; lemma 7–8 mm. long, acute, minutely 2-toothed, the awn exserted above the middle, 12–15 mm. long, geniculate; callus-hairs 1 mm. long; anthers 3 mm. long.——July. Grassy slopes in high mountains; Honshu (Shinano Prov.); very rare.

18. TRISETUM Pers. KANI-TSURIGUSA ZOKU

Perennials, or rarely annuals; leaf-blades linear; panicles medium-sized, 2- to 6-flowered, articulate between the florets; spikelets compressed; glumes unequal, membranous, keeled, acute; lemmas longer than to as long as the glumes, keeled, bidentate, green to brownish, chartaceous, with hyaline margins, the awn usually elongate, from upper part of the back, geniculate and twisted; callus minute, short-hairy; rachilla-joints short-hairy on the back; palea membranous, 2-keeled; stamens 3; ovary glabrous or pubescent at tip.——About 60 species, chiefly in temperate regions.

1A. Axis of panicle and pedicels more or less pubescent; panicle contracted, spikelike, 3–10(–12) cm. long; anthers 0.5–1 mm. long.
　　1. *T. spicatum*
1B. Axis of panicle and pedicels glabrous.
　　2A. Lemmas nearly smooth, membranous; spikelets 4–6 mm. long; panicles 5–12 cm. long, 1–1.5 cm. wide, rather spikelike; branches smooth; internodes of culm usually long-pubescent; anthers 0.5–0.8 mm. long. .... 2. *T. koidzumianum*
　　2B. Lemmas scabrous, chartaceous; spikelets 6–10 mm. long; panicles loose or contracted, 10–20 cm. long, 2–10 cm. wide.
　　　　3A. Anthers 0.6–1 mm. long, much shorter than the lemma; lemmas strongly scabrous, attenuate at apex into 2 short awns.
　　3. *T. bifidum*
　　　　3B. Anthers 2–3 mm. long, ½–⅔ as long as the lemma; lemmas scabrous, minutely bidentate. .... 4. *T. sibiricum*

1. **Trisetum spicatum** (L.) Richt. *Aira spicata* L.; *A. subspicata* L.; *T. subspicatum* (L.) Beauv.——Rishiri-kani-tsuri. Low tufted perennial; culms 15–30 cm. long, erect, usually densely pubescent in upper part; leaf-blades soft, flat, 5–10 cm. long, 2–5 mm. wide, usually densely white-villous; ligules 1–2 mm. long; panicles cylindric, spikelike, 3–10 cm. long, 7–15 mm. wide, the branches short, white-pubescent; spikelets 2- to 3-flowered, pale yellowish green, sometimes somewhat purple-tinged, 5–6 mm. long; glumes unequal, acuminate, the first 3–4 mm., the second 4–5 mm. long; lemmas 4–5 mm. long, obscurely 5-nerved, very acute, the awn arising slightly below the tip, exserted, geniculate, 4–7 mm. long; ovary glabrous.——July–Aug. Grassy places in alpine regions; Hokkaido, Honshu (centr. distr.); very rare.—— Cooler regions and at high elevations in the Tropics.

Var. **kitadakense** (Honda) Ohwi. *Trisetum kitadakense* Honda——Kitadake-kani-tsuri. Glabrescent, only sparingly pilose on axis of panicles and pedicels.——Honshu (Mounts Kitadake and Senjo in Kai).

2. **Trisetum koidzumianum** Ohwi. *T. agrostideum* sensu Koidz., non Fries; *T. flavescens* var. *variegatum* sensu Honda, non Asch. & Graebn.——Miyama-kani-tsuri. Perennial with very short rhizomes; stolons absent; culms somewhat tufted, 15–20(–40) cm. long, slender, with long, spreading, white hairs except in upper part; leaf-blades flat, thin, 5–15 cm. long, 2–4 mm. wide, pilose or glabrous; sheaths pubescent; ligules 1–2 mm. long; panicles 5–7(–10) cm. long, rather dense, glabrous, 8–15 mm. wide, the branches 3- to 5-nate, short, smooth; spikelets 2-flowered, 4.5–5.5 mm. long, pale yellowish green, sometimes variegated with purple; glumes 3–3.5 mm. long, 3-nerved; lemmas about 4 mm. long, nearly smooth, membranous, acuminate and minutely bidentate, the awn slightly twisted, about 5 mm. long; callus appressed-puberulent.——July–Aug. Alpine meadows; Honshu (centr. distr.); very rare.

3. **Trisetum bifidum** (Thunb.) Ohwi. *Bromus bifidus* Thunb.; *T. flavescens* sensu auct. Japon., non Beauv.; *T. flavescens* var. *papillosum* Hack. and var. *macranthum* Hack.; *T. biaristatum* Nakai; *T. taquetii* Hack.——Kani-tsurigusa. Rather weak perennial with short innovation shoots at base; culms 40–80 cm. long; leaf-blades flat, thin, soft, 10–20 cm. long, 3–5 mm. wide, often whitish pubescent; ligules 0.5–1.5 mm. long; panicles nodding, 10–20 cm. long, 2–5 cm. wide, the branches geminate; spikelets 6–8 mm. long, lustrous, yellowish brown or pale green, sometimes somewhat purplish, 2- to 3-flowered; glumes lanceolate, the first 2–3 mm., the second 5–7 mm. long; lemmas 5–7 mm. long, rigid-chartaceous, very scabrous, with hyaline margins, attenuate at apex into 2 awn-like points, the awn from below the tip, 6–10 mm. long, geniculate, twisted; callus short-bearded; anthers oblong.—— May–June. Hills and lowlands; Hokkaido (w. distr.), Honshu, Shikoku, Kyushu; very common.——Korea, China, and Formosa.

4. **Trisetum sibiricum** Rupr. *T. flavescens* var. *genuinum* Hack., pro parte; *T. flavescens* var. *purpurascens* sensu auct. Japon., non Arcang.; *T. flavescens* var. *sibiricum* (Rupr.) Ostenf.; *T. homochlamys* Honda——Chishima-kani-tsuri. Culms rather tall, 50–100 cm. long; leaf-blades 15–25 cm. long, 4–10 mm. wide, sometimes sparingly pilose, flat; ligules 1–3 mm. long; panicles rather broad, sometimes slightly nodding, the branches subverticillate; spikelets yellowish brown or rarely pale green, usually purple-variegated, 6–10 mm. long, 3- to 5-flowered; glumes somewhat unequal, acuminate, lanceolate, the first 3–4 mm., the second 5–7 mm. long; lemmas broadly lanceolate, 6–7 mm. long, short-acuminate, chartaceous; the awn 7–12 mm. long, prominently geniculate and twisted; anthers linear; ovary glabrous.——July–Aug. Grassy places; lowlands and low mountains in Hokkaido, high mountains of Honshu, Shikoku, Kyushu.——e. Europe, Siberia, China, Manchuria, Kamchatka, Sakhalin, and Alaska.

19. KOELERIA Pers. Mino-boro Zoku

Tufted perennials or annuals; leaf-blades flat or involute, linear; panicles narrow; spikelets 2- to 4-flowered, compressed, lustrous, articulate above the glumes; glumes membranous, the second slightly broader, 1- to 3-nerved, keeled, acute; lemmas thick-membranous, obscurely 5-nerved, obtusely keeled, acute to bidentate, awnless or short-awned just below the apex; callus glabrous; rachilla nearly glabrous; palea 2-keeled, membranous; stamens 3; ovary glabrous; caryopsis free, not adherent to the lemma and palea.——Thirty to forty species in temperate regions.

1. **Koeleria cristata** (L.) Pers. *Aira cristata* L.; *K. gracilis* Pers.; *K. tokiensis* Domin; *K. gracilis* var. *tokiensis* (Domin) Honda; *K. cristata* var. *tokiensis* (Domin) Ohwi ——Mino-boro. Densely tufted perennial; culms 20–40 cm. long, pubescent; leaf-blades flat, thin, 5–15 cm. long, 1.5–3 mm. wide; ligules 0.2–0.5 mm. long, ciliolate; panicles erect, compact, cylindric, 5–15 cm. long, 6–15 mm. wide, the branches short; spikelets silvery green, lustrous, 4–5 mm. long, compressed, 3- to 4-flowered; glumes rather unequal, lanceolate, acuminate, sometimes pilose on the keel; lemma broadly lanceolate, 3.5–5 mm. long; callus glabrous; rachilla puberulent; anthers 1.5–2 mm. long.——May–July. Lowlands and mountains; Hokkaido, Honshu, Shikoku, Kyushu; rather common.——Widely distributed in the temperate regions of the N. Hemisphere.

Koeleria phleoides (Vill.) Pers. *Festuca phleoides* Vill. ——Mino-boro-modoki. Annual with much denser panicles and awned lemmas.——European introduction recently brought to our area.

20. DESCHAMPSIA Beauv. Kome-susuki Zoku

Slender, rather tall perennials, rarely annuals; leaf-blades linear to filiform; panicles open; spikelets 2- to 4-flowered, articulate above the glumes and between the florets; glumes equal, membranous, 1- to 3-nerved, keeled, as long as or longer than the florets; lemmas membranous, lustrous, ovate to oblong, rounded on the back, rather truncate and with 2–4 hyaline teeth at apex, the lateral nerves not reaching to the margin, the mid-nerve excurrent into a geniculate or erect awn; callus hairy; rachilla-joints pilose; palea membranous, 2-keeled; stamens 3; ovary glabrous; caryopsis free within the lemma and palea, shallowly grooved ventrally.——More than a dozen species in the cooler and alpine regions of the world.

1A. Lemma membranous, nearly as long as the glumes; anthers 1.5–2.5 mm. long; leaf-blades somewhat involute to rather flat, glabrous.
2A. Awn distinctly geniculate and twisted; leaf-blades filiform. .... 1. *D. flexuosa*
2B. Awn suberect; leaf-blades folded and acuminate or rather flat. .... 2. *D. caespitosa*
1B. Lemma rather rigid, lustrous, much shorter than the thin glumes; anthers about 0.8 mm. long; leaf-blades short, flat, puberulent.
　　　3. *D. atropurpurea* var. *paramushirensis*

1. **Deschampsia flexuosa** (L.) Trin. *Aira flexuosa* L. ——KOME-SUSUKI. Densely tufted perennial with short rhizomes; culms slender, 20–60 cm. long, smooth; leaves mostly radical, the blades filiform, folded, 5–15 cm. long, 0.5–1 mm. wide, smooth, glabrous, acute; ligules 1–1.5 mm. long; panicles very loose and open, nearly as broad as long, 5–12 cm. long, the branches 2- to 3-nate, capillary, spreading, nearly smooth; spikelets lustrous, 2-flowered, 4–6 mm. long, purplish; glumes equal, hyaline toward the margin, acuminate, narrowly ovate, 1-nerved, nearly smooth, glabrous; lemmas ovate, obscurely 5-nerved, slightly shorter than to as long as the glumes, 4-denticulate at apex, the awn arising near the base, slightly twisted and geniculate, 5–7 mm. long; callus-hairs about 1 mm. long; rachilla-joints appressed-puberulent; anthers 2–2.5 mm. long.——July–Aug. Bare ground of volcanoes and alpine regions; Hokkaido, Honshu (centr. and n. distr. and on Mount Omine in Yamato and Mount Daisen in Hoki), Shikoku, Kyushu; common.——Alpine and boreal regions of Eurasia and N. America.

2. **Deschampsia caespitosa** (L.) Beauv. var. **festucaefolia** Honda. *D. caespitosa* sensu auct. Japon., non Beauv.; *D. brevifolia* sensu auct. Japon., non R. Br.; *D. caespitosa* subsp. *orientalis* Hult.——HIROHA-NO-KOME-SUSUKI. Densely tufted, glabrous perennial without stolons; culms slender, 15–70 cm. long; leaf-blades folded to rather flat, 5–15 cm. long, 1–3 mm. wide, acute, strongly nerved above, acute; ligules 3–7 mm. long, hyaline; panicles erect or nodding, 5–20 cm. long, the branches spreading to ascending, somewhat scabrous; spikelets 2- to 3-flowered, 5–7 mm. long, lustrous; glumes lanceolate, acuminate or acute, nearly smooth, membranous, the first 2.5–3 mm. long, 1-nerved, slightly shorter than the lemma, the second 3-nerved, 3–4.5 mm. long; lemmas 3.5–4.5 mm. long, obscurely 5-nerved, truncate and 4-toothed at apex, the awn exserted about the middle of the lemma, nearly erect, slightly or scarcely exserted, 2–4 mm. long; callus-hairs 2/5–1/4 as long as the lemma; anthers 1.5–2.5(–3) mm. long. ——July–Aug. High mountains; Hokkaido, Honshu (centr.

distr. and northw.), Kyushu; rather common and very variable. This variety differs from the typical phase of Eurasia and N. America in the shorter culms, more scabrous panicle-branches, and slightly larger spikelets, but numerous intermediates are known.

Var. **levis** (Takeda) Ohwi. *Trisetum leve* Takeda; *D. takedana* Honda——YŪBARI-KANI-TSURI. Woodland form with taller slender culms, nodding panicles and linear-lanceolate second glume not exceeded by the tips of the florets.—— Reported from Mount Yubari in Hokkaido.

Var. **macrothyrsa** Ohwi & Tatew. *D. macrothyrsa* Tatew. & Ohwi——ONI-KOME-SUSUKI. Distinct phase of moist ground, taller, with culms about 1 m. long, with larger and more effuse panicles about 30 cm. long, the second glumes linear-lanceolate, the florets not longer than the tip of the glumes.——Lowlands near Nemuro in Hokkaido.

3. **Deschampsia atropurpurea** Wahlenb. var. **paramushirensis** Kudō. *D. atropurpurea* sensu auct. Japon., non Wahl.; *D. pacifica* Tatew. & Ohwi; *Vahlodea atropurpurea* subsp. *paramushirensis* (Kudo) Hult.; *V. flexuosa* (Honda) Ohwi; *Erioblastus flexuosus* Honda, non *Aira flexuosa* L.; *V. paramushirensis* (Kudo) Roshev.——TAKANE-KOME-SUSUKI, YUKI-WARIGAYA. Whitish-pubescent, soft perennial; culms slender, 15–40 cm. long, tufted; leaf-blades flat, thin, 5–10 cm. long, 2–4 mm. wide, glaucescent, short-acuminate; ligules 1–2 mm. long; panicles very loose, sometimes nodding, 7–12 cm. long, 5–7 cm. wide, the branches few, geminate, filiform, with few spikelets toward apex; spikelets 2-flowered, about 5 mm. long, flattened, reddish purple, glaucescent, dull; glumes equal, narrowly ovate, folded, acuminate, finely puberulent; lemmas broadly ovate, 2–2.5 mm. long, included, lustrous, awned from the middle, the awn longer than the lemma, twisted and slightly geniculate; anthers short.——July–Aug. Grassy slopes of alpine regions; Hokkaido (Mount Taisetsu); very rare.—— N. Kuriles, Kamchatka, and Aleutians. The typical phase occurs in n. Eurasia and N. America.

21. HOLCUS L.　　SHIRAGEGAYA ZOKU

Perennials with flat leaf-blades; spikelets in contracted panicles, 2-flowered, compressed, articulate above the pedicel and falling entire; lower floret bisexual, awnless, the upper staminate and awned; glumes unequal, folded, membranous, longer than the florets; lemmas chartaceous, lustrous, 5-nerved, glabrous, the awn from the back, sometimes curved, short; callus glabrous; rachilla slender, glabrous, not prolonged beyond the upper floret.——About 10 species, in Europe and Africa.

1. **Holcus lanatus** L. *Nothoholcus lanatus* (L.) Nash ——SHIRAGEGAYA. Densely grayish velvety-pubescent, soft perennial; culms 30–80 cm. long; panicles erect, 8–15 cm. long, 1.5–3 cm. wide; spikelets 4–5 mm. long, pale glaucous-green, sometimes tinged reddish purple, villous; anthers about 2 mm. long.——European introduction, naturalized and locally abundant.

22. ACHNATHERUM Beauv.　　HANEGAYA ZOKU

Perennials with short rhizomes and flat or involute leaf-blades; panicles erect, elongate; spikelets bisexual, terete, broadly lanceolate to narrowly ovoid, 1-flowered; glumes equal, lanceolate or narrowly ovate, acute to acuminate, membranous with hyaline margins; lemma nearly as long as the glumes, involute, obscurely 5-nerved, usually pubescent, the awn exserted between the minute terminal teeth, slightly geniculate, twisted, scabrous to plumose; callus short, obtuse; rachilla not produced beyond the floret; palea enclosed in the involute lemma, 2-nerved; lodicules 3; stamens 3, the anthers glabrous or pilose at apex; ovary glabrous.——More than a dozen species, in Eurasia and N. America.

1. **Achnatherum pekinense** (Hance) Ohwi. *Stipa pekinensis* Hance; *S. effusa* (Maxim.) Nakai, non Mez; *S. sibirica* var. *effusa* Maxim.; *S. extremiorientalis* Hara——HANE-GAYA. Rather rigid erect perennial; culms 80–150 cm. long; leaf-blades slightly involute, 30–60 cm. long, 7–15 mm. wide, scabrous on the margin, glabrous; sheaths pilose on the collar; ligule 0.5–1 mm. long, truncate; panicles erect, ovate, open at anthesis, 20–40 cm. long, the branches subverticillate, scabrous; spikelets lanceolate, 8–12 mm. long, sometimes purplish, lustrous; glumes membranous, green, with broad hyaline margins, 3-nerved, smooth, not keeled; lemma 6–8 mm. long, sparingly pubescent, the awn 20–25 mm. long, erect, slightly flexuous at tip; anthers linear, 4–6 mm. long, short-pilose at apex.——Aug.–Sept. Edges of woods and grassy slopes in mountains; Hokkaido, Honshu; rather rare.——e. Siberia, n. China, Korea, Sakhalin, and s. Kuriles.

23. ORTHORAPHIUM Nees   HIROHA-NO-HANEGAYA ZOKU

Rather rigid, tall perennial with flat or nearly flat leaf-blades and narrow erect panicles; spikelets bisexual, lanceolate, terete, articulate above the glumes; glumes lanceolate, green, scarcely hyaline-margined, 5- to 7-nerved, the nerves connected by few transverse veinlets; lemma coriaceous, brownish, involute, terete, lanceolate, usually pubescent, the awn exserted between 2 minute teeth at apex, erect, rather stout, slightly flexuous at tip, obscurely jointed at base; callus obtuse, short-pilose; palea as long as the lemma, 2-nerved; stamens 3; lodicules 3.——Two species, in India, China, Korea, and Japan.

1. **Orthoraphium coreanum** (Honda) Ohwi var. **kengii** (Ohwi) Ohwi. *Achnatherum coreanum* var. *kengii* Ohwi; *Stipa coreana* var. *kengii* Ohwi as a synonym of *A. coreanum* var. *kengii* Ohwi——HIROHA-NO-HANEGAYA. Perennial; culms 60–100 cm. long, erect, with 5- to 7-nodes; leaf-blades grayish green, rather flat, 20–40 cm. long, 7–15 mm. wide; sheaths short-pilose on the collar; ligules truncate, about 1 mm. long; panicles erect, linear, exserted, 15–30 cm. long, dense, rather few-flowered, the branches geminate; spikelets 12–15 mm. long; glumes green, equal, broadly lanceolate, acute, smooth, glabrous, rounded on back; lemma about 12 mm. long, the awn stout, 25–30 mm. long, scabrous, 2-grooved on inner side; anthers linear, 7–8 mm. long, glabrous.——Mountain woods; Hokkaido, Honshu, Shikoku, Kyushu.——The typical phase occurs in Korea and China.

24. MILIUM L.   IBUKI-NUKABO ZOKU

Rather soft perennials with flat, linear leaf-blades and open panicles; spikelets 1-flowered, ovate, not compressed, articulate above the glumes; glumes equal, obscurely 3-nerved, green, not keeled; lemma ovate, as long as the glumes, coriaceous, lustrous, obscurely 5-nerved, smooth, awnless, the callus minute, glabrous; palea simulating the lemma in texture and length; anthers and ovary glabrous.——A few species in the cooler regions of the N. Hemisphere.

1. **Milium effusum** L. IBUKI-NUKABO. Glabrous, green, soft perennial; culms erect, sometimes slightly decumbent at base, 60–120 cm. long; leaf-blades flat, linear, 10–20 cm. long, 10–15 mm. wide, abruptly acute; ligules hyaline, 5–10 mm. long, obtuse, smooth; panicles open, loose, 15–25 cm. long, ovate, the branches 2- to 5-nate, scabrous; spikelets about 3 mm. long, ovate; glumes ovate, rather obtuse, smooth, green; lemma coriaceous and glossy, smooth, scarcely shorter than the glumes, rounded on back, slightly dorsiventrally compressed; anthers about 2 mm. long.——June–July. Wet grassy places in mountain woods; Hokkaido, Honshu, Shikoku, Kyushu; rather common.——Formosa, Korea, s. Kuriles, Sakhalin, and cooler regions of Eurasia and N. America.

25. PHALARIS L.   KUSA-YOSHI ZOKU

Annuals or perennials with linear, usually flat leaves and contracted panicles; spikelets laterally compressed, articulate above the glumes; glumes membranous, navicular, 3-nerved, acute or cuspidate, the keel often winged; fertile with 2 (rarely 1), minute, appressed scales corresponding to sterile lemmas, attached at base; lemma of fertile floret ovate, not longer than the glumes, usually pubescent, obscurely 5-nerved, coriaceous to chartaceous, acute, lustrous, laterally compressed; callus very short, glabrous; rachilla not prolonged; palea lanceolate, faintly 2-nerved; stamens 3; anthers and ovary glabrous.——About 20 species, in the temperate zones of Eurasia and N. America.

1. **Phalaris arundinacea** L. *Digraphis arundinacea* (L.) Trin.; *P. japonica* Steud.; *Baldingera arundinacea* (L.) Dumort.; *P. arundinacea* var. *genuina* Hack. and var. *japonica* (Steud.) Hack.——KUSA-YOSHI. Gregarious perennial; culms rather stout, 70–180 cm. long, with 6–10 nodes; leaf-blades flat, glaucous-green, scabrous, 20–30 cm. long, 8–15 mm. wide, glabrous; ligules truncate, 2–3 mm. long; panicles 10–15 cm. long, 1–3 cm. wide, erect, very dense, sometimes spikelike, pale green or sometimes slightly purplish, the branches erect or ascending, single or in pairs, the branchlets short, densely spiculose; spikelets 4–5 mm. long, ovate, flattened, appressed to the branchlets, very acute; glumes equal, short-acuminate, folded, scabrous, the keel with a very narrow wing above; florets lanceolate-ovate, 3–3.5 mm. long, the attached sterile lemmas minute, villous; anthers 1.5–2 mm. long.——May–June. Wet sunny places; Hokkaido, Honshu, Shikoku, Kyushu; common.——Cooler regions of the N. Hemisphere.

Var. **picta** L. RIBONGURASU. Leaves variegated. Sometimes cultivated as an ornamental.

26. ANTHOXANTHUM L.   HARUGAYA ZOKU

Slender annuals or perennials, fragrant; leaf-blades linear, flat or loosely involute; panicles usually contracted, sometimes spikelike; spikelets with 1 terminal perfect floret and 2 sterile lemmas falling together, both or at least one of them neuter and epaleate; lemmas oblong, obtuse, usually pubescent, bifid, shorter than to as long as the second glume, usually awned; fertile floret smaller, lustrous, ovate, smooth; palea 1-nerved, folded; stamens 2; ovary glabrous; stigmas long and slender.——More than a dozen species, in temperate regions of the N. Hemisphere and in high mountains of the Tropics.

1A. Panicles spikelike; sterile florets 2/5–3/5 as long as the second glume; tufted perennial without rhizomes. .... 1. *A. odoratum*
1B. Panicles densely spiculose, sometimes spikelike; sterile florets as long as the second glume or slightly shorter; perennial with short rhizomes. .... 2. *A. japonicum*

1. Anthoxanthum odoratum L. HARUGAYA. Erect, slender, tufted perennial or annual; culms 20–50 cm. long, sometimes loosely pubescent; leaf-blades flaccid, flat, 5–10 cm. long, 3–6 mm. wide, usually loosely pubescent; ligules ovate, 2–4 mm. long; panicles long-exserted, spikelike, sometimes interrupted, broadly lanceolate-cylindric, 4–7 cm. long, 7–15 mm. wide, the branches short, scabrous, usually sparsely pilose; spikelets lanceolate, lustrous, 8–10 mm. long, fulvous-yellow or yellowish green; glumes strongly unequal, folded, sometimes sparingly long-pilose, the first narrowly ovate, acuminate, 1-nerved, 3–6 mm. long, the second broadly lanceolate, 3-nerved, 8–10 mm. long; sterile lemmas narrowly oblong, brown-pubescent except the tip, the lower one with a short erect awn slightly below the apex, the upper with a geniculate, twisted awn about 7–9 mm. long from the middle; fertile floret brown, about 2 mm. long, lustrous; anthers 4–5.5 mm. long. ——May–July. Introduced and naturalized; Hokkaido, Honshu, Shikoku, Kyushu; common.——Europe and Siberia, now naturalized widely in temperate regions.

Var. **furumii** (Honda) Ohwi. *A. nipponicum* Honda, incl. var. *furumii* Honda——MIYAMA-HARUGAYA. Plant glabrous; panicles cylindric, not broadened in lower part, glabrous; sterile florets 3.5–4 mm. long, the lemmas glabrous on the upper third.——July–Aug. Indigenous in alpine regions; Hokkaido (Mount Rishiri), Honshu (Akaishi Mountain Range); very rare.——Korea.

2. Anthoxanthum japonicum (Maxim.) Hack. *Hierochloë japonica* Maxim.——TAKANE-KŌBŌ. Perennial; culms slender, 20–60 cm. long; leaf-blades 10–20 cm. long, 4–10 mm. wide, flat or with slightly involute margins, smooth or nearly so, glabrous beneath, sparsely pilose above; ligules truncate or acute, 2–6 mm. long; panicles 5–10 cm. long, usually nodding, lustrous, rather densely spiculose, the branches geminate, smooth, often with a few hairs just below the spikelets; spikelets oblong, rather compressed, 5–7 mm. long, brownish green; glumes unequal, membranous, folded, glabrous, the first ovate, 3–4 mm. long, abruptly acute, 1-nerved, the second broadly lanceolate, 5–7 mm. long, gradually acuminate, 3-nerved; sterile lemmas chartaceous, brown, nearly as long as the upper glume, with long, brown hairs except at the bifid apex, the lower with a short erect awn from above the middle, the upper with a geniculate, twisted awn 4–5 mm. long, inserted near the base; anthers 2–2.5 mm. long.——June–July. In mountains; Honshu (centr. and n. distr. and Mount Omine in Yamato).

Var. **sikokianum** (Ohwi) Ohwi. *A. sikokianum* Ohwi——ISHIZUCHI-KŌBŌ. Leaf-blades loosely involute, 2–6 mm. wide; panicles denser, the second glume oblong-ovate, abruptly acuminate.——July–Aug. High mountains; Shikoku, Kyushu (Yakushima).

27. HIEROCHLOË R. Br. KŌBŌ ZOKU

Stoloniferous fragrant perennials; radical leaves elongate, the cauline short, the upper with slightly inflated sheaths; spikelets paniculate, slightly flattened, 3-flowered, articulate above the glumes; lower 2 florets staminate, disarticulating with the fertile floret, the uppermost bisexual, fertile; lemma of sterile floret chartaceous to coriaceous, pubescent, bifid, often awned; callus short, glabrous or short-hairy; palea 2-keeled; stamens 3; fertile lemma glabrous or ciliate toward the summit, lustrous, awnless; palea 3-nerved; caryopsis glabrous, free within the lemma and palea.——About 10 species, in temperate regions.

1A. Lemma of upper staminate floret with a long geniculate awn, that of the lower awnless; ligules truncate, ciliolate, less than 0.5 mm. long. .... 1. *H. alpina*
1B. Lemmas of both staminate florets awnless or with a short erect awn; ligules 1–4 mm. long.
  2A. Lemmas of staminate florets membranous, with hairy callus; leaf-blades smooth, glabrous. .... 2. *H. pluriflora*
  2B. Lemmas of staminate florets coriaceous, with hairy or glabrous callus; leaf-blades and sheaths usually scabrous to short-pilose.
    3. *H. odorata* var. *pubescens*

1. Hierochloë alpina (Sw.) Roem. & Schult. *Aira alpina* Liljebl., non L.; *Holcus alpinus* Sw.——MIYAMA-KŌBŌ. Glabrous perennial with ascending, branching rhizomes; culms slender, 15–30 cm. long, covered with few, whitish, lustrous sheaths of the old leaves at base; leaf-blades 1–2 mm. wide, less than 10 mm. long, flat or nearly so, the basal blades elongate, to 30 cm. long, involute, lustrous, 1–2 mm. wide; ligules 0.2–0.5 mm. long; sheaths lustrous, often tinged with red-purple; panicles 1.5–3 cm. long, 1–1.5 cm. wide, oblong, with 10–12 spikelets, the branches short, geminate; spikelets 5–6 mm. long, lustrous, yellowish brown, narrowly obovate, with an exserted awn; glumes 3-nerved, papery, smooth, glabrous, obtuse, as long as or slightly longer than the florets. ——July–Aug. Alpine regions; Hokkaido, Honshu (centr.

distr.).——Arctic regions and high mountains of the N. Hemisphere.——Forma **monstruosa** (Koidz.) Ohwi. *H. alpina* var. *monstruosa* Koidz.; *H. monstruosa* (Koidz.) Honda——Ō-MIYAMA-KŌBŌ. Viviparous phase.——Hokkaido.

2. Hierochloë pluriflora Koidz. *H. pauciflora* sensu auct. Japon., non R. Br.——EZO-KŌBŌ. Rhizomes short, creeping; culms loosely tufted, 20–40 cm. long, slender; radical leaves 30–40 cm. long, the blades involute, 1–4 mm. wide, or 5–6 mm. wide when flattened; culm-blades lanceolate, very short, on slightly inflated sheaths; ligules deltoid, ciliolate; panicles open, 2–3 cm. long, 1–1.5 cm. wide, pedicels usually with a few hairs at apex; spikelets usually several, ovate, 5–6 mm. long, pale yellowish brown, lustrous; glumes equal, ovate, papery, 3-nerved, slightly longer than the florets, obtuse;

lemmas of the staminate florets membranous, awnless or nearly so, 4.5–5.5 mm. long.——July–Aug. Alpine regions; Hokkaido (Mount Yubari); very rare.

Var. **intermedia** (Hack.) Ohwi. *H. alpina* var. *intermedia* Hack.; *H. intermedia* (Hack.) Kawano——Ezo-yama-kōbō. Lemmas of both bisexual and staminate florets distinctly short-awned.——June–July. Hokkaido (w. distr.); very rare.

3. **Hierochloë odorata** (L.) Beauv. var. **pubescens** Krylov. *H. bungeana* Trin.; *H. odorata* sensu auct. Japon., non Beauv.; *H. borealis* sensu auct. Japon., non Roem. & Schult.; *H. odorata* var. *sachalinensis* Printz; *H. odorata* subsp. *pubescens* (Krylov) Hara——Kōbō. Culms 20–50 cm. long, rather slender; radical leaves elongate, 20–40 cm. long, loosely involute to nearly flat, scabrous, 2–5 mm. wide, the cauline lanceolate, 1–4 cm. long, rather flat; sheaths usually short-pilose; ligules truncate to obtuse, glabrous, 1.5–3 mm. long; panicles broadly ovate, open, 4–8 cm. long, lustrous, the branches 2- to 3-nate, spreading; spikelets broadly obovate, 4–6 mm. long, slightly flattened, yellowish brown, awnless; glumes smooth, papery, ovate, acute, 1-nerved, the upper with a pair of short lateral nerves; lemmas of staminate florets scabrous above, rounded on back, ciliate, obtuse, usually awnless; callus glabrous, rarely with a tuft of short hairs; fertile floret smooth.——Apr.–June. Dry, grassy places; Hokkaido, Honshu, Shikoku, Kyushu; common.——Kuriles, Sakhalin, Korea, and Siberia. The typical phase occurs in temperate regions of the N. Hemisphere.

28. BRACHYPODIUM Beauv.   Yama-kamojigusa Zoku

Perennials or annuals with linear leaf-blades and nearly sessile, many-flowered spikelets solitary at each node of an elongate axis, subcylindric, not twisted, articulate above the glumes and between the florets; glumes several-nerved, not keeled, acuminate; lemmas firm, rounded on back, 5(–7)-nerved, acuminate and with a straight awn at apex; palea usually as long as the body of the lemma, 2-keeled; stamens 3; ovary with a pubescent beak; caryopsis adherent to the lemma and palea.——About 20 species, in temperate regions and in high mountains of the Old World Tropics.

1. **Brachypodium sylvaticum** (Huds.) Beauv. *Festuca sylvatica* Huds.; *F. misera* Thunb.; *B. miserum* (Thunb.) Koidz.——Yama-kamojigusa.   More or less pilose, tufted perennial with short rhizomes; culms rather rigid, 48–80 cm. long; leaf-blades flat or loosely involute, 10–20 cm. long, 5–10 mm. wide, narrowed at both ends; racemes nodding or pendulous, 7–13 cm. long; spikelets about 10, erect, pale green, 15–30 mm. long, 4- to 10-flowered; pedicels about 1 mm. long; glumes unequal, lanceolate, acuminate, the lower 5–7 mm. long, 3- to 4-nerved, the upper 8–11 mm. long, 7- to 8-nerved; lemmas lanceolate, 10–12 mm. long, acute, slightly scabrous, the apical awn straight, 5–10 mm. long; palea slightly shorter than the lemma; anthers 2–3 mm. long.——June–July. Woods; Hokkaido, Honshu, Shikoku, Kyushu; very common. ——Temperate regions of the Old World.

29. AGROPYRON Gaertn.   Kamojigusa Zoku

Perennials (few annuals) sometimes with creeping rhizomes; leaf-blades linear; inflorescence loosely to densely spicate, the spikelets solitary and alternate on a continuous rachis, very rarely in 2's or 4's, never twisted, 3- to 10-flowered, green, sessile; glumes firm, flat or keeled, linear to lanceolate; lemmas somewhat coriaceous to chartaceous, scabrous to pilose, rounded on back, 5-nerved, awned from the apex or awnless; palea 2-keeled; stamens 3; ovary with a hairy beak; caryopsis adherent to the lemma and palea.——About 100 species, in all cooler regions.

1A. Plants with long creeping rhizomes; lemmas awnless or short-awned; anthers 4–6 mm. long. . . . . 1. *A. repens*
1B. Plants without creeping rhizomes; lemmas commonly long-awned; anthers 1–3 mm. long.
  2A. Glumes narrowly oblong, acute, sometimes 2-toothed and awn-pointed; lemmas 1- or 2-toothed at apex, the awn recurved when dried; palea distinctly shorter than the lemma, rounded to emarginate, the internerve oblanceolate, slightly broadened above.
                                           2. *A. ciliare*
  2B. Glumes lanceolate to broadly so, narrowed at tip, sometimes awned, entire; palea as long as the lemma or nearly so, the internerve lanceolate to linear-lanceolate, not broadened above.
    3A. Callus prominently scabrous or appressed scabrous-pilose; woods in mountains.
      4A. Spike erect; awns more or less recurved. . . . . 3. *A. gmelinii* var. *tenuisetum*
      4B. Spike nodding, slender; awns erect. . . . . 4. *A. yezoense*
    3B. Callus smooth, glabrous or nearly so; grassy places in low mountains and in cultivated fields.
      5A. Spikelets always appressed to the rachis; spikes usually erect; keels of palea scarcely winged, ciliate with rather rigid, short hairs, the internerve linear-oblong; leaf-sheaths not ciliate; wet, sunny places. . . . . 5. *A. humidum*
      5B. Spikelets ascending at anthesis, appressed to the rachis before and after anthesis; spikes nodding; keels of palea winged, minutely serrulate-scabrous, the internerve lanceolate, usually narrowed above; leaf-sheaths usually ciliate; waste places and low mountains. . . . . 6. *A. tsukushiense*

1. **Agropyron repens** (L.) P. Beauv. *Triticum repens* L.; *Elytrigia repens* (L.) Desv.; *A. sachalinense* Honda—— Shiba-mugi.   Erect perennial, with long creeping rhizomes; culms rather rigid, 40–80 cm. long; leaf-blades nearly flat, green or glaucous-green, 3–8 mm. wide; ligules very short; spikes erect, 7–15 cm. long, the spikelets green or glaucous-green, 10–17 mm. long, 5- to 7-flowered; glumes equal, lanceolate, 7–12 mm. long, acuminate, 5- to 7-nerved; lemmas lanceolate, 7–11 mm. long, glabrous, nearly awnless or with a short, straight awn 2–4 mm. long; anthers linear, 4–6 mm. long.——Naturalized in Hokkaido and Honshu; rather rare. ——Europe, Mediterranean region, Siberia, n. China, Sakhalin, and N. America.

2. **Agropyron ciliare** (Trin.) Franch. var. **minus** (Miq.) Ohwi. *Brachypodium japonicum* var. *minor* Miq.; *A. ciliare* sensu auct. Japon., non Franch.; *Bromus racemifer* Steud.——

AO-KAMOJIGUSA. Tufted perennial without stolons; culms 30–100 cm. long; leaf-blades flat, green, 15–25 cm. long, 4–10 mm. wide, scabrous; spikes nodding, 10–20 cm. long, pale green; spikelets numerous, erect, 4- to 7-flowered, 10–15 mm. long; glumes narrowly oblong, 5- to 7-nerved, 5–8 mm. long; lemmas 10–15 mm. long; glumes narrowly oblong, 5- to 7-nerved, 5–8 mm. long; lemmas narrowly oblong, 7–10 mm. long, with long, spreading, rigid hairs or rarely glabrous, prominently 5-nerved above, the awn 15–20 mm. long, from an entire or notched apex, rarely wanting; callus nearly smooth and glabrous; palea 2/3–4/5 as long as the lemma, the keels ciliate, at least above.——May–July. Waste grounds and cultivated fields; Hokkaido, Honshu, Shikoku, Kyushu; very common.

Var. **pilosum** (Korsh.) Honda. *A. ciliare* f. *pilosum* Korsh.; *A. amurense* Drobov; *A. japonicum* Honda; *A. japonense* Honda; *A. mite* Honda; *A. ciliare* var. *hackelianum* (Honda) Ohwi; *A. japonicum* var. *hackelianum* Honda; *A. hackelianum* (Honda) Beetle——TACHI-KAMOJI. Spikes erect; lemmas glabrous, rarely strigose toward the margins.——Low mountains; Honshu, Shikoku, Kyushu; rather common.——China, Korea, Ussuri, and Manchuria. The typical phase occurs in n. China.

3. **Agropyron gmelinii** (Ledeb.) Scribn. & Smith var. **tenuisetum** (Ohwi) Ohwi. *A. turczaninovii* var. *tenuisetum* Ohwi——INU-KAMOJIGUSA. Green perennial; culms solitary or few together, 80–120 cm. long, usually with villous nodes; leaf-blades nearly flat, green, 15–25 cm. long, 5–8 mm. wide; spikes erect, 10–15 cm. long; spikelets numerous, green to slightly purple-tinged, appressed, 15–20 mm. long, 5- to 7-flowered; glumes broadly lanceolate, 6- to 8-nerved, gradually acute, 9–11 mm. long; lemmas 9–11 mm. long, scabrous, entire; awn rather stout, elongate, slightly recurved above; palea as long as the lemma.——June–July. Mountains; Honshu (Mount Kirigamine, in Shinano Prov.); rare.——Korea and s. Kuriles. The typical phase occurs in Siberia and Manchuria.

Agropyron caninum (L.) Beauv. *Triticum caninum* L. ——IBUKI-KAMOJIGUSA. Glumes 3-nerved, awns slender, erect.——European introduction naturalized on Mount Ibuki in Honshu.

4. **Agropyron yezoense** Honda. EZO-KAMOJIGUSA. Culms 70–100 cm. long, rather slender; leaf-blades flat, 15–25 cm. long, 3–5 mm. wide, green; spikes nodding, slender, 10–15 cm. long, green, the numerous spikelets loosely arranged, 15–20 mm. long, 5- to 7-flowered, appressed; glumes lanceolate, 3- to 5-nerved, short-acuminate, 6–8 mm. long; lemmas prominently 5-nerved above, rigidly ciliate toward the margin, 7–9 mm. long, entire, the awn straight, slender, 15–25 mm. long; palea

as long as the lemma, ciliate on the keels.——July. Woods; Hokkaido, Honshu (mts. in Shinano); rare.——e. Siberia, Manchuria, and n. Korea.

Var. **tashiroi** (Ohwi) Ohwi. *A. tashiroi* Ohwi. Lemmas glabrous.——Honshu (Chūgoku Distr.); rare.——Korea (Mount Kongo).

5. **Agropyron humidorum** Owhi & Sakamoto. *A. mayebaranum* Honda pro parte. Ō-TACHI-KAMOJI-GUSA. Culms rather stout, tufted, 40–80 cm. long; leaf-blades 10–20 cm. long, 3–7 mm wide, flat, glabrous, green; spikes erect, 10–20 cm. long; spikelets loosely arranged, pale green, sometimes tinged with dull purple, becoming brownish below when mature, 17–22 mm. long, 4- to 7-flowered, appressed; glumes broadly lanceolate, 3(–5)-nerved, 6–8 mm. long; lemmas glabrous, nearly smooth, with very narrow, hyaline margins, the awn rather stout, straight, 2.5–3 cm. long.——May–July. Wet grassy places in lowlands; Honshu, Kyushu; rather common.—— China.

Agropyron × mayebaranum Honda. *A. hatusimae* Ohwi;; *A mayebaranum* var. *intermedium* Hatus. TARIHONO-Ō-TACHI-KAMOJI. Spike nodding; lemmas scabrous on nerves above. Hybrid of *A. humidorum × A. tsukushiense* var. *transiens*.——Kyushu (n. distr.).

6. **Agropyron tsukushiense** (Honda) Ohwi var. **transiens** (Hack.) Ohwi. *A. semicostatum* var. *transiens* Hack.; *A. semicostatum* sensu auct. Japon., non Nees; *A. komoji* Ohwi ——KAMOJIGUSA. Tufted perennial; culms 40–100 cm. long; leaf-blades green to glaucous-green, flat, 20–30 cm. long, 5–10 mm. wide; spikes nodding to pendulous, 15–25 cm. long; spikelets erect, rather loosely arranged, pale green to glaucous-green, sometimes purple-tinged, 15–25 mm. long, 5- to 10-flowered; glumes oblanceolate or broadly lanceolate, much smaller than the lemmas, 3- to 5-nerved, short-acuminate and short-awned or awnless, 5–8 mm. long; lemmas 9–12 mm. long, broadly lanceolate, entire, glabrous, with hyaline margins, awn straight, scabrous, 2–3 cm. long; palea as long as the lemma, the keels winged, serrulate-scabrous.——May–July. Grassy places and waste grounds in lowlands; Hokkaido, Honshu, Shikoku, Kyushu; very common.——Korea, Manchuria, China, and Ryukyus.

Var. **tsukushiense**. *Elymus tsukushiensis* Honda; *A. semicostatum* var. *tsukushiense* (Honda) Ohwi——ONI-KAMOJI. Lemmas long-pubescent.——Kyushu (n. Distr.); very rare.

Agropyron nakashimae Ohwi. ZARAGE-KAMOJI. An alleged hybrid of *A. ciliare* var. *minus* and *A. tsukushiense* var. *transiens,* with the aspect of No. 5, but the spikes erect or only slightly nodding, the lemmas pilose with long rigid hairs toward margin.——Honshu, Kyushu; rare.

30. LOLIUM L. DOKU-MUGI ZOKU

Annuals or perennials with flat leaf-blades; spikes slender, often nodding; spikelets solitary, placed edgewise on the continuous rachis, several-flowered, the rachilla disarticulating above the glumes and between the florets; first glume wanting (except on the terminal spikelet), the second glume strongly 3- to 5-nerved, equaling or exceeding the second floret; lemmas rounded on the back, obtuse, 5- to 7-nerved, obtuse, acute, or awned.——About 8 species in Eurasia, two widely distributed.

1A. Glumes shorter than the spikelet.
  2A. Short-lived perennial, usually producing sterile shoots at base; lemmas quite or nearly awnless. .... 1. *L. perenne*
  2B. Annual or biennial, without sterile shoots at base; lemmas, at least the upper, awned. .... 2. *L. multiflorum*
1B. Glumes as long as to longer than the spikelet; annual or biennial.
  3A. Spikes flat; spikelets much broader than the rachis, with 6–10 florets. .... 3. *L. temulentum*
  3B. Spikes nearly cylindric; spikelets scarcely wider than the rachis, with 3–6 florets. .... 4. *L. loliaceum*

1. **Lolium perenne** L. Hoso-MUGI. Smooth, glabrous, lustrous, short-lived perennial; culms 30–60 cm. long; auricles at summit of the leaf-sheath minute; spikes mostly 15–20 cm. long; spikelets mostly 6- to 10-flowered; lemmas awnless or nearly so.——Naturalized in waste places and cultivated fields; Honshu, Shikoku, Kyushu; rather rare.——Introduced from Europe.

2. **Lolium multiflorum** Lam. NEZUMI-MUGI. Resembles the preceding but usually stouter; culms to 1 m. long, pale or straw-colored at base; auricles at summit of leaf-sheath prominent; spikelets 7- to 20-flowered, 1–2 cm. long; lemmas 6–8 mm. long, at least the upper awned.——May–June. Naturalized in Honshu, Shikoku, Kyushu; common.——Introduced from Europe.

3. **Lolium temulentum** L. var. **temulentum**. DOKU-MUGI. Rather robust annual; culms 60–90 cm. long; auricle at summit of leaf-sheath well-developed; spikes flat, 50–80 cm.

long; spikelets mostly 5- to 7-flowered; glumes equaling or longer than the spikelet; florets 6–10, plump, awned.——May–July. Naturalized in waste places and cultivated fields; Honshu, Shikoku, Kyushu; common.

Var. **leptochaeton** A. Br. *L. arvense* With. Awnless.——Rather common in our area.——An introduction from Europe.

4. **Lolium loliaceum** (Bory & Chaub.) Hand.-Mazz. *Rottboellia loliacea* Bory & Chaub.; *L. subulatum* Vis.——Bō-MUGI. Rather stout low annual or biennial; culms usually tufted, 10–30 cm., rarely to 40 cm. long, leaves not prominent, the blade short, rather flat; spikes firm, stout, often slightly curved, nearly cylindric, 10–15 cm. long; spikelets sunken in the excavations of the rachis, partly hidden by the appressed strongly nerved obtuse glumes; florets 3–6; lemmas oblong, about 5 mm. long, usually awnless.——May–June. Naturalized; Honshu (Kantō to Chūgoku Distr.); rather rare.——Introduced from Europe.

31. ELYMUS L. Ezo-MUGI ZOKU

Perennials with or without creeping rhizomes; inflorescence loosely to densely spicate; spikelets sessile, in pairs or in 3's alternate or opposite on a continuous rachis, slightly twisted, bringing the florets more or less dorsiventral on the rachis, 2- to 8-flowered, articulate above the glumes; glumes lanceolate to broadly linear, smaller than the lemmas; lemmas rather coriaceous, broadly lanceolate, attenuate-awned, slightly dorsally compressed, scabrous or pubescent, 5-nerved; palea as long as the lemma, 2-keeled; stamens 3; ovary with a hairy beak.——About 50 species, Eurasia and in N. and S. America.

1A. Plants with long, stout rhizomes; spikes thick, pubescent; spikelets awnless; anthers 5–8 mm. long. .... 1. *E. mollis*
1B. Plants tufted, without rhizomes; spikes glabrous; spikelets awned; anthers 1.5–2 mm. long.
  2A. Spikes erect; glumes slightly or scarcely shorter than the lemma. .... 2. *E. dahuricus*
  2B. Spikes nodding to pendulous; glumes much shorter than the lemma.
    3A. Glumes attenuate into a short awn; lemmas scabrous, with a slightly recurved awn. .... 3. *E. sibiricus*
    3B. Glumes abruptly acuminate, often dentate at apex, awnless; lemmas nearly smooth, with a rather stout, distinctly recurved awn.
                4. *E. yubaridakensis*

1. **Elymus mollis** Trin. *E. arenarius* var. *villosus* E. Mey.; *E. arenarius* sensu auct. Japon., non L.; *E. arenarius* var. *coreënsis* Hack.; *Leymus mollis* (Trin.) Pilger; *E. arenarius* var. *mollis* Trin.——TENKIGUSA, KUSADŌ, HAMA-NINNIKU. Robust perennial with long, stout rhizomes; culms 50–100 cm. long, firm, smooth, densely pubescent below the spike; leaf-blades firm, glaucous-green, flat, 20–40 cm. long, 7–12 mm. wide, smooth and glabrous beneath, scabrous on the elevated nerves on upper surface; spikes erect, dense, 10–25 cm. long, whitish green; spikelets 2(–5)-nate, erect, 1–2.5 cm. long, 3- to 5-flowered; glumes chartaceous, broadly lanceolate, acuminate, 3- to 7-nerved, as long as the florets or slightly exceeding them; lemmas awnless, 12–20 mm. long.——June–July. Gregarious on sandy beaches; Hokkaido, Honshu, Kyushu; common.——Korea, Sakhalin, Kuriles, Kamchatka to N. America.

2. **Elymus dahuricus** Turcz. *E. excelsus* Turcz.; *E. dahuricus* var. *cylindricus* Franch.; *E. dahuricus* var. *excelsus* (Turcz.) Roshev.; *E. cylindricus* (Franch.) Honda——HAMA-MUGI. Tufted perennial; culms 50–100 cm. long, glabrous; leaf-blades flat, 10–20 cm. long, 4–8 mm. wide, usually glabrous; spikes erect, obscurely 1-sided, dense; spikelets numerous, appressed, usually geminate, 1–1.5 cm. long, 3- to 4-flowered, green; glumes lanceolate, 8–11 mm. long, 3(–5)-nerved, scabrous on nerves, acuminate, with a short, terminal awn; lemmas 7–11 mm. long, lanceolate, scabrous, the awn 10–20 mm. long, straight or nearly so; palea as long as the lemma.——June–Aug. Grassy places, especially near the sea; Hokkaido, Honshu, Kyushu; common.——e. Siberia,

n. China, Korea, Manchuria, and Mongolia.

Var. **villosulus** (Ohwi) Ohwi. *E. villosulus* Ohwi; *E. osensis* Ohwi——YAMA-MUGI. More slender; nodes villous.——Mountain woods; Hokkaido, Honshu (centr. distr.); rare.——s. Kuriles.

3. **Elymus sibiricus** L. *E. yezoensis* Honda——EZO-MUGI. Perennial; culms 40–80 cm. long; leaf-blades 7–15 cm. long, 5–8 mm. wide, flat, scabrous; spikes pendulous, 8–15 cm. long, rather loosely spiculose; spikelets geminate or solitary at both ends of the spike, sometimes slightly purplish, 3- to 4-flowered, 12–14 mm. long; glumes linear-lanceolate, scabrous, 3–5 mm. long, 3-nerved, attenuately short-awned; lemmas 8–11 mm. long, 5-nerved, scabrous, the awn rather slender, slightly recurved, 1–2 cm. long; palea as long as the lemma.——July–Aug. Grassy places and woods; Hokkaido, Honshu (Shinano Prov.); rare.——e. Europe, Siberia, China, Manchuria, and Kamchatka.

4. **Elymus yubaridakensis** (Honda) Ohwi. *Clinelymus yubaridakensis* Honda——TAKANE-EZO-MUGI. Tufted perennial; culms 40–80 cm. long; leaf-blades 7–15 cm. long, 5–8 mm. wide, flat, scabrous; spikes pendulous, 8–12 cm. long, on slender peduncles; spikelets in pairs or, near both ends, solitary, 12–15 mm. long; glumes linear-lanceolate or linear-oblanceolate, 7–8 mm. long, 3-nerved, scabrous, acute, often with a tooth at apex; lemmas broadly lanceolate, about 10 mm. long, the awn prominently recurved, rather thick, 2.5–3 cm. long.——July. Alpine regions; Hokkaido (Mount Yubari); very rare.

32. ASPERELLA Humb.　Azumagaya Zoku

Perennials with flat, rarely involute leaf-blades; spikes slender, loosely flowered; spikelets geminate, rarely solitary, spreading to ascending, 2- to 4- or sometimes 1-flowered; glumes minute or wanting; lemmas broadly lanceolate, 5-nerved, hispid-scabrous, coriaceous, green, attenuate into long awns; rachilla scabrous, fragile; stamens 3; ovary with a hairy beak; caryopsis adherent to the lemma and palea.——Few species in temperate regions of the N. Hemisphere, exclusive of Europe.

1A.　Upper part of the culm and axis short-pubescent; spikelets usually geminate, 1- or 2-flowered; anthers about 3 mm. long
　　　1. *A. longearistata*
1B.　Culms and spikes glabrous; spikelets solitary, 1-flowered; anthers 4 mm. long. . . . . 2. *A. japonica*

1.　Asperella longearistata (Hack.) Ohwi.　*A. sibirica* var. *longearistata* Hack.; *Hystrix longearistata* (Hack.) Honda ——Azumagaya.　Rhizomes short; culms 70–100 cm. long, slender, short-pubescent below the spike and on nodes; leaf-blades flat, thin, broadly linear to lanceolate, 10–25 cm. long, 1–2 cm. wide, green; spikes exserted, pendulous, 10–15 cm. long, slender, the axis densely pubescent; spikelets sessile, 10–12 mm. long, 1- or 2-flowered, green; florets readily deciduous; glumes subulate, 4–6 mm. long, sometimes wanting; lemmas about 10 mm. long, sparsely hispid toward the apex, the awn straight, 15–20 mm. long; rachilla slender; palea as long as the lemma, lanceolate.——June–July.　Mountain woods; Hokkaido, Honshu, Shikoku, Kyushu; rather rare.——Korea.

2.　Asperella japonica Hack.　*Hystrix japonica* (Hack.) Ohwi; *H. hackelii* Honda——Iwa-take-sō.　Rhizomes slender; culms in loose tufts, slender, 60–80 cm. long, smooth; leaf-blades deep green, thin, flat, broadly linear, 10–20 cm. long, 8–15 mm. wide; spikes slender, pendulous or prominently nodding, about 10 cm. long, the axis glabrous; spikelets solitary, sessile, 1-flowered, slightly ascending; florets readily deciduous; glumes wholly suppressed or short-subulate, to 4 mm. long; lemmas about 10 mm. long, lanceolate, long-hispid near the margin, the awn erect, slender, 15–25 mm. long; rachilla slender; palea as long as the lemma.——June–July.　Woods; Honshu (Shinano and westw.), Shikoku, Kyushu; rare.

33. HORDEUM L.　Ō-mugi Zoku

Tufted annuals or perennials; leaf-blades flat; spikelets 1-flowered, 3 together, alternate on opposite sides at each node; middle spikelet sessile, the glumes subulate, distorted at base, standing at the sides of the 1 fertile floret; rachilla disarticulating above the glumes and, in the central spikelet, prolonged behind the palea as a bristle; palea as long as the body of the lemma, adhering to the caryopsis; stamens 3; lateral spikelets pedicelled, reduced, the florets sterile; caryopsis adherent to the lemma and palea. ——More than 20 species, chiefly in temperate regions.

1A.　Glumes of the fertile florets not ciliate, about 12 mm. long; lemma about 5 mm. long. . . . . 1. *H. hystrix*
1B.　Glumes of the fertile florets ciliate below, about 2.5 cm. long; lemma 8–10 mm. long. . . . . 2. *H. murinum*

1.　Hordeum hystrix Roth.　*H. gussoneanum* Parl.—— Hime-mugi-kusa.　Annual; culms 20–40 cm. long; blades linear, usually pilose below; sheaths often pilose; spike 2–4 cm. long, erect, pale green; glumes of fertile spikelets scabrous, narrowly subulate; lemma broadly lanceolate, 5–6 mm. long, the awn nearly erect, scabrous, about 10 mm. long.—— Naturalized in waste places; Honshu.——Europe and Asia Minor.

2.　Hordeum murinum L.　Mugi-kusa.　Annual; culms 10–50 cm. long, glabrous; leaf-blades flat, smooth, glabrous; spike 4–7 cm. long, pale green; glumes of fertile florets linear-lanceolate, 3-nerved, 2.5–3 cm. long, 0.3–1 mm. wide, ciliate below the middle.——May–July.　Naturalized in waste places and cultivated fields; Honshu, Shikoku, Kyushu.——Europe.

34. BROMUS L.　Suzume-no-cha-hiki Zoku

Perennials with short rhizomes or annuals, with linear leaf-blades and closed, cylindric sheaths; panicles loose to compact; spikelets 3- to many-flowered, usually rather large, slightly flattened laterally, articulate above the glumes and between the florets; glumes 1- to 9-nerved, membranous, sharply to obtusely keeled; lemmas green to pale green, 5- to 7-nerved, rounded or sometimes keeled, 2-toothed, membranous or chartaceous, awned between the minute teeth or awnless; palea 2-keeled; stamens 3; ovary hairy at apex, the styles usually lateral; caryopsis adherent to the palea.——Many species, chiefly in temperate and cooler regions, also in high mountains of the Tropics.

1A.　Lower glume 1-nerved, the second 3-nerved; lemmas usually narrow.
　2A.　Perennials.
　　3A.　Lemmas broadly lanceolate, nearly awnless or the awn less than half as long as the lemma; spikelets pale green or whitish green, sometimes brownish purple.
　　　4A.　Lemmas long-ciliate, the awn 2–4 mm. long; anthers 1–1.2 mm. long; panicles nodding, loose, the branches rather long.
　　　1. *B. yezoensis*
　　　4B.　Lemmas glabrous, the awn to 1 mm. long; anthers 4–5 mm. long; panicles erect, rather dense, the branches rather short.
　　　2. *B. inermis*
　　3B.　Lemmas narrowly lanceolate, the awn half as long to as long as the lemma; spikelets deep green. . . . . 3. *B. pauciflorus*
　2B.　Annuals; naturalized species.
　　5A.　Upper glume not more than 10 mm. long; lemmas 10–12 mm. long, with a straight awn 12–14 mm. long. . . . . 4. *B. tectorum*
　　5B.　Upper glume more than 10 mm. long; lemmas 17–23 mm. long, with an awn 2–5 cm. long.

6A. Lower glume about 8 mm. long; lemmas 17–20 mm. long, with an awn 2–3 cm. long. . . . . 5. *B. sterilis*
6B. Lower glume about 15 mm. long; lemmas 25–30 mm. long, with an awn 3.5–5 cm. long. . . . . 6. *B. rigidus*
1B. Lower glume 3- to 5-nerved, the second 7- to 9-nerved; annuals with rather broad lemmas.
  7A. Spikelets cylindric while young, slightly flattened before anthesis; lemmas rounded on back, not keeled.
    8A. Lemmas narrowly oblong, scabrous, with a distinct, sometimes spreading awn. . . . . 7. *B. japonicus*
    8B. Lemmas broadly oblong, the awn short, erect, or wanting.
      9A. Panicle open, the branches spreading; lemmas chartaceous, glabrous. . . . . 8. *B. secalinus*
      9B. Panicle contracted, usually dense, the branches erect; lemmas membranous, densely pubescent. . . . . 9. *B. mollis*
  7B. Spikelets prominently flattened throughout; lemmas sharply keeled on back. . . . . 10. *B. catharticus*

1. Bromus yezoensis Ohwi. *B. ciliatus* sensu auct. Japon., non L.——KUSHIRO-CHA-HIKI. Sparingly pilose perennial with short rhizomes; culms 70–120 cm. long, glabrous, nodes about 6, pilose; leaf-blades membranous, linear, 20–30 cm. long, 6–10 mm. wide, flat; ligules about 1 mm. long, truncate; panicles nodding, broadly ovate, 15–25 cm. long, the branches 2- to 3-nate; spikelets pale green, tinged with yellowish brown, 6- to 7-flowered, 15–20 mm. long; glumes narrow, the lower 6–7 mm., upper 8–9 mm. long; lemmas 10–12 mm. long, appressed-pubescent especially near margin, obtuse and bidentulate, with a short awn between the teeth; palea 8–9 mm. long; anthers elliptic, yellow, 1–1.2 mm. long.——June–July. Grassy places in lowlands; Hokkaido.——s. Kuriles and s. Sakhalin.

2. Bromus inermis Leyss. *Zerna inermis* (Leyss.) Lindm.; *B. glabrescens* Honda; *B. tatewakii* Honda——KO-SUZUME-NO-CHA-HIKI. Perennial with creeping rhizomes; culms 40–60 cm. long, glabrous; leaf-blades 10–15 cm. long, 3–8 mm. wide, nearly flat, somewhat firm; ligules less than 2 mm. long, truncate; panicles erect, 10–15 cm. long, rather dense, the branches subverticillate, short; spikelets narrowly oblong, 12–25 mm. long, pale green, sometimes slightly purple-tinged, 6- to 8-flowered; glumes acute, the lower 4–5 mm., upper 6–7 mm. long; lemmas narrowly oblong, about 10 mm. long, membranous, obtuse, with 2 minute teeth, the awn wanting or to 1 mm. long between the teeth; anthers linear, orange-yellow, 4–5 mm. long.——Hokkaido, Kyushu; very rare; possibly not indigenous.——Europe, Siberia, Mongolia, Manchuria, and N. America.

3. Bromus pauciflorus (Thunb.) Hack. *Festuca pauciflora* Thunb.; *F. remotiflora* Steud.; *Bromus remotiflorus* (Steud.) Ohwi——KITSUNEGAYA. Rhizomes short, culms 60–100 cm. long, retrorsely short-pubescent; leaf-blades deep green, nearly flat, 25–40 cm. long, 4–7 mm. broad; sheaths pilose; ligules semirounded, 1–2 mm. long; panicles nodding, deep green, 20–30 cm. long, very loose, the branches geminate, elongate, scabrous; spikelets loosely 6- to 10-flowered, 3–4 cm. long; glumes linear-lanceolate, acuminate, the lower 5–7 mm., the upper 8–12 mm. long; lemmas narrowly lanceolate, 12–15 mm. long, the awn erect, as long as to half as long as the lemma; anthers 1.5–2 mm. long.——June–July. Woods; Hokkaido, Honshu, Shikoku, Kyushu; very common.——s. Korea and China.

4. Bromus tectorum L. *Schedonorus tectorum* (L.) Fries; *Anisantha tectorum* (L.) Nevski——UMA-NO-CHA-HIKI. Annual; culms 15–40 cm. long, soft-pubescent above; leaf-blades pubescent, thin, 3–5 mm. wide; sheaths pubescent; panicles 10–15 cm. long, somewhat nodding, the branches subverticillate; spikelets nodding, broadly lanceolate, 12–20 mm. long, 5- to 8-flowered, pale green; glumes linear-lanceolate, sparingly pubescent, gradually acuminate, the lower 4–6 mm., the upper 8–10 mm. long; lemmas narrowly lanceolate, pubescent, with 2 lanceolate teeth 2–3 mm. long, the awn erect; palea about ¾ as long as the lemma; anthers about 0.7 mm. long, orange-yellow, elliptic.——Naturalized in Honshu.——Introduction from Europe.

5. Bromus sterilis L. *Zerna sterilis* (L.) Panzer; *Anisantha sterilis* (L.) Nevski——ARECHI-NO-CHA-HIKI. Allied to, but larger than the preceding; culms 30–70 cm. long; leaf-blades and sheaths pubescent; panicles 10–20 cm. long, loose; spikelets larger, 2.5–3.5 mm. long, 5- to 10-flowered; glumes nearly linear; lemmas linear-lanceolate, scabrous, gradually attenuate above into 2 teeth about 2 mm. long, long-awned; anthers about 0.7 mm. long.——Naturalized in Honshu.——Europe and w. Asia.

6. Bromus rigidus Roth. *B. villosus* Forsk., non Scop. ——HIGE-NAGA-SUZUME-NO-CHA-HIKI, Ō-KITSUNEGAYA. Pubescent annual; culms 30–60 cm. long; leaf-blades 4–5 mm. wide; panicles 10–20 cm. long, nodding or nearly erect, branches nearly simple, with 1 or 2 spikelets, each 3–4 cm. long, 6- to 8-flowered; glumes linear-lanceolate, the lower 15–20 mm. long; lemmas scabrous, with 2 teeth about 3 or 4 mm. long, the awn rising from between the teeth; anthers 0.7–1 mm. long.——Naturalized in Honshu and Shikoku.——Europe.

7. Bromus japonicus Thunb. *B. vestitus* Schrad.; *B. patulus* Mert. & Koch——SUZUME-NO-CHA-HIKI. Densely pubescent annual; culms 30–70 cm. long; leaf-blades flat, 15–30 cm. long, 3–6 mm. wide; ligules subrounded, 1–2.5 mm. long; panicles 10–25 cm. long, broadly ovate, somewhat nodding to secund, branches 4- to 6-nate; spikelets rather dense, oblong, 15–25 mm. long, 6–8 mm. wide, rather densely 6- to 10-flowered, yellowish green; glumes scabrous, the lower broadly lanceolate, 5–7 mm. long, acute, 3-nerved, the upper 7- to 9-nerved, narrowly oblong, 7–8 mm. long, acute; lemmas thinly chartaceous, narrowly ovate-oblong, 9–11 mm. long, obtuse, slightly inflated above the middle on margin, the awn inserted below the apex, 2–3 mm. long, straight in the lower floret, to 12 mm. long, recurved in the upper florets; palea 7–8 mm. long; anthers 1 mm. long.——May–July. Waste places and cultivated fields; Hokkaido, Honshu, Shikoku, Kyushu; very common.——Temperate regions of the N. Hemisphere.

8. Bromus secalinus L. KARASU-NO-CHA-HIKI. Glabrous annual; culms erect, 30–60 cm. long; leaf-blades 3–6 mm. wide; panicles 7–15 cm. long, nodding, the branches subverticillate; spikelets narrowly oblong, 15–20 mm. long, slightly inflated, 6–8 mm. wide, 5- to 15-flowered; lower glume lanceolate, 3- to 5-nerved, upper glume oblong, 6–7 mm. long, obtuse, 7-nerved; lemmas elliptic, 7-nerved, obtuse, chartaceous, lead-green, nearly smooth, glabrous, 2-toothed at apex, the awn somewhat flexuous, shorter than the lemma, those of the lower florets often absent; anthers about 2 mm. long.——Naturalized in Hokkaido and Honshu (n. and centr. distr.).——Europe and Siberia.

9. Bromus mollis L. *B. hordeaceus* sensu auct. Japon., non L.——HAMA-CHA-HIKI. Annual; culms 10–60 cm. long; leaf-blades linear, flat, 3–5 mm. wide, short-pubescent; panicles erect, 5–10 cm. long, the branches short, erect, short-pubescent; spikelets oblong, densely 6- to 10-flowered, 1–1.5 cm. long; glumes obtuse, puberulent, the lower 3- to 5-nerved, broadly lanceolate, 5–6 mm. long, the upper 5- to 7-nerved, oblong, 6–8 mm. long; lemmas membranous, elliptic, 7–8 mm. long, obtuse, 7-nerved, short-pubescent, the awn 6–9 mm. long, between the teeth; palea shorter than the lemma; anthers about 2 mm. long.——Naturalized in Hokkaido and Honshu.——Europe and Siberia.

10. Bromus catharticus Vahl. *Festuca unioloides* Willd.; *Ceratochloa unioloides* (Willd.) Beauv.; *B. unioloides* (Willd.) HBK.——INU-MUGI.   Annual; culms tufted, rather firm, stout, 40–100 cm. long, glabrous; leaf-blades 20–30 cm. long, 4–10 mm. wide, sparingly pilose, nearly flat, the sheaths whitish pilose; ligules whitish, 3–5 mm. long, ovate; panicles 10–25 cm. long, ovate, rather firm, the branches geminate, scabrous, spreading; spikelets compressed, yellowish green, narrowly ovate, 3- to 6-flowered, lustrous, 2–3 cm. long; glumes broadly lanceolate, acuminate, keeled, the lower 3- to 5-nerved, 10–12 mm. long, the upper 7- to 9-nerved, 12–15 mm. long; lemmas 14–18 mm. long, broadly lanceolate, chartaceous, folded and strongly keeled, very acute, terminated by an awn about 1 mm. long; palea 8–9 mm. long; anthers usually not exserted, 0.5 mm. long, oblong.——June–July. Naturalized in Honshu, Shikoku, Kyushu; very common.—— S. America.

35. BRACHYELYTRUM Beauv.   Kōyazasa Zoku

Slender perennials with flat leaf-blades and narrow panicles; spikelets narrow, 1-flowered, articulate above the glumes and between the florets; glumes minute, unequal, green; lemma narrowly lanceolate, slightly flattened dorsally, green, minutely scabrous, 5-nerved, entire at apex, long-awned; callus oblique, short; rachilla produced behind the upper floret as a slender bristle, glabrous; palea as long as the lemma, 2-keeled; stamens 3; ovary with a hairy beak; styles terminal; caryopsis adherent to the lemma and palea.——Two species, one in Japan, the other in N. America.

1. Brachyelytrum japonicum Hack. *B. erectum* var. *japonicum* Hack.——KŌYAZASA.   Rhizomes short, slender, branched; culms very slender, 50–70 cm. long, glabrous; leaf-blades green, flat, thin, broadly linear, 7–12 cm. long, 4–7 mm. wide, sparingly pilose above and on margin; ligules 2–3 mm. long, truncate; panicles 10–15 cm. long, green, branches short, erect or appressed to the axis; spikelets several, short-pedicelled, 1-flowered, 8–9 mm. long; glumes linear-lanceolate, 1-nerved, acute, the lower 1–1.5 mm., the upper 2–3 mm. long; lemma narrowly lanceolate, minutely scabrous, gradually narrowed to the awned apex, the awn straight, delicate, 10–15 mm. long; callus minutely pilose on both sides; rachilla prolonged as a slender bristle about 4 mm. long, appressed to the palea.—— July–Aug. Woods in mountains; Honshu (centr. distr. and westw.), Shikoku, Kyushu.——Korea (Quelpaert Isl.).

36. AULACOLEPIS Hack.   Hiroha-no-konukagusa Zoku

Slender or low perennials with flat or involute leaf-blades; spikelets in panicles, 1-flowered, flattened, articulate below the floret; glumes unequal, the lower minute or smaller than the upper, the upper shorter than the lemma, keeled, acute, lanceolate; lemma green, dull, rather strongly keeled, broadly lanceolate, acute, thinly chartaceous to membranous, glabrous, awnless; callus short, short-hairy; rachilla prolonged beyond the floret as a slender bristle, glabrous; palea as long as the lemma, with 2 closely parallel nerves; stamens 3; ovary glabrous; caryopsis lanceolate, free within the lemma and palea.——Several species, in e. Asia and Malaysia, extending to the Himalayas.

1. Aulacolepis treutleri (Kuntze) Hack. var. **japonica** (Hack.) Ohwi. *A. japonica* Hack.——HIROHA-NO-KONUKA-GUSA.   Slender, glabrous perennial; culms 80–120 cm. long; leaf-blades flat, thin, green, scabrous, 20–30 cm. long, 10–22 mm. wide; ligules hyaline, 2–5 mm. long; panicles ovate, erect, green, 20–30 cm. long, 10–15 cm. wide, very loose and open, the branches subverticillate, spreading, scabrous; spikelets ap-pressed to the branchlets, about 3.5 mm. long, flattened; glumes broadly lanceolate, keeled, 1-nerved, the lower 1–1.5 mm., the upper 2–2.5 mm. long; lemmas about 3.5 mm. long, acute, green, 5-nerved, intermediate nerves weak; anthers about 1.2 mm. long.——June–July. Mountain woods; Honshu (centr. distr. and westw.); rare. The typical phase of the species occurs in India to s. China and Formosa.

37. DACTYLIS L.   Kamogaya Zoku

Coarse perennials with linear leaf-blades and interrupted panicles; spikelets densely arranged on one side toward the ends of the panicle-branches, 3- or 4-flowered, flattened; glumes unequal, keeled, acute, 1- to 3-nerved, short-ciliate on the keel; lemmas longer than the glumes, keeled, folded, rather chartaceous, green, 5-nerved, setulose-scabrous, mucronate; palea 2-keeled; callus glabrous; rachilla short, glabrous; stamens 3; ovary glabrous.——Few species, in Eurasia and N. Africa.

1. Dactylis glomerata L. KAMOGAYA.   Tufted, stout perennial; culms erect, in large tussocks, 80–120 cm. long, smooth, glabrous; leaf-blades flat, scabrous, green or glaucous-green, 30–60 cm. long, 5–10 mm. wide; ligules deltoid, 7–12 mm. long, hyaline; panicles erect, lobed, 8–20 cm. long, the branches distant, solitary, stiff, spreading to ascending, sca-brous, the spikelets in dense 1-sided fascicles borne at the ends of the branches; spikelets 7–8 mm. long, glaucous-green, 2- to 4-flowered; glumes lanceolate, acuminate, scabrous, the lower 1-nerved, 3–4 mm. long, the upper 3-nerved, 5–6 mm. long; lemmas 6–7 mm. long; anthers 2–3 mm. long.——July–Aug. Naturalized in pastures and waste places; Hokkaido, Honshu, Shikoku, Kyushu; common.——Eurasia and w. Asia.

38. FESTUCA L. Ushi-no-kegusa Zoku

Perennials or annuals of various habits; leaf-blades linear, flat or convolute; spikelets in panicles, 2- to many-flowered, more or less laterally flattened, articulate above the glumes; glumes unequal, shorter than the lemmas; lemmas 5-nerved, rounded or slightly keeled on the back, glabrous or pubescent, awnless or with a straight apical awn; callus not hairy; palea 2-keeled; stamens 1–3; ovary glabrous or puberulent at apex; styles terminal; caryopsis often adherent to the lemma and palea.—— Cosmopolitan genus with many species especially in cooler regions, and in high mountains of the Tropics.

1A. Perennials with short or elongate rhizomes; stamens 3, protruding.
  2A. Leaf-blades less than 5 mm. wide, awned, if wider the lemmas awnless.
    3A. Anthers 0.5–2.5 mm. long; lemmas usually awned, rounded on back.
      4A. Lower glume more than ⅓ as long as the lowest lemma, lanceolate, acute.
        5A. Innovation shoots intravaginal; stolons wanting. .... 1. *F. ovina*
        5B. Innovation shoots partly extravaginal; stolons often present. .... 2. *F. rubra*
      4B. Lower glume shorter, ovate, obtuse.
        6A. Lowest lemma 4.5–6 mm. long, commonly awned; anthers elliptic, 0.5–0.8 mm. long. .... 3. *F. parvigluma*
        6B. Lowest lemma 3–4 mm. long, always awnless; anthers linear, 1.5 mm. long. .... 4. *F. japonica*
    3B. Anthers 3–4.5 mm. long; lemmas awnless, more or less keeled on back.
      7A. Culms 15–30 cm. long; alpine regions. .... 5. *F. takedana*
      7B. Culms 40–150 cm. long; naturalized introductions.
        8A. Glumes rather obtuse, the lower 2–3 mm., the upper 3.5–4.5 mm. long; anthers 3 mm. long. .... 6. *F. elatior*
        8B. Glumes acuminate, the lower 5–7 mm., the upper 6–7 mm. long; anthers 4–4.5 mm. long. .... 7. *F. arundinacea*
  2B. Leaf-blades 6–12 mm. wide, flat; lemmas long-awned, keeled on back; anthers about 1 mm. long. .... 8. *F. extremiorientalis*
1B. Annuals, without rhizomes; stamens not protruding.
  9A. Spikelets 3- to 5-flowered, pale green.
    10A. Lemmas not ciliate. .... 9. *F. myuros*
    10B. Lemmas long-ciliate toward tip. .... 10. *F. megalura*
  9B. Spikelets 5- to 13-flowered, often purplish. .... 11. *F. octoflora*

1. Festuca ovina L. var. **ovina**. *F. ovina* var. *vulgaris* Koch——Ushi-no-kegusa, Shin-ushi-no-kegusa. Low densely tufted perennial without stolons; culms 20–40 cm. long, slender but firm, glabrous or scabrous below the panicle; radical leaves elongate, 5–20 cm. long, rather firm, involute, glabrous, green or glaucous-green, the blades 0.4–0.6 mm. wide, weakly 3- to 7-nerved, scarcely sulcate; culm-leaves short; ligules less than 0.5 mm. long; panicles erect, narrow, 5–8 cm. long, lanceolate to narrowly ovate, the branches solitary or geminate; spikelets 5–7 mm. long, green to glaucous-green, sometimes purplish, 3- to 6-flowered; glumes lanceolate, acute, the upper longer, 3–3.5 mm. long, 3-nerved with short lateral nerves; lemmas narrowly oblong, rather coriaceous to chartaceous, 3.5–4.5 mm. long, scabrous, rounded on back, with a straight apical awn to 2 mm. long, or awnless; anthers 1.5–2.5 mm. long; ovary glabrous.——June–Aug. Mountains; Hokkaido, Honshu (centr. distr. and northw.); rather rare. The species is widespread in Eurasia and N. America.

Var. **duriuscula** (L.) Koch. *F. duriuscula* L.——Kōrai-ushi-no-kegusa. Blades firm, 0.7–1 mm. wide, otherwise simulating the typical phase.——Shikoku; rare.——Eurasia.

Var. **coreana** St. Yves. *F. ovina* var. *pubiculmis* Ohwi, non Asch. & Graebn., 1900; *F. ovina* var. *nipponica* Ohwi—— Ao-ushi-no-kegusa. Culms 20–40 cm. long, puberulent or pubescent below the panicle; leaf-blades less firm, sometimes puberulent; panicles green to glaucous-green, 4–8 cm. long. ——Low mountains; Honshu, Shikoku, Kyushu; common. ——Korea.

Var. **tateyamensis** Ohwi. Takane-ushi-no-kegusa. Rather soft, smooth and glabrous alpine phase; culms 10–20 cm. long; leaf-blades filiform, sulcate; panicles 2–4 cm. long, purplish; lemmas 3–4 mm. long; anthers 1.5 mm. long.—— Alpine regions; Honshu (centr. distr.); rare.

Var. **chiisanensis** Ohwi. Chiisan-ushi-no-kegusa. Allied to var. *coreana*, but the leaves more soft, sulcate.—— Mountains; Shikoku; rare.

2. Festuca rubra L. var. **rubra**. *F. rubra* var. *baicalensis* Griseb.; *F. rubra* var. *muramatsui* Ohwi; *F. rubra* var. *genuina* Hack.——Ō-ushi-no-kegusa. Tufted perennial with short rhizomes, sometimes also stoloniferous; culms slender, 15–50 cm. long; radical leaves elongate, the blades loosely involute, 10–20 cm. long, 1–2 mm. wide, glabrous or nearly so; sheaths often partly reddish; ligules very short; panicles 5–12 cm. long; spikelets 5–10 mm. long, 3- to 7-flowered; glumes broadly lanceolate, the upper larger, 3–6 mm. long; lemmas narrowly ovate, 4–7 mm. long, glabrous or sparingly pubescent, scabrous, the awn to 3 mm. long; anthers 2–2.5 mm. long.—— June–Aug. Common in mountains, sometimes on rocks near the seashore; Hokkaido, Honshu (centr. distr. and northw.). ——Sakhalin, Kuriles, and Korea, to the temperate and boreal regions of Europe and N. America.

Var. **pacifica** Honda. *F. rubra* var. *planifolia* sensu auct. Japon., non Trautv.——Hiroha-no-ō-ushi-no-kegusa. Leaf-blades loosely involute, about 2.5 mm. wide; panicles pale green, otherwise plants simulating the preceding var.—— Rocks near the seashore; Honshu (centr. distr. and westw.), Shikoku, Kyushu.

Var. **hondoensis** Ohwi. *F. hondoensis* (Ohwi) Ohwi—— Yama-ō-ushi-no-kegusa. Innovation shoots mostly intravaginal; lemmas violascent, awnless, otherwise plants simulating var. *rubra*.——Mountains; Honshu (centr. distr.); rare.

Var. **musashiensis** (Honda) Ohwi. *F. musashiensis* Honda ——Asakawa-sō. Slender, glabrous, green; leaves elongate; spikelets green; glumes and lemmas ciliate.——Honshu (Asakawa near Tokyo); rare.

3. Festuca parvigluma Steud. Toboshigara. Green, slender perennial with elongate slender rhizomes and stolons; culms in loose tufts, 30–60 cm. long, slender, with brownish hyaline sheaths at base; leaf-blades deep green, elongate, loosely involute, 1.5–3 mm. wide; ligules 0.2–0.3 mm. long, truncate; panicles nodding, very loose, 8–15 cm. long, the branches elongate, spreading, solitary, scabrous; spikelets pale green, somewhat lustrous, 7–10 mm. long, 3- to 5-flowered; glumes ovate, obtuse, membranous, 1–1.5 mm. long; lemmas broadly lanceolate, acute, 5–7 mm. long; awn terminal, very slender, 5–7 mm. long, erect, slightly flexuous; anthers elliptic. ——May–June. Woods in low mountains; Hokkaido, Honshu, Shikoku, Kyushu; very common.——Formosa and China.

Var. **breviaristata** Ohwi. IBUKI-TOBOSHIGARA. Lemmas awnless or nearly so.——Honshu (Mount Ibuki in Omi and Mount Omine in Yamato) and Shikoku (Mount Kamegamori); rare.

4. **Festuca japonica** Makino. *F. fauriei* Hack.——YAMA-TOBOSHIGARA. Closely allied to the preceding species; culms 30–60 cm. long, tufted; panicles very loose, branches slender, geminate; spikelets 3- to 4-flowered, 4–6 mm. long; lemmas 3–4 mm. long, rather acute, awnless; anthers about 1.5 mm. long. ——May–July. Mountains; Honshu; rare.——Korea and Formosa.

5. **Festuca takedana** Ohwi. (?) *Poa nuda* Hack.; *Leiopoa nuda* (Hack.) Ohwi——TAKANE-SOMOSOMO. Glabrous low perennial with slender rhizomes and short creeping stolons; culms few together, slender, 15–30 cm. long, smooth, glabrous; leaf-blades linear, of the radical somewhat elongate, of the culm-leaves 5–10 cm. long, 3–4 mm. wide, nearly smooth, loosely involute, glaucous above; ligules truncate, less than 1 mm. long; panicles nodding, 4–7 cm. long, ovate, loose, the branches geminate, smooth; spikelets 7–8 mm. long, dull, pale green with a brown and purple hue, 3- or 4-flowered; glumes nearly equal, 3-nerved, lanceolate, acute, scaberulous, keeled, 3–4 mm. long; lemmas longer than the glumes, slightly keeled, narrowly ovate, 6–7 mm. long, rather herbaceous, glabrous, acute to very acute, 3(–5)-nerved; palea as long as the lemma; callus and rachilla glabrous; anthers (2.5–) 3–3.5 mm. long, linear; ovary puberulent at apex.——July–Aug. Alpine regions; Honshu (centr. distr.); very rare.

6. **Festuca elatior** L. *F. pratensis* Huds.; *F. elatior* subsp. *pratensis* (Huds.) Hack.——HIROHA-NO-USHI-NO-KEGUSA. Rhizomes short, plant rarely stoloniferous; culms 30–100 cm. long; leaf-blades narrowly linear, elongate, rather flat, 3–7 mm. wide; ligules very short; panicles narrowly ovate, 10–20 cm. long, the branches ascending, geminate, scabrous; spikelets pale green, 8–12 mm. long, 5- to 8-flowered; glumes lanceolate, rather acute to obtuse, the lower 2–3 mm., the upper 3.5–4.5 mm. long; lemmas chartaceous, narrowly ovate, 5–7 mm. long, nearly acute, glabrous, obscurely 5-nerved, nearly as long as the palea; anthers about 3 mm. long; ovary glabrous.——June–Aug. Naturalized; Hokkaido (common), Honshu, Shikoku, Kyushu.——Europe and Siberia.

7. **Festuca arundinacea** Schreb. *F. elatior* var. *arundinacea* (Schreb.) Wimm.; *F. elatior* subsp. *arundinacea* (Schreb.) Hack.——ONI-USHI-NO-KEGUSA. Closely allied to the preceding but stouter; culms 50–150 cm. long; leaf-blades flat or nearly so, 5–7 mm. wide; panicles 10–20 cm. long, rather loose, open, the branches geminate, with numerous spikelets; spikelets 15–18 mm. long, sometimes slightly purple-tinged; glumes lanceolate, acuminate to very acute, the lower 5–7 mm., the upper 6–7 mm. long; lemmas 8–9 mm. long;

anthers 4–4.5 mm. long.——June–Aug. Naturalized; Honshu. ——Europe.

8. **Festuca extremiorientalis** Ohwi. *F. gigantea* sensu auct. Japon., non Vill.; *F. subulata* var. *japonica* Hack.; *F. iwamotoi* Honda——Ō-TOBOSHIGARA, TŌ-TOBOSHIGARA, ONI-TOBOSHIGARA. Perennial with short rhizomes and ascending innovation shoots; culms 80–120 cm. long, glabrous, smooth; leaf-blades flat, green, 20–30 cm. long, 5–12 mm. wide, scabrous, sometimes pilose above; sheaths often retrorsely scabrous; ligules 2–3 mm. long, truncate, brownish; panicles loose, open, nodding, pale green and sometimes slightly purple-tinged, the branches 2- or 3-nate, scabrous; spikelets 5–7 mm. long, 4- or 5-flowered; glumes unequal, lanceolate, 1- to 3-nerved, the lower about 3 mm., the upper 4.5–5 mm. long; lemmas lanceolate, 5–6 mm. long, scaberulous; awn terminal or between 2 minute teeth, slender, flexuous, 4–7 mm. long; anthers 1 mm. long; caryopsis puberulent at apex.——June–Aug. Mountain woods; Hokkaido, Honshu (centr. distr. and northw.); rather common.——e. Siberia, n. China, and Korea.

9. **Festuca myuros** L. *Vulpia myuros* (L.) Gmel.——NAGINATAGAYA. Tufted annual; culms erect, slender, rather firm; leaf-blades glaucous-green, involute, filiform, 5–15 cm. long, 0.5–1 mm. wide; ligules nearly truncate, about 1 mm. long; panicles erect, narrow, very dense, the branches solitary, short, appressed to the axis; spikelets short-pedicelled, 6–8 mm. long, pale green, 3- to 5-flowered; lower glume broadly lanceolate, 1–2 mm. long, 1-nerved, the upper glume linear-lanceolate, 5–6 mm. long, acuminate, 3-nerved, keeled; lemmas narrowly lanceolate, acuminate, 5–6 mm. long, not keeled, prominently scabrous; awn erect, delicate, about 15 mm. long; stamens 1–3; anthers about 1.2 mm. long.——May–June. Naturalized in grassy places in lowlands and waysides; Honshu, Shikoku, Kyushu; common.——Europe, N. Africa, and w. Asia; naturalized in e. Asia and America.

10. **Festuca megalura** Nutt. *Vulpia megalura* (Nutt.) Rydb.——Ō-NAGINATAGAYA. Closely allied to the preceding; culms 30–60 cm. long; ligules 0.2–0.3 mm. long; panicles 15–30 cm. long, the branches solitary or geminate, appressed to the axis; spikelets 7–10 mm. long, 3- to 5-flowered; lower glume 0.5–1 mm., the upper 3 mm. long; lemmas with spreading long hairs on upper margins, the awn about 15 mm. long; anthers about 0.2 mm. long; caryopsis glabrous.——May–June. Naturalized in grassy places in lowlands; Honshu (Kinki Distr. and westw.), Shikoku, Kyushu.——N. America.

11. **Festuca octoflora** Walt. *Vulpia octoflora* (Walt.) Rydb.——MURASAKI-NAGINATAGAYA. Tufted annual; culms slender, erect, 15–30 cm. long; leaf-blades narrow, soft, involute or rather flat, 2–10 cm. long; ligules about 0.5 mm. long; panicles narrow, dense, erect, 5–10 cm. long, the branches solitary, ascending, scabrous; spikelets elliptic, 5–8 mm. long, 5- to 10-flowered; glumes linear-lanceolate, about 4 mm. long, the lower 1-nerved, the upper 3-nerved; lemmas lanceolate, scabrous, 4–5 mm. long, pale green and purple-tinged; awn slender, elongate.——June. Naturalized; Honshu (Harima Prov.); rare.——N. America.

39. **PUCCINELLIA** Parl.   CHISHIMA-DOJŌ-TSUNAGI ZOKU

Perennials or annuals with linear leaf-blades; spikelets in panicles, slightly flattened laterally, 3- to 10-flowered, articulate below the florets; florets bisexual except the uppermost; glumes shorter than the lemmas, 1- to 3-nerved; lemmas chartaceous, nearly smooth, 5-nerved, not keeled, obtuse to rather acute, awnless, with short, appressed hairs at base; callus short, glabrous;

rachilla glabrous; palea as long as the lemma, 2-keeled; stamens 3; ovary glabrous, the styles short; caryopsis rather plump, free within the lemma and palea.——About 100 species, in temperate and cooler regions of the N. Hemisphere, especially on alkaline soils and near seashores.

1A. Culms 30–100 cm. long; panicles pale green, 10–40 cm. long, the branches 3- to 6-nate, appressed to slightly ascending, scabrous; verticils spikelet-bearing from the base. .... 1. *P. nipponica*
1B. Culms 10–40 cm. long; panicles more or less purple-tinged, 5–15 cm. long, the branches solitary or geminate, rarely ternate, some of them spreading to deflexed, smooth or nearly so; verticils without spikelets near base. .... 2. *P. pumila*

1. **Puccinellia nipponica** Ohwi. *P. adpressa* Ohwi—— Tachi-dojō-tsunagi. Rhizomes almost wanting, innovation shoots short; culms tufted, 30–100 cm. long, smooth, glabrous; leaf-blades 10–20 cm. long, 2–3 mm. wide when flat, or when involute about 1 mm. wide, minutely papillose, glabrous, rather soft, glaucous-green; ligules hyaline, ovate, 2–3 mm. long, glabrous; panicles lanceolate, partly enclosed within the uppermost sheath, glaucous-green, branches erect or appressed to the axis, subverticillate, very unequal; spikelets 4–6 mm. long, 3- or 4-flowered; glumes lanceolate, very acute, slightly longer than the lowest lemma, the lower 1-nerved, 2–2.5 mm. long, the upper 3-nerved, about 3 mm. long; lemmas 3–3.5 mm. long, rather acute, sparingly short-pubescent at base; anthers narrowly oblong, about 0.7 mm. long.——June–Aug. Seashores; Honshu (Rikuzen Prov.); very rare.——Korea.

2. **Puccinellia pumila** (Vasey) Hitchc. *Glyceria pumila* Vasey; *P. kurilensis* (Takeda) Honda; *Atropis convoluta* sensu auct. Japon., non Griseb.; *Festuca thalassica* sensu auct. Japon., non Kunth; *P. distans* var. *convoluta* Honda excl. syn.; *Atropis kurilensis* Takeda——Chishima-dojō-tsunagi. Glabrous tufted perennial with short innovation-shoots; culms 10–40 cm. long, rather thick, smooth, ascending at base; leaf-blades soft and rather thick-membranous, glaucous-green, slightly involute, 5–10 cm. long, 2–3 mm. wide, glabrous; ligules semi-rounded, hyaline, smooth, 2–3 mm. long; panicles finally exserted, erect, pale green and sometimes slightly purple-tinged, 5–15 cm. long, the branches unequal, nearly smooth, the lower deflexed; spikelets 5–8 mm. long, 2–3 mm. wide, 5- to 7-flowered; glumes 1-nerved, narrowly oblong, rather acute, the lower 1.5–2 mm., the upper 2.5–3 mm. long; lemmas narrowly oblong, rather obtuse, about 3 mm. long, smooth, short-appressed-puberulent near base; anthers about 0.7 mm. long, oblong.——June–Oct. Seashores; Hokkaido, Honshu (near Tokyo), Kyushu (n. distr.).——Kuriles, Sakhalin, Ussuri, Kamchatka, Aleutians to N. America.

40. POA L. Ichigo-tsunagi Zoku

Perennials, rarely annuals, sometimes stoloniferous; spikelets in panicles, 2- to many-flowered, more or less laterally flattened, articulate below the florets; glumes keeled, 1- to 3-nerved, acute, shorter than the lemma; lemmas keeled, ovate, acute, membranous, awnless, 5-nerved, usually white-pubescent on keel and nerves below; callus short, often with a tuft of cobwebby hairs; palea 2-keeled; stamens 3; ovary glabrous; caryopsis free within the lemma and palea.——More than 200 species, in the temperate and cooler regions of the world, and on high mountains of the Tropics.

1A. Palea appressed-pubescent on the keel; annuals or perennials; anthers 0.5–1 mm. long and 1/7–1/3 as long as the lemma.
  2A. Lemmas 2–3.5 mm. long; panicle-branches with many spikelets.
    3A. Panicle-branches smooth.
      4A. Lemmas oblong-ovate, with slender or obscure intermediate nerves, soft-pubescent on margins below; anthers 0.7–1 mm. long.
                  1. *P. annua*
      4B. Lemmas narrowly oblong, with thickened, densely appressed-pubescent intermediate nerves, and glabrescent on margins; anthers 0.5–0.8 mm. long. .... 2. *P. crassinervis*
    3B. Panicle-branches scabrous.
      5A. Leaf-blades 1.5–5 mm. wide; culms slender, 30–80 cm. long; anthers 0.6–1 mm. long.
        6A. Panicle-branches spreading at anthesis; lemmas 2–3 mm. long, obtuse, green, appressed-pubescent on both sides at base; anthers 0.7–1 mm. long, 1/4–1/3 as long as the lemma. .... 3. *P. acroleuca*
        6B. Panicle-branches erect; lemmas 2.5–3.5 mm. long, acute, glabrous on both sides, yellowish-green or vivid green, anthers 0.4–0.7 mm. long, 1/7–1/5 as long as the lemma. ....4. *P. hisauchii*
      5B. Leaf-blades 4–7 mm. wide; culms less slender, 30–50 cm. long; anthers 0.5–0.8 mm. long and 1/5–1/4 as long as the lemma; panicle-branches thicker. .... 5. *P. nipponica*
  2B. Lemmas 3.5–4.5 mm. long; panicles loose and open, the branches spreading, with 1–3 spikelets toward the ends; culm-base with bulbously thickened internodes; anthers 0.7–1 mm. long and 1/6–1/4 as long as the lemma. .... 6. *P. tuberifera*
1B. Palea scabrous on the keel; perennials.
  7A. Anthers 1/8–1/3 as long as the lemma; callus scarcely cobwebby.
    8A. Leaf-sheaths terete; spikelets glaucescent; anthers 0.8–1.5 mm. long. .... 7. *P. hakusanensis*
    8B. Leaf-sheaths slightly compressed, sometimes retrorsely scaberulous on the keel; spikelets green, sometimes partly dull purplish; anthers 0.5–1 mm. long. .... 8. *P. radula*
  7B. Anthers 1/3–2/3 as long as the lemma.
    9A. Culms stout, 4–5 mm. across at base; panicles erect, contracted, with erect branches; lemmas 5–6 mm. long; anthers 2–2.5 mm. long. .... 9. *P. eminens*
    9B. Culms slender, less than 3 mm. across at base.
      10A. Culms terete, sometimes slightly compressed, but not bifacial.
        11A. Glumes and lemmas soft, rounded, erose-dentate near tip in upper glumes and lower lemmas; innovation-shoots much-elongate, first erect, later procumbent, rooting and bearing secondary shoots from the nodes; panicles 2–8 cm. long, nodding; anthers 2–3 mm. long. .... 10. *P. fauriei*

11B.   Glumes and lemmas entire, membranous, acute to obtuse; innovation-shoots not elongate.
   12A.   Lemmas not cobwebby at the base.
      13A.   Plant estoloniferous; culms erect, slender but not soft; anthers 1.2–2 mm. long; panicles erect. .... 11. *P. glauca*
      13B.   Plant short-stoloniferous; culms ascending at base, rather stout and soft; anthers 1.8–2.2 mm. long, panicles nodding.
                                                                   12. *P. hayachinensis*
   12B.   Lemmas cobwebby on the callus.
      14A.   Spikelets glaucescent or purplish; lemmas of the lower florets usually more than 4 mm. long.
         15A.   Glumes lanceolate, acuminate, the lateral nerves prominent, almost reaching the tip of the glumes; anthers
             1.5–2.2 mm. long. .... 13. *P. macrocalyx*
         15B.   Glumes broadly lanceolate, the lateral nerves about ⅔ the length of the lower glume and about ¾ the length of the
             upper glume.
            16A.   Culms 30–80 cm. long; panicles nodding; lemmas 4–5 mm., rarely 3.5 mm. long; anthers 1.5–2.2 mm. long.
                                                  14. *P. sachalinensis*
            16B.   Culms usually less than 30 cm. long, stouter; panicles erect or nearly so; lemmas 5–6 mm. long; anthers 1.6–2 mm.
             long. .... 15. *P. komarovii* var. *shinanoana*
      14B.   Spikelets green; lemmas usually 2–3.5 mm. long, sometimes variegated with purple.
         17A.   Ligules 3–6 mm. long, if less than 3 mm. long, acute at tip.
            18A.   Lemmas with prominent intermediate nerves, the marginal nerves glabrous or short-pubescent near base; panicles
             ovate, with spreading branches; anthers 1–1.5 mm. long. .... 16. *P. trivialis*
            18B.   Lemmas with weak or obsolete intermediate nerves, the marginal nerves short-pubescent at base.
               19A.   Culms smooth; ligules usually obtuse; panicles broad, effuse; lemmas with obsolete intermediate nerves.
                                                17. *P. palustris*
               19B.   Culms scabrous below the panicles and on upper half of the internodes; ligules acute; panicles usually narrow
               and contracted; lemmas with weak but distinct intermediate nerves. .... 18. *P. sphondylodes*
         17B.   Ligules 0.5–1.5 (–2) mm. long, obtuse to truncate.
            20A.   Culms densely tufted, erect, the basal sheaths sometimes tinged red-purple; plants estoloniferous.
               21A.   Ligules 1–2 mm. long; uppermost culm-leaves shorter than the sheath. .... 19. *P. viridula*
               21B.   Ligules scarcely developed; uppermost culm-leaves mostly longer than the sheath. .... 20. *P. nemoralis*
            20B.   Culms solitary or loosely tufted; plant stoloniferous; anthers 1.2–1.5 mm. long.
               22A.   Sheaths slightly compressed; panicles open, loose; spikelets prominently compressed, pale or yellowish green.
                                                21. *P. matsumurae*
               22B.   Sheaths scarcely compressed; panicles dense and contracted; spikelets moderately compressed, green.
                                              22. *P. pratensis*
10B.   Culms strongly compressed, 2-edged, erect from long creeping base; anthers 0.8–1.2 mm. long, about half as long as the
   lemma; panicles green, dense; lemmas 1.5–2.5 mm. long. .... 23. *P. compressa*

1. Poa annua L. SUZUME-NO-KATABIRA.   Soft annual, 10–30 cm. high, tufted, glabrous; leaf-blades smooth, flat, 4–10 cm. long, 1.5–3 mm. wide; ligules semirounded, 3–6 mm. long; panicles broadly ovate, usually 4–5 cm. long, the branches geminate, rather thick, smooth, spreading; spikelets ovate, pale green, 3–5 mm. long, 3- to 5-flowered; glumes 1- to 3-nerved, nearly smooth, the lower 1.5 mm. long, the upper narrowly ovate, 2–2.5 mm. long, acute; lemmas about 3 mm. long, sometimes white-pubescent on lower half, the nerves white-pubescent except near the tip, the intermediate nerves often glabrous, the web wanting or nearly so.——Mar.–Nov. Cultivated fields and along roadsides; Hokkaido, Honshu, Shikoku, Kyushu; very common. A cosmopolitan weed.

Var. **reptans** Hausskn. *P. supina* sensu auct. Japon., non Schrad.——TSURU-SUZUME-NO-HIE.   With elongate creeping innovation-shoots, bearing new ones from the nodes.——Rare in our area.——Europe.

2. Poa crassinervis Honda. TSUKUSHI-SUZUME-NO-KA-TABIRA.   Closely allied to the preceding; ligules 2–3 mm. long; panicles oblong, the branches in pairs, rather thick; spikelets oblong, 4–6 mm. long, 3- to 6-flowered, pale green; glumes narrowly oblong, the upper 3-nerved, 3 mm. long, the lower 2 mm. long; lemmas narrowly oblong, obtuse, with glabrous sides, the intermediate nerves strong, white-villous below the middle, the web obscure.——Apr.–June. Honshu (Izumo Prov.), Shikoku, Kyushu; rare.

3. Poa acroleuca Steud. *P. psilocaulis* Steud.; *P. familiaris* Steud.; *P. acroleuca* var. *psilocaulis* (Steud.) Munro——MIZO-ICHIGO-TSUNAGI.   Green, soft, glabrous annual or short-lived perennial; culms smooth, slender, 30–80 cm. long, weak; leaf-blades flat, flaccid, 10–15 cm. long, 1.5–3 mm. wide, nearly smooth; sheaths slightly compressed, nearly smooth; ligules hyaline; panicles 10–20 cm. long, narrowly ovate, nodding, green, the branches geminate, scabrous, slender, spreading; spikelets ovate, compressed, 3- to 5-flowered, 3–5 mm. long; glumes lanceolate, acute, unequal, 1- to 3-nerved, with scabrous keel, the lower 1.5 mm., the upper 2.5 mm. long; lemmas narrowly ovate, about 2.5 mm. long, appressed-pubescent on the sides below and on the nerves, obtuse.——May–July. Shaded places in lowlands and low mountains; Hokkaido, Honshu, Shikoku, Kyushu; very common.——Formosa, Korea, and China.

Var. **submoniliformis** Makino. TAMA-MIZO-ICHIGO-TSUNAGI.   A few basal internodes of culms bulbously thickened.——Occurs less frequently with the typical phase.

4. Poa hisauchii Honda. *P. acroleuca* var. *spiciformis* Honda——YAMA-MIZO-ICHIGO-TSUNAGI.   Closely allied to the preceding; leaves vivid green; panicles narrow with rather erect branches, lanceolate, nodding; lemmas broadly lanceolate, about 3 mm. long, vivid or yellowish green, glabrous, rather acute, the intermediate nerves slender but distinct, glabrous, the midrib and marginal nerves pubescent on lower half, the web distinct; anthers short.——May–July. Shaded places in mountains; Hokkaido, Honshu, Shikoku, Kyushu; common.——Korea.

5. Poa nipponica Koidz. Ō-ICHIGO-TSUNAGI.   Allied to No. 1, but more robust and taller; culms 40–50 cm. long; leaf-blades flat, thin, 10–20 cm. long, 3–7 mm. wide; ligules 1–2 mm. long; panicles open, ovate, 10–20 cm. long, branches scabrous; spikelets vivid green, broadly ovate, 4–5 mm. long;

glumes unequal, acute, scabrous on the keel, the upper narrowly ovate; lemmas 2.5–3 mm. long.——May–July. Shaded waste places and parks; Hokkaido, Honshu, Shikoku, Kyushu; common.——Korea. Possibly a hybrid of *P. acroleuca* × *P. annua.*

6. Poa tuberifera Faurie ex Hack. MUKAGO-TSUZURI. Weak perennial without stolons or distinct rhizomes; culms somewhat tufted, 20–50 cm. long, slender, smooth, the basal 1 or 2 internodes short, bulbously thickened, globose to ellipsoidal; leaf-blades flaccid, smooth, 5–15 cm. long, 2–4 mm. wide; ligules truncate, 1–2 mm. long; sheaths weakly keeled, the margins connate to the top; panicles very loose, 8–15 cm. long, with rather few spikelets, the branches in pairs, spreading, slender, slightly scabrous; spikelets pale green, long-pedicelled, 5–6 mm. long, 2- to 4-flowered, prominently flattened; glumes unequal, acute, the upper broadly lanceolate, 3–4 mm. long, 3-nerved; lemmas 3.5–4.5 mm. long, acute, green, glabrous or short-pubescent on the sides at base, the midrib and marginal nerves appressed-pubescent at base, the intermediate nerves very weak, the web wanting.——May–June. Woods; Honshu, Shikoku, Kyushu; rare.

7. Poa hakusanensis Hack. *P. yezomontana* Honda—— HAKUSAN-ICHIGO-TSUNAGI. Rhizomes short, with ascending short innovations; culms 40–70 cm. long, terete, smooth; leaf-blades flat, 15–20 cm. long, 3–6 mm. wide, short-acuminate, soft, nearly smooth; sheaths without a keel, the lower terete with fused margins; ligules about 1 mm. long, truncate; panicles broadly ovate, open, diffuse, 8–15 cm. long, the branches 2- to 4-nate, nearly smooth, spreading, spikelet-bearing on the upper parts, the verticils naked at base; spikelets 5–7 mm. long, 2- or 3-flowered, glaucous-green; glumes broadly lanceolate, acute, the lower 1- to 3-nerved, about 3.5 mm. long, the upper 3-nerved, 4–5 mm. long, scabrous on the nerves; lemmas narrowly ovate, rather acute, about 5 mm. long, glabrous on the sides, the midrib and marginal nerves short-pubescent near base, the web scanty and short.——July–Aug. High mountains; Hokkaido, Honshu (centr. distr.).

8. Poa radula Fr. & Sav. *P. sudetica* sensu auct. Japon., non Haenke; *P. ibukiana* Koidz.——IBUKI-SOMOSOMO, CHI-SHIMA-SOMOSOMO. Rhizomes short; culms rather tufted or solitary, 80–150 cm. long, retrorsely scabrous on the upper part of the internodes and below the panicle; leaf-blades thin, flat, green, 20–30 cm. long, 5–10 mm. wide, scabrous; sheaths connate to above the middle, somewhat keeled on back, slightly retrorsely scabrous; ligules 1–3 mm. long; panicles large and open, to 30 cm. long, loosely flowered, the branches sub-verticillate, spreading, prominently scabrous; spikelets green, 4–7 mm. long, prominently flattened, 4- to 8-flowered; glumes scabrous, very acute, lanceolate, the lower 1-nerved, 2.5–3 mm. long, the upper 3-nerved, 3–3.5 mm. long; lemmas broadly lanceolate, keeled, 3.5–4.5 mm. long, acute, glabrous on the sides, with glabrous, distinct intermediate nerves, the keel and marginal nerves appressed-pubescent near base, the web short and scanty.——June–Aug. Woods in mountains; Hokkaido, Honshu (as far west as Mount Ibuki in Ōmi Prov. in centr. distr.).——s. Kuriles, Kamchatka, Sakhalin, and Ussuri.

9. Poa eminens Presl. *P. glumaris* Trin.; *P. kurilensis* Hack.; *P. glumaris* var. *kurilensis* (Hack.) Kudō; *P. trinii* Scribn. & Merr.——ONI-ICHIGO-TSUNAGI. Stout, glaucescent perennial with long creeping rhizomes; culms 4–5 mm. thick above the base, 40–100 cm. long, erect, smooth, terete; leaf-blades nearly flat, rather thick, 10–30 cm. long, 5–10 mm.

wide, smooth, glabrous, very acute; sheaths terete; ligules 1–2 mm. long, truncate; panicles erect, 10–25 cm. long, 3–5 cm. wide, narrowly oblong, densely spicate, the branches 3- to 5-nate, nearly smooth, rather thick; spikelets ovate, glaucous-green, sometimes with a purple hue, 6–8 mm. long, 3- or 4-flowered; glumes lanceolate, 3-nerved, acute, 5–6 mm. long; lemmas 5-nerved, 5–6 mm. long, glabrous on the sides, the keel pubescent on lower half, the marginal nerves pubescent near base, the web scanty; anthers 2–2.5 mm. long.——July–Aug. Sandy shores; Hokkaido; rare.——Sakhalin, Kuriles, Ussuri, Kamchatka, and nw. N. America.

10. Poa fauriei Hack. *P. tateyamensis* Honda——AINU-SOMOSOMO. Tufted perennial, with elongate innovation-shoots becoming decumbent and rooting, also with innovations at the nodes; culms slender, soft, smooth, 20–60 cm. long, with several nodes; leaf-blades flat, soft, thin, green, smooth, 4–10 cm. long, 2–3 mm. wide, acute; sheaths much shorter than the internodes; ligules about 1 mm. long; panicles broadly ovate, nodding, 2–8 cm. long, the branches geminate, smooth, with 1 to 4 spikelets; spikelets 5–7 mm. long, 4- to 6-flowered, pale green, very soft; glumes equal, broadly lanceolate, thin-margined, smooth, 1- to 3-nerved, 3.5–4.5 mm. long; lemmas oblong, 4–5 mm. long, 5(–6)-nerved, rounded to obtuse, with few irregular teeth at apex, short-pubescent near base and on nerves below, the web wanting or very scanty.——Grassy mountain slopes; Hokkaido (w. distr.), Honshu (Japan Sea side, as far west as Echizen Prov.); rare.

11. Poa glauca Vahl. *P. misera* var. *alpina* Koidz.; *P. sphondylodes* var. *alpina* (Koidz.) Honda; *P. extremiorientalis* Ohwi——TAKANE-TACHI-ICHIGO-TSUNAGI. Rhizomes short; culms somewhat tufted, erect, with 1 or 2 nodes toward base, 20–30 cm. long; leaf-blades glaucous-green, rather flat, 5–10 cm. long, 1.5–2.5 mm. wide, smooth; sheaths longer than the internodes, sometimes tinged reddish purple; ligules truncate, about 0.5 mm long; panicles erect, densely spicate, lanceolate, 3–7 cm. long, the branches short, scabrous, 3- to 5-nate, with few spikelets; spikelets short-pedicelled, lustrous, pale or glaucous-green with a purple hue, 3–4 mm. long, 2- or 3-flowered; glumes lanceolate, very acute, 3-nerved, the lower about 2.5 mm., the upper about 3 mm. long; lemmas 2.5–3 mm. long, narrowly ovate, 5-nerved, rather acute, usually bronzed toward the margin, glabrous on the sides, the keel short-pubescent on lower 2/3, the marginal nerves pubescent on lower half, the intermediate nerves not distinct, the web wanting or nearly so; anthers 1.2–1.5 mm. long.——June–Aug. Alpine regions; Honshu (centr. distr.); rare.——Sakhalin, n. Korea, and generally in high mountains of the N. Hemisphere.

Var. **kitadakensis** (Ohwi) Ohwi. *P. kitadakensis* Ohwi ——KITADAKE-ICHIGO-TSUNAGI. Culms 40–70 cm. long; panicles 10–15 cm. long, with elongate branches to 8 cm. long, the verticils widely naked at base; spikelets 5–6 mm. long; glumes acuminate, 4–5 mm. long; lemmas 4–4.5 mm. long, acute to very acute; anthers 1.5–2 mm. long.——July–Aug. Alpine regions; Honshu (Akaishi mountain range); rare.

12. Poa hayachinensis Koidz. NANBU-SOMOSOMO. Rhizomes short with innovation-shoots; culms 30–50 cm. long, smooth, ascending at base; leaf-blades rather flat, 7–15 cm. long, 4–6 mm. wide, short-acuminate, smooth, rather thick and soft, glaucous above; sheaths connate on the lower half, smooth, terete; ligules truncate, 0.5–1 mm. long; panicles nodding, ovate, 10–15 cm. long, open, the branches 2- or 3-nate, spreading, smooth, rather thick; spikelets rather long-pedicelled, narrowly oblong, 7–8 mm. long, 3- to 5-flowered, glaucous-green with a purple hue; glumes unequal, lanceolate, smooth, very acute, the lower 1-nerved, about 4 mm. long, the upper 3-nerved, about 5 mm. long; lemmas lanceolate, acuminate, 5-nerved, glabrous on the sides, the intermediate nerves obscure, pubescent on the keel on the lower half and on the marginal nerves near base, the web wanting; anthers about 3 mm. long.——July–Aug. Grassy slopes in alpine regions; Hokkaido (Mount Taisetsu and Mount Yubari), Honshu (Mount Hayachine in Rikuchu).

13. Poa macrocalyx Trautv. & Mey. *P. stenantha* var. *japonica* Hack.——KARAFUTO-ICHIGO-TSUNAGI. Rhizomes short, slender, long-creeping; culms 30–70 cm. long, smooth; leaf-blades flat, 10–20 cm. long, 2.5–5 mm. wide, rather soft; sheaths terete, smooth, or slightly compressed on the innovation-shoots; ligules 2–3 mm. long; panicles nodding or pendulous, sometimes nearly erect, ovate, 10–20 cm. long, the branches geminate or subverticillate, nearly smooth; spikelets often purplish, sometimes glaucous, ovate, 6–10 mm. long, 3- to 5-flowered; glumes lanceolate, 4–6 mm. long, acuminate, the lateral nerves prominent, reaching almost to the apex; lemmas narrowly ovate, very acute, 4–6 mm. long, 5-nerved, glabrous or with pubescent intermediate nerves, the keel and the marginal nerves pubescent on the lower half, the web elongate, prominent; anthers 1.5–2.5 mm. long.——Grassy places in lowlands, especially near seashores; Hokkaido; rather common and very variable.——Around the Okhotsk Sea, Sakhalin, Kuriles, Kamchatka, and the Aleutians.

Var. **fallax** (Hack.) Ohwi. *P. stenantha* var. *fallax* Hack.——WATAGE-SOMOSOMO. Lemmas appressed-pubescent on the lower half.——Near seashores; Hokkaido; rare.

Var. **scabriflora** (Hack.) Ohwi. *P. scabriflora* Hack.——ZARABANA-SOMOSOMO. Lemmas scabrous with minute, hairlike aculeoli.——Near seashores; Hokkaido.——Sakhalin.

Var. **tatewakiana** Ohwi. *P. tatewakiana* Ohwi——HOSOBANA-SOMOSOMO. Culms 30–60 cm. long; panicles pendulous or strongly nodding; spikelets glaucous-green; lemmas narrow, gradually acuminate.——Near seashores; Hokkaido (e. distr.).——s. Kuriles.

14. Poa sachalinensis (Koidz.) Honda. *P. macrocalyx* var. *sachalinensis* Koidz.; *P. yezoensis* Ohwi; *P. sachalinensis* var. *yezoensis* Ohwi——HIME-KARAFUTO-ICHIGO-TSUNAGI. Allied to and possibly not distinct from the preceding, but more slender and with slender stolons; culms 30–80 cm. long, smooth; leaf-blades 20–30 cm. long, 3–6 mm. wide; panicles 7–20 cm. long; spikelets 5–8 mm. long, pale or glaucous-green, sometimes partly purplish; glumes unequal, the lower 1- to 3-nerved, 2.5–4 mm. long, the upper 4–4.5 mm. long, 3-nerved, the lateral nerves ½–¾ the length of the glume; lemmas 4–4.5 mm. long.——July–Aug. Grassy places in lowlands, especially near seashores; Hokkaido.——Sakhalin.

Var. **yatsugatakensis** (Honda) Ohwi. *P. yatsugatakensis* Honda——TANI-ICHIGO-TSUNAGI. Lemmas rather broad, acute.——Along rivers in high mountains; Honshu (Mount Yatsugatake in Shinano); rare.

15. Poa komarovii Roshev. var. **shinanoana** (Ohwi) Ohwi. *P. alpina* sensu auct. Japon., non L.; *P. shinanoana* Ohwi——MIYAMA-ICHIGO-TSUNAGI, TAKANE-ICHIGO-TSUNAGI. Low, smooth perennial with short ascending innovation-shoots; culms loosely tufted, 10–25 cm. long, rather stout, erect from ascending base; leaf-blades rather thick and soft, flat or with slightly involute margins, 4–6 cm. long in the cauline, slightly longer in the radical, 4–5 mm. wide, glaucous above; ligules 2–4 mm. long, semirounded; panicles broadly ovate, open, erect or slightly nodding, 5–7 cm. long, the branches geminate, smooth, spreading, rather thick; spikelets short-pedicelled, 6–8 mm. long, 2- to 4-flowered, purplish, rarely pale green; glumes slightly unequal, acuminate, with very short lateral nerves, the lower 3–4 mm., the upper 4–5 mm. long; lemmas broadly lanceolate, 5–6 mm. long, very acute, weakly 5-nerved, sometimes pubescent below, the midrib and marginal nerves pubescent on the lower half, the web rather short; anthers 1.5–2 mm. long.——July–Aug. Grassy slopes and on rocks in alpine regions; Honshu (centr. distr.).——n. Korea.——Some plants have viviparous spikelets. The typical phase occurs in Kamchatka and n. Kuriles.

16. Poa trivialis L. *P. uda* Honda——Ō-SUZUME-NO-KATABIRA. Culms 40–100 cm. long, decumbent at base, rather slender, often retrorsely scabrous below the panicle; leaf-blades green, flat, soft, 10–20 cm. long, 3–5 mm. wide; sheaths slightly compressed, often somewhat retrorsely scabrous; ligules hyaline, deltoid, 3–6 mm. long; panicles ovate, open, 10–20 cm. long, the branches subverticillate, scabrous; spikelets green, 2.5–3.5 mm. long, 2- or 3-flowered; glumes lanceolate, about 2.5 mm. long, acute, the lower 1- to 3-nerved, the upper 3-nerved; lemmas 2.5–3 mm. long, narrowly ovate, acute, 5-nerved, glabrous on the sides, the keel and marginal nerves short-pubescent near base, the web elongate; anthers 1.2–1.7 mm. long.——May–June. Lowlands, possibly naturalized; Hokkaido, Honshu (centr. distr. and northw.).——Europe and w. Asia.

17. Poa palustris L. NUMA-ICHIGO-TSUNAGI. Culms 50–100 cm. long, slender, erect from ascending to short-creeping base, smooth; leaf-blades linear, 10–20 cm. long, 1–2 mm. wide; ligules 3–5 mm. long, obtuse; panicles erect, 12–25 cm. long, very loose and open, ovate, the branches subverticillate, scabrous, spreading; spikelets 2- or 3-flowered, ovate, 3–5 mm. long, acute, pale green; glumes lanceolate, 2–2.5 mm. long, acute, 3-nerved, with a scabrous keel; lemmas acute, often bronzed toward margin, 2.5–3 mm. long, glabrous, the keel and marginal nerves pubescent on lower 1/3–1/2 of the entire length; anthers 1–1.5 mm. long.——May–June. Naturalized; Hokkaido, Honshu (centr. distr. and northw.).——Cooler regions of the N. Hemisphere.

18. Poa sphondylodes Trin. *P. linearis* Trin.; *P. diantha* Steud.; *P. strictula* Steud.; *P. sphondylodes* var. *diantha* (Steud.) Munro; *P. palustris* var. *strictula* (Steud.) Hack.; *P. misera* Koidz., excl. syn.; *P. sphondylodes* var. *strictula* (Steud.) Koidz.——ICHIGO-TSUNAGI, ZARA-TSUKI-ICHIGO-TSUNAGI. Tufted, glaucous-green perennial; culms slender, somewhat firm, erect, 30–60 cm. long, sometimes reddish at the base, scabrous on the upper part of the internodes and below the panicle; leaf-blades glaucous-green, slightly involute, scabrous, 5–15 cm. long, 1.5–3 mm. wide; ligules lanceolate, acute, 3–8 mm. long; panicles erect, 5–15 cm. long, lanceolate, contracted, the branches 3- to 6-nate, scabrous, erect to ascending; spikelets pale- or glaucous-green, ovate, 3–5 mm. long, 3- to 6-flowered; glumes unequal, lanceolate, acute, the lower 1.5–2 mm. long, 1-nerved, the upper 3-nerved, 2–2.5(–3) mm. long; lemmas about 3 mm. long, rather acute, glabrous on both sides, pubescent on the lower ½–¾ of the keel and on 1/3–2/3 the length of the marginal nerves, the web rather copious; anthers 1.2–1.5 mm. long.——May–July. Sandy, sunny places; Hokkaido, Honshu, Shikoku, Kyushu; very common.——China, e. Siberia, Korea, and Formosa.

Var. **subtrivialis** Ohwi. HIROHA-ICHIGO-TSUNAGI. Plant larger, with open panicles.——Honshu (Mino Prov. and Chūgoku Distr.), Shikoku.

19. **Poa viridula** Palib. *P. nemoralis* sensu auct. Japon., pro maxim. parte, non L.——AO-ICHIGO-TSUNAGI. Tufted green perennial; culms erect, 30–60 cm. long, usually smooth; leaf-blades flat or loosely involute, 5–15 cm. long, 1.5–3 mm. wide, green, the uppermost shorter than the sheath; ligules semirounded to truncate, 1–2 mm. long; panicles narrow, erect, dense, green, 8–15 cm. long, 1.5–3 cm. wide, the branches 3- to 5-nate, scabrous; spikelets 4–5 mm. long, 2- or 3-flowered; glumes rather unequal, 3-nerved, acuminate, about 3 mm. long; lemmas green, bronzed toward the margin, 3 mm. long, acute, glabrous, with delicate intermediate nerves, the keel pubescent on lower half, the marginal nerves pubescent near base, the web rather short, not copious; anthers about 1.5 mm. long.——May–June. Hills and low mountains; Hokkaido, Honshu (Kinki Distr. and northw.).——Korea, Manchuria, Sakhalin, and s. Kuriles.

20. **Poa nemoralis** L. TACHI-ICHIGO-TSUNAGI. Closely allied to the preceding; culms erect, slender, tufted, 20–50 cm. long; leaf-blades green, 1–3 mm. wide; ligules truncate, about 0.5 mm. long; panicles erect, open, loose, the branches 3- to 5-nate; spikelets 3–4 mm. long, ovate; glumes lanceolate, acuminate; lemmas oblong, acute, glabrous on the sides, with obscure intermediate nerves, 2–3 mm. long, pubescent on lower half of the keel and marginal nerves, the web sparse; anthers 1.2–1.5 mm. long.——Mountain woods; Honshu (Kantō and centr. Distr.); rare.——n. Korea, Kuriles, and cooler parts of the N. Hemisphere.

21. **Poa matsumurae** Hack. *P. tomentosa* Koidz.; *P. iwateana* Ohwi; *P. chosenensis* Ohwi; *P. trivialis* var. *tomentella* Honda; *P. iwayae* Honda——ITO-ICHIGO-TSUNAGI. Slender, green or vivid green perennial with slender, elongate rhizomes; culms 50–80 cm. long, smooth, solitary or few; leaf-blades thin, flat, 10–15 cm. long, 1.2–3 mm. wide, often sparsely pilose above; sheaths slightly compressed, the lower sometimes puberulent; ligules 1–2 mm. long, truncate; panicles loose, open, green or yellowish green, 8–15 cm. long, 3–5 cm. wide, the branches subverticillate, slender, scabrous, spreading; spikelets 4–6 mm. long, prominently compressed, 4- or 5-flowered; glumes narrowly lanceolate, 2.5–3 mm. long, very acute, 3-nerved, scabrous on the keel; lemmas broadly lanceolate, about 3.5 mm. long, glabrous, with distinct intermediate nerves, the keel and marginal nerves pubescent on lower ½–¾ of the entire length, the web copious; anthers 1.2–1.5 mm. long.——May–July. Mountain woods; Honshu (centr. distr. and northw.), perhaps also in Hokkaido; rare. ——Korea.

22. **Poa pratensis** L. NAGAHAGUSA. Very variable, rhizomatous perennial; culms erect, 30–80 cm. long, smooth, rather firm; leaf-blades nearly flat to loosely involute, 7–20 cm. long, 2–4 mm. wide, usually glabrous; ligules truncate, 1–2 mm. long; panicles narrowly ovate, 8–15 cm. long, contracted, densely spiculose, erect, the branches subverticillate, scabrous; spikelets ovate, 3–6 mm. long, 3- to 5-flowered; glumes slightly unequal, acute, scabrous on the keel, the lower lanceolate, 1-nerved, 1.5–2 mm. long, the upper broadly lanceolate, 3-nerved, 2–2.5 mm. long; lemmas about 3 mm. long, acute, glabrous on the sides, the intermediate nerves distinct, the keel and marginal nerves pubescent on lower ½–¾ of the entire length, the web copious; anthers 0.8–1.2 mm. long.——May–July. Spontaneous in mountains, although believed to be naturalized, very common in lowland pastures and parks; Hokkaido, Honshu, Shikoku, Kyushu.——Eurasia.

23. **Poa compressa** L. KO-ICHIGO-TSUNAGI. Rhizomatous perennial; culms 20–60 cm. long, erect, strongly flattened, smooth; leaf-blades nearly flat, green, 7–15 cm. long, 1.5–3 mm. wide; ligules truncate to semirounded, short; panicles erect, rather firm, densely spiculose from near the base, deep green, lanceolate, 3–7 cm. long, the branches 2- or 3-nate, scabrous, 1–2 cm. long; spikelets 3–5 mm. long, 3- to 6-flowered, short-pedicelled; glumes about equal, 3-nerved, broadly lanceolate, rather acute; lemmas 2–3 mm. long, obtuse, bronzed toward the margin, glabrous on the sides, with rather obscure intermediate nerves, the keel and marginal nerves pubescent only near the base, the web short, not copious; anthers 0.8–1.2 mm. long.——May–July. Naturalized; Hokkaido, Honshu; rather rare.——Europe and Siberia.

41. GLYCERIA R. Br. DOJŌ-TSUNAGI ZOKU

Glabrous perennials with flat leaf-blades, terete closed sheaths and narrow to broad panicles; spikelets 3- to many-flowered, linear to ovate, articulate below the florets; glumes hyaline, 1- (3-) nerved, usually shorter than the florets; lemmas rounded on back, without a keel, awnless, 7- to 9-nerved, scabrous, with a hyaline margin; callus minute, glabrous; rachilla slender, glabrous; palea 2-keeled, as long as the lemma or slightly longer, the keels sometimes narrowly winged; stamens 3; ovary glabrous; caryopsis free within the lemma and palea, plump.——About 40 species, in the temperate and cooler regions of the N. Hemisphere.

1A. Spikelets cylindric, 8- to 15-flowered, 15–50 mm. long; palea with winged keels; panicle-branches usually smooth.
  2A. Lemmas 7–11 mm. long, lanceolate, acute; anthers 1–1.2 mm. long. . . . . 1. *G. acutiflora*
  2B. Lemmas 2.5–5.5 mm. long, narrowly oblong; anthers about 0.7 mm. long. . . . . 2. *G. depauperata*
1B. Spikelets linear to broadly elliptic, slightly compressed, 3- to 7-flowered, 4–10 mm. long; palea with wingless keels; panicle-branches usually scabrous.
  3A. Lemmas about 2.2 mm. long; keels of palea and joints of rachilla strongly curved; anthers 0.5–0.6 mm. long.
                                                           3. *G. ischyroneura*
  3B. Lemmas 2.5–4 mm. long; keels of palea and joints of rachilla nearly straight.
    4A. Ligules about 1 mm. long; plant rather firm and stout; anthers 0.5–0.6 mm. long. . . . . 4. *G. leptolepis*
    4B. Ligules 2–3 mm. long; plant soft or slender with open, loose, nodding panicles.
      5A. Lower glume about 1 mm. long, the upper 1.3–1.5 mm. long; anthers 0.5–0.6 mm. long; leaf-blades thin.
                                                        5. *G. lithuanica*
      5B. Lower glume 2–2.5 mm. long, the upper 2.5–3 mm. long; anthers 0.7–1 mm. long; leaf-blades very thin.
                                                        6. *G. alnasteretum*

1. Glyceria acutiflora Torr. *Hemibromus japonicus* Steud.; *G. japonica* (Steud.) Miq.——MUTSU-OREGUSA, MINO-GOME. Soft perennial; culms 20–70 cm. long, erect from a long, creeping base with sterile ascending innovations; leaf-blades thin, flat, 10–30 cm. long, 3–6 mm. wide, nearly smooth, glaucescent above, short-acuminate; ligules 4–7 mm. long, narrowly deltoid; panicles linear, 10–30 cm. long, partly enclosed within the uppermost leaf-sheath, with 5 to 20 spikelets, the branches geminate, strongly unequal, appressed to the axis, smooth; spikelets 25–50 mm. long, pale green, 9- to 15-flowered, short-pedicelled; glumes unequal, 1-nerved, lanceolate, rather acute, the lower about 2 mm., the upper 4–5 mm. long; lemmas lanceolate, minutely scabrous, 7–9 mm. long, 7-nerved, with short, parallel veinlets between the nerves; palea 0.7–1 mm. longer than the lemma, bicuspidate; caryopsis 3–4 mm. long. ——May–June. Ponds, paddy fields; Honshu, Shikoku, Kyushu; common.——Korea, Ryukyus, China, and N. America.

2. Glyceria depauperata Ohwi. *Glyceria fluitans* var. *leptorrhiza* sensu auct. Japon., non Maxim.——HIME-UKIGAYA. Similar to but much smaller than the preceding; leaf-blades 2–4 mm. wide; ligules 2–5 mm. long; panicles only slightly exserted, 10–25 cm. long; spikelets rather many, very short-pedicelled, 10–25 mm. long, 7- to 15-flowered; glumes narrowly ovate, the lower 1 mm., the upper about 2 mm. long; lemmas 2.5–3.5 mm. long.——Ponds and riverbanks; Hokkaido, Honshu.

Var. **infirma** (Ohwi) Ohwi. *G. fluitans* sensu auct. Japon., non R. Br.; *G. leptorrhiza* var. *infirma* Ohwi——UKIGAYA. Plant with larger lemmas, 4–5.5 mm. long.——Occurs with the typical phase.

3. Glyceria ischyroneura Steud. *G. tonglensis* sensu auct. Japon., non C. B. Clarke; *G. caspica* sensu auct. Japon., non Trin.; *G. tonglensis* var. *honshuana* Kelso——DOJŌ-TSUNAGI. Slender to rather stout perennial; culms erect from an ascending base, 80–120 cm. long; leaf-blades flat, somewhat membranous, 30–60 cm. long, 3–7(–10) mm. wide; sheaths terete; ligules short, truncate; panicles lanceolate to ovate, 15–40 cm. long; spikelets narrowly oblong, 5–7 mm. long, 3- to 6-flowered, pale green, sometimes tinged pale purple; glumes oblong, obtuse, weakly 1-nerved, the 1st about 1 mm. long, the 2d 1.5–2 mm. long; lemmas ovate, obtuse, about 2.2 mm. long, 7-nerved; palea with conspicuously curved keels, slightly inflated below.——May–June. Ponds and river-banks; Hokkaido, Honshu, Shikoku, Kyushu; common.——Korea.

4. Glyceria leptolepis Ohwi. *G. aquatica* sensu auct. Japon., non Wahl.; *G. ussuriensis* Komar.——HIROHA-NO-DOJŌ-TSUNAGI. Stout, pale green perennial; culms 80–150 cm. long; leaf-blades rather thick, firm, scabrous, 40–60 cm. long, 5–12 mm. wide, short-acuminate, with loosely involute margins; sheaths terete; ligules truncate, about 1 mm. long; panicles large, 20–30 cm. long, ovate, nodding, the branches subverticillate, scabrous; spikelets numerous, 6–8 mm. long, 4- to 6-flowered, pale green, rarely slightly brownish; glumes obtuse, narrowly oblong, 1-nerved, 1.5–2 mm. long; lemmas narrowly oblong, 3–4 mm. long, punctate-scaberulous, 7-nerved, obtuse, rather thin; palea as long as the lemma; caryopsis narrowly obovoid, about 1.5 mm. long.——July–Aug. Ponds and riverbanks; Hokkaido, Honshu, Kyushu; rather rare.——Korea, Ussuri, Manchuria, China, and Formosa.

5. Glyceria lithuanica (Gorski) Lindm. *Poa lithuanica* Gorski; *G. remota* (Forselles) Fries——KARAFUTO-DOJŌ-TSUNAGI. Stoloniferous perennial; culms 80–100 cm. long; leaf-blades thin, 20–30 cm. long, 5–8 mm. wide, rather flat, vivid-green, scabrous above; panicles nodding to pendulous, loose, 15–25 cm. long, the branches slender, 2- to 5-nate, slightly scabrous; spikelets numerous, ovate or oblong, 4–7 mm. long, usually purpurascent, 3- to 5-flowered; glumes narrowly oblong, obtuse, hyaline, often somewhat brown- and purple-tinged, 1-nerved, the lower about 1 mm., the upper 1.3–1.5 mm. long; lemmas narrowly oblong, 3–3.5 mm. long, thin, 7-nerved, with hyaline margins.——July–Aug. Wet places in woods; Hokkaido, Honshu (centr. and n. distr.); rather common.——Korea, Kuriles, Sakhalin, Manchuria, Kamchatka, e. Siberia, and w. Europe.

6. Glyceria alnasteretum Komar. *G. remota* var. *japonica* Hack.; *G. arundinacea* sensu auct. Japon., non Kunth——MIYAMA-DOJŌ-TSUNAGI. Closely allied to the preceding; leaf-blades thinner, vivid- to deep-green, flat, 4–7 mm. wide, nearly smooth; panicle branches geminate, smooth or slightly scabrous; glumes lanceolate, rather acute, the lower 2–2.5 mm., the upper 2.5–3 mm. long; lemmas narrowly oblong, 3.5–4 mm. long; anthers narrowly oblong.——July–Aug. Wet places in cool, coniferous woods; Hokkaido, Honshu (centr. distr. and northw.); rather common.——Sakhalin, Kuriles, and Kamchatka.

41a. TORREYOCHLOA Church   HAI-DOJŌ-TSUNAGI ZOKU

Resembles *Glyceria* in general aspect; culms often decumbent at base; leaf-sheaths open, not fused on margin; panicles open, branches ascending to spreading, flexuous; spikelets oblong to ovate, few-flowered; lemmas 7- to 9-nerved; lodicules truncate, united.——Few species in the temperate and northern regions of e. Asia and N. America.

1A. Lemmas 2–2.5 mm. long; anthers less than 0.5 mm. long; spikelets 4–6 mm. long; panicles less than 10 cm. long. ...... 1. *T. natans*
1B. Lemmas about 3.5 mm. long; anthers about 0.5 mm. long; spikelets 6–9 mm. long; panicles larger. .... 2. *T. viridis*

1. Torreyochloa natans (Komar.) Church. *Glyceria natans* Komar.; *G. pallida* sensu auct. Japon., non Trin.——HOSOBA-DOJŌ-TSUNAGI. Slender glabrous perennial; culms ascending, long-creeping at base, 20–40 cm. long; leaf-blades 5–10 cm. long, 1.5–3 mm. wide, flat, thin, glaucescent; ligules narrowly deltoid, 3–6 mm. long; panicles narrowly ovate, rather loose and open, 6–10 cm. long, the branches 2- to 4-nate, scabrous, slender, ascending; spikelets narrowly oblong, 4–6 mm. long, 3- or 4-flowered, pale green; glumes broadly lanceolate, rather obtuse, the 1st 1–1.2 mm. long, the 2d with a pair of short lateral nerves, about 1.5 mm. long; lemmas narrowly ovate, 2–2.5 mm. long, 7-nerved, scaberulous.——July–Aug. Ponds; Hokkaido, Honshu (Kamikochi in Shinano); rare.——s. Kuriles, Sakhalin, Ussuri, and Kamchatka.

2. Torreyochloa viridis (Honda) Church. *Glyceria viridis* Honda——HAI-DOJŌ-TSUNAGI. Allied to the preceding but larger; culms 30–50 cm. long; leaf-blades 10–15 cm. long, 4–6 mm. wide; panicles exserted, 10–20 cm. long, 3–5 cm. wide, the branches subverticillate, ascending, slender, scabrous; spikelets 6–9 mm. long, green, 7- to 9-flowered, the pedicels appressed to the branchlets; glumes 1-nerved, oblong, obtuse, about 1.5 to 2 mm. long; lemmas about 3.5 mm. long, ovate, 7-nerved, scaberulous.——June–July. Ponds; Hokkaido, Honshu (Kantō Distr. and northw.).

42. MELICA L. Komegaya Zoku

Perennials with slender culms, linear leaf-blades and closed sheaths; panicles loose and open, or compact and spikelike, sometimes loose and nearly simply racemose; pedicels usually pubescent under the spikelet; spikelets 3- to 5-flowered, nodding to pendulous, articulate below the florets; florets bisexual, the upper often sterile and convolute into a clavate mass; glumes papery, 1- to 9-nerved, smaller than or as long as the lowest floret; lemmas rather chartaceous or membranous, 7- to 13-nerved, sometimes long-pilose, obtuse, with hyaline margins usually awnless, rounded on back; callus minute, glabrous; rachilla glabrous; palea 2-keeled; stamens 3; ovary glabrous; caryopsis plump.——About 30 species, in temperate regions of the N. Hemisphere.

1A. Spikelets narrowly lanceolate, pale green; lemmas 4–6 mm. long, lanceolate; culms 100–150 cm. long; panicles large, much-branched.
.... 1. *M. onoei*

1B. Spikelets elliptic, sometimes purple-tinged; lemmas 5–8 mm. long, oblong; culms slender, 20–50 cm. long; panicles scarcely branched, racemelike. .... 2. *M. nutans*

1. Melica onoei Fr. & Sav. *M. matsumurae* Hack.; *M. kumana* Honda——Michi-shiba, Hanabigaya. Rather firm, pale green perennial without stolons; culms erect, slender, 100–150 cm. long, with 10 or more nodes; leaf-blades loosely involute, thin but rather firm, scabrous, 15–30 cm. long, 4–10 mm. wide, sheaths longer than the internodes, sometimes pubescent or retrorsely scabrous; ligules truncate, 0.2–0.3 mm. long; panicles at length exserted, erect, 25–50 cm. long, the branches subverticillate, spreading, slender but firm; pedicels short-pilose; spikelets 7–10 mm. long, 3- or 4-flowered, the florets all nearly equal; glumes lustrous, rather acute, the first narrowly ovate, 2.5–3 mm. long, 1-nerved, the second broadly lanceolate, 4.5 mm. long, with a pair of slender lateral nerves; lemmas pale green with hyaline margins, obtuse, 7- to 9-nerved, glabrous, 4–6 mm. long; caryopsis broadly lanceolate, 3 mm. long.——Aug.–Sept. Thin woods; Honshu (Kantō Distr. and westw.), Shikoku, Kyushu; rather rare.——Korea and China.

Var. **pilosella** Honda. Usuge-michi-shiba. Lemmas with a few long appressed setose hairs.——Kyushu (Higo Prov.).

2. Melica nutans L. *M. grandiflora* Koidz.; *M. nutans* var. *argyrolepis* Komar.——Komegaya. Slender perennial with slender creeping rhizomes and elongate stolons; culms erect, 20–50 cm. long, obtusely angled; leaf-blades thin, scabrous, rather flat, 5–15 cm. long, 2–5 mm. wide, sometimes sparsely pilose; lower sheaths tinged red-purple; ligules truncate, very short; panicles exserted, narrow, nearly simple and racemose, loosely spiculose, 8–15 cm. long; spikelets 5–15, 2-flowered exclusive of the 2 sterile lemmas, pendulous, elliptic, 6–8 mm. long, very obtuse to rounded at apex, pale green, sometimes purple-tinged, lustrous; glumes oblong, papery, smooth, lustrous, the lower 3-nerved, 3–4 mm. long, the upper 5-nerved, 5–6 mm. long; lemmas chartaceous, many-nerved, scaberulous; palea rather broad.——May–July. Grassy slopes of hills and mountains; Hokkaido, Honshu, Shikoku, Kyushu; rather common.——Kuriles, Sakhalin, Korea, China to Siberia, and Europe.

43. SCHIZACHNE Hack. Fōrii-gaya Zoku

Slender perennial with linear leaf-blades and rather few-spiculose, nodding panicles; spikelets slightly flattened, 4- or 5-flowered, articulate below the florets; florets bisexual; glumes papery, scarcely keeled, rather acute, smooth, 1- to 5-nerved; lemmas longer than the glumes, chartaceous to thick-membranous, 7- to 13-nerved, not keeled, pale green, with hyaline, smooth margins, awned just below the apex, callus pilose; rachilla glabrous; palea 2-keeled; stamens 3; ovary glabrous; caryopsis plump, lustrous.——One species, in e. Asia and N. America.

1. Schizachne purpurascens (Torr.) Swallen. *Trisetum purpurascens* Torr.; *Avena callosa* Turcz.; *S. fauriei* Hack.; *Melica callosa* (Turcz.) Ohwi; *S. callosa* (Turcz.) Ohwi—— Fōrii-gaya, "Faurie"-gaya. Loosely tufted, glabrous perennial; culms slender, 40–70 cm. long; leaf-blades 1–2 mm. wide; ligules 1.5–2.5 mm. long, hyaline; panicles exserted, nodding, 5–8 cm. long, 5- to 10-flowered, the branches geminate, scabrous, slender and short, usually with 1 spikelet; spikelets narrowly oblong, 12–15 mm. long, 3- to 5-flowered, pale green with a brown-purple hue; glumes broadly lanceolate, rather acute, lustrous, the lower 1-nerved, 3–4 mm. long, the upper 5-nerved with weak, short lateral nerves, 6–8 mm. long; lemmas broadly lanceolate, 8–11 mm. long, glabrous, scaberulous, 7-nerved, slightly bidentate, the awn inserted about 1.5 mm. below the apex, straight or slightly recurved, scabrous, slender, 1.5–2 times as long as the lemma; callus-hairs 1.5–2 mm. long.——June–July. Coniferous woods; Hokkaido, Honshu (centr. distr.); rare.——Korea, Sakhalin, Ussuri, Kamchatka, and s. Kuriles, and N. America.

44. BRYLKINIA F. Schmidt Hogaerigaya Zoku

Slender perennial with elongate rhizomes, solitary culms, linear leaf-blades, and closed sheaths; inflorescence a simple raceme; pedicels solitary, short, alternate on a 4-angled axis, jointed at base, falling with the spikelet, pendulous, short-pilose, slightly and gradually thickened above; spikelets pendulous, much-flattened laterally, 3-flowered, the lower 2 florets neutral and without a palea, the uppermost bisexual and fertile; glumes and sterile lemmas green, chartaceous, keeled, the upper gradually larger, acuminate to awn-pointed; fertile lemma green, chartaceous, with a narrow wing on back above, the awn excurrent from the apex, long, scabrous; callus slightly thickened, glabrous, very short; palea 2-keeled; stamens 3; caryopsis plump, free, not adherent to the lemma and palea.——One species, in e. Asia.

1. Brylkinia caudata (Munro) F. Schmidt. *Ehrharta caudata* Munro ex A. Gray; *Brylkinia schmidtii* Ohwi——Ho-GAERIGAYA. Green, glabrous, perennial; culms 20–40 cm. long; leaf-blades flat, smooth, 15–20 cm. long, 2–5 mm. wide; ligules obsolete; raceme 8–20 cm. long, 1-sided with 10–15 spikelets; pedicels 4–6 mm. long, jointed at base; spikelets 12–15 mm. long; glumes narrowly lanceolate, acuminate; sterile lemma similar, gradually larger, epaleate; fertile lemma largest, broadly lanceolate, nearly 7-nerved, 10–13 mm. long, minutely 2-toothed at apex, the awn 1 or 2 times as long as the lemma; anthers about 3 mm. long.——May–July. Mountain woods; Hokkaido, Honshu, Shikoku, Kyushu; rather common.——Sakhalin, s. Kuriles, and Manchuria.

45. LEERSIA Sw. SAYANUKAGUSA ZOKU

Perennials growing in wet places; culms rather slender; leaf-blades short-linear, flat, scabrous; spikelets short-pedicelled, laterally flattened, 1-flowered; glumes wanting; florets articulate at base; lemma chartaceous to membranous, oblong and folded, keeled, 5-nerved, uniformly scabrous or only on nerves; palea 3-nerved; stamens 6 or 3, not protruded in cleistogamous spikelets. ——About 10 species, in tropical and warm regions.

1A. Panicles narrow, 5–10 cm., rarely to 15 cm. long, the branches thick, erect to ascending, flowered nearly to base; stamens 6.
1. *L. japonica*
1B. Panicles usually 10–20 cm. long, the branches slender, spreading, naked on lower ⅓–½; stamens 3. . . . . 2. *L. oryzoides*

1. Leersia japonica (Honda) Makino ex Honda. *Homalocenchrus japonicus* Honda——ASHI-KAKI. Stoloniferous, scabrous perennial; culms slender, 30–50 cm. long, erect from a decumbent, branching base; leaf-blades broadly linear, striately nerved, flat, 5–15 cm. long, 5–8 mm. wide, glabrous; ligules 2–2.5 mm. long; panicles narrow, 5–15 cm. long, exserted, the branches simple, erect to ascending, with about 10 spikelets, glabrous, slightly flattened, puberulent on axils; spikelets whitish green, sometimes slightly red-tinged, flattened, 4.5–6 mm. long, 1.5–1.7 mm. wide, acute, appressed to the branches, obtuse at base, dimidiate-ovate, bristly-ciliate; stamens 6, the anthers 2.5–3.5 mm. long; ovary glabrous.——Aug.–Oct. Ponds and paddy-fields; Honshu, Shikoku, Kyushu.——China, Korea, and Ryukyus.

2. Leersia oryzoides (L.) Sw. *Phalaris oryzoides* L.; *Homalocenchrus oryzoides* (L.) Poll.; *Oryza oryzoides* (L.) Dalla Torre & Sarnth.——EZO-NO-SAYANUKAGUSA. Slender green perennial; culms 50–80 cm. long, with retrorsely pilose nodes; leaf-blades thin, broadly linear, flat, 15–25 cm. long, 8–12 mm. wide, abruptly acuminate, scabrous; ligules 1–1.5 mm. long; panicles 10–20 cm. long, partly or at length wholly exserted, sometimes wholly inclosed within the sheath, the branches single, slender, spreading, loosely spiculose; spikelets oblong, flat, whitish green, rather acute, 4.5–6 mm. long, 1.5–2 mm. wide, bristly ciliate, loosely short-setose on nerves; stamens 3, the anthers linear, to 1.5 mm. in chasmogamic flowers, minute in the cleistogamic.——Aug.–Oct. Wet places; Hokkaido, Honshu, Shikoku, Kyushu; common.——Widely distributed in temperate regions.

Var. **japonica** Hack. *Homalocenchrus oryzoides* var. *japonicus* (Hack.) Honda; *L. sayanuka* Ohwi——SAYANUKAGUSA. Spikelets linear-oblong, short-ciliate, green, 5–6.6 mm. long, about 1.5 mm. wide; leaf-blades not ciliate.——Aug.–Oct. Wet places; Hokkaido (sw. distr.), Honshu, Shikoku, Kyushu; rather common.——Forma **latifolia** (Honda) Ohwi. *Leersia oryzoides* var. *latifolia* Honda——HIROHA-NO-SAYANUKAGUSA. A larger phase of var. *japonica* with leaves 10–25 mm. wide, and panicles 20–30 cm. long.——Kyushu.

46. CHIKUSICHLOA Koidz. TSUKUSHIGAYA ZOKU

Tall aquatic perennials with linear flat leaf-blades; panicles large, erect; spikelets perfect, 1-flowered, terete, the pedicels elongate, jointed about the middle, the upper part falling with the spikelet, forming a stipe; glumes wanting; lemmas elliptic, thin-membranous, brownish, scabrous, 5- to 7-nerved, not keeled on back; awned or acuminate; palea as long as the lemma; stigma laterally exserted; stamen 1, the anther linear; ovary glabrous; caryopsis ellipsoidal, plump.——Two species in e. Asia.

1. Chikusichloa aquatica Koidz. TSUKUSHIGAYA. Tufted perennial with short-branched rhizomes and ascending innovations; culms 100–120 cm. long, erect; leaf-blades thin and firm, 30–70 cm. long, 8–12 mm. wide, long-acuminate, vivid green, scabrous; ligules truncate, 2–3 mm. long; panicles erect, 40–50 cm. long, about 20 cm. wide, rather open, the numerous very slender branches fasciculate, capillary, ascending to spreading; spikelets 3 mm. long, ellipsoidal, scabrous, with a terminal erect awn 5–6 mm. long; pedicels very slender, elongate, jointed about the middle, the upper part scabrous, falling with the spikelet, forming a stipe 5–6 mm. long; anther 1.3–1.5 mm. long.——Aug.–Oct. Honshu (Uzen and Yamato Prov.), Kyushu; very rare.——China.

47. ZIZANIA L. MAKOMO ZOKU

Large, stout, aquatic perennials or annuals, with flat or slightly involute leaf-blades and monoecious spikelets in panicles; spikelets rather large, 1-flowered, jointed at base and readily deciduous from the pedicels, somewhat angular, not flattened, scabrous; glumes wholly reduced; lemma thinly membranous, 5-nerved, awned from the apex or awnless; callus short; rachilla not prolonged; palea 3-nerved, the marginal nerves closely aligned to the marginal nerves of the lemma; stamens 6; caryopsis cylindrical.——Three or four species, in e. Asia and N. America.

1. Zizania latifolia Turcz. *Limnochloa caduciflora* Turcz. ex Trin. in syn.; *Z. caduciflora* (Turcz.) Hand.-Mazz.; *Hydropyrum latifolium* Griseb.; *Z. aquatica* sensu auct. Japon., non L.; *Z. aquatica* var. *latifolia* (Turcz.) Komar.——MA-KOMO. Stout aquatic perennial with thick creeping rhizomes; culms 1–2.5 m. long, glabrous; leaf-blades glaucous-green, 50–100 cm. long, 2–3 cm. wide, short-acuminate with a short caudate tip, tapering to the base, the midrib beneath thickened in lower part; sheaths spongy-thickened, obtuse on back; ligules white, elongate-deltoid, acute; panicles at length exserted, 40–60 cm. long, narrowly pyramidal, the branches ascending, scabrous, subverticillate, with a tuft of long white hairs in axils, pedicels short, the staminate spikelets on the lower part, lanceolate, 8–12 mm. long, usually purplish, acute or short-awned, stamens 6, the anthers linear, 6–10 mm. long; pistillate spikelets on upper part pale green, linear, 18–25 mm. long, very scabrous; awns erect, 2–3 cm. long, scabrous.——Aug.–Oct. Ponds and riverbanks; Hokkaido, Honshu, Shikoku, Kyushu; very common.——Formosa, Ryukyus, Indochina, China, Korea, and e. Siberia.

48. LOPHATHERUM Brongn. SASA-KUSA ZOKU

Tall perennials with flat, broadly lanceolate to ovate-oblong leaf-blades; spikelets subsessile, on one side of racemelike panicle branches, with 1 perfect floret and few short-awned sterile lemmas in a group at the summit of the rachilla; spikelets more or less laterally flattened, falling entire; glumes chartaceous, short; lemmas ovate, chartaceous, rounded on back, 5- to 7-nerved, the fertile lemma awnless, the sterile ones short-awned; callus short, fused with the rachilla, glabrous; palea membranous, 2-keeled; stigma with short branchlets.——Few species, in se. Asia, China, and Japan.

1A. Spikelets lanceolate to narrowly so, scarcely flattened; fertile lemma narrowly oblong, not bowed out on back, not keeled; caryopsis narrowly oblong. .... 1. *L. gracile*
1B. Spikelets narrowly ovate, distinctly flattened; fertile lemma broadly ovate, obtuse and bowed out on back; caryopsis oblong.
2. *L. sinense*

1. Lophatherum gracile Brongn. *L. elatum* Zoll. & Moritzi; *Acroelytrium japonicum* Steud.; *L. japonicum* (Steud.) Steud.; *L. pilosulum* Steud.; *L. humile* Miq.; *L. annulatum* Fr. & Sav.——SASA-KUSA. Erect, tufted perennial, some of the roots often slenderly tuberous; culms 40–80 cm. long; leaf-blades green, flat, broadly lanceolate, 10–30 cm. long, 2–5 cm. wide, short-acuminate, glabrous or sparsely pilose, reticulate beneath, rounded to a short petiolelike base; sheaths glabrous or often pilose around the mouth; ligules short, truncate; panicles long-exserted, 15–30 cm. long, the branches distant, single or in pairs, angular, simple, spikelike, glabrous or sparingly pilose; spikelets 7–8 mm. long, green, glabrous or sparsely pilose, with a tuft of short hairs at base; fertile lemma awnless; palea subhyaline, 2-keeled; sterile lemmas 3–6, the awns 1–1.5 mm. long, forming a minute 1-sided tuft; caryopsis 3.5 mm. long.——Aug.–Oct. Woods; Honshu (Kantō Distr. and westw.), Shikoku, Kyushu; locally abundant, and very variable.——Ryukyus, Formosa, s. Korea, China, India, and Malaysia.

2. Lophatherum sinense Rendl. TŌ-SASA-KUSA. Allied to the preceding; spikelets 7–8 mm. long, green, lustrous, distinctly flattened, narrowly ovate; glumes with obtuse keel, scabrous above; fertile lemma ovate to broadly so, rather acute, bowed out on back below the middle; anthers 1.5–2 mm. long; caryopsis glabrous.——Aug.–Oct. Honshu (Kinki Distr. and westw.), Shikoku, Kyushu; rare.——China.

49. PHRAGMITES Adans. YOSHI ZOKU

Tall leafy perennials with creeping rhizomes and long stolons; blades flat; panicles large, the branches ascending, rather densely spiculose; rachilla with copious long silky hairs, disarticulating above the glumes and at the base of each segment between the florets, the segments forming plumelike points below the florets; lowest floret usually staminate; glumes delicately membranous, unequal; lemmas much longer than the glumes, delicately membranous, 3-nerved, glabrous, loosely involute above as if short-awned at apex; palea short, 2-keeled; stamens 3; caryopsis glabrous, rather plump.——Few species.

1A. Stoloniferous, hirsute on nodes; first glume 1/2–3/5 as long as the lowest lemma, gradually narrowed to tip; upper leaf-sheaths usually dull purplish above; spikelets 8–12 mm. long. .... 1. *P. japonica*
1B. Rhizomatous, also with long stolons; first glume less than half as long as the lowest lemma; sheaths usually not purplish.
2A. Spikelets 12–17 mm. long; glumes 4–5, 6–8 mm. long; panicles 15–40 cm. long. .... 2. *P. communis*
2B. Spikelets 5–8 mm. long; glumes 3–4 mm. long; panicles 30–70 cm. long. .... 3. *P. karka*

1. Phragmites japonica Steud. *P. prostrata* Makino; *P. communis* var. *pumila* Hack., excl. syn.——TSURU-YOSHI, JI-SHIBARI. Tall robust stoloniferous perennial with glaucescent leaves, the stolons hairy on the nodes; culms 1.5–3 m. long, pubescent below the panicle and on the nodes; leaf-blades rather coriaceous, glaucous-green, elongate, 2–3 cm. wide, flat, gradually narrowed above; sheaths often somewhat purplish; ligules a series of hairs; panicles nodding, to 30 cm. long, broadly ovate, the branches subverticillate, scabrous, long-pilose on axils and on pedicels; spikelets 8–12 mm. long, 3- or 4-flowered; glumes 1- to 3-nerved; lowest lemma staminate, narrowly lanceolate, 6–10 mm. long, the fertile lemmas linear-lanceolate, 3-nerved; callus with long hairs on both sides; palea 2–3 mm. long; anthers about 1.5 mm. long.——Aug.–Oct. Sandy places along rivers; Honshu, Shikoku, Kyushu; common.——Korea, Ryukyus, Formosa, China, and Ussuri.

2. Phragmites communis Trin. *Arundo phragmites* L.; *A. vulgaris* Lam.; *P. longivalvis* Steud.; *P. communis* var. *longivalvis* (Steud.) Miq.; *P. nakaiana* Honda——YOSHI, KITA-YOSHI, ASHI. Culms 1–3 m. long, leafy, firm, with glabrous or sparsely pilose nodes mostly covered by the sheaths, from long-creeping rhizomes; leaf-blades 20–50 cm. long, 2–4 cm. wide, gradually narrowed and declined in upper half, sometimes sparingly pilose; panicles broadly ovate, 15–40 cm. long, the branches subverticillate, elongate, sometimes nodding, scabrous, sparsely long-hairy on axils and pedicels; spikelets

(10–)12–17 mm. long, 2- to 4-flowered; glumes unequal; lowest lemma 10–15 mm. long, gradually acuminate, the fertile lemmas involute at tip and apparently short-awned; anthers about 2 mm. long.——Aug.–Oct. Gregarious in wet lowlands; Hokkaido, Honshu, Shikoku, Kyushu; very common. ——Widespread in temperate regions.

3. Phragmites karka (Retz.) Trin. *Arundo karka* Retz.; *A. roxburghii* Kunth; *P. roxburghii* (Kunth) Steud.; *Sericura japonica* Steud.——SEI-TAKA-YOSHI, SEIKO-NO-YOSHI. Allied to the preceding; culms taller, 2–4 m. long, as much as 2 cm. in diameter at base; leaf-blades more ascending than in *P. communis,* 40–70 cm. long, 2.5–4 cm. wide, firm, flat, usually not drooping; panicles 30–70 cm. long, the spikelets 5–8 mm. long; lemmas 5–7 mm. long.——Aug.–Oct. Gregarious in wet lowlands; Honshu, Shikoku, Kyushu; locally common. ——Korea, China, Ryukyus, Formosa, India, Malaysia, Micronesia, and Australia.

50. ARUNDO L. DAN-CHIKU ZOKU

Stout, robust, tall perennials; culms often biennial or triennial and branched; leaf-blades flat, broadly linear; panicles large, spikelets 2- to 5-flowered; florets bisexual, jointed at base; glumes rather unequal, delicately membranous, 3-nerved, shorter than to as long as the florets; lemmas lanceolate, acute, delicately membranous, 3- to 5-nerved, with long white hairs on back, the lateral nerves ending in minute teeth, the midrib excurrent from apex into a short, erect awn; callus minute, short-hairy, rarely glabrous; rachilla with short internodes; palea shorter than the lemma, 2-keeled; stamens 3; ovary glabrous.——A few species, in tropical and warmer regions of the Old World.

1. Arundo donax L. *A. benghalensis* Retz.; *A. bifaria* Retz.; *A. donax* var. *benghalensis* (Retz.) Makino——DAN-CHIKU. Stout perennial with thick, short, branched rhizomes; culms 2–4 m. long, terete, 2–4 cm. in diameter, smooth, glabrous; leaf-blades flat, 50–70 cm. long, 2–5 cm. wide, glaucous-green, rather thick, gradually narrowed to tip, rounded at base; ligules truncate, 1–2 mm. long, short-ciliate; panicles erect, 30–70 cm. long, whitish with a purple hue, slightly lustrous, the branches scabrous; spikelets 8–12 mm. long, 3- to 5-flowered; glumes equal, narrowly lanceolate, acuminate, 3-nerved, slightly longer than the florets; lemmas lanceolate, 7–10 mm. long, 3- to 5-nerved, with shorter veinlets between the nerves, 2-toothed at apex, with long white hairs on back; the awn between the teeth at apex 1–3 mm. long, slender, erect; callus broadly ovate, with short hairs 1.5–2 mm. long on both sides; palea 1/2–2/3 as long as the lemma; anthers 2.5–3 mm. long.——Aug.–Nov. Sand-dunes near seashores; Honshu (Kantō Distr. and southw.), Shikoku, Kyushu; common.——Ryukyus, Formosa, China, India, and the Mediterranean region.

51. PHAENOSPERMA Munro TAKI-KIBI ZOKU

Rather large perennial, glabrous throughout, with elongate leaf-blades and large, open, erect panicles; spikelets 1-flowered, nodding or pendulous on a short jointed pedicel; glumes shorter than the floret, membranous or rather hyaline, not keeled, the margins short–connate at base; lemma chartaceous, green, lustrous, awnless, smooth, not keeled on back, convolute while young; callus fused with the base of the glumes; rachilla not prolonged; lodicules 3; stamens 3; caryopsis rather large, globose, glabrous, exserted from between the lemma and palea, mucronate, with a persistent style-base.——One species, in e. Asia.

1. Phaenosperma globosum Munro. *Garnotia japonica* Hack.——TAKI-KIBI, KASHIMAGAYA, Ō-TATSU-NO-HIGE. Rhizomes short; culms 1–1.5 m. long, erect, smooth; leaf-blades rather thin, narrowly lanceolate, 30–50 cm. long, the radical to 60 cm. long, 2–3 cm. wide, gradually narrowed to the base, petiolate, rather flat, green beneath, glaucous above; ligules lanceolate, 5–10 mm. long, smooth; panicles 30–50 cm. long, at length open and effuse, the branches single or geminate, the base globosely thickened, with 2–8 nearly simple short branches; spikelets deciduous from the very short pedicel, about 4 mm. long, pendulous, nearly truncate at base; glumes hyaline, 1- to 3-nerved, margins connate at base, the lower 2.5–3 mm., the upper 3.5–4 mm. long; lemmas about 4 mm. long, obtuse, lustrous, smooth; caryopsis 2.5–3 mm. long and as broad.——Aug.–Oct. Honshu (centr. distr. and westw.), Shikoku, Kyushu; rare.——s. Korea, China, and Formosa.

52. DIARRHENA Beauv. TATSU-NO-HIGE ZOKU

Erect, slender perennials; panicles open or narrow; spikelets 2- to 4-flowered, cylindric; florets easily deciduous from the joints at base; glumes shorter than the lemma, rather obtuse; lemmas ovate, chartaceous, convolute while young, acute to mucronate, green, usually smooth, lustrous, 3-nerved and often with a pair of short, accessory nerves outside rather close to the lateral nerve, not at all or slightly keeled above; callus short, glabrous; rachilla glabrous; palea 2-keeled; stamens 2; ovary glabrous; caryopsis ellipsoidal, exserted between the lemma and palea at maturity, beaked.——A few species, in e. Asia and N. America.

1A. Anthers about 1 mm. long; panicle-branches slender, divaricately branched; lowest lemma about 3 mm. long; palea mostly smooth on the keels; plant slender. . . . . 1. *D. japonica*
1B. Anthers about 2 mm. long; panicle-branches erect to ascending, with few, erect branchlets; lowest lemma 3.5–4.5 mm. long; palea scabrous on the keels; plant more robust. . . . . 2. *D. fauriei*

1. Diarrhena japonica Fr. & Sav. *Onoea japonica* Fr. & Sav. nom. nud.——TATSU-NO-HIGE. Culms slender, smooth, erect, from short rhizomes, 50–80 cm. long, with short lanceolate innovation-buds covered by coriaceous scales at base; leaf-blades flat, narrowly lanceolate, 20–30 cm. long, 8–15 mm. wide, narrowed at both ends, glabrous or rarely sparsely pilose above, the ligules 0.5–1 mm. long, truncate, thick-membranous; panicles exserted, 10–20 cm. long, very effuse and open, the branches single or in pairs, divaricately branched, very slender, few-flowered; spikelets green, 1- to 3-flowered, 3–5 mm. long; glumes nearly hyaline, 1–1.5 mm. long; lemmas narrowly ovate, smooth, glabrous, obtuse; keels of palea mostly smooth; anthers about 1 mm. long; caryopsis ovoid, with a whitish, mammilliform beak.——Aug.–Sept. Mountain woods; Hokkaido, Honshu, Shikoku, Kyushu; rather rare.——Korea (Quelpaert Isl.).

2. Diarrhena fauriei (Hack.) Ohwi. *Molinia fauriei* Hack.; *Neomolinia fauriei* (Hack.) Honda——HIROHA-NUMA-GAYA. Allied to but plants slightly larger than the preced-ing, and the culms stiffer, 60–80 cm. long; leaf-blades flat, 25–30 cm. long, 7–18 mm. wide, often sparingly pilose above; panicles at length exserted, narrow, the branches erect or nearly so, stiffer; spikelets 5–7 mm. long, 2- rarely 1-flowered, broadly lanceolate while young, narrowly ovate at maturity; glumes about 1.5 and 2 mm. long, respectively; lemmas narrowly ovate, 3.5–4.5 mm. long, smooth, the keels of palea scabrous above; anthers about 2 mm. long.——Aug.–Sept. Mountain woods; Honshu (Shinano Prov.); very rare.——Korea, Manchuria, and Ussuri.

53. MOLINIOPSIS Hayata   NUMAGAYA ZOKU

Erect perennials, from short, creeping, ligneous rhizomes, a portion of the thickened basal part of culms persistent for several years; spikelets 2- to 8-flowered, rather flattened, articulate below the florets; florets bisexual; glumes slightly distant, thinly chartaceous, keeled, shorter than the lemmas; lemmas chartaceous, smooth, glabrous, lustrous, convolute while young, acute, awnless, not keeled, 3-nerved; callus slightly elongate, with a tuft of hairs; rachilla nearly glabrous; palea as long as the lemma, 2-keeled, ovate; stamens 3; ovary glabrous; caryopsis oblong.——Few species, in e. Asia. Differing from the European *Molinia,* in the hairy callus.

1. Moliniopsis japonica (Hack.) Hayata. *Molinia japonica* Hack.; *Moliniopsis spiculosa* Honda, excl. syn.; *Fluminea spiculosa* Honda, excl. syn.; *Graphephorum nipponicum* Honda; *Moliniopsis nipponica* (Honda) Honda——NUMA-GAYA. Densely tufted perennial from short rhizomes; culms firm, smooth, 30–100 cm. long, the few basal short, thickened internodes persistent for several years; leaf-blades linear, 20–50 cm. long, 2–10 mm. wide, with loosely involute margins, glaucous above, obscurely jointed at base with the sheath, the sheaths with a ring of short hairs; panicles 10–40 cm. long, narrow to open, branches single or geminate, sometimes ternate, with few long hairs in the axils while young; spikelets 8–12 mm. long, lustrous, 2- to 6-flowered; glumes narrowly deltoid, acute, 1- to 3-nerved, 3–5 mm. and 4–5 mm. long, respectively; lemmas broadly lanceolate, rounded on back, 4–5.5 mm. long; callus-hairs about 1 mm. long; anthers 2–3 mm. long; stigma often purplish.——Aug.–Oct. Wet places in lowlands to alpine regions, sometimes in sphagnum bogs; Hokkaido, Honshu, Kyushu; rather common.——Korea and s. Kuriles.

54. HAKONECHLOA Makino ex Honda   URAHAGUSA ZOKU

Tufted perennial with elongate culms, not thickened at base; leaf-blades broadly linear, glaucous above; spikelets in loose panicles, 5- to 10-flowered, slightly flattened; glumes thinly chartaceous, 1- to 3-nerved, acute, scarcely keeled, slightly distant; lemmas convolute when young, chartaceous, lustrous, narrowed into a slightly bidentate apex, the midnerve excurrent between minute teeth into an erect awn about 3 mm. long, the internerves glabrous, the margins appressed-pilose; rachilla elongate, pilose, disarticulating at the base of each segment, forming a hairy stipe below the floret; palea slightly shorter than the lemma, sparsely long-pilose on the margins; stamens 3; ovary glabrous; stigma deep yellow or deep purple when dried; caryopsis oblong, scarcely flattened.——One species in Japan.

1. Hakonechloa macra (Munro) Makino. *Phragmites macra* Munro——URAHAGUSA. Creeping rhizomes and stolons covered with short, lustrous, rather coriaceous, white scales; culms slender, 40–70 cm. long, ascending to spreading, smooth; leaf-blades broadly linear, 10–25 cm. long, 4–8 mm. wide, flat or with loosely involute margins, glabrous, glaucous above, narrowed at both ends; sheaths sometimes dull purplish with a minute ring of short hairs on the collar; ligules nearly obsolete; panicles open, ovate, 5–15 cm. long, the branches geminate, scabrous; spikelets yellowish green, linear-oblong, 1–2 cm. long; glumes broadly lanceolate, acute, 3–4 mm. and 4–5 mm. long, respectively; lemmas broadly lanceolate, 6–7 mm. long, the awn slender, erect, 1/2–4/5 as long as the lemma; rachilla joints about 2 mm. long, densely clothed with white hairs 1–1.5 mm. long; anthers 2–3 mm. long; caryopsis about 2 mm. long, glabrous, with a shallow groove on ventral side at base.——Aug.–Oct. Wet rocky cliffs in mountains; Honshu (Tōkaidō Distr. from Sagami to Kii Prov.); rather rare. Variegated forms are cultivated as ornamental pot-plants.

55. ERAGROSTIS Beauv.   SUZUMEGAYA ZOKU

Glabrous to pilose annuals or perennials with linear, flat or involute leaf-blades and paniculate inflorescence; spikelets few- to many-flowered, flat; florets bisexual; rachilla articulate below the florets, sometimes not disarticulate until after the fall of the lemmas and paleas with the caryopsis; glumes 2, shorter than the floret; lemmas membranous or chartaceous, ovate, rather acute to obtuse, imbricate, folded, glabrous or very rarely puberulent, 3-nerved, entire, awnless, sometimes falling with the caryopsis; palea attached to the persistent rachilla with slightly curved keels; stamens 3 or fewer; ovary glabrous; caryopsis ellipsoidal, not adherent to the lemma and palea.——More than 100 species, chiefly in warm-temperate and tropical regions.

1A. Spikelets less than 2 mm. long; florets 0.7–0.8 mm. long; rachilla jointed, deciduous; anthers 1/6–1/4 as long as the lemma; culms strict; panicles narrow and elongate, 10–60 cm. long. .... 1. *E. japonica*
1B. Spikelets more than 2 mm. long; florets 1 mm. long or more; rachilla rather persistent.
  2A. Spikelets, pedicels, sheaths and underside of leaf-blades sometimes, at least in part, with glandular depressions.
    3A. Annuals; anthers less than 1 mm. long; lemmas rather obtuse.
      4A. Spikelets about 2.5 mm. wide; lemmas and glumes with few glands on the keel. .... 2. *E. cilianensis*
      4B. Spikelets 1.5–2 mm. wide; lemmas and glumes with obscure glands or glandless. .... 3. *E. poaeoides*
    3B. Perennial; anthers about 1 mm. long; spikelets without glands. .... 4. *E. ferruginea*
  2B. Spikelets and other parts not glandular.
    5A. Annuals; spikelets less than 1.5 mm. wide.
      6A. Mouth of leaf-sheaths and axils of panicles with tufts of few long hairs. .... 5. *E. pilosa*
      6B. Leaf-sheaths, axils of panicles and other parts without long hairs. .... 6. *E. multicaulis*
    5B. Perennials; spikelets 1.5–2 mm. wide, 5–20 mm. long.
      7A. Spikelets dull green to brownish, not purplish, lateral ones on short to rather long, ascending pedicels. ..... 7. *E. bulbillifera*
      7B. Spikelets dull grayish purple, lateral ones on short appressed pedicels. .... 8. *E. aquatica*

1. Eragrostis japonica (Thunb.) Trin. *Poa japonica* Thunb.; *E. tenella* sensu auct. Japon., non Beauv.; *E. aurea* Steud.; *E. interrupta* var. *tenuissima* sensu Matsum. & Hayata, non Stapf——KOGOME-KAZE-KUSA. Annual; culms erect, branched at base, 50–120 cm. long; leaf-blades 10–20 cm. long, 2–5 mm. wide, glabrous; panicles straight, 20–60 cm. long, sometimes shorter, narrow, erect, the branches single, numerous, 3–6 cm. long, slender, branching from base; spikelets rather densely arranged, ovate or broadly so, 1–1.5 mm. long, 1–1.2 mm. wide, rather flat, 3- to 5-flowered, pale green and slightly red-tinged; glumes 1-nerved, obtuse, about 0.5 mm. long; lemmas elliptic, very obtuse, about 0.7 mm. long; palea as long as the lemma, obtuse, the keels scaberulous; caryopsis ellipsoidal, nearly terete, smooth, about 0.3 mm. long.—— Aug.–Sept. Honshu, Shikoku, Kyushu.——s. Korea, China, Ryukyus, and Formosa.

2. Eragrostis cilianensis (All.) Lutati. *Poa cilianensis* All.; *Briza eragrostis* L.; *Poa megastachya* Koeler; *E. major* Host; *E. megastachya* (Koeler) Link——SUZUMEGAYA. Tufted annual; culms 20–60 cm. long, usually ascending at base, with a ring of glands below each node; leaf-blades flat or loosely involute, 10–20 cm. long, 2–6 mm. wide, green, with indistinct glands beneath, on the margin, and on the back of the sheath, long-hairy at the mouth; panicles 5–20 cm. long, the branches single, spreading, with few long hairs on the axils; pedicels with a prominent gland; spikelets 5–15 mm. long, 2.5–3 mm. wide, 10- to 30-flowered, pale lead-green; glumes about 2 mm. long; lemmas ovate, rather obtuse, 2–2.5 mm. long, smooth; palea about ¾ as long as the lemma; caryopsis broadly ovate, 0.6 mm. long.——Aug.–Oct. Waste grounds and along roadsides; Honshu, Shikoku, Kyushu; rather common.——Widespread in warm-temperate and tropical regions.

3. Eragrostis poaeoides Beauv. ex Roem. & Schult. *Poa eragrostis* L.; *E. eragrostis* (L.) Beauv.; *E. minor* Host ——KO-SUZUMEGAYA. Smooth, rather tufted annual; culms rather slender, 10–40 cm. long, often with a ring of glands below each node; leaf-blades 3–10 cm. long, 2–5 mm. wide, often sparsely pilose; panicles ovate, 5–20 cm. long, the branches spreading; pedicels with a nodelike sessile gland on upper half; spikelets 3–8 mm. long, 1.2–1.7 mm. wide, pale green to lead-green, 5- to 12-flowered; glumes acute; lemmas ovate, rather obtuse, smooth, 1.7–2 mm. long; caryopsis ellipsoidal, slightly laterally compressed.——Aug.–Oct. Waste grounds and along roadsides; Honshu, Shikoku, Kyushu; rather common.——Widespread in warm-temperate and tropical regions.

4. Eragrostis ferruginea (Thunb.) Beauv. *Poa ferruginea* Thunb.; *P. barbata* Thunb.; *E. orientalis* Trin.; *E. pogonia* Steud.; *E. thunbergii* Koidz., non Franch., 1892—— KAZE-KUSA. Tufted perennial; culms 30–80 cm. long; leaf-blades rather firm, 30–40 cm. long, 2–6 mm. wide, with loosely involute margins, long white hairs on upper side near base, and a ring of short hairs around the mouth outside; panicles narrowly ovate, 20–40 cm. long, the branches rather numerous, spreading, scabrous, glabrous, with a sessile, yellowish gland on pedicels above the middle; spikelets lanceolate to narrowly oblong, flat, 6–10 mm. long, acute, 5- to 10-flowered, dull purplish, rarely pale green; glumes narrowly lanceolate, acute, 1-nerved, the first 1–2 mm., the second 2–2.5 mm. long; lemmas membranous, lustrous, narrowly ovate, 2.5–3 mm. long, rather acute, slightly shorter than the palea; anthers 0.8–1.2 mm. long; caryopsis slightly compressed laterally, ellipsoidal, about 1 mm. long.——Aug.–Oct. Waste grounds and along roadsides; Honshu, Shikoku, Kyushu; common.—— Korea, Manchuria, China, and the Himalayas.

5. Eragrostis pilosa (L.) Beauv. *Poa pilosa* L.; *P. verticillata* Cav.; *E. verticillata* (Cav.) Beauv.; *E. filiformis* Link ——Ō-NIWA-HOKORI. Annual; culms erect from ascending base, tufted, slender, 30–70 cm. long; leaf-blades 10–20 cm. long, 2–4 mm. wide with loosely involute margins; sheath glabrous, with long hairs at the mouth; panicles 10–25 cm. long, narrowly ovate, the branches single or subverticillate and rather many, spreading, with long hairs in axils; spikelets lanceolate, rather flat, 3–5 mm. long, 0.7–1 mm. wide, 5- to 10-flowered, lead-green, often tinged with reddish purple; glumes broadly lanceolate, 1-nerved, rather obtuse, the first 0.5–0.7 mm., the second about 1.5 mm. long; lemmas ovate, rather acute, 3-nerved, about 1.5 mm. long, deciduous with the palea; anthers about 0.25 mm. long; caryopsis 0.6 mm. long, smooth. ——Aug.–Oct. Honshu, Shikoku, Kyushu; very common in lowlands.——Widespread in warm-temperate and tropical regions.

6. Eragrostis multicaulis Steud. *Glyceria airoides* Steud.; *E. pilosa* var. *nana* Miq.; *E. pilosa* var. *imberbis* Franch.; *E. pilosa* var. *condensata* Hack.; *E. damiensiana* (E. Bonn.) E. Bonn. ex Thell.; *E. peregrina* Wieg.; *E. niwahokori* Honda——NIWA-HOKORI. Allied to the preceding but smaller; culms 7–30(–50) cm. long; leaf-blades and sheaths glabrous, smooth; panicles 6–10 cm. long, without long hairs in the axils; spikelets ovate to oblong, rather dense, 4- to 8-flowered, 2–3.5 mm. long, 1–1.5 mm. wide; stamens 2 or 3; anthers elliptic, about 0.25 mm. long.——Aug.–Oct. Waste grounds and cultivated fields; Hokkaido, Honshu, Shikoku,

Kyushu; very common.——Sakhalin, Korea, Ussuri, Amur, China, Ryukyus, Formosa, Malaysia; naturalized in Europe and America.

7. Eragrostis bulbillifera Steud. *E. brownii* sensu Miq., non Nees; *E. bahiensis* sensu auct. Japon., non Schrad.; *E. atrovirens* sensu auct. Japon., non Trin.——ITO-SUZUMEGAYA. Perennial with short rhizomes and lanceolate, short innovations; culms 30–60 cm. long, slender but rather stiff, smooth; leaf-blades 10–20 cm. long, 1–3 mm. wide, with loosely involute margins, rather firm, nearly glabrous, with long hairs near base, the ligules very short; panicles broadly ovate, 10–20 cm. long, 8–12 cm. wide, the branches with a few hairs in the axils, branchlets short, ascending or nearly erect; spikelets lead-green to brownish green, dull, linear, 5–10 mm. long, 1.7–2 mm. wide, 15- to 25-flowered; glumes rather equal, 1.2–1.5 mm. long, acute; lemmas ovate to broadly so, about 2 mm. long, rather acute, deciduous; palea at length deciduous, slightly shorter than the lemma; anthers less than 1 mm. long; caryopsis ellipsoidal, about 0.6 mm. long, lustrous.——Aug.-Oct. Grassy places and waste grounds in lowlands; Honshu (Tōkaidō and westw.), Shikoku, Kyushu; rare.——Ryukyus, Formosa, China, and Indochina.

8. Eragrostis aquatica Honda. NUMA-KAZE-KUSA. Somewhat resembles the preceding; culms 30–50 cm. long; leaf-blades 5–20 cm. long, 2–4 mm. wide; sheaths sparsely long-hairy near the mouth; panicles 10–20 cm. long, loosely spiculose, the branches slender but rather stiff; pedicels short, appressed to the branches or branchlets; spikelets linear, 8–15 mm. long, 1.3–1.5 mm. wide, flattened, grayish brown to purplish, dull, rather densely (5–)10- to 20-flowered; glumes lanceolate, acute, 1–1.5 mm. and 1.5–2 mm. long respectively; lemmas deciduous, ovate, 1.7–2 mm. long, acute, rather weakly but distinctly 3-nerved; palea persistent at first, 1.2–1.5 mm. long; anthers less than 1 mm. long; caryopsis oblong, less than 1 mm. long.——Aug.-Nov. Honshu (w. distr.); rare.

56. ELEUSINE Gaertn.   O-HI-SHIBA ZOKU

Annuals; leaf-blades linear, inflorescence digitate, with few branches, somewhat laterally flattened; spikelets densely arranged in 2 rows on one side of the branches, nearly sessile, 3- to many-flowered, jointed below the florets; florets bisexual; glumes 2, shorter than the lemmas, keeled; lemmas narrowly oval to ovate, keeled, rather obtuse, with 3 closely parallel nerves, glabrous, awnless; rachilla glabrous; palea 2-keeled, slightly shorter than the lemma; caryopsis with a thin, loose pericarp, or with a naked seed early breaking away from the pericarp, dark brown, with fine oblique cross lines.——Few species, in warm-temperate and tropical regions.

1. Eleusine indica (L.) Gaertn. *Cynosurus indicus* L.; *E. japonica* Steud.; *E. indica* var. *oligostachya* Honda——O-HI-SHIBA. Annual; culms tufted, 30–80 cm. long, smooth; leaf-blades green, slightly folded or flat, 15–40 cm. long, 3–7 mm. wide, usually sparingly long-hairy near base; ligules less than 1 mm. long, hyaline, denticulate; branches of digitate inflorescence 2–6, very rarely solitary, ascending, 7–15 cm. long, 3–4 mm. wide, the rachis about 1 mm. wide, glabrous except the axils; spikelets laterally flattened, pale green, ovate, 4–5 mm. long, 4- to 5-flowered, slightly lustrous; glumes lanceolate, obtuse, with a scabrous, rather acute keel on back, the first 1.5–2 mm. long, the second slightly longer with 1 or 2 pairs of short lateral nerves; lemmas broadly lanceolate, 3–3.5 mm. long, acutely keeled on back; anthers elliptic, about 0.7 mm. long; caryopsis ovoid, 1.5 mm. long, with 3 obtuse angles, enclosed in a loose paricarp.——Aug.-Oct. Waste grounds and cultivated fields; Honshu, Shikoku, Kyushu; one of the most common weeds in Japan.——Widespread in warm temperate and tropical regions.

Eleusine coracana (L.) Gaertn. *Cynosurus coracanus* L.; *E. indica* var. *coracana* (L.) Fiori——SHIKOKU-BIE. A stout grass with branches about 1 cm. wide, brownish when mature, and with larger seeds.——Rarely cultivated on sterile soils.

57. KENGIA Packer   CHŌSEN-GARIYASU ZOKU

Somewhat tufted perennials with many-noded culms and short, flat to involute leaf-blades; inflorescence often cleistogamous within the sheaths; panicles rather few-flowered, with few, ascending to spreading, simple branches; spikelets short-pedicelled, slightly flattened laterally, 2- to 10-flowered, articulate below the florets; florets bisexual; glumes shorter than the lemma, membranous, 1- to 5-nerved; lemmas lanceolate, thinly chartaceous, 3-nerved, slightly keeled, pilose toward the margin, nearly entire or with 2 minute teeth at apex, awnless or with a short straight apical awn; callus obtuse, short-hairy; rachilla scabrous; palea as long as the lemma, 2-keeled; caryopsis lanceolate, glabrous, nearly terete.——Few species, in temperate regions of Asia and e. Europe.

1. Kengia hackelii (Honda) Packer. *Cleistogenes hackelii* (Honda) Honda; *C. serotina* var. *aristata* (Hack.) Y. L. Keng; *Diplachne serotina* var. *chinensis* Maxim., pro parte; *D. serotina* var. *aristata* Hack.; *D. hackelii* Hondo——CHŌSEN-GARIYASU. Erect perennial with short rigid rhizomes and small, ovoid, rather flattened, rigid innovation-buds covered at base with few, yellowish, lustrous scales; culms 40–100 cm. long, smooth, sometimes with a cleistogamous inflorescence within the upper sheath; leaf-blades flat or with loosely involute margins, somewhat spreading, linear, 4–10 cm. long, 3–6(–10) mm. wide, gradually acute, sometimes with a few long hairs on both sides; sheaths short, sparsely long-hairy near the mouth and on upper margin; ligules very short, minutely ciliate; panicles 4–8 cm. long, more or less one-sided, with 2 or 3 nearly simple, ascending branches; pedicels short, appressed to the branches; spikelets narrowly lanceolate, 2- to 4- sometimes 1-flowered, lead-green with reddish purple markings; glumes unequal, membranous, 1-nerved, the first narrowly ovate, obtuse, about 1 mm. long, the second broadly lanceolate, acute, 1.5–2 mm. long; lemmas lanceolate, 4–5 mm. long, with a slender awn 2–4 mm. long in cleistogamous florets, the awn longer than the lemma; anthers 1.5–2 mm. long in chasmogamic florets.——Aug.-Oct. Honshu, Shikoku, Kyushu.——Korea and China, Ussuri, and Mongolia.

58. DIPLACHNE Beauv.　HAMAGAYA ZOKU

Culms sometimes branched; leaf-blades linear, with involute margins; inflorescence a panicle or compound spike; spikelets short-pedicelled, somewhat laterally flattened, narrowly oblong, 3- to 8-flowered, rather densely arranged in 2 rows on the branches, articulate below the florets; glumes membranous, slightly folded, unequal, 1-nerved, shorter than the lemmas; lemmas oblong, keeled, chartaceous, 3-nerved, 2- to 4-toothed at apex, the midrib often shortly excurrent at apex; callus short; rachilla glabrous; palea as long as the lemma, 2-keeled; stamens 3; ovary glabrous; caryopsis free within the lemma and palea.——Cosmopolitan, with many species in warmer and tropical regions.

1. Diplachne fusca (L.) Beauv. *Festuca fusca* L.; *Bromus polystachyus* Forsk.; *Leptochloa fusca* (L.) Kunth; *D. fascicularis* sensu auct. Japon., non Beauv.; *Leptochloa fascicularis* sensu auct. Japon., non A. Gray; *D. polystachya* (Forsk.) Backer——HAMAGAYA, MITSUBAGAYA, TAKAO-BARENGAYA. Annual; culms erect or decumbent, branched at base, 30–60 cm. long; leaf-blades rather flat, 20–30 cm. long, 2–5 mm. wide, glaucous-green; ligules hyaline, 3–4 mm. long, denticulate, smooth, glabrous; panicles 15–25 cm. long, narrowly ovate, rather dense, pale green, sometimes lead-gray, rarely dull red, the branches 5–10 cm. long, numerous, nearly simple; pedicels short, appressed to the branch; spikelets lanceolate, 7–10 mm. long, densely 8- to 14-flowered; glumes unequal, broadly lanceolate, acuminate, the lower about 2 mm., the upper 3–3.5 mm. long; lemmas narrowly oblong, minutely appressed-puberulent on the margin below the middle, the midrib excurrent into an awn 1–2 mm. long; anthers elliptic, about 0.3 mm. long; caryopsis lanceolate, biconvex, about 2 mm. long.——Near seashores; Honshu (Ise, Totomi, and Kazusa Prov.), Kyushu (Chikuzen Prov.); rare.——Formosa, China, Malaysia, India, Australia, and Africa.

59. LEPTOCHLOA Beauv.　AZEGAYA ZOKU

Annuals or perennials; leaf-blades linear; panicles with an elongate axis and nearly simple branches; spikelets sessile or short-pedicelled, crowded in 2 rows on side of the rachis, 2- to 7-flowered, rather flat, articulate below the florets; florets bisexual; glumes 2, membranous, 1-nerved; lemmas small, membranous, obtuse or nearly so, folded, usually awnless, 3-nerved, appressed-puberulent; callus minute; rachilla glabrous; palea 2-keeled; caryopsis smooth, glabrous, ovate, obsoletely rugose.——Widespread in temperate and warmer regions.

1A. Leaf-sheaths glabrous; spikelets 5- to 7-flowered, 2.5–3 mm. long; lemmas 1–1.2 mm. long. .... 1. *L. chinensis*
1B. Leaf-sheaths sparsely papillose-pilose with long hairs; spikelets 2- or 3-flowered, about 1.5 mm. long; lemmas 0.6–0.8 mm. long.
　　　　　　　　　　　　　　　　　　　　　　　　　　　　　　　　　　　2. *L. panicea*

1. Leptochloa chinensis (L.) Nees. *Poa chinensis* L.; *L. eragrostoides* Steud.; *L. tenerrima* sensu auct. Japon., non Roem. & Schult.——AZEGAYA. Annual; culms ascending from branching base, 30–70 cm. long; leaf-blades membranous, green and slightly glaucous, 7–15 cm. long, 3–8 mm. wide, rather flat; ligules about 1 mm. long; panicles at length exserted, narrowly ovate, 15–40 cm. long, the branches rather numerous and simple, irregular on the main-axis, spreading, 4–10 cm. long, filiform, compressed-triangular, scabrous, spikelet-bearing from near base; spikelets short-pedicelled, appressed or erect, pale green and slightly reddish; glumes broadly lanceolate, acute, the first 1-nerved, 0.7–1 mm. long, the second 3-nerved, 1.2–1.5 mm. long; lemmas membranous, 1–1.2 mm. long, obtuse to rather rounded at apex, with sparse, short, appressed hairs; anthers minute, 0.15–0.2 mm. long; caryopsis ellipsoidal, 0.8 mm. long, dorsiventrally slightly compressed, with obsolete, linear-lanceolate areolae caused by minute longitudinal striations and short transverse lines connecting them.——Aug.–Oct. Honshu, Shikoku, Kyushu; rather common.——China, Formosa, Malaysia, India, and Australia.

2. Leptochloa panicea (Retz.) Ohwi. *Poa panicea* Retz.; *L. filiformis* sensu auct. Japon., non Beauv.——ITO-AZEGAYA. More slender than the preceding; culms 50–100 cm. long from ascending base; leaf-blades 10–20 cm. long, 3–6 mm. wide, flat, thin, with spreading, long, whitish hairs callose at base; sheaths sparingly long-pilose; ligules hyaline, truncate, 0.5–1 mm. long; panicles at length exserted, 20–30 cm. long, the branches spreading, slender, 5–10 cm. long; spikelets about 1.5 mm. long, rather flat; glumes unequal, 1-nerved, membranous, broadly lanceolate, 0.7–1 mm. long; lemmas ovate, 0.6–0.8 mm. long, white puberulent; anthers about 0.1 mm. long, elliptic; caryopsis ovoid, obsoletely rugulose, about 0.5 mm. long.——Kyushu; rather rare.——Ryukyus, China, Formosa, and tropical Asia.

60. TRIPOGON Roth　TORIKOGUSA ZOKU

Densely tufted perennials; leaf-blades very narrow, often filiform with involute margins; culms slender, often delicate; inflorescence a simple terminal spike; spikelets somewhat flattened laterally, in 2 rows on one side of the axis, nearly sessile, 3- to 10-flowered, jointed below the florets; glumes 2, unequal, 1-nerved or the upper 3-nerved, narrow, rather membranous; lemmas 3-nerved, thinly chartaceous, narrow, truncate, keeled, the midrib excurrent into a straight or geniculate awn, the lateral nerves ending in a short tooth or mucro; callus obtuse, short-hairy; rachilla glabrous; palea as long as the lemma, 2-keeled; stamens 3; caryopsis glabrous.——With more than a dozen species in warm regions.

1A. Glumes 1-nerved; lemma with a geniculate awn extending from the midrib; anthers 1–1.5 mm. long. .... 1. *T. japonicus*
1B. Glumes unequal, the first 1-nerved, the second 3-nerved; lemma with a short straight awn extending from the midrib; anthers 0.7–1 mm. long. .... 2. *T. chinensis* var. *coreensis*

1. **Tripogon japonicus** (Honda) Ohwi. *T. longearistatus* Honda var. *japonicus* Honda——FUKURODAGAYA. Rhizomes short; culms 15–25 cm. long, densely tufted, slender, smooth, with old sheaths bearing intravaginal innovations from the base; leaf-blades narrowly linear or subulate, involute or with involute margins, the radical to 20 cm. long, with long white hairs near base on upper side; ligules very short; spikes at length exserted, erect, 10–15 cm. long, narrow; spikelets in 2 rows on the axis, secund, sessile, appressed, linear-oblong, 6- to 8-flowered, 5–7 mm. long, pale lead-green, lustrous; glumes lanceolate, 1-nerved, the first acute to mucronate at apex, nearly 3 mm. long, the second gradually narrowed to tip, ending in a short awn about 1 mm. long; lemmas narrowly oblong, 2.5–3 mm. long, obtuse with 2 hyaline teeth at apex; awn 4–5 mm. long, geniculate; anthers 1–1.5 mm. long.——Honshu (Hitachi and Shimotsuke Prov.); very rare.

2. **Tripogon chinensis** Hack. var. **coreënsis** Hack. *T. coreënsis* (Hack.) Ohwi——NEZUMIGAYA. Densely tufted, strict perennial; culms 20–30 cm. long, with few nodes at base, striate, very slender but rather stiff, glabrous, simple; leaf-blades filiform, 3–5 cm. long, 0.5–0.7 mm. wide, scarcely thicker than the culms, erect, tightly convolute to loosely involute, smooth outside, long-pilose around the mouth inside and often sparsely so or glabrous on margin; ligules very short, hyaline, 0.5 mm. long; spikes solitary on the top of the culms, 7–10 cm. long, about 2 mm. across; rachis slender, smooth; spikelets sessile, nearly cylindric, 3- or 4-flowered, 4–6 mm. long, 1.2 mm. wide, lead-green; glumes linear-lanceolate, acuminate, scarious, scaberulous above; lemmas narrowly oblong, about 3 mm. long, smooth or obsoletely punctate, shallowly bifid, with 2 hyaline, semirounded erose tips, the lateral nerve excurrent as a minute mucro, the midnerve excurrent as an erect scaberulous delicate awn 1–2 mm. long; callus with hairs about 1/6–1/5 as long as the lemma; rachilla-joints slender, glabrous, about half as long as the lemma.——Oct. Kyushu (Hirado Isl. in Hizen); very rare.——Korea.

61. CYNODON L. C. Rich. GYŌGI-SHIBA ZOKU

Firm, low perennials with long stolons, short leaves, and digitate inflorescence; spikelets nearly sessile, in 2 rows on one side of flattened simple branches of the inflorescence, strongly laterally flattened, 1-flowered, awnless, articulate below the floret; glumes 2, 1-nerved, folded, acute, shorter than the lemma, unequal or barely so, membranous; lemmas ovate, strongly folded, flattened, 3-nerved, chartaceous, smooth, pubescent on the keel; rachilla prolonged behind the floret as a bristle, glabrous; palea 2-keeled, lanceolate, as long as the lemma; stamens 3; ovary glabrous.——Few species, widely distributed, chiefly in Africa and Australia.

1. **Cynodon dactylon** (L.) Pers. *Panicum dactylon* L.; *Capriola dactylon* (L.) O. Kuntze——GYŌGI-SHIBA. Culms rather firm, 15–40 cm. long, from axils of long creeping, leafy stolons; leaf-blades pale green, rather flat or with involute margins, 5–8 cm. long, 1.5–4 mm. wide, short-linear, sparsely long-pilose especially near the very short ligule; inflorescence digitate; spikes 3–7, ascending to spreading, 2.5–5 cm. long, sessile; spikelets 2–3 mm. long, rather dense, appressed to the axis, pale green, sometimes with a pale reddish hue; glumes unequal, lanceolate, 1-nerved, acute, the first incurved and appressed to the lemma, about 1.5 mm. long, the second ascending, about 2 mm. long, slightly shorter than the lemma; lemmas ovate, folded, obtuse or mucronate, 2.5–3 mm. long, the keel slightly curved, appressed-pubescent; anthers 1–1.5 mm. long——June–Aug. Sunny places; Hokkaido, Honshu, Kyushu, Shikoku; common.——Widespread in all temperate and tropical regions.

Var. **nipponica** Ohwi. Ō-GYŌGI-SHIBA. A robust phase with spikes 5–10 cm. long; spikelets 3–3.5 mm. long.——Honshu and westward; connected by intermediates with the typical phase.

62. CHLORIS Sw. O-HIGE-SHIBA ZOKU

Annuals or perennials; leaf-blades flat, sometimes folded; inflorescence digitate, the branches 2–8; spikelets sessile, rather flattened, with 1 fertile floret and 1 to few sterile ones above, articulate below the fertile floret; glumes 2, narrow, acute, membranous, folded, keeled; lemmas chartaceous, usually 3-nerved, broadly ovate, folded, pubescent on the keel and marginal nerves, 2-toothed at apex, awned from between the teeth; sterile lemmas usually truncate at apex, often awned; callus short-hairy; rachilla glabrous; palea slightly shorter than the lemma, 2-keeled; stamens 3; caryopsis glabrous, smooth, the hilum minute.——About 40 species, chiefly in the Tropics.

1. **Chloris virgata** Sw. *C. caudata* Trin.——O-HIGE-SHIBA. Rather stout, firm annual; culms 30–50 cm. long; leaf-blades 10–25 cm. long, 4–7 mm. wide, flat; ligules nearly obsolete; uppermost sheath slightly inflated, elongate, the blade short; branches of inflorescence about 10, sparsely pilose, erect or ascending, 3–7 cm. long, sessile; spikelets numerous on one side of the branches, dense, pale green and sometimes dull purple; glumes narrowly lanceolate, the first 1.5 mm. long, the second awn-pointed, 3–4 mm. long; lemma obtuse, 3.5–4 mm. long, whitish hairy on upper margins, the keel obtuse, pubescent, bowed out about the middle; the awn delicate, erect, 10–15 mm. long; anthers 0.3 mm. long; sterile floret single, clavate, 1.5–2 mm. long, long-awned.——Naturalized (?); Kyushu; very rare.——China, Korea, centr. Asia, Africa, and America.

63. COELACHNE R. Br. HINAZASA ZOKU

Delicate grasses with flat, lanceolate, short leaf-blades, and open to narrow, small panicles; spikelets pedicelled, or 1- or 2-flowered, the lower bisexual, the upper, when present, pistillate; glumes 2, hyaline, with 1–3 obscure nerves, not keeled; lemmas longer than the glumes, scarcely keeled, hyaline, glabrous, awnless, nearly nerveless; callus and rachilla glabrous; palea as long as the lemma, with 2 faint keels; loosely involute; stamens 2 or 3; caryopsis lustrous, slightly dorsiventrally flattened, glabrous, the hilum minute.——Two or three species, in e. and s. Asia, Australia, and Africa.

1. Coelachne japonica Hack. *C. pulchella* sensu auct. Japon., non R. Br.——HINAZASA.   Annual; culms 5–20 cm. long, smooth, from a decumbent or ascending base, branching below, with puberulent nodes; inflorescence scarcely exserted from the sheaths of the upper nodes; leaf-blades flat, soft-herbaceous, spreading, lanceolate, 1–3 cm. long, 2–6 mm. wide, short-acuminate, glabrous, nearly rounded at base, the nerves prominent on upper surface; sheaths short, sparsely short-pubescent near the apex; ligules wanting; panicles short-exserted, open, loose, ovate, 15–30 mm. long, 1–2 cm. wide, the branches few, spreading; spikelets 1 to 3 on the branches, usually pale green, about 2.5 mm. long; glumes rather persistent, membranous, glabrous, not keeled, the first ovate, with a nearly rounded summit, about 0.7 mm. long, obsoletely 1-nerved, the second obsoletely 3-nerved, with few short bristles on upper part of the nerves; lemmas about 2.5 mm. long; stamens 2, the anthers about 0.25 mm. long, elliptic; caryopsis thick-lenticular, ovate, glabrous, lustrous, the hilum indistinct. ——Aug.–Oct. Wet places; Honshu, Kyushu; rather rare.

64. SPOROBOLUS R. Br.   NEZUMI-NO-O ZOKU

Annuals or perennials; leaf-blades often involute, long-linear to short-linear; panicles sometimes effuse, usually contracted; spikelets 1-flowered, nearly terete, bisexual; glumes thinly chartaceous to membranous, 1-nerved, sometimes faintly so, lustrous, the first usually somewhat shorter than the floret; lemmas usually glabrous, loosely involute, awnless, 1-nerved; rachilla not produced; palea with 2 faint keels; stamens 2 or 3; caryopsis with the pericarp early deciduous, exposing the seed, falling free at maturity.——Many species in warm-temperate and tropical regions.

1A. Low annual 5–20 cm. high; leaves chiefly cauline; leaf-blades narrowly lanceolate, long-ciliate, the cilia callose at base; spikelets brown.
　　　1. *S. japonicus*
1B. Tall perennial 20–80 cm. high; leaves chiefly basal; leaf-blades linear, elongate, glabrous or sparsely hairy above; spikelets lead-gray to purple-tinged.
　　2A. Plant tufted, without creeping rhizomes; panicles 15–40 cm. long; first glume about ⅓, the second ½ to ⅔ as long as the lemma.
　　　2. *S. elongatus*
　　2B. Plant with extensive creeping rhizomes; panicles 4–8 cm. long; glumes and lemmas about equal in length. . . . . 3. *S. virginicus*

1. Sporobolus japonicus (Steud.) Maxim. *Agrostis japonica* Steud.; *S. ciliatus* sensu auct. Japon., non Presl; *S. ciliatus* var. *japonicus* (Steud.) Hack.; *S. piliferus* sensu auct. Japon., non Kunth——HIGE-SHIBA.   Small erect tufted annual; culms 5–30 cm. long, simple or branching at base, geniculate and ascending at base, glabrous; leaf-blades flat or with loosely involute margins, narrowly lanceolate, 4–10 cm. long, 2–5 mm. wide, spreading, papillose-pilose toward the base, long-ciliate, the cilia callose at base; ligules a series of dense, short hairs; panicles erect, linear, 3–7 cm. long, about 5 mm. wide, the branches erect, short, glabrous, rather thick; spikelets 2–2.2 mm. long, at first broadly lanceolate, becoming ovoid in maturity, lustrous, deep-brown; first glume about 1.5 mm. long, acute, involute, the second slightly shorter than the floret, narrowly ovate; lemma ovate to narrowly so, obtuse; stamens 3, the anthers elliptic, about 0.3 mm. long; caryopsis ovoid, 2/3–3/4 as long as the lemma.——Aug.–Oct. Wet places; Honshu, Shikoku, Kyushu; rather common.——China.

2. Sporobolus elongatus R. Br. *S. indicus* sensu auct. Japon., non R. Br.; *Cinna japonica* Steud.; *Agrostis fertilis* Steud.——NEZUMI-NO-O.   Tufted perennial; culms erect, 30–80 cm. long, firm, glabrous; leaves chiefly radical, the blades firm, linear, 20–60 cm. long, 1.5–5 mm. wide, loosely involute or with involute margins, gradually acuminate, glabrous or with short hairs above; sheaths white-hairy near the summit; ligule a ring of short hairs; panicles narrow and rather spikelike, rarely slightly interrupted, erect, 15–40 cm. long, 5–10 mm. wide, the branches numerous, appressed, densely spicu-lose; spikelets 2–2.5 mm. long, at first lanceolate, obovate when mature, lustrous, pale lead-green; glumes unequal, the first ovate, obtuse, 1/4–1/3 as long as the spikelet, the second broadly lanceolate, rather obtuse, 1–1.5 mm. long, 1/2–2/3 as long as the spikelet; lemma narrowly ovate, 2–2.2 mm. long, the nerve faint, smooth, obtuse; stamens 3; anthers 0.5–0.7 mm. long, pale yellow.——Sept.–Nov. Waste grounds, riverbanks, and sunny places in lowlands; Honshu, Shikoku, Kyushu; very common.——Warmer parts of Asia and Australia.

Var. **purpureosuffusus** Ohwi. *S. indicus* var. *purpureosuffusus* (Ohwi) T. Koyama——MURASAKI-NEZUMI-NO-O. Plant larger with an elongate inflorescence and pale reddish purple spikelet.——Grows with the typical phase.

3. Sporobolus virginicus (L.) Kunth. *Agrostis virginica* L.; *Vilfa virginica* (L.) Beauv.——SONARE-SHIBA. Rather firm, glabrous perennial with much-branched creeping rhizomes; culms 10–40 cm. long, erect from ascending base, many-noded; leaf-sheaths short, usually 3–5 cm. long, overlapping, slightly pilose on the mouth, blades loosely involute, nearly smooth, 3–5 cm. long, 1–3 mm. wide or somewhat more than 10 cm. long in the innovation shoots, gradually narrowed to a fine point, ligule a hairy rim; panicle linear, spikelike, erect, pale, 4–8 cm. long, spikelets broadly lanceolate, about 2.5 mm. long, barely acute, glumes lanceolate, the first a little shorter than the second, lemma acute, palea as long as the lemma.——Oct. On seashores; Kyushu (Mageshima near Tanegashima); rare.——Pantropic.

65. MUHLENBERGIA Schreb.   NEZUMIGAYA ZOKU

Annuals or perennials; culms usually slender; leaf-blades linear, usually flat; panicles narrow or sometimes open; spikelets 1-flowered, bisexual, somewhat laterally flattened, articulate below the floret; glumes 2, rather equal, membranous, shorter than the lemma; lemma broadly lanceolate or narrowly ovate, slightly keeled, 3-nerved, chartaceous or rather membranous, pubescent toward the margins below, with a straight awn between 2 minute teeth at apex; callus short, usually short-hairy; rachilla not produced; palea as long as the lemma, 2-keeled; caryopsis narrow, terete, smooth, glabrous.——Many species, chiefly in America, but a few in e. Asia, India, and Malaysia.

1A. Culms creeping and branching at base, without elongate rhizomes; glumes ½–⅔ as long as the lemma; anthers 0.7–1 mm. long.
.... 1. *M. japonica*

1B. Culms erect from base, with elongate rhizomes covered with smooth, lustrous, whitish, bladeless scales.
  2A. Glumes ovate, obtuse or nearly so, often denticulate at summit, ¼–⅓ as long as the lemma; anthers 0.7–1 mm. long; culms 50–120 cm. long; awns 8–10 mm. long. .... 2. *M. longistolon*
  2B. Glumes lanceolate, acuminate or very acute, entire, 3/5–4/5 as long as the lemma.
    3A. Spikelets 2.5–3 mm. long; glumes 3/5–2/3 as long as the lemma; culms much-branched in upper part; scales on rhizomes rather soft, appressed, scarcely inflated on back; anthers 0.5–0.7 mm. long. .... 3. *M. ramosa*
    3B. Spikelets 3–4 mm. long; glumes 2/3–4/5 as long as the lemma; culms not or scarcely branched in upper part; scales on rhizomes coriaceous, inflated on back; anthers 0.8–2 mm. long.
      4A. Spikelets 3–3.5 mm. long; anthers about 1 mm. long; leaf-blades 3–7 mm. wide; rhizomes, including the scales, 3–4 mm. across; panicle-branches ascending. .... 4. *M. curviaristata*
      4B. Spikelets 4–4.5 mm. long; anthers 1.5–2 mm. long; leaf-blades 2–4 mm wide; rhizomes about 2 mm. in diameter; panicle-branches appressed to the axis. .... 5. *M. hakonensis*

1. Muhlenbergia japonica Steud. NEZUMIGAYA. Slender, loosely tufted perennial; culms 15–40 cm. long, ascending to erect from creeping, branching base; leaf-blades soft, thin, glaucous-green, 5–15 cm. long, 2–4 mm. wide, gradually narrowed to the apex; sheaths rather flat; ligules very short, ciliolate; panicles somewhat exserted, nodding or erect, 8–15 cm. long, lanceolate, rather dense, pale lead-green, with a purplish glaucous hue, the branches single, slender, scabrous; spikelets 2.5–3 mm. long; glumes broadly lanceolate, 1.5–2 mm. long, acute, white with a green keel; lemmas rather acute, with a delicate erect awn 4–8 mm. long; stamens 3.——Aug.-Oct. Woods in lowlands and low mountains; Hokkaido, Honshu, Shikoku, Kyushu; common.——Ussuri, China, and Korea.

2. Muhlenbergia longistolon Ohwi. *M. huegelii* sensu auct. Japon., non Trin.——Ō-NEZUMIGAYA. Rhizomes to 15 cm. long, 3–4 mm. in diameter, with white, coriaceous, lustrous scales; culms erect, 50–120 cm. long, slightly flattened, sparingly branched; leaf-blades 10–20 cm. long, 3–8 mm. wide, flat; panicles slightly nodding, 10–30 cm. long, pale green with a purple hue, the branches ascending, scabrous; spikelets lanceolate, about 3 mm. long; glumes membranous, white, obtuse, obscurely dentate, faintly nerved, the first 0.5–0.7 mm., the second 0.7–1 mm. long; lemma pale lead-green, the awn straight, delicate, 8–10 mm. long; caryopsis lanceolate, 1.5–2 mm. long, free within the lemma and palea, not adherent.——Aug.-Sept. Woods in mountains; Hokkaido, Honshu, Shikoku, Kyushu; rather common.——Korea, China, and Amur.

3. Muhlenbergia ramosa (Hack.) Makino. *M. japonica* var. *ramosa* Hack.; *M. incumbens* Honda——KIDACHI-NO-NEZUMIGAYA. Rhizomes to 10 cm. long, with rather chartaceous, appressed, not inflated scales; culms erect or declinate above, 40–100 cm. long, freely branching above; leaf-blades 8–15 cm. long, 2–5 mm. wide; panicles 10–15 cm. long, narrow, rather densely spiculose, the branches single or geminate, nearly erect; spikelets 2.5–3 mm. long; glumes broadly lanceolate, 1.5–2 mm. long, 1-nerved, acute; awn of the lemma 5–8 mm. long; anthers 0.5–0.7 mm. long; caryopsis narrowly oblong, 1–1.5 mm. long, the hilum punctate.——Aug.-Oct. Woods in low mountains; Honshu (centr. distr. and westw.), Kyushu.

4. Muhlenbergia curviaristata (Ohwi) Ohwi. *M. ramosa* var. *curviaristata* Ohwi.——KOSHINO-NEZUMIGAYA. Rhizomes about 10 cm. long, 3–4 mm. in diameter, covered with coriaceous, white, ovate, slightly keeled, inflated scales about 5–10 mm. long; culms erect, 60–100 cm. long, simple or sparingly branched near the middle; leaf-blades 10–15 cm. long, 3–7 mm. wide; panicles 15–35 cm. long, the branches 2- to 4-nate, ascending; spikelets lanceolate, 3–3.5 mm. long; glumes 2/3–4/5 as long as the spikelets, very acute; lemma acuminate, the awn 3–4 mm. long, slightly curved at tip; anthers 0.8–1.2 mm. long.——Aug.-Oct. Woods in mountains; Honshu (Japan Sea side of centr. distr.); rather rare.

Var. **nipponica** Ohwi. MIYAMA-NEZUMIGAYA. Phase with erect awns 5–10 mm. long.——Hokkaido, Honshu (Tajima Prov. and eastw.).——s. Kuriles.

5. Muhlenbergia hakonensis (Hack.) Makino. *M. japonica* var. *hakonensis* Hack.——TACHI-NEZUMIGAYA. Rhizomes to 10 cm. long, about 2 mm. in diameter, covered with rather coriaceous, many-nerved, inflated, white scales; culms erect, slender, 40–80 cm. long; nearly simple; leaf-blades 10–20 cm. long, 2–4 mm. wide; panicles linear, erect, 10–15 cm. long, 5–10 mm. wide; the branches single or geminate, appressed; spikelets rather few, narrowly lanceolate, 4–4.5 mm. long; glumes narrowly lanceolate, 3–3.5 mm. long, acuminate; awn of the lemma 6–10 mm. long, slender, erect; anthers 1.5–2 mm., rarely 1.3 mm. long; caryopsis narrowly lanceolate, about 2 mm. long.——Aug.-Sept. Woods in mountains; Honshu (Kantō Distr. and westw.), Shikoku, Kyushu; rather rare.——Korea (Quelpaert Isl.).

66. LEPTURUS R. Br. HAI-SHIBA ZOKU

Rather small, firm annuals or perennials; leaf-blades linear, flat or involute, ligules short; inflorescence a simple terminal cylindrical spike; spikelets sessile, solitary, alternate, embedded in the hollows on opposite sides of the jointed spike, the axis disarticulate at the joints at maturity; lower glume suppressed except in the terminal spikelet, upper glume well-developed, longer than the florets, narrow, acute to acuminate or awned, rigid, with 5–9 nerves; florets 1 or 2, the upper fertile or frequently reduced; lemmas membranous or hyaline, finely 3-nerved.——Few species in the tropical and warm regions of the Old World.

1. Lepturus repens (G. Forst.) R. Br. *Rottboellia repens* G. Forst.; *Monerma repens* (G. Forst.) Beauv.——HAI-SHIBA. Firm perennial with extensively creeping stolons; culms 15–30 cm. long or sometimes more, erect or ascending, branched below; leaf-sheaths 2–5 cm. long, usually overlapping, the ligule truncate, short, ciliolate, the blades fragile, linear, 5–15 cm. long, gradually narrowed to a fine point, usually loosely involute, glabrous, pale green; spikelets 7–15 mm. long, the upper glume linear-lanceolate, gradually narrowed to tip, rigid.——Oct. Seashores, Kyushu (Tanegashima); rare.——Africa, India to Malaysia, North Australia, Polynesia, and Japan.

67. ZOYSIA Willd.   SHIBA ZOKU

Perennials with long rhizomes and stolons; leaf-blades short, linear to subulate; inflorescence a spikelike raceme, branchlets short; spikelets 1-flowered, bisexual, laterally compressed, articulate above the pedicel, glabrous; first glume usually wanting, rarely developed as a small hyaline membrane, the second coriaceous, strongly flattened, lustrous, oblong, awnless; lemma small, membranous, included within the second glume, faintly 3-nerved, smooth, glabrous; palea usually obsolete; caryopsis smooth, glabrous; styles slender, sometimes colored, protruding above the glumes and lemma.——Few species, in e. Asia, Australia, and Africa.

1A. Plants stoloniferous; spikelets 2.5–3.5 mm. long, 0.8–1.5 mm. wide.
  2A. Spikelets ovate, 2–2.5 times as long as wide; leaf-blades 2–5 mm. wide. . . . . 1. *Z. japonica*
  2B. Spikelets narrowly oblong, 3–4 times as long as wide; leaf-blades less than 1 mm. wide. . . . . 2. *Z. tenuifolia*
1B. Stolons absent; spikelets 4–8 mm. long, 1.2–2.2 mm. wide.
  3A. Spikelets 1.2–1.5 mm. wide, lanceolate. . . . . 3. *Z. sinica*
  3B. Spikelets 1.8–2.2 mm. wide, narrowly oblong. . . . . 4. *Z. macrostachya*

1. Zoysia japonica Steud. *Z. pungens* sensu auct. Japon., non Willd.; *Z. pungens* var. *japonica* (Steud.) Hack.; *Osterdamia japonica* (Steud.) Hitchc.——SHIBA. Stolons elongate, with shoots at the nodes; culms 10–20 cm. long, erect, firm, leafy only near base; leaf-blades 5–10 cm. long, 2–5 mm. wide, flat or with loosely involute margins, sparsely long-pilose while young; racemes long-exserted, 3–5 cm. long, linear-lanceolate; spikelets obliquely oblong-ovate, about 3 mm. long, 1.2–1.5 mm. wide, lustrous; second glume coriaceous, smooth, glabrous, rather obtuse or with a mucro about 1 mm. long, faintly 5-nerved; lemma slightly shorter than the second glume, folded, lanceolate, chartaceous, rather obtuse, 1-nerved; anthers about 1.5 mm. long.——May–June. Sunny slopes in mountains and hills; Hokkaido, Honshu, Shikoku, Kyushu; common and frequently cultivated as a lawn grass.——Korea and China.

2. Zoysia tenuifolia Willd. *Z. pungens* var. *tenuifolia* (Willd.) Dur. & Schinz; *Osterdamia tenuifolia* (Willd.) O. Kuntze——KŌRAI-SHIBA. Much slenderer than the preceding; culms 7–20 cm. long; leaf-blades filiform-subulate, involute or strongly folded, 0.3–0.7 mm. wide, ascending to erect, long-hairy at the mouth; racemes 1–3 cm. long, 2–4 mm. wide, exserted, pale yellow; spikelets rather few; second glume broadly lanceolate, rather acute, coriaceous, 2.5–3.5 mm. long, about 0.7 mm. wide, faintly 5-nerved, rarely with a mucro at apex; lemma lanceolate, 2–2.5 mm. long, rather obtuse, 1-nerved, keeled; anthers about 1 mm. long.——Honshu, Shikoku, Kyushu; possibly not indigenous; widely cultivated as a lawn grass in southern countries.——Ryukyus, Formosa, China, and s. Asia.

3. Zoysia sinica Hance var. **sinica.** *Z. matrella* var. *macrantha* Nakai; *Z. sinica* var. *macrantha* (Nakai) Ohwi; *Osterdamia liukiuensis* Honda; *Z. liukiuensis* (Honda) Honda ——KO-ONI-SHIBA, Ō-HARI-SHIBA. Spikelets slightly shorter than in var. *nipponica*, 4–5 mm. long.——Sandy seashores; Kyushu (Tanegashima).——Ryukyus, Formosa, and China.

Var. **nipponica** Ohwi. *Z. sinica* sensu auct. Japon., non Hance——NAGAMI-NO-ONI-SHIBA. Rhizomatous; culms erect or ascending, rather slender, firm, 10–25 cm. long; leaf-blades ascending or nearly erect, flat or involute, 3–7 cm. long, 2–3 mm. wide, long-hairy at the mouth; racemes slightly exserted, linear-lanceolate, 3–5 cm. long, 4–7 mm. wide, densely flowered; spikelets lanceolate, appressed, 5–7 mm. long, 1.2–1.5 mm. wide; second glume coriaceous, lustrous; lemma 3–4 mm. long, lanceolate; anthers linear, 1.5–2 mm. long.——Sand dunes near seashores; Honshu (Kantō Distr. and westw.), Shikoku, Kyushu; rather common.——Korea and Manchuria.

4. Zoysia macrostachya Fr. & Sav. *Osterdamia macrostachya* (Fr. & Sav.) Honda——ONI-SHIBA. Rhizomes long-creeping; culms 10–20 cm. long; leaf-blades firm, short-linear or narrowly lanceolate, spreading, nearly flat to involute, 3–5 cm. long, 2–4 mm. wide, long-hairy at the mouth; uppermost sheath slightly inflated above; racemes partly enclosed within the sheath, 3–4 cm. long, 6–8 mm. wide, densely flowered; spikelets rather appressed, narrowly oblong; second glume 6–8 mm. long, 1.8–2.2 mm. wide, biconvex; lemma oblanceolate, about 4 mm. long, rather acute, folded; stamens 3, the anthers 2–2.5 mm. long; caryopsis obovate-oblong, slightly less than 2 mm. long, somewhat laterally flattened, glabrous, the hilum linear.——June–Aug. Sand dunes along seashores; Hokkaido (sw. distr.), Honshu, Shikoku, Kyushu; common. ——Ryukyus.

Zoysia hondana Ohwi. *Z. sinica* var. *robusta* Honda—— SUNA-SHIBA. An alleged hybrid of *Z. japonica* and *Z. macrostachya*.

68. ARUNDINELLA Raddi   TODA-SHIBA ZOKU

Rather small to stout grasses; leaf-blades linear; panicles narrow to open; spikelets pedicelled, acute to acuminate, articulate above the pedicel, 2-flowered, commonly geminate on branchlets, the lower floret sterile; first glume short, awnless, the second acuminate or with a short, erect awn, rather large, few-nerved; sterile lemma awnless, rather thin, with a hyaline palea; fertile lemma membranous or chartaceous, smaller, entire or with 2 minute awnlike points, the awn terminal, slender, geniculate, sometimes wanting; palea membranous to chartaceous; stamens 3; styles divided from base, the stigma feathery; caryopsis oblong, free but closely packed within the lemma and palea.——About 50 species, in tropical and subtropical regions of Asia and America.

1. Arundinella hirta (Thunb.) C. Tanaka var. **hirta.** *Poa hirta* Thunb.; *Agrostis ciliata* Thunb.; *Arundinella anomala* Steud.; *Arundinella murayamae* Honda; *Arundinella oleagina* Honda; *Arundinella smaragdina* Koidz.; *Agrostis thunbergii* Steud.; *Panicum mandshuricum* Maxim., incl. var. *pekinense* Maxim.——KE-TODA-SHIBA. Tufted, rather stout, strongly pilose throughout, with firm creeping rhizomes; culms 30–120 cm. long, erect, sometimes scarcely pilose; leaf-blades linear, somewhat spreading, 15–40 cm. long, 5–15 mm. wide, flat or involute; ligules very short, long-ciliate; sheaths usually longer than the internodes, often pilose; panicles 8–30 cm. long, dense to rather loose, the branches short to elongate, sometimes spikelike, usually hairy in the axils; spikelets usually geminate, pedicelled, green and often dull purplish, sometimes glaucescent, ovate to narrowly so, 3.5–4.5 mm. long; glumes acuminate, firmly membranous, the first slightly shorter than the spikelet, 3-nerved, sometimes with 1 or 2 pairs of accessory short nerves, the second longer, 5- to 7-nerved; sterile lemma slightly shorter, but similar to the second glume, with a hyaline palea and stamens; fertile lemma 2.5–3 mm. long, awnless or rarely with a short, straight awn from the entire apex and a tuft of hairs at base 1/4–1/3 as long as the lemma.——Aug.-Oct. Lowlands and mountains; Hokkaido, Honshu, Shikoku, Kyushu; very common and variable.——Ussuri, Manchuria, and China. Varieties based on pubescence are as follows:

Var. **hondana** Koidz. ONI-TODA-SHIBA. Similar to the typical phase, the spikelets pilose.

Var. **ciliata** Koidz. USUGE-TODA-SHIBA. Plant thinly pilose.

Var. **riparia** (Honda) Ohwi. *A. riparia* Honda——MIGIWA-TODA-SHIBA. Culms usually geniculate at base; fertile lemma with the awn about 3 mm. long, slightly geniculate.——Honshu (Doro-kyo in Kii); rare.

69. PSEUDORAPHIS Griff.   UKI-SHIBA ZOKU

Glabrous grasses of wet places or in water; leaf-blades flat, linear to lanceolate; panicles simple, with rather numerous sub-equal branches, the very slender main axis and the branches ending in a bristle; spikelets 1 to few, not or obsoletely jointed at base, nearly sessile; first glume minute, the second gradually narrowed to tip, flat, many-nerved; sterile lemma slightly shorter than the glumes, acute, with a palea and 3 stamens; fertile lemma still shorter, chartaceous or herbaceous with a pistil and 2 sterile stamens.——Few species, in e. Asia, India, Malaysia, and Australia.

1. Pseudoraphis ukishiba Ohwi. *Chamaeraphis depauperata* sensu Hack., non Nees nec Balansa; *C. spinescens* var. *depauperata* sensu auct. Japon., non Hook. f.——UKI-SHIBA. Rather soft, glabrous perennial; culms tufted, about 20 cm. long in terrestrial plants, greatly elongate in water, ascending at base, leafy; leaf-blades thin, flat, glaucous-green, linear, 3–5 cm. long, 2–4 mm. wide, abruptly acuminate; ligules truncate, about 1 mm. long, ciliolate; panicles racemose, 3–6 cm. long, terminal, soon appearing lateral by the elongation of the lateral bud, often reflexed from the base, partly enclosed in the uppermost sheath, the axis smooth, the branches numerous, simple, ascending, scabrous, about 2.5 cm. long with 1 spikelet slightly below the middle; spikelets nearly sessile, appressed to the branchlet, lanceolate, 4–5 mm. long, gradually acuminate; first glume nearly truncate, about 0.7 mm. long, the second longer, 7-nerved, membranous, pale green, loosely flowered; sterile lemma slightly shorter than the glumes, glabrous; fertile lemma ovate, 1.2 mm. long, smooth.——June–Aug. Ponds and wet places; Honshu, Shikoku, Kyushu.——s. Korea and China (?).

70. PENNISETUM L. C. Rich.   CHIKARA-SHIBA ZOKU

Annuals or perennials; culms often branched; leaf-blades linear, flat or involute; spikelets 1–3, in involucrate fascicles on a thick axis, forming a feathery or bristly spike, deciduous together with the involucre, the involucre of separate or partly connate, simple or branched, sometimes pinnate bristles; first glume small, the second sometimes as long as the spikelet; sterile lemma usually staminate and paleate, membranous to chartaceous; fertile lemma chartaceous, bisexual, paleate; stamens 3; styles 2 or fused at base; caryopsis dorsally compressed, the hilum small.——About 130 species, in warmer and tropical regions.

1A. At least the inner involucral bristles pinnate at base; culms often branched, bearing few spikes. .... 1. *P. orientale* var. *triflorum*
1B. Involucral bristles scabrous, not pinnate; culms simple.
  2A. Leaf-sheaths slightly compressed; leaf-blades loosely involute to flat, 2–6 mm. wide, firmly herbaceous; pedicels of involucre spreading, 2–3 mm. long; second glume 3.5–4 mm. long, about half as long as the spikelet, 3- to 6-nerved. .. 2. *P. alopecuroides*
  2B. Leaf-sheaths terete; leaf-blades tightly convolute, terete, about 1 mm. in diameter, coriaceous; pedicels of involucre ascending, 0.7–1 mm. long; second glume about 1.5 mm. long, 1/4–1/3 as long as the spikelet, nerveless. .... 3. *P. sordidum*

1. Pennisetum orientale L. C. Rich. var. **triflorum** (Nees) Stapf. *P. triflorum* Nees——EDA-UCHI-CHIKARA-SHIBA. Tufted perennial, about 80–120 cm. high; culms glabrous, puberulent above, branching below; leaf-blades flat, slightly pilose, 40–60 cm. long, 9–15 mm. wide, acuminate; spikes cylindric, 15–25 cm. long, about 15 mm. wide, densely flowered, dull purpurascent, the involucre much-spreading, with 3 or 4 spikelets; pedicels about 1 mm. long, villous; bristles unequal, 4–13 mm. long, the longest to 20 mm. long, the inner pinnate at base; spikelets about 5 mm. long; glumes membranous, lanceolate, acuminate, the first 1/4 as long as the spikelet, nerveless, obtuse; sterile lemma as long as the fertile.——Aug.-Nov. Naturalized; Honshu (near Tokyo); rare.——N. Africa, Asia Minor, and India.

2. Pennisetum alopecuroides (L.) Spreng. *Panicum alopecuroides* L.; *Cenchrus purpurascens* Thunb.; *Pennisetum* *japonicum* Trin.; *Pennisetum purpurascens* (Thunb.) Kuntze, non HBK.——CHIKARA-SHIBA.   Tufted, stout perennial 30–80 cm. high; culms erect, densely white-pubescent in upper part; leaf-blades 30–60 cm. long, 5–8 mm. wide, gradually acuminate, glaucous and scabrous above, often loosely convolute, long-hairy near the mouth; ligules very short, ciliolate; sheaths slightly keeled, smooth; spike terminal, solitary, cylindric, 10–15 cm. long, about 2 cm wide; involucres with 1 spikelet, the pedicels 2–3 mm. long, the bristles 5–30 mm. long, unequal, scabrous; spikelets about 7 mm. long, lanceolate or broadly so, acute; glumes rather chartaceous, the first minute, the second about half as long as the spikelet, narrowly oblong, acute; sterile lemma as long as the fertile, both broadly lanceolate and acute.——Aug.–Nov. Grassy places and waste grounds in lowlands; Hokkaido (sw. distr.), Honshu, Shikoku, Kyushu; very common.——Korea, China, Formosa to Philippines. Our material consists of three phases based on the color of the involucral bristles:——Forma **purpurascens** (Thunb.) Ohwi.   CHIKARA-SHIBA.   Bristles dark purple; the commonest form.——Forma **viridescens** (Miq.) Ohwi.

Gymnothrix japonica var. *viridescens* Miq.——AO-CHIKARA-SHIBA.   Bristles pale green.——Forma **erythrochaetum** Ohwi.   BENI-CHIKARA-SHIBA.   Bristles red.

3.   **Pennisetum sordidum** Koidz.   SHIMA-CHIKARA-SHIBA. Glabrous, densely tufted perennial; culms 40–60 cm. long, smooth, covered by the leaf-sheaths; leaf-blades often exceeding the culms in length, strongly convolute, 1–1.5 mm. wide, 3–4 mm. wide when flat, scabrous above, gradually narrowed to a filiform tip, hairy at the mouth; sheaths terete; ligule a series of fine hairs; spikes 8–12 cm. long, dense, the axis scabrous, bearing ascending involucres, each with 1 spikelet; bristles unequal, scabrous, whitish, 1–2 cm. long; pedicels 0.7–1 mm. long, hairy; spikelets lanceolate, about 6 mm. long, acuminate; first glume small, ovate, 1/5–1/4 as long as the spikelet, nerveless, the second 1/4–1/3 as long as the spikelet, about 1.5 mm. long; sterile lemma acuminate, 3- to 5-nerved; fertile lemma as long as the sterile one, awn-pointed.——Oct.–Nov. Near seashores; Kyushu (s. distr.).——Bonins, Ryukyus, and Amami-Oshima.

71. SETARIA Beauv.   ENOKORO-GUSA ZOKU

Annuals or perennials, with flat or plicate leaf-blades and simple or branched culms; panicle open or contracted, often cylindric and spikelike, the branchlets ending in a bristle; spikelets ovate, obtuse, jointed at the base; glumes membranous, the first small, the second usually shorter than the floret; sterile lemma membranous, paleate or almost epaleate, sometimes staminate; fertile lemma coriaceous, ovate, usually rugulose, sometimes smooth; stamens 3; styles not connate.——About 100 species, widespread in warmer regions.

1A. Panicles loose and open, with rather distant branches, the bristles sparse, the axis visible; leaf-blades 1.5–7 cm., rarely 5–17 mm. wide.
  2A. Leaf-blades plicate, 1.5–7 cm. wide; tufted perennials; florets finely rugose.
    3A. Leaf-blades 3–7 cm. wide; florets ovate, slightly rugulose. . . . . 1. *S. palmifolia*
    3B. Leaf-blades 1.5–3 cm. wide; florets narrowly ovate, distinctly rugulose. . . . . 2. *S. plicata*
  2B. Leaf-blades flat, not plicate, 5–15 mm. wide; perennials with much-elongated rhizomes; florets smooth. . . . . 3. *S. chondrachne*
1B. Panicles cylindric, spikelike, the branches and branchlets very dense, concealing the axis; leaf-blades flat or slightly folded, 3–30 mm. wide.
  4A. Palea of sterile floret lanceolate, nerveless, hyaline, much smaller than the fertile floret, or obsolete; leaf-sheaths ciliate.
    5A. Spikelets 2–2.7 mm. long; second glume nearly as long as the floret; floret not visible. . . . . 4. *S. viridis*
    5B. Spikelets about 3 mm. long; second glume 2/3–3/4 as long as the floret; floret naked on back; leaf-blades usually pilose above.
                                                                       5. *S. faberi*
  4B. Palea of sterile floret nearly as large as the floret, ovate, hyaline, sometimes brown-tinged, often staminate, 2-keeled; leaf-sheaths not ciliate.
    6A. Spikelets about 3 mm. long, somewhat whitish; bristles golden-yellow; plants annual. . . . . 6. *S. glauca*
    6B. Spikelets 2–2.8 mm. long; bristles brown, brownish purple or rarely golden-yellow or whitish; plants annual. . . 7. *S. pallidefusca*

1. **Setaria palmifolia** (Koenig) O. Stapf. *Panicum palmifolium* Koenig; *P. plicatum* Willd., non Lam.——SASA-KIBI. Large, tufted, strigose-hirsute perennial 80–200 cm. high; leaf-blades oblanceolate or lanceolate, plicate, 30–60 cm. long, 3–7 cm. wide, scabrous, hirsute or nearly glabrous, sessile or gradually narrowed to a petiolelike base; ligules very short, long bristly-hairy; sheaths and sheath-margins long bristly-hairy; panicles 20–40 cm. long, somewhat nodding at maturity; branches rather spikelike, nearly simple or short-branched, the branchlets partly ending in a short bristle; spikelets ovate, green, about 3 mm. long, abruptly acuminate; first glume 2/5–1/2 as long as the spikelet, the second slightly shorter than the spikelet, 7-nerved, ovate; sterile lemma largest, 5-nerved, acuminate; floret nearly as long as the sterile lemma, transversely finely rugose, lustrous, mucronate.——Aug.–Nov. Kyushu (s. distr.).——Tropical Asia, including the Ryukyus and Formosa.

2. **Setaria plicata** (Lam.) T. Cooke.   *Panicum plicatum* Lam.; *P. excurrens* Trin.; *S. excurrens* (Trin.) Miq.; *P. paucisetum* Steud.; *S. excurrens* var. *pauciseta* (Steud.) Ohwi—— KO-SASA-KIBI.   Nearly glabrescent tufted perennials; culms 80–130 cm. long, glabrous; leaf-blades thin, narrowly lanceolate, plicate, glabrous, 20–30 cm. long, 1–3 cm. wide, acuminate, the lower ones narrowed to a petiolelike base; sheaths hairy on mouth and margins; panicles 20–30 cm. long, loosely branched, scabrous, the branches slender, spikelike, ascending; spikelets green, sometimes slightly purplish, ovate-oblong or narrowly ovate, 3.3–3.5 mm. long, acute; first glume ovate-rounded, about 1/3 as long as the spikelet, the second 2/3–3/4 as long, oblong, obtuse, 5- to rarely 7-nerved; sterile lemma narrowly ovate, 3.3–3.5 mm. long, rather acute, 5- rarely 7-nerved; fertile lemma as long as the sterile, lustrous, ovate-oblong, acute, distinctly rugulose.——Sept.–Oct.   Kyushu.——Ryukyus, Formosa, s. China, India, and Malaysia.

3. **Setaria chondrachne** (Steud.) Honda. *Panicum chondrachne* Steud.; *P. matsumurae* Hack.; *S. matsumurae* (Hack.) Hack.——INU-AWA. Rather glabrous perennial; rhizomes elongate, covered by coriaceous, short scales; culms 50–80 cm. long, branching below, glabrous; leaf-blades thin, flat, broadly linear, 20–40 cm. long, 5–15 mm. wide, acuminate, scabrous; ligules very short, ciliolate; sheaths ciliate; panicles 15–30 cm. long, narrow, linear to narrowly lanceolate, scabrous, the branches short, spreading, loosely flowered; spikelets ovate, 2–2.2 mm. long, acute, green or yellowish green; first glume 2/5–1/2 as long as the spikelet, the second 2/3 as long, 5-nerved, ovate, acute; sterile lemma 5-nerved, acute; fertile lemma as long as the sterile one, 2–2.2 mm. long, narrowly ovate, mucronately acute, smooth and lustrous.——Aug.–Oct. Thickets and woods in lowlands; Honshu (Uzen, Kantō, Etchu Distr. and westw.), Shikoku, Kyushu; rather common. ——s. Korea and China.

4. **Setaria viridis** (L.) Beauv. *Panicum viride* L.; *S. italica* subsp. *viridis* (L.) Thell.——ENOKOROGUSA. Weedy annual; culms erect, glabrous, branching at base, 20–70 cm. long; leaf-blades linear to broadly so, 5–20 cm. long, 5–15 (rarely to 20) mm. wide, glabrous, flat, the sheaths and ligules ciliate; panicles cylindric, terete, spikelike, nearly erect, 2–5 cm. long, pale green, sometimes purpurascent, the axis spreading-hairy, the bristles numerous, densely arranged, spreading, short, scabrous, 6–8 mm. long; spikelets about 2 mm. long, ovate, obtuse; first glume 2/5–1/3 as long as the spikelet, the second nearly as long, ovate, obtuse, 5-nerved; sterile lemma similar, slightly longer than the second glume, the fertile lemma ovate-elliptic, as long as the sterile, obtuse, minutely rugulose; anthers dark brown.——Aug.–Nov. Waste grounds and cultivated fields; Hokkaido, Honshu, Shikoku, Kyushu; very common.——Widespread in temperate and subtropical regions.

Var. **pachystachys** (Fr. & Sav.) Makino & Nemoto. *Panicum pachystachys* Fr. & Sav., pro parte; *S. pachystachys* (Fr. & Sav.) Fr. & Sav. ex Matsumura——HAMA-ENOKORO. A shorter and more branched annual; blades slightly thicker; panicles 1–4 cm. long; spikelets and bristles very dense and numerous, the latter rarely purpurascent; anthers yellowish brown.——Grassy places near seashores; Hokkaido, Honshu, Shikoku, Kyushu; very common.

5. **Setaria faberi** Herrm. *S. autumnalis* Ohwi——AKI-NO-ENOKOROGUSA. Similar to but larger than the preceding species; culms 40–100 cm. long, branching below, slightly scabrous above; leaf-blades flat, 10–30 cm. long, 8–20 mm. wide, often pilose above, the sheath-margins and ligules ciliate; panicles cylindric, spikelike, contracted, 5–10 cm. long, nodding, green, sometimes slightly purpurascent, the bristles numerous; spikelets densely arranged, ovate to broadly so, obtuse, 2.8–3 mm. long; first glume about 1/3 as long as the spikelet, the second 2/3–3/4 as long, broadly ovate; sterile lemma as long as the spikelet; fertile lemma ovate, rather obtuse, punctate and minutely rugulose, rather lustrous.——Sept.–Nov. Waste grounds and thickets; Hokkaido, Honshu, Shikoku, Kyushu; very common; grows in more shaded places than the preceding species.——China; naturalized in N. America.

6. **Setaria glauca** (L.) Beauv. *Panicum glaucum* L.; *P. lutescens* Weigel; *S. lutescens* (Weigel) F. T. Hubb.; *S. pumila* sensu auct., non Roem. & Schult.——KIN-ENOKORO. Smooth, nearly glabrous annual; culms 20–50 cm. long, branching at base, puberulent below the panicle; leaf-blades linear, 10–25 cm. long, gradually acute, smooth, glabrous except for a few long hairs near the mouth; ligule a series of short hairs; sheaths slightly flattened, glabrous; panicles cylindric, spikelike, erect, 3–10 cm. long, golden-yellow, the axils pilose-scabrous, with the bristles spreading, scabrous, 7–10 mm. long; spikelets broadly ovate, rather acute, 2.8–3 mm. long, whitish; first glume 3-nerved, about half as long as the spikelet, the second 3/5–2/3 as long, broadly ovate, 5-nerved; sterile lemma as long as the floret, 5-nerved; fertile lemma broadly ovate, finely rugulose, rather obtuse, lustrous.——Aug.–Oct. Waste grounds and cultivated fields in lowlands; Hokkaido, Honshu, Shikoku, Kyushu; common northw.——Cosmopolitan in temperate areas.

7. **Setaria pallidefusca** (Schumach.) Summerh. & C. E. Hubb. *Panicum pallidefuscum* Schumach.; *S. geniculata* sensu Ohwi, non Beauv.——KO-TSUBU-KIN-ENOKORO. Similar to the preceding; panicles slightly narrower, with slightly shorter bristles; spikelets pale green, 2–2.8 mm. long; bristles brown to purplish brown, rarely golden-yellow or white.—— Aug.–Oct. Waste grounds and cultivated fields in lowlands; Honshu, Shikoku, Kyushu; common and more variable than the preceding.——Warmer parts of Africa, Asia, and Northern Australia.

72. SACCIOLEPIS Nash   NUMERIGUSA ZOKU

Annuals or perennials; leaf-blades linear, flat or involute; culms often branched at base; panicles dense and spikelike, green or purplish; spikelets small, deciduous from the apex of the short thickened pedicel, ovate to broadly lanceolate, terete or somewhat laterally compressed, slightly gibbous on one side at base; glumes prominently nerved, fragile, membranous; first glume small, the second saccate at base; sterile lemma erect, with a reduced palea within; fertile lemma rather small, crustaceous, smooth, ovoid, lustrous, substipitate; stigma long; hilum minute.——About 30 species, widespread in wet places in warm-temperate and tropical regions.

1. **Sacciolepis indica** (L.) Chase. *Panicum indicum* L.; *Aira spicata* L.; *S. spicata* (L.) Honda——HAI-NUMERI. Glabrous, soft annual; culms tufted, erect, smooth, 10–40 cm. long, from a branching, ascending or decumbent base; leaf-blades linear, 4–10 cm. long, 2–4 mm. wide, glabrous; ligules very short, hyaline; sheaths rather short, glabrous; panicles spikelike, long-exserted, erect, cylindric, 1–6 cm. long, pale green, 4–6 mm. across, densely flowered, the axis smooth; pedicels short, slightly dilated and concave at apex; spikelets about 3 mm. long, acute, numerous, narrowly conical-ovoid, slightly oblique, often sparsely white-pilose; first glume ovate, about half as long as the spikelet, the second longer, narrowly ovate-oblong, prominently several-nerved; sterile lemma slightly shorter, not inflated at base; fertile lemma crustaceous, lustrous, narrowly ovoid-oblong, rather acute, pale yellow, about 1.5 mm. long.——Aug.–Oct. Wet paddy fields and riverbanks; Honshu, Shikoku, Kyushu; common.——India, China, Malaysia, Korea, Ryukyus, Formosa, and Australia.

Var. **oryzetorum** (Makino) Ohwi. *Panicum indicum* var. *oryzetorum* Makino; *P. oryzetorum* (Makino) Makino; *S. oryzetorum* (Makino) Honda; *S. angusta* sensu auct. Japon., non Stapf——NUMERIGUSA. Culms erect, 30–60 cm. long; leaves 10–20 cm. long, 3–6 mm. wide; spike purplish, 3–12 cm. long, erect, densely flowered.——Paddy fields and along river-banks; Honshu, Shikoku, Kyushu; common.——Ryukyus, Formosa, and China.

73. PANICUM L.   KIBI ZOKU

Annuals or perennials; leaf-blades linear or short linear, sometimes convolute; culms sometimes stout; panicles usually loose and open, much branched; spikelets pedicellate, lanceolate to broadly ovate, terete or slightly compressed; glumes membranous, the first usually shorter than the spikelet; sterile lemma similar to the second glume, often paleate and staminate; fertile lemma coriaceous, usually lustrous, smooth or rarely rugulose; stamens 3.——About 500 species, widely distributed in warm-temperate and tropical regions.

1A. Spikelets about 2 mm. long; culms much branched; panicles very loose and open; annual. .... 1. *P. bisulcatum*
1B. Spikelets 2.5–3 mm. long.
  2A. Annual. .... 2. *P. dichotomiflorum*
  2B. Perennial with creeping rhizomes. .... 3. *P. repens*

1. Panicum bisulcatum Thunb. *P. acroanthum* Steud. ——NUKA-KIBI. Glabrous, slender, soft annual; culms 30–120 cm. long, erect, from a decumbent weak base, much branched, straggling, glabrous; leaf-blades linear, flat, slightly scabrous, 5–30 cm. long, 4–12 mm. wide; ligules hyaline, truncate, to 0.5 mm. long, glabrous; sheaths smooth, ciliolate; panicles large and effuse, loose, rounded-ovate, erect, 12–30 cm. long and as wide, the branches very slender, much branched, spreading, intricate, very loosely spiculose; spikelets pedicelled, elliptic, deep-green and purplish, rather acute, 1.8–2 mm. long; first glume deltoid, acute or rather obtuse, 1/3–2/5 as long as the spikelet, the second and sterile lemma longer, elliptic, sometimes short-puberulent, weakly 5-nerved, thinly membranous; fertile lemma elliptic, 1.5–1.8 mm. long, obtuse, lustrous, smooth, gray-brown when mature.——July–Oct. Wet lowlands; Hokkaido, Honshu, Shikoku, Kyushu; very common.——Ussuri, Korea, Manchuria, China, Formosa, and India.

2. Panicum dichotomiflorum Michx. Ō-KUSA-KIBI. Rather large tufted annual; culms 40–100 cm. long, branched, glabrous, erect from an ascending base; leaf-blades linear, 20–40 cm. long, 8–15 mm. wide, flat or sometimes with loosely involute margins, glabrous; ligule a series of short hairs; sheaths glabrous, often somewhat purplish; panicles rather spherical, the branches spreading, rather slender, elongate, scabrous; spikelets loosely arranged, on rather long pedicels, ovate-oblong, acute, slightly compressed, about 2.5 mm. long, pale green, often purplish, glabrous; first glume truncate or obtuse, 1/5–1/4 as long as the spikelet, the second and sterile lemma longer, acute, 5- to 7-nerved, membranous; fertile lemma oblong, rather obtuse, smooth, lustrous, 1.7–2 mm. long.——Sept.–Oct. Naturalized in waste grounds; Honshu. ——N. America.

3. Panicum repens L. HAI-KIBI. Rather firm, pale green perennial; rhizomes stout, long-creeping; culms 40–100 cm. long, erect, usually simple, glabrous; leaf-blades linear, with involute margins or nearly flat, coarsely and sparsely pilose above, 8–20 cm. long, 5–8 mm. wide; ligules hyaline, truncate, ciliolate; sheaths ciliate; panicles at length exserted, erect, 15–25 cm. long, the branches elongate, ascending at anthesis, scabrous, 10 to 15 cm. long; spikelets on upper part of branches and branchlets pedicelled, pale green above, about 3 mm. long, scarcely lustrous, narrowly ovate-oblong; first glume truncate, 1/5–1/4 as long as the spikelet, the second and the sterile lemma equal, larger, weakly 7-nerved, acute, thick-membranous; fertile lemma white, lustrous, ovate-oblong, rather acute, nearly 2 mm. long.——Sept.–Oct. Grassy places especially near seashores; Shikoku, Kyushu.——Ryukyus, Formosa, and in all warm and tropical regions.

74. DIGITARIA Hall.   ME-HI-SHIBA ZOKU

Perennials and annuals with linear, usually flat, membranous leaf-blades; inflorescence digitate, the branches usually simple, in fascicles, terminal on the culm or slightly distant, the rachis slender, angular or flat, often narrowly wing-margined; spikelets usually geminate, rarely solitary, lanceolate to ovate, dorsally compressed, unequally pedicellate or nearly sessile, usually puberulent; first glume abaxial, minute or absent, the second membranous; sterile lemma nearly as long as the fertile, membranous, 5- to 9-nerved, with a minute palea; fertile lemma lanceolate to ovate, chartaceous with membranous margins.——More than 100 species, in warm-temperate and tropical regions.

1A. Spikelets lanceolate, 2.2–3 mm. long, acuminate; pedicels triangular; fertile lemma chartaceous, usually pale, acuminate.
  2A. Perennial; racemes or branches of inflorescence erect, fasciculate even after anthesis; spikelets 2.2–2.5 mm. long; culms long-decumbent, slender, from short rhizomes. .... 1. *D. henryi*
  2B. Annuals; racemes spreading after anthesis; spikelets 2.5–3 mm. long; culms erect from a decumbent base, without rhizomes.
    3A. Plant slender, nearly glabrous; culms often purplish; leaf-blades 4–7 cm. long; rachis of racemes smooth on margins; first glume obsolete or nearly so. .... 2. *D. timorensis*
    3B. Plant usually pilose; culms usually green; leaf-blades 8–20 cm. long; rachis of racemes scabrous on margins; first glume minute but distinct. .... 3. *D. adscendens*
1B. Spikelets ovate-elliptic, 1.5–2.5 mm. long; pedicels terete; fertile lemma usually black or dark brown, barely acute.
  4A. Spikelets 1.5–2 mm. long, rather flat on the outer face; hairs of spikelets crisped, not thickened at tip. .... 4. *D. violascens*
  4B. Sipkelets 2–2.5 mm. long, rather convex on the outer face; hairs of spikelets at least somewhat thickened at tip. .. 5. *D. ischaemum*

1. **Digitaria henryi** Rendle. *Syntherisma henryi* (Rendle) Newbold; *Panicum henryi* (Rendle) Makino & Nemoto ——HENRI-ME-HI-SHIBA. Culms few, erect, slender, glabrous, 10–20 cm. long, from a long-creeping, branching base; leaf-blades membranous, green, flat, linear, 4–7 cm. long, 3–5 mm. wide, short-acuminate, glabrous except for a few long, spreading hairs on margins near base; ligules hyaline, 1.5–2 mm. long, glabrous; sheaths smooth, glabrous, rather short; racemes 3–6, short, erect, pale green, 5–8 cm. long, slender, the rachis scabrous on margins; spikelets broadly lanceolate, acute, long-pubescent on both sides, 2.2–2.5 mm. long; first glume minute, deltoid, the second linear-lanceolate, 1/2–2/3 as long as the spikelet, pubescent; sterile lemma broadly lanceolate, acute, 7-nerved, the midrib somewhat distant from the others, glabrous on the margins, long-pubescent on both faces; fertile lemma very slightly shorter than the sterile one, lanceolate, short-acuminate, whitish.——Sept.–Nov. Grassy places near seashores; Kyushu (s. distr.); rather rare.——Ryukyus, Formosa, and s. China.

2. **Digitaria timorensis** (Kunth) Balansa. *D. chinensis* auct., non Hornem.; *Panicum sanguinale* var. *timorense* (Kunth) Hack.; *P. timorense* Kunth.——KO-ME-HI-SHIBA. Deep green slender annual, nearly glabrous throughout; culms long-creeping, branching, erect in upper part, smooth, often purplish; leaf-blades thin, flat, broadly linear, 4–7 cm. long, 4–10 mm. wide, with few long hairs near the base; ligules 1–1.5 mm. long; sheaths smooth, usually glabrous; racemes 2–4, digitate, slender, spreading, the rachis 4–7 cm. long, smooth on margins; spikelets lanceolate, 2.8–3 mm., rarely 2.5 mm. long, acuminate, green; first glume nearly absent, the second membranous, lanceolate, 3-nerved, pubescent, 3/5–2/3 as long as the spikelet; sterile lemma lanceolate, acuminate, 5-nerved, pubescent toward the margin; fertile lemma herbaceous, lanceolate, acuminate, pale green, as long as the sterile one.——July–Oct. Shaded places near dwellings; Honshu (Kantō Distr. and westw.), Shikoku, Kyushu; rather common.——s. Korea, Ryukyus, Formosa, China, Indochina, Malaysia, and Micronesia.

3. **Digitaria adscendens** (HBK.) Henr. *Panicum adscendens* HBK.; *D. marginata* Link; *D. sanguinalis* var. *marginata* (Link) Fern.; *P. sanguinale* sensu auct. Japon., non L.; *Syntherisma sanguinalis* sensu auct. Japon., non Dulac ——ME-HI-SHIBA. Diffuse, hirsute or rarely glabrescent annual; culms erect, 30–80 cm. long, ascending to creeping from branching base, glabrous; leaf-blades flat, linear, 8–20 cm. long, 5–12 mm. wide, glabrous to sparingly hairy, often somewhat glaucous; ligules hyaline, 1–3 mm. long, glabrous; sheaths mostly hirsute; racemes 3–8, ascending to spreading, pale green, often somewhat purplish, 5–15 cm. long; rachis scabrous on margins; spikelets appressed, about 3 mm. long, rather acu-

minate, lanceolate; first glume deltoid, minute, the second lanceolate, 3/5–2/3 as long as the spikelet, 3-nerved, acuminate, pubescent; sterile lemma longer, broadly lanceolate, membranous, pubescent especially toward the margins; fertile lemma lanceolate, herbaceous, acuminate, lead-gray to white, as long as the sterile one.——July–Nov. Waste grounds and cultivated fields; Hokkaido, Honshu, Shikoku, Kyushu; very common and variable.——Widespread in warmer regions. Many specimens with long-ciliate spikelets often considered to be *D. fimbriata* Link. KUSHIGE-ME-HI-SHIBA, but this variant scarcely merits nomenclatural recognition.

4. **Digitaria violascens** Link. *Panicum violascens* (Link) Kunth; *Paspalum chinense* Nees ex Hook. & Arn.; *Paspalum filiculme* Nees ex Miq.; *D. ischaemum* var. *asiatica* Ohwi; *Syntherisma ischaemum* sensu auct. Japon., non Nash; *D. ropalotricha* sensu auct. Japon., non Buse; *D. filiculmis* (Nees ex Miq.) Ohwi——AKI-ME-HI-SHIBA. Glabrescent, tufted annual; culms 20–50 cm. long, erect from a short decumbent or ascending base, glabrous; leaf-blades flat, 6–12 cm. long, 5–8 mm. wide, slightly glaucescent above, slightly scabrous, glabrous or often with a few, spreading, long hairs near the base; ligules hyaline, pale-brown, 1–1.5 mm. long; sheaths glabrous; racemes 4–10, digitate, ascending, 4–10 cm. long, densely flowered; rachis less than 1 mm. wide, scabrous on margins; spikelets ovate-elliptic, rather acute, 1.5–2 mm. long, white-green and sometimes slightly reddish purple tinged; first glume usually wanting, rarely hyaline and truncate, early withering, the second slightly shorter than the spikelet, 3-nerved, puberulent; sterile lemma 7-nerved, ovate-elliptic, grayish crisped-puberulent between the nerves; fertile lemma as long as the sterile one, ovate, coriaceous to herbaceous, dark brown, rather acute.——Aug.–Oct. Grassy waste places and cultivated fields; Hokkaido, Honshu, Shikoku, Kyushu; very common and variable.——China, Formosa, Ryukyus, Korea, and S. America (naturalized).

Var. **lasiophylla** (Honda) Tuyama. *Syntherisma ischaemum* var. *lasiophylla* Honda——ARAGE-AKI-ME-HI-SHIBA. Plant slender, erect, with 1–3 erect racemes and hairy leaf-blades.

Var. **intersita** (Ohwi) Ohwi. *D. ischaemum* var. *intersita* Ohwi——USUGE-AKI-ME-HI-SHIBA. Similar to the preceding variety but less hairy. Both varieties grow in rather wet places.

5. **Digitaria ischaemum** (Schreb.) Schreb. ex Muhl. *Panicum ischaemum* Schreb.; *D. humifusa* Pers.; *Syntherisma glabra* Schrad.; *D. glabra* (Schrad.) Beauv.——KITA-ME-HI-SHIBA. Similar to the preceding; spikelets slightly longer, 2–2.5 mm. long, somewhat convex on the upper side; second glume and sterile lemma puberulent, at least some of the hairs clavately thickened at tip.——Aug.–Sept. Hokkaido, Honshu; rare.——Widespread in temperate regions.

75. PASPALUM L.　SUZUME-NO-HIE ZOKU

Perennials or annuals with digitate or racemose inflorescences; spikelets geminate, densely arranged in 2 rows on one side of the rachis, 1-flowered, ovate to rather orbicular, plano-convex; first glume wanting, the second next to the rachis, similar to the sterile lemma, 3- to many-nerved, rounded on the back; sterile lemma flat; fertile lemma coriaceous, broadly ovate to nearly orbicular, usually smooth, rounded on back.——About 200 species in warm-temperate and tropical regions, especially abundant in America.

1A. Spikelets oblong, acute, more than twice as long as broad; racemes 2.
  2A. Spikelets puberulent, about 3 mm. long; fertile lemma nearly as long as the spikelet; leaf-blades flat. .... 1. *P. distichum*
  2B. Spikelets glabrous, 3.5–4.5 mm. long; fertile lemma distinctly shorter than the spikelet; leaf-blades tapering to an involute point.
                                                                  2. *P. vaginatum*

1B. Spikelets broadly elliptic to rather orbicular, slightly longer than broad.
   3A. Spikelets with hairs 1–2 mm. long on margins.
      4A. Racemes 3–7, spreading; spikelets 3–3.5 mm. long; culms usually geniculate at base. . . . . 3. *P. dilatatum*
      4B. Racemes 10–20, ascending; spikelets 2–3 mm. long; culms erect. . . . . 4. *P. urvillei*
   3B. Spikelets glabrous or minutely pubescent.
      5A. Leaf-blades with spreading hairs; spikelets rather loosely arranged on the rachis; fertile lemma pale when mature. 5. *P. thunbergii*
      5B. Leaf-blades usually glabrous, if pubescent, spikelets closely arranged on the rachis.
         6A. Spikelets glabrous or nearly so; racemes 2–10; florets brown at maturity. . . . . 6. *P. orbiculare*
         6B. Spikelets puberulent; racemes 7–20; florets pale at maturity. . . . . 7. *P. longifolium*

1. Paspalum distichum L.   KISHŪ-SUZUME-NO-HIE.
Culms 20–40 cm. long, glabrous, from a long-creeping base; leaf-blades soft, linear, 5–10 cm. long, 3–7 mm. wide, glabrous, pale green; ligules hyaline, truncate, minutely lacerate on margin, 2–3 mm. long, glabrous; sheaths smooth, long-ciliate; racemes 2, geminate, ascending, at length spreading, 3–6 cm. long; rachis 1.5–2 mm. wide, flat; spikelets nearly sessile, pale green, obovate-oblong, acute, about 3 mm. long, loosely puberulent; first glume wanting, the second as long as the spikelet, 4(–6)-nerved, membranous; sterile lemma simulating the glumes, 3(–5)-nerved, acute, the fertile lemma nearly as long, rather coriaceous, pale, acute and with few minute hairs at apex.——July–Sept. Wet places near seashores; Honshu (Kinki Distr. and westw.), Shikoku, Kyushu.——Tropical regions of Asia and America.

2. Paspalum vaginatum Sw.   SAWA-SUZUME-NO-HIE.
Perennial, glabrous except for white hairs at mouth of sheaths; culms erect from an ascending base, rather thick, smooth; leaf-blades soft, linear, almost filiform, 6–15 cm. long, strongly involute, smooth, 2–3 mm. wide when flat, nerves not prominent; ligules truncate, 0.5–1 mm. long, hyaline, entire; sheaths slightly folded, smooth; racemes 2 or 3, erect; rachis 1.5–2 mm. wide; spikelets very short-pedicelled, appressed, ovate-oblong, 3.5–4.5 mm. long, acute, pale yellow, slightly lustrous, glabrous; second glume thinly membranous, acute, as long as the spikelet, with a pair of weak nerves close to the margins; sterile lemma similar, with a distinct midnerve; fertile lemma slightly shorter than the spikelet, slightly flattened, pale, acute, with few fine hairs at apex.——June–Sept. Kyushu (Yakushima).——Ryukyus, Formosa, and in tropical regions generally.

3. Paspalum dilatatum Poir.   SHIMA-SUZUME-NO-HIE.
Culms 80–100 cm. long, tufted, erect, glabrous; leaf-blades linear, 20–40 cm. long, flat, slightly scabrous and with long hairs near the mouth, gradually narrowed to both ends, glaucous; ligules 2–4 mm. long, glabrous, truncate; sheaths glabrous; racemes 5–10, spreading, 5–10 cm. long, with a tuft of long hairs in the axils; rachis rather slender; spikelets short-pedicelled, ovate-orbicular, 3–3.5 mm. long, acute, pale green, sometimes slightly dull reddish, rounded at base, with ascending to appressed white hairs on margin; second glume membranous, resembling the sterile lemma; fertile lemma orbicular, rounded at apex, about 2 mm. long.——Naturalized; Honshu.——N. America.

4. Paspalum urvillei Steud.   TACHI-SUZUME-NO-HIE.
Culms densely tufted, 75–100 cm. long, erect, glabrous; leaf-sheaths usually coarsely hirsute, the blades linear, 15–40 cm.

long, 3–15 mm. wide, flat, pilose at base; ligules 3–5 mm. long; inflorescence 10–40 cm. long; racemes 10–20, rather crowded, ascending, to 15 cm. long; spikelets 2–3 mm. long, broadly ovate, abruptly acute, pale green or slightly purplish; upper glume and lower lemma fringed with long silky hairs.—— Naturalized in Kyushu (Kokura City); rare.——S. America, S. Africa, Australia, and warmer parts of Asia; naturalized in N. America.

5. Paspalum thunbergii Kunth.   *P. mollipilum* Steud. ——SUZUME-NO-HIE.   Tufted, hirsute perennial; culms 40–90 cm. long, erect, with pilose nodes; leaf-blades flat, linear, 10–30 cm. long, 5–8 mm. wide, pilose; ligules membranous, truncate, 1–2 mm. long; sheaths rather elongate, pilose, the radical slightly folded; racemes 3–5, 5–10 cm. long, with a tuft of long hairs in the axils; spikelets short-pedicelled, broadly elliptic or nearly orbicular, 2.5–2.7 mm. long, rounded and short-mucronate at apex, glabrous or with minute, spreading hairs near margins; second glume and sterile lemma alike, 3-nerved, thinly membranous, pale green; fertile lemma orbicular-elliptic, coriaceous, slightly lustrous, rounded to very obtuse, pale.——Aug.–Oct. Waste places and cultivated fields in lowlands; Honshu, Shikoku, Kyushu; common.——Ryukyus, China, and Korea.

6. Paspalum orbiculare Forst.   *P. scrobiculatum* var. *orbiculare* (Forst.) Hack.; *P. thunbergii* var. *minor* Makino ——SUZUME-NO-KOBIE.   Rather tufted perennial; culms usually erect, 30–80 cm. long; leaf-blades flat, linear, 15–30 cm. long, 5–10 mm. wide; ligules truncate, 0.5–1.5 mm. long, glabrous; sheaths slightly folded or nearly terete, usually with a few long hairs on the mouth; racemes 2–10, 3–10 cm. long, densely flowered, ascending, with a few long hairs in the axils; spikelets pale green, elliptic to broadly so, or obovate-elliptic, 2–2.5 mm. long, glabrous or spreading-puberulent on the margins, rounded to very obtuse; second glume and sterile lemma thinly membranous, 3-nerved; fertile lemma as long as the sterile one, coriaceous, lustrous, brown when mature.—— Aug.–Oct. Lowlands; Honshu (Tōkaidō and westw.), Shikoku, Kyushu.——Ryukyus, Formosa, China, and in tropical regions generally.

7. Paspalum longifolium Roxb.   NAGABA-SUZUME-NO-HIE.   Rather glabrous, tufted perennial; culms erect, 80–130 cm. long, ascending and branching at base; leaf-blades 10–20 cm. long, 5–8 mm. wide; racemes 7–20, loosely arranged, spreading to ascending, 5–8 cm. long; spikelets geminate, in 2 rows on one side of the rachis, obovate, 2–2.5 mm. long, pale green, with spreading, minute hairs.——Naturalized; Honshu.——India, s. China, Philippines, and Malaysia.

76. ERIOCHLOA HBK.   NARUKOBIE ZOKU

Usually villous or pubescent perennials or annuals, with linear, flat to involute leaf-blades and a racemose inflorescence; spikelets short-pedicelled or sessile, in 2 rows on one side of the raceme, solitary or rarely geminate, slightly dorsally compressed, ovate-elliptic or ovate, soft-pubescent; first glume reduced to a ring with the pedicel, adaxial; second and the sterile lemma alike, membranous, the latter usually paleate, sometimes staminate; fertile lemma coriaceous, often short-pointed at apex.—— About 20 species, in warm-temperate and tropical regions.

1. **Eriochloa villosa** (Thunb.) Kunth. *Paspalum villosum* Thunb.; *Panicum tuberculiflorum* Steud.——NARUKOBIE. Tufted, villous perennial; culms 50–100 cm. long, villous; leaf-blades flat, linear, 10–25 cm. long, 7–15 mm. wide; ligules very short, ciliate; racemes 4–7, loosely arranged on the axis, 2–5 cm. long, white-villous, spreading, rachis about 1 mm. wide, 3-angled; spikelets slightly compressed, broadly ovate, 4.5–5 mm. long inclusive of the basal ring, acute; second glume and sterile lemma as long as the spikelet, broadly ovate, rather acute, membranous, softly pubescent, pale, with 5, green, weak nerves; fertile lemma as large as the sterile one, subobtuse, coriaceous, punctate, yellow.——July–Oct. Grassy places in lowlands; Honshu, Shikoku, Kyushu; rather common.—— Ussuri, China, Manchuria, and Korea.

77. OPLISMENUS Beauv. CHIJIMIZASA ZOKU

Annuals or perennials with rather broad, lanceolate leaf-blades; panicles spikelike or with racemosely arranged simple branches; spikelets ovate, short-pedicelled; first glume abaxial, membranous, 3- to 5-nerved, apically awned, the second similar, usually awnless; sterile lemma longer, 5- to 7-nerved, sometimes paleate and staminate; fertile lemma ovate, awnless, rather coriaceous to chartaceous, smooth.——More than a dozen species, widespread in all warm-temperate and tropical regions.

1A. Branches of inflorescence not elongate, rarely to 1.5 cm. long, very slender and densely flowered. .... 1. *O. undulatifolius*
1B. Branches (racemes) of inflorescence elongate, 2–5(–7) cm. long, rather thick, loosely flowered. .... 2. *O. compositus*

1. **Oplismenus undulatifolius** (Arduino) Roem. & Schult. *Panicum undulatifolium* Arduino; *P. hirtellum* sensu Thunb., non L.——KE-CHIJIMIZASA.   Coarsely pubescent perennial; culms 10–30 cm. long, branching and long-creeping at base; leaf-blades flat, broadly lanceolate to narrowly ovate, 3–7 cm. long, 1–1.5 cm. wide, acuminate, slightly oblique and cuneate to obtusely rounded at base; ligules very short; sheaths short; panicles simple and spikelike, exserted, erect, linear or lanceolate, 6–12 cm. long, loosely and interruptedly racemose, the branches 6–10, very short, rarely to 15 mm. long, slender, densely flowered; spikelets short-pubescent, narrowly ovate, about 3 mm. long, acute, subsessile; first glume broadly lanceolate, obtuse, about half as long as the spikelet, with a straight, antrorsely scaberulous awn, the second narrowly ovate, about 2/3 as long as the spikelet, long-awned, 3-nerved; sterile lemma as long as the spikelet, broadly ovate, acute with a short awn, 5- to 7-nerved; fertile lemma pale, smooth, lustrous, acute, about 2.7 mm. long.——Aug.–Oct. Woods in lowlands; Hokkaido, Honshu, Shikoku, Kyushu; very common and variable.——Korea, China, India to s. Europe.

Var. **japonicus** Steud. *Panicum japonicum* Steud.; *O. japonicus* (Steud.) Honda; *O. tsushimensis* Honda——KO-CHIJIMIZASA.   Glabrescent. Nearly as common as the typical phase.

Var. **microphyllus** (Honda) Ohwi. *O. microphyllus* Honda——CHABO-CHIJIMIZASA.   Plant smaller and less pilose; leaf-blades 1–3 cm. long; branches of the inflorescence very short with a few spikelets.

2. **Oplismenus compositus** (L.) Beauv. *Panicum compositum* L.; *O. polliniaefolius* Honda——EDA-UCHI-CHIJIMIZASA.   Larger than the preceding; culms 20–40 cm. long, long-creeping and loosely branching at base; leaf-blades flat, broadly lanceolate to semi-ovate, 3–10 cm. long, 1–2 cm. wide, acuminate, sometimes pubescent; ligules very short, truncate; sheaths glabrous to pilose; panicles exserted, 10–20 cm. long, the racemes 6–10, loosely arranged on the axis, ascending to spreading, 3–8 cm. long, loosely spiculose, the rachis triangular; spikelets green, rarely slightly dull purplish, narrowly ovate, 3–3.5 mm. long, usually glabrous; awn slightly thicker than in the preceding species.——Oct.–Nov. Woods in lowlands and mountains; Honshu (Izu Isls.), Kyushu; rare. ——Ryukyus, Formosa, s. China, and in tropical regions generally.

Var. **patens** (Honda) Ohwi. *O. patens* Honda——ŌBA-CHIJIMIZASA.   Plant stouter, glabrescent; leaf blades 10–15 cm. long, 2–3.5 cm. wide; spikelets slightly larger; fertile lemma terminated with an apical straight mucro.——Kyushu (Yakushima).——Ryukyus and Formosa.

78. ECHINOCHLOA Beauv. HIE ZOKU

Annuals or perennials, usually rather stout; leaf-blades linear, flat; panicles pyramidal, with spikelike branches densely spiculose on one side; spikelets ovoid, subsessile, acuminate or awned, spinulose-scabrous; glumes and sterile lemma membranous, the first smaller, acuminate, the second nearly as long as the spikelet, inflated, 5- to 7-nerved, awned or acuminate; sterile lemma similar to the second glume, sometimes rather coriaceous and lustrous, often long-awned, sometimes paleate and staminate; fertile lemma ovate, smooth, lustrous, coriaceous, acute, convex on back.——More than a dozen species, in the warmer regions of the world.

1. **Echinochloa crusgalli** (L.) Beauv. var. **crusgalli.** *Panicum crusgalli* L.; *E. crusgalli* (L.) Beauv.; *E. caudata* Rashev.; *E. crusgalli* var. *caudata* (Rashev.) Kitag.; *Oplismenus crusgalli* (L.) Kunth——INUBIE.   Coarse annual; culms 80–120 mm. long, rather thick, branching at base; leaf-blades linear, flat, 30–50 cm. long, 1–2 cm. wide, scabrous, scarcely thickened on margin; ligules wanting; sheaths smooth, often reddish in lower ones; panicles exserted, more or less nodding, 10–25 cm. long, rather densely branched, the branches (racemes) to 3–5 cm. long, ascending, sessile; spikelets 3–4 mm. long, densely arranged on branches, ovoid, often long-awned, pale green and often dull purplish, short bristly on nerves; first glume about 2/5 as long as the spikelet, deltoid, the second as long as the spikelet, short-awned; sterile lemma membranous, with a straight scabrous awn 2–4 cm. long, or awnless; fertile lemma ovate-elliptic, acute, pale yellow, lustrous, smooth, 3–3.5 mm. long.——Aug.–Oct.   Wet paddy fields and ponds; Honshu, Shikoku, Kyushu; very common and variable.——Widespread in all warmer regions.——Forma **kanashiroi** Ohwi. Leaf-blades and sheaths pilose.

Var. **praticola** Ohwi. *Panicum crusgalli* var. *muticum* sensu auct. Japon., non Wirtgen——HIME-INUBIE.   An awnless, more slender plant.——July. Very common in (not wet) lowlands; Hokkaido, Honshu, Shikoku, Kyushu.—— Formosa, Ryukyus, to Korea.

Var. **formosensis** Ohwi. *E. crusgalli* var. *kasaharae* Ohwi

——HIME-TA-INUBIE. Intermediate between the typical phase and the following variety; leaf-blades narrower; spikelets pale green; first glume about 2/5 as long as the spikelet; sterile lemma thickened, lustrous, glabrous, at least on the back.——Honshu, Shikoku.——Formosa, China, and India.

Var. **oryzicola** (Vasing.) Ohwi. *E. oryzicola* Vasing.; *E. crusgalli* var. *hispidula* Honda, excl. syn.——TA-INUBIE.

Culms nearly erect from base; leaf-blades thickened on margins, scabrous; panicles usually pale green, erect, spikelets about 5 mm. long, awnless or short-awned; first glume 1/2–3/5 as long as the spikelet; sterile lemma often thickened and coriaceous, glabrous, lustrous.——Aug.–Oct. Wet places; common in paddy fields; Honshu, Shikoku, Kyushu; a troublesome weed.——China, Korea, and India.

79. BRACHIARIA Griseb.   NIKU-KIBI ZOKU

Perennials or annuals; leaf-blades linear to lanceolate, usually flat; racemes loosely arranged on axis of inflorescence; spikelets geminate or solitary, short-pedicellate, in 2 rows on rachis, oblong to elliptic or ovate; first glume adaxial, usually short, the second and the sterile lemma similar, membranous, 5- to 7-nerved; fertile lemma oblong or elliptic, coriaceous, sometimes mucronate, usually rugulose.——About 50 species, chiefly in Africa, few in Asia and America.

1. **Brachiaria villosa** (Lam.) A. Camus. *Panicum villosum* Lam.; *P. coccospermum* Steud.——BIRŌDO-KIBI. Tufted, densely pubescent annual; culms erect, branching at base, pubescent, 15–20 cm. long; leaf-blades lanceolate to broadly linear, flat, 3–5 cm. long, 5–8 mm. wide, silvery-green, short-acuminate, rounded at base; ligule a series of short hairs; sheaths rather short; inflorescence exserted, erect, slightly one-sided, 4–6 cm. long, the branches 6–10, ascending, to 2–3 cm. long; spikelets short-pedicelled, appressed to the rachis, ovate-elliptic, 2.5–2.7 mm. long, acute, pubescent, pale green, becoming pale yellow at maturity; first glume deltoid, 2/5 as long as the spikelet, acute, the second nearly as long as the spikelets, acute, 7- rarely 5-nerved, membranous; sterile lemma as long as the spikelets, acute, 5-nerved; fertile lemma slightly shorter, pale yellow, rather acute, narrowly ovate, finely rugulose, not lustrous.——Honshu, Shikoku, Kyushu; rare.——Ryukyus, Formosa, China, Malaysia, and India.

80. ISACHNE R. Br.   CHIGOZASA ZOKU

Small or rather large perennials or annuals; culms often branching at base; leaf-blades lanceolate to narrowly ovate, usually with raised parallel nerves; panicles loose to contracted; spikelets ovate, obovate, or globose, 2-flowered; florets all bisexual; first and second glumes alike, membranous, nearly as long as the florets; lemmas nearly alike, usually coriaceous to chartaceous, elliptic to orbicular, often puberulent.——About 60 species, widespread in warmer regions of the world, abundant in tropical Asia.

1A. Culms 20–50 cm. long, erect; leaves broadly linear; spikelets often dull purplish; pedicels with a sessile, obsolete gland. .. 1. *I. globosa*
1B. Culms 5–20 cm. long, long-decumbent at base; leaves broadly lanceolate; spikelets pale green; pedicels glandless.
  2A. Spikelets about 1.5 mm. long; plant very small, 5–10 cm. high, delicate; glumes glabrous at base, scabrous and with a few short bristles toward tip. . . . . 2. *I. nipponensis*
  2B. Spikelets about 2 mm. long; plant rather firm, 10–20 cm. high, with thicker leaves, glumes uniformly scabrous. .... 3. *I. kunthiana*

1. **Isachne globosa** (Thunb.) O. Kuntze. *Milium globosum* Thunb.; *I. australis* R. Br.; *Panicum lepidotum* Steud.; *Helopus globosum* Steud.; *Eriochloa japonica* Kunth ex Steud.——CHIGOZASA. Perennial; culms erect, ascending or shortly decumbent at base, 30–60 cm. long, glabrous; leaf-blades firmly membranous, flat, broadly linear, 4–7 cm. long, 3–7 mm. wide, gradually acuminate, scabrous, with slightly thickened margins, nerves not prominently raised; ligule a series of short hairs; sheaths rather short, glabrous; panicles erect, open, 3–6 cm. long, the branches slender, elongate, loosely branching, the pedicels with a pale yellow, obscure gland, thickened at apex; spikelets broadly obovate to obovate-globose, 2–2.2 mm. long, rounded at apex, pale green, sometimes slightly purplish, slightly lustrous; glumes equal, broadly elliptic, obtuse, weakly nerved, smooth or slightly scabrous above, firmly membranous; lemmas as long as to slightly longer than the glumes, elliptic, unequal, somewhat coriaceous, pale yellow, rounded at apex, glabrous to puberulent on margin.——June–Aug. Wet places; Hokkaido, Honshu, Shikoku, Kyushu.——Ryukyus, Formosa, China, se. Asia, Malaysia, Micronesia, and Australia.

2. **Isachne nipponensis** Ohwi. *I. myosotis* sensu auct. Japon., non Nees——HAI-CHIGOZASA. Slender, delicate perennial; culms erect, long-creeping at base, glabrous, 5–10 cm. long, with pubescent nodes; leaf-blades membranous, flat, broadly lanceolate, 15–30 mm. long, 4–8 mm. wide, short-acuminate, cuneate at base, sparsely long-pilose on both sides, many-nerved; sheaths 7–15 mm. long, long-ciliate; ligule a series of short hairs; panicles loosely flowered, broadly ovate, 3–5 cm. long, the branches slender, ascending; spikelets green, about 1.5 mm. long, broadly elliptic to nearly globose; glumes equal, 3- to 7-nerved, rounded at apex, with few bristles of unequal length on upper half; lemmas equal, elliptic, 1.2–1.3 mm. long, loosely puberulent.——Sept.–Oct. Wet places; Honshu (Kantō Distr. and westw.), Shikoku, Kyushu; rather rare.——s. Korea.

3. **Isachne kunthiana** (Wight & Arn.) Nees. *Panicum kunthianum* Wight & Arn.; *I. schmidtii* Hack.; *I. commelinifolia* Warb.; *I. firmula* sensu auct. Japon., non Buse; *I. myosotis* var. *nudiglumis* Hack.; *I. kunthiana* var. *nudiglumis* (Hack.) T. Koyama——ATSUBA-HAI-CHIGOZASA. Culms long-creeping, 10–20 cm. long, and glabrous; leaf-blades herbaceous, broadly lanceolate, 3–6 cm. long, 6–15 mm. wide, gradually acute, obtuse at base, often puberulent beneath, thickened on margins, with raised closely parallel slender nerves above; ligules nearly reduced to a series of fine hairs; sheaths rather short, ciliate; panicles at length exserted, open, 3–4 cm. long, the branches rather thick, ascending, few-flowered; spikelets

short-pedicelled, elliptic, grayish green, 2–2.3 mm. long, rounded at apex, dull; glumes alike, thick-membranous, with 9–11 raised, prominently scabrous nerves; lemmas alike, broadly ovate-elliptic, rather prominently compressed, coria-ceous, rounded at apex, sparsely puberulent on margins.——Reported to occur in Kyushu.——Ryukyus, Formosa, s. China, Indochina, Malaysia, and Micronesia.

81. SPINIFEX L.   Tsuki-ige Zoku

Stout dioecious perennials; culms rather woody; leaf-blades linear, pungent; inflorescence capitate, the branches radiate, sessile, pungent, each with one pistillate spikelet, the staminate spikelets several, 2-flowered, subsessile, the glumes 5- to 7-nerved, shorter than the florets, chartaceous; pistillate spikelets solitary at base of long naked branches, the glumes and sterile lemmas alike or the first somewhat longer; sterile lemma sometimes paleate; fertile lemma 5- to 7-nerved, involute, acute.——Several species, of sandy seashores in Australia, Malaysia, Polynesia, India, and e. Asia.

1. Spinifex littoreus (Burm. f.) Merr. *Stipa spinifex* L.; *Stipa littorea* Burm. f.; *Spinifex squarrosus* L.——Tsuki-ige. Stout perennials; culms decumbent and branching at base, 30–50 cm. long; leaf-blades crowded at the nodes, coriaceous, subulate-linear, 5–20 cm. long, 2.5–3 mm. wide, glabrous, spine-pointed, terete with a groove above, smooth but scabrous on margin, pale green, often glaucous; ligules very short, densely white-hairy; sheaths often crowded and overlapping, glabrous, at first ciliate; staminate inflorescence 4–8 cm. across; racemes many, ending in a spine; staminate spikelets alternate on upper half of the raceme, 8–12 mm. long, the glumes herbaceous, nerved, acuminate; pistillate spikelets single, near the base of long spinelike branches of the inflorescence; glumes many-nerved; fertile lemma rather coriaceous, broadly lanceolate, about 10 mm. long, slightly shorter than the sterile lemma, acuminate, glabrous, few-nerved only in upper part.——July–Aug. Sandy seashores; Kyushu (Yakushima and Tanega-shima); very rare.——Ryukyus, Formosa, s. China, Malaysia, and India.

82. DIMERIA R. Br.   Karimata-gaya Zoku

Slender annuals or perennials; leaf-blades linear, acute; racemes 1 to several, digitate; rachis persistent; spikelets alternate, in 2 rows on one side of a continuous rachis, solitary, sessile, lanceolate, laterally flattened, deciduous from the base; glumes folded and flattened, the first narrowly linear; sterile lemma hyaline; fertile lemma hyaline, keeled, 1-nerved, 2-toothed, usually awned; palea minute or wanting; stamens 2; caryopsis linear; laterally compressed.——About 20 species, Madagascar to India, e. Asia, Malaysia, and Australia.

1. Dimeria ornithopoda Trin. var. **tenera** (Trin.) Hack. *D. tenera* Trin.; *Andropogon stipaeformis* Steud.; *D. stipaeformis* (Steud.) Miq.; *D. higoensis* Honda; *D. mikii* Honda; *D. neglecta* Tzvel.——Karimata-gaya. Annual; culms erect or from ascending base, cespitose or solitary, 7–40 cm. long; leaf-blades linear, the margins often slightly recurved, 4–7 cm. long, 3–5 mm. wide; ligules ovate, obtuse, hyaline, glabrous, 1–2.5 mm. long; sheaths rather short, glabrous or sparsely long-hairy, smooth; inflorescence terminal and from axils of upper leaves; racemes 1–5, commonly 2 or 3, erect, sessile, 2–8 cm. long, pale green to pale brown; rachis not jointed; spikelets compressed, appressed to the rachis, nearly sessile, lanceolate, acute, 2.5–4 mm. long, with a tuft of very short hairs at base; first glume linear-lanceolate, membranous, with a scabrous, purplish keel, the second as long, lanceolate; fertile lemma hyaline, about 2/3 as long as the glumes, 2-toothed, 1-nerved, the awn between the teeth, 2–10 mm. long, geniculate and twisted, minutely scabrous.——Aug.–Oct. Wet places in lowlands; Hokkaido, Honshu, Shikoku, Kyushu; common.——India, Malaysia, and Australia.——Subvar. **microchaeta** Hack. Hime-karimata-gaya. Awnless.——Occurs with the species.

83. ECCOILOPUS Steud.   Abura-susuki Zoku

Rather tall tufted perennials; leaf-blades linear, often petiolate and sagittate at base; panicles exserted; racemes pedunculate, with a jointed persistent rachis; spikelets geminate, lanceolate to ovate, nearly terete, with a tuft of short hairs at base, deciduous from the pedicels; glumes nearly equal, herbaceous, nerved; sterile lemma paleate or epaleate; fertile lemma hyaline, bifid, the awn perfect or imperfect.——Several species, in e. Asia.

1. Eccoilopus cotulifer (Thunb.) A. Camus. *Andropogon cotulifer* Thunb.; *Spodiopogon cotulifer* (Thunb.) Hack.; *E. andropogonoides* Steud.——Abura-susuki. Tall tufted perennial; culms erect, simple, 80–120 cm. long, rather stout, resinously lustrous below the panicle; radical leaf-blades flat, linear, 40–60 cm. long, 1–1.5 cm. wide, acuminate, gradually narrowed to the petiole, sparsely pilose or glabrous; ligules 2–4 mm. long, appressed-pubescent dorsally; sheaths smooth, the upper cauline not petiolate; panicles 20–30 cm. long, nodding; racemes loosely arranged, rather many, pendulous or nodding on rather long, nearly smooth, slender pedun-cles, 3–5 cm. long, sometimes branched at base, loosely spic-ulose; rachis persistent, with a tuft of short hairs at each node; joints and pedicels slightly thickened above; spikelets equal, lanceolate, terete, about 6 mm. long, scabrous, acuminate, white-pilose, pale green, rarely dull purple; glumes herbaceous, with raised nerves; fertile lemma with an exserted awn.——Sept.–Oct. Grassy places and thickets in low mountains; Hokkaido, Honshu, Shikoku, Kyushu; common and variable.——Korea, Formosa, China, and India.——Forma **sagittiformis** Ohwi. Radical leaves with blades sagittate at base.

84. IMPERATA Cyr.　Chigaya Zoku

Perennials with long-creeping rhizomes; culms erect; panicles dense, with an elongate continuous axis; racemes numerous, rather short, the rachis persistent, not disarticulating; spikelets alike, geminate, lanceolate, awnless, long-sericeous, deciduous from pedicels of various length; glumes membranous; sterile lemma epaleate, hyaline; fertile lemma hyaline; palea nerveless; stamens 1 or 2.——About 10 species, in warm-temperate to tropical regions.

1. **Imperata cylindrica** (L.) Beauv. var. **koenigii** (Retz.) Durand & Schinz. *Saccharum koenigii* Retz.; *I. koenigii* (Retz.) P. Beauv.; *I. arundinacea* var. *koenigii* (Retz.) Benth.; *I. pedicellata* Steud.——Chigaya, Fushige-chigaya. Rhizomes elongate; culms 30–80 cm. long, rather slender, firm, long-hairy on nodes; leaf-blades flat, linear, 20–50 cm. long, 7–12 mm. wide, acuminate, gradually narrowed to a petiole-like base, scabrous; ligules truncate, very short; sheaths smooth or often long-pilose; panicles spikelike, nearly erect, silvery white, 10–20 cm. long, cylindric, the racemes less than 3 cm. long, sparsely long-hairy; spikelets lanceolate, 3.5–4.5 mm. long, glabrous, terete, acute, with a tuft of silvery white hairs about 12 mm. long at the base; glumes lanceolate, acuminate, membranous, few-nerved; stamens 2, 2.5–3 mm. long; stigma blackish purple, lanceolate in outline.——May–June. Waste grounds in lowlands; Hokkaido, Honshu, Shikoku, Kyushu; very common.——In temperate regions.——Forma **pallida** Honda. *I. cylindrica* var. *genuina* sensu auct. Japon., non Durand & Schinz——Kenashi-chigaya. Culm-nodes glabrous; common within our area.——China, Korea, and Manchuria.

Imperata cylindrica var. **major** (Nees) C. E. Hubb. The tropical phase, which differs from ours in the larger panicles with smaller, narrower stigmas.

85. MISCANTHUS Anderss.　Susuki Zoku

Robust perennials; leaf-blades elongate-linear; racemes few to rather many on the short axis of the panicle; rachis slender, many-jointed; perfect spikelets usually awned, the secondary spikelet and its pedicel usually wanting, or the pedicel (rarely a spikelet) developed only at the lower joints of the articulate rachis; perfect spikelet bisexual, usually awned, lanceolate, deciduous from the pedicel; callus-hairs longer or shorter than the spikelets; glumes chartaceous; sterile lemma hyaline, epaleate; fertile lemma hyaline, 2-toothed; palea short, nerveless, stamens 2 or 3.——About 20 species, in India, e. Asia, Malaysia, and Polynesia.

1A. Rhizomes elongate; spikelets dull gray-brown, usually awnless; callus-hairs 2–4 times as long as the spikelet, silvery white; plants gregarious, in wet places. . . . . 1. *M. sacchariflorus*
1B. Rhizomes short, thick; spikelets usually awned, sometimes awnless; callus-hairs nearly as long or shorter than the spikelet, white to purplish; plants growing in dry fields.
　2A. Inflorescence paniculate, the axis elongate, much longer than the racemes. . . . . 2. *M. floridulus*
　2B. Inflorescence corymbose, the axis shorter than or slightly longer than the racemes.
　　3A. Leaf-blades linear, very long, usually coriaceous or thick-herbaceous; racemes usually many. . . . . 3. *M. sinensis*
　　3B. Leaf-blades, at least some of cauline ones, narrowly lanceolate, herbaceous; racemes usually 2–12.
　　　4A. Spikelets awned; first glume gradually narrowed above and bidentate at apex; leaf-blades pilose beneath.
　　　　5A. Awn perfect, about 15 mm. long; racemes 2–5; leaf-blades 8–12 mm. wide. . . . . 4. *M. oligostachyus*
　　　　5B. Awn imperfect, 5–8 mm. long, slightly exserted; racemes 6–10; leaf-blades 1–2 cm. wide, rounded at base.

5. *M. intermedius*
　　　4B. Spikelets awnless; first glume acute, bidentate; leaf-blades usually glabrous. . . . . 6. *M. tinctorius*

1. **Miscanthus sacchariflorus** (Maxim.) Benth. *Imperata sacchariflora* Maxim.; *I. eulalioides* Miq.; *Triarrhena sacchariflora* (Maxim.) Nakai; *M. ogiformis* Honda——Ogi. Stout, coarse perennial with stout rhizomes; culms 1–2.5 m. long, glabrous; leaf-blades flat, linear, 40–80 cm. long, 1–3 cm. wide, acuminate, rather firm, glabrous except at base above, very scabrous on margins, glaucescent beneath; ligules very short; sheaths glabrous or the lower coarsely hairy; panicles large, open, exserted, 25–40 cm. long, rather digitately many-racemed, the central axis densely pubescent at the base; racemes nearly sessile, often branched at base, 20–40 cm. long, slender; spikelets 5–6 mm. long, lanceolate, acuminate; callus-hairs 10–15 mm. long; glumes lanceolate, herbaceous, pale grayish brown, with hyaline margins, sparsely long-hairy, the first 2-keeled; fertile lemma awnless or with a short, included awn.——Sept.–Oct. Wet places in lowlands; Hokkaido, Honshu, Shikoku, Kyushu; common.——Manchuria, Ussuri, Korea, and n. China.

2. **Miscanthus floridulus** (Labill.) Warb. *Saccharum floridulum* Labill.; *M. formosanus* A. Camus; *M. japonicus* Anderss.——Tokiwa-susuki.　Robust, tufted, evergreen perennial; culms 1–2.5 m. long, stout, smooth; leaf-blades elongate, 1.5–3 cm. wide, pale glaucous-green, scabrous on margin, pubescent above near base; ligules truncate, to 2 mm. long; sheaths smooth; panicles pyramidal, 30–50 cm. long, erect, white, the axis nearly reaching to the summit of the panicle, puberulent-scaberulous, with axillary tufts of hairs; racemes numerous, nearly sessile, branched at base, slender, 10–20 cm. long; spikelets on pedicels of various length, obliquely spreading, lanceolate, 3–3.5 mm. long, acute, with a tuft of hairs 4–6 mm. long at base; glumes herbaceous, glabrous; awn of fertile lemma 8–10 mm. long.——July–Aug. Lowlands; Honshu (s. Kantō, Tōkaidō Distr. and westw.), Shikoku, Kyushu.——Ryukyus, Formosa, and the Pacific Islands.

3. **Miscanthus sinensis** Anderss. Susuki.　Tufted stout perennial, with thick, short rhizomes; culms 1–2 m. long; leaves radical and cauline, the blades flat, 1–2 cm. wide, much-elongate, pale green beneath, coarsely scabrous on margins, midrib thickened toward base; panicles corymbose, 20–30 cm. long, the axis short, less than half as long as the racemes; racemes rather numerous and dense, 15–30 cm. long; spikelets

geminate, ascending or obliquely spreading, lanceolate, 5–7 mm. long, acuminate, yellowish, awned, with a tuft of white or purplish hairs 7–12 mm. long, pedicellate; glumes rather coriaceous, acuminate, with hyaline apex and margins; awn of fertile lemma 8–15 mm. long.——Aug.–Oct. Slopes in lowlands and mountains; Hokkaido, Honshu, Shikoku, Kyushu; common and variable.——s. Kuriles, Korea, China, Ryukyus, and Formosa.——Cv. **Gracillimus.** *M. sinensis* var. *gracillimus* Hitchc.——ITO-SUSUKI. A narrow-leaved cultivated phase.——Cv. **Zebrinus.** *M. sinensis* forma *zebrinus* (Nichols.) Nakai; *Eulalia japonica zebrina* Nichols.; *M. sinensis* var. *zebrinus* (Nichols.) Beal——TAKANOHA-SUSUKI. Cultivated phase with narrow, very scabrous, intermittent transverse whitish bands across the leaf blades.

Var. **condensatus** (Hack.) Makino. *M. condensatus* Hack.; *M. hidakanus* Honda——HACHIJŌ-SUSUKI. Stouter; leaf-blades 1.5–4 cm. wide, glaucous beneath, less scabrous, panicles much denser.——Near seashores, sometimes in mountains; Hokkaido, Honshu, Shikoku, Kyushu; rather common. ——Korea, China, Indochina, and the Pacific Islands.

4. Miscanthus oligostachyus Stapf. *M. matsumurae* Hack.——KARI-YASU-MODOKI. Tufted perennial; culms 60–80 cm. long, with pilose nodes; leaf-blades flat, linear to broadly so, 20–40 cm. long, 8–12 mm. wide, scabrous on margins, sparsely long-pilose especially beneath; ligules semirounded, obtuse, glabrous, membranous, 2–3 mm. long; sheaths rather short, glabrous, striate; panicles corymbose, the axis very short; racemes 2–5, erect, 7–12 cm. long, green or purplish; spikelets erect, lanceolate, 7–8 mm. long, acuminate, sparsely long-hairy; callus with a tuft of hairs 3–5 mm. long; glumes herbaceous, acuminate, scabrous; awn of fertile lemma

about 15 mm. long.——Aug.–Oct. Mountains; Honshu, Shikoku, Kyushu.

5. Miscanthus intermedius (Honda) Honda. *M. longiberbis* var. *intermedius* Honda; *M. tinctorius* var. *intermedius* (Honda) Ohwi——Ō-HIGENAGA-KARI-YASU-MODOKI. Rhizomes short; culms tufted, 120–180 cm. long, rather stout; leaf-blades flat, narrowly to broadly lanceolate, 40–60 cm. long, 1–2 cm. wide, long-acuminate, glaucescent beneath, sparsely pilose, rounded at base; ligules truncate, brownish, 1–2 mm. long; sheaths smooth, short-pubescent above; corymbs exserted, with a very short axis; racemes 6–10, erect, 10–15 cm. long; spikelets erect to ascending, 7–8 mm. long, acuminate, with a tuft of hairs 5–7 mm. long at base; first glume gradually narrowed, 2-costate at tip, sparsely white-pilose; awn of fertile lemma imperfect, slightly exserted, 5–8 mm. long.—— Aug.–Sept. Mountains; Honshu (Japan Sea side of centr. and n. distr.).

6. Miscanthus tinctorius (Steud.) Hack. *M. sieboldii* Honda; *Saccharum tinctorium* Steud.; *Imperata tinctoria* (Steud.) Miq.——KARI-YASU. Tufted perennial; culms 80–100 cm. long; leaf-blades flat, broadly linear, 20–40 cm. long, 8–12 mm. wide, glabrous, scabrous on margins, pilose on both sides near base, midrib rather slender; ligules semirounded or truncate, 2–3 mm. long; sheaths sometimes sparsely pilose; corymbs exserted, erect, with a very short axis; racemes 2–12, 10–15 cm. long; spikelets 5–6 mm. long, acute, with a tuft of hairs at base about half as long as the spikelet, awnless; glumes rather coriaceous, brownish, acute, long-pilose, the first bidentate, scabrous on the 2 ribs above.——Aug.–Oct. Mountains; Honshu (centr. distr.).

86. SPODIOPOGON Trin. Ō-ABURA-SUSUKI ZOKU

Rather coarse perennials; leaf-blades linear to narrowly lanceolate; racemes 1–3, peduncled; rachis articulate at nodes; spikelets geminate, ovate, terete, the first pedicelled, deciduous from apex of the pedicel, the second nearly sessile, deciduous with the joint and pedicel; glumes coriaceous, nerved, acute, the first scarcely keeled, the second nearly as long; sterile lemma hyaline, paleate and staminate; fertile lemma hyaline; stamens 3.——More than a dozen species, widespread in temperate regions.

1A. Leaf-blades linear, 25–40 cm. long, pilose to glabrous; rachis with 3 to 5 joints and 7 to 11 spikelets. .... 1. *S. sibiricus*
1B. Leaf-blades lanceolate to linear-lanceolate, less than 20 cm. long, thinly membranous, glabrous except at the mouth; rachis with 1 or 2 joints and 3 to 5 spikelets. .... 2. *S. depauperatus*

1. Spodiopogon sibiricus Trin. *Andropogon sibiricus* (Trin.) Steud.——Ō-ABURA-SUSUKI. Perennials with elongate, rather thick rhizomes covered with small scales; culms erect, 80–120 cm. long, glabrous; leaf-blades flat, linear to linear-lanceolate, usually 25–40 cm. long, 1–1.5 cm. wide, pilose or glabrescent, narrowed to the obtuse base; ligules hyaline, 1–1.5 mm. long, brownish; sheaths shorter than the leaf-blades; panicles erect, 15–25 cm. long, the axis and branches terete, smooth, resinously lustrous in life, the branches erect or ascending, to 4–6 cm. long, usually slightly branched, often falsely verticillate; racemes linear, 2–3.5 cm. long, erect, the joints and pedicels 2–3 mm. long, thickened above with a tuft of spreading hairs about 1 mm. long at apex; spikelets narrowly ovate, terete, 4.5–5.5 mm. long, acute, with spreading hairs about 2 mm. long, pale green, awned; glumes rather coriaceous, with raised nerves; awn of fertile lemma perfect, 7–12 mm. long.——Aug.–Oct. Slopes in mountains; Hokkaido, Honshu, Shikoku, Kyushu; rather common.—— Korea, Manchuria, China, and Siberia.

2. Spodiopogon depauperatus Hack. MIYAMA-ABURA-SUSUKI. Glabrous or glabrescent perennial with short rhizomes; culms 60–80 cm. long, with few, short, sheathing leaves at base; cauline blades lanceolate to linear-lanceolate, completely flat, usually glabrous, membranous, 15–20 cm. long, 1–1.5 cm. wide, cuneate at base; ligules semirounded to truncate, glabrous, 0.5–1.5 mm. long; sheaths shorter than the blades, glabrous; panicles exserted, 8–12 cm. long, the branches falsely verticillate or solitary, sometimes slightly branched, glabrous, ascending, to 2–3 cm. long; racemes 6–8 mm. long, with 3 to 5 spikelets, the joints and pedicels 2–3 mm. long, thickened above, with a tuft of spreading, white or reddish hairs less than 1 mm. long; spikelets narrowly ovate, 4–5 mm. long, acute, terete, nerved, sparsely white-pilose; glumes firmly herbaceous, pale green, or slightly purplish, scaberulous, the first 2-denticulate, the second awn-pointed; awn of fertile lemma perfect, 7–8 mm. long.——June–Aug. Mountains; Honshu (Japan Sea side of centr. and n. distr.); rather rare.

87. SACCHARUM L. SATŌ-KIBI ZOKU

Usually stout, tufted perennials; culms terete, solid; leaf-blades linear, with a thick midnerve; panicles large, dense, with an elongate axis, pubescent; racemes slender, many-jointed, on numerous branches of panicles; spikelets geminate, one sessile, the other pedicellate, perfect, awnless, disarticulating below the spikelets; callus hairs long; glumes hyaline at tip, the first narrow, the second 1- to 5-nerved; sterile lemma hyaline, epaleate; fertile lemma hyaline, obtuse, sometimes reduced; palea minute, nerveless.——Less than 10 species, in warmer regions of Malaysia, e. Asia, India, and N. Africa.

1. **Saccharum spontaneum** L. var. **arenicola** (Ohwi) Ohwi. *Saccharum spontaneum* sensu auct. Japon., non L.—— WASE-OBANA.   Tufted perennial; culms erect, firm, to 1.5 m. high, 6 mm. in diameter, sheathed at base, long-villous in upper part; leaf-blades coriaceous, linear, 40–60 cm. long, 4–6 mm. wide, gradually long-acuminate, narrowed at base, with recurved margins, glaucous beneath, scabrous on upper surface and margins, the midrib thickened toward the base; mouth of sheath long-hairy; ligules 1.5–3 mm. long, ciliate, ovate or ovate-deltoid; panicle narrow, erect, about 30 cm. long, the branches erect; racemes many-jointed, the joints 4–6 mm. long, sparsely long-hairy toward the apex; pedicels 2–3 mm. long; sessile spikelet broadly lanceolate, 4.5–5 mm. long; callus with a tuft of silky hairs about 10 mm. long.——Aug.– Sept. Sandy places near seashores; Honshu (s. Kantō and Tōkaidō Distr.); locally common.——Ryukyus.

88. MICROSTEGIUM Nees ASHIBOSO ZOKU

Slender, branched annuals or perennials, decumbent to ascending at base; leaf-blades thin, lanceolate; inflorescence racemose-corymbose; racemes several, rarely solitary, slender, many-jointed; spikelets geminate, one of them pedicelled, the other sessile or short-pedicelled, lanceolate to narrowly ovate; callus-hairs short; glumes rather coriaceous to herbaceous, the first sometimes with a groove between the 2 keels, the second navicular, with a keel on back; sterile lemma hyaline, sometimes paleate; fertile lemma short, hyaline, bidentate, the awn perfect or imperfect; stamens 1–3.——About 30 species, in warmer regions.

1A. Stamens 2; rachilla-joints slender, glabrous except at both ends; keels of the first glume scaberulous; sterile lemma slightly shorter than the spikelet.
2A. One of the pair of spikelets sessile, the other pedicelled. . . . . 1. *M. nudum*
2B. Both of the spikelets pedicelled, the pedicels of different length. . . . . 2. *M. japonicum*
1B. Stamens 3; rachilla-joints usually ciliate on the angles, gradually thickened above; keels of the first glume scabrous above; sterile lemma less than half as long as the spikelet or wanting.
3A. Cleistogamous spikelets wanting; anthers the same length, linear; spikelets all alike; first glume 2-keeled. . . . . 3. *M. ciliatum*
3B. Cleistogamous spikelets present; anthers of 2 forms, linear in the chasmogamic spikelets, minute in the cleistogamous; chasmogamic spikelets of 2 forms, the pedicelled one compressed laterally, with a keeled first glume. . . . . . . 4. *M. vimineum* var. *polystachyum*

1. **Microstegium nudum** (Trin.) A. Camus. *Pollinia nuda* Trin.; *Leptatherum royleanum* Nees; *M. mayebaranum* Honda; *P. arisanensis* Hayata——MIYAMA-SASAGAYA.   Slender, freely branched perennial without distinct rhizomes; culms 20–40 cm. long, erect from a long decumbent base, the nodes pubescent; leaf-blades 4–8 cm. long, 7–10 mm. wide, glabrous, slightly lustrous; ligules truncate, about 0.5 mm. long; sheaths ciliate; inflorescence exserted, the axis 2–3 cm. long; racemes 4–7, slender, pale green, 4–8 cm. long, the joints 6–7 mm. long, very slender, about twice as long as the pedicel, with a tuft of short hairs at apex; spikelets geminate, one of them sessile, the other short-pedicelled, about 4 mm. long, lanceolate, rather acute, minutely scabrous, with a broad, shallow groove above, and a tuft of short hairs at base; awn delicate, 15–20 mm. long, flexuous.——Sept.–Oct. Woods in low mountains; Honshu (Tōkaidō Distr. and westw.), Shikoku, Kyushu.——Formosa, China, Malaysia, India, and the Caucasus.

2. **Microstegium japonicum** (Miq.) Koidz. *Pollinia japonica* Miq.; *P. nuda* sensu auct. Japon., non Trin.——SASA-GAYA.   Very slender, glabrous, freely branching perennial without distinct rhizomes; culms smooth, 20–70 cm. long, erect from a decumbent base, the nodes glabrous; leaf-blades thin, flat, broadly lanceolate to narrowly ovate, glabrous, 3–7 cm. long, 7–10 mm. wide, abruptly acuminate, acute to rounded at the oblique base; sheaths 2–4 cm. long, the uppermost longest; ligules very short; inflorescence long-exserted; racemes 3 to 6, very slender, spreading to deflexed, sessile, 4–6 cm. long, many-jointed, green, the joints 3–5 mm. long,

scaberulous, with a ring of very short hairs at apex; spikelets geminate, 3 mm. long, alike, broadly lanceolate, acute, with a tuft of short hairs at base, the pedicels 1–1.5 mm. and 2.5–3 mm. long, respectively; first glume scaberulous, with a broad, shallow groove between the 2 slender keels above, the second as long, 3-nerved, rather acute; fertile lemma about 2 mm. long, with a delicate, minutely scaberulous, flexuous awn 6–8 mm. long; anthers 2.——Aug.–Oct. Woods in low mountains; Hokkaido, Honshu, Shikoku, Kyushu; common.—— s. Korea and China.

Var. **boreale** (Ohwi) Ohwi. *M. boreale* Ohwi——KITA-SASAGAYA.   Culm-nodes short-pubescent.——Shaded places in mountains; Hokkaido, Honshu (centr. distr. and northw.).

3. **Microstegium ciliatum** (Trin.) A. Camus. *Pollinia ciliata* Trin.; *P. ciliata* subsp. *wallichiana* (Nees) Hack.; *P. wallichiana* Nees——Ō-SASAGAYA.   Culms long-decumbent and rooting at base, freely branched, slender, 50–80 cm. long, green, glabrous; leaf-blades flat, broadly linear, 5–8 cm. long, 4–7 mm. wide, sparsely pilose above, glabrous beneath, narrowed at base; ligules ovate-deltoid, obtuse, reddish brown, hyaline, glabrous, 2–3 mm. long; sheaths glabrous, often with a few long hairs near apex; inflorescence exserted, the axis 1–3 cm. long; racemes 4–7, ascending, slender, pale green with a brown hue, 7–10 cm. long, rather densely flowered; rachis joints subterete, 3–4 mm. long, reddish brown ciliate, slightly thickened above; pedicels about half as long as the joint; spikelets lanceolate, about 3.5 mm. long, those with pedicels barely smaller than the sessile ones; first glume with a distinct groove between the ciliate keels, second glume awn-pointed;

fertile lemma with a delicate awn about 2 cm. long.——Oct.-Nov. Woods in low mountains; Kyushu (Yakushima and Tanegashima).——Ryukyus, Formosa, s. China, Malaysia, and India.

4. Microstegium vimineum (Trin.) A. Camus var. **polystachyum** (Fr. & Sav.) Ohwi. *Pollinia imberbis* Nees; *M. vimineum* var. *imberbe* (Nees) Honda; *P. japonica* var. *polystachya* Fr. & Sav.——ASHIBOSO. Slender annual; culms 40–100 cm. long from a decumbent, branched, rooting base, glabrous; leaf-blades thin, flat, lanceolate, 4–10 cm. long, 8–15 mm. wide, short-acuminate, obliquely cuneate at base; ligules truncate, about 0.5 mm. long, puberulent on back; sheaths pubescent at least on outer margin above, shorter than the leaf-blades and internodes of culms; inflorescence at length exserted; racemes 1–3, nearly erect to ascending, straight, green, 5–7 cm. long, the rachis joints rather thick, 3-angled, 3.5–5 mm. long, thickened above, short-pilose on the angles; pedicels about 2/3 as long as the spikelets; spikelets green, 5–8 mm. long, the pedicelled ones somewhat flattened laterally, the sessile ones with a shallow groove on the face between the 2 ciliate keels; awn long-exserted, to 15 mm. long.——Sept.-Oct. Woods in lowlands and in low mountains; Hokkaido, Honshu, Shikoku, Kyushu; common.——Korea, China, Ryukyus, Formosa, Malaysia, India, and the Caucasus.

89. EULALIA Kunth   UN-NU-KE ZOKU

Erect perennials or annuals; leaf-blades elongate, linear; racemes elongate, many-jointed, long-hairy, sessile on the short axis of the inflorescence; spikelets geminate, dorsiventrally slightly compressed, usually awned, the sessile spikelet disarticulating with the joint and pedicel; first glume herbaceous to coriaceous, 2-keeled, hairy, the second navicular; sterile lemma hyaline, obtuse, usually epaleate; fertile lemma bifid, with a perfect awn from the sinus.——About 20 species, in Africa, India, e. Asia, Malaysia, and Australia.

1A. Culms at base covered by the yellow-brown densely villous sheath-bases; nerves between the keels of first glume very weak, not fused.
1. *E. speciosa*
1B. Culms at base not covered by villous sheath-bases; nerves between the keels of first glume distinct, anastomosing above.
2. *E. quadrinervis*

1. Eulalia speciosa (Deb.) O. Kuntze. *Pseudopogonatherum speciosum* (Deb.) Ohwi; *Erianthus speciosus* Deb.; *Pollinia tanakae* (Makino) Makino; *Eulalia tanakae* (Makino) Honda——UN-NU-KE. Culms tufted, erect, firm, 80–120 cm. long, covered at base with yellow-brown densely villous sheath-bases; leaves radical and cauline, the blades linear, 30–50 cm. long, 5–7 mm. wide, though appearing as if 2–3 mm. in diameter resulting from the loosely revolute margins, glabrous or nearly so, often with short hairs near base; ligules truncate, short; sheaths white-pilose above; inflorescence exserted; racemes few, erect, 12–15 cm. long, yellow-brown with a purple hue, the joints villous, about 4 mm. long, oblique at both ends; pedicels slightly shorter than the joints; spikelets lanceolate, about 5 mm. long, rather acute, slightly flattened above, white-pilose; awn about 2 cm. long.——Sept.-Oct. Honshu (Owari and Mikawa Prov.).——Korea, China, and India.

2. Eulalia quadrinervis (Hack.) O. Kuntze. *Pollinia quadrinervis* Hack.; *Pseudopogonatherum quadrinerve* (Hack.) Ohwi——UN-NU-KE-MODOKI. Culms tufted, 60–100 cm. long; leaf-blades rather flat, 20–40 cm. long, 3–8 mm. wide, usually white-pilose; ligules short, truncate; sheaths with spreading white hairs; racemes 3–7, on a short axis 8–12 cm. long, the joints villous with purple-brown hairs; spikelets broadly lanceolate, 5–6 mm. long, rather acute, sparsely hairy; awn 12–17 mm. long.——Sept.-Nov. Honshu (Tōkaidō Distr. and westw.), Shikoku, Kyushu.——Ryukyus, India, and China.

90. POGONATHERUM Beauv.   ITACHIGAYA ZOKU

Tufted perennials; leaf-blades linear; racemes solitary on the slender culms; rachis slender, articulate, the segments long-ciliate; spikelets small, geminate, one of them sessile, disarticulating with the rachis joint and the pedicel, the other short-pedicellate, 1- or 2-flowered, with a tuft of long hairs at base; first glume membranous, obtuse, ciliate, convex on back, the second navicular, awn-pointed; sterile lemma hyaline, sometimes paleate and staminate; fertile lemma hyaline, bifid, with a long awn from the sinus; palea nerveless; stamens 1 or 2; sterile lemma wanting in the pedicellate spikelet.——Few species, in India, Malaysia, Polynesia, N. Queensland, and e. Asia.

1. Pogonatherum crinitum (Thunb.) Kunth. *Andropogon crinitus* Thunb.; *A. monandrus* Roxb.; *P. saccharoideum* var. *monandrum* (Roxb.) Hack.——ITACHIGAYA. Culms slender, densely tufted, 10–30 cm. long, simple or slightly branched at base, glabrous, short-pilose on nodes; leaf-blades thin, broadly linear to narrowly lanceolate, 4–6 cm. long, 3–5 mm. wide, acuminate, glabrous, often with loosely involute margins, minutely papillose-scaberulous; ligules very short, long-pilose; sheaths short, ciliate above; raceme solitary, 2–3 cm. long, the joints very short, densely white-pilose; spikelets broadly lanceolate, about 1.5 mm. long; second glume with a delicate awn about 10 mm. long at apex; awn of fertile lemma about 15 mm. long, brown.——Aug.-Nov. Rocky cliffs and roadbanks in lowlands; Honshu (Kii Prov. and Chūgoku Distr.), Shikoku, Kyushu; rather common.——Ryukyus, Formosa, China, India, and Malaysia.

91. SORGHUM Moench   MOROKOSHI ZOKU

Tall annuals or perennials; leaf-blades flat; panicles large, with an elongate axis and slender fascicled branches of 1 to few racemes; joints of racemes rather few; spikelets geminate, one of them sessile and fertile, bisexual, 1-flowered, ovate, compressed dorsally, the other pedicelled and sterile but well-developed; first glume coriaceous; sterile lemma hyaline, epaleate; fertile lemma oblong, bifid, usually awned from the sinus; palea small or wanting.——More than 100 species in warmer regions, especially abundant in Africa.

1. Sorghum nitidum (Vahl) Pers. var. **majus** (Hack.) Ohwi. *Andropogon serratus* var. *genuinus* subvar. *major* Hack.; *A. dichroanthus* Steud.; *S. nitidum* var. *dichroanthum* (Steud.) Ohwi; *S. dichroanthum* (Steud.) Ohwi; *A. serratus* Thunb.——MOROKOSHIGAYA. Tufted perennial; culms 50–120 cm. long, with a ring of long white hairs on nodes; leaf-blades flat, linear, 5–10 mm. wide, sometimes loosely involute, glabrous or pilose on upper surface, glaucous beneath; ligules truncate to semirounded, 1–2.5 mm. long; sheaths pilose arourd mouth and on margin; panicles exserted, 10–20 cm. long, the branches very slender, terete, with a single raceme, 2–6 cm. long, short-pilose in axils, the lower falsely verticillate; racemes rather numerous, 1–1.5 cm. long, few-jointed, the joints reddish ciliate; pedicelled spikelets staminate, awnless or short-awned, the pedicels similar to and as long as the rachis joints; sessile spikelets broadly lanceolate, brown-pilose, acute, about 5 mm. long; first glume coriaceous, lustrous, dark brown, with a pale green, membranous tip; fertile lemma reduced to a geniculate awn 2–2.5 cm. long.——Aug.–Oct. Thickets and open places in lowlands; Honshu (Kii Prov. and westw.), Shikoku, Kyushu.——s. Korea, China, and Malaysia. The typical phase occurs in warmer parts of Asia and Australia.

92. BOTHRIOCHLOA O. Kuntze MONTSUKIGAYA ZOKU

Perennials or annuals; leaf-blades linear; panicles often corymbose, simple or branched, bearing 1 to many racemes, the joints of racemes few to rather many, with a translucent thin line in the center; spikelets geminate, the pedicelled staminate or neuter, without a fertile lemma, awnless; sessile spikelet bisexual, disarticulating with the joint and pedicel; callus short-hairy; first glume herbaceous to thick-membranous, 2-keeled, the second navicular; sterile lemma hyaline, nerveless, epaleate; fertile lemma small, narrow, with a perfect awn at apex; palea wanting.——About 30 species, in warmer regions.

1. Bothriochloa parviflora (R. Br.) Ohwi. *Holcus parviflorus* R. Br.; *Andropogon micranthus* Kunth; *Capillipedium parviflorum* (R. Br.) Stapf——HIME-ABURA-SUSUKI. Glabrous to hirsute perennial; culms erect, 50–100 cm. long, tufted, often branched at base, with ciliate nodes; leaf-blades flat, linear, 8–30 cm. long, 5–8 mm. wide, scabrous; ligules truncate, less than 1 mm. long; sheaths glabrous, sometimes spreading-hirsute to -villous; panicles exserted, erect, 5–15 cm. long, the branches very slender, divaricately forked, with short white hairs on axils; racemes peduncled, 5–8 mm. long, bearing a few reddish brown spikelets, the joints ciliate; pedicelled spikelets lanceolate, staminate, awnless; sessile spikelets lanceolate, acuminate, 2.7–3.2 mm. long, bisexual, flattened, scaberulous; callus with short, white hairs; first glume herbaceous, pale green and purplish, slightly lustrous; fertile lemma reduced to a slender, geniculate awn about 18 mm. long.——July–Oct. Open places and thickets in lowlands; Honshu (Kantō Distr. and westw.), Shikoku, Kyushu; variable.——Ryukyus, Formosa, China, Manchuria, Korea, India, Malaysia, and Australia.

93. ARTHRAXON Beauv. KOBUNAGUSA ZOKU

Low, decumbent annuals or perennials with broad, flat leaf-blades clasping at base, and usually a flabellate panicle; racemes few to rather many, on the short axis of the panicle; rachis slender, many-jointed; spikelets geminate, the pedicelled one staminate or reduced to a mere pedicel, the sessile one bisexual, usually awned, lanceolate; first glume chartaceous, rounded on back, 5- to 10-nerved, scabrous, the second keeled, obtuse; sterile lemma short, hyaline, epaleate; fertile lemma hyaline, nearly entire, rising from base, awnless; palea wanting or minute; stamens 2 or 3.——About 15 species, in warmer parts of Asia and Africa.

1. Arthraxon hispidus (Thunb.) Makino. *Phalaris hispida* Thunb.; *A. ciliaris* Beauv.; *Pleuroplitis langsdorffii* Trin.; *A. langsdorffianus* (Trin.) Hochst.; *A. pauciflorus* Honda; *A. kobuna* Honda——KOBUNAGUSA. Culms long-creeping and branched at base, erect to ascending, 20–50 cm. long, slender, leafy, the nodes pilose; leaf-blades flat, narrowly ovate, 2–6 cm. long, 1–2.5 cm. wide, short-acuminate, cordate and clasping at base, ciliate, glabrous or pilose on both surfaces; ligules truncate, lacerate, 1–2 mm. long; sheaths short, ciliate, usually hirsute; inflorescence flabellate, long-exserted, with 3 to 20 racemes, these often branched at base, nearly sessile, 3–5 cm. long; rachis-joints usually glabrous; spikelets pale green to purple, lanceolate, acute, the strong nerves aculeate-scaberous; sterile lemma awnless or with a geniculate awn; anthers 0.5–1 mm. long.——Sept.–Nov. Low mountains and lowlands; Hokkaido, Honshu, Shikoku, Kyushu; very common and variable.——e. and tropical Asia.

Arthraxon lanceolatus (Roxb.) Hochst. *Andropogon lanceolatus* Roxb.; *Andropogon serrulatus* (Link) Link; *Arthraxon serrulatus* Link——ONI-KOBUNAGUSA. Culms appressed-pubescent; perfect spikelets aculeolate on both sides, the pedicelled spikelets well developed, staminate, larger than the sessile.——Once found naturalized in Honshu (near Ōtsu in Ōmi).——China, Malaysia to India and Africa.

94. CYMBOPOGON Spreng. OGARUKAYA ZOKU

Aromatic perennials, with rather tall, branched culms; racemes geminate, subtended by spathes, aggregate into a compound inflorescence; spikelets geminate, the lowermost pair of one or both racemes sterile or staminate and similar to the pedicelled spikelets above; sessile spikelets bisexual, dorsally compressed; glumes nearly equal, the first flat, faintly nerved, the upper margins keeled; fertile lemma hyaline, 2-lobed, usually with a geniculate awn; palea wanting; pedicelled spikelets awnless; stamens 3.——About 30 species in tropical and warmer regions.

1. **Cymbopogon tortilis** (Presl) A. Camus. *Anthistiria tortilis* Presl; *Andropogon hamatulus* Nees; *C. hamatulus* (Nees) A. Camus; *C. tortilis* var. *goeringii* (Steud.) Hand.-Mazz.; *Andropogon goeringii* Steud.; *Andropogon nardus* subsp. *marginatus* var. *goeringii* (Steud.) Hack.; *C. goeringii* (Steud.) A. Camus——OGARUKAYA. Perennial with short rhizomes; culms tufted, erect, rather firm, 60–100 cm. long, glabrous; leaf-blades flat, linear, 15–40 cm. long, 3–5 mm. wide, with slightly recurved margins; ligules deltoid, 1–3 mm. long, glabrous, obtuse to rounded; culm-sheaths rather short; inflorescence compound, narrow, cylindric, interrupted, erect, 20–40 cm. long, the branches rather short, the spathes 1–2 cm. long; racemes geminate, scarcely exserted from the spathe, spreading or reflexed, 15–20 mm. long, pale green to dull purplish; pedicelled spikelets broadly lanceolate, 4–6 mm. long, staminate, awnless; fertile spikelets broadly lanceolate, 5–6 mm. long, with an awn about 10 mm. long.——Aug.–Nov. Open places in hills; Honshu, Shikoku, Kyushu.——Formosa, Philippines, Indochina, China, and Manchuria.

95. ANDROPOGON L.   USHI-KUSA ZOKU

Annuals or perennials; leaf-blades linear; racemes 1 to few, digitate, spathed at base, forming a racemose or corymbose inflorescence; spikelets geminate, the pedicelled staminate or much reduced, the sessile bisexual, usually awned; callus-hairs short; first glume coriaceous or chartaceous, prominently inflexed on margins, 2-keeled, flat or grooved on back, the second navicular, often awn-pointed; sterile lemma hyaline, 2-nerved; fertile lemma usually bifid and awned from the sinus, hyaline; palea small or wanting.——With about 150 species, widespread in warm-temperate to tropical regions.

1. **Andropogon brevifolius** Sw. *Schizachyrium brevifolium* (Sw.) Nees——USHI-KUSA. Slender, erect annual; culms 10–40 cm. long, glabrous, branched above and near base; leaf-blades soft, thin, flat, broadly linear or linear-oblong, 2–4 cm. long, 2–5 mm. wide, abruptly acute, nearly glabrous, rounded at base; ligules very short; sheaths compressed and keeled, glabrous, short; racemes simple, peduncled, terminal and axillary from upper culm-nodes, scarcely exserted from the sheathed spathe, linear, 1–2 cm. long, the joints thickened gradually toward the apex; pedicels slender; sessile spikelets lanceolate, about 3 mm. long, the first glume lanceolate; fertile lemma hyaline, bifid, awned from the sinus, the awn perfect, about 8 mm. long, delicate, geniculate, the pedicellate spikelet reduced to an awn about 3 mm. long.——Aug.–Nov. Wet places; Honshu, Shikoku, Kyushu; common.——Widespread in warm-temperate to tropical regions.

96. THEMEDA Forsk.   MEGARUKAYA ZOKU

Rather large annuals or perennials with culms usually branching above; racemes solitary on upper branches, short, spathed at base; spikelets geminate, the lower 2 pairs alike, staminate or neuter, forming an involucre, sessile or nearly so, subtending the upper spikelets, one of the other pairs of spikelets bisexual, sessile, terete, awned; callus acute, villous; first glume coriaceous, involute, the second keeled, with a groove on each side; sterile lemma hyaline, nerveless; fertile lemma small, with a long, thick apical awn; palea minute or absent; involucral spikelets flattened dorsally, awnless; first glume herbaceous, 2-keeled; fertile lemma with its palea sometimes wanting; pedicelled spikelet similar to the involucral ones but smaller and narrower.——About 10 species, in warm-temperate to tropical regions.

1. **Themeda japonica** (Willd.) C. Tanaka. *Anthistiria japonica* Willd.; *Themeda forskalii* var. *major* subvar. *japonica* (Willd.) Hack.; *T. triandra* var. *japonica* (Willd.) Makino ——MEGARUKAYA. Culms tufted from short rhizomes, erect, 70–100 cm. long, glabrous; leaves radical and cauline, the blades flat with recurved margins, 30–50 cm. long, 3–8 mm. wide, scabrous, glaucous beneath, with long hairs near base; ligules hyaline, truncate, lacerate, 1–3 mm. long; sheaths slightly compressed, usually with spreading, coarse hairs; inflorescence a false panicle, 20–40 cm. long, leafy; racemes solitary, spathed at base; spikelets of the lower 2 pairs alike, staminate, the first glume herbaceous, oblanceolate, 8–10 mm. long, obtuse, sparsely long-hairy on back; sessile spikelet bisexual, fusiform-cylindric, 8–10 mm. long inclusive of the rigid, oblique callus densely beset with rusty-brown bristly hairs 2–4 mm. long, the first glume coriaceous, convolute, obtuse, with ascending, rigid, short hairs on upper portion, the second glume as long as the first; awn of fertile lemma perfect, thick, geniculate, about 5 cm. long, with short, soft hairs.——Sept.–Oct. Low mountains and lowlands; Honshu, Shikoku, Kyushu; common.——Korea, Manchuria, China, and India.

97. ISCHAEMUM L.   KAMO-NO-HASHI ZOKU

Branched annuals or perennials; leaf-blades flat, lanceolate; racemes 2(–10), fasciculate at summit of branches, sessile; rachis thickened, 3-angled, flat or slightly concave on the face, the joints disarticulating with the sessile spikelet; spikelets geminate, alike or nearly so; pedicellate spikelet bisexual, staminate or neuter, the sessile spikelet bisexual; first glume coriaceous, 2-keeled with inflexed margins, flat or convex on face, the second navicular; sterile lemma membranous or thinly herbaceous, paleate and staminate; fertile lemma bifid, usually awned from the sinus; palea hyaline; stamens 3.——About 50 species, in warm-temperate to tropical regions, abundant in s. Asia.

1A. Culm-nodes glabrous; leaf-blades glabrous or with spreading hairs; spikelets broadly lanceolate to narrowly obovate, narrowly winged.
    1. *I. aristatum*
1B. Culm-nodes ciliate; leaf-blades usually appressed-pubescent; spikelets obovate to broadly so, usually broadly winged, commonly long-hairy on the face. . . . . 2. *I. anthephoroides*

1. Ischaemum aristatum L.   *I. crassipes* (Steud.) Thell.; *Andropogon crassipes* Steud.; *I. sieboldii* Miq.; *I. sieboldii* var. *formosanum* Hack.; *I. crassipes* var. *aristatum* Nakai; *I. ikomanum* Honda——KAMO-NO-HASHI.   Culms 30–70 cm. long, branched and decumbent or ascending at base, glabrous; leaf-blades flat, linear, 15–30 cm. long, 5–10 mm. wide, narrowed at the ends, glabrous, rarely with spreading hairs; ligules short, truncate or bifid; sheaths glabrous or ciliate; racemes 2, geminate, sessile, erect and appressed to each other, 4–7 cm. long, joints of the rachis thick, nearly glabrous; spikelets alike, broadly lanceolate to narrowly obovate, usually glabrous, 5–6 mm. long, rather acute, awned or awnless; first glume narrowly winged at tip.——July–Nov. Sandy seashores; Honshu, Shikoku, Kyushu; common and extremely variable.——Korea, Manchuria, Formosa, and China.

2. Ischaemum anthephoroides (Steud.) Miq.   *Rottboellia anthephoroides* Steud.; *I. eriostachyum* Hack.; *I. hokianum* Honda; *Andropogon anthephoroides* (Steud.) Steud.——KE-KAMO-NO-HASHI.   Culms 30–80 cm. long, from a branched, decumbent or ascending base, glabrous, hairy on upper parts and nodes; leaf-blades flat, broadly linear, 15–30 cm. long, 8–12 mm. wide, usually appressed-pubescent on both sides; ligules truncate, often bifid, sometimes lacerate; sheaths usually appressed-pubescent; racemes 2, geminate, erect and appressed to each other, 6–12 cm. long, sessile, with long white hairs, the joints thick, 3-angled, white-ciliate on the angles; sessile spikelet obovate to broadly so, 7–8 mm. long, obtuse, usually hirsute on back; first glume with broad, more or less serrulate wings; awn of fertile lemma perfect, exserted.——July–Sept. Sandy seashores; Hokkaido, Honshu, Shikoku, Kyushu; common.——Korea and China.

98. PHACELURUS Griseb.   AI-ASHI ZOKU

Rather large, stout perennials; culms leafy; leaf-blades flat, with a thick midnerve; racemes 2 to several, congested, terminal on the peduncle, the joints 3-angled, thick; spikelets alike, 2-flowered, the first glume of sessile spikelets coriaceous, lanceolate, rather acute, 2-keeled, with inflexed margins, the second navicular; sterile lemma lanceolate, paleate and staminate; fertile lemma slightly shorter; pedicelled spikelets acute to acuminate, the first glume with 1 keel.——Several species, in e. Asia, India, and the Mediterranean region.

1. Phacelurus latifolius (Steud.) Ohwi.   *Rottboellia latifolia* Steud.; *Ischaemum latifolium* (Steud.) Miq., non Kunth, 1829; *R. foliata* Steud.——AI-ASHI.   Rhizomes elongate, covered with short scales; culms stout, 80–120 cm. long, erect, simple, glabrous; leaf-blades flat, broadly linear to narrowly lanceolate, 20–40 cm. long, 1–4 cm. wide, long-acuminate, with a rounded base; ligules truncate, 1–2 mm. long, ciliolate; sheaths terete, ciliate and pilose around the mouth; inflorescence digitate, terminal, exserted, glaucescent, sometimes dull purplish, erect, sessile, 10–25 cm. long; racemes 5–12, rarely reduced to 1, the axis very short, the segments 3-angled, glabrous, slightly thickened above, nearly as long as the pedicel; sessile spikelets appressed in the depression between the joint and pedicel, 3-angled, lanceolate, glabrous, acute, about 10 mm. long, awnless; first glume nearly coriaceous, inflexed on margins, flat but somewhat concave above between the keels.——June–Oct. Near seashores; Hokkaido, Honshu, Shikoku, Kyushu; rather common and gregarious. ——Korea, Manchuria, and China.

99. HEMARTHRIA R. Br.   USHI-NO-SHIPPEI ZOKU

Perennials; culms decumbent, often creeping at base, rather slender, compressed, branching toward the ends; leaf-blades flat, linear; racemes solitary, compressed, with a narrow spathe at base, the peduncle and rachis obtusely angled; spikelets awnless, geminate on the nodes of the thickened rachis, one sessile and perfect, the other pedicelled and sterile, the pedicels thickened, appressed and partly adnate to the slightly grooved side of the rachis-joint; racemes compressed-cylindric, persistent until after maturity; sessile spikelets narrowly ovate-oblong; first glume coriaceous, flat on back, tapering to a blunt tip, keeled and inflexed near the margins; pedicelled spikelet similar, smaller.——Several species, in the Tropics and warmer regions.

1A.   Culms long-creeping, rarely ascending or nearly erect at base; leaf-blades 7–15 cm. long, 2–5 mm. wide, slightly narrowed and obtuse at apex; sessile spikelets 4–6 mm. long including the basal callus; first glume of sessile spikelets narrowly oblong, abruptly narrowed above the middle to the obtuse apex. .... 1. *H. compressa*
1B.   Culms erect or rarely short-creeping at base; leaf-blades 20–30 cm. long, 3–7 mm. wide, gradually narrowed and acuminate at apex; sessile spikelets 5–8 mm. long including the basal callus; first glume lanceolate to broadly so, usually gradually narrowed from the base to the obtuse apex. .... 2. *H. sibirica*

1. Hemarthria compressa (L. f.) R. Br.   *Rottboellia compressa* L. f.——KOBANO-USHI-NO-SHIPPEI.   Rhizomes well developed; culms creeping to ascending at base, erect above, glabrous, 70–100 cm. long, branching; leaf-blades glaucous, 7–15 cm. long, 2–5 mm. wide, slightly narrowed to the obtuse apex, glabrous, rather rounded at base; ligules truncate, ciliolate, about 0.5 mm. long; sheaths short, ciliate above; racemes solitary, peduncled, subulate, slightly curved, 5–8 cm. long, included in the sheath at base; rachis joints 3–3.5 mm. long, the spikelets adherent to the rachis joints; first glume of sessile spikelets 4–6 mm. long including the basal callus, narrowly oblong, narrowed above the middle to the obtuse apex; first glume of pedicelled spikelets broadly lanceolate, gradually narrowed near base to the obtuse apex.——July–Sept. Wet places in lowlands and near seashores; Kyushu.——Ryukyus, China, Formosa, Indochina, and India.

2. Hemarthria sibirica (Gand.) Ohwi.   *Rottboellia sibirica* Gand.; *R. compressa* var. *japonica* Hack.; *H. japonica* (Hack.) Roshev.; *R. japonica* (Hack.) Honda——USHI-NO-SHIPPEI.   Perennial; culms erect, ascending or short-creeping at base, 80–120 cm. long, branching, glabrous; leaf-blades flat, linear to broadly so, glaucous, 20–30 cm. long, 3–7 mm. wide; ligules truncate, about 0.5 mm. long, ciliolate; sheaths short, glabrous; racemes 5–8 cm. long, narrow, subulate, slightly curved; rachis joints 4–6 mm. long, fused with the pedicel; sessile and pedicelled spikelets similar in shape; first glume 5–8 mm. long inclusive of the basal callus, lanceolate to broadly so, gradually narrowed from base to the obtuse apex. ——July–Sept. Wet places in lowlands; Honshu, Shikoku, Kyushu; common.——Korea, China, Manchuria, Ussuri, and e. Siberia.

100. COIX L. JUZUDAMA ZOKU

Tall, stout annual or perennial; culms branched; leaf-blades rather broad, flat; inflorescences numerous, on long peduncles clustered in the axils of the upper leaves, monoecious, the pistillate enclosed in a globose to ovoid, bead-like involucral bract from which protrudes a staminate raceme; pistillate spikelets 3 together, 1 fertile, 2 reduced and sterile, the first glume broad, many-nerved, the second keeled; fertile lemma and palea hyaline; staminate spikelets geminate or in 3's, 2-flowered, lanceolate, the first glume herbaceous, with a narrow wing on each side, many-nerved, the stamens 3.——Few species, in tropical Asia.

1. **Coix lacryma-jobi** L. JUZUDAMA. Perennial; culms tufted, branched at base, about 1 m. long, rather thick; leaf-blades 30–60 cm. long, 2–4 cm. wide; involucral bracts ovate, hard, lustrous, about 1 cm. long; first glume of staminate spikelets herbaceous, oblong, obtuse, pale green.——Frequently cultivated and sometimes naturalized in our area.

Var. **mayuen** (Romain) Stapf. *C. mayuen* Romain—— HATO-MUGI. Similar to the typical phase but annual, the inflorescence often drooping and with broader bracteal leaves; involucral bracts crustaceous, elliptical or oblong.——Frequently cultivated in our area, possibly native of Indochina or China.

Fam. 42. CYPERACEAE KAYATSURI-GUSA KA Sedge Family

Grasslike or rushlike herbs; culms usually triquetrous, solid; leaves with a closed sheath at base, linear, more or less scabrous; flowers in spikelets, hermaphroditic or unisexual, sessile, solitary, subtended by few to many spirally imbricate or distichous bracts or scales (squamae); perianth wanting or represented by hypogynous bristles or rarely by scales; stamens 1–6, usually 3; anthers basifixed, 2-locular, the filaments distinct; ovary 1-locular; style 2- to 3(–4)-fid; ovule 1, anatropous; fruit a small achene, lenticular or trigonous in transverse section; seeds with abundant mealy or fleshy endosperm.——Cosmopolitan with about 70 genera, and about 3,500 species.

1A. Flowers bisexual or sometimes unisexual, not inclosed in a perigynium.
  2A. Spikelets many-flowered; flowers always bisexual.
    3A. Scales distichous; spikelet flat; perianth wanting; style-base continuous with the achene. ....1. *Cyperus*
    3B. Scales spirally arranged, except in some species of *Fimbristylis*.
      4A. Style-base not swollen, gradually passing into the achene.
        5A. Perianth-segments 0–6, bristle-shaped, petaloid, or membranous.
          6A. Perianth-segments or at least some of them broadly dilated, petaloid or membranous.
            7A. Perianth-segments 2; plants usually glabrous. ....2. *Lipocarpha*
            7B. Perianth-segments 6; plants hairy. ....3. *Fuirena*
          6B. Perianth-segments bristle-shaped, rarely wanting. ....4. *Scirpus*
        5B. Perianth of numerous silky bristles, much-elongate after anthesis. ....5. *Eriophorum*
      4B. Style-base bulbously thickened, distinct from the achene.
        8A. Perianth wanting; blade of leaves elongate; spikelets 1 to numerous.
          9A. Style-base deciduous from the achene. ....6. *Fimbristylis*
          9B. Style-base more or less persistent on the achene. ....7. *Bulbostylis*
        8B. Perianth of bristles, rarely wanting; achene crowned by a persistent style-base; leaves bladeless; spikelet solitary, terminal.
                  8. *Eleocharis*
  2B. Spikelets few-flowered; flowers bisexual or unisexual.
    10A. Flowers or most of them bisexual.
      11A. Spikelets fertile, except the uppermost.
        12A. Style 3-fid; achenes trigonous; scales distichous.
          13A. Achene not crowned with a style-base. ....9. *Schoenus*
          13B. Achene crowned with the persistent style-base. ....10. *Carpha*
        12B. Style 2-fid or scarcely divided; achenes lenticular; scales spirally imbricate. ....11. *Rhynchospora*
      11B. Lower flowers of the spikelet sterile; style of fertile flowers 3-fid; achene trigonous.
        14A. Leaves dorsiventrally flattened; inflorescence corymbose, flat or convex; achenes somewhat drupelike, beakless. . 12. *Cladium*
        14B. Leaves bilaterally compressed or terete; inflorescence a slender panicle; achenes coriaceous, beaked. .. .. 13. *Machaerina*
    10B. Flowers always unisexual; spikelets often androgynous (i.e., the upper part staminate, the lower pistillate).
      15A. Inflorescence paniculate or globose, composed of rather small spikelets; achene falling separately from the scales. .... 14. *Scleria*
      15B. Inflorescence glomerate, subsessile, composed of very small spikelets; achene falling with the 2 surrounding scales.
                  15. *Diplacrum*
1B. Flowers unisexual, the pistillate enclosed in a perigynium (prophyllum).
  16A. Achene surrounded by a perigynium, open on one side at least above the middle. ....16. *Kobresia*
  16B. Achene inclosed in a perigynium, open only at the tip. ....17. *Carex*

1. CYPERUS L. KAYATSURI-GUSA ZOKU

Annuals or perennials; culms erect, leafy only at base; leaves usually elongate, sometimes bladeless, sheathing at base; inflorescence simple or compound, umbelliform or capitate, with few to numerous spikelets, usually subtended by 1 to several leaflike bracts; spikes usually peduncled, sometimes sessile; spikelets sessile, 1- to many-flowered; flowers bisexual; scales usually 2-ranked; perianth none; stamens 1–3; style continuous with the ovary, not enlarged at base, deciduous; stigmas 2 or 3; achenes lenticular or 3-angled.——About 700 species, mostly in tropical and warm-temperate regions of both hemispheres.

1A.   Stigmas 2; achenes lenticular.
  2A.   Achenes dorsiventrally compressed, the ventral surface facing the rachilla. . . . . 1. *C. serotinus*
  2B.   Achenes laterally compressed, the angle facing the rachilla.
    3A.   Spikelets many-flowered, not jointed at base.
      4A.   Surface cells of the achene longitudinally oblong, thus the achene transversely wrinkled. . . . . 2. *C. diaphanus*
      4B.   Surface cells of the achene hexagonal, thus the achene puncticulate.
        5A.   Culms decumbent or ascending at base, often producing roots at the nodes; scales distinctly sulcate. . . . . 3. *C. sanguinolentus*
        5B.   Culms erect; scales not sulcate.
          6A.   Scales 3.5–5 mm. long, with a curved keel. . . . . 4. *C. unioloides*
          6B.   Scales 1.5–2.5 mm. long, with a straight keel.
            7A.   Spikelets usually almost erect, gradually tapering to the apex, partially tinged with dark red; achenes narrowly oblong, subtruncate. . . . . 5. *C. polystachyos*
            7B.   Spikelets usually spreading, linear (parallel-sided), brownish or dark brown; achenes oblong-obovate, not truncate.
                      6. *C. globosus*
    3B.   Spikelets 1-flowered, jointed at base, in a dense terminal globose head; rhizomes long-creeping. . . . . 7. *C. brevifolius*
1B.   Stigmas 3; achenes trigonous (Nos. 25–28, sometimes the stigmas 2 and achenes lenticular).
  8A.   Rachilla articulate; spikelets hardly flattened.
    9A.   Rachilla articulate only at base; spikelets 4–5 mm. long. . . . . 8. *C. cyperoides*
    9B.   Rachilla articulate below each scale; spikelets 1–1.5 cm. long. . . . . 9. *C. odoratus*
  8B.   Rachilla not articulate; spikelets usually flattened.
    10A.   Spikelets spicate, the rachis distinct.
      11A.   Rhizomes long-creeping; culms not tufted.
        12A.   Culms long-leaved at base; subtending bracts longer than or nearly as long as the inflorescence.
          13A.   Rachis hispid; scales broadly ovate, 2 mm. long; plant of wet places. . . . . 10. *C. pilosus*
          13B.   Rachis not hispid; scales narrowly ovate, 3–3.5 mm. long; plant of sandy places. . . . . 11. *C. rotundus*
        12B.   Culms with bladeless or short-bladed sheaths at base; subtending bracts shorter than the inflorescence. . . 12. *C. monophyllus*
      11B.   Rhizomes thick and very short, or absent; culms tufted.
        14A.   Culms stout, thickened at base; perennials or apparently annuals.
          15A.   Scales elliptic or ovate, acute or mucronate; achenes elliptic or narrowly ovate.
            16A.   Spikelets scarcely flattened, 4–6 mm. long; scales acute. . . . . 13. *C. ohwii*
            16B.   Spikelets compressed, 5–8 mm. long; scales with a short recurved mucro at apex. . . . . . . 14. *C. exaltatus* var. *iwasakii*
          15B.   Scales narrowly oblong, obtuse; achenes narrowly oblong. . . . . 15. *C. glomeratus*
        14B.   Culms slender, not thickened at base; annuals or biennials.
          17A.   Scales 1–1.5 mm. long, rounded and emarginate or mucronate, nearly as long as the achene.
            18A.   Rachis ciliate; scales sanguineous, entire. . . . . 16. *E. orthostachyus*
            18B.   Rachis glabrous; scales light yellow to reddish brown, mucronate.
              19A.   Rachis and rachilla nearly wingless; scales light yellow, minutely mucronate. . . . . 17. *C. iria*
              19B.   Rachis and rachilla winged; scales distinctly mucronate.
                20A.   Scales light yellow to tawny with a straight mucro. . . . . 18. *C. microiria*
                20B.   Scales reddish brown, the mucro of the scale awnlike, more or less recurved. . . . . 19. *C. amuricus*
        17B.   Scales, including the short awn, 3–3.5 mm. long, rather obtuse, nearly 3 times as long as the achene. . . 20. *C. compressus*
    10B.   Spikelets fascicled or digitate, the rachis wanting; small annuals or perennials.
      21A.   Spikelets distinctly compressed; scales acutely keeled, always 2-ranked.
        22A.   Achenes nearly as long as the scale, acutely 3-angled; scales dark sanguineous or blackish brown, rounded and slightly emarginate, not mucronate. . . . . 21. *C. difformis*
        22B.   Achenes ¼–⅓ as long as the scale, obtusely angled, whitish; scales pale to reddish brown, mucronate.
          23A.   Scales obtuse with an extremely short mucro, often partially reddish.
            24A.   Scales straight; rachilla quite or almost hidden by the densely imbricate scales. . . . . 22. *C. haspan*
            24B.   Scales ultimately recurved; rachilla distinctly visible. . . . . 23. *C. tenuispica*
          23B.   Scales rounded or truncate, the midrib pale green, excurrent into a long recurved awn. . . . . 24. *C. hakonensis*
      21B.   Spikelets terete or slightly flattened; scales 2-ranked or spiral, obtusely keeled, pale green.
        25A.   Achenes ellipsoidal or obovoid, obtusely angled; scales ovate or ovate-orbicular, several-nerved.
          26A.   Scales ovate-orbicular, obtuse. . . . . 25. *C. niigatensis*
          26B.   Scales ovate, rather acute or gradually pointed. . . . . 26. *C. nipponicus*
        25B.   Achenes oblong or lanceolate, acutely angled or winged; scales oblong to lanceolate, hyaline, 3- to 5-nerved.
          27A.   Scales 2-ranked, with a smooth or sparsely spinulose mucro; achenes acute but not winged. . . . . 27. *C. extremiorientalis*
          27B.   Scales spirally arranged or inconspicuously 2-ranked, with a smooth or minutely papillose mucro; achenes narrowly winged. . . . . 28. *C. pacificus*

1.   Cyperus serotinus Rottb. *C. japonicus* Miq.; *Juncellus serotinus* (Rottb.) C. B. Clarke; *Duval-jouvea serotina* (Rottb.) Palla; *C. makinoi* Nakai——MIZU-GAYATSURI. Rhizomes elongate, terminated by an oblong tuber; culms 50–100 cm. long, stout, usually smooth; leaves 5–8 mm. wide; corymbs compound, large, with 5–8 erect-spreading smooth rays, subtended by 3 or 4 spreading leaflike bracts; spikes 2–4 cm. long and as broad, the rachis usually appressed-setulose; spikelets 1–2 cm. long, 2–2.5 mm. wide, turgid-biconvex, dark sanguineous to pale green; scales broadly ovate, 2–2.5 mm. long, obtuse, not keeled, entire, involute; achenes 1.5 mm. long, nearly orbicular, biconvex, dorsiventrally compressed, brownish; stigmas 2, rarely 3.——Aug.–Oct. Marshy places; Hokkaido, Honshu, Shikoku, Kyushu.——Korea, Manchuria, Formosa, China, India, and Europe.

2. Cyperus diaphanus Schrad. *C. setiformis* Korsh.; *Pycreus setiformis* (Korsh.) Nakai; *P. gratissimus* Kitag.; *C. latespicatus* var. *setiformis* (Korsh.) T. Koyama——TACHI-GAYA-TSURI, HITORI-GAYATSURI. Somewhat resembling No. 3; culms 10–30 cm. long; leaves very slender, 1–1.5 mm. wide; inflorescence of 1 or 2 sessile spikes bearing several spikelets; spikelets oblong, mostly flattened, 8–15 mm. long, pale to partly sanguineous; scales narrow-ovate, about 3 mm. long, obtuse, entire, 3-nerved; achenes 1/3 as long as the scale, obovate-orbicular, biconvex, dark brown; stigmas 2.——Aug.–Oct. Honshu (Aomori); rare.——Bonin Isls., Korea, Manchuria, Amur, centr. Asia, India, and Malaysia.

3. Cyperus sanguinolentus Vahl. *C. eragrostis* Vahl non Lam.; *Pycreus sanguinolentus* (Vahl) Nees; *P. eragrostis* (Vahl) Palla; *P. rubromarginatus* E. G. Camus——KAWARA-SUGANA. Culms tufted, 10–40 cm. long, decumbent at base; leaves 1–3 mm. wide; inflorescence corymbiform (often a head in forma *nipponicus* T. Koyama); bracts 2 or 3, much longer than the inflorescence; spikelets spicate, divergent, compressed, narrowly oblong, 1–2 cm. long, 2.5–3.5 mm. wide, (rarely to 4 mm. wide in forma *spectabilis* (Makino) Ohwi), sanguineous; scales broadly ovate, 2.5–3.5 mm. long, obtuse, entire; achenes 2/5–1/2 as long as the scale, biconvex, dark brown, broadly obovate; stigmas 2.——July–Oct. Wet places; Hokkaido, Honshu, Shikoku, Kyushu; common.——Warmer parts of the E. Hemisphere.

4. Cyperus unioloides R. Br. *C. angulatus* Nees; *Pycreus angulatus* (Nees) Nees; *C. tosaensis* Makino——MUGIGARA-GAYATSURI. Culms 30–80 cm. long, slender, stiff; leaves 2–3 mm. wide; corymbs simple, with 0 to 4 rays, to 8 cm. long; bracts 2 or 3, the lowest one much exceeding the inflorescence; spikelets compressed, 1–2.5 cm. long, 4–5 mm. wide, broadly lanceolate, acute, divergent, pale green to yellowish brown; scales narrowly ovate, 3.5–5 mm. long, acute; achenes to 1.5 mm. long, inflated-biconvex, nearly orbicular, dark brown; stigmas 2.——July–Oct. Shikoku, Kyushu.——Pantropic.

5. Cyperus polystachyos Rottb. *Pycreus polystachyos* (Rottb.) Beauv.——IGA-GAYATSURI. Rhizomes very short; culms tufted, 10–50 cm. long, obtusely 3-angled, stiff: leaves 1–3 mm. wide; inflorescence simple, corymbose or capitate; bracts 3 to 5, much longer than the inflorescence; rays of inflorescence to 7 cm. long; spikelets fascicled, nearly erect, linear, laterally compressed, 1–2.5 cm. long, 1.5 mm. wide, usually somewhat reddish; scales nearly erect, narrowly ovate, 1.5–2 mm. long, rather obtuse; achenes half as long as the scale, narrowly obovate, compressed-biconvex, brown, somewhat truncate at apex; stigmas 3, rarely 2.——Aug.–Oct. Sandy places near seashores; Honshu (Kantō Distr. and westw.), Shikoku, Kyushu.——Widespread in warmer parts of the world.

6. Cyperus globosus All. *C. flavidus* Retz.; *Pycreus globosus* (All.) Reichenb.; *C. nilagiricus* Hochst. ex Steud.; *C. globosus* var. *nilagiricus* (Steud.) C. B. Clarke; *C. fuscoater* Meinsh.——AZE-GAYATSURI. Culms slender, obtusely angled, smooth, 10–50 cm. long; leaves 1–2 mm. wide; corymbs simple, with 1–5 rays to 8 cm. long; bracts 2 to 4, the lower ones longer than the inflorescence; spikelets linear, 1–2.5 cm. long, 2–2.5 mm. wide, yellowish brown, strongly compressed;

scales narrowly ovate, 1.5–2 mm. long, obtuse, entire; achenes 1/3–2/5 as long as the scale, biconvex, dark brown, obovate; stigmas 2.——Aug.–Oct. Wet places; Honshu, Shikoku, Kyushu; common.——Korea, Manchuria, China, Formosa, Malaysia, India, Australia, s. Europe, and tropical Africa.

7. Cyperus brevifolius (Rottb.) Hassk. var. **brevifolius.** *Kyllinga brevifolia* Rottb.; *K. intermedia* var. *oligostachya* C. B. Clarke——AIDA-KUGU, TAIWAN-HIME-KUGU. Rhizomes elongate–creeping; culms solitary, erect, rather soft, 7–30 cm. long; leaves soft, 2–3 mm. wide; head solitary (very rarely 2 or 3), globose or ovoid-globose, sessile, 5–10 mm. long, pale green; bracts 3, leaflike, elongate; spikelets numerous, jointed at base, densely disposed, broadly lanceolate, 3–3.5 mm. long, compressed, 1-flowered, acute; scales membranous, pale green or with reddish brown flecks, the keel spinulose or sometimes smooth, the apex prolonged to a slightly recurved mucro; achenes half as long as the scale, obovate, brownish, laterally compressed; stigmas 2.——July–Oct. Honshu, Shikoku, Kyushu.——Ryukyus, Formosa, China, and other warmer regions.

Var. **leiolepis** (Fr. & Sav.) T. Koyama. *Kyllinga monocephala* var. *leiolepis* Fr. & Sav.; *K. gracillima* Miq.; *K. brevifolia* var. *gracillima* (Miq.) Kuekenth.; *C. brevifolius* var. *gracillimus* (Miq.) Kuekenth.; *K. brevifolia* var. *leiolepis* (Fr. & Sav.) Hara——HIME-KUGU. Spikelets slightly broader than in the typical phase, oblong, distinctly turgid, rather obtuse, lustrous, the keel of scale smooth, the mucro not recurved.——July–Oct. Hokkaido, Honshu, Shikoku, Kyushu. ——Korea and China.

8. Cyperus cyperoides (L.) O. Kuntze. *Scirpus cyperoides* L.; *Mariscus sieberianus* Nees; *C. sieberianus* (Nees) K. Schum.; *C. umbellatus* sensu Miq., non alior.; *M. cyperinus* sensu auct. Japon., non Vahl——INU-KUGU, KUGU. Rhizomes hard, abbreviated; culms 30–80 cm. long, thickened at base; leaves 3–6 mm. wide; inflorescence umbelliform or capitate, simple, with 5 to 15 rays to 8 cm. long; bracts 4 or 5, leaflike, longer than the inflorescence; spikes cylindrical; spikelets spreading, linear-lanceolate, terete or obscurely 3-angled, 4–5 mm. long, pale green, in part slightly yellowish, acuminate, 0.5–0.7 mm. wide, 1- or 2-flowered; scales erect, narrowly oblong, about 3 mm. long, involute, obtuse; achenes 2/3 as long as the scale, narrowly oblong, obtusely 3-angled; stigmas 3.——Aug.–Oct. Honshu (Izu Isls., s. Kinki, Chūgoku Distr.).——s. Korea, China, Ryukyus, Formosa, Malaysia, India, Australia, and Africa.

9. Cyperus odoratus L. *C. ferax* L. C. Rich.; *Maricus ferax* (L. C. Rich.) C. B. Clarke; *Torulinium confertum* Hamilt.; *T. ferax* (L. C. Rich.) Urban; *C. speciosus* Makino, non Vahl——KIN-GAYATSURI, MUTSUORE-GAYATSURI. Culms (10–) 20–60 cm. long, stout, slightly thickened at base; leaves 5–10 mm. wide; corymbs frequently compound, large, with 5 to 10 stiff rays to 15 cm. long, the bracts 5–8, the lower much longer than the inflorescence; spikelets spreading, linear, terete, 1–1.5 cm. long, 1–1.2 mm. thick, more or less flexuous, 5- to 16-flowered, the rachilla thick; scales loosely arranged, appressed, 3 mm. long, elliptic, obtuse, nearly keelless, partly yellowish; achenes half as long as the scale, narrowly obovate, 3-angled, slightly compressed, brownish; stigmas 3.——Honshu (Awa); rare.——Ryukyus, Formosa, Indochina, and Malaysia and the Tropics generally.

10. Cyperus pilosus Vahl. *C. subulatus* Steud.; *C. piptolepis* Steud.; *C. marginellus* Nees——ONI-GAYATSURI. Somewhat resembling No. 1, but rhizomes slender, elongate, without a terminal tuber, the spikelets more numerous and dense; culms stout, 30–80 cm. long, scabrous above; leaves 5–10 mm. wide; spikelets linear, 7–20 mm. long, 2–3 mm. wide, often somewhat reddish brown, scales broadly ovate, about 2 mm. long, acute, 2-nerved on both sides; achenes 3/5 as long as the scale, broadly ellipsoidal, 3-angled, brownish, the faces slightly concave; stigmas 3.——July–Oct. Wet places; Honshu (centr. distr. and westw.), Shikoku, Kyushu.——Ryukyus, Formosa, China, Malaysia, India, N. Australia, and tropical West Africa.

11. Cyperus rotundus L. var. **rotundus.** *C. laevissimus* Steud.——Hama-suge. Rhizomes long and slender, firm; culms slender, 20–40 cm. long, smooth, the base bulbously enlarged and covered with brown fibers; leaves 2–6 mm. wide; corymbs simple or compound, with 1–7 rays, to 10 cm. across; the bracts 1 or 2, as long as or slightly longer than the inflorescence; spikelets spreading, 1.5–3 cm. long, 1.5–2 mm. wide, lustrous, sanguineous, loosely 20- to 40-flowered; scales narrowly ovate, ascending, 3–3.5 mm. long, somewhat obtuse; achenes 2/5 as long as the scale, oblong, compressed-triangular, dark brown; stigmas 3.——July–Oct. Abundant in sandy places; Honshu, Shikoku, Kyushu.——Widely distributed in tropical and warm-temperate regions of both hemispheres.

Var. **yoshinagae** (Ohwi) Ohwi. *C. yoshinagae* Ohwi——Tosa-no-hama-suge. Spikelets lanceolate, short, 2.5 mm. wide; scales somewhat densely arranged.——Shikoku; rare.

12. Cyperus monophyllus Vahl. *C. tegetiformis* sensu auct. Japon., non Roxb.; *C. malaccensis* var. *brevifolius* Boecklr.——Shichitō-i, Ryukyu-i. Rhizomes stout, long-creeping; culms 1–1.5 m. long, 3-angled, 2.5–4 mm. thick above, naked, with 2 or 3 bladeless or short-bladed sheaths at base; corymbs usually compound, with 5–10 rays to 7 cm. long, the bracts 2–4, leaflike, usually shorter than the inflorescence, to 7 cm. long; spikelets linear, hardly compressed, 1–3 cm. long, 1–1.5 mm. thick, light ferruginous, loosely 20- to 40-flowered; scales oblong, 2–2.5 mm. long, very obtuse, nearly erect, not keeled; achenes slightly shorter than the scale, compressed-trigonous, linear-oblong; stigmas 3.——Aug.–Oct. Occasionally cultivated, also wild in the warmer parts; Honshu, Shikoku, Kyushu.——Ryukyus, Formosa, China, and Indochina.

13. Cyperus ohwii Kuekenth. Tsukushi-ō-gayatsuri. Coarse herb; rhizomes thick, short; culms stout, 1–1.5 m. long, thickened at base; leaves 10–15 mm. wide; corymbs large, 10–20 cm. long and as broad, compound to decompound with erect to spreading rays to 7 cm. long, the bracts 3–5, leaflike, much longer than the inflorescence; spikes cylindrical, bearing numerous densely arranged spikelets; spikelets obliquely spreading, 4–6 mm. long, slightly compressed, light yellowish; scales elliptic, about 2 mm. long, acute; achenes ¾ as long as the scale, grayish brown, concave-convex; stigmas 3.——Sept.–Oct. Kyushu (Fukuoka Park).——Indochina and Malaysia.

14. Cyperus exaltatus Retz. var. **iwasakii** (Makino) T. Koyama. *C. iwasakii* Makino; *C. exaltatus* sensu auct. Japon., non Retz.; *C. tokiensis* C. B. Clarke——Kan'en-gayatsuri. Culms loosely tufted, stout, 80–120 cm. long; leaves 8–15 mm. wide; corymbs large, decompound, 10–30 cm. long and as broad, with 5–10 stiff rays to 20 cm. long, the bracts 4 or 5, leaflike, much longer than the corymb; spikes 10–15 mm. thick, cylindric, very dense with numerous spikelets; spikelets spreading, 5–10 mm. long, flattened, often yellowish brown in front; scales ovate, 1.7–2 mm. long, the keel green, ending in

a recurved awnlike mucro; achenes half as long as the scale, elliptic, stramineous; stigmas 3.——Sept.–Oct. Swamps and shallow ponds in lowlands; Honshu (Hitachi and Tokyo).——Korea.——The typical phase occurs in se. Asia.

15. Cyperus glomeratus L. *Chlorocyperus glomeratus* (L.) Palla——Numa-gayatsuri. Annual or sometimes a short-lived perennial; culms loosely tufted, 20–70 cm. long; leaves 2–8 mm. wide; corymbs rather large, simple or compound, 3–10 cm. long, with 3–5 rays to 10 cm. long, the bracts 3 or 4, leaflike, much exceeding the inflorescence; spikes oblong-cylindric, very dense, with numerous spikelets; spikelets ascending to nearly erect, linear, compressed, 5–10 mm. long, 1.5 mm. wide, ferrugineous; scales narrowly oblong, slightly obtuse, somewhat acutely keeled; achenes half as long as the scale, narrowly oblong, 3-angled, slightly compressed, grayish brown; stigmas 3.——Swampy places; Honshu (Kantō and Yechigo Distr.)——Korea, China, Amur, Manchuria, Ussuri, India, and Europe.

16. Cyperus orthostachyus Fr. & Sav. *C. truncatus* Turcz., non L. C. Rich.; *C. fimbriatus* sensu Miq., non Nees; *C. truncatus* var. *orthostachyus* (Fr. & Sav.) C. B. Clarke——Ushi-kugu. Annual; culms tufted, soft, 20–70 cm. long; leaves soft, 2–6 mm. wide; corymbs large, compound, 5–20 cm. long, with 5–7 oblique, spreading to ascending rays to 20 cm. long, the bracts 3–5, the lowest one longer than the inflorescence; spikelets linear, slightly compressed, 5–10 mm. long, 1.5 mm. wide, dark reddish purple; scales broadly elliptic, 1.2 mm. long, rounded-truncate, entire; achenes slightly shorter than the scale, obovate, 3-angled, dark brown; stigmas 3.——Aug.–Oct. Wet places; Hokkaido, Honshu, Shikoku, Kyushu; common.——Korea, China, Manchuria, Ussuri, and Dahuria.

17. Cyperus iria L. *C. paniciformis* Fr. & Sav.; *C. iria* var. *paniciformis* (Fr. & Sav.) C. B. Clarke——Kogome-gayatsuri. Annual; culms loosely tufted, 20–60 cm. long; leaves 2–6 mm. wide; corymbs simple or compound, to 15 cm. long, with 3–5 rays, the bracts 4 or 5, the lower 2 or 3 longer than the inflorescence; spikelets spreading or ascending, numerous, densely 10- to 30-flowered, linear, 5–10 mm. long, 1.5–2 mm. wide, compressed, yellowish; scales broadly obovate, 1–1.5 mm. long, emarginate and minutely mucronate, the midrib green; achenes slightly shorter than the scale, obovate, 3-angled, brownish; stigmas 3.——Cultivated fields and wastelands; Honshu, Shikoku, Kyushu; common.——Ryukyus, Formosa, Korea, China, Indochina, Malaysia, India, Africa, and Australia.

18. Cyperus microiria Steud. *C. textori* Miq.; *C. iria* var. *microiria* (Steud.) Fr. & Sav.; *C. japonicus* Makino, non Miq.; *C. amuricus* var. *textori* (Miq.) Kuekenth.; *C. amuricus* var. *japonicus* Kuekenth., excl. basionym; *Chlorocyperus franchettii* Palla——Kayatsurigusa, Ki-gayatsuri. Annual, somewhat resembling No. 17; leaves 2–5 mm. wide; spikelets linear, 7–12 mm. long, 1.5 mm. wide, yellowish, often slightly reddish brown; scales 1.5 mm. long, rounded at apex with the green excurrent midrib ending in a distinct erect mucro; achenes slightly shorter than the scale; stigmas 3.——Aug.–Oct. Waste grounds and cultivated fields; Honshu, Shikoku, Kyushu; very common.——Korea, Manchuria, and China.

19. Cyperus amuricus Maxim. *C. amuricus* var. *japonicus* Miq.; *C. krameri* Fr. & Sav.; *C. textori* var. *laxa* Fr. & Sav.; *C. pterygorrhachis* C. B. Clarke ex Lév.; *C. amuricus* var. *pterygorrhachis* (C. B. Clarke) Ohwi——Cha-gayatsuri. Annual, resembling No. 18; culms 20–60 cm. long; spikelets linear, compressed, 7–12 mm. long, 1.5–2 mm. wide, reddish brown; scales broadly obovate, 1.5 mm. long, rounded, the midrib green, excurrent as a slightly recurved awnlike mucro. ——Aug.–Oct. Cultivated fields and waste grounds; Honshu, Shikoku, Kyushu; rather common.——Korea, Ussuri, Amur, China, and Formosa.

20. Cyperus compressus L. KUGU-GAYATSURI. Annual; culms tufted, 10–40 cm. long; leaves 1–3 mm. wide; inflorescence simple, to 10 cm. long, with 1–5 ascending rays about 5 cm. long, rarely the rays abbreviated and the inflorescence capitate, the bracts 2 or 3, longer than the inflorescence; spikelets linear, 1–2.5 cm. long, 2.5–3 mm. wide, strongly compressed, pale and somewhat yellowish at maturity, slightly lustrous; scales broadly ovate, 3–3.5 mm. long, obtuse, inconspicuously several-nerved, the keel thick, green, with a slightly recurved awnlike mucro about 0.7 mm. long at apex; achenes 1/3 as long as the scale, lustrous, 3-angled; stigmas 3.——Aug. -Oct. Waste grounds, roadsides, seashores, quite common; Honshu (Kantō Distr. and westw.), Shikoku, Kyushu.—— Ryukyus, Formosa, China, India, and throughout the Tropics.

21. Cyperus difformis L. TAMA-GAYATSURI. Soft annual; culms tufted, 15–60 cm. long; leaves 2–5 mm. wide; inflorescence a simple or compound corymb to 7 cm. across, with 1–6 rays to 5 cm. long, or sometimes capitate, the bracts 2 or 3, the lowest longer than the inflorescence; spikelets very densely disposed in subglobose spikes, linear, 3–10 mm. long, 1 mm. wide, compressed or slightly turgid, dark reddish brown to pale green, dull; scales obovate-orbicular, 0.5 mm. long, entire, rounded-emarginate, the keel rather acute, green; achenes nearly as long as the scale, 3-angled, pale, obovate; stigmas 3.——Aug.–Oct. Common in wet places or in paddy fields; Hokkaido, Honshu, Shikoku, Kyushu.——Ryukyus, Formosa, Korea, China, India, Malaysia, Australia, s. Europe, and Africa.

22. Cyperus haspan L. KO-AZE-GAYATSURI, MIZU-HANABI. Rhizomes elongate, slender; culms solitary, nearly naked, 20–60 cm. long; leaves flaccid, short, often bladeless, 2–6 cm. long; corymbs compound or simple, to 15 cm. across, with many spreading slender rays, the bracts 1 or 2, leaflike, longer or shorter than the inflorescence; spikelets digitate in groups of 4 or 5, linear, flat, 5–15 mm. long, 1.5–2 mm. wide, sanguineous; scales oblong to narrowly so, about 1.5 mm. long, emarginate and mucronulate with the green midrib ending as an erect slightly excurrent mucro; achenes 1/3 as long as the scale, white, 3-angled, elliptic; stigmas 3.——Aug.–Oct. Swampy places and rice paddies; Honshu, Shikoku, Kyushu. ——Ryukyus, Formosa, China, Indochina, India, Malaysia, and Australia.

23. Cyperus tenuispica Steud. *C. flavidus* sensu auct. plur., non Retz.; *C. pseudohaspan* Makino——HIME-GAYATSURI, MIZU-HANABI. Closely akin to the preceding, but rhizomes absent; culms tufted, flaccid; leaves 2–4 mm. wide; spikelets 3–8 mm. long, 1–1.5 mm. wide; scales oblong to oblong-lanceolate, nearly 1 mm. long, obtuse and emarginate, with a green keel excurrent as a slightly recurved mucro; achenes 2/5–1/2 as long as the scale, white, broadly obovate, obtusely 3-angled; stigmas 3.——Aug.–Nov. Wet places; Honshu, Shikoku, Kyushu.——Ryukyus, Formosa, s. Korea, China, Indochina, India, Malaysia, tropical Australia, and tropical Africa.

24. Cyperus hakonensis Fr. & Sav. *C. trinervis* var. *flaccidus* Kuekenth., pro parte——HINA-GAYATSURI. Small green annual; culms flaccid, tufted, 5–25 cm. long; leaves usually reduced to short-bladed or bladeless hyaline sheaths; corymbs simple or subcompound, rather loose, with 3–5 short rays, rarely to 10 cm. long, the bract solitary, decurrent on the culm, nearly as long as the inflorescence; spikelets digitate, in groups of 2–6, strongly flattened, light green, 5–12 mm. long, 2 mm. wide; scales spreading, broadly ovate, 1 mm. long, the keel conspicuous, more or less winglike, excurrent at apex as a recurved awn 1/5–1/4 as long as the scale; achene 1/3 as long as the scale (including the awn), broadly obovate, 3-angled, light yellow; stigmas 3.——Aug.–Oct. Wet grassy places in low grounds; Honshu, Shikoku, Kyushu.——Korea and China.

25. Cyperus niigatensis Ohwi. NIIGATA-GAYATSURI. Annual; culms tufted, about 10 cm. long; leaves 1–2 mm. wide; inflorescence capitate, globular, 1–1.5 cm. wide, the bracts 4 or 5, spreading, to 10 cm. long; spikelets numerous in the head, ovate, 2–3 mm. long, 1.5 mm. wide, slightly flattened, obtuse, yellowish green; scales ascending, 2-ranked, ovate-orbicular, obtuse, entire, hyaline, 5- to 7-nerved; achenes 1/2–2/3 as long as the scale, obovate, brownish, obtusely angled, lunate in cross section, the midrib sometimes slightly conspicuous; stigmas 2 or 3.——Honshu (Niigata).

26. Cyperus nipponicus Fr. & Sav. var. **nipponicus.** *C. michelianus* var. *nipponicus* (Fr. & Sav.) Kuekenth.; *Juncellus nipponicus* (Fr. & Sav.) C. B. Clarke; *J. pygmaeus* sensu auct. Japon., pro parte, non C. B. Clarke; *Dichostylis nipponica* (Fr. & Sav.) Palla——AO-GAYATSURI, Ō-TAMA-GAYATSURI. Soft green annual; culms 5–25 cm. long; leaves 1–2.5 mm. wide; inflorescence capitate, globose, dense, 1–2.5 cm. across, occasionally umbelliform with 1–5 rays to 5 cm. long, the bracts leaflike; spikelets slightly flattened, lanceolate or narrowly ovate, 3–7 mm. long, 1.5–2 mm. wide, pale green; scales 2-ranked, dense, ovate, 1.7–2 mm. long, several-nerved, hyaline, obtuse with a broad tip; achenes 2/5–1/2 as long as the scale, obovate to elliptic, lunate in cross section, obtusely angled, light brown; stigmas 2 or 3.——Aug.–Oct. Honshu, Shikoku, Kyushu.——Korea, China, and Manchuria.

Var. **spiralis** Ohwi. *Scirpus stauntonii* C. B. Clarke; *C. stauntonii* (C. B. Clarke) Ohwi——Ō-SHIRO-GAYATSURI. Spikelets not flattened, obscurely angled; scales spirally arranged. ——Honshu, Shikoku.——China.

27. Cyperus extremiorientalis Ohwi. *Juncellus pygmaeus* sensu auct. Japon., non C. B. Clarke——HIME-AO-GAYATSURI. Soft green annual; culms 3–15 cm. long; leaves 1–2 mm. wide; inflorescence capitate, globose, 1–1.5 cm. in diameter, densely many-spiculose, the bracts leaflike, longer than the inflorescence; spikelets mostly flat, lanceolate to broadly so, 2–6 mm. long, 1–1.3 mm. wide, acute, pale green; scales broadly lanceolate, 1.5–2 mm. long, 2-ranked, hyaline, faintly 3- to 5-nerved, very obtuse and mucronate, the mucro extremely short, erect, sparsely spinulose or smooth; achenes 1/2–2/3 as long as the scale, narrowly oblong, planconvex, light brown, acutely angled; stigmas 2.——Aug.–Oct. Honshu, Shikoku.

28. Cyperus pacificus (Ohwi) Ohwi. *C. michelianus* sensu auct. Japon., non Link; *Scirpus michelianus* sensu auct. Japon., non Roem. & Schult.; *C. michelianus* var. *pacificus* Ohwi; *Isolepis micheliana* sensu auct. Japon., non Roem. & Schult.——SHIRO-GAYATSURI. Soft green annual; culms 3–

20 cm. long; leaves 1–2 mm. wide; inflorescence capitate, globular, 5–15 mm. in diameter, the bracts 3–6, 4–10 cm. long; spikelets numerous, dense, 3–5 mm. long, 1.5 mm. wide, somewhat compressed; scales spirally arranged or inconspicuously 2-ranked, broadly lanceolate, 1.5 mm. long, hyaline, slenderly 3- to 5-nerved, scarcely keeled, very acute and mucronate at tip, the mucro extremely short, erect or nearly so, smooth or bearing a few papillose processes; achenes 3/5–1/2 as long as the scale, narrowly ovate-oblong, unequally biconvex, acutely angled and narrowly winged.——Aug.–Oct. Hokkaido, Honshu (Kinki Distr. and eastw.)——Korea.

2. LIPOCARPHA R. Br.　Hinji-gayatsuri Zoku

Annuals or perennials; culms leafy only at base; leaves elongate, linear; spikelets several, many flowered, in a headlike cluster; inflorescence terminal, subtended by leaflike bracts; flowers bisexual; scales spirally imbricate; perianth-segments (or prophylla) 2, hyaline, one of them placed between the scales of the spikelet and the achene, the other between the achene and the axis (rachilla) of the spikelet; stamens 1 or 2; style continuous with the ovary, scarcely enlarged at base, deciduous; stigmas 2 or 3; achenes narrowly oblong, beakless, surrounded by the perianth-segments.——About 15 species, in tropical to warm-temperate regions.

1.　**Lipocarpha microcephala** (R. Br.) Kunth. *Hypaelyptum microcephalum* R. Br.; *L. zollingeriana* Boecklr.; *Cyperus zollingerianus* (Boecklr.) T. Koyama; *Isolepis squarrosa* sensu Miq., non Roem. & Schult.——Hinji-gayatsuri.　Flaccid green annual; culms tufted, erect, obtusely 3-angled, smooth; leaves 1–2 mm. wide, often filiform; head solitary, small, 5–8 mm. across, deltoid, usually composed of 3 spikelets, the bracts 2, divergent to reflexed, leaflike; spikelets ovoid-globular, obtuse, 3–5 mm. long, sessile; scales numerous, broadly oblanceolate, 1–1.3 mm. long, pale green, sometimes brown-flecked, with an apical recurved short awn-like mucro; achenes 1 mm. long, lanceolate, obtusely angled, light yellow, covered by 2 hyaline scales; stigmas 3.——Aug.–Oct. Wet places and waste grounds in lowlands; Honshu, Shikoku, Kyushu.——Korea, Manchuria, China, Formosa, India, Malaysia, and Australia.

3. FUIRENA Rottb.　Kuro-tama-gayatsuri Zoku

Hairy annuals or perennials; culms nodose; leaves slender, flat; inflorescences terminal and lateral, short-peduncled; spikelets hairy; flowers bisexual; scales spirally imbricate; perianth of 3 stipitate, cordate or elliptic scales with as many alternating short bristles; stamens 3; style continuous with the ovary, scarcely thickened at base, deciduous; stigmas 3; achenes 3-angled, beakless. ——About 30 species, in tropical and subtropical regions of the E. and W. Hemispheres, one in our area.

1.　**Fuirena ciliaris** (L.) Roxb. *Scirpus ciliaris* L.; *F. glomerata* Lam.——Kuro-tama-gayatsuri.　Annual; culms 10–40 cm. long, with 2 or 3 nodes; leaves flat, short-linear, 5–15 cm. long, 3–7 mm. wide, flaccid, short-pointed, the ligule ferrugineous, 1–2 mm. long; inflorescences 1–2; spikelets 3–10, oblong, 4–7 mm. long, 3 mm. wide, dark gray-green; scales elliptic, 1.5 mm. long, loosely pilose, grayish, hyaline, rounded at apex, the awn about 1 mm. long, recurved; achenes acutely 3-angled, obovate-orbicular, about 0.7 mm. long, light brown, mucronate; perianth-segments 6, the inner 3 nearly as long as the achene, squarrose, long-stiped, glabrous, the outer 3 bristle-shaped, very short.——Wet places and paddy fields; Honshu (Shimosa Prov. and Chūgoku Distr.), Shikoku, Kyushu.——s. Korea, Ryukyus, Formosa, China, India, and Malaysia.

4. SCIRPUS L.　Hotaru-i Zoku

Perennials, rarely annuals, usually glabrous, sometimes small, sometimes stout; culms naked or with a few nodes; leaves linear, sometimes reduced to a truncate sheath; inflorescence terminal or pseudolateral, headlike, corymbose, or somewhat umbelliform, subtended by 1–several bracts, sometimes consisting of a solitary spikelet; spikelets usually many-flowered; flowers bisexual; scales spirally arranged (or all except the lowest) 1-flowered; perianth of 0–8 usually scabrous bristles or rarely of flat, smooth, elongated fibers; stamens 1–3; style continuous with the ovary, not enlarged at base, deciduous; stigmas 2 or 3; achenes obovate, lenticular or 3-angled.——Cosmopolitan, with about 200 species.

1A.　Inflorescence terminal; bracts several or none, sometimes 1.
　　2A.　Bracts none or 1, scalelike; spikelet solitary, terminal.
　　　3A.　Leaves bladeless or nearly so.
　　　　4A.　Culms nearly terete, smooth; basal sheaths coriaceous, lustrous; perianth-segments of bristles, hardly exceeding the achene.
　　　　　　　　　　　　　　　　　　　　　　　　　　　　　1. S. caespitosus
　　　　4B.　Culms sharply 3-angled, scabrous on the angles; basal sheaths hyaline, dull; perianth-segments filiform, silky, elongate after anthesis, much exceeding the achene, to 2 cm. long. .... *2. S. hudsonianus*
　　　3B.　Leaves, at least some of them, with elongate blades. .... *3. S. pseudofluitans*
　　2B.　Bracts several, flat, leaflike; culms more or less leafy, tall, stout; leaves linear.
　　　5A.　Spikelets several to numerous, small; achenes 0.7–1.3 mm. long.
　　　　6A.　Stigmas 3; achenes compressed-triangular to trigonal.
　　　　　7A.　Culms slenderly striate, distinctly 3-angled above, often scabrous on the angles; leaf-sheaths loosely surrounding the culm; spikelets dark gray.
　　　　　　8A.　Culms 50–120 cm. long; cauline leaves at least 20 cm. long; scales less than 2 mm. long.
　　　　　　　9A.　Spikelets 5–7 mm. long, solitary or few together; bristles frequently prominently flexuous.

10A. Bristles slightly longer than the achene, retrorsely scabrous nearly to the base; rays of inflorescence and branchlets scabrous near tip; sterile shoots not elongate. . . . . 4. *S. sylvaticus* var. *maximowiczii*

10B. Bristles prominently flexuous, 3–4 times as long as the achene, with spreading to ascending spines only on apical part; rays of inflorescence and branchlets smooth; sterile shoots much elongate. . . . . 5. *S. radicans*

9B. Spikelets 2–2.5 mm. long, densely congested in groups of 8–40; bristles erect, shorter than the achene.

6. *S. atrovirens* var. *georgianus*

8B. Culms 15–40 cm. long; cauline leaves 3–7 cm. long; scales 3.5–4 mm. long. . . . . 7. *S. maximowiczii*

7B. Culms smooth, lustrous, stiff, obtusely 3-angled; leaf-sheaths tightly surrounding the culm.

11A. Leaves not septate-nodose, prominently scabrous on margin; rays of inflorescence smooth or slightly scabrous at tip; spikelets dark gray.

12A. Cauline leaves 4–8 mm. wide; scales lanceolate, 0.7 mm. wide at base. . . . . 8. *S. mitsukurianus*

12B. Cauline leaves 3–6 mm. wide; scales ovate or narrowly so, 1–1.3 mm. wide at base.

13A. Lateral inflorescence with 5–10 capitate clusters of spikelets, the terminal twice compound, with slender rays.

9. *S. karuizawensis*

13B. Lateral inflorescence with 1 or 2 capitate clusters of spikelets, the terminal once-compound, with 3–6 stiff rays.

10. *S. fuirenoides*

11B. Leaf-sheaths and under surface of the blade septate-nodose; rays and branchlets of inflorescence prominently scabrous; spikelets reddish brown. . . . . 11. *S. wichurae*

6B. Stigmas 2; achenes biconvex; spikelets reddish brown.

14A. Sheath of cauline leaves partially brownish; rays and branchlets of inflorescence smooth; spikelets in groups of 4–10, 2.5–3.5 mm. thick; scales concolorous, 1-nerved or very obscurely veined. . . . . 12. *S. ternatanus*

14B. Sheath of cauline leaves pale green throughout; rays and branchlets of inflorescence scabrous at apex; spikelets in groups of 2–5, about 1.5 mm. thick; scales rusty brown with a broad green midrib. . . . . 13. *S. kiushuensis*

5B. Spikelets several, sometimes solitary or numerous, relatively large; achenes 3–4 mm. long.

15A. Spikelets 1–3; inflorescence simple; scales short-awned; stigmas 2; achenes with more or less concave faces, about 3 mm. long, broadly obovate, compressed. . . . . 14. *S. planiculmis*

15B. Spikelets usually rather numerous; inflorescence compound; scales with a recurved awn; stigmas 3; achenes 3-angled, with flat faces, 3.5–4 mm. long, cuneate at base. . . . . 15. *S. fluviatilis*

1B. Inflorescence appearing lateral, the lowest bract culmlike as a continuation of the culm; spikelets 7–20 mm. long; leaves bladeless, except No. 23.

16A. Inflorescence usually capitate; scales membranous to more or less coriaceous, faintly nerved; rhizomes almost wanting, except in No. 16.

17A. Rhizomes slender, elongate; culms solitary; spikelets usually 1. . . . . 16. *S. lineolatus*

17B. Rhizomes wanting or very short; culms tufted; spikelets 3–12.

18A. Culms slender, terete or nearly so.

19A. Achenes 2 mm. long; scales nearly coriaceous.

20A. Spikelets acute, pale green; bristles 4, much exceeding the achene. . . . . 17. *S. wallichii*

20B. Spikelets rather obtuse, more or less stained ferruginous when dried; bristles 5–6, shorter than or slightly longer than the achene. . . . . 18. *S. juncoides*

19B. Achenes 1.2–1.5 mm. long, the bristles 1.5 times as long as the achene; scales relatively soft.

21A. Spikelets 4–4.5 mm. wide; scales with an insignificant midrib on back; achenes compressed trigonous. . . 19. *S. hondoensis*

21B. Spikelets 3–4 mm. wide; scales with a very broad green midrib; achenes planoconvex. . . . . 20. *S. komarovii*

18B. Culms acutely 3-angled

22A. Culms thick; spikelets long, not angled; achenes indistinctly rugose; style and stigma 4 mm. long; anthers 2.5 mm. long.

21. *S. triangulatus*

22B. Culms relatively slender, the faces distinctly concave; spikelets angled; achenes distinctly rugose; style and stigma 2.5 mm. long; anthers 0.7 mm. long. . . . . 22. *S. mucronatus*

16B. Inflorescence corymbose, rarely capitate; scales hyaline, emarginate, mucronate, nerveless; achenes usually smooth; rhizomes creeping or elongate.

23A. Radical leaves with an elongate blade, linear, often overtopping the inflorescence. . . . . 23. *S. nipponicus*

23B. Radical leaves bladeless or nearly so.

24A. Culms acutely 3-angled; rhizomes slender, elongate. . . . . 24. *S. triqueter*

24B. Culms terete; rhizomes short, stout. . . . . 25. *S. tabernaemontani*

1. Scirpus caespitosus L. *Trichophorum caespitosum* (L.) Hartm.——MINE-HARI-I.   Rhizomes densely tufted, densely covered with coriaceous, decaying sheaths; culms 5–30 cm. long, terete, erect, leafless, smooth, enveloped by a few coriaceous, lustrous, brownish sheaths 7–20 mm. long; spikelet solitary, terminal, narrowly ovoid, 3–5 mm. long, 2- to 5-flowered, tawny, bractless; achenes broadly obovate, compressed-trigonous, mucronate; bristles 6, erect, 1.5 times as long as the achene, nearly smooth; anthers 3, linear.——July–Aug. Wet slopes and moors in high mountains; Hokkaido, Honshu (n. and centr. distr.).——Sakhalin, Kuriles, and circumpolar.

2. Scirpus hudsonianus (Michx.) Fern. *Eriophorum alpinum* L.; *Trichophorum alpinum* (L.) Pers.; *S. tricho-*

phorum Asch. & Graebn.; *Eriophorum hudsonianum* Michx. ——HIME-WATA-SUGE, MIYAMA-SAGI-SUGE.   Resembles No. 1, but slightly larger; culms slender, 10–30 cm. long, acutely angled, scabrous, enveloped in hyaline bladeless sheaths at base; spikelet solitary, terminal, broadly lanceolate, 5–7 mm. long, narrowly ovoid when mature, bractless, 5- to 10-flowered, reddish brown; achenes 1.3 mm. long, narrowly obovate, compressed-trigonous, mucronate; bristles 6, filiform, much elongate and exserted, to 2 cm. long, white, smooth; anthers linear.——June–July. Sphagnum moors; Hokkaido——n. Korea, Kuriles, and circumpolar.

3. Scirpus pseudofluitans Makino. *S. fluitans* subsp. *pseudofluitans* (Makino) T. Koyama——BYAKKO-I.   Pale green, flaccid; sterile shoots elongate, the branches aggregated, elongate, leafy; leaves smooth, thick, linear, 5–10 cm. long, 1–2 mm. wide; fertile culms leafless except at base, 5–15 cm. tall; spikelet solitary, terminal, oblong, more or less acute, 5–7 mm. long, bractless, many-flowered, pale green; scales ovate, 4–4.5 mm. long, obtuse; achenes narrowly obovate, 1.5 mm. long, planoconvex; bristles wanting; stamens 3, the anthers linear.——Rare, in streams and around springs; Honshu (Iwaki Prov.).

4. Scirpus sylvaticus L. var. **maximowiczii** Regel. *S. orientalis* Ohwi——KURO-ABURAGAYA, YAMA-ABURAGAYA. Short-rhizomatous; culms 80–120 cm. long, 6- to 8-noded, obtusely 3-angled at base, acutely angled above, scabrous on the angles below the inflorescence; cauline leaves flat, linear, 20–40 cm. long, 5–10 mm. wide, long-acuminate, the sheaths loosely surrounding the culm, 5–10 cm. long; inflorescence a solitary, terminal, large, compound umbelliform panicle subtended by 2–5 leaflike bracts much exceeding the inflorescence, the rays numerous, the longer ones to 15 cm. long, scabrous above; spikelets solitary or in 3's, numerous, dark gray, narrowly ovoid, 4–7 mm. long; achenes 1 mm. long, pale; bristles 5–6, slightly longer than the achene, retrorsely scabrous.——July–Aug. Swampy places in woods; Hokkaido, Honshu (n. and centr. distr. especially in mountains)——Korea, Sakhalin, Manchuria, and Ussuri.——The typical phase occurs in Europe.

5. Scirpus radicans Schk. *S. hokkaidoensis* Beetle; *S. sylvaticus* var. *radicans* (Schk.) Willd.——TSURU-ABURAGAYA, KE-NASHI-ABURAGAYA. Culms 1–1.5 m. long, with 7–10 nodes; leaves 7–10 mm. wide, 20–35 cm. long; inflorescence 10–20 cm. long and as broad, the rays and branchlets glabrous; spikelets solitary on each branchlet, narrowly oblong-ovoid, 5–7 mm. long; scales dark gray, about 2 mm. long; bristles 6, 3–4 times as long as the achene, prominently flexuous, nearly glabrous; anthers 1 mm. long, otherwise with the characters of No. 4.——Swampy places; Hokkaido, Honshu (Uzen); rare.——Sakhalin, Korea, Siberia, and Europe.

6. Scirpus atrovirens Willd. var. **georgianus** (Harper) Fern. *S. georgianus* Harper; *S. hattorianus* Makino——IWAKI-ABURAGAYA. Culms stout, more than 1 m. long, glabrous, leafy; cauline leaves flat, linear, 20–30 cm. long, 8–10 mm. wide, the sheaths 4–5 cm. long; inflorescence terminal, compound, 5–10 cm. long, the lower 2 or 3 bracts longer than the others, leaflike, narrow, exceeding the inflorescence, the rays smooth, sometimes with short branchlets; spikelets sessile, disposed in capitate clusters of 8–10 spikelets, broadly ovoid, obtuse, 2–2.5 mm. long, dark gray; scales about 1 mm. long, very obtuse; achenes slightly shorter than the scale, ellipsoid, pale, compressed-trigonous; bristles 4–5, slender, erect, with a few spines at tip; anthers 0.8 mm. long.——Honshu (Iwaki); rare.——N. America.

7. Scirpus maximowiczii C. B. Clarke. *Eriophorum japonicum* Maxim.; *S. japonicum* (Maxim.) Fern., non Fr. & Sav.; *E. maximowiczii* (C. B. Clarke) Beetle——TAKANE-KURO-SUGE, MIYAMA-WATA-SUGE. Rhizomes short; culms solitary, with 1–3 nodes, 15–40 cm. tall; radical leaves flat, broadly linear, 3–6 mm. wide, long-attenuate, obtuse at tip, the cauline 3–7 cm. long, the sheath 3–4 cm. long, loosely surrounding the culm at base; inflorescence terminal, compound, 3–5 cm. long, nodding on one side, the bracts 1 or 2, short, the rays scabrous; spikelets oblong, rarely in capitate clusters of 1–3, 7–10 mm. long, 3–4 mm. thick, dark gray; scales membranous, 3.5–4 mm. long, very obtuse; achenes broadly obovate, 1.3 mm.

long, the bristles 6, flexuous, 5–6 mm. long, sparsely spinulose at tip; anthers 2–3 mm. long.——July–Aug. Wet places in high mountains; Hokkaido, Honshu (n. & centr. distr.).——s. Kuriles, Sakhalin, Korea, and Ussuri.

8. Scirpus mitsukurianus Makino. MATSU-KASA-SUSUKI. Culms 1–1.5 m. long, stout, terete, with 5–7 distant nodes; radical leaves shorter than the culm, the cauline 4–8 mm. wide, elongate, long-attenuate, the sheath 3–10 cm. long, tightly surrounding the culm; inflorescence of 2–5 interrupted subcompound panicles, the terminal one largest, 5–10 cm. long, the subtending bracts several, 2 or 3 of them longer than the others, longer than the inflorescence; spikelets sessile, densely congested in globular-capitate clusters of 10–20, ovate-oblong, 4–6 mm. long, dark brownish gray; scales lanceolate, 3 mm. long, 0.7 mm. wide, acute; achenes obovate, compressed-trigonous, 1 mm. long, mucronate, pale, the bristles 5–6, about 5 mm. long, sparsely ascending-spinulose at tip.——Aug.–Oct. Swamps in low grounds; Honshu, Shikoku, Kyushu.

9. Scirpus karuizawensis Makino. *S. coreanus* Palla; *S. fuirenoides* var. *jaluanus* Komar.; *S. jaluanus* (Komar.) Nakai; *S. fuirenoides* var. *karuizawensis* (Makino) Hara——HIME-MATSU-KASA-SUSUKI. Differs from the following in the lateral branches of the panicle with 5–10 clusters of spikelets, the terminal panicle decompound, and by the more slender rays.——Aug.–Oct. Swamps; Honshu (Shinano Prov., Karuizawa); rare.——Korea and Manchuria.

10. Scirpus fuirenoides Maxim. KO-MATSU-KASA-SUSUKI. Rhizomes short; culms 80–120 cm. long, stout, with 4 or 5 nodes; radical leaves linear, the cauline 3–4 mm. wide, gradually attenuate; partial panicles 4–6, the lateral composed of 1 or 2 clusters of spikelets, the terminal with 3–6 stiffly erect simple rays, the bracts leaflike, narrow; spikelets sessile, oblong, 5–7 mm. long, 2.5–3 mm. thick, dark grayish brown when mature, in globose-capitate clusters of 10–20 spikelets; scales ovate-deltoid, 2.5–3 mm. long, 1–1.2 mm. wide, acute; achenes obovate, mucronate, compressed-trigonous, about 1 mm. long, pale, the bristles 6, flexuous, about 5 mm. long, sparsely ascending-spinulose at tip.——Aug.–Oct. Swamps; Honshu, Kyushu.

11. Scirpus wichurae Boecklr. var. **wichurae**. *S. eriophorum* sensu auct. Japon., non Michx.; *S. cyperinus* sensu auct. Japon., non Kunth; *S. concolor* Maxim.——ABURA-GAYA. Culms 1–1.5 m. long, stout, obtusely 3-angled, with 5–8 distant nodes; cauline leaves flat, 30–40 cm. long, 5–15 mm. wide, acute, rather stiff, gradually narrowed at tip; partial panicles compound, 1–4, the lateral ones small, the terminal often decompound, the bracts 2 or 3, longer or shorter than the partial inflorescence; spikelets peduncled or sessile, sometimes in clusters of 2–5 (forma **concolor** Ohwi. ABURA-GAYA), sometimes solitary (forma **wichurae**. AIBA-sō), oblong or ellipsoid, 4–8 mm. long, 3–4 mm. thick, obtuse, reddish brown; scales narrow-ovate, 2–2.5 mm. long; achenes compressed-trigonous, 0.8–1 mm. long, obovate, pale, mucronate, the bristles 6, about 4 mm. long, slender, flexuous, ascending, scabrous at tip.——Aug.–Oct. Wet places in hills and lowlands; Hokkaido, Honshu, Shikoku, Kyushu; common and variable.

Var. **asiaticus** (Beetle) T. Koyama. *S. borealis* T. Koyama, excl. basionym; *S. wichurae* var. *borealis* Ohwi, excl. typus ——HIGE-ABURA-GAYA. Lateral partial panicles usually wanting, the terminal one large, with numerous spikelets; spikelets dense, globular; scales ovate to broadly so.——Honshu, Shikoku, Kyushu.——Korea, China to the Himalayas.

12. Scirpus ternatanus Reinw. *S. chinensis* Munro, non Osbeck——Ō-ABURA-GAYA. Culms stout, 60–100 cm. long, leafy; cauline leaves elongate, stiff, 8–15 mm. wide, long-attenuate, the sheaths loosely enveloping the culm, usually brownish and somewhat lustrous at base; inflorescence terminal, decompound, very large, with smooth stiff rays to 10 cm. long, the bracts 3 or 4, leaflike; spikelets sessile, ovoid, obtuse, reddish brown, in clusters of 4–10; scales membranous, 1–1.2 mm. long, ovate-deltoid, obtuse; achenes slightly shorter than the scale, flattened, obovate, pale, the bristles 0–3, slightly longer than the achene, erect, slender, ascending-scabrous above the middle.——Kyushu (Hyuga Prov.; Tanegashima)——Ryukyus, Formosa, China, and Malaysia.

13. Scirpus rosthornii Diels. *S. kiushiuensis* Ohwi; *S. ternatanus* var. *kiushiuensis* (Ohwi) T. Koyama——TSUKUSHI-ABURA-GAYA. More slender than the preceding; culms 60–100 cm. long, glabrous, with 5–7 nodes; cauline leaves flat, 15–20 cm. long, 6–8 mm. wide, gradually attenuate, the sheaths rather loosely enveloping the culm; inflorescence terminal, large, to 10 cm. wide, the rays stiff, to 5 cm. long, scabrous in upper part; spikelets sessile, oblong, 3–4 mm. long, 1.5 mm. thick, in clusters of 2–5; scales ovate-orbicular, 1 mm. long, very obtuse, brownish with a broad green midrib; achenes slightly shorter than the scale, pale, compressed, 0.7 mm. long; bristles none.——Aug.-Oct.——s. Kyushu.——China.

14. Scirpus planiculmis F. Schmidt. *S. maritimus* var. *affinis* sensu auct. Japon., non C. B. Clarke; *S. maritimus* var. *compactus* sensu auct. Japon., non Mey.; *S. biconcavus* Ohwi——EZO-UKI-YAGARA, KO-UKI-YAGARA. Rhizomes elongate; culms 40–100 cm. long, 3-angled, without nodes except at base, more or less thickened at base; leaves elongate, flat, 2–5 mm. wide; inflorescence capitate or with 1 or 2 short rays, the bracts 1–3, much longer than the inflorescence, leaflike; spikelets 1–6, sessile, ovoid, 8–15 mm. long, 6–8 mm. wide, lustrous, reddish or yellowish brown; scales 3–6 mm. long, elliptic, hyaline, nerveless on both sides, pubescent, emarginate, the midrib prolonged into a slightly recurved awn 1–2 mm. long; achenes flattened, broadly obovate, 3 mm. long, tawny and lustrous, obtusely angled, the faces more or less concave, bristles none.——July–Oct. Swampy places especially near seashores; Hokkaido, Honshu, Shikoku, Kyushu.——Sakhalin, Manchuria, Korea, and Ryukyus.

15. Scirpus fluviatilis (Torr.) A. Gray. *S. maritimus* var. *fluviatilis* A. Gray; *S. yagara* Ohwi; *S. fluviatilis* var. *yagara* (Ohwi) T. Koyama——UKI-YAGARA, YAGARA. Rhizomes elongate; culms 70–150 cm. long, stout, 3-angled, leafy below, globosely thickened at base; leaves flat, 5–10 mm. wide, gradually narrowed upward; inflorescence terminal, usually simple, with 3–8 rays to 7 cm. long, the bracts 2–4, leaflike, elongate; spikelets oblong, 2 or 3 on each ray, 10–20 mm. long, 6–8 mm. across; scales oblong, hyaline, brownish, nerveless, 5–6 mm. long, puberulent, emarginate, the midrib ending in a recurved awn 2–3 mm. long; achenes obovate, 3.5–4 mm. long, cuneate at base, 3-angled, at first whitish, finally dark brown, lustrous, with equal flat sides, the bristles 6, shorter than the scale, retrorsely scabrous.——July–Oct. Marshy places and shallow waters; Hokkaido, Honshu, Shikoku, Kyushu.——e. Asia, Australia, and N. America.

16. Scirpus lineolatus Fr. & Sav. HIME-HOTARU-I. Rhizomes slender, elongate, forming a tuber at apex in autumn; culms loose, solitary, terete, 7–30 cm. long, soft, naked, the basal sheaths 1 or 2, hyaline, bladeless, obliquely truncate at apex; spikelet solitary, pseudolateral, 7–10 mm. long, 3 mm. across, broadly lanceolate, acute, sessile, tawny to reddish brown, more or less lustrous; bract 1, terete, 1–4 cm. long, erect, continuous with the culm; scales obtuse, hyaline, 4–5 mm. long, faintly nerved, oblong; stigmas 2; achenes broadly obovate, 2 mm. long, dark brown, lustrous, compressed, glabrous; bristles 4–5, twice as long as the achene, retrorsely scabrous; anthers 2.5 mm. long.——July–Oct. Marshy places; Hokkaido, Honshu, Shikoku, Kyushu.——Formosa.

17. Scirpus wallichii Nees. *S. sasakii* Hayata——TAI-WAN-YAMA-I. Culms tufted, green, slender, 10–40 cm. long, obtusely 4- or 5-angled, naked; basal sheaths cylindrical, obliquely truncate at tip, bladeless; inflorescence capitate, pseudolateral, dense, composed of 2–5 spikelets; bract 1, erect, 5–13 cm. long, continuous with the culm; spikelets narrowly ovoid, 3–17 mm. long, 3–3.5 mm. wide, acute, sessile, pale green; scales elliptic, 3.5–4 mm. long, obtuse; stigmas 2; achenes broadly obovate, about 2 mm. long, planoconvex, slightly rugose; bristles 4, much longer than the achene, retrorsely scabrous.——Aug.-Oct. Swampy places; Honshu, Shikoku, Kyushu.——Korea, China, Formosa, and India.

18. Scirpus juncoides Roxb. var. **hotarui** (Ohwi) Ohwi. *S. hotarui* Ohwi; *S. juncoides* sensu auct. Japon., saltem pro parte, non Roxb.——HOTARU-I. Rhizomes nearly wanting; culms 30–60 cm. long, slender, smooth; basal sheaths obliquely truncate at apex, bladeless; inflorescence of 1–3(–5) radiate sessile spikelets, pseudolateral, headlike; bract 1, decurrent on the culm, erect, 5–8 cm. long, shallowly sulcate in front; spikelets ovoid to narrowly ovoid, 8–12 mm. long, 5–6 mm. across; scales ovate-orbicular, rounded at apex; achenes dark brown, lustrous, compressed-trigonous, broadly obovate, 2 mm. long, rugose; stigmas 3; bristles shorter or slightly longer than the achene, 6, retrorsely scabrous, brownish.——Rather common in wet places; Hokkaido, Honshu, Shikoku, Kyushu.——Korea and Manchuria.

Var. **ohwianus** (T. Koyama) T. Koyama. *S. ohwianus* T. Koyama; *S. juncoides* sensu auct. Japon., pro parte, non Roxb.——INU-HOTARU-I. Culms terete, rather soft, opaque, indistinctly obtuse-angled; inflorescence of 3–9(–12) short-cylindric spikelets; achenes lenticular, unequally convex; style bifid; otherwise with the characters of var. *hotarui*.——Honshu, Shikoku, Kyushu.——Ryukyus, Formosa, and China. The typical phase occurs in India.

Scirpus trapezoideus Koidz. *S. juncoides* Roxb. var. *triangulatus* (Honda) Ohwi; *S. erectus* var. *triangulatus* Honda; *? S. quadrangulus* Makino——SHIKAKU-HOTARU-I, SAN-KAKU-HOTARU-I. May be a natural hybrid between Nos. 17 and 20, distinguished from No. 17 by the distinctly 3–5 angled stout culms.——Honshu, Kyushu.

19. Scirpus hondoensis Ohwi. MIYAMA-HOTARU-I. Rhizomes abbreviated; culms soft, 15–40 cm. long; spikelets 2–4, broadly ovoid, 5–6 mm. long, 4–4.5 mm. wide, obtuse, light green; scales somewhat hyaline, 3 mm. long; stigmas 3; achenes 1.5 mm. long, compressed-triangular; bristles 5–6, 1½ times as long as the achene; otherwise with the characters of the preceding species.——July–Sept. Marshy places on edge of lakes or ponds in high mountains; Honshu (n. and centr. distr.).

20. Scirpus komarovii Roshev.   *S. supinus* var. *leiocarpus* Komar.; *S. okuyamae* Ohwi——Ko-hotaru-i.   Rhizomes scarcely evident; culms tufted, 20–50 cm. long, soft, terete, slender; basal sheaths cylindric, the mouth hyaline-translucent, obliquely truncate, sometimes cuspidate; inflorescence capitate, pseudolateral, of 3–10 sessile spikelets; bract 1, 7–10 cm. long, continuous with the culm; spikelets broadly ovoid, 5–8 mm. long, 3–4 mm. wide, obtuse, light green; scales elliptic, membranous, 2.5–3 mm. long; stigmas 2; achenes broadly obovate, 1.2–1.5 mm. long, flattened, dark brown, nearly smooth; bristles 4 or 5, 1½ times as long as the achene, retrorsely scabrous; anthers 0.3 mm. long.——July–Sept. Hokkaido, Honshu (Kai, Mutsu).——Korea, Manchuria, and Ussuri.

21. Scirpus triangulatus Roxb.,   *S. mucronatus* auct. Japon., non L.; *S. preslii* Dietr.; *S. mucronatus* var. *robustus* Miq.; *S. mucronatus* var. *subleiocarpa* Fr. & Sav.——Kangare-i.   Rhizomes abbreviated; culms tufted, 50–120 cm. long, acutely 3-angled, stout, naked; basal sheaths 3-angled, obliquely truncate at apex; inflorescence capitate, pseudolateral, of 4–20 sessile spikelets; bracts 1, 3-angled, 3–10 cm. long, erect or ascending, continuous with the culm; spikelets oblong, 1–2 cm. long, 4–6 mm. wide, light green to light brown, terete; scales oblong or ovate, 4–5 mm. long, somewhat firm; stigmas 3; achenes 2–2.5 mm. long, compressed-triangular, indistinctly rugose or nearly smooth; bristles 5–6, 1.5–2 times as long as the achene, retrorsely scabrous; anthers 2.5 mm. long, linear. ——Aug.–Oct. Marshes and ponds; common; Hokkaido, Honshu, Shikoku, Kyushu.——Korea, China, Formosa, Malaysia, and India.

22. Scirpus mucronatus L.   *S. abactus* Ohwi——Hime-kangare-i.   Smaller than the preceding; culms 40–70 cm. long, distinctly concave on the faces, 3–5 mm. wide at top; spikelets 5–10, ovoid, 6–10 mm. long, 4–5 mm. wide, slightly angled; bract solitary, 3–6 cm. long, erect, or spreading; scales ovate-orbicular, 3–5 mm. long, rather loosely imbricate; stigmas 3; achenes 2 mm. long, distinctly rugose; bristles 6, shorter than the achene, retrorsely scabrous; anthers lanceolate, 0.7 mm. long; otherwise as in the preceding species.——Aug.–Oct. Marshy places and shallow ponds; Honshu (Kii Prov.), Shikoku, Kyushu; rare.——Korea to Europe.

23. Scirpus nipponicus Makino.   *S. etuberculatus* subsp. *nipponicus* (Makino) T. Koyama; *S. depauperatus* Komar., non Poir.——Shizui, Te-ga-numai.   Rhizomes slender, terminated by a small tuber; culms 40–60 cm. long, 3-angled, about 2 mm. wide in upper part, without nodes, smooth, leafy at base; leaves slightly longer than the culm, flaccid, 3-angled, glabrous, 2–3 mm. wide, subobtuse; inflorescence corymbose, pseudolateral, 1–3 times dichotomously branched, with 5–8 loosely disposed spikelets; bract 1, continuous with the culm,

erect, 10–20 cm. long; spikelets narrowly oblong to broadly lanceolate, 10–15 mm. long, 5–6 mm. thick, subacute, reddish brown; scales thin, narrowly oblong-ovate, 4–5 mm. long, obtuse, hyaline; stigmas 2; achenes obovate, biconvex, 2 mm. long, dark brown, dull; bristles 4, twice as long as the achene, retrorsely scabrous.——July–Oct. Ponds and ditches; Hokkaido, Honshu, Shikoku, Kyushu; rare.——Manchuria.

24. Scirpus triqueter L.   *S. pollichii* Godr. & Gren.——Sankaku-i.   Rhizomes slender, long-creeping; culms rather distant, 50–100 cm. long, acutely 3-angled, 2–7 mm. thick at base; basal sheaths 3-angled, with short blades 0.5 cm. long; inflorescence pseudolateral, umbelliform with a few rays, or capitate; bract 1, continuous with the culm, 3-angled, 2–5 cm. long; spikelets oblong to ovate, 7–17 mm. long, 5–7 mm. thick, obtuse, ferrugineous, in clusters of 2 or 3; scales oblong, about 4 mm. long, hyaline, dark brown striate, ciliate, shallowly bifid, the midrib excurrent into a short, erect mucro; achenes broadly obovate, biconvex, 2–2.5 mm. long, nearly smooth, lustrous; bristles 3–5, nearly as long as the achene, retrorsely scabrous; stigmas 2.——July–Oct. Marshy places and shallow ponds; Hokkaido, Honshu, Shikoku, Kyushu; common.——Ussuri, Korea, Manchuria, China, Ryukyus, Malaysia, India, and s. Europe.

25. Scirpus lacustris L. var. **creber** (Fern.) T. Koyama. *S. tabernaemontani* sensu auct. Japon., non Gmel.; *S. validus* Vahl; *S. validus* Vahl var. *creber* Fern.; *S. lacustris* sensu auct. Japon., non L.; *S. lacustris* var. *tabernaemontani* (Gmel.) Doell; *S. ciliatus* Steud.——Futo-i.   Rhizomes stout, creeping, short; culms thick, terete, glaucous-green, 1–2 m. long; basal sheaths cylindric, with a short subulate blade to 10 cm. long; inflorescence pseudolateral, umbelliform, once to twice compound, with 4–7 rays; bract 1, continuous with the culm; spikelets in groups of 2 or 3 or solitary, ovoid, reddish brown, 5–10 mm. long; scales membranous, elliptic to broadly ovate, about 3 mm. long, ciliate, with reddish brown flecks, shallowly bifid and mucronate; stigmas usually 2, rarely 3; achenes 2 mm. long, biconvex, dark grayish brown, smooth; bristles as long as the achene, retrorsely scabrous.——July–Oct. Ponds and lake margins; Hokkaido, Honshu, Shikoku, Kyushu; common.——Sakhalin, Kuriles, Korea to Europe. Very variable.——Forma **creber.** Kita-futo-i.   Spikelets usually aggregated, dark colored.——Occurring chiefly in northern areas.——Forma **luxurians** Miq. Nami-futo-i.   Spikelets numerous, usually solitary, light colored, scarcely flecked.—— Common in central areas.——Forma **australis** Ohwi. Nagabo-futo-i.   Spikelets densely disposed, 10–15 mm. long, light colored, not flecked.——Common in southern areas.—— Cv. **Zebrinus.**   *S. lacustris* forma *zebrinus* Hort. ex Asch. & Graebn. Shima-futo-i.   Culms transversely white-banded. Cultivated.——Cv. **Pictus.**   *S. lacustris* forma *pictus* Honda ——Tatejima-futo-i.   Culms longitudinally white-lined. Cultivated.

5. ERIOPHORUM L.   Wata-suge Zoku

Tufted, rhizomatous or stoloniferous perennials; culms few-leaved; radical leaves slender, elongate, flat or 3-angled; cauline leaves usually reduced to a sheath; inflorescence of 1 to several spikelets, simple, terminal, bracteate or not; scales spirally arranged, many, the lower one empty; flowers bisexual; perianth of many elongate flat fibrous filaments smooth or nearly so, much longer than the scale; stamens 3; achenes obovate, compressed-trigonous, rarely scabrous on margin at tip; style slender, continuous with the achene and not thickened at base, the stigmas 3.——About 15 species, in the subarctic and cool-temperate regions of the N. Hemisphere.

1A. Spikelets several; cauline leaves with distinct blades; rhizomes long-creeping; anthers 2–3 mm. long. .... 1. *E. gracile*
1B. Spikelet solitary, terminal; cauline leaves reduced to an inflated sheath.
  2A. Densely tufted, without stolons; radical leaves compressed, 3-angled, scabrous on margin; anthers 2.5–3 mm. long.
                                                   2. *E. vaginatum*
  2B. Stoloniferous, hardly tufted; radical leaves terete, sulcate on upper surface, smooth; anthers 0.8–1 mm. long.
                                     3. *E. scheuchzeri* var. *tenuifolium*

1. Eriophorum gracile Koch. *E. coreanum* Palla; *E. gracile* subsp. *coreanum* (Palla) Hult.; *Scirpus ardea* T. Koyama, incl. var. *coreanus* (Palla) T. Koyama——SAGI-SUGE. Scarcely caespitose; rhizomes elongate; culms slender, rather soft, 20–50 cm. long, obtusely 3-angled; radical leaves compressed, 3-angled, sometimes over-topping the culm, abruptly obtuse, the cauline 1–2, short, subulate, sheathing at base; inflorescence simple, the lowest bracts elongate, 1–2 cm. long; spikelets 2–5, oblong in flower, 5–10 mm. long, the peduncle papillose-pilose; scales narrowly oblong, obtuse, gray, faintly nerved; perianth-segments many, white, linear, elongate after anthesis and much exceeding the scale, to 2 cm. long; achenes narrowly oblong, sometimes mucronate, 3–3.5 mm. long.—— June–Aug. High moors; Hokkaido, Honshu (Kinki Distr. and northeast).——Widely distributed in the cooler regions of the N. Hemisphere.

2. Eriophorum vaginatum L. *E. scheuchzeri* sensu Matsum., non Hoppe; *E. fauriei* E. G. Camus; *E. scabridum* Ohwi; *Scirpus fauriei* (E. G. Camus) T. Koyama——WATA-SUGE, SUZUME-NO-KE-YARI. Densely caespitose in large clumps; culms erect, firm, obtusely 3-angled, 20–50 cm. long; basal leaves compressed-trigonous, with acute angles, 1–1.5 mm. wide, abruptly acute, slightly scabrous on margin, the cauline 1 or 2, reduced to a bladeless inflated sheath, blackish above, acute; spikelet solitary, terminal, narrowly ovoid, 1–2 cm. long at anthesis, nearly globose in fruit; scales numerous, imbricate, lanceolate-deltoid, dark gray, gradually narrowed and acute, 1-nerved; achenes obovate, 2–2.5 mm. long; perianth-segments white, flat, filiform, much-elongate, to 2–2.5 cm. long in fruit; anthers 2.5–3 mm. long.——June–Aug. High moors; Hokkaido, Honshu (n. and centr. part); rather common.——Korea, Kuriles, Sakhalin, Siberia, and Europe.

3. Eriophorum scheuchzeri Hoppe var. **tenuifolium** Ohwi. *Scirpus leucocephalus* (Boecklr.) T. Koyama forma *tenuifolius* (Ohwi) T. Koyama——EZO-WATA-SUGE. Rhizomes elongate; culms 10–30 cm. long, rather soft, obscurely angled, 1- or 2-leaved at base; radical leaves at length elongate, terete, sulcate on upper surface, smooth, 0.7 mm. wide, abruptly obtuse; sheaths of cauline leaves tubular, slightly inflated in upper part, sometimes terminated by a short setaceous blade; spikelet solitary, terminal, obovoid in flower, 1–1.5 cm. long, nearly hemispherical in fruit; scales deltoid-lanceolate, gradually narrowed to tip, acute, dark gray, 1-nerved; perianth-segments white, filiform, much elongate, to 2–2.5 cm. long in fruit; achenes obovate, about 2 mm. long, mucronate; anthers 0.8–1 mm. long.——July–Aug. High moors; Hokkaido (Mount Daisetsu); rare.——The typical phase with larger parts, occurs in Europe to Siberia and N. America.

6. FIMBRISTYLIS Vahl   TEN-TSUKI ZOKU

Tufted annuals or perennials; culms usually leafless, often more or less bifacially compressed; leaves radical, fascicled, rarely bladeless; inflorescence usually umbelliform, sometimes capitate or consisting of a solitary spikelet, usually subtended by a leafy bract or bracts at base; spikelets few- to many-flowered; flowers bisexual; scales spirally arranged or 2-ranked; perianth none; stamens 1–3; style-base somewhat thickened, articulate with the ovary and deciduous when mature; stigmas 2–3; achenes obovate, biconvex or 3-angled, smooth to tuberculate, rarely verrucose.——Cosmopolitan, with about 150 species, in tropical and warm-temperate regions.

1A. Spikelets compressed; scales at least in young spikelets distinctly distichous; stigmas 3, styles slender.
  2A. Scales glabrous, lustrous; spikelets 1 or 2, ovoid, 4–6 mm. wide, pale. .... 1. *F. monostachya*
  2B. Scales puberulent, dull; spikelets many, lanceolate, 2–2.5 mm. wide, reddish brown. .... 2. *F. fusca*
1B. Spikelets not compressed; scales wholly spiral; stigmas 2 or 3.
  3A. Achenes narrowly oblong.
    4A. Achenes with clavate marginal tubercles; spikelets pale green; leaves filiform, 0.25–0.5 mm. wide. .... 3. *F. verrucifera*
    4B. Achenes without processes; spikelets somewhat tawny; leaves 2–2.5 mm. wide. .... 4. *F. stauntonii*
  3B. Achenes obovoid to broadly so.
    5A. Stigmas 3; achenes trigonous.
      6A. Culms not surrounded at base by a bladeless sheath above the tuft of long-bladed radical leaves; leaf-blades flattened dorsiventrally; spikelets lanceolate or oblong-lanceolate.
        7A. Ligule a dense fringe of short hairs; rather soft herbs with flattened culms.
          8A. Annual herb without rhizomes; scales of spikelets 1.5–2 mm. long; stamens 1 or 2; anthers 0.25 mm. long.
                                      5. *F. autumnalis*
          8B. Perennial herb with short rhizomes; scales of spikelets about 3 mm. long; stamens 3; anthers about 1.5 mm. long.
                                        6. *F. complanata*
        7B. Ligule absent; rather firm perennial herbs; culms not compressed.
          9A. Rhizomes creeping, covered by short scales; scales of spikelets 5–6 mm. long. .... 7. *F. pierotii*
          9B. Rhizomes not creeping, forming a dense tuft; scales of spikelets 1.5–2 mm. long. .... 9. *F. cymosa*
      6B. Culms surrounded at base by a bladeless leaf-sheath; leaf-blades laterally compressed, equitant; spikelets subglobose, terete, not angled. .... 10. *F. miliacea*
    5B. Stigmas 2; achenes lenticular.
      10A. Scales thinly membranous, ciliate, with a raised green excurrent midrib on back. .... 12. *F. kadzusana*
      10B. Scales chartaceous, not ciliate, the midrib not prominently raised.
        11A. Achenes smooth or nearly so.

12A.   Spikelets angled; scales more or less keeled.
     13A.   Leaves more or less silky hairy beneath; spikelets 6–10 mm. long, about 4 mm. in diameter. . . . . 8. *F. sericea*
     13B.   Leaves glabrous; spikelets 3–5(–6) mm. long, about 2 mm. in diameter. . . . . 9. *F. cymosa*
12B.   Spikelets terete; scales not keeled on back.
     14A.   Culms surrounded at base by a bladeless leaf-sheath; spikelets many to very many, small. . . . . 11. *F. diphylloides*
     14B.   Culms not surrounded at base by a bladeless leaf-sheath; spikelets 1 to few.
         15A.   Scales 1-nerved, minutely puberulent. . . . . 13. *F. ferruginea* var. *sieboldii*
         15B.   Scales faintly many-nerved, glabrous. . . . . 14. *F. subbispicata*
11B.   Achenes reticulate.
     16A.   Spikelets 2–7 mm. wide, terete; scales flat or nearly so on back, chartaceous.
         17A.   Plants hairy, at least partially. . . . . 16. *F. dichotoma*
         17B.   Plants glabrous. . . . . 15. *F. longispica*
     16B.   Spikelets about 1.5 mm. wide, angled; scales more or less keeled on back, membranous.
         18A.   Style-base with long hairs covering the upper part of the achene; achenes nearly smooth.
             19A.   Scales with a long recurved apical awn. . . . . 17. *F. squarrosa*
             19B.   Scales with a short straight apical awn. . . . . 18. *F. velata*
         18B.   Style-base without long hairs; achenes smooth. . . . . 19. *F. æstivalis*

1. Fimbristylis monostachya (L.) Hassk. *Cyperus monostachyos* L.; *Abildgaardia monostachya* (L.) Vahl——YARI-TEN-TSUKI. Rhizomes short; culms 15–40 cm. long; leaves shorter than the culm, 0.7–1 mm. wide, glabrous; spikelet terminal, solitary, very rarely 2, ovoid, more or less compressed, lustrous, 8–15 mm. long, 4–6 mm. wide, with 1 or 2 scaly bracts at base; scales 2-ranked or finally partially subspiral owing to the torsion of the rachilla, pale or grayish stramineous, somewhat coriaceous, broadly ovate, 4–6 mm. long, acute and mucronate, glabrous, keeled, indistinctly 3-nerved; achenes 2.5–3 mm. long, pale, broadly obovate, coarsely verrucose; anthers 1.5–2 mm. long.——Sept. Honshu (Sagami and Kii Prov.), Kyushu.——s. Korea, Formosa, China, India, Malaysia, Australia, and Africa.

2. Fimbristylis fusca (Nees) C. B. Clarke. *Abildgaardia fusca* Nees——ONOE-TEN-TSUKI. Culms slender, 20–40 cm. long; leaves shorter than the culms, 1–2 mm. wide, slightly setulose-scabrous, abruptly pointed at apex; inflorescence usually compound, with rather many loosely disposed nearly simple spikelets, the bracts setaceous, to 5 cm. long; spikelets lanceolate, 7–10 mm. long, 2–2.5 mm. thick, compressed, dull, dark brown, subacute, 3- to 8-flowered; scales few, 2-ranked, keeled, the lower 2 or 3 small, empty, the middle ones 4–5 mm. long, gradually narrowed to apex, minutely puberulent; achenes broadly obovoid, 1 mm. long, 3-angled, whitish, verruculose; anthers linear, about 1.5 mm. long.——Sept.–Oct. Shikoku, Kyushu; rare.——China, India, and Malaysia.

3. Fimbristylis verrucifera (Maxim.) Makino. *Isolepis verrucifera* Maxim.; *F. dipsacea* C. B. Clarke var. *verrucifera* (Maxim.) T. Koyama; *F. dipsacea* sensu auct. Japon., non C. B. Clarke——AO-TEN-TSUKI. Culms tufted, 5–15 cm. long, soft; leaves 3–5 cm. long, filiform, flaccid, 0.25–0.5 mm. wide; inflorescence umbelliform, nearly simple, bearing to 20 spikelets, the bracts 3–10, filiform, the lower 1 or 2 longer than the inflorescence; spikelets ovoid-globular to oblong, pale green, 3–6 mm. long, 2–2.5 mm. across, densely many-flowered; scales oblong, thin-membranous, about 1 mm. long, obtuse and awned; achenes narrowly oblong, 0.7 mm. long, short-stiped, with 4–6 clavate-globular tubercles along the margins, the reticulations oblong-hexagonal.——Wet places in lowlands; Honshu, Shikoku, Kyushu.——s. Korea, Manchuria, Amur, and Ussuri.

4. Fimbristylis stauntonii Deb. & Fr. var. **stauntonii.** HATAKE-TEN-TSUKI. Culms many, tufted, slender, 7–40 cm. long; leaves shorter than the culm, glabrous, 2–2.5 mm. wide,

short-ciliate around the mouth of leaf-sheaths; inflorescence erect, loosely umbelliform, simple or branched, 5–10 cm. long, with numerous spikelets, the bracts 2 or 3, somewhat leaflike; spikelets ovoid to oblong, 3–5 mm. long, 2–2.5 mm. across, obtuse, light brown; scales thin-membranous, broadly lanceolate, 1.5–2 mm. long, abruptly cuspidate; achenes narrow-oblong, obscurely 3-angled or inflated-biconvex, 0.7 mm. long, with transversely oblong reticulations; style and stigmas 1.5–2 mm. long, persistent, the stigmas spreading, rather conspicuous; stamens 1 or 2, the anthers 0.3 mm. long.——Aug.–Oct. Kyushu; rare.——Korea, Manchuria, and China.

Var. **tonensis** (Makino) Ohwi. *F. tonensis* Makino——TONE-TEN-TSUKI. Scales 2–2.5 mm. long; style and stigmas 2.5–3 mm. long, the stigmas persistent; achenes 0.8–1 mm. long.——Honshu (Kazusa, Shimosa, Yamato and Settsu); rare.

5. Fimbristylis autumnalis (L.) Roem. & Schult. *Scirpus autumnalis* L.; *F. complanata* var. *microcarya* sensu auct. Japon., non C. B. Clarke——HIME-HIRA-TEN-TSUKI, HIME-TEN-TSUKI, KUSA-TEN-TSUKI. Culms tufted, 10–30 cm. long, flaccid, rather flat, glabrous; leaves shorter than the culm, flaccid, flat, 1.5–2.5 mm. wide, abruptly acute, glabrous, the sheaths distichous, keeled, compressed, minutely ciliate along the dorsal margin of the mouth; inflorescence umbelliform, compound, with many spikelets, the bracts 2 or 3, linear, rather short; spikelets rather loosely arranged, lanceolate, angled, 7- to 16-flowered, 3–6 mm. long, 1.5 mm. wide, brownish red; scales membranous, narrowly ovate, 1.5–2 mm. long, acute, keeled; achenes 0.7 mm. long, whitish, with transversely oblong reticulations; anthers 0.2 mm. long.——July–Oct. Common in wet places in lowlands; Hokkaido, Honshu, Shikoku, Kyushu.——Korea, Ryukyus, and N. America.

6. Fimbristylis complanata (Retz.) Link. *Scirpus complanatus* Retz.; *F. complanata* var. *kraussiana* sensu auct. Japon., non C. B. Clarke——NO-TEN-TSUKI, HIRA-TEN-TSUKI. Rhizomes abbreviated; culms distinctly compressed to ancipital above, glabrous, 20–80 cm. long; leaves flat, 1.5–3 mm. wide, glabrous, abruptly acuminate, the sheath distichous, minutely ciliate along the dorsal margin of the mouth; inflorescence umbelliform, compound, 2 to 7(–10) cm. long, the bracts linear, shorter than the inflorescence; spikelets numerous, broadly lanceolate to narrowly ovate, 5–8 mm. long, 1.5–2 mm. wide, acute, angled; scales membranous, oblong, about 3 mm. long, acute, reddish brown, keeled; achenes about 1 mm. long, whitish, smooth or sparsely verrucose, the reticulations transversely oblong, not elevated; stamens 3, the anthers linear, about 1.5 mm. long.——July–Oct. Honshu, Shikoku, Kyushu. ——Korea, China, Ryukyus, Formosa, India, and Malaysia.

7. Fimbristylis pierotii Miq. NOHARA-TEN-TSUKI, BU-ZEN-TEN-TSUKI. Rhizomes rather hard, creeping, to 5 cm. long, covered with small deltoid scales; culms solitary, 20–60 cm. long, slender, glabrous; leaves several, basal, shorter than the culm, 1–2 mm. wide, glabrous, the mouth of sheath not hairy; inflorescence 2–4 cm. long, with 3–10 loosely disposed spikelets, the bracts 2 or 3, subulate or scaly, short; spikelets broadly lanceolate, 7–15 mm. long, 3–4 mm. wide, acute, brownish; scales narrowly ovate, 5–6 mm. long; achenes about 1.2 mm. long, whitish, with transversely oblong reticulations, the areoles slightly depressed in the center; stamens 3, the anthers about 2.5 mm. long, linear.——July–Oct. Grassy places in mountains; Honshu (Kinki, Chūgoku Distr.), Shikoku, Kyushu; rather rare.——Korea, Philippines, and India.

8. Fimbristylis sericea R. Br. *Scirpus sericeus* Poir.; *F. velutina* Franch.——BIRŌDO-TEN-TSUKI. Rhizomes creeping, covered with brown withered leaves; culms 10–30 cm. long, firm and thick, appressed-sericeous while very young; leaves flat, thick, 1.5–2 mm. wide, slightly shorter than the culm, often more or less revolute, the mouth of the sheath not hairy; inflorescence simple, umbelliform, with a few rather thick spreading rays, the bracts 1 or 2, short; spikelets 3–10, clustered at the apices of the rays, narrowly ovoid, 6–10 mm. long, about 4 mm. thick, grayish brown; scales broadly ovate, 3–4 mm. long, keeled, many-nerved; achenes 1.5 mm. long, broadly obovate, biconvex, dark brown, the reticulations minute and indistinct; stamens 2 or 3, the anthers 1.5 mm. long, linear-oblong.——Aug.–Oct. Sandy places near seashores; Honshu (Hitachi, Etchu and westw.), Shikoku, Kyushu.——Ryukyus, Formosa, China, Indochina, India, Malaysia, and Australia.

9. Fimbristylis cymosa R. Br. *F. spathacea* Roth; *F. wightiana* Nees; *Scirpus glomeratus* Retz., non L.; *F. glomerata* (Retz.) Nees——SHIO-KAZE-TEN-TSUKI, SHIBA-TEN-TSUKI. Rhizomes thick, short, densely covered with withered leaves; culms 15–40 cm. long, erect, firm, sulcate, many-leaved at base; leaves stiff, flat, 7–20 cm. long, 1.5–3 mm. wide, abruptly acuminate, recurved, the mouth of the sheath not hairy; inflorescence compound, umbelliform, the bracts 1–3, much shorter than the inflorescence; spikelets many to very many, short-peduncled or sessile, in small clusters, oblong, 3–5 mm. long, 2 mm. thick; scales broadly ovate, ferruginous, obtuse, 1.5–2 mm. long; achenes broadly obovate, biconvex, rarely 3-angled, about 1 mm. long, dark brown, smooth; stamens 1 or 2, the anthers 0.8 mm. long.——Aug.–Oct. Near seashores; Shikoku, Kyushu.——Ryukyus, Formosa, Malaysia, India, and Australia.

Var. **depauperata** (T. Koyama) T. Koyama. KUJŪKURI-TEN-TSUKI. Apparently annual; rhizomes wanting; leaves softer, fewer, 1 mm. wide; inflorescence with rather few loosely disposed spikelets.——Sept.–Oct. Near seashores; Honshu (Kadzusa Prov.).

10. Fimbristylis miliacea (L.) Vahl. *Scirpus miliaceus* L.; *F. littoralis* Gaudich.——HIDERI-KO. Annual or short-lived perennial; culms tufted, 10–60 cm. long, compressed-tetraquetrous, surrounded at base by a bladeless sheath; leaves laterally compressed, distichous, soft, 1.5–2.5 mm. wide, long-attenuate, the sheath keeled, strongly compressed; inflorescence compound to decompound, loose to rather dense, the bracts 2–4, short, setaceous; spikelets ovoid-globular to globular, 2.5–4 mm. long, 1.5–2 mm. across, obtuse or rounded, reddish brown; scales ovate, about 1.5 mm. long, obtuse, hyaline, somewhat cymbiform; achenes obovate, whitish, lustrous, 0.6 mm. long, obtusely 3-angled, with transversely oblong reticulations; stamens 1 or 2, the anthers 0.5 mm. long.——July–Oct. Common in wet places along rivers or in paddy fields; Honshu, Shikoku, Kyushu.——Pantropic.

11. Fimbristylis diphylloides Makino. *F. brevicollis* Kuekenth.; *F. globulosa* var. *torresiana* sensu auct. Japon., non C. B. Clarke; *F. campylophylla* Tuyama——KURO-TEN-TSUKI. Rhizomes very short or nearly absent; culms 10–50 cm. long, slightly compressed, surrounded below by a bladeless sheath; leaves flat, 1.5–3 mm. wide; inflorescence simple or subcompound; the bracts 4–6, erect, 1–2 cm. long; spikelets many, ovoid, 4–6 mm. long, 2.5–3 mm. thick, dark brown to reddish brown; scales membranous, narrowly ovate, about 2 mm. long, obtuse; achenes biconvex, lustrous, stramineous, obovate, 0.8 mm. long, with distinct, transversely oblong reticulations; stamens 1 or 2, the anthers linear, 0.5 mm. long.——Aug.–Oct. Honshu (Shimotsuke, Noto and westw., including Sado Isl.), Shikoku, Kyushu.——Korea, Ryukyus, and China.

12. Fimbristylis kadzusana Ohwi. ISSUN-TEN-TSUKI. Annual; culms 4–15 cm. long; leaves few, flat, aggregated at the base of culm, 2–6 cm. long, nearly 1 mm. wide, the sheaths brown; inflorescence simple, the bracts 1 or 2, scalelike; spikelets 1–3, oblong, chestnut-brown, 5–7 mm. long, dull, obtuse; scales thin, membranous, appressed, oblong, slightly ciliate, emarginate, the green midrib prolonged into a short upright awn; achenes biconvex, broadly obovate, about 1 mm. long, dark brown, more or less lustrous, nearly smooth; stamens 3, the anthers 0.5–0.8 mm. long.——Sept.–Oct. Honshu (Kadzusa); rare.

13. Fimbristylis ferruginea (L.) Vahl var. **sieboldii** (Miq.) Ohwi. *F. sieboldii* Miq.; *F. leiocarpa* Miq., non Maxim.——ISO-YAMA-TEN-TSUKI. Rhizomes hard, short, ascending; culms erect, 15–40 cm. long, firm, slightly thickened at base, few-leaved at base; lower leaves bladeless, the upper elongate but shorter than the culms, 1–1.5 mm. wide, stiff, glabrous; the basal sheaths coriaceous, dark brown; inflorescence simple, the bracts 1–3, linear, the lower one slightly longer or as long as the inflorescence; spikelets 1–5, narrowly ovoid or narrowly oblong, 7–13 mm. long, about 3 mm. thick, acute, terete, dull, dark brown or dark ferrugineous; scales very obtuse and short-mucronate, 1-nerved, whitish puberulent; achenes biconvex, dark brown, nearly smooth, 1–1.2 mm. long; stamens 3, the anthers linear, 0.8–1 mm. long.——Aug.–Oct. Near seashores; Honshu (Kadzusa, Noto Prov. and westw.), Shikoku, Kyushu.——Korea, Ryukyus.——The typical phase, SHIMA-TEN-TSUKI, with more robust parts and many spikelets, is pantropic.

14. Fimbristylis subbispicata Nees & Meyen var. **subbispicata**. *F. japonica* Sieb. & Zucc.; *F. gynophora* C. B. Clarke; *F. crassipes* Palla; *F. tristachya* var. *subbispicata* (Nees & Meyen) T. Koyama——YAMAI. Rhizomes very short or nearly absent; culms erect, slightly compressed, 10–60 cm. long; leaves rather firm, 0.7–1 mm. wide, abruptly obtuse, glabrous; inflorescence of a single spikelet, rarely to 3, the bract wanting or 1, short, rarely to 5 cm. long; spikelet oblong to ovoid, 8–25 mm. long, 4–7 mm. thick, terete, lustrous, yellowish stramineous, partially reddish brown; scales rather firm, not keeled, elliptic, 4–6 mm. long, obtuse and mucronate; achenes obovate-orbicular, biconvex, dark brown at maturity, nearly smooth, distinctly stiped, the body 1–1.2 mm. long; stamens 3, the anthers linear, 1.2–1.8 mm. long.——July–Oct. Rather common in wet places in mountains or in lowlands; Hokkaido, Honshu, Shikoku, Kyushu.——Korea, Manchuria, Ryukyus, Formosa, China, India, and Malaysia.

Var. **pacifica** (Ohwi) Ohwi. *F. pacifica* Ohwi; *F. tristachya* var. *pacifica* (Ohwi) T. Koyama——Iso-TEN-TSUKI. Plant very slender; spikelet 7–15 mm. long, 2.5–3 mm. wide, paler, stramineous; scales membranous, 3–4 mm. long; stamens 2; otherwise as in the typical phase.——Cliffs along seashores and in wet places in lowlands near the sea; Honshu (Hachijô Isl. in Izu), Shikoku, Kyushu; rare.——Ryukyus.

15. **Fimbristylis longispica** Steud. *F. buergeri* Miq.; *F. koreënsis* C. B. Clarke——NAGABO-TEN-TSUKI. Rhizomes short-creeping, thick; culms stout, 40–60 cm. long; leaves flat, 2–4 mm. long, abruptly acute, rather firm; inflorescence compound, with rather many spikelets, 4–6 cm. long, the bracts 2–5, much exceeding the inflorescence, to 30 cm. long; spikelets rather densely arranged, narrowly ovoid to narrowly oblong, 7–15 mm. long, 3–4 mm. thick, acute, terete, lustrous, yellowish brown; scales rather firm, 3.5 mm. long, broadly ovate, obtuse, not keeled; achenes obovate-orbicular, biconvex, tawny, lustrous, with distinct, elevated, rectangular reticulations; stamens 3, the anthers nearly 1 mm. long.——Aug.–Oct. Marine, wet places; Honshu, Shikoku, Kyushu.——Korea, Manchuria, and China.

16. **Fimbristylis dichotoma** (L.) Vahl var. **dichotoma**. *Scirpus dichotomus* L.; *S. annuus* All.; *S. diphyllus* Retz.; *F. diphylla* (Retz.) Vahl——TEN-TSUKI. Rhizomes very short or nearly absent; culms 15–50 cm. long, compressed, glabrous or hairy; leaves flat, 1.5–5 mm. wide, abruptly acute, hairy, rarely nearly glabrous, the sheath frequently hairy; inflorescence umbelliform, compound, usually somewhat hairy, the lowest bract linear, longer or shorter than the inflorescence; spikelets usually rather numerous, ovoid or narrowly so, 5–8 mm. long, 2.5–3 mm. thick, terete, acute, sessile or short-peduncled, fulvous to tawny, frequently somewhat reddish brown, lustrous; scales ovate-orbicular, 2–3 mm. long, obtuse and often short-mucronate, not keeled, glabrous; achenes broadly obovate, 0.8–1.2 mm. long, biconvex, light yellow, with elevated, transversely oblong reticulations; stamens 2 or 3, the anthers 0.5–0.8 mm. long.——Rather wet places in lowlands; Hokkaido, Honshu, Shikoku, Kyushu; in all warmer regions of the world. Very common and widely variable, the following forms noted from this area:——Forma **floribunda** (Miq.) Ohwi. *F. diphylla* var. *floribunda* Miq.——KUGU-TEN-TSUKI. Robust; spikelets numerous; culms slightly thickened at base; nearly glabrous throughout.——Forma **tomentosa** (Vahl) Ohwi. *F. tomentosa* Vahl; *F. diphylla* var. *tomentosa*

(Vahl) Benth.——KE-TEN-TSUKI. Leaf-sheaths densely tomentose, otherwise as in *F. floribunda*.——Forma **depauperata** (C. B. Clarke) Ohwi. *F. diphylla* var. *depauperata* C. B. Clarke——HOSOBA-TEN-TSUKI. Slender; leaves filiform; spikelets few, usually 1 or 2.

Var. **tashiroana** (Ohwi) Ohwi. *F. tashiroana* Ohwi——TSUKUSHI-TEN-TSUKI. Leaves 1–2 mm. wide, glabrous; spikelets narrow, 8–15 mm. long, the lateral ones on long peduncles 2–7 cm. long; scales oblong, 4 mm. long; achenes obovate-deltoid; anthers 1.2–1.8 mm. long.——Aug.–Oct. Near mountain hot springs; Kyushu.——Forma **cincta** (Ohwi) Ohwi. *F. tashiroana* var. *cincta* Ohwi; *? F. dichotoma* f. *tomentosa* × *F. dichotoma* var. *tashiroana*——KE-TSUKUSHI-TEN-TSUKI. Leaves hairy, 2–3 mm. wide.

17. **Fimbristylis squarrosa** Vahl. *Pogonostylis squarrosus* (Vahl) Bertol.——AZE-TEN-TSUKI. Culms densely tufted, slender, 10–20 cm. long, filiform; leaves filiform, with slightly involute margins, 0.2–0.8 mm. wide, hairy, acute, the sheaths pale, prominently hairy; inflorescence compound, 3–5 cm. long, the bracts 3–5, the lower 1 or 2 nearly as long as the inflorescence; spikelets numerous, densely many-flowered, lanceolate or broadly so, 4–10 mm. long, 1.5 mm. in diameter, acute, ferrugineous; scales thinly membranous, oblong, 1.5–2 mm. long, keeled, obtuse, with a recurved awn 0.5–1 mm. long; achenes light brown, biconvex, obovate, 0.8 mm. long, smooth; style with long, pendent hairs at base; stamen 1, the anther oblong, 0.2 mm. long.——Wet places in lowlands; Hokkaido, Honshu; common.——Korea, Manchuria, China, India, Africa, and s. Europe.

18. **Fimbristylis velata** R. Br. *F. squarrosa* var. *esquarrosa* Makino; *F. makinoana* Ohwi——ME-AZE-TEN-TSUKI. Allied to the preceding; culms 10–25 cm. long; leaves and leaf-sheaths hairy; spikelets lanceolate, 4–7 mm. long, 1–1.5 mm. wide, acute; scales thinly membranous, oblong, 1.5–2 mm. long, the awn straight, 1/10–1/5 as long as the body of the scale, smooth; achenes 0.5 mm. long, smooth, the style-base with long, pendent hairs; stamen 1, the anther linear, 0.2 mm. long.——Aug.–Oct. Wet places in lowlands; Honshu, Shikoku, Kyushu; rather common.——Korea, Manchuria, Ussuri, Malaysia, and Australia.

19. **Fimbristylis aestivalis** (Retz.) Vahl. *Scirpus aestivalis* Retz.; *F. leiocarpa* Maxim.; *F. tokyoensis* Makino——KO-AZE-TEN-TSUKI. Very closely resembling the preceding; culms 5–15 cm. long, slender; leaves and leaf-sheaths hairy; spikelets lanceolate, 3–7 mm. long, 1–1.5 mm. across, acute; scales oblong or ovate, to 1.5 mm. long, brownish, obtuse and mucronate; achenes obovate, 0.6 mm. long, lustrous, smooth, stramineous; style-base often puberulent but not long-hairy; stamen 1, the anther small, 0.2–0.3 mm. long.——Aug.–Oct. Honshu; rare.——Amur, Formosa, China, Indochina, India, Malaysia, and tropical Australia.

7. BULBOSTYLIS Kunth   HATAGAYA ZOKU

Annuals or perennials; culms slender to capillary; leaves filiform, aggregated at the base of the culm; inflorescence terminal, often umbelliform, simple to compound, sometimes capitate, usually with bracts at base; spikelets rather many-flowered; flowers bisexual; scales spirally arranged; perianth absent; stamens 1–3; style slender, the base thickened and bulbous, persistent; stigmas 3; achenes obovate, 3-angled.——In all tropical and warm-temperate regions, with about 90 species.

1A. Spikelets and scales ferrugineous; scales with a recurved awn; achenes smooth. .... 1. *B. barbata*
1B. Spikelets and scales chestnut-brown; scales entire; achenes undulate-corrugate. .... 2. *B. densa*

1. Bulbostylis barbata (Rottb.) Kunth. *Scirpus barbatus* Rottb.; *Isolepis barbata* (Rottb.) R. Br.; *Fimbristylis barbata* (Rottb.) Benth.——HATAGAYA. Tufted annual; culms numerous, erect, 5–30 cm. long, filiform, glabrous, longitudinally several-ribbed, leafy only at base; leaves much shorter than the culm, filiform, glabrous, narrower than the culm; sheaths 5–20 mm. long, ferrugineous, glabrous, with few long white hairs on the mouth; inflorescence a dense head of 2–15 sessile spikelets, 5–12 mm. across, with few filiform bracts at base longer than the inflorescence; spikelets lanceolate, angled, 3–8 mm. long, 2 mm. thick; scales ovate, about 2 mm. long, with a recurved short awn; achenes obovate-orbicular, pale, smooth, 0.6–0.7 mm. long; style-base minute, depressed.——Aug.–Oct. Abundant in waste grounds and cultivated fields; Honshu, Shikoku, Kyushu.——Korea, Ryukyus, Formosa, China, India, Malaysia, and Australia.

2. Bulbostylis densa (Wall.) Hand.-Mazz. var. **densa.** *Scirpus densus* Wall.; *B. trifida* Kunth; *B. capillaris* var. *trifida* (Kunth) C. B. Clarke; *B. capillaris* sensu auct. Japon., non Kunth——ITO-HANABI-TEN-TSUKI. Annual; culms many, tufted, filiform, 5–40 cm. long; leaves aggregated at the base of the culm, glabrous, much shorter than the culm, the sheaths glabrous, with few long white hairs on the mouth; inflorescence with few to rather many spikelets, and 1–5 slender smooth rays 2–5 cm. long, the bracts setaceous or scaly, short; spikelets oblong, 3–5 mm. long, angled, chestnut-brown, the lateral peduncled, the terminal sessile; scales ovate, about 2 mm. long, subacute, keeled; achenes pale, obovate-orbicular, 0.8 mm. long, 3-angled, inconspicuously undulate-corrugate and puncticulate; style-base bulbose.——Aug.–Oct. Waste grounds and cultivated fields; Hokkaido, Honshu, Shikoku, Kyushu.——Korea, Formosa, China, and India.

Var. **capitata** (Miq.) Ohwi. *Isolepis capillaris* var. *capitata* Miq.; *B. japonica* C. B. Clarke; *B. capillaris* var. *capitata* (Miq.) Makino——ITO-TEN-TSUKI, KURO-HATAGAYA. Spikelets or nearly all of them sessile, in a head.——Aug.–Oct. Honshu (centr. distr. and westw.), Shikoku, Kyushu.——Ryukyus.

8. ELEOCHARIS R. Br.   HARI-I ZOKU

Annuals or perennials, sometimes stoloniferous; culms tufted, naked; leaves basal, bladeless; spikelets solitary, many- to few-flowered, terminal, without subtending bracts; scales usually spirally imbricate; flowers bisexual; bristles 6 or fewer or sometimes absent; stamens 1–3; achenes 3-angled or lenticular, crowned by the thickened style-base; stigmas 2 or 3.——Cosmopolitan, with more than 100 species.

1A. Spikelet cylindrical, scarcely broader than the culm; scales herbaceous, appressed.
  2A. Culms sharply 3-angled, not septate; achenes reticulate with raised longitudinal and transverse lines. . . . . 1. *E. fistulosa*
  2B. Culms terete, septate; achenes not longitudinally lined, the reticulations 4- to 6-angled.
    3A. Scales broadly elliptical, rounded or truncate; base of tubercle (or style-base) indistinctly discoid, thus the tubercle continuous with the achene; connective of anther projecting and deltoid at the tip. . . . . 2. *E. dulcis*
    3B. Scales narrowly oblong, obtuse; base of tubercle distinctly discoid; connective of anther projecting and linear at the tip.
                                                               3. *E. kuroguwai*
1B. Spikelet conspicuously broader than the culm; scales usually membranous, not closely appressed.
  4A. Stigmas 3; achenes 3-angled.
    5A. Achene continuous, not constricted below the style base.
      6A. Body of achene 3 mm. long, tipped by a prominent disc; culms 25–50 cm. tall; spikelets about 10 mm. long. 4. *E. margaritacea*
      6B. Body of achene 1 mm. long, without disc at tip; culms slender, flaccid, 3–5 cm. tall; spikelets about 3 mm. long. . . 5. *E. parvula*
    5B. Achene distinctly constricted below the style-base.
      7A. Delicate plant; scales of spikelets few, 2-ranked; achenes with raised transverse lines and longitudinal ribs.
                                                            6. *E. acicularis* var. *longiseta*
      7B. Slender to coarse plants; scales of spikelets numerous, spirally ranked; achenes smooth, without raised transverse lines.
        8A. Culms relatively slender, often rhizomatous at base, angled; spikelets 1–2 cm. long; style-base large, spongy.
          9A. Bristles retrorsely spinose; scales more or less coriaceous; culms acutely 4-angled. . . . . 7. *E. tetraquetra*
          9B. Bristles plumose; scales soft, obtuse; culms 3–6 angled or nearly terete. . . . . 8. *E. wichurae*
        8B. Culms very slender, not sharply angled, usually densely tufted; spikelets less than 1 cm. long.
          10A. Style-base not more than ⅓ as wide as the achene. . . . . 9. *E. congesta*
          10B. Style-base as broad as the achene or nearly so. . . . . 10. *E. attenuata*
  4B. Stigmas 2; achenes lenticular.
    11A. Spikelets ovoid, 3–8 mm. long; culms slender; achenes lustrous; style-base spongy; tufted.
      12A. Achenes tawny when mature; style-base compressed, lamelliform; spikelets 4–8 mm. long, 3–4 mm. thick. . . . . . . 11. *E. ovata*
      12B. Achenes black, lustrous; style-base depressed, hardly compressed; spikelets 3–5 mm. long, 2–2.5 mm. thick.
                                                      12. *E. atropurpurea*
    11B. Spikelets lanceolate to short-cylindric, usually 1–3 cm. long; achenes lusterless, turgid-biconvex; style-base spongy; culms more or less stout, rhizomatous.
      13A. Style-base large, spongy, nearly as broad as the achene; only the lowest scale empty. . . . . 13. *E. kamtschatica*
      13B. Style-base small, much narrower than the achene; lower 2 scales empty.
        14A. Bristles 5–6; culms thick, soft; spikelets ferrugineous. . . . . 14. *E. mamillata* var. *cyclocarpa*
        14B. Bristles 0–4; culms firmer.
          15A. Bristles firm, straight, nearly 3 times as long as the achene, densely retrorse-scabrous; achenes 1–1.2 mm. long; scales ferrugineous. . . . . 15. *E. parvinux*
          15B. Bristles slender, 1.5 to 2 times as long as the achene, more or less flexuous, loosely retrorse-scabrous.
            16A. Culms with elevated ribs; scales sanguineous, very obtuse. . . . . 16. *E. valleculosa*
            16B. Culms almost smooth; scales dark purplish red, obtuse to slightly acute. . . . . 17. *E. intersita*

1. **Eleocharis fistulosa** Link ex Spreng. *Scirpus acutangulus* Roxb.; *E. acutangula* (Roxb.) Schult.; *S. fistulosus* Poir., non Forsk.——MISUMI-I.   Rhizomes sometimes elongate; culms soft, acutely 3-angled, 40–80 cm. tall, 2.5–4 mm. thick, smooth, the uppermost sheath 5–10 cm. long, partially reddish brown; spikelets 2–3 cm. long, 3–4 mm. thick, cylindric, rather acute, glaucous-green; scales appressed, broadly ovate, 4–5 mm. long, very obtuse; achenes broadly obovate, 1.5–2 mm. long, light yellow, lustrous, biconvex, the bristles 6, retrorsely scabrous, slightly exceeding the style base; stigmas 3. ——Shallow water on edge of ponds; Honshu (Kii Prov.), Kyushu; rare.——Ryukyus, Formosa, China, India, and Australia.

2. **Eleocharis dulcis** (Burm. f.) Trin. *Cyperus dulcis* Rumph.; *Andropogon dulce* Burm. f.; *Scirpus plantaginoides* Rottb.; *E. plantaginea* (Retz.) Roem. & Schult.; *S. plantagineus* Retz.——INU-KURO-GUWAI, SHIRO-GUWAI.   Rhizomes elongate, usually terminated by a tuber; culms terete, 40–80 cm. long, 2–5 mm. thick, glaucous-green, smooth, septate-nodose within, the sheaths 5–20 cm. long, frequently partially reddish; spikelets cylindric, 4 cm. long, 3–4 mm. thick; scales broadly elliptic, 5–6 mm. long; achenes obovate-orbicular, 2 mm. long, tawny, lustrous, smooth, the bristles with short spines at tip, the spinulae shorter toward apex, the style-base short-deltoid, with a strongly depressed inconspicuous basal disc.——July–Oct.   Shallow water; Honshu (s. part of Kinki Distr.), Kyushu; rare.——Ryukyus, Formosa, China, India, and Malaysia.

3. **Eleocharis kuroguwai** Ohwi. *Scirpus plantagineus* sensu Fr. & Sav. non Retz.; *E. plantaginea* sensu auct. Japon., pro parte——KURO-GUWAI.   Resembles the preceding; spikelets cylindric, 2–4 cm. long, 3–4 mm. thick, pale green; scales narrowly oblong, obtuse, 6–8 mm. long; achenes 2 mm. long, very abruptly contracted at apex and with a prominent apical discoid tubercle, the bristles spinulose at tip, with spinulae longer toward the apex; anther connective with a linear mucro at apex.——July–Oct.   Ponds and ditches; Honshu (Kantō, Hokuriku Distr. and westw.), Shikoku, Kyushu.——Korea.

4. **Eleocharis margaritacea** (Hult.) Miyabe & Kudo. *Scirpus margaritaceus* Hult.——SHIROMI-NO-HARI-I.   Rhizomes short; culms slightly tufted, 25–50 cm. long, sulcate, slender, the upper sheaths 3–10 cm. long, stramineous; spikelets narrowly ovoid to broadly lanceolate, 7–12 mm. long, 3–4 mm. thick, acute to acuminate, dark golden-brown, lustrous; scales narrowly ovate, 5–6 mm. long; achenes obovate, obtusely 3-angled, 3 mm. long, white, narrowed at base, puncticulate, with an annular apical disc, the style-base depressed-trigonous, the bristles 6, longer than the achene, retrorsely scabrous; stigmas 3.——June–July.   Moors; Hokkaido (Sarobetsu, Hamatombetsu), Honshu (Rikuchu Prov.), Yanagisawa; rare. ——s. Kuriles and Kamchatka.

5. **Eleocharis parvula** (Roem. & Schult.) Link. *Scirpus parvulus* Roem. & Schult.; *S. pollicaris* Delile——CHABO-I.   Rhizomes extremely slender, terminated by a small tuber; culms soft, slender, 3–5 cm. long, light green, the sheaths thin-membranous, soon withering, inconspicuous; spikelets ovoid to oblong, 3 mm. long, compressed, light green, few-flowered; scales ovate, hyaline, lustrous, 1.5 mm. long; achenes obovoid, about 1 mm. long, stramineous, smooth, with an indistinct apical annular disc, the bristles 4, slender, slightly longer or shorter than the achene, minutely retrorsely scabrous.——Shikoku and Kyushu.——Europe, Siberia, and N. Africa.

6. **Eleocharis acicularis** (L.) Roem. & Schult. var. **longiseta** Svenson. *E. acicularis* sensu auct. Japon., non Roem. & Schult.; *Scirpus yokoscensis* Fr. & Sav.; *E. svensonii* Zinserl.——MATSUBA-I.   Rhizomes nearly absent; culms filiform to capillary, 3–10 cm. long, sulcate, deep green, the sheaths 2–10 mm. long, often partially reddish; spikelets few-flowered, narrowly ovoid to broadly lanceolate, 2–4 mm. long, compressed, acute, pale green and usually partly sanguineous; scales narrowly ovate, hyaline; achenes narrowly obovate, 1 mm. long, raised-reticulate with several longitudinal ribs and very many transverse lines, the style base depressed, triangular, conical, the bristles 3–4, unequal, much longer than the achene.——June–Sept.   Common in wet places especially in paddy fields; Hokkaido, Honshu, Shikoku, Kyushu.——e. Siberia, Manchuria, Korea, China, Ryukyus, and Formosa.——The typical phase, with 1–3 bristles shorter than the achene, occurs in the Ryukyus, China, Korea, Manchuria, e. Siberia, and N. America.

7. **Eleocharis tetraquetra** Nees var. **tetraquetra**. *E. erythrochlamys* Miq.——MA-SHIKAKU-I.   Rhizomes short; culms 30–50 cm. long, firm, acutely 4-angled, 1–2 mm. thick; sheaths often somewhat reddish; spikelets narrowly ovoid to broadly lanceolate, 8–17 mm. long, 3–5 mm. thick, often slightly bent, densely many-flowered, ferrugineous; scales oblong, obtuse, dull, 3–4 mm. long; achenes obovate, 1.5–2 mm. long, compressed-trigonous, tawny, nearly smooth, the style base compressed, trigonous, spongy, rather large, the bristles 6, firm, erect, densely retrorsely spinulose, with spinulae 1 to 2 times as wide as the bristle.——June–Oct.   Honshu (Chūgoku), Shikoku, Kyushu.——Ryukyus, Formosa, China, India, Malaysia, and Australia.

Var. **tsurumachii** (Ohwi) Ohwi. *E. tsurumachii* Ohwi ——KADO-HARI-I.   Culms slightly softer, the scales obtuse at apex.——Honshu (Hitachi Prov.).

8. **Eleocharis wichurae** Boecklr. *Scirpus hakonensis* Fr. & Sav.; *E. tetraquetra* var. *wichurae* (Boecklr.) Makino——SHIKAKU-I.   Resembles the preceding; culms rather soft, 30–50 cm. long, typically 4-angled, or 3-, 5-, or 6-angled in forms (see below), sometimes rhizomatous; scales rather soft, elliptic or oblong, very obtuse, ferrugineous-sanguineous; bristles 6, as long as the style-base, with long, soft, plumose, whitish, spreading to slightly recurved hairs.——July–Oct.   Common in wet places in lowlands and mountains; Hokkaido, Honshu, Shikoku, Kyushu.——Korea, Manchuria, and Ussuri.—— Forma **petasata** (Maxim.) Hara. *Scirpus petasatus* Maxim. Culms 3-angled.——Forma **teres** Hara.   Culms nearly terete or 5- or 6-angled.

Eleocharis × **yezoensis** Hara. *E. tetraquetra* var. *yezoensis* Hara; *E. wichurae* var. *yezoensis* (Hara) Ohwi——HIME-SHIKAKU-I.   Alleged hybrid of *E. congesta* × *E. wichurae*. Culms short, slender; scales rounded, hyaline.——Hokkaido (Mount Apoi).

9. **Eleocharis congesta** D. Don var. **japonica** (Miq.) T. Koyama. *E. pellucida* Presl; *E. afflata* Steud.; *E. japonica* Miq.; *E. afflata* var. *japonica* (Miq.) C. B. Clarke; *Scirpus japonicus* (Miq.) Fr. & Sav.——HARI-I.   Culms 5–40 cm. long, often densely tufted, green, capillary or filiform, ribbed, 0.2–1 mm. wide, the sheaths somewhat reddish; spikelets lanceolate to narrowly ovoid, 3–8 mm. long, 1.5–2.5 mm. thick, sometimes proliferous at base; scales membranous, oblong to elliptic, 1.5–2.5 mm. long, obtuse, reddish; achenes obovoid, 0.7–0.8 mm. long, obtusely 3-angled, yellowish

green, lustrous, nearly smooth, the style base compressed, trigonous, 1/3–1/2 as broad as and 1/5–1/3 as long as the achene, the bristles 6, usually slightly longer than the achene, retrorsely scabrous.——June–Oct. Common in wet places; Hokkaido, Honshu, Shikoku, Kyushu.——Korea, Manchuria, China, Formosa, and Philippines.

Var. **congesta.** *E. subprolifera* Steud.; *E. pellucida* forma *attenuata* Ohwi, excl. syn.——Ō-HARI-I.   Plant more robust, achenes larger, 1–1.2 mm. long, with firm bristles more or less densely scabrous, 1.5 times as long as the achene.—— Honshu, Kyushu.——India and Indochina.

Var. **thermalis** (Hult.) T. Koyama. *E. pellucida* var. *thermalis* (Hult.) Hara; *Scirpus japonicus* var. *thermalis* Hult.; *E. japonica* var. *thermalis* (Hult.) Hara——Ezo-HARI-I. Culms short, spikelets very rarely proliferous, rather loosely flowered, scales somewhat keeled, dark sanguineous-fuscous; achenes olive-colored, 1.2 mm. long.——Hokkaido, Honshu, Shikoku, Kyushu; rather rare.——Kamchatka.

Var. **nipponica** (Makino) Ohwi. *E. nipponica* Makino ——YARI-HARI-I.   Spikelets linear, sharp-tipped, 7–17 mm. long, 1.5–2.5 mm. thick; scales more obtuse than in the typical phase; achenes 1–1.3 mm. long, the style base ovate-deltoid, ½–¾ as wide as the body of the achene.——Honshu, Kyushu; rare.

10. Eleocharis attenuata (Fr. & Sav.) Palla var. **leviseta** (Nakai) Hara. *E. leviseta* Nakai——CHŌSEN-HARI-I. Resembles the preceding species; culms 30–50 cm. long, densely tufted, ribbed, about 1 mm. thick; spikelets scarcely proliferous, ovoid to narrowly so, densely many-flowered, 4–12 mm. long, 2.5–4 mm. thick; scales membranous, elliptic or oblong, about 2 mm. long, rounded at apex, pale and reddish ferrugineous; achenes obovate, 1.2 mm. long, tawny, lustrous, smooth, obtusely 3-angled, the style base depressed, 2/3 to nearly as wide as the body of the achene, rather acute on the basal angles, the bristles 6, smooth, about as long as the body of the achene.——July–Oct. Wet places; Honshu (Shinano Prov.; Kutsukake); rare.——Korea.

Var. **attenuata.** *Scirpus attenuatus* Fr. & Sav.; *E. pellucida* forma *attenuata* (Fr. & Sav.) Ohwi, pro parte; *E. leviseta* var. *major* (Hara) Hara; *E. major* Hara——SEI-TAKA-HARI-I. Bristles retrorsely scabrous, usually slightly longer than the body of the achene.——Wet places; Honshu, Shikoku, Kyushu.——Ryukyus.

11. Eleocharis ovata (Roth) Roem. & Schult. *Scirpus ovatus* Roem. & Schult.; *S. soloniensis* var. *nipponica* Hara ——MARU-HO-HARI-I.   Culms slender, tufted, sulcate, 6–40 cm. long; sheaths light green, the mouth transversely truncate and short-mucronate; spikelets ovoid, densely many-flowered, obtuse, 4–8 mm. long, 3–4 mm. thick, ferrugineous; scales ovate, 2–2.5 mm. long, obtuse; achenes obovoid, pale, tawny, 1 mm. long, lustrous, smooth, turgid-lenticular, the style base compressed, deltoid, 1/2–2/3 as wide as the body of the achene, the bristles 6, twice as long as the achene, retrorsely scabrous, light ferrugineous.——July–Oct. Wet places; Hokkaido, Honshu (n. and centr. distr.); rather rare. ——Manchuria, Ussuri, Dahuria, India, and Europe.

12. Eleocharis atropurpurea (Retz.) Presl. *E. atropurpurea* var. *hashimotoi* Ohwi——KURO-MINO-HARI-I.   Delicate herb; culms tufted, nearly capillary, 5–15 cm. long, green, sulcate, the sheaths 5–15 mm. long, membranous, obliquely truncate at apex; spikelets ovoid to narrowly so, 3–5 mm. long, 2–2.5 mm. thick, reddish purple; scales membranous, elliptic,

obtuse, 1.5 mm. long; achenes deltoid-obovate, about 0.5 mm. long, biconvex, black, lustrous, smooth, cuneate at base, abruptly contracted at apex, the style-base minute, depressed, 1/5–1/4 as wide as the body of the achene, the bristles 4–5, delicate, white, longer than the body of the achene, slightly scabrous.——July–Oct. Rare; Honshu (Omi Prov.), Kyushu. ——Formosa, India, Africa, Europe, and N. America.

13. Eleocharis kamtschatica (C. A. Mey.) Komar. *Scirpus kamtschaticus* C. A. Mey.; *S. sachalinensis* Meinsh.; *E. savatieri* C. B. Clarke; *E. sachalinensis* (Meinsh.) Komar.; *Scirpus mitratus* Fr. & Sav.; *E. mitrata* (Fr. & Sav.) Makino; *E. uniglumis* sensu auct. Japon., non Schult.——HIME-HARI-I. Rhizomes sometimes elongate; culms slender, 20–50 cm. long, 1–1.5 mm. thick, terete, obsoletely striate, the sheaths 5–10 cm. long, somewhat reddish, transversely truncate at apex, sometimes mucronate; spikelets ovoid to lanceolate, 7–20 mm. long, 3–5 mm. thick, acute, dark sanguineous and purplish brown; only the lowest scale empty and clasping, the others narrowly ovate, 4–5 mm. long, erect; achenes tawny, turgid-biconvex, 1–1.5 mm. long, nearly smooth, dull, the style base large, spongy, ovate to deltoid, 1–2 mm. long, nearly as broad as the achene, the bristles 5, slender, slightly shorter to as long as the achene, retrorsely scabrous, or in forma **reducta** (Ohwi) Ohwi. KURO-HARI-I, the bristles rudimentary or lacking.—— July–Oct. Wet places especially near the seacoast; Hokkaido, Honshu, Kyushu; rare.——Korea, Ussuri, Sakhalin, Kuriles, Kamchatka, and northern N. America.

14. Eleocharis mamillata Lindb. f. var. **cyclocarpa** Kitag. *E. ussuriensis* Zinserl.; *E. palustris* sensu auct. Japon., saltem pro parte maj.; *E. mamillata* sensu auct. Japon., vix Lindb. f.——Ō-NUMA-HARI-I, NUMA-HARI-I.   Rhizomes elongate; culms soft, terete, 30–70 cm. long, 1.5–5 mm. thick, strongly compressed when dried, the sheaths hyaline, truncate at apex, sometimes reddish at base; spikelets lanceolate to ovoid, 1–3 cm. long, 3–6 mm. thick, rather obtuse at apex, ferrugineous, densely many-flowered; scales broadly lanceolate to narrowly ovate, 4–5 mm. long, slightly obtuse to acute; achenes yellow to tawny, turgid-biconvex, slightly lustrous, obovate-orbicular, 1.5–2 mm. long, nearly smooth, the style base depressed-deltoid, spongy, half as wide as the achene, the bristles 5–6, twice as long as the achene, retrorsely scabrous. ——July–Oct. Shallow ponds and wet places; Hokkaido, Honshu, Kyushu; rather common.——Korea, Manchuria, and Ussuri.   The typical phase is widely distributed in the northern part of the N. Hemisphere.

15. Eleocharis parvinux Ohwi. KO-TSUBU-NUMA-HARI-I. Resembles the preceding, but smaller; culms 30–60 cm. long, 1–2 mm. thick; spikelets ellipsoid to broadly lanceolate, 7–15 mm. long, 3–4 mm. thick, ferrugineous; scales many, membranous, erect, deltoid-lanceolate, 3–4 mm. long, acute; achenes obovate, inflated-biconvex, 1–1.2 mm. long, the style-base deltoid, half as broad as the achene or narrower, the bristles 4, pale, rather firm, nearly 3 times as long as the body of the achene, not flexuous, somewhat densely scabrous.——July–Oct. Wet places and shallow ponds in lowlands; Honshu (Kantō Distr.).

16. Eleocharis valleculosa Ohwi. SUJI-NUMA-HARI-I. Rhizomatous; culms 30–50 cm. long, rather firm, slender, distinctly several ribbed, 1–2 mm. thick, the sheaths 5–10 cm. long, hyaline, frequently somewhat reddish; spikelets narrow-ovate to lanceolate, 7–15 cm. long, 3–3.5 mm. thick; lower 2 scales empty and clasping, the others fertile, oblong, erect, 3–4 mm. long, reddish brown, obtuse; achenes 1–1.3 mm. long, turgid-biconvex, yellowish, the style-base depressed, spongy, 1/3 as broad as the achene, the bristles 4 in forma **setosa** Kitag., slender, ferrugineous, retrorsely scabrous, often flexuous, longer than the style base, sometimes none in forma **valleculosa.**——July–Oct.　Wet sandy places; Honshu (Kai, Ugo), Kyushu; rare.——Korea, Manchuria, and China.

17. Eleocharis intersita Zinserl. *E. palustris* sensu auct. Japon., pro parte, non Roem. & Schult.——KURO-NUMA-HARI-I, NUMA-HARI-I.　Rhizomatous; culms terete, 30–60 cm. tall,

1.5–3 mm. thick; spikelets narrowly to broadly ovate, rarely broadly lanceolate, 7–15 mm. long, 3–5 mm. wide; scales narrowly ovate, 3–4 mm. long, dark purplish red, subacute to obtuse; achenes tawny, obovate or broadly so, 1.2–1.7 mm. long, turgid-biconvex, the style base deltoid, 1/3 to 1/2 as wide as the achene, the bristles 4, slender, ferrugineous, retrorsely scabrous, longer than the style base, more or less flexuous.—— July–Oct.　Marshy places and shallow ponds in mountains; Hokkaido, Honshu (n. distr.).——Sakhalin, Kuriles, Manchuria, and e. Siberia.

9.　SCHOENUS L.　NOGUSA ZOKU

Culms solitary or tufted, few-leaved; radical leaves aggregated, slender, rather stiff, sheathing at base; spikelets narrow, 2- to 5-flowered, in a fascicle or panicle; flowers bisexual; scales 2-ranked, few, deciduous, the lower smaller and empty, the rachilla slightly flexuous, rather thick; bristles 6 to few or 0; stamens 3; achenes somewhat sunken in the rachilla, 3-angled, ovate, beakless; style deciduous from the achene, hardly thickened at base; stigmas 3.——Australia, s. Asia, and Africa.

1.　Schoenus apogon Roem & Schult. *S. albescens* (Fr. & Sav.) Matsum.; *Chaetospora albescens* Fr. & Sav.; *C. japonica* Fr. & Sav.——NOGUSA.　Culms tufted, many, slender, erect, 10–15 cm. long, smooth, 1- or 2-leaved; radical leaves erect, aggregated, about 0.5 mm. wide, the sheaths somewhat sanguineous, the mouth glabrous; corymbs 2 or 3, small, umbelliform or capitate, bearing 2–10 spikelets, the bracts seta-

ceous or leaflike, sheathing below; spikelets compressed, lanceolate, 4–6 mm. long, acute, stramineous, usually somewhat dark red-tinged, somewhat lustrous, with 5 or 6 scales; achenes about 1 mm. long, obovoid-globular, whitish, turgid, 3-angled, minutely reticulate, the bristles 6, nearly 2 mm. long, antrorsely scabrous, caducous.——Honshu, Shikoku, Kyushu; rather rare.——Ryukyus, Malaysia, and Australia.

10.　CARPHA Banks & Soland.　INU-NOGUSA ZOKU

Low perennials; leaves basal; inflorescences corymbose or nearly capitate, with numerous spikelets, the terminal large, the lateral smaller; bracts 1 or 2, leaflike; spikelets narrow, 3- to 6-scaled, 1- or 2-flowered; flowers bisexual; scales firmly membranous, 2-ranked, the lower 2 or 3 empty, the fertile scales larger; bristles 6, usually plumose, scabrous at tip; stamens 3; style thickened at base and continuous with the 3-angled achene; stigmas 3.——Mainly Australian.

1.　Carpha aristata Kuekenth. INU-NOGUSA.　Roots fibrous; culms slender, leafy, 40 cm. long; leaves about 6, shorter than the culm, soft, 1.5–2 mm. wide, the sheaths long, inflated; inflorescence divided in 2 parts, spikelike, oblong, 8–12 mm. long; spikelets rather numerous, oblong-lanceolate, 5

mm. long, 1-flowered, with 4 or 5 scales; scales stramineous, the fertile prominently awned at apex; bristles 6, slightly longer than the achene, antrorsely scabrous; achenes oblong, 3-angled, ferrugineous, minutely reticulate throughout.——Reported in Kyushu (Nagasaki) by Kuekenthal.

11.　RHYNCHOSPORA Vahl　MIKAZUKIGUSA ZOKU

Perennials; culms solitary, more or less leafy; leaves radical and also often cauline, linear; inflorescence simple or compound, the partial inflorescence 1 to several, terminal and lateral, capitate, corymbose, or fascicled; spikelets lanceolate, brown or whitish, few-scaled and few-flowered; scales 1-nerved, membranous, the lower empty; flowers bisexual, perianth usually of 6, sometimes rather many, smooth or scabrous bristles; achenes biconvex, beaked at apex with a thick persistent style-base; stigmas 2, sometimes almost undivided; stamens 2 or 3.——Fifty to sixty species, mostly in tropical and temperate regions of both hemispheres, a few in cooler zones.

1. **Rhynchospora rubra** (Lour.) Makino. *Schoenus ruber* Lour.; *R. wallichiana* Kunth——IGA-KUSA. Rather firm perennial; culms 20–40 cm. long, slender, obtusely angled, erect, smooth, rather firm, leafy only at base; leaves stiff, 1.5–3 mm. wide; inflorescence a solitary, terminal, globular head 15 mm. across, with numerous sessile spikelets; involucral bracts leaflike, spreading or reflexed; spikelets 6–7 mm. long, somewhat distichously 5–6 scaled, tawny-stramineous; achenes obovate, dark brown, 1.5–1.8 mm. long, smooth, short-setulose on upper margin, the beak small, conical, the style long, undivided; bristles 6, half as long as the achene, antrorsely scabrous.——Aug.–Oct. Wet sunny grassy places; Honshu (Kadzusa Prov. and westw.), Shikoku, Kyushu.——Ryukyus, Formosa, China, Indochina, India, and Malaysia.

2. **Rhynchospora malasica** C. B. Clarke. *R. nipponica* Makino——MIKURIGAYA. Rhizomes elongate, the sterile innovations elongate; culms 40–100 cm. long, stout; leaves on the upper 2/3 of the culm, flat, rather firm, deep green, 5/8 mm. wide; basal sheaths bladeless, rusty-brown; heads 5–8, in an interrupted spike, globose, 1.5 cm. across, with numerous sessile spikelets; lower bracts leaflike, sheathless; spikelets lanceolate-ovoid, lustrous, acuminate, 6–7 mm. long, 1-flowered, reddish stramineous; achenes broadly obovate, shining, 2 mm. long, smooth and often minutely wrinkled, the beak slender, the style slender, the stigmas 2 mm. long, 2-cleft; bristles 6, smooth, twice as long as the achene.——Sept.–Oct. Marshy places; Honshu (Tōkaidō, Kinki Distr.), Kyushu; rare.——Ryukyus, Formosa, and Malaysia.

3. **Rhynchospora chinensis** Nees & Mey. *R. glauca* var. *chinensis* (Nees & Mey.) C. B. Clarke; *R. japonica* Makino; *R. longisetigera* Hayata——INU-NO-HANA-HIGE. Culms slender, 30–60 cm. long, smooth, very loosely few-leaved; leaves 2–3.5 mm. wide; corymbs 3–5, erect, interrupted, the peduncles unequal; spikelets broadly lanceolate, 7–8 mm. long, with 4 or 5 scales, usually 2-flowered, dark brown; achenes broadly obovate, glabrous, 2–2.2 mm. long, with inconspicuous transverse wrinkles, the beak slightly shorter than the achene, deltoid, flattened, glabrous; bristles 6, slender, 4–5.5 mm. long, antrorsely scabrous.——July–Oct. Wet and swampy places; Honshu (centr. distr. and westw.), Shikoku, Kyushu.——Korea, Ryukyus, Formosa, China, India, and Malaysia.

4. **Rhynchospora fauriei** Franch. Ō-INU-NO-HANA-HIGE. Rhizomes abbreviated; culms 20–60 cm. long, slender, smooth, very loosely few-leaved; leaves 1.5–2.5 mm. wide; corymbs 2 or 3, the branches short; spikelets broadly lanceolate, 7–8 mm. long, dark reddish brown; achenes broadly obovate, glabrous, 2 mm. long, faintly transversely wrinkled, the beak slightly shorter than the achene, deltoid, glabrous; bristles 6, slender, 3–4 times as long as the achene, retrorsely scabrous.——July–Oct. Wet places; Hokkaido, Honshu (Mimasaka Prov. and eastw.), Kyushu.

5. **Rhynchospora brownii** Roem. & Schult. *R. glauca* sensu auct. Asiat., non Vahl——TORA-NO-HANA-HIGE. Culms slender, 60–80 cm. long, very loosely few-leaved; leaves chiefly radical, rather firm, 1.5–2.5 mm. wide; corymbs 2–3, the branches to 4 cm. long, nodding; spikelets narrowly ovate, dark brown, 3–4.5 mm. long, few-flowered; achenes 2 mm.

long, broadly ovate, glabrous, faintly transversely rugose, the beak about half as long as the achene, compressed, deltoid; bristles 6 or 7, reddish brown, slightly shorter than the achene, antrorsely scabrous.——Aug.–Oct. Wet places; Honshu (s. part of Kinki, Chūgoku Distr.), Shikoku, Kyushu.——Ryukyus, Formosa, and widely distributed in tropical and subtropical regions of the Old World.

6. **Rhynchospora yasudana** Makino. *R. franchetiana* C. B. Clarke, pro parte——MIYAMA-INU-NO-HANA-HIGE. Mostly tufted; culms slender, 15–30 cm. long; leaves 1–2 mm. wide; corymbs 4–6, narrow, short, erect, with a few spikelets; spikelets lanceolate, dark brown, 5–6 mm. long; achenes narrowly oblong, 2–2.5 mm. long, light tawny, nearly smooth, the beak nearly 2 mm. long, glabrous; bristles 6, slightly longer than the achenes, erect, slender, retrorsely scabrous.——High moors and wet slopes in high mountains; sw. Hokkaido, Honshu (Mount Hyonosen in Tajima Prov. and eastward).

7. **Rhynchospora fujiiana** Makino. *R. fauriei* var. *leviseta* C. B. Clarke; *R. franchetiana* C. B. Clarke, pro parte; *R. coreana* Palla——KO-INU-NO-HANA-HIGE. Tufted in small clumps, often forming a large colony; culms slender, 30–100 cm. long, few-leaved; radical leaves linear, 1–1.5 mm. wide; corymbs 4–5, interrupted, dense, with a few spikelets; spikelets lanceolate, 5–6 mm. long, dark reddish brown; achenes broadly obovate, 2 mm. long, faintly transversely wrinkled, the beak slightly shorter than the achene; bristles 6, slightly longer than the achene, smooth.——Wet places in lowlands; Hokkaido, Honshu, Shikoku, Kyushu; common. The following are based on variations of the bristles: Forma **scabriseta** (Makino) T. Koyama. *R. fujiiana* var. *scabriseta* Makino. Bristles antrorsely scabrous.——Forma **retrososcabra** ? (Takeda) T. Koyama. *R. fujiiana* var. *retrososcabra* ? Takeda. Bristles retrorsely scabrous.——Korea.

8. **Rhynchospora faberi** C. B. Clarke. *R. miyakeana* Makino; *R. umemurae* Makino; *R. hattoriana* Makino; *R. breviseta* Palla; *R. umemurae* var. *yakushimensis* Masam.; *R. yakushimensis* Masam.——ITO-INU-NO-HANA-HIGE, HIME-INU-NO-HANA-HIGE. Slender, loosely tufted herb; culms 10–40 cm. long, filiform; leaves 0.5–1 mm. wide, nearly filiform; corymbs 3–4, small, very distant, with 2–5 spikelets; spikelets narrowly ovate, dark brown; achenes broadly obovate, 1.5–2 mm. long, transversely wrinkled, dark reddish brown, the beak elongate-deltoid; bristles 6, slightly longer than the achene, retrorsely scabrous.——July–Oct. Wet places in lowlands; Hokkaido, Honshu, Shikoku, Kyushu; common.——Korea, China, and Ussuri.

9. **Rhynchospora alba** (L.) Vahl. *Schoenus albus* L.; *R. alba* var. *kiushiana* Makino——MIKAZUKIGUSA. Slender herbs; culms 10–60 cm. long, smooth; leaves filiform, involute, 0.5–1.5 mm. wide; corymbs 1–3, loosely arranged, small; spikelets lanceolate, whitish, 4–6 mm. long; achenes obovate, 2–2.5 mm. long, with faint transverse wrinkles, the beak half as long as the achene, compressed, glabrous; bristles 9–15, longer than the achene, retrorsely scabrous, with minute ascending ciliae near base.——July–Oct. Wet places and bogs; Hokkaido, Honshu, Kyushu (rare).——Europe, Asia, and e. N. America.

12. CLADIUM R. Br.　HITO-MOTO-SUSUKI ZOKU

Rather large perennials; culms usually sparsely leafy; radical leaves narrow, dorsiventrally compressed, usually thick and rigid, the ligule inconspicuous; inflorescence a compound panicle, the partial inflorescences corymbose, flat or convex at apex;

spikelets few-scaled, 1- to 7-flowered, the lower scales small and empty, the upper gradually enlarged, the uppermost empty or sterile; achenes 3-angled, coriaceous or spongy, drupe-like, glabrous, beakless; stigmas 3.——Few species, mainly tropical.

1. Cladium chinense Nees. *C. japonicum* Steud.; *C. mariscus* sensu auct. Japon., non R. Br.; *C. jamaicense* sensu auct. Japon., non Crantz——HITO-MOTO-SUSUKI, SHISHI-KIRI-GAYA. Culms stout, 1–2 m. long, with several nodes, obtusely 3-angled, smooth, sometimes with axillary tufts of leaves at nodes; leaves flat, with loosely incurved margin below, 8–10 mm. wide, very coarse and firm, coarsely serrulate-scabrous on margins, caudately acuminate in a long, subulate, somewhat 3-angled apex; partial inflorescences 5–7, corymbose, 4–8 cm.

across, with very numerous, dense spikelets; spikelets oblong when mature, brown, 3 mm. long; scales broadly ovate, obtuse, about the 7 lower ones small and empty, the 3 upper large and nearly equal; achenes broadly ovoid, 2.5 mm. long, tawny, truncate, excavated at base.——Aug.–Oct. Wet places near the seacoast; Honshu (Awa and Noto Prov. and westw.), Shikoku, Kyushu.——Ryukyus, Formosa, s. Korea, China, India, Malaysia, and Australia.

13. MACHAERINA Vahl ANPERA-I ZOKU

Moderate to large perennial herbs, often with long stolons; culms obscurely trigonous, leafy at base; leaves slender, terete or laterally flattened, rarely bladeless, long-sheathing; ligule almost wanting; inflorescence a slender compound panicle, partial panicles spicate or paniculate, not corymbose, with many spikelets; spikelets bisexual, with 2-ranked few scales, the lower scales small and empty, the upper (usually 1) perfect; achenes obscurely trigonous, coriaceous, glabrous, crowned by a conical or sometimes depressed usually hispidulous beak; stigmas 3; hypogynous bristles 3 or none.——About 70 species, chiefly in Oceania.

1. Machaerina nipponensis (Ohwi) Ohwi & T. Koyama. *M. rubiginosa* var. *nipponensis* T. Koyama; *Cladium nipponense* Ohwi; *C. glomeratum* sensu auct. Japon., non R. Br.; *Chapelliera glomerata* sensu Franch. & Sav., non Nees ——NEBIKIGUSA, ANPERA-I. Rhizomes elongate; culms 60–100 cm. long, smooth, nearly terete, 1-leaved near the middle; leaves radical, fascicled, smooth, linear, 3–4 mm. across, gradually narrowed to the obtuse apex, somewhat cinereous; cauline

leaves sheathlike, 10–15 cm. long, with a subulate apex; inflorescences 3–5, loosely disposed in slender compound panicles, bearing fascicles of 5–8 spikelets, the bracts nearly bladeless; spikelets 5–6 mm. long, red-brown, 6- to 7-flowered; achenes oblong, obtusely 3-angled, glabrous, with a densely hairy style-base.——June–Oct. Wet places; Honshu (Tōkaidō and westw.), Shikoku, Kyushu.——Ryukyus and (?) China.

14. SCLERIA Berg. SHINJUGAYA ZOKU

Usually perennial; culms stout to slender, 3-angled, leafy; leaves linear, sheathing at base, the ligule usually distinct; spikelets unisexual or androgynous, in several terminal and lateral paniculate partial inflorescences, few-scaled; pistillate spikelets 1-flowered; achenes globular, sometimes obscurely 3-angled, rugose, clathrate or smooth, usually lustrous, glabrous or puberulent; hypogynium (disc surrounding ovary at base) attached below the achene, 3-lobed or rarely entire, usually depressed; achene with attached hypogynium falling from the receptacle.——More than 100 species, in all tropical regions.

1A. Rhizomes stout, short-creeping; partial inflorescences large, with many spikelets; achenes nearly smooth.
  2A. Hypogynium orbicular or very obscurely 3-lobed; leaf-sheaths wingless or 3-winged. .... 1. *S. terrestris*
  2B. Hypogynium distinctly 3-lobed, the lobes ovate-deltoid, acute; leaf-sheaths always 3-winged. .... 2. *S. levis*
1B. Rhizomes absent; partial inflorescences smaller, 1–3 cm. long, with few spikelets; bracts nearly erect; achenes reticulate.
  3A. Hypogynium 3-lobed, the lobes acute.
    4A. Achenes with surfaces of the reticulation dull, the depressed pits lustrous; leaf-sheaths wingless. .... 3. *S. mikawana*
    4B. Achenes wholly lustrous; leaf-sheaths 3-winged. .... 4. *S. parvula*
  3B. Hypogynium 3-lobed, the lobes very obtuse; leaf-sheaths wingless. .... 5. *S. rugosa*

1. Scleria terrestris (L.) Fassett. *Zizania terrestris* L.; *S. elata* Thw.; *S. doederleiniana* Boecklr.; *S. luzonensis* Palla; *S. scrobiculata* sensu auct. Japon., non Nees & Mey.——Ō-SHIN-JUGAYA. Rather coarse perennial; culms 60–100 cm. long, acutely angled, retrorsely scabrous; leaves stiff, 5–12 mm. wide, gradually acuminate, scabrous, the sheath winged, or in depauperate specimens wingless; partial inflorescences 3–5, 4–8 cm. long, the branches spreading, hispid-scabrous; spikelets numerous, 3–5 mm. long; achenes globular, about 2.5 mm. across, white to gray, lustrous, loosely and minutely puberulent; hypogynium appressed, orbicular, scarcely lobed, rather thickened on margin.——July–Oct. (Kyushu (Yakushima). ——Ryukyus, Formosa, China, India, Malaysia, and Australia.

2. Scleria levis Retz. *S. hebecarpa* Nees; *S. zeylanica* Poir.; *S. japonica* Steud.; *S. pubescens* Steud.——SHINJUGAYA. Culms rather stiff, 60–80 cm. long, acutely angled, usually retrorsely scabrous on the angles; leaves rather firm, scabrous,

5–8 mm. wide, gradually acuminate, the sheaths broadly 3-winged; partial inflorescences usually 2, sometimes only 1, the terminal larger, 4–5 cm. long, with obliquely spreading branches; spikelets 3–4 mm. long; achenes globular, white, lustrous, sparsely puberulent, about 2.5 mm. across; hypogynium 3-lobed, the lobes ovate-deltoid, acute, flat.——July–Oct. Wet slopes; Honshu (Kii Prov.), Shikoku, Kyushu.——Bonins, Ryukyus, Formosa, China, India, Malaysia, and Australia.

3. Scleria mikawana Makino.——MIKAWA-SHINJUGAYA. Culms 30–50 cm. long, acutely angled, smooth; leaves 2.5–4 mm. wide, nearly smooth, glabrous, gradually narrowed and obtuse at apex, the sheaths wingless; partial inflorescences 2–3, erect, 1.5–3 cm. long, the branches smooth; spikelets 4–5 mm. long; achenes white, tawny at maturity, nearly globose, about 2 mm. across, glabrous, lustrous in the hollows, dull on the raised reticulations; hypogynium 3-lobed, the lobes flat, ovate, abruptly narrowed and acute, abruptly thin-margined.——July–Oct. Wet places in lowlands; Honshu (Kadzusa, Mikawa, Ōgami, and Tamba Prov.) Kyushu; rare.——Africa, India, and Malaysia.

4. **Scleria parvula** Steud. *S. fenestrata* Fr. & Sav.; *S. tessellata* sensu auct. Japon., non Willd.; *S. koreana* Palla——Ko-SHINJUGAYA. Slender, glabrescent perennial; culms 30–50 cm., smooth; leaves flat, 3–5 mm. wide, long-attenuate, the sheaths broadly winged on the angle; partial panicles 4–6, distant, erect, 1.5–3 cm. long, the branches smooth, suberect; spikelets 4–5 mm. long; achenes nearly globose, whitish with indistinct brown flecks, lustrous, 2 mm. across, glabrous or slightly puberulent, with elevated reticulations; hypogynium appressed, the lobes ovate-deltoid, abruptly contracted at apex, with a projection on the center.——July–Oct. Wet places; Honshu, Shikoku, Kyushu; rather common.——Korea, China, India to Africa.

5. **Scleria rugosa** R. Br. var. **rugosa.** *S. pubigera* Makino; *S. onoei* var. *pubigera* (Makino) Ohwi; *S. lateriflora* Boecklr.——KE-SHINJUGAYA. Rather soft perennial; culms 10–30 cm. long, slightly curved, smooth, densely pilose; leaves flat, 2–3 mm. wide, abruptly acute, pilose, the sheaths acutely angled, wingless, pilose; partial panicles 3–5, bearing rather few spikelets, 7–15 mm. long, on arcuate or declined peduncles; spikelets 2–4 mm. long; achenes whitish, lustrous, globose, 1.5 mm. across, glabrous, with coarse somewhat imperfect reticulations; hypogynium appressed, the lobes deltoid, very obtuse, slightly thickened on margin, flat, with a somewhat elevated longitudinal line in the center.——July.–Oct. Wet places; Honshu (Kadzusa Prov. and westw.), Shikoku, Kyushu; rare.——Ryukyus, Formosa, China, India, Ceylon, Australia, and Malaysia.

Var. **glabrescens** (Koidz.) Ohwi & T. Koyama. *S. onoei* Fr. & Sav.; *S. onoei* var. *glabrescens* Koidz.——MANEKI-SHINJUGAYA. Nearly glabrous throughout.—July–Oct. Wet places; Honshu, Shikoku, Kyushu; very rare.——s. Korea.

15. DIPLACRUM R. Br. KAGASHIRA ZOKU

Small green annuals; culms leafy; leaves short, linear, short-sheathing at base; spikelets minute, in a short-peduncled, axillary, densely capitate inflorescence; pistillate spikelets terminal, 2-scaled, 1-flowered, the scales all alike, few-nerved, 3-toothed, the style 3-fid, the hypogynium (disc subtending the achene) inconspicuous; staminate spikelets lateral, 1- or 2-flowered, about 3-scaled, the scales narrow, membranous; stamens 1 to 3; achenes globose, obtusely mucronate, lustrous, falling together with the scales.——About 2 species, in tropical regions of Africa, Asia, and Australia.

1. **Diplacrum caricinum** R. Br. *Scleria caricina* (R. Br.) Benth.; *S. onoei,* sensu auct. Japon., non Fr. & Sav.——KAGASHIRA, HIME-SHINJUGAYA. Glabrous annual; roots fibrous, purplish; culms simple or branching at base, 5 to 20 cm. long, several-leaved; leaves flat, short, broadly linear, 15–25 mm. long, 2–3 mm. wide, rather abruptly pointed at apex, the sheaths 3–10 mm. long; partial inflorescences several, capitate, axillary, short-peduncled, 3–5 mm. across, light green; pistillate spikelet 2–3 mm. long, short-peduncled, the scales oblong, 5- to 8-nerved, 3-toothed at apex, the central tooth longer than the others; achenes scarcely 1 mm. long, globose, indistinctly 3-ribbed, coarsely reticulate, more or less puberulent on the net.——July–Oct. Wet places; Honshu (Kadzusa Prov. and westw.), Shikoku, Kyushu; rare——Ryukyus, Formosa to India, and Australia.

16. KOBRESIA Willd. HIGE-HARI-SUGE ZOKU

Perennials; culms 3-angled; leaves slender; spikelets in a terminal simple or branched spike, 1- to several-flowered, the terminal spikelet staminate, the lateral androgynous or pistillate; flowers unisexual, the perianth wanting, the staminate consisting of 3 stamens and subtending bract (scale); pistillate flowers of a single pistil surrounded by a prophyllum (perigynium) and bract (scale); perigynium open on one side at least above the middle; style single, hardly thickened at base; stigmas 2–3; achenes biconvex or 3-angled.——About 20 species, Europe, Asia, and N. America.

1. **Kobresia bellardii** (All.) Degl. *K. myosuroides* (Vill.) Fiori & Paoletti; *Carex myosuroides* Vill.; *C. bellardii* All.; *K. scirpina* Willd.; *Elyna bellardii* (All.) C. Koch——HIGE-HARI-SUGE. Densely tufted glabrous perennial; culms slender, erect, 10–25 cm. long, obtusely angled, smooth; basal sheaths broad, lustrous, dark chestnut-brown; leaves as long as the culm, filiform, sulcate; spike 1.5–3 cm. long, linear, with few loose spikelets; terminal spikelet staminate, several-flowered, the others androgynous, 2-flowered; scales thin-membranous, broadly ovate, obtuse, chestnut-brown, lustrous; perigynia ovate, 4 mm. long, free on margin, chestnut-brown; achenes obovate-oblong, 3 mm. long, brown, lustrous, obsoletely 3-angled; stigmas 3.——July–Aug. Dry alpine slopes; Hokkaido, Honshu (centr. distr.); rare.——Kuriles, n. Korea, Siberia to Europe, and N. America.

17. CAREX L. SUGE ZOKU

Perennials (rarely annuals or biennials), commonly with naked culms; leaves usually radical or subradical, narrow, elongate; inflorescence often spicate; spikes, unisexual, androgynous (staminate above and pistillate below), or gynecandrous (pistillate above and staminate below), the flowers without a perianth, each subtended by a scale; scales spirally imbricate; staminate flowers of 2 or 3 stamens; pistillate flower of a single ovary enclosed in a perigynium; style 1, protruding from the beak of the perigynium, 2- or 3-fid; achene enclosed in a perigynium, lenticular or trigonous.——Cosmopolitan, with about 2,000 species, especially abundant in temperate and cooler regions, and in mountains of the Tropics.

1A. Spike solitary and terminal.

  2A. Style 2-fid; achenes lenticular; spikes dioecious or sometimes androgynous.

    3A. Leaves 1–1.5 mm. wide; perigynia distinctly longer than the scales, obliquely spreading when mature, with a rather long beak.

                                 2. *C. kabanovii*

    3B. Leaves 0.5–0.7 mm. wide; perigynia as long as the scales, short-beaked, horizontally spreading when mature. .... 3. *C. gynocrates*

  2B. Style 3-fid; achenes triangular.

    4A. Scales of the pistillate flowers dark brown, lustrous; perigynia brownish or partially so; species of alpine regions.

      5A. Leaves narrow, rather rigid; culms obtusely angled, nearly smooth; perigynia reflexed at maturity. .... 75. *C. pyrenaica*

      5B. Leaves flat, rather flaccid; culms sharply angled, scabrous; perigynia nearly erect at maturity. .... 77. *C. hakkodensis*

    4B. Scales of the pistillate flowers pale to light yellowish brown or sometimes sanguineous.

      6A. Scales of the pistillate flowers sanguineous; perigynia pubescent; species of woods.

        7A. Spikelets unisexual (plants dioecious); leaves about 1 mm. wide. .... 125. *C. grallatoria*

        7B. Spikelets androgynous; leaves 1–2.5 mm. wide. .... 126. *C. heteroclita*

      6B. Scales of the pistillate flowers pale or light yellowish brown; perigynia glabrous.

        8A. Perigynia thin-membranous, erect or spreading.

          9A. Spikes loosely flowered; perigynia 5–6 mm. long. .... 155. *C. rhizopoda*

          9B. Spikes densely flowered; perigynia 1.5–4 mm. long.

            10A. Spikes small, 3–6 mm. long, rather few-flowered; perigynia not inflated; achene compressed-trigonous.

              11A. Leaves involute, less than 1 mm. wide; culms obtusely angled, almost smooth; perigynia 2 mm. long, oval, nerveless or nearly so. .... 149. *C. hakonensis*

              11B. Leaves flat, 1.5–3 mm. wide; culms 3-angled, scabrous in the upper part; perigynia 2.5–3 mm. long, narrowly ovoid, finely nerved. .... 150. *C. onoei*

            10B. Spikes terete or ellipsoidal, 5–20 mm. long, rather many-flowered; perigynia somewhat inflated, loosely enclosing the 3-angled and not compressed achene.

              12A. Culms densely scabrous on the angles; spikelets pale, 1 or 2 of the scales in lower part often rather elongate and somewhat bractlike. .... 151. *C. fulta*

              12B. Culms smooth or slightly scabrous only on the angles in upper part.

                13A. Culms sharply angled, smooth, flaccid; perigynia narrowly ovoid, 3–3.5 mm. long, gradually attenuate above into a rather long beak. .... 152. *C. uda*

                13B. Culms somewhat obtusely angled; perigynia oval or ovoid, abruptly rather short beaked.

                  14A. Spikes 1–2 cm. long; perigynia 1.5–1.8 mm. long. .... 153. *C. biwensis*

                14B. Spikes 5–10 mm. long; perigynia 2.5–4 mm. long. .... 154. *C. capillacea*

        8B. Perigynia subcoriaceous, reflexed when mature, 6–6.5 mm. long, narrowly lanceolate; sphagnum bogs. .... 188. *C. pauciflora*

1B. Spikes 2 to numerous.

  15A. Spikes usually bisexual, sessile, without prophylla at base; bracts not sheathing; style 2-fid, rarely 3-fid.

    16A. Spikes androgynous, rarely unisexual and plants dioecious.

      17A. Style 2-fid; perigynia less than 7 mm. long.

        18A. Rhizomes long-creeping.

          19A. Perigynia wingless.

            20A. Perigynia planoconvex, crescent-shaped in transverse section, margins acute; along sandy coasts or in meadows.

                                  1. *C. arenicola*

            20B. Perigynia biconvex, inflated, oblong in transverse section, margins obtuse; moist places. .... 4. *C. disperma*

          19B. Perigynia narrowly winged.

            21A. Perigynia slightly pubescent at least when young; pistillate scales yellowish brown or pale; forest species.

                                    5. *C. pallida*

            21B. Perigynia glabrous; pistillate scales ferrugineous or reddish brown.

              22A. Stoloniferous; wet areas. .... 6. *C. pseudocuraica*

              22B. Rhizomatous; grassy river banks. .... 7. *C. lithophila*

        18B. Rhizomes short; culms tufted.

          23A. Perigynia membranous, with acute or winged margins, pale to pale brown on both sides.

            24A. Upper half of perigynia prominently winged; bracts foliaceous, elongate; meadows. .... 8. *C. neurocarpa*

            24B. Perigynia not winged or narrowly winged from base to apex.

              25A. Perigynia many-nerved, with several tubercles scattered on the back. .... 9. *C. paxii*

              25B. Perigynia smooth on back.

                26A. Longitudinal outer band of leaf-sheaths thinly membranous, transversely wrinkled, projected and longer than the mouth of sheath at apex.

                  27A. Scales brownish ferrugineous; perigynia slightly scabrous on margins near the tip; culms slender, rigid; grassy places. .... 10. *C. laevissima*

                  27B. Scales pale or tawny; perigynia distinctly scabrous-margined; culms rather stout but soft; swampy areas.

                                  13. *C. stipata*

                26B. Outer band of leaf-sheaths not wrinkled, with truncate apex not longer than the mouth of sheath; wet places in mountainous regions.

                28A. Culms scabrous on the angles; perigynia lanceolate-ovate, gradually attenuate-beaked. .... 11. *C. albata*

                28B. Culms glabrous; perigynia ovate or oval, rather short-beaked. .... 12. *C. nubigena* var. *franchetiana*

         23B. Perigynia coriaceous, lustrous, dark chestnut brown, biconvex, with obtuse margins; swamps. .... 14. *C. diandra*

      17B. Style 3-fid; perigynia nearly 10 mm. long; species of dune sands.

        29A. Culms obtusely angled, smooth or slightly scabrous on one angle; pistillate scales herbaceous, sulfur-colored, central portion broad, many-nerved, the midrib projecting as a broad awnlike mucro; perigynia straight, longer than or as long as the scales. .... 15. *C. kobomugi*

29B. Culms acutely angled, scabrous on the angles; pistillate scales membranous, chestnut-brown, the central portion green, 3-nerved, the midrib excurrent as a needlelike, smooth awn; perigynia strongly divergent, usually longer than the scales.
16. *C. macrocephala*

16B. Spikes gynecandrous.
  30A. Stigmas 3; achenes moderately compressed; along roadsides or in fields. .... 17. *C. gibba*
  30B. Stigmas 2.
    31A. Perigynia not white-puncticulate, gradually attenuate-beaked.
      32A. Perigynia reflexed when mature, with wingless but sharp margins; swamps.
        33A. Perigynia deltoid-ovate, slightly convex on back, scabrous on the margins near apex. .... 18. *C. echinata*
        33B. Perigynia lanceolate-ovate, strongly convex on back, nearly smooth on the margins. .... 19. *C. omiana*
      32B. Perigynia erect at maturity, somewhat winged on the margins; wet meadows or forests.
        34A. Spikes densely aggregated in a terminal hemispherical head; annual or biennial herb with fibrous roots; perigynia narrowly lanceolate, long-stipitate, 7–10 mm. long. .... 20. *C. cyperoides*
        34B. Spikes somewhat loosely disposed; perennials with short rhizomes; perigynia broadly lanceolate to broadly ovate, sessile, 3–3.5 mm. long.
          35A. Spikes loosely arranged; bracts conspicuous.
            36A. Lower spikes with a very long leafy bract.
              37A. Perigynia oval, much compressed, with a rather broad beak. .... 21. *C. planata*
              37B. Perigynia broadly lanceolate to ovate, with a very narrow beak.
                38A. Leaf-blades 2–4 mm. wide; spikelets 8–10, narrowly oblong, 8–15 mm. long; perigynia 4–4.5 mm. long.
22. *C. rochebrunii*
                38B. Leaf-blades 1–2 mm. wide; spikelets 4–7, ovoid-globose, 4–6 mm. long; perigynia about 3 mm. long.
23. *C. remotiuscula*
            36B. Lower 1 or 2 spikes with a short setaceous bract. .... 24. *C. deweyana* var. *senanensis*
          35B. Spikes approximate, usually without elongate bracts. .... 25. *C. maackii*
    31B. Perigynia densely white-puncticulate, not winged on the margins, abruptly beaked or beakless.
      39A. Pistillate scales chestnut-brown, reddish brown, or yellowish brown, or if pale, the perigynia distinctly beaked.
        40A. Pistillate scales chestnut-brown to yellowish brown; perigynia usually smooth.
          41A. Pistillate scales chestnut-brown; perigynia at maturity longer than the scales; culms 20–30 cm. long; grassy slopes in alpine regions. .... 26. *C. lachenalii*
          41B. Pistillate scales reddish brown to yellowish brown; swampy areas.
            42A. Culms 20–40 cm. long; leaves glaucous-green and minutely papillose; pistillate scales ovate-orbicular, ferrugineous or nearly so; perigynia grayish. .... 27. *C. mackenziei*
            42B. Culms (20–)40–80 cm. long, slender; pistillate scales brown, ovate; perigynia yellowish green or somewhat brownish.
              43A. Culms loosely tufted; leaf-blades 3–4 mm. wide, nearly smooth; perigynia elliptic, longer than scales, the scales brown, nearly obtuse. .... 28. *C. traiziscana*
              43B. Culms densely tufted; leaf-blades 2.5–3 mm. wide, scabrous; perigynia ovate, more or less thickened, as long as or slightly shorter than the chestnut-brown scales. .... 29. *C. nemurensis*
        40B. Pistillate scales pale or light yellow; perigynia slightly scabrous on the margins at apex; swamps or moist places in alpine regions.
          44A. Plants glaucous-green; leaf-blades 1.5–4 mm. wide; perigynia glaucous-green, becoming brownish at maturity.
30. *C. curta*
          44B. Plants bright green; leaves 1–2 mm. wide; perigynia light green at first, becoming brownish at maturity.
31. *C. brunnescens*
      39B. Pistillate scales pale; perigynia almost beakless, smooth; moist places.
        45A. Culms not slender; leaf-blades 1.5–3 mm. wide; perigynia 3.5–4 mm. long. .... 32. *C. pseudololiacea*
        45B. Culms slender; leaves 1–2 mm. wide; perigynia 2.5–3.5 mm. long.
          46A. Leaves green, blades flat, 1–2 mm. wide; head 2–3 cm. long, the lower spikes distant; perigynia about twice as long as the scales, with elevated thick nerves. .... 33. *C. loliacea*
          46B. Leaves glaucous-green, slightly folded, blades 1–1.5 mm. wide; head densely capitate, 5–8 mm. long, composed of 2 or 3 closely approximate spikes; perigynia as long as the scales, with weak nerves. .... 34. *C. tenuiflora*
15B. Spikes bisexual or unisexual, peduncled or sometimes sessile with prophylla at base, bracts sheathing at base or not; style 3-fid, rarely 2-fid.
  47A. Bracts not sheathing at base; style 2-cleft; achenes compressed.
    48A. Perigynia smooth.
      49A. Stigmas extremely slender and long, persistent on the ripe achene; mountain streams. .... 35. *C. sadoensis*
      49B. Stigmas distinctly shorter than the perigynia, falling off immediately after anthesis.
        50A. Culms aphyllopodic (i.e., the outer sheaths of culm-base bladeless), the outer sheaths cinnamon-colored, membranous, not keeled dorsally, nor ventrally reticulate; pistillate scales pale or light yellowish.
          51A. Culms 40–80 cm. long, stout; pistillate scales ovate, acute. .... 36. *C. shimidzensis*
          51B. Culms 20–50 cm. long, slender; pistillate scales obovate, emarginate, short mucronate-tipped. .... 37. *C. incisa*
        50B. Leaf-sheaths of culm-base somewhat reddish purple, more often brown and coriaceous, keeled dorsally, fibrously reticulate ventrally.
          52A. Rhizomes elongate.
            53A. Leaves with incurved margins.
              54A. Leaves strongly flaccid; culms acutely angled; perigynia faintly nerved; swampy places and paddy fields.
38. *C. thunbergii*
              54B. Leaves slightly flaccid; culms obtusely angled; perigynia nerveless; costal swamps. .... 39. *C. subspathacea*

53B.   Leaves with recurved or reflexed margins.
  55A.   Perigynia densely puncticulate, nerveless; sheaths of culm-base at least somewhat reddish purple; usually in costal swamps.
    56A.   Pistillate spikes pendulous on long capillary peduncles. .... 41. *C. lyngbyei*
    56B.   Pistillate spikes sessile or short-peduncled, erect. .... 40. *C. ramenskii*
  55B.   Perigynia smooth or nearly so.
    57A.   Leaf-sheaths of culm-base completely bladeless, reddish purple; perigynia with slender nerves.
      58A.   Perigynia ovate-oval, abruptly short-beaked; alpine or subalpine regions. .... 42. *C. aphyllopus*
      58B.   Perigynia narrowly ovate, gradually attenuate-beaked; mountainous areas of Chugoku District. 43. *C. impura*
    57B.   Leaf-sheaths of the culm-base, at least some of them, bladed, light to dark brown, rarely somewhat reddish purple; perigynia nerveless.
      59A.   Leaf-blades 2–3 mm. wide; perigynia nearly beakless, with entire mouth; wet places in lowlands.
                    44. *C. cinerascens*
      59B.   Leaf-blades 3–5 mm. wide; perigynia short-beaked, the beak minutely bidentate; along streams and valleys in mountainous regions. .... 45. *C. heterolepis*
52B.   Rhizomes very short; culms tufted.
  60A.   Leaf-sheaths of the culm-base not fibrously reticulate, membranous; species of alpine regions in Hokkaido.
    61A.   Leaves flaccid; terminal spike usually gynecandrous; perigynia weakly nerved. .... 46. *C. eleusinoides*
    61B.   Leaves rigid; terminal spike usually staminate; perigynia nerveless. .... 47. *C. bigelowii*
  60B.   Leaf-sheaths of the culm-base splitting ventrally into reticulate fibers, subcoriaceous dorsally.
    62A.   Perigynia not puncticulate, smooth or papillose.
      63A.   Perigynia papillose, if smooth spikelets 8–10 mm. wide when mature; species of damp meadows.
        64A.   Pistillate spikes short-cylindric, 10–14 mm. wide; perigynia distinctly inflated, loosely inclosing the achene.
                    49. *C. maximowiczii*
        64B.   Pistillate spikes cylindric, 5–7 mm. wide; perigynia not inflated and tightly inclosing the achene.
          65A.   Terminal spikes gynecandrous; sheaths of the culm-base dark brown.
            66A.   Pistillate scales acute or subacute, mucronate; perigynia 4- or 5-nerved on both sides. .. 50. *C. subcernua*
            66B.   Pistillate scales emarginate and short-awned at apex; perigynia nerveless. ...... 51. *C. dimorpholepis*
          65B.   Terminal spikes staminate, rarely androgynous or bearing the pistillate flowers in middle part; lateral spikelets pistillate or rarely androgynous; sheaths of the culm-base brown or yellowish brown.
                    52. *C. phacota*
      63B.   Perigynia smooth or somewhat sparsely resinous-puncticulate.
        67A.   Sheaths of the culm-base brownish, somewhat reddish; perigynia vivid green when dried; bordering streams and mountain ravines. .... 53. *C. otaruensis*
        67B.   Sheaths of the culm-base dark brown to chestnut brown.
          68A.   Perigynia much-inflated, brownish when mature; achene loosely inclosed in the perigynium; spikelets erect.
                    54. *C. aequialta*
          68B.   Perigynia tightly inclosing the achene or only slightly turgid, light green or yellowish when dried; spikelets nodding or somewhat pendulous; species of damp forests or on rocks.
            69A.   Culms smooth; leaves nearly smooth; perigynia few-nerved and slightly turgid. ...... 55. *C. flabellata*
            69B.   Culms and leaves scabrous; perigynia nearly nerveless and not turgid. .... 56. *C. kiotensis*
    62B.   Perigynia densely puncticulate; species of sphagnum-moors.
      70A.   Perigynia prominently few-nerved, 3.5–4.5 mm. long, grayish green. .... 57. *C. middendorffii*
      70B.   Perigynia nerveless, 2–2.5 mm. long.
        71A.   Sheaths of the culm-base dark brown; lowest bract leaflike; pistillate scales broadly lanceolate; perigynia turgid, loosely inclosing the achene. .... 58. *C. schmidtii*
        71B.   Sheaths of the culm-base dark reddish; lowest bract setaceous or scalelike; pistillate scales oblong; perigynia elliptic and tightly inclosing the achene. .... 59. *C. caespitosa*
48B.   Perigynia scabrous on the margins or at least so on beak; sheaths of the culm-base bladeless.
  72A.   Sheaths of the culm-base dark brown, splitting ventrally into reticulate fibers; stream sides in mountains. .. 48. *C. forficula*
  72B.   Sheaths of the culm-base reddish purple or sanguineous.
    73A.   Perigynia oval, much compressed, thin-membranous, abruptly short beaked; sandy places in subalpine regions.
                    60. *C. angustisquama*
    73B.   Perigynia ovate to lanceolate, gradually attenuate-beaked.
      74A.   Perigynia narrowly ovate, the long beak bifurcate into 2 long recurved lobes; stigmas very long, persistent; achenes tightly inclosed in the perigynium; rocky areas in alpine regions. .... 61. *C. doenitzii*
      74B.   Perigynia lanceolate, with a sharply bidentate beak; stigmas not very long, deciduous; achenes small.
        75A.   Pistillate spikes oblong-cylindric; perigynia ovate to broadly elliptic, about 5 mm. long, short-stipitate.
                    62. *C. scitaeformis*
        75B.   Pistillate spikes ovoid to globose, large; perigynia linear-lanceolate, 12–15 mm. long, with a long, sparsely hairy stipe; fields. .... 63. *C. podogyna*
47B.   Lowest bract sheathing at base, or, if bracts not sheathing, then achenes trigonous with a 3-cleft style.
  76A.   Tips of perigynia membranous, bidentate or entire.
    77A.   Lowest bract not sheathing at base, leaflike to scalelike; style 3-fid.
      78A.   Perigynia hairy.
        79A.   Perigynia nerveless; ligule obsolete.
          80A.   Perigynia compressed-trigonous, rounded and not angled dorsally, gradually tapering into a beak; leaves sparsely hairy; staminate scales minutely ciliate; pine forests. .... 128. *C. mira*
          80B.   Perigynia obtusely, sometimes indistinctly trigonous, not compressed, abruptly beaked; leaves glabrous.

81A. Culms rigid, erect, 5–15 cm. long; leaves stiff; pistillate scales ciliolate; perigynia 2 mm. long; gravelly places in alpine regions. .... 129. *C. melanocarpa*
81B. Culms weak, reclining, often longer, to 50 cm. long; scales not ciliate.
  82A. Scales purple-brown.
    82AA. Perigynia obscurely angled, globose-obovoid, sparsely hispidulous, often nearly glabrous when mature; alpine meadows. .... 130. *C. vanheurckii*
    82BB. Perigynia obtusely trigonous, narrowly ovate to elliptic, rarely (in var. *lanceata*) narrowly lanceolate, pubescent; subalpine woods. .... 131. *C. oxyandra*
  82B. Scales tawny or pale. .... 131a. *C. chinoi*
79B. Perigynia conspicuously nerved; ligule prolonged, scarious, dorsally 2-lobed; leaves pilose; forest hillsides.
127. *C. gifuensis*
78B. Perigynia glabrous or scabrous, sometimes densely papillose.
  83A. Perigynia grayish, densely papillose, compressed-trigonous; scales cupreus or dark reddish purple; swamps.
    84A. Pistillate spikes pendulous on long peduncles.
      85A. Perigynia 2.5–3 mm. long, almost beakless; pistillate scales long-acuminate; pistillate spikes 2 or 3.
133. *C. paupercula*
      85B. Perigynia 3.5–4 mm. long, short-beaked; pistillate scales sharply mucronate; pistillate spikes 1 or 2. . 132. *C. limosa*
    84B. Pistillate spikes erect, sessile.
      86A. Plant densely tufted; sheaths of the culm-base chestnut brown; pistillate scales obtuse. .... 73. *C. meyeriana*
      86B. Plant long-rhizomatous; sheaths of the culm-base reddish purple; pistillate scales abruptly short-awned.
74. *C. buxbaumii*
  83B. Perigynia not grayish, smooth or scabrous.
    87A. Perigynia compressed or compressed-trigonous, loosely inclosing the achene; spikelets or scales usually dark brown to dark reddish purple; pistillate spikes 1–3 cm. long.
      88A. Pistillate scales 1–1.5 mm. long, obtuse; perigynia slightly to scarcely compressed.
        89A. Pistillate spikes 6–8 mm. long, elliptic to oblong; perigynia orange-yellow, firm-membranous; alpine regions.
64. *C. lehmannii*
        89B. Pistillate spikes 1–2(–4) cm. long, short-cylindric; perigynia light green, thin-membranous; banks along mountain streams.
          90A. Culms scabrous above; perigynia longer than the scales, ovate to ovate-lanceolate, with a short recurved beak; stream-banks in high mountains. .... 66. *C. augustinowiczii*
          90B. Culms smooth; perigynia 2 or 3 times as long as the scales, lanceolate, long-beaked; along streams in low mountains. .... 67. *C. curvicollis*
      88B. Pistillate scales 2–2.5 mm. long, acute to acuminate; perigynia distinctly compressed.
        91A. Terminal spike gynecandrous, if entirely staminate, the perigynia strongly nerved.
          92A. Perigynia thin-membranous, nerveless; species of alpine meadows.
            93A. Sheaths of the culm-base cinnamon colored; spikes with staminate flowers at base; perigynia thin-membranous. .... 68. *C. mertensii* var. *urostachys*
            93B. Sheaths of the culm-base purplish red; basal spikes pistillate; perigynia thick-membranous.
69. *C. atrata* var. *japonalpina*
          92B. Perigynia somewhat coriaceous, strongly nerved; sandy seashores. .... 70. *C. gmelinii*
        91B. Terminal spike entirely staminate.
          94A. Perigynia smooth on the margins. .... 71. *C. flavocuspis*
          94B. Perigynia setulose-scabrous on margins. .... 72. *C. scita*
    87B. Perigynia not compressed; pistillate spikes 1–1.5 cm. long.
      95A. Bracts scalelike; lateral spikes sessile, spreading at right angles; inflorescence a dense panicle 3–8 cm. long; sandy places. .... 166. *C. satsumensis*
      95B. Bracts leaflike; lateral spikes nearly erect or on erect peduncles; inflorescence a loose raceme.
        96A. Perigynia brownish or blackish when dried; species of wet places.
          97A. Perigynia nerveless or few-nerved, obtusely angled, beak often reflexed.
            98A. Pistillate scales long-awned at apex, pale; perigynia scabrous. .... 183. *C. nemostachys*
            98B. Pistillate scales awnless, sometimes cuspidate, often purple-brown on both sides; perigynia smooth.
184. *C. dispalata*
          97B. Perigynia distinctly nerved, inflated, loosely inclosing the achene, glabrous.
            99A. Pistillate scales short-awned; leaves whitish beneath; pistillate spikes very densely many-flowered.
179. *C. olivacea* var. *angustior*
            99B. Pistillate scales awnless; leaves pale green beneath; pistillate spikes rather densely many-flowered.
              100A. Perigynia 10–12 mm. long, oblong-ovoid. .... 177. *C. idzuroei*
              100B. Perigynia about 6 mm. long, ovoid-oval. .... 178. *C. hymenodon*
        96B. Perigynia somewhat lustrous, green when dried, smooth; forests.
          101A. Culm-base dark purplish; terminal spike gynecandrous, clavate; perigynia ascending. .... 65. *C. peiktusanii*
          101B. Culm-base stramineous; terminal spikes nearly always staminate, linear; perigynia spreading to slightly recurved.
            102A. Leaves whitish beneath; perigynia light green, beak erect to somewhat recurved.
              103A. Pistillate spikes oblong to short-cylindric, 1–3 cm. long; leaves 2–4 mm. wide, slightly whitish beneath.
                104A. Pistillate spikes peduncled; perigynia gradually long-beaked; stigmas slightly longer than the perigynium, slender, persistent. .... 161. *C. japonica*

104B. Pistillate spikes usually sessile; perigynia strongly inflated, abruptly short-beaked; stigmas short, deciduous after anthesis. .... 162. *C. aphanolepis*

103B. Pistillate spikes cylindric, 3–7 cm. long; leaves 5–10 mm. wide, distinctly whitish beneath.

163. *C. doniana*

102B. Leaves thin, fresh green, somewhat paler beneath; perigynia fresh green, widely spreading or somewhat reflexed, beak erect.

105A. Spikes contiguous or approximate, the lowest one peduncled, the lowest bract leaflike, nearly erect; pistillate scales slightly yellowish when immature. .... 164. *C. planiculmis*

105B. Spikes contiguous to overlapping, sessile, the lowest bract reflexed; pistillate scales pale.

165. *C. mollicula*

77B. Lowest bract sheathing at base; style 2- or 3-fid.

106A. Style 2-fid; achenes compressed-lenticular.

107A. Leaves 8–12 mm. wide; spikes light green, unisexual; perigynia glabrous; beak of achenes with an annulate base; flowering in spring; seacoasts. .... 114. *C. matsumurae*

107B. Leaves 1–4 mm. wide; spikes reddish brown, mostly androgynous; perigynia scabrous; achenes without any appendix at apex; flowering in late summer.

108A. Stigmas much-elongate, 6–8 mm. long, much longer than the perigynia, persistent; pistillate scales narrowly oblong.

173. *C. teinogyna*

108B. Stigmas shorter than the perigynia, deciduous; pistillate scales oblong to ovate.

109A. Pistillate part of spikelets densely flowered; perigynia strongly nerved, scabrous on both sides.

110A. Leaves yellowish green to bright green; spikes narrowly cylindric, 2–3 mm. thick, stramineous; perigynia small, 2.5–2.7 mm. long; open dry woods near the seacoast. .... 168. *C. brunnea*

110B. Leaves very stiff, yellowish to dark green; spikes cylindric to oblong, 3.5–4 mm. wide; perigynia 3–3.5 mm. long; woods of inland hills. .... 169. *C. lenta*

109B. Pistillate part of spikes somewhat loosely flowered; perigynia faintly nerved, scabrous on the margins only.

111A. Terminal spike staminate, the others androgynous. .... 170. *C. autumnalis*

111B. Spikes all androgynous.

112A. Rather stout herb to 1 m. high; leaves 3–4 mm. wide; perigynia about 1.8 mm. wide, with a very short, rather flat stipe. .... 171. *C. nachiana*

112B. Slender herb to 40 cm. high; leaves 1–3 mm. wide; perigynia less than 1.5 mm. wide, with an elongate stipe.

172. *C. sacrosancta*

106B. Style 3-fid; achenes trigonous to compressed-trigonous.

113A. Achenes with an annular or beaklike appendage at apex, not compressed; perigynia membranous, light green, obtusely 3-angled.

114A. Perigynia small, 2–5(–6) mm. long; achenes crowned by a sessile annular appendage.

115A. Perigynia lageniform; annular appendage at apex of achenes sessile, continuous with the achene-body, 0.7 mm. wide, concave at apex, the short straight style from the depressed center; forests or grassy places.

116A. Pistillate scales obtuse, mucronate; perigynia ovoid-fusiform, 3–3.5 mm. long, shallowly constricted above the middle. .... 83. *C. formosensis*

116B. Pistillate scales not mucronate; perigynia more plump, broader, scarcely constricted above the middle.

84. *C. genkaiensis*

115B. Perigynia ovoid, ellipsoidal, or rhomboidal; appendage of achenes discoid, about ⅓ as wide as the body.

117A. Plants small, low, usually the pistillate spikes densely flowered, oblong or narrowly so, the lowest bract usually short-sheathing at base.

118A. Staminate scales ferrugineous, tawny or reddish brown, awnless.

119A. Basal sheaths brownish or pale, nearly entire; perigynia pubescent or glabrous.

120A. Pistillate spikes wholly radical, except the uppermost, hidden in the tuft of leaves. .... 85. *C. pudica*

120B. Pistillate spikes wholly cauline, except the lowest one.

121A. Basal sheaths yellowish brown, lustrous; leaves about 2 mm. wide; staminate spike filiform, inconspicuous. .... 86. *C. mitrata*

121B. Basal sheaths brownish to pale; leaves 2.5–4 mm. wide; staminate spike lanceolate, conspicuous.

122A. Plant tufted; innovations short, ascending, extravaginal; perigynia sparsely puberulent or more often nearly glabrous, nerveless, the stipe much thickened, white when mature.

87. *C. subumbellata*

122B. Plant with stolons; perigynia glabrous to sparsely pubescent, the stipe not thickened.

123A. Plant sometimes stoloniferous, sometimes the perigynia glabrous. .... 100. *C. sachalinensis*

123B. Plant without stolons; perigynia pubescent.

124A. Pistillate scales pale tawny; perigynia pubescent, strongly nerved; beak conspicuous.

88. *C. nervata*

124B. Pistillate scales castaneous; perigynia sparsely puberulent, faintly nerved; beak very short.

89. *C. caryophyllea* var. *microtricha*

119B. Basal sheath splitting into brown fibers; perigynia hispidulous, rarely glabrate. ...... 90. *C. sabynensis*

118B. Staminate spike pale green; staminate scales entire, acuminate or awned at apex.

125A. Perigynia pubescent; pistillate scales short-awned. .... 91. *C. breviculmis*

125B. Perigynia glabrous or nearly so.

126A. Plant green when dry; lower 1 or 2 spikes always radical; perigynia oblong-fusiform. .... 92. *C. jacens*

126B. Plant blackish when dried; spikes all cauline; perigynia broadly ovoid, obtusely angled, transversely wrinkled near the middle. .... 93. *C. rugata*

117B.   Plants relatively tall and stout, usually the lowest bract long-sheathing at base; pistillate spikes cylindric to short-cylindric, rather loosely to densely flowered.

  127A.   Staminate scales acuminate to obtuse, sometimes gradually pointed.

    128A.   Leaves 0.5–3(–4) mm. wide, not coriaceous.

      129A.   Leaves glabrous.

        130A.   Basal sheaths castaneous to brown perigynia; pilose.

          131A.   Plants densely tufted.

            132A.   Pistillate spikes oblong-cylindric, rather densely flowered, the scales brown. .. 94. *C. stenostachys*

            132B.   Pistillate spikes terete, rather loosely flowered, the scales pale.

              133A.   Basal sheaths dark brown; perigynia broadly obovoid. .... 95. *C. polyschoena*

              133B.   Basal sheaths light brown; perigynia oblong-ovoid. .... 96. *C. clivorum*

          131B.   Plants with slender long-creeping rhizomes; basal sheaths brownish; pistillate spikes cylindric, rather loosely flowered. .... 97. *C. pisiformis*

        130B.   Basal sheaths pale, when castaneous the perigynia glabrous or pubescent.

          134A.   Leaves mostly flat, 1.5–4 mm. wide.

            135A.   Plants densely tufted.

              136A.   Perigynia glabrous. .... 98. *C. tenuinervis*

              136B.   Perigynia pubescent.

                137A.   Perigynia rhomboid or ovoid-fusiform, about 3 mm. long, twice as long as the scales; pistillate scales entire or mucronate at apex; staminate scales connate ventrally, infundibuliform. .... 104. *C. tristachya*

                137B.   Perigynia obovate-elliptic, 2 mm. long, nearly as long as the scales; pistillate scales short-awned; staminate scales not as above. .... 99. *C. tashiroana*

            135B.   Plants with creeping rhizomes; perigynia usually glabrous. .... 100. *C. sachalinensis*

          134B.   Leaves often involute, less than 1 mm. wide; perigynia glabrous. .... 101. *C. fernaldiana*

      129B.   Leaves, sheaths, or bracts hairy.

        138A.   Blades and sheaths hairy; perigynia short-beaked, hairy. .... 102. *C. duvaliana*

        138B.   Blades and sheaths glabrous; beak of perigynia nearly as long as the body, glabrous.

                                                      103. *C. mayebarana*

    128B.   Leaves 3–10 mm. wide, rather thick and stiff.

      139A.   Leaves smooth or nearly so; perigynia ovoid-trigonous or lenticular, thick, with strong elevated nerves.

                                                   115. *C. hachijoensis*

      139B.   Leaves scabrous; perigynia oblong, obtusely 3-angled, thin, faintly nerved.

        140A.   Basal sheaths splitting into dark brown fibers.

          140AA.   Perigynia pubescent, 3.5–4 mm. long; culms 30–50 cm. long. .... 105. *C. daisenensis*

          140BB.   Perigynia glabrous, 3–3.5 mm. long; culms less than 20 cm. long; leaves abruptly acute.

                                      110a. *C. omurae*

        140B.   Basal sheaths entire or nearly so.

           141A.   Basal sheaths purplish brown or reddish fuscous; perigynia oblong to obovate-oblong.

                                  106. *C. dolichostachya*

           141B.   Basal sheaths castaneous; perigynia obovoid. .... 107. *C. atroviridis*

  127B.   Staminate scales emarginate and short-awned to abruptly pointed; perigynia spreading or ascending.

    142A.   Perigynia pubescent, with a short beak, toothed or nearly entire on the mouth.

      143A.   Scales light green.

        144A.   Pistillate spikes fastigiate, solitary, simple; perigynia many-ribbed. .... 108. *C. tsushimensis*

        144B.   Pistillate spikes not fastigiate, binate or ternary; perigynia faintly nerved. .... 109. *C. sociata*

      143B.   Scales dark brown, yellow-brown or dark reddish purple.

        145A.   Perigynia hispidulous, with an erect beak. .... 110. *C. oshimensis*

        145B.   Perigynia sparsely puberulent or nearly glabrous, often with a recurved beak. ...... 111. *C. conica*

    142B.   Perigynia glabrous, often with a scabrous beak, somewhat rigidly 2-toothed at apex.

      146A.   Leaves very stiff, many-veined on upper surface. .... 112. *C. morrowii*

      146B.   Leaves stiff to rather soft, strongly 2-ribbed on upper surface. .... 113. *C. foliosissima*

114B.   Perigynia rather large, 5–8 mm. long, usually long-beaked; appendage of achene beaklike or annulate, rarely nearly wanting.

  147A.   Achenes constricted, annulate at apex; pistillate scales mostly pale.

    148A.   Leaves hairy. .... 116. *C. laticeps*

    148B.   Leaves glabrous. .... 117. *C. insaniae*

  147B.   Achenes not constricted, not annulate; pistillate scales reddish brown to yellowish brown.

    149A.   Leaves stiff, 5–10 mm. wide; spikes 3–6, the lateral ones androgynous, short-cylindric, many-flowered.

                                               118. *C. boottiana*

    149B.   Leaves rather flaccid, 2–3 mm. wide; spikes usually 2, the lateral one entirely pistillate, ovoid to globose, few-flowered. .... 119. *C. longerostrata*

113B.   Achenes without any appendage.

  150A.   Perigynia appressed-trigonous, usually thinly membranous; scales dark brown or yellowish brown.

    151A.   Spikes nodding to pendulous; perigynia strongly lustrous, nearly glabrous; species of gravelly places, alpine.

      152A.   Plant loosely tufted; leaves nearly flat, 2–3 mm. wide; spikes sparsely flowered, the pistillate long-peduncled, dark reddish brown, 2–3 cm. long, the terminal one staminate. .... 78. *C. stenantha*

      152B.   Plant densely tufted; leaves folded, scarcely flattened, 1–2.5 mm. wide; spikes densely flowered, the pistillate oblong-clavate, 1–2 cm. long, dark brown, the terminal one gynecandrous or staminate. .. 76. *C. siroumensis*

151B. Culms and spikes erect or nearly so; perigynia less lustrous, hairy or slightly so; species of mountains.
   153A. Lateral spikes androgynous, solitary or ternary, 1.5–3 cm. long; forests. .... 82. *C. reinii*
   153B. Lateral spikes pistillate, single; rocky places.
     154A. Terminal staminate spike 4–10 cm. long, narrowly linear. .... 79. *C. makinoensis*
     154B. Terminal staminate spike 0.5–3 cm. long, linear-clavate.
       154AA. Perigynia glabrous except the faintly scabrous beak. .... 79a. *C. phaeodon*
       154BB. Perigynia more or less pubescent or hispidulous.
         155A. Culms nearly smooth; sheaths of bract tight around the culm; staminate scales golden-brown; perigynia sparsely appressed-hispidulous. .... 80. *C. chrysolepis*
         155B. Culms scabrous above; sheaths of bract loose around the culm; staminate scales castaneous; perigynia pubescent. .... 81. *C. blepharicarpa*
150B. Perigynia scarcely compressed, if distinctly so then with tawny pale scales.
  156A. Sipkes 2–10, in a raceme or spikelike inflorescence.
   157A. Leaves flat, linear-lanceolate, 1–3 cm. wide; lateral spikes always androgynous; species of forests.
    158A. Spikes short-cylindric; pistillate flowers very loose, stigmas long, slender.
     159A. Leaves scabrous on the margins; perigynia glabrous. .... 144. *C. siderosticta*
     159B. Leaves long-ciliate on the margins; perigynia pilose. .... 145. *C. ciliatomarginata*
    158B. Spikes globose-ovoid; pistillate flowers dense; stigmas short, thickened. .... 146. *C. pachygyna*
   157B. Leaves linear to filiform.
    160A. Perigynia hairy, with a thick stipe at base.
     161A. Culms loosely tufted; perigynia acutely trigonous, nerveless or faintly nerved, with an erect beak; pistillate scales truncate, obtuse to rounded, sometimes mucronate or short-awned.
      162A. Culms, leaves, and scales pilose; perigynia 2-ribbed; dry rocky places. .... 120. *C. lasiolepis*
      162B. Culms, leaves, and scales glabrous.
       163A. Principal leaves cauline; culms 40–50 cm. long, sharply angled, scabrous; perigynia faintly nerved.
         198. *C. poculisquama*
       163B. Leaves all radical; culms slender, 10–20 cm. long, nearly smooth; perigynia nerveless except for the 2 ribs.
        164A. Plant wholly greenish except the pale scales. .... 121. *C. hashimotoi*
        164B. Basal sheaths and scales reddish brown. .... 122. *C. quadriflora*
     161B. Culms densely tufted; perigynia obtusely trigonous, ribbed, with a short recurved beak; pistillate scales pointed or gradually acuminate; open woods.
      165A. Leaves filiform, 0.5–1.5 mm. wide, hispidulous-ciliate; culms smooth, 3–6 cm. long, hidden among the leaves. .... 123. *C. humilis*
      165B. Leaves linear, 1–2 mm. wide, scabrous; culms slightly scabrous, 10–40 cm. long. .... 124. *C. lanceolata*
    160B. Perigynia glabrous, smooth or scabrous above, sometimes densely papillose.
     166A. Lateral spikes androgynous.
      167A. Culms 40–80 cm. long; spikes 2–4 cm. long, sparsely flowered; lower bracts elongate, leafy; perigynia 9–11 mm. long, light green, long-beaked; forests. .... 157. *C. dissitiflora*
      167B. Culms 15–30 cm. long; spikes 1–2 cm. long, loosely flowered; perigynia 3 mm. long, gray-green, short-beaked. .... 143. *C. tumidula*
     166B. Lateral spikes usually pistillate, seldom gynecandrous.
      168A. Perigynia nerveless or faintly nerved.
       169A. Perigynia minutely puncticulate, dull, glaucous-green.
        170A. Scales or basal sheaths dark reddish brown or cupreous brown.
         171A. Culms long- to short-rhizomatous.
          172A. Basal sheaths light yellowish to pale.
           173A. Pistillate spikes pendulous, rather densely flowered; pistillate scales reddish cupreous; perigynia ascending, glaucous, abruptly short-beaked; swamps. .... 135. *C. laxa*
           173B. Pistillate spikes erect, rather loosely flowered; pistillate scales purplish cupreous.
            174A. Bracts leafy, short-sheathing at base; perigynia erect, glaucous, nearly beakless; peat bogs.
             134. *C. livida*
            174B. Bracts with short blades, long-sheathing at base; perigynia ascending to spreading, abruptly short- to long-beaked; forests or meadows. .... 136. *C. vaginata* var. *petersii*
          172B. Basal sheaths at least partially dark purplish brown.
           175A. Blades and sheaths usually hairy; pistillate spikes cylindric, loosely flowered; forests.
            137. *C. pilosa*
           175B. Blades and sheaths glabrous; pistillate spikes oblong-ellipsoidal; rather densely flowered.
            176A. Leaves of fruiting plants not yet fully elongated, flaccid, abruptly pointed; perigynia obovoid, with a deflexed beak sometimes minutely scabrous on keels; Kyushu.
             138. *C. kujuzana*
            176B. Leaves of fruiting plants fully elongated, as long as or slightly longer than the culm, gradually tapering to apex; perigynia ovoid-fusiform, with a smooth, slightly recurved beak; Honshu. .... 139. *C. dissitispicula*
         171B. Culms tufted; forests. .... 140. *C. filipes*
        170B. Scales pale green to pale tawny; basal sheaths pale.
         177A. Pistillate spikes pendulous, the scales ferrugineous, elliptic, rounded and mucronate; wet places.
          141. *C. papulosa*

177B. Spikes erect, the upper contiguous; pistillate scales pale green, ovate; forests or grassy places.
142. *C. parciflora*

169B. Perigynia smooth or scabrous at tip, more or less lustrous, not grayish.
178A. Perigynia linear-lanceolate, pale green, appressed, 7–8 mm. long; stigmas much-elongate, persistent.
156. *C. bostrychostigma*

178B. Perigynia ovoid, yellowish to somewhat brownish when mature, nearly erect, 3–7 mm. long.
179A. Perigynia more than 5 mm. long.
180A. Perigynia about 7 mm. long, loosely inclosing the achene, inflated, scabrous at tip; tufted herb with solitary to ternary spikelets at the nodes; leaves 3–6 mm. wide. 174. *C. metallica*
180B. Perigynia 5–6 mm. long, tightly inclosing the achene, acutely 3-angled, sparsely pilosulous; loosely to scarcely tufted, rhizomatous herb; spikes 3 or 4; leaves 2–3 mm. wide.
175. *C. macrandrolepis*

179B. Perigynia 3–5 mm. long, tightly inclosing the achene.
181A. Culms 10–25 cm. long, slender; spikes 0.5–2.5 cm. long; perigynia brownish; stigmas deciduous; species of alpine meadows.
182A. Terminal spike not exceeding the proximate pistillate one; leaves 1–1.5 mm. wide.
158. *C. capillaris*
182B. Terminal spike long-peduncled, much longer than the proximate one; leaves 2.5–3(–4) mm. wide. .... 159. *C. tenuiformis*
181B. Culms 50–70 cm. long; pistillate spikes 2–5 cm. long; perigynia light yellowish green; stigmas very long, persistent; open woods in *Fagus* zone. .... 176. *C. hondoensis*

168B. Perigynia conspicuously many-nerved, dark brown when dried.
183A. Perigynia compressed-trigonous, prominently papillose; plant wholly glaucous, not reddish when living; wet places. .... 148. *C. maculata*
183B. Perigynia not compressed, glabrous; basal sheaths sanguineous to reddish purple.
184A. Sheaths and scales sanguineous; perigynia lustrous; terminal spike often gynecandrous, the pistillate bearing few staminate flowers at the base; forests. .... 160. *C. alliiformis*
184B. Bracts and scales greenish; perigynia dull; plants of grassy places.
185A. Pistillate scales awned; perigynia spreading, abruptly beaked.
186A. Perigynia about 3 mm. long, abruptly short-beaked. .... 180. *C. brownii*
186B. Perigynia 5–6 mm. long, long-beaked. .... 181. *C. transversa*
185B. Pistillate scales entire, mostly obtuse, occasionally slightly acute, awnless; perigynia erect-patent, gradually rather long-beaked; open forests. .... 147. *C. ischnostachya*

156B. Spikes numerous, in compound panicles; cladoprophyllum utricle-shaped. .... 167. *C. cruciata*
76B. Tips of perigynia firm, sharply bifid or 2-toothed; style 3-fid; achenes trigonous.
187A. Plants wholly glabrous.
188A. Perigynia obtusely 3-angled, sometimes slightly inflated.
189A. Perigynia ovoid to oblong-conical, several-nerved.
190A. Perigynia thick-membranous, 2.5–3 mm. long; spikes closely contiguous; staminate spike solitary; wet places.
182. *C. viridula*
190B. Perigynia somewhat ligneous, 6–8 mm. long; staminate spikes 2–4, the pistillate usually distant.
191A. Culms usually 10–20 cm. long; pistillate spikes approximate, the lowest bract short-sheathing; leaves 2–4 mm. wide; sandy places along seacoasts. .... 185. *C. pumila*
191B. Culms 30–100 cm. long; pistillate spikes distant; tidal swamps.
192A. Leaves 1.5–2.5 mm. wide; pistillate spikes 1 or 2, oblong, 1–2 cm. long; bracts sheathless. .. 186. *C. scabrifolia*
192B. Leaves 5–10 mm. wide; pistillate spikes 2–4, cylindric, 2.5–5 cm. long; lowest bract often short-sheathing.
187. *C. rugulosa*

189B. Perigynia lanceolate, many-nerved; wet or swampy places.
193A. Bracts all sheathing; perigynia 10–13 mm. long, with a shallowly 2-toothed beak; pistillate spikes ovoid-globose, several-flowered; scales entire, obtuse. .... 189. *C. michauxiana* var. *asiatica*
193B. Bracts, except the lowermost, not sheathing; perigynia 4–9 mm. long, the beak 2-fid; pistillate spikes oblong to cylindric, densely many-flowered; scales awned.
194A. Pistillate spikes cylindric, 2–5 cm. long, 6–8 mm. across, nodding; perigynia 4–5 mm. long, the beak bifid, with erect teeth. .... 190. *C. pseudocyperus*
194B. Pistillate spikes oblong, 1.5–3 cm. long, about 15 mm. across, erect-patent; perigynia 6–9 mm. long, the beak with recurved teeth. .... 191. *C. capricornis*
188B. Perigynia oval, strongly inflated, lustrous, loosely inclosing the achenes.
195A. Perigynia about 10 mm. long, with a long slender beak. .... 192. *C. dickinsii*
195B. Perigynia 5–8 mm. long, with a short or moderately long beak.
196A. Leaves thick, sulcate above, about 1.5 mm. wide; perigynia short-beaked. .... 193. *C. oligosperma*
196B. Leaves rather flat, 3–15 mm. wide, recurved on the margins; perigynia rather long beaked.
197A. Culms scabrous on the angles; leaves 3–6 mm. wide; pistillate spikes 2 or 3, cylindric, 3–7 cm. long; perigynia 6–8 mm. long. .... 194. *C. vesicaria*
197B. Culms nearly glabrous; leaves 8–15 mm. wide; pistillate spikes 2 to 5, long-cylindric, 5–10 cm. long; perigynia 5–6 mm. long. .... 195. *C. rhynchophysa*
187B. Leaves, culms, bracts or perigynia hairy.
198A. Longer leaves on the upper half of the culm gradually reduced below into bladeless sheaths; all bracts sheathing; in forests.

199A. Inflorescence spiciform, with at least the lower spikes distant; upper leaves approximate but not overlapping.
.... 196. *C. ligulata*

199B. Inflorescence capitate, with densely fastigiate spikes; longer leaves 8–13 mm. wide, aggregated on the apical part of the culm. .... 197. *C. phyllocephala*

198B. Leaves all radical or culms with only 1 or 2 cauline leaves.

200A. Perigynia coriaceous; leaves 1.5–3 mm. wide, involute or flat, canaliculate in upper part; peat bogs.
.... 199. *C. lasiocarpa* var. *occultans*

200B. Perigynia thick-membranous or herbaceous; leaves 3–6 mm. wide.

201A. Bracts sheathing at base; leaves, ligule, and perigynia usually hairy; forests. ... 200. *C. drymophila* var. *abbreviata*

201B. Bracts, except the lowest one, not sheathing.

202A. Leaves glabrous; perigynia hispid; staminate spikes 2–4; wet places along rivers. .... 201. *C. miyabei*

202B. Leaves pilose; perigynia glabrous; staminate spikes always solitary; wet places in woods. .... 202. *C. latisquamea*

1. Carex arenicola F. Schmidt. *C. chaetorhiza* Fr. & Sav.; *C. yedoensis* Boecklr.——KURO-KAWAZU-SUGE. Rhizomes long-creeping; culms distant, 10–30 cm. long, scabrous; leaf-blades, 2–3 mm. wide, greenish; spikes several, androgynous, oval, 5–8 mm. long, densely aggregated into a narrow, ovate, bractless head; pistillate scales narrowly ovate, ferrugineous-castaneous, acuminate; perigynia slightly longer than the subtending scales, ovoid, 3–4 mm. long, spreading, plano-convex, crescent-shaped in transverse section, faintly nerved, tightly inclosing the achene, margins acute, spongy-thickened at base, the long beak scabrous-margined, the tip membranous. ——May–June. Sandy grassy places in lowlands; Hokkaido, Honshu, Kyushu; common.——Korea, Ussuri, Sakhalin, and Kuriles.

2. Carex kabanovii V. Krecz. *C. yezoalpina* Akiyama ——YARI-SUGE. Rhizomes short and slender; culms 12–17 cm. long, obtusely angled, smooth; leaf-blades short, sulcate, 1–1.5 mm. wide, rather thick, somewhat obtuse; spike solitary, terminal, unisexual or rarely androgynous, 15–25 mm. long, somewhat lustrous; pistillate scales ovate, yellowish brown, obtuse, 1-nerved; perigynia conspicuously longer than the scales, about 2.5 mm. long, obliquely spreading at tip, yellowish brown, faintly nerved, biconvex, gradually rather long beaked, the tip hyaline.——June–July. High moors; Hokkaido (Daisetsu-zan); rare.——Sakhalin.

3. Carex gynocrates Wormsk. *C. dioica* var. *gynocrates* (Wormsk.) Ostenf.——KANCHI-SUGE. Rhizomes slender, rather short; culms 10–20 cm. long, smooth; leaf-blades thick, 0.5–0.7 mm. wide, sulcate above; spike solitary, terminal, unisexual or rarely androgynous, 7–14 mm. long, lustrous; pistillate scales oval, acute, castaneous-cupreous, 3-nerved; perigynia nearly as long as the scales, 3 mm. long, deflexed when mature, ellipsoidal, biconvex, coriaceous, dark yellowish brown, slightly lustrous, smooth or scabrous on the margins above, several-nerved on both sides, the beak short, emarginate; achenes tightly inclosed, about 2 mm. long; style bifid.—June–July. High moors; Honshu (Rikuchu Prov., Yakeishi-dake); rare.——Kuriles, Sakhalin, Kamchatka, Siberia, and Canada.

4. Carex disperma Dewey. *C. tenella* Schk., non Thuill.; *C. misera* Franch., non Phil.; *C. nakaii* Lév. & Van't.; *C. tenella* var. *brachycarpa* Kuekenth.; *C. tenella* var. *nakaii* (Lév. & Van't.) Lév.——HOSO-SUGE. Rhizomes slender, elongate; culms 20–50 cm. long, slender, acutely angled, scabrous; leaf-blades thin, 1–1.5 mm. wide; spikelets 3–5, androgynous, distant, 2–3 mm. long, few-flowered, ebracteate or the lowest one with a setaceous bract; pistillate scales pale, ovate, acute; perigynia ovoid, slightly longer than the scales, 2.5–3 mm. long, pale yellowish, lustrous, biconvex, inflated, faintly nerved, subcoriaceous, abruptly attenuate and very short, smooth beaked, the mouth entire; achenes tightly inclosed;

style bifid.——June–July. Wet mossy places in woods; Hokkaido; rare.——Subarctic regions of the N. Hemisphere.

5. Carex pallida C. A. Mey. *C. siccata* sensu auct. Japon., non Dewey; *C. siccata* subsp. *pallida* (C. A. Mey.) Kuekenth.; *C. accrescens* Ohwi——USU-IRO-SUGE, EZO-KAWAZU-SUGE. Rhizomes slender, long-creeping; culms 30–60 cm. long, loose, scabrous, acutely angled; leaf-blades flat, 3–5 mm. wide, shorter than the culm; spikes several, unisexual or androgynous, oblong, 5–10 mm. long, aggregated into a short, cylindric, ebracteate spike often interrupted below; pistillate scales ovate, acute, yellowish brown, with a broad green midrib; perigynia longer than the scales, 4–4.5 mm. long, reflexed at tip, flat, ovate, broadly winged and scabrous on the margin, sparsely appressed-pubescent, at least while young; style bifid. ——Coniferous woods; Hokkaido, Honshu (Rikuchu Prov.) ——s. Kuriles, Sakhalin, Korea, and e. Siberia.

6. Carex pseudocuraica F. Schmidt. *C. curaica* var. *major* Boecklr.——TSURU-SUGE, TSURU-KAWAZU-SUGE. Sterile culms ultimately leafy, much elongated, decumbent, bearing scapes from the nodes the following year; fertile culms solid, 20–40 cm. long, acutely angled, slightly scabrous above; leaf-blades flaccid, 2–4 mm. wide; inflorescence oblong-cylindric, 15–30 mm. long, ebracteate, the 5–8 loosely aggregated spikes androgynous (or unisexual), ovoid, 5–10 mm. long, light castaneous; perigynia ovate-elliptic, slightly longer than the ovate, acutely pointed scales, compressed, more or less spongy toward the narrowly-winged margins, scabrous at tip, short-beaked; style bifid.——June–July. Swamps; Hokkaido, Honshu (Mutsu, Yechigo).——s. Kuriles, Sakhalin, Korea, and Ussuri.

7. Carex lithophila Turcz. *C. disticha,* sensu auct. Japon., non Huds.; *C. intermedia,* sensu auct. Japon., non Retz. nec Gooden.; *C. distichoidea* Lév. & Van't.——ASAMA-SUGE, TAIRIKU-KAWAZU-SUGE. Rhizomes long-creeping; culms 20–40 cm. long, acutely 3-angled, very scabrous above; leaf-blades flat, 2.5–3 mm. wide; inflorescence dense, 1.5–2 (–3) cm. long, ovoid to oblong-ovoid, ebracteate, the spikes 8 to 10, androgynous or unisexual, ovoid, 5–8 mm. long, reddish brown; pistillate scales ovate, reddish brown; perigynia slightly shorter than the scales, oval, 4–5 mm. long, compressed, spongy, several-nerved on both sides, scabrous, winged on the margins, short-beaked; style bifid.——May–June. Honshu (Kantō and ne. Shinano); rather rare.——Korea, Manchuria, and e. Siberia.

8. Carex neurocarpa Maxim. MIKOSHIGAYA. Plant usually wholly ferrugineous-puncticulate; culms tufted, 30–60 cm. long, obtusely angled, smooth; leaf-blades flat, 2–3 mm. wide; inflorescence dense, ovoid to ovoid-cylindric, 3–6 cm. long, with 2 or 3 spreading, elongate, foliaceous bracts at base, the spikes numerous, androgynous, ovoid-globose, 4–8 mm.

long; pistillate scales membranous, oval, short-awned, ferruginous-spotted; perigynia longer than the scales, 4 mm. long, nearly erect, ovate, slightly turgid, planoconvex, membranous, many-nerved, with broad serrate wings on upper half, beak long, 2-toothed; stigmas 2.——May-June. Grassy places in lowlands; Honshu (n. and centr. part, including Kinki Distr.); locally common.——Amur, Korea, and China.

9. Carex paxii Kuekenth. *C. succedanea* Nakai; *C. kengii* Kuekenth.; *C. paxii* var. *succedanea* (Nakai) Ohwi——KIBI-NO-MINOBORO-SUGE. Rhizomes short; culms tufted, 40–60 cm. long, obtusely angled, smooth; leaf-blades 1.5–2.5 mm. wide, flat; inflorescence ovate-cylindric, 2–6 cm. long, dense, with scalelike bracts, the spikes androgynous, ovate-orbicular, 4–6 mm. long; scales pale, oval; perigynia longer than the scales, spreading, 3.5–4 mm. long, ovate-conic, many-nerved, glaucous, greenish yellow, lustrous, planoconvex, with several large tubercles on the back, narrowly winged and scabrous on the margins, gradually beaked, the tip 2-toothed; achenes loosely inclosed, biconvex; style bifid.——May. Honshu (Chūgoku Distr.); rare.——Korea and China.

10. Carex laevissima Nakai. HIME-MIKOSHIGAYA. Rhizomes short; culms tufted, 15–40 cm. long, acutely angled, scabrous above, slender, firm; the blades 2–3 mm. wide; inflorescence dense, short-cylindric, 2–5 cm. long, nearly bractless, the spikes numerous, usually androgynous, sometimes nearly unisexual, oval, 5–8 mm. long; pistillate scales narrowly ovate, membranous, brownish ferrugineous, acute; perigynia slightly longer than the scales, 3–3.5 mm. long, narrowly ovate, faintly nerved, planoconvex, gradually long-beaked, the sharp margins usually nearly smooth, sometimes very slightly scabrous near the minutely 2-toothed apex; achenes small, about 1.2 mm. long, nearly orbicular, biconvex; style slender, brown, bifid.—May. Honshu (Bitchu, Settsu); rare.——Korea and Ussuri.

11. Carex albata Boott. *C. nubigena* var. *albata* (Boott) Kuekenth.——MINOBORO-SUGE. Rhizomes short, tufted; culms 20–60 cm. long, scabrous on the angles; leaf-blades 2–3 mm. wide, rather firm, deep green, sometimes slightly curved upward, gradually long-attenuate; inflorescence a terminal spike 3–5 cm. long, dense, ovate-oblong, the bracts all scalelike, or the lower 1–3 nearly foliaceous; spikes numerous, androgynous; globular-ovoid, 5–8 mm. long; pistillate scales ovate, acute, midrib green; perigynia longer than the scales, 4–4.5 mm. long, ovate-lanceolate, nerved, planoconvex, the margins acute, slightly winged above the middle, scabrous toward tip, gradually long-beaked, the beak flat, 2-toothed; achenes small; stigmas rather long, slender.——May-July. Wet places in mountains; sw. Hokkaido, Honshu (ne. part s. to centr. distr.); rather common.

12. Carex nubigena D. Don var. **franchetiana** Ohwi. *C. phaeoleuca* Ohwi; *C. fallax* var. *franchetiana* (Ohwi) Ohwi——TSUKUSHI-MINOBORO-SUGE. Resembles the preceding, differing in the somewhat stouter, glabrous, obtusely angled culms, and the shorter bracts not longer than the culm; perigynia ovate or oval, about 4 mm. long, the beak somewhat shorter, the margins scabrous only at tip.——Apr.-June. Wet places in mountains; Honshu (Chūgoku Distr.), Kyushu.——The typical phase occurs in Formosa to sw. China, India, and Malaysia.

13. Carex stipata Muhl. Ō-KAWAZU-SUGE. Rhizome short; culms tufted, soft, 30–60 cm. long, scabrous on the prominent, acute angles; leaf-blades 3–7 mm. wide, flaccid, bright green; inflorescence rather dense, ovoid-cylindric, 3–6

cm. long, the bracts scalelike or the lower elongate, to 3 cm. long; spikes numerous, androgynous, ovoid-globular, 6–10 mm. long; scales pale, 1-nerved, acuminate; perigynia longer than the scales, spreading, broadly lanceolate, 5 mm. long, thick, planoconvex, pale green, nerved, spongy and prominently thickened at the base, gradually long-beaked, the margins scabrous except at base.——June-July. Moors; Hokkaido, Honshu (centr. distr. northw.).——N. America.

14. Carex diandra Schrank. *C. teretiuscula* Gooden.——KURI-IRO-SUGE. Rhizomes short-creeping; culms tufted, slender, 50–80 cm. long, obtusely angled, scabrous above; leaf-blades flat, 2–2.5 mm. wide, scabrous on the margins; inflorescence ovate to oblong, 2–3.5 cm. long, ebracteate; spikes numerous, crowded, androgynous, ovoid-globose, 5–7 mm. long; pistillate scales oval, acute, castaneous, with whitish, somewhat translucent margins; perigynia coriaceous, lustrous, longer than the scales, broadly ovoid, 3 mm. long, spreading, dark chestnut-brown at maturity, thick-biconvex, indistinctly 3- to 5-nerved at base, obtusely margined, stipitate, abruptly beaked, the beak moderately long, setulose-scabrous on margins, the tip almost entire; style bifid.——June-July. Swamps; Hokkaido, Honshu (Shimokita Penin. in Mutsu and Izunanohara in Shinano Prov.); rather rare.——Sakhalin, s. Kuriles, and widely dispersed in the cooler regions of the N. Hemisphere.

15. Carex kobomugi Ohwi. *C. macrocephala* sensu auct. Japon., pro parte; *C. macrocephala* var. *kobomugi* (Ohwi) Miyabe & Kudo——KŌBŌ-MUGI, FUDE-KUSA. FIG. 1. Rhizomes stout, extensively creeping, covered with brown fibrous remains of leaf-sheaths; culms stout, stiff, obtusely angled, nearly smooth, 10–20 cm. long; leaf-blades rather few, 4–6 mm. wide, coriaceous, light green, somewhat lustrous above, prominently scabrous on the margins; heads much-crowded, usually unisexual, ovoid to oblong-ovoid, 4–6 cm. long in the pistillate, nearly bractless or with a few short bracts; spikes unisexual (rarely androgynous), the pistillate ovoid, about 15 mm. long; pistillate scales narrowly ovate, sulfur-yellow, with broad herbaceous margins, the broad many-nerved keel prolonged into a scabrous, awnlike mucro; perigynia longer than or as long as the scales, coriaceous, appressed-erect, ovate, about 10 mm. long, planoconvex to somewhat concave-convex, rounded on the back, many-nerved, with a narrow irregularly serrate wing on the margins, the beak long, incurved, bifid at tip; achenes compressed-trigonous; stigmas 3.——Apr.-June. Common on sand dunes along the seacoast; w. Hokkaido, Honshu, Shikoku, Kyushu.——Korea, Formosa, Ussuri, and Manchuria; introduced in e. N. America.

16. Carex macrocephala Willd. ex Spreng. *C. anthericoides* Presl——EZO-NO-KŌBŌ-MUGI. FIG. 1. Resembles the preceding, differs in the acutely angled culms distinctly scabrous above; head solitary, 4–6 (–8) cm. long; pistillate scales ovate, membranous, chestnut-brown, 3-nerved, the green midrib excurrent into a needlelike smooth awn; perigynia longer than the scales, strongly divergent, many-nerved, conspicuously wing-margined, the wings dentate-serrate, the beak much longer than in No. 15, and the tip more deeply bifid; achenes compressed-triangular; stigmas 3.——May-July. Sandy places along seacoasts; Hokkaido (except the sw. part).——s. Kuriles, Sakhalin, Kamchatka, and w. coast of N. America.

17. Carex gibba Wahlenb. MASU-KUSA. Rhizomes short; culms tufted, 30–70 cm. long, obtusely angled, smooth; leaf-blades rather soft, 2–4 mm. wide; heads 4–10 cm. long,

FIG. 1.—*Carex kobomugi* Ohwi (1–19). 1, Male plant at anthesis; 2, staminate spikelet with a bract at base; 3, staminate scale; 4, upper part of anther; 5, pistillate plant in fruiting stage; 6, part of leaf; 7, upper part of sheath; 8, part of a culm; 9, staminate flower and scale; 10, pistillate spikelet; 11, bract of lower spikelet; 12, bract of upper spikelet; 13, pistillate scale; 14, 15, utricles; 16, 17, beak of utricles; 18, cross section of utricle; 19, achene. *Carex macrocephala* Willd (20–29), pistillate plant; 21, part of culm; 22, pistillate spikelet with a bract at base; 23, pistillate scale; 24, 25, utricles; 26, 27, beak of utricle; 28, cross section of utricle; 29, achene; 30, diagram of staminate flower; 31, diagram of pistillate flower.

Tetsuo Koyama, delin.

interrupted below, the bracts divergent, elongate; spikes 5–8, gynecandrous, oblong or obovoid, 5–10 mm. long, green; pistillate scales narrowly obcordate, whitish, nearly truncate at apex, the green midrib prolonged into a mucro; perigynia nearly orbicular, 3–3.5 mm. long, much longer than the scales, planoconvex, nerveless, membranous, winged on margins, scabrous above the middle, the beak short, 2-toothed at tip; achenes biconvex with a discoid appendage at apex; stigmas 3.——May–June(–July). Waste grounds and roadsides; Honshu, Shikoku, Kyushu; very common.——Korea and China.

18. Carex echinata Murr. *C. stellulata* Gooden.; *C. leersii* Willd.; *C. muricata* sensu auct., non L.; *C. basilata* Ohwi ——KITA-NO-KAWAZU-SUGE. Rhizomes short; culms tufted, 20–50 cm. long, sharply angled, scabrous above; leaf-blades 1–2 mm. wide; head loose, the bracts scalelike to setaceous; spikes 3 or 4, gynecandrous, globose, 4–6 (terminal spike sometimes to 10) mm. long; pistillate scales ovate, brownish; perigynia about 3 mm. long, about twice as long as the scales, spreading to divaricate, deltoid-ovate, planoconvex, brownish when mature, scabrous on the acute margins near tip, thickened and spongy at base, gradually long-beaked; stigmas 2.——June–July. Wet places; Hokkaido.——Kuriles and Korea, e. Siberia to Europe.

19. Carex omiana Fr. & Sav. *C. stellulata* sensu auct. Japon., non Gooden.; *C. stellulata* var. *omiana* (Fr. & Sav.) Kuekenth.——YACHI-KAWAZU-SUGE. Resembles the preceding; culms 30–50 cm. long, nearly smooth; leaf-blades 1.5–2.5 mm. wide; spikes 3–5 sometimes 2, somewhat approximate or rather loose, gynecandrous, globose, the terminal with a slightly longer staminate portion than in those of the lateral; perigynia nearly twice as long as the scales, 4–5 mm. long, strongly spreading to somewhat deflexed, lanceolate-ovate, strongly convex on back, faintly nerved, gradually long-beaked, minutely bifid at tip, prominently thickened at the spongy base; achenes tightly inclosed; stigmas 2.——May–July. Wet places; Hokkaido, Honshu, Kyushu.——s. Kuriles.

Var. **monticola** Ohwi. KAWAZU-SUGE. Culms 10–30 cm. long; spikes slightly smaller; perigynia 3.5–4 cm. long, broadly ovate-lanceolate, with a slightly longer beak.——Wet places usually in mountains; Hokkaido, Honshu (centr. distr. and northw.).

Var. **yakushimana** Ohwi. *C. ohwii* Masam.——CHABO-KAWAZU-SUGE. Culms short, to 10 cm. long, leaf-blades somewhat involute, narrow; perigynia 4 mm. long, nearly smooth, short-beaked.——Wet places in high mountains; Kyushu (Yakushima).

20. Carex cyperoides Murray. KAYA-TSURI-SUGE. Annual or biennial with fibrous roots; culms loosely tufted, soft, smooth, 15–30 cm. long; leaf-blades 1.5–2.5 mm. wide, glabrous, spikes numerous, densely crowded in a terminal hemispherical head 15–20 mm. long and as wide, the bracts 2–3, leafy, much longer than the inflorescence; pistillate scales lanceolate, awn-tipped, membranous, pale; perigynia much longer than the scales, 7–10 mm. long, erect, narrowly lanceolate, scabrous and narrowly winged on the margins, planoconvex, pale green, membranous, long-stipitate, with a flat, long, deeply bifid beak; achenes small; stigmas 2.——Wet sandy soil; Hokkaido, Honshu (Lake Kawaguchi in Kai); rare.——Siberia and Europe.

21. Carex planata Fr. & Sav. *C. ponmoshirensis* Miyabe & Kudo——TAKANE-MASU-KUSA. Rhizomes abbreviated; culms densely tufted, to 60 cm. long, obtusely angled, smooth;

leaf-blades 1.5–2.5 mm. wide, soft, deep green; spikes 2–5(10), pistillate with a few staminate flowers at the base, ovoid to ovoid-globular, 6–10 mm. long, loosely aggregated into a rather short head 3–8 cm. long, the lower bracts leafy, elongate, much longer than the head; pistillate scales pale, membranous, ovate; perigynia appressed, much longer than the scales, compressed, oval, 4 mm. long, light green, rather broad, the wings broad, scabrous-margined, the beak short, rather broad, sulcate in front, the tip shallowly bifid; achenes small, tightly inclosed; stigmas 2.——May–June. Forests; Hokkaido (Ishikari Prov.), Honshu, Shikoku, and Kyushu.

22. Carex rochebrunii Fr. & Sav. *C. remota* sensu auct. Japon., non L.; *C. remota* subsp. *rochebrunii* (Fr. & Sav.) Kuekenth.——YABU-SUGE. Rhizomes short; culms tufted, 40–60 cm. long, obtusely angled, smooth; leaf-blades 2–4 mm. wide; spikes 8–10, gynecandrous, narrowly oblong, pale, densely flowered, 8–15 mm. long, mostly aggregated into a slender head 5–10 cm. long, but the lower ones separate, the lower bracts elongated, leafy, much longer than the culm; pistillate scales pale, membranous, oblong; perigynia longer than the scales, 4–4.5 mm. long, compressed-planoconvex, faintly nerved, scabrous and narrowly winged on the margins, gradually narrowed into a flat, very narrowly winged beak, the tip deeply 2-fid; stigmas 2.——May. Forests; Honshu (Kinki Distr. and eastw.), Shikoku.——China, Formosa, and Malaysia.

23. Carex remotiuscula Wahlenb. *C. remotaeformis* Komar.; *C. remota* subsp. *rochebrunii* var. *remotaeformis* (Komar.) Kuekenth.; *C. rochebrunii* var. *remotaeformis* (Komar.) Akiyama——ITO-HIKI-SUGE. Resembles the preceding, but more slender; culms tufted, 30–50 cm. long, sharply angled, scabrous; leaf-blades thin, 1–2 mm. wide; inflorescence often more or less flexuous, 3–10 cm. long; spikes 4–7, gynecandrous, ovoid-globose, 4–6 mm. long, pale green, the lower ones widely separate; pistillate scales ovate, thin, pale; perigynia longer than the scales, nearly erect, about 3 mm. long, ovate-lanceolate, planoconvex, faintly nerved, the margins narrowly winged, not spongy, gradually long-beaked, deeply 2-toothed at the tip; stigmas 2.——June–July. Forests; Hokkaido, Honshu (Mount Kirigamine in Shinano); rare.——Korea, Ussuri, and Sakhalin.

24. Carex deweyana Schwein. var. **senanensis** (Ohwi) T. Koyama. *C. senanensis* Ohwi; *C. deweyana* sensu auct. Japon., non Schwein.; *C. hondae* Akiyama——HO-SUGE. Rhizomes short; culms tufted, sharply angled, scabrous; leaf-blades 2–4 mm. wide, thin, flat, bright green; inflorescence rather dense, looser toward base, the lowest bract setaceous; spikes 7-9, gynecandrous, each with a few staminate flowers at base, oblong to slenderly so, 3–4 mm. across, pale green; pistillate scales membranous, ovate, faintly nerved, acute and mucronate; perigynia slightly longer than the scales, planoconvex, ovate-lanceolate, thin, faintly nerved, about 4 mm. long, gradually long-beaked, scabrous and 2-toothed at apex; stigmas 2.——June–July. Wet grassy places in alpine and subalpine zones in the middle part of Honshu. The typical phase occurs in N. America.

25. Carex maackii Maxim. *C. nipponica* Franch.——YAGAMI-SUGE. Rhizomes short; sterile culms elongate, with nodes and many leaves; fertile culms tufted, 40–60 cm. long, acutely angled, scabrous; leaf-blades 2–3 mm. wide, soft, flat; head cylindric, 3–5 cm. long, the bract usually small, scalelike; spikes several to numerous, approximate, gynecandrous, ovoid-globose, 5–8 mm. long, light green, densely flowered; pistillate scales pale, ovate, obtuse, the midrib broad, green; perigynia twice as long as the scales, ovate, 3.5 mm. long, slightly recurved at tip, concavo-convex, inconspicuously nerved, prominently spongy toward the periphery, margins scabrous near tip and narrowly winged, the beak distinct, minutely 2-toothed at tip; stigmas 2.——May–June. Wet grassy places along rivers in lowlands; Hokkaido, Honshu, Kyushu; locally common.——Amur, Ussuri, Korea, and China.

26. Carex lachenalii Schk. *C. bipartita* auct.; *C. lagopina* Wahlenb.——Takane-yagami-suge. Rhizomes short; culms somewhat tufted, erect, 20–30 cm. long, acutely angled, scabrous at tip; leaf-blades 1.5–2.5 mm. wide, deep green; head short, the bracts reduced to scales, the spikes 3 or 4, approximate or contiguous, gynecandrous or the lateral often pistillate, oblong, 5–10 mm. long, dark brown; pistillate scales ovate-oval, shining, chestnut-brown; perigynia slightly longer than the scales, obovate, 3 mm. long, planoconvex, smooth, light castaneous, rather abruptly short-beaked, with an entire hyaline tip; stigmas 2.——July–Aug. Dry alpine slopes; Hokkaido (Mount Daisetsu), Honshu (Kai and Shinano Prov.); rare.——Korea and n. Kuriles to Arctic regions and high mountains of the N. Hemisphere.

27. Carex mackenziei V. Krecz. *C. norvegica* Willd. ex Schk., non Retz.——Noruge-suge, Karafuto-suge. Rhizomes short; culms loosely tufted, 20–40 cm. long; leaf-blades 2–3 mm. wide, soft, flat, glaucous-green, densely papillose; spikes 3–5, mostly gynecandrous, contiguous or the lower ones more distant, the terminal sometimes staminate, usually clavate, the pistillate oblong, 7–12(–15) mm. long; pistillate scales ovate-orbicular, very obtuse, ferrugineous; perigynia as large as the scales, oval, 3 mm. long, at first glaucous, becoming glaucous-brown at maturity, faintly nerved, the margins obtuse and nearly smooth, the beak very short with an entire tip; achenes tightly invested by the perigynium; stigmas 2. ——June–July. Salt marshes near seashores; Hokkaido (Nemuro); rare, locally abundant.——Sakhalin, n. Siberia, n. Europe, and Canada.

28. Carex traiziscana F. Schmidt. Hiroha-Ozenuma-suge. Rhizomes short; culms loosely tufted, 40–60 cm. long, acutely angled, sparsely scabrous in upper part; leaf-blades 3–4 mm. wide, flat, glaucous-green, nearly smooth; spikes 6 or 7, gynecandrous, ovoid-globose, 5–8 mm. long, the upper ones approximate, the lower distant, forming a more or less flexuous elongate head, the bracts scalelike; pistillate scales rather obtuse, red-castaneous, paler toward the margins; perigynia longer than the scales, elliptic, 3 mm. long, tawny, ferrugineous at maturity, scabrous in upper part, strongly compressed, with acute margins, several-nerved, very short-beaked and minutely 2-toothed at tip; stigmas 2.——June–July. Wet grassy places; Hokkaido, Honshu (Ozegahara Moor in Kodzuke); rare.——Sakhalin.

29. Carex nemurensis Franch. *C. traiziscana,* sensu auct. Japon., saltem pro parte, non F. Schmidt——Hosoba-Oze-numa-suge. Resembles the preceding; culms densely tufted, 40–70 cm. long, acutely angled, scabrous; leaf-blades 2–3 mm. wide, deep green, scabrous on the margin; spikes gynecandrous, obovoid to globose-ellipsoidal, 5–7 mm. long, the terminal one with a slightly longer staminate portion; pistillate scales acute, castaneous, lustrous; perigynia about as long as the scales, ovate, nearly 3 mm. long, plano-convex, veiny, pale green, becoming dark brown at maturity, slightly scabrous on the margins near tip, abruptly short-beaked and entire at tip. ——June–July. Sphagnum bogs; Hokkaido, Honshu (Oze-gahara in Kotzuke Prov., and Mount Kirigamine in Shinano Prov.); rather common.——Sakhalin and Kuriles.

30. Carex curta Gooden. *C. canescens* sensu auct. plur., non L.——Hakusan-suge. Rhizomes short; culms somewhat tufted, 20–60 cm. long, 3-angled, slightly scabrous in upper part; leaf-blades 1.5–4 mm. wide, flat, glaucous-green; head naked, 2–6 cm. long; spikes 4–7, gynecandrous, oblong to nearly globose, 4–10 mm. long, glaucous-green to pale green; pistillate scales oval, pale yellowish; perigynia longer than the scales, broadly ovate or obovate, 2–2.2 mm. long, nearly obtuse, glaucous- or gray-green, becoming brownish at maturity, several-nerved, slightly scabrous on margins above, the beak very short, 2-toothed; stigmas 2.——June–July. Bogs and wet meadows; Hokkaido, Honshu (alpine regions of centr. distr., and northw.); rather common.——Arctic areas and alpine regions of the N. Hemisphere.

31. Carex brunnescens (Pers.) Poir. *C. curta* var. *brunnescens* Pers.; *C. vitilis* Fries; *C. laeviculmis* sensu auct. Japon., non Meinsh.; *C. sphaerostachya* Dewey; *C. brunnescens* var. *sphaerostachya* (Dewey) Kuekenth.——Hime-kawazu-suge. Rhizomes short; culms tufted, slender, 15–40 cm. long, acutely angled, slightly scabrous above; leaf-blades 1–2 mm. wide, bright green, flaccid; spikes 2–5, gynecandrous, or the lateral pistillate, nearly globose, 4–7 mm. long, light green, sometimes yellowish tawny, lustrous, the lower spikes loose, occasionally with setaceous basal bracts; pistillate scales pale or light tawny, with a green midrib; perigynia slightly exceeding the scales, ovate or ovate-elliptic, about 2 mm. long, indistinctly nerved, plano-convex, minutely scabrous on the margins above, the beak short, shallowly cleft in front, minutely toothed at apex.——June–July. Wet places in alpine or subalpine regions; Hokkaido, Honshu (n. and centr. distr.).—— Sakhalin, Kuriles, Siberia, Europe, and N. America.

32. Carex pseudololiacea F. Schmidt. Hiroha-ippon-suge, Ō-tsuru-suge. Rhizomes short, sparsely branched, culms loosely tufted, 20–40 cm. long, acutely angled, slightly scabrous below the apex; leaf-blades 1.5–3 mm. wide, glaucous-green, flat, soft; head short, ovoid, the bracts scalelike or the lowest one often short-setaceous; spikes 3 or 4, gynecandrous, 4–7 mm. long, contiguous; pistillate scales oval, acute, pale to slightly tawny; perigynia oblong-elliptic, 3.5–4 mm. long, much longer than the scales, glaucous-green, prominently nerved, thick-membranous, abruptly contracted and beakless or nearly so, minutely 2-toothed at tip; stigmas 2. ——Sphagnum bogs; Hokkaido, Honshu (Mount Azuma in Uzen Prov.).——Sakhalin and Kuriles.

33. Carex loliacea L. Akan-suge. Tufted; culms 20–50 cm. long, slender; leaf-blades 1–2 mm. wide, thin, flat, nearly smooth; head solitary, slenderly cylindric, 2–3 cm. long, ebracteate; spikes 3 or 4(–5), gynecandrous, obovoid, sub-globose at maturity, few-flowered, 3–5 mm. long, the lateral shorter than the terminal, the upper ones contiguous, the lower distant; pistillate scales pale, obtuse; perigynia about twice as long as the scales, ovate-oblong, 2.5–3 mm. long, unequally biconvex, at first glaucous-green, brownish at maturity, rather thin-membranous, with elevated thick nerves, obtusely margined, beakless, with an entire smooth tip; stigmas 2.—— June–July. Akan volcanic group in Hokkaido; rare.—— Korea, Sakhalin, Siberia, Europe, and Alaska.

34. Carex tenuiflora Wahlenb. *C. arrhyncha* Franch.;

C. tenuiflora var. *arrhyncha* (Franch.) Kuekenth.——IPPON-SUGE. Culms slender, in small tufts, erect, 20–60 cm. long, nearly smooth; leaf-blades 1–1.5 mm. wide, glaucous-green, narrow, nearly flat; head densely capitate, deltoid, 5–8 mm. long, the bracts scalelike; spikes 2 or 3, gynecandrous, obovoid-globose, 3–6 mm. long, few-flowered; pistillate scales obtuse, pale, more or less translucent, somewhat lustrous, persistent; perigynia ovate, planoconvex, as long as the scales, 3–3.5 mm. long, glaucous-green to gray, faintly many-nerved, smooth, beakless, with an entire tip; stigmas 2.——May–July. Sphagnum bogs; Hokkaido, Honshu (Nikko in Shimotsuke Prov.).——Sakhalin, Kuriles, Korea, and widely distributed in the N. Hemisphere.

35. Carex sadoensis Franch. SADO-SUGE. Rhizomes creeping; culms 30–70 cm. long, loosely tufted in small clumps, 3-angled, somewhat scabrous; leaf-blades 3–4 mm. wide, flat, soft, somewhat glaucous beneath, the basal leaf-sheaths with blades, scarcely filamentous, stramineous, not keeled; spikes 5–8, contiguous or the upper ones sometimes overlapping, erect, the terminal staminate, 2–3 cm. long, the lateral pistillate, cylindric, 3–5 cm. long, usually sessile; lowest bract leaflike; pistillate scales narrowly oblong, dark reddish brown, acute; perigynia broader and shorter than the scales, elliptic, 2–2.5 mm. long, compressed-biconvex, often brown-spotted, nerveless, with a rather abrupt, long, sometimes scabrous beak minutely 2-toothed at the tip; stigmas 2, very long, slender, persistent, reddish brown.——May–July. Abundant in wet places along mountain streams; Hokkaido, Honshu (n. and centr. distr., and on Mount Daisen in Hōki Prov.).——s. Sakhalin, and s. Kuriles.

36. Carex shimidzensis Franch. *C. nervulosa* Franch.; *C. sorachensis* Lév. & Van't.——AZUMA-NARUKO, MIYAMA-NARUKO. Rhizomes short and thick, short-creeping; culms loosely tufted, rather thick, 40–80 cm. long; leaf-blades rather soft and thin, broadly linear, 4–10 mm. wide, slightly glaucous, flat, the basal sheaths bladeless, cinnamon-colored, not keeled, soft, not splitting into fibers; terminal spike staminate or gynecandrous, the lateral 2–5, approximate, pendulous, terete, very densely many-flowered, yellowish green, sometimes the lowest one distant and long-peduncled, the lower 1–3 bracts leaflike, longer than the culm; pistillate scales ovate, pale, acute; perigynia longer than the scales, 2.5–3 mm. long, ascending, biconvex, somewhat turgid, glabrous, nearly nerveless, abruptly short-beaked, the beak slightly recurved; stigmas 2.——May–June. Wet mountain slopes; Hokkaido, Honshu, kyushu (n. distr.).——s. Kuriles and Korea (Ullung Isl.).

37. Carex incisa Boott. *C. textori* Miq.——KAWARA-SUGE, TANI-SUGE. Rhizomes loosely tufted, short-creeping; culms nodding, 20–50 cm. long; leaf-blades 3–6 mm. wide, soft, slightly glaucous, the basal sheaths bladeless or short-bladed, not disintegrating into fibers; spikes 4–6, nearly fastigiate, nodding to pendulous, the terminal one staminate or gynecandrous, 2–4 cm. long, the lateral pistillate, linear-cylindric, pale green, somewhat loosely many-flowered, 3–7 cm. long, the lower peduncled, the lowest bract leaflike, slightly longer to nearly as long as the culm; pistillate scales obovate, pale with brown flecks, emarginate, the broad green midrib prolonged into a short mucro; perigynia nearly twice as long as the scales, 3 mm. long, nerveless, biconvex, very short-beaked with an entire tip; stigmas 2.——May–June. Wet places in lowlands and low mountains; Hokkaido, Honshu; common.

38. Carex thunbergii Steud. *C. gaudichaudiana* var. *thunbergii* (Steud.) Kuekenth.——AZE-SUGE. Rhizomes long-creeping; culms 20–80 cm. long, erect; leaves somewhat flaccid; basal leaf-sheaths with blades, pale, partially brownish or purplish, the upper leaf-blades 1.5–4 mm. wide, soft, glaucous above, with slightly involute margins; staminate spikes solitary or sometimes 2, rarely 3, linear, 2–6 cm. long, the pistillate 2 or 3, long-cylindric to oblong, 2–5 cm. long, erect, usually subsessile; pistillate scales oblong, dark reddish purple to purplish brown, paler toward the margins; perigynia 3–3.5 mm. long, oblong to elliptic, planoconvex, faintly nerved, pale green, obsoletely puncticulate, abruptly very short-beaked with an entire tip; stigmas 2, rather stout.——May–June. Common in rice paddies and in swampy river flats; Hokkaido, Honshu. Very polymorphic.

Var. **appendiculata** (Trautv.) Ohwi. *C. acuta* var. *appendiculata* Trautv.; *C. semiplena* var. *tenuinervis* Kuekenth.; *C. vulpicaudata* Akiyama——Ō-AZE-SUGE, EZO-AZE-SUGE. Rhizomes rarely creeping; culms densely tufted; leaves less soft.——May–July. Wet places, especially in bogs; Hokkaido, Honshu (Senjōgahara in Nikko).——Sakhalin, s. Kuriles, Kamchatka, Amur, Ussuri, and n. Korea.

39. Carex subspathacea Wormsk. *C. salina* var. *minor* Boott; *C. descendens* sensu auct. Japon., non Kuekenth.——HIME-USHIO-SUGE. Rhizomes long-creeping; culms 3–30 cm. long, smooth; basal leaf-sheaths with blades, somewhat purplish, the upper leaf-blades 1–2 mm. wide, smooth, soft, slightly involute; terminal spike staminate, linear, 5–20 mm. long, the lateral 1 to 3, pistillate, erect, approximate, oblong, 5–15 mm. long, sessile, rather loosely 7- to 15-flowered, the lower 1 or 2 bracts leaflike, involute; pistillate scales ovate, dark purplish brown, obtuse; perigynia longer than the scales, suberect, ovate, 3.5–4 mm. long, thick, planoconvex, glaucous-green, nerveless, thick, densely papillose, abruptly very short-beaked with a rather firm entire tip; stigmas 2.——June–July. Salt marshes; Hokkaido.——Kamchatka, Kuriles, Sakhalin, and circumpolar areas.

40. Carex ramenskii Komar. *C. salina* sensu Ohwi, non Wahlenb.——USHIO-SUGE, UMIBE-SUGE. Rhizomes stout, creeping; culms 30–50 cm. long, smooth; leaf-blades 2–5 mm. wide, flat, the margins slightly recurved; basal leaf-sheaths bladeless, somewhat dark reddish purple, not splitting into fibers; upper 1 to 3 spikes staminate, linear, the lower 2 or 3 pistillate, short-cylindric, sometimes with a few staminate flowers toward the apex, erect, 2–3 cm. long, sessile or short-peduncled, the lower 1 or 2 bracts leaf-like, about as long as the inflorescence; pistillate scales narrowly ovate, dark brownish purple, acute; perigynia slightly longer than the scales, broadly ovate, thick plano-convex, glaucous-brown, densely papillose, rather thick, inconspicuously nerved, abruptly very short beaked, with a firm entire tip; stigmas 2.——June–July. Salt-marshes; Hokkaido (Nemuro).——Sakhalin and Kamchatka.

41. Carex lyngbyei Hornem. *C. cryptocarpa* C. A. Mey.; *C. prionocarpa* Franch.; *C. lyngbyei* var. *cryptocarpa* (C. A. Mey.) Kuekenth.——YARAME-SUGE. Rhizomes thick, stout, long-creeping, forming a large mat; culms 30–100 cm. long, scabrous on the angles above; leaf-blades 3–8 mm. wide, flat, grayish green, somewhat glaucous, the basal leaf-sheaths partly reddish or reddish purple, reticulate-fibrous; upper 1 to 3 spikes staminate, 2–4 cm. long, the lower 2–4 pistillate, sometimes androgynous or partially staminate, cylindric, 2–6 cm. long, pendulous on long capillary peduncles, the lower 1 or 2 bracts leaf-like; pistillate scales lanceolate to ovate, purplish brown; perigynia shorter than the scales, elliptic, 3–3.5 mm. long, unequally biconvex, glaucous, obscurely nerved, rather firm, abruptly very short beaked, with an entire tip; stigmas 2.——Abundant in wet places along seacoasts in Hokkaido, and in shallow ponds of subalpine regions in Honshu (Echigo, Rikuchu, and Ugo Prov.)——Sakhalin, Kuriles to Siberia and N. America.

42. Carex aphyllopus Kuekenth. TATEYAMA-SUGE. Rhizomes loosely tufted and short-creeping, forming a large loose mat; culms 30–100 cm. long, scabrous on the angles above, rather soft; leaf-blades 3–5 mm. wide, soft, flat, the basal sheaths bladeless, reddish purple, reticulate-fibrous; staminate spikes 1 to 3, broadly linear, the pistillate 2 to 4, sometimes with a short staminate portion above, 2–5 cm. long, erect or nodding, sessile or peduncled, dark reddish purple, the lower 1 or 2 bracts leaf-like, nearly as long as the inflorescence; pistillate scales ovate, dark purplish crimson to dark blood-red, awnless or sometimes mucronate or awned (forma *aristata* Ohwi. NAGABO-TATEYAMA-SUGE); perigynia about 3 mm. long, broadly ovate-oval, planoconvex, sometimes with 2 or 3 very slender nerves on each side, rather abruptly short-beaked, subentire; stigmas 2.——June–July. Wet sunny slopes in high mountains; Honshu (centr. and n. distr. on the Japan Sea side); locally common.

43. Carex impura Ohwi. *C. aphyllopus* var. *impura* (Ohwi) T. Koyama——HIRUZEN-SUGE. Related to the preceding; culms 50–80 cm. long, nearly smooth; basal leaf-sheaths brownish red; spikes 4 or 5, approximate, erect, sessile, the upper 1 or 2 staminate, brownish purple, the lower 1–3(–4) pistillate, short-cylindric, 1–2 cm. long, the lowest bract leaf-like; pistillate scales ovate, brownish purple, long-cuspidate; perigynia slightly longer than the scales, ovate, 3–5 mm. long, brownish, somewhat turgid, faintly nerved, slightly recurved toward apex, gradually narrowed to a rather conspicuous beak, entire at the tip; stigmas 2.——June–July. Honshu (Mount Hiruzen in Mimasaka Prov.).

44. Carex cinerascens Kuekenth. *C. micrantha* Kuekenth.; *C. ouensanensis* Ohwi——NUMA-AZE-SUGE. Rhizomes long; culms loosely tufted, about 60 cm. long; leaf-blades rather firm, 2–3 mm. wide, the lower leaf-sheaths bladeless, brownish; spikes 3 to 5, the upper ones continuous, the terminal staminate, 2–4 cm. long, the lateral pistillate, short-cylindric, 1.5–4 cm. long, erect, sessile, the lowest bract leaflike, about as long as the inflorescence; pistillate scales oblong, obtuse, dark brown, with white margins, 3-nerved; perigynia slightly longer than the scales, 2 mm. long, pale green, planoconvex, nerveless, nearly beakless, with entire mouth; stigmas 2.——May–June. Wet river banks in lowlands; Honshu (Tōhoku and Kantō Distr.); rare.——Korea.

45. Carex heterolepis Bunge. *C. latinervia* Lév. & Van't;. *C. periculosa* Honda; *C. tobae* Honda——YAMA-AZE-SUGE. Rhizomes stout, extensively long-creeping; culms 20–60 cm. long, acutely angled, the angles scabrous in upper part; leaf-blades 3–5 mm. wide, rather firm, the basal sheaths (at least the lower ones) bladeless, brownish stramineous, subrigid, splitting somewhat into fibers; spikes 3 to 7, erect, the terminal staminate, the others pistillate, cylindric, contiguous to approximate, sessile, 2–6 cm. long, densely many-flowered, the bracts leaf-like; pistillate scales oblong, dark brown or dark purplish brown; perigynia slightly longer than the scales,

2.5–3 mm. long, nerveless, often with yellow resinous spots, smooth, abruptly contracted into a short minutely 2-toothed beak; stigmas 2.——May–June. Wet places along streams in low mountains; Hokkaido (sw. distr.), Honshu, Kyushu (n. distr.); rather common.——Korea and n. China.

46. Carex eleusinoides Turcz. HIME-AZE-SUGE. Resembles somewhat *Carex thunbergii;* culms 10–30 cm. long; leaves flaccid, the blades about 2 mm. wide, the basal sheaths in part purplish brown; spikes 3–5, contiguous, short-cylindric to oblong, the terminal gynecandrous or staminate, the lateral pistillate, 1–1.5 cm. long, the lower 1 or 2 bracts leaf-like, usually longer than the culm; pistillate scales oblong, obtuse, dark purple; perigynia longer than the scales, 2.5 mm. long, oval, pale green, minutely punctate, weakly nerved, short-beaked, with an entire tip; stigmas 2.——Alpine regions; Hokkaido; rare.——n. Kuriles, n. Korea, and e. Siberia.

47. Carex bigelowii Torrey. *C. concolor* sensu auct., non R. Br.; *C. hyperborea* Drejer; *C. vulgaris* var. *hyperborea* (Drejer) Boott; *C. rigida* sensu auct. plur., non Gooden.——OHAGURO-SUGE. Rhizomes ascending; culms 10–30 cm. long, acutely angled, smooth; leaves rigid, the blades 3–4 mm. wide, flat, somewhat firm, the basal sheaths sometimes bladeless, firm, castaneous, shining; spikes 4 or 5, fastigiate, the terminal staminate, 10–15 mm. long, the lateral pistillate, oblong, 1–2 cm. long, the upper ones sessile, the lowest one short-peduncled, the lowest bract short, leaflike; pistillate scales elliptic, dark brown, rounded at the apex, narrowly to scarcely white-hyaline on margins; perigynia as large as or slightly larger than the scales, obovate, 2.5 mm. long, green, becoming brownish at maturity, turgid, densely puncticulate, nerveless, abruptly very short-beaked, with an entire tip; stigmas 2.——Alpine slopes; Hokkaido (Mount Daisetsu); rare.——n. Korea, Kamchatka, Siberia, n. Europe, and N. America.

48. Carex forficula Fr. & Sav. var. **forficula.** TANIGAWA-SUGE. Densely tufted in large clumps; culms 30–50 cm. long, scabrous above on the angles; leaf-blades 2–4 mm. wide, rather stiff, the basal sheaths bladeless, dark brown, splitting ventrally into reticulate fibers; spikes 3 to 6, erect, contiguous except the lower, the terminal staminate, linear, the lateral pistillate, erect, cylindric, 2–5 cm. long, sessile except the often short-peduncled lowest one, the lowest bract leaflike, longer than the inflorescence; pistillate scales dark purplish brown, with a green midrib, 3-nerved, short-mucronate; perigynia slightly longer than the scales, 3.5–4 mm. long, moderately appressed, nerveless, abruptly rather long-beaked, coarsely scabrous on the margins, the apex 2-toothed; stigmas 2.——May–June. Along mountain streams; sw. Hokkaido, Honshu, Shikoku, Kyushu.——Korea and n. China.

Var. **scabrida** Kuekenth. *C. diamantina* Lév. & Van't.——Ō-TANIGAWA-SUGE. Spikelets slightly larger; perigynia scattered-spinose.——Kyushu (Tsushima).——Korea.

49. Carex maximowiczii Miq. var. **maximowiczii.** *C. pruinosa* var. *picta* Franch.; *C. pruinosa* subsp. *maximowiczii* (Miq.) Kuekenth.——GŌ-SO. Tufted, with ascending short rhizomes; culms 40–70 cm. long, acutely angled, the angles scabrous in upper part; leaves 4–6 mm. wide, rather soft, flat, the basal sheaths bladeless, soft, cinnamon-colored, scarcely splitting ventrally; spikes 2 to 4, approximate, the terminal staminate, linear, ferrugineous, the lateral pistillate, short-cylindric, 2–3.5 cm. long, 10–14 mm. wide, the lowest bract leaflike, longer than the inflorescence; pistillate scales ovate, obtuse, ferrugineous, hyaline on the margins, with a green midrib, mucronate; perigynia longer than the scales, oval-elliptic, 3.5–4.5 mm. long, inflated, biconvex, grayish green to grayish brown, densely papillose, abruptly very short-beaked, entire at tip; stigmas 2.——Wet places; Hokkaido, Honshu, Kyushu; quite common.——Ryukyus, China, Korea, and s. Kuriles.

Var. **levisaccus** Ohwi. *C. pruinosa* var. *levisaccus* (Ohwi) Makino & Nemoto——HOSHI-NASHI-GŌ-SO. Perigynia vivid green, smooth, not papillose.——Rarely occurs with the typical phase.

50. Carex subcernua Ohwi. TSUKUSHI-NARUKO. Culms 40–60 cm. long, tufted, acutely angled, scabrous above on the angles; leaf-blades 3–4 mm. wide, firm, the basal sheaths (at least the lower ones) bladeless, dark brown, more or less fibrous; spikes 4 or 5, contiguous, the terminal gynecandrous, about 3 cm. long, the lateral pistillate, with few staminate flowers at the base, 2.5–4 cm. long, the lower ones distant and peduncled, pendulous, the lower bracts 1 or 2, leaflike, elongate; pistillate scales oblong, ferrugineous, acute or subacute, mucronate; perigynia slightly longer and broader than the scales, 3–3.5 mm. long, brownish green, broadly elliptic, densely papillose, prominently 4- or 5-nerved, compressed-biconvex, abruptly very short-beaked, entire at the tip; stigmas 2.——May–June. Wet places; Kyushu; rare.

51. Carex dimorpholepis Steud. *C. cernua* Boott, non J. F. Gmel.; *C. rubescens* Boecklr.——AZE-NARUKO. Plant densely tufted; culms 40–80 cm. long, acutely angled, scabrous on the angles above; leaves with flat blades 4–10 mm. wide, the basal sheaths bladeless, dark cinnamon-brown, slightly to scarcely splitting into fibers ventrally; spikes 4 to 6, approximate, pendulous, cylindric, 3–6 cm. long, peduncled, all or the upper ones gynecandrous, the others pistillate, the lower 2 or 3 bracts leaflike, much longer than the culms; pistillate scales narrowly obovate, with short ferrugineous striations, emarginate, the midrib 3-nerved, prolonged into an erect awn; perigynia longer than the scales, broadly ovate, 2.5–3 mm. long, compressed, nerveless, densely papillose, brownish when dry, abruptly very short-beaked, with a ferrugineous entire tip; stigmas 2.——May–June. Wet fields in lowlands; Honshu, Shikoku, Kyushu; common.——Korea, China, Ryukyus, and Indochina.

52. Carex phacota Spreng. *C. gracilipes* Miq., pro parte; *C. jauriei* Franch.; *C. cincta* Franch.; *C. cincta* var. *subphacota* Kuekenth.; *C. subphacota* (Kuekenth.) Nakai; *C. shichiseitensis* Hayata——HIME-GŌ-SO, AO-GŌ-SO. Tufted plant; culms 20–60 cm. long, acutely angled, scabrous on the angles; leaf-blades 2–6 mm. wide, cinereous to glaucous-green, rather firm, the basal sheaths nearly bladeless, brown to yellowish, sparingly fibrous; spikes 3 to 5, the terminal usually staminate, the lateral pistillate, very often with a short staminate portion at apex, cylindric, 2–6 cm. long, erect, declinate or sometimes pendulous, approximate to contiguous, usually long-peduncled, the lower bracts leaf-like, much longer than the culm; pistillate scales narrowly obovate, ferrugineous, truncate to emarginate, 3-nerved, the green midrib prolonged into a long slightly recurved awn; perigynia nearly twice as long as the body of the scales, elliptic, 2.5–3.5 mm. long, compressed-biconvex, almost nerveless, densely papillose, cinereous-green to cinnamon-colored, or becoming dark brown when dried, abruptly beaked, very short-emarginate at the tip; stigmas 2.——May. Wet places in lowlands; Hokkaido, Hon-

shu, Shikoku, Kyushu; common.——Korea, Ryukyus, Formosa, China, India, and Malaysia.

53. Carex otaruensis Franch. *C. cardioglochis* Lév. & Van't.; *C. mitoensis* Ohwi; *C. prescottiana* var. *otaruensis* (Franch.) Kuekenth.——OTARU-SUGE, HIME-TEKIRI-SUGE, MITO-SUGE. Tufted plant; culms 30–60 cm. long, acutely angled, scabrous above on the angles, firm; leaf-blades 3–5 mm. wide, rather flat, the basal sheaths bladeless, brownish to reddish, conspicuously fibrous-reticulate ventrally; spikes 4 or 5, approximate or slightly distant, the terminal staminate, elongate-linear, 3–6 cm. long, the lateral pistillate, cylindric, 3–5 cm. long, the lower ones nodding to pendulous on long peduncles, the lower bracts leaflike; pistillate scales pale, often slightly brownish, narrowly oblong, acute; perigynia slightly longer than the scales, 2.5–3 mm. long, obovate, biconvex, nerveless, vivid green, glabrous, rather abruptly short-beaked with a hyaline emarginate tip; stigmas 2.——May–June. Bordering streams and in mountain ravines; Hokkaido (sw. distr.), Honshu, Shikoku, Kyushu; occasional.

54. Carex aequialta Kuekenth. *C. prescottiana* var. *fuscocinnamomea* Kuekenth.——TODA-SUGE, AWA-SUGE. Rhizomes relatively thick; culms few, tufted, 30–60 cm. long, acutely angled, nearly smooth; leaf-blades 3–5 mm. wide, nearly smooth, glaucous beneath, the basal sheaths bladeless, dark brown, fibrous; spikes 3 or 4, contiguous, erect, fastigiate, the terminal staminate, linear, ferrugineous, usually shorter than the succeeding ones, the lateral spikes pistillate, often with a short staminate portion at the apex, cylindric, 3–5 cm. long, the lowest one peduncled, the lowest bract leaflike, much longer than the inflorescence; pistillate scales pale ferrugineous, narrowly oblong, membranous, rounded or mucronate; perigyna nearly as long as and much broader than the scales, 2.5–3 mm. long, much inflated, ovate-orbicular, brownish, membranous, faintly nerved, abruptly very short-beaked, entire at the tip; style flexuous, stigmas 2.——Apr.–May. Wet river banks; Honshu (Kantô Distr. and Owari Prov.), Kyushu (n. distr.); rare.——China.

55. Carex flabellata Lév. & Van't. *C. prescottiana* sensu auct. Japon., non Boott; *C. prescottiana* var. *flabellata* (Lév. & Van't.) Ohwi——YAMA-TEKIRI-SUGE. Rhizomes tufted; culms 40–60 cm. long, slender, glabrous, acutely angled; leaf-blades 4–8 mm. wide, cinereous-green, nearly smooth, the basal sheaths bladeless, castaneous, fibrous; spikes 3 to 5, nodding or somewhat pendulous, approximate, the terminal staminate, linear, the lateral pistillate, cylindric, 2–5 cm. long, the lower slightly nodding on rather long, nearly smooth peduncles, the lowest bract leaflike; pistillate scales oblong, truncate, pale or somewhat yellowish, the 3-nerved midrib prolonged into a short awn; perigynia longer than the scales, elliptic, 2.5–3 mm. long, somewhat turgid, biconvex, light green or yellowish, few-nerved, abruptly very short-beaked, with an emarginate tip.——May–June. Wet slopes; Yezo, Honshu; rather rare.

56. Carex kiotensis Fr. & Sav. *C. fuscescens* Boecklr.; *C. jizogatakensis* Lév. & Van't.; *C. prescottiana* var. *kiotensis* (Fr. & Sav.) Kuekenth. & var. *fuscescens* (Boeckl.) Kuekenth.——TEKIRI-SUGE. Closely resembles the preceding but the culms scabrous on the angles, at least in the upper part, and the leaves distinctly scabrous, the basal sheaths dark brown; spikes 5 to 7, 3–10 cm. long, on scabrous peduncles; pistillate scales obovate, pale, truncate or rounded, the 3-nerved midrib ending in a small mucro; perigynia nearly twice as long as the scales, 2–2.5 mm. long, not turgid, nearly nerveless.——May–June. Wet slopes; Hokkaido, Honshu, Shikoku, Kyushu; rather common.

57. Carex middendorffii F. Schmidt. *C. levicaulis* Franch.; *C. grandilimosa* Akiyama——Tomari-suge, Kuro-suge, Horomui-suge. Rhizomes ascending, stout, firm, branched, tufted; culms stout, rigid, 30–70 cm. long, acutely 3-angled; leaf-blades firm, conduplicate or flat, 2–4 mm. wide, grayish green, the basal sheaths rigid, bladeless, stramineous, rarely partly reddish, fibrous; spikes 3–5, the upper 2 or 3 staminate, clavate or lanceolate, the lower 1 or 2 pistillate or with a short staminate part at tip, oblong to ellipsoidal, 1.5–3 cm. long, only the lowest one long-peduncled and nodding, the lowest bract short, leaf-like; pistillate scales narrowly ovate, dark purplish brown, rather obtuse; perigynia broadly elliptic, 3.5–4.5 mm. long, grayish green, planoconvex, thick-membranous, densely punctate, prominently few-nerved, abruptly very short-beaked, with a firm, entire tip; stigmas 2.——June–July. Sphagnum bogs; Hokkaido, Honshu (n. and centr. distr.); common.——Kuriles, Sakhalin, Kamchatka, and Ussuri.

Var. **kirigaminensis** (Ohwi) Ohwi. *C. leiogona* Franch.; *C. caulorrhiza* Lév. & Van't.; *C. semiplena* Kuekenth.; *C. caulorrhiza* var. *kirigaminensis* Ohwi——Oni-aze-suge, Kita-aze-suge, Kirigamine-suge. Differs from the typical phase in the erect cylindric spikes 2–4 cm. long, the upper ones sessile, the lowest one short-peduncled.——Occasionally found with the typical phase.

58. Carex schmidtii Meinsh. *C. aperta* Boott, pro parte; *C. subvaginata* Meinsh.; *C. vladimiroviensis* Lév.——"Schmidt"-suge, Shumitto-suge, Kōrai-aze-suge. Rhizomes densely tufted; culms 50–70 cm. long, acutely 3-angled, prominently scabrous on the angles above; leaf-blades 2–4 mm. wide, the basal sheaths bladeless, lustrous, dark brown, fibrous; spikes 3 or 4, approximate, the upper 1 or 2 staminate, linear, the lower 1 or 2 pistillate, cylindric, sessile, 1–3 cm. long, the lowest bract leaflike, not longer than the inflorescence; pistillate scales broadly lanceolate, dark castaneous, rather obtuse; perigynia broader than and nearly as long as the scales, 2.5 mm. long, turgid, biconvex, nerveless, minutely punctate, brownish, abruptly and very shortly beaked, with an entire tip; stigmas 2.——Sphagnum bogs; Hokkaido; rare.——e. Siberia, n. Korea, Sakhalin, Kamchatka, and Kuriles.

59. Carex caespitosa L. var. **caespitosa**. *C. usta* Franch., non Bailey; *C. rubra* Lév. & Van't.; *C. caespitosa* var. *rubra* (Lév. & Van't.) Lév.——Kabu-suge, Kuro-o-suge. Plant densely tufted in large clumps; culms rather slender, 40–70 cm. long, scabrous, acutely angled; leaf-blades rather firm, scabrous, 2–3 mm. wide, the basal sheaths bladeless, dark reddish, prominently reticulate-fibrous; spikes 2–4, approximate toward the top of the culm, the terminal staminate, linear, the lateral pistillate, sessile, short-cylindric, 1.5–3 cm. long, about 4 mm. thick, erect, the bracts short, the lowest setaceous or scalelike; pistillate scales oblong, rather obtuse, dark purplish brown; perigynia longer than the scales, elliptic, 2–2.5 mm. long, minutely punctate, nerveless, abruptly very short-beaked, the tip entire; stigmas 2.——June–July. Wet places, especially in sphagnum bogs; Hokkaido.——Cooler regions of Eurasia.

Var. **minuta** (Franch.) Kuekenth. *C. minuta* Franch.——Kuro-me-suge. Basal sheaths dark reddish; pistillate spikes shorter, 7–15 mm. long, oblong; perigynia smaller.——Wet places, especially in sphagnum bogs; Hokkaido.——Sakhalin.

60. Carex angustisquama Franch. Yama-tanuki-ran. Culms tufted, 30–50 cm. long, scabrous above; leaf-blades 3–5 mm. wide, the basal sheaths bladeless, brownish crimson; spikes 4–6, approximate, the terminal staminate or androgynous, the lateral pistillate or with a few staminate flowers at the apex, oblong or ovoid, 1.5–2.5 cm. long, densely many-flowered, erect or the lower nodding on a rather long peduncle, the lower bracts leaflike; pistillate scales narrowly ovate, dark purplish brown, acute; perigynia longer and broader than the scales, 4–5 mm. long, oval, much appressed, thin-membranous, glabrous, with 2 spinulose-scabrous, slender marginal nerves, otherwise nerveless, abruptly and very short-beaked, the tip minutely 2-toothed; achenes very small; stigmas 2.——June–Aug. Subalpine regions, especially in wet gravelly places in volcanoes; Honshu (n. distr.).

61. Carex doenitzii Boecklr. var. **doenitzii**. *C. plocamo-styla* Maxim.; *C. dicuspis* Franch.; *C. dicraea* C. B. Clarke; *C. nagatadakensis* Masam.——Ko-tanuki-ran. Rhizomes densely tufted, short, firm, ascending, with stout, yellow, tomentose roots; culms 30–60 cm. long; leaf-blades 3–5 mm. wide, stiffish, glaucous beneath, the basal sheaths bladeless, lustrous, dark red or brown, firm, reticulate-fibrous; spikes 2 or 3, the upper 1 or 2 staminate, linear-clavate, the others pistillate or sometimes with a short staminate portion at the apex, oblong to narrowly obovate, densely flowered, 1.5–3 cm. long, nodding, the lower ones long-peduncled, the bracts 1 or 2, leaflike, not sheathing at base; pistillate scales lanceolate, acute, dark sanguineous, the green, 3-nerved midrib prolonged into an erect awn; perigynia shorter than the scales, 4 to 6(–9) mm. long, erect, narrowly ovate, biconvex, brown-striate, sparsely spinulose, gradually long-beaked, scabrous, deeply bifid or bifurcate at apex, the lobes short, bristlelike; stigmas 2, very long, persistent.——June–July. Rocks in mountains; Honshu (n. to centr. distr. and s. Kinki), Kyushu (Yakushima). Very polymorphic.

Var. **okuboi** (Franch.) Kuekenth. *C. okuboi* Franch.——Shima-tanuki-ran. Spikelets 4–7, contiguous, thick, 1.5–2 cm. long, sessile.——Dry sandy mountain slopes; Honshu (Idzu Isls.).

62. Carex scitaeformis Kuekenth. Ō-tanuki-ran. Plant rather large, tufted; culms stout, 40–70 cm. long, soft; leaf-blades broadly linear, flat, rather soft, the basal sheaths reddish to cinnamon-colored, sparingly reticulate-fibrous, bladeless or nearly so; spikes 4 to 7, the upper 2 or 3 staminate, the others pistillate or rarely androgynous, short-cylindric to oblong, pendulous or nodding on capillary peduncles, the lower 1 or 2 bracts leaflike, not sheathing at base; pistillate scales broadly lanceolate, dark purplish sanguineous, acute and awned; perigynia ovate or broadly elliptic, appressed, about 5 mm. long, thin-membranous, rather flat, sparsely setulose on the margins, short-stipitate, gradually narrowed into a beak with a minutely 2-toothed apex; achenes never maturing; stigmas 2.——Rocky places in mountains; Honshu (Japan Sea side in n. and centr. distr.); rare. This is probably a natural hybrid of *C. aphyllopus* × *C. podogyna*.

63. Carex podogyna Fr. & Sav. *C. trichopoda* Franch.; *C. doenitzii* var. *trichopoda* (Franch.) Kuekenth.——Tanuki-ran. Culms stout, tufted, 30–100 cm. long; leaf-blades 5–10 mm. wide, soft, partially cinnamon-colored, the ligules thin, the basal-sheaths bladeless, reddish to cinnamon-colored; spikes 3–6, somewhat distant, the upper 1–3 staminate, lanceolate, the others pistillate, pendulous on long capillary peduncles, ovoid to nearly globose, densely many-flowered, 2–4 cm. long, the lower bracts leaflike, not sheathing; pistillate scales lanceolate, dark purplish brown, obtuse and short-awned; perigynia longer than the scales, linear-lanceolate, 12–15 mm. long, compressed, light green, often dark brown variegated, thin-membranous, nerveless, long-stipitate, hairy on the margins and sometimes sparsely so on the faces, gradually long beaked, with a 2-toothed apex; stigmas 2.——June–July. Wet cliffs and rocky banks of ravines; Hokkaido (sw. distr.), Honshu (n. and centr. distr. especially on the Japan Sea side).

64. Carex lehmannii Drejer. *C. hidewoi* Ohwi——SEN-JŌ-SUGE. Plant loosely tufted, short rhizomatous; culms 20–30 cm. long, glabrous; leaf-blades 3–4 mm. wide, deep green, thin, flat, rather soft, the basal sheaths bladeless, soft, dark reddish, somewhat fibrous-reticulate ventrally; spikes 3 or 4, contiguous, the terminal gynecandrous, the others pistillate, densely flowered, erect, elliptic to oblong, 6–8 mm. long, short-peduncled (the lowest one rather long-peduncled), the lowest bract leaflike, much longer than the inflorescence, not sheathing; pistillate scales small, broadly ovate, dark purplish brown; perigynia twice as long as the scales, ellipsoidal, about 2 mm. long, spreading, inflated, inconspicuously 3-angled, orange-yellow, firm-membranous, faintly nerved, abruptly short-beaked, with an entire, dark brown tip; stigmas 3, short. ——July–Aug. Alpine forests; Honshu (Senjo-dake in Kai Prov.); very rare.——India, w. China, s. Siberia, and n. Korea.

65. Carex peiktusanii Komar. *C. hancockiana* var. *peiktusanii* (Komar.) Kuekenth.——MANSHŪ-KURO-KAWA-SUGE. Culms tufted, 40–60 cm. long, rather slender, scabrous above on the angles, few-leaved at purplish base; leaf-blades 2–4 mm. wide, slightly paler beneath, soft, scabrous on margin, the basal sheaths with blades, dark purplish, slightly lustrous, sparingly reticulate-fibrous; spikes 3 or 4, nodding, the terminal gynecandrous, clavate, the lateral pistillate, oblong to oblong-cylindric, 10–25 mm. long, on long setaceous peduncles, the lowest bract foliaceous, as long as or longer than the inflorescence; pistillate scales ovate, ascending, pale yellowish brown, membranous, acute and short-awned; perigynia nearly equaling the scales, ovoid, 3–3.5 mm. long, ascending, inflated-trigonous, pale green, membranous, slenderly nerved, short-beaked, the tip ferruginous, acutely bidentate; achenes loosely invested by the perigynium; stigmas 3, short, slender.——July–Aug. Calcareous rocks in mountains; Honshu (Mount Toyokuchi in Shinano Prov.); very rare.——Korea and Manchuria.

66. Carex augustinowiczii Meinsh. *C. bidentula* Franch.; *C. soyaeensis* Kuekenth.; *C. infirma* C. B. Clarke; *C. flaccidior* (F. Schmidt) Miyabe & Kudo; *C. eleusinoides* var. *flaccidior* F. Schmidt——HIRAGISHI-SUGE, EZO-AZE-SUGE. Rhizomes densely tufted, short-creeping; culms 30–50 cm. long, soft, scabrous above; leaf-blades soft, flat, 2–4 mm. wide, the lower basal sheaths bladeless, very short, lustrous, castaneous, the upper sheaths stramineous, soft, elongate, bladed; spikes 4–6, sessile, approximate, the terminal staminate or rarely with a few pistillate flowers at the apex, the others pistillate or sometimes with a few staminate flowers at base, oblong to short cylindric, 1–3 cm. long, erect, the lowest bract leaflike, not sheathing; pistillate scales narrowly ovate, dark purplish brown or dark sanguineous, rather small; perigynia longer than the scales, ovate to ovate-lanceolate, 2.5–3(–4) mm. long, somewhat turgid, obtusely 3-angled, pale green, thin-membranous, faintly several-nerved, with an abrupt, short recurved beak, with an entire tip; stigmas 3.——June–July. Moist mossy rocks along torrents and in ravines in subalpine re-gions; Hokkaido, Honshu (centr. and n. distr.).——Sakhalin. Polymorphic.

Var. **sharensis** (Franch.) Ohwi. *C. sharensis* Franch.—— SHARI-SUGE. Perigynia 4 mm. long, slightly narrower than in the typical phase, with a longer beak.——June–July. Hokkaido.

67. Carex curvicollis Fr. & Sav. *C. viridula* Fr. & Sav., non Michx.——NARUKO-SUGE. Plant densely tufted, short-rhizomatous; culms 20–40 cm. long, glabrous, slender, nodding above; leaf-blades soft, thin, flat, 2.5–3.5 mm. wide, the basal sheaths pale or the lower ones small, scalelike and dark brown; spikes 3–5, the terminal staminate, linear, the lateral pistillate, short-cylindric, 15–40 mm. long, sessile, or the lowest one peduncled, nodding, the lowest bract not over-topping the culm, not sheathing; pistillate scales small, ovate, dark brown or dark purplish brown; perigynia 2 or 3 times as long as the scales, lanceolate, 4–5 mm. long, obsoletely angled, light or pale green, thinly membranous, faintly nerved, gradually long slightly recurved-beaked, with an entire membranous tip; achenes small, obovate-orbicular; style rather long, the stigmas short, 3-fid.——Along streams in low mountains; Hokkaido (sw. distr.), Honshu, Shikoku, Kyushu; common.

68. Carex mertensii Presc. var. **urostachys** (Franch.) Kuekenth. *C. urostachys* Franch.; *C. mertensii* sensu auct. Japon., non Presc.——KINCHAKU-SUGE, IWAKI-SUGE. Culms tufted, 30–60 cm. long, nodding above; leaf-blades 4–8 mm. wide, soft, flat, the ligule ferrugineous, the basal leaf-sheaths bladeless, soft, cinnamon-colored; spikes 5–8, contiguous, nodding, densely many-flowered, short-cylindric or oblong, 2–3 cm. long, peduncled, all pistillate but with a few staminate flowers at base, the lower 1 or 2 bracts leaflike, longer than the inflorescence; pistillate scales narrowly ovate, dark purple, gradually narrowed to the short-awned tip; perigynia longer than the scales, appressed, 4 mm. long, ovate-elliptic, very much compressed, glabrous, thin-membranous, nearly white, slenderly 2-nerved on the dorsal side, abruptly very short beaked, with a dark-colored entire tip; achenes very small; stigmas 3, short.——July. Sandy and gravelly slopes in sub-alpine regions, especially on volcanoes; Hokkaido, Honshu (n. and centr. distr. mainly on the Japan Sea side).——s. Kuriles. The typical phase occurs in N. America.

69. Carex atrata L. var. **japonalpina** T. Koyama. *C. japonalpina* (T. Koyama) T. Koyama——KUROBO-SUGE. Culms 20–50 cm. long, in small loose tufts, slightly nodding above; leaf-blades 3–5 mm. wide, the basal sheaths dark purplish red; spikes 3 to 5, contiguous, the terminal gynecandrous, the others pistillate, approximate, oblong to narrowly so, 10–25 mm. long, nodding, peduncled, the lowest bract long-bristleform, as long as the inflorescence; pistillate scales ovate, dark brown, acute; perigynia nearly as long as and slightly broader than the scales, elliptic, 3–3.5 mm. long, membranous, tawny-flecked, nerveless, abruptly short-beaked, slightly emarginate; achenes loosely inclosed, small; stigmas 3, short.—— July–Aug. Alpine slopes; Honshu (centr. mountains); very rare.——n. Korea. A variable widespread species, Siberia to Europe and N. America.

70. Carex gmelinii Hook. & Arn. NEMURO-SUGE. Rhizomes tufted; culms 30–70 cm. long, slender, often nodding above; leaf-blades 3–5 mm. wide, the basal sheaths dark sanguineous; spikes 3–5, contiguous, oblong to narrowly so, 1.5–3 cm. long, the terminal gynecandrous, the others pistillate, sessile, or the lower ones peduncled, the bracts nearly leaflike, not sheathing; pistillate scales ovate-elliptic, dark brown, the yellowish midnerve excurrent into a short erect awn; perigynia somewhat coriaceous, longer than the scales, elliptic, 4–5 mm. long, turgid, planoconvex, strongly nerved, stramineous to light cinnamon-brown, abruptly very short-beaked, with a nearly entire dark colored tip; stigmas 3.——June–July. Sandy places along seacoasts; Hokkaido, Honshu (Mutsu Prov.).——Sakhalin, Kuriles, Ussuri, Kamchatka, and n. Korea.

71. Carex flavocuspis Fr. & Sav. *C. gansuensis* Franch.; *C. macrochaeta* subsp. *flavocuspis* (Fr. & Sav.) Kuekenth., incl. var. *denticulata* Kuekenth. and var. *platycarpa* Kuekenth.; *C. tolmiei* var. *denticulata* (Kuekenth.) Ohwi.——MIYAMA-KURO-SUGE, CHA-IRO-TANUKI-RAN. Rhizomes short-creeping; culms few, 10–50 cm. long, rather stout, glabrous; leaf-blades relatively stiff, glabrous, 3–5 mm. wide, the basal sheaths partly dark brown, the lower ones short and nearly bladeless; spikes 3–5, approximate, the terminal staminate, the others pistillate, oblong to short-cylindric, 1.5–3 cm. long, the lower ones peduncled, sometimes nodding, the lowest bract leaflike; pistillate scales narrowly ovate, dark brown, awned or pointed; perigynium 4–5 mm. long, compressed, elliptic, faintly nerved, smooth, somewhat membranous, abruptly very short-beaked, the apex entire or minutely 2-toothed.——July–Aug. Dry alpine slopes; Hokkaido, Honshu (n. and centr. distr.); rather abundant.——Kuriles, Sakhalin, and Kamchatka (var. *paramushirensis* Ohwi).

72. Carex scita Maxim. Culms loosely tufted, 20–70 cm. long; leaf-blades 3–5 mm. wide, the basal sheaths bladeless, reddish purple, slightly reticulate-fibrous; spikes 3–6, the terminal staminate, linear-oblong, the lateral pistillate, short-cylindric, 1–3 cm. long, peduncled, usually nodding or pendulous, the lower bracts leaflike; pistillate scales broadly lanceolate, dark purplish brown, awned; perigynia slightly longer than the scales, about 4 mm. long, membranous, brown-flecked, compressed-trigonous, several-nerved, setulose-scabrous on the margins and nerves; achenes loosely inclosed; stigmas 3.——Kuriles, Sakhalin, Kamchatka and the Okhotsk Sea region. A rather polymorphic species, with the following variants in our area:

Key to Varieties

1A. Leaves smooth, not papillose beneath; pistillate scales nearly as long as the perigynia, gradually narrowed toward the apex.
  2A. Culms glabrous; plants of Honshu.
    3A. Perigynia transversely wrinkled in upper part. . . Var. *scita*
    3B. Perigynia not wrinkled. . . . . Var. *brevisquama*
  2B. Culms scabrous; plants of Hokkaido.
    4A. Lower spikes distinctly peduncled. . . . . . Var *riishirensis*
    4B Spikes shot-peduncled or sessile. . . . . . Var. *scabrinervia*
1B. Leaves papillose beneath; pistillate scales half as long as the perigynia, abruptly contracted toward the apex.
                        Var. *parvisquama*

Var. **scita**. MIYAMA-ASHIBOSO-SUGE. Perigynia oblong-lanceolate to oblong, the upper part transversely wrinkled, the beak erect, somewhat immersed in the upper part of the perigynium.——Alpine slopes; Honshu (s. part of centr. distr.); rather rare.

Var. **brevisquama** (Koidz.) Ohwi. *C. tenuiseta* Franch.; *C. tenuiseta* var. *brevisquama* Koidz.——ASHIBOSO-SUGE, SHIRO-UMA-SUGE. Pistillate scales narrower than in var. *scita;* perigynia lanceolate, not wrinkled, the beak gradually narrowed,

not immersed.——July–Aug. Alpine slopes; Honshu (Japan Sea side of centr. and n. distr.).

Var. **riishirensis** (Franch.) Kuekenth. *C. ciliolata* Franch.; *C. riishirensis* Franch.——RISHIRI-SUGE. Culms slender, slightly scabrous above; lower spikes distinctly peduncled; perigynia oblong to ovate, not transversely wrinkled, abruptly short-beaked.——June–Aug. Alpine regions; Hokkaido; rather common.——Sakhalin, Kuriles, and Kamchatka.

Var. **scabrinervia** (Franch.) Kuekenth. *C. scabrinervia* Franch.; *C. xanthoathera* Franch.; *C. urolepis* Franch.; *C. scita* sensu auct. Japon., pro parte——SHIKOTAN-SUGE. Plant somewhat stouter; spikes short-peduncled or sessile, the upper 1 or 2 staminate; perigynia elliptic to broadly so, strongly compressed.——Rocks near seashore; Hokkaido (Rebun Isl. and Nemuro).——Sakhalin and Kuriles.

Var. **parvisquama** T. Koyama. DAISEN-ASHIBOSO-SUGE. Leaf-blades somewhat stiff, densely papillose and whitish beneath, long-acuminate; spikes approximate to contiguous; pistillate scales half as long as the elliptic perigynia.——Honshu (Mount Daisen in Hōki Prov.).

73. Carex meyeriana Kunth. *C. funicularis* Franch.; *C. crassinervia* Franch.——NUMA-KUROBO-SUGE, SHIRAKAWA-SUGE. Plant very densely tufted; culms 30–50 cm. long, firm, slender, scabrous; leaf-blades 1–1.5 mm. wide, firm, glaucous-green, conduplicate, the basal sheaths rigid, bladeless, chestnut-brown, lustrous, fibrous-reticulate; spikes 2 or 3, on the apical part of the culm, the terminal staminate, linear, 2–3 cm. long, the others pistillate, nearly globose, approximate, 5–10 mm. long, sessile, the bracts usually scalelike, short, not sheathing; pistillate scales ovate-oblong, dark purplish brown, obtuse; perigynia slightly broader than the scales, ellipsoidal, 3–3.5 mm. long, compressed-trigonous, cinereous-glaucous, densely papillose, several-nerved, abruptly short-beaked, the apex dark brown; stigmas 3, relatively thick.——May–July. Wet boggy places and high moors; Honshu (n. and centr. distr.), Kyushu.——Korea and e. Siberia.

74. Carex buxbaumii Wahlenb. *C. polygama* Schk., non J. F. Gmel.; *C. tarumensis* Franch.——TARUMAI-SUGE. Rhizomes long-creeping; culms few, 30–40 cm. long, acutely angled, scabrous above; leaf-blades grayish green, glaucescent, 2–3 mm. wide, the basal sheaths bladeless, reddish purple, fibrous; spikes 3 or 4, nearly approximate, the terminal gynecandrous or staminate or with a few pistillate flowers in the middle, the others pistillate, sessile, erect, oblong or short-cylindric, 1–2.5 cm. long, densely many-flowered, the bracts short; pistillate scales narrowly ovate-lanceolate, dark brownish purple, abruptly short-awned; perigynia longer than and slightly broader than the scales, elliptic, 3 mm. long, glaucous-green, densely and minutely papillose, somewhat compressed, several-nerved, abruptly very short-beaked, shallowly 2-toothed at apex; stigmas 3, rather thick.——June–July. Boggy places; Hokkaido; rare.——s. Kuriles, Europe, N. America, and Australia.

75. Carex pyrenaica Wahlenb. KIN-SUGE, SEI-TAKA-KIN-SUGE. Rhizomes short; culms densely tufted, 10–40 cm. long, obtusely angled, glabrous; leaf-blades 1–1.5 mm. wide, glabrous, rather thick; spikes solitary, terminal, androgynous, broadly lanceolate at anthesis, oblong when mature, 1–2 cm. long, densely rather many-flowered; pistillate scales narrowly ovate, deciduous, dark brown, obtuse; perigynia longer than the scales, 4–5.5 mm. long including the rather prominent stipe, spreading to reflexed when mature, broadly lanceolate, slightly compressed, nerveless, moderately long beaked; stigmas 3, slender.——July–Aug. Alpine slopes; Hokkaido, Honshu (n. and centr. distr.).——Europe, Siberia, and N. America.

76. Carex siroumensis Koidz. TAKANE-NARUKO. Culms slender, densely tufted, nodding above, 20–30 cm. long; leaf-blades 1–2.5 mm. wide, folded, the basal sheaths dark brown, sparsely fibrous; spikes 3 or 4, densely flowered, contiguous except the rather long-peduncled lowest one, the terminal gynecandrous and obovate or staminate and linear, 1–1.5 cm. long, the others pistillate, oblong-clavate, 1–2 cm. long, dark brown, subsessile, the lower bracts bristleform, sheathing at base; pistillate scales oblong-ovate, dark brown, acute; perigynia nearly twice as long as the scales, 5–6 mm. long, broadly lanceolate, membranous, compressed-trigonous, frequently brown-flecked, very faintly nerved, sparsely setulose-scabrous, with appressed to ascending short hairs, gradually tapering at both ends, short-beaked, with a hyaline, shallowly 2-toothed tip; stigmas 3.——July–Aug. Dry rocky places in alpine regions; Honshu (centr. distr.); rare.——n. Korea.

77. Carex hakkodensis Franch. ITO-KIN-SUGE. Rhizomes loosely tufted, with short ascending innovations; culms densely matted, 10–50 cm. long, slender, acutely angled, very scabrous; leaf-blades flat, rather thin and soft, 1.5–3 mm. wide, the basal sheaths membranous, not fibrous, pale; spike solitary, terminal, usually androgynous, rarely monoecious, usually oblong to narrowly so, the staminate part very short; pistillate scales oblong, castaneous, emarginate, the green midrib excurrent as a mucro; perigynia 6–8 mm. long including the stipe, nearly twice as long as the scales, suberect, moderately compressed, tawny, faintly or obsoletely nerved, lustrous, the beak very shallowly 2-toothed, with a hyaline tip; stigmas 3. ——July–Aug. Wet alpine slopes; Hokkaido, Honshu (n. and centr. distr.); rather rare.——Kuriles.

78. Carex stenantha Fr. & Sav. var. **stenantha**. *C. stenantha* var. *yatsugatakensis* Akiyama——IWA-SUGE. Plant loosely tufted, with short ascending innovations; culms slender, 15–40 cm. long, glabrous, 3-angled, nodding above; leaf-blades rather short, flat, 2–3 mm. wide, the basal sheaths partially dark red; spikes 3–5, distant, sparsely flowered, the lower ones often very long, on filiform peduncles, the terminal staminate, linear, 2–3 cm. long, the lateral pistillate, narrowly cylindric, long-peduncled, pendulous, dark reddish brown, 2–3 cm. long, the bracts bristleform, long-sheathing at base; pistillate scales oblong, rounded at apex, dark brown; perigynia longer than the scales, 6–8 mm. long, erect, membranous, lustrous, compressed-trigonous, distinctly but slenderly nerved, stramineous and often fuscous in upper part, slightly pubescent especially along the margins, short-stipitate, gradually long-beaked, with an obliquely truncate, hyaline tip; stigmas 3.—— July–Aug. Rocky and gravelly places in alpine regions; Honshu (n. and centr. distr.); common.

Var. **daisetsuensis** Akiyama. *C. ktausipalii* Meinsh.; *C. stenantha* sensu auct. Hokkaido, non Fr. & Sav.——TAISETSU-IWA-SUGE. Differs in the broader, oblong-lanceolate perigynia with a shorter beak.——Alpine regions; Hokkaido. ——Kuriles and Sakhalin (Mt. Ktausipal).

79. Carex makinoensis Franch. IWA-KAN-SUGE. Rhizomes densely tufted, the neck densely covered with dark brown comose fibrous remains of old leaf-sheaths; culms 30–50 cm. long, slightly scabrous above; leaf-blades 2–3 mm. wide,

stiff; spikes 3–5, approximate except the lowest, the terminal staminate, narrowly linear, 4–10 cm. long, the lateral pistillate, cylindric or short-cylindric, 2–4 cm. long, erect, the bracts long-sheathing at base; staminate scales dark chestnut-brown, minutely ciliate; pistillate scales yellowish-golden, ovate, acute; perigynia longer than the scales, 4–6 mm. long, nearly erect, oblong-fusiform, compressed-trigonous, membranous, stramineous, sparsely hairy, nerves slender, the beak acutely 2-toothed; stigmas 3.——Apr.–May. Mountains; Shikoku, Kyushu.——Ryukyus and Formosa.

79a. Carex phaeodon T. Koyama. HASHI-NAGA-KAN-SUGE. Perennial; rhizomes loosely tufted, long-creeping; leaves linear, plicate, 3–6 mm. wide, to 25 cm. long, rather rigid; basal sheaths short-bladed, light purple-brown to red-brown; culms 20–30 cm. long, smooth; spikelets 3–6, the terminal staminate, linear-clavate, brownish, 2–3 cm. long, the lateral pistillate or sometimes androgynous, oblong-cylindrical, 10–25 mm. long, about 7 mm. wide, densely many-flowered, peduncled, with sheathing bracts; pistillate scales oblong-elliptic to oblong-ovate, pale with brownish margins; perigynia erect-spreading, oblanceolate or linear-oblong, 4.8–5.3 mm. long, trigonous, smooth, thinly membranous, nearly nerveless except for lateral 2 keels, the beak long, with a bidentate mouth; achenes tightly inclosed; stigmas 3.——Wet rocks along streams; Honshu (Suruga Prov.); rare.

80. Carex chrysolepis Fr. & Sav. var. **chrysolepis**. *C. picea* var. *asensis* Lév. & Van't.; *C. odontostoma* var. *variegata* Kuekenth.; *C. kiusiuana* Ohwi——KO-IWA-KAN-SUGE. Rhizomes densely tufted; culms 15–30 cm. long, obtusely angled, nearly glabrous; leaf-blades flat, 1.5–3 mm. wide, the basal sheaths pale brown, fibrous when withered; spikes 3 or 4, approximate, the terminal staminate, linear-clavate, 1.5–3 cm. long, golden-brown, the others pistillate, slenderly oblong, erect, 8–20 mm. long, the lower ones usually peduncled, the lowest bract short, sheathing at base; pistillate scales narrow-ovate, golden-castaneous, truncate and mucronate; perigynia 4–5 mm. long, nearly erect, oblong to somewhat fusiform, membranous, light yellow-green, inconspicuously nerved, sparsely appressed-hispidulous, abruptly to gradually rather long-beaked, with a 2-toothed apex; stigmas 3.——Apr.–June. High mountains; Shikoku, Kyushu.

Var. **odontostoma** (Kuekenth.) Ohwi. *C. odontostoma* Kuekenth.——MIYAMA-IWA-SUGE. Perigynia narrow, 6–7 mm. long, the beak longer, 2-toothed.——Kyushu.——Formosa.

Var. **glabrior** (Ohwi) Ohwi. *C. odontostoma* var. *glabrior* Ohwi; *C. glabrior* (Ohwi) Akiyama——KANSAI-IWA-SUGE. Perigynia extremely narrow, nearly linear, sparsely appressed-hirsute only along the margins, the beak very long with a deeply 2-toothed apex.——Rocky places in high mountains; Honshu (Yamato and Tajima Prov.); rare.

81. Carex blepharicarpa Franch. var. **blepharicarpa**. *C. hayatae* Lév.; *C. yezomontana* Akiyama——SHŌJŌ-SUGE. Rhizomes densely to somewhat loosely tufted, the neck densely covered with brown fibers; culms 10–50 cm. long, angled, scabrous; leaf-blades rather stiff, 2–4 mm. wide, the basal sheaths castaneous or brown; spikes 2–5, approximate or contiguous, the lower ones distant or rarely the lowest radical, the terminal staminate, clavate, 1–3 cm. long, the lateral pistillate, oblong or oblong-cylindric, 1–3 cm. long, the bracts short, loosely long-sheathing at base; pistillate scales obovate-elliptic, ferrugineous to castaneous, rounded-truncate and mucronate; perigynia 4–6 mm. long, fusiform, nearly nerveless, pubescent, the beak very short with a 2-toothed apex; stigmas 3.——Common in mountains; Hokkaido, Honshu, Shikoku, Kyushu.—— Forma **dueensis** (Meinsh.) T. Koyama. *C. dueensis* Meinsh.; *C. blepharicarpa* var. *dueensis* (Meinsh.) Akiyama——TAKANE-SHŌJŌ-SUGE.   A rather dwarf alpine phase, with short-creeping, sparsely branching rhizomes forming a loose tuft.——Sakhalin, s. Kuriles, Korea (Utsuryoto Isl.).

Var. **stenocarpa** Ohwi. *C. hirtifructus* Kuekenth.; *C. blepharicarpa* var. *hirtifructus* (Kuekenth.) Ohwi——NAGAMI-SHŌJŌ-SUGE, TSUKUBA-SUGE.   Slender plant; perigynia long and narrow, long-beaked.——Usually in shallow soil over rocks; occurs within the area of the typical phase, especially common in the southern part.

82. Carex reinii Fr. & Sav.   *C. nambuensis* Franch.; *C. ogawae* Akiyama——KO-KAN-SUGE, NAMBU-SUGE.   Tufted and rhizomatous; culms 30–60 cm. long, glabrous; leaves fascicled, the blades 3–5 mm. wide, flat, deep green, stiff, coriaceous, lustrous above, the margins prominently antrorsely scabrous on the upper half and retrorsely so on the lower half, the basal sheaths brown, dark brown fibrous; inflorescence much interrupted; spikes 4–10, on simple or forked peduncles, solitary or ternate at each node, often all androgynous, 1.5–3 cm. long, the pistillate slightly shorter than the staminate, few-flowered, the bracts short-bladed, long-sheathing at base; pistillate scales ovate, dark brown, obtuse; perigynia 5–6 mm. long, twice as long as the scale, obliquely spreading, obovate-fusiform, plano-convex, many-nerved, sparsely hairy, gradually narrowed at both ends, with a short, recurved 2-toothed beak; stigmas 3.——Apr.–May. Woods in hills and low mountains; Honshu, Shikoku, Kyushu.

83. Carex formosensis Lév. & Van't. *C. ligata* var. *formosensis* (Lév. & Van't.) Kuekenth.; *C. formosensis* var. *kabashimensis* Ohwi; *C. shimotsukensis* Honda——TAIWAN-SUGE, Ō-MIYAMA-KAN-SUGE.   Rhizomes tufted; culms 30–50 cm. long, obtusely triquetrous, nearly glabrous; leaf-blades rather stiff, 2–6 mm. wide, the basal sheaths dark to pale brown, sparsely fibrous; spikes 3–7, erect, the terminal staminate, linear, 1–2 cm. long, the lateral pistillate, short-cylindric, 1–4 cm. long, loosely flowered, approximate to somewhat contiguous, the peduncles inclosed in the leaf-sheath and not exserted; staminate scales tubular-infundibuliform with the margins connate at base; pistillate scales oblong, pale green, obtuse and mucronate; perigynia 3–3.5 mm. long, exserted, erect, ovoid-fusiform, many-nerved, sparsely pubescent, obtuse on the angles, with a shallow constriction above the middle, the beak short with an emarginate tip.——Apr.–May. Woods; Honshu (Mount Takadate in Shimotsuke Prov.), Kyushu. ——s. Korea, Ryukyus, and Formosa.

84. Carex genkaiensis Ohwi. *C. formosensis* var. *vigens* (Kuekenth.) Ohwi; *C. mitrata* var. *vigens* Kuekenth.—— GENKAI-MOEGI-SUGE.   Resembles the preceding, the pistillate scales not mucronate; perigynia more plump, scarcely constricted above the middle, with a shorter beak; achenes very faintly constricted.——Apr.–May. Kyushu (Hizen and Chikuzen Prov. and Tsushima).——s. Korea.

85. Carex pudica Honda. *C. iseana* Akiyama——MAME-SUGE.   Culms tufted, hidden among the leaves, short, to about 15 cm. long; leaf-blades rather short, flat, much longer than the culms, 2–3 mm. wide, deep green, somewhat soft,

nearly glabrous, the basal sheaths stramineous; spikes 2–5, the terminal staminate, lanceolate, about 5 mm. long, few-flowered, long-peduncled, the lateral pistillate, all radical except sometimes the uppermost one which is often approximate to the terminal staminate spikelet, erect, usually oblong, 5–10 mm. long, peduncled, loosely few-flowered, the bracts leaflike, short-sheathing at base; pistillate scales partly reddish tawny, obovate, obtuse to nearly rounded and mucronate; perigynia longer than the scales, about 3 mm. long, nearly erect, ovoid-fusiform, obtusely 3-angled, sparsely puberulent, many-nerved, extremely short-beaked, with an entire tip.——Apr.–May. Wet open woods on hillsides; Honshu (Kinki Distr. eastw. to Kantō Distr.); uncommon.

86. Carex mitrata Franch. var. **mitrata**. *C. kingiana* Lév. & Van't.——NUKA-SUGE.   Rhizomes densely tufted, forming a large clump; culms 10 to 30 cm. long, slender, glabrous; leaf-blades 1.5–2 mm. wide, yellow-green, somewhat lustrous, overtopping the culms, the basal sheaths yellowish brown, lustrous; spikes 3 or 4, erect, approximate to contiguous, the terminal staminate, filiform, 5–10 mm. long, inconspicuous, the others pistillate, 5–12 mm. long, narrowly oblong to short-cylindric, the lowest one conspicuously peduncled, the bracts bladeless, rather long-sheathing; pistillate scales obovate, usually rounded-truncate at the apex, pale or pale brown with pale margins, the midrib usually not reaching the apex; perigynia slightly exserted, 2.3 mm. long, obovate, faintly nerved, sparsely puberulent to nearly glabrous, abruptly short-beaked, entire at apex; stigmas 3.——Apr.–May. Woods in low mountains; Honshu, Shikoku, Kyushu; rather common.——s. Korea.

Var. **aristata** Ohwi.   NOGE-NUKA-SUGE.   Lowest bract foliaceous; pistillate scales ending in a mucro or short awn. With the typical phase.——Formosa.

87. Carex subumbellata Meinsh. var. **subumbellata**. *C. depressa* var. *subumbellata* (Meinsh.) Kuekenth.——MI-YAKE-SUGE.   Plant tufted, with short ascending extravaginal innovations; culms 20–30 cm. long, glabrous; leaf-blades 2–3 mm. wide, the basal sheaths brownish, nearly entire; spikes 3–5, approximate, the terminal staminate, 10 mm. long, the others pistillate, oblong-cylindric, 1–2.5 cm. long, the lowest one distinctly peduncled and remote from the others, sometimes radical, the lower 1 or 2 bracts nearly leaflike and rather short-sheathing; pistillate scales obovate, castaneous, with a green midrib usually prolonged as a short mucro; perigynia nearly erect, 3 mm. long, obovate-oblong, sparsely puberulent or almost glabrous, with a 2-nerved midrib and a thickened, white, prominent stipe when mature, abruptly very short-beaked, with an emarginate tip; stigmas 3, short.——June–Aug. Alpine slopes; Hokkaido (Mount Yūbaridake).—— Sakhalin.

Var. **verecunda** Ohwi.   *C. heribaudiana* Lév. & Van't., pro parte; *C. artinux* C. B. Clarke, pro parte.——KUMOMA-SHIBA-SUGE.   Rhizomes very abbreviated; spikes all contiguous on the apical part of the culm, the lowest bract longer than the spike.——Alpine slopes; Honshu (centr. distr., including Iwashiro, Kozuke, and Shimotsuke Prov.); rather rare.

88. Carex nervata Fr. & Sav.   *C. praecox* sensu auct. Japon., non Jacq.; *C. praecox* var. *vidalii* Fr. & Sav.; *C. homoiolepis* Fr. & Sav.; *C. vidalii* Fr. & Sav. ex Franch.; *C. caryophyllea* subsp. *nervata* (Fr. & Sav.) Kuekenth.——SHIBA-SUGE.   Rhizomes very loosely tufted, long-creeping; culms few, 10–30 cm. long, glabrous; leaf-blades flat, 2–3 mm. wide, much shorter than the culm, the basal sheaths pale, splitting into brownish fibers when withered; spikes 2–4, contiguous, erect, the terminal staminate, clavate-linear, 10–15 mm. long, light greenish yellow, the others pistillate, oblong, 7–12 mm. long, short-peduncled, the lowest bract setaceous, usually very short-sheathing; pistillate scales obovate, pale to tawny, acute; perigynia 2–2.5 mm. long, obliquely spreading, strongly nerved, short-pubescent, abruptly short-beaked, emarginate at tip; stigmas 3.——Apr.–May. Sunny fields in lowlands and hills; Hokkaido (rare), Honshu, Shikoku, Kyushu; common. ——s. Korea.

89. Carex caryophyllea var. **microtricha** (Franch.) Kuekenth. *C. verna* var. *microtricha* (Franch.) Ohwi; *C. microtricha* Franch.; *C. squamoidea* Akiyama——CHA-SHIBA-SUGE, HAMA-SHIBA-SUGE. Rhizomes long-creeping, rarely loosely tufted, the neck sparsely covered with brown fibers; culms 10–40 cm. long, nearly glabrous; leaf-blades rather short, 2–3 mm. wide; spikes 2–4, approximate, or the lower ones distant, the terminal staminate, clavate-lanceolate, 1–2 cm. long, the others pistillate, oblong, 7–20 mm. long, peduncled, the lowest bract nearly setaceous, short-sheathing; pistillate scales ovate, castaneous, acute to mucronate; perigynia slightly longer than the scales, about 3 mm. long, obliquely spreading, obovoid, sparsely puberulent, sometimes glabrescent, faintly nerved, abruptly short-beaked, with an emarginate tip; stigmas 3.——May–July. Sunny slopes near the seacoast and in mountains; Hokkaido, Honshu (n. and centr. distr.).——Sakhalin, Kuriles, Kamchatka, Korea, and widely distributed in the northern part of e. Asia. The typical phase occurs in Europe and Siberia and is naturalized in N. America.

90. Carex sabynensis Less. *C. pediformis* var. *obliqua* Turcz.; *C. pediformis* var. *caespitosa* F. Schmidt; *C. kamikawensis* Franch.; *C. recticulmis* Franch.; *C. umbrosa* subsp. *sabynensis* (Less.) Kuekenth.——KAMIKAWA-SUGE. Rhizomes tufted; the neck thickly covered with dark brown fibers; culms slender, 20–50 cm. long, nearly glabrous; leaf-blades elongate, 2–3 mm. wide; spikes 2 or 3, erect, approximate, the terminal staminate, clavate, 1–1.5 cm. long, the lateral pistillate, ovoid to oblong, 5–15 mm. long, the upper ones nearly sessile, the lowest bract setaceous, short-sheathing; pistillate scales oblong, brown, obtuse or rounded; perigynia slightly longer than the scales, about 3 mm. long, obliquely spreading, obovoid, faintly nerved, hispidulous, abruptly very short-beaked, 2-toothed at apex; stigmas 3, short.——May–July. Sunny fields or open woods; Hokkaido, Honshu (n. distr.).——e. Siberia, Sakhalin, s. Kuriles and Korea.

Var. **rostrata** (Maxim.) Ohwi. *C. pediformis* var. *rostrata* Maxim.; *C. lucidula* Franch.; *C. umbrosa* subsp. *sabynensis* var. *stolonifera* Kuekenth.; *C. subbracteata* Ohwi; *C. praestabilis* Ohwi——TSURU-KAMIKAWA-SUGE. Rhizomes tufted and creeping; leaves elongate and much overtopping the culms after flowering.——May–July. Wet grassy places; Honshu (n. and centr. distr.), Kyushu (n. distr.).——e. Siberia and Korea.

91. Carex breviculmis R. Br. *C. leucochlora* Bunge; *C. lonchophora* Ohwi; *C. royleana* Nees ex Wight; *C. langsdorffii* Boott; *C. filiculmis* Fr. & Sav.——AO-SUGE. Culms 5–40 cm. long, slender, tufted; leaves narrowly linear, 1–5 mm. wide, shorter than to nearly as long as the culm; spikelets 2–6, erect, the terminal staminate, 4–20 mm. long, light green, the others pistillate, oblong to short-cylindric, nearly sessile or the lower ones short-pedunculate, 5–30 mm. long, densely flowered, the lowest bracts leaflike, sheathing at base; pistillate scales ob-

ovate, light green, acute or short-awned; perigynia slightly longer than the scales, obovoid, green, 1.5–3 mm. long, pubescent, many-nerved, abruptly short-beaked, minutely 2-toothed at tip; stigmas 3.——Apr.–July. Abundant in open fields; Hokkaido to Kyushu.——Korea, China, India, and Australia.

Var. **puberula** (Boott) T. Koyama. *C. puberula* Boott ——KO-AO-SUGE. Plant low; leaves shorter than the culms; spikelets rather few-flowered, the lowest one often almost radical.——Shady hillsides; Hokkaido, Honshu (centr. part and north.).

Var. **discoidea** Boott. *Carex discoidea* Boott; *C. aphanandra* Fr. & Sav.; *C. perangusta* Ohwi; *C. leucochlora* var. *aphanandra* (Fr. & Sav.) T. Koyama——HIME-AO-SUGE. Plant slender, loosely tufted with long slender creeping rhizomes; spikelets small, crowded at the top of the culm; bracts barely sheathing; perigynia about 1.5 mm. long, less hairy.——Honshu, Kyushu.——Ryukyus.

Var. **fibrillosa** (Fr. & Sav.) Kuekenth. *C. fibrillosa* Fr. & Sav.; *C. breviculmis* var. *pluricostata* Kuekenth.; *C. breviculmis* subsp. *royleana* forma *fibrillosa* (Fr. & Sav.) Kuekenth.; *C. tosaensis* Akiyama——HAMA-AO-SUGE. Plant robust; stolons tough, elongate; lowest bracts longer than the inflorescence; pistillate spikelets densely many-flowered; pistillate scales easily falling off in fruit; perigynia about 3 mm. long, strongly ribbed, yellowish when mature.——Sandy coasts; Honshu, Shikoku, Kyushu.——s. Korea.

92. Carex jacens C. B. Clarke. *C. geantha* Ohwi—— HAGAKURE-SUGE. Rhizomes very short; culms slender, many, tufted, 7–15 cm. long, glabrous; leaf-blades about 2 mm. wide, thin, flat, soft, overtopping the culms; spikes 5–7, the terminal staminate, pale, 6–7 mm. long, the others pistillate, narrowly oblong, 6–10 mm. long, the upper one contiguous, the lower 1 or 2 always radical or nearly so, the bracts leaflike, sheathing at base; pistillate scales obovate, pale green, very obtuse and mucronate, thin-membranous; perigynia about 3 mm. long, nearly erect, oblong-fusiform, thin-membranous, faintly nerved, rather abruptly short-beaked, with an emarginate tip; stigmas 3.——June–July. Coniferous woods; Hokkaido, Honshu (n. and centr. distr.).——s. Kuriles.

93. Carex rugata Ohwi. *C. kingiana* Lév. & Van't., pro minima parte; *C. breviculmis* subsp. *royleana* var. *kingiana* (Lév. & Van't.) Kuekenth.——KUSA-SUGE. Soft, green, loosely tufted herb, blackish when dried; culms 15–30 cm. long, glabrous, soft; leaf-blades soft, thin, vivid green, 2–3 mm. wide, the basal sheaths pale; spikes 3 or 4, cauline, the terminal staminate, few-flowered, nearly sessile, 5–10 mm. long, the others pistillate, oblong, 5–10 mm. long, approximate, the lower ones distant, the peduncles scarcely exserted from the sheaths of short-bladed bracts; pistillate scales pale green, narrowly obovate, obtuse, the green midrib exserted and mucronate; perigynia longer than the scales, broadly ovoid, 2.5–3 mm. long, thin-membranous, several-nerved, with few transverse wrinkles on the sides, abruptly short-beaked with an emarginate tip.——Apr.–May. Open woods in hillsides and low mountains; Hokkaido (rare), Honshu, Kyushu.

94. Carex stenostachys Fr. & Sav. var. **stenostachys**. *C. pisiformis* sensu auct. Japon., pro parte, non Boott—— NISHI-NO-HOMMONJI-SUGE. Densely tufted, forming large clumps; culms many, 30–50 cm. long, slender but rather firm, slightly scabrous; leaf-blades rather stiff, 2–3 mm. wide, the basal sheaths castaneous, entire and scarcely fibrous, rigid; spikes nearly always 3, approximate, erect, the terminal staminate, linear, brown, lustrous, 2–3 cm. long, the lateral pistillate, oblong–cylindric, 1.5–2 cm. long, the lowest one often separate from the others, the bracts with a short leaflike or setaceous blade, sheathing at base, the peduncles scarcely exserted from the sheath; pistillate scales obovate, brown or brownish, rounded at the apex; perigynia ovate-oblong, 3 mm. long, nerved, pubescent, abruptly short-beaked, the tip minutely 2-toothed; stigmas 3.——Apr.–May. Common in open woods; Honshu (centr. distr. westw. to Iwami Prov.).

Var. **cuneata** (Ohwi) Ohwi & T. Koyama. *C. cuneata* Ohwi——MICHI-NO-KU-HOMMONJI-SUGE. Plant more loosely tufted, with short ascending rhizomes; culms 15–40 cm. long; leaf-blades flat, 2.5–4 mm. wide, the basal sheaths dark brown; spikes 4 or 5, all or the lower ones distant, the terminal staminate, 1.5-2 cm. long, the others pistillate, 1–2 cm. long, the bracts short; pistillate scales often mucronate; perigynia as in the typical phase.——Apr.–June. Thickets and thin woods; Honshu (Tōhoku, n. Kantō and Echigo Distr.).

Var. **ikegamii** T. Koyama. KOSHI-NO-HOMMONJI-SUGE. Intermediate between var. *cuneata* and the typical phase; rhizomes densely tufted in large clumps; leaves broader, softer; spikes usually 3, the lower ones distant.——Japan Sea side of centr. Honshu.

95. Carex polyschoena Lév. & Van't. *C. albomas* C. B. Clarke; *C. pisiformis* forma *polyschoena* (Lév. & Van't.) Kuekenth.——SHIRO-HOMMONJI-SUGE. Rhizomes densely tufted; culms 30–50 cm. long; leaf-blades 1.5–3 mm. wide, the basal sheaths dark brown, somewhat fibrous; spikes 2–4, erect, 1.5-2.5 cm. long, the terminal staminate, linear, pale, the others pistillate, distant, somewhat loosely flowered, the lowest bract leaflike, longer than the spike, sheathing at base; pistillate scales pale green, obovate, rounded and short-awned; perigynia broadly obovoid, 3–3.5 mm. long, suberect, nerved, pubescent, abruptly short-beaked, obtusely 2-toothed at tip; stigmas 3.——Apr.–May. Kyushu (Tsushima).——Korea and Manchuria.

96. Carex clivorum Ohwi. YAMA-Ō-ITO-SUGE. Plant densely tufted; culms 20–40 cm. long, slender; leaf-blades flat, green, 2–3 mm. wide, the basal sheaths light brown, somewhat lustrous, fibrous, the fibers sparsely covering the neck of the rhizome; spikes usually 3, distant, the terminal staminate, linear, long-peduncled, 2.5–3 cm. long, pale greenish yellow, the others pistillate, mostly loosely flowered, 2–3 cm. long, broadly linear, the bracts short, rather long-sheathing; pistillate scales narrowly oblong, truncate or emarginate and mucronate; perigynia longer than the scales, 4 mm. long, suberect, oblong-ovoid, faintly but distinctly nerved, sparsely pubescent, rather long-beaked, minutely 2-toothed at apex; stigmas 3.——Apr.–May. Open woods on hillsides and in low mountains; Honshu (Musashi, Kai, Mikawa, Owari, and Mino Prov.).

97. Carex pisiformis Boott. *C. amphora* Fr. & Sav. ——HOMMONJI-SUGE. Rhizomes long-creeping; culms 30–40 cm. long, glabrous, slender; leaf-blades flat, about 3 mm. wide, the basal sheaths brownish to stramineous, slightly fibrous; spikes 3, distant, the terminal staminate, 2.5–3 cm. long, the others pistillate, short-cylindric, peduncled, rather loosely flowered, 1.5–2 cm. long, the bracts short, long-sheathing; pistillate scales obovate, pale green, rounded to truncate and mucronate at apex; perigynia slightly longer than the scales, 3.5 mm. long, suberect, obovate-ellipsoid, nerved, pubescent, abruptly short-beaked, the apex 2-toothed; stigmas 3.

——Apr.–May. Open woods in hills; Honshu (sw. Kantō Distr.); rather common.

98. Carex tenuinervis Ohwi. *C. pseudostrigosa* Lév. & Van't., pro parte; *C. alterniflora* var. *tenuinervis* (Ohwi) Ohwi——TSURU-NASHI-Ō-ITO-SUGE. Rhizomes densely tufted; culms 20–30 cm. long, slender; leaf-blades flat, 2–2.5 mm. wide, the basal sheaths pale or stramineous; spikes 3, the terminal staminate, linear, 2–3 cm. long, pale, the others pistillate, broadly linear, 1–1.5 cm. long, erect, somewhat loosely flowered, the bracts leaflike, sheathing at base; pistillate scales obovate-elliptic, pale green, rounded and cuspidate; perigynia longer than the scales, 2.5 mm. long, nearly erect, obovoid, glabrous, faintly nerved, abruptly short-beaked, the beak scabrous, minutely 2-toothed; stigmas 3.——Apr.–May. Kyushu.

99. Carex tashiroana Ohwi. NO-SUGE. Rhizomes densely tufted; culms 20–30 cm. long, glabrous; leaf-blades 1.5–2.5 mm. wide; spikes 2 or 3, slightly distant, erect, the terminal staminate, 1.5–2 cm. long, pale green, the others pistillate, broadly linear, loosely 5- to 10-flowered, 1–2.5 cm. long, the bracts setaceous, sheathing at base; pistillate scales pale, short-awned; perigynia nearly as long as the scales, 2 mm. long, obovate-elliptic, faintly nerved, short pubescent, abruptly short-beaked, the beak recurved, 2-toothed at apex; stigmas 3.——Apr. Honshu (Aki Prov.); rare. This may be an aberrant form of *C. tenuinervis*.

100. Carex sachalinensis F. Schmidt. Rhizomes long-creeping; culms slender, 15–30 cm. long, nearly glabrous; leaf-blades flat, soft, 2–3 mm. wide; spikes 2–4, distant, erect, the terminal staminate, linear, 1–2 cm. long, the others pistillate, short-cylindric, on long-exserted peduncles 7–15 mm. long, the lowest one sometimes nearly radical, the bracts short, long-sheathing at base; pistillate scales obovate, pale, usually mucronate; perigynia longer than the scales, erect, obovoid to obovoid-oblong, usually glabrous, usually with an abrupt erect beak, often scabrous on the margins, hyaline and minutely 2-toothed at apex; stigmas 3.——Very polymorphic, with the following varieties:

Var. **sachalinensis**. *C. pseudoconica* Fr. & Sav.; *C. pisifomis* var. *sachalinensis* (F. Schmidt) Kuekenth.; *C. korsakoviensis* Lév.——GONGEN-SUGE. Bracts long-sheathing at base, with blades shorter than the spikes; perigynia longer than the scales, 3 mm. long, erect, glabrous, nerved.——May–July. Damp woods in mountains; Hokkaido, Honshu (n. centr. distr.).——Sakhalin and s. Kuriles.

Var **iwakiana** Ohwi. KO-ITO-SUGE. Closely resembles var. *sachalinensis* but the perigynia distinctly 3-angled, slightly hairy.——Honshu (Iwaki Prov.).

Var. **longiuscula** Ohwi. *C. nikomontana* Akiyama——MIYAMA-AO-SUGE. Spikes loosely flowered; perigynia to 4 mm. long, gradually long-beaked.——Coniferous woods in mountains; Honshu (centr. distr.).

Var. **conicoides** (Honda) Ohwi. *C. conicoides* Honda——WATARI-SUGE. Differs from var. *arimaensis* in the broader leaves, 2.5 to 3 mm. wide, and long-sheathing bracts shorter than the spikes; plant with long leafy stolons.——Apr.–May. Kyushu; rare.

Var. **pineticola** (Ohwi) Ohwi. *C. pineticola* Ohwi——MATSU-KAZE-SUGE. Akin to var. *conicoides;* pistillate spikes densely flowered and somewhat tawny; perigynia 2.5 mm. long and often sparsely pubescent.——Pine woods near seacoasts; Honshu (Shimōsa Prov.); rare.

Var. **fulva** (Ohwi) Ohwi. *C. alterniflora* var. *fulva* Ohwi; *C. heribaudiana* Lév. & Van't., pro parte; *C. artinux* C. B. Clarke, pro parte——KI-ITO-SUGE. Plant relatively low and small; culms usually 10–20 cm. long; spikes tawny, lustrous.——Under shrubs in high mountains; Honshu (Ugo, Shimotsuke, Yechigo, Kaga, Shinano Prov. and elsewhere).

Var. **arimaensis** (Ohwi) Ohwi. *C. alterniflora* var. *arimaensis* Ohwi——ARIMA-ITO-SUGE. Plant stoloniferous, with short, green leaves; culms low; intermediate between var. *alterniflora* and var. *conicoides.*——Honshu (Kinki Distr.).

Var. **alterniflora** (Franch.) Ohwi. *C. alterniflora* Franch.; *C. pseudostrigosa* Lév. & Van't.; *C. scabroaristata* Akiyama ——Ō-ITO-SUGE. Resembles var. *sikokiana;* culms usually taller; pistillate spikes loosely flowered; scales and sheaths pale, not reddish.——Woods in hills and mountains; Hokkaido, Honshu, Shikoku, Kyushu; rather abundant.——Formosa.

Var. **elongatula** (Ohwi) Ohwi. *C. alterniflora* var. *elongatula* Ohwi——KUJŪ-SUGE. Resembles the preceding variety; rhizomes loosely tufted; perigynia long-beaked.——Kyushu (Mount Kujū); rare.

Var. **sikokiana** (Fr. & Sav.) Ohwi. *C. sikokiana* Fr. & Sav.; *C. tenuissima* var. *sikokiana* (Fr. & Sav.) Kuekenth.—— BENI-ITO-SUGE, KANSAI-Ō-ITO-SUGE, SHIKOKU-ITO-SUGE. Culms 20–50 cm. long; leaf-blades about 2 mm. wide, the basal sheaths entire, at least partially reddish; lateral spikes 1.5–4 cm. long; the bracts leaflike, slightly longer than the spikes; staminate and/or pistillate scales frequently reddish.—— Woods on hillsides and in mountains; Honshu (Kinki Distr.), Shikoku, Kyushu, s. Korea.

Var. **aureobrunnea** (Ohwi) Ohwi. *C. alterniflora* var. *aureobrunnea* Ohwi——CHA-ITO-SUGE. Basal leaf-sheaths yellowish brown, slightly lustrous.——Apr.–May. Shikoku, Kyushu.

101. Carex fernaldiana Lév. & Van't. *C. mariesii* C. B. Clarke; *C. ischne* C. B. Clarke; *C. tenuissima* sensu auct. Japon., non Boott——ITO-SUGE. Closely related to the preceding species; rhizomes tufted; culms 15–30 cm. long, capillary, smooth; leaf-blades 0.3–1 mm. wide, sometimes flattish, at least the outer always filiform-involute, the sheaths stramineous green; spikes 2 or 3, erect, the terminal staminate, linear, often slightly brownish, 1–1.5 cm. long, the others pistillate, short-cylindric, loosely few-flowered, 7–15 mm. long, on short usually included peduncles, the lowest one sometimes slightly exserted, the lowest bract somewhat leaflike, longer than the spike, long-sheathing; perigynia 2.5–3.5 mm. long, glabrous. ——Apr.–June. Woods in hills and mountains; Hokkaido, Honshu, Shikoku, Kyushu; rather common.——Formosa.

102. Carex duvaliana Fr. & Sav. *C. hololasius* Lév. & Van't.; *C. tenuissima* var. *duvaliana* (Fr. & Sav.) Kuekenth. ——KE-SUGE. Related to *C. sachalinensis* var. *alterniflora;* pilose throughout; rhizomes loosely tufted, creeping; culms 30–50 cm. long, pilose; leaf-blades 1.5–2.5 mm. wide, the basal sheaths pale, entire; bracts leaflike, long-sheathing; perigynia 3–3.5 mm. long, short-beaked.——Apr.–May. Thickets in hills; Honshu (Kantō Distr. and westw.), Shikoku, Kyushu (rare).

103. Carex mayebarana Ohwi. KE-KUSA-SUGE, KE-HIE-SUGE. Rhizomes slender, creeping; culms 20–30 cm. long, glabrous; leaf-blades soft, glabrous, flat, 2–3 mm. wide, the basal sheaths pale; spikes 3 or 4, distant, the terminal staminate, linear, 1–1.5 cm. long, sometimes with a few pistillate flowers intermixed, the others wholly pistillate, loosely few-flowered, 6–15 mm. long, broadly linear, the bracts leaflike, longer than the spikes, thinly pubescent, long-sheathing, the peduncles mostly included in the sheaths of the bracts; pistillate scales narrowly oblong, pale, glabrous, mucronate; perigynia slightly longer than the scales, about 6 mm. long, erect, glabrous, light green, nerveless, the beak nearly as long as the body, glabrous, minutely 2-toothed and hyaline at tip; stigmas 3.——Kyushu, Shikoku; rare.

104. Carex tristachya Thunb. var. **tristachya.** *C. monadelpha* Boott——MOEGI-SUGE. Rhizomes densely tufted; culms erect, 20–40 cm. long; leaf-blades flat, stiffish, 3–5 mm. wide, the sheaths pale, splitting into hairlike brown fibers, these covering the neck of the rhizome; spikes 3–5, contiguous, fastigiate, the terminal staminate, filiform, 1–3 cm. long, the lateral pistillate, erect, short-cylindric, 1–3 cm. long, the lowest one sometimes remote from the others and long-peduncled, the lowest bract bladed, sheathing at base; staminate scales small; pistillate scales elliptic, pale green, entire or mucronate at the apex, the midrib not reaching the margin; perigynia twice as long as the scales, 3 mm. long, erect, rhomboid or ovoid-fusiform, about 3 mm. long, puberulent, obtusely 3-angled, many-nerved, the beak short, somewhat recurved, minutely 2-toothed; stigmas 3, short.——Dry fields and rocky slopes in hills and in low mountains; Honshu (Kantō Distr. and westw.), Shikoku, Kyushu; common.——Korea and China.

Var. **pocilliformis** (Boott) Kuekenth. *C. pocilliformis* Boott——KOPPU-MOEGI-SUGE. Staminate scales connate ventrally, infundibuliform; perigynia slightly smaller than in the typical phase.——Honshu, Shikoku, Kyushu.——Ryukyus, Formosa, and Malaysia.

105. Carex daisenensis Nakai. DAISEN-SUGE. Rhizomes densely tufted, the neck rather densely covered with dark brown fibers; culms 30–50 cm. long, slender; leaf-blades 4–6 mm. wide, flat, stiff; spikes 3 or 4, distant, erect, the terminal staminate, linear, 2–2.5 cm. long, the lateral pistillate, broadly linear, loosely flowered, 1.5–2 cm. long, peduncled, the bracts setaceous, sheathing at base; pistillate scales obovate, pale green, obtuse, the green midrib sometimes excurrent as a mucro; perigynia longer than the scales, 3.5–4.5 mm. long, oblong-fusiform, the beak moderately elongate, minutely 2-toothed.——Apr.–June. Honshu (Kinki Distr. and westw., mainly on the Japan Sea side), Kyushu (n. distr.).

106. Carex dolichostachya Hayata. *C. multifolia* Ohwi; *C. foliosissima* Franch., non F. Schmidt——MIYAMA-KAN-SUGE. Rhizomes sometimes short-creeping (forma *stolonifera* Ohwi): culms many, 20–50 cm. long; leaves densely fascicled, the blades 3–8 mm. wide, subcoriaceous to rather soft, slightly scabrous, flattened, the basal sheaths purplish brown or reddish fuscous, more or less lustrous, scarcely fibrous; spikes 3–5, distant, the terminal staminate, linear, 2–4 cm. long, the others pistillate, broadly linear, 1.5–3 cm. long, loosely many-flowered, long-peduncled, the bracts long-sheathing, short-setaceous; pistillate scales obovate, pale or tawny to purplish brown, rounded or truncate at the apex and mucronate; perigynia longer than the scales, erect, oblong or obovate-oblong, nerved, 3–4 mm. long, usually sparsely pubescent, rather abruptly short-beaked, with a minutely 2-toothed, membranous apex; stigmas 3.——May—July. Woods in mountains; Hokkaido, Honshu, Shikoku, Kyushu; common and variable.—— Formosa.

Var. **imbecillis** Ohwi. AO-MIYAMA-KAN-SUGE. Plant

wholly green, with thinner, softer, broader leaves, stoloniferous.——Kyushu.——Forma **pallidisquama** (Ohwi) Ohwi. *C. multiflora* var. *pallidisquama* Ohwi. Pistillate scales pale. ——Forma **glaberrima** (Ohwi) Ohwi. *C. multiflora* var. *glaberrima* Ohwi. Perigynia glabrous.

107. Carex atroviridis Ohwi. *C. yakushimensis* Masam. ——YAKUSHIMA-SUGE. Culms to 25 cm. long, tufted, glabrous; leaf-blades 5–6 mm. wide, rather thick, flat, deep green, the basal sheaths castaneous, scarcely fibrous; spikes 3–5, the terminal staminate, broad-linear, 2–3 cm. long, the others pistillate, distant, 1–2 cm. long, on a slightly exserted peduncle, the bracts bladeless or with a setaceous blade, long-sheathing; pistillate scales obovate, brown, acute; perigynia slightly longer than the scales, obovoid, many-nerved, glabrous or sparsely short-pubescent, abruptly beaked, the apex 2-toothed; stigmas 3.——Kyushu (Yakushima).

108. Carex tsushimensis (Ohwi) Ohwi. *C. chinensis* var. *tsushimensis* Ohwi——TSUSHIMA-SUGE. Culms tufted, 30–40 cm. long; leaf-blades stiff, flat, 2–4 mm. wide, the basal sheaths sparingly fibrous; spikes 3 or 4, loosely arranged, the terminal staminate, lanceolate, 1.5 cm. long, the others pistillate, erect, with a few staminate flowers at the base, oblong-cylindric, 15–25 mm. long, densely many-flowered, on a long-exserted peduncle, the bracts short-bladed or setaceous, with a long sheath at base; pistillate scales obovate, pale green, truncate or emarginate, with a scaberulous apical awn, giving a comose aspect to the spikes; perigynia longer than the scales, 2.5 mm. long, obliquely spreading to divergent, broadly rhomboid, 3-angled, many-ribbed, loosely hispidulous, cuneately narrowed at both ends, the beak short, erect, emarginate; stigmas 3, short.——Kyushu (Tsushima); rare.

109. Carex sociata Boott. *C. chinensis* sensu auct. Japon., non Retz.; *C. legendrei* Lév. & Van't.; *C. nexa* var. *strictior* Kuekenth.; *C. ligata* var. *strictior* Kuekenth.; *C. atronucula* Hayata; *C. uraiensis* Hayata——TASHIRO-SUGE, KUMIAI-SUGE. Rhizomes densely tufted; culms many, 20–50 cm. long, obtusely angled; leaf-blades stiff, flat or nearly so, 3–6 mm. wide, the sheaths pale, often with brown nerves, fibrous when withered and old; spikes 4–8, the terminal staminate, 1.5–3 cm. long, lanceolate, the others pistillate, often with a few staminate flowers at base, short-cylindric, densely many-flowered, 1.5–3 cm. long, long-peduncled, the lower ones usually binate to ternate at each node, the lower bracts leaflike, loosely long-sheathing; pistillate scales pale, oblong, short-awned; perigynia longer than the scales, 2.5 mm. long, obliquely spreading, rhomboid, obtusely 3-angled, faintly nerved, puberulent, rather abruptly short-beaked, with a minutely 2-toothed hyaline apex; stigmas 3, very slender.——Mar.–May. Shikoku, Kyushu.——Ryukyus and Formosa.

Var. **uber** (Ohwi) Ohwi. *C. uber* Ohwi——TSUKUSHI-SUGE. Leaves thin and soft.——Shikoku, Kyushu.——Ryukyus and Formosa.

110. Carex oshimensis Nakai. ŌSHIMA-KAN-SUGE. Rhizomes densely tufted; culms 20–50 cm. long, obtusely 3-angled, glabrous; leaves fascicled, the blades 3–6 mm. wide, flat, stiff, deep green, the basal sheaths splitting into fibers, these thinly covering the neck of the rhizome; spikes 3–5, erect, the terminal (rarely the upper 2 or 3) staminate, pale to dark brown, clavate, 1.5–2.5 cm. long, the others pistillate (sometimes the upper ones androgynous), distant, short-cylindric, on long, exserted peduncles, 1.5–5 cm. long, densely many-flowered, the bracts short-bladed with inflated sheaths green throughout

or dark brown at the base; pistillate scales oblong, abruptly awned; perigynia slightly longer than the scales, 3 mm. long, obliquely spreading, obovoid, hispidulous, nerved, abruptly short-beaked, entire at tip.——Apr.–May. Dry woods and rocky slopes; Honshu (Izu Isls.); common.

110a. Carex omurae T. Koyama. SURUGA-SUGE. Rhizomes decumbent to ascending, rather loosely tufted, covered with brown fibers; leaves spreading, very hard, linear-ensiform, to 16 cm. long, 2.5–4 mm. wide, abruptly short-pointed, prominently scabrous on both margins; culms slender, 1–3 to a clump, 15–20 cm. long, smooth; spikelets 3–4, the upper 2 usually close, upright; terminal spikelet staminate, linear, 1–1.5 cm. long, brown to red-brown; lateral spikelets pistillate, cylindrical, 1.5–2 cm. long, loosely several-flowered, with an exserted peduncle; bracts sheathlike, almost bladeless, light red-brown; pistillate scales lanceolate-deltoid, acute, half as long as the perigynium, pale; perigynia patent, ellipsoidal, 3–3.5 mm. long, obscurely 3-angled, membranous, greenish tawny, faintly many-nerved; beak relatively long, somewhat curved when mature, the mouth minutely bidentate; achene tightly enclosed, elliptical, crowned by a depressed indistinct disc at apex; stigmas 3.——Coniferous forest. Honshu (Tō-kaido Distr.).

111. Carex conica Boott. *C. excisa* Boott; *C. digama* Nakai; *C. doiana* Akiyama; *C. yoshinoi* Ohwi; *C. okushirensis* Akiyama——HIME-KAN-SUGE. Rhizomes tufted, sometimes with elongate ascending innovations, the neck thinly covered with brown fibers; culms 20–50 cm. long, obtusely angled, smooth; leaf-blades 2–4 mm. wide, flat, stiff, dark green, scabrous; spikes 3–5, erect, the terminal staminate, clavate, dark brown, 1.5–2.5 cm. long, the others pistillate (or very rarely the upper ones androgynous), distant, short-cylindric, 1–2.5 cm. long, rather densely flowered, the bracts short, with a long inflated sheath, green throughout or purplish brown at the base; pistillate scales obovate, dark purplish brown, sometimes pale green, abruptly mucronate; perigynia slightly longer than the scales, ellipsoidal, 2.5–3 mm. long, usually light green, sparsely puberulent or almost glabrous, abruptly recurved-beaked, entire at the tip; stigmas 3.—Apr.–June. Open woods on hillsides and in low mountains; Hokkaido (sw. distr.), Honshu, Shikoku, Kyushu; very common.——s. Korea.

112. Carex morrowii Boott var. **morrowii**. KAN-SUGE, IZU-KAN-SUGE. Rhizomes short, tufted and forming a large clump; culms many, 20–40 cm. long, obtusely 3-angled, smooth; leaves fascicled, the blades 5–10 mm. wide, flat, thick and very stiff, deep green (white-striped in a widely grown garden cultivar), lustrous, scabrous on margin, many veined on upper side, the basal sheaths dark castaneous, dull, sparsely fibrous when withered or old; spikes 4–6, distant, the terminal staminate, linear to narrowly clavate, 2–4 cm. long, the others pistillate, short-cylindric, on long exserted erect peduncles, the lower bracts short-bladed, inflated and long-sheathing; pistillate scales ovate, brownish, sometimes pale, abruptly cuspidate; perigynia broadly ovoid, 3–3.5 mm. long, obliquely spreading to somewhat divergent, stramineous or yellowish green, glabrous, nerved, turgid, abruptly tapering into a moderately long, recurved beak with a firm sharply 2-toothed tip; stigmas 3. ——Apr.–May. Woods in low mountains; Honshu (Iwaki Prov. and westw., chiefly on the Pacific side), Shikoku, Kyushu.

Var. **laxa** Ohwi. YAKUSHIMA-KAN-SUGE. Leaves 3–5 mm. wide; pistillate scales rather loosely flowered; perigynia narrowly ovoid, gradually long-beaked.——Kyushu (Yaku-shima); rare.

Var. **temnolepis** (Franch.) Ohwi. *C. temnolepis* Franch.; *C. foliosissima* var. *temnolepis* (Franch.) Kuekenth.; *C. kin-pokusanensis* Akiyama——HOSOBA-KAN-SUGE. Leaves 3–5 mm. wide; pistillate spikes densely flowered; beak of the perigynia usually scabrous on the margins.——May–June. Woods in mountains; Honshu (Ugo Prov. to Tanba, mainly on the Japan Sea side).

113. Carex foliosissima F. Schmidt var. **foliosissima**. *C. morrowii* sensu auct. Japon., pro parte, non Boott; *C. yesanensis* Franch.; *C. crassicaulis* Franch.; *C. niigatensis* Koidz.; *C. morrowii* subsp. *foliosissima* (F. Schmidt) Ohwi ——OKU-NO-KAN-SUGE. Rhizomes short-creeping, in loose tufts; culms smooth, obtusely angled, 15–40 cm. long; leaf-blades flat, stiff to rather soft, 5–10 mm. wide, scabrous or slightly so, 2-ribbed on upper surface, the basal sheaths dark brown to dark reddish brown, dull; spikes 3–5, loosely ar-ranged, on erect exserted peduncles, the terminal staminate, linear, 1.5–3 cm. long, the others pistillate, short-cylindric, 2–3 cm. long, densely many-flowered, the bracts very short-bladed, with prolonged inflated sheaths; pistillate scales ovate, abruptly cuspidate, reddish brown to pale green; perigynia spreading, broadly obovoid, light yellow-green, 2.5–3.5 mm. long, glabrous, nerved, with a rather long recurved beak, acutely 2-toothed at tip; stigmas 3.——Woods in mountains; Hokkaido, Honshu, Kyushu; common.——Sakhalin.

Var. **latissima** (Ohwi) Akiyama. *C. morrowii* subsp. *foliosissima* var. *latissima* Ohwi——HABABIRO-SUGE. Leaves 15–20 mm. wide.——Especially abundant in mountains on the Japan Sea side of Honshu.

114. Carex matsumurae Franch. *C. taquetii* Lév.; *C. viridissima* Nakai——KI-NO-KUNI-SUGE, KISHŪ-SUGE. Plant densely tufted forming large clumps; culms many, 30-40 cm. long, glabrous; leaves evergreen, with flat, thick, deep green, lustrous, smooth blades, 8–12 mm. wide, the basal sheaths pale green with brown nerves; spikes 4 or 5, distant, light green, the upper somewhat approximate, the terminal staminate, broadly linear, 3–5 cm. long, the lateral pistillate, cylindric, 25–35 mm. long, densely many-flowered, erect on exserted peduncles, the bracts short, with a long sheath; pistillate scales ovate, pale green; perigynia much longer than the scales, 4–5 mm. long, nearly erect, obovate, pale green, glabrous, bi-convex, several-ribbed, short-beaked, with a membranous emarginate tip; beak of achenes with annulate base; stigmas 2. ——Mar.–May. Evergreen woods near seacoasts; Honshu (Etchū, Shima Prov., and westw.), Shikoku, Kyushu.——s. Korea.

115. Carex hachijoensis Akiyama. HACHIJŌ-KAN-SUGE. Rhizomes tufted, sometimes creeping; culms 20–40 cm. long, 3-angled, almost smooth; leaves longer than the culms, the blades 6–7 mm. wide, rather thick, green, nearly glabrous, long-attenuate, the basal sheaths light brown, sparingly fibrous when withered or old; spikes 3 or 4, the terminal staminate, linear, light brown, 2.5–3 cm. long, 2–2.5 mm. across, the lateral pistillate, erect, linear-cylindric, 1.5–2 cm. long, with a few staminate flowers at the apex, 5 mm. thick, the peduncles nearly included, the bracts spathaceous, 7–15 mm. long, pale green, with a setaceous short blade; pistillate scales oblong, very obtuse and mucronate, pale and slightly tinged with red, the midrib green, 3-nerved; perigynia about 3.5 mm. long, ovoid to ellipsoid, trigonous or lenticular, glabrous, pale green,

many-ribbed, the beak smooth, short, hyaline, subentire; achenes narrowed above and crowned by a depressed disc about 0.7 mm. across at apex; stigmas (2–) 3.——June. Hon-shu (Hachijō Isl. in Izu).

116. Carex laticeps C. B. Clarke. *C. hancei* C. B. Clarke ——Ō-MUGI-SUGE. Prominently pilose throughout; rhizomes short-creeping; culms 30–50 cm. long; leaf-blades 3–5 mm. wide, rather soft, cinereous-green, with whitish hairs, the basal sheaths stramineous, dull; spikes 2 or 3, distant, the terminal staminate, linear-clavate, 2–3 cm. long, rusty-brown, on a long-exserted peduncle, the others pistillate, oblong to oblong-cylindric, 2–3 cm. long, about 10 mm. thick, the bracts short, leaflike, long-sheathing, the peduncles nearly included; pistillate scales oblong, cuspidate, whitish; perigynia longer than the scales, 5–6 mm. long, obliquely spreading, yellowish green, pilose, nerved, the beak long, erect, acutely 2-toothed; achene constricted on the angles near the middle, obliquely beaked; stigmas 3.——Honshu (Bingo and Bitchu Prov.); rare.——Korea and China.

117. Carex insaniae Koidz. *C. fauriei* Franch.—— HIROBA-SUGE. Rhizomes loosely tufted; culms 5–40 cm. long; leaf-blades flat, rather thick, glabrous, deep green, broadly linear, 8–12 mm. wide, the basal sheaths pale; spikes 2–4, the terminal linear-clavate, 1–2 cm. long, the others pistil-late, oblong, peduncled, rather densely flowered, the lowest bract leaflike, long-sheathing; pistillate scales elliptic, rounded at the apex and mucronate, pale green; perigynia longer than the scales, 5–6 mm. long, obliquely spreading, broadly ellipsoid, loosely pubescent, inconspicuously 3-angled, green, abruptly short-beaked, with a hyaline emarginate tip; achenes ellipsoid, with a short annulate beak, 3-angled, each angle with a con-striction near the middle; stigmas 3.——May–June. Woods and thickets in low mountains; Hokkaido, Honshu (Japan Sea side in n. and centr. distr.).——s. Kuriles.

Var. **papillaticulmis** (Ohwi) Ohwi. *C. papillaticulmis* Ohwi——AOBA-SUGE. Leaves 4–8 mm. wide; terminal spike linear; perigynia rather long-beaked.——Honshu (Pacific side of centr. part and Kinki Distr.), Shikoku, Kyushu.

Var. **subdita** (Ohwi) Ohwi. *C. subdita* Ohwi; *C. nan-kaiensis* Honda; *C. kiyozumiensis* Akiyama; *C. tosana* Ma-kino——AO-HIE-SUGE. Plant slender; leaf-blades 2–4 mm. wide, paler green; perigynia 5–6 mm. long.——Woods on hillsides; Honshu (Tōkaidō and s. Kinki Distr.), Shikoku.

118. Carex boottiana Hook. & Arn. *C. bongardii* Boott; *C. stupenda* Lév. & Van't.; *C. oahuensis* var. *boottiana* (Hook. & Arn.) Kuekenth.——HIGE-SUGE, Ō-HIGE-SUGE, Iso-SUGE. Rhizomes densely tufted, forming large clumps, with ascending short innovations; culms few to rather many, 30–50 cm. long; leaf-blades stiff, coriaceous, lustrous, 5–10 mm. wide, the margins revolute and prominently scabrous, the basal sheaths castaneous, splitting into brown fibers; spikes 3–6, the terminal staminate, thick-clavate, 3–6 cm. long, densely many-flowered, castaneous, the lateral androgynous, short-cylindric, densely many-flowered, peduncled, the lower bracts leaflike, long-sheathing; pistillate scales oblong, dark or yellowish brown, emarginate and awned; perigynia 5–6 mm. long, obliquely spreading, oblong-elliptic, somewhat coriaceous, indistinctly 3-angled, glabrous, many-nerved, with a rather long deeply bifid beak, the teeth slightly curved; achenes with a curved beak; stigmas 3.——Rocks near sea-coasts; Honshu (Noto and Awa Prov. westw.), Shikoku, Kyushu.——s. Korea, Ryukyus, China, and Bonin Isls.

119. Carex longerostrata C. A. Mey. *C. longerostrata* var. *recurvifolia* Kuekenth.——HIE-SUGE, MATSUMAE-SUGE. Rhizomes tufted, with ascending short innovations; culms slender, 20–40 cm. long; leaves somewhat flaccid, with flat blades 2–3 mm. wide, much elongate after flowering, the basal sheaths stramineous, early splitting into brown fibers; spikes usually 2, the terminal staminate, clavate, 10–15 mm. long, densely many-flowered, brownish, the lateral one pistillate, ovoid to globose, about 10 mm. long and as wide, few-flowered, the bracts short-bladed, short-sheathing, the peduncle scarcely exserted from the sheath; pistillate scales narrowly ovate, abruptly short-awned, yellowish brown; perigynia longer than the scales, obliquely spreading, 7–8 mm. long, obovate-ellipsoid, yellowish green, inconspicuously nerved, obsoletely 3-angled, abruptly long-beaked, bifid at apex; achenes without appendage at apex.——May–July. Grassy slopes and thickets in mountains; Hokkaido, Honshu.——e. Siberia, Korea, Sakhalin, and s. Kuriles.——Forma **pallida** (Kitag.) T. Koyama. *C. tenuistachya* Nakai; *C. tenuistachya* var. *pallida* Kitag.; *C. longerostrata* var. *pallida* Ohwi——CHŪZENJI-SUGE. With long-creeping rhizomes.——Honshu, Kyushu.

120. Carex lasiolepis Franch. *C. adumana* Makino——AZUMA-SUGE. Plant pilose throughout; rhizomes tufted, with ascending, short innovations; culms 5–15 cm. long, slender, pilose; leaf-blades rather short, 3–5 mm. wide, flat, soft, pilose, yellowish or pale green, the basal sheaths pilose, pale, partially dark reddish, scarcely fibrous; spikes 3 or 4, long-peduncled, 5–7 mm. long, the terminal staminate, narrowly obovate, rather many-flowered, the others pistillate, densely rather few-flowered, the lower ones usually radical and nodding on elongate peduncles, the bract setaceous, sheathing, pale; pistillate scales oblong, broadly rounded and mucronate at the apex, dark reddish brown; perigynia much longer than the scales, 4–4.5 mm. long, nearly erect, acutely 3-angled, nerveless except for 2 costal nerves, with a thickened, prominent stipe, abruptly short-beaked, the apex hyaline, minutely 2-toothed.——Apr.–May. Shallow soil over rocks in mountains; Hokkaido (Mount Apoi), Honshu (n. and centr. distr.), Kyushu (n. distr.).

121. Carex hashimotoi Ohwi. SAYAMA-SUGE. Rhizomes tufted; culms hidden among the leaves, 5–10 cm. long, slender; leaf-blades fresh green, soft, flat, 3–5 mm. wide, the sheaths pale; spikes 3 or 4, the terminal staminate, peduncled, pale green, the others pistillate, short-cylindric to oblong, 5- to 10-flowered, all but the uppermost radical, the bract on the culm sheathing, short-pointed; pistillate scales obovate-elliptic, broadly rounded, pale; perigynia twice as long as the scales, 4 mm. long, acutely 3-angled, very weakly nerved, pale green, membranous, sparsely puberulent, gradually narrowed to a stipelike base nearly as long as the body, rather abruptly short-beaked, with a minutely 2-toothed hyaline apex; stigmas 3, short.——Apr.–May. Honshu (Ōmi, Mino, Hida and s. Shinano Prov.); rare.

122. Carex quadriflora (Kuekenth.) Ohwi. *C. digitata* var. *pallida* Meinsh.; *C. digitata* subsp. *quadriflora* Kuekenth. ——AKA-SUGE. Rhizomes short-creeping to ascending, forming loose tufts; culms 10–20 cm. long, slender, soft, with a reddish brown bladeless sheath at base; leaf-blades flat, soft, glabrous, deep green, 2–4 mm. wide, the sheaths dark red; spikes 3 or 4, the terminal staminate, short-linear, 7–10 mm.

long, few-flowered, the upper lateral ones longer, the others pistillate, approximate, 1–2 cm. long, very loosely few-flowered on a flexuous axis, the peduncles nodding above, the bracts bladeless, the sheath elongate, tubular, reddish brown; pistillate scales obovate, truncate, reddish, clasping at base; perigynia much longer than the scales, 4.5–5 mm. long, acutely 3-angled, ellipsoidal, sparsely short-pubescent, nerveless except for 2 costal nerves, with a prominent, thickened stipe shorter than the body, abruptly short-beaked, with a ferrugineous, entire, membranous mouth; stigmas 3, slender, short.——Hokkaido (Mount Kitami-fuji); rare.——Korea, Manchuria, and Ussuri.

123. Carex humilis Leyss. var. **nana** (Lév. & Van't.) Ohwi. *C. humilis* sensu auct. As. Orient., non Leyss.; *C. lanceolata* var. *nana* Lév. & Van't.; *C. nanella* Ohwi——HOSOBA-HIKAGE-SUGE, HIME-HIKAGE-SUGE, HINATA-SUGE. Rhizomes densely tufted; culms smooth, hidden among the leaves, 3–6 cm. long; leaf-blades filiform, ciliate-hispidulous, elongating after anthesis, 0.5–1.5 mm. wide, the basal sheaths partly reddish, splitting into brown fibers, these thinly covering the neck of the rhizome; spikes 2–4, erect, the terminal staminate, 5–10 mm. long, few-flowered, longer than the lateral pistillate ones, the pistillate ovoid, few-flowered, 5–7 mm. long, the bracts spathaceous, bladeless, broadly hyaline-margined; pistillate scales ovate, acute, hyaline, pale to partially reddish, clasping the rachis at base, broadly scarious-margined; perigynia slightly shorter than the scales, nearly orbicular, 3 mm. long, indistinctly trigonous, obsoletely thick-nerved, rather densely pubescent, with a rather short oblique thick stipe, abruptly very short-beaked, with a ferrugineous entire tip.——Mar.–May. Shallow soils over rocks or in open woods in hills and low mountains; Hokkaido, Honshu, Shikoku, Kyushu.——Korea. The typical phase occurs in Europe and Siberia.

Var. **callitrichos** (V. Krecz.) Ohwi. *C. callitrichos* V. Krecz.——ITO-HIKAGE-SUGE. Leaves involute, capillary, less than 0.5 mm. wide; staminate spike many-flowered, ellipsoidal. ——Kyushu (Tsushima).——s. Kuriles, Korea to Ussuri and Manchuria.

124. Carex lanceolata Boott. *C. longisquamata* Meinsh.; *C. delicatula* C. B. Clarke; *C. yesoensis* Koidz.; *C. subpediformis* Suto & Suzuki——HIKAGE-SUGE. Rhizomes densely tufted, the neck with brown fibers; culms 10–40 cm. long, slightly scabrous above; leaf-blades linear, flat, 1.5–2 mm. wide, scabrous, the basal sheaths partly reddish brown, scaberulous, splitting into parallel fibers; spikes 3–6, erect, the terminal pistillate, linear-clavate, 10–15 mm. long, rather few-flowered, shorter or slightly longer than the lower ones, the lateral spikes pistillate, short-cylindric, 1–2 cm. long, the upper approximate, the lower distant, loosely flowered, the lower bracts spathaceous, bladeless, broadly hyaline-margined; peduncles scarcely exserted from the sheath; pistillate scales ovate, acute, ferrugineous to castaneous, rarely pale, with broadly scarious margins; perigynia shorter than the scales, obovoid, 3 mm. long, many-ribbed, densely pubescent, with a thick oblique stipe, the beak extremely short, ferrugineous, slightly recurved with an entire tip.——Apr.–June. Open woods in hills and mountains; Hokkaido, Honshu, Shikoku, Kyushu; common.——Korea, Manchuria, and Ussuri.

125. Carex grallatoria Maxim. HINA-SUGE. Rhizomes slender, short, with ascending innovations; culms slender,

5–20 cm. long; leaf-blades about 1 mm. wide, overtopping the culm after anthesis, the basal sheaths dark reddish, rarely pale; spikes unisexual (plant dioecious), solitary, terminal, bractless, the pistillate loosely 3- to 6-flowered, broadly linear, about 1 cm. long, the staminate loosely few-flowered, linear, reddish brown; pistillate scales narrowly ovate, pale to partially reddish brown, slightly lustrous, clasping at base; perigynia nearly as long as or slightly longer than the scales, ellipsoidal, nerveless except for 2 scabrous costal nerves, pale green, membranous, the beak short, recurved, with an entire hyaline mouth; stigmas 3, rather long.——Apr.–May. Rocks in woods in mountains; Honshu, Shikoku, Kyushu; gregarious and matted.

126. Carex heteroclita Franch. *C. grallatoria* var. *heteroclita* (Franch.) Kuekenth.——SANAGI-SUGE. Closely allied to the preceding; leaf-blades flat, 1–2.5 mm. wide, short; spikes androgynous, the staminate part shorter than the loosely few-flowered pistillate part.——Rocks in woods in mountains; Honshu (Kantō, westw. to Kinki Distr.), Shikoku, Kyushu.——Formosa.

127. Carex gifuensis Franch. *C. argyrostachys* Lév. & Van't.——KURO-HINA-SUGE. Rhizomes slightly woody, short-creeping, much-branched, forming mats; culms 10–30 cm. long, slender, pilose-scabrous; leaves elongating after anthesis, pilose, the blades rather flat, 1.5–2.5 mm. wide, light to yellowish green, at first pilose-scabrous above, the basal sheaths partially dark reddish, prominently reticulate-fibrous, the ligules prolonged, pale, scarious, dorsally 2-lobed; spikes 2 or 3, contiguous, erect, the terminal staminate, narrowly lanceolate, 10–15 mm. long, usually reddish, the lateral pistillate, sessile, oblong, 5–10 mm. long, densely rather few-flowered, the lowest bract scalelike, clasping, short-setaceous at tip; pistillate scales ovate, abruptly short-cuspidate, dark castaneous, the margins broadly hyaline; perigynia longer than the scales, 4 mm. long, pilose, several-ribbed, with a thick stipe, abruptly short-beaked, with a dark purplish brown entire tip; stigmas 3.——Apr.–June. Dry open woods; Honshu (Shimotsuke, Mino, and Ise Prov.); rare.

128. Carex mira Kuekenth. SAWA-HIME-SUGE. Rhizomes densely tufted, forming large clumps; culms 20–40 cm. long, slender, decumbent when the achenes are ripe; old leaves persistent and surrounding the neck at base, circinately curved, the new leaves erect, flat, about 2 mm. wide, sparsely hairy, the basal sheaths partially dark red, the ligules scarcely developed; spikes 2–4, erect, approximate, the terminal staminate, broadly linear, dark reddish brown, with minutely ciliate scales, the lateral pistillate, oblong, 5–10 mm. long, sessile, the lowest bract auriculate, clasping, not sheathing, short-awned; pistillate scales elliptic, dark reddish brown, minutely ciliate; perigynia as long as and narrower than the scales, ovoid-fusiform, 3–3.5 mm. long, compressed-trigonous, loosely appressed-pilose, gradually narrowed at each end, short-beaked, with a dark brown minutely 2-toothed mouth; stigmas 3.——Pine woods in mountains; Honshu (Mikawa Prov. to Kinki and Chūgoku Distr.); rare.——Korea.

129. Carex melanocarpa Cham. *C. brachyphylla* Turcz.; *C. ericetorum* subsp. *melanocarpa* (Cham.) Kuekenth. ——TAKANE-HIME-SUGE. Rather firm tufted herb; culms 5–15 cm. long, erect, smooth, rigid; leaf-blades stiff, lustrous, spreading, 1.5–2.5 mm. wide, the sheaths partially reddish purple; spikes 2 or 3, approximate, erect, the terminal staminate, linear, 7–13 mm. long, the lateral pistillate, narrowly oblong to nearly globose, densely few-flowered, subsessile, 4–8 mm. long, the bracts very short, auriculate, not sheathing, sometimes with a setaceous tip; pistillate scales broadly elliptic, very obtuse to nearly rounded, ciliolate, dark reddish brown; perigynia slightly longer than the scales, 2 mm. long, ellipsoidal, obtusely angled, with 2 costal nerves, sparsely short-setulose mainly on the upper part, abruptly short-beaked, the tip hyaline, emarginate; stigmas 3.——July–Aug. Dry alpine slopes; Hokkaido (Mount Yubaridake); very rare.——Sakhalin, and e. Siberia.

130. Carex vanheurckii Muell.-Arg. *C. amblyolepis* Trautv. & Mey., non Peterm.; *C. pensylvanica* var. *amblyolepis* (Trautv. & Mey.) Kuekenth.; *C. pseudowrightii* Honda; *C. vanheurckii* var. *pseudowrightii* (Honda) Ohwi——NUIO-SUGE, SHIROUMA-HIME-SUGE. Rhizomes loosely branched, with short-creeping innovations; culms slender, slightly scabrous, 10–40 cm. long, ultimately decumbent; leaf-blades flat, 1.5–2 mm. wide, rather soft, the basal sheaths dark reddish purple, splitting into fibers; spikes 2 or 3, erect, the terminal staminate, many-flowered, linear, 1–2 mm. long, with obtuse, margined, thin scales broadly scarious, the others pistillate, contiguous, 4–7 mm. long, ovate-ellipsoidal, rather few-flowered, the bracts scalelike, dark reddish purple, 4–6 mm. long; pistillate scales ovate-elliptic, acute, dark reddish purple; perigynia nearly as long as the scales, about 2.7 mm. long, globose-obovoid, with 2 costal nerves, obscurely 3-angled, sparsely hispidulous, often glabrate, abruptly short-beaked, the beak purplish with an entire hyaline mouth; stigmas 3.——July–Aug. Alpine slopes; Hokkaido, Honshu (n. and centr. distr.); rather rare.——e. Siberia, Sakhalin, Kuriles, and Kamchatka.

131. Carex oxyandra (Fr. & Sav.) Kudo var. **oxyandra.** *C. montana* var. *oxyandra* Fr. & Sav.; *C. wrightii* Franch., non Dewey; *C. pilulifera* sensu auct. Japon., non L.——HIME-SUGE. Plant loosely tufted, with short ascending rhizomes; culms 10–50 cm. long, slender, scabrous; leaf-blades flat, 2–3 mm. wide, soft, the basal sheaths reddish, sparsely fibrous; spikes 3–6, all but the lowest approximate, the terminal staminate, 5–8 mm. long, dark red, linear, with acute, scarious-margined scales, the lateral spikes pistillate, ovate-globose, 5–7 mm. long, few-flowered, the bracts scalelike, sheathless, sometimes ending in a short awnlike bristle; pistillate scales dark reddish purple, ovate-oblong, acute; perigynia longer than the scales, narrowly ovate to elliptic, 2.5–3.5 mm. long, obtusely trigonous, with 2 costal nerves, sparsely short-pubescent, abruptly short-beaked, with a minutely 2-toothed tip.——May–July. Coniferous woods, sometimes in sandy places in volcanoes, often in alpine regions; Hokkaido, Honshu, Shikoku, Kyushu; rather common.——Sakhalin, Kuriles, and Formosa.

Var. **lanceata** (Kuekenth.) Ohwi. *C. wrightii* var. *lanceata* Kuekenth.——NAGAMI-NO-HIME-SUGE. Pistillate scales broadly lanceolate, long-attenuate; perigynia 4–5 mm. long, narrowly lanceolate, sparsely puberulent, long-beaked.——Woods; Honshu (centr. and Kinki Distr.); rather rare.

131a. Carex chinoi Ohwi ex T. Koyama. TATESHINA-HIME-SUGE. Rhizomes covered with yellow-brown fibrous remains of leaf-sheaths; leaves narrowly linear, nearly as long as the culms; basal sheaths yellow-brown; terminal staminate spikelet 5–10 mm. long; lateral pistillate spikelets globose, 3–6 mm. across; bracts setaceous; pistillate scales elliptic- to oblong-ovate, yellow-brown with a lighter colored margin, short-cuspidate; perigynia narrow-obovate, 2.5–3 mm. long, trigonous, almost nerveless except the lateral 2 keels, short-stipitate at base; achenes elliptical; otherwise with the characters of *C. oxyandra*.——Honshu (Mount Tateshina, Shinano Prov.); rare.

132. Carex limosa L. *C. limosa* var. *fuscocuprea* Kuekenth.; *C. fuscocuprea* (Kuekenth.) V. Krecz.——YACHI-SUGE. Rhizomes elongate, long-creeping, with yellow-tomentose, thick roots; culms 20–40 cm. long; leaf-blades 1.5–2.5 mm. wide, cinereous-green, the sheaths reddish brown; spikes 2 or 3, distant, the terminal staminate, linear, 2–2.5 cm. long, on an erect peduncle, dark tawny, the others pistillate or rarely with a short terminal staminate portion, ovoid or oblong, 15–20 mm. long, pendulous on long setaceous peduncles, the lowest bract bristlelike, erect; pistillate scales ovate, acute, sharply mucronate, copper-brown; perigynia narrower and slightly shorter than the scales, elliptic, 3.5–4 mm. long, glaucous-green, compressed-triangular, several-nerved, densely puncticulate, abruptly very short-beaked, with a ferrugineous emarginate tip; achene compressed-trigonous; stigmas 3.——June–Aug. Sphagnum bogs; Hokkaido, Honshu (Tajima Prov. and eastw.); rather common.——Widely distributed in bogs in the N. Hemisphere.

133. Carex paupercula Michx. *C. limosa* var. *irrigua* Wahlenb.; *C. irrigua* (Wahlenb.) Smith; *C. gentiliana* Lév.; *C. magellanica* sensu auct. Japon., non Lam.——DAKE-SUGE. Somewhat resembles the preceding; culms 20–40 cm. long; leaf-blades flat, bright green, 2–3 mm. wide, the basal sheaths brown, rarely reddish; spikes 2–4, approximate, the terminal staminate (rarely gynecandrous), linear, 7–15 mm. long, few-flowered, the lateral 2 or 3 pistillate, oblong, 10–15 mm. long, pendulous on long setaceous peduncles, the lowest bract leaflike, erect, scarcely sheathing; pistillate scales broadly lanceolate, long-acuminate, dark copper-red; perigynia suberect, shorter and broader than the scales, broadly elliptic, 2.5–3 mm. long, compressed-trigonous, densely puncticulate, slenderly few-nerved, nearly beakless, the tip ferrugineous, entire; stigmas 3.——June–July. Sphagnum bogs in mountains; Honshu (Shinano and Uzen Prov.).——n. Kuriles, and widely distributed in the N. Hemisphere.

134. Carex livida (Wahlenb.) Willd. *C. limosa* var. *livida* Wahlenb.; *C. fujitae* Kudo——MUSEN-SUGE. Rhizomatous, the roots smooth, white, fibrous; culms 20–30 cm. long, glabrous; leaf-blades erect, cinereous-green, rather flat, 2–3 mm. wide, the sheaths partially grayish brown; spikes 2–4, erect, the terminal staminate, lanceolate, reddish brown, 1–1.5 cm. long, the others pistillate, short-cylindric, few-flowered, short-peduncled, 1–2 cm. long, the bract leaflike, longer than the spikes, very short-sheathing; pistillate scales elliptic, obtuse, light brown, deciduous; perigynia slightly longer than the scales, ovoid, 4 mm. long, erect, 3-angled, glaucous, densely puncticulate, inconspicuously nerved, nearly beakless, with a ferrugineous entire tip; stigmas 3.——July. Bogs in high mountains; Hokkaido (Mount Taisetsu).——Korea, Kuriles, and widely occurring in northern part of the N. Hemisphere.

135. Carex laxa Wahlenb. *C. macrochlamys* Franch.——ITO-NARUKO-SUGE. Rhizomatous; culms 20-40 cm. long, slender, glabrous; leaf-blades soft, 1.5–2.5 mm. wide, the sheaths pale or light brownish; spikes 2 or 3, loose, the terminal staminate, linear, 1.5–2.5 cm. long, tawny, the lateral pistillate, oblong to narrowly so, rather densely flowered, pendulous on long peduncles, the lowest bract short, leaflike,

long-sheathing; pistillate scales oval, very obtuse, reddish cupreus; perigynia as large as the scales, ascending, ovoid, more or less compressed-trigonous, glaucous, densely puncticulate, indistinctly nerved, abruptly short-beaked, with a ferrugineous entire tip; stigmas 3.——Bogs; Hokkaido, Honshu (Rikuchu Prov.); rare.——n. Korea, Kuriles, and Siberia to Europe.

136. Carex vaginata Tausch var. **petersii** (C. A. Mey.) Akiyama. *C. falcata* Turcz.; *C. petersii* C. A. Mey.; *C. sparsiflora* var. *petersii* (C. A. Mey.) Kuekenth.——SAYA-SUGE, KE-YARI-SUGE. Rhizomes elongate, slender; culms 20–50 cm. long, glabrous; leaf-blades soft, flat, bright green, 2–5 mm. wide, the sheaths partially dark reddish purple; spikes 2–4, distant, the terminal staminate, broadly linear, 1.5–2 cm. long, dark purplish brown, the lateral pistillate, short-cylindric, rather densely many-flowered, long-peduncled, 1.5–2 cm. long, the bracts short-bladed, long-sheathing; pistillate scales ovate, dark purplish brown or dark reddish purple; perigynia longer than the scales, ascending to spreading, obtusely 3-angled, ovoid, membranous, glabrous, faintly nerved, abruptly short to rather long-beaked, with an obsoletely 2-toothed apex; stigmas 3.——June–July. Woods in mountains; Hokkaido, Honshu (Mount Yakeishidake in Rikuchu Prov.); rare.—— e. Siberia, Sakhalin, Kamchatka, s. and centr. Kuriles, and Korea. The typical phase occurs in Canada, Siberia, and Europe.

137. Carex pilosa Scop. *C. auriculata* Franch.; *C. hakodatensis* Lév. & Van't.; *C. pilosa* var. *auriculata* (Franch.) Kuekenth.——SAPPORO-SUGE, HANA-MAGARI-SUGE, MIMI-SUGE. Rhizomes slender, elongate; culms 30–60 cm. long, usually loosely pilose; leaf-blades 5–10 mm. wide, flat, abruptly acuminate, soft, loosely pilose or sometimes glabrous, deep green, the basal sheaths (at least the lower ones) bladeless and dark purplish sanguineous; spikes 3 or 4, distant, the terminal staminate, linear-clavate, 10–20 mm. long, long-peduncled and much overtopping the next one, the lateral pistillate, cylindric, 2–3 cm. long, loosely flowered, the lower ones long-peduncled, the bracts leaflike, usually pilose, long-sheathing; pistillate scales ovate, acute, dark reddish purple; perigynia longer than the scales, ovoid, 4–5 mm. long, obtusely 3-angled, light green, membranous, glabrous, weakly nerved, abruptly rather long-beaked, with an obliquely truncate, emarginate, membranous apex; stigmas 3.——June–July. Woods; Hokkaido, Honshu (n. distr. south to Yechigo Prov.); rare.——Sakhalin, Korea, and westw. to n. Europe.

138. Carex kujuzana Ohwi. KUJŪ-TSURI-SUGE. Rhizomes creeping, rather short; culms solitary to few, soft, 50–60 cm. long, acutely angled, glabrous; leaf-blades 3–4 mm. wide, soft, flat, flaccid, abruptly pointed; spikes 2 or 3, widely distant, the terminal staminate, long-pedunculate, lanceolate, 2–2.5 mm. long, dark reddish, the lateral pistillate, 1–1.5 cm. long, oblong, 5- to 8-flowered, pendulous on very long capillary peduncles, the bracts leaflike, long-sheathing; pistillate scales acute, pale tawny, partially dark reddish; perigynia longer than the scales, 6 mm. long, spreading, broadly obovoid, light yellowish green, slenderly nerved, the beak long, deflexed, with an obliquely truncate hyaline tip; stigmas 3.——May. Grassy slopes in mountains; Kyushu (Mount Kuju); very rare.

139. Carex dissitispicula Ohwi. *C. kunioi* T. Koyama——KARUIZAWA-TSURI-SUGE. Rhizomes elongate, creeping; culms solitary to few, slender, soft, 25–60 cm. long, glabrous; leaf-blades of the sterile fascicles elongate and overtopping the culm, 3–4 mm. wide, soft, flattish, the basal sheaths bladeless or short-bladed, partially dark purple or purplish brown, the nerves pale green, areolate with transverse veinlets; spikes 2 or 3, widely distant, the terminal staminate, long-pedunculate, 15–20 mm. long, linear, purplish brown, the lateral pistillate, oblong, 10–20 mm. long, loosely 5- to 13-flowered, pendulous on long capillary peduncles, the bracts short-bladed, long-sheathing; pistillate scales ovate-elliptic, acuminate, pale; perigynia longer than the scales, 5.5–6.5 mm. long, ovoid-fusiform, 3-angled, slenderly nerved, glabrous, yellowish green, the beak long, slightly recurved, with an obliquely truncate hyaline tip; stigmas 3.——May–June. Wet grassy places; Honshu (Karuisawa in Shinano Prov.); rare.

140. Carex filipes Fr. & Sav. Rhizomes short-creeping; culms tufted, 30–50 cm. long, soft, acutely angled, glabrous; leaf-blades thin and soft, flat, nearly glabrous, glaucous to bright green, rather abruptly narrowed toward apex, the basal sheaths bladeless; spikes 3 or 4, distant, the terminal staminate, the others pistillate, short-cylindric, very loosely several-flowered, spreading to pendulous, the bracts leaflike, long-sheathing; pistillate scales ovate, acuminate, pale, partially dark reddish or brownish; perigynia nearly as long as the scales, 5–6 mm. long, broadly ovoid-fusiform, glabrous, pale green, membranous, weakly nerved, rather long-beaked, with a truncate hyaline tip; stigmas 3.

Var. **filipes**. TAMA-TSURI-SUGE. Leaf-blades 2–4 mm. wide, the sheaths dark reddish purple; staminate spike dark purplish brown or sometimes pale, linear, 10–15 mm. long, short- or rather long-peduncled, few-flowered, the pistillate loosely several-flowered, the upper contiguous, the lower distant.——Apr.–June. Woods or grassy places in hills and mountains; Honshu (n. distr., westw. to Kinki Distr.), Shikoku; rather common.

Var. **tremula** (Ohwi) Ohwi. *C. arisanensis* var. *tremula* Ohwi; *C. tremula* (Ohwi) Ohwi——HIME-JUZU-SUGE. Plant smaller; leaf-blades narrow, short, abruptly acuminate to acute; staminate spike broadly linear, very short, few-flowered, inconspicuous.——Apr.–May. Shikoku, Kyushu.

Var. **rouyana** (Franch.) Kuekenth. *C. rouyana* Franch. ——Ō-TAMA-TSURI-SUGE. Plant larger; leaf-blades 3–7 mm. wide, the sheaths pale or brownish, rarely partially purplish brown; staminate spike elongate, always long-peduncled and surpassing the adjacent one.——Apr.–June. Woods in mountains and hills; Honshu (Kantō and westw. to Kinki Distr.).

Var. **arakiana** (Ohwi) Ohwi. *C. rouyana* var. *arakiana* Ohwi; *C. arakiana* (Ohwi) Ohwi——HIRO-HA-NO-Ō-TAMA-TSURI-SUGE. Resembles the preceding variety, the leaves usually living two years, 8–12 mm. wide, deep green, the basal leaf-sheaths at least partially dark sanguineous-purple.——May. Honshu (Tango and Aki Prov.).

141. Carex papulosa Boott. *C. flectens* Boott; *C. grandisquama* Franch.; *C. sekimotoi* Honda——EZO-TSURI-SUGE. Plant loosely tufted; culms soft, 30–50 cm. long, glabrous; leaf-blades flat, flaccid, 3–7 mm. wide, grayish green, the basal sheaths stramineous, short-bladed; spikes 2 or 3, widely distant, the terminal staminate, narrowly lanceolate, 1.5–3 cm. long, long-peduncled, ferruginous, the lateral pistillate, oblong, long-peduncled, pendulous, the bracts short-bladed, long-sheathing; pistillate scales elliptic, rounded and mucronate, ferruginous; perigynia longer than the scales, 5–6 mm. long, erect, ovoid-fusiform, 3-angled, grayish brown, punctticulate, faintly nerved, dull, glabrous, rather gradually long-beaked,

with a truncate tip; stigmas 3.——June–July. Wet places; Hokkaido, Honshu (n. distr. southw. to Shimotsuke, Hitachi, and Shinano Prov.), Kyushu; rare.——Korea and Ussuri.

142. Carex parciflora Boot var. **parciflora**. *C. glehnii* F. Schmidt; *C. jackiana* subsp. *parciflora* (Boott) Kuekenth.; *C. kamikochiana* Nakai——GUREN-SUGE, "GLEHN"-SUGE. Rhizomes abbreviated or very shortly elongate, loosely to densely tufted; culms 50–70 cm. long, soft, glabrous, acutely 3-angled; leaf-blades flat, 5–10 mm. wide, soft, bright or yellowish green, the sheaths pale; spikes 4 or 5, erect, the terminal staminate, linear, pale, 1–1.5 cm. long, peduncled (sessile in forma *ochrolepis* (Franch.) T. Koyama), the lateral pistillate, distant, narrowly oblong, 1.5–3 cm. long, the lower long-peduncled, the bracts leaflike, long-sheathing; pistillate scales ovate, pale green; perigynia longer than the scales, 4–4.5 mm. long, obliquely spreading, ovate-oblong, obtusely 3-angled, membranous, glabrous, gradually long-beaked, with an obliquely truncate hyaline tip; stigmas 3.——June–July. Wet places in lowlands and mountains; Hokkaido, Honshu (n. distr.).——Sakhalin, and s. Kuriles.

Var. **macroglossa** (Fr. & Sav.) Ohwi. *C. macroglossa* Fr. & Sav.; *C. jackiana* subsp. *parciflora* var. *macroglossa* (Fr. & Sav.) Kuekenth.; *C. filipes* var. *oligostachys* Kuekenth. pro parte, excl. synon.——KO-JUZU-SUGE. Plant more slender; leaf-blades glaucous-green, usually narrower; pistillate spikes oblong, rather few-flowered, 1–1.5 cm. long; stigmas 3.——Forma **subsessilis** Ohwi. MUGI-SUGE. Perigynia to 8 mm. long.——Wet places in lowlands; Hokkaido, Honshu, Kyushu. ——s. Korea.

Var. **vaniotii** (Lév.) Ohwi. *C. vaniotii* Lév.; *C. gagaensis* Akiyama——NAGABO-NO-KO-JUZU-SUGE. Plant slender; leaves vivid green, 2–5 mm. wide; culms 20–30 cm. long; perigynia suberect, obtusely angled, ovate-fusiform, narrower than in the typical phase.——June–July. Wet places in subalpine regions; Honshu (n. and centr. distr., chiefly on the Japan Sea side).

143. Carex tumidula Ohwi. IWAYA-SUGE. Rhizomes slender, long-creeping; culms 15–30 cm. long, glabrous; leaf-blades linear, longer than the culms, 2–3 mm. wide, soft, flat, the basal sheaths membranous, pale, usually indistinctly punctate-puberulent, the lower ones bladeless and ferrugineous, the ligules oblique, ferrugineous; spikes 7–9, erect, 1–2 cm. long, pedunculate, androgynous except the terminal one, frequently binate to ternate at each node, sometimes sparingly branching at base, the pistillate loosely 1- to 3-flowered, the bracts leaf-like, with an inflated sheath at base, at least the lowest one much longer than the inflorescence; scales of both sexes nearly alike, oblong, very obtuse, pale ferrugineous; perigynia 3 mm. long, ovate-rhomboid, 3-angled, membranous, gray-green, inconspicuously nerved, abruptly narrowed at each end, short-beaked, with an emarginate tip; stigmas 3, rather thick.——Apr.–May. Mountains; Shikoku (Iyo Prov.); rare.

144. Carex siderosticta Hance. TAGANE-SŌ. Rhizomes creeping, elongate, slender; culms few, 10–40 cm. long, with only a few sheaths at the base; leaves of sterile fascicles with broadly lanceolate to lanceolate blades 1–3 cm. wide, thin, soft, glabrous or sparsely puberulent, scabrous on margins; spikes 4–8, androgynous, erect, short-cylindric, few-flowered, 1–2 cm. long, the bracts spathaceous, the sheath ampliate above, usually bladeless; pistillate scales oblong, obtuse, frequently brown-spotted; perigynia 3 mm. long, ellipsoidal, 3-angled, pale green, membranous, glabrous, slenderly nerved, the beak extremely short, with an entire tip; stigmas 3, slender, elongate.——Apr.–May. Woods in mountains; Hokkaido, Honshu, Shikoku, Kyushu; rather common.——Korea, Ussuri, Manchuria, and China.

145. Carex ciliatomarginata Nakai. *C. siderosticta* var. *pilosa* Lév.——KE-TAGANE-SŌ.   Much-resembling the preceding, the leaf-blades slightly smaller and prominently long-ciliate at least on the lower margins; terminal spike usually staminate only, ovate-oblong; perigynia pilose.——Apr.–May. Dry shaded grassy places in mountains; Honshu (centr. distr. and westw.), Shikoku, Kyushu; rather rare.——Korea.

146. Carex pachygyna Fr. & Sav. SASA-NO-HA-SUGE. Rhizomes creeping, elongate, thickened at the nodes, somewhat ligneous; culms soft, 15–30 cm. long; leaf-blades lanceolate to broadly so, 1–2 cm. wide, flat, soft, glabrous or sparsely pilose, the basal sheaths pale cinnamon, the lower ones bladeless; spikes rather many, globose-ovoid, androgynous, binate to ternate at the nodes, on unequal erect peduncles, densely few-flowered, 4–6 mm. long and as wide, the staminate part very short and inconspicuous, the bracts bladeless, spathaceous, the sheaths pale or yellowish green, ampliate above; pistillate scales ovate-orbicular, rounded at the apex, pale; perigynia longer than the scales, 2–2.5 mm. long, ascending, ellipsoidal, 3-angled, many-nerved, pale green, glabrous, rounded at the apex and nearly beakless; stigmas 3, thick, short.——Apr.–May. Woods in low mountains; Honshu (Kinki Distr. and westw.), Shikoku.

147. Carex ischnostachya Steud. *C. ringgoldiana* Boott ——JUZU-SUGE.   Rhizomes densely tufted; culms erect, 30–60 cm. long, rather firm; leaf-blades flat, deep green, 5–10 mm. wide, the basal sheaths bladeless, dark reddish; spikes 3–6, contiguous except the lower ones, the terminal one staminate, pale, filiform, 2–3 cm. long, shorter than the pistillate, the pistillate cylindric, rather loosely to rather densely many-flowered, erect, 2–5 cm. long, the bracts leaflike, longer than the culm, long-sheathing; pistillate scales small, broadly ovate, obtuse, entire, pale; perigynia 3–4 times longer than the scales, narrowly ovoid, 3.5–5 mm. long, inflated, erect-patent, obsoletely 3-angled, glabrous, olive-colored, prominently nerved, gradually rather long-beaked, obliquely truncate and hyaline at tip; stigmas 3, slender, short.——Apr.–June. Wet places in lowlands and hills, especially along streams; Hokkaido, Honshu, Shikoku, Kyushu; common.——Ryukyus, Korea, and China.

148. Carex maculata Boott. *C. micans* Boott; *C. maculata* forma *viridans* Kuekenth.——TACHI-SUGE.   Rhizomes tufted; culms slender, erect, 20–60 cm. long, glabrous; leaf-blades glaucous, somewhat soft, nearly glabrous, 3–5 mm. wide, densely puncticulate, the sheaths light ferrugineous, thinly membranous, the ligule light ferrugineous and very thin; spikes 3 or 4, erect, the terminal staminate, linear, sessile, 1–3 cm. long, light ferrugineous, the lateral pistillate, cylindric, densely many-flowered, erect, 1–4 cm. long, the upper ones nearly sessile, the lower long-peduncled, the lower bracts leaf-like, longer than the culm, long-sheathing; pistillate scales narrowly ovate, obtuse, pale ferrugineous; perigynia much longer than the scales, 2.5 mm. long, broadly ellipsoidal, ascending, compressed-trigonous, grayish green, frequently brownish when dried, densely papillose, with several raised nerves, rounded and abruptly short-beaked, the mouth ferrugineous, hyaline, entire; stigmas 3, short.——May–June. Wet places in lowlands; Honshu, Shikoku, Kyushu.——s. Korea, Ryukyus, Formosa, China, and India.

149. Carex hakonensis Fr. & Sav. *C. krameri* Fr. & Sav.; *C. onoei* var. *krameri* (Fr. & Sav.) Kuekenth.——KO-HARI-SUGE, KOKE-SUGE.   Rhizomes extremely short, densely tufted; culms very slender, obtusely angled, nearly glabrous, 10–20 cm. long; leaf-blades capillary, involute, to 1 mm. wide; spike solitary, terminal, androgynous, ovoid, 3–5 mm. long, the staminate few-flowered, short, inconspicuous, the bracts wanting; pistillate scales broadly ovate, reddish or dark brown; perigynia slightly longer than the scales, about 2 mm. long, ascending, elliptic-ovoid, compressed-trigonous, not turgid, with 2 costal nerves, abruptly short-beaked, emarginate at apex; stigmas 3. ——May–Aug. Wet places in mountains; Hokkaido, Honshu, Shikoku, Kyushu.——Korea.

150. Carex onoei Fr. & Sav. *C. capituliformis* Meinsh.; *C. hakonensis* var. *onoei* (Fr. & Sav.) Ohwi——HIKAGE-HARI-SUGE.   Somewhat resembling the preceding; culms 15–30 cm. long, soft, acutely angled, scabrous on the angles; leaves flat, soft, 1.5–3 mm. wide; perigynia 2.5–3 mm. long, ovate, compressed-triangular, faintly nerved, abruptly very short-beaked, 2-toothed at tip.——June–July. Wet shaded places; Hokkaido (Hakodate), Honshu (Nikko, Mount Kirigamine in Shinano and elsewhere)——Korea and Ussuri.

151. Carex fulta Franch. NIKKŌ-HARI-SUGE, HIME-TAMA-SUGE.   Rhizomes short, tufted; culms 20–40 cm. long, soft, acutely angled, 0.6–1 mm. thick, prominently scabrous on the angles; leaf-blades 2–3 mm. wide, flat, soft, thin; spike solitary, terminal, androgynous, ovoid-globose, 5–7 mm. long, 4–5 mm. in diameter, the staminate inconspicuous, the pistillate with rather many flowers; pistillate scales ovate-elliptic, pale, sometimes the lower 1 or 2 ending in a short awn; perigynia longer than the scales, 2–2.5 mm. long, spreading, broadly ovoid, slightly compressed, 3-angled, few-nerved on each side, abruptly short-beaked, with a membranous entire tip; stigmas 3, short.——June–July. Wet places in mountains; Honshu (n. and centr. distr. and Bitchu Prov.); rather rare.

152. Carex uda Maxim. EZO-HARI-SUGE, Ō-HARI-SUGE. Culms 15–50 cm. long, soft, about 1 mm. thick, in dense tufts, acutely angled, glabrous; leaf-blades 2–3 mm. wide, flat, soft, thin; spike solitary, terminal, 7–10 mm. long, androgynous, ovate to oblong, the staminate 2–3 mm. long, shorter than the pistillate; pistillate scales oblong-lanceolate, acute, pale ferrugineous; perigynia longer than the scales, somewhat reflexed, ovate-lanceolate, inflated, 3-angled, membranous, pale green, faintly nerved, gradually long-beaked, emarginate; stigmas 3.——Wet places in mountains; Hokkaido, Honshu (Mount Kirigamine in Shinano Prov.); rare.——Sakhalin, Korea, Amur, and Ussuri.

153. Carex biwensis Franch. *C. capillacea* sensu auct. Japon., non Boott; *C. rara* subsp. *capillacea* Kuekenth., pro parte; *C. rara* var. *biwensis* (Franch.) Kuekenth.——MATSUBA-SUGE.   Plant tufted; culms 10–40 cm. long, obtusely angled, glabrous; leaf-blades 1.5 mm. wide; spike solitary, terminal, 1–2 cm. long, the staminate linear, longer or shorter than the pistillate, about 1 mm. across, the pistillate oblong, about 3 mm. across, densely many-flowered; pistillate scales elliptic, very obtuse, rusty-brown; perigynia about as long as the scales, 1.5–2 mm. long, spreading, broadly ovoid, inflated, trigonous, nerved, abruptly very short-beaked with an entire tip.—— Apr.–June. Wet places; Honshu, Shikoku, Kyushu; rather common.——Korea, Manchuria, and Ussuri.

154. Carex capillacea Boott var. **capillacea**. *C. ontakensis* Franch.; *C. rara* subsp. *capillacea* (Boott) Kuekenth. ——HARIGANE-SUGE.   Culms tufted, slender, sulcate, glabrous,

10–30 cm. long; leaf-blades filiform, about 1 mm. wide, glabrous, flat or slightly involute, erect; spike solitary, androgynous, 5–10 mm. long, the staminate lanceolate, 3- to 5-flowered, 3–5 mm. long, the pistillate densely flowered, 5 mm. across, longer or shorter than the staminate part; pistillate scales ovate, very obtuse, ferrugineous; perigynia slightly longer than the scales, 2.5–3 mm. long, spreading, broadly ovoid, inflated, trigonous, faintly nerved, abruptly short-beaked, submarginate; stigmas 3, short.——Apr.–May. Wet places; Hokkaido (rare), Honshu, Kyushu; rather common.——Korea, China, India, Malaysia, and Australia.

Var. **sachalinensis** (F. Schmidt) Ohwi. *C. aomorensis* Franch.; *C. uda* var. *sachalinensis* F. Schmidt; *C. capillacea* var. *aomorensis* (Franch.) Ohwi; *C. nana* Boott, non Cham.; *C. rara* subsp. *capillacea* var. *nana* (Boott) Kuekenth.——MICHI-NO-KU-HARI-SUGE, AOMORI-HARI-SUGE. Perigynia 3–4 mm. long, slightly larger than in the typical phase.——June–July. Hokkaido, Honshu (n. and centr. distr.).——Kuriles, Sakhalin, Korea, and Ussuri.

155. Carex rhizopoda Maxim. *C. krebsiana* Boecklr.——SHIRAKO-SUGE. Culms 20–50 cm. long, tufted, acutely angled, scabrous; leaf-blades 2–3(–5) mm. wide, soft, flat, bright green, scabrous; spike solitary, androgynous, narrowly oblong, 1.5–4 cm. long, the pistillate many-flowered, longer than to nearly as long as the staminate; pistillate scales ovate, acute, pale, the midrib green; perigynia nearly twice as long as the scales, 5–6 mm. long, suberect, ovate, 3-angled, light green, weakly many-nerved, thinly membranous, gradually long-beaked, obliquely truncate and hyaline at tip; stigmas 3.——Apr.–June. Wet places along streams in low mountains; Hokkaido, Honshu, Shikoku, Kyushu; common.——s. Kuriles (?)

156. Carex bostrychostigma Maxim. *C. explens* Kuekenth.——YAMAJI-SUGE. Rhizomes short-creeping, loosely tufted, dark brown fibrous; culms 10–30 cm. long, glabrous, often nodding; leaf-blades flat, soft, nearly glabrous, 3–4 mm. wide, the sheath often yellow-tawny; spikes 5–10, the terminal staminate, broadly linear, loosely rather many-flowered, 1–2.5 cm. long, ferrugineous-tawny, the lateral pistillate, broadly linear, loosely rather many-flowered, 2–4 cm. long, peduncled, the upper approximate, the lower 2 or 3 distant, the lower bracts leaflike, sheathing; pistillate scales lanceolate-ovate, acute, fulvous, with a green 3-nerved midrib; perigynia longer than the scales, 7–8 mm. long, appressed, linear-lanceolate, 3-angled, thinly membranous, pale green, nerveless, glabrous, slightly lustrous, gradually long-beaked, with an obliquely truncate hyaline mouth; stigmas 3, persistent, elongate, slender, brownish.——Apr.–June. Honshu (Kinki and Chūgoku Distr.), Shikoku, Kyushu; rather rare.——Korea, Manchuria, and Ussuri.

157. Carex dissitiflora Franch. *C. subdissitiflora* Kuekenth. ex Matsum. nom. nud.——MIYAMA-JUZU-SUGE. Rhizomes short, tufted; culms 40–80 cm. long, soft, glabrous, few-leaved; leaf-blades flat, soft, bright green, 3–7 mm. wide, the basal sheaths pale, ultimately dark brown fibrous; spikes 4–6, very loosely arranged, 2–4 cm. long, solitary or binate at each node, androgynous, the upper lateral ones on a peduncle arising from a fertile perigynium, the lower ones on long sometimes sparingly branched peduncles, the lower bracts leaflike, sheathing; pistillate scales pale green to light yellowish brown; perigynia twice as long as the scales or longer, 9–11 mm. long, appressed, ovoid-fusiform, light green, faintly nerved, thinly membranous, 3-angled, long-beaked, obliquely truncate at the

hyaline tip; stigmas 3.——May–July. Damp woods and wet grassy places along streams in mountains; Hokkaido, Honshu, Shikoku, Kyushu; rather rare.——s. Kuriles and Formosa (var.).

158. Carex capillaris L. TAKANE-SHIBA-SUGE. Rhizomes short; culms 10–20 cm. long, slender, glabrous, nodding at tip; leaf-blades short, flat, 1–1.5 mm. wide, the sheaths partially dark red; spikes 3 or 4, peduncled, nodding, small, the terminal staminate, linear-lanceolate, 5–10 mm. long, the lateral pistillate, narrowly oblong, 8–15 mm. long, loosely few-flowered, the upper 1 or 2 longer than the staminate; pistillate scales obovate, obtuse to acute, light castaneous; perigynia longer than the scales, 3–3.5 mm. long, suberect, ovoid, brownish green, membranous, glabrous, with 2 costal nerves, somewhat lustrous, the beak rather long, glabrous or slightly scabrous, with hyaline entire tip; stigmas 3, short.——July–Aug. Rather dry alpine slopes; Hokkaido (Mount Yubari-dake), Honshu (Mount Shirouma and Mount Hayachine); very rare.——n. Korea, Sakhalin, Kuriles, and the cooler zones and alpine regions of the N. Hemisphere.

159. Carex tenuiformis Lév. & Van't. *C. koreana* Komar., non Bailey——ONOE-SUGE, REBUN-SUGE. Plant somewhat tufted; culms slender, 15–40 cm. long, nodding in upper part; leaf-blades deep green, 2.5–4 mm. wide, the basal sheaths dark sanguineous in part; spikes 2–4, loosely arranged, the terminal staminate, long-peduncled, clavate-linear, 1–1.5 cm. long, longer than the others, the pistillate linear-oblong, 1–2.5 cm. long, loosely flowered, long-peduncled, nodding, the bracts short-bladed, long-sheathing; pistillate scales acute, reddish brown; perigynia longer than the scales, 3–4.5 mm. long, suberect, ovoid-fusiform, obtusely angled, brownish green, slightly lustrous, glabrous or sparsely spinulose at tip, with 2 costal nerves, moderately long-beaked, hyaline at tip; stigmas 3, slender.——June–Aug. Rather dry alpine slopes; Hokkaido, Honshu (n. and centr. distr.).——Korea, Ussuri, Sakhalin, and s. Kuriles.

160. Carex alliiformis C. B. Clarke. *C. purpurascens* Kuekenth.——RYŪKYŪ-SUGE. Culms 20–40 cm. long, glabrous; leaf-blades flat, soft, dark green, 6–12 mm. wide, the basal sheaths sanguineous; spikes 3–7, erect, approximate, the lower ones distant, the upper 1–3 staminate and linear or gynecandrous and clavate, the others mostly all pistillate, often with a few staminate flowers at base, narrowly oblong, densely many-flowered, peduncled, the bracts leaflike, the sheaths often dark purplish red at base; pistillate scales ovate, acute, dark reddish purple; perigynia longer than the scales, 3.5–4 mm. long, slightly turgid, membranous, lustrous, glabrous, green, many-nerved, rather abruptly short-beaked, with an entire hyaline tip; stigmas 3.——Apr.–May. Dense woods; s. Kyushu; rare.——Ryukyus, Formosa, China, and Indochina.

161. Carex japonica Thunb. *C. japonica* var. *minor* Boott; *C. motoskei* Miq.; *C. trichostyles* Fr. & Sav.; *C. japonica* var. *gracilis* Miq.; *C. japonicaeformis* Nakai——HIGO-KUSA. Rhizomes elongate, slender; culms 20–40 cm. long, scabrous, often 1-leaved; leaf-blades flat, 2.5–4 mm. wide, somewhat glaucous beneath, the basal sheaths pale; spikes 2–4, loosely arranged, the terminal staminate, linear, pale green, 1.5–3 cm. long, peduncled, the lateral pistillate, ellipsoidal to narrowly oblong, 1–2 cm. long, densely many-flowered, nodding on slender peduncles, the lower bracts leaflike, longer than the culm, sheathless; pistillate scales narrowly ovate, acuminate, pale; perigynia longer than the scales, 3.5–4 mm. long, spreading, ovoid, nerved, slightly inflated, membranous, pale green, glabrous, gradually long-beaked, minutely 2-toothed and hyaline at tip; stigmas 3, persistent, brown, slender, slightly longer than the perigynium.——Apr.–June. Woods in lowlands; Hokkaido, Honshu, Shikoku, Kyushu; rather common.——Korea and China.

162. Carex aphanolepis Fr. & Sav. *C. japonica* var. *humilis* Franch.; *C. japonica* var. *aphanolepis* (Fr. & Sav.) Franch. ex Lév. & Van't.; *C. vernicosa* C. B. Clarke——E-NASHI-HIGO-KUSA, SAWA-SUGE. Closely related to *C. japonica;* culms 20–40 cm. long; leaf-blades 2–4 mm. wide; spikes 2–4, the pistillate globose to oblong, usually sessile, 7–12 mm. long, erect; perigynia much longer than the scales, 3 mm. long, spreading, ellipsoidal, spongy-membranous, pale, strongly inflated, abruptly short-beaked; stigmas short, deciduous after anthesis.——Apr.–June. Woods on hillsides and in lowlands; Hokkaido, Honshu, Shikoku, and Kyushu.——Korea.

163. Carex doniana Spreng. *C. chlorostachys* D. Don, non Stev.; *C. zollingeri* Kunze; *C. consocialis* Steud.; *C. alopecuroides* var. *chlorostachys* (D. Don) Kuekenth.——SHIRA-SUGE. Culms rather stout, tufted, 50–70 cm. long; leaf-blades flat, light green, whitish beneath, 5–10 mm. wide, the basal sheaths pale; spikes 4–6, the terminal staminate, elongate-linear, light green, 3–6 cm. long, the others pistillate, cylindric, densely many-flowered, erect or nodding, 3–7 cm. long, the upper ones sessile, the lower bracts leaflike, sheathless; pistillate scales narrowly ovate, very acute, whitish with a green midrib; perigynia longer than the scales, 3 mm. long, spreading, narrowly ovoid, somewhat inflated, membranous, nerved, pale green, glabrous, rather abruptly short-beaked, the tip hyaline, minutely 2-toothed; stigmas 3.——Apr.–June. Woods and thickets in lowlands and in low mountains; Hokkaido, Honshu, Shikoku, Kyushu; common.——Korea, China, Ryukyus, Formosa, Malaysia, and India.

164. Carex paniculmis Komar. *C. japonica* var. *naipiangensis* Lév. & Van't.; *C. paniculigera* Nakai; *C. macromollicula* Nakai——HIKAGE-SHIRA-SUGE. Resembles the preceding; leaf-blades to 10 mm. wide, bright green, not glaucous beneath; spikes 4 or 5, erect, contiguous or approximate, the terminal staminate, 3–5 cm. long, slightly ferrugineous, the pistillate 2–5 cm. long, the lower occasionally nodding, the lowest bract leaflike, nearly erect; pistillate scales ovate, pale or often obscurely tawny; perigynia 3.5–4 mm. long.——June–July. Coniferous woods; Hokkaido, Honshu (centr. distr.); rather rare.——Korea, Manchuria, Ussuri, and Sakhalin.

165. Carex mollicula Boott. *C. arcuata* Franch.——HIME-SHIRA-SUGE. Culms 15–30 cm. long, soft but rather stout, scabrous, acutely angled; leaf-blades flat, soft, rather thin, bright green, 4–8 mm. wide, the basal sheaths pale; spikes 3–6, nearly contiguous, erect, sessile, the upper ones almost fastigiate, the terminal staminate, linear, pale green, 1.5–3 cm. long, the lateral pistillate, short-cylindric, densely many-flowered, sessile or nearly so, 1.5–3 cm. long, the lowest one often somewhat distant and short-peduncled, the lowest bract leaflike, spreading or slightly deflexed, sheathless; pistillate scales narrowly ovate, long-acuminate, pale; perigynia longer than the scales, 3–4 mm. long, spreading or slightly reflexed, narrowly ovoid, inflated, trigonous, pale green, glabrous, few-nerved, gradually long-beaked, with a minutely 2-toothed hyaline tip; stigmas 3, short, slender.——May–July. Damp woods in mountains; Hokkaido, Honshu, Shikoku, Kyushu; rather rare.——s. Korea and s. Kuriles.

166. Carex satsumensis Fr. & Sav. *C. nikoensis* Fr. & Sav.——ABURA-SHIBA. Rhizomes elongate, slender; culms glabrous, obtusely angled, 10–30 cm. long; leaf-blades 2–3 mm. wide, flat, rather stiff; inflorescence an acute, ovoid-conical to oblong-conical dense panicle 3–8 cm. long, with numerous spikes; spikes androgynous, ovate, 5–12 mm. long, spreading, the lateral sessile, branching at base, the bracts scalelike or short, sheathless; pistillate scales lanceolate-ovate, obscurely tawny; perigynia longer than the scales, about 2.5 mm. long, spreading, narrowly ovoid, obtusely angled, faintly nerved, brownish, glabrous, rather abruptly long-beaked, with a hyaline obliquely truncate tip; stigmas 3, long.——Apr.–July. Honshu (Iwashiro Prov. and westw.); Shikoku, Kyushu.——Formosa and Philippines.

167. Carex cruciata Wahlenb. *C. hakkuensis* Hayata ——HANABI-SUGE. Large stout herb; rhizomes thick, short-creeping; culms 40–100 cm. long, stout, glabrous, obtusely angled; leaf-blades flat, 6–12 mm. wide, coriaceous, the basal sheaths pale or partially sanguineous; inflorescence a large compound panicle, with numerous peduncled spikes, the branches hispid-scabrous, the bracts leaflike, much longer than the inflorescence, long-sheathing; spikes androgynous, oblong or ovoid, 5–8 mm. long, light tawny; bractlets scalelike, awned; cladoprophyllum utricle-shaped; pistillate scales broadly ovate, cuspidate, lustrous, pale and ferrugineous-striolate, slenderly nerved; perigynia longer than the scales, 3.5–4 mm. long, ovoid, inflated, glabrous or hairy at tip, rather abruptly long-beaked, with an obliquely truncate hyaline tip; stigmas 3.——Kyushu (Nagaura in Hizen Prov.); rare.——Ryukyus, Formosa, China, and India.

168. Carex brunnea Thunb. var. **brunnea.** *C. gentilis* var. *oshimensis* Kuekenth.; *C. amamioshimensis* Akiyama——KOGOME-SUGE, KOGOME-NAKIRI-SUGE. Plant tufted, forming a large clump, culms slender, rather firm, 40–80 cm. long; leaves with rather stiff yellowish green to bright green blades (2–) 3–4 mm. wide, the basal sheaths brown; spikes numerous, androgynous, narrowly cylindric, 1–3 cm. long, 2–3 mm. thick, often branched at the base, binate to ternate, stramineous, nodding; peduncles slender, branched, scabrous, exserted from the sheath of the bracts, the lower bracts leaflike, long-sheathing; pistillate scales ovate, acute, ferrugineous; perigynia longer than the scales, 2.5–2.7 mm. long, nearly erect, elliptic, brownish, membranous, distinctly nerved, planoconvex or unequally biconvex, sparsely hispidulous on the nerves, abruptly short-beaked, with a minutely 2-toothed hyaline tip; stigmas 2, shorter than the perigynium.——Aug.–Sept. Dry open woods near the seacoast; Honshu (s. Kantō, westw. to s. Kinki, along the Pacific side), Shikoku (?), Kyushu.——Ryukyus, Formosa, s. China, and Philippines.

Var. **abscondita** T. Koyama. SHIO-KAZE-NAKIRI. Rhizomes decumbent; culms very scabrous, short, mostly hidden among the leaves; spikes not branched; perigynia very short-beaked or nearly beakless.——Honshu (Pacific coast in Mikawa Prov.).

169. Carex lenta D. Don. var. **sendaica** (Franch.) T. Koyama. *C. brunnea* var. *sendaica* (Franch.) Kuekenth.; *C. sendaica* Franch.; *C. franchetiana* Lév. & Van't.——SENDAI-SUGE. Rhizomes decumbent, short to long-creeping; culms 10–30 cm. long, slender, as long as the leaves; leaf-blades flat, yellowish to dark green, very stiff; spikes 3 or 4, androgynous or the lower ones pistillate, cylindric to oblong, 7–15 mm. long, 3.5–4 mm. wide, approximate, simple, peduncled, ascending to nearly erect, the bracts bristle-shaped, short-sheathing; pistillate scales ovate, pale, reddish brown, hyaline; perigynia much larger than the scales, 3–3.5 mm. long, obovate-elliptic, prominently nerved, abruptly short-beaked; stigmas 2, shorter than the perigynium.——Aug.–Oct. Rocky open woods or sandy places; Honshu (Sendai and southwestw.), Shikoku; rare. ——Korea and China.

Var. **lenta.** *C. lenta* D. Don; *C. sendaica* var. *nakiri* (Ohwi) T. Koyama; *C. brunnea* var. *nakiri* Ohwi——NAKIRI-SUGE. Rhizomes short, forming a large clump; basal sheaths dark brown, lustrous; leaves very stiff, deep green, conduplicate; spikes numerous, androgynous, oblong-cylindric, 5–20 mm. long, 3.5–4 mm. thick, unbranched or with branching peduncles solitary to binate on each node, the bracts (at least the lower ones) elongate, leaflike, over-topping the inflorescence; perigynia broadly elliptic, prominently nerved.——Aug.–Nov. Open woods on hillsides; Honshu (Rikuzen, Etchu, and westw.), Shikoku, Kyushu; quite common.——s. Korea, China, and the Himalayas.

Var. **pseudosendaica** T. Koyama. SENDAI-SUGE-MODOKI. Intermediate between the typical phase and var. *sendaica;* rhizomes long-creeping; spikes usually 5–8, unbranched, all androgynous.——Honshu (Mikawa Prov.).

170. Carex autumnalis Ohwi. Ō-NAKIRI-SUGE. Somewhat resembles *C. lenta* var. *lenta;* rhizomes loosely tufted; culms 70–80 cm. long; leaf-blades stiff, 2.5–3 mm. wide, the basal sheaths dark brown; spikes about 10, the terminal staminate, nearly sessile, 15–20 mm. long, the lateral androgynous, broadly linear, loosely flowered, 1–3 cm. long, the upper ones nearly sessile, solitary, the lower frequently paired on unequal long peduncles, the bracts long-sheathing, the lowest one with a short bristlelike blade at the tip; perigynia about twice as long as the scales, about 3 mm. long, elliptic, faintly nerved, very sparsely hispidulous especially on the upper part, or nearly glabrous, the beak short, emarginate; stigmas 2, about 2 mm. long.——Sept.–Oct. Honshu (Yamato Prov. and Chūgoku Distr.), Shikoku (Awa); rare.——Ryukyus.

171. Carex nachiana Ohwi. KISHU-NAKIRI-SUGE. Resembles *C. lenta* var. *lenta,* but larger; culms tufted, to 1 m. long; leaves 3–4 mm. wide, deep green; spikes numerous, androgynous, solitary to ternate at each node, often sparsely branching below, loosely rather many-flowered, 1.5–3 cm. long, the lower bracts leaflike, long-sheathing; pistillate scales acute; perigynia nearly as long as the scales, suberect, elliptic to broadly ovate, 3.5–4 mm. long, about 1.5 mm. wide, glabrous, minutely hispidulous-scabrous on the margins, short-stipitate, moderately long-beaked, with a minutely 2-toothed hyaline tip; stigmas 2, about 3 mm. long.——Sept.–Oct. Woods; Honshu (Kii Prov. and Oki Isl.), Kyushu (Kagoshima in Satsuma Prov. and Yakushima); rare.——China.

172. Carex sacrosancta Honda. *C. kasugayamensis* Akiyama——JINGŪ-SUGE, HIME-NAKIRI-SUGE. Somewhat resembles but more slender than *C. lenta* var. *lenta;* culms 20–40 cm. long, slender; leaf-blades 1–3 mm. wide; spikes 4–6, androgynous, linear-cylindric, loosely flowered, simple, solitary or paired at each node, 1–2.5 cm. long, the lower bracts leaflike, sheathing; pistillate scales oblong, rather obtuse, ferruginous; perigynia longer than the scales, elliptic, 4 mm. long, less than 1.5 mm. wide, faintly nerved, loosely his-

pidulous on the nerves, stipitate, rather long-beaked, minutely 2-toothed at tip; stigmas filiform, shorter than the perigynia.——Sept.–Oct. Honshu (Miyakejima, Ise, and Yamoto Prov.), Shikoku, Kyushu; rare.

173. Carex teinogyna Boott. *C. scabriculmis* Ohwi; *C. teinogynia* var. *scabriculmis* Kuekenth.——FUSA-NAKIRI-SUGE. Rhizomes tufted; culms 40–60 cm. long, nodding above; leaf-blades rather flat, 2–3.5 mm. wide, the basal sheaths brown; spikes numerous, dense, androgynous, 2- to 5-nate, linear, loosely flowered, peduncles often sparsely branching, the lower bracts leaflike, sheathing; pistillate scales narrowly oblong, acuminate, ferrugineous; perigynia nearly as long as the scales, 3.5–4 mm. long, erect, elliptic, faintly nerved, minutely hispidulous on the margins, rather long-beaked, with a minutely 2-toothed hyaline tip; stigmas persistent, very slender and much-elongate, 6–8 mm. long.——Aug.–Oct. Wet rocks along ravines and streams; Honshu (s. Kinki and Chūgoku Distr.), Shikoku, Kyushu; locally abundant.——s. Korea and China to India and Malaysia.

174. Carex metallica Lév. *C. pachinensis* Hayata——FUSA-SUGE, SHIRAHO-SUGE. Plant densely tufted, forming large clumps; culms 30–60 cm. long, glabrous, nodding above; leaf-blades flat, stiff, deep green, lustrous above, 3–6 mm. wide, the basal sheaths pale, ultimately splitting into copious brown fibers covering the neck of the rhizomes; spikes 5–10, approximate, the terminal gynecandrous and clavate-cylindric, or merely staminate and broadly linear, the lateral pistillate, the upper ones with a few staminate flowers at base thick-cylindric, 2–5 cm. long, peduncled, solitary or paired, nodding, the lower bracts leaflike, much longer than the culm, sheathing; pistillate scales narrow-ovate, acuminate and cuspidate, pale; perigynia twice as long as the scales, about 7 mm. long, ascending, narrowly ovoid, light yellowish green, lustrous, inflated and loosely inclosing the achene, slenderly nerved, scabrous at tip, rather long-beaked, with a hyaline tip; stigmas 3, short.——Apr.–May. Near seashores; Honshu (Harima Prov.), Shikoku, Kyushu.——s. Korea, Ryukyus, and Formosa.

175. Carex macrandrolepis Lév. *C. distantiflora* Nakai; *C. sharyotensis* Hayata——KATA-SUGE. Rhizomes loosely to scarcely tufted, the neck covered with dark brown straight fibers; culms 15–40 cm. long, slender, glabrous; leaf-blades 2–3 mm. wide, flat, soft, bright to yellowish green; spikes 3 or 4, distant, or the upper ones approximate, the terminal staminate, linear, long-peduncled, 1–3 cm. long, pale brownish yellow, the lateral pistillate, oblong, about 1 cm. long, somewhat loosely few-flowered, peduncled, the lowest one sometimes radical, the bracts leaflike, short- to somewhat long-sheathing; pistillate scales obovate, rounded or emarginate and cuspidate, pale; perigynia longer than the scales, 5–6 mm. long, ascending, rhomboid, acutely 3-angled, yellowish green, lustrous, sparsely hispidulous, subcoriaceous, with 2 costal nerves, the beak short, erect, with a 2-toothed, hyaline tip.——Dry grassy places on hills and in lowlands; Honshu (Suruga Prov. to Kinki Distr.), Kyushu; rather rare.——Formosa and Korea.

176. Carex hondoensis Ohwi. *C. arnellii* sensu auct. Japon., non Christ; *C. arnellii* var. *hondoensis* (Ohwi) T. Koyama——AIZU-SUGE. Rhizomes decumbent or ascending, densely covered with pale brown fibers; culms 50–70 cm. long, nodding above; leaf-blades 3–4 mm. wide, flat, soft; spikes 4 or 5, the upper 2 or 3 staminate, approximate, narrowly lanceolate, 1–2 cm. long, obscurely tawny, the others pistillate, 2–5 cm. long, long-peduncled, pendulous, cylindric, the lower ones distant, the bracts leaflike, short-sheathing; pistillate scales narrowly ovate, acuminate and short-awned, pale tawny; perigynia longer than the scales, 4.5–5 mm. long, ellipsoidal, distinctly 3-angled, light yellowish green, somewhat lustrous, membranous, with 2 costal nerves, long-beaked, with a hyaline, 2-toothed tip; stigmas 3, very long, persistent.——May–July. Grassy places and open woods in mountains; Honshu (centr. distr.).

177. Carex idzuroei Fr. & Sav. *C. pseudovesicaria* Lév. & Van't.——UMA-SUGE. Rhizomes elongate; culms 40–60 cm. long; leaf-blades flat, soft, 4–8 mm. wide, the basal sheaths pale, partially reddish purple; spikes 4 or 5, the upper 1 or 2 staminate, linear, 2–4 cm. long, the others pistillate, oblong, 2–3 cm. long, rather densely flowered, erect, distant, sessile, the lower peduncled, the lower bracts leaflike, scarcely sheathing; pistillate scales lanceolate-ovate, cuspidate, pale to stramineous; perigynia nearly twice as long as the scales, 10–12 mm. long, obliquely spreading, oblong-ovoid, inflated, thick but rather soft, light green but becoming dark grayish brown when dried, glabrous, ribbed, long-beaked, with a hyaline 2-toothed tip; stigmas 3.——May–June. Wet muddy places along rivers and ditches in lowlands; Honshu (Kantō Distr. and westw.), Shikoku, Kyushu.——China.

178. Carex hymenodon Ohwi. *C. aequibilirostris* Suto & Suzuki——YAMA-KUBO-SUGE, HIME-MIKURI-SUGE. Rhizomes long-creeping; culms 40–50 cm. long, glabrous; leaf-blades flat, fresh-green, soft, 4–7 mm. wide, the basal sheaths pale, partially dark reddish purple; spikes 3 or 4, erect, the upper 1 or 2 staminate, linear, 3–4 cm. long, long-peduncled, the others pistillate, approximate, short-cylindric, 1.5–3 cm. long, the lower ones short-peduncled, the bracts leaflike, nearly sheathless; pistillate scales ovate, acute, pale; perigynia longer than the scales, about 6 mm. long, spreading, ovoid-oval, inflated, obsoletely 3-angled, light green, glabrous, many-nerved, the beak short, erect, with 2-toothed hyaline tip; stigmas 3.——May–June. Swampy places in hills and low mountains; Honshu (Shimotsuke Prov.); rare.

179. Carex olivacea Boott var. **angustior** Kuekenth. *C. confertiflora* Boott; *C. olivacea* sensu auct. Japon., non Boott——MIYAMA-SHIRA-SUGE. Rhizomes stout, creeping; culms stout, 30–80 cm. long, with 1 or 2 nodes; leaf-blades flat, soft, 8–15 mm. wide, prominently whitish beneath, the basal sheaths pale; spikes 3–6, erect, the terminal (rarely the upper 2) staminate, linear, 3–7 cm. long, pale yellow, the lateral pistillate, very densely many-flowered, cylindric, 2.5–5 cm. long, 7–9 mm. thick, nearly sessile, the bracts leaflike, longer than the culm, sheathless or the sheath extremely short; pistillate scales narrowly oblong, obtuse and short-awned, pale or light tawny; perigynia slightly longer than the scales, 4 mm. long, spreading, broadly obovate, glabrous, inflated, light green, grayish brown when dried, wrinkled, slenderly many-nerved, the beak short, recurved, with a hyaline 2-toothed tip; stigmas 3.——May–July. Wet muddy places in mountains; Hokkaido, Honshu, Shikoku, Kyushu; rather abundant.—— The typical phase occurs in India, Indochina, and Malaysia.

180. Carex brownii Tuckerm. *C. nipposinica* Ohwi ——AWABO-SUGE. Plant loosely tufted; culms 30–70 cm. long, acutely 3-angled, with 1 node; leaf-blades flat, rather stiff, deep green, 3–5 mm. wide, the basal sheaths bladeless, brownish red; spikes 3 or 4, 1.5–3 cm. long, all but the lowest one approximate or nearly fastigiate, the terminal staminate,

linear, pale to light tawny, usually overtopped by the lower pistillate ones, 1.5–3 cm. long, erect, densely flowered, cylindric, peduncled, the bracts leaflike, the lower longer than the inflorescence, long-sheathing; pistillate scales elliptic, pale, with a broad green 3-nerved midrib projecting as a long awn; perigynia as long as or slightly shorter than the awns of the scales, about 3 mm. long, spreading, obovoid-globose, much inflated, thick-membranous, obsoletely 3-angled, light green, brownish when dried, abruptly short-beaked, with a hyaline 2-toothed tip; stigmas 3, short.——Wet sunny grassy places or open woods; Hokkaido (sw. distr.), Honshu, Shikoku, Kyushu; rather rare.——Korea, China, Formosa to Australia.

181. Carex transversa Boott. *C. brownii* var. *transversa* (Boott) Kuekenth.——YAWARA-SUGE. Closely allied to the preceding, but more densely tufted; culms smaller; leaves slightly flaccid; perigynia 5–6 mm. long, the beak longer than the awn of the scale.——Apr.–June. Sandy wet places in lowlands and hillsides; Honshu, Shikoku, Kyushu; common.—— Korea, China, and Ryukyus.

Carex × furusei T. Koyama. Hybrid of *C. brownii* × *C. transversa*——HASHI-NAGA-AWABO-SUGE. Differs from *Carex transversa* in the rather short-beaked perigynia and stouter culms decumbent at base.——Honshu (Awa, Shinano Prov.).

182. Carex viridula Michx. *C. oederi* var. *viridula* (Michx.) Kuekenth.; *C. flava* var. *viridula* (Michx.) Bailey ——EZO-SAWA-SUGE. Rhizomes densely tufted, forming small clumps; culms 10–30 cm. long, slender, firm, glabrous; leaf-blades erect, stiff, yellowish green, 1.5–2.5 mm. wide, gradually acuminate, the basal sheaths stramineous; spikes 3 or 4, erect, closely contiguous, the lower ones sometimes distant and peduncled, the terminal staminate, linear, 7–15 mm. long, light ferrugineous, the lateral pistillate, oblong-globose, 5–8 mm. long, densely flowered, subsessile, the lowest bract leaflike, much longer than the inflorescence, short-sheathing; pistillate scales broadly ovate, rather acute, ferrugineous; perigynia longer than the scales, 2.5–3 mm. long, broadly obovoid, turgid, 3-angled, thick-membranous, glabrous, strongly several-nerved, yellowish green, abruptly short-beaked, the tip rigid, deeply 2-toothed; stigmas 3, short, slender.——June–July. Wet places in lowlands and highlands; Hokkaido, Honshu (Shinano to Hitachi Prov. and northw.); rather rare.——Sakhalin, s. Kuriles, Kamchatka, and N. America.——Forma **oederioides** (Tatew. & Akiyama) T. Koyama. *C. oederioides* Tatew. & Akiyama; *C. oederi* var. *oederioides* (Tatew. & Akiyama) Ohwi. A much depauperate phase.——Hokkaido.

183. Carex nemostachys Steud. *C. excurva* Boott; *C. zollingeri* Kunze, ex Boecklr.——AKI-KASA-SUGE. Rhizomes loosely tufted, bearing stout, elongate stolons; culms 30–60 cm. tall; leaves with flat, stiff, elongate blades 4–5 mm. wide, the basal sheaths pale-stramineous; spikes 5–7, erect, densely arranged, the terminal one staminate, linear-cylindric, 4–8 cm. long, pale to pale-stramineous, the lateral spikes pistillate, narrowly cylindric, densely many-flowered, 4–10 cm. long, sessile, the lower ones short-peduncled, the lower 1 or 2 bracts leaflike, sheathless; pistillate scales lanceolate, long-awned, pale; perigynia longer than the scales, 3–3.5 mm. long, obliquely spreading, obovoid, pale yellowish green, membranous, few-nerved, lusterless, scabrous, abruptly contracted into a long recurved beak with a hyaline obscurely 2-toothed tip; stigmas 3, slender, deciduous.——July–Sept. Swampy places along streams in low mountains; Honshu (Yamato Prov. and w. Chūgoku Distr.), Shikoku, Kyushu.——Ryukyus, China, and Indochina.

184. Carex dispalata Boott var. **dispalata.** *C. subanceps* Boecklr.; *C. coronata* Lév.——KASA-SUGE. Rhizomes clustered, long-creeping; culms rather stout, 40–100 cm. long; leaf-blades 4–8 mm. wide, flat, rather stiff, the basal sheaths partially dark reddish purple, splitting into fibers, the terminal staminate, linear-cylindric, 4–7 cm. long, dark reddish purple, the others pistillate, cylindric, sessile or peduncled, erect or nodding, densely many-flowered, 3–10 cm. long, the bracts leaflike, sheathless; pistillate scales ovate to oblong, sometimes cuspidate, purple-brown; perigynia longer than the scales, 3–4 mm. long, spreading to divergent, glabrous, thick-membranous, inconspicuously nerved or nerveless except for 2 costal nerves, greenish, dark brown when dried, dull, the beak long, recurved, with an obliquely truncate hyaline tip; stigmas 3, rather thick, deciduous.——Apr.–July. Wet places around ponds and along rivers in lowlands; Hokkaido, Honshu, Shikoku, Kyushu; common.——Korea, Ussuri, Manchuria, China, s. Kuriles, and Sakhalin.

Var. **takeuchii** (Ohwi) Ohwi. *C. persistens* Ohwi; *C. persistens* var. *takeuchii* Ohwi——KINKI-KASA-SUGE. Culms 30–70 cm. long; leaf-blades softer, the basal sheaths pale; perigynia thinner, light yellowish, slightly brownish when dried, closely investing the achene, the beak slightly elongate, erect; stigmas 3, slender, persistent, longer than the perigynium. ——May–June. Wet places along streams and ravines in mountains; Honshu (Kinki, Chūgoku Distr.).

185. Carex pumila Thunb. *C. littorea* Labill.; *C. platyrhyncha* Fr. & Sav.; *C. nutans* var. *japonica* Fr. & Sav.; *C. nutans* var. *platyrhyncha* (Fr. & Sav.) Kuekenth.——KŌBŌ-SHIBA. Rhizomes extensively creeping, branching; culms 10–20(–30) cm. long, obtusely angled, glabrous, firm, the basal sheaths reddish brown, splitting into fibers; leaf-blades flat, stiff, 2–4 mm. wide; upper 2 or 3 (rarely 1) spikes staminate, linear, 2–3 cm. long, reddish brown, the lower 1 or 2 pistillate, approximate, oblong to short-cylindric, 1.5–3 cm. long, short-peduncled, erect, the bracts leaflike, short-sheathing; pistillate scales narrowly ovate, acute, cuspidate, somewhat reddish brown; perigynia longer than the scales, 6–8 mm. long, ascending, ovoid-conical, ligneous, brownish green, obsoletely 3-angled, slenderly nerved, glabrous, abruptly short-beaked, with a firm bifurcate tip; stigmas 3.——May–July. Sandy places along seacoasts; Hokkaido, Honshu, Shikoku, Kyushu; common.——Sakhalin, s. Kuriles, Korea, China, Ryukyus, and Formosa to Australia.

186. Carex scabrifolia Steud. *C. pierotii* Miq.; *C. yabei* Lév. & Van't.——SHIO-KUGU. Rhizomes short-creeping; culms 3-angled, sparingly scabrous above, 30–50 cm. long; leaves overtopping the culm, blades 1.5–2.5 mm. wide, slightly involute, sulcate, the basal sheaths bladeless, partially dark reddish purple, splitting into fibers; spikes 3 or 4, the upper 2–4 staminate, approximate, linear, 2–4 cm. long, pale to reddish brown, the lower 1 or 2 pistillate, oblong, 1–2 cm. long, nearly sessile, the lowest bract leaflike, longer than the inflorescence, sheathless; pistillate scales deltoid-ovate, acute, pale ferrugineous; perigynia longer than the scales, 6–8 mm. long, oblong-ellipsoidal, ligneous, slenderly nerved, glabrous, the beak short, erect, thick, shallowly bilobed; stigmas 3, short. ——Apr.–July. Wet places near tidal zone at mouth of rivers or in lagoons, below high tide level; Hokkaido, Honshu, Shi-

koku, Kyushu; common.——Korea, Manchuria, Ussuri, Ryukyus, Formosa, and China.

187. Carex rugulosa Kuekenth. *C. riparia* var. *rugulosa* (Kuekenth.) Kuekenth.——Ō-KUGU. Somewhat resembles the preceding, but larger; rhizomes long-creeping; culms stout, glabrous, 40–70 cm. long, somewhat thickened at base; leaf-blades flat, 5–10 mm. wide, the basal sheaths bladeless, partially dark reddish brown, sparsely fibrous; upper 3–5 spikes staminate, sessile, linear-cylindric, 2–4 cm. long, dark brown, approximate, the lower 2–4 pistillate, cylindric, erect, 2.5–5 cm. long, short-peduncled, the bracts leaflike, the lowest sometimes sheathing; pistillate scales very obtuse, short-awned, castaneous to ferrugineous; perigynia longer than the scales, 6–7 mm. long, oblong, ligneous, faintly nerved, glabrous, gradually short-beaked, the tip bifurcate; stigmas 3, short.——May–July. Wet places at mouth of rivers; Hokkaido, Honshu, Kyushu; rare.——Korea, Manchuria to Ussuri.

188. Carex pauciflora Lightf. *C. microglochin* sensu auct. Japon., non Wahlenb.——TAKANE-HARI-SUGE, MIGAERI-SUGE. Rhizomes slender, long-creeping; culms 10–15 cm. long, erect, glabrous, slender, somewhat firm; leaf-blades sulcate above, about 1 mm. wide, shorter than the culm; spike solitary, terminal, androgynous, oblanceolate at anthesis, 6–8 mm. long, loosely flowered, composed of 2–3 staminate and 2–4 pistillate flowers; pistillate scales caducous at maturity, lanceolate, rather acute, light ferrugineous; perigynia longer than the scales, 6–6.5 mm. long, at first appressed, reflexed at maturity, linear-lanceolate, slenderly many-nerved, light green, thick membranous, gradually long-beaked, glabrous, entire, truncate at tip; stigmas 3, short.——June–July. High moors; Hokkaido, Honshu (Mount Naeba in Echigo, Mount Adzuma in Iwashiro, Oze in Shimotsuke and elsewhere); rare.——Kuriles, Sakhalin, and widely distributed in the N. Hemisphere.

189. Carex michauxiana Boecklr. var. **asiatica** (Hult.) Ohwi. *C. michauxiana* sensu auct. Japon., non Boecklr.; *C. michauxiana* subsp. *asiatica* Hult.——MITAKE-SUGE. Rhizomes tufted; culms erect, sometimes briefly ascending at base, somewhat firm, 20–50 cm. long, nearly glabrous, 1- to 3-leaved on the lower half; leaf-blades flat, stiff, 3–5 mm. wide, shorter than the culm, the basal sheaths pale, sparsely grayish brown fibrous; spikes 3–5, the terminal staminate, linear, pale tawny, 1–1.5 cm. long, somewhat inconspicuous, the others pistillate, ovoid-globose, several-flowered, the upper approximate and subsessile, the lower widely distant, long-peduncled, erect, 1–1.5 cm. long, the bracts leaflike, long-sheathing; pistillate scales ovate, obtuse, entire, pale tawny; perigynia 2 or more times as long as the scales, 10–13 mm. long, spreading, narrowly lanceolate, rather coriaceous, yellowish green, glabrous, slightly lustrous, many-nerved, gradually long-beaked and shallowly 2-toothed; stigmas 3, short.——June–July. High moors; Hokkaido, Honshu (n. and centr. distr.).——Kuriles, Kamchatka. The typical phase occurs in N. America.

190. Carex pseudocyperus L. KUGU-SUGE. Culms 30–70 cm. tall, tufted; leaf-blades rather stiff, 5–7 mm. wide, the basal sheaths pale; spikes 3–5, contiguous, the terminal staminate, erect, linear, 2–3 cm. long, the others pistillate, cylindric, 2–5 cm. long, nodding, peduncled, densely many-flowered, the bracts leaflike, longer than the inflorescence, the lowest one subvaginate; pistillate scales narrowly ovate, long-awned, pale, ciliolate; perigynia more than twice as long as the scales, 4–5 mm. long, spreading or reflexed when fully mature, broadly lanceolate, compressed-trigonous, light yellowish green, many-nerved, gradually long-beaked, with a deeply bifid firm tip, the lobes needlelike, erect; stigmas 3, short.——Swamps; Hokkaido, Honshu (Shinano and Mutsu Prov.); rare.——Siberia, N. America, Europe, and N. Africa.

191. Carex capricornis Meinsh. *C. pseudocyperus* var. *brachystachya* Regel & Maack; *C. brachystachya* (Regel & Maack) Akiyama, non Schrank——Jōrō-suge. Culms tufted, 40–70 cm. long, firm, erect; leaf-blades flat, rather stiff, 4–6 mm. wide, the basal sheaths (at least the lower ones) bladeless, partially reddish brown; spikes 4–6, approximate, the terminal staminate, linear, rarely gynecandrous and clavate, 1.5–3 cm. long, somewhat inconspicuous, the others pistillate, oblong, 1.5–3 cm. long, 15–18 mm. across, erect-patent, densely many-flowered, subsessile, the lowest one sometimes long-peduncled and pendulous, the lowest bract leaflike, elongate, not sheathing; pistillate scales narrowly oblong, more or less emarginate and long-awned, pale, minutely ciliate at apex; perigynia 4 to 5 times as long as the scales, 6–9 mm. long, spreading, linear-lanceolate, compressed, yellowish green, glabrous, many-nerved, gradually very long-beaked, deeply bifurcate at tip, the lobes needlelike, 1.5 mm. long, strongly divergent; stigmas 3.——May–July. Swampy places; Hokkaido, Honshu (Kantō Distr.); local and rather rare.——Korea, Manchuria, and Ussuri.

192. Carex dickinsii Fr. & Sav. *C. coreana* Bailey——Oni-suge, Mikuri-suge. Culms few, 20–50 cm. long; leaf-blades flat, rather firm, light green, 4–8 mm. wide, the basal sheaths pale, with short blades; spikes 3(2–4), the terminal staminate, erect, long-pedunculate, linear, 2–3 cm. long, stramineous-green, the others pistillate, ellipsoidal to nearly globose, densely flowered, 15–20 mm. long, about 15 mm. wide, closely approximate, subsessile, the bracts leaflike, spreading to reflexed; pistillate scales broadly ovate, tawny or pale; perigynia much longer than the scales, about 10 mm. long, spreading to divergent, broadly ovoid-conical, strongly inflated, lustrous, bright green, nerved, abruptly long-beaked, the beak slender, acutely 2-toothed at tip; stigmas 3.——May–July. Shallow water in ditches and edges of ponds; Hokkaido, Honshu, Shikoku, Kyushu; rather common.——Korea.

193. Carex oligosperma Michx. *C. tsuishikarensis* Koidz. & Ohwi; *C. rotundata* sensu Miyabe & Kudo, non Wahlenb.——Horomui-kugu. Culms 20–50 cm. long, glabrous; leaf-blades glaucous-green, thick, sulcate above, nearly smooth, about 1.5 mm. wide, the basal sheaths light gray; spikes 2 or 3, distant, erect, the terminal staminate, linear, 2–3 cm. long, the lateral pistillate, ovoid-globose to oblong, rather densely flowered, 1–1.5 cm. long; lowest bract leaflike, sheathing; pistillate scales broadly ovate, acute, castaneous, paler and scarious toward margin; perigynia longer than the scales, 5–5.5 mm. long, obliquely spreading, broadly ovoid, slightly turgid, rather coriaceous, grayish green and partially dark brown, glabrous, nerved, abruptly short-beaked, 2-toothed at tip; stigmas 3, short.——June–Aug. Bogs and swampy places in lowlands; Hokkaido, Honshu (Rikuchu and Shinano Prov.); rare.——s. Kuriles and e. N. America.

194. Carex vesicaria L. *C. vesicaria* var. *monile* sensu auct. Japon., non Boecklr.; *C. vesicaria* var. *tenuistachya* Kuekenth.; *C. vesicata* Meinsh.——Oni-naruko-suge. Plant loosely tufted; culms 30–100 cm. long, scabrous on the angles; leaf-blades flat, rather stiff, 3–6 mm. wide, the basal leaf-sheaths bladeless, partially reddish purple, sparingly fibrous; upper 2 or 3 spikes staminate, broadly linear, 3–5 cm. long, contiguous, the lower 2 or 3 pistillate, cylindric, 3–7 cm. long,

7–10 mm. wide, distant, densely flowered, the lower one or more peduncled, erect or often nodding, the bracts leaflike, sheathless; pistillate scales ovate-lanceolate, acute, sanguineous or light tawny; perigynia longer than the scales, 6–8 mm. long, spreading, ovoid-conical, inflated-trigonous, light yellowish green, sometimes partially dark brown, lustrous, nerved, abruptly rather long-beaked, inflated, bifurcate at tip; stigmas 3, short.——May–July. Swamps or marshy places along rivers and around ponds; Hokkaido, Honshu, Kyushu; rather common northward.——Widely distributed in the cooler zones of the N. Hemisphere.

195. Carex rhynchophysa C. A. Mey. *C. laevirostris* Blytt; *C. ventricosa* Franch., non Curt.——Ō-kasa-suge. Rhizomes thickened, creeping; culms 60–100 cm. long, stout, spongy-thickened at base, nearly glabrous; leaf-blades flat, rather thick, 8–15 mm. wide, the basal sheaths pale, sometimes partially sanguineous; upper 3–7 spikes staminate, linear, 3–6 cm. long, contiguous, the lower 2–5 pistillate, distant, long-cylindric, 5–10 cm. long, about 1 cm. across, densely many-flowered, nearly sessile except the lowest one usually long-peduncled and sometimes nodding, the bracts leaflike; pistillate scales oblong-lanceolate, castaneous or ferrugineous, sometimes pale, scarious toward the margins especially broadly so above; perigynia longer than the scales, 5–6 mm. long, spreading, broadly ovoid, firmly membranous, strongly inflated, stramineous-green, sometimes partially dark brownish, lustrous, nerved, abruptly long-beaked, bifurcate at tip; stigmas 3.——June–Aug. Swamps and shallow water around ponds; Hokkaido, Honshu (n. and centr. distr.).——Korea, Ussuri, Sakhalin, Kuriles, Siberia, and N. America.

196. Carex ligulata Nees. *C. keiskei* Miq.; *C. hebecarpa* var. *ligulata* (Nees) Kuekenth.; *C. bakanensis* Lév. & Van't.; *C. hebecarpa* sensu Matsum., non C. A. Mey.——Satsuma-suge. Rhizomes thick, loosely tufted; culms erect, ciliate-scabrous on the angles, nearly always hidden under the leaf-sheaths, 3-angled, 40–70 cm. long; lower leaf-sheaths bladeless, dark sanguineous-purple, the normal leaf-blades on the upper 2/3 of the culm, closely approximate, flat, linear-lanceolate, 4–8 mm. wide, the sheaths sharply 3-angled, sometimes hispid, the ligules conspicuous, ferrugineous, scarious; spikes 5–7, contiguous, the terminal staminate, linear, 1–3 cm. long, pale-ferrugineous, the lateral pistillate, cylindric, 15–40 mm. long, erect, densely many-flowered, peduncled, the bracts leaflike, at least the lower ones overtopping the culm, long-sheathing; pistillate scales ovate, obscurely ferrugineous; perigynia longer than the scales, 4–5 mm. long, suberect, broadly obovoid, 3-angled, ferrugineous-green, densely whitish tomentose, rather long beaked; stigmas 3.——Woods; Honshu (s. Kantō Distr. and westw.), Shikoku, Kyushu; rare.——s. Korea (variety), Formosa, China, and India.

197. Carex phyllocephala T. Koyama. *C. hebecarpa* var. *maubertiana* forma *latifolia* Makino——Tenjiku-suge. Culms erect, 37–45 cm. long, obtusely angled, glabrous, about 2.5 mm. thick, nearly always hidden under the leaf-sheaths; lower leaf-sheaths bladeless, dark sanguineous purple, the normal leaves densely aggregated on the apical part of the culm, the blades spreading, 10–20 cm. long, 8–13 mm. wide, the sheaths loosely surrounding the culm, hispidulous on the angles, the ligules glabrous, ferrugineous, about 2 mm. long; spikes 8–10, forming a capitate inflorescence, the terminal staminate, linear, to 2 cm. long, tawny, the others pistillate, cylindric, 25–30 mm. long, 3–3.5 mm. thick, densely many-flowered; pistillate scales rather small, ovate-elliptic, cuspidate, membranous, ferrugineous and flecked with dark brown, hyaline toward the margins; perigynia much longer than the scales, 2.8–3 mm. long, obliquely spreading, ellipsoidal or ovoid-ellipsoidal, obtusely 3-angled, thick-membranous, ferrugineous, almost nerveless; costal nerves 2, densely whitish setulose, abruptly rather long-beaked, minutely 2-toothed; stigmas 3.——Cultivated in Honshu (Mikawa Prov.), Kyushu (Chikugo Prov.); rare.——China.

198. Carex poculisquama Kuekenth. AKANE-SUGE. Rhizomes thickened, short; culms few, loosely tufted, slender, sharply 3-angled, firm, scabrous, 40–50 cm. long, nearly always hidden under the leaf-sheaths; lower leaf-sheaths scalelike, reddish brown, the normal leaf-blades all cauline, stiff, 2.5–3.5 mm. wide, with slightly recurved margins, somewhat cinereous-green, the sheaths long, flecked with purplish brown, scarious in front, scabrous on the angles; spikes about 4, approximate, or only the lowest one remote from the others, the terminal staminate, narrowly linear, almost equaling the adjacent one in length, about 1.5 cm. long, 1 mm. thick, brownish, the bracts leaflike, overtopping the inflorescence, sheathing; staminate scales infundibuliform, connate ventrally; pistillate spikes cylindric, 1.5–2.5 cm. long, rather loosely flowered, peduncled, erect; pistillate scales ovate-deltoid, pale to partially ferrugineous, clasping the axis of the spike at base, the midrib broad, green, ending in a short recurved awn; perigynia longer than the scales, about 4 mm. long, rhomboid-ellipsoid, acutely 3-angled, faintly nerved, light green, sparsely pubescent, abruptly narrowed at both ends, the beak short, erect, 2-toothed; stigmas 3, short, slightly thickened, recurved. ——May–June. Limestone mountains; Honshu (Mount Iwafune in Shimotsuke and Akiyoshidai in Nagato); very rare. ——China.

199. Carex lasiocarpa Ehrh. var. **occultans** (Franch.) Kuekenth. *C. filiformis* var. *occultans* Franch.; *C. occultans* (Franch.) V. Krecz.; *C. striata* var. *japonica* Koidz.; *C. koidzumii* Honda——MUJINA-SUGE. Rhizomes rather stout and elongate; culms 70–100 cm. long, slender, somewhat firm; leaves with narrow stiff blades, canaliculate above, 1.5–3 mm. wide, involute or flat, the basal sheaths bladeless, dark sanguineous-purple; upper 1–4 spikes staminate, contiguous, reddish purple, 2–5 cm. long, linear, erect, the lower 1 or 2 pistillate, oblong to short-cylindric, densely flowered, 2–4 cm. long, the lowest short-peduncled, the bracts setaceous, with or without a short sheath; pistillate scales narrowly ovate, brownish or reddish brown; perigynia nearly as long as or slightly longer than the scales, ovoid, 4–5 mm. long, coriaceous, brown-hairy to nearly glabrous, gradually short-beaked, deeply bifurcate at tip, the lobes erect; stigmas 3.——June–Aug. High moors; Hokkaido, Honshu (Kozuke, Kaga, Rikuchu Prov.); rather rare.——Sakhalin, s. Kuriles, n. Korea, and e. Siberia. The typical phase occurs in the cooler zones of Eurasia and N. America.

200. Carex drymophila Turcz. var. **abbreviata** (Kuekenth.) Ohwi. *C. akanensis* Franch.; *C. drymophila* sensu auct. Japon., non Turcz.; *C. drymophila* var. *akanensis* (Franch.) Kuekenth.; *C. amurensis* Kuekenth., incl. var. *abbreviata* Kuekenth.——AKAN-KASA-SUGE. Culms 60–80 cm. long; leaf-blades flat, somewhat soft, often somewhat hairy, 4–6 mm. wide, the basal sheaths bladeless, dark reddish and purplish brown, splitting into fibers; upper 2 or 3 spikes staminate, linear, 1–3 cm. long, the lower 3 or 4 pistillate, cylindric, 2–5 cm. long, rather loosely flowered, the lower one or more short-peduncled, the lower bracts leaflike, sheathing, the ligules in front and the upper part of the sheaths at least somewhat hairy; pistillate scales narrowly ovate, short-awned, sometimes reddish; perigynia longer than the scales, 4–5 mm. long, spreading, broadly ovoid-conical, inflated, rather membranous, brownish green, nerved, abruptly rather long-beaked, reddish bifurcate at tip; stigmas 3. ——July. Wet meadows and wet places in woods; Hokkaido; rare.——Sakhalin, e. Siberia, n. Korea, and Manchuria. The typical phase occurs in Korea, Manchuria, and e. Siberia.

201. Carex miyabei Franch. *C. saruensis* Franch.; *C. wallichiana* var. *miyabei* (Franch.) Kuekenth.; *C. jedia* var. *miyabei* (Franch.) T. Koyama——BIRŌDO-SUGE. Rhizomes elongate, rather stout; culms 30–60 cm. long; leaf-blades flat, 3–5 mm. wide, glabrous, the basal sheaths bladeless, somewhat reddish brown, splitting into fibers; upper 2–4 spikes staminate, linear, 15–30 mm. long, reddish brown, the lower 2 or 3 pistillate, short-cylindric, 2–4 cm. long, the lower one or more peduncled, the lower bracts leaflike, sheathless; pistillate scales narrowly ovate, acute, sanguineous; perigynia longer than the scales, 3–4 mm. long, ascending, obtusely 3-angled, broadly obovoid, membranous, hispid, abruptly rather long-beaked, acutely 2-toothed; stigmas 3.——May–June. Sandy or wet clay soil along rivers and streams; Hokkaido, Honshu (Kinki Distr. and eastw.), Kyushu.

202. Carex latisquamea Komar. *C. villosa* Boott, non Stokes; *C. villosa* Boott var. *wrightii* Fr. & Sav.; *C. villosa* var. *latisquamea* (Komar.) Kuekenth.; *C. villosa* Boott var. *straminea* Akiyama——HATABE-SUGE. Rhizomes short; culms 1 or few together, slender, 40–75 cm. long, pilose; leaf-blades flat, thinly pilose, 3–6 mm. wide, the basal sheaths bladeless, light brown, pilose; spikes 3 or 4, the terminal staminate, linear, 15–25 mm. long, the lateral pistillate, oblong, 1–2 cm. long, distant, rather densely flowered, the lowest long-peduncled, the lowest bract leaflike, somewhat long-sheathing; pistillate scales ovate, acute, tawny; perigynia much longer than the scales, 5–6 mm. long, spreading, ovoid, obtusely 3-angled, firmly membranous, brownish, glabrous, nerved, moderately long-beaked, bidentate at tip; stigmas 3.——June–July. Hokkaido, Honshu (Shinano Prov.), Kyushu (Mount Kuju and Kojōmura in Higo); rare.——Korea, Manchuria, and Ussuri.

Fam. 43. **PALMAE** SHURO KA Palm Family

Arborescent plants with woody stems, or rarely acaulescent; stem or trunk simple, rarely branched, naked or covered with the persistent basal part of old leaves, occasionally very spiny, rarely clambering; leaves alternate, pinnately or palmately divided, plicate in bud, the petiole usually expanding into a broad sheath and surrounding the stem at base; flowers unisexual or bisexual, small, usually densely aggregated in a bracteate, paniculate or spicate inflorescence; bracts 3; perianth-segments 6, in 2 series, usually free, imbricate or valvate in bud; stamens 3 or 6, with versatile anthers; ovary of 3 free or connate carpels each terminated by a stigma; ovules 1 or 2 in each carpel, anatropous; fruit a drupe or berry, soft or bony; embryo small.——About 150 genera with about 1,500 species, chiefly of tropical regions.

1A. Leaves pinnate, with a well-developed rachis; carpels connate, forming a single berry not subtended by a scale. .... 1. *Arenga*
1B. Leaves palmate, without a well-developed rachis; carpels free, each becoming an independent berry.
   2A. Petioles stout, prominently spiny on margin; rachis of leaf-blades very short; ovary of 3 carpels, connate below; styles connate or distinct. .... 2. *Livistona*
   2B. Petioles slenderer, with small teeth on margin toward the base; rachis not developed; ovary of 3 carpels, distinct or connate below; styles distinct. .... 3. *Trachycarpus*

1. ARENGA Labill.   KURO-TSUGU ZOKU

Caulescent or acaulescent; leaves pinnate, the pinnae many, elongate, irregularly toothed or incised near tip, 1-ribbed, with 1 or 2 auricles at base; inflorescence much branched, on a peduncle surrounded at base with numerous bracts; flowers usually unisexual; stamens many; pistillate flowers globose; sepals accrescent after flowering; ovary 3-locular, with a conical stigma; fruit ovoid-globose; seeds compressed or planoconvex.——More than 10 species, in the Tropics of Asia and Australia.

1. **Arenga engleri** Becc. *Didymosperma engleri* (Becc.) Warb.; *A. saccharifera* Miq. sensu auct. Japon., non Labill.; *A. tremula* sensu auct. Japon., non Becc.——KURO-TSUGU. Acaulescent or with a pseudostem consisting of the compact petioles; leaves radical, large, suberect, firm, with numerous leaflets to 60 cm. long, 3 cm. wide, plicate and gradually narrowed at base, obtuse, deeply dentate at apex, deep green above, light grayish, with an elevated midrib beneath, the upper leaflets gradually smaller, ascending, the terminal one broader, rounded or obtuse at apex; leaf-sheaths disintegrated into black fibers; inflorescence unisexual; staminate flowers with a small calyx; petals oblong, yellowish, 1.5–2 cm. long; stamens numerous, the anthers slender; pistillate flowers with deltoid petals; fruit globose, 1.5 cm. across.——Kyushu (Yakushima); possibly not indigenous.——Ryukyus and Formosa.

2. LIVISTONA R. Br.   BIRŌ ZOKU

Trunk tall, marked with ringlike scars; leaves large, palmate, conduplicate in bud, parted or divided into many linear segments, the petiole long, spiny on margin; inflorescence from leaf-axils, long-peduncled, loosely paniculate, the bracts many, tubular, sheathing; flowers small, bisexual; sepals 3, imbricate in bud; petals 3, connate at base, the lobes valvate in bud, coriaceous; stamens 6, the filaments connate at base into a ring, the anthers cordate; ovary of 3, nearly free to connate carpels with a short common style at apex; seeds erect with a concave spot on ventral side.——More than 10 species, in tropical Asia and Australia.

1. **Livistona chinensis** R. Br. var. **subglobosa** (Hassk.) Becc. *L. subglobosa* Martius; *Saribus subglobosus* Hassk.; *Corypha japonica* Kittlitz——BIRŌ. Trunk straight, to 5 m. high, stout, simple; leaves large, orbicular-flabellate, palmately parted, to 1 m. across, the axis extremely short, the segments linear, plicate, 1-ribbed, transversely veined, gradually tapering to a bifid pendulous apex, the petiole 1–2 m. long, stout, retrorsely spiny on margins; inflorescence elongate, loosely branching; flowers ovoid in bud, about 4 mm. long; sepals broadly ovate; petals ovate, woody, subacute; stamens 6; carpels 3, approximate and terminated by a single common style; ovule 1 in each locule; fruit nearly globose, about 1.5 cm. long.——Rather rare in warmer coastal areas; Shikoku, Kyushu.——Ryukyus and Formosa.

3. TRACHYCARPUS H. Wendl.   SHURO ZOKU

Trunk straight and erect, treelike; leaves orbicular, flabellate, large, palmately parted into linear plicate segments, the petiole spineless; panicles from leaf-axils, much branched, with numerous multibracteate flowers, bractlets small; flowers unisexual (plant dioecious), with 3 sepals and 3 petals; stamens 6, with short basifixed anthers; carpels 3, free, each with a single stigma and a single basal ovule; fruiting carpels 1–3, globose to ellipsoidal; seeds erect, furrowed on the ventral side, with an embryo on the dorsal side.——Few species, in northern India, China, and Japan.

1A. Leaf-segments pendulous at tip; petioles rather long, spreading; inflorescence densely flowered. .... 1. *T. fortunei*
1B. Leaf-segments stiff, not pendulous; petioles shorter; inflorescence very densely flowered. .... 2. *T. wagnerianus*

1. **Trachycarpus fortunei** (Hook.) H. Wendl. *T. excelsus* H. Wendl., excl. syn.; *Chamaerops fortunei* Hook.——SHURO, WA-JURO. An evergreen tree with a straight thick simple trunk, 3–7 m. long, covered with dark brown fibrous remains of the petiole bases; leaves aggregated at the top of the stem, large, 50–80 cm. across, orbicular-flabellate, the segments linear, plicate, somewhat obtuse and bifid at apex, 1.5–3 cm. wide, the petioles nearly 1 m. long, tapering to a broad sheath surrounding the stem at base; inflorescence peduncled, deflexed, subtended by large bracts; flowers small, yellowish; fruiting carpels 1–3, globose, about 1 cm. across, black-indigo, bloomy when mature.——May–June. Kyushu (s. distr.); widely planted in centr. and s. Japan.

2. **Trachycarpus wagnerianus** Becc. *T. fortunei* sensu auct. Japon., non H. Wendl.——TŌ-JURO. Resembles the preceding but leaves on shorter petioles, the segments stiff, darker colored, slightly smaller; inflorescence very densely flowered.——Native of China, planted in s. Japan.

Plate 7

Themedia japonica (Willd.) C. Tanaka. (Gramineae). Narashino near Tokyo, Honshu. (Photo J. Ohwi.)

Themeda japonica (Willd.) C. Tanaka. (Gramineae). Narashino near Tokyo, Honshu. (Photo J. Ohwi.)

Plate 8

Miscanthus sinensis Anderss. (Gramineae). Narashino near Tokyo, Honshu. (Photo J. Ohwi.)

Mount Shibutsu from Ozegahara Moor, showing alpine zone on upper part of slope; floating islands in foreground. (Photo Midori Kogure.)

Plate 9

Livistona subglobosa Mart. (Palmae). Island of Birojima, Osumi Prov., Kyushu. (Photo M. Tatewaki.)

Crinum asiaticum var. *japonicum* Bak. (Amaryllidaceae). Edge of shore, Osezaki in Idzu Prov., Honshu. (Photo J. Ohwi.)

Fam. 44. **ARACEAE** Tennan-shō Ka   Aroid Family
(Contributed by Dr. Tetsuo Koyama.)

Perennial herbs; leaves radical or alternate on elongate, often twining stems, frequently with transversely reticulate nervules; flowers unisexual or bisexual, sessile, densely aggregated in a spadix subtended by a spathe; perianth wanting or small and scaly; anthers 2- or 3-locular; ovary sessile, 1- to 3-locular, with 1 to several ovules; fruit berrylike, few-seeded.——About 100 genera, with more than 500 species, mainly in the Tropics.

1A. Perianth absent; flowers unisexual, except in *Calla*.
2A. Spadix with an appendage at apex.
3A. Ovules anatropous; pistillate part of spadix continuous with the staminate; inflorescence and normal leaves appearing at different times. .... 1. *Amorphophallus*
3B. Ovules orthotropous; pistillate and staminate parts of spadix distinct; inflorescence on the peduncle arising from the axil of a normal leaf, arising together.
4A. Neuter flowers absent between the pistillate and staminate parts.
5A. Ovary with a single ovule; pistillate part of spadix adnate to the inner side of the spathe. .... 2. *Pinellia*
5B. Ovary always with 2 or more ovules in each locule; spadix not adnate to the spathe. .... 3. *Arisaema*
4B. Neuter flowers present between the pistillate and staminate parts. .... 4. *Typhonium*
2B. Spadix not appendaged at apex.
6A. Stamens connate. .... 5. *Alocasia*
6B. Stamens free. .... 6. *Calla*
1B. Perianth of 4 to 8 free tepals; flowers bisexual.
7A. Leaves broad; spathe broad, surrounding or inclosing the spadix, not leaflike, usually colored or white.
8A. Ovules orthotropous; ovary 2-celled and 1- or 2-ovuled in each cell. .... 7. *Lysichiton*
8B. Ovules pendulous; ovary 1(–2)-celled, with 1 or 2 ovules in each cell. .... 8. *Symplocarpus*
7B. Leaves linear; spadix leaflike, green, narrow, appearing like a continuation of the stem, neither surrounding nor covering the spadix. .... 9. *Acorus*

1. **AMORPHOPHALLUS** Bl.   Konnyaku Zoku

Tuber usually depressed-globose; leaf solitary, radical, large, long-petioled, ternately to digitately compound; scape usually elongate; spathe large, ovate to broadly so, the blade deciduous or persistent; inflorescence (spadix) erect, terete, densely flowered, the pistillate part continuous with the staminate, without a neutral intermediate area; anthers 2 to 4, sessile, opening by pores at tip; ovary globose, 1- to 4-locular; stigma entire or 2- to 4-fid; ovule single in each locule, arising nearly from the base, anatropous; berry globose; seeds without endosperm, embryo similar to seed in shape.——About 100 species in the Old World Tropics.

1A. Appendage of spadix 30–50 cm. long including the inflorescence, much longer than the spathe, 20–30 cm. long; style about 1 mm. long including the inconspicuously 3-lobed stigma; larger in all aspects than the following. .... 1. *A. konjac*
1B. Appendage of spadix 15–20 cm. long including the inflorescence, about as long as the spathe; style extremely short, terminated by an inconspicuously 2-lobed stigma; smaller in all aspects than the preceding. .... 2. *A. kiusianus*

1. Amorphophallus konjac K. Koch. *Conophalus konjac* Schott ex Miq., nom. nud.; *Dracontium polyphyllum* sensu Thunb., non L.; *A. rivieri* var. *konjac* (K. Koch) Engler——Konnyaku. Tuber large, depressed-globose, to 25 cm. across; leaf solitary, arising from the center of the tuber, more than 1 m. tall, the petiole long, spotted with purple, the blade ternate, the primary leaflets again binate or ternate or 2- or 3-parted, the ultimate leaflets or segments pinnately parted on a winged axis, the lobes ovate to narrowly so, caudate-acuminate, 4–8 cm. long, entire, glabrous; scape arising from the center of the tuber, about 1 m. long, with several scales at base; inflorescence 30–50 cm. long; spathe broadly ovate, tubular-infundibuliform, with dark purple variegations, glabrous outside, with small protuberances at base inside; spadix with the pistillate flowers below and continuous with the upper staminate flowers, the appendage erect, much longer than the fertile part,

long-exserted from the spathe, dark purple, obtuse; berries yellowish red when mature.——May. Native of Indochina, widely cultivated in our area for "Konnyaku" or mannan from the tubers.

2. Amorphophallus kiusianus (Makino) Makino. *A. konjac* var. *kiusianus* Makino——Yama-konnyaku. Related to the preceding, but slightly smaller; petioles with yellowish brown spots, the leaf-blades ternate, the leaflets dichotomously forked and pinnately lobed, the ultimate lobes oblong, caudate-acuminate, 3–20 cm. long; spathe 15–20 cm. long, glabrous, the tubular part short; spadix erect, terminated by an obtuse terete appendage; ovary usually 2-locular; style very short, the stigma indistinctly 2-lobed; berry at first greenish, becoming reddish purple, then deep blue at maturity.——May–June. Kyushu (s. distr., including Yakushima and Tanegashima); the tuber not edible.

2. **PINELLIA** Tenore   Hange Zoku

Perennials with small subterranean tubers; leaves petioled, simple or 3- to 7-lobed or -parted; scape solitary; spathe persistent, the tubular part incurved, cylindric, the blade longer than the tubular part, oblong, concave; spadix surrounded by the tubular part of the spathe, the pistillate part adnate to the tubular part of spathe, separated from the staminate by the apical constriction of the tubular spathe, the appendage filiform, slenderly elongate and long-exserted from the spathe; flowers naked, the staminate with 1 or 2 anthers, the pistillate with a single 1-locular ovary containing a solitary orthotropous ovule; berry inclosing a single seed with copious endosperm.——Few species, in e. Asia.

1A. Leaves 3-foliolate, the leaflets oblong to lanceolate, acute and mucronate at apex, with small bulbils on the petiole. .... 1. *P. ternata*
1B. Leaves deeply 3-parted, the segments broadly or narrowly ovate, short-acuminate; plant slightly larger, without bulbils. 2. *P. tripartita*

1. Pinellia ternata (Thunb.) Breitenb. *Arum ternatum* Thunb.; *P. tuberifera* Tenore; *A. macrourum* Bunge; *Typhonium tuberculigerum* Schott; *Arisaema loureiri* Bl.——KA-RASU-BISHAKU. Leaves few from a small tuber, with small bulbils borne at the middle and on the uppermost part of the petiole; leaflets 3, sessile, ovate-elliptic to oblong, 3–12 cm. long, 1–5 cm. wide, entire, glabrous; scape solitary, 20–40 cm. long; spathe typically green, 6–7 cm. long, the tubular part 1.5–2 cm. long, the blade lanceolate, rounded at apex, curved at tip, glabrous outside, puberulent within; spadix with the appendage filiform, erect, 6–10 cm. long, glabrous; berry small, green.——May–July. Cultivated fields and roadsides; Hokkaido, Honshu, Shikoku, Kyushu; quite common.——Korea and China.——Forma **angustata** (Schott) Makino. *P. an-*

gustata Schott.——SHIKA-HANGE. Leaflets linear.

2. Pinellia tripartita (Bl.) Schott. *Antherurus tripartitus* Bl.; *Arisaema tripartitum* (Bl.) Engler——Ō-HANGE. Tuber to 3 cm. across, covered with brown fibers; leaves petioled, deeply 3-lobed to nearly 3-parted, the segments 8–20 cm. long, 2–12 cm. wide, abruptly short-caudate; scape 20–50 cm. long; spathe green outside, typically light purplish within, 6–10 cm. long, the tubular part 2–3 cm. long, the blades broadly lanceolate, obtuse, slightly incurved at tip, smooth outside, densely papillose inside; spadix with a filiform appendage 15–25 cm. long.——June–Aug. Honshu (Mino Prov. and westw.), Shikoku, Kyushu.——Forma **atropurpurea** (Makino) Ohwi.——MURASAKI-Ō-HANGE. Spathe dark purple within.

3. ARISAEMA Martius   TENNAN-SHŌ ZOKU

(Synopsis by Dr. Tetsuo Koyama.)

Perennial herbs with a depressed-globose tuber or corm, rarely rhizomatous; normal leaves 1 or 2, rarely 3, petioled or nearly sessile, with 3–17 sessile or short-petioluled, reticulate-nervulose leaflets, the petiole arising from within a cluster of scapelike sheaths (pseudostem), these surrounded at base by 2 or 3 scalelike leaves (cataphylls); inflorescence solitary, terminal or pseudo-lateral; spathe persistent, the tube cylindric, involute, the blade galeate or leaflike; spadix bisexual or unisexual (plants paradioecious, i.e., dioecious, but the sex reversible according to conditions), the staminate part above the pistillate, the appendage somewhat elongate, terete, clavate, or sometimes filiform; staminate flower with 1 to 5 stamens; pistillate flower with a single 1-locular ovary, with 1–9 orthotropous ovules; berries 1- to few-seeded; seeds ovoid-globose.——About 150 species, chiefly of tropical and cooler regions of e. Asia, a few in Africa and N. America.

1A. Spadix-appendage filiform, long-exserted from the spathe.
  2A. Normal leaves 2; spadix-appendage 10–15 cm. long, 2.5 to 3.5 times as long as the tube of spathe; terminal leaflet not smaller than lateral ones. .... 1. *A. negishii*
  2B. Normal leaf solitary; spadix-appendage longer.
    3A. Terminal leaflet much shorter than the adjacent lateral ones; pseudostem more than 40 cm. long; spathe nearly green.
                                                 2. *A. heterophyllum*
    3B. Terminal leaflet as large as or larger than the adjacent lateral ones; pseudostem nearly wanting; spathe dark reddish purple.
      4A. Spathe-blade with a T-shaped white mark inside; scape arising near base of petiole. .... 3. *A. kiushianum*
      4B. Spathe-blade without a T-shaped mark; scape arising above base of petiole.
        5A. Enlargement at base of spadix-appendage densely wrinkled. .... 4. *A. thunbergii*
        5B. Enlargement at base of spadix-appendage smooth. .... 5. *A. urashima*
1B. Spadix-appendage terete or clavate, obtuse or sometimes capitate at apex, shorter than the blade of the spathe.
  6A. Spadix-appendage gradually passing into the floriferous part; spathe and spadix-appendage different in staminate and pistillate plants. .... 6. *A. heterocephalum*
  6B. Spadix-appendage truncate at base; spathe and spadix-appendage equal or nearly alike in staminate and pistillate plants.
    7A. Spathe-blade saccate-galeate; leaflets always 3, filiform at apex, rather thick, lustrous above, entire. .... 7. *A. ringens*
    7B. Spathe-blade spreading, declined, or ascending anteriorly.
      8A. Leaflets always 3, with many subtranslucent papillae on margin; spathe-blade acuminate. .... 8. *A. ternatipartitum*
      8B. Leaflets 3–15, pedate, the margins entire or loosely undulate to toothed; spathe-blade gradually or abruptly acute.
        9A. Normal leaf solitary (rarely 2).
          10A. Spadix-appendage rather stout, 3–10 mm. across, terete, or gradually thickened toward apex; spathe-blade deltoid-ovate, 3–5 cm. wide.
            11A. Scape 3–10 cm. long, apparently lateral; leaflets somewhat palmate, without a rachis; Kyushu. .... 9. *A. nanum*
            11B. Scape 15–50 cm. long.
              12A. Lateral leaflets 2(–3) on each side, nearly radiate; rachis slightly developed.
                13A. Spathe-blade obliquely ascending, 5–6 cm. long. .... 10. *A. robustum*
                13B. Spathe-blade strongly declined over the spathe, 8–10 cm. long. .... 11. *A. ovale*
              12B. Lateral leaflets 3–8 on each side; rachis, at least the lower part, well developed.
                14A. Leaflets obovate, abruptly short-acuminate; Izu Peninsula. .... 12. *A. izuense*
                14B. Leaflets lanceolate to oblong-obovate, tapering toward apex; Kinki District and westward.

15A. Spathe-blade oblong-ovate, abruptly long-acuminate, strongly declined over the spadix; spadix-appendage cylindric-clavate, slightly thickened at apex. .... 13. *A. suwoense*

15B. Spathe-blade deltoid, abruptly narrowed from the top of the tubular part, erect, the apex gradually elongate to a linear tip; spadix-appendage narrowly cylindric, terete. .... 14. *A. simense*

10B. Spadix-appendage thickened at both ends or at least at base, the median and sometimes the apical part narrow, 1–2 mm. thick; spathe-blade broadly deltoid-lanceolate to narrowly deltoid-ovate, 2–3 cm. wide; leaflets 7–19.

16A. Spathe entirely green, papillose on margin. .... 15. *A. yoshiokae*

16B. Spathe usually purplish, entire.

17A. Spathe-blade 7–10 cm. long.

18A. Leaflets 9–11; spathe-blade linear, ovate, gradually acuminate. .... 16. *A. iyoanum*

18B. Leaflets 7–9.

19A. Spathe-blade ovate-lanceolate, long-acuminate, dark purple with 4 white stripes; Honshu (Harima Prov.).

17. *A. seppikoense*

19B. Spathe-blade oblong-lanceolate, subabruptly acute, light purple-brown; Aki. .... 18. *A. akiense*

17B. Spathe-blade 3–6 cm. long.

20A. Spathe-blade, at least in part, dark purple, 5–6 cm. long, deltoid-lanceolate. .... 19. *A. monophyllum*

20B. Spathe-blade greenish, ovate, 3–4 cm. long. .... 20. *A. maximowiczii*

9B. Normal leaves 2, rarely 1 in depauperate individuals.

21A. Leaflets 3–5; spathe-blade abruptly long-acuminate; spadix-appendage capitately thickened at apex. ... 21. *A. sikokianum*

21B. Leaflets (5–)7–17; spadix-appendage cylindric, slender or clavately thickened at apex.

22A. Spathe-blade filiform-elongate.

23A. Spadix-appendage thick-cylindric; spathe-blade dark purple with vertical white stripes. .... 22. *A. kishidae*

23B. Spadix-appendage clavate, thickened at apex; spathe-blade green. .... 23. *A. tosaense*

22B. Spathe-blade acuminate to long-acuminate, not filiform-elongate.

24A. Spathe-blade broadly lanceolate, gradually long-acuminate; spadix-appendage slenderly cylindric.

25A. Spathe-blade bent forward horizontally. .... 24. *A. longilaminum*

25B. Spathe-blade declined and arching over the spadix.

26A. Spadix-appendage erect. .... 25. *A. angustifoliatum*

26B. Spadix-appendage slightly bent forward. .... 26. *A. shinanoense*

24B. Spathe-blade narrowly to broadly ovate.

27A. Spathe-blade densely yellowish papillose inside. ....27. *A. yamatense*

27B. Spathe-blade smooth inside.

28A. Spathe-blade obliquely ascending or bent horizontally, subabruptly short-acute.

29A. Plant loosely tufted with numerous young plants; spathe-blade oval to orbicular-ovate. .... 28. *A. proliferum*

29B. Plant solitary; spathe-blade ovate.

30A. Spadix-appendage thick-clavate. .... 29. *A. speirophyllum*

30B. Spadix-appendage linear or cylindric. .... 30. *A. peninsulae*

28B. Spathe-blade declined above, long-acuminate.

31A. Spathe-blade oblong-deltoid, tapering from base; appendage of spadix cylindric, the apex as broad as the other part.

32A. Spadix-appendage slenderly cylindric or linear.

33A. Spadix-appendage linear; spathe-blade obliquely declined only at tip; leaflets 9–18, often serrulate; Honshu. .... 31. *A. angustatum*

33B. Spadix-appendage cylindric; spathe-blade declined; leaflets 11–13 in the lower one, entire; Kyushu.

32. *A. koshikiense*

32B. Spadix-appendage cylindric, rather thick, to 8 mm. across at base; tubular part of spathe rather long.

33. *A. solenochlamys*

31B. Spathe-blade broadly ovate to ovate-lanceolate, tapering from below the middle but not from base.

34A. Leaflets palmate on the scarcely developed rachis; scape longer than the leaves.

35A. Leaflets 5. .... 34. *A. nikkoense*

35B. Leaflets more than 5.

36A. Leaflets 5–7, not variegated, narrowly oblong; entire or serrate; auricles of spathe very distinct.

35. *A. stenophyllum*

36B. Leaflets 8–14, with a broad white line along the midrib, linear-lanceolate, always undulate on margin; auricles of spathe narrow and rather inconspicuous. .... 36. *A. undulatifolium*

34B. Leaflets pedate on the well-developed rachis at least between the lower ones.

37A. Spathe-blade broadly deltoid-ovate, the rachis of leaflets spirally curved on apical part, the auricles almost wanting. .... 37. *A. takedae*

37B. Spathe-blade narrower, gradually tapering toward both ends from the middle, the auricles developed.

38A. Auricles of spathe very broad; Kantō to Ōshima in Izu Prov. ....38. *A. limbatum*

38B. Auricles of spathe narrow.

39A. Spadix-appendage bulbously thickened at apex, smooth or rugose.

40A. Spathe-blade elongate, nearly as long as the tubular part, spreading nearly at right angle over the mouth; Kyushu. .... 39. *A. mayebarae*

40B. Spathe-blade always shorter than the tubular part, strongly declined at tip over the mouth; eastern Honshu. .... 40. *A. serratum*

39B. Spadix-appendage slightly or scarcely thickened, smooth.

41A. Tuber globose, not depressed; Izu Islands. .... 41. *A. hachijoense*

41B. Tuber depressed-globose; western Honshu. .... 42. *A. japonicum*

1. **Arisaema negishii** Makino.　Shima-tennan-shō, Hengodama.　Tuber depressed-globose, more or less irregular in shape, 3–7 cm. across; scalelike leaves 2, whitish, tunicate; leaves 2, the petioles terete, 5–20 cm. long; leaflets 9–15, lanceolate, 8–20 cm. long, 1.5–5 cm. wide, undulate, cuneate at base; peduncle erect, elongate after anthesis, 4–8 mm. in diameter; spathe light or pale green, the tube cylindric, slightly ampliate above, the blade obliquely ascending forward, narrowly ovate, 7–8 cm. long, gradually acuminate; appendage of spadix linear, 10–15 cm. long, purplish, somewhat thickened, long-exserted, gradually thickened at base, the floriferous part often with rather loose hornlike tubercles.——Mar.–May. Honshu (Hachijo and Miyake Isls. in Izu).

2. **Arisaema heterophyllum** Bl.　*A. thunbergii* var. *heterophylum* (Bl.) Engl.——Maizuru-tennan-shō.　Monoecious; tuber hemispherical, 2–4 cm. across; scalelike leaves 4 or 5, tunicate; normal leaf solitary, on a long erect stemlike sheath to 1 m. tall and 1–2 cm. thick, the petiole erect-spreading, 6–14 cm. long; leaflets 13–19, oblanceolate or oblong, subabruptly contracted at apex, undulate on margin, the terminal leaflet much shorter than the others, the first lateral ones 10–25 cm. long, 2–6.5 cm. wide; spathe on a long erect scape, green, the tube 3.5–8 cm. long, dark purplish on margin, the blade ovate, 7–12 cm. long, 3–7 cm. wide, usually green, bent forward and arching over the spadix, abruptly short-acuminate; appendage of spadix filiform, erect, 15–30 cm. long, slightly thickened at base and passing into the floriferous part.——May–July. Wet places in lowlands; Honshu (Kantō Distr. and westw.), Shikoku, Kyushu; rather rare.——s. Korea, Formosa, and China.

3. **Arisaema kiushianum** Makino.　Hime-urashima-sō.　Tuber nearly globose, with several small lateral tubers; leaf always solitary, the petiole terete, 15–40 cm. long; leaflets 7–13, lanceolate, long-acuminate, 10–20 cm. long and 2–4 cm. wide in the terminal one; scape arising near base of petiole, 5–15 cm. long; spathe uniformly dark purple, the blade broadly ovate, strongly arching over the spadix, abruptly narrowed to a long cusp, with a white T-shaped mark inside; spadix-appendage slender, long-exserted, 15–18 cm. long, pendulous at tip, thickened below and passing into the floriferous part.——May–June.——Kyushu.

4. **Arisaema thunbergii** Bl.　*Flagellarisaema thunbergii* (Bl.) Nakai——Nangoku-urashima-sō.　Tuber depressed-globose, usually bearing a few small lateral tubers; leaf solitary, the petiole nearly erect, terete, 30–60 cm. long; leaflets 9–17, linear- to broad-lanceolate, acuminate, 10–25 cm. long and 1–4 cm. broad in the terminal one; peduncle arising from the lower part of the petiole, 10–20 cm. tall; spathe uniformly dark purple or reddish purple, sometimes purplish bronze, the blade ovate, strongly arching over the spadix, the upper part semipendulous and long-filiform tapering, 8–10 cm. long; spadix-appendage filiform, 30–50 cm. long, erect, pendulous at tip, gradually thickened below forming a densely wrinkled, fusiform enlargement and further narrowed toward the floriferous part.——Apr.–May. Honshu (Chūgoku Distr.), Shikoku, Kyushu.

5. **Arisaema urashima** Hara.　*A. thunbergii* sensu auct. Japon., pro parte; *A. thunbergii* var. *urashima* (Hara) Makino; *Flagellarisaema urashima* (Hara) Nakai——Urashima-sō.　Closely resembles the preceding; tuber depressed-globose, bearing several to rather many small lateral tubers; leaf solitary (very rarely 2), the petiole terete, nearly erect, 30–50

cm. long, the leaflets 11–15, oblanceolate to broadly lanceolate, acuminate, the terminal one 10–18 cm. long, 2–3.5 cm. wide; peduncle 10–20 cm. long, arising 7–15 cm. above base of petiole; spathe dark to bronze-purple, sometimes reddish purple, the blade ovate to broadly so, abruptly long taillike at tip, strongly arching over the spadix; spadix-appendage filiform, 40–60 cm. long, long-exserted, the upper half bent abruptly downward and pendulous, gradually thickened below forming a smooth oblong enlargement then gradually passing into the floriferous part.——Apr.–May. Hokkaido (Oshima and Hidaka Prov.), Honshu, Shikoku.

6. **Arisaema heterocephalum** Koidz.　Amami-tennan-shō, Hosoba-tennan-shō.　Pseudostem 30–70 cm. long, from a hemispherical tuber; leaves 2, the petiole 7–20 cm. long; leaflets 11–19, pedate, linear or linear-oblanceolate, the terminal one 8–17 cm. long, 15–20 mm. wide; spathes dimorphic, the pistillate with a green tube about 8 cm. long, ampliate in upper part, the blade broadly ovate, 7 cm. long, short-acuminate, the spadix-appendage 6.5 cm. long, narrowly clavate-cylindric, an erect stipe about 6 cm. long; staminate spathe tubular, 7 cm. long, 5 mm. across, the spadix-appendage capitate at apex, on a stipe 7 cm. long.——Reported to occur in Kyushu (Yakushima)——Ryukyus, from Amami-Oshima to Okinawa.

7. **Arisaema ringens** (Thunb.) Schott.　*Arum ringens* Thunb.; *Arisaema triphyllum* sensu Thunb., non L.; *A. sieboldii* de Vriese; *A. praecox* de Vriese; *A. ringens* var. *sieboldii* (de Vriese) Engl. and var. *praecox* (de Vriese) Engl.; *Ringentiarum ringens* (Thunb.) Nakai——Musashi-abumi.　Tuber depressed-globose, bearing a few small lateral tubers; leaves 2, the petiole terete, erect, 15–25 cm. long; leaflets 3, rather thick, ternate, sessile, broadly rhomboidal-ovate, 8–20 cm. long, 4–10 cm. wide, entire, prominently lustrous above, subabruptly narrowed into a filiform tail 5–15 mm. long; peduncle short, 3–10 cm. long; spathe greenish (forma *praecox* T. Koyama) or uniformly dark purple (forma *sieboldii* T. Koyama), with a large auricle on each side on the margin of the tubular part, the blade saccate-galeate, the lower half obliquely ascending, the upper half strongly bent forward, 3.5–4.5 cm. long, short-caudate; spadix-appendage cylindric, the base thickened and abruptly contracted above the floriferous part.——Mar.–May. Woods near seacoast; Honshu (Kantō Distr. and westw.), Shikoku, Kyushu.——Ryukyus, China, and s. Korea.

8. **Arisaema ternatipartitum** Makino.　Mitsuba-tennan-shō.　Tuber depressed-globose or hemispherical; leaves 2, the petiole erect-ascending, 7–20 cm. long; leaflets 3, rhomboid-ovate, 7–15 cm. long, 3–8 cm. wide, abruptly mucronate, cuneate and sessile at base in the terminal one, rounded at base in the lateral ones, bearing subtranslucent dense papillae on margin; peduncle 10–15 cm. long, shorter than the petiole; spathe dark purple, broadly auricled on the upper margin of the tube, the blade long-deltoid, about 3 cm. wide, 7 cm. long, acuminate, arching over the spadix; spadix-appendage cylindric, rather stout, 2.5–3 mm. across, thickened and truncate at base.——Apr.–May. Honshu (Suruga Prov.), Shikoku, Kyushu.

9. **Arisaema nanum** Nakai.　Hime-tennan-shō.　Tuber hemispheric; leaf solitary, the petiole terete, 10–25 cm. long; leaflets 5(–7), somewhat palmate, elliptic to oblong, abruptly short-cuspidate, cuneate at base, the terminal one short-petiolulate, 8–15 cm. long, 4–7 cm. wide, the adjacent

lateral ones slightly smaller, usually decurrent on one side at base; scape 3–10 cm. long; peduncle 2–5 cm. long; spathe dark purple, without auricles, the blade oblong, scarcely narrowed at base, 8–13 cm. long, arching over the spadix and pendulous at tip, acuminate; spadix-appendage thick, clavate-cylindric, thickened below and truncate at base.——May. Kyushu.

10. Arisaema robustum (Engl.) Nakai. *A. amurense* Maxim. var. *robustum* Engl.; *A. amurense,* sensu auct. Japon., non Maxim.; *A. sadoense* Nakai——Hiro-ha-tennan-shō, Sado-tennan-shō. Pseudostem 20–35 cm. long, from a depressed-globose tuber bearing small tubers on the sides; leaf solitary, the petiole terete, nearly erect, 10–25 cm. long, the blade palmately 5-foliolate, the rachis not developed, the leaflets obovate or ovate, 10–15 cm. long, 4–7 cm. wide, abruptly short-acuminate; peduncle 2–5 cm. long; spathe green with longitudinal white stripes or dark purple (forma *atropurpureum* T. Koyama), somewhat auriculate on the upper margin of the tube, the blade ovate, 5–6 cm. long, long-acuminate, obliquely ascending, scarcely narrowed at base; spadix-appendage cylindric, thickened and truncate at base.——May–June. Woods; Hokkaido, Honshu, Kyushu.——Korea and Sakhalin.

Arisaema hakonecola Nakai. Kamiyama-tennan-shō, is poorly known and may be the same as *A. robustum,* differing allegedly in the larger spathe with a uniformly dark purple blade arching over the spadix, and a very stout appendage of the spadix.——Honshu (Hakone).

11. Arisaema ovale Nakai. Ashū-tennan-shō. Tuber subdepressed-globose, 3–5 cm. across; scalelike leaves 3, tunicate; normal leaf solitary, the petiole thick, about 20 cm. long, 5–10 mm. in diameter, the leaflets 5, pedate or nearly palmate, oblong to broadly ovate-elliptic, entire, 10–15 cm. long, 3–6 cm. wide; peduncle very short; spathe 5–6 cm. long, dark purple with longitudinal white stripes, auricled on the upper margins of the tube, the blade ovate, 8–10 cm. long, strongly arching over the spadix; spadix-appendage whitish, clavate-cylindric, stout, 4–6 cm. long, longitudinally rugose, on a short stipe 5–8 mm. long.——Honshu (Tanba, Ōmi, Echizen, Etchu, Noto, and elsewhere).

12. Arisaema izuense Nakai. Izu-tennan-shō. Leaflets to 17, obovate, to 8 cm. long, 3.5–4 cm. wide, abruptly short-acuminate; spathe dark purple, without stripes, the blade deltoid-ovate, gradually tapering to apex, scarcely auricled on the upper margins of the long cylindrical tube; spadix-appendage prominently thickened at tip; otherwise with the characters of *A. ovale* Nakai.——Honshu (Izu Penin.).

13. Arisaema suwoense Nakai. Yamaguchi-tennan-shō. Resembles the preceding; tuber rather small, about 15 mm. across; leaf solitary, the petiole terete, about 20 cm. long; leaflets 7–9, oblong to obovate-oblong, abruptly acuminate, cuneate at base, the terminal one 10–12 cm. long, 4–5 cm. wide; peduncle about 20 cm. long; spathe purplish, the auricles rather narrow, the blade 12–14 cm. long, oblong-ovate, abruptly long-acuminate, strongly declined over the spadix, the spadix-appendage stout, cylindric-clavate, slightly thickened at apex, truncate at base, about 8 mm. in diameter at apex.——May. Honshu (Suwo).

14. Arisaema simense Nakai. Shima-mamushigusa. Tuber depressed-globose, without lateral tubers; leaf solitary, the petiole terete, about 10 cm. long, leaflets 7–13, lanceolate to broadly so, acuminate, the terminal one 6–20 cm. long,

about 3 cm. wide; spathe pale green, the upper margins of the tube slightly recurved, the blade abruptly narrowed from a deltoid base to a long linear tip, erect, 6–8 cm. long; spadix-appendage narrowly cylindric, terete, not thickened at apex.——Honshu (Kinki Distr.), Kyushu.

15. Arisaema yoshiokae Nakai. Inugatake-tennan-shō. Resembles the preceding; pseudostem about 50 cm. long; leaf solitary, the petiole terete, about 10 cm. long; leaflets 15–19, oblong to lanceolate, the terminal one 12–16 cm. long, 2–5 cm. wide; peduncle 2–5 cm. long; spathe greenish, the margin of the tube scarcely recurved, the blade bent forward, ovate-lanceolate, gradually long taillike at apex, papillose on the margin; spadix-appendage slender, about 2 mm. in diameter, the base thickened to about 5 mm., truncate.——May. Kyushu (Inugatake in Buzen Prov.).

16. Arisaema iyoanum Makino. Omogo-tennan-shō. Tuber depressed-globose, bearing a few small additional tubers; leaf solitary, long-petioled; leaflets 9–11, lanceolate, long-acuminate, cuneate at base; spathe light green to purplish, the blade 7–9 cm. long, linear-ovate, gradually acuminate, broadened at base, the upper margin of the tubular part recurved; spadix-appendage narrowly cylindric, rounded at apex, somewhat thickened and truncate at base.——Shikoku (Omogokei in Iyo Prov.).

17. Arisaema seppikoense Kitam. Seppiko-tennan-shō. Tuber depressed-globose; pseudostem 17–25 cm. long, purple-spotted; normal leaf solitary, long-petioled, pedately 7- to 9-foliolate; leaflets oblong-lanceolate, acuminate, narrowed toward base, sparsely puberulent beneath, the terminal one 17–23 cm. long, 4–5 cm. wide; peduncle short; spathe dark purplish, the tube 4–4.5 cm. long, dark purple, the blade ovate-lanceolate, 7–8.5 cm. long, about 2.5 cm. wide, dark purple with 4 white longitudinal stripes, long-acuminate; spadix-appendage purplish, rounded at tip, 2–4 mm. in diameter.——May. Honshu (Harima Prov.).

18. Arisaema akiense Nakai. Aki-tennan-shō. Tuber depressed-globose, covered with the fibrous remains of old sheaths, scalelike leaves 4, the outer 2 whitish, the others light brown or greenish; normal leaf solitary, the petiole terete, about 25 cm. long including the sheath; leaflets 7, rather loosely pedate, lanceolate-oblong, 6–13 cm. long, 2–3.5 cm. wide, acuminate, acute at base, undulate on margin; peduncle shorter than the leaf; spathe large, the tube to 4 cm. long, about 10 mm. across, with longitudinal whitish purplish stripes above the middle, the mouth ampliate, the blade longer than the tube, oblong-lanceolate, subabruptly acute, purple-brown, bent down and arching over the spadix, recurved on margin; spadix-appendage cylindric-clavate, whitish but purplish at base.——May. Honshu (Aki Prov.).

19. Arisaema monophyllum Nakai var. **monophyllum.** *A. akitense* Nakai——Hitotsuba-tennan-shō, Akita-tennan-shō. Tuber depressed-globose; pseudostem 20–80 cm. long; leaf solitary, the petiole terete, 7–15 cm. long; leaflets 7–11, ovate to oblong, sometimes obovate, abruptly cuneate-acuminate at tip, entire (forma **integrum** Nakai) or serrulate (forma **serrulatum** Nakai), the terminal one as large as or sometimes slightly smaller than the next lateral ones, 10–20 cm. long, 3–7 cm. wide; peduncle 3–10 cm. long at anthesis, later elongating; spathe light green, marked with a dark purple transverse band inside, the upper margins of the tube scarcely recurved, the blade deltoid-lanceolate, 5–6 cm. long, gradually long-acuminate, erect or slightly oblique, in part dark purple;

spadix-appendage narrowly cylindric, slightly bent forward and somewhat thickened at apex, slightly thickened and truncate at base.——Apr.–May. Honshu (Shimotsuke, Musashi, Sagami, Idzu, Shinano, Ugo and elsewhere).

Var. **atrolinguum** (F. Maekawa) Kurata. *A. atrolinguum* F. Maekawa——Kuro-hashi-tennan-shō. Blade of spathe dark purple, slightly arching over the spadix.——Honshu (Sagami, Idzu).

20. Arisaema maximowiczii (Engl.) Nakai. *A. serratum* var. *maximowiczii* Engl.——Tsukushi-mamushigusa. Related to *A. monophyllum;* tuber depressed-globose; pseudostem 12–37 cm. long; leaf single, on a petiole about 6.5 cm. long; leaflets 7–11, oblong-lanceolate, acuminate, the terminal one larger, 9–10 cm. long, 17–35 mm. wide; peduncle 5–6 cm. long; spathe greenish, the upper margins of the tube strongly recurved, the blade ovate, 3–4 cm. long, about 2 cm. wide, acuminate, greenish; spadix-appendage narrowly cylindric, rounded and about 1 mm. across at apex, gradually thickened to 5 mm. in diameter at base.——Apr.–June. Kyushu.

21. Arisaema sikokianum Fr. & Sav. *A. sazensoo* Makino; *A. magnificum* Nakai——Yuki-mochi-sō, Kanki-sō. Tuber depressed-globose; pseudostem 5–20 cm. long; leaves 2, the petiole terete, 7–12 cm. long; leaflets 3–5, obovate, oblong, or elliptic, 5–20 cm. long, 3–10 cm. wide, abruptly short-acuminate; peduncle 5–12 cm. long; spathe dark purple, the upper margins of the tube not recurved, the blade obliquely ascending, narrowed toward base, elliptic, 8–13 cm. long, 4–5 cm. wide, abruptly long-acuminate; spadix-appendage thick-cylindric, whitish, 6 mm. wide, capitately thickened at apex forming a globose head 10–18 mm. across.——May–June. Honshu (Yamato), Shikoku, Kyushu.

22. Arisaema kishidae Makino. Murou-mamushigusa. Tuber depressed-globose; pseudostem 17–35 cm. long; leaves 2, the petioles 5–15 cm. long; leaflets 5–9, oblong to ovate, abruptly long-acuminate, the terminal one 5–20 cm. long, 2–8 cm. wide; peduncle 5–8 cm. long; spathe purplish, the upper margins of the tube slightly recurved, the blade obliquely declined, broadly deltoid-lanceolate, 10–13 cm. long, gradually elongate at tip, forming a rather short taillike apex, dark purple with vertical white stripes; spadix-appendage thick-cylindric, not or scarcely thickened at apex, slightly thickened and truncate at base.——Apr.–May. Honshu (s. Kinki Distr.).

23. Arisaema tosaense Makino. Ao-tennan-shō. Tuber depressed-globose; pseudostem 10–30 cm. long; leaves 2, the petioles terete, 5–10 cm. long; leaflets 7–15, narrowly ovate to broadly oblanceolate, abruptly tapering to a rather short taillike apex, the terminal one 15–25 cm. long, 3–12 cm. wide; peduncle 3–30 cm. long; spathe dark green with white longitudinal stripes, the upper margins of the tubular part recurved, the blade green, 15–25 cm. long, broadly deltoid-lanceolate and obliquely declined, scarcely narrowed toward the base, gradually tapering toward tip to a tail nearly as long as the body of the blade; spadix-appendage usually thickened and clavate in upper part, rounded at apex, 5–10 mm. in diameter, slightly narrowed and truncate at base.——June–July. Honshu (Kinki and Chūgoku Distr.), Shikoku, Kyushu.

24. Arisaema longilaminum Nakai. Murou-tennan-shō. Tuber depressed-globose; leaves 2, petioled; leaflets 7–15, lanceolate, the terminal one 5–25 cm. long; peduncle 10–25 cm. long; spathe dark purple, the blade deltoid-lanceolate, 7–15 cm. long, 2–4 cm. wide, gradually tapering at tip, erect and

bending forward horizontally; spadix-appendage narrowly cylindric, the upper part bent slightly forward and rather thick, thickened and truncate at base.——Honshu (s. Kinki Distr.).

25. Arisaema angustifoliatum (Miq.) Nakai. *A. japonicum* var. *angustifoliatum* Miq.——Nagaha-mamushigusa. Tuber depressed-globose; leaves usually 2, sometimes 1, petiolate; leaflets 9–12, lanceolate, acuminate, about 13 cm. long; spathe usually purple, the upper margins of the tube slightly recurved, the blade deltoid-lanceolate, 4.5–6 cm. long, obliquely declined, gradually tapering toward the apex; spadix-appendage erect, thickened at both ends, especially at base.——Mar.–May.——Kyushu.

26. Arisaema shinanoense Nakai. Karuizawa-tennan-shō. Tuber globose; pseudostem 15–30 cm. long; leaves 2, the lower one on a petiole 7–12 cm. long; leaflets 7–9, broadly lanceolate, abruptly acuminate, the terminal one 12–15 cm. long, 2.5–5 cm. wide; peduncle 10–20 cm. long; spathe dark purple, the upper margin of the tube slightly recurved, truncate, the blade deltoid-lanceolate, 5–9 cm. long, gradually long-acuminate, scarcely narrowed toward the base, obliquely declined; spadix-appendage purple, narrowly cylindric, slightly bent forward, 2–3 mm. thick, slightly thickened at both ends.——June–July. Honshu (near Karuizawa in Shinano).

27. Arisaema yamatense (Nakai) Nakai. *A. japonicum* var. *yamatense* Nakai——Yamato-tennan-shō. Pseudostem rather long; leaves 2, the petioles 2–3 and 5–12 cm. long respectively; leaflets 7–11, pedate, broadly oblanceolate, 7–12 cm. long, 2–3 cm. wide, long-acuminate; peduncle 2–3 cm. long; spathe green, the upper margins of tube recurved, the blade ovate to broadly so, 4–7 cm. long, abruptly short-acuminate, scarcely narrowed toward base, obliquely ascending, densely yellowish, papillose inside; spadix-appendage nearly erect, narrowly cylindric, the upper part gradually narrowed to the abruptly thickened capitellate apex 2–3 mm. wide.——Apr.–May. Woods; Honshu (Kinki Distr. and Kaga Prov.).

Var. **sugimotoi** (Nakai) Kitam. *A. sugimotoi* Nakai——Suruga-tennan-shō. Spathe-blade abruptly long-acuminate, nearly as long as the tubular part; spadix-appendage usually recurved above.——Honshu (Tōkaidō Distr.).

28. Arisaema proliferum Nakai. Ko-mochi-tennan-shō. Tuber hemispherical, 3–7 cm. across, with rather many supernumerary young tubers and young shoots; pseudostem 30–70 cm. long, 2–3 cm. in diameter at base, 7–15 mm. above; scalelike leaves 4; normal leaves 2; leaflets 5–13, the terminal leaflet oblong or narrowly elliptic, petiolulate or nearly sessile, the lateral leaflets oblanceolate or oblanceolate-oblong; scape equaling or shorter than the leaves; spathe greenish, the tube 4.5–8.5 cm. long, whitish at base, 1.5–2 cm. in diameter, slightly recurved on upper margins, the blade dark green with white longitudinal stripes, orbicular-ovate or oval, 5–6 cm. long, 4–4.5 cm. wide, acuminate; spadix-appendage 4–6.5 cm. long including the short stipe 3–10 mm. long, clavate, green, thickened at apex, about 5 mm. in diameter.——Hokkaido, Honshu (centr. and n. distr.).

29. Arisaema speirophyllum Nakai. Uzumaki-tennan-shō. Tuber depressed-hemispheric, without lateral tubers; pseudostem 30–60 cm. long, 2–4 cm. in diameter at base; leaves 2, the lower one with a petiole 5–7 cm. long; leaflets 16–22, pedate, the median and adjacent laterals petiolulate, the others sessile or nearly so and spirally disposed, broadly ovate, oval, or obovate; peduncle slightly longer than the leaves; spathe-tube 5–9 cm. long, 15–25 mm. in diameter, glaucous-green with longitudinal white stripes, shortly recurved on upper margins, the spathe-blade broadly ovate, abruptly acuminate, dark green outside, dark green with longitudinal white stripes inside; spadix-appendage clavate or subobconical-cylindric, long-stipitate at base, longitudinally many-furrowed toward base, rounded and smooth at apex.——Honshu (Iwaki, Rikuchu Prov. and elsewhere).

Arisaema boreale Nakai. KOSHIMA-TENNAN-SHŌ. Very closely resembles *A. speirophyllum* and differs from it in the smaller habit and smaller number of broadly elliptic leaflets.——Hokkaido (Koshima in Oshima Prov.).

30. **Arisaema peninsulae** Nakai. *A. angustatum* var. *peninsulae* Nakai ex Miyabe & Kudo——KŌRAI-TENNAN-SHŌ. Tuber depressed-globose, 15–50 mm. in diameter; basal scalelike leaves 3, membranous; normal leaves 2, the petioles ascending, 3–16 cm. long; leaflets 5–14, the median one on a petiolule 1–4 cm. long, oblong, ovate, obovate, or elliptic, acuminate, green (often white-variegated in forma *variegatum* T. Koyama); peduncle 6–20 cm. long; spathe green with longitudinal white stripes, the tube about 5 cm. long, the blade ovate, subabruptly narrowed toward apex, straight, forwardly ascending; spadix-appendage linear or cylindric, erect or slightly bent forward at apex.——Hokkaido, Honshu, Kyushu (Tsushima).——Korea, and Manchuria.

31. **Arisaema angustatum** Fr. & Sav. HOSOBA-TENNAN-SHŌ. Tuber depressed-globose; pseudostem 20–80 cm. long; leaves 2, the lower one larger, the petiole 5–15 cm. long; leaflets 9–18, lanceolate, linear-lanceolate, or oblong, acuminate, serrulate or entire (forma *integrum* Nakai); scape 5–15 cm. long; spathe green or purplish above, upper margins of the tube recurved, the blade deltoid-ovate, 5–7 cm. long, subabruptly acute, tip obliquely declined over the spadix; spadix-appendage linear, rounded, about 2 mm. in diameter at apex, to 6 mm. in diameter at base.——May–June. Honshu (Kantō and n. distr. on the Pacific side).

32. **Arisaema koshikiense** Nakai. KOSHIKIJIMA-TENNAN-SHŌ. Tuber depressed-globose, about 4 cm. across; pseudostem 20–40 cm. long; leaves 2, the lower one larger, 11- to 13-foliolate, with an erect spreading petiole about 10 cm. long, the upper one smaller, with a petiole about 5 cm. long, 5- to 9-foliolate; leaflets narrowly oblong or narrowly lanceolate, entire; peduncle 6–17 cm. long, longer than the leaves; spathe-tube 6–6.5 cm. long, 1–1.5 cm. in diameter, purplish, with longitudinal pale stripes, the spathe blade broadly ovate or deltoid-ovate, acuminate, declined over the spadix, purple to dark purple with longitudinal white stripes, 3–5 cm. wide; spadix-appendage narrowly cylindric.——Mar. Kyushu (Satsuma, Ōsumi, Hizen Prov. and elsewhere).

33. **Arisaema solenochlamys** Nakai ex F. Maekawa. YAMAJI-NO-TENNAN-SHŌ. Differing from No. 28 in the solitary not tufted stem without lateral young shoots, and from No. 31 by the thick spadix-appendage to 8 mm. across at base and the tubular part of the spathe longer; resembles also No. 32, but the leaflets usually minutely toothed on the margin.——May–June. Honshu (Ugo, Shinano, Musashi, Shimotsuke Prov.).

34. **Arisaema nikkoense** Nakai. YUMOTO-MAMUSHIGUSA, YAMAGONNYAKU. Tuber depressed-globose; pseudostem 15–20 cm. long, to 1 cm. thick; leaves 2, the petioles 5–10 cm. long; leaflets 5, nearly palmate, oblong to narrowly so, usually dentate, acuminate, the terminal one 8–12 cm. long, 2–4 cm.

wide; peduncle 10–20 cm. long, longer than the leaves; spathe-tube ampliate above, upper margins slightly recurved, the spathe-blade ovate to narrowly so, 6–8 cm. long, declined over the spadix, greenish or partly purplish; spadix-appendage thick-cylindric or clavate, slightly thickened at tip, truncate at base.——May–July. Woods; Honshu (Kantō Distr.)——Related to *A. robustum*.

Arisaema alpestre Nakai. TAKANE-TENNAN-SHŌ. Possibly conspecific with *A. nikkoense*, said to differ in the lanceolate, entire leaflets and reddish to purplish scalelike basal leaves.——Honshu (Mount Yatsugatake).

35. **Arisaema stenophyllum** Nakai & F. Maekawa. HAUCHIWA-TENNAN-SHŌ. Tuber depressed-globose, 15–30 mm. in diameter; basal scalelike leaves 4 or 5, crimson or light purplish; normal leaves 2, palmate-pedate, 7–11 cm. long; leaflets 5–7, narrowly oblong, entire or serrate, long-acuminate; peduncles longer than the leaves; spathe with a cylindric tube 3.8–4 cm. long, with large auricles on the recurved upper margins of the tube, the blade bent forward and arching over the spadix, elongate- to broadly ovate, acuminate, dark purple with longitudinal white stripes; spadix-appendage thick, clavate, whitish, or with purplish spots.——May–June. Honshu (Mount Hakone).——Allied to *A. limbatum*.

36. **Arisaema undulatifolium** Nakai. NAGABA-MAMUSHIGUSA. Tuber depressed-globose or hemispherical, sometimes with 1–3 small tuberlets; basal scalelike leaves 3; normal leaves 2, palmate-pedate, the petiole terete, 3–10 cm. long; leaflets 8–14, sometimes the middle one short-petioluled, linear-lanceolate, 3–25 cm. long, 5–40 mm. wide, declined at apex in vigorous specimens, with a broad white line along the midrib, the margin prominently undulate, entire, or serrulate (forma *serrulatum* Nakai); peduncle longer than the leaves; spathe with the tube 3–4 cm. long, the blade ovate, bent down, recurved on the upper margins, reddish or brownish purple with longitudinal white stripes, auricles narrow, not prominent.——Apr. Honshu (Idzu).

37. **Arisaema takedae** Makino. *A. amplissimum*, sensu auct. Japon., non Bl.——Ō-MAMUSHIGUSA. Tuber depressed-globose; pseudostem 20–80 cm. long; leaves 2, the lower one larger, with a petiole 8–11 cm. long; leaflets 11–17, arranged on 2 spirally curved rachises, ovate-oblong, the terminal one 12–20 cm. long, 5–10 cm. wide; peduncle longer or shorter than the petiole; spathe with the tube ampliate above, recurved on upper margin, the blade large, broadly deltoid-ovate, declined over the spadix, 8–12 cm. long, abruptly acuminate, dark purple; spadix-appendage thick, cylindric, rounded at apex, slightly wrinkled, 10–15 mm. in diameter, furrowed longitudinally in lower part, truncate at base.——May–July. Hokkaido, Honshu (n. and centr. distr. as far westw. as Kinki Distr.).

38. **Arisaema limbatum** Nakai & F. Maekawa. MIMIGATA-TENNAN-SHŌ. Tuber depressed-globose; pseudostem 20–60 cm. long; leaves 2, the petioles 7–20 cm. long; leaflets 7–11, oblong to narrowly so, abruptly acuminate, the terminal one 10–20 cm. long, 2.5–6 cm. wide; peduncles 10–15 cm. long; spathe with the upper margin of the tube broadly auricled, the blade ovate, acuminate, bent forward and arching over the spadix, dark purple with longitudinal white stripes; spadix-appendage cylindric, slightly thickened at both ends, rounded at apex, truncate at base.——Apr.–May. Honshu (Kantō Distr. and northw., incl. Ōshima in Izu).

Arisaema aequinoctiale Nakai & F. Maekawa. KIYOZUMI-

TENNAN-SHŌ, HIGAN-MAMUSHIGUSA.   Closely related to *A. limbatum*. A stout plant with elliptic leaflets gradually attenuate below and occasionally strongly toothed on margin. ——Honshu (Boso Penin.).

39.   Arisaema mayebarae Nakai.   HITO-YOSHI-TENNAN-SHŌ.   Tuber depressed-globose, about 3 cm. across; pseudostem about 30 cm. long; basal scalelike leaves 4, normal leaves 2, the lower one larger, with a petiole about 10 cm. long, the upper one smaller, with a petiole 4 cm. long; leaflets about 5, oblong, the terminal one petioluled, 11 cm. long; peduncle about 12 cm. long; spathe with the tube about 9 cm. long, 16 mm. across at base, ampliate above, the blade ovate to broadly so, bent nearly at right angles or more downward, about 8.5 cm. long, nearly as long as the tubular part, acuminate, dark purple; spadix-appendage clavate, 6 cm. long, purplish, thickened above, about 7 mm. in diameter at the verrucose apex.——Kyushu. Allied to *Arisaema serratum*, but differs chiefly in the longer blade of the spathe bent forward horizontally.

40.   Arisaema serratum (Thunb.) Schott.   *Arum serratum* Thunb.; *Arisaema japonicum* var. *serratum* (Thunb.) Engl.; *A. capitellatum* Nakai; *A. niveum* Nakai; *A. niveum* var. *viridescens* Nakai——KANTŌ-MAMUSHI-SŌ, MURASAKI-MA-MUSHI-SŌ.   Tuber depressed-globose, without lateral tubers; pseudostem 30–60 cm. long, terete, dark purple-spotted; leaves 2, the lower one larger, with an ascending petiole 10–20(–30) cm. long; leaflets 7–13, the terminal oblong to elliptic, 10–20 cm. long, 3–8 cm. wide, acuminate; peduncle erect, 10–20 cm. long; spathe to 10 cm. long, the upper margin of the tube narrowly recurved, the blade arching over the spadix, often elongate, ovate to broadly so, 7–10 cm. long, acuminate, shorter than the tubular part, dark purple with white longitudinal stripes, the tip strongly declined over mouth; spadix-appendage stout, cylindric to clavate, truncate at base, usually thickened in upper part, globose, apex 6–10 mm. in diameter, wrinkled. ——May–June. Honshu (Kantō, Shinano Prov.); very polymorphic.——Forma **capitellatum** (Nakai) T. Koyama.   *A. capitellatum* Nakai——NOBIDOME-TENNAN-SHŌ.   Globose en-largement of spadix-appendage to 12 mm. across.

41.   Arisaema hachijoense Nakai.   *A. japonicum* sensu Matsumura, pro parte, non Bl.; *A. serratum* forma *blumei* Makino, pro parte, excl. syn.——HACHIJŌ-TENNAN-SHŌ.   Tuber globose, not depressed; pseudostem 10–30 cm. long; basal scales 3–4; normal leaves 2; leaflets 5–9, the terminal oblong to elliptic, 4–14 cm. long, 2–5 cm. wide, entire; peduncle as long as the leaves, 7–11 cm. long; spathe erect, with white longitudinal stripes, the tube 3–6 cm. long, 18–22 mm. in diameter, margins on the upper part recurved, the blade ovate-oblong, 7–10 cm. long, acuminate, longitudinally white-striped; spadix-appendage gradually and slightly narrowed toward the apex.——Mar.–Apr.   Honshu (Izu Isl.).

42.   Arisaema japonicum Bl.   *A. pseudojaponicum* Nakai; *A. serratum* forma *blumei* Makino and forma *japonicum* Makino, pro parte.——MAMUSHI-SŌ, JATŌ-SŌ, CHINZEI-TENNAN-SHŌ.   Tuber depressed-globose, with few lateral tubers; pseudostem 30–50 cm. long; leaves 2, the lower one larger, the petiole nearly erect, 10–25 cm. long; terminal leaflet oblong to broadly lanceolate, 10–30 cm. long, 3–8 cm. wide, abruptly acuminate; peduncle 10–30 cm. long; spathe with the upper margin of the tube narrow, recurved, the blade narrowly ovate or oblong, 6–9 cm. long, obliquely bent downward, straight, long-acuminate, dark purple (forma **japonicum**) or green (forma **viridans** T. Koyama), longitudinally white-striped; spadix-appendage cylindric, 1–3(–5) mm. in diameter, not thickened above, smooth, slightly thickened and truncate at base.——Apr.–May.   Honshu (Echizen, Mino, Ōmi Prov. and westw.), Shikoku, Kyushu.

Var. **brachyspatha** T. Koyama.   TŌKAI-JATŌ-SŌ.   Spathe-blade considerably shorter, bent forward horizontally, abruptly narrowed at apex, short-acuminate, otherwise with the characters of the typical phase.——Honshu (Suruga, Mikawa, Owari Prov. and elsewhere).

Var. **akitense** Nakai.   YAMASE-MAMUSHI-SŌ.   Leaflets larger, oval to obovate-oval, to 10 cm. wide, otherwise with the characters of the typical phase.——Honshu (Ugo Prov.).

4.   TYPHONIUM   Schott   RYŪKYŪ-HANGE ZOKU

Perennial herbs with tubers; leaves palmately 3- to 5-parted, petiolate; tubular part of the spathe short, persistent, constricted at the mouth, the blade narrowly ovate to linear, deciduous; spadix exserted from the spathe, neuter between the pistillate and staminate parts, the appendage elongate, smooth; ovary 1-locular, with 1 or 2 orthotropous erect ovules; stigma sessile; berries ovoid, 1- or 2-seeded; endosperm present.——More than 20 species, in tropical regions of the Old World.

1.   Typhonium divaricatum (L.) Decne.   *Arum divaricatum* L.; *A. trilobatum* Thunb.; *T. trilobatum* (Thunb.) Masam.——RYŪKYŪ-HANGE.   Tuber nearly globose; leaves radical, on slender petioles 10–20 cm. long, broadly cordate-sagittate, 3-lobed, 5–10 cm. long and as wide, the lobes acute, entire, the middle one broadly ovate, the lateral smaller, divaricate; scape radical, 4–8 cm. long; spathe 2.5–3 cm. long, the tube narrowly ovate, 15–18 mm. long, the blade elongate-deltoid, 8–12 cm. long, dark purple, gradually broadened below and rounded at base, gradually narrowed above to a subacute apex; neuter flowers linear, the upper ones ascending, the others reflexed, the appendage of spadix linear, erect, purplish, nearly uniformly wide from base to acute apex, somewhat obliquely truncate at base.——May–Aug.   Kyushu (s. distr.).——Ryukyus, Formosa, India, Malaysia, and s. China.

5.   ALOCASIA   Neck.   KUWAZU-IMO ZOKU

Perennials; stem short, hypogeous or epigeous; leaves sheathing at base, long-petioled, peltate at least when young, ovate-cordate, simple, entire or sometimes pinnately parted, with ovate segments; peduncles axillary, several; spathe with oblong persistent tube shorter than the deciduous, sometimes cucullate blade; spadix with the neuter part slender, the appendage thick-cylindric, not longer than the spathe, continuous with the staminate part; stamens 3–8; ovary 1-locular, with a very short style, the stigma 3- or 4-lobed; ovules several, erect, orthotropous; berries usually reddish, the seeds almost globose, the endosperm copious.——Rather numerous species in the Tropics of Asia.

1. Alocasia macrorrhiza (L.) Schott. *Arum macrorrhizon* L.; *Colocasia macrorrhiza* (L.) R. Br.; *Arum peregrinum* L.——KUWAZU-IMO. Stems epigeous, stout, terete, sometimes more than 1 m. tall; leaves bright green, on a long stout sheathing petiole, the blade large, broadly ovate, to 60 cm. long, 50–60 cm. wide, acute, entire, cordate-sagittate at base, nearly peltate, with the basal sinus of lower margins slightly connate at base, the lateral nerves 9–12 pairs, the lowest pair nearly straight-reflexed; peduncles 15–20 cm. long; spathe greenish, the tube light green, oblong, 4–8 cm. long, the blade oblong, concave, acute, longer than the tube; spadix nearly as long as the spathe, the appendage narrowly conical. ——Woods in low mountains; Shikoku (s. distr.), Kyushu (s. distr.).——Formosa, s. China, se. Asia, and Australia.

6. CALLA L.   HIME-KAIU ZOKU

Rather soft perennial herbs of swampy places, with long-creeping rhizomes; leaves with long-sheathing petiole, the blades cordate, entire, rounded to mucronate at apex; scape radical; spathe more or less cuplike, without tubular part, the limb broadly ovate to ovate-oblong, abruptly acute, abruptly narrowed and more or less decurrent at base, white at anthesis, greenish in fruit, persistent; spadix peduncled, cylindric; flowers mostly bisexual, the upper ones staminate, without a perianth; stamens about 6, with broad flat filament; ovary 1-locular, with 6–9 anatropous ovules; stigma sessile; berries globose, the seeds several, oblong, with abundant endosperm.——One widely distributed species in the boreal zone of the N. Hemisphere.

1. Calla palustris L. HIME-KAIU, MIZU-ZAZEN, MIZU-IMO. Rhizomes stout, 1–1.5 cm. in diameter, decumbent; leaves on a long terete petiole 10–25 cm. long, cordate at base, entire, 7–12 cm. long and as wide, nerves slender, many, inconspicuous, curved inward at tip; scapes 15–30 cm. long; spathe 4–6 cm. long, 3–5 cm. wide, rather flat, ending in a mucro to 7 mm. long; spadix 1.5–3 cm. long, narrowly oblong at anthesis, to 5 cm. long, and ellipsoidal in fruit, densely flowered, the peduncle 7–12 mm. long.——Shallow water and swampy places around ponds in moors; Hokkaido, Honshu (n. distr.).——Europe, Siberia, n. Asia, and N. America.

7. LYSICHITON Schott   MIZU-BASHŌ ZOKU

Rather soft, large, glabrous perennials with stout erect rhizomes; leaves several, radical, tufted, rosulate, the petiole spongy, nearly planoconvex in cross section, the blade large, simple, entire, elliptic to narrowly oblong, subobtuse, decurrent at base, the midrib thick, the lateral nerves curved inward at tip, not reaching the margin; scape elongate after anthesis; spathe withered in fruit, the basal part elongate, tightly enclosing the peduncle, inrolled but not connate on margin, the blade ovate, abruptly acuminate; spadix without appendage, cylindric, densely many-flowered; flowers bisexual; tepals 4, narrowly oblong, incurved below the apex; stamens 4, the filaments somewhat flat; ovary conical-ovoid, completely or incompletely 2-locular; stigma sessile; ovules 1 or 2 to each locule; berries 2-seeded; endosperm absent.——Two species, 1 in N. America and another in e. Asia.

1. Lysichiton camtschatcense (L.) Schott. *Dracontium camtschatcense* L.; *Pothos camtschaticus* Spreng.; *Arctiodracon japonicum* A. Gray; *A. camtschaticus* (Spreng.) A. Gray—MIZU-BASHŌ. Large, soft perennial with thick, white, short, erect rhizomes; leaves elongate after anthesis, 40–80 cm. long, 15–30 cm. wide, elliptic to narrowly oblong, entire, gradually narrowed at base to a flat petiole, the lateral nerves very slender; scape 10–30 cm. long, tightly covered by the narrow-tubular basal part of the spathe; spathe-blade white, elliptic to ovate, 8–12 cm. long, abruptly acute; spadix terete, cylindric, 4–8 cm. long at anthesis, oblong to short-cylindric, to 12 cm. long and 5 cm. across in fruit; flowers bisexual, perianth-segments small, rounded at apex; anthers yellow, ovoid.—— May–July. Common in swampy places; Hokkaido, Honshu (centr. and n. distr.).——Kuriles, Kamchatka, Sakhalin, and Ussuri.

8. SYMPLOCARPUS Salisb.   ZAZEN-SŌ ZOKU

Large stout perennials with very stout erect rhizomes; leaves radical, rosulate, tufted, the petioles thick, long-sheathing at base, the blade cordate to ovate-elliptic, abruptly decurrent on the petiole, with several pairs of lateral nerves, incurved above; scape short at anthesis, elongate but shorter than the leaves in fruit; spathe navicular-hemispherical, thick and fleshy, dark purple or purplish brown, rarely green, acute, without a basal tubular part; spadix ellipsoidal, distinctly peduncled, much smaller than the spathe, densely flowered, without an appendage; flowers bisexual; perianth-segments 4, thick, incurved at tip; stamens 4, the filaments flattened; ovary 1-locular, 1-ovuled, elongate; style conical, stigma small.——Two or three species, e. Asia and e. N. America.

1A. Leaf-blades orbicular-cordate, acute; flowering before the leaves; fruits ripening in summer of the same year. . . . . 1. *S. renifolius*
1B. Leaf-blades ovate to narrowly ovate, obtuse; flowering after the leaves; fruits ripening the second spring. . . . . 2. *S. nipponicus*

1. Symplocarpus renifolius Schott. *Symplocarpus foetidus* sensu auct. Asiat., non Salisb. ex Nutt.; *Spathyema foetida* forma *latissima* Makino; *Symplocarpus foetidus* forma *latissimus* Makino in syn.; *S. foetidus* var. *latissimus* (Makino) Hara——ZAZEN-SŌ. Large, fleshy, glabrous, foetid perennial; leaves tufted, long-petiolate, the blade orbicular, acute, deeply cordate at base, 30–40 cm. long and as wide; flowering before the leaves, the peduncle short; spathe 8–20 cm. long, 5–12 cm. wide, purplish to green, navicular-hemispherical gradually acuminate; spadix ellipsoidal, about 2 cm. long, densely flowered; fruit ripening in summer of the same year.——Wet places; Hokkaido, Honshu (centr. and n. distr.). ——Amur, Sakhalin, and Ussuri.

2. **Symplocarpus nipponicus** Makino. *Spathyema nipponica* (Makino) Makino——Hime-zazen-sō. Resembles No. 1 but smaller; leaves several, rather long-petioled, the blades narrowly ovate to narrowly ovate-oblong, 10–20 cm. long, 7–12 cm. wide, obtuse, cordate or nearly so, usually green, rarely variegated (forma **variegata** T. Koyama. Fuiri-hime-zazen-sō); flowering after the leaves; spathe broadly elliptic, tinged with dark brown-purple; fruit ripening the second spring.——July. Wet places; Hokkaido, Honshu (Chūgoku Distr. and eastw.).——n. Korea.

9. ACORUS L.　Shōbu Zoku

Perennials with decumbent, rather thick, aromatic, branching rhizomes; leaves tufted, linear, equitant, ensiform, not narrowed below, sheathing and surrounding the neck of the rhizome at base; scapes radical, erect, elongate; spathe consisting of a leaflike sessile bract subtending the spadix and apparently continuous with the stem or peduncle; spadix cylindrical without an appendage; flowers bisexual; perianth-segments 6, 2-ranked, small, thick, incurved and truncate at apex; stamens 6, the filaments linear; ovary oblong, 2- or 3-locular, with several ovules in each locule; style very short, the stigma small; berries reddish, the seeds few, oblong, with a fleshy endosperm.——Two species in temperate and warmer parts of Eurasia and N. America.

1A. Leaves with a distinct midrib; spadix 4–7 cm. long, 6–10 mm. thick; plant robust, large. . . . . 1. *A. calamus*
1B. Leaves without a midrib; spadix 5–10 cm. long, 3–5 mm. thick; plant slightly smaller. . . . . 2. *A. gramineus*

1. **Acorus calamus** L. *A. calamus* var. *angustatus* Bess.; *A. spurius* Schott; *Orontium cochinchinense* Lour.; *A. cochinchinensis* (Lour.) Schott; *A. asiaticus* Nakai——Shōbu. Rhizomes 8–12 mm. in diameter, creeping; leaves 50–80 cm. long, 6–15 mm. wide, flat, acuminate, smooth, deep green, ensiform, with a prominent raised midrib, scarcely narrowed below, the sheathes flattened laterally; peduncle leaflike, shorter than the leaves, slightly flattened; bract (spathe) leaflike, 20–40 cm. long, 5–8 mm. wide; spadix sessile, terete, cylindric, densely flowered, 4–7 cm. long, 6–10 mm. thick, ascending, light yellow; anthers light yellow, slightly longer than the perianth. ——May–Aug. Shallow water along rivers; Hokkaido, Honshu, Shikoku, Kyushu.——Siberia, e. Asia, N. America, and Europe (introduced).

2. **Acorus gramineus** Soland. Sekishō. Rhizomes creeping, 5–8 mm. in diameter, much branched; leaves deep green, linear-ensiform, gradually narrowed to an acuminate apex, 30–50 cm. long, 2–6 (rarely 8) mm. wide, shining, without a midrib, smooth; peduncle 10–30 cm. long, compressed-triangular, 3–5 mm. wide; bract (spathe) leaflike, rather short, 7–15 cm. long, 2–5 mm. wide, erect; spadix ascending to nearly erect, narrowly cylindric, terete, 5–10 cm. long, 3–4 mm. thick, densely flowered, yellow.——Apr.–May. Common in wet places along streams and around ponds; Honshu, Shikoku, Kyushu.——Formosa, China, and India.

Var. **pusillus** Engl. Arisugawa-zekishō, Kōrai-zekishō. Smaller in all respects. Much-prized as an ornamental pot-plant, with numerous cultivars. Introduced from China.

Fam. 45. LEMNACEAE　Uki-kusa Ka　Duckweed Family

Minute, free-floating, aquatic, stemless herbs; plant body a green scalelike or leaflike "frond" solitary or few in a group, with or without roots, new individuals produced from the lateral gemmae, or reproductive pouches, these overwintering in cold areas at bottom of ponds or ditches; inflorescence, when present, of 1 to several unisexual flowers, each with a single stamen or pistil; ovules 1 to several, erect or pendulous.——About 4 genera, with about 25 species, widespread in temperate and warm regions.

1A. Fronds with 1 or more roots; flowers marginal.
　2A. Roots 2 to several; fronds 5- to 12-nerved. . . . . 1. *Spirodela*
　2B. Roots single; fronds 1- to 5-nerved. . . . . 2. *Lemna*
1B. Fronds without roots, very minute; flowers on upper side of the frond at the center. . . . . 3. *Wolffia*

1. SPIRODELA Schleid.　Uki-kusa Zoku

Fronds flat, free-floating, 5- to 12-nerved; roots with a thin root-cap and a single vascular bundle; spathe saclike; anthers 2-locular; ovary with 2 anatropous ovules; fruit globose, winged on margin.——About 3 species, widely distributed in warmer and tropical regions.

1A. Roots about 10; fronds broadly obovate, producing winter buds in the form of bulblets. . . . . 1. *S. polyrhiza*
1B. Roots 3 to 5; fronds smaller, oblong to obovate, not forming the winter buds. . . . . 2. *S. oligorrhiza*

1. **Spirodela polyrhiza** (L.) Schleid. *Lemna polyrhiza* L.——Uki-kusa. Fronds flat, broadly obovate, 5–8 mm. long, 4–6 mm. wide, rounded to a very obtuse apex, purplish below, with 6–11 descending roots 3–5 cm. long, palmately 5- to 11-nerved, the lateral gemmae (daughter fronds) produced from the edges near the attachment of the root; autumnal fronds or winter buds orbicular, thickish; flowers globose, surrounded by a small spathe; inflorescence composed of 3 flowers, 2 staminate and 1 pistillate.——Common in ditches, ponds, and paddy fields; Hokkaido, Honshu, Shikoku, Kyushu.——Cosmopolitan.

2. **Spirodela oligorrhiza** (Kurz) Hegelm. *Lemna oligorrhiza* Kurz; *S. melanorhiza* F. Muell.; *S. pleiorhiza* F. Muell. ex Kurz——Shima-uki-kusa, Hime-uki-kusa. Fronds

membranous, narrowly obovate to oblong, palmately 3- to 5-nerved, 4–6 mm. long, 1.5–2.5 mm. wide, purplish beneath; roots 3–5; winter buds absent; plants growing throughout the year in the summer condition.——Shikoku, Kyushu (according to S. Miki).——Formosa, China, Malaysia, and Australia.

2. LEMNA L.   Ao-uki-kusa Zoku

Fronds small, flat, free-floating or submersed, 1- to 5-nerved; root 1, with a thin root-cap, without a vascular bundle; ovary with 1–6 orthotropous or anatropous ovules; fruit ovoid.——About 8 species, in warm-temperate and tropical regions.

1A.   Fronds submersed, with a slender stipe at one end, ovate-oblong, sagittate or truncate at base. . . . . 1. *L. trisulca*
1B.   Fronds floating, broadly obovate, not stiped. . . . . 2. *L. paucicostata*

1. Lemna trisulca L. Hinji-mo.   Fronds submersed, flat, green, 8–10 mm. long, 2–4 mm. wide, narrowly deltoid to ovate-oblong or broadly deltoid-lanceolate, rather obtuse to subacute, minutely serrulate on the upper margin, 3-nerved, truncate or rather sagittate and short-stiped at base, with 1 or few daughter fronds; root single, sometimes wanting.——Ponds, ditches, and still waters; Hokkaido, Honshu, Shikoku.——Europe, Asia, Africa, Australia, and N. America.

2. Lemna paucicostata Hegelm. Ao-uki-kusa.   Fronds floating, obliquely ovate-elliptic, 3–5 mm. long, 2–4 mm. wide, rounded at tip, rounded or very obtuse at base, entire, green, 3-nerved, sessile; root single, with an acute root-cap; ovules orthotropous, erect.——Common in ponds and paddy fields; Hokkaido, Honshu, Shikoku, Kyushu. Widely distributed in tropical regions.

Lemna minor L. Ko-uki-kusa is a cosmopolitan species allied to No. 2, differing in the obtuse root-cap, horizontal semianatropous ovules and fronds purplish beneath.——Reported from Hokkaido.

3. WOLFFIA Horkel   Mijinko-uki-kusa Zoku

Fronds minute, free-floating, globose to hemispherical or ovoid, rootless and nerveless, producing lateral gemmae from one edge of the frond; ovary with a single orthotropous ovule; fruit globose, smooth.——A few species, in tropical and temperate regions.

1. Wolffia arrhiza (L.) Wimmer. *Lemna arrhiza* L.; *L. globosa* Roxb.; *W. microscopica* sensu Miki, an etiam Kurz(?)——Mijinko-uki-kusa, Ko-tsubu-uki-kusa, Kona-uki-kusa.   Fronds ovoid-hemispherical, nearly flat above, convex beneath, more or less margined, about 0.7 mm. long, producing lateral gemmae from one end of the frond, without roots and nerves.——Honshu (vicinity of Tokyo).——Cosmopolitan.

Fam. 46. ERIOCAULACEAE   Hoshi-kusa Ka   Pipewort Family

(Contributed by Dr. Tetsuo Koyama.)

Herbs or semishrubby plants; stems short or nearly wanting; leaves linear, cauline or radical; peduncles surrounded below by a tubular bladeless sheath; inflorescence a terminal involucrate head; flowers unisexual; perianth segments arranged in 2 series, each consisting of 2 to 3 segments, the outer sepaloid, usually connate, the inner petaloid, always free or rarely minute to wanting; stamens in 2 or a single series; ovary superior, (1-) 2- or 3-locular, the ovule single in each locule, orthotropous, pendulous, the style solitary, 2- or 3-fid; capsule loculicidally dehiscent; embryo short, at the apical end of mealy albumen.——Eleven genera, with about 1,000 species, in both hemispheres.

1. ERIOCAULON L.   Hoshi-kusa Zoku

Herbs: stems usually very short; leaves radical, rosulate, linear; flowers dimerous or trimerous, monoecious, both sexes in the same head, subtended by a scalelike (floral) bract; staminate flowers: calyx of 2 to 3, free or connate sepals, the corolla tubular, distinctly or indistinctly 2- or 3-lobed at apex, the stamens in two series; pistillate flowers: sepals free or connate, the petals free, rarely none, bearing a small black gland at the apex.——About 250 species, in wet places in the tropical and temperate regions of both hemispheres; all of our species flowering in autumn.

1A.   Flowers wholly dimerous. . . . . 1. *E. decemflorum*
1B.   Flowers at least partially trimerous.
  2A.   Sepals of pistillate flowers 2–3, free.
    3A.   Pistillate flowers without petals; sepals 2; anthers white to gray.
      4A.   Bracts white, chartaceous, gradually narrowed above to a short awn; pistillate sepals winged on back.
                                                 2. *E. echinulatum* var. *seticuspe*
      4B.   Bracts hyaline, membranous, obtuse to subacute; pistillate sepals not winged on back. . . . . . . 3. *E. cinereum* var. *sieboldianum*
    3B.   Pistillate flowers petaliferous; sepals 2–3; anthers black.
      5A.   Involucral bracts much longer than the disc flowers; calyx white. . . . . 4. *E. zyotanii*
      5B.   Involucral bracts shorter than the disc flowers; calyx bluish black.
        6A.   Stem distinct, elongate; leaves narrow, 1-nerved. . . . . 5. *E. cauliferum*
        6B.   Stem almost wanting; leaves several- to many-nerved.

7A. Receptacle densely pilose; leaves 1–2 (–3) mm. wide, long-acuminate. .... 6. *E. parvum*
    7B. Receptacle glabrous; leaves 5–7 mm. wide, gradually narrowed above and obtuse at tip. .... 7. *E. senile*
2B. Sepals of pistillate flowers 3, connate into a 3-lobed spathe open on one side, bractlike.
    8A. Anthers white; peduncles 8- to 10-grooved; caudex obovoid, erect. .... 8. *E. heleocharoides*
    8B. Anthers blackish; peduncles 3- to 6-ribbed; caudex wanting or obconical and very short.
        9A. Heads of numerous flowers (see No. 9); involucral bracts usually longer than the disc flowers (except in Nos. 13–15); floral bracts whitish, glabrous or white-bearded on the upper margin; calyx whitish.
            10A. Petals of pistillate flowers wholly glabrous.
                11A. Heads few-flowered; involucral bracts few, much longer than the disc flowers; floral bracts and calyces minutely toothed on the upper margin. .... 9. *E. takae*
                11B. Heads many-flowered.
                    12A. Involucral bracts much exceeding the disc flowers, acute; calyces minutely toothed on the upper margin.
                                                            10. *E. omuranum*
                    12B. Involucral bracts slightly longer than the disc flowers, rather obtuse at apex; calyces entire-margined. 11. *E. japonicum*
            10B. Petals of pistillate flowers pilose inside.
                13A. Involucral bracts obtuse.
                    14A. Involucral bracts longer than the disc flowers; locules of ovary 2; stigmas 2 or sometimes 3. .... 12. *E. perplexum*
                    14B. Involucral bracts not exceeding the disc flowers; locules of ovary and stigmas 3.
                        15A. Petals of pistillate flowers glabrous at apex; receptacle glabrous. .... 13. *E. robustius*
                        15B. Petals of pistillate flowers white-bearded at apex; floral bracts and calyces usually also bearded on the upper margin; receptacle pilose.
                              16A. Calyx puberulous only on the upper margin; leaves broad. .... 14. *E. buergerianum*
                              16B. Calyx densely puberulous both on the upper margin and on the dorsal part of the lobes; leaves linear; heads densely white puberulent. .... 15. *E. nudicuspe*
            13B. Involucral bracts acuminate, conspicuously exceeding the disc flowers.
                17A. Petals and calyx of pistillate flowers perfectly glabrous; a relatively stout herb. .... 16. *E. hondoense*
                17B. Petals and calyx bearded on the upper margin; floral bracts also bearded on back.
                    18A. Involucral bracts lanceolate, about twice as long as the disc flowers; floral bracts of the pistillate flowers white-puberulous on the dorsal side; receptacle pilose. .... 17. *E. sikokianum*
                    18B. Involucral bracts 2 to 3 times longer than the disc flowers; floral bracts of pistillate flowers puberulous on the apical part outside; receptacle glabrous. .... 18. *E. miquelianum*
        9B. Heads few-flowered (except in No. 22); involucral bracts as long as or slightly shorter than the disc flowers (see Nos. 20–22); floral bracts blackish or whitish, usually glabrous, rarely loosely puberulous on the upper margin; calyx blackish, rarely white.
            19A. Calyx white, thin.
                20A. Heads 2–3 mm. long; receptacle glabrous. .... 19. *E. pallescens*
                20B. Heads 4–6 mm. long; receptacle pilose. .... 20. *E. dimorphoelytrum*
            19B. Calyx blackish; involucral bracts also often blackish.
                21A. Involucral bracts acute, exceeding the disc flowers.
                    22A. Floral bracts acute; staminate corolla lobes puberulous at apices. .... 21. *E. atroides*
                    22B. Floral bracts obtuse; staminate corolla lobes glabrous on margins. .... 22. *E. sekimotoi*
                21B. Involucral bracts rounded to obtuse, rarely the outer ones somewhat acute, as long as or slightly shorter than the disc flowers.
                    23A. Heads cyathiform, very few-flowered, higher than wide; very slender herbs with leaves about 1 mm. wide.
                      24A. Calyces of pistillate flowers glabrous outside.
                        25A. Calyx of pistillate flowers with 2 deltoid lobes; petals of pistillate flowers 2 or 3.
                                          23. *E. sachalinense* var. *kushiroense*
                      25B. Calyx of pistillate flowers 3-lobed; flowers wholly trimerous. .... 24. *E. nanellum*
                  24B. Calyces of pistillate flowers densely pilose outside. .... 25. *E. ozense*
              23B. Heads hemisphaerical, the height less than the width; leaves 2–5 mm. wide, several-nerved. .... 26. *E. atrum*

---

1. Eriocaulon decemflorum Maxim. var. **decemflorum.** KO-INU-NO-HIGE. Stemless annual 5–15 cm. high; leaves linear, gradually long-acuminate, yellowish green, 3- to 5-nerved; peduncles slender, longer than the leaves, 4-ribbed; heads turbinate, white to pale green, 3–4 mm. long, about 10-flowered; involucral bracts ovate-lanceolate, longer than the disc flowers; floral bracts white.——Wet places in lowlands and mountains; Honshu, Shikoku, Kyushu.——China (?).
Var. **nipponicum** (Maxim.) Nakai. *E. nipponicum* Maxim.——ITO-INU-NO-HIGE. Larger than the typical variety; peduncles to 30 cm. tall; heads to 5 mm. long; floral bracts, sepals and petals densely white-bearded.——Wet places in lowlands; Hokkaido, Honshu, Shikoku, Kyushu; quite common.——Korea and China.

2. Eriocaulon echinulatum Martius var. **seticuspe** (Ohwi) Ohwi. *E. seticuspe* Ohwi——HYŪGA-HOSHI-KUSA.

Stemless annual; leaves radical, erect-ascending, cinereous green, 2–4 cm. long, 2.5–3 mm. wide, 5- to 7-nerved; peduncles 8–10 cm. tall, 5-angled; heads globose, many-flowered, 4–5 mm. in diameter; involucral bracts many, ovate-deltoid, distinctly short-awned, about 2.5 mm. long including the awn; flowers wholly dimerous, the staminate apetalous with gray anthers, the pistillate with 2 free winged sepals, apetalous; stigmas 2, very rarely 3.——Kyushu (Hyuga Prov.); rare. The typical phase occurs in s. China, Indochina, and India.

3. Eriocaulon cinereum R. Br. var. **sieboldianum** (Sieb. & Zucc.) T. Koyama. *E. sexangulare* sensu auct. Japon., non L.; *E. heteranthum* Benth.; *E. formosanum* Hayata——HOSHI-KUSA. Stemless annual; leaves short, rosulate, narrowly linear, long-acuminate, 2.8–8 cm. long, 1–2 mm. wide; peduncles 5–15 cm. long, scarcely ribbed; heads ovoid-globose, 3–4 mm. long, 4 mm. across, sordid white or brownish white, densely many-flowered; involucral bracts about 1.5 mm. long, obtuse-tipped, slightly shorter than the disc flowers; floral bracts obovate-oblong, 1.5–2 mm. long, rather acute; staminate flowers about 2 mm. long, with spathiform calyx and tubular 3-lobed corolla; pistillate flowers without petals, the sepals 2, free, linear, blackish above the middle, pilose, the stigmas 3. ——Wet lowlands; Honshu, Shikoku, Kyushu.——Korea, Ryukyus, Formosa, China, Philippines, Malaysia, Indochina, India, and Africa. The typical phase is Australian.

4. **Eriocaulon zyotanii** Satake. IZU-NO-SHIMA-HOSHI-KUSA. Stemless; leaves linear, flat, gradually narrowed toward apex, 2.5–5 cm. long, 1.5–2 mm. wide, 5-nerved; peduncles 7–9 cm. tall, 4-ribbed; heads turbinate, 3–4 mm. across, light yellow, more than 10-flowered; involucral bracts several (5 or 6), lanceolate, acute, 4–6 mm. long, greenish, 3-nerved, exceeding the disc flowers; floral bracts obovate-oblong, sparsely puberulous on the upper margin, slightly shorter than the flowers; flowers wholly trimerous; staminate flowers to 2 mm. long, the sepals oblong, obtuse-tipped, connate below, sparsely ciliolate on the upper margin, the corolla tubular, trilobate at apex; pistillate flowers shortly pedicellate, 2 mm. long, the sepals oblong, acutish, whitish, nearly glabrous, the petals 3, free, oblong, pilose inside, the stigmas 3. ——Honshu (only in Kodzu-shima, Izu Prov.).

5. **Eriocaulon cauliferum** Makino. TAKANO-HOSHI-KUSA. Annual; stems erect, elongate, stout, terete, soft, 4–20 cm. long, 3.5–5 mm. thick, densely many-leaved; leaves subulate-linear, 1-nerved, 3–9 cm. long, 0.3 mm. wide; peduncles arising from the apex of the stem, 8–20 cm. long, slender, 6-ribbed; heads depressed-globose, 2–3 mm. long, 3–4 mm. across, bluish black, loosely puberulous; involucral bracts 7–9, hyaline, ovate or orbicular, 1 mm. long, blackish; receptacle glabrous; flowers many, trimerous; floral bracts obovate-cuneate, acuminate, nearly as long as the flowers; staminate flowers 1.5 mm. long, with a spathelike calyx deeply trilobate above and a tubular corolla; pistillate flowers 1 mm. long, pedicellate, the sepals 3, free, ovate-elliptic, rounded at apex, the petals 3, free, linear-oblanceolate, sparsely ciliate, the stigmas 3, longer than the style.——Honshu (only in Tatara-numa, Kozuke Prov.); very rare.

6. **Eriocaulon parvum** Koern. KURO-HOSHI-KUSA. Stemless or nearly so; leaves tufted, narrowly linear, long-attenuate above to a filiform point, 3- to 5-nerved, 4–10 cm. long, 1–3 mm. wide; peduncles slender, 10–20 cm. long, 5- or 6-ribbed; heads globose, 4–5 mm. in diameter, dark gray, white-puberulous; involucral bracts shorter than the disc flowers, broadly ovate, acute, 1–6 mm. long, to 1 mm. wide; receptacle pilose; floral bracts obovate-cuneate, acute, 2 mm. long, white-bearded along upper margin and on the dorsal side above; flowers trimerous, 1.5–1.8 mm. long; staminate flowers with a spathelike calyx shallowly trilobate at apex; pistillate flowers with 3 free oblong-ovate acute sepals and 3 free linear-oblong petals, the stigmas 3.——Honshu (Kantō, Etchu, and westw.), Shikoku, Kyushu; occasional.——s. Korea.

7. **Eriocaulon senile** Honda. GOMASHIO-HOSHI-KUSA. Acaulescent annual; leaves many, tufted, broadly linear, gradually narrowed to an obtuse tip, flat, 5–7 mm. wide, 9- to 13-nerved; peduncles many, 10–25 cm. long, 6-ribbed; heads globose to depressed-globose, 4–5 mm. long, 5–6 mm. across, blackish, loosely white-pilose; involucral bracts about 9, obovate-cuneate, obtuse or rounded, 2 mm. long, shorter than the disc flowers; flowers numerous, trimerous; floral bracts

obovate-oblong, obtuse, blackish, puberulent with 2-celled hairs along the upper margin and on the back; receptacle glabrous; staminate flowers 2 mm. long, with a spathelike trilobate calyx white-bearded at apex; pistillate flowers about 2 mm. long, the sepals 3, free, oblong, olive-black, acute, puberulent on the upper margin and on the back, the petals 3, free, lanceolate, pilose, with a black gland inside above, the stigmas 3, shorter than the style.——Wet lowlands; Honshu (Tōkaidō Distr. and westw.), Shikoku, Kyushu; relatively rare.

8. **Eriocaulon heleocharoides** Satake. KOSHIGAYA-HOSHI-KUSA. Caudex erect, obovoid, 1–1.5 cm. long, 7–10 mm. thick; leaves tufted, linear, long-attenuate, 7–15 cm. long, 3–4 mm. wide; peduncles numerous, 12–28 cm. long, very shallowly 9- or 10-grooved; heads obovoid-conical, 6–7 mm. long, 5 mm. across, nearly glabrous; involucral bracts about 14, broadly obovate, 2 mm. long and as wide, rather obtuse, shorter than the disc flowers; receptacle glabrous with numerous flowers; floral bracts acute, oblanceolate, nearly as long as the flower, puberulent with unicellular hairs along the upper margin; staminate flowers 2 mm. long with a spathelike minutely trilobed and glabrous calyx; pistillate flowers 2 mm. long, the calyx spathelike, glabrous, minutely 3-lobed, the petals 3, free, oblanceolate, pilose inside, the stigmas 3, slightly longer than the style.——Honshu (wet sandy banks of the Motoara River, Musashi Prov.); very local.

9. **Eriocaulon takae** Koidz. AZUMA-HOSHI-KUSA. Tufted acaulescent annual; leaves several, linear, 1–4.5 cm. long, 0.5–1.5 mm. wide, prominently acuminate, 3-nerved; peduncles 1–3, tufted, filiform, 8–15 cm. long, 4-ribbed; heads turbinate, about 3 mm. in diameter; involucral bracts 3–5, lanceolate-oblong, rather obtuse, minutely toothed on the upper margin, slightly longer than the disc flowers; receptacle glabrous; floral bracts ovate-oblong, acute, glabrous; flowers few, 2 mm. long; staminate flowers with a spathelike, glabrous calyx minutely toothed along the upper margin; pistillate flowers with spathelike indistinctly toothed calyx sparsely ciliate on the upper margin, the petals 3, free, lanceolate, with a black gland inside extending upward, the stigmas 3, longer than the style.——Honshu (Iwashiro Prov.).

10. **Eriocaulon omuranum** T. Koyama. SHINANO-INU-NO-HIGE. Slender acaulescent annual; leaves slenderly linear, few, 5–12 cm. long, to 1 mm. wide, long-acuminate, 4-nerved; peduncles 1–4, 10–15 cm. long, 4-ribbed; heads turbinate, 4.5–5.5 mm. long, 4–6 mm. across, the receptacle glabrous; involucral bracts 5 or 6, lanceolate, 2.5–6 mm. long, gradually acuminate, much exceeding the disc flowers; flowers 9–17 to a head; floral bracts lanceolate, hyaline, short-acute, slightly shorter than the flower; staminate flowers with a spathelike rounded to shallowly emarginate calyx; pistillate flowers 3.7 mm. long, the sepals connate into a hyaline oval quite glabrous spathe, minutely 3-toothed at apex, the petals 3, free, narrowly ovate, stipitate, glabrous, with a small gland inside extending upward, the stigmas 3, shorter than the style.——Honshu (Lake Shirakaba, Shinano Prov.).

11. **Eriocaulon japonicum** Koern. YAMATO-HOSHI-KUSA. Acaulescent annual; leaves tufted, linear, 6–12 cm. long, 2–3 mm. wide, gradually narrowed to an acute apex, 5- to 7-nerved; peduncles numerous, 12–16 cm. long, 4-ribbed; heads hemisphaerical, 3–4 mm. long, 5–6 mm. in diameter, wholly glabrous; involucral bracts about 20, 4–5 mm. long, 1.5–2 mm. wide, the outer lanceolate, rather obtuse, the inner small and acute, slightly longer than the disc flowers; receptacle gla-

brous; floral bracts hyaline, oblong, acute to acuminate, nearly glabrous; pistillate flowers almost 2 mm. long, the calyx glabrous, spathelike, trilobate, the petals 3, free, lanceolate, cuneate at base, the stigmas 3, almost as long as the style.——Honshu (Kadzusa Prov.); very local and scarce.

12. Eriocaulon perplexum Satake & Hara. EZO-INU-NO-HIGE. Acaulescent annual; leaves broadly linear, 2–6 cm. long, 1–3 mm. wide, 5- to 8-nerved; peduncles 5–14 cm. long, 4- or 5-ribbed; heads hemisphaerical, 3–6 mm. across; involucral bracts 6–8, obovate or oblong-obovate, obtuse-tipped, 1.2–2 mm. wide, 3–4 mm. long, white, slightly longer than the disc flowers; receptacle pilose; flowers many, both dimerous and trimerous together in the same head; staminate flowers 2.5 mm. long, with a bearded spathiform 2- or 3-lobed calyx; pistillate flowers with obovate, acute bracts pubescent along the upper margin, the calyx spathiform, long-ciliate and 2- or 3-lobed on the upper margin, the petals 2 or 3, oblanceolate, the stigmas 2 or 3, as long as to exceeding the style.——Hokkaido (Hidaka Prov.); local.

13. Eriocaulon robustius (Maxim.) Makino. *E. alpestre* var. *robustius* Maxim.——HIROHA-INU-NO-HIGE. Relatively robust, usually stemless annual or sometimes with an obconical caudex; leaves broadly linear, gradually attenuate to a rather obtuse apex, flat, 5–17 cm. long, 5–8(–10) mm. wide, 9- to 17-nerved; peduncles numerous, usually 5-ribbed; heads hemisphaerical or obconical-globose, 4 mm. long, 4–6 mm. in diameter; involucral bracts 10–12, ovate-oblong, 2–2.5 mm. long, 1.5 mm. wide, rounded at apex, shorter than the disc flowers; receptacle glabrous; flowers numerous, 1.5–2 mm. long; floral bracts obovate, obtuse-tipped; staminate calyx obovate, spathelike, trilobed at tip, obtuse, glabrous; pistillate flowers with obovate, spathelike, trilobed calyx, nearly glabrous inside, the petals 3, free, oblong-lanceolate, pilose inside, glabrous at apex, the stigmas 3, slightly shorter than the style.——Wet places and paddy fields in lowlands; Hokkaido, Honshu, Shikoku, Kyushu; quite common.——Korea, Manchuria, and e. Siberia. Forma **perpusillum** (Nakai) Satake. *E. alpestre* var. *perpusillum* Nakai; *E. robustius* var. *perpusillum* Nakai) Satake——CHABO-INU-NO-HIGE. A dwarf form.

Var. **nigrum** Satake. KURO-HIROHA-INU-NO-HIGE. Floral bracts and calyx dark brown, relatively small, otherwise almost as in the typical variety.——Honshu.

14. Eriocaulon buergerianum Koern. *E. pachypetalum* Hayata——Ō-HOSHI-KUSA. Rather large, stemless annual; leaves many, broadly linear, gradually tapering above toward an obtuse apex, 13- to 17-nerved, 8–20 cm. long, 4–8 mm. wide; peduncles numerous, densely tufted, 15–30 cm. long, 5- or 6-ribbed; heads hemisphaerical, 4–5 mm. high, 6 mm. in diameter, densely white-pilose; involucral bracts broadly obovate to orbicular-obovate, 2.5 mm. long, shorter than the disc flowers; receptacle pilose; floral bracts membranous, cuneate, 2.2 mm. long, 1.5 mm. wide, densely puberulent on the upper margin and on the back, the pistillate bracts deltoid at apex; flowers numerous; pistillate flowers with an elliptic spathelike calyx open on one side, hyaline, minutely 3-lobed at apex, ciliate, pilose inside, the petals 3, free, lanceolate, almost glabrous at the tips, the stigmas 3, as long as the style.——Paddy fields; Honshu (Kinki Distr. and westw.), Shikoku, Kyushu.——China, Ryukyus, and Formosa.

15. Eriocaulon nudicuspe Maxim. SHIRATAMA-HOSHI-KUSA, KONPEITŌ-SŌ. Stemless plant; leaves tufted, ascending, linear, 3–15 cm. long, less than 3 mm. wide, subulate at apex,

more or less flattened; peduncles 1–3(–5), erect, 20–40 cm. long, 4-ribbed, loosely twisted; heads globose, 6–8 mm. long, 5–7 mm. across, densely white-pilose; involucral bracts 6–10, broadly lanceolate, obtuse; floral bracts rhomboid-lanceolate, more or less deltoid above, obtuse, 3 mm. long, 1.5–2 mm. wide, with 2- or 3-celled hairs outside and on the upper margin; flowers trimerous, numerous; receptacle pilose; staminate flowers about 3 mm. long, with an obovate, hyaline spathe, truncate and 3-lobed at the densely villose apex; pistillate flowers slightly longer than the staminate, the calyx spathelike, ovate-rhomboid, the apex densely villose and 3-lobed, pilose inside and outside, the petals 3, free, oblanceolate, pilose inside, pubescent at the apex, the stigmas 3, almost as long as the style.——Wet places around springs in low hills; Honshu (Mikawa, Owari, Mino, and Ise Prov.); locally abundant.

16. Eriocaulon hondoense Satake. *E. miquelianum* sensu auct. Japon., pro parte, non Koern.; *E. hondoense* var. *stellatum* Satake——NIPPON-INU-NO-HIGE, HOSHIZAKI-INU-NO-HIGE. Relatively robust, stemless annual; leaves many, tufted, linear to broadly linear, 10–20 cm. long, 5–8 mm. wide, flat, gradually attenuate to a somewhat firm obtuse apex, about 13-nerved; peduncles numerous, 15–22 cm. tall, 5-ribbed, loosely twisted; heads broadly turbinate to somewhat obconical-globose, 6–8 mm. across, quite glabrous; outer involucral bracts 8 or 9, lanceolate, acute, 3-nerved, 7–9 mm. long or more, about 2 times longer (rarely to 3 times in forma **stellatum** (Satake) T. Koyama) than the disc flowers; receptacle glabrous; floral bracts narrowly obovate, shallowly navicular, shortly cuspidate at apex, as long as or slightly shorter than the flowers, glabrous or sparsely puberulent on the upper margin; pistillate flowers with a hyaline spathelike calyx, 3-lobed at apex, glabrous outside, pilose inside, glabrous or short-ciliolate on the upper margin, the petals lanceolate, cuneate at base, quite glabrous at apex, pilose inside, the stigmas 3, shorter than the style.——Wet lowlands; Hokkaido, Honshu, Shikoku, Kyushu; common.——Korea (Quelpaert Isl.).

17. Eriocaulon sikokianum Maxim. SHIRO-INU-NO-HIGE. Stemless annual; leaves linear, 12–18 cm. long, 3–5 mm. wide, gradually attenuate to a somewhat firm acute tip; peduncles numerous, 15–40 cm. tall, prominently 5- or 6-ribbed, twisted; heads hemisphaerical or turbinate, about 1 cm. across; involucral bracts about 12–14, broadly lanceolate to lanceolate, 5–8 mm. long, 2–2.5 mm. wide, 1.5 times longer than the disc flowers or shorter; floral bracts obovate, rather acute, puberulent, slightly shorter than the flowers; pistillate flowers with spathelike, 3-lobed, puberulent calyx, the petals 3, free, lanceolate, cuneate at base, rather acute and puberulent at apex, glabrous outside, pilose inside. A variable species.

Var. **sikokianum.** SHIRO-INU-NO-HIGE. Receptacle pilose; floral bracts and calyces rather densely white-puberulent on the upper margin; ovary 3-locular; stigmas 3.——Honshu, Shikoku, Kyushu.——Korea.

Var. **lutchuense** (Koidz.) Satake. *E. lutchuense* Koidz. ——OKINAWA-HOSHI-KUSA. Receptacle quite glabrous; floral bracts and calyces also not bearded; otherwise almost as in the typical variety.——Honshu (Shimotsuke Prov.), Kyushu. ——Ryukyus.

Var. **piliphorum** (Satake) Satake. *E. piliphorum* Satake ——NAGATO-HOSHI-KUSA. Receptacle pilose; some flowers with the ovary 2-locular and the stigmas 3.——Honshu (Nagato Prov.).

Var. **matsumurae** (Nakai) Satake. *E. matsumurae* Nakai ——MATSUMURA-INU-NO-HIGE, Ō-INU-NO-HIGE. Receptacle glabrous; some pistillate flowers with the ovary 2-locular and with 2 stigmas.——Honshu (Bitchu Prov.).

Var. **mikawanum** (Satake & T. Koyama) T. Koyama. *E. mikawanum* Satake & T. Koyama——MIKAWA-INU-NO-HIGE. Plants slender, small; receptacle pilose; pistillate flowers with unilocular ovary and single stigma.——Honshu (Tsukude Moor, Mikawa Prov.).

18. Eriocaulon miquelianum Koern. var. **miquelianum.** INU-NO-HIGE. Acaulescent annual; leaves linear, 6–20 cm. long, 3–4 mm. wide; peduncles numerous, exceeding the leaves, slender, twisted, 4- to 5-ribbed; heads turbinate, 6–10 mm. across; involucral bracts many, spreading, the outer lanceolate, acute, twice as long or much longer than the disc flowers, the inner ovate; receptacle usually glabrous; flowers numerous; floral bracts ovate-cuneate, nearly as long as the flowers, puberulent on the back above; staminate flowers 2.5 mm. long; pistillate flowers 2.5–3 mm. long, the calyx an oval 3-lobed spathe, white-puberulent on the upper margin, glabrous outside and pilose inside, the petals 3, free, ovate-lanceolate, narrowed at base, long-pilose inside, the stigmas 3, longer than the style.——Honshu, Shikoku, Kyushu.——China.

Var. **atrosepalum** Satake. TAKAYU-INU-NO-HIGE. Heads few-flowered; receptacle pilose; pistillate calyces blackish.——Honshu (Uzen).

Var. **monococcon** (Nakai) T. Koyama. *E. monococcon* Nakai——EZO-HOSHI-KUSA. Very slender; receptacle glabrous; pistillate flowers with unilocular ovary and single stigma.——Peaty soil; Hokkaido, Honshu (Yamato Prov.); scarce.

19. Eriocaulon pallescens (Nakai) Satake. *E. sachalinense* var. *pallescens* Nakai——SHIRO-EZO-HOSHI-KUSA. Slender, small, stemless annual; leaves narrowly linear, to 7 cm. long, 1 mm. wide, gradually narrowed to an acute apex, 1- to 3-nerved; peduncles few, 5–6 cm. tall, capillary, 3-ribbed; heads turbinate, 2–3 mm. long, 2 mm. across, few-flowered, the receptacle glabrous; involucral bracts 2–3, narrowly ovate, rather obtuse, 2.5–3 mm. long, 1.5 mm. wide; floral bracts elliptic, obtuse-tipped, ciliolate; pistillate flowers about 2 mm. long, the calyx pale, thin, deeply 3-lobed, white-pubescent on the upper margin, the petals 3, free, lanceolate, glabrous outside, pilose inside, the stigmas 3.——Hokkaido (Iburi Prov.).

20. Eriocaulon dimorphoelytrum T. Koyama. YUKI-INU-NO-HIGE. Small stemless annual; leaves few, slenderly linear, 1–4 cm. long, 0.3–1 mm. wide, acuminate; peduncles 1–3, 5–7 cm. long; heads lageniform-turbinate, 3.5–6 mm. long, 2.5–3.5 mm. wide, white, 3- to 6-flowered; involucral bracts pale white, ovate-lanceolate. 3.2–5.5 mm. long, cuspidate, exceeding the flowers; receptacle pilose; floral bracts oblong-elliptic, white, short-puberulent on back above; pistillate flowers 2.7–3 mm. long, the calyx white, thin, orbicular-oval, distinctly 3-lobed, white-pubescent on the upper margin, glabrous on both sides, the petals ovate-lanceolate, cuneate and short-stipitate at base, pilose inside, the stigmas 3.——Peat bogs; Honshu (Ozegahara Moor).

21. Eriocaulon atroides Satake. KURO-INU-NO-HIGE-MODOKI. Stemless annual; leaves few to several, linear, long-attenuate to an acute tip, 2–8 cm. long, 1–4 mm. wide, 3- to 5-nerved; peduncles slender, 10 cm. long or more, 4- to 5-ribbed, loosely twisted; heads short-turbinate, 3–4 mm. long, about 5 mm. across, 4- to 10-flowered, blackish, the receptacle pilose; involucral bracts 5 or 6, lanceolate or broadly so, 3.5–4 mm. long, 1–1.2 mm. wide, rather acute, exceeding the disc flowers; floral bracts lanceolate, rather acute, with 2-celled hairs along the upper margin; pistillate flowers to 2.5 mm. long, with an oval blackish hyaline spathelike calyx 3-toothed and white-puberulent at apex, pilose inside, the petals 3, free, lanceolate, pilose inside, the stigmas 3, as long as the style.——Honshu (Kodzuke, Shimotsuke, and Uzen Prov.).——Forma **nanum** Satake. OZE-INU-NO-HIGE. A small form with a lighter colored pistillate calyx and slightly shorter involucre.——Honshu (Ozegahara Moor).

22. Eriocaulon sekimotoi Honda. INU-NO-HIGE-MODOKI. Relatively robust stemless annual; leaves tufted, linear, 10–13 cm. long, 5–6 mm. wide, gradually attenuate to a somewhat firm obtuse tip, 9- to 11-nerved; peduncles numerous, 10–18 cm. long, twisted, 5-ribbed; heads turbinate, 5 mm. long, 6–7 mm. across, but to 1–1.2 cm. wide including the involucre; involucral bracts 10–12, spreading, lanceolate, about 2 mm. wide, acute, the outer 5–7 mm. long, 3-nerved, much exceeding the disc flowers; receptacle more or less pilose; floral bracts ovate-oblong, glabrous except for the ciliolate upper margin, blackish above; pistillate flowers 2.7–3 mm. long, the calyx a blackish spathe almost glabrous except pilose inside, 3-lobed, the petals 3, free, lanceolate, glabrous at apex, pilose inside, the stigmas 3, slightly longer than the style.——Honshu (Shimotsuke Prov.); local. Possibly this is a natural hybrid between *E. atrum* Nakai × *E. hondoense* Satake——Forma **glabrum** Satake. YASHŪ-INU-NO-HIGE. Receptacle almost glabrous; involucre shorter, nearly as long as the disc flowers. Occurs with the typical phase.

23. Eriocaulon sachalinense Miyabe & Nakai var. **kushiroense** (Miyabe & Kudo) T. Koyama. *E. kushiroense* Miyabe & Kudo——KUSHIRO-HOSHI-KUSA. Slender stemless annual; leaves narrowly linear, 5–7 cm. long, 2 mm. wide, gradually narrowed to an acute apex, 5-nerved; peduncles few to several, about 10 cm. long, soft, 3-ribbed; heads obovoid, 2 mm. long, 2–3 mm. across, few-flowered; involucral bracts 4 or 5, broadly elliptic to obovate, obtuse-tipped, nearly as long as the disc flowers, olive-black above; receptacle glabrous; pistillate flowers with obovate, glabrous, rounded bracts, the calyx an olive-black spathe, 2- or 3-lobed, ciliolate on the upper margin, glabrous outside, pilose inside, the petals 3, free, lanceolate, glabrous outside, pilose inside, the stigmas 2 or 3. ——Peat bogs; Hokkaido (Kushiro Prov.). The typical phase always has 2-lobed pistillate calyces and 2 stigmas.

24. Eriocaulon nanellum Ohwi var. **nanellum.** MI-YAMA-HINA-HOSHI-KUSA. Delicate stemless annual; leaves few, filiform, 1.5–3 cm. long, 1 mm. wide, 3- to 5-nerved; peduncles filiform, few, 2–8 cm. long; heads dark brown, about 2 mm. long, cyathiform, few-flowered; involucral bracts 3 or 4, oblong, obtuse or rounded at apex, 2 mm. long, 1 mm. wide; receptacle glabrous; floral bracts oblong, slightly shorter than the flowers, blackish; pistillate flowers with the calyx spathe-like, wholly glabrous except sometimes short-ciliolate on the upper margin, olive-black, irregularly 3- or sometimes 4-lobed, the petals 3, free, oblanceolate, glabrous, the stigmas 3, shorter than the style.——High mountains; Honshu (n. distr.).

Var. **albescens** Satake. SHIROBANA-MIYAMA-HINA-HOSHI-KUSA. Heads greenish, not blackish; pistillate calyces also greenish.——Honshu (Ugo Prov.).

Var. **filamentosum** (Satake) Satake. *E. filamentosum* Satake——ITO-HOSHI-KUSA. Staminate calyces deeply 3-fid, otherwise with the characters of the typical phase.——Honshu (n. distr.).

Var. **nosoriense** (Ohwi) Ohwi & T. Koyama. *E. nosoriense* Ohwi——NOSORI-HOSHI-KUSA. Pistillate flowers with irregular deeply 3-fid calyces ciliolate on margin, the petals sparsely pilose inside.——Honshu (Nosori-ike Moor, Kodzuke Prov.).

25. Eriocaulon ozense T. Koyama. HARA-INU-NO-HIGE. Slender, stemless annual; leaves 4–7, erect-spreading, narrowly linear, 1–6 cm. long, 0.3–1 mm. wide, gradually acuminate; peduncles few, 6–12 cm. long, slender, 4-ribbed; heads obovoid, 2.8–3.5 mm. long, 2.3–3 mm. wide, grayish white, few-flowered; involucral bracts 3–5, ovate-oval or ovate-elliptic, thin, whitish, deltoid-acute at apex; receptacle pilose; floral bracts oval to ovate-oval, thin, the upper ¼ blackish tinged and short-puberulent outside; pistillate flowers 2–2.3 mm. long, the calyx ovate-orbicular, spathelike, sharply 3-lobed at apex, densely pilose on both sides, the lobes quite glabrous and blackish, the petals ovate-rhomboid, densely pilose inside, rounded and white pubescent at tip, the stigmas 3, shorter than the style.——Sphagnum bogs; Honshu (Oze Moor, Kodzuke Prov.).

26. Eriocaulon atrum Nakai var. **atrum.** KURO-INU-NO-HIGE. Stemless annual; leaves linear, 2–10 cm. long, 2–4 mm. wide, acute, 7- to 9-nerved; peduncles several, 7–10 cm. long, soft, 4- or 5-ribbed; heads hemisphaerical, 2 mm. long, 4 mm. across; involucral bracts ovate-oblong, obtuse, about 2 mm. long, shorter than the disc flowers, usually the upper half blackish; receptacle pilose; floral bracts broadly obovate, 2 mm. long, rather obtuse, glabrous except minutely ciliate on the margin; pistillate flowers exceeding the bracts; calyx an olive-black 3-lobed spathe, glabrous outside, minutely ciliolate on the upper margin, pilose inside; petals 3, free, lanceolate, cuneate at base, subbifid at apex, long-pilose inside; stigmas 3, longer than the style.——Hokkaido, Honshu, Shikoku, Kyushu.—— Korea.

Var. **nakasimanum** (Satake) T. Koyama. *E. nakasimanum* Satake——TSUKUSHI-KURO-INU-NO-HIGE. Differs from the typical variety in the glabrous receptacle, not blackish involucre, and quite glabrous petals.——Kyushu.

Var. **intermedium** Nakai ex Satake. SAIKOKU-KURO-INU-NO-HIGE. A transitional phase between the typical variety and var. *nakasimanum*, distinguishable from the former by the not blackish involucral bracts.——Honshu (w. distr.) and Kyushu.

Var. **hananoegoense** (Masam.) T. Koyama. *E. hananoegoense* Masam.——YAKUSHIMA-HOSHI-KUSA. A dwarf phase of the typical variety known only on Yakushima (Kyushu).

Var. **glaberrimum** (Satake) T. Koyama. *E. glaberrimum* Satake——NEMURO-HOSHI-KUSA. Flowers of both sexes wholly glabrous except the pistillate petals which are pilose inside.——Hokkaido (Nemuro).

Fam. 47. COMMELINACEAE   TSUYU-KUSA KA   Spiderwort Family

Perennial or annual herbs; leaves flat, parallel-nerved, with a midrib, the base forming a membranous closed sheath about the stem; flowers bisexual, often ephemeral, actinomorphic or zygomorphic; perianth-segments of 2 whorls, the sepals 3, usually green, often persistent, the petals 3, free or connate below; stamens 6, in 2 series, often some of them sterile, the filament often bearded; ovary superior, 2- or 3-locular, the stigmas small; ovules few, on the inner angle of the locule, orthotropous; capsule loculicidal when dehiscent; seeds angled.——About 25 genera, with about 350 species, in all tropical regions, a few in temperate zones.

1A. Capsule not dehiscent, fragile; inflorescence a terminal paniculate cyme; leaves large. .... 1. *Pollia*
1B. Capsule dehiscent.
  2A. Androecium of 6 perfect stamens; stem twining. .... 2. *Streptolirion*
  2B. Androecium of 3 perfect and 1 to 3 sterile stamens.
    3A. Cyme surrounded by a spathe. .... 3. *Commelina*
    3B. Cyme not surrounded by a spathe. .... 4. *Aneilema*

1. POLLIA Thunb.   YABU-MYŌGA ZOKU

Large perennial herbs with flat, broadly lanceolate, alternate leaves on simple stems; cymes several to numerous in a terminal panicle; sepals 3, persistent; petals smaller, equal, white, obovate; stamens 6, all or only 3 perfect, the filaments glabrous, the anthers with 2 parallel locules; ovary 3-locular, the ovules few in each locule; capsules globose, berrylike, crustaceous and fragile. ——Fifty to sixty species, in e. and s. Asia and tropical Africa.

1. Pollia japonica Thunb. *Commelina japonica* Thunb.; *Aneilema japonicum* (Thunb.) Kunth; *P. japonica* var. *minor* Hayata; *P. minor* Honda——YABU-MYŌGA. Perennial; stem terete, erect from long-creeping slender rhizomes, 30–80 cm. long, rather stout; leaves about 10, loosely alternate, flat, narrowly oblong to broadly oblanceolate, 20–30 cm. long, 3–6 cm. wide, acute at both ends, scabrous above, sparsely puberulent beneath, the lower ones reduced to a sheath; peduncles erect, 10–30 cm. long, white-puberulent; inflorescence 5–15 cm. long, the pedicels about 4 mm. long, the bracts narrowly ovate, thinly membranous, acute; flowers white, about 7 mm. across, with 6 perfect stamens; fruit about 5 mm. in diameter.—— Aug.–Sept. Honshu (Kantō Distr. and westw.), Shikoku, Kyushu.——Ryukyus.

2. STREPTOLIRION Edgew.   AOI-KAZURA ZOKU

Slender scandent herbs; stem branched, much elongate; leaves long-petioled, the blade ovate-cordate, acute, with a tubular sheath at base; flowers few in a short cyme subtended by a small leaflike spathe; sepals oblong, free; petals linear; stamens 6, with hairy filaments and transversely fixed anthers; ovary 3-locular, the style slender, the ovules 2 in each locule; capsule 3-angled, ellipsoidal, prominently beaked; seeds tubercled and wrinkled.——Three species, in India and e. Asia.

1. Streptolirion volubile Edgew. Aoi-kazura. Nearly glabrous herb; leaves long-petioled, the petiole 3–6 cm. long, the sheath usually truncate at apex, ciliate, 1–2 cm. long, the blade 5–8 cm. long, 3–5 cm. wide, abruptly long-acuminate, deeply cordate at base, minutely ciliate on margin, sometimes with long hairs above; flowers 2 or 3, white, 5–6 mm. across; sepals 1- to 3-nerved, obtuse, about 4 mm. long; filaments curled-pubescent; capsule 8–11 mm. long including the terminal beak, glabrous.——Aug.–Oct. Honshu (Bitchu and Bingo Prov.).——China and India.

3. COMMELINA L. Tsuyu-kusa Zoku

Small to large herbs; stems usually procumbent and rooting at the nodes; leaves lanceolate to ovate, the petiole forming a sheath around the stem; cymes usually dichotomous, several-flowered, subtended by a leaflike bract folded once to form a large spathe; sepals 3, membranous, the lateral 2 connate at base; petals 3, the lateral 2 large, often clawed at base; perfect stamens 3, the staminodes 2 or 3; ovary 2- or 3-locular, 2 of the locules with 1 or 2 ovules, the other one, if present, sterile or with 1 ovule; seeds wrinkled or pitted.——About 180 species, chiefly in tropical regions, a few in warm-temperature areas.

1A. Spathe connate on lower margin.
  2A. Leaves ovate or broadly ovate, obtuse; capsule 5-seeded. .... 1. *C. benghalensis*
  2B. Leaves broadly lanceolate, acute; capsule 3-seeded. .... 2. *C. auriculata*
1B. Spathe not connate on margin; leaves lanceolate to narrowly lanceolate, acute to gradually acuminate at apex.
  3A. Spathe narrowly ovate, gradually tapering from the rounded base to an acute apex. .... 3. *C. diffusa*
  3B. Spathe broadly cordate, with a cordate base, rounded or nearly so at apex. .... 4. *C. communis*

1. Commelina benghalensis L. Maruba-tsuyu-kusa. A more or less pubescent annual; stem creeping or ascending, branching; leaves ovate to broadly ovate, 3–5 cm. long, 1.5–3 cm. wide, obtuse at apex, abruptly contracted at base; spathe obtriangular, more or less ampliate above, 1–1.5 cm. long and as wide, somewhat puberulent and sparsely pilose externally; flowers rather small, 8–10 mm. across, pale blue; seeds transversely short-wrinkled.——July–Oct. Honshu (s. Kantō Distr. and westw.), Shikoku, Kyushu.——Widely distributed in tropical regions of Asia and Africa.

2. Commelina auriculata Bl. *C. blumei* Dietr.——Hōrai-tsuyu-kusa. Nearly glabrous or sparsely pilose annual; stems creeping, branched below; leaves broadly lanceolate, acute, 3–6 cm. long, 1–2 cm. wide, more or less oblique at base; spathe obliquely obtriangular, usually more or less pilose, 7–10 mm. long and as wide; flowers blue; seeds nearly smooth but densely puberulent.——Aug. Kyushu (Yakushima)——s. Asia.

3. Commelina diffusa Burm. *C. nudiflora*, sensu Burm. f.; *C. ochreata* Schauer; *C. salicifolia* sensu Benth., non Roxb.——Shima-tsuyu-kusa. Nearly glabrous annual herb resembling the preceding species, but the spathe longer, the margins not connate in front; stem creeping, ascending above, branching below; leaves lanceolate to broadly lanceolate, 4–6 cm. long, 1–2 cm. wide, gradually acute to acuminate; spathe broad, rounded or shallowly cordate at base, gradually tapering above to a rather acute apex, 2–3 cm. long, 1.5–2 cm. wide when unfolded, glabrous; flowers blue, small; capsule 3-locular, 5-seeded; seeds elevate-reticulate.——Aug.–Oct. Reported to occur in Kyushu (Yakushima).——Ryukyus (Amami-Oshima and southw.), Formosa; widely distributed in the Tropics.

4. Commelina communis L. Tsuyu-kusa. Nearly glabrous green annual; stem creeping and branching below, ascending above, 15–50 cm. long; leaves soft but rather thick, ovate-lanceolate, abruptly acute at apex, subabruptly narrowed below to a membranous sheath with long-pubescent mouth, 5–7 cm. long, 1–2.5 cm. wide, glabrous; spathe broadly cordate, rounded or with a short cusp at apex, nearly 2 cm. long, glabrous or loosely pubescent outside; flowers blue, about 12 mm. across, the 2 lateral petals large and orbicular, clawed at base, the central one small and usually white.——June–Sept. Rather common in partly shaded grassy places, cultivated fields, and roadsides in lowlands and hills; Hokkaido, Honshu, Shikoku, Kyushu.——Sakhalin, Ussuri, Korea, Manchuria, China to Siberia, and the Caucasus.

Var. **ludens** (Miq.) C. B. Clarke. *C. ludens* Miq.; *C. communis* var. *angustifolia* Nakai——Hosoba-tsuyu-kusa. Leaves narrow, puberulent beneath; spathe also more or less narrow and puberulent.——Mountainous districts.

Cv. **Hortensis.** *C. communis* var. *hortensis* Makino——Ō-bōshibana. A cultivated phase with large showy flowers yielding a blue dye from the corolla.

4. ANEILEMA R. Br. Ibo-kusa Zoku

Soft, simple or branching herbs, often with a bulbous root; flowers in terminal and lateral paniculate inflorescences, usually ephemeral, subtended by nonspathaceous bracts and bracteoles; sepals 3, free, membranous; petals 3, obovate, unequal, or all equal in length and width; stamens 2 or 3, the filaments sometimes bearded, the anthers oblong; staminodes 2–4; ovary sessile, 2- or 3-locular, the ovules few to rather numerous in each locule.——About 60 species, chiefly in tropical regions.

1A. Annual herb without rosulate leaves at base; flowers solitary, rarely 2 or 3, axillary; sepals 4–6 mm. long; leaf-blades glabrous.
                                                       1. *A. keisak*
1B. Perennial herb with rosulate leaves at base; flowers several, on a one-sided, short-peduncled cyme; sepals 3–4 mm. long; leaves sparsely ciliate. .... 2. *A. nudiflorum*

1. Aneilema keisak Hassk. *Murdannia keisak* (Hassk.) Hand.-Mazz.; *A. japonicum* sensu Nakai, excl. syn.; *A. oliganthum* Fr. & Sav.——Ibo-kusa. Juicy, soft annual without rosulate leaves; stem 10–30 cm. long, decumbent and branching below, with a puberulent longitudinal stripe on one side; leaves lanceolate, more or less spreading, flat to slightly folded, gradually attenuate above to a rather obtuse apex, usually glabrous, the sheaths 5–10 mm. long, coarsely pubescent in front;

flower usually solitary, axillary, the peduncles 1.5–3 cm. long, usually with a linear bract sometimes bearing an additional flower in the axil, recurved in fruit; sepals lanceolate, acute; petals pink, slightly longer than the calyx; stamens 3, the filaments bearded at base; staminodes 3; capsule 10 mm. long, with several seeds in each locule.——Sept.–Oct. Common in wet places; Honshu, Shikoku, Kyushu.——China, Ryukyus, and e. N. America.

2. Aneilema nudiflorum R. Br. *Commelina nudiflora* L., pro parte; *Murdannia nudiflora* (R. Br.) Brenan; *A. malabaricum* sensu auct. Japon., non Merr.——SHIMA-IBO-KUSA. Perennial soft herb with rosulate radical leaves; leaves rather

thick, lanceolate to narrowly so, 4–7 cm. long, 4–10 mm. wide, acute, the sheaths 5–10 mm. long, coarsely pubescent in front as on the margin of the blade; peduncles terminal and lateral, ascending or decumbent, glabrous, 3–7 cm. long, sometimes with a single bract; cymes densely several-flowered, one-sided, glabrous, the pedicels slightly recurved, 3–4 mm. long; sepals elliptic, rounded at apex, 3–4 mm. long; petals slightly longer than the sepals, pale rose-purple; stamens 3, the filaments bearded; staminodes 3; capsule ellipsoidal, slightly longer than the sepals with 2 seeds in each locule, these about 1.2 mm. long. ——Aug.–Oct. Kyushu (Yakushima and Tanegashima).—— Ryukyus, China, Philippines, Malaysia, and India.

Fam. 48. PONTEDERIACEAE   MIZU-AOI KA   Pickerelweed Family

Aquatic or marsh herbs with or without rhizomes; leaves radical and cauline, broad, parallel-nerved; flowers bisexual, ephemeral, in a spike arising from the sheath of the uppermost leaf; perianth-segments 6, in 2 whorls, free, similar, usually withering away very early after anthesis; stamens 1–6, with basifixed or versatile anthers; ovary superior, 1- to 3-locular, with 3 parietal placentae, the style slender, terminated by an entire or 3-fid stigma; ovules anatropous; capsule membranous, loculicidally dehiscent into 3 valves; seeds small.——About 5 genera, with about 40 species, chiefly in the Tropics.

1. MONOCHORIA Presl   MIZU-AOI ZOKU

Aquatic or growing in wet places; leaves all radical, except one on the apical part of the stem, long-petioled, the blade ovate, lanceolate, linear, or sagittate; inflorescence a sessile or peduncled raceme, with a membranous sheath at the base of the peduncle; perianth of 6, nearly free segments, campanulate, withering away early after flowering; stamens 6, similar, or the central one of the inner series larger than the others, the filaments glabrous, with a tooth on one side at the base, the anthers basifixed; ovary 3-locular, with numerous ovules.——Several species, chiefly in the Old World Tropics.

1A. Stems 20–40 cm. long; leaves cordate; inflorescence many-flowered, long-peduncled, exceeding the uppermost leaf; flowers 2.5–3 cm. across. . . . . 1. *M. korsakovii*
1B. Stems shorter; leaves broadly lanceolate to deltoid-ovate, sometimes shallowly cordate at base; inflorescence few-flowered, sessile or nearly so, much shorter than the uppermost leaf. . . . . 2. *M. vaginalis* var. *plantaginea*

1. Monochoria korsakovii Regel & Maack. *M. vaginalis* var. *korsakovii* (Regel & Maack) Solms-Laub.——MIZU-AOI. Glabrous, soft, somewhat fleshy annual; stems 20–40 cm. long; radical leaves long-petioled, the blade cordate, reniform-cordate, or rarely ovate-cordate, 5–10 cm. long and as wide, very abruptly attenuate to an obtuse apex, the cauline leaf similar, but short-petioled; inflorescence terminal, rather many-flowered, finally elongate and 5–15 cm. long in fruit; flowers on obliquely ascending pedicels, 2.5–3 cm. across, purplish blue; perianth-segments elliptic, about 15 mm. long; anthers versatile, the larger one single, about 4 mm. long, light blue, the others about 3 mm. long and yellowish; style curved; capsules narrowly ovoid-conical, about 10 mm. long.——Sept.–Oct. Common in ponds and ditches; Hokkaido, Honshu, Shikoku, Kyushu.——Korea, Ussuri, and China.

2. Monochoria vaginalis (Burm. f.) Presl var. **plan-**taginea (Roxb.) Solms-Laub. *Pontederia plantaginea* Roxb.; *Monochoria plantaginea* (Roxb.) Kunth——KO-NAGI, SASA-NAGI, MIZU-NAGI. Similar to the preceding species but smaller; radical leaves long-petioled, the blade broadly lanceolate to deltoid-ovate, 3–7 cm. long, 1.5–3 cm. wide, rounded to shallowly cordate at base, subabruptly narrowed to an acute or obtuse apex, the cauline leaves on shorter petioles 3–10 cm. long; inflorescence much shorter than the leaves, one-sided, 3- to 7-flowered, abruptly deflexed at the base in fruit; flowers purplish blue, 1.5–2 cm. across; perianth-segments oblong; capsule ellipsoidal, about 10 mm. long; seeds ellipsoidal, about 1 mm. long, longitudinally ribbed and transversely lineolate.—— Sept.–Oct. Common in paddy fields and shallow ponds in lowlands; Honshu, Shikoku, Kyushu.——Korea and China. The typical phase occurs in se. Asia.

Fam. 49 PHILYDRACEAE   TANUKI-AYAME KA   Philydrum Family

Erect perennials with 2-ranked, laterally flattened, linear leaves; inflorescence a spike or panicle; flowers small, subtended by bracts, bisexual, zygomorphic; perianth-segments 4 in 2 series, persistent, petaloid; stamen solitary, arising from the base of the anterior perianth-segment, with a flat filament and straight or twisted anther; ovary superior, 3- or 1-locular with 3 parietal placentae, the ovules anatropous, numerous; style terminal, with an entire stigma; capsule loculicidally dehiscent into 3 valves; seeds numerous, minute.——Three genera with few species, in Asia, Australia, and the Pacific Islands.

1. PHILYDRUM Banks ex Gaertn.   TANUKI-AYAME ZOKU

Rather soft perennial herb; stamen not adnate to the perianth, with a twisted anther; ovary 1-locular.——Only one species known.

1. **Philydrum lanuginosum** Banks ex Gaertn. *Garciana cochinchinensis* Lour.——TANUKI-AYAME. Stems herbaceous, 50–100 cm. long, with long woolly white hairs especially on the inflorescence; leaves 2-ranked, linear, ensiform, gradually narrowed above, smooth, 30–70 cm. long, 1–2 cm. wide at base; scape terminal, usually simple, with several short leaves; spike erect, 20–50 cm. long, rather many-flowered, the bracts oblong, attenuate above, nearly erect, 2–6 cm. long; flowers 1 or 2 in axil of each bract, sessile, yellow; perianth-segments ovate, membranous, withering but persistent and enclosing the fruit, 15–18 mm. long; stamen about 2/3 as long as the perianth-segments; inner perianth segments membranous, glabrous, cuneate, rather acute, slightly shorter than the fertile stamens; capsule oblong, about 12 mm. long, with long white tomentose hairs, the seeds narrowly cocoon-shaped, 0.7 mm. long, minutely papillose on the upper half.——Aug.–Oct. Wet places in lowlands; Kyushu.——Ryukyus, Formosa, China, Indian, Malaysia, and Australia.

Fam. 50. **JUNCACEAE** IGUSA KA Rush Family

(Contributed by Dr. Tetsuo Koyama.)

Annual or perennial herbs, rarely shrubs; stems erect, usually simple; leaves terete, flat, ensiform or linear, sometimes scale-like, more or less sheathing at base; inflorescence corymbose, or cymose, sometimes capitate or reduced to a single flower; flowers small, usually with a bracteole at base, bisexual; perianth-segments 6, free, 2-ranked, persistent, glumaceous; stamens 3 or 6, with basifixed anthers; ovary superior, imperfectly to perfectly 3-locular, the stigmas 3, the ovules anatropous, ascending; capsule loculicidally dehiscent, 3-valved, the seeds few to numerous, with an erect embryo.——About 8 genera, with more than 300 species, widely distributed in both hemispheres.

1A. Leaves more or less pilose on the margin; flowers always bearing a bracteole at base; capsule 1-locular, 3-seeded. . . . . 1. *Luzula*
1B. Leaves glabrous; flowers sometimes with a bracteole at base; capsule 1- to 3-locular, many-seeded. . . . . 2. *Juncus*

1. **LUZULA** DC. SUZUME-NO-YARI ZOKU

Perennials; stems usually few-leaved; leaves basal and cauline, flat or sulcate above, the sheath cylindrical, not auricled at apex; inflorescence cymose, umbellike or capitate; flowers solitary on the pedicels or densely aggregated in a capitate cluster, always with a bracteole at base; perianth-segments 6, glumaceous, 2-ranked, similar; filaments linear or filiform, bearing oblong to linear anthers; ovary 1-locular, the style terminal, single, with a twisted 3-fid stigma; ovules 3, anatropous; capsule 3-valved and 3-seeded; seeds ovoid to ellipsoidal with or without a caruncle at one end, sometimes very sparsely pilose at base, the embryo small, the endosperm mealy.——About 60 species, cosmopolitan, in temperate and subarctic regions.

1A. Leaves narrow, 2–3 mm. wide, sulcate above, acute at apex; flowers solitary or few together on the slender, recurved ultimate branches of the inflorescence; seeds not caruncled but sparsely pilose at one end. . . . . 1. *L. wahlenbergii*
1B. Leaves terminated by a callose obtuse tip, flattened, not sulcate above.
  2A. Flowers solitary on the slender ultimate branches of the umbellike inflorescence; caruncle rather large; plants stoloniferous.
    3A. Perianth 3.5–4 mm. long; anthers broadly linear, as long as or longer than the filament. . . . . 2. *L. plumosa* var. *macrocarpa*
    3B. Perianth 2–3 mm. long; anthers oblong, conspicuously shorter than the filaments. . . . . 3. *L. rostrata*
  2B. Flowers sessile, more or less clustered and forming a capitate inflorescence or glomerule; caruncle small or absent; plants without stolons.
    4A. Caruncle conspicuous, 1/4–1/2 as long as the seed.
      5A. Inflorescence a single head, very rarely up to 3 heads. . . . . 4. *L. capitata*
      5B. Inflorescence of several, peduncled, small heads. . . . . 5. *L. multiflora*
    4B. Caruncle very short to nearly wanting.
      6A. Flowers brownish, dark brown, or black-brown; anthers nearly half as long as the filaments. . . . . 6. *L. oligantha*
      6B. Flowers tawny or greenish brown; anthers nearly as long as the filaments. . . . . 7. *L. pallescens*

1. **Luzula wahlenbergii** Rupr. *L. spadicea* var. *kunthii* E. Mey.; *L. spadicea* var. *wahlenbergii* (Rupr.) Buchen.; *? L. parviflora* var. *yezoensis* Satake——KUMOMA-SUZUME-NO-HIE. Rhizomes short; scape slender, declined above after flowering, 15–20 cm. long, 2- or 3-leaved; radical leaves narrowly linear, 5–10 cm. long, 2–3 mm. wide, acute, more or less plicate, sulcate above toward apex; inflorescence a terminal, nodding, compound cyme, with a short bract; flowers rather numerous, solitary or few on the ultimate branches of inflorescence, the bracteoles hyaline, ciliate; perianth-segments lanceolate, acute, equal in length, about 2 mm. long, dark reddish brown; stamens 2/3 as long as the perianth, the anthers oblong, shorter than the filaments; capsule nearly as long as the perianth, chestnut-brown.——July–Aug. Alpine, on dry, grassy slopes and rocky places; Hokkaido, Honshu (centr. distr.). ——Kuriles, Kamchatka, Siberia, Europe, and N. America.

2. **Luzula plumosa** E. Mey. var. **macrocarpa** (Buchen.) Ohwi. (?) *L. jimboi* var. *integra* Satake; *L. pilosa* sensu auct. Japon., non Willd.; *L. plumosa* sensu auct. Japon., non E. Mey.; *L. japonica* Buchen.; *L. rufescens* sensu auct. Japon., non Fisch.; (?) *Luzula rufescens* var. *brevipes* Fr. & Sav.; *L. brachycarpa* Satake; *L. rufescens* var. *macrocarpa* Buchen.—— NUKABŌSHI-SŌ, KURO-BOSHI-SŌ, MIYAMA-KUROBŌSHI-SŌ. A very polymorphic perennial, more or less tufted, often stoloniferous; radical leaves linear to broadly linear, 8–15 cm. long, 3–8 mm. wide, the cauline 2 or 3; inflorescence a terminal, rather loosely flowered, umbellate cyme; flowers solitary, pedicelled; perianth-segments lanceolate, acute, light green, partially red-brown, nearly equal in length, 3.5–4 mm. long; stamens slightly shorter than the perianth, the anthers broadly linear, as long as or longer than the filaments; capsule longer than the perianth, light yellow at maturity; seeds with a curved caruncle nearly as large as the body of the seed.——Apr.–July. Rather common in hills and mountains; Hokkaido, Honshu,

Shikoku, Kyushu.——China, Korea, and Amur. The typical phase, with slightly smaller capsules nearly as long as the perianth, occurs in India and sw. China.

3. Luzula rostrata Buchen.   Miyama-nuka-bōshi-sō. Very much like the preceding but more slender; leaves 3–6 mm. wide; flowers reddish to dark red-brown; perianth-segments lanceolate, acute, 2.5–3 mm. long; stamens 2/3 as long as the perianth-segments, the anthers oblong, conspicuously shorter than the filaments; capsule longer than the perianth; seeds with a caruncle as long as the body of the seed.——June–Aug. Alpine regions; Hokkaido, Honshu (n. and centr. distr.).

4. Luzula capitata (Miq.) Miq. ex Komar. L. campestris var. capitata Miq.——Suzume-no-yari, Suzume-no-hie. Tufted perennial; radical leaves broadly linear to linear, 7–15 cm. long, 2–6 mm. wide; stems 10–30 cm. long; inflorescence a capitate cluster of many flowers or sometimes composed of 2 to 3 peduncled clusters; flowers greenish to red-brown; perianth-segments 2.5–3 mm. long, broadly lanceolate, sharply acute; stamens 2/3 as long as the perianth-segments, the filaments very short, terminated by a narrowly oblong anther; caruncle half as long as the body of the seed.——Apr.–June. Quite common in grassy places in mountains and lowlands; Hokkaido, Honshu, Shikoku, Kyushu.——China, Korea, Sakhalin, e. Siberia, and Kamchatka.

5. Luzula multiflora (Retz.) Lej.   L. intermedia (Thuill.) Spenner; Juncus intermedius Thuill.; J. multiflorus Retz.; L. campestris var. multiflora (Retz.) Celak.; L. campestris sensu auct. Japon., non DC.——Yama-suzume-no-hie. Related to the preceding; radical leaves 6–15 cm. long, 2–5 mm. wide; inflorescence of several peduncled heads; flowers tawny to red-brown; perianth-segments about 2.5 mm. long, acute to cuspidate at apex; anthers narrowly oblong, as long as or slightly longer than the perianth; caruncle half as long as the seed.——Hokkaido, Honshu, Shikoku, Kyushu.—— Widely distributed in temperate regions of the N. Hemisphere and Australia.

Var. **kjellmanniana** (Miyabe & Kudo) G. Sam. L. kjellmanniana Miyabe & Kudo——Chishima-suzume-no-hie. Somewhat stouter, with dark to black-brown flowers; caruncle 1/3 as long as the seed, reported from Hokkaido (Ishikari Prov.).——n. Kuriles, and Kamchatka.

6. Luzula oligantha G. Sam.   L. campestris var. pauciflora Buch.; L. campestris var. sudetica sensu auct. Japon., non Celak.; L. sudetica var. nipponica Satake and var. microstachya Satake——Takane-suzume-no-hie.   Perennial herb in small, tufted clumps; leaves 5–10 cm. long, 2–3 mm. wide; stem 10–20 cm. long; heads several, somewhat loosely disposed in a terminal cyme; flowers dark brown; perianth-segments lanceolate, about 2 mm. long, acute at apex, nearly equal in length; stamens 2/3 as long as the perianth, the anthers oblong, half as long as the filaments; caruncle very short and inconspicuous.——June–Aug. Alpine slopes; Hokkaido, Honshu (n. and centr. distr.), Shikoku.——China, Korea, Kuriles, and Kamchatka.

7. Luzula pallescens (Wahlenb.) Bess. Juncus pallescens Wahlenb.; L. campestris var. pallescens (Wahlenb.) Wahlenb.——Oka-suzume-no-hie.   Closely resembling Luzula multiflora, but smaller; leaves 2–3.5 mm. wide; flowers including the capsule 2–2.5 mm. long, tawny or brown; perianth-segments lanceolate, acute, all equal in length; stamens 2/3 as long as the perianth, the anthers oblong, as long as or longer than the filaments; capsule nearly as long as the perianth; caruncle minute and indistinct.——Honshu, Shikoku, Kyushu; rare.——Korea, China, Kuriles, Sakhalin to Siberia, n. Europe, and Alaska.

2. JUNCUS L.   Igusa Zoku

Perennials, rarely annuals, always glabrous; leaf-blades various, grasslike, terete, stemlike or scalelike, the sheaths open on one side, usually auriculate at top; flowers small, disposed in cymes or in heads, sometimes solitary, bracteoles sometimes present; perianth-segments 6, in 2 series, glumelike; stamens 3 or 6; ovary superior, 1-locular, with parietal placentae, or the placentae reaching the center and the ovary apparently 3-locular, the ovules numerous, anatropous, the style 1, the stigmas 3, twisted; capsule 3-valved, loculicidally dehiscent; seeds numerous, sometimes with a taillike appendage on one or both ends, the embryo erect, the endosperm mealy.——Nearly cosmopolitan with about 300 species.

1A. Flowers prophyllate at base or each flower subtended by a pair of bracteoles.
  2A. Stems bearing normal leaves at least at base; leaves grasslike, flat or sulcate above; inflorescence terminal.
    3A. Annual; capsule perfectly 3-locular . . . . 1. J. bufonius
    3B. Perennial; capsule imperfectly 3-locular, or the placentae not quite reaching the center.
      4A. Perianth-segments obtuse, shorter than the capsule; anthers nearly as long as the filaments; auricle of the sheath small.
                                                                                   2. J. gracillimus
      4B. Perianth-segments acute, longer than the capsule; auricles longer, scarious. . . . . 3. J. tenuis
  2B. Stems without normal leaves, naked; leaves scalelike or sheathlike at the base of the stems; inflorescence pseudolateral, the lowest bract being stemlike and continuing above the base of the inflorescence.
    5A. Inflorescence few-flowered; lowest bract as long as or only slightly longer than the inflorescence. . . . . 4. J. beringensis
    5B. Inflorescence many-flowered; lowest bract much longer than the inflorescence.
      6A. Capsule perfectly 3-locular, brownish green; stamens 3, rarely 6 . . . . 5. J. effusus var. decipiens
      6B. Capsule imperfectly 3-locular.
        7A. Stamens 6; capsule dark brown to straw-colored.
           8A. Anthers 1/3 to 1/2 as long as the filaments; capsules straw-colored; rhizomes slender, with short internodes. . . 6. J. filiformis
           8B. Anthers longer than the filaments; capsules reddish brown or blackish brown; rhizomes thick, with elongate internodes.
              9A. Capsule elongate-ovoid, 4–5 mm. long; anthers 3 times as long as the filaments; stems twisted. . . . . 7. J. yokoscensis
              9B. Capsules oblong-obovoid, 6–7 mm. long; anthers as long as the filaments. 8. J. haenkei
        7B. Stamens 3; capsules greenish. . . . . 9. J. setchuensis var. effusoides
1B. Flowers not prophyllate.
  10A. Leaves grasslike, flattened, not septate; perianth-segments densely puncticulate, minutely scaberulous-papillose. . . 10. J. prominens

10B. Leaves terete, laterally compressed, or filiform.
  11A. Leaves filiform or slenderly cylindric; inflorescence of a single head or of several small few-flowered heads.
    12A. Rhizomes prominent, elongate; leaves compressed-cylindric; stems sparsely leafy to the top; heads 2 or 3, rather large, the lowest bract always exceeding the inflorescence. .... 11. *J. triceps*
    12B. Rhizomes very short or absent; leaves needlelike, mostly confined to the lower part of the stem; heads solitary or 2, the lowest bract shorter than the inflorescence.
      13A. Rather firm; flowers dark brown to black-brown; valves of capsule firm. .... 12. *J. triglumis*
      13B. Soft herbs; flowers light green; capsules brownish to straw-colored at maturity, the valves membranous.
        14A. Perianth-segments linear; stamens exserted, the filaments longer than the perianth. .... 13. *J. maximowiczii*
        14B. Perianth-segments narrowly lanceolate; stamens slightly exserted, the filaments slightly shorter than the perianth.
          14. *J. potaninii*

  11B. Leaves terete or laterally compressed; stamens always shorter than the perianth.
    15A. Seeds with taillike appendages at both ends.
      16A. Heads few; stamens 6; capsules rather large. .... 15. *J. kamtschatcensis*
      16B. Heads many; stamens 3; capsules rather small.
        17A. Inner perianth-segments obtuse, the outer acute. .... 16. *J. fauriensis*
        17B. Inner perianth-segments acute, the outer obtuse. .... 17. *J. tokubuchii*
    15B. Seeds not tailed.
      18A. Stem terete, always wingless.
        19A. Heads solitary; perianth-segments castaneous; flowers conspicuously pedicellate. .... 18. *J. mertensianus*
        19B. Heads many; perianth-segments greenish.
          20A. Stamens 6; capsules abruptly contracted to a mucronate apex. .... 24. *J. krameri*
          20B. Stamens 3; capsules gradually acuminate.
            21A. Inner perianth-segments longer than the outer; heads few-flowered. .... 23. *J. papillosus*
            21B. Inner perianth-segments nearly as long as the outer; heads many-flowered. .... 22. *J. wallichianus*
      18B. Stems compressed, usually winged.
        22A. Heads black-brown; rhizomes creeping, with more or less elongate internodes. .... 19. *J. ensifolius*
        22B. Heads greenish brown or brownish, 2- to several-flowered; internodes of rhizomes very short.
          23A. Heads usually 2-flowered; auricles rather large, oblong; stamens 3. .... 20. *J. yakeishidakensis*
          23B. Heads few- to many-flowered; auricles small; stamens 3 or 6.
            24A. Stamens 6; heads usually many-flowered; stems broadly winged. .... 21. *J. alatus*
            24B. Stamens 3; stems narrowly or scarcely winged.
              25A. Inner perianth-segments longer than the outer; capsules lanceolate or narrowly lanceolate, distinctly exceeding the perianth. .... 25. *J. diastrophanthus*
              25B. Inner perianth-segments nearly as long as the outer; capsules ovoid- or pyramidal-lanceolate, as long as the perianth or nearly so. .... 26. *J. leschenaultii*

1. Juncus bufonius L. HIME-KŌGAI-ZEKISHŌ. Annual; stems tufted, 10–30 cm. long, terete, slender; leaves more or less compressed, channeled; flowers solitary, 5–6 mm. long, forming a terminal large concave cyme, the bract leafy, much shorter than the inflorescence; perianth-segments whitish green, narrowly lanceolate, acuminate, the outer caudate; stamens 6, about half as long as the perianth-segments, the anthers 1/3–1/2 as long as the filaments; capsule oblong, tawny, shorter than the perianth; seeds obovate-elliptic.——June–Oct. Wet sandy places; Hokkaido, Honshu, Shikoku, Kyushu.——Cosmopolitan.

2. Juncus gracillimus (Buchen.) V. Krecz. & Gontsch. *J. compressus* var. *gracillimus* Buchen.——DORO-I, MIZU-I. Rhizomes firm, creeping, the internodes short; stems 40–70 cm. long, terete, rather firm, about 2 mm. across at base; leaves linear, grasslike, rather flat, whitish green, shorter than the stem, the auricles small; flowers solitary, in a terminal compound cyme; bracts leafy, longer or shorter than the inflorescence; perianth-segments ovate-oblong, 2.2 mm. long, obtuse, dark brown-purple on back, red-brown on both sides; stamens 6, 2/3 as long as the perianth-segments, the anthers nearly as long as the filaments; capsule ellipsoidal, brownish, lustrous; seeds ellipsoidal.——Wet or swampy places; Hokkaido, Honshu, Shikoku, Kyushu.——n. China, Korea, Ussuri, Amur, and Sakhalin.

3. Juncus tenuis Willd. *J. dudleyi* sensu auct. Japon., non Wieg.——KUSA-I, SHIRANE-I. Green; rhizomes very short; stems tufted, 30–60 cm. long, terete, slender; leaves flattened, grasslike, shorter than the stem, slightly involute, the

auricle elliptic, 2–3 mm. long, grayish white, scarious; inflorescence terminal, cymose, the bracts leafy, longer than the inflorescence; flowers solitary; perianth-segments nearly equal in length, lanceolate, acute, light green, 3.5–4 mm. long; stamens 6, half as long as the perianth, the anthers half as long as the filaments; capsule light green, ovoid-ellipsoidal, as long as to slightly shorter than the perianth; seeds obliquely obovoid.——July–Sept. Waste places along roadsides in mountains; Hokkaido, Honshu, Shikoku, Kyushu; quite common and rather variable.——Europe, centr. Asia, N. and S. America, and Australia.

4. Juncus beringensis Buchen. *J. fauriei* Lév. & Van't., pro parte——MIYAMA-I, TATEYAMA-I. Rhizomes short-creeping, with very short internodes; culms tufted, naked, terete, inconspicuously striate, 15–40 cm. tall; leaves at the basal part of the culm scalelike, more or less lustrous; inflorescence pseudolateral, 2- to 5-flowered, the lowest bract terete, appearing like a continuation of the culm, 2–4 cm. long, as long as or longer than the inflorescence; flowers 5–6 mm. long including the capsule, long-pedicelled; perianth-segments lanceolate, acute, black-brown, equal in length; stamens 6, slightly shorter than the perianth-segments, the anthers linear, 2–3 mm. long, the filaments very short; capsule ellipsoidal, dark brown, longer than the perianth, obtuse; seeds oblong, 1 mm. long, long-tailed at both ends, to 3 mm. long.——July–Aug. Wet alpine slopes; Hokkaido, Honshu (centr. distr.).——Kamchatka and Bering Sea coast.

5. Juncus effusus L. var. **decipiens** Buchen. *J. decipiens* (Buchen.) Nakai——I, TŌSHIN-SŌ. Rhizomes short-creeping with very short internodes; culms terete, inconspicuously striate, 25–60 cm. long; leaves few, basal, scalelike, brownish, lustrous; inflorescence pseudolateral, many-flowered, the lowest bract stemlike, appearing like a continuation of the culm, 10–20 cm. long; flowers many, solitary, 2–2.5 mm. long inclusive of the capsule, light green; perianth-segments equal in length, acute; stamens 3, slightly shorter than the perianth, the anthers slightly shorter than the filaments; capsules ovoid or ellipsoidal, brownish; seeds obliquely obovoid, 0.5 mm. long.——Aug.–Oct. Wet places in lowlands and mountains; Hokkaido, Honshu, Shikoku, Kyushu; very common and rather variable.——Ussuri, China, Korea and N. America. The typical phase occurs in Eurasia and N. America.

6. Juncus filiformis L. *J. filiformis* var. *curvatus* (Buchen.) Kudo; *J. curvatus* Buchen.; *J. brachyspathus* var. *curvatus* (Buchen.) Satake——Ezo-hoso-i, Rishiri-i, Karafuto-hoso-i. Rhizomes short-creeping with abbreviated internodes; culms slender, nearly terete, 30–90 cm. long inclusive of the culmlike lowest bract of the inflorescence, the basal sheaths few, scalelike; inflorescence a pseudolateral, few-flowered cyme, the lowest bract appearing like a continuation of the culm, 10–20 (–40) cm. long, sometimes longer than the culm; flowers about 3.5 mm. long including the capsule, perianth-segments narrowly lanceolate, the outer acute, longer than the somewhat obtuse inner segments; stamens 6, half as long as the inner perianth-segments, anthers half as long as the filaments; capsule oblong, obtuse, yellowish, nearly as long as the inner perianth-segments; seeds ellipsoidal, 0.7 mm. long.——Wet or swampy places in alpine or subalpine regions; Hokkaido, Honshu (centr. distr.).——Sakhalin and the Kuriles to Siberia, Europe, and N. America.

7. Juncus yokoscensis (Fr. & Sav.) Satake. *J. fauriei* Lév. & Van't., pro parte; *J. balticus* var. *japonicus* Buchen.; *J. glaucus* var. *yokoscensis* Fr. & Sav.——Inu-i, Hira-i, Neji-i. Deep green perennial; rhizomes horizontally creeping, firm, the internodes slightly elongate or short; culms erect, distinctly compressed, 20–50 cm. tall, more or less twisted; the basal sheaths scalelike; inflorescence pseudolateral, the lowest bract having the appearance of a continuation of the culm, sometimes longer than the inflorescence; flowers solitary, numerous, dense, 4–5 mm. long including the capsule; perianth-segments broadly lanceolate, acute, brownish green dorsally, blackish brown on both margins, the inner slightly shorter than the outer; stamens 6, half as long as the perianth, the anthers linear, 1.5 mm. long, the filaments very short; capsules oblong-ovoid, longer than the perianth, abruptly acute, dark brown; seeds obovoid, 0.8 mm. long.——Wet sandy places, especially near seashores; Hokkaido, Honshu (n. and centr. distr.).

Var. **laxus** Satake. Tsukushi-inu-i. Inflorescence loose; capsules gradually narrowed above and terminated by a mucro; perianth-segments all equal in length.——Kyushu.

8. Juncus haenkei E. Mey. *J. balticus* sensu auct. Japon., non Willd.; *J. balticus* var. *haenkei* (E. Mey.) Buchen.——Hama-i, Ō-inu-i. Rhizomes firm, creeping horizontally; culms 30–50 cm. long, rather thick, terete, the basal sheaths few, scalelike; inflorescence pseudolateral, 6- to 10-flowered, the lowest bract culmlike, having the appearance of a continuation of the culm, 10–20 cm. long; flowers solitary, 6–7 mm. long inclusive of the capsule; perianth-segments lanceolate, acute, greenish brown dorsally, reddish brown on both margins; stamens 6, about half as long as the perianth-segments, the anthers as long as or slightly longer than the filaments;

capsules obovoid-oblong, somewhat acute at apex, exceeding the perianth, dark brown; seeds obovoid-oblong, nearly 1 mm. long.——Aug.–Oct. Wet sandy places along seacoasts; Hokkaido.——Kuriles, Sakhalin, Korea, e. Siberia, Kamchatka, and Alaska.

9. Juncus setchuensis Buchen. var. **effusoides** Buchen. *J. pauciflorus* sensu auct. Japon., non R. Br.——Hoso-i. Cinereous-green; rhizomes short-creeping, branching; culms erect, 25–50 cm. long, slender, longitudinally striate, the basal sheaths few, scalelike; inflorescence pseudolateral, cymose, many-flowered, the lowest bract having the appearance of a continuation of the culm, sometimes up to 20 cm. long; flowers solitary, 3 mm. long inclusive of the capsule, light green; perianth-segments acute, equal in length; stamens 3, slightly shorter than the perianth; capsule exceeding the perianth, almost spherical, very obtuse; seeds broadly ovoid, about 0.6 mm. long.——Aug.–Oct. Wet places; Honshu, Shikoku, Kyushu.——China and Korea.

10. Juncus prominens (Buchen.) Miyabe & Kudo. *J. falcatus* var. *prominens* Buchen.——Sekishō-i, Ezo-no-mikuri-zekishō. Rhizomes long-creeping; stems terete, 15–30 cm. long, minutely papillose and scaberulous; leaves grasslike, flattened, 2–3 mm. wide, shorter than the culms, the cauline 1 to 2; inflorescence of 3 or 4, few-flowered heads, the lowest bract short; flowers about 4 mm. long inclusive of the capsule, pedicellate; perianth-segments nearly equal in length, the outer broadly lanceolate, acute, the inner narrowly ovate, obtuse, both dark reddish brown, papillose and scaberulous outside; stamens 6, half as long as the perianth–segments, the anthers slightly longer than the filaments; capsules ellipsoidal, slightly longer than the perianth, brown, emarginate; seeds obovoid.——July–Sept. Peaty soils in lowlands; Hokkaido (Nemuro), Honshu (Mutsu).——Kuriles, Kamchatka, and N. America.

11. Juncus triceps Rostk. *J. castaneus* sensu auct. Japon., non Smith; *J. satakei* Kitag.——Kuro-kōgai-zekishō. Stems terete, 25–40 cm. long; basal leaves scalelike, the cauline 3 or 4, chiefly on the lower part of the stem, rather flat, slightly involute, pluritubulose but smooth, 10–12 cm. long, about 1 mm. wide, somewhat obtuse; heads 1 or 2, hemisphaerical, 1–1.5 cm. in diameter, 4- to 6-flowered, the lowest bract somewhat leaflike, usually exceeding the inflorescence; flowers large, short-pedicelled, dark brown, 7–8(–9) mm. long inclusive of the capsule; perianth-segments lanceolate, acute; stamens 6, nearly as long as the perianth, the anthers nearly half as long as the filaments; capsule narrowly oblong, acute, about twice as long as the perianth; seeds ellipsoidal, the body about 1 mm. long, but 3–4 mm. long inclusive of the long taillike appendage at both ends.——Hokkaido (Mount Daisetsu).——Kuriles, Korea, and e. Siberia.

12. Juncus triglumis L. Takane-i, Shirouma-zekishō. Rhizomes extremely short; stems terete, 6–15 cm. long; leaves aggregated at the lower part of the culm, much shorter than the culm, somewhat terete, 1–5 cm. long, 1 mm. wide, the auricles rather large, ovate, obtuse; head solitary, terminal, usually 3-flowered, the bract scalelike, reddish brown, nearly as long as the head; flowers 4 mm. long or 6 mm. long inclusive of the capsule, short-pedicelled; perianth-segments equal in length, ovate-lanceolate, obtuse, somewhat membranous, reddish brown; stamens 6, usually as long as the perianth, the anthers much shorter than the filaments; capsules ovoid, obtuse; seeds with a white taillike appendage on each end, 2 mm. long.——Aug.–Sept. Alpine regions; rare; Hokkaido (Mount

Daisetsu), Honshu (Mount Shirouma in Shinano).——Korea, Kamchatka, Siberia, China, Himalayas, and Europe.

13. Juncus maximowiczii Buchen. *J. cupreus* Lév. & Van't.——Ito-i.   Soft slender herb with very short rhizomes; stems tufted, 10–15 cm. long, usually bearing 1 leaf; leaves all filiform, more or less compressed, deeply channeled above, 0.5–1 mm. thick, the radical leaves slightly shorter to longer than the stem; heads solitary, terminal, 1- to 4-flowered, the lowest bract short, leaflike; flowers 5–7 mm. long including the capsule, pedicellate; perianth-segments equal in length, hyaline, light green, linear to broadly linear, rather obtuse; stamens 6, longer than the perianth-segments and nearly as long as the mature fruit, the anthers much shorter than the filaments, exserted from the perianth; capsules narrowly obovoid-ellipsoidal, longer than the perianth, yellowish or light brown, somewhat membranous; seeds with a white tail on each end, 2 mm. long.——Aug.–Sept. Wet rocks in forest zone; Honshu (centr. distr.).——Korea.

14. Juncus potaninii Buchen. *J. luzuliformis* var. *potaninii* (Buchen.) Buchen.——Ezo-ito-i.   Related to the preceding but more delicate; stems 5–12 cm. long; flowers 4–5 mm. long including the capsule; perianth-segments lanceolate, rather acute, the outer slightly longer than the inner; stamens as long as or slightly longer than the perianth, the filaments slightly shorter than the perianth; capsule slightly longer than the perianth, oblong.——Hokkaido, Honshu (Mount Yatsugatake in Shinano); rare.——n. Korea and China.

15. Juncus kamtschatcensis (Buchen.) Kudo. *J. fauriensis* var. *kamtschatcensis* Buchen.——Miyama-hoso-kōgai-zekishō.   Rhizomes short; stems erect, nearly terete, 10–20 cm. long; cauline leaves 2 or 3, shorter than the stems, slightly compressed, rather clearly septate, the auricles large, elliptical, 1–2 mm. long; inflorescence terminal, bearing (2–)5 heads each 3- to 6-flowered, the bracts short, leaflike, shorter than or slightly longer than the inflorescence; flowers reddish brown, about 5 mm. long including the capsule; perianth-segments narrowly lanceolate, acute, the inner somewhat obtuse, nearly as long as the outer; stamens 6, 2/3 as long as the perianth-segments, the anthers ovoid, shorter than the white filaments; capsules oblong, obtuse, exceeding the perianth-segments, dark brown; seeds oblong, with a tail at each end, about 2 mm. long.——Aug.–Sept. Swampy places in alpine or subalpine regions; Hokkaido, Honshu (n. distr.).——Kuriles and Kamchatka.

16. Juncus fauriensis Buchen. Hoso-kōgai-zekishō. Related to the preceding species, but the stems 20–40 cm. long; inflorescence terminal, with rather numerous heads, the bracts much shorter than the inflorescence; flowers 3–4 mm. long including the capsule; perianth-segments lanceolate, reddish brown; stamens 3, slightly shorter than the outer perianth-segments, the anthers nearly as long as the filaments.——Aug.–Sept. In moist places in alpine or subalpine regions; Hokkaido, Honshu (n. and centr. distr.).

17. Juncus tokubuchii Miyabe & Kudo. Horomui-kō-gai.   Perennial; stem about 40 cm. long; basal leaves scale-like, the cauline linear, usually shorter than the stem, terete, unitubulose, 10–20 cm. long, perfectly septate, acuminate; inflorescence terminal, compound, cymose, with many, loosely disposed, 2- to 10-flowered heads; flowers 3 mm. long, or 4.5–5 mm. long including the capsule; perianth-segments nearly equal, the outer ovate-lanceolate, obtuse, the inner lanceolate, rather acute; stamens 3, slightly shorter than the perianth-

segments; capsule elongate-ovoid, exceeding the perianth, castaneous or reddish brown; seeds with a filiform appendage on each end.——Hokkaido.

18. Juncus mertensianus Bong. *J. ensifolius* sensu auct. Japon., non Wikstr.——Ezo-no-mikuri-zekishō, Kumoma-mikuri-zekishō.   Rhizomes creeping; stem loosely tufted, 10–25 cm. long, about 1 mm. across, slightly compressed; cauline leaves 2 or 3, cylindrical, 7–13 cm. long, 1 mm. thick, more or less compressed, the septa perfect but not visible, the auricles ovate, obtuse, hyaline, stramineous; heads usually solitary, terminal, 1–1.3 cm. in diameter, 10- to 25-flowered, the lowest bract short, leaflike; flowers distinctly pedicellate, about 4 mm. long, castaneous or blackish brown; perianth-segments lanceolate, acute, the inner slightly shorter than the outer; stamens 6 (rarely 3), 2/3 as long as the perianth, the anthers shorter than the filaments; capsule ellipsoidal or obovoid, obtuse, not exceeding the perianth, castaneous, rather thick; seeds obovoid, 0.5–0.6 mm. long.——Alpine regions; Hokkaido (Mount Daisetsu).——N. America and Aleutian Isls.

19. Juncus ensifolius Wikstr. *J. xiphioides* sensu auct. Japon., non E. Mey.; *J. oligocephalus* Satake & Ohwi——Mikuri-zekishō, Kuro-mikuri-zekishō.   Rhizomes short, creeping; stems loosely tufted, erect, 30–50 cm. tall, compressed, narrowly 2-winged; leaves cauline, few, linear-ensiform, compressed, slightly shorter or longer than the stem, pluritubulose, acuminate, the septa clear; inflorescence terminal, of 2–5 globose heads, each 8–10 mm. in diameter, densely many flowered, the lowest bract leaflike, shorter than the inflorescence; flowers pedicellate, about 3 mm. long in fruit, mostly black-brown; perianth-segments lanceolate, acute, the inner slightly shorter than the outer; stamens 3, 2/3 as long as the inner perianth-segments, the anthers half as long as the filaments; capsule oblong, as long as the perianth-segments, rather thin; seeds obovoid, about 0.6 mm. long.——Aug.–Sept. Moors; Hokkaido, Honshu (n. and centr. distr.).——Kuriles and N. America.

20. Juncus yakeishidakensis Satake. Miyama-zehishō. Rhizomes indistinct, the internodes very short; stems tufted, 10–20 cm. long, compressed, broadly 2-winged, 1–1.5 mm. wide including the wings, 3-leaved; leaves all cauline, linear or linear-ensiform, 4–10 cm. long, 1–2 mm. wide, acute, pluritubulose, the septa not prominent, the auricles large, white, ovate; inflorescence of 4–6 heads, terminal, the lowest bract leaflike, shorter than the inflorescence, the heads 4- to 8-flowered; flowers pedicellate, about 4 mm. long when mature; perianth-segments linear-lanceolate or lanceolate, greenish, nearly equal in length, acute and tipped with an awn; stamens 3, 2/3 as long as the outer perianth-segments, the anthers half as long as the filaments; capsules ovoid-conical, brownish, slightly longer than the perianth, thin; seeds obovoid, 0.6 mm. long.——Honshu (n. distr.); rare.

21. Juncus alatus Fr. & Sav. Hanabi-zekishō.   Rhizomes very short; stems tufted, 25–40 cm. long, strongly compressed, broadly 2-winged; leaves cauline, few, linear-ensiform, compressed, shorter than the inflorescence, 4–6 mm. long, gradually attenuate to an acute tip, pluritubulose, the septa distinct; inflorescence of rather numerous 4- to 7-flowered heads, terminal, the lowest bract leaflike, much shorter than the inflorescence; flowers 4–5 mm. long when mature, perianth-segments equal in length or the inner slightly longer than the outer, lanceolate, cuspidate, greenish; stamens 6, 2/3 as long as the perianth, the anthers much shorter than the filaments, oblong; capsules elongate, conical-ovoid, exceeding the perianth, reddish brown, lustrous, rather thick; seeds obovoid, 0.5 mm. long.——June–Sept. Wet places; Honshu, Shikoku, Kyushu.——China.

22. Juncus wallichianus Laharpe. *J. prismatocarpus* var. *leschenaultii* subvar. *unitubulosus* Buchen.; *J. leschenaultii* var. *radicans* Fr. & Sav.——HARI-KŌGAI-ZEKISHŌ. Rhizomes very short; stems terete or slightly compressed, not winged, 10–50 cm. long, 2- or 3-leaved; leaves all cauline, shorter than the stem, linear-subulate, unitubulose, perfectly septate; inflorescence terminal, cymose, the heads usually many, the lowest bract leaflike, short; flowers 4–5 mm. long including the capsule; perianth-segments lanceolate to narrowly lanceolate, cuspidate; stamens 3, about half as long as the outer perianth-segments, the anthers much shorter than the filaments; capsules narrowly oblong, 3-angled, gradually acuminate at apex, exceeding the perianth, thin; seeds obovoid, 0.6 mm. long.——Aug.–Sept. Swampy places; Hokkaido, Honshu, Shikoku, Kyushu; common.——Sakhalin, Ussuri, Ryukyus, Formosa, China to India.

23. Juncus papillosus Fr. & Sav. *J. nipponensis* Buchen.; *J. umbellifer* Lév. & Van't.——AO-KŌGAI-ZEKISHŌ. Rhizomes elongate; stems terete, 20–30 cm. long, 2- or 3-leaved; leaves all cauline, slightly compressed, unitubulose, shorter than the stem, 1–2 mm. wide, perfectly septate; inflorescence rather large, terminal, with many loosely disposed small 2- or 3-flowered heads; flowers 4–5 mm. long including the capsule, the perianth-segments narrowly lanceolate, cuspidate, greenish, the inner longer than the outer; stamens 3, 2/3 as long as the outer perianth-segments, the anthers much shorter than the filaments; capsules lanceolate, mucronate, much longer than the perianth, brown, thin; seeds obovoid, about 0.6 mm. long. ——Aug.–Sept. Swampy places; Hokkaido, Honshu, Shikoku, Kyushu.——China, Korea, Manchuria to e. Siberia.

24. Juncus krameri Fr. & Sav. TACHI-KŌGAI-ZEKISHŌ. Rhizomes creeping; stem terete, slightly thickened, 30–60 cm. long, 2- or 3-leaved; leaves all cauline, shorter than the inflorescence, terete, unitubulose, perfectly septate; inflorescence a relatively dense terminal cyme of rather numerous, 3- to 10-flowered, small heads, the lowest bract leaflike, exceeding

the inflorescence; flowers sessile, 3–4 mm. long when mature; perianth-segments broadly lanceolate, acute, greenish, the inner slightly longer than the outer; stamens 6, shorter than the perianth-segments, the anthers ovoid, much shorter than the filaments; capsules ellipsoidal, 3-angled, rather obtuse, slightly longer than the perianth, rather thick; seeds obovoid, 0.5 mm. long.——Aug.-Oct. Wet or marshy places; Hokkaido, Honshu, Kyushu.——Kuriles, Korea, Ryukyus, and n. China.

25. Juncus diastrophanthus Buchen. HIROHA-NO-KŌGAI-ZEKISHŌ. Rhizomes very short; stems compressed, 2-winged, 20–40 cm. long, 2–3 mm. wide, usually 3-leaved; leaves all cauline, 10–20 cm. long, 3–5 mm. wide, linear-ensiform, strongly compressed, pluritubulose, the septa inconspicuous; inflorescence a compound cyme of rather numerous globose heads, the lowest bract leaflike, short; flowers 5–6 mm. long including the capsule; perianth-segments linear, greenish, cuspidate, equal or the inner slightly longer than the outer; stamens 3, half as long as the perianth, the anthers oblong, shorter than the filaments; capsules linear-lanceolate to 3-angled-cylindric, cuspidate, much exceeding the perianth, thin; seeds obovoid, 0.6 mm. long.——Aug.–Sept. Wet places; Hokkaido, Honshu, Shikoku, Kyushu; common.——Korea and China.

26. Juncus leschenaultii J. Gay. *J. prismatocarpus* var. *leschenaultii* subvar. *pluritubulosus* Buchen.——KŌGAI-ZEKISHŌ, HIRA-KŌGAI-ZEKISHŌ. Stems more or less tufted, 20–40 cm. tall, compressed, narrowly 2-winged, about 2 mm. wide, few-leaved; leaves all cauline, 10–20 cm. long, 2–3 mm. wide, strongly compressed, pluritubulose, ensiform; inflorescence a compound cyme of usually numerous heads, large or rather small, the lowest bract leaflike, shorter than the inflorescence; flowers 4–5 mm. long including the capsule; perianth-segments narrowly lanceolate, cuspidate; stamens 3, 1/3–1/2 as long as the perianth-segments, the anthers shorter than the filaments; capsule 3-angled, lanceolate to elongate-ovoid, acute, brownish, slightly exceeding the perianth, rather membranous; seeds obovoid, about 0.6 mm. long.——Aug.–Nov. Wet places; Hokkaido, Honshu, Shikoku, Kyushu; very common.——Kamchatka, Korea, Formosa, China, and India; highly variable.

Fam. 51 **STEMONACEAE** BYAKUBU KA Stemona Family

Perennial herbs with rhizomes or bulbs; stems erect, ascending or scandent; leaves alternate or opposite, longitudinally nerved, with connecting transverse veinlets; flowers on axillary peduncles, 1 to several, actinomorphic, bisexual; perianth-segments 4, superior or semisuperior, in 2 ranks; stamens 4, inserted at the base of the perianth-segments or of the ovary, the anthers dorsally attached; ovary 1-locular, the stigmas 1 or 3, the ovules more than 2, erect or pendant, anatropous; capsules dehiscent, 2-valved; seeds oblong, with an elongate embryo, the endosperm hardened.——Three genera, with about 30 species, chiefly in the Tropics and warmer regions of e. and s. Asia, Australia, and N. America.

1A. Stems twining, leafy, elongate; connective of anthers attenuated at tip into a linear appendage. .... 1. *Stemona*
1B. Stems erect, simple, few-leaved at top; anthers inappendiculate. .... 2. *Croomia*

1. **STEMONA** Lour. BYAKUBU ZOKU

Roots tuberous; rhizomes short; stems elongate, twining; leaves alternate or more often verticillate in 4's or 3's, lanceolate to cordate, 3- to many-nerved, petiolate, with very many and prominent transverse close veinlets; peduncles axillary, solitary or racemosely few-flowered; flowers bisexual; perianth segments 4, free, in 2 series, many-nerved, acuminate; filaments short, more or less connate at base into a ring, anthers linear, erect, the connective attenuated at tip of the anther-locule into a linear appendage; ovary superior, 1-locular; stigma sessile; capsules ovoid or oblong, dehiscent, 2-valved; seeds oblong, striate, with a fascicle of fleshy filaments on the funicle.——Few species, India to Australia and China.

1. Stemona japonica Miq. BYAKUBU. Perennial herb with short rhizomes and several thickened roots; stems elongate, scandent in upper portion; leaves verticillate in 4's or 3's, spreading, petiolate, ovate, long-acuminate, rounded to broadly cuneate at base, slightly undulate on margin, 5-nerved and with closely parallel, transverse veinlets between the nerves; peduncles axillary, 1- to 2-flowered, often partly fused to the petiole; flowers erect, half-open, perianth-segments 4, about 12 mm. long, pale green, lanceolate.——June. Introduced and planted for medicinal purposes.——China.

2. CROOMIA Torr. & Gray NABEWARI ZOKU

Rhizomes horizontally creeping; stems erect, declined above, simple, unbranched, surrounded at base by 1 or 2 scales; leaves several, alternate, petioled, oblong-cordate, several-ribbed, with transverse parallel veinlets between the ribs; peduncles axillary, shorter than the leaves, 1- to several-flowered; flowers rather small, the bract small; perianth-segments 4, 2-ranked, free, spreading; stamens with the filament slightly thickened; ovary broad at base, sessile, the ovules several, pendant from the top of the locule; capsules beaked, 2-valved; seeds with a tuft of fleshy setose appendages on the funicle.——Three species, the following in Japan and one in e. N. America.

1A. One of the outer perianth-segments larger and broader than the others. . . . . 1. *C. heterosepala*
1B. Perianth-segments nearly equal in size, more or less recurved on margin. . . . . 2. *C. japonica*

1. Croomia heterosepala (Bak.) Okuyama. *C. japonica* var. *heterosepala* Bak.; *C. japonica* sensu auct. Japon., non Miq. ——NABEWARI. Smooth glabrous perennial; stems 30–60 cm. long, with several leaves on the upper half; leaves flat, oblong-ovate or ovate, 6–15 cm. long, 3–8 cm. wide, abruptly acute to abruptly short-acuminate, rounded-truncate to shallowly cordate at base, 5- to 9-ribbed, the 3 median ribs extending to the apex, the peduncles decumbent, 2.5–5 cm. long, with a small bract below the middle, the pedicels 5–8 mm. long, slightly thickened upward; flowers yellowish green, pendulous, one of the 4 perianth-segments larger than the others, 8–10 mm. long, the smaller broadly ovate, 5–7 mm. long, both flat and spreading, obtuse, sparsely and minutely papillose inside; filaments nearly smooth.——Apr.–June. Mountain forests; Honshu (s. Kantō Distr. and westw.), Shikoku, Kyushu.

2. Croomia japonica Miq. *C. kiushiana* Makino—— HIME-NABEWARI. Resembling the preceding but larger; flowers 1 to 4 on one peduncle, smaller, the perianth-segments oblong, narrowly ovate or elliptic, all equal in shape and size, recurved on margin, minutely papillose inside; filaments papillose; seeds broadly obovoid, 4 mm. long, with prominent longitudinal grooves.——Apr.–June. Honshu (Chūgoku Distr.), Kyushu.

Fam. 52. LILIACEAE YURI KA Lily Family

Usually perennial herbs, mostly from bulbs and rhizomes, rarely shrubs or scandent tendril-bearing subshrubs; leaves alternate, rarely opposite or verticillate, usually entire; inflorescence 1- to many-flowered, racemose, umbellate, spicate, paniculate, or fasciculate; flowers bisexual, rarely unisexual, actinomorphic or zygomorphic, the perianth-segments (tepals) 6, rarely 4 or many, usually inferior, in 2 whorls, free or connate; stamens usually 6, the anthers usually longitudinally dehiscent; ovary usually superior, 3-locular, sometimes 1-locular with 3 parietal placentae, the ovules 1 to many in each locule; endosperm copious. ——About 230 genera, with about 3,000 species, cosmopolitan.

1A. Herbs (in ours) without tendrils; inflorescence various, but not an axillary umbel.
  2A. Ovary superior or nearly so; tepals free or connate.
    3A. Carpel valves accrescent and persistent after anthesis and enclosing the seeds.
      4A. Fruit a capsule.
        5A. Plants with rhizomes.
          6A. Capsules septicidal (loculicidal in *Narthecium* and *Metanarthecium*); anthers extrorse or introrse.
            7A. Tepals persistent or tardily deciduous; stigmas 3-lobed or entire.
              8A. Anthers linear to ovate, not peltate, the locules distinct; stems scaly.
                9A. Plants saprophytic, without chlorophyll. . . . . 1. *Protolirion*
                9B. Plants with chlorophyll.
                  10A. Stigmas 3 or style 3-lobed.
                    11A. Leaves spirally arranged, dorsiventrally flattened.
                      12A. Leaves oblong, with a petiolelike base; flowers zygomorphic, sessile; perianth partially suppressed.
                            2. *Chionographis*
                      12B. Leaves linear, slightly keeled on back; flowers actinomorphic, pedicelled. . . . . 3. *Japonolirion*
                  11B. Leaves 2-ranked, laterally flattened, linear, sessile; flowers actinomorphic. . . . . 4. *Tofieldia*
                  10B. Stigma 1, small, on a slender style.
                    13A. Leaves 2-ranked, rigid, laterally flattened, linear, sessile; flowers yellow; filaments villous. . . . . 5. *Narthecium*
                    13B. Leaves spirally arranged, thin, dorsiventrally flattened, broadly linear to oblong.
                      14A. Ovary ovoid to globose; capsules loculicidal. . . . . 6. *Metanarthecium*
                      14B. Ovary 3-lobed; capsules with inflated valves, septicidal on inner (upper) side. . . . . 7. *Heloniopsis*
              8B. Anthers somewhat extrorse, globose-reniform, finally peltate, with confluent locules; stems usually leafy.
                15A. Leaves cauline; inflorescence pubescent; plant usually stout and robust; tepals not gland-bearing at base.
                            8. *Veratrum*
                15B. Leaves mostly radical; inflorescence glabrous; plant usually slender; tepals with 1–2 glands near base.
                            9. *Zigadenus*

7B.   Tepals deciduous; stigmas 3, deeply 2-cleft; flowers large. .... 10. *Tricyrtis*
   6B.   Capsules loculicidal; anthers introrse.
       16A.   Tepals connate, corollas infundibuliform; leaves dorsiventrally flattened, basal; seeds glabrous.
          17A.   Leaves flat, petiolate, broad; flowers white to purplish. ....11. *Hosta*
          17B.   Leaves strap-shaped, usually sulcate; flowers yellowish. .... 12. *Hemerocallis*
       16B.   Tepals free; leaves laterally flattened, evergreen; seeds 6, with long hairs at base. .... 13. *Alectorurus*
   5B.   Plants with bulbs.
       18A.   Tepals persistent or tardily deciduous; stems naked, the leaves basal.
          19A.   Inflorescence a many-flowered raceme. .... 14. *Scilla*
          19B.   Inflorescence umbellate-cymose with 1 or 2 large bracts at base.
             20A.   Inflorescence umbellate-cymose with 2 slender green bracts at base; flowers few.
                21A.   Flowers yellowish; tepals without glands. .... 15. *Gagea*
                21B.   Flowers white; tepals usually with a gland or depression near base. .... 16. *Lloydea*
             20B.   Inflorescence umbellate, enveloped while young by a membranous bract.
                22A.   Tepals free. .... 17. *Allium*
                22B.   Tepals connate toward the base. .... 18. *Nothoscordum*
       18B.   Tepals deciduous; stems usually leafy.
          23A.   Anthers versatile. .... 19. *Lilium*
          23B.   Anthers basifixed.
             24A.   Tepals narrow, strongly recurved above. .... 20. *Erythronium*
             24B.   Tepals broader, not strongly recurved.
                25A.   Tepals often maculate near base but not pitted; flowers campanulate to infundibuliform. .... 21. *Tulipa*
                25B.   Tepals with a depression or nectariferous pit near base; flowers campanulate. .... 22. *Fritillaria*
  4B.   Fruit a berry; plants with rhizomes.
     26A.   Tepals all alike; leaves cauline and alternate or radical.
       27A.   Rhizomes sympodially elongate, the aerial stems arising apically.
          28A.   Leaves scalelike, bearing in the axils leaflike or needlelike cladodes. .... 23. *Asparagus*
          28B.   Leaves larger, leaflike.
             29A.   Filaments partly thickened; leaves linear, evergreen. .... 24. *Dianella*
             29B.   Filaments linear; leaves broader, not evergreen.
                30A.   Tepals connate, tubular; stamens adnate below on the tube. .... 25. *Polygonatum*
                30B.   Tepals free or nearly so; stamens free or nearly so.
                  31A.   Flowers supra-axillary. .... 26. *Streptopus*
                  31B.   Flowers or inflorescence terminal.
                    32A.   Leaves radical. .... 27. *Clintonia*
                    32B.   Leaves cauline.
                      33A.   Stems simple, unbranched; inflorescence a raceme or panicle.
                         34A.   Tepals 6; leaves few to several. .... 28. *Smilacina*
                         34B.   Tepals 4; leaves 2. .... 29. *Maianthemum*
                      33B.   Stems often forked; flowers 1 to several, umbellate; perianth deciduous. .... 30. *Disporum*
       27B.   Rhizomes monopodially elongate, aerial stems or branches arising from the axils.
          35A.   Styles columnar, the stigma simple.
             36A.   Flowers in spikes, erect; corolla with a short tube, the lobes much longer than the tube. .... 31. *Reineckea*
             36B.   Flowers in recemes, nodding; corolla globose-campanulate, with short recurved lobes. .... 32. *Convallaria*
          35B.   Styles at the apex, dilated into broad stigma-lobes.
             37A.   Scapes short, 1-flowered. .... 33. *Aspidistra*
             37B.   Scapes bearing a terminal dense many-flowered spike. .... 34. *Rohdea*
     26B.   Inner tepals usually smaller than the outer; leaves ternate or verticillate at the top of stem.
       38A.   Leaves 4 to many; flowers 4-merous to polymerous. .... 35. *Paris*
       38B.   Leaves 3; flowers 3-merous. .... 36. *Trillium*
  3B.   Carpel-valves soon dehiscent and deciduous after anthesis and exposing large fleshy globose seeds; tepals free or nearly so.
     39A.   Flowers erect; ovary superior; filaments distinct; anthers oblong, obtuse at both ends; seeds commonly purple-black.
                                                             37. *Liriope*
     39B.   Flowers nodding; ovary partly superior; filaments very short; anthers acute, sagittate at base; seeds blue-black or blue.
                                                        38. *Ophiopogon*
2B.   Ovary partly inferior, adnate to the tepals; tepals connate and tubular in lower portion; fruit a capsule. .... 39. *Aletris*
1B.   Subshrubs or herbs often with tendrils; leaves flat, the 3–7 parallel nerves joined with fine transverse veinlets; flowers in axillary umbels, small, unisexual; ovules orthotropus, pendulous, 1–2 in each locule.
  40A.   Tepals free; stamens free. .... 40. *Smilax*
  40B.   Tepals connate, tubular; stamens 3, connate. .... 41. *Heterosmilax*

1. **PROTOLIRION** Ridl.   S<small>AKURAI-SŌ</small> Z<small>OKU</small>

Small pale yellow saprophytic herbs from slender rhizomes; stems slender, usually simple, the leaves small, scalelike; flowers in a terminal corymbose raceme, small, bisexual; tepals 6, connate at base, the outer smaller than the inner, the inner ovate, ascending-spreading, with a small gland at base; stamens 6, the filaments subulate, the anthers ovate; capsule of 3 obliquely spreading ovoid carpels, connate toward the base, dehiscent on upper margin, the ovules numerous; styles short, the stigma capitellate; seeds small, brown.——Three species, in Malaya, s. China, and Japan.

1. **Protolirion sakuraii** (Makino) Dandy. *Miyoshia sakuraii* Makino; *P. miyoshia-sakuraii* (Makino) Makino——Sakurai-sō. Low glabrous saprophytic herb, pale brown when dried; stems 7–12 cm. long, rather firm, erect; scalelike leaves rather numerous, membranous, contiguous on lower part and loose on upper part of the stems, broadly ovate to narrowly deltoid, 1-nerved, subobtuse, 2–5 mm. long; racemes 1.5–3 cm. long, several- to more than 10-flowered, loose, the pedicels ascending, 2–3 mm. long, the bracts lanceolate, shorter to slightly longer than the pedicel; flowers 3.5–4 mm. wide, the tube infundibuliform, slightly shorter than the inner tepals, the outer tepals deltoid-ovate, about half as long as the inner ones, the inner about 1.5 mm. long, subobtuse.——July. Woods; Honshu (Mino Prov.); very rare.——Formosa.

2. CHIONOGRAPHIS Maxim.   Shiraito-sō Zoku

Glabrous perennial herbs from short stout rhizomes; leaves radical, oblong, entire or the margin minutely undulate; scapes erect, simple, with a few ascending small linear leaves; flowers in a dense to loose terminal spike, bisexual, zygomorphic, sessile, small, the tepals 6, the upper 3 or 4 linear to filiform, the lower 2 or 3 very short or suppressed; stamens 6, the filaments short, inserted at the base of the tepals, the anthers small, subglobose; ovary globose, the ovules 2 in each carpel; styles 3, stigmatic on inner side; seeds fusiform.——Several species, in Japan, s. Korea, and s. China.

1A. Upper tepals linear, slightly broadened above, 0.5–0.7 mm. wide, white; lower tepals small, nearly as long as the stamens; anther locules 2, distinct; spike usually densely flowered. .... 1. *C. japonica*
1B. Upper tepals filiform, not broadened above, 0.2–0.4 mm. wide, usually greenish white, sometimes with a purple hue; lower tepals completely suppressed; anther locules 2, confluent; spikes usually loosely flowered. .... 2. *C. koidzumiana*

1. **Chionographis japonica** Maxim. *Melanthium luteum* Thunb., excl. syn.; *Chionographis lutea* (Thunb.) Baill.——Shiraito-sō. Radical leaves several, narrowly obovate-oblong or oblong, long- or short-petioled, 3–8 cm. long, 1.5–3 cm. wide, obtuse to subacute, the margins slightly undulate; scapes 15–45 cm. long, the cauline leaves linear-lanceolate to linear, green, sessile, ascending to subappressed; inflorescence elongating after anthesis, 4–20 cm. long, rather densely many-flowered, the flowers white, the upper 3–4 tepals 7–12 mm. long, obtuse, spreading; capsules oblong, erect, 3–3.5 mm. long, the seeds lanceolate, 2 in each carpel, with a short tail at one end, about 3 mm. long.——May–June. Honshu (Kantō Distr., Echigo Prov. and westw.), Shikoku, Kyushu.——s. Korea.

Var. **hisauchiana** Okuyama. Azuma-shiraito-sō. A smaller phase with shorter tepals about 3–5 mm. long.——Honshu (Kantō Distr.).

2. **Chionographis koidzumiana** Ohwi. *C. sparsa* F. Maekawa; *C. koidzumiana* var. *mikawana* Ohwi & Okuyama——Chabo-shiraito-sō. Glabrous green perennial herb from short sparingly fibrillose rhizomes; radical leaves long-petioled, ovate-oblong to ovate, sometimes broadly ovate, subacute to rather obtuse, broadly cuneate to rounded at base, 15–60 mm. long, 10–25 mm. wide, the margin undulate; scapes 10–30 cm. long, slender, simple, erect, the cauline leaves 1–3 cm. long, green, lanceolate, subobtuse, sessile, ascending to appressed; inflorescence elongating after anthesis, very slender, the flowers relatively numerous, the upper tepals filiform, scarcely dilated above, about 10 mm. long, pale greenish, the lower ones suppressed; stamens 6, the 3 outer slightly longer than the inner.——Moist places along streams in mountains; Honshu (Mikawa and Kii Prov.), Shikoku, Kyushu (Yakushima); rare.

3. JAPONOLIRION Nakai   Oze-sō Zoku

Rhizomes short-creeping, scaly; leaves radical, tufted, linear, scabrous on margin; scapes arising from the axils of radical leaves, erect, terete, loosely scaled, simple, the racemes solitary; flowers small, regular, greenish, short-pedicelled; bracteoles absent, the tepals 6, membranous, broadly lanceolate, obtuse, 1-nerved without a gland or depressed pit; stamens 6, the filaments subulate, the anthers ovate, 2-locular, basifixed, introrse, the locules parallel, distinct; ovary superior, the carpels 3, connate except at the apex; styles 3, short, recurved, stigmatic on the inner surface; capsules septicidally dehiscent.——Monotypic.

1. **Japonolirion osense** Nakai. *J. saitoi* Makino & Tatew.; *J. osense* var. *saitoi* (Makino & Tatew.) Ohwi——Oze-sō. Fig. 2 Glabrous green perennial; rhizomes about 2 mm. in diameter, with obtuse membranous, many-nerved scales 4–6 mm. long; radical leaves sometimes longer than the scape, glabrous, linear, 5–15 cm. long, 2–4 mm. wide, acuminate to gradually so, somewhat narrowed at base, the margin scabrous, with few parallel nerves; scapes 10–30 cm. long, the scales membranous, lanceolate, 1-nerved, subobtuse; inflorescence dense, racemose, simple, erect, several- to relatively many-flowered, the bracts similar to the scales, the pedicels 2–4 mm. long, ascending; tepals 2–2.5 mm. long, ascending to spreading, persistent; capsules oblong-ovate, about 3 mm. long.——July–Aug. Alpine meadows; Hokkaido, Honshu (Mount Shibutsu and Mount Tanigawa in Kotsuke); rare.

4. TOFIELDIA Huds.   Chishima-zekishō Zoku

Green glabrous perennials from short creeping rhizomes; leaves mostly radical, 2-ranked, laterally flattened, linear; scapes slender, few-leaved or naked; racemes sometimes spikelike, the flowers small, on short pedicels, in axils of bracts, solitary or in 3's, bracteolate; tepals 6, persistent, linear-oblong to oblanceolate, white, greenish, or brownish red; stamens 6, the filaments linear-subulate, the anthers ovate, introrse, 2-locular; ovary superior, sessile, ovoid, 3-lobed at apex, the ovules numerous; styles short, the stigma introrse; capsules septicidal, 3-locular, the seeds small, narrowly oblong, caudate at one end or without appendage.——About 20 species, in the temperate and northern regions of the N. Hemisphere.

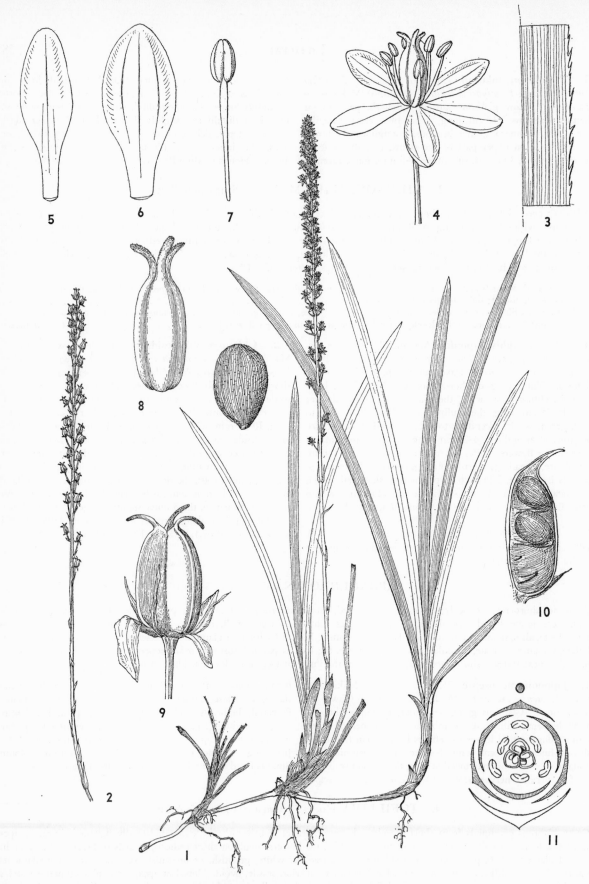

Fig. 2.—*Japonolirion osense* Nakai. 1, Habit; 2, fruiting scape; 3, part of leaf; 4, flower; 5, 6, outer and inner tepals; 7, stamens; 8, pistil; 9, capsule; 10, capsule with seeds; 11, floral diagram.

1A. Scapes and racemes glabrous, smooth; pedicels solitary; seeds not appendaged.
  2A. Leaves scabrous on margin.
    3A. Pedicels erect; capsules about twice as long as the tepals; leaves abruptly acute, mucronate, or short-acuminate. .... 1. *T. okuboi*
    3B. Pedicels arcuately spreading; flowers nodding to pendulous; capsules slightly longer to equaling the length of the tepals; leaves acuminate. .... 2. *T. coccinea*
  2B. Leaves smooth on margin; pedicels slender, obliquely spreading, straight. .... 3. *T. nuda*
1B. Scapes and pedicels with viscid glandular tubercles above; pedicels ternate; seeds caudate. .... 4. *T. japonica*

1. Tofieldia okuboi Makino. HIME-IWA-SHŌBU. Rhizomes short-creeping; leaves linear, laterally flattened, broadest above the middle, 3–6 cm. long, 2–5 mm. wide, abruptly acute and mucronate or short-acuminate, scabrous on margin; scapes 6–15 cm. long, erect, with 1 or 2 short leaves; racemes loosely about 10-flowered, the pedicels 2–6 mm. long, erect, rather stout; flowers erect, greenish white, about 3 mm. long, the bracteoles 3, minute, nearly free; capsules oblong, about 4 mm. long, erect; style very short; stigma small, discoid.—July–Aug. Alpine meadows and on rocks; Hokkaido, Honshu (centr. and n. distr.); rare.

2. Tofieldia coccinea J. Rich. *T. nutans* Willd.—CHISHIMA-ZEKISHŌ. Rhizomes short; leaves rather firm, laterally flattened, linear, 2.5–5 cm. long, 2–4 mm. wide, scabrous on margin, acuminate; scapes 6–12 cm. long, with 1–2 short leaves; racemes 1–2 cm. long, oblong to ellipsoidal, densely many-flowered; pedicels arcuate, 2–3 mm. long; flowers nodding to pendulous, about 3 mm. long, white to brownish, the subtending bracteole 3-lobed; anthers usually yellowish; capsules obovoid-globose, slightly longer than the tepals; styles about 0.5 mm. long.—July–Aug. Alpine slopes and rocky places; Hokkaido, Honshu (n. and centr. distr.).—Korea, Sakhalin, Ussuri, Amur, Ochotsk Sea region, and N. America. Very variable.

Var. **fusca** (Miyabe & Kudo) Hara. *T. nutans* var. *fusca* (Miyabe & Kudo) Ohwi; *T. fusca* Miyabe & Kudo—KURO-MINO-IWA-ZEKISHŌ. Tepals and capsules dark brown.—Occurs with the typical phase.

Var. **kondoi** (Miyabe & Kudo) Hara. *T. kondoi* Miyabe & Kudo; *T. yezoensis* Miyabe & Kudo—APOI-ZEKISHŌ. Plants slender; leaves 3–10 cm. long, 1–2.5 mm. wide; racemes loosely flowered; flowers white; style about 1 mm. long; capsules broadly obovoid.—Occurs with the typical phase.—Hokkaido.

Var. **gracilis** (Fr. & Sav.) Ohwi. *T. gracilis* Fr. & Sav.; *T. sordida* Fr. & Sav.; *T. stenantha* Fr. & Sav.—CHABO-ZEKISHŌ. Plants slender; leaves gradually long-acuminate, 3–8 cm. long; scapes 8–20 cm. long; racemes loosely flowered, to 6 cm. long, the pedicels slightly longer; flowers white, the anthers purplish; capsules broadly obovoid; style about 0.6 mm. long, slender.—Honshu (n. and centr. distr.).

Var. **kiusiana** (Okuyama) Hara. *T. nutans* var. *kiusiana* (Okuyama) Ohwi; *T. kiusiana* Okuyama—MIYAMA-ZEKISHŌ. Differs from var. *gracilis* in the longer scapes, 20–30 cm. long, the pedicels 4–6 mm. long; flowers 4.5–5 mm. long, the anthers pale; style erect, about 1 mm. long, slender.—Kyushu (Mount Dogatake in Hyuga Prov.).

3. Tofieldia nuda Maxim. *T. yoshiiana* Makino—HANA-ZEKISHŌ, IWA-ZEKISHŌ. Plants glabrous, sometimes with filiform or capillary stolons, the rhizomes short; leaves laterally flattened, linear, arcuate, 3- to 7-nerved, 5–12(–20) cm. long, long-acuminate, the margin smooth; scapes 12–30 cm. long, 2- or 3-leaved; racemes 3–6 cm. long, densely to loosely many-flowered, the pedicels slender, straight, spreading, 5–12 mm. long, the bracts lanceolate, acute, or subobtuse, 1.5–2 mm. long, the bracteoles 3-fid; flowers white, 3.5–4 mm. long, the anthers pallid or pale purple; capsules ovoid-ellipsoid, slightly longer than the tepals; style slender, 0.6–1.5 mm. long.—Rocks in mountains; Honshu (Kantō, w. Tōkaidō, Kinki, and Chūgoku Distr.), Kyushu.—Korea.

4. Tofieldia japonica Miq. *Triantha japonica* (Miq.) Makino—IWA-SHŌBU. Rhizomes short-creeping, branching, ascending; leaves linear, laterally flattened, 10–40 cm. long, 4–8 mm. wide, acuminate, the margin scabrous, broadest above the middle; scapes 20–40(–60) cm. long, 1- or 2-leaved, with viscid glandular-tubercles on upper portion; racemes 2–7 cm. long, erect, usually simple, the pedicels straight, obliquely spreading, ternate, 6–12 mm. long, the bracts ovate to lanceolate, the bracteoles 3, connate on lower half; flowers white, 5–7 mm. long, glabrous, the anthers cordate, dark purple; capsules broadly ovoid-ellipsoid, longer than the tepals; style 1.5–2 mm. long, the stigma punctiform; seeds ellipsoidal, with a filiform curved appendage.—Aug.–Sept. Bogs and moist slopes in mountains; Honshu (n. and centr. distr., westw. to Ōomi and Bingo Prov.).

5. NARTHECIUM Moehr. KINKŌKA ZOKU

Perennial herbs from creeping rhizomes; leaves nearly all radical, 2-ranked, laterally flattened, linear, rather firm, scapes simple, with few small leaves; racemes solitary, rarely branched below; flowers small, yellow, the bracts narrow, the pedicels with a linear bracteole near the middle; tepals 6, in 2 whorls, persistent, equal, linear, few-nerved; stamens 6, slightly shorter than the tepals, the filaments white-villous, the anthers linear, basifixed; ovary lanceolate, 3-locular, the ovules many; style simple, the stigma punctiform; capsules lanceolate, narrowed above, loculicidal; seeds many, small, caudate at both ends.—About 4 species in temperate regions of the N. Hemisphere.

1. Narthecium asiaticum Maxim. KINKŌKA. Glabrous; leaves linear, slightly arcuate, 10–25 cm. long, 5–10 mm. wide, long-acuminate, smooth, few-nerved; scapes erect, 20–50 cm. long, with few short green linear-lanceolate leaves; racemes 6–12 cm. long, rather many-flowered, glabrous, smooth, the bracts linear, usually slightly shorter than the pedicels; pedicels ascending, 7–15 mm. long, with a linear bracteole near the middle; tepals spreading, yellow in anthesis, erect and green in fruit, subacute, 8–10 mm. long, linear; anthers pale yellow, the filaments white-villous; capsules lanceolate, acuminate, 12–15 mm. long, about 2.5 mm. wide; seeds lanceolate, about 2 mm. long, the caudate tails about 10 mm. long.—July–Aug. Bogs, wet rocks, and moist places along streams in mountains; Hokkaido (Horomui), Honshu (n. to centr. distr. and Ise Prov.).

6. METANARTHECIUM Maxim.　NOGI-RAN ZOKU

Rhizomes short and stout; leaves radical, broadly oblanceolate, flat, abruptly acuminate to acute, with a slender midrib; scapes naked, simple or sparingly branched above; spikes elongate, glandular-pilose, the bracts and bracteoles small, the pedicels very short; tepals linear, persistent, spreading above, 1-nerved; stamens 6, inserted near the base of the tepals, the filaments linear, glabrous, the anthers oblong; ovary 3-locular, ovoid, the ovules many; style columnar, the stigma shallowly 3-lobulate; capsules narrowly ovoid-conical, loculicidal; seeds small, brown, ovoid, with few longitudinal striations, not appendaged.——One or two species, in Japan and Formosa.

1. Metanarthecium luteoviride Maxim. *Aletris luteoviridis* (Maxim.) Franch.——NOGI-RAN.　Radical leaves spreading, yellowish green, glabrous, oblanceolate to broadly so, gradually narrowed at base, abruptly acuminate to acute at apex, about 10-nerved, 8–20 cm. long, 1–4 cm. wide; scapes 20–40 cm. long, often with 1 or 2 branches above, naked; spikes 3–20 cm. long, usually minutely puberulent, many- or sometimes several-flowered, the bracts linear, longer than the very short pedicels, the bracteoles solitary, small; flowers yellow-green, the tepals subobtuse, 5–8(–10) mm. long, the filaments slightly broadened below; capsules oblong, acute, shorter than the tepals; style about 2 mm. long.——June–Aug.　Mountain meadows; Hokkaido, Honshu, Shikoku, Kyushu.

Var. **nutans** Masam.　YAKUSHIMA-NOGI-RAN.　Pedicels to 7 mm. long.——Kyushu (Yakushima).

7. HELONIOPSIS A. Gray　SHŌJŌBAKAMA ZOKU

Glabrous perennial herb from short stout rhizomes; leaves radical, rosulate, evergreen, narrowly oblong to oblanceolate, narrowed gradually below to a petiolelike base; scapes arising from the center of the rosette, erect, simple, with few lanceolate scalelike leaves; racemes few, rarely 1-flowered, often abbreviated and umbellate, the pedicels sometimes bracteate, bracteoles absent; flowers nodding in anthesis, erect in fruit, the tepals 6, persistent, spathulate or oblong, spreading, pinkish, purplish, or white, becoming greenish in fruit, obtuse; stamens 6, the filaments filiform; style simple, elongate, the stigma capitate; ovules numerous; capsules 3-lobed, septicidal on upper margin; seeds linear, caudate at both ends.——Few species, in Japan, Formosa, and Korea.

1. Heloniopsis orientalis (Thunb.) C. Tanaka. *Scilla orientalis* Thunb.; *H. pauciflora* A. Gray; *Sugerokia orientalis* (Thunb.) Koidz.; *H. japonica* Maxim.; *H. grandiflora* Fr. & Sav.——SHŌJŌBAKAMA.　Rhizomes short and stout, the roots firm; leaves oblanceolate, or subspathulate, 7–15 cm. long, 1.5–4 cm. wide, acute, short-mucronate, gradually narrowed below, green and slightly lustrous above; scapes 10–60 cm. long, elongate after flowering, with several small, greenish appressed sessile lanceolate leaves, especially below; racemes 3- to 10-flowered, the pedicels 1.5–2 cm. long in fruit; flowers rose-purple, changing to purple- or brown-green in fruit, the tepals 1–1.5 cm. long, the anthers usually narrowly oblong; styles about 2 cm. long.——Apr.–June.　Mountain thickets and meadows, sometimes ascending to the alpine zone; Hokkaido, Honshu, Shikoku, Kyushu (rare); rather common especially northward.——Korea and Sakhalin.

Var. **flavida** (Nakai) Ohwi. *H. japonica* var. *flavida* Nakai; *Sugerokia nipponica* Ohwi; *H. nipponica* (Ohwi) Nemoto; *H. japonica* var. *albiflora* Honda; *H. japonica* var. *tessellata* Nakai.——SHIROBANA-SHŌJŌBAKAMA.　Plants rather slender, the leaves slightly thinner, the margin undulate, often violascent at base; flowers white, pale green in fruit, the tepals gradually narrowed below.——Apr.–June.　Honshu (Kantō Distr. and westw.).

Var. **breviscapa** (Maxim.) Ohwi. *H. breviscapa* Maxim.; *H. japonica* var. *breviscapa* (Maxim.) Honda; *Scilla japonica* Thunb.; *Sugerokia japonica* Miq.; *H. orientalis* var. *yakusimensis* (Masam.) Ohwi; *H. japonica* var. *yakusimensis* Masam.; *H. breviscapa* var. *yakusimensis* (Masam.) Hara——TSUKUSHI-SHŌJŌBAKAMA.　The tepals slightly shorter, white to pale rose, cuneate below, the fruit reddish to yellowish green.——Mountains; Kyushu.

8. VERATRUM L.　SHURO-SŌ ZOKU

Rhizomes short, thickened, the roots stout; stems erect, rather stout, leafy, thickened below, usually with fibers disintegrated from the sheaths at base; leaves narrow to broad, plicately nerved, narrowed and usually sheathed at base; flowers polygamous, green, purple, brown, or white, numerous, on racemelike branches of the terminal panicle, short-pedicelled or nearly sessile, the floral bracts becoming increasingly foliar toward the upper leaves, the tepals 6, persistent, spreading, oblong to spathulate, many-nerved, narrowed below, not glandular-pitted; stamens 6, inserted at base of the tepals, the filaments filiform, the anthers small, cordate-orbicular, extrorse, the locules confluent; ovary ovoid, trigonous, slightly 3-lobed at apex; the ovules usually many; styles 3, short, stigmatic on the inside at apex.——About 50 species, in the temperate to cooler regions of the N. Hemisphere.

1A. Cauline leaves well-distributed, usually with some hairs on blades or sheaths; tepals white to greenish, usually denticulate.
  2A. Ovary glabrous; tepals slightly shorter than the stamens; pedicels longer than the tepals. .... 1. *V. stamineum*
  2B. Ovary hairy or with hairlike tubercles; tepals much longer than the stamens; pedicels shorter than the tepals.
    3A. Tepals broadly lanceolate, subacute; inflorescence very densely flowered. .... 2. *V. oxysepalum*
    3B. Tepals narrowly ovate to oblong, obtuse to subrounded at apex; inflorescence rather loosely flowered. .... 3. *V. grandiflorum*
1B. Cauline leaves confined to lower part of stem, glabrous; tepals green or dark brown-purple, entire.
  4A. Tepals ascending in fruit; inflorescence 10–20 cm. long, the bracts of main axis as long as to longer than the branches, rarely shorter; flowers greenish, rarely dark purple. .... 4. *V. longebracteatum*
  4B. Tepals reflexed in fruit; inflorescence 15–50 cm. long, the bracts of main axis much shorter than the branches; flowers greenish to dark purple. .... 5. *V. maackii*

1. **Veratrum stamineum** Maxim. KO-BAIKEI-SŌ. Plants rather stout, erect, simple, 50–100 cm. high; leaves alternate, broadly elliptic, the upper oblong to narrowly so, 10–20 cm. long, (2–)5–15 cm. wide, glabrous on both sides or nearly so, sheathed at base; inflorescence 15–25 cm. long, with short papillose hairs, the several branches rather stout, ascending, the ultimate bracts broadly lanceolate, 3–6 mm. long, the pedicels obliquely spreading, 7–12 mm. long; tepals white, spreading or more or less erect, oblong, 4–7.5 mm. long; stamens slightly longer than the tepals; capsules erect, narrowly ovate, about 2 cm. long, tapered above.——July–Sept. Wet meadows or bogs in mountains; Hokkaido, Honshu (n. and centr. distr.).

Var. **lasiophyllum** Nakai. URAGE-KO-BAIKEI. Leaves pilose beneath, especially on the nerves. Occurs with the typical phase.

Var. **micranthum** Satake. MIKAWA-BAIKEI-SŌ. Leaves with papillose hairs beneath; inflorescence rather large, loosely flowered; tepals distinctly shorter than the stamens.——May. Honshu (Mikawa Prov.).

2. **Veratrum oxysepalum** Turcz. *V. album* var. *oxysepalum* (Turcz.) Miyabe & Kudo; *V. album* subsp. *oxysepalum* (Turcz.) Hult.——EZO-BAIKEI-SŌ. Plants stout, erect, 50–150 cm. high, leafy, papillose-hairy above; leaves elliptic, acute to obtuse, 15–30 cm. long, 5–13 cm. wide, usually glabrescent; panicles 25–40 cm. long, the ultimate bracts 6–10 mm. long, ovate, the pedicels 1–2(–3) mm. long; flowers 8–12(–15) mm. in diameter, the tepals oblong-lanceolate, subacute, 8–12 mm. long, 2–4 mm. wide, denticulate, green; capsules ovate, narrowed above.——June–Aug. Bogs; Hokkaido.——Kamchatka, Ussuri, Dahuria, and Sakhalin.

3. **Veratrum grandiflorum** (Maxim.) Loes. f. *V. album* var. *grandiflorum* Maxim.; *V. patulum* sensu Ohwi, non Loes.; *V. sikokianum* Nakai——BAIKEI-SŌ. Plants stout, 1–1.5 m. high, papillose hairy above; leaves rather many, broadly elliptic to narrowly oblong, 20–30 cm. long, 7–20 cm. wide, the upper ones narrowed, the lower surface more or less papillose hairy, especially on the nerves, longitudinally plaited; panicles 20–50 cm. long, the branches suberect to ascending, loosely flowered, the ultimate bracts ovate to oblong, acute to obtuse, 3–10 mm. long, the pedicels 1–8 mm. long; flowers 15–25 mm. across, greenish white, the tepals spreading, oblong, obtuse to rounded, denticulate, about twice as long as the stamens; capsules narrowly ovate-oblong, narrowed at both ends especially above, 2–2.5 cm. long.——June–Aug. Wet places and bogs in mountains; Hokkaido, Honshu, Shikoku, Kyushu.

4. **Veratrum longebracteatum** Takeda. TAKANE-AO-YAGI-SŌ. Plants 20–60 cm. high, papillose hairy above; leaves few, the radical and lower cauline broadly lanceolate to narrowly oblong, glabrous, acuminate, 15–20 cm. long, 2–4 cm. wide, gradually narrowed below, the upper ones linear; flowers green, rarely brown-purple, 12–15 mm. in diameter, the pedicels 5–12 mm. long, the ultimate bracts lanceolate, acuminate, about twice as long as the pedicels, the tepals narrowly oblong, ascending in fruit; capsules elliptic, about 12 mm. long.——July–Aug. Alpine slopes; Honshu (n. and centr. distr.).

5. **Veratrum maackii** Regel var. **maackii**. *V. nigrum* var. *maackii* (Regel) Maxim.; *Zigadenus japonicus* Miq.——NAGABA-SHURO-SŌ, HOSOBA-SHURO-SŌ. Resembling var. *japonicum*, the lower leaves broadly linear, 20–40 cm. long, 1–1.5 (–2) cm. wide; capsules narrowly obovate-oblong, to 20 mm. long.——July–Sept. Woods in mountains; Honshu (centr. distr. and westw.), Shikoku, Kyushu.——Ussuri, Manchuria, and Korea.

Var. **parviflorum** (Miq.) Hara & Mizushima. *Veratrum album* var. *parviflorum* Miq.; *V. maacki* var. *maximowiczii* Nakai; *V. angustisepalum* Loes. f.——AO-YAGI-SŌ. Plants 50–100 cm. high, few-leaved; lower leaves oblong to ovate-oblong, 20–30 cm. long, 6–10 cm. wide, ascending, acute, narrowed at base, the upper linear; panicles 30–50 cm. long, the lower branches ascending, rather slender, the bracts of the main axis shorter than the branches, the ultimate bracts broadly lanceolate, obtuse, 2–5 mm. long, papillose-hairy; flowers rather loosely arranged, green, 8–10 mm. wide, the pedicels 6–10 mm. long, papillose-hairy; tepals spreading to slightly reflexed, broadly oblanceolate to oblong-lanceolate, subacute, more than twice as long as the stamens; capsules elliptic, the spreading to subreflexed tepals adnate at base.——June–Aug. Woods in mountains; Hokkaido, Honshu (n. and centr. distr.).——Korea.

Var. **japonicum** (Baker) Shimizu. *V. nigrum* var. *japonicum* Baker; *V. nigrum* var. *reymondianum* Loes. f.; *V. maackii* var. *reymondianum* (Loes. f.) Hara——SHURO-SŌ. Leaves radical and lower cauline, glabrous, narrowly oblong to linear-lanceolate, 20–35 mm. wide, gradually tapered toward the base; inflorescence 15–30 cm. long, branched, slender, the bracts shorter than the branches; flowers loosely arranged, brown-purple, about 10 mm. across, the pedicels 8–12 mm. long; tepals broadly lanceolate, subobtuse, about twice as long as the stamens; capsules elliptic, 12–15 mm. long.——July–Sept. Woods in mountains; Hokkaido, Honshu.

9. ZIGADENUS Michx. RISHIRI-SŌ ZOKU

Rhizomes horizontal, or the plants bulbous; scapes erect, simple; leaves radical or nearly so, linear; inflorescence racemose, often branching and subpaniculate; flowers bisexual or polygamous, the tepals 6, persistent, in 2 whorls, obliquely spreading, with 2 distinct or connate glands near base; stamens 6, inserted at base and slightly shorter than the tepals, the filaments filiform, the anthers small, suborbicular, the locules confluent; styles 3; ovules many; capsules ovate or oblong, 3-lobed, septicidal, the seeds oblong or sublanceolate.——About 10 species, mainly in N. America, 1 species in Asia.

1. **Zigadenus sibiricus** (L.) A. Gray. *Melanthium sibiricum* L.; *Z. japonicus* Makino, non Miq.; *Anticlea japonica* (Makino) Gates; *A. makinoana* (Miyabe & Kudo) Tatew. & Ohwi——RISHURI-SŌ. Bulbs slender; scapes 12–25 cm. long; leaves few, radical, linear, 10–20 cm. long, 4–10 mm. wide, subacute to obtuse, tapered below, the cauline 0–2, short; racemes simple or with 1 or 2 short branches, loosely few-flowered, 3–12 cm. long, the bracts narrowly lanceolate to broadly ovate, 3–25 mm. long; flowers about 10 mm. in diameter, pale yellowish green or purplish outside, the tepals ascending, spreading at anthesis, erect in fruit, 7–8 mm. long, 2.5–4 mm. wide, oblong or narrowly rhombic-ovate, obtuse, the glands obcordate; capsules conical, narrowed above.——Aug. Alpine zone; Hokkaido (Rishiri Isl.).——Northern Asia.

10. TRICYRTIS Wall.   Hototogisu Zoku

Rhizomes short-creeping; stems erect, sometimes branched, often reclinate above; leaves alternate, sessile, all cauline, oblong to ovate, sometimes clasping or perfoliate; flowers solitary or in axillary and terminal cymes or in fascicles, bisexual, showy, rather large, infundibuliform; pedicels sometimes small-bracteate; tepal 6, 2-seriate, free to base, early deciduous, recurved above, the outer ones saccate or short-spurred; stamens 6, the filaments slightly flattened, the anthers oblong, extrorse; ovary oblong, 3-angled, 3-locular; style columnar, stigmas 3, spreading, bifid, tuberculate on the inner side, the ovules numerous; capsules narrowly oblong, or linear-trigonous, septicidal; seeds ovate or orbicular, flat.——More than 10 species, in e. Asia to India.

1A. Outer tepals short-spurred; flowers nodding; stems declined; leaves clasping.
  2A. Flowers fasciculate or solitary in upper leaf-axils.
    3A. Leaves shallowly cordate at base, auriculate on one side. .... 1. *T. macrantha*
    3B. Leaves deeply cordate at base, auriculate on both sides. .... 2. *T. macranthopsis*
  2B. Flowers in a terminal corymb; auricles at base of leaves on each side of the stem. .... 3. *T. ishiiana*
1B. Outer tepals saccate; flowers erect.
  4A. Tepal ground-color yellow.
    5A. Flowers fasciculate or solitary in leaf-axils.
      6A. Leaves merely clasping.
        7A. Pedicels distinct, nearly equaling to longer than the capsules and flowers.
          8A. Tepals narrowly obovate, scarcely spotted, rounded, the mucro reflexed, 1.5–2 mm. long; anthers about 4 mm. long.
                4. *T. ohsumiensis*
          8B. Tepals oblong-spathulate, the outer ones acute, with an erect or slightly recurved mucro 3–4 mm. long, the inner tepals obtuse to subretuse, the mucro scarcely 1 mm. long; anthers 2.5–3 mm. long. .... 5. *T. flava*
        7B. Pedicels very short, much shorter than the flowers and capsules; tepals oblong-spathulate, the outer ones acute, with a short, erect to ascending mucro 1.5–2 mm. long, the inner ones obtuse, the mucro ascending to spreading, about 0.5 mm. long; anthers about 2 mm. long; stems very short. .... 6. *T. nana*
      6B. Leaves perfoliate, broadly lanceolate to ovate-oblong, long-acuminate at apex. .... 7. *T. perfoliata*
    5B. Flowers distinctly cymose, few; cymes peduncled. .... 8. *T. latifolia*
  4B. Tepal ground-color white.
    9A. Tepals spreading-recurved from near the base, about 2 cm. long.
      10A. Flowers in terminal or axillary, 1- or 2-flowered pedunculate cymes; stems hispid-hirsute to glabrous; pedicels densely pilose with rather soft curved short hairs sometimes gland-tipped at apex. .... 9. *T. macropoda*
      10B. Flowers in axillary and terminal fascicles or solitary; common peduncles usually not developed; stems setose to glabrous; pedicels densely hispidulous with short spreading hairs. .... 10. *T. affinis*
    9B. Tepals ascending, short-spreading toward apex, 2.5–3 cm. long; flowers solitary or fascicled; stems hirsute. .... 11. *T. hirta*

1. Tricyrtis macrantha Maxim. *Brachycyrtis macrantha* (Maxim.) Koidz.——Jōrō-hototogisu, Tosa-jōrō-hototogisu. Closely resembles 2 and 3 but the stems usually smaller, with pale brown coarse ascending hairs; leaves ovate-oblong to narrowly ovate, gradually acuminate, rounded to shallowly cordate at base, the basal auricles turned on one side of stem.—— July–Aug. Shikoku; rare.

2. Tricyrtis macranthopsis Masam. *Brachycyrtis macranthopsis* (Masam.) Honda; *T. macrantha* var. *macranthopsis* (Masam.) Okuyama & T. Koyama——Kii-jōrō-hototogisu. Rhizomes short; stems 40–80 cm. long, reclined above; leaves distichous, narrowly oblong-ovate, 7–17 cm. long, 2.5–5 cm. wide, lustrous above, glabrous, gradually acuminate, cordate and clasping with 2 basal auricles or the lower edges placed on each side of the stems, pale green beneath and subglabrous to sparsely hirsute on the nerves; flowers axillary and terminal, solitary to few, pendulous, 3–4 cm. long, tubular-campanulate, glabrous; tepals oblanceolate, clear yellow, with brownish purple dots inside, the outer ones with a short spur at base, short-mucronate.——Aug.–Oct. Mountains; Honshu (Kii Prov.); rare. Occasionally planted in gardens.

3. Tricyrtis ishiiana (Kitag. & T. Koyama) Ohwi & Okuyama. *T. macrantha* var. *ishiiana* Kitag. & T. Koyama. ——Sagami-jōrō-hototogisu. Closely resembles the preceding; stems 20–50 cm. long; leaves oblong-ovate to lanceolate, 4.5–10 cm. long, nearly glabrous, the basal edges clasping the stem; flowers 3–5 in a terminal corymb, bracts ovate, up to 8 mm. long, the spur of the outer tepals much broader, 5–6 mm. wide.——Honshu (Sagami Prov.); rare.
Var. **surugensis** Yamazaki. Suruga-jōrō-hototogisu.

Leaves pubescent on costas and nerves beneath; bracts short; spur small, 1–1.5 mm. wide; anthers dark brown.——Mountains; Honshu (Mount Tenshigatake in Suruga Prov.); rare.

4. Tricyrtis ohsumiensis Masam. Takakuma-hototogisu. Stems 20–50 cm. long, nearly glabrous; lower leaves oblong-lanceolate, acute, the upper elliptic to oblong, clasping, abruptly acute, 5–20 cm. long, 2–6 cm. wide, glabrous; flowers axillary and terminal, one or 2, yellow, nearly spotless, 2.5–3.5 cm. long, the pedicels slightly shorter to longer than the flowers, the tepals narrowly obovate or obovate-oblong, about 1 cm. wide, the inner segments slightly narrower.——Sept.–Oct. Kyushu (Ōsumi Prov.).

5. Tricyrtis flava Maxim. *T. yatabeana* Masam.—— Kibana-no-hototogisu. Stems 30–50 cm. long, pilose, purplish; leaves rather densely arranged, oblanceolate to broadly elliptic, 7–15 cm. long, acuminate, sometimes spotted above; flowers 1 or 2, axillary and terminal, about 2.5 cm. long, with dark purple spots inside, the tepals narrowly oblong-obovate to oblong-spathulate, the outer segments long-mucronate.—— Sept.–Oct. Woods in mountains; Honshu (s. Kinki Distr.), Shikoku, Kyushu.

6. Tricyrtis nana Yatabe. *T. flava* var. *nana* (Yatabe) Makino; *T. flava* sensu Masam., non Maxim.——Chabo-hototogisu. Stems 5–15 cm. long, short-pubescent, few-leaved; leaves acuminate, 6–12 cm. long, 1.5–4 cm. wide; flowers 1 or 2, axillary and terminal, yellow, with brown-purple spots, the tepals oblanceolate, about 2 cm. long, acute.—— Aug.–Sept. Woods in mountains; Honshu (s. Kinki Distr.), Shikoku, Kyushu.

7. Tricyrtis perfoliata Masam. Kibana-no-tsuki-nuki-

HOTOTOGISU. Stems simple, 50–70 cm. long, declined above, glabrous; lower leaves narrowly ovate, the upper broadly lanceolate to narrowly ovate-oblong, 7–18 cm. long, 2.5–4 cm. wide, perfoliate, gradually narrowed above and long-acuminate, glabrous; flowers solitary in the axil of median leaves, yellow, the pedicels shorter than the flowers, the tepals 2.5–3 cm. long, narrowly oblong, ascending, slightly purple-spotted inside, acute.——Oct. Kyushu; rare.

8. Tricyrtis latifolia Maxim. TAMAGAWA-HOTOTOGISU. Plants nearly glabrous, except the inflorescence; stems 40–80 cm. long; leaves obovate, 8–15 cm. long, 4–9 cm. wide, rather thin, abruptly acuminate, deeply cordate and prominently clasping, short-ciliate while young; inflorescence cymose, at summit of stems and in upper axils, few- to several-flowered, pedunculate; flowers yellow, with purple spots inside, the tepals obliquely spreading, 2–2.5 cm. long, subacute, the outer segments broadly oblanceolate, the inner linear-oblong.—— July–Sept. Woods in mountains; Hokkaido, Honshu, Shikoku, Kyushu.

Var. **makinoana** (Tatew.) Hiyama. *T. makinoana* Tatew.; *T. latifolia* var. *nikkomontana* Hiyama——HAGOROMO-HOTOTOGISU. The stems and leaf veins hairy beneath. Occurs with the typical phase.

9. Tricyrtis macropoda Miq. YAMA-HOTOTOGISU. Stems 40–70 cm. long; leaves ovate to oblong, 8–13 cm. long, 3–6 cm. wide, acuminate, subcordate and clasping except a few lower ones; inflorescence axillary and terminal, few- to several-flowered, densely soft-pubescent, the peduncles distinct; flowers white with purple spots, the tepals 1.5–2 cm. long, flared on upper half, acute, the outer segments broadly lanceolate, the inner narrowly oblanceolate.——July–Sept. Honshu, Shikoku,

Kyushu.——Forma **hirsuta** Masam. Plants hirsute.—— Forma **glabrescens** Masam. Plants glabrescent.

Var. **chugokuensis** (Koidz.) Ohwi. *T. chugokuensis* Koidz.; *T. affinis* var. *chugokuensis* (Koidz.) Ohwi.——CHŪ-GOKU-HOTOTOGISU. Cymes terminal and axillary, several-flowered, on longer pilose pedicels and peduncles.——Honshu (Chūgoku Distr.); rare.

10. Tricyrtis affinis Makino. *T. japonica* sensu Masam., non Miq.——YAMAJI-NO-HOTOTOGISU. Stems 30–60 cm. long, setose, with slightly reflexed hairs or glabrous; leaves narrowly oblong, ovate-oblong, or broadly oblanceolate, 8–18 cm. long, 2.5–5 cm. wide, loosely hairy, acuminate, the upper leaves clasping; flowers 1 to several in axils and at summit of stems, about 2 cm. long, the pedicels nearly as long to slightly shorter than the flowers, the tepals white, with dark purple spots inside, subacute, the outer segments broadly oblanceolate, the inner lanceolate, spreading-recurved.——Aug.–Oct. Hokkaido, Shikoku, Kyushu.

11. Tricyrtis hirta (Thunb.) Hook. *Uvularia hirta* Thunb.; *T. japonica* Miq.——HOTOTOGISU. Perennial; stems 40–80 cm. long, hairy; leaves narrowly ovate-oblong to broadly lanceolate, 8–15 cm. long, 2–5 cm. wide, acuminate, those above usually clasping, the lower ones free; flowers (1–) 2–3, axillary and terminal, the pedicels usually shorter than the flowers, the tepals obliquely ascending, 2.5–3 cm. long, white with purple spots inside, acute, the outer segments oblanceolate, the inner ones narrowly so.——Aug.–Oct. Honshu, Shikoku, Kyushu; rather common.

Var. **masamunei** (Makino) Masam. *T. masamunei* Makino ——SATSUMA-HOTOTOGISU. Plants glabrate.——Kyushu (Satsuma).

11. HOSTA Tratt. GIBŌSHI ZOKU

(Treatment based mainly on monograph by Dr. F. Maekawa.)

Perennial glabrous herbs from short rhizomes, often clothed at base with fibers of disintegrated leaf-sheaths; leaves radical, tufted, flat, petiolate; scapes terminal, sometimes few-scaled, the racemes simple; flowers pedicellate, bracteate, the tepals connate, tubular-campanulate, the lobes 6, in 2 series, imbricate in bud, white to purplish; stamens 6, free, the filaments filiform, all ascending above on one (upper) side, the anthers 2-locular, versatile, parallel; ovary 3-locular, superior; style filiform, the stigma capitate; capsules loculicidal, 3-valved; seeds many.——More than 40 species, all but 1 or 2 in Japan.

1A. Bracts 2; flowers nocturnal, fragant, white, about 11.5 cm. long; stamens inserted on the perianth near the base. ... 1. *H. plantaginea*
1B. Bract solitary; flowers diurnal, not fragant, smaller; stamens free to the base.
  2A. Bracts navicular, densely imbricate in bud; perianth thin, the upper dilated portion often with translucent lines.
    3A. Scapes erect, without leaflike bracts; perianth-tube with strongly impressed translucent lines on upper dilated portion, and 6 distinct grooves on the lower narrower portion.
      4A. Scapes longitudinally striate; leaves chartaceous.
        5A. Scapes fistulose; racemes elongate, spikelike; bracts greenish. .... 2. *H. venusta*
        5B. Scapes solid; racemes capitately congested; bracts whitish purple.
          6A. Leaves abruptly cordate. .... 3. *H. capitata*
          6B. Leaves more or less truncate. .... 4. *H. nakaiana*
      4B. Scapes not longitudinally striate; bracts not leaflike.
        7A. Lower part of the perianth-tube much shorter than the upper dilated portion, colored in bud, the translucent lines very much impressed.
          8A. Leaves yellow on margin. .... 5. *H. opipara*
          8B. Leaves green.
            9A. Lower narrower portion of perianth-tube purple; leaves decurrent on the petiole, acute to acuminate.
              10A. Leaves glaucescent; flowers deep purple. .... 6. *H. atropurpurea*
              10B. Leaves green; flowers pale purple. .... 7. *H. rectifolia*
            9B. Lower narrower portion of the perianth-tube whitish; leaves distinctly petioled, elliptic, very obtuse. .. 8. *H. decorata*
        7B. Lower portion of the perianth-tube nearly as long as the dilated upper portion, nearly white in bud, the translucent lines sometimes obsolete.
          11A. Leaves erect, narrow, strongly lustrous above; flowers small and few. .... 9. *H. longissima*
          11B. Leaves somewhat spreading, dull, flowers many, larger.

12A. Leaves narrow, widely spreading. .... 10. *H. helonioides*
12B. Leaves obliquely spreading; bracts clasping.
  13A. Flowering in autumn; leaves slightly lustrous. .... 11. *H. okamii*
  13B. Flowering in summer; leaves dull.
    14A. Leaves decurrent on the petiole. .... 12. *H. rhodeifolia*
    14B. Leaves abruptly petioled.
      15A. Flower-bud clavate, obtuse. .... 13. *H. clavata*
      15B. Flower-bud narrow, subacute. .... 14. *H. albomarginata*
3B. Scapes ascending above, with leaflike bracts; perianth angled on the lower narrow portion but obsoletely grooved, the translucent lines not impressed in anthesis. .... 15. *H. undulata*
2B. Bracts flat or the midrib impressed sometimes only at base, spreading at anthesis, rarely navicular or imbricate; broad upper portion of the perianth equally purplish or white.
16A. Broad upper portion of perianth-tube obconical, with impressed translucent lines.
  17A. Perianth-lobes linear-lanceolate, about 5 mm. wide. .... 16. *H. gracillima*
  17B. Perianth-lobes wider.
    18A. Rather large plant with thin, lusterless leaves; bracts withering after anthesis. .... 17. *H. tardiva*
    18B. Small plant with rather coriaceous, lustrous leaves; scapes short; bracts persistent. .... 18. *H. cathayana*
16B. Broad upper portion of perianth-tube narrow, campanulate-infundibuliform, with incomplete translucent lines, the narrower lower portion angled, not grooved; bracts often whitish or purplish.
19A. Bracts very thinly membranous, loosely imbricate, elongate; perianth thinner; flowering in early autumn; leaves not coriaceous.
  20A. Leaves narrow, spathulate or ovate; scapes few-flowered.
    21A. Leaves narrower; flowers, fewer, narrow.
      22A. Leaves spathulate, flat, long-petioled. .... 19. *H. tardiflora*
      22B. Leaves very narrow, often undulate and tortuous. .... 20. *H. tortifrons*
    21B. Leaves ovate; flowers more in number, broader. .... 21. *H. longipes*
  20B. Leaves broad, cordate, prominently coriaceous; scapes densely many-flowered. .... 22. *H. rupifraga*
19B. Bracts herbaceous, green or whitish; flowering in summer; leaves usually large, firmly coriaceous.
23A. Bracts flat.
  24A. Bracts completely flat, radiating in a stellate manner, often whitish; perianth somewhat thick.
    25A. Plants stout; leaves firm, coriaceous, flat or undulate, distinctly glaucous while young; scapes elongate; filaments thick.
      26A. Leaves completely flat, yellowish green, the margin involute below. .... 23. *H. nigrescens*
      26B. Leaves undulate, deep green. .... 24. *H. fluctuans*
    25B. Plants relatively slender; leaves herbaceous, rugose or with impressed nerves above, minutely papillose or papillose and pilose on raised nerves beneath.
      27A. Scapes shorter or not more than twice as long as the leaves; broader upper portion of the perianth-tube obconical, spreading at an angle of 160–180°.
        28A. Leaves nearly orbicular, spreading from the base, distinctly rugulose above, firm and blue-green; flowers not fully opening. .... 25. *H. tokudama*
        28B. Leaves ovate, elliptic, or orbicular, gradually spreading above; flowers fully opening. .... 26. *H. sieboldiana*
      27B. Scapes more than twice as long as the leaves; bracts usually white to green; broad upper portion of perianth-tube narrowly obconical, sometimes slightly inflated at base, spreading at an angle of 90–150°.
        29A. Leaves 18–23 cm. long, long-petioled, green. .... 27. *H. montana*
        29B. Leaves about 13–16 cm. long, thinner in texture. .... 28. *H. crispula*
  24B. Bracts flat, colored, imbricate to form an elongate beak; perianth often thin.
    30A. Perianth white; broader upper portion of perianth-tube slightly inflated at base. .... 29. *H. kikutii*
    30B. Perianth sordid purplish on both sides; broader upper portion of perianth-tube abruptly infundibuliform.
      31A. Scapes erect. .... 30. *H. tosana*
      31B. Scapes declined. .... 31. *H. caput-avis*
23B. Bracts often boat-shaped, green or slightly colored, imbricate.
32A. Leaves large, flat; stamens shorter than the perianth. .... 32. *H. sacra*
32B. Leaves rather large to moderate in size; stamens longer than the perianth.
  33A. Broad upper portion of the perianth-tube abruptly broadened, the lobes recurved. .... 33. *H. hippeastrum*
  33B. Broad upper portion of the perianth-tube infundibuliform, the lobes erect to spreading.
    34A. Bracts ovate-lanceolate, entirely green, short. .... 34. *H. kiyosumiensis*
    34B. Bracts oblong-lanceolate, imbricate, white.
      35A. Perianth about 5 cm. long, clavately infundibuliform, the upper broad portion longer than wide. .. 35. *H. densa*
      35B. Perianth about 4 cm. long, the upper broad portion shorter than wide. .... 36. *H. pachyscapa*

1. Hosta plantaginea (Lam.) Asch. *Hemerocallis plantaginea* Lam.; *Hemerocallis alba* Andr.; *Funkia subcordata* Spreng.; *Niobe plantaginea* (Lam.) Nash——MARUBA-TAMA-NO-KANZASHI. Large rank herb from stout rhizomes; leaves cordate, long-petioled, 15–22 cm. long, 10–17 cm. wide, green, ovate-orbicular, abruptly acute, with 8 or 9 pairs of nerves, smooth, glabrous, the margin slightly undulate; scapes 40–65 cm. long with 1–2(–4) scales; racemes elongate, the bracts 1(–2), 3–8 cm. long, narrowly ovate, or ovate-lanceolate, spreading, green; perianth infundibuliform, about 11.5 cm.

long; stamens as long as the perianth; capsules cylindric, 3-angled, pendulous, about 6.5 cm. long, 7–8 mm. wide; seeds winged on margin.——Aug. Introduced from China, long cultivated in our gardens.

Var. **japonica** Kikuchi & F. Maekawa. TAMA-NO-KANZASHI. Leaves elongate, rather loose, the flowers narrower, usually sterile.——China. Long cultivated in our gardens.

2. Hosta venusta F. Maekawa. *H. venusta* var. *decurrens* F. Maekawa——OTOME-GIBŌSHI. Scapes about 20 cm. long, slender, angled, about twice as long as the leaves; leaves

3–4 cm. long, about 2 cm. wide, petiolate, ovate or ovate-elliptic, mucronate, abruptly decurrent on the petiole, with 3–4 pairs of nerves; flowers pale purple, 2.5–3 cm. long, the pedicels 6–8 mm. long, spreading, the bracts 3–4.5 mm. long.——June–July. Honshu (centr. distr.). Sometimes cultivated in our gardens.——Korea (Quelpaert Isl.).

Hosta minor Nakai. KEIRIN-GIBŌSHI. Small herb with broadly ovate-orbicular leaves short-cordate to truncate at base; scapes much longer than the leaves, longitudinally lined; flowers 5–5.5 cm. long. Poorly known.——Reported from Kyushu (Tsushima).——Korea.

3. **Hosta capitata** (Koidz.) Nakai. *H. caerulea* var. *capitata* Koidz.——IYA-GIBŌSHI. Scapes to 65 cm. long, about twice as long as the leaves, loosely 3-scaled, terete; leaves cordate-ovate or broadly ovate, 8–12 cm. long, 5–7.7 cm. wide, long-petioled, abruptly acuminate, green, dull, with 7–9 pairs of impressed nerves above, the margins undulate; inflorescence 3.5 cm. long, dense, the flowers 4.5–5 cm. long, purplish, the bracts about 2 cm. long, elliptic, purplish, acute, the pedicels 4–6 mm. long.——Shikoku.

4. **Hosta nakaiana** F. Maekawa. KANZASHI-GIBŌSHI. Leaves long-petioled, oblong-ovate, recurved above, acuminate, cordate-truncate, 6–7.5 cm. long, 2.8–4.5 cm. wide, the margins slightly incurved near base, minutely undulate, green above, membranously chartaceous, with 5–7 pairs of nerves; scapes much longer than the leaves, 35–45 cm. long, angled, with a scalelike leaf near the top; racemes capitate, the bracts imbricate, elliptic, 14–17 mm. long, 6–10 mm. wide, strongly boat-shaped, white to purplish, obliquely spreading, acute or mucronate, the pedicels about 5 mm. long, obliquely spreading; perianth purple, 4.5–5 cm. long.——Honshu (Kinki Distr.), Kyushu.

5. **Hosta opipara** F. Maekawa. NISHIKI-GIBŌSHI. Leaves horizontally spreading, flat, long-petioled, elliptic or ovate-elliptic, slightly undulate, acuminate, abruptly narrowed at base, 13–18 cm. long, 7–12 cm. wide, with about 9 nerves, the margin whitish or yellowish; scapes about 80 cm. long, erect, the bracts broadly ovate or ovate-elliptic, 9–15 mm. long, acute; flowers about 4 cm. long, purplish.——Planted in Honshu (Mutsu Prov.).

6. **Hosta atropurpurea** Nakai. KUROBANA-GIBŌSHI. Leaves broadly lanceolate to ovate-oblong, petioled, 5–20 cm. long, glaucescent; scapes 45–60 cm. long, 1-scaled; racemes erect, the bracts lanceolate, purplish; flowers deep purple.——High mountains; Hokkaido.

7. **Hosta rectifolia** Nakai. TACHI-GIBŌSHI. Leaves firm, long-petioled, oblong-lanceolate, 17–20 cm. long, 6–7 cm. wide, slightly undulate, with 7 pairs of nerves, acute, gray-green, decurrent below on the petiole; scapes about 80 cm. long, the bracts loose; flowers slightly nodding, 3.7–5 cm. long, whitish, the outer perianth-lobes ovate, subacute, about 12 mm. wide, the pedicels 9–12 mm. long, the bracts elliptic, subacute, green with purplish striae; stamens shorter than the perianth, the anthers whitish.——Hokkaido, Honshu (centr. and n. distr.).——s. Kuriles, Sakhalin, and Ussuri.

8. **Hosta decorata** L. H. Bailey. OTAFUKU-GIBŌSHI. Leaves ascending, long-petioled, elliptic, 9–12 cm. long, 5–6 cm. wide, short-acuminate to subobtuse, narrowed below and rounded at base, flat, dull, 5-nerved, the margins white; scapes about 50 cm. long, median scales clasping, the bracts involute, broadened at base, the pedicels very short; flowers nodding, about 5.2 cm. long, pale purple, the outer perianth-lobes ob-

long, acuminate, about 2 cm. long, 1 cm. wide.——Cultivated in Honshu.——Forma **normalis** Stearn. MIDORI-OTAFUKU-GIBŌSHI. Leaves green.——Spontaneous in Honshu (Shinano Prov.).

9. **Hosta longissima** Honda. *H. japonica* var. *longifolia* Honda; *H. lancifolia* var. *longifolia* (Honda) Honda——NAGABA-MIZU-GIBŌSHI. Leaves linear-oblanceolate, 17–19 cm. long, 1.7–2 cm. wide, subobtuse, long-tapering to the petiole, spreading-recurved, flat above, dull; scapes about 50 cm. long; racemes usually 3-flowered; flowers small, spreading, pale rose-purple, the outer perianth-lobes oblong-lanceolate, erect, acute, about 12 cm. long, 6 mm. wide, the pedicels 6–9 mm. long, shorter than the bract, purplish, curved; capsules nodding, about 3.3 cm. long.——Honshu (w. distr.).

Var. **brevifolia** F. Maekawa. MIZU-GIBŌSHI, SAJI-GIBŌSHI. Leaves erect, spreading, 17–19 cm. long, 1.7–2 cm. wide; racemes 3- to 5-flowered, the bracts 8–9 mm. long, very acute. ——Wet places; Honshu (centr. distr. and westw.).

10. **Hosta helonioides** F. Maekawa. HAKAMA-GIBŌSHI. Leaves spreading, linear-oblanceolate, acuminate, gradually narrowed to the petiole, with 3–4 pairs of lateral nerves, green and dull above; scapes erect, the scales rather numerous, oblong-lanceolate, boat-shaped, incurved above; bracts boat-shaped, involute on margin, 17–32 mm. long, oblong-lanceolate, acuminate, green; pedicels 5–10 mm. long; flowers pendulous, about 5 cm. long, with pale-purple striae, the perianth-lobes usually spreading or reflexed at tip.——Honshu (n. distr.).

11. **Hosta okamii** F. Maekawa. MURASAMO-GIBŌSHI. Leaves rather thick, obliquely spreading, petiolate, lanceolate to oblong-lanceolate, 7–10 cm. long, 22–28 mm. wide, acute, obsoletely glandular-dotted, cuneate below and decurrent on the petiole, flat except the sometimes undulate margin, rather lustrous; scapes about 50 cm. long, slender, terete, green; racemes loose, one-sided, about 12-flowered, the bracts thinly membranous, ovate, boat-shaped, 7–12 mm. long, translucent, equaling to slightly longer than the spreading pedicel; perianth about 4 cm. long, pale purple, the outer lobes spreading, narrowly elliptic, about 16.5 mm. long; stamens long-exserted. ——Honshu (Kinki Distr.), Shikoku.

12. **Hosta rhodeifolia** F. Maekawa. OMOTO-GIBŌSHI. Leaves lanceolate to oblong-lanceolate, green to yellowish and slightly lustrous above, flat with 5 or 6 pairs of nerves, the margin slightly involute, narrowed below; scapes long-exserted, 75–120 cm. long, erect, with 5 or 6 scales, the bracts ovate, subobtuse, boat-shaped, with yellow-white margins, clasping, 1.5–3 cm. long; flowers nodding, about 4.5 cm. long, pale purple, with purple striae, the pedicels abruptly curved below. ——Cultivated in Honshu (w. distr.).——Forma **viridis** F. Maekawa.——AOBA-OMOTO-GIBŌSHI. Leaves green.——Spontaneous in Honshu (Kinki Distr.).

13. **Hosta clavata** F. Maekawa. KO-GIBŌSHI, MUSASHINO-GIBŌSHI. Leaves lanceolate or oblong-lanceolate, 7–11 cm. long, 2–3 cm. wide, acute, undulate on lower margin and narrowly decurrent on the petiole, dull above, with 3–4 pairs of nerves; scapes 40–45 cm. long, erect, slender, the bracts green, very acute, the pedicels very short, curved; flower-buds clavate, obtuse, the flowers at right angles to the scape, about 4 cm. long, white to pale purplish.——Mountains; Honshu (Kantō Distr.).

14. **Hosta albomarginata** (Hook.) Ohwi. *Funkia albomarginata* Hook.; *Aletris japonica* Thunb.; *Hemerocallis*

japonica (Thunb.) Thunb.; *Hemerocallis lancifolia* Thunb.; *Funkia lancifolia* (Thunb.) Spreng.; *Funkia ovata* var. *lancifolia* (Thunb.) Miq.; *Niobe japonica* (Thunb.) Nash; *Hosta japonica* (Thunb.) Voss, non Tratt.; *Funkia ovata* var. *albomarginata* (Hook.) Miq.; *Hosta lancifolia* (Thunb.) Engler; *Hosta lancifolia* var. *albomarginata* (Hook.) Stearn——Koba-gibōshi.　Leaves obliquely spreading, narrowly ovate, elliptic, or ovate-elliptic, 10–16 cm. long, 5–8 cm. wide, acute to acuminate, narrowed and obtuse at base, dull and gray-green above, flat or slightly undulate, 5- to 6-nerved; scapes erect, about 45 cm. long, loosely 2- to 5-scaled; racemes erect, about 30-flowered, the bracts 1.5–3 cm. long, green, narrowly ovate, acuminate; flowers spreading in anthesis, later nodding, about 5 cm. long, the pedicels about 6 mm. long, the perianth pale purple. ——Honshu, Shikoku, Kyushu. A variegated-leaved phase of this species is cultivated.

15. Hosta undulata (Otto & Dietr.) L. H. Bailey. *Funkia undulata* Otto & Dietr.; *F. ovata* var. *undulata* (Otto & Dietr.) Miq.; *H. lancifolia* var. *undulata* (Otto & Dietr.) L. H. Bailey——Suji-gibōshi.　Leaves rather large, ovate, with a broad white line down the center, the margins undulate; scapes much longer than the leaves, 80–100 cm. long, leafy or sometimes naked; racemes about 10-flowered, the bracts 2–2.5 cm. long; perianth 5.5–6 cm. long, white or pale purple.——Cultivated in our area.

Var. **erromena** (Stearn) F. Maekawa. *H. erromena* Stearn——Ō-ha-tsuki-gibōshi.　Leaves about 20 cm. long, 13 cm. wide, long-petioled, lustrous, green, the margins undulate, with about 10 pairs of nerves; scapes about 100 cm. long, with 2 or 3 leaflike scales.——Known only in cultivation.

16. Hosta gracillima F. Maekawa. Hime-iwa-gibōshi.　Leaves spreading, petiolate, lanceolate to ovate-lanceolate, 2.5–6 cm. long, 1–2 cm. wide, gradually caudate-acuminate, lustrous and green above, obtuse at base, with 3(–4) pairs of nerves, the margins crisped or flat; scapes erect or nearly so, slender, terete, 20–25 cm. long; flowers about 10, loosely arranged, disposed on one side of the scape, the bracts linear, spreading, 1–1.8 cm. long, firm, the pedicels 2–5 mm. long, purplish; perianth 3–3.5 cm. long, pale purple.——Shikoku.

17. Hosta tardiva Nakai. Nankai-gibōshi.　Leaves obliquely spreading, petiolate, ovate, 10–16 cm. long, 4.5–8 cm. wide, gradually acuminate, attenuated and subobtuse at base, lustrous above, flat, with 5–6 pairs of nerves, the margin slightly undulate; scapes about twice as long as the leaves, 50–60 cm. long; flowers disposed on one side of the scape, about 3.8 cm. long, pale purple, the pedicels about 6 mm. long, purplish, the bracts greenish.——Honshu (se. Kinki Distr.), Shikoku.

18. Hosta cathayana Nakai. Akikaze-gibōshi.　Closely resembles the preceding, the leaves smaller and lustrous, the bracts wholly green, persistent.——Honshu (Kinki and Chūgoku Distr.).——China.

19. Hosta tardiflora (Irving) Stearn. *Funkia tardiflora* Irving; *H. sparsa* Nakai; *Funkia japonica* var. *tardiflora* Hort. ——Aki-gibōshi.　Leaves erect, long-petioled, thick and firm, lanceolate, 10–15 cm. long, 3–4 cm. wide, narrowed at base, deep green and slightly lustrous, 5-nerved, entire, flat; scapes nearly as long as the leaves, erect, terete, 1-scaled; flower buds suboptuse, the bracts narrowly ovate, 12–16 mm. long, purplish to white, withering after anthesis, spreading, the pedicels nearly as long as the bract, arcuate, spreading;

flowers 4–4.5 cm. long, purple.——Cultivated in Honshu (near Nagoya).

20. Hosta tortifrons F. Maekawa. Kogarashi-gibōshi.　Closely resembles the preceding; leaves smaller, shorter, undulate and twisted.——Cultivated in Honshu (Tokyo).

21. Hosta longipes (Fr. & Sav.) Matsum. *Funkia longipes* Fr. & Sav.——Iwa-gibōshi.　Leaves chartaceous, ovate-cordate or elliptic-ovate, 12–13 cm. long, 8–9 cm. wide, abruptly mucronate, truncate or subcordate at base, flat, dull above, with 7–9 pairs of nerves, the margin with 4 or 5 undulations on each side; scapes ascending, about 30 cm. long; flowers pale purple, about 4 cm. long, the pedicels 4–11 mm. long, the bracts membranous, purplish to white, as long as or slightly shorter than the pedicels, withering after anthesis.——Honshu, Kyushu.

22. Hosta rupifraga Nakai. Hachijō-gibōshi.　Leaves rather coriaceous, broadly ovate, 8–13 cm. long, 4–8 cm. wide, with 6 to 8 pairs of nerves, nearly flat, the margins barely undulate, acuminate, lustrous; scapes stout, many-flowered, the bracts subcoriaceous, boat-shaped; flowers dense, purplish, 3–4 cm. long, the pedicels 10–13 mm. long, pendulous.——Honshu (Hachijo Isl.).

23. Hosta nigrescens (Makino) F. Maekawa. *H. sieboldiana* var. *nigrescens* Makino——Kuro-gibōshi.　Leaves long-petioled, broadly ovate to ovate-orbicular, 20–30 cm. long, 13–21 cm. wide, mucronate, cordate, coriaceous, flat, glaucous, becoming yellowish green in age, slightly lustrous, with 12–13 pairs of impressed nerves; scapes much longer than the leaves, terete, erect, slightly curved above, 75–140 cm. long, 2- to 4-scaled; flowers 20–40, rather dense, 4.5–5 cm. long, pale purple, the bracts 2–3 cm. long, spreading, ovate-lanceolate to linear, greenish to white, with a purplish margin.——July–Aug. Honshu (n. distr.); commonly cultivated.

24. Hosta fluctuans F. Maekawa. Kuro-nami-gibōshi.　Leaves long-petioled, ovate to ovate-elliptic, 20–25 cm. long, 10–18 cm. wide, spreading, abruptly acute, truncate to obtuse at base, with 9–10 pairs of nerves, lustrous above, glaucous beneath, smooth; scapes much longer than the leaves, often arcuate and recurved, 1–1.3 m. long; flowers purplish, many, dense, 4–6 cm. long, the pedicels more than twice as long as the bracts, recurved and arcuate.——July–Aug. Honshu (ne. distr.).

25. Hosta tokudama F. Maekawa. *Funkia sieboldiana* var. *condensata* Miq.; *H. sieboldiana* var. *glauca* Makino—— Tokudama.　Leaves spreading on the erect petiole, cordate-orbicular, abruptly acuminate, glaucous and rugose above; scapes 30–45 cm. long, as long as to slightly longer than the leaves; racemes dense, short, one-sided, recurved, the bracts firm, purplish, narrowly ovate to lanceolate; flowers 4–4.3 cm. long, white to pale purple.——June–July. Commonly cultivated in our gardens and said to be spontaneous in Honshu (Inaba Prov.).

26. Hosta sieboldiana (Lodd.) Engler. *Hemerocallis sieboldiana* Lodd.; *Funkia sieboldiana* (Lodd.) Hook.; *Niobe sieboldiana* (Lodd.) Nash——Tō-gibōshi.　Leaves glaucescent or green, spreading, elliptic or ovate-elliptic, 25–35 cm. long, 14–23 cm. wide, acuminate, cordate and abruptly decurrent on the petiole, prominently undulate, scabrous on nerves beneath, 13- to 14-nerved; scapes erect, equaling to twice as long as the leaves, 50–60 cm. long, the bracts lanceolate, acuminate, spreading, persistent, 2.5–6 cm. long, 1–1.7

cm. wide, greenish to white, often purplish, the pedicels spreading, 11–15 mm. long; flowers 5–5.7 cm. long, white. ——June. Honshu (Hokuriku and n. Kinki Distr.); much planted in our gardens.

Var. **hypophylla** F. Maekawa. *Funkia glauca* Sieb.; *H. glauca* (Sieb.) Stearn——HAGAKURE-GIBŌSHI. Cultivated selection with larger ovate-orbicular abruptly acuminate leaves 24–33 cm. long; flowers pale purple, about 6 cm. long; scapes shorter.——Occurs with the typical phase.

Var. **fortunei** (Bak.) Asch. & Graebn. *Funkia fortunei* Bak.; *H. fortunei* (Bak.) L. H. Bailey.——RENGE-GIBŌSHI. Smaller plant with orbicular-depressed leaves.——Occurs with the typical phase.

27. **Hosta montana** F. Maekawa. *H. fortunei* var. *gigantea* L. H. Bailey; *H. cucullata* Koidz., pro parte——ŌBA-GIBŌSHI. Leaves broadly to narrowly ovate or ovate-cordate, 18–23 cm. long, 10–13 cm. wide, abruptly cuspidate, green, slightly rugulose and papillose on the 10–13 pairs of nerves beneath; scapes erect, much longer than the leaves, 1- to 3-scaled, the bracts stellately spreading, ovate to lanceolate, 18–33 mm. long, 7–14 mm. wide, acute; flowers disposed horizontally on the scape, pale purple to white, 4.5–5 cm. long, the pedicels 10–13 mm. long.——July–Aug. Hills and mountains, Hokkaido, Honshu (n. and centr. distr.).

28. **Hosta crispula** F. Maekawa. *H. fortunei* var. *marginato-alba* L. H. Bailey; *H. latifolia* var. *albomarginata* Wehrh.——SAZANAMI-GIBŌSHI. Leaves arcuate-spreading, ovate, 13–16 cm. long, 6–10 cm. wide, gradually cuspidate, rounded at base, green, the white margin undulate, dull above, with 7–8 pairs of nerves, strigillose on the nerves beneath; scapes much longer than the leaves, terete, 50–80 cm. long; racemes 30- to 40-flowered, the bracts ovate to lanceolate or ovate-elliptic, whitish, spreading, 1.5–2 cm. long, 7–12 mm. wide; flowers about 4 cm. long.——July. Planted in our gardens.

29. **Hosta kikutii** F. Maekawa. HYŪGA-GIBŌSHI. Leaves elliptic, elliptic-ovate to lanceolate, 15–20 cm. long, 7.5–14 cm. wide, flat or undulate, acuminate to abruptly cuspidate, narrowly decurrent on the petiole or cordate to truncate, green, with 8–10 pairs of nerves; flowers dense, white, tardily expanding, about 4.5 cm. long, the bracts white, lanceolate, 15–20 mm. long, the pedicels about 25 mm. long.——Kyushu (Hyuga Prov.).

Var. **yakusimensis** (Masam.) F. Maekawa. *H. sieboldiana* var. *yakusimensis* Masam.; *H. polyneuron* F. Maekawa—— SUDARE-GIBÔSHI, HIME-HYŪGA-GIBŌSHI. Leaves lanceolate, many-nerved; perianth softer and thinner than the typical form, with the upper portion campanulate-dilated; stamens exserted.——Kyushu (Yakushima).

30. **Hosta tosana** F. Maekawa. TOSA-NO-GIBŌSHI. Resembles the preceding; flowers subcapitate, the bracts densely imbricate, green, boat-shaped, attenuate above, the pedicels short, purple-dotted, the perianth about 48 mm. long, the narrower portion white, 6-grooved, the dilated upper portion gradually obconical, pale purple.——Shikoku.

31. **Hosta caput-avis** (F. Maekawa) F. Maekawa. *H. tosana* var. *caput-avis* F. Maekawa——UNAZUKI-GIBŌSHI. Leaves 3–6, rather firm and thick, obliquely ovate, oblong to lanceolate, 8–12 cm. long, 5–12 cm. wide, cordate, long-attenuate, the petioles 15–25 cm. long; scapes declined from near base, terete; flowers 4–4.5 cm. long, the pedicels purplish, about 1 cm. long, the bracts whitish.——June. Rocks along ravines and valleys in mountains; Honshu (Kii Prov.), Shikoku.

32. **Hosta sacra** F. Maekawa. SAKURA-GIBŌSHI. Leaves ovate-orbicular to ovate-elliptic, 18–20 cm. long, 12–20 cm. wide, abruptly cuspidate, slightly cordate to obtuse at base, with about 13 pairs of nerves, green above, glaucous beneath; scapes 75–100 cm. long, erect; racemes about 20-flowered, the bracts ovate to lanceolate or narrowly elliptic, imbricate, boat-shaped, green or rose-purple, 18–28 mm. long; flowers pale purple, 4.5–4.8 cm. long.——Honshu (Hirosaki).

33. **Hosta hippeastrum** F. Maekawa. RAPPA-GIBŌSHI. Leaves thin, horizontally spreading on long ascending petioles, ovate, 13–20 cm. long, 7–12 cm. wide, abruptly acuminate, cordate, deep green, flat, with 8–9 pairs of nerves, the margins slightly undulate; scapes erect, firm, about 50 cm. long, the bracts obliquely spreading, about 22 mm. long, 12 mm. wide; flowers about 15, disposed horizontally on the scape, 3.5–4.5 cm. long, pale purple.——Cultivated in Honshu (Iga Prov.).

34. **Hosta kiyosumiensis** F. Maekawa. KIYOSUMI-GIBŌ-SHI. Leaves petiolate, ovate to elliptic-ovate, 8.5–11 cm. long, 4.5–7 cm. wide, caudate, truncately narrowed to the petiole, yellowish green and dull above, with 5–6 pairs of nerves; scapes much longer than the leaves, 28–36 cm. long; flowers 4 or 5, disposed laterally on the scape, white, about 5 cm. long, the bracts 15–25 mm. long, 4–6 mm. wide, ovate-lanceolate, green, boat-shaped, the pedicels 1–1.2 cm. long, arcuate, shorter than the bracts.——Honshu (Awa Prov.).

Var. **petrophila** F. Maekawa. IWAMA-GIBŌSHI. Plants smaller with flowers often purplish, the dilated upper portion of the perianth narrower.——Honshu (Kinki Distr.).

35. **Hosta densa** F. Maekawa. KEYARI-GIBŌSHI. Leaves spreading, ovate to ovate-elliptic, 16–22 cm. long, 9–12 cm. wide, acuminate, obtuse at base, with 8–10 pairs of nerves, deep green and dull above; scapes longer than the leaves, about 60 cm. long; racemes ascending, many-flowered; flowers about 40, dense, pale purple, 4.7–5 cm. long, the bracts oblong-lanceolate, acute, green, 2–2.5 cm. long, 7–9 mm. wide, the pedicels horizontally spreading, 10–13 mm. long.——Honshu (Kinki Distr.).

36. **Hosta pachyscapa** F. Maekawa. BENKEI-GIBŌSHI. Leaves horizontally spreading, deep green, elliptic to ovate-elliptic, sometimes oblong, 13–17 cm. long, 6–11 cm. wide, acuminate, slightly undulate, lustrous, abruptly narrowed below and obtuse at base, about 7-nerved on each side; scapes ascending, 70–95 cm. long, with leaflike scales; flowers dense, about 4.5 cm. long, pale purple, the bracts imbricate, oblong-lanceolate, green, the margins purplish, acuminate, the pedicels 5–8 mm. long, arcuate.——Honshu (Tōkaidō, Kinki Distr.).

12. HEMEROCALLIS L. WASUREGUSA ZOKU

Rhizomes short, roots sometimes thickened and fusiform; leaves radical, elongate, narrow, flat, 2-ranked, linear; scapes naked or nearly so; inflorescence racemose, the flowers few to several, large, erect or horizontal, on short pedicels, bracteate; perianth infundibuliform, the lobes narrowly oblong, much longer than the tube, pale lemon yellow to orange or orange-red, the venation reticulate; stamens 6, inserted at the summit of tube; ovary sessile, 3-locular; style filiform, erect, the stigma small; capsules 3-angled, transversely rugulose, loculicidally dehiscent; seeds plump.——About 20 species in Eurasia.

1A. Flowers lemon-yellow; roots not thickened; perianth-tube 2–4 cm. long. . . . . 1. *H. vespertina*
1B. Flowers orange-yellow to orange-red; roots usually partially thickened and fusiform.
  2A. Flowers orange-yellow; in May and June, or July in high mountains.
    3A. Inflorescence simple, very short.
      4A. Flowers 8–10 cm. long; inner tepals 2–2.5 cm. wide. . . . . 2. *H. middendorffii*
      4B. Flowers 5–7 cm. long; inner tepals about 1.2 cm. wide. . . . . 3. *H. dumortieri*
    3B. Inflorescence elongate, sometimes forked. . . . . 4. *H. aurantiaca*
  2B. Flowers orange-red, rarely orange-yellow; in July and Aug., sometimes until Nov.
    5A. Leaves 7–15 mm. wide, rarely more.
      6A. Leaves rather thin; scapes and inflorescence without fascicles of axillary leaves. . . . . 5. *H. longituba*
      6B. Leaves rather firm and thick; scapes and inflorescence often bearing fascicles of axillary leaves. . . . . 6. *H. littorea*
    5B. Leaves 25–40 mm. wide; flowers usually double. . . . . 7. *H. fulva* var. *kwanso*

1. Hemerocallis vespertina Hara. *H. thunbergii* Baker?; *H. citrina* sensu Nakai, non Baroni——ASAMA-KISUGE, YŪSUGE.   Leaves suberect, green, 5–15 mm. wide, smooth; scapes erect, slender, 50–100 cm. long; inflorescence simple or forked, rather loosely 3- to 15-flowered, the lower bracts linear-lanceolate, those above ovate-caudate, the pedicels 2–15 mm. long; flowers nocturnal, lemon-yellow, fragrant, the perianth-tube 2–4.5 cm. long, the inner segments obtuse, about 7–8 cm. long, 18–25 mm. wide, cross-veins few.——July–Aug. Mountains; Honshu (centr. and w. distr.), Kyushu.——Korea.

Hemerocallis yezoensis Hara. *H. citrina* sensu Miyabe & Kudo, non Baroni; *H. coreana* Nakai, pro parte——EZO-KISUGE.   Allied to No. 1; reported to have prominently arcuate declined leaves; flowers yellow, diurnal.——Seashore; Hokkaido.

2. Hemerocallis middendorffii Trautv. & Meyer. EZO-ZENTEI-KA.   Roots rarely somewhat fusiform; leaves soft, vivid green, flat, 1.5–2.5 cm. wide, smooth, reclined above; scapes erect, 40–70 cm. long, simple; inflorescence very short, the axis scarcely distinct; flowers few, 8–10 cm. long, orange-yellow, scarcely pedicelled, the bracts broadly ovate to ovate-cordate, acute, the perianth-tube 1–1.5 cm. long, the inner perianth-segments narrowly obovate to narrowly oblong, about 2.5 cm. wide, obtuse, with few cross-veinlets.——June–July. Meadows in high mountains; Hokkaido.——Kuriles, Sakhalin, Amur, Ussuri, Korea, Manchuria, and n. China.

Var. **esculenta** (Koidz.) Ohwi. *H. middendorffii* sensu auct. Japon., non Trautv. & Meyer; *H. esculenta* Koidz.——ZENTEI-KA, NIKKŌ-KISUGE.   Pedicels 3–12 mm. long, the bracts narrowly ovate; perianth-tube 1–2 cm. long.——Meadows in high mountains of Honshu (centr. and n. distr.).

Hemerocallis exaltata Stout. TOBISHIMA-KANZŌ.   A robust insular plant with tall stems and broader leaves. Occurs only in Tobishima, a small islet lying on the Japan Sea side near Ugo Prov. (Akita Pref.) in Honshu.

3. Hemerocallis dumortieri Morr. HIME-KANZŌ.   Resembles the preceding but smaller; leaves green, flat, 1–1.5 cm. wide, reclined above; scapes 25–50 cm. long, as long as to slightly longer than the leaves; racemes simple, very short, densely few-flowered, the bracts ovate to oblong-ovate, subacute, membranous, the pedicels very short; flowers 5–7 cm. long, orange-yellow, the perianth-tube about 1 cm. long, the inner segments oblanceolate, obtuse, sometimes with transverse veinlets; anthers black.——May–June. Mountains; Hokkaido, Honshu (centr. and n. distr.).——e. Siberia, Manchuria, and Korea.

4. Hemerocallis aurantiaca Bak. *H. aurantiaca* var.

major Bak.——WASUREGUSA, NAMBAN-KANZŌ, KANZŌ.   Leaves rather thick, 60–80 cm. long, 1.5–2.5 cm. wide, flat, smooth, glaucous, ascending, arcuate above; scapes nearly as long as the leaves, slightly thickened, often with few linear subfoliaceous bracts; inflorescence short, forked, few-flowered, the bracts linear-lanceolate, 2–8 cm. long, the upper bracts membranous on margin, the pedicels 1–2 cm. long, the lower portion adnate to the rachis; flowers orange-yellow, 10–13 cm. long, the perianth-tube 1–2 cm. long, the inner segments narrowly oblong, subobtuse, 3–3.5 cm. wide, with prominent cross-veins, membranous and undulate on margin.——May–July. Chinese plant rarely cultivated in Japan.

5. Hemerocallis longituba Miq. *H. fulva* var. *longituba* (Miq.) Maxim.; *H. sendaica* Ohwi——NO-KANZŌ, BENI-KANZŌ.   Roots somewhat thickened; leaves flat, slightly folded below, 50–70 cm. long, 7–12 mm. wide, green, smooth, arcuate above, rather thin; scapes 50–60 cm. long, often with few scales; inflorescence forked, loosely about 10-flowered, the pedicels very short, the bracts ovate, 3–8 mm. long; flowers orange-yellow to orange-red, about 12 cm. long, the perianth-tube slender, 2–4 cm. long, about 2/5 to 1/3 as long as the lobes, the inner segments broadly lanceolate, gradually acute to obtuse, with few cross-veins.——July–Aug. Honshu, Shikoku, Kyushu; rather common.

Hemerocallis exilis Satake. MUSASHINO-KANZŌ.   An alleged hybrid between *H. longituba* and *H. vespertina*. Leaves elongate, spreading, narrower; perianth-tube about 3.5 cm. long, the ovary sterile.——July.

6. Hemerocallis littorea Makino. *H. aurantiaca* var. *littorea* (Makino) Nakai——HAMA-KANZŌ.   Leaves flat, smooth, broad, deep green, relatively broad, thick and firm; scapes stout, often bearing a fascicle of relatively short leaves in leaf-axils throughout the inflorescence; inflorescence branched, elongate, few- to many-flowered, the bracts ovate to narrowly deltoid-ovate, 5–20 mm. long, the pedicels 1–10 mm. long; flowers orange-yellow, 10–12 cm. long, the tube 1.5–2 cm. long, the inner perianth-segments broadly lanceolate, subacute, about 2 cm. wide, slightly thicker than the outer ones, the margin membranous.——Aug.–Oct. Near the seashore; Honshu (Kantō Distr. and westw.), Kyushu.

7. Hemerocallis fulva L. var. **kwanso** Regel. *H. disticha* var. *kwanso* (Regel) Nakai——YABU-KANZŌ, ONI-KANZŌ.   Roots somewhat thickened; leaves green, flat, smooth, 40–60 cm. long, 2.5–4 cm. wide, arcuate above; scapes 80–100 cm. long, loosely few-scaled; inflorescence forked, loosely several-flowered, the pedicels to 2 cm. long, often adnate to the axis at base, the bracts ovate-deltoid, membranous, 4–10 mm. long;

flowers about 10 cm. long, orange-red, double, sterile, the perianth-tube about 2 cm. long, the inner segments oblong, with cross veinlets.——July–Aug. Hills and low mountains; Hok-kaido, Honshu, Shikoku, Kyushu; common.——The typical phase with single fertile flowers occurs from Europe to Siberia.

13. ALECTORURUS Makino   KEIBI-RAN ZOKU

Dioecious perennial, smooth, glabrous; leaves evergreen, radical, laterally flattened, 2-ranked, gladiate, obsoletely jointed at base with a densely imbricate sheath; scapes erect, flat, naked; inflorescence a terminal many-flowered panicle, the bracts small, the branches subracemose; flowers small, white, jointed at base with the short pedicel; perianth campanulate, the segments 6, persistent, oblong, 1-nerved; stamens 6, attached at base of the segments, the anthers introrse; ovary sessile, globose, 3-locular; style solitary, erect, the stigma slightly thickened; ovules 2 in each locule, anatropous, ascending; capsules coriaceous, loculicidal; seeds oblong, with obsolete longitudinal lines and long white hairs on one end; embryo imbedded within the endosperm.——A single species in Japan.

1. **Alectorurus yedoensis** (Maxim.) Makino. *Anthericum yedoense* Maxim.; *Bulbinella yedoensis* (Maxim.) Matsum.——KEIBI-RAN.   Rhizomes short, the roots stout; leaves broadly linear, 10–30 cm. long, 1–2.5 cm. wide near the middle, with 3–6 pairs of longitudinal nerves, gladiate, rather thick, falcate, acuminate, the apex usually obtuse; scapes 15–40 cm. long, flat, the margin slightly winged; panicle 7–15 cm. long, the branches relatively few, obliquely spreading; perianth 4–5 mm. long; stamens exserted; capsules depressed-globose, about 3.5 mm. wide; seeds about 2.2 mm. long, brown, with a tuft of white hairs on one end.——July–Aug. Rocky cliffs in mountains; Shikoku, Kyushu; rare.

14. SCILLA L.   TSURUBO ZOKU

Bulbous herbs; radical leaves linear; scapes naked, simple; flowers relatively small, blue, pinkish, or white, pedicelled, in a terminal raceme, the bracts small, the tepals persistent, free or slightly connate at base, ascending or the corolla rarely campanulate; filaments inserted at base or near the middle on the tepals, the anthers elliptic or ovate; ovary 3-locular, with 1–2(–10) ovules in each locule; style filiform, the stigma small; capsules membranous, subglobose; seeds obovoid or globose, black, sometimes angled.——About 90 species, in temperate and tropical mountain regions of the Old World.

1. **Scilla scilloides** (Lindl.) Druce. *Ornithogalum japonicum* Thunb.; *O. sinense* Lour.; *Barnardia scilloides* Lindl.; *S. chinensis* Benth.; *S. japonica* Bak., non Thunb.; *S. thunbergii* Miyabe & Kudo——TSURUBO.   Bulbs ovoid-globose, 2–3 cm. long; leaves linear, usually 2, 15–25 cm. long, 4–6 mm. wide, rather thick, smooth, soft, concave above, abruptly acuminate, the apex obtuse; scapes erect, 20–40 cm. long; racemes many-flowered, 7–12 cm. long, rather dense, the bracts narrowly lanceolate, thinly membranous, 1–2 mm. long, the pedicels ascending, 5–12 mm. long, flowers rose-purple, the tepals barely spreading, narrowly oblong, about 4 mm. long; filaments filiform, the lower half broadened, the margin soft-pubescent; style 1.5–2 mm. long; ovary short pubescent on the angles; capsules obovoid-globose, about 5 mm. long; seeds broadly lanceolate, about 4 mm. long.——Aug.–Sept. Hills and lowlands; Hokkaido, Honshu, Shikoku, Kyushu; common.——Korea, China, Formosa, Ryukyus, Manchuria, and Ussuri.

15. GAGEA Salisb.   KIBANA-NO-AMANA ZOKU

Bulbous herbs usually with a single radical leaf; leaves linear; scapes slender, the inflorescence umbellate, the subtending bracts 1–3, the pedicels unequal, with a small bract at base; flowers yellow or yellow-green, the tepals 6, free, in 2 series, 3- to 5-nerved, oblong, persistent; stamens 6, the filaments filiform, slightly broadened at base, the anthers rounded to elliptic; ovary obovoid, 3-locular, the ovules numerous; style single, columnar; capsules obovoid-globose to elliptic, loculicidal, 3-angled; seeds narrowly obovate or relatively flat.——About 50 species, in Europe, the Mediterranean region, and temperate regions of Asia.

1A. Outer coat of bulbs yellowish; leaves 5–7 mm. wide; tepals 12–15 mm. long; plants relatively large. .... 1. *G. lutea*
1B. Outer coat of bulbs dark brown; leaves about 2 mm. wide; tepals 7–9 mm. long; plants relatively small. .... 2. *G. japonica*

1. **Gagea lutea** (L.) Ker-Gawl. *Ornithogalum luteum* L.; *G. sylvatica* Loud.; *G. fascicularis* Salisb.; *G. pratensis* sensu Nakai, non Dumort.; *G. coreana* Nakai, pro parte——KIBANA-NO-AMANA.   Bulbs ovoid, about 1.5 cm. long; radical leaves linear, sessile, soft, smooth, rather fleshy, flat, slightly involute, gradually narrowed to the obtuse apex, 15–30 cm. long, 5–7(–9) mm. wide; scapes 15–25 cm. long; inflorescence bracts 2, the lower one 4–8 cm. long, the upper one shorter; flowers 3–10, umbellate, the pedicels 1–4 cm. long, the tepals linear-oblong, 2.5–3.5 mm. wide, obtuse; capsules subglobose, 3-angled, about 7 mm. long and as wide.——Apr.–May. Hokkaido, Honshu.——Sakhalin, Kuriles, Korea, China, and e. Siberia to Europe.

2. **Gagea japonica** Pascher. *G. nipponensis* Makino; *G. pusilla* sensu auct. Japon., non Roem. & Schult.——HIME-AMANA.   Bulbs ovoid, 8–10 mm. long, with a dark brown outer coat; radical leaves solitary, linear, loosely involute, 10–20 cm. long, about 2 mm. wide, soft and rather thick, obtuse; scapes 8–15 cm. long; inflorescence bracts 2, the lower one linear, 2–3 cm. long, the upper one 7–15 mm. long; flowers (1–)2–5, the pedicels 1–2.5 cm. long, slender, the tepals yellow, changing to greenish yellow in fruit, narrowly oblong or broadly oblanceolate, obtuse, 7–9 mm. long; capsules globose, about 5 mm. wide.——Apr. Meadows along rivers in lowlands; Hokkaido, Honshu; rare. Plants of the s. Kuriles and Hokkaido are distinguished as **Gagea vaginata** Pascher. EZO-HIME-AMANA.

16. **LLOYDEA** Salisb.　Chishima-amana Zoku

Small bulbous herbs; leaves linear, radical; scapes with few short leaves; flowers 1–6, white, the tepals ascending, free, persistent, lanceolate, 3- to 5-nerved, usually with a gland inside near base; stamens 6, shorter than the tepals; ovary sessile, 3-angled, 3-locular, many-ovuled; style short, the stigma shortly 3-lobed; capsules membranous, obovoid or subglobose; seeds oblong, small, slightly flattened.——About 18 species, in the Mediterranean region, e. Asia, European alps, and N. America.

1A.　Bulbs cylindric, 4–7 cm. long, with a chartaceous coat; tepals with a transverse gland near base. . . . . 1. *L. serotina*
1B.　Bulbs broadly ellipsoidal, small, less than 1 cm. long, with a thinly membranous coat; tepals without a gland. . . . . 2. *L. triflora*

1. Lloydea serotina (L.) Reichenb. *Bulbocodium serotinum* L.; *L. alpina* Salisb.——Chishima-amana. Bulbs cylindric, the outer coat pale yellow-brown, barely split longitudinally, with fine vertical lines; radical leaves usually 2, 7–20 cm. long, about 1 mm. wide, abruptly subobtuse, flat, thick, keeled on back, the margin barely roughened and uneven; scapes 7–15 cm. long, usually decumbent, the leaves 1–3 cm. long, nearly flat, with involute margin; flowers white, nodding, 10–13 mm. long, the tepals narrowly oblong, obtuse, white, several-nerved, the nerves often dark purplish outside; stamens half as long as the tepals; anthers elliptic; style 2–2.5 mm. long, the stigma obsoletely 3-lobed; capsules 6–7 mm. long, broadly obovoid.——June–Aug. Rocky slopes and cliffs in alpine zone; Hokkaido, Honshu (n. and centr. distr.).——Circumboreal.

2. Lloydea triflora (Ledeb.) Bak. *Ornithogalum triflorum* Ledeb.; *Gagea triflora* (Ledeb.) Roem. & Schult.——Hosoba-no-amana. Bulbs broadly ellipsoidal, 6–10 mm. long, the outer coat thinly membranous, entire, without vertical lines; radical leaves usually solitary, 10–20 cm. long, 1.5–3 mm. wide, gradually obtuse, smooth, linear, 3-angled, soft; scapes 10–25 cm. long, few-leaved, the lowest leaf narrowly lanceolate, 3–6 cm. long, 4–6 mm. wide; inflorescence 2- to 4(–6)-flowered, the bracts linear, 7–12 mm. long, about 1 mm. wide; flowers 1–1.5 cm. long, the tepals white, oblanceolate, obtuse, with green striations; stamens about 3/5 as long as the tepals, the anthers small, broadly elliptic; style about 4 mm. long, obscurely 3-lobed.——May–June. Thickets and meadows in mountains; Hokkaido, Honshu, Shikoku.——Siberia, Kamchatka, Ussuri, n. and centr. China, Sakhalin, Kuriles, and Korea.

17. **ALLIUM** L.　Negi Zoku

Perennials with tunicate bulbs, solitary or tufted, sometimes with a short rhizome at base; leaves linear, cylindric, or rarely flat and oblong; scapes naked above, the lower part enclosed by the sheaths of the radical leaves; inflorescence umbellate, terminal, entirely enveloped while young by 2-lobed membranous or hyaline spathe, the pedicels slender; flowers small, bulbils sometimes produced in place of the flowers, the tepals free or slightly connate below, usually rose or white, rarely yellowish, usually 1-nerved, obliquely spreading to suberect; filaments usually dilated below and connate at base, the inner often toothed on each side; ovary 3-locular, the ovules usually 2, but sometimes more in each locule; style single, filiform, sometimes shallowly 3-lobed; capsules membranous; seeds black, angled, flat.——About 300 species, in the N. Hemisphere.

1A.　Leaves flat, oblong to broadly lanceolate, 3–10 cm. wide; bulbs with a disintegrated netted fibrous outer coat.
　　1. *A. victorialis* var. *platyphyllum*
1B.　Leaves terete to flat, less than 1 cm. wide.
　2A.　Inflorescence 1- or 2-flowered, with a small spathe at base; leaves flat, broadly linear, 3–8 mm. wide; outer bulb coat transversely wrinkled, the cells transversely elongate; style 3-fid. . . . . 2. *A. monanthum*
　2B.　Inflorescence many-flowered, the spathe 2-parted; leaves linear or terete; style entire.
　　3A.　Flowers white; tepals ascending from the base, spreading above; bulbs and rhizomes present, usually tufted, the outer fibrous coat yellowish-brown.
　　　4A.　Rhizomes well-developed; tepals 5–6 mm. long, acute; stamens not exserted; filaments entire. . . . . 3. *A. tuberosum*
　　　4B.　Rhizomes poorly developed; tepals 3.5–4 mm. long, obtuse; stamens exserted; filaments broadened below, often with an obsolete tooth on each side. . . . . 4. *A. togasii*
　　3B.　Flowers rose-purple to rose; tepals spreading from the base; bulbs solitary to several, rhizomes sometimes present.
　　　5A.　Tepals oblong to ovate-oblong, sometimes elliptic, obtuse to rounded; leaves semiterete, 3-angled, or flat.
　　　　6A.　Bulbs subglobose, with thin entire outer coat; umbels partially or wholly consisting of bulbils; spathe prominently beaked.
　　5. *A. grayi*
　　　　6B.　Bulbs ovoid, with firm and usually fibrous outer coat; umbels not consisting of bulbils; spathe beakless or only short-beaked.
　　　　　7A.　Tepals ovate-oblong, obtuse or subobtuse; leaves flat; outer bulb coat reticulate mesh-fibrous; pedicels 4–8 mm. long.
　　6. *A. splendens*
　　　　　7B.　Tepals elliptic to oblong, rounded, the outer ones slightly shorter and concave; outer bulb coat entire or only slightly reticulate mesh-fibrous; pedicels 12–30 mm. long.
　　　　　　8A.　Outer bulb coat slightly thickened, reticulate mesh-fibrous, the cells longitudinal; leaves dying in winter; pedicels 12–15 mm. long. . . . . 7. *A. thunbergii*
　　　　　　8B.　Outer bulb coat entire, thin, membranous; leaves evergreen.
　　　　　　　9A.　Leaves hollow. . . . . 8. *A. chinense*
　　　　　　　9B.　Leaves solid, very narrow. . . . . 9. *A. virgunculae*
　　5B.　Tepals lanceolate to broadly lanceolate, gradually acuminate; leaves terete, sometimes filiform.
　　　10A.　Bulbs ovoid, attached to obsolete rhizomes, the outer coat rather firm. . . . . 10. *A. schoenoprasum*
　　　10B.　Bulbs lanceolate-cylindric to narrowly ovoid, attached to a short distinct rhizome, the outer coat thin, membranous.
　　11. *A. maximowiczii*

1. **Allium victorialis** L. var. **platyphyllum** (Hult.) Makino. *A. victorialis* sensu auct. Japon., non L.; *A. victorialis* subsp. *platyphyllum* Hult.; *A. latissimum* Prokhan.——GYŌJA-NINNIKU. Plants with strong onion odor; bulbs lanceolate, 4–7 cm. long, the outer coat consisting of netted fibers, brown-yellow; leaves 2–3, oblong to broadly oblanceolate, 20–30 cm. long, 3–10 cm. wide, subobtuse to subacute, entire, narrowed to the petiolelike base below, slightly glaucous, somewhat fleshy, the midrib slightly raised beneath; scape 40–70 cm. long, the spathe ovoid, enveloping the lower portion, 2-parted in anthesis; flowers white or creamy-white, the pedicels 1.5–3 cm. long, the tepals ascending, oblong, 5–6 mm. long, obtuse; stamens and style longer than the tepals, the anthers yellow-green.——June–July. Moist woods in high mountains; Hokkaido, Honshu (centr. and n. distr. and Yamato Prov.).——Sakhalin, Kuriles, Kamchatka, Korea, China, and e. Siberia. The typical phase occurs from Siberia to Europe and India; also in N. America.

2. **Allium monanthum** Maxim. HIME-NIRA, HIME-BIRU, HIME-AMANA. Small delicate plant with slight odor; bulbs broadly ovoid, 6–10 mm. long, the outer coat transversely undulate, firmly membranous; leaves 1–2, nearly radical, linear, 10–20 cm. long, 3–8 mm. wide, lunate in cross section, soft, narrowed at both ends, subobtuse, 9- to 13-nerved, smooth, the midrib obscure; scapes 5–12 cm. long, delicate, the spathe thinly membranous, ovoid, 6–7 mm. long, entire; flowers solitary, rarely 2, white to pinkish, 4–5 mm. long, campanulate, the pedicels rather short, the tepals oblong to narrowly ovate, subobtuse; style short, 3-lobed nearly to the middle; capsules globose.——Apr. Woods and thickets in hills and the lower mountains; Hokkaido, Honshu, Shikoku.——Korea, Manchuria, and Ussuri.

3. **Allium tuberosum** Rottl. *A. odorum* sensu auct. Japon., non L.; *A. yezoense* Nakai——NIRA. Plants with strong odor; bulbs tufted, attached to short rhizomes, the outer fibrous bulb coat yellowish; leaves few, linear, 20–30 cm. long, 3–4 mm. wide, flat, soft, obtuse; scapes 30–50 cm. long, the spathe acute; flowers white, spreading, the pedicels slender, 1.5–3.5 cm. long, the tepals 5–6 mm. long, narrowly oblong, acute; stamens slightly shorter than the tepals, the anthers yellow, the filaments broadened below, entire; style not exserted.——June–Sept. Naturalized and said to be spontaneous in Honshu and Kyushu; widely cultivated as a vegetable.——China and India.

4. **Allium togasii** Hara. KANKAKEI-NIRA. Bulbs oblong, on short rhizomes, the outer fibrous coat brownish and finely netted; leaves 3–4, narrowly linear, 12–20 cm. long, 1.2–2 mm. wide, flattened, fleshy, weakly 3-ribbed beneath, green; scapes terete, 12–25 cm. long; flowers white, the pedicels filiform, 6–12 mm. long, the tepals of the inner whorl about 4 mm. long; stamens about 6 mm. long, exserted, the filaments of the inner whorl abruptly dilated and often obsoletely toothed at base; capsules globose, 4–4.5 mm. long. ——July. Shikoku (Kankakei, Azuki Isl. in Sanuki Prov.); very rare.

5. **Allium grayi** Regel. *A. nipponicum* Fr. & Sav.——No-BIRU. Plants odoriferous; bulbs subglobose, white, about 15 mm. in diameter, the outer coat thinly membranous, entire; leaves few, nearly 3-angled in cross section, shorter than the scape, shallowly grooved on upper side, 2–3 mm. wide, smooth; scapes erect, rather firm, terete, 40–60 cm. long, the spathe long beaked at apex; inflorescence pale pinkish, 4–6 mm. long, the flowers usually partially or almost wholly replaced by sessile bulbils, the pedicels slender, about 15 mm. long, the tepals narrowly ovate-oblong, subobtuse, the midrib tinged red-purple on back; filaments exserted, not toothed; style elongate.——May–June. Lowland meadows to mountain foothills; Hokkaido, Honshu, Shikoku, Kyushu.——Ryukyus.

6. **Allium splendens** Willd. *A. lineare* sensu auct. Japon., non Schrad.——MIYAMA-RAKKYŌ. Bulbs few or solitary, lanceolate, the outer coat thick, netted-fibrous; leaves 3–4, linear, 3–5 mm. wide, flat, shorter than the scapes, the margin more or less uneven; scapes 15–40 cm. long; flowers rose-purple, about 4 mm. long, the pedicels 4–7 mm. long, the tepals narrowly ovate to ovate-oblong, obtuse, the midrib slightly deeper colored on back; filaments about equal to or slightly longer than the tepals, the inner ones with a deltoid to ovate tooth on each side at base.——July–Aug. Alpine meadows; Hokkaido, Honshu (centr. distr.).——Kuriles, Kamchatka, Sakhalin, n. Korea, and e. Siberia.

7. **Allium thunbergii** G. Don. *A. odorum* sensu Thunb., non L.; *A. japonicum* Regel; *A. chinense* sensu auct. Japon., non Don; *A. pseudojaponicum* Makino——YAMA-RAKKYŌ. Bulbs few or solitary, narrowly ovoid, the outer coat slightly thickened, brown-tinged, entire or slightly netted-fibrous; leaves usually slightly shorter than the scape, obtusely trigonous, 2–5 mm. wide; scapes 30–60 cm. high, the spathe abruptly acute, broadly ovoid; flowers rose-purple, numerous, the pedicels 1–1.5 cm. long, the tepals 4–5 mm. long, elliptic, rounded at apex, the inner segments slightly longer than the outer; filaments much-exserted, broadened at base, the teeth obsolete.——Sept.–Nov. Low mountains; Honshu (Iwaki Prov., Kantō Distr. and westw.), Shikoku, Kyushu.——s. Korea.

8. **Allium chinense** G. Don. *A. bakeri* Regel; *A. splendens* sensu Miq., non Willd.——RAKKYŌ. Bulbs narrowly ovoid, the outer coat whitish; leaves evergreen, linear, flat, rounded on back, soft; scapes 30–40 cm. long; flowers rose-purple, campanulate, the pedicels 15–30 mm. long, slightly arcuate, the tepals broadly ovate-elliptic, rounded at apex, about 5 mm. long, the inner segments slightly longer; filaments much-exserted, the inner 3 with a lanceolate tooth on each side at base; style long-exserted.——Sept.–Nov. Chinese species much cultivated for the bulbs and sometimes established around cultivated fields.

9. **Allium virgunculae** F. Maekawa & Kitam. ITO-RAKKYŌ. Bulbs gregarious, narrowly oblong, 5–7 mm. in diameter, about 2 cm. long, the outer coat membranous, pale rose; leaves 3–5, spreading, linear, dark green, densely white-puncticulate, 10–20 cm. long, 1 mm. wide, solid; scapes erect, 8–22 cm. long, slender, about 0.8 mm. in diameter, terete, striate; spathe whitish, membranous, broadly ovoid, about 5 mm. long, acuminate; flowers 2–12, rose, about 5.5 mm. long, the pedicels 12–15 mm. long; tepals broadly ovate, 3–4 mm. wide, obtuse; filaments about 6 mm. long, the inner ones with an acute tooth on each side at base; style long-exserted; capsules about 4 mm. long, 5 mm. wide.——Kyushu (Hirato Isl. in Hizen).

10. **Allium schoenoprasum** L. *A. sibiricum* L.; *A. schoenoprasum* var. *sibiricum* (L.) Regel——EZO-NEGI. Bulbs narrowly ovoid, the outer coat rather firm, gray-brown, entire or nearly so, obscurely striate with fine longitudinal cells; leaves 2–3, shorter than the scape, terete, hollow, slender; scapes 40–50 cm. long, often purplish below, terete; spathe

broadly ovoid, abruptly acute; flowers many, rose, narrowly campanulate, the pedicels 7–10 mm. long, the tepals lanceolate, acuminate above, 9–12(–14) mm. long; stamens 1/2–2/3 as long as the tepals, not toothed; style not exserted.——May–June. Hokkaido, Honshu.——Siberia to E. Europe, and N. America.

Var. **foliosum** Regel. *A. ledebourianum* Roem. & Schult.——Asatsuki. Closely resembles the typical phase; bulbs ovoid; tepals 5–8(–10) mm. long, distinctly shorter than the pedicels, filaments equaling to slightly shorter than the tepals.——May–June. Mountains; Hokkaido, Honshu, Shikoku.——Siberia.

Var. **yezomonticola** Hara. Hime-Ezo-negi. Scapes 10–20 cm. long; leaves slender; umbels 2–3 cm. wide, the pedicels 3–8 mm. long, the tepals 6–8 mm. long, 1.5–2.5 mm. wide; stamens 4–5 mm. long.——July–Aug. Hokkaido.

Var. **caespitans** Ohwi. Kabu-asatsuki. Closely resembles var. *foliosum,* the bulbs broadly ovoid, densely caespitose, semi-epigeal.——Honshu (Shimotsuke Prov.); rare.

11. Allium maximowiczii Regel. *A. schoenoprasum* var. *orientale* Regel——Shirouma-asatsuki. Allied to the preceding species; bulbs lanceolate-cylindric, the coat thinly membranous, attached to a short but distinct rhizome; scapes 30–50 cm. long, relatively soft and stout; pedicels 8–15 mm. long; tepals 6–8 mm. long, equaling to slightly longer than the stamens.——(July)–Aug. Alpine meadows; Honshu (centr. distr.).——The typical phase occurs in Dahuria to Ussuri and the high mountains of n. Korea.

Var. **shibutsuense** (Kitam.) Ohwi. *A. schoenoprasum* var. *shibutsuense* Kitam.——Shibutsu-asatsuki. Scapes and leaves slender and relatively firm; flowers and umbels smaller.——Alpine regions; Honshu (Mount Shibutsu in Kotsuke Prov.).

18. NOTHOSCORDUM Kunth Sutego-biru Zoku

Inodorous bulbous perennials; leaves linear, flat; scapes solitary, naked, the umbel solitary, terminal, the spathe 2-parted, the bracts small or reduced; tepals persistent, connate below the middle, 1-nerved; stamens 6, the filaments adnate to the tepals, more or less dilated below, the anthers oblong; ovary 3-locular, the ovules 6–12 in each carpel; style filiform, the stigma small; capsules membranous, 3-valved.——About 30 species, mainly in N. and S. America, few in e. Asia.

1. Nothoscordum inutile (Makino) Kitam. *Allium inutile* Makino——Sutego-biru. Bulbs tunicate, subglobose, 1–1.5 cm. in diameter, the outer coat membranous; leaves produced in autumn, dying the following summer, linear, to 30 cm. long, flat, rounded on the back, the midrib distinct; scapes appearing before the leaves, 20–30 cm. long, soft, the spathe thinly membranous; flowers few to several, white, in a terminal umbel, broadly campanulate, the pedicels 1.5–2 cm. long, the tepals subulate-linear, 7–8 mm. long, short connate at base, subobtuse at apex; stamens half as long as the tepals, the filaments inserted on the upper portion of the perianth-tube, not toothed; style short, 3-lobed.——Sept.–Oct. Meadows in lowlands; Honshu; rare.

19. LILIUM L. Yuri Zoku

Bulbs consisting of densely imbricate fleshy scales; stems erect or recurved above; leaves linear to linear-lanceolate, rarely ovate-cordate, alternate, sometimes more or less irregularly whorled; flowers large, 1 to more than 10, in loose corymbose racemes or dense spikes, erect or nodding, white, yellow-orange, or reddish, the perianth infundibuliform or campanulate, the tepals deciduous, free, narrowed at base, more or less recurved above, with a nectary groove inside near the base; stamens 6, the filaments usually filiform, the anthers linear or broadly lanceolate, versatile; ovary sessile, 3-locular, the ovules many; styles elongate, simple, the stigma sometimes 3-lobed; capsules loculicidal, oblong; seeds flat.——About 70 species, in the temperate regions of the N. Hemisphere, especially abundant in e. Asia.

1A. Leaves long-petiolate, ovate-cordate, convolute in bud; flowers narrowly infundibuliform, 10–15 cm. long, short-pedicelled, in dense terminal spikes; tepals saccate at base; styles elongate after anthesis; bulb-scales few. .... 1. *L. cordatum*
1B. Leaves sessile or very short-petiolate, linear to lanceolate, flat in bud; flowers broadly infundibuliform; tepals recurved toward the tip, scarcely saccate at the base; bulb-scales usually many, lanceolate to ovate.
  2A. Flowers erect; stamens obliquely spreading.
    3A. Stems angled or narrowly winged, usually more or less woolly, commonly papillose-scabridous on the angles; tepals 7–10 cm. long; style longer than the ovary. .... 2. *L. maculatum*
    3B. Stems terete, nearly glabrous; tepals 3–4 cm. long; style not longer than the ovary. .... 3. *L. concolor* var. *partheneion*
  2B. Flowers horizontal or nodding.
    4A. Stamens obliquely spreading; tepals often spotted, obliquely spreading, recurved above.
      5A. Leaves sparse, alternate.
        6A. Leaves sessile; flowers yellow to orange-red; nectary grooves pubescent.
          7A. Style not longer than the ovary, erect; plants without stolons and axillary bulbils. .... 4. *L. callosum*
          7B. Style distinctly longer than the ovary, ascending-arcuate above; plants with stolons or axillary bulbils.
            8A. Plants with stolons, axillary bulbils absent. .... 5. *L. leichtlinii*
            8B. Plants with bulbils in the leaf-axils, without stolons. .... 6. *L. lancifolium*
        6B. Leaves distinctly petioled; flowers white to rose; nectary grooves glabrous.
          9A. Tepals strongly recurved or somewhat revolute from the base, rose to darker spotted, rarely unspotted. .... 7. *L. speciosum*
          9B. Tepals obliquely spreading, recurved gradually on upper half, white, usually with reddish or yellowish spots.
                                                   8. *L. auratum*
      5B. Leaves whorled in one to several series.
        10A. Tepals orange-red, not thickened, much recurved near base; bulb-scales often with a joint near middle. 9. *L. medeoloides*

10B. Tepals strongly thickened, orange-yellow, gradually recurved above; bulb-scales not jointed. . . . . 10. *L. hansonii*
4B. Stamens nearly erect; tepals not spotted, slightly recurved above.
  11A. Leaves sessile; flowers white or partially purplish. . . . . 11. *L. longiflorum*
  11B. Leaves short-petioled; flowers pale rose.
    12A. Tepals 5–7 cm. long, slightly recurved only near the tip, the outer segments nearly as broad as the inner, abruptly obtuse
      at apex; anthers yellow; leaves rather soft, usually broadly lanceolate. . . . . 12. *L. rubellum*
    12B. Tepals 10–13 cm. long, clearly recurved above, the outer segments much narrower than the inner, gradually narrowed
      above to an obtuse apex; anthers deep brown; leaves relatively firm, usually narrower than in the preceding species.
                                                    13. *L. japonicum*

1. Lilium cordatum (Thunb.) Koidz. *Hemerocallis cordata* Thunb.; *L. cordifolium* Thunb.; *Cardiocrinum cordatum* (Thunb.) Makino——UBA-YURI. Stout glabrous perennial; stems 50–100 cm. long, pale green, hollow; leaves loosely whorled on lower half of the stem, long-petioled, oblong-ovate to broadly ovate, cordate, 7–10 cm. long, 7–15 cm. wide, abruptly acuminate, the upper surface lustrous, pale to vivid green, the veins often variegated, the upper leaves bractlike; racemes simple, 3–7 cm. long, few-flowered, the bracts narrowly lanceolate, deciduous; flowers horizontal, greenish white, tubular-infundibuliform, 7–10 cm. long, the pedicels very short, the tepals narrowly oblanceolate.——July–Aug. Moist woods in lowlands to foothills; Honshu (Kantō Distr. and westw.), Shikoku, Kyushu.

Var. **glehnii** (F. Schmidt) Woodc. *Cardiocrinum glehnii* (F. Schmidt) Makino; *L. glehnii* F. Schmidt; *C. cordatum* var. *glehnii* (F. Schmidt) Hara——Ō-UBA-YURI. Plants stouter; leaves broadly ovate; flowers often up to 20, on a rather long rachis, 10–15 cm. long.——Honshu (n. and centr. distr.), Hokkaido.——Sakhalin and s. Kuriles.

2. Lilium maculatum Thunb. *L. elegans* Thunb.; *L. davuricum* subsp. *thunbergianum* (Schult. f.) Wils.; *L. thunbergianum* Schult. f.; *L. venustum* Kunth——SUKASHI-YURI. Bulb-scales compactly imbricate, entire or partially jointed; stems usually 20–80 cm. long, relatively stout, erect, usually with papillose tubercles below and on angles, more or less white-woolly while young; leaves sessile, lanceolate to linear-lanceolate, 4–10 cm. long, 1–2 cm. wide, 3- to 5-nerved, spreading; tepals 8–10 cm. long, usually orange-red, sometimes yellow or nearly scarlet in cultivated races, spotted, oblanceolate, densely hairy on margin of the nectary groove; anthers reddish; style erect; capsules erect, 4–5 cm. long.——June–Aug. Rocks along seashores; Honshu, Shikoku. Widely grown with numerous horticultural variants.

Var. **davuricum** (Ker-Gawl.) Ohwi. *L. pennsylvanicum* Ker-Gawl.; *L. davuricum* Ker-Gawl.——EZO-SUKASHI-YURI. Bulb-scales more often jointed; stems less densely papillose-tubercled but densely woolly.——June–Aug. Rocks and sandy meadows along seashores; Hokkaido.——Kuriles, Kamchatka, Sakhalin, Korea, Manchuria, Amur, and Dahuria.

Var. **bukosanense** (Honda) Okuyama & T. Koyama. *L. bukosanense* Honda——MIYAMA-SUKASHI-YURI. Stems slender, ascending to reclining, erect only toward the top, papillose-tubercled on lower portion; leaves broadly linear; flowers solitary, rarely 2, erect, spotted.——July. Rocky cliffs in mountains; Honshu (Mount Buko in Musashi).

3. Lilium concolor Salisb. var. **partheneion** (Sieb. & de Vriese) Bak. *L. partheneion* Sieb. & de Vriese; *L. cordion* Sieb. & de Vriese——HIME-YURI. Bulbs small, ovoid, with few scales; stems nearly smooth, 30–80 cm. long, terete, green; leaves linear, 3–7 cm. long, 3–6 mm. wide, sessile, smooth except for minute semirounded tubercles on margin, the nerves slender; flowers 1–5, erect, orange-yellow to reddish,

the tepals oblanceolate, loosely woolly outside, usually spotted, 3–4 cm. long, the nectary grooves papillose on both sides; anthers the same color as the tepals; styles as long as the ovary.——June–July. Meadows in mountains; Honshu, Shikoku, Kyushu; rare.——Korea, Manchuria, Amur. The typical phase of China has purplish, more densely papillose-tubercled stems.

4. Lilium callosum Sieb. & Zucc. *L. tenuifolium* var. *stenophyllum* Bak.——NO-HIME-YURI. Bulbs small, with few scales; stems 30–100 cm. long, terete, glabrous, erect; leaves sessile, linear, 5–13 cm. long, 3–6 mm. wide, narrowed at both ends, glabrous, the margin often with semirounded short tubercles; flowers few, nodding, orange-red or yellowish, with obsolete spots, the tepals 3–4 cm. long, oblanceolate, recurved from below the middle, the nectary grooves slightly hairy on margin; style shorter than the ovary, erect; capsules narrowly oblong, 3–4 cm. long.——July–Aug. Honshu, Shikoku, Kyushu.——Ryukyus, Formosa, China, Manchuria, Korea, and Amur.

5. Lilium leichtlinii Hook. f. var. **leichtlinii**. *L. leichtlinii* var. *majus* G. F. Wils.——KIBANA-NO-ONI-YURI. Flowers yellow.——Said to occur in Honshu (Suruga and Izu Prov.) and in Amami-Oshima in the Ryukyus; frequently cultivated in our gardens.

Var. **tigrinum** (Regel) Nichols. *L. pseudotigrinum* Carr.; *L. maximowiczii* Regel; *L. leichtlinii* var. *maximowiczii* (Regel) Bak.; *L. maximowiczii* var. *tigrinum* Regel——KO-ONI-YURI. Bulbs subglobose, with slender hypogeal stolons; stems 1–2 m. long, terete, erect, usually papillose-tubercled; leaves linear, 8–15 cm. long, 5–12 mm. wide, glabrous or slightly white-hairy; flowers few, nodding, orange-red, spotted, often white-woolly outside, the tepals gently revolute, lanceolate, 6–8 cm. long, hairy on margin of the nectary groove.——July–Aug. Honshu, Shikoku, Kyushu.——Korea and Manchuria.

6. Lilium lancifolium Thunb. *L. tigrinum* Ker-Gawl.——ONI-YURI. Bulbs broadly ovoid-globose, 4–8 cm. across; stems stout, erect, 1–2 m. long, purplish, white-woolly while young; leaves broadly linear to lanceolate, 5–18 cm. long, 5–15 mm. wide, sessile, with few fleshy scaly bulbils in axils; flowers odorless, 4–20, orange-red, with darker spots inside, the tepals revolute, lanceolate, or the inner segments broadly lanceolate, 7–10 cm. long, short-hairy on margin of the nectary groove; styles elongate, upwardly curved.——July–Aug. Hokkaido, Honshu, Shikoku, Kyushu; commonly cultivated in gardens.——Korea, Manchuria, and China.

7. Lilium speciosum Thunb. KANOKO-YURI. Bulbs 7–10 cm. long and as much in diameter; stems 1–1.5 m. long, smooth, glabrous, terete; leaves short-petioled, broadly lanceolate to oblong, 10–18 cm. long, 2–6 cm. wide, smooth, glabrous, 5- to 7-nerved; flowers few to more than 10, reddish pink and usually white above, commonly dark red-spotted, sometimes pure white, nodding, the pedicels elongate, rather stout, the bracts leaflike, the tepals geniculately reflexed on lower half, broadly lanceolate, 8–10 cm. long, the nectary gland glabrous on margin; style elongate, curved upward; capsules oblong or ovate-oblong, with 3 broad grooves and 6 shortly winged angles.——Aug. Shikoku, Kyushu; rare. Frequently cultivated.

8. Lilium auratum Lindl. *L. dexteri* Hovey——Yama-yuri. Bulbs depressed-globose, 6–10 cm. in diameter; stems 1–1.5 m. long, slightly purple-tinged, smooth, glabrous; leaves short-petioled, spreading, broadly to narrowly lanceolate, 10–15 cm. long, 1.5–3.5 cm. wide; flowers usually few, sometimes up to 20, broadly infundibuliform, nodding or at right angles, on stout often bracteate pedicels, white with dark red spots, strongly fragrant, the tepals obliquely spreading, broadly lanceolate to narrowly ovate, 10–18 cm. long, recurved above, the inner segments broader, yellowish and glabrous on margin along the nectary groove; style elongate, curved upward at tip; capsules oblong, 5–8 cm. long.——July–Aug. Hills and mountains; Honshu. Naturalized in Hokkaido.

Var. **platyphyllum** Bak. *L. auratum* var. *hamaoanum* Makino; *L. platyphyllum* (Bak.) Makino——Saku-yuri. Taller and stouter than the typical phase; leaves 3.5–6 cm. wide; flowers larger, more numerous, the tepals broader, with fewer yellowish spots to nearly spotless.——Honshu (Izu Isls.). Said to grow with the typical phase on the Izu Peninsula.

9. Lilium medeoloides A. Gray. *L. avenaceum* Fisch. ex Maxim.; *L. maculatum* Bak. non Thunb.; *L. sadoinsulare* Masam. & Satomi; *L. medeoloides* var. *sadoinsulare* (Masam. & Satomi) Masam. & Satomi——Kuruma-yuri. Bulbs 2–2.5 cm. in diameter, the scales jointed toward the middle and base; stems smooth, 30–70 cm. long, terete; leaves mostly approximate below the middle of stem, verticillate, lanceolate to oblanceolate or narrowly ovate, 5–12 cm. long, 3–5 cm. wide, upper leaves alternate, small, few; flowers 1–5, orange-red, foetid, rather small, the pedicels bracteate, erect, the tepals much-recurved from near base, lanceolate, 3–4.5 cm. long, with few to rather numerous dark spots, the nectary glands glabrous or with short-hairs on margin near base; capsules obovate.——July–Aug. Damp woods and meadows, sometimes ascending to alpine regions; Hokkaido, Honshu (centr. and n. distr.).——Kuriles, Kamchatka, Sakhalin, s. Korea, and China.

10. Lilium hansonii Leichtl. *L. maculatum* Hook. f., non Thunb.; *L. medeoloides* var. *obovata* Fr. & Sav.——Take-shima-yuri. Bulbs 3.5–7 cm. in diameter, the scales imbri-

cate, narrowly ovate, inarticulate, acute on margin; stems 1–1.5 m. long; leaves verticillate, oblanceolate to narrowly oblong, 10–18 cm. long, 2–4 cm. wide; flowers 4–12, yellow, fragrant, nodding to horizontal, the tepals very thick, spreading, gently recurved, broadly lanceolate to oblanceolate, 3–4 cm. long, with glabrous nectary glands; capsules subglobose, 2.5–3.5 cm. long and as much in diameter.——June–July. Said to be spontaneous in Hokkaido.——Korea and Ussuri.

11. Lilium longiflorum Thunb. Teppō-yuri. Bulbs depressed-globose, 5–6 cm. in diameter; stems 30–100 cm. long, smooth; leaves sessile, narrowly lanceolate, 10–18 cm. long, 5–15 mm. wide; flowers few, sometimes solitary, white, the midrib and marginal zone often with a purple hue, fragrant, tubular-infundibuliform, the tepals oblanceolate, unspotted, 13–18 cm. long, with slightly recurved tip, the nectary glands glabrous; styles elongate, slightly curved upward at tip; capsules oblong.——May–June. Kyushu (Yakushima and Tanegashima).——Ryukyus.

12. Lilium rubellum Bak. *L. japonicum* var. *rubellum* (Bak.) Makino——Hime-sa-yuri. Bulbs 2.5–3 cm. in diameter, ovoid-globose; stems 30–80 cm. long, smooth, naked below; leaves broadly lanceolate to narrowly oblong, 5–10 cm. long, 3–3.5 cm. wide, abruptly narrowed at both ends, short-petioled, smooth; flowers 1–3(–6), fragrant, rose-colored, infundibuliform, the tepals oblanceolate, 5–7 cm. long, unspotted, with glabrous nectary glands, recurved slightly near the summit; style elongate, curved slightly upward at tip.——June–Aug. Low mountains, sometimes also on alpine slopes; Honshu (Japan Sea side of centr. and n. distr.).

13. Lilium japonicum Thunb. *L. krameri* Teutschell; *L. belladonna* Leichtl.; *L. japonicum* var. *angustifolium* Makino; *L. makinoi* Koidz.; *L. abeanum* Honda; *L. japonicum* var. *abeanum* (Honda) Kitam.——Sasa-yuri. Bulbs broadly ovoid, 2–4 cm. in diameter, the scales ovate; stems rather slender, 50–100 cm. long, glabrous, slightly glaucous; leaves broadly lanceolate to narrowly oblong, 5–10 cm. long, 3–3.5 cm. wide, abruptly narrowed at each end, short-petioled, smooth; flowers 1–3(–5), fragrant, rose-colored, infundibuliform, nodding or horizontal, the tepals broadly oblanceolate, 12–15 cm. long, the inner segments broader, the outer narrowed above, the nectary gland greenish, glabrous; style elongate, slightly curved above; capsules obovoid, about 4 cm. long.——July–Aug. Thickets in hills and low mountains; Honshu (centr. distr. and westw.), Shikoku, Kyushu.

20. ERYTHRONIUM L.　　Katakuri Zoku

Perennial bulbous herbs; scapes single, simple, 2-leaved on lower half; leaves petiolate, flat, ovate to broadly lanceolate; flowers nodding, solitary or several on slender peduncles, relatively large, the tepals 6, free, in 2 series, lanceolate, 3- to 5-nerved, recurved above; stamens 6, shorter than the tepals, the filaments subulate, slightly flattened at base, often unequal in length, the anthers oblong to linear; ovary sessile, 3-locular, the ovules many; style filiform or slightly broadened above, shortly 3-lobed; capsules membranous to coriaceous, 3-angled, globose to elliptic.——About 15 species, in the temperate regions of the N. Hemisphere, most abundant in w. N. America.

1. Erythronium japonicum Decne. *E. dens-canis* var. *japonicum* Bak.——Katakuri. Bulbs cylindric-lanceolate, 5–6 cm. long, about 1 cm. in diameter, with two bulblets subtending at the base; scapes 20–30 cm. long, with a pair of leaves below the middle; leaves petiolate, narrowly ovate or oblong, 6–12 cm. long, 2.5–5 cm. wide, obtuse, mucronate or acute, entire; flowers solitary, terminal, nodding, long-pedun-

culate, rose-purple, the tepals lanceolate, 5–6 cm. long, 5–10 mm. wide, with a 3-lobed dark purple spot at base; stamens slightly unequal, the anthers broadly linear, 6–8 mm. long; capsules suborbicular, 3-angled; stigma short, 3-lobed.——Apr. Woods in lowlands and low mountains; Hokkaido, Honshu, Shikoku (rare); commoner northward.——Korea.

21. TULIPA L. AMANA ZOKU

Bulbous perennial herbs; scapes few-leaved; leaves oblong to linear, radical or lower cauline; flowers erect, the tepals 6, deciduous, free, eglandular, blotched inside near base; filaments filiform or subulate, shorter than the perianth, the anthers linear-oblong; ovary oblong, many-ovuled; style often very short; capsules oblong to subglobose, loculicidal, 3-valved; seeds flat, often narrowly winged on margin.——About 50 species, in the temperate regions of the Old World, especially abundant in centr. Asia.

1A. Leaves 15–25 cm. long, 5–10 mm. wide, glaucous, usually gradually narrowed to a small obtuse apex; bracts 2–3. .... 1. *T. edulis*
1B. Leaves 10–15 cm. long, 7–15 mm. wide, abruptly contracted to an obtuse apex, brownish green with a broad white band along the midrib above. .... 2. *T. latifolia*

1. Tulipa edulis (Miq.) Bak. *Orithya edulis* Miq.; *O. oxypetala* sensu A. Gray non Kunth; *Amana edulis* (Miq.) Honda——AMANA. Bulbs broadly ovoid, 3–4 cm. long, with dark brown outer coat; leaves 2, relatively thick, attached near base of the scape, paired, linear, 15–25 cm. long, 5–10 mm. wide, gradually narrowed to an obtuse point, glaucous; scapes slender, 15–30 cm. long, often decumbent, the bracts 2(–3), linear, 2–3 cm. long; flowers 1–3, 2–2.5 cm. long, erect, the pedicels 2–4 cm. long, the tepals lanceolate, subobtuse or subacute, white with dark purple striations inside; style 4–5 mm. long; capsules nearly globose and angled, about 1.2 cm. long and as broad.——Mar.–Apr. Meadows in lowlands along rivers; Honshu (Kantō Distr. and westw.); Shikoku, Kyushu; common.——China.

2. Tulipa latifolia (Makino) Makino. *T. edulis* var.

latifolia Makino; *Amana latifolia* (Makino) Honda——HIRO-HA-AMANA. Bulbs ovoid, 3–4 cm. long, the outer coat dark brown, long-pubescent inside; scapes 15–20 cm. long, simple, rarely forked above the leaves; leaves 2, essentially paired, soft, 10–15 cm. long, 7–15 mm. wide, brownish green to purplish, with a broad white longitudinal band above, abruptly contracted to the obtuse apex; bracts 3, linear, 2–3 cm. long, 1–2 mm. wide, equaling to shorter than the pedicels; flowers solitary, 2–2.5 cm. long, on slender pedicels, the tepals lanceolate, subobtuse, white with several dark purple nerves inside, green near base; stamens less than half the length of tepals, the filaments flat, broadly lanceolate, greenish, 1.5–2 times as long as the anthers.——Mar.–Apr. Wooded hills; Honshu (Musashi and Ise Prov.); rare.

22. FRITILLARIA L. BAIMO ZOKU

Bulbous herbs with few to many, sometimes jointed scales; scapes simple, erect, leafy; leaves alternate, opposite, or verticillate, rarely attenuate above into a curved point; flowers relatively large, nodding, solitary or few, campanulate, the tepals 6, oblong or ovate, with a depressed nectary gland above the base; filaments filiform or slightly flattened; ovary 3-carpellate, many-ovuled; style filiform or columnar, entire or 3-lobed; capsules globose to obovoid, angled, loculicidal; seeds flattened, narrowly winged on margin.——About 60 species, in the temperate regions of the N. Hemisphere.

1A. Bulbs with many jointed scales; flowers 1 to several, brown-purple, sometimes greenish. .... 1. *F. camtschatcensis*
1B. Bulbs with 2 large inarticulate scales.
  2A. Leaves many, the upper elongate to a hooked point; flowers 1 to several. .... 2. *F. verticillata* var. *thunbergii*
  2B. Leaves in 2 whorls, not elongate above, the lower whorl 2-leaved, the upper 3-leaved; flowers solitary.
    3A. Flowers narrowly companulate; tepals oblanceolate or linear-oblong, glandular at base, colored net inside absent; filaments and style densely papillose. .... 3. *F. amabilis*
    3B. Flowers broadly campanulate; tepals obovate-oblong, glandular on lower third, colored net inside present. .... 4. *F. japonica*

1. Fritillaria camtschatcensis (L.) Ker-Gawl. *Lilium camtschatcense* L.——KURO-YURI. Bulbs depressed-globose (a seashore race exists with short stolons), 1.5–3 cm. in diameter, the scales many, white short-stiped or jointed; stems 10–50 cm. long, erect, with few to several whorls of leaves; leaves in verticils of (2–)3–5, sessile, 3–7 cm. long, 1–3 cm. wide, subobtuse to subacute, lanceolate to ovate-oblong; flowers 1 to several, on short pedicels, nodding, dark brown-purple, broadly campanulate, foetid, the tepals rather thick, 2–3 cm. long, 7–10 mm. wide, obtuse, papillose-hairy at apex, the glands nearly basal; style 3-parted nearly to the base; capsules obovoid-globose, 2–2.5 cm. long.——(June) July–Aug. Hokkaido (rocky places along seashores in eastern distr. and alpine regions in centr. distr.), Honshu (centr. and n. alpine regions). ——Kamchatka, Kuriles, Sakhalin, Ussuri, and N. America.

2. Fritillaria verticillata Willd. var. **thunbergii** (Miq.) Bak. *Uvularia cirrhosa* Thunb., non *F. cirrhosa* Don; *F. thunbergii* Miq.——BAIMO. Bulbs 1.5–3 cm. in diameter, scales semiglobose, white, thick, fleshy; stems 30–80 cm. long, erect, many-leaved; leaves opposite or ternate, linear-lanceolate,

gradually narrowed toward the tip, sessile, 7–15 cm. long, 5–10 cm. wide, whitish beneath, the upper cauline leaves bractlike and producing a hooked tendrillike point; flowers several or sometimes solitary, the pedicels 1–2 cm. long, the tepals pale yellow, 2.5–3 cm. long, about 1 cm. wide, obsoletely reticulate inside, oblong, obtuse, with a nectary gland near the base; style longer than the stigmas; fruit squarrose with 6 broad wings, the tip retuse.——Apr.–May.——Chinese plant long cultivated in our gardens for medicinal and ornamental purposes, sometimes naturalized.——The typical phase occurs in the Altai Mts. and centr. Asia.

3. Fritillaria amabilis Koidz. HOSOBANA-KO-BAIMO. Bulbs globose, 6–8 mm. in diameter, the scales 2, thickened, semiglobose; stems 15–20 cm. long; leaves cauline, consisting of a pair of opposite leaves near the middle and 3 verticillate leaves (bracts) above, the lower pair lanceolate or linear-lanceolate, 4–6 cm. long, 4–10 mm. wide, subacute, sessile, the upper 3 usually smaller, commonly linear, the pedicels about 1 cm. long; flowers tubular-campanulate, the tepals oblanceolate to linear-oblong, not reticulate, 1.5–2 cm. long, glandular near the base; filaments and style rather densely papillose.——Mar.–Apr. Honshu (Chūgoku Distr.), Shikoku, Kyushu; rare.

4. Fritillaria japonica Miq. var. **koidzumiana** (Ohwi) Hara & Kanai. *F. koidzumiana* Ohwi——Ko-BAIMO, KOSHINO-KO-BAIMO.   Bulbs globose, about 1 cm. in diameter, the scales 2, thickened, semiglobose; stems 15–20 cm. long, erect; leaves lanceolate to broadly linear, opposite below, the upper verticillate, 4–6 cm. long, 5–18 mm. wide, sessile, obtuse; flowers solitary, broadly campanulate, about 2 cm. long and as wide, the pedicels about 1 cm. long, recurved, the tepals narrowly oblong, 6–7 mm. wide, pale yellow, on the inside reticulate, dark purple, loosely fimbriate on margin near middle and on midrib inside.——Apr.–May. Honshu (Izu, Suruga, Echigo, and Iwashiro Prov.); rare.

Var. **japonica.** *F. japonica; F. muraiana* Ohwi.——AWA-KO-BAIMO, KO-BAIMO.   Closely resembles the preceding but slightly smaller; tepals entire, 16–18 mm. long, 4–5 mm. wide.——Apr. Honshu (Ise and Mino Prov.), Shikoku; rare.

23. ASPARAGUS L.   KUSA-SUGI-KAZURA ZOKU

Rhizomatous dioecious herbs or subshrubs; stems erect or sometimes scandent, branching; leaves reduced to scales, the branchlets (cladodes) in their axils functioning as leaves; flowers small, solitary to several in umbels or racemes, bisexual (or plants dioecious), campanulate to tubular, the tepals free or connate below; stamens 6; ovary 3-locular; style simple, the stigmas 3; fruit a globose berry; seeds 1 to few, globose.——About 300 species in the temperate to subtropical regions of the Old World, especially in drier areas.

1A.   Flowers 1–2 in axils, campanulate or tubular-campanulate, 5–7 mm. long; pedicels 8–12 mm. long, slender, jointed at or above the middle; cladodes slender, straight; fruit red.
  2A.   Branches nearly terete, slender, smooth; flowers 5–6 mm. long; anthers as long as the filaments; cladodes 1–2 cm. long, slender.
                                                                      1. *A. officinalis*
  2B.   Branches 3-angled, with semiround tubercles on the angles; flowers 6–7 mm. long; anthers longer than the filaments; cladodes 1–3 cm. long. . . . . 2. *A. oligoclonos*
1B.   Flowers several, fasciculate in axils, rarely 1–2, broadly campanulate or infundibuliform, 2–5 mm. long; pedicels 2–5 mm. long; cladodes often arcuate or falcate.
  3A.   Pedicels jointed near the middle.
    4A.   Cladodes 3–6 in a fascicle, straight, 3-angled, acute; roots elongate; flowers about 5 mm. long; fruit red. . . . . 3. *A. kiusianus*
    4B.   Cladodes 1 or 2(–3) in a fascicle, loosely falcate, lustrous, flattened, 3-angled, acuminate with a short awn at the tip; roots fusiform; flowers about 2.5 mm. long; fruit whitish. . . . . 4. *A. cochinchinensis*
  3B.   Pedicels jointed immediately beneath the flowers; roots slender; cladodes 3–7 in a fascicle, 3-angled, acuminate with a short awn at apex, loosely falcate; fruit red. . . . . 5. *A. schoberioides*

1. Asparagus officinalis L.   ORANDA-KIJI-KAKUSHI. Rhizomes short-creeping with stout fibrous roots; stems 1–1.5 m. long, smooth, much branched, terete; cladodes slender, 5–8 in a fascicle, filiform, acuminate, 1–2 cm. long, straight; flowers tubular-campanulate, yellowish green, the pedicels solitary or paired, slender; anthers narrowly ovate, obtuse, nearly as long as the filaments; fruit red, globose.——May–June. Europe. Cultivated as a vegetable in our gardens.

2. Asparagus oligoclonos Maxim.   *A. tamaboki* Yatabe ——TAMABŌKI, TSUKUSHI-TAMABŌKI.   Rhizomes short, with elongate roots; stems 50–100 cm. long, much branched above, erect, striate, angled, minutely scabrous when dried; cladodes 1–8 in a fascicle, linear, acute, straight, angled, 1–3.5 cm. long, deep green; flowers paired or solitary, 6–7 mm. long, tubular-campanulate, yellowish green, the pedicels jointed at or above the middle, 8–9 mm. long; anthers narrowly ovate-lanceolate, mucronate, longer than the filaments; fruit red.——May–June. Mountains; Kyushu.——Korea, Manchuria, Amur, and Ussuri.

3. Asparagus kiusianus Makino.   HAMA-TAMABŌKI. Rhizomes short, with elongate roots; stems angled, 50–80 cm. long, much branched, the branchlets angled, scaberulous; cladodes (1–)3–6 in a fascicle, straight, deep green, 3-angled, acute, 1–1.5(–1.8) cm. long; flowers yellow-green, 2–6 in axils, about 5 mm. long, campanulate-infundibuliform, the pedicels 2.5–4 mm. long, jointed near the middle; anthers oblong, mucronulate, longer than the filament; fruit red.——Near seashores; Kyushu (Hizen and Chikuzen Prov.).

4. Asparagus cochinchinensis (Lour.) Merr. *Melanthium cochinchinense* Lour.; *A. lucidus* Lindl.——KUSA-SUGI-KAZURA, TEMMONDŌ.   Rhizomes short, the adventitious roots fusiform; stems subscandent, 1–2 m. long, the branches slender, acutely angled, nearly smooth; cladodes 1–2 in a fascicle, linear, flattened, 3-angled, acuminate with a short awn at apex, 1–2 cm. long, 1–1.2 mm. wide, somewhat falcate, lustrous; flowers 1–3, pale yellow-green, campanulate-infundibuliform, about 3 mm. long, the pedicels 2–5 mm. long, jointed near the middle; anthers ovate-elliptic, obtuse, shorter than the filaments; fruit whitish.——May. Near seashores; Hokkaido (naturalized), Honshu, Shikoku, Kyushu.——Ryukyus, Formosa, China, and Korea.

Var. **pygmaeus** (Makino) Ohwi. *A. lucidus* var. *pygmaeus* Makino; *A. pygmaeus* (Makino) Makino——TACHI-TEMMON-DŌ.   Stems erect, 15–20 cm. long.——Cultivated in our gardens.

5. Asparagus schoberioides Kunth. *A. wrightii* A. Gray; *A. parviflorus* Turcz.; *A. sieboldii* Maxim.; *A. rigidulus* Nakai——KIJI-KAKUSHI.   Roots elongate; stems 50–100 cm. long, much branched above, the branchlets acutely angled, papillose on the angles; cladodes 3–7 in a fascicle, narrowly linear, flattened with a raised rib on one side, somewhat falcate, 1–2 cm. long, acuminate, short-awned at the tip; flowers 2–4 in the axils, broadly campanulate, 2.5–3 mm. long, pale greenish yellow, the pedicels short, 1–2 mm. long, jointed near the top; anthers cordate, much shorter than the filaments; fruit red.——May–July. Hokkaido, Honshu, Shikoku, Kyushu.——Sakhalin, Korea, Ussuri, Dahuria, Manchuria, China, and Formosa.

Var. **subsetaceus** Franch. HOSOBA-KIJI-KAKUSHI.   The cladodes more slender.——Occurs with the typical phase.

24. DIANELLA Lam. Kikyō-ran Zoku

Evergreen perennial herbs from stout branching rhizomes; leaves linear, 2-ranked, radical, subcoriaceous, flat, with a raised costa beneath; scapes relatively tall, with few, short, linear leaves; inflorescence paniculate, loosely branched, bearing many short terminal racemes, the bracts small, the pedicels jointed at top; flowers blue, sometimes pale yellow, relatively small, nodding, the tepals free, narrowly oblong, 3- to 7-nerved; stamens 6, the filaments usually partially thickened, the anthers opening by terminal pores; ovary 3-locular, the ovules 3–8 in each locule; style filiform, the stigma small; fruit a berry, sub-globose, blue; seeds ovate, often flattened, lustrous, black.——About 20 species, from India to the Pacific Islands, also in n. Australia and Madagascar.

1. Dianella ensifolia (L.) DC. *Dracaena ensifolia* L.; *Dianella nemorosa* Lam.——Kikyō-ran.　Evergreen glabrous herb from stout creeping rhizomes, the old decayed leaf sheaths covering the base; leaves 50–60 cm. long, 1–2 cm. wide, rather firm and lustrous above, broadly linear, gradually tapering, the tip obtuse, also gradually narrowed and loosely folded below to a densely imbricated sheath, the midrib impressed above and keeled beneath; scapes 50–100 cm. long, with few small loose leaves; panicle 15–30 cm. long, loosely branched, the flowers on upper part, the pedicels arcuate, 7–15 mm. long, the bracts small, membranous, the tepals spreading, blue, about 6 mm. long, narrowly oblong; filaments relatively stout above, geniculate near the middle; berry ovoid-globose to globose, blue-purple, about 4 mm. in diameter.——Apr.-Aug. Near seashores; Honshu (Kii Prov.), Shikoku, Kyushu.——Ryukyus, Formosa, and Malaysia.

25. POLYGONATUM Adans. Amadokoro Zoku

Rhizomes thickened, creeping; stems simple, leafy; leaves alternate or rarely verticillate, entire, lanceolate to broadly elliptic, many-nerved; inflorescence an axillary cyme sometimes reduced to a fascicle or single flower, the pedicels jointed below the flower, the bracts sometimes developed; flowers pendulous and tubular, the tepals 6, tubular-connate; filaments adnate to the corolla-tube, filiform or flattened, the anthers oblong to lanceolate, bifid at base, introrse; ovary 3-locular, the ovules 4–6 in each locule; style slender, the stigma small; fruit a globose few-seeded berry.——About 30 to 40 species, in the temperate regions of the N. Hemisphere.

1A. Bracts absent.
  2A. Filaments with dense multicellular long hairs; stems terete, finely striate; leaves smooth, usually whitish beneath; peduncles elongate, horizontally spreading; rhizomes with thickened internodes. .... 1. *P. lasianthum*
  2B. Filaments smooth, glabrous or papillose-tuberculate; peduncles strongly arcuate-recurved from the base.
    3A. Stems angled except at base, erect or ascending above; rhizomes cylindrical, long-creeping, with long internodes.
      4A. Stems erect, 15–30 cm. long; filaments, leaves on lower side, and the leaf-margins papillose; rhizomes 2–4 mm. in diameter; leaves pale green beneath. .... 2. *P. humile*
      4B. Stems ascending above, 30–100 cm. long; filaments and leaves smooth or nearly so; rhizomes 5–20 mm. in diameter; leaves glaucous beneath. .... 3. *P. odoratum*
    3B. Stems nearly terete, prominently arcuate-ascending above; rhizomes with short thickened internodes, submoniliform; perianth-tube slightly elongate and stipitate at the base.
      5A. Flowers 25–35 mm. long, pale green; leaves glaucescent beneath, smooth; stems terete; filaments 7–10 mm. long, much thickened at base, papillose except at apex; anthers 4.5–5.5 mm. long. .... 4. *P. macranthum*
      5B. Flowers 18–23(–25) mm. long; leaves usually papillose on nerves beneath; filaments thickened below, smooth or nearly so, 4.5–7 mm. long; anthers 2.5–3 mm. long.
        6A. Leaves lanceolate-ovate to narrowly ovate, acute; filaments smooth. .... 5. *P. trichosanthum*
        6B. Leaves linear-lanceolate to oblong-lanceolate. .... 6. *P. falcatum*
1B. Bracts distinct.
  7A. Bracts membranous, 1-nerved, as long as to shorter than the pedicels; peduncles 1.5–3 cm. long; flowers greenish, 22–25 mm. long; perianth pubescent inside. .... 7. *P. inflatum*
  7B. Bracts herbaceous, several- to many-nerved, much longer than the pedicels; peduncles 5–15(–20) mm. long; flowers 12–25 mm. long; perianth-tube whitish to pale yellow-green, glabrous.
    8A. Peduncles, pedicels, bracts, and frequently the underside of leaves prominently papillose; flowers yellow-green, about 13 mm. long; pistil shorter than the stamens; style twice as long as the ovary .... 8. *P. cryptanthum*
    8B. Peduncles and pedicels smooth; bracts and leaves beneath rarely papillose; flowers whitish or pale green, 20–25 mm. long; pistil longer than the stamens; style 2.5–4 times as long as the ovary.
      9A. Bracts lanceolate, usually papillose on the margin, usually inserted at or above the middle on the pedicels.
        10A. Filaments glabrous. .... 9. *P. desoulavyi* var. *yezoense*
        10B. Filaments mealy-puberulent. .... 10. *P. miserum*
      9B. Bracts ovate, smooth, inserted at base of pedicels. .... 11. *P. involucratum*

1. Polygonatum lasianthum Maxim. Miyama-naruko-yuri.　Slender herbaceous perennial from creeping knotty rhizomes 3–8 mm. in diameter; stems 30–60 cm. long, ascending above, weakly striate; leaves firm, narrowly oblong to broadly elliptic, short-petiolate, smooth, often glaucescent beneath, thinly chartaceous; peduncles 1–2 cm. long, 1- to 3-flowered, horizontally spreading, the pedicels shorter than the common peduncle; flowers white, the tips greenish, 15–18 mm. long, sometimes with a very short stipe at the base.——May-June. Thin woods, low mountains and hills; Hokkaido, Honshu, Shikoku, Kyushu.——Forma **amabile** (Yatabe) Makino. *P. amabile* Yatabe——Hime-naruko-yuri.　Leaves with purplish variegation.

2. **Polygonatum humile** Fisch. *P. officinale* var. *humile* (Fisch.) Bak.——HIME-IZUI. Rhizomes elongate, slender, 3–4 mm. in diameter; stems erect, 15–30 cm. long, angled; leaves oblong-lanceolate, 4–7 cm. long, 1.5–3 cm. wide, sessile, the nerves loosely papillose beneath; flowers 1 to 2, yellowish green, 15–18 mm. long, the pedicels 7–15 mm. long, the filaments slightly papillose, the anthers deltoid-lanceolate, 3–3.5 mm. long, slightly shorter than the filaments.——June–July. Meadows and thin woods in lowlands from sea level to the foot of the mountains; Hokkaido, Honshu (centr. and n. distr.), Kyushu.——Sakhalin, Kuriles, Korea, Manchuria to Siberia.

3. **Polygonatum odoratum** (Mill.) Druce var. **pluriflorum** (Miq.) Ohwi. *P. officinale* var. *pluriflorum* Miq.; *P. odoratum* var. *japonicum* Hara, excl. syn.; *P. japonicum* sensu auct. Japon., non Morr. & Decne.; *P. officinale* sensu auct. Japon., non All.——AMADOKORO. Rhizomes elongate, 4–7 mm. in diameter, the internodes long; stems angled, 30–60 cm. long, ascending above; leaves oblong to narrowly oblong, nearly sessile, 5–10 cm. long, 2–5 cm. wide, glabrous, usually glaucous beneath; flowers 1–2, axillary, white, greenish near the tip, 15–20 mm. long, the pedicels 0.8–2 cm. long; filaments minutely papillose, the anthers about 4 mm. long, equaling to slightly longer than the filaments.——Apr.–May. Meadows and thin woods in lowlands and foothills; Hokkaido, Honshu, Shikoku, Kyushu; common.——Korea and China. The typical phase occurs elsewhere in the temperate regions of the Old World.

Var. **thunbergii** (Morr. & Decne.) Hara. *P. thunbergii* Morr. & Decne.; *P. giganteum* var. (?) *thunbergii* (Morr. & Decne.) Maxim.; *P. japonicum* Morr. & Decne.; *P. officinale* var. *japonicum* (Morr. & Decne.) Miq.; *P. odoratum* var. *japonicum* (Morr. & Decne.) Hara——YAMA-AMADOKORO. Plant rather stout; stems 50–100 cm. long; leaves 8–15 cm. long, with minute papillae on the nerves beneath; flowers 2–2.5 cm. long.——Apr.–May. Mountains; Honshu.

Var. **maximowiczii** (F. Schmidt) Koidz. *P. maximowiczii* F. Schmidt; *P. officinale* var. *maximowiczii* (F. Schmidt) Maxim.; *P. japonicum* var. *maximowiczii* (F. Schmidt) Nakai; *P. hondoense* Nakai——Ō-AMADOKORO. Plant stout, 60–100 cm. high; leaves 10–20 cm. long, 3–8 cm. wide, the nerves beneath with minute papillae; flowers (1–)2–4, axillary, 22–25 mm. long.——May–June. Hokkaido, Honshu (n. distr.).——s. Kuriles, Sakhalin, and Ussuri.

4. **Polygonatum macranthum** (Maxim.) Koidz. *P. giganteum* var. *macranthum* Maxim.; *P. silvicolum* Makino; *P. sadoense* Nakai; *P. iyoense* Nakai——Ō-NARUKO-YURI, YAMA-NARUKO-YURI, ONI-NARUKO-YURI. Rhizomes stout; stems erect, ascending above, stout, slightly glaucous, (40–)80–130 cm. long, greenish, terete; leaves usually narrowly oblong to broadly lanceolate, glabrous, 15–30 cm. long, 5–12 cm. wide, glaucescent beneath; flowers (1–)3–4, axillary, pale green, (25–)30–35 mm. long, tubular, gradually narrowed at base; filaments minutely papillose, the anthers 4.5–5.5 mm. long, shorter than the filaments.——May–July. Mountains, woods; Hokkaido, Honshu, Shikoku, Kyushu.

5. **Polygonatum trichosanthum** Koidz. MARUBA-ŌSEI. Stems more than 60 cm. high, terete, glabrous; leaves ovate, 7.5–13.5 cm. long, 2.5–5.5 cm. wide, glaucescent beneath, the nerves slightly scabrous; flowers 3–4, glabrous, about 18 mm. long, the pedicels glabrous; filaments longer than the anthers. ——Cultivated in Tokyo, but said to occur in Kyushu (Cape Nomo in Hizen Prov.).

6. **Polygonatum falcatum** A. Gray. *P. canaliculatum* var. *sublanceolatum* Miq.; *P. giganteum* var. *falcatum* (A. Gray) Maxim.; *P. falcatum* var. *tenuiflorum* (Koidz.) Ohwi; *P. tenuiflorum* Koidz.; *P. petiolatum* Lév.——NARUKO-YURI. Rhizomes stout, creeping, with approximate nodes; stems terete, 50–80 cm. long, ascending above; leaves lanceolate to narrowly so, sessile, 8–13 cm. long, 1.8–2.5 cm. wide, tapered at both ends, the apex obtuse, glabrous above often with a white longitudinal band in the center, slightly scabrous beneath on the nerves, papillose; flowers (1–)3–5 in an axillary umbel or corymb, greenish white, about 2 cm. long, usually contracted into a short stipe at base, the pedicels slender, shorter than the peduncle; filaments nearly smooth, the anthers about 3 mm. long, shorter than the filaments.——May–June. Hills and low mountains; Hokkaido (according to Miyabe and Kudo), Honshu, Shikoku, Kyushu.——Korea.

7. **Polygonatum inflatum** Komar. *P. nipponicum* Makino——MIDORI-YŌRAKU. Glabrous glaucescent herb from elongate creeping rhizomes 4–5 mm. in diameter, the internodes elongate; stems 30–60 cm. long, rather stout, ascending and angled above; leaves 5–9, oblong, rather thin, glabrous, short-petiolate, 10–15 cm. long, 4–7 cm. wide, glaucous beneath; peduncles 3- to 7-flowered, 1.5–3 cm. long, with 3–7 bracts at the apex, the bracts thinly membranous, broadly lanceolate, acuminate, 1-nerved, slightly folded, 4–8 mm. long, pale green; flowers pale green, 22–25 mm. long, urceolate-tubular, the pedicels nearly as long as the bracts, the perianth inside partially pubescent; anthers about 4 mm. long, about half as long as the filaments.——June. Mountains; Kyushu; rare.——Korea and Manchuria.

8. **Polygonatum cryptanthum** Lév. & Van't. *P. fauriei* Lév. & Van't.; *P. taquetii* Lév. & Van't.——USUGI-WANI-GUCHI-SŌ. Rhizomes creeping, the internodes relatively long; stems 20–40 cm. long, terete below, angled above; leaves 5–6, subsessile, ovate to elliptic, 3–4.5 cm. long, 2–3.5 cm. wide, prominently papillose on margin, on the nerves beneath, or sometimes on both sides; peduncles arcuately recurved, 3–6 mm. long, flattened, densely papillose, the bracts 2(–3), 7–10 mm. long, attached at the top of the peduncle, ovate, many-nerved, papillose; flowers yellow-green, 10–13 mm. long, glabrous, the pedicels 1–3 mm. long; anthers about 2.5 mm. long, the filaments smooth, nearly filiform, shorter than the anthers. ——May. Kyushu (Tsushima).——s. Korea.

9. **Polygonatum desoulavyi** Komar. var. **yezoense** (Miyabe & Tatew.) Satake. *P. mediobracteatum* var. *yezoense* Miyabe & Tatew.; *P. desoulavyi* var. *mediobracteatum* (Ohwi) Satake; *P. mediobracteatum* Ohwi——KŌRAI-WANIGUCHI-SŌ, EZO-WANIGUCHI-SŌ. Rhizomes elongate, 3–4 mm. in diameter, with relatively long internodes; stems terete, ascending and angled above, 20–40 cm. long, 5- to 8-leaved; leaves oblong, 5–10 cm. long, 2–4 cm. wide, sessile, papillose only slightly on nerves beneath; peduncles arcuate and reclined, the pedicels 8–12 mm. long, nearly as long as the peduncle; bracts 2, attached to the pedicels about the middle or lower, lanceolate or narrowly ovate, 2–2.5 cm. long, 5–8 mm. wide, many-nerved, the margin usually papillose; flowers about 2 cm. long, white with a green tip, glabrous; filaments about 4 mm. long, nearly smooth, as long as the anthers.——June. Hokkaido (Makomanai in Ishikari Prov.); rare.——Korea. The typical phase occurs in Ussuri and Korea.

10. **Polygonatum miserum** Satake. KO-WANIGUCHI-SŌ. Rhizomes slender, much elongate, 2–3 mm. in diameter, with

long internodes; stems terete, 15–20 cm. long, angled and ascending above, 6- to 7-leaved; leaves oblong to narrowly so, 4–5 cm. long, 1–1.7 cm. wide, subsessile, slightly paler beneath, the nerves obscurely papillose; peduncles about 5 mm. long, the pedicels 7–8 mm. long, bracteate near the middle, the bracts lanceolate, with semirounded tubercles on margin, 7–12 mm. long, 1.5–3 mm. wide; flowers 17–19 mm. long, greenish white, glabrous; filaments mealy-puberulent, the anthers about 4 mm. long, slightly longer than the filaments.——June. Honshu (Azusayama in Shinano); rare.

11. Polygonatum involucratum (Fr. & Sav.) Maxim. *Periballanthus involucratus* Fr. & Sav.; *P. periballanthus* Makino; *? P. periballanthus* var. *ibukiense* Makino; *? P. ibukiense*

(Makino) Makino——Waniguchi-sō. Rhizomes slender, 3–4 mm. in diameter, with elongate internodes; stems 20–40 cm. long, terete, ascending and angled above; leaves 4–7, narrowly ovate to ovate-elliptic, 5–10 cm. long, 2.5–4 cm. wide, subsessile to short-petiolate, glabrous, subglaucous beneath; peduncles 1–2 cm. long, smooth, bracts 2(–3), at the top of peduncle, ovate to ovate-orbicular, smooth, many-nerved, 1.5–3 cm. long, 1–2.5 cm. wide, the pedicels paired, 2–4 mm. long; flowers white, greenish at the tip, urceolate-tubular, about 2.5 cm. long, glabrous; filaments laterally flattened, granular-papillose, slightly longer than the anthers.——May–June. Meadows in mountains; Hokkaido, Honshu, Shikoku, Kyushu; rather rare.——Korea, Manchuria, and Ussuri.

26. STREPTOPUS Michx. Takeshima-ran Zoku

Perennial herbs from creeping rhizomes; stems simple or sometimes branched; leaves ovate to lanceolate; flowers 1–2, white, pale pink, or yellowish brown, nodding, campanulate or subrotate, on slender peduncles often partially adnate to the stem; tepals 6, free, broadly lanceolate; stamens 6, inserted at base of the tepals, the filaments broadened at base; ovary ovoid, 3-locular, the ovules many, the style columnar; fruit a globose, many-seeded berry; seeds oblong, often slightly curved, usually striate, with a thin testa.——About 7 species, in the temperate regions of the N. Hemisphere; most abundant in N. America.

1A. Flowers campanulate; style columnar, elongate; pedicels jointed; plants relatively stout with glaucescent clasping leaves.
1. *S. amplexifolius* var. *papillatus*
1B. Flowers rotate; style nearly absent; pedicels not jointed; plants relatively slender, the leaves neither glaucous nor clasping.
2. *S. streptopoides*

1. Streptopus amplexifolius (L.) DC. var. **papillatus** Ohwi. *S. amplexifolius* sensu auct. Japon., non DC.——Ōba-takeshima-ran. Rhizomes usually short, creeping; stems leafy, erect, 50–100 cm. long, often sparingly hirsute below, ascending and branched below the middle, rather stout; leaves thinly membranous, ovate to ovate-elliptic, 6–12 cm. long, 3–6 cm. wide, abruptly acuminate, glabrous, the margin usually papillate, rounded-cordate and clasping at base, glaucous beneath; peduncles axillary, adnate on the stem to the insertion of the next leaf, the free portion 2–4 cm. long, pendulous on upper half, geniculate at the middle; flowers solitary on the peduncles, greenish white, broadly campanulate, pendulous, the tepals lanceolate, recurved on upper half, 8–10 mm. long; anthers lanceolate, acuminate, about 4 mm. long, smooth, glabrous, longer than the filaments, the style 4–5 mm. long; fruit red, about 1 cm. long, ovoid-globose.——June–Aug. Damp coniferous woods in mountains; Hokkaido, Honshu (n. and centr. distr.); rather rare.——China, Sakhalin, Kuriles, Kamchatka, and Amur. The typical phase occurs in the cooler regions of the N. Hemisphere.

2. Streptopus streptopoides (Ledeb.) Frye & Rigg. *Smilacina streptopoides* Ledeb.; *Kruhsea tilingii* Regel; *Streptopus ajanensis* Tiling; *K. streptopoides* (Ledeb.) Kearney——Hime-takeshima-ran. Rhizomes slender, creeping, usually elongate; stems slender, 15–30 cm. long, unbranched or with

1 or 2 branches, slightly ascending above; leaves thinly membranous, ovate-oblong to oblong, 4–8 cm. long, 1.5–3 cm. wide, green, papillose on the margin, abruptly acuminate, sub-rounded, sessile, the pedicels solitary, 1–2 cm. long, adnate to the base of the leaf above, the free portion slender, not jointed; flowers solitary, rotate, 5–6 mm. in diameter, nodding, the tepals lanceolate, slightly reflexed above, about 3 mm. long, caudate; anthers transversely elliptic, slightly scabrous on lower half; stigma nearly sessile; fruit red, globose.——June–July. Damp coniferous woods in mountains; Hokkaido.——e. Siberia, Sakhalin, and w. N. America.

Var. **japonicus** (Maxim.) Fassett. *S. ajanensis* var. *japonicus* Maxim.; *S. japonicus* (Maxim.) Ohwi; *S. streptopoides* sensu auct. Japon., non Frye & Rigg——Takeshima-ran. More robust, the stems stouter, 20–60 cm. long, with 2 to 6 branches, rarely simple; leaves broadly lanceolate to narrowly ovate, 5–10 cm. long, 1–3 cm. wide, acuminate, rarely abruptly so, the margins essentially smooth, cuneate to rounded at base; anthers scabrous on lower half with short hairlike processes.——May–July. Damp coniferous woods in high mountains; Honshu (n. and centr. distr.); rare.——Forma **atrocarpus** (Koidz.) Ohwi. *S. streptopoides* var. *atrocarpus* Koidz.; *S. ajanensis* var. *japonicus* forma *atrocarpus* (Koidz.) Makino & Nemoto——Kuro-mi-no-takeshima-ran. A rare phase with black fruit.——Honshu (centr. distr.).

27. CLINTONIA Raf. Tsubame-omoto Zoku

Perennial herbs from short rhizomes; leaves several, radical, obovate to lanceolate, entire; scapes erect, usually simple, often short-pubescent; flowers 1 to several, rather small, in short terminal racemes or umbels, the tepals free, oblong to lanceolate, ascending to spreading; stamens 6, the filaments filiform, the anthers oblong, semi-extrorse; ovary 3-locular, the ovules 2–12 in each carpel, the style columnar; berry globose, the seeds shiny brown.——About 6 species; Himalayas, e. Asia, and N. America.

1. Clintonia udensis Trautv. & C. A. Mey. *C. alpina* var. *udensis* (Trautv. & C. A. Mey.) Macbride——Tsubame-omoto. Rhizomes short-creeping; radical leaves few, narrowly oblong to narrowly obovate-oblong, abruptly cuspidate

to acute, soft and relatively thick, narrowed toward the base, the margin pubescent when young; scapes naked, erect, 20–70 cm. long, rarely short-branched above, soft-pubescent; flowers several, racemose, loose, erect, the bracts caducous, broadly linear, 6–8 mm. long, the pedicels pubescent, 1–4 cm. long, elongating slightly after anthesis, ascending to obliquely spreading, the tepals white, somewhat spreading, 12–15 mm. long, narrowly oblong; anthers oblong, the style deeply 3-fid; berry globose, deep blue to nearly black, about 1 cm. in diameter, the seeds ovoid, about 3 mm. long, shining brown.——May–July. Damp coniferous woods in high mountains; Hokkaido, Honshu (centr. and n. distr. and Mount Ōmine in Yamato).——Kuriles, Sakhalin, Manchuria, China, and e. Siberia.

28. SMILACINA Desf.   YUKIZASA ZOKU

Perennial herbs from creeping rhizomes; stems simple, erect at base, few-leaved, usually ascending above; leaves flat, usually oblong or elliptic, with several longitudinal nerves; inflorescence terminal, racemose or paniculate, the flowers small, bisexual, or the plants sometimes dioecious, the tepals ascending to spreading, free or nearly so, persistent, the anthers ovate, introrse; ovary globose, 3-locular, the ovules 2 in each locule; style short, the stigma 3-fid or nearly entire; berry globose, few seeded.—— About 20 species, in the temperate regions of the Himalayas, e. Asia, and N. America, 1 in Siberia.

1A. Flowers bisexual; stigma entire or obsolete, 3-lobed.
  2A. Rhizomes long-creeping, with elongate slender internodes; stems 30–70 cm. long. .... 1. *S. japonica*
  2B. Rhizomes short-creeping, with short thickened internodes; stems stout, 80–150 cm. long. .... 2. *S. robusta*
1B. Flowers unisexual, the plants dioecious; stigma in the pistillate flowers deeply 3-cleft.
  3A. Stigma-lobes ovate, spreading, scarcely recurved; pedicels pilose; pistillate perianth whitish in fruit; stems not striate.
                                                                      3. *S. hondoensis*
  3B. Stigma-lobes lanceolate, recurved above; pedicels nearly glabrous; pistillate perianth purple-brown in fruit; staminate perianth greenish; stems distinctly striate. .... 4. *S. yesoensis*

1. Smilacina japonica A. Gray. *S. hirta* Maxim.; *S. trinervis* Miyabe & Kudo——YUKIZASA.   Rhizomes creeping, elongate, slender, 4–6 mm. in diameter; stems erect, simple, sparsely hirsute, ascending above; leaves 5–7, narrowly oblong to elliptic, sometimes broadly ovate, 6–15 cm. long, 2–5 cm. wide, the tip abruptly narrowed, obtuse, rounded at the base, sessile or the lower ones short-petioled, pilose especially beneath; inflorescence paniculate, terminal, rather densely hirsute, the flowers whitish; style nearly as long as the ovary, the stigma entire or very shallowly 3-lobed.——May–July. Woods and thickets in mountains; Hokkaido, Honshu, Shikoku, Kyushu; rather common.——Korea, Manchuria, China, Ussuri, and Amur.

Var. **mandshurica** Maxim. Ō-YUKIZASA.   Plants larger, more leafy with about 10 narrowly oblong leaves.——Kyushu (Tsushima).——Korea and Manchuria.

2. Smilacina robusta Makino & Honda. *S. japonica* var. *robusta* (Makino & Honda) Ohwi——HARUNA-YUKIZASA. Stout perennial from densely noded moniliform rhizomes 5–10 cm. long and 7–15 mm. in diameter; stems erect, ascending above, stout, 80–150 cm. long, simple, to 2 cm. in diameter near the base, glabrescent and pale purplish in lower part, leafy except at base; leaves lanceolate-oblong to ovate-oblong, 15–20 cm. long, 6–9 cm. wide, abruptly narrowed to an obtuse tip, rounded and short-petioled at the base, green and short-pubescent above, rather densely pubescent and glaucescent beneath, the petioles 5–7 mm. long; inflorescence paniculate, densely pilose, 5–15 cm. long, 3–10 cm. wide, erect, many-flowered, the bracts minute, the flowers bisexual, white, 7 mm. in diameter, short-pedicelled, the tepals narrowly oblong or spathulate-oblong, 3.5–4 mm. long, 1-nerved; stamens 2.5 mm.

long, the filaments white, subulate, the anthers about 0.7 mm. long; style very short, the stigma lobes short and thick.—— June–Aug. Mountains; Honshu (Kotsuke and Shimotsuke Prov.); rather rare.

3. Smilacina hondoensis Ohwi. YAMATO-YUKIZASA, ŌBA-YUKIZASA.   Dioecious perennial; rhizomes stout, creeping, about 7 mm. in diameter; stems stout, leafy, erect at base, ascending above, 8- to 11-leaved, the indument short-spreading; leaves oblong to broadly ovate, short-petioled, 10–15 cm. long, 3–6 cm. wide, abruptly narrowed to an subobtuse tip, rounded at the base, short-pubescent on both sides especially beneath, or glabrescent above; inflorescence paniculate, rather densely short-pubescent, the tepals in the staminate flowers nearly as long as the stamens, the anthers small and abortive in the pistillate flowers; style short, the stigmas 3.——June–July. Woods in mountains; Honshu (centr. to n. distr. and Yamato Prov.).

4. Smilacina yesoensis Fr. & Sav. *S. viridiflora* Nakai ——HIROHA-NO-YUKIZASA, MIDORI-YUKIZASA.   Dioecious perennial; rhizomes creeping, fleshy, 4–7 mm. in diameter; stems relatively stout, 7- to 11-leaved, glabrous or scattered pilose above; leaves oblong, rounded at the base, short-petioled, 7–20 cm. long, 2.5–8 cm. wide, glabrous above, scattered pilose beneath; staminate inflorescence a many-flowered panicle, the pistillate a simple raceme or with 1 or 2 short branches below, the axis often glabrous and angled; tepals twice as long as the stamens, greenish in anthesis, in the pistillate flowers changing to purple-brown in fruit; style very short, the stigma-lobes lanceolate, recurved and appressed to the ovary.——June–July. Damp coniferous woods; Honshu (centr. and n. distr.), also reported from Hokkaido.

29. MAIANTHEMUM Weber   MAIZURU-SŌ ZOKU

Low herbs from slender long-creeping rhizomes; stems 2-leaved; leaves usually ovate-cordate, petiolate, the sterile radical leaves solitary, long-petioled; inflorescence an erect terminal raceme, the flowers bisexual, white, small, pedicellate, the tepals 4 in 2 series, free, obliquely spreading, oblong; ovary 2-locular, the ovules 2 in each carpel; style solitary, columnar, relatively short, the stigma small; berry globose, 1- to 3-seeded; seeds globose to ovoid, pale brown.——About 3 species, in the temperate regions of the N. Hemisphere.

1A. Plants 8–15 cm. high, usually with short, hairlike papillae on the leaves beneath, on margin, and on the raceme-axis. ... 1. *M. bifolium*
1B. Plants taller, 10–25 cm. high, glabrous. .... 2. *M. dilatatum*

1. Maianthemum bifolium (L.) F. W. Schmidt. *Convallaria bifolia* L.; *Smilacina bifolia* (L.) Desf.; *M. convallaria* Wiggers——HIME-MAIZURU-SŌ. Rhizomes slender, long-creeping; stems 8–15 cm. long, glabrous to scattered-papillose; leaves 2, cauline, petiolate, deltoid-ovate, cordate, 2–5 cm. long, 1.5–4 cm. wide, glabrous to scattered papillose, the marginal cells columnar or ovate-deltoid; racemes about 20-flowered, 2–3 cm. long, the pedicels slender, 3–8 mm. long, the flowers white, the tepals elliptic, 1-nerved, spreading or slightly recurved, about 2 mm. long; style short, the stigma shallowly 2-lobed; fruit globose, red, about 5 mm. in diameter.——May–July. Damp coniferous woods in mountains; Hokkaido, Honshu (centr. and n. distr.); rare.——Kuriles, Sakhalin, n. Korea, also in Europe and North America.

2. Maianthemum dilatatum (Wood) A. Nels. & Macbr. *Convallaria bifolia* var. *kamtschatica* Gmel. ex Cham.; *M. bifolium* var. *dilatatum* Wood; *Unifolium dilatatum* (Wood) Howell; *M. kamtschaticum* (Gmel.) Nakai; *M. bifolium* var. *kamtschaticum* (Gmel.) Trautv. & C. A. Mey.——MAIZURU-SŌ. Closely resembles the preceding, but usually slightly larger, deeper green and more lustrous, 10–25 cm. high, glabrous; leaves cordate, ovate to deltoid-cordate, 3–10 cm. long, 2.5–8(–10) cm. wide, the marginal cells semirounded, the pedicels 3–7 mm. long; style rather stout.——May–July. Coniferous woods; Hokkaido, Honshu, Shikoku, Kyushu; rather common.——Kuriles, Sakhalin, e. Siberia, Manchuria, Korea, Kamchatka, and N. America.

30. DISPORUM Salisb. CHIGO-YURI ZOKU

Perennial herbs from creeping rhizomes; stems leafy, usually sparingly branched; leaves oblong to lanceolate, sessile or short-petiolate; inflorescence 1- to few-flowered, umbellate, terminal, sessile, ebracteate, the pedicels simple; flowers tubular or rotate, pendulous, the tepals 6, free, deciduous, sometimes saccate at base; stamens 6, inserted at base of the perianth, the anthers lanceolate or oblong, semi-extrorse; ovary globose to obovoid, 3-locular, the ovules 2(–6) in each locule; berry globose, the seeds ovoid-globose, brown.——More than 10 species, in Malaysia, Himalayas, e. Asia, and N. America.

1A. Flowers tubular, slightly broadened above; tepals oblanceolate, saccate at base; filaments and style more than twice as long as the anthers and ovary. .... 1. *D. sessile*
1B. Flowers cup-shaped; tepals lanceolate, the base not saccate; filaments and style as long as to twice as long as the anthers and ovary.
  2A. Tepals glabrous, white or greenish; leaves abruptly acuminate.
    3A. Flowers white; style including stigma 2 to 3 times as long as the ovary. .... 2. *D. smilacinum*
    3B. Flowers pale green; style including stigma 1 to 1½ times as long as the ovary. .... 3. *D. viridescens*
  2B. Tepals distinctly papillose inside and on margin near base, yellowish; leaves caudate-acuminate. .... 4. *D. lutescens*

1. Disporum sessile Don. *Uvularia sessilis* Thunb.; *Tovaria hallaisanensis* Lév.; *D. hallaisanense* (Lév.) Ohwi——HŌCHAKU-SŌ. Plants from relatively short rhizomes; stems 30–60 cm. long, ascending and sparingly branched above; leaves oblong, sometimes broadly lanceolate to elliptic, 5–15 cm. long, 1.5–4 cm. wide, abruptly acuminate, subsessile, mammillate on the nerves beneath and the margin, green on both sides, 3- to 5-nerved; flowers 1–3 at the end of the branches, pendulous, about 3 cm. long, white, greenish near the tip, the pedicels 1.5–3 cm. long, the tepals abruptly acute and sometimes cuspidate, gradually narrowed toward the saccate base, the inside and lower margin short soft pubescent; filaments glabrous, about 2 cm. long, the anthers linear, 5–6 mm. long; style glabrous, about 15 mm. long, 3-cleft; ovary ellipsoidal; berry blue-black, about 1 cm. in diameter, globose.——Apr.–May. Woods in hilly country and foothills; Hokkaido, Honshu, Shikoku, Kyushu; common.——Sakhalin and Ryukyus (?).

2. Disporum smilacinum A. Gray. *D. smilacinum* var. *album* Maxim.——CHIGO-YURI. Glabrous perennial herb from slender, elongate rhizomes; stems 15–40 cm. long, simple or sparingly branched; leaves oblong, elliptic, or elliptic-ovate, abruptly acuminate, subsessile, 4–7 cm. long, 1.5–3.5 cm. wide, rounded at the base, the margin smooth with semi-rounded cells; flowers 1–2, white, terminal, pendulous on slender pedicels 1–2 cm. long, the tepals thin, obliquely spreading, lanceolate, 15–18 mm. long, acuminate, glabrous; filaments glabrous, 5–6 mm. long, the anthers narrowly oblong, 2–3 mm. long; style 5–7 mm. long, smooth, 3-cleft; berry

black, globose, about 7 mm. in diameter.——Apr.–May. Woods in hilly country and foothills; Hokkaido, Honshu, Shikoku, Kyushu.——Korea.

3. Disporum viridescens (Maxim.) Nakai. *Uvularia viridescens* Maxim.; *D. smilacinum* var. *viridescens* (Maxim.) Maxim.——Ō-CHIGO-YURI. Plant relatively larger, 30–70 cm. high, usually branched; leaves 6–12 cm. long, 2–5 cm. wide, the margins and nerves beneath with semirounded mammillae; flowers usually larger than in No. 2, pale green, the tepals 13–18 mm. long; filaments 4–5 mm. long, the anthers 3–4 mm. long; style about 4 mm. long, usually as long as the ovary.——Mountains; Honshu (centr. and Kantō Distr.).——Korea, Manchuria, and Ussuri.

4. Disporum lutescens (Maxim.) Koidz. *D. viridescens* var. *lutescens* Maxim.——KIBANA-CHIGO-YURI. Glabrous green perennial; stems 25–50 cm. high, simple or forked above, erect, ascending above; leaves several, loosely disposed, ovate-oblong, 7–12 cm. long, 3–5 cm. wide, rather thin, smooth, 5- to 7-nerved, the upper ones larger, caudately acuminate, rounded at base, sessile, the nerves slender, with about 3 veinlets between them; flowers 1–3, in terminal sessile umbels, the pedicels 1–1.5 cm. long, slender, the tepals yellowish, about 1.5 cm. long, obliquely spreading, lanceolate, acute with an obtuse tip, 12–15 mm. long, 3–4 mm. wide, distinctly papillose inside and on margin near base; stamens about ½ as long as the tepals, the anthers about 3 mm. long, the filaments about 4 mm. long; ovary obovoid, the style about twice as long as the ovary.——May–June. Mountains; Honshu (Mt. Koyasan in Kii), Kyushu; rare.

31. REINECKEA Kunth   Kıchıjō-sō Zoku

Perennial evergreen herbs; leaves tufted on the summit of the stems, broadly linear to lanceolate, narrowed below to a petiolelike base; scapes naked, shorter than the leaves, simple, erect; flowers in a terminal spike, sessile, the bracts membranous, the tepals connate at base forming a short tube; stamens 6, the filaments filiform, the anthers oblong; ovary 3-locular, the ovules 2 in each locule; stigma 3-lobed; berry globose, few-seeded.——A single species, in e. Asia.

1. Reineckea carnea (Andr.) Kunth. *Sansevieria carnea* Andr.; *S. sessiliflora* Ker-Gawl.——Kıchıjō-sō. Stems prostrate, stout, greenish, elongate, with a tuft of leaves at the summit; leaves broadly linear to lanceolate, to narrowly oblanceolate, 10–40 cm. long, 5–20 mm. wide, deep green, glabrous, acute, gradually narrowed below; inflorescence spicate, 3–8 cm. long, densely many-flowered, glabrous, the scapes 8–12 cm. long, the bracts membranous, ovate-deltoid, 4–7 mm. long; flowers pale rose, the perianth 8–12 mm. long, the segments slightly longer than the tube, reflexed, narrowly oblong, obtuse, slightly fleshy; stamens shorter than to as long as the style, the anthers about 2 mm. long.——Aug.–Oct. Woods in lowlands and foothills; Honshu (Kantō Distr. and westw.), Shikoku, Kyushu. Often planted in our gardens.——China.

32. CONVALLARIA L.   Suzu-ran Zoku

Rhizomes slender, elongate; leaves paired, radical, oblong, erect; scapes naked, erect, axillary from the radical scale; racemes terminal, loosely few-flowered, one-sided, bracts linear, deciduous; flowers white or pinkish, nodding, pedicelled, relatively small, fragrant, globose-campanulate, the tepals connate, the free lobes at the tip short, recurved; stamens inserted at base of the perianth, the filaments narrowed above, the anthers oblong, slightly introrse; ovary 3-locular, the ovules several in each carpel; style columnar, the stigma small; fruit a globose berry; seeds small.——About 4 species, in the temperate regions of the N. Hemisphere.

1. Convallaria keiskei Miq. *C. majalis* sensu auct. Japon., non L.; *C. japonica* Greene; *C. majalis* var. *manshurica* Komar.; *C. majalis* var. *keiskei* (Miq.) Makino——Kımı-kage-sō, Suzu-ran. Plants glabrous; leaves 2, 12–18 cm. long, 3–7 cm. wide, oblong to ovate-elliptic, glabrous, acuminate, acute at base, green above, glaucescent beneath, enveloped by an involute sheath at base; scapes 20–35 cm. long, shorter than the leaves; racemes 5–10 cm. long, loosely about 10-flowered, one-sided, the bracts membranous, broadly linear to lanceolate, shorter than to as long as the pedicel; flowers 6–8 mm. long, the pedicels 6–12 mm. long, curved above, the perianth-lobes recurved, ovate-deltoid, acute; anthers broadly deltoid-lanceolate, about 1.5 mm. long, as long as the subulate filaments; style 3–4 mm. long; berry globose, about 6 mm. in diameter, red when mature.——Apr.–June. Highlands and mountain meadows; Hokkaido, Honshu, Kyushu.——Sakhalin, Korea, China, e. Siberia.

33. ASPIDISTRA Ker-Gawl.   Ha-ran Zoku

Rhizomes rather stout, creeping, covered with scalelike leaves; scapes short, stout, with few scales, 1-flowered, in the axils of rhizome scales; perianth tubular-campanulate; tepals 8 or 9, connate, fleshy; filaments nearly absent, the anthers narrowly ovate; ovary 4-carpellate, with 2 to 6 ovules in each locule; stigma very large, peltate, covering the throat; berry relatively large, fleshy, few-seeded.——Six to seven species, e. Asia, Formosa, and the Himalayas.

1. Aspidistra elatior Bl. *Plectogyne variegata* Link—— Ha-ran, Ba-ran. Evergreen perennial herb from stout creeping rhizomes, the roots stout; leaves radical, long-petioled, smooth, erect, 30–50 cm. long, deep green, lustrous, oblong-lanceolate, acuminate at both ends, the midrib rather thick; scapes 2–7 cm. long, stout, glabrous, with 4–5 membranous scalelike bracts; flowers 2–2.5 cm. long, brown-purple, the perianth-lobes 8–9, narrowly deltoid, obtuse, about 1 cm. long, thick; anthers nearly sessile, inserted on the tube below the middle.——Feb.–May. Introduced Chinese plant long cultivated in our gardens.

34. ROHDEA Roth   Omoto Zoku

Rhizomes ascending, stout; leaves tufted at the summit of the rhizomes, coriaceous and slightly lustrous, lanceolate to oblong, dilated at base; scapes erect, axillary, simple, much shorter than the leaves, naked; flowers in a dense terminal simple spike, sessile, the bracts short, ovate, the tepals 6, highly connate, forming a broadly campanulate corolla, the free portion short, incurved; stamens 6, the filaments almost wholly adnate to the tepals, the anthers ovate; ovary globose, 3-locular, with 2 ovules in each locule; style very short, simple, the stigma 3-lobed; berry globose, red, usually with a solitary seed.——Two species, in Japan and sw. China.

1. Rohdea japonica (Thunb.) Roth. *Orontium japonicum* Thunb.——Omoto. Dark green glabrous evergreen perennial from stout creeping to ascending rhizomes and thick adventitious roots; leaves tufted, 2-ranked, lanceolate to oblanceolate, coriaceous, 30–50 cm. long, 3–5 cm. wide, gradually narrowed to both ends, subacuminate at the apex, abruptly dilated at the base, the midrib slightly elevated beneath; scapes 10–20 cm. long, stout, naked or 1-scaled; spike 2–3.5 cm. long, narrowly oblong, densely many-flowered; flowers pale yellow, sessile, about 5 mm. across, the bracts membranous, relatively small, broadly deltoid, obtuse, the perianth-lobes relatively short, incurved; fruit globose, red, rarely yellow, about 8 mm. in diameter.——May–June. Thickets and woods in the warmer parts; Honshu (Tōkaidō Distr. and westw.), Shikoku, Kyushu.——China.

35. PARIS L. TSUKUBANE-SŌ ZOKU

Rhizomes creeping, elongate, rarely short and stout; stems simple, erect, naked except the few scales at base; leaves 4 to many, in a terminal whorl, lanceolate to ovate, with 3 distinct main nerves and anastomosing veinlets; flowers solitary, terminal, pedunculate, the tepals free, 4–10, the outer whorl green or white, ovate to lanceolate, the inner ones narrow or absent; stamens 8–20, the filaments narrow, flat, the anthers linear, often with a connective at apex; ovary subglobose, 4- to 10-carpellate, or the placentae parietal; styles 4–10, linear; berry indehiscent or loculicidal, the seeds subglobose.——About 20 species, in Europe and Asia.

1A. Rhizomes slender, creeping; leaves 4–12 cm. long, usually in verticils of 4 or 8; outer tepals usually 4, green, the inner tepals filiform or absent.
    2A. Connective extending beyond the anther; leaves usually 6–8; inner tepals filiform. . . . . .1. *P. verticillata*
    2B. Connective not extending beyond the anthers; leaves usually 4; inner tepals absent. . . . . 2. *P. tetraphylla*
1B. Rhizomes stout and short; leaves 20–30 cm. long, usually in verticils of 8–10; outer tepals usually 8, white, the inner tepals minute or absent. . . . . 3. *P. japonica*

1. Paris verticillata Bieb. *P. quadrifolia* sensu auct. Japon., non L.; *P. dahurica* Fisch.; *P. quadrifolia* var. *obovata* sensu auct. Japon., non Regel & Tiling; *P. hexaphylla* Cham. ——KURUMABA-TSUKUBANE-SŌ. Glabrous perennial herb from slender creeping rhizomes 3–4 mm. in diameter; stems 20–40 cm. long; leaves lanceolate to narrowly oblong, 6–8, 7–12 cm. long, 1–4 cm. wide, acuminate, narrowed to a subsessile base; peduncles 1-flowered, 5–15 cm. long, erect; outer tepals 4(–5), broadly lanceolate to narrowly ovate, green, spreading, (2–) 3–4 cm. long, (0.5–)1–1.5 cm. wide, acuminate, the inner ones 4(–5), filiform, yellow, reflexed after anthesis, 1.5–2 cm. long; stamens 8–10, slightly longer than the inner tepals, the filaments nearly filiform, 5–7 mm. long, the anthers linear, 5–8 mm. long; the free portion of the connective prominent, filiform, about 5–7 mm. long; styles usually 4; ovary dark purplish brown——June–July. Woods in mountains; Hokkaido, Honshu, Shikoku, Kyushu.——Sakhalin, Korea, China, and Siberia.

2. Paris tetraphylla A. Gray. TSUKUBANE-SŌ. Rhizomes slender, elongate, 3–4 mm. in diameter; stems 20–40 cm. long, glabrous; leaves oblong or elliptic, rarely broadly lanceolate, 4(–5), 4–10 cm. long, 1.5–4 cm. wide, acuminate, acute at base, sessile, 3-nerved; flowers solitary, the peduncle 3–10

(–15) cm. long, erect; outer tepals green, 1–2 cm. long, 3–8 mm. wide, acuminate; stamens 8, the filaments 3–4 mm. long, as long as the anthers, the connective not extended beyond the anther; styles 4.——May–Aug. Woods in mountains; Hokkaido, Honshu, Shikoku, Kyushu.

Var. **penduliflora** Murata & Yamanaka. UNAZUKI-TSUKU-BANE-SŌ. Differs from the typical phase in the flowers nodding at anthesis and the leaves narrower, 8–24 mm. wide.——Shikoku.

3. Paris japonica (Fr. & Sav.) Franch. *Trillidium japonicum* Fr. & Sav.; *Trillium japonicum* (Fr. & Sav.) Matsum.; *Kinugasa japonica* (Fr. & Sav.) Tatew. & Suto——KINUGASA-SŌ. Rhizomes stout and very short, ascending, 1.5–2 cm. in diameter; stems stout, terete, 30–80 cm. long, glabrous or sometimes scattered pubescent; leaves 8–10, broadly lanceolate to narrowly obovate-oblong, 20–30 cm. long, 3–8 cm. wide, abruptly acuminate, gradually narrowed below to the subsessile base; peduncles solitary, 1-flowered, 3–8 cm. long; outer tepals 8–10, white, spreading, narrowly oblong, narrowly ovate to broadly lanceolate, 3–5 cm. long, acuminate; anthers 5–8 mm. long, nearly as long as the filaments, the connective not extended beyond the anther; styles about 8.——June–Aug. High mountains; Honshu (n. and centr. distr.).

36. TRILLIUM L. ENREI-SŌ ZOKU

Rhizomes stout, short; stems erect, simple, naked except for the few scalelike leaves at base; leaves 3, relatively large, in a terminal whorl, reticulate-veined, usually rhombic-orbicular, sessile or short-petioled; flowers solitary, terminal, pedunculate or sessile, white, rose-purple, or purple-brown; tepals 6, in 2 series, free, the outer whorl (sepals) green to brownish, persistent, relatively small, the inner whorl (petals) larger, petallike, withering, or absent; stamens 6, the filaments short, flat, the anthers linear; ovary ovoid to globose, 3-locular, the ovules many; styles simple, deeply 3-lobed, the lobes stigmatic on inner side; berry globose to ovoid; seeds many, ovoid.——About 30 species, in the temperate regions of the Himalayas, e. Asia, and N. America.

1A. Petals usually absent, sometimes 2 or 3 smaller than the sepals, brown-purple, obtuse; anthers about 3 mm. long, slightly shorter than the filaments; sepals 12–20 mm. long; ovary globose. . . . . 1. *T. smallii*
1B. Petals 3, larger than the sepals, white or pale rose-purple; anthers 6–15 mm. long; sepals 2–3 cm. long; ovary conical-ovoid.
    2A. Anthers 6–8 mm. long, as long as to slightly longer than the filaments; sepals and petals subacute to acute. . . . . . . 2. *T. tschonoskii*
    2B. Anthers 12–15 mm. long, much longer than the filaments; sepals and petals obtuse. . . . . 3. *T. kamtschaticum*

1. Trillium smallii Maxim. *T. apetalon* Makino; *T. tschonoskii* var. *smallii* (Maxim.) T. Itō; *T. smallii* var. *apetalon* (Makino) Takeda——ENREI-SŌ. Rhizomes stout, short; stems simple, erect, glabrous, 20–40 cm. long, terete, smooth; leaves 3, sessile, depressed rhombic-orbicular, 7–17 cm. long and as wide, glabrous, 3-nerved, abruptly short-acuminate, entire; peduncles 2–4 cm. long; flowers ascending to horizontal in anthesis, erect in fruit, the sepals 3, persistent, green or purple-brown, ovate to ovate-oblong, subobtuse, 15–20 mm. long, rather thick, the petals absent, or 1(–3), small, purple-

brown; anthers oblong, 2.5–3.5 mm. long, the filaments slightly longer than the anthers, broadened below; stigma very short; berry globose, 1–1.2 cm. in diameter, the seeds many, about 2.5 mm. long, arcuate-oblong, brown, with obscure longitudinal wrinkles.——Apr.–May. Woods in foothills; Hokkaido, Honshu, Shikoku, Kyushu. The fruit is edible. Plants from the western part of the range usually have narrower leaves slightly longer than wide.

Trillium amabile Miyabe & Tatew. *T. smallii* Maxim., pro parte——KOJIMA-ENREI-SŌ. An alleged hybrid of No. 1

and No. 3 with rather larger sepals, well-developed purple orbicular or ovate-orbicular petals about 2 cm. long and as wide; anthers 6–7 mm. long, the filaments 1–2 mm. long.——sw. Hokkaido.

2. Trillium tschonoskii Maxim.　Shirobana-enrei-sō. Leaves rhombic-orbicular to depressed-orbicular, 7–17 cm. long and as wide; sepals broadly lanceolate to narrowly ovate, subacute, green, herbaceous, 2–2.5 cm. long; petals white, ovate, subacute, slightly longer than the sepals; anthers (4–) 6–8 mm. long, the connective slightly produced above, the filaments 3–8 mm. long; ovary conical-ovoid; stigma rather thick; berry about 15 mm. in diameter.——Apr.–June. Mountains; Honshu, Shikoku.——Korea.——Forma **violaceum** Ma-

kino.　Murasaki-enrei-sō.　Petals rarely pale purple or rose-purple.

3. Trillium kamtschaticum Pall.　*T. obovatum* sensu auct. Japon., non Pursh; *T. pallasii* Hult.——Ōbana-no-enrei-sō.　Leaves ovate to rhombic-orbicular, sometimes depressed, 7–15 cm. long and as wide, 3- to 5-nerved; sepals broadly lanceolate to ovate-oblong, 2.5–4 cm. long, obtuse, sometimes subacute, herbaceous, green, the petals white, ovate or elliptic, 2.5–4.5 cm. long, obtuse; anthers linear, (10–)12–15 mm. long, the filaments 3–5 mm. long; ovary conical; stigma rather thick.——May–June. Hokkaido.——Kuriles, Sakhalin, Korea, Manchuria, Ussuri, and Kamchatka.

37. LIRIOPE Lour.　Yabu-ran Zoku

Glabrous perennial herbs from short rhizomes, stolons sometimes also present; leaves radical, linear, tufted; scapes naked, terete, angled or slightly compressed, simple; inflorescence racemose, terminal; flowers relatively small, or short, jointed with the pedicels, pale purple or white, semiglobose to campanulate, bracteate; tepals 6, free, equal, ascending, membranous, persistent but withering in fruit; stamens 6, the filaments linear, the anthers oblong, obtuse; ovary superior, 3-locular with 2 ovules in each locule; style solitary, columnar, the stigma small; carpels soon splitting and exposing the seeds; seeds 1 to few, globose, rather fleshy, usually dark purple.——Few species, in the warmer regions of e. Asia.

1A. Leaves 2–3 mm. wide; scapes 10–15(–20) cm. long, relatively few-flowered. .... 1. *L. minor*
1B. Leaves (3–)4–12 mm. wide; scapes 25–50 cm. long, many-flowered.
  2A. Leaves (3–)4–7 mm. wide; plants usually stoloniferous; flowers relatively loose. .... 2. *L. spicata*
  2B. Leaves (7–)8–12 mm. wide; plants usually without stolons; flowers many, dense. .... 3. *L. platyphylla*

1. Liriope minor (Maxim.) Makino.　*Ophiopogon spicatus* var. *minor* Maxim.; *L. graminifolia* var. *minor* (Maxim.) Bak.; *L. spicata* var. *minor* (Maxim.) Wright; *Mondo tokyoense* Nakai——Hime-yabu-ran.　Plants rhizomatous and with slender stolons; leaves narrowly linear, 10–20 cm. long, 2–3 mm. wide, 5- to 7-nerved, obtuse, subcoriaceous, slightly arcuate above; scapes 10–15 cm. long, slender, slightly flattened and obtusely angled; racemes 1–5 cm. long, about 10-flowered, the bracts and bracteoles membranous, small, the pedicels 2–4 mm. long, 1- to 3-nate, jointed below the flower or above the middle; tepals pale purple, about 4 mm. long; seeds globose, purple-black, 4–5 mm. in diameter.——June–Sept. Thickets in lowlands and foothills; Hokkaido, Honshu, Shikoku, Kyushu; rather common.——Ryukyus and China.

2. Liriope spicata Lour.　*L. gracilis* Nakai (excl. syn. ?); *Convallaria spicata* Thunb.; *L. spicata* var. *koreana* Palib.; *L. koreana* (Palib.) Nakai——Ko-yabu-ran, Ryūkyū-yabu-ran. Plants stoloniferous, the roots sometimes fusiform; leaves deep green, obtuse, gradually narrowed below, 7- to 11-nerved, 30–

40 cm. long, 4–7 mm. wide; scapes obtusely angled, 25–40 cm. long; racemes 8–12 cm. long, many-flowered; flowers 2–5 in the axil of a bract, pale purple, the pedicels 3–6 mm. long, jointed at or above the middle, the tepals about 4 mm. long; seeds blackish.——July–Sept. Honshu, Shikoku, Kyushu.——Ryukyus and China.

3. Liriope platyphylla F. T. Wang & Tang.　*Ophiopogon spicatus* var. *communis* Maxim.; *L. graminifolia* var. *densiflora* Bak.; *L. spicata* var. *densiflora* (Bak.) Wright; *L. muscari* L. H. Bailey, excl. syn. Decne.——Yabu-ran.　Rhizomes thick, short, fasciculate; leaves deep green, flat, 30–50 cm. long, 8–12 mm. wide, obtuse, 11- to 15-nerved, narrowed gradually below; scapes 30–50 cm. long, subterete; racemes 8–12 cm. long, many-flowered; flowers few to several in a group, fascicled, the pedicels 2–5 mm. long, jointed at the middle or beneath the flower, the tepals pale purple, about 4 mm. long.——Aug.–Oct. Honshu (Kantō Distr. and westw.), Shikoku, Kyushu.——Ryukyus, Formosa, and China.

38. OPHIOPOGON Ker-Gawl.　Ja-no-hige Zoku

Rhizomes short and rather stout; plants often with slender stolons; leaves radical, tufted, linear to oblong-linear, coriaceous; scapes simple, the racemes terminal; flowers small, pale purple to white, broadly campanulate, usually nodding, the pedicels jointed at the middle or beneath the flower, the tepals 6, rather thick; stamens 6, the filaments very short, the anthers lanceolate, acute; ovary subinferior, 3-locular, the ovules 2 in each locule; style simple, the stigma small; seeds early exposed, globose, usually blue at maturity.——More than 10 species, India to e. Asia.

1A. Leaves (7–) 10–15 mm. wide; pedicels 1–2 cm. long; plants tufted. .... 1. *O. jaburan*
1B. Leaves 2–6 mm. wide; pedicels 4–10 mm. long.
  2A. Leaves 4–6 mm. wide; plants stoloniferous. .... 2. *O. planiscapus*
  2B. Leaves 2–3(–4) mm. wide.
    3A. Leaves 10–20(–30) cm. long; plants usually stoloniferous; pedicels less than 8 mm. long. .... 3. *O. japonicus*
    3B. Leaves 30–40 cm. long, prominently tufted; plants without stolons, pedicels 8–12 mm. long. .... 4. *O. ohwii*

1. Ophiopogon jaburan (Kunth) Lodd.　*Slateria jaburan* Sieb.; *Convallaria japonica* var. *major* Thunb.; *Flueggea jaburan* Kunth——Noshi-ran.　Tufted, estoloniferous peren-

nial; leaves linear, 30–80 cm. long, (7–)10–15 mm. wide, relatively thick and lustrous, deep green, 9- to 13-nerved, gradually narrowed above, obtuse, slightly scabridulous on margin, gradually narrowed at base; scapes flattened, broader at the top, narrowly winged on margin, 30–50 cm. long, 4–7 mm. wide; racemes 7–10 cm. long, rather densely flowered, the lower bracts linear, membranous and dilated at base, the pedicels 3–8 in a group, fascicled, 1–2 cm. long, jointed at or above the middle; flowers nodding, 7–8 mm. long, pale purple; anthers lanceolate, 4–5 mm. long; seeds blue.——Aug. Honshu (w. distr.), Shikoku, Kyushu. Often planted, and there is a garden cultivar with white-variegated leaves.——Ryukyus.

2. **Ophiopogon planiscapus** Nakai. *O. japonicus* var. *wallichianus* Maxim., excl. syn.; *O. wallichianus* sensu auct. Japon., non Hook. f.——Ōba-ja-no-hige. Plants with partially fusiform-thickened roots and often with slender stolons; leaves 30–50 cm. long, 4–6 mm. wide, deep green, flat, several-nerved, gradually narrowed above, obtuse, scabridulous on the margin especially above; scapes 20–30 cm. long, 3-angled, compressed, 1.5–2 mm. wide; racemes 5–7 cm. long; flowers pale purple or white, 6–7 mm. long, nodding, the bracts linear-lanceolate, the pedicels 1–3 in a group, 5–10 mm. long, jointed at or above the middle; anthers 2.5–3 mm. long; seeds dull blue.——July–Aug. Woods and thickets in lowlands and foothills; Honshu, Shikoku, and Kyushu.

3. **Ophiopogon japonicus** (L. f.) Ker-Gawl. *Convallaria japonica* L. f.; *C. japonica* var. *minor* Thunb.——Ja-no-hige. Perennial herb with slender stolons; roots partially thickened; leaves fasciculate, somewhat arcuate above, 10–20 (–30) cm. long, 2–3(–4) mm. wide, few-nerved, obtuse; scapes 7–12 cm. long, compressed, 1–1.5 mm. wide, acute on margin; racemes solitary, 1–3 cm. long, about 10-flowered, the flowers pale purple to nearly white, 4–5 mm. long, nodding, the pedicels 2–6 mm. long, jointed at the middle or just below the flower; anthers about 2.5 mm. long; seeds globose, deep blue. ——July–Aug. Shaded places in lowlands to the foothills; Hokkaido, Honshu, Shikoku, Kyushu; often planted in our gardens.——Korea and China.

Var. **caespitosus** Okuyama. Kabudachi-ja-no-hige. Rhizomes tufted, the plants scarcely stoloniferous.——Honshu (Awa Prov.).

4. **Ophiopogon ohwii** Okuyama. *Mondo longifolium* Ohwi; *O. longifolium* Ohwi, non Decne.; *Mondo gracile* var. *brevipedicellatum* Koidz.; *O. gracilis* var. *brevipedicellatus* (Koidz.) Nemoto; *O. japonicus* var. *umbrosus* Maxim.——Nagaba-ja-no-hige. Rhizomes short, densely tufted; stolons absent; leaves suberect, 30–40 cm. long, 1.5–2.5 mm. wide, 3- to 5-nerved, scabrous above, obtuse; scapes about half as long as the leaves, flat, acutely angled, about 1.5 mm. across; racemes 6–8 cm. long, rather densely flowered; flowers white, 4–5 mm. long, nodding, the pedicels 2–4 in a group, 8–12 mm. long, jointed near the middle; anthers about 2.5 mm. long.——Aug. Honshu, Shikoku, Kyushu.

39. ALETRIS L. Sokushin-ran Zoku

Rhizomes short; leaves mainly radical, tufted, linear to lanceolate, vivid to pale green; scapes usually simple, sometimes naked; inflorescence terminal, racemose, the bracts relatively small, lanceolate; flowers subsessile or short-pedicelled, small, tubular-campanulate, white or yellow-green, the perianth tubular, becoming suburceolate in fruit, the segments ascending, persistent in fruit; stamens 6, the filaments short, the anthers ovate; ovary adnate to the perianth-tube; style simple; seeds many, small. ——About 10 species, in e. Asia, Malaysia, Tibet, Himalayas, and N. America.

1A. Leaves linear, 3–7 mm. wide, 3-nerved; scapes and exterior of perianth with short curly farinaceous hairs; flowers white. .. 1. *A. spicata*
1B. Leaves lanceolate to oblanceolate, 1–2 cm. wide, 7- to 11-nerved; scapes and exterior of the perianth glandular-viscid. .. 2. *A. foliata*

1. **Aletris spicata** (Thunb.) Franch. *Hypoxis spicata* Thunb.; *H. farinosa* Thunb.; *A. japonica* Lamb.——Sokushin-ran. Rhizomes stout, short; leaves linear, pale green, 3-ribbed, 15–30 cm. long, 3–7 mm. wide, slightly folded; scapes erect, few to several, 30–50 cm. long, sparingly leafy, with short white curled farinaceous hairs; spike 15–25 cm. long, many-flowered, the bracts linear, longer to shorter than the flowers, the pedicels nearly absent; flowers erect, 5–6 mm. long, white and sometimes slightly tinged dull red; fruit obovoid, about 3 mm. long.——Apr.–May. Hills and low mountains; Honshu (Kantō Distr. and westw.), Shikoku, Kyushu; commoner westw.——Ryukyus and Formosa.

2. **Aletris foliata** (Maxim.) Bureau & Franch. *Metanarthecium foliatum* Maxim.; *A. dickinsii* Franch.——Nebari-nogi-ran. Rhizomes stout, short; leaves lanceolate or oblanceolate, sometimes linear-lanceolate, 10–20 cm. long, 1–2 cm. wide, acute to acuminate, vivid green, glabrous, chartaceous, about 10-nerved, spreading; scapes 20–40 cm. long, glandular-viscid above, with few small linear leaves; spike solitary, 5–20 cm. long, loosely flowered below, the bracts lanceolate, shorter to slightly longer than the flowers; flowers rather many, short-pedicelled, 6–8 mm. long, yellow-green, glandular-viscid on the exterior; capsules 3–4 mm. long, enveloped by the persistent perianth-tube.——June–Aug. Mountain meadows and lower alpine regions; Hokkaido, Honshu (n. and centr. distr.), Shikoku.

40. SMILAX L. Shiode Zoku

Scandent or erect, branched, dioecious shrubs or herbs usually with tendrils; leaves 2-ranked, alternate, often evergreen, lanceolate to elliptic or cordate-orbicular, 3- to 7-nerved, petiolate; umbels few- to many-flowered, the flowers yellowish green, the tepals free; stamens 6(–15) in the staminate flowers, the staminodia (1–)6 in the pistillate flowers; ovary ovoid, 3-locular, the ovules 1–2 in each locule; stigmas 3; berry globose; seeds few.——About 300 species, abundant in the Tropics, several in e. Asia, N. America, and the Mediterranean region.

1A. Unarmed herbs, the stems dying in winter; leaves herbaceous to membranous; anthers coiled after anthesis.
2A. Leaves relatively thin, membranous, vivid green above, pale white and glaucescent beneath, rounded to truncate at base, the petioles 1–4 cm. long; anthers narrowly oblong, about 0.7 mm. long, 1/5 to 1/3 as long as the filaments. ...... 1. *S. nipponica*
2B. Leaves relatively thick, green and lustrous above, pale green beneath, usually cordate, the petioles 5–25 mm. long; anthers linear, nearly 2 mm. long, ⅓ to ½ as long as the filaments. .... 2. *S. riparia* var. *ussuriensis*

1. Smilax nipponica Miq. *S. herbacea* var. *nipponica* (Miq.) Maxim.; *S. oldhamii* sensu auct. Japon., non Miq.; *S. higoensis* Miq.——Tachi-shiode. Herbaceous perennial, 1–2 m. high, erect while young, becoming somewhat scandent; leaves relatively thin and membranous, ovate-oblong to nearly ovate, 5–15 cm. long, 2.5–7 cm. wide, 5- to 7-nerved, abruptly acute to acuminate, rounded to subtruncate at base, vivid green above, glaucescent and usually with papillae beneath, the petioles 1–4 cm. long, the upper ones with subtending tendrils; inflorescence simple, many-flowered, the peduncles in lower stem axils, 4–10 cm. long, the staminate tepals spreading, broadly oblanceolate, about 4 mm. long; stamens 1/2–2/3 as long as the tepals, the anthers about 0.7 mm. long; pistillate tepals nearly boat-shaped, appressed to the ovary; fruit globose, black, bloomy.——May–June. Meadows and thickets in mountains; Honshu, Shikoku, Kyushu.——Formosa, and a variety in Korea and China.

2. Smilax riparia A. DC. var. **ussuriensis** (Regel) Hara & T. Koyama. *S. ovatorotunda* var. *ussuriensis* (Regel) Hara; *S. herbacea* var. *oldhamii* (Miq.) Maxim., excl. syn.; *S. nipponica* sensu auct. Japon., non Miq.; *S. maximowiczii* Koidz.; *S. higoensis* var. *maximowiczii* (Koidz.) Kitag.——Shiode. Plant herbaceous, much branched; stems striate; leaves relatively thick, rather firm, slightly lustrous above, paler beneath, ovate or ovate-oblong, 5–15 cm. long, 2.5–7 cm. wide, short-acuminate, usually cordate or sometimes rounded at base, 5- to 7-nerved, glabrous, rarely with papillae on slightly raised veinlets beneath, the petioles 5–25(–30) mm. long, tendril-bearing; umbels many-flowered, simple, axillary, the pedicels 7–12 mm. long, the staminate tepals recurved, lanceolate, about 4 mm. long, the stamens 2/3–4/5 as long as the tepals; anthers about 1.5 mm. long; staminodes absent in the pistillate flowers.——July–Aug. Hokkaido, Honshu, Shikoku, Kyushu.——Forma **stenophylla** (Hara) T. Koyama. *S. ovatorotunda* forma *stenophylla* Hara; *S. higoensis* auct., non Miq.; *S. oldhamii* var. *higoensis* Ohwi, excl. syn.——Hosoba-shiode. A narrow-leaved phase.

3. Smilax vaginata Decne. var. **stans** (Maxim.) T. Koyama. *S. stans* Maxim.——Maruba-sankirai. Stems erect, angled, unarmed, firm, green, occasionally to rather profusely branched; leaves thinly chartaceous, deltoid-ovate to broadly ovate, 4–7 cm. long, 3–6 cm. wide, short-acuminate, rounded to shallowly cordate, green above, whitish beneath, (3–)5-nerved, the petioles 8–15 mm. long, without tendrils; umbels pedunculate, axillary, several-flowered; flowers yellow-green, the pistillate tepals elliptic, about 3 mm. long, the inner whorl slightly narrower than the outer; staminodia present in pistillate flowers; berry black, globose, about 6 mm. in diameter. ——May–June. Honshu, Shikoku, Kyushu; rare.

4. Smilax biflora Sieb. ex Miq. *S. china* var. *biflora* (Sieb.) Makino——Hime-kakara, Hime-saru-tori-ibara. Rhizomes creeping; stems 15–30 cm. long, densely branched, flexuous, the branchlets angled, often scattered-prickly; leaves broadly ovate to orbicular, mucronate, usually rounded at both ends, subcoriaceous, green, whitish beneath, 5–15 mm. long, 3-nerved, the petioles 2–3 mm. long, the tendrils 0–1.5 mm. long, subulate; umbels axillary, short-pedunculate, about 2-flowered; flowers yellow-green, the staminate tepals elliptic, obliquely spreading, recurved above, about 3 mm. long, the inner whorl narrowly oblong; stamens 1/2–3/5 as long as the tepals, the anthers elliptic, about 0.5 mm. long; berry red, globose, about 5 mm. long.——Apr. Mountains; Kyushu (Yakushima).——Ryukyus (Amami-oshima).

Var. **trinervula** (Miq.) Hatusima. *S. trinervula* Miq.; *S. sarumame* Ohwi; *S. china* var. *trinervula* (Miq.) Makino ——Saru-mame. Stems erect, 30–50 cm. long, loosely branched; leaves elliptic to oblong or suborbicular, 1.5–4 cm. long, 1–2.5 cm. wide, glaucous beneath.——May–June. Mountains; Honshu (Kantō Distr. and westw.).

5. Smilax china L. *Coprosmanthus japonicus* Kunth; *S. japonica* (Kunth) A. Gray——Saru-tori-ibara. Rhizomes woody, stout, creeping; stems firm, 0.5–2 m. long, profusely branched, ascending to spreading, the branches green, usually more or less prickly; leaves coriaceous, lustrous, ovate to ovate-orbicular or elliptic, 3–12 cm. long, 2–10 cm. wide, mucronate, rounded or sometimes shallowly cordate or cuneate at base, 3- to 5-nerved, smooth, sometimes glaucescent beneath; petioles 7–20 mm. long, with a long tendril near the middle on each side; umbels axillary, many-flowered, pedunculate; staminate tepals ascending, elliptic, recurved above, about 4 mm. long, the inner whorl narrower; filaments about 2/3 as long as the outer tepals, the anthers elliptic, about 0.6 mm. long; fruit red, globose, about 7 mm. in diameter.——Apr. Hills and mountains; Hokkaido, Honshu, Shikoku, Kyushu; common. ——China and Korea.

6. Smilax sebeana Miq. *S. iriomotensis* Masam.; *S. maritima* Hatus.——Hama-saru-tori-ibara. Evergreen scandent liana with terete, smooth or loosely prickly stems; leaves chartaceous to thickly so, orbicular-ovate, ovate, oblong-ovate or elliptic, 3–12 cm. long, 1.3–10 cm. wide, the upper surface lustrous, the lower surface very glaucous and waxy, cuneate to rounded at base, shortly acute to rounded at apex, nerves 3 to 7, the lateral veins reticulate, prominently elevated on the lower surface, the petioles short, curved, the lower 2/3 to 1/2 sheathing; tendrils shorter than to almost as long as the blade; umbel solitary, with 17–56 smooth slender rays to 15 mm. long; peduncle moderately compressed, smooth; perianth segments slightly recurved in flowering, the staminate 3.5–5 mm. long, the pistillate 3–3.5 mm. long; stamens 6; berry globose, (5–)7–8 mm. in diameter, blue-black to dark purple; seeds 1 to 2 to a berry, obovoid, dark red.——Occasional on rocky slopes along the seacoast; Kyushu (Ohsumi, Satsuma).——Ryukyus and Formosa.

7. Smilax sieboldii Miq. *S. oldhamii* Miq.——Yama-kashŪ, Saikachi-ibara. Scandent subshrub with rather prickly angled branches; leaves chartaceous, lustrous above, duller beneath, broadly ovate to cordate or deltoid-ovate, 5–12 cm. long, 3–9 cm. wide, short-acuminate, cordate or sometimes rounded at base, 5(–7)-nerved, the margin slightly uneven, the petioles 5–15 mm. long, with tendrils; umbels simple, solitary, relatively many-flowered, the peduncles 1–2 cm. long; staminate tepals 4–5 mm. long, ascending, subrecurved on upper half, the outer ones broadly lanceolate, the inner ones narrower; stamens slightly shorter than the tepals, the anthers linear-oblong, about 1 mm. long; fruit globose, blue-black,

about 6 mm. in diameter.——May–June. Honshu, Shikoku, Kyushu.——Korea, China, and Formosa.

8. Smilax bracteata Presl. *S. stenopetala* A. Gray——Satsuma-sankirai. Plants scandent; branches striate, sometimes loosely prickly; leaves coriaceous, rather lustrous, 3- to 7-nerved, oblong, ovate to orbicular, 5–8 cm. long, 3–7 cm. wide, acute to abruptly cuspidate, cuneate to rounded at base, the margin slightly thickened and firm, the petioles 8–12 mm. long, with tendrils near the middle; inflorescence corymbose, often branched, many-flowered, short-pedunculate; staminate tepals much-recurved, about 5 mm. long, the outer ones linear-oblong, the inner ones linear-lanceolate; stamens nearly as long as the tepals, the anthers broadly linear, 1.2–1.5 mm. long, coiled.——Dec.–Feb. Kyushu (s. distr.).——Ryukyus to Philippines.

41. HETEROSMILAX Kunth   Karasukiba-sankirai Zoku

Scandent dioecious shrubs or subshrubs with elongate branching stems; leaves ovate, sometimes cordate, 3- to 5-nerved, the petioles with a tendril on each side; inflorescence umbellate, axillary, pedunculate; flowers relatively small, pedicelled, the tepals connate, forming an urceolate tube; stamens 3, inserted at the base of the perianth or forming a connate tube; staminodes 1–3, filiform in the pistillate flowers; ovary ovoid, 3-locular, with 2 ovules in each locule; fruit a globose berry; seeds 1–3.——Few species in the warmer regions of India, Malaysia, and e. Asia.

1. Heterosmilax japonica Kunth. Karasukiba-san-kirai. Glabrous scandent subshrub with elongate, nearly terete branches; leaves firmly chartaceous, narrowly deltoid-ovate to broadly ovate, 5–10 cm. long, 3–8 cm. wide, 5(–7)-nerved, abruptly acuminate mucronate, rounded to cordate at base, slightly lustrous, green, with raised netted-veinlets on

both sides, the petioles 1–2.5 cm. long, with a solitary long tendril on each side; umbels many-flowered, the pistillate perianth 3–4 mm. long, with small obtuse teeth at apex.——Aug.–Oct. Kyushu (Yakushima).——Ryukyus, Formosa, and China.

Fam. 53. AMARYLLIDACEAE   Higanbana Ka   Amaryllis Family

Mostly herbs with tunicate bulbs or rarely with short rhizomes; leaves usually radical, linear; flowers actinomorphic or slightly zygomorphic; tepals 6, in 2 series, usually equal; stamens 6, the anthers usually introrse; ovary inferior, 3-locular; ovules anatropous, in 2 series, the embryo small, erect, enclosed in a fleshy endosperm; fruit a loculicidal capsule, rarely a berry, seeds few.——About 86 genera and about 1000 species, chiefly in tropical and warmer regions, relatively few in the temperate regions.

1A. Plants with tunicate bulbs; scapes naked; inflorescence subtended by a spathelike bract at base.
  2A. Corona absent; bulbs subcylindric and elongate (in ours); leaves rather large, 4–8 cm. wide (in ours). . . . . 1. *Crinum*
  2B. Corona present; bulbs ovoid and much shorter (in ours).
    3A. Filaments between the very minute corona-lobes. . . . . 2. *Lycoris*
    3B. Filaments within the large cup-shaped or tubular corona. . . . . 3. *Narcissus*
1B. Plants rhizomatous; stems usually leafy; inflorescence usually without a spathe.
  4A. Ovary with a filiform beak at apex; fruit fleshy. . . . . 4. *Curculigo*
  4B. Ovary beakless; fruit a capsule. . . . . 5. *Hypoxis*

1. CRINUM L.   Hama-omoto Zoku

Bulbs tunicate; radical leaves elongate, relatively broad; flowers large, in terminal umbels on a simple naked scape, sessile or short-pedicelled, the involucral bracts 2, membranous, the bracts many, linear; tepals connate below, forming a long narrow tube, the segments narrowly oblong to linear; stamens inserted on the throat of the tube, the filaments free, filiform, the anthers versatile; ovary 3-locular, the style simple, filiform, the stigma small; ovules few to many; capsules irregularly globose and dehiscent; seeds few, large.——About 130 species, pantropical, few in temperate regions.

1. Crinum asiaticum L. var. **japonicum** Bak. *C. asiaticum* var. *declinatum* sensu auct. Japon., non Miq.; *C. maritimum* Sieb.——Hama-omoto. Bulbs subcylindric, 30–50 cm. long, 3–5 cm. in diameter; leaves rather fleshy, 30–70 cm. long, 4–8 cm. wide, hyaline on the margin, glabrous, slightly lustrous, gradually narrowed above; scapes erect, 50–80 cm. long, naked, decumbent in fruit; umbels many-flowered, the subtending bracts 2, linear-oblong, abruptly acuminate, 6–8 cm. long;

flowers white, the perianth-tube slender, the segments broadly linear, acute, 6–7 cm. long, 5–6 mm. wide, shorter than the tube; anthers linear, 17–22 mm. long; capsules subglobose; seeds globose or rounded with obtuse angles, 2–2.5 cm. long and as wide, the testa gray-white and spongy.——Aug.–Sept. Sandy sea beaches; Honshu (s. Kantō Distr. and westw.), Shikoku, Kyushu. The typical phase is widely distributed in tropical Asia.

2. LYCORIS Herb.   HIGANBANA ZOKU

Bulbs tunicate, ovoid; leaves radical, flat, linear or bandlike, usually appearing after anthesis; scapes solid, naked, erect; flowers pedicelled, in a terminal umbel, involucral bracts 2, linear or lanceolate, the perianth infundibuliform, the tube rather short, the segments sometimes strongly recurved, equal, the corona-segments usually minute and scalelike on the throat; filaments elongate, erect or upwardly curved, inserted on the throat of the perianth tube; anthers oblong, versatile; ovary 3-locular; style slender, simple, filiform, the stigma minute; ovules many in each locule, 2-seriate; capsules globose to ovoid, often beaked at apex; seeds rather large, globose, black-brown.——Few species, in e. Asia.

1A. Perianth-segments distinctly undulate on margin; flowers in September and October.
  2A. Flowers clear yellow, fertile; leaves yellowish green, about 15 mm. wide. . . . . 1. *L. aurea*
  2B. Flowers red or white, sterile; leaves deep green, with a white longitudinal band above.
    3A. Flowers white; leaves 10–14 mm. wide. . . . . .2. *L. albiflora*
    3B. Flowers red; leaves 6–8 mm. wide. . . . . 3. *L. radiata*
1B. Perianth-segments not undulate on margin; flowers in July and August.
  4A. Flowers orange-red, 5–9 cm. long; leaves to 15 mm. wide. . . . . 4. *L. sanguinea*
  4B. Flowers pale rose-purple, 9–10 cm. long; leaves about 2 cm. wide. . . . . 5. *L. squamigera*

1. Lycoris aurea (L'Hérit.) Herb. *Nerine aurea* (L'-Hérit.) Bury; *Amaryllis aurea* L'Hérit.——SHŌKI-ZUISEN, SHŌ-KI-RAN.   Bulbs broadly ovoid, 5–6 cm. in diameter; leaves strap-shaped, to 60 cm. long, 12–18 mm. wide, glaucous, rather fleshy, lustrous; scapes erect, about 60 cm. long, the involucral bracts 3–5 cm. long, membranous, lanceolate, gradually narrowed above, with a minute obtuse apex; flowers few large, clear yellow, not fragrant, the perianth-segments equal, oblanceolate, acute, ascending, about 8 cm. long, the tube 1.5–2 cm. long, the pedicels 8–15 mm. long; fiilaments slightly exserted, the anthers 3.5–4 mm. long.——Sept.–Oct. Warmer parts of Shikoku and Kyushu.——Ryukyus, Formosa, and China.

2. Lycoris albiflora Koidz.   SHIROBANA-MANJU-SHAGE, Resembles species No. 1, but the flowers smaller, white; leaves 10–13 mm. wide, green; flowers about 5 cm. long, sterile, the perianth-segments only slightly recurved above; stamens shorter than in No. 3, only slightly exserted, the pedicels shorter.——Sept. Reported to be spontaneous in Kyushu; rarely planted. A reputed hybrid of Nos. 1 and 3.

3. Lycoris radiata (L'Hérit.) Herb. *Amaryllis radiata* L'Hérit.; *Nerine japonica* Miq.——HIGANBANA.   Bulbs broadly ellipsoidal, 2.5–3.5 cm. in diameter; leaves appearing in autumn, narrowly strap-shaped, deep green, 30–40 cm. long, 6–8 mm. wide; scapes 30–50 cm. long; umbels terminal, several-flowered, the involucral bracts broadly linear to lanceolate, thinly membranous, 2–3 cm. long; flowers red, inodorous, the pedicels 6–15 mm. long; perianth-tube 6–8 mm. long, the segments narrowly oblanceolate, about 4 cm. long, 5–6 mm. wide, prominently recurved, undulate on margin; stamens 7–8 cm.

long, much exserted.——Sept. Around cultivated fields and meadows in lowlands and hills; Honshu, Shikoku, Kyushu; common.——Ryukyus and China.

4. Lycoris sanguinea Maxim. KITSUNE-NO-KAMISORI. Bulbs tunicate, 1.5–2 cm. in diameter; leaves strap-shaped, 10–12 mm. wide, pale green, soft, appearing in spring, somewhat glaucescent; scapes 30–50 cm. long, the involucral bracts narrowly lanceolate, membranous, 2–4 cm. long; flowers several, orange-red, without odor, 5–6 cm. long, the pedicels 2–5 cm. long; perianth-tube 12–15 mm. long, greenish, the segments narrowly oblanceolate, subobtuse, ascending, scarcely or slightly recurved at apex; stamens slightly shorter than the perianth, the anthers pale yellow, about 2.5 mm. long; style as long as the perianth; fruit globose.——Aug. Wooded hills and low mountains; Hokkaido (naturalized), Honshu, Shikoku, Kyushu; common.——China.

Var. **kiushiana** Makino. *L. kiushiana* Makino——Ō-KI-TSUNE-NO-KAMISORI.   Flowers larger, 7–9 cm. long, the pedicels longer, to 8 cm. long, the stamens exserted and longer.——July. Honshu and Kyushu.

5. Lycoris squamigera Maxim. NATSU-ZUISEN.   Bulbs broadly ovoid, 4–5 cm. in diameter; leaves appearing in late autumn, elongating in spring, soft, pale green, 18–25 mm. wide, strap-shaped; scapes rather stout, 50–70 cm. long; flowers large, several, the involucral bracts broadly lanceolate, 2–4 cm. long, the pedicels 1–2 cm. long, the perianth 9–10 cm. long, pale rose-purple, the tube about 2.5 cm. long, the segments obliquely spreading, oblanceolate, about 15 mm. wide; stamens not exserted.——Aug.–Sept. Honshu, Shikoku, Kyushu; often planted in our gardens.

3. NARCISSUS L.   SUISEN ZOKU

Bulbs tunicate; leaves radical, strap-shaped or linear; flowers in a terminal umbel on a naked solitary scape, subtended by a single spathelike involucral membranous bract, tubular toward the base; perianth-segments 6, ovate, spreading, forming a connate tube below, the corona cup-shaped or tubular; stamens inserted on the tube, the filaments short; ovary 3-locular, inferior; style simple, filiform, the stigma small; ovules many in each locule; capsules loculicidal; seeds black.——About 40 species, mainly in the Mediterranean region, 1 species in e. Asia.

1. Narcissus tazetta L. var. **chinensis** Roem. SUISEN. Bulbs ovoid-globose; leaves strap-shaped, rather fleshy, glaucous, rounded to very obtuse, 20–40 cm. long, 8–15 mm. wide; scapes 20–40 cm. long, the bract scarious, 3–5 cm. long; flowers few to several, white, the pedicels rather unequal; perianth-tube pale green, about 2 cm. long, narrow, the segments

spreading, about 1.5 cm. long, ovate-orbicular to broadly elliptic, with a minute mucro at the apex, the corona yellow, cup-shaped, about 1 cm. across at the margin.——Jan.–Apr. Naturalized (?) along seashores; Honshu (Kantō Distr. and westw.), Kyushu.——China. The typical phase occurs mainly along the Mediterranean.

4. CURCULIGO Gaertn.   KIMBAIZASA ZOKU

Rhizomes short; radical leaves usually narrowly lanceolate, longitudinally plicate; inflorescence spicate or racemose, few-flowered, concealed between the leaves or long-pedunculate and densely flowered, the bracts linear; perianth-segments equal, spreading, tubular below; stamens 6, inserted at the base of the perianth-segments, the filaments short, the anthers linear; ovary 3-locular, often beaked at apex; style columnar, short; stigma 3-lobed, oblong, erect; ovules anatropous, 2 to many in each locule; fruit rather fleshy, indehiscent; seeds subglobose, black, crustaceous, the hilum often with an appendage; embryo small, the endosperm fleshy.——Few species in Africa, India, se. Asia, and tropical America.

1. Curculigo orchioides Gaertn. *C. ensifolia* R. Br.; *Hypoxis orchioides* (Gaertn.) Kurz——KIMBAIZASA.   Rhizomes perpendicular, with rather stout adventitious roots; leaves lanceolate-acuminate, 20–30 cm. long, 1–2 cm. wide, tapering at base to an elongate petiole, becoming sheathlike at base, loosely pilose, the nerves slender, several; flowers few, on very short peduncles, the bracts membranous, lanceolate, 2–4 cm. long; flowers long-pilose outside, the tepals yellow, lanceolate, few-nerved, about 8 mm. long, the perianth-stipe filiform, 2–2.5 cm. long, becoming a beak in fruit; fruit elliptic.——May–Aug. Honshu (Chūgoku Distr.), Shikoku, Kyushu.——China, Ryukyus, Formosa, India, Malaysia, and Australia.

5. HYPOXIS L.   KO-KIMBAIZASA ZOKU

Rhizomes tuberlike or cormlike, relatively small; radical leaves linear or slightly broader, longitudinally nerved; flowers 1 to many, umbellate or racemose on relatively short or elongated peduncles, the bracts linear, small, or sometimes absent; perianth 6-parted, the segments spreading; stamens 6, inserted at the base of the perianth-segments, the filaments short, the anthers erect, linear to ovate; ovary not beaked at apex, 3-locular; style short, columnar, the stigma lobes 3, erect, oblong, thickened; ovules many, 2-seriate in each locule; capsules globose to ellipsoidal, supporting the perianth at the apex while young; seeds small, subglobose, muricate.——About 100 species, mainly in Africa, few in India, Malaysia, e. Asia, N. and S. America.

1. Hypoxis aurea Lour. *H. minor* D. Don——KO-KIMBAIZASA.   Rhizomes tuberous, rather irregular, 3–8 mm. in diameter; leaves linear, 10–25 cm. long, usually broadest above the middle, 2–4 mm. wide, 3–5(–7)-nerved, long-acuminate, membranous at base, with scattered long yellowish white hairs; scapes slender, 3–10 cm. long, 1- to 2-flowered, long-pilose, the bracts filiform, 5–8 mm. long; flowers short-pedicelled, long-pilose outside, yellow, the perianth-segments deltoid-lanceolate, subacute, few-nerved, 4–6.5 mm. long; anthers about 1.3 mm. long, obtuse, shallowly 2-lobed at base; capsules oblong, narrowed below, 6–8 mm. long; seeds globose, black-brown, about 1.2 mm. across, lusterless, densely muricate with subrounded mammillae, the appendage short.——Apr.–May. Sunny slopes and thickets in the foothills; Honshu (Rikuzen Prov. and southw.), Shikoku, Kyushu; rather rare.——China, India, and Malaysia.

Fam. 54. DIOSCOREACEAE   YAMA-NO-IMO KA   Yam Family

Usually scandent herbs; leaves alternate or opposite, flat, petiolate, entire or palmately lobed; flowers unisexual (plants dioecious or monoecious), small, actinomorphic, the tepals 6, 2 seriate, free or connate at base; stamens 6, fertile or sometimes with 3 staminodia and 3 fertile; the anthers small; ovary 3-angled, inferior, 3-locular or with 3 imperfect septae; styles 3, distinct; fruit a winged capsule or sometimes a berry.——About 10 genera, with more than 650 species, mainly tropical, some in e. Asia, the Mediterranean region, the Pyrenees, N. and S. America.

1. DIOSCOREA L.   YAMA-NO-IMO ZOKU

Scandent perennial dioecious herbs from a fleshy horizontal or vertical rhizome, or the tuberous stems much-elongate and branched; leaves alternate or opposite, sometimes lobed; flowers relatively small, erect or nodding, the tepals free or connate at base; stamens free or sometimes connate and forming a column (sometimes partially suppressed or reduced to staminodia); ovary inferior, 3-angled; style short; ovules 2 in each locule; capsules 3-winged, loculicidal; seeds much flattened, winged or wingless; endosperm fleshy or cartilaginous; embryo oblong, dividing the endosperm, the cotyledons suborbicular.——About 600 species, widely distributed in the Tropics and subtropics, e. Asia, the Mediterranean region, N. and S. America.

1A. Tepals rather fleshy, white or purplish, ascending or erect, not spreading; rhizomes obsolete, the roots fleshy, tuberous, cylindric or flabellate, perpendicular; leaves with bulbils in the axils.
  2A. Inflorescence erect in the staminate, pendulous in the pistillate; tuber whitish, covered with fine scattered roots; leaves usually opposite.
    3A. Plants uniformly green; leaves broadly lanceolate to narrowly deltoid-ovate, cordate at base. .... 1. *D. japonica*
    3B. Stems and petioles purplish; leaves deltoid-ovate to deltoid, broadly cordate, the basal margin sometimes auriculately dilated.
                                                 2. *D. batatas*
  2B. Inflorescence of both sexes pendulous; tuber dark grayish brown, covered with coarse thickened roots, warty-thickened at base; leaves alternate, depressed-cordate to deltoid. .... 3. *D. bulbifera*
1B. Tepals membranous, yellowish green, usually spreading; rhizomes long-creeping, with slender fibrous roots; leaves without bulbils in the axils.
  4A. Fertile stamens 3.
    5A. Staminodia 3 in the staminate flowers; leaves broadly ovate to narrowly deltoid-ovate, 5–8(–10) cm. long, broadly cordate; staminate flowers sessile. .... 4. *D. gracillima*

5B. Staminodia absent in the staminate flowers; leaves broadly deltoid-lanceolate to narrowly ovate-deltoid, 8–15 cm. long, deeply cordate, the basal margins often imbricate; staminate flowers pedicellate. .... 5. *D. asclepiadea*
4B. Fertile stamens 6.
  6A. Seeds winged on one side.
    7A. Leaves entire, glabrous; flowers fully expanding. .... 6. *D. tokoro*
    7B. Leaves palmately lobed, usually pilose; flowers campanulate, not fully expanding. .... 7. *D. nipponica*
  6B. Seeds winged all the way around.
    8A. Leaves entire; staminate flowers rather loose, distinctly pedicelled. .... 8. *D. tenuipes*
    8B. Leaves palmately lobed; staminate flowers dense, sessile to short-pedicelled.
      9A. Leaves usually pilose; staminate flowers short-pedicelled. .... 9. *D. quinqueloba*
      9B. Leaves glabrous, sometimes minutely papillose on the nerves beneath; staminate flowers sessile. .... 10. *D. septemloba*

1. Dioscorea japonica Thunb. YAMA-NO-IMO. Root fleshy, cylindric, elongate, perpendicular; stems elongate, branched; leaves opposite, sometimes alternate, cordate at base, long-petioled, narrowly oblong-deltoid, 5–10 cm. long, 2–5 cm. wide, long-acuminate, green, glabrous; inflorescence 1- to 3-nate in leaf axils, spicate, short-peduncled; flowers white, the pistillate pendulous, loosely few-flowered, the staminate erect, rather loosely many-flowered, sessile; stamens 6; capsules broadly elliptic, 1.2–1.4 cm. long, about 2 cm. wide; seeds broadly wing-margined.——Aug. Wooded foothills; Honshu, Shikoku, Kyushu.

2. Dioscorea batatas Decne. *D. polystachya* sensu auct. Japon., non Turcz.——NAGA-IMO. Root fleshy, cylindric to flabellate, perpendicular; stems usually purplish; leaves long-petiolate, nearly deltoid-ovate, acuminate, broadly cordate and often auriculate at base, purplish on petioles and nerves, with bulbils in axils; inflorescence spicate, 1- to 3-nate, in leaf axils; flowers sessile, white, the staminate erect, rather loosely many-flowered, the pistillate pendulous, loosely few-flowered; stamens 6; capsules obovate-orbicular.——Aug.-Sept. Hokkaido, Honshu, Shikoku, Kyushu; frequently planted.——Korea, China, and Formosa.

3. Dioscorea bulbifera L. *D. sativa* sensu auct. Japon., non L.——NIGA-KASHŪ. Tuberous root depressed-globose, 5–8 cm. across; stems elongate, branched; leaves alternate, long-petiolate, depressed-cordate or deltoid, glabrous, abruptly acuminate, entire, with bulbils in the axils; staminate spikes 1- to 3-nate, in leaf axils, short-pedunculate, rather densely many-flowered; flowers sessile, the tepals linear-lanceolate, ascending, rather thick, 2.5–3 mm. long; stamens 6.——Aug.-Sept. Honshu (Chūgoku Distr.), Shikoku, Kyushu.——Forma **domestica** (Makino) Makino & Nemoto. *D. sativa* f. *domestica* Makino ——KASHŪ-IMO. Cultivated for the edible bulbils and tuberous roots.——Forma **spontanea** (Makino) Makino & Nemoto. *D. sativa* forma *spontanea* Makino—— NIGA-KASHŪ. A wild phase with bitter, nonedible bulbils and tuberous roots.

4. Dioscorea gracillima Miq. TACHIDOKORO. Rhizomes long-creeping; stems elongate, branched; leaves ovate, deltoid-ovate to elliptic, 5–10 cm. long, 2.5–7 cm. wide, acuminate, deeply cordate, the margins somewhat uneven, entire or shallowly lobed, glabrous; inflorescence spicate; flowers sessile with spreading tepals, the staminate erect, often branched, the pistillate loosely arranged; stigma 3-fid, entire; capsules depressed ovate-orbicular; seeds wing-margined.—— May–June. Honshu, Shikoku, Kyushu.

5. Dioscorea asclepiadea Prain & Burkill. TSUKUSHI-TACHIDOKORO. Plants glabrous; stems slightly elongate, green; leaves thin, green, broadly deltoid-lanceolate to narrowly ovate-deltoid, 8–15 cm. long, 4–7 cm. wide, deeply cordate, mucro-nate, entire or sometimes undulate, the basal margins imbricate; staminate inflorescence branched, the flowers loose, yellow-green, short-pedicelled, the tepals spreading.——Apr.-May. Kyushu.

6. Dioscorea tokoro Makino. *D. yokusaii* Prain & Burkill——ONIDOKORO, TOKORO. Plants glabrous; stems elongate, branched; leaves thin, cordate or deltoid-cordate, 5–10 cm. long, 5–10 cm. wide, acuminate, entire; inflorescence somewhat elongate, erect, the staminate often branched; staminate flowers yellowish green, short-pedicelled, the pistillate pendulous; tepals minutely toothed on upper margin, flat, spreading; capsules broadly obovate-elliptic, about 15 mm. long; seeds with a broad wing on one side.——July–Aug. Thickets in lowlands to the foothills; Hokkaido, Honshu, Shikoku, Kyushu; common.

7. Dioscorea nipponica Makino. UCHIWADOKORO. Stems elongate; leaves thin, long-petioled, broadly ovate to ovate-cordate, 7–15 cm. long, 4–12 cm. wide, usually with short coarse hairs, palmately lobed or rarely entire, long-acuminate; staminate inflorescence spicate, elongate, sometimes branched, the flowers few, in fascicles, short-pedicelled, the pistillate usually simple, pendulous; capsules broadly obovate; seeds with a broad wing on one side.——Aug. Hokkaido, Honshu (n. and centr. distr.).——Korea and China.

8. Dioscorea tenuipes Fr. & Sav. HIMEDOKORO. Stems elongate; leaves thin, deltoid-cordate, or sometimes lanceolate with the basal margins prominently auriculate, glabrous, 5–10 cm. long, 2.5–7 cm. wide, long-acuminate, sometimes with minute papillae on the nerves beneath, entire, the petioles usually with a small auriclelike appendage at base; inflorescence pendulous, the staminate flowers loose, slenderly pedicelled, the tepals entire, slightly recurved on margin; capsules orbicular, somewhat angular; seeds wing-margined.——July-Aug. Honshu (Kantō Distr. and westw.), Shikoku, Kyushu.

9. Dioscorea quinqueloba Thunb. KAEDEDOKORO. Plants elongate, branched, usually with short coarse hairs; leaves relatively large, green, cordate, 6–12 cm. long, 4–10 cm. wide, palmately 5- to 9-lobed, the median lobes narrowly ovate, acuminate, the lateral lobes rounded to obtuse, smaller; staminate inflorescence sometimes branching, the flowers short-pedicelled, becoming reddish brown when dry; tepals spreading, rather thick; capsules obovate-orbicular; seeds wing-margined.——July–Sept. Honshu (centr. distr. and westw.), Shikoku, Kyushu.——Korea, China, and Ryukyus.

10. Dioscorea septemloba Thunb. KIKUBADOKORO. Plants glabrous; leaves becoming dark when dry, 5–12 cm. long and as wide, usually cordate, usually palmately 5- to 7-lobed, the median lobe deltoid-ovate to narrowly ovate, long-acuminate, the lateral lobes deltoid, acuminate to rounded, the petioles without a basal appendage; inflorescence pendulous, the staminate flowers sessile, the tepals membranous, yellowish green, spreading; capsules obovate-orbicular, about 20 mm. long.——June–Aug. Honshu, Shikoku, Kyushu.

Var. **sititoana** (Honda & Jotani) Ohwi. *D. sititoana* Honda & Jotani——SHIMA-UCHIWADOKORO.   With larger obovate-elliptic capsules 18–28 mm. long.——Izu Isls.

Fam. 55. IRIDACEAE   AYAME KA   Iris Family

Usually perennial rhizomatous or cormose herbs; leaves narrow, often 2-ranked, frequently laterally flattened; stems simple or branched, flowers 2-bracteate, bisexual, actinomorphic or zygomorphic, usually large and showy; tepals 6, often connate at base, 2-seriate, ephemeral, petaloid, the tube adnate at base to the ovary; stamens 3, alternate with the inner tepals, the filaments filiform, free or connate, the anthers 2-locular, extrorse; ovary inferior, usually 3-locular, the ovules usually many in each locule, anatropous; style 3-fid, the lobes sometimes lobulate; capsules loculicidal; seeds many, in 1 or 2 rows in each locule, the endosperm fleshy or horny, the embryo small, straight.——About 57 genera, with about 1,500 species, cosmopolitan.

1A. Style-branches petaloid, broad, spreading and covering the anthers; tepals dimorphic. .... 1. *Iris*
1B. Style-branches slender, filiform, alternate with and not covering the anthers; tepals monomorphic. .... 2. *Belamcanda*

1. IRIS L.   AYAME ZOKU

Perennial herbs from stout branching rhizomes or corms; leaves mostly radical, 2 ranked, usually ensiform and laterally flattened, parallel-nerved; stems or scapes erect, simple or sparingly short-branched, 1- to many-flowered; flowers large and showy, ephemeral, 2 to several in the axil of a bract, the pedicels short or longer than the bract; perianth-segments in 2 series, usually dimorphic, tubular-connate below, the outer tepals clawed below, the limb (fall) spreading to reflexed, the inner whorl usually erect, narrowed below; style-branches 3, spreading, covering the stamens, the style crests petaloid, bifid or fimbriate; capsules oblong to globose, 3- or 6-angled, often subterete; seeds many.——About 150 species, in the temperate regions of the N. Hemisphere.

1A. Plants relatively low and slender, the scapes less than 30 cm. high.
  2A. Leaves 2–5 mm. wide, the nerves not prominent, the sheath persistent, fibrous; bracteoles absent; perianth-tube about twice as long
      as the segments, the outer segments without a raised ridge inside. .... 1. *I. rossii*
  2B. Leaves 5–15 mm. wide, 5- to 7-nerved, the sheath not fibrous; inflorescence sparsely branched, few-flowered; bracteoles brownish, scarious, about 2 cm. long; perianth-tube shorter than the segments, the outer segments with a crested ridge on the middle inside.
        2. *I. gracilipes*
1B. Plants relatively tall, the scapes 40–70 cm. long.
  3A. Outer perianth-segments (falls) with a median crested ridge inside; leaves 2–3 cm. wide; inflorescence more or less branched.
    4A. Pedicels as long as to slightly longer than the bract; leaves pale beneath; flowers rather numerous on each branch; bracteoles
      about 15 mm. long; inflorescence multibranched. .... 3. *I. japonica*
    4B. Pedicels nearly absent; leaves concolorous; flowers 2–3 on each branch; bracteoles about 4 cm. long; inflorescence usually with
      a single branch. .... 4. *I. tectorum*
  3B. Outer perianth-segments (falls) without a crested ridge.
    5A. Stems solid.
      6A. Inner perianth-segments very small, not more than 2 cm. long, lanceolate, acuminate-setose; flowers blue-purple; leaves without
        a prominent midrib. .... 5. *I. setosa*
      6B. Inner perianth-segments larger, petaloid, usually more than half as long as the outer segments.
        7A. Leaves flat without a prominent midrib; flowers blue-purple in wild form; inner perianth-segments oblanceolate; ovary
          3-angled toward the base; capsules not beaked. .... 6. *I. laevigata*
        7B. Leaves with a prominent raised midrib; flowers reddish-purple in wild form; ovary 6-angled; capsules beaked. .. 7. *I. ensata*
    5B. Stems hollow; leaves 5–12 mm. wide; claw of the outer perianth-segments yellow, suffused with blue. .... 8. *I. sanguinea*

1. Iris rossii Bak. EHIME-AYAME, TARE-YUE-SŌ.   Rhizomes short, somewhat tufted, branching, ascending, brown fibrous; leaves elongating after anthesis, to 30 cm. long, 2–5 mm. wide, as long as the scape in anthesis, thin, glaucous beneath, midrib not prominent, minutely papillose on upper margin; scapes 5–15 cm. long, 3- to 4-bracteate, 1-flowered, the bracts linear, green, long-acuminate, 4–6 cm. long, the pedicels about 8 mm. long; flowers blue-purple, 3.5–4 cm. across, the perianth-tube slender, 4–6 cm. long, the outer segments narrowly obovate, without a crested ridge in the center, clawed, the inner segments erect, slightly shorter than the outer ones; anthers yellow, about 4 mm. long, shorter than the filaments, the stigma-appendage deeply bifid; capsules globose.——Apr.–May. Honshu (Chūgoku Distr.), Shikoku, Kyushu.—— Korea, Manchuria, and n. China.

2. Iris gracilipes A. Gray. HIME-SHAGA.   Rhizomes short, branching, scarcely fibrous; leaves pale green, somewhat glaucous beneath, 20–40 cm. long, 5–15 mm. wide, thin, several-nerved, slightly papillose on upper margin; scapes 15–30 cm. long, 2- to 3-flowered, the green bracts 2–3, 5–10 cm. long, ensiform, the branches 3–7 cm. long, the bracteoles connate at base, 1.5–2 cm. long, the pedicels 1–2 mm. long, much shorter than the ovary; flowers 4–5 cm. across, pale purple, the perianth-tube about 12 mm. long, the outer segments obovate, very shortly clawed, with a median crest toward the lower half, the inner segments ascending, spathulate-oblong; stigma appendages bifid, fringed; fruit globose; seeds obovate, about 3 mm. long, keeled.——May–June. Mountains; Hokkaido (sw. distr.), Honshu (Kinki Distr. and eastw.), Kyushu.

3. Iris japonica Thunb. SHAGA.   Plants long-stoloniferous the rhizomes creeping; leaves evergreen, 30–60 cm. long, 2.5–3 cm. wide, green above, somewhat paler beneath, glabrous, the midrib not prominent; scapes 30–70 cm. long, the branches 3–10 cm. long, rather many-flowered, the bracts conduplicate, green, acuminate, the lower ones ensiform, the bracteoles 3–10 cm. long, ovate-oblong, subobtuse, the pedicels as long as to longer than the bracteoles; flowers about 5 cm. across, pale blue-purple, the outer perianth-segments obovate, denticulate, crested, the inner segments relatively small, obliquely ascending; stigma-appendage fimbriate; fruit very rarely maturing.——Apr.–May. Wooded hills; Honshu, Shikoku, Kyushu; common.——China.

4. Iris tectorum Maxim. ICHI-HATSU. Rhizomes short, stout, tufted; leaves 30–60 cm. long, 2.5–3.5 cm. wide, pale green, acuminate, glabrous, the midrib not prominent; scapes 30–50 cm. long, with few ensiform bracts, the branches 1–2, 2- to 3-flowered, the bracteoles obovate-oblong, membranous, acute or mucronate, 4–5 cm. long, the pedicels nearly absent; flowers purple, about 10 cm. across, the perianth-tube about 2.5 cm. long, the outer perianth-segments obovate, short-clawed, densely fringe-crested on lower half, the inner segments nearly as large as the outer, obliquely spreading; stigma-appendages irregularly toothed.——May. Chinese plant long cultivated in Japan, frequently grown on the roof of straw-thatched houses.

5. Iris setosa Pall. *I. setosa* var. *hondoensis* Honda——HIŌGI-AYAME. Rhizomes ascending, brown, fibrous; leaves 20–40 cm. long, 1–2 cm. wide, acuminate, the midrib obscure; scapes 30–70 cm. long, branched, the bracts green, ensiform, the bracteoles broadly lanceolate, essentially membranous, subobtuse, 2–3 cm. long, the pedicels 3–4 cm. long; flowers blue-purple, about 8 cm. across, the perianth-tube short, the outer segments broadly obovate, the claw yellowish with red-purple veins, the inner segments much smaller, lanceolate, acuminate-setose; anthers purple, the stigma-appendage dentate.——July–Aug. Bogs in mountains; Honshu (centr. distr. and northw.), Hokkaido.——Korea, Kuriles, Sakhalin, e. Siberia, and Aleutian Isls.

6. Iris laevigata Fischer. KAKITSUBATA. Rhizomes fibrous, branched; leaves glaucous, 40–60 cm. long, 1.5–3 cm. wide, without a prominent midrib; scapes 50–70 cm. long, with few short leaves, usually 3-flowered, the bracteoles broadly lanceolate, sometimes narrowly oblong, greenish, 5–7 cm. long, the pedicels shorter than the ovary; flowers blue-purple, some-

times white, about 12 cm. across, the perianth-tube short, the outer segments clawed, broadly ovate-elliptic, yellow in the median portion near the base, the inner ones oblanceolate, small, erect; anthers white; stigma-appendages nearly rounded.——May–June. Wet places; Hokkaido, Honshu; much-cultivated in gardens.——Manchuria, Korea, and China.

7. Iris ensata Thunb. var. **ensata.** *I. ensata* Thunb.; *I. ensata* var. *hortensis* Makino & Nemoto; *I. kaempferi* Sieb.; *I. laevigata* var. *kaempferi* (Sieb.) Maxim.——HANA-SHŌBU. The cultivated selections often have double flowers with larger often spreading inner perianth-segments.

Var. **spontanea** (Makino) Nakai. *I. kaempferi* var. *spontanea* Makino——NO-HANA-SHŌBU. Rhizomes branched, brown-fibrous; leaves 20–60 cm. long, 5–12 mm. wide, the midrib prominent; scapes 40–80 cm. long, unbranched, few-bracteate and -flowered, the uppermost bract green, linear-lanceolate, acuminate, 5–8 cm. long, the pedicels 3–5 cm. long; flowers red-purple, about 10 cm. across, the outer perianth-segments elliptic, clawed, the limb yellow at base, the inner segments smaller, erect; anthers yellow, the stigma-appendages subentire.——June–July. Hokkaido, Honshu, Kyushu.

8. Iris sanguinea Hornem. *I. sibirica* sensu Thunb., non L., nec Thunb. (1784); *I. sibirica* var. *sanguinea* Ker-Gawl.; *I. nertschinskia* Lodd.; *I. sibirica* var. *orientalis* Maxim. ——AYAME. Rhizomes creeping, branched, dark brown fibrous; leaves glaucescent, 30–50 cm. long, 5–12 mm. wide, the midrib slender, not prominent; scapes 30–60 cm. long, hollow, unbranched, 1- to 3-bractate, the uppermost bract linear-lanceolate, green, acuminate, 5–6 cm. long, the bracteoles similar but sometimes longer than the bracts; flowers 2 or 3, red-purple, about 8 cm. across, the pedicels 2–4 cm. long, shorter than the bracteoles, much longer than the ovary, the outer perianth-segments broadly obovate, the claw yellow with transverse blue lines, the inner segments erect, small; anthers dark purple; stigma-appendages dentate.——May–July. Mountain meadows; Hokkaido, Honshu, Kyushu.——Korea and e. Siberia.

Var. **violacea** Makino. *I. kamayama* Makino——KAMA-YAMA-SHŌBU. Leaves narrower and firmer, slightly tortuous, deeper green; flowers deep violet, the inner perianth-segments slightly longer.——Said to have been introduced from Korea and sometimes to be planted in our area.

2. **BELAMCANDA** Adans.   HI-ŌGI ZOKU

Tall rhizomatous perennial; leaves 2-ranked, ensiform, mostly radical and lower cauline, laterally flattened, linear; panicle terminal; flowers pedicelled, orange-red, the tepals 6, nearly equal, connate below; stamens 3, inserted at the base of the perianth-segments; anthers basifixed; ovary 3-locular; style 3-fid; capsules many-seeded, loculicidal.——A single species in e. Asia.

1. Belamcanda chinensis (L.) DC. *Ixia chinensis* L.; *B. punctata* Moench; *Gemmingia chinensis* (L.) O. Kuntze ——HI-ŌGI. Plants stoloniferous, the rhizomes short; stems 1–1.5 m. long, densely leafy toward the base; leaves laterally flattened, broadly linear, 30–50 cm. long, 2–4 cm. wide, glaucous, acuminate; inflorescence terminal, once or twice branched, the bracts few at the apex, narrowly ovate, mem-

branous, obtuse, about 1 cm. long, the pedicels 1–4 cm. long; flowers 5–6 cm. across, orange-red, dark-spotted, the tepals narrowly oblong, obtuse, spreading, narrowed and connate at base; anthers about 1 cm. long; capsules obovate-ellipsoidal, about 3 cm. long; seeds shining black.——Aug.–Sept. Honshu (w. distr.), Shikoku, Kyushu.——China and n. India.

Fam. 56. **ZINGIBERACEAE**   SHŌGA KA   Ginger Family

Perennial herbs with thickened rhizomes; stems usually simple, erect, tightly enveloped by the sheaths of radical leaves; leaves often large, long-sheathed below, linear to oblong, with a ligule at top of the sheath inside, midrib usually stout, nerves pinnately arranged; flowers 1 to few in the axil of a bract, arranged in a raceme or spike, bisexual, zygomorphic; calyx or outer perianth tubular, often splitting on one side, the inner perianth or corolla infundibuliform, 3-lobed, the posterior segment longer than the others; stamens 6, 1 fertile, the others reduced; lip placed anteriorly, larger than the petals, 2- to 3-lobed, the anthers introrse, 2-locular; ovary inferior, the ovules many, anatropous; style slender; fruit a capsule or berrylike.——About 47 genera, with about 1,400 species, mainly in the Tropics, few in temperate regions.

1A. Lip 3-lobed; appendage of the anther-connective subulate, involute, forming a tube around the style; flowers in a headlike spike.
　　1. *Zingiber*
1B. Lip 2-lobed; appendage of the anther-connective absent; flowers in a spike or panicle. . . . . 2. *Alpinia*

1. ZINGIBER Boehm. Shōga Zoku

Rhizomes thickened, aromatic; scapes usually with scalelike leaves; inflorescence a densely flowered spike; flowers one to few in the axils of bracts, the bracteoles spathlike, the calyx membranous, tubular, 3-toothed, usually splitting on one side; tube of the inner perianth whorl slender, dilated above, the segments often unequal, the posterior one largest, erect, the lateral staminodes adnate to the recurved lip; fertile stamen 1, the filament short, deeply grooved; style slender, the stigma small, subglobose, glands 2, subulate; ovary 3-locular, many-ovuled; capsules subglobose to ellipsoidal; seeds relatively large, with a lobed aril.——More than 50 species, e. Asia to India and Malaysia.

1. **Zingiber mioga** (Thunb.) Rosc. *Amomum mioga* Thunb.——Myōga. Glabrous rhizomatous perennial; leaves radical, the sheaths closely involute and forming a pseudostem, the blade lanceolate to narrowly oblong, 20–35 cm. long, 3–6 cm. wide, acuminate, narrowed to a short petiolelike base; scapes 5–15 cm. long, with few scalelike leaves at base; spike 5–7 cm. long, erect, oblong, acute; bracts acuminate, the outer ones narrowly ovate, the inner ones lanceolate; flowers pale yellow, about 5 cm. across, the perianth-tube elongate, exserted from the bracts, the segments lanceolate, the lip obovate, with a small lobule on each side.——Aug.–Oct. Woods; Honshu, Shikoku, Kyushu; often cultivated for the edible shoots.

Zingiber officinale (Willd.) Rosc. *Amomum officinale* Willd.——Shōga. Plants herbaceous; scapes 20–25 cm. long; bracts nearly equal, oblong, rounded at apex; flowers yellow, the lip purple, yellow-spotted.——Introduced from tropical Asia; widely cultivated in our area for the pungent rhizomes.

2. ALPINIA Roxb. Hana-myōga Zoku

Rhizomes stout, branching; leaves lanceolate to oblong, large, the sheaths involute and forming a pseudostem at base; inflorescence racemose or paniculate, the scapes arising from the pseudostem, the bracts deciduous or caducous; calyx tubular, 3-toothed; corolla-tube short, the segments oblong or lanceolate, the posterior one broadest, the lateral staminodia small or absent, the lip spreading, obovate, often with a small appendage on each side at base, the connective scarcely produced; ovary 3-locular; style filiform, the stigma subglobose; fruit globose, dry or fleshy, usually indehiscent; seeds globose or angled.—— About 150 species in the Tropics of the Old World.

1A. Leaves pilose at least beneath; bracts small, caducous; flowers nearly sessile, 1 or 2 from each node, about 2.5 cm. long; inflorescence simple, unbranched. . . . . 1. *A. japonica*
1B. Leaves glabrous on both sides; inflorescence branched.
　　2A. Bracts 5–7 mm. long, flat; flowers about 2 cm. long. . . . . 2. *A. intermedia*
　　2B. Bracts 1.5–2.5 cm. long, enveloping the calyx; flowers 2.5–5 cm. long.
　　　　3A. Inflorescence erect, the axis glabrous; bracts about 1.5 cm. long; flowers 2.5–3 cm. long. 3. *A. formosana*
　　　　3B. Inflorescence nodding, the axis hairy; bracts 2–2.5 cm. long; flowers 4–5 cm. long. . . . . 4. *A. speciosa*

1. **Alpinia japonica** (Thunb.) Miq. *Globba japonica* Thunb.——Hana-myōga. Plants 40–60 cm. high; rhizomes branching; pseudostems ascending, several-leaved; leaves broadly lanceolate to narrowly oblong, 20–40 cm. long, 5–7 cm. wide, sessile or short-petiolate, rather densely pilose on both sides especially beneath, the ligules 2–4 mm. long, puberulent, the sheath pilose; spike solitary, terminal, densely short-pilose, 10–15 cm. long, the bracts narrowly oblong, caducous, shorter than the calyx; flowers nearly sessile, 1 or 2 at each node of the inflorescence, white with red striations, about 2.5 cm. long; calyx tubular, split inside, puberulent, with 3 obtuse teeth on upper margin, the lip ovate; fruit broadly ellipsoidal, 1–1.7 cm. long, puberulent, red.——May–June. Woods in foothills; Honshu (s. Kantō Distr. and westw.), Shikoku, Kyushu.——Formosa and China.

Var. **kiushiana** Kitam. *A. kiushiana* Kitam.——Tsuku-shi-hana-myōga. Calyx-tube, ovary, perianth, and fruit nearly glabrous.——Kyushu (Ibusuki in Satsuma Prov.).

2. **Alpinia intermedia** Gagnep. *A. chinensis* sensu auct. Japon., non Rosc.——Ao-no-kumatake-ran. Glabrous perennial herb; leaves narrowly oblong, 30–50 cm. long, 6–12 cm. wide, glabrous, the ligules 5–6 mm. long, loosely ciliolate, the petioles short; inflorescence 10–20 cm. long, erect, the branches 0.8–1.3 cm. long, slender, 3- to 4-flowered at the summit, the bracts pale, membranous, 7–10 mm. long, obtuse; calyx-tube 3.5–4.5 mm. long, with small teeth, sometimes loosely ciliolate, the lip ovate, narrowed below.——July. Woods; Honshu (Izu Isls. and Kii Prov.), Shikoku, Kyushu.——Ryukyus and Formosa.

3. **Alpinia formosana** K. Schum. *A. kumatake* Makino; *A. satsumensis* Gagnep.——Kumatake-ran. Glabrous tufted perennial, 1–2 m. high, from stout rhizomes; leaves narrowly oblong or broadly lanceolate, 50–70 cm. long, 8–12 cm. wide, acuminate at both ends, sessile or short-petiolate, ciliolate, deep green above, the ligules 5–7 mm. long, ciliate; scapes erect; inflorescence 15–20 cm. long, racemose to paniculate, the axis glabrous to slightly puberulous, the branches 1–1.5 cm. long, the bracts membranous and enveloping the calyx, rounded and mucronate, scarious on margin; flowers white, tinged red, 2.5–3 cm. long; calyx-tube 8–10 mm. long, shallowly toothed,

ciliolate, the lip broadly ovate; fruit broadly ellipsoidal, vermilion.——July–Aug. Woods in lowlands to foothills; Kyushu (s. distr., Yakushima and Tanegashima).——Ryukyus and Formosa.

4. Alpinia speciosa (Wendl.) K. Schum. *Zerumbet speciosum* Wendl.——Gettō. Stout tufted glabrous perennial, 1.5–2.5 m. high; leaves lanceolate to broadly so, 50–70 cm. long, 7–15 cm. wide, acuminate at both ends, short-petiolate, rather densely ciliate, the ligules 8–10 mm. long, with brown hairs on back; inflorescence nodding, 20–30 cm. long, with brown hairs on axis, the branches 1–2 cm. long, the bracts 2–2.5 cm. long, broadly elliptic, covering the flower-bud, subcoriaceous, nerveless, often 2-lobed; calyx about 1.5 cm. long, tubular, split on one side, with deltoid teeth; inner perianth-segments white to pinkish, obtuse, the lip broadly ovate, subtrilobed, creamy white with red striations; ovary densely pubescent with brown hairs; fruit globose-ovoid, about 2 cm. long, longitudinally ribbed, yellow-vermilion, crowned with the persistent perianth base.——July–Aug. Woods in lowlands to the foothills; Kyushu (Cape Sata), Yakushima, and Tanegashima.——Ryukyus and Formosa.

Fam. 57. BURMANNIACEAE   Hina-no-shakujō Ka   Burmannia Family

Delicate saprophytic herbs without chlorophyll or rarely green; leaves cauline, alternate, scalelike, small; flowers bisexual, usually actinomorphic, the perianth tubular to prismatic, sometimes with 3 or 6 longitudinal wings, the lobes 6 or 3, the outer whorl alike, the inner segments sometimes very small or nearly obsolete; stamens 3 or 6, inserted on the tube, the filaments very short, sometimes absent, the anthers 2-locular, transversely or longitudinally split; ovary inferior, 3-locular, the placentae axile or 1-locular with parietal placentae; style simple, stigmas 3; capsules many-seeded, the seeds minute; endosperm absent, the embryo simulating the seed in shape.——About 16 genera, with about 120 species, mainly in the Tropics.

1A. Perianth-tube persistent; style as long as the tube; stamens 3; anthers transversely split; outer perianth-lobes larger, the inner whorl smaller or absent. .... 1. *Burmannia*
1B. Perianth-tube transversely splitting and deciduous after anthesis; style very short; stamens 6; anthers longitudinally split; outer perianth-lobes not connate above, the inner segments often connate or imbricate and forming a 3-pored depressed-conical body bearing 3 filiform or linear appendages at apex. .... 2. *Glaziocharis*

1. BURMANNIA L.   Hina-no-shakujō Zoku

Saprophytes without chlorophyll or greenish delicate annual or perennial herbs; stems simple or branched; leaves usually reduced to lanceolate scales; flowers 1 to several on a terminal cyme; perianth tubular at base, often 3-angled, the lobes 6, the outer segments larger, the inner ones minute or absent; anthers 3, nearly sessile, inserted at the base of the inner lobe on the throat of the tube; style filiform, the stigmas 3, short; ovary triangular, 3-locular, the placentae axile; capsules irregularly dehiscent, crowned with the persistent base of the perianth-tube; seeds many, ellipsoidal.——About 60 species, mainly in the Tropics.

1A. Perianth-tube wingless; inflorescence headlike, usually several-flowered; outer perianth-lobes with a narrow involute lateral lobule on each side. .... 1. *B. championii*
1B. Perianth-tube winged; inflorescence umbellate or cymose, sometimes 1-flowered.
  2A. Inflorescence a few-flowered umbel; flowers 7–10 mm. long.
    3A. Flowers white; wings relatively narrow above; the inner perianth-lobes absent; connective without a spur at base.
                                        2. *B. cryptopetala*
    3B. Flowers blue-purple; wings relatively broad, subtruncate at apex; the inner perianth-lobes 3; connective with a pendulous spur at base. .... 3. *B. itoana*
  2B. Inflorescence a rather loose cyme; flowers 4–5 mm. long; connective with a pendulous spur at base. .... 4. *B. liukiuensis*

1. Burmannia championii Thw. *B. japonica* Maxim. ex Makino——Hina-no-shakujō. Slender white saprophytic herb, 5–15 cm. high from subglobose rhizomes; stems simple; scalelike leaves lanceolate, acute, appressed, 1.5–4 mm. long, the bracts about 3 mm. long, resembling the scalelike leaves; flowers 2–10, nearly sessile, in a dense headlike inflorescence, white, 3-ribbed, 6–10 mm. long, the outer perianth-lobes erect, about 1.5 mm. long, deltoid, acute, the inner lobes erect, spathulate, rounded at apex; capsule obovate-orbicular, about 2.5 mm. long.——Aug.–Oct. Woods in mountains; Honshu (centr. distr. and westw.), Shikoku, Kyushu.——s. China, Malaysia to Ceylon.

2. Burmannia cryptopetala Makino. Shiro-shakujō. Delicate white saprophytic herb, 5–15 cm. high; stems usually simple; scalelike leaves lanceolate or narrowly ovate, 3–4 mm. long, acute; inflorescence simple or a once-branched umbel, 1- to 12-flowered, dense, the bracts simulating the scalelike leaves; flowers erect, 8–10 mm. long, white, with a yellow hue above, short-pedicelled, winged, the outer perianth-lobes ovate, acute, 1.5–2 mm. long, the inner perianth-lobes absent.——Aug.–Oct. Honshu (Kinki Distr.), Kyushu.——China (Hainan Isl.).

3. Burmannia itoana Makino. Ruri-shakujō. Purplish delicate herb 5–12 cm. high; stems simple or with few branches from the base; scalelike leaves and bracts few, small, lanceolate to ovate, 2–3 mm. long, membranous; flowers 1–2, erect, short-pedicelled, purplish, 8–10 mm. long, 3-winged, the outer perianth-lobes erect, slightly incurved, about 1.5 mm. long, the inner ones orbicular, minute.——Aug.–Sept. Kyushu (Yakushima).——Ryukyus and China (Hainan Isl.).

4. Burmannia liukiuensis Hayata. *B. coelestis* sensu auct. Japon., non Don; *B. nepalensis* sensu auct. Japon., non Hook. f.——Kirishima-shakujō. Very delicate annual, 5–15 cm. high; stems simple, filiform; scalelike leaves lanceolate, appressed, 1–3 mm. long; flowers (1)2–8, whitish, in a loose terminal cyme, the bracts simulating the scalelike leaves, the pedicels slender; perianth 4–5 mm. long, the outer lobes broadly deltoid, obtuse, about 0.5 mm. long, the inner ones minute,

orbicular; capsules subglobose, about 2 mm. across.——Sept.–Oct. Woods in the foothills; s. Kyushu (Tanegashima).——Ryukyus.

2. GLAZIOCHARIS Taub. ex Warming   TANUKI-NO-SHOKUDAI ZOKU

Delicate saprophytic herbs, from slender elongate branched rhizomes; stems simple, with small scalelike leaves; flowers relatively large, solitary and terminal on the scape, the perianth-tube urceolate, the lobes 6, the outer whorl free, the inner segments free or connate at apex, with a filiform or linear, sometimes apically thickened appendage at apex, the tube with a ringlike thickened margin; stamens 6, inserted on the upper margin of the tube, the filaments broad; ovary inferior, with 3 parietal placentae.——Two species, one in Japan and the other in S. America.

1. Glaziocharis abei Akasawa. TANUKI-NO-SHOKUDAI. Delicate glabrous low saprophytic perennial herb; rhizomes filiform, elongate, sparingly branched; scapes solitary, about 1 cm. long, few-scaled; flowers solitary, about 15 mm. long, 6–7 mm. across, the perianth-tube ovoid-campanulate, about 10 mm. long, the upper portion deciduous after anthesis on the transverse split above the base, the outer perianth-lobes

about 3 mm. long, broadened at base, linear and spreading above, the inner lobes obovate-cuneate, erect, incurved above, 5–6 mm. long, each bearing an erect linear rather short appendage on back, the upper portion overlapping but free from each other; stigma glabrous.——July–Aug. Shikoku (Awa Prov.), very rare.

Fam. 58. **ORCHIDACEAE**   RAN KA   Orchid Family

Perennial herbs of various habits, rarely saprophytic and leafless, sometimes with thick corms or rhizomes; aerial stems more or less elongate, rarely scandent; leaves simple, entire, flat, plicate, or rarely equitant, with parallel nerves, usually sheathed at the base; flowers zygomorphic, perfect, rarely imperfect, bracteate, minute to large and showy, solitary or in spikes or racemes; perianth superior, with 6 segments in 2 whorls, the 3 outer (sepals) nearly equal, usually free, the 2 inner lateral ones (petals) usually similar to the sepals, or smaller, the median one (lip or labellum) adaxial, usually larger, variously shaped, often spurred or saccate at the base; stamens 1 or 2, adnate to the style and forming a column (gynostemium); anthers 2-locular, the pollen commonly collected in 2–8 waxy or granular masses (pollinia) attached to a gland; stigma viscid or rough, placed under the rostellum or in a cavity between the anther-locules; ovary inferior, usually twisted (so that the posterior flower-parts are in front), commonly unilocular; placentae 3, parietal, with very numerous ovules; fruit a capsule with 3 or 6 valves, rarely berrylike; seeds minute, without endosperm; embryo undifferentiated.——A large cosmopolitan family comprising about 600 genera, with more than 20,000 species, especially abundant in the Tropics.

1A. Fertile stamens 2.
  2A. Flowers nearly actinomorphic; sepals, petals and lip similar in shape. .... 1. *Apostasia*
  2B. Flowers distinctly zygomorphic; lip strongly different from the sepals and petals, large, saccate. .... 2. *Cypripedium*
1B. Fertile stamen 1.
  3A. Anthers attached to the column by a broad base, persistent; pollinia with caudicles attached to a viscid disc or glands at the base.
    4A. Stigma 1, entire, within the cavity under the anther-locules, or 2 and projecting from the cavity.
      5A. Stigma 1, undivided.
        6A. Glands contained in a bursicule of the rostellum; flowers relatively large, showy, rose-purple or white. .... 3. *Orchis*
        6B. Glands naked, not in a bursicule; flowers mostly greenish.
          7A. Stigma not elevated, not thickened; lip commonly simple. .... 4. *Platanthera*
          7B. Stigma elevated, thickened.
            8A. Lip 3-lobed near the apex, the terminal lobe minute, simulating the lateral ones. .... 5. *Coeloglossum*
            8B. Lip subsimple or 3-lobed nearly to base, the terminal lobe elongate, the lateral lobes smaller. .... 6. *Tulotis*
      5B. Stigmas 2, more or less prominent and projecting from the cavity.
        9A. Lip without a spur. .... 7. *Herminium*
        9B. Lip spurred.
          10A. Rostellum not armed, beaked or angled.
            11A. Stems 1- to 3-leaved, mostly 1- or 2- leaved; plants mostly 5–20 (–25) cm. high. .... 8. *Amitostigma*
            11B. Stems (2-) 3- to 7-leaved; plants mostly (10-) 20–60 cm. high. .... 9. *Gymnadenia*
          10B. Rostellum with 2 arms, not beaked or angled. .... 10. *Habenaria*
    4B. Stigmas 2, divided into 2 horns, not in cavity under the anther-locules; flowers small; lip entire, without a spur. .. 11. *Androcorys*
  3B. Anther easily detached from the column or early withering; pollinia with a caudicle and gland at the apex.
    12A. Pollinia granular, soft; anther commonly persistent; inflorescence terminal.
      13A. Anthers reclined on the column, incumbent.
        14A. Column very short, with a tooth or wing on each side at the apex. .... 12. *Microtis*
        14B. Column distinct, somewhat elongate, without wings.
          15A. Plant with short slender rhizomes; leaves if present cauline.
            16A. Lip narrow and flat, more or less retuse or bifid.
              17A. Plant with 2 opposite green leaves. .... 13. *Listera*
              17B. Plant saprophytic, without green leaves. .... 14. *Neottia*

16B.   Lip concave at base, sometimes loosely enveloping the column; leaves alternate.
  18A.   Lip divided by a constriction into an upper and lower portion; leaves if present more or less plicate.
    19A.   Plant saprophytic, without green leaves. .... 15. *Aphyllorchis*
    19B.   Plant not saprophytic, with green parts.
      20A.   Lip with a chin at base, distinctly jointed between the upper and lower portions. .... 16. *Cephalanthera*
      20B.   Lip without a distinct chin at base, not jointed between the upper and lower portions. .... 17. *Epipactis*
  18B.   Lip not divided into an upper and lower portion; leaves if present flat, not plicate.
    21A.   Plant with a single green cauline leaf; flowers solitary, with a green bract at base. .... 18. *Pogonia*
    21B.   Plant saprophytic, without green parts; flowers more than 1.
      22A.   Plant with branching stem; perianth without a calyculus at base. .... 19. *Galeola*
      22B.   Plant small with unbranched stem; perianth with 3 small calyculae at base. .... 20. *Lecanorchis*
15B.   Plant with corms or fleshy to coralloid rhizomes.
  23A.   Leaves jointed at base. .... 21. *Bletilla*
  23B.   Leaves not jointed, soft, or reduced to scales in saprophytic plants.
    24A.   Perianth-segments not connate.
      25A.   Leaf solitary, radical, petiolate, broad; flowers before the leaves (solitary in our species). .... 22. *Nervilia*
      25B.   Leaves cauline, reduced to scales or sheaths.
        26A.   Lip spurred; column without an appendage. .... 23. *Epipogium*
        26B.   Lip spurless; column with a tooth at base. .... 25. *Stigmatodactylus*
    24B.   Perianth-segments connate; plant saprophytic, without green leaves. .... 26. *Gastrodia*
13B.   Anthers more or less erect; rostellum erect or nearly so.
  27A.   Leaves flat, not plicate, thinnish to more or less fleshy.
    28A.   Stems erect, the roots fasciculate at base. .... 27. *Spiranthes*
    28B.   Stems creeping or decumbent at base and rooting at intervals.
      29A.   Flowers with a definite oblong stipe between the pollinia and the viscid disc of the rostellum on the column.
        30A.   Sepals free; column without distinct prolongations. .... 34. *Zeuxine*
        30B.   Sepals connate to the middle; column with 2 upright narrow arms. .... 30. *Cheirostylis*
      29B.   Flowers without a definite narrow stipe between the pollinia and the viscid disc of the rostellum, on the column.
        31A.   Sepals more or less connate, often up to the middle.
          32A.   Lip ventricose at base, spurless. .... 33. *Myrmechis*
          32B.   Lip with a bilobulate spur. .... 31. *Vexillabium*
        31B.   Sepals free or nearly so.
          33A.   Lip with a dentate or fimbriate claw. .... 32. *Odontochilus*
          33B.   Lip without a definite claw.
            34A.   Stigma simple; base of lip with hairs. .... 28. *Goodyera*
            34B.   Stigma bilobed; base of lip variously callose. .... 29. *Hetaeria*
  27B.   Leaves plicate, thin, many-nerved; stems slender, but firm. .... 35. *Tropidia*
12B.   Pollinia waxy or cartilaginous; anther easily detached; inflorescence terminal or lateral.
  35A.   Inflorescence terminal.
    36A.   Lip slipper-shaped, with bifid or obtuse spur toward the apex.
      37A.   Leaf solitary, radical; flowers solitary and terminal. .... 36. *Calypso*
      37B.   Leaves reduced to scales; flowers racemose, fleshy; saprophytic herbs without green parts. .... 24. *Yoania*
    36B.   Lip not slipper-shaped.
      38A.   Leaves distichous, equitant, evergreen; flowers minute, numerous, in spikes; lip not spurred. .... 37. *Oberonia*
      38B.   Leaves dorsi-ventrally flattened.
        39A.   Lip spurred. .... 38. *Tipularia*
        39B.   Lip not spurred.
          40A.   Lip rather similar to the sepals and petals. .... 39. *Didiciea*
          40B.   Lip different from sepals and petals.
            41A.   Rhizomes slender, elongate, not thickened. .... 40. *Ephippianthus*
            41B.   Rhizomes short, thickened.
              42A.   Flowers small; column very short.
                43A.   Anthers deciduous. .... 41. *Microstylis*
                43B.   Anthers not deciduous, persistent, withering in place. .... 42. *Malaxis*
              42B.   Flowers small to large; column elongate.
                44A.   Flowers 1 (rarely paired); leaves solitary. .... 43. *Eleorchis*
                44B.   Flowers in racemes, smaller; leaves more than 1. .... 44. *Liparis*
  35B.   Inflorescence lateral.
    45A.   Leaves convolute in bud.
      46A.   Pollinia 4 or 8, with caudicle but without stipe; leaves jointed or not jointed at base.
        47A.   Leaves not jointed at base.
          48A.   Lip nearly free from the column. .... 45. *Phajus*
          48B.   Lip adnate to the column at base. .... 46. *Calanthe*
        47B.   Leaves jointed at base. .... 47. *Tainia*
      46B.   Pollinia 2 or 4, without a caudicle; leaves jointed at base.
        49A.   Lip saccate or spurred at base.
          50A.   Lip broad; perianth parts spreading. .... 48. *Eulophia*
          50B.   Lip narrow, linear; perianth parts subparallel. .... 49. *Cremastra*
        49B.   Lip not spurred at base.

51A. Flowers in racemes; stems thickened at base; leaves 1–3 cm. wide. . . . . 50. *Oreorchis*
51B. Flower solitary; stems not thickened at base; leaves 3–4 cm. wide. . . . . 51. *Dactylostalix*
45B. Leaves conduplicate in bud.
  52A. Stems sympodial.
    53A. Column produced into a distinct foot; pollinia unappendaged or with a caudicle.
      54A. Stem or pseudobulb with several joints.
        55A. Pollinia 4, without caudicles; stems 5–40 cm. long. . . . . 52. *Dendrobium*
        55B. Pollinia 8, with caudicles; stems usually shorter. . . . . 53. *Eria*
      54B. Stem with a single joint or pseudobulb; plants commonly with a long-creeping rhizome. . . . . 54. *Bulbophyllum*
    53B. Column without a foot; pollinia 2, with a caudicle; stems abbreviated, concealed by a tuft of leaves; leaves long linear.
      55. *Cymbidium*
  52B. Stems monopodial; inflorescence lateral.
    56A. Lip not spurred; leaves terete. . . . . 56. *Luisia*
    56B. Lip spurred.
      57A. Plant without leaves (in our species); stems very abbreviated. . . . . 60. *Taeniophyllum*
      57B. Plant with green leaves.
        58A. Column with a definite foot.
          59A. Spur straight. . . . . 62. *Sarcochilus*
          59B. Spur incurved under the lip. . . . . 61. *Aërides*
        58B. Column without a foot.
          60A. Spur divided within by a longitudinal septum. . . . . 57. *Sarcanthus*
          60B. Spur not divided by a longitudinal plate.
            61A. Lip with a saccate (mostly short) spur. . . . . 58. *Saccolabium*
            61B. Lip with a very slender, elongate, curved spur. . . . . 59. *Neofinetia*

1. APOSTASIA Bl.   Yakushima-ran Zoku

Terrestrial herbs with short rhizomes; stems erect, loosely leaved; leaves rather narrow, nerved; inflorescence a spike, terminal and axillary from upper nodes, sometimes branched; flowers small, nearly actinomorphic, subtended by a narrow bract; perianth-segments nearly equal, free; column short; anthers 2, stalked, erect, on both sides of rostellum, the locules parallel, contiguous; pollinia viscid; rostellum apical, erect; ovary 3-locular; capsule linear.——Few species, in the Tropics of India, China, Malaysia, and Australia.

1. Apostasia nipponica Masam. Yakushima-ran. Glabrous herb about 6 cm. high; leaves few, cauline, lanceolate, long-acuminate, about 2 cm. long, 5 mm. wide; spikes 2–3; flowers 2–3, the bracts ovate, about 4 mm. long, 1.5 mm. wide, acuminate to acute; perianth-segments lanceolate, about 3.5 mm. long, 1 mm. wide, acuminate.——July. Kyushu (Yakushima); very rare.

2. CYPRIPEDIUM L.   Atsumori-sō Zoku

Terrestrial herbs with short, creeping rhizomes and rather stout fibrous roots; stems (in ours) leafy, 1- to few-flowered; leaves few, sometimes opposite, many-nerved, sometimes plicate; flowers usually large, showy, zygomorphic; sepals spreading, free or the 2 lateral united; petals spreading, narrow; lip large, saccate or pouch-shaped; column incurved, with a short-stalked, 2-loculed anther on each side, and a dilated, rather petallike staminode adnate to the back at tip; stigma terminal, obsoletely 3-lobed.——About 40 species, in the temperate regions of the N. Hemisphere, one in the American Tropics.

1A. Leaves 2, ovate-rounded to rounded-cordate, subopposite, not sheathed, flat, not plicate, 2- to 5(-7)-nerved; flowers 2–2.5 cm. across; bracts broadly linear; stems and ovaries glabrous. . . . . 1. *C. debile*
1B. Leaves oblong-elliptic to flabellate, or orbicular, 2 to several, alternate or opposite, many-nerved, plicate; flowers larger; bracts broader; stems and ovaries usually pubescent.
  2A. Leaves 2, nearly opposite, not at all or slightly sheathed at base; flowers yellowish brown.
    3A. Leaves rounded-flabellate, with parallel nerves not converging at tip. . . . . 2. *C. japonicum*
    3B. Leaves oblong-elliptic, nerves converging at tip. . . . . 3. *C. yatabeanum*
  2B. Leaves 3–5, oblong-elliptic, alternate; flowers rose to rose-purple, rarely white. . . . . 4. *C. macranthum*

1. Cypripedium debile Reichenb. f. *C. cardiophyllum* Fr. & Sav.——Ko-atsumori-sō.   Small, glabrous perennial with slender, short-creeping rhizomes and elongate roots covered with prominent root-hairs; stems 10–20 cm. long, glabrous, with 2 or 3, thinly membranous scaly sheaths at base; leaves 2, opposite, sessile, not sheathed, flat, rounded-cordate, 3–5 cm. long and as wide, glabrous except on the margins, 2- to 5(-7)-nerved, acute to short-acute, rounded to slightly cordate at base; peduncle solitary, terminal, strongly declined, glabrous, 2–3 cm. long; bracts broadly linear, 1.5–2 cm. long; flowers 2–2.5 cm. across, yellowish green with a purple hue, pendulous, glabrous; perianth-segments acute with an obtuse tip, about 1.5 cm. long; dorsal sepal narrowly ovate, the lateral ones wholly united to the tip; petals broadly lanceolate; lip slightly shorter than the petals.——May–June. Woods in mountains; Hokkaido (s. distr.), Honshu, Shikoku.

2. Cypripedium japonicum Thunb. Kumagai-sō. Rhizomes long-creeping, with rather stout roots from the nodes; stems 20–40 cm. long, with spreading, crisped hairs and few membranous sheathing scales at base; leaves 2, nearly opposite, near apex, many-nerved, plicate, pilose on the nerves, flabellate-suborbicular, 10–15 cm. long, 12–20 cm. wide, truncately rounded and abruptly acute to cuspidate at apex, cuneate to rather truncate at base, undulately toothed; peduncles solitary, 1-flowered, 10–15 cm. long, pubescent; bract oblong, acuminate, 3–6 cm. long, many-nerved; flowers yellowish brown; ovary densely pubescent; lateral sepals narrowly oblong, about 5 cm. long, wholly connate; petals nearly as long as the sepals, with long crisped hairs inside at base; lip slightly longer than the perianth, hairy inside near base.——Apr.–May. Woods and bamboo-forests in lowlands; Hokkaido (Oshima Prov.), Honshu, Shikoku, Kyushu; rather common.——China.

3. Cypripedium yatabeanum Makino. *C. guttatum* var. *yatabeanum* (Makino) Pfitz.——Kibana-no-atsumori-sō. Loosely pilose perennial with long, creeping rhizomes; stems 10–25 cm. long, with few sheathing scales at base; leaves 2, approximate, nearly on the summit of the stem, oblong to elliptic, 7–15 cm. long, 5–8 cm. wide, obtuse with an acute apex, obtuse and slightly clasping at base, many-nerved, these converging at apex, plicate, slightly pilose on both sides and on margins; peduncles solitary and terminal, 10–20 cm. long, densely pubescent; bract large, broadly lanceolate, leaflike, acuminate, 2–3 cm. long; flowers yellowish brown; dorsal sepal erect, ovate, 2–2.5 cm. long, acute, short-pubescent outside toward the base, the lateral sepals connate into one lamina,

slightly bifid at the tip, 13–18 mm. long; petals obliquely ovate, attenuate, ending in a rounded apex; lip obliquely descending, longer than the rest of the perianth; ovary villous.——July. Woods in high mountains; Hokkaido, Honshu (centr. distr.); very rare.——Kuriles and Kamchatka.

4. Cypripedium macranthum Sw. *C. thunbergii* Bl.; *C. speciosum* Rolfe; *C. atsmori* Morr.; *C. calceolus* var. *atsmori* (Morr.) Pfitz.; *C. macranthum* var. *speciosum* (Rolfe) Koidz.——Atsumori-sō. Rhizomes short-creeping, with rather thick firm roots from the nodes; stems 25–40 cm. long, 3- to 5-leaved, white-hirsute; leaves alternate, oblong–elliptic, 10–20 cm. long, 5–8 cm. wide, abruptly acuminate, shortly sheathed at base, many-nerved, plicate; flowers rose to rose-purple, rarely white, 4–6 cm. across; dorsal sepal ovate, acuminate, 4–5 cm. long, the lateral sepals wholly connate; petals narrowly ovate, acuminate, crisped-pubescent inside at base.——May–July. Grassy mountain slopes; Hokkaido, Honshu; rather rare.——Korea, Manchuria, n. China, Kamchatka, Siberia, and e. Europe.——Forma **albiflorum** (Makino) Ohwi. *C. speciosum* var. *albiflorum* Makino; *C. macranthum* var. *leucanthum* Hatus.——Shirobana-atsumori-sō. Flowers white; rare.——Forma **rebunense** (Kudo) Ohwi. *C. rebunense* Kudo; *C. macranthum* var. *rebunense* (Kudo) Miyabe & Kudo——Rebun-atsumori-sō. Flowers pale yellow; rare.

3. ORCHIS L. Hakusan-chidori Zoku

Terrestrial, rarely epiphytic herbs, rarely growing on rocks or tree-trunks, with or without tuberous roots; stems few-leaved; leaves linear to elliptic-oblong, sometimes nearly radical; flowers rather small, showy, in a spike, the bracts small or conspicuous; sepals free, spreading, the dorsal often forming a hood with the petals; lip slightly adnate to the column, usually 3-lobed, spurred at base; column short with parallel anther-locules; pollinia with a caudicle attached to glands, these in a bursicule or pouch.——About 100 species, in temperate and cooler regions of the world.

1A. Roots thickened, tuberous or digitately divided; flowers usually rose-purple; lip distinctly 3-lobed, with an elongate, cylindrical spur at base.
  2A. Roots digitately divided; leaves 3–6; sepals and petals long-acuminate. . . . . 1. *O. aristata*
  2B. Roots tuberous, undivided; leaves 1–3 (–4); sepals and petals acute to obtuse.
    3A. Sepals and petals 1-nerved; leaves 2 or 3, linear or broadly so. . . . . 2. *O. graminifolia*
    3B. Sepals and petals 3-nerved; leaves lanceolate to narrowly oblong.
      4A. Leaves 2 or 3, lanceolate. . . . . 3. *O. jooiokiana*
      4B. Leaves 1, oblong to narrowly so.
        5A. Leaves near the middle of stem; peduncles ascending. . . . . 4. *O. chidori*
        5B. Leaves nearly radical; peduncles erect. . . . . 5. *O. curtipes*
1B. Roots linear, slender, slightly if at all thickened; flowers white to pale rose; lip cuneate or obsoletely 3-lobed, with a short or very short spur at base.
  6A. Leaves 1; flowers usually 2, pale rose; perianth-segments acuminate; spur linear, 7–10 mm. long. . . . . 6. *O. cyclochila*
  6B. Leaves 2; flowers 2–6, white; perianth-segments obtuse; spur ellipsoidal, 3–4 mm. long; lip cuneate. . . . . 7. *O. fauriei*

1. Orchis aristata Fisch. *O. latifolia* var. *beeringiana* Cham.; *O. beeringiana* (Cham.) Kudo——Hakusan-chidori. Roots somewhat thickened, digitately 2- to 3-lobed; stems 10–40 cm. long, erect; leaves 3–6, oblanceolate to lanceolate, blades of the lower leaves broadly linear, 7–15 cm. long, 7–30 mm. wide, obtuse, the upper blades acute, often with dark-purple spots; spike or raceme few-flowered, 4–10 cm. long, the bracts linear-lanceolate, long-acuminate, as long as the flower; flowers rather densely arranged, rose-purple; sepals lanceolate, long-acuminate, 8–13 mm. long, 3-nerved, the dorsal sepal erect, the lateral sepals spreading upward; petals narrowly ovate, obtuse, 2- to 3-nerved, slightly shorter than the sepals; lip obovate-triangular, spreading, slightly longer than the sepals, papillose inside, 3-lobed at tip, the lateral lobes depressed-deltoid, rounded at apex, shorter than the median; spur linear-cylindric, 1–1.5 cm. long, straight.——June–Aug. Grassy alpine

slopes; Hokkaido, Honshu (centr. distr. and northw.).——Kuriles, Sakhalin, Korea, Aleutians, and Alaska.——Forma **albiflora** (Koidz.) Tatew. *Orchis aristata* var. *albiflora* Koidz.——Shirobana-hakusan-chidori. Flowers white.

2. Orchis graminifolia (Reichenb. f.) Tang & Wang. *Ponerorchis graminifolia* Reichenb. f.; *Gymnadenia graminifolia* (Reichenb. f.). Reichenb. f.; *G. rupestris* Miq; *O. rupestris* (Miq.) Schltr.——Uchō-ran. Terrestrial herb with ovoid tuberous roots; stems slender, 8–15 cm. long, suberect; leaves 2 or 3, rarely 4, linear to broadly so, 7–12 cm. long, 3–8 mm. wide, acute to acuminate, gradually arcuate toward tip, nearly flat; racemes one-sided, few-flowered; bracts narrowly lanceolate, 7–12 mm. long; flowers rose-purple; sepals oblong, obtuse, 1-nerved, about 6 mm. long, the dorsal sepal navicular, the lateral ones oblique; petals obliquely ovate, obtuse, erect, as long as the sepals, 1- rarely 2-nerved, forming a hood with the median sepal; lip ascending, longer than the sepals, rather deeply 3-lobed, the lobes ovate, the terminal lobe slightly larger, entire or retuse; spur 1–1.5 cm. long, straight or nearly so.——June–Aug. Shaded rocky cliffs in mountains; Honshu (Kantō Distr. and westw.), Shikoku, Kyushu; rather rare.—— Korea.

3. Orchis jooiokiana Makino. *Ponerorchis pauciflora* var. *jooiokiana* (Makino) Ohwi; *O. matsumurana* Schltr.—— NYOHŌ-CHIDORI. Terrestrial herb with obovoid tuberous roots; stems 12–30 cm. long; leaves usually 2, sometimes 1 or 3, lanceolate, rather acute, 4–8 cm. long, 1–1.5 cm. wide, blades of upper leaves broadly linear, acuminate; flowers 3–8, in slightly one-sided spikes, rose-purple; bracts lanceolate, 1–2 cm. long; sepals 3-nerved, broadly lanceolate, rather acute, 8–10 mm. long, the lateral sepals oblique; petals erect, narrowly ovate, oblique, obtuse, indistinctly 2-nerved, shorter than the sepals; lip spreading, broadly 4-angled, cuneate, longer than the sepals, 3-lobed above, 13–15 mm. long and as broad, the lateral lobes ovate, the terminal lobe larger, retuse; spur straight, 15–17 mm. long.——July and Aug. High mountains; Honshu (Kantō and centr. distr.); rare.

4. Orchis chidori (Makino) Schltr. *Gymnadenia chidori* Makino; *O. rupestris* var. *chidori* (Makino) Soó; *Ponerorchis chidori* (Makino) Ohwi—HINA-CHIDORI. Roots tuberous; stems slender, 7–12 cm. long, 1-leaved near the middle; leaf ascending, broadly lanceolate or narrowly oblong, 7–12 cm. long, 12–25 mm. wide, short-acuminate, flat, spreading; spikes one-sided; flowers 1–6, rose-purple, the bracts narrowly lanceolate, obliquely ascending, acuminate, 1–1.5 cm. long; sepals 3-nerved, 5–6 mm. long, subobtuse, the dorsal sepal oblong, slightly navicular, erect, forming a hood with the petals, the lateral sepals subovate; petals slightly shorter than the sepals, obtuse; lip ascending, 8–10 mm. long, rather deeply 3-lobed, the lobes ovate, 4-angled; spur 13–17 mm. long, straight or slightly curved.——July–Aug. Shaded rocks in mountains; Honshu, Shikoku; very rare.

5. Orchis curtipes Ohwi. CHABO-CHIDORI. Small epiphyte on tree-trunks; stems about 7 cm. long, erect; leaves 1, near the base, narrowly ovate-oblong, spreading, about 6 cm. long, 2 cm. wide, flat, with a short sheath at base; spikes one-sided, few-flowered, the lowermost bract at base of spike leaf-like, to 4 cm. long, rounded at base, the others narrowly lanceolate, nearly erect, gradually acuminate, 1–2 cm. long, 2–3 mm. wide; flowers rose-purple, the dorsal sepal erect, nar-

rowly ovate, slightly navicular, rather obtuse, weakly 3-nerved, 5–6 mm. long, forming a hood with the petals, the lateral sepals spreading, broadly falcate-lanceolate, 3-nerved, 6–7 mm. long, rather obtuse; petals obliquely ovate, 3-nerved, subobtuse, as long as the dorsal sepal; lip about 1 cm. long, flat, spreading, 3-lobed, the terminal lobe broadly lanceolate, entire, obtuse, about 6 mm. long, 2.5 mm. wide, the lateral lobes shorter than the terminal, about 4.5 mm. long, oblong-ovate, obtuse, entire; spur straight, gradually narrowed to tip, obtuse, about 15 mm. long.——Aug. Hokkaido (Kushiro); very rare.

6. Orchis cyclochila (Fr. & Sav.) Maxim. *Habenaria cyclochila* Fr. & Sav.; *Gymnadenia cyclochila* (Fr. & Sav.) Korsh.; *Galeorchis cyclochila* (Fr. & Sav.) Nevski——KAMOME-RAN. Small terrestrial herb with slightly thickened roots; stems 7–15 cm. long, 1-leaved at base; leaves radical, petiolate, oblong to broadly ovate, or broadly elliptic, 4–6 cm. long, 2–5 cm. wide, obtuse to rounded at apex; flowers usually 2, pale rose with small deeper colored dots on lip, rarely white; bracts narrowly oblong to narrowly ovate, subacute, 1–2 cm. long; sepals broadly lanceolate, 8–10 mm. long, acuminate, 3-nerved, gradually narrowed to a somewhat obtuse tip, the dorsal sepal erect, the lateral sepals ascending-spreading; petals lanceolate, rather obtuse, slightly shorter than the sepals; lip broadly ovate, mucronate, about 1 cm. long, slightly contracted above the middle and obsoletely 3-lobed; spur linear, slightly curved, slender, 7–10 mm. long, subacute, as long as or slightly shorter than the ovary.——May–June. Mossy places in woods in mountains; Hokkaido (Kushiro), Honshu (centr. and n. distr.), Shikoku; rare.——Korea and Ussuri.

7. Orchis fauriei Finet. *Chondradenia yatabei* Maxim. ex Makino; *O. chondradenia* Makino; *C. fauriei* (Finet) Sawada——ONOE-RAN. Terrestrial herb with slightly thickened roots; stems 8–15 cm. long, erect, 2-leaved; leaves nearly radical, oblong, elliptic, or broadly oblanceolate, 6–10 cm. long, 1.5–4 cm. wide, rounded to acute, shortly sheathed at base; flowers 2–6, white; bracts broadly lanceolate, 1–2 cm. long; sepals somewhat spreading, oblong, obtuse, 5-nerved, 8–10 mm. long; petals narrowly oblong, very obtuse, with obsolete minute teeth on upper margins, as long as the sepals but slightly thicker; lip ascending, cuneate, as long as but thicker than the sepals, slightly broadened and spreading at tip, about 3.5 mm. wide near the retuse apex; spur ellipsoidal, 3–4 mm. long.——July–Aug. Grassy places in mountains; Honshu (Kantō and n. distr.); rare.

4. PLATANTHERA L. C. Rich. TSURE-SAGI-SŌ ZOKU

Terrestrial, with elongate, slightly thickened roots; leaves usually cauline, 1 to several, linear to elliptic; racemes few- to many-flowered, bracteate; flowers rather small, greenish, rarely white; sepals subequal, ovate, the dorsal one erect, sometimes forming a hood with the petals, the lateral sepals spreading; petals similar to the sepals, the lip usually simple, linear, sometimes ovate or ligulate; anther-locules parallel or divergent, usually widely separated; rostellum not extended, low and broad; stigma depressed.——About 100 species, in temperate and cooler regions of the N. Hemisphere.

1A. Sepals about 2 mm. long; lip ovate-orbicular, with a very short spur; leaves 2, approximate, broadly elliptic or nearly orbicular.
　　　　　　　　　　　　　　　　　　　　　　　　　　　　　　　　　　　　　　　1. P. chorisiana
1B. Sepals 2–12 mm. long; lip linear-lanceolate to broadly lanceolate, sometimes ligulate, with an elongate or short spur.
　2A. Lower 1 or 2 leaves largest, the upper ones abruptly reduced and scalelike; dorsal sepal flat, not navicular.
　　3A. Spur 2–3 mm. long; sepals white, about 4 mm. long; flowers greenish white, the lip ligulate. . . . . *2. P. brevicalcarata*
　　3B. Spur 5–30 mm. long, if shorter then the flowers greenish with rather thick petals.
　　　4A. Flowers white, rather large; leaves nearly radical, usually paired, narrowly oblong.
　　　　5A. Anther-locules approximate, divergent at base; dorsal sepal ovate, 5.5–6 mm. long; spur gradually narrowed toward the obtuse apex; bracts lanceolate. . . . . *3. P. metabifolia*

1. Platanthera chorisiana (Cham.) Reichenb. f. *Habenaria chorisiana* Cham.; *Pseudodiphyllum chorisianum* (Cham.) Nevski; *Platanthera matsudae* Makino; *Platanthera ditmariana* Komar.——Takane-tombo. Small green herb with slender, slightly thickened roots; stems 8–15 cm. long (to 40 cm. long in var. **elata** Finet); leaves 2, approximate, on the lower part of stems, elliptic to broadly so, or nearly orbicular, 2–6 cm. long, 2–5 cm. wide, rounded to mucronate at apex, abruptly narrowed at base; scalelike leaves few, linear-lanceolate; racemes 2–10 cm. long, many to rather few-flowered; bracts as long as to slightly longer than the flowers, rarely shorter, linear, broadened at base; flowers minute, green; sepals nearly equal, oblong to elliptic, rounded at apex, 1-nerved, about 2 mm. long, the lateral sepals spreading, slightly oblique; petals ovate to broadly so, slightly shorter than the sepals; lip ovate-orbicular, obtuse, as long as the sepals, somewhat thickened; spur 1–1.3 mm. long, slightly incurved and thickened above, obtuse.——July–Sept. High mountains; Hokkaido, Honshu (centr. and n. distr.); rare.——Along Bering Sea, Kamchatka, Kuriles, and Sakhalin.

2. Platanthera brevicalcarata Hayata. *P. yakumontana* Masam.——Niitaka-chidori, Tsukushi-chidori. Small herb with somewhat thickened, elongate roots; stems 10–15 cm. long; leaves 1–2, nearly radical, oblong to elliptic, 2.5–4 cm. long, 1.5–2 cm. wide, rounded at apex, abruptly narrowed at base; scalelike leaves 1–3, rather large, broadly lanceolate, acute, ascending; racemes 2–5 cm. long, 5- to 20-flowered; bracts lanceolate, acute, 4–7 mm. long; flowers greenish white; sepals white, nearly equal, 1-nerved, narrowly oblong, the dorsal sepal about 3 mm. long, the lateral sepals about 4 mm. long, obtuse, spreading; petals rather thick, broadly semi-ovate, about 3 mm. long, subobtuse, erect, forming a hood with the dorsal sepal; lip oblong-ligulate, somewhat carnose, obtuse, 3.5–4 mm. long; spur 2–3 mm. long, slightly curved, thickened at tip, obtuse.——July. High mountains; Kyushu (Mount Kirishima and high mountains of Yakushima); rare.——Formosa.

3. Platanthera metabifolia F. Maekawa. *P. chlorantha* sensu auct. Japon., non Cust.; *P. bifolia* sensu auct. Japon., non L. C. Rich.——Ezo-chidori, Futaba-tsure-sagi. Roots somewhat fusiform-thickened; stems 20–50 cm. long; leaves 2, radical, oblong to elliptic, 8–15 cm. long, 3–5 cm. wide, obtuse, narrowed to base; racemes 8–16 cm. long, rather densely flowered, the bracts lanceolate; flowers white; dorsal sepal ovate to ovate-elliptic, 5.5–6 mm. long, the lateral sepals about 8 mm. long, broadly lanceolate, obtuse, spreading; petals broadly lanceolate, erect, rather thick, slightly shorter than the sepals, forming a hood with the dorsal sepal; lip broadly linear, 1–1.3 cm. long, subcarnose, obtuse; spur 2–2.7 cm. long, gradually narrowed toward the obtuse apex.——July–Aug. Hokkaido.——Kuriles and Sakhalin.

4. Platanthera okuboi Makino. Hachijō-tsure-sagi. Roots fusiform; stems 20–45 cm. long; leaves 1–2, nearly radical, oblong to narrowly so, 10–20 cm. long, 3–6 cm. wide, obtuse, narrowed at base; scalelike leaves rather large, erect, broadly lanceolate, 2–6 cm. long; racemes rather densely many-flowered, 5–10 cm. long; bracts broadly lanceolate, 15–25 mm. long; flowers white to slightly green; dorsal sepal broadly ovate, obtuse, 5-nerved, 6–8 mm. long, the lateral sepals broadly lanceolate, 8–10 mm. long, obtuse, 3- to 5-nerved; petals erect, shorter than the dorsal sepal, slightly thickened, oblique, narrowly ovate, 3-nerved, about 6 mm. long; lip broadly linear, 10–13 mm. long, obtuse; spur 2.5–3 cm. long.——May–June. Honshu (Izu Isls.).

5. Platanthera florenti Fr. & Sav. *P. listeroides* Takeda ——Jimbai-sō. Roots elongate; stems 20–40 cm. long, rather slender; leaves 2, radical, oblong to elliptic, 5–12 cm. long, 3–5 cm. wide, obtuse to abruptly acute, lustrous, abruptly narrowed to base, somewhat spreading, the upper leaves small, several, ascending, loose, broadly lanceolate, 5–10 mm. long, acute; racemes rather loosely 5- to 10-flowered; bracts spreading; flowers pale green; dorsal sepal ovate to ovate-elliptic, 3-nerved, obtuse, 5–6.5 mm. long, the lateral sepals falcate, lanceolate, 3-nerved, obtuse, 7–8 mm. long, reflexed; petals deltoid-ovate, about 7 mm. long, slightly shorter than the median sepal, oblique, abruptly narrowed at tip, only slightly thickened at apex, 2-nerved; lip broadly linear, obtuse, 7–10 mm. long; spur 1.5–2 cm. long, incurved, gradually narrowed toward the apex; rostellum broad.——Aug.–Sept. Woods in mountains; Honshu, Shikoku, Kyushu; rather rare.

6. Platanthera ophrydioides F. Schmidt. *P. amabilis* Koidz.; *P. reinii* Fr. & Sav.; *P. platycorys* Schltr.; *P. ophrydioides* var. *platycorys* (Schltr.) F. Maekawa——KISO-CHIDORI. Roots somewhat thickened; stems 15–30 cm. long, rather slender; larger leaves solitary on lower part of stem, spreading, elliptic to oblong, 3–6 cm. long, 2–4 cm. wide, obtuse to rounded or abruptly acute at apex, rounded to subcordate and slightly clasping at base, the upper leaves 1–2, small, lanceolate; racemes 3–8 cm. long, 5- to 15-flowered; bracts 5–15 mm. long; flowers pale green; dorsal sepal narrowly triangular-ovate, 3-nerved, 5–6 mm. long, slightly mucronate, the lateral sepals reflexed, lanceolate to linear-lanceolate, 6–7 mm. long; petals erect, the lower half triangular-ovate, membranous, abruptly narrowed and elongate at tip, with the upper half broadly linear, only slightly thickened near apex; lip broadly linear, 6–8 mm. long, subobtuse; spur 6–10 mm. long, narrowed toward the rather acute apex, incurved.——Aug. Coniferous woods in mountains; Hokkaido, Honshu, Shikoku, Kyushu; rather common.——Sakhalin and s. Kuriles. Forma **australis** (Ohwi) Makino. *P. ophrydioides* var. *australis* Ohwi——NAGABA-KISO-CHIDORI. Southern phase with linear-oblong or oblanceolate leaves.——Honshu (Kii Prov. and westw.), Shikoku, Kyushu.

Var. **takedae** (Makino) Ohwi. *P. takedae* Makino; *P. ophrydioides* subsp. *takedae* (Makino) Soó——MIYAMA-CHIDORI, NIKKŌ-CHIDORI. Flowers slightly smaller than in the typical phase, the spur conical, barely 1 mm. long.——Coniferous woods in mountains; Hokkaido, Honshu (centr. distr. and northw.); rather rare.

Var. **uzenensis** Ohwi. GASSAN-CHIDORI. Flowers smaller, the dorsal sepal ovate-rounded, 2–2.5 mm. long, the lateral sepals 3–3.5 mm. long; petals obliquely ovate, with a thickened shorter upper half; lip 3–4 mm. long; spur narrowly oblong, 2–2.5 mm. long.——Honshu (Japan Sea side of centr. and n. distr.).

7. Platanthera mandarinorum Reichenb. f. var. **macrocentron** (Fr. & Sav.) Ohwi. *P. oreades* var. *macrocentron* Fr. & Sav.——HASHI-NAGA-YAMA-SAGI-SŌ. Roots somewhat thickened; stems 30–40 cm. long, 1-leaved below the middle; principal leaves lanceolate to oblong, sometimes linear-oblong, 6–12 cm. long, 1–3.5 cm. wide, ascending, obtuse, acute or rounded, rounded to abruptly narrowed at base, the upper reduced leaves 2–5, 1–2 cm. long, erect, lanceolate to narrowly so; racemes 5–12 cm. long; bracts lanceolate, 5–20 mm. long; flowers pale green; dorsal sepal broadly ovate to cordate-rounded, 3-nerved, 5–7 mm. long, the lateral sepals linear-lanceolate, 8–12 mm. long, gradually falcate, 3-nerved; petals obliquely ovate, obtuse, suddenly narrowed above the middle, the upper portion thickened; lip broadly linear, obtuse, 12–15 mm. long; spur rather obtuse, 2–3 cm. long; rostellum broad. ——Mountains and hills; Honshu, Shikoku, Kyushu; highly variable.——China.

Var. **maximowicziana** (Schltr.) Ohwi. *P. minor* sensu auct. Japon., non Reichenb. f.; *P. maximowicziana* Schltr.—— TAKANE-SAGI-SŌ. Stems 10–20 cm. long; leaves broadly lanceolate to narrowly oblong, obliquely spreading, the upper reduced leaves 1 or 2 or none; flowers smaller than in the typical phase, the spur shorter.——July–Aug. High mountains; Hokkaido, Honshu (centr. and n. distr.)——s. Kuriles.

Var. **brachycentron** (Fr. & Sav.) Koidz. *P. oreades* var. *brachycentron* Fr. & Sav.——YAMA-SAGI-SŌ. Stems 20–40 cm. long; leaves linear-oblong to oblong, the upper leaves 2 or

3, small, erect; flowers smaller than in the typical phase; dorsal sepal 4–5 mm. long, the lateral sepals 6–8 mm. long; petals somewhat elongate; lip 10–15 mm. long; spur 12–20 mm. long.——May–July. Grassy places in hills and mountains; Hokkaido, Honshu, Shikoku, Kyushu; rather common.—— Sakhalin, Korea, and China.

Var. **hachijoensis** (Honda) Ohwi. *P. hachijoensis* Honda ——HACHIJŌ-CHIDORI. Stems rather thick, 20–30 cm. long; leaves slightly clasping at base, the upper leaves reduced, the bracts larger than in the typical phase.——May–June. Honshu (Izu Isls.).

8. Platanthera minor (Miq.) Reichenb. f. *P. japonica* var. *minor* Miq.; *P. interrupta* Maxim.——ŌBA-NO-TOMBO-SŌ. Whole plant blackened when dry; roots fusiform; stems 30–60 cm. long; larger leaves 1 or 2, alternate, the lowest oblong or narrowly oblong, 7–12 cm. long, 2.5–3.5 cm. wide, subobtuse, narrowed to base, the upper gradually smaller, narrower, acute to acuminate, the scalelike leaves several, acuminate, lanceolate; racemes 10–25 cm. long, loosely 10- to 25-flowered; bracts broadly lanceolate, as long as the ovary, the margins usually papillose; flowers greenish white; dorsal sepal broadly ovate, obtuse, 3- to 5-nerved, 4–5 mm. long, the lateral markedly spreading, narrowly oblong, 3- to 5-nerved, 6–7 mm. long; petals erect, semi-ovate, obtuse, 3- to 4-nerved, slightly shorter than the dorsal sepal; lip broadly linear, obtuse, 5–7 mm. long; spur 12–15 mm. long, linear, pendulous, obtuse.——June–July. Low mountains; Honshu, Shikoku, Kyushu; rather common.

9. Platanthera tipuloides Lindl. *Habenaria keiskei* Miq.; *P. keiskei* (Miq.) Fr. & Sav.; *P. sororia* Schltr.; *P. tipuloides* var. *sororia* (Schltr.) Soó——HOSOBA-NO-KISO-CHIDORI. Roots elongate, thickened near base; stems slender, 20–30 cm. long, 1-leaved below the middle; leaves narrowly oblong to linear-oblong, ascending, 3–7 cm. long, 1–2 cm. wide, obtuse, sessile and abruptly contracted at base, the upper leaves 1–3, small; racemes 4–6 cm. long, rather densely many-flowered; bracts smooth on margin, nearly as long as the ovary; flowers small, yellowish green; dorsal sepal ovate to narrowly so, 2–3 mm. long, the lateral sepals reflexed, narrowly elliptic, 3.5–4.5 mm. long, obtuse; petals semi-ovate, thickened, obtuse, as long as the dorsal sepal; lip 5–6 mm. long, broadly linear; spur pendulous to gently incurved, slender, slightly thickened toward the obtuse apex, 12–17 mm. long; anther-locules parallel, approximate.——July–Aug. Coniferous woods or grassy places in mountains; Hokkaido, Honshu, Shikoku (Mount Tsurugi); rather common northward.——e. Siberia, Kamchatka, Kuriles, and Sakhalin.

Var. **nipponica** (Makino) Ohwi. *P. nipponica* Makino; *P. matsumurana* Schltr.——KOBA-NO-TOMBO-SŌ. Plant 20–40 cm. high; leaves broadly linear, rather erect; spikes loosely few-flowered; spur ascending.——June–Aug. Bogs in mountains; Hokkaido, Honshu, Shikoku, Kyushu; rather rare.

Var. **linearifolia** Ohwi. NAGABA-TOMBO-SŌ. Leaves linear, to 15 cm. long, 5–6 mm. wide; flowers few, loosely arranged.——Aug.–Sept. Kyushu (Yakushima).

10. Platanthera sachalinensis F. Schmidt. Ō-YAMA-SAGI-SŌ. Roots more or less thickened; stems 40–60 cm. long, rather stout; larger leaves 1–3, alternate, the lowest ascending, obovate-oblong, 10–20 cm. long, 4–7 cm. wide, obtuse, narrowed below into a sheath, the upper leaves gradually smaller, acute to acuminate, the upper 2 or 3 scalelike; racemes 10–30 cm. long, densely many-flowered; bracts usually slightly longer than the flowers, ascending; flowers greenish, small;

dorsal sepal narrowly ovate, 3–3.5 mm. long, the lateral sepals obliquely ovate, 4–5.5 mm. long, obtuse, reflexed; petals semi-ovate, thickened, about 3 mm. long, slightly shorter than the dorsal sepal; lip broadly linear, 5–7 mm. long; spur slender, 15–20 mm. long.——Grassy places in mountains; Hokkaido, Honshu, Shikoku, Kyushu.——Sakhalin and s. Kuriles.

Var. **hondoensis** Ohwi. OBANA-Ō-YAMA-SAGI-SŌ. Racemes more loosely flowered; flowers 10–25, the dorsal sepal about 5 mm. long, the lateral sepals about 7 mm. long; spur 2.2–2.5 cm. long.——Honshu (Mount Mitsumine in Musashi, and Mount Fuji in Suruga).

11. Platanthera japonica (Thunb.) Lindl. *Orchis japonica* Thunb.; *P. manubriata* Kraenzl.——TSURE-SAGI-SŌ. Roots horizontal, slightly thickened; stems 40–60 cm. long, stout, 5- to 8-leaved; lower 3–5 leaves narrowly oblong, 12–20 cm. long, 4–7 cm. wide, acute to obtuse or acuminate, short-sheathed at base, the upper leaves 2–4, becoming bractlike, broadly linear, racemes 10–20 cm. long, densely many-flowered; bracts usually slightly longer than the flowers, acuminate; flowers white; dorsal sepal elliptic, 7–8 mm. long, rather obtuse, navicular, keeled, the lateral sepals obliquely ovate, obtuse, about 8 mm. long; petals subdeltoid, obtuse, erect, somewhat thickened, slightly shorter than the sepals; lip broadly linear, 13–15 mm. long; spur pendulous, linear, 3–4 cm. long. ——May–June. Hokkaido, Honshu, Shikoku, Kyushu; rather rare.——China.

12. Platanthera hologlottis Maxim. *Habenaria neuropetala* Miq.; *P. neuropetala* (Miq.) Fr. & Sav.; *Limnorchis hologlottis* (Maxim.) Nevski——MIZU-CHIDORI. Roots horizontally spreading, thickened; stems stout, 50–80 cm. long; leaves 12–16, alternate, the upper ones gradually reduced, the lower 4–6 larger, linear-lanceolate to broadly linear, 10–20 cm. long, 1–2 cm. wide, acute to acuminate; racemes 10–20 cm. long, densely many-flowered; bracts linear-lanceolate, long-acuminate; flowers white, fragrant; dorsal sepal elliptic, flat, 5- to 7-nerved, very obtuse, 4–5 mm. long, the lateral sepals pendulous, falcate, narrowly oblong, 6–7 mm. long, obtuse; petals narrowly ovate, obtuse, thickened, about 7-nerved, slightly shorter than the dorsal sepal; lip ligulate, narrowly oblong, carnose, rounded at apex; spur longer than the sepals, 1–1.2 cm. long, linear, slender.——June–July. Wet places and bogs; Hokkaido, Honshu, Shikoku; locally abundant.——s. Kuriles, Korea, Manchuria, and e. Siberia.

13. Platanthera hyperborea (L.) Lindl. *Orchis hyperborea* L.; *P. convallariaefolia* Lindl.; *P. makinoi* Yabe; *P. hyperborea* var. *makinoi* (Yabe) Takeda——SHIROUMA-CHI-DORI, YŪBARI-CHIDORI. Stems about 30 cm. long, rather thick; leaves several, alternate, narrowly oblong, sheathed at base, the lower to 7 cm. long, 2 cm. wide, the upper gradually reduced, obtuse to acute; racemes about 7 cm. long, densely several- to many-flowered; bracts nearly as long as the flowers or longer, ovate-lanceolate; flowers yellowish green; dorsal sepal broadly ovate, obtuse, 4–5 mm. long, the lateral sepals oblique, narrowly ovate, obtuse, about 5 mm. long, 6-nerved; petals obliquely ovate or narrowly deltoid, acute; lip oblong-ovate, obtuse, about 5 mm. long; spur about as long as the sepals.—— Grassy places in alpine regions; Hokkaido (Mount Yubari), Honshu (Mount Shirouma in Shinano); very rare.——Kuriles, Kamchatka, N. America, and Iceland.

5. COELOGLOSSUM Hartm.   AO-CHIDORI ZOKU

Terrestrial herbs with alternate leaves; inflorescence spicate, the flowers small, greenish, the bracts green; sepals connivent in upper part and forming a hood with the petals; lip cuneate, 3-lobed, the terminal lobe minute; spur very short; anther-locules distinct; pollinia oblong, partly hidden by a small bursicule; caudicles rather elongate; glands small, scarcely broader than the caudicle; rostellum bifid, very low; stigma depressed.——Few species, in the temperate regions of the N. Hemisphere.

1. Coeloglossum viride (L.) Hartm. var. **bracteatum** (Willd.) Richt. *Orchis bracteata* Willd.; *C. bracteatum* (Willd.) Parl.; *Habenaria viridis* var. *bracteata* (Willd.) A. Gray; *Platanthera viridis* var. *bracteata* (Willd.) Reichenb. f.; *P. bracteata* (Willd.) Torr.——AO-CHIDORI. Roots thickened, conical, more or less lobed; stems erect, 15–30 cm. long, 2- or 3-leaved on lower half; leaves narrowly obovate-oblong or oblong, 4–8 cm. long, 1.5–3.5 cm. wide, rounded to acute at apex, sheathing at base; spikes 4–12 cm. long, rather dense, many-flowered; bracts usually longer than the flower, narrowly lanceolate, 1–4 cm. long; flowers greenish brown, small; sepals narrowly ovate, 5–7 mm. long, obtuse, 5- to 7-nerved, ascending; petals linear-lanceolate, 1-nerved, shorter than the sepals; lip pendulous, oblanceolate, rather thick, longer than the sepals, shallowly 3-lobed, the lateral lobes about 2 mm. long, erect, broadly lanceolate, obtuse, the terminal lobe subrounded, about 0.5 mm. long; spur narrowly ovoid, about 3 mm. long. ——May–July. High mountains; Hokkaido, Honshu (centr. and n. distr.); rather rare.——Korea, China, e. Siberia, Europe, and N. America.

6. TULOTIS Raf.   TOMBO-SŌ ZOKU

Terrestrial herbs with slender slightly thickened roots; leaves 1–3, alternate; inflorescence spicate, with small, green bracts; flowers small, green or yellowish green; sepals and petals nearly equal, narrow; lip 3-lobed, the lateral lobes nearly basal, small, the median lobe elongate, with a callosity at the base; anther-locules parallel or divergent, the pollinia granular, with a caudicle at the elongated base.——About 8 species, in e. Asia, and N. America.

1A. Spur linear, 5–7 mm. long; lip obtuse. .... 1. *T. ussuriensis*
1B. Spur oblong, about 1 mm. long; lip subacute. .... 2. *T. iinumae*

1. Tulotis ussuriensis (Regel) Hara. *Perularia ussuriensis* (Regel) Schltr.; *Platanthera ussuriensis* (Regel) Maxim.; *Platanthera tipuloides* var. *ussuriensis* Regel; *Platanthera herbiola* var. *japonica* Finet; *Perularia fuscescens* sensu auct. Japon., non Lindl.——TOMBO-SŌ. Roots elongate, slightly thickened; stems 20–50 cm. long, slender, 2-leaved on the lower part, with scalelike leaves toward the top; larger leaves sub-approximate, oblong to broadly oblanceolate, 8–13 cm. long,

1–5 cm. wide, rounded to acute, narrowed to the base; spikes 3–15 cm. long, the bracts narrowly lanceolate, long-acuminate, nearly as long as the flowers; flowers greenish, rather numerous, the dorsal sepal ovate, 3-nerved, obtuse, 2–3.5 mm. long, the lateral sepals spreading, narrowly oblong, obtuse, 3–4 mm. long; petals narrowly ovate, obtuse, thickened; lip 3–4 mm. long, obtuse, 3-lobed, the lateral lobes deltoid, small, the terminal lobe rather thick, linear-oblong, obtuse, slightly reflexed at tip; spur linear, 5–7 mm. long.——July–Aug. Grassy places and woods in low mountains; Hokkaido, Honshu, Shikoku, Kyushu.——Korea, China, and Ussuri.

2. Tulotis iinumae (Makino) Hara. *Perularia iinumae* (Makino) Ohwi; *Habenaria iinumae* Makino; *Platanthera iinumae* Makino; *Gymnadenia iinumae* (Makino) Miyabe & Kudo——Iinuma-mukago. Roots slender, slightly thickened; stems slender, 25–35 cm. long, 2-leaved near the middle, with several reduced leaves above; larger leaves subapproximate, oblong, 8–15 cm. long, 2–4 cm. wide, rounded to abruptly acuminate at apex, sheathed at base; spikes 5–10 cm. long, many-flowered, the bracts linear-lanceolate, spreading above; flowers small, greenish; sepals nearly erect, 1-nerved, obtuse, ovate, about 2 mm. long; petals slightly shorter than the sepals; lip slightly longer than the sepals, ovate, subacute, thick, ascending, spreading toward tip, with an ascending acute minute lobe on each side above the base; spur oblong, 1–1.5 mm. long.——July–Aug. Grassy places and thickets in mountains; Hokkaido (Oshima Prov.), Honshu, Shikoku, Kyushu.

7. HERMINIUM L.   Mukago-sō Zoku

Small terrestrial herbs with tuberous roots; leaves few, sheathing at base; spikes terminal; flowers small, greenish; sepals equal, free; petals equal, or smaller than the sepals; lip spreading or pendulous, adnate to the column at base, 3-lobed, the terminal lobe smaller; spur absent; column very short, rostellum short; anther-locules parallel; glands naked.——More than a dozen species in Europe and Asia.

1A. Leaves 3–5, linear to broadly so; lip much longer than the sepals, 3-lobed from the middle, the lateral lobes linear, nearly entire, the terminal lobe much shorter than the lateral lobes. . . . . 1. *H. longicrure*
1B. Leaves 2, ovate to broadly lanceolate; lip 3-lobed from the base, slightly longer than the sepals, the lateral lobes ascending, slightly smaller than the terminal lobe, both lanceolate. . . . . 2. *H. monorchis*

1. Herminium longicrure (Wright) Wang & Tang. *Aceras longicruris* Wright; *H. angustifolium* var. *longicrure* (Wright) Makino——Mukago-sō. Slender herb with ovoid roots; stems rather slender, 20–40 cm. long, loosely 3- to 5-leaved; leaves linear to broadly so, 8–15 cm. long, 4–8 mm. wide, acuminate; spikes 5–15 (–20) cm. long, the bracts ovate-triangular, 1-nerved, long-acuminate, shorter or slightly longer than the flowers; flowers green, not fully open; sepals oblong, very obtuse, 1-nerved, 2–2.5 mm. long, equal; petals slightly shorter than the sepals, linear-lanceolate; lip 5–8 mm. long, linear, pendulous, slightly broadened at base, 3-lobed from the middle, the lateral lobes linear, obtuse, ascending, the terminal lobe very short; spur absent; column very short.——June–Aug. Grassy places in lowlands; Hokkaido (Oshima Prov.), Honshu, Shikoku, Kyushu; rather rare.——Ryukyus, Formosa, China, Manchuria, and Korea.

2. Herminium monorchis (L.) R. Br. *Ophrys monorchis* L.——Kushiro-chidori. Small herb with thickened roots; stems 10–35 cm. long, slender; leaves 2, radical, ovate or broadly lanceolate, 4–10 cm. long, 1–2 cm. wide, acute to acuminate, the cauline usually 1, near middle of stem, small; spikes densely flowered, elongate, to 20 cm. long in fruit; bracts green, as long as the ovary, lanceolate, caudate-acuminate; flowers somewhat secund, green; sepals ascending, 1-nerved, the dorsal ovate, about 2 mm. long, obtuse, the lateral sepals slightly narrower and longer; petals oblong, 3–4 mm. long, obtuse; lip 4–5 mm. long, ascending, somewhat broadened at base, inflated, 3-lobed from the base, the lateral lobes short, ascending, the terminal lobe linear, obtuse; ovary 3–4 mm. long.——June–July. Hokkaido (Kushiro); very rare.——Korea, China, n. India, Siberia, and Europe.

8. AMITOSTIGMA Schltr.   Hina-ran Zoku

Terrestrial herbs, with tuberous roots; stems slender, 5–20 cm. long, 1- to 4-leaved; leaves narrow, sometimes nearly radical; flowers in spikes, rose-purple or pale rose, the bracts green, lanceolate, often shorter than the ovary; sepals free, oblong or ovate, 1- to 3-nerved; petals similar to the sepals, often forming a hood with the dorsal sepal; lip rather large, 3-lobed, the terminal lobe larger, often shallowly bifid, spur relatively short; column very short, adnate to the base of lip; staminodes developed, the anther-locules parallel, approximate, not elongate at base; rostellum small; stigma transversely divided into 2 lobes; ovary often short-stalked.——About 20 species, in e. Asia and India.

1A. Flowers relatively small; lip about 3.5 mm. long; perianth-segments 1-nerved, about 2.2 mm. long; leaves solitary, broadly lanceolate, subacute. . . . . 1. *A. gracile*
1B. Flowers relatively large; lip 7–13 mm. long; perianth-segments 3–6 mm. long.
  2A. Leaves solitary (rarely 2 in No. 2).
    3A. Leaves linear to broadly so; plant of bogs; flowers white to pale rose; lobes of lip lanceolate. . . . . 2. *A. kinoshitae*
    3B. Leaves narrowly oblong to lanceolate; plant of shaded rocky cliffs; flowers rose-purple; lobes of lip ovate. . . . . 3. *A. keiskei*
  2B. Leaves 2–4; flowers rose-purple; lobes of lip broadly cuneate to nearly rounded; plants of shaded rocks.
    4A. Leaves approximate at base of stem, the lower 2 oblong to broadly oblanceolate; bracts nearly hyaline; lip with a rather deeply bifid terminal lobe. . . . . 4. *A. lepidum*
    4B. Leaves alternate, cauline, linear to broadly so; bracts green; lip with entire terminal lobe. . . . . 5. *A. kurokamianum*

1. **Amitostigma gracile** (Bl.) Schltr. *Mitostigma gracile* Bl.; *Gymnadenia gracilis* (Bl.) Miq.; *G. tryphiaeformis* Reichenb. f.; *Orchis gracilis* (Bl.) Soó——Hina-ran. Roots somewhat tuberous-thickened; stems 5–15 cm. rarely to 20 cm. long, slender, 1-leaved near base; leaves broadly lanceolate to linear-oblong, ascending, 4–8 cm. long, 1–2 cm. wide, subacute, rounded at base, the scalelike leaves usually 1, on upper part of stem, small; spikes few- to rather many-flowered, secund, 1–4 cm. long; bracts small, ovate to narrowly so, acuminate, 3–5 mm. long, 1-nerved; flowers rose-purple; sepals 2.2–2.5 mm. long, 1-nerved, obtuse, elliptic, the lateral sepals obliquely ovate, ascending; petals oblique, broadly ovate, about 2.2 mm. long, obtuse, 1-nerved, as long as the sepals; lip suberect, about 3.5 mm. long, 3-lobed to below the middle, the lateral lobes narrow, short, the terminal longest, spathulate-cuneate, rounded at apex, 3-nerved at base; spur cylindric, 1–1.5 mm. long; capsules short-pedicelled, 5–6 mm. long.——June–July. Honshu (w. distr.), Shikoku, Kyushu.——Korea.

2. **Amitostigma kinoshitae** (Makino) Schltr. *Gymnadenia kinoshitae* Makino; *G. gracilis* var. *kinoshitae* (Makino) Finet; *G. keiskei* var. *kinoshitae* (Makino) Makino; *Orchis kinoshitae* (Makino) Soó; *A. hisamatsui* Miyabe & Tatew.—— Koani-chidori. Roots narrowly oblong; stems slender, erect, 12–20 (rarely to 25) cm. long, 1- or 2-leaved below the middle; leaves linear to broadly so, 4–8 cm. long, 4–8 mm. wide, subacute; spikes 2- to 5-flowered, secund; bracts broadly lanceolate, 3–8 mm. long, subacute; flowers white to pale rose; sepals 3.5–4.5 mm. long, obtuse, 3-nerved, the dorsal sepal elliptic, the lateral sepals obliquely ovate; petals about as long as the sepals, obliquely ovate, 2-nerved; lip 8–10 mm. long, 3-lobed, cuneate, the lateral lobes ascending, lanceolate, the terminal lobe slightly longer, spathulate-cuneate, shallowly bifid; spur 1–1.5 mm. long, obtuse; capsules short-pedicelled.——June–Aug. Bogs, rarely wet rocks along streams in mountains; Hokkaido, Honshu (Kantō Distr. and northw.).——s. Kuriles.

3. **Amitostigma keiskei** (Maxim.) Schltr. *Gymnadenia keiskei* Maxim.; *Orchis keiskei* (Maxim.) Soó——Iwa-chidori. Roots somewhat fusiform; stems 8–15 cm. long, slender, erect, 1-leaved below the middle; leaves lanceolate to narrowly ob-

long, 3–5 cm. long, 8–15 mm. wide, ascending, obtuse; spikes rather densely few-flowered, secund; bracts narrowly lanceolate to broadly linear, 4–10 mm. long, obtuse; flowers rose-purple, rarely white; sepals obtuse, 1- to 3-nerved, 3.5–4.5 mm. long, the dorsal sepal elliptic, the lateral sepals obliquely oblong; petals obliquely ovate, obtuse, 2-nerved, as long as the sepals; lip ascending, about 10 mm. long, rather deeply 3-lobed, the lateral lobes ascending, ovate, the terminal lobe somewhat longer, slightly broader, bifid; spur 2–2.5 mm. long, obtuse. ——May–June. Shaded rocky cliffs in mountains; Honshu (centr. distr.).

4. **Amitostigma lepidum** (Reichenb. f.) Schltr. *Gymnadenia lepida* (Reichenb. f.) Soó——Okinawa-chidori. Roots mostly oblong; stems 8–15 cm. long; leaves 3, sometimes 2, nearly radical and approximate, spreading, oblong to broadly oblanceolate, 4–8 cm. long, 1.5–2.5 cm. wide, obtuse, mucronate, the upper leaves narrower and shorter; flowers few; bracts nearly hyaline, broadly lanceolate, acuminate, 1-ribbed, 5–10 mm. long; sepals oblong, 1- or sometimes 2-nerved, 4.5–6 mm. long, obtuse; petals oblique, broadly ovate, obtuse, 2-nerved, about 3 mm. long; lip 8–10 mm. long, rather deeply 3-lobed, the lateral lobes ascending, the terminal lobe larger, broadly cuneate, deeply bifid; spur 4–5 mm. long, obtuse.——Mar.–Apr. Kyushu (s. distr.).——Ryukyus.

5. **Amitostigma kurokamianum** (Ohwi & Hatus.) Ohwi. *Orchis kurokamiana* Ohwi & Hatus.——Kuro-kami-ran. Roots somewhat ellipsoidal; stems erect, 5–15 cm. long, slender, 2-leaved; leaves alternate, cauline, linear to broadly so, 5–10 cm. long, 5–8 mm. wide, arcuate-recurved at the acuminate tip; spikes 2–3 cm. long, rather densely 2- to 10-flowered, secund; bracts green, linear, 5–10 mm. long, acuminate, broadened at base; flowers rose-purple; sepals oblong, obtuse, 1- or sometimes 3-nerved, the lateral sepals spreading, 6–6.5 mm. long; petals obliquely ovate, about 5 mm. long; lip 8–10 mm. long, 3-lobed, the lateral lobes ascending, the terminal lobe larger, entire, elliptic, 4–5 mm. long; spur short-cylindric, about 3 mm. long; capsules short-pedicelled. ——June–July. Rocks in mountains; Kyushu (Mount Kurokami).

9. GYMNADENIA R. Br. Tegata-chidori Zoku

Terrestrial herbs often with thickened, sometimes lobed roots; stems leafy; leaves linear to oblong; spikes terminal, mostly many-flowered, often secund, sometimes twisted, the bracts small; flowers relatively small, rose-purple to pale rose; dorsal sepal often forming a hood with the petals, the lateral sepals ascending to spreading; lip 3-lobed; anther-locules parallel, with a narrowly grooved rostellum between the locules; glands slender, naked, distinct; stigma not distinct.——About 10 species, mostly in boreal regions.

1A. Stems 20–60 cm. long; leaves 3–7, cauline, 7–20 cm. long; lateral lobes of lip spreading.
  2A. Spur 15–20 mm. long, about 1.5 times as long as the ovary, 3 to 4 times longer than the perianth; roots digitately thickened; leaves broadly linear, flat. .... 1. *G. conopsea*
  2B. Spur 2–5 mm. long, much shorter than the ovary, and somewhat shorter than the perianth; roots not as above; leaves elliptic to narrowly oblong, slightly plicate. .... 2. *G. camtschatica*
1B. Stems 10–20 cm. long, slender; leaves usually 2, radical, 3–6 cm. long; petals forming a hood with the dorsal sepal. .... 3. *G. cucullata*

1. **Gymnadenia conopsea** (L.) R. Br. *Orchis conopsea* L.; *G. conopsea* var. *ussuriensis* Regel; *G. ibukiensis* Makino ——Tegata-chidori. Roots digitate, thickened; stems 30–60 cm. long, rather stout; leaves 3–7, disposed mostly on lower half of stems, alternate, broadly linear, sometimes lanceolate or linear, 10–20 cm. long, 1–2.5 cm. wide, obtuse, becoming acute or acuminate toward the top and smaller, linear-lanceolate,

long-acuminate; spikes rather densely many-flowered, 7–15 cm. long; bracts broadly lanceolate, long-acuminate; flowers rose-purple; sepals ovate, obtuse, spreading, 4–5 mm. long, 2- or 3-nerved; petals oblique, broadly ovate, obtuse, shorter than the sepals; lip 6–8 mm. long, with 3 equally rounded lobes; spur linear, 15–20 mm. long.——July–Aug. Grassy places in high mountains; Hokkaido, Honshu (centr. distr.

and northw.); rather common.——Korea, Sakhalin, Kuriles, China, Siberia, and Europe.

2. Gymnadenia camtschatica (Cham.) Miyabe & Kudo. *Orchis camtschatica* Cham.; *Neolindleya camtschatica* Nevski; *Platanthera decipiens* Lindl.; *G. vidalii* Fr. & Sav.——NOBINE-CHIDORI. Roots slightly thickened, slenderly cylindric; stems 20–60 cm. long, 4- to 7-leaved; leaves alternate, elliptic to narrowly oblong, 7–15 cm. long, 2–6 cm. wide, rounded at apex, slightly plicate, the upper 1 or 2 leaves small, broadly lanceolate, acuminate; spikes mostly many-flowered, 5–15 cm. long; bracts lanceolate, green, usually conspicuously longer than the flowers; flowers pale rose; sepals 3-nerved, narrowly ovate, rather obtuse, ascending, about 5 mm. long; petals obliquely ovate, shorter than the sepals; lip broadly cuneate-ovate, slightly longer than the sepals, 3-lobed at apex, the lateral lobes narrowly ovate, obtuse, the terminal lobe rather short; spur prominently curved, 2–5 mm. long, narrowed toward the obtuse apex.——May–June. Woods in mountains; Hokkaido, Honshu (centr. distr. and northw.), Shikoku.——Kamchatka and Sakhalin.

3. Gymnadenia cucullata (L.) L. C. Rich. *Orchis cucullata* L.; *Neottianthe cucullata* (L.) Schltr.——MIYAMA-MOJIZURI. Roots globose; stems slender, 10–20 cm. long, with few scalelike linear leaves; radical leaves usually 2, spreading, oblong to broadly lanceolate, 3–6 cm. long, 1–2.5 cm. wide, acute to acuminate; spikes secund, more or less many-flowered; bracts linear, slightly longer or shorter than the ovary; flowers pale rose; sepals and petals alike, narrowly lanceolate, 1-nerved, 6–8 mm. long, gradually narrowed to an acute apex, nearly erect, slightly connivent above; petals forming a hood with the dorsal sepal; lip narrowly ovate, erect, 7–8 mm. long, 3-lobed, the lobes nearly erect, parallel, linear, the terminal lobe slightly broader; spur curved inward, longer than the perianth.——July–Aug. Woods in high mountains; Hokkaido, Honshu (centr. distr. and northw.).——Cooler regions of Eurasia.

10. HABENARIA Willd. MIZU-TOMBO ZOKU

Terrestrial herbs sometimes with tuberous roots; inflorescence spicate; leaves linear to oblong; flowers green to white; sepals alike or the dorsal broader; petals simple or bifid, sometimes fimbriate; lip slightly adnate to base of the column, usually 3-lobed, spurred; column short; anther-locules parallel or divergent at base, sometimes tubular at base; glands naked; staminodes rarely present; stigma 2-lobed; rostellum usually small, 2-armed, erect, between the anther-locules.——About 700 species, widely distributed in temperate and tropical regions.

1A. Arms of rostellum very short; anther-locules erect, closely parallel; flowers greenish.
  2A. Spikes 4–5 cm. long; sepals 3-nerved; spur short-cylindric, about 6 mm. long; leaves radical. 1. *H. iyoensis*
  2B. Spikes 7–25 cm. long; sepals 1-nerved; spur ovoid to clavate, 1.5–4 mm. long.
    3A. Larger leaves mostly basal; spur ovoid, about 1.5 mm. long; plant brown when dry. . . . . 2. *H. formosana*
    3B. Larger leaves on lower half of the stems; spur clavate, 3–4 mm. long; plant blackened when dry. . . . . 3. *H. flagellifera*
1B. Arms of rostellum relatively long; anther-locules ascending, usually divergent; flowers white to pale green.
  4A. Lateral lobes of lip filiform to linear, entire or obsoletely toothed; flowers white to pale green.
    5A. Lateral sepals subreniform, strongly reflexed; lateral lobes of lip obliquely spreading; flower-buds somewhat strongly 4-angled in lateral view; flowers greenish, with a spur rather abruptly inflated at the apex. . . . . 4. *H. sagittifera*
    5B. Lateral sepals obliquely ovate, spreading to slightly deflexed; lateral lobes of lip horizontally spreading, straight or scarcely recurved; flower buds obliquely 3-angled in lateral view; flowers white, with the spur gradually inflated.
      6A. Flowers relatively large; lateral sepals 7–8 mm. long; lip about 15 mm. long; spur 3–4 cm. long. . . . . 5. *H. linearifolia*
      6B. Flowers relatively small; lateral sepals about 5 mm. long; lip about 10 mm. long; spur rather short, 4–15 mm. long.
                                                 6. *H. yezoensis*
  4B. Lateral lobes of lip flabellate or cuneate, with laciniate upper margins; flowers white.
    7A. Leaves broadly linear or narrowly lanceolate, 3–6 mm. wide; flowers 1–3, rarely 4; petals obliquely ovate, few-nerved, longer than the sepals, obsoletely toothed or shallowly laciniate on lower margin; lateral lobes of lip deeply laciniate. . . 7. *H. radiata*
    7B. Leaves broadly lanceolate to narrowly ovate-oblong, 2–4 cm. wide; flowers rather densely arranged and numerous; petals lanceolate, 1- to 2-nerved, shorter than the sepals, entire; lateral lobes of lip shallowly laciniate. . . . . 8. *H. miersiana*

1. Habenaria iyoensis Ohwi. *Peristylus iyoensis* Ohwi ——IYO-TOMBO. Glabrous, erect perennial about 20 cm. high, with ovoid tuberous roots; leaves radical, several, broadly oblanceolate, 3–6 cm. long, 1–1.5 cm. wide; scape or stem erect, with few, appressed, scalelike leaves; spikes 4–5 cm. long, 7- to 12-flowered; bracts lanceolate, acuminate, 5–15 mm. long; sepals 3-nerved, about 3 mm. long, the dorsal sepal broadly ovate, erect, obtuse, the lateral sepals obliquely ovate, reflexed; lip 3-lobed nearly to the base, about 3.5 mm. long, the lateral lobes spreading, filiform, flexuous, about 4 mm. long, the terminal lobe 2.5 mm. long, broadly linear; spur nearly straight, short-cylindric, about 6 mm. long, obtuse; ovary about 1 cm. long.——Oct. Honshu (s. Kantō Distr.), Shikoku, Kyushu; very rare.

2. Habenaria formosana Schltr. *H. tentaculata* var. *acutifolia* Hayata——TAKASAGO-SAGI-SŌ. Rhizomes slender, short, loosely sheathed; stems erect, 30–50 cm. long; leaves 2–4, nearly radical, rosulate, narrowly oblong to broadly lanceolate, 7–20 cm. long, 1.5–4 cm. wide, ascending, abruptly acute, the upper leaves 6–8, appressed to the stem, broadly linear; spikes 7–25 cm. long, mostly rather many-flowered; bracts lanceolate, long-acuminate, 1–1.5 cm. long, 3-nerved, appressed to the ovary; flowers sessile; sepals narrowly ovate, 1-nerved, obtuse, about 3.5 mm. long; petals slightly shorter than the sepals, 1-nerved, ovate; lip 3-lobed from the oblong-cuneate, rather carnose base, the lateral lobes filiform-linear, ascending, spreading, flagelliform, the median lobe about 2 mm. long, obtuse, the spur ovoid, apiculate, about 1.5 mm. long; column very short.——Kyushu (Yakushima); very rare.——Ryukyus and Formosa.

3. Habenaria flagellifera Makino. *Coeloglossum flagelliferum* Maxim. ex Makino; *Peristylus flagelliferus* (Makino) Ohwi; *P. hiugensis* Ohwi; *P. satsumanus* Ohwi——MUKAGO-TOMBO. Glabrous, erect perennial, blackened when dry, with ovoid tuberous roots; stems 20–50 cm. long; principal leaves 3–5, on lower part of stem, alternate, broadly lanceolate,

5–10 cm. long, 1.5–2.5 cm. wide, short-acuminate, the reduced upper leaves few, appressed to the stem; spike 10–20 cm. long, rather many-flowered; bracts broadly lanceolate, 5–10 mm. long; flowers green; dorsal sepal narrowly ovate, rather obtuse, the lateral sepals obliquely oblong, obtuse, reflexed at anthesis; petals narrowly ovate, nearly as long as the lateral sepals, erect; lip 3-lobed nearly to the base, about 2 mm. long, the lateral lobes 6–7 mm. long, the terminal lobe short, obtuse, lanceolate; spur clavate, 3–4 mm. long; ovary 5–7 mm. long.——Sept.–Oct. Grassy places in lowlands; Honshu (Kantō Distr. and westw.), Shikoku, Kyushu.

4. Habenaria sagittifera Reichenb. f. *H. oldhamii* Kraenzl.; *H. tosaensis* Makino; *H. sagittifera* var. *reichenbachiana* Takeda——MIZU-TOMBO. Slender perennial with tuberous roots; stems erect, 40–70 cm. long, with few leaves on the lower part; leaves linear, 5–15 cm. rarely to 20 cm. long, 3–6 mm. wide, gradually narrowed at tip, sheathing at base, the upper scalelike leaves few, erect, linear, broadened at base; spike rather densely flowered, the bracts narrowly ovate, 8–15 mm. long, abruptly narrowed at tip; flowers rather numerous, greenish, 8–10 mm. across; sepals unequal, the dorsal rounded-cordate, 3–4 mm. long, 5-nerved, the lateral sepals prominently oblique, subreniform, 6–7 mm. long and as wide, strongly reflexed; petals subdeltoid, much dilated on lower margins, as long as the dorsal sepal, the anterior margin appressed to the column; lip about 2 cm. long, 3-lobed, the lobes linear, the lateral lobes obliquely spreading, about 10 mm. long, sometimes obsoletely toothed, the terminal lobe about 15 mm. long, entire; spur about 15 mm. long, slender, with an abruptly inflated apex, nearly as long as to slightly shorter than the ovary. ——July–Sept. Bogs and wet grassy places in lowlands; Hokkaido, Honshu, Shikoku, Kyushu.

5. Habenaria linearifolia Maxim. *H. sagittifera* var. *linearifolia* (Maxim.) Takeda——SAWA-TOMBO. Erect perennial with tuberous roots; stems 40–70 cm. long, few-leaved on lower part; leaves linear, 10–20 cm. long, 3–6 mm. wide, gradually narrowed to the tip, slightly plicate or flat, 1-costate on back, sheathing at base, the upper reduced leaves few, appressed, linear, broadened at base; spikes 7–15 cm. long, many-flowered; bracts 1–1.5 cm. long; flowers white, 1–1.5 cm. across; dorsal sepal erect, ovate, 6–7 mm. long, 5-nerved, the lateral sepals spreading, obliquely ovate, about 7 mm. long; petals erect, subdeltoid, about 6 mm. long, with dilated lower margins; lip about 15 mm. long, about 2 mm. wide, 3-lobed to below the middle, the lobes linear, the lateral lobes horizontally spreading, sometimes declined at tip, usually toothed on one side near the apex; spur 3–4 cm. long, gradually inflated toward apex; ovary 2–2.5 cm. long.——Aug. Bogs and wet grassy places; Hokkaido, Honshu (n. distr.).——Korea, Manchuria, Ussuri, and Amur.

6. Habenaria yezoensis Hara. *H. linearifolia* var. *brachycentra* Hara——HIME-MIZU-TOMBO. Slender, erect herb with tuberous roots; stems 25–50 cm. long, 6- to 7-leaved on lower part; leaves 6–11 cm. long, 4–6 mm. wide, acute, 3-nerved, sheathing at base, the upper reduced leaves lanceolate; spikes loosely 2- to 8-flowered, to 10 cm. long; bracts lanceolate,

broadened at base; flowers white, suboblong (in bud) in lateral view; dorsal sepal erect, broadly ovate, about 4 mm. long and as wide, 3-nerved, the lateral sepals spreading, greenish, obliquely ovate, about 5 mm. long; petals erect, white, obliquely ovate-triangular, about 4 mm. long; lip greenish, about 10 mm. long, the lateral lobes spreading, gently recurved, minutely toothed near apex, about 6 mm. long, the terminal lobe about as long; spur greenish, about 4 mm. long, curved forward, inflated toward apex; ovary about 1.5 cm. long.——Aug. Bogs; Hokkaido.

Var. **longicalcarata** Miyabe & Tatew. *H. osensis* Makino ——OZE-NO-SAWA-TOMBO. Spur about 15 mm. long, gradually inflated upward.——July–Aug. Bogs; Honshu (Oze in Kotsuke); rare.——s. Kuriles.

7. Habenaria radiata (Thunb.) Spreng. *Orchis radiata* Thunb.; *Platanthera radiata* (Thunb.) Lindl.; *Pecteilis radiata* (Thunb.) Raf.; *Hemihabenaria radiata* (Thunb.) Finet—— SAGI-SŌ. Slender, erect perennial with nearly spherical, tuberous roots; stems 20–40 cm. long, with 3–5 leaves on lower part and few scalelike leaves above; leaves ascending, broadly linear to narrowly lanceolate, 5–10 cm. long, 3–6 mm. wide, acuminate, sheathing at base, the upper scalelike leaves appressed to the stem, distant, linear, gradually acuminate; bracts green, 7–12 mm. long; flowers 1–3, rarely 4, about 3 cm. across, white; sepals 5- to 7-nerved, narrowly ovate, obtuse, green, 8–10 mm. long, the lateral ones spreading, oblique; petals white, obliquely ovate, erect, forming a hood with the dorsal sepal, 10–12 mm. long, subacute, toothed or shallowly laciniate on the outer margin; lip white, short-cuneate, 3-lobed, about 15 mm. long, the lateral lobes obliquely flabellate, deeply laciniate on outer margin, the terminal lobe linear, rather obtuse, entire, about 1 cm. long, 2 mm. wide, 3-nerved; spur 3–4 cm. long, narrowly cylindric, gradually thickened toward the apex; ovary about 1.5 cm. long.——Aug. Wet grassy places in lowlands; Honshu, Shikoku, Kyushu.

8. Habenaria miersiana Champ. *H. geniculata* sensu auct. Japon., non D. Don; *H. sieboldiana* Miq.——DAI-SAGI-SŌ. Terrestrial, erect perennial with broadly ovoid, tuberous roots; stems 30–60 cm. long; leaves 4–5, broadly lanceolate to narrowly ovate-oblong, 8–15 cm. long, 2–4 cm. wide, acute, short-sheathing at base, with narrow translucent margins, the upper leaves scalelike, appressed to the stem, narrowly lanceolate, 3–5 cm. long, with a filiform point; inflorescence rather densely flowered; bracts broadly linear; flowers white, 2–2.5 cm. across; sepals narrowly ovate, 10–13 mm. long, acute, the lateral sepals oblique and slightly deflexed; petals lanceolate, 8–10 mm. long, acuminate, entire, 1- to 2-nerved, forming a hood with the dorsal sepal; lip 15–18 mm. long, 3-lobed, the lateral lobes spreading, broadly cuneate, shallowly laciniate, the terminal lobe linear, 1-nerved, 5–8 mm. long; spur about 3 cm. long, gradually thickened toward the obtuse apex; ovary about 2 cm. long.——Aug.–Oct. Thickets; Honshu (Awa Prov.), Shikoku, Kyushu; rare.——Ryukyus, Formosa, and s. China.

11. ANDROCORYS Schltr. MISUZU-RAN ZOKU

Small terrestrial plants with a single, erect, narrow, radical leaf, the scape naked; spikes loose; flowers small, few; dorsal sepal erect, concave, ovate, adnate to the base of petals, the lateral sepals reflexed, oblong; petals spreading, concave, oblique, ovate-rounded, forming a hood with the dorsal sepal; lip small, entire, ligulate, narrowed toward the obtuse apex, adnate to base of the lateral sepals, spurless; anthers large, the locules small, lateral, incurved, the connective large; rostellum deltoid, erect, with spreading lateral lobes; stigmas 2, papillose, stalked, the caudicles attached to the rostellum; ovary sessile, slightly twisted.——Two species, one in w. China and one in Japan.

1. Androcorys japonensis F. Maekawa. MISUZU-RAN. Roots tuberous, small; leaf membranous, solitary, basal, erect, oblanceolate-oblong, 15–22 mm. long, 5–7 mm. wide, obtuse, long-attenuate at base; scapes 7.5–9 cm. long, erect, covered at base with leaf-sheaths, the bracts minute; flowers 2–3, 2–2.5 mm. across; dorsal sepal elliptic, obtuse, about 1 mm. long, the lateral sepals slightly reflexed, falcate, oblong, obtuse, irregularly denticulate, about 1.5 mm. long; petals oblique, ovate-rounded, 0.8 mm. long, incurved at tip, narrowed to the base, with minute irregular teeth on margins, forming an open hood with the dorsal sepal; lip ligulate, not spurred, lanceolate, 1.5 mm. long, obtuse, broader at base; column about 1 mm. long. ——Aug. High mountains; Honshu (Yatsugatake); very rare.

12. MICROTIS R. Br. NIRABA-RAN ZOKU

Small terrestrial plants with tuberous roots; stems low, 1-leaved; leaf elongate, linear, terete, sheathing the lower part of the stem; inflorescence densely spicate, minutely bracteate; flowers minute; dorsal sepal erect, broad, incurved, the lateral sepals lanceolate or oblong, free, spreading or reflexed; petals similar to the lateral sepals or smaller; lip attached to the base of the column, sessile, spreading, oblong, entire or bifid, often with callosities at base; spur absent; column very short, with 2 auricles at apex; stigma short, obtuse, attached at base of rostellum; anther erect; pollinia granular.——About 10 species, chiefly in Australia, a few in e. Asia and Malaysia.

1. Microtis formosana Schltr. *M. parviflora* sensu auct. Japon., non R. Br.——NIRABA-RAN. Tubers globose, 5–8 mm. across; stems 10–30 cm. long, 1-leaved near the middle, enveloped by a sheath on lower half, with few leafless membranous sheaths at the base, naked on upper half; leaves nearly radical, terete, 2–2.5 mm. across, longer than or as long as the stem, gradually narrowed to the tip, obtuse, long-sheathing at base; spikes 2–8 cm. long, rather densely flowered; bracts membranous, narrowly ovate-deltoid, 2–4 mm. long, nearly nerveless, mostly acuminate, mucronate; flowers green, about 2.5 mm. across; dorsal sepal elliptic, about 2 mm. long, erect, concave, slightly incurved at tip, obtuse, the lateral sepals gently reflexed, narrowly oblong, obtuse, slightly shorter than the dorsal; petals narrowly oblong, obtuse, about 2/3 as long as the dorsal sepal; lip spreading, somewhat fleshy, quadrangular-oblong, nearly as long as the dorsal sepal, truncate at apex, slightly broadened at base, with 2 obscure callosities at the base; spur absent; ovary short-pedicelled, capsules 3–4 mm. long.——Apr.–May. Grassy places in lowlands; Honshu (Izu Isls., Kazusa, Kii, and Suwo Prov.), Shikoku, Kyushu; rare. ——China, Ryukyus, and Formosa.

13. LISTERA R. Br. FUTABA-RAN ZOKU

Terrestrial, slender, rhizomatous perennials; stems erect, simple, with a few scales or scalelike leaves at base, the principal leaves 2, opposite, near middle of the stem, sessile, spreading, not sheathing, ovate-rounded; inflorescence racemose, the flowers small, usually pedicelled, yellowish green, the bracts small; sepals equal, free, reflexed to spreading; petals resembling the sepals; lip longer than the sepals, attached to base of column, spreading, often pendulous at tip, entire or bifid, usually auriculate near base; column very short, without a foot, wingless; rostellum terminal, sometimes reclined; stigma broad; anther-locules contiguous, the pollinia often bifid in each locule, granular; capsules pedicelled, obovoid or globose, erect or spreading.——About 30 species, in Europe, Siberia, n. India, e. Asia, and N. America.

1A. Leaves slightly above the base of the stem, elliptic to broadly ovate; scalelike leaves on the scape 4–10, with a filiform slightly reflexed apex; lip without auricles. . . . . 1. *L. makinoana*
1B. Leaves nearly at the middle of the stem; scalelike leaves 0–2.
   2A. Lip 5–8.5 mm. long, with obtuse lobes, with or without auricles.
      3A. Lip auriculate near base, glabrous; perianth segments reflexed.
         4A. Lobes of lip linear-oblong, the auricles loosely embracing the column; leaves deltoid or deltoid-cordate. . . . . 2. *L. japonica*
         4B. Lobes of lip ovate, the auricles ascending, not embracing the column. . . . . 3. *L. nipponica*
      3B. Lip without auricles, ciliolate, the lobes ovate; perianth segments spreading. . . . . 4. *L. yatabei*
   2B. Lip 2.5–3.5 mm. long, with linear lobes, the auricles spreading outward; scapes naked. . . . . 5. *L. cordata*

1. Listera makinoana Ohwi. *L. savatieri* sensu auct. Japon., non Maxim. ex Komar.; *L. patentifolia* Nevski—— AO-FUTABA-RAN. Rhizomes slender, creeping, with filiform roots; stems 10–25 cm. long, slender, 2-leaved near the base, glabrous below the leaves, glandular-puberulous on upper portion above the leaves; leaves 2, spreading, elliptic-ovate to ovate-rounded, somtimes deltoid-ovate, 12–40 mm. long, 10–30 mm. wide, obtuse to subacute, broadly truncate to shallowly cordate at base, 3-nerved; scalelike leaves 4–10, loosely alternate, ascending, narrowly linear, 2–5 mm. long, with a filiform slightly reflexed apex; flowers 5–20, greenish, loosely arranged, the bracts spreading, small; sepals oblong-lanceolate, rather obtuse, 2–2.7 mm. long; petals linear, obtuse; lip 2–2.5 times as long as the sepals, narrowly obovate-cuneate, not auricled, the lobes ovate, rounded at apex.——Aug. Woods in mountains; Honshu, Shikoku, Kyushu.

2. Listera japonica Bl. *L. sikokiana* Makino——HIME-FUTABA-RAN, MURASAKI-FUTABA-RAN. Rhizomes slender, creeping; stems erect, 5–18 cm. long, slender, usually 2-leaved above the middle, the lower portion glabrous, the upper portion loosely glandular-puberulent; leaves 2, deltoid to deltoid-cordate, 1–2 cm. long, 0.7–2 cm. wide, sessile, acute or mucronate, broadly truncate to shallowly cordate at base; scalelike leaves absent; flowers 2–5, very loose, purplish green, with reflexed perianth-segments 2.3–3 mm. long; sepals obtuse, narrowly ovate, the lateral ones narrowly oblong; petals linear-oblong, obtuse, the lip 2.5–3 times as long as the sepals, 5.5–8.5 mm. long, cuneate, bilobed, with rather acute auricles embracing the column near the base, the lobes linear-oblong, 3–5 mm. long; capsules 3 mm. long, long-pedicelled.——Apr.–May. Woods in mountains; Honshu, Shikoku, Kyushu; rather rare.——Ryukyus.

3. **Listera nipponica** Makino. *L. savatieri* Maxim. ex Komar., pro parte; *L. brevidens* Nevski——Miyama-futaba-ran. Rhizomes slender, short; stems 10–20 cm. long, erect, 2-leaved near the middle, the upper part with short, spreading, glandular-pubescence, the lower part glabrous; leaves reniform to cordate, 15–25 mm. long, 10–30 mm. wide, abruptly mucronate, truncate-cordate to truncate-rounded at base, the scalelike leaf broadly lanceolate or wanting, racemes 3- to 10-flowered, loose, the bracts ascending, broadly lanceolate, 1–2 mm. long; flowers brownish green; perianth-lobes reflexed; sepals 3–4 mm. long, strongly reflexed from the base, narrowly lanceolate, obtuse; petals as long as the sepals, narrowly oblong; lip twice as long as the sepals, broadly cuneate-obovate, bifid, auriculate near the base, the lobes ovate; capsules ellipsoidal, 3–4 mm. long, the pedicels slightly longer.——July–Aug. Coniferous woods; Hokkaido, Honshu (centr. and n. distr.), Kyushu (Mount Sobosan); rare.——Ussuri.

4. **Listera yatabei** Makino. *L. major* Nakai; *L. savatieri* Maxim. var. *major* (Nakai) Beauverd; *L. convallarioides* subsp. *yatabei* (Makino) Beauverd——Takane-futaba-ran. Rhizomes slender; stems 15–20 cm. long, slender, 2-leaved near

the middle, the lower part glabrous, the upper part glandular-pubescent; leaves reniform to cordate, 15–30 mm. long, 20–30 mm. wide, obtuse or sometimes acute, 3-nerved, the scalelike leaves 2, narrowly linear, acuminate, 1–3 mm. long; racemes loosely 5- to 10-flowered; flowers greenish brown, the bracts ascending, lanceolate to elliptic, 1–2 mm. long; perianth segments spreading; sepals narrowly oblong, about 2 mm. long, obtuse; petals linear; lip 6–8 mm. long, narrowly cuneate, ciliolate, the lobes ovate, nearly erect, obtuse; ovary about 3 mm. long, slightly shorter than the pedicels.——Aug. Coniferous woods; Hokkaido, Honshu (centr. distr.); rare.——s. Kuriles, Sakhalin, Korea, Ussuri, and e. Siberia.

5. **Listera cordata** (L.) R. Br. *Ophrys cordata* L.——Futaba-ran, Ko-futaba-ran. Rhizomes short, slender; stems 10–20 cm. long, slender, 2-leaved near the middle, the lower portion glabrous, the upper part glandular-puberulent; leaves deltoid-reniform or deltoid-cordate, 8–20 mm. long, 1–2.5 cm. wide, obtuse, mucronate, slightly cordate to truncate at base; scalelike leaves absent; racemes loosely 4–10 flowered; flowers greenish yellow or purplish; bracts spreading, deltoid-ovate, rather acute, about 1 mm. long; sepals spreading, narrowly oblong, obtuse, 1.5–2 mm. long, the petals narrowly ovate, nearly as long as the sepals; lip spreading, cuneate, 2.5–3.5 mm. long, deeply bifid, toothed at the base, the lobes linear; ovary pedicelled; capsules ellipsoidal, 3–4 mm. long, slightly shorter than the pedicels.——July–Aug. Woods in high mountains; Hokkaido, Honshu (centr. and n. distr.), Shikoku; rather rare.——Europe, Siberia, and N. America. The Japanese phase has relatively smaller flowers and broader leaves than the plant in Europe and is sometimes called var. **japonica** Hara.

14. NEOTTIA L.   Sakane-ran Zoku

Saprophytic rhizomatous herbs without chlorophyll; stems simple, erect; leaves scalelike, sheathing, scattered on the stems; inflorescence racemose, the flowers small, short-pedicelled, the bracts small; sepals nearly alike, free, concave, spreading; petals similar to the sepals but narrower; lip spreading from the base of the column, longer than the sepals, flat or concave, entire or bifid; column rather long, wingless, nearly terete, without a foot; rostellum incurved; stigma broad; anther erect or incurved, the locules contiguous; pollinia granular; capsules ovoid or ellipsoidal, erostrate.——About 10 species in Eurasia, especially in India and e. Asia.

1A. Stems stout, 25–45 cm. long; upper part of plant inclusive of inflorescence and ovaries brownish crisped-pubescent; flowers 6–7 mm. across; perianth-segments very obtuse; lip deeply bifid, 10–12 mm. long, 2.5–3 times as long as the sepal.
　　1. *N. nidus-avis* var. *manshurica*
1B. Stems slender, 10–25 cm. long; plant glabrous; flowers 3–4 mm. across; perianth-segments gradually acuminate; lip entire, 2–3 mm. long, nearly as long as the sepals. . . . . 2. *N. asiatica*

1. **Neottia nidus-avis** (L.) L. C. Rich. var. **manshurica** Komar. *N. nidus-avis* sensu auct. Orient. Asiat., non L. C. Rich.; *N. papilligera* Schltr.——Sakane-ran. Rhizomes thick, short; stems stout, fleshy, 25–45 cm. long, brownish crisped-pubescent above, erect, simple; leaves scalelike, tubular, membranous, alternate, loose, glabrous, 1.5–4 cm. long; racemes 10–15 cm. long, loosely or somewhat densely flowered, erect, with short, crisped, brown, glandular hairs above; bracts deltoid-lanceolate to deltoid-ovate, membranous, 1-ribbed, gradually narrowed to the tip, 4–7 mm. long; flowers yellow, 6–7 mm. across, with a glandular-pubescent pedicel; sepals and petals obovate, 1-nerved, 5–6 mm. long, very obtuse, concave; lip 10–12 mm. long, deeply bifid, the limb oblong, papillose inside, often pubescent without, the lobes ascending, narrowly oblong, falcate, obtuse or rounded, some-

times truncate at apex; column about 3.5 mm. long.——May–June. Woods in mountains; Hokkaido, Honshu (centr. distr.); very rare.——Sakhalin, s. Kuriles, Manchuria, and e. Siberia. The typical phase occurs in Europe and Siberia.

2. **Neottia asiatica** Ohwi. *N. micrantha* Lindl., non La Llave & Lex.; *N. subsessilis* Ohwi——Hime-muyō-ran. Rhizomes short; stems 10–25 cm. long, rather slender, glabrous, 3- to 4-scaled; scalelike leaves loose, sheathing, tubular, membranous, 2–3 cm. long, very obtuse, few-nerved; racemes 5–10 cm. long, loosely-flowered, glabrous, the bracts membranous, ovate, rather acute, 1–1.5 mm. long, spreading; flowers 3–4 mm. across; sepals and petals ovate-lanceolate, gradually acuminate, 1-nerved, spreading, recurved at tip, about 3 mm. long; petals slightly shorter than the sepals; lip nearly as long as the sepals, deltoid-ovate, 2–3 mm. long, very acute,

spreading, 3-nerved, entire, with slightly involute margins at tip; ovary glabrous, short-pedicelled; capsules about 3 mm. long.——June–Aug. Coniferous woods; Hokkaido, Honshu (centr. and n. distr.); very rare.——Sakhalin, Ussuri, Kamchatka, and Korea.

15. APHYLLORCHIS Bl.   TANEGASHIMA-MUYŌ-RAN ZOKU

Saprophytic herb with fleshy roots; stems simple, with loose alternate scales; inflorescence loosely racemose, the flowers rather small, sessile or pedicelled, the bracts small; sepals nearly equal, free, erect at first, later spreading; petals as long as the sepals but narrower; lip sessile, somewhat spreading, short-clawed, with 2 broad auricles, the limb entire or 3-lobed; column rather long, nearly terete; stigma ovate, depressed, under the short rostellum; anther short-stalked, ovate, the locules contiguous; pollinia granular.——About 15 species, in India, Malaysia, and e. Asia.

1. Aphyllorchis tanegashimensis Hayata. TANEGASHI-MA-MUYŌ-RAN. Stems 40–60 cm. long, erect; scalelike leaves loose, broadly ovate, 1–1.5 cm. long, 8–10 mm. wide, slightly clasping, very obtuse, membranous, appressed, the upper scales narrowly ovate, obtuse, 2 cm. long; racemes rather densely many-flowered, 15–20 cm. long, the bracts lanceolate, acute with an obtuse tip, membranous, 8–15 mm. long, 1.5–3 mm. wide; flowers ascending at first, later spreading or slightly deflexed, pedicelled; sepals oblong-lanceolate, obtuse, 3-nerved, ascending, 1–1.2 cm. long; petals nearly as long as the sepals but slightly narrower, 1-nerved; lip nearly as long as the sepals, erect, auriculate, slightly recurved at tip, the claw about 2 mm. long, the limb ovate, obsoletely 3-lobed, the lateral lobes small, obscure, the terminal deltoid, obtuse, the auricles broader than long, obliquely deltoid, acute, about 3 mm. long; column slightly shorter than the sepals; capsules deflexed, 3–3.5 cm. long, 7–10 mm. wide, straight, linear, terete, with the base of the column persistent at the apex.——Sept. In broad-leaved evergreen forests; Kyushu (s. distr. and Tanegashima).——Ryukyus.

16. CEPHALANTHERA L. C. Rich.   KIN-RAN ZOKU

Terrestrial, usually leafy, erect herb with short rhizomes and elongate roots; leaves cauline, usually plicate, lanceolate to oblong or elliptic, with prominent nerves; flowers rather small, in spikes, erect, white or yellow, the bracts largest in lower part of spike, leaflike; sepals free, not fully open, nearly alike; petals slightly smaller than the sepals, often broader; lip with a chin at base, erect, distinctly jointed between the upper and lower portions, inflated at base, the lateral lobes erect, embracing the column, the terminal lobe larger; column rather long, not winged; rostellum scarcely surpassing the broad stigma; anther broadly ovate, the locules contiguous; pollinia bifid in each locule, granular; capsules erect.——More than a dozen species, in the temperate regions of the N. Hemisphere.

1A. Flowers yellow; sepals narrowly ovate-oblong, obtuse, 14–17 mm. long; bracts deltoid, about 2 mm. long; lip 5- to 7-striate. 1. *C. falcata*
1B. Flowers white; sepals broadly lanceolate, gradually narrowed to the tip, 8–12 mm. long; lip 3-striate.
  2A. Stems and leaves glabrous; leaves narrowly oblong, 3–8.5 cm. long, rarely much-reduced; bracts narrowly deltoid, 1–3 mm. long, the lower 1–2 sometimes rather large, commonly not surpassing the flower; lip with a somewhat acute basal chin. ... 2. *C. erecta*
  2B. Stems, leaves, and ovaries with short, papillose hairs; leaves lanceolate to linear-lanceolate, 7–15 cm. long, acuminate; bracts linear, the lower 1–2 elongate, commonly longer than the spike; lip with a short, obtuse basal chin. .... 3. *C. longibracteata*

1. Cephalanthera falcata (Thunb.) Bl. *Serapias falcata* Thunb.; *Epipactis falcata* (Thunb.) Sw.; *C. japonica* A. Gray; *C. platycheila* Reichenb. f.——KIN-RAN. Smooth, terrestrial herb; stems 40–80 cm. long, 6- to 8-leaved, covered at base by a few membranous sheaths; leaves broadly lanceolate, 8–15 cm. long, 2–4 cm. wide, gradually acute, prominently nerved, narrowed to the base; spikes 3- to 12-flowered, glabrous, smooth, the bracts membranous, deltoid, about 2 mm. long; flowers yellow, rather large, half-open; sepals ovate-oblong, 14–17 mm. long, obtuse, narrowed to the base, the dorsal sepal narrower, concave on back; lip shortly dilated, 3-lobed, longitudinally 5- to 7-striate on the inside, included within the sepals, recurved at tip; column 8–10 mm. long.——Apr.–June. Thickets and thin woods in low mountains; Honshu (Kantō Distr. and westw.), Shikoku, Kyushu; common.——Korea and China.

2. Cephalanthera erecta (Thunb.) Bl. *Serapias erecta* Thunb.; *Epipactis erecta* (Thunb.) Wettst.; ? *C. elegans* Schltr.——GIN-RAN. Slender, glabrous perennial; stems 20–40 cm. long, 3- to 6-leaved, with few sheaths at base; leaves narrowly oblong, 3–8.5 cm. long, 1–2.5 cm. wide, short-acuminate, narrowed to the sheathing base, prominently several-nerved; spikes nearly smooth, 4–8 cm. long, several- to about 10-flowered; bracts narrowly deltoid to broadly lanceolate, 1–3 mm. long, the lower 1 or 2 often somewhat larger, or rarely to 4 cm. long; flowers white; sepals lanceolate, gradually narrowed at tip, obtuse, 7–9 mm. long; petals broadly lanceolate, obtuse; lip about 2/3 as long as the perianth, with a distinct acute chin at base, the terminal lobe transversely elliptic, short-acute; column as long as the lip; ovary slightly shorter than the perianth.——May. Thickets, thin forests, and grassy slopes in low mountains; Hokkaido, Honshu, Shikoku, Kyushu; common.

Var. **subaphylla** (Miyabe & Kudo) Ohwi. *C. subaphylla* Miyabe & Kudo——YŪSHUN-RAN. Green leaves smaller, the bracts well developed.——Hokkaido, Honshu, Kyushu.

Var. **shizuoi** (F. Maekawa) Ohwi. *C. shizuoi* F. Maekawa ——KUGENUMA-RAN. Chin low, obtuse.——Near seashores; Honshu (Kugenuma in Sagami).

3. Cephalanthera longibracteata Bl. *Epipactis longibracteata* (Bl.) Wettst.——SASABA-GIN-RAN. Slender, erect perennial; stems 30–50 cm. long, rather firm, white-scabrous on the angles, alternately 6- to 8-leaved and with a few leafless sheaths at the base; leaves oblong-lanceolate to linear-lanceolate, 7–15 cm. long, 1.5–3 cm. wide, sometimes long-acuminate, shortly sheathed at base, scabrous on margins and on nerves beneath, prominently nerved; spikes loosely several-flowered; bracts linear to broadly so, 4–7 mm. long, the lower 1 or 2 elongate, leaflike, as long as the spike, 5–10 mm. wide; flowers white; sepals erect, about 12 mm. long, lanceolate, acute with an obtuse tip; petals short and broad; lip 2/3 as long as the sepals, the basal chin obtuse, short, not prominent, the termi- nal lobe cordate, abruptly acute; ovary as long as the perianth. ——May–June. Thickets and grassy places in low mountains; Hokkaido, Honshu, Shikoku, Kyushu; rather common.—— Korea.

17. EPIPACTIS R. Br.   Kaki-ran Zoku

Terrestrial erect herbs with creeping rhizomes; stems leafy; leaves ovate to lanceolate, plicate; flowers in racemes, short-pedicelled, nodding, greenish or yellowish, sometimes purplish, the bracts herbaceous, narrow, often longer than the flowers; sepals free, somewhat spreading, equal; petals similar to the sepals; lip sessile at the base of the column, broad, with a dilated limb, the lateral lobes erect, the terminal lobe spreading, jointed at the base with the lower portion, often undulate on the margins; column short; stigma transverse under the broad prominent rostellum; pollinia bilobed in each locule, granular; capsules spreading or pendulous.——About 25 species in temperate regions of Europe, Asia, and N. America.

1A. Lower portion of lip obcordate, the upper portion broadly ovate, obtuse; plant nearly glabrous throughout; bracts scarcely longer than the flowers; perianth segments 12–15 mm. long; petals obtuse. . . . . 1. E. thunbergii
1B. Lower portion of lip semirounded, truncate, the upper portion deltoid, acute; stems, leaves, and axis of inflorescence prominently scabrous; bracts usually longer than the flowers; perianth segments 9–12 mm. long; petals acute. . . . . 2. E. papillosa

1. **Epipactis thunbergii** A. Gray. *Serapias longifolia* sensu Thunb., non L.; *E. longifolia* Bl. sensu auct. Japon., excl. syn.; *E. gigantea* sensu auct. Japon., non Dougl.; *E. gigantea* var. *manshurica* Maxim. ex Komar.——Kaki-ran, Suzu-ran. Rhizomes slender; stems 30–70 cm. long, smooth, 6- to 12-leaved, purplish below and covered with few leafless sheaths at base; leaves narrowly ovate to broadly lanceolate, 7–12 cm. long, rounded and short-sheathed at base, gradually acuminate, glabrous except for a few mammillae on the nerves above; bracts scarcely longer than the flowers; petals ovate, obtuse, nearly as long as the sepals, rounded on lower margin; lip as long as the petals, the lower portion obcordate, the up-per portion broadly ovate, with 2 elevated lines near the base, obtuse; column about 6 mm. long; ovary pedicelled, spreading to pendulous; capsules about 2 cm. long, narrowly oblong.—— June–Aug. Bogs and moist places; Hokkaido, Honshu, Shi-koku, Kyushu; rather common.——Korea, Ussuri, and Man-churia.
2. **Epipactis papillosa** Fr. & Sav. *E. latifolia* var. *papil-losa* (Fr. & Sav.) Maxim. ex Komar.——Ezo-suzu-ran, Ao-suzu-ran. Rhizomes short; stems 30–70 cm. long, promi-nently pilose with brown, crisped, hairlike protuberances, with few leafless sheaths at base; leaves 5–7, elliptic-ovate to broadly lanceolate, 7–12 cm. long, 2–4 cm. wide, prominently short-acuminate, the upper blades long-acuminate, with whitish hairlike protuberances on nerves and on margin, the lower blades short-sheathed at base; racemes 10–20 cm. long, many-flowered; bracts usually longer than the flowers; flowers spreading or nodding, greenish; sepals ascending with re-curved tips, narrowly ovate, acute, 9–12 mm. long; petals rather short, ovate, as long as the sepals, acute; lip pale green, as long as the petals, the lower portion semirounded or de-pressed-obovate, truncate, brown within, the upper portion deltoid, acute; column as long as the lower lobe of the lip; ovary pedicelled; capsules about 1 cm. long, ellipsoidal.—— Aug. Mountains; Hokkaido, Honshu, Kyushu; rare.

Var. **sayekiana** (Makino) T. Koyama & Asai. *Epipactis sayekiana* Makino——Hama-kaki-ran. Flowers pale green and purple spotted.——Near seashores; Honshu (Sagami and Kazusa Prov.).

18. POGONIA Juss.   Toki-sō Zoku

Small, terrestrial herbs with slender, elongate roots; stems slender with a single, narrow, flat leaf at the middle and a bract at apex; basal bracts narrower and smaller, green; flowers solitary, rather large, with a green bract at base; sepals and petals nearly alike, free, suberect; lip erect from base of the column, free, not spurred, undivided or 3-lobed, or often with 2–4 elevated lines and sometimes fleshy-hairy toward tip inside; column elongate, not winged, slightly broadened; rostellum short; stigma flat, placed under the rostellum; anther mobile, the locules parallel; pollinia one in each locule, granular, without a caudicle.——Few species, in e. Asia and N. America.

1A. Flowers half open, rose-purple, 2–2.5 (rarely 1.5) cm. long; sepals oblong-oblanceolate, 3–5 mm. wide, obtuse; lip slightly longer than the perianth segments, the lateral lobes deltoid, the terminal one distinctly recurved; bogs. . . . . 1. P. japonica
1B. Flowers scarcely open, pale rose-purple, 1–1.5 cm. long; sepals linear-oblanceolate, 1.5–3 mm. wide, narrowed at tip and obtuse; lip as long as the petals, the lateral lobes small, the terminal one slightly recurved; grassy slopes. . . . . 2. P. minor

1. **Pogonia japonica** Reichenb. f. *P. similis* Bl.; *P. ophio-glossoides* sensu auct. Japon., non Nutt.——Toki-sō. Rhi-zomes short, slender; stems 20–40 cm. long, with a few blade-less sheaths at base; leaf suberect, lanceolate or linear-oblong, 4–10 cm. long, 7–12 mm. wide, obtuse, decurrent and forming two narrow wings on stem; flowers 2–2.5 cm. long, solitary, terminal, rose-purple; bract leaflike, 2–4 cm. long, 3–6 mm. wide, usually longer than ovary; sepals 3–5 mm. wide, obtuse, the dorsal broadly oblong-oblanceolate, the lateral narrower, as long as the dorsal; petals narrowly oblong, very obtuse, slightly shorter than the sepals, minutely and irregularly toothed on upper margin; lip slightly longer than the perianth segments, 3-lobed, the terminal lobe obovate, distinctly recurved, with fleshy hairs within and on margin, the lateral lobes deltoid; ovary about 1.5 cm. long.——May–July. Bogs; Hokkaido, Honshu, Shikoku, Kyushu.——Korea and China.
2. **Pogonia minor** (Makino) Makino. *P. japonica* var. *minor* Makino——Yama-toki-sō. Very similar to the pre-ceding but smaller, 15–25 cm. high; leaves usually oblanceolate to broadly so, sometimes narrowly oblong, 3–7 cm. long, 4–12

mm. wide; flowers pale rose-purple, 1.5–3 mm. wide, narrowed to an obtuse tip; dorsal sepal narrowly oblanceolate, the lateral ones linear-oblanceolate; lip as long as the petals, nearly erect, with smaller lateral lobes, the terminal lobe slightly recurved; ovary 1–1.5 cm. long.——June–July. Grassy slopes in mountains; Hokkaido, Honshu, Shikoku, Kyushu; rare.

19. GALEOLA Lour.   TSUCHI-AKEBI ZOKU

Saprophytic, leafless herbs without chlorophyll; rhizomes often greatly elongate; stems erect or scandent, elongate, branched; inflorescence racemose, the flowers rather large, the bracts usually small; sepals equal, concave, spreading; petals similar to the sepals; lip sessile, broad, inflated on back, loosely enveloping the column, as long as the sepals, the lateral lobes absent or broad, glabrous or pilose inside; column rather long, sometimes narrowly winged, without a foot; stigma under the rostellum transversely broadened; anther reclined, the locules separated; pollinia bilobed in each locule, granular (sometimes slightly waxy); fruit relatively large, fleshy, indehiscent or occasionally dehiscent; seeds often winged.——More than a dozen species, in e. Asia, Malaysia, India, and e. Australia.

1A. Plant glabrous, scandent, to 20 m. long or sometimes more, much branched; flowers about 12 mm. long; perianth-segments linear-oblong; lip navicular. . . . . 1. G. altissima
1B. Plant brown-pubescent, erect, 50–100 cm. high, branched above; flowers about 2.5 cm. long; perianth-segments narrowly oblong; lip obovate. . . . . 2. G. septentrionalis

1. **Galeola altissima** Reichenb. f. TSURU-TSUCHI-AKEBI, TAKA-TSURU-RAN.   Roots firm, elongate; plant glabrous, scandent, to 20 m. long; stems rather slender, firm, much-branched, lustrous, flexuous, reddish brown, with slightly thickened nodes, glabrous; racemes densely flowered; flowers about 12 mm. long, glabrous; perianth-segments linear-oblong; sepals obtuse, 5-nerved, petals narrower than the sepals, 3-nerved, rounded at apex; lip navicular, erose and undulate on upper margins, with a broad fleshy puberulent disc reaching to the middle and a bilobed woolly mass above it; column rather long; capsules linear, acuminate.——Kyushu (Tanega-shima); very rare.——Ryukyus, Formosa, India, and Malaysia.

2. **Galeola septentrionalis** Reichenb. f. TSUCHI-AKEBI. Rhizomes elongate, thickened; plant brown-pubescent, 50–100 cm. high; stems fleshy, firm, erect, loosely scaly, branched above; scales deltoid- to deltoid-lanceolate, acute to acuminate, slightly fleshy and firm, striate, concave, 1–1.5 cm. long; branches ascending, 5–10 cm. long, interspersed with small scales; racemes densely flowered; bracts 3–5 mm. long; flowers about 2.5 cm. across, yellowish brown; sepals narrowly oblong, obtuse, ascending, fleshy, densely brown-puberulent outside; petals glabrous; lip erect, fleshy, obovate, rounded to obtuse, with elevated cristate lines within, finely toothed on margin, slightly shorter than the sepals; column rather long, slightly incurved; ovary brown-puberulent; fruit pendulous, narrowly cylindric-oblong, 6–8 cm. long, fleshy, red when mature; seeds winged.——June–Aug. Hokkaido, Honshu, Shikoku, Kyushu; rare.

20. LECANORCHIS Bl.   MUYŌ-RAN ZOKU

Leafless, unbranched saprophytic herbs without chlorophyll, blackening when dried; rhizomes creeping, with short-sheathing loosely imbricate scales; stem erect, simple, slender, short-scaled; leaves relatively small; racemes loosely flowered, the bracts scalelike; flowers somewhat tubular, not fully open, white to pale purple; perianth with 3 small appressed calyculae at the base; sepals and petals narrow, alike, free; lip adnate to the column, narrow, erect, broadened at tip and loosely involute, often hairy on the inner surface, sometimes 3-lobed; column elongate, subterete, clavate, thickened at tip; rostellum short, with the rounded, prominent stigma under it; anther on margin of clinandrium, erect or divergent, the locules distinct; pollinia granular.——Few species, in e. Asia, Malaysia.

1A. Lip 3-lobed, the terminal lobe villous inside; flowers loosely arranged.
 2A. Flowers white, 1.5–2 cm. long; lip adnate high on the column, the terminal lobe wholly villous inside inclusive of the margin, the lateral lobes small, semi-orbicular, inserted at base on each side of the terminal lobe or at the top of the limb. . . . . 1. L. japonica
 2B. Flowers pale yellow, about 1 cm. long; lip adnate to the lower half of column, the limb broad, the terminal lobe with a large villous area inside, the lateral lobes glabrous, smaller, only slightly lower than the terminal and long-decurrent on the limb.
2. L. kiusiana
1B. Lip entire, sparingly pubescent inside toward tip, cucullate at apex; flowers pale purple, rather densely arranged. . . . . . . 3. L. nigricans

1. **Lecanorchis japonica** Bl. MUYŌ-RAN.   Rhizomes somewhat thickened, firm, elongate, rather densely scaly; stems erect, 30–40 cm. long, brownish black when dried, with a few loose scales; scalelike leaves rather membranous, 5–8 mm. long, short-sheathing, obliquely truncate, obtuse or subacute, the upper sometimes not sheathing at base; flowers several, 1.5–2 cm. long, not opening widely, white, the bracts deltoid-ovate, subacute, 2–4 mm. long, the caliculus about 1 mm. long, slightly connate at base, deltoid; sepals and petals oblanceolate, obtuse; lip with an oblanceolate limb, adnate high on the column, 3-lobed, the lateral lobes suborbicular, at base of the terminal lobe, the terminal lobe orbicular, villous inside.——July–Aug. Woods in mountains; Honshu, Shikoku, Kyushu; rare.

2. **Lecanorchis kiusiana** Tuyama. USUKI-MUYŌ-RAN. Saprophytic erect perennial, about 10 cm. high; stems turning black when dried, with a few remote scales; flowers few, loosely arranged, about 1 cm. long, pale yellow, not fully opening; sepals and petals oblanceolate; lip adnate to the lower half of the column, the free portion involute, dilated, 3-lobed, the terminal lobe semirounded, undulate-lobulate, loosely papillose on margin, with a large villous area inside, the lateral lobes smaller, glabrous.——May. Kyushu; rare.

3. Lecanorchis nigricans Honda. *L. purpurea* Masam., nom. seminudum——Kuro-muyō-ran. Similar to the preceding; stems more slender, firm, 20–40 cm. long; flowers few, pale purple, rather densely arranged; bracts 2–3 mm. long, deltoid, spreading; lip entire, adnate to the column on the lower half, the free portion obovate, mitre-shaped, obtuse, cucullate at apex, sparsely pubescent inside near tip.——July-Aug. Woods in mountains; Honshu (s. Kinki Distr.), Kyushu; rare.

21. BLETILLA Reichenb. f. Shi-ran Zoku

Terrestrial herbs with pseudobulbs; leaves thin, few from the nodes of the bulbs, jointed at base, plicate, elliptic-lanceolate to oblong; inflorescence spicate, the bracts caducous; flowers rose-purple, rarely white, relatively large; sepals and petals alike, free, ascending; lip 3-lobed, with elevated cristate lines on the inner surface, the lobes loosely embracing the column, the terminal lobe slightly recurved; column with a narrow wing on each side; pollinia 4 in each locule, the caudicle short.——Few species, in e. Asia.

1. Bletilla striata (Thunb.) Reichenb. f. *Limodorum striatum* Thunb.; *Epidendrum tuberosum* Lour.; *Cymbidium hyacinthinum* J. E. Smith; *Bletia hyacinthina* (J. E. Smith) R. Br.——Shiran. Pseudobulbs ovoid-globose, about 4 cm. long; stems 30–70 cm. long, slender, covered at base with several leaf-sheaths, and with 1 or 2 bladeless sheaths near the middle; leaves elliptic-lanceolate to oblong, 20–30 cm. long, 2–5 cm. wide, acuminate, glabrous, gradually narrowed to the sheathing base; inflorescence spicate, loosely 3- to 7-flowered; bracts green, oblong-lanceolate, 2–3 cm. long, 6–8 mm. wide, acute, several-nerved, the margins scarious; flowers rose-purple; sepals and petals alike, 2.5–3 cm. long, 6–8 mm. wide, subacute, several-nerved, ascending; lip as long as the perianth, cuneate-obovate, with loosely involute margins, 3-lobed, the terminal lobe suborbicular, undulate on margin, with 5 elevated lines inside; column about 2 cm. long.——May. Grassy slopes in foothills; Honshu (Kantō Distr. westw.), Shikoku, Kyushu; rare; frequently cultivated in gardens.——Ryukyus and China.——Forma **gebina** (Lindl.) Ohwi. *Bletia gebina* Lindl.; *Bletilla striata* var. *gebina* (Lindl.) Reichenb. f.—— Shirobana-shi-ran. Flowers white.

22. NERVILIA Gaudich. Mukago-saishin Zoku

Terrestrial herbs with bulblike rhizomes; leaves solitary, radical, developing after anthesis, rounded-flabellate to cordate, plicate, petiolate, parallel-nerved; scapes with a few membranous bladeless sheaths near the base; inflorescence racemose, 1- to many-flowered; perianth-segments narrow, nearly alike, subparallel; lip sessile or clawed at base of the column, narrow, entire or 2- to 3-fid; spur scarcely developed; column elongate, clavate; stigma oblong or broader; rostellum short; anther substipitate; pollinia 1 or 2 in each cell; capsules erect or pendulous.——About 60 species, in Africa, India, Malaysia, e. Asia, and n. Australia.

1. Nervilia nipponica Makino. *N. punctata* Makino, excl. syn.——Mukago-saishin. Leaves solitary, long-petioled, cordate-rounded, subacute, 7- to 9-nerved, 3.5–4.5 cm. wide; scapes about 10 cm. long, slender, pale purple, loosely sheathed, 1-flowered, the sheaths (reduced leaves) tubular, obliquely truncate; flowers solitary, terminal, at first ascending, nodding later, about 10 mm. long, purplish; bract thinly membranous, oblanceolate, about 5 mm. long; perianth-segments narrowly oblanceolate, subobtuse, thinly membranous, nearly erect; petals slightly shorter than the sepals; lip narrow, as long as the petals, 3-lobed, the lateral lobes elliptic, rugulose, white, purple-spotted inside, the median lobe very small, obtuse, white; ovary obovoid, prominently pedicelled.——June. Honshu (Izu Isls., Musashi and Izumi Prov.), Kyushu (Tanegashima); very rare.

23. EPIPOGIUM Sw. Torakichi-ran Zoku

Saprophytic herbs without chlorophyll; rhizomes coralloid or tuberous; stems loosely scaly; flowers racemose, short-pedicelled, spreading to pendulous; bracts membranous, usually shorter than the flowers; sepals equal, free, narrow, ascending or reflexed; petals similar to the sepals; lip sessile, ovate, inflated below forming a saccate spur, the lateral lobes absent or small, the terminal spreading, with 2–4 series of papillae within; column short, broadened at tip; rostellum broad and short; stigma prominent, broad, separated from the rostellum; pollinia 1 in each locule, granular, often with filiform caudicles attached to the glands; capsules ovoid, pendulous.——About 7 species, in Europe, w. Africa, Asia, Malaysia, and ne. Australia.

1A. Rhizomes coralloid, much branched and thickened; lip on upper side of flower auriculate. .... 1. *E. aphyllum*
1B. Rhizomes tuberous, not branched; lip entire.
  2A. Sepals and petals narrowly ovate; spur as long as the lip; racemes 1- to 7-flowered. .... 2. *E. japonicum*
  2B. Sepals narrowly lanceolate; petals oblong-lanceolate; spur much shorter than the lip; racemes 8- to 27-flowered. ...... 3. *E. rolfei*

1. Epipogium aphyllum (F. W. Schmidt) Sw. *Satyrium epipogium* L.; *Orchis aphylla* F. W. Schmidt; *E. gmelinii* L. C. Rich.; *E. epipogium* (L.) Karst.——Torakichi-ran. Glabrous, saprophytic, pale brown herb with short, thickened, much-branched coralloid rhizomes; stems 7–20 cm. long, fleshy, loosely sheathed, slightly thickened at base, the sheaths short, thinly membranous, nerveless, obliquely truncate, 5–10 mm. long; racemes loosely 2- to 8-flowered, the bracts narrowly ovate, thinly membranous, erect, nearly nerveless, acuminate, 6–10 mm. long; flowers brownish, about 18 mm. across, pedicelled; sepals narrowly lanceolate, acute, reflexed, 12–14 mm. long, 1-nerved; petals slightly shorter than the sepals, lanceolate, acute; lip nearly as long as the petals, spreading, rose-colored and purple-spotted within, with 4–6 cristate, papillose lines in the center, the lateral lobes (auricles) small, ovate, the terminal lobe large, slightly concave, obtuse, with undulate margins at tip; spur oblong, rounded at apex, slightly shorter than the lip, 6–8 mm. long, 3–4 mm. wide; ovary pedicelled, nearly globose.——Aug.–Oct. Coniferous woods; Hokkaido, Honshu; very rare.——Europe, Caucasus, Himalayas, Siberia, Korea, Sakhalin, Ussuri, and Kamchatka.

2. Epipogium japonicum Makino. *Galera japonica* (Makino) Makino——AOKI-RAN. Rhizomes tuberous, 10–18 mm. long, transversely lined, ovoid; stems erect, fleshy, yellowish, with 2 to 5 loose, thinly membranous, appressed sheaths; racemes erect, loosely 1- to 7-flowered, the bracts ovate, thinly membranous, longer than the pedicels; flowers about 2 cm. long, pendulous; sepals and petals ascending, narrowly ovate, acute, pale yellow with purple spots; lip inferior, somewhat spreading, deltoid, entire, strongly concave, papillate within, whitish with purple spots; spur thick, oblong, obtuse, nearly as long as the lip; column rather short, broadened at tip; ovary globose.——Sept.–Oct. Woods in mountains; Honshu; very rare.

3. Epipogium rolfei (Hayata) Schltr. *Galera rolfei* Hayata; *E. makinoanum* Schltr.; *E. nutans* sensu auct. Japon., non Reichenb. f.——TASHIRO-RAN. Rhizomes tuberous, ellipsoidal, 3–4 cm. long; stems erect, 25–55 cm. long, the sheaths 5–6, loose, membranous; racemes erect, 6–25 cm. long, 8- to 27-flowered, the bracts membranous, reflexed, broadly lanceolate, gradually narrowed to tip, nerveless, 8–12 mm. long; flowers pedicelled, pendulous, about 12 mm. across; sepals narrowly lanceolate, 8–10 mm. long, acuminate, 3-nerved, thin, nearly erect; petals oblong-lanceolate, nearly as long as the sepals; lip inferior, as long as the sepals, broadly ovate, entire, prominently concave, with 2 cristate glandular lines within; spur oblong, much shorter than the lip, about 4 mm. long; ovary ellipsoidal, 6–7 mm. long in fruit, slightly longer than the pedicel.——May–July. Honshu (s. Kantō Distr.), Kyushu (s. distr. incl. Yakushima).——Ryukyus and Formosa.

24. YOANIA Maxim. SHŌKI-RAN ZOKU

Stout fleshy saprophytic herbs with branched rhizomes; scapes simple, scaly; flowers relatively large, fleshy, purplish rose to rose-yellow, in loose or rather dense racemes; sepals free, spreading, oblong, the petals somewhat shorter, ovate; lip suberect, oblong, as long as the petals, concave, broadened at the base, with a large saccate, porrect, obtuse spur below the limb, bearded at the mouth; column slightly shorter than the lip, erect, flat, quadrangular-oblong, auriculate; stigma depressed, transversely broadened; rostellum obsolete; anther persistent, cuspidate; pollinia 2 in each locule, sectile, granular, oblong; capsules spreading to nodding, on long pedicels.——About 3 species, in Japan, Formosa, and the Himalayas.

1A. Flowers 3–7, rather loose, rose-colored. . . . . 1. *Y. japonica*
1B. Flowers 6–15, rather dense, yellow. . . . . 2. *Y. amagiensis*

1. Yoania japonica Maxim. SHŌKI-RAN. Rhizomes elongate, fleshy; stems fleshy, stout, 15–25 cm. long, terete; scales 6–8 mm. long, somewhat spreading, rounded at the apex, fleshy, with membranous margins, convex on back; flowers 3–7, rather loose, rose-colored, the bracts ovate, 6–8 mm. long; sepals fleshy, spreading, oblong, about 22 mm. long, 10–12 mm. wide, obtuse; petals slightly shorter than the sepals; lip as long as the petals, erect, fleshy, the limb subtrapezoid, 1.6 cm. long and as wide, with a broad, papillose line within; spur porrect, oblong, rounded at apex, about 7mm. long, 3–4 mm. wide, as high as the limb; column about 12 mm. long, the lower edges winged and continuous with the margins of the lip; ovary about 7–10 cm. long inclusive of the pedicel. ——July–Aug. Woods in mountains; Hokkaido, Honshu, Kyushu; rare.

2. Yoania amagiensis Nakai & F. Maekawa. KIBANA-NO-SHŌKI-RAN. Rhizomes much branched, thickened, puberulent, membranous-scaly at the nodes; stems erect, 20–30 cm. long, fleshy, stout, terete, with scattered deltoid scales; flowers 6–15, rather dense, yellow; sepals oblong, 25–27 mm. long, rounded at apex, about 9-nerved; petals ovate, obtuse, minutely serrate; lip about 22 mm. long, inflated; spur porrect, slightly longer than the limb, incurved, 6–8 mm. long, 4–5 mm. across, rounded at the apex, the limb broadly ovate, obtuse, longitudinally ridged; ovary, inclusive of long pedicel, 8–10 cm. long.——July. Woods in foothills; Honshu (Izu, sw. Kantō and Yamato Prov.), Kyushu; rare.

25. STIGMATODACTYLUS Maxim. KŌROGI-RAN ZOKU

Greenish white, delicate, saprophytic herbs with small fleshy rhizomes; stems with one scalelike leaf near the middle and several more near the base; flowers few, racemose, small, the bracts ovate, small; sepals and petals alike, spreading, linear, acute; lip spreading, rounded, thin, flat, with a small bilobed appendage near the base; column erect, slightly incurved in upper part, narrowly wing-margined, with a small fingerlike appendage on the center within, toothed at base; pollinia 4, ellipsoidal, nearly waxy; ovary short-pedicelled.——About 4 species, in India, Japan, and Malaysia.

1. Stigmatodactylus sikokianus Maxim. KŌROGI-RAN. Pale green, glabrous, delicate, leafless herb, with tuberous rhizomes 2–3 mm. in diameter; stems 5–10 cm. long, simple, the basal scales small, the median one ovate, 3–5 mm. long, acuminate, spreading, nerveless; flowers 2 or 3, pale purplish green, loosely racemose, the bracts similar to the median cauline scale; sepals linear, membranous, obtuse, 1-nerved, scattered-ciliate near the base, the dorsal sepal 4 mm. long, the lateral ones 2.5 mm. long; petals linear, acute, about 3.5 mm. long; lip rounded, about 4 mm. long, with obsolete teeth on upper margin, the appendage about half as long as the lip, lanceolate and deeply bifid; column erect, about 3.5 mm. long, 2-toothed; ovary pedicelled, 5–7 mm. long.——Aug.–Oct. Woods in mountains; Shikoku, Kyushu; very rare.

26. GASTRODIA R. Br.   Oni-no-yagara Zoku

Saprophytic, leafless herbs without chlorophyll, with thickened, rather short rhizomes; stems simple, erect, loosely scaly; flowers erect or pendulous, in loose racemes, bracteate, rather small; sepals and petals connate, urceolate or short-tubular, 5-fid, the lobes short, ovate, equal, or the inner (petals) smaller; lip adnate to the base of the column, erect, more or less adnate to but shorter than the perianth, somewhat involute, flat or with 1 or 2 pairs of elevated lines within; column subterete, the stigma barely perceptible; rostellum small; pollinia granular, without caudicles; capsules erect, ellipsoidal, the pedicels short or sometimes much elongate after anthesis.——About 25 species, in Africa, India, Malaysia, Australia, and e. Asia.

1A. Rhizomes glabrous; stems 60–100 cm. long, rarely shorter; flowers many, 7–8 mm. long; pedicels not elongating after anthesis.
   2A. Pedicels shorter than the ovary. .... 1. *G. elata*
   2B. Pedicels longer than the ovary. .... 2. *G. gracilis*
1B. Rhizomes with unicellular hairs; stems 3–10 cm. long; flowers few, 1–2 cm. long; pedicels much elongating after anthesis.
   3A. Stems 3–4 cm. long; flowers usually 2, about 2 cm. long, tubular; lip with 4 crispate ridges within, and a spherical appendage on the margin on each side near the base. .... 3. *G. nipponica*
   3B. Stems 3.5–10 cm. long; flowers 2–7, about 11 mm. long, broadly tubular-campanulate; lip with an elevated bilobed ridge above the center, and a 4-angled short-cylindric appendage on each side near the base. .... 6. *G. confusa*

1. **Gastrodia elata** Bl. Oni-no-yagara. Tall, glabrous, pale brown herb; rhizomes oblong, somewhat fleshy, glabrous, about 10 cm. long, 3.5 cm. across; stems yellowish brown, 60–100 cm. long, terete, the scales loose, membranous, 1–2 cm. long, the lower scales short-sheathing; racemes 10–30 cm. long, many-flowered, the bracts lanceolate to linear-oblong, obtuse, membranous, weakly nerved, 7–12 mm. long, about 2 mm. wide; flowers yellowish brown, the pedicels shorter than the ovary; perianth-tube broadly ovoid, oblique at the mouth, 7–8 mm. long, slightly inflated at base, the lobes small, deltoid; lip about 2/3 as long as the tube, adnate to the under side near base, the claw short, the limb broadly ovate, obsoletely 3-lobed, the terminal lobe larger, short-fimbriate; column subterete, slightly broadened at tip, 6–7 mm. long; capsules oblong-obovoid, 12–15 mm. long, short-pedicelled.——Woods in mountains; Hokkaido, Honshu; rather rare.——Forma **viridis** (Makino) Makino. *Gastrodia viridis* Makino. Plant pale green; occurs with the typical phase.——Formosa and China.

2. **Gastrodia gracilis** Bl. Nayo-temma. Rhizomes smooth, glabrous; stems 10–30 cm. long, slender, with 6–7 loosely arranged, membranous basal sheaths; racemes 2–5 cm. long, loosely 6- to 14-flowered; bracts much shorter than the pedicels, ovate or elliptic-ovate, 2–3.5 mm. long, obtuse, rarely subacute; pedicels slender, longer than the ovary, 7–10 mm. long, to 2 cm. long in fruit; flowers spreading to nodding; perianth-tube about 8 mm. long, slightly inflated at base on the lower side, the outer lobes nearly equal, very obtuse; capsules broadly ellipsoidal, about 1 cm. long.——June–July. Honshu (Awa and Aki Prov.); very rare.

3. **Gastrodia nipponica** (Honda) Honda & Tuyama. *Didymoplexis pallens* sensu auct. Japan., non Griff.; *D. nipponica* Honda——Haruzaki-yatsushiro-ran. Rhizomes fusiform, 2–3.5 cm. long, to 5 mm. across, cross-lined, with unicellular hairs; stems terete, 3–4 cm. long, with 2 or 3 scales; flowers usually 2, about 2 cm. long, tubular, the bracts broadly ovate, slightly clasping, about 4 mm. long; perianth tube violet-brown, about 20 mm. long, 12 mm. across, narrowed to the base, broadest above the middle, the outer lobes ovate, the terminal lobe narrower, slightly retuse, the lateral lobes acuminate, obsoletely rugulose, the inner lobes oblong, about 3 mm. long, obtuse; lip about 9 mm. long, 5.5 mm. wide, the claw short, adnate to base of the perianth-tube, with a globose appendage on each side at the apex, the limb broadly ovate, obtuse, with 4 crispate ridges within, the inner ridges disappearing on the lower half, the outer ridges disappearing on the upper half; column about 7 mm. long, lanceolate.——May. Woods; Honshu (Kii Prov.), Shikoku, Kyushu; very rare.

4. **Gastrodia confusa** Tuyama. Akizaki-yatsushiro-ran. Rhizomes pyriform, 1.5–4 cm. long, 0.7–1.5 cm. across, with unicellular hairs; stems 3.5–10 cm. long, whitish, the scales few, ovate, not at all or very short-sheathing, membranous; racemes short; flowers 2–7, about 11 mm. long, broadly tubular-campanulate, the bracts ovate, gradually narrowed to tip, 4–7 mm. long, perianth-tube about 11 mm. long, 8 mm. across, slightly inflated at base on the lower side, slightly contracted at the mouth, divided to the middle on the ventral side, the outer lobes slightly incurved, the terminal lobe shorter, subtruncate and retuse, the lateral lobes ovate, subacute, the inner lobes rounded, 3–4 mm. wide at base; lip about 5.5 mm. long, 4 mm. wide, the claw short, with a depressed, tubular, 4-angled protuberance on each side at apex, the limb broadly ovate, short-cuspidate, obtuse, with a median bilobed ridge within; column about 6 mm. long, auricled at the apex, wing-margined, obovate, slightly incurved; stigma in a depression of the column on the lower portion of the ventral side; rostellum small, subtruncate; pedicels to 30 cm. long following anthesis.——Oct. Woods; Honshu (Sagami and Kii Prov.), Shikoku, Kyushu; very rare.

27. SPIRANTHES L. C. Rich.   Nejibana Zoku

Plants terrestrial, erect; rhizomes fasciculate, short; leaves narrow, mostly radical; inflorescence spicate, one-sided, spirally twisted; flowers small, sessile, white or rose; sepals free, nearly equal, the dorsal sepal frequently erect and forming a hood, the lateral sepals obliquely attached on the top of the ovary, sometimes decurrent; lip sessile or clawed, erect, concave, loosely embracing the column, flat above, entire or 3-lobed, with a callose tubercle on each side; column subterete, often decurrent on the ovary, without a foot; stigma broad; rostellum erect, short and obtuse or elongate, more or less bifid; anther erect, ovate, the locules separate; pollinia granular, sometimes with caudicles; capsules erect, oblong.——About 100 species, widely distributed in warmer and tropical regions.

1. Spiranthes sinensis (Pers.) Ames. *Aristotelea spiralis* Lour.; *S. spiralis* Makino, non C. Koch; *Neottia sinensis* Pers.; *S. australis* sensu auct. Asiat., non Lindl.; *N. amoena* Bieb.; *S. amoena* (Bieb.) Spreng.——NEJIBANA, MOJIZURI. Roots slightly thickened; stems 10–40 cm. long, erect, slender, few-leaved, densely white-pubescent in upper part; radical leaves linear to narrowly oblanceolate, 5–20 cm. long, 3–10 mm. wide, acute to acuminate, narrowed to the base, the stem-leaves few, smaller, appressed, lanceolate, acuminate, sheathed at base; spikes 5–15 cm. long, densely white-pubescent, densely many-flowered, the bracts narrowly ovate or oblong, 4–8 mm. long, long-acuminate, appressed to the ovary; flowers pale rose, rarely white, 4–5 mm. across, slightly open; sepals lanceolate, 4–5 mm. long, gradually narrowed, obtuse; petals erect, slightly shorter than the sepals, linear-lanceolate; lip obovate, white, slightly longer than the sepals, minutely dentate on the margins, slightly recurved at tip, with a small callosity on each side; ovary sessile; capsules erect, ellipsoidal, 6–7 mm. long.——Apr.–Aug. Grassy places in lowlands and hills; Hokkaido, Honshu, Shikoku, Kyushu; very common.——Sakhalin, Kuriles, Siberia, China, Manchuria, Korea, Ryukyus, Formosa, India, Malaysia, and Australia.

28. GOODYERA R. Br. SHUSU-RAN ZOKU

Terrestrial, rarely subepiphytic perennials, mostly with fleshy roots; leaves mostly basal, alternate, ovate to lanceolate, petiolate; inflorescence spicate, the flowers relatively small, bracteate; sepals nearly equal, the dorsal erect, forming a hood with petals, the lateral ones spreading to suberect; lip sessile, ascending, concave, entire, acute, recurved, hairy at base; column short, nearly terete, not appendaged; stigma simple, broad; rostellum erect; anther-locules separate; pollinia oblong; capsules erect, erostrate.——About 100 species, in Europe, N. America, especially abundant in the warmer part of Asia.

1A. Flowers many, 15–100 in dense spikes; sepals 3–4 mm. long.
  2A. Stems stout, 40–80 cm. long, erect from a very short creeping base; spikes 7–25 cm. long; leaves acuminate, 8–15 cm. long.
           1. *G. procera*
  2B. Stems 10–20 cm. long, declined to ascending from a long creeping base; spikes 4–7 cm. long; leaves acute, often abruptly so, 2–6 cm. long.
    3A. Stems ascending; leaves ovate, 3–6 cm. long, 2–3.5 cm. wide; sepals obtuse, ovate, the lateral sepals narrowly ovate-oblong.
           2. *G. hachijoensis*
    3B. Stems declined; leaves broadly lanceolate to ovate or narrowly so; sepals subobtuse, narrowly ovate. .... 3. *G. pendula*
1B. Flowers 1–12, usually in loose spikes; sepals 4–30 mm. long.
  4A. Inflorescence 3- to 12-flowered, rarely fewer; flowers obliquely ovate in lateral view; sepals narrowly ovate, 4–10 mm. long.
    5A. Inflorescence pedunculate, upper part of stem and ovary densely pubescent; sepals sparingly puberulent.
      6A. Sepals 4–5 mm. long. .... 4. *G. repens*
      6B. Sepals 7–10 mm. long.
        7A. Flowers 2–3; sepals about 10 mm. long, acuminate. .... 5. *G. ogatae*
        7B. Flowers 4–12; sepals 7–10 mm. long, obtuse.
          8A. Inflorescence with minute straight or crisped hairs; bracts ascending, not appressed to the ovary; sepals 7–8 mm. long; leaves with the midrib white-striate. .... 6. *G. velutina*
          8B. Inflorescence with rather long, crisped hairs; bracts erect, appressed to the ovary; sepals 8–10 mm. long; leaves white-reticulate on upper surface, rarely green. .... 7. *G. schlechtendaliana*
    5B. Inflorescence sessile, glabrescent or sparingly papillose-pilose. .... 8. *G. maximowicziana*
  4B. Inflorescence 1- to 3-flowered; flowers long-tubular; sepals linear, 2.5–3 cm. long. .... 9. *G. macrantha*

1. Goodyera procera (Ker-Gawl.) Hook. *Neottia procera* Ker-Gawl.; *G. lancifolia* Fr. & Sav.——KINGIN-SŌ. Stems erect from a short-creeping base, stout, fleshy, 40–80 cm. long, rather densely leafy on the lower half; leaves broadly lanceolate to oblong, 8–15 cm. long, 2–6 cm. wide, acuminate, acute to nearly rounded at base, 5- to 7-nerved, the upper leaves few, reduced, the petioles 3–7 cm. long, broadened into a short sheath at base; spikes densely many-flowered, 7–25 cm. long, slightly pubescent; sepals 3–4 mm. long, ovate, obtuse; lip broadly ovate, rounded at apex, about 2 mm. long, pilose, the inner side with 2 callose protuberances; column nearly as long as the lip; ovary glabrous.——Kyushu (Yakushima).——Ryukyus, Formosa, China, India, and Malaysia.

2. Goodyera hachijoensis Yatabe. *Peramium hachijoense* (Yatabe) Makino——HACHIJŌ-SHUSU-RAN. Stems about 15 cm. long, ascending from a long-creeping base; leaves ovate, 3–6 cm. long, 2–3.5 cm. wide, acute, rounded at base, with a white line along the midnerve, the petioles 1–1.5 cm. long inclusive of the short sheaths; spike short, densely many-flowered, slightly one-sided, short-pubescent, the scape with a few scales, about 5 cm. long; bracts linear-lanceolate, acuminate, 6–12 mm. long, ciliate, longer than the ovary, embracing the lower part; flowers small, white changing to yellow; sepals ovate, about 4 mm. long, obtuse, the dorsal sepal ovate, the lateral sepals narrowly oblong-ovate, slightly greenish; petals forming a hood with the dorsal sepal, white, oblanceolate; lip broadly ovate, rather obtuse, as long as the sepals, saccate-dilated at base, pale yellow, pubescent within; column short, about 2/3 as long as the lip; rostellum short, bifid; pollinia bilobed in each locule; ovary about 6 mm. long, glabrous.——Sept.–Oct. Woods in mountains; Honshu (Hachijō); rare.

Var. **yakushimensis** (Nakai) Ohwi. *G. yakushimensis* Nakai——YAKUSHIMA-SHUSU-RAN. Leaves ovate to oblong-ovate, with white spots along the midnerve, the bracts pubescent below.——Woods in mountains; Kyushu (Yakushima and Tanegashima); rather rare.

Var. **leuconeura** (F. Maekawa) Ohwi. *G. leuconeura* F. Maekawa——SHIRAITO-SHUSU-RAN. Leaves with white nerves, the margins undulate; spur and limb of the lip slender.——Woods in mountains; Kyushu (Yakushima and Tanegashima).

Var. **matsumurana** (Schltr.) Ohwi. *G. matsumurana* Schltr.; *Peramium matsumuranum* (Schltr.) Makino——KAGOME-RAN. Slightly larger than the preceding varieties; leaves broader, sometimes subcordate, 4–6 cm. long, 2–3 cm.

wide, white-reticulate on upper side along the nerves and veinlets, the bracts lanceolate.——Kyushu (Yakushima and Tanegashima) ; rare.——Ryukyus and Formosa.

3. Goodyera pendula Maxim. TSURI-SHUSU-RAN. Stems short-creeping, declined above, 10–15 cm. long; leaves few, alternate, broadly lanceolate to ovate or narrowly so, 2–3.5 cm. long, 5–10 mm. wide, acuminate, acute to cuneate at base, 3- to 5-nerved, the petioles rather broad, shortly sheathing at base, the upper 3 or 4 leaves linear-lanceolate; inflorescence with arcuate peduncles, densely many-flowered, one-sided, 4–6 cm. long, scattered crisped-pubescent, the bracts lanceolate, membranous, 4–7 mm. long, 1-nerved, spreading; flowers white, not fully open; sepals about 4 mm. long, narrowly ovate, subobtuse, slightly crisped-pubescent without, 1-nerved; petals narrowly oblanceolate, closely appressed to the dorsal sepal; lip slightly shorter than the sepals, saccate at base, smooth within, the limb narrowly ovate.——Aug. Tree-trunks and rocks in mountain forests; Hokkaido, Honshu, Shikoku, Kyushu; rare.

Var. **brachyphylla** F. Maekawa. HIROHA-TSURI-SHUSU-RAN. A northern variant, mostly with shorter and broader leaves. ——Occurs with the typical phase.

4. Goodyera repens (L.) R. Br. *Satyrium repens* L. ——HIME-MIYAMA-UZURA. Stems erect, 10–20 cm. long, from creeping base; leaves chiefly basal, ovate, 1–2.5 cm. long, 7–18 mm. wide, acute to subobtuse, rounded to acute at base, the petioles about 1 cm. long, short-sheathing; inflorescence 3–7 cm. long, 5- to 15-flowered, slightly one-sided, long crisped-pubescent, with 1 or 2 remote scales; flowers white, the bracts lanceolate, appressed to the ovary, shorter than the flowers; sepals deltoid-ovate, 4–5 mm. long, somewhat elongate at tip, obtuse; petals oblanceolate, closely appressed to the dorsal sepal; lip as long as the sepals, inflated below, the limb short, deltoid-ovate, recurved, glabrous within; column about 2/5 as long as the sepals.——Aug. Coniferous woods in mountains; Hokkaido, Honshu, (centr. and n. distr.); rare.——Europe, Siberia, Sakhalin, s. Kuriles, Korea, Himalayas, and N. America.

5. Goodyera ogatae Yamamoto. SHIMA-SHUSU-RAN. Stems 10–15 cm. long, creeping at base; leaves few, on the lower portion of the stems, ovate, acute, rounded at base, the petioles short, short-sheathing at the base; peduncles with few scalelike leaves, pubescent; flowers 2–3, pale reddish brown, the bracts linear-lanceolate, about 12 mm. long, long-acuminate, 3-nerved, ciliate; sepals broadly lanceolate, about 10 mm. long, acuminate, the lateral sepals spreading; petals as long as the sepals; lip nearly as long as the sepals, inflated, elliptic, the limb rather short, deltoid, reflexed; column about 10 mm. long.——Woods; Kyushu (Yakushima); very rare.——Formosa.

6. Goodyera velutina Maxim. *Peramium velutinum* (Maxim.) Makino——SHUSU-RAN, BIRŌDO-RAN. Stems 10–15 cm. long from an elongate creeping base; leaves several, oblong–ovate or ovate, 2–4 cm. long, 1–2 cm. wide, acute, deep green and purplish, white-striate on the midnerve, rounded at base, the petioles sheathing below; peduncles short, whitish pilose, the scalelike leaves 2–3, linear; inflorescence 4- to 10-flowered, secund, minutely hairy, the bracts ascending, linear-

lanceolate, 6–12 mm. long, long-acuminate; sepals 7–8 mm. long, narrowly ovate, subobtuse, whitish, 1-nerved; petals closely appressed to the dorsal sepal, broadly oblanceolate; lip as long as the sepals, saccate at base, pubescent within, the limb ovate, obtuse; capsule 7–10 mm. long in fruit.——Aug.- Sept. Woods; Honshu (centr. distr. and westw.), Shikoku, Kyushu; rather rare.——Korea.

7. Goodyera schlechtendaliana Reichenb. f. *G. japonica* Bl.——MIYAMA-UZURA. Stems 12–25 cm. long from a creeping base; leaves ovate to narrowly so, 2–4 cm. long, 1–2.5 cm. wide, acute, flat, entire, rounded or subcordate at base, white-reticulate on upper side, rarely green, the petioles 1–2 cm. long, with a membranous short sheath at base; scapes elongate, densely soft crisped-pubescent, the scalelike leaves 2–3, erect, linear, often short-sheathing; inflorescence 5–10 cm. long, 7- to 12-flowered, rather long crisped-hairy; flowers rose to white, half-open; bracts erect, appressed to the ovary; sepals 8–10 mm. long, narrowly ovate to broadly lanceolate, obtuse, 1-nerved; petals broadly oblanceolate, appressed to the median sepals; lip nearly as long as the sepals, saccate at base, semi-rounded, pilose within, the limb ovate, subobtuse; column about 3/5 as long as the sepals; capsules 8–12 mm. long.—— Aug.–Sept. Woods; Honshu, Shikoku, Kyushu; rare.—— Korea and China.——Forma **similis** (Bl.) Makino. *G. similis* Bl.——AO-MIYAMA-UZURA. Leaves green throughout; occurs with the typical phase.

8. Goodyera maximowicziana Makino. *G. bifida* sensu Maxim., non Bl.; *G. foliosa* var. *laevis* Finet; *Peramium maximowiczianum* (Makino) Makino——AKEBONO-SHUSU-RAN. Stems 7–10 cm. long, from a long-creeping base; leaves green, oblong-ovate to ovate, 2–3.5 cm. long, 1.2–2 cm. wide, acute, acute to rounded at base, often oblique, the petioles about 1 cm. long, short-sheathing at base; inflorescence sessile, rather densely 3- to 7-flowered, secund, glabrescent or sparingly papillose-pilose, the bracts lanceolate, 1–1.5 cm. long, nearly erect, ciliate; flowers half-open, white; sepals narrowly ovate, obtuse, 8–10 mm. long; petals broadly oblanceolate, closely appressed to the median sepal; lip nearly as long as the sepals, saccate-inflated at base, pilose within, the limb ovate, obtuse; column about 3/5 as long as the sepals.——Woods; Hokkaido, Honshu, Shikoku, Kyushu; rather rare.——Korea and Ryukyus (Amami Oshima).

9. Goodyera macrantha Maxim. *Peramium macranthum* (Maxim.) Makino——BENI-SHUSU-RAN. Stems 4–8 cm. long from a long-creeping base; leaves ovate to narrowly so, 2–4 cm. long, 1–2.2 cm. wide, acute to obtuse, usually rounded at the base, white-reticulate on upper side, the petioles 1–2 cm. long, short-sheathing at base; inflorescence 1- to 3-flowered, loosely long crisped-pubescent, the bracts broadly linear, suberect, 1.5–2 cm. long, long-acuminate, slightly pubescent; flowers long-tubular; sepals broadly linear, 2.5–3 cm. long, obtuse, loosely pubescent without; petals linear, narrowed toward the base; lip 17–20 mm. long, saccate at base, sparsely pubescent within, the limb elongate, lanceolate, subacute, with a recurved tip; column about 1/3 as long as the sepals; capsules 15–18 mm. long.——July–Aug. Woods; Honshu (centr. distr. and westw.), Shikoku, Kyushu; rare.—— Korea (Wangtao).

29. HETAERIA Bl.   HIME-NO-YAGARA ZOKU

Terrestrial, sometimes saprophytic herbs from creeping, rarely somewhat thickened rhizomes; stems erect; leaves ovate to lanceolate, sometimes reduced to scales; flowers small, in spikes; sepals free or connate at base, the lateral sepals broadest; petals narrow, often closely appressed to and forming a hood with the dorsal sepal; lip erect, from the base of the column, saccate at base, with callose protuberances within, the margin often adnate to the lower margin of the column, the limb entire or bifid; column short, with 2 appendages in front; stigma broad, bilobed, depressed, with protuberances on the lower margins; anther-locules separate; pollinia sectile, with caudicles; capsules erect.——About 65 species, in Malaysia, Polynesia, India, and se. Asia.

1A. Stems with green leaves.
  2A. Lateral sepals about 3 mm. long, ovate, semi-open at anthesis. .... 1. *H. yakusimensis*
  2B. Lateral sepals about 5 mm. long, lanceolate, fully open at anthesis. .... 2. *H. xenantha*
1B. Aphyllous saprophyte, without chlorophyll, wholly glabrous. .... 3. *H. sikokiana*

1. Hetaeria yakusimensis (Masam.) Masam.

Zeuxine yakusimensis Masam.——YAKUSHIMA-AKA-SHUSU-RAN. Terrestrial, red-brown when dry; stems erect from a short-creeping base, 10–15 cm. long, 2- to 5-leaved near the base, glabrous toward the base, short-pilose on the upper part; leaves obovate or ovate, about 3 cm. long, 1.5 cm. wide, acute, glabrous, rounded at base, the petioles about 2 cm. long, sheathing at base; inflorescence about 10 cm. long, loosely 6- to 10-flowered; bracts 1-nerved, membranous, ciliate, as long as the ovary; median sepal broadly ovate, acute, about 3 mm. long, the lateral sepals oblique-oblong; petals subovate, obtuse; lip cucullate-ovate, as long as the sepals, with 2 callose protuberances within; column with 2 winglike ridges within.——Sept. Woods in mountains; Kyushu (s. distr., incl. Yakushima); rare.

2. Hetaeria xenantha Ohwi & T. Koyama.

ŌSUMI-KINU-RAN. Stem arising from a long creeping base, terete, glabrous, including the inflorescence 8–15 cm. tall; leaves 3 to 7 on a stem, broadly ovate or elliptic, to 4 cm. long, abruptly acute, upper surface deep green, velvetlike, the petiole to 1 cm. long, the lower half sheathing; scape wiry, wholly pubescent, with 2 or 3 scales below the middle, terminated by a 4–5 cm. long spikelike raceme of 3 to 6 remote flowers, the bracts scalelike, lanceolate, as long as or slightly shorter than the ovary, ciliate on the upper margin; sepals glabrous, the median one ovate, 5 mm. long, acute, reddish or brownish green outside, the lateral lanceolate, 5 mm. long, gradually attenuate, olive-green outside; petals forming a hood with the central sepal, 4.5 mm. long, thin, elliptic-ovate, white; labellum conical-ovoid, ascending, 4.5 mm. long, 2 mm. wide, white, glabrous, with a pair of hooked processes inside the saclike base, the limb 3-lobed, the lateral lobes papillose, involute; column including rostellum 2.8 mm. long, with an enlargement on each side.——Aug.–Sept. Wet evergreen woods; Kyushu (Ohsumi).

3. Hetaeria sikokiana (Makino & F. Maekawa) Tuyama.

Chamaegastrodia sikokiana Makino & F. Maekawa——HIME-NO-YAGARA. Plants pale brown, saprophytic, without green leaves; rhizomes rather thick, creeping, branching, with small loose scales; stems erect, rather fleshy, 5–15 cm. long, glabrous, with loose scales; leaves scalelike, 4–10 mm. long, thinly membranous, subobtuse, 1-nerved, sheathing at base; inflorescence 2–3.5 cm. long, 5- to 10-flowered; bracts ovate-oblong or ovate, 5–8 mm. long, erect; sepals obtuse, 1-nerved, 3–4 mm. long; petals narrowly oblong, the lip about 6 mm. long, saccate-inflated at base, with 2 globose callosities within, the limb T-shaped; column with 2 horn-shaped appendages on the inner surface.——June–July. Honshu (Rikuzen Prov. and southw.), Shikoku, Kyushu; rare.

30. CHEIROSTYLIS Bl. KAIRO-RAN ZOKU

Small, terrestrial herbs from creeping rhizomes; stems leafy at base, with scalelike leaves in the upper part; inflorescence racemose, loosely flowered, bracteate; flowers rather small, usually short-pedicelled; sepals at the base more or less united into a tube becoming 3-lobed above the middle; petals 2, free, closely appressed to the inner side of the dorsal sepal; lip saccate-inflated, with two protuberances on the inside, the limb bifid, entire or toothed; column short, with an appendage near the rostellum on each side; pollinia caudiculate, bilobed in each locule; stigma lateral, bilobed.——About 20 species, in tropical and subtropical regions of Asia and Africa.

1. Cheirostylis okabeana Tuyama.

TANEGASHIMA-KAI-RO-RAN. Stems 10–20 cm. long, slightly fleshy, from creeping rhizomes, 4- to 6-leaved toward the base; leaves oblong-ovate to broadly ovate, 7–20 mm. long, 5–12 mm. wide, acute, mucronate, glabrous, red-brown when dry, slightly fleshy, the petioles 5–10 mm. long, broadened and short-sheathing at base; peduncles with white recurved hairs, the scales 2 or 3, remote, appressed, sheathing at base; racemes short, 2- to 5-flowered, the bracts 5–7 mm. long, lanceolate; ovary white-pubescent, subsessile; flowers small; sepals slightly united below into a tube, the lobes or segments to 3.7 mm. long, the dorsal lobe slightly shorter; lip nearly included, with 2 callosities on inner side of inflated lower part, the limb T-shaped, the lobes ovate, entire, obtuse.——Woods in mountains; Kyushu (Tanegashima); very rare.

31. VEXILLABIUM F. Maekawa HAKUUN-RAN ZOKU

Small terrestrial herbs; stems long-creeping; leaves small, ovate, the petioles sheathing at base; inflorescence scapose, erect, spicate, few-flowered; flowers small, white, the bracts membranous; calyx campanulate, slightly pubescent outside; petals as long as and appressed to the dorsal sepal; lip slightly exserted, saccate-inflated at base, with 2 protuberances within, the claw slender, short and entire, the limb triangular to quadrangular or oblong; spur bilobulate; column erect, nearly terete, the rostellum unequally bifid, minutely papillose; pollinia sectile.——About 4 species, in Japan and Korea.

1A. Sepals distinctly connate toward base.
  2A. Limb of lip quadrangular-oblong, truncate or retuse; leaves less than 7 mm. long.
    3A. Leaves 3–7 mm. long; spikes 1- to 4-flowered. .... 1. *V. nakaianum*

1.   Vexillabium nakaianum F. Maekawa.   HAKUUN-RAN. Stems 5-12 cm. long, long-creeping at base, with 2–4 leaves in lower part, pubescent above; leaves ovate, 3–7 mm. long, 2.5–7 mm. wide, acute, the petioles 3–6 mm. long, sheathing at base; scalelike leaves 2; spikes 1- to 4-flowered; flowers white; sepals forming a campanulate tube below the middle, pubescent outside, the dorsal lobe 3.2–3.5 mm. long, acute, the lateral lobes 4.8 mm. long; petals obovate-oblong, the lip about 4.5 mm. long, retrorsely saccate at base, forming a short, retuse spur, the claw entire, narrow, the limb about 2 mm. long, 4.5 mm. wide, retuse.——Aug.   Woods; Kyushu (Mount Kurodake in Bungo); very rare.

2.   Vexillabium fissum F. Maekawa.   Ō-HAKUUN-RAN. Stems about 13 cm. long, ascending, from a short-creeping base, glabrous at base, pubescent above; leaves about 6, rounded to ovate-elliptic, 1–1.5 cm. long, 8–11 mm. wide, rounded at both ends or obtuse at apex, 3-nerved, the petioles 7–8 mm. long, dilated at base and sheathing for half the length; spikes scattered-pubescent, about 7-flowered; bracts linear-lanceolate, as long as the ovary, acute, about 6 mm. long; flowers several; sepals pubescent outside, connate below the middle, the dorsal lobe about 4 mm. long, narrowly ovate, the lateral lobes oblong, 4.5 mm. long; petals appressed to the dorsal sepal; lip about 6 mm. long, with an obtuse, saccate, retrorse base, the claw entire, the limb about 2 mm. long, 5 mm. wide.——July.   Honshu (Miyake Isl. in Izu); very rare.

3.   Vexillabium yakushimense (Yamamoto) F. Maekawa.   *Anoectochilus yakushimensis* Yamamoto——YAKU-SHIMA-HIME-ARIDŌSHI-RAN.   Stems 4–10 cm. long, long-creep-ing at base, 3- to 5-leaved in the lower part, glabrous, with scattered white multicellular hairs above; leaves broadly ovate, 7–15 mm. long, rounded at base, 3- to 5-nerved, the petioles short, sheathing at base; upper reduced leaves 1–3, lanceolate, 5–8 mm. long, appressed to the stem; spikes 1- to 3-flowered; flowers white, the bracts scarious, pubescent, 3–6 mm. long, triangular-lanceolate, acute; calyx 3.5–4 mm. long on dorsal side, 5–5.5 mm. on lateral side, tubular-campanulate, loosely pubescent outside, the lobes acute; petals appressed to the dorsal sepal; lip 7–8 mm. long, the spur saccate, the claw slen-der, sometimes 1- to 2-toothed on the upper margins, the limb triangular, retuse, about 4 mm. long, 6 mm. wide.——Aug. Woods; Kyushu (Yakushima); very rare.

4.   Vexillabium inamii Ohwi.   ISE-RAN.   Small peren-nial creeping herb; stems 5–6 cm. long, 3- to 4-leaved toward top; leaves orbicular-ovate, 6–10 mm. long, 5–8 mm. wide, short-petiolate; scape 2–3 cm. long, with white multicellular hairs, 1-bracteate below the middle; flowers 1–3, rose-colored, in loose 1-sided racemes, the bracts broadly lanceolate, 5–6 mm. long, appressed to the ovary, loosely ciliate; sepals only very briefly connate at base, the median deltoid-ovate, about 3.5 mm. long, erect, obtuse, 1-nerved, loosely pubescent out-side at base, the lateral oblong, erect, about 5 mm. long; petals as long as the median sepal, the lip exserted from the perianth, the claw lanceolate, about twice as long as the limb, with 2 minute tubercles on the sac, the limb spreading, obcordate-trapezoid, about 3.5 mm. long, 5 mm. wide, bifid at apex. ——Aug.   Evergreen forest; Honshu; rare.

32.   ODONTOCHILUS Bl.   INABA-RAN ZOKU

Terrestrial rhizomatous erect herbs, leafy toward the base; leaves ovate to lanceolate, petiolate; inflorescence spicate, the flowers relatively small, sessile; dorsal sepal erect, forming a hood with the petals, the lateral sepals forming a chin with the foot of the column; lip with protuberances inside the saccate base, the claw spreading, fimbriate or dentate, the limb abruptly dilated upward and deeply bifid, conduplicate in bud; column very short, without frontal appendages; stigma often bifid; anther-locules distinct; pollinia sectile, with caudicles; capsules erect.——About 10 species, in India, Malaysia, and Formosa.

1.   Odontochilus inabae Hayata.   *Anoectochilus inabae* Hayata——INABA-RAN.   Stems rather stout, 10–20 cm. long, glabrous and leafy below; leaves alternate, ovate-oblong to ob-long, about 4.5 cm. long, 2 cm. wide, acute, rounded to obtuse at base, the petioles about 2 cm. long, sheathing at base; pe-duncles with few scales, pubescent; inflorescence loosely few-flowered; flowers rose; dorsal sepal ovate to broadly so, about 5.5 mm. long, slightly caudate and obtuse, the lateral sepals oblong, spreading, about 11 mm. long; petals lanceolate, ap-pressed to the dorsal sepal; lip about 23 mm. long, saccate at base, the claw slender, fimbriate, the limb V-shaped, bifid, the lobes ascending, cuneate; column about 0.8 mm. long; ovary pubescent.——Reported to occur in Kyushu (Yaku-shima).——Formosa.

2.   Odontochilus hatusimanus Ohwi & T. Koyama. HATSUSHIMA-RAN.   Stem slenderly terete, arising from a long-creeping base, 10–15 cm. tall, including the inflorescence; leaves 4–6 on a stem, the blades ovate to ovate-elliptic, to 4 cm. long, 2 cm. wide, 3-nerved, fresh green, abruptly short-acute, rounded at base, the petiole patent, nearly 15 mm. long, sheathing below; scape erect, slenderly terete, pubescent, with few scales below the middle, terminated by a spikelike raceme of 2 to 7 flowers; flowers white and pink-tinged, about 4 mm. long; bracts lanceolate, light reddish, 2/3 as long as the ovary; sepals pubescent outside, the central one ovate-acute, the lat-eral ovate, 3.2 mm. long; petals appressed on the central sepal, thin, elliptic, 3 mm. long; labellum patent, much ex-serted, T-shaped, 5.5 mm. long, the hypochilus rounded-ciliate, with a pair of horns inside, the mesochilus slenderly cylindric, the epichilus 2-lobed, the lobes ovate-acute, 1.8 mm. long; column 1.5 mm. long, with 2-fid rostellum and 2 appendages at apex.——July–Aug.   Kyushu (Ohsumi).

33. MYRMECHIS Bl.   ARIDŌSHI-RAN ZOKU

Small, low, terrestrial herbs from long-creeping rhizomes; leaves few, on lower part of stem, ovate, short-petiolate; flowers relatively small, spicate, few, sometimes solitary; sepals connate at base, mostly erect, the lateral sepals forming a short chin with the foot of the column; petals broadly lanceolate; lip attached to the base of the column, with 2 protuberances within, ventricose at base, saccate, spurless, the claw narrow, entire, the limb short, broad, bifid; column very short; stigmas 2, distinct, on each side of the short erect rostellum; pollinia sectile; caudicles very short; capsules erect, sessile.——Few species in e. Asia.

1A. Lip slightly longer than the perianth, distinctly dilated in upper part on the bifid obtriangular limb. . . . . 1. *M. japonica*
1B. Lip slightly shorter than the perianth, not dilated, subtruncate. . . . . 2. *M. tsukusiana*

1. Myrmechis japonica (Reichenb. f.) Rolfe. *Rhamphidia japonica* Reichenb. f.; *M. gracilis* sensu Fr. & Sav., non Bl.——ARIDŌSHI-RAN.   Stems 3–8 cm. long, long-creeping at base, glabrous in lower part, the upper part with scattered, white, crisped multicellular hairs; leaves 5–7, green, broadly ovate, 5–12 mm. long, 4–8 mm. wide, obtuse, the petioles 5–10 mm. long, briefly dilated below with a short membranous sheath at base; flowers 1 or 2, rarely 3, white, the bracts scarious, 1-nerved, narrowly 3-angled, appressed to the ovary, acute, 4–6 mm. long; sepals lanceolate, 6–7 mm. long, gradually narrowed to the obtuse tip, the dorsal sepal recurved, the lateral ones ascending; petals linear; lip slightly longer than the perianth segments, saccate at base, distinctly dilated in upper part to the bifid obtriangular limb; column very short. ——Woods in mountains; Hokkaido, Honshu (centr. and n. distr.); rare.

2. Myrmechis tsukusiana Masam. TSUKUSHI-ARIDŌSHI-RAN.   Stems 4–5 cm. long, creeping at base; leaves flat, entire, ovate-rounded to ovate, 5–10 mm. long, 5–7 mm. wide, obtuse to acute, short-petiolate; peduncles 15–25 mm. long, pubescent; bracts obovate-lanceolate, embracing the base of the ovary, ciliate, membranous; flowers 1 or 2; sepals ovate-lanceolate, about 4 mm. long, 2.5 mm. wide, slightly narrowed to the obtuse tip, slightly united at base; petals lanceolate, 4.5 mm. long, 2 mm. wide; lip slightly shorter than the perianth segments, spreading, lanceolate, about 4 mm. long, 2 mm. wide at base, saccate, subtruncate at apex; column very short, erect.——July–Aug. Woods; Kyushu (Yakushima); very rare.

34. ZEUXINE Lindl.   KINU-RAN ZOKU

Terrestrial, erect, unbranched herbs from a creeping base; leaves linear to ovate, petiolate, sheathing at base; flowers small, sessile, in spikes, the bracts membranous; sepals nearly equal, free, the dorsal erect, forming a hood with the narrow closely appressed petals, the lateral sepals spreading; lip slightly adnate to the base of the column, erect, included, concave, smooth or with 2 callosities within, the limb entire or bifid; column not appendaged, short, the stigma with 2 protuberances; anther-locules contiguous; pollinia granular; capsules erect, ovate.——About 40 species, in India, Malaysia, tropical Africa, and e. Asia.

1. Zeuxine strateumatica (L.) Schltr. *Orchis strateumatica* L.; *Z. sulcata* Lindl.; *Adenostylis strateumatica* (L.) Ames——KINU-RAN, HOSOBA-RAN.   Stems 5–10 cm. long, rather densely leafy at base; leaves fleshy, linear, 1.5–4 cm. long, 3–4 mm. wide, acuminate, sheathing at base; spikes 2–4 cm. long, densely flowered, the bracts scarious, triangular-ovate, 7–12 mm. long, long-acuminate, 1-nerved, longer than the flowers; flowers sessile, half-open, abruptly recurved above the ovary; sepals ovate-oblong, 5–6 mm. long, obtuse, 1-nerved; petals and lip as long as the sepals; capsules ellipsoidal, about 6 mm. long.——Apr. Kyushu (s. distr. incl. Yakushima and Tanegashima).——India, Malaysia, China, Formosa, and Ryukyus.

35. TROPIDIA Lindl.   NETTAI-RAN ZOKU

Terrestrial, slender, sometimes branched herbs; leaves thin, plicate, many-nerved; flowers relatively small, in terminal spikes; bracts herbaceous; sepals not fully open, the dorsal narrowest, the lateral forming a short chin below the lip at base; petals as long as the lateral sepals; lip sessile, erect, grooved, saccate or spurred at base, the limb entire, as long as the sepals, usually acuminate; column short; rostellum elongate, acuminate, erect, with a membranous margin embracing the anther on each side; anther-locules contiguous; pollinia granular; capsules ovoid, acutely angled.——More than 20 species, in tropical Asia and the Pacific Islands.

1. Tropidia nipponica Masam. YAKUSHIMA-NETTAI-RAN. Stems 10–30 cm. long, slender, erect, stiff, glabrous, the lower portion with 3–7 membranous sheaths; leaves 1–4, on the upper part of the stem, thin, firm, ovate-oblong to lanceolate, 5–10 cm. long, 2–3.5 cm. wide, caudate-acuminate, narrowed to the petiole, many-nerved, plicate; uppermost leaf at the base of the peduncle broadly linear, about 4 cm. long, 3–5 mm. wide; spikes about 10-flowered, the peduncle about 6 cm. long, the bracts lanceolate, 1–3 mm. long, acuminate; flowers yellow; dorsal sepal narrowly obovate, 6–8 mm. long, 3–3.5 mm. wide, the lateral sepals completely united into an obovate, navicular blade; petals lanceolate, about 6.5 mm. long, 3 mm. wide; lip ovate-lanceolate, about 5 mm. long, 3.5 mm. wide, saccate at base, abruptly recurved at apex; column about 3 mm. long. ——July–Aug. Honshu (Hachijō Isl. in Izu), Shikoku (Tosa Prov.), Kyushu (Yakushima and Koshiki); very rare.

36. CALYPSO Salisb.　HOTEI-RAN ZOKU

Terrestrial herb with a single elliptic, plicate, petiolate, radical leaf; flowers rather large, showy, solitary, terminal, nodding; sepals and petals narrow, equal, free; lip longer than the perianth lobes, spreading, broadly ovate, strongly inflated and saccate above, the lateral lobes flat, the upper margins united to the dorsal lobe which is entire or shallowly bifid; column shorter than the sepals, winged, broad and without a foot; pollinia 2 in each locule, sessile, waxy, ovoid; capsules oblong-fusiform, 6-ribbed.——A single species, in the cooler regions of Eurasia and N. America.

1. Calypso bulbosa (L.) Reichenb. f. var. **speciosa** (Schltr.) Makino. *C. speciosa* Schltr.; *C. bulbosa* var. *japonica* Makino, excl. syn; *Cytherea bulbosa* var. *speciosa* (Schltr.) Makino——HOTEI-RAN.　Glabrous perennial from a short, fusiform or narrowly ovoid pseudobulb; leaves ovate-elliptic, obtuse to acute, 2.5–5 cm. long, 1.5–3 cm. wide, slightly plicate, deep green above, purple beneath, the petiole 1.5–4 cm. long; scape 6–15 cm. long, erect, slender, with 2 membranous, sheathing scales near the base; flowers solitary, terminal, the bract broadly linear, 1.2–2.5 cm. long, acuminate; sepals and petals spreading, brownish rose, linear-lanceolate, 2–3 cm. long, 3–4 mm. wide, acuminate; lip pendulous, broadly ovate, 2.5–3 cm. long, white with pale brown spots inside, the pouch antrorse, bifid, softly long-pubescent within, the limb contracted, rounded at apex, with undulate margins; column flattened, ovate-elliptic, 1.3–1.5 cm. long, about 1 cm. wide; capsules clavate, cylindric, about 2.5 cm. long, long-pedicelled, ascending.——May–June.　Coniferous woods in mountains; Honshu (north and centr. distr.); very rare. The Japanese phase nearly corresponds with the N. American var. *occidentalis* Holz, which has the limb of the lip shorter than the apical appendages of the pouch.

37. OBERONIA Lindl.　YŌRAKU-RAN ZOKU

Evergreen epiphytes; stems elongate; leaves coriaceous to herbaceous, distichous, equitant, more or less fleshy, laterally flattened or rarely terete, sheathed at base; flowers minute, numerous, in spikes, or racemes, the bracts very small; sepals free, nearly equal; petals shorter and narrower than the sepals; lip sessile, spurless, inflated on the back, fimbriate or 2- to 3-lobed, broadened at base and embracing the column in natural position; column very short, terete, narrowed at base, without a foot; pollinia 4, in 2 locules, waxy; capsules small.——About 150 species, in tropical Asia, Malaysia, Pacific Islands, and Australia.

1. Oberonia japonica (Maxim.) Makino. *Malaxis japonica* Maxim.; *O. makinoi* Masam.——YŌRAKU-RAN.　Small, pendulous, caespitose epiphyte; stems 1–4 cm. long, densely leafy; leaves 4–10, laterally flattened, linear-oblong, 1–4 cm. long inclusive of the short sheath, 2–6 mm. wide, short-acute to obtuse, mucronate; inflorescence terminal, densely flowered, 2–8 cm. long, about 2.5 mm. across, the peduncle slender, 1–2 cm. long; bracts deltoid, membranous, 0.5–2 mm. long, acuminate, spreading; flowers yellow to greenish yellow, about 1 mm. across; sepals spreading, broadly ovate, obtuse; petals ovate; lip obovate, 5-lobed, the terminal lobe shorter than the others; capsules short-pedicelled, obovoid, about 2.5 mm. long. ——Apr.–June.　Tree-trunks and rocks; Honshu (s. Kantō and westw.), Shikoku, Kyushu.

38. TIPULARIA Nutt.　HITOTSU-BOKURO ZOKU

Small terrestrial herbs; leaves solitary, basal, ovate, rather long-petiolate; scapes from the pseudobulb, with few sheathing scales at base; flowers small, short-pedicelled, in loose racemes, the bracts minute or absent; sepals and petals equal, spreading, free, narrow; lip erect at the base of column, sessile, as long as the sepals, slenderly spurred, 3-lobed, the lateral lobes small, the terminal lobe flat, oblong, entire or retuse; column shorter than the sepals, erect, narrowly winged above; anther terminal, pollinia 4 in 2 cells, waxy, unappendaged; capsules clavate-fusiform or oblong-cylindric, pendulous.——About 4 species, in N. America, e. Asia, and the Himalayas.

1. Tipularia japonica Matsum. HITOTSU-BOKURO.　Pseudobulbs conical-ovoid, 1–1.5 cm. long; leaves solitary, the petioles 3–7 cm. long, the blades deltoid- to elliptic-ovate, 3.5–7 cm. long, 1.5–3 cm. wide, acuminate, rounded at base, green above, purplish beneath; scapes erect, slender, 20–30 cm. long, with 2 or 3 sheathing scales on lower part; inflorescence loosely 5- to 10-flowered, 4–6 cm. long, the bracts minute; flowers yellowish green; sepals and petals narrowly oblanceolate, about 4 mm. long, obtuse; lip obovate, 3-lobed, about 3 mm. long, the terminal lobe obtuse, entire, the lateral ones minutely toothed; spur slender, about 5 mm. long; capsules fusiform, short-pedicelled, pendulous, about 1 cm. long.——May–June. Woods in mountains; Honshu, Shikoku, Kyushu.

39. DIDICIEA King & Prain　HITOTSU-BOKURO-MODOKI ZOKU

Small terrestrial herbs with pseudobulbs; leaves solitary; inflorescence racemose; flowers small, yellowish green or dark purple, the bracts minute; sepals and petals nearly alike, erect, narrow; lip without a foot at the base of the column, erect, fleshy, oblanceolate, sometimes inflated, spurless; column short, winged or nearly terete with a groove on the inner side; anthers 2-locular, deciduous; pollinia 4, without caudicles; rostellum short, truncate.——Two species, one in Japan and another in Sikkim Himalaya.

1. Didiciea japonica Hara.　HITOTSU-BOKURO-MODOKI. Pseudobulbs globose, about 1 cm. long; leaf ovate, about 4.5 cm. long, 2.3 cm. wide, acute, cordate, brownish purple beneath, 5-nerved, the petiole about 3 cm. long; scape slender, about 15 cm. long, with 2 membranous sheaths at base; inflorescence about 4 cm. long, loosely about 8-flowered, the bracts deltoid, delicate; sepals narrowly oblong, obtuse, about 3 mm. long, 1 mm. wide, 3-nerved, brown-purple, ascending; petals slightly shorter than the sepals; lip simulating the petals, completely spurless; column about 1.5 mm. long, nearly terete, with a groove on the ventral side, obliquely truncate; ovary about 4 mm. long, short-pedicelled.——May. Kyushu (Mount Tara in Hizen); very rare.

40. EPHIPPIANTHUS Reichenb. f.　　Ko-ichiyō-ran Zoku

Delicate terrestrial, glabrous herbs with slender, elongate rhizomes; leaves solitary; scapes erect with few scales on the lower portion; racemes loosely few-flowered; bracts minute, nerveless; flowers small, yellowish green; sepals and petals narrowly oblong, somewhat spreading, obtuse; lip as long as the sepals, short-clawed, narrowly oblong, spreading, short-auriculate, the limb or lobe obtuse, entire; column slender, erect, shorter than the sepals, slightly incurved in upper part, angled on each side; stigma large; anthers terminal; pollinia 4, in 2 locules, globose; ovary clavate-cylindrical, short-pedicelled.——Two species in e. Asia.

1A. Column slender, not flattened or appendaged; lip entire. .... 1. E. schmidtii
1B. Column flattened, with an auriculate wing on each side on the upper portion; lip denticulate. .... 2. E. sawadanus

1. **Ephippianthus schmidtii** Reichenb. f.　*E. sachalinensis* Reichenb. f.; *Liparis schmidtii* (Reichenb. f.) Benth.
——Ko-ichiyō-ran.　Slender glabrous perennial with creeping rhizomes; leaves solitary, broadly ovate, flat, 1.5–3 cm. long, 1–2.5 cm. wide, obtuse, rounded to subcordate at base, the petioles 2–5 cm. long; scapes slender, 10–20 cm. long, with 2 sheathing scalelike leaves in lower part; racemes 1–3 cm. long, loosely 2- to 7-flowered, the bracts membranous, about 1 mm. long; flowers horizontally spreading or nodding, yellowish green; sepals narrowly oblong, about 5 mm. long, obtuse, 3-nerved; petals slightly shorter than the sepals; lip short-clawed at base, abruptly dilated above, the limb about as long as the sepals, flat, entire, oblong-ovate, slightly recurved, with a fleshy, callose protuberance on each side at base; column about 2/3 as long as the sepals; ovary pedicelled; capsules pendulous, clavate-fusiform, 8–10 mm. long.——July–Aug.

Mossy places in coniferous woods in mountains; Hokkaido, Honshu (centr. and n. distr.).——Sakhalin, and s. Kuriles.

2. **Ephippianthus sawadanus** (F. Maekawa) Ohwi. *Hakoneaste sawadana* F. Maekawa——Hakone-ran.　Leaves oblong, 1.5–3 cm. long, 1–2.5 cm. wide, obtuse, acute to subrounded at base, the petioles 1.5–2.5 cm. long; scapes 8–15 cm. long, with 1 or 2 remote membranous sheathing scales 5–12 mm. long; racemes 1–4 cm. long; flowers 3–5, yellowish green, the bracts membranous, 1 mm. long, obtuse, toothed; sepals 4.5–5 mm. long, obtuse, 3-nerved; petals slightly shorter than the sepals, denticulate; lip 4–5 mm. long, 2–2.2 mm. wide, slightly involute and brownish purple on inner side at tip, denticulate; column 2.5 mm. long, flattened, cuneate-obovate, with an auriculate wing on each side; ovary about 3 mm. long. ——June. Woods in mountains; Honshu (Hakone, Fuji, Chichibu, Yamato); very rare.

41. MICROSTYLIS Nutt.　　Hozaki-ichiyō-ran Zoku

Plants small, terrestrial or epiphytic; stems thickened at base; leaves few, chiefly radical, membranous, broad, flat, sheathed at base; flowers small or minute, in terminal racemes, the bracts small, narrow; sepals free, spreading; petals usually narrower than the sepals, sometimes filiform; lip sessile, erect or spreading, inflated on the back, entire, 2- to 3-lobed or fimbriate, cordate or auriculate; column very short, terete, emarginate, auriculate; anthers erect, deciduous, the locules separate, persistent; pollinia 4, in 2 locules, waxy, ovoid; capsules small, ovoid.——About 200 species, widely distributed, especially abundant in the Tropics.

1. **Microstylis monophyllos** (L.) Lindl.　*Ophrys monophyllos* L.; *Malaxis monophyllos* (L.) Sw.; *Microstylis diphylla* Lindl.; *Acroanthes monophylla* (L.) Greene; *Liparis inconspicua* Makino——Hozaki-ichiyō-ran.　Pseudobulb ovoid, covered with old leaf-sheaths; leaves 1 or rarely 2, thin, broadly ovate to elliptic, 4–8 cm. long, 3–5 cm. wide, subacute to obtuse, rounded to abruptly narrowed at base, the sheaths 3–8 cm. long; peduncles 15–30 cm. long, naked; racemes 8–15 cm. long, rather densely flowered, the bracts deltoid-lanceolate, 1–2 mm. long, acuminate; flowers 2.5–3 mm. across, pale green; sepals slightly recurved, linear, 2.5 mm. long; petals lanceolate, as long as the sepals, subobtuse, 1-nerved; lip as long as the sepals, the lower half reniform, abruptly narrowed at tip, subacute, 3-nerved; column very short; capsules obovoid, pedicelled, erect, 5–6 mm. long, crowned with the remnants of the decayed perianth.——July. Coniferous woods; Hokkaido, Honshu (centr. and n. distr.), Shikoku; rather rare.——Cooler regions of N. America and Eurasia.

42. MALAXIS Sw.　　Yachi-ran Zoku

Stems short; leaves 2 or 3, basal or nearly so, small, short-sheathing at base; flowers minute, green, racemose, short-pedicelled; sepals free, spreading, the petals much smaller, spreading; lip ascending, without a foot, longer than the petals, cordate and loosely embracing the column; column very short, terete, retuse, toothed at the apex; anthers in a depression on the column, nearly erect, persistent, the locules separate; pollinia 4, in 2 locules, waxy, obovoid-oblong.——A single species widely distributed in Eurasia and N. America.

1. **Malaxis paludosa** (L.) Sw.　*Ophrys paludosa* L.; *Hammarbya paludosa* (L.) O. Kuntze——Yachi-ran.　Stems 5–10 cm. long, from a creeping base; leaves 1–3, with small tubers in the axils, narrowly oblong to broadly lanceolate, 1–2.5 cm. long, 4–10 mm. wide, obtuse, gradually narrowed to a short-sheathing base; peduncles slender, naked; racemes 3–5 cm. long, rather densely flowered, the bracts broadly lanceolate, appressed to the ovary, 1.5–2.5 mm. long; flowers pale green,

2.5–3 mm. across; sepals narrowly ovate, 1-nerved, suboptuse, about 2 mm. long, the dorsal sepal recurved, the lateral ones spreading; petals ovate, about 1 mm. long, 1-nerved; lip on upper side of flower, erect, deltoid-ovate, 1.5 mm. long, with lower margins slightly involute; capsules ellipsoid-globose, pedicelled, about 2.5–3 mm. long.——Aug. Bogs; Hokkaido, Honshu (centr. and n. distr.); very rare.——s. Kuriles, Sakhalin, Siberia, Europe, and N. America.

43. ELEORCHIS F. Maekawa   Sawa-ran Zoku

Pseudobulbs small, globose; scapes erect, slender; leaves solitary, linear-lanceolate, sheathing toward the base and enveloping the scape; flowers 1 (rarely paired), rose-purple, the bracts minute, thinly membranous; sepals and petals alike, free, oblanceolate, erect, gibbous; lip narrowly obovate or oblanceolate, adnate below to the base of column; column narrowly cylindric, ridged on each side; rostellum flat, truncate; anthers terminal, reclined; pollinia 4, in 2 locules, granular, the caudicles not developed.—— One species in Japan.

1. **Eleorchis japonica** (A. Gray) F. Maekawa. *Arethusa japonica* A. Gray; *Bletilla japonica* (A. Gray) Schltr.——Sawa-ran, Asahi-ran.   Pseudobulbs about 6 mm. across; scapes 20–30 cm. long, slender, terete, naked, with 1 or 2 sheathing scales at base; leaf radical, erect, linear-lanceolate or broadly linear, 6–15 mm. long, 4–8 mm. wide, gradually narrowed at both ends; flowers 1 (rarely paired), the bracts deltoid, thinly membranous, 2–3 mm. long; sepals oblanceolate, subacute, rose-purple, nearly erect, 2–2.5 cm. long, 5-nerved; petals nearly as long but slightly narrower than the sepals; lip as long as the sepals, obovate-oblong, denticulate, obsoletely 3-lobed, rounded to truncate at apex; column about 2 cm. long; ovary linear, 1–1.5 cm. long, (to 2.5 cm. in fruit), nearly sessile, erect.——July. Bogs; Hokkaido, Honshu (centr. and n. distr.).——s. Kuriles.

Var. **conformis** (F. Maekawa) F. Maekawa. *E. conformis* F. Maekawa——Kirigamine-asahi-ran.   Flowers erect, tubular; sepals 13–17 mm. long; lip flat, entire, broadly oblanceolate.——July. Bogs; Honshu; rare.

44. LIPARIS L. C. Rich.   Kumo-kiri-sō Zoku

Terrestrial or epiphytic herbs; stems leafy on the pseudobulb; leaves sometimes jointed at the base, flat or conduplicate, evergreen or deciduous; flowers small, white, purple or brownish, racemose, the bracts small; sepals free, spreading, as wide as or wider than the petals; lip attached to the base of the column, the lateral lobes inconspicuous or rather prominent, embracing the column in a natural position, the terminal lobe larger, with a globose tubercule on each side at base; anthers terminal; pollinia 4, in 2 locules, waxy; capsules globose to oblong-cylindric.——About 300 species in Eurasia and N. America, especially abundant in the Tropics.

1A. Leaves usually solitary, evergreen, narrowly oblong or linear-oblong, with a distinct joint within the sheath; peduncles compressed, recurved in upper part; flowers one-sided; column winged. .... 1. *L. plicata*
1B. Leaves usually 2 or 3, broadly lanceolate to broadly ovate-elliptic, without a distinct joint within the sheath; peduncles erect, 3- to 5-angled.
  2A. Pseudobulbs cylindric, 3–10 cm. long; lip cuneate-oblong.
    3A. Lip rounded at apex, the appendages at base elliptic, obtuse; bracts reflexed. .... 2. *L. formosana*
    3B. Lip retuse, the appendages at base subulate, acute; bracts spreading. .... 3. *L. nervosa*
  2B. Pseudobulbs ovoid-globose, 1–2 cm. long.
    4A. Lip oblong-cuneate or cuneate, with slightly recurved margins, truncate to rounded at apex.
      5A. Leaves broadly lanceolate, acuminate; veinlets closely approximate, longitudinal. .... 4. *L. odorata*
      5B. Leaves narrowly ovate to ovate-rounded, rather obtuse to abruptly acute; veinlets subremote.
        6A. Leaves broadly ovate to orbicular ovate, abruptly acute, rounded to cordate at base; flowers whitish; sepals linear-oblong, acute; lip straight, with 2 callosities at base. .... 5. *L. auriculata*
        6B. Leaves elliptic to oblong, obtuse at both ends; flowers greenish or greenish purple; sepals oblong, obtuse; lip abruptly reflexed above the middle, the callosities absent at base or obscure. .... 6. *L. kumokiri*
  4B. Lip obovate to narrowly obovate-oblong, rounded to obtuse, sometimes mucronate or caudate, flat, abruptly narrowed at base.
    7A. Lip about ⅔ as long as the petals, caudate. .... 7. *L. krameri*
    7B. Lip about as long as the petals, broadly obovate, short-pointed.
      8A. Lip 7–8 mm. long, 4.5–5.5 mm. wide, pointed; plant 20–40 cm. high. .... 8. *L. japonica*
      8B. Lip 12–18 mm. long, 10–15 mm. wide, rounded, mucronate; plant 15–30 cm. high. .... 9. *L. makinoana*

1. **Liparis plicata** Fr. & Sav. *Cestichis plicata* (Fr. & Sav.) F. Maekawa——Chi-kei-ran.   Evergreen herb; pseudobulbs elongate-conical, 1–3 cm. long, few-scaled while young; leaves usually solitary, narrowly oblong or linear-oblong, conduplicate, ribbed, 10–18 cm. long, 1.5–3 cm. wide, acuminate at both ends, distinctly jointed with the sheath; peduncles 10–18 cm. long, terminal on the pseudobulbs, flat, narrowly wing-margined, arcuately recurved in upper part; racemes loosely 5- to 10-flowered, ascending; bracts subulate-lanceolate, 3–5 mm. long; flowers pale yellowish green, one-sided; dorsal sepal linear-oblong, obtuse, 6–7 mm. long, the lateral ones slightly shorter, spreading, arcuate in upper part; petals linear, 6–7 mm. long, obtuse; lip cuneate-obovate, about 5 mm. long, rounded-truncate, smooth, gradually recurved at tip, with 2 indistinct callosities at the base inside; column winged near the tip; capsules pedicelled, clavate-fusiform, 8–10 mm. long.——Oct.–Dec. Shikoku, Kyushu (s. distr.). Ryukyus and Formosa.

2. **Liparis formosana** Reichenb. f. Yūkoku-ran. Pseudobulbs elongate, cylindric, terete, the internodes to 5 cm. long; leaves 3, obliquely ovate to elliptic, 5–15 cm. long, 3–8 cm. wide, abruptly acuminate, short-sheathed at base; peduncles terminal, 20–30 cm. long, erect; racemes terminal, many-flowered; bracts narrowly ovate, 1-nerved, membranous, reflexed, 2–3 mm. long, acute; flowers brownish purple; sepals linear-oblong, obtuse, the lateral gradually spreading, 8–10 mm. long; petals linear-oblong, as long as the lateral sepals; lip cuneate-obovate, rounded at apex, gradually recurved in

upper part, 8–10 mm. long, 5–6 mm. wide, shallowly grooved at tip, entire, the appendages at base elliptic, obtuse; column about 5 mm. long, narrowly winged at apex.——June. Woods in mountains; Kyushu.——Ryukyus, Formosa, and China.

Var. **hachijoensis** (Nakai) Ohwi. *L. hachijoensis* Nakai; *L. bicallosa* Schltr., quoad pl. Japon.; *L. bituberculata* sensu Makino, non Lindl.——SHIMA-SASABA-RAN. Flowers smaller than in the typical phase.——June–Aug. Honshu (Izu-shi-chitō Isls.).

3. **Liparis nervosa** (Thunb.) Lindl. *Ophrys nervosa* Thunb.; *Sturmia nervosa* (Thunb.) Reichenb. f.; *L. cornicaulis* Makino; *L. bambusaefolia* Makino——KOKU-RAN. Terrestrial herb with cylindrical pseudobulbs, the internodes to 4 cm. long; leaves 2 or 3, obliquely ovate to elliptic, 5–12 cm. long, 2.5–5 cm. wide, acute to abruptly acuminate, acute and short-sheathed at base; peduncles 15–30 cm. long; racemes loosely 5- to 10-flowered, 2/5–1/2 the length of the peduncles; bracts deltoid, spreading, 1-nerved, membranous, 1–2 mm. long, acute; flowers dark purple; sepals narrowly oblong, obtuse, the lateral ones spreading, about 5 mm. long; petals linear-oblanceolate, obtuse, about 5 mm. long; lip cuneate-obovate, gradually recurved at tip, shallowly grooved on the upper surface, retuse, the appendages at base subulate; column about 3 mm. long, narrowly winged; capsules pedicelled, erect, clavate-fusiform, 15–25 mm. long.——June–July. Woods; Honshu (Tōkaidō, Kinki, and westw.), Shikoku, Kyushu.

4. **Liparis odorata** (Willd.) Lindl. *Malaxis odorata* Willd.; *Leptorchis odorata* (Willd.) O. Kuntze——SASABA-RAN. Pseudobulbs ovoid-globose, about 2 cm. long; leaves 3–5, membranous, imbricate at the base, broadly lanceolate to narrowly oblong, 8–15 cm. long, 2–4 cm. wide, acuminate at both ends, plicate, sheathing at base; peduncles 20–30 cm. long, erect; racemes 5–15 cm. long; bracts lanceolate, acuminate, 1-nerved, obliquely ascending, 3–6 mm. long; flowers yellowish green, with a light purple hue; sepals about 6 mm. long, lanceolate, obtuse, the lateral ones narrowly oblong, spreading under the lip; petals linear, slightly broadened toward tip, subacute, about 7 mm. long; lip obovate-cuneate, recurved at tip, nearly truncate, shallowly grooved, as long as the lateral sepals; column narrowly wiinged.——July. Shikoku, Kyushu.——Ryukyus, Formosa, s. China, and India.

5. **Liparis auriculata** Bl. *L. yakushimensis* Masam.——GIBŌSHI-RAN. Pseudobulbs short, fusiform-ovate, 1–2 cm. long; leaves 2, broadly ovate to orbicular-ovate, 5–10 cm. long, 3–7 cm. wide, abruptly acute, rounded to cordate at base, the sheaths 2–7 cm. long; peduncles 15–30 cm. long, erect; racemes 3–7 cm. long; bracts lanceolate, ascending, acuminate, 1.5–2.5 mm. long; flowers whitish; sepals linear-oblong, 5–6 mm. long, acute, the lateral ones spreading; petals linear, slightly broadened at tip, obtuse, about 5 mm. long; lip cuneate-obovate, about 5 mm. long, rounded at apex, nearly straight, with 2 callosities at base; column about 3.5 mm. long, narrowly winged.——July–Aug. Woods; Honshu, Shikoku, Kyushu; rather rare.

6. **Liparis kumokiri** F. Maekawa. *L. auriculata* sensu auct. Japon., pro parte, non Bl.——KUMO-KIRI-SŌ. Pseudobulbs 1–1.2 cm. long; leaves 2, elliptic to oblong, 5–12 cm. long,

2.5–5 cm. wide, obtuse or subacute, acute and short-sheathing at base; peduncles 15–25 cm. long, erect; racemes 3–7 cm. long, loosely 5- to 15-flowered; flowers pale green to purplish; bracts spreading, ovate-deltoid, 1–1.5 mm. long, acute; sepals narrowly oblong, 5.5–6.5 mm. long, obtuse, spreading; petals narrowly linear, as long as the sepals; lip cuneate-obovate, abruptly reflexed above the middle, shallowly grooved on the upper surface, rounded-truncate at the apex; column about 3 mm. long, narrowly winged with a low ridge above on each side; capsules pedicelled, erect, clavate, 10–15 mm. long.——June–Aug. Woods; Hokkaido, Honshu, Shikoku, Kyushu.——Korea.

7. **Liparis krameri** Fr. & Sav. *Leptorchis krameri* (Fr. & Sav.) O. Kuntze——JIGABACHI-SŌ. Pseudobulbs globose, 7–12 mm. long; leaves 2, ovate, 3–8 cm. long, 2–4 cm. wide, obtuse or subacute, rounded at base, the petioles sheathing at base; peduncles 8–20 cm. long, erect; racemes 3–8 cm. long, 10- to 20-flowered; flowers pale green or dark purple; bracts deltoid, subacute, spreading, 1–1.5 mm. long; sepals linear, 10–12 mm. long, subacute, the lateral ones spreading; petals reflexed, filiform, 8–10 mm. long, acute; lip 6–8 mm. long, about 2/3 as long as the petals, abruptly reflexed near the base, the limb spreading, narrowly ovate-oblong, with slightly recurved margins, caudate; column about 2 mm. long, almost wingless.——May–July. Woods; Hokkaido, Honshu, Shikoku, Kyushu.——Korea.

Var. **shichitoana** Ohwi. *? Liparis nikkoensis* Nakai——HIME-JIGABACHI-SŌ. Plant smaller; sepals about 7 mm. long; lip 6 mm. long.——Honshu.

8. **Liparis japonica** (Miq.) Maxim. *Microstylis japonica* Miq.——SEITAKA-SUZUMUSHI-SŌ. Plants 20–40 cm. high; pseudobulbs ovoid-globose, 6–12 mm. long; leaves 2, narrowly ovate or ovate-oblong, 6–10 cm. long, 2–3.5 cm. wide, acute, cuneate or acute at base, the petioles 4–8 cm. long, sheathing at base; peduncles 20–40 cm. long, erect; racemes 10–20 cm. long, loosely many-flowered; bracts spreading, deltoid-ovate, 1–1.5 mm. long, acute; flowers pale green or purplish; sepals linear-lanceolate, 8–9 mm. long, subobtuse; petals as long as the sepals, filiform; lip obovate, flat, 7–8 mm. long, 4.5–5.5 mm. wide, pointed, erect, slightly recurved at tip; column about 2.5 mm. long, narrowly deltoid-winged.——June–Aug. Woods; Hokkaido, Honshu, Shikoku, Kyushu.——Korea, Manchuria, and Amur.

9. **Liparis makinoana** Schltr. *L. lilifolia* sensu auct. Japon., non L. C. Rich.——SUZUMUSHI-SŌ. Plant 15–30 cm. high; pseudobulbs ovoid-globose, 8–12 mm. long; leaves 2, elliptic to oblong, 5–10 cm. long, 2.5–3.5 cm. wide, obtuse at tip, abruptly narrowed at base to the short-sheathing petiole; peduncles 15–30 cm. long; racemes 10–20 cm. long, rather loosely many-flowered; bracts spreading, deltoid-ovate, acute, 1–1.5 mm. long; flowers pale green or purplish; sepals linear-lanceolate, 8–9 mm. long, subobtuse; petals as long as the sepals, filiform; lip obovate, flat, erect, 12–18 mm. long, 10–15 mm. wide, rounded, with a slightly recurved mucronate tip; column about 2.5 mm. long, narrowly winged at tip.——May–July. Woods; Hokkaido, Honshu, Shikoku, Kyushu.——Korea.

45. PHAJUS Lour. GANZEKI-RAN ZOKU

Rather large terrestrial herbs with pseudobulbs; leaves relatively large, plicate, narrowed at both ends, the petiole attached to the pseudobulb; scapes from the base of the pseudobulbs; flowers racemose, relatively large, yellowish, rarely white or purple, pedicelled, the bracts minute or leaflike, sometimes caducous; sepals free, ascending; petals narrower than the sepals; lip erect, dilated at tip, usually short-spurred at base, the lateral lobes large and loosely embracing the column in natural position, the terminal lobe broad; column broadened in upper part, with a ridge or wing on both sides, without a foot; anthers incurved; pollinia 8, 4 in each locule, waxy; capsules oblong.—About 50 species, in w. Africa, tropical and e. Asia, and Polynesia.

1A. Leaves to 17 cm. wide; sepals and petals narrowly oblong, acuminate, about 5 cm. long; lip with the lobes undulate near tip.
.... 1. *P. tankervilleae*
1B. Leaves 5–8 cm. wide; sepals and petals oblong, obtuse, about 3.5 cm. long; lip with the lobes strongly papillose, crispate-dentate near tip. .... 2. *P. minor*

1. Phajus tankervilleae (Banks ex L'Hérit.) Bl. *Limodorum tankervilleae* Banks ex L'Hérit.; *Bletia tankervilleae* (Banks ex L'Hérit.) R. Br.; *P. grandifolius* Lour.; *Calanthe speciosa* Vieill.——Kaku-ran, Kwaku-ran, Kwa-ran. Glabrous, terrestrial, tufted herb; leaves 2 or 3 from the pseudobulb, ovate-lanceolate or oblong, to 70 cm. long, 17 cm. wide, plicate, attenuate, acuminate, gradually narrowed at base to the long petiole; scapes 60–70 cm. long, erect, terete, the bladeless sheaths 4–7 cm. long; racemes erect, the bracts deciduous; sepals and petals narrowly oblong, about 5 cm. long, 1–1.5 cm. wide, nearly equal, acuminate, spreading; lip slightly shorter than the sepals, broadly rhombic-obovate, loosely embracing the column in natural position, shallowly 3-lobed, 3-keeled on the inside, the lobes short, rounded, with undulate margins; spur terete, slender, about 1 cm. long, slightly broadened toward the base; column about 2 cm. long. ——Kyushu (Tanegashima); very rare; sometimes cultivated. ——Ryukyus, Formosa, China, Malaysia, Polynesia, and Australia.

2. Phajus minor Bl. *P. maculatus* var. *minor* (Bl.) Fr. & Sav.——Ganzeki-ran. Pseudobulbs narrowly ovoid-conical, angled, 3–5 cm. long; leaves 3–5 on the pseudobulb, narrowly oblong to broadly oblanceolate, 20–40 cm. long, 5–8 cm. wide, acuminate at both ends, plicate, the petioles 20–30 cm. long; scapes 40–60 cm. long, erect, rather stout, with few membranous, sheathing scales; racemes 15–25 cm. long, 5- to 18-flowered, the bracts membranous, oblong, subacute, nerved, 15–20 mm. long; flowers yellow, becoming indigo-blue when dry; sepals and petals narrowly oblong, 3–3.5 cm. long, 1–1.5 cm. wide, obtuse, nerved, obliquely spreading; lip cuneate-obovate, slightly shorter than the other perianth lobes, loosely embracing the column in natural position, 3-lobed, with strongly papillose, crispate-dentate margins near tip, the lateral lobes semirounded, the terminal lobe broader and denticulate; spur 7–8 mm. long, lanceolate; column about 2 cm. long, long-pubescent on the inner surface; capsules pedicelled, pendulous, narrowly oblong, 3–4 cm. long.——June. Honshu (Izu-Shichitō Isls. and Kii Prov.), Shikoku, Kyushu; rather rare. Forma **punctatus** Ohwi. Leaves white-maculate. Occurs with the typical phase.

46. CALANTHE R. Br.   Ebine Zoku

Plants terrestrial, with pseudobulbs; leaves few, radical, relatively large, plicate; scapes usually leafless; inflorescence simple, racemose, the bracts small or prominent; sepals equal, free, spreading; petals similar to the sepals or narrower; lip clawed, adnate to the base of the column, spurred or spurless, the limb spreading, often 3-lobed, the terminal lobe sometimes bifid; column short, erect, without a foot, winged or rarely wingless; anthers nearly terminal; pollinia 8, in 2 locules, waxy; capsules oblong, usually pendulous.——About 200 species, in tropical and temperate regions of Asia, 1 in Central America.

1A. Bracts caducous; scapes from lower nodes of elongate sterile shoots; flowers small, pale yellow; spur absent. .... 1. *C. venusta*
1B. Bracts persistent; scapes radical.
  2A. Scapes from the fascicles of young leaves; flowers in elongate loose racemes.
    3A. Flowers spurred.
      4A. Flowers yellowish green; sepals 7–9 mm. long; spur 2–8 mm. long.
        5A. Leaves 1–2 cm. wide; bracts 1–2 cm. long, linear-lanceolate; spur narrowly cylindric, about 8 mm. long. .... 2. *C. bungoana*
        5B. Leaves 3–6.5 cm. wide; bracts about 4 mm. long, short-setaceous; spur oblong, about 2 mm. long. ...... 3. *C. oblanceolata*
      4B. Flowers white, rose, brownish or yellow; sepals 15–30 mm. long.
        6A. Spur shorter than the sepals; leaves glabrous.
          7A. Leaves oblanceolate to narrowly oblong; petals linear-lanceolate; bracts lanceolate, 1–2 cm. long. .... 4. *C. nipponica*
          7B. Leaves obovate-oblong; petals obovate-spathulate or narrowly ovate; bracts 5–10 mm. long. .... 5. *C. discolor*
        6B. Spur longer than or as long as the sepals.
          8A. Leaves puberulent beneath; lip 3-lobed; spur 15–18 mm. long; ovary puberulent. .... 6. *C. aristulifera*
          8B. Leaves glabrous beneath; lip entire, minutely fimbriate; spur 2–2.3 cm. long; ovary glabrous. .... 7. *C. schlechteri*
    3B. Flowers spurless.
      9A. Petals broadly linear; lip flat, cuspidate. .... 8. *C. reflexa*
      9B. Petals oblanceolate; lip with cristate ridges on inner side, deeply retuse. .... 9. *C. torifera*
  2B. Scapes from the fascicles of full-grown leaves; flowers in dense short racemes; spur slender; leaves sparsely pilose beneath.
    10A. Leaves narrowly oblong to obovate-oblong, acuminate at both ends; flowers white to rose-purple; lip with 3 series of protuberances near base within; spur longer than the sepals. .... 10. *C. furcata*
    10B. Leaves elliptic to ovate, abruptly acuminate, rounded at base; flowers white; lip with a relatively large 5-lobed callosity; spur shorter than the sepals. .... 11. *C. japonica*

1. Calanthe venusta Schltr. *C. gracilis* sensu auct. Japon., non Lindl.——Tokusa-ran. Plant indigo-black when dried; sterile shoots elongate, 50–100 cm. long, leafy at apex, the nodes conspicuous; leaves narrowly oblong, 20–40 cm. long, 4–7 cm. wide, acuminate at both ends, glabrous, the lower ones smaller and sheathlike; flowering stems 40–60 cm. long, from the lower nodes of the sterile shoots, with several reduced sheathing leaves in the lower part; racemes 30–40 cm. long, spreading white-pubescent, loosely many-flowered, the bracts broadly linear, acuminate, 1.5–3 cm. long, caducous;

flowers pale yellow; spur absent; sepals reflexed-spreading, oblong-ovate, 10–13 mm. long, about 3 mm. wide, caudate-acuminate, 3-nerved; petals 7–8 mm. long, obtuse, oblong-ovate; lip as long as the sepals, quadrangular-oblong, 3-lobed, the lateral lobes ascending, ovate, the terminal lobe broadly clawed, the limb transversely oblong, truncate, crispate-denti-culate; ovary pedicelled.——Nov. Woods in mountains; Kyushu (s. distr.); rather common.——Ryukyus and Formosa.

2. Calanthe bungoana Ohwi. Tagane-ran. Rhizomes short; leaves linear-oblanceolate, 30–50 cm. long, 1.5–2 cm. wide, acuminate, gradually narrowed at base, glabrous, short-petiolate; scapes about 65 cm. long, white-puberulent, with 2 sheathing scales on the lower half; racemes rather densely many-flowered, about 10 cm. long, the bracts linear-lanceolate, 1–2 cm. long, green; flowers yellowish green; sepals ovate, puberulent without, about 7 mm. long, acute, reflexed-spread-ing; petals oblanceolate, about 6 mm. long, acute; lip slightly longer than the sepals, 3-lobed, with 3 crested merging veins on the inner side, the lateral lobes 4–5 mm. long, elliptic, the terminal lobe broadened at tip, bifid; spur about 8 mm. long, straight; ovary pedicelled, puberulent.——Kyushu (Bungo Prov.); very rare.

3. Calanthe oblanceolata Ohwi & T. Koyama. Saku-rajima-ebine. Fig. 3. Terrestrial perennial herb with thick, rather short rhizomes; leaves in fascicles of 3 or 4, oblanceolate, 40–50 cm. long, 3–6.5 cm. wide, 3- to 7-nerved, densely puberu-lent on under side; scape 40–70 cm. long, minutely puberulent; racemes densely many-flowered, about 20 cm. long; flowers greenish yellow, half-open; median sepal ovate-oblong, about 9 mm. long, 3-nerved, pubescent outside, the lateral elliptic-oblong; petals linear-oblanceolate, about 1.3 mm. wide, the lip straight, about 9 mm. long, 3-lobed, with 3 irregular raised ridges inside, the lateral lobes small, the median obtriangular, rounded to emarginate at apex, the spur short, straight, about 2 mm. long.——May. Kyushu (s. distr.); rare.

4. Calanthe nipponica Makino. *C. trulliformis* var. *hastata* Finet.—Kinsei-ran. Rhizomes short; leaves 3–5, oblanceolate to narrowly oblong, 15–30 cm. long, 1.5–3.5 cm. wide, glabrous, acuminate, narrowed to a short sheath at base; scapes 30–60 cm. long, puberulent in upper part; racemes loosely 5- to 12-flowered, 15–20 cm. long, the bracts lanceolate, acuminate, 1–2 cm. long, green; sepals spreading, broadly lanceolate, 15–20 mm. long, acuminate; petals linear-lanceo-late, 15–17 mm. long, acuminate, narrowed toward the base; lip deltoid-ovate, as long as the sepals, the limb abruptly broadened, shallowly 3-lobed, with crested ridges in the center, the lateral lobes short, ascending-spreading, the terminal lobe quadrangular-ovate, cuspidate, undulate on margins; spur about 5 mm. long.——June. Woods; Hokkaido, Honshu (centr. and n. distr.); rather rare.

5. Calanthe discolor Lindl. *C. lurida* Decne.——Ebine. Rhizomes short and thick; leaves 2 or 3, narrowly obovate-ob-long, 15–25 cm. long, 4–6 cm. wide, acute, petiolate, glabrous; scapes 30–50 cm. long, puberulent in upper part, with 1 or 2 sheathing scales; racemes 5–15 cm. long, loosely about 10-flow-ered, the bracts lanceolate, 5–10 mm. long, scarious; sepals spreading, narrowly ovate, 15–20 mm. long, acute or mucro-nate, narrowed toward the base, dark brown to greenish; petals slightly smaller than the sepals, obovate-spathulate or narrowly ovate; lip flabellate, as long as the sepals, rose or white, deeply 3-lobed, with 3 median ridges, the lateral lobes ascending, broadly cuneate, entire, the terminal lobe broadly

cuneate, bifid; spur 5–10 mm. long.——Woods in foothills; Hokkaido (sw. distr.), Honshu, Shikoku, Kyushu; common.

Var. **bicolor** (Lindl.) Makino. *C. striata* R. Br. ex Miq.; *C. striata* var. *bicolor* (Lindl.) Maxim.; *C. bicolor* Lindl.—— Takane. Flowers larger, not fully expanding; terminal lobe of lip not prominently lobed.——Apr.–May. Kyushu. ——Forma **sieboldii** (Decne.) Ohwi. *C. sieboldii* Decne.; *C. striata* var. *sieboldii* (Decne.) Maxim.; *C. striata* forma *sieboldii* (Decne.) Ohwi——Ki-ebine. Flowers clear yellow.

6. Calanthe aristulifera Reichenb. f. *C. kirishimensis* Yatabe——Kirishima-ebine. Rhizomes short; leaves 2 or 3, narrowly obovate-oblong, 15–20 cm. long, 3–5 cm. wide, acute to short-acuminate, narrowly cuneate, puberulent beneath, rather long-petiolate; scapes 30–40 cm. long, puberulent in upper part, with a few scalelike leaves; racemes 8–12 cm. long, loosely about 10-flowered, brownish puberulent, the bracts linear-lanceolate, membranous, acuminate, 4–8 mm. long; flow-ers half-opened at anthesis, white, sometimes with a rose tint; sepals narrowly ovate-oblong, about 15 mm. long, acuminate, puberulent outside; petals narrower but as long as the sepals; lip 3-lobed, slightly shorter or as long as the sepals, deltoid-flabellate, the inner side with 3 elevated ridges, the lateral lobes ascending, the terminal lobe broadly cuneate, truncate, mucronate; spur 15–18 mm. long; ovary puberulent.——May. Honshu (Hachijo Isl.), Kyushu.

7. Calanthe schlechteri Hara. *C. alpina* sensu auct. Japon., non Hook. f.——Kiso-ebine. Rhizomes short; leaves 2 or 3, narrowly obovate-oblong, 15–20 cm. long, 3–4 cm. wide, acute, narrowly cuneate, glabrous; scapes 25–30 cm. long, glabrous, with 1 scale; racemes 5–7 cm. long, 3- to 8-flowered, glabrous, the bracts lanceolate, 8–15 mm. long, acuminate; flowers pendulous; sepals broadly lanceolate to narrowly ovate, 15–17 mm. long, 4–6 mm. wide, acuminate, mucronate, purplish; petals lanceolate, 12–15 mm. long, about 3 mm. wide, acuminate, purplish; lip erect, slightly shorter than the sepals, nearly rounded, entire, minutely fimbriate, pale yellow with darker striations; spur narrowed above, 2–2.3 cm. long; ovary glabrous.——July–Sept. Honshu (centr. distr.), Shikoku (Mount Ishizuchi); very rare.

8. Calanthe reflexa Maxim. *Paracalanthe reflexa* (Maxim.) Kudo; *C. okushiriensis* Miyabe & Tatew.; *C. reflexa* var. *okushiriensis* (Miyabe & Tatew.) Ohwi——Natsu-ebine. Rhizomes short; leaves 3–5, narrowly oblong to broadly ob-lanceolate, 10–30 cm. long, 3–6 cm. wide, acuminate, usually puberulent beneath, narrowed toward the base; scapes 20–40 cm. long, puberulent above, with a solitary green scale; ra-cemes 10–15 cm. long, loosely 10- to 20-flowered, the bracts lanceolate, acuminate, green, 1–2 cm. long, puberulent; flowers pale purple; sepals gradually acuminate, 15–20 mm. long, the dorsal sepal narrowly ovate, the lateral ones ovate, reflexed; petals 12–15 mm. long, 1.5–2 mm. wide, broadly linear, acumi-nate; lip flat, broadly ovate, as long as the sepals, cuspidate, deeply 3-lobed, the terminal lobe relatively large, broadly cuneate-elliptic, with undulate margins, rounded, mucronate; spur absent.——Aug. Honshu, Shikoku, Kyushu.

9. Calanthe torifera Schltr. *C. tricarinata* sensu auct. Japon., non Lindl.——Sarumen-ebine. Rhizomes short; leaves 2–4, narrowly obovate-oblong, 15–25 cm. long, 6–8 cm. wide, abruptly acuminate, narrowed toward the base, glabrous or scattered puberulent beneath; scapes 30–50 cm. long, with 1 scale about the middle; racemes 15–30 cm. long, loosely 7- to 15-flowered, the bracts membranous, narrowly triangular,

Fig. 3.—*Calanthe oblanceolata* Ohwi & T. Koyama. 1, Habit; 2, fruiting raceme; 3, tepals; 4, labellum, seen from below and from above.

5–8 mm. long, acuminate; sepals spreading, narrowly ovate-oblong, 2–2.5 cm. long, 7–10 mm. wide, acute, 5-nerved, brownish green, tinted purple; petals broadly oblanceolate, 17–22 mm. long, 4–5 mm. wide, acute; lip as long as the sepals, deeply retuse, with cristate ridges on inner side, the lateral lobes small, the terminal lobe rounded, about 15 mm. wide, purplish brown, strongly crisped-undulate, with 3 crested incised ridges within; spur absent.——May–June. Hokkaido, Honshu, Shikoku, Kyushu; rather rare.

10. Calanthe furcata Batem. *C. veratrifolia* sensu auct. Japon., non R. Br.; *C. triplicata* Ames, excl. syn.; *C. matsumurana* Schltr.; *C. textori* Miq.——Tsuru-ran, Kwa-ran. Rhizomes short; leaves 3–6, rather long-petiolate, narrowly oblong to obovate-oblong, 20–40 cm. long, acuminate at both ends, puberulent beneath; scapes 40–80 cm. long, arising from the fascicle of fully grown leaves, densely white-puberulent; racemes 5–10 cm. long, densely many-flowered, the bracts spreading, recurved, narrowly ovate, 7–20 mm. long, acute, few-nerved, green; flowers white to rose-purple; sepals obovate or elliptic, spreading, 12–15 mm. long, obtuse, mucronate; petals as long

as the sepals and slightly narrower; lip ascending, longer than the sepals, 3-lobed, with 3 series of yellow protuberances near the base within, the lateral lobes ascending, oblong, the terminal lobe large, broadly obtriangular, deeply bifid, the lobules rounded, ascending, cuneate; spur slender, 1.5–2 cm. long, curved upward, longer than the sepals.——July–Oct. Kyushu (s. distr., incl. Yakushima and Tanegashima).——Ryukyus, Formosa, China, Malaysia, Bonins, and Micronesia.

11. Calanthe japonica Bl. *C. austrokiusiuensis* Ohwi ——Hiro-ha-no-kwa-ran. Rhizomes short; leaves few, elliptic to ovate, to 20 cm. long, 10 cm. wide, abruptly acuminate, rounded at base, puberulent beneath, the petioles 7–10 cm. long; scapes 30–40 cm. long, densely puberulent above, with 1 or 2 scales; racemes 8–10 cm. long, many-flowered, densely puberulent, the bracts spreading or reflexed, narrowly ovate, few-nerved; flowers white; sepals broadly ovate to broadly obovate, about 8 mm. long, subobtuse, spreading, with minute brown pubescence outside; petals rounded-rhombic, about 6 mm. long, obtuse; lip with a relatively large 5-lobed callosity; spur shorter than the sepals.——Kyushu (s. distr.).

47. TAINIA Bl. Hime-token-ran Zoku

Rhizomes creeping, the sterile shoots with pseudobulbs; leaves solitary and terminal on the pseudobulb, jointed at base, long-petiolate, lanceolate or oblong; scape with a few sheathing scales at base; racemes simple; flowers pedicelled, loosely arranged, relatively large; sepals narrow, the dorsal free, the lateral ones adnate to the base of the column; petals as long as the sepals, sometimes narrower; lip erect, from the base of the column, spurless or short-spurred, the lateral lobes erect, the terminal lobe spreading, broader, entire; column rather elongate, slightly incurved, narrowly winged; anthers terminal, 2-locular; pollinia 8, 4 in each locule, ovate, waxy.——About 15 or more species, in India, e. Asia, and Malaysia.

1. Tainia laxiflora Makino. *Oreorchis laxiflora* T. Ito ex Makino, in syn.——Hime-token-ran. Rhizomes slender, the conical pseudobulbs about 10 mm. long, interspersed while young with sheathing scales; leaves rather thick, deep green, oblong-lanceolate, gradually acute, glabrous, conspicuously nerved, the petioles 2–3 cm. long; scapes erect, slender, 20–30 cm. long, with 1 or 2 sheaths at base; racemes about 10 cm. long, loosely flowered, the bracts linear-lanceolate, 4–5 mm. long, acuminate; flowers brownish white; sepals yellowish

brown, obliquely spreading, linear-lanceolate, 12–15 mm. long, about 2 mm. wide, acute; petals white, ascending-spreading, similar to and contiguous with the dorsal sepal; lip 8–10 mm. long, white, tinged rose-purple, cuneate, with a short chin at base, 3-lobed, the lateral lobes ascending, short, obtuse, the dorsal lobe transversely oblong, narrowed at base, truncate and entire at apex, spurless; ovary pedicelled.——May. Woods; Honshu (Izu Shichitō Isl.), Kyushu; rather rare.——Ryukyus.

48. EULOPHIA R. Br. Imo-ran Zoku

Terrestrial herb with pseudobulbs; leaves distichous, radical, usually from one side of pseudobulb; scapes erect, scaly, rising laterally from the base of leafy stems; racemes simple or rarely branched, the flowers relatively small, pedicelled, the bracts membranous; sepals equal, free, the lateral spreading, sometimes adnate below to the base of the column; petals spreading, often connivent above; lip erect from the base of the column, saccate or spurred at the base, 3-lobed, the lateral lobes erect, loosely embracing the column or obsolete, the terminal lobe spreading to recurved, entire or bifid, with elevated lines or crests on the center within; column short, thick, without a foot, usually winged; anthers terminal, subglobose; pollinia 2 in each cell, ovoid, waxy; capsules ovoid to oblong, not beaked, pendulous.——More than 50 species, in Africa, Asia, and Australia, a few in South America.

1. Eulophia ochobiensis Hayata. Imone-yagara. Stout, glabrous, terrestrial herb about 50 cm. high; scapes terete, the scales few, thinly membranous, appressed, ovate, about 3 cm. long, acute, weakly nerved, short-sheathing; racemes to 30 cm. long, erect, loosely many-flowered, the bracts greenish, flat, broadly linear, 1.5–2.5 cm. long, 1.5–2.5 mm. wide, 3- or 5-nerved, acuminate; flowers slightly expanded; sepals obovate-oblong, about 2 cm. long, 8 mm. wide, abruptly acuminate, 5- to 7-nerved; petals obovate, obtuse, 3/4 as long

as the sepals; lip suberect, slightly shorter than the sepals, obovate, 3-lobed, papillose on the nerves inside, with 2 low ridges on the center, the lateral lobes rounded, 4–5 mm. long and as wide, the terminal lobe ascending, ovate-rounded, 7 mm. long and as wide, obtuse, entire, the spur compressed-conical, obtuse, about 4 mm. long and as wide; column plano-convex, about 6 mm. long, 4 mm. wide.——July. Kyushu (Tanegashima and Ohsumi Prov.); very rare.——Ryukyus and Formosa.

49. CREMASTRA Lindl.　Saihai-ran Zoku

Terrestrial herbs with creeping rhizomes and pseudobulbs; leaves solitary, large, slightly plicate, narrowed to the petiole; scapes loosely sheathed; racemes simple, the flowers slender, nearly tubular, short-pedicelled, pendulous or nodding, the bracts linear, short; sepals and petals free, narrow, erect; lip from the base of the column, free, linear, erect, slightly saccate, the lateral lobes small, the terminal lobe oblong, entire; column linear, erect, essentially terete; anthers terminal; pollinia 4 in each cell, waxy; capsules ellipsoidal.——About 5 species, in e. Asia and the Himalayas.

1A. Leaves 25–40 cm. long, 4–5 cm. wide; flowers densely arranged, 3–3.5 cm. long, with erect sepals; terminal lobe of lip not geniculate.
1. C. variabilis

1B. Leaves 10–12 cm. long, 3–5 cm. wide; flowers loosely arranged, 18–20 mm. long, with ascending sepals; terminal lobe of lip geniculate.
2. C. unguiculata

1. Cremastra variabilis (Bl.) Nakai. *Hyacinthorchis variabilis* Bl.; *C. wallichiana* sensu auct. Japon., non Lindl.; *C. mitrata* A. Gray; *C. appendiculata* Makino, excl. syn.——Saihai-ran. Terrestrial herb; pseudobulbs ovoid, about 3 cm. long; leaves solitary, rarely 2, terminal on the pseudobulb, narrowly oblong, 25–45 cm. long, 4–5 cm. wide, long-acuminate, narrowed below to the petiole, often white- or pale yellow-maculate, with 3 main nerves and intervening veinlets; scapes erect, 30–50 cm. long, terete, loosely few-sheathed; racemes 10–20 cm. long, 10- to 20-flowered, sulcate, rather dense, the bracts thinly membranous, linear-lanceolate, 7–10 mm. long, acuminate; flowers pendulous, rose to rose-purple with a brownish tinge; sepals and petals linear-oblanceolate, 3–3.5 cm. long, 4–5 mm. wide, acuminate; lip about 3 cm. long, slightly broadened at base, erect, 3-lobed, the lateral lobes lanceolate, small, erect, the terminal lobe oblong, about 9 mm. long, truncate and slightly recurved at tip; column about 2.5 cm. long, slightly broadened at tip; capsules 2–2.5 cm. long, nearly sessile, pendulous.——May–June. Woods in hills and low elevations in the mountains; Hokkaido, Honshu, Shikoku, Kyushu; rather common.——s. Sakhalin, s. Kuriles, and s. Korea.

2. Cremastra unguiculata (Finet) Finet. *Oreorchis unguiculata* Finet——Token-ran. Pseudobulbs ovoid-globose, about 10 mm. long; leaves 2, sometimes solitary, terminal on the pseudobulb, oblong, 10–12 cm. long, 3–5 cm. wide, acute to abruptly acuminate, 3-nerved and with intervening veinlets, deep green with rather large dark purple spots above, purplish beneath, the petioles 4–6 cm. long; scapes slender, erect, 25–40 cm. long, with 2 sheathing scales; racemes loosely few-flowered, 8–10 cm. long, glabrous; flowers ascending, not fully open, 18–20 mm. long, the bracts lanceolate, membranous, 4–6 mm. long; sepals ascending, linear-oblanceolate, 18–20 mm. long, acute, narrowed at base, yellowish brown with purple striations; petals linear, 14–16 mm. long, acute, closely contiguous with the dorsal sepal; lip as long as the petals, linear, embracing the lower part of the column, barely inflated at base, white, 3-lobed, the lateral lobes erect, lanceolate, much smaller than the terminal one, the terminal lobe obovate, about 5 mm. long, rounded at apex, undulate on margin; column 12–13 mm. long.——May–June. Woods in foothills; Hokkaido, Honshu, Shikoku; rare.

50. OREORCHIS Lindl.　Kokei-ran Zoku

Terrestrial herbs; pseudobulbs with 1 or 2 narrow plicate leaves; scape erect, simple, with a few sheathing scales at base; flowers relatively small, in terminal racemes, pedicelled, the bracts linear, minute; sepals and petals equal, free, narrow, ascending or with incurved tips; lip as long as the sepals, inserted at the base of the column, spurless, the claw erect, the limb 3-lobed, the lateral lobes erect, short, narrow, the terminal lobe broad; column elongate, nearly terete, subclavate in upper part, wingless, without a foot; anther terminal, pollinia 4, distinct, nearly globose; capsules oblong or fusiform, reflexed.——About 10 species, in n. India, China, Formosa, Korea, Japan and e. Siberia.

1. Oreorchis patens (Lindl.) Lindl. *Corallorhiza patens* Lindl.; *O. lancifolia* A. Gray; *O. gracilis* Fr. & Sav.; *O. patens* var. *gracilis* (Fr. & Sav.) Makino——Kokei-ran. Pseudobulbs ovoid-globose, 1.5–2 cm. long; leaves 1 or 2, narrowly to broadly lanceolate, 20–30 cm. long, 1–3 cm. wide, gradually acuminate at both ends, short-petiolate; scapes 30–40 cm. long, with 2 membranous sheaths on the lower part; racemes 10–20 cm. long, relatively many-flowered; flowers yellowish brown; bracts membranous, linear, 4–6 mm. long, acuminate; sepals and petals ascending, lanceolate, 8–10 mm. long, subobtuse; petals closely contiguous on the dorsal sepal; lip ascending, as long as the sepals, white with dark purple spots, 3-lobed from the base, the lateral lobes lanceolate, 4–5 mm. long, obtuse, erect, the terminal lobe cuneate-obovate, denticulate, 2-ridged on the inside; column about 9 mm. long; capsules pendulous, short-pedicelled, narrowly fusiform, about 20 mm. long.——Woods in mountains; Hokkaido, Honshu, Shikoku, Kyushu; rather rare.——China, Ussuri, Sakhalin, s. Kuriles, and Kamchatka.

51. DACTYLOSTALIX Reichenb. f.　Ichiyō-ran Zoku

Terrestrial herbs with short rhizomes; leaves solitary, ovate, petiolate; scapes 1- to 2-sheathed; flowers terminal, solitary, relatively large, short-pedicelled; sepals and petals narrow, alike, ascending; lip slightly concave, short-clawed at base, spurless, 3-lobed, the lateral lobes short, erect, the terminal obovate, recurved at tip, undulate on margins; column rather long, winged, with 2 appendages at the apex; anther with 4 pollinia.——One species, in Japan.

1. **Dactylostalix ringens** Reichenb. f. *Pergamena uniflora* Finet; *Calypso japonica* Maxim. ex Komar.; *D. maculosa* Miyabe & Kudo——ICHIYŌ-RAN. Rhizomes creeping, slender; leaves solitary, radical, ovate, 3–6 cm. long, 3–4 cm. wide, subobtuse, obtuse to rounded at base, petiolate; scapes 10–20 cm. long, with 2 or 3 short sheaths on lower half; flowers solitary, the bracts thinly membranous, quadrangular-elliptic to ovate, 2–3 mm. long; sepals and petals pale green, oblanceolate to linear-lanceolate, 2–2.5 cm. long, subobtuse, ascending, 3-nerved; lip erect, ovate, cuneate, 3-lobed, the lateral lobes broadly ovate, 2–3 mm. long and nearly as broad, obtuse, purplish on upper half, the terminal lobe obovate-rounded, recurved, about 7 mm. long and as wide, white with purple spots, more or less undulate on the margins, with 2 short ribs near the base; spur absent; column about 10 mm. long, flat, narrowly cuneate-oblong, rather broadly winged; ovary 3-angled; capsules clavate-fusiform, short-pedicelled, about 2 cm. long, ascending.——May–July. Woods in mountains; Hokkaido, Honshu, Shikoku.

52. DENDROBIUM Sw. SEKKOKU ZOKU

Tufted epiphytic herbs; stems usually elongate, leafy, terete, nodose, fleshy, with tight and persistent leaf-sheaths; leaves subcoriaceous, flat; flowers 1 to several, relatively large, in racemes or fascicles, the bracts minute or absent; sepals more or less alike, the dorsal free, the lateral ones oblique at the base and adnate to the base of the column, often forming a spurlike chin; petals similar to the dorsal sepal; lip attached to the base of the column-foot, narrowed at base, often 3-lobed, the limb often with 2 to 3 series of elevated ribs; column short, usually winged; anther terminal, 2-locular, pollinia 4 in each cell, waxy; capsules ovoid to oblong.——About 1,000 species, in India, Malaysia, s. Asia, and Australia.

1A. Flowers 1 or 2, white or pale rose, fasciculate; dorsal sepal and petals 2–2.5 cm. long. . . . . 1. *D. moniliforme*
1B. Flowers 3 to 8, pale greenish yellow, loosely racemose; dorsal sepal and petals 12–15 mm. long. . . . . 2. *D. tosaense*

1. **Dendrobium moniliforme** (L.) Sw. *Epidendrum moniliforme* L.; *E. monile* Thunb.; *Oncidium japonicum* Bl.; *D. monile* (Thunb.) Kraenzl.——SEKKOKU, SEKIKOKU. Stems tufted, erect, 5–25 cm. long, terete, the internodes 2–4 cm. long, 3–6 mm. in diameter, covered by a tight membranous sheath; leaves lanceolate, 4–7 cm. long, 7–15 mm. wide, obtuse, rounded at the base; flowers 1 or 2 in the upper nodes of 3-year old stems, fasciculate, pedicelled, white or pale rose, fragrant, the minute bracts basal; dorsal sepal lanceolate, 2–2.5 cm. long, 5–7 mm. wide, acute, the lateral ones obliquely broadened at base; petals similar to the dorsal sepal, slightly shorter; lip cuneate, erect to ascending, the limb slightly recurved at tip, about as long as the petals, narrowly ovate-triangular, acute; column very short, about 2 mm. long on the back, the chin in front about 8 mm. long.——May–June. Epiphytic on tree-branches and rocks in mountains; Honshu (centr. and s. distr. incl. Izu Shichitō Isls.), Shikoku, Kyushu; rather rare.——s. Korea and China.

2. **Dendrobium tosaense** Makino. KIBANA-NO-SEKKOKU. Stems tufted, 25–40 cm. long, terete, the internodes 2–6 cm. long; leaves lanceolate, 2–7 cm. long, 7–17 mm. wide, acute to subobtuse, rounded to obtuse at the base and jointed at the attachment with the sheath; racemes 4–7 cm. long, from the upper nodes of 2- to 3-year old stems, short-pedunculate, the axis slender, flexuous, the bracts linear-lanceolate, 3–6 mm. long, acute; flowers 3–8, pale greenish yellow, loosely racemose; dorsal sepal broadly lanceolate, 12–15 mm. long, acute; petals similar, 10–12 mm. long, the chin about 10 mm. long; lip ascending, the limb nearly as long as the sepals, slightly recurved at the tip, ovate, acute, narrowed to the base, with a few short dark purple striations at the center below; column pale green, rounded on the back, 3–4 mm. long, with dark purple spots within.——July–Nov. Epiphytic on tree branches; Shikoku, Kyushu; rather rare.——Ryukyus.

53. ERIA Lindl. OSA-RAN ZOKU

Epiphytic herbs with short rhizomes and pseudobulbs (stems); stems 2- to several-leaved; scapes 1- to many-flowered; flowers relatively small, in racemes, the bracts narrow; sepals equal, the dorsal lobe free or united with the lateral lobes at base, the lateral sepals forming a spurlike chin at the base and adnate with the column-foot; petals similar to the dorsal sepal; lip joined with the foot, recurved at tip, entire or with a small lateral lobe on each side; column short, concave in front, often wing-margined, the base elongate, simulating a foot in front; anther terminal, the locules 2, distinct, incompletely 4-septate; pollinia 8, 4 in each locule, on 2 glands, with caudicles; capsules oblong to linear.——About 400 species, in e. Asia, Malaysia, and India.

1A. Pseudobulbs 1–2.5 cm. long, 5–10 mm. in diameter; leaves 4–8 cm. long; flowers 1 or 2, the peduncle naked, scapose; terminal lobe of lip 3-ribbed. . . . . 1. *E. reptans*
1B. Pseudobulbs 1–6 cm. long, 5–25 mm. in diameter; leaves 8–30 cm. long; flowers 10–16, the peduncle few-scaled, scapiform; terminal lobe of lip 5-ribbed. . . . . 2. *E. yakushimensis*

1. **Eria reptans** (Fr. & Sav.) Makino. *Dendrobium reptans* Fr. & Sav.; *E. japonica* Maxim.; *E. hosokawae* Hawkes & Heller——OSA-RAN. Pseudobulbs in a series on the short, creeping rhizomes, narrowly oblong-cylindrical to quadrangular-oblong, slightly flattened, 1–2.5 cm. long, 5–10 mm. in diameter, red-brown, sheathed while young; leaves 2, narrowly oblong to broadly lanceolate, 4–8 cm. long, 1.2–2 cm. wide, acuminate, narrowed to the base and jointed with the short sheaths, several-nerved, deciduous in autumn; scapes 1 or 2, 2–6 cm. long, with scattered short brown crisped hairs; flowers 1 or 2, the bracts narrowly ovate, 3–4 mm. long; flowers white; dorsal sepal and petals free, broadly lanceolate, 8–10 mm. long, subobtuse, slightly incurved at tip, the lateral sepals forming a chin 3–4 mm. long at base; lip rather short, broadly cuneate, 3-ribbed within, pale reddish brown in the center, retuse, the lateral lobes relatively small, the terminal trans-

versely quadrangular-oblong, truncate; column terete, about 4 mm. long.——July–Aug. Epiphyte on trees, sometimes also occurring on mossy rocks; Honshu (sw. distr. and Izu-shi-chitō Isls.), Shikoku, Kyushu.——Ryukyus (Amami-Oshima).

2. **Eria yakushimensis** Nakai. Ō-OSA-RAN. Plant evergreen; pseudobulbs 1–6 cm. long, 5–25 mm. in diameter, green, slightly compressed and angular, 3-sheathed while young; leaves 2, linear-lanceolate to oblanceolate, 8–30 cm. long, 15–40 mm. wide, acute, lustrous above; scapes with few scales; racemes 5–8 cm. long, the bracts minute or absent; flowers 10–16; dorsal sepal oblong-lanceolate, 10–11 mm. long, 3–4 mm. wide, subobtuse, pale yellowish green; petals slightly shorter than the sepals, linear-oblong; lip recurved, the lateral lobes obtuse, purplish at base, the terminal lobe 5-keeled, with 1 pair disappearing toward the base; column about 3 mm. long.——Nov. Epiphyte on trees, also occurring on mossy rocks; Kyushu (Yakushima and Tanegashima); rare.

54. BULBOPHYLLUM Thouars   MAMEZUTA-RAN ZOKU

Plants epiphytic, with scaly long-creeping rhizomes, the pseudobulbs 1- or 2-leaved; stem with a single joint or pseudobulb; peduncles lateral, simple, leafless; inflorescence umbellate to racemose, the flowers sometimes solitary, the bracts minute; dorsal sepal free, as long as or longer than the lateral ones, joined with the base of the column and forming a short chin; petals relatively small; lip narrowed at base and attached to the base of the column, versatile, the lateral lobes small, the terminal prominent; column erect, short, winged in upper part; pollinia waxy, usually 4.——About 700 species in e. Africa, India, Malaysia, Australia, e. Asia, and tropical America.

1A. Lateral sepals free; leaves obtuse; rhizomes without fibers of the disintegrated scales.
  2A. Pseudobulbs absent; leaves elliptic to ovate, 7–13 mm. long, not prominently nerved; petals much shorter than the dorsal sepal, glabrous. .... 1. *B. drymoglossum*
  2B. Pseudobulbs ovoid, 5–8 mm. long; leaves usually oblanceolate to narrowly oblong, 1–3.5 cm. long, 7- to 9-nerved; petals nearly as long as the dorsal sepal, ciliate. .... 2. *B. inconspicuum*
1B. Lateral sepals much longer than the dorsal, the lower margins sometimes partly connate; leaves subacute; rhizomes with persistent fibers of the disintegrated scales.
  3A. Pseudobulbs 6–8 mm. long; leaves lanceolate, 4–8 cm. long, 6–10 mm. wide, acute; flowers about 10 mm. long. .... 3. *B. japonicum*
  3B. Pseudobulbs 1–2 cm. long; leaves narrowly ovate-oblong, 10–17 cm. long, 3–5 cm. wide, subobtuse; flowers about 3.5 cm. long.
                                        4. *B. makinoanum*

1. **Bulbophyllum drymoglossum** Maxim. MAMEZUTA-RAN. Plants evergreen, epiphytic, with slender elongate rhizomes; pseudobulbs absent; leaves solitary at the nodes, rather fleshy, flat, elliptic to ovate, 7–13 mm. long, 5–10 mm. wide, obtuse, nearly sessile, the nerves inconspicuous; scapes 1-flowered, 7–10 mm. long, delicate, with few small scales at the base, the bracts ovate, thinly membranous, about 1.5 mm. long, obtuse; flowers pale yellow; sepals broadly lanceolate, gradually acuminate, 7–8 mm. long; petals oblong, glabrous, much shorter than the dorsal sepal; lip broadly lanceolate, obtuse, slightly recurved, shorter than the sepals.——June. Tree-trunks; Honshu (centr. distr. and westw.), Shikoku, Kyushu; rather rare.——s. Korea.

2. **Bulbophyllum inconspicuum** Maxim. MUGI-RAN. Plants evergreen, epiphytic, with slender rhizomes; pseudobulbs ovoid, 5–8 mm. long, relatively remote; leaves oblanceolate to narrowly oblong, 1–3.5 cm. long, 6–8 mm. wide, rounded to obtuse or retuse, 7- to 9-nerved, narrowed to a short petiole; flowers 1 or 2 on short lateral peduncles, the bracts thinly membranous, oblong, about 2 mm. long; sepals ovate-elliptic, 3–3.5 mm. long, the dorsal somewhat shorter; petals nearly as long as the dorsal sepal, ciliate, spreading; lip narrowly ovate, recurved at tip.——June. Tree-trunks; Honshu (centr. and western), Shikoku, Kyushu.

3. **Bulbophyllum japonicum** (Makino) Makino. *Cirrhopetalum japonicum* Makino——MIYAMA-MUGI-RAN. Evergreen epiphyte with rather slender, creeping fibrous rhizomes; pseudobulbs ovoid-globose, 6–8 mm. long; leaf solitary, somewhat coriaceous, lanceolate to nearly so, 4–8 cm. long, 6–10 mm. wide, obtuse, acute at base; flowers 2 or 3, about 10 mm. long, rose-purple, in umbels, the bracts membranous, lanceolate, 1–2 mm. long, acuminate; scape 1.5–2 cm. long; sepals ascending, the dorsal oblong, about 2.5 mm. long, the lateral ones broadly lanceolate, about 7 mm. long, gradually acuminate, the outer margins contiguous; petals ovate, shorter than the dorsal sepal; lip reflexed, rather thick, rounded; column short; capsules ellipsoidal.——June–July. Tree-trunks; Honshu (centr. and western), Shikoku, Kyushu.

4. **Bulbophyllum makinoanum** (Schltr.) Masam. *Cirrhopetalum makinoanum* Schltr.——SHIKŌ-RAN. Evergreen epiphyte with short, creeping, fibrous rhizomes; pseudobulbs somewhat tufted, narrowly ovoid-conical, 1–2 cm. long, green, angled or only slightly so; leaves solitary, thick and slightly lustrous, narrowly ovate-oblong, 10–17 cm. long, 3–5 cm. wide, subacute to obtuse, acute at the base, the petioles 1–1.5 cm. long; scapes slender, nodding to pendulous, 10–15 cm. long, with 1 or 2 small scales; flowers 2–4, about 3.5 cm. long, in umbels, pedicelled, yellowish brown, the bracts lanceolate, acuminate, 4–5 mm. long; sepals ascending, the dorsal one narrowly oblong, 8–10 mm. long, the lateral ones linear-lanceolate, 25–30 mm. long, gradually acuminate, slightly connate; petals about half as long as the dorsal sepal.——July–Sept. Tree-trunks in mountains; Kyushu (Yakushima and Tanegashima).——Ryukyus.

55. CYMBIDIUM Sw.   SHUN-RAN ZOKU

Terrestrial or epiphytic herbs with short rhizomes (rarely saprophytic without green leaves); stems abbreviated, concealed by a tuft of leaves; leaves fasciculate, linear; scapes few-scaled, erect or declined, 1- to many-flowered; flowers rather large, solitary or in racemes, the bracts short; sepals equal, free, spreading; petals similar to the sepals or slightly shorter; lip without a foot, the lateral lobes rather broad, loosely embracing the column in natural position, the terminal lobe undivided, with 2 elevated ridges within; column rather long, without a foot, subterete; anther terminal, the pollinia 2, with a caudicle; capsules ovoid-oblong, often with a remnant of the perianth and column at the apex.——About 70 species, in Madagascar, India, Malaysia, e. Asia, and Australia.

1A. Plant with green leaves; rhizomes short.
  2A. Leaves linear.
    3A. Scapes 1-flowered. .... 1. *C. goeringii*
    3B. Scapes more than 1-flowered.
      4A. Racemes erect; sepals broadly linear, acuminate; terrestrial. .... 2. *C. kanran*
      4B. Racemes pendulous; sepals oblanceolate, acute to obtuse; epiphytic. .... 3. *C. dayanum* var. *austrojaponicum*
  2B. Leaves narrowly oblong to broadly oblanceolate.
    5A. Leaves minutely and distinctly serrulate toward apex; flowers in July. .... 4. *C. lancifolium*
    5B. Leaves entire; flowers in October and November. .... 5. *C. javanicum* var. *aspidistrifolium*
1B. Plant saprophytic, without green leaves; rhizomes elongate. .... 6. *C. nipponicum*

1. Cymbidium goeringii (Reichenb. f.) Reichenb. f. *Maxillaria goeringii* Reichenb. f.; *C. virescens* Lindl.——SHUN-RAN. Rhizomes short; stems tufted; leaves evergreen, linear, 20–35 cm. long, 6–10 mm. wide, acuminate, minutely toothed, the short basal sheaths becoming fibrous with age; scapes erect, terete, fleshy, 10–25 cm. long, 1-flowered, with hyaline, sheathing scales; bracts lanceolate, 3–4 cm. long, acute; flowers greenish; sepals rather fleshy, oblanceolate, subobtuse, 3–3.5 cm. long, 7–10 mm. wide; petals simulating the sepals but slightly shorter, ascending; lip slightly shorter than the sepals, white with reddish purple spots, recurved above, papillose within, the lateral lobes small, obtuse; column about 15 mm. long.——Mar.–Apr. Grassy places in foothills; Hokkaido (Okushiri Isl.), Honshu, Shikoku, Kyushu.——China.

2. Cymbidium kanran Makino. KAN-RAN. Terrestrial herb; leaves evergreen, tufted, linear, arcuate, 20–70 cm. long, 6–17 mm. wide, acuminate, scabridulous on the margins; scapes 25–60 cm. long, loosely sheathed; racemes loosely 5- to 12-flowered, erect, the bracts herbaceous, linear, 8–30 mm. long, acuminate; flowers fragrant, purplish green; sepals spreading, broadly linear, acuminate, the dorsal one 3–4 cm. long, 3.5–4.5 mm. wide, longer than the lateral ones; petals slightly shorter than the sepals, linear-lanceolate, 2–3 cm. long, 4.5–5.5 mm. wide; lip slightly shorter than the petals, about half as long as the dorsal sepal, smooth, glabrous, spreading, the lateral lobes small; column 10–14 mm. long.——Dec.–Jan. Honshu (Kii Prov.), Kyushu; very rare.

3. Cymbidium dayanum Reichenb. f. var. **austrojaponicum** Tuyama. *C. alborubens* Makino; *C. simonsianum* sensu auct. Japon., non King & Pantling——HETSUKA-RAN. Evergreen epiphytic herb; leaves tufted, arcuate, 30–50 cm. long, 10–13 mm. wide, acuminate, slightly narrowed at base, nearly glabrous; scapes lateral, pendulous, with few sheathing scales on the lower part; racemes few- to relatively many-flowered, 10–20 cm. long, pendulous, the bracts lanceolate, 5–8 mm. long, acuminate; sepals spreading, oblanceolate, 3–3.2 cm. long, 5–7 mm. wide, acute to obtuse, white, brownish purple near the base; petals slightly smaller, the lip about 2 cm. long, erect, the lobes ovate, the lateral lobes small, obtuse, the terminal one cuspidate; column about 13 mm. long; capsules fusiform, about 6 mm. long inclusive of the short pedicel.——Oct. Epiphyte; Kyushu (s. distr.).——Ryukyus and Formosa.

4. Cymbidium lancifolium Hook. *C. nagifolium* Masam.——NAGI-RAN. Rhizomes short, sparsely scaly, few-leaved, the roots elongate, rather thick, puberulent; leaves jointed at the sheathing base, the blade narrowly oblong to broadly oblanceolate, 4–15 cm. long, 1.5–3 cm. wide, short-acuminate, narrowed at base, scabridulous on the upper margins; scapes lateral, 10–15 cm. long, few-sheathed; racemes loosely 2- to 4-flowered, pendulous, glabrous, with a slightly flexuous axis, the bracts membranous, linear-lanceolate, 8–15 mm. long, acuminate; sepals linear-lanceolate, 22–25 mm. long, 2.5–3 mm. wide, acuminate, white with a greenish flush, the lateral sepals slightly reflexed; petals narrowly oblong, about 2 cm. long, 5–6 mm. wide, subacute, white with a brown-purple band; lip white, brown-purple spotted within, obovate-oblong, 13 mm. long, about 10 mm. wide, subobtuse, fleshy; column about 13 mm. long; anther yellow, minutely mamillate; capsules erect, about 5 cm. long inclusive of the short pedicel.——July–Aug. Honshu (s. Kantō, Tōkaidō, Izu-Shichitō Isls. and westw.), Shikoku, Kyushu; rare.——Ryukyus and Formosa.

5. Cymbidium javanicum Bl. var. **aspidistrifolium** (Fukuyama) F. Maekawa. *C. aspidistrifolium* Fukuyama——AKIZAKI-NAGI-RAN. Closely resembles No. 4, the leaves entire; scapes shorter than the leaves; flowers grass-green, with thicker, more fleshy sepals; petals oblong, broader and shorter, about 17 mm. long, 6.6 mm. wide.——Kyushu (Tsushima); very rare.——Formosa. A variable species occurring also in the Himalayas and Malaysia.

6. Cymbidium nipponicum (Fr. & Sav.) Makino. *Bletia nipponica* Fr. & Sav.; *C. pedicellatum* Finet; *Pachyrhizanthe nipponicum* (Fr. & Sav.) Nakai——MAYA-RAN. Plant leafless, saprophytic; rhizomes to 15 cm. long, slightly puberulent, scaly; scapes erect, 15–20 cm. long, the sheaths 7–20 mm. long, membranous; racemes 2- to 5-flowered, the bracts membranous, broadly lanceolate, 5–10 mm. long, long-acuminate; flowers white with a brownish purple hue; sepals oblanceolate, about 2 cm. long, 3–4 mm. wide, acuminate; petals narrowly oblong, slightly shorter than the sepals; lip about 15 mm. long, cuneate, obsoletely 3-lobed, the terminal lobe slightly recurved, minutely undulate on the margins; column 8–10 mm. long, densely papillose.——July–Aug. Honshu, Kyushu; rare.

Cymbidium sinense (Andr.) Willd. *Epidendrum sinense* Andr.; *C. hoosai* Makino——HŌSAI-RAN. Leaves 15–32 mm. wide; sepals linear-oblanceolate, 25–35 mm. long, 5–6.5 mm. wide.——Reported to occur in Kyushu (Yakushima).——Ryukyus, Formosa, and China.

56. LUISIA Gaudich.   BŌ-RAN ZOKU

Stout epiphytic herbs with elongate leafy stems and closely appressed sheaths; leaves jointed, terete, evergreen, fleshy; inflorescence very short, spicate; flowers sessile, relatively small, the bracts short, thick; sepals free, equal or the dorsal smaller;

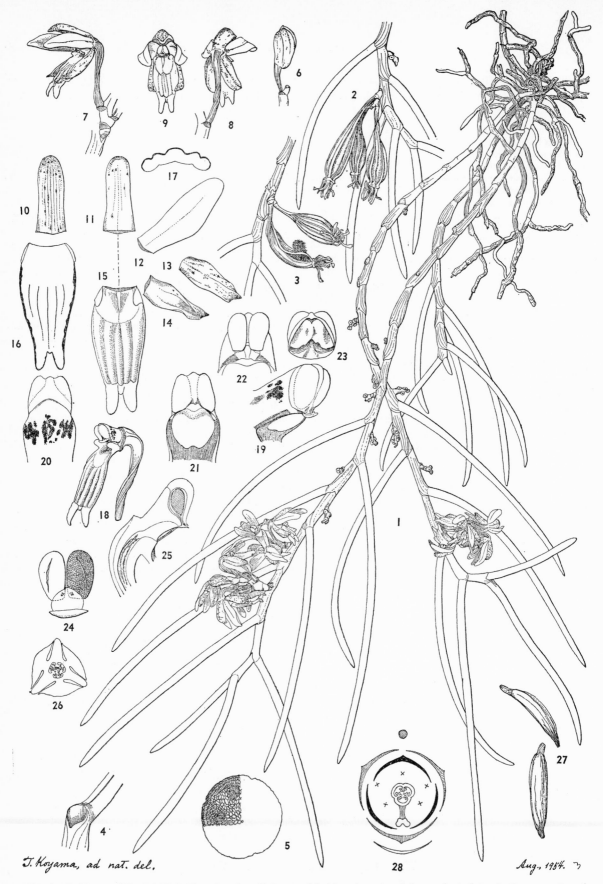

FIG. 4.—*Luisia teres* (Thunb.) Bl. 1, Flowering plant; 2, 3, part of fruiting plant; 4, apical part of a leaf sheath; 5, cross section of stem; 6, flower bud; 7, 8, 9, flowers; 10, 11, 12, 13, 14, tepals; 15, 16, inner and outer side of labellum; 17, cross section of the same; 18, labellum and a gynostemium; 19, 20, 21, 22, 23, apical part of gynostemium; 24, pollinia; 25, longitudinal section of a gynostemium; 26, cross section of ovary; 27, seeds; 28, floral diagram.

petals similar to the sepals; lip sessile, from the base of the column, spurless, often longer than the sepals, nearly flat, scarcely inflated at base, with a callosity or small lateral lobe on each side near base, the limb flat, entire or bifid; column very short, thick; anther terminal, 2-locular; pollinia 2, waxy; capsules narrowly oblong, not beaked, erect.——About 40 species, in India, Malaysia. e. Asia, and the Pacific islands.

1. Luisia teres (Thunb.) Bl. *Epidendrum teres* Thunb.; *L. Jauriei* Schltr.——Bō-RAN. FIG. 4.

Evergreen epiphyte; stems simple or sparsely branched, tufted, 10–40 cm. long, about 3–4 mm. in diameter, erect or pendulous, the internodes sheathed; leaves obliquely spreading, linear, 6–12 cm. long, 3.5–4 mm. wide, terete, abruptly obtuse, scarcely narrowed at base; spikes sessile, 1- to 5-flowered, dense, the axis short, 1.5–2 mm. thick, the bracts deltoid, 2–3 mm. long; flowers greenish yellow with a brown hue; sepals oblong, obtuse, pale green, relatively thick, the dorsal sepal about 8 mm. long, ascending, rounded on the back, the lateral ones about 10 mm. long, spreading, obtusely keeled on the back, cucullate, contiguous with the margin of the lip; petals flat, pale green, ascending, about 10 mm. long, narrowly spathulate-oblong, rounded; lip thick, flat, spreading, pale green outside, dark brown-purple within, oblong, about 15 mm. long, ascending, auriculate, shallowly bifid, the lobes ascending, lanceolate, 2–3 mm. long; column thick, very short; capsules clavate-fusiform, 25–30 mm. long inclusive of the thick pedicel.——July–Aug. Tree-trunks; Honshu (Ise and Kii Prov.), Shikoku, Kyushu.

57. SARCANTHUS Lindl. MUKADE-RAN ZOKU

Evergreen epiphytic herbs with leafy stems; leaves terete or flat, coriaceous, joined with the nodal sheaths; inflorescence spicate, axillary, simple or branched, sometimes reduced to a single flower; flowers small, short-pedicelled, usually yellowish green; sepals and petals spreading, subequal; lip 3-lobed, the lateral lobes inserted at the mouth of the spur, small, the terminal lobe ovate to lanceolate, the limb with a thin longitudinal septum above the spur, callose below; the column wingless, the foot wanting; anther terminal, 2-locular; pollinia usually 4, sometimes 2.——About 100 species, in India, e. Asia, Malaysia, and Polynesia.

1. Sarcanthus scolopendrifolius Makino. MUKADE-RAN.

Stems slender, creeping, branched, about 1.5 mm. in diameter; leaves distichous, 7–10 mm. long, slightly recurved, subulate, obtuse, grooved above; sheaths tubular, closely appressed to the internodes; peduncles axillary, 1-flowered, 2–3 mm. long, 1- or 2-bracteate; sepals spreading, pale rose, spathulate-oblong, obtuse, slightly connate at the base, the petals similar but slightly shorter; lip saccate-spurred, the terminal lobe deltoid-ovate, obtuse, entire; column short; capsules clavate-ovoid, 6–7 mm. long, nearly sessile.——June–July. Trees or rocks; Honshu (Kantō Distr. and westw.), Shikoku, Kyushu; rare.——s. Korea.

58. SACCOLABIUM Bl. MATSU-RAN ZOKU

Evergreen epiphyte without pseudobulbs; stems leafy; leaves spreading, distichous, coriaceous or fleshy, flat or rarely terete, the sheaths persistent, closely clasping; peduncles simple or branched; flowers in racemes, the bracts diminutive; sepals and petals alike, free; lip sessile, saccate or spurred, the lateral lobes relatively small, erect, the terminal one larger; column short, wingless, sometimes angled; the foot wanting; anther terminal, 1- or imperfectly 2-locular; pollinia 2, rarely bifid; capsules oblong-cylindric.——About 20 species in India, Malaysia, and e. Asia.

1A. Leaves 3–6 cm. long, 6–15 mm. wide, obscurely 7- to 9-nerved. .... 1. S. japonicum
1B. Leaves 7–20 mm. long, 3–5 mm. wide, obscurely 3-nerved.
  2A. Leaves uniformly green; petals broadly elliptic; lip ciliate. .... 2. S. ciliare
  2B. Leaves with purple spots; petals narrowly oblong; lip not ciliate.
    3A. Leaves narrowly oblong, 7–20 mm. long, 3–5 mm. wide; lip entire, with a saccate spur. .... 3. S. matsuran
    3B. Leaves elliptic to ovate-elliptic, 5–11 mm. long, 2.5–5 mm. wide; lip 3-lobed, with a short-cylindric spur. ...... 4. S. toramanum

1. Saccolabium japonicum Makino. *Gastrochilus japonicus* (Makino) Schltr.——KASHI-NO-KI-RAN.

Stems 1–4 cm. long, densely leafy; leaves evergreen, oblanceolate, 3–6 cm. long, 6–15 mm. wide, subacute, somewhat falcate, obscurely 7- to 9-nerved, the midrib raised beneath; racemes 1–3 cm. long, 4- to 10-flowered, the bracts deltoid, 1–2 mm. long; flowers pale yellow; sepals and petals about 4 mm. long, spathulate-oblong, obtuse, rose-purple spotted on the inside; lip relatively large spurred, the spur about 3 mm. across at the mouth, about 4 mm. in depth, the lateral lobe on the margin of the mouth, the limb depressed deltoid, obtuse; column short; capsules sessile, narrowly oblong-cylindric, about 2 cm. long. ——July–Aug. Shikoku, Kyushu; rare.——Ryukyus.

2. Saccolabium ciliare (F. Maekawa) Ohwi. *Gastrochilus ciliaris* F. Maekawa——MATSUGE-KAYA-RAN.

Stems slender; leaves dimorphic, uniformly green, numerous, spreading, obsoletely 3-nerved, the midrib depressed above, some of them elliptic, 9–10 mm. long, about 4.5 mm. wide, acute to subacute, others linear-oblong, 16–20 mm. long, 3.5–4 mm. wide, the bracts ovate, small; flowers 2–4, on an abbreviated axis, pale yellowish green, partially open; sepals narrowly oblong, 2–2.5 mm. long, obtuse; petals broadly elliptic, about 2 mm. long; lip whitish, saccate, the limb reniform, ciliate; column short.——Kyushu (Yakushima); very rare.

3. Saccolabium matsuran Makino. *Gastrochilus matsuran* (Makino) Schltr.——MATSU-RAN, BENI-KAYA-RAN.

Stems slender, 1–3 cm. long; leaves evergreen, dense, narrowly oblong to linear-oblong, 7–20 mm. long, 3–5 mm. wide, slightly incurved at tip, purple-spotted, the midrib depressed above, elevated beneath; inflorescence 8–10 mm. long, 2- to 4-flowered, the bracts deltoid, 0.5–1 mm. long, obtuse; flowers yellowish green; sepals and petals spreading, narrowly oblong, 3–3.5 mm. long, obtuse; lip rather large, saccate-spurred, entire, the limb reniform, obtuse-truncate; column very short; capsules obovoid-oblong, short-pedicelled, about 10 mm. long.——Tree-trunks; Honshu, Shikoku. Kyushu.

4. Saccolabium toramanum Makino. *Gastrochilus toramanus* (Makino) Schltr.——Momi-ran.   Epiphytic; stems to 7 cm. long, creeping; leaves small, equal, spreading, very dense, elliptic to ovate-elliptic, 5–11 mm. long, 2.5–5 mm. wide, acute to acuminate, retuse, 3-nerved, green with purple spots, grooved above; flowers few, in short racemes on very short peduncles; sepals elliptic, 2.5–2.7 mm. long, yellowish green with a purple center, petals similar; lip 3-lobed, cylindric-spurred, the dorsal lobe larger, transversely oblong with rounded margins, about 1.5 mm. long, 3 mm. wide, retuse to truncate at apex, the lateral lobes minute, erect, on the mouth of the spur, the column short; capsules oblong, about 9 mm. long, pedicelled.——Oct.   Honshu (Kii, s. Shinano and Iwaki), Shikoku; very rare.

59. NEOFINETIA Hu   Fū-ran Zoku

Epiphytic, evergreen herb with short, densely leafy, erect stems; leaves linear, distichous, conduplicate, firm, jointed, short-sheathed at base; peduncles lateral, short, 1-scaled at base; flowers long-pedicelled, rather small, the bracts small; sepals spreading, linear-lanceolate, alike, free, the petals similar; lip lingulate, erect, long-spurred, the lateral lobes small, the terminal lobe ovate, flat, entire; spur elongate, curved; column short, broad, wingless, without a foot; rostellum broad, short, bifid; capsules cylindric, long-pedicelled.——A single species in Japan.

1. Neofinetia falcata (Thunb.) Hu.   *Orchis falcata* Thunb.; *Angraecum falcatum* (Thunb.) Lindl.; *Oeceoclades lindleyana* Regel; *Nipponorchis falcata* (Thunb.) Masam.; *Finetia falcata* (Thunb.) Schltr.; *Angraecopsis falcata* (Thunb.) Schltr.——Fū-ran.   Stems tufted, 1–3 cm. long, flattened; leaves evergreen, conduplicate, coriaceous, thick, lustrous, gradually arcuate, 5–10 cm. long, 7–8 mm. wide, obtuse, the sheaths short, jointed at apex; peduncles 3–10 cm. long, with 1 or 2 small lanceolate scales on the lower part; flowers 3–5, white, the bracts ascending, lanceolate, 4–7 mm. long; sepals and petals linear-lanceolate, about 10 mm. long, obtuse; lip slightly recurved, 7–8 mm. long, lingulate, fleshy, 3-lobed, the lateral lobes subrounded, with long-decurrent lower margins, the terminal lobe narrowly ovate, 3–4 mm. long, obtuse, the spur linear, curved, about 4 cm. long; capsules cylindric, about 3 cm. long, the pedicel about the same length.——July.   Honshu (Izu Prov. and westw.), Shikoku, Kyushu; rare.——Ryukyus.

60. TAENIOPHYLLUM Bl.   Kumo-ran Zoku

Small epiphytic herbs with aerial roots; stems very short or wanting; leaves linear (in our species entirely suppressed); peduncles very short, simple; flowers minute, short-pedicelled, the bracts minute; sepals ascending or incurved at tip, nearly equal, often slightly connate at base, the petals similar; lip saccate or spurred, 3-lobed, the lateral lobes attached near the mouth of spur or sometimes absent, the terminal lobe small, spreading, fleshy; column very short; anther terminal, 2-locular; pollinia 4, waxy; capsules oblong-linear.——About 120 species, in India, Malaysia, Australia, Polynesia, and e. Asia.

1. Taeniophyllum aphyllum (Makino) Makino. *Sarcochilus aphyllus* Makino——Kumo-ran.   Minute leafless, green, epiphytic herb with fasciculate roots, 2–3 cm. long, 1–1.2 mm. in diameter; stems nearly wanting; leaves absent; peduncles few, tufted, slender, 4–7 mm. long, 1- to 3-flowered, the bracts deltoid, acute, about 1 mm. long; flowers minute; sepals pale green, not expanded, lanceolate, about 2 mm. long, adnate to the base of the petals, gradually acuminate; petals slightly shorter than the sepals, broadly lanceolate; lip lanceolate, the margin involute, gradually acuminate toward the inflexed apex; spur saccate; column very short, adnate to the base of the lip; capsules oblong-ovoid, 4–5 mm. long, ascending, nearly sessile.——Tree-trunks; Honshu (Kantō Distr. and westw.), Shikoku, Kyushu; very rare.

61. AËRIDES Lour.   Nago-ran Zoku

Evergreen epiphyte; stems short, not thickened, leafy; leaves distichous, coriaceous or fleshy, jointed with the embracing sheaths at base; inflorescence simple or branched, racemose; flowers usually rather small, the bracts small; sepals equal, spreading, the lateral lobes broadest, adnate to the foot of the column, the petals similar; lip adnate to the foot of column, erect or incurved, spurred, the lateral lobes small, erect, inserted near the mouth of the spur, the terminal lobe spreading, ovate; column short, extended to form a broad foot, wingless; anther terminal, 2-locular; pollinia 2; capsules oblong or clavate.——About 20 species in India, Malaysia, and e. Asia.

1. Aërides japonicum Reichenb. f. Nago-ran.   Stems ascending, 3- to 5-leaved; leaves narrowly oblong, 8–15 cm. long, 1.5–2 cm. wide, obtuse or retuse, rather fleshy, the midrib depressed above; peduncles cauline, declined, 5–12 cm. long, few-flowered, the bracts ovate, obtuse, spreading, 4–5 mm. long; flowers pale greenish white; sepals ascending, oblong, about 12 mm. long, obtuse; petals similar to the sepals; lip as long as the sepals, 3-lobed, the lateral lobes small, the terminal lobe spathulate, rounded, undulate on the upper margin; spur saccate, inserted below the lip and bent forward; capsules short-pedicelled.——Honshu (Izu Prov. and westw.), Shikoku, Kyushu; rare.——Ryukyus.

62. SARCOCHILUS R. Br.   Kaya-ran Zoku

Epiphytic; stems leafy; leaves distichous, evergreen, coriaceous or fleshy, linear to oblong; peduncles lateral, cauline, simple or rarely branched; sepals spreading, free, alike or the lateral lobes broader and adnate to the base of the column; petals similar to the dorsal sepal; lip inserted on the foot of the column, sometimes forming a short inflated chin at base, the lateral lobes erect; spur straight; column erect, subterete, wingless, extended to form a basal foot; anther terminal, 2-locular; pollinia 2, globose, waxy.——About 30 species, in India, Malaysia, e. Asia, and Australia.

1. Sarcochilus japonicus (Reichenb. f.) Miq. *Thrix-spermum japonicum* Reichenb. f.——KAYA-RAN. Stems rather slender; leaves 10–20, flat, distichous, relatively dense, lanceolate, 2–4 cm. long, 4–6 mm. wide, obtuse, mucronate, gradually narrowed toward the base; peduncles slender, 2–4 cm. long, 2- to 5-flowered, with a small scale near the middle, the bracts broadly ovate, spreading, about 3 mm. long; flowers pale yellow; sepals narrowly oblong, 7–8 mm. long, obtuse; petals narrowly spathulate-oblong, slightly shorter than the sepals; lip shallowly 3-lobed, the lateral lobes spreading, the terminal lobe obsolete; column short; capsules cylindric, nearly sessile, 4–5 cm. long, about 3.5 mm. wide.——Apr.–May. Honshu (Kantō Distr. and westw.), Shikoku, Kyushu.

SUBCLASS 2. DICOTYLEDONEAE   DICOTYLEDONS   SōSHIYō RUI

1. Archichlamydeae   RIBENKA RUI

Fam. 59. SAURURACEAE   DOKUDAMI KA   Lizard's-tail Family

Perennial erect herbs with simple, petiolate, entire, alternate leaves; inflorescence racemose or spicate; flowers small, bisexual, bracteolate, the perianth absent; stamens 6–8, hypogynous or adnate to the ovary and essentially epigynous; anthers 2-locular, longitudinally dehiscent; pistil with 3 or 4 free or united carpels, the ovules orthotropous, 1 or 2 in each locule; fruit a capsule or berrylike, indehiscent or incompletely dehiscent; seeds ovoid or globose, the testa membranous, the endosperm copious, mealy, the embryo minute.——About 3 genera, with 4 or 5 species, in e. Asia and N. America.

1A. Stamens free from the ovary; carpels free; inflorescence without involucral bracts. . . . . 1. *Saururus*
1B. Stamens with the lower part of filaments adnate to the carpels; carpels connate; inflorescence with 4–6 petaloid involucral bracts at base. . . . . 2. *Houttuynia*

1. SAURURUS L.   HAN-GE-SHō ZOKU

Glabrous perennials (except inflorescence); leaves alternate, ovate-cordate, the stipules membranous, adnate to the petiole; inflorescence spicate, terminal, bracteate on the pedicel; flowers small, the perianth absent; stamens 6 or 8, sometimes fewer, the filaments filiform, free, the anthers erect; pistil of 3 or 4 free or nearly free carpels, the ovules 2 to 4 in each carpel; fruit subglobose, folliclelike; mature seed globose, usually solitary in each carpel.——Two species, one in N. America and one in e. Asia.

1. Saururus chinensis (Lour.) Baill. *Spathium chinense* Lour.; *Saururus loureiri* Decne.; *Saururopsis chinensis* (Lour.) Turcz.——HAN-GE-SHō, KATA-SHIRO-GUSA. Foetid perennial with thick creeping rhizomes; stems 50–100 cm. long, erect, sparsely branched; leaves narrowly to broadly ovate, 5–15 cm. long, 3–8 cm. wide, acute, 5- to 7-nerved, entire, paler beneath, cordate-auriculate, the uppermost white on the lower half, the petioles rather broad, with a ridge beneath, 1–5 cm. long, slightly clasping at base; racemes 10–15 cm. long, densely many-flowered, erect, nodding while young, sparsely pubescent with crisped hairs, the bracts ovate-rounded, about 1.5 mm. wide, the pedicels 2–3 mm. long; stamens 6 or 7; carpels 3–5, glabrous.——June–Aug. Around ponds and along rivers; Honshu, Shikoku, Kyushu; common.——Korea, Ryukyus, China, and Philippines.

2. HOUTTUYNIA Thunb.   DOKUDAMI ZOKU

Erect perennial herbs with alternate, petiolate, entire, cordate leaves; stipules rather large, membranous, adnate to the petiole at base; spikes terminal, pedunculate, with 4–6 whitish involucral bracts at base; flowers sessile, bisexual, small, naked; bracts small; stamens 3–8, the filaments adnate to the ovary; carpels 3 or 4, with parietal placentae; styles 3 or 4, simple; fruit subglobose, dehiscent below the style inside; seeds rather numerous, globose.——A single species, in e. Asia.

1. Houttuynia cordata Thunb. *Polypara cordata* (Thunb.) Buek; *P. cochinchinensis* Lour.——DOKUDAMI. Foetid, glabrous perennials with elongate, slender rhizomes; stems 20–50 cm. long, usually simple; leaves broadly ovate-cordate, 3–8 cm. long, 3–6 cm. wide, deep green, paler beneath, the stipules oblong, obtuse, persistent; peduncles 2–3 cm. long, the involucral bracts usually 4, elliptic or oblong, spreading, persistent, 1.5–2 cm. long, white above; spikes 1–3 cm. long, short-cylindric, densely many-flowered, the flowers sessile.——June–July. Shaded places in lowlands; Honshu, Shikoku, Kyushu; very common.——Ryukyus, Formosa, China, Himalayas, and Java.

Fam. 60. PIPERACEAE   KOSHō KA   Pepper Family

Herbs, shrubs or rarely small trees, sometimes scandent, often fleshy; stems and branches sometimes jointed; leaves alternate, verticillate, rarely opposite, entire, the stipules sometimes absent; spikes or racemes densely many-flowered; flowers small, the perianth absent, bisexual or unisexual, bracteate; stamens 2–10; ovary unilocular, with a solitary basal orthotropous ovule; stigmas 1–6, sessile or the styles short; fruit a berry, 1-seeded, the embryo small, the endosperm fleshy.——About 5 genera, with more than 1,000 species, chiefly in the Tropics.

1A. Stigma solitary; erect sometimes fleshy herbs, often creeping at base. . . . . 1. *Peperomia*
1B. Stigmas (2–)3–4(–5); commonly scandent woody plants, sometimes herbs. . . . . 2. *Piper*

1. PEPEROMIA Ruiz & Pav. SADA-SŌ ZOKU

Annuals or perennials; stipules absent; leaves alternate, opposite, or verticillate, entire, usually fleshy, commonly with pellucid dots; spikes densely or loosely flowered; flowers minute, bracteate, sometimes sunken in the axis, bisexual, sessile, the perianth absent; stamens 2, the filaments usually short, the anthers globose or depressed-globose, 2-locular and often confluent; stigma often penicillate; berry small, scarcely fleshy, the seeds similar to the fruit in shape.——About 400 species, pantropic.

1. Peperomia japonica Makino. SADA-SŌ. Fleshy, densely soft-puberulent, green perennial herb; stems 10–30 cm. long, erect, terete, from ascending base, sparsely branched; leaves spreading, verticillate in 3's to 5's, slightly fleshy, elliptic, obovate or broadly elliptic, 1–2.5 cm. long, 7–15 mm. wide, entire, rounded to broadly obtuse, the petioles 3–8 mm. long; spike erect, short-pedunculate, 2–4 cm. long, rather densely many-flowered, glabrous except the peduncle; flowers minute, sessile; fruit globose, about 0.4 mm. long.——Shikoku, Kyushu (s. distr., Yakushima and Tanegashima).

2. PIPER L. KOSHŌ ZOKU

Scandent, woody or rarely herbaceous plants or sometimes small trees, with jointed branches; leaves alternate, entire or rarely 3-lobed, 3- to many-nerved, the stipules rarely absent; spikes terminal; flowers bisexual or unisexual, usually sessile, sometimes sunken in the axis, adnate to a peltate bract; perianth absent; stamens 2–4, rarely to 8, the anthers ovoid or globose, sometimes oblong; stigmas 2–5; berry small, globose or ovoid; seeds globose.——More than 700 species, pantropic.

1. Piper kadzura (Choisy) Ohwi. *Ipomoea kadzuru* Choisy; *P. futokadzura* Sieb.——FUTO-KAZURA. Scandent over trees and rocks, often radicant; leaves ovate (sometimes rounded-cordate in young specimens), 5–8 cm. long, usually 3–4 cm. wide, long-acuminate, rounded at base, entire, rather thick, dark green, paler beneath and often scattered soft hairy, 5-nerved or nearly so; spikes 3–8 cm. long, pedunculate, nodding to pendulous; flowers unisexual (plants dioecious), the bracts peltate; fruit globose, red, 3–4 mm. in diameter.——May–June. Woods near seashores; Honshu (Kantō Distr. and westw.), Shikoku, Kyushu.——s. Korea and Ryukyus.

Fam. 61. CHLORANTHACEAE SENRYŌ KA Chloranthus Family

Herbs or shrubs, sometimes aromatic; leaves opposite, usually serrate, pinnately veined, the petioles joined by a horizontal ridge on stem or by a short sheath, the stipules small, linear, subtending the petiole or inserted on the margin of the sheath; inflorescence a simple or branched spike (rarely capitate), essentially terminal; flowers unisexual or bisexual, the perianth absent in the bisexual ones; stamens 1 or 3, sometimes 2, connate below, sometimes adnate to the ovary; ovary 1, 1-locular; stigma sessile or rarely with a short stalk, truncate or short-linear, undivided; ovules solitary, orthotropous; fruit a drupe, ovoid or globose, the exocarp fleshy or juicy, the endocarp hard; seeds with copious endosperm, the embryo small.——Four genera, with about 40 species, chiefly in the Tropics and warmer temperate regions.

1A. Stamen 1; evergreen shrubs. .... 1. *Sarcandra*
1B. Stamens 3, connate below; perennial herbs or shrubs. .... 2. *Chloranthus*

1. SARCANDRA Gardn. SENRYŌ ZOKU

Shrubs with jointed nodes; leaves evergreen, opposite or verticillate in 3's or 4's, petiolate, serrate, pinnately nerved, the stipules short, with 2 marginal projections, green; spikes terminal, sometimes branched; flowers small, naked, sessile; stamen 1, adnate on the back of ovary and jointed at base with it, the anther 2-locular, introrse; stigma sessile, truncate; ovules pendulous, ventral.——One species.

1. Sarcandra glabra (Thunb.) Nakai. *Bladhia glabra* Thunb.; *Chloranthus glaber* (Thunb.) Makino; *Chloranthus brachystachyus* Bl.; *Ascaria serrata* Bl.——SENRYŌ. Glabrous evergreen shrub 30–80 cm. high; stems terete, green, spreading, the nodes slightly swollen; leaves oblong or ovate-oblong, 10–15 cm. long, 4–6 cm. wide, deep green, lustrous, sharply serrate, acuminate, the petioles 5–15 mm. long; spikes terminal, short-pedunculate, simple to few-branched below, 15–30 mm. long, erect; flowers small, white; stamen 1, the anther 2-locular, introrse; fruit fleshy, globose, orange–scarlet, 5–7 mm. in diameter.——Woods; Honshu (Tōkaidō and Kinki Distr.), Shikoku, Kyushu; rather common.——s. Korea, Ryukyus, Formosa, China, India, and Malaysia.

Cv. **Flava.** *S. glabra* forma *flava* (Makino) Ohwi; *Chloranthus glaber* var. *flavus* Makino——KIMINO-SENRYŌ. Fruit yellow.——Cultivated.

2. CHLORANTHUS Sw. CHA-RAN ZOKU

Herbs or shrubs with swollen, jointed nodes; leaves simple, petiolate, serrate, the petioles connected by a transverse ridge on stem; inflorescence a terminal spike, simple or branched; flowers bisexual, the perianth absent; stamens 3, sometimes partly reduced and sterile, the anthers 1- or 2-locular, introrse or extrorse, the filaments connate, forming a dilated, 3-lobed body, often attached on one side of the ovary; stigma sessile, truncate.——More than 10 species, in s. and e. Asia.

1A. Evergreen shrub with leaves in evenly spaced pairs. . . . . 1. *C. spicatus*
1B. Herbs with leaves in few pairs near the apex.
   2A. Fertile stamens 2, the median one sterile; filaments elongate, linear, short-connate at base, the connective extending beyond the anther; spike usually solitary. . . . . 2. *C. japonicus*
   2B. Stamens 3, all fertile; filaments connate below, dilated, shortly 3-lobed above; spikes 1–3, usually 2. . . . . 3. *C. serratus*

1. Chloranthus spicatus (Thunb.) Makino. *Nigrina spicata* Thunb.; *C. inconspicuus* Sw.——CHA-RAN. Small glabrous shrub, the branches ascending or spreading toward the base; stipules membranous, linear, 2–3 mm. long, mucronate; leaves oblong-ovate to ovate-elliptic, 4–8 cm. long, 2.5–4 cm. wide, rather obtuse, acute at base, shallowly undulate-serrate, the petioles 5–12 mm. long; inflorescence terminal, spicate, pedunculate, glabrous, the spikes 10–18, ascending, 2–3 cm. long; stamens 3, the filaments connate.——Chinese shrub long cultivated in Japan.

2. Chloranthus japonicus Sieb. *Tricercandra quadrifolia* A. Gray; *C. mandshuricus* Rupr.; *Tricercandra japonica* (Sieb.) Nakai——HITORI-SHIZUKA. Glabrous perennial from short, branching, creeping rhizomes; stems erect, usually simple, 20–30 cm. long; leaves small, membranous and scalelike toward the base, the uppermost consisting of 2 ordinary pairs simulating a whorl of 4 leaves, ovate or elliptic, 4–12 cm. long, 2–6 cm. wide, abruptly acute, sharply serrate, the petioles 10–15 mm. long; spikes usually simple, solitary, 2–3 cm. long, erect, the peduncle 2–5 cm. long; flowers white, sessile; fila-

ments white and conspicuous, shortly connate below and adnate to the ovary, linear, obtuse, spreading, 4–5 mm. long; fruit obliquely obovate, narrowed towards the base, 2.5–3 mm. long.——Apr.–May. Woods in low mountains; Hokkaido, Honshu, Shikoku, Kyushu; common.——s. Kuriles, Sakhalin, Manchuria, and Korea.

3. Chloranthus serratus (Thunb.) Roem. & Schult. *Nigrina serrata* Thunb.——FUTARI-SHIZUKA. Erect glabrous perennial from branching rhizomes; stems 30–50 cm. long, usually simple; lower leaves reduced to membranous connate sheaths, the ordinary leaves 4 or 6, paired, approximate, ovate-oblong to elliptic or ovate, 8–15 cm. long, 4–7 cm. wide, acuminate, sharply serrulate, abruptly narrowed toward the base, the petioles 8–12 mm. long; spikes erect, usually 2, sometimes to 5, or solitary, 3–5 cm. long; flowers small, white; filaments 3, 2.5–3 mm. long, connate into a broadly obovate, 3-lobed body, the median lobe slightly longer than the lateral ones; anthers introrse; fruit broadly obovoid, about 3 mm. long.——Apr.–June. Woods in low mountains; Hokkaido, Honshu, Shikoku, Kyushu; common.——China.

Fam. 62. SALICACEAE   YANAGI KA   Willow Family

Trees and shrubs, usually dioecious; leaves alternate, rarely subopposite, simple, usually stipulate, petiolate, usually serrate; inflorescence of aments; flowers unisexual, appearing before or with the leaves, in the axils of bracts; perianth wanting; stamen 1 to many; ovary 1-locular, the style 1, stigmas 2–4; fruit a 2- to 4-valved capsule, the 2–4 placentae parietal; seeds few to numerous, with a tuft of long hairs (coma), the endosperm absent.——Few genera, with about 350 species, circumboreal, in the temperate to cooler regions (few at high elevations in the Tropics); flowering mostly early in spring, but some species in summer.

1A. Bud-scales numerous, imbricate; spikes pendulous with laciniate to toothed bracts; flowers with a cup-shaped gland. . . . . . . 1. *Populus*
1B. Bud-scales solitary; spikes with entire or denticulate bracts; flowers with 1 or 2 globose to cylindric glands or glandless.
   2A. Spikes pendulous; stigma deciduous with the upper part of the style.
      3A. Flowers anemophilous, without a gland. . . . . 2. *Chosenia*
      3B. Flowers entomophilous, with 2 dorsi-ventral glands in the staminate ones, the glands lateral in the pistillate ones. . . . 3. *Toisusu*
   2B. Spikes erect; flowers entomophilous or anemophilous, with 1 or 2 glands; stigma persistent. . . . . 4. *Salix*

1. POPULUS L.   HAKO-YANAGI ZOKU

Dioecious trees; buds with imbricate, glandular or sometimes hairy bud-scales; leaves alternate, deciduous, usually ovate to ovate-rhomboid or subdeltoid, petiolate, entire or serrulate, involute in bud; flowers unisexual, with an oblique, cup-shaped gland, borne in the axils of laciniate, incised, or rarely entire bracts on pendulous, precocious aments; stamens 4 to many; ovary 1-locular, the style solitary, short, the stigma 2- to 4-fid; capsules 2- to 4-valved; seeds small, ovoid, brown; cotyledons elliptic.——About 30 species, in the temperate regions of the N. Hemisphere.

1A. Petioles laterally compressed above.
   2A. Leaves deltoid or rhomboid-deltoid, glabrous, paler beneath; buds without hairs. . . . . 1. *P. nigra* cv. Italica
   2B. Leaves broadly ovate or rhomboid-rounded, slightly glaucous beneath, hairy while young; bud-scales pubescent. . . . . 2. *P. sieboldii*
1B. Petioles not obviously compressed.
   3A. Leaves densely silvery lanate beneath, incised on margins. . . . . 3. *P. alba*
   3B. Leaves pale white and usually puberulous on nerves beneath, obtusely serrate. . . . . 4. *P. maximowiczii*

1. Populus nigra L. cv. **Italica**. *P. nigra* var. *italica* Koehne; *P. italica* Moench——SEIYŌ-HAKO-YANAGI, POPURA. Deciduous tree of fastigiate habit, staminate; branchlets glabrous, orange-red while young, the buds reddish, glabrous, viscid; leaves broadly deltoid or broadly rhombic-deltoid, abruptly short-acuminate, broadly cuneate to nearly truncate at the base, obtusely serrulate, glabrous, often with reddish nerves, the petioles 2–5 cm. long; staminate aments precocious,

3–5 cm. long, the flowers with rather numerous stamens.——Apr. The Lombardy poplar, a cultivar of European origin, often cultivated in parks and gardens.

2. Populus sieboldii Miq. *P. tremula* var. *villosa* sensu Fr. & Sav., non Wessmael; *P. tremula* var. *sieboldii* (Miq.) Kudo——YAMA-NARASHI, HAKO-YANAGI. Deciduous tree with rather thick, purplish gray branches, white-pubescent while young, the bud-scales white-pubescent; leaf-blades ovate to broadly so, 4–8 cm. long, 3–7 cm. wide, acute to abruptly acute, serrate, deep green, with a small gland on each side of the rounded to broadly cuneate base, slightly glaucous beneath and white-pubescent while young; aments precocious, the staminate 4–7 cm. long, 8–10 mm. across, stamens 5–7; the pistillate 5–8 cm. long, about 7 mm. across.——Apr.–May. Mountains; Hokkaido, Honshu, Shikoku, Kyushu.

3. Populus alba L. Urajiro-hako-yanagi, Hakuyō, Gin-doro. Deciduous tree; branches and bud-scales white-woolly while young; leaf-blades ovate, elliptic or narrowly so, 4–10 cm. long, acute to obtuse, rounded to shallowly cordate at base, sinuately toothed or incised, deep green above, densely white-woolly beneath; pistillate aments about 5 cm. long; staminate flowers with 6–10 stamens.——Sometimes cultivated in Japan, native of Europe and centr. Asia.

4. Populus maximowiczii Henry. *P. suaveolens* var. *latifolia* Gombócz; *P. suaveolens* sensu auct. Japon., non Fisch.——Doro-yanagi, Doro-no-ki, Doro. Deciduous tree with thick branches, the branchlets at first pubescent, grayish; leaves subcoriaceous, ovate, elliptic or oblong, sometimes broadly ovate, 6–10 cm. long, abruptly short-acuminate, obtuse to shallowly cordate at base, obtusely serrulate, deep green above, puberulous on the nerves and whitish beneath, the petioles 1–4 cm. long, puberulous; staminate aments 5–10 cm. long, the stamens rather numerous; pistillate aments elongate after flowering; capsules rather loosely arranged, nearly sessile.——Apr.–June. Sunny fluvial plains and riverbanks; Hokkaido, Honshu (centr. distr. and northw.).——Korea, Manchuria, Ussuri, Amur, Sakhalin, and Kamchatka.

The following are poorly known and reported to occur in Hokkaido:

(1) **Populus jesoensis** Nakai. Ezo-yama-narashi. Plant glabrous; leaves larger and with longer petioles than in *P. sieboldii* Miq.

(2) **Populus davidiana** Dode. Chōsen-yama-narashi. Leaves smaller than in *P. jesoensis* Nakai and without glands at base of blade.——Korea, Manchuria, and Ussuri.

(3) **Populus koreana** Rehd. Chirimen-doro. Young branches glabrous, viscid; nerves of leaves impressed above.

2. CHOSENIA Nakai   Keshō-yanagi Zoku

Deciduous dioecious tree; buds provided with a single bud-scale; leaves relatively narrow, alternate, petiolate, usually serrate, sometimes entire, the stipules absent; flowers precocious, anemophilous; staminate aments pendulous, the bracts thinly membranous, adnate at base to the filaments; stamens 5, slightly unequal, the filaments filiform, the anthers 2-locular; pistillate aments pendulous, the bracts thinly membranous, caducous, hairy; ovary 1, 1-locular, stipitate; capsules 2-valved, the ovules 4, the endosperm absent.——A single species, in e. Asia.

1. Chosenia arbutifolia (Pall.) B. V. Skvortz. *Salix arbutifolia* Pall.; *C. bracteosa* (Turcz.) Nakai; *Salix macrolepis* Turcz., pro parte; *C. macrolepis* (Turcz.) Komar.; *S. splendida* Nakai; *C. splendida* (Nakai) Nakai; *S. bracteosa* Turcz.——Keshō-yanagi. Upright deciduous tree; branchlets glabrous, usually glaucous, sometimes reddish; leaves with glaucous petioles 2–5(–7) mm. long, the blades oblanceolate or oblong-lanceolate, 6–8 cm. long, 1–2 cm. wide, obscurely serrate or sometimes entire, acute at both ends, quite glabrous, rather thick; staminate aments 1–2.5 cm. long, 4–5 mm. in diameter, the axis glabrous, the bracts 3- to 5-nerved, broadly ovate, about 2.5 mm. long, obtuse, loosely hairy on back; pistillate aments 1–2 cm. long in flower, 4–5 cm. long in fruit, the bracts caducous; ovary glabrous, stipitate, the styles 2, free, jointed above, the stigma bifid, lobes linear.——Alluvial plains; Hokkaido, Honshu (Kamikochi in Shinano); locally abundant.——Korea, Sakhalin, e. Siberia, and the area surrounding the Ochotsk Sea.

3. TOISUSU Kimura   Ōba-yanagi Zoku

Deciduous dioecious trees with a single overlapping bud-scale, the margins free; leaves narrowly ovate, usually serrate, alternate, stipulate in vigorous shoots; aments cylindric, pendulous in anthesis, the pistillate bracts deciduous; flowers entomophilous, the staminate with 2 dorsi-ventral glands and 5–10 stamens, the pistillate with a gland on each side; ovary stipitate, the styles 2, free or connate to the middle, the stigmas bifid, slender, becoming deciduous with the upper part of the style; ovules 4.—— Few species in e. Asia.

1. Toisusu urbaniana (Seemen) Kimura. *Salix cardiophylla* sensu auct. Japon., non Trautv. & Mey.; *S. urbaniana* Seemen; *T. cardiophylla* var. *urbaniana* (Seemen) Kimura——Ōba-yanagi. Tree; branchlets dark reddish brown, sparsely short-pubescent while young; leaves narrowly ovate to broadly lanceolate, 5–13 cm. long, 1–4.5 cm. wide, acuminate, minutely serrate, obtuse to rounded or sometimes shallowly cordate at the base, fulvous especially on the underside when young, short-puberulous on the midrib above, glabrescent in age, glaucous beneath; aments appearing with the leaves, terminal on short leafy spurs, pendulous, 5–10 cm. long, the staminate densely flowered, about 8 mm. across, the bracts broadly ovate, obtuse, ciliate, densely silky on lower half, the pistillate elongate, to 10 cm. long; ovary densely grayish white-pubescent. ——May–June. Hokkaido, Honshu (centr. distr. and northw.).——Kuriles.

Var. **schneideri** (Miyabe & Kudo) Kimura. *Salix urbaniana* var. *schneideri* Miyabe & Kudo——Tokachi-yanagi. Ovary glabrous.——Hokkaido, Honshu (n. distr.).——Sakhalin.

× **Toisochosenia kamikochica** (Kimura) Kimura. *Salix kamikochica* Kimura——Kamikochi-yanagi. A bigeneric hybrid of *Chosenia arbutifolia* × *Toisusu urbaniana*. Occurs with the parents.

4. SALIX L.   Yanagi Zoku

Usually deciduous, dioecious trees or shrubs; leaves alternate, rarely subopposite, petiolate or sessile, lanceolate, sometimes rounded-cordate, entire to serrate, those on vigorous young shoots often stipulate; aments appearing before or with the leaves; flowers solitary in the axils of bracts, with 1 or 2 glands; staminate flowers with 1–12, usually 2, stamens, the filaments elongate sometimes united below; ovary solitary, 2 carpellate, sessile or stipitate, the style slender or absent, the stigmas 2, each usually bifid; capsules 1-locular, 2-valved; seeds usually many, silky-comose.——More than 300 species, chiefly in the temperate and cooler regions of the N. Hemisphere, a few in S. America but none elsewhere in the S. Hemisphere.

1A. Staminate flowers with 2 glands.
  2A. Stamens 3–5; bracts yellowish, deciduous in the pistillate flowers.
    3A. Bud-scales with free imbricate margins. .... 1. *S. chaenomeloides*
    3B. Bud-scales united on the margins. .... 2. *S. subfragilis*
  2B. Stamens 2.
    4A. Erect shrubs or trees; leaves 4–15 cm. long, lanceolate, acuminate.
      5A. Pistillate flowers with 2 glands; leaves glabrous.
        6A. Spikes about 2 cm. long; leaves and branchlets green when dried; branchlets slender, often twisted; cultivated.
                3. *S. matsudana* Cv. Tortuosa
        6B. Spikes not over 1.5 cm. long, densely flowered; leaves and branchlets often dark brown when dried; branchlets rather thick; indigenous. .... 4. *S. eriocarpa*
      5B. Pistillate flowers with 1 gland; leaves usually pubescent.
        7A. Ovary glabrous.
          8A. Leaves linear-lanceolate; branchlets slender, drooping; styles short; cultivated. .... 5. *S. babylonica*
          8B. Leaves lanceolate to broadly so; branchlets not drooping; styles nearly absent; indigenous. .... 6. *S. serissaefolia*
        7B. Ovary pubescent.
          9A. Leaves pale green beneath. .... 7. *S. yoshinoi*
          9B. Leaves glaucous beneath.
            10A. Petioles 2–5 mm. long; branchlets densely puberulent. .... 8. *S. jessoensis*
            10B. Petioles 5–10 mm. long; branchlets glabrous or glabrescent. .... 9. *S. koreënsis*
    4B. Dwarf alpine subshrub; leaves obovate to rounded, rounded at apex, 6–12 mm. long, subentire. .... 10. *S. pauciflora*
1B. Staminate flowers with 1 gland.
  11A. Stamens free.
    12A. Dwarf alpine shrubs; branchlets short, creeping, much-branched; aments terminal on short leafy branchlets.
      13A. Styles long; leaves 2–4 cm. long, the petioles 5–25 mm. long.
        14A. Leaves with prominently raised reticulate veinlets beneath. .... 11. *S. hidakamontana*
        14B. Leaves without prominently raised veinlets beneath.
          15A. Leaves elliptic to oblong; bracts narrowly ovate, subobtuse, pubescent on both sides; alpine in Honshu. 12. *S. nakamurana*
          15B. Leaves elliptic to ovate-rounded; bracts elliptic, rounded at apex, densely pubescent; alpine in Hokkaido. 13. *S. yesoalpina*
      13B. Styles short; leaves 1–3 cm. long, the petioles 2–5 mm. long. .... 18. *S. paludicola*
    12B. Shrubs or trees of mountains and lowlands, with erect or ascending stems and branches.
      16A. Styles very short or absent.
        17A. Much-branched large shrubs; leaves 4–8 cm. long, 2–4 cm. wide, the petioles 1–2.5 cm. long.
          18A. Sapwood of young branches with short, elevated, longitudinal striations; leaves pubescent beneath. .... 14. *S. bakko*
          18B. Sapwood of young branches without striations.
            19A. Adult leaves pubescent beneath. .... 15. *S. hultenii*
            19B. Adult leaves glabrous on both sides. .... 16. *S. taraikensis*
        17B. Small shrubs; the branches erect, slightly branched; leaves 2.5–4 cm. long, 5–12 mm. wide, the petioles 1–5 mm. long.
                17. *S. subopposita*
      16B. Styles distinct, longer than the stigma.
        20A. Spikes terminal on short leafy branchlets.
          21A. Leaves rounded with a mucro or very short-acute at apex, usually distinctly toothed; ovary glabrous or white-pubescent.
                19. *S. reinii*
          21B. Leaves rounded to obtuse, minutely and sometimes obscurely toothed; ovary yellow-brown villous. ...... 20. *S. hidewoi*
        20B. Spikes lateral, sessile or nearly so.
          22A. Leaves oblong to broadly elliptic, acute or short-acuminate.
            23A. Leaves 10–20 cm. long, narrowly oblong to elliptic, often yellowish pubescent; ovary nearly sessile. ...... 21. *S. futura*
            23B. Leaves 3–8 cm. long, elliptic, oblong, or obovate-elliptic, rarely lanceolate.
              24A. Young branchlets and the underside of leaves yellow-brown pubescent; leaves elliptic or oblong; ovary stiped.
                22. *S. vulpina*
              24B. Young branchlets and other parts white-pubescent or glabrous.
                25A. Leaves broadly cuneate to shallowly cordate at base; ovary long-stiped. .... 23. *S. shiraii*
                25B. Leaves acute to obtuse at base; ovary short-stiped.
                  26A. Leaves ovate-oblong; fruiting spikes 3–5 cm. long; staminate spikes 2–2.5 cm. long. .... 24. *S. rupifraga*
                  26B. Leaves lanceolate; fruiting spikes 6–12 cm. long; staminate spikes 3–8 cm. long. .... 25. *S. japonica*
         22B. Leaves lanceolate, gradually narrowed and acuminate.
          27A. Leaves glabrous from the beginning; branchlets and leaves glaucous. .... 26. *S. rorida*
          27B. Leaves pubescent at least at first.
            28A. Leaves silky-pubescent beneath.
              29A. Branchlets thick, villous; flower-buds large, densely arranged on the branchlets; leaves silky-villous beneath.
                27. *S. kinuyanagi*
              29B. Branchlets more slender, silvery-pubescent; flower-buds rather loosely arranged; leaves silvery-pubescent.
                28. *S. petsusu*
            28B. Leaves appressed-puberulent beneath. .... 29. *S. sachalinensis*
  11B. Stamens with the filaments united at base.
    30A. Stamens 2 with filaments connate beneath.
      31A. Styles absent or very short and less than half the length of the ovary; filaments pilose; leaves lanceolate to narrowly so.

32A. Leaves sessile or nearly so, scarcely exceeding 4 cm. in length, usually opposite and entire, rounded to shallowly cordate and slightly clasping the branches, rounded to obtuse at apex. .... 30. *S. integra*

32B. Leaves distinctly petiolate, 6–16 cm. long, serrate, obtuse to acute and not clasping, acuminate to acute.

    33A. Petioles 2–5 mm. long; leaves linear-lanceolate, sometimes subopposite, obtuse to rounded at base; shrub cultivated and frequently naturalized. .... 31. *S. koriyanagi*

    33B. Petioles 2–8 mm. long; leaves always alternate.

        34A. Styles absent, the stigma entire. .... 32. *S. miyabeana*

        34B. Styles short but distinct, the stigma bifid. .... 33. *S. gilgiana*

31B. Styles longer than the ovary, 2–3 mm. long; leaves white-pubescent beneath, with raised lateral nerves.

    35A. Adult leaves silky-pubescent beneath; bracts densely hairy. .... 34. *S. gracilistyla*

    35B. Adult leaves glabrous; bracts glabrous except at base. .... 35. *S. graciliglans*

30B. Stamen 1, or the flowers with 1 or 2 stamens in one ament.

    36A. Staminate aments 4–7 cm. long, about 4 mm. across; glands cylindric; bracts dark colored; pistillate aments 3–4 cm. long.

                                  36. *S. buergeriana*

    36B. Staminate aments 2.5–4 cm. long, 6–10 mm. across; glands ovoid; bracts darkened above; pistillate aments 2–2.5 cm. long.

        37A. Stamen 1; bracts relatively short-villous; stipe of the ovary as long as the gland; leaves elliptic or narrowly ovate.

                                  37. *S. harmsiana*

        37B. Stamens 1 or 2; bracts long-villous; stipe of the ovary longer than the gland; leaves narrowly oblong. ...... 38. *S. sieboldiana*

1. Salix chaenomeloides Kimura. *S. glandulosa* Seemen, non Raf.——Akame-yanagi. Tree; young branches hairy, soon glabrate, yellowish, rather slender; leaves with petioles 5–12 mm. long with a small discoid gland near apex on each side, the blades oblong or narrowly so, sometimes elliptic, 5–12 cm. long, 1.5–5 cm. wide, short-acuminate, obtuse to acute at base, serrate with incurved teeth, glabrous and brownish red above while young, glaucous and glabrous or slightly pubescent only near base beneath, the lateral nerves many, rather prominent on both sides; staminate aments 2–6 cm. long, rather loosely flowered, the filaments glabrous, the pistillate aments 2–4 cm. long in anthesis, 6–10 cm. in fruit, the bracts 1.5 mm. long, short-pubescent; styles very short, the stigma bifid; capsules about 3 mm. long, ovoid, glabrous, the stipe 1–1.5 mm. long.——Honshu, Shikoku, Kyushu.——Korea and China.

2. Salix subfragilis Anderss. *S. nipponica* Fr. & Sav.; *S. triandra* var. *nipponica* (Fr. & Sav.) Seemen; *S. amygdalina* var. *nipponica* (Fr. & Sav.) C. K. Schn.; *S. triandra* sensu auct. Japon., non L.; *S. hamatidens* Lév.——Tachi-yanagi. Small tree or shrub; leaves with petioles 3–10 mm. long, the blades ovate- to elliptic-lanceolate, sometimes broadly lanceolate, 5–12 cm. long, 5–25 mm. wide, long-acuminate, acute at base, serrulate, whitish beneath, becoming glabrous, the lateral nerves slender, rather many; staminate aments 2–5 cm. long, about 8 mm. across, rather densely flowered; stamens 3, the filaments hairy below, the pistillate aments 2–3.5 cm. long; ovary glabrous, stipitate.——Wet lowlands; Hokkaido, Honshu, Shikoku, Kyushu.——Sakhalin, Ussuri, Manchuria, n. China, and Korea.

3. Salix matsudana Koidz. cv. **Tortuosa.** *S. matsudana* var. *tortuosa* Vilm.——Unryū-yanagi. Glabrous tree with slender tortuous, flexuous branchlets; leaves with petioles 2–8 mm. long, the blades linear-lanceolate, 5–8 cm. long, 4–10 mm. wide, long-acuminate, serrulate, cuneate to rounded at base, glaucous and slightly silky-pubescent beneath, becoming glabrate; aments 1–2 cm. long, the peduncles with entire leaves; ovary nearly sessile, the styles almost absent.——Occasionally cultivated as a garden ornamental. The typical phase of the species occurs in n. China, Korea, Manchuria, and e. Siberia.

4. Salix eriocarpa Fr. & Sav. *S. dolichostyla* Seemen ——Ja-yanagi. Trees; branches fulvous, short-pubescent; leaves with petioles 4–7 mm. long, the blades lanceolate, 7–12 cm. long, 8–20 mm. wide, acuminate, acute at base, serrulate, glabrous and glaucous beneath, slightly short-pilose above, the lateral nerves slender, rather many; aments appearing with the leaves, the staminate nearly sessile, with few small leaves at base, 8–20 mm. long, about 5 mm. across, densely flowered, the bracts oblong-ovate, obtuse; ovary densely white-pubescent; style short, glabrous.——Wet lowlands; Honshu, Shikoku, Kyushu.

5. Salix babylonica L. *S. pendula* Moench——Shidare-yanagi. Tree with the branches slender, pendulous, brownish when old, short-pubescent to glabrate while young; leaves with petioles 3–6 mm. long, the blades linear-lanceolate, 8–15 cm. long, 5–12 mm. wide, acuminate, serrulate, acute at base, glabrous or with scattered appressed pubescence, glaucous beneath, the lateral nerves many; aments nearly sessile, 2–4 cm. long in the staminate, 1–2 cm. in the pistillate, densely leafy at base, the bracts narrowly ovate, with long hairs only at base, 1.2–1.5 mm. long; ovary glabrous, nearly sessile; style very short, the stigma bifid, rather thickened.——Cultivated in Japan, but spontaneous in centr. China. Introduced and cultivated species related to *S. babylonica* are:

Salix ohsidare Kimura. Ō-shidare. Branchlets thicker; leaves longer (10–20 cm.), 2.3–4 cm. wide; peduncles 5–8 mm. long; style slightly longer.——Uncommon; Honshu (n. distr.).

Salix elegantissima K. Koch. Rokkaku. Leaves lanceolate, to 2 cm. wide, pale green beneath, the petioles 1–1.5 cm. long; pistillate aments to 5 cm. long, pedunculate; flowers with 2 glands, the ovary short-stiped, pubescent on the lower portion.

6. Salix serissaefolia Kimura. Kogome-yanagi. Tree with ascending grayish-brown branches, the branchlets short grayish pubescent soon glabrate; leaves with short-pilose petioles 2–6 mm. long, the blades thinly coriaceous, lanceolate to broadly lanceolate, 4–7 cm. long, 8–12 mm. wide, gradually long-acuminate, obtuse at base, serrulate, lustrous and yellowish green above, whitish beneath, the lateral nerves slender, spreading, paler colored and prominent above; pistillate aments rather slender, 1–1.7 cm. long, short-pedunculate, rather densely flowered, the bracts yellowish green, elliptic, nearly truncate or rounded at apex, about 2 mm. long, hairy only at base; ovary nearly sessile, with silky hairs near base; style very short, the stigma linear; capsules ovoid, about 2.5 mm. long.——Honshu (Kantō to Kinki Distr.).

7. Salix yoshinoi Koidz. Yoshino-yanagi. Small tree with grayish branchlets densely white-pubescent while young;

leaves with white-villous petioles about 5 mm. long, the blades lanceolate, to 10 cm. long and 2.2 cm. wide, short-acuminate, serrulate, rounded at base, sparingly pubescent above at first, becoming glabrous, densely pubescent beneath at first, later sparingly so, the lateral nerves many; aments appearing with the leaves, the pistillate 1–1.5 cm. long, short-pedunculate, the bracts ovate, sparsely hairy on outer side, rounded at apex; ovary white-villous, sessile; style elongate, glabrous; staminate aments 1.5–2.5 cm. long, about 7 mm. across, the filaments long-hairy at base.——Honshu (Kinki and Chūgoku Distr.).

8. Salix jessoensis Seemen. *S. hondoensis* Koidz.—— SHIRO-YANAGI. Tree with brownish branches, the branchlets densely short-pubescent with grayish hairs; leaves with pubescent petioles 2–5 mm. long, the blades lanceolate or narrowly so, 5–10 cm. long, 8–15 mm. wide, long-acuminate, serrulate, the juvenile ones glaucous and densely appressed silky-pubescent but only sparingly so when mature; aments short-pedunculate, the staminate 2–3 cm. long, rather loosely flowered, the bracts oblong, rather densely hairy near base, the filaments hairy below, the pistillate 3–4.5 cm. long, densely flowered, the bracts retuse to obtuse; ovary pubescent; style very short or almost absent, the stigma oblong, entire.—— Hokkaido, Honshu.

9. Salix koreënsis Anderss. *S. pseudojessoensis* Lév. ——KŌRAI-YANAGI. Trees with grayish brown or brownish green branches, the branchlets glabrous or short-pubescent; leaves with glabrous petioles 5–10 mm. long, the blades lanceolate, 8–13 cm. long, 12–25 mm. wide, acuminate, serrulate, rounded to acute at base, glabrous on both sides except the midrib, glaucous beneath; staminate aments 1–3 cm. long, 6–7 mm. across, densely flowered, the bracts ovate-oblong, obtuse, about 2 mm. long, long white-hairy; anthers reddish brown, the filaments hairy below, the pistillate aments 7–15 mm. long, densely flowered; ovary villous, sessile, the style relatively long, the stigma bifid.——Honshu (western distr.), Kyushu. ——Korea and Manchuria.

10. Salix pauciflora Koidz. EZO-MAME-YANAGI. Small prostrate shrub with slender branchlets, short-pubescent, changing to red-brown in age; leaves with pubescent petioles 1–4 mm. long, the blades thinly coriaceous, obovate to elliptic, sometimes nearly rounded, 6–12 mm. long, 3–10 mm. wide, entire, subacute to obtuse or rounded at base, lustrous and glabrous above, pale green and sparsely long-pubescent on midrib beneath, glabrous or short-pubescent on margin; staminate aments globose, 5- to 30-flowered, terminal on short lateral branchlets, leafy below, 3–10 mm. long, the bracts obovate, rounded at apex, reddish on upper half, hairy on margin and inner side, glabrescent outside, 1.2–2.2 mm. long; filaments free, glabrous; pistillate aments 5- to 30-flowered, 13–17 mm. long in fruit; capsules about 5 mm. long, glabrous, stipitate; style short, the stigmas very short, bifid.——July–Aug. Alpine slopes; Hokkaido (Mount Daisetsu).——Forma **hebecarpa** Kimura. Ovary pubescent. Occurs with the typical phase.

11. Salix hidakamontana Hara. HIDAKA-MINE-YANAGI. Low shrub 10–20 cm. high, with decumbent branches, the branchlets silky-pubescent while young; leaves with petioles 1–2.5 cm. long, the blades rounded to elliptic, 2–4 cm. long, 1.5–3.5 cm. wide, sometimes retuse at apex, rounded to shallowly cordate at base, entire or with obsolete undulate teeth, glabrous, glaucous, the veins and veinlets raised reticulate beneath, impressed above; pistillate aments terminal on short

leafy branchlets, densely flowered, 2–5 cm. long in fruit, the bracts ovate or oblong, obtuse or subacute, 1–1.5 mm. long, silky villous; capsules narrowly ovoid, 4–6 mm. long, glabrous, subsessile.——Alpine regions; Hokkaido (Hidaka mt. range).

12. Salix nakamurana Koidz. RENGE-IWA-YANAGI, TAKANE-IWA-YANAGI. Small prostrate shrub with rooting branches, the branchlets at first slightly silky, soon glabrate, changing to dark purple-brown in age; leaves with glabrous petioles, the blades elliptic or obovate, sometimes oblong, 2–4 cm. long, 1–3 cm. wide, rounded or sometimes obtuse at apex, obtuse to acute at base, entire or with remote obsolete teeth, glaucous beneath, the lateral nerves prominently arcuate above; aments 2.5–4 cm. long, on short lateral branches, the bracts narrowly ovate or oblong, subobtuse to nearly rounded at apex, black-brown above, hairy on both sides; filaments short-connate below; ovary glabrous, short-stipitate; style distinct, the stigma short-linear, bifid.——July. Alpine slopes; Honshu (high mountains of Kai and Shinano Prov.); rare.

Var. **eriocarpa** Kimura. Ovary hairy. Occurs with the typical phase.

13. Salix yezoalpina Koidz. *S. cyclophylla* Seemen, non Rydb.——MARUBA-YANAGI. Low shrub with prostrate, rooting branches, the branchlets glabrous or sometimes slightly pubescent while young, becoming purple-brown in age; leaves with glabrous petioles 5–15 mm. long, the blades variously elliptic or rounded, 2–4.5 cm. long, 1–3.5 cm. wide, essentially entire, slightly silky while young, glaucous beneath, rounded, barely mucronate or retuse, broadly cuneate to shallowly cordate, the lateral nerves and reticulate veinlets slightly raised beneath; staminate aments 15–30 mm. long, densely flowered, the pistillate slightly elongate after anthesis, to 2–5 cm. long, the bracts elliptic, nearly half as long as the capsule, rounded at apex, densely long-hairy; capsules subsessile, glabrous; style slender, the stigmas rather slender, bifid, the filaments glabrous.——July. Alpine slopes; Hokkaido.

Var. **neoreticulata** (Nakai) Kimura. *S. neoreticulata* Nakai——INU-MARUBA-YANAGI. Distinguished from the typical phase by the leaves lustrous above and the prominently raised reticulate veinlets beneath, and silky ovaries.——Hokkaido (Mount Daisetu and Mount Rishiri).

14. Salix bakko Kimura. *S. caprea* sensu auct. Japon., non L.——BAKKO-YANAGI. Tree with grayish brown branches pubescent at first, the sapwood with elevated, short, longitudinal striations; leaves with petioles 1–2 cm. long, the blades oblong or elliptic, 6–13 cm. long, 3–4 cm. wide, acuminate to acute, undulate-serrate or nearly entire, acute to rounded at base, vivid green and glabrous above, glaucous and densely white crisped-pubescent; staminate aments 3–5 cm. long, rather broad, short-pedunculate, very densely flowered, the bracts narrowly oblong to lanceolate, 3–4 mm. long, subacute, densely long-hairy on both surfaces; stamens 2, the filaments free, hairy below; pistillate aments to 10 cm. long in fruit, 15–20 mm. across; ovary villous, long-stipitate; style and stigmas very short, glabrous.——Mountains; Hokkaido (western distr.), Honshu (Kinki Distr. and eastw.).

Salix leucopithecia Kimura. FURI-SODE-YANAGI. An alleged hybrid of *S. bakko* × *S. gracilistyla*; sometimes cultivated for ornament in early spring; only the staminate plant is known.

15. Salix hultenii Floderus var. **angustifolia** Kimura. *S. caprea* sensu auct. Japon., pro parte, non L.——EZO-NO-BAKKO-YANAGI, EZO-NO-YAMANEKO-YANAGI. Small tree or large shrub closely allied to the preceding species, the second year branches purple-castaneous, the sapwood not or scarcely ridged; leaves with petioles 1–2 cm. long, pubescent above, the blades chartaceous, narrowly obovate or oblong, 6–12 cm. long, 3–5 cm. wide, short-acuminate, with obsolete undulate teeth, acute to obtuse at base, glabrous except the midrib above, glaucous and densely crisped-pubescent.——Hokkaido, Kuriles, Sakhalin, and Kamchatka.

Salix × yoitiana Kimura is an alleged hybrid of No. 15 possibly crossed with *S. gracilistyla*.

16. Salix taraikensis Kimura. *S. hastata* sensu auct. Japon., non L.——TARAIKA-YANAGI. Large shrub with the juvenile branchlets silky pubescent, soon becoming glabrous; leaves with nearly glabrous petioles 1–1.5 cm. long, the blades chartaceous, narrowly obovate or rhomboid-elliptic, 6–10 cm. long, 2–3.5 cm. wide, acuminate to acute, obsoletely serrulate, acute at base, usually broadest at or above the middle, glabrous except for silky pubescence on nerves beneath and on the upper surface while young, glaucous beneath, deep green and lustrous above; staminate aments elliptic or short-cylindric, densely flowered, 2.5–3.8 cm. long, 18–27 mm. wide; pistillate aments 2–2.7 cm. long, short-cylindric, pedunculate, the bracts oblong, obtuse or nearly so, densely hairy on both surfaces, 2.3–2.6 mm. long; ovaries white-villous, stipitate; style short, the stigma bifid, the lobes short, rather thick; capsules about 9 mm. long.——Hokkaido (Kushiro and Nemuro Prov.).——Sakhalin.

17. Salix subopposita Miq. *S. repens* var. *subopposita* (Miq.) Seemen; *S. sibirica* var. *subopposita* (Miq.) C. K. Schn. ——NO-YANAGI, HIME-YANAGI. Small shrub, the branches rather slender, erect, deep brown, densely short grayish pubescent while young; leaves with petioles 1–5 mm. long, the blades broadly lanceolate, broadly oblong-lanceolate to narrowly oblong, 2.5–4 cm. long, 5–12 mm. wide, acute, entire, acute at base, glabrous or nearly so above, glaucous and sparsely gray-yellow appressed-pubescent beneath, densely so while young, the nerves slightly impressed above, slightly raised beneath; staminate aments to 2.7 cm. long and 7 mm. across, densely flowered, the bracts oblong, rounded at apex, silky; pistillate aments short-cylindric, short-pedunculate, 1–2 cm. long, 8–10 mm. across in fruit, densely flowered; capsules rather densely appressed grayish pubescent, about 4 mm. long; style nearly absent, stigma bifid, short.——Honshu (Chūgoku Distr.), Kyushu.——s. Korea (Quelpaert Isl.).

18. Salix paludicola Koidz. MIYAMA-YACHI-YANAGI. Dwarf procumbent shrub about 5 cm. high with rooting narrow branches and yellowish brown, slightly lustrous, glabrous branchlets; leaves with glabrous petioles 2–5 mm. long, the blades glabrous, obovate, 1–3 cm. long, 5–17 mm. wide, lustrous above, glaucous beneath, obtuse to rounded, broadly cuneate at base, entire or undulate-serrulate; staminate aments densely flowered, about 3 cm. long, about 10 mm. wide; pistillate aments terminal on short lateral few-leaved branchlets, densely flowered, 3–4 cm. long in fruit, the rachis hairy, the bracts broadly ovate, yellowish brown, rounded at apex, densely hairy outside; ovary short-stipitate, densely pubescent; style short, the stigmas shortly bifid; capsules lanceolate, 7–8 mm. long, loosely pubescent.——Hokkaido (alpine zone on Mount Daisetsu); rare.

Salix × pseudopaludicola Kimura is an alleged hybrid between Nos. 18 and 20.

19. Salix reinii Fr. & Sav. *S. glabra* Fr. & Sav., non

Scop.; *S. kakista* C. K. Schn.——MIYAMA-YANAGI, MINE-YANAGI. Glabrous shrub, with relatively thick dark brown branches; leaves with petioles 5–13 mm. long, the blades obovate or elliptic, 2–6 cm. long, 1.5–3.5 cm. wide, rounded and mucronate to abruptly acute, with incurved, undulate-serrate teeth, rounded to obtuse at base, deep green above, glaucous beneath, the lateral nerves slender; staminate aments short-pedunculate, slender, 2.5–4 cm. long, rather loosely flowered, the axis short-pilose, the bracts broadly lanceolate to narrowly ovate, obtuse to acute, slightly pubescent below and near margins; pistillate aments narrowly cylindric, 3–5 cm. long; capsules stipitate, glabrous; style rather slender, the stigmas shortly bifid.—— Hokkaido, Honshu (centr. and n. distr.); rather common. ——s. Kuriles.

Var. **eriocarpa** Kimura. KE-MIYAMA-YANAGI. Branchlets and leaves silky hairy while young; ovaries densely white silky-hairy.——Alpine mountains; Honshu (centr. distr.).

20. Salix hidewoi Koidz. EZO-MIYAMA-YANAGI. Similar to the preceding; shrub with glabrous branchlets at first dark purple, changing to dark brown; leaves with petioles 4–10 mm. long, the blades oblong, rarely obovate, 1–5 cm. long, 7–25 mm. wide, obtuse to rounded, obtuse at base, slightly undulate-serrate, soon glabrous, not lustrous above, pale or slightly glaucous beneath; pistillate aments pedunculate, densely flowered, about 3 cm. long at anthesis, the axis villous, the bracts oblong, rounded at apex, yellowish brown villous; style short, stigmas bifid; fruiting aments 3.5–4.5 cm. long, about 1 cm. across; capsules 6–7 mm. long, puberulous, becoming glabrate.——Hokkaido.——s. Kuriles.

21. Salix futura Seemen. *S. vulpinoides* Koidz.; *S. vulpina* var. *tomentosa* Koidz.——Ō-NEKO-YANAGI. Large shrub, the branchlets dark brown, rather thick, at first brown-villous; leaves with petioles 8–15 mm. long, the blades narrowly oblong to elliptic, 8–20 cm. long, 3–6 cm. wide, acuminate, with margins slightly recurved, depressed-serrate, obtuse to acute at base, chartaceous, deep green and appressed pubescent above while young, glaucous and with pale to yellowish brown pubescence beneath especially on the nerves; staminate aments long-cylindric, densely flowered, sessile, about 5 mm. across, the bracts ovate, very obtuse, about 1.2 mm. long, with relatively long dense hairs on both surfaces; ovary silky-pubescent, nearly sessile; style somewhat less than 1 mm. long, the stigma short, bifid.——Honshu (centr. distr. and northw.).

Var. **psilocarpa** Kimura. Ovary glabrescent.——Occurs with the typical phase.

Salix × sirakawensis Kimura is a hybrid of No. 21 with *S. integra*.

22. Salix vulpina Anderss. *S. miquelii* Anderss.; *S. miquelii* var. *vulpina* (Anderss.) Anderss.; *S. ignicoma* Lév. & Van't.; *S. daiseniensis* Seemen, pro parte.——KITSUNE-YANAGI. Shrub with lustrous dark brown branches, whitish and yellowish brown pubescent when young; leaves with petioles 6–10 mm. long, the blades oblong to elliptic, 3–8 cm. long, 2–3.5 cm. wide, acute to rather obtuse, acute to obtuse at base, irregularly obtusely serrate, not lustrous above, more or less glaucous and usually yellow-brown pubescent along the midrib beneath; staminate aments short-pedunculate, 2.5–4 cm. long, 5–6 mm. across, densely flowered, the bracts small, reddish brown villous; filaments free or slightly connate and more or less pubescent below; capsules stipitate, glabrous; style slender, the stigmas short.——Hokkaido, Honshu, Shikoku, Kyushu.——s. Kuriles. In western Honshu (Kinki Distr. and westw.), Shikoku, and Kyushu this plant is sometimes differentiated as *S. alopechroa* Kimura with aments having much more reduced leaves at base, the staminate shorter, thicker, and larger.

23. Salix shiraii Seemen. SHIRAI-YANAGI. Shrub with dark brown branches, gray-appressed pubescent only when young; leaves with petioles 6–10 mm. long, the blades ovate to narrowly so, 2.5–5.5 cm. long, acuminate to caudately acuminate, sometimes acute, vivid green, acutely appressed-toothed, rounded to shallowly cordate, rarely subacute at base, glaucous and glabrous, sometimes gray-pubescent beneath when young; staminate aments nearly sessile, about 4 cm. long, 5 mm. wide, densely flowered, the bracts small, broadly ovate, rounded at apex, densely white-hairy; stamens 2, the filaments free, glabrous; pistillate aments rather loosely flowered; capsules long-stipitate, glabrous, 2.5–3 mm. long; style slender, the stigmas short.——Mountains; Honshu (near Nikko).

24. Salix rupifraga Koidz. KOMA-IWA-YANAGI. Shrub, with grayish brown glabrous branches, often with long grayish appressed-pubescence while young; leaves with petioles 4–10 mm. long, the blades ovate-oblong to oblong, 4–6.5 cm. long, 1.5–2.5 cm. wide, acuminate to very acute, loosely and acutely appressed-toothed, acute to obtuse or sometimes nearly rounded at base, silky pubescent on both surfaces when young, becoming glabrate except the puberulent midrib above, glaucous beneath; staminate aments cylindric, 2–2.5 cm. long, about 6 mm. across, short-pedunculate, the bracts elliptic or ovate, very obtuse, long white-hairy on both surfaces, about 1.3 mm. long; pistillate aments to 5 cm. long in fruit, densely flowered; ovaries glabrous, short-stipitate; styles and stigmas very short, stigmas bifid.——Mountains; Honshu (Kai and se. Shinano Prov.).

Var. **eriocarpa** Kimura. Distinguished by the pubescent ovaries. Occurs with the typical phase.

25. Salix japonica Thunb. *S. babylonica* var. *japonica* Anderss.; *S. fauriei* Seemen——SHIBA-YANAGI. Shrub with brownish slender branches pubescent while young; leaves with petioles 3–5 mm. long, silky at first, soon glabrous, the blades lanceolate to broadly so, 5–8 cm. long, 1.5–2.5 cm. wide, glabrous, appressed-toothed, obtuse to rounded at base, vivid green above, glaucous or grayish beneath; staminate aments 3–8 cm. long, about 5 mm. across, rather loosely flowered; filaments pilose below; pistillate aments short-pedunculate, linear, rather loosely flowered, 6–12 cm. long, the axis scattered short-pilose, the bracts ovate, rounded at apex, sparsely short-pilose; capsules stipitate, glabrous; style short, the stigmas oblong.——Hills and mountains; Honshu (centr. distr.); common.

26. Salix rorida Lacksch. *S. daphnoides* sensu auct. Japon., non Vill.; *S. lackschewitziana* Toepfer——EZO-YANAGI. Tree with glabrous, reddish, slightly glaucous branches; leaves with petioles 2–8 mm. long, the blades glabrous, lanceolate, 5–12 cm. long, 7–30 mm. wide, long-acuminate, acute to obtuse or rounded at base, glaucous beneath, serrulate; staminate aments nearly sessile, 1.5–3.5 cm. long, 12–13 mm. across, very densely flowered, the bracts narrowly obovate, 2–2.5 mm. long, glandular on margin, densely long silvery-hairy, the filaments glabrous; pistillate aments 2–4 cm. long, 1–1.5 cm. across, very densely flowered; ovaries glabrous, stipitate; style slender, rather elongate, the stigmas short, bifid.——Hokkaido, Honshu.——Sakhalin, Amur, Ussuri, and Korea.

Var. **roridaeformis** (Nakai) Ohwi. *S. roridaeformis* Nakai; *S. lackschewitziana* var. *roridaeformis* (Nakai) Kimura; *S. rorida* var. *eglandulosa* Kimura——KO-EZO-YANAGI. Bracts eglandular.——Honshu (Kamikochi in Shinano).——Korea and Sakhalin.

27. Salix kinuyanagi Kimura. *S. viminalis* sensu auct. Japon., non L.——KINU-YANAGI. Tree or shrub with grayish villous, rather thick branchlets; leaves with petioles 10–18 mm. long, the blades chartaceous, narrowly lanceolate, 10–20 cm. long, 1–2 cm. wide, caudately acuminate, obtuse to acute at base, with obsoletely undulate and slightly recurved margins, appressed puberulent above while young, long-silvery appressed-villous beneath; staminate aments ellipsoidal or ovoid, densely flowered, sessile, 2.5–3.5 cm. long, 15–20 mm. in diameter, the bracts broadly lanceolate, acute, villous outside, 3–3.2 mm. long; filaments free, glabrous.——Widely cultivated in Honshu, Shikoku, and Kyushu; pistillate plant unknown to us.

Salix × thaumasta Kimura is a hybrid of No. 27 with *S. gracilistyla*.

28. Salix petsusu Kimura. *S. viminalis* sensu auct. Japon., non L.; *S. viminalis* var. *yezoensis* C. K. Schn.; *S. yezoensis* (C. K. Schn.) Kimura——EZO-NO-KINU-YANAGI. Tree resembling No. 27, differing in the rather slender branches, these less densely hairy with shorter pubescence, the branchlets grayish brown, densely white-pubescent while young; leaves with pubescent petioles 8–15 mm. long, the blades lanceolate to oblong-lanceolate, 10–20 cm. long, 1.5–2 cm. wide, long-acuminate, acute to obtuse at base, entire, usually glabrous above, with dense silvery appressed hairs beneath; staminate aments short-cylindric or ovoid-cylindric, 2–3.5 cm. long, 1–1.5 cm. in diameter, sessile; pistillate aments cylindric, 3.5–6(–10) cm. in fruit, the bracts 2–2.5 mm. long, dark brown above, white-hairy on back.——Hokkaido, Honshu (centr. distr. and northw.).——Sakhalin.

29. Salix sachalinensis F. Schmidt. *S. opaca* Anderss. ex Seemen; *S. sikokiana* Makino; *S. makinoana* Seemen, pro parte; *S. pilgeriana* Seemen; *S. sachalinensis* var. *pilgeriana* (Seemen) C. K. Schn.——ONOE-YANAGI. Tree; branches glabrous, brownish, slightly pubescent while young, the hairs short and grayish; leaves with petioles 2–6 mm. long, the blades lanceolate to narrowly so, 6–10 cm. long, 8–20 mm. wide, gradually acuminate at both ends, entire or obsoletely undulate with slightly recurved margins, short-appressed silky pubescent beneath especially while young, pale and scarcely glaucous on under side; staminate aments short-pedunculate, densely flowered, cylindric, 2–4 cm. long, 5–6 mm. in diameter, the bracts lanceolate, obtuse, with long whitish hairs especially on margin, the filaments glabrous; pistillate aments densely flowered, 2.5–5 cm. long; capsules with short whitish pubescence, stipitate; style elongate, slender, to 1.5 mm. long, the stigma short, oblong, entire.——Valleys in mountains; Hokkaido, Honshu, Shikoku; rather common.——Amur, Ussuri, Sakhalin, Kamchatka, and Kuriles.

The following hybrids are known: **S. × euerata** Kimura (*S. gilgiana* × *S. sachalinensis*); **S. × ampherista** Kimura (*S. sachalinensis* × *S. vulpina*); **S. × ikenoana** Kimura (*S. integra* × *S. sachalinensis*).

30. Salix integra Thunb. *S. purpurea* sensu auct. Japon., non L.; *S. multinervis* Fr. & Sav.; *S. purpurea* var. *multinervis* (Fr. & Sav.) Matsum.; *S. savatieri* Camus; *S. purpurea* subsp. *amplexicaulis* sensu auct. Japon., non Bory &

Chaub.——INU-KŌRI-YANAGI. Shrub with glabrous lustrous branches; leaves nearly sessile, usually opposite, narrowly oblong to oblong, 3–6 cm. long, 7–20 mm. wide, rather thin, glabrous, rounded to subacute, obsoletely serrulate or entire, rounded to shallowly cordate, whitish beneath; staminate aments 15–25 mm. long, about 7 mm. in diameter, densely flowered, the bracts broadly obovate, rounded at apex, about 2 mm. long, loosely long-ciliate, the filaments glabrous; pistillate aments 15–30 mm. long, cylindric and densely flowered at first; ovary silky, sessile; style short, the stigmas very short, bifid; capsules about 3 mm. long, grayish white, with rather dense and appressed short hairs.——Honshu, Kyushu.—— Korea.

The following hybrids are recorded: **S. × hapala** Kimura (*S. gilgiana × S. integra*); **S. × sirakawensis** Kimura (*S. futura × S. integra*); **S. × isikawae** Kimura (*S. vulpina × S. sachalinensis*).

31. Salix koriyanagi Kimura. *S. purpurea* var. *japonica* Nakai; *S. purpurea* sensu auct. Japon., non L.——KŌRI-YANAGI. Shrub with glabrous, yellowish brown to brown branches; leaves alternate or sometimes opposite, with petioles 2–5 mm. wide, the blades linear-lanceolate, 6–8 cm. long, 5–10 mm. wide, gradually acuminate at the apex, obtuse to rounded at base, remotely serrate, thinly coriaceous, glabrous, deep green above, glaucescent beneath, the lateral nerves many, slender, rather spreading; staminate aments cylindric, densely flowered, 2–3 cm. long, the bracts obovate, long-hairy; anthers deep purple; pistillate aments 2–3 cm. long, nearly sessile, densely flowered, the bracts narrowly obovate, obtuse, about 1.5 mm. long, loosely long-hairy; ovary sessile, white-villous; style and stigma very short, stigma bifid.——Widely cultivated for making baskets and furniture.——Korea.

32. Salix miyabeana Seemen. *S. sapporoensis* Lév.—— EZO-NO-KAWA-YANAGI. Shrub or small tree with glabrous, pale brown branches; leaves with petioles 3–8 mm. long, the blades lanceolate or narrowly so, 5–15 cm. long, acuminate at both ends, undulate-serrate, glabrous, glaucescent beneath; staminate aments 3–5.5 cm. long, sessile, densely flowered, the filaments completely connate; anthers yellow; pistillate aments 4–8 cm. long, short-pedunculate, the bracts ovate, densely long-pubescent; ovary villous, sessile; stigma sessile, very short, undivided.——Hokkaido, Honshu (n. distr.).

33. Salix gilgiana Seemen. *S. gymnolepis* Lév. & Van't.; *S. purpurea* var. *sericea* Seemen, non Wimm.; *S. purpurea* subsp. *gymnolepis* (Lév. & Van't.) Koidz.——KAWA-YANAGI. Shrub or small tree with elongate branches, at first short silky-pubescent, soon glabrate; leaves with petioles 2–8 mm. long, the blades lanceolate to narrowly so, or nearly linear, 5–12 cm. long, 5–10 mm. wide, acuminate, acute at base, glabrous, glaucous beneath, serrate; staminate aments 2–5 cm. long, sessile, the bracts obovate, rounded at apex, long-hairy, about 2 mm. long, the filaments connate, hairy on lower half, the anthers purple; pistillate aments 2–4 cm. long at anthesis, 3–5 cm. long in fruit; ovary appressed-pubescent, very short stipitate; style about 0.3 mm. long, the stigma short, bifid.——Hokkaido (Oshima), Honshu, Shikoku, Kyushu.——Korea.

34. Salix gracilistyla Miq. *S. thunbergiana* Bl. ex Anderss.; *S. brachystachys* sensu auct. Japon., non Benth.—— NEKO-YANAGI. Shrub with rather thick erect branches, the young branchlets with long dense whitish pubescence, glabrescent; leaves with densely pubescent petioles 5–10 mm. long, the blades linear-oblong to oblanceolate, 5–12 cm. long, 1.5–3

cm. wide, acute, obtuse to rounded at base, densely appressed grayish white pubescent beneath, serrulate or nearly entire, the lateral nerves prominently curved upward and elevated as in the veinlets; staminate aments 3–4 cm. long, 10–12 mm. in diameter, densely flowered, the bracts ovate, acuminate, densely long-hairy, about 2 mm. long; pistillate aments 2–5 cm. long; ovary long-hairy; capsules long-pubescent; style slender, about 2 mm. long, the stigma very short, bifid.—— Hokkaido, Honshu, Shikoku, Kyushu.——Korea, Manchuria, and China.

Var. **melanostachys** (Makino) C. K. Schn. *S. thunbergiana* var. *melanostachys* Makino; *S. gracilistyla* subsp. *melanostachys* Makino——KURO-YANAGI, KURO-ME. Aments and bracts dark brown; rarely cultivated; possibly only a cultivar.

The following hybrids are reported: **S. × thaumasta** Kimura (*S. gracilistyla × S. koriyanagi*); **S. × yoitiana** Kimura (*S. gracilistyla × S. hultenii* var. *angustifolia*); **S. × koiei** Kimura (*S. gilgiana × S. gracilistyla*); **S. × iwahisana** Kimura (*S. gracilistyla × S. rorida*); **S. × cremnophila** Kimura (*S. gracilistyla × S. japonica*); **S. × turumatii** Kimura (*S. bakko × S. gracilistyla*). **Salix × Hatusimana** Kimura (*S. gracilistyla × S. integra* ? or × *S. koriyanagi*) is an alleged hybrid.

35. Salix graciliglans Nakai. *S. nakaii* Kimura—— CHŌSEN-NEKO-YANAGI. Shrub; branches glabrous, appressed-silky while young; leaves with petioles 6–10 mm. long, pubescent or glabrous, the blades linear-oblanceolate or linear-oblong, 6–10 cm. long, 1–3 cm. wide, acuminate, acute or narrowly cuneate at base, glabrous above, glabrate but usually pubescent beneath while young, serrulate; staminate aments 15–25 mm. long, 7–8 mm. in diameter, sessile, the bracts about 2 mm. long, oblong, acuminate to obtuse, glabrous except the long whitish hairs at base; pistillate aments 3–5 cm. long; ovary densely appressed pubescent; capsules densely pubescent; style 2–3 mm. long, the stigma very short, bifid. ——Reported to occur in Honshu (Tōkaidō Distr.) and Shikoku.——Korea.

36. Salix buergeriana Miq. HASHI-KAERI-YANAGI, NA-GABO-NO-YAMA-YANAGI. Shrub or small tree; branches glabrous, white-pubescent while young; leaves subrhombic-oblong to rhombic-obovate, sometimes broadly elliptic, acute, with incurved undulate teeth on margins, obtuse to acute at base, white villous or with brownish hairs while young, glabrate or nearly so except the midrib, glaucous beneath; staminate aments 4–7.5 cm. long, about 4 mm. in diameter, short-pedunculate, the filaments pubescent below, the bracts villous with long grayish fulvous hairs on both surfaces, broadly ovate; pistillate aments 3–4 cm. long, about 5 mm. in diameter, to 6–8 cm. long in fruit; ovary densely pubescent; capsules slightly pubescent, stipitate; style mostly elongate, the stigma bifid.——Kyushu.

37. Salix harmsiana Seemen. TSUKUSHI-YAMA-YANAGI. Shrub or tree; branches glabrous, white-pubescent while young; leaves oblong, elliptic or broadly so, acute, with incurved teeth on margin, sparsely pubescent; staminate aments 2.5–4 cm. long, about 6 mm. in diameter, short-pedunculate; stamen 1, pubescent below, the bracts ovate, obtuse, with long yellow-brown hairs on both sides; pistillate aments to 27 mm. long in anthesis, about 5 mm. in diameter; ovary densely pubescent, stipitate; style short, the stigma bifid.——Kyushu.

38. Salix sieboldiana Bl. YAMA-YANAGI. Shrub or small tree with dark brown branches; leaves with petioles

1–1.5 cm. long, the blades lanceolate-oblong to oblong, sometimes elliptic, 3–10 cm. long, 15–35 mm. wide, short-acuminate, obtuse at base, with incurved-undulate obtuse teeth on margins, white-villous while young, soon glabrate above, glaucous and glabrate or pubescent only on the midrib beneath; staminate aments 3–3.5 cm. long, about 7 mm. in diameter, short-pedunculate, very densely flowered, the bracts ovate, obtuse, long-villous; filaments short-pubescent below; style short, the stigma bifid.——Kyushu.

Fam. 63. MYRICACEAE   YAMA-MOMO KA   Sweet-gale Family

Shrubs and trees sometimes aromatic; leaves evergreen or deciduous, simple, entire or serrate, penninerved, without stipules; flowers unisexual (the plants monoecious or dioecious), in axils of densely imbricate scales on aments, without a perianth, bracteoles absent or many; stamens 2 to many, usually 4–6; ovary sessile, 1-locular; style short, the stigmas 2; ovule solitary, basal, erect, orthotropous; fruit a small or rather large, globose to ovoid drupe, often with a heavy waxy coating, the exocarp juicy or fleshy, the endocarp hard; seed erect, without endosperm, the embryo erect, with fleshy, flat cotyledons.——Two genera, with about 35 species, chiefly in the Tropics, with 1 species widely distributed in boreal regions.

1. MYRICA L.   YAMA-MOMO ZOKU

Staminate aments oblong to cylindric, sometimes short; pistillate aments short-cylindric to ovoid; ovary with 2–4 small bracteoles at base; leaves without stipules, entire or serrate.——More than 30 species, chiefly in the tropics, few widely distributed temperate species.

1A. Deciduous shrub of bogs; pistillate flowers with 2 winglike bracteoles adnate at the base; fruit small, resinous-scaly, about 2 mm. long.
1. *M. gale* var. *tomentosa*
1B. Evergreen tree of forests; pistillate flowers subtended by 2–4 small free bracteoles; fruit relatively large, exceeding the scales, tuberculate, about 15 mm. long. .... 2. *M. rubra*

1. Myrica gale L. var. **tomentosa** C. DC.   *M. tomentosa* (C. DC.) Aschers. & Graebn.; *Gale japonica* Cheval.—— YACHI-YANAGI.   Shrub with dark brown branches, soft-pubescent while young; leaves oblanceolate to narrowly obovate, 2–4 cm. long, 7–15 mm. wide, somewhat obtuse, long-cuneate toward the base, more or less short-pubescent, with few depressed teeth on upper margins, pale green beneath, with yellow sessile glands on both surfaces, sessile or barely petiolate; pistillate aments sessile, oblong to ellipsoidal, 1–1.5 cm. long, 5–6 mm. wide at maturity; fruit about 2 mm. long, resinous, broadly ovoid, the bracteoles 2, winglike, adnate at the base to the fruit.——Apr. Bogs; Hokkaido, Honshu (Ise and eastw.); rather rare.——Kuriles, Sakhalin, n. Korea, to e.

Siberia. The typical phase occurs widely in bogs in northern areas of the N. Hemisphere.

2. Myrica rubra Sieb. & Zucc.   *M. nagi* sensu DC., non Thunb.——YAMA-MOMO.   Evergreen glabrous tree; leaves coriaceous, oblanceolate to broadly so, 6–12 cm. long, 1–3.5 cm. wide, obtuse or acute with an obtuse tip, long-cuneate at base, entire, sometimes with few appressed teeth on upper margins, the raised nerves reddish on under side, the petioles 5–10 mm. long; pistillate aments sessile, narrow, 8–12 mm. long, with small scales; fruit globose, tuberculate, 12–15 mm. across, dark red when mature.——Mar.–Apr. Honshu (sw. Kantō Distr., and Wakasa Prov. and westw.), Shikoku, Kyushu.—— s. Korea, Ryukyus, Formosa, and China.

Fam. 64. JUGLANDACEAE   KURUMI KA   Walnut Family

Frequently foetid monoecious trees or shrubs; leaves alternate, imparipinnate, the leaflets sessile or short-petiolate, the stipules none; flowers unisexual, the staminate in long, lateral, usually pendulous aments, the perianth irregularly lobed or absent, the stamens 3 to many, the filaments short, the anthers oblong; pistillate flowers terminal, solitary or in racemes, 2-bracteolate at base, the perianth 3- to 6-lobed; ovary inferior, 1- or incompletely 2- to 4-locular; ovules erect, orthotropous; styles 2, stigmatic on the inner side; fruit a drupe or nut, 2- to 4-lobed; seeds large, the endosperm absent, the embryo usually corrugate, oily.—— Six genera, with about 50 species, in the temperate regions of the N. Hemisphere.

1A. Fruit a winged nut.
  2A. Nuts borne in axils of persistent bracts of upright conelike aments. .... 1. *Platycarya*
  2B. Nuts borne on pendulous, loose aments. .... 2. *Pterocarya*
1B. Fruit a large drupe. .... 3. *Juglans*

1. PLATYCARYA Sieb. & Zucc.   NOGURUMI ZOKU

Deciduous tree; leaves alternate, large, the leaflets doubly serrate; aments erect, the staminate slender, short-pedunculate, usually few, below the pistillate, the scales lanceolate, the flowers naked; stamens 8–10; pistillate aments solitary, ovoid-oblong, with imbricate, narrowly lanceolate, persistent, thick bract; bracteoles 2, adnate to the ovary; styles 5, short, thick; fruit an axillary winged nutlet on the conelike aments.——One species.

1. Platycarya strobilacea Sieb. & Zucc.   *Petrophiloides strobilacea* (Sieb. & Zucc.) Reid & Chandler——NOGURUMI, NOBU-NO-KI, YAMAGURUMI.   Deciduous tree with terete

branches brown-pubescent while young; leaves rather large, the leaflets 7–15, sessile, lanceolate to narrowly ovate, 4–10 cm. long, 1–3 cm. wide, long-acuminate, doubly serrate, loosely brown pubescent at first, soon glabrous except in the axils of the nerves beneath; aments few to several at the summit of the young branchlets, erect, the terminal aments pistillate, 3–4 cm. long, 2–3 cm. across, ellipsoidal, with dark brown, linear-lanceolate, coriaceous, acuminate scales; fruit about 5 mm. long, glabrous, obtriangular inclusive of the wing; staminate aments borne laterally below the pistillate.——June–July. Mountains; Honshu, Shikoku, Kyushu.——Korea, China, and Formosa.

2. PTEROCARYA Kunth   SAWAGURUMI ZOKU

Deciduous monoecious trees; leaves alternate, pinnate, without stipules; flowers unisexual, on pendulous aments, the staminate with 2 bracteoles adnate to the bract, the perianth 1- to 4-parted, the stamens 6–18; pistillate flowers with 2 bracteoles at base, the perianth adnate to the ovary, the limb free, shortly 4-lobed; style short, bifid; fruit rather small, 2-winged; seed 4-locular at base; cotyledons 4-lobed, green, epigeal.——About 10 species, chiefly in China, 1 species in Japan, 1 in Asia Minor.

1. Pterocarya rhoifolia Sieb. & Zucc. SAWAGURUMI. Tree; branches rather thick, terete, the branchlets short-pubescent or nearly glabrous; leaves large, clustered on the upper part of branchlets, petiolate, imparipinnate, the leaflets 11–21, ovate-oblong to lanceolate or broadly oblanceolate, 6–12 cm. long, 1.5–4 cm. wide, acuminate, serrulate, with yellowish sessile glands and brown pubescence in axils beneath; stami-nate aments short-pedunculate; stamens 10; pistillate aments pedunculate, very loosely flowered in the axils of scales, elongate after anthesis, 20–30 cm. long in fruit; nut conical, about 8 mm. long, 12–18 mm. wide inclusive of the winglike, obtusely toothed bracteoles.——May. Valleys in mountains; Hokkaido, Honshu, Shikoku, Kyushu; rather common.

3. JUGLANS L.   KURUMI ZOKU

Deciduous trees or rarely shrubs; leaves alternate, imparipinnate, foetid, without stipules, the leaflets opposite, entire or ser-rate; staminate aments axillary and pendulous, elongate; flowers in the axils of scales, with 2 bracteoles, the perianth segments 1–4; stamens 3–40; pistillate aments terminal, loosely few to rather many flowered; flowers with a 3-lobed involucre consisting of a bract and 2 bracteoles, the perianth 4-lobed; style bifid, the stigma pinnate; nut large, drupelike, indehiscent, incompletely 2- or 4-locular, with a ligneous wall; cotyledons 2- to 4-lobed, retained within the nut in germination.——About 15 species, in the temperate regions of the N. Hemisphere.

1. Juglans ailanthifolia Carr. *J. sieboldiana* Maxim., non Goeppert; *J. allardiana* Dode; *J. coarctata* Dode; *J. lavallei* Dode; *J. sachalinensis* (Miyabe & Kudō) Komar.; *J. mirabunda* Koidz.——ONIGURUMI.   Tall erect tree; branches grayish brown, densely glandular-pubescent when young; leaves large, petiolate, the leaflets 9–21, ovate-oblong, 8–12 cm. long, 3–4 cm. wide, abruptly acute to acuminate, appressed-serrulate, mi-nutely stellate-pubescent above while young, rather densely stellate-pubescent beneath, sessile and obliquely truncate at base; petioles and rachis densely glandular; staminate aments 10–30 cm. long; pistillate aments 10- to 20-flowered, peduncu-late, densely brown-pubescent with crisped hairs; nut pubes-cent, with a hard shell, broadly ovoid to nearly globose, 2.5–3.5 cm. long, mucronate, rugose, with raised sutures.——May. Hokkaido, Honshu, Shikoku, Kyushu; very variable.

Var. cordiformis (Maxim.) Rehd. *J. cordiformis* Maxim., non Wangenh.; *J. subcordiformis* Dode——HIMEGURUMI. Nut cordate or cordate-ovoid, rather depressed, the shell rela-tively thin, nearly smooth, with a shallow groove on each side.——Cultivated.

Juglans regia L. var. **orientis** (Dode) Kitam. *J. orien-tis* Dode; *J. regia* var. *sinensis* sensu auct. Japon., non DC.——CHŌSEN-GURUMI.   A widely cultivated Chinese tree with glabrous leaves and branchlets, the leaflets 3–9, obtuse, entire except in the young tree; nut relatively thin-shelled.

Juglans avellana Dode and **Juglans notha** Rehd. are al-leged hybrids of *J. ailanthifolia* Carr. × *J. regia* var. *orientis* (Dode) Kitam.

Fam. 65. BETULACEAE   KABA-NO-KI KA   Birch Family

Deciduous trees or shrubs; leaves alternate, toothed, simple, the stipules usually caducous; staminate aments pendulous, elongate, the flowers 1–3 in axils of scales, the perianth 2- to 4-lobed, sometimes absent; stamens 2–15, the filaments free, the anther-locules distinct or connate; pistillate aments erect or pendulous, elongate or short, the flowers composed of a single 2-locular ovary; tepals present or absent; styles 2 or solitary and deeply bifid; ovules 1 or 2 in each locule, anatropous; fruit a nut, sometimes winged; endosperm absent; cotyledons fleshy.——About 7 genera, with about 100 species, chiefly in the N. Hemis-phere, usually flowering in early spring.

1A. Staminate flowers solitary on each scale, without perianth; pistillate flowers with perianth; nut attached to a leaflike bract.
  2A. Staminate flowers without bracteole; pistillate flowers in spikelike aments; fruit small, subtended by a large bract.
    3A. Bract of the nut flat, toothed or incised. .... 1. *Carpinus*
    3B. Bract of the nut connate at base, forming a tube. .... 2. *Ostrya*
  2B. Staminate flowers 2-bracteolate; pistillate flowers 2–4, capitate; nut large, enveloped by a large foliaceous involucre. .... 3. *Corylus*
1B. Staminate flowers 3–6 on each scale, with a perianth; pistillate flowers without a perianth; nut small, on inner side of closely im-bricated bracts or scales, sometimes winged on margin.
  4A. Stamens 2, the filaments bifid; scales of pistillate aments deciduous when mature. .... 4. *Betula*
  4B. Stamens 4, the filaments not lobed; scales persistent. .... 5. *Alnus*

1. CARPINUS L. KUMA-SHIDE ZOKU

Deciduous trees or large shrubs; leaves more or less 2-seriate on the branchlets, petiolate, toothed, penninerved, the lateral nerves 7- to 24-paired, straight; staminate aments pendulous, clothed with bud-scales in winter, the flowers naked, the stamens 3–15; pistillate aments terminal, the scales with 2 flowers in the axils; flowers 2-bracteolate, the perianth adnate to the ovary, with 6–10 teeth at the top; style short, the stigmas linear; nutlets surrounded by a rather large leaflike longitudinally striate scale or bract.——About 20 species in the temperate to warmer regions of the N. Hemisphere.

1A. Leaves with 7–15 pairs of lateral nerves, rounded to cuneate at base; staminate scales broadly ovate, subsessile; fruiting scales ovate; bark smooth.
  2A. Leaves 2–5(–6) cm. long, with 10–13 pairs of lateral nerves, acute to obtuse; fruiting aments few-scaled, spreading to ascending; small tree or large shrub. .... 1. *C. turczaninovii*
  2B. Leaves 4–10 cm. long, with 12–15 pairs of lateral nerves, acute to acuminate; fruiting aments many-scaled, pendulous; trees.
    3A. Leaves acute to abruptly acuminate, pilose, the petioles pilose; fruiting scales loosely long-hairy, 20–25 mm. long, toothed only on one side. .... 2. *C. tschonoskii*
    3B. Leaves caudately acuminate, glabrescent, the petioles slender, reddish, glabrate; fruiting scales short-hairy, 10–18 mm. long, with a lobule at base on each side, coarsely toothed on both sides or on one side. .... 3. *C. laxiflora*
1B. Leaves with 15–24 pairs of lateral nerves, usually more or less cordate or sometimes rounded at base; staminate scales stipitate; fruiting scales ovate, toothed, membranous, closely imbricate; bark scaly.
  4A. Leaves narrowly ovate to ovate-oblong, 2.5–4 cm. wide, barely cordate to rounded at base, with 20–24 pairs of lateral nerves. .... 4. *C. japonica*
  4B. Leaves ovate, 4–7 cm. wide, prominently cordate, with 15–20 pairs of lateral nerves. .... 5. *C. cordata*

1. Carpinus turczaninovii Hance. *C. paxii* Winkl.; *C. stipulata* Winkl.; *C. tanakaeana* Makino; *C. turczaninovii* var. *makinoi* Winkl.——IWA-SHIDE. Small tree or large shrub with pilose branchlets and petioles; stipules persistent; leaves ovate, 2.5–5(–6) cm. long, 18–20 mm. wide, acute to sub-obtuse, double-toothed, rounded at base, nearly lusterless, glabrous and with slightly raised reticulate veinlets above, appressed-pubescent beneath, with tufts of axillary hairs, the lateral nerves of 10–13 pairs; petioles 5–12 mm. long; fruiting aments ascending to spreading, 4- to 8-scaled, the peduncles pubescent, 1–2 cm. long, gradually curved; scales ovate to narrowly so, 10–18 mm. long, oblique, coarsely toothed; nuts broadly ovoid, about 4 mm. long, sparingly pilose above.——Honshu (Chūgoku Distr.), Shikoku, Kyushu.——Korea and China.

2. Carpinus tschonoskii Maxim. *C. yedoensis* Maxim.——INU-SHIDE. Tree with the branchlets, young leaves and petioles pubescent; leaves ovate to ovate-oblong, 4–8 cm. long, acute to short-acuminate, doubly serrulate with mucronate teeth, rounded at base, slightly appressed-pilose above, pilose beneath especially on the nerves and in the axils, the lateral nerves of 12–15 pairs, the petioles 8–12 mm. long; fruiting aments pendulous, pubescent, the peduncles 1.5–3 cm. long, loosely pubescent; scales subovate, 20–25 mm. long, pilose, coarsely toothed on one side; nuts ovoid-orbicular, about 4 mm. long, few-nerved, glabrous or thinly hairy above.——Honshu, Shikoku.——Korea and China.

3. Carpinus laxiflora (Sieb. & Zucc.) Bl. *Distegocarpus laxiflora* Sieb. & Zucc.——AKA-SHIDE. Tree; branches loosely pilose while young; leaves ovate or ovate-elliptic, 4–7 cm. long, 2.5–3.5 cm. wide, brownish when unfolded, caudate, doubly serrulate, rounded or rarely shallowly cordate at base, scattered long hairy while young, later appressed-pilose and with axillary tufts beneath, the petioles 8–12 mm. long, slender, glabrescent; peduncles 1.5–2 cm. long, glabrescent, sometimes minutely puberulous above; ament scales narrowly subovate or ovate-subdeltoid, 10–18 mm. long, with a lobule on each side at base, the margin toothed on one or both sides; nut broadly deltoid-ovate, striate, glabrous or puberulous above.——Hokkaido, Honshu, Shikoku, Kyushu.——Korea.

4. Carpinus japonica Bl. *Distegocarpus carpinus* Sieb. & Zucc.; *D. carpinoides* Sieb. & Zucc.; *C. carpinus* (Sieb. & Zucc.) Sarg.; *C. distegocarpus* Koidz.; *C. carpinoides* (Sieb. & Zucc.) Makino——KUMA-SHIDE. Tree with pubescent young branchlets; leaves narrowly ovate to ovate-oblong, 6–10 cm. long, 2.5–4 cm. wide, long-acuminate, rounded to shallowly cordate at base, doubly serrate with acute mucronate teeth, glabrescent above, with tufts of axillary hairs and long brownish hairs on nerves beneath, the lateral nerves of 20 to 24 pairs, the petioles 8–15 mm. long, puberulent above; fruiting aments pendulous, narrowly oblong, the peduncles 2–4 cm. long, puberulent, the scales 15–22 mm. long, narrowly ovate, with short appressed hairs, coarsely toothed; nuts oblong, about 4 mm. long, with few striations.——Honshu, Shikoku, Kyushu.

Var. **cordifolia** Winkl. *C. carpinoides* var. *cordifolia* (Winkl.) Makino.——Ō-KUMA-SHIDE. A large-leaved phase, occurring with the species.

5. Carpinus cordata Bl. *C. erosa* Bl.; *C. cordata* var. *pseudojaponica* Winkl.——SAWA-SHIBA. Tree, the branchlets soon glabrate; leaves ovate to broadly so or elliptic-ovate, 7–13 cm. long, 4–7 cm. wide, abruptly acuminate, cordate, irregularly serrulate, with the teeth ending in a short awn, nearly glabrous or with scattered hairs on the nerves and axillary tufts of brownish short hairs on lower side, the lateral nerves of 15 to 20 pairs, the petioles 1.5–3 cm. long; fruiting aments narrowly oblong, pendulous, the peduncles glabrous or puberulous, 2–4 cm. long, the scales 2–2.5 cm. long, oblong, toothed; nuts oblong, about 5 mm. long, glabrous, with about 10 striations on each side.——Hokkaido, Honshu, Shikoku, Kyushu.——Korea and China.

2. OSTRYA Scop. ASADA ZOKU

Deciduous trees with scaly scabrous bark; leaves petiolate, ovate or narrowly so, doubly toothed; staminate aments elongate, pendulous (naked in winter), the perianth absent; stamens 3–14, the filaments bifid above; pistillate aments slender, erect, 2-flowered in axils of the scales, the scales imbricate, deciduous in fruit; ovary with an adnate perianth surrounded by the involucre which consists of the adnate bract and bracteole; stigmas 2, linear; nuts striate, enveloped by an accrescent rather membranous involucre, coarsely hirsute below.——Seven to eight species, in the N. Hemisphere.

1. **Ostrya japonica** Sarg.   *O. virginica* var. *japonica* Maxim. ex Sarg.; *O. japonica* var. *homochaeta* Honda——ASADA.   Tree with pubescent and glandular-pilose young branchlets; leaves rather thin, narrowly ovate or ovate-oblong, 7–12 cm. long, 3–5 cm. wide, acuminate to very acute with irregular awn-tipped teeth, pubescent and glandular-pilose on both sides or becoming glabrate except on nerves beneath, the lateral nerves of 9 to 13 pairs, the petioles 4–8 mm. long, pubescent and glandular-pilose; fruiting aments narrowly ovoid, nodding, pedunculate, the scales ovate-elliptic, 13–17 mm. long, entire, acute, short-pubescent, connate and clasping the nut below; nuts oblong-ovate, 5–6 mm. long, glabrous or with short hairs above.——May–June. Hokkaido, Honshu, Shikoku, Kyushu.——Korea and China.

3. CORYLUS L.   HASHIBAMI ZOKU

Deciduous trees or shrubs; leaves alternate, simple, scattered-pilose, doubly toothed, petiolate, induplicate in bud, the stipules caducous; staminate aments elongate, pendulous (naked in winter), the flowers naked, each scale with 4–8 axillary stamens, the filaments bifid above, the anthers pilose at apex; pistillate aments capitate, the flowers paired in the axils of small bracts, the perianth short; ovary 2-locular, the ovules 1 or 2 in each locule; styles 2, the stigma exserted; nuts rather large, enveloped by an involucre, the cotyledons thick, hypogeal.——About 20 species, in the N. Hemisphere.

1A. Involucre leaflike, deeply incised, not tubular, not bristly; leaves broadly obovate, obovate-rounded or nearly obversed deltoid.
           1. *C. heterophylla*
1B. Involucre connate and tubular above the nut, densely bristly; leaves obovate to broadly so. .... 2. *C. sieboldiana*

1. **Corylus heterophylla** Fisch.   Ō-HASHIBAMI, OHYŌ-HASHIBAMI.   Shrub or small tree, the branchlets pubescent and glandular-pilose while young; leaves deltoid-obovate to orbicular-obdeltoid, 5–12 cm. long and as wide, truncate and abruptly acuminate, incised and doubly toothed, cordate, pubescent above while young, with short spreading hairs beneath and loose ascending hairs on nerves, the petioles 5–20 mm. long, pubescent and glandular-pilose; nuts 1–3 in a cluster, the involucre campanulate, 2.5–3.5 cm. long, puberulent, glandular-pilose below, striate, deeply toothed on margin; nuts subglobose, about 1.5 cm. across.——Honshu (centr. distr.).——Korea, Manchuria, Ussuri, and Amur.
    Var. **thunbergii** Bl.   *C. heterophylla* var. *yezoensis* Koidz.; *C. yezoensis* (Koidz.) Nakai——HASHIBAMI.   Leaves obovate-orbicular to broadly obovate, abruptly short-acuminate, rarely glandular-pilose; involucres scarcely glandular-pilose.——Hokkaido, Honshu, Kyushu.
2. **Corylus sieboldiana** Bl.   *C. rostrata* var. *sieboldiana* (Bl.) Maxim.; *C. rostrata* var. *mitis* Maxim.——TSUNO-HA-SHIBAMI.   Large shrub with pubescent branches; leaves obovate to broadly so, 5–10 cm. long, abruptly acuminate, ob-soletely incised and acutely serrulate, rounded or rarely shallowly cordate at base, often with a large brown-purple blotch on the center above, glabrous to slightly appressed-pilose on upper side, becoming yellowish brown when dry and loosely ascending-pilose on the nerves beneath, the petioles 1–1.5 cm. long, pubescent; nuts 1–3 in a cluster, short-conical, 6–8 mm. long, the involucre tubular, 1.5–5 cm. long, narrowed above the nut, shortly lobed or deeply dentate, puberulent and densely bristly-hairy.——Hokkaido, Honshu, Shikoku, Kyushu.
    Var. **brevirostris** C. K. Schn.   *C. brevirostris* (C. K. Schn.) Miyabe.——TOKKURI-HASHIBAMI.   A phase with short involucres 1–2 cm. long.——Occurs with the typical phase.——Korea.
    Var. **mandshurica** (Maxim.) C. K. Schn.   *C. mandshurica* Maxim.; *C. rostrata* var. *mandshurica* (Maxim.) Regel——Ō-TSUNO-HASHIBAMI.   Leaves relatively large, distinctly incised, often cordate, pale green beneath, rather thin; involucres gradually narrowed above, with spreading bristlelike hairs; nuts larger.——Hokkaido, Honshu (n. and centr. distr.).——Korea, Manchuria, Amur, and Ussuri.

4. BETULA L.   SHIRA-KAMBA ZOKU

Deciduous shrubs and trees; leaves petiolate, usually ovate, serrate-toothed, rarely incised, penninerved, the stipules deciduous; staminate aments elongate, pendulous (naked in winter), the scales usually 3-flowered in the axils, the bracteoles 2, adnate to the scale; perianth 4-parted; stamens 2, the filaments 2-fid above; pistillate aments short, sometimes cylindric, 3-flowered in the axils of bracts, the perianth absent; nuts usually wing-margined; deciduous with the scales.——About 40 species, in the temperate and cold regions of the N. Hemisphere.

1A. Fruiting aments 2–4, racemose, long-cylindric, (2–) 3–7 cm. long; leaves 8–14 cm. long, cordate. .... 1. *B. maximowicziana*
1B. Fruiting aments solitary, globose to short-cylindric, 1.5–5 cm. long; leaves less than 10 cm. long.
    2A. Leaves distinctly glaucous beneath; fruiting aments erect. .... 2. *B. corylifolia*
    2B. Leaves pale green and not glaucous beneath.
        3A. Fruiting aments erect or pendulous, short-cylindric or oblong-cylindric, the peduncles 3–12 mm. long.
            4A. Fruiting aments pendulous; bark white and smooth; leaves broadly deltoid-ovate to subdeltoid. .... 3. *B. platyphylla*
            4B. Fruiting aments erect; bark brownish.
                5A. Leaves narrowly ovate-oblong to broadly ovate, dull above, with upward pointing teeth; petioles 5–15 mm. long.
                    6A. Trees; leaves 4–8 cm. long.
                        7A. Midlobe of fruiting scales slightly shorter than the lateral lobes; leaves irregularly toothed, the lateral nerves 6- to 8-paired; petioles 5–15 mm. long. .... 4. *B. davurica*
                        7B. Midlobe of fruiting scales longer than the lateral; leaves serrulate, the lateral nerves 9- to 18-paired; petioles 4–10 mm. long.
                            8A. Branchlets distinctly glandular; leaves with 9–10 pairs of lateral nerves. .... 5. *B. schmidtii*
                            8B. Branchlets scarcely glandular; leaves with 14 to 18 pairs of lateral nerves. .... 6. *B. chichibuensis*

1. Betula maximowicziana Regel. *B. candelae* Koidz. ——Udai-kamba, Saihada-kamba. Tree with gray or orange-gray bark peeling off into papery thin pieces, the branchlets red-brown, glabrous, lustrous; leaves broadly ovate to nearly ovate-orbicular, 8–14 cm. long, 6–10 cm. wide, short-acuminate, deeply cordate at base, serrate, with glandular tipped teeth, scattered pilose while young above, slightly appressed pilose on nerves and later often becoming glabrous beneath, the lateral nerves of 10 to 12 pairs, the petioles 25–35 mm. long; fruiting aments 2–4 in a cluster, racemose, pedicellate, pendulous, long-cylindric, (2–)3–7 cm. long, 6–7 mm. in diameter; fruiting scales glabrous, about 3 mm. long, 3-lobed, the lateral lobes ascending, shorter than the midlobe; nuts about 3 mm. long, the wings 2–3 times as wide as the body.——May–June. Hokkaido, Honshu (centr. and n. distr.).——s. Kuriles.

2. Betula corylifolia Regel & Maxim. Neko-shide, Urajiro-kamba. Tree with grayish or whitish bark, the branchlets glabrous or nearly so, dark purple-brown to dark red-brown; leaves elliptic, obovate or ovate, 4–8 cm. long, 2.5–5 cm. wide, rather thin, acute, coarsely doubly toothed, the base obtuse to rounded, vivid green and glabrous above, glaucous and slightly appressed-pilose on nerves beneath, the lateral nerves of 8 to 14 pairs, the petioles 7–15 mm. long, white-silky while young; fruiting aments erect, short-pedunculate, 3–4 cm. long, oblong-cylindric, the fruiting scales thinly long-pilose, 12–15 mm. long, 3-lobed, the lateral lobes ascending, about half as long as the midlobe; nuts ovoid-orbicular, about 3 mm. long, narrowly winged on both sides.——Mountains; Honshu (Kinki Distr. and eastw.).

3. Betula platyphylla Sukatschev var. **japonica** (Miq.) Hara. *B. japonica* Sieb., non Thunb.; *B. alba* var. *japonica* Miq.; *B. alba* var. *tauschii* Regel, pro parte; *B. tauschii* (Regel) Koidz.——Shira-kaba, Shira-kamba. Tree with white bark, the branchlets dark purple-brown, with resinous glands while young; leaves broadly deltoid-ovate, 5–7 cm. long, 4–6 cm. wide, truncate at base, acuminate, doubly toothed, glabrous or sparsely short-pilose beneath, usually with tufts of axillary hairs, rather dull above, the lateral nerves weak, of 6 to 8 pairs, the petioles 1–3 cm. long, rather slender; fruiting aments short-cylindric, pendulous, pedunculate, 3–4.5 cm. long, 8–10 mm. in diameter, the scales 3.5–4.5 mm. long, puberulous outside, the lateral lobes spreading, obovate, 2–3 times as long as the midlobe; nuts narrowly obovate to ovate, 1.5–2.2 mm. long, puberulous, the wing 1.5–2 times wider than the nuts.——Highlands; Hokkaido, Honshu (centr. and n. distr.); locally abundant.

Var. **kamtschatica** (Regel) Hara. *B. alba* var. *kamtschatica* Regel——Ezo-no-shira-kamba. Leaves often broadly cuneate at base.——Occurs with var. *japonica*.——The typical phase occurs in the cooler areas of e. Asia.

Betula × avaczensis Komar. Oku-Ezo-shira-kamba, is a hybrid between No. 3 and No. 6. Reported from Hokkaido, originally described from Kamchatka.

4. Betula davurica Pall. *B. maackii* Rupr.——Ko-ono-ore, Yaegawa-kamba. Tree with gray-brown to gray bark, the branchlets very resinous, pilose; leaves rhombic-ovate to narrowly ovate, 4–8 cm. long, acute or acuminate, irregularly toothed, broadly cuneate at base, pilose on nerves beneath, the lateral nerves of 6 to 8 pairs, the petioles 5–15 mm. long, slightly pilose; fruiting aments erect, short-pedunculate, 2–2.5 cm. long, narrowly oblong, the scales glabrous, lustrous, firm, the midlobe deltoid, the lateral ones spreading, rounded, as long as the midlobe; wings about half as wide as the nut.——Hokkaido, Honshu (centr. and n. distr.).——Korea, Manchuria, Ussuri, and Amur.

Var. **okuboi** Miyabe & Tatew. Hidaka-yaegawa. Leaves with 7–9 pairs of lateral nerves; midlobe of fruiting scales longer than the lateral ones.——Reported to occur in Hokkaido.

5. Betula schmidtii Regel. Ono-ore-kamba. Tree with dark gray bark peeling off into small thick pieces, the branchlets dark-brown, resinous-glandular and pilose while young; leaves narrowly ovate-elliptic to ovate, acuminate, with irregular suberect small teeth, rounded to broadly cuneate at base, 4–8 cm. long, rather dull, appressed-pilose on nerves beneath, the lateral nerves of 9 to 10 pairs, the petioles 5–10 mm. long, pilose; fruiting aments short-pedunculate, oblong-cylindric, 2–3 cm. long, 8–9 mm. in diameter, erect, the scales 5–6 mm. long, slightly resinous, short-ciliolate, the midlobe linear-lanceolate, the lateral ones obliquely ascending, about half as long as the midlobe, subacute; nuts ovoid, about 2 mm. long, with very narrow wings.——Honshu (centr. and n. distr.).——Korea, Manchuria, and Ussuri.

6. Betula chichibuensis Hara. Chichibu-minebari. Tree; branchlets densely villous while young, scarcely glandular; leaves ovate or oblong-ovate, 3–6 cm. long, 1.5–3.2 cm. wide, short-acuminate, minutely and unequally duplicate-serrate, rounded to broadly cuneate at base, densely white-villous at least on the nerves beneath, lateral nerves of 14 to 18 pairs, the petioles 4–7 mm. long, white-villous; pistillate aments cylindric, short-peduncled, erect, 1.5–2.5 cm. long, 7–10 mm. wide, the scales 3.5–5 mm. long, deeply 3-lobed, pubescent and eglandular on back, the lateral lobes slightly shorter than the median; nuts ovate, 2–3 mm. long, scarcely winged. ——Honshu (Chichibu).

7. Betula ermanii Cham. *B. incisa* Koidz.; *B. shikokiana* Nakai——Dake-kamba, Take-kamba. Tree or large shrub with gray-brown to gray-white bark peeling off into thin pieces, the branchlets resinous when young, soon becoming brownish purple; leaves deltoid-ovate to broadly so, 5–10 cm. long, 4–7 cm. wide, acuminate, irregularly mucronate-toothed, rounded to shallowly cordate, thinly pilose on nerves beneath or glabrous except for axillary tufts of brown hairs, the lateral nerves of 7 to 12 pairs, the petioles 1–3.5 cm. long, usually glabrous; fruiting aments 2–3.5 cm. long, 8–10 mm. in diameter, short-cylindric to oblong, erect, the scales 6–8 mm. long, ciliolate, the midlobe linear-lanceolate, the lateral ones nearly orbicular, ascending, 1/3–1/2 as long as the midlobe; nuts broadly obovate, puberulous above, 2–3 mm. long, the wing narrower than the nut.——June–July. Mountains; Hokkaido, Honshu (centr. and n. distr.), Shikoku; extremely variable.——Kuriles, Sakhalin, Kamchatka, and Korea.

Var. **japonica** (Shirai) Koidz. *B. bhojpatra* var. *japonica* Shirai; *B. nikoensis* Koidz.——NAGABA-NO-DAKE-KAMBA, NAGABA-NO-SHIRA-KAMBA, MA-KAMBA.　　Leaves deltoid-ovate, the lateral nerves of 14 to 15 pairs, the fruiting scales with a narrow midlobe and rather spreading lateral lobes.——Honshu (centr. distr. and Kantō Distr.).

Var. **subcordata** Koidz. AKA-KAMBA.　　Leaves shallowly cordate.——Occurs with the typical phase.

8. Betula tatewakiana M. Ohki & S. Watanabe. YACHI-KAMBA.　　Branching shrub about 1 m. high; branchlets pubescent while young, soon becoming glabrous, very densely glandular; leaves coriaceous, ovate or obovate, 1.2–3 cm. long, 1–2 cm. wide, acute at both ends, acutely serrate except on the lower margin, glandular and hairy beneath, especially on the nerves, the petioles 2–6 mm. long, pilose; fruiting aments erect, axillary, oblong-cylindric, 1–1.5 cm. long, 3.5–7.5 mm. across; bracts sparsely pilose on back, 3-lobed, ciliate, 2.8–4.2 mm. long, 2.7–3.4 mm. wide, the lobes subequal; nutlets obovate or elliptic, 3 mm. long, winged.——Bogs; Hokkaido (Tokachi Prov.); rare.

9. Betula apoiensis Nakai. *B. miyoshii* Nakai; *B. miijimae* Nakai——APOI-KAMBA.　　Much-branched shrub about 1 m. high, the branchlets densely resinous-glandular, short-pubescent while young; leaves ovate to broadly so to ovate-orbicular, 1.5–4 cm. long, 1–3 cm. wide, acute, usually rounded at base, irregularly acute-toothed, with silky-pubescence beneath while young, glabrescent, sometimes persisting only on the nerves, the petioles 2–10 mm. long; fruiting aments erect, subglobose to short-cylindric, 1–3 cm. long, 6–12 mm. in diameter, short-pedunculate, 2–5 mm. long, the scales 3–4 mm. long, glabrous, ciliolate, the midlobe lanceolate to ovate, the lateral lobes obovate, shorter than the midlobe.——Hokkaido (Mount Apoi).

10. Betula globispica Shirai.　JIZŌ-KAMBA, INUBUSHI. Tree with whitish peeling bark, the branchlets grayish or yellowish brown, scattered pilose while young; leaves broadly ovate, 4–7 cm. long, 3–5 cm. wide, short-acuminate, irregularly mucronate-toothed, rounded to broadly cuneate at base, smooth, slightly lustrous, soon becoming glabrous above, long-pilose on nerves beneath, the lateral nerves of 8 to 10 pairs, the petioles 5–15 mm. long, slightly pilose; fruiting aments ellipsoidal to subglobose, 2.5–3.5 cm. long, the scales 13–15 mm. long, the lobes narrowly lanceolate, obtuse, ciliolate, the lateral lobes ascending, 1/2–2/3 as long as the midlobe; nuts 3–4 mm. long, broadly obovoid, with very narrow wings.——Mountains; Honshu (Kantō and centr. distr.); rare.

11. Betula grossa Sieb. & Zucc. *B. ulmifolia* Sieb. & Zucc.; *B. carpinifolia* Sieb. & Zucc., non Willd.; *B. solennis* Koidz.——YOGUSO-MINEBARI, AZUSA, MIZUME.　　Tree, with dark gray smooth bark, the branchlets yellow-brown and scattered long-pilose while young, becoming glabrous, changing to chestnut-brown; leaves rather thin, ovate, 5–10 cm. long, 3–6 cm. wide, acuminate, shallowly cordate to nearly rounded at base, acutely toothed, dull, nearly glabrous or scattered long-pilose above, pilose on nerves beneath, the lateral nerves of 8 to 14 pairs, the petioles 1–2.5 cm. long, pilose; fruiting aments sessile or nearly so, ellipsoidal to oblong, 2–2.5 cm. long, 12–15 mm. in diameter, the scales ciliate, 6–7 mm. long, the midlobe narrowly oblong, obtuse, the lateral lobes ascending, elliptic, about half as long as the midlobe; nuts ovate, about 2 mm. long, the wings slightly narrower than the nut.——Mountains; Honshu, Shikoku, Kyushu. The branches have a characteristic odor when broken.

5. ALNUS Mill.　HAN-NO-KI ZOKU

Deciduous trees or shrubs; leaves alternate, pinnately nerved, toothed, rarely entire, sometimes shallowly lobed; staminate aments elongate, usually pendulous, with 3 flowers in the axil of each bract, the perianth-segments usually 4; stamens 4; pistillate aments short, globose to short-cylindric, conelike, the densely imbricate scales slightly woody, persistent and with 2 flowers in the axil of each scale, the perianth absent, the bracteoles adnate to the bract (scale); ovary sessile, 2-locular, the ovules solitary in each locule; styles 2, cylindric; nuts flat, sometimes laterally winged.——More than 30 species, in the N. Hemisphere and S. America.

1A. Winter-buds sessile, with 2 or more unequal scales; pistillate aments with protective winter bud-scales; staminate aments sessile, solitary; flowering with the leaves.
　2A. Leaves narrowly ovate, with 12–26 pairs of lateral nerves.
　　3A. Aments usually from the lateral buds, the pistillate solitary; branches glabrous, rather stout; leaves with 12–15 pairs of lateral nerves. .... 1. *A. sieboldiana*
　　3B. Aments from the terminal and lateral buds near the top, the pistillate lateral, 1–6 in a group; branches usually pilose while young.
　　　4A. Leaves with 13–17 pairs of lateral nerves; pistillate aments paired, sometimes solitary, erect. .... 2. *A. firma*
　　　4B. Leaves with 20–26 pairs of lateral nerves; pistillate aments 3–6 in a group, nodding. .... 3. *A. pendula*
　2B. Leaves broadly ovate, with 10–12 pairs of lateral nerves. .... 4. *A. maximowiczii*
1B. Winter-buds short-stiped, with 2 or 3 nearly equal scales; pistillate aments naked during the winter; staminate aments more or less pedunculate; flowering usually before the leaves.
　5A. Leaves plicate in bud, broadly elliptic to ovate-orbicular or obovate-orbicular, shallowly incised on the upper margin to emarginate, the lateral nerves nearly parallel and straight.
　　6A. Leaves ovate-orbicular to broadly elliptic, obtuse to acute, rounded to subtruncate at base, often glaucescent beneath.
　　　5. *A. hirsuta*

6B. Leaves obovate-orbicular to nearly obcordate, truncate to emarginate or truncate-rounded at apex, broadly cuneate at base, without teeth on lower margin.
    7A. Leaves glaucous beneath. . . . . 6. *A. matsumurae*
    7B. Leaves pale green and not glaucous beneath.
        8A. Leaves membranous, pale green beneath, irregularly undulate-toothed, with 6 or 7 pairs of weak lateral nerves, with axillary tufts of brown hairs beneath; fruiting aments oblong or short-cylindric, 15–25 mm. long, 6–8 mm. in diameter.
                7. *A. fauriei*
        8B. Leaves firm, pale brownish green beneath, the teeth minute, depressed mucronate, the lateral nerves of 7–9 pairs prominently raised beneath; fruiting aments ovate-ellipsoid, 1.5–2 cm. long, 8–12 mm. in diameter. . . . . 8. *A. serrulatoides*
5B. Leaves not plicate in bud, narrowly ovate, ovate-elliptic or narrowly oblong, acuminate, the lateral nerves more or less arcuate above.
    9A. Leaves dull above, cuneate or acute at base, the lateral nerves of 7–9 pairs; petioles 1.5–3 cm. long. . . . . 9. *A. japonica*
    9B. Leaves smooth and slightly lustrous above, becoming reddish when dry especially on nerves, shallowly cordate, the lateral nerves 9–12 pairs, rather prominently raised beneath; petioles 1–1.5 cm. long. . . . . 10. *A. trabeculosa*

1. Alnus sieboldiana Matsum. *A. firma* var. *sieboldiana* (Matsum.) Winkl.——ŌBA-YASHABUSHI. Large shrub or small tree with rather stout yellow-brown to gray-brown, glabrous branches; leaves ovate or narrowly so to deltoid-ovate, 6–10 cm. long, 3–6 cm. wide, appressed pilose on the nerves beneath while young, gradually acute to short-acuminate, rounded at base, doubly serrate with mucronate teeth, pale green beneath, the lateral nerves of 12 to 15 pairs, the petioles 10–15 mm. long, glabrous; pistillate inflorescence with peduncles 1–2 cm. long, pubescent while young, ascending to spreading, rather thick; fruiting aments broadly ellipsoidal, 2–2.5 cm. long.——Mar. Lowlands and foothills, especially abundant near the sea; Honshu (Kantō Distr. and westw. to Kii Prov.).

2. Alnus firma Sieb. & Zucc. *A. yasha* Matsum.; *A. firma* var. *yasha* (Matsum.) Winkl.——YASHABUSHI. Much-branched small tree or large shrub with gray-brown often puberulous branchlets often scattered pilose while young; leaves ovate or narrowly so to narrowly ovate-deltoid, 5–10 cm. long, 2.5–4.5 cm. wide, gradually acuminate, rounded at base, mucronate-toothed, with a mucro between the teeth, the nerves beneath with appressed hairs, glabrescent, the hairs sometimes also on the upper side, the lateral nerves 13–17 pairs, the petioles 7–12 mm. long; staminate inflorescence from the terminal and upper lateral buds; pistillate aments 1 or 2, the peduncles pubescent while young, erect or ascending, 5–15 mm. long, the mature aments ovoid-ellipsoid, 15–20 mm. long. ——Apr. Kyushu; common in mountains.
Var. **hirtella** Fr. & Sav. *A. hirtella* (Fr. & Sav.) Koidz. ——MIYAMA-YASHABUSHI. Branches and leaves pubescent. ——Honshu; common.

3. Alnus pendula Matsum. *A. firma* var. *multinervis* Regel; *A. multinervis* (Regel) Callier——HIME-YASHABUSHI. Much-branched small tree or large shrub with dark gray-brown, slender, evanescent pilose branches; leaves narrowly ovate to broadly lanceolate, long-acuminate, with short, mucronate, double teeth, broadly cuneate at base, pale green and appressed pilose on nerves beneath, with 20–26 pairs of lateral nerves, the petioles pilose, 3–8 mm. long; staminate aments from the terminal and upper lateral buds, the pistillate on the lower part, 3–6 in a group, nodding or subpendulous; fruiting aments broadly ellipsoidal, about 15 mm. long.——Apr. Mountains; Hokkaido, Honshu; common.——e. Asia.

4. Alnus maximowiczii Callier. *A. viridis* var. *sibirica* sensu auct. Japon., non Regel; *A. sinuata* var. *kamtschatica* Callier; *A. viridis* subsp. *maximowiczii* (Callier) Hult.—— MIYAMA-HAN-NO-KI. Much-branched, large to small shrub or small tree, with rather thick, gray-brown to dark gray glabrous to glabrescent branches; leaves broadly ovate to ovate-orbicular, 6–10 cm. long, 4–7 cm. wide, abruptly short-acuminate to short-acute, rounded to shallowly cordate at base, with minute mucronate teeth, pale green and glabrous or nearly so beneath except for axillary tufts of hairs, the lateral nerves of 10 to 12 pairs, the petioles glabrous, 1–3 cm. long; staminate inflorescence from the terminal and upper lateral buds, the pistillate aments 3–5 in a group, loosely racemose on a short axis, arising from the buds below the staminate, the peduncles ascending, the pistillate aments broadly ellipsoidal, 1–1.5 cm. long.——June. High mountains, often ascending to the alpine zone; Hokkaido, Honshu (centr. and n. distr.); common and very variable.——Kamchatka, Sakhalin, Kuriles, Ussuri, and Korea.

5. Alnus hirsuta Turcz. *A. incana* var. *hirsuta* Spach; *A. tinctoria* Sarg.; *A. sibirica* var. *hirsuta* (Turcz.) Koidz.—— KE-YAMA-HAN-NO-KI. Tree with densely pubescent branchlets and inflorescence; leaves broadly ovate-orbicular to broadly elliptic, 7–12 cm. long and as wide, obtuse to acute, sometimes nearly rounded at apex, rounded to subtruncate at base, shallowly pinnately incised, irregularly toothed, usually short-pilose but often becoming glabrate above, densely pubescent beneath, the lateral nerves of 6 to 8 pairs, the petioles 1.5–3 cm. long; staminate inflorescence with 2–4 aments, borne from the apical and upper lateral buds, the pistillate aments 3–5, borne immediately below; fruiting aments ellipsoidal to ovate-oblong, 15–25 mm. long.——Apr. Mountains and hills; Hokkaido, Honshu, Shikoku, Kyushu; very variable.——Sakhalin, Kamchatka, Korea, and e. Siberia.
Var. **sibirica** (Fischer) C. K. Schn. *A. sibirica* Fisch.; *A. incana* var. *sibirica* Spach; *A. incana* var. *glauca* Regel—— YAMA-HAN-NO-KI. With nearly glabrous parts; leaves prominently glaucous beneath; a common variant in our area, occurring with the typical phase.

6. Alnus matsumurae Callier. *A. incana* var. *emarginata* Matsum.; *A. emarginata* (Matsum.) Shirai——YAHAZU-HAN-NO-KI. Small tree or rarely large shrub with dark gray-purple glabrous branches; leaves obcordate-orbicular, 5–10 cm. long and as wide, glabrous or nearly so above, appressed-pilose on the nerves beneath while young, gray-glaucous beneath, rounded and distinctly emarginate, irregularly toothed, broadly cuneate at base, the lateral nerves of 6 to 9 pairs, the petioles glabrous, 1–3 cm. long, dark purple-brown when dried; staminate aments solitary or paired, terminal and on upper part of branches, the pistillate aments 2–5 in a group, the fruiting ones elliptic, 15–18 mm. long.——May–June. High mountains; Honshu (centr. and n. distr.); rather common.

7. Alnus fauriei Lév. *A. glutinosa* var. *cylindrostachya* Winkl.; *A. schneideri* Callier; *A. cylindrostachya* (Winkl.)

Makino——Miyama-kawara-han-no-ki.   Small tree with dark purple-brown or slightly grayish glabrous branches; leaves cuneately obovate or cuneately obcordate-orbicular, 5–12 cm. long, 4–11 cm. wide, rounded or sometimes slightly retuse, undulately toothed, pale green beneath, glabrous except the axillary tufts of brownish hairs or thinly pilose on nerves beneath, pale green beneath, broadly cuneate to cuneate at base, the lateral nerves of 6 to 7 pairs, the petioles glabrous, 5–15 mm. long; staminate and pistillate inflorescence each bearing 4 or 5 aments, the fruiting aments ellipsoidal to cylindric-oblong, 15–25 mm. long, 6–8 mm. in diameter.—— Mountains; Honshu (n. and centr. distr. along the Japan Sea).

8. Alnus serrulatoides Callier.   *A. maritima* var. *obtusata* Fr. & Sav.; *A. glutinosa* var. *japonica* Matsum.; *A. glutinosa* var. *obtusata* (Fr. & Sav.) Winkl.; *A. obtusata* (Fr. & Sav.) Makino——Kawara-han-no-ki.   Small tree with glabrous dark brown to dark purple-brown or sometimes slightly grayish branches; leaves obovate-cuneate to broadly obovate or obovate-orbicular, 6–10 cm. long, 4–9 cm. wide, rounded, truncate or emarginate, cuneate at base, glabrous above, often slightly pilose on nerves, with spreading hairs near axils beneath, serrulate, the lateral nerves of 7 to 9 pairs, slightly raised and rather reddish beneath, the petioles 8–15 mm. long; staminate inflorescence terminal, consisting of 4 or 5 aments, the pistillate from the upper axils, consisting of 1–4 aments; fruiting aments ovoid-ellipsoid, 15–20 mm. long, 8–12 mm. in diameter.——Mar.–Apr. Along rivers; Honshu (Tōkaidō, Kinki, and Chūgoku Distr.), Shikoku.

9. Alnus japonica (Thunb.) Steud.   *Betula japonica* Thunb.——Han-no-ki.   Tree with gray-brown, glabrescent to loosely brown-pubescent branches; leaves oblong-ovate, ovate-elliptic to broadly lanceolate, 6–13 cm. long, 2.5–5 cm. wide, acuminate, acute to cuneate at base, glabrous except for axillary tufts of red-brown hairs beneath or slightly pubescent on both sides, pale green beneath, with 7–9 pairs of rather slender arcuate lateral nerves, the short marginal teeth mucronate, the petioles 1–3.5 cm. long; staminate inflorescence consisting of 2 to 5 aments borne on the upper part of the branches, the pistillate consisting of 1–5 aments borne just below, the fruiting aments ovoid-ellipsoid, 15–20 mm. long. ——Mar. Wet lowlands; Hokkaido, Honshu, Shikoku, Kyushu; common and frequently planted around paddy fields; variable.——Korea, Ussuri, and Manchuria.

Alnus mayrii Callier.   Usuge-hiro-ha-han-no-ki, and **Alnus borealis** Koidz., Hiro-ha-han-no-ki, are alleged hybrids between No. 9 and No. 5. Both are allied to *Alnus japonica* but with broader, elliptic to oblong, short-acute, indistinctly incised leaves.

10. Alnus trabeculosa Hand.-Mazz.   *A. nagurae* Inokuma——Sakuraba-han-no-ki.   Small tree with gray-brown branchlets, glabrous or yellowish pilose while young; leaves obovate-elliptic to oblong, 6–9 cm. long, 3–5 cm. wide, short-acuminate to abruptly acute, rather regularly serrulate, glabrous or slightly pilose on nerves beneath, reddish when dried, obtusely rounded to shallowly cordate, the lateral nerves of 9 to 12 pairs, raised beneath, the petioles 5–15 mm. long, more or less pilose on upper side; staminate inflorescence terminal, consisting of 4 or 5 aments, the pistillate of 3 to 5 aments borne just below, the fruiting aments short-pedunculate, ovoid-ellipsoid, 1.5–2 cm. long.——Wet places in lowlands and hills; Honshu (Kantō, Tōkaidō, Kinki, and Chūgoku Distr.); rather rare.——China.

Fam. 66. **FAGACEAE**   Buna Ka   Oak Family

Deciduous or evergreen monoecious trees or shrubs; leaves alternate, pinnately nerved, entire, toothed or dentate, sometimes pinnatifid, the stipules usually deciduous; flowers usually in axillary spikes on the twigs; perianth 4- to 7-merous; staminate spikes (aments) slender and elongate, the flowers in axils of scales; stamens as many or twice as many as the perianth-segments or rarely more, the filaments usually slender; pistillate flowers in 2's or 3's or solitary, fasciculate or in short spikes, sometimes disposed at the base of the staminate aments; ovary 3- to 6-locular, the ovules 2 in each locule; styles 3, sometimes 6; nuts 1-seeded (sometimes 2- to 3-seeded), wholly or partially enveloped by an involucre; seeds lacking endosperm, the cotyledons fleshy.—— About 10 genera, with about 600 species, in tropical and temperate regions.

1A.   Staminate flowers many in heads; pistillate flowers in pairs on a common peduncle; involucres 2- to 4-lobed; nuts trigonous; cotyledons plicate, epigeal; flowers anemophilous. ....1. *Fagus*
1B.   Staminate flowers many in elongate, slender aments; pistillate flowers solitary or rather many, in spikes or aments; involucres entire or irregularly incised; nuts spherical; cotyledons hypogeal.
    2A.   Staminate aments pendulous; stigma flat; plants anemophilous. ....2. *Quercus*
    2B.   Staminate aments erect; stigma punctiform; plants entomophilous.
        3A.   Deciduous trees, without terminal buds; ovary 6-locular; styles 6, free; pistillate flowers 1–3 in an involucre. ......3. *Castanea*
        3B.   Evergreen trees with terminal buds; ovary 3-locular; styles 3, connate below; pistillate flowers (in ours) solitary in an involucre.
            4A.   Involucre entirely enclosing the nut; leaves usually in 2 series; buds flat, the scales distichous. ....4. *Castanopsis*
            4B.   Involucre cuplike, exposing the nut above; leaves and scales not in 2 series; buds not flattened. ....5. *Pasania*

1. **FAGUS** L.   Buna Zoku

Deciduous trees with smooth bark; leaves alternate, pinnately nerved, dentate or nearly entire; staminate flowers many, in long more or less erect pedunculate heads, the perianth 4- to 7-parted; stamens 8–16; bracts of pistillate aments many, forming a connate involucre, the flowers usually geminate in the involucre; styles 3, slender, reflexed; nuts ovoid-trigonous, enveloped by a 4-lobed, spinose or scaly involucre; seeds solitary, the cotyledons rather thick, flat and plicate, epigeal.——About 10 species, in the temperate regions of the N. Hemisphere.

1A.   Leaves nearly glabrous, pale green beneath, with 7 to 11 pairs of lateral nerves; fruiting peduncles 5–15 mm. long, pubescent.
                                                         1. *F. crenata*
1B.   Leaves slightly pubescent and slightly glaucous beneath, with 10–14 pairs of lateral nerves; fruiting peduncles 3–4 cm. long, glabrous.
                                                        2. *F. japonica*

1. **Fagus crenata** Bl. *F. sieboldii* Endl.; *F. sylvatica* var. *sieboldii* (Endl.) Maxim.——BUNA. Tree with grayish bark; young branchlets slightly long-pubescent, soon glabrous; leaves ovate or broadly so, or rhombic-ovate, 5–8 cm. long, 3–5 cm. wide, short-acuminate, broadly cuneate to obliquely rounded at base, undulately and obsoletely toothed, long-pubescent while young especially on nerves beneath and on margin, the lateral nerves of 7 to 11 pairs, ascending, straight or nearly so, the petioles 3–10 mm. long, slightly pubescent; involucre about 15 mm. long, nearly as long as the nuts, provided with rather dense linear awnlike fleshy appendages 3–7 mm. long, the peduncles rather thick, pubescent.——May. Important forest-tree of temperate regions; Hokkaido, Honshu, Shikoku, Kyushu.

2. **Fagus japonica** Maxim. INU-BUNA. Young branchlets soon becoming glabrous; leaves ovate or ovate-elliptic, 5–8 cm. long, 2.5–5 cm. wide, short-acuminate, obsoletely undulate, rounded to obtusely rounded at base, long-pubescent on both sides while young, soon glabrate above, the hairs persistent beneath, the lateral nerves of 10 to 14 pairs, the petioles 8–10 mm. long, soon glabrate; involucre 6–8 mm. long, about 1/2–2/3 the length of the nuts, with a short ovate-deltoid appendage, the peduncles slender, glabrous, 3–4 cm. long.——Apr.–May. Mountains; Honshu, Shikoku, Kyushu.

2. QUERCUS L. KO-NARA ZOKU

Evergreen or deciduous trees, rarely shrubs; leaves alternate, short-petiolate or subsessile, pinnately nerved, simple, toothed or entire, sometimes pinnatifid; staminate flowers in slender, pendulous aments, the perianth 4- to 7-merous; stamens 4–12; pistillate flowers solitary or paired, usually in short spikes; ovary 3(–5)-locular; styles short or elongate, broadened above, the stigma on the upper side; nuts (acorns) globose to subcylindric, enveloped at base or nearly completely by a cup-shaped involucre (cupule) sculptured with ringlike annuli or scales.——More than 200 species, in the temperate to subtropical regions of the N. Hemisphere.

1A. Involucres with ringlike annuli; evergreen trees.
  2A. Adult leaves glabrous, green beneath, often slightly brown-tinged when dried, usually decurrent on the petiole.
    3A. Petioles 2–4 cm. long; leaves entire, loosely arranged on branchlets. .... 1. *Q. acuta*
    3B. Petioles less than 2 cm. long.
      4A. Leaves vivid to pale green when dry, thinly coriaceous, the midnerve not impressed above. .... 2. *Q. hondae*
      4B. Leaves dark brownish when dry, coriaceous, densely disposed toward the tip of branchlets; midrib impressed above.
                                                   3. *Q. sessilifolia*
  2B. Adult leaves glaucous or densely pubescent beneath.
    5A. Underside of adult leaves, branchlets, and petioles with dense yellow-brown tomentum. .... 4. *Q. gilva*
    5B. Underside of leaves glaucescent, glabrous or with scattered appressed hairs.
      6A. Leaves with persistent hairs thinly dispersed beneath, oblong or obovate-elliptic, the teeth prominent, the lateral nerves prominently raised beneath; nuts maturing in autumn of first year. .... 6. *Q. glauca*
      6B. Leaves glabrate beneath, lanceolate or broadly lanceolate, the teeth low, the lateral nerves only slightly raised beneath.
        7A. Costa not sunken on upper side; nuts ripening in autumn of the first year; branches rather stout, dark purple-brown.
                                                   5. *Q. myrsinaefolia*
        7B. Costa sunken on upper side; nuts ripening in autumn of second year; branches slender, gray-brown. .... 7. *Q. salicina*
1B. Fruiting involucres with imbricate scales; deciduous or evergreen trees.
  8A. Evergreen small tree or large shrub with leaves less than 6 cm. long. .... 8. *Q. phillyraeoides*
  8B. Deciduous trees with leaves 6–25 cm. long.
    9A. Leaves obovate to narrowly so, or elliptic, usually prominently toothed or dentate, the teeth not awn-tipped, rarely only mucronate.
      10A. Branches relatively slender, glabrous or slightly pubescent while young; teeth of leaves usually ascending, obtuse or mucronate; scales of fruiting involucres short, appressed or slightly adnate and scalelike.
        11A. Petioles 0–5 mm. long; leaf-blade long-pilose while young, becoming entirely glabrous, pale green or sometimes slightly glaucescent beneath. .... 9. *Q. mongolica*
        11B. Petioles 5–30 mm. long; leaf-blades with long thin deciduous hairs, usually glaucous, sometimes with persistent stellate-pubescence beneath.
          12A. Leaves 6–15 cm. long, thinly appressed-pubescent beneath, with 7–12 pairs of lateral nerves; involucres distinctly stipitate; branches slender, pubescent while young. .... 10. *Q. serrata*
          12B. Leaves 10–25(–35) cm. long, short stellate-pubescent beneath, sometimes appressed-pilose only while young, with 10–15 pairs of lateral nerves; involucres nearly sessile; branches rather stout, soon becoming glabrous. .... 11. *Q. aliena*
      10B. Branches stout, densely short yellowish villous at least while young; leaves with large spreading rounded teeth; scales of fruiting involucres linear, elongate, spreading to somewhat recurved. .... 12. *Q. dentata*
    9B. Leaves oblong-lanceolate, with awn-tipped teeth.
      13A. Leaves whitish and densely stellate-pubescent beneath. .... 13. *Q. variabilis*
      13B. Leaves glabrous or very thinly pilose except for tufts of axillary hairs beneath. .... 14. *Q. acutissima*

1. **Quercus acuta** Thunb. *Cyclobalanopsis acuta* (Thunb.) Oerst.; *Q. buergeri* Bl.; *Q. laevigata* Bl.; *Q. marginata* Bl.; *C. buergeri* (Bl.) Oerst.; *C. laevigata* (Bl.) Oerst.; *C. marginata* (Bl.) Oerst.——AKAGASHI. Evergreen; branches stout, dark brown, densely long red-brown pubescent while young, soon becoming glabrous; leaves coriaceous, ovate-elliptic to narrowly so, or oblong, 8–20 cm. long, 3–6 cm. wide, abruptly long-acuminate, pubescent while young, soon becoming glabrous, entire, often slightly undulate, abruptly decurrent on the petiole, lustrous and deep green above, slightly paler and often becoming reddish beneath when dried, with 8–13 pairs of lateral nerves, the midrib not impressed above, the petioles slightly flattened, not grooved, 2–4 cm. long; staminate aments pendulous, the bracts scarious, 3–8 mm.

long; stamens 10–12; pistillate aments few-flowered, the fruiting involucres densely appressed-puberulent; nuts ellipsoidal, about 2 cm. long.——May. Honshu, Shikoku, Kyushu; common and often planted.——Korea and China.

Var. **acutaeformis** Nakai. Hiro-ha-akagashi. With broader thicker leaves.——Reported from Kyushu.

Quercus × idzuensis Makino, a hybrid allegedly between No. 1 and No. 6.——Honshu (Idzu Prov.).

2. **Quercus hondae** Makino. *Cyclobalanopsis hondae* (Makino) Schottky——Hanagagashi. Branches slender, dark red-brown, glabrous; leaves thinly coriaceous, rather densely disposed toward the top of the branches, lanceolate or oblanceolate, 7–13 cm. long, 2–3 cm. wide, long-acuminate, vivid or rather pale green, slightly paler beneath, subentire or with teeth undulate-ascending and awn-tipped, decurrent to an obscure or very short petiole, the midrib slightly raised or flat above, reddish beneath, the lateral nerves 8 to 13 pairs, very slender.——Mountains; Kyushu.

3. **Quercus sessilifolia** Bl. *Cyclobalanopsis sessilifolia* (Bl.) Nakai; *Q. paucidentata* Franch.; *C. paucidentata* (Franch.) Kudō & Masam.——Tsukubanegashi. Evergreen; branches dark gray-brown, pale yellow-brown villous, soon becoming glabrous; leaves coriaceous, obovate to broadly oblanceolate, 6–10 cm. long, 2–4 cm. wide, densely pale yellow-brown tomentose while young but later becoming glabrous beneath, reddish when dry, abruptly or short-acuminate, sparsely toothed or subentire, acute or subcuneate at base, the midrib impressed above, with 8–13 pairs of rather slender lateral nerves, the petioles 5–20 mm. long; stamens 10–20; fruiting involucres densely short-pubescent; nuts ellipsoidal, about 15 mm. long.——May. Honshu (Kantō Distr. and westw.), Shikoku, Kyushu.

Quercus × takaoyamensis Makino. A hybrid of No. 3 and No. 1.

4. **Quercus gilva** Bl. *Cyclobalanopsis gilva* (Bl.) Oerst. ——Ichiigashi. Evergreen; branches dark purple-brown, much-branched, the branchlets, petioles, and undersides of leaves yellow-brown stellate-villous; leaves coriaceous, oblanceolate to broadly so, 5–12 cm. long, 1.5–3 cm. wide, abruptly acuminate, acute or sometimes obtuse at base, with acute short-awned teeth on upper margin, stellate-pubescent but soon glabrate and deep green above, the midrib impressed above, the lateral nerves 13 to 18 pairs, somewhat raised beneath, the petioles 5–15 mm. long; stamens 8–10; fruiting involucres short-villous; nuts ellipsoidal, about 2 cm. long.—— May. Honshu (s. Kantō Distr. and westw.), Shikoku, Kyushu.——Formosa and China.

5. **Quercus myrsinaefolia** Bl. *Cyclobalanopsis myrsinaefolia* (Bl.) Oerst.; *Q. vibrayana* Fr. & Sav.; *Q. glauca* var. *nudata* Bl.——Shira-kashi. Glabrous sparsely leaved evergreen tree, with dark purple-brown branches; leaves coriaceous, lanceolate, sometimes narrowly oblong-ovate, 5–12 cm. long, 15–25(–30) mm. wide, gradually long-acuminate, subacute to rounded at base, with short mucronate teeth on upper half, lustrous above, prominently glaucous beneath, the midrib scarcely impressed above, the lateral nerves very slender, 10 to 16 pairs, only slightly raised beneath, the petioles 1–2 cm. long; fruiting involucres stipitate.——Apr.–May. Warmer parts of Honshu, Shikoku, Kyushu.

6. **Quercus glauca** Thunb. *Cyclobalanopsis glauca* (Thunb.) Oerst.; *Q. glauca* var. *caesia* Bl.——Ara-kashi. Evergreen; branches rather stout, dark brown, yellowish pubescent while young; leaves coriaceous, oblong, elliptic, to obovate-oblong, 7–13 cm. long, 2.5–6 cm. wide, abruptly acuminate, rounded to subacute at base, with acute teeth on upper half, lustrous and glabrous above, glaucous beneath, densely appressed brownish hairy while young, becoming sparsely so when fully developed, the midrib slightly impressed above, the lateral nerves 8 to 11 pairs, prominent beneath, the petioles 1.5–2(–3) cm. long, glabrous; stamens 10–15; fruiting involucres densely puberulous; nuts ellipsoidal, about 2 cm. long. ——Apr.–May. Honshu, Shikoku, Kyushu.——Formosa and China; common and variable; frequently planted, with many cultivars. Grown in gardens.

7. **Quercus salicina** Bl. *Cyclobalanopsis salicina* (Bl.) Oerst.; *Q. glauca* var. *stenophylla* Bl.; *Q. stenophylla* (Bl.) Makino; *C. stenophylla* (Bl.) Schottky; *Q. miyagii* sensu auct. Japon. pro parte, non Koidz.——Urajirogashi. Evergreen; branches rather slender, gray-brown or pale red-brown, much-branched, pale yellow-brown villous while young; leaves rather loosely arranged, lanceolate to broadly so, sometimes narrowly oblong, long-acuminate, obtuse to acute at base, with short upright rather long mucronate teeth on upper half, glabrous and lustrous above, yellow-brown pubescent while young becoming glaucous and glabrous beneath, the midrib impressed above, the lateral nerves 10 to 13 pairs, slightly raised beneath, the petioles 10–15 mm. long, glabrous; fruiting involucres densely puberulous; nuts ellipsoidal to oblong-ovoid, 15–22 mm. long.——May. Warmer regions; Honshu, Shikoku, Kyushu.——s. Korea.

8. **Quercus phillyraeoides** A. Gray. *Q. ilex* var. *phillyraeoides* (A. Gray) Franch.——Ubamegashi. Evergreen shrub or small tree with dense, dark gray-brown branches, the branchlets densely pale red-brown stellate-pubescent; leaves coriaceous, dark green, obovate-oblong to broadly so, 3–6 cm. long, 1.5–3 cm. wide, slightly lustrous, obtuse to subacute, rounded to shallowly cordate at base, undulately toothed except near base, the midrib scarcely impressed above, the lateral nerves 6 to 9 pairs, not prominent, the petioles 2–5 mm. long; stamens 4–5; fruiting involucres densely puberulous; nuts ovoid, 13–20 mm. long.——Apr.–May. Thickets and open woods near seashores; Honshu (Kantō Distr. and westw.), Shikoku, Kyushu.——Ryukyus, China. Variable; much planted in parks and for hedges.——The following are cultivated: Forma **crispa** (Matsum.) Kitam. & Horikawa. *Q. phillyraeoides* var. *crispa* Matsum.——Biwabagashi, Chirimegashi. Leaves crisped.——Forma **wrightii** (Nakai) Makino. *Q. wrightii* Nakai; *Q. phillyraeoides* var. *wrightii* (Nakai) Masam.——Ke-ubamegashi. Leaves densely stellate-hairy on under side.

9. **Quercus mongolica** Fisch. Mongori-nara. Deciduous tree with rather thick, glabrous branches; leaves obovate or broadly so, 10–17 cm. long, 6–12 cm. wide, densely aggregated toward the tip of branchlets, obtuse to subacute with an obtuse tip, auriculate at base, glabrous and deep green above, pale green and sometimes slightly glaucescent beneath with scattered hairs on nerves or nearly glabrous, with coarse deeply undulate obtuse teeth, the petioles 0–5 mm. long; fruiting involucres subsessile, with keeled incurved and appressed scales; nuts ovoid-ellipsoid.——June–July. Woods in the temperate and cooler regions; Hokkaido, Honshu (Tanba Prov. and eastw., centr. and n. distr.).——s. Kuriles, Sakhalin, Manchuria, Korea, e. Mongolia, and e. Siberia.

Var. **grosseserrata** (Bl.) Rehd. & Wils. *Q. crispula* Bl.;

Q. grosseserrata Bl.; *Q. crispula* var. *grosseserrata* (Bl.) Miq. ——Mizu-nara. Leaves obovate to narrowly so, acute to subacute, with rather acute teeth; involucral scales less prominently keeled and strongly appressed.——May–June. Mountains; Hokkaido, Honshu, Shikoku, Kyushu.——Sakhalin and s. Kuriles.

10. Quercus serrata Thunb. *Q. glandulifera* Bl.; *Q. canescens* Bl.; *Q. urticaefolia* Bl.——Ko-nara, Hahaso, Nara. Large or small deciduous tree with rather slender branchlets pubescent while young; leaves obovate to narrowly ovate, sometimes oblong, rarely broadly lanceolate, 6–15 cm. long, 2.5–7 cm. wide, acute, cuneate to acute or subrounded at base, coarsely mucronate-toothed, with grayish white appressed-pubescence especially beneath, deep green and glabrate above, the petioles 5–15 mm. long; fruiting involucres with short appressed-hairs; nuts ellipsoidal or oblong-ovoid, 15–20 mm. long.——Apr.–May. Hokkaido, Honshu, Shikoku, Kyushu; very common and highly variable.——s. Kuriles, Korea, and China.

Var. **donarium** (Nakai) Kitam. & Horikawa. *Q. donarium* Nakai——Teriha-ko-nara. Leaves narrower and prominently lustrous above.——Occurs with the typical phase.

Quercus neostuxbergii Koidz. Ao-nara. Poorly known and possibly synonymous with *Q. serrata*.

11. Quercus aliena Bl. *Q. acutidentata* (Maxim.) Koidz.; *Q. aliena* var. *acutidentata* Maxim.——Naragashiwa. Deciduous tree with rather stout glabrous branches; leaves oblong-obovate or obovate, 10–25(–35) cm. long, obtuse to subacute, undulate-toothed, glabrous above, densely brownish or grayish stellulate-pubescent beneath, the petioles 1–3 cm. long, glabrous; fruiting involucres axillary, whitish puberulent; nuts ellipsoidal or ovoid, 2–2.5 cm. long.——Apr. Honshu (Kinki Distr. and westw.), Shikoku, Kyushu.

Hybrids of Nos. 9 and 10 are: **Quercus × pellucida** (Bl.) Nakai. Ō-mizu-nara; and **Quercus × major** Nakai. Ōba-ko-nara.

12. Quercus dentata Thunb. *Q. obovata* Bunge——Kashiwa. Deciduous tree with stout short gray-yellow villous branches; leaves obovate to broadly so, 10–30 cm. long, 6–18 cm. wide, obtuse, narrowed and auriculate below, with few coarse rounded teeth, pubescent at first on the nerves above, densely pale grayish with simple or sometimes stellate hairs beneath, the lateral nerves 4 to 10 pairs, the petioles 2–5 mm. long, not prominent; fruiting involucres fascicled, subsessile, enveloping more than half the nut, the scales broadly linear, dense, the upper ones to 12 mm. long, white-puberulous outside; nuts ovoid-globose, 15–20 mm. long.——May. Sunny slopes and thickets, lowlands to mountains; Hokkaido, Honshu, Shikoku, Kyushu.——s. Kuriles, Korea, and China.

Hybrids of Nos. 10 and 12 are **Quercus × nipponica** Koidz. *Q. angustelepidota* Nakai——Kashiwa-ko-nara; and **Quercus × takatorensis** Makino. Ko-gashiwa.

13. Quercus variabilis Bl. *Q. bungeana* Forbes——Abe-maki. Deciduous tree, the branches soon becoming glabrous; leaves narrowly ovate-oblong, 8–15 cm. long, 2.5–5 cm. wide, short-acuminate with a short awn at apex, rounded to broadly cuneate at base, obsoletely undulate-toothed, with 9 to 16 pairs of lateral nerves, the marginal teeth awn-tipped, glabrous above, densely pale yellow-brown or grayish stellate-pubescent beneath, the petioles glabrous, 10–25 mm. long; fruiting involucres nearly sessile, slightly shorter than the nut, the scales linear, elongate, dense; nuts ovoid-globose, 15–20 mm. long.——Apr.–May. Honshu (centr. distr. and westw.), Shikoku, Kyushu.——Korea and China.

14. Quercus acutissima Carruth. *Q. serrata* sensu auct. Japon., olim, non Thunb.——Kunugi. Deciduous tree, the branches pubescent, becoming glabrate; leaves narrowly linear-oblong, 8–15 cm. long, 2–4 cm. wide, acute, rounded to broadly cuneate at base, pubescent on both sides at first, soon becoming glabrous above, pale green and nearly glabrate beneath except for axillary tufts of hairs, the lateral nerves 12 to 16 pairs, marginal teeth awn-tipped, the petioles 1–2 cm. long; fruiting involucres sessile, enveloping more than half the nut, the scales broadly linear, slightly reflexed, dense; nuts subglobose, about 2 cm. long and as wide.——May. Honshu, Shikoku, Kyushu.——Korea and China.

3. CASTANEA Mill. Kuri Zoku

Monoecious deciduous trees, rarely shrubs, the branches without terminal buds; leaves petiolate, alternate, simple, toothed, with parallel lateral nerves; staminate flowers on erect or spreading slender aments; perianth 6-merous; stamens 10–20; pistillate flowers usually subtending the staminate aments, usually in fascicles of 3; ovary 6-locular; styles 6, linear; ovules 2 in each locule; nuts 1–3(–7), brown, enclosed in spiny 2- to 4-lobed involucres.——About 10 species, in the temperate regions of the N. Hemisphere.

1. Castanea crenata Sieb. & Zucc. *C. stricta* Sieb. & Zucc.; *C. vesca* var. *pubinervis* Hassk.; *C. pubinervis* (Hassk.) C. K. Schn.; *C. japonica* Bl.——Kuri. Tree, the branches densely gray-white short-hairy while young; leaves oblong-lanceolate or narrowly oblong, 8–15 cm. long, 3–4 cm. wide, acuminate, obtuse to shallowly cordate at base, undulately toothed with 15 to 20 pairs of lateral nerves that project beyond the margin as short awns, puberulous on nerves above, glabrescent to densely pale yellowish brown soft hairy, with discoid sessile glands beneath; staminate aments 10–15 cm. long, densely flowered, yellowish white, erect, or suberect; fruiting involucres with long subglabrous spines; nuts 2–3, the hilum occupying the whole basal area; styles about 3 mm. long.——July. Foothills; Hokkaido (sw. distr.), Honshu, Shikoku, Kyushu; much planted for the nuts.

Castanea mollissima Bl. *C. bungeana* Bl.——Shinaguri. Leaves coarsely toothed; sessile discoid glands absent beneath; hilum smaller; young branchlets and petioles puberulent and spreading pilose; sometimes cultivated in our area.——Korea and China.

4. CASTANOPSIS Spach Shii-no-ki Zoku

Evergreen monoecious trees with entire or toothed, coriaceous leaves; staminate flowers in erect aments, the perianth 5- to 6-merous; stamens 10 or 12; pistillate flowers 1–3 in an involucre; ovary 3-locular; styles 3, the stigmas minute; fruit wholly enclosed in an ovoid to globose, sometimes fissured, spinose or tuberculate, or transversely striate involucre, the nut maturing the second year.——About 30 species, chiefly in e. and s. Asia, 1 species in western N. America.

1. Castanopsis cuspidata (Thunb.) Schottky. *Quercus cuspidata* Thunb.; *Pasania cuspidata* (Thunb.) Oerst.; *P. cuspidata* var. *thunbergii* Makino; *Lithocarpus cuspidata* (Thunb.) Nakai; *Pasaniopsis cuspidata* (Thunb.) Kudo; *Shiia cuspidata* (Thunb.) Makino——TSUBURA-JII. Much branched evergreen tree, the branchlets with scurfy appressed hairs while young; leaves rather coriaceous, ovate-oblong, broadly lanceolate or oblong, sometimes ovate, usually 5–10 cm. long, 2–3 cm. wide, long-acuminate, acute to subrounded at base, undulately toothed on upper margin, smooth, glabrous and deep green above, gray-brown to brownish beneath with appressed brownish scales, the petioles 5–10 mm. long, soon becoming smooth; staminate aments 5–10 cm. long; pistillate aments on the upper part of the branches; fruiting involucre sub-globose, sessile, with small protuberances arranged in 1 or 2 series of transverse lines, densely soft grayish puberulent; nuts subglobose to ovoid-globose, 8–10 mm. in diameter.——May-June. Warmer parts; Honshu (Kantō Distr. and westw.), Shikoku, Kyushu.——Korea.

Var. **sieboldii** (Makino) Nakai. *Pasania cuspidata* var. *sieboldii* Makino; *Pasania sieboldii* (Makino) Makino; *Lithocarpus cuspidata* var. *sieboldii* (Makino) Nakai; *Shiia sieboldii* (Makino) Makino; *Pasaniopsis sieboldii* (Makino) Kudo——SUDA-JII. Branches thicker; leaves thicker and the nuts larger, ovoid-oblong; bark fissured in younger trees.——Warmer areas; Honshu (Kantō Distr. and westw.), Shikoku, Kyushu; common in environs of Tokyo.——Korea.

5. PASANIA Oerst.   MATEBA-SHII ZOKU

Evergreen trees with entire or toothed, coriaceous, petiolate leaves; staminate flowers in slender erect aments, the perianth 4- to 6-merous; stamens 6–12, longer than the perianth; pistillate flowers 1–3 or 5 together, subtending the staminate aments or borne separately, involucrate; ovary 3-locular; styles 3, erect, cylindric with a punctiform stigma; nuts wholly or partially involucrate, maturing the second year, the involucres with annular rings or scales.——About 100 species, in e. and s. Asia.

1A. Young branches nearly glabrous; leaves yellowish green beneath, with about 12 pairs of lateral nerves. . . . . 1. *P. edulis*
1B. Young branches densely short-pubescent; leaves silvery beneath, with about 7 pairs of lateral nerves. . . . . 2. *P. glabra*

1. Pasania edulis Makino. *Quercus edulis* Makino; *Lithocarpus edulis* (Makino) Nakai; *Q. glabra* var. *micrococca* Bl.; *Q. glabra* var. *sublepidota* Bl.; *L. sublepidota* (Bl.) Koidz.——MATEBA-SHII. Branches stout, glabrous or nearly so; leaves coriaceous, oblanceolate to obovate-oblong, sometimes oblong, 8–16 cm. long, 3–7 cm. wide, acute with an obtuse tip, acute at base, glabrous except the ascending pubescence on midrib beneath and on petioles while young, lustrous above, obsoletely scaly and somewhat brownish beneath, with 10 to 12 pairs of lateral nerves, the petioles 1.5–3 cm. long; staminate aments linear, erect; nuts narrowly ovoid to oblong, 2–2.5 cm. long, the involucre shallow, with small imbricate scales.——June. Honshu (s. Kantō Distr. and westw.), Shikoku, Kyushu.——Ryukyus.

2. Pasania glabra (Thunb.) Oerst. *Quercus glabra* Thunb.; *Q. sieboldiana* Bl.; *Lithocarpus glabra* (Thunb.) Nakai; *Kuromatea glabra* (Thunb.) Kudo; *P. sieboldiana* (Bl.) Nakai——SHIRIBUKAGASHI. Branches rather stout, with densely grayish or yellowish brown pubescence while young; leaves coriaceous, oblong, sometimes broadly oblanceolate, 10–15 cm. long, 4–6 cm. wide, abruptly acute to acuminate with an obtuse tip, acute at base, pubescent on both surfaces while very young, glabrescent and lustrous above, silvery beneath with obscure lepidote scales, sometimes with few obscure undulate teeth on upper margin, the lateral nerves 6 to 8 pairs, the petioles pubescent, 8–10 mm. long; staminate aments erect, densely flowered, linear; nuts broadly ovoid or ellipsoidal, 1.5–2 cm. long, depressed at base, the involucre shallow, with small adnate scales.——Oct.–Nov. Honshu (Kinki and Chūgoku Distr.), Shikoku, Kyushu.——Ryukyus and Formosa.

Fam. 67. ULMACEAE   NIRE KA   Elm Family

Monoecious or hermaphroditic shrubs or trees sometimes with short spine-tipped spurs; leaves simple, petiolate, alternate, the stipules deciduous; flowers bisexual or unisexual, axillary and fasciculate, solitary or cymose; perianth 4- to 5-(–8) parted; stamens as many as the perianth lobes and opposite them, the filaments erect; ovary superior, 1-locular, the ovules anatropous, pendulous; styles solitary and bifid or 2; fruit a samara, a drupe or resembling an achene; seeds lacking endosperm, the embryo straight or curved.——About 15 genera, with about 200 species, chiefly in the N. Hemisphere.

1A. Leaves pinnately nerved, the lowest pair much shorter than the median ones; perianth-segments connate below; fruit dry.
  2A. Flowers bisexual; fruit a samara. . . . . 1. *Ulmus*
  2B. Flowers unisexual; fruit achenelike. . . . . 2. *Zelkova*
1B. Leaves palmately 3-nerved from base; perianth-segments nearly free or very shortly connate below; fruit a drupe.
  3A. Flowers cymose, unisexual or polygamous; leaves serrulate, with lateral nerves not ending in the marginal teeth. . . . . . . 3. *Trema*
  3B. Flowers unisexual, the pistillate solitary or few, in axillary fascicles, the staminate cymose.
    4A. Lateral nerves of leaves arcuate near the margin, not ending in the teeth. . . . . 4. *Celtis*
    4B. Lateral nerves of leaves straight, ending in the teeth. . . . . 5. *Aphananthe*

1. ULMUS L.   NIRE ZOKU

Deciduous or semievergreen trees and shrubs; leaves simple, toothed, petiolate, sometimes shallowly 2- to 3-fid at the apex, usually oblique at base, the stipules lateral, free; flowers bisexual, small, precocious or in leaf axils in autumn, fasciculate or racemose, the perianth campanulate, 4- to 9-merous; stamens as many as the perianth-lobes; styles short, deeply bifid; fruit a broadly winged pedicelled pendulous samara; seeds flat, endosperm absent, the embryo straight.——About 20 species, in the temperate regions of the N. Hemisphere.

1A. Leaves 2–5 cm. long, 1–2 cm. wide, with simple rather obtuse teeth; samara 8–13 mm. long; flowers in autumn in leaf-axils; perianth deeply lobed. .... 1. *U. parvifolia*
1B. Leaves 3–13 cm. long, 2–7 cm. wide, with usually acute, sometimes double teeth; flowers precocious, before the leaves; perianth shallowly lobed.
2A. Seeds on upper portion of the samara; leaves unlobed, the petioles 4–12 mm. long. .... 2. *U. davidiana* var. *japonica*
2B. Seeds at the center of the samara; leaves usually shallowly 3- to 5-(–9) lobed above, the petioles 3–5 mm. long. .... 3. *U. laciniata*

1. Ulmus parvifolia Jacq. *U. sieboldii* Daveau; *U. shirasawana* Daveau——Aki-nire. Small much-branched tree with puberulent branches; leaves rather firm, lustrous, obliquely oblong, obovate or narrowly ovate, 2–5 cm. long, 1–2 cm. wide, obtuse to acute, obliquely obtuse at base, simply toothed, glabrous except puberulent on the midrib above, with 7 to 15 pairs of lateral nerves, the petioles 4–12 mm. long; samaras elliptic or nearly orbicular, short-pedicellate, glabrous, 8–13 mm. long, the seed near the center.——Sept. Honshu (centr. distr. and westw.), Shikoku, Kyushu.——Korea, Formosa, and China.

2. Ulmus davidiana Planch. var. **japonica** (Rehd.) Nakai. *U. campestris* sensu auct. Japon., non L.; *U. campestris* var. *japonica* Rehd.; *U. japonica* Sarg., non Sieb.; *U. propinqua* Koidz.——Haru-nire, Nire. Deciduous tree with densely rusty-pubescent branches; leaves obovate, sometimes narrowly so, or elliptic, 3–10 cm. long, 2–6 cm. wide, doubly-toothed, abruptly acuminate, usually obliquely cuneate-obtuse at base, scabrous above, pilose beneath at least on nerves, with 7–13 pairs of lateral nerves, the petioles 4–12 mm. long; sam-

aras obovate, glabrous, 10–16 mm. long, the seeds on upper portion close to the base of style.——Apr.–June. Hokkaido, Honshu, Shikoku, Kyushu; rare.——Forma **suberosa** Nakai has corky branches.——Forma **kijimae** (Makino) Sugimoto. *U. kijimae* Makino——Tsukushi-nire. Leaves glabrous. ——Kyushu. The typical variety occurs in Sakhalin, s. Kuriles, Korea, e. Siberia, and n. China.

3. Ulmus laciniata (Trautv.) Mayr. *U. montana* var. *laciniata* Trautv.——Ohyō, Atsuni, Atsushi. Tree with pale brown branches puberulous when young; leaves rather thin, obovate to broadly obovate-cuneate, 8–13 cm. long, 5–7 cm. wide, usually shallowly 3- to 9-lobed toward the tip, abruptly acuminate, obtuse or cuneately narrowed to an obtuse base, acutely double-toothed, scabrous and short appressed-puberulent above, pilose beneath, with 10 to 17 pairs of lateral nerves; samaras elliptic, glabrous, about 15 mm. long, the seed nearly at the center.——May–June. Temperate forests; Hokkaido, Honshu, Kyushu.——Korea, n. China, e. Siberia, and Kamchatka.

2. ZELKOVA Spach Keyaki Zoku

Deciduous trees or shrubs; leaves short-petiolate, alternate, stipulate, penninerved, simple-toothed; flowers polygamous, 4- to 5-merous, short-pedicelled, the staminate in the lower, the bisexual ones in the upper axils of young branchlets; stamens 4–5; style lateral; drupe short-pedicelled, small, oblique, usually rather flattened, the cotyledons broad.——Few species in e. and w. Asia.

1. Zelkova serrata (Thunb.) Makino. *Corchorus serratus* Thunb.; *Ulmus keakii* Sieb.; *Planera acuminata* Lindl.; *Z. acuminata* (Lindl.) Planch.; *Z. keakii* (Sieb.) Maxim.—— Keyaki. Tall tree with gray-brown smooth bark and much divided slender branches, the branchlets brownish red, puberulent, the lateral ones deciduous in autumn; leaves membranous, narrowly ovate or narrowly ovate-oblong, 3–7 cm. long (sometimes to 12 cm. long in young vigorous shoots), 1–2.5 (–5) cm. wide, long-acuminate, shallowly cordate to rounded at base, slightly scabrous or nearly smooth and dull above,

acutely mucronate-toothed, thinly pilose on nerves beneath while young, the petioles 1–3 mm. long, puberulous on upper side; fruit sessile, nerved, about 5 mm. across, obliquely ovate-orbicular, glabrous.——Apr.–May. Lowlands and mountains; Honshu, Shikoku, Kyushu; frequently planted around houses. ——Korea, China, and perhaps Formosa.

Var. **stipulacea** Makino. *? Zelkova schneideriana* Hand.-Mazz.——Me-geyaki. Leaves pilose on both surfaces.—— Honshu.

3. TREMA Lour. Urajiro-enoki Zoku

Shrubs or trees; leaves alternate, rather thin, serrulate, palmately 3- to 7-nerved, the stipules lateral, free, deciduous; flowers unisexual or polygamous, small, in small axillary cymes, 4- to 5-merous; stamens erect in bud; ovary sessile; style terminal, bifid, the stigma lobes linear; ovules pendulous; drupe small, erect, ovoid to nearly globose, with thick endocarp, the endosperm fleshy, the embryo curved.——About 20 species in the Tropics and subtropics.

1A. Leaves whitish and densely sericeous-pubescent beneath. .... 1. *T. orientalis*
1B. Leaves green beneath. .... 2. *T. cannabina*

1. Trema orientalis (L.) Bl. *Celtis orientalis* L.; *Sponia orientalis* (L.) Planch.——Urajiro-enoki. Small tree with densely appressed-pilose elongate branches; leaves narrowly ovate-oblong to broadly lanceolate, 7–15 cm. long, 1.5–5 cm. wide, gradually acuminate, obliquely cordate at base, obtusely serrulate, scaberulous and appressed short-hairy above, white and densely silky-pubescent beneath, palmately 3- to 5-nerved, the petioles 8–12 mm. long; cymes longer or shorter than the petioles, rather many-flowered; fruit ovoid-globose, black, 3–4

mm. across.——Kyushu (lowlands in Yakushima and Tanegashima).——Ryukyus, Formosa, s. China to India, Malaysia, and Australia.

2. Trema cannabina Lour. *Celtis amboinensis* Willd.; *T. amboinensis* (Willd.) Bl.; *T. virgata* Bl.——Kiri-enoki. Closely resembles the preceding; leaves green, glabrous to hirsute beneath.——Kyushu (s. distr. to Tanegashima and Yakushima).——Ryukyus, Formosa, s. China, Malaysia, India, and Australia.

4. CELTIS L.　　ENOKI ZOKU

Usually deciduous trees; leaves alternate, entire or toothed, often oblique at base, penninerved, with 3 main nerves from the base, the stipules free, lateral; inflorescence cymose, staminate or polygamous, axillary or from the base of young branchlets, the flowers 5(or 4)-merous, the bisexual 1–3, axillary, pedicelled; perianth-segments and stamens usually 5; ovary sessile; style bifid, the lobes subplumose and often twice bilobed; ovule pendulous, anatropous; drupe juicy, ovoid to globose, the endocarp bony, often rugose or reticulate; seeds with scanty endosperm, the embryo curved.——About 70 species, in temperate to tropical regions of the N. Hemisphere.

1A.　Fruit black when mature, the pedicels 20–25 mm. long; leaves with prominent teeth. . . . . 1. *C. jessoensis*
1B.　Fruit brown or red-brown when mature, the pedicels 5–15 mm. long; leaves with shallow not prominent teeth.
　　2A.　Leaves 5–10 cm. long, 2.5–5 cm. wide, acute or short-acuminate, relatively thin, glabrous or nearly so above.
　　　2. *C. sinensis* var. *japonica*
　　2B.　Leaves 3–7 cm. long, 2–3(–3.5) cm. wide, caudately long-acuminate, relatively thick, firm, more or less scabrous and appressed-
　　　　hairy above. . . . . 3. *C. leveillei*

1.　Celtis jessoensis Koidz. *C. aphananthoides* Koidz.; *C. bungeana* var. *jessoensis* (Koidz.) Kudo; *C. hashimotoi* Koidz.——EZO-ENOKI. Tree with puberulous branches; leaves narrowly or sometimes broadly ovate, 6–10 cm. long, 3–5 cm. wide, acuminate, obliquely rounded to obtuse at base, with incurved teeth except near base, scaberulous or nearly glabrous above, glaucous or in young individuals pale green beneath, short-hirsute especially on nerves, the petioles 2–7 mm. long, puberulous; drupe black when mature, 6–7 mm. across, the pedicels 2–2.5 cm. long, glabrous.——Hokkaido, Honshu, Shikoku, Kyushu.——Korea.

2.　Celtis sinensis Pers. var. **japonica** (Planch.) Nakai. *C. willdenowiana* Schult.; *C. japonica* Planch.——ENOKI. Tree with pale gray-brown branches and appressed-pilosulous branchlets; leaves oblique, broadly ovate to ovate-oblong, 5–10 cm. long, 3.5–6 cm. wide, acute to short-acuminate, obtuse at base, with short incurved teeth toward the tip, the lower margin entire, smooth or slightly scabrous and glabrous above, paler and thinly yellow-brown pilose to glabrous beneath; drupe red-brown, broadly ellipsoidal, 6–7 mm. across, the pedicels 6–15 mm. long.——Apr.–May. Lowlands and hills; Honshu, Shikoku, Kyushu; very common.——Korea and China. The typical phase occurs in s. China.

3.　Celtis leveillei Nakai. KOBA-NO-CHŌSEN-ENOKI. Tree with gray-brown branches densely yellow-brown appressed-hairy while young; leaves rather thick and firm, obliquely oblong, obovate or elliptic, 3–7 cm. long, 2–3.5 cm. wide, caudately long-acuminate, cuneate and acute or rarely rounded at base, with rather coarse appressed teeth on upper half, both surfaces appressed-pilose, scabrous above, pale-white beneath; drupe subglobose, brown at maturity, about 6 mm. long, the pedicels pilose, 5–10 mm. long.——Honshu (Kinki Distr. and westw.), Kyushu; rather rare.——China and Korea.

Var. **holophylla** Nakai. CHŪGOKU-ENOKI. Leaves obsoletely or not prominently toothed.——Honshu (Chūgoku Distr.), Shikoku.

5. APHANANTHE Planch.　　MUKU-NO-KI ZOKU

Deciduous trees; leaves petiolate, alternate, toothed, with the largest lateral nerves from the base, the stipules free and lateral; staminate inflorescence cymose, axillary near base of young branchlets, rather many-flowered; flowers 4- or 5-merous; pistillate flowers solitary in upper axils, pedicelled, the segments of pistillate perianth narrower; ovary sessile; style bifid, the stigmas (style-branches) linear, rather thick; ovules pendulous; drupe ovoid to subglobose, the pericarp juicy, the endocarp hard, the endosperm scanty or absent, the embryo incurved.——Few species, in e. Asia, Malaysia, and Australia.

1.　Aphananthe aspera (Thunb.) Planch. *Prunus aspera* Thunb.; *Homoioceltis aspera* (Thunb.) Bl.——MUKU-NO-KI. Tall tree with scabrous branches when young; leaves rather thin, ovate to narrowly so, 5–10 cm. long, 3–6 cm. wide, long-acuminate, obtusely rounded at base, acutely toothed, 3-nerved from the base, scabrous and green above, short appressed-pilose beneath, the lateral nerves 7 to 12 pairs, ending in a tooth, the petioles 5–10 mm. long; flowers green, small; drupe ovoid-globose, short appressed-pilose, black when mature, 7–8 mm. across, the pedicels 7–8 mm. long.——May. Lowlands and hills; Honshu, Shikoku, Kyushu; common.——Korea and China.

Fam. 68.　MORACEAE　　KUWA KA　　Mulberry Family

Evergreen or deciduous, dioecious or monoecious trees, shrubs or rarely herbs, often with colored juice; leaves alternate, entire or toothed, sometimes lobed, the stipules prominent, deciduous; inflorescence basically cymose; flowers small, actinomorphic, spicate, capitate, or on the inner side of a fleshy urceolate receptacle (syconium of *Ficus*); staminate flowers 2- to 4(–6)-merous, the stamens opposite the segments, the filaments erect or incurved in bud; pistillate perianth 4-lobed or -parted; stigmas 1–2; ovary superior or inferior, with a solitary pendulous ovule; fruit a small achene or drupe, the perianth and axis sometimes aggregated or connate, forming a syncarp, the embryo curved.——About 55 genera, with more than 1,000 species, chiefly in the Tropics, few in the temperate regions of the world.

1A.　Trees or erect herbs; leaves alternate.
　　2A.　Flowers variously arranged but not on the inner side of a fleshy urceolate receptacle.
　　　　3A.　Stamens incurved in bud; branches unarmed.
　　　　　　4A.　Herbs; inflorescence cymose. . . . . 1. *Fatoua*
　　　　　　4B.　Trees or shrubs; inflorescence in spikes or heads.

1. FATOUA Gaudich.   KUWA-KUSA ZOKU

Annual monoecious herbs with alternate, petiolate, simple, toothed leaves, the stipules deciduous; cymes densely flowered, capitate, nearly axillary; staminate flowers 4-merous, the filaments incurved in bud; pistillate perianth usually 6-lobed; ovary oblique; style sublateral, filiform, unbranched, dentate, or branched; ovules pendulous; achenes obliquely globose, slightly flattened, enclosed in a persistent perianth, the pericarp crustaceous, the embryo curved.——Few species, in Australia, Malaysia, and e. Asia.

1. Fatoua villosa (Thunb.) Nakai. *Urtica villosa* Thunb.; *U. japonica* Thunb., non L. f.; *F. aspera* Sieb. & Zucc.; *F. japonica* (Thunb.) Bl.——KUWA-KUSA.   Erect, branched, short-pilose annual 30–80 cm. high; leaves membranous, broadly ovate, 5–10 cm. long, 3–5 cm. wide, acuminate, subtruncate at base, obtusely toothed, coarsely appressed-pilose above, short-pilose beneath, often slightly scabrous on both surfaces, the petioles rather slender; flowers small, greenish, rather numerous, in axillary glomerules; achenes about 1 mm. across, compressed-trigonous, dispersed with minute raised points.——Sept.–Oct. Grassy places and cultivated fields in lowlands; Honshu, Shikoku, Kyushu; common.——Ryukyus and China.

2. MORUS L.   KUWA ZOKU

Dioecious or monoecious deciduous shrubs or trees, unarmed, with milky juice; leaves often 3- to 5-lobed, toothed; staminate aments rather long, many-flowered; stamens 4, filaments incurved in bud; pistillate aments elongate or short, the perianth 4-parted; ovary sessile; style terminal, bifid, the stigma lobes linear; ovules pendulous; fruit an achene, enclosed by a persistent juicy perianth; seeds subglobose, with a membranous testa, the endosperm fleshy, the embryo straight.——More than 10 species, in the temperate to warmer regions of the N. Hemisphere.

1. Morus tiliaefolia Makino. KEGUWA.   Tree with rather stout, coarsely pilose, gray-brown branches; leaves broadly ovate or ovate-orbicular, 10–22 cm. long, 6–18 cm. wide, abruptly caudate-acuminate, cordate at base, obtusely toothed, often 3-lobed, deep green, scabrous above, prominently short-pilose beneath especially on nerves, the petioles densely soft-puberulent, 3–5 cm. long; spikes short-cylindric, the pistillate 15–20 mm. long.——Apr.–May. Honshu (Chūgoku Distr.), Shikoku, Kyushu.——Korea.

2. Morus bombycis Koidz. *M. japonica* L. H. Bailey, non Sieb.; *M. kagayamae* Koidz.——YAMAGUWA.   Tree or large shrub with glabrous or puberulous branches while young; leaves ovate or broadly so, 8–20 cm. long, 5–12 cm. wide, short-acuminate, rounded to shallowly cordate at base, nearly glabrous and smooth or slightly scabrous above, puberulous beneath especially on nerves, irregularly toothed, often 3-lobed, vivid to deep green above, the petioles nearly glabrous, 1–4 cm. long; staminate spikes short-cylindric, 2–3 cm. long; fruiting spikes ellipsoidal, purple-black.——Apr.–May. Mountains; Hokkaido, Honshu, Shikoku, Kyushu.——Korea, Sakhalin, and s. Kuriles. Much used in feeding silk worms.

3. Morus australis Poir. *M. acidosa* Griff.——SHIMAGUWA.   Tree with nearly glabrous branches; leaves ovate or broadly so, 8–18 cm. long, 5–15 cm. wide, rounded to shallowly cordate at base, larger in vigorous shoots, with short rather regular teeth, sometimes 3-lobed, glabrescent on both sides or puberulous beneath on nerves, slightly scabrous above, the petioles 2–6 cm. long, nearly glabrous; staminate spikes short-cylindric; fruiting spikes ellipsoidal.——Kyushu (Yakushima and Tanegashima).——China, Formosa, and Ryukyus.

3. BROUSSONETIA Vent.   KŌZO ZOKU

Dioecious or monoecious trees or shrubs, sometimes scandent; leaves alternate, more or less scabrous, toothed, sometimes 3- to 5-lobed, the stipules free, membranous, deciduous; staminate flowers 4-merous, the perianth-lobes valvate in bud, the filaments incurved in bud; pistillate perianth ovoid or tubular, 3- to 4-toothed, entirely surrounding the ovary; style simple, filiform, with stigmatic hairs from near base; fruiting spikes capitate; fruit stipitate, small, fleshy, exserted from the persistent perianth at maturity, the endocarp seedlike and crustaceous.——Few species, in e. and se. Asia.

1. Broussonetia papyrifera (L.) Vent. *Morus papyrifera* L.——KAJI-NO-KI.   Small tree with stout gray-brown coarsely spreading hirsute branches while young; leaves thick-herbaceous, oblique, broadly ovate, 7–18 cm. long, 5–15 cm. wide, short-acuminate, shallowly cordate at base, coarsely hirsute, dull, toothed, often deeply 3- to 5-lobed, deep green and scabrous above, pale green beneath, the stipules broadly lanceolate or obliquely narrow-ovate, acuminate, deciduous; staminate spikes cylindric, densely many-flowered, 4–8 cm. long, the peduncles 2–4 cm. long, pendulous, the perianth coarsely hairy; pistillate spikes densely many-flowered, globose, short pedunculate, about 2 cm. across, villous; style slender, filiform; drupe exserted from the perianth when mature, red, oblanceolate, the style persistent.——May–June.   Honshu (warmer part of centr. distr. and westw.), Shikoku, Kyushu; cultivated for making paper.——Ryukyus, Formosa, China, and Malaysia.

2. Broussonetia kazinoki Sieb.   *B. sieboldii* Bl.—— Kōzo.   Monoecious shrub with brownish bark, the branches slender, elongate, purplish, short-pilose while young; leaves thinly membranous, oblique, ovate or broadly so, 5–15 cm. long, 3–8 cm. wide, long-acuminate, rounded to shallowly cordate at base, slightly scabrous above, short-pilose while young beneath, toothed, the petioles 3–20 mm. long, short-pilose, the stipules narrowly lanceolate; staminate spikes 1–1.5 cm. long, the peduncles about 1 cm. long; pistillate spikes globose, rather many-flowered, the peduncles nearly as long as the petioles; styles filiform, elongate; fruiting heads globose, red, the fruits stipitate, with a persistent style.——Apr.–May. Honshu, Shikoku, Kyushu; often cultivated for paper making. ——Ryukyus, Formosa, China, and Korea.

3. Broussonetia kaempferi Sieb.   TSURU-KŌZO.   Closely resembles the preceding; dioecious, with much-elongate and scandent branches; leaves ovate to narrowly so, acuminate, toothed, rounded at base, subglabrous above, only sparingly short-pilose beneath.——Apr.–May.   Shikoku, Kyushu.

4. CUDRANIA Tréc.   HARIGUWA ZOKU

Dioecious, branching, sometimes scandent shrubs, the branchlets often reduced to spines; leaves alternate, short-petiolate, entire or sometimes 3-lobed, penninerved, the stipules small; inflorescence axillary, glomerate, short-peduncled, the bracteoles many; staminate flowers 4-merous, the perianth segments oblong, obtuse; pistillate perianth broader, imbricate in bud, enveloping the ovary; styles simple, undivided or bifid, the stigmatose branches elongate, filiform or rather thick; ovules pendulous; fruiting perianth and bracts accrescent, rather fleshy, forming a false headlike syncarp; fruit ovoid, rather flat, with crustaceous pericarp. ——Few species in tropical Asia.

1A. Leaves oblong to obovate, rather coriaceous, glabrescent, the midrib impressed above especially toward the base; petioles becoming glabrous; peduncles of staminate heads 4–6 mm. long. .... 1. *C. cochinchinensis* var. *gerontogea*
1B. Leaves ovate to broadly so, rather chartaceous, more or less pilose on nerves, the midrib scarcely impressed above; petioles puberulous on upper side; peduncles of staminate heads 10–12 mm. long. .... 2. *C. tricuspidata*

1. Cudrania cochinchinensis (Lour.) Kudo & Masam. var. **gerontogea** (Sieb. & Zucc.) Kudo & Masam.   *Maclura gerontogea* Sieb. & Zucc.——KA-KATSUGA-YU.   Evergreen shrub; branches elongate, scandent, with erect spines in axils, woolly-pubescent on buds and stipules; leaves oblong to obovate, 4–7 cm. long, 2–3.5 cm. wide, entire, subcoriaceous, very obtuse to rounded with a minute awn at apex, slightly recurved on margin, the petioles 5–20 mm. long; flower heads axillary, the peduncles 3–6 mm. long, appressed-puberulent. ——May–July. Honshu (Kinki and Chūgoku Distr.), Shikoku, Kyushu.——Formosa and China. The typical phase occurs in e. Africa to India, Malaysia, and Australia.

2. Cudrania tricuspidata (Carr.) Bureau.   *Maclura tricuspidata* Carr.; *Cudrania triloba* (Hance) Forbes & Hemsl.; *Cudranus trilobus* Hance——HARIGUWA.   Small puberulent tree; leaves membranous to chartaceous, ovate to broadly so, 6–10 cm. long, 3–6 cm. wide, entire or 3-lobed, abruptly acute to acuminate, with an obtuse tip, puberulous especially on midrib above and on nerves beneath; petioles 15–25 mm. long, puberulous on upper side; peduncles of staminate heads 10–12 mm. long, densely soft-pubescent.——June. Rarely planted in our area.——Korea and China.

5. FICUS L.   ICHIJIKU ZOKU

Evergreen or deciduous, erect or scandent trees or shrubs, with milky juice; leaves alternate, entire or toothed, often lobed, the stipules caducous; receptacle fleshy, axillary or from the nodes of denuded branches, globose to ovate, usually enveloping the pistillate and staminate flowers, often with 3 bracts at base; staminate perianth 2- to 6-lobed, the stamens 1–6, the filaments short, erect; pistillate perianth small or reduced; ovary erect; styles solitary, somewhat lateral, the stigma peltate or filiform; achenes small, with crustaceous pericarp.——About 800 species in all tropical and warmer regions.

1A. Scandent woody plants; leaves evergreen, coriaceous, entire, with prominently elevated and finely reticulated veinlets beneath; receptacles axillary.
    2A. Leaves broadly lanceolate to ovate-oblong, long-acuminate, 5–12 cm. long; receptacles sessile, globose, not more than 10 mm. across.
                                                                                                                            1. *F. nipponica*
    2B. Leaves oblong or elliptic to ovate or broadly so, obtuse or subacute, 1–10 cm. long; receptacles 1.5–4 cm. across; peduncles 7–12 mm. long.
        3A. Leaves oblong or elliptic, obtuse, 4–10 cm. long, the petioles 1.5–2 cm. long; receptacles 3–4 cm. across, pyriform. .. 2. *F. pumila*
        3B. Leaves broadly to narrowly ovate, subacute with an obtuse tip, 1–5 cm. long, the petioles 5–10 mm. long; receptacles globose, 15–17 mm. across. .... 3. *F. stipulata*
1B. Erect trees or shrubs.
    4A. Large deciduous shrub with herbaceous leaves. .... 4. *F. erecta*
    4B. Evergreen trees with entire coriaceous leaves.
        5A. Petioles 7–20 mm. long; leaves 5–8(–10) cm. long, cuneately obtuse at base; receptacles sessile. .... 5. *F. microcarpa*
        5B. Petioles 3–6 cm. long; leaves 7–12 cm. long, rounded at base; receptacles on short peduncles 3–7 mm. long. .... 6. *F. wightiana*

1. **Ficus nipponica** Fr. & Sav. *F. foveolata* sensu auct. Japon., non Wall.; *? F. wrightii* Benth.; *F. foveolata* var. *nipponica* (Fr. & Sav.) King——ITABI-KAZURA. Woody scandent shrub, the branches attached to rocks or tree-trunks by means of aerial roots, the young branches puberulous; leaves evergreen, pale green when dried, narrowly ovate-oblong, narrowly ovate or broadly lanceolate, 5–12 cm. long, 2.5–4 cm. wide, long-acuminate, rounded at base, glabrous, whitish with finely elevated prominently reticulate veinlets beneath, the lateral nerves 5- to 6-paired, prominently arcuate short of the margin, the petioles 1–2 cm. long, puberulent while young, the stipules linear-lanceolate, about 10 mm. long; receptacles axillary, sessile, globose, glabrous, 7–10 mm. across, the basal bracts ovate-orbicular, obtuse, persistent, membranous, appressed-puberulent.——Honshu (Kantō Distr. and westw.), Shikoku, Kyushu.——Ryukyus, Formosa, China, and s. Korea.

2. **Ficus pumila** L. *F. hanceana* Maxim.——Ō-ITABI. Scandent shrub, the young branchlets appressed pubescent; leaves evergreen, coriaceous, lustrous, 4–10 cm. long, oblong to ovate-elliptic, obtuse, very obtuse to rounded at base, sometimes subobovate, glabrous or puberulous on nerves beneath, recurved on margin, the lateral nerves 3 to 5 pairs, more or less ascending, straight, the veinlets very prominently raised beneath, the petioles 1.5–2 cm. long, with brown appressed hairs while young; receptacles pyriform or subglobose, sometimes ellipsoidal, 3–4 cm. across, slightly pubescent only while young, the subtending bracts deciduous, the peduncles thick, 8–15 mm. long.——Honshu (Tōkaidō Distr. and westw.), Shikoku, Kyushu.——Ryukyus, Formosa, and China.

3. **Ficus stipulata** Thunb. *F. thunbergii* Maxim.; *F. foveolata* var. *thunbergii* (Maxim.) King——HIME-ITABI. Scandent evergreen much branched shrub, the branchlets and petioles red-brown pubescent; leaves narrowly to broadly ovate, 1–5 cm. long, 1–2 cm. wide, subacute with an obtuse tip, subrounded at base, often loosely soft-pubescent above while young, brownish spreading-puberulent on the nerves especially beneath, the lateral nerves 5 to 6 pairs, arcuate short of the margin, the veinlets prominent beneath, the petioles 5–10 mm. long; receptacles globose, 15–17 mm. across.——Warmer parts; Honshu (w. Kantō Distr. and westw.), Shikoku, Kyushu.——Ryukyus.

4. **Ficus erecta** Thunb. var. **erecta**. *F. japonica* Bl.

——INU-BIWA. Deciduous, erect, loosely branching shrub to 4 m. high, the branches rather thick, glabrous; leaves obovate or narrowly so, 10–20 cm. long, 4–10 cm. wide, acuminate, rounded or sometimes shallowly cordate to subobtuse at base, glabrous, entire, pale green beneath, the petioles 1–4 cm. long, glabrous; receptacles axillary, solitary, globose, abruptly narrowed below, 3-bracteate at base, glabrous, 1–1.3 cm. across, the peduncles glabrous, rather slender, 1–2 cm. long.——May–Aug. Thickets and woods, low mountains and hills; Honshu (Kantō Distr. and westw.), Shikoku, Kyushu.

Var. **sieboldii** (Miq.) King. *F. sieboldii* Miq.——HOSOBA-INU-BIWA. Leaves narrower, lanceolate.——Occurs with the typical phase.

Var. **yamadorii** Makino. KE-INU-BIWA. Coarsely pilose on branches, leaves and receptacles.——Occurs with the typical phase.——s. Korea, Ryukyus, and Formosa.

5. **Ficus microcarpa** L. f. *F. retusa* sensu auct., non L.——GAJU-MARU. Evergreen much branched glabrous tree with pendant aerial roots from the trunks and branches; leaves coriaceous, oblong or obovate, 5–8 cm. long, 3–5 cm. wide, slightly cuspidate with an obtuse tip, obtuse or cuneately obtuse at base, entire, the lateral nerves not strong, rather many, parallel except the lowest pair, the petioles 7–20 mm. long; receptacles sessile, depressed-globose, about 8 mm. across, with 3 ovate-orbicular bracts at base.——Kyushu (Yakushima and Tanegashima).——Ryukyus, Formosa, China to India, Malaysia, and Australia.

6. **Ficus wightiana** Wall. *F. superba* var. *japonica* Miq.——AKO. Glabrous evergreen tree with rather stout gray-brown branches; leaves coriaceous, oblong or narrowly so, 7–12 cm. long, 3.5–5 cm. wide, rounded at both ends, with an obtusely cuspidate apex, entire, the lateral nerves 6 to 9 pairs, slender, ascending, parallel, arcuate near the margin, the petioles 3–6 cm. long, rather stout; receptacles globose, 6–8 mm. across, the peduncles 3–7 mm. long, puberulous.——Honshu (Kii Prov.), Shikoku, Kyushu; rather common on lowlands near the sea.——Ryukyus, Formosa, and China.

Ficus elastica L. INDO-GOMU-NO-KI. Subscandent tree with lustrous leaves to 30 cm. long, with numerous slender parallel lateral nerves; often planted for ornamental purposes in the warmer parts of our area.——India.

6. HUMULUS L. KARA-HANA-SŌ ZOKU

Scandent dioecious retrorsely scabrous herbs with opposite, petiolate, membranous, palmately lobed, toothed leaves, the stipules lateral, free; staminate inflorescence a loosely branched large axillary panicle, the perianth-segments 5, imbricate in bud, the stamens 5; pistillate inflorescence a conelike, few-flowered, bracteate, pedunculate spike, the bracts accrescent after anthesis, the perianth thinly membranous, closely appressed to the sessile ovary, the style bifid, the branches deciduous; ovules pendulous; achenes broadly ovate, slightly compressed, often glandular-dotted, enveloped by the accrescent perianth, the pericarp crustaceous, the endosperm fleshy, the embryo coiled.——Two species, in the temperate regions of the N. Hemisphere.

1A. Leaves palmately 5- to 7-cleft; pistillate bracts ovate-orbicular in fruit, green, partly purplish brown, herbaceous, 7–10 mm. long; annual. .... 1. *H. japonicus*
1B. Leaves 3-lobed or sometimes unlobed; pistillate bracts ovate to ovate-oblong in fruit, papery-membranous, pale yellow-green or slightly brown-tinged, pale inside, 15–20 mm. long; rhizomatous perennial. .... 2. *H. lupulus* var. *cordifolius*

1. **Humulus japonicus** Sieb. & Zucc. KANA-MUGURA. Much branched scandent, green annual, retrorsely spinose-scabrous on stems and petioles; leaves membranous, long-petiolate, palmately 5- to 7-cleft, 5–12 cm. long and as wide, cordate, regularly serrulate, the lobes acuminate, slightly narrowed toward base, with yellow sessile discoid glands beneath, scabrous on both surfaces; staminate flowers in loosely arranged erect panicles 15–25 cm. long, pedicelled, nodding, the anthers pale yellow, pendulous on slender filiform filaments; pistillate spike conelike, broadly ovoid, the bracts ovate-orbicular, 7–10 mm. long, imbricate, herbaceous, accrescent after anthesis, scabrously pilose on back and especially on mar-

gin; achenes ovate-orbicular, inflated-lenticular, yellow-brown, puberulous near the top, glandless, 4–5 mm. long and as wide. ——Sept.–Oct.   Thickets and roadsides in lowlands; Hokkaido, Honshu, Shikoku, Kyushu; very common.——Ryukyus, Formosa, and China.

2. **Humulus lupulus** L. var. **cordifolius** (Miq.) Maxim. *H. cordifolius* Miq.——KARA-HANA-SŌ.   Much-branched, scandent, perennial herb, retrorsely prickly on stems and petioles; leaves green, long-petiolate, ovate-orbicular, 5–12 cm. long and as wide, acuminate, cordate at base, scabrous on both surfaces, rather irregularly toothed, often 3-lobed, the terminal

lobe larger than the lateral ones; pistillate spikes pendulous, short-pedunculate, pale yellow-green, ovoid-globose in fruit, the bracts obtuse, 15–20 mm. long and as wide, concave, papery-membranous, accrescent, ovate; achenes inflated-lenticular, 2.5–3 mm. long and as wide, covered with minute impressed points and yellow discoid glands; staminate inflorescence paniculate, erect, 10–15 cm. long, the flowers short-pedicelled.——Aug.–Sept. Thickets in mountains; Hokkaido, Honshu (centr. and n. distr.).——China. The typical phase occurs in Europe.

Fam. 69. **URTICACEAE**   IRA-KUSA KA   Nettle Family

Monoecious or dioecious herbs, rarely trees, often with stinging hairs; leaves opposite (often unequally so) or alternate, simple, toothed or incised, petiolate, commonly with cystoliths in the epidermal cells, the stipules free or connate, often caducous; flowers in glomerate cymes, the perianth 2- to 5-cleft, sometimes utriclelike; stamens as many as the perianth-segments and opposite them, the filaments incurved in bud; ovary superior, 1-locular; style simple or nearly absent, the stigma capitate, penicillate or pinnate, sometimes filiform; ovules solitary, usually orthotropous; fruit an achene, the endosperm oily.——About 40 genera, with about 500 species, cosmopolitan.

1A. Stinging hairs present.
  2A. Leaves opposite; stigma penicillate-capitate. .... 1. *Urtica*
  2B. Leaves alternate; stigma filiform. .... 2. *Laportea*
1B. Stinging hairs absent.
  3A. Pistillate perianth not utriclelike in fruit, usually shorter than the achene or deeply lobed to parted.
    4A. Leaves opposite. .... 3. *Pilea*
    4B. Leaves alternate.
      5A. Leaves equilateral, petiolate. .... 4. *Nanocnide*
      5B. Leaves not equilateral, sessile or short-petiolate.
        6A. Pistillate perianth suppressed, not accrescent; bracts of the pistillate inflorescence connate, forming a disclike receptacle at base. .... 5. *Elatostema*
        6B. Pistillate perianth 4- to 5-parted, covering part of the achene; bracts of pistillate inflorescence not connate. .... 6. *Pellionia*
  3B. Pistillate perianth connate and wholly investing the achene in fruit.
    7A. Leaves toothed; herbs or shrubs.
      8A. Pistillate perianth membranous, not adnate to the achene; suffrutescent herbs with alternate or opposite leaves. .. 7. *Boehmeria*
      8B. Pistillate perianth adnate to the achene, fleshy or juicy in fruit; woody plants with alternate leaves.
        9A. Pistillate perianth fleshy in fruit; stigma discoid or peltate. .... 8. *Villebrunea*
        9B. Pistillate perianth juicy in fruit; stigma capitate. .... 9. *Debregeasia*
    7B. Leaves entire; herbs or essentially so (in ours).
      10A. Glomerules involucrate at base; stigma short-linear, penicillate; delicate herbs. .... 10. *Parietaria*
      10B. Glomerules without involucral bracts; stigma filiform, pilose, suffrutescent herbs.
        11A. Lateral nerves of leaves much shorter than the midrib; perianth-segments of the staminate flowers without a transverse fold or gibbosity on back. .... 11. *Pouzolzia*
        11B. Lateral nerves reaching nearly to the apex of leaves; perianth-segments of staminate flowers with a transverse fold or gibbosity on back. .... 12. *Gonostegia*

1. **URTICA** L.   IRA-KUSA ZOKU

Dioecious or monoecious annuals or perennials with stinging hairs; leaves opposite, petiolate, dentate or serrate, 3- to 5-(7-) nerved, the stipules free or connate between the petioles; flowers in glomerate cymes, the staminate 4-merous; pistillate perianth 4-merous, the segments small, unequal; ovary erect; stigma penicillate-capitate; ovules orthotropous; achenes ovate or oblong, more or less flattened, enclosed by 2 accrescent inner perianth-segments.——About 40 species, in the temperate and warmer regions of the N. Hemisphere.

1A. Stipules 4, connate, in a pair.
  2A. Leaves broadly ovate or cordate-orbicular, with 7–11 coarse, narrowly deltoid, often double teeth on each side, the petioles slightly shorter than the leaves. .... 1. *U. thunbergiana*
  2B. Leaves narrowly to broadly ovate, rounded to shallowly cordate at base, with 10–25 usually simple teeth on each side; petioles much shorter than the leaves. .... 3. *U. platyphylla*
1B. Stipules 4, free, lanceolate or broadly linear.
  3A. Leaves narrowly ovate-oblong to lanceolate, 6–12 cm. long. .... 4. *U. angustifolia*
  3B. Leaves ovate to broadly so, 5–10 cm. long. .... 2. *U. laetevirens*

1. **Urtica thunbergiana** Sieb. & Zucc. IRA-KUSA.   Perennial; stems erect, simple or slightly branched, 40–80 cm. long, retrorsely puberulous; leaves broadly ovate to ovate-

orbicular, 5–12 cm. long, 4–10 cm. wide, acuminate, cordate at base, coarsely incised-serrate, with scattered coarse appressed hairs above, puberulent beneath especially on nerves, the petioles 3–10 cm. long, white appressed-puberulent; inflorescence axillary, narrowly pyramidal or subspicate, the pistillate above the staminate; flowers small, pale green, the 2 inner segments accrescent after anthesis and investing the achene; achenes flat, green, ovate.——Oct.–Nov. Woods in mountains; Honshu (Kantō Distr. and westw.), Shikoku, Kyushu.

2. Urtica laetevirens Maxim.　Koba-no-ira-kusa. Erect perennial herb 50–100 cm. long; stems thinly ascending-puberulous; leaves ovate or sometimes rather narrowly ovate, 5–10 cm. long, 2.5–5 cm. wide, acuminate, rounded to broadly cuneate at base, coarsely toothed, vivid green, scattered-pilose above, often sparingly short-pilose beneath; staminate inflorescence on upper portion of stems above the pistillate, the glomerules rather loose.——July–Oct. Woods in mountains; Hokkaido, Honshu.——Korea and China.

3. Urtica platyphylla Wedd.　*U. takedana* Ohwi——Ezo-ira-kusa.　Tall, often gregarious perennial; stems erect, 50–180 cm. long, obtusely angled, simple or slightly branched, ascending- or slightly recurved-puberulent; leaves deep green, ovate, 8–15 cm. long, 4–8 cm. wide, acuminate, rounded or shallowly cordate at base, short-pilose on both sides, especially on nerves beneath, coarsely toothed, the petioles 2–5 cm. long, much shorter than the blade, ascending-puberulent; inflorescence narrowly pyramidal, the pistillate toward the upper portion of the plant.——July–Oct. Moist woods and riverbanks in lowlands and foothills; Hokkaido, Honshu (centr. and n. distr.).——Kuriles, Kamchatka, Sakhalin, and e. Siberia.

4. Urtica angustifolia Fisch.　*U. dioica* var. *angustifolia* Fisch.; *U. foliosa* Bl.——Hosoba-ira-kusa.　Resembles the preceding; stems 50–150 cm. long, erect, simple or slightly branched, obtusely angled, rather sparingly appressed-pilosulous; leaves deep green, narrowly oblong to lanceolate, 6–12 cm. long, 1.5–5 cm. wide, acuminate, rounded at base, coarsely toothed, nearly glabrous or puberulous beneath especially on nerves, the petioles 1–3 cm. long, slightly puberulous or nearly glabrous; inflorescence narrowly pyramidal.——Aug.–Sept. Woods in mountains; Hokkaido, Honshu, Shikoku, Kyushu.——Korea, China, and e. Siberia.

Var. **sikokiana** (Makino) Ohwi.　*U. sikokiana* Makino——Nagaba-ira-kusa.　A western phase with slender stems, the leaves narrower and more glabrate, thinner.

2. LAPORTEA Gaudich.　Mukago-ira-kusa Zoku

Dioecious or monoecious herbs, shrubs or trees with stinging hairs; leaves alternate, large, simple, toothed or entire, penninerved, rarely 3-nerved, the stipules distinct, free or connate, deciduous; panicles or spikes axillary or terminal, the bracts small or absent; staminate flowers 4- to 5-merous; perianth-segments equal or unequal, the outer ones rarely absent; ovary erect when young, ascending afterward; stigma linear, usually elongate; ovules orthotropous; achenes flat, ascending.——Abundant in the Tropics, few in temperate e. Asia and N. America; our plants are herbs with pinnately 3-nerved leaves.

1A. Pistillate inflorescence a one-sided panicle; leaves narrowly ovate, rounded to obtuse at base; plant with bulbils in leaf-axils; roots thickened fusiform; achenes 2.5–3 mm. long. .... 1. *L. bulbifera*
1B. Pistillate inflorescence a linear spike; leaves broadly ovate, shallowly cordate to rounded at base; plant sometimes with bulbils; roots slender, not thickened; achenes 1.7–1.8 mm. long. .... 2. *L. macrostachya*

1. Laportea bulbifera (Sieb. & Zucc.) Wedd.　*Urtica bulbifera* Sieb. & Zucc.; *Fleurya bulbifera* (Sieb. & Zucc.) Bl.——Mukago-ira-kusa.　Green perennial 40–70 cm. high with rather stout erect stems; leaves rather long-petioled, narrowly ovate, 8–15 cm. long, 4–7 cm. wide, acuminate, obtuse to rounded at base, coarsely mucronate-toothed, puberulous on both surfaces especially on nerves; staminate inflorescence paniculate, 4–7 cm. long, peduncled, axillary; pistillate inflorescence on the upper part of the stems, 7–15 cm. long inclusive of the long peduncle, ascending, the branches spreading-puberulent on one side, the 2 lateral perianth-segments oblong or obliquely narrow-obovate, obtuse, about 2.5 mm. long, accrescent after anthesis; achenes smooth, pale green, obliquely ovate-orbicular, 2.5–3 mm. long.——Aug.–Sept. Woods in mountains; Hokkaido, Honshu, Shikoku, Kyushu; rather common.——China.

2. Laportea macrostachya (Maxim.) Ohwi.　*Sceptrocnide macrostachya* Maxim.; *L. grossedentata* Wright——Miyama-ira-kusa.　Green erect perennial 40–80 cm. high, the stems slightly retrorsely puberulent; leaves petiolate, broadly ovate or orbicular, 8–20 cm. long, 5–15 cm. wide, more or less pilose on both surfaces, the stipules small; staminate inflorescence paniculate, 5–10 cm. long, puberulous, the pistillate spicate, nearly terminal on the stems, peduncled, 20–30 cm. long, ascending, puberulous, the glomerules rather loosely several-flowered, the 2 lateral perianth-segments obliquely ovate, obtuse, puberulous, nearly as long as the achene, accrescent after anthesis; achenes obliquely elliptic, flat, 1.7–1.8 mm. long, smooth.——Woods in mountains; Hokkaido, Honshu.

3. PILEA Lindl.　Mizu Zoku

Dioecious or monoecious, soft, juicy annual or perennial herbs, rarely subshrubs; leaves opposite, equal or unequal, entire or toothed, 3-nerved, rarely penninerved or nearly nerveless, the stipules connate, often caducous; inflorescence cymose, axillary, loosely or densely flowered, subcapitate, the bracts small; staminate flowers 2- to 5-merous, perianth-segments valvate in bud; pistillate perianth unequally 3-lobed or equally 5-lobed, the reduced stamens scalelike, opposite the segments, sometimes absent; ovary erect; stigma sessile, penicillate; achenes ovate, slightly flattened, with a membranous testa.——More than 100 species, in the Tropics except Australia; few in temperate regions.

1A. Leaves orbicular, less than 10 mm. long, very obtuse, undulately toothed or nearly entire, with brown spots beneath, the cystoliths more or less transversely arranged above; achenes smooth. .... 1. *P. peploides*
1B. Leaves longer than broad, 1.5–13 cm. long, acute to caudate at apex, distinctly toothed, without spots beneath, the cystoliths rather irregularly arranged above.
  2A. Pistillate perianth unequally 3-lobed.
    3A. Pistillate perianth divided nearly to base, the longer segments lanceolate or linear-lanceolate, obtuse; achenes more or less redbrown spotted; leaves broadly rhombic-ovate, 1.5–10 cm. long.
      4A. Pistillate perianth-segments shorter than the achene, linear-lanceolate; achenes about 1.2 mm. long; leaves 3–10 cm. long.
        2. *P. mongolica*
      4B. Pistillate perianth-segments, or at least one of them, as long as to slightly longer than the achenes; achenes 1.8–2 mm. long; leaves 1.5–5 cm. long, acute with an obtuse tip. .... 3. *P. hamaoi*
    3B. Pistillate perianth spathelike, with 3 acute-deltoid teeth or with as many ovate lobes at apex; achenes smooth or verrucose; leaves oblong to ovate, 6–15 cm. long.
      5A. Leaves oblong, acuminate; achenes 1.3–1.5 mm. long, smooth. .... 4. *P. petiolaris*
      5B. Leaves oblong-ovate, broadest slightly below the middle, long caudate-acuminate; achenes about 0.6 mm. long, strongly verrucose. .... 5. *P. pseudopetiolaris*
  2B. Pistillate perianth with 5 nearly equal segments, parted to the base; achenes smooth, not brown-spotted.
    6A. Peduncles of pistillate inflorescence mammillate-puberulous on one side; leaves 2–12 cm. long, narrowly ovate or oblong, with 7–13 coarse teeth on each side. .... 6. *P. kiotensis*
    6B. Peduncles of pistillate inflorescence glabrous; leaves 1–3(–8) cm. long, broadly ovate or ovate-orbicular, with 3–6 coarse teeth on each side. .... 7. *P. japonica*

1. **Pilea peploides** (Gaudich.) Hook. & Arn. *Dubreulia peploides* Gaudich.——Koke-mizu. Glabrous somewhat tufted monoecious annual, 7–15 cm. high, the stems branching below; leaves orbicular or depressed-ovate, 6–10 mm. long and as wide, very obtuse to rounded at the tip, usually sparingly undulate-toothed, broadly cuneate and 3-nerved at base, the cystoliths linear, more or less transversely arranged on the upper surface, brown-spotted beneath, the petiole shorter or longer than the blade; flowers more or less capitate in leaf-axils, sessile, the perianth-segments broadly oblanceolate, shorter than the achenes; achenes smooth, pale brown, broadly lenticular-ovate, about 0.6 mm. long.——Mar.–Aug. Honshu (Kantō Distr. and westw.), Shikoku, Kyushu.——Ryukyus, Formosa, China, Korea, Manchuria, Hawaii, Galapagos Isl., Malaysia, and India.

2. **Pilea mongolica** Wedd. *P. viridissima* Makino——Ao-mizu. Glabrous, soft, fleshy, monoecious annual; stems erect, terete, often branched, 30–50 cm. long, pale green; leaves deep green, broadly rhombic-ovate, 1.5–10 cm. long, 1–7 cm. wide, short-acuminate or subacute with an obtuse tip, broadly cuneate at base, coarsely deltoid-toothed, 3-nerved, glabrous or scattered pubescent above and on margin, the petioles unequally paired, 1.5–10 cm. long; inflorescence densely cymose, 1–3 cm. long, the pistillate perianth-segments unequal, the longer ones linear-lanceolate, obtuse, with a prominent slightly raised midrib, ½–¾ as long as the achene; achenes broadly ovate-lenticular, 1–1.2 mm. long, with scattered slightly raised brown spots.——July–Oct. Shaded places in lowlands and mountains; Hokkaido, Honshu, Shikoku, Kyushu.——Korea, China, and e. Siberia.

3. **Pilea hamaoi** Makino. Mizu. Glabrous, soft, juicy monoecious annual; stems erect, sparingly branched, green or purplish, 20–40 cm. long; leaves deep green, broadly rhombic to rhomboid-ovate, 1.5–5 cm. long, 1–3.5 cm. wide, acute with an obtuse tip, broadly cuneate at the 3-nerved base, glabrous, ciliate, sparingly toothed; cystoliths short; inflorescence cymose, 5–20 mm. long, axillary; pistillate perianth-segments unequal, the larger ones lanceolate or linear-oblong, as long as to slightly longer than the achene, very obtuse, with a prominent midrib; achenes broadly ovate-lenticular, pale brown-spotted, 1.8–2 mm. long.——July–Oct. Shaded places; Hokkaido, Shikoku, Kyushu.——China.

4. **Pilea petiolaris** (Sieb. & Zucc.) Bl. *Urtica petiolaris* Sieb. & Zucc.; *P. strangulata* Fr. & Sav.——Miyama-mizu. Soft, juicy, monoecious perennial, 40–80 cm. high, glabrous, sometimes sparingly branched below; leaves ovate to narrowly so, or oblong, 6–15 cm. long, 3–5 cm. wide, acuminate with a rather obtuse tip, obtuse or cuneate and 3-nerved at base, shallowly toothed, the stipules rather large, deciduous; inflorescence densely cymose, 1–4 cm. long, the pistillate perianth split to the base on one side, broadly ovate, 3-toothed or -lobed, about half as long as the achene; achenes broadly ovate, usually greenish brown, lenticular, not spotted, 1.3–1.5 mm. long, broadly ovate.——Aug.–Oct. Shaded wet places in mountains; Honshu (Kantō Distr. and westw.), Shikoku, Kyushu.

5. **Pilea pseudopetiolaris** Hatus. Ko-miyama-mizu. Slightly succulent perennial herb; rhizomes elongate, creeping; stems erect, 40–80 cm. long; leaves narrowly ovate to ovate-elliptic, 6–10 cm. long, 2.5–4 cm. wide, caudately long-acuminate, rounded to obtuse at base, obtusely serrate, trinerved, scattered pilose above, the petioles 2–3 cm. long; female cymes axillary, solitary, about 1 cm. long, short-peduncled, densely flowered; male cymes larger, 2–3.5 cm. long, nearly sessile, glabrous; achenes ovate-fusiform, acute, about 0.6 mm. long, densely verrucose.——Aug.–Oct. Honshu, Shikoku, Kyushu.

6. **Pilea kiotensis** Ohwi. *Achudemia iseana* Makino; *P. iseana* Makino——Miyako-mizu. Glabrous perennial (?) herb; stems 30–40 cm. long, rather juicy and soft, sometimes slightly branched; leaves ovate to narrowly so, sometimes oblong, 2–12 cm. long, 1–5 cm. wide, caudately acuminate with an acute tip, cuneate at base, coarsely toothed, nearly glabrous on both sides, 3-nerved, the stipules about 5 mm. long, deciduous; pistillate inflorescence cymose, the peduncles short-puberulous on one side or subsessile, the perianth-segments 5, elliptic to oblong, nearly equal, obtuse, the midrib tuberculate on back; achenes about 1 mm. long, 1.5–2 times longer than the perianth, broadly ovate, lenticular, smooth, pale, not punctate.——Oct. Honshu (Kinki Distr.), Kyushu (n. distr.).

7. **Pilea japonica** (Maxim.) Hand.-Mazz. *Achudemia japonica* Maxim.——Yama-mizu. Glabrous monoecious annual; stems branched, 10–20 cm. long, slightly red-brown, rather slender, soft, juicy; leaves broadly ovate or ovate-orbicular, 1–3(–8) cm. long, obtuse or acute with an obtuse tip, 3-nerved, the pairs unequal, ciliate, with 1–4 coarse teeth on

each side, the petioles slender, shorter to slightly longer than the blades; flowers cymose, the peduncles glabrous or sessile, the pistillate perianth-segments 5, nearly equal, broadly lanceolate, obtuse; achenes broadly ovate, pale, sublenticular, about 1 mm. long, not punctate, smooth.——Sept.–Oct. Woods in lowlands and foothills; Honshu (Kantō Distr. and westw.), Shikoku, Kyushu.

4. NANOCNIDE Bl. KATEN-SŌ ZOKU

More or less pubescent small monoecious herbs; leaves alternate, petiolate, ovate-orbicular or depressed, toothed or incised, 3- to 5-nerved, the stipules free; staminate inflorescence in peduncled cymes, in the upper axils, the perianth 4- to 5-cleft, the segments with a transverse ridge on back; pistillate inflorescence a dense sessile axillary cyme or glomerule, the bracts free, the perianth 4-cleft, the segments often with a small apical bristle, the outer ones slightly broader, keeled on back; ovary erect; achenes oblong, covered by the perianth but dehiscing from it when mature.——Three species in Japan, China, and Ryukyus.

1. **Nanocnide japonica** Bl. KATEN-SŌ. Gregarious perennial herb 10–30 cm. high; stems slightly pubescent above, soft, green; leaves deltoid or rhombic-ovate to orbicular, obtuse, deeply obtuse-toothed, subtruncate at base, slightly pilose on both sides, the stipules ovate, obliquely spreading, 1–2 mm. long; staminate inflorescence distinctly peduncled, the perianth white; pistillate flowers small, green, subsessile in short dense peduncled cymes in upper axils, the segments lanceolate, slightly longer than the achene; achenes lenticular, broadly ovate, pale, punctulate, about 1.3 mm. long; stigma mammillate, densely pubescent.——Apr.–May. Shaded places in lowlands and foothills; Honshu, Shikoku, Kyushu.——Formosa and China.

5. ELATOSTEMA Forst. UWABAMI-SŌ ZOKU

Herbs and subshrubs; stems rather thick; leaves alternate, sometimes with an odd minute opposite one, subsessile, toothed, usually oblique at base, penninerved or 3-nerved; flowers small, unisexual, in involucrate dense headlike cymes, involucral bracts free or connate, staminate perianth 4- to 5-merous; stamens opposite the perianth-segments; pistillate perianth minute, 3- to 5-parted, persistent; stigma penicillate; achenes small, ovate, usually longitudinally striate.——About 60 species, in Africa and Asia, especially abundant in the Tropics.

1A. Staminate glomerules pedunculate; leaves long-caudate; some of the nodes thickened in autumn. .... 1. *E. umbellatum*
1B. Staminate and pistillate glomerules sessile or nearly so; nodes never thickened.
    2A. Stems spreading-pilose; leaves brownish green when dried, slightly caudate-acuminate with an obtuse tip; bracts setulose.
        2. *E. densiflorum*
    2B. Stems nearly glabrous; leaves vivid yellow-green when dried, acute with an obtuse tip, not elongate above; bracts densely puberulent.
        3. *E. laetevirens*

1. **Elatostema umbellatum** Bl. *Procris umbellata* (Bl.) Sieb. & Zucc.; *Pellionia umbellata* (Bl.) Wedd.; *E. japonicum* Wedd.——HIME-UWABAMI-SŌ. Soft, green, glabrous, perennial herb; stems 15–30 cm. long, ascending to erect and unbranched below, often declined and branched above, rather fleshy or juicy, terete, short-decumbent at base; leaves 2-seriate, alternate, rather thin and soft, oblique, narrowly ovate, 1.5–7 cm. long, 8–20 mm. wide, elongate and long-caudate with an obtuse tip, sessile, penninerved, with 2–4(–5) coarse teeth; staminate glomerules pedunculate, 1–3 cm. long, the pistillate ones small, axillary, sessile, the bracts slightly puberulous above; achenes pale, about 0.6 mm. long, loosely warty.——May–July. Moist shaded places in mountains; Honshu (Kantō Distr. and westw.), Shikoku, Kyushu.

Var. **majus** Maxim. *E. umbellatum* var. *involucratum* (Fr. & Sav.) Makino; *E. involucratum* Fr. & Sav.——UWABAMI-SŌ. Plants rather large; stems 20–50 cm. long; leaves 5–12 cm. long, 2–4 cm. wide, with 5–12 coarse teeth on each side, the bracts broadly lanceolate to elliptic; achenes about 0.8 mm. long, nearly smooth.——May–July. Moist shaded places in mountains; Hokkaido, Honshu, Shikoku, Kyushu.

2. **Elatostema densiflorum** Fr. & Sav. *E. nipponicum* Makino——TOKI-HOKORI. Annual; stems erect below, ascending above, 10–25 cm. long, juicy, often branched, spreading-pubescent; leaves oblique, cuneate-oblong, 4–8 cm. long, 2–3 cm. wide, acute with an obtuse tip, cuneate at base on one side, rounded on the other, slightly pubescent on nerves beneath, coarsely toothed above, subtrinerved; inflorescence sessile, the bracts short-setulose.——Sept.–Oct. Shaded places in lowlands; Honshu (n. and centr. distr.).

3. **Elatostema laetevirens** Makino. *E. sessile* var. *cuspidatum* sensu Maxim., non Wedd.——YAMA-TOKIHOKORI. Perennial rhizomatous herb with stolons, resembling the preceding but glabrous; stems fleshy, terete, ascending above, decumbent below, 15–40 cm. long; leaves vivid green, 2-seriate, slightly elongate above and caudate with an obtuse tip; inflorescence sessile, the bracts densely short-pubescent outside. ——Aug.–Oct. Woods in mountains; Hokkaido, Honshu, Shikoku, Kyushu.

6. PELLIONIA Gaudich. SANSHŌ-SŌ ZOKU

Dioecious or monoecious herbs sometimes from a woody base; leaves alternate, nearly sessile, 2-seriate, oblique at base, entire or toothed, the stipules small or obscure; cymes axillary, sessile or peduncled, often densely flowered, capitate; flowers 4- to 5-merous; ovary erect; stigma penicillate; achenes small.——About 20 species, in the Tropics of Asia.

1A. Stems prostrate, herbaceous.
   2A. Leaves short curved-puberulent on nerves and midrib beneath, obliquely obovate, obtuse at apex, 8–20 mm. long. .... 1. *P. minima*
   2B. Leaves glabrous or minutely papillose-puberulent on midrib beneath, subovate or narrowly oblique-ovate, slightly elongate above.
                                                                                                          2. *P. radicans*
1B. Stems woody, ascending to erect, 20–40 cm. long. .... 3. *P. scabra*

1. Pellionia minima Makino. SANSHŌ-SŌ. Deep green perennial herb; stems slender, slightly branched, elongate, creeping prostrate, 10–30 cm. long, minutely puberulous, brownish green; leaves obliquely obovate, auriculate-rounded on outer margin at base, pale green, minutely puberulent on nerves beneath, the petioles very short; staminate glomerules cymose, the peduncles 1–1.5 cm. long, puberulous; pistillate glomerules sessile, densely flowered, small, the perianth-segments linear; achenes elliptic, pale, about 0.7 mm. long, prominently raised punctate.——Apr.–June. Woods; Honshu (Tōkaidō Distr. and westw.), Shikoku, Kyushu.

2. Pellionia radicans (Sieb. & Zucc.) Wedd. *Procris radicans* Sieb. & Zucc.; *Elatostema radicans* (Sieb. & Zucc.) Wedd.——Ō-SANSHŌ-SŌ. Resembling the preceding but larger; stems terete, prostrate, branched, elongate, brownish green, nearly glabrous; leaves alternate, 2-seriate, subovate or obliquely ovate in smaller ones, elongate above in the larger ones, 2–5 cm. long, 1–2 cm. wide, deep green, sometimes appressed pilose above, pale green and nearly glabrous beneath, the petioles very short, 1–3 mm. long; staminate cymes 1–1.5

cm. across, the peduncles papillose-puberulous; pistillate glomerules small, subglobose; achenes about 0.8 mm. long, minutely raised punctate.——June–July. Woods; Honshu (Kinki Distr. and westw.), Shikoku, Kyushu.

Pellionia yosiei Hara. *Elatostema yosiei* Hara. This may be a larger phase of *P. radicans;* leaves to 9 cm. long and 3 cm. wide, the staminate inflorescence with peduncles 1–5.5 cm. long.

3. Pellionia scabra Benth. *P. pellucida* Merr., excl. syn. ——KI-MIZU. Shrub 20–40 cm. high, branches ascending to nearly erect, rather slender, grayish, slightly puberulent; leaves thin, sessile, oblique, oblanceolate or subobovate, 6–10 cm. long, 1–3 cm. wide, caudately acuminate, cuneate and obtuse at base, short-toothed, coarsely scattered-hairy above, paler and the nerves puberulent beneath; staminate glomerules cymose, pedunculate, the pistillate glomerules subsessile or short-peduncled, densely flowered; achenes minutely punctate.—— Woods; Honshu (Izu Isls., Tōkaidō and s. Kinki Distr.), Shikoku, Kyushu.——Ryukyus and China.

7. BOEHMERIA Jacq.   KARA-MUSHI ZOKU

Monoecious herbs or shrubs, more or less hairy; leaves opposite or alternate, petiolate, equilateral or slightly unequal, toothed, sometimes 2- to 3-lobulate, the stipules free, rarely connate, deciduous; inflorescence spicate or paniculate, the flowers small, unisexual, the glomerules sessile; staminate flowers short-pedicelled or sessile, 4-merous, rarely 3- or 5-merous; pistillate flowers usually sessile, the perianth tubular or utriclelike, the orifice with 2–4 minute teeth; ovary entirely invested by the perianth; stigma elongate, short-puberulent on one side; ovules erect; achenes enclosed within the perianth.——About 100 species, pantropical and difficult to determine; the following list is a conservative estimate of those occurring in our area.

1A. Leaves alternate, usually densely white-tomentose beneath; inflorescence a loose panicle.
   2A. Branches and petioles densely coarse-pubescent; leaves ovate-orbicular. .... 1. *B. nivea*
   2B. Branches and petioles densely short-puberulent; leaves broadly ovate. .... 2. *B. nipononivea*
1B. Leaves opposite, pilose or pubescent beneath but without white tomentum.
   3A. Pistillate perianth-tube puberulent throughout; plants slightly hairy.
      4A. Subshrub with woody branching base; leaves 4–8 cm. long, rhombic-ovate. .... 3. *B. spicata*
      4B. Herbaceous, usually with simple stems; leaves 8–20 cm. long, ovate-orbicular. .... 4. *B. tricuspis*
   3B. Pistillate perianth-tube above uniformly short appressed-puberulent or the hairs longer on the upper portion than on the lower portion.
      5A. Pistillate perianth-tube glabrous on lower portion.
         6A. Stigma very short, about 0.5 mm. long; achenes about 1.2 mm. long. .... 5. *B. formosana*
         6B. Stigma slender, 1.5–2 mm. long; achenes about 2 mm. long. .... 6. *B. sieboldiana*
      5B. Pistillate perianth-tube uniformly hairy, the hairs on lower part short and ascending, clearly shorter than those above.
         7A. Leaves with the teeth equal.
            8A. Leaves bullate-rugose, prominently scabrous above, more or less unequal, often 2- to 3- lobulate, the teeth small and many; maritime plants with relatively thick leaves. .... 7. *B. biloba*
            8B. Leaves not bullate-rugose above, equilateral, with larger teeth.
               9A. Leaves rhombic-ovate or ovate, broadly cuneate at base, short-pubescent beneath. .... 8. *B. kiusiana*
               9B. Leaves ovate-orbicular or orbicular, rounded to cordate at base, often densely short-pubescent beneath. ...... 9. *B. pannosa*
         7B. Leaves with the teeth unequal, often doubly toothed in upper part.
            10A. Pistillate glomerules densely arranged and approximate; leaves membranous to rather thick, rounded to cordate or obtuse at base.
               11A. Leaves velvety short-pubescent beneath, usually broadly ovate, obtusely rounded at base, the teeth usually simple.
                                                                                                  10. *B. holosericea*
               11B. Leaves short-pubescent, the upper teeth often double. .... 11. *B. longispica*
            10B. Pistillate glomerules rather loosely arranged; leaves thinly membranous, usually truncate or very broadly cuneate at base.
                                                                                                  12. *B. platanifolia*

1. Boehmeria nivea (L.) Gaudich. *Urtica nivea* L. ——NAMBAN-KARA-MUSHI. Densely hirsute perennial herb; stems simple or sparingly branched, to 1.5 m. long, grayish

white, coarse, spreading-pubescent; leaves membranous, ovate-orbicular, 10–15 cm. long, 6–12 cm. wide, slightly elongate to the caudate tip, cuneate-rounded at base, regularly dentate-toothed, green and slightly hirsute above, densely white-woolly beneath with spreading-hairs on the nerves, the petioles slender, as long as to slightly shorter than the blade; panicles 5–10 cm. long, loosely glomerate; pistillate perianth with short ascending hairs; stigma rather short; achenes about 1 mm. long.——Sept. Cultivated and occasionally naturalized in Honshu, Shikoku, Kyushu.——Malaysia, China, and Indochina.

Var. **candicans** Wedd. *B. utilis* Bl.——RAMII. Parts larger; leaves broadly cuneate at base, the petioles usually longer than the blade; frequently cultivated in our area for fiber.——s. China, Philippines, and Malaysia.

2. **Boehmeria nipononivea** Koidz. *B. frutescens* sensu auct. Japon., non Thunb.——KARA-MUSHI. Resembles No. 1 but more slender, the stems and petioles short-hirsute; leaves broadly ovate, membranous, caudately acuminate, rounded or sometimes broadly cuneate at base, the petioles shorter than the blade; inflorescence shorter; stigma short.——Aug.–Sept. Honshu, Shikoku, Kyushu.——China and Formosa.

Var. **concolor** (Makino) Ohwi. *B. nivea* var. *concolor* Makino——AO-KARA-MUSHI. Leaves scarcely white-woolly beneath.——Occurs with the typical phase.

3. **Boehmeria spicata** (Thunb.) Thunb. *Urtica spicata* Thunb.; *U. japonica* L. f.; *B. japonica* Miq., quoad syn.——KO-AKA-SŌ. Subshrub with branched woody base; stems 50–100 cm. long, erect, the branches rather slender; leaves membranous, rhomboid-ovate, 4–8 cm. long, 2.5–4 cm. wide, caudate at apex, coarsely toothed, cuneate to broadly cuneate at base, slightly pilose on both sides especially on the nerves beneath, the petioles usually reddish; glomerules of staminate flowers loosely arranged, small, the pistillate perianth minutely puberulous; achenes about 1 mm. long; stigma slender, rather elongate.——Aug.–Oct. Mountains; Honshu, Shikoku, Kyushu; common.——China.

Boehmeria minor Satake. KO-YABU-MAO. Intermediate between *B. spicata* and *B. biloba,* differing from the former in the relatively larger parts, with leaves slightly bullate and rugulose above and prominently short-pilose beneath, especially on the nerves.

4. **Boehmeria tricuspis** (Hance) Makino. *B. platyphylla* var. *tricuspis* Hance; *B. japonica* var. *tricuspis* (Hance) Maxim.——AKA-SŌ. Erect perennial herb about 50–80 cm. high; stems simple, usually reddish, nearly glabrous or sparingly short-pilose; leaves thin, ovate-orbicular, 8–20 cm. long, 5–15 cm. wide, coarsely toothed, broadly cuneate to truncate at base, usually deeply incised at apex, long caudate between the incision, slightly short pilose above and on the nerves beneath; achenes about 1.5 mm. long, enclosed in the minutely puberulent perianth; stigma rather short.——July–Sept. Hills and mountains; Hokkaido, Honshu, Shikoku, Kyushu; rather common.——China.

Var. **unicuspis** Makino. *B. paraspicata* Nakai——KUSA-KO-AKA-SŌ, MARUBA-KO-AKA-SŌ. Leaves scarcely incised and not caudate.——Hokkaido, Honshu, Shikoku.

5. **Boehmeria formosana** Hayata. TAIWAN-TORI-ASHI. Stems erect, often branched below, somewhat woody at base, sparingly short appressed-pilose; leaves rather membranous, ovate-oblong, 8–12 cm. long, 3–6 cm. wide, short-acuminate, cuneate to broadly cuneate at base, toothed, nearly glabrous above, sparingly short pilose on nerves beneath, the petioles 1–5 cm. long; glomerules of pistillate flowers small, forming a slender uninterrupted spike in fruit, the perianth slightly appressed-pilose; achenes about 1.5 mm. long; stigma short.——

Kyushu (Yakushima).——Ryukyus, Formosa, and China.

6. **Boehmeria sieboldiana** Bl. *B. platyphylla* var. *sieboldiana* (Bl.) Wedd.——NAGABA-YABU-MAO. Erect perennial herb; stems 1–2 m. long, scattered short-pilose above; leaves rather thin, oblong-ovate to ovate, 10–20 cm. long, 4–10 cm. wide, caudately acuminate at apex, broadly cuneate to rounded at base, regularly toothed, slightly pilose to nearly glabrous on both sides, the petioles rather slender; glomerules of pistillate flowers rather contiguous, the perianth slightly appressed-pilose on upper half; achenes 2–2.5 mm. long; stigma slender, rather elongate.——Aug.–Oct. Hills and lowlands; Honshu (s. Kantō Distr. and westw.), Shikoku, Kyushu. Poorly known are:

Boehmeria egregia Satake. SHIMA-NAGABA-YABU-MAO. Closely allied to *B. sieboldiana* but with larger parts, rather short stigmas and usually 3 stamens.——Honshu (Izu Isls. and Awa Prov.).

Boehmeria pseudosieboldiana Honda. INU-YABU-MAO. With thicker leaves slightly obtuse at base.——Kyushu. Similar to but may not be conspecific with *B. sieboldiana.*

7. **Boehmeria biloba** Wedd. *B. bifida* Bl.; *Splitgerbera japonica* Miq.; *B. splitgerbera* (Miq.) Koidz.——RASEITA-SŌ. Coarse perennial herb; stems 30–70 cm. long, rather stout, scabrous, brownish red; leaves thick, broadly ovate, obovate-orbicular or ovate-orbicular, 6–15 cm. long, 4–10 cm. wide, short-acuminate, acute to cuneate at base, often shallowly 2- to 3-lobed toward the apex, rather minutely toothed, often unequal, prominently bullate and scabrous above, impressed and short-pilose on nerves and veinlets beneath; glamerules of pistillate flowers contiguous, forming a rather short thick spike in fruit, the perianth with short ascending hairs; achenes narrowly cuneate, about 1.5 mm. long.——Aug.–Oct. Waste ground and rocky cliffs near seashores; Hokkaido (s. distr.), Honshu. Poorly known are:

Boehmeria arenicola Satake. HAMA-YABU-MAO; **B. kiyozumensis** Satake. KIYOSUMI-YABU-MAO; and **B. tenuifolia** Satake. USUBA-RASEITA-SŌ. Reported from the seashore area of Kantō District, Honshu; all are intermediate between *B. biloba* and *B. longispica.*

8. **Boehmeria kiusiana** Satake. TSUKUSHI-YABU-MAO. Resembles *B. sieboldiana,* but with leaves very hairy beneath and the perianth of the pistillate flowers hairy; intermediate between *B. biloba* and *B. longispica.*——Aug.–Sept. Honshu (Nagato Prov.), Kyushu (Chikuzen Prov.).

9. **Boehmeria pannosa** Nakai & Satake. SAIKAI-YABU-MAO. Large, stout perennial herb; stems robust, densely short-pubescent above; leaves broadly ovate to rounded-cordate, 10–20 cm. long and as wide, short-acuminate, rounded to shallowly cordate at base, rather regularly toothed, with short coarse hairs above, densely short-pubescent beneath; glomerules of pistillate flowers in fruit loosely arranged in axillary spikes, the perianth rather densely short-hairy above.——Aug. Honshu (Nagato Prov.), Kyushu.

Boehmeria gigantea Satake. NIŌ-YABU-MAO. Large poorly understood plant with coarsely toothed leaves; perhaps closely allied to *B. pannosa.*——Honshu (Suwo Prov.).

10. **Boehmeria holosericea** Bl. *B. platyphylla* var. *holosericea* (Bl.) Wedd.——ONI-YABU-MAO. Large stout perennial; stems 1–1.5 m. long, rather densely short-pubescent; leaves rather thick, ovate-orbicular, 10–15 cm. long and as wide, obtuse to shallowly truncate-cordate at base, hairy above, densely short-pubescent beneath, the upper ones with coarse acute double teeth; pistillate glomerules rather thick in fruit,

forming a contiguous elongate spike; achenes about 1.5 mm. long, the enveloping perianth pilosulous.——Aug.–Sept. Honshu, Shikoku, Kyushu.——China and Indochina.

Var. **izuosimensis** (Satake) Satake & Mizushima. *B. izuosimensis* Satake——ŌSHIMA-YABU-MAO. Leaves as long as or shorter than wide, shallowly cordate to truncate at base.—— Izu Isls. (Oshima Isl.).

11. Boehmeria longispica Steud. *B. grandifolia* Wedd.; *B. miqueliana* T. Tanaka——YABU-MAO. Large perennial herb with erect, rather densely short-pubescent stems 80–100 cm. long; leaves rather thick, ovate-orbicular, acuminate, rounded at base, with coarse double teeth in the upper ones, scabrous above, prominently short-pilose beneath; glomerules on elongate axillary spikes, rather contiguous; achenes about 2 mm. long, the enveloping perianth pilose.——Aug.–Oct. Hokkaido, Honshu, Shikoku, Kyushu; common.

Boehmeria robusta Satake. MARUBA-YABU-MAO. Resembles the preceding but with thinner leaves. Poorly known.—— Honshu (s. Kantō Distr.).

12. Boehmeria platanifolia Fr. & Sav. *B. japonica* var. *platanifolia* Maxim.; *B. maximowiczii* Nakai & Satake——ME-YABU-MAO. Stems erect, simple, to 1 m. long, short-hairy above; leaves membranous, orbicular to depressed ovate-orbicular, often incised and long-caudate at apex, rounded to truncate at base, with unequal coarse teeth below and larger often double teeth on the upper margin, pilose on both sides, especially above and on nerves beneath; pistillate glomerules small, rather loosely arranged on slender axillary spikes; achenes about 1.5 mm. long, the enveloping perianth pilose.——Aug.– Oct. Lowlands and hills; Hokkaido, Honshu, Shikoku, Kyushu; common.——Korea and China.

8. VILLEBRUNEA Gaudich. HADO-NO-KI ZOKU

Dioecious trees or shrubs; leaves alternate, long-petiolate, entire or undulately toothed, penninerved or 3-nerved, the stipules usually connate on the petioles; inflorescence mostly cymose, the glomerules small; staminate flowers 4-merous, the stamens opposite the perianth-segments; pistillate perianth connate, tubular, wholly adnate to the ovary, toothed on margin; stigma discoid or peltate, hairy on margin; achenes wholly enclosed in the fleshy perianth; seeds erect.——About 10 species, in India, Malaysia, and e. Asia.

1A. Branchlets and petioles with long and short hairs; leaves membranous, white-woolly beneath at least while young, pistillate and staminate glomerules sessile; bracteoles subacute. .... 1. *V. frutescens*
1B. Branchlets and petioles with short appressed hairs; leaves relatively firm, dark green, without woolly hairs; pistillate glomerules pedunculate, the staminate sessile; bracteoles rather rounded at apex. .... 2. *V. pedunculata*

1. Villebrunea frutescens (Thunb.) Bl. *Urtica frutescens* Thunb.; *Oreocnide frutescens* (Thunb.) Miq.; *Nirwamia pellucida* Raf.; *Boehmeria fruticosa* Gaudich.; *Morocarpus microcephalus* Benth.; *V. fruticosa* (Gaudich.) Nakai; *V. microcephala* (Benth.) Nakai; *Pellionia pellucida* (Raf.) Merr. ——IWAGANE. Erect deciduous shrub with slender elongate branches; leaves oblong, sometimes ovate or elliptic, 6–12 cm. long, 2.5–5 cm. wide, acutely caudate-acuminate, cuneate with an obtuse end or rounded at base, dentate-serrate, coarsely hirsute above and on nerves beneath, the petioles elongate, slender, the stipules free, deciduous; fruit ovoid, about 1.5 mm. long.——Mar.–May. Kyushu.——China.

2. Villebrunea pedunculata Shirai. *Oreocnide pedunculata* (Shirai) Masam.——HADO-NO-KI. Dark green shrub closely resembling the preceding; leaves somewhat thicker, 5–10 cm. long, 2–4 cm. wide, obtuse to rounded at base, glabrous on both sides or sparsely appressed-pilose above and short appressed-puberulent on both sides, obtusely toothed, the petioles 1–3 cm. long; staminate glomerules sessile, axillary, the peduncles of the pistillate ones 3–5 mm. long; fruit about 1.5 mm. long.——Feb.–Apr. Honshu (Aogashima in Izu Isls. and Kii Prov.), Shikoku, Kyushu.——Ryukyus and Formosa.

9. DEBREGEASIA Gaudich. YANAGI-ICHIGO ZOKU

Monoecious or dioecious shrubs; leaves alternate, petiolate, obtusely toothed, 3-nerved, often white-woolly beneath, the stipules connate above the petioles, bifid; glomerules globose, axillary, sessile or pedunculate, the inflorescence sometimes compound and cymose; staminate flowers 4(3–5)-merous; pistillate perianth connate, ovoid to obovoid, toothed at apex, becoming juicy in fruit; ovary erect, enclosed in the perianth; stigma capitate; achenes enclosed in the perianth.——Ethiopia to India, Malaysia, and e. Asia.

1. Debregeasia edulis (Sieb. & Zucc.) Wedd. *Morocarpus edulis* Sieb. & Zucc.; *Morocarpus japonicus* Miq.; *D. japonica* (Miq.) Koidz.——YANAGI-ICHIGO. Deciduous shrub with elongate branches, the branchlets with short appressed hairs; leaves lanceolate to linear-oblong, 7–15 cm. long, 12–25 mm. wide, acuminate, obtuse at base, subtrinerved, dark green and slightly lustrous above, with appressed minute teeth on

margin, appressed-puberulent on the nerves above, densely white-woolly beneath, the petioles 1–3 cm. long; glomerules 1–3, axillary, globose, short-pedunculate, rather many-flowered; fruit orange-yellow, about 1 mm. long.——Mar.–May. Honshu (s. Kantō Distr. through Tōkaidō to s. Kinki Distr.), Shikoku, Kyushu.——Ryukyus, Formosa, and China.

10. PARIETARIA L. HIKAGE-MIZU ZOKU

Slender often delicate annuals or perennials; leaves alternate, petiolate, small, trinerved, without stipules; flowers in small cymes, sometimes in axillary glomerules, the outer bracts sometimes connate below the glomerules; perianth of bisexual and staminate flowers (3-) 4-parted, in pistillate flowers connate into a tube; ovary free from the perianth; stigma linear, curved, bearded; achenes small, inclosed in the persistent perianth.——Seven or eight species in temperate regions.

1. **Parietaria debilis** Forst. var. **micrantha** (Ledeb.) Wedd. *P. micrantha* Ledeb.——HIKAGE-MIZU. Delicate soft-pubescent annual 10–15 cm. high; leaves alternate, long-petioled, ovate to broadly so, 8–25 mm. long, 5–15 mm. wide, obtuse, deep green; flowers few, axillary, the bracts linear; achenes broadly ovate, black, lustrous, about 1.5 mm. long.——Aug.–Oct. Honshu, Kyushu; rather rare.——Siberia, N. China, and Korea. The typical phase occurs in warmer areas from Asia to Australia, Africa, and S. America.

11. POUZOLZIA Gaudich. ŌBA-HIME-MAO ZOKU

Glabrescent or hirsute monoecious shrubs or herbs; leaves opposite or alternate, equilateral, usually entire, 3-nerved, the lateral nerves not reaching to the tip of the blade, the cystoliths punctiform, the stipules axillary, free, persistent; flowers in axillary glomerules or in terminal spikes, the bracts small, scarious; staminate perianth 3- to 5-parted, the segments ovate, convex on back; stamens 3-5; pistillate perianth tubular, often ovoid, nerved, sessile, contracted and 2- to 4-toothed at apex; stigma filiform, deciduous, villous; achenes black or ivory, enclosed in the sometimes persistent costate perianth, shining, smooth.——More than 20 species in the Tropics, abundant in Asia.

1. **Pouzolzia zeylanica** (L.) J. Benn. *Parietaria zeylanica* L.; *Parietaria indica* L.; *Parietaria alienata* L.; *Parietaria cochinchinensis* Lour.; *Pouzolzia indica* (L.) Gaudich.; *Pouzolzia microphylla* Wight; *Pouzolzia indica* var. *alienata* (L.) Wedd.; *P. indica* var. *alienata* subvar. *microphylla* (Wight) Wedd.——ŌBA-HIME-MAO, TSURU-MAO-MODOKI, ARIE-HIME-MAO. Ascending much-branched subshrub with elongate, slender, terete, pilose branches; leaves membranous, ovate to narrowly so, or sometimes broadly lanceolate in the upper ones, 2.5–4 cm. long, 5–20 mm. wide, acute to subobtuse, rounded to subcuneate at base, opposite or alternate, entire, weakly 3-nerved with the lateral nerves reaching 2/3–3/4 the entire length, the cystoliths white-punctiform, callose, the petioles 3–15 mm. long, the stipules deltoid, hyaline; flowers few, in axillary glomerules, nearly sessile; staminate perianth-segments 4, ovate, coarsely pilose; pistillate perianth ovoid in fruit, about 1.2 mm. long, several-ribbed, coarsely pilose; achenes ovoid, scarcely compressed, shining, smooth, black, 1 mm. long, acute at apex; stigma brown, about 1 mm. long, densely pilose, deciduous.——Aug.–Oct. Honshu (Aogashima in Izu Isls.), Kyushu (Yakushima).——Ryukyus, Formosa, China to Malaysia, and India.

12. GONOSTEGIA Turcz. TSURU-MAO ZOKU

Glabrescent to pubescent, monoecious, rarely dioecious, perennial herbs or subshrubs; leaves opposite or ternate, often alternate in the upper ones, usually equilateral, entire, 3-nerved, the lateral nerves reaching nearly to apex, the cystoliths punctiform, the stipules axillary, free or connate (interpetiolar); flowers with a scarious bract at base, in axillary glomerules or in spikes; staminate perianth 3- to 5-lobed, the segments ovate, short-acuminate, with a transverse fold on the upper half or infracted, the stamens as many as the segments; pistillate perianth tubular, often ovoid, enclosing the achene in fruit, toothed at the mouth; stigma filiform, hairy; achenes small, lustrous.——About 15 species in India, Malaysia, and se. Asia.

1. **Gonostegia hirta** (Bl.) Miq. *Urtica hirta* Bl.; *Pouzolzia hirta* (Bl.) Hassk.; *Memorialis hirta* (Bl.) Miq.——TSURU-MAO. Perennial suffrutescent herb with much-elongated procument or declined branches 30–50 cm. long, short-pubescent above; leaves membranous, ovate to lanceolate, 3–7 cm. long, 1.5–3 cm. wide, acute to acuminate, rounded to shallowly cordate at base, 3- to 5-nerved, entire, with appressed hairs on both sides or nearly glabrous, the petioles 1–3 mm. long; staminate flowers short-pedicellate, pilose; fruit ovoid, about 1 mm. long, the enveloping perianth with short white hairs especially above; stigma pilose.——Sept.–Oct. Honshu (Kii Prov. and Chūgoku Distr.), Kyushu.——China, Ryukyus, Formosa, Malaysia, and India.

Fam. 70. PODOSTEMACEAE KAWAGOKE-SŌ KA Riverweed Family

Submersed herbs attached to rocks in running water, often by frondlike or mosslike, flat, often pinnately divided photosynthetic roots; stems simple or scarcely developed, or sometimes much elongate and branched; flowers bisexual or unisexual, solitary or cymose, actinomorphic, often enclosed in a spathe when young; perianth membranous, lobed or absent; stamens 1–4, free or the filaments connate, the anthers 2-locular, longitudinally split; ovary superior, usually 2-locular, with a central placenta; styles 2–3, short or slender; ovules many; fruit a septicidal capsule; seeds minute, without endosperm.——About 43 genera, with about 140 species, mostly in the Tropics.

1A. Frondlike body rounded, obsoletely and irregularly lobulate; leaves unlobed; stamens 2, the filaments connate; capsules ribbed, 2-valved, the valves equal. .... 1. *Hydrobryum*
1B. Frondlike body pinnately branching; leaves palmately lobed; stamens 1; capsules dehiscing with 2 unequal valves, the smaller valve deciduous. .... 2. *Cladopus*

1. HYDROBRYUM Endl. KAWAGOROMO ZOKU

Submersed herbs, the frondlike body flat, rounded, irregularly shallowly lobulate, green; flowers borne on the frond, solitary on a very short simple scape covered with scalelike or filiform green leaves, the spathe entirely enveloping the flower in bud, splitting on one side in anthesis, the perianth-segments 2, subulate, adnate at base of the filament; stamens 2, the filaments connate below; ovary 2-locular, ovoid; stigmas 2, ovate, linear or cuneate, entire or lobed; capsules ovoid, 2-valved, ribbed, many-seeded.——About 10 species in India and e. Asia.

1A.   Frond thin, 0.1–0.3 mm. thick; subulate leaves 2–8 mm. long.
  2A.   Frond 0.1–0.15 mm. thick; subulate leaves 5–8 mm. long; stigma broadly subulate, 0.3–0.5 mm. long. . . . . 1. *H. floribundum*
  2B.   Frond 0.2–0.3 mm. thick; subulate leaves 2–4 mm. long; stigma cylindric, 0.4–0.5 mm. long. . . . . 2. *H. puncticulatum*
1B.   Frond about 0.5 mm. thick; subulate leaves 8–12 mm. long; stigma elongate-linear, 1.5–2 mm. long. . . . . 3. *H. japonicum*

1.   Hydrobryum floribundum Koidz.   *H. griffithii* var. *floribundum* (Koidz.) Koidz.; *Hydroanzia floribunda* (Koidz.) Koidz.——Usu-kawagoromo.   Frond flat, fixed by means of rootlets on rocks in running water, irregularly orbicular and shallowly lobulate, green, rarely to 30 cm. across; fascicles of leaves scattered on the upper surface of the frond, the subulate leaves 5–8 mm. long, in groups of 3–6 together, deciduous from the frond before anthesis (in autumn), their minute bases becoming accrescent, the scapes clothed with elliptic navicular bracts; flowers solitary, erect, on a very short scape, completely covered with a narrowly ovate spathe when young, soon splitting down one side; stamens 2, the filaments connate below, often to above the middle, the staminodes or perianth segments 2, subulate, as long as the ovary; stigmas 2, broadly subulate; capsules oblong-ellipsoidal, longitudinally 11- to 19-ribbed.——Nov.–Dec. Kyushu (s. distr.).

2.   Hydrobryum puncticulatum Koidz.   *Hydroanzia puncticulata* (Koidz.) Koidz.——Yakushima-kawagoromo. Closely resembles the preceding but the frond fleshy, green, irregularly orbicular, rarely more than 10 cm. across, 0.2–0.3 mm. thick; subulate leaves 2–4 mm. long, in scattered fascicles of 2's to 5's; flowers smaller and fewer on a frond than in the preceding, the spathe smaller; stamens 2, the filaments variously connate; stigma cylindric to long-subulate, 0.4–0.5 mm. long, rarely to 0.8 mm. long; capsules 6- to 12-ribbed.——Dec. Kyushu (Yakushima).

3.   Hydrobryum japonicum Imamura.   *H. griffithii* var. *japonicum* (Imamura) Koidz.; *Hydroanzia japonica* (Imamura) Koidz.——Kawagoromo.   The fronds nearly orbicular, lobulate, about 0.5 mm. thick, deep green; subulate leaves 8–12 mm. long, the bracts larger and the scapes longer; stigma 1.5–2 mm. long, long-subulate or linear; capsules dark brown, 11- to 22-ribbed, the valves 2, more or less unequal, the smaller valve often deciduous.——Nov.–Dec. Kyushu (s. distr.).

2.   CLADOPUS Moeller   Kawagoke-sō Zoku

Submersed perennial herbs on rocks in running water attached by rootlets, with flat green pinnately parted frondlike bodies bearing fascicles of small, palmately-lobed leaves and very short scapes on the upper side near the frond-sinus; flowers solitary and terminal on the scape, short-pedicelled, entirely enveloped while young by an ovoid spathe, splitting at the time of anthesis; stamen 1, attached to the ovary laterally, with a subulate segment on each side near base; ovary ovoid, oblique, 2-locular, the placenta central, many-ovuled; stigmas 2, spathulate or subulate; capsules globose, 2-valved, the valves unequal, the smaller one deciduous.——Few species in Malaysia, China, and Japan.

1A.   Fronds profusely parted; autumnal leaves of nearly equal size.
  2A.   Leaves deciduous. . . . . 1. *C. japonicus*
  2B.   Leaves evergreen; fronds very densely pinnatisect, the segments relatively thick, short, obtuse. . . . . 2. *C. austrosatsumensis*
1B.   Fronds loosely pinnatiparted; upper autumnal leaves larger than the lower. . . . . 3. *C. doianus*

1.   Cladopus japonicus Imamura.   *Hemidistichophyllum japonicum* (Imamura) Koidz.; *Lawiella kiusiana* Koidz.——Kawagoke-sō.   Fronds flat, deciduous, tightly adhering by dense rootlets on rocks in running water, regularly and alternately pinnately parted, deep green, the segments 2–4 mm. wide, 0.2–0.4 mm. thick, thinner toward the margin; fascicles of leaves on the upper side near the sinuses of the fronds; summer leaves 3–15, usually 7–8 in a fascicle, subulate or linear, 4–8 mm. long, about 0.3 mm. wide, 0.1 mm. thick, obtuse, usually deciduous before anthesis; autumnal leaves much shorter and developing in the place of the summer ones, 6–12, 2-seriate, depressed-orbicular, bifid and palmately lobed, densely arranged on the scape and concealing it; scapes very short, erect, 2–3 mm. long; flowers enclosed in a spathe while young, erect, solitary, the spathe about 1.5 mm. long, pinkish at tip; stamen 1, the filament subulate, subtended by a subulate segment (staminode or perianth-segment) on each side; stigmas 2, shortly connate below, usually obovate-spathulate; capsules globose, 1.5 mm. long, the stipe 4–5 mm. long.——Oct.–Dec. Kyushu (s. distr.).

2.   Cladopus austrosatsumensis (Koidz.) Ohwi.   *Lawiella austrosatsumensis* Koidz.——Tokiwa-podosutemon. Fronds evergreen, very densely pinnatisect, the segments 3–6 mm. wide, 0.4–0.6 mm. thick; summer leaves persistent over winter; autumnal leaves larger, stouter.——Kyushu (s. distr.).

3.   Cladopus doianus (Koidz.) Koriba.   *Lawiella doiana* Koidz.; *Hemidistichophyllum doianum* (Koidz.) Koidz.——Tsukushi-podosutemon.   Fronds evergreen, loosely pinnatiparted, the segments about 0.5 mm. thick, 1–2 mm. broad; summer leaves densely fascicled on upper surface near the base of the leaf sinus, 3–7 mm. long, about 0.3 mm. wide, subulate, acute to obtuse, sessile; autumnal leaves rhombic-orbicular or obovate-orbicular, sessile, bifid, 4- to 8-lobulate, larger than the foregoing species, the scape longer, to 10 mm. long; flowers solitary, subtended by an ovoid-elliptic, often reddish spathe, longitudinally split, the pedicel erect; stamen 1, often bifid, the filament with a minute subulate segment (staminode or perianth-segment) on each side at base; capsules globose to ellipsoidal, about 1.5 mm. long, unequally 2-valved, the smaller valve deciduous, the stipe about 3 mm. long.——Kyushu (s. distr.).

Fam. 71.   PROTEACEAE   Yama-mogashi Ka   Protea Family

Trees or shrubs, rarely herbs; leaves usually alternate, coriaceous, entire or pinnatiparted, exstipulate; flowers bisexual, the perianth uniseriate, regular or irregular, tubular, segments 4, hypogynous, gibbous, free, recurved above; stamens 4, adnate to the perianth-segments, the glands or scales 4, sometimes connate, alternate with the stamens or sometimes absent; ovary 1-locular, usually 1- to many-ovuled; fruit a drupe, follicle, samara, or achene; seeds without endosperm.——About 55 genera, with about 1,200 species, mostly in the drier regions of the S. Hemisphere.

1. HELICIA Lour. YAMA-MOGASHI ZOKU

Trees or shrubs; leaves alternate, entire, sometimes toothed; flowers in terminal or axillary racemes, bisexual, the perianth-segments 4, regular, each with a narrow contiguous claw that forms a narrow tube, the limb narrow, reflexed; anthers oblong, the connective extended at tip; ovary superior, sessile; style slender, elongate, the stigma terminal; ovules 2; nut ellipsoidal.——About 30 species, in tropical Asia and Australia.

1. **Helicia cochinchinensis** Lour. *H. lancifolia* Sieb. & Zucc.——YAMA-MOGASHI. Small evergreen tree with purple-brown branches and green glabrous branchlets; leaves coriaceous, oblanceolate to oblong, alternate, 5–12 cm. long, 1.5–5 cm. wide, abruptly acuminate, narrowed below, entire or with coarse teeth toward the tip, glabrous, the petioles 6–10 mm. long; inflorescence racemose, axillary, about 10 cm. long, short-peduncled, glabrous, densely many-flowered, the flowers in pairs, short-pedicelled, the perianth slender-tubular, 12–14 mm. long just before anthesis, slightly thickened above, the segments 4, white, linear, revolute in anthesis; stamens inserted on the upper part of the perianth-segments, the anthers broadly linear, 2–2.5 mm. long; fruit berrylike, ellipsoidal, purple-black, 10–12 mm. long.——Aug.–Oct. Warmer parts; Honshu (Tō-kaidō and s. Kinki to Chūgoku Distr.), Shikoku, Kyushu.——Ryukyus, Formosa, China and Indochina.

Fam. 72. OLACACEAE BOROBORO-NO-KI KA Olax Family

Trees, shrubs or rarely vines; leaves alternate, rarely opposite, usually entire, exstipulate; inflorescence a few-flowered usually axillary cyme or spike, the flowers relatively small, greenish, yellowish or white, rarely purplish, actinomorphic, bisexual or unisexual; sepals small, 4–6; corolla 3–6, free or connate into a tube; stamens equal in number or 2 to 3 times the number of corolla lobes; ovary imperfectly 2- to 5-locular or 1-locular, the ovules few, pendulous, placentation axile; fruit drupelike, 1-locular, 1-seeded, the seeds with copious endosperm.——About 40 genera, with about 230 species, mainly in the Tropics.

1. SCHOEPFIA Schreb. BOROBORO-NO-KI ZOKU

Shrubs or trees, glabrous, often blackened when dried; leaves entire, coriaceous; flowers yellow or white, in short axillary spikes or racemes; calyx adnate to the ovary, accrescent in fruit, the lobes small; petals 4, sometimes to 6, connate into a tube; stamens as many as and opposite the corolla-lobes, inserted on the tube; ovary 3-locular, inferior; stigma 3-lobed; ovules 3, pendulous; drupe berrylike.——More than 10 species, in the Tropics of Asia and America.

1. **Schoepfia jasminodora** Sieb. & Zucc. *Schoepfiopsis jasminodora* (Sieb. & Zucc.) Miers——BOROBORO-NO-KI. Small deciduous tree, glabrous, the branches yellowish gray, the slender branchlets deciduous with the leaves; leaves rather coriaceous, ovate or broadly deltoid-ovate, 4–6 cm. long, 2–4 cm. wide, caudately acuminate, rounded and petiolate at base, entire, the petioles rather broad, 4–7 mm. long; spikes loosely 3- to 4-flowered, peduncled, the flowers sessile, yellow, the perianth-tube 6–7 mm. long, tubular, the lobes ovate, slightly recurved above, about 3 mm. long; ovary about 2 mm. long, inferior; drupe ellipsoidal, about 8 mm. long.——Apr. Kyushu.——Ryukyus.

Fam. 73. SANTALACEAE BYAKU-DAN KA Sandalwood Family

Trees, shrubs, or herbs; sometimes parasitic; leaves alternate or opposite, sometimes reduced to scales, the stipules absent; flowers small, green, usually with bracts and bracteoles, actinomorphic, bisexual or unisexual, the perianth 3- to 8-toothed or -divided; stamens 3–6, adnate to and opposite the perianth-lobes or teeth; ovary 1-locular, inferior, rarely superior; style usually short, the stigma entire or 3- to 6-lobed; ovules 2–3, on central placentae; fruit a nut or drupe; seeds subglobose, with copious endosperm.——About 30 genera, with about 400 species, mostly tropical, a few in temperate regions.

1A. Herbs; leaves alternate; flowers bisexual; bracts solitary, free from the ovary. .... 1. *Thesium*
1B. Shrubs; leaves usually opposite; flowers unisexual; bracts 4, adnate to the ovary. .... 2. *Buckleya*

1. THESIUM L. KANABIKI-SŌ ZOKU

Green semiparasitic herbs; leaves alternate, linear, sometimes scalelike; inflorescence a spike, raceme, or compound cyme, the flowers sometimes solitary in axils, the bracts usually leaflike, subtending the flower or on the pedicel, the bracteoles 2, sometimes absent; flowers bisexual, small, the perianth-tube adnate below to the ovary, tubular to campanulate, (4–)5-lobed above; stamens 4–5, inserted at the base of the perianth lobes; ovary inferior, 2- to 3-ovuled; fruit nutlike, small, enclosed in the perianth; seeds solitary.——About 200 species, abundant in Africa and the Mediterranean region, few in Asia and S. America.

1A. Pedicels ascending, very short, to 4 mm. long; fruit with raised reticulate veinlets. .... 1. *T. chinense*
1B. Pedicels spreading, curved, 3–15 mm. long; fruit nearly smooth except for the longitudinal nerves. .... 2. *T. refractum*

1. **Thesium chinense** Turcz. *T. decurrens* Bl.——KANA-BIKI-SŌ. Slightly glaucescent glabrous perennial herb; stems few or solitary, erect, 10–25 cm. long, sometimes branched, obtusely angled to terete; leaves rather thick, linear, 2–4 cm. long, 1–3 mm. wide, acute, entire; flowers small, green, solitary on short pedicels, the bract solitary, rather small, similar to the leaves, the bracteoles 2, linear, 2–6 mm. long, the perianth 2.5–3 mm. long, tubular; fruit ellipsoidal-globose, 2–2.3 mm. long, green, with raised reticulations, crowned with the persistent perianth-lobes.——Apr.–June. Meadows in low-

lands and mountains; Honshu, Shikoku, Kyushu; common. ——Korea, China, and Ryukyus.

2. **Thesium refractum** C. A. Mey. *T. longifolium* sensu auct. Japon., non Turcz.; *T. repens* sensu auct. Japon., non Ledeb.——Kama-yari-sō. Resembles the preceding but larger; flowers on relatively long pedicels, the bracteoles shorter, the perianth 4–5 mm. long; fruit nearly smooth, longitudinally nerved.——June–July. Hokkaido.——e. Siberia, Manchuria, and Korea.

2. BUCKLEYA Torr. Tsukubane Zoku

Dioecious semiparasitic deciduous shrubs; leaves opposite, short-petioled, entire; staminate flowers small, 4-merous, in small terminal, axillary umbels, with a disc at base of the perianth, the bracts absent, the stamens inserted at the base of the perianth-segment; pistillate flower solitary and terminal, short-pedicelled, the bracts 4, spreading, leaflike, adnate below to the ovary, the perianth-tube adnate to ovary; staminodes absent; ovary inferior; style short; fruit ovoid or ellipsoidal, with 4 large persistent apical bracts.——Two or three species, in N. America and Japan.

1. **Buckleya lanceolata** (Sieb. & Zucc.) Miq. *Quadriala lanceolata* Sieb. & Zucc.; *Calycopteris joan* Sieb., nom. seminud; *B. joan* (Sieb.) Makino——Tsukubane. More or less short-pubescent shrub; leaves ovate, caudately acuminate, 4–8 cm. long, 1.5–3 cm. wide; flowers pale green; fruit 7–10 mm. long, the apical bracts oblanceolate, spreading, 2.5–3 cm. long.—— June. Woods at low elevations in mountains; Honshu (Kinki Distr. and eastw.).

Fam. 74. LORANTHACEAE Yadorigi Ka Mistletoe Family

Shrubs parasitic on shrubs and tree branches; leaves usually opposite, entire, sometimes minute and scalelike, without stipules; flowers bisexual or unisexual, spicate or cymose (sometimes fascicled in axils), usually bracteate and bracteolate, the calyculus (a modified calyx) inserted at base of the ovary, truncate on margin, or absent, the perianth of 1 or 2 whorls with 4–8 free or connate lobes valvate in bud; stamens as many as and opposite the perianth lobes; ovary inferior, 1-locular, 1-ovuled; stigma simple; ovules erect and adnate to the wall of the ovary; fruit a 1-seeded juicy, viscid berry, the seeds adnate to the pericarp, the endosperm fleshy, the embryo straight.——About 30 genera, with about 1,500 species, mainly in the Tropics, a few in temperate regions.

1A. Flowers bisexual.
  2A. Sepals free; flowers small, in spikes. . . . . 1. *Hyphear*
  2B. Sepals connate below to a tube; flowers rather large, in cymes. . . . . 2. *Taxillus*
1B. Flowers unisexual, small.
  3A. Leaves prominent; sepals deciduous. . . . . 3. *Viscum*
  3B. Leaves scalelike; sepals persistent. . . . . 4. *Korthalsella*

1. HYPHEAR Danser Hozaki-yadorigi Zoku

Parasitic shrubs with dichotomously branched stems and branches; leaves opposite, deciduous; flowers spicate, small, bisexual; sepals 4–6, free, 2-seriate, as many as the stamens; ovary inferior, the calyculus small; style short; fruit a berry.——Few in Eurasia.

1. **Hyphear tanakae** (Fr. & Sav.) Hosokawa. *Loranthus tanakae* Fr. & Sav.——Hozaki-yadorigi. Green glabrous shrubs with dichotomously forked dark-brown glabrous branches parasitic on deciduous trees; leaves nearly opposite, cuneate-lanceolate to oblong, 2–3.5 cm. long, 1–1.5 cm. wide, rounded at apex, narrowed at base, slightly nerved on both sides, glabrous, entire, the petiole 1–4 mm. long; spikes terminal on short branchlets, 3–4 cm. long, loosely several- to rather many-flowered; flowers sessile, the calyculus about 1 mm. long, subcampanulate, the sepals small, 4–6; fruit ellipsoidal, 5–6 mm. long.——July. Mountains; Honshu (centr. and n. distr.); rare.——s. Korea.

2. TAXILLUS Van Tiegh. Matsu-gumi Zoku

Parasitic shrub often with stellate hairs; leaves subopposite; flowers in short axillary cymes, sometimes fasciculate, pedicelled, bracteate at base; sepals 4–5, reflexed above, connate below and often deeply split on the lower side; anthers 4-locular; fruit ovoid to ellipsoidal, coriaceous.——More than 60 species; abundant in s. Asia and S. Africa.

1A. Leaves oblanceolate, 1.5–3 cm. long, 3–7 mm. wide, the midrib scarcely distinct; flowers glabrous, about 15 mm. long.
                                                   1. *T. kaempferi*
1B. Leaves broadly ovate, the midribs distinct; flowers pubescent, longer. . . . . 2. *T. yadoriki*

1. **Taxillus kaempferi** (DC.) Danser. *Viscum kaempferi* DC.; *Loranthus kaempferi* (DC.) Maxim.——Matsu-gumi. Evergreen much branched shrub, the young branchlets dark brown-pubescent; leaves nearly opposite, thick, coriaceous, oblanceolate, 1.5–3 cm. long, 3–7 mm. wide, obtuse to rounded at apex, narrowed at base, nearly nerveless, dark brown pubescent beneath while very young, glabrous above, the petiole about 1 mm. long; inflorescence cymose, umbelliform, 1- to 4-flowered, short-pedunculate, the pedicels nearly as long as the ovary; flowers about 15 mm. long, deep red, the perianth-tube slender, glabrous, the lobes 4, reflexed, about 4 mm. long, brown-yellow; fruit globose, red, about 5 mm. across.——Aug.

29B. Leaves solitary; perianth-lobes decurrent on the limb, the limb without a distinct ring on the throat.
  29. *A. variegatum*

22B. Flowers large; perianth-tube about 20 mm. long. . . . . 30. *A. megacalyx*

1. Asarum caulescens Maxim. *Japonasarum caulescens* (Maxim.) F. Maekawa——Futaba-aoi. Rhizomes elongate, the internodes 5–15 mm. long; leaves membranous, deciduous, long-petioled, usually 2, approximate, cordate, 6–15 cm. wide, abruptly acuminate, deeply cordate at base, loosely pilose on margin and on both sides; flowers solitary, terminal, the pedicels 3–4 cm. long, longer than the flowers, loosely pubescent with multicellular hairs, the perianth-segments ovate, slightly united at base, forming a short false tube on lower half, reflexed on upper half.——Mar.–May. Woods in mountains; Honshu, Shikoku, Kyushu.

2. Asarum sieboldii Miq. *Asiasarum sieboldii* (Miq.) F. Maekawa——Usuba-saishin. Rhizomes with short internodes; leaves usually in pairs, long-petioled, cordate or reniform-cordate, 5–10 cm. wide, usually abruptly acute, scattered short-pilose on both sides especially while young; flowers solitary, on short pedicels, glabrous, the perianth-tube depressed-globose, the lobes broadly deltoid-ovate, often recurved on margin; styles 6; stamens 12.——Mar.–Apr. Honshu, Kyushu (n. distr.).

3. Asarum dimidiatum F. Maekawa. *Asiasarum dimidiatum* (F. Maekawa) F. Maekawa——Kurofune-saishin. Closely resembling the preceding but the leaves thicker, 4–6 cm. long; styles 3; stamens 6.——Apr.–May. Shikoku, Kyushu (centr. distr.).

4. Asarum heterotropoides F. Schmidt. *Asiasarum heterotropoides* (F. Schmidt) F. Maekawa——Oku-Ezo-saishin. Closely resembling No. 2, but the leaves yellowish green, usually obtuse, 4–6 cm. wide; perianth-lobes fleshy, flat, obtuse, recurved above after anthesis.——May–June. Hokkaido.——Sakhalin, s. Kuriles.

5. Asarum sakawanum Makino. *Heterotropa sakawana* (Makino) F. Maekawa——Sakawa-saishin. Leaves rather thick, long-petioled, broadly to narrowly ovate, 6–10 cm. long, 4–7 cm. wide, acute, deeply cordate at base, loosely short-pilose on nerves above, ciliate; flowers glabrous, relatively large, the perianth-tube globose, about 12 mm. long, the lobes obtuse, much longer than the tube, the reduced petals 3, short-tubular, the styles ascending, incurved.——Apr.–May. Shikoku (w. half of Tosa Prov.).

6. Asarum costatum (F. Maekawa) F. Maekawa. *Heterotropa costata* F. Maekawa——Tosa-no-aoi. Leaves ovate-elliptic, subhastate, 10–15 cm. long, scattered-hairy on both sides, ciliate; flowers glabrous, the perianth tubular, 6-angled, the lobes obtuse, ascending, the petals absent; styles ascending, recurved.——May. Shikoku (e. half of Tosa Prov.).

7. Asarum hexalobum (F. Maekawa) F. Maekawa. *Heterotropa hexaloba* F. Maekawa——Sanyō-aoi. Leaves ovate to broadly so, deeply cordate, 5–10 cm. long, 4–8 cm. wide, subobtuse or sometimes acute, long-petioled, glabrous beneath; flowers glabrous, short-pedicelled; perianth-tube depressed-globose to obconical-globose, prominently constricted at the throat, with 6 longitudinal inflations, and 6 prominently longitudinal raised and inconspicuous transverse lines inside; perianth lobes nearly spreading, ovate-orbicular to broadly ovate, about 10 mm. long, glabrous; stamens 6, the staminodes 6.——Apr. Honshu (w. Chūgoku Distr.), Kyushu (n. distr.).

Asarum perfectum (F. Maekawa) F. Maekawa. *A. hexalobum* var. *perfectum* F. Maekawa——Kinchaku-aoi. A

close relative of No. 7, differing chiefly in the oblong leaves and rather small flowers short-pilose inside, the lobes not connivent above; stamens 12, staminodes absent. Poorly known.——Kyushu and Shikoku (w. distr.).

8. Asarum crassum F. Maekawa. *Heterotropa crassa* (F. Maekawa) F. Maekawa——Nangoku-aoi. Leaves broadly ovate or ovate-cordate, 11–17 cm. long, very thick, glabrous, lustrous and with impressed veinlets above; flowers densely white-puberulent, the perianth-tube semi-globose, the lobes obliquely spreading.——Feb. Cultivated in Kyushu.

9. Asarum parviflorum (Hook.) Regel. *Heterotropa parviflora* Hook.; *A. elegans* Duchart.——Kobana-kan-aoi. Leaves ovate, acute, long-petioled; flowers small, shorter than the bracts, the perianth-tube broadly ellipsoidal, prominently constricted at apex, the limb shallowly tubulose, the lobes broadly ovate, nearly spreading.——Reported to be spontaneous in Honshu (Kantō Distr.).

10. Asarum constrictum F. Maekawa. *Heterotropa constricta* (F. Maekawa) F. Maekawa——Tsukubane-aoi. Leaves 1 or 2, ovate, 7–10 cm. long, flat, acute, cordate, lustrous, the nerves not impressed or raised, glabrous beneath; flowers on very short pedicels, the perianth-tube 7–8 mm. long, trapezoidal in lateral view, truncate and much-constricted at apex, with about 15 longitudinal ribs and cross-veins inside, the lobes spreading, ovate, about 10 mm. long.——Mar. Rarely cultivated in Honshu (near Tokyo).

11. Asarum asperum F. Maekawa. *Heterotropa aspera* (F. Maekawa) F. Maekawa——Miyako-aoi. Leaves solitary, ovate-orbicular, 6–8 cm. long, subhastate at base, dull, subobtuse, glabrous beneath; perianth-tube trapezoidal in lateral view, about 6 mm. long, 9 mm. across, prominently constricted at apex, with 15 longitudinal ribs and 2 or 3 transverse veins inside, the lobes spreading, 8–10 mm. long.——Apr. Honshu (w. distr.), Shikoku.

12. Asarum tamaense Makino. *Heterotropa tamaensis* (Makino) F. Maekawa——Tama-no-kan-aoi. Leaves solitary, broadly elliptic to ovate-orbicular, obtuse, 5–10 cm. long, lustrous, with impressed veinlets above, glabrous beneath; flowers short-pedicelled, glabrous, the perianth short-tubular, slightly constricted at apex, subtruncate at base, the limb spreading, the lobes cordate-orbicular, densely crisped-pubescent inside.——Apr. Honshu (w. Kantō Distr.).

13. Asarum muramatsui Makino. *Heterotropa muramatsui* (Makino) F. Maekawa——Amagi-kan-aoi. Closely resembling No. 12; leaves ovate to ovate-elliptic, with prominently impressed veinlets above, nearly glabrous, strongly lustrous; styles more than half the length of the perianth-tube, strongly recurved above.——May. Honshu (Izu Prov.).

14. Asarum curvistigma F. Maekawa. *Heterotropa curvistigma* (F. Maekawa) F. Maekawa——Kagigata-aoi. Leaves ovate to ovate-elliptic, 5–10 cm. long, 4–7 cm. wide; the veinlets slightly impressed above, glabrous and smooth beneath; flowers short-pedicelled, the perianth-tube broadly campanulate, slightly constricted at apex, 7–13 mm. long, 10–13 mm. wide, with 15–18 longitudinal ribs and cross-veinlets inside, the lobes spreading, usually short-pubescent inside.——Sept.–Oct. Honshu (Tōkaidō Distr.).

15. Asarum asaroides (Morr. & Decne.) Makino. *Heterotropa asaroides* Morr. & Decne.; *A. thunbergii* A.

Braun——TAI-RIN-AOI. Leaves broadly ovate to nearly deltoid-ovate, 8–12 cm. long, 5–10 cm. wide, long-petioled, abruptly obtuse, short-pubescent on nerves above, glabrous beneath; flowers very shortly pedicelled, relatively large, glabrous, the perianth-tube depressed-pyriform, prominently constricted at apex, 2–2.5 cm. long, finely net-veined inside, the lobes orbicular-cordate or deltoid-orbicular, spreading, undulate on margin, 1–1.5 cm. long.——Apr.–June. Honshu (w. Chūgoku Distr.), Kyushu (n. distr.).

16. Asarum satsumense F. Maekawa. *Heterotropa satsumensis* (F. Maekawa) F. Maekawa——SATSUMA-AOI. Resembling No. 15; leaves broadly ovate-elliptic; flowers smaller, the perianth-lobes prominently undulate on margin; styles with a short, narrow auriclelike wing on each side near top.——May. Kyushu (s. distr.).

17. Asarum blumei Duchart. *A. albivenium* Regel; *A. leucodictyon* Miq.; *Heterotropa blumei* (Duchart.) F. Maekawa——RAN'YŌ-AOI. Leaves solitary, broadly hastate-ovate to broadly ovate, 6–10 cm. long, 4–8 cm. wide, acute to obtuse, the basal lobes more or less spreading, short-pilose on both sides especially on nerves and on margin, long-petioled; flowers glabrous, rather distinctly pedicelled, the perianth-lobes inflated, obtusely 4-angled, in lateral view, 1–1.2 cm. long and as wide, slightly constricted at apex, net-veined inside, the lobes spreading, ovate-orbicular, slightly incurved at apex, the throat narrow.——Mar.–May. Honshu (Izu, Sagami, Suruga Prov.).

18. Asarum kiusianum F. Maekawa. *Heterotropa kiusiana* (F. Maekawa) F. Maekawa——TSUKUSHI-AOI. Leaves usually oblong, 6–10 cm. long, acute, deeply cordate to slightly hastate-cordate at base, smooth and scarcely lustrous above; flowers prominently pedicelled, erect, straight, pale purple or dull purple, the perianth nearly tubular, 10–15 mm. long, slightly broadened above and somewhat constricted at apex, with about 15 longitudinal ribs and lateral veinlets inside, the lobes ovate-deltoid, spreading.——Apr. Kyushu (nw. distr.).

Var. **tubulosum** (F. Maekawa) F. Maekawa. *A. tubulosum* F. Maekawa——AKEBONO-AOI. Flowers white, the perianth-tube scarcely broadened above.——Kyushu.

Var. **melanosiphon** (F. Maekawa) F. Maekawa. *A. melanosiphon* F. Maekawa——KIKYŌ-KAN-AOI. Perianth-tube dark purple, the limb thinner, smooth inside, white to paler purple.——Kyushu.

Asarum unzen F. Maekawa. *Heterotropa unzen* F. Maekawa——UNZEN-KAN-AOI. Intermediate between No. 18 and No. 15; flowers about 2.5 cm. across, the perianth-tube obovoid-cylindric, the lobes similar to No. 15; styles simulating No. 18.——Kyushu (w. distr.).

19. Asarum subglobosum F. Maekawa. *Heterotropa subglobosum* F. Maekawa——MARUMI-KAN-AOI. Resembling No. 18, but the perianth-tube roundish, the limb nearly flat, smooth.——Kyushu (Hyuga Prov.).

20. Asarum trigynum (F. Maekawa) Araki. *Heterotropa trigyna* F. Maekawa——SANKO-KAN-AOI. Resembling No. 18, but with 3 styles and 6 stamens.——Apr. Kyushu (Koshiki Isl.).

21. Asarum yakusimense Masam. *Heterotropa yakusimensis* (Masam.) F. Maekawa——YAKUSHIMA-AOI. Leaves about 8 cm. long, deltoid-ovate, thinly pubescent beneath and on petioles; flowers nearly sessile, 1.5–2 cm. across, dark purple, the perianth-tube short-cylindric, pubescent outside, the limb often partially yellowish.——Nov.–Feb. Kyushu (Yakushima and Tanegashima).

22. Asarum nipponicum F. Maekawa. *Heterotropa nip-*
ponica (F. Maekawa) F. Maekawa; *A. blumei* sensu auct. Japon., non Duchart.——KAN-AOI. Leaves flat, ovate to broadly so, 6–10 cm. long, 4–7 cm. wide, glabrous beneath, subobtuse or subacute; flowers short-pedicelled, sometimes subsessile, the perianth-tube campanulate, not constricted at apex, with 9, rarely 12 longitudinal ribs and lateral veinlets inside, the lobes spreading.——Oct.–Feb. Honshu (Kantō and centr. Distr.).

Var. **brachypodion** F. Maekawa——SUZUKA-KAN-AOI. Perianth-lobes larger, much longer than the tube; leaves with more or less impressed veinlets above.——Honshu (Tōkaidō to Kinki Distr. and w. Hokuriku Distr.).

Var. **kooyanum** (Makino) F. Maekawa. *A. kooyanum* Makino——KŌYA-KAN-AOI. Perianth-lobes short, about half as long as the tube.——Honshu (Kii Prov.).

Asarum rigescens F. Maekawa. *Heterotropa rigescens* (F. Maekawa) F. Maekawa——ATSUMI-KAN-AOI. Perianth fleshy, strongly net-veined inside. Poorly known.——Jan.–Feb. Honshu (Kinki and centr. Distr.).

23. Asarum nankaiense F. Maekawa. *Heterotropa nankaiensis* (F. Maekawa) F. Maekawa——NANKAI-AOI. Resembles *A. nipponicum* but without impressed veinlets on leaves above; flowers in bud with 3 impressions on perianth-lobes, slightly constricted at base.——Honshu (Kii Prov.), Shikoku.

24. Asarum kumageanum Masam. *Heterotropa kumageanum* (Masam.) F. Maekawa——KUWA-IBA-KAN-AOI. Leaves long-petioled, thick, lustrous, cordate-oblong, 3–4 cm. wide, acute, cordate at base, appressed-pilose only on nerves above; flowers pedicelled, slightly pubescent, the perianth-tube obconical, the lobes deltoid-orbicular, about 1 cm. long and as wide, rounded at apex.——Feb.–Mar. Kyushu (Yakushima).

25. Asarum savatieri Franch. *Heterotropa savatieri* (Franch.) F. Maekawa——OTOME-AOI. Leaves ovate-orbicular to broadly ovate, 5–7 cm. long; flowers about 1.5 cm. across, the perianth tubular-globose, slightly constricted at apex, about 1 cm. long, with 15–21 rows of delicate reticulations, the lobes deltoid-ovate, spreading, the limb with a small mouth on the center.——July–Aug. Honshu (Sagami, Izu, and Suruga Prov.).

26. Asarum fauriei Franch. *Heterotropa fauriei* (Franch.) F. Maekawa——MICHI-NO-KU-SAISHIN. Rhizomes elongate, with long internodes; leaves 1–3, rounded to reniform-orbicular, about 3 cm. across, deep green and lustrous above, not variegated; flowers small, 1–1.5 cm. across, the perianth-tube short; styles exserted from the tube.——Apr.–May. Honshu (centr. and n. distr.).——cv. **Serpens**. *A. fauriei* var. *serpens* F. Maekawa.——SONOU-SAISHIN. A cultivar of gardens with reniform-orbicular retuse lustreless gray-green leaves with pale variegation above.

Var. **nakaianum** (F. Maekawa) Ohwi. *Asarum nakaianum* F. Maekawa——MIYAMA-AOI. Perianth-tube very short, cup-shaped, the leaves dull.——Honshu (centr. distr.).

27. Asarum viridiflorum Regel. *Heterotropa viridiflora* (Regel) F. Maekawa——MOEGI-KAN-AOI. Leaves pale bluegreen, orbicular, relatively thin, flat, about 5 cm. wide; flowers about 1.5 cm. across, the perianth-tube broadly tubulose, 5–9 mm. long, about 9 mm. across, very finely reticulate inside, the lobes broadly deltoid-ovate, spreading.——Mar.–Apr. Cultivated, but possibly indigenous in our area.

28. Asarum takaoi F. Maekawa. *Heterotropa takaoi* (F. Maekawa) F. Maekawa——HIME-KAN-AOI. Leaves often paired, ovate-orbicular to ovate, broadly obtuse to subobtuse,

5–8 cm. long, 4–7 cm. wide; flowers about 1.5 cm. across, the perianth broadly tubular, with about 18 longitudinal ribs and accompanying veinlets inside, the lobes nearly spreading, ovate, obtuse; styles nearly as long as the tube.——Feb.–Mar. Honshu (w. Tōkaidō and Hokuriku to e. Kinki Distr.).

Var. **hisauchii** (F. Maekawa) F. Maekawa. *A. hisauchii* F. Maekawa——Zeniba-saishin. Leaves usually orbicular and small.——Cultivated in Kinki Distr. of Honshu (spontaneous in Tōkaidō and Kinki Distr. incl. Kiso region).

Var. **dilatatum** (F. Maekawa) F. Maekawa. *Heterotropa dilatata* F. Maekawa——Suehiro-aoi. Perianth-tube much inflated, deeply cup-shaped, with 3 shallow wrinkles on the limb at base on the throat.——Honshu (Kinki Distr.).

29. **Asarum variegatum** A. Braun & Bouché. *Heterotropa variegata* (A. Braun & Bouché) F. Maekawa——Koba-

no-kan-aoi. Small herb; leaves solitary, reniform-orbicular, about 3 cm. wide, slightly retuse, flat, dull, long-petioled; flowers resembling No. 28 but thicker, the perianth-lobes decurrent to the limb, without a distinct limit between the throat.——Mar.–Apr. Known only in cultivation.

30. **Asarum megacalyx** F. Maekawa. *Heterotropa megacalyx* (F. Maekawa) F. Maekawa——Koshi-no-kan-aoi. Leaves long-petioled, ovate-hastate or hastate, 9–12 cm. long, 6.5–8 cm. wide, acute, lustrous, glabrous beneath; perianth tubular-globose, about 2 cm. long and as wide, subtruncate at base, with 15 longitudinal ribs and accompanying transverse veinlets inside, the lobes spreading, broadly ovate, about 12 mm. long, 14 mm. wide.——Apr.–May. Honshu (Hokuriku and w. part of n. distr.).

2. ARISTOLOCHIA L. Uma-no-suzu-kusa Zoku

Scandent or rarely erect perennial herbs or shrubs; leaves alternate, petiolate, entire or 3- to 5-lobed, usually cordate, 5- to 7-nerved; flowers zygomorphic, solitary or in axillary fascicles, the calyx (perianth) adnate below to the ovary, the tube narrow, usually inflated around the style, the throat slightly narrowed, the limb open or recurved, entire or 3- to 6-lobed; stamens usually 6, the anthers sessile, adnate to the style or stigma, 2-locular, the locules longitudinally split; ovary inferior, usually 6-locular, the placentae parietal; style short, 3- to 6-lobed; capsules 6-valved; seeds numerous, horizontal, flat, or recurved on margin, compressed.——About 600 species, mainly in the Tropics and subtropics, a few in the temperate regions.

1A. Plants pubescent; calyx-tube prominently curved, the limb orbicular and shallowly 3-lobed.
  2A. Hairs rather dense and long; leaves subcoriaceous, usually orbicular-cordate; calyx-limb about 1.5 cm. across, yellow-green, brown-striate inside, the tube 2.5–3.5 cm. long. . . . . 1. *A. kaempferi*
  2B. Hairs loose and relatively short; leaves thick-membranous or chartaceous, usually deeply 3-lobed; calyx-limb larger, yellow, changing to black-purple, the tube 4–5.5 cm. long. . . . . 2. *A. onoei*
1B. Plants glabrous; calyx-tube slightly curved, the limb obliquely cut, narrowly deltoid, entire.
  3A. Leaves narrowly deltoid-ovate, rather thick-membranous, 1.5–3 times as long as wide, the petioles 8–20 mm. long; flowers solitary in axils; calyx-limb short-attenuate, acute. . . . . 3. *A. debilis*
  3B. Leaves cordate, thinly membranous, nearly as long as wide, the petioles 1–7 cm. long; flowers few in axils; calyx-limb with a long filiform apex. . . . . 4. *A. contorta*

1. **Aristolochia kaempferi** Willd. *A. lineata* Duchart.; *Hocquartia kaempferi* (Willd.) Nakai——Ōba-uma-no-suzu-kusa. Scandent shrub, 2–3 m. high, the young branches and leaves densely soft-pubescent while young; leaves often polymorphic, petiolate, cordate-orbicular, 8–15 cm. long and as wide, sometimes 3-lobed, obtuse to subacute, cordate, subcoriaceous, glabrate above, gray-white pubescent beneath; flowers solitary in axils, pedicels 4–5 cm. long, densely brownish pubescent; calyx-tube recurved near the middle, the 3 lobes orbicular, glabrous inside, green-yellow, brown-striate, valvate and covering the throat in bud; capsules oblong, 6-ribbed, 5–7 cm. long; seeds elliptic, about 5 mm. long, rounded on back, strongly recurved, with a raised longitudinal line on ventral face.——May–June. Honshu (Kantō Distr. and westw.), Shikoku, Kyushu.——Ryukyus and China.

2. **Aristolochia onoei** Fr. & Sav. *A. arimaensis* Makino; *A. kaempferi* var. *trilobata* Fr. & Sav.——Hosoba-uma-no-suzu-kusa, Arima-uma-no-suzu-kusa. Resembling the preceding but more slender and less hairy; leaves thinner, shorter-pubescent beneath, broadly ovate to lanceolate, usually 3-lobed; flowers rather large, the calyx-limb yellow, soon changing to black-purple.——June. Honshu (centr. distr. and westw.), Shikoku, Kyushu.

3. **Aristolochia debilis** Sieb. & Zucc. Uma-no-suzu-kusa.

Scandent glabrous perennial from stout creeping elongate rhizomes; young branches dark purple, soon glaucescent; stems slender, 30–100 cm. long, branching; leaves petiolate, narrowly ovate-deltoid, obtuse, cordate, 4–7 cm. long, 2–5 cm. wide, the petioles 8–20 mm. long; flowers solitary in axils, the pedicels 2–4 cm. long, the calyx-tube yellowish green, slender, slightly recurved above, inflated and globose at base, the limb obliquely truncate, narrowly deltoid-acuminate, slightly recurved on margin, brown-purple inside; capsules globose, about 1.5 cm. long.——June–Aug. Thickets and meadows in lowlands; Honshu (Kantō Distr. and westw.), Shikoku, Kyushu.——China.

4. **Aristolochia contorta** Bunge. *A. nipponica* Makino——Maruba-uma-no-suzu-kusa, Ko-uma-no-suzu-kusa. Resembles the preceding; plant glabrous; leaves cordate or broadly ovate-cordate, 4–10 cm. long, 3.5–8 cm. wide, acute or obtuse, cordate, thinly membranous, glaucescent, the petioles 1–7 cm. long; flowers few in axils, fascicled, the pedicels 1–4 cm. long; calyx-tube inflated and globose at base, the limb dilated, obliquely truncate, narrowly deltoid, long-acuminate to a filiform point; fruit obovoid-globose, about 3 cm. long.——July–Aug. Honshu.——Korea, n. China, Manchuria, and Ussuri.

Fam. 76. RAFFLESIACEAE Yakko-sō Ka Rafflesia Family

Fleshy parasitic herbs with or without reduced scalelike leaves; stems short or nearly absent, scaly or naked; flowers terminal, solitary, or in a terminal raceme, actinomorphic, bisexual or unisexual, the perianth of one whorl, fleshy, globose, or campanulate, the tube adnate below to the ovary, the limb 3- to 10-lobed, the segments imbricate or valvate in bud; stamens 8 to many, the anthers 2-locular, sessile around a fleshy genital column or inserted on the throat of the perianth-tube, free or connate; ovary superior, very rarely half-superior, 1-locular or imperfectly many-loculed by intricate lamellate placentae; stigma undivided or lobed; ovules many, orthotropous or anatropous; fruit fleshy, indehiscent or irregularly ruptured; seeds minute, many, with reticulate outer coat.——About 10 genera, with about 60 species, mainly pantropical.

1. MITRASTEMON Makino　　Yakko-sō Zoku

Glabrous herbs without chlorophyll, with short thickened scabrous rhizomes; stems erect, simple, short, fleshy, unbranched, scaly; scalelike leaves decussately opposite, ovate, obtuse, entire, the upper ones becoming gradually larger; flowers solitary, terminal, erect, bractless, bisexual; perianth of a single connate whorl, truncate, tubular, fleshy, persistent; stamens completely connate, the anthers extrorse, dehiscing by pores, the connective connate and forming a depressed-conical body over the stigma; ovary superior, relatively large, ovoid-globose, sessile, 1-locular, with lamellate placentae; style short, stout, the stigma thick, depressed-conical, subbilobed; testa 1-layered.——Few species, in e. and se. Asia, Central America.

1. **Mitrastemon yamamotoi** Makino. Yakko-sō. Stems 5–7 cm. high, white, black-brown when dry; scalelike leaves thick, few-paired, 10–20 mm. long, ovate-deltoid or deltoid, ascending, subcoriaceous, rounded on back, incurved on margin; flowers 15–20 mm. long, the perianth 5–8 mm. long; stamens 14–20 mm. long, the filaments connate and surrounding the ovary, the anthers 4–5 mm. long; stigma 4–6 mm. across.——Nov. Shikoku, Kyushu; parasitic on roots of *Castanopsis;* rare.

Fam. 77. BALANOPHORACEAE　　Tsuchi-tori-mochi Ka　　Balanophora Family

Variously colored monoecious or dioecious root-parasitic herbs without chlorophyll; rhizomes tuberous, often lobed, sometimes scaly, often scabrous; stems erect, terete, usually scaly; leaves reduced to scales, alternate; spikes terminal, ovoid, clavate or globose, with stout fleshy axis, the bracts absent or many; flowers unisexual, the pistillate sessile, minute, densely arranged on the face of the axis, many, sometimes somewhat hairy, the perianth absent or small and adnate to the ovary, rarely free, bilabiate or tubular; staminate flowers larger, interspersed among the pistillate or at one end of the spike, naked or with a 3- to 8-lobed perianth; stamens 1 or 2, or as many as and opposite the perianth-lobes; ovary 1- to 3-locular, small, the ovules solitary in each locule; fruit small, indehiscent, 1-seeded.——About 15 genera, with about 100 species, mainly in the Tropics of the Old World.

1. BALANOPHORA Forst.　　Tsuchi-tori-mochi Zoku

Fleshy glabrous parasitic herbs; scapes with persistent alternate scalelike leaves; staminate flowers dispersed among the pistillate ones or on one end of the spike, rarely on different spikes, the perianth-tube cylindric, solid, the lobes ovate, 3–6, imbricate in bud; stamens 3–8, the filaments absent or united in a column; pistillate flowers naked; ovary ellipsoidal, 1-locular, sometimes stipitate; style single, slender; ovules solitary, anatropous; fruit indehiscent; seeds globose, the endosperm copious, oily.—— About 70 species, in India, Australia, and s. Asia, few in e. Asia.

1A. Plants dioecious; staminate flowers unknown; rhizomes lenticellate; spikes blood-red or yellow-red, ovoid, or cylindric-ovoid. .... 1. *B. japonica*

1B. Plants monoecious; rhizomes not lenticellate; spikes pale yellow or whitish, ovoid to narrowly so. .... 2. *B. tobiracola*

1. **Balanophora japonica** Makino. *Balania japonica* (Makino) Van Tiegh.——Tsuchi-tori-mochi. Stout glabrous reddish herb 5–10 cm. high; rhizomes irregularly globose, lobed, pale brown, 2–3 cm. across, with raised whitish lenticels; scapes terete, stout, erect, few-scaled; scales suberect, fleshy, reddish, broadly ovate to oblong, entire; spikes terminal, solitary, stout, narrowly ovoid to ovoid, 3–6 cm. long, 2–3 cm. wide, blood-red; pistillate flowers minute, fleshy, with minute ovoid-globose bodies between them; ovary stipitate, ellipsoidal, small, the style linear.——Oct.–Dec. On roots of evergreen trees, especially *Symplocos;* Honshu, Shikoku, Kyushu.—— Amami-Oshima.

Var. **nipponica** (Makino) Ohwi. *B. nipponica* Makino; *Balania nipponica* (Makino) Masam.——Miyama-tsuchi-tori-mochi. More slender; spikes ovoid-cylindric to narrowly ovoid, orange- to reddish yellow, sometimes reddish.——July–Oct. On roots of deciduous trees; Honshu (centr. and n. distr.), Kyushu; rare.

2. **Balanophora tobiracola** Makino. *Balaneikon tobiracola* (Makino) Setchell——Kiire-tsuchi-tori-mochi. Resembling the preceding species; scapes 10–15 cm. long; rhizomes not lenticellate; scapes and spikes pale yellow; scales ovate; spikes monoecious, the staminate flowers larger, dispersed among the pistillate.——Oct.–Nov. On roots of *Pittosporum, Raphiolepis,* and *Ligustrum;* Kyushu (s. distr.)—— Ryukyus and Formosa.

Fam. 78. POLYGONACEAE　　Tade Ka　　Buckwheat Family

Herbs, rarely shrubs; leaves alternate, rarely opposite, simple or lobed, the petioles often dilated below into a stipular sheath (ochrea); flowers usually small, bracteate, solitary or fasciculate in axils or in spikes or racemes, rarely in a compound panicle, bisexual, sometimes unisexual, actinomorphic, the pedicels usually jointed at apex; perianth of 4–6 free, frequently highly colored sepals sometimes connate at base, imbricate in bud (petals absent), sometimes accrescent after anthesis; stamens 6–9, rarely many, inserted at the base of perianth-segments; ovary superior, trigonous or lenticular, 1-locular, 1-ovuled; styles 2 or 3, often connate at base; fruit an achene, indehiscent, enclosed in the persistent perianth, the pericarp hard, the testa membranous; endosperm copious, mealy, the embryo often curved.——About 30 genera, with about 800 species, cosmopolitan.

1A. Stigma fimbriate; inner perianth-segments usually accrescent after anthesis.
  2A. Perianth-segments (sepals) 6; styles 3; achenes trigonous. .... 1. *Rumex*
  2B. Perianth-segments (sepals) 4; styles 2; achenes lenticular. .... 2. *Oxyria*
1B. Stigma capitellate or 2-lobed, rarely fimbriate but the inner perianth-segments unchanged after anthesis. .... 3. *Polygonum*

1. RUMEX L.   Gishi-gishi Zoku

Perennial or annual herbs, rarely subshrubs; leaves radical and cauline, alternate, the stipules sheathed; flowers bisexual or unisexual, small, fascicled at the nodes, ultimately disposed in terminal racemes or panicles; perianth-segments biseriate, 6, rarely 4, nearly equal, herbaceous or coriaceous, entire or dentate to spine-margined, the midrib sometimes swollen at the middle, the outer 3 unchanged in fruit, the inner 3 usually accrescent after anthesis; stamens 6; styles 3, the stigma laciniate-fimbriate; achenes trigonous, enclosed in the accrescent inner perianth-segments; embryo lateral.——Many species, cosmopolitan.

1A. Leaves hastate, with acute auricles on each side; flowers usually unisexual (plants dioecious); styles lateral.
  2A. Perianth-segments not accrescent in fruit. .... 1. *R. acetosella*
  2B. Perianth-segments of the inner whorl accrescent and winglike in fruit.
    3A. Leaves broadly lanceolate or oblong; sheath dentate or lobed from the earlier stage. .... 2. *R. acetosa*
    3B. Leaves narrowly ovate to broadly so; sheath entire while young. .... 3. *R. montanus*
1B. Leaves cuneate, rounded or cordate at base, without auricles on lower margin; flowers bisexual; styles terminal; inner perianth-segments accrescent.
  4A. Inner perianth-segments orbicular to orbicular-cordate in fruit, the midrib not thickened or only slightly swollen in one sepal.
    5A. Verticils or fascicles of fruit dense; leaves glabrous, the lower oblong or narrowly so, the upper linear-oblong.
      6A. Leaves cordate; inner perianth-segments truncate to rounded at base. .... 4. *R. aquaticus*
      6B. Leaves narrowed to truncate or rounded at base; inner perianth-segments cordate. .... 5. *R. longifolius*
    5B. Verticils or fascicles of fruit interrupted; leaves with hairlike papillae especially on nerves beneath.
      7A. Lower leaves ovate or broadly so.
        8A. Inner perianth-segments shallowly toothed in fruit. .... 6. *R. madaio*
        8B. Inner perianth-segments with hooked spines on margin in fruit. .... 7. *R. nepalensis*
      7B. Lower leaves deltoid-cordate; inner perianth-segments entire. .... 8. *R. gmelinii*
  4B. Inner perianth-segments with a prominent tubercle at the center.
    9A. Inner perianth-segments orbicular to orbicular-cordate, 4–7 mm. wide, acutely toothed to nearly entire.
      10A. Radical leaves cordate; inner perianth-segments orbicular-cordate or ovate-cordate, more or less distinctly toothed.
          9. *R. japonicus*
      10B. Radical leaves rounded to truncate or cuneate at base; inner perianth-segments ovate-orbicular, entire or nearly so.
          10. *R. crispus*
  9B. Inner perianth-segments ovate or narowly so.
    11A. Inner perianth-segments entire, about 1.5 mm. wide; stems usually dark purplish; leaves often partially blood-purple variegated.
          11. *R. conglomeratus*
    11B. Inner perianth-segments prominently toothed or spined on margin.
      12A. Inner perianth-segments 2.5–3 mm. long, narrower than the spines on margin; verticils of flowers many-flowered, often globose. .... 12. *R. maritimus*
      12B. Inner perianth-segments 4–6 mm. long, broader than the length of marginal teeth or spines; verticils of flowers few- to rather many-flowered.
        13A. Lower leaves oblong-ovate or broadly ovate, cordate, with hairlike papillae beneath; verticils of flowers rather contiguous above, separated in the lower ones. .... 13. *R. obtusifolius*
        13B. Lower leaves oblong-lanceolate, smooth, rounded at base; verticils of flowers rather widely separated. .. 14. *R. nipponicus*

1. Rumex acetosella L. Hime-suiba. Dioecious perennial herb, acid to taste, from slender creeping branched rhizomes, glabrous or with scattered hairlike papillae; stems slender, erect, few-leaved, sometimes branching; radical leaves long-petioled, hastate, 3–6 cm. long, 1–2 cm. wide, acute to obtuse, the cauline rather small, the sheaths scarious; inflorescence paniculate, the flowers small, about 2 mm. long, the sepals unchanged in fruit, not warty; achenes broadly elliptic, trigonous, brown, slightly lustrous, about 1.2 mm. long, the angles rather obtuse.——May–July. Waste grounds and meadows in lowlands and foothills; Hokkaido, Honshu, Shikoku, Kyushu; naturalized and very common locally.——Europe and w. Asia.

2. Rumex acetosa L. Suiba. Dioecious perennial herb, acid tasting, glabrous or with obsolete granular papillae on the inflorescence; rhizomes short, with thick adventitious roots; stems erect, 30–80 cm. long, usually simple, few-leaved; radical leaves petioled, broadly lanceolate to oblong, hastate, 5–10 cm. long; panicles terminal, erect, 10–30 cm. long, narrow, the branches erect, slender; staminate flowers with the sepals all alike, 2.5–3 mm. long, the outer sepals of the pistillate flowers elliptic, reflexed, to 1 mm. long, the inner ones strongly accrescent, often pinkish, orbicular-cordate, membranous, net-veined, about 5 mm. long and as wide; achenes elliptic, black-brown, lustrous, trigonous, about 2 mm. long, the angles acute.——May–June. Meadows; Hokkaido, Honshu, Shikoku, Kyushu; very common.——Temperate regions of the N. Hemisphere.

3. Rumex montanus Desf. *R. arifolius* All., non L. f. ——Takane-suiba. Closely resembles the preceding; radical and lower cauline leaves narrowly to broadly ovate, 6–15 cm. long, 3–7 cm. wide, subacute at apex, the auricles obliquely deflexed, the sheath usually entire; upper cauline leaves with longer auricles.——July–Aug. High mountains and alpine regions; Hokkaido, Honshu (centr. and n. distr.).——Europe and Siberia.

4. Rumex aquaticus L. NUMA-DAIŌ. Stems 80–150 cm. high, often purplish, sulcate; radical and lower cauline leaves oblong-ovate or elongate-deltoid, smooth, to 35 cm. long, broadest near base; inner sepals 5–8.5 mm. long, 4.5–7 mm. wide in fruit, ovate-deltoid, truncate to rounded at base, entire or nearly so.——Reported to be spontaneous in Honshu (n. and centr. distr.).——Temperate regions of the Old World.

5. Rumex longifolius DC. *R. domesticus* Hartm.; *? R. suzukianum* Rechinger f.——NO-DAIŌ. Smooth or with scattered hairlike papillae; stems 80–120 cm. long, rather stout; radical leaves long-petiolate, oblong-ovate to oblong, 20–35 cm. long, 10–20 cm. wide, obtuse, cordate, slightly undulate on margin, the upper leaves linear-lanceolate to narrowly oblong, rather small, shallowly cordate or broadly cuneate at base and short-petioled; inflorescence erect, large, narrow, densely many-flowered, the pedicels slender, rather elongate; outer sepals in fruit spreading, oblong, about 1.8 mm. long, the inner whorl orbicular-cordate, membranous, weakly net-veined, 5–6(–7) mm. long and as wide, entire or with shallow obsolete teeth on margin; achenes dark-brown, lustrous, acutely 3-angled, about 3 mm. long.——June–Aug. Wet meadows in lowlands; Hokkaido, Honshu (centr. and n. distr.).—— Temperate regions of Eurasia.

6. Rumex madaio Makino. *R. daiwoo* Makino——MA-DAIŌ. Resembles No. 5; leaves beneath with distinct hairlike papillae especially on nerves; stems stout, 80–100 cm. long; radical leaves long-petioled, ovate or broadly so, sometimes ovate-oblong, 20–30 cm. long, 12–20 cm. wide, subobtuse, cordate; verticils of flowers rather distant; lobes of the outer sepals spreading in fruit, the inner whorl orbicular-cordate, slightly toothed on margin.——May–July. Hokkaido (?), Honshu, Shikoku, Kyushu.

7. Rumex nepalensis Spreng. *R. andreaeanus* Makino; *R. daiwoo* var. *andreaeanus* (Makino) Makino——KIBUNE-DAIŌ. Resembles No. 6; perianth-segments of the inner whorl ovate-orbicular, slightly cordate at base, rather firm, 5–6 mm. long and as wide, rather prominently netted-veined, the margin with slender hooked spreading spines 2–3 mm. long; achenes gray-brown, lustrous, ovate, acutely 3-angled, 3–3.5 mm. long.——May–June. Stream banks and ravines in mountains; Honshu (Yamashiro Prov.); rare; possibly introduced from China.——China to the Himalayas.

8. Rumex gmelinii Turcz. KARAFUTO-NO-DAIŌ. Perennial herb, densely pilose on leaves beneath and with whitish hairlike papillae; rhizomes stout; stems 50–100 cm. long, few-leaved; radical leaves deltoid-cordate, 10–20 cm. long and as wide, very obtuse, long-petioled and deeply cordate, the cauline smaller, the upper short-petioled, narrowly deltoid-ovate, more or less cordate; inflorescence usually not leafy, rather large, erect, the outer sepals ascending, the lobes lanceolate; 1.5–1.8 mm. long, boat-shaped, the lobes of the inner whorl broadly ovate or broadly elliptic, membranous, obtuse, rounded at base, 4–4.5 mm. long, weakly net-veined, nearly entire; achenes ovate, acutely 3-angled, dark brown, lustrous, about 3 mm. long.——July–Sept. Wet places in high mountains;

Hokkaido; rare.——n. Korea, s. Kuriles, Sakhalin, Manchuria, and the Ochotsk Sea area.

9. Rumex japonicus Houtt. *R. crispus* var. *japonicus* Makino——GISHI-GISHI. Smooth green or scattered papillose-hairy perennial herb; stems erect, 40–100 cm. long, rather stout; radical leaves long-petioled, narrowly ovate-oblong to narrowly oblong, 10–25 cm. long, 4–10 cm. wide, subobtuse, cordate, undulate on margin, the upper cauline leaves small, narrow, short-petioled, cuneate at base; inflorescence narrow, dense, leafy; segments of the outer sepal-whorl spreading, oblanceolate, boat-shaped, the lobes of the inner whorl rather firm, cordate-orbicular, dentate, the tubercles narrowly ovoid, 2–2.5 mm. long; achenes broadly ovate, dark brown, acutely trigonous, lustrous, about 2.5 mm. long.——June–Aug. Wet meadows and along ditches in lowlands; Hokkaido, Honshu, Shikoku, Kyushu; common.——Sakhalin, Korea, and Kamchatka.

Var. **yezoensis** (Hara) Ohwi. *R. yezoensis* Hara; *? R. nikkoensis* Makino——Ō-GISHI-GISHI. Inner sepal-lobes 6–7 mm. wide, larger, with prominent teeth on lower margin; achenes 3–4 mm. long.——Hokkaido.

10. Rumex crispus L. NAGABA-GISHI-GISHI. Resembling No. 9; stems sometimes purplish, 30–80 cm. long; radical leaves lanceolate to oblong-lanceolate, 12–30 cm. long, 4–6 cm. wide, distinctly undulate on margin; inner sepals broadly ovate, sometimes subcordate, 4–5 mm. long, 3.5–4.5 mm. wide, usually entire, the tubercles 1.5–2 mm. long.——June–Oct. Naturalized in our area.——Europe, Asia, and North Africa; naturalized also in N. America.

11. Rumex conglomeratus Murr. ARECHI-GISHI-GISHI. Glabrous perennial; stems branched, often purplish, 30–100 cm. long; radical leaves oblong-lanceolate, 10–20 cm. long, 3–7 cm. wide, subacute, rounded to shallowly cordate, long-petioled, undulate on margin, the upper ones smaller, short-petioled; inflorescence much-branched, leafy, diffuse, the verticils small, many-flowered, the pedicels short, the outer sepals ascending-appressed, about 0.5 mm. long, the inner ones narrowly deltoid-ovate, about 3 mm. long, 1.5 mm. wide, entire, with few weak netted veinlets, the tubercles ovoid, rather prominent, about 1.2 mm. long; achenes broadly ovate, trigonous, dark brown, lustrous, about 1.5 mm. long.——June–Oct. Naturalized and common around cities in our area.——Europe.

12. Rumex maritimus L. *R. ochotkius* Rechinger f.——KOGANE-GISHI-GISHI. Low, rather stout, smooth or sometimes papillose-pilose annual or biennial herb; stems 10–50 cm. long, with ascending branches; lower leaves narrowly to broadly lanceolate, 7–15 cm. long, 1–3.5 cm. wide, subacute, obtuse to acute at base; inflorescence leafy, the verticils densely many-flowered, subglobose, widely separated, or upper ones contiguous; perianth in fruit yellow-brown, the outer sepals nearly spreading, about 1 mm. long, the inner deltoid-ovate, acute, 2.5–3 mm. long, 1–2 mm. wide, sparingly net-veined, the marginal spines 2–3 mm. long, the tubercles ovoid, 1–1.5 mm. long; achenes yellow-brown, oblong, about 1.2 mm. long, lustrous, acutely trigonous.——July–Aug. Waste ground near seashores; Hokkaido, Honshu (n. distr.).——Ochotsk Sea area to Siberia, India, and Europe.

13. Rumex obtusifolius L. EZO-NO-GISHI-GISHI. Perennial herb with hairlike papillae on the leaves beneath especially on the nerves; stems stout, 30–80 cm. long; radical leaves long-petioled, oblong, ovate, or broadly so, 15–20 cm. long, 8–12 cm. wide, obtuse to acute, cordate; inflorescence leafy, with ascending branches; outer sepals spreading, about 1.5 mm. long, the inner narrowly deltoid-ovate or ovate-orbicular, subacute, 4–6 mm. long, 3–5 mm. wide, with rather prominent netted veinlets, short-spined or toothed on margin, the tubercles ovoid, 1.5–2 mm. long; achenes dark brown, ovate, acutely 3-angled, lustrous, about 2.5 mm. long.——June–Sept. Hokkaido, Honshu; very variable.——Eurasia and N. Africa.

14. Rumex nipponicus Fr. & Sav. *R. obtusifolius* var. *nipponicus* (Fr. & Sav.) Nakai; *R. dentatus* subsp. *nipponicus* (Fr. & Sav.) Rechinger f.——Ko-GISHI-GISHI. Smooth perennial herb; branches ascending to nearly spreading; lower leaves oblong-lanceolate to lanceolate, obtuse, rounded at base, long-petioled, 4–6 cm. long, 1–2 cm. wide; inflorescence leafy, the verticils widely separated, the outer sepals ascending, the inner deltoid-ovate, 4–5 mm. long, 2–3 mm. wide, with few marginal spines about 2 mm. long; achenes similar to the preceding.——May–June. Lowlands; Honshu (Kantō Distr. and westw.), Shikoku, Kyushu.——China and Korea.

2. OXYRIA Hill   Maruba-gishi-gishi Zoku

Low glabrous perennial herbs; radical leaves long-petioled, reniform-orbicular, entire, the margin undulate, the stipules scarious; scapes arising from short stout rhizomes, erect, naked or 1-leaved; inflorescence paniculate, the flowers few at each node, small, bisexual, the pedicels jointed at apex; perianth-segments or sepals 4, in 2 series, the outer whorl reflexed in fruit, the inner whorl slightly accrescent after anthesis, appressed to the achene; stamens 6; ovary compressed; styles 2, the stigma broad, fimbriate; achenes flat, winged on margin; embryo lateral.——Two species in the arctic and alpine regions of the N. Hemisphere.

1. Oxyria digyna (L.) Hill. *Rumex digynus* L.; *O. reniformis* Hook.——Maruba-gishi-gishi, Jin'yō-suiba. Glabrous, sometimes slightly reddish perennial herb from stout rhizomes; scapes 10–30 cm. long, naked or with a small subtending leaf below the inflorescence; leaves succulent, somewhat tufted, long-petioled, reniform-orbicular, rounded to subretuse, 1.5–2.5 cm. long, 2–4 cm. wide; inflorescence 4–10 cm. long, the branches erect; inner sepal-lobes spathulate-obovate, about 2 mm. long, rounded at apex, membranous; achenes ovate, flat, about 2 mm. long, with a membranous wing on margin, orbicular-cordate, retuse.——July–Aug. Wet barren alpine slopes; Hokkaido, Honshu (centr. distr.); rare.——Arctic and alpine regions of the N. Hemisphere.

3. POLYGONUM L.   Tade Zoku

Herbs, rarely subshrubs, often scandent; leaves alternate, with a sheathlike membranous stipule (ochrea) at base; flowers in fascicles or solitary, more often in spiciform panicles (spikes), racemes or panicles, the pedicels jointed; perianth-segments usually 5, 2-seriate, one of the whorls sometimes accrescent after anthesis; ovary lenticular or trigonous; styles 2 or 3, free or connate at base, the stigma capitellate or rarely fimbriate; fruit enclosed within the persistent perianth; embryo curved.——About 200 species, cosmopolitan, especially abundant in the N. Hemisphere.

1A. Styles hooked at apex, persistent, firm; perianth-segments 4. .... 1. *P. filiforme*
1B. Style not hooked, deciduous, not hardened; perianth-segments 5, rarely 4.
  2A. Flowers fasciculate in leaf-axils, not forming a spike or raceme; leaves small, linear to oblong, entire.
    3A. Leaf-sheath nerveless or with short nerves near base; leaves with obsolete lateral nerves; achenes smooth, lustrous.
                                                                           2. *P. plebeium*
    3B. Leaf-sheath nerved except at apex; leaves with more or less raised lateral nerves at least beneath; achenes punctulate and scarcely lustrous.
      4A. Achenes exserted; plant glaucescent, becoming dark brown when dried. .... 3. *P. polyneuron*
      4B. Achenes scarcely exserted; plant green, scarcely glaucescent, not becoming dark brown when dried. .... 4. *P. aviculare*
  2B. Flowers in spikes or racemes, except sometimes in Nos. 44–46.
    5A. Stems scapose, simple; radical leaves conspicuous; rhizomes thickened.
      6A. Spike solitary, always terminal.
        7A. Stem-leaves 1 or 2, small, or none. .... 5. *P. tenuicaule*
        7B. Stem-leaves usually more than 2, foliaceous.
          8A. Spikes bearing small bulbils on the lower part. .... 6. *P. viviparum*
          8B. Spikes not bearing bulbils.
            9A. Petioles of radical leaves wingless. .... 7. *P. hayachinense*
            9B. Petioles of radical leaves winged. .... 8. *P. bistorta*
      6B. Spikes axillary and terminal. .... 9. *P. suffultum*
    5B. Stems leafy, not scapose, often branched; radical leaves not prominent.
      10A. Stems with retrorse prickles.
        11A. Stipules, at least some of them, with a marginal dilated green limb.
          12A. Leaves deltoid, nearly as long as wide.
            13A. Stems much elongate, scandent.
              14A. Plant glabrous; leaves peltate; spikes with a rather prominent leaflike orbicular bract at base; perianth becoming slightly fleshy and bluish in fruit; achenes black and lustrous. .... 10. *P. perfoliatum*
              14B. Plant loosely puberulous; leaves basifixed; spikes without prominent leaflike bracts; perianth not becoming fleshy; achenes dull. .... 11. *P. senticosum*

13B. Stems 10–30 cm. long, not scandent. .... 12. *P. debile*
12B. Leaves ovate-deltoid or narrower, distinctly longer than wide.
15A. Leaflike limb of stipules deeply toothed; midlobe of leaves narrow; achenes lustrous; stellate-pubescent.
13. *P. maackianum*
15B. Leaflike limb of stipules entire or with a few deep incisions.
16A. Petioles more or less winged; achenes dull, deep brown. .... 14. *P. thunbergii*
16B. Petioles not winged; achenes gray and lustrous. .... 15. *P. oreophilum*
11B. Stipules without dilated limb.
17A. Leaves gradually narrowed toward the petiole, entire; erect herb, the stems sparingly prickly. ...... 16. *P. bungeanum*
17B. Leaves truncate to cordate, rarely rounded at base, often 3-lobed.
18A. Stipules 3–4 mm. long, pilose; spikes 1- to 3-flowered. .... 17. *P. breviochreatum*
18B. Stipules 1–2 cm. long, glabrous or sparsely pilose, sometimes ciliate.
19A. Inflorescence once or twice forked. .... 18. *P. hastatoauriculatum*
19B. Inflorescence terminal and axillary, not forked.
20A. Stipules obliquely truncate, scarcely ciliate, rather acute at the tip; leaves deeply cordate at base, glabrous or loosely ciliate on margin; pedicel shorter than the bract. .... 19. *P. sieboldii*
20B. Stipules nearly rectangularly truncate, ciliate; leaves truncate to shallowly cordate at base, setulose-scabrous on margin.
21A. Leaves with a rather distinct lobule on each side near base; pedicels longer than the bract.
20. *P. hastatosagittatum*
21B. Leaves without distinct lobule on each side; pedicels shorter than the bract. .... 21. *P. nipponense*
10B. Stems without prickles, glabrous or hairy.
22A. Stems not scandent; perianth-segments essentially unchanged in fruit.
23A. Annuals or rarely perennials; inflorescence a solitary spike or raceme, or compound; achenes small, lenticular or trigonous, the pericarp coriaceous or crustaceous.
24A. Spikes globose or capitate.
25A. Leaves with appressed sessile discoid glands beneath; stems to 30 cm. long .... 22. *P. nepalense*
25B. Leaves without sessile discoid glands; stems longer. .... 23. *P. chinense*
24B. Spike cylindric to linear.
26A. Stipules tubular, at least partly with a green dilated limb; leaves ovate or broadly so, cordate. .... 24. *P. orientale*
26B. Stipules without dilated limb.
27A. Peduncles with stipitate glands. .... 25. *P. viscosum*
27B. Peduncles without stipitate glands.
28A. Stipules not ciliate.
29A. Petioles adnate to the stipules nearly the whole length; aquatic or terrestrial perennial; spikes 1–2.
26. *P. amphibium*
29B. Petioles adnate to the stipules near the base; annual terrestrial herbs without floating leaves; spikes few to many.
30A. Stems 50–120 cm. long, stout, with thickened nodes; achenes small, about 2 mm. across; spikes to 10 cm. long, often nodding or pendulous. .... 27. *P. lapathifolium*
30B. Stems to 60 cm. long, without thickened nodes; achenes 2–3 mm. across; spikes scarcely more than 5 cm. long, erect. .... 28. *P. scabrum*
28B. Stipules ciliate.
31A. Perennials from long-creeping rhizomes; flowers rather large, loosely arranged, white or pink.
32A. Flowers white; achenes lustrous. .... 29. *P. japonicum*
32B. Flowers pink; achenes dull. .... 30. *P. conspicuum*
31B. Annuals or biennials without rhizomes.
33A. Perianth-segments distinctly glandular-dotted.
34A. Achenes lenticular; leaves and other parts very acrid. .... 31. *P. hydropiper*
34B. Achenes trigonous; leaves not acrid. .... 32. *P. pubescens*
33B. Perianth-segments scarcely glandular-dotted.
35A. Peduncles and upper internodes of stems and branchlets viscid. .... 33. *P. viscoferum*
35B. Peduncles and upper internodes not viscid.
36A. Cilia of stipules 1–2 mm. long; spikes thick. .... 34. *P. persicaria*
36B. Cilia of stipules 3–10 mm. long; spikes rather slender.
37A. Leaves ovate to lanceolate; ciliae of stipules elongate, usually as long as the stipules; plants of woods, thickets and grassy places.
38A. Plant of woods, with vivid green leaves abruptly narrowed above, the tip often caudate; spikes loosely flowered, flowers pale rose-colored. .... 35. *P. caespitosum* var. *laxiflorum*
38B. Plant of sunny grassy places, thickets, or cultivated fields, with deep green leaves gradually narrowed above, the tip obtuse; spikes densely flowered, pale-red, rarely white. .. 36. *P. longisetum*
37B. Leaves narrowly lanceolate to broadly linear, sometimes lanceolate; ciliae of stipules short or rather elongate, usually shorter than the stipule; plants of wet grassy places along rivers.
39A. Leaves rather thick-membranous, often with indistinct discoid glands beneath.
40A. Achenes lenticular.
41A. Spikes densely flowered, erect. .... 37. *P. kawagoeanum*
41B. Spikes loosely flowered, often nodding. .... 38. *P. foliosum*
40B. Achenes trigonous. .... 39. *P. trigonocarpum*

39B. Leaves thinly membranous, green or pale green, without sessile discoid glands beneath.
    42A. Spikes densely flowered, erect; achenes trigonous. . . . . 40. *P. erectominus*
    42B. Spikes loosely flowered, nodding above; achenes trigonous or lenticular. . . . . . . 41. *P. taquetii*
23B. Perennials with rhizomes; inflorescence a densely flowered panicle; achenes rather large, often exserted from the perianth, the pericarp membranous or chartaceous.
    43A. Plants less than 50 cm. high; achenes included or only slightly exserted from the perianth; leaves subsessile or very short-petioled.
        44A. Leaves lanceolate to broadly so, gradually narrowed below, not rounded at base. . . . . 42. *P. ajanense*
        44B. Leaves ovate to broadly so, rounded to very broadly cuneate at base. . . . . 43. *P. nakaii*
    43B. Plants 40–100 cm. high; achenes 3–4 times as long as the perianth; lower leaves distinctly petioled. . . 44. *P. weyrichii*
22B. Stems scandent, or if not the outer perianth-segments winged on back.
    45A. Plants bisexual; flowers in loose spikes or axillary fascicles.
        46A. Perianth-segments not winged. . . . . 45. *P. convolvulus*
        46B. Outer perianth-segments winged on the back in fruit.
            47A. Fruit inclusive of perianth obovate, about 10 mm. long, prominently decurrent to the pedicel. . . 46. *P. dentatoalatum*
            47B. Fruit inclusive of perianth elliptic or nearly orbicular, 5–7 mm. long, only slightly decurrent on the pedicel.
                                                                   47. *P. dumetorum*
    45B. Plants dioecious; flowers in dense much branched panicles.
        48A. Stems scandent. . . . . 48. *P. multiflorum*
        48B. Stems erect, ascending above.
            49A. Leaves 6–12(–15) cm. long, scarcely glaucous, often papillose-pilose to papillose-scabrous beneath; stems 50–150 cm. long. . . . . 49. *P. cuspidatum*
            49B. Leaves usually 20–30 cm. long, glaucous, glabrous beneath; stems about 2.5 m. long. . . . . 50. *P. sachalinense*

1. Polygonum filiforme Thunb. *P. virginianum* var. *filiforme* (Thunb.) Merr.; *Tovara filiformis* (Thunb.) Nakai; *T. virginiana* var. *filiformis* (Thunb.) Steward; *Sunania filiformis* (Thunb.) Rafin.——MIZU-HIKI. Coarsely appressed-strigose perennial herb from short, rather thick rhizomes; stems 40–80 cm. long; leaves elliptic to oblong, rarely obovate or ovate, 7–15 cm. long, 4–9 cm. wide, abruptly acute with an obtuse tip, acute at the base, the petiole 5–30 mm. long; stipules 5–10 mm. long, short-ciliate; spikes few, elongate, 20–40 cm. long, loosely flowered, narrow; flowers short-pedicelled, rose-pink, rarely white, the perianth to 3 mm. long in fruit; achenes inflated-biconvex, brown, elliptic, lustrous, about 2.5 mm. long; styles 2, free, persistent, hooked at apex.——Aug.–Oct. Woods and thickets in lowlands and hills; Hokkaido, Honshu, Shikoku, Kyushu; common; variable.

Var. **neofiliforme** (Nakai) Ohwi. *P. neofiliforme* Nakai; *Tovara neofiliformis* (Nakai) Nakai; *Sunania neofiliformis* (Nakai) Hara——SHIN-MIZU-HIKI. Plant thinly pilose with shorter hairs; leaves usually short-acuminate.——Honshu; common.

Var. **smaragdinum** (Nakai) Ohwi. *Tovara smaragdina* Nakai ex F. Maekawa; *Sunania filiformis* forma *smaragdina* (Nakai) Hara——NAGABA-MIZU-HIKI. Leaves narrower, ovate-oblong to broadly lanceolate; flowers and fruit slightly smaller.——Honshu (Izu).

2. Polygonum plebeium R. Br. YAMBARU-MICHI-YANAGI. Glabrous herb; stems procumbent, often rooting at the nodes, the branches with short internodes; leaves linear-oblong to lanceolate, 8–15(–20) mm. long, 1.5–3.5 mm. wide, rather thick, obtuse, subsessile, the sheaths hyaline, scarious, 2–2.5 mm. long, nearly nerveless, laciniate; perianth-segments about 2 mm. long; achenes trigonous, rhombic-ovoid, smooth, lustrous, dark brown, not exserted from the perianth.——Reported to occur in Kyushu (Yakushima).——Ryukyus, Formosa, s. China, and Malaysia to Australia.

3. Polygonum polyneuron Fr. & Sav. *P. gymnolepis* Fr. & Sav.; *P. propinquum* sensu auct. Japon., non Ledeb.—— AKI-NO-MICHI-YANAGI, HAMA-MICHI-YANAGI. Glabrous annual; stems prominently striate, suberect, much-branched, with the branches rather thick; leaves oblanceolate to broadly so, 5–30 mm. long, 2–8 mm. wide, sometimes slightly glaucous, rather thick, subacute, slightly narrowed toward the base, subsessile, the upper leaves often much smaller and deciduous, the inflorescence then loosely spicate; sheaths 5–10 mm. long, rather many-nerved, laciniate; perianth-segments about 3 mm. long; achenes trigonous, exserted above, punctulate, dull.——Sept.–Oct. Sand dunes near seashores; Hokkaido, Honshu, Shikoku, Kyushu.

4. Polygonum aviculare L. *P. heterophyllum* Lindm. ——MICHI-YANAGI. Glabrous annual; stems much-branched, prostrate to ascending, rarely suberect, 10–40 cm. long, slenderly striate; leaves narrowly oblong to linear-lanceolate, 1.5–4 cm. long, 3–12 mm. wide, membranous, usually green when dry, obtuse; sheaths laciniate, 5–10 mm. long, weakly nerved; perianth 2.5–3 mm. long, green, red at the apex; achenes trigonous, scarcely exserted or included, punctulate, dull.—— May–Oct. Meadows and riverbanks in lowlands; Hokkaido, Honshu, Shikoku, Kyushu; very common and variable.—— Temperate regions of the N. Hemisphere.

5. Polygonum tenuicaule Biss. & Moore. *Bistorta tenuicaulis* (Biss. & Moore) Nakai——HARU-TORA-NO-O. Glabrous, frequently stoloniferous perennial with short thick rhizomes; radical leaves few, long-petioled, oblong-ovate, 3–8 cm. long, 2–3 cm. wide, acuminate to acute, truncate to subrounded or shallowly cordate and abruptly decurrent to the petiole, membranous, sometimes slightly whitish beneath; scapes erect, 7–15 cm. long, 1- or 2-leaved or leafless; cauline leaves short-petioled or sessile, ovate or broadly so, sometimes ovate-cordate; sheaths brown, scarious; spike 2–3.5 cm. long, rather densely flowered, cylindric; perianth about 3 mm. long, white, slightly shorter than the stamens, as long as the pedicel; achenes slightly exserted, trigonous, lustrous; styles slender. ——Apr.–June. Honshu, Shikoku, Kyushu.

6. Polygonum viviparum L. MUKAGO-TORA-NO-O. Usually glabrous erect perennial from short thick rather large rhizomes; stems 10–30 cm. long, with few leaves; radical leaves rather thick, long-petioled, linear-lanceolate to oblong-ovate, 2–10 cm. long, 8–25 mm. wide, subacute to obtuse, shallowly cordate to abruptly cuneate at base, slightly lustrous above and deep green, with fine veinlets, often glaucous beneath; spikes solitary, 2–5 cm. long, cylindric, densely flowered, with bulbils on lower portion; flowers rose, about 3 mm. long, with slightly exserted stamens.——July–Oct. Alpine slopes; Hokkaido, Honshu (centr. and n. distr.).——Arctic and alpine regions of the N. Hemisphere.

7. Polygonum hayachinense Makino. *Bistorta hayachinensis* (Makino) Gross——NAMBU-TORA-NO-O. Glabrous perennial; stems erect, 10–30 cm. long, sparsely leafy, the sheaths to 3.5 cm. long; radical leaves rather firm, ovate to narrowly so, 3.8 cm. long, 15–35 mm. wide, obtuse to subacute, shallowly cordate to rounded at base, not decurrent to the petiole, rather lustrous above, often glaucous beneath; spikes solitary and terminal, ellipsoidal, densely flowered, 15–30 mm. long; perianth about 3.5 mm. long, about as long as the pedicel; stamens slightly exserted.——Aug. Alpine slopes; Honshu (Mount Hayachine in Rikuchu); very rare.

8. Polygonum bistorta L. *Bistorta vulgaris* Hill; *B. major* S. F. Gray——IBUKI-TORA-NO-O. Glabrous perennial, sometimes white-pilose on the leaves underneath; rhizomes thick; stems 30–80 cm. long; radical leaves long-petioled, lanceolate to broadly so, 8–20 cm. long, 2–5 cm. wide, acuminate, the tip obtuse, truncate and broadly decurrent to the petiole, with rather prominent veinlets, deep green above, paler to glaucous beneath; the lower cauline leaves similar to the radical, but smaller, the upper ones cordate, with longer sheaths; spikes densely many-flowered, cylindric, 3–6 cm. long, the perianth pale pink or white, about 3 mm. long, the pedicels relatively long; stamens exserted; achenes about 3 mm. long, trigonous, lustrous.——July–Sept. Grassy slopes in highlands and subalpine regions; Hokkaido, Honshu, Shikoku, Kyushu; rather common.——High mountains and temperate regions of the Old World.

9. Polygonum suffultum Maxim. *Bistorta suffulta* (Maxim.) Greene——KURIN-YUKI-FUDE. Perennial, glabrous or white-pilose on underside of leaves; rhizomes thickened, stems 20–40 cm. long; radical leaves membranous, long-petioled, broadly ovate, 5–10 cm. long, 3–7 cm. wide, abruptly acute to abruptly acuminate, cordate, glaucous beneath; cauline leaves 3–5, ovate-cordate, sessile and clasping, rather small, short-spicate from the axils; terminal spikes cylindric, 1–3 cm. long, rather densely flowered; perianth 2.5–3 mm. long, white; stamens exserted; achenes lustrous, slightly exserted.——Mountains; Honshu, Shikoku, Kyushu.——Korea and China.

10. Polygonum perfoliatum L. *Persicaria perfoliata* (L.) Gross——ISHIMI-KAWA. Scandent glabrous annual; stems much elongate, branched, 1–2 m. long, retrorsely prickly; leaves deltoid, thinly membranous, retrorsely prickly on the nerves beneath, glaucous or pale green, 3–6 cm long and as wide, acute to subacute, truncate to shallowly cordate, the margins sometimes minutely retrorsely scabrous, the petioles long, retrorsely prickly; sheaths scarcely tubular, the dilated leaflike limb orbicular, perfoliate, green; spikes short, 1–2 cm. long, subtended by an orbicular leaflike bract, the pedicels short; perianth 3–4 mm. long, pale greenish white, the segments broadly elliptic, becoming fleshy and blue in fruit; achenes inflated, obsoletely trigonous, black, lustrous, about 3 mm. long and as wide.——July–Oct. Wet thickets and along rivers in lowlands; Hokkaido, Honshu, Kyushu; common.——Korea, China, Malay Peninsula, and India.

11. Polygonum senticosum (Meissn.) Fr. & Sav. *Truellum japonicum* Houtt.; *Chylocalyx senticosus* Meissn.; *Persicaria senticosa* (Meissn.) Gross——MAMAKO-NO-SHIRI-NU-GUI. Slender, puberulent, much-branched, scandent annual; stems 1–2 m. long, scabrous; leaves deltoid, 4–8 cm. long and as wide, acuminate, obtuse or subobtuse at the tip, broadly cordate, green, puberulous on both sides, the petioles retrorsely prickly; sheaths very short, membranous, the marginal limb clasping, reniform or subcordate; inflorescence somewhat capitate, densely several-flowered, reddish, long-peduncled, the pedicels very short, the perianth 3.5–4 mm. long, the segments elliptic; achenes ovoid-globose, inflated, obsoletely trigonous, dull, black, about 3 mm. long and as wide.——June–Sept. Thickets in lowlands and foothills; Hokkaido, Honshu, Shikoku, Kyushu; common.——Korea and China.

12. Polygonum debile Meissn. *P. debile* var. *triangulare* Meissn.; *Persicaria debilis* (Meissn.) Gross; *Persicaria triangularis* (Meissn.) Nakai——MIYAMA-TANI-SOBA. Soft glabrescent annual; stems ascending to erect, decumbent at base, branched, 10–30 cm. long, the nodes sometimes with soft very short retrorse prickles on the nodes, glabrous or puberulous; leaves cauline, deltoid, 2–4 cm. long and as wide, petiolate, the upper subsessile, thin, green, scattered coarsely pilose above and on the nerves beneath, short-ciliolate, acuminate with an obtuse tip, truncate at base, the petioles not prickly; sheaths short, with a small dilated limb; inflorescence spicate or subcapitate, terminal and axillary, often forked, short-peduncled; flowers few, white, about 3 mm. long, the perianth-segments elliptic; achenes ovoid, trigonous, lustrous, brown, about 3 mm. long.——July–Oct. Damp woods in mountains; Honshu, Shikoku, Kyushu; rather common.

13. Polygonum maackianum Regel. *P. thunbergii* var. *maackianum* (Regel) Maxim.; *Persicaria maackiana* (Regel) Nakai——SADE-KUSA. Scandent or straggling annual, the entire plant minutely stellate-pubescent; stems ascending or sometimes suberect, decumbent at the base, branched, 40–100 cm. long, retrorsely prickly; leaves divided, petiolate, hastate, 4–7 cm. long, 3–7 cm. wide, cordate, the median lobe narrowly ovate, acuminate to acute, obtuse at tip, the lateral lobes similar but small, rather deflexed, the petioles and the nerves on underside of leaves retrorsely prickly; sheaths short, the orbicular green limb 5–15 mm. across; inflorescence pedunculate, few-flowered, cymose, the peduncles usually with stipitate glands, the pedicels very short; flowers reddish, 4–5 mm. long, the perianth-segments elliptic; achenes ovoid, trigonous, about 3.5 mm. long, lustrous, dark brown.——Aug.–Oct. Wet places along ditches, riverbanks, and paddy fields on lowlands; Honshu, Shikoku, Kyushu; rather common.——Korea, China and Ussuri.

14. Polygonum thunbergii Sieb. & Zucc. *P. debile* var. *hastatum* Meissn.; *P. hastatotrilobum* Meissn.; *P. stoloniferum* F. Schmidt; *Persicaria thunbergii* (Sieb. & Zucc.) Gross; *Persicaria stellatotomentosa* Nakai——MIZO-SOBA. Glabrous or pilose, very rarely stellate-pilose annual; stems elongate, erect from the decumbent base, 30–100 cm. long, retrorsely prickly, glabrous; leaves petiolate, the upper ones sessile, retrorsely prickly on the nerves beneath, hastate, coarsely pilose on both sides, 4–10 cm. long, 3–7 cm. wide, truncate to shallowly cordate, the midlobe ovate to broadly so, acuminate, subobtuse at the tip, the lateral (basal) lobes smaller, acute to obtuse, the petioles often narrowly winged, usually retrorse-prickly; sheaths 5–8 mm. long, obliquely truncate at apex, membranous, short-ciliate, rarely with a minute limb; inflorescence cymose, 10- to 20-flowered, the peduncles short-pubescent and stipitate-glandular, the pedicels very short; flowers reddish or rarely pale green, 5–6 mm. long; achenes included, ovoid, trigonous, acute, yellow-brown, dull.——Aug.–Oct. Wet grassy places in lowlands and mountains; Hokkaido, Honshu, Shikoku, Kyushu; very common and variable; possibly this species hybridizes with No. 13.——China, Korea, and Ussuri.

15. Polygonum oreophilum (Makino) Ohwi. *Polygonum thunbergii* var. *oreophilum* Makino; *Persicaria thunbergii* var. *oreophila* (Makino) Nemoto; *Persicaria oreophila* (Makino) Hiyama——YAMA-MIZO-SOBA.   Stems slender, elongate, diffuse, loosely branched, sparsely retrorse-prickly, the internodes elongate; leaves thinly hairy on both sides, hastate, broadly ovate, truncate at base, the midlobe large, not contracted at base, the lateral ones spreading, deltoid or ovate-deltoid, acute or rather obtuse; petioles rather long, not winged; inflorescence terminal and axillary; achenes gray, lustrous.——Mountains; Honshu.

16. Polygonum bungeanum Turcz. *Persicaria bungeana* (Turcz.) Nakai; *Polygonum chanetii* Lév.——HARI-TADE. Annual; stems erect, branched, 20–60 cm. long, scantily short-prickly; leaves lanceolate, 5–10 cm. long, 1–2.5 cm. wide, acute to subobtuse, acute at base, short-pilose on the underside especially on nerves, the petioles 5–12 mm. long; sheaths 3–8 mm. long, membranous, slightly pilose; spikes 2–5 cm. long, very loosely flowered, the peduncles with stipitate glands; flowers pale green, slightly red-tinged, about 3 mm. long, on very short pedicels, the perianth-segments elliptic; achenes subglobose, much inflated, biconvex, black, scarcely lustrous, about 2.5 mm. long and as wide.——July–Sept. Sparsely naturalized in Honshu.——Korea, China, Ussuri, and Manchuria.

17. Polygonum breviochreatum Makino. *Persicaria breviochreata* (Makino) Ohki; *Persicaria ramosa* Nakai—— NAGABA-NO-YA-NO-NEGUSA.   Slender annual, 30–50 cm. high; stems elongate, branched, long-decumbent at the base, with rather hairlike retrorse prickles; leaves thin, broadly lanceolate, 3–7 cm. long, 1–2 cm. wide, acute to acuminate, subobtuse at tip, sagittate, green, glabrous, or scattered-strigose above, often loosely stellate-pubescent beneath on lower leaves, the basal lobules deltoid and obliquely deflexed, the petioles 2–10 mm. long, the sheaths tubular, slightly pilose, 3–4 mm. long, long-ciliate; spikes terminal and lateral, remotely 1- to 3-flowered, the peduncles short ascending-setose, often glandular; flowers pale green with a pale reddish tinge, 3–4 mm. long; achenes broadly ovoid-ellipsoidal, inflated-trigonous, about 3 mm. long, yellow-brown, lustrous.——Aug.–Oct. Shaded grassy places in foothills; Honshu (Kantō Distr. and westw.), Shikoku, Kyushu.——Korea.

18. Polygonum hastatoauriculatum Makino. *P. auriculatum* Makino, non Meissn.; *Persicaria hastatoauriculata* (Makino) Nakai——HOSOBA-NO-UNAGI-TSUKAMI.   Annual; stems elongate, 30–80 cm. long, ascending or erect from long-decumbent slender branching base, striate, with loose retrorse prickles; lower leaves petiolate, the upper thinly chartaceous, sessile to short-petiolate, hastate, lanceolate to narrowly so, 4–8 cm. long, 5–15 mm. wide, gradually narrowed above to an obtuse tip, cordate, the lobules deflexed auriculate on each side, the margin scabrous, the sheaths scarious, glabrous, 1–1.5 cm. long, often lobed on one side; spikes 1–3 cm. long, interruptedly 3- to 5-flowered, the peduncles stipitate-glandular; flowers reddish, 2.5–3 mm. long; achenes orbicular, inflated-biconvex or obtusely trigonous, about 2 mm. long; achenes

orbicular, inflated-biconvex, or obtusely trigonous, about 2 mm. long, pale, dull.——Aug.–Nov. Wet places; Honshu (Kantō Distr. and westw.), Shikoku, Kyushu.——s. Korea and Ryukyus.

19. Polygonum sieboldii Meissn. *P. sagittatum* sensu auct. Japon., non L.; *P. sagittatum* var. *sieboldii* (Meissn.) Maxim.; *P. anguillanum* Koidz.; *Persicaria sieboldii* (Meissn.) Ohki——UNAGI-TSUKAMI.   Glabrous annual; stems slender, angled, ascending to erect, decumbent and branched at the base, often prominently retrorse-prickly, 20–100 cm. long; leaves rather thin, broadly lanceolate or narrowly ovate-oblong, 4–8 cm. long, 1.5–3 cm. wide, obtuse to subacuminate and obtuse at the tip, slightly glaucous, glabrous on both sides, deeply cordate; stipules scarious, 5–10 mm. long, glabrous, obliquely cut; spikes globose to ovoid-globose, densely flowered, about 10 mm. long, the peduncles slender, glabrous, the pedicels very short; flowers slightly reddish, 3–3.5 mm. long; achenes obtusely trigonous, ovoid-globose, nearly lustreless, black, about 3 mm. long.——June–Sept. Wet grassy places in lowlands.——Hokkaido, Honshu.——Korea, China, Manchuria, and e. Siberia.

Var. **sericeum** Nakai. *Persicaria sieboldii* var. *tomentosa* Hara——KE-AKI-NO-UNAGI-TSUKAMI.   Leaves and sometimes the peduncles pilose.——Hokkaido, Honshu; rare.——Korea.

20. Polygonum hastatosagittatum Makino. *Persicaria hastatosagitta* (Makino) Nakai; *Polygonum strigosum* var. *hastatosagittatum* (Makino) Steward.——NAGABA-NO-UNAGI-TSUKAMI.   Annual; stems slender, elongate, from decumbent branching base, 30–80 cm. long, scattered retrorse-prickly; leaves narrowly ovate-oblong to lanceolate, subacuminate, truncate to shallowly cordate at base, the lower corners (lobes) rounded or deltoid to subhastate, the nerves beneath retrorse-prickly, antrorsely scabrous-setulose on upper margin, retrorsely so on lower margin, the petioles usually retrorse-prickly, the sheaths scarious, tubular, 2–3.5 cm. long, truncate, ciliolate; spikes globose to broadly ovoid, few toward the ends of the branches, densely rather many-flowered, 1–1.5 cm. long, the peduncles short-pubescent and glandular-pilose; flowers about 3 mm. long, the pedicels 3–5 mm. long, often glandular-pilose, longer than the bracts; achenes ovoid, trigonous, lustrous, deep brown, about 3 mm. long.——Sept.–Oct. Wet grassy places in lowlands; Honshu, Kyushu; rather common.

21. Polygonum nipponense Makino. *P. hastatosagittatum* var. *latifolium* Makino; *Persicaria nipponensis* (Makino) Gross——YA-NO-NEGUSA.   Closely allied to the preceding; leaves broader, oblong to elliptic-ovate, abruptly acute, 5–8 cm. long, 2.5–4.5 cm. wide, truncate to shallowly cordate at base without a distinct lobe on lower margin; flowers with pedicels 1–2 mm. long, shorter than the flowers.——Sept.–Oct. Wet grassy places in lowlands; Hokkaido, Honshu, Shikoku, Kyushu; common.——Korea.

22. Polygonum nepalense Meissn. *P. alatum* var. *nepalense* (Meissn.) Hook. f.; *Persicaria nepalensis* (Meissn.) Miyabe; *Polygonum quadrifidum* Hayata; *P. lyratum* Nakai ——TANI-SOBA.   Annual; stems branched and decumbent below, 30–50 cm. long, glabrous, the nodes with fleshy retrorse hairs; leaves petiolate or sessile, ovate-deltoid, gradually acute, 1.5–5 cm. long, 1–3 cm. wide, rounded to truncate and abruptly decurrent on the winged petiole, scattered pilose and with a yellowish sessile discoid gland beneath, glabrous above, the sheaths thinly membranous, 4–8 mm. long, nearly nerveless; spikes globose, terminal and axillary, densely flowered, with a leaflike bract at base, the peduncles short, often with stipitate glands on the upper parts; flowers white, 2.5–3 mm. long; achenes black, biconvex, orbicular, about 1.5 mm. across, impressed-punctulate, dull.——Aug.–Oct. Damp woods in mountains; Hokkaido, Honshu, Shikoku, Kyushu; common. ——Korea, China, Formosa, and India.

23. Polygonum chinense L. *P. chinense* var. *thunbergii* Meissn.; *Rumex umbellatus* Houtt.; *Persicaria umbellata* (Houtt.) Nakai——Tsuru-soba. Perennial herb; stems elongate, much branched, glabrous, sometimes scandent; leaves broadly ovate to oblong-ovate, 4–10 cm. long, 3–6 cm. wide, glabrous or with few rather fleshy hairs on nerves beneath, abruptly acuminate to acute, truncate at base, the petioles short, the sheaths membranous, obliquely truncate, nerved, 1–1.5 cm. long; inflorescence globose, an erect terminal panicle, densely 10- to 20-flowered, 6–10 mm. across, the peduncles often with scattered fleshy hairs; flowers white, the perianth-segments about 3 mm. long, slightly accrescent, to 4 mm. long after flowering, slightly fleshy; achenes glabrous, black, trigonous, ovoid-globose, about 2.5 mm. long, dull.——Thickets and hedges in lowlands and hills; Honshu (Izu Isls., Kii Prov. and Chūgoku Distr.), Shikoku, Kyushu; common southw. ——China, Ryukyus, and Formosa.

24. Polygonum orientale L. *P. orientale* var. *pilosum* (Roxb.) Meissn.; *P. pilosum* Roxb.; *Amblygonum pilosum* (Roxb.) Nakai——Ō-ke-tade. Prominently spreading-pilose annual; stems rather stout, erect, much branched, 1–1.5 m. long; leaves broadly ovate to ovate-cordate, 10–20 cm. long, 7–15 cm. wide, short-acuminate, cordate or the upper ones subrounded at base, long-petiolate, the sheaths tubular, membranous, pilose, 7–30 mm. long, somewhat wing-margined; inflorescence branched, the flower spikes rather dense, peduncled, nodding to pendulous, 5–12 cm. long; flowers red to pink, 3–4 mm. long; achenes orbicular, about 3 mm. long, dull, compressed.——Cultivated and often naturalized.—— China, India, and Malaysia.

25. Polygonum viscosum Hamilt. *Persicaria viscosa* (Hamilt.) Gross——Nioi-tade. Annual; stems erect, branched above, 40–120 cm. long, coarsely long-hairy and short stipitate-glandular; leaves short-petiolate, lanceolate to deltoid-lanceolate, gradually acuminate, cuneate to rather abruptly narrowed at base, 7–15 cm. long, 1.5–4 cm. wide, often with sessile discoid glands on the upper side or on both sides, the sheaths tubular, pilose, 7–15 mm. long; spikes 3–5 cm. long, densely many-flowered, the peduncles and bracts long-hairy and short-stipitate glandular; flowers reddish, 2.5–3 mm. long; achenes black-brown, trigonous, 2.5–3 mm. long, broadly ovoid.——Aug.–Oct. Honshu, Shikoku, Kyushu.——Korea, China, and India.

26. Polygonum amphibium L. *Persicaria amphibia* (L.) S. F. Gray; *Persicaria amurensis* (Korsh.) Nieuwl.; *Polygonum amphibium* var. *amurense* Korsh. and var. *vestitum* Hemsl.——Ezo-no-mizu-tade. Perennial herb with creeping rhizomes; stems elongate, glabrous, slightly branched in the floating form, short and ascending in the terrestrial form; floating leaves oblong, 5–15 cm. long, 2–6 cm. wide, glabrous, lustrous above, red-brown when dried, obtuse to rounded with a mucro at apex, shallowly cordate at base, the lateral nerves rather numerous, spreading, the petioles rather long, the sheaths membranous, glabrous; spikes axillary, 2–4 cm. long, pedunculate, often branched below, cylindric, erect, densely many-flowered; flowers reddish, 3–4 mm. long, the

pedicels longer than the bracts; achenes black-brown, biconvex, suborbicular, about 2.5 mm. long, slightly lustrous.——June–Sept. Ponds and shallow lakes; Hokkaido, Honshu (Mutsu and Ugo Prov.).——Temperate regions of the N. Hemisphere.

27. Polygonum lapathifolium L. *P. nodosum* Pers.; *Persicaria nodosa* (Pers.) Opiz; *Polygonum tenuiflorum* Presl; *P. komarovii* Lév.——Ō-inu-tade. Annual; stems rather stout, much-branched, erect, glabrous, 50–120 cm. long, reddish, the nodes swollen; leaves lanceolate, 7–20 cm. long, 1.5–5 cm. wide, acuminate, acute at base, often red-brown when dried, with minute strigose hairs on both sides especially on nerves and on margin, the lateral nerves 20 to 30 pairs, the petioles short; sheath tubular, slenderly nerved, glabrous or rarely short-ciliate, 1–2 cm. long; spikes many, cylindric, nodding or pendulous, densely many-flowered, 4–10 cm. long; flowers about 2.5 mm. long, white or reddish, the perianth-segments 4, the lateral nerves of the 2 outer segments distinct, dichotomously forked in 2 recurved branches; achenes flat, suborbicular, lustrous, deep brown, 1.5–2 mm. across.—— June–Nov. Sunny grassy places and waste grounds in lowlands; Hokkaido, Honshu, Shikoku, Kyushu; very common. ——Temperate regions of the N. Hemisphere.

28. Polygonum scabrum Moench. *P. lapathifolium* L. pro parte; *P. tomentosum* Schrank, non Willd.; *P. incanum* F. W. Schmidt——Sanae-tade. Closely resembling No. 27, but smaller; stems 20–60 cm. long, slightly branched; leaves narrowly lanceolate to narrowly ovate, 5–12 cm. long, 0.8–3.5 cm. wide, obtuse to subacute, scabrous, also appressed pilosulous on midrib on both sides and on margin, the lateral nerves 7 to 15 pairs, ascending, arcuate, the petioles short; sheaths membranous, tubular, 7–15 mm. long; spikes ellipsoidal to short-cylindric, 1–4 cm. long, erect, densely many-flowered; flowers white or reddish, about 3 mm. long, the perianth-segments nearly as long as the achenes; achenes flat, orbicular, lustrous, dark brown, 2–3 mm. across.——May–Oct. Grassy sunny places and waste grounds in lowlands; Hokkaido, Honshu, Shikoku, Kyushu; common.——Temperate and warmer regions of the N. Hemisphere.

29. Polygonum japonicum Meissn. *Persicaria japonica* (Meissn.) Gross——Shirobana-sakura-tade. Dioecious perennial herb with long creeping rhizomes; stems erect, 60–100 cm. long, sparingly branched, glabrescent to slightly appressed-pilose; leaves rather thick and firm, lanceolate, 7–12 cm. long, 1–2 cm. wide, acuminate at both ends, strigose above especially toward the margin and on the nerves beneath, the lateral nerves 15 to 25 pairs, the petioles very short; sheaths tubular, nerved, truncate at apex, membranous, with rather firm bristly cilia 8–15 mm. long; spikes terminal, few to several, linear or narrowly cylindric, 7–12 cm. long, nodding, loosely to rather densely many-flowered, the pedicels distinct, the perianth white, sometimes greenish below in fruit, 3–4 mm. long; stamens shorter than the style in pistillate flowers; styles 2 or 3; achenes lenticular or compressed-trigonous, lustrous, elliptic-ovate, 2–2.5 mm. long.——Aug.–Oct. Wet grassy places in lowlands; Hokkaido, Honshu, Shikoku, Kyushu; rather common.——s. Korea, Ryukyus, and China.

30. Polygonum conspicuum (Nakai) Nakai. *P. japonicum* var. *conspicuum* Nakai; *Persicaria conspicua* (Nakai) Nakai; *Polygonum sterile* Nakai; *Persicaria sterilis* (Nakai) Nakai——Sakura-tade. Closely resembles the preceding; stems less branched; spikes solitary to few; flowers larger, more loosely arranged, reddish, the perianth-segments 4–6 mm. long, more widely spreading; achenes compressed-trigonous, dull. ——Aug.–Oct. Wet sunny places in lowlands; Honshu, Shikoku, Kyushu; rather common.——Korea.

31. Polygonum hydropiper L. *Persicaria hydropiper* (L.) Spach; *Persicaria vernalis* Nakai——Yanagi-tade, Ma-tade. Annual; stems often profusely branched, erect, 40–80 cm. long, glabrous; leaves lanceolate to narrowly ovate, 3–12 cm. long, 1–3 cm. wide, acuminate to obtuse, acute at base, short-petiolate, essentially glabrous, slightly scabrous on margin, the underside with obscure impressed discoid glands; sheaths short-tubular, membranous, glabrous, ciliate; spikes slender, narrow, nodding, 5–10 cm. long, sometimes shorter and included within the sheath, loosely flowered, the bracts usually sparingly short-ciliate; perianth green-yellow, reddish on the upper portion, 2.5–4 mm. long; achenes flattened, black, punctulate, dull, ovate to ovate-orbicular, 2–3 mm. long. ——Aug.–Oct. Wet places in lowlands and hills, especially along rivers and streams; Hokkaido, Honshu, Shikoku, Kyushu; very common and exceedingly variable.——Widely distributed in the N. Hemisphere.——Cv. **Fastigiatum.** *P. hydropiper* var. *fastigiatum* Makino; *P. fastigiatoramosum* Makino; *P. maximowiczii* Regel; *P. hydropiper* var. *maximowiczii* (Regel) Makino——Azabu-tade. Slender cultivar with narrow, smaller leaves and flowers; achenes about 1.5 mm. long; used as flavoring.——Forma **angustissimum** Makino. Ito-tade. Very slender phase with nearly linear leaves.—— Forma **gramineum** (Meissn.) Ohwi. *P. gramineum* Meissn. ——Satsuma-tade. Allied to f. *angustissimum;* the petioles 1–2.5 cm. long, distinct.

32. Polygonum pubescens Bl. *P. flaccidum* sensu auct., non Roxb.; *Persicaria rottleri* Hara, excl. syn.; *Persicaria flaccida* Nakai, excl. syn.——Bontoku-tade. Annual; stems erect, branched, 40–80 cm. long, glabrous or appressed short-pubescent; leaves lanceolate to broadly so, 5–10 cm. long, 1–2.5 cm. wide, acuminate, gradually narrowed below to the short petiole, appressed short-pubescent, at least on the nerves beneath, often dark blotched on the middle above, and with scattered, obscure, impressed glands, the sheaths tubular, membranous, 8–13 mm. long, short-pilose, with cilia 3–8 mm. long; spikes narrowly linear, nodding, loosely flowered, 5–10 cm. long, the bracts strigose-ciliate; flowers about 3 mm. long, the perianth-segments green on the lower half, reddish on upper half, glandular-dotted; achenes about 2 mm. long, trigonous, ovoid, black, punctulate, lustreless.——Sept.–Oct. Wet places; Honshu, Shikoku, Kyushu; common.——Korea, China, Formosa, Ryukyus, Malaysia, and India.

33. Polygonum viscoferum Makino. *Persicaria viscofera* (Makino) Gross——Nebari-tade. Annual with long ascending coarse hairs; stems simple or branched at base, erect, rather thick, 40–80 cm. long; leaves lanceolate or broadly so, 4–10 cm. long, 1–2 cm. wide, acuminate, obtuse to subacute at the subsessile base; sheaths tubular, 5–10 mm. long, ciliate; spikes several to more than 10, erect, 3–5 cm. long, rather densely many-flowered, the peduncles on the upper part and the upper internodes viscid; flowers greenish or rarely with a reddish tinge, pedicellate, about 2 mm. long; achenes black, lustrous, trigonous, obovoid, 1.8–2 mm. long.——July–Sept. Waste ground in hills and foothills; Hokkaido, Honshu, Shikoku, Kyushu.——Ryukyus and Korea.

Var. **robustum** Makino. *P. makinoi* Nakai; *Persicaria makinoi* (Nakai) Nakai; *Polygonum excurrens* Steward—— Ō-nebari-tade. Relatively large, erect, less hairy, loosely

pilose and appressed short-strigose on leaves above or toward margin; flowers greenish.——Waste ground in hills and low mountains; Hokkaido, Honshu, Kyushu.——Korea.

34. Polygonum persicaria L. *Persicaria vulgaris* Webb & Moq.; *Persicaria mitis* auct. Japon., non Gilib.——Haru-tade. Erect annual 20–50 cm. high; stems rather soft, the branches ascending at base, glabrous or often pubescent; leaves lanceolate or broadly so, 6–12 cm. long, 1–3 cm. wide, obtuse, acute at base, loosely short-pilose or glabrous except on the nerves and margin, short-petioled, the lateral nerves not prominent, 7 to 10 pairs, the sheaths thin-membranous, 3–10 mm. long, glabrescent, with cilia 1–2 mm. long; spikes few, terminal, cylindric, erect, 2–5 cm. long, densely many-flowered, the peduncles sometimes loosely glandular-pilose; flowers reddish, 2.5–3 mm. long, the perianth-segments with very slender nerves; achenes ovate-orbicular, compressed or trigonous, black-brown, lustrous, about 2 mm. long.——May–June (Oct.). Hokkaido, Honshu, Shikoku, Kyushu; rather common.—— Eurasia and N. America.

35. Polygonum caespitosum Bl. var. **laxiflorum** Meissn. *P. yokusaianum* Makino; *Persicaria yokusaiana* (Makino) Nakai; *Polygonum caespitosum* subsp. *yokusaianum* (Makino) Danser——Hana-tade. Slender, decumbent, branched, glabrescent annual; leaves thinly membranous, ovate to narrowly so, sometimes broadly lanceolate, 3–7 cm. long, 1.5–3 cm. wide, abruptly acuminate, caudate at tip, vivid green, loosely ascending-pilose above and on nerves beneath, short-petiolate; sheaths 3–8 mm. long, tubular, the cilia about as long as the sheath; spikes several, linear, 3–10 cm. long, loosely flowered, often interrupted; flowers clear pinkish, 2–3 mm. long; achenes dark brown, lustrous, trigonous, ovoid, about 2 mm. long.——Aug.–Sept. Woods in foothills; Hokkaido, Honshu, Shikoku, Kyushu; rather common.——Korea and China. The typical phase occurs in Formosa and Malaysia.

36. Polygonum longisetum de Bruyn. *P. blumei* Meissn.; *P. kinashii* Lév. & Van't.; *Persicaria longiseta* (de Bruyn) Kitag.; *Persicaria blumei* (Meissn.) Gross——Inu-tade. Annual, usually more or less pubescent; stems erect or ascending, branched, decumbent at base, glabrous; leaves broadly lanceolate to narrowly oblong-ovate, deep green, 4–8 cm. long, 1–2.5 cm. wide, gradually narrowed and obtuse at tip, acute at base, glabrous or appressed short-pilose above and on nerves beneath, short-petiolate; sheaths 5–10 mm. long, the cilia about as long as the sheath; spikes several, terminal, cylindric, rather densely many-flowered, erect; flowers reddish, 2–2.5 mm. long; achenes black-brown, lustrous, trigonous, ovoid, 1.8–2.2 mm. long.——July–Nov. Wet, grassy, sunny places in lowlands; Hokkaido, Honshu, Shikoku, Kyushu; very common.——Korea, China, Ryukyus, and Malaysia.—— Forma **albiflorum** (Makino) Ohwi. *P. blumei* var. *albiflorum* Makino——Shirobana-inu-tade. A rare phase with white flowers.

37. Polygonum kawagoeanum Makino. *Persicaria kawagoeana* (Makino) Nakai; *Polygonum minus* subsp. *procerum* Danser——Shima-hime-tade. Annual, 30–60 cm. high; stems somewhat tufted, erect from branched decumbent base, usually red-brown, slender, with slightly thickened nodes; leaves linear-lanceolate or broadly linear, 4–8 cm. long, 4–8 mm. wide, acute or gradually acuminate, obtuse to acute at base, reddish brown tinged, glabrescent or appressed-pilose on both sides, often with sessile discoid glands beneath, sessile or short-petiolate; sheaths 5–13 mm. long, appressed-pilose, the cilia rather elongate; spikes cylindric, 2–3 cm. long, densely many-flowered, erect; flowers about 1.5 mm. long, green with a red tinge; achenes broadly obovoid, black-brown, lustrous, lenticular, about 1.2 mm. long.——Aug.–Oct. Wet places; Honshu (Kazusa Prov.), Kyushu; rather rare.——Ryukyus.

38. Polygonum foliosum H. Lindb. var. **paludicola** (Makino) Kitam. *Persicaria foliosa* var. *paludicola* (Makino) Hara; *Persicaria paludicola* (Makino) Nakai——YANAGI-NUKABO.   Usually a more or less reddish annual, 30–60 cm. high; stems slender, elongate, branched above, decumbent at base; leaves broadly linear or narrowly lanceolate, 4–8 cm. long, 3–7 mm. wide, gradually acuminate to acute, nearly sessile or very short-petiolate, sparsely short-appressed pilose on upper side, on the margins and on nerves beneath, often with sessile discoid glands beneath; sheaths 5–10 mm. long, short-pilose, the cilia rather short; spikes rather many, linear, terminal and lateral, loosely flowered, 3–5 cm. long, nodding; flowers green with a red tinge, 1.5–1.8 mm. long; achenes lenticular, elliptic, about 1.5 mm. long, lustrous, black-brown.——Sept.–Oct. Wet places in abandoned paddy fields, on riverbanks and along ditches; Hokkaido, Honshu, Kyushu.——Korea. The typical phase occurs in Eurasia.

Var. **nikaii** (Makino) Kitam.   *P. nikaii* Makino; *P. paludicola* var. *nikaii* Makino; *Persicaria foliosa* var. *nikaii* (Makino) Hara; *P. nikaii* (Makino) Nakai——SAIKOKU-NUKABO. Spikes very loosely flowered; flowers compressed, 2–2.5 mm. long.——Sept.–Oct. Wet places; Honshu (Kantō Distr. and westw.), Shikoku, Kyushu.——China.

39. Polygonum trigonocarpum (Makino) Kudo & Masam.   *P. minus* forma *trigonocarpum* Makino; *Persicaria trigonocarpa* (Makino) Nakai——HOSOBA-INU-TADE.   Slender annual; stems loosely branched, 30–50 cm. long, reddish brown, glabrous; leaves narrowly lanceolate to linear, 3–7 cm. long, 4–8 mm. wide, gradually narrowed and acute at apex and at base, short appressed-pilose, especially on nerves and on margin, with sessile discoid glands beneath, the petioles very short; sheaths 5–12 mm. long, appressed-pilose and rather long-ciliate; spikes narrowly cylindric, 2–3 cm. long, densely rather many-flowered; flowers 2–2.2 mm. long, reddish, the perianth often with a few sessile glands; achenes trigonous, ovoid, lustrous, about 1.8 mm. long.——Sept.–Oct. Wet places; Hokkaido, Honshu.

40. Polygonum erectominus Makino. *Persicaria erecto-minor* (Makino) Nakai——HIME-TADE.   Annual; stems loosely branched, erect from decumbent or ascending, geniculate base, loosely rather few-leaved; leaves broadly linear to narrowly lanceolate, 3–8 cm. long, 5–10 mm. wide, subacute or acuminate, obtuse at the tip, vivid to pale green, sometimes purplish beneath, glabrous on upper side, the margin pilose, loosely short-pilose on the nerves beneath, short-petiolate; sheaths 5–12 mm. long, appressed-pilose and short-ciliate, pale green, thinly membranous; spikes erect, short-cylindric, 1.5–2 cm. long, many-flowered, interrupted below; flowers glaucous, red, about 2 mm. long; achenes broadly ovoid-ellipsoidal, black, lustrous, trigonous, 1.5–2 mm. long.——May–Oct. Wet places along rivers and margins of ponds in lowlands; Honshu, Shikoku, Kyushu; rather rare.——Korea and China.

41. Polygonum taquetii Lév. *Persicaria taquetii* (Lév.) Koidz.; *Polygonum minutulum* Makino; *Persicaria minutula* (Makino) Nakai; *Polygonum taquetii* var. *minutulum* (Makino) Ohwi——NUKABO-TADE.   Slender ascending to decumbent annual 20–40 cm. high; stems branched especially

below, sometimes scattered appressed-pilose, pale green or pale brown-yellow; leaves thin, lanceolate to broadly linear, 2–5 cm. long, 3–7 mm. wide, obtuse to subacute, narrowed to a subsessile base, usually pale green, nearly glabrous to scattered appressed-pilose on both sides; sheaths 3–8 mm. long, pale green, appressed pilose and rather short-ciliate; spikes linear or filiform, often branching below, nodding or pendulous, terminal and axillary, very loosely or remotely flowered, 2–4 cm. long, on capillary peduncles; flowers pale green and rose tinged, 1.5–2 mm. long; achenes ovate, trigonous or lenticular, deep brown, 1.5–1.7 mm. long.——Sept.–Oct. Wet sandy places along rivers in lowlands; Honshu, Shikoku, Kyushu.——Korea.

42. Polygonum ajanense (Regel & Tiling) Grig. *P. polymorphum* var. *ajanense* Regel & Tiling; *Pleuropteropyrum ajanense* (Regel & Tiling) Nakai——HIME-IWA-TADE.   Glabrous or pubescent perennial; stems flexuous, erect, sometimes branched, 10–30 cm. long, with ascending branches; leaves lanceolate or broadly so, 3–6 cm. long, 4–10 mm. wide, obtuse or acute, gradually narrowed below to the sessile base; sheaths scarious; inflorescence rather densely flowered, forming a terminal nodding panicle; flowers pale green, 3–4 mm. long; achenes included in the perianth, acutely trigonous, brown, lustrous, ovoid, about 3.5 mm. long.——July–Aug. Alpine gravelly and sandy slopes; Hokkaido; rare.——s. Kuriles, Sakhalin, n. Korea, and e. Siberia.

43. Polygonum nakaii (Hara) Ohwi. *Pleuropteropyrum nakaii* Hara; *P. japonicum* Nakai, excl. syn.——OYAMA-SOBA.   Rhizomatous loosely tufted perennial; stems erect, spreading, flexuous, 20–40 cm. long, glabrous, rather stout, profusely branched; leaves ovate to broadly so, 4–12 cm. long, 2–5 cm. wide, acute to acuminate, rounded to broadly cuneate at base, slightly pubescent, soon glabrate, very shortly petiolate or subsessile; sheaths membranous, long scattered-pubescent or glabrate; spikes erect, terminal, densely many-flowered, paniculate, much-branched below; flowers pale green, 3–4 mm. long; achenes slightly longer to as long as the perianth, acutely trigonous, dark yellow-brown, broadly ovoid, 3–3.5 mm. long, lustrous.——July–Sept. Alpine gravelly to sandy slopes; Hokkaido (Hidaka Prov.), Honshu (centr. distr. and northw.).

44. Polygonum weyrichii F. Schmidt. *Pleuropteropyrum weyrichii* (F. Schmidt) Gross——URAJIRO-TADE.   Rhizomes rather thick; stems slightly flexuous, erect, stout, 30–100 cm. long, soft-pubescent, branched above; leaves broadly to narrowly ovate, 8–17 cm. long, acuminate, rounded to cuneately rounded at base, appressed-pubescent above, densely white-tomentose beneath, the petioles distinct, usually 1–4 cm. long in the lower ones; sheaths pubescent, rather long; inflorescence paniculate, large, soft-pubescent, much-branched; flowers 2–2.5 mm. long, pale yellow-green; achenes somewhat membranous, broadly ellipsoidal, lustrous, brown, about 8 mm. long, 3 or 4 times as long as the perianth.——July–Aug. Hokkaido (gravelly places near seashores and on alpine slopes), Honshu (alpine slopes of n. and centr. distr.).——s. Kuriles and Sakhalin.

Var. **alpinum** Maxim.   *P. paniculatum* sensu Fr. & Sav., non Bl.; *P. savatieri* Nakai; *Pleuropteropyrum savatieri* (Nakai) Koidz.; *P. alpinum* (Maxim.) Koidz.——ON-TADE.   Less densely pubescent, green beneath.——Alpine gravelly or sandy slopes; Hokkaido, Honshu (n. and centr. distr.).——s. Kuriles and Sakhalin.

45. Polygonum convolvulus L. *Bilderdykia convol-*

vulus (L.) Dumort.——SOBA-KAZURA. Minutely pilose-papillose annual; stems slender, scandent, 40–100 cm. long; leaves sagittate-cordate, 3–4.5 cm. long, 2.5–4.5 cm. wide, acute to acuminate, cordate at the base, the petioles 1–5 cm. long; sheaths obliquely truncate, 2–5 mm. long; flowers in loose terminal panicles and in small axillary fascicles, greenish, about 3.5 mm. long in fruit, wingless; achenes ovoid, dull, punctulate, enclosed in the perianth, acutely trigonous, about 3 mm. long.——June–Sept. Naturalized; Hokkaido, Honshu, Kyushu.——Eurasia.

46. Polygonum dentatoalatum F. Schmidt. *P. scandens* var. *dentatoalatum* (F. Schmidt) Maxim.; *Bilderdykia scandens* var. *dentatoalata* (F. Schmidt) Nakai——Ō-TSURU-ITADORI. Minutely pilose-papillose annual; stems scandent, slender, elongate, sometimes more than 1 m. long; leaves 3–6 cm. long, 2.5–4 cm. wide, acute to short-acuminate, cordate, the basal lobes rounded, the petioles 1–6 cm. long; sheaths short; flowers in short terminal and lateral spikes, or in fascicles in leaf-axils, rather long-pedicellate; fruit obovoid inclusive of the wing, decurrent on the pedicel, about 10 mm. long, pale green, accrescent after anthesis, the outer 3 perianth-segments winged on the back; achenes black, trigonous, cordate, about 4 mm. long, rather lustrous.——Aug.–Oct. Hokkaido, Honshu (Kinki Distr. and eastw.).——Sakhalin.

47. Polygonum dumetorum L. *Bilderdykia dumetorum* (L.) Dumort.; *B. pauciflora* Nakai, excl. syn.——TSURU-TADE. Closely resembles the preceding; leaves subsagittate-ovate, often long-acuminate, cordate, the lobes subacute; fruit ellipsoidal to subglobose inclusive of the wings, 5–7 mm. long, slightly and abruptly decurrent below on the relatively short pedicel; achenes black, slightly lustrous, about 2.7 mm. long.——July–Sept. Hokkaido, Honshu, Kyushu (Tsushima).——Eurasia.

48. Polygonum multiflorum Thunb. *Pleuropterus cordatus* Turcz.——TSURU-DOKUDAMI. Scandent dioecious perennial herb from relatively thick rhizomes, minutely pilose-papillose; stems much-elongate, branched, scandent, 1–2 m. or longer; leaves ovate-cordate, 3–6 cm. long, 2.5–4.5 cm. wide, short-acuminate, with rounded lower margins, rather long-petiolate; sheaths rather short; spikes branched, terminal, paniculate; flowers white, small, 1.5–2 mm. long, accrescent after anthesis; fruit broadly obovate to suborbicular, 7–8 mm.

long, the 3 prominent wings abruptly decurrent below on the pedicel; achenes dark brown, lustrous, acutely trigonous, about 2.5 mm. long.——Sept.–Oct. Naturalized in our area.——China.

49. Polygonum cuspidatum Sieb. & Zucc. *Reynoutria japonica* Houtt.; *P. confertum* Hook. f.; *P. zuccarinii* Small; *Pleuropterus zuccarinii* (Small) Small; *Polygonum reynoutria* Makino; *Pleuropterus cuspidatus* (Sieb. & Zucc.) Gross; *R. yabeana* Honda; *R. uzenensis* (Honda) Honda; *R. japonica* var. *uzenensis* Honda; *R. hastata* Nakai; *Polygonum sieboldii* de Vries, non Meissn.——ITADORI. Rather stout dioecious perennial herb with thick rhizomes; stems ascending from an erect base, 50–150 cm. long, hollow, branched above; leaves broadly ovate to ovate-elliptic, 6–15 cm. long, short acuminate, pale green, usually papillose to short-pilose beneath, usually truncate at base, the petioles 1–3 cm. long; sheaths thinly membranous; spikes branched, numerous, terminal, paniculate; flowers densely arranged, white, 2.5–3 mm. long, the pistillate accrescent after anthesis; fruit with the outer perianth-segments winged on back, 6–10 mm. long, subcuneate at base; achenes ovate-elliptic, 2–2.5 mm. long, black-brown, lustrous, acutely trigonous.——July–Oct. Sunny places in hills and high mountains; Hokkaido, Honshu, Shikoku, Kyushu; very common and variable.——Korea, China, and Formosa.

Var. **compactum** (Hook. f.) Bailey. *P. compactum* Hook. f.; *P. sieboldii* var. *compactum* (Hook. f.) Bailey——MEIGE-TSU-SŌ. A dwarf alpine phase often tinged with red.

Var. **terminale** (Honda) Ohwi. *Reynoutria japonica* var. *terminalis* Honda; *R. hachidyoensis* var. *terminalis* Honda——HACHIJO-ITADORI. Leaves larger, lustrous.——Honshu (Izu Isls.).

50. Polygonum sachalinense F. Schmidt. *Reynoutria sachalinensis* (F. Schmidt) Nakai; *R. sachalinensis* var. *brachyphylla* Honda; *R. brachyphylla* (Honda) Nakai——Ō-ITADORI. Resembles the preceding but stouter, glaucous, not papillose; stems 1–2 m. long; leaves 15–30 cm. long, 10–20 cm. wide, glaucous beneath, glabrous, cordate; inflorescence densely pubescent; fruit cuneate-obovate; achenes ovate, about 3 mm. long.——Aug.–Sept. Along ravines and streams in mountains; Hokkaido, Honshu (Japan Sea side of centr. and n. distr.); rather common.——Sakhalin and s. Kuriles.

Fam. 79. **CHENOPODIACEAE** AKAZA KA Goosefoot Family

Herbs, rarely shrubs (sometimes fleshy); leaves alternate, occasionally opposite, simple, the stipules absent, rarely minute; inflorescence various, in axillary fascicles, or the fascicles arranged on panicle branches; flowers small, bracteate or bractless, bisexual or unisexual (plants sometimes polygamous), greenish, actinomorphic; perianth-segments 2–5, free or connate at base, sometimes absent; stamens 2–5, opposite the segments; ovary usually superior, free from the perianth, 1-locular; ovules solitary, amphitropous; styles 1–3; fruit a nutlet enclosed by an utricle, the embryo annular or spirally coiled.——About 100 genera, with about 1,500 species, cosmopolitan, especially abundant in drier regions.

1A. Leaves flat; embryo annular, usually with copious endosperm in the center.
  2A. Flowers bisexual, in fascicles or in panicles; sepals distinct, persistent; testa crustaceous.
    3A. Sepals without wings. .... 1. *Chenopodium*
    3B. Sepals with a horizontal wing on back. .... 2. *Kochia*
  2B. Flowers unisexual, the pistillate without a perianth, with 2 large bracts. .... 3. *Atriplex*
1B. Leaves linear, terete or scalelike; embryo spirally coiled; endosperm absent or scanty.
  4A. Stems jointed; leaves opposite, scalelike; embryo conduplicate. .... 4. *Salicornia*
  4B. Stems not jointed; leaves alternate, linear to terete; embryo spiral.
    5A. Leaves not spine-pointed; sepals not winged on back. .... 5. *Suaeda*
    5B. Leaves spine-pointed; sepals transversally winged in fruit. .... 6. *Salsola*

1. CHENOPODIUM L.　Akaza Zoku

Pilose to puberulent annual or perennial herbs; leaves petiolate, flat, entire or pinnately lobed; flowers small, bisexual, sessile, ebracteate, in glomerules, these often arranged in spikes or panicles; perianth-segments 2 or 3, free or slightly connate, deciduous with the fruit; stamens 1–5; styles 2–3; fruit indehiscent, an achene or utricle, the seed solitary, horizontal or vertical, the endosperm mealy, embryo annular.——About 80 species, cosmopolitan, especially abundant in temperate regions.

1A. Branchlets of inflorescence short-spinose; flowers cymose; leaves entire, not mealy. .... 1. *C. aristatum*
1B. Branchlets of inflorescence not short-spinose; flowers fasciculate, spicate or in panicles.
　2A. Plants not glandular-pilose; embryo in a perfect ring.
　　3A. Stems decumbent or ascending; sepals 2–5; seeds oriented horizontally and vertically. .... 5. *C. glaucum*
　　3B. Stems erect; sepals 5; seeds oriented horizontally.
　　　4A. Leaves entire, sometimes with dilated lower margin.
　　　　5A. Leaves narrowly lanceolate to ovate-oblong, sometimes ovate-orbicular, rather thick; inflorescence-axis rather stout, densely flowered, rather densely reddish farinaceous while young. .... 4. *C. virgatum*
　　　　5B. Leaves deltoid or rhombic-ovate, sometimes ovate, thin; inflorescence-axis slender, rather loosely flowered, often whitish mealy while young.
　　　　　6A. Sepals broadly obovate to elliptic; nutlets smooth, the seeds keeled on margin, slightly depressed on the faces.
　　　　　　　　　　　　　　　　　　　　　　　　　　　　　　　　　　　　　2. *C. bryoniaefolium*
　　　　　6B. Sepals oblong to narrowly obovate; nutlets slightly granular, the seeds lenticular, not keeled on margin. .. 3. *C. koraiense*
　　　4B. Leaves more or less toothed.
　　　　7A. Seeds about 1 mm. across; leaves white-mealy at least while young beneath, not elongate above.
　　　　　8A. Leaves rhombic, deltoid-ovate, or narrowly ovate, sometimes broadly lanceolate, irregularly toothed; seeds keeled on margin; flowering and fruiting in September and October. .... 6. *C. album*
　　　　　8B. Leaves oblong-ovate to lanceolate, deeply undulate-toothed, with lowest pair of teeth lobelike, ascending; seeds not keeled; flowering and fruiting in July and August in our area. .... 7. *C. ficifolium*
　　　　7B. Seeds about 2 mm. across; leaves green, relatively large, elongate above. .... 8. *C. hybridum*
　2B. Plants glandular-pilose especially on leaves beneath; embryo imperfectly ringed. .... 9. *C. ambrosioides*

1. Chenopodium aristatum L. *Teloxys aristata* (L.) Moq.——Hari-sembon.　Glabrous, smooth or loosely papillose annual with erect branched stems 10–40 cm. long; leaves broadly linear to linear-lanceolate, 1.5–3 cm. long, 2–4 mm. wide, obtuse or subacute, rather thick, gradually decurrent toward the sessile or short-petioled base; inflorescence rather large, densely flowered while young, more expanded and rather loose in fruit, pale green, the branchlets 2–4 mm. long, short-spinose; sepals oblong, about 0.5 mm. long, obtuse, green on back, rather broadly white-membranous on margin; seeds horizontal, discoid, black, about 0.5 mm. across, the embryo imperfectly annular.——July–Aug. Honshu (centr. distr.); rare.——Korea, s. China, and s. Siberia.

2. Chenopodium bryoniaefolium Bunge. Midori-akaza.　Annual; stems erect, branched, 50–100 cm. long, the branches glabrous, elongate, ascending; leaves thin, hastate-ovate to deltoid or ovate, 3–5 cm. long, 2.5–4.5 cm. wide, obtuse or acute, with a large tooth on each side or 3-lobed, with spreading acute to obtuse lateral teeth, broadly cuneate to subtruncate at base, slightly glaucous, mealy beneath while young, the petioles 1–4 cm. long; inflorescence loosely flowered; sepals 5, elliptic or broadly obovate, obtuse, greenish on back, 1–1.3 mm. long; utricles orbicular, flat, smooth, the seeds black, about 1.3 mm. across, discoid, lustrous, indistinctly striolate, keeled on margin.——Aug. Mountains; Honshu (Kantō Distr.); rare.——e. Siberia and Manchuria.

3. Chenopodium koraiense Nakai. *C. acuminatum* var. *ovatum* sensu Nakai, non Fenzl——Iwa-akaza, Yama-akaza. Glabrous slightly mealy annual with elongate ascending branches; leaves thinly membranous, deltoid-ovate, rhombic-ovate, or narrowly rhombic-ovate, 1–4 cm. long, 0.8–3 cm. wide, acute, broadly cuneate and often dilated toward the base, slightly glaucescent, the petioles 5–20 mm. long; inflorescence elongate, loosely flowered; sepals 5, 0.8–1 mm. long, oblong, obtuse, green on back; utricles horizontal, deep brown, slightly granular-tubercled, the seeds lenticular, flat, 1–1.2 mm. across,

lustrous, black, rugulose, obtuse and not keeled on margin. ——Aug.–Oct. Honshu (Chūgoku Distr.), Kyushu (n. distr.).——Korea.

4. Chenopodium virgatum Thunb.　*C. acuminatum* var. *japonicum* Fr. & Sav.; *C. acuminatum* var. *virgatum* sensu auct. Japon., non Moq.——Kawara-akaza, Maruba-akaza. Annual herb, densely red-brown puberulent while young; stems erect, 30–60 cm. long, the branches ascending to suberect; leaves rather thick, white-mealy while young especially beneath, narrowly lanceolate to ovate-orbicular, 1–4 cm. long, 0.6–2 cm. wide, obtuse to acute, often short-awned at apex, cuneate to acute at base, the petioles 5–25 mm. long; inflorescence a terminal densely flowered panicle; sepals obovate, about 1 mm. long, with a broad green band on back and faint transverse ridges near the middle; utricles smooth, the seeds black, about 1 mm. across, horizontal, planoconvex, with an obtuse margin.——June–Oct. Waste ground along rivers and seashores; Honshu, Shikoku, Kyushu; common.——Ryukyus and Formosa.

5. Chenopodium glaucum L. Urajiro-akaza.　Glabrous annual; stems decumbent to ascending, rather fleshy, much-branched, 15–30 cm. long; leaves oblong-ovate, sometimes broadly lanceolate, 2–4 cm. long, 5–15 mm. wide, obtuse, acuminate at base, slightly fleshy, deep green above, whitish mealy beneath, with few deeply undulate teeth, the petioles 5–20 mm. long; inflorescence spicate, axillary and terminal, short, rather densely flowered; sepals 2–3 or 5, about 0.5 mm. long, membranous, obtuse, with a slender green rib on back; utricles horizontal and vertical, the seeds smooth, dark brown, shining, flat, about 0.8 mm. across, obtuse on margin.——June–Sept. Chiefly in waste ground along seashores; Hokkaido, Honshu, Shikoku.——Eurasia.

6. Chenopodium album L. Shiro-akaza, Shiroza. Glabrous annual, white-mealy on young parts and underside of leaves; stems much branched, striate, 60–150 cm. long; leaves deltoid-ovate to rhombic-ovate, narrowly ovate in the upper ones, 3–6 cm. long, 1.5–3.5 cm. wide, obtuse to subacute, with small deeply undulate teeth, the petioles 1–4 cm. long; flowers in fascicles or glomerules, these arranged in panicles; sepals obovate, about 1 mm. long, obtuse, with a broad green band, low-keeled in the middle; seeds flat, black, 1–1.3 mm across, shining.——Waste ground and cultivated fields in lowlands and hills; very common.——Temperate regions of Eurasia.

Var. **centrorubrum** Makino. *C. centrorubrum* (Makino) Nakai——AKAZA. Leaves often deltoid-ovate, reddish mealy while young, the panicles somewhat looser.——Rarely cultivated, frequently naturalized in our area.——India (?) or China (?).

Chenopodium purpurascens Jacq. *C. album* var. *purpurascens* Makino; *C. elegantissimum* Koidz.——MURASAKI-AKAZA. Tall purplish stout plant with leaves sometimes more than 10 cm. long; fascicles of flowers dispersed on the much-branched pendulous panicle; seeds usually pale brownish. Sometimes cultivated in the warmer parts of our area.

7. Chenopodium ficifolium Smith. KO-AKAZA. Annual, slightly white-mealy on young parts and underside of leaves; stems 30–60 cm. long; leaves green, deltoid-oblong or narrowly deltoid-ovate, 2–5 cm. long, 1–3 cm. wide, obtuse, cuneate at base, deeply undulate-toothed, the lowest pair of teeth lobelike, the petioles slender; fascicles of flowers subinterrupted, forming a dense panicle; sepals obovate, about

1 mm. long, obtuse, green on back, the midrib raised; seeds black, about 1 mm. across, discoid, with obtuse margin.——June–Aug. Hokkaido, Honshu, Kyushu; rather common in waste ground and along roadsides in lowlands.——Europe to w. Siberia.

8. Chenopodium hybridum L. *C. bonus-henricus* sensu Makino & Nemoto, non L.——USUBA-AKAZA. Glabrous green rather tall annual with angled stems; leaves broadly ovate, 6–13 cm. long, 3–7 cm. wide, much elongate above, acuminate, the few large lobes deltoid; spikes rather loosely flowered; flowers rather large; sepals ovate, obtuse, about 2 mm. long, green, ribbed on the back; seeds discoid, flat, about 2 mm. across, minutely pitted.——Sept. Hokkaido.——Europe.

9. Chenopodium ambrosioides L. *C. ambrosioides* var. *pubescens* Makino——KE-ARITA-SŌ. Green or yellow-green annual with pale white glandular-hairs, not farinose; stems much-branched, ascending, 60–80 cm. long; leaves narrowly ovate to lanceolate, 2–5 cm. long, 5–15 mm. wide, acute to obtuse, undulate to sinuate, short-petioled; spikes leafy, short or elongate; flowers small; sepals usually 3, about 0.8 mm. long, green; seeds horizontal or vertical, shining, thick-lenticular, about 0.6 mm. across.——July–Nov. Naturalized in Honshu, Shikoku, Kyushu; common in waste ground and along roadsides.——Tropical America.

2. KOCHIA Roth HŌKIGI ZOKU

Usually silky-pilose herbs often woody at base; leaves alternate, sessile, narrow, entire; flowers small, sessile, solitary or in axillary fascicles, bisexual and pistillate, the bracts and bracteoles absent; perianth subglobose, urceolate, or depressed-urceolate, 5-parted, the segments incurved, with a horizontal wing on back; stamens 5; ovary broadly ovoid; styles slender; stigmas 2 or 3, capillary, tubercled; fruit utriculate, enclosed in the perianth, depressed-globose, the pericarp membranous, persistent, loosely enclosing the seed; seeds horizontal, orbicular, flattened, the testa membranous, the embryo nearly annular.——About 80 species, Europe, Asia, Australia, and Africa.

1. Kochia scoparia (L.) Schrad. *Chenopodium scoparia* L.——HŌKIGI. Much-branched annual; stems erect, 50–100 cm. long, soft-pubescent above; leaves narrowly oblanceolate, lanceolate, or narrowly so, 2–5 cm. long, 2–8 mm. wide, subacuminate to acute, gradually narrowed below and petioled in the lower ones, subtrinerved, more or less long-pilose; flowers few, in axillary fascicles, sessile, pale green, often forming a terminal spike by the reduction of the upper leaves; perianth urceolate-globose, depressed, 5-merous, the segments subdeltoid, the wing on back horizontal, spreading, subchartaceous;

styles very short, the stigmas 2, slender; seeds about 1.5 mm. across, broadly obovate, enclosed in the slightly accrescent perianth.——Aug.–Oct. Cultivated for making brooms.——Europe to s. Asia and China.

Var. **littorea** Makino. *Kochia littorea* (Makino) Makino——ISO-HŌKIGI. Stems rather strongly flexuous, with ascending to nearly spreading branches.——Sept.–Oct. Near seashores; Honshu (Tōkaidō Distr. and westw.), Shikoku, Kyushu.——Korea.

3. ATRIPLEX L. HAMA-AKAZA ZOKU

Monoecious or dioecious herbs or shrubs often with scurfy indument; stems leafy; leaves alternate, rarely opposite, petioled or sessile; fascicles of flowers unbranched or paniculate; staminate flowers ebracteate, the sepals 3–5, the stamens of the same number; pistillate flowers surrounded by 2 accrescent bracts, the perianth absent, the ovary globose, the stigmas 2; utricles enveloped by 2 bracts, the pericarp often membranous, the seeds erect or horizontal, the embryo annular.——More than 100 species, cosmopolitan.

1A. Leaves ovate-deltoid to lanceolate, 2–4 cm. wide; pistillate bracts 6–10 cm. long and wide; seeds dark brown, 3–4 mm. across, slightly lustrous; stems with ascending branches. .... 1. *A. subcordata*
1B. Leaves lanceolate to linear, 3–15 mm. wide; pistillate bracts 3–4 mm. long and as wide in fruit; seeds black, about 1.5 mm. across, lustrous; stems with suberect branches. .... 2. *A. gmelinii*

1. Atriplex subcordata Kitag. *A. tatarica* sensu auct. Japon., non L.; *A. patulus* sensu auct. Japon., non L.; *A. littoralis* var. *dilatata* Fr. & Sav.——HAMA-AKAZA, KO-HAMA-AKAZA. Scurfy annual; stems erect, 40–60 cm. long, with ascending branches; leaves ovate-deltoid, narrowly ovate to

lanceolate, 3–8 cm. long, 2–4 cm. wide, acute to subobtuse, slightly glaucescent, broadly cuneate to gradually narrowed at base, often with few teeth; fascicles of flowers axillary, forming a terminal spike; pistillate bracts subsessile in fruit, subdeltoid, truncate to shallowly cordate at base, subentire, acute to subacuminate, nearly smooth, 6–10 mm. long, 5–9 mm. wide, with shallow grooves at base; seeds dark brown, suborbicular, flat, 3–4 mm. across.——Aug.–Oct. Sandy seashores; Hokkaido, Honshu.——Kuriles and Sakhalin.

Atriplex hastata L. of Europe with opposite lower leaves is sometimes naturalized in our area.

2. Atriplex gmelinii C. A. Mey. *A. littoralis* var. *japonica* Koidz.; *A. subcordata* var. *japonica* (Koidz.) Honda; *A. littoralis* var. *angustissima* sensu auct. Japon., non Moq.—— Hosoba-no-hama-akaza. Annual; stems erect, branched, glabrous, 30–50 cm. long, with suberect branches; leaves alternate, lanceolate to narrowly so, sometimes linear, 5–10 cm. long, 3–15 mm. wide, acute to subacute, narrowed at base, entire or few-toothed, glabrous, white-scurfy while young, the petioles 5–15 mm. long; flowers fasciculate in upper axils, forming a terminal spike; pistillate bracts sessile in fruit, broadly ovate-deltoid, acute, 3-nerved, not tubercled, entire or subentire, 3–4 mm. long and as wide; seeds black, orbicular, flat, lustrous, about 1.5 mm. across.——Sandy seashores; Hokkaido, Honshu.——Kuriles, Kamchatka, Sakhalin, and Korea.

4. SALICORNIA L.   Akkeshi-sō Zoku

Fleshy glabrous herbs or small shrubs with opposite, terete, jointed branches; leaves opposite, scalelike, usually connate at base; inflorescence spicate, the flowers 3–7, fascicled and impressed in upper leaf-axils, bisexual or staminate in the lateral ones, sessile, the perianth obconical or rhomboidal, fleshy, shallowly 3- to 4-toothed or truncate on margin, spongy in fruit; stamens 2 or 1; ovary ovoid; styles or stigmas 2; utricles enclosed in the perianth; seeds erect, flat, the embryo conduplicate.——About 30 species, cosmopolitan in saline soil.

1. Salicornia europaea L. *S. europaea* var. *herbacea* L.; *S. herbacea* L.——Akkeshi-sō. Fleshy, usually reddish annual, 10–30 cm. high, the branches opposite, erect; leaves scalelike, small, connate at the base in a short tube; spikes 1–5 cm. long, 3–4 mm. across, terete, joints about 3 mm. long, the perianth rhomboid-ovoid, small.——Aug.–Oct. Saline marshes; Hokkaido, Honshu (n. distr.), Shikoku; rare southward.—— Wet saline places especially along seashores in the temperate regions of the N. Hemisphere.

5. SUAEDA Forsk.   Matsu-na Zoku

Usually glabrous herbs or shrubs; leaves alternate, fleshy, narrow or flat, entire; flowers small, axillary, sessile or nearly so, solitary or fascicled, bisexual or unisexual, with 1 bract and 2 bracteoles; perianth 5-fid to -divided, the segments rather fleshy, rarely appendaged on back; stamens 5; ovary sessile; stigmas 2–5, short; utricles globose, ovate or flat, enveloped by and sometimes adnate below to the perianth; seeds horizontal, oblique or vertical, globose, often flat, the embryo spiral.——About 100 species, in saline soils along seashores, cosmopolitan.

1A. Leaves linear to linear-oblong or oblanceolate, slightly flattened, very obtuse to subacute; perianth 2–4 mm. across in fruit; testa pale, membranous.
  2A. Plant and flowers purplish in fruit; leaves rounded to very obtuse; utricles 2.5–4 mm. across. .... 1. *S. japonica*
  2B. Plant green, the flowers changing to red in fruit; leaves acute to subobtuse; utricles 2–2.5 mm. across. .... 2. *S. malacosperma*
1B. Leaves narrowly linear, obtuse to subacute.
  3A. Utricles 2.5–3 mm. across; flowers 1 or 2, axillary, the upper ones short-pedicelled. .... 3. *S. asparagoides*
  3B. Utricles 1.2–1.5 mm. across; flowers 3–5, axillary, sessile. .... 4. *S. maritima*

1. Suaeda japonica Makino. Shichimen-sō. Glabrous annual; stems erect, much branched, later becoming slightly woody, 20–40 cm. long; leaves many, fleshy, sessile, terete, linear to oblanceolate, very obtuse, 5–35 mm. long, 2–4 mm. wide; flowers few in leaf-axils, sessile, the perianth rather fleshy, to 4 mm. across in fruit, the segments ovate-orbicular, subobtuse; utricles depressed-globose, the pericarp tightly appressed around the seeds; stigmas 2; seeds lenticular, horizontal, 1.5–2 mm. across, the testa membranous.——Sept.–Oct. Seashores; Kyushu (n. distr.).——Korea.

2. Suaeda malacosperma Hara. Hiro-ha-matsu-na. Glabrous annual; stems 10–30 cm. long, usually branched; leaves broadly linear, to 35 mm. long, 4 mm. wide, fleshy, slightly flattened, acute to subobtuse and minutely cuspidate, sessile, bluish green; flowers 1–4, axillary, sessile, usually pistillate, the bracteoles minute, hyaline, the perianth depressed-globose, the segments rather fleshy in fruit, thickened deltoid on the back; utricles depressed-globose; seeds white, the testa membranous.——Oct.–Nov. Seashores; Honshu (Bingo Prov.), Kyushu.

3. Suaeda asparagoides (Miq.) Makino. *Salsola* (?) *asparagoides* Miq.; *Schoeberia maritima* var. *asparagoides* Fr. & Sav.; *Suaeda glauca* sensu auct. Japon., non Bunge—— Matsu-na. Glabrous annual; stems erect, 50–80 cm. long, much branched, later becoming woody, the branches ascending; leaves green, narrowly linear, 1–3 cm. long, subobtuse to subacute; inflorescence loosely arranged on an elongated branching axis, aphyllous toward upper part; flowers small, green, in 1- to 2-flowered fascicles, very short pedicelled or sessile, the bracts and bracteoles deltoid, membranous, 0.5–0.8 mm. long; perianth-segments 5, connate at base, narrowly ovate; utricles globose or slightly depressed, 2.5–3 mm. across. ——Aug.–Sept. Seashores; Honshu (Kantō Distr. and westw.), Shikoku, Kyushu.

4. Suaeda maritima (L.) Dumort. *Chenopodium maritimum* L.——Hama-matsu-na. Glabrous annual; stems much-branched, erect, 30–50 cm. long; leaves narrowly linear, acute or subacute, 1–3 cm. long; inflorescence in 3- to 5-flowered axillary fascicles, the flowers small, sessile; perianth-segments 5, elliptic, not accrescent in fruit; utricles 1.2–1.5 mm. across, depressed-globose; seeds black, shining, discoid or lenticular, about 1 mm. across.——Sept.–Oct. Sandy seashores; Honshu (Kantō Distr. and westw.), Shikoku, Kyushu.——Along seashores in the N. Hemisphere.

6. SALSOLA L. Oka-hijiki Zoku

Herb or shrubs, glabrous or short-pilose; branches not jointed; leaves alternate, rarely opposite, narrow, sessile, sometimes scalelike, usually spine-pointed; flowers small, solitary or in fascicles in leaf-axils, often accrescent in fruit, bisexual, 2-bracteolate; perianth 4- to 5-merous, the segments oblong or lanceolate, concave, often incurved at apex, thickened and becoming horizontally winged on back; stamens 5 or fewer, staminodes absent; ovary globose; stigmas 2(–3), linear; utricles enclosed by the lower portion of perianth, broadly ovoid to globose, depressed at the apex or mucronate, the pericarp not adnate to the seed; seeds horizontal, the embryo spiral.——About 100 species, mainly in saline soils and along seashores, especially abundant in the N. Hemisphere.

1. Salsola komarovii Iljin. *S. soda* sensu auct. Japon., non L.——Oka-hijiki. Glabrous annual; stems decumbent or ascending, spreading, much branched, 10–30 cm. long; leaves alternate, green, fleshy, linear-terete, spine-pointed, 1–2.5 cm. long; flowers solitary in axils, sessile, the bracteoles 2, narrowly ovate, 4–5 mm. long; tepals 5, cartilaginous in fruit, transversely ridged on back, inflexed in upper portion; fruit depressed-obconical, about 2 mm. across; seeds as large as the fruit, the testa white, membranous, easily separable.——July–Oct. Sandy places along sea beaches; Hokkaido, Honshu, Shikoku, Kyushu.——Korea, Manchuria, n. China, and e. Siberia.

Fam. 80. **AMARANTHACEAE** Hiyu Ka Amaranth Family

Herbs, sometimes shrubs; leaves opposite or alternate, simple, entire or with obsolete teeth, the stipules absent; flowers small, greenish, white, rose, or rarely yellow, bisexual, rarely unisexual, actinomorphic; inflorescence densely cymose, disposed in spikes or panicles; perianth-segments 4–5, free or connate at base, regular, scarious; stamens 1–5, opposite the perianth-segments, more or less connate at base, often with a membranous appendage between them; carpels 2–3, connate; ovary superior, 1-locular; ovules 1 to many, campylotropous, erect; fruit a circumscissile capsule, or frequently a utricle, rarely a drupe or berry; seeds lenticular, the testa shining; endosperm mealy, the embryo annular.——About 64 genera, with about 800 species, especially abundant in the Tropics.

1A. Leaves alternate; anthers 2-locular.
  2A. Flowers unisexual or polygamous; ovules solitary. .... 1. *Amaranthus*
  2B. Flowers bisexual; ovules 2 to many. .... 2. *Celosia*
1B. Leaves opposite.
  3A. Anthers 2-locular; flowers in elongate spikes, deflexed in fruit. .... 3. *A. caudatus*
  3B. Anthers 1-locular; flowers in heads, not deflexed in fruit.
    4A. Stigma sessile, usually capitellate, entire or nearly so. .... 4. *Alternanthera*
    4B. Stigmas 2, subulate, on a very short style. .... 5. *Philoxerus*

1. AMARANTHUS L. Hiyu Zoku

Erect or decumbent, often slightly pubescent, monoecious or polygamous annuals; leaves alternate, flat, petiolate, ovate, lanceolate or rhombic, uually entire; flowers small, in axillary glomerules, forming a terminal spike or panicle, green, sometimes white or reddish, the bracts often spine-tipped, sometimes minute, the bracteoles 2; perianth-segments 5, sometimes 3, membranous, persistent; stamens 1–3 or –5, free; ovary ovoid; style very short, the stigmas 2–3; ovules solitary, erect; utricle indehiscent or circumscissile; seeds erect, orbicular, flat, shining, smooth, the embryo annular.——About 50 species, cosmopolitan.

1A. Utricles transversely or imperfectly dehiscent; bracts usually prominent.
  2A. Perianth-segments and stamens 5; spikes elongate; tall herbs with rather stout thickened stems.
    3A. Leaf-axils without spines; utricles transversely dehiscent.
      4A. Spikes erect or ascending, greenish; bracts with an acute green spine at apex; perianth-segments broadly spathulate; seeds black.
        5A. Bracts 4–6 mm. long inclusive of the longer spine; perianth 3 mm. long, about half as long as the bract; spikes thicker, pale green. .... 1. *A. retroflexus*
        5B. Bracts 2–4 mm. long; perianth 1.5–2 mm. long, about ⅔ as long as the bract; spikes narrower, green. ...... 2. *A. patulus*
      4B. Spikes narrow, pendulous above, reddish or white; bracts with a slender pale rather short spine at apex; perianth-segments obovate; seeds usually white, reddish tinged toward the margin. .... 3. *A. caudatus*
    3B. Leaf-axils with spines; utricles imperfectly dehiscent. .... 4. *A. spinosus*
  2B. Perianth-segments 3; stamens 2–3; spikes usually not elongate.
    6A. Stems 80–150 cm. long, stout; leaves 5–20 cm. long.
      7A. Spikes not elongate; leaves linear to rhombic-ovate. .... 5. *A. tricolor*
      7B. Spikes elongate in the terminal ones; leaves broadly ovate-deltoid. .... 6. *A. mangostanus*
    6B. Stems 10–60 cm. long, rather slender, much-branched; leaves 1–4 cm. long.
      8A. Stems and branches decumbent; bracts narrowly ovate, green; utricles smooth. .... 7. *A. graecizans*
      8B. Stems and branches erect; bracts spine-tipped; utricles rugose. .... 8. *A. albus*
1B. Utricles indehiscent; bracts not prominent, shorter than the perianth.
  9A. Utricles prominently rugose. .... 9. *A. viridis*
  9B. Utricles smooth.
    10A. Leaves 2.5–4 cm. wide; stems erect, with ascending base. .... 10. *A. lividus*
    10B. Leaves 1–1.5(–2) cm. wide; stems ascending to decumbent. .... 11. *A. deflexus*

1. Amaranthus retroflexus L. AOGEITŌ, AOBIYU. Rather coarse annual; stems rather stout, green, soft-pubescent, 1–2 m. long, branched; leaves rhombic-ovate, 5–10 cm. long, 3–6 cm. wide, acute or subobtuse with a short mucro, cuneate at base, glabrous above, soft-pubescent on the nerves beneath, the petioles 3–8 cm. long; inflorescence paniculate, elongate, erect or ascending, often short-branched below, the axis densely soft-pubescent; flowers pale green (plants polygamous), densely disposed, the bracts subulate, 4–6 mm. long, spine-pointed, the lower margins membranous; perianth-segments about 3 mm. long, oblanceolate, obtuse to rounded, short-awned, white-membranous on margin, slightly longer than the utricle.——Aug.–Oct. A weed in lowlands; Hokkaido, Honshu, Kyushu; very common.——A pantropic weed.

2. Amaranthus patulus Bertol. *A. hybridus* subsp. *cruentus* var. *patulus* (Bertol.) Thell.——HOSO-AOGEITŌ. Resembles the preceding; spikes green, the bracts 2–4 mm. long; perianth 1.5–2 mm. long, the segments obtuse with a minute point, slightly shorter than the utricle; stigma very short.——Aug.–Oct. A weed in lowlands; Honshu, Shikoku, Kyushu.——Europe.

Amaranthus paniculatus L. SUGIMORIGEITŌ. Annual with slender red to yellow spikes, the bracts with shorter spines.——Sometimes cultivated as an ornamental in our area.

3. Amaranthus caudatus L. SENNIN-KOKU, HIMOGEITŌ. Erect, slightly pubescent annual to 1 m. high; leaves rhombic-ovate, 5–10 cm. long, 3–6 cm. wide, acute, mucronulate, nearly glabrous above, scattered soft-pubescent on the nerves beneath; inflorescence much elongate, branched below, pendulous or nodding, rose to white, the bracts 2–3 mm. long, membranous, rounded at apex, slightly toothed, slightly shorter to as long as the utricle.——Commonly cultivated in our area as an ornamental.——Tropics.

4. Amaranthus spinosus L. HARIBIYU. Scattered short-pubescent to nearly glabrous annual, 40–80 cm. high, with angled stems; leaves narrowly to sometimes broadly ovate, 3–8 cm. long, 1.5–4 cm. wide, obtuse with a small mucro at tip, cuneate at base, the petioles 1–8 cm. long, subtended by spreading spines about 1 cm. long; lower spikes axillary and globose, the upper elongate, the bracts narrowly lanceolate; perianth-segments 5, about 1.3 mm. long, nearly as long as the utricle, narrowly oblong, membranous, cuspidate; fruit imperfectly circumscissile; stigma rather long, about 1 mm. long, filiform.——Aug.–Oct. Honshu, Kyushu.——Tropical America, now widely naturalized as a weed in the warmer regions of the world.

5. Amaranthus tricolor L. *A. melancholicus* L.; *A. gangeticus* L.——HAGEITŌ. Annual, often slightly soft-pubescent while young, later glabrescent; stems 80–150 cm. long, erect, coarse; leaves linear-lanceolate to rhombic-ovate, 7–20 cm. long, 2–7 cm. wide, acute or subacute, short-mucronate, cuneate at base, often variously colored and blotched with red and yellow, the petioles 3–10 cm. long; spike axillary, globose, densely flowered, pale green, the bracts narrowly ovate, membranous, awn-tipped, slightly shorter than the perianth; perianth-segments 3, broadly lanceolate, short-awned, about 3 mm. long; fruit about 2/3 as long as the perianth.——Aug.–Oct. Widely cultivated in our gardens.——Tropical Asia.

6. Amaranthus mangostanus L. *A. inamoenus* Willd. ——HIYU. Nearly glabrous annual; stems erect, 80–150 cm. long; leaves broadly rhombic-ovate, or broadly deltoid-ovate, 5–12 cm. long, 2–7 cm. wide, obtuse, retuse at apex, broadly cuneate and decurrent on the petiole; petioles 3–10 cm. long; lower spikes globose, the upper ones elongate, interrupted, pale green, the bracts ovate, awn-pointed, shorter than the perianth, membranous; perianth-segments narrowly ovate, membranous, awn-tipped, about 3 mm. long; stamens 3; utricles shorter than the perianth.——Aug.–Oct. Cultivated as a vegetable.——India.

7. Amaranthus graecizans L. *A. blitoides* S. Wats.—— INU-HIME-SHIROBIYU. Annual, glabrous or nearly so; stems branched, ascending to procumbent, 10–60 cm. long; leaves narrowly obovate, 8–25 mm. long, 2.5–15 mm. wide, obtuse, mucronate or subacute, acute at base, the petioles 5–15 mm. long; spikes small, axillary, subglobose, the bracts narrowly ovate or broadly lanceolate, green, short-awned at apex; perianth-segments shorter than the bracts, narrowly oblong, cuspidate, about 1 mm. long; utricles nearly smooth, slightly longer than the perianth.——Sept.–Oct. North American weed naturalized in Honshu.

8. Amaranthus albus L. HIME-SHIROBIYU. Resembles *A. graecizans;* leaves oblong or subspathulate, 8–13 mm. long, 2–7 mm. wide, obtuse with a short mucro at tip, decurrent below on the petiole, 3–8 mm. long; spikes axillary, rather few-flowered, the bracts linear-subulate, recurved above, 2–3 mm. long, the perianth-segments much shorter than the bracts, membranous, acute; utricles rugose, longer than the perianth. ——Aug.–Sept. Naturalized in Honshu (Kinki Distr.).—— N. America.

9. Amaranthus viridis L. *Euxolus viridis* (L.) Moq.; *Chenopodium caudatum* Jacq.; *E. caudatus* (Jacq.) Moq.; *A. gracilis* Desf.——AOBIYU, HONAGA-INUBIYU. Nearly glabrous annual; stems erect, sparsely branched, 50–80 cm. long; leaves broadly deltoid-ovate, 4–8 cm. long, 2.5–6 cm. wide, retuse, broadly cuneate to subtruncate at base; spikes on upper part of the stems, to 8 cm. long, the bracts membranous, narrowly ovate, minute, 0.8 mm. long, much shorter than the perianth, minutely awned at apex; perianth-segments 3, broadly oblanceolate, acute, with a green midrib, 1–1.2 mm. long; utricles globose, slightly longer than the perianth, prominently rugose, indehiscent.——July–Oct. Honshu, Kyushu. ——Tropical American weed widely naturalized in the warmer regions of the world.

10. Amaranthus lividus L. *A. blitum* L., pro parte; *A. ascendens* sensu auct. Japon., non Loisel.; *Euxolus ascendens* (Loisel.) Hara; *E. viridis* var. *ascendens* (Loisel.) Moq.—— INU-BIYU. Resembles *A. viridis;* leaves deeply retuse, broadly cuneate at base; spikes rather stout, the lateral ones usually not elongate, the perianth-segments lanceolate; utricles nearly smooth.——June–Oct. Hokkaido, Honshu, Kyushu.—— Cosmopolitan weed of European (?) origin.

11. Amaranthus deflexus L. *Euxolus deflexus* (L.) Raf.——HAIBIYU. Scattered short-pubescent; stems much branched, ascending or decumbent, 10–30 cm. long; leaves ovate or narrowly ovate-deltoid, 1–4 cm. long, 5–20 mm. wide, obtuse, retuse or subacute, cuneate at base, nearly glabrous, the petioles 1–3 cm. long; inflorescence terminal, subelongate, 2–4 cm. long, sparsely branched below, the bracts narrowly ovate, short-awned at tip, about 1 mm. long; perianth-segments 2–3, oblong, acuminate, with a green midrib on back, about 2 mm. long; utricles about 1.5 times as long as the perianth, finely few-nerved.——Aug.–Oct. Naturalized near Tokyo.——Europe, N. America, N. Africa, and S. America.

2. CELOSIA L. Keitō Zoku

Mostly annuals; leaves alternate, petiolate, linear to broadly ovate, entire; spikes axillary and terminal, densely many-flowered; flowers rather small, white or reddish, lustrous, bisexual, sessile, with a bract and 2 bracteoles at base, the perianth-segments 5, scarious, oblong to lanceolate, subobtuse, finely nerved, erect; stamens 5, connate below into a cup; ovary ovoid; style single, short to elongate, the stigmas 2–3; ovules 2–8; utricles globose or ovoid, circumscissile; seeds erect, lenticular, black, shining, smooth, the embryo annular.——About 60 species, in the Tropics.

1A. Sipkes long-peduncled, the flowers not fasciated; flowers 8–10 mm. long. . . . . 1. *C. argentea*
1B. Spikes short-peduncled or sessile, often fasciated; flowers about 4 mm. long. . . . . 2. *C. cristata*

1. Celosia argentea L. Nogeitō. Glabrous annual; stems erect, 40–80 cm. long; leaves lanceolate to narrowly ovate, 5–8 cm. long, 1–2.5 cm. wide, acute or gradually acuminate, decurrent on the petiole or sessile; spikes lanceolate or cylindric, 5–8 cm. long, 1–2.5 cm. wide, densely many-flowered, the bracts and bracteoles broadly lanceolate, about 4 mm. long, white, scarious, acuminate; perianth-segments lanceolate, 8–10 mm. long, acuminate, white, 1-nerved above, with few weak nerves below; utricles globose, about half as long as the peri-anth; style about 3 mm. long, erect; seeds about 1.5 mm. across.——July–Nov. Honshu (w. distr.), Shikoku, Kyushu; naturalized.——A pantropic weed.

2. Celosia cristata L. Keitō. Coarse annual, the stems and leaves reddish; leaves rather large, ovate, ovate-deltoid or lanceolate; spikes often fasciated, sessile or short-peduncled, red, yellow, or white, the perianth-segments about 4 mm. long.——Aug.–Oct. Widely cultivated in gardens in our area.——Pantropic.

3. ACHYRANTHES L. Inokozuchi Zoku

Herbs, rarely woody at base; stems branched, thickened on nodes; leaves opposite, petiolate, ovate to lanceolate, entire; inflorescence an elongate terminal spike, greenish or white, the bracts persistent, membranous, reflexed in fruit, the bracteoles 2, erect, subulate, deciduous with the fruit; flowers bisexual, sessile, deflexed in fruit, the perianth-segments 4–5, subcoriaceous, nearly equal, lanceolate, connivent; stamens 5, rarely 4, connate at base, the staminodes entire or fimbriate; ovary oblong; style single, the stigma capitellate; ovules solitary; utricles membranous, enclosed in the persistent perianth, indehiscent; seeds solitary, oblong.——About 20 species, mainly tropical.

1A. Appendage of bracteoles nearly orbicular, about 0.7 mm. long; staminodes slightly toothed or shallowly bifid; roots scarcely thickened; plant nearly glabrous or glabrescent. . . . . 1. *A. japonica*
1B. Appendage of bracteoles obtusely deltoid to suborbicular, 0.3–0.5 mm. long; staminodes subrounded or nearly quadrangular, slightly toothed or subentire; roots thickened.
2A. Leaves glabrescent, thinly membranous, lanceolate to broadly so. . . . . 2. *A. longifolia*
2B. Leaves usually prominently pubescent, thickly membranous, elliptic to ovate. . . . . 3. *A. fauriei*

1. Achyranthes japonica (Miq.) Nakai. *A. bidentula* var. *japonica* Miq.——Hikage-inokozuchi. Nearly glabrous to slightly pubescent perennials with scarcely thickened roots; stems 50–100 cm. long, obtusely quadrangular, branched, the nodes thickened; leaves oblong, elliptic or obovate, 10–20 cm. long, 4–10 cm. wide, acuminate; inflorescence elongate, usually loosely flowered in fruit, the bracts membranous, ovate-deltoid, with a green long-protruded midrib, to ½ as long as the perianth; flowers horizontal in anthesis, deflexed in fruit, the bracteoles subulate, shorter than the perianth, the basal auricles membranous, whitish, about 0.8 mm. long, nerveless, orbicular, the perianth 4–5 mm. long, the segments linear-lanceolate, acuminate, the inner ones slightly shorter, the staminodes sessile, denticulate or obsoletely 2-lobulate; utricles oblong, about 2.5 mm. long, about 1 mm. across, thinly membranous, 1-seeded, the style slender, erect, about 1 mm. long. ——Aug.–Sept. Woods in lowlands and hills; Honshu, Shikoku, Kyushu; common.
Var. **hachijoensis** Honda. Hachijo-inokozuchi. A maritime phase with thicker glabrescent lustrous leaves.——Honshu (Hachijo Isl. and westw.), Shikoku, Kyushu.——Ryukyus.

2. Achyranthes longifolia (Makino) Makino. *A. biden-tula* var. *longifolia* Makino——Yanagi-inokozuchi. Resembles the preceding, the roots thickened; leaves lanceolate to broadly so, 10–20 cm. long, 2–5 cm. wide, gradually acuminate at both ends, nearly glabrous; bracts often with shallow teeth; style about 1.5 mm. long.——Aug.–Sept. Honshu (centr. distr. and westw.), Shikoku, Kyushu.——China and Formosa.

3. Achyranthes fauriei Lév. & Van't. Hinata-inokozuchi. Resembles the preceding two species but rather prominently pubescent; leaves elliptic, broadly ovate or oblong, 10–15 cm. long, 4–10 cm. wide, short-acuminate, usually prominently pubescent; bracts ovate, with a protruding midrib, the auricles of the bracteoles thinly membranous, about 0.5 mm. long, orbicular; perianth 5–5.5 mm. long; staminodes quadrangular, with obsolete teeth on upper margin or entire; style about 1 mm. long.——Aug.–Sept. Thickets and roadsides in lowlands to hills; Hokkaido (?), Honshu, Shikoku, Kyushu; common.——China.

4. ALTERNANTHERA Forsk. Tsuru-nogeitō Zoku

Prostrate or decumbent, rarely erect, branched, glabrous to tomentose herbs; leaves opposite, sessile or short-petiolate, obovate to linear, entire or obsoletely toothed; glomerules small, sometimes congested, the flowers small, usually white, the bracts, bracteoles, and perianth-segments obtuse to long-acuminate, the perianth-segments 5, sessile, dorsally compressed, unequal;

stamens 2–5, connate below, the filaments interposed between the staminodes or the staminodes absent, the anthers 1-locular; ovary orbicular; style short or absent, the stigma usually capitellate; ovules solitary, pendulous on an elongate filiform funicle; utricles compressed, ovoid to orbicular-obcordate; seeds lenticular, with a smooth, coriaceous testa, the embryo annular.—— About 170 species, mostly in the New World Tropics, several widely dispersed as pantropical weeds.

1A. Leaves 5–15 mm. wide; perianth-segments narrowly ovate, glabrous, acute, with a prominent rather stout midrib on back.
1. *A. sessilis*

1B. Leaves 3–6 mm. wide; perianth-segments lanceolate, acuminate, with a weaker midrib on back, usually with few long soft dorsal hairs.
2. *A. nodiflora*

1. Alternanthera sessilis (L.) R. Br. *Gomphrena sessilis* L.; *Illecebrum sessile* (L.) L.——Tsuru-nogeitō. Prostrate or decumbent somewhat fleshy densely branched annual; stems and branches terete, the nodes with long white curved hairs, the internodes with two longitudinal rows of hairs while young; leaves spreading, glabrous, lanceolate to oblanceolate, 3–7 cm. long, 5–15 mm. wide, obtuse, sparsely crenate-toothed or subentire, gradually narrowed to a short-petioled or subsessile base, the midrib rather prominent; glomerules solitary or few, many-flowered, axillary, sessile, globose, the bracts and bracteoles minute; perianth-segments 5, white-scarious, more or less lustrous, smooth, glabrous, 2–2.5 mm. long, narrowly ovate, acute, with a prominent midrib especially toward the top; stamens 2–3, less than half as long as the perianth, the anthers minute, yellow; utricles obtusely margined, slightly shorter than the perianth, obcordate; seeds reddish brown, about 1 mm. across, shining.——July–Sept. Naturalized in the warmer parts of our area. A pantropic weed.

2. Alternanthera nodiflora R. Br. *A. denticulata* R. Br. ——Hosoba-tsuru-nogeitō. Much-branched prostrate annual with long soft slender white hairs on nodes while young; branches terete, often rooting at the nodes, with two longitudinal lines of short white crisped hairs on the internodes while very young; leaves broadly linear to linear-lanceolate, soft, 2.5–6 cm. long, 3–6 mm. wide, subobtuse, obsoletely mucronate-toothed or subentire, sessile or short-petioled; glomerules one to few, axillary, densely many-flowered, sessile, about 4 mm. across, the bracts and bracteoles less than half as long as the perianth, white-scarious; perianth 2–2.5 mm. long, the segments lanceolate to broadly so, long–acuminate, sparsely soft white-hairy, the midrib faint on back; utricles gray-brown, obcordate-reniform, retuse; seeds brownish yellow, shining, about 0.7 mm. across.——June–Nov. Naturalized in lowlands in the warmer parts.——Tropical Asia to Australia and Africa.

5. PHILOXERUS R. Br. Iso-fusagi Zoku

Decumbent to erect, glabrous or densely hairy mostly fleshy herbs; leaves opposite, obovate-spathulate, narrowly oblong or linear, entire; inflorescence globose to cylindric, axillary and terminal, white or reddish, sessile or short-peduncled, the bract solitary, papyraceous, the bracteoles 2, keeled; flowers bisexual, short-pedicelled, the perianth dorsally flattened, the segments 5, oblong, obtuse, nerved, chartaceous; stamens 5, the filaments slightly connate at base, the anthers 1-locular; ovary flattened; style very short, the stigmas 2; ovules solitary, pendulous; utricles indehiscent, the seeds lenticular, smooth, the embryo annular.——About 15 species, along seashores, e. N. America, w. Africa, Australia, and s. Japan.

1. Philoxerus wrightii Hook. f. Iso-fusagi. Rather fleshy tufted perennial herb forming dense mats on seashores; stems glabrous, decumbent to ascending, branched below, 2–5 cm. long; leaves glabrous, narrowly spathulate-obovate, 4–8 mm. long, 2–3 mm. wide, very obtuse, gradually narrowed to the base; inflorescence a terminal very short-peduncled few-flowered head; flowers rose-colored, the bracts and bracteoles small, membranous, ovate, about half as long as the perianth; perianth-segments elliptic, 3–3.5 mm. long, obtuse, scarious; seeds dull brown.——July–Oct. Rocks along seashores usually where washed by waves; Kyushu (Yakushima); rather rare.

Fam. 81. PHYTOLACCACEAE Yama-gobō Ka Pokeweed Family

Herbs, rarely shrubs or trees; leaves alternate, entire, usually without stipules; inflorescence usually racemose; flowers bisexual or polygamous, sometimes unisexual, the sepals 4–5, free or connate below, imbricate in bud, the petals absent; stamens as many as or more than and alternate with the sepals; ovary superior, several locular, or composed of several free carpels; styles short or absent, the stigmas linear, filiform or subulate, usually papillose inside; ovules solitary in each carpel, amphitropous or campylotropous; fruit usually a berry, capsule or samaralike.——About 22 genera, with about 120 species, mostly pantropic.

1. PHYTOLACCA L. Yama-gobō Zoku

Tall herbs, shrubs, or trees with erect or scandent glabrous or pulverulent branches; leaves alternate, sessile or petiolate, large, entire, exstipulate; racemes terminal, the pedicels bracteate and with 1–3 bracteoles at the base; sepals (4–)5, persistent; stamens 5–30, inserted at the base of the sepals; ovary subglobose; carpels 5–15, free or connate; fruit a globose berry; seeds flattened, the embryo annular, the endosperm mealy.——About 35 species, in the Tropics and subtropics, especially abundant in America.

1A. Inflorescence erect in fruit, the peduncles 1–3 cm. long.
  2A. Carpels free, 8; seeds smooth. . . . . 1. *P. esculenta*
  2B. Carpels connate, 7–10; seeds with slender concentric striations. . . . . 2. *P. japonica*
1B. Inflorescence nodding in fruit, the peduncles 4–12 cm. long; carpels 10, connate; seeds smooth. . . . . 3. *P. americana*

1. **Phytolacca esculenta** Van Houtte. *P. acinosa* var. *esculenta* Maxim.; *P. kaempferi* A. Gray; *P. acinosa* var. *kaempferi* (A. Gray) Makino——YAMA-GOBŌ.   Glabrous perennial; stems stout, to 1 m. long, branched, green; leaves suborbicular to ovate-elliptic, 10–15 cm. long, abruptly acuminate, acute at base, entire, the petioles 1.5–2.5 cm. long; racemes 5–12 cm. long, rather densely flowered, short-peduncled, slightly scurfy, the pedicels 10–12 mm. long; flowers white, about 8 mm. across, the sepals elliptic; berry depressed-globose, purple-black; seeds reniform-orbicular, rather compressed, black, shining, smooth, about 3 mm. long.——June–Sept. Rarely cultivated as a vegetable; sometimes naturalized.—— China.

2. **Phytolacca japonica** Makino. MARUMI-NO-YAMA-GOBŌ.   Resembles the preceding; leaves oblong to ovate-ob-long; racemes erect, distinctly scurfy; flowers pale rose-colored; berry globose; seeds reniform-orbicular, rather flattened, black, shining, about 3 mm. across, with fine concentric lines.—— June–Sept. Honshu (Kantō Distr. and westw.), Shikoku, Kyushu.

3. **Phytolacca americana** L.   *P. decandra* L.——YŌSHU-YAMA-GOBŌ.   Resembles No. 1; stems 1–1.5 m. long, reddish purple, glabrous; leaves ovate-elliptic to oblong, 10–30 cm. long, 5–16 cm. wide, short-acuminate, abruptly acute at base, the petioles 1–4 cm. long; racemes nodding in fruit, rather densely many-flowered, 10–15 cm. long; flowers white with a rosy tinge, rather small; berry depressed-globose; seeds reniform-orbicular, rather flat, about 3 mm. across, shining, black, smooth.——June–Nov. Naturalized in lowlands and hills; common.——N. America.

Fam. 82. AIZOACEAE   ZAKURO-SŌ KA   Mesembryanthemum Family

Herbs, rarely shrubs; leaves opposite or verticillate, stipulate or exstipulate; flowers small, actinomorphic, bisexual, solitary or fasciculate, sometimes cymose, the sepals 4 or 5, free or connate below, the petals 4 or 5, small or absent, free or connate below; stamens 4 or 5 or fewer, rarely more; ovary superior or inferior, the carpels 3–5, sometimes connate, the ovules 1 to many; capsules loculicidal or circumscissile; seeds amphitropous, the embryo slender, curved.——About 100 genera, with more than 600 species, mostly in the Tropics, especially abundant in Africa.

1A. Calyx-tube adnate to the ovary; fleshy plant with exstipulate alternate leaves. .... 1. *Tetragonia*
1B. Calyx-lobes free from the ovary; slender herbs with falsely verticillate leaves and small stipules. .... 2. *Mollugo*

1. TETRAGONIA L.   TSURU-NA ZOKU

Herbs or subshrubs, glabrous, hairy or tubercled, rather fleshy; leaves alternate, flat, entire, exstipulate; flowers axillary, sessile or pedicelled, green or yellow-green; calyx 3- to 5-lobed, the tube adnate to the ovary, often angled; petals absent; stamens 1 to many, inserted on the calyx-tube; ovary inferior, 3- to 8- rarely 1- or 2-locular; styles as many as the locules of the ovary, the ovules solitary in each locule, pendulous; fruit nutlike, indehiscent, the seeds subreniform, the embryo curved, subcylindric.—— About 100 species, mainly in the Tropics, especially in Africa.

1. **Tetragonia tetragonoides** (Pall.) O. Kuntze. *Desmovia tetragonoides* Pall.; *T. expansa* Murr.; *T. japonica* Thunb.——TSURU-NA.   Fleshy, glabrous, mealy, perennial herb; stems branched, decumbent below, 40–60 cm. long; leaves thick, ovate-deltoid, 4–6 cm. long, 3–4.5 cm. wide, obtuse, broadly cuneate to subtruncate at base, the petiole about 2 cm. long; flowers 1 or 2 in leaf-axils, the pedicels very short, rather stout, the calyx-tube 3–4 mm. long, accrescent and becoming 6–7 mm. long in fruit, with 4 or 5 large spinelike tubercles on the shoulder, the calyx-teeth broadly ovate; petals absent; nut several-seeded, indehiscent.——Apr.–Nov. Sandy seashores; Hokkaido (sw. distr.), Honshu, Shikoku; common; sometimes cultivated as a vegetable.——China, s. Asia, S. America, and Australia.

2. MOLLUGO L.   ZAKURO-SŌ ZOKU

Glabrous to pilose herbs; leaves alternate to falsely verticillate, flat, the stipules membranous; flowers green, in terminal, fasciculate, axillary cymes; sepals 5, equal, membranous, persistent; petals absent; stamen 3–5, rarely many, usually alternate with the sepals; ovary ovoid, 3- to 5-locular; styles 3–5; ovules many, sometimes few in each locule; capsules membranous, loculicidally dehiscent.——About 15 species, mostly in the Tropics.

1A. Sepals obsoletely 1-nerved; capsules globose; seeds minutely warty; leaves in verticils of 3–5. .... 1. *M. pentaphylla*
1B. Sepals 3-nerved; capsules ovoid-ellipsoid; seeds with few raised lines; leaves in verticils of 4–7. .... 2. *M. verticillata*

1. **Mollugo pentaphylla** L.   *M. stricta* L.——ZAKURO-SŌ. Delicate glabrous annual; stems much-branched, 10–30 cm. long, ascending to diffuse, angled; leaves in verticils of 3–5, slightly lustrous above, lanceolate to oblanceolate, 1.5–3 cm. long, 3–7 mm. wide, obtuse to subacute, acute at base, entire, 1-nerved; inflorescence loosely flowered, the bracts delicate, membranous, the pedicels 1–4 mm. long; sepals 5, elliptic, obtuse to rounded, obsoletely 1-nerved, about 1.5 mm. long; capsules globose, about 2 mm. long; seeds rather numerous, dark brown, nearly lusterless, reniform-orbicular, flat, about 0.5 mm. across, minutely warty.——July–Oct. Waste ground and cultivated fields; Honshu, Shikoku, Kyushu; common in lowlands and hills.——China, Malay Peninsula, and India.

2. **Mollugo verticillata** L. KURUMABA-ZAKURO-SŌ.   Slender glabrous diffuse annual; stems branched, 10–20 cm. long, nearly terete; leaves in verticils of 4–7, oblanceolate to broadly linear, 12–25 mm. long, 2–7 mm. wide, obtuse or subacute, 1-nerved, gradually narrowed to a short petiole; flowers sev-

eral, in axillary fascicles, the pedicels 2–5 mm. long, the sepals oblong, obtuse or rounded, 1.5–2 mm. long, 3-nerved; capsules ovoid-ellipsoid, about 3 mm. long; seeds yellow-brown, lustrous, reniform-orbicular, about 0.3 mm. across, with few raised concentric lines.——July–Aug. Naturalized in our area.

Fam. 83. **PORTULACACEAE** SUBERI-HIYU KA Purselane Family

Often fleshy herbs or small shrubs, glabrous to long-pubescent; leaves alternate or opposite, entire, the stipules scarious or sometimes changing to a tuft of hairs or absent; flowers solitary, racemose or paniculate, actinomorphic, bisexual; sepals usually 2, rarely 5, free or adnate to the base of the ovary, imbricate in bud; petals 4–5, rarely many, inferior or perigynous, free or connate below, imbricate in bud; stamens as many as the petals, sometimes fewer or more; ovary 1-locular; styles 2–8, connate below; ovules 2 to many, amphitropous, or basal on a free-central placenta; capsules membranous or crustaceous, with as many valves as the styles, circumscissile or rarely indehiscent; seeds many, small, rarely 1 or 2, sometimes strophiolate, the testa crustaceous, the embryo curved.——About 16 genera with about 500 species, mainly in S. America, Australia, and S. Africa, a few in the N. Hemisphere.

1A. Ovary partially united with the calyx; capsules circumscissile. .... 1. *Portulaca*
1B. Ovary free from the calyx; capsules longitudinally dehiscent. .... 2. *Montia*

1. **PORTULACA** L. SUBERI-HIYU ZOKU

Diffuse to ascending, glabrous or rarely pilose herbs; leaves simple; flowers terminal, the sepals 2, connate below and adnate partially to the ovary, the petals 4 or 5, sometimes many; stamens 7–20, inserted on the sepals; ovary many-ovuled, partially united with the calyx; styles deeply 3- to 8-fid; capsules circumscissile, membranous.—About 20 species, mainly in America, a few in the Tropics.

1. Portulaca oleracea L. SUBERI-HIYU. Fleshy glabrous annual; stems decumbent to ascending, much branched, terete, 10–30 cm. long, reddish brown; leaves alternate, fasciculate, cuneate-obovate, 15–25 mm. long, 5–15 mm. wide, rounded at apex, short-petiolate, flat; flowers sessile, yellow; sepals 2, keeled on back; petals obovate, retuse, about 4 mm. long; stamens 7–12; seeds obliquely globose, about 0.5 mm. across, black, sparsely tubercled toward margin.——July–Sept. Waste ground and cultivated fields in lowlands and foothills; Hokkaido, Honshu, Shikoku, Kyushu; common.——Cosmopolitan.

2. **MONTIA** L. NUMA-HAKOBE ZOKU

Delicate glabrous annual; leaves opposite, slightly fleshy; flowers very small, solitary or in few-flowered racemes; sepals 2, ovate-orbicular, herbaceous, persistent; petals 3–5, slightly connate below; stamens 3–5, inserted near base of the petals; ovary free, 3-ovuled; styles 3-fid; capsules globose, 3-valved, longitudinally dehiscent; seeds suborbicular, nearly flat.——About 5 species, in the cold regions of both hemispheres.

1. Montia lamprosperma Cham. *M. fontana* var. *lamprosperma* (Cham.) Fenzl; *M. fontana* sensu auct. Japon., non L.; *M. rivularis* sensu auct. Japon., non Gmel.——NUMA-HA-KOBE. Stems 3–8 cm. long, decumbent at base; leaves opposite, spathulate-oblanceolate to narrowly obovate, 5–10 mm. long, 1.5–4 mm. wide, obtuse, narrowed below and petiolelike; flowers 2 or 3, terminal, the pedicels 3–10 mm. long, the bracts very small, membranous; sepals 2, orbicular, about 1 mm. long; petals 5, small, slightly unequal; capsules globose, 2- to 3-seeded; style very short; seeds obovate-orbicular, slightly flattened, black, slightly lustrous, about 1.2 mm. across, nearly smooth, superficial cells slightly impressed on margin.——June–Aug. Wet places along rivulets in mountains; Hokkaido, Honshu (Nikko).——Europe, Asia, N. America, Africa, and New Zealand.

Fam. 84. **CARYOPHYLLACEAE** NADESHIKO KA Chickweed Family

Herbs, rarely shrubby below, branches often thickened and jointed at the nodes; leaves opposite, gradually reduced to small bracts above, entire, nerveless or 1- to 3-nerved, usually shortly connate at the nodes, sometimes with scarious or bristly stipules; flowers bisexual, rarely unisexual, sometimes partially cleistogamous, solitary or cymose or in panicles, actinomorphic; sepals 4–5 (–7), persistent, free or connate below into a tube, imbricate in bud; petals as many as the sepals or absent, entire or lobed; stamens 4–10, rarely fewer, receptacle (gynophore) sometimes prominent; ovary 1-locular, rarely imperfectly 2- to 5-locular at the base, the ovules 2 to many, inserted at base or on the free-central placenta; styles 2–5; capsules dehiscing loculicidally and/or septicidally with as many valves or twice the number of the styles, rarely indehiscent; seeds many, rarely 1; endosperm mealy.——About 80 genera, with about 2,000 species, cosmopolitan.

1A. Sepals free.
2A. Stipules present.
3A. Styles connate at base; stipules of few bristles. .... 1. *Drymaria*

3B. Styles free; stipules scarious.
 4A. Styles and carpels 5. . . . . 2. *Spergula*
 4B. Styles and carpels 3. . . . . 3. *Spergularia*
2B. Stipules absent.
 5A. Petals entire or absent.
  6A. Styles as many as the sepals and alternate with them. . . . . 4. *Sagina*
  6B. Styles fewer than the sepals.
   7A. Valves of the capsule as many as the style, entire.
    8A. Rather fleshy seashore plants, with flat leaves; disc prominent, 8- to 10-lobed; fruit fleshy. . . . . 5. *Honkenya*
    8B. Small tufted alpine plants, with subulate leaves; disc not prominent; fruit a dry capsule. . . . . 6. *Minuartia*
   7B. Valves of the capsule bifid.
    9A. Seeds not strophiolate. . . . . 7. *Arenaria*
    9B. Seeds strophiolate. . . . . 8. *Moehringia*
 5B. Petals retuse to 2-parted, rarely entire or absent.
  10A. Cleistogamic flowers present; roots at least some of them thickened. . . . . 9. *Pseudostellaria*
  10B. Cleistogamic flowers absent; roots not thickened (in ours).
   11A. Styles 5, opposite the sepals, rarely 3 or 4; capsules cylindric. . . . . 10. *Cerastium*
   11B. Styles 3, rarely 5 but alternate with the sepals; capsules ovoid to globose. 11. *Stellaria*
1B. Sepals connate below into a prominent tube; petals distinctly clawed; stipules absent.
 12A. Styles 2; calyx-tube with many longitudinal nerves, with 1 to few pairs of bracts at base; seeds peltately attached. . . 12. *Dianthus*
 12B. Styles 3–5; calyx-tube 5- to 10-nerved, sometimes up to 20-nerved, bractless or the bracts not at the base; seeds laterally attached.
  13A. Fruit a berry becoming crustaceous and irregularly dehiscent. . . . . 13. *Cucubalus*
  13B. Fruit a capsule, dehiscent by teeth or valves.
   14A. Capsules and ovary 1-locular.
    15A. Teeth of capsules usually 5, as many as the styles. . . . . 14. *Lychnis*
    15B. Teeth of capsules usually 6, twice as many as the styles. . . . . 15. *Melandrium*
   14B. Capsules and ovary 3- to 5-locular at base. . . . . 16. *Silene*

1. DRYMARIA Willd. Yambaru-hakobe Zoku

Diffuse, rarely erect, dichotomously branched herbs; leaves flat, the stipules small, bristly or fugaceous; flowers pedicellate, small, in spreading terminal axillary cymes; sepals 5, herbaceous or scarious on the margin, free; petals 5, 2- to 6-fid; stamens 5 or fewer; ovary 1-locular, few to many-ovuled; styles 3-fid; capsules 3-valved; seeds reniform-globose or laterally flattened, with a lateral hilum.——About 40 species, mainly in tropical America, a few widely dispersed in the Tropics of the Old World.

1. Drymaria cordata Willd. ex Roem. & Schult. var. **pacifica** Mizushima. Omunagusa. Much-branched, slender, glabrous herb with obtusely angled branches; leaves thinly membranous, depressed cordate-orbicular, 8–15 mm. wide, slightly shorter than broad, rounded, entire, truncately rounded at base, weakly 3-nerved, the petioles nearly absent to 3 mm. long, the stipules consisting of a few whitish setae; cymes terminal, soon becoming lateral, glabrous, loosely several-flowered, the peduncles slender, 3–5 cm. long, the bracts scarious, broadly lanceolate, acute, greenish on the back, the pedicels smooth; sepals 5, about 3 mm. long, narrowly oblong, rounded and pale green on back, weakly 3-nerved, the margins scarious; petals white, slightly shorter than the sepals, oblanceolate, cleft to the middle, the lobes linear-lanceolate, acutish; stamens slightly more than half as long as the sepals, the anthers pale yellow; styles 3, slightly connate at base; capsules ovoid, about 2 mm. long, the valves 3, ovate, entire, membranous; seeds 4 or 5, dark brown, about 0.5 mm. long, mammillate.——Apr.–May. Lowlands; Honshu (Hachijo Isl.).——Bonins, Hawaii, S. America, and Africa. The typical phase occurs in S. America and Africa.

2. SPERGULA L. Ō-tsume-kusa Zoku

Annual or biennial herbs, dichotomously or fasciculately branched; leaves subulate, apparently verticillate, the stipules scarious, small; flowers pedicelled, the cymes racemose; sepals 5; petals 5, entire; stamens 10, rarely 5; ovary 1-locular, many-ovuled; styles 5, alternate with the sepals; capsules 5-valved, entire, the valves opposite the sepals; seeds flat, acutely margined or winged.——Two or three species in the Old World, spread elsewhere as a weed.

1. Spergula arvensis L. Nohara-tsume-kusa. Tufted annual or biennial, glabrous or slightly short-pubescent above; stems ascending, 20–50 cm. long, slender, the stipules small, about 1 mm. long; leaves narrowly linear or subulate, 1.5–4 cm. long, 0.5–1 mm. wide, obtuse, verticillate in 12's to 20's; inflorescence loosely flowered, often once to twice branched, 5–10 cm. long, the bracts small, membranous, the pedicels 1.5–4 cm. long, deflexed in fruit; sepals 3–4 mm. long, ovate, acute, slightly longer or as long as the petals; capsules slightly longer than the sepals; seeds orbicular, slightly flattened, about 1.2 mm. across, black, narrowly winged with scattered pale mammillae on both surfaces.——July–Aug. Waste grounds and cultivated fields in lowlands; Hokkaido, Honshu (Kantō and centr. distr.).——Naturalized from Europe.

Var. **sativa** (Boenn.) Koch. *S. sativa* Boenn.——Ō-tsume-kusa. Seeds smooth.——Occurs with the typical phase.

Var. **maxima** Koch. Ō-tsume-kusa-modoki. With large mammillate seeds.——Naturalized in our area from Europe.

3. SPERGULARIA Pers.   Ushio-tsume-kusa Zoku

Spreading diffuse herbs mainly growing near the seashore and in saline soils; leaves linear to subulate, the axillary tufts of leaves often appearing verticillate, the stipules scarious; flowers pedicelled, white or pinkish, axillary, the cymes often racemose; sepals 5; petals 5, entire, rarely fewer or absent; stamens 10 or fewer; ovary 1-locular, many-ovuled; styles 3, free; capsules 3-valved; seeds reniform-globose or laterally flattened, often winged on margin.——About 30 species.

1A. Stipules free, lanceolate or narrowly deltoid; flowers pale pinkish; capsules slightly longer than the sepals. .... 1. *S. rubra*
1B. Stipules connate at base, the free portion broadly deltoid, often shallowly toothed; flowers white; capsules 1½ times longer than the sepals. .... 2. *S. marina*

1. **Spergularia rubra** (L.) Presl. *Arenaria rubra* L.; *Tissa rubra* (L.) Britt.——Usu-beni-tsume-kusa. Annual or biennial herb often glandular-pilose on upper part and on sepals; stems decumbent and branching below, 5–15 cm. long, usually tufted; stipules 3–3.5 mm. long, thinly scarious, white, lustrous; leaves linear, 8–15 mm. long, 0.5–1 mm. wide, ending in a minute awn; flowers axillary, the pedicels 3–8 mm. long; sepals broadly lanceolate, obsoletely nerved, subobtuse, 3–4 mm. long, narrowly scarious on margin; petals oblanceolate, 2–3 mm. long, obtuse; capsules 4–5 mm. long, ovoid-globose; seeds subovate-globose, about 0.3 mm. across, with scattered obsolete tubercles on back.——June–Oct. Seashores; Hokkaido.——Seashores in the cold temperate to frigid regions of the N. Hemisphere.

2. **Spergularia marina** (L.) Griseb. *Arenaria rubra* var. *marina* L.; *A. marina* (L.) All.; *S. salina* J. & C. Presl; *S. marina* var. *asiatica* (Hara) Hara; *S. salina* var. *asiatica* Hara——Ushio-tsume-kusa. Resembles the preceding species but larger; stems 10–20 cm. long, the upper parts and sepals often slightly glandular-pilose; stipules broadly deltoid or broadly ovate, 1.5–2 mm. long, thinly membranous, white, lustrous, connate on lower half, often 2- to 3-toothed on margin; leaves linear; flowers axillary, the pedicels 3–8 mm. long; sepals narrowly ovate or narrowly ovate-oblong, obtuse, obsoletely 3-nerved, white-scarious on margins, 3–4 mm. long; petals narrowly obovate, about 2 mm. long, white, obtuse; capsules 5–6 mm. long; seeds ovate, slightly flattened, minutely mammillate to nearly smooth, 0.5–0.7 mm. long, barely winged.——July–Sept. Wet saline habitats near the sea; Hokkaido, Honshu, Kyushu (n. distr.).——Sakhalin, Korea, Manchuria, and cooler regions of the N. Hemisphere.

4. SAGINA L.   Tsume-kusa Zoku

Small annuals, biennials or perennials; leaves subulate, opposite, exstipulate; flowers small, solitary in leaf axils or in terminal cymes, pedicelled, white; sepals 4 or 5; petals 4 or 5 or absent, entire, rarely retuse; stamens as many or fewer than the sepals, sometimes twice as many; ovary 1-locular; styles as many as and alternate with the sepals; capsules 4- or 5-valved, the valves opposite the sepals; seeds minute.——More than 10 species, in the N. Hemisphere.

1A. Plant glabrous throughout; pedicels recurved while young, straight at maturity. .... 1. *S. saginoides*
1B. Plant usually glandular-pilose in upper part or sometimes glabrous; pedicels straight from the beginning.
  2A. Superficial cells of seeds with a wart in the center. .... 2. *S. japonica*
  2B. Superficial cells of seeds smooth, not warty. .... 3. *S. maxima*

1. **Sagina saginoides** (L.) Karst. *Spergula saginoides* L.; *Sagina linnaei* Presl——Chishima-tsume-kusa. Small glabrous perennial; leaves linear, mucronate or very short-awned; pedicels 1–2.5 cm. long, slender, ascending or erect, recurved near the tip in young fruit, strict in fruit; sepals 5, rarely 4, ovate-oblong, elliptic, or ovate, 1.5–2 mm. long, obtuse to rounded; petals nearly rounded to obovate, about half to nearly as long as the calyx; stamens 5–10; capsules as long as to slightly longer than the calyx; seeds light brown, 0.3–0.4 mm. across, nearly smooth, distinctly grooved dorsally.—— July–Aug. Alpine; Hokkaido, Honshu (centr. distr.); rare. ——Sakhalin, Kuriles, Kamchatka, Siberia, Caucasus, Europe, and N. America.

2. **Sagina japonica** (Sw.) Ohwi. *Spergula japonica* Sw.; *Spergella japonica* Sw. ex Steud.; *Sagina maxima* sensu auct. Japon., non A. Gray; *S. sinensis* Hance; *S. taquetii* Lév.—— Tsume-kusa, Takano-tsume. Small tufted glabrous annual or biennial, the pedicels, sepals, and sometimes the upper part of the stems glandular-pilose; stems 2–20 cm. long, ascending and branched below; leaves subulate, slightly flattened, 7–18 mm. long, 0.8–1.5 mm. wide, entire, acutely and minutely spine-tipped; flowers axillary, solitary, sometimes forming a terminal leafy cyme; sepals 5, elliptic, rounded at apex, about 2 mm. long; petals narrowly ovate, slightly shorter than to nearly as long as the sepals, entire, rounded at apex, white; stamens 5–10; capsules ovoid-globose, slightly longer than the sepals, deeply 5-parted; seeds minute, ovoid, about 0.4 mm. long, minutely warty, dark brown.——Mar.–Oct. Lowlands to mountains; Hokkaido, Honshu, Shikoku, Kyushu; very common especially around dooryards.——China and Korea to India.

3. **Sagina maxima** A. Gray. *S. taquetii* auct., non Lév. ——Hama-tsume-kusa. Resembles the preceding but slightly stouter, often with a rosette of basal leaves in winter and the leaves slightly broader and thicker, the pedicels and sepals glandular-pilose; seeds yellow-brown, without warts, the cells slightly convex, scarcely impressed.——June–Aug. Near seashores; Hokkaido, Honshu.——Korea and Formosa.

Var. **crassicaulis** (S. Wats.) Hara. *S. crassicaulis* S. Wats. ——Ezo-hama-tsume-kusa. Pedicels and sepals glabrous. ——Hokkaido.——Kamchatka and western N. America.

5. HONKENYA Ehrh.   HAMA-HAKOBE ZOKU

Slightly fleshy glabrous perennial herb; stems decumbent and much-branched at base; leaves ovate to oblanceolate; flowers small, bisexual or staminate, whitish, solitary in axils or cymose; sepals 5; petals 5, entire, small, shorter than the sepals; stamens 10; ovary 1-locular; styles 3, sometimes 4 or 5; capsules rather fleshy, globose, 3- to 5-valved, the valves entire; seeds many, rather large, not strophiolate.——One species, on seashores in the cooler regions of the N. Hemisphere.

1. **Honkenya peploides** (L.) Ehrh. var. **major** Hook. *Ammodenia major* (Hook.) Kudo; *H. peploides* var. *oblongifolia* (Torr. & A. Gray) Fenzl; *H. oblongifolia* Torr. & A. Gray; *A. peploides* var. *oblongifolia* (Torr. & A. Gray) Maxim. ——HAMA-HAKOBE.   Lustrous fleshy green perennial from slender creeping rhizomes; stems tufted, much branched, decumbent at base, 10–30 cm. long; leaves oblong, ovate or broadly oblanceolate, 1.5–4 cm. long, 4–20 mm. wide, acute to obtuse, mucronate, sessile; flowers in upper axils, sometimes leafy-cymose, short-pedicelled, yellowish green; sepals 4–5 mm. long, narrowly ovate, subacute; petals white, obovate, smaller than the sepals, obtuse; capsules globose, about 8 mm. across; seeds 3–4 mm. long.——June–Sept. Gravelly slopes along seashores; Hokkaido, Honshu (n. distr.).——Korea, Sakhalin, Kuriles, and the Ochotsk Sea area.——Alaska. The typical phase occurs in the Arctic regions of n. Europe to Spain, and N. America.

6. MINUARTIA L.   TAKANE-TSUME-KUSA ZOKU

Small tufted perennial herbs; leaves subulate, densely disposed on the short stems and branches, 1- to 3-nerved, often with a short awn; flowers solitary or in cymes, white, rarely pinkish; sepals 5, 1- to 3-nerved; petals entire, longer to shorter than the sepals; stamens 10; ovary 1-locular; styles 3; capsules narrowly ovoid to ovoid-cylindric, shallowly 3-cleft, the valves nearly entire.——About 100 species, in high latitudes and mountainous regions of the N. Hemisphere.

1A. Sepals acuminate or acute, minutely awn-tipped; petals slightly longer than to as long as the sepals; capsules 3–3.5 mm. long, slightly longer than the sepals. .... 1. *M. verna* var. *japonica*
1B. Sepals obtuse, not awn-tipped; petals always longer than the sepals; capsules 6–15 mm. long, 1.5–2.5 times as long as the sepals.
  2A. Leaves 3-nerved, more or less ciliate; seeds with long tubercles on margin. .... 2. *M. macrocarpa*
  2B. Leaves 1-nerved, glabrous or nearly so except near base.
    3A. Seeds glabrous. .... 3. *M. hondoensis*
    3B. Seeds minutely tuberculate. .... 4. *M. arctica*

1. **Minuartia verna** (L.) Hiern var. **japonica** Hara. *M. verna* sensu auct. Japon., non Hiern——HOSOBA-TSUME-KUSA, KOBA-NO-TSUME-KUSA.   Stems much branched near the base, tufted, erect or ascending, 4–12 cm. long, slender, short-pubescent on upper parts; leaves subulate, 3–10 mm. long, 0.2–0.3 mm. wide, acute to subacute, minutely awn-tipped, glabrous, persistent; flowers 1–5; sepals narrowly ovate or broadly lanceolate, about 3 mm. long, 3-nerved, glabrous or sparingly short-pilose; petals obovate, obtuse, white, as long as to slightly longer than the sepals; seeds reniform-orbicular, 0.6 mm. long, the superficial cells scarcely elevated.——July–Aug. Rocky cliffs and slopes in alpine regions; Hokkaido, Honshu (n. and centr. distr.).——The species occurs widely over Eurasia to N. Africa.

2. **Minuartia macrocarpa** (Pursh) Ostenf. var. **jooi** (Makino) Hara. *Alsine jooi* Makino; *A. macrocarpa* sensu auct. Japon., non Fenzl——MIYAMA-TSUME-KUSA.   Low spreading perennial herb, the stems creeping, much branched, the branches erect, 3–5 cm. long, spreading-pubescent above; leaves dense, numerous, subulate, somewhat flattened, often slightly arcuate, yellowish green, subobtuse to subacute, 7–12 mm. long, about 1 mm. wide, sparingly ciliate; flowers solitary, terminal, erect, short-pedicellate; sepals linear-oblong to lanceolate, 6–7 mm. long, obsoletely 3-nerved, obtuse; petals narrowly obovate, white, 7–8 mm. long, obtuse; capsules 10–15 mm. long; seeds about 1.5 mm. across, tuberculate.——July–Aug. Dry alpine slopes; Honshu (centr. distr.); rare.

Var. **minutiflora** Hult. *M. subfalcata* Nakai, excl. syn.—— EZO-MIYAMA-TSUME-KUSA.   Leaves hairy with scattered cellular hairs on the margin.——Alpine slopes; Hokkaido.—— Kamchatka.

3. **Minuartia hondoensis** Ohwi. *M. arctica* var. *hondoensis* Ohwi; *Alsine arctica* Makino, non Fenzl——TAKANE-TSUME-KUSA.   Low tufted perennial herb; stems much branched below, 3–7 cm. high, slender, ascending below, erect above, the upper internodes glandular short-hairy; leaves dense, subulate, 8–15 mm. long, 0.5–0.7 mm. wide, obtuse, green, slightly purplish red at base, sometimes loosely short-pilose on margin near base, 1-nerved; flowers usually solitary, the pedicels with glandular and short eglandular hairs; sepals linear-oblong, obtuse, slightly pilose, 5–6 mm. long, 3-nerved; petals white, narrowly obovate, rounded at apex, 7–9 mm. long; capsules 8–10 mm. long; seeds reniform-orbicular, 0.6–1 mm. across, smooth, superficial cells scarcely convex.——Dry slopes and on rocks in high mountains; Honshu (n. and centr. distr.).

4. **Minuartia arctica** (Stev.) Asch. & Graebn. *Arenaria arctica* Stev.——EZO-TAKANE-TSUME-KUSA.   Closely resembles the preceding; seeds tuberculate.——July–Aug. Alpine slopes; Hokkaido.——Cold regions of the N. Hemisphere.

7. ARENARIA L.   NOMI-NO-TSUZURI ZOKU

Annual or perennial herb; leaves flat, rarely linear to subulate; inflorescence paniculate-cymose; flowers rarely solitary, white; sepals 5, herbaceous; petals as many and alternate with the sepals, entire, or sometimes absent; stamens 10; styles usually 3; capsules ovoid, enclosed by the persistent calyx, dehiscing or with 3 lobes, each 2-cleft or with 6 lobes; seeds many to few, subglobose to discoid, the hilum naked.——Circumboreal.

1A. Annual or biennial herb; petals shorter than the sepals. .... 1. *A. serpyllifolia*
1B. Perennial herbs; petals slightly longer than the sepals.
  2A. Stems with minute papillose hairs, more or less in 2 series; leaves nearly glabrous, broadly lanceolate to narrowly ovate, 3–7 mm. long, 1.5–3 mm. wide; sepals 3–4 mm. long, acuminate. .... 2. *A. katoana*
  2B. Stems with crisped hairs all around; leaves soft-pubescent especially on margin, oblong to broadly ovate, 1–2 cm. long, 5–10 mm. wide; sepals 6–8 mm. long, subobtuse. .... 3. *A. merckioides*

1. **Arenaria serpyllifolia** L.   *A. serpyllifolia* var. *leptoclados* auct. Japon., non Reichenb., and var. *tenuior* auct. Japon., non Mert. & Koch——NOMI-NO-TSUZURI. Short-pubescent annual or biennial; stems slender, much-branched, 5–20 cm. long; leaves small, narrowly to broadly ovate or elliptic, sometimes lanceolate in upper ones, 3–7 mm. long, 1–5 mm. wide, acuminate to acute, mucronate, sessile; flowers in leafy cymes, axillary, solitary, terminal; pedicels spreading or ascending, 5–15 mm. long, slender; sepals broadly lanceolate, acuminate, 3–4 mm. long; petals about half as long as the sepals, obovate-orbicular, obtuse, white; capsules slightly longer than the sepals; seeds obliquely elliptic, 0.3–0.5 mm. long, dark brown, thick-discoid, the superficial cells slightly convex-elliptic.——Mar.–June. Sunny slopes and waste ground in lowlands and hills; Hokkaido, Honshu, Shikoku, Kyushu.——Europe and Asia.

2. **Arenaria katoana** Makino. KATŌ-HAKOBE. Perennial; stems tufted, branched below, terete, 5–8 cm. long, rather densely leaved, with 2 longitudinal rows of minute hairs; leaves sessile, ovate or narrowly so to broadly lanceolate, 3–7 mm. long, 1.5–3 mm. wide, acuminate, glabrous except a few short hairs on margin near base; flowers 1 or 2, the pedicels minutely pubescent, 5–12 mm. long; sepals broadly lanceolate, glabrous, acuminate, 3–4 mm. long; petals ovate, white, 5–6 mm. long, subobtuse; styles slender, erect, about 2 mm. long; capsules conical-ovoid, about 5 mm. long; seeds reniform-orbicular, minutely mammillate, nearly 1 mm. long.——July-Aug. Alpine; Hokkaido (Mount Yubari), Honshu (Mount Hayachine in Rikuchu, and Mount Shibutsu in Kotsuke); rare.

Var. **lanceolata** Tatew. *A. tatewakii* Nakai.——APOI-TSUME-KUSA. Leaves narrower, lanceolate to narrowly so.——Hokkaido (Mount Apoi in Hidaka Prov.).

3. **Arenaria merckioides** Maxim. *A. chokaiensis* Yatabe——CHŌKAI-FUSUMA, MEAKAN-FUSUMA. Rhizomes slender, elongate, much branched; stems tufted, angled, 5–15 cm. long, nearly simple; leaves oblong to elliptic or ovate, 1–2 cm. long, 5–10 mm. wide, acute with an obtuse tip, pubescent while young, sparsely so later and confined to the margins, rather fleshy, sessile; flowers 1–4, the pedicels soft-pubescent, 10–15 mm. long; sepals lanceolate, obtuse, slightly pubescent; petals narrowly ovate; seeds reniform-orbicular, red-brown, about 1.5 mm. across inclusive of the rather thickish marginal wing.——July–Aug. Alpine; Hokkaido (Mount Meakan), Honshu (Mount Chokai and Mount Gassan); rare.

8. MOEHRINGIA L.   Ō-YAMA-FUSUMA ZOKU

Rather low annuals or perennials; leaves flat, thin, sometimes linear, sessile or short-petiolate; flowers axillary, solitary or cymose, small, white; sepals and petals 4 or 5; stamens 8–10; styles 2 or 3; capsules ellipsoidal, the valves 4–6; seeds rather few, usually lustrous, with a membranous strophiole on the hilum.——About 20 species, temperate regions of the N. Hemisphere.

1A. Leaves ovate to broadly so, distinctly petioled; sepals acuminate, broadly lanceolate, longer than the capsule. .... 1. *M. trinervia*
1B. Leaves oblong to narrowly so, sessile; sepals very obtuse, elliptic, about half as long as the capsule. .... 2. *M. lateriflora*

1. **Moehringia trinervia** (L.) Clairv. *Arenaria trinervia* L.; *M. platysperma* Maxim.; *M. trinervia* var. *platysperma* (Maxim.) Makino——TACHI-HAKOBE. Short-pubescent biennial; stems erect or ascending, mostly decumbent and branched below, 7–20 cm. long; leaves ovate or broadly so, 1–2.5 cm. long, 5–12 mm. wide, acute or cuspidate, abruptly narrowed to subrounded at base, subtrinerved, subsessile or distinctly petiolate in the lower ones, short-ciliate; flowers solitary in axils, white, the pedicels 1–4 cm. long; sepals broadly lanceolate, 1-nerved, acuminate, short-pilose on the nerves, broadly white and glabrous on the margin, 3.5–5 mm. long; petals obovate, entire, 1/3–1/2 as long as the sepals; capsules 2.5–3 mm. long, the valves 6, recurved above; seeds reniform-orbicular, lenticular, about 0.8 mm. across, with a small strophiole at base.——Mar.–July. Foothills; Hokkaido, Honshu, Shikoku, Kyushu; rather rare.——Europe, Siberia, and Formosa.

2. **Moehringia lateriflora** (L.) Fenzl. *Arenaria lateriflora* L.——Ō-YAMA-FUSUMA. Perennial herb; stems slender, 10–15 cm. long, branched, slender; leaves oblong to narrowly so, sometimes elliptic, 1–2.5 cm. long, 3–10 mm. wide, obtuse or subobtuse, sessile, short-pilose on margin and on nerves beneath; inflorescence axillary or terminal, 1- to 3-flowered, 1–3 cm. long, puberulent, often with a pair of minute bracts near the middle; sepals elliptic, obtuse to subrounded at apex, 2–3 mm. long, obsoletely 1- to 3-nerved; petals entire, longer or nearly as long as the sepals; capsules 3.5–5.5 mm. long, 3-valved, the valves bifid, erect; seeds about 1 mm. long.——Meadows in mountains; Hokkaido, Honshu, Shikoku, Kyushu.——Temperate regions of the N. Hemisphere.

9. PSEUDOSTELLARIA Pax   WACHIGAI-SŌ ZOKU

Low perennial herbs usually with thickened roots; stems erect or ascending, sometimes long-creeping after anthesis; leaves flat, 1-nerved; flowers dimorphic, the chasmogamic flowers 1 to several on or near the top of stems, in cymes; sepals 5, sometimes 6 or 7; petals as many as the sepals, white, slightly longer than the sepals, entire or bifid; stamens usually 8 or 10; styles 2–4; ovary globose, the ovules rather numerous, the placenta nearly basal; capsules globose, 2- to 4-valved, the valves bifid; cleistogamous flowers usually on lower part of stems, short-pedicelled or subsessile, the sepals 4 or 5, the petals absent; stamens reduced; seeds slightly compressed, tubercled.——More than 10 species, in e. Asia to the Altai mts. and the Himalayas to Europe.

1A. Upper 2 pairs of leaves approximate; flowers 1–5, terminal.
  2A. Chasmogamic flowers solitary, terminal, usually 6- to 8-merous, the pedicels glabrous; roots few-fascicled, only slightly thickened.
        1. *P. palibiniana*
  2B. Chasmogramic flowers 1 or 2(–5), terminal, 5-merous, the pedicels puberulous on one side; main root solitary, fusiform.
        2. *P. heterophylla*
1B. Leaf pairs not approximate; flowers terminal and lateral.
  3A. Leaves glabrous except on margin at base, linear to ovate, narrowed below.
    4A. Leaves nearly all alike, linear to linear-lanceolate; flowers usually few to several in a terminal cyme. . . . . 3. *P. sylvatica*
    4B. Leaves heteromorphic, the upper ovate or narrowly ovate, the lower spathulate; flowers 1–3, solitary in axils, terminal.
        4. *P. heterantha*
  3B. Leaves ciliate, ovate to broadly ovate, rounded at base. . . . . 5. *P. japonica*

1. Pseudostellaria palibiniana (Takeda) Ohwi. *Krascheninnikovia palibiniana* Takeda; *K. palibiniana* var. *polymera* Nakai——HIGENE-WACHIGAI-SŌ. Roots 1–4, elongate, slightly thickened; stems 10–20 cm. long, the internodes with 2 longitudinal series of soft hairs; lower leaves spathulate-oblanceolate, 2 to 4 pairs, obtuse, slightly pubescent toward the base, the upper leaves approximate, broadly lanceolate to ovate or broadly so, 3–4 cm. long, 1.5–2.5 mm. wide, abruptly acute to obtuse, rather broadly cuneate at base, glabrous, sessile; flowers solitary, white, the pedicels 1.5–2.5 cm. long, glabrous; sepals 5–8, sometimes sparsely ciliate at base; petals broadly oblanceolate, entire, 6–8 mm. long, obtuse; cleistogamous flowers near base of stems, short-petioled.——Apr.–May. Woods in mountains; Honshu (Iwaki Prov. southw. to centr. distr.).——Korea.

2. Pseudostellaria heterophylla (Miq.) Pax. *Krascheninnikovia heterophylla* Miq.; *K. koidzumiana* Ohwi; *Pseudostellaria koidzumiana* (Ohwi) Ohwi——WADA-SŌ. Resembles the preceding; main root thickened, fusiform near base; stems 10–15 cm. long, with 2 longitudinal rows of short hairs on the internodes; upper leaves ovate to broadly so or rhombic-ovate, usually acute, 3–6 cm. long, 1–4 cm. wide, often slightly pilose on the nerves beneath, sessile; flowers 1–5, fasciculate, terminal, white, the pedicels 1–2 cm. long, short-pubescent; sepals 5, slightly pilose on back; petals 5, narrowly oblong, bifid, 7–8 mm. long; cleistogamous flowers short-pedicellate, on nodes below the chasmogamic ones.——Apr.–May. Honshu, Kyushu.——Korea, China, and Manchuria.

3. Pseudostellaria sylvatica (Maxim.) Pax. *Krascheninnikovia sylvatica* Maxim.——KUSHIRO-WACHIGAI-SŌ. Roots 1 to few, thickened and fusiform; stems simple, usually solitary, 4-angled, 15–30 cm. long, with 2 longitudinal rows of short hairs on the internodes; leaves linear to linear-lanceolate, 3–7 cm. long, 2–7 mm. wide, acuminate, subsessile, glabrous except for a few hairs at base; flowers few (sometimes solitary), usually in a loose terminal cyme, white, the pedicels short-pubescent on one side, 15–30 mm. long; sepals 5, glabrous; petals narrowly obovate, 5–6 mm. long, shallowly 2-lobulate; cleistogamous flowers pedicelled, borne on the lower nodes.——May–June. Hokkaido, Honshu (Rikuchu Prov.); rare.——Korea, China, and e. Siberia.

4. Pseudostellaria heterantha (Maxim.) Pax. *Krascheninnikovia heterantha* Maxim.; *K. maximowicziana* Fr. & Sav.; *P. musashiensis* Hiyama——WACHIGAI-SŌ. Delicate herb, roots 1 or 2, thickened, fusiform; stems 1 or 2, often decumbent at base, 8–12 cm. long; leaves narrowly oblanceolate to subspathulate, 10–25(–40) mm. long, 2–4 mm. wide, acute or subacute, glabrous except for a few soft hairs at base, the upper leaves often ovate or narrowly so, short-petiolate, 15–30 mm. long, 7–12 mm. wide, abruptly acuminate to acute; flowers 1 or 2(–3), white, solitary in leaf axils, terminal, the pedicels 2–3 cm. long, with a longitudinal series of short hairs; sepals 5; petals 5, obovate, about 6 mm. long, rounded to obtusely mucronate; cleistogamous flowers few, pedicelled, borne near the base.——Apr.–June. Woods in the foothills; Honshu (Kantō Distr. and westw.), Shikoku, Kyushu.——China and India.

5. Pseudostellaria japonica (Korsh.) Pax. *Krascheninnikovia japonica* Korsh.; *K. ciliata* Honda ——NAMBU-WACHIGAI-SŌ. Roots thickened, fusiform; stems erect, 15–20 cm. long, with 2 longitudinal rows of short hairs; lower leaves few-paired, nearly linear to lanceolate, 1.5–2.5 cm. long, 2–3 mm. wide, the upper leaves 4-paired, ovate or broadly so, 1.5–3 cm. long, 1–2 cm. wide, acute to short-acuminate, rounded at base, sessile, loosely pilose on both sides, rather long-pilose on margin and on nerves beneath; flowers 1–3, terminal and axillary, the pedicels about 15 mm. long, pubescent; sepals 5; petals obovate, 4–6 mm. long, white.——Honshu (n. distr.).——Manchuria.

10. CERASTIUM L. MIMINAGUSA ZOKU

Annuals, biennials or perennials usually more or less pubescent or tomentose; cymes terminal, leafy or bracteate; sepals 5, rarely 4; petals 5, rarely 4, retuse to bifid, sometimes fimbriate or entire, white; stamens 10 or fewer; ovary 1-locular, many-ovuled; styles 5 and opposite the sepals, rarely 3 or 4; capsules cylindrical, with twice as many teeth as the styles; seeds reniform-orbicular or broadly obovate, slightly flattened.——About 70 species, cosmopolitan, except lowlands of the Tropics.

1A. Petals bifid or fimbriate; teeth of capsules erect, with revolute margins.
  2A. Petals shorter to barely longer than the sepals; biennials or short-lived perennials.
    3A. Pedicels shorter than to as long as the sepals; sepals pale green; inflorescence small, densely flowered; biennial or annual.
        1. *C. viscosum*
    3B. Pedicels at least some of them longer than the sepals; sepals often purplish on margin; inflorescence rather loose; short-lived perennials. . . . . 2. *C. vulgatum*
  2B. Petals distinctly longer than the sepals; perennials.
    4A. Seeds 1–1.2 mm. wide, mammillate; growing on rocks or in meadows near seashores. . . . . 3. *C. fischerianum*
    4B. Seeds less than 1 mm. wide, with low semirounded warts; plants of mountains.

5A.   Stems 25–35 cm. long, loosely tufted; radical leaves spathulate, the cauline broadly lanceolate, 15–20 mm. long, 3–6 mm. wide.
                                                                                  4. *C. furcatum*

5B.   Stems 10–15(–20) cm. long, densely tufted; radical and cauline leaves linear-lanceolate, 8–20 mm. long, 1.5–3 mm. wide.
                                                                             5. *C. schizopetalum*

1B.   Petals entire; teeth of capsules flat, reflexed; leaves 4–8 cm. long, 1–2 cm. wide. .... 6. *C. pauciflorum* var. *oxalidiflorum*

1. Cerastium viscosum L. *C. glomeratum* Thuill.—— Oranda-miminagusa.   Biennial herb, short spreading-pilose and glandular-pubescent; stems 10–60 cm. long, somewhat tufted; radical leaves broadly spathulate or broadly lanceolate, the cauline narrowly oblong to ovate, 7–20 mm. long, 4–12 mm. wide, obtuse, mucronate; inflorescence dense, the bracts green, small, the pedicels shorter than to as long as the sepals; sepals lanceolate, acute, 4–5 mm. long, with long ascending hairs on back, slightly membranous on margin; petals shorter than the sepals; seeds about 0.5 mm. wide.——Apr.–May. Meadows and roadsides in lowlands and foothills; Honshu, Shikoku, Kyushu; very common.——Introduced from Europe.

2. Cerastium vulgatum L. *C. caespitosum* Gilib.; *C. holosteoides* Fries; *C. triviale* Link——Ō-miminagusa.   Perennial herb, 15–30 cm. high, pilose and often glandular-hairy on upper parts; stems erect, tufted; leaves ovate to narrowly ovate-oblong, 1–4 cm. long, 4–12 mm. wide, acute to subobtuse, mucronate; inflorescence spreading in fruit, the pedicels 5–25 mm. long, erect in anthesis, abruptly curved near top in fruit; flowers white; sepals broadly lanceolate to narrowly ovate, pubescent, purplish toward the scarious margin, subacute to obtuse, 5.5–6.5 mm. long; petals nearly as long as the sepals; capsules 8–10 mm. long; seeds 0.6–0.7 mm. wide, low-tubercled.——June–July. Lowlands to foothills; Hokkaido.—— Europe, Siberia, n. Korea, and Sakhalin.

Var. **hallaisanense** Nakai. *C. caespitosum* var. *ianthes* (Williams) Hara; *C. ianthes* Williams; *C. holosteoides* var. *hallaisanense* (Nakai) Mizushima——Miminagusa.   Slightly smaller in all parts than in the typical phase; sepals 4–5(–5.5) mm. long.——May–June. Lowlands to foothills; Hokkaido, Honshu, Shikoku, Kyushu; common.——Korea and China.

3. Cerastium fischerianum Ser. *C. alpinum* var. *fischerianum* (Ser.) Regel; *C. schmidtianum* Takeda; *C. robustum* Williams; *? C. rishiriense* Miyabe & Tatew.——Ōbana-no-miminagusa.   Rather tufted, spreading-pubescent perennial; stems 15–60 cm. long; leaves oblong-lanceolate, narrowly ovate or elliptic, 1–5 cm. long, 3–20 mm. wide; inflorescence loose, the pedicels 1–3 cm. long; sepals ovate-oblong or narrowly oblong, 6–8 mm. long, obtuse, with scarious margins; petals cuneate-obovate, 10–12 mm. long, bifid, rarely 2-lobed or twice toothed; capsules 12–15 mm. long; seeds mammillate, 1–1.2 mm. wide.——June–Aug. Lowlands, especially near seashores; Hokkaido, Honshu, Kyushu (n. distr.); variable.—— Kamchatka, Kuriles, Sakhalin, Ochotsk Sea region, and n. Korea.

4. Cerastium furcatum Cham. & Schltdl. var. **tetraschistum** (Takeda) Ohwi. *C. takedae* Hara; *C. takedae* var. *tetraschistum* Takeda; *? C. mitsumoriense* Miyabe & Tatew. ——Hosoba-miminagusa.   Slightly pubescent, loosely tufted perennial; stems 25–35 cm. long, with a single longitudinal line of hairs on the internodes; radical leaves spathulate, obtuse, 5–15 mm. long, sometimes spreading-hairy, the cauline lanceolate to ovate-oblong, glabrescent, 15–20 mm. long, 3–6 mm. wide; inflorescence loosely few-flowered, the pedicels 1–2 cm. long, short-pubescent; sepals 3.5–5 mm. long, ovate-oblong, the margins broadly scarious; petals about 8 mm. long, spathulate-oblong, bifid, rarely twice-lobed.——July–Aug. Alpine; Honshu (centr. distr.).

Var. **ibukiense** Ohwi. *C. vulgatum* var. *alpinum* sensu Fr. & Sav., non Koch——Koba-no-miminagusa.   Stems ascending and branching below; flowers smaller than in the preceding variety; petals about 7 mm. long.——May–June. Mountains; Honshu (Mount Ibuki in Oomi); rare.——The typical phase occurs in e. Siberia, Manchuria, Korea, and n. China.

5. Cerastium schizopetalum Maxim. Miyama-miminagusa.   Densely tufted, pubescent perennial; stems 10–20 cm. long, branched at base, rather elongate; leaves linear-lanceolate or narrowly lanceolate, 8–20 mm. long, 1.5–3 mm. wide, acute to subacute; inflorescence loosely few-flowered, glandular-pilose, the pedicels 1–2 cm. long; sepals ovate-oblong, 3–5 mm. long, glabrous and scarious on margin, obtuse; petals 10–12 mm. long, bifid, the lobes again shallowly 2- to 3-lobed; capsules 7–8 mm. long; seeds about 0.6 mm. long, minutely tubercled.——July–Aug. Alpine; Honshu (centr. distr.).

Var. **bifidum** Takeda. *C. rupicola* Ohwi. Kumoma-miminagusa.   Petals simply bifid.——Alpine regions of Honshu (Mount Shirouma).

6. Cerastium pauciflorum Stev. var. **oxalidiflorum** (Makino) Ohwi. *C. oxalidiflorum* Makino——Tagasode-sō. Densely tufted pubescent perennial herb; stems 30–40 cm. long; leaves broadly lanceolate, 4–8 cm. long, 1–2 cm. wide, subacuminate to acute with an obtuse tip; inflorescence rather loosely more than 10-flowered, glandular-pilose, the pedicels 2–5 cm. long; sepals elliptic to ovate-oblong, obtuse, 5–6 mm. long, scarious on margin; petals cuneate, entire, rounded at apex, about 15 mm. long; capsules 11–17 mm. long, the teeth reflexed; seeds tubercled, about 1 mm. across.——May–June. Mountains; Honshu (centr. distr.); rare.——The typical phase occurs in Siberia and Manchuria.

11. STELLARIA L.   Hakobe Zoku

Erect to decumbent, often tufted herbs, rarely with branched hairs; flowers solitary in the leaf axils or in cymes, white to green; sepals 5; petals 5, bifid, sometimes lacerate or retuse, rarely absent; ovary 1-locular, many- or rarely few-ovuled; styles 3, rarely 5, alternate with the sepals; capsules 3- (5-) valved, the valves bilobed; seeds reniform-globose or flattened, rarely winged or tubercled on the margin.——About 100 species, cosmopolitan in temperate and cold regions.

1A.   Leaves at least the lower ones distinctly petioled, ovate or deltoid-ovate; hairs if present simple and not branched.
    2A.   Styles 5. .... 1. *S. aquatica*
    2B.   Styles 3.
        3A.   Pedicels and stems with a longitudinal series of soft hairs.
            4A.   Leaves 4–8 cm. long, ciliate; sepals obsoletely nerved. .... 2. *S. bungeana*

4B. Leaves 1–4 cm. long, nearly glabrous; sepals 1-nerved.
  5A. Sepals acute, distinctly 1-ribbed; petals usually longer than the sepals. . . . . 3. *S. sessiliflora*
  5B. Sepals obtuse, 1-nerved; petals as long as to shorter than the sepals.
    6A. Stamens 1–7; seeds 1–1.2 mm. wide, with semirounded obtuse tubercles. . . . . 4. *S. media*
    6B. Stamens 5–10; seeds about 1.5 mm. wide, with conical subacute tubercles. . . . . 5. *S. neglecta*
3B. Pedicels and stems glabrous; leaves often with a few long multicellular appressed hairs on the upper side. . . . . 6. *S. diversiflora*
1B. Leaves sessile or nearly sessile, with branched hairs.
  7A. Leaves broadly ovate, with branched hairs.
    8A. Petals longer than the sepals. . . . . 7. *S. uchiyamana*
    8B. Petals absent. . . . . 8. *S. tomentella*
  7B. Leaves linear to oblong, rarely ovate, with simple hairs.
    9A. Petals bifid or absent; leaves and stems glabrous or glabrescent.
      10A. Leaves usually not more than 5 cm. long and 1.2 cm. wide; capsules dehiscent, many-seeded.
        11A. Leaves linear or broadly so, 1–2.5 mm. wide.
          12A. Stems slightly scabrous; petals as long as to slightly longer than the sepals. . . . . 9. *S. longifolia*
          12B. Stems smooth; petals about twice as long as the sepals.
            13A. Leaves ascending, linear, scarcely broadened below; stems 20–50 cm. long; plant of wet places in lowlands.
10. *S. filicaulis*
            13B. Leaves horizontally spreading, slightly broadened below; stems 10–20 cm. long; plant of gravelly slopes in alpine zone.
              14A. Seeds tubercled on margin. . . . . 11. *S. nipponica*
              14B. Seeds winged on margin. . . . . 12. *S. pterosperma*
        11B. Leaves lanceolate, oblong or sometimes ovate.
          15A. Leaves ovate to lanceolate, acuminate to very acute, rather firm.
            16A. Plant glaucous; leaves 1–3 cm. long; flowers solitary in upper axils, ebracteate; petals 1½ times as long as the sepals.
13. *S. ruscifolia*
            16B. Plant yellow-green or pale green; leaves 2–4 cm. long; flowers cymose, the bracts small; petals shorter than the sepals or absent. . . . . 14. *S. fenzlii*
          15B. Leaves oblong to oblong-lanceolate, obtuse to subacute, 1–2.5 cm. long, herbaceous or membranous.
            17A. Bracts thinly membranous, white; plant usually glaucous. 15. *S. alsine* var. *undulata*
            17B. Bracts green; flowers axillary and solitary.
              18A. Petals longer than the sepals. . . . . 16. *S. humifusa*
              18B. Petals absent. . . . . 17. *S. calycantha*
      10B. Leaves 4–12 cm. long, 1.5–3 cm. wide; capsules indehiscent, 1-seeded. . . . . 18. *S. monosperma* var. *japonica*
    9B. Petals fimbriate; stems and leaves silky-pubescent. . . . . 19. *S. radians*

1. Stellaria aquatica (L.) Scop. *Cerastium aquaticum* L.; *Malachium aquaticum* (L.) Fries——USHI-HAKOBE. Biennial or perennial, decumbent at base, erect above, slightly glandular-pubescent on upper parts, 20–50 cm. high; leaves ovate or broadly so, glabrescent, 1–6 cm. long, 0.8–3 cm. wide, rounded to shallowly cordate, the lower ones petiolate, those above sessile; flowers in leafy cymes, the pedicels 5–15 mm. long, deflexed in fruit; sepals narrowly ovate, obtuse, 4–5.5 mm. long, short-glandular; petals nearly as long as the sepals, deeply 2-parted, white; styles 5, short; capsules 5-valved, the valves bifid; seeds ellipsoidal, about 0.8 mm. long, slightly flattened, mammillate.——Apr.–Oct. Banks and roadsides in lowlands; Hokkaido, Honshu, Shikoku, Kyushu; common. ——Temperate regions of the N. Hemisphere and N. Africa.

2. Stellaria bungeana Fenzl. *S. nemorum* var. *bungeana* (Fenzl) Regel——EZO-NO-MIYAMA-HAKOBE, Ō-HAKOBE. Stems 50–80 cm. long, branched below, ascending to suberect above, the internodes with a longitudinal line of soft pubescence; leaves ovate-oblong or ovate, 4–8 cm. long, 2–3 cm. wide, abruptly acuminate, rounded at base, ciliate, the midrib often slightly pubescent on both sides, pale green above, the upper ones sessile, the lower petioled; cymes relatively large, many-flowered, the bracts small, green, leaflike, the pedicels with a line of soft hairs; sepals ovate, subobtuse, 4–5 mm. long, pubescent; petals slightly longer than the sepals, 2-fid; capsules globose; seeds reniform-orbicular, about 1.2 mm. across, densely mammillate, dark brown, slightly flattened.——May–Sept. Hokkaido.——e. Siberia, Manchuria, n. Korea, and Sakhalin.

3. Stellaria sessiliflora Yabe. *S. nemorum* var. *japonica*
Fr. & Sav.; *S. japonica* (Fr. & Sav.) Makino, non Miq.; *S. franchetii* Honda——MIYAMA-HAKOBE. Perennial herb; stems 20–30 cm. long, usually ascending to decumbent at base, with a longitudinal line of short soft hairs; leaves petioled, ovate or broadly so, or sometimes cordate-orbicular, 1–4 cm. long, 7–25 mm. wide, acuminate to acute, rounded to shallowly cordate at base, often mucronate, green on both sides, glabrous above, the midrib beneath slightly pubescent; flowers axillary, the pedicels 0–4 cm. long, with a longitudinal line of short soft hairs; sepals lanceolate, 1-nerved, long-pubescent, acute, 4–6 mm. long; petals usually slightly longer than the sepals, deeply bifid; capsules ovoid-globose, 6-valved; seeds dark brown, obliquely reniform-orbicular, 1–1.2 mm. long, densely mammillate with semirounded tubercles.——May–July. Along streams in woods in uplands; Hokkaido (sw. distr.), Honshu, Shikoku, Kyushu.

4. Stellaria media (L.) Vill. *Alsine media* L.; *S. media* var. *minor* Makino; *S. minor* (Makino) Honda——KO-HA-KOBE. Biennial or annual; stems 10–20 cm. long, decumbent at base, much branched, often purplish, with a longitudinal line of soft hairs; leaves ovate, 1–2 cm. long, 8–15 mm. wide, acute, rounded to obtuse at base, glabrous, often slightly pubescent on margin near base, petiolate in the lower, sessile in the upper ones; flowers in upper axils, the bracts green, leaflike, the pedicels 5–40 mm. long, with a line of short hairs; sepals narrowly oblong, obtuse, obsoletely 1-nerved, 3–4 mm. long, soft-pubescent; petals white, deeply bifid, slightly shorter than the sepals; stamens 1–7, usually 3–5; seeds slightly flattened, with obtuse mammillae, 1–1.2 mm. across, reniform-orbicular.——Mar.–Sept. Sunny waste places, cultivated fields and roadsides in lowlands and hills; Hokkaido, Honshu, Shikoku, Kyushu; common.——Nearly cosmopolitan.

5. Stellaria neglecta Weihe. *S. media* sensu auct. Japon., pro parte, non Vill.; *S. media* var. *neglecta* (Weihe) Weihe; *S. media* var. *procera* Klett & Richt.——Midori-hakobe. Closely resembles the preceding, but slightly larger and paler green; stamens 8–10, sometimes 5–6; seeds about 1.5 mm. across, more or less distinctly flattened, mammillate, especially on the margin.——Mar.–Sept. Hokkaido, Honshu, Shikoku, Kyushu.——Eurasia.

6. Stellaria diversiflora Maxim. *S. diandra* Maxim.; *S. diversiflora* var. *diandra* (Maxim.) Makino——Sawa-hakobe, Tsuru-hakobe. Perennial; stems 5–30 cm. long, often decumbent at base, branched, glabrous, terete; leaves petiolate, broadly ovate to deltoid-ovate, 1–3.5 cm. long, 0.8–2.5 cm. wide, acute to acuminate and often mucronate, glabrous beneath, with scattered appressed multicellular hairs above, green on both sides, the petioles usually slightly hairy; pedicels axillary, glabrous, 5–30 mm. long; sepals lanceolate, 1-nerved, glabrescent, acute, 4–6 mm. long; petals shorter to as long as the sepals, 2-fid, sometimes absent; stamens 2–10; capsules globose, 6-valved; seeds ellipsoidal, slightly flattened, about 1.3 mm. long, dark brown, mammillate.——May–July. Woods in mountains; Honshu, Shikoku, Kyushu.

7. Stellaria uchiyamana Makino. Yama-hakobe. Perennial herb; stems, leaves and sepals stellate-pubescent, the hairs 2- to 6-branched; stems long-creeping and branched below, ascending above, 20–30 cm. long; leaves ovate-orbicular to broadly ovate, 1–2.5 cm. long, 0.8–2.5 cm. wide, the petioles 1–2 mm. long, 1-nerved; flowers axillary, solitary, the pedicels stellate-pilose, 2–4 cm. long; sepals lanceolate, acuminate, subtrinerved, 4–5 mm. long; petals white, slightly longer than the sepals, deeply bifid; capsules narrowly ovoid; seeds ellipsoidal, 1.5 mm. long, with semirounded mammillae.——Apr.–June. Low elevations; Honshu (Kinki Distr. and westw.), Shikoku, Kyushu.

8. Stellaria tomentella Ohwi. *S. tomentosa* Maxim., non Link; *S. iinumae* Ohwi——Ao-hakobe. Resembles the preceding, but rather smaller and the stellate hairs slightly larger; leaves scarcely petioled; sepals lanceolate, subobtuse, 3–4 mm. long, subuninerved; petals absent.——Apr.–June. Woods in mountains; Honshu (Kinki Distr. and westw.), Shikoku, Kyushu.

9. Stellaria longifolia Muhl. *S. friesiana* Ser.; *S. mosquensis* Bieb.——Nagaba-tsume-kusa. Slender, mostly erect, glabrous, usually tufted perennial; stems 20–40 cm. long, branched below, 4-angled, minutely scabrous; leaves linear to broadly so, 1.5–2.5 cm. long, 1.5–2.5 mm. wide, acute, minutely scabrous, the midrib impressed above, raised beneath; inflorescence pedunculate, several-flowered, loose, the pedicels smooth, the bracts small, scarious; sepals ovate, glabrous, obsoletely 3-nerved, obtuse to subacute, 2.5–3 mm. long; petals slightly longer than the sepals, deeply 2-cleft; capsules oblong, 4–5 mm. long, brownish; seeds obliquely obovate, slightly flattened, about 0.8 mm. long, nearly smooth, dull.——June–Aug. Woods in mountains; Hokkaido.——Cold regions of the N. Hemisphere.

10. Stellaria filicaulis Makino. Ito-hakobe. Glabrous perennial, 30–50 cm. high; stems slender, erect, sparsely branched, 4-angled; leaves linear, 2–3 cm. long, 1–2 mm. wide, acute or acuminate, the midrib impressed above, raised beneath; flowers white, solitary, the pedicels 2–5 cm. long, slen-

der, deflexed in fruit; sepals lanceolate or narrowly so, acuminate, 1-ribbed, 4–5 mm. long; petals about 1.7 times as long as the sepals, deeply bifid; capsules oblong, yellowish, 1.5 times as long as the sepals; seeds dark brown, about 0.7 mm. long, ellipsoidal, obscurely rugulose.——May–June. Wet places in lowlands; Honshu (Kantō Distr.).——Korea and Manchuria.

11. Stellaria nipponica Ohwi. *S. florida* var. *angustifolia* Maxim.——Iwa-tsume-kusa. Perennial herb essentially glabrous except slightly hairy on the lower portion; stems tufted, erect, branched below, 10–20 cm. long; leaves linear, horizontally spreading, gradually acuminate, 2–3.5 cm. long, 1–2 mm. wide, the midrib raised beneath; flowers 1–3, the pedicels erect, 2–6 cm. long, the bracts rather small; sepals lanceolate, acuminate, mostly 3-nerved, 4–5.5 mm. long; petals white, 1.5–2 times as long as the sepals, deeply bifid; capsules ellipsoidal, slightly shorter than to as long as the sepals; seeds flat, brown, reniform-orbicular, about 1 mm. long, tubercled on margin.——July–Sept. Rocky slopes in mountains; Honshu (alpine regions in centr. distr.).

Var. **yezoensis** Hara. Ō-iwa-tsume-kusa. With a firmer, stouter habit than in the typical phase, the sepals more distinctly 3-nerved and rather larger, the seeds more densely tuberculate with longer mammillae on margin.——Alpine regions; Hokkaido (Mount Apoi in Hidaka).

12. Stellaria pterosperma Ohwi. Ezo-iwa-tsume-kusa. Closely resembles the preceding; leaves rather thicker, firmer, 1–2 cm. long, 1.5–2.5 mm. wide, loosely soft-pubescent on margin; flowers 1–3, the pedicels 2–5 cm. long, straight, the bracts small, membranous to subfoliaceous; capsules narrowly oblong, as long as to slightly longer than the sepals; seeds reniform-orbicular, about 2 mm. long inclusive of the winged margin.——July–Sept. Alpine; Hokkaido (Mount Daisetsu).

13. Stellaria ruscifolia Willd. Shikotan-hakobe. Glabrous, coriaceous, glaucous perennial herb; stems 8–15 cm. long, usually unbranched; leaves ovate to narrowly so, 1–3 cm. long, 7–12 mm. wide, acuminate, rounded to shallowly cordate, somewhat cartilaginous on the margin, slightly clasping, 1-nerved; flowers few, solitary in axils, the pedicels 4–6 cm. long; sepals broadly lanceolate, acute, 5–7 mm. long, obsoletely 3-nerved; petals bifid, 1.5–2 times as long as the sepals; capsules as long as the sepals; seeds flat, pale brown, reniform-orbicular, about 1 mm. across, scabrous, with rather prominent tubercles on margin.——July–Aug. Alpine regions of Hokkaido and Honshu (centr. distr.) and rocks near the sea in Hokkaido (e. distr.).——Ochotsk Sea region, inclusive of the Kuriles.

14. Stellaria fenzlii Regel. *S. yezoensis* Maxim.; *S. pilosula* Franch.; *S. sachalinensis* Takeda——Shiraoi-hakobe. Pale green perennial herb; stems 15–30 cm. long, usually simple, nearly glabrous except for short hairs on nodes; leaves lanceolate to broadly so, 2–4 cm. long, 6–10 mm. wide, acuminate, rounded at base, often with few soft hairs at base and on midrib beneath; inflorescence terminal, loose, the bracts small, membranous on margin, the pedicels 1–4 cm. long; sepals broadly lanceolate, 2.5–4 mm. long, very acute, mostly 3-nerved; petals about half as long as the sepals, deeply 2-parted or sometimes absent; capsules 1½–2 times as long as the sepals, narrowly oblong, pale yellow; styles usually 3; seeds dark brown, obliquely elliptic, slightly flattened, about 0.7 mm. long, nearly smooth, dull.——June–Aug. Coniferous woods; Hokkaido, Honshu (centr. and n. distr.).——Kamchatka, Amur, Sakhalin, and Kuriles.

15. Stellaria alsine Grimm var. **undulata** (Thunb.)

Ohwi. *S. undulata* Thunb.; *S. uliginosa* sensu auct. Japon., non Murr.; *S. uliginosa* var. *undulata* (Thunb.) Fenzl; *S. japonica* Miq.; *S. alsine* var. *phaenopetala* Hand.-Mazz.——NOMI-NO-FUSUMA. Glabrous, usually glaucescent annual or biennial; stems tufted, 5–20 cm. long; leaves oblong to broadly lanceolate or elliptic, sometimes broadly oblanceolate, 8–13 mm. long, 2.5–4 mm. wide, obtuse and mucronate or acute, 1-nerved; inflorescence few-flowered, the bracts minute, white-membranous, the pedicels 5–25 mm. long, deflexed in fruit; sepals broadly lanceolate, acuminate, 3–4 mm. long, 1-ribbed, the lateral nerves obsolete; petals white, deeply 2-fid, slightly longer than the sepals in the earlier flowers, absent in the later ones; styles very short; capsules ellipsoidal, nearly as long as the sepals; seeds reniform-orbicular, dark brown, about 0.5 mm. long, the superficial cells convex on the body, sinuate on the margin.——Apr.–Oct. Wet waste grounds and sometimes woods in foothills; Hokkaido, Honshu, Shikoku; common.——China and Korea. The typical phase with smaller petals and longer styles occurs widely in temperate regions of the N. Hemisphere.

16. Stellaria humifusa Rottb. *S. humifusa* var. *oblongifolia* Fenzl——EZO-HAKOBE. Glabrous perennial; stems decumbent and branched at base, 5–10 cm. long; leaves narrowly oblong to elliptic or broadly lanceolate, slightly fleshy, 10–15 mm. long, 2.5–4 mm. wide, acute to subobtuse, obsoletely 1-nerved; flowers axillary, the pedicels 1–3 cm. long; sepals narrowly ovate to broadly lanceolate, 4–5 mm. long, acute, weakly 3-nerved; petals white, slightly longer than the sepals, deeply bifid; capsules slightly shorter than the sepals, ovoid; seeds about 0.8 mm. long, nearly smooth.——July–Aug. Wet sandy beaches; Hokkaido (e. distr.).——Cooler regions of the N. Hemisphere.

17. Stellaria calycantha (Ledeb.) Bong. *Arenaria calycantha* Ledeb.; *S. yezoalpina* Nakai——KANCHI-YACHI-HAKOBE.

Tufted, essentially glabrous perennial herb; stems 5–15 cm. long, branched and decumbent at base; leaves narrowly oblong to broadly lanceolate, acute, 5–15 mm. long, 2–4 mm. wide; flowers solitary, axillary, the pedicels 1–2 cm. long; sepals broadly lanceolate, acute, weakly 3-nerved, 2–3 mm. long; petals absent; seeds nearly smooth.——Wet places in alpine regions; Hokkaido, Honshu (centr. distr.); rare.——Kamchatka, n. Kuriles, N. America, and N. Europe.

18. Stellaria monosperma Hamilt. var. **japonica** Maxim. *S. paniculigera* Makino——Ō-YAMA-HAKOBE. Perennial; stems suberect, branched, 40–80 cm. long, with a longitudinal line of hairs; leaves oblong to broadly lanceolate or broadly oblanceolate, 4–12 cm. long, 1.5–3 cm. wide, glabrescent, short-petioled or subsessile, acuminate, acute at base; inflorescence loose, the pedicels 3–12 mm. long, short-glandular, the bracts small, foliaceous; sepals membranous, lanceolate, acuminate, short glandular-hairy, 3.5–5 mm. long; petals shorter than the sepals, bifid, white; capsules ovoid-globose, 2.5–3 mm. long, membranous, indehiscent; seeds solitary, about 2.5 mm. across, nearly smooth.——Aug.–Oct. Woods in mountains; Honshu (Kantō Distr. and westw.), Shikoku, Kyushu.——The typical phase occurs from China to India.

19. Stellaria radians L. EZO-Ō-YAMA-HAKOBE. Rather loosely appressed, silky-pubescent perennial; stems erect, 50–80 cm. long, mostly simple or sparsely branched; leaves lanceolate to oblanceolate, 6–12 cm. long, 1–2.5 cm. wide, short-acuminate, acute at base, subsessile, the lateral nerves rather distinct; cymes relatively large, loose, the pedicels 1–3 cm. long, deflexed in fruit, the bracts foliaceous; sepals ovate or narrowly so, subobtuse, 6–8 mm. long, 1-nerved; petals broadly obovate, 8–10 mm. long, laciniate.——July–Aug. Meadows in lowlands; Hokkaido.——Kuriles, Kamchatka, Sakhalin, Manchuria and e. Siberia.

12. DIANTHUS L. NADESHIKO ZOKU

Perennial herbs or rarely subshrubs; leaves flat, mostly linear; inflorescence cymose, paniculate or sometimes fasciculate; flowers solitary, red, rose, rarely white or yellowish; calyx tubular, 5-toothed, parallel-veined, subtended by 2 to many bracts; petals 5, the claw slender, the limb toothed or laciniate, rarely entire, hairy above or glabrous; stamens 10; gynophore often stipelike; ovary 1-locular, the styles 2; capsules cylindric or ovoid, 4-toothed or -valved; seeds discoid, flat, peltately attached.——About 100 species, in Europe, Asia, N. Africa, and n. N. America.

1A. Petals toothed on upper margin; inflorescence a dense many-flowered cyme; leaves deep green, elliptic to lanceolate or spathulate, the upper bracts often elongate.
  2A. Innovation-shoots slightly elongate, with a tuft of leaves at top; cauline leaves spathulate to elliptic, usually obtuse; stems glabrous.
    3A. Inflorescence dense, many-flowered; stems 20–50 cm. long, erect from the base. . . . . 1. *D. japonicus*
    3B. Inflorescence loose, few-flowered; stems 15–30 cm. long, often decumbent at base. . . . . 2. *D. kiusianus*
  2B. Innovation-shoots not elongate; cauline leaves lanceolate or narrowly so, acute to acuminate; stems often with short papillose hairs below the nodes. . . . . 3. *D. shinanensis*
1B. Petals finely laciniate on upper margin; inflorescence loosely 1- to 8-flowered; leaves glaucous, linear to narrowly lanceolate, the upper bracts short and broad. . . . . 4. *D. superbus*

1. Dianthus japonicus Thunb. *D. nipponicus* Makino; *D. ellipticus* Turcz.; *D. barbatus* var. *japonicus* (Thunb.) Yatabe——FUJI-NADESHIKO, HAMA-NADESHIKO. Stout glabrous perennial; stems erect, ascending from the base, 20–50 cm. long, terete; radical leaves oblanceolate-oblong, papillose on margin, gradually narrowed below to a short petiole, the cauline leaves rather thick, sessile or nearly so, narrowly ovate to oblong or narrowly so, 4–8 cm. long, 1–2.5 cm. wide, obtuse to rounded or subacute, lustrous above; inflorescence densely many-flowered, rather flat-topped, the bracts elliptic, long-caudate; calyx-tube 1.5–2 cm. long; petal-limb obtusely deltoid,

6–7 mm. long, rose or rose-purple, toothed on upper margin; capsules slightly exserted beyond the calyx-tube; seeds about 2 mm. long, black, punctate-scabrous.——July–Oct. Hokkaido (?), Honshu, Shikoku, Kyushu.——Ryukyus and China.

2. Dianthus kiusianus Makino. *D. hachijoensis* Nakai——HIME-HAMA-NADESHIKO. Resembling the preceding species but smaller; stems often decumbent at the base, 15–30 cm. long; cauline leaves narrowly oblong to oblanceolate, 1–4 cm. long, 4–12 mm. wide, rounded to obtuse, mucronate; flowers few, loosely racemose, the bracts elliptic, subulate at apex; calyx-tube 15–20 mm. long, the teeth nearly awn-tipped; petals rose-purple, about 10 mm. long, obtusely triangular, toothed on upper margin.——June–Oct. Near the sea; Kyushu.

3. Dianthus shinanensis (Yatabe) Makino. *D. barbatus* var. *shinanensis* Yatabe; *D. takenakae* Honda——SHINA-NO-NADESHIKO, MIYAMA-NADESHIKO. Perennial herb; stems erect, 20–40 cm. long, rather stout, often with short papilloselike hairs below the nodes; cauline leaves linear-lanceolate to broadly lanceolate, 3–6 cm. long, 3–8 mm. wide, deep green, glabrous, papillose on margin, acuminate to acute, narrowed below to a short petiole; inflorescence densely flat-topped, the bracts oblong, long-linear to subulate-appendaged; calyx-tube 1.5–2 cm. long; petals rose-purple, obtusely triangular, 7–9 mm. long, scattered-pubescent on the inside near base; capsules slightly exceeding the calyx; seeds about 2 mm. long.——July–Aug. Valleys and ravines in mountains; Honshu (centr. distr.).

4. Dianthus superbus L. EZO-KAWARA-NADESHIKO. Stems 30–50 cm. long, erect, often branching at base, glabrous; leaves broadly linear, narrowly oblanceolate or linear-lanceolate, 4–7 cm. long, 2–10 mm. wide, papillose on margin, acuminate, glaucous, shortly clasping at base; flowers 3–8, pale rose or rarely white, the bracts 2-paired, the lower pair rarely becoming larger, concealing the upper ones; calyx-tube 2–3 cm. long, 2–3 times as long as the upper bracts; petal limb 2–2.5 cm. long, broadly obovate, deeply laciniate, hairy at base above.——June–Sept. Mountains; Hokkaido, Honshu (centr. and n. distr.).——Europe, Siberia, Manchuria, and Sakhalin.

Var. **speciosus** Reichenb. *D. superbus* var. *monticola* Makino——TAKANE-NADESHIKO. Dwarf alpine phase.——Honshu (centr. distr. and northw.), Hokkaido.

Var. **longicalycinus** (Maxim.) Williams. *D. superbus* forma *longicalycinus* Maxim.; *D. oreadum* Hance; *D. longicalyx* Miq.——KAWARA-NADESHIKO, NADESHIKO. Stems 30–80 cm. long; bracts 3- to 4-paired, the upper ones somewhat larger, short-cuspidate; calyx-tube 2.5–4 cm. long, rather slender, 3–4 times as long as the upper bracts.——July–Oct. Meadows, lowlands and mountains; Honshu, Shikoku, Kyushu.——Formosa, Korea, and China.

Dianthus chinensis L. SEKICHIKU. Plant with long-cuspidate bracts and white to deep red laciniate petals.——Cultivated.——China and Korea.

Dianthus barbatus L. AMERIKA-NADESHIKO. Frequently cultivated in our gardens.——Europe to Siberia, Manchuria, and China.

13. CUCUBALUS L.   NAMBAN-HAKOBE ZOKU

Short-pilose perennials; stems elongate, branched; leaves ovate; inflorescence paniculate, loose, the flowers solitary, terminal; calyx at anthesis nearly cylindric, accrescent, cup-shaped in fruit with recurved upper margin, 10-nerved, 5-fid; petals 5, white, long-clawed, the limb bifid, with 2 small scales inside at base; stamens 10; receptacle stipelike; ovary 1-locular, many-ovuled; styles 3, slender; capsules globose, slightly fleshy and berrylike; seeds black, lustrous.——One variable species, in the temperate regions of the Old World.

1. Cucubalus baccifer L. var. **japonicus** Miq. NAMBAN-HAKOBE. Stems much elongate and branched, more than 1 m. long; leaves ovate to narrowly so, 2–5 cm. long, 7–20 mm. wide, acuminate, abruptly short-petiolate, nearly glabrous above, short-pilose beneath especially on the nerves, ciliolate, the petioles 1–4 mm. long; calyx 10–12 mm. long, green, accrescent, nearly glabrous, the lobes narrowly ovate, acute; petals about 15 mm. long, the limb cuneate, bifid, as long as the claw; capsules globose, about 8 mm. across, the gynophore 2.5–3 mm. long; seeds about 1.3 mm. long.——July–Oct. Thickets in mountains and foothills; Hokkaido, Honshu, Shikoku, Kyushu; rather common.——Korea, Manchuria, s. Kuriles, and Sakhalin.

14. LYCHNIS L.   SENNŌ ZOKU

Perennial herbs with flat leaves; calyx slightly inflated, clavate, ovoid or tubular, 5-toothed, 10-nerved; petals 5, clawed below, the limb variously toothed or lobed, rarely entire, with 2 small scales inside at base; stamens 10; styles 5, opposite the calyx-teeth, rarely 4 or 5; ovules many; capsules often short-stiped, with as many teeth or lobes as the styles; seeds often tubercled.——Thirty to forty species, in the temperate or cold regions of the N. Hemisphere.

1A. Flowers white; pedicels often longer than the calyx; calyx 7–12 mm. long. .... 1. *L. gracillima*
1B. Flowers vermilion or deep red, rarely white; pedicels short, nearly as long as the calyx or shorter, sometimes nearly absent; calyx 15–30 mm. long.
  2A. Petals entire; flowers vermilion. .... 2. *L. miqueliana*
  2B. Petals toothed or laciniate.
    3A. Bracts ascending, linear to lanceolate, 5–10 mm. long, 1–2 mm. wide; limb of petals laciniate.
      4A. Leaves lanceolate to linear-lanceolate, 5–10 mm. wide; flowers light red. .... 3. *L. kiusiana*
      4B. Leaves ovate to narrowly so, 1–3 cm. wide; flowers deep red.
        5A. Calyx 15–18 mm. long. .... 4. *L. wilfordii*
        5B. Calyx 25–30 mm. long. .... 5. *L. senno*
    3B. Bracts appressed to the calyx, lanceolate to broadly so, 1–1.5 cm. long, 5–7 mm. wide; limb of petals toothed.
      6A. Petals deeply bifid, the lobes narrowly oblong; flowers deep red; calyx 15–17 mm. long. .... 6. *L. fulgens*
      6B. Petals unlobed or very shallowly 2-lobed.
        7A. Plant glabrous; flowers yellowish red; claw of petals longer than the calyx. .... 7. *L. coronata*
        7B. Plant pubescent; flowers deep red, rarely white; claw of petals not exceeding the calyx. .... 8. *L. sieboldii*

1. Lychnis gracillima (Rohrb.) Makino. *Silene gracillima* Rohrb.; *L. stellarioides* Maxim.; *L. ugoensis* Makino——SENJU-GAMPI, SHIRANE-GAMPI. Caudex short; stems 40–100 cm. long, rather slender, scattered white-pubescent; leaves thin, lanceolate or broadly so, 5–13 cm. long, 1–3.5 cm. wide, sessile, nearly glabrous, long-acuminate, slightly narrowed below; inflorescence rather loose, the pedicels 2–6 cm. long, glabrous; calyx campanulate-tubular in anthesis, clavate-campanulate in fruit, membranous, glabrous, 7–12 mm. long, the teeth short-ciliate; petals white, the limb 12–15 mm. long, bifid and toothed; capsules about 10 mm. long, short-stipitate, narrowly ovoid; seeds tuberculate.——July–Aug. Woods and thickets in mountains; Honshu (n. and centr. distr.).

2. Lychnis miqueliana Rohrb. FUSHIGURO-SENNŌ. Stems 50–80 cm. long, slightly branched below, scattered-pubescent or nearly glabrous, with dark thickened nodes; leaves ovate, obovate or oblong, 5–12 cm. long, 2.5–5 cm. wide, sessile, pubescent on nerves and on margin while young, short-acuminate to acute; bracts subulate-linear, 5–10 mm. long, 1–2 mm. wide, recurved above; pedicels 3–7 mm. long; calyx glabrous, 2.5–3 cm. long, narrowly clavate, the teeth short-ciliate; limb of petals 2.5–3 cm. long, vermilion, very rarely white, cuneate-obovate, nearly entire; seeds black, tuberculate.—— July–Oct. Woods in mountains; Honshu (Kantō Distr. and westw.), Shikoku, Kyushu.

3. Lychnis kiusiana Makino. OGURA-SENNŌ. Stems 60–80 cm. long, simple, glabrescent or slightly recurved-scabrous; leaves lanceolate to linear-lanceolate, 5–10 cm. long, 6–12 mm. wide, gradually acuminate, subacute at base, loosely scabrous; bracts linear, 3–8 mm. long, about 1 mm. wide, ascending; pedicels short-pilose, 3–12 mm. long; calyx membranous, narrowly tubular, slightly thickened above, nearly glabrous, 2–2.5 cm. long, the teeth short-ciliate; claw of petals slightly longer than the calyx, the limb red, laciniate.——July–Aug. Honshu (w. distr.), Kyushu; rare.

4. Lychnis wilfordii (Regel) Maxim. *L. laciniata* var. *japonica* Maxim.; *L. fulgens* var. *wilfordii* Regel; *L. laciniata* Maxim., non Lam.——EMBI-SENNŌ. Stems simple, 50–80 cm. long, glabrescent or scattered soft-pubescent; leaves narrowly ovate to oblong-ovate, 3–7 cm. long, 1–2 cm. wide, acuminate, rounded and slightly clasping at base, short-ciliate; bracts spreading, linear or linear-lanceolate, 3–5 mm. long; pedicels 3–10 mm. long, yellowish brown pubescent; calyx nearly glabrous, tubular at anthesis, short-clavate in fruit, 12–18 mm. long, membranous, the teeth short-ciliate; limb of petals deep red, about 2 cm. long, deeply and finely laciniate; capsules about 13 mm. long, narrowly oblong; seeds with elongate tubercles.——July–Aug. Hokkaido, Honshu (centr. distr.); rare.——Ussuri, Manchuria, and n. Korea.

5. Lychnis senno Sieb. & Zucc. SENNŌ. Plant short-pubescent; stems 50–70 cm. long; leaves ovate, 4–6 cm. long,

2–3 cm. wide, short-acuminate, acute at base; bracts lanceolate or narrowly so, 8–12 mm. long, 2–3 mm. wide, ascending; calyx 2.5–3 cm. long, clavate, membranous, long-pubescent; limb of petals usually deep red, rarely white, 15–20 mm. long, cleft with few lobes.——July–Aug. Chinese plant cultivated in old gardens.

6. Lychnis fulgens Fischer. EZO-SENNŌ. Roots slightly thickened at base; stems 50–70 cm. long, with scattered long white hairs; leaves narrowly ovate, 4–6 cm. long, 1.5–2.5 cm. wide, acuminate, rounded at base, sessile; bracts broadly lanceolate, 10–15 mm. long, 5–7 mm. wide, with long white hairs, erect; flowers deep red; calyx tubular, 15–17 mm. long, membranous; limb of petals broadly cuneate, about 2 cm. long, deeply bifid, the lobes narrowly oblong, obtuse, shallowly toothed on upper margin.——Honshu (Shinano Prov.), also reported to occur in Hokkaido.——n. Korea, Manchuria, and Ussuri.

7. Lychnis coronata Thunb. *L. grandiflora* Jacq.; *Agrostemma banksia* Meerb.——GAMPI. Plant nearly glabrous; stems green, 40–80 cm. long, thickened at the nodes; leaves ovate or elliptic, 5–8 cm. long, 2–4 cm. wide, slightly scabrous on margin, rounded and narrowed toward the base, sessile; bracts erect, broadly lanceolate, 10–15 mm. long, 5–8 mm. wide, short-ciliate; calyx 2.5–3 cm. long, glabrous, rather thick, gradually broadened above, long-tubular, the teeth short, short-ciliate; claw of petals slightly longer than the calyx, the limb 2–2.5 cm. long, yellowish red, with rather deep irregular teeth on upper margin.——June–July. Chinese plant cultivated in our gardens.

8. Lychnis sieboldii Van Houtte. *L. coronata* var. *sieboldii* (Van Houtte) L. H. Bailey; *L. speciosa* Carr.——MATSUMOTO, MATSUMOTO-SENNŌ. Roots slightly thickened at base; stems long-pubescent, 40–70 cm. long; leaves narrowly to broadly ovate, 5–8 cm. long, 2.5–4.5 cm. wide, acuminate to short-acuminate, scattered scabrous above, pilose beneath and on margin, rounded to abruptly narrowed at the sessile base; inflorescence rather dense, the bracts broadly lanceolate, 15–20 mm. long, 6–10 mm. wide, sometimes slightly recurved above, long-ciliate; calyx 2.5–3 cm. long, submembranous, long-pubescent on upper portion of ribs, the teeth rather short, acuminate, the tip subobtuse; limb of petals deep red, rarely white, 2–2.5 cm. long, broadly cuneate, shallowly 2-lobed, the lobes shallowly toothed.——June–July. Honshu (Shinano Prov.), Kyushu; very rare.

15. MELANDRIUM Fries FUSHIGURO ZOKU

Annual, biennial, or perennial herbs; leaves flat; inflorescence cymose; calyx tubular or tubular-campanulate, 10- or 20-nerved, often inflated, 5-toothed; petals clawed, the limb bifid, with 2 small scales inside at base; stamens 10; capsules 3- to 6-valved, often stipitate; styles 3–6.——Many species in he N. Hemisphere and in S. America.

1A. Styles 5 or 6; capsules 5- or 6-toothed. .... 1. *M. apetalum*
1B. Styles 3; capsules 3-lobed or -toothed.
  2A. Biennial; limb of petals much shorter than the calyx-tube.
    3A. Leaves and calyx rather densely short-puberulent; seeds about 0.5 mm. across. .... 2. *M. apricum*
    3B. Leaves and calyx glabrous or scattered-pubescent; seeds about 0.7 mm. across. .... 3. *M. firmum*
  2B. Perennial; limb of petals as long as to longer than the calyx-tube.
    4A. Calyx tubular, subtruncate at base; flowers pale rose, rarely white; leaves lanceolate or narrowly so. .... 4. *M. keiskei*
    4B. Calyx campanulate, narrowed below; flowers white; leaves ovate to narrowly so. .... 5. *M. yanoei*

1. **Melandrium apetalum** (L.) Fenzl. *Lychnis apetala* L.; *M. apetalum* forma *okadae* Makino——TAKANE-MANTEMA. Tufted perennial with stout rhizomes; stems 10–20 cm. long, pubescent, with 2–4 pairs of leaves; radical and cauline leaves narrowly lanceolate, 3–8 cm. long, 2–10 mm. wide, short-pubescent especially on margin, acuminate, 1-nerved, gradually narrowed to the somewhat clasping base; flowers solitary, terminal; calyx narrowly tubular-ovoid in anthesis becoming ovoid in fruit, 12–15 mm. long, 5-toothed, 10-nerved, blackened on nerves, white-pubescent; limb of the petals about 2 mm. long, slightly retuse; capsules narrowly ovoid, 5- to 6-toothed, the teeth bidentate; seeds flat, red-brown, broadly winged, smooth, about 2 mm. across inclusive of the wing.—— July–Aug. Alpine slopes; Honshu (centr. distr.); rare.—— N. Europe and Siberia to the Kuriles and N. America.

2. **Melandrium apricum** (Turcz.) Rohrb. *Silene aprica* Turcz.——HIME-KE-FUSHIGURO. Densely gray-puberulent biennial; stems 20–50 cm. long; leaves linear-lanceolate to lanceolate or oblanceolate, sometimes narrowly obovate, 2–8 cm. long, 2–15 mm. wide, acute; inflorescence bracts herbaceous, the pedicels 5–30 mm. long; calyx narrowly ovoid in anthesis, ovoid in fruit, 8–10 mm. long, 5-toothed; limb of petals about 2 mm. long, retuse, pale rose; capsules ovoid, 6–8 mm. long, very short-stipitate, shallowly 6-lobed; styles 3; seeds reniform, about 0.5 mm. across, with rather small elongate tubercles on margin.——May–Aug. Honshu (w. distr.), Kyushu.——Korea, n. China, and e. Siberia.

3. **Melandrium firmum** (Sieb. & Zucc.) Rohrb. *Silene firma* Sieb. & Zucc.; *S. melandryformis* Maxim.; *M. apricum* var. *firmum* (Sieb. & Zucc.) Rohrb.——FUSHIGURO. Erect biennial; stems glabrous, 30–80 cm. long; leaves oblong or ovate, sometimes broadly lanceolate, 3–10 cm. long, 1–3 cm. wide, short-acuminate, narrowed at base, ciliate, often loosely pubescent on both sides; cymes terminal and axillary, the bracts often membranous, the pedicels glabrous, 1–3 cm. long; flowers erect; calyx ovoid, glabrous or nearly so, 7–10 mm. long, 10-nerved; limb of the petals 2–3 mm. long, white, 2-lobed; capsules ovoid, 7–8 mm. long, short-stipitate; seeds reniform, about 0.6 mm. long, small-tubercled.——June–Sept. Hokkaido, Honshu, Shikoku, Kyushu.——Forma **pubescens** (Makino) Makino. *M. firmum* var. *pubescens* (Makino) Makino; *M. apricum* var. *firmum* forma *pubescens* Makino—— KE-FUSHIGURO. Plant pubescent.——e. Siberia, China, Korea, and Ryukyus.

4. **Melandrium keiskei** (Miq.) Ohwi. *Silene keiskei* Miq.; *S. maximowicziana* Rohrb.——Ō-BIRANJI. Perennial herb with short, stout, branched rhizomes; stems tufted, slender, mostly erect or ascending, sometimes decumbent, nearly glabrous or slightly short-pubescent; leaves lanceolate or narrowly so, acuminate, sessile, 1.5–6 cm. long, 3–10 mm. wide, glabrescent; cymes loosely few-flowered; calyx tubular, 10–13 mm. long, glabrous or nearly so, subtruncate at base; limb of petals 10–15 mm. long, obovate, bifid; capsules oblong; seeds flat, reniform, about 1 mm. across, with cylindric hairlike tubercles.——July–Sept. Mountains; Honshu (centr. distr.).

Var. **minus** (Takeda) Takeda. *Silene keiskei* forma *minor* Takeda——BIRANJI. Soft hairy on leaves, pedicels, and calyx.——Honshu (Kantō and centr. distr.).

Var. **akaisialpinum** Yamazaki. TAKANE-BIRANJI. Tufted; stems low; flowers 1–2, rose-colored, large, 3–4 cm. across, the pedicels and calyx short-hairy.——Alpine; Honshu (centr. distr.).

5. **Melandrium yanoei** (Makino) Williams. *Silene yanoei* Makino——TEBAKO-MANTEMA. Somewhat tufted perennial; stems erect, 20–40 cm. long, short-pubescent above; leaves ovate to narrowly so, 2–4 cm. long, 1–1.5 cm. wide, rather long-acuminate, nearly glabrous, slightly scabrous on margin, sessile or short-petioled; cymes 1- to few-flowered, loose, the pedicels slender, often with a pair of small linear bractlike leaves; calyx campanulate, glabrous, green, 6–7 mm. long, 5-lobed to the middle; limb of petals cuneate, 6–7 mm. long, white, bifid; capsules ovoid; styles 3; seeds reniform, about 1 mm. across, with diminutive tubercles.——Aug. Shikoku; rare.

16. SILENE L. MANTEMA ZOKU

Annuals, biennials, or perennials, rarely suffrutescent; leaves flat; inflorescence cymose; flowers white, rose or rose-purple; calyx tubular, ovoid or campanulate, 5-toothed, 10- or more-nerved; petals 5, the claw rather long, the limb usually bifid, with 2 small scales inside at base; stamens 10; styles 3(–5); ovary 1-locular, 3- to 5-septate at base; capsules ovoid or oblong, the teeth twice as many as the styles.——About 200 species.

1A. Biennial; petals white to rose-purple, the limb subentire; calyx ovoid in fruit. . . . . 1. *S. gallica*
1B. Perennial; petals pale rose or white, the limb bifid; calyx obovoid or clavate in fruit.
  2A. Stems with tufts of short leaves at the nodes; calyx glabrous, 6–7 mm. long; limb of petals white, deeply bifid; stamens nearly as long as the petals, long-exserted. . . . . 2. *S. foliosa*
  2B. Stems without tufts of leaves at the nodes; calyx pubescent, 12–15 mm. long; limb of petals pale rose to white, shallowly bifid; stamens shorter than the petals, only slightly exserted. . . . . 3. *S. repens*

1. **Silene gallica** L. *S. anglica* L.——SHIROBANA-MANTEMA. Biennial with coarse spreading hairs; stems erect, 30–50 cm. long, often branched at base; leaves oblanceolate to broadly so, 2–4 cm. long, 5–10 mm. wide, acute to rounded, mucronate, sessile above, the radical ones subpetioled; flowers solitary in axils, forming a terminal bracteate spikelike raceme, the pedicels 1–8 mm. long; calyx toothed, inflated after anthesis, ovoid, 8–10 mm. long, puberulent and pilose; limb of petals 3–4 mm. long, white, with 2 small oblong obtuse scales inside at base; capsules ovoid; seeds reniform, black, about 0.8 mm. across, with slightly raised concentric lines and minute tubercles.——May–July. Naturalized in Honshu.——Introduced from Europe.

Var. **quinquevulnera** (L.) Rohrb. *S. quinquevulnera* L. ——MANTEMA. Limb of petals with a large dark red-purple blotch, the scales lanceolate, acute.——Waste ground, along railroads and sandy beaches; naturalized in Honshu, Shikoku, Kyushu; common.——Introduced from Europe.

2. **Silene foliosa** Maxim. EZO-MANTEMA. Perennial herb; stems tufted, erect, 20–40 cm. long, shortly recurved-hairy, thickened on nodes; leaves at base rosulate, petiolate, narrowly oblanceolate, 5–7 cm. long, 4–8 mm. wide, acute,

ciliolate, the cauline leaves narrowly oblanceolate, 2–5 cm. long, 2–6 mm. wide, gradually narrowed at base, subsessile; flowers few at a node, the pedicels short, glabrous, the bracts membranous; calyx tubular in anthesis, clavate-obovate in fruit, glabrous, 6–7 mm. long, the teeth deltoid or broadly ovate-deltoid; limb of petals white, cuneate, 5–6 mm. long, deeply bifid; styles and stamens long-exserted; capsules 4–5 mm. long, the teeth recurved; seeds about 1 mm. long, with raised superficial cells.——July–Aug. Hokkaido.——Korea, Amur, and Ussuri.

3. Silene repens Pers. KARAFUTO-MANTEMA. Perennial herb; stems erect, branched at base, 10–30 cm. long, with recurved short hairs; leaves lanceolate to narrowly so, 3–5 cm.

long, 3–8 mm. wide, sometimes oblanceolate, acute, sessile, glabrescent to short-pubescent, especially on margin; flowers few, terminal, the pedicels very short, puberulent; calyx tubular, 12–15 mm. long, puberulent; limb of petals cuneate-obovate, bifid, 6-7 mm. long, pale rose to white; stamens slightly exserted; gynophore 6–7 mm. long.——July–Aug. Meadows in lowlands, especially near the sea; Hokkaido.—— Europe, Siberia, Korea, and Manchuria.

Var. **latifolia** Turcz. CHISHIMA-MANTEMA. More densely long-hairy; leaves broadly oblanceolate to narrowly oblong, 5–15 mm. wide.——Perhaps in Hokkaido (e. distr.).

Fam. 85. NYMPHAEACEAE   SUIREN KA   Waterlily Family

Aquatic herbs with creeping rhizomes; leaves long-petioled, involute in bud, often peltate, usually orbicular to cordate; flowers usually solitary from the axils of radical leaves, usually rather large; sepals 3–5; petals 3 to many; stamens 6 to many, free, the anthers erect, introrse or extrorse, longitudinally split, the locules adnate to the connective; carpels 3–8, free or united, sometimes embedded in the upper surface of an enlarged obconical receptacle; stigma distinct, often disc- or ring-shaped; ovules solitary in each locule, orthotropous, pendulous or many; carpels free and indehiscent or united to form a fleshy or spongy fruit; cotyledons fleshy.——About 8 genera, with about 60 species, cosmopolitan.

1A. Plant with a gelatinous sheath; flowers small, reddish purple; sepals and petals 3; stamens 12–18. .... 1. *Brasenia*
1B. Plants without a gelatinous sheath; flowers relatively large; sepals 4–6; petals 5 to many; stamens many.
  2A. Leaves floating or slightly elevated above water level; carpels united; ovules many.
    3A. Flowers yellow to orange-yellow; ovary inserted on the receptacle, superior; sepals, petals, and stamens inferior, free; sepals 5.
                                                                       2. *Nuphar*
    3B. Flowers white or rose-purple; ovary more or less adnate to the receptacle; sepals, petals, and stamens attached to the periphery of the ovary; sepals 4.
      4A. Leaves prickly, 30–120 cm. across; stigma-disc depressed, adnate to the inner wall of an enlarged, cup-shaped receptacle.
                                                                             3. *Euryale*
      4B. Leaves unarmed, smaller; stigma-disc free from the receptacle; ovary semi-inferior. .... 4. *Nymphaea*
  2B. Leaves elevated above water level often 1 m. or more on long petioles; flowers deep pink or rarely white; carpels free, embedded in the flat-topped spongy receptacle; ovules 1 or 2. .... 5. *Nelumbo*

1. BRASENIA Schreb.   JUNSAI ZOKU

Aquatic herb with transparent gelatinous sheath; stems elongate, branched, leafy; leaves alternate, floating, peltate, elliptic, palmately nerved, entire; flowers small, reddish purple, axillary; sepals 3, resembling the 3 petals; stamens 12–18, the filaments subulate, the anthers laterally split; carpels 6–18, free; style short, the stigma lateral; ovules 2 or 3, pendulous, attached to the dorsal wall; mature carpels coriaceous, indehiscent, 1- to 2-seeded; seeds with endosperm.——One species.

1. Brasenia schreberi J. F. Gmel. *B. peltata* Pursh; *Hydropeltis purpurea* Michx.; *B. purpurea* (Michx.) Casp. ——JUNSAI. Perennial herb with branched creeping rhizomes; stems much elongated, slender, sparingly branched; leaves long-petioled, 6–10 cm. long, 4–6 cm. wide, purplish beneath; flowers long-peduncled, about 2 cm. across; sepals broadly lanceolate, obtuse, about 10 mm. long; petals linear-oblong, obtuse, about 1.5 cm. long; anthers linear, about 4 mm. long; style about 8 mm. long.——May–Aug. Lakes and ponds; Hokkaido, Honshu, Shikoku, Kyushu.——e. Asia, India, Australia, w. Africa, and N. America.

2. NUPHAR Smith   KŌ-HONE ZOKU

Perennial herbs with stout thickened creeping rhizomes; leaves long-petiolate, radical, floating or briefly elevated above the water, rounded-cordate to narrowly ovate, entire, sagittate; peduncles 1-flowered; flowers yellow or orange-yellow; sepals 5 or 6, concave, petaloid, coriaceous, persistent, often becoming green in fruit; petals many, linear to spathulate or narrowly oblong; stamens many, shorter than the sepals, inferior, the filaments short, flat, the anthers introrse; carpels many, united into a compound ovary; stigma linear, the stigma-disc peltate; ovules many; fruit ovoid, berrylike; seeds with endosperm.——Ten or more species, in the Old World and N. America.

1A. Leaves narrowly ovate, raised above the water; stigma-disc dentate. .... 1. *N. japonicum*
1B. Leaves floating.
  2A. Flowers broadly cup-shaped; filaments 1 or 2 times as long as the anthers; petioles terete. .... 2. *N. subintegerrimum*
  2B. Flowers ellipsoidal; filaments 3–10 times as long as the anthers; petioles slightly flattened.
    3A. Petioles hollow; stigma-disc without a raised center. .... 3 *N. oguraense*
    3B. Petioles solid; stigma-disc with a raised center. .... 4. *N. pumilum*

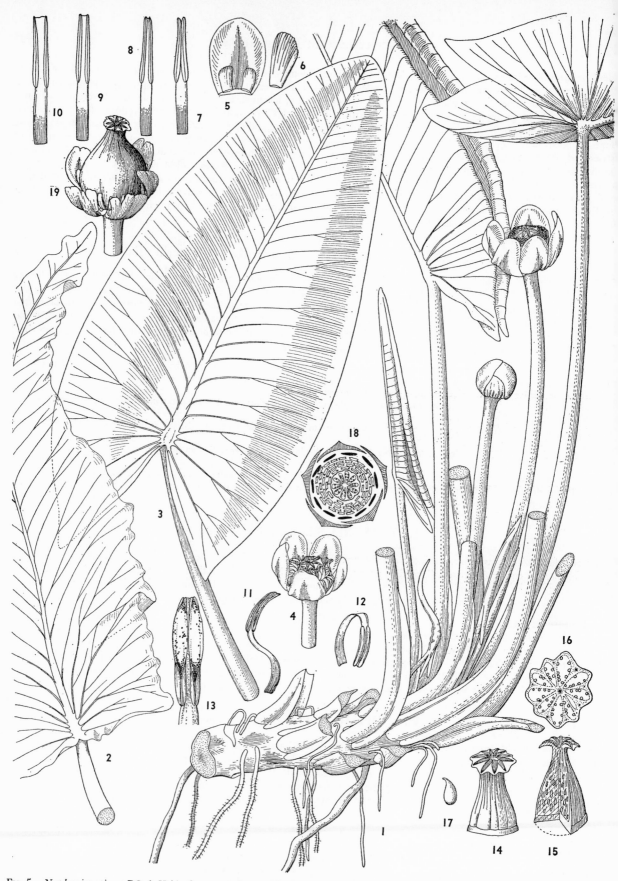

FIG. 5.—*Nuphar japonicum* DC. 1, Habit; 2, young submersed leaf; 3, aerial leaf; 4, flower; 5, sepal with 2 petals inside; 6, petal; 7, 8, 9, 10, 11, 12, stamens; 13, anther; 14, pistil; 15, the same, cut longitudinally; 16, ovary, cut transversely; 17, ovule; 18, floral diagram; 19, fruit.

1. Nuphar japonicum DC. Kô-hone. Fig. 5. Rhizomes stout, creeping; leaves radical, the submersed ones narrow, membranous, with undulate margin, the aerial ones subcoriaceous, narrowly ovate to oblong, 20–30 cm. long, 7–12 cm. wide, obtuse to rounded, sagittate, glabrous above, slightly pubescent beneath while young, entire; peduncles raised above the water; flowers yellow, 4–5 cm. across; sepals obovate-orbicular, about 2.5 cm. long, subcoriaceous, becoming green in fruit, rounded at apex; petals obovate-cuneate, about 8 mm. long; anthers 4–6 mm. long; stigma-disc 6–8 mm. across, dentate.——June–Sept. Ponds, lakes, and shallow streams; Hokkaido (sw. distr.), Honshu, Shikoku, Kyushu; common.

Var. **rubrotinctum** (Casp.) Ohwi. *N. japonicum* var. *subintegerrimum* forma *rubrotinctum* Casp.; *N. subintegerrimum* forma *rubrotinctum* (Casp.) Makino——Beni-kô-hone. Flowers orange-red.——Known only in cultivation.

2. Nuphar subintegerrimum (Casp.) Makino. *N. japonicum* var. *subintegerrimum* Casp.——Hime-kô-hone. Resembles the preceding; floating leaves broadly ovate, 4–8 cm. long, 5–8 cm. wide; flowers 3–4 cm. across; sepals 15–20 mm. long; anthers 3–5 mm. long; stigma-disc 4–5 mm. across, shallow-toothed; stigma broader, contiguous below.——June–Sept. Honshu, Shikoku.

3. Nuphar oguraense Miki. Ogura-kô-hone. Rhi-zomes stout; floating leaves broadly ovate to ovate-orbicular, 5–8 cm. long, 4–6 cm. wide, rounded, deeply cordate-sagittate at base, entire, the petioles flattened, slightly hollow; flowers yellow, 3–4 cm. across; sepals obovate, about 2 cm. long; petals broadly spathulate, about 6 mm. long, shallowly retuse; anthers 2–2.5 mm. long, oblong; stigma-disc 6–7 mm. across, shallowly dentate.——July. Ponds and shallow lakes; Honshu (w. distr. and westw.), Kyushu.

4. Nuphar pumilum (Timm) DC. *Nymphaea lutea* var. *pumila* Timm; *N. subpumilum* Miki——Nemuro-kô-hone. Rhizomes stout, creeping; floating leaves broadly ovate to ovate-orbicular, sometimes elliptic, 6–8 cm. long, 6–9 cm. wide, the petioles slightly flattened; flowers about 2.5 cm. across, yellow, floating; sepals broadly obovate or broadly elliptic, 12–20 mm. long, rounded at apex; petals spathulate-obovate, 5–7 mm. long, rounded at apex, often obsoletely toothed; anthers elliptic, 2.5–3 mm. long; stigma-disc 6–8 mm. across, pale yellow or partially reddish; stigma-disc 10–20 mm. across.——July–Aug. Ponds and shallow lakes; Hokkaido (e. distr.), Honshu (n. distr.).——Europe to Siberia.

Var. **ozeënse** (Miki) Hara. *N. ozeënse* Miki——Oze-kô-hone. With dark red stigma-disc.——Honshu (Oze in Kotsuke).

3. EURYALE Salisb. Onibasu Zoku

Large, prickly, acaulescent aquatic annual herb; leaves floating, orbicular, rugulose especially while young, prominently nerved and net-veined beneath; peduncles radical, 1-flowered; flowers rose-purple; sepals 4, united below into an ellipsoidal tube; petals many, smaller than the calyx-lobes; stamens many, inserted on the inner side of the petals, all fertile, the filaments short, the anthers oblong, the connective truncate; ovary inferior, the 8 carpels syncarpous; stigma discoid, depressed, the margin adnate to the inner wall of the receptacle; ovules few in each locule; fruit berrylike, spongy, densely prickly, irregularly dehiscent; seeds with endosperm.——One species.

1. Euryale ferox Salisb. Onibasu. Leaves peltate, long-petioled, orbicular, lustrous above, purplish and short-pubescent beneath, 30–120 cm. across, flat, subentire, slightly emarginate at base; calyx-lobes 15–20 mm. long, narrowly deltoid, subacute to obtuse, persistent, erect in fruit; petals broadly lanceolate or oblanceolate, 6–10 mm. long, obtuse; anthers 3–3.5 mm. long, the filaments unequal in length; fruit ellipsoidal or subglobose, 5–7 cm. long, beaked; seeds subglobose, about 1 cm. across, with a fleshy aril.——Aug.–Oct. Ponds and lakes in lowlands; Honshu, Kyushu.——India, China, and Formosa.

4. NYMPHAEA L. Suiren Zoku

Perennial aquatic, acaulescent herbs with bulbous or stout short-creeping rhizomes; leaves orbicular, cordate-sagittate, basifixed or peltate; peduncles 1-flowered; flowers white, blue, rose or yellowish; sepals 4; petals many; stamens many, the outer filaments often petaloid, the anthers introrse; ovary syncarpous, the carpels many, adnate below to the receptacle; styles free, short, the stigma linear; fruit spongy, irregularly ruptured; seeds with endosperm.——Cosmopolitan, with about 50 species.

1. Nymphaea tetragona Georgi var. **tetragona**. *N. tetragona* var. *lata* Casp.; *N. pygmaea* (Salisb.) Aiton; *Castalia pygmaea* Salisb.——Ezo-no-hitsujigusa. Rhizomes stout and fleshy; leaves long-petioled, ovate-orbicular to ovate-elliptic, 5–12 cm. long, 8–15 cm. wide, basifixed, entire, rounded, deeply cordate-sagittate; lobes of leaves shorter than the length from the attachment of petiole to apex, the inner margin often overlapping; petals usually 8, white; sepals green, broadly lanceolate to narrowly ovate, 3–3.5 cm. long, 10–15 mm. wide, subacute; outer filaments broadly oblanceolate, the inner ones nearly linear, 3–5 mm. long, the connective not produced above; fruit ovoid-globose, covered by the persistent calyx. ——July–Oct. Ponds and shallow lakes; Hokkaido, Honshu, Shikoku, Kyushu.——e. Siberia, Manchuria, China, Korea, India, and N. America.

Var. **angusta** Casp. *N. tetragona* var. *angusta* subvar. *orientalis* Casp.——Hitsujigusa. Lobes of leaves longer than the length from the attachment of petiole to apex, with more or less distant inner margin; petals usually 12–15.——Ponds and lakes at low elevations; Honshu, Shikoku, Kyushu.

5. NELUMBO Adans. Hasu Zoku

Coarse acaulescent, aquatic herbs; rhizomes long-creeping; leaves long-petiolate, raised 1 m. or more above water, orbicular, peltate, radiately nerved; peduncles radical, elongate, 1-flowered; flowers deep pink, white or yellowish; sepals 4 or 5, free, caducous petals many, inferior, early deciduous; stamens numerous, the anthers introrse, the connective with a clavate appendage; carpels many, free, embedded in the flat-topped spongy receptacle; style short, the stigma terminal; ovules 1 or 2, pendulous; nut ellipsoidal, the pericarp hard, without endosperm; cotyledons thickened.——Two species, one in N. America, another in Australia and e. Asia.

1. Nelumbo nucifera Gaertn. *Nymphaea nelumbo* L.; *Nelumbo nelumbo* (L.) Karst.; *Nelumbium speciosum* Willd.; *Nelumbium nelumbo* (L.) Druce——Hasu.   Large, glaucous, perennial herb; leaves depressed-orbicular, 20–50 cm. across, slightly retuse, the petioles and peduncles with short fleshy prickles; flowers 15–20 cm. across, deep pink or sometimes white; sepals caducous, green; petals obovate, obtuse, nervulose, 8–12 cm. long, 3–7 cm. wide; anthers linear, yellow, 15–20 mm. long, the filaments linear, 7–25 mm. long; receptacle spongy, in fruit to 10 cm. high and as wide, flat, the seeds embedded within; nuts ellipsoidal, about 20 mm. long.—— July–Aug.   Much planted in ponds and shallow lakes, with several cultivated races.——India, China, Iran, and Australia.

Fam. 86. CERATOPHYLLACEAE   Matsu-mo Ka   Hornwort Family

Slender submersed monoecious or dioecious aquatic herbs; stems slender, elongate, branching; leaves dichotomously forked into filiform, slightly and obsoletely toothed segments; flowers unisexual, solitary in axils, sessile, the involucral bracts multifid, the segments entire or toothed, linear, the perianth absent; staminate flowers with 10-20 stamens, the anthers nearly sessile, extrorse, the connective ending in a thickened appendage; pistillate flowers with a single superior ovary, the ovules orthotropous, pendulous, the styles filiform, the stigma terminal; fruit an achene, the embryo of 4 verticillate cotyledons, without endosperm. ——A single genus of 2 or 3 species, widely distributed in fresh waters of the world.

1. CERATOPHYLLUM L.   Matsu-mo Zoku

Characters of the family.

1. Ceratophyllum demersum L.   Matsu-mo, Kingyo-mo.   Dark green fragile aquatic herb; stems 20–100 cm. long, branched, slender; leaves verticillate, in 5's to 12's, sessile, 1.5–2.5 cm. long, much-divided, the ultimate segments filiform, with scattered minute spiny teeth; anthers oblong; achenes narrowly ovoid, sessile, 4–5 mm. long, with 2 spines near base; style persistent, spinelike, often longer than the body.——June– July.   Ponds, lakes, and shallow ditches; Hokkaido, Honshu, Shikoku, Kyushu; common.——Cosmopolitan.

Var. **quadrispinum** Makino. *C. oryzetorum* Komar.; *C. pentacanthum* Hayata, non Haynald; *C. demersum* var. *pentacorne* Kitag.——Yotsubari-kingyo-mo.   Achenes with 2 pairs of spines, one near base, the other on the shoulder.—— Honshu, Kyushu.——Manchuria, Ussuri, China, and Formosa.

Fam. 87. TROCHODENDRACEAE   Yamaguruma Ka   Trochodendron Family

Evergreen trees; leaves alternate, petiolate, rhombic-obovate to narrowly obovate, coriaceous, undulate toothed, penninerved; stipules absent; inflorescence racemose, terminal; flowers small, bisexual, the perianth absent; stamens many, the filaments filiform, the anthers elliptic; carpels 5–10, united at base, many-ovuled; styles short, recurved in fruit; stigma narrowly oblong; fruit partially sunken in the receptacle, dehiscent along the inner carpel margins; seeds linear.——A single genus and species.

1. TROCHODENDRON Sieb. & Zucc.   Yamaguruma Zoku

Characters of the family.

1. Trochodendron aralioides Sieb. & Zucc.   Yama-guruma.   Glabrous tree with rather stout branches, the branchlets greenish; leaves usually crowded at the ends of branchlets, broadly to narrowly obovate, rhombic-obovate, 6–12 cm. long, 2–7 cm. wide, obtuse, subcuneate at base, deep green and lustrous above, slightly paler beneath; inflorescence short-pedunculate, 7–12 cm. long; flowers 10–20, 10–12 mm. across, the pedicels 2–4 cm. long; anthers 1.2–1.7 mm. long; styles persistent; fruit depressed-globose, 7–10 mm. across; seeds linear, 4–6 mm. long.——May–June.   Mountains; Honshu, Shikoku, Kyushu.——Formosa and s. Korea.——Forma **longifolium** (Maxim.) Ohwi. *T. aralioides* var. *longifolium* Maxim.——Nagaba-no-yamaguruma.   Phase with narrower, broadly oblanceolate leaves, often passing into the typical phase.——Occurs with the typical phase.

Fam. 88. EUPTELEACEAE   Fusazakura Ka   Euptelea Family

Large deciduous nearly glabrous tree; leaves petiolate, alternate, depressed-orbicular, penninerved, irregularly acute-toothed; stipules absent; flowers before the leaves, bisexual, short-pedicelled, the perianth absent; stamens many, pendulous on capillary filaments; carpels rather many, 1- to 4-ovuled; stigma sessile, broadly linear, decurrent inside; fruit stipitate, indehiscent, wing-margined; seeds 1–4, pendulous on the inner angle of the locule.——A single genus with perhaps 2 species, in e. Asia.

1. EUPTELEA Sieb. & Zucc.   FUSAZAKURA ZOKU

Characters of the family.

1. Euptelea polyandra Sieb. & Zucc. FUSAZAKURA. Branches red-brown; petioles 3–7 cm. long, rather slender, hairy on upper side while young; leaves 6–12 cm. long and as wide, abruptly long-acuminate, subtruncate to rounded at base, irregularly coarse-toothed, pale green and sometimes slightly glaucous beneath, slightly pilose in axils; anthers linear, 6–7 mm. long, the connective long-mucronate; fruit usually 1-seeded, glabrous, obliquely obovate, 5–7 mm. long, the stipe as long as the body; stigma linear, slightly concave in fruit. ——Mar.–May. Mountains; Honshu, Shikoku, Kyushu; rather common.

Fam. 89. CERCIDIPHYLLACEAE   KATSURA KA   Cercidiphyllum Family

Deciduous dioecious trees with dimorphic branches (short spurs and longer vegetative shoots); buds 2-scaled; leaves petiolate, simple, palmately nerved, the stipules connate inside the petiole; flowers appearing before the leaves, small-scaled, the perianth absent; stamens many, nearly sessile, the filaments filiform, the anthers linear, red, the connective extended at the apex; pistillate flowers short–pedicelled; carpels 3–5, follicular, free; style very slender, elongate, rose-purple; fruit a many-seeded follicle, dehiscent along the inner suture; seeds small, winged; embryo small.——A single genus of 2 or 3 species in e. Asia.

1. CERCIDIPHYLLUM Sieb. & Zucc.   KATSURA ZOKU

Characters of the family. Two species in Japan.

1A. Seeds winged on one end; bark longitudinally fissured while still young; leaves of short spurs orbicular-cordate, slightly narrowed above and rounded at apex, shallowly cordate to nearly truncate at base; leaves of long shoots usually narrowed above.
       1. C. japonicum
1B. Seeds winged on both ends; bark not fissured until very old; leaves larger, those of short spur nearly orbicular, broadly rounded at apex, usually distinctly cordate at base, those of long shoots usually obtuse at apex. .... *2. C. magnificum*

1. Cercidiphyllum japonicum Sieb. & Zucc. *C. ovale* Maxim.——KATSURA. Trunk often branched above the base; leaves vivid green on upper side, glaucous beneath, obtusely serrulate, 3–7 cm. long and as wide, the petioles reddish, 2–2.5 cm. long; staminate flowers very short-pedicelled, with few small bracts; stamens many, the anthers linear, red, 3–4 mm. long; pistillate flowers short-petioled; style very long; follicles nearly sessile, cylindric, slightly curved, about 15 mm. long; style very slender; seeds flat, small, winged on one end, 5–6.5 mm. long, inclusive of the wing.——Apr.–May. Hokkaido, Honshu, Shikoku, Kyushu.——China.——Forma **pendulum** (Miyoshi) Ohwi. *C. japonicum* var. *pendulum* Miyoshi—— SHIDARE-KATSURA. Rare form with pendulous branches.—— Honshu; sometimes cultivated.

2. Cercidiphyllum magnificum (Nakai) Nakai. *C. japonicum* var. *magnificum* Nakai——HIRO-HA-KATSURA. Closely resembles the preceding, but the bark not fissured until the tree becomes very old; leaves larger, orbicular, 7–10 cm. long and as wide, deeply cordate; seeds 6–7 mm. long, inclusive of the wings.——Mountains at higher elevations than the preceding; Honshu (centr. and n. distr.).

Fam. 90. RANUNCULACEAE   KIM-PŌGE KA   Buttercup Family

Annuals or perennials, rarely shrubs, sometimes scandent; leaves often divided, alternate, sometimes opposite; flowers actinomorphic or zygomorphic, bisexual or sometimes unisexual, often large; sepals free, often petaloid, 3 to many, usually imbricate in bud; petals free, 3 to many, sometimes absent, well developed or sometimes reduced to nectaries; stamens many, free; anthers introrse, longitudinally split; carpels many or sometimes few, 1-locular, usually apocarpous, or sometimes united to form a compound ovary; styles distinct, the stigma terminal or oblique; ovules anatropous, ascending or pendulous; fruit usually a follicle or an achene, rarely berrylike and indehiscent, 1- to many-seeded; embryo small.——About 40 genera, with about 1,500 species, abundant in the temperate and frigid regions of both hemispheres, few in the Tropics.

1A. Sepals valvate in bud; leaves opposite, styles much elongate, taillike in fruit, often plumose-hairy. .... 1. *Clematis*
1B. Sepals imbricate in bud; leaves usually alternate or radical, rarely opposite.
    2A. Fruit an achene.
       3A. Flowers with involucral leaves.
          4A. Styles elongate and taillike in fruit. .... 2. *Pulsatilla*
          4B. Styles not accrescent after anthesis, short. .... 3. *Anemone*
       3B. Flowers not involucrate.
          5A. Petals present.
             6A. Petals with a nectary gland at the base.
                7A. Ovules ascending, funicles ventral; terrestrial or aquatic herbs not glaucous in our species. .... 4. *Ranunculus*
                7B. Ovules pendulous, funicles dorsal; terrestrial glaucous herbs. .... 5. *Callianthemum*
             6B. Petals without glands. .... 6. *Adonis*
          5B. Petals absent.
             8A. Leaves ternately compound. .... 7. *Thalictrum*
             8B. Leaves simple, palmately cleft to lobed. .... 8. *Trautvetteria*
    2B. Fruit a follicle, rarely a berry.

9A. Petals spurred.
    10A. Flowers actinomorphic; leaves ternately compound. .... 9. *Aquilegia*
    10B. Flowers zygomorphic; leaves palmately lobed or divided. .... 10. *Aconitum*
9B. Petals, when present, not spurred.
    11A. Petals smaller than the petaloid sepals, often reduced to a nectary or absent.
        12A. Leaves ternately or pinnately compound.
            13A. Sepals many. .... 11. *Anemonopsis*
            13B. Sepals 3 to 5.
                14A. Fruit a berry. .... 12. *Actaea*
                14B. Fruit a follicle.
                    15A. Follicles and carpels stipitate, evergreen herbs. .... 13. *Coptis*
                    15B. Follicles and carpels usually sessile; summer-green herbs.
                        16A. Flowers in racemes or spikes. ....14. *Cimicifuga*
                        16B. Flowers solitary, in umbels or panicles. .... 15. *Isopyrum*
        12B. Leaves orbicular or palmately cleft.
            17A. Petals present; leaves palmately cleft or dissected.
                18A. Petals entire; carpels sessile; scapes not involucrate. .... 16. *Trollius*
                18B. Petals cleft; carpels stipitate; scapes involucrate. .... 17. *Eranthis*
            17B. Petals absent; leaves merely toothed. .... 18. *Caltha*
    11B. Petals well developed; sepals 3 to 5, greenish, much shorter than the petals.
        19A. Leaves palmately cleft; sepals caducous; seeds flat, winged; follicles 2, connate inside, flat. .... 19. *Glaucidium*
        19B. Leaves ternately compound; sepals persistent; seeds subglobose, not winged; follicles 3–5. .... 20. *Paeonia*

1. CLEMATIS L.   SENNIN-SŌ ZOKU

Scandent or rarely erect bisexual or dioecious perennial herbs or shrubs; leaves opposite, petioled, pinnate or ternate, rarely simple, entire or toothed; flowers in panicles, corymbs, or solitary, white, purple or yellowish; sepals 4, sometimes 5 or more, petaloid, free, deciduous; petals absent or small, linear to spathulate, much smaller than the sepals; stamens many, the filaments glabrous or pubescent; carpels many; achenes 1-seeded; style elongate, often much so after anthesis, persistent, pilose.——About 250 species, widely distributed in the temperate zones, mainly in the N. Hemisphere.

1A. Sepals erect, usually recurved at the tip; flowers tubular or campanulate, nodding.
    2A. Petals present; flowers solitary; stamens densely pubescent. .... 1. *C. ochotensis*
    2B. Petals absent.
        3A. Leaflets entire, often lobed or divided; flowers solitary, usually brown-villous; filaments densely pubescent. .... 2. *C. fusca*
        3B. Leaflets toothed, often lobed to divided.
            4A. Erect herbs slightly woody at base.
                5A. Lateral leaflets broadly cuneate or subrounded on both halves at base; flowers 15–20 mm. long. .... 3. *C. stans*
                5B. Lateral leaflets subcordate on outer half at base, broadly cuneate to subrounded on inner half; flowers 20–25 mm. long.
                                4. *C. speciosa*
            4B. Scandent herbs with woody base.
                6A. Filaments villous; flowers tubular-campanulate.
                    7A. Flowers in fascicles or solitary in axils of leaves of young branchlets.
                        8A. Sepals purplish, thick, ovate-lanceolate, scarcely dilated above, acute, densely short-tomentose near margin.
                            9A. Bracts remote from the flowers, small, green. .... 5. *C. japonica*
                            9B. Bracts subtending the flowers, often large and colored. .... 6. *C. obvallata*
                      8B. Sepals creamy white, membranous, spathulate-oblong, dilated above, obtuse, densely pubescent on back. .. 7. *C. tosaensis*
                  7B. Flowers solitary, terminal and axillary toward the top of young branchlets.
                      10A. Young branchlets glabrous or nearly so; leaves green; peduncles 2–8 cm. long, glabrous; sepals 12–18 mm. long, rather thick, narrowly oblong, white-tomentose near the margin, abruptly acute; achenes several, with a tail 2.5–3 cm. long.
                                  8. *C. lasiandra*
                      10B. Young branchlets scattered pubescent; leaves glaucous; peduncles 10–14 cm. long, pubescent; sepals 2.5–3 cm. long, membranous, oblong-ovate, gradually long-acuminate, loosely pubescent on both sides; achenes many, the tail 4.5–5 cm. long. .... 9. *C. serratifolia*
                6B. Filaments glabrous or nearly so; flowers broadly campanulate. .... 10. *C. williamsii*
1B. Sepals spreading from the base; flowers rotate, erect in anthesis, petals absent.
    11A. Flowers large, solitary, 5–15 cm. across; spring.
        12A. Peduncles with a pair of leaflike bracts; sepals 6 in the normal form. .... 11. *C. florida*
        12B. Peduncles not bracteate; sepals 8 in the normal form. .... 12. *C. patens*
    11B. Flowers small, 10–25 mm. across; summer to autumn.
        13A. Flowers 1 to 3 in leaf-axils; anthers oblong, about 1 mm. long, blackened when dry. .... 13. *C. pierotii*
        13B. Flowers 3 to many, in terminal and axillary panicles; anthers linear, 2–3 mm. long, yellow.
            14A. Leaflets toothed. .... 14. *C. apiifolia*
            14B. Leaflets entire or lobed to divided, not toothed.
                15A. Petiolules ultimately jointed above; achenes lanceolate. .... 15. *C. uncinata* var. *ovatifolia*
                15B. Petiolules not jointed; achenes lanceolate to broadly ovate.
                  16A. Leaves trifoliolate or reduced to 1 or 2 leaflets, the petiolules of equal length.
                    17A. Leaflets broadly ovate to broadly elliptic, obtuse, cuneate to rounded at base; anthers about 1/10 as long as the filaments.
                                16. *C. crassifolia*

17B. Leaflets narrowed above, acuminate, rounded to shallowly cordate; anthers ⅓ to ½ as long as the filaments.
17. *C. meyeniana*

16B. Leaves pinnately compound, or if trifoliolate, the petiolules of the terminal leaflet much longer than the others.
18A. Plant gray-green to sordid-green when dry; sepals oblanceolate; achenes 7–10 mm. long. .... 18. *C. maximowicziana*
18B. Plant purplish-brown when dry; sepals oblong-lanceolate; achenes oblong or rhombic-oblong, about 5 mm. long.
19 *C. fujisanensis*

1. **Clematis ochotensis** (Pall.) Poir. *Atragene ochotensis* Pall.; *C. alpina* var. *ochotensis* (Pall.) O. Kuntze; *A. alpina* var. *ochotensis* (Pall.) Regel & Tiling——Mɪʏᴀᴍᴀ-ʜᴀɴsʜō-ᴢᴜʀᴜ. Slender scandent herb, woody at base, the young shoots loosely pubescent, the flowering branches usually short, with a pair of leaves; leaves usually twice ternate, petioled, the leaflets broadly lanceolate to ovate-orbicular, 2.5–8 cm. long, 1.5–4 cm. wide, acuminate, petioluled, sometimes 2- to 3-parted, coarsely toothed, membranous; flowers solitary, terminal, pendulous, pubescent, campanulate, 2.5–3 cm. long, usually purplish, on ebracteate peduncles 7–15 cm. long; sepals 4, broadly lanceolate to narrowly ovate, narrowed above; petals few, spathulate-linear, about half as long as the sepals; stamens many, slightly shorter than the petals, the filaments pubescent, the anthers yellow, oblong; achenes broadly ovate, 3–4 mm. long, spreading-pilose; style 2.5–3 cm. long.——June–Aug. Alpine; Hokkaido, Honshu (centr. and n. distr.).——Kuriles, Sakhalin, Manchuria, Korea, e. Siberia, and Kamchatka.

Var. **fauriei** (H. Boiss.) Tamura. *C. alpina* var. *fauriei* H. Boiss.; *C. ochotensis* var. *ternata* Nakai——Kᴏᴍɪʏᴀᴍᴀ-ʜᴀɴsʜōᴢᴜʀᴜ. Phase with once-ternate leaves.——Honshu (n. distr.).

2. **Clematis fusca** Turcz. *C. kamtschatica* Bong.; *C. fusca* var. *kamtschatica* Regel & Tiling; *C. fusca* var. *ajanensis* Regel & Tiling; *C. fusca* var. *yezoensis* Miyabe——Kᴜʀᴏʙᴀɴᴀ-ʜᴀɴsʜōᴢᴜʀᴜ. Rather firm perennial herb, woody at the base, sometimes scandent above, more or less brownish pubescent on young parts; leaves pinnate, 5- to 9-foliolate, petioled, the leaflets firm-chartaceous, narrowly to broadly ovate, 2–7 cm. long, 1–4 cm. wide, entire, petioluled, those at the base often 2- to 3-lobed; flowers terminal and axillary, broadly campanulate, nodding, 15–20 mm. long, the terminal ones bractless, the peduncle 2–7 cm. long, the lateral flowers with a pair of minute bracts, the peduncles 5–15 cm. long; sepals relatively thick, narrowly ovate, dark purple, dark brown-villous, slightly recurved at apex; anthers linear, about 4 mm. long, the filaments villous; achenes broadly ovate, with obliquely spreading hairs; style about 2.5 cm. long, pale yellowish brown, feathery.——July–Aug. Grassy places; Hokkaido.——e. Siberia, Kamchatka, Sakhalin, and Kuriles.

Var. **glabricalyx** Nakai. Phase with less hairy reddish purple sepals.

3. **Clematis stans** Sieb. & Zucc. *C. kousabotan* Decne.; *C. lavallei* Decne. and var. *foliosa* Decne.; *C. savatieri* Decne.; *C. heracleifolia* var. *stans* (Sieb. & Zucc.) O. Kuntze——Kᴜsᴀ-ʙᴏᴛᴀɴ. Erect dioecious short-pubescent herb often woody at the base; stems 50–100 cm. long, sometimes branched; leaves petiolate, ternate, the leaflets thick-chartaceous, ovate to depressed-orbicular, 5–10 cm. long, 4–10 cm. wide, often 2- to 3-cleft, short-acuminate, with coarse deltoid teeth, the lateral leaflets subsessile; cymes axillary, often forming a large terminal panicle, the pedicels short white-tomentose; flowers 15–20 mm. long, narrowly campanulate, nodding, pale purple-blue, short white-tomentose outside; sepals recurved on upper half, lanceolate; stamens of staminate flowers 10–12 mm. long, the anthers linear, 4–5 mm. long, as long as to slightly shorter than the loosely pubescent filaments; achenes elliptic, about 3.5 mm. long, with short ascending hairs; style about 2 cm. long in fruit.——Aug.–Oct. Sunny places in mountains; Honshu.

Var. **austrojaponensis** (Ohwi) Ohwi. *C. austrojaponensis* Ohwi——Tsᴜᴋᴜsʜɪ-ᴋᴜsᴀ-ʙᴏᴛᴀɴ. Anthers shorter, about 3 mm. long, the filaments 2–3 times as long as the anthers.——Shikoku, Kyushu.

Clematis urticifolia Kitag. Tᴀᴄʜɪ-ᴋᴜsᴀ-ʙᴏᴛᴀɴ. Closely related to *C. stans,* but differs in the urceolate flowers. Reported from Honshu (Sado Isl.).——Korea.

4. **Clematis speciosa** (Makino) Makino. *C. heracleifolia* var. *speciosa* Makino; *C. heracleifolia* var. *hookeri* Makino, excl. syn.——Ō-ᴋᴜsᴀ-ʙᴏᴛᴀɴ. Closely resembles No. 3; stems stout; leaves chartaceous, petiolate, somewhat larger, ternate, the leaflets ovate to broadly so, 10–18 cm. long, 4–10 cm. wide, short-acuminate, sometimes 2- to 3-lobed, short-toothed, the lateral leaflets usually petioluled, prominently oblique at base; flowers 2–2.5 cm. long; sepals linear-lanceolate, dilated above; stamens about 1/3 as long as the sepals, the anthers shorter than the filaments, about 3 mm. long.——Oct. Shikoku, Kyushu.

5. **Clematis japonica** Thunb. *C. ternata* Makino——Hᴀɴsʜō-ᴢᴜʀᴜ. Scandent, sparsely pubescent subshrub; leaves petiolate, ternate, the leaflets ovate or elliptic, 5–8 cm. long, 2–4 cm. wide, subsessile, 3- to 5-nerved, acuminate to acute, cuneate to obtuse at base, with coarse teeth especially on upper half; flowers 1 to few, between the bud-scales at the base of young shoots, nodding, campanulate, 2.5–3 cm. long, purplish red or brownish red, the peduncles 5–10 cm. long, with a pair of small bracts above the middle; sepals 4, broadly lanceolate to narrowly ovate, slightly fleshy, nearly glabrous except the short white-tomentose margin; stamens about half as long as the sepals, the filaments flat, pubescent above; achenes broadly lanceolate, about 6 mm. long, glabrous or sparingly pilose; style 3–3.5 cm. long.——May–June. Thickets in mountains; Honshu, Kyushu.——Forma **purpureofusca** Hisauchi. Peduncles shorter, bibracteate near base.——Forma **villosula** Ohwi. Sepals yellow-villous externally, sometimes divided.

6. **Clematis obvallata** (Ohwi) Tamura. *C. japonica* var. *obvallata* Ohwi; *C. japonica* forma *obvallata* (Ohwi) Ohwi——Kōʏᴀ-ʜᴀɴsʜōᴢᴜʀᴜ. Closely resembles the preceding; bracts at the top of the peduncles, larger, often colored and involucrelike.——Mountains; Honshu (s. Kinki Distr.), Shikoku.

Var. **shikokiana** Tamura. Sʜɪᴋᴏᴋᴜ-ʜᴀɴsʜōᴢᴜʀᴜ. Bracts smaller.——Shikoku.

7. **Clematis tosaensis** Makino. *C. japonica* var. *brevipedicellata* Makino——Tᴏʀɪɢᴀᴛᴀ-ʜᴀɴsʜōᴢᴜʀᴜ, Aᴢᴜᴍᴀ-ʜᴀɴsʜōᴢᴜʀᴜ. Scandent subshrub closely resembling *C. japonica;* young shoots elongate, pubescent; leaflets 3; peduncles 1–8 cm. long, ebracteate; flowers creamy white, 2–3 cm. long; sepals membranous, spathulate-oblong, slightly dilated above, obtuse, white-pubescent outside, glabrous to scattered pubes-

cent inside; anthers much shorter than the filaments, the filaments linear, slightly pubescent; achenes 5–6 mm. long, broadly lanceolate.——Apr.–June. Thickets in mountains; Honshu, Shikoku.——Forma **cremea** (Makino) Tamura. *C. japonica* var. *cremea* Makino——Shiro-hanshōzuru. Peduncles longer, bracteate near the middle.——Honshu (Nikko in Shimotsuke).

8. Clematis lasiandra Maxim. Takane-hanshōzuru. Nearly glabrous scandent herb with slender woody branches; leaves petiolate, once to twice ternately compound, thin, the leaflets broadly lanceolate to broadly ovate, 4–8 cm. long, 1.5–4 cm. wide, acuminate, acute to shallowly cordate at base, with coarse deltoid, mucronate teeth, petiolate or subsessile; flowers solitary or 2 or 3, campanulate, 12–18 mm. long, the terminal peduncles naked, the lateral ones 2–8 cm. long, with a pair of small broadly linear bracts on lower part; sepals 4, rather thick, narrowly oblong to broadly lanceolate, nearly glabrous except the short white densely tomentose margin; stamens slightly shorter to nearly as long as the sepals, the filaments long-villous, the anthers broadly linear, about 2 mm. long; achenes many, ovate, about 3 mm. long, with spreading hairs; style slender, 2.5–3 cm. long, white.——Aug.–Oct. Honshu (Chūgoku Distr.), Shikoku, Kyushu.

9. Clematis serratifolia Rehd. *C. orientalis* var. *serrata* Maxim.; *C. orientalis* var. *wilfordii* Maxim.; *C. sibiricoides* Nakai——Ō-waku-note. Scandent subshrub, the nodes and young parts short-pubescent; leaves pinnately biternate, the petioles 4–5 cm. long, the leaflets 3–5 cm. long, 1–3 cm. wide, petiolulate, often irregularly 2- to 3-fid, coarsely toothed, sparingly pubescent on both sides, glaucous, acuminate; peduncles erect, 10–14 cm. long, white-pubescent, with small bracts near base; flowers nodding, broadly campanulate; sepals 2–3 cm. long, about 1 cm. wide, narrowly ovate, gradually narrowed above and acuminate, membranous; stamens 5–7 mm. long, the filaments lanceolate-subulate, silky especially on margin, the anthers 2–2.5 mm. long, yellowish, obtuse; achenes ovate, about 2.5 mm. long, pilose, the tail 4.5–5 cm. long, slender, long white-pilose.——Hokkaido (near Wakkanai in Kitami Prov.).——Manchuria, n. Korea, and Ussuri.

10. Clematis williamsii A. Gray. *C. montana* var. *williamsii* (A. Gray) O. Kuntze; *C. montana* var. *bissetii* O. Kuntze——Shirobana-no-hanshōzuru. Scandent, gray-brown pubescent; leaves ternate, the leaflets short-petiolulate, rather thin, often nearly glabrous above, ovate to oblong, 3–8 cm. long, 1–3.5 cm. wide, few-toothed, cuneate to rounded at base, usually 3-fid, the median lobes acuminate; flowers white, cup-shaped or broadly campanulate, 1.5–5 cm. long, nodding, solitary, the peduncles axillary with a pair of minute bracts on lower half; sepals 4, broadly elliptic, membranous, about 2 cm. long, glabrous inside, appressed-pubescent outside; stamens shorter than the sepals, the filaments linear, glabrous, becoming black when dried, the anthers oblong, about 2 mm. long, yellow, the connective also blackened when dried; achenes ovate, about 5 mm. long, pilose; style about 2.5 cm. long, pilose, pale yellowish-brown.——May–June. Honshu (Tōkaidō Distr. and westw.), Shikoku, Kyushu; rather rare.

Clematis eriopoda Maxim. Hime-botanzuru. Species attributed to Japan by Maximowicz, but obscure to us. Described as having ternate leaves, the leaflets ovate, 3-sected, incise-toothed, the median lobes linear, the lateral ones ovate, 2–2.5 cm. long, 1–2 cm. wide, the peduncles solitary, 1-

flowered, bracteate, villous above; sepals 4, nearly spreading, elliptic, densely silky outside, slightly longer than the stamens; filaments nearly glabrous, much longer than the anthers.

11. Clematis florida Thunb. *Anemone japonica* Houtt.; *C. hakonensis* Fr. & Sav.——Tessen. Slightly pubescent, often glabrate woody herb; stems scandent, slender; leaves petiolate, 1 or 2 times ternate, the median primary petiolules longer than the lateral, the ultimate leaflets ovate to narrowly so, usually entire, rather chartaceous, 2–4 cm. long, 1–2 cm. wide, acute; peduncles 1-flowered, solitary on the axils, about 10 cm. long, with a pair of leaflike simple bracts near middle; flowers white, large, 5–8 cm. across; sepals usually 6, spreading, ovate, acute, with a longitudinal band of soft short dense pubescence on each side externally; anthers linear, 3–5 mm. long, the filaments linear, about twice as long as the anthers, glabrous, blackened when dry; carpels silky.——May–June. Chinese plant long grown as an ornamental in our area; several cultivars are known.

12. Clematis patens Morr. & Decne. Kazaguruma. Resembles the preceding; leaves 3- to 5-foliolate, the leaflets ovate, chartaceous, often 3-fid, 3–10 cm. long, 2–5 cm. wide, acute, rounded to shallowly cordate; peduncles solitary and terminal on short young shoots, 1-flowered, ebracteate, 10–15 cm. long; flowers white to pale blue-purple, 10–15 cm. across; sepals usually 8, caudate, narrowly ovate to narrowly oblong, with a broad band of short white hairs on each side externally; anthers linear, 6–10 mm. long, the filaments glabrous, as long as to slightly longer than the anthers; achenes ovate; style long yellowish-brown pilose.——May–June. Honshu, Shikoku, Kyushu. Several cultivars are grown in this country.——China.

13. Clematis pierotii Miq. Koba-no-botanzuru, Yama-botanzuru. Slightly short-pubescent climber; leaves petiolate, twice ternate, the leaflets membranous, petioluled or sessile, ovate to broadly so, 1.5–5 cm. long, 8–20 mm. wide, usually 3-fid, with few irregular teeth, the midlobes of terminal leaflets often elongate, acuminate, the other lobes acute to subobtuse, linear-lanceolate to narrowly ovate; peduncles axillary, solitary, 1- to 3-flowered, 5–10 cm. long, leafy-bracteate; flowers white, about 2.5 cm. across, erect; sepals oblong, white-pubescent outside especially toward the margin; stamens slightly shorter than the sepals, glabrous, the anthers narrowly oblong, 1–1.5 mm. long; achenes broadly lanceolate, about 5 mm. long, spreading-pilose; style yellowish, about 2 cm. long, feathery.——Aug.–Sept. Shikoku, Kyushu.——Ryukyus.

14. Clematis apiifolia DC. Botanzuru. Slightly pubescent climber; stems rather thick below, woody, the shoots elongate, much branched, the inflorescence terminal; leaves petiolate, ternate, the leaflets nearly membranous, narrowly to broadly ovate, 4–8 cm. long, 2–5 cm. wide, acuminate, often 3-lobed, sparsely coarse-toothed, rounded to shallowly cordate at base; inflorescence pedunculate, 5–10 cm. long, rather many-flowered, cymose-paniculate; flowers white, 15–20 mm. across; sepals 4, broadly oblanceolate, spreading, short white-puberulent outside; stamens slightly shorter than the sepals, glabrous, the anthers linear, about 2 mm. long; achenes ovate, with spreading hairs, about 4 mm. long; style 1–1.2 cm. long.——Aug.–Sept. Thickets in lowlands and foothills; Honshu, Shikoku, Kyushu; common.——Korea and China.

Var. **biternata** Makino. *C. brevicaudata* sensu auct. Japon., non DC.——Me-botanzuru, Ko-botanzuru. Leaves often partially twice ternate, the achenes often glabrous.——Honshu (centr. distr.).

Clematis × takedana Makino. MURASAKI-BOTANZURU. Hybrid of *C. apiifolia* × *C. stans*. Stems only slightly scandent, the leaflets firmer; flowers pale purple, the sepals spreading and recurved above; filaments often long-pubescent.——Honshu (n. and centr. distr.); rare.

15. Clematis uncinata Champ. var. **ovatifolia** (T. Ito) Ohwi. *C. ovatifolia* T. Ito; *C. longiloba* sensu Maxim., non DC.——KII-SENNIN-SŌ. Glabrous, somewhat glaucous climber; leaves petiolate, somewhat coriaceous, twice ternate or the lower ones pinnately 5-foliolate, the lowest leaflets once again ternate, the ultimate leaflets ovate to ovate-oblong, 4–7 cm. long, 2–3 cm. wide, 3- or obsoletely 5-nerved, entire, acute to abruptly acuminate, rounded at base, the petiolules jointed below the top; panicles loosely many-flowered, 10–25 cm. long, the pedicels 2–2.5 cm. long, bracteate; flowers 2.5–3 cm. across; sepals 4, white, oblanceolate, subacuminate, the margin densely puberulent; anthers about 3 mm. long; achenes lanceolate, truncate, about 6 mm. long, glabrous; style 1.8–2.2 cm. long, feathery.——Aug. Honshu (s. Kinki Distr.), Kyushu (Higo Prov.). The typical phase occurs in s. China and Formosa.

16. Clematis crassifolia Benth. YAMA-HANSHŌZURU. Glabrous woody climber; stems prominently fissured and corky, the new shoots terete; leaves petiolate, ternate, the leaflets thick-coriaceous, elliptic or broadly ovate, 8–12 cm. long, 6–8 cm. wide, obtuse, broadly cuneate to rounded at base, entire, petiolules distinct, 3–4 cm. long, stout; inflorescence axillary, many-flowered, the bracts small; sepals 4, spreading, linear- to narrowly oblong, 15–18 mm. long, 3–5 mm. wide, white-woolly on margin and inside; anthers much shorter than the filaments.——Kyushu (s. distr. incl. Yakushima and Tanegashima).——s. China and Formosa.

17. Clematis meyeniana Walp. *C. lutchuensis* Koidz. ——YAMABARU-SENNIN-SŌ, TERIHA-NO-SENNIN-SŌ. Glabrous woody climber, the young shoots often yellowish brown pubescent in leaf axils, changing to dark brown when dried; leaves petiolate, ternate, the leaflets rather coriaceous, petioluled, ovate to elongate-ovate, sometimes deltoid-ovate, 5-nerved, 4–10 cm. long, 2.5–5 cm. wide, acuminate, rounded to shallowly cordate at base; panicles axillary, pyramidal, 7–20 cm. long, peduncled, rather many-flowered, the bracts subulate, small, the pedicels rather short, the flowers white, 15–20 mm. across; sepals 4, spreading, with white tomentum on margin;

stamens about 2/3 as long as the sepals, glabrous, the anthers linear, yellow, 1/2–2/3 as long as the filaments; achenes yellowish brown villous; style feathery except at apex, yellowish brown.——July–Aug. Kyushu (Tanegashima and Yakushima).——Ryukyus, Formosa, s. China, and Philippines.

18. Clematis maximowicziana Fr. & Sav. *C. paniculata* Thunb., non J. F. Gmel.; *C. recta* var. *paniculata* (Thunb.) O. Kuntze; ? *C. terniflora* DC.; *C. dioscoreifolia* Lév. & Van't.; *C. flammula* var. *robusta* Carr.; *C. dioscoreifolia* var. *robusta* (Carr.) Rehd.; *C. terniflora* var. *robusta* (Carr.) Tamura—— SENNIN-SŌ. Branched suffrutescent woody climber, short-pubescent only while young, becoming grayish green when dry; leaves petiolate, pinnately 5-foliolate, the leaflets petiolulate, narrowly ovate to ovate-orbicular, sometimes broadly lanceolate or deltoid-ovate, 3–7 cm. long, (1–)2–4(–6) cm. wide, 3- to 5-nerved, usually entire, sometimes 2- to 3-lobed, obtuse, mucronate or acute, slightly cordate to rounded at base; inflorescence a rather many-flowered terminal panicle, 5–15 cm. long, the bracts subulate; flowers white, 2–2.8 cm. across, erect; sepals 4, spreading, oblanceolate, white-puberulent on margin; stamens 1/2–2/3 as long as the sepals, glabrous, the anthers linear, yellow, 2.5–3 mm. long; achenes rather few, ovate, 7–10 mm. long, slightly appressed-puberulent; style 2.5–3 cm. long, white, feathery.——Aug.–Oct. Hokkaido, Honshu, Shikoku, Kyushu; very common.——Korea and China.

19. Clematis fujisanensis Hisauchi & Hara. *C. kyushuensis* Tamura——FUJI-SENNIN-SŌ. Nearly glabrous climber, to 5 m. high; leaves 20–25 cm. long, 10–13 cm. wide, 3- to 5-foliolate, the petioles 5–7 cm. long, the leaflets deep green, chartaceous, petiolulate, ovate, ovate-lanceolate to lanceolate, 2–9 cm. long, very acute, entire, rarely unequally bifid, 5-nerved, white-pubescent while young, soon becoming glabrous, the petiolules 1–4 cm. long; panicles corymbose, few to many-flowered, nearly glabrous to sparsely pubescent, with minute subulate-deltoid bracts, the pedicels 2–3 cm. long; flowers 1.5–3 cm. across, white; sepals oblong-lanceolate, rounded, mucronate, sparingly pilose on outer side near tip, white-tomentulose on margin; stamens many, the anthers linear, 2–3 mm. long; achenes oblong or rhombic-oblong, about 5 mm. long, 3 mm. wide, appressed white-puberulent, the style 2.5–4 cm. long, with ascending-spreading white hairs. ——Aug.–Sept. Honshu (Kantō Distr. and westw.), Kyushu; rare.

2. PULSATILLA Mill. OKINAGUSA ZOKU

Rhizomatous perennial herbs usually with long silky hairs at least while young; leaves radical, petiolate, palmately, ternately or pinnately compound, rarely simple; scapes erect, naked except for a whorl of involucral leaves, 1-flowered, the involucral leaves ternate; flowers nodding, campanulate, long-peduncled; sepals 5–12, deciduous, petaloid; petals absent; stamens many; carpels many, sessile, usually pilose; style elongate, becoming taillike after anthesis, feathery.——About 30 species, in temperate and arctic regions of the N. Hemisphere.

1A. Radical leaves pinnately 5-foliolate, the lateral leaflets subsessile; sepals dark reddish brown inside; tail of the achenes with hairs about 3–4 mm. long. .... 1. *P. cernua*
1B. Radical leaves ternately compound, the linear segments 1–2 mm. wide; sepals pale yellow inside; tail of the achenes with hairs 0.5–1 mm. long. .... 2. *P. nipponica*

1. Pulsatilla cernua (Thunb.) Spreng. *Anemone cernua* Thunb.——OKINAGUSA. Perennial silky-pubescent herb with thickened rhizomes; radical leaves petiolate, tufted, broadly ovate, 5-foliolate, 6–10 cm. long, the leaflets subcoriaceous, ovate or rhombic-orbicular, 2- to 5-parted, 3–5 cm. long,

acutely incised, often nearly glabrate above; scapes 10–40 cm. long; involucral leaves sessile, palmately parted, 3–5 cm. long, sometimes again 3- to 5-cleft, the segments linear; peduncles 15–25 cm. long, the flowers nodding, solitary, tubular-campanulate, 2.5–3 cm. long; sepals 6, silky white-villous outside, glabrous and dark reddish brown to dark reddish purple inside, narrowly oblong; achenes narrowly ovate, about 3 mm. long, white-silky, the tail white, feathery, about 4 cm. long, the spreading hairs 3–4 mm. long; stigma punctate.——Apr.–May. Grassy slopes in foothills; Honshu, Shikoku, Kyushu; rather common.——Korea and China.

2. **Pulsatilla nipponica** (Takeda) Ohwi. *Anemone taraoi* var. *nipponica* Takeda——Tsukumogusa. Silky-villous when young; rhizomes thickened, with a tuft of leaves at top; radical leaves long-petiolate, deltoid-orbicular, 2–6 cm. long and as wide, ternately compound, the leaflets petiolulate, 1.5–3 cm. long, finely dissected into linear acute segments 1–2 mm. wide; scapes 10–30 cm. long; involucral leaves sessile, 15–25 mm. long, deeply parted; peduncles elongate after anthesis, to 5–10 cm. long; flowers broadly campanulate or broadly tubular, one-sided, about 2.5 cm. long; sepals 6, oblong, white-silky outside, glabrous and pale yellow inside, membranous; achenes hairy, narrowly ovate, about 4 mm. long, the tail elongate, slightly thickened below, firm, 3–3.5 cm. long, densely pilose with brownish-yellow hairs 0.5–1 mm. long.——June–July. Alpine slopes; Hokkaido, Honshu (centr. distr.); rare.

3. ANEMONE L. Ichirin-sō Zoku

Perennial herbs usually with creeping rhizomes; radical leaves petioled, the cauline usually involucrelike, commonly ternate, rarely opposite, all palmately to ternately divided or lobed; flowers 1 to many, usually on pedicels, usually large and showy; sepals 5 to many, petaloid, deciduous, imbricate in bud, rarely absent; petals absent; stamens many; carpels many or rather few, free, each with a pendulous ovule; achenes sessile or stiped, the style short, pilose or glabrous.——About 100 species, in temperate to cold regions of the world.

1A. Flowers sessile, the involucral leaves simple, entire; radical leaves 3-lobed, the lobes entire. . . . . 1. *A. hepatica*
1B. Flowers pedunculate, the involucral leaves variously lobed; radical leaves incised and toothed, usually compound.
  2A. Achenes sessile; involucral leaves usually dissimilar in shape to the radical or cauline ones.
    3A. Achenes ovate or oblong, scarcely flattened.
      4A. Involucral leaves petiolate.
        5A. Flowers usually solitary; plants nonstoloniferous; rhizomes without persistent base of petioles; radical leaves absent or single at base of the solitary scape.
          6A. Primary leaflets petioluled; scapes 10–30 cm. long; flowers large; sepals 15–30 mm. long.
            7A. Plants slightly glaucous, the long silky hairs deciduous; petioles of involucral leaves scarcely dilated. . . . . 2. *A. raddeana*
            7B. Plants green, not glaucous, the hairs on the peduncle persistent; petioles of involucral leaves dilated and winged on the margin.
              8A. Sepals 8–12, linear-oblong, 5–8 times as long as wide; anthers ellipsoidal, less than 1 mm. long. . . 3. *A. pseudoaltaica*
              8B. Sepals 5(–6), elliptic, 1.5–3 times as long as wide; anthers oblong, 1–1.5 mm. long. . . . . 4. *A. nikoensis*
          6B. Primary leaflets sessile; scapes 5–15 cm. long; flowers small; sepals 5–12 mm. long.
            9A. Sepals 5, 5–7 mm. long; leaflets of involucral leaves lanceolate or linear-lanceolate. . . . . 5. *A. debilis*
            9B. Sepals 5–7, 10–12 mm. long; leaflets of involucral leaves ovate-oblong to oblong. . . . . 6. *A. yezoensis*
        5B. Flowers 1 to 3 on a scape; plants often stoloniferous; rhizomes with persistent fibers disintegrated from the decayed petioles; radical leaves few; scapes 1–3. . . . . 7. *A. stolonifera*
      4B. Involucral leaves sessile.
        10A. Leaflets deltoid-ovate; involucral leaves rather narrow, pinnately incised; flowers solitary, pale purple, the peduncles 3–5 cm. long; sepals 12–15, linear-lanceolate, 5–10 times as long as wide. . . . . 8. *A. keiskeana*
        10B. Leaflets rhombic-obovate, the lateral ones 2- to 3-parted; involucral leaves 3-parted; flowers 1–3, white, or slightly pinkish, the peduncles 5–10 cm. long; sepals 5, elliptic, 2–3 times as long as wide. . . . . 9. *A. flaccida*
    3B. Achenes elliptic or ovate, prominently flattened, often with a thick wing on margin.
      11A. Rhizomes slender, long-creeping, bearing scapiform stems at intervals with few pairs of opposite sessile leaves; stems dichotomously branched above, naked below, with few pairs of scales at base; flowers solitary in the dichotomous axils.
        . . . . 10. *A. dichotoma*
      11B. Rhizomes thick and rather short, densely covered with decayed petiole-bases; scapes simple, rarely once-branched, with 1 or 2 pairs of sessile involucral leaves, surrounded at base with a tuft of radical petioled leaves; flowers in umbels.
        12A. Peduncles unbranched. . . . . 11. *A. narcissiflora*
        12B. Peduncles or some of them branched. . . . . 12. *A. sikokiana*
  2B. Achenes stipitate; involucral leaves few-paired, leaflike. . . . . 13. *A. hupehensis* var. *japonica*

1. **Anemone hepatica** L. var. **japonica** (Nakai) Ohwi. *Hepatica nobilis* var. *japonica* Nakai——Misumi-sō. White-silky while young; rhizomes elongate; radical leaves long-petioled, depressed-deltoid, cordate, 2–3 cm. long, 3–5 cm. wide, 3-lobed, the lobes broadly ovate-deltoid, acute, rather coriaceous; stems 10–15 cm. long, the involucral leaves 3, narrowly ovate, subobtuse; sepals 6–9, white or slightly pinkish, 8–10 mm. long, slightly longer than the involucres, narrowly oblong, subobtuse at apex; carpels hairy, the stigma slightly thickened.——Feb.–May. Honshu.——Forma **variegata** (Makino) Hara. *Hepatica triloba* var. *obtusa* forma *variegata* Makino; *A. hepatica* var. *japonica* forma *nipponica* (Nakai) Ohwi; *H. nobilis* var. *nipponica* Nakai——Suhama-

sō. Leaves sometimes variegated, with obtuse lobes. Occurs with var. *japonica*. The typical form occurs in Eurasia.

Var. **pubescens** Hiroe. Ke-suhama-sō. Leaves often hairy on both sides, the lobes roundish.——Honshu (west. distr.), Shikoku.

2. **Anemone raddeana** Regel. Azuma-ichige. Rhizomes creeping, 1.5–3 cm. long, slenderly fusiform, with scattered, small, scarious scales at apex; leaves long-petiolate, 1 or 2 times ternate, the leaflets 3-parted; stems solitary, 15–20 cm. long, erect, scattered long-pilose, the involucral leaves 3, short-petiolate, once-ternate, the leaflets oblong, 15–35 mm. long, 5–15 mm. wide, obtuse, with irregularly toothed upper half; peduncles 2–3 cm. long; flowers white with a purple hue;

sepals about 2 cm. long; anthers elliptic, about 1 mm. long; ovary puberulous.——Mar.–May. Damp woods in mountains; Hokkaido, Honshu, Shikoku.——Amur, Ussuri, Korea, and Sakhalin.

3. Anemone pseudoaltaica Hara. *A. altaica* sensu auct. Japon., non Fisch.——KIKUZAKI-ICHIGE. Rhizomes slender, 2–10 cm. long, long-creeping; radical leaves twice-ternate, long-petiolate, the leaflets ovate, 3- to 5-parted, irregularly toothed; stems solitary, 10–30 cm. long, glabrous, the involucral leaves 3, petiolate, ternate, the peduncles pubescent; sepals pale purple or white, linear-elliptic, about 20 mm. long; anthers elliptic; achenes broadly lanceolate, about 4 mm. long, densely short, spreading, white-pubescent; styles slender, rather short.——Apr.–June. Woods in mountains; Hokkaido, Honshu (n. and centr. distr.).

4. Anemone nikoensis Maxim. ICHIRIN-SŌ. Rhizomes slender, creeping, 1–6 cm. long, slightly thickened at intervals; radical leaves 1–2 times ternate, long-petiolate, the leaflets ovate, subacute, pinnately parted; stems 20–30 cm. long, solitary, the involucral leaves 3, petiolate, once-ternate, the leaflets ovate, subacute, 2–5 cm. long, pinnately parted, irregularly toothed, the peduncles solitary, pubescent; sepals 5, white, usually with a pinkish hue, elliptic, 2–3 cm. long; anthers oblong; style short; carpels densely short-pilose.——Apr.–May. Woods in foothills; Honshu, Shikoku, Kyushu.

5. Anemone debilis Fisch. *A. ranunculoides* var. *gracilis* Schltdl.; *A. gracilis* (Schltdl.) F. Schmidt——HIME-ICHIGE. Delicate herb, 5–15 cm. high, sparsely pubescent; rhizomes short-creeping, 7–15 mm. long; involucral leaves ternate, petiolate, sometimes only 3-parted, with the leaflets linear-lanceolate or lanceolate, 2–5 cm. long, 3–12 mm. wide, usually coarsely toothed, sessile; peduncles solitary; flowers white; sepals 5, oblong, 5–7 mm. long; anthers elliptic, about 0.4 mm. long; achenes narrowly ovate, about 3 mm. long, rather densely grayish white-pubescent; style narrow, rather short.——June–July. Damp coniferous woods in mountains; Hokkaido, Honshu (n. and centr. distr. and Yamato Prov.).——Kuriles, Sakhalin, Kamchatka, Manchuria, and e. Siberia.

6. Anemone yezoensis (Miyabe) Koidz. *A. umbrosa* var. *yezoensis* Miyabe; *A. debilis* var. *soyensis* Makino, excl. syn.——EZO-ICHIGE. Resembles the preceding; rhizomes slender; leaflets of involucral leaves ovate-oblong or oblong, 2–3 cm. long, 7–15 mm. wide, coarsely toothed; sepals 5–7, narrowly ovate, 10–12 mm. long; anthers elliptic, about 0.7 mm. long.——May–July. Hokkaido.——Sakhalin.

7. Anemone stolonifera Maxim. *A. siuzevii* Komar. ——SANRIN-SŌ. Sparsely pubescent stoloniferous herb 15–30 cm. high; radical leaves long-petiolate, 3-foliolate, the lateral leaflets 2-parted, the terminal one rhombic-obovate, 2.5–4 cm. long, 3-parted, incised and toothed, short-petiolulate or subsessile; scapes 1 to 3, the involucral leaves ternate, short-petiolate, 3-parted, the segments cuneate to broadly oblanceolate, 3-cleft and coarsely toothed, 2–3 cm. long; flowers 1–3, white, on erect peduncles 3–10 cm. long; sepals 5, elliptic, 7–8 mm. long; achenes rather few, broadly ovate, puberulent, about 3 mm. long; style slender and rather short.——May–July. Woods in mountains; Hokkaido, Honshu (centr. and n. distr.).——Formosa, Manchuria, Korea, and (?) China.

8. Anemone keiskeana T. Ito. YUKI-WARI-ICHIGE, RURI-ICHIGE. Rhizomes rather slender, creeping, to 10 cm. long; radical leaves long-petiolate, nearly glabrous, deltoid, 3-foliolate, the leaflets 3–5 cm. long, sessile, deltoid-ovate, acuminate,

obtusely toothed or incised, the involucral leaves narrowly ovate or narrowly oblong, sessile, 10–35 mm. long, toothed or pinnately lobulate, the peduncles 3–5 cm. long, solitary, loosely ascending-pubescent; sepals 12–15, linear-oblong, 15–20 mm. long, white to purplish toward the base; anthers elliptic, 1–2 mm. long.——Feb.–Apr. Honshu (w. distr.), Shikoku, Kyushu.

9. Anemone flaccida F. Schmidt. *A. baicalensis* var. *laevigata* A. Gray; *A. laevigata* (A. Gray) Koidz.; *A. soyensis* H. Boiss.; *A. tagawae* Ohwi; *A. amagisanensis* Honda——NIRIN-SŌ. Slightly pubescent perennial with stout creeping rhizomes; radical leaves long-petiolate, 3-parted, the lateral segments 2-parted, the terminal one rhombic-obovate, 2–4 cm. long, 1.5–3.5 cm. wide, 3-fid and coarsely toothed; stems 1 or 2, 20–30 cm. long, the involucral leaves sessile, rather unequal, cuneate-ovate or oblong, 3-fid, coarsely toothed, 2–5 cm. long; peduncles 1–3, 5–10 cm. long; flowers white; sepals 5, elliptic, 8–15 mm. long; anthers elliptic, about 1 mm. long; achenes ovate, about 2.5 mm. long, short white-pubescent; style very short.——Moist shaded places along streams and ravines in foothills; very common.——Sakhalin, Amur, and n. China.

10. Anemone dichotoma L. FUTAMATA-ICHIGE. Rather stout, white-pubescent herb with slender, elongate rhizomes; stems erect, rather firm, 40–80 cm. long, naked on lower half, with few scales at base, dichotomously branching above, with a single flower on each branch; leaves opposite, sessile, 3-parted, the segments cuneate-oblanceolate, 4–8 cm. long, 1–2 cm. wide, acute, coarsely toothed on upper half and sometimes again 3-lobed, nearly glabrous above; peduncles 3–7 cm. long; flowers white; sepals 5, broadly elliptic, 8–15 mm. long, appressed-puberulent externally; anthers elliptic, about 0.8 mm. long; achenes glabrous, ovate, flat, 4–5 mm. long, the style about 1.5 mm. long.——June–July. Grassy places in lowlands, especially near the sea; Hokkaido.——Europe, Siberia, Manchuria, and Korea.

11. Anemone narcissiflora L. HAKUSAN-ICHIGE. Perennial herb, usually densely white-pilose; rhizomes stout, covered with old leaf bases; radical leaves long-petiolate, orbicular-cordate, 3-parted, the lateral segments deeply 2-lobed, each segment further dissected into linear subacute ultimate lobes; stems 1 to 4, 15–40 cm. long, the involucral leaves 2–4 cm. long, dissected into linear lobes, the peduncles 1–7 cm. long; flowers white; sepals 5–7, ovate or elliptic, 12–15 mm. long; anthers oblong, about 1 mm. long; achenes flat, glabrous, with a rather thick wing on the margin, broadly elliptic, 6–7 mm. long, about 5 mm. wide, the style short and slightly incurved.——July–Aug. Wet alpine slopes; Hokkaido, Honshu (centr. and n. distr.).——Sakhalin, Kuriles, n. Korea to Himalayas, Siberia to Europe, and N. America.

Var. **sachalinensis** Miyabe & Miyake. KARAFUTO-SENKA-SŌ. Plants more densely pubescent.——Honshu (n. distr.), Hokkaido.——Sakhalin and s. Kuriles.

12. Anemone sikokiana Makino. *A. narcissiflora* var. *sikokiana* Makino——SHIKOKU-ICHIGE. Resembles the preceding; radical leaves 5-angled, cordate-orbicular, 3-parted, the lateral segments 2-parted, 3–5 cm. long, each segment 2- to 3-fid, coarsely incised and lobed; some of the peduncles branched, bearing smaller involucral leaves and few flowers. ——Alpine slopes; Shikoku (Mount Ishizuchi in Iyo Prov.); rare.

13. Anemone hupehensis Lemoine var. **japonica** (Thunb.) Bowles & Stearn. *Atragene japonica* Thunb.; *Anemone japonica* (Thunb.) Sieb. & Zucc., non Houtt.; *A. vitifolia* var. *japonica* (Thunb.) Finet & Gagnep.; *A. hybrida* var. *japonica* (Thunb.) Ohwi; *A. nipponica* Merr.; *A. sieboldii* Honda; *Eriocapitella japonica* (Thunb.) Nakai——SHŪMEI-GIKU, KIBUNEGIKU. Rather tall, appressed-pubescent perennial herb 50–80 cm. high, with erect dichotomously forked stems; radical leaves petiolate, ternate, the leaflets petiolulate, ovate, 5–12 cm. long, 3–7 cm. wide, often 3-fid, irregularly toothed, acute, rounded to cordate, the cauline of few pairs, sessile or short-petioled, ternate or 3-fid, the upper ones gradually reduced; flowers long-pedunculate, 5–7 cm. across; sepals about 30, slightly unequal, spreading, some of the outer ones green and small, the inner ones petaloid, oblanceolate, pale rose-purple, silky-pubescent outside; carpels very numerous, forming a rather large head.——Sept.–Oct. Cultivated and often naturalized in our area.——China.

4. RANUNCULUS L.   KIMPŌGE ZOKU

Annual or perennial sometimes aquatic herbs; leaves entire, palmate or ternate, sometimes finely dissected, the cauline alternate; flowers yellow, white or sometimes orange-red, solitary or in terminal panicles; sepals 3–5 or more, caducous or deciduous, greenish; petals larger than the sepals, normally 5, sometimes few or many, flat, petaloid, with a nectariferous spot often covered by a small scale near base inside; stamens usually many, sometimes rather few; ovules solitary, ascending; achenes many, glabrous or puberulous, forming a globose to ellipsoidal head, on a short or slightly elongate receptacle, smooth or striate, sometimes spiny, terminated by a short beaklike style.——About 400 species, in the temperate and cold regions of both hemispheres and in high mountains of the Tropics.

1A. Achenes not transversely nerved; terrestrial herbs with yellow flowers.
  2A. Achenes longitudinally striate; plants stoloniferous, growing near seashores; leaves elliptic to oblong, 3–5 toothed on upper margin; flowers small, yellow, the petals 2.5–3 mm. long. .... 1. *R. kawakamii*
  2B. Achenes not striate.
    3A. Achenes not spiny, smooth, glabrous or puberulent.
      4A. Achenes plump or lenticular, without a flat face on each side, 1.2–2.5 mm. long; radical leaves simple to 3-foliolate.
        5A. Radical leaves linear or linear-lanceolate, simple, entire, 1–1.5 mm. wide; flowers small, 5–7 mm. across; plant long-stoloniferous. .... 2. *R. reptans*
        5B. Radical leaves cordate-orbicular or depressed-orbicular, usually palmately lobed or toothed.
          6A. Sepals villous with dark brown woolly hairs; plant of alpine regions. .... 3. *R. shinanoalpinus*
          6B. Sepals glabrous or with pale hairs.
            7A. Annual or biennial herbs; achenes in an oblong, cylindric head on an elongate receptacle. .... 4. *R. sceleratus*
            7B. Perennial herbs; achenes in a globose to ellipsoidal head, on a short receptacle.
              8A. Flowers small, 4–8 mm. across; low alpine plants with small, palmately 5- to 9-parted radical leaves.
                9A. Stems 3–7 cm. long; leaves 5- to 7-cleft. .... 5. *R. pygmaeus*
                9B. Stems 8–20 cm. long; leaves 5- to 9-parted.
                  10A. Anthers narrowly oblong; plant deep green, slightly pubescent; petals narrowly oblong, scarcely longer than the sepals. .... 6. *R. kitadakeanus*
                  10B. Anthers ellipsoidal; plants vivid green, pubescent; petals broadly obovate, longer than the sepals.
                                           7. *R. yatsugatakensis*
              8B. Flowers rather large, 12–25 mm. across.
                11A. Radical leaves at least partially ternately divided; stems 5–25 cm. long; plant of grassy places along rivers, with partially fusiform roots. .... 8. *R. ternatus*
                11B. Radical leaves palmately parted to cleft; plants of mountains and hills with slender or slightly thickened roots.
                  12A. Achenes lenticular, glabrous.
                    13A. Stems 3–8 cm. long, solid. .... 9. *R. yakushimensis*
                    13B. Stems 15–80 cm. long, hollow.
                      14A. Plants not stoloniferous; radical leaves usually rather many.
                        15A. Hairs spreading. .... 10. *R. japonicus*
                        15B. Hairs appressed. .... 11. *R. acris*
                    14B. Plants stoloniferous; radical leaves rather few. .... 12. *R. grandis*
                12B. Achenes very plump, puberulent. .... 13. *R. franchetii*
  4B. Achenes rather flat, 3–4 mm. long, often with an indistinct ridge near margin on face; radical leaves ternate.
    16A. Plants stoloniferous; flowers 15–22 mm. across; petals obovate-orbicular; sepals not reflexed in anthesis. ...... 14. *R. repens*
    16B. Plants without stolons; flowers 6–12 mm. across; petals narrowly obovate; sepals reflexed in anthesis.
      17A. Receptacle elongate, 6–10 mm. long in fruit; heads ellipsoidal to oblong; leaf-segments narrow. ...... 15. *R. chinensis*
      17B. Receptacle scarcely elongate, 1–5 mm. long; heads globose.
        18A. Stems decumbent; flowers solitary, cauline, the pedicels 1–5 cm. long. .... 16. *R. sieboldii*
        18B. Stems erect or decumbent only at base; flowers in a terminal corymbose panicle.
          19A. Style rather slender, distinctly recurved; achenes without or with an indistinct ridge along upper margin; leaf-segments broadly ovate.
            20A. Achenes somewhat ridged near margin; style strongly recurved; stems erect or rarely decumbent.
                                             17. *R. quelpaertensis*
            20B. Achenes without a ridge near margin; style slightly recurved; stems prostrate. .... 18. *R. hakkodensis*
         19B. Style rather thick and short, scarcely recurved; achenes with an indistinct ridge all around the margin; leaf-segments ovate to cuneate-lanceolate, coarsely hirsute.
          21A. Leaves once ternate, the ultimate leaf segments ovate or oblong. .... 19. *R. cantoniensis*
          21B. Leaves twice ternate; ultimate leaf segments cuneate-oblanceolate. .... 20. *R. tachiroei*

1. Ranunculus kawakamii Makino. *Halerpestes kawakamii* (Makino) Tamura——HIME-KIMPŌGE, TSURU-HIKI-NO-KASA. Small glabrous leafy stoloniferous herb; leaves ovate-oblong to ovate-orbicular, 8–20 mm. long, 7–12 mm. wide, rounded at apex, rounded to subcuneate at base, 3(–5) short-toothed on upper margin; petioles of radical leaves 3–10 cm. long, rather shorter in those of the stolons; stems erect, 5–12 cm. long, 1- to 3-flowered, slender, the bracts 1–2 mm. long, usually lanceolate, entire, the pedicels erect, 2–5 cm. long; flowers yellow, about 6 mm. across; sepals broadly ovate, spreading, glabrous, concave, about 2.5 mm. long; petals narrowly ovate or narrowly oblong, 1.5–2 times as long as the sepals; anthers reniform-orbicular, very short; heads of carpels globose, about 4 mm. across, the receptacles sparsely short-hairy; achenes 1–1.2 mm. long, glabrous, longitudinally nerved; style short, slightly recurved.——Aug. Wet sandy places near seashore; Honshu (Kantō Distr. and northw.); rare.

2. Ranunculus reptans L. *R. repens* var. *flagellifolius* (Nakai) Ohwi; *R. flagellifolius* Nakai——ITO-KIMPŌGE. Delicate nearly glabrous perennial stoloniferous herb; radical leaves linear to filiform, 3–10 cm. long, about 1 mm. wide, sometimes slightly broader above, often with a callose point at apex; leaves of stolons 1–5 cm. long; flowers solitary, 5–6 mm. across, the pedicels 2–5 cm. long, radical or from the nodes of stolons, slightly pubescent above while young; sepals ovate or broadly so, glabrous, spreading, concave; petals 5, sometimes 4 or 6, obovate, slightly longer than the sepals; anthers very short, ellipsoidal.——Aug.–Sept. Wet places in streams and around ponds; Hokkaido, Honshu (Oze in Shimotsuke Prov. and Nozori Lake in Kotsuke, Nikko and elsewhere); rare.——s. Kuriles, Sakhalin and throughout the cooler regions of the N. Hemisphere.

3. Ranunculus shinanoalpinus Ohwi. *R. altaicus* var. *minor* Nakai, pro parte——TAKANE-KIMPŌGE. Rhizomes short; stems 8–12 cm. long, erect, unbranched, 1- to 2-leaved, usually 1-flowered; radical leaves obovate-orbicular to broadly cuneate, sometimes depressed-orbicular, relatively thick, 5- to 7-toothed on upper margin, rounded at apex, rounded to broadly cuneate, rarely truncate at base, nearly glabrous on both sides, slightly long-pubescent on margin, the petioles 1–3 cm. long, the cauline leaves sessile, with 2-stipulelike membranes at base, 3-parted, the segments lanceolate, entire, obtuse; pedicels 2–5 cm. long, brown-pubescent; sepals 5, 5–7 mm. long, spreading, concave, oblong, brown woolly-villous; petals 5, yellow, 6–8 mm. long, broadly obovate; anthers narrowly oblong; heads of carpels globose, the receptacle glabrous; achenes ovate, glabrous.——July–Aug. Alpine regions; Honshu (Mount Shirouma in Shinano); rare.

4. Ranunculus sceleratus L. TAGARASHI. Nearly glabrous annual or biennial herb; stems 10–50 cm. long, often branched, erect; leaves lustrous, the lower cauline petiolate, reniform-orbicular, 1.2–4 cm. long, 1.5–5 cm. wide, 3-parted to

cleft, cordate, the lateral segments 2-cleft, the median segments cuneate, obtuse, both slightly lobed and with coarse obtuse teeth, the upper ones subsessile, dilated and with a membrane on each side at base, 3-sected, the segments lanceolate, obtuse; flowers rather numerous, 6–8 mm. across; sepals somewhat reflexed, slightly pubescent, elliptic; petals as long as the sepals, obovate; anthers oblong; heads of carpels ellipsoidal to oblong-cylindric, the receptacle 7–12 mm. long, glabrous or with scattered short white hairs; achenes many, broadly obovate, glabrous, 1–1.2 mm. long, with a very short style.——Apr.–June. Wet places in lowlands, especially abundant in paddy fields; Hokkaido, Honshu, Shikoku, Kyushu; very common.——Eurasia, N. America, and n. Africa.

5. Ranunculus pygmaeus Wahlenb. KUMOMA-KIMPŌGE. Small, glabrous or scattered pubescent perennial herb with short rhizomes; stems 3–7 cm. long, 1- to 2-flowered; radical leaves petiolate, reniform-orbicular, 5–7 mm. long, 7–12 mm. wide, truncate to broadly cordate, 5- to 7-cleft, the segments ovate, obtuse, the cauline leaves 1–2, subsessile, 3-parted, the segments narrowly ovate, obtuse; flowers about 5 mm. across, pedicellate, elongate after anthesis; sepals elliptic, scattered-pubescent, concave; petals nearly as long as the sepals, yellow, obovate; heads of carpels ellipsoid-globose, about 3 mm. long, the receptacle glabrous; achenes broadly obovate, glabrous, about 1 mm. long.——July. Alpine regions; Honshu (Mount Shirouma in Shinano); rare.——Northern regions and high mountains of Eurasia and N. America.

6. Ranunculus kitadakeanus Ohwi. KITADAKE-KIMPŌGE. Glabrescent to sparingly pubescent perennial; rhizomes short, sparsely fibrous at neck; scapes 2- to 4-leaved, 1- to 2-flowered; radical and lower cauline leaves petiolate, 12–20 mm. wide, 3- to 5-parted, the median segments obovate to cuneate, often 3-fid, the middle and upper cauline leaves subsessile or sessile, 3- to 5-sected into linear, subacute segments, the pedicels much elongate after anthesis; flowers about 7 mm. across, yellow; sepals spreading, slightly pubescent, concave, elliptic; petals 5, only slightly longer than the sepals, narrowly spathulate-oblong; heads of carpels globose, 5–6 mm. across, the receptacle short-pilose, about 4 mm. long; achenes obovate, glabrous, about 2 mm. long, the styles slender, short, suberect.——July–Aug. Alpine regions; Honshu (Mount Kitadake in Kai Prov.); rare.

7. Ranunculus yatsugatakensis Honda & Kumazawa. YATSUGATAKE-KIMPŌGE. Loosely appressed-pubescent perennial with short rhizomes; stems 10–30 cm. long, erect, usually simple, 2- to 3-leaved; radical leaves 1–3, petiolate, orbicular-cordate, 8–15 mm. long and as wide, palmately 5- to 9-parted, the segments subobtuse, incised, the median ones obovate, the cauline leaves sessile or short-petiolate, 3-sected, the lateral segments 3-parted, linear-lanceolate, subobtuse, 0.7–1.2 mm. wide; flowers usually 1, long-pedicellate, 8–10 mm. across; sepals broadly elliptic, white-pubescent, slightly concave;

petals 5, broadly obovate, 1.5 times as long as the sepal; anthers small, elliptic; achenes glabrous.——Aug. Alpine regions; Honshu (Mount Yatsugatake in Shinano Prov.); rare.

8. Ranunculus ternatus Thunb. *R. zuccarinii* Miq.; *R. extorris* Hance——HIKI-NO-KASA.   Scattered pubescent to glabrescent perennial herb, often with the roots thickened, narrowly fusiform; stems 10–30 cm. long, few-flowered, few-leaved; radical leaves petiolate, simple, 3-parted or 3-sected, the segments reniform-orbicular to cuneate-obovate, often petiolulate, 1–2 cm. long, 5–15 mm. wide, sometimes again 3-lobed, with few obtuse teeth or incised, the cauline leaves 1–4, sessile, 3-sected, the segments nearly linear, obtuse; flowers long-pedicellate, yellow, 12–17 mm. across; sepals 5, spreading, broadly elliptic, sparsely pubescent, concave; petals 5, obovate-orbicular, 2–2.5 times as long as the sepals; anthers oblong; heads of carpels ellipsoidal to globose, 5–7 mm. long, the receptacles glabrous; achenes obovate-orbicular, about 1.2 mm. long, glabrous, the style slender, rather short, scarcely recurved.——Apr. Grassy places along rivers in lowlands; Honshu, Shikoku, Kyushu; rather common.——Ryukyus, Formosa, and China.

9. Ranunculus yakushimensis (Makino) Masam. *R. acris* var. *japonicus* subvar. *yakushimensis* Makino——HIME-UMA-NO-ASHIGATA.   Glabrous or pubescent perennial herb; rhizomes short; stems ascending, rooting from the lower nodes, 3–8 cm. long, 1- to 2-flowered, few-leaved; radical leaves cordate-orbicular or depressed-cordate, 4–8 mm. long, 5–10 mm. wide, 3-parted, the segments often 3-lobed, subacute, the petioles 1–4 cm. long, the cauline leaves petiolate or sessile, lanceolate, simple or 3-parted into lanceolate segments; flowers long-pedicellate, 12–15 mm. across; sepals oblong or elliptic, slightly concave, glabrous to slightly pubescent; petals obovate, about twice as long as the sepals; anthers oblong.——July–Aug. Kyushu (Yakushima); rare.

10. Ranunculus japonicus Thunb. *R. acris* var. *japonicus* (Thunb.) Maxim.; *R. acris* sensu auct. Japon., saltem pro parte, non L.; *R. propinquus* var. *hirsutus* A. Gray——UMA-NO-ASHIGATA, KIMPŌGE.   Spreading-pubescent perennial herb; rhizomes short, slightly tufted; radical leaves long-petiolate, orbicular-cordate, obtusely 5-angled, 2.5–7 cm. long, 3–10 cm. wide, 3-parted, coarsely toothed and then shallowly incised, the median segments cuneate-obovate, the lateral ones 2-fid, the cauline leaves few, similar to the radical but short-petiolate or sessile, the segments slightly narrower, the uppermost ones 3-parted to simple and bractlike; flowers 12–20 mm. across, rather long-pedicellate; sepals 5, elliptic, pubescent, spreading, concave; petals 5, 2–2.5 times as long as the sepals, obovate-orbicular; anthers oblong; heads of carpels globose, the receptacle globose, glabrous; achenes obovate-orbicular, rather flat, glabrous, 2–2.5 mm. long, the style very short, slightly recurved.——Apr.–June. Grassy places in lowlands and mountains; Hokkaido (sw. distr.), Honshu, Shikoku, Kyushu; common and rather variable.——Korea, China, Ryukyus, and Formosa.

11. Ranunculus acris L. var. **nipponicus** Hara. *? R. novus* Lév. & Van't.——MIYAMA-KIMPŌGE.   Resembles the preceding, the hairs appressed to ascending, sparse; rhizomes rather tufted; radical leaves much divided, the segments narrower, acute; scapes 1–2; flowers few, 20–25 mm. across; style very shortly recurved.——June–Aug. Alpine regions; Hokkaido, Honshu (n. and centr. distr.); rather common.——? n. Korea.

Var. **subcorymbosus** (Komar.) Tatew. *R. subcorymbosus*

Komar.——KENASHI-MIYAMA-KIMPŌGE.   Stems sparsely pilose. ——Alpine areas; Hokkaido, Honshu (n. distr.).——Sakhalin and Kamchatka.

12. Ranunculus grandis Honda. Ō-UMA-NO-ASHIGATA. Closely resembles *R. japonicus;* stems 30–100 cm. long, with spreading or somewhat reflexed hairs; rhizomes short; plants short-stoloniferous; leaf-segments slightly narrower, subacute; flowers long-pedicellate, about 20 mm. across, yellow; sepals ovate-orbicular, concave; petals broadly obovate, about twice as long as the sepals; flower heads globose.——Mountains; Honshu (Rikuchu Prov.).

Var. **mirissimus** (Hisauchi) Hara. *R. mirissimus* Hisauchi ——GUNNAI-KIMPŌGE.   Stems 20–30 cm. long, the stolons elongate, the hairs spreading, denser; leaves rather deeply cordate.——May–July. Mountains; Honshu (Kai Prov.).

Var. **austrokurilensis** (Tatew.) Hara. *R. acris* var. *austrokurilensis* Tatew.; *R. transochotensis* Hara; *R. grandis* var. *transochotensis* (Hara) Hara; *R. grandis* var. *ozensis* Hara ——SHIKOTAN-KIMPŌGE.   Plant rather slender, 25–50 cm. high, the hairs appressed, the pedicels elongate, the style short, recurved.——June–Aug. Lowlands to the mountains; Hokkaido, Honshu (Ozegahara in Shimotsuke).——A variable species, occurring also in the Kuriles, Kamchatka, n. Korea, and Manchuria.

13. Ranunculus franchetii H. Boiss. EZO-KIMPŌGE. Nearly glabrous perennial, 15–20 cm. high, with numerous horizontally spreading roots; radical leaves few, long-petiolate, orbicular-cordate, 2–2.5 cm. long, 2.5–4 cm. wide, palmately parted, the median segments cuneate, the lateral ones again 2- to 3-lobed, these deeply incised or sometimes coarsely toothed, the ultimate lobes subobtuse; stems 1- to 3-flowered, the cauline leaves sessile, 3-sected, the lateral segments 2-parted, all the segments broadly linear to lanceolate, entire or few-toothed, subobtuse; flowers yellow, about 20 mm. across; sepals 5, narrowly ovate, concave, glabrous or scattered-pubescent; petals obovate, about twice as long as the sepals; anthers narrowly oblong; achenes plump, puberulent.——May. Hokkaido.——Manchuria, n. Korea, Amur, and Sakhalin.

14. Ranunculus repens L. HAI-KIMPŌGE.   Stoloniferous perennial herb, scattered-pubescent; rhizomes short; radical leaves long-petiolate, ternate, the leaflets petiolulate, broadly ovate, 4–8 cm. long and as wide, 3- to 5-parted, the segments acute, again 2- to 3-fid, incised and coarsely toothed; stems 20–50 cm. long; cauline leaves short-petiolate or sessile, with a stipulelike membrane at base of the petiole on each side; flowers yellow, 15–22 mm. across; sepals ovate, spreading, slightly concave, nearly glabrous to appressed-pubescent; petals 5, about twice as long as the sepals, broadly obovate; heads of carpels globose, the receptacle short, scattered short-pilose; achenes obovate, about 3 mm. long, with an indistinct ridge along the margin, the style rather thick, short, slightly recurved.——May–June. Wet places in lowlands, especially near seashores; Hokkaido, Honshu (centr. and n. distr.).——Cold regions of Eurasia and N. America.

15. Ranunculus chinensis Bunge. KO-KITSUNE-NO-BO-TAN.   Perennial herb with spreading coarse hairs; rhizomes short; stems rather stout, erect, branched, 40–80 cm. long; radical leaves few, long-petiolate, 2- to 3-parted, the segments cuneate, subacute, 2–4 cm. long, 5–10 mm. wide, again 2- to 3-fid, incised and coarsely toothed; flowers 6–8 mm. across, several to rather many, the pedicels appressed-pubescent; sepals narrowly ovate, reflexed; petals 5, nearly as long as the sepals, narrowly obovate; heads of carpels oblong or narrowly so, 10–15 mm. long, 8–10 mm. across, the receptacle elongate, 6–10 mm. long, short-pilose; achenes elliptic, 3–3.5 mm. long, with an indistinct ridge along both margins, the style very short, scarcely recurved.——May–June. Wet places; Honshu. ——Korea, China, and e. Siberia.

16. Ranunculus sieboldii Miq. *R. asiaticus* sensu Thunb., pro parte; *R. ternatus* sensu DC., non Thunb.—— SHIMA-KITSUNE-NO-BOTAN. Perennial with coarse spreading hairs; rhizomes short; stems ascending or procumbent, 30–50 cm. long, rather stout; radical and lower cauline leaves petiolate, ternate, the leaflets ovate-orbicular to elliptic, petiolulate, 3–5 cm. long, 2- to 3-fid, toothed and incised, the segments obovate; pedicels 1-flowered, 1–5 cm. long, ebracteate; flowers yellow, 8–12 mm. across; sepals reflexed, ovate, slightly concave, thinly hirsute; petals 5, obovate, slightly larger than the sepals; heads of carpels globose, 10–12 mm. across, the receptacle short, loosely pilose; achenes broadly obovate, about 4 mm. long, flat, with an indistinct ridge along the margin on both sides, the style prominent, slightly recurved.——Mar.–June. Honshu (Chūgoku Distr.), Shikoku, Kyushu.——Ryukyus and Formosa.

17. Ranunculus quelpaertensis (Lév.) Nakai var. **glaber** (H. Boiss.) Hara. *R. hakkodensis* var. *glaber* (H. Boiss.) Ohwi & Okuyama; *R. ternatus* var. *glaber* H. Boiss.; *R. vernyi* var. *glaber* Nakai; *R. glaber* Makino——KITSUNE-NO-BOTAN. Nearly glabrate or sparsely hairy perennial herb; stems suberect to ascending, 15–80 cm. long, rather stout, branched; radical and lower cauline leaves petiolate, ternate, the leaflets petiolulate, broadly ovate to ovate-orbicular, 2–6 cm. long, 1.5–4 cm. wide, acute to subobtuse, 2- to 3-cleft, incised and acutely toothed, the upper ones 3-parted, incised and toothed, sessile; flowers 8–12 mm. across; sepals reflexed, ovate, concave; petals 5, obovate-oblong, slightly longer than the sepals; heads of carpels globose, 8–10 mm. across, the receptacle short, short-pilose; achenes obovate, flat, glabrous, about 3.5 mm. long, with an indistinct ridge along the upper margin, the style prominent, distinctly recurved.——Apr.– July. Wet places in lowlands and mountains; Hokkaido, Honshu, Shikoku, Kyushu; common.——s. Kuriles, Korea, China, and Ryukyus.

Var. **quelpaertensis**. *R. hakkodensis* var. *quelpaertensis* (Lév.) Ohwi and Okuyama; *R. repens* var. *quelpaertensis* Lév.; *R. quelpaertensis* (Lév.) Nakai; *R. vernyi* var. *quelpaertensis* (Lév.) Nakai; *R. ternatus* var. *quelpaertensis* (Lév.) Ohwi——YAMA-KITSUNE-NO-BOTAN. Plants short, densely hirsute, growing in mountains.

Var. **yaegatakensis** (Masam.) Ohwi & Okuyama. *R. yaegatakensis* Masam.; *R. ternatus* var. *yaegatakensis* (Masam.) Ohwi——HIME-KITSUNE-NO-BOTAN. A dwarf variant.—— Kyushu (Yakushima).

18. Ranunculus hakkodensis Nakai. TSURU-KITSUNE-NO-BOTAN. Sparsely hairy perennial; stems prostrate, 40–60 cm. long, glabrescent; radical leaves petiolate, 4–5.5 cm. wide, 3-parted, the segments obovate, irregularly incised and obtusely toothed, the petiolules 3–8 mm. long; petals 5, elliptic, 3.5–4 mm. long; achenes rather few, broadly obovate, about 3 mm. long.——July–Aug. Wet places in mountains; Honshu (n. distr.).

19. Ranunculus cantoniensis DC. *Hecatonia pilosa* Lour.; *R. polycephalus* Makino; *R. silerifolius* Lév.——KE-KITSUNE-NO-BOTAN. Perennial herb; stems 30–80 cm. long, erect, branched, with coarse spreading long hairs at base, appressed-strigose above; radical and lower cauline leaves once

ternate, petiolate, the leaflets ovate to broadly so, 2.5–5 cm. long, 2–3.5 cm. wide, petiolulate, subacute to acute, 2- to 3-fid, the lobes oblong, incised and acutely coarse-toothed; flowers about 10 mm. across; sepals reflexed, ovate, slightly concave, sparingly pilose outside; petals elliptic or oblong, slightly longer than the sepals; heads of carpels globose, 8–10 mm. across, the receptacle ellipsoidal, short white-hairy; achenes obovate-orbicular, about 3 mm. long, glabrous, with an indistinct ridge along both margins, the style rather short, scarcely recurved.——May–July. Wet places in lowlands; Honshu, Shikoku, Kyushu.——Korea, Formosa, and s. China.

20. Ranunculus tachiroei Fr. & Sav. OTOKOZERI. Perennial herb; stems erect, rather stout, branched above, 50–100 cm. long, with spreading coarse long hairs toward the base, appressed-hairy above; radical and lower cauline leaves rather long-petiolate, twice ternate, the leaflets petiolulate or subsessile, 2- to 3-parted, the segments cuneate to oblanceolate or broadly linear, 3–6 cm. long, 7–12 mm. wide, acute, entire and subcuneate at base, coarsely acute-toothed or incised above; flowers rather numerous, long-pedicellate, 10–12 mm. across; sepals reflexed, ovate, slightly pubescent outside; petals oblong, yellow, slightly longer than the sepals; heads of carpels globose, about 8 mm. across, the receptacle short, ellipsoidal, white-pilose; achenes about 3 mm. long, obovate-orbicular, glabrous, with an indistinct ridge along both margins; style rather prominent, scarcely recurved.——July–Aug. Wet places; Honshu.

21. Ranunculus muricatus L. TOGEMI-NO-KITSUNE-NO-BOTAN. Biennial, nearly glabrous; stems branched from the base, ascending, 15–40 cm. long; radical and lower cauline leaves long-petiolate, orbicular-cordate, 4–10 cm. long, 3(–5)-fid, coarsely toothed and often incised, the median lobes broadly cuneate; flowers 8–12 mm. across; sepals slightly reflexed, shorter than the petals; petals 5–8 mm. long, narrowly obovate; heads of carpels globose, 12–15 mm. across; achenes elliptic, obscurely ridged near the margin and spiny, about 5 mm. long, the style slightly recurved.——Mar.–Apr. Naturalized in our area.——Introduced from Europe.

22. Ranunculus arvensis L. ITO-KITSUNE-NO-BOTAN. Biennial, sparingly pubescent herb; stems erect, 20–40 cm. long; lower leaves petiolate, the upper ones sessile or short-petiolate, both 3-foliolate and again trisected, the segments 2- to 3-parted, 1.5–3 cm. long, the ultimate segments linear, 1–3 mm. wide, subacute; flowers 10–12 mm. across, relatively long-pedicellate; sepals narrowly ovate, shorter than the petals; heads of carpels globose, 10–12 mm. wide, achenes broadly obovate, obsoletely ridged along the margin, mammillate-spinose on both sides, about 5 mm. long; style prominent, scarcely recurved.——Apr. Rarely naturalized in our area. ——Introduced from Europe.

23. Ranunculus nipponicus (Makino) Nakai. *R. aquatilis* var. *nipponicus* Makino, nom. seminud.——ICHŌBA-BAIKA-MO. Aquatic herb; stems elongate; submerged leaves several times dissected into filiform segments, short-petiolate, 3–4 cm. long, the sheath usually dilated above, thinly membranous, appressed-puberulent; floating leaves (when present) palmately cleft or lobed; flowers about 15 mm. across, white with a yellow center; heads of carpels globose, the receptacle hairy; achenes glabrous or with sparse stiff hairs.——July–Aug. Mountains; Honshu (centr. distr.).

Var. **major** Hara. *R. aquatilis* var. *flaccidus* forma *drouetii* sensu auct., Japon., non Hiern; *R. pantothrix* var. *submersus* Nakai, excl. syn.——BAIKA-MO, UMEBACHI-MO. Floating leaves absent; flowers 15–20 mm. across, the peduncles 3–10 cm. long; common.——Ponds and streams; Hokkaido, Honshu.

Var. **japonicus** (Nakai) Hara. *R. aquatilis* var. *japonicus* Nakai——MISHIMA-BAIKA-MO. Resembles the typical phase but with palmately parted floating leaves.——Ponds and streams; Honshu (Mishima in Izu Prov.).——Korea.

24. Ranunculus kazusensis Makino. HIME-BAIKA-MO. Slender, glabrous, aquatic, perennial herb; stems slender, glabrous, elongate, loosely leafy; leaves short-petiolate, the upper ones sessile, 3–4.5 cm. long, finely dissected, the stipules thinly membranous, appressed-puberulent on the outside; pedicels slender, glabrous, 1–3 cm. long; flowers white with a yellow center, about 6 mm. across; sepals elliptic, glabrous; petals obovate, 1.2–4 mm. long, 1.5–2 times as long as the sepals; receptacle short-pilose; heads of carpels globose, 3–3.5 mm. across; achenes about 1.2 mm. long, glabrous or with few appressed hairs on the back.——June–Aug. Honshu (Kazusa Prov.), Kyushu (Higo Prov.).——Korea.

25. Ranunculus yesoensis Nakai. CHITOSE-BAIKA-MO. Leaves petiolate or sessile in the upper ones, several times dissected into filiform segments, glabrous, the stipules oblong, thinly membranous, 5–6 mm. long, glabrous; pedicels 3–6.5 cm. long, slender; flowers 6–7 mm. across, white with a yellow center; sepals 5, greenish, about 3 mm. long, oblong; petals obovate, about 4 mm. long, spreading; stamens about 15; achenes obovate, about 1 mm. long, glabrous, the receptacles glabrous.——Hokkaido.

5. CALLIANTHEMUM C. A. Mey.   UMEZAKI-SABA-NO-O ZOKU

Low, glaucescent, glabrous perennial herbs with short, stout rhizomes; radical leaves long-petiolate, ternately or pinnately compound, the cauline leaves few, reduced in size; stems usually simple; flowers 1 to 3, white or creamy white with a dark orange-red center; sepals 5, deciduous, greenish, smaller than the petals; petals 5–15, with a depressed nectary near base inside; carpels many, sessile, free, each with a pendulous ovule; achenes ovate, capitate; style short, persistent.——Few species, in the alpine regions and far northern areas of the Old World.

1A. Terminal primary petiolules 1–2 times as long as the lateral ones; cauline leaves dilated at base; flowers opening before the leaves; petals 8–10, linear- to narrowly oblong; sepals ovate to broadly elliptic .... 1. *C. miyabeanum*
1B. Terminal primary petiolules 1.5–4 times as long as the lateral ones; cauline leaves not dilated at base; flowers opening after the leaves; petals 6 or 7, oblong, elliptic, or obovate. .... 2. *C. insigne* var. *hondoense*

1. Callianthemum miyabeanum Tatew. HIDAKA-SŌ. Glaucous perennial herb, 10–25 cm. high; radical leaves petiolate, ternate, the leaflets ovate-deltoid, 2–4 cm. long, 3-sected or 3-foliolate, deeply parted; stems few-leaved; flowers 1 or 2, long-pedunculate, creamy white, about 2 cm. across; sepals ovate to broadly elliptic, 4–5 mm. long; petals linear- to narrowly oblong, obtuse or subrounded, retuse, creamy white, dark orange-red near base; styles prominently recurved.——May–June. Alpine regions; Hokkaido (Mount Apoi in Hidaka Prov.); rare.

2. Callianthemum insigne (Nakai) Nakai var. **hondoense** (Nakai & Hara) Ohwi. *C. hondoense* Nakai & Hara ——KITADAKE-SŌ. Closely resembling the preceding species; petiolules of the primary terminal division of the radical leaves much longer than the lateral ones; stems usually 1-flowered; flowers 2–2.5 cm. across; sepals obovate to oblong, 7–8 mm. long; petals obovate to elliptic, often retuse.——June. Alpine regions; Honshu (Mount Kitadake in Kai); rare.

6. ADONIS L.   FUKUJU-SŌ ZOKU

Perennial or sometimes annual herbs with branched stems; leaves alternate, pinnately dissected, the lower often reduced to a sheath or scale; flowers terminal, rather large, yellow, sometimes red; sepals several, green, deciduous; petals 5–40, petaloid, deciduous, without a nectariferous gland; stamens many; carpels many, free, each with a pendulous ovule; fruiting heads globose to cylindric; achenes with a short style.——Few species, in the N. Hemisphere.

1. Adonis amurensis Regel & Radde. FUKUJU-SŌ. Rhizomes short and thick, with numerous stout fibrous roots; stems 10–30 cm. long, often branched, usually papillose-puberulent above, sometimes pubescent; lower leaves membranous, sheathlike, the normal cauline leaves alternate, deltoid-ovate, 3–10 cm. long, petiolate, pinnately dissected, the ultimate segments lanceolate, glabrous or pubescent underneath; stipules small, lobed, green; flowers yellow, 3–4 cm. across; sepals several, ovate; petals longer than the sepals, 20–30, spreading, broadly oblanceolate; anthers 1–2 mm. long; achenes puberulent, 3–3.5 mm. long, on receptacles about 1 cm. long, the heads ovoid-globose; style short and recurved.——Mar.–May. Hokkaido, Honshu, Kyushu.——Korea, Manchuria, and e. Siberia.

7. THALICTRUM L.   KARAMATSU-SŌ ZOKU

Perennial herbs; leaves ternately compound, radical and cauline, usually alternate, often stipulate and stipellate; inflorescence paniculate, corymbose or racemose; flowers small, bisexual or unisexual (plants dioecious), rarely polygamous, greenish, white, sometimes rose-colored; sepals 3–5, greenish and inconspicuous or sometimes white or rose-colored and more or less petaloid, caducous or deciduous; petals absent; stamens many, prominent, the filaments capillary or clavate, the anthers small; carpels few to many, the ovules solitary, pendulous, the receptacle minute; achenes often stipitate, commonly laterally flattened, sometimes narrowly winged, usually nerved; style beaklike, elongate, sometimes circinnate, or short, with a deltoid to linear stigma on inner side.——About 100 species, in the cooler regions of the N. Hemisphere, few in S. America and S. Africa.

1A. Inflorescence paniculate or racemose, more or less pyramidal; flowers usually green or yellow-green, rarely purplish; anther-connective mucronate; filaments filiform, narrower than the anthers.
  2A. Inflorescence racemose, loosely about 10-flowered; stems 10–15 cm. long, naked or with a single cauline leaf.
                                                                1. *T. alpinum* var. *stipitatum*
  2B. Inflorescence paniculate, many-flowered; stems 20–150 cm. long, more or less distinctly leafy.
    3A. Achenes 2–6, sessile; sepals small, caducous; flowers greenish yellow to pale yellow.
      4A. Stems usually prominently ridged, simple, erect, scarcely branched; leaves once to twice pinnately ternate, rather longer than broad. .... 2. *T. simplex* var. *brevipes*
      4B. Stems striate or nearly terete, often flexuous; leaves 2- to 3-ternate, nearly as broad as long.
        5A. Stems rather flexuous; nerves and veinlets of leaflets prominently impressed above, raised beneath; stigma lanceolate-deltoid, acute; stipels absent. .... 3. *T. foetidum* var. *glabrescens*
        5B. Stems nearly erect; nerves and veinlets of leaflets not prominently raised beneath, nearly flat above; stigma ovate-deltoid; stipels often present. .... 4. *T. minus*
    3B. Achenes 8–20, short-stipitate; sepals rather prominent, deciduous after anthesis; flowers purplish. .... 5. *T. rochebrunianum*
1B. Inflorescence corymbose, with a flat or only slightly convex top; flowers white or rose-colored; anther-connectives not mucronate; filaments broadened above and clavate, as broad as or broader than the anther.
  6A. Achenes 3- to 4-winged, distinctly stipitate. .... 6. *T. aquilegifolium*
  6B. Achenes laterally flattened or scarcely so, not winged.
    7A. Achenes sessile or very short-stipitate, not flattened; stems 30–100 cm. long, many-leaved.
      8A. Achenes sessile; stems 30–60 cm. long, rather firm, slightly flexuous below. .... 7. *T. actaefolium*
      8B. Achenes very short-stipitate; stems 50–100 cm. long.
        9A. Stipules prominently lacerate on margin; stipels absent; achenes with 3 or 4 longitudinal nerves on each side; styles scarcely hooked at apex. .... 8. *T. baicalense*
        9B. Stipules obsoletely toothed or subentire; stipels present at base of primary division of leaves; achenes with 3 or 4 raised ribs on each side; style slender, prominently hooked at apex. .... 9. *T. sachalinense*
    7B. Achenes distinctly stipitate, flattened laterally, with 1–4 slender longitudinal nerves on each side; stems 10–70 cm. long, slender or delicate, 1–2(–3)-leaved.
      10A. Style including the stigma thick, beaklike, 0.3–1.5 mm. long, straight or slightly recurved, not flexuous, much shorter than the body of the achene.
        11A. Beak of achene very short, about 0.5 mm. long; filaments usually much-dilated above.
          12A. Achenes about 3 mm. long; radical leaves usually much shorter than the stem; leaflets ovate to oblong, rarely rhombic, shallowly cordate to rounded at base; stipes rather short. .... 10. *T. filamentosum*
          12B. Achenes about 2 mm. long; radical leaves nearly as long as the stem; leaflets rhombic-elliptic, usually cuneate at base; stipes elongate. .... 11. *T. microspermum*
        11B. Beak of achene rather elongate, about 1 mm. long; filaments usually gradually and slightly dilated above.
          13A. Leaflets linear-lanceolate to linear-oblong, usually entire or sometimes few-toothed. .... 12. *T. integrilobum*
          13B. Leaflets orbicular to rhombic-elliptic, few-toothed on upper margin.
            14A. Leaflets rhombic-elliptic to orbicular-rhombic, broadly cuneate to truncate at base; flowers few to many; achenes oblong-lanceolate, with a straight beak 1–1.2 mm. long. .... 13. *T. watanabei*
            14B. Leaflets orbicular to orbicular-elliptic, shallowly cordate to rounded at base; flowers few; achenes broadly oblanceolate, with a slightly recurved beak 0.7–1 mm. long. .... 14. *T. nakamurae*
      10B. Style including the stigma slender, flexuous, about 2 mm. long, as long as the body of the achene. .... 15. *T. toyamae*

1. Thalictrum alpinum L. var. **stipitatum** Yabe. *T. nipponoalpinum* Honda——HIME-KARAMATSU. Small glabrous perennial herb with short-branched, creeping rhizomes; radical leaves petiolate, 2–3 times ternate-pinnate, 3–5 cm. long, deltoid in outline, exstipellate, the leaflets 2- to 3-lobed, petiolulate or sessile, orbicular to cuneate-obovate, 2–3 mm. long and as wide, slightly recurved on margin, the lobes rounded to very obtuse, entire or 2- to 3-lobulate; stems 10–15 cm. long, 1-leaved or leafless, erect, nodding above; racemes occupying 1/2–2/3 of the entire stem, loosely about 10-flowered, the bracts small, the pedicels recurved, 1–4 cm. long; flowers small, pale greenish yellow; sepals 4, narrowly oblong, 2.5–3 mm. long, 3-nerved; anthers linear, about 2 mm. long, mucronate; achenes 3–5, subobovate, about 3 mm. long, short-stipitate; style very short, the stigma deltoid, puberulous.——July-Aug. Rather dry alpine slopes; Honshu (centr. distr.); very rare.——China (?) and Himalayas (?). The species is widely distributed throughout northern regions and mountainous areas of the N. Hemisphere.

2. Thalictrum simplex L. var. **brevipes** Hara. *T. simplex* auct. Japon., non L.; *T. affine* auct Japon., non Ledeb.; *T. simplex* var. *affine* auct. Japon., non Regel——NO-KARA-MATSU. Erect, glabrous perennial herb 60–100 cm. high; stems prominently acute-angled; lower leaves twice, the upper ones once to twice ternate-pinnate, short-petiolate or sessile, the stipules minutely toothed, the stipels sometimes developed on the lower portion, the leaflets cuneate to narrowly obovate or lanceolate, 2–4 cm. long, 4–5 mm. wide, acute to acuminate, sometimes 2- to 3-lobed; panicle large, narrowly ovoid, the pedicels short; flowers pale greenish yellow or pale yellow; sepals about 3 mm. long, about 3-nerved; anthers 1.5–2 mm. long, mucronate; achenes 2–6, sessile, scarcely compressed, ellipsoidal, about 3 mm. long, with 8–10 raised longitudinal ribs; style very short, the stigma ovate.——June–Aug. Grassy places in lowlands along rivers; Honshu, Kyushu.——Korea, Manchuria, China, and e. Asia. The species occurs across northern Asia to Europe.

3. Thalictrum foetidum L. var. **glabrescens** Takeda. *T. foetidum* sensu auct. Japon., non L.; *T. yesoense* Nakai—— CHABO-KARAMATSU. Minutely glandular-puberulent to nearly glabrous perennial; stems 15–40 cm. long, slightly flexuous toward base, branched above; lower leaves petiolate, 3–6 times ternate, 5–15 cm. long and as wide, the stipules undulately toothed or subentire, exstipellate, the leaflets broadly obovate-cuneate, ovate or depressed-ovate, 3–7(–10) mm. long, 3–8 mm. wide, green above, glaucous beneath, 2- to 3-lobed, often 2- to 3-toothed, the nerves and veinlets prominently impressed above, elevated beneath, the upper leaves smaller and sessile;

inflorescence loose, the pedicels 2–5 cm. long; anthers broadly linear, 2–2.5 mm. long, mucronate; achenes 2–4, sessile, about 3 mm. long, few-ribbed on each side, slightly compressed laterally, the style very short, the stigma lanceolate-deltoid. ——May–July. Rocky cliffs near the sea and in mountains; Hokkaido, Honshu (Rikuchu Prov.).——The species occurs across northern Asia to Europe.

4. **Thalictrum minus** L. var. **stipellatum** (C. A. Mey.) Tamura. *T. kemense* Fries; *T. kemense* var. *stipellatum* C. A. Mey. ex Maxim.; *T. minus* sensu auct. Japon., non L.; *T. minus* var. *pseudosimplex* H. Boiss.; *T. simplex* var. *divaricatum* Huth; *T. chionophyllum* Nakai——Ō-KARAMATSU. Mostly glabrous perennial 30–100 cm. high; stems usually striate, leafy; lower leaves rather large, short-petiolate or sessile, 2–4 times ternate, the stipules undulately toothed, usually stipellate, the leaflets oblong, obovate or flabellate-obovate, 1–3 cm. long, 8–20 mm. wide, rounded to sometimes cordate, 2- to 3-lobed, green above, glaucous beneath; panicles rather large; flowers yellowish green, the pedicels usually rather long; sepals caducous, 3-nerved, oblong; anthers broadly linear, about 3 mm. long, mucronate; achenes 2–6, narrowly obovate, about 3 mm. long, sessile, few-nerved on each side, the style very short, the stigma ovate-deltoid.——July–Sept. Hills and mountains; Hokkaido, Honshu; very variable.——Kuriles, Sakhalin, Korea, Manchuria, China, Siberia to Europe, and N. America.

Var. **hypoleucum** (Sieb. & Zucc.) Miq. *T. hypoleucum* Sieb. & Zucc.; *T. thunbergii* DC.; *T. thunbergii* var. *hypoleucum* (Sieb. & Zucc.) Nakai; *T. minus* var. *elatum* sensu auct. Japon., non Lecoyer; *T. minus* var. *majus* Miq.; *T. yamamotoi* Honda——AKI-KARAMATSU. Panicles looser and larger; pedicels short; anthers slightly shorter and the stigma narrower.——A more southern variety occurring widely in Hokkaido, Honshu, Shikoku, Kyushu.——s. Kuriles, Sakhalin, Korea, Manchuria, Amami-oshima, and China.

5. **Thalictrum rochebrunianum** Fr. & Sav. SHIKIN-KARAMATSU. Glabrous perennial; stems erect, slightly branched above, 70–100 cm. long, striate; leaves deltoid in outline, petiolate, 3–4 times ternate, the upper ones sessile, the stipules subentire, the stipels absent, the leaflets obovate to elliptic, obtuse, mucronate, 2–3 cm. long, 1.5–2.5 cm. wide, rounded to cordate at base, often 2- to 3-lobed; panicles sometimes branched below, large, the pedicels 1–2 cm. long, nodding above, the bracts obcordate, small; flowers pale purple, the sepals rather large, elliptic, deciduous after anthesis, 6–8 mm. long; anthers lanceolate, yellow, about 1.2 mm. long, short-mucronate; achenes about 4 mm. long, 8–20, narrowly oblong, with several prominent longitudinal ribs, the stipe 1–1.5 mm. long; style very short, the stigma small, ovate.—— July–Aug. Grassy places in highlands; Honshu (centr. distr.); rather rare.

6. **Thalictrum aquilegifolium** L. *T. contortum* L.; *T. rubellum* Sieb. & Zucc.; *T. aquilegifolium* var. *sibiricum* Regel & Tiling; *T. aquilegifolium* vars. *asiaticum* Nakai, var. *intermedium* Nakai, and var. *japonicum* Nakai; *T. mitsinokuense* Koidz.; *T. daisenense* Nakai; *T. nipponense* Huth; *T. anomalum* Lév. & Van't.; *T. taquetii* Lév.——KARAMATSU-SŌ. glabrous, more or less glaucescent perennial herb; stems erect, rather stout, 50–120 cm. high, slightly striate, leafy; leaves 3–4 times ternate, the lower petiolate, the upper sessile; stipules membranous, nearly entire, reflexed, the stipels smaller, the leaflets obovate to cordate-orbicular, 1.5–3.5 cm. long, 1–3 cm. wide, often 3-lobed, obtuse; panicles densely rather many-

flowered, flat-topped; flowers white or sometimes slightly reddish; sepals elliptic, about 3–4 mm. long, 5- to 7-nerved, caducous; stamens many, about 10 mm. long, the filaments dilated above, white, the anthers whitish yellow, broadly linear, 1–1.2 mm. long; achenes 5–10, pendulous, 6–8 mm. long, 3- to 4-winged, broadly obovate to elliptic, the stipe 4–5 mm. long; style and stigma short.——July–Sept. Grassy places in highlands and alpine meadows; Hokkaido, Honshu, Shikoku, Kyushu; more common northward.——Cooler regions of the N. Hemisphere.

7. **Thalictrum acteaefolium** Sieb. & Zucc. SHIGIN-KARA-MATSU. Glabrous perennial with thick fibrous roots; stems 30–70 cm. long, striate, somewhat flexuous; leaves deltoid in outline, 2 or 3 times ternate, often aggregated toward the lower portion of stem, petiolate, the upper ones smaller and sessile, the stipules narrow, undulate on margin, the stipels usually absent, the leaflets thinly chartaceous, slightly unequal, broadly ovate, rhombic-elliptic or ovate-orbicular, glaucescent beneath, 1–4 cm. long, 1–3 cm. wide, rather acute, cuneate to truncate at base, irregularly coarse-toothed; flowers white; sepals elliptic, 3–4 mm. long; stamens many, the filaments white, very slightly dilated above, the anthers linear, 1.5–1.8 mm. long; achenes 3–5, sessile, few-ribbed, narrowly ovate; style slender, rather short, about 1.5 mm. long, strongly recurved above, the stigma on inner side of the style above.—— July–Oct. Woods in hills and mountains; Honshu (Kantō Distr. and westw.), Shikoku, Kyushu.

8. **Thalictrum baicalense** Turcz. *T. franchetii* Huth; *T. baicalense* var. *japonicum* H. Boiss.——HARU-KARAMATSU. Glabrous herb; stems 50–100 cm. long, striate, erect; leaves 2 to 3 times ternate, petiolate in lower ones, sessile in upper ones, deltoid in outline, glaucescent, exstipellate, the stipules membranous, lacerate, the leaflets broadly obovate to orbicular, membranous, 2–3.5 cm. long and as wide, rounded to obtuse, often shallowly 3-lobed and with few rounded teeth, truncate, to cuneate at base; inflorescence corymbose, densely rather many-flowered, with the flowers creamy white; sepals obovate, 3- to 5-nerved, 3–4 mm. long; stamens many, the filaments slightly dilated above, the anthers 0.8–1 mm. long, narrowly oblong, sometimes slightly mucronate; achenes 4–10, 2.5–3 mm. long, ellipsoidal or broadly obovoid, rather plump, with 3–5 weak nerves on each side, the stipe about 0.3 mm. long; style rather thick, 0.3–0.5 mm. long, the stigma obovate. ——June–July. Mountains; Hokkaido, Honshu (centr. distr. and northw.).——Korea, Manchuria, and e. Siberia.

9. **Thalictrum sachalinense** Lecoyer. *T. akanense* Huth——Ezo-KARAMATSU, MIYAMA-AKI-KARAMATSU. Glabrous perennial herb; stems erect, 50–80 cm. long, striate, leafy; leaves deltoid in outline, 2–3 times ternate, the upper sessile, the lower ones petiolate, the stipules rather broad, membranous, undulate-toothed or subentire, the stipels membranous, broadly cuneate-obovate or rhombic-orbicular, rounded at apex, shallowly cordate to broadly cuneate at base, shallowly 2- to 3-lobed on upper margin, the lobes rounded, often with few small rounded teeth; inflorescence flat-topped, rather densely flowered; flowers white; sepals obovate, 3- to 7-nerved; stamens white, many, the filaments slightly dilated above, the anthers broadly linear, about 1 mm. long; achenes 8–15, about 3 mm. long, ovoid, longitudinally ridged, the stipe about 0.5 mm. long; styles slender, about 1.2 mm. long, prominently hooked, the stigma oblanceolate, about half as long as the style.——June–July. Hokkaido.——n. Korea.

10. Thalictrum filamentosum Maxim. var. **tenerum** (Huth) Ohwi. *T. tenerum* Huth; *T. tuberiferum* var. *tenerum* (Huth) H. Boiss.; *T. tuberiferum* Maxim.; *T. clavatum* sensu auct. Japon., non DC.; *T. tuberiferum* var. *glabrescens* Honda; *T. tuberiferum* var. *pubescens* Honda——Miyama-karamatsu. Rather slender, sometimes pubescent, rhizomatous, perennial herb; stems erect, 20–70 cm. long; radical leaves usually solitary, long-petiolate, 2- to 3-ternate, thinly membranous, the stipels absent, the leaflets ovate to ovate-oblong or ovate-orbicular, glaucescent beneath, 1.5–8 cm. long, 1–5 cm. wide, usually gradually narrowed above, obtuse, cuneate to shallowly cordate at base, often 2- to 3-lobulate and toothed, the cauline leaves 1–3, smaller, sessile, the stipules nearly absent; inflorescence densely flowered, rather flat-topped; flowers many, white; sepals elliptic; stamens rather many, the filaments much dilated above, the anthers oblong, about 1 mm. long; achenes 2–6, oblanceolate to narrowly obovate, 3–5 mm. long, flattened laterally, 1- to 4-nerved on each side, the stipe 3–4 mm. long; style rather thick and short, about 0.5 mm. long, nearly straight, the stigma narrowly oblong.——May–Aug. Woods and thickets in mountains; Hokkaido, Honshu, Shikoku, Kyushu.——Korea, Manchuria, and s. Kuriles.

Var. **yakusimense** (Koidz.) Tamura. *T. yakusimense* Koidz.——Yakushima-karamatsu. Plant small with the leaflets cuneate to rounded at base, the stipes on the achenes shorter.——Kyushu (Yakushima).——Korea, Manchuria, and s. Kuriles. The typical phase occurs in Amur, Manchuria, and Korea.

11. Thalictrum microspermum Ohwi. *T. microspermum* var. *sikokianum* Honda; *T. filamentosum* var. *sikokianum* (Honda) Ohwi, excl. syn. nonnul.——Kogome-karamatsu. Slender glabrous perennial herb; stems to 20 cm. long, very slender, few-leaved, erect, simple; radical leaves 3-ternate, the petioles 3–5 cm. long, the stipules entire (stipels absent), the leaflets petiolulate, rhombic to deltoid-elliptic, 1–3.5 cm. long, glaucous beneath, thinly membranous, very obtuse, cuneate to subtruncate at base, coarsely toothed on upper margin; inflorescence corymbose-paniculate, about 10 cm. long and as wide, loosely many-flowered, the branches spreading; flowers white, on long pedicels, the bracts linear, about 1 mm. long; achenes 6–12, oblong-lanceolate, about 2 mm. long inclusive of the style, flattened laterally, straight, few-nerved on each side, the stipe 3–4 mm. long, very slender, straight, the style straight, very short.——Apr.–May. Shikoku, Kyushu.

12. Thalictrum integrilobum Maxim. *T. filamentosum* var. *integrilobum* (Maxim.) Ohwi——Nagaba-karamatsu. Slender glabrous rhizomatous perennial herb; radical leaves solitary, sometimes nearly as long as the stem, 3- to 4-ternate, the leaflets petiolulate, linear-lanceolate or linear-oblong, 1.5–3 cm. long, 3–5 mm. wide, obtuse, entire to obsoletely few-toothed; corymbs few- to rather many-flowered, somewhat convex, loose; flowers white, the pedicels 5–15 mm. long; filaments linear-oblanceolate, slightly dilated above, often twice as broad as the anther; achenes few, oblanceolate, 3–4 mm. long, the stipe 1.5–2 mm. long, the beak (style and stigma) 0.8–1 mm. long, straight or nearly so, linear, with a linear stigma inside.——Hokkaido (Mount Apoi in Hidaka Prov.); rare.

13. Thalictrum watanabei Yatabe. *T. capillipes* Nakai ——Tama-karamatsu. Slender glabrous rhizomatous perennial herb, some of the roots thickened near middle; stems 20–30 cm. long; radical leaves solitary, petiolate, 3- to 4-ternate, the leaflets petiolulate, elliptic- to orbicular-rhombic, thinly membranous, 1–2 cm. long, 7–15 mm. wide, obtuse to broadly cuneate at base, sparsely coarse-toothed on the upper half; cauline leaves 1–2, much reduced and sessile, ternate or biternate; corymbs loosely few- to rather many-flowered, very loose; flowers white, the pedicels 5–20 mm. long, obliquely spreading, the filaments linear-oblanceolate, usually broader than the anthers; achenes 3–6, lanceolate to oblanceolate, 3–3.5 mm. long, the stipe 1.5–2 mm. long, the beak 1–1.5 mm. long, linear, straight, the stigma linear on inner side.——June–Sept. Honshu (Totomi and Yamato Prov.), Shikoku, Kyushu.

14. Thalictrum nakamurae Koidz. Hime-miyama-karamatsu. Closely resembles the preceding; roots short, thickened near base; stems 20–30 cm. long, very slender; leaflets orbicular to elliptic-orbicular, very rarely rhombic-orbicular, rounded to shallowly cordate; corymbs loosely few-flowered; flowers white; filaments slightly broadened above; young achenes oblanceolate to narrowly obovate, on slender stipes; style about 1 mm. long or less, slightly recurved at apex.——July–Aug. Wet rocks in mountains; Honshu (Mount Komagatake in Echigo and Mount Tanigawa in Kotsuke Prov.); rare.

15. Thalictrum toyamae Ohwi. Hire-furi-karamatsu. Glabrous, slender perennial herb (sometimes stoloniferous), some of the roots thickened; stems solitary, slender, 10–20 cm. long, loosely 1- to 3-leaved, with few divaricate branches above; leaves without stipules and stipels, the radical ones solitary, as long as the stems, 3-ternate, deltoid in outline, 7–15 cm. long and wide, the petioles 3–6 cm. long, the leaflets thinly membranous, ovate- to orbicular-rhombic, 12–30 mm. long and as wide, usually cuneate or rarely subcordate at base, glaucescent beneath; cauline leaves ternate or simple, short-petiolate or sessile; achenes 5–10, oblong-fusiform, 1.7–2 mm. long, flattened laterally, weakly 2- to 3-nerved on each side, the stipe 1–1.2 mm. long; style about 2 mm. long, slender and slightly flexuous, stigmatic on inner side.——June. Kyushu (Mount Kurokami in Hizen); rare.

8. TRAUTVETTERIA Fisch. & Mey. Momiji-karamatsu Zoku

Rather tall perennial herbs; leaves alternate, petiolate, palmately cleft, irregularly toothed; inflorescence paniculate-corymbose; flowers small, white; sepals 3–5, caducous; petals absent; stamens many, the filaments slightly dilated above; carpels many, 1-ovuled, glabrous, sessile; achenes ovate, rhombic in transverse section, ridged, slightly flattened laterally, rather inflated, terminated by a short style.——Two species, one in N. America and one in e. Asia.

1. Trautvetteria japonica Sieb. & Zucc. *T. palmata* var. *japonica* (Sieb. & Zucc.) Huth——Momiji-karamatsu. Rhizomes short-creeping, the plants sometimes stoloniferous; stems erect, 25–60 cm. long, 2- to 3-leaved, short appressed-pubescent above; radical and lower cauline leaves long-petiolate, glabrous, cordate-orbicular or reniform-orbicular, 5–12 cm. long, 6–15 cm. wide, 7- to 9-parted to -cleft, the segments rhombic-ovate, acute to acuminate, often 1- to 2-lobuled, coarsely acute-toothed and incised, the upper leaves smaller, sessile; inflorescence flat-topped, densely many-flowered, often lobed; anthers elliptic, 0.5–0.7 mm. long; achenes rather many, forming a globose head on a short small receptacle, broadly ovate, sometimes with few appressed hairs, 2.5–3.5 mm. long; style slender, short, slightly hooked at apex.——July–Aug.

Wet places in mountains; Hokkaido, Honshu (n. and centr. distr.).——s. Kuriles.

Var. **borealis** Hara. *T. palmata* Fisch. &. Mey.; *Ranunculus pleurocarpus* Maxim.——OKU-MOMIJI-KARAMATSU. Leaves pubescent beneath; style shorter.——Hokkaido (Kitami Prov.).——s. Kuriles, Sakhalin, Amur, and Ochotsk Sea region.

9. AQUILEGIA L.   ODAMAKI ZOKU

Perennial herbs; stems erect, sometimes branched; leaves radical and cauline, alternate, petiolate, 2- to 3-ternate; flowers actinomorphic, usually rather large; sepals 5, petallike, colored, flat; petals slightly smaller than the sepals, spurred at base; stamens many, the inner ones reduced to membranous staminodes; carpels 5, sessile, free; ovules many; fruit of 5 dehiscent follicles; seeds black, usually smooth and shining, sometimes rugulose.——About 50 species, in the temperate regions of the N. Hemisphere.

1A. Stamens many; flowers large; spur much exserted in normal forms.
  2A. Carpels glabrous; spur incurved above; flowers blue, white or bluish purple. . . . . 1. *A. flabellata*
  2B. Carpels pubescent; spur erect, sometimes incurved above; flowers usually yellowish brown to purplish brown. . . . . 2. *A. buergeriana*
1B. Stamens 9–14; flowers small; spur very short, scarcely exserted. . . . . 3. *A. adoxoides*

1. Aquilegia flabellata Sieb. & Zucc.   *A. akitensis* Huth; *A. sibirica* var. *flabellata* (Sieb. & Zucc.) Finet & Gagnep.——ODAMAKI.   Glaucescent perennial herb with short, stout, branched rhizomes; stems erect, 20–50 cm. long, few-leaved, often slightly puberulent above; radical leaves few, petiolate, ternate, the petioles sometimes slightly pubescent, the leaflets 3-parted, the segments flabellate, 1.5–4 cm. long and as wide, 3-lobulate, obtusely incised on upper margin, glaucous beneath; flowers few, large, 3–4 cm. across; sepals about 2.5 cm. long, usually bluish purple, rarely white, ovate to elliptic, obtuse to rounded; petals about 1.5 cm. long, erect, pale yellow, sometimes white, the spur nearly as long as the body of the petal, incurved above, rarely absent in some cultivated forms; carpels glabrous; style rather long.——Apr.–May. Frequently cultivated as a pot plant, believed to have originated from the wild phase var. *pumila;* the plant was first described on the basis of the cultivated phase.

Var. **pumila** Kudo.   *A. buergeriana* var. *pumila* Huth; *A. akitensis* sensu auct. Japon., non Huth; *A. fauriei* Lév. & Van't., non Lév.; *A. japonica* Nakai & Hara; *A. flabellata* var. *prototypica* Takeda——MIYAMA-ODAMAKI.   Plants smaller than in the cultivated phase; stems to 30 cm. high, 1- to 2(–3)-flowered, the leaves less prominently glaucous.——Alpine regions in Honshu (centr. and n. distr.), Hokkaido. ——s. Kuriles, Sakhalin, and n. Korea.

2. Aquilegia buergeriana Sieb. & Zucc.   YAMA-ODA-MAKI.   Stems 30–60 cm. long, branched, scattered soft-pubescent; leaves twice ternate, long-petiolate, the leaflets rather thin, broadly rhombic to broadly obovate-cuneate, 1.5–4

cm. long, 2- to 3-cleft, obtusely incised, sparingly soft-pubescent beneath, glaucescent especially below; flowers few, 3–3.5 cm. across, the pedicels spreading-puberulent often with glandular hairs; sepals narrowly ovate, acute, 2–2.5 cm. long, brownish purple to yellowish, ascending to obliquely spreading; petals 12–15 mm. long, yellowish, the spur erect, slender, 17–20 mm. long, often very slightly inflated at the end; carpels soft-pubescent; seeds black, punctulate.——June–Aug. Honshu, Shikoku, Kyushu.

Var. **oxysepala** (Trautv. & Mey.) Kitam.   *A. oxysepala* Trautv. & Mey.; *A. vulgaris* var. *oxysepala* (Trautv. & Mey.) Regel——EZO-YAMA-ODAMAKI.   Spurs incurved.——Hokkaido and Honshu (n. and Hokuriku Distr.).——Korea, Manchuria, e. Siberia, and n. China.

3. Aquilegia adoxoides (DC.) Ohwi.   *Isopyrum adoxoides* DC.; *I. japonicum* Sieb. & Zucc.; *I. tuberosum* Lév.; *Semiaquilegia adoxoides* (DC.) Makino——HIME-UZU.   Slender perennial herb, the rhizomes globose, tuberous; scapes sparingly branched, scattered-puberulent, few-leaved; radical leaves few, 2–5 cm. long and as wide, long-petiolate, sparingly puberulent, slightly glaucescent and somewhat purplish beneath, ternate, the leaflets 1–2.5 cm. long, 2- to 3-cleft or parted, obtusely incised, the cauline ones smaller; flowers nodding, pale rose; sepals narrowly oblong, 5–6 mm. long; petals 2.5–3 mm. long, tubular below and with a very short spur at base; stamens 9–14, some of the inner ones reduced to flat staminodes; carpels 3–4(–5), sessile, 5–6 mm. long in fruit; seeds slightly roughened.——Mar.–May. Mountains; Honshu (Kantō and westw.), Shikoku, Kyushu; common.——s. Korea and China.

10. ACONITUM L.   TORI-KABUTO ZOKU

Erect or sometimes scandent perennial herbs with thickened roots; leaves petiolate, alternate, palmately 3- to 7-cleft or divided; flowers racemose or paniculate, zygomorphic, blue, pale purple, white or yellow; sepals 5, petallike, the upper or median one clearly hooded or galeate, the others flat, the lower 2 narrower than the others; petals 2, small, hidden under the helmet, nectariferous, stipitate; stamens many; follicles 3–5, the seeds many, often winged or rugose.——About 200 species, in temperate and northern regions of the N. Hemisphere. Common in our area.

1A. Upper sepals with elongate-conic suberect to recurved tip; roots perennial, often branched; flowers pale purple or pale yellow.
  2A. Flowers pale yellow; upper sepals slightly narrowed toward the tip and rounded, suberect; leaves deeply parted to divided, the segments and teeth acute. . . . . 1. *A. gigas*

2B. Flowers pale purple to nearly white; upper sepals distinctly narrowed and slightly recurved; leaves palmately cleft to shallowly parted, the segments and teeth acute to obtuse. .... 2. *A. loczyanum*
1B. Upper sepals erect or slightly incurved; roots yearly renovated, obconical or elongate-conical, undivided; flowers blue or bluish purple, rarely white.
    3A. Stems scandent; carpels glabrous. .... 3. *A. volubile*
    3B. Stems erect or ascending, not scandent.
        4A. Stems ascending to arcuate.
            .... 4. *A. japonicum*
        4B. Stems erect or nearly so, sometimes flexuous above.
            5A. Inflorescence relatively loose, usually few-flowered, the pedicels relatively elongate; plant of alpine regions.
                6A. Carpels and follicles glabrous or sparsely hairy. .... 5. *A. senanense*
                6B. Carpels and follicles uniformly and rather densely puberulent. .... 6. *A. yuparense*
            5B. Inflorescence relatively dense, usually many-flowered, the pedicels relatively short; plants of mountains and hills.
                7A. Leaves palmately divided, the median segments petioluled.
                    8A. Ultimate lobes of leaves linear. .... 7. *A. sachalinense*
                    8B. Ultimate lobes of leaves lanceolate. .... 8. *A. yezoense*
                7B. Leaves palmately parted. .... 9. *A. chinense*

1. Aconitum gigas Lév. & Van't. *Lycoctonum gigas* (Lév. & Van't.) Nakai; *A. pallidum* sensu auct. Japon., non Reichenb.——Ō-REIJIN-SŌ. Vivid green perennial herb; stems 50–100 cm. long, rather stout, slightly angled, with short incurved hairs on upper portion; radical and lower cauline leaves long-petiolate, reniform-orbicular, 10–20 cm. wide, 7- to 9-parted or nearly divided, usually again 3-cleft and incised, acutely toothed, thinly pilose on both sides especially on the nerves beneath, sometimes nearly glabrous on both sides; racemes terminal and axillary, to 30 cm. long, the flowers many, pedicellate, about 2.5 cm. long, pale yellow; carpels 3, glabrous.——July–Aug. Wet shaded places in high mountains; Hokkaido, Honshu (n. and centr. distr.).——Sakhalin.

2. Aconitum loczyanum R. Raymund. REIJIN-SŌ. Short-pubescent perennial herb; stems often angled below, to 70 cm. long, sparsely branched; radical and lower cauline leaves long-petiolate, reniform-orbicular or orbicular-cordate, 6–15 cm. wide, deeply to shallowly 5(–7)-cleft, incised and toothed; racemes terminal and axillary, 5–15 cm. long, the pedicels about 1 cm. long, pilose; flowers 2–2.5 cm. long; upper sepals pilose outside; carpels 3, with spreading hairs, becoming glabrescent.——Aug.–Oct. Wet shaded places in woods and along streams in mountains; Honshu, Shikoku, Kyushu.

Var. **pterocaule** (Koidz.) Ohwi. *A. pterocaule* Koidz. ——AZUMA-REIJIN-SŌ. Stems narrowly winged on lower parts; carpels with incurved hairs or glabrous; pedicels and outer side of sepals with incurved hairs.——Honshu (centr. and n. distr.).

3. Aconitum volubile Pall. HANA-KAZURU, HANA-ZURU, TSURUBUSHI. Stems scandent above, with short curved hairs on upper part; leaves 3-sected, the segments sessile or very short-petiolulate, narrow, 4–8 cm. long, 2- or 3-parted, sometimes with few incisions on margin; inflorescence terminal and lateral, rather loosely few-flowered, the flowers about 3 cm. long, bluish purple; carpels usually 5, glabrous.——Sept.–Oct. Kyushu (Kirishima Mts.).——Korea, Manchuria, and e. Siberia.

4. Aconitum japonicum Thunb. YAMA-TORI-KABUTO. Stems 30–100 cm. long, erect, more or less arcuate above, rarely slightly scandent on upper portion, puberulent above; leaves 5–10 cm. wide, 3(–5)-cleft or -divided, glabrous, the segments obovate-rhombic or rhombic-ovate, more or less incised and coarsely toothed, the median segment often petiolulate; inflorescence corymbose, terminal and lateral, rather small, usually few-flowered, the axillary branches usually not

elongate, the pedicels puberulent, spreading-hairy or sometimes glabrous; flowers bluish purple, about 3 cm. long; carpels usually 3, sometimes 4 or 5, glabrous or loosely few-haired.—— Sept.–Nov. Hills and mountains; Hokkaido (sw. distr.), Honshu, Shikoku, Kyushu; common and very variable.

5. Aconitum senanense Nakai. ?*A. zigzag* Lév. & Van't.; *A. matsumurae* Nakai——HOSOBA-TORI-KABUTO, TA-KANE-TORI-KABUTO. Erect perennial herb often flexuous and short-pubescent above, usually unbranched; leaves 6–10 cm. wide, 3(–5)-parted or -divided, the segments deeply cleft and narrowly lobulate; inflorescence usually a terminal corymbiform panicle, rather loosely few-flowered, the lateral branches when present few and rather elongate; flowers bluish purple, 2.5–3 cm. long; carpels 3–5, glabrous.——July–Sept. Alpine regions; Honshu (south-centr. distr.).

Var. **incisum** (Nakai) Ohwi. *A. hakusanense* Nakai; *A. metajaponicum* Nakai; *A. zigzag* forma *incisum* Nakai—— HAKUSAN-TORI-KABUTO. Stems slightly taller, 40–80 cm. long; leaves cleft to parted, the segments shallowly lobed; lateral inflorescence often ascending, rather many, the terminal branches nearly pyramidal, rather many-flowered.——Mountain regions of Honshu (Japan Sea side of centr. and n. distr.).

6. Aconitum yuparense Takeda. EZO-HOSOBA-TORI-KA-BUTO. Stems 40–50 cm. long, erect, with incurved short hairs on upper portion; leaves 5–11 cm. wide, 5- to 7-parted, the segments deeply lobed, narrowly caudate; inflorescence few-flowered, terminal and lateral, the lateral branches few and erect; flowers bluish purple, about 2.5 cm. long; carpels 3–5, densely white-puberulent.——Aug. Alpine slopes; Hokkaido (Mount Yubaridake); rare. Closely allied to *Aconitum kurilense* Takeda of the s. Kuriles.

Aconitum ito-seiyanum Miyabe & Tatew. SEIYA-BUSHI. Allied to *A. yuparense* but differs in the 5-sected leaves.—— Reported to occur in Hokkaido (Teshio Prov.).

7. Aconitum sachalinense F. Schmidt. KARAFUTO-BUSHI. Erect perennial herb, puberulent on upper portion; stems 50–100 cm. long, rather stout; leaves rather dense, palmately 5-sected, the median segments 6–8 cm. long, deeply cut, short-petioluled, the ultimate lobes 2–4 mm. wide, broadly linear; inflorescence a dense rather many-flowered terminal panicle, the lateral branches mostly erect; flowers bluish purple, about 2.5 cm. long; carpels 3, glabrous.——Hokkaido. ——Sakhalin.

8. Aconitum yezoense Nakai. EZO-TORI-KABUTO. Stems erect, slightly flexuous above, short recurved-pubescent; leaves 3-sected, the median segments petiolate, 7–10 cm. long, the lateral ones 2-parted, both cut into narrow lobes; inflorescence terminal and axillary, rather densely flowered, the axillary ones on elongate peduncles; flowers bluish purple, about 3 cm. long; carpels glabrous, usually 3.——Aug.–Oct. Hokkaido.

9. **Aconitum chinense** Sieb. & Zucc. Tori-kabuto. Stems 80–120 cm. long, firm, recurved short-hairy on upper portion; leaves rather dense, somewhat thick, 3-parted, the segments 10–15 cm. long, deeply incised and coarsely toothed,

the lateral segments deeply 2-cleft; branches of inflorescence rather many-flowered, erect; flowers deep bluish purple, large, 3–4 cm. long; carpels usually 5, glabrous.——Sept.–Oct. Cultivated for cut flowers and planted in gardens.

Aconitum zuccarinii Nakai. Aizu-tori-kabuto. Closely allied to and doubtfully distinct from *A. chinense,* differing usually in having 3 carpels.——Honshu (Iwashiro and Shimotsuke Prov.). Poorly known.

11. ANEMONOPSIS Sieb. & Zucc.　Renge-shōma Zoku

Erect perennial rhizomatous herbs; leaves alternate, mostly basal and on lower portion of stems, 2- to 3-ternate, petiolate; flowers in loose racemiform panicles, long-pedicelled, rather large, nodding, pale purple; sepals 7–10, flat, petaloid, oblong, deciduous; petals about 10, much smaller than the sepals, nectariferous at base; stamens many; carpels 2–4, sessile, free, rather many-ovuled; style elongate, slender; fruit of 2–4 follicles; seeds scaly.——A single species, in Japan.

1. **Anemonopsis macrophylla** Sieb. & Zucc. *Xaveria macrophylla* (Sieb. & Zucc.) Endl.——Renge-shōma. Plant glabrous; stems 40–80 cm. long; leaves large, the petioles dilated and membranous on both sides at base, the leaflets rhombic-ovate to broadly ovate or oblong, 4–8 cm. long, 2–5 cm. wide, acuminate, often 3-lobed, incised and acutely

toothed; inflorescence 15–30 cm. long, more than 10-flowered; flowers 3–3.5 cm. across; sepals oblong, about 2 cm. long; petals 10–12 mm. long, obovate, rather thick; stamens many; style elongate, filiform, the stigma punctate; follicles about 15 mm. long.——July–Sept. Woods in mountains; Honshu (centr. distr.); rare.

12. ACTAEA L.　Ruiyō-shōma Zoku

Rhizomatous perennials; leaves radical and cauline, alternate, ternately compound; inflorescence racemose, terminal, the bracts small, linear to lanceolate, persistent; flowers small, white, pedicellate; sepals 3–5, petaloid or sepallike, caducous; petals 4–10, unguiculate, flat, caducous; stamens many, the filaments filiform, longer than the sepals; carpels solitary, ellipsoidal, several-ovuled; stigma sessile, shallowly 2-lobed, discoid; fruit a berry.——Few species, in the temperate regions of the N. Hemisphere.

1A. Pedicels rather thickened in fruit; fruit black; inflorescence relatively short. .... 1. *A. asiatica*
1B. Pedicels slender and not thickened in fruit; fruit red; inflorescence rather elongate. .... 2. *A. erythrocarpa*

1. **Actaea asiatica** Hara. *A. spicata* var. *nigra* sensu auct. Japon., non Willd.; *A. acuminata* sensu Komar., non Wall.——Ruiyō-shōma.　Stems 40–70 cm. long, 2- to 3-leaved, few-scaled at base, puberulent above; leaves radical and lower cauline, 2- to 4-ternate, the leaflets membranous, ovate to narrowly so, 4–10 cm. long, 2–6 cm. wide, acuminate, incised and acutely toothed, sometimes 3-lobed, scattered short-pubescent when young; racemes densely flowered, 3–5 cm. long, the pedicels 10–15 mm. long in fruit, dark red, horizontally spreading; sepals about 3 mm. long; petals broadly ovate, 2–2.5 mm. long; fruit globose, black, about 6

mm. across.——May. Woods in mountains; Hokkaido, Honshu, Shikoku, Kyushu.——China, Manchuria, Ussuri, and Korea.

2. **Actaea erythrocarpa** Fisch. *A. spicata* var. *erythrocarpa* (Fisch.) Ledeb.——Akami-no-ruiyō-shōma.　Closely resembling the preceding species; inflorescence dense in anthesis, 5–10 cm. long in fruit, the pedicels often deflexed in fruit, much slenderer than the axis; fruit red.——June. Woods in mountains; Hokkaido, Honshu (n. distr.).——Europe, Siberia, Manchuria, Korea, Sakhalin, and s. Kuriles.

13. COPTIS Salisb.　Ō-ren Zoku

Small evergreen perennial herbs with creeping rhizomes; leaves radical, ternate or palmately 5-parted, sometimes 2- to 4-ternate, petiolate; scapes 1- to few-flowered; flowers white, rather small, bisexual or unisexual (plants often dioecious); sepals 5–6, petaloid, deciduous; petals 5–6, small, nectariferous; stamens many; carpels few to many, stipitate, free, several-ovuled; follicles verticillately arranged on a receptacle, chartaceous; seeds shining.——About 10 species, in temperate regions of the N. Hemisphere.

1A. Follicles without nerves on each side, obovate in cross section; style slender, rather elongate, the stigma narrow, decurrent on inner side of the style; leaves always 3-foliolate. .... 1. *C. trifolia*
1B. Follicles with 1 longitudinal nerve on each side, obtriangular in cross section; style short, the stigma essentially punctate; leaves 3 or more foliolate.
　2A. Flowers solitary, rarely 2; sepals obovate or elliptic; petals obovate; leaflets 3 or 5, sessile, ternately or palmately arranged.
　　3A. Leaves 3-foliolate, rather firm, with rather prominent veinlets beneath. .... 2. *C. trifoliolata*
　　3B. Leaves 5-foliolate, firmly chartaceous, with faintly raised veinlets beneath. .... 3. *C. quinquefolia*
　2B. Flowers 1 to 3; sepals lanceolate; petals spathulate; leaves once to thrice ternately compound, if simply ternate the leaflets distinctly petiolulate. .... 4. *C. japonica*

1. **Coptis trifolia** (L.) Salisb. *Helleborus trifolius* L. ——Mɪᴛsᴜʙᴀ-ō-ʀᴇɴ, Mɪᴛsᴜʙᴀ-ᴡō-ʀᴇɴ. Slender nearly glabrous perennial with very slender creeping rhizomes; leaves long-petiolate, the leaflets 3, sessile, obovate or obovate-orbicular, 1–2.5 cm. long, 1–2 cm. wide, obtuse, shallowly incised and irregularly mucronate-toothed, sometimes shallowly lobulate; scapes 5–10 cm. long; flowers white, 7–10 mm. across, the limb of petals elliptic, yellow; follicles ovoid, 3–5 mm. long, on a stipe 5–7 mm. long; style recurved above.——June–July. Coniferous woods and sphagnum-moors; Hokkaido, Honshu (centr. and n. distr.).——Cooler regions of the N. Hemisphere.

2. **Coptis trifoliolata** (Makino) Makino. *C. quinquefolia* var. *trifoliolata* Makino; *C. trifoliolata* var. *oligodonta* F. Maekawa; *C. oligodonta* (F. Maekawa) Satake——Mɪᴛsᴜʙᴀ-ɴᴏ-ʙᴀɪᴋᴀ-ō-ʀᴇɴ. Closely resembles the preceding in habit but slightly coarser and the rhizomes less elongate; leaflets firmer and thicker, the petioles grooved above; scapes 7–15 cm. long, sometimes with a small scale; flowers 12–15 mm. across; sepals broader, 6–8 mm. long; petals about half as long as the stamens, the limb slightly shorter or as long as the claw; follicles 6–8 mm. long, with a longitudinal nerve on each side, the stipe about 5 mm. long; style short, the stigma nearly punctate.——June–Aug. Wet alpine slopes and high mountains; Honshu (Japan Sea side of centr. and n. distr.).

3. **Coptis quinquefolia** Miq. *C. quinquefolia* var. *stolonifera* Makino——Bᴀɪᴋᴀ-ō-ʀᴇɴ, Bᴀɪᴋᴀ-ᴡō-ʀᴇɴ, Gᴏᴋᴀʏō-ō-ʀᴇɴ. Rhizomes short-creeping (plants often stoloniferous); leaflets 5, evergreen, irregularly mucronate-toothed, 1–2.5 cm. long, 7–20 mm. wide, the lateral ones smaller; scapes 7–15 cm. long, often 1-scaled; flowers white, 12–18 mm. across; sepals obovate; petals nearly as long as the stamens, the limb much shorter than the claw; follicles similar to those of the preceding.——Apr.–May. Woods in mountains; Honshu (centr. distr. and westw.), Shikoku.

Var. **ramosa** (Makino) Ohwi. *C. quinquefolia* forma *ramosa* Makino; *C. ramosa* (Makino) Tamura; *C. quinquefolia* var. *pedatoquinquefolia* Koidz.——Ō-Gᴏᴋᴀʏō-ō-ʀᴇɴ. Plant without stolons; leaflets pinnately incised; scapes often 2-flowered.——Kyushu (Yakushima).

4. **Coptis japonica** (Thunb.) Makino. *Thalictrum japonicum* Thunb.; *C. orientalis* Maxim.; *C. anemonaefolia* Sieb. & Zucc.; *C. occidentalis* var. *japonica* Huth——Ō-ʀᴇɴ, Wō-ʀᴇɴ. Perennial herb with short, rather thick rhizomes; radical leaves ternate, the leaflets simple or shallowly lobulate, 2–5 cm. long, broadly ovate, acute, petiolulate, irregularly acute-toothed; scapes 10–25 cm. long, 1- to 3-flowered, nearly glabrous or short-pubescent above; flowers about 10 mm. across, white, bisexual or polygamous; sepals lanceolate; follicles about 10 mm. long, the stipe 6–8 mm. long; style short.——Mar.–Apr. Woods in mountains; Hokkaido, Honshu, Shikoku.

Var. **japonica**. Kɪᴋᴜʙᴀ-ō-ʀᴇɴ, Kɪᴋᴜʙᴀ-ᴡō-ʀᴇɴ. Leaves once ternate, the leaflets pinnately cleft to parted or sometimes nearly divided.

Var. **dissecta** (Yatabe) Nakai. *C. anemonaefolia* var. *dissecta* Yatabe——Sᴇʀɪʙᴀ-ō-ʀᴇɴ, Sᴇʀɪʙᴀ-ᴡō-ʀᴇɴ. Leaves twice ternate.——Occurs with the typical phase.

Var. **major** (Miq.) Satake. *C. brachypetala* Sieb. & Zucc.; *C. brachypetala* var. *major* Miq.——Kᴏ-sᴇʀɪʙᴀ-ō-ʀᴇɴ, Hᴏsᴏʙᴀ-sᴇʀɪʙᴀ-ō-ʀᴇɴ, Kᴏ-sᴇʀɪʙᴀ-ᴡō-ʀᴇɴ. Leaves thrice ternate.——Occurs with the typical phase.

14. CIMICIFUGA L.   Sᴀʀᴀsʜɪɴᴀ-sʜōᴍᴀ Zᴏᴋᴜ

Erect perennial herbs; leaves radical and cauline, the lower ones long-petiolate, ternately compound; inflorescence racemose or spicate, terminal, often loosely branched, the flowers small, white, pedicelled or sessile; sepals 2–5, petaloid, caducous; petals 0–8, unguiculate, often 2-fid; stamens many, the filaments filiform or slightly flattened above; carpels 1–8, sessile or stipitate, many-ovuled; styles very short or slender; follicles 1–8; seeds smooth or scaly.——About 10 species, in the temperate regions of the N. Hemisphere.

1A. Flowers distinctly pedicelled; carpels 2–7, stipitate; style circinnate, with a narrow oblique stigma inside; plant of mountains and alpine meadows. ....1. C. simplex
1B. Flowers nearly sessile; carpels solitary, very rarely 2, scarcely stipitate; style very short, the stigma short, terminal.
    2A. Leaflets acute to obtuse or short-acuminate, with short callose hairs on nerves above. .... 2. C. japonica
    2B. Leaflets acuminate to long-acuminate, glabrous above except for short hairs sometimes near margin. .... 3. C. acerina

1. **Cimicifuga simplex** Wormsk. *C. foetida* var. *simplex* (Wormsk.) Regel; *C. yesoensis* (Nakai) Kudo; *C. simplex* var. *yesoensis* Nakai——Sᴀʀᴀsʜɪɴᴀ-sʜōᴍᴀ. Perennial herb with short-creeping rhizomes; stems white-puberulent above, 40–100 cm. long, few-leaved; radical and lower cauline leaves long-petiolate, 2- to 3-ternate, the leaflets ovate to narrowly so, 3–8 cm. long, 1.5–5 cm. wide, acuminate, sometimes 3-lobed, irregularly incised and toothed; spikes elongate, 20–30 cm. long, often branched below, densely many-flowered, the pedicels 5–10 mm. long; sepals white, about 4 mm. long; petals slightly smaller, shallowly 2-lobed; stamens about 10 mm. long, white; carpels 2–7, the stipe elongate after anthesis, slightly puberulent to glabrous.——Aug.–Oct. Mountain meadows and alpine regions; Hokkaido, Honshu, Shikoku, Kyushu.——Sakhalin, Kuriles, and Kamchatka.

2. **Cimicifuga japonica** (Thunb.) Spreng. *Actaea japonica* Thunb.; *Pityrosperma obtusilobum* Sieb. & Zucc. var. *biternatum* Sieb. & Zucc.; *C. obtusiloba* (Sieb. & Zucc.) Miq.; *C. biternata* (Sieb. & Zucc.) Miq.——Iɴᴜ-sʜōᴍᴀ. Perennial herb 60–80 cm. high; radical leaves long-petiolate, 1- to 2-ternate, the leaflets chartaceous, broadly ovate to orbicular-cordate, 6–10 cm. long, 5–10 cm. wide, subobtuse to acuminate, palmately lobulate and irregularly toothed, shallowly cordate, with short callose hairs on nerves above, and more or less spreading hairs on nerves beneath; stems scapelike, usually leafless; inflorescence 20–35 cm. long, short-pubescent, often branched below; flowers many, white, sessile or nearly so; sepals about 5 mm. long; petals 2-cleft; stamens about 8 mm. long, the filaments filiform, the anthers broadly elliptic; carpels solitary, rarely 2.——Aug.–Oct. Honshu (centr. distr. and westw.).

3. **Cimicifuga acerina** (Sieb. & Zucc.) T. Tanaka. *Pityrosperma acerina* Sieb. & Zucc.; *C. japonica* var. *acerina* (Sieb. & Zucc.) Huth; *C. acerina* var. *typica* Hara, var. *macro-*

phylla (Koidz.) Hara, and var. *intermedia* Hara; *C. macrophylla* Koidz.——Ōba-shōma.   Rhizomatous perennial herb; radical leaves 1- to 2-ternate, the leaflets deep green above, orbicular-cordate, 7–20 cm. long, 6–18 cm. wide, acuminate to long-acuminate, glabrous above except short-pilose near margin, short-hairy on nerves beneath, palmately cleft to lobulate, irregularly acute-toothed; scapes 80–120 cm. long, usually leafless; inflorescence often branched below, 15–30 cm. long,

puberulent, many-flowered; flowers white; sepals 4–5 mm. long; stamens about 8 mm. long, the filaments filiform, the anthers oblong; carpels solitary, glabrous.——Sept.–Oct. Honshu, Shikoku, Kyushu.

Var. **peltata** (Makino) Hara.   *C. japonica* forma *peltata* Makino; *C. peltata* (Makino) Koidz.——Kiken-shōma.   Leaflets peltate.——Honshu (centr. distr.).

15. ISOPYRUM L.   Shiro-kane-sō Zoku

Slender rather soft perennial herbs; leaves ternately compound, the cauline alternate; inflorescence cymose-paniculate or the flowers solitary, yellowish or white; sepals 5–6, petaloid, deciduous; petals 5, small, nectariferous or absent; stamens many, sometimes 5 or 10; carpels 2–10, free, several-ovuled; follicles 2–10; seeds smooth or granular-roughened.——About 20 species, in the temperate regions of the N. Hemisphere.

1A. Petals absent; carpels more than 2, ascending in fruit, the receptacle not thickened; leaves 1- to 2-ternate, the leaflets all alike.
                                                                                                                              1. *I. raddeanum*
1B. Petals present, small; carpels 2, horizontally spreading in fruit, the receptacle thickened and recurved on margin; leaves ternate, the terminal leaflets simple, the lateral ones 2-nate, 3-nate or pedately compound.
   2A. Stems and lower leaf surface soft-pubescent. . . . . 2. *I. dicarpon*
   2B. Stems and leaves glabrous.
      3A. Seeds granular-roughened; rhizomes and stolons absent. . . . . 3. *I. trachyspermum*
      3B. Seeds smooth; rhizomes or stolons present.
         4A. Flowers 1.2–1.5 cm. across; rhizomes long-creeping; leaflets rhombic-ovate to broadly so. . . . . 4. *I. stoloniferum*
         4B. Flowers 7–8 mm. across; rhizomes short-creeping.
            5A. Lower 2 cauline leaves not closely contiguous; leaflets broadly ovate. . . . . 5. *I. numajirianum*
            5B. Lower 2 cauline leaves closely contiguous.
               6A. Rhizomes densely scaly, the plants rarely stoloniferous; Japan Sea side. . . . . 6. *I. nipponicum*
               6B. Rhizomes loosely scaly, the plants sometimes stoloniferous; Hakone and Izu in Honshu. . . . . 7. *I. hakonense*

1.   Isopyrum raddeanum (Regel) Maxim.   *Enemion raddeanum* Regel; *E. raddeanum* var. *japonicum* Fr. & Sav. ——Chichibu-shiro-kane-sō.   Perennial herb with short rhizomes; stems 20–30 cm. long, with few scaly leaves at base; cauline leaves 2–3, alternate, on upper part of the stem, short-petiolate or sessile, 1- to 2-ternate, the leaflets broadly cuneate-ovate, 2–4 cm. long, 1–3 cm. wide, often 2- to 3-lobed and irregularly toothed, glaucous and soft short-pubescent beneath or glabrous on both sides; flowers few, terminal, umbellate, about 12 mm. across, white, the pedicels to 3 cm. long, erect; petals absent; filaments slightly broadened above; carpels 3–5, glabrous, 4–5 mm. long in fruit, slightly ascending, broadly ovate; seeds granular-roughened.——May.   Woods in mountains; Honshu (n. and centr. distr.); rare.——Korea, Manchuria, and Ussuri.

2.   Isopyrum dicarpon Miq.   *I. stipulaceum* Fr. & Sav. ——Saba-no-o.   Scattered soft-pubescent perennial; rhizomes much abbreviated; radical leaves ternate, petiolate, the petioles prominently dilated at base, the terminal leaflets simple, ovate, 1–3 cm. long, 7–20 mm. wide, subacute with an obtuse tip, often lobed, toothed, the lateral leaflets again dissected with 2–5 smaller leaflets, the lower cauline leaves subopposite; flowers to 10, white with purple striations, about 8 mm. across, the petals stipitate, nectariferous, the limb 2-lobed, the outer lobe larger, inflexed; carpels 2; seeds smooth.—— Mar.–Apr.   Shikoku, Kyushu.

Var. **univalve** Ohwi.   Saikoku-saba-no-o.   Petals without the inner lobe.——Mar.–Apr.   Honshu (Kinki Distr.).

3.   Isopyrum trachyspermum Maxim.   Tōgoku-saba-no-o.   Resembling the preceding but quite glabrous; stems rather tufted, 5–20 cm. long; terminal leaflets broadly ovate to broadly obovate, 5–15 mm. long, 4–15 mm. wide, obtuse to

subrounded; flowers 5–8 mm. across, pale yellowish-green; sepals obliquely ascending; limb of petals divided into an inner and outer lobe; carpels 2; seeds granular-roughened.——Mar.– May.   Woods in foothills and mountains; Honshu (Rikuzen Prov. and southw.), Shikoku, Kyushu; rather common.

4.   Isopyrum stoloniferum Maxim.   Tsuru-shiro-kane-sō, Shiro-kane-sō.   Resembles *I. dicarpon,* but quite glabrous, 10–20 cm. high; rhizomes slender, elongate; petioles slightly dilated at base, the terminal leaflets rhombic-ovate or broadly so, 1–2 cm. long, subacute with an obtuse tip; flowers creamy yellow, 12–15 mm. across, expanded; sepals spreading; limb of petals bilabiate; carpels 2; seeds smooth.——May– June.   Woods in mountains; Honshu (Tōkaidō to s. Kinki Distr.).

5.   Isopyrum numajirianum Makino.   Kōya-shiro-kane-sō.   Resembles species Nos. 4 and 6; rhizomes short; stems about 20 cm. long, without radical leaves; lower 2 cauline leaves distant, the terminal leaflets broadly ovate, retuse, 10– 25 mm. long, 7–20 mm. wide; flowers yellowish white, broadly campanulate; seeds smooth.——May.   Woods in mountains; Honshu (s. Kinki Distr.), Shikoku; rare.

6.   Isopyrum nipponicum Franch.   *I. makinoi* Nakai; *I. pterigionocaudatum* Koidz.——Azuma-shiro-kane-sō.   Rhizomes short, branched, creeping, with broad fleshy scales; stems without radical leaves, 10–25 cm. long, the cauline leaves few, the terminal leaflets usually cuneate-obovate or cuneate-oblong, obtuse to subrounded, 2–4 cm. long, 10–25 mm. wide, glaucescent beneath; flowers 7–10 mm. across, few to more than 10, pale greenish yellow, sometimes with a purple hue, not fully expanded; limb of petals simple, inflexed; seeds smooth.——May–July.   Wet places in mountains; Honshu (Japan Sea side, Uzen Prov. to Echizen and Tajima Prov.).

Var. **sarmentosum** Ohwi. *I. ohwianum* Koidz.——SAN'IN-SHIRO-KANE-SŌ. Plants sarmentose.——Honshu (Japan Sea side, Echizen Prov. to Chūgoku Distr.).

7. Isopyrum hakonense F. Maekawa & Tuyama. HA-KONE-SHIRO-KANE-SŌ. Glabrous perennial; rhizomes branched, short-creeping, loosely scaly; stems 10–20 cm. long, rather slender, the cauline leaves 1 pair, subsessile, the radical leaves long-petiolate, the terminal leaflets broadly ovate or broadly rhombic-ovate, 1.5–2 cm. long, subobtuse, slightly narrower, broadly cuneate to abruptly narrowed below, 3-lobulate; flowers few, 6–8 mm. across, white, not fully expanded; sepals about 4 mm. long; seeds smooth.——May. Honshu (Izu and Sagami Prov.).

16. TROLLIUS L. KIMBAI-SŌ ZOKU

Erect glabrous perennial herbs with thick short rhizomes; leaves radical and cauline, alternate, palmately divided to lobed; flowers solitary or few, yellow, bisexual, erect, pedicelled; sepals 5 to rather numerous, petaloid, deciduous; petals 5–10, linear, short-clawed and with a nectariferous pit at base; stamens many; carpels 3 to many, free, sessile; fruit of few to rather many follicles; seeds many, usually angled, smooth.——More than 10 species, in temperate and northern regions of the N. Hemisphere.

1A. Petals longer than the stamens. . . . . 1. *T. hondoensis*
1B. Petals shorter than to as long as the stamens.
  2A. Sepals 5–7, spreading; anthers usually 2.5–5 mm. long. . . . . 2. *T. riederianus*
  2B. Sepals 10–16, ascending and connivent above; anthers usually 1.5–2 mm. long. . . . . 3. *T. pulcher*

1. Trollius hondoensis Nakai. *T. asiaticus* var. *lede-bourii* Maxim.; *T. ledebourii* var. *macropetalus* sensu auct. Japon., non Regel——KIMBAI-SŌ. Stems erect, few-leaved; radical and lower cauline leaves long-petiolate, 6–12 cm. long and as wide, 3-sected, the terminal segments rhombic-elliptic to rhombic-obovate, pinnately cleft on the upper half, irregularly acute-toothed, the lateral segments 2- to 3-parted, green above, paler beneath, firmly chartaceous; upper cauline leaves sessile or short-petiolate, rather small; flowers 2.5–4 cm. across; sepals elliptic, spreading; petals subacute or obtuse; anthers 3–5 mm. long; styles 2.5–3.5 mm. long.——June–Aug. Wet meadows in mountains especially along streams; Honshu (Uzen to Oomi Prov.).

2. Trollius riederianus Fisch. & Mey. *T. patulus* var. *brevistylus* Regel; *T. membranostylis* sensu Miyabe, non Hult. ——CHISHIMA-KIMBAI-SŌ. Erect perennial herb with short rhizomes; radical and lower cauline leaves long-petiolate, cordate-orbicular, 4–12 cm. long and as wide, 3-sected, the segments obovate to rhombic-elliptic, the lateral divisions 2- to 3-parted, pale and dull beneath, pinnately parted or cleft, incised and toothed; upper cauline leaves small, sessile; flowers yellow, 3–4 cm. across; sepals 5–7, broadly elliptic; petals linear, rather thick, obtuse to rounded; anthers 2.5–5 mm. long; style 1.5–2.5 mm. long.——July–Aug. Alpine meadows; Hokkaido.——Kuriles, Sakhalin, and Kamchatka.

Var. **japonicus** (Miq.) Ohwi. *T. japonicus* Miq.; *T. nishidae* Miyabe; *T. patulus* sensu auct. Japon., non Salisb.—— SHINANO-KIMBAI-SŌ. Leaves more deeply cut; styles rather slender, 2.4–4 mm. long.——July–Aug. Alpine meadows; Hokkaido, Honshu (n. and centr. distr.).

Trollius citrinus Miyabe, with lemon-colored flowers and spreading petals, is reported as growing with *T. riederianus* var. *japonicus* in Hidaka Prov., Hokkaido. Possibly it is just a variant, but its proper status can not now be determined with certainty.

3. Trollius pulcher Makino. BOTAN-KIMBAI-SŌ. Glabrous rather stout perennial herb; stems 20–60 cm. long, tufted, usually 3-leaved, 1- to 2-flowered; radical leaves long-petiolate, reniform, to 12 cm. long, 19 cm. wide, 3-sected, the segments cuneate, incised-toothed, the lateral segments broader, deeply 2-parted, the terminal segment elliptic, 3-cleft; cauline leaves gradually becoming smaller above, the petioles dilated below; flowers orange-yellow, hemispheric, 3–5.5 cm. across, erect; sepals many, concave, connivent above, imbricate, orbicular, about 2.5 cm. long and as wide in the outer ones, narrower in inner ones, minutely toothed on upper margin, rounded; petals 5–7, 8.5–10 mm. long, shorter than the stamens, erect, linear-spathulate, obtuse, short-clawed, entire; anthers short; carpels many; style erect, straight.——Hokkaido (Rishiri Isl.).

17. ERANTHIS Salisb. SETSUBUN-SŌ ZOKU

Small tuberous-rooted perennials; radical leaves palmately dissected, the involucral leaves sessile; flowers solitary and terminal, short-peduncled, yellow or white; sepals 5–8, petallike, deciduous; petals reduced to small 2- to 4-fid nectaries; stamens numerous; carpels follicular, few, free, short-stiped.——Few species in the Old World.

1. Eranthis pinnatifida Maxim. *E. keiskei* Fr. & Sav.; *Shibateranthis pinnatifida* (Maxim.) Satake & Okuyama; *S. keiskei* (Fr. & Sav.) Nakai——SETSUBUN-SŌ. Small perennial herb 5–15 cm. high, nearly glabrous, the peduncle minutely puberulous; radical leaves long-petioled, reniform-orbicular, 3–4 cm. wide, somewhat 5-angled in outline, 3-parted, the lateral divisions deeply bifid, the lobes and central division rhombic-ovate, obtuse, pinnatifid; involucral leaves sessile; flowers white, about 2 cm. across; sepals 5, ovate, 10–15 mm. long; petals yellow, 2- to 4-fid, shorter than the stamens; carpels 1–5, about 10 mm. long, glabrous, short-stiped, the style about 2–3 mm. long; seeds brownish, rounded, smooth.——Mar.–Apr. Woods in mountains; Honshu (Kantō and westw.); rather rare.

18. CALTHA L.　Ryū-kinka Zoku

Glabrous, often stoloniferous perennial herbs with short rhizomes; radical leaves simple, palmately nerved, entire or toothed, petiolate, cordate, the cauline ones few or absent; flowers yellow or creamy yellow, few to rather numerous; sepals 5 to rather many, yellow, petaloid, large, deciduous; petals absent; stamens many or sometimes few; carpels sessile; style short; fruit of few to several follicles; seeds many, smooth.——More than 10 species, in temperate and northern regions of the N. Hemisphere.

1.　**Caltha palustris** L. var. **membranacea** Turcz. *C. palustris* var. *sibirica* Regel; *C. membranacea* (Turcz.) Schipcz.; *C. sibirica* (Regel) Makino; *C. palustris* var. *nipponica* Hara——Ryū-kinka.　Rhizomes short, thick-rooted; radical leaves long-petiolate; reniform-orbicular or ovate-cordate, 5–10 cm. long and as wide, undulate-toothed; scapes 15–50 cm. long, erect, few-leaved; flowers golden-yellow, 2–3 cm. across; follicles about 1 cm. long.——Apr.–July. Wet places in mountains; Honshu, Kyushu.——e. Siberia, China, and Korea.——Forma **decumbens** Makino. *C. palustris* var. *enkoso* Hara——Enkō-sō.　Stems decumbent.——Wet places; Honshu.

Var. **pygmaea** Makino. Kobano-ryū-kinka.　Dwarf-growing phase, recently found wild.——Honshu (Hirugano in Hida).——Long cultivated.

Var. **barthei** Hance. *C. barthei* (Hance) Koidz.; *C. fistulosa* Schipcz.; *C. palustris* forma *gigas* Lév.——Ezo-no-ryū-kinka.　Plant 50–80 cm. high; radical leaves 10–25 cm. wide, often deltoid-toothed; flowers large, rather many; style short.——May–July.　Wet places; Hokkaido, Honshu (n. distr.).——Sakhalin and Kuriles.

19. GLAUCIDIUM Sieb. & Zucc.　Shirane-aoi Zoku

Perennial herbs, white-pubescent while young; rhizomes stout; cauline leaves 2, palmately lobed, the basal leaves long-petiolate; flowers solitary, terminal, pale blue-purple; sepals 4, petaloid, spreading, deciduous; petals absent; stamens many; carpels 2, sometimes solitary, sessile, slightly connate below, many-ovuled; follicles flattened laterally, somewhat 4-angled; seeds obovate, flat, many, broadly winged on margin.——A single species, in Japan.

1.　**Glaucidium palmatum** Sieb. & Zucc. *G. paradoxum* Makino——Shirane-aoi.　Stems rather stout, naked below, 2-leaved above; radical (sterile) leaves membranous, long-petiolate, the cauline reniform or cordate-orbicular, 8–20 cm. long and as wide, palmately 7- to 11-lobed, soft-pubescent while young, the lobes acuminate, irregularly acute-toothed, often incised, the petioles 10–15 cm. long; flowers terminal, solitary, 5–8 cm. across, pedunculate; sepals 4, broadly ovate, thin, pale blue-purple, rarely white; stamens many; carpels usually 2; follicles flattened, about 15 mm. long; seeds narrowly oblong, about 10 mm. long inclusive of the wing.——May–July.　Woods in high mountains; Hokkaido, Honshu (n. and centr. distr.).

20. PAEONIA L.　Botan Zoku

Perennial herbs or suffruticose subshrubs; leaves ternately or pinnately compound; flowers large, terminal and axillary, pale pink, yellow or purple-red; sepals 3–5, herbaceous, greenish; petals 5–20, petaloid, without nectary glands, deciduous; stamens many; carpels 2–5, free; follicles rather large, several- to many-seeded; seeds black.——About 30 species in the Old World, mainly in Asia, 2 in N. America.

1A.　Plants herbaceous, the flowers with a low disc below the carpels.
  2A.　Leaflets rather thin, not shining, oblong, usually not decurrent below, abruptly acute to subobtuse, whitish beneath; flowers not fully expanded; filaments pale red; stigma purplish; plant of mountains.
    3A.　Flowers white; stigma short, slightly recurved; follicles reflexed. .... 1. *P. japonica*
    3B.　Flowers rose; stigma rather elongate, prominently circinnate; follicles spreading. .... 2. *P. obovata*
  2B.　Leaflets firm, chartaceous, shining above, narrowly obovate to lanceolate, often decurrent on the petiolules, usually acuminate, pale green beneath; flowers fully expanded; cultivated. .... 3. *P. lactiflora*
1B.　Plant shrubby, the flowers with the disc dilated and enveloping the carpels; cultivated. .... 4. *P. suffruticossa*

1.　**Paeonia japonica** (Makino) Miyabe & Takeda. *P. obovata* var. *japonica* Makino——Yama-shakuyaku.　Stems 40–50 cm. long, with few scalelike leaves at base; rhizomes stout, short; cauline leaves 3–4, petiolate, large, usually twice ternate, the leaflets oblong to obovate, 4–12 cm. long, 3–7 cm. wide, abruptly acute to obtuse, acuminate at base, often petiolulate, glaucous beneath; flowers white, 4–5 cm. across, solitary and terminal; sepals ovate; petals 5–7, ascending, connivent above, obovate, 3–4 cm. long; anthers 5–7 mm. long; stigma short, slightly recurved.——Apr.–June. Thickets and open woods in mountains; Honshu, Shikoku, Kyushu.——Korea and Manchuria.

Var. **pilosa** Nakai. Ke-yama-shakuyaku.　Leaves pubescent beneath.——Occurs with the typical phase.

2.　**Paeonia obovata** Maxim. Benibana-yama-shakuyaku.　Closely allied to the preceding; leaves scattered soft-pubescent beneath; flowers pale rose; stigma elongate, circinnate.——Apr.–June.　Woods and thickets in mountains; Hokkaido, Honshu, Shikoku.——Korea, Manchuria, China, Amur, and Sakhalin.

Var. **glabra** Makino. Ke-nashi-benibana-yama-shakuyaku.　Leaflets glabrous beneath.——Occurs with the typical phase.

3.　**Paeonia lactiflora** Pall. *P. albiflora* Pall.——Shaku-yaku.　Erect perennial herb, 50–80 cm. high, the roots thickened, narrowly fusiform; leaves once to twice pinnately ternate, the leaflets often 3-parted, the segments lanceolate to narrowly obovate, shining above, pale green beneath; flowers large, white to rose or red-purple, often double.——May.

Long cultivated in our area for ornament and medical purposes.——e. Siberia, Manchuria, Korea, China, and Tibet.

4. Paeonia suffruticosa Andr. *P. moutan* Sims——Bo-TAN. Sparsely branched small shrub 1–1.5 m. high; leaflets usually 3- to 5-cleft, often whitish beneath; flowers large and showy, the disc thick-membranous, enveloping the carpels; carpels densely brown-pubescent.——Long cultivated in our area and many cultivars of it are grown.——nw. China, Tibet, and Bhutan.

Fam. 91. **LARDIZABALACEAE**   AKEBI KA   Lardizabala Family

Usually scandent woody climbers; leaves alternate, palmately or ternately compound, exstipulate; flowers polygamous or unisexual, actinomorphic, usually racemose; sepals 6, rarely 3, petallike; petals absent or small, reduced to a nectary; stamens 6; ovary superior; carpels 3 to many, free, dehiscent or indehiscent; ovules many; fruit berrylike; seeds with copious endosperm. ——About 8 genera with about 20 species, in e. Asia, India, and S. America.

1A. Sepals 3; leaves deciduous; stamens free. .... 1. *Akebia*
1B. Sepals 6; leaves evergreen; stamens monadelphous. .... 2. *Stauntonia*

1. AKEBIA Decne.   AKEBI ZOKU

Deciduous woody climbers; leaves palmately 3- to 5-foliolate, petiolate; racemes axillary; flowers unisexual (plants monoecious), the pistillate larger, on the lower part of the raceme, purplish; sepals 3, free; petals absent; stamens in staminate flowers 6, free; carpels 3–9, many-ovuled; ovules on parietal placentae; berry large, oblong.——Few species, in e. Asia.

1A. Leaflets 5, entire, oblong; pistillate pedicels declined. .... 1. *A. quinata*
1B. Leaflets 3(–5), undulate-toothed or subentire, ovate or broadly so; pistillate pedicels horizontally spreading. .... 2. *A. trifoliata*

1. Akebia quinata (Thunb.) Decne. *A. sempervirens* Nakai; *Rajania quinata* Thunb.——AKEBI. Glabrous climber; leaflets 5, oblong or oblong-obovate, 3–5 cm. long, 1–2 cm. wide, somewhat retuse, obtuse to rounded at base, entire, slightly glaucescent beneath; flowers pale purple, the pistillate 2.5–3 cm. across, long-pedicellate; sepals broadly elliptic, 15–20 mm. long, the staminate 12–16 mm. across, on upper part of raceme, the pedicels slender; sepals 7–8 mm. long; fruit 5–8 cm. long, oblong, becoming dark purplish when mature.—— Apr.–May. Thickets in hills and mountains; Honshu, Shikoku, Kyushu; common.——Korea and China.

2. Akebia trifoliata (Thunb.) Koidz. *Clematis trifoliata* Thunb.; *A. lobata* Decne.; *A. quercifolia* Sieb. & Zucc.; *A. clematifolia* Sieb. & Zucc.——MITSUBA-AKEBI. Glabrous woody climber; leaves trifoliolate, the leaflets ovate to broadly so, 3–6 cm. long, 1–2.5 cm. wide, shallowly retuse, rounded at base, usually undulate-toothed; sepals of the pistillate flowers 7–10 mm. long; staminate flowers rather numerous, 4–5 mm. across, short-pedicelled; sepals about 2 mm. long; fruit oblong, becoming dark purplish.——Apr.–May. Thickets in hills and low mountains; Hokkaido, Honshu, Shikoku, Kyushu; common.——China.

Akebia pentaphylla Makino. GOYŌ-AKEBI. An alleged hybrid of the two native species, having 5 leaflets and the flower characters of No. 2.

Akebia pentaphylla var. **integrifolia** Y. Kimura. KUWA-ZOME-AKEBI. An Alleged hybrid having the leaves of No. 1, but paler flowers rather larger than in those of *A. pentaphylla*. ——Known from Honshu (Shinano Prov.).

2. STAUNTONIA DC.   MUBE ZOKU

Evergreen woody climber; leaves palmately 5- to 7-foliolate; racemes axillary, short, few-flowered; flowers unisexual (plants monoecious); sepals 6, the outer 3 somewhat broader; stamens 6, connate; pistillate flowers with 6 staminodia; carpels 3, free; stigma capitate; ovules many; berry ovoid-globose, usually indehiscent; seeds many, ovoid or oblong, embedded in edible fleshy pulp.——Few species, in e. Asia.

1. Stauntonia hexaphylla (Thunb.) Decne. *Rajania hexaphylla* Thunb.——MUBE. Glabrous climber; leaflets (3–)5–7, coriaceous, oblong, ovate or obovate, 6–10 cm. long, 2–4 cm. wide, entire, pale green and prominently net-veined beneath; flowers few, pale yellow with dark red-purple striations, the outer sepals lanceolate, obliquely ascending, about 20 mm. long, gradually narrowed above, subacute, the inner sepals linear; berry purplish, ovate-globose, about 5 cm. long. ——Apr.–May. Thickets in lowlands and foothills in warmer regions; Honshu (Kantō Distr. and westw.), Shikoku, Kyushu.——Ryukyus and s. Korea.

Fam. 92. **BERBERIDACEAE**   MEGI KA   Barberry Family

Herbs or shrubs; leaves alternate, simple, ternately or sometimes pinnately compound, the stipules absent or present; flowers bisexual, in cymes, racemes, or panicles or solitary; sepals and petals imbricate, biseriate or the latter sometimes reduced to nectaries; stamens as many as the petals and opposite them, the anthers usually opening by 2 valves hinged at top; ovary superior, 1-locular, few- to many-ovuled; style short or absent, the stigma usually peltate; fruit a berry or capsule; seeds with endosperm.——About 10 genera with more than 250 species, mainly in temperate regions of the N. Hemisphere, few in S. America.

1A. Woody plants; petals reduced to nectaries; fruit a berry.
  2A. Leaves simple, often reduced to spines, deciduous (in ours); anthers opening by valves; sepals 6. 1. *Berberis*
  2B. Leaves pinnate or ternate, evergreen.
    3A. Leaves tripinnate; leaflets entire; flowers white; anthers longitudinally split. . . . . 2. *Nandina*
    3B. Leaves simply pinnate; leaflets spiny on margins; flowers yellow; anthers opening by valves. . . . . 3. *Mahonia*
1B. Herbs.
  4A. Sepals and petals present.
    5A. Petals with a nectary gland; leaves ternate, rarely binate.
      6A. Carpels soon deciduous after anthesis, exposing 2 seeds. . . . . 4. *Caulophyllum*
      6B. Carpels persistent and enveloping the seeds.
        7A. Fruit a dehiscent follicle. . . . . 5. *Epimedium*
        7B. Fruit an indehiscent berry. . . . . 6. *Ranzania*
    5B. Petals without a nectary gland; leaves simple, often peltate; fruit a berry. . . . . 7. *Diphylleia*
  4B. Sepals and petals absent; leaves ternate; fruit a follicle. . . . . 8. *Achlys*

1. BERBERIS   Megi Zoku

Deciduous or rarely evergreen shrubs with yellowish wood; leaves simple, entire or spine-toothed, penni-nerved, the leaves of the shoots reduced usually to 3-parted spines; flowers in axillary racemes, solitary, or fasciculate in axils, rather small, yellow; sepals 6, subtended by 2 or 3 bracteoles; petals 6, in 2 series, usually with 2 nectariferous glands at base; stamens 6, free, the anthers opening by 2 small terminal valves; carpels solitary; stigma peltate; ovules few; berry indehiscent, 1- to several-seeded.
——About 200 species, abundant in c. Asia, S. America, and e. Asia, few in N. America, Europe, and n. Africa.

1A. Leaves spiny-toothed.
  2A. Branches angled; leaves oblanceolate, 1–2 cm. wide, acute or obtuse; flowers 3–7, in umbels or short racemes. . . . . . . 1. *B. sieboldii*
  2B. Branches obsoletely striate; leaves obovate or oblong, 1.5–3 cm. wide, very obtuse or nearly rounded; flowers more than 10, in
    racemes. . . . . 2. *B. amurensis* var. *japonica*
1B. Leaves entire.
  3A. Branches terete or nearly so; leaves 3–8 cm. long; flowers 3–5, often in short racemes. 3. *B. tschonoskyana*
  3B. Branches prominently grooved; leaves 1–3.5 cm. long; flowers 2–4, fasciculate in axils. . . . . 4. *B. thunbergii*

1. Berberis sieboldii Miq. Hebi-noborazu. Glabrous, spiny, deciduous shrub; leaves oblanceolate to broadly so, 4–8 cm. long, 1–2 cm. wide, acute, sometimes obtuse, narrowed below, pale beneath, densely spiny-toothed; racemes much shorter than the leaves, 2–4 cm. long; flowers yellow, about 6 mm. across; fruit globose, dark red.——May. Honshu (sw. part of centr. distr. and Kinki Distr.).

2. Berberis amurensis Rupr. var. **japonica** (Regel) Rehd. *B. vulgaris* var. *japonica* Regel; *B. japonica* (Regel) C. K. Schn., non R. Br.; *B. regeliana* Koehne; *B. amurensis* sensu auct. Japon., non Rupr.——Hiro-ha-no-hebi-noborazu. Glabrous spiny shrub, branches longitudinally grooved, the spines spreading; leaves obovate to oblong, usually narrowly obovate, 3–10 cm. long, 1.5–3 cm. wide, not lustrous, very obtuse to rounded, narrowed below, pale beneath, the veinlets rather prominent; racemes nodding, as long as to slightly longer than the leaves, usually more than 10-flowered; flowers yellow, about 6 mm. across; berry oblong, red, about 10 mm. across.——June. Hokkaido, Honshu (centr. and n. distr.), Kyushu.——Korea, Manchuria, and Amur.——Forma **bretschneideri** (Rehd.) Ohwi. *B. bretschneideri* Rehd.; *B. amur-* ensis var. *bretschneideri* (Rehd.) Hara——Aka-jiku-hebi-no-borazu. Branches and petioles red-brown.——Forma **brevifolia** (Nakai) Ohwi. *B. amurensis* var. *brevifolia* Nakai——Maruba-hebi-noborazu. Leaves obovate-orbicular to broadly obovate.——Honshu (n. Kantō Distr.).

3. Berberis tschonoskyana Regel. *B. sikokiana* Yatabe ——Ōba-megi, Miyama-megi. Shrub with branches sometimes obsoletely grooved, often spiny; leaves elliptic to oblong, 3–8 cm. long, 1–2 cm. wide, obtuse, narrowed below, whitish or paler beneath; flowers 3–8, axillary, often on a short axis slightly longer than the leaves, yellow, about 6 mm. across; fruit oblong, 8–10 mm. long, red.——June. Mountains; Honshu (Kantō to Kinki Distr.), Shikoku, Kyushu.

4. Berberis thunbergii DC. Megi. Spiny shrub with branches prominently grooved, the spines rarely 3-parted; leaves obovate, elliptic or oblong, 1–3.5 cm. long, obtuse, mucronate or rounded, narrowed below, whitish beneath; flowers 2–4, umbellate, yellow, about 6 mm. across; the peduncles short or wanting; fruit ellipsoidal, 7–10 mm. across, red.——Apr. Honshu (Kantō Distr. and westw.), Shikoku, Kyushu; common.

2. NANDINA Thunb.   Nanten Zoku

Evergreen unarmed shrub; leaves alternate, short-petiolate, 2- to 3-pinnately compound, the axis jointed at base, the leaflets entire, the petioles dilated at base; stipules absent; panicles terminal, many-flowered; flowers white, rather small; sepals many, 3-seriate, the inner ones gradually larger, becoming petallike, the nectaries 3 or 6; anther-locules longitudinally split; ovary 1, 2-ovuled; fruit a globose, 2-seeded berry.——A single species, India to e. Asia.

1. Nandina domestica Thunb. Nanten. Erect, glabrous, simple to sparingly branched shrub; leaves rather densely arranged especially toward the top of stems and branchlets, the leaflets narrowly ovate to broadly lanceolate, 3–7 cm. long, 1–2.5 cm. wide, subcoriaceous, acuminate, lustrous above, entire; panicles large, pedunculate, erect; flowers white, about 6 mm. long; fruit red, sometimes white, rarely pale purple, 6–7 mm. across.——June. Ravines and valleys in mountains in warmer parts; Honshu (Tōkaidō to Kinki Distr. and westw.), Shikoku, Kyushu; long cultivated, with many cultivars.——Centr. China and India.

3. MAHONIA Nutt.   HIIRAGI-NANTEN ZOKU

Evergreen spineless shrubs or rarely small trees with yellow wood; leaves alternate, pinnately compound, the leaflets coriaceous, spinose-dentate; stipules minute; flowers in many-flowered racemes or panicles, yellow; sepals 9; petals 6; filaments dilated, usually with 2 toothlike appendages at the apex; stigma peltate; fruit a berry, dark blue, bloomy; seeds solitary or few.——About 50 species in N. America and e. Asia.

1. Mahonia japonica (Thunb.) DC. *Ilex japonica* Thunb.; *Berberis japonica* (Thunb.) R. Br.——HIIRAGI-NAN-TEN.   Shrub to small tree; leaves 30–50 cm. long, the leaflets 9–15, coriaceous, rigid, usually ovate-oblong, 4–10 cm. long, with 1–3 teeth on inner side and 5–6 teeth on outer side, dark bluish green above, yellowish green beneath, the lateral ones very oblique, the terminal leaflet stalked, truncate or subcordate at the base, larger; flowers yellow, in fascicled, long-drooping racemes; fruit bluish black.——Introduced and long cultivated in our area,——China and Formosa.

4. CAULOPHYLLUM Michx.   RUIYŌ-BOTAN ZOKU

Glabrous perennial rhizomatous herbs; leaves 1–2, membranous, pinnately 2- to 3-ternate, the leaflets sometimes 3-lobed; inflorescence cymose, terminal and axillary; flowers yellow, 3- to 6-bracteate at base; sepals 6; petals 6, very short, reduced to nectaries, opposite the sepals; stamens 6, the anthers opening by a valve on each side; ovary solitary, 1-carpellate; stigma lateral; ovules 2, orthotropous; fruits berrylike, globose, blue-black, often bloomy.——Two species, one in N. America and one in e. Asia.

1. Caulophyllum robustum Maxim. *C. thalictroides* var. *robustum* (Maxim.) Regel; *Leontice robusta* (Maxim.) Diels——RUIYŌ-BOTAN.   Glaucescent glabrous perennial herb, 40–80 cm. high, with stout rhizomes; stems solitary, erect, simple, few-scaled at base, 1- to 2-leaved on upper half; leaves petiolate or sessile, 2- to 3-ternate, the leaflets membranous, ovate-oblong, 4–8 cm. long, 2–4 cm. wide, subacuminate to acute, often 2- to 3-lobed to -divided, glaucous beneath; flowers yellow, 10–12 mm. across, pedicellate; sepals obovate, narrowed below, 6–8 mm. long; fruits berrylike, 6–7 mm. across, globose, the stipe 4–6 mm. long, thickened.——May-July. Woods in mountains; Hokkaido, Honshu, Shikoku, Kyushu.

5. EPIMEDIUM L.   IKARI-SŌ ZOKU

Perennial herbs with short-creeping rhizomes and much-branched roots; leaves binate or ternate, the leaflets entire or spine-margined, the cauline ones 1 or 2; racemes simple or slightly branched, terminal; flowers white, purplish, or yellowish; sepals 8, 2-seriate, the outer ones smaller, the inner ones petallike; petals 4, often spurred at base; stamens 4, the anthers opening by a valve on each side; ovary 1, the ovules in 2 series along the ventral suture; capsules slender, 2-valved, the valves unequal, the smaller one deciduous, the larger one persistent; seeds few, the raphe arillate.——More than 20 species, in Asia and Europe.

1A. Petals scarcely spurred.
  2A. Stamens longer than the perianth; filaments about 2 mm. long, slightly shorter than the anthers; leaves 1- or 2-ternate, evergreen; pedicels nearly as long as the flowers. .... 1. *E. sagittatum*
  2B. Stamens shorter than the perianth; filaments 0.5 mm. long, much shorter than the anthers; leaves binate; pedicels 1–3 times as long as the flowers.
    3A. Racemes many-flowered, subpaniculate; leaves summer-green; leaflets ovate-oblong, 6–10 cm. long, acute. .... 2. *E. setosum*
    3B. Racemes 4- to 15-flowered; leaves evergreen; leaflets ovate to broadly so, 2.5–5 cm. long, obtuse. .... 3. *E. diphyllum*
1B. Petals long-spurred.
  4A. Primary division of cauline leaves paired; leaflets obtuse to subobtuse. .... 4. *E. trifoliatobinatum*
  4B. Primary division of cauline leaves in 3's; leaflets usually acute to acuminate.
    5A. Leaves evergreen, rather firm; leaflets prominently glaucous beneath. .... 5. *E. sempervirens*
    5B. Leaves usually summer-green, thinly chartaceous; leaflets pale to slightly glaucous beneath.
      6A. Leaflets 3–6 cm. long; flowers white, rose, or pale purple. .... 6. *E. grandiflorum*
      6B. Leaflets 5–12 cm. long; flowers creamy yellow. .... 7. *E. koreanum*

1. Epimedium sagittatum (Sieb. & Zucc.) Maxim. *Aceranthus sagittatus* Sieb. & Zucc.; *E. ikariso* Sieb. ex Regel; *E. sinense* Sieb.——HOZAKI-NO-IKARI-SŌ.   Evergreen perennial herb; cauline leaves solitary, 3- to 9-foliolate, the leaflets ovate-oblong, 6–10 cm. long, 3–4 cm. wide, spine-margined; flowers white, about 6 mm. across; inner sepals about 4 mm. long, ovate.——A Chinese plant long cultivated for medical use, now becoming rare.

2. Epimedium setosum Koidz. Ō-BAIKA-IKARI-SŌ. Summer-green herb, 25–40 cm. high; cauline leaves solitary, binately 2- to 4-foliolate, the leaflets narrowly ovate to narrowly ovate-oblong, 6–10 cm. long, 3–4 cm. wide, acute, sagittate, more or less spine-margined, brownish long-pubescent beneath while young; inflorescence rather many-flowered; flowers white, 10–12 mm. across, the inner sepals 6–7 mm. long.——Apr.–May. Honshu (Chūgoku Distr.); rare.

3. Epimedium diphyllum (Morr. & Decne.) Lodd. *Aceranthus diphyllus* Morr. & Decne.; *E. japonicum* Sieb. ——BAIKA-IKARI-SŌ.   Evergreen perennial herb, 20–30 cm. high; leaves 1 or 2 times binate or sometimes ternate in the last division, the cauline solitary, the leaflets rather coriaceous, ovate to broadly so, 2.5–5 cm. long, 2–3.5 cm. wide, entire or slightly spine-margined, obtuse, cordate, scattered-puberulent beneath; flowers white, 10–12 mm. across.——Apr.–May. Shikoku, Kyushu.

Epimedium × youngianum Fisch. & Mey. UMEZAKI-IKARI-SŌ. A hybrid of *E. diphyllum* × *E. grandiflorum* Garden origin.

4. **Epimedium trifoliatobinatum** Koidz. HIME-IKARI-SŌ. Perennial, 30–50 cm. high; leaves binate, the primary ternate, the leaflets broadly ovate to narrowly so, 4–7 cm. long, 3–4 cm. wide, obtuse or abruptly acute, obliquely sagittate to cordate, somewhat coriaceous, usually glaucous and thinly brownish puberulent beneath, entire or scarcely spine-toothed; flowers about 10, white, rather large.——Honshu (Chūgoku Distr.), Shikoku, Kyushu.

Epimedium longifolium Decne. KO-IKARI-SŌ. This may be a hybrid of No. 4 and No. 5.

5. **Epimedium sempervirens** Nakai. *E. macranthum* var. *hypoleucum* Makino; *E. hypoleucum* (Makino) F. Maekawa; *E. sempervirens* var. *hypoleucum* (Makino) Ohwi—— TOKIWA-IKARI-SŌ. Evergreen herb, 40–60 cm. high, with rather thick rhizomes; leaflets coriaceous, spine-margined, broadly ovate to ovate-elliptic, 6–10 cm. long, 3–5 cm. wide, abruptly acuminate, deeply cordate, glaucous beneath, and with fine appressed hairs beneath, flowers white or purplish, rather large, the pedicels often forked.——Apr.–May. Honshu (Japan Sea side from Kinki Distr. to n. distr.).

Var. **hypoglaucum** (Makino) Ohwi. *E. macranthum* var. *hypoglaucum* Makino; *E. hypoglaucum* (Makino) F. Maekawa——URAJIRO-IKARI-SŌ. Leaflets persistent, with fine hairs beneath; flowers purplish to white.——Often found growing with the typical phase in Honshu (Japan Sea side from centr. distr. to Kinki).

6. **Epimedium grandiflorum** Morr. *E. macranthum* Morr. & Decne.; *E. grandiflorum* var. *violaceum* (Morr.) Stearn; *E. violaceum* Morr.; *E. macranthum* var. *thunbergianum* Miq.; *E. macranthum* var. *violaceum* (Morr.) Franch. ——IKARI-SŌ. Summer-green perennial herb, 20–40 cm. high; stems few-scaled at base; leaves 2- to 4-ternate, the leaflets ovate to deltoid-ovate, 3–6 cm. long, 2–4 cm. wide, subacute, spinulose-margined, cordate, glabrous above, glaucescent and glabrate beneath; flowers white to pale purple, with an elongated exserted spur.——Mar.–May. Honshu, Shikoku, Kyushu.

Var. **higoense** Shimizu. *E. multifoliolatum* Koidz.—— YACHIMATA-IKARI-SŌ, SOHAYAKI-IKARI-SŌ. Adult leaflets with scattered minute spreading hairs above.——Calcareous rocks; Kyushu (Higo Prov.); rare.

7. **Epimedium koreanum** Nakai. *E. cremeum* Nakai; *E. sulphurellum* Nakai; *E. coelestre* Nakai——KIBANA-IKARI-SŌ, SHIROBANA-IKARI-SŌ. Rhizomes short, loosely tufted, sparingly branched; leaves usually 2-ternate, the leaflets ovate to broadly so, 5–12 cm. long, 3–9 cm. wide, oblique, spinulose-toothed, green and glabrous above, abruptly short-acuminate; stems brown-pubescent above when young, 1-leaved; sepals 5–7 mm. long; petals 7–8 mm. long, the spur spreading, about 15 mm. long.——Apr. Hokkaido (sw. distr.), Honshu (Japan Sea side of centr. and n. distr.).——Korea.

6. RANZANIA T. Itō   TOGAKUSHI-SHŌMA ZOKU

Soft glabrous glaucescent perennial herb with creeping branched rhizomes; stems simple, 2-leaved at the summit, naked below; leaves ternate, petiolate, petiolulate; stipules absent; flowers in a terminal fascicle, pale purple, nodding, with 3 caducous sepallike bracts outside; sepals 6, petallike; petals 6, small, bearing 2 nectariferous glands at base; stamens 6; anthers dehiscing from a small valve on each side; ovary solitary, many-ovuled, the placentae parietal; fruit a berry.——A single species, in Japan.

1. **Ranzania japonica** (T. Itō) T. Itō. *Podophyllum japonicum* T. Itō; *Yatabea japonica* (T. Itō) Maxim.——TOGA-KUSHI-SHŌMA. Stems 30–50 cm. long, terete, erect; leaflets 3, petiolulate, orbicular to ovate-orbicular, 3-lobed or -incised, subacute, 8–12 cm. long and as wide, thinly membranous, cordate, glaucous beneath; pedicels fasciculate at the summit of the stem, 4–8 cm. long, nodding in anthesis, to 15 cm. long and erect in fruit; flowers about 2.5 cm. across; sepals ovate, obliquely spreading, obtuse, about 15 mm. long; fruit ellipsoidal.——June–July. Woods in mountains; Honshu (Japan Sea side of centr. and n. distr.); rare.

7. DIPHYLLEIA Michx.   SANKA-YŌ ZOKU

Soft perennial herbs with stout creeping rhizomes; stems simple, 2-leaved above, terete; leaves petiolate or sessile, orbicular or sometimes peltate; inflorescence cymose or umbellate, the flowers white; sepals 6, caducous; petals 6, flat, white; stamens 6; ovary solitary, the ovules in 2 series on one side within the ovary; berry indehiscent, ellipsoidal; seeds oblong, curved.——About 3 species, in N. America and e. Asia.

1. **Diphylleia grayi** F. Schmidt. *D. cymosa* sensu auct. Japon., non Michx.; *D. cymosa* var. *grayi* (F. Schmidt) Maxim. ——SANKA-YŌ. Rhizomes stout, knotty; stems 30–60 cm. long, terete, rather stout, scattered curled white-pubescent, remotely 2(–3)-leaved on upper half, with few membranous scales at base; leaves petiolate, orbicular or reniform-orbicular, the lower one peltate, to 25 cm. long, 30 cm. wide, the upper one sessile with a deep sinus at each end, basifixed, toothed and incised, white-pubescent beneath; flowers 3–10, pedicelled, about 2 cm. across; petals white, spreading, obovate, about 10 mm. long; berry ellipsoidal, blue, bloomy, 10–13 mm. long, several-seeded.——June–July. Woods in high mountains; Hokkaido, Honshu (centr. and n. distr., and the mountains in Yamato Prov.).——Sakhalin.

8. ACHLYS DC. NAMBU-SŌ ZOKU

Glabrous herbs with creeping rhizomes; leaves radical, palmately nerved, 3-sected to -parted, long-petiolate, the segments sessile, entire, sometimes obtusely lobulate; scapes erect, simple; inflorescence terminal, spicate, the flowers minute, bractless, the perianth absent; stamens 9–15, the filaments filiform, dilated above, the outer ones longer; anthers short, 2-locular, opening by a small valve on each side; ovary solitary, the carpels 1; stigma broad, sessile; ovules solitary, erect from the base of ovary; follicles small, oblique, slightly laterally compressed.——Two species, one in western N. America and one in e. Asia.

1. Achlys japonica Maxim. *A. triphylla* var. *japonica* (Maxim.) T. Itō——NAMBU-SŌ. Slender, glabrous, rather firm perennial herb with slender, creeping rhizomes; leaves radical, usually solitary, 3-sected, the petioles 15–20 cm. long, the segments sessile, with the median one cuneate-obovate or flabellate, 5–8 cm. long, 3–6 cm. wide, 3-lobed or undulately incised on upper margin, obtuse, the lateral segments subcordate; flowers rather many, white, small; stamens unequal, to about 5 mm. long; fruit 3–4 mm. long, subobcordate, ascending.——Hokkaido, Honshu (n. distr.).

Fam. 93. MENISPERMACEAE TSUZURA-FUJI KA Moonseed Family

Dioecious herbs or scandent woody vines, rarely erect; leaves alternate, often peltate, entire or palmately cleft; stipules absent; inflorescence paniculate or cymose; sepals free or rarely connate, 1 to several; petals free or connate; stamens many or few, free or connate; carpels 3 to 6, usually 3, free, the ovules 2, 1 ovule soon aborting; stigma entire or lobed; fruit a drupe, the stones usually incurved, sometimes reniform; endosperm present or absent.——About 70 genera, with about 400 species, mainly in the Tropics.

KEY TO FLOWERING MATERIAL

1A. Carpels 2–6.
  2A. Sepals and petals spirally arranged; stamens 12–24. .... 1. *Menispermum*
  2B. Sepals and petals verticillate; stamens 3–12.
    3A. Stamens 9–12. .... 2. *Sinomenium*
    3B. Stamens 3–6. .... 3. *Cocculus*
1B. Carpels 1.
  4A. Perianth of pistillate flowers actinomorphic. .... 4. *Stephania*
  4B. Perianth of pistillate flowers not regular. .... 5. *Cissampelos*

KEY TO STERILE MATERIAL

1A. Erect shrub; leaves evergreen, oblong-obovate, 3-nerved. .... *Cocculus laurifolius*
1B. Scandent shrubs or herbs; leaves ovate to orbicular.
  2A. Leaves peltate.
    3A. Leaves scattered-pubescent on upper side, more densely so beneath. .... *Cissampelos insularis*
    3B. Leaves glabrous above, glabrous or rarely pubescent beneath.
      4A. Leaves broadly ovate, not angled, rounded at base. .... *Stephania japonica*
      4B. Leaves orbicular, 5- to 9-angled or shallowly lobed, barely cordate or truncate at base. .... *Menispermum dauricum*
  2B. Leaves basifixed.
    5A. Branches and upper side of leaves glabrous; inflorescence large. .... *Sinomenium acutum*
    5B. Branches and leaves soft-pubescent; inflorescence rather short. .... *Cocculus trilobus*

1. MENISPERMUM L. KŌMORI-KAZURA ZOKU

Scandent, dioecious, herbaceous or woody perennials; leaves orbicular-angulate or palmately lobed, peltate; inflorescence racemose or paniculate; staminate sepals 4–10, 2-seriate, the inner whorl larger; petals 6–9, suborbicular, shorter than the sepals, inflexed on margin; stamens 12–24, free, the anthers 4-locular; perianth of the pistillate flowers similar to the staminate, the stamens reduced to staminodes; carpels 2–4; stigma dilated; drupes 2–3, subglobose, oblique, the stones orbicular-reniform, subcristate on back, concave on the sides.——Two species, one in N. America and one in e. Asia.

1. Menispermum dauricum DC. KŌMORI-KAZURA. Suffrutescent climber, the young branches elongate, glabrous or slightly pilose on upper part; leaves reniform to cordate-orbicular, 7–13 cm. long and as wide, shallowly 5- to 9-lobulate or angled, membranous, glaucous beneath, usually glabrous, the petioles 5–15 cm. long; inflorescence racemose, glabrous, pedunculate; flowers green-yellow; drupe reniform-globose, 8–10 mm. long, black.——May–June. Hokkaido, Honshu, Shikoku, Kyushu.——Korea, Manchuria, n. China, and e. Siberia.

2. SINOMENIUM Diels TSUZURA-FUJI ZOKU

Deciduous woody dioecious climbers with simple, sometimes palmately lobed leaves; panicles rather large; sepals and petals 6, the latter inflexed and embracing the filaments; stamens 9–12, the anthers 4-locular; pistillate flowers with 9 staminodia; carpels 3; the style recurved, the stigma lobulate; drupe flat, oblique, the stone curved, with a tuberculate dorsal rib and numerous transverse ribs; seeds lunate.——A single species, in e. Asia.

1. Sinomenium acutum (Thunb.) Rehd. & Wils. *Menispermum acutum* Thunb.; *Cocculus diversifolius* Miq., non DC.; *Cebatha miqueliana* O. Kuntze; *S. diversifolium* (Miq.) Diels; *Cocculus heterophyllus* Hemsl. & Wils.——TSUZURA-FUJI, Ō-TSUZURA-FUJI. Scandent shrub, the branches nearly glabrous; leaves orbicular to ovate-orbicular, 6–10 cm. long, 3–12 cm. wide, 5- to 7-angled, lobed or cleft, subcordate to truncate at base, with 2 or 3 pairs of nerves, glaucescent and pubescent beneath while young, glabrous and green above; petioles 5–10 cm. long; inflorescence 10–20 cm. long, many-flowered, more or less short-pilose, the branches nearly racemose; flowers pale green, the perianth short-pubescent; drupe oblique, black, 6–7 mm. long, 4–5 mm. wide, flattened, orbicular.——July. Honshu (s. Kantō Distr. and westw.), Shikoku, Kyushu.——China.

3. COCCULUS DC. Ao-TSUZURA-FUJI ZOKU

Scandent or erect shrubs; leaves simple, or sometimes lobed; cymes or panicles axillary; sepals 6, often pilose, the outer 3 smaller; petals 6, often bifid, auriculate and inflexed at base; stamens 6–9, free; carpels 3 or 6; style cylindric, the stigma lateral; drupes obovoid or subglobose, the stones muricate on back, compressed laterally with shallowly concave sides; cotyledons linear, flat.——More than 10 species, in e. and s. Asia, Africa, and N. America.

1A. Erect glabrous evergreen shrub; leaves oblong-obovate, deep green and lustrous above. .... 1. *C. laurifolius*
1B. More or less pubescent deciduous scandent shrub; leaves broadly ovate to ovate-cordate, sometimes shallowly 3-lobed. .. 2. *C. trilobus*

1. Cocculus laurifolius DC. KŌSHŪ-UYAKU, ISOYAMA-AOKI. Glabrous evergreen erect shrub, the branches slightly flattened, green, multistriate; leaves coriaceous-chartaceous, deep green, lustrous, oblong-obovate or obovate, sometimes elliptic, 5–12 cm. long, 2–4.5 cm. wide, acuminate, entire, 3-ribbed and with transverse parallel veinlets; petioles 5–10 mm. long, 1-ribbed beneath and obtriangular in cross section; inflorescence axillary and terminal, the flowers small, yellow; drupes depressed-globose, black, about 6 mm. across, slightly compressed.——July. Kyushu (s. distr.).——Ryukyus, Formosa, China, and se. Asia.

2. Cocculus trilobus (Thunb.) DC. *Menispermum trilobum* Thunb.; *C. thunbergii* DC.; *Cebatha orbiculata* O. Kuntze——Ao-TSUZURA-FUJI. Deciduous scandent pubescent shrub; branches elongate, terete, obsoletely striate; leaves firmly membranous to chartaceous, broadly ovate to ovate-cordate, 2–12 cm. long, 2–10 cm. wide, obtuse, mucronulate, sometimes shallowly 3-lobed, with 1 or 2 pairs of lateral nerves, the petioles 1–3 cm. long; inflorescence axillary and terminal, narrowly paniculate; flowers small, glabrous, yellow-white; drupe black, bloomy, 6–7 mm. across, the stones U-shaped, with small tubercles on back, finely rugulose.——July–Aug. Thickets in lowlands and foothills; Hokkaido, Honshu, Shikoku, Kyushu; common.——China, Formosa, Malaysia, Philippines, Indochina to the Himalayas.

4. STEPHANIA Lour. HASU-NO-HA-KAZURA ZOKU

Plants scandent, herbaceous or woody, dioecious; leaves simple, deciduous or evergreen, often peltate; inflorescence a simple or compound umbel, or sometimes paniculate; staminate sepals 6–10, usually equal, 2-seriate; petals 3–4, shorter than the sepals, suborbicular or obovate, thickened, rarely absent; stamens connate into a column, peltate, the anthers marginal at the top of the column, confluent and forming a marginal ring; pistillate sepals 3–5; petals 2–4, similar to the staminate ones; carpels solitary, with a parted style; drupe compressed, oblique, the stones compressed laterally, tuberculate on back, excavated on both sides; embryo linear.——About 35 species, in the Tropics.

1. Stephania japonica (Thunb.) Miers. *Menispermum japonicum* Thunb.; *Cocculus japonicus* (Thunb.) DC.——HASU-NO-HA-KAZURA. Glabrous scandent herb with elongate-striate branches; leaves peltate, thinly chartaceous or membranous, often glaucous beneath, broadly ovate to ovate-orbicular, 6–12 cm. long, 5–10 cm. wide, obtuse, rounded to slightly cordate; inflorescence an axillary, pedunculate, compound umbel or cyme; flowers pale green; style several-parted; drupe globose, red, about 6 mm. across; seeds U-shaped, coarsely rugose on back.——July–Sept. Warmer parts; Honshu (Tō-kaidō to Chūgoku Distr.), Shikoku, Kyushu.——India, Malaysia, and China.

5. CISSAMPELOS L. MIYAKOJIMA-TSUZURA-FUJI ZOKU

Woody climbers; leaves often peltate, glabrous or pubescent; staminate inflorescence corymbiform, paniculate or cymose, axillary, the pistillate cymose, racemose, or rarely in axillary fascicles or similar to the staminate; staminate sepals 4–5, the petals 2–5, free or connate; stamens united into a column, the anthers 2–4, marginal; carpels solitary; style 3-fid or dentate, the lobes entire or bilobed; drupe subglobose, oblique, the stones compressed laterally, tuberculate on back, excavated on each side; seeds U-shaped.——About 20 species, in the Tropics.

1. Cissampelos insularis Makino. *Paracyclea insularis* (Makino) Kudo & Yamamoto——MIYAKOJIMA-TSUZURA-FUJI. Scandent shrubs with striate branches, pubescent while young; leaves ovate-cordate or broadly ovate, 3–10 cm. long and as wide, subobtuse, pubescent on both sides especially beneath, green above, paler beneath; inflorescence paniculate or racemose, sessile in the pistillate, the staminate short-peduncu-late, 3–8 cm. long; staminate petals 4–5, free; fruit subglobose, about 5 mm. long, thinly long-pilose.——Aug.–Oct. Warmer parts; Honshu (s. Kinki and w. Chūgoku Distr.), Shikoku, Kyushu; rare.——Ryukyus.

Fam. 94. MAGNOLIACEAE Mokuren Ka Magnolia Family

Evergreen or deciduous, often aromatic trees or shrubs or sometimes scandent woody vines; leaves entire, simple, rarely lobed or toothed, the stipules large, rarely absent; flowers usually bisexual, sometimes unisexual, actinomorphic: sepals 3, often petaloid; petals 6 to many; stamens many, rarely connate; carpels usually spirally arranged on an elongate often woody torus, rarely verticillate, each with 1 to few usually anatropous ovules, dehiscing along dorsal or ventral sutures or berrylike and indehiscent; endosperm oily.——About 10 genera, with about 100 species, in temperate and subtropical regions of Asia and America.

1A. Erect woody plants; mature fruit a woody, dehiscent follicle; flowers bisexual.
  2A. Stipules prominent, deciduous; carpels spirally arranged, the seeds dorsally dehiscent at maturity.
    3A. Fruit stipitate. .... 1. *Michelia*
    3B. Fruit sessile. .... 2. *Magnolia*
  2B. Stipules absent; carpels in a single simple verticil, the seeds dehiscent on inner side when mature. .... 3. *Illicium*
1B. Woody climbers; mature carpels fleshy, indehiscent, berrylike; flowers unisexual; plants dioecious or monoecious.
  4A. Receptacles much elongate after anthesis, the carpels spicate. .... 4. *Schisandra*
  4B. Receptacles not elongate, the carpels capitate. .... 5. *Kadsura*

1. MICHELIA L. Ogatama-no-ki Zoku

Trees; leaves simple, penninerved, stipulate; flowers solitary, axillary, rather small; sepals and petals (tepals) similar, 9 to many, 3- to many-seriate, imbricate; anthers linear, the locules introrse; carpels many, 2- to many-ovuled, loosely arranged on an elongate conelike axis, coriaceous, persistent; seeds at first suspended by a filiform funicle.——More than 10 species, in mountains of tropical and subtropical regions of Asia.

1. **Michelia compressa** (Maxim.) Sarg. *Magnolia compressa* Maxim.——Ogatama-no-ki. Erect evergreen tree, brownish appressed-pubescence on young branches, buds, and young leaves beneath; leaves oblong-obovate or oblong, 5–10 cm. long, 2–4 cm. wide, coriaceous, lustrous and deep green above, paler beneath, acute with an obtuse tip; petioles 2–3 cm. long, appressed-pubescent; flowers solitary in upper axils, about 3 cm. across, white, with a purplish red center; perianth-segments 12, oblanceolate, 15–25 mm. long; carpels short-pubescent while young, 2- to 3-seeded, the fruit 5–10 cm. long.——Feb.–Apr. Warmer regions; Honshu (s. Kantō, Tōkaidō, and s. Kinki Distr.), Shikoku, Kyushu.——Ryukyus and Formosa.

2. MAGNOLIA L. Mokuren Zoku

Evergreen or deciduous trees or shrubs with rather stout branches, the winter buds large, enveloped by a single outer scale; leaves usually entire, simple, petiolate; stipules membranous, adnate to the base of petiole and enclosing the young successive leaves, the scar encircling the stem; flowers large, solitary, terminal, sessile or short-pedunculate, bisexual, the bud enclosed in a stipular spathe; sepals 3, often petaloid; petals 6–12, in 2–4 series; stamens many, the anthers linear, the locules introrse; carpels many, 2-ovuled, on a conelike, woody receptacle; seeds with an aril.——About 35 species, in temperate to warmer regions of eastern N. America, Mexico, West Indies, e. Asia, and the Himalayas.

1A. Flowers after or with the leaves; fruit conelike, usually symmetrical, subglobose to oblong.
  2A. Evergreen tree; leaves thick-coriaceous; cultivated. .... 1. *M. grandiflora*
  2B. Deciduous trees and shrubs.
    3A. Leaves 20–40 cm. long, mostly at the tip of the branches; branches and winter buds glabrous; flowers erect on a short stout peduncle. .... 2. *M. obovata*
    3B. Leaves 6–18 cm. long, alternate on the branches.
      4A. Flowers erect, 12–15 cm. across, on a short stout peduncle. .... 3. *M.* × *watsonii*
      4B. Flowers nodding, 8–10 cm. across, on a rather long slender peduncle. .... 4. *M. sieboldii*
1B. Flowers appearing before the leaves, subsessile; fruit cylindric, usually curved and unsymmetric.
  5A. Sepals 3, much shorter and narrower than the petals.
    6A. Flowers expanding, white or partially reddish; indigenous trees and shrubs.
      7A. Leaves broadly lanceolate to ovate-oblong, acute, glaucous and white-puberulent beneath; branches rather slender; flowers without a small leaf at base of the peduncle. .... 5. *M. salicifolia*
      7B. Leaves obovate, abruptly acuminate, pale green beneath; branches rather stout; flowers usually with a small leaf at base of the peduncle. .... 6. *M. kobus*
    6B. Flowers not expanding, purplish, with ascending sepals and petals; leaves broadly obovate, 6–12 cm. wide; cultivated.
                                                7. *M. liliflora*
  5B. Perianth segments similar in shape and color.
    8A. Perianth-segments 9, obovate; leaves broadly obovate, 8–15 cm. long. .... 8. *M. denudata*
    8B. Perianth-segments 12–18, oblanceolate; leaves narrowly obovate or broadly oblanceolate, 5–8 cm. long. .... 9. *M. stellata*

1. **Magnolia grandiflora** L. Taisan-boku. Evergreen tree, the buds and young branches densely soft brown-tomentose; leaves coriaceous, oblong to narrowly so, or ovate, sometimes ovate-oblong, 12–23 cm. long, 5–10 cm. wide, subobtuse to subacute, lustrous and deep green above, brown-tomentose beneath, the petioles about 2 cm. long; flowers cup-shaped, erect, 15–20 cm. across, white, fragrant, on a short stout peduncle; fruit ovoid, 7–10 cm. long, brown-hairy.——May–July. Commonly planted in parks and gardens.——se. United States.

2. **Magnolia obovata** Thunb. *M. hypoleuca* Sieb. & Zucc.; *M. glauca* sensu Thunb., non L.——Hō-NO-KI. Deciduous tree; branches stout, glabrous, purplish; winter buds glabrous; leaves obovate to narrowly cuneate-obovate, 20–40 cm. long, 13–25 cm. wide, usually rounded and broadly mucronate, obtuse to narrowly rounded at base, glabrous above, glaucous and pubescent beneath, the petioles glabrous or slightly pubescent on upper side, 2–3 cm. long; flowers cup-shaped, white, about 15 cm. across, erect, fragrant; sepals 3, similar to the petals but smaller; petals 6–9, obovate, slightly fleshy; anthers 16–18 mm. long; fruit narrowly oblong.——May–June. Mountains; Hokkaido, Honshu, Shikoku, Kyushu.——China.

3. **Magnolia × watsonii** Hook. f. UKEZAKI-ŌYAMA-RENGE. Small tree with glabrous rather stout branches; leaves obovate or oblong-obovate, 10–18 cm. long, 6–13 cm. wide, obtuse, nearly glabrous above, glaucous and pubescent beneath, the petioles 1.5–2 cm. long; flowers erect, cup-shaped, 12–15 cm. across, the peduncles 2–3 cm. long.——Reported to be a hybrid of No. 2 and No. 4.

4. **Magnolia sieboldii** C. Koch. *M. parviflora* Sieb. & Zucc., non Bl.; *M. oyama* Kort; *M. verecunda* Koidz.——ŌYAMA-RENGE. Large shrub with glabrous, rather stout branches, the buds scattered brown-hairy; leaves obovate or broadly so, 7–15 cm. long, 5–10 cm. wide, green and scattered-pubescent on nerves above, whitish to glaucous and sparsely long-pubescent beneath, the petioles pubescent, 2–3.5 cm. long; flowers nodding, 8–10 cm. across, fragrant, white, cup-shaped, the peduncle rather long and pubescent; sepals ovate, pale rose, 1/2–2/3 as long as the petals; petals white, obovate, usually 6; fruit ellipsoidal, 2.5–3.5 cm. long, glabrous.——May–July. Woods in mountains; Honshu (Kantō to Kinki Distr.), Shikoku, Kyushu; rare.——Korea.

5. **Magnolia salicifolia** (Sieb. & Zucc.) Maxim. *Buergeria salicifolia* Sieb. & Zucc.——TAMU-SHIBA. Small deciduous tree with rather slender glabrous branches; leaves broadly lanceolate to ovate-oblong, 6–12 cm. long, 2–5 cm. wide, thinly chartaceous, acute, glabrous above, glaucous and white-puberulent beneath, acute at base, the petioles glabrous, 1–1.5 cm. long; flowers white, 7–10 cm. across; sepals 3, lanceolate, 20–30 mm. long, 4–7 mm. wide, about half as long as the petals; petals 6(–12), narrowly obovate, to broadly oblanceolate, 17–25 mm. wide, spreading; fruit cylindric, 5–7 cm. long, glabrous.——Apr.–May. Mountains; Honshu (mainly Japan Sea side), Kyushu.

6. **Magnolia kobus** DC. *M. thurberi* Hort.; *M. praecocissima* Koidz.——KOBUSHI. Deciduous tree with rather stout glabrous branches; leaves thinly chartaceous, obovate to broadly so, 6–13 cm. long, 3–6 cm. wide, abruptly acuminate, the tip subobtuse, glabrous above, paler green and slightly pubescent on nerves beneath; flowers white, sometimes with a rose-purple center, about 10 cm. across, usually with a small leaf at base of the very short peduncle; sepals 3, linear, 15–18

mm. long, 3–4 mm. wide; petals narrowly obovate, 2–3.5 cm. broad, spreading, usually reddish near the base; fruit short-cylindric, 7–10 cm. long.——Apr.–May. Hills and base of mountains; Hokkaido, Honshu, Shikoku, Kyushu; rather common.

Var. **borealis** Sarg. *M. praecocissima* var. *borealis* (Sarg.) Koidz.——KITA-KOBUSHI. Leaves larger, 10–17 cm. long, 6–8 cm. wide; flowers slightly larger, often pink tinged.——Hokkaido and Honshu (Japan Sea side of n. and centr. distr.).——s. Korea.

7. **Magnolia liliflora** Desr. *M. discolor* Vent.; *M. purpurea* Curt.; *Yulania japonica* Spach; *M. obovata* sensu Willd., non Thunb.——MOKUREN, SHI-MOKUREN. Large deciduous shrub with rather stout erect nearly glabrous branches; leaves obovate or broadly so, 8–18 cm. long, 6–12 cm. wide, abruptly short-acuminate, narrowed below, scattered short-pilose above, pale green and short-pilose on nerves beneath, the petioles 10–15 cm. long; flowers tubular-campanulate, not expanding, 8–10 cm. long; sepals lanceolate, about 3 cm. long, 7–8 mm. wide, recurved below, with incurved margin above; petals 6, dark purple, usually whitish inside, 3–4 cm. wide.——Apr.–May. Widely cultivated in our area.——centr. China.

Var. **gracilis** (Salisb.) Rehd. *M. gracilis* Salisb.——TŌ-MOKUREN. Shrub with rather slender, erect branches, the leaves narrower; sepals linear-lanceolate, incurved on margin, rather straight, about 4 mm. wide; petals slightly smaller, narrower, dark red-purple, with whitish inner side.——Chinese shrub cultivated in gardens.

8. **Magnolia denudata** Desr. *M. conspicua* Salisb.; *M. yulan* Desf.——HAKU-MOKUREN. Deciduous tree, resembling the preceding, but the flowers white and open-campanulate; leaves broadly ovate, 8–15 cm. long; perianth-segments (tepals) 9, narrowly obovate, 7–8 cm. long, 3–4 cm. wide; fruit glabrous.——Apr. Ornamental tree from centr. China, cultivated in our area.

Var. **purpurascens** (Maxim.) Rehd. & Wils. *M. purpurascens* (Maxim.) Makino; *M. conspicua* var. *purpurascens* Maxim.——SARASA-RENGE. Perianth-segments pale red-purple outside and white inside.——Cultivated in our area; flowering slightly later than the typical phase.——centr. China.

9. **Magnolia stellata** (Sieb. & Zucc.) Maxim. *Buergeria stellata* Sieb. & Zucc.——SHIDE-KOBUSHI. Small tree or large shrub with densely pubescent young branches; leaves thinly chartaceous, narrowly obovate or broadly oblanceolate, 5–8 cm. long, 1–3 cm. wide, obtuse to rounded, cuneate at base, glabrous above, pale green and glabrous or slightly short-pubescent only on nerves beneath, the petioles 2–5 mm. long, usually appressed short-pubescent; flowers white to rose, about 8 cm. across; perianth-segments (tepals), 12–18, oblanceolate, 8–12 mm. wide, spreading, subrecurved before withering; carpels glabrous.——Apr. Mountains; Honshu (western Tōkaidō Distr.); often planted in gardens.

3. **ILLICIUM** L. SHIKIMI ZOKU

Glabrous, often aromatic, evergreen trees and shrubs; leaves simple, alternate, glandular-dotted, usually entire, petiolate; flowers solitary, pedicelled, white, yellowish or purplish; sepals 3 or 6, free, membranous, in 1 or 2 series, imbricate; petals many, spreading, usually linear-oblong; stamens many, the filaments thickened, the anther-locules introrse; carpels many, verticillate, 1-ovuled; follicles coriaceous or ligneous.——About 20 species, in e. Asia and N. America.

1. **Illicium religiosum** Sieb. & Zucc. *I. anisatum* L. pro parte; *I. japonicum* Sieb.——SHIKIMI. Small evergreen tree; leaves coriaceous, lustrous, narrowly obovate or oblong, 4–10

cm. long, 1.5–3.5 cm. wide, abruptly short-acuminate, the tip obtuse, entire, the lateral nerves and veinlets not prominent; petioles 10–15 mm. long; flowers pale yellow-white, fragrant,

Plate 10

Betula platyphylla var. *japonica* Hara (Betulaceae). Okuyukiusu in Nemuro Prov., Hokkaido. (Photo M. Tatewaki.)

Callianthemum miyabeanum Tatew. (Ranunculaceae). Mount Appi, Hidaka, Hokkaido. (Photo M. Tatewaki.)

Plate 11

Machilus thunbergii Sieb. & Zucc. (Lauraceae). Enoshima Island, Sagami Prov., Honshu. (Photo J. Ohwi.)

Stauntonia hexaphylla (Thunb.) Decne. (Lardizabalaceae). Near Kobe, Honshu. (Photo J. Ohwi.)

2.5–3 cm. across, the peduncles about 2 cm. long, the bracts caducous; petals broadly linear or lanceolate; seeds yellow-brown.——Mar.–May. Thickets and woods in foothills; Honshu (Kantō Distr. and westw.), Shikoku, Kyushu; common and much planted in temples and used as a decoration in Buddhist cemeteries.——Ryukyus, Formosa, and China.

4. SCHISANDRA Michx. MATSUBUSA ZOKU

Scandent dioecious woody climbers; leaves alternate, membranous, deciduous, exstipulate; flowers solitary, from the lower axils of young shoots, small, nodding, long-pedunculate, pale yellow or rose, the perianth-segments 6–12, petallike; stamens 5–15, connate; carpels many, indehiscent, 2-ovuled, loosely arranged in a spike on a much-elongate receptacle, berrylike, red or black; seeds reniform.——More than 10 species, 1 in N. America, the others in e. and s. Asia.

1A. Leaves narrowly obovate to obovate-orbicular, with 5–10 short teeth, the midrib impressed on upper side; petioles half or less than half as long as the blade; fruit red; seeds smooth. . . . . 1. *S. chinensis*
1B. Leaves ovate-orbicular, with 3–5 short teeth, the midrib not impressed on upper side; petioles about as long as to half as long as the blade; fruit blue-black; seeds tubercled. . . . . 2. *S. repanda*

1. **Schisandra chinensis** (Turcz.) H. Baill. *Kadsura chinensis* Turcz.; *Sphaerostemma japonica* Sieb. & Zucc.; *Maximowiczia chinensis* (Turcz.) Rupr.——CHŌSEN-GOMISHI. Small scandent shrub; leaves membranous, narrowly obovate to obovate-orbicular, rarely elliptic, 4–7 cm. long, 3–5 cm. wide, glabrous or sometimes papillose on nerves beneath, the petioles 2–2.5 cm. long, pale green; flowers creamy white, pendulous on slender peduncles; petals oblong, about 10 mm. long, ascending; fruit globose, red; seeds smooth.——May–June. Mountains; Hokkaido, Honshu (n. and centr. distr.).——Korea, China, Amur, and Sakhalin.

2. **Schisandra repanda** (Sieb. & Zucc.) Radlk. *Trochostigma repanda* Sieb. & Zucc.; *S. nigra* Maxim.——MATSUBUSA. Closely resembles the preceding; leaves slightly thicker, broader, flat above, glabrous, 2–6 cm. long, 3.5–5 cm. wide, rounded and abruptly petiolate at base, paler beneath, the nerves not impressed, the petioles 2–5 cm. long, often reddish, the petals creamy white; fruit globose, blue-black; seeds tubercled.——May–June. Mountains; Hokkaido, Honshu, Shikoku, Kyushu.——s. Korea.——Forma **hypoleuca** (Makino) Ohwi. *S. nigra* var. *hypoleuca* Makino; *S. discolor* Nakai——URAJIRO-MATSUBUSA. Leaves whitish beneath.——Occurs with the typical phase.

5. KADSURA Juss. SANE-KAZURA ZOKU

Evergreen scandent dioecious shrubs; leaves alternate, petiolate, simple, rather thick, lustrous, scattered-toothed; stipules absent; flowers axillary, solitary, creamy white; perianth-segments 9–15, the outer gradually smaller, the inner petaloid; stamens in the staminate flowers many, free or connate in a head, the connective broad; carpels many, in heads, 2- to 3-ovuled, the fruiting ones berrylike, red, on a globose incrassate fleshy receptacle; seeds reniform.——About 10 species, in the tropical and temperate regions of Asia.

1. **Kadsura japonica** (Thunb.) Dunal. *Uvaria japonica* Thunb.——SANE-KAZURA, BINAN-KAZURA. Leaves elliptic, oblong or ovate, 5–10 cm. long, 3–5 cm. wide, rather thick and soft, deep green and lustrous above, paler and often purplish beneath, the lateral nerves and veinlets very weak and obsolete; peduncles with few small scalelike bracts on lower part; flowers about 1.5 cm. across, pendulous, the inner petals elliptic, about 10 mm. long; seeds about 5 mm. across.——Aug. Honshu (Kantō Distr. and westw.), Shikoku, Kyushu.——China and Formosa.

Fam. 95. LAURACEAE KUSU-NO-KI KA Laurel Family

Usually aromatic, sometimes dioecious erect trees or shrubs, rarely twining epiphytic parasites; leaves alternate, simple or palmately cleft, penninerved, sometimes 3-nerved at base; stipules absent; flowers bisexual or unisexual, actinomorphic, rather small, in axillary umbels, racemes, or panicles, the perianth-segments 4 or 6, often connate at base; stamens 6–12, the anthers 2- to 4-locular, the locules opening upward by a flaplike valve; ovary 1-locular, with a solitary pendulous ovule; fruit a berry or dry, often enclosed by the persistent perianth base; seeds without endosperm.——About 40 genera, with about 1,500 species, mainly in the Tropics, relatively few in temperate regions.

1A. Flowers bisexual; inflorescence without involucral bracts.
  2A. Twining parasitic herbs; leaves scalelike, small; anthers 2-locular. . . . . 1. *Cassytha*
  2B. Trees with large leaves.
    3A. Fruit entirely enclosed within the perianth-tube; anthers 2-locular. . . . . 2. *Cryptocarya*
    3B. Fruit naked; anthers 4-locular.
      4A. Perianth-segments deciduous; leaves usually 3-nerved. . . . . 3. *Cinnamomum*
      4B. Perianth-segments persistent, reflexed in fruit; leaves usually penninerved. 4. *Machilus*
1B. Flowers unisexual (plants dioecious); inflorescence involucrate at base.
  5A. Anthers (except in *Lindera citriodora*) 2-locular; perfect stamens 9; leaves mostly deciduous (evergreen in *Lindera strychnifolia*).
    6A. Fruit an indehiscent berry. . . . . 5. *Lindera*
    6B. Fruit dry, irregularly splitting, 5- to 6-lobed, the seed exposed. . . . . 6. *Parabenzoin*
  5B. Anthers 4-locular; leaves evergreen.
    7A. Perianth-segments 4; stamens 6(–8); leaves usually 3-nerved. . . . . 7. *Neolitsea*

7B. Perianth-segments 6; stamens 9(–12); leaves usually penninerved.
    8A. Lower surface of leaves tomentose; perianth-tube developed into a cup enclosing the base of the fruit. .... 8. *Litsea*
    8B. Lower surface of leaves glabrous or nearly so; perianth-tube not developed into a cup enclosing the base of fruit.
                                                         9. *Actinodaphne*

1. CASSYTHA L.   SUNAZURU ZOKU

Epiphytic twining parasitic herbs; stems filiform, much-branched, adhering to the host by holdfasts; leaves small, scalelike, alternate; inflorescence spicate, racemose, or in heads, in axils of scalelike leaves; flowers sessile or short-pedicelled, bisexual, sometimes dimorphic, the bracteoles 2, minute; perianth-segments 6, persistent, the inner 3 connate below into an ovoid tube; perfect stamens 9, the 6 outer ones introrse, without glands, the inner 3 extrorse, each with a gland on each side; staminodes subsessile or stipitate; fruit enclosed in a fleshy perianth-tube.——About 30 species, in the tropics.

1. Cassytha filiformis L. SUNAZURU. Glabrous, slender, herbaceous, much branched; scales, bracts, bracteoles and 3 outer perianth-segments ovate-cordate, about 0.5 mm. long, ciliolate; flowers sessile, in rather loose axillary spikes; inner perianth-segments elliptic or ovate-deltoid, 1.5–2 mm. long; fruit globose, 6–7 mm. long and as wide, enclosed by the perianth-tube.——Oct.–Nov. Near seashores; Kyushu (Yakushima). Widely distributed in the Tropics.

2. CRYPTOCARYA R. Br.   SHINA-KUSU-MODOKI ZOKU

Trees or shrubs; leaves alternate or rarely subopposite, 3-nerved or penninerved; inflorescence a panicle; flowers small, bisexual; perianth-segments 6, nearly equal, the tube nearly ovoid, contracted at apex after anthesis; perfect stamens 9, in 3 series, those of outer 2 series not glanduliferous, introrse, the inner with a gland on each side, extrorse; staminodes ovate; ovary sessile; berry entirely enclosed in the perianth-tube, globose or ellipsoidal, the perianth-segments deciduous.——More than 100 species, in the Tropics.

1. Cryptocarya chinensis (Hance) Hemsl. *Beilschmiedia chinensis* Hance——SHINA-KUSU-MODOKI. Tree; leaves narrowly elliptic, 7–10 cm. long, 3–5 cm. wide, acuminate, 3-nerved, the petioles 7–12 mm. long; panicles axillary and terminal, 2–3 cm. long; flowers 2–3 mm. long; fruit about 10 mm. across, 13–15 mm. long.——Kyushu (Hyuga Prov. and Tanegashima).——Formosa and s. China.

3. CINNAMOMUM Bl.   KUSU-NO-KI ZOKU

Aromatic trees; leaves evergreen, coriaceous, subopposite or alternate, 3-nerved; inflorescence an axillary or terminal panicle, sometimes a raceme or umbel; flowers bisexual, the perianth-segments 6, equal, the tube short; perfect stamens 9 or fewer, in 3 series, the outer 2 series without glands, the anthers introrse, 4-locular, the innermost series with 2 glands on each filament, the anthers extrorse; staminodes capitellate and stipitate; ovary sessile; berry with a persistent perianth-tube below.——About 100 species, in tropical Asia, few in Australia.

1A. Leaves with impressed axillary glands beneath. .... 1. *C. camphora*
1B. Leaves without axillary glands beneath.
    2A. Leaves obovate-oblong, 3–6 cm. long, rounded to very obtuse, with pale yellowish silky appressed-hairs beneath. .. 2. *C. daphnoides*
    2B. Leaves lanceolate to ovate-orbicular, 6–15 cm. long, acute to obtuse, glabrous or scattered-hairy.
        3A. Inflorescence glabrous; leaves acute. .... 3. *C. japonicum*
        3B. Inflorescence appressed-pubescent; leaves short-caudate. .... 4. *C. sieboldii*

1. Cinnamomum camphora (L.) Sieb. *Laurus camphora* L.——KUSU-NO-KI. Glabrous evergreen tree with yellow-brown branches; buds many-scaled; leaves subcoriaceous, ovate or broadly so or elliptic, 6–10 cm. long, 3–6 cm. wide, abruptly acuminate, lustrous above, often glaucescent beneath, 3-nerved or subpenninerved, with 2 impressed glands in the lower nerve axils beneath, the petioles 15–25 mm. long; inflorescence axillary, paniculate, glabrous, pedunculate, shorter than the leaves, the bracts caducous, membranous, linear, soft-pubescent, the perianth-segments about 1.5 mm. long, yellow-green, soft-pubescent inside, elliptic; fruit globose, 7–8 mm. across, black.——May–June. Warmer parts; Honshu (Kantō Distr. and westw.), Shikoku, Kyushu; widely planted. ——Ryukyus, Formosa, and China.

2. Cinnamomum daphnoides Sieb. & Zucc. *C. sericeum* Sieb. ex Miq.——KŌCHI-NIKKEI. Evergreen tree, the young branchlets pale yellowish and densely appressed silky-pubescent; leaves coriaceous, obovate to oblong, 3–6 cm. long, 1.2–3 cm. wide, rounded at apex, obtuse at base, appressed-pubescent above while young, densely yellowish silky appressed-tomentose beneath, deep green above, the margin slightly recurved, the midrib 3-parted from the base, the petioles 5–8 mm. long, silky-pubescent; inflorescence densely appressed silky-pubescent; perianth-segments 3–3.5 mm. long, appressed-pubescent on both sides, broadly ovate-elliptic; fruit ellipsoidal, about 10 mm. long.——Kyushu (s. distr.); common near the sea.—— Ryukyus.

3. Cinnamomum japonicum Sieb. ex Nakai. *C. pedunculatum* Nees, excl. syn. Thunb.——YABU-NIKKEI. Glabrous evergreen tree; leaves subcoriaceous, narrowly ovate or oblong, 6–12(–15) cm. long, 2–5 cm. wide, obtuse, glaucous beneath, the midrib 3-parted near base, the petioles 8–18 mm. long; inflorescence shorter than the leaves, glabrous; peduncles prominent, the pedicels rather short; perianth-segments about 2.5 mm. long, broadly ovate, puberulent inside; fruit ellipsoidal, 10–12 mm. long, black.——June. Honshu (Kantō Distr. and westw.), Shikoku, Kyushu.——Korea, China, Ryukyus, and Formosa.

Cinnamomum brevifolium Miq. *C. maruba* Jungh.——
Maruba-yabu-nikkei. Leaves nearly orbicular, 3-nerved
from the base.——This may be a variant of No. 3.
4. Cinnamomum sieboldii Meissn. *C. loureiri* sensu
auct. Japon., non Nees——Nikkei. Evergreen tree; branches
with scattered short appressed hairs while young; leaves coriaceous, narrowly ovate-oblong, rarely broadly lanceolate, 8–15

cm. long, (2–)2.5–5 cm. wide, glabrous (young leaves appressed white-pubescent), caudate, glaucous beneath, the midrib 3-parted near base, the petioles 15–20 mm. long; inflorescence much shorter than the leaves, rather densely appressed
white-hairy; perianth-segments about 5 mm. long, oblong, appressed-puberulent on both sides.——May–June. Planted in
warmer districts.——Ryukyus (?).

4. MACHILUS Nees   Tabu-no-ki Zoku

Evergreen trees with alternate, penninerved, petiolate leaves; panicles axillary; flowers bisexual, small, the perianth-segments 6,
equal or the outer 3 slightly smaller, persistent, the tube short; perfect stamens 9, in 3 series, the anthers 4-locular, the outer 6
without glands, introrse, the inner 3 each with 2 glands at base, extrorse; staminodes 3, capitate, stipitate; ovary sessile; berry ellipsoidal or globose, with persistent reflexed perianth-segments at base.——About 60 species, in s. and e. Asia.

1A. Leaves obovate to broadly oblanceolate, sometimes oblong, glabrous, abruptly obtuse. . . . . 1. *M. thunbergii*
1B. Leaves obovate-oblong to lanceolate or oblanceolate, usually tapered and sometimes caudate. 2. *M. japonicus*

1. Machilus thunbergii Sieb. & Zucc. Tabu-no-ki, Inu-
gusu. Large evergreen tree with stout, terete, glabrous
branches, the bud-scales many, the inner ones yellow-brown,
woolly; leaves coriaceous, obovate, elliptic or broadly oblanceolate, 8–15 cm. long, 3–7 cm. wide, abruptly obtuse, usually
acute at base, lustrous and deep green above, penninerved,
pale green and with slightly raised veinlets beneath, the petioles 2–3 cm. long, stout; inflorescence glabrous; flowers pedicelled, the perianth-segments 5–7 mm. long, narrowly oblong,
brown-pubescent inside; fruit globose, about 10 mm. across,
black-purple.——Mar.–May. Hills, lowlands, and foothills;
Honshu, Shikoku, Kyushu; common.——s. Korea, Ryukyus,
Formosa, and China.

2. Machilus japonicus Sieb. & Zucc. *M. thunbergii* var.
japonicus (Sieb. & Zucc.) Yatabe——Hosoba-tabu, Aogashi.
Evergreen tree with glabrous branches, the bud-scales many,
ciliate; leaves lanceolate or obovate-oblong, 8–20 cm. long, 2–4
cm. wide, sometimes oblanceolate, somewhat lustrous, whitish
with slightly raised veinlets beneath, gradually acuminate or
caudate, acute at base, the petioles 15–20 mm. long; inflorescence glabrous; flowers pedicelled, the perianth-segments 4–5
mm. long, narrowly oblong, puberulent inside and on margin;
fruit globose, about 10 mm. long.——May–June. Honshu
(Kinki Distr. and westw.), Shikoku, Kyushu.——s. Korea and
Ryukyus (?).

5. LINDERA Thunb.   Kuro-moji Zoku

Dioecious trees and shrubs; bud-scales few to many or absent; leaves alternate or subopposite, deciduous, sometimes evergreen,
penninerved, sometimes 3-nerved; inflorescence umbellate, sessile or peduncled, with usually 4(2–5) involucral bracts in bud;
flowers unisexual, the perianth-segments 6, deciduous, nearly equal, the tube short; stamens usually 9, in 3 series, the outer
6 without glands, the inner each with 2 glands, the anthers introrse, 2-locular (4-locular in No. 5); staminodes absent; berry
globose or ovoid.——About 100 species, in warmer and tropical regions of the N. Hemisphere, few in N. America, abundant in
Asia.

1A. Leaves 3-nerved from the base.
    2A. Leaves often 3-cleft, deciduous, ovate-orbicular. . . . . 1. *L. obtusiloba*
    2B. Leaves unlobed, evergreen, elliptic or broadly so. . . . . 2. *L. strychnifolia*
1B. Leaves penninerved.
    3A. Branchlets yellowish brown; winter-buds sessile.
        4A. Petioles 3–4 mm. long; leaves oblong to elliptic; fruit black, 6–7 mm. across. . . . . 3. *L. glauca*
        4B. Petioles 7–20 mm. long; leaves oblanceolate; fruit red, 5–6 mm. across. . . . . 4. *L. erythrocarpa*
    3B. Branchlets dark green; flowers opening before or with the leaves; fruit black.
        5A. Leaves glabrous; pedicels 4–6 mm. long, glabrous; winter-buds sessile. . . . . 5. *L. citriodora*
        5B. Leaves long-pubescent at least while young; pedicels 15–20 mm. long, pubescent; winter-buds stipitate.
            6A. Petioles 10–15 (–20) mm. long; leaves without prominently raised veinlets beneath. . . . . 6. *L. umbellata*
            6B. Petioles 5–10 (–15) mm. long; leaves with prominently raised veinlets beneath. . . . . 7. *L. sericea*

1. Lindera obtusiloba Bl. *Benzoin obtusilobum* (Bl.)
O. Kuntze——Dan-kōbai. Large shrub; leaves alternate,
ovate-orbicular or broadly so, 5–15 cm. long, 4–13 cm. wide,
usually 3-lobed, green and slightly pubescent above while
young, whitish and often long-pubescent becoming glabrate
beneath, subobtuse to subacute, truncate to shallowly cordate
at base, the lobes ovate-deltoid, the lateral ones erect, obtuse,
the petioles 1.5–3 cm. long; flowers opening before the leaves,
the peduncles absent, the pedicels 12–15 mm. long, pubescent,
slightly thickened at apex; fruit globose, about 8 mm. across,

red.——Apr.–May. Mountains; Honshu (Kantō Distr. and
westw.), Shikoku, Kyushu.——Korea and China.
2. Lindera strychnifolia (Sieb. & Zucc.) F. Vill. *Daphnidium strychnifolium* Sieb. & Zucc.; *Benzoin strychnifolium*
(Sieb. & Zucc.) O. Kuntze——Tendai-uyaku. Evergreen
shrub; leaves subcoriaceous, elliptic to broadly obovate, 4–6
cm. long, 2.5–4 cm. wide, caudate, with an obtuse tip, glaucous
beneath, densely pale brown-pubescent while young, glabrescent, except the midrib, the veinlets rather prominent, the
petioles 5–10 mm. long; peduncles absent, the pedicels 3–4

mm. long at anthesis, pubescent.——Apr. Chinese shrub often planted in warmer parts and sometimes naturalized.

3. Lindera glauca (Sieb. & Zucc.) Bl. *Benzoin glaucum* Sieb. & Zucc.——YAMA-KŌBASHI. Large deciduous shrub; leaves oblong to elliptic, often ovate or obovate-oblong, acute to subacute at both ends, whitish, with deciduous hairs beneath, green, glabrous, the midrib puberulous above, the petioles short, 3–4 mm. long; peduncles very short, the pedicels 13–15 mm. long, thickened at apex; fruit globose, black, 6–7 mm. across.——Apr. Honshu (Kantō Distr. and westw.), Shikoku, Kyushu.——Korea, China, and Formosa.

4. Lindera erythrocarpa Makino. *Benzoin erythrocarpa* (Makino) Rehd.; *L. thunbergii* Makino, excl. syn.——KANA-KUGI-NO-KI. Large deciduous shrub; leaves oblanceolate to oblong, 6–13 cm. long, 1.5–2.5 cm. wide, acute or slightly obtuse, glabrous above, glaucescent and with sparse deciduous hairs beneath, the petioles 6–10 mm. long; peduncles 3–5 mm. long, the pedicels pubescent, 10–12 mm. long in fruit; fruit globose, 5–6 mm. across, red.——May. Honshu (s. part of centr. distr. and westw.), Shikoku, Kyushu.——Korea and China.

5. Lindera citriodora (Sieb. & Zucc.) Hemsl. *Aperula citriodora* (Sieb. & Zucc.) Bl.; *Benzoin citriodorum* Sieb. & Zucc.; *L. citrata* Koidz., excl. syn.; *Litsea citriodora* (Sieb. & Zucc.) Hatusima——AO-MOJI. Small deciduous glabrous tree; branches dark green; leaves alternate, thinly chartaceous, broadly lanceolate to narrowly oblong-ovate, 7–15 cm. long, 2–4.5 cm. wide, long-acuminate, glaucous beneath, the veinlets slightly raised above, the petioles 12–25 mm. long; peduncles 10–12 mm. long, the pedicels 4–6 mm. long, slightly thickened at apex; anthers 4-locular; fruit about 5 mm. across, globose, black.——Mar. Low elevations; Kyushu.——Ryukyus and Formosa.

6. Lindera umbellata Thunb. *Benzoin umbellatum* (Thunb.) O. Kuntze; *L. hypoglauca* Maxim.; *Benzoin hypoglaucum* (Maxim.) O. Kuntze; *L. umbellata* var. *hypoglauca* (Maxim.) Makino; *Benzoin thunbergii* Sieb. & Zucc.; *L. obtusa* Fr. & Sav.——KURO-MOJI. Erect deciduous shrub; leaves thinly chartaceous, narrowly oblong to ovate-oblong, 5–9 cm. long, 1.5–3.5 cm. wide, whitish hairy beneath, glabrescent, the lateral nerves 4 to 6 pairs, the veinlets weak and not raised beneath, the petioles slender, 10–15 mm. long; inflorescence pubescent, the peduncles 4–8 mm. long, the pedicels 15–22 mm. long; perianth-segments elliptic, 3–4 mm. long; fruit about 6 mm. across, black.——Honshu, Shikoku, Kyushu; rather common.——China.

Var. **membranacea** (Maxim.) Momiyama. *L. membranacea* Maxim.; *Benzoin membranaceum* (Maxim.) O. Kuntze ——ŌBA-KURO-MOJI. Plant larger than in the typical phase, the leaves narrowly oblong to obovate, 6–12 cm. long, 3–5 cm. wide, the lateral nerves 5 to 8 pairs.——Hokkaido, Honshu (n. distr.).

Var. **lancea** Momiyama. HIME-KURO-MOJI. Leaves lanceolate-oblong to oblanceolate, acuminate, the silky hairs beneath persistent, the petioles shorter, pubescent, the peduncles shorter, with yellowish velvety persistent hairs.——Honshu (Suruga Prov. to Kii and Kawachi Prov.), Shikoku, Kyushu.

7. Lindera sericea (Sieb. & Zucc.) Bl. *Benzoin sericeum* Sieb. & Zucc.; *L. umbellata* var. *sericea* (Sieb. & Zucc.) Makino; *Benzoin umbellatum* var. *sericeum* (Sieb. & Zucc.) Rehd.——KE-KURO-MOJI. Shrub; leaves chartaceous, narrowly obovate, 6–15 cm. long, 2–6 cm. wide, acuminate to very acute, cuneate, short-pubescent on both sides and with scattered long hairs beneath especially on nerves and on midrib above, whitish beneath, the lateral nerves 6 to 9 pairs, the veinlets prominent and raised beneath; inflorescence few-flowered, the pedicels slightly thickened above in fruit, 15–18 mm. long; fruit globose, about 7 mm. across, black.——Honshu (Kii Prov.), Shikoku, Kyushu.

Var. **glabrata** Bl. *L. sericea* var. *tenuis* (Nakai) Momiyama; *L. subsericea* Makino; *Benzoin sericeum* var. *tenue* Nakai——USUGE-KURO-MOJI, MIYAMA-KURO-MOJI. Leaves somewhat smaller and thinner, not short-pubescent on both surfaces but scattered long-hairy beneath.——Honshu (Kantō to centr. distr. and westw.), Shikoku, Kyushu.

6. PARABENZOIN Nakai   SHIRO-MOJI ZOKU

Deciduous dioecious shrubs; leaves alternate, petiolate, simple or 3-fid, penninerved or 3-nerved; inflorescence umbellate, short-pedunculate, with 4 involucral bracts when young; flowers similar to those of *Lindera*, the pedicels thickened; fruit large, globose, irregularly splitting into 5–6 segments and exposing the single yellow-brown seed.——Japan and China.

1A. Leaves usually 3-cleft, 3-nerved from the base. .... 1. *P. trilobum*
1B. Leaves simple, undivided, not 3-nerved from the base. .... 2. *P. praecox*

1. Parabenzoin trilobum (Sieb. & Zucc.) Nakai. *Benzoin trilobum* Sieb. & Zucc.; *Lindera triloba* (Sieb. & Zucc.) Bl.——SHIRO-MOJI. Deciduous shrub; leaves thinly chartaceous, broadly deltoid-obovate, often 3-lobed, 8–12 cm. long, 7–10 cm. wide in the divided ones, 3–5 cm. wide in the simple ones, glaucescent and glabrous beneath except for spreading hairs on the nerves, 3-nerved from the base, the lobes ovate, acuminate, the petioles 10–20 mm. long; inflorescence short-peduncled, the pedicels thickened in fruit, 8–12 mm. long; fruit globose, about 12 mm. across.——Apr. Honshu (centr. distr. and westw.), Shikoku, Kyushu; rather common in mountains.——China.

2. Parabenzoin praecox (Sieb. & Zucc.) Nakai. *Benzoin praecox* Sieb. & Zucc.; *Lindera praecox* (Sieb. & Zucc.) Bl.——ABURA-CHAN. Deciduous shrub to small tree; leaves chartaceous, ovate to narrowly so or elliptic, 5–7 cm. long, 2–4 cm. wide, acuminate, glabrous, whitish and often soft-pubescent on the nerves beneath, the petioles slender, 10–20 mm. long; flowers opening before the leaves, the pedicels thickened, 8–10 mm. long, slightly ligneous; fruit globose, brownish yellow, about 15 mm. across.——Mar.–Apr. Mountains; Honshu, Shikoku, Kyushu; rather common.

7. NEOLITSEA Merr. SHIRO-DAMO ZOKU

Evergreen dioecious trees; leaves alternate, petiolate, entire, usually 3-nerved from above the base or rarely penninerved, often silky-pubescent while young; inflorescence umbellate, subsessile, enclosed while young by involucral bracts; flowers unisexual, the perianth-segments 4, deciduous, the tube very short; stamens in the staminate flowers 6(-8), all fertile, in 3 series, the outer 4 without glands, the inner 2 each with a gland on each side, the anthers introrse, 4-locular; fruit a berry, red or black.——About 60 species, in e. Asia, Malaysia, and India.

1A. Leaves 8–18 cm. long; petioles 2–3 cm. long; flowers yellow, in autumn; fruit red. . . . . 1. *N. sericea*
1B. Leaves 5–12 cm. long; petioles 8–15 mm. long; flowers reddish, in spring; fruit black. . . . . 2. *N. aciculata*

1. Neolitsea sericea (Bl.) Koidz. *Laurus sericea* Bl.; *Litsea glauca* Sieb.; *Malapoenna sieboldii* O. Kuntze; *N. glauca* (Sieb.) Koidz.; *N. sieboldii* (O. Kuntze) Nakai; *N. latifolia* Koidz., non S. Moore——SHIRO-DAMO. Evergreen tree with greenish branchlets; leaves coriaceous, oblong or ovate-oblong, 8–18 cm. long, 4–7 cm. wide, obtuse, 3-nerved from above the base, white beneath, densely yellow-brown silky while young, glabrescent or the hairs becoming scattered, the petioles 2–3 cm. long; inflorescence umbellate, sessile, yellow-brown villous, the pedicels 7–10 mm. long in fruit, slightly thickened above; fruit ellipsoidal, 12–15 mm. long, red.——Sept.–Oct. Honshu, Shikoku, Kyushu.——Korea, Formosa, and China.

2. Neolitsea aciculata (Bl.) Koidz. *Litsea foliosa* Sieb. & Zucc., non Nees; *Litsea aciculata* Bl.; *Malapoenna aciculata* (Bl.) O. Kuntze——INUGASHI, MATSURA-NIKKEI. Evergreen tree; leaves coriaceous, oblong to obovate-oblong, 5–12 cm. long, 2–4 cm. wide, obtuse, 3-nerved from above the base, white, glabrous or slightly appressed-pubescent beneath, the petioles 8–15 mm. long; flowers reddish, the pedicels hairy, 7–8 mm. long; fruit ellipsoidal, slightly longer than the pedicel, with a small cup-shaped perianth-tube at base, black.——Mar.–Apr. Honshu (Kantō Distr. and westw.), Shikoku, Kyushu.——Ryukyus and Korea.

8. LITSEA Lam. HAMA-BIWA ZOKU

Evergreen dioecious trees; leaves alternate, rarely subopposite, usually penninerved; inflorescence umbellate, sessile or pedunculate, subtended by 4–6 involucral bracts; perianth-segments 6, the tube short or ovoid; stamens 9–12, the first and second series without glands, the anthers introrse, 4-locular; fruit a berry often surrounded by an inflated perianth-tube at base.——About 150 species, abundant in tropical Asia and Australia, few in Africa and America.

1. Litsea japonica (Thunb.) Juss. *Tomex japonica* Thunb.; *Fiwa japonica* (Thunb.) J. F. Gmel.; *Tetranthera japonica* (Thunb.) Spreng.——HAMA-BIWA. Evergreen tree; branches yellowish velvety; leaves coriaceous, oblong to narrowly so, 7–15 cm. long, 2–5 cm. wide, slightly recurved on margin, obtuse, glabrous above, yellowish velvety beneath, the lateral nerves 8 to 12 pairs, prominently raised beneath, the petioles 15–40 mm. long, densely woolly; inflorescence short-pedunculate, often branched; flowers very short-pedicelled, the perianth-segments deciduous; stamens 9; fruit ellipsoidal, blue-purple, 15–18 mm. long, about 12 mm. across, with a cup-shaped perianth-tube at base.——Oct.–Nov. Thickets and woods near the sea; Honshu (Nagato Prov.), Shikoku, Kyushu.——Ryukyus and s. Korea.

9. ACTINODAPHNE Nees KAGO-NO-KI ZOKU

Evergreen dioecious trees; leaves alternate, sometimes subverticillate, coriaceous, penninerved; inflorescence umbellate, enveloped by deciduous imbricate bracts while young, pedunculate or sessile; flowers small, the perianth-segments 6, sometimes persistent, the tube short; perfect stamens 9, in 3 series, the outer 6 without glands, the anthers introrse, 4-locular; fruit a berry with a persistent perianth at base.——About 100 species, in the Tropics of Asia and Malaysia.

1A. Leaves broadly oblanceolate to obovate-oblong, 5–10 cm. long, glabrous, obtuse; inflorescence subsessile; fruit globose, red.
. . . . 1. *A. lancifolia*
1B. Leaves lanceolate, sometimes oblanceolate, 10–25 cm. long, scattered puberulous beneath, long-acuminate; inflorescence a compound umbel, short pedunculate; fruit ellipsoidal, black. . . . . 2. *A. longifolia*

1. Actinodaphne lancifolia (Sieb. & Zucc.) Meissn. *Daphnidium lancifolium* Sieb. & Zucc.; *Iozoste lancifolia* (Sieb. & Zucc.) Bl.; *Litsea lancifolia* (Sieb. & Zucc.) F. Vill. ——KAGO-NO-KI. Evergreen tree; leaves coriaceous, obovate-oblong to broadly oblanceolate, 5–10 cm long, obtuse, glabrous, whitish beneath, the lateral nerves 7 to 10 pairs, the petioles 8–15 mm. long, scattered-pubescent; inflorescence subsessile, the pedicels 5–10 mm. long in fruit, thickened; fruit obovoid-globose, 7–8 mm. long, 6–7 mm. across, with a 6–parted persistent perianth at base.——Sept. Honshu (Kantō Distr. and westw.), Shikoku, Kyushu.——s. Korea and Ryukyus.

2. Actinodaphne longifolia (Bl.) Nakai. *Machilus* (?) *longifolia* Bl.; *Iozoste acuminata* Bl.; *A. acuminata* (Bl.) Meissn.; *Fiwa longifolia* (Bl.) Nakai——BARIBARI-NO-KI, AO-KAGO-NO-KI. Evergreen tree with stout glabrous branches; leaves thinly coriaceous, lanceolate to narrowly so, sometimes oblanceolate, 10–25 cm. long, 15–30 mm. wide, long-acuminate, glaucescent and slightly puberulous beneath, the lateral nerves 10 to 15 pairs, the petioles 10–30 mm. long; inflorescence on a peduncle 5–10 mm. long, the perianth-segments deciduous; fruit ellipsoidal, about 15 mm. long, black.——Aug. Honshu (s. Kantō Distr. and westw.), Shikoku, Kyushu.——Ryukyus.

Fam. 96. **PAPAVERACEAE**   Keshi Ka   Poppy Family

Usually herbs, glabrous or coarsely hairy, often glaucescent, usually with colored juice; leaves alternate, simple or pinnate, stipules absent; flowers solitary, subumbellate or racemose, actinomorphic or zygomorphic; sepals 2(–4), free, usually caducous; petals 4(–12); stamens many and indefinite or few and definite, sometimes with partially connate filaments; ovary superior, 1- to 8-locular, the placentae parietal, nervelike or inflexed and septalike; style short, the stigma various, often discoid; fruit a capsule, dehiscent by valves or opening by pores, rarely indehiscent; seeds sometimes arillate.——About 30 genera, with about 500 species, mostly in the subtropics and cooler regions of the N. Hemisphere.

1A. Stamens many; petals if present not spurred; carpels 2–8; sepals usually large.
  2A. Placentae 4–8; stigma discoid; capsules globose to obovoid, opening by pores; flowers usually large and showy. .... 1. *Papaver*
  2B. Placentae 2; stigma 2-fid; capsules compressed or linear, valves dehiscent to base.
    3A. Petals 4; capsules linear; ovules rather numerous. .... 2. *Chelidonium*
    3B. Petals absent; capsules cuneate-oblanceolate, flattened; ovules few. .... 3. *Macleaya*
1B. Stamens 4–6; carpels 2; sepals minute.
  4A. Stamens 6; filaments usually connate below; flowers spurred.
    5A. Outer 2 petals spurred or saccate at base. .... 4. *Dicentra*
    5B. Upper petal only spurred or saccate at base. .... 5. *Corydalis*
  4B. Stamens 4; filaments free; petals not spurred nor saccate at base; leaves radical, pectinately pinnate. .... 6. *Pteridophyllum*

1. PAPAVER L.   Keshi Zoku

Annuals, biennials, or perennials often with coarse hairs, usually glaucous, with milky juice; leaves usually pinnately lobed; flowers large, showy, yellow, white, or red, long-pedunculate, nodding in bud, erect at anthesis; sepals 2, green, caducous; petals large, 4; stamens many; ovary ovoid to globose or obovoid, 1- to many-locular, many-ovuled; stigma broadly discoid, radiately striate; capsules globose to obovoid, opening by apical pores; seeds minutely impressed-punctate.——About 50 species, mostly in Eurasia, few in N. America.

1A. Acaulescent; leaves and peduncles radical; flowers yellow; capsules hispid. .... 1. *P. fauriei*
1B. Caulescent, the stems leafy; peduncles cauline, axillary.
  2A. Perennial herb with long coarse hairs on stems and leaves; capsules glabrous. .... 2. *P. orientale*
  2B. Annuals or biennials.
    3A. Hairy; leaves pinnatisect, not clasping. .... 3. *P. rhoeas*
    3B. Glabrous; leaves pinnately cleft to incised, clasping. .... 4. *P. somniferum*

1. Papaver fauriei Fedde. *P. nudicaule* var. *fauriei* Fedde; *P. nudicaule* sensu auct. Japon., non L.——Rishirihinageshi. Hispid perennial herb, 10–20 cm. high with short-tufted rhizomes; leaves radical, long-petiolate, ovate to narrowly so, pinnately divided, the segments often 2- to 4-lobed, the ultimate lobes oblong, acute; scapes 10–20 cm. high, erect, with spreading hairs below, the hairs appressed above; sepals about 10 mm. long, ovate, densely rusty-brown hairy; petals greenish yellow, about 2 cm. long; capsules broadly ellipsoidal, coarsely appressed-hairy, 7–11 mm. long, 6–10 mm. across.——July–Aug. Alpine gravelly slopes; Hokkaido (Rishiri Isl.); very rare.

2. Papaver orientale L. Onigeshi. Coarsely long-hairy perennial herb, 60–100 cm. high; stems simple, erect, 1-flowered; leaves pinnately divided, the segments broadly lance-olate, coarsely incised; calyx densely coarse-hairy, to 3 cm. long; petals deep red, 5–6 cm. long, with a rather large black spot at base; capsules broadly obovoid.——May–June. A Mediterranean plant often cultivated.

Papaver bracteatum Lindl. *P. orientale* var. *bracteatum* (Lindl.) Ledeb.——Botangeshi. Flowers subtended by 2 rather large bracts.——Cultivated in our area.

3. Papaver rhoeas L. Hinageshi. Biennial 30–80 cm. high, with coarse spreading hairs; stems erect, loosely branched, slender; leaves pinnately divided, toothed; peduncles with coarse spreading hairs; sepals about 2 cm. long, hairy; petals white to deep red, 3–4 cm. long; capsules glabrous, broadly obovoid, about 10 mm. long.——May–July. European plant commonly cultivated as an ornamental.

4. Papaver somniferum L. Keshi. Large, coarse, glabrous, slightly branched, erect biennial, 50–150 cm. high; leaves glaucous, narrowly ovate to narrowly oblong, clasping, irregularly incised and toothed, the lower ones pinnatifid; sepals glabrous, 1.5–2 cm. long; petals 4(–8), white or red-purple, 5–7 cm. long; capsules ovoid-globose, glabrous, 4–6 cm. long, 3.5–4 cm. across.——May–June. European plant cultivated as a medicinal herb.

2. CHELIDONIUM L.   Kusa-no-ō Zoku

Soft sometimes glaucescent annuals, biennials, or perennials with orange juice; stems leafy or acaulescent; leaves pinnately divided; flowers solitary in leaf-axils or few in axillary umbels, rather large, yellow, pedicelled; sepals 2, green, caducous; petals 4; stamens many; placentae 2, parietal; style short, the stigma 2-lobed; capsules linear, glabrous, erect, dehiscent; seeds lustrous, small, arillate.——Few species, in Eurasia.

1A. Biennial; stems leafy, 30–80 cm. long; umbels several-flowered, axillary, pedunculate; petals 10–12 mm. long.
                                                         1. *C. majus* var. *asiaticum*
1B. Perennial, rhizomatous; stems few-leaved above, 30–40 cm. long; flowers solitary or paired in axils; petals 2–2.5 cm. long.
                                                          2. *C. japonicum*

1. **Chelidonium majus** L. var. **asiaticum** (Hara) Ohwi. *C. Majus* var. *hirsutum* Trautv. & Mey., excl. syn.; *C. majus* sensu auct. Japon., non L.; *C. majus* subsp. *asiaticum* Hara—— KUSA-NO-Ō. Erect, branched glaucescent biennial herb with multicellular curled hairs at first, later becoming glabrescent; stems 30–80 cm. long; leaves 1- to 2-pinnately parted or divided, 7–15 cm. long, 5–10 cm. wide, obtuse, glabrous to pubescent beneath, obtusely toothed and incised, the lower leaves petiolate; umbels pedunculate, axillary, few-flowered; flowers pedicelled; sepals 6–8 mm. long; petals 10–12 mm. long.——May–July. Meadows and thickets in lowlands; Hokkaido, Honshu, Shikoku, Kyushu; common.——The temperate regions of e. Asia. The typical phase occurs in Europe and Siberia.

2. **Chelidonium japonicum** Thunb. *Hylomecon japonicum* (Thunb.) Prantl; *C. uniflorum* Sieb. & Zucc.; *Stylophorum japonicum* (Thunb.) Miq.——YAMABUKI-SŌ. Soft perennial herb with short rhizomes, with scattered multicellu- lar curled hairs; radical leaves long-petioled, pinnate, the leaf- lets 5–7, broadly ovate to elliptic, 1.5–5 cm. long, 1.2–3 cm. wide, short-petiolulate, acute, incised and irregularly toothed; scapes 30–40 cm. long, with few, small, dissected leaves above; flowers 1 to few, axillary, the pedicels 4–6 cm. long; sepals narrowly ovate, acute, 15–17 mm. long; petals 2–2.5 cm. long; capsules linear, 3–5 cm. long, about 3 mm. across, many- seeded; stigma 2-lobed.——Apr.–June. Woods in lowlands and hills; Honshu.——Forma **lanceolatum** (Yatabe) Ohwi. *Stylophorum lanceolatum* Yatabe; *Hylomecon japonicum* var. *lanceolatum* (Yatabe) Makino——HOSOBA-YAMABUKI-SŌ. Leaf- lets of cauline leaves minutely toothed, broadly lanceolate.—— Forma **dissectum** (Fr. & Sav.) Ohwi. *Stylophorum japoni- cum* var. *dissectum* Fr. & Sav.; *Hylomecon japonicum* var. *dissectum* (Fr. & Sav.) Makino——SERIBA-YAMABUKI-SŌ. Leaf- lets of cauline leaves rhombic-ovate, deeply pinnatifid and incised.

3. MACLEAYA R. Br. TAKENIGUSA ZOKU

Large erect perennial more or less glaucous herbs with yellow juice; stems stout, hollow, leafy; leaves petiolate, large, pinnati- fid; flowers white, small, pedicelled, in large erect panicles; sepals 2, white, caducous; petals absent; stamens many; ovary 1; style short, the stigma thickened, 2-lobed; ovules few; capsules broadly oblanceolate, pendulous, flat, the valves membranous, glabrous; seeds few.——Two species, in Japan and China.

1. **Macleaya cordata** (Willd.) R. Br. *Bocconia cordata* Willd.; *B. jedoensis* Carr.; *B. japonica* André; *M. yedoensis* André; *M. cordata* var. *yedoensis* (André) Fedde——TAKENI- GUSA, CHAMPAGIKU. Stems stout, 1–2 m. high, erect, usu- ally simple, branched above, terete; leaves radical and cauline, broadly ovate, 12–25 cm. long, obtuse, subcordate, pinnately incised, white and rather densely short-pubescent beneath; sepals oblanceolate, about 10 mm. long; stamens many, the anthers linear, yellow, 3–4 mm. long; capsules flat, glabrous, obtuse, cuneate, about 20 mm. long, about 5 mm. wide, gray- brown.——June–Aug. Waste grounds and open meadows in lowlands and foothills; Honshu, Shikoku, Kyushu; common. ——China and Formosa.

Var. **thunbergii** (Miq.) Miq. *Bocconia cordata* var. *thunbergii* Miq.——KE-NASHI-CHAMPAGIKU. Leaves glabrous beneath.——Occurs with the typical phase.

4. DICENTRA Bernh. KOMA-KUSA ZOKU

Glabrous, glaucous perennial herbs with pinnately or ternately dissected leaves; inflorescence racemose; flowers rose or yellow- ish, nodding, the pedicels bracteate; sepals 2, scalelike; petals 4, the outer 2 boat-shaped, reflexed above in anthesis, saccate or short-spurred at base, the inner 2 rather small, clawed, connate at apex, ridged on back stamens 6; ovary 1, the placentae 2; style slender, the stigma flat, 2- to 4-lobed; capsules lanceolate, dehiscent, 2-valved; seeds usually arillate.——More than 10 species, in N. America and Asia.

1A. Tall leafy-stemmed perennial herb; racemes relatively long, many-flowered; flowers about 30 mm. long; sepals caducous.

1. *D. spectabilis*

1B. Small scapose perennial herb; racemes very short, few-flowered; flowers about 20 mm. long; sepals persistent. .... 2. *D. peregrina*

1. **Dicentra spectabilis** (L.) Lem. *Fumaria spectabilis* L.; *Dielytra spectabilis* (L.) G. Don; *Eucapnos spectabilis* (L.) Sieb. & Zucc.——KEMAN-SŌ. Rhizomes rather stout, elongate, tufted; stems leafy, 40–80 cm. long, angled, juicy, reddish; leaves long-petiolate, twice ternate, the leaflets 3- to 5-parted to -divided, 3–6 cm. long, the segments broadly ob- lanceolate, usually incised on upper margin, slightly glauces- cent; racemes solitary, axillary, pedunculate, 20–30 cm. long, many-flowered; flowers nodding, flattened, deeply cordate, 27–32 mm. long, 18–20 mm. wide, rose, the inner petals white; sepals thinly membranous, lanceolate, caducous, 6–7 mm. long. ——Apr.–June. Cultivated in gardens.——Korea, Manchuria, and China.

2. **Dicentra peregrina** (Rudolph) Makino. *Fumaria peregrina* Rudolph; *D. pusilla* Sieb. & Zucc.; *D. peregrina* var. *pusilla* (Sieb. & Zucc.) Makino——KOMA-KUSA. Glaucous, glabrous perennial herb with short erect rhizomes; leaves radical, petiolate, finely dissected, the lobes linear-oblong, the segments obtuse, 2–6 mm. long, about 1 mm. wide, glaucous; scapes 5–15 cm. long, the raceme very short, few-flowered; flowers rose, flat, slightly cordate, about 20 mm. long, 10–12 mm. wide; sepals broadly ovate, persistent, about 3 mm. long, green; capsules narrowly oblong, 8–10 mm. long, about 4 mm. wide, enveloped by the persistent withered perianth; style slender, as long as the capsule, the stigma transversely oblong, flat; seeds black, shining.——July–Aug. Gravelly to sandy slopes mainly in volcanoes; Hokkaido, Honshu (centr. and n. distr.); rare.——Sakhalin, Kuriles, Kamchatka, and e. Siberia.

5. CORYDALIS DC.   Ki-keman Zoku

Usually glabrous and glaucous, sometimes sparingly papillose biennial or perennial herbs; leaves ternately or pinnately divided, alternate or radical; inflorescence racemose, terminal or axillary, bracteate; flowers pedicelled, rose-purple, white or yellow; sepals 2, small or absent; petals 4, free or partially connate, the outer 2 larger, the upper one spurred at base, the inner 2 connate at apex, ridged on the back; stamens 6, united in pairs; ovary 1-locular, with 2 nervelike placentae, 1- to many-ovuled; styles slender, the stigma flat, often dentate on margin; capsules linear to ovoid, sometimes torulose; seeds usually arillate.——About 200 species, in Africa, Eurasia, and America, especially abundant in centr. Asia and China.

1A. Plants with tubers; cauline leaves few; flowers rose-purple or blue, rarely white.
  2A. Tubers usually 1-stemmed; radical leaves present only when the tuber not scapose; lowest cauline leaf reduced to a scale.
    3A. Capsules linear or oblong-linear; tuber with more or less yellow flesh; bracts usually entire. . . . . 1. *C. ambigua*
    3B. Capsules lanceolate to narrowly ovate; tuber with white flesh; bracts often incised.
      4A. Flowers 15–25 mm. long. . . . . 2. *C. lineariloba*
      4B. Flowers 10–13 mm. long. . . . . 3. *C. capillipes*
  2B. Tubers with few radical leaves and scapes; lowest cauline leaf not reduced to a scale.
    5A. Stems slender, 5–10 cm. long; capsules linear, 17–20 mm. long. . . . . 4. *C. decumbens*
    5B. Stems stout, up to 1 m. long; capsules broadly oblanceolate, about 25 mm. long. . . . . 5. *C. curvicalcarata*
1B. Plants without tubers; cauline leaves often rather numerous; flowers yellow, rarely rose-purple or whitish.
  6A. Upper petals connate to the claw of the inner petals; seeds smooth.
    7A. Stems much elongate, sometimes more than 1 m. long; flowers pale yellow, in loose racemes, with a spur about as long as the
        body at base; sepals subentire. . . . . 6. *C. ochotensis*
    7B. Stems erect, 20–50 cm. long; flowers rose-purple, rarely rose-white, in dense racemes, with a spur shorter than the body at base;
        sepals finely dissected into capillary segments. . . . . 7. *C. incisa*
  6B. Upper petals free; seeds muricate, papillose or impressed-punctate.
    8A. Flowers 6–13 mm. long; stigma with 4 distinct lobes on upper margin; aril clavate, not appressed to the seed.
      9A. Flowers green-yellow, 6–13 mm. long, sometimes with a purple hue; capsules strongly flexuous; seeds with impressed dots.
                                                      8. *C. ophiocarpa*
      9B. Flowers pale yellow, 6–7 mm. long; capsules straight; seeds with conical papillae. . . . . 9. *C. racemosa*
    8B. Flowers 15–23 mm. long (12–18 mm. long in *C. tashiroi*); stigma with small teeth on margin; aril appressed to the seed.
      10A. Leaves pinnate.
        11A. Capsules straight; flowers 12–18 mm. long, with a spur 1–4 mm. long at base. . . . . 10. *C. tashiroi*
        11B. Capsules more or less curved, slightly torulose; flowers 15–23 mm. long, with a longer spur at base.
          12A. Racemes 3–10 cm. long, densely many-flowered, the flowers 20–23 mm. long; seeds with impressed dots.
                                                     11. *C. speciosa*
          12B. Racemes 2–5 cm. long, 2- to 8-flowered, the flowers 18–20 mm. long; seeds tuberculate. . . . . 12. *C. pallida*
      10B. Leaves ternate. . . . . 13. *C. heterocarpa*

1. Corydalis ambigua Cham. & Schltdl.   Ezo-engo-saku.   Glabrous, glaucescent, smooth or slightly papillose perennial herb, the tuber globose, 1–2 cm. across; stems solitary, 10–30 cm. long, 2-leaved above, the scalelike leaf near the base sometimes bearing a branch from the axil; leaves petiolate, 1- to 2(–3)-ternate, the ultimate segments thin, linear to ovate-orbicular, 10–30 mm. long, 2–25 mm. wide, entire or variously lobed to parted, obtuse to rounded; racemes terminal, densely many-flowered, erect, the bracts broadly lanceolate to ovate-orbicular, usually entire, the pedicels 7–12 mm. long; flowers blue-purple, rarely white, 17–25 mm. long; capsules linear or oblong-linear, 15–23 mm. long, 2.5–3 mm. across; seeds dark brown, shining, smooth.——Apr.–May. Woods and meadows in lowlands to mountains; Hokkaido, Honshu (n. distr.); very variable in leaf-shape and in the papillosity of leaf-margins and racemes.——Sakhalin, Kuriles, and Ochotsk Sea region.

2. Corydalis lineariloba Sieb. & Zucc.   *C. orthoceras* Sieb. & Zucc.; *C. senanensis* Fr. & Sav.; *C. vernyi* Fr. & Sav.; *C. laxa* Fr. & Sav.; *C. hallaisanensis* Lév.; *C. remota* sensu auct. Japon., non Fisch.——Yama-engo-saku, Sasaba-engo-saku. Glabrous, smooth or slightly papillose perennial herb, the tuber globose, 7–15 mm. across; stems 10–20 cm. long, 2-leaved, the scalelike leaf at the base often with a small axillary bulblet; leaves 2- to 3-ternate, the leaflets or ultimate segments thin, linear to ovate-orbicular, often deeply 3-cleft to -parted, 15–25 mm. long, 3–15 mm. wide; flowers in a dense terminal raceme, the bracts lanceolate to flabellate-cuneate, usually toothed to in-cised, the pedicels 5–15 mm. long; flowers 15–25 mm. long, blue-purple; capsules lanceolate to ovate-oblong, 7–13 mm. long, 2.5–4 mm. wide; seeds smooth, shining.——Apr.–May. Woods and meadows in hills and mountains; Honshu, Kyushu; rather common.——Korea and Manchuria.

Var. **capillaris** (Makino) Ohwi.   *C. bulbosa* var. *capillaris* Makino; *C. capillaris* (Makino) Takeda——Hime-engo-saku. Differs from the typical phase in being very delicate; leaves 3(–4)-ternate, the leaflets or ultimate segments ovate-orbicular to oblong, 5–8 mm. long, 3–5 mm. wide, thinner; flowers rather few.——Shikoku, Kyushu.

Var. **papilligera** (Ohwi) Ohwi.   *C. papilligera* Ohwi——Kinki-engo-saku.   Differs from the preceding variety in the ovate-oblong capsules, 5–10 mm. long, 4–5 mm. wide, and the seeds with minute papillose-tubercles toward the margin.——Honshu.

3. Corydalis capillipes Franch.   Michinoku-engo-saku, Hime-yama-engo-saku.   Glabrous slender perennial herb, the tuber globose, 5–10 mm. across; scapes 10–20 cm. long, 2-leaved, with a single subterranean scale and axillary bulblet; leaves petiolate, 2- to 3-ternate, the leaflets or ultimate segments linear to oblong, thin, 1–3 cm. long, 2–10 mm. wide, often deeply parted, sometimes entire; racemes many- or rather few-flowered, the bracts narrowly ovate to flabellate-cuneate, entire to incised; flowers 10–13 mm. long; capsules ovate-oblong, about 3 mm. long, 2 mm. wide.——Apr.–May. Woods and meadows in mountains; Honshu (Japan Sea side of n. and centr. distr.).

4. Corydalis decumbens (Thunb.) Pers. *Fumaria decumbens* Thunb.——JIROBŌ-ENGO-SAKU. Slender glabrous glaucescent perennial herb, the globose tuber renewed each season; radical leaves few, 2- to 3-ternate, long-petiolate, the leaflets usually 2- to 3-parted to -cleft, the segments oblanceolate to obovate, 1–2 cm. long, 3–7 mm. wide, thin, glaucous; scapes decumbent, erect above, 10–30 cm. long, without a scalelike leaf below ground; cauline leaves 2, petiolate, 2-ternate; racemes rather few-flowered, the bracts ovate, entire, subacute, the pedicels 5–10 mm. long; flowers rose to blue-purple, 15–22 mm. long; capsules linear, 15–22 mm. long, about 1.5 mm. wide, slightly torulose when dried; seeds dark brown, about 1.2 mm. across, minutely mammillate.——Apr.–May. Meadows along rivers in lowlands; Honshu (Kantō Distr. and westw.), Shikoku, Kyushu.——Formosa and China.
——Forma **albescens** (Takeda) Ohwi. *C. decumbens* lusus *albescens* Takeda. Flowers whitish.——Occurs with the typical phase.

5. Corydalis curvicalcarata Miyabe & Kudo. Ezo-Ō-KEMAN. Stems erect, stout, glabrous, about 1 m. long; cauline leaves 2- to 3-ternate-pinnate, the ultimate segments pinnate, the lobes broadly lanceolate to elliptic, acute, glaucous beneath; racemes short, nearly simple, about 4.5 cm. long, the pedicels about 1 cm. long, the bracts linear; flowers white, about 16 mm. long, the upper petals slightly longer than the upwardly curved rounded spur; capsules clavate, about 2 cm. long, 6–7 mm. wide; seeds 3–3.5 mm. across.——June. Hokkaido.

6. Corydalis ochotensis Turcz. TSURU-KEMAN. FIG. 6. Glabrous glaucescent biennial herb; stems branched, elongate, soft and juicy, decumbent or ascending, acutely angled, sometimes more than 1 m. long; leaves petiolate, nearly deltoid in outline, 7–12 cm. long and as wide, 2- to 3-ternate, the leaflets often 3-cleft, the segments oblong or obovate, 10–15 mm. long, 5–10 mm. wide, subrounded, the petiole winged; racemes terminal, loose, the bracts ovate to broadly so, usually entire, 3–10 mm. wide, weakly veined, the pedicels 4–7 mm. long; flowers pale yellow, 15–20 mm. long; sepals subentire; capsules obovate-cuneate to elongate-obovate, 12–15 mm. long, 3.5–4.5 mm. wide; seeds in 2 series, smooth.——Aug.–Sept. Mountains; Honshu (Kantō and centr. distr.).——e. Siberia and Ochotsk Sea region.

Var. **raddeana** (Regel) Nakai. *C. raddeana* Regel; *C. ochotensis* forma *raddeana* (Regel) Nakai——NAGAMI-NO-TSURU-KEMAN. Bracts ovate to lanceolate, 2–5 mm. wide; capsules linear-oblanceolate, 2–2.5 mm. wide; seeds in one series.——Aug.–Oct. Mountains; Hokkaido, Honshu, Kyushu.——Manchuria, Korea, and e. Siberia.

7. Corydalis incisa (Thunb.) Pers. *Fumaria incisa* Thunb.——MURASAKI-KEMAN. Soft glabrous slightly glaucescent biennial; tap-root somewhat fusiform in winter; stems 20–50 cm. long, tufted, angular, slightly branched, leafy; radical leaves long-petiolate, ovate-orbicular, 3–8 cm. long, more or less triangular in outline, 2-ternate, the leaflets ternately or pinnately lobed, the lobes cuneate, incised, 1–2 cm. long, the cauline petiolate; racemes erect, 4–12 cm. long, the bracts flabellate-cuneate to oblanceolate, incised, the pedicels 10–15 mm. long; flowers rose-purple, 12–18 mm. long; sepals finely dissected into capillary segments; capsules linear-oblong, nodding, about 15 mm. long, 3–3.5 mm. wide, straight.——Apr.–June. Thickets and bamboo forests in lowlands and foothills; Hokkaido, Honshu, Shikoku, Kyushu; common.——Forma

pallescens Makino. SHIRO-YABU-KEMAN. Flowers nearly white.——Occurs with the typical phase.

8. Corydalis ophiocarpa Hook. f. & Thoms. *C. japonica* Makino; *C. makinoana* Matsum.——YAMA-KI-KEMAN. Glabrous, glaucous, perennial herb; stems somewhat tufted, nearly erect, 40–80 cm. long, angled, greenish; leaves petiolate, ovate, 8–15 cm. long, 4–8 cm. wide, pinnately compound, the leaflets pinnately divided or parted, the segments ovate or narrowly so, often incised, the petioles winged; racemes terminal, short-pedunculate, rather densely many-flowered, the bracts lanceolate, with a filiform point, the pedicels 2–6 mm. long, curved, much shorter than the bracts; flowers greenish yellow, sometimes with a dark purple hue, 8–13 mm. long, the spur short; capsules linear, strongly flexuous, about 3 cm. long; seeds impressed-punctate.——May–July. Mountains; Honshu (Kantō Distr. and westw.), Shikoku.——Formosa, China, and India.

9. Corydalis racemosa (Thunb.) Pers. *Fumaria racemosa* Thunb.——HOZAKI-KI-KEMAN. Glabrous biennial herb; stems spreading, 20–60 cm. long; leaves petiolate, ovate, 5–10 cm. long, 3–5 cm. wide, 2-ternate, the leaflets broadly ovate; racemes 5–10 cm. long, the bracts linear; flowers 6–7 mm. long; capsules linear, straight, 17–30 mm. long, 2–2.5 mm. wide; seeds about 1 mm. across, with conical tubercles.——Apr.–May. Kyushu, Shikoku.——China.

10. Corydalis tashiroi Makino. *C. formosana* Hayata ——SHIMA-KI-KEMAN. Glabrous glaucescent biennial; stems 15–40 cm. long; leaves petiolate, broadly ovate, 8–20 cm. long, bipinnate, the leaflets pinnately parted to divided, the segments broadly ovate, incised; racemes bracteate; flowers 12–18 mm. long, the spur very short; capsules linear, straight, 3.5–4.5 cm. long, 2.5–3 mm. wide; seeds with impressed dots.——Mar.–Apr. Kyushu.——Ryukyus, Formosa, and s. China.

11. Corydalis speciosa Maxim. *C. pallida* var. *speciosa* (Maxim.) Komar.; *C. maximowicziana* Nakai; *C. pterophora* Ohwi——EZO-KI-KEMAN. Glabrous glaucescent biennial herb; stems 20–40 cm. long, terete, erect, often ascending at base; leaves petiolate or subsessile, narrowly ovate, 10–15 cm. long, 4–6 cm. wide, the leaflets pinnate or pinnatiparted, incised; racemes densely many-flowered, elongate in fruit, 3–10 cm. long; flowers 20–23 mm. long; capsules linear, 2–3 cm. long, torulose, gently curved; seeds with impressed dots.——May–June. Hokkaido, Honshu (Mount Hayachine in Rikuchu). ——e. Siberia, Manchuria, n. China, and Sakhalin.

12. Corydalis pallida (Thunb.) Pers. *Fumaria pallida* Thunb.; *C. satsumensis* T. Ito; *C. triflora* Ohwi; *C. triflora* var. *arakiana* Ohwi——FŪRO-KEMAN. Glabrous, glaucescent, biennial herb; stems 15–50 cm. long, terete, several-leaved, erect or ascending; leaves petiolate or subsessile, ovate, 1- to 2-pinnate, the leaflets broadly ovate, incised; racemes 2–5 cm. long, 2- to 8-flowered; flowers 18–20 mm. long; capsules linear, 1–2 cm. long, slightly torulose; seeds with conical tubercles. ——Mar.–May. Honshu (centr. distr. and westw.), Shikoku, Kyushu.

Var. **tenuis** Yatabe. *C. pallida* sensu auct. Japon., pro parte Maxim., non Pers.; *C. hondoensis* Ohwi——MIYAMA-KI-KE-MAN. Plant slightly larger; racemes longer, 3–10 cm. long, rather many-flowered; flowers 20–23 mm. long; capsules 2–3 cm. long, prominently torulose.——Honshu (Kinki Distr. and eastw.); common.

13. Corydalis heterocarpa Sieb. & Zucc. *C. wilfordii* Regel——TSUKUSHI-KI-KEMAN. Glabrous glaucescent herb;

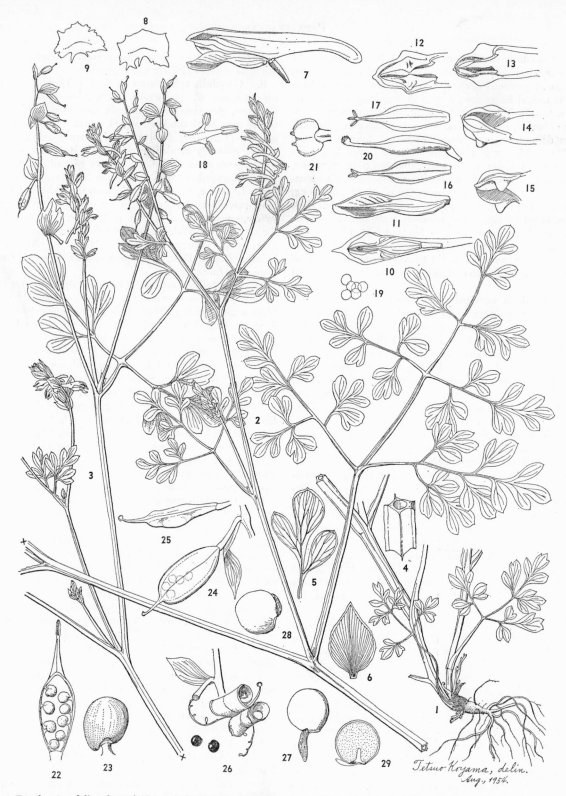

Fig. 6.—*Corydalis ochotensis* Turcz. 1, 2, 3, Habit; 4, portion of stem; 5, part of a leaf; 6, bract; 7, flower; 8, 9, sepals; 10, flower; 11, lower petal; 12, 13, 14, 15, apical part of inner petal; 16, 17, stamens; 18, stamens enlarged; 19, pollen; 20, pistil; 21, stigma; 22, longitudinal section of ovary; 23, ovule; 24, 25, capsules; 26, the same at dehiscence; 27, 28, seeds; 29, the same, cut longitudinally.

stems rather stout, often reddish, terete, 40–60 cm. long; leaves usually petiolate, broadly ovate-deltoid, 10–25 cm. long and as wide, 2- to 3-ternate, the leaflets incised to parted; racemes 5–10 cm. long, many-flowered; flowers 15–25 mm. long, the spur short; capsules broadly linear, 25–35 mm. long, 3–4 mm. wide, torulose; seeds in one series, densely papillose with short-cylindric tubercles.——Mar.–June. Honshu (Chūgoku Distr.), Shikoku, Kyushu.——s. Korea.

Var. **japonica** (Fr. & Sav.) Ohwi. *C. wilfordii* var. *japonica* Fr. & Sav.; *C. pallida* var. *platycarpa* Maxim. ex Palib., pro parte; *C. platycarpa* (Maxim.) Makino——KI-KEMAN.

Capsules narrowly lanceolate, scarcely torulose; seeds in two series or nearly two series.——Apr.–May. Thickets and thin woods in lowlands, especially near the sea; Honshu (Kantō Distr. and westw.), Shikoku, Kyushu; rather common.——Ryukyus.

Var. **brachystyla** (Koidz.) Ohwi. *C. brachystyla* Koidz.——MUNIN-KI-KEMAN. Closely resembles the preceding variety; capsules lanceolate, 2–3 cm. long, 4–5 mm. wide, not torulose; seeds with short conical tubercles.——Honshu (Izu Isls.).——Bonins and Ryukyus.

6. PTERIDOPHYLLUM Sieb. & Zucc. OSABAGUSA ZOKU

Stemless perennial herbs with short rather thick rhizomes; leaves radical, pectinate, deep green, petiolate; scapes simple, naked, erect; racemes solitary, terminal; flowers nodding, rather small, the pedicels solitary or ternate; sepals 2, deciduous, small; petals 4, white, deciduous, equal, obliquely ascending, oblong; stamens 4, free, the filaments short, erect, the anthers linear; ovary 1-locular; styles slender, the stigma bifid; capsules obovate-globose, 2-valved, 2-seeded.——A single species, in Japan.

1. Pteridophyllum racemosum Sieb. & Zucc. OSABA-GUSA. Nearly glabrous perennial herb; leaves pectinate, broadly oblanceolate in outline, 10–20 cm. long, 2.5–3 cm. wide, obtuse, gradually narrowed below, the segments many, spreading, linear-oblong, about 3 mm. wide, short-toothed, sparsely hispid above, slightly auriculate on upper side near base; scapes erect, 15–25 cm. long; racemes loose, 5–15 cm. long, the pedicels slender, solitary to ternate, 10–15 mm. long, spreading; sepals 2, broadly ovate, 1–1.2 mm. long; petals about 5 mm. long; anthers linear, 2–2.2 mm. long; capsules obovoid, about 3.5 mm. across and as long; style 2.5–3 mm. long; seeds ellipsoidal, 1.5 mm long.——June–Aug. Coniferous woods; Honshu (centr. and n. distr.); rare.

Fam. 97. CAPPARIDACEAE FŪCHŌ-SŌ KA Caper Family

Herbs or woody plants; leaves simple or palmately compound, 3- to 9-foliolate, alternate, rarely opposite; stipules 2 or absent, often becoming spiny; inflorescence racemose or corymbose; flowers actinomorphic, usually bisexual; sepals 4, sometimes connate; petals 4, 2, or absent; stamens 4 or more, the disc often much elongate; ovary stipitate or sessile, usually 1-locular, the placentae parietal; carpels 2 or more; fruit a capsule or berry.——About 35 genera, with about 500 species, mainly tropical and warm-temperate.

1. CRATAEVA L. GYOBOKU ZOKU

Glabrous trees or shrubs with lenticellate branches; leaves 3-foliolate, petiolate; flowers in corymbs, showy, often polygamous; calyx 4-parted, the segments imbricate in bud, deciduous; petals 4, long-clawed, not overlapped in bud; receptacle hemispherical, lobed; stamens 8–20, inserted on the margin of receptacle, the filaments filiform, elongate; ovary ovoid, long-stipitate, 1- to 2-locular, the placentae 2, parietal; ovules many; stigma sessile, discoid; fruit a globose or ovoid berry; seeds few or many.——Few species, in the Tropics.

1. Crataeva religiosa Forst. *C. falcata* DC.——GYO-BOKU. Small tree with pale lenticellate gray-brown branches; leaves long-petiolate, the leaflets rather thin, ovate to narrowly so, 6–10 cm. long, 2–5 cm. wide, usually acuminate, entire, glaucescent beneath; inflorescence a terminal corymb, the flowers large, pale yellow, the pedicels about 3 cm. long; petals ovate, clawed, 15–20 mm. long; stamens exserted, about 4 cm. long; ovary narrowly ovoid, 4–5 mm. long, the stipe nearly as long as the pedicel; berry fleshy, nodding, about 2.5 cm. long.——June–July. Near the sea; Kyushu (s. distr.).——s. China, India, Malaysia to Australia, and Africa.

Fam. 98. CRUCIFERAE ABURA-NA KA (JŪJI KA) Mustard Family

Mostly herbs; leaves cauline or radical, alternate, simple to pinnate; stipules absent; flowers in racemes, sometimes solitary in leaf-axils, rarely bracteate; sepals 4, deciduous; petals 4 or absent, clawed, deciduous; stamens 6 (rarely 4), tetradynamous, in 2 series; disc with 2, 4, or 6 glands; ovary 2-locular, with parietal placentae; style usually persistent, simple, the stigma discoid, sometimes bifid; fruit a silique or silicle, rarely indehiscent and jointed; seeds without endosperm; cotyledons large, incumbent or accumbent.——About 200 genera, with about 2,000 species, mostly in temperate regions and high latitudes of both hemispheres.

1A. Hairs simple, not branched, without glands, or hairs absent.
  2A. Capsule a flattened long-stipitate silique; stamens longer than the petals. .... 1. *Macropodium*
  2B. Capsules sessile or nearly so, not distinctly stipitate at the base; stamens not longer than the petals.
    3A. Capsules indehiscent.
      4A. Capsules 1-seeded, flat; flowers yellow. .... 2. *Isatis*
      4B. Capsules few-seeded, torulose; flowers white or rose-purple. .... 3. *Raphanus*

3B. Capsules dehiscent.
  5A. Capsules laterally flattened; septa narrow; valves navicular, sometimes winged on the keel.
    6A. Seeds solitary in each locule of capsule; cotyledons incumbent. .... 4. *Lepidium*
    6B. Seeds 2 to several in each locule; cotyledons accumbent. .... 5. *Thlaspi*
  5B. Capsules terete or dorsally flattened.
    7A. Flowers white.
      8A. Valves elastic and coiled at maturity. .... 6. *Cardamine*
      8B. Valves not elastic or coiled at maturity.
        9A. Flowers bracteate; seeds rather large, few; cotyledons incumbent. .... 7. *Wasabia*
        9B. Flowers not bracteate; seeds small, many; cotyledons accumbent.
          10A. Capsule a silique. .... 8. *Nasturtium*
          10B. Capsule an inflated silicle. .... 9. *Cochlearia*
    7B. Flowers yellow.
      11A. Cotyledons incumbent; fruit terete or flattened; valves often 3-nerved; plants hairy. .... 10. *Sisymbrium*
      11B. Cotyledons accumbent; valves costate or keeled or 1-nerved; plants glabrous.
        12A. Siliques 3–4 cm. long, costate or keeled. .... 11. *Barbarea*
        12B. Siliques shorter, weakly 1-nerved or nerveless. .... 12. *Rorippa*
1B. Hairs bifid or stellate, rarely simple and gland-tipped.
  13A. Capsules flattened laterally, a silicle; valves navicular. .... 13. *Capsella*
  13B. Capsules dorsally flattened or terete, a silique, rarely a silicle; valves not navicular.
    14A. Capsules a silicle; cotyledons accumbent. .... 14. *Draba*
    14B. Capsule a silique.
      15A. Cotyledons accumbent; flowers white, rarely purple. .... 15. *Arabis*
      15B. Cotyledons incumbent.
        16A. Flowers white or rose-purple.
          17A. Siliques narrowly linear.
            18A. Valves thinly membranous; stamens free. .... 16. *Arabidopsis*
            18B. Valves firmly membranous; longer stamens connate in pairs. .... 17. *Dontostemon*
          17B. Siliques gradually narrowed above. .... 18. *Berteroella*
        16B. Flowers yellow to yellowish.
          19A. Stigma entire; leaves pinnately dissected; valves thinly membranous. .... 19. *Descurainia*
          19B. Stigma bifid; leaves simple, toothed or incised; valves firmly membranous. .... 20. *Erysimum*

1. MACROPODIUM R. Br.   HAKUSEN-NAZUNA ZOKU

Nearly glabrous perennial herbs; leaves alternate, toothed, the lower petiolate; inflorescence a densely flowered spikelike raceme, elongate, ebracteate; flowers many, white, rather large; sepals narrow, loose, equal; petals narrow, longer than the sepals; stamens exserted, the anthers elongate, contorted; siliques elongate, flattened, long-stipitate, the valves flat, with a weak nerve; style very short, the stigma punctiform; seeds in 1 or 2 series, flat, winged.——Two species, in the Altai Mountains and e. Asia.

1. Macropodium pterospermum F. Schmidt. HAKU-SEN-NAZUNA. Rhizomes creeping, short; stems erect, 40–100 cm. long, usually simple, leafy, with simple minute hairs above; leaves narrowly ovate to broadly lanceolate, 7–12 cm. long, 1.5–5 cm. wide, acuminate, glabrous on both sides or often minutely hairy on nerves beneath, toothed, the lower ones petiolate, the upper ones subsessile; inflorescence elongate, many-flowered, often to 40 cm. long in fruit, the pedicels spreading, 7–13 mm. long; petals slightly longer than the sepals, linear, 6–7 mm. long; stamens 13–15 mm. long; siliques 4–6 cm. long, about 4 mm. wide, the stipe 15–30 mm. long; seeds about 10, winged on margin.——July–Aug. Wet grassy places in alpine regions; Hokkaido, Honshu (n. and centr. distr.); rare.——Sakhalin.

2. ISATIS L.   TAISEI ZOKU

Usually glabrous, sometimes pubescent or tomentose, often glaucescent, branched annual or biennial herbs; leaves entire or obsoletely toothed, the cauline sagittate; flowers in loose ebracteate racemes, small, usually yellow; silique oblong to linear, indehiscent, 1-seeded, flat, woody or bony on the center, coriaceous and thickened or foliaceous on margin; stigma sessile; seed solitary; cotyledons incumbent.——About 20 to 30 species, in Europe, N. Africa, and Asia.

1. Isatis tinctoria L. var. **yezoensis** (Ohwi) Ohwi. *I. oblongata* sensu auct. Japon., non DC.; *I. japonica* sensu auct. Japon., non Miq.; *I. yezoensis* Ohwi——HAMA-TAISEI, EZO-TAISEI. Glaucous, glabrous; stems erect, 30–40 cm. long; lower cauline leaves narrowly ovate, 10–12 cm. long, 2–3 cm. wide, obtuse, shallowly toothed, sagittate; petals small, yellow; siliques cuneate-obovate, about 15 mm. long, 4–5 mm. wide, subacute.——June–July. Seashores; Hokkaido.——Korea and Ussuri.

3. RAPHANUS L.   DAIKON ZOKU

Glabrous or hispidulous branched annuals or biennials; tap-root often thickened; lower leaves lyrate; racemes elongate, bractless; flowers white or purplish rose, rather large, pedicelled; sepals erect, the lateral ones subsaccate at base; stamens free; siliques elongate, terete, torulose, coriaceous or spongy, the locellus 1-seeded; style elongate, the stigma emarginate; seeds globose; cotyledons conduplicate.——Few species, in the Old World, especially in the Mediterranean region.

1. **Raphanus sativus** L. forma **raphanistroides** Makino. *R. raphanistroides* (Makino) Nakai; *R. macropodus* var. *spontaneus* Nakai; *R. acanthiformis* Morel; *R. macropodus* Lév.; *R. sativus* var. *raphanistroides* (Makino) Makino——HAMA-DAIKON. Annual or biennial with a slenderly thickened tap-root; stems 40–80 cm. long, branched, terete; leaves lyrate, slightly glaucescent; flowers pale rose-purple; petals about 2 cm. long, obovate, clawed; siliques erect, 5–8 cm. long, 5–6 mm. across, the pedicels 1–2 cm. long, ascending, glabrous, somewhat coriaceous, torulose, not dehiscent, terete, 2- to 5-seeded.——Apr.–June.——A degenerated phase of cv. **Hortensis**. *R. sativus* var. *hortensis* Backer, the "Daikon" or Japanese radish.

4. LEPIDIUM L. MAME-GUMBAI-NAZUNA ZOKU

Glabrous or pubescent erect or diffuse often branched herbs or subshrubs; flowers in racemes, ebracteate, small, white; sepals short, equal; petals short or absent; stamens 2–4; silicles flat, oblong, ovate, obovate or obcordate, rarely subglobose, often laterally flattened, keeled, emarginate, sometimes winged, with a narrow membranous septum; style absent or filiform, the stigmas emarginate; cotyledons accumbent or incumbent.——About 60 species, in temperate to warmer regions of the world.

1A. Cauline leaves cordate, entire, strongly clasping. . . . . 1. *L. perfoliatum*
1B. Cauline leaves petiolate or sessile, not clasping.
    2A. Silicles orbicular, 2.5–3 mm. long and as wide. . . . . 2. *L. virginicum*
    2B. Silicles elliptic, about 5 mm. long, 3 mm. wide. . . . . 3. *L. sativum*

1. **Lepidium perfoliatum** L. KOSHIMINO-NAZUNA. Erect biennial; lower cauline and radical leaves twice pinnately divided, the upper leaves cordate, entire, acute; flowers small.——Apr. (–Oct.). Naturalized in our area.——s. Europe and w. Asia.
2. **Lepidium virginicum** L. MAME-GUMBAI-NAZUNA. Erect biennial herb; stems branched, 20–50 cm. long, minutely puberulent; cauline leaves cuneate-oblanceolate to linear, 1.5–5 cm. long, 2–10 mm. wide, acute, narrowed at base, deeply toothed or entire, racemes elongate after anthesis; flowers many, small, white; silicles orbicular, with a narrow wing on upper margin, the seeds narrowly winged on upper margin.——May–June. Naturalized in our area.——N. America.
3. **Lepidium sativum** L. KOSHŌ-SŌ. Nearly glabrous annual or biennial; cauline leaves pinnately parted, the upper ones entire and linear; flowers rather small, white; silicles elliptic, entire, narrowly winged on upper margin.——May–Aug. Very rarely cultivated.——Europe.

5. THLASPI L. GUMBAI-NAZUNA ZOKU

Usually glabrous, often glaucescent annuals, biennials, or perennials; stems erect, sometimes scapose; radical leaves rosulate, entire or toothed, the cauline oblong or narrowly ovate, auriculate; flowers in ebracteate racemes, white, rose, or pale purple; sepals erect, equal; petals obovate; stamens free; silicles short, laterally flattened, oblong, obcordate or obcuneate, emarginate or rarely acute, the valves keeled or winged, the septum narrow, membranous; style short or elongate, the stigma emarginate; seeds 2 to several in each locule; cotyledons accumbent.——About 30 species, in temperate to Arctic regions and mountains of the N. Hemisphere, rare in the S. Hemisphere.

1A. Biennial; silicles obcordate-orbicular, prominently wing-margined, retuse; style very short. . . . . 1. *T. arvense*
1B. Perennial; silicles cuneate-oblong, wingless, subtruncate; style slender, about 1.5 mm. long. . . . . 2. *T. japonicum*

1. **Thlaspi arvense** L. GUMBAI-NAZUNA. Glabrous glaucous biennial; stems erect, 20–40 cm. long, often branched; leaves broadly lanceolate to narrowly ovate, 3–6 cm. long, 1–2.5 cm. wide, subobtuse, short-toothed, clasping; racemes elongate, 10–20 cm. long; flowers white; petals about 3.5 mm. long; silicles retuse to obcordate-orbicular, 1.5 cm. long, 1–1.2 cm. wide, including the winged margin, slightly shorter than the pedicel; seeds about 1.2 mm. long, wrinkled.——Apr.–June. Waste grounds and cultivated fields in lowlands; Hokkaido, Honshu, Shikoku, Kyushu; naturalized and rather common.——Europe to n. Africa, n. and w. Asia to India.
2. **Thlaspi japonicum** H. Boiss. TAKANE-GUMBAI. Tufted glaucescent glabrous perennial with short much-branched caudex; stems 8–15 cm. long, firm, terete, erect, few-leaved; radical leaves spathulate-elliptic, entire, obtuse to rounded at apex, the cauline narrowly ovate to elliptic, 1–2 cm. long, 7–12 mm. wide, clasping, auriculate, subacute; racemes elongate after flowering, 2–3 cm. long, 10- to 30-flowered, the pedicels spreading, rather stout, as long as to slightly longer than the silicle; silicles cuneate-oblong, wingless, about 8 mm. long, subtruncate; style slender, about 1.5 mm. long, the valves folded, wingless; seeds wingless, smooth.——Hokkaido; rare.

6. CARDAMINE L. TANETSUKE-BANA ZOKU

Glabrous or pubescent perennials or biennials, sometimes rhizomatous or stoloniferous; leaves simple or pinnately divided; inflorescence racemose or corymbose; flowers white or rose-purple, ebracteate; sepals equal; petals clawed; siliques elongate, linear, the valves flat, coiled, the septum membranous, hyaline; style short or elongate; stigma simple or shallowly 2-lobed; seeds in 1 series, flat, winged or wingless; cotyledons accumbent.——About 100 species in temperate to frigid regions of the world and in mountains, especially abundant in Europe and Asia.

1A. Rhizomes scarcely thickened; scapes or stems leafy.
  2A. Stolons if present short; sparsely pubescent biennials or perennials; flowers 2.5–6(–7) mm. long.
    3A. Cauline leaves 10–20, usually distinctly auriculate; petals oblanceolate or absent. .... 1. *C. impatiens*
    3B. Cauline leaves few, not or scarcely auricled; petals cuneate-obovate to obovate.
      4A. Rhizomes scarcely developed or branched; leaflets 3–20 mm. long.
        5A. Siliques and pedicels pilose; terminal leaflets or radical leaves reniform or cordate; petals about 5 mm. long.
          6A. Stems 7–20 cm. long; racemes few-flowered, not elongate. .... 2. *C. tanakae*
          6B. Stems 20–40 cm. long; racemes rather many-flowered, elongate after anthesis. .... 3. *C. dentipetala*
        5B. Siliques and pedicels glabrous; petals 3–4 mm. long.
          7A. Siliques 17–22 mm. long, about 1 mm. wide, the stigma only slightly thickened.
            8A. Leaflets of radical leaves 7–17, the terminal one slightly larger than the others; leaflets of upper leaves usually linear
                 to narrowly oblanceolate. .... 4. *C. flexuosa*
            8B. Leaflets of radical leaves (3–)5–11, the terminal one much larger than the others; leaflets of upper leaves oblanceolate to
                 oblong. .... 5. *C. scutata*
          7B. Siliques 25–30 mm. long, 1.2–1.5 mm. wide, the stigma distinctly thickened. .... 6. *C. longifructus*
      4B. Rhizomes much branched; leaflets not more than 10 mm. long; alpine plant with stems 5–10 cm. long. ...... 7. *C. nipponica*
  2B. Stolons elongate after anthesis, leafy; glabrous perennials; flowers 6–10 mm. long.
    9A. Stems erect from the base, sometimes decumbent after anthesis; plants stoloniferous; rhizomes not prominent. ...... 8. *C. lyrata*
    9B. Stems usually ascending from the creeping rhizomes; stolons elongate after anthesis.
      10A. Leaflets 5–11.
        11A. Leaflets of cauline leaves sessile.
          12A. Leaflets of upper leaves linear, acute. .... 9. *C. schinziana*
          12B. Leaflets elliptic to lanceolate, usually obtuse.
            13A. Siliques gradually narrowed above to the stigma. .... 10. *C. yezoensis*
            13B. Siliques abruptly and shortly narrowed above to the stigma. .... 11. *C. torrentis*
        11B. Leaflets of cauline leaves petiolulate. .... 12. *C. kiusiana*
      10B. Leaflets 3(–5). .... 13. *C. fauriei*
1B. Rhizomes more or less thickened.
  14A. Plants short-rhizomatous; stems 10–20(–30) cm. long; leaves 3-foliolate or if 5-foliolate, the leaflets of the lowest pair very small;
       racemes short, rather few-flowered.
    15A. Radical leaves absent; leaflets of cauline leaves rhombic-lanceolate to linear-lanceolate; petals 8–10 mm. long.
                                                    14. *C. anemonoides*
    15B. Radical leaves present; leaflets orbicular-reniform or ovate; petals 4–5 mm. long. .... 15. *C. arakiana*
  14B. Plants stoloniferous; stems 30–70 cm. long; leaflets 5–7; racemes rather elongate.
    16A. Cauline leaves auriculate; leaflets obtuse or subobtuse, rather thick, 3–5 cm. long, short-pilose on upper side; racemes glabrous.
                                                    16. *C. appendiculata*
    16B. Cauline leaves not auriculate; leaflets acuminate to long-acute, thin and membranous, 4–10 cm. long, puberulent beneath, scattered
       short-hairy above; racemes usually puberulent. .... 17. *C. leucantha*

1. Cardamine impatiens L. JA-NINJIN.   Slender, sparsely short-pubescent annual or biennial; stems erect, often branched above; leaves pinnately divided, the radical petiolate, the cauline short-petiolate or sessile, usually auriculate, the pinnae 7–20, thinly membranous, pinnately lobed or deeply incised; racemes many-flowered, 10–15 cm. long, erect, the pedicels 3–6 mm. long, slender; flowers greenish white; sepals oblanceolate, slightly pilose; petals white, broadly oblanceolate, 3–3.5 mm. long, 1.5–2 times as long as the sepals, sometimes smaller or absent; siliques glabrous or scattered-pilose.——Wet shaded places in hills and mountains; Hokkaido, Honshu, Shikoku, Kyushu; rather common.——The temperate regions of Eurasia.

2. Cardamine tanakae Fr. & Sav. *C. tenuis* Koidz., excl. basionym; *C. chelidonioides* S. Moore——MARUBA-NO-KONRON-SŌ.   Biennial or perennial (?), usually rather densely white-pilose; stems 7–20 cm. long; radical leaves petiolate, 1- to 7-foliolate, the leaflets petiolulate, orbicular-cordate or ovate-orbicular, coarsely toothed, rounded at apex, the terminal leaflet 1–3 cm. long and as wide, the lateral leaflets smaller, the cauline 1–5, petiolate, 3- to 7-foliolate, sometimes auriculate; racemes rather few-flowered, densely white-pilose, the pedicels 7–15 mm. long; petals cuneate-obovate, white, 5–7 mm. long; siliques ascending, 2–2.5 cm. long, 1–1.2 mm. wide, scattered-pilose; stigma slightly thickened; seeds about 1.2 mm. long.——Wet places; Honshu, Shikoku, Kyushu.

3. Cardamine dentipetala Matsum. *C. drakeana* H.

Boiss.——Ō-KE-TANETSUKE-BANA.   Resembles No. 5, but scattered-pilose on all parts; siliques pilose.——June. Wet woods in mountains; Honshu (Japan Sea side of n. and centr. distr.); rather rare.

4. Cardamine flexuosa With. *C. scutata* subsp. *flexuosa* (With.) Hara——TANETSUKE-BANA.   Annual or biennial herb; stems 10–30 cm. long, usually spreading-pilose on lower part, sometimes branched above; lower leaves 7- to 17-foliolate, often very thinly pilose, the leaflets ovate to broadly so, petiolulate, sometimes 3- to 5-lobed, the terminal lobe slightly larger, 3–15 mm. long, 6–15 mm. wide, the leaflets of the upper leaves 3–11, lanceolate, sometimes denticulate or incised, sessile or subsessile; racemes 10- to 20-flowered; flowers white; petals cuneate-obovate, 3–4 mm. long; siliques glabrous, about 2 cm. long, 1 mm. wide; seeds about 1 mm. wide.——Mar.–June. Wet places in lowlands and paddy fields; Hokkaido, Honshu, Shikoku, Kyushu; common.——Temperate regions of the N. Hemisphere.

Var. **fallax** O. E. Schulz. *C. fallax* (O. E. Schulz) Nakai; *C. brachycarpa* Franch.; *C. parviflora* sensu auct. Japon., non L.; *C. koshiensis* Koidz.——TACHI-TANETSUKE-BANA.   More densely pubescent; stems slender and erect; growing in dryer places than the typical phase.

5. Cardamine scutata Thunb. *C. regeliana* Miq.; *C. angulata* var. *kamtschatica* Regel; *C. flexuosa* var. *kamtschatica* (Regel) Matsum.; *C. autumnalis* Koidz.; *C. scutata* subsp. *regeliana* (Miq.) Hara——ŌBA-TANETSUKE-BANA.   Resembles the preceding, but slightly larger, softer and nearly glabrous; leaflets 3–11, the terminal leaflets much larger than the lateral; petals about 4 mm. long; siliques about 2–2.5 cm. long.——Wet places in lowlands and in the mountains; Hokkaido, Honshu, Shikoku, Kyushu; common.——e. Siberia, Kamchatka, Sakhalin, Korea, and Manchuria.

6. Cardamine longifructus Ohwi. *C. flexuosa* var. *latifolia* Makino——MIZU-TANETSUKE-BANA. Glabrous perennial herb with short branched rhizomes; stems tufted, striate, green, sparsely branched, few-leaved; radical leaves pinnately 3- to 7-foliolate, with petioles 4–6 cm. long, the terminal leaflets reniform, few-toothed, 1.5–2 cm. long, 2–2.5 cm. wide, truncate to shallowly cordate at base, the lateral leaflets much smaller, the lower cauline leaves simulating the radical, the median and upper leaflets sessile, the terminal rhombic-orbicular to obovate-cuneate, obtusely and irregularly toothed, 2–2.5 cm. long, 1.5–2 cm. wide, the lateral leaflets narrowly oblong; racemes erect; flowers several to rather many, about 6 mm. long; sepals 2 mm. long; petals obovate-cuneate; siliques suberect to ascending, nearly straight, 2.5–3 cm. long; style very short; stigma slightly thickened; seeds rather many, wingless.——Apr.–May. Wet places in hills and low mountains; Honshu (Chūgoku Distr.), Shikoku, Kyushu (?).

7. Cardamine nipponica Fr. & Sav. MIYAMA-TANE-TSUKE-BANA. Glabrous perennial herb, the stem much-branched; stems or scapes tufted, 5–10 cm. long; radical leaves few, 1.5–3 cm. long, 3- to 5(–7)-foliolate, the leaflets elliptic, obovate or oblong, subsessile, 2–6 mm. long, 1.5–5 mm. wide, entire, the terminal leaflets slightly larger to nearly as large as the lateral ones, the cauline leaves few, short-auriculate, clasping; inflorescence densely few-flowered; petals 5–6 mm. long, cuneate-obovate; siliques erect, 2–3 cm. long, 1–1.2 mm. wide, gradually narrowed above.——July–Aug. Alpine slopes; Hokkaido, Honshu (centr. and n. distr.).

8. Cardamine lyrata Bunge. MIZU-TAGARASHI. Glabrous perennial stoloniferous herb; stems 30–50 cm. long, leafy, erect in anthesis, decumbent in fruit; lower cauline leaves 11- to 13-foliolate, sessile, the terminal leaflets 1–2 cm. long and as wide, orbicular-cordate, the lateral ones smaller, subsessile, ovate, obtuse, 3–10 mm. long, the upper ones 5- to 7- foliolate; leaves of the stolons often reduced to a simple terminal leaflet; racemes many-flowered; petals white, 8–10 mm. long, obovate; siliques 2–3 cm. long, about 1 mm. wide, nearly erect, the pedicels 1–2 cm. long, obliquely spreading; seeds winged.——May. Wet places and paddy fields in lowlands and foothills; Honshu (centr. distr. and westw.), Shikoku, Kyushu.——Korea, n. China, and e. Siberia.

9. Cardamine schinziana O. E. Schulz. *C. nasturtii-folia* H. Boiss., non Steud.——EZO-NO-JA-NINJIN. Glabrous perennial; stems 20–30 cm. long, leafy, often short-creeping below; leaflets 5–11, oblong to lanceolate, often slightly incised, the terminal one 15–25 mm. long, 8–15 mm. wide, the lateral ones sessile, 5–20 mm. long, 3–8 mm. wide; flowers white; petals 7–8 mm. long.——June. Wet places; Hokkaido (Hidaka Prov.).

10. Cardamine yezoensis Maxim. *C. prorepens* Fisch. forma *valida* Takeda; *C. valida* (Takeda) Nakai——AINU-WASABI. Perennial stoloniferous herb, mostly glabrous or thinly pilose on the stems near base; stems 30–50 cm. long, often sparsely branched above; leaves petiolate, 5–12 cm. long, without auricles at base, the leaflets 5–9, ovate to ovate-oblong, 1–3 cm. long, 7–15 mm. wide, glabrous, sessile, scattered

short-toothed; racemes elongate, many-flowered; petals 8–10 mm. long.——July–Sept. Hokkaido.——Sakhalin.

11. Cardamine torrentis Nakai. *C. nipponica* Nakai, non Fr. & Sav.——OKUYAMA-GARASHI. Closely resembling the preceding species; stems rather stout, branched, often decumbent; leaflets usually ovate to suborbicular, 1.5–4 cm. long, 1–3.5 cm. wide; racemes rather short; petals 5–6 mm. long; siliques 25–35 mm. long, about 1.5 mm. wide; seeds wingless, about 1.5 mm. long.——May–July. Wet places along streams and ravines in mountains; Honshu (centr. distr.).

12. Cardamine kiusiana Hara. TAKACHIHO-GARASHI. Closely resembles the two preceding species; stems to 60 cm. long; lateral leaflets of the cauline leaves petiolulate, ovate to oblong, the uppermost leaves lanceolate to oblong; petals 8–10 mm. long; siliques 15–25 mm. long.——Shikoku (Mount Tsurugi ?), Kyushu (Hyuga Prov.).

13. Cardamine fauriei Franch. *C. geifolia* Koidz.——EZO-WASABI. Glabrous stoloniferous perennial; stems 30–40 cm. long, few-leaved; leaves petiolate, the lower ones 3(–5)-foliolate, the terminal leaflet ovate-orbicular or orbicular-cordate, obtusely toothed, 2.5–4 cm. long and as wide, the lateral lobes cordate-orbicular to oblong, petiolulate or sessile, entire or toothed, 1–3 cm. long, 7–15 mm. wide, each of the upper leaves of the stolons reduced to a terminal leaflet only, often incised-toothed; racemes elongate after anthesis, on long pedicels; petals 7–9 mm. long; siliques 15–20 mm. long, 1–1.2 mm. wide.——May–July. Streams and ravines in mountains; Hokkaido, Honshu (?).

14. Cardamine anemonoides O. E. Schulz. *Dentaria corymbosa* Matsum.; *C. matsumurana* Nemoto——MITSUBA-KONRON-SŌ. Perennial herb, scattered short-pubescent on the leaves above and on margin; stems 10–20 cm. long, 4- to 5-leaved above, with 1–2 reduced leaves at base; rhizomes slightly thickened, fusiform, scaly; radical leaves absent, the cauline leaves petiolate, 1- to 3-foliolate, the leaflets broadly rhombic-lanceolate to linear-lanceolate, toothed or incised, often connate below and forming a 3-parted leaf, 1–8 cm. long, 8–20 mm. wide, the petioles to 20 mm. long; flowers 2–7, in subcorymbose racemes; petals ascending, 8–10 mm. long, white, sometimes reddish, cuneate-obovate; siliques erect, 3–4 cm. long, 1–1.2 mm. wide, the pedicels about 20 mm. long; seeds wingless, about 1.5 mm. long.——Apr.–May. Honshu (Kantō Distr. and westw.), Shikoku, Kyushu.

15. Cardamine arakiana Koidz. *Dentaria arakiana* Koidz.——Ō-MARUBA-KONRON-SŌ. Allied to species No. 2; rhizomes short, thickened, scaly; stems 10–30 cm. long, simple, few-leaved, with spreading hairs; radical leaves ternate or simple, thin, the terminal leaflets orbicular-reniform, or reniform-cordate, 15–30 mm. long, 15–40 mm. wide, short-pilose above, undulate-toothed, the lateral leaflets very small, short-petiolulate, obliquely cordate; cauline leaves smaller, the leaflets slightly narrower; racemes several-flowered, short, glabrous; sepals glabrous; petals narrowly cuneate-oblong, 6–7 mm. long, slightly shorter than the pedicels in anthesis; siliques 2–3 cm. long, about 1.8 mm. wide, gradually narrowed above to the stigma, the style about 2 mm. long; seeds about 1.5 mm. long, wingless.——Apr.–May. Honshu (Tanba Prov.).

16. Cardamine appendiculata Fr. & Sav. *Dentaria appendiculata* (Fr. & Sav.) Matsum.; *C. macrophylla* var. *appendiculata* (Fr. & Sav.) Yatabe——HIRO-HA-KONRON-SŌ, TADE-NO-UMI-KONRON-SŌ. FIG. 7. Rhizomes slender, creeping, slightly

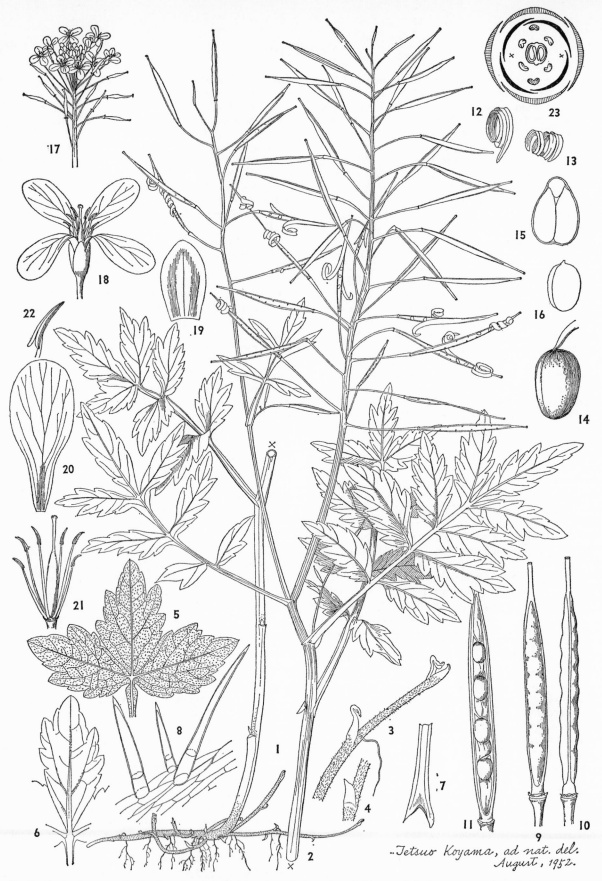

Fig. 7—*Cardamine appendiculata* Fr. & Sav. 1, 2, Fruiting plant; 3, 4, portion of rhizome; 5, leaflets, upper surface; 6, the same, lower surface; 7, base of petiole; 8, hairs on upper surface of leaf; 9, 10, siliques; 11, the same, with valve taken off; 12, 13, valves; 14, 15, 16, seeds; 17, raceme; 18, flower; 19, sepal; 20, petal; 21, flower, with sepals and petals removed; 22, anther; 23, floral diagram. Specimen from Lake Tadenoumi in Shimotsuke, Honshu.

thickened at apex; stems 30–50 cm. long, several-leaved; leaves petiolate, the radical and lower cauline long-petiolate, the leaflets 5–7, nearly equal, ovate-oblong to oblong, 3–5 cm. long, 1–3 cm. wide, obtuse, short-puberulent above, incised and obtusely toothed, the petioles of the cauline leaves slightly auriculate; racemes solitary or few, about 10-flowered, the pedicels 1–2 cm. long; petals about 8 mm. long, white, cuneate-obovate; siliques 2–3 cm. long, 1.5–1.8 mm. wide, gradually narrowed above; seeds about 2 mm. long, wingless.——May–July. Wet places along streams and ravines in mountains; Honshu (centr. and n. distr.).

17. Cardamine leucantha (Tausch) O. E. Schulz.

Dentaria leucantha Tausch; *C. dasyloba* (Turcz.) Komar.; *D. dasyloba* Turcz.; *D. macrophylla* var. *dasyloba* (Turcz.) Makino——KONRON-SŌ. Rhizomatous; stems 40–60 cm. long, erect, short-pubescent; cauline leaves few, petiolate, short-pubescent beneath, the leaflets 5–7, thinly membranous, broadly lanceolate to ovate-oblong, 4–8 cm. long, 1–2.5(–3) cm. wide, acuminate, toothed; racemes many-flowered, elongate after anthesis, short-pubescent; sepals usually pilose; petals 8–10 mm. long; siliques slightly pilose while young, long-pedicelled.——Apr.–June. Wet places along streams and ponds in mountains; Hokkaido, Honshu, Shikoku, Kyushu.——Korea, Manchuria, n. China, and e. Siberia.

7. WASABIA Matsum. WASABI ZOKU

Glabrous perennial herbs; radical leaves long-petiolate, simple, cordate-orbicular, undulate-toothed; stems leafy, simple; flowers bracteate, in racemes, white; sepals ascending; petals clawed, obovate; siliques linear-oblong, rather short, torulose, cylindric, the valves membranous, without a midrib, the septum perfect; style rather elongate, the stigma simple; seeds few in 1 series, rather large, plump; cotyledons incumbent.——Two species, in Japan.

1A. Rhizomes stout, 1–2 cm. across; style about 2 mm. long; petals 8–9 mm. long; leaves to 15 cm. across. .... 1. *W. japonica*
1B. Rhizomes short and narrow, 1–2 mm. across; style 0.5–0.7 mm. long; petals 5–7 mm. long; leaves to 5 cm. across. ...... 2. *W. tenuis*

1. Wasabia japonica (Miq.) Matsum. (?) *Cochlearia wasabi* Sieb.; *Eutrema wasabi* (Sieb.) Maxim.; *Alliaria wasabi* (Sieb.) Prantl; *W. wasabi* (Sieb.) Makino; *W. pungens* Matsum.; (?) *Lunaria japonica* Miq.; *E. japonica* (Miq.) Koidz.; (?) *E. okinosimensis* Takenouchi——WASABI. Rhizomes creeping, stout; stems ascending, 20–40 cm. long, several-leaved; radical leaves long-petiolate, cordate-reniform, 8–15 cm. long and as wide, rounded at apex, undulate-toothed, the petioles dilated at base; cauline leaves petiolate, broadly ovate, 2–4 cm. long, shallowly cordate; inflorescence simple, solitary, many-flowered, the bracts similar to the cauline leaves but small, the pedicels slender, 1–3(–5) cm. long; sepals and stamens about 4 mm. long; petals cuneate-obovate; siliques on stipes 3–5 mm. long; style slender.——Mar.–May. Wet places along streams in mountains and often cultivated in such places for the pungent rhizomes; Honshu, Shikoku, Kyushu.——Sakhalin.

2. Wasabia tenuis (Miq.) Matsum. *Nasturtium tenue* Miq.; *Cardamine tenuis* (Miq.) Koidz.; *C. bracteata* S. Moore; *Eutrema bracteata* (S. Moore) Koidz.; *W. bracteata* (S. Moore) Hisauchi; *E. hederaefolia* Fr. & Sav.; *W. hederaefolia* (Fr. & Sav.) Matsum.——YURI-WASABI. Slender glabrous perennial herb with narrow short rhizomes and slender fibrous roots; stems ascending, 10–30 cm. long, few-leaved; radical leaves long-petiolate, reniform-orbicular, cordate, 2–5 cm. long and as wide, undulate-toothed, mucronate, the nerves palmate; cauline leaves petiolate, ovate, cordate, 1–2.5 cm. long, 0.8–2.5 cm. wide; flowers in loose racemes, few, the bracts similar to the cauline leaves but small, the pedicels 5–15(–40) mm. long, slender, spreading to deflexed; siliques nodding, broadly linear, torulose, 8–12 mm. long, about 2 mm. across, sessile or nearly so; seeds 4–8, in 1 series, 2–2.5 mm. long, obsoletely punctulate.——Apr.–May. Streams in mountain woods; Honshu, Shikoku, Kyushu.

8. NASTURTIUM R. Br. ORANDA-GARASHI ZOKU

Perennial herbs of wet places; leaves pinnate; racemes elongate, many-flowered, without bracts; flowers white; siliques linear; style stout, short; valves membranous, nerveless; seeds in 2 series, many, small; cotyledons accumbent.——A single species.

1. Nasturtium officinale R. Br. *N. nasturtium-aquaticum* (L.) Karst.; *Sisymbrium nasturtium-aquaticum* L.; *Rorippa nasturtium-aquaticum* (L.) Schinz & Thell.; *Radicula nasturtium-aquaticum* (L.) Britten & Rendle——ORANDA-GARASHI. Glabrous perennial rhizomatous herb; stems 30–40 cm. long, rather stout; leaves petiolate, pinnate, the leaflets 3–7(–9), ovate or broadly so, undulately few toothed, the terminal leaflet larger than the lateral, 2–3 cm. long, 15–25 mm. wide; racemes rather elongate, the pedicels often deflexed; siliques 1–2 cm. long, 1.5–2 mm. wide, with a short, stout style; seeds in 2 series.——Widely naturalized in our area, growing in water or in very wet places; common.——Eurasia.

9. COCHLEARIA L. TOMOSHIRI-SŌ ZOKU

Glabrous annuals or perennials; leaves alternate, sometimes rosulate, entire or pinnately parted; flowers white, rarely purple or yellow, ebracteate, in racemes or rarely solitary on naked scapes; petals shortly clawed; stamens straight or geniculate; silicles sometimes short-stiped, oblong or globose, the valves usually ventricose, loosely reticulate or veiny; style short or elongate, the stigma simple or capitellate; seeds many to few, usually in 2 series, small; cotyledons accumbent or rarely incumbent.——About 25 species, in the temperate and far northern regions of the N. Hemisphere, especially along the sea and in high mountains.

1. Cochlearia oblongifolia DC. *C. officinalis* subsp. *oblongifolia* (DC.) Hult.——TOMOSHIRI-SŌ. Lustrous glabrous biennial; stems 10–20 cm. long, branched at base, few-leaved; leaves rather thick, the radical rosulate, rather many, usually withering before anthesis, the lower cauline petiolate, the upper leaves sessile, ovate or oblong, 1–2.5 cm. long, 8–15 mm. wide, obtuse, entire or few-toothed, narrowed at base; flowers white, small, the pedicels 7–15 mm. long; silicles ovoid-oblong, 5–7 mm. long, 3–3.5 mm. wide; style slender, short, the stigma capitellate.——Rocks by the sea; Hokkaido (Nemuro Prov.).——Kamchatka, Kuriles, Sakhalin, and Ochotsk Sea region.

10. SISYMBRIUM L. Kakane-garashi Zoku

Annuals, biennials, or perennials usually with coarse simple hairs; cauline leaves alternate, variously divided to entire; racemes usually elongate; flowers yellow or yellowish; petals often with a slender claw at base; siliques linear to narrowly so, sometimes narrowed above, terete or angled, the valves often 3-nerved; style short, the stigma usually bifid; seeds in 1 series; cotyledons incumbent.——Many species, mainly in the Old World.

1A. Siliques about 12 mm. long, appressed to the axis, on very short pedicels, gradually narrowed from the base to the apex; flowers small; biennial. . . . . 1. *S. officinale*
1B. Siliques 8–10 cm. long, erect or obliquely spreading, narrowly linear, uniformly thickened from base to the apex.
  2A. Annual or biennial herbs; siliques as broad as the stout pedicels.
    3A. Pedicels 3–4 mm. long; sepals pilose, obtuse. . . . . 2. *S. orientale*
    3B. Pedicels 7–10 mm. long; sepals glabrous with a mucro below the apex. . . . . 3. *S. altissimum*
  2B. Perennial herb; siliques twice as broad as the pedicel; flowers rather large. . . . . 4. *S. luteum*

1. Sisymbrium officinale (L.) Scop. *Erysimum officinale* L.——Kakine-garashi. More or less coarsely hispid-pilose annual or biennial herb, 30–80 cm. high with obliquely ascending branches; leaves pinnately cleft to parted, petiolate, the segments 2 or 3 pairs, unequal, spreading, the terminal ones relatively large and toothed, the upper leaves nearly sessile; flowers small; petals about 3 mm. long; siliques 10–12 mm. long, acuminate, appressed to the axis, 1–1.2 mm. across at base, shortly soft-pubescent, terete, the pedicels appressed to the axis, slenderer than the silique, 2–3 mm. long.——May–July. Naturalized; Honshu (near Tokyo).
Var. **leiocarpum** DC. Hama-garashi. Siliques glabrous——Naturalized in Hokkaido and Honshu (near Tokyo).——Europe, w. Asia, and N. America.
2. Sisymbrium orientale L. *S. columnae* var. *orientale* (L.) DC.——Inu-kakine-garashi. Resembles the preceding; pedicels erect, stout; petals about 6 mm. long, narrowly spathulate; siliques 7–8 cm. long, narrowly linear.——May–July. Naturalized in our area.——Eurasia.

3. Sisymbrium altissimum L. Hatazao-garashi. Resembles the preceding; pinnae of leaves 4 or 5 pairs, the pedicels ascending, 7–10 mm. long; sepals glabrous; siliques 8–10 cm. long.——May–July. Widely naturalized.——Europe and w. Asia.
4. Sisymbrium luteum (Maxim.) O. E. Schulz. *Hesperis lutea* Maxim.——Kibana-hatazao. Coarsely pilose branched perennial herb, 80–120 cm. high; leaves petiolate, the lower ones pinnate, the lateral segments 1 to 3 pairs, the terminal ones narrowly deltoid-ovate, 8–12 cm. long, 3–5 cm. wide, acute, short-toothed; racemes elongate after anthesis, the pedicels 12–15 mm. long, ascending; sepals 8–9 mm. long, more or less coarsely pilose; petals yellow, 12–13 mm. long; siliques narrowly linear, 8–10 cm. long; stigma shallowly bilobed.——May–June. Honshu (centr. distr.), Kyushu (Tsushima); rare.——Korea and Manchuria.

11. BARBAREA R. Br. Yama-garashi Zoku

Erect branched glabrous biennials or perennials; leaves lyrate to sinuately pinnatifid or entire; flowers yellow, sometimes bracteate; sepals equal; petals clawed, obovate; siliques linear, compressed-tetragonous or subcompressed, the valves costate or keeled; style short, the stigma capitate or bifid; seeds in 1 series, oblong, elliptic, sometimes margined; cotyledons accumbent.—— About 10 species, in temperate and cold regions.

1. Barbarea orthoceras Ledeb. *B. cochlearifolia* H. Boiss.; *B. patens* H. Boiss.; *Sisymbrium japonicum* H. Boiss.; *B. vulgaris* lusus *sibirica* Regel; *B. sibirica* (Regel) Nakai; *B. hondoensis* Nakai; *B. vulgaris* var. *stricta* sensu auct. Japon., non Regel——Yama-garashi, Miyama-garashi. Deep green glabrous perennial herb, 20–40 cm. high; stems striate, branched above; radical leaves petiolate, lyrate, the terminal segments large, ovate to cordate, 1–3 cm. long, 1–2.5 cm. wide, the lateral segments to 5 pairs, smaller, the cauline leaves nearly equal, decurrent and auriculate at base, clasping; racemes elongate after anthesis, erect, many-flowered, the pedicels slender; petals yellow, 7–8 mm. long, obovate; siliques erect or ascending, 3–4 cm. long, about 1.5 mm. wide, flat, straight; seeds thinly margined, 1.5–2 mm. long.——May–Aug. Wet places especially along streams and around ponds in high mountains; Hokkaido, Honshu (centr. and n. distr.). ——e. Siberia, Kamchatka, Manchuria, Korea, Sakhalin, Kuriles, and N. America.

12. RORIPPA Scop. Inu-garashi Zoku

Branched annuals or perennials, glabrous or with simple hairs; leaves entire, lobed or parted; flowers yellow, sometimes solitary in upper axils, usually in racemes; sepals short, equal, spreading; petals slightly clawed, spreading or rarely absent; stamens 6 or fewer; siliques often curved, nearly terete, on spreading or arcuate sometimes very short pedicels, the valves membranous, weakly 1-nerved or nerveless; style short or rather long, the stigma simple or 2-lobed; seeds in 2 series, small, turgid; cotyledons accumbent.——About 50 species, in temperate and warmer regions of the N. Hemisphere, few in the S. Hemisphere.

1. Rorippa dubia (Pers.) Hara. *Sisymbrium dubium* Pers.; *R. sinapis* Ohwi & Hara, excl. syn.; *Nasturtium heterophyllum* Bl.; *Cardamine sublyrata* Miq.; *N. sublyratum* (Miq.) Fr. & Sav.; *R. sublyrata* (Miq.) Hara; *N. indicum* forma *apetalum* Makino——MICHIBATA-GARASHI. Glabrous perennial with short rhizomes; stems branched at base, 10–20 cm. long, often ascending-spreading, few-leaved; radical leaves narrowly ovate, the lower cauline more or less lyrate, petiolate, 4–10 cm. long, irregularly dentate-serrate; racemes 3–5 cm. long, erect or ascending, straight; flowers about 10 in a raceme, small, the pedicels 4–6 mm. long; petals absent; sepals 2–2.5 mm. long; siliques narrowly linear, straight, 20–25 mm. long, 1–1.2 mm. wide; seeds minute, about 0.5 mm. long.—— May–Aug. In shaded places around houses and on roadsides in lowlands; Honshu, Shikoku, Kyushu.——India, China, Malaysia, Ryukyus, and Formosa.

2. Rorippa indica (L.) Hiern. *Sisymbrium indicum* L.; *S. sinapis* Burm.; *R. sinapis* (Burm.) Ohwi & Hara; *Nasturtium indicum* (L.) DC.; *S. atrovirens* Hornem.; *N. atrovirens* (Hornem.) DC.; *R. atrovirens* (Hornem.) Ohwi & Hara; *N. montanum* Wall. ex Benth.; *N. montanum* var. *obtusulum* Miq.; *N. obtusulum* (Miq.) Koidz.——INU-GARASHI. Resembles the preceding; stems often more branched, 20–40 cm. long; terminal segment of radical leaves often larger, 1–2 cm. wide, irregularly dentate-serrate, the upper leaves sessile, lanceolate; racemes 4–6 cm. long, often branched below; petals yellow, spathulate, slightly longer than the petals, about 3 mm. long; siliques narrowly linear, 15–22 mm. long, about 1.2 mm. wide, slightly incurved, the pedicels 5–8 mm. long; style thick and short.——May–Sept. Rather wet places in lowlands, especially in paddy fields and along roadsides; Hokkaido, Honshu, Shikoku, Kyushu; common.——China, Ryukyus, Formosa, Malaysia, India, and Africa.

3. Rorippa islandica (Oed.) Borb. *Sisymbrium islandicum* Oed.; *Nasturtium palustre* (L.) DC.; *R. palustris* (L.) Bess.; *S. amphibium* var. *palustre* L.; *Radicula palustris* (L.) Moench——SUKASHI-TAGOBŌ. Nearly glabrous biennial herb; stems spreading, branched, 30–50 cm. long; radical leaves petiolate, lyrate, 7–15 cm. long, 15–30 mm. wide, toothed, the cauline leaves sessile, auriculate, toothed, lobed or undivided; racemes 5–15 cm. long, many-flowered, the pedicels spreading or reflexed, 5–7 mm. long; petals cuneate-obovate, about 3 mm. long, slightly longer than the sepals, yellow; siliques short, oblong or oblong-cylindric, 4–6 mm. long, 2–2.5 mm. wide; style very short; seeds minute.——Apr.–Oct. Wet places in lowlands especially in paddy fields and along river margins; Hokkaido, Honshu, Shikoku, Kyushu; common.——Temperate and warmer regions of the N. Hemisphere.

4. Rorippa nikkoensis Hara. *Nasturtium amphibium* sensu auct. Japon., non R. Br.——MIGIWA-GARASHI. Glabrous perennial herb; stems to 50 cm. long, branched; lower leaves lyrate, the upper ones parted to cleft, 7–15 cm. long, 15–50 mm. wide, auriculate, the terminal segments narrowly ovate to oblong, irregularly toothed, the lateral segments 2–10, rather small; racemes rather many-flowered, the pedicels spreading, 4–6 mm. long; petals nearly as long as the sepals, 1.5–2 mm. long, broadly spathulate; siliques globose or broadly ellipsoidal, 3–4 mm. long, 2–3 mm. wide; style rather slender; seeds about 0.8 mm. long.——July–Aug. Mountains; Honshu (Nikko).

5. Rorippa cantoniensis (Lour.) Ohwi. *Ricotia cantoniensis* Lour.; *Nasturtium microspermum* DC.; *N. sikokianum* Fr. & Sav.——KO-INU-GARASHI. Glabrous branched annual or biennial, 10–40 cm. high; leaves cleft, 3–6 cm. long, 1–2 cm. wide, irregularly incised and toothed, auriculate, the upper leaves (or bracts) broadly lanceolate, 5–20 mm. long, incised, sessile; flowers solitary in leaf-axils, nearly sessile; petals broadly oblanceolate, about 3 mm. long, slightly longer than the sepals; siliques short-cylindric, 6–10 mm. long, 2–3 mm. wide; style short and thick; seeds minute.——Apr.–June. Edge of paddy fields and along margins of rivers in lowlands; Honshu (Kantō Distr. and westw.), Shikoku, Kyushu; common.——China and Manchuria.

13. CAPSELLA Medic. NAZUNA ZOKU

Annual or biennial herbs, stellate-pubescent; radical leaves rosulate, lyrate; stems erect, few-leaved; flowers small, white, bractless; sepals equal; siliques cuneate or obtriangulate, flattened laterally or subterete, the valves navicular, flattened, keeled on back, the septum narrow, membranous; style short, the stigma sessile; seeds many, wingless; cotyledons incumbent or sometimes accumbent.——Few species, in both hemispheres.

1. Capsella bursa-pastoris (L.) Medic. *Thlaspi bursa-pastoris* L.——NAZUNA. Annual or biennial herb with simple and stellate hairs; radical leaves lyrate, the cauline lanceolate, auriculate and sessile, toothed; stems 10–40 cm. long, often branched below; flowers many, in racemes, the pedicels slender; petals obovate, clawed, about 2–2.5 mm. long; silicles obtriangular, glabrous, flattened, 6–7 mm. long, 5–6 mm. wide, shallowly and broadly retuse to emarginate; style short; seeds 20–25, obovate, about 0.8 mm. long.——Mar.–June. Cultivated fields and along roadsides in lowlands; Hokkaido, Honshu, Shikoku, Kyushu; very common.——Temperate regions of the N. Hemisphere.

14. DRABA L. INU-NAZUNA ZOKU

Usually densely tufted annuals, biennials, or perennials with stellate or simple hairs; leaves entire or incised-toothed, the radical rosulate; stems naked or few-leaved; racemes short or elongate; flowers bractless, small, white or yellow, rarely rose or purple; silicles elliptic to broadly linear, flat, many- to few-seeded, the valves nearly flat, rarely nerved, deciduous, the septum membranous; styles short or elongate, the stigma simple; seeds in 2 series, wingless, sometimes with a membranous appendage; cotyledons accumbent.——About 270 species, mostly in mountains and northern regions of the N. Hemisphere.

1A. Biennial; flowers yellow; style very short; silicles much shorter than the pedicels. .... 1. *D. nemorosa* var. *hebecarpa*
1B. Perennials; innovation-shoots present; style distinct.
  2A. Flowers yellow. .... 2. *D. japonica*
  2B. Flowers white.
    3A. Petals 7–8 mm. long; style 2.5–4 mm., rarely 1.5 mm. long; silicles pilose. .... 3. *D. sachalinensis*
    3B. Petals 3–6 mm. long; style less than 2 mm. long.
      4A. Silicles puberulent, prominently contorted; petals 5–6 mm. long; style 0.2–0.5 mm. long; rocks by the sea. ...... 4. *D. borealis*
      4B. Silicles glabrous, not or slightly contorted; petals 3–4 mm. long; rocks and dry slopes in mountains and alpine regions.
        5A. Seeds appendaged on one end. .... 5. *D. sakuraii*
        5B. Seeds not appendaged.
          6A. Leaves glabrous on both sides, ciliate. .... 6. *D. shiroumana*
          6B. Leaves minutely stellate-pubescent on both-sides. .... 7. *D. kitadakensis*

1. Draba nemorosa L. var. **hebecarpa** Lindbl. *D. hirta* sensu Miq. non L.——INU-NAZUNA. Rather prominently stellate-pilose biennial; stems erect, simple or branched at base, 10–20 cm. long, slender, leafy; radical leaves rosulate, spathulate, 2–4 cm. long, 8–15 mm. wide, slightly toothed, nearly sessile, the cauline leaves few, narrowly ovate to narrowly oblong, 1–3 cm. long, 5–15 mm. wide, broadly cuneate, sessile; racemes many-flowered, the pedicels 10–20 mm. long, spreading; petals about 3 mm. long; silicles narrowly oblong, flat, 5–8 mm. long, 2–2.2 mm. wide, short-pilose; style very short or nearly absent; seeds not appendaged, about 0.4 mm. long.——Mar.–May. Sunny places in lowlands; Hokkaido, Honshu, Shikoku, Kyushu; common.——Temperate regions of the N. Hemisphere.

Var. **leiocarpa** Lindbl. ME-INU-NAZUNA. Silicles glabrous.——Rare. Occurs with var. *hebecarpa*.

2. Draba japonica Maxim. *D. yezoensis* Nakai——NAMBU-INU-NAZUNA. Perennial herb, stellate-pubescent, the caudex branched, tufted, the innovation-shoots slightly elongate; stems 5–10 cm. long, few-leaved; rosulate leaves oblanceolate to narrowly so, 5–15 mm. long, 1.5–3 mm. wide, entire or with few coarse teeth, subobtuse, gradually narrowed below, the cauline leaves sessile, 8–10 mm. long, 2–6 mm. wide; racemes short-pilose; petals obovate, clawed below, yellow, 4–4.5 mm. long; silicles narrowly ovate, glabrous, 4–6 mm. long, 2.5–3.5 mm. wide; style 0.2–0.5 mm. long, the stigma discoid; seeds not appendaged, about 1.5 mm. long.——June–Aug. Alpine regions; Hokkaido (Mount Yubari), Honshu (Mount Hayachine); rare.

3. Draba sachalinensis Trautv. *D. grandiflora* Franch., non C. A. Mey.; *D. franchetii* O. E. Schulz; *D. moiwana* Koidz.; *D. shinanomontana* Ohwi; *D. sachalinensis* var. *shinanomontana* (Ohwi) Okuyama——MOIWA-NAZUNA. Perennial herb, densely grayish stellate-puberulent; stems 10–25 cm. long, 2- to 4-leaved on lower half, the sterile innovation shoots slightly elongate, with a tuft of rosulate leaves at the tip; rosulate leaves oblanceolate, few-toothed or subentire, 2–3 cm. long, 5–8 mm. wide, subacute, gradually narrowed below, the cauline leaves ovate to broadly lanceolate, 1.5–2 cm. long, 3–10 mm. wide, toothed, rarely entire, cuneate; racemes rather many-flowered, stellate-puberulent, the pedicels ascending, 15–20 mm. long; petals 7–8 mm. long; silicles broadly lanceolate, stellate-puberulent, 8–10 mm. long, 2.5–3 mm. wide;

stigma capitellate; seeds not appendaged, 1–1.2 mm. long.——May–June. Rocky places in mountains; Hokkaido, Honshu (Shinano Prov.).——Sakhalin.

4. Draba borealis DC. SHIROBANA-NO-INU-NAZUNA, EZO-INU-NAZUNA. Rather densely stellate-puberulent perennial, the innovation shoots somewhat elongate and sparingly branched; stems few-leaved, 6–20 cm. long; rosulate leaves spathulate, 15–30 mm. long, 5–8 mm. wide, obtuse, few-toothed or sometimes entire, the margin with scattered simple hairs, the cauline leaves ovate to oblong, 8–25 mm. long, toothed, usually acute, the innovation-leaves usually without simple hairs; racemes rather many-flowered, puberulent, the pedicels about 10 mm. long; petals obovate, clawed, 5–6 mm. long; silicles broadly lanceolate, usually prominently contorted, 5–8 mm. long, 2.5–3 mm. wide, puberulent, rarely glabrous; style 0.2–0.4 mm. long, the stigma capitellate; seeds without appendage, about 1 mm. long.——June–July. Rocky cliffs near the sea; Hokkaido (Nemuro and Hidaka Prov.).——Kuriles, Sakhalin, and Ochotsk Sea region.

5. Draba sakuraii Makino. *D. sinanensis* Makino——TOGAKUSHI-NAZUNA. Loosely tufted perennial, rather densely stellate-pubescent; stems 7–20 cm. long, 2- to 5-leaved, with scattered stellate and simple hairs, the innovation-shoots rarely elongate; rosulate leaves oblanceolate or broadly so, 1–2.5 cm. long, 2–8 mm. wide, toothed or subentire, subobtuse, often with simple hairs on margin, the cauline leaves narrowly ovate to oblong, usually coarsely few-toothed, 8–15 mm. long, 5–10 mm. wide, puberulent, the pedicels puberulent; petals 3.5–4(–5) mm. long, cuneate-obovate; silicles narrowly lanceolate, flat; style 1–2 mm. long, the stigma capitellate; seeds with a membranous tail.——June–July. Rocky dry places in high mountains; Honshu (n. part of centr. distr.).

Var. **nipponica** (Makino) Takeda. *D. nipponica* Makino; *D. ontakensis* Makino; *D. nikoensis* Koidz.; *D. sakuraii* var. *ontakensis* (Makino) Takeda; *D. sakuraii* var. *rigidula* Takeda; *D. linearis* Satake——KUMOMA-NAZUNA. Plant more slender, less puberulent, with glabrous more slender pedicels and shorter-tailed seeds.——Honshu (centr. distr. and Nikko).

6. Draba shiroumana Makino. SHIROUMA-NAZUNA. Perennial with short slightly branched caudex; stems 5–10 cm. long, 1- to 4-leaved, glabrous; rosulate leaves linear-spathulate, 5–12 mm. long, 1.5–2 mm. wide, entire or rarely few-toothed, glabrous on both sides, ciliate, gradually narrowed below, the cauline sessile, broadly linear, 5–12 mm. long, 1.5–2 mm. wide; racemes glabrous, rather few-flowered; petals about 3.5 mm. long, cuneate-obovate; silicles broadly linear, often slightly contorted; style minute; seeds with obtuse margin, not appendaged, 1–1.5 mm. long.——July–Aug. Dry gravelly or rocky slopes in alpine regions; Honshu (centr. distr., on Mount Shirouma and in Akaishi Mts.); rare.

Draba okamotoi Ohwi is a sterile, alleged hybrid possibly between *D. shiroumana* and *Arabis tanakana*.

7. Draba kitadakensis Koidz. *D. nakaiana* Hara; *D. kamtschatica* var. *yezoensis* Nakai; *D. oiana* Honda——KITA-DAKE-NAZUNA, SŌUN-NAZUNA.   Perennial gray-green herb, densely appressed stellate-puberulent; stems 10–15 cm. long, short-branched and with short sterile shoots at base, erect, 3- to 10-leaved; rosulate leaves oblanceolate or narrowly so, 6–12 mm. long, 1.5–3 mm. wide, acute, narrowed at base, few-toothed or entire, the cauline narrowly ovate or broadly lanceolate, 8–20 mm. long, acute, toothed; racemes rather densely more than 10-flowered, the pedicels 3–7(–15) mm. long, densely minute-puberulent; petals about 3 mm. long, narrowly cuneate-obovate; silicles lanceolate, contorted, 7–10 mm. long, about 2 mm. wide; style 0.2–1.5 mm. long, the stigma scarcely thickened, truncate; seeds about 1 mm. long, obtuse on margin, not tailed.——May–June. Rocky cliffs in mountains; Hokkaido, Honshu (e. Shinano and Kai Prov.); rare.

15. ARABIS L.   HATAZAO ZOKU

Annuals, biennials, or perennials, nearly glabrous or with bifurcate, stellate, or simple hairs; radical leaves usually spathulate, toothed or entire, the cauline sessile, often toothed or lobed; racemes bractless; flowers white or rose-purple; sepals obtuse, equal or the lateral ones gibbous; petals clawed, entire, longer than the sepals; siliques linear, sessile, the valves flat, sometimes with a midrib, the septum membranous; style short, the stigma simple or shallowly bilobed; seeds in 1 or sometimes 2 series, flat, sometimes narrowly winged on margin; cotyledons accumbent.——Many species, mainly in temperate to northern regions of the N. Hemisphere.

1A. Cauline leaves sessile and clasping or when not clasping, the plants erect, 50–80 cm. high; radical leaves entire or toothed.
  2A. Cauline leaves auriculate and clasping, at least in the lower ones.
    3A. Petals 5–10 mm. long, cuneate-obovate, white.
      4A. Siliques nearly appressed to the axis; style 0.3–0.7 mm. long, stout. . . . . 1. *A. stelleri* var. *japonica*
      4B. Siliques spreading to ascending, rarely suberect; style usually about 2 mm. long, slender. . . . . 2. *A. serrata*
    3B. Petals 4–5 mm. long, cuneate-oblanceolate, white or creamy white; siliques obliquely ascending, erect, or appressed.
      5A. Seeds narrowly winged; plant coarsely hirsute; petals white; seeds 1-seriate. . . . . 3. *A. hirsuta*
      5B. Seeds not winged; plant glaucous, coarsely pubescent only on lower half; petals creamy white; seeds imperfectly 2-seriate.
                                           4. *A. glabra*
  2B. Cauline leaves not auriculate or clasping.
    6A. Stems 50–80 cm. long, often branched above, hispid; cauline leaves hispid; petals cuneate-oblanceolate, 3–4 mm. long; siliques nodding or pedulous. . . . . 5. *A. pendula*
    6B. Stems scapelike, 3–25 cm. long, simple, appressed stellate-pubescent; cauline leaves appressed stellate-pubescent; petals cuneate-obovate; siliques spreading to ascending.
      7A. Style 3–5 mm. long; plants stoloniferous; leaves toothed, 1–7 cm. long, 8–12 mm. wide; petals 7–12 mm. long.
                                           6. *A. flagellosa*
      7B. Style less than 3 mm. long; stolons absent; leaves entire, 3–10 mm. long, 1–3 mm. wide; petals 2.5–3 mm. long.
                                         7. *A. tanakana*
1B. Lower cauline leaves often petioled; radical leaves usually lyrate.
  8A. Siliques (2.5–)3–4 cm. long, not torulose or scarcely inflated; style stout, about 0.5 mm. long; stems erect, bearing no innovation-shoots on the upper part.
    9A. Green or only slightly glaucescent perennial; seeds rather many, 1.5 mm. long, with a narrow wing at the end.
                                       8. *A. lyrata* var. *kamtschatica*
    9B. Glaucescent biennial; seeds many, about 1 mm. long, wingless. . . . . 9. *A. kawasakiana*
  8B. Siliques 10–20 mm. long, slightly torulose and somewhat inflated; style slender, about 1 mm. long; stems usually decumbent after anthesis and bearing innovation-shoots. . . . . 10. *A. gemmifera*

1. Arabis stelleri DC. var. **japonica** (A. Gray) F. Schmidt. *A. stelleri* sensu auct. Japon., non DC.; *A. alpina* var. *japonica* A. Gray; *A. japonica* (A. Gray) A. Gray; *A. yokoscensis* Fr. & Sav.——HAMA-HATAZAO.   Rather densely hairy with usually 2- to 3-branched hairs; stems erect, rather stout, 20–40 cm. long; radical leaves oblong to oblanceolate, 3–7 cm. long, 8–25 mm. wide, few-toothed, gradually narrowed to a rather broad petiole at base; racemes often slightly branched; flowers rather large and many; siliques appressed to the axis, dense, 4–6 cm. long, 1.5–2 mm. wide; style short and stout; seeds narrowly winged, about 1.5 mm. long, in 1 series.——Apr.–June. Dunes along the sea; Hokkaido, Honshu, Shikoku, Kyushu; common.——Korea, Amur, and Sakhalin.

2. Arabis serrata Fr. & Sav. *A. amplexicaulis* var. *ser-*

rata (Fr. & Sav.) Makino; *A. serrata* var. *platycarpa* Ohwi——FUJI-HATAZAO.   Usually more or less tufted perennial with 2- to 4-branched short hairs; stems 10–30 cm. long, few-leaved; radical leaves rosulate, 1.5–7 cm. long, 8–15 mm. wide, prominently coarse-toothed, narrowed to the relatively long petiole; flowers white, few to rather many; petals cuneate-obovate, 6–10 mm. long; siliques 3.5–6 cm. long, 1.5–2 mm. wide; style 1.5–2 mm. long; seeds 1.2 mm. long, narrowly winged.——July–Aug. Rocky or sandy places in mountains; Honshu (Mount Fuji); very variable.

Var. **sikokiana** (Nakai) Ohwi. *A. boissieuana* var. *sikokiana* Nakai; *A. sikokiana* (Nakai) Honda——SHIKOKU-HATAZAO.   Radical leaves long-attenuate at base; siliques 7–9 cm. long.——Mountains; Honshu (Tōkaidō and Kinki Distr. and westw.), Shikoku, Kyushu.——Korea (Quelpaert Isl.).

Var. **japonica** (H. Boiss.) Ohwi. *A. amplexicaulis* var. *japonica* H. Boiss.; *A. boissieuana* Nakai; *A. serrata* var. *japonica* f. *grandiflora* (Nakai) Ohwi; *A. fauriei* var. *grandiflora* Nakai; *A. kishidae* Nakai; *A. serrata* var. *japonica* f. *glabrescens* (Ohwi) Ohwi; *A. serrata* var. *glabrescens* Ohwi; *A. serrata* var. *japonica* f. *fauriei* (H. Boiss.) Ohwi; *A. fauriei* H. Boiss.; *A. iwatensis* Makino——IWA-HATAZAO. Radical leaves gradually narrowed to a rather short petiole, the teeth usually less prominent.——Rather common in mountains, often alpine; Honshu (centr. and n. distr.).

Var. **glauca** (H. Boiss.) Ohwi. *A. glauca* H. Boiss.; *A. pseudoauriculata* H. Boiss.; *A. boissieuana* var. *glauca* (H. Boiss.) Koidz.——EZO-NO-IWA-HATAZAO. Resembles var. *Japonica,* but the radical leaves pilose with coarser, 2-parted, distinctly stipitate hairs above.——Hokkaido, Honshu (n. distr.).——s. Kuriles and Sakhalin.

3. Arabis hirsuta (L.) Scop. *Turritis hirsuta* L.; *A. sagittata* DC.; *A. nipponica* H. Boiss.——YAMA-HATAZAO. Erect perennial with 2- to 4-branched hairs; stems erect, usually simple, leafy, 30–80 cm. long, often with spreading hairs on lower part; radical leaves oblanceolate, 5–10 cm. long, 1–2 cm. wide, obsoletely few-toothed, gradually narrowed to the petiole, the cauline sessile, 2–7 cm. long; petals cuneate-oblanceolate, 3–5 mm. long; siliques appressed to the axis or erect, 4–6 cm. long, 1–1.2 cm. wide, the axis much-elongate; style very short, stout; seeds about 1.2 mm. long, very narrowly winged.——May–June. Sunny hills and low mountains; Hokkaido, Honshu, Kyushu.——Cool to warm regions of the N. Hemisphere.

4. Arabis glabra (L.) Bernh. *Turritis glabra* L.; *A. perfoliata* Lam.——HATAZAO. Glaucescent erect biennial, hairy only on the lower half; stems usually simple, to 80 cm. long, terete; radical leaves oblong-oblanceolate or narrowly oblong, hirsute, the cauline sagittate, entire or nearly so, gradually narrowed and subobtuse to obtuse at apex; flowers on an elongate raceme, the petals creamy white; siliques appressed to the axis; seeds about 0.8 mm. long, not winged, nearly 2-seriate.——Apr.–June. Sunny places in lowlands to foothills; Hokkaido, Honshu, Shikoku, Kyushu.——Temperate regions of the N. Hemisphere.

5. Arabis pendula L. *A. subpendula* Ohwi——EZO-HATAZAO. Erect hirsute biennial, with simple spreading and 2- to 5-branched short hairs; stems 50–100 cm. long, erect, branched above, hirsute; lower cauline leaves oblong, obovate or broadly lanceolate, 3–10 cm. long, 1–5 cm. wide, acute to acuminate, sessile and slightly auriculate, toothed, coarsely hirsute above, usually hirsute on the nerves beneath; racemes loosely many-flowered; sepals glabrous to sparingly pilose, about 1.8 mm. long; petals white, oblanceolate, about 3 mm. long; siliques flat, nodding or deflexed, 7–10 cm. long, 1.5–2 mm. wide; style very short; seeds narrowly winged on margin, 1-seriate.——June–Aug. Woods and thickets in mountains; Hokkaido, Honshu (n. and centr. distr.).——Europe, Siberia, Manchuria, and Sakhalin.

6. Arabis flagellosa Miq. SUZU-SHIRO-SŌ. Perennial stoloniferous herb with (2–)3- to 4-branched hairs; rhizomes slender, branched; stems simple, 10–25 cm. long, few-leaved; radical leaves petiolate, 3–7 cm. long, 8–20 mm. wide, spathulate, toothed, the cauline sessile, 1–2 cm. long, ovate; leaves of the stolons broadly obovate to narrowly ovate, 1–3.5 cm. long, 1–1.5 cm. wide, toothed, cuneate; siliques 4–10, in a loose raceme, ascending to spreading, 2–3 cm. long, about 1.5

mm. wide, glabrous; style slender; seeds several to more than 10, very narrowly winged.——Mar.–Apr. Foothills; Honshu (Kinki Distr. and westw.), Shikoku, Kyushu.——Ryukyus.——Forma **lasiocarpa** (Matsum.) Ohwi. *A. flagellosa* var. *lasiocarpa* Matsum.——KE-SUZU-SHIRO-SŌ. Siliques pilose.——Occurs with the typical phase.

7. Arabis tanakana Makino. KUMOI-NAZUNA. Stellate-pubescent perennial herb; stems 3–8 cm. long, erect, few-leaved; radical leaves rosulate, linear-spathulate, entire, 5–10 mm. long, 1.5–3 mm. wide; racemes usually solitary, 3- to 10-flowered; petals 2.5–3 mm. long; siliques flat, spreading, about 10 mm. long, 1 mm. wide, linear; seeds nearly in 2 series, wingless.——July–Aug. Dry alpine slopes and rocky cliffs; Honshu (centr. distr. on Mount Shirane in Kai and Mount Shirouma in Shinano); very rare.

8. Arabis lyrata L. var. **kamtschatica** Fisch. *A. kamtschatica* Fisch.; *A. lyrata* sensu auct. Japon., non L.; *A. petraea* var. *kamtschatica* (Fisch.) Regel——MIYAMA-HATAZAO. Green to glaucous perennial, nearly glabrous to sparingly pilose; stems few, erect or ascending, 10–30 cm. long, few-leaved, sometimes sparingly branched, soft-hairy toward the base; radical leaves lyrate, abruptly petiolate, 2–5 cm. long, 5–8 mm. wide, with simple and branched stipitate hairs, the cauline leaves broadly lanceolate to broadly linear, 1–4 cm. long, 1–10 mm. wide, the lowermost dentate-serrate, the upper entire; racemes loosely few-flowered; petals cuneate-obovate, white, sometimes with a rosy hue, 3.5–4.5 mm. long; siliques obliquely spreading, 3–4 cm. long, 1–1.2 mm. wide; style short; seeds about 1.5 mm. long with a very narrow wing on the upper end.——June–Aug. Rocky dry slopes in high mountains; Hokkaido, Honshu (centr. and n. distr. and Mount Daisen in Hōki Prov.).——ne. Asia and N. America.

9. Arabis kawasakiana Makino. TACHI-SUZU-SHIRO-SŌ. Resembles the preceding; glaucous biennial herb; stems erect, 25–40 cm. long, few-leaved, slightly branched below, usually glabrous; flowers several to more than 10; petals 4–6 mm. long; siliques 2–4 cm. long, 1–1.2 mm. wide, straight; style very short; seeds about 1 mm. long, wingless.——Apr.–June. Sunny sandy places in lowlands especially near the sea; Honshu (Kinki Distr. and westw.), Shikoku.

10. Arabis gemmifera (Matsum.) Makino. *Cardamine gemmifera* Matsum.; *A. axillaris* Komar.; *A. maximowiczii* N. A. Busch; *C. greatrexii* Miyabe & Kudo; *A. greatrexii* (Miyabe & Kudo) Miyabe & Tatew.; *A. halleri* sensu auct. Japon., non L.; *A. senanensis* Makino——HAKUSAN-HATAZAO. Perennial; stems 10–30 cm. long, solitary or tufted, often spreading-pubescent, few-leaved, decumbent after anthesis and bearing innovation-shoots in the leaf-axils; radical leaves lyrate, 2–6 cm. long, 8–15 mm. wide, with scattered bifurcate hairs or nearly glabrous, the lower cauline leaves often petiolate, the upper ones oblanceolate to narrowly lanceolate, 1–5 cm. long, acute, few-toothed, glabrous; racemes elongate and loose after anthesis; petals 5–6 mm. long; siliques obliquely spreading, 10–20 mm. long, 0.8–1 mm. wide, somewhat torulose, compressed, glabrous; style slender, about 1 mm. long, the stigma capitellate; seeds scarcely winged, about 1 mm. long, in 1 series.——Mar.–May. Hokkaido (sw. distr.), Honshu, Shikoku, Kyushu.——Forma **alpicola** (Hara) Ohwi. *A. gemmifera* var. *alpicola* Hara——IBUKI-HATAZAO. Densely hairy on all parts.——Honshu (Mount Ibuki in Oomi Prov.).——Korea.

16. ARABIDOPSIS Heynh.　SHIRO-INU-NAZUNA ZOKU

Biennials or perennials with the habit of *Arabis kawasakiana,* the hairs bifurcate; radical leaves lyrate; flowers white, small, in racemes; style very short, the stigma bilobed; siliques narrowly linear; seeds in 1 or 2 series; cotyledons incumbent.——Few species, in the N. Hemisphere.

1. **Arabidopsis thaliana** (L.) Heynh. *Arabis thaliana* L.; *Stenophragma thaliana* (L.) Celak.; *Arabis pubicalyx* Miq.; *Arabis thaliana* var. *pubicalyx* (Miq.) Makino——SHIRO-INU-NAZUNA. Biennial herb; stems 20–30 cm. long, branched below, often with simple spreading hairs on the lower portion; radical leaves rosulate, 1–5 cm. long, broadly oblanceolate, 3–15 mm. wide, gradually narrowed below, with few obsolete teeth, the hairs bifurcate, the cauline leaves few or absent; petals oblanceolate, about 3 mm. long; stamens slightly shorter than the petals; siliques obliquely spreading to sub-erect, 15–20 mm. long, about 0.7 mm. wide; seeds about 0.5 mm. long, the margin obtuse.——Mar.–May. Sunny grassy places in lowlands, especially near the sea; Hokkaido, Honshu, Shikoku, Kyushu.——Europe, Africa, centr. Asia, China, Korea, and N. America.

17. DONTOSTEMON Andrz.　HANA-HATAZAO ZOKU

Erect annual, biennial, or perennial branched herbs, the hairs simple, often glandular-tipped; leaves entire to pinnately cleft; flowers in bractless racemes, rather small, pale rose to white; sepals equal; stamens of the longer 4 with the filaments united in pairs; siliques linear-cylindric, angled, the valves inflated; style short, stout, the stigma simple; seeds many, in 1 series, sometimes margined; cotyledons linear, incumbent.——Few species, in e. Asia.

1. **Dontostemon dentatus** (Bunge) Ledeb. *Andreoskia dentata* Bunge.——HANA-HATAZAO. Biennial with short in-curved puberulence; stems erect, rather firm, 25–50 cm. long, often branched, puberulent; leaves rather many, petiolate in the lower ones, sessile in the upper, lanceolate, 2–8 cm. long, 3–10 mm. wide, sparingly toothed, narrowed at both ends, sparsely puberulent; flowers rather many; petals 8–10 mm. long, obovate, clawed, pale rose-purple, about 2.5 times as long as the sepals; siliques 4–5 cm. long, about 1.5 mm. across, the pedicels 7–10 mm. long, rather stout, obliquely spreading, glabrous, straight or slightly curved; style stout and short; seeds narrowly margined, about 1.5 mm. long.——June–July. Honshu (centr. distr. and s. part of n. distr.); rare.

Var. **glandulosus** Maxim. NEBARI-HANA-HATAZAO. Hairs glandular.——Occurs with the typical phase.——e. Asia.

18. BERTEROELLA O. E. Schulz　HANA-NAZUNA ZOKU

Erect branched stellate-pubescent annuals; cauline leaves rather numerous, obovate-oblong or oblanceolate, narrowed below, nearly sessile, entire, rounded and mucronate at apex; petals pale rose; filaments of the inner stamens dilated at base; siliques sessile, stellate-puberulent, linear, the valves 3-nerved; style rather long, glabrous, the stigma small, truncate; seeds wingless; cotyledons incumbent.——A single species.

1. **Berteroella maximowiczii** (Palib.) O. E. Schulz. *Sisymbrium maximowiczii* Palib.——HANA-NAZUNA. Gray-green biennial, densely stellate-pubescent; stems erect, 20–40 cm. long, often branched above; leaves 1.5–3 cm. long, 5–13 mm. wide; racemes many-flowered, the pedicels erect, slender, 3–8 mm. long; petals 3–4 mm. long, cuneate-oblong; siliques white-pubescent, erect, subulate-linear, 10–12 mm. long; style about 2.5 mm. long; seeds few, about 1.5 mm. long.——May–Oct. Honshu (Chūgoku Distr.), Kyushu (Tsushima Isl.); rare.——Korea, Manchuria, and n. China.

19. DESCURAINIA Webb & Berthel.　KUJIRAGUSA ZOKU

Annuals, biennials, or perennials, the minute hairs stellate or bifurcate; stems usually branched, leafy; leaves pinnate or bipinnate; flowers small, white or yellow, in bractless racemes; siliques linear, on long slender pedicels, the valves 1-nerved; style very short, the stigma simple; seeds small, in 1 or 2 series; cotyledons incumbent.——About 40 species, in the N. Hemisphere and S. America.

1. **Descurainia sophia** (L.) Prantl. *Sisymbrium sophia* L.——KUJIRAGUSA. White-pubescent annual or biennial; stems erect, 30–60 cm. long, sometimes branched; leaves sessile, 3–5 cm. long, 2–2.5 cm. wide, narrowly ovate, bipinnate, the ultimate segments oblanceolate, subentire, 3–5 mm. long, 1–1.5 mm. wide; flowers small, the pedicels slender, 10–15 mm. long; petals oblanceolate, white, about 2.5 mm. long, slightly longer than the sepals; siliques narrowly linear, 15–25 mm. long, about 1 mm. across, ascending, slightly curved; seeds about 1 mm. long, unappendaged, in 1 series.——June–Aug. Honshu (centr. distr., naturalized in Kantō Distr.).——Europe and Asia.

20. ERYSIMUM L.　EZO-SUZU-SHIRO ZOKU

Annuals, biennials, or perennials, the hairs appressed, 2- or 3-branched or stellate, rarely tomentose; stems erect, branched; leaves linear to oblong, entire, sinuate, toothed or rarely pinnatifid; racemes bractless; flowers yellow, orange-yellow or rarely purple; lateral sepals gibbous; stamens free, entire; siliques linear, 4-angled, terete or sometimes flat, the valves linear, often 1-nerved, the replum often keeled on back; style short or elongate, the stigma 2-lobed, capitate or emarginate; seeds in 1 series, oblong; cotyledons incumbent.——About 100 species, in the temperate regions of the N. Hemisphere.

1. Erysimum cheiranthoides L. var. **japonicum** H. Boiss. *E. japonicum* (H. Boiss.) Makino——EZO-SUZU-SHIRO-SŌ.   Erect biennial, uniformly clothed with appressed 2- or 3-branched hairs; stems 50–100 cm. long, densely many-leaved, terete, obsoletely striate, often slightly purplish red near base, simple or short-branched above; leaves membranous, ascending, linear-lanceolate, 4–6 cm. long, 6–8 mm. wide, obtuse, gray-green, sessile, subentire, slightly recurved on margin; racemes rather densely many-flowered, pilose, the pedicels oblique, 5–7 mm. long; flowers 4–5 mm. long; sepals rather densely pilose, 2–2.5 mm. long, obtuse and scarious at the apex; petals cuneate; siliques erect, obtusely 4-angled, sparsely pilose, 12–15 mm. long, about 1 mm. across, the valves obtuse, 1-nerved; style very short, the stigma distinctly discoid, dilated, obsoletely 2-lobed; seeds about 15 on each side of a white membranous septum.——By the sea; Hokkaido (Tokoro in Kitami); rare.——The typical phase occurs in the temperate regions of Eurasia and N. America.

Fam. 99. DROSERACEAE   Mōsengoke Ka   Sundew Family

Perennials or rarely annuals, mostly terrestrial or rarely aquatic, usually glandular-hairy; leaves circinnate in bud; stipules sometimes absent; flowers solitary or in racemes, bisexual, actinomorphic; sepals 4–5; petals as many, free; stamens 4–5(–20); ovary 1-locular; carpels 3–5; styles 3–5, the stigma simple or lobed; fruit a capsule; seeds many or few; endosperm copious.—— Six genera, with more than 100 species, all insectivorous.

1A. Terrestrial; leaves alternate or rosulate, entire, with glandular hairs on upper side; flowers in racemes. .... 1. *Drosera*
1B. Aquatic; leaves verticillate, with sensitive hispid hairs; flowers solitary. .... 2. *Aldrovanda*

1. DROSERA L.   Mōsengoke Zoku

Perennials or rarely annuals with short rhizomes; stems absent or elongate and leafy; leaves alternate, cauline or radical, circinnate in bud, with long glandular-tipped bristlelike hairs; stipules absent or scarious, adnate to base of the petiole; inflorescence circinnate while young; flowers in radical or axillary racemes; sepals 4–8, persistent; petals 4–8, white or rose, persistent; stamens 4–8; ovary superior, free from the calyx; ovules on parietal placentae, many; capsules loculicidally dehiscent.——About 100 species in both hemispheres.

1A. Stemless, the plants scapose; stipules scarious, adnate to the base of the petiole, laciniate.
 2A. Leaves 3–15 cm. long; racemes glabrous or nearly so; sepals 4–6 mm. long, subobtuse or obtuse, as long as to shorter than the oblong capsule.
  3A. Leaves depressed-obovate to depressed-orbicular, abruptly long-petiolate; seeds nearly linear. .... 1. *D. rotundifolia*
  3B. Leaves linear-oblanceolate or oblanceolate, gradually long-petiolate; seeds lanceolate, impressed-punctulate. .... 2. *D. anglica*
 2B. Leaves 10–20 mm. long, closely appressed to the ground; racemes with dense minute glandular hairs; sepals 2.5–3 mm. long, acute, longer than the globose capsules. .... 3. *D. spathulata*
1B. Stems elongate; stipules absent.
 4A. Annual herb, not tuberous; leaves narrowly linear, 4–6 cm. long, to 2.5 mm. wide, the petiole 7–15 mm. long; sepals short-glandular, lanceolate. .... 4. *D. indica*
 4B. Perennial herb with tubers; leaves transversely lunate, broader than long, 2–3 mm. long, 4–6 mm. wide, the petioles 8–15 mm. long, distinct; sepals laciniate, ovate. .... 5. *D. peltata* var. *nipponica*

1. Drosera rotundifolia L. Mōsengoke.   Perennial usually with a reddish hue and short rhizomes; leaves rosulate, ascending or spreading, depressed-obovate to depressed-orbicular, 5–10 mm. long and as wide, abruptly petiolate, with long gland-tipped bristlelike hairs on upper side, the petioles 3–13 cm. long; scapes 6–30 cm. long, glabrous, the inflorescence racemose, terminal, few-flowered, one-sided; sepals narrowly oblong, with short glandular-hairs on margin; petals about twice as long as the sepals, 4–6 mm. long; styles 3, the stigma 2-parted.——June–Aug. Wet places in mountains; Hokkaido, Honshu, Shikoku, Kyushu; rather common.—— Temperate to cold regions of the N. Hemisphere.
2. Drosera anglica Huds. *D. longifolia* L. pro parte; *D. intermedia* sensu auct. Japon., non Heyne——NAGABA-NO-MŌSENGOKE.   Usually reddish perennial with slightly elongate rhizomes; leaves erect, 2–4 cm. long, 1.5–4 mm. wide, rounded at apex, gradually narrowed to the petiole; scapes 10–20 cm. long, the inflorescence terminal, racemose, one-sided; petals 5–7 mm. long, white; seeds lanceolate, obtuse at both ends.——July–Aug. Bogs; Hokkaido, Honshu (Oze in Kotsuke); rare.——Bogs in the colder regions of the N. Hemisphere.
Drosera × obovata Mert. & W. Koch. SAJIBA-MŌSEN-GOKE.   Hybrid of Nos. 1 and 2; leaves broadly oblanceolate to narrowly obovate, 1–2 cm. long, 3–6 mm. wide.——Occurs where the two parent species grow together.
3. Drosera spathulata Labill. *D. loureiri* Hook. & Arn.; *D. burmannii* DC., non Vahl——KO-MŌSENGOKE.   Usually reddish perennial with very short rhizomes; radical leaves rosulate, rather many, spreading, obovate-spathulate, 10–20 mm. long, 2.5–3.5 mm. wide, rounded at the apex with long gland-tipped bristlelike hairs on the upper side, cuneate and decurrent to a rather broad short-petiole; scapes few, 5–15 cm. long, slender; flowers few to more than 10, in a terminal unbranched, one-sided glandular-hairy raceme; flowers about 4 mm. across; sepals lanceolate; petals rose, 1.5–2 times as long as the sepals; style 2-parted, the stigma capitellate; capsules 1.5 mm. long; seeds minute, fusiform, obtuse and mucronulate at both ends.——June–Sept. Shaded places in wet bare soils in lowlands and foothills; Honshu (Rikuzen and southw.), Shikoku, Kyushu; commoner near the sea.—— Ryukyus, Formosa, China, to s. Asia and Australia.
4. Drosera indica L. *D. makinoi* Masam.——NAGABA-NO-ISHI-MOCHI-SŌ.   Glandular-hairy annual; stems 7–15 cm. long, loosely leafy; leaves narrowly linear, 4–6 cm. long, 2–2.5 mm. wide, gradually narrowed to a filiform point, with long gland-tipped bristlelike hairs on the upper side, the petioles 7–15 mm. long, as wide as the lower portion of the blade,

without long glandular-tipped hairs on upper side; racemes short-pedunculate, 5–10 cm. long, loosely 3- to 10-flowered, rather densely short glandular-hairy, the pedicels spreading, 10–15 mm. long; flowers white or rose; sepals lanceolate, acute; petals 6–8 mm. long; capsules subglobose, about 3.5 mm. long; styles 3, bilobed, the stigma capitellate; seeds about 0.5 mm. long, broadly ovate-elliptic, black, slightly acute at one end, longitudinally ribbed and cross-lined.——July–Aug. Wet grounds in lowlands; Honshu (Kantō and Tōkaidō Distr.).——China, Malaysia, Australia, India, and Africa.

5. Drosera peltata Smith var. **nipponica** (Masam.) Ohwi. *D. nipponica* Masam.; *D. peltata* var. *lunata* sensu auct. Japon., non Clarke; *D. lunata* sensu auct. Japon., non Hamilt.——Ishi-mochi-sō. Tuberous perennial herb, mostly glabrous except for glandular hairs on upper side of the leaves; stems erect, 10–30 cm. long, slender, sparingly branched above; radical leaves withering before anthesis, the cauline loosely alternate, lunate, 2–3 mm. long, 4–6 mm. wide, with long gland-tipped bristlelike hairs on the upper side, acute on the angles, the petioles slender; racemes 2- to 10-flowered, 2–6 cm. long including the short peduncle, the pedicels 4–6 mm. long; sepals ovate, laciniate, the segments glandular-tipped; petals 6–8 mm. long; capsules subglobose, about 2.5 mm. long; styles 4-parted; seeds broadly elliptic, slightly attenuate at both ends, about 0.5 mm. long with fine longitudinal ribs and obsolete cross-lines.——May–June. Wet places; Honshu (Kantō Distr. and westw.), Shikoku, Kyushu.——China. The typical phase occurs in India, Australia, and e. Asia.

2. ALDROVANDA L.   Mujina-mo Zoku

Floating aquatic herb, green, glabrous, without roots; stems jointed, slightly elongate, sparingly branched, 15–20 mm. across, inclusive of the verticillate leaves; leaves 6–8 at each node, the petioles cuneate, spreading, with few bristles above, the blade cochleate, with a hinge-joint; flowers solitary on an axillary pedicel, bractless, pale green, 5-merous, the calyx 5-sected; petals 1.5 times as long as the calyx, narrowly oblong, thinly membranous; stamens 5; ovary superior, ovoid, with 5 parietal placentae; ovules in 2 series; styles 5, rather elongate, the stigma flabellate, irregularly lobed; capsules ovoid-globose, slightly longer than the petals, crowned with the withered corolla; seeds minute, ellipsoidal, black, shining, with a mucro at base.——A single species.

1. Aldrovanda vesiculosa L. Mujina-mo.——July–Aug. Shallow ponds; Honshu; rare.——centr. and s. Europe, India, and Australia.

Fam. 100. CRASSULACEAE   Benkei-sō Ka   Stonecrop Family

Usually fleshy herbs or shrubs; leaves opposite, alternate or verticillate, without stipules; flowers in terminal cymes, rarely solitary or in spikes, bracteate or ebracteate; sepals 3–5 or rarely more, as many as the petals; petals free or connate at base, imbricate in bud; stamens as many or more often twice as many as the petals; nectariferous glands opposite the carpels or rarely absent; carpels 3–5, rarely more, free or rarely connate at base; ovules many or sometimes few, inserted on the suture on the ventral side, ascending or pendulous; fruit follicular, the seeds small, usually oblong, with fleshy endosperm.——About 20 genera, with about 1,300 species, nearly cosmopolitan, except Australia, especially abundant in S. Africa and centr. Asia.

1A. Plant fleshy; carpels free or nearly so.
    2A. Stamens 3–5, as many as the sepals, petals, and carpels. . . . . 1. *Tillaea*
    2B. Stamens 8–10, twice as many as the sepals, petals, and carpels. . . . . 2. *Sedum*
1B. Plant not fleshy; carpels connate below and forming a 5-loculed capsule. . . . . 3. *Penthorum*

1. TILLAEA L.   Azuma-tsume-kusa Zoku

Small, glabrous, rather fleshy herbs; leaves opposite, linear, entire, often terete; flowers axillary, solitary or in cymes or terminal panicles; sepals 3–5, often connate at base; petals 3–5, linear; stamens 3–5, the filaments filiform, the glands scalelike, 3–5 or absent; carpels 3–5, free, the ovules 1 to many in each locule.——Wet places or in water, with more than 20 species, in both hemispheres.

1. Tillaea aquatica L. *Bulliarda aquatica* (L.) DC.; *Tillaeastrum aquaticum* (L.) Britt.; *Elatine tetrandra* Maxim. ——Azuma-tsume-kusa. Glabrous, sparsely branched annual 2–6 cm. high; leaves broadly linear, opposite and connate at the base, 5–8 mm. long, about 1 mm. wide; flowers sessile and solitary in the axils, 4-merous; petals whitish, about 1.5 mm. long, obtuse, ovate or narrowly so, membranous, about twice as long as the calyx; stamens shorter than the petals; carpels erect, narrowly ovate, slightly shorter than the persistent petals, the glands terete; seeds few, narrowly oblong.——May–June. Wet places in lowlands; Hokkaido, Honshu.——Eurasia, n. Africa, and N. America.

2. SEDUM L.   Kirin-sō Zoku

Fleshy and usually glabrous annuals or perennials; leaves alternate, simple, opposite or verticillate, rarely rosulate, entire or toothed, rarely spine-tipped, usually sessile; flowers in terminal cymes or solitary, rarely in spikes, yellow, white, or rose-purple, bisexual (plants sometimes dioecious); sepals 4–5, free; petals as many, free, persistent; stamens twice as many as the sepals and petals, the filaments subulate or filiform; nectariferous glands 5 or 4, minute; ovary superior; carpels 5, sometimes 4, free or shortly connate below; styles short, the ovules rather many; fruit a follicle.——About 350 species, abundant in the temperate regions of the Old World, also in N. and S. America.

1A. Plants dioecious, the flowers usually 4-merous, rarely 5-merous.
    2A. Leaves elliptic, obovate or narrowly oblong, glaucous, obsoletely serrate, 5–15 mm. wide, smooth on margin. .... 1. *S. rosea*
    2B. Leaves oblanceolate not glaucous, coarsely serrate, 3–5 mm. wide, uneven on margin. .... 2. *S. ishidae*
1B. Plants bisexual, the flowers usually 5-merous.
    3A. Leaves not spine-tipped; inflorescence cymose.
       4A. Rhizomes stout; leaves basifixed, usually toothed.
          5A. Flowers greenish white to rose-purple.
            6A. Flowers rose-purple.
              7A. Stems tufted, 5–25 cm. or rarely to 30 cm. long, densely leafy; leaves 1–2.5 cm. or rarely to 3 cm. long.
                 8A. Leaves ternate, flabellate-orbicular, manifestly shorter than broad, sparingly undulate-toothed. .... 3. *S. sieboldii*
                 8B. Leaves opposite or alternate, ovate-orbicular to oblanceolate, as long as to narrower than wide.
                    9A. Leaves short-petiolate, usually few-toothed. .... 4. *S. cauticolum*
                    9B. Leaves sessile, entire. .... 5. *S. pluricaule* var. *yezoense*
              7B. Stems few or solitary, 30–50 cm. long, loosely leaved; leaves 4–8 cm. long.
                10A. Stamens conspicuously longer than the petals; leaves verticillate or opposite. .... 6. *S. spectabile*
                10B. Stamens as long as or slightly shorter than the petals; leaves alternate or opposite. .... 7. *S. telephium* var. *purpureum*
            6B. Flowers greenish white or greenish.
              11A. Leaves sessile, rounded or obtuse at base; flowers 7–10 mm. across. .... 8. *S. tsugaruense*
              11B. Leaves more or less narrowed to a petiole or petiolelike base; flowers usually smaller.
                12A. Leaves verticillate, subacute to obtuse, dentate, impressed-punctate with scattered minute black spots.
                                                              9. *S. verticillatum*
                12B. Leaves alternate or opposite, rarely subverticillate, rounded to subobtuse, undulate-toothed to nearly entire.
                  13A. Plants glaucous; leaves 5–8 cm. long.
                    14A. Stems rather stout, erect; lowest pair of branches on inflorescence rather elongate. .... 10. *S. erythrostictum*
                    14B. Stems rather slender, ascending; lowest pair of branches of inflorescence rather short. .... 11. *S. okuyamae*
                13B. Plants brownish green.
                    15A. Leaves usually alternate, rarely opposite, decurrent to a short petiole. .... 12. *S. sordidum*
                    15B. Leaves opposite, abruptly and distinctly petiolate. .... 13. *S. viride*
          5B. Flowers yellow.
            16A. Stems solitary or few, usually simple, 20–50 cm. long; leaves oblanceolate, 3–6 cm. long. .... 14. *S. aizoon*
            16B. Stems tufted, often branched at base, 7–20 cm. long; leaves obovate or broadly oblanceolate, 2–3.5 cm. long.
                                            15. *S. kamtschaticum*
    4B. Plants without stout rhizomes; leaves more or less attached to the stem above the base, entire.
       17A. Leaves terete or only slightly flattened, linear to narrowly oblong.
          18A. Leaves of fertile stems verticillate.
            19A. Leaves linear, gradually acute and scarcely flattened at apex; anthers conical-ovoid, 0.5–0.6 mm. long. .... 16. *S. lineare*
            19B. Leaves narrowly oblanceolate, narrowed and flattened toward the obtuse apex; anthers narrowly ovoid, 0.8–1 mm. long.
                                          17. *S. sarmentosum*
         18B. Leaves of fertile stems alternate.
            20A. Flowers solitary, terminal; leaves oblong-cylindric, 3–5 mm. long. .... 18. *S. uniflorum*
            20B. Flowers few to many, in terminal cymes.
              21A. Petals 1.5–2 times as long as the sepals, spotless or with few brown short striations; leaves cylindric-oblong to lanceolate-cylindric, 5–10 mm. or sometimes to 12 mm. long, obtuse.
                22A. Inflorescence on lateral stems; leaves obovate-cylindric, rounded at apex; carpels ascending; on rocks near seashores.
                                            19. *S. oryzifolium*
                22B. Inflorescence terminal on the main axis; leaves rather narrower, obtuse; carpels spreading; on rocks in lowlands and high mountains. .... 20. *S. japonicum*
              21B. Petals 2.5–4 times as long as the sepals, usually brown-striated; leaves linear, 10–20 mm. long.
                23A. Flowers 4-merous; leaves linear, obtuse; plant of shaded places on tree-trunks. .... 21. *S. hakonense*
                23B. Flowers 5-merous; leaves linear to oblanceolate-linear, obtuse to subobtuse.
                  24A. Leaves broadly linear, very many and densely disposed on the stem. .... 22. *S. polytrichoides*
                  24B. Leaves linear-spathulate, to about 4.5 mm. wide. .... 23. *S. yabeanum*
       17B. Leaves distinctly flattened, at least the sterile ones spathulate.
          25A. Perennial stoloniferous herbs; leaves entire or emarginate, smooth on margin; stems without bulbils in leaf-axils.
            26A. Leaves dimorphic, linear or only slightly broadened above on the flowering shoots, obovate on the vegetative shoots; petals as long as to slightly longer than the sepals. .... 24. *S. subtile*
             26B. Leaves nearly all alike, spathulate; petals 2–5 times as long as the sepals.
              27A. Carpels erect in fruit. .... 25. *S. formosanum*
              27B. Carpels obliquely spreading or ascending in fruit.
                28A. Leaves on flowering shoots opposite. .... 26. *S. makinoi*
                28B. Leaves on flowering shoots alternate.
                  29A. Leaves obtuse to rounded at apex; carpels 3–5. .... 27. *S. tricarpum*
                  29B. Leaves emarginate; carpels 5. .... 28. *S. tosaense*
          25B. Annual or biennial herbs without stolons; leaves entire, somewhat uneven on the upper margin; reproducing by bulbils.
                                            29. *S. bulbiferum*
    3B. Plant either with spine-tipped leaves or with spicate inflorescence.
       30A. Inflorescence a few-flowered cyme; leaves 7–12 mm. long, 1.5–2.5 mm. wide. .... 30. *S. leveilleanum*
       30B. Inflorescence a many-flowered dense spike; leaves 2–8 cm. long, 4–40 mm. wide.
          31A. Leaves callose on upper margin, ending in a short spine, broadly linear, slightly flattened. .... 31. *S. erubescens*

1. Sedum rosea (L.) Scop. *Rhodiola rosea* L.; *R. tachiroei* (Fr. & Sav.) Nakai; *R. maxima* Nakai; *R. hideoi* Nakai; *R. rosea* var. *vulgare* (Regel & Tiling) Hara; *R. sachalinensis* A. Boreau; *S. rhodiola* DC.; *S. rhodiola* var. *vulgare* Regel & Tiling; *S. rhodiola* var. *tachiroei* Fr. & Sav.; *S. roseum* var. *vulgare* (Regel & Tiling) Maxim.; *S. roseum* var. *tachiroei* (Fr. & Sav.) Takeda——IWA-BENKEI. Dioecious glaucous perennial herb with stout brown-scaly caudex; stems tufted, 7–30 cm. long, erect; leaves fleshy, flat, obovate to elliptic, sometimes narrowly oblong, 1–3 cm. long, 5–15 mm. wide, usually rounded with an obtuse tip, sessile, shallowly toothed especially on upper half; the lower leaves smaller; inflorescence dense, many-flowered; staminate flowers yellowish; stamens longer than the petals; pistillate flowers rather small, often purplish; carpels usually erect, 6–7 mm. long, the style short-spreading.——June–Aug. Hokkaido (rocks in alpine regions and along the sea), Honshu (alpine regions of n. and centr. distr.).——Alpine and cooler regions of the N. Hemisphere.

2. Sedum ishidae Miyabe & Kudo. *Rhodiola ishidae* (Miyabe & Kudo) Hara; *R. stephanii* sensu auct. Japon., non Trautv. & Mey.——HOSOBA-IWA-BENKEI, AO-NO-IWA-BENKEI. Resembles the preceding; leaves green, not glaucous, oblanceolate, rather prominently narrowed at each end, deeply toothed, the margin translucent, slightly uneven; petals slightly longer; carpels 7–10 mm. long.——June–Aug. Alpine regions; Hokkaido, Honshu (n. distr. and Nikko in Shimotsuke Prov.).

3. Sedum sieboldii Sweet. MISEBAYA. Rhizomes stout; stems tufted, 15–30 cm. long, declined above; leaves ternate, glaucous, often more or less purplish (white-variegated in some garden cultivars), flabellate–orbicular, 10–15 mm. long, 13–20 mm. wide, fleshy, sessile, with obsolete undulate teeth; inflorescence densely flowered, rather small; flowers about 6 mm. across, rather numerous, on short pedicels; sepals narrowly ovate, subacute; petals spreading, acute, about 4 mm. long, narrowly ovate; stamens nearly as long as the petals; anthers red-purple; carpels erect, ovate.——Oct.–Nov. Frequently cultivated as a pot plant; recently discovered in Shikoku (Shōdoshima in Sanuki Prov.).

Sedum kagamontanum Maxim. KAGA-NO-BENKEI-SŌ. Stems branched, stouter, the lower leaves ternate, to 3 cm. long, the upper leaves alternate; may be a larger form of *S. sieboldii*. Reported to be spontaneous in Honshu (Kaga Prov.).

4. Sedum cauticolum Praeg. HIDAKA-MISEBAYA. Rhizomes branched; stems 10–15 cm. long, tufted, ascending to suberect; leaves opposite, sometimes alternate, glaucous, ovate-orbicular to elliptic, 10–25 mm. long, 7–20 mm. wide, obtuse or subrounded, fleshy, flat, broadly cuneate at base, usually with few undulate teeth, the petioles 2–7 mm. long; inflorescence densely flowered; flowers rose-purple; sepals linear-lanceolate, sometimes ovate; petals ovate-lanceolate, 5–7 mm. long; stamens nearly as long as the petals, the anthers red-purple.——Sept. Hokkaido (Hidaka and Tokachi Prov.).

5. Sedum pluricaule (Maxim.) Kudo var. **yezoense** (Miyabe & Tatew.) Tatew. & Kawano. *S. telephium* var. *pluricaule* auct., non Maxim.; *S. yezoense* Miyabe & Tatew.;

(?) *S. sieboldii* var. *erectum* Makino——EZO-MISEBAYA. Glaucous perennial; stems tufted, 5–8 cm. long; leaves narrowly oblong to oblanceolate or obovate, 1–2.5 cm. long, 4–6 mm. wide, obtuse, entire; inflorescence densely flowered; flowers rose-purple, short-pedicelled; sepals lanceolate; petals broadly lanceolate, 4–6 mm. long, about twice as long as the sepals; stamens slightly shorter than the petals, the anthers dark purple.——July–Sept. Reported to occur in Hokkaido.——The typical phase occurs in Sakhalin and Amur.

6. Sedum spectabile Boreau. Ō-BENKEI-SŌ. Rather stout, glaucous perennial; stems erect, 30–70 cm. long, terete, thick; leaves verticillate or opposite, flat, fleshy, ovate to narrowly so, 4–10 cm. long, 2–5 cm. wide, obtuse, sessile, sparsely undulate-toothed; inflorescence large, densely many-flowered; flowers about 10 mm. across; sepals lanceolate, acute; petals broadly lanceolate, acuminate, 5–6 mm. long, about 3 times as long as the sepals; stamens 6–7 mm. long, prominently longer than the petals, the anthers purplish.——July–Sept. Cultivated in gardens and for cut flowers.——Korea and Manchuria.

7. Sedum telephium L. var. **purpureum** L. *S. purpureum* (L.) Link——MURASAKI-BENKEI-SŌ. Stems often solitary, erect, 30–50 cm. long; leaves glaucous, flat, more or less fleshy, alternate or opposite, narrowly ovate to oblong, 3–6 cm. long, 12–25 mm. wide, subobtuse, narrowed to the sessile base, short-toothed; inflorescence densely many-flowered, rather large; sepals broadly lanceolate; petals lanceolate, acute, 4–5 mm. long, 3–4 times as long as the sepals, brown-striated when dried, about as long as the stamens; carpels erect, rather shorter than the petals.——Aug.–Sept. Hokkaido, Honshu, Kyushu.——Europe and Siberia.

8. Sedum tsugaruense Hara. TSUGARU-MISEBAYA. Erect glaucous perennial herb, 10–40 cm. high, from a short thick caudex; leaves opposite or alternate, or in lower ones sometimes ternate, spreading, ovate to elliptic, 2–4 cm. long, 1.2–3.5 cm. wide, irregularly crenate; cymes 3–5 cm. across, densely many flowered, sometimes branched; flowers pedicelled, ovate-oblong in bud; sepals deltoid-lanceolate, about 1 mm. long; petals 5, oblong-lanceolate, pale green, 5 mm. long, spreading; stamens exserted; carpels 5, rarely 4, the fruiting ones suberect, 4.5–5 mm. long.——Sept. Honshu (Mutsu Prov.).

9. Sedum verticillatum L. MITSUBA-BENKEI-SŌ. Stems erect, 30–60 cm. long; leaves somewhat glaucous, verticillate, narrowly ovate, oblong or broadly lanceolate, 3–13 cm. long, 1–3 cm. wide, subobtuse, toothed, often partly dark brown spotted, the veinlets becoming brown and prominent when dry, the petioles 2–10 mm. long; inflorescence densely many-flowered, sometimes lobed; flowers yellow-green; sepals narrowly ovate, acute; petals oblong-lanceolate, acute, about 4 mm. long, 4–5 times as long as the sepals; stamens about as long as the petals, the anthers dark-colored; carpels erect in fruit, about 3 mm. long, ovoid.——Aug.–Oct. Hokkaido, Honshu, Shikoku, Kyushu.——Europe and Siberia.

10. Sedum erythrostictum Miq. *S. alboroseum* Baker——BENKEI-SŌ. Rather stout glaucous perennial; stems 30–60 cm. long; leaves opposite or alternate, fleshy, elliptic or oblong-ovate, 6–10 cm. long, 3–4 cm. wide, obtuse, obsoletely toothed, short-petiolate; inflorescence large, very many-flowered; sepals narrowly deltoid; petals oblong-lanceolate, 5–6 mm. long, acute, 3–4 times as long as the sepals; stamens about as long as the petals, the anthers purplish, narrowly ovoid.——July–Oct. Planted in our area, unknown as a wild plant.

11. Sedum okuyamae Ohwi. KO-BENKEI-SŌ. Resembles the preceding, but more slender, and the upper part of the stems declined, the lowest branches of the inflorescence shorter; glaucous glabrous perennial; stems solitary, terete, about 30 cm. long; leaves opposite, obtuse, toothed, 3–9 cm. long, 2–4 cm. wide, yellowish brown spotted, the veinlets slender, obscure; flowers many, pale green; sepals narrowly deltoid, about 2 mm. long; petals acuminate, about 5 mm. long; carpels suberect, 4–5 mm. long in fruit.——Sept. Honshu (n. distr.).

12. Sedum sordidum Maxim. CHICHIPPA-BENKEI-SŌ. Stems 20–40 cm. long; leaves alternate, sometimes opposite, ovate or ovate-orbicular, 2–4 cm. long, 15–30 mm. wide, obtuse, tinged with brown-purple, remotely undulate-toothed, the petiole often to 10 mm. long; inflorescence many-flowered, often lobed; flowers pale yellow-green; sepals ovate, acute; petals narrowly ovate, about 4 mm. long, 2–3 times as long as the sepals, acute; stamens about as long as the petals, the anthers brown-purple; carpels ovate in fruit, erect.——Aug.–Oct. Honshu (Uzen, Echigo, and Shinano Prov.); rather rare in mountains.

13. Sedum viride Makino. AO-BENKEI-SŌ. Closely allied to the preceding; leaves usually opposite, rather loosely arranged, with shallower depressed marginal teeth, the petioles distinct, to 15 mm. long; inflorescence densely flowered, usually not lobed.——Aug.–Oct. Shikoku, Kyushu.

14. Sedum aizoon L. *S. maximowiczii* Regel——HOSO-BA-NO-KIRIN-SŌ. Rhizomes stout; stems 1–2, erect, 20–50 cm. long, rather stout; leaves fleshy, usually alternate, oblanceolate to broadly so, rarely narrowly oblong, 3–6 cm. long, 5–7 mm. wide, subacute, toothed, sessile; inflorescence many-flowered; flowers yellow, 10–13 mm. across; sepals broadly linear, obtuse, slightly broadened below; petals lanceolate to broadly so, long-acuminate, 6–7 mm. long, 1.5–2 times as long as the sepals; stamens shorter than the petals, the anthers yellow; carpels ovate, ascending, briefly connate at base.——July–Sept. Hokkaido, Honshu.——Siberia.

15. Sedum kamtschaticum Fisch. *S. sikokianum* Maxim.; *S. aizoon* subsp. *kamtschaticum* (Fisch.) Froederstr.; *S. ellacombianum* Praeg.——KIRIN-SŌ. Resembles the preceding; rhizomes stout; stems tufted, 5–30 cm. long, sometimes branched at base; leaves alternate, rarely opposite, broadly oblanceolate to obovate, 2–4 cm. long, 1–2 cm. wide; flowers many to few; carpels ascending to spreading; anthers yellow.——July–Sept. Hokkaido, Honshu, Shikoku, and Kyushu; rather common on rocks in mountains especially northward.——Sakhalin, Kuriles, Kamchatka, and Ussuri.

16. Sedum lineare Thunb. O-NO-MANNENGUSA. Stems 10–20 cm. long, fleshy; sterile innovation-shoots ascending, later creeping; leaves linear, 20–25 mm. long, about 2 mm. wide, ternate, rarely subopposite, rather flat and broadened at base, gradually narrowed and terete above, acute, with an obtuse tip; inflorescence rather large, lobed, loosely rather many-flowered; flowers sessile in the axils of linear bracts; sepals broadly linear to lanceolate; petals lanceolate, 6–7 mm. long, acuminate, 1.5 times as long as the sepals; stamens shorter than the petals, the anthers oblong, cordate, mucronate following anthesis; carpels 5, suberect.——May–June. Honshu, Shikoku, Kyushu.——Ryukyus.

17. Sedum sarmentosum Bunge. *S. lineare* var. *contractum* Miq.——TSURU-MANNENGUSA. Resembles the preceding; leaves oblanceolate, 15–25 mm. long, 3–5(–6) mm. wide, obtuse, flat; sepals linear-lanceolate, 1/2–3/4 as long as the petals; anthers narrowly ovoid to oblong, cordate.——May–June. Frequently planted; reported to be spontaneous in Kyushu.——Korea, Manchuria, and n. China.

18. Sedum uniflorum Hook. & Arn. KOGOME-MANNEN-GUSA. Stems creeping, tufted and much-branched, 3–10 cm. long, fleshy, densely leafy; leaves alternate, narrowly cylindric-oblong, 3–5 mm. long, 1.5–2 mm. wide, rounded at the apex, slightly narrowed at base, sessile; flowers terminal, solitary; sepals narrowly oblong, slightly narrowed at base, 2.5–3 mm. long; petals oblong, subobtuse, longer than the sepals; stamens shorter than the petals.——Rocks near the sea; Kyushu (Yakushima); rare.——Ryukyus.

19. Sedum oryzifolium Makino. TAITŌGOME. Stolons long-creeping; stems erect, 5–12 cm. long; sterile shoots densely leafy on upper part; leaves alternate, cylindric-obovoid to -oblong, 3–6 mm. long, 1–1.5 mm. wide, rounded to very obtuse, sessile; inflorescence often trichotomously branched, rather densely many-flowered; sepals ovate-oblong, rounded at the apex; petals broadly lanceolate, acuminate, 4–5 mm. long, at most 1.5 times as long as the sepals; stamens shorter than the petals, the anthers ovate-globose, cordate; carpels slightly ascending.——May–July. Rocks near the sea; Honshu (Kantō Distr. and westw.), Shikoku, Kyushu.

20. Sedum japonicum Sieb. ME-NO-MANNENGUSA. Flowering stems erect, terete, 10–15 cm. long, bearing sterile shoots at base; leaves nearly terete to slightly flattened, 5–10 mm. long, about 2 mm. wide, broadest slightly above the middle, rounded to very obtuse; inflorescence usually 3 or 4 times branched, many-flowered; flowers nearly sessile; sepals unequal in length, linear-oblong to oblong, obtuse to subrounded; petals lanceolate, acuminate, 6–7 mm. long; stamens shorter than the petals, the anthers ovoid-globose, cordate; carpels spreading.——May–June. Rocks in thin woods and sunny slopes in foothills; Honshu, Shikoku, Kyushu; common.

Var. **senanense** (Makino) Makino. *S. senanense* Makino ——MIYAMA-MANNENGUSA. Slightly smaller in all parts, and frequently reddish, the leaves narrowly oblong, slightly narrowed above, obtuse.——May–Aug. Alpine slopes; Honshu (centr. distr.).

21. Sedum hakonense Makino. MATSU-NO-HA-MANNEN-GUSA. Stems 5–10 cm. long, erect, ascending or creeping, much branched at the base; leaves linear, 1–2.5 cm. long, (1–)2–3 mm. wide, fleshy, slightly flattened, somewhat rounded at the apex; inflorescence often trichotomously branched, rather loosely many-flowered; flowers sessile; sepals 4, rather unequal, deltoid to oblong, very obtuse; petals 4, yellow, broadly lanceolate, about 4 mm. long, acute, much longer than the sepals; stamens 8, as long as the petals, the anthers cordate-globose.——Aug. Honshu (Sagami, Izu, and Suruga Prov.); rare.

Var. **rupifragum** (Koidz.) Ohwi. *Sedum rupifragum* Koidz.——Ō-ME-NO-MANNEGUSA. Flowers 4- to 5-merous.——Honshu (Izumo).

22. Sedum polytrichoides Hemsl. *S. kiusianum* Makino——UNZEN-MANNENGUSA. Stems densely tufted, erect, branching at base, about 10 cm. long, usually simple above, densely many-leaved, dark brown below; leaves broadly linear, 6–15 mm. long, 1–2.5 mm. wide, acute, entire, slightly narrowed below; cymes rather small, the branches obliquely spreading, few-flowered; flowers sessile, about 1 cm. across; sepals 5, slightly unequal, linear to subulate-linear, about 2 mm. long; petals 5, acuminate, yellow, 5–6 mm. long; fruiting carpels 5, shortly connate at base.——Rocks in mountains; Honshu (w. distr.), Kyushu; rare.——Korea and China.

23. Sedum yabeanum Makino.——TSUSHIMA-MANNENGUSA. Stems 5–15 cm. long, erect, ascending to short-creeping and branched at the base; leaves linear to narrowly oblanceolate, 1–2 cm. long, 1.5–3(–4) mm. wide, subacute, entire, ascending; inflorescence often 3-lobed; flowers sessile, yellow; sepals broadly lanceolate, obtuse; petals narrowly lanceolate, acuminate, 5–6 mm. long, 3 or 4 times as long as the sepals; stamens shorter than the petals, the anthers globose, slightly cordate; carpels 5, obliquely spreading.——June–July. Kyushu.

24. Sedum subtile Miq. *S. zentaro-tashiroi* Makino ——HIME-RENGE, KO-MANNENGUSA. Green slender perennial; stems rather slender, 4–10 cm. long, short-creeping from the base; sterile shoots loosely leaved with a tuft of rosulate leaves at the apex; leaves of sterile shoots opposite, rarely verticillate in 3's to 5's, obovate, 5–15 mm. long, 3–5 mm. wide, rounded to very obtuse, narrowed below to a slender petiolelike base; leaves of flowering shoots alternate, rarely opposite, linear, 5–20 mm. long, 1–2 mm. wide, obtuse; inflorescence loosely flowered, the bracts linear; sepals narrowly lanceolate to broadly linear, obtuse, slightly shorter to as long as the petals; petals thin, lanceolate, about 5 mm. long, yellow; stamens shorter than the petals, the anthers ovoid; carpels 5, ascending.——Apr.–June. On mossy rocks in mountain forests; Honshu, Shikoku, Kyushu.

25. Sedum formosanum N. E. Br. HAMA-MANNEGUSA. Green fleshy perennial; stems 10–15 cm. long, ascending, rather stout, densely leaved, sparsely branched; leaves alternate, obovate to nearly orbicular, 15–22 mm. long, 8–12 mm. wide, broadly rounded at apex, narrowed to the cuneate base; cymes rather large, 5–8 cm. across, many-flowered; flowers sessile, about 1 cm. across, yellow, the bracts leaflike; sepals unequal, linear-lanceolate, 2–3 mm. long, obtuse; petals narrowly lanceolate, acuminate, 6–7 mm. long; stamens much shorter than the petals; carpels 5, the fruiting ones erect, broadly lanceolate, 5–6 mm. long, connivent above, shortly acuminate, terminated by a slender style about 1 mm. long.——Kyushu.——Ryukyus, Formosa, and China.

26. Sedum makinoi Maxim. MARUBA-MANNENGUSA. Stems 8–20 cm. long, ascending from a creeping branched base; leaves opposite, rarely partly verticillate, alternate in the inflorescence, spathulate, 7–15 mm. long, 3–8(–10) mm. wide, flat, fleshy; inflorescence large, much branched, the lower bracts leaflike, ternate; flowers sessile, loosely arranged; sepals linear-spathulate, ascending, spreading and rounded at apex, unequal in length; petals narrowly lanceolate, 5–6 mm. long, 3–5 times as long as the sepals; stamens slightly shorter than the petals, the anthers broadly ovoid, cordate; carpels ascending.——June–July. Honshu, Shikoku, Kyushu.

27. Sedum tricarpum Makino. TAKANE-MANNENGUSA. Flowering stems 10–25 cm. long, rather stout, the sterile basal shoots more or less creeping; leaves alternate, spathulate, 20–25 mm. long, 5–10 mm. wide, rounded at apex, flat, fleshy, dense and disposed mainly toward the top of the sterile shoots,

otherwise loosely arranged; inflorescence large, much branched; flowers rather loosely arranged, the lower ones with an oblanceolate or narrowly spathulate, leaflike bract; sepals linear-spathulate to lanceolate, obtuse; petals narrowly lanceolate, about 5 mm. long, much longer than the sepals; anthers broadly ovoid; carpels 3–5, ascending.——May–July. Honshu (Chūgoku Distr.), Shikoku, Kyushu.

28. Sedum tosaense Makino. YAHAZU-MANNENGUSA. Closely allied to No. 27; leaves spathulate, fleshy, rounded and retuse; carpels consistently 5, spreading in fruit.——Shikoku (Tosa Prov.).

29. Sedum bulbiferum Makino. *S. alfredii* sensu auct. Japon., non Hance; *S. lineare* var. *floribundum* Miq.; *S. subtile* var. *obovatum* Fr. & Sav.——KO-MOCHI-MANNENGUSA. Biennial or perennial with axillary bulbils; stems 7–20 cm. long, short-decumbent and branched at base, without sterile shoots; bulbils consisting usually of 2 pairs of small fleshy scalelike leaves; lower cauline-leaves opposite, ovate-spathulate, abruptly narrowed below, the upper ones alternate, spathulate-oblanceolate, 10–15 mm. long, 2–4 mm. wide, obtuse; inflorescence often branched; sepals broadly lanceolate to oblanceolate, obtuse; petals lanceolate, 4–5 mm. long, yellow, 1.5–2.5 times as long as the sepals; stamens shorter than the petals, the anthers ovoid-ellipsoidal or ovoid-oblong; carpels 5, 4–5 mm. long.——May–June. Shaded places in lowlands; Honshu, Shikoku, Kyushu; common.

Sedum alfredii Hance var. **nagasakianum** Hara is allied to *S. bulbiferum,* but the axillary bulbils are lacking; cauline leaves opposite to ternate.——Reported from Kyushu.

Sedum rosulatobulbosum Koidz. KŌRAI-KO-MOCHI-MANNENGUSA. Differs from *S. bulbiferum* in the obovate leaves, the tufted stems, and the late-appearing many-leaved bulbils. ——Reported to occur in Hachijo Isl. in Izu Prov.——Amami-oshima and Korea.

30. Sedum leveilleanum Hamet. *Cotyledon sikokiana* Makino; *S. orientoasiaticum* Makino; *Orostachys sikokianus* (Makino) Ohwi; *Meterostachys sikokianus* (Makino) Hara ——CHABO-TSUME-RENGE. Caudex short, rather stout; stems 3–7 cm. long; basal leaves rosulate, linear to narrowly oblanceolate, only slightly flattened, fleshy, glabrous, 7–12 mm. long, about 2 mm. wide, more or less callose and short spined at apex; cauline leaves obsoletely or short spine-tipped, the lowermost ternate, the upper ones alternate; flowers few, in cymes, white to pinkish; sepals 1/3–1/2 as long as the petals; petals 3–4 mm. long; ovules 2 in each locule.——July–Sept. Rocks in mountains; Honshu (w. distr.), Shikoku, Kyushu. ——s. Korea.

31. Sedum erubescens (Maxim.) Ohwi. *Umbilicus erubescens* Maxim.; *Cotyledon erubescens* (Maxim.) Fr. & Sav.; *Orostachys erubescens* (Maxim.) Ohwi——TSUME-RENGE. Monocarpic perennial; stems 6–15 cm. long, densely linear-lanceolate leaved; winter basal leaves narrowly spathulate, fleshy, flat, 1.5–3 cm. long, 4–7 mm. wide, callose, few-toothed on the upper margin, short spine-tipped; summer cauline and radical leaves scarcely callose-margined; spikes solitary, terminal, 4–10 cm. long, 15–20 mm. wide, densely many-flowered; flowers white to pinkish, 6–8 mm. long, very short-pedicellate; anthers reddish at first, dark purple following pollen dehiscence.——Sept.–Oct. Rocks and dry slopes; Honshu (Kantō Distr. and westw.), Shikoku, Kyushu.——Korea, Manchuria, Ussuri, and n. China.

Var. **polycephalum** (Makino) Ohwi. *Cotyledon poly-cephala* Makino; *S. polycephalum* (Makino) Makino; *Oro-stachys polycephalus* (Makino) Hara——YATSUGASHIKA. Plants with sterile innovation-shoots at base.

Var. **japonicum** (Maxim.) Ohwi. *Cotyledon japonica* Maxim.; *S. japonicola* Makino; *Orostachys japonicus* (Maxim.) Berger——HIRO-HA-TSUME-RENGE. Radical leaves ovate-oblong to narrowly so, often to 10 mm. wide, the cauline leaves oblong-linear.——Rocks near the seashore; Honshu.

32. Sedum iwarenge (Makino) Makino. *Cotyledon iwarenge* Makino; *Orostachys iwarenge* (Makino) Hara; *C. malacophylla* var. *japonica* Fr. & Sav.——IWA-RENGE. Monocarpic perennial; stems erect, densely leaved on lower half, 10–25 cm. long; caudex and stems densely many-leaved; leaves glaucous, fleshy, flat, oblong-spathulate, 3–7 cm. long, 7–28 mm. wide, obtuse to rounded; spike solitary and terminal, erect, very densely many-flowered, 5–20 cm. long; flowers short-pedicellate, white; petals 5–7 mm. long, about twice as long as the sepals.——Sept.–Nov. Not known wild in our area but frequently planted.——China (?).

33. Sedum furusei (Ohwi) Ohwi. *Orostachys furusei* Ohwi——REBUN-IWA-RENGE. Glaucous perennial; sterile stems slender, ascending, short, densely leaved at the apex, the basal innovation-shoots somewhat elongate, ascending to short-creeping; radical leaves fleshy, flat, obovate, 10–20 mm. long, 5–10 mm. wide, cuneate, entire; spikes many-leaved, 5–10 cm.

long, simple, many-flowered, the flowers dense, sessile; sepals 5, ascending, 3.5 mm. long, lanceolate-spathulate, subequal; petals 5, greenish white, slightly longer than the sepals, narrowly ovate, 4.5–5 mm. long, ascending in anthesis, closely appressed to the carpels in fruit; carpels 5, erect, oblong, as long as the petals; styles erect, about 1 mm. long.——Oct.–Nov. Rocks near the sea; Hokkaido (Momoiwa in Rebun Isl.); rare.

34. Sedum aggregeatum (Makino) Makino. *Cotyledon malacophylla* sensu auct. Japon., non Pall.; *C. aggregeata* Makino; *Orostachys aggregeatus* (Makino) Hara——AO-NO-IWA-RENGE, KO-IWA-RENGE. Closely resembles No. 32, but the leaves smaller, obtusely rounded, green, not glaucous, 2–4 cm. long, 1–2 cm. wide.——June–Oct. Rocks near the sea; Honshu (Pacific Ocean side from Mutsu to Iwaki Prov.).

Var. **genkaiense** (Ohwi) Ohwi. *Orostachys genkaiensis* Ohwi. GENKAI-IWA-RENGE. Rosulate leaves larger, to 8 cm. long.——Kyushu (n. distr.); rare.

Var. **boehmeri** (Makino) Ohwi. *Cotyledon malacophylla* var. *boehmeri* Makino; *C. boehmeri* (Makino) Makino; *Orostachys boehmeri* (Makino) Hara; *S. boehmeri* (Makino) Makino——KO-MOCHI-RENGE. Innovation-shoots stout, horizontally spreading, with a tuft of rosulate leaves at the apex. ——Seashores; Hokkaido, Honshu (Japan Sea side from Mutsu to Ugo Prov.).

3. PENTHORUM L. TAKO-NO-ASHI ZOKU

Perennials with stolons; stems erect; leaves alternate, membranous, lanceolate, long-acuminate at both ends, sessile, serrulate; cymes terminal, the branches scorpioid; flowers yellow-green, small; calyx-tube short, the lobes 5; petals 5 or absent; stamens 10, nectary glands absent; carpels 5, united on the lower half; style short; capsules depressed, 5-beaked, dehiscent from the base of the free portion of the carpels, the upper portion deciduous; seeds many, ovoid.——Two species, one in e. Asia, the other in N. America.

1. Penthorum chinense Pursh. *P. sedoides* var. *chinense* (Pursh) Maxim.; *P. sedoides* sensu auct. Japon., non L.; *P. sedoides* forma *angustifolium* Miq.——TAKO-NO-ASHI. Stems 30–70 cm. long, terete; inflorescence branched, many-flowered; capsules about 7 mm. across; seeds ovoid.——Aug.–Sept. Wet muddy places along rivers in lowlands; Honshu (Kantō Distr. and westw.), Shikoku, Kyushu; locally common.——Korea, China, Manchuria, and e. Siberia.

Fam. 101. **SAXIFRAGACEAE** YUKI-NO-SHITA KA Saxifrage Family

Herbs or shrubs, rarely trees; leaves alternate or opposite, often stipulate; flowers bisexual, sometimes unisexual, the marginal ones sometimes sterile; calyx often gamosepalous and adnate to the ovary, the perianth lobes imbricate or valvate in bud; stamens as many as the calyx-lobes or twice as many, sometimes more, staminodia often present; disc frequently well developed; ovary superior or inferior; carpels 2–5, nearly free or united; fruit a capsule, sometimes a berry.——About 110 genera, with about 1,200 species, most abundant in the N. Hemisphere but also in the S. Hemisphere.

1A. Fruit a capsule.
  2A. Herbs; carpels usually 2; stamens as many or twice as many as the petals.
    3A. Staminodia absent.
      4A. Ovary 2-locular with axile placentae.
        5A. Leaves pinnately, ternately, or palmately compound, rarely simple; seeds irregularly shaped.
          6A. Flowers ebracteate, apetalous; carpels connate below; leaves palmately divided; stipules scarious. .... 1. *Rodgersia*
          6B. Flowers bracteate.
            7A. Carpels nearly free; leaves ternately compound, rarely simple; stipules brown-scarious. .... 2. *Astilbe*
            7B. Carpels connate; leaves simple, evergreen, coriaceous; stipules indistinct. .... 3. *Tanakaea*
        5B. Leaves simple; seeds globose.
          8A. Calyx-tube usually slender, completely adnate to the ovary; stamens 10. .... 4. *Saxifraga*
          8B. Calyx-tube campanulate, adnate to the ovary only at base; petals and stamens inserted on the margin of the calyx-tube.
            9A. Stamens 10; petals deciduous; leaves peltate. .... 5. *Peltoboykinia*
            9B. Stamens 5; petals rather persistent; leaves palmately cleft, petiolate, basifixed. .... 6. *Boykinia*
      4B. Ovary 1-locular with parietal placentae.
        10A. Flowers 4-merous, apetalous, in cymes; bracts rather leaflike. .... 7. *Chrysosplenium*
        10B. Flowers 5-merous, usually petaloid, in racemes; bracts not leaflike.

1. RODGERSIA A. Gray   Yaguruma-sō Zoku

Large rather thick perennial herbs with short, stout creeping rhizomes; indument of two forms, hairlike and paleaceous; leaves palmately compound, large, toothed, often trilobulate, the stipules scarious; stems erect, few-leaved; inflorescence paniculate, large, terminal, the ultimate branches scorpioid-cymose; flowers white, small, bractless, bisexual; calyx-tube shallow, short, the lobes 5, spreading, white in anthesis; petals narrowly linear, small, sometimes absent; stamens 10; capsules subsuperior, 2-locular, dehiscent between the 2 persistent styles; seeds many, small.——Few species, in e. Asia.

1. Rodgersia podophylla A. Gray. *R. japonica* A. Gray ex Regel; *Astilbe podophylla* (A. Gray) Franch.——Yaguruma-sō. Stems 20–130 cm. long, terete, few-leaved, rather stout; radical leaves long-petiolate, the leaflets 5, chartaceous, 15–35 cm. long, 10–25 cm. wide, acuminate, cuneate, sessile, irregularly toothed, 3- to 5-lobulate on the upper portion, the lobules deltoid, the petioles long-hairy at top; inflorescence 20–40 cm. long, the branches densely many-flowered, densely pubescent, the hairs scalelike; calyx-lobes ovate, acute, 2–4 mm. long; filaments exserted; capsules broadly ovoid, subsuperior, about 5 mm. long; styles 2, persistent.——June–July. Woods in mountains; Hokkaido, Honshu.——Korea.

2. ASTILBE Hamilt.   Chidake-sashi Zoku

Perennial rhizomatous herbs, with scattered long brown scalelike hairs; leaves radical and cauline, alternate, ternately compound, rarely simple, the stipules scarious, the leaflets lanceolate to ovate-orbicular, toothed; inflorescence a terminal panicle; flowers small, white or rose-purple, bracteate, bisexual or unisexual (plants sometimes dioecious); calyx-tube short, lobes 5, sometimes 7–11; petals persistent, linear or absent; stamens 5, 10, or 8; ovary subsuperior; carpels 2, connate at the base; capsules 2-locular, dehiscent between the styles; seeds small.——Few species, in e. Asia, Himalayas, and N. America.

1. Astilbe platyphylla H. Boiss. Momiji-shōma. Dioecious stoloniferous perennial; stems to 50 cm. long; radical leaves 2-ternate, the terminal leaflets broadly rhombic-ovate, 8–17 cm. long, 6–15 cm. wide, long-acuminate, 2- to 5-lobulate, doubly acute-toothed; inflorescence 10–20 cm. long, the branches simple or the lower ones branched, the axils minutely glandular-hairy and with scalelike hairs; flowers yellow-green; calyx-lobes 7–11, lanceolate, subacute; petals absent; stamens 8–10; capsules 4–5 mm. long.——June–July. Hokkaido (Hidaka, Tokachi, and Oshima Prov.); rare.

2. **Astilbe simplicifolia** Makino. Hɪᴛᴏᴛsᴜʙᴀ-ꜱʜōᴍᴀ. Rhizomes short; stems 10–30 cm. long, usually with 1 leafless node; radical leaves simple, thinly chartaceous, broadly to narrowly ovate, often 3- to 5-lobulate, 3–8 cm. long, 2–5 cm. wide, doubly incised-toothed, the petioles 5–15 cm. long; inflorescence 5–20 cm. long, the axis with short glandular hairs, the branches spreading; flowers in loose panicles, white, the pedicels 1–5 mm. long; calyx-lobes narrowly ovate to broadly lanceolate, white in anthesis, spreading; petals linear-spathulate, as long as the stamens; anthers pale yellow; capsules 3–4 mm. long.——Honshu (Suruga and Sagami Prov.); rare.

3. **Astilbe chinensis** (Maxim.) Franch. var. **davidii** Franch. *A. davidii* (Franch.) Henry——Ō-ᴄʜɪᴅᴀᴋᴇ-ꜱᴀsʜɪ. Rhizomes stout; stems 30–70 cm. long; radical leaves 2- to partially 3-ternate, the terminal leaflets ovate, 2–8 cm. long, 1–4 cm. wide, short-acute, doubly acute-toothed, chartaceous; inflorescence 15–30 cm. long, erect, the branches densely flowered, the axis densely long brown-hairy, the pedicels very short; flowers rose-purple; petals linear, about 5 mm. long, twice as long as the stamens; anthers purple; capsules 3–4 mm. long.——July–Aug. Reported from Kyushu (Tsushima).—— Korea, Manchuria, n. China, Amur, and Ussuri. The typical phase does not occur in our area.

4. **Astilbe microphylla** Knoll. *A. chinensis* var. *japonica* Maxim.——Cʜɪᴅᴀᴋᴇ-ꜱᴀsʜɪ.   Stems 30–80 cm. long; radical leaves 2–4 times ternately compound, long-petiolate, the terminal leaflets chartaceous, broadly ovate, obovate or elliptic, 1–5 cm. long, 5–30 mm. wide, usually obtuse, irregularly doubly acute-toothed; inflorescence narrow, 10–20 cm. long, rather loosely short-branched, the axis glandular-hairy, the ultimate branches racemose; flowers pale rose-colored; petals linear-spathulate, obtuse, 3–5 mm. long, longer than the stamens; capsules 3–4 mm. long.——July–Aug.   Mountains, rather common in sunny grassy places; Honshu, Shikoku, Kyushu.

5. **Astilbe japonica** (Morr. & Decne.) A. Gray. *Hoteia japonica* Morr. & Decne.——Aᴡᴀᴍᴏʀɪ-ꜱʜōᴍᴀ.   Perennial herb; stems 50–80 cm. long; radical leaves 2-ternate, thick-chartaceous, the terminal leaflets rhombic-lanceolate, oblanceolate or rhombic-oblong, 3–7 cm. long, 1–2 cm. wide, deeply and doubly acute-toothed; panicles 10–20 cm. long, short-glandular, densely flowered; flowers white; petals narrowly spathulate, 3–4 mm. long, usually longer than the stamens; capsules 3–4 mm. long.——May–June. Rocks in mountain ravines; Honshu (Kinki Distr. and westw.), Shikoku, Kyushu.

Var. **terrestris** (Nakai) Murata. *A. glaberrima* Nakai; *A. glaberrima* forma *terrestris* Nakai——Yᴀᴋᴜsʜɪᴍᴀ-ꜱʜōᴍᴀ. Stems 5–50 cm. long, the terminal leaflets often 3-lobed, broadly lanceolate, long-acuminate, thinly chartaceous.—— Kyushu (Yakushima).

6. **Astilbe thunbergii** (Sieb. & Zucc.) Miq. *A. perplexi-*

pexa Koidz.; *Hoteia thunbergii* Sieb. & Zucc.; *A. kiusiana* Hara; *A. thunbergii* var. *kiusiana* Hara——Aᴋᴀ-ꜱʜōᴍᴀ. Rhizomes stout; stems 40–80 cm. long; radical leaves 3-ternate, the terminal leaflets chartaceous, ovate or narrowly so, 4–12 cm. long, 2–5 cm. wide, caudate, doubly acute-toothed, obtuse to shallowly cordate at base; panicles 10–30 cm. long, slenderly branched, the lower branches simple, elongate, densely glandular; flowers white; petals linear-spathulate, 3–4 mm. long, longer than the stamens; capsules 2–3 mm. long.——May–July. Sunny grassy slopes; Honshu, Shikoku, Kyushu; common and very variable.

Var. **sikokumontana** (Koidz.) Murata. *A. sikokumontana* Koidz.——Hɪᴍᴇ-ᴀᴋᴀ-ꜱʜōᴍᴀ. Leaflets doubly incised-toothed, sometimes narrower than in the typical variety.——Mountains; Shikoku.

Var. **shikokiana** (Nakai) Ohwi. *A. shikokiana* Nakai—— Sʜɪᴋᴏᴋᴜ-ᴛᴏʀɪᴀsʜɪ-ꜱʜōᴍᴀ.   Rhizomes sometimes long-creeping.——Shikoku.

Var. **fujisanensis** (Nakai) Ohwi. *A. fujisanensis* Nakai ——Fᴜᴊɪ-ᴀᴋᴀ-ꜱʜōᴍᴀ.   Resembles the typical phase, but the leaflets firmer, smaller, and doubly incised-toothed, and the petals rather short, 2–2.5 mm. long, nearly as long as the stamens, the capsules 3–4 mm. long.——Mountains; Honshu (Sagami, Suruga, and Izu Prov.).

Var. **congesta** H. Boiss. *A. odontophylla* Miq.; *A. senanensis* Matsum.; *A. pedunculata* (H. Boiss.) Nakai——Tᴏʀɪ-ᴀsʜɪ-ꜱʜōᴍᴀ.   Resembles the typical phase, but the terminal leaflets ovate to broadly so, 5–12 cm. long, 4–10 cm. wide, and shallowly cordate, the panicles broadly pyramidal, the lower branches paniculately compound.——Hokkaido, Honshu (n. and centr. distr.).——Forma **bandaica** (Honda) Ohwi. *A. congesta* var. *bandaica* Honda; *A. odontophylla* var. *bandaica* (Honda) Hara; *A. bandaica* (Honda) Koidz.——Bᴀɴᴅᴀɪ-ꜱʜōᴍᴀ. An alpine phase.

Var. **formosa** (Nakai) Ohwi. *A. chinensis* var. *formosa* Nakai; *A. formosa* (Nakai) Nakai; *A. nipponica* Nakai—— Hᴀɴᴀ-ᴄʜɪᴅᴀᴋᴇ-ꜱᴀsʜɪ.   Leaflets short-acuminate, the panicle rachis densely glandular, the petals narrow, 4–6 mm. long.—— Grassy slopes and thickets in mountains; Honshu (Shinano and Etchu Prov.).

Var. **okuyamae** (Hara) Ohwi. *A. okuyamae* Hara; *A. odontophylla* var. *okuyamae* Hara——Mɪᴋᴀᴡᴀ-ꜱʜōᴍᴀ.   Resembles var. *congesta,* but the leaflets ovate-cordate, regularly doubly shallow-toothed, the panicle-branches rather distant, not much elongate, the petals narrow, nearly as long as the stamens.——Honshu (Mikawa Prov.).

Var. **hachijoensis** (Nakai) Ohwi. *A. hachijoensis* Nakai ——Hᴀᴄʜɪᴊō-ꜱʜōᴍᴀ.   Leaflets firmly chartaceous and lustrous above, the terminal ones broadly to narrowly ovate, often 3-lobed, the cauline leaves usually 3-foliolate, the bracts and bracteoles prominent, 2–4(–10) mm. long, the petals white, about 3 mm. long.——Honshu (Izu Isls.).

3. TANAKAEA Fr. & Sav.   Iᴡᴀ-ʏᴜᴋɪ-ɴᴏ-sʜɪᴛᴀ Zᴏᴋᴜ

Evergreen dioecious perennial with creeping rhizomes and stolons; radical leaves tufted, petiolate, oblong, irregularly coarse-toothed, firm and thick, flat, exstipulate; stems scapose, erect, usually naked; inflorescence paniculate, the bracts small, linear; flowers white; calyx-tube short, shallow, the lobes 5(4–7); petals absent; stamens 10; ovary nearly superior; styles short, capitellate; capsules nearly superior; seeds acute at both ends.——One species in China and Japan.

1. Tanakaea radicans Fr. & Sav. IWA-YUKI-NO-SHITA. Radical leaves subcoriaceous, green above, paler beneath, obtuse or subacute, 2–8 cm. long, 1–5 cm. wide, rounded or sometimes cordate at base, the petioles loosely coarse-hairy; scapes 10–30 cm. long, loosely long coarse-hairy; inflorescence densely flowered, erect or often declined or nodding in fruit, 5–15 cm. long; flowers small, whitish, apetalous; capsules 3–4 mm. long; carpels 2, connate nearly the whole distance, dehiscent.——May–June. Wet shaded rocks; Honshu (Tōkaidō Distr.), Shikoku, Kyushu; rare.

4. SAXIFRAGA L.   YUKI-NO-SHITA ZOKU

Perennials, sometimes annuals or biennials; leaves simple, often toothed or lobed; flowers 1 to many, usually in cymes, white or yellowish, bisexual; calyx-tube prominent or shallow and short, the lobes or segments 5; petals 5, equal or unequal; stamens 10, the filaments filiform or sometimes dilated above; carpels 2, connate at base; styles 2, short or rather elongate, free, the stigma small; capsules dehiscent between the styles; seeds many, small, sometimes tuberculate, sometimes caudate.——More than 300 species, especially abundant in the N. Hemisphere.

1A. Calyx-tube well developed; ovary semisuperior; leaves reniform-orbicular, palmately lobed.
  2A. Inflorescence with bulbils; petals oblanceolate, 6–8 mm. long, 3–4 times as long as the calyx-lobes; alpine. . . . . 1. *S. cernua*
  2B. Inflorescence without bulbils; petals broadly obovate, about 5 mm. long, 1.5–2 times as long as the calyx-lobes; wet rocks near the sea.
                                                         2. *S. bracteata*
1B. Calyx-tube short and shallow; ovary superior or nearly so.
  3A. Stems scapose, simple or only sparsely branched at base; leaves distinctly petiolate (except nos. 6–7).
    4A. Flowers actinomorphic; petals all alike and equal.
      5A. Radical leaves reniform-orbicular, long-petiolate.
        6A. Flowers red-brown or greenish brown; petals twice to several times as long as the stamens; disc prominently raised.
                                             3. *S. fusca*
        6B. Flowers white or sometimes purplish red; petals slightly longer than the stamens; disc narrow.
          7A. Scapes 15–80 cm. long, naked or with a single small leaf; radical leaves cordate, 3–10 cm. long and nearly as wide, irregularly toothed; seeds long-tailed at each end. . . . . .4. *S. japonica*
          7B. Scapes 5–25 cm. long, naked; radical leaves reniform, 2–4 cm. long, 3–6 cm. wide, regularly toothed; seeds not tailed.
                                           5. *S. reniformis*
      5B. Radical leaves ovate to oblong-cuneate, sessile or short-petiolate.
        8A. Scapes 10–40 cm. long; radical leaves 4–10 cm. long, with many teeth; hairs with 1 longitudinal row of cells.
                                        6. *S. sachalinensis*
        8B. Scapes 5–10 cm. long; radical leaves 1–3 cm. long, coarsely 5- to 11-toothed; hairs with many rows of cells. .. 7. *S. laciniata*
    4B. Flowers zygomorphic; petals of two forms.
      9A. Aerial stems stout, usually simple, 3–20 cm. long, erect; leaves ovate-orbicular. . . . . 8. *S. sendaica*
      9B. Aerial stems absent; leaves radical, usually reniform-orbicular.
        10A. Upper petals nearly lanceolate, gradually narrowed at base, not spotted; seeds smooth, not tubercled; summer and autumn flowering. . . . . 9. *S. fortunei*
        10B. Upper petals broadly ovate, abruptly narrowed below to a claw, spotted; seeds tubercled.
          11A. Leaves palmately cleft to lobed, the lobes broadly ovate to oblong; autumn flowering. . . . . 10. *S. cortusaefolia*
          11B. Leaves shallowly and obsoletely lobulate, the lobules depressed; spring flowering.
             12A. Stolons present; leaves deep green and obsoletely variegated on upper side, purplish red beneath; upper petals with yellow and small red spots. . . . . 11. *S. stolonifera*
             12B. Stolons absent; leaves green and not variegated on upper side; upper petals with yellow spots only near base.
                                      12. *S. nipponica*
  3B. Stems well-developed, short, much-branched and tufted, densely leafy; leaves small, ciliate.
    13A. Petals narrowed at base, not clawed, lanceolate, with small spots.
      14A. Leaves 3-lobed, glandular on the margin. . . . . 13. *S. nishidae*
      14B. Leaves undivided, ciliate and bristle-margined. . . . . 14. *S. cherlerioides* var. *rebunshirensis*
    13B. Petals broadly ovate, distinctly clawed at base, not spotted; leaves 6–25 mm. long, often shallowly 3-lobed, with long glandular hairs on margin. . . . . 15. *S. merkii*

1. Saxifraga cernua L. MUKAGO-YUKI-NO-SHITA. Perennial with short rhizomes and numerous small scalelike fleshy bulbils; scapes erect, simple, 5–25 cm. long, long-pubescent, naked except for few leaves on lower half; radical leaves few, reniform-orbicular, palmately 5- to 7-cleft, 5–15 mm. long, 7–17 mm. wide, slightly fleshy, the petioles slightly dilated at base, with soft curled hairs; cauline leaves short-petioled, broadly cuneate to shallowly cordate, the lobes acute, the upper leaves gradually becoming bracteate; flowers few, loosely spicate (bulbils appearing in place of the lowermost flowers), the bracts lanceolate to elongate-ovate, sessile, the rachis with curled glandular hairs, the terminal flowers usually well-developed, white; calyx-lobes broadly lanceolate, obtuse, glandular-pilose, about 2.5 mm. long; petals ascending, spreading above, oblanceolate, 3-nerved, 6–8 mm. long; capsules usually not developed.——Aug. Wet shaded rocks in high mountains; Honshu (alpine regions in Shinano Prov.); rare.——Mountains and northern regions of the N. Hemisphere.

2. Saxifraga bracteata D. Don. *S. laurentiana* Ser.; *S. rivularis* var. *laurentiana* (Ser.) Engl.; *S. fauriei* H. Boiss.; *S. exilis* sensu Yabe & Yendo, non Steph.——KIYOSHI-SŌ. Resembles the preceding, but larger and less pubescent; stems stouter, often shortly branched; leaves 1–4 cm. wide, the shallowly lobed bracts leaflike; inflorescence cymose, loose, few-flowered, without bulbils; calyx-lobes broadly ovate; petals broadly obovate, about 5 mm. long, 1.5–2 times as long as the calyx-lobes; capsules subinferior, broadly obovoid, about 8 mm. long, shallowly 2-lobed; seeds ovoid, smooth.——July–Aug. Wet rocks near the sea; Hokkaido (Nemuro Prov.).——Kuriles, Sakhalin, Ochotsk Sea region, Alaska, and Kamchatka.

3. Saxifraga fusca Maxim. *S. fusca* var. *divaricata* Fr. & Sav.——Ezo-kuro-kumo-sō. Perennial herb with short creeping rhizomes; radical leaves long-petiolate, reniform-orbicular, 3–6 cm. long, 4–10 cm. wide, regularly coarse-toothed, nearly glabrous; scapes 20–30 cm. high, glabrous below; panicles terminal; flowers many, loose, brown, the pedicels slender, 5–12 mm. long, sparingly glandular-hairy; calyx-lobes broadly ovate, obtuse, about 1 mm. long, recurved; petals narrowly ovate, 2.5–3 mm. long, obtuse, spreading, the disc prominent; stamens short, the filaments linear; capsules superior, erect, 4–6 mm. long, the carpels connate at base; seeds oblong, about 0.8 mm. long, 1-ribbed, short-mucronate at both ends, with minute tubercles in longitudinal rows.——July–Sept. Streams and ravines; Hokkaido, Honshu (n. distr.).——s. Kuriles.——Forma **kurilensis** Ohwi. *S. fusca* var. *kurilensis* Ohwi. Chishima-kuro-kumo-sō. Without glandular hairs on the inflorescence.——Forma **intermedia** Hara. Usuge-kuro-kumo-sō. With longer, flexuous glandular-hairs on the inflorescence.

Var. **kikubuki** Ohwi. *S. fusca* sensu auct. Japon., non Maxim.; *S. kikubuki* Ohwi; *S. fusca* var. *kiusiana* Hara——Kuro-kumo-sō, Kikubuki. Hairs eglandular; scapes long-pubescent at base, the inflorescence short-pubescent, the pedicels 2–7 mm. long, the flowers dark purple-brown, sometimes greenish brown.——July–Sept. Wet places along streams and ravines in mountains; Honshu (centr. and n. distr.), Shikoku, Kyushu.

4. Saxifraga japonica H. Boiss. Fuki-yuki-no-shita. Rhizomes stout, short-creeping; radical leaves tufted, ovate-cordate, the blades 3–10(–15) cm. long and as wide, irregularly acute-toothed, nearly glabrous, the petioles 10–35 cm. long; scapes 15–80 cm. long, often with 1(–3) small leaves, loosely curly-pubescent; panicles 15–40 cm. long, rather loosely many-flowered; flowers white, the pedicels 5–20 mm. long; calyx-lobes ovate, recurved, obtuse; petals obliquely spreading, ovate, obtuse, 3–4 mm. long; stamens slightly longer than the petals, the filaments nearly linear, the disc surrounding the base of the ovary; capsules 10–12 mm. long, the carpels erect, obliquely ascending on the upper part; seeds fusiform, about 1 mm. long, longitudinally mammillate-ribbed, the tail at each end as long as the body.——July–Aug. Wet rocks, stream banks and ravines in mountains; Hokkaido, Honshu (n. and centr. distr.), Shikoku.

5. Saxifraga reniformis Ohwi. *S. punctata* sensu auct. Japon., non L.; *S. ohwii* Tatew., quoad plant.; *S. punctata* subsp. *reniformis* (Ohwi) Hara——Chishima-iwabuki. Rhizomes creeping, rather slender, short; radical leaves long-petiolate, reniform or reniform–orbicular, 5–10 cm. long, 3–6 mm. wide, regularly coarse-toothed, nearly glabrous; scapes 5–25 cm. high; inflorescence corymbose, densely flowered, the peduncles and axis soft short-pubescent; calyx-lobes recurved, narrowly ovate, about 2 mm. long, obtuse; petals obliquely spreading, white, sometimes with a reddish purple hue, oblong, about 3 mm. long, shortly clawed below; stamens as long as to slightly shorter than the petals, the filaments slightly flattened and dilated toward tip; capsules 4–5 mm. long; carpels oblong, erect, with a spreading apex; style very

short; seeds broadly fusiform, subacute, about 0.7 mm. long.——July. Wet places along streams and ravines in mountains and alpine regions; Hokkaido (Rishiri Isl. and Mount Daisetsu); rare.——Sakhalin, Kuriles, and Kamchatka.

6. Saxifraga sachalinensis F. Schmidt. *S. virginiensis* var. *yesoensis* Franch.; *S. yesoensis* (Franch.) Engl.; *S. reflexa* sensu auct. Japon., non Hook.——Yama-hana-sō. Perennial rhizomatous herb, rather long glandular-hairy on the vegetative parts, shorter-pubescent on the inflorescence; radical leaves rosulate, ovate to oblong, 2–10 cm. long, 1–5 cm. wide, subobtuse to rounded at apex, rather fleshy, abruptly narrowed below and decurrent, dentate, often purplish beneath, the petioles 0.5–2.5 cm. long; scapes 10–40 cm. long; panicle loosely many-flowered, the pedicels 5–15 mm. long; calyx-lobes recurved, narrowly ovate, about 2 mm. long; petals white, ovate, obtuse, 3–4 mm. long, 3-nerved; stamens nearly as long as the petals, the filaments broadened above; capsules 4–6 mm. long; carpels ovoid, ascending, shortly reflexed at apex; seeds fusiform-ellipsoidal, about 0.7 mm. long, mucronate at both ends, longitudinally ribbed.——May–June. Hokkaido.——Sakhalin and s. Kuriles.

7. Saxifraga laciniata Nakai & Takeda. Kumoma-yuki-no-shita, Hime-yama-hana-sō. Rhizomes short; radical leaves rosulate, oblong-cuneate, 2–3 cm. long, 8–12 mm. wide, sessile, with few teeth on the upper margin, glandular-ciliate, slightly fleshy, yellowish green; scapes 5–10 cm. long, glandular-pubescent; inflorescence cymose, terminal, rather few-flowered; calyx-lobes recurved, ovate-oblong, 2–3 mm. long; petals white, narrowly ovate, 4–5 mm. long, clawed at base, with 2 yellow spots on lower portion, 3-nerved; stamens shorter than the petals, the filaments subulate; capsules superior; carpels connate on lower half, ovoid, 5–6 mm. long; seeds ovoid, small, tuberculate on the longitudinal ribs.——July–Aug. Alpine; Hokkaido (Mount Daisetsu and Mount Yubari); rare.——Sakhalin and n. Korea.

Saxifraga takedana Nakai. Tsuru-kumomagusa, with elongate rhizomes.——Occurs with and probably only an ecological variant of the preceding.

8. Saxifraga sendaica Maxim. Sendai-sō. Rather fleshy perennial herb; stems simple, terete, stout, 3–20 cm. long, erect, 3–8 mm. in diameter, fleshy, leafy toward the summit, with scarious scalelike leaves below; leaves long-petiolate, ovate-orbicular, 4–10 cm. long, 3–7 cm. wide, shallowly 7- to 9-lobed, irregularly toothed, rather fleshy, lustrous above, glabrous, glandular-hairy on the margins, petioles 2–10 cm. long; scapes 5–8 cm. long; panicle corymbose, the pedicels short glandular-puberulent; calyx-lobes ovate-oblong, obtuse, about 4 mm. long; petals lanceolate, acute, white, of two lengths, the longest to 20 mm. long; stamens about 5 mm. long.——Oct. On shaded rocks in mountains; Honshu (Ise and Yamato Prov.), Shikoku, Kyushu; rare.——Forma **laciniata** (Nakai) Ohwi. *S. sendaica* var. *laciniata* Nakai. A rare phase with deeply lobed leaves.

9. Saxifraga fortunei Hook. f. var. **incisolobata** (Engl. & Irmsch.) Nakai. *S. cortusaefolia* sensu auct. Japon., non Sieb. & Zucc.; *S. mutabilis* Koidz.; *S. cortusaefolia* var. *typica* subvar. *incisolobata* Engl. & Irmsch.——Dai-monji-sō. Slightly fleshy perennial with short rhizomes; radical leaves long-petiolate, stipulate, reniform-orbicular, 3–15 cm. long, 4–20 cm. wide, cordate, rarely truncate at base, palmately lobed and dentate, nearly glabrous to coarsely hirsute, the petioles 3–30 cm. long; scapes 5–35 cm. long, naked or rarely with a small leaf, glabrous to thinly glandular-pubescent; inflorescence 1–25 cm. long, broadly pyramidal, the pedicels 3–20 mm. long, scattered glandular-pubescent; calyx-lobes spreading, ovate-oblong, 2–3 mm. long; petals spreading, white, sometimes reddish, the 3 upper ones broadly lanceolate, 3–4 mm. long, the lower 2 narrowly lanceolate, 5–15 mm. long, rarely few-toothed; stamens 4–7 mm. long, the filaments scarcely broadened above; styles short, slightly elongate after anthesis; capsules superior, ovoid, 4–6 mm. long; carpels connate nearly to the top; seeds narrowly fusiform, acute, 0.8 mm. long, smooth.——July–Oct. Wet rocks in mountains; Hokkaido, Honshu, Shikoku, Kyushu; common and variable.——s. Sakhalin, s. Kuriles, Korea, Manchuria, Ussuri, and China.——Forma **partita** (Makino) Ohwi. *S. cortusaefolia* var. *partita* Makino; *S. fortunei* var. *partita* (Makino) Nakai——Kaede-dai-monji-sō. With cleft to parted leaves.

Var. **obtusocuneata** (Makino) Nakai. *S. cortusaefolia* var. *obtusocuneata* Makino; *S. mutabilis* var. *obtusocuneata* (Makino) Koidz.——Uchiwa-dai-monji-sō. Leaves ovate-orbicular or flabellate-orbicular, broadly cuneate to nearly truncate at base.——Along streams in mountains; Honshu, Shikoku, Kyushu.

Var. **crassifolia** (Engl. & Irmsch.) Nakai. *S. cortusaefolia* var. *typica* subvar. *crassifolia* Engl. & Irmsch.; *S. cortusaefolia* var. *typica* subvar. *brevifolia* Engl. & Irmsch.; *S. jotanii* Honda; *S. cortusaefolia* var. *jotanii* (Honda) Makino——Izu-no-shima-dai-monji-sō. Prominently long-hirsute, fleshier, the leaves shallowly lobed; scapes and pedicels with longer glandular hairs.——Oct.–Jan. Honshu (Awa and Izu Isls.).

10. Saxifraga cortusaefolia Sieb. & Zucc. *S. cortusaefolia* var. *madida* Maxim.; *S. madida* (Maxim.) Makino; *S. cortusaefolia* var. *maximowiczii* (Engl. & Irmsch.) Hara; *S. madida* var. *maximowiczii* Engl. & Irmsch.——Jinji-sō. Resembles the preceding but with longer and denser hairs; leaves often deeply cleft, sometimes with needle-shaped cystoliths (under magnification); 3 upper petals broadly ovate, distinctly clawed and with a yellow spot near the base, about 4 mm. long; capsules broad; seeds narrowly fusiform, about 0.7 mm. long, minutely tuberculate.——Sept.–Oct. Wet shaded places especially on rocks and along streams in mountains; Honshu (Kantō Distr. and westw.), Shikoku, Kyushu; rather rare.

Var. **stolonifera** (Makino) Koidz. *S. madida* var. *stolonifera* Makino——Tsuru-jinji-sō. Plant stoloniferous.——Kyushu (Hizen and Higo Prov.).

11. Saxifraga stolonifera Curt. *S. sarmentosa* L.; *S. ligulata* Murr.; *S. chinensis* Lour.; *Sekika sarmentosa* (L.) Moench——Yuki-no-shita. Rather fleshy stoloniferous perennial with coarse brownish red hairs; rhizomes short; leaves fleshy, reniform-orbicular, 3–5 cm. long, 3–9 cm. wide, shallowly lobulate, toothed, pale variegated above, purplish red beneath, the petioles 3–10 cm. long, dilated and long-ciliate at base; scapes 20–40 cm. long; inflorescence a broad panicle 10–20 cm. long, short glandular-pubescent; calyx-lobes narrowly ovate, 3–4 mm. long, recurved above; petals spreading, the upper 3 ovate, acute, about 3 mm. long, short-clawed, spotted, the lower 2 unequal, lanceolate, 10–20 mm. long, spotless; stamens 4–7 mm. long, the filaments slightly broadened above; capsules ovoid-globose, 4–5 mm. long, shallowly 2-lobed; styles rather long; seeds ovoid, warty, about 0.5 mm. long.——May–July. Wet soils and rocks; Honshu, Shikoku, Kyushu.——China. Frequently planted in gardens.——Forma **aptera** (Makino) Hara. *S. sarmentosa* var. *aptera* Makino

——Hoshizaki-yuki-no-shita. Abnormal phase with 5 equal petals.

12. Saxifraga nipponica Makino. Haru-yuki-no-shita. Sparsely long-pilose; rhizomes rather long, branched; leaves reniform-orbicular or orbicular, 2–6 cm. long, 2–7 cm. wide, deeply cordate, toothed, very shallowly lobulate, green, with needle-shaped cystoliths; scapes 20–40 cm. long; cymes paniculate, 10–30 cm. long, with soft glandular-hairs; calyx-lobes recurved after anthesis, 2–4 mm. long; petals spreading, white, the upper 3 broadly ovate, subacute, short-clawed at base, spotted on the lower portion, 3–4 mm. long, the lower 2 larger, unequal, lanceolate, acute, 10–25 mm. long, not spotted; stamens 3–5 mm. long, the filaments slightly broadened above; capsules broadly ovoid, 4–5 mm. long, shallowly 2-fid; styles ascending; seeds small, granular-tuberculate.——Apr.–May. Mountains; Honshu (n. and centr. distr. incl. Kantō); rare.

13. Saxifraga nishidae Miyabe & Kudo. Ezo-no-kumo-magusa. Stems slender, short, much-branched, tufted; leaves lanceolate-cuneate, 4–10 mm. long, 2–3 mm. wide, sessile, 3-lobed, the lobes narrowly ovate, short-awned, scattered glandular-hairy on margin; stems 3–7 cm. long, scattered glandular-hairy, loosely leafy; inflorescence 1- to 3-flowered, the pedicels short glandular-hairy; calyx-lobes ovate-oblong, 3–4 mm. long; petals narrowly oblong, subobtuse, 3-nerved, 6–7 mm. long, with small spots; stamens about 5 mm. long, the filaments subulate; styles slender.——Rocky places in alpine regions; Hokkaido (Mount Yubari); rare.

14. Saxifraga cherlerioides D. Don var. **rebunshirensis** (Engl. & Irmsch.) Hara. *S. bronchialis* sensu auct. Japon., non L.; *S. bronchialis* var. *genuina* forma *rebunshirensis* Engl. & Irmsch.; *S. bronchialis* var. *cherlerioides* sensu auct. Japon., non Engl.; (?) *S. firma* Litv.——Shikotan-sō. Stems slender, much-branched, densely many-leaved; leaves evergreen, thick, firm, spathulate-lanceolate or narrowly lanceolate, 6–15 mm. long, 1.5–3 mm. wide, short spine-tipped, spreading bristly on the margin; stems 3–12 cm. long, scattered glandular-puberulent, sometimes with curled hairs at the base, loosely few-leaved; cymes 3- to 10-flowered; calyx-lobes ascending, narrowly ovate, 2–3 mm. long, subacute; petals narrowly oblong or oblanceolate, 5–7 mm. long, ascending, white to cream-colored, with small brown-red spots; stamens 4–5 mm. long, the filaments subulate; capsules broadly ovoid, 4–5 mm. long, 2-fid; seeds about 0.8 mm. long, granular-tuberculate.——July–Aug. Rocks and rocky slopes in alpine regions; Hokkaido, Honshu (centr. distr.); rare.——Forma **togakushiensis** (Hara) Ohwi. *S. bronchialis* var. *pseudoburseriana* sensu auct. Japon., non F. Schmidt; *S. cherlerioides* var. *togakushiensis* Hara——Hime-kumomagusa. Smaller phase with leaves prominently spine-tipped and ciliate.——Honshu (Mount Togakushi) and Hokkaido (Sounbetsu in Ishikari).——The typical phase and several distinct variants occur in Siberia from the Ural Mts. eastw., also in N. America.

15. Saxifraga merkii Fisch. var. **merkii**. *S. merkii* var. *robusta* Takeda; *S. merkii* var. *typica* Engl. & Irmsch.——Chishima-kumomagusa. Stems short, densely branched, densely leafy; leaves rather fleshy, green, broadly lanceolate to obovate, 6–17(–20) mm. long, 3–8(–10) mm. wide, rounded to subacute at the tip, long glandular-hairy on margin, entire, scattered-pilose above, very rarely shallowly 3-lobed; scapes 3–10 cm. long, 1- to 8-flowered, densely glandular-pilose; calyx-lobes spreading, broadly ovate, 2–3 mm. long, rounded at apex; petals broadly ovate, 6–7 mm. long, spreading, white, spotless, abruptly narrowed to a prominent claw at base; stamens about 4 mm. long, the filaments linear; capsules broadly ovoid, 6–7 mm. long, bifid, superior.——July–Aug. Gravelly and sandy slopes in alpine regions; Hokkaido.—— Kuriles and Kamchatka.

Var. **idsuroei** (Fr. & Sav.) Engl. *S. idsuroei* Fr. & Sav. ——KUMOMAGUSA. Leaves often shortly 3-lobed, the flowers 1–4.——Alpine regions; Honshu (centr. distr.).

5. PELTOBOYKINIA Hara   YAWATA-SŌ ZOKU

Rather large perennial herbs with short creeping rhizomes; stems few-leaved; radical leaves long-petioled, large, peltate, palmately lobed, the stipules membranous; cymes terminal; calyx-tube shallowly campanulate, adnate to the ovary on lower half, the lobes 5; petals 5, pale yellow, toothed, glandular-dotted, ascending, deciduous; stamens 10; ovary 2-locular, subsuperior; carpels connate nearly all the way, the placentae axile; styles 2, free; capsules subsuperior, enclosed in the somewhat inflated calyx-tube; seeds minute, longitudinally tuberculate.——Two species, in Japan.

1A. Leaves 7- to 13-lobed, the lobes ovate-deltoid, abruptly acute, broader than long. .... 1. *P. tellimoides*
1B. Leaves 8- to 10-cleft, the lobes narrowly cuneate-ovate, acuminate, 2–3 times as long as wide. .... 2. *P. watanabei*

1. Peltoboykinia tellimoides (Maxim.) Hara. *Saxifraga tellimoides* Maxim.; *Boykinia tellimoides* (Maxim.) Engl.——YAWATA-SŌ. Rhizomes stout and short; radical leaves 1 or 2, long-petioled, orbicular, peltate, cordate, 10–25 cm. long and as wide, shallowly palmate-lobed, toothed, obsoletely incised, glabrous or slightly glandular-pilosulous, lustrous above; stems 30–60 cm. long, rather stout, 2- to 3-leaved, the cauline leaves smaller, short-petioled or subsessile; inflorescence several-flowered, the bracts small; calyx-lobes narrowly deltoid, erect, acute, 4–5 mm. long; petals creamy white, broadly oblanceolate, 10–12 mm. long, nearly acute, ascending, few-toothed toward tip; capsules 10–13 mm. long, broadly ovoid, enclosed within the calyx-tube.——May–July. Woods in mountains; Honshu (n. and centr. distr.); rare.

2. Peltoboykinia watanabei (Yatabe) Hara. *Saxifraga watanabei* Yatabe; *S. tellimoides* var. *watanabei* (Yatabe) Makino; *Boykinia tellimoides* var. *watanabei* (Yatabe) Engl.; *B. watanabei* (Yatabe) Makino; *P. tellimoides* var. *watanabei* (Yatabe) Hara——WATANABE-SŌ. Closely resembles the preceding species; leaves palmately cleft below the middle, the lobes narrow, acuminate, slightly narrowed at base; rhizomes short; radical leaves long-petioled, about 25 cm. long and as wide, 8- to 10-cleft, the lobes much longer than wide, usually 2- to 3-incised, glabrous, irregularly toothed; calyx-lobes deltoid, acute, about 4 mm. long; petals pale yellow, short-glandular, 8–10 mm. long, about 2.5 mm. wide.——June–July. Woods in mountains; Shikoku, Kyushu; rare.

6. BOYKINIA Nutt.   ARASHIGUSA ZOKU

More or less pubescent perennial herbs; leaves petiolate, orbicular-cordate, palmately lobed; scapes few-leaved; cymes paniculate, terminal, the bracts small; flowers small, greenish-yellow or white, bisexual; calyx-tube campanulate or subglobose, adnate to the ovary most of the way, the lobes 5; petals 5, deciduous or somewhat persistent, as many as the stamens; ovary nearly inferior; carpels adnate to the calyx-tube; capsules 2-locular, 2-beaked; styles free; seeds minute.——About 10 species, in N. America, 1 in Japan.

1. Boykinia lycoctonifolia (Maxim.) Engl. *Saxifraga lycoctonifolia* Maxim.; *Therofon lycoctonifolia* (Maxim.) Takeda; *Neoboykinia lycoctonifolia* (Maxim.) Hara——ARASHIGUSA. Rhizomes slender, creeping, rather elongate; radical leaves solitary, long-petiolate, reniform-orbicular, 3–10 cm. long, 5–15 cm. wide, palmately 7- to 9- lobed or -cleft, incised and toothed, dull and nearly glabrous on upper side, long-

pubescent on the nerves beneath; stems 15–40 cm. long, few-leaved, long glandular-hairy; inflorescence glandular-pilose; flowers small, yellow-green; calyx-lobes narrowly deltoid; petals slightly shorter than the sepals, obovate, about 2 mm. long; capsules ellipsoidal, 7–8 mm. long; seeds longitudinally and minutely tuberculate.——July–Aug. Wet grassy slopes in high mountains; Hokkaido, Honshu (centr. and n. distr.).

7. CHRYSOSPLENIUM L.   NEKO-NO-ME-SŌ ZOKU

Small more or less succulent perennials or biennials; leaves alternate or opposite, simple, short-toothed, petiolate, stipules absent; radical leaves often few; stems scapose, the bracts subfoliaceous often colored; flowers small, in cymes, green, yellow, or white; calyx-tube cup-shaped or infundibuliform, the limb 4-lobed; petals absent; stamens 8 or sometimes 4; ovary 1-locular; styles 2, short, free; capsules subinferior or nearly superior, depressed or laterally compressed, bifid; seeds elliptic, sometimes fusiform, small, smooth or tuberculate, often longitudinally ribbed.——Principally in China and Japan, few in Europe and Siberia, also a few in N. and S. America.

1A. Leaves opposite or very rarely partially alternate in the cauline ones.
  2A. Stolons hypogeal, leafless, bearing a bulbil at apex; nearly glabrous. .... 1. *C. maximowiczii*
  2B. Stolons epigeal, with few pairs of leaves and without bulbils.
    3A. Plants brownish or whitish pilose.
      4A. Calyx-lobes erect at anthesis; seeds longitudinally warty; stamens as long as to slightly longer than the calyx-lobes; capsules with 2 ascending or nearly erect beaks.

5A. Calyx-lobes oblong or narrowly ovate, sometimes elliptic, usually white at anthesis; stamens exserted; hairs on upper side of leaves rather long; anthers dark purple. . . . . 2. *C. album*

5B. Calyx-lobes rounded to ovate-rounded, yellow or pale green at anthesis; anthers yellow.

6A. Calyx-lobes yellow to greenish yellow at anthesis; stamens slightly shorter than the calyx-lobes.
3. *C. pilosum* var. *sphaerospermum*

6B. Calyx-lobes pale green at anthesis; stamens as long as to slightly longer than the calyx-lobes. . . . . . . 4. *C. rhabdospermum*

4B. Calyx-lobes spreading at anthesis; seeds nearly smooth; stamens much shorter than the calyx-lobes; capsules with 2 horizontally spreading beaks. . . . . 5. *C. ramosum*

3B. Plants glabrous, sometimes sparsely hairy in the leaf-axils.

7A. Seeds with few longitudinal ribs; stamens 8.

8A. Ribs of seeds nearly smooth; calyx-lobes spreading; stamens very short. . . . . 6. *C. kamtschaticum*

8B. Ribs of seeds tubercled.

9A. Stamens much shorter than the calyx-lobes, about 0.5 mm. long; disc prominent. . . . . 7. *C. echinus*

9B. Stamens longer than to slightly shorter than the calyx-lobes, 1–3 mm. long; disc not prominent.

10A. Rosulate leaves short-petioled or nearly sessile; seeds ribbed, the ribs wrinklelike with low tubercles. . . . . . . 8. *C. fauriei*

10B. Rosulate leaves distinctly petioled; seeds with prominently tubercled ribs. . . . . 9. *C. macrostemon*

7B. Seeds with a single longitudinal rib on one side, minutely papillose; stamens 4. . . . . 10. *C. grayanum*

1B. Leaves always alternate.

11A. Seeds minutely mammillate; flowers green or greenish yellow.

12A. Leaves with rounded teeth.

13A. Stolons epigeal, leafy. . . . . 11. *C. flagelliferum*

13B. Stolons hypogeal, naked. . . . . 12. *C. tosaense*

12B. Leaves with depressed teeth; stolons absent; plants bearing bulbils at base. . . . . 13. *C. japonicum*

11B. Seeds quite smooth, not tubercled; flowers clear yellow; leaves short-toothed. . . . . 14. *C. alternifolium* var. *sibiricum*

1. Chrysosplenium maximowiczii Fr. & Sav. MUKAGO-NEKO-NO-ME, TAMA-NEKO-NO-ME. Stems erect, 3–15 cm. long, sparsely pubescent while young, nearly glabrous later; stolons with a small resting bulbil at the tip from which the new plant arises each year; leaves all cauline, 2 to 3 pairs, ovate-orbicular, 4–18 mm. long, 5–15 mm. wide, short-toothed, petiolate; flowers yellow-green; calyx-lobes erect, ovate-orbicular, about 1 mm. long; stamens 8, to 1 mm. long, shorter than the calyx-lobes, the disc indistinct; capsule-valves suberect, unequal, slightly compressed laterally; seeds ovoid-globose, about 0.6 mm. long, mammillate on the longitudinal ribs.——Mar.–Apr. Wet places; Honshu (Sagami, Awa, Izu, Shimotsuke Prov.); rare.

2. Chrysosplenium album Maxim. *C. stamineum* Fr. & Sav.; *C. beauverdii* Terrac.; *C. naminoense* Honda——SHIRO-BANA-NEKO-NO-ME, HANA-NEKO-NO-ME. Stems 5–15 cm. long, long-pubescent; stolons elongate, leafy; leaves opposite, flabellate-suborbicular, 4–10 mm. long, 5–12 mm. wide, often slightly variegated on the upper surface, obsoletely 5- to 9-toothed, subtruncate and abruptly decurrent to a petiole; flowers rather few, dense, white in anthesis, pale green in fruit; calyx-lobes erect, 3–5 mm. long, broadly lanceolate to narrowly oblong; stamens erect, the anthers dark purple.——Mar.–May. Woods in foothills; Honshu (Iwaki Prov. and southw.), Shikoku, Kyushu.

Var. **stamineum** (Fr. & Sav.) Hara. A smaller plant with leaves usually 5-toothed. Western part of the distribution.

Var. **flavum** Hara. KIBANA-HANA-NEKO-NO-ME. Flowers yellow; sepals ovate, obtuse to shortly acute; anthers yellowish, often tinged with red.——Honshu (Tōkaidō Distr.); rare.

3. Chrysosplenium pilosum Maxim. var. **sphaerospermum** (Maxim.) Hara. *C. pilosum* var. *fulvum* (Terrac.) Hara; *C. multicaule* Fr. & Sav.; *C. fulvum* Terrac.; *C. vidalii* Fr. & Sav.——KOGANE-NEKO-NO-ME. Soft pubescent herb; stems 5–10 cm. long, with well-developed leafy stolons; radical leaves usually withering before anthesis, the lower cauline short-petiolate, flabellate, 4–15(–20) mm. long, 5–12(–20) mm. wide, with few short teeth; flowers few, yellow or rarely greenish yellow at anthesis, pale green in fruit; calyx-lobes ovate-orbicular or depressed-rounded, erect, 1–1.5 mm. long; stamens 8, slightly shorter than the calyx-lobes; capsules and seeds nearly the same as in the preceding species.——Mar.–May. Wet places; Honshu, Shikoku, Kyushu.

4. Chrysosplenium rhabdospermum Maxim. *C. doianum* Ohwi——TSUKUSHI-NEKO-NO-ME. Radical leaves petiolate, depressed-orbicular, 5–10 mm. long, 5–15 mm. wide, undulate-toothed; scapes 5–10 cm. long, loosely few-flowered; calyx-tube cup-shaped, the lobes pale green, erect, ovate to depressed-rounded; stamens slightly longer than to as long as the calyx-lobes, the anthers yellow; capsules compressed, erect, the valves unequal; seeds mammillate on the ribs.——Mar.–Apr. Kyushu.

5. Chrysosplenium ramosum Maxim. *C. ramosum* var. *atrodiscum* (Suto) Hara; *C. yesoense* Fr. & Sav.; *C. crenulatum* var. *atrodiscum* Suto; *C. microphyllum* Tatew. & Suto——MARUBA-NEKO-NO-ME. Thinly long-pubescent; stems 7–15 cm. long; stolons very elongate, branched, with widely spaced pairs of orbicular leaves; radical leaves withering before anthesis, the cauline petiolate, ovate-orbicular or suborbicular, 5–15 mm. long, 6–15 mm. wide, broadly cuneate at base, with subrounded teeth; flowers green, purple-green or purple-brown; calyx-lobes horizontally spreading, depressed-rounded; stamens inserted on the margin of a prominent disc, much shorter than the calyx-lobes; capsules subinferior, the valves horizontally spreading, nearly equal; seeds narrowly ovoid-ellipsoidal, about 1 mm. long, obsoletely ribbed or nearly smooth.——May–June. Hokkaido, Honshu (Kinki Distr. and northw.).——s. Sakhalin, Korea, Manchuria and Amur.

6. Chrysosplenium kamtschaticum Fisch. *C. costulatum* Franch.; *C. corrugatum* Franch.; *C. nodulosum* Franch.; *C. aomorense* Franch.; *C. flabellatum* Terrac.; *C. kamtschaticum* var. *costulatum* (Franch.) Suto; *C. kamtschaticum* var. *aomorense* (Franch.) Hara——CHISHIMA-NEKO-NO-ME. Slender glabrous herb; stems 5–15 cm. long; stolons much elongate, often with a pair of leaves at wide intervals, with a tuft of rosulate leaves at apex; radical leaves rosulate, ovate-orbicular, short-petiolate, undulately toothed; leaves on stolons orbicular, 5–15 mm. long and as wide, abruptly decurrent at base, short-petiolate, sometimes nearly sessile; flowers few to more than 10, yellowish green or brownish green; calyx-lobes spreading at anthesis, orbicular, about 1 mm. in diameter, the disc prominent; stamens marginal, shorter than the calyx-lobes; capsules ascending, the valves unequal; seeds longitudinally ribbed and finely cross-barred on the ribs.——Apr.–May. Hokkaido, Honshu (Kinki Distr. and northw.).——Kuriles, Sakhalin, and Kamchatka.

7. Chrysosplenium echinus Maxim. *C. echinulatum* Fr. & Sav.——IWA-NEKO-NO-ME. Glabrous stoloniferous herb without radical leaves in anthesis; stems 5–12 cm. high; leaves distinctly petioled, orbicular, 5–20 mm. in diameter, subtruncate at base, with undulate incurved teeth; flowers green; calyx-lobes spreading, depressed-orbicular or broadly ovate, 1–1.5 mm. long; stamens about 0.5 mm. long, on the margin of the prominent disc; capsule-valves ascending, unequal; seeds about 1 mm. long, longitudinally ribbed, with elongate terete tubercles on the ribs.——Apr.–May. Honshu (centr. distr. and westw.), Shikoku, Kyushu.

8. Chrysosplenium fauriei Franch. HOKURIKU-NEKO-NO-ME. Glabrous stoloniferous herb; radical leaves ovate-orbicular, 1–2.5 cm. long and as wide, nearly sessile or short-petiolate; leaves of stolons orbicular to flabellate-orbicular, short-petiolate or sessile, with incurved teeth; flowers rather many, yellow; calyx-lobes erect, elliptic to orbicular, 1.5–2 mm. long; stamens 8, 2.5–3 mm. long, longer than the calyx-lobes; capsules ascending, the valves unequal; seeds about 0.8 mm. long, longitudinally ribbed with low mammillae on the ribs.——May. Honshu (Japan Sea side from Echigo to Izumo Prov.).

Var. **kiotense** (Ohwi) Ohwi. *C. kiotense* Ohwi——BO-TAN-NEKO-NO-ME. Stamens shorter than the calyx-lobes, and the seeds with rugose ribs.——Honshu (centr. distr. and westw.).

9. Chrysosplenium macrostemon Maxim. *C. discolor* Fr. & Sav.; *C. calcitrapa* Franch.; *C. macrostemon* var. *calcitrapa* (Franch.) Hara——IWA-BOTAN. Stems 5–15 cm. long; stolons elongate after anthesis; radical leaves often persistent at anthesis, long-petiolate, broadly ovate, 1–2.5 cm. long, 0.8–2 cm. wide, with incurved teeth; leaves of stolons ovate-orbicular to ovate-elliptic, 5–40 mm. long, 4–35 mm. wide; flowers rather numerous, yellow-green; calyx-lobes nearly erect at anthesis, elliptic, 1–1.8 mm. long; stamens longer than the calyx-lobes, 2–3 mm. long, the anthers yellow; capsules ascending, the valves unequal; seeds about 0.8 mm. long, longitudinally ribbed with a series of clavate tubercles on the ribs.——Mar.–May. Honshu (Kantō Distr. and westw.), Shikoku, Kyushu.

Var. **atrandrum** Hara. YOGORE-NEKO-NO-ME. Flowers with dark-colored anthers.——Occurs with the typical phase.

Var. **shiobarense** (Franch.) Hara. *C. discolor* var. *shiobarense* (Franch.) Suto; *C. viridescens* Suto; *C. discolor* var. *viridescens* Suto——NIKKŌ-NEKO-NO-ME. Radical leaves usually withering before anthesis; stolons much-elongate at anthesis, with smaller, orbicular leaves; anthers red-purple, rarely yellow.——Apr.–May. Honshu (Kantō Distr. and westw.), Shikoku, Kyushu.

10. Chrysosplenium grayanum Maxim. *C. nipponicum* Fr. & Sav.; *C. dickinsii* Fr. & Sav.; *C. grayanum* var.

polystichum Fr. & Sav.; *C. grayanum* var. *nipponicum* (Fr. & Sav.) Fr. & Sav.; *C. grayanum* var. *dickinsii* (Fr. & Sav.) Fr. & Sav.——NEKO-NO-ME-SŌ, MIZU-NEKO-NO-ME. Glabrous, without radical leaves; stems 5–20 cm. long; stolons elongate after anthesis, sparsely leafy, without a tuft of rosulate leaves at apex; leaves short-petiolate, broadly ovate to ovate-orbicular, 5–20 mm. long, 5–18 mm. wide, with incurved teeth; flowers more than 10, yellowish green; calyx-lobes suborbicular, about 1 mm. long, erect, concave on back; stamens 4, about 0.5 mm. long; capsules ascending, the valves unequal; seeds broadly ovoid, about 0.7 mm. long, 1-ribbed, minutely mammillate.——Apr.–May. Hokkaido, Honshu, Shikoku, Kyushu.——Korea, s. Kuriles, and China.

11. Chrysosplenium flagelliferum F. Schmidt. TSURU-NEKO-NO-ME. Green sparsely pubescent herb; radical leaves sometimes persistent until anthesis; stems 5–20 cm. long; stolons leafy; leaves alternate, cordate-orbicular, 3–10 mm. long, 4–12 mm. wide, with rounded teeth, slightly cordate; leaves on stolons similar to the radical ones but larger, distinctly cordate, to 4 cm. long and as wide, long-petiolate; flowers yellowish green; calyx-lobes spreading, depressed-orbicular or elliptic, 1–2 mm. long; stamens short; capsules subinferior, subtruncate; seeds ovoid, about 0.6 mm. long, 1-ribbed, minutely mammillate.——Apr.–May. Hokkaido, Honshu (n. Kinki Distr. and northw.), Shikoku.——Korea, Manchuria, Amur, Ussuri, Sakhalin, and Kuriles.

12. Chrysosplenium tosaense (Makino) Makino. *C. flagelliferum* var. *tosaense* Makino——TACHI-NEKO-NO-ME. Closely resembling the preceding species; stolons subterranean; radical leaves long-petioled, reniform-orbicular, 5–17 mm. long, 5–25 mm. wide, 5- to 9-toothed, slightly whitish beneath; stems erect, 5–12 cm. high, 3- to 4-angled, leafless or 1- to 2-leaved; flowers pale green; sepals spreading, broadly ovate, obtuse; stamens short; ovary inferior, embedded in the disc; capsules convex; seeds 1-ribbed, minutely mammillate.——Mar.–May. Honshu (Kantō Distr. and westw.), Shikoku, Kyushu; rather common.

13. Chrysosplenium japonicum (Maxim.) Makino. *C. alternifolium* sensu auct. Japon., non L.; *C. alternifolium* var. *japonicum* Maxim.; *C. alternifolium* var. *papillosum* Fr. & Sav. ——YAMA-NEKO-NO-ME. Thinly soft-pubescent; stems 10–20 cm. long, thickened and bearing bulbils at base; stolons absent; radical leaves prominent, long-petioled, reniform-orbicular, 5–20 mm. long, 8–30 mm. wide, cordate, very shallowly toothed; flowers rather many, pale green; calyx-lobes spreading, depressed-orbicular, about 1 mm. long; stamens very short, about 0.5 mm. long; capsules subinferior, cup-shaped, subtruncate; seeds broadly ovoid, about 0.6 mm. long, 1-ribbed, minutely mammillate.——Mar.–May. Hokkaido (sw. distr.), Honshu, Shikoku, Kyushu; rather common.——Korea and Manchuria.

14. Chrysosplenium alternifolium L. var. **sibiricum** (Steph.) Ser. *C. sibiricum* Steph.; *C. alternifolium* sensu auct. Japon., non L.——EZO-NEKO-NO-ME. Thinly pubescent; stems 5–12 cm. long; stolons subterranean; radical leaves long-petioled, reniform-orbicular, 3–6 mm. long, 5–12 mm. wide, shallowly toothed; flowers yellow in anthesis; calyx-lobes spreading, subrounded, 1–1.5 mm. long; stamens 8, about 0.5 mm. long; capsules nearly inferior; seeds ellipsoidal, glabrous, smooth.——Apr.–June.——Bogs; Hokkaido (ne. distr.).——Sakhalin, Kuriles, n. China, Siberia, Himalayas, and N. America.

8. TIARELLA L. ZUDA-YAKUSHU ZOKU

Perennial herbs with short-creeping rhizomes; radical leaves long-petiolate, simple, sometimes 3-foliolate, the cauline leaves few, with small stipules; racemes or cincinni simple or sparsely branched, the bracts minute; calyx 5-lobed, subpetaloid, the calyx-tube cup-shaped, adnate to the ovary at base; petals 5, small; stamens 10, filaments filiform; ovary nearly superior, 1-locular, with 2 parietal placentae; styles slender, the stigma punctiform; capsules nearly superior, thinly chartaceous; carpels very unequal; seeds few, smooth.——Few species, in N. America, one in e. Asia and the Himalayas.

1. **Tiarella polyphylla** D. Don. ZUDA-YAKUSHU. Rhizomes creeping, branched; stolons subterranean; radical leaves few, cordate-orbicular, 2–7 cm. in diameter, shallowly 5-lobed, obtusely toothed, long-hairy and glandular-puberulent, the stipules membranous, brown, the petioles 2–10 cm. long; scapes 10–40 cm. long, glandular-puberulent, with 2 or 3 short-petioled cauline leaves; racemes often with a short branch near base, the pedicels spreading, 3–7 mm. long; flowers rather numerous, white, nodding; calyx-lobes 1–2 mm. long, narrowly ovate; petals subulate-linear, slightly longer than the calyx-lobes; stamens exserted; capsule-valves very unequal, 7–12 mm. long.——June–Aug. Damp woods in mountains; Hokkaido, Honshu (Kinki Distr and northw.), Shikoku.——China, Formosa, and Himalayas.

9. MITELLA L. CHARUMERU-SŌ ZOKU

Low usually hirsute perennial herbs with creeping rhizomes; radical leaves long-petiolate, simple, ovate, incised to lobulate, the stipules scarious; stems leafless or few-leaved; inflorescence racemose, terminal, bracteate; flowers small; calyx 5-lobed, the calyx-tube cup-shaped or campanulate, more or less adnate to the ovary; petals 5, on the margin of the calyx-tube, often pinnatiparted, sometimes simple or absent; stamens 5 or 10; ovary 1-locular, superior or subinferior; styles 2; seeds ovoid, small, often tubercled.——About 20 species, in N. America, Siberia, Japan, and Formosa.

1A. Ovary nearly superior, nearly free from the calyx-tube.
  2A. Stamens 5; petals subulate-linear, entire; stems 1- to 3-leaved. . . . . 1. M. integripetala
  2B. Stamens 10; petals pinnatiparted; stems scapose or 1-leaved. . . . . 2. M. nuda
1B. Ovary inferior, adnate at least partially to the calyx-tube.
  3A. Petals absent; scapes 2–7 cm. long, 1- to 3-flowered; calyx-tube campanulate. . . . . 3. M. doiana
  3B. Petals present; scapes 10–50 cm. long, often many-flowered; calyx-tube cup-shaped to broadly obconical.
    4A. Stamens attached to the disc; leaves wider than to as wide as long; racemes few-flowered; style undivided. . . . . . 4. M. pauciflora
    4B. Stamens attached to the base of the petal; leaves usually longer than wide; racemes rather many-flowered; style often lobed.
      5A. Petioles and the underside of leaves glabrous. . . . . 5. M. acerina
      5B. Petioles and leaves hirsute.
        6A. Calyx-lobes erect with an ascending to slightly recurved tip.
          7A. Petals entire, smooth, subulate-linear; calyx-lobes not ciliate; seeds papillose. 6. M. leiopetala
          7B. Petals pinnatiparted, glandular-roughened; calyx-lobes ciliate; seeds smooth or papillose.
            8A. Leaves obtuse or subobtuse; seeds not papillose. . . . . 7. M. stylosa
            8B. Leaves acute to subacute; seeds papillose. . . . . 8. M. makinoi
        6B. Calyx-lobes spreading in anthesis, often reflexed at apex.
          9A. Rhizomes short, thickened, with a tuft of leaves at the apex.
            10A. Leaves broadly ovate-orbicular, the petioles eglandular-hirsute; stipules usually glabrous.
              11A. Calyx-lobes not ciliate; petals mostly 7-parted. . . . . 9. M. koshiensis
              11B. Calyx-lobes ciliate; petals 9- to 11-parted. . . . . 10. M. furusei
            10B. Leaves usually elongate-ovate, the petioles glandular-hirsute; stipules long-ciliate. . . . . 11. M. japonica
          9B. Rhizomes long-creeping, rather slender, loosely leaved.
            12A. Seeds papillose; leaves broadly ovate; calyx-lobes deltoid; styles 2, 2-lobed. 12. M. kiusiana
            12B. Seeds not papillose; leaves narrowly ovate; calyx-lobes depressed-deltoid; styles 2, undivided. . . . . . . 13. M. yoshinagae

1. **Mitella integripetala** H. Boiss. EZO-NO-CHARUMERU-SŌ. Stoloniferous; radical leaves membranous, long-petiolate, deltoid-ovate, cordate, shallowly 3- to 5-lobed, toothed, scattered-pilose on upper side, glabrescent beneath; flowering stems with 1–3 short-petiolate leaves; racemes rather many-flowered, the pedicels less than 3 mm. long, minutely glandular-puberulent; calyx-tube cup-shaped, the lobes lanceolate, about 2 mm. long, acute; petals linear, 4–5 mm. long, entire, minutely glandular; styles 2, subulate, entire.——Hokkaido (sw. distr.), Honshu (n. distr.); rare.

2. **Mitella nuda** L. MARUBA-CHARUMERU-SŌ. Rhizomes slender, elongate; stolons filiform; radical leaves few, ovate-orbicular, 15–35 mm. long and as wide, often subcordate, obtuse, obsoletely incised and short-toothed, scattered-pilose on both sides, the petioles 5–10 cm. long, slender, glandular-hairy, the stipules short; flowering stems 15–25 cm. long, short glandular-pubescent, often with a single small leaf; ra-

cemes few-flowered; calyx-lobes deltoid-ovate, acute, spreading, subrecurved at the tip; petals 5, pinnatiparted, not glandular; seeds glabrous, smooth, black.——June. Woods in mountains; Hokkaido, Honshu (centr. distr.); rare.——n. Korea, Manchuria, Ussuri to Siberia, and N. America.

3. **Mitella doiana** Ohwi. HIME-CHARUMERU-SŌ. Rhizomes creeping; radical leaves tufted, ovate-cordate, 8–13 mm. long, 7–10 mm. wide, shallowly 3-lobed, acute, deeply cordate, hirsute on both sides, the petioles 1–3 cm. long, hirsute, the stipules scarious, entire, glabrescent; flowering stems 2–7 cm. high, hirsute and glandular-pubescent, 1- to 3-flowered toward the top; calyx-tube subcampanulate, the lobes ovate-deltoid, erect in anthesis, with a slightly recurved tip; petals absent; styles 2, terete, about 0.8 mm. long, scarcely thickened at apex; seeds papillose.——May–July. Kyushu (Yakushima); rare.

4. **Mitella pauciflora** Rosend. KO-CHARUMERU-SŌ. Rhizomes rather slender, elongate; leaves all radical, broadly

ovate to ovate-orbicular, 2–5 cm. long, 2.5–6 cm. wide, deeply cordate, hirsute and minutely glandular-puberulous on both sides, subacute to acute, shallowly 5-lobed, incised and toothed; scapes 15–25 cm. long, glandular-puberulent, coarsely hairy at the base, 2- to 10-flowered, the pedicels short; calyx-lobes depressed-deltoid, subacute, spreading, subrecurved at the tip; petals pinnatiparted, minutely glandular, the disc prominent; stamens inserted on the margin of the disc, very short; style very short, entire, the stigma punctiform; seeds often glandular.——Apr.–June. Along streams and ravines in mountains; Honshu, Shikoku, Kyushu.

5. Mitella acerina Makino. Momiji-charumeru-sō. Rhizomes elongate, branched, creeping; leaves broadly ovate, 4–12 cm. long, 4–10 cm. wide, acute to acuminate, deeply cordate, acutely 5- to 7-lobed, incised and toothed, thinly pilose above, glabrous beneath, the petioles 15–30 cm. long, glabrous, the stipules glabrous; scapes glabrous, 20–40 cm. long; racemes 10–20 cm. long, rather densely short-glandular, many-flowered; calyx-lobes erect in anthesis, recurved at the tip, ovate-deltoid; petals 3-parted, minutely glandular; styles 2, very short, the stigma thickened, indistinctly 2-lobed; seeds not papillose.——May–June. Wet places along ravines and streams in mountains; Honshu (Japan Sea side of centr. and Kinki Distr.).

6. Mitella leiopetala Ohwi & Okuyama. *M. japonica* var. *integripetala* Makino; *M. stylosa* var. *integripetala* (Makino) Ohwi; *M. stylosa* forma *integripetala* (Makino) Hara ——Hariben-charumeru-sō, Takimi-charumeru-sō. Rhizomes somewhat thickened, creeping, with a tuft of leaves at the apex; radical leaves ovate, 4–5 cm. long, 3–4 cm. wide, cordate, subacute to obtuse, obsoletely lobulate and irregularly toothed, hirsute on upper side, sparsely papillose-pilose beneath, the petioles 7–10 cm. long, purplish, hirsute, the stipules white-scarious, oblong; scapes terete, about 25 cm. long, whitish papillose-pilosulous, white-hairy at base; racemes about 7 cm. long, many-flowered, the bracts minute, the pedicels shorter than the calyx; calyx obconical, brown-purple, 3.5–4 mm. long, the lobes about 1.2 mm. long, deltoid, erect, the tip recurved, glandular externally; petals linear-subulate, 2.5–3 mm. long, entire or very rarely with a short lobe on one or both sides; styles very short, the stigma 4-lobed; seeds papillose.——Apr. Wet places along ravines and streams in mountains; Honshu (Ise and Mino Prov.); rare.

7. Mitella stylosa H. Boiss. *M. longispica* Makino, pro parte; *M. japonica* Makino, pro parte, non Maxim.——Charumeru-sō. Rhizomes rather short, thickened; radical leaves tufted, broadly ovate, 4–8 cm. long, 3–5 cm. wide, obtuse to subacute, deeply cordate, shallowly and irregularly toothed, long-hirsute and minutely glandular-puberulent on both sides; scapes 30–50 cm. high, hirsute below; racemes 15–30 cm. long, many-flowered, short glandular-puberulent; calyx-lobes erect, ovate-deltoid; petals 3-parted, the segments narrowly linear, glandular-dotted; styles very short, thickened, the stigma shallowly 2- to 4-lobed; seeds not papillose.——Apr.–May. Wet places along streams and ravines in foothills; Honshu (Kantō Distr. and westw.), Kyushu.

8. Mitella makinoi Hara. *M. japonica* Makino, pro parte, non Maxim.——Shikoku-charumeru-sō. Rhizomes ascending, thickened, with a tuft of radical leaves at the tip; leaves broadly ovate, 2–7 cm. long, 2–6 cm. wide, acute, cordate, 5- to 7-lobulate, toothed, coarsely hirsute on upper side, with rather short spreading glandular hairs beneath; scapes 20–40 cm. long, glandular-hairy; racemes 15–25 cm. long,

many-flowered; calyx-lobes erect; petals pinnately parted, glandular-dotted; style short, the stigma thickened, retuse; seeds papillose.——Apr.–May. Shikoku.

9. Mitella koshiensis Ohwi. Koshi-no-charumeru-sō. Rhizomes thickened; leaves tufted, broadly ovate to orbicular-ovate, 3–6 cm. long and nearly as wide, cordate, acute, shallowly 5-lobed, incised, long-hirsute on both sides, the petioles 3–10 cm. long, with coarse reddish hairs, the stipules scarious, glabrous; scapes long-hirsute; racemes glandular-puberulous, rather many-flowered; calyx-lobes depressed-deltoid, subacute, spreading in anthesis, with a slightly recurved tip; petals reddish, pinnately parted, the segments linear, glandular-dotted; stamens very short; styles very short, the stigma shallowly 2-lobed, the lobes retuse; seeds not papillose.——May. Honshu (Japan Sea side, Etchu and Echigo Prov.).

10. Mitella furusei Ohwi. Mikawa-charumeru-sō. Rhizomes creeping, densely leaved near the top; leaves broadly ovate, 3–6 cm. long and nearly as wide, slightly lobulate, irregularly toothed, acute, pilose, the petioles with long multi-cellular hairs, 5–10 cm. long; scapes 30–45 cm. long, pilose; racemes simple, loosely many-flowered, glandular on the axis and pedicels; flowers greenish; calyx-lobes 5, deltoid, spreading at anthesis, erect in fruit, uniformly glandular-pilose on back and on margin, acute, often shallowly lobulate; petals 5, purplish, spreading, 5–6 mm. long, glandular-roughened, pinnately divided, with 4–5 pairs of filiform lobes; stamens 5, anthers yellow, nearly sessile, obsoletely lobulate; seeds fusiform, about 1.2 mm. long, not papillose.——May. Honshu (Mikawa Prov.).

11. Mitella japonica (Sieb. & Zucc.) Maxim. *Mitellopsis japonica* Sieb. & Zucc.; *M. triloba* Miq.——Ō-charumeru-sō. Rhizomes thickened; leaves tufted, ovate to narrowly so, 5–10 cm. long, 3–7 cm. wide, acute to short acuminate, cordate, usually shallowly 5-lobed and incised, long-hirsute on upper side, glandular-hairy beneath, the nerves sometimes without glandular hairs, the petioles with spreading glandular hairs, the stipules ciliate; scapes 20–35 cm. long, glandular pilose; racemes many-flowered; calyx-lobes depressed-deltoid, acute; petals with 5–7 linear-subulate glandular-dotted segments; styles very short, the stigma thickened, the lobes retuse; seeds not papillose.——Apr.–May. Wet places along streams and ravines in mountains; Honshu (Kinki Distr. and westw.), Shikoku, Kyushu.

12. Mitella kiusiana Makino. Tsukushi-charumeru-sō. Rhizomes elongate, branched; leaves rather loosely arranged, broadly ovate, 2–7 cm. long, 1–6 cm. wide, acute, cordate, acutely 5-lobulate, and irregularly toothed, long-hirsute above and on nerves beneath, the petioles 4–10 cm. long, brown-hirsute, the stipules glabrescent; scapes 15–20 cm. long, hirsute below, glandular-pilosulous above; racemes rather many-flowered; calyx-lobes deltoid, acute, spreading, with a slightly recurved tip; petals pinnately parted into linear-subulate glandular-dotted segments; styles 2, bifid, the lobes entire or somewhat retuse; stigma not thickened; seeds papillose on back.——Apr.–May. Along ravines and streams in mountains; Kyushu.

13. Mitella yoshinagae Hara. Tosa-charumeru-sō. Rhizomes elongate, branched; leaves rather loosely arranged, ovate, 3–9 cm. long, 2–7 cm. wide, acute, cordate, 3- to 4-lobulate on each side and incised, long-hirsute on upper side and on nerves beneath, minutely glandular-pilose beneath; scapes hirsute on upper side and on nerves beneath, minutely glandular-pilose beneath; scapes 20–30 cm. long, glandular-pilosulous, many-flowered; calyx-lobes depressed-deltoid, acuminate, spreading, with a slightly recurved tip; petals pinnately parted into linear-subulate glandular-dotted segments; styles 2, short, the stigmas spreading, depressed-globose, sometimes retuse; seeds not papillose.——Apr. Shikoku.

10. PARNASSIA L.   Umebachi-sō Zoku

Glabrous perennial herbs; leaves radical, long-petiolate, orbicular-cordate, entire; stems scapiform, unbranched, angled, 1-flowered, naked to several leaved; flowers white, erect; calyx-lobes 5; petals 5; stamens 5, alternate with the staminodes; ovary superior or subinferior, 1-locular; style very short; capsules loculicidally dehiscent, 3- to 4-valved; seeds many.——More than 10 species, in the temperate to northern regions of the N. Hemisphere.

1A. Stems 2- to 8-leaved; petals fimbriate; staminodes 3-fid, the lobes terminating in a globose gland. 1. *P. foliosa*
1B. Stems 1-leaved; petals entire.
  2A. Flowers large, 2–2.5 cm. across; staminodes digitately cleft into many filiform segments, each ending in a globose gland; anthers
    grayish white; pollen pale yellow. . . . . 2. *P. palustris*
  2B. Flowers small, 8–10 mm. across; staminodes 3- to 5(–7)-cleft, the lobes subulate, obtuse, not gland-tipped; anthers dark gray; pollen
    dark brown. . . . . 3. *P. alpicola*

1. Parnassia foliosa Hook. f. & Thoms. var. **nummularia** (Maxim.) T. Itō. *P. nummularia* Maxim.——Shira-hige-sō. Glabrous green perennial; stems several, tufted, 15–30 cm. long; leaves deeply cordate, rounded, mucronate, 15–30 mm. in diameter; radical leaves long-petiolate, the cauline sessile and clasping; flowers 2–2.5 cm. across; petals ovate, fimbriate, distinctly clawed at base; anthers about 1.5 mm. long, the filaments spreading in anthesis.——Aug.–Sept. Wet places in mountains; Honshu (centr. distr. and westw.), Shikoku, Kyushu.

Var. **japonica** (Nakai) Ohwi. *P. japonica* Nakai; *P. yudzuruana* T. Itō——Ō-shira-hige-sō. Plants larger than in the preceding with leaves to 6 cm. across, the flowers 3–3.5 cm. across, the petals about 15 mm. long, the staminodes about 5 mm. long.——Aug.–Sept. Wet places in mountains; Honshu (Mount Togakushi and Mount Ontake in Shinano).

2. Parnassia palustris L. *P. mucronata* Sieb. & Zucc.——Umebachi-sō. Glabrous perennial with short rhizomes; stems 10–50 cm. long, 1-leaved, usually few together; leaves orbicular-cordate, sometimes broadly ovate, 15–35 mm. long and nearly as wide, rounded to obtuse, sometimes subacute, the radical leaves long-petiolate, the cauline sessile and clasp-ing; flowers 2–2.5 cm. across, white; petals broadly ovate to elliptic, scarcely clawed at base, rounded at apex; anthers broadly ovoid, 2–3 mm. long, the staminodes 12- to 22-cleft; capsules 10–12 mm. long, broadly ovoid.——Aug.–Oct. Wet places; Hokkaido, Honshu, Shikoku, Kyushu.——Temperate and northern regions of the N. Hemisphere.

Var. **tenuis** Wahlenb. *P. palustris* var. *alpina* Drude——Ko-umebachi-sō. Small alpine phase with 9- to 11-cleft staminodia.——Alpine regions of Hokkaido and Honshu (centr. distr.).——Cold regions and high mountains of the Old World.

3. Parnassia alpicola Makino. *P. simplex* Hayata——Hime-umebachi-sō. Small perennial; stems 1-leaved, 5–15 cm. long, solitary or few together; leaves reniform-orbicular, 7–15 mm. long and as wide, the radical ones long-petiolate, the cauline sessile and clasping; flowers white, 8–10 mm. across; petals 4–6 mm. long, ovate or oblong, entire, with distinct ascending claws at base; anthers orbicular, about 0.5 mm. long, the staminodes shortly 3- to 8-cleft; capsules broadly ovoid, 5–6 mm. long.——Aug. Wet places in alpine regions; Honshu (centr. and n. distr.); rare.

11. SCHIZOPHRAGMA Sieb. & Zucc.   Iwagarami Zoku

Scandent deciduous woody vines climbing by means of aerial roots; leaves opposite, long-petiolate, simple, usually toothed, exstipulate; inflorescence corymbose, some of the marginal flowers neutral and one of the sepals large and petallike, the bisexual flowers small, many, 4- to 5-merous; calyx-teeth small, deltoid; petals 5, caducous; stamens 10; style columnar, short, thickened, the stigma 4- to 5-lobed, capitate; ovary nearly inferior; capsules narrowly obconical, 10-ribbed, splitting between the ribs; seeds small, linear, many.——Few species, in e. Asia.

1. Schizophragma hydrangeoides Sieb. & Zucc. Iwagarami. Stems or branches to 10 m. or more long; leaves suborbicular to broadly ovate, 5–12 cm. long and as much wide, abruptly short-acuminate, rounded to cordate, coarsely acute-toothed, glabrous above, glaucous and with tufts of axillary hairs or glabrous beneath, the petioles 3–10 cm. long; inflorescence rather large, densely many-flowered, the enlarged neutral calyx-lobes deltoid-ovate, 2–3 cm. long; petals of bi-sexual flowers narrowly ovate, about 2 mm. long, caducous; capsules 6–8 mm. long inclusive of a rather elongate style, longitudinally ribbed.——June–July. Woods and thickets in mountains; Hokkaido, Honshu, Shikoku, Kyushu; rather common.

Var. **concolor** Hatus. Teriha-iwagarami. Leaves green and lustrous on both sides.——Kyushu.

12. DEINANTHE Maxim.   Gimbai-sō Zoku

Coarse perennial herbs with creeping woody rhizomes; stems erect, leafy; leaves opposite, rather large, petiolate, coarsely hirsute, broadly obovate, toothed, membranous, usually bilobed; inflorescence a terminal corymb, the marginal flowers often neutral, the bisexual flowers white, the bracts persistent; calyx-tube short, obconical, the lobes rather large, persistent; petals 5, imbricate in bud, obovate, deciduous; stamens very many, inserted on the margin of the disc; ovary subinferior, imperfectly 5-locular; style solitary, grooved; capsules subsuperior, subglobose, deflexed; seeds many, small, with a short taillike wing at each end.——Two species, in e. Asia.

1. Deinanthe bifida Maxim. GIMBAI-SŌ.   Stems 40–70 cm. long, simple; lower leaves small, scarious, scalelike, the normal leaves 10–20 cm. long, 6–12 cm. wide, toothed and sometimes obscurely incised, shallowly 2-fid, the petiole 1–10 cm. long; inflorescence glabrescent, few- to more than 20-flowered; sterile flowers with 2 or 3 enlarged calyx-lobes; calyx-lobes of bisexual flowers elliptic, 6–8 mm. long, obtuse, green; petals obovate-orbicular, about 10 mm. long; capsules 8–10 mm. across, the style conical.——July–Aug. Woods in mountains in warmer regions; Honshu (s. Kantō Distr. and southwestw.), Shikoku, Kyushu.

13. PLATYCRATER Sieb. & Zucc.   BAIKA-AMA-CHA ZOKU

Deciduous shrubs; leaves opposite, petiolate, toothed, exstipulate; inflorescence loose, cymose, the marginal neutral flowers with enlarged 3- to 5-lobed calyx; bisexual flowers rather large, 4-merous; calyx-tube narrowly obconical, the calyx lobes lanceolate, persistent, acute; petals ovate, white; stamens very many; styles 2, free; ovary inferior, 2-locular; capsules narrowly obconical; seeds small, many, with a taillike wing at each end.——Two species, in Japan and China.

1. Platycrater arguta Sieb. & Zucc.   *P. serrata* Makino, excl. basionym——BAIKA-AMA-CHA.   Sparsely and coarsely hairy; leaves rather thinly membranous, oblong to broadly lanceolate, 10–15 cm. long, 3–5(–7) cm. wide, short-toothed, acuminate, paler beneath, the petioles 5–30 mm. long, winged above; inflorescence terminal, several-flowered, the pedicels slender; calyx in neutral flowers shallowly 3- to 5-lobed, the calyx-lobes in bisexual flowers 4–7 mm. long; petals about 10 mm. long; capsules 5–8 mm. long; styles persistent, 7–8 mm. long.——July–Aug. Honshu (Totomi, Yamato, and Kii Prov.), Shikoku, Kyushu.

14. CARDIANDRA Sieb. & Zucc.   KUSA-AJISAI ZOKU

Sparsely hirsute, suffrutescent herb; leaves alternate, rarely nearly opposite, broadly lanceolate to oblong, acutely toothed, exstipulate; inflorescence a terminal corymb with few neutral peripheral flowers; neutral flowers with 3, often colored, petaloid calyx-lobes; bisexual flowers small, many, white or rose-colored; calyx-tube ellipsoidal, the teeth 4–5, deltoid; petals 5, small, imbricate in bud; stamens many; ovary nearly inferior, imperfectly 3-locular; styles 3; capsules ovoid; seeds many, small, with a small narrow wing at each end.——Few species, in e. Asia.

1. Cardiandra alternifolia Sieb. & Zucc. KUSA-AJISAI.   Stems 20–70 cm. long, usually simple; leaves membranous, acuminate at both ends, 10–20 cm. long, 2.5–6 cm. wide; calyx-teeth deltoid, obtuse; petals broadly obovate, about 3 mm. long; capsules broadly ovoid, about 4 mm. long inclusive of the styles; styles 3; seeds small, elliptic, with a very short wing at each end.——July–Sept. Woods in mountains; Honshu (Kantō Distr. and westw.), Shikoku, Kyushu.

15. HYDRANGEA L.   AJISAI ZOKU

Erect or scandent deciduous shrubs; leaves opposite, simple, petiolate, usually toothed, exstipulate; inflorescence a terminal cyme or panicle; flowers small, bisexual, some of the marginal ones neutral with enlarged petallike sepals; calyx-lobes in bisexual flowers 4–5, small, toothlike; petals 4–5, usually valvate in bud; stamens 8–20; ovary inferior or subsuperior; styles 2–5, small; capsules 2- to 5-locular; seeds small, many, often winged.——Species 20–25, in temperate and warmer regions of e. Asia, India, and N. America.

1A. Scandent shrub; petals connate above, calyptrately deciduous at anthesis; seeds flat, winged on margin; stamens 15–20. . . 1. *H. petiolaris*
1B. Erect shrubs; petals usually not connate, spreading in anthesis; seeds winged or wingless.
  2A. Inflorescence a pyramidal panicle; seeds with a taillike wing at each end. . . . . 2. *H. paniculata*
  2B. Inflorescence a flat-topped cyme.
    3A. Seeds with a taillike wing at each end; ovary inferior; styles usually 2.
      4A. Leaves toothed, with few incisions; inflorescence not enveloped by large bracts in bud. . . . . 3. *H. sikokiana*
      4B. Leaves minutely toothed, not incised; inflorescence enveloped by few large bracts in bud. . . . . 4. *H. involucrata*
    3B. Seeds entire or scarcely winged; ovary subsuperior; styles 2–3.
      5A. Petals ovate-oblong, not narrowed at base. . . . . 5. *H. macrophylla*
      5B. Petals obovate or broadly spathulate, narrowed at base.
        6A. Inflorescence destitute of neutral flowers; more or less pilose shrub, with compact many-flowered inflorescence.   6. *H. hirta*
        6B. Inflorescence at least partially bearing neutral flowers; glabrescent shrubs with rather loose inflorescence.
          7A. Leaves acuminate, 6–12 cm. long, rather many-toothed; petals obtuse.
            8A. Leaves coarsely dentate; anthers ellipsoidal; branches purple-brown. 7. *H. angustipetala*
            8B. Leaves short-toothed; anthers subglobose; branches gray-brown. . . . . 8. *H. scandens*
          7B. Leaves acute to short-acuminate, 3–5(–6) cm. long, with few coarse teeth; petals mucronate or subacute.
                                                                       9. *H. luteovenosa*

1. Hydrangea petiolaris Sieb. & Zucc.   *H. scandens* sensu Maxim., non Ser. (1830) mec Poeppig. (1830); *H. cordifolia* Sieb. & Zucc.; *H. anomala* D. Don subsp. *petiolaris* (Sieb. & Zucc.) McClintock.——GOTŌ-ZURU, TSURU-AJISAI. Scandent shrub, the branches with aerial roots; leaves ovate-cordate or broadly ovate, 4–10 cm. long, 3–10 cm. wide, acutely toothed, nearly glabrous, abruptly acute, paler and with tufts of axillary hairs beneath, the petioles 3–8 cm. long; inflorescence rather large, cymose; enlarged calyx-lobes of neutral flowers sometimes toothed; petals narrowly ovate, not narrowed at base;

stamens 15–20, the anthers subglobose; capsules globose, truncate at apex.——June–July. Woods in mountains; Hokkaido, Honshu, Shikoku, Kyushu; rather common.——Sakhalin, s. Kuriles, and s. Korea.

2. Hydrangea paniculata Sieb. NORI-UTSUGI. Erect spreading branched shrub, the branches rather stout, usually scattered-pubescent when young; leaves elliptic, ovate or oblong, 5–12 cm. long, 3–8 cm. wide, abruptly acuminate, slightly pilose above, paler and sometimes coarsely hirsute on nerves beneath, the petioles 1–3 cm. long; inflorescence relatively large, 8–30 cm. long, pyramidal, more or less pubescent, the neutral flowers usually white, sometimes pale rose; calyx-teeth in bisexual flowers deltoid; petals narrowly ovate; anthers depressed-globose; capsules ovoid, 4–5 mm. long inclusive of the short style; seeds 3–4 mm. long inclusive of the tail-like wing at the apices.——July–Sept. Sunny slopes and open woods in mountains; Hokkaido, Honshu, Shikoku, Kyushu; common and variable.——s. Kuriles, Sakhalin, and China.—— Cv. **Grandiflora.** *H. paniculata* forma *grandiflora* (Sieb.) Ohwi; *H. paniculata* var. *grandiflora* Sieb.; *H. paniculata* var. *hortensis* Maxim.——MINAZUKI. Cultivated phase with all or most of the flowers neutral.

3. Hydrangea sikokiana Maxim. YAHAZU-AJISAI. Shrub with stout terete branches; leaves broadly ovate, 10–20 cm. long, 7–15 cm. wide, sharply serrate, sparingly incised toward the tip, coarsely hirsute on both sides especially on the nerves beneath, and paler beneath, the petioles 5–15 cm. long; inflorescence a rather large cyme, hirsute, with large neutral marginal flowers; calyx-teeth in bisexual flowers short; petals narrowly ovate; anthers globose-cordate; capsules globose, truncate, about 3.5 mm. across; styles 2.——July. Honshu (w. distr.), Shikoku, Kyushu.

4. Hydrangea involucrata Sieb. *H. cuspidata* Makino, excl. basionym, non Miq.——TAMA-AJISAI. Coarsely hirsute and scabrous shrub with rather stout branches; leaves oblong, elliptic or obovate, sometimes broadly ovate, 10–20 cm. long, 5–10 cm. wide, with short awn-shaped teeth, acuminate, sometimes subcordate, coarsely hairy beneath, the petioles 1–8 cm. long; inflorescence rather small, enveloped at first by a few large whitish bracts; petals in bisexual flowers narrowly ovate; anthers subglobose; styles 2.——July–Sept. Sunny rocky places or in open woods in mountains; Honshu (n. and centr. distr.), Shikoku, Kyushu.——Cv. **Hortensis.** *H. involucrata* forma *hortensis* (Maxim.) Ohwi; *H. involucrata* var. *hortensis* Maxim.——YAE-NO-GYOKUDAN-KA. Cultivated phase with the sterile and bisexual flowers double.

5. Hydrangea macrophylla (Thunb.) Ser. var. **macrophylla**——AJISAI. Nearly glabrous slightly fleshy shrub with stout branches; leaves lustrous above, broadly elliptic to broadly elliptic-ovate or broadly obovate, 10–15 cm. long, 7–10 cm. wide, abruptly acute to abruptly acuminate, dentate, yellowish to fresh-green, sparsely short-pilose on nerves above while young, paler and sparsely short-pilose while young beneath, the petioles 2–4 cm. long; inflorescence rather large, flat-topped, short-pilose, with neutral flowers on the periphery; calyx-teeth in bisexual flowers small; petals narrowly ovate; anthers globose; capsules ovoid, 6–8 mm. long inclusive of the styles, enveloped for 2/3 of the length by the calyx-tube.—— June–July.——Forma **normalis** (Wils.) Hara. *H. hortensis* var. *azisai* (Sieb.) A. Gray; *H. azisai* Sieb. Inflorescence with neutral flowers only on the periphery.——Sunny places near the sea; Honshu (Sagami and Idzu Prov. inclusive of Idzu

Isls.); sometimes cultivated in gardens.——Forma **macrophylla.** *Viburnum macrophyllum* Thunb.; *H. macrophylla* (Thunb.) Ser.; *H. otaksa* Sieb. & Zucc.; *H. macrophylla* forma *otaksa* (Sieb.) Wils.——AJISAI. The most commonly cultivated phase with a large globose inflorescence and most of the flowers neutral.

Var. **megacarpa** Ohwi. MUTSU-AJISAI. Intermediate between the typical phase and var. *acuminata,* with thin leaves and the capsules about 6 mm. long, inclusive of the styles.—— Hokkaido, Honshu (n. distr.).

Var. **acuminata** (Sieb. & Zucc.) Makino. *H. acuminata* Sieb. & Zucc.; *H. hortensis* var. *acuminata* (Sieb. & Zucc.) A. Gray; *Viburnum cuspidatum* Thunb.; *H. cuspidata* (Thunb.) Miq.; *H. belzonii* Sieb. & Zucc.; *V. serratum* Thunb.; *H. serrata* (Thunb.) Ser.; *H. macrophylla* subsp. *serrata* (Thunb.) Makino——YAMA-AJISAI, SAWA-AJISAI. Branches more slender than in the typical phase, the leaves thinner, lanceolate to elliptic, 5–15 cm. long, 2–10 cm. wide, often caudate, green to vivid green, often loosely hirsute, the nerves short-pilose beneath, the capsules shorter, 3–4 mm. long inclusive of the styles.——June–Aug. Mountains; Honshu, Shikoku, Kyushu; common and variable.——Korea.——Cultivars of var. *acuminata* are frequently grown in gardens of which the following are the best known: Cv. **Rosalba.** *H. japonica* Sieb. & Zucc.; *H. japonica rosalba* Van Houtte; *H. serrata* forma *rosalba* (Van Houtte) Wils.; *H. macrophylla* forma *japonica* Ohwi, excl. syn.——BENIGAKU. The neutral and bisexual flowers deep rose-colored.——Cv. **Prolifera.** *H. macrophylla* var. *acuminata* forma *prolifera* (Regel) Ohwi; *H. macrophylla* forma *stellata* (Sieb. & Zucc.) Makino & Nemoto; *H. sitsitan* Sieb.; *H. stellata* Sieb. & Zucc.; *H. hortensia* var. *stellata* (Sieb. & Zucc.) Maxim.; *H. stellata* forma *prolifera* Regel; *H. serrata* forma *prolifera* (Regel) Rehd.——SHICHIDANKA. The neutral and bisexual flowers double.

Var. **thunbergii** (Sieb.) Makino. *H. thunbergii* Sieb., excl. syn.——AMA-CHA. Closely resembles var. *acuminata* but differs in the sugary taste of the dried leaves, and the calyx-lobes of the neutral flowers usually orbicular.——Spontaneous in Honshu (centr. distr.); often cultivated.

Var. **angustata** (Fr. & Sav.) Hara. *H. hortensis* var. *angustata* Fr. & Sav.; *H. macrophylla* var. *amagiana* Makino; *H. serrata* var. *amagiana* Makino——AMAGI-AMA-CHA. Fresh leaves sugary to taste, lanceolate to broadly so, the calyx-lobes in the neutral flowers ovate to ovate-orbicular, obtuse to rounded at the apex.——Spontaneous in Honshu (Izu Prov.).

6. Hydrangea hirta (Thunb.) Sieb. *Viburnum hirtum* Thunb.——KO-AJISAI. Small much-branched shrub with slender purple-brown branches; leaves thinly membranous, ovate to broadly so or obovate, 4–8 cm. long, 3–5 cm. wide, acute, dentate, the petioles 1–3 cm. long; inflorescence many-flowered, rather compact, without neutral flowers; flowers blue-purple; calyx-teeth small, ovate-deltoid; petals narrowly ovate or narrowly obovate; anthers subglobose; capsules subglobose, about 3 mm. long inclusive of the persistent styles.—— June–July. Honshu (Kantō and westw.), Shikoku.

Hydrangea amagiana Makino. AMAGI-KO-AJISAI. Alleged hybrid of *H. macrophylla* × *H. scandens.* Leaves narrow, coarsely dentate on upper half.——Honshu (sw. Kantō Distr.) and Mount Amogi in Izu Prov.

7. Hydrangea angustipetala Hayata. *H. umbellata* Rehd. & Wils.; *H. kawagoeana* var. *grosseserrata* (Engl.) Hatus.; *H. grosseserrata* Engl.——YAKUSHIMA-AJISAI. Glabrescent shrub with slender, purple-brown branches; leaves broadly lanceolate, 8–15 cm. long, 2–4 cm. wide, long-acuminate, coarsely toothed, paler beneath, the petioles 8–15 mm. long; inflorescence rather small, with few neutral marginal flowers; calyx-teeth of bisexual flowers narrowly deltoid; petals broadly spathulate; anthers elliptic; capsules broadly ovoid, 4–5 mm. long inclusive of the rather prominent styles.——Woods along streams and ravines at low elevations in the mountains; Kyushu (Yakushima).——Ryukyus, Formosa, and China.

8. Hydrangea scandens (L. f.) Ser. *Viburnum scandens* L. f.; *V. virens* Thunb.; *H. virens* (Thunb.) Sieb.——GAKU-UTSUGI, KONTERIGI. Sparingly short-pilose or nearly glabrous shrub with slender gray-brown branches; leaves ovate-oblong, 5–8 cm. long, 2–3.5 cm. wide, narrowly ovate or oblong, somewhat chartaceous, acuminate, vivid-green above, short-toothed, paler beneath, the petioles 3–10 mm. long; inflorescence rather small, with few neutral flowers; calyx-teeth of the bisexual flowers small, deltoid; petals obovate; anthers subglobose; capsules broadly ellipsoidal, 4–5 mm. long, inclusive of slightly spreading styles.——May–June. Honshu (Tōkaidō and Kinki Distr.), Shikoku, Kyushu.

9. Hydrangea luteovenosa Koidz. *H. virens* sensu Sieb. & Zucc., excl. syn., non Sieb.——KO-GAKU-UTSUGI. Closely allied to the preceding but somewhat smaller, with purple-brown branches, the leaves often tinged dark purplish, 3–5 cm. long, 1–2 cm. wide, acute or short-acuminate, the teeth prominent but fewer, the petioles 2–5 mm. long; petals slightly larger, acute to mucronate.——Apr.–June. Honshu (Izu Prov. and westw.), Shikoku, Kyushu.

16. PHILADELPHUS L.   BAIKA-UTSUGI ZOKU

Deciduous shrubs; leaves short-petioled, opposite, toothed or sometimes entire, 3- to 5-nerved; inflorescence a raceme, cyme, or panicle; flowers rarely single, white, 4-merous; calyx-tube obconical, the lobes valvate in bud; petals rather large, imbricate in bud, white; stamens many; styles 4, connate below; ovary 4-locular, nearly inferior; capsules 4-valved, dehiscent; seeds small, many.——Temperate regions of the N. Hemisphere.

1. Philadelphus satsumi Sieb. ex Lindl. & Paxt. *P. satsumanus* Sieb. ex Miq.; *P. matsumuranus* Koehne; *P. acuminatus* Lange; *P. coronarius* var. *satsumi* Maxim.; *P. schrenkii* sensu auct. Japon., non Rupr.——BAIKA-UTSUGI. Thinly pubescent shrub, the second-year branches brownish, the bark longitudinally split and peeling; leaves ovate, to narrowly so or oblong, 5–8 cm. long, 2–3.5 cm. wide, long-acuminate, sparsely short-toothed, sometimes slightly pilose above, paler and nearly glabrous beneath except tufts of axillary hairs, the petioles 5–10 mm. long; racemes about 10-flowered; flowers white; calyx-tube and pedicels often puberulous; calyx-lobes ovate, acute, 4–5 mm. long; petals broadly ovate, 12–15 mm. long; styles pilose to glabrous.——May–July. Honshu, Shikoku, Kyushu.——Forma **nikoensis** (Rehd.) Ohwi. *P. satsumanus* var. *nikoensis* Rehd.——KE-BAIKA-UTSUGI. Leaves prominently pubescent beneath.

Var. **lancifolius** (Uyeki) Murata. *P. shikokianus* Nakai; *P. shikokianus* var. *lancifolius* Uyeki——SHIKOKU-UTSUGI. Styles pilose near the base.——Honshu, (Tōkaidō and Kinki Distr.), Shikoku, Kyushu.

17. DEUTZIA Thunb.   UTSUGI ZOKU

Deciduous or sometimes semi-evergreen stellate-pubescent shrubs; leaves opposite, short-petioled, exstipulate, usually dentate; flowers in racemes, cymes or sometimes solitary, white; calyx-teeth 5; petals 5; stamens 10, the filaments winged, usually with a tooth on each side at tip; ovary inferior; styles 3–5, free; capsules 3- to 5-valved; seeds minute, many.——e. Asia, Himalayas, and Mexico. Our species have the petals valvate in bud.

1A. Flowers in panicles or racemes.
  2A. Leaves all alike, petiolate, sometimes subsessile; filaments toothed.
    3A. Leaves grayish brown on upper side, densely stellate-pubescent beneath.
      4A. Leaves rather thick, pale green beneath. . . . . 1. *D. crenata*
      4B. Leaves membranous, densely stellate-pubescent and whitish beneath. . . . . 2. *D. maximowicziana*
    3B. Leaves vivid-green, glabrous to thinly stellate-pubescent beneath. . . . . 3. *D. gracilis*
  2B. Leaves slightly dimorphic, sessile and barely clasping in those immediately below the inflorescence, very short-petiolate in the others; filaments of the longer stamens without teeth. . . . . 4. *D. scabra*
1B. Flowers 1, very rarely 2, from lateral buds of the previous year, usually without basal leaves. . . . . 5. *D. uniflora*

1. Deutzia crenata Sieb. & Zucc. *D. scabra* Thunb., pro parte——UTSUGI. Shrub; leaves ovate to broadly lanceolate, 3–6 cm. long, 1.5–3 cm. wide, acuminate to subobtuse, with 4- to 6-rayed stellate hairs above, pale green and with 10- to 15-rayed stellate hairs beneath; petioles 2–5 mm. long; inflorescence a simple raceme sometimes branched at base; calyx with simple and stellate hairs; petals about 15 mm. long; styles 3–4; capsules 3.5–6 mm. across.——May–June. Hokkaido, Honshu, Shikoku, Kyushu; common and very variable.

Var. **heterotricha** (Rehd.) Hara. *D. heterotricha* Rehd.; *D. hebecarpa* Nakai; *D. ferruginea* Satake——BIRŌDO-UTSUGI. Leaves beneath, pedicels and calyx-tube with spreading hairs.——Honshu, Kyushu.

Var. **nakaiana** (Engl.) Hara. *D. floribunda* Nakai, non Lemoine; *D. nakaiana* Engl.——KO-UTSUGI. Leaves relatively narrower, with shorter teeth, the flowers smaller and more numerous, the petals 5–8 mm. long, the capsules about 3 mm. across.——Honshu (w. distr.), Shikoku, Kyushu.

2. Deutzia maximowicziana Makino (Feb. 1892). *D. discolor* Maxim., non Hemsl.; *D. hypoleuca* Maxim. (Nov. 1892)——URAJIRO-UTSUGI. Closely allied to the preceding, but the branches more slender; leaves green, acute to acuminate, minutely toothed, membranous, white stellate-pubescent beneath.——May–June. Honshu (Shinano Prov. and westw.), Shikoku.

3. Deutzia gracilis Sieb. & Zucc. *D. nagurae* Makino;

D. gracilis var. *nagurae* (Makino) Makino——HIME-UTSUGI. Young branches glabrous or glabrescent; leaves membranous, vivid-green, ovate to broadly lanceolate, 3.5–10 cm. long, 2–4 cm. wide, acuminate, minutely toothed, scattered stellate-pubescent above, somewhat paler beneath with 5- to 8-rayed stellate hairs or subglabrous, the petioles 3–8 mm. long; racemes often branched at base; flowers white, the pedicels slender; calyx-teeth deltoid, glabrous; petals about 10 mm. long; capsules about 4 mm. across.——May–June. Mountains; Honshu (Kantō Distr. and westw.), Shikoku, Kyushu.

Var. **zentaroana** (Nakai) Hatus. *D. zentaroana* Nakai ——BUNGO-UTSUGI. Leaves larger and prominently stellate-pubescent beneath.——Occurs with the typical phase.

Var. **ogatae** (Koidz.) Ohwi. *D. ogatae* Koidz.——AO-KO-UTSUGI. Closely resembles the preceding variety, but the inflorescence is paniculate with more numerous smaller flowers.——Shikoku.

4. Deutzia scabra Thunb. *D. sieboldiana* Maxim.; *D. sieboldii* Koehne; *D. kiusiana* Koidz.; *D. subvelutina* Nakai; *D. taradakensis* Nakai; *D. microcarpa* Nakai; *D. reticulata* Koidz.; *D. sieboldiana* var. *dippeliana* C. K. Schn.; *D. sieboldii*

forma *dippeliana* (C. K. Schn.) Nakai——MARUBA-UTSUGI, TSUKUSHI-UTSUGI. Shrub; leaves subtending the inflorescence sessile, the others short-petiolate, green, thinly chartaceous, ovate to oblong, 3–8 cm. long, 2–4 cm. wide, acuminate, minutely dentate-serrulate, the stellate hairs 3- to 4-rayed on upper side, 4- to 6-rayed beneath, also with scattered simple and stipitate stellate hairs on nerves below, the petioles 2–4 mm. long; panicles rather many-flowered, densely stellate and spreading-hairy; petals 6–7 mm. long; capsules about 3 mm. across.——Apr.–May. Honshu (Kantō Distr. and westw.), Shikoku, Kyushu.

5. Deutzia uniflora Shirai. UME-UTSUGI. Shrub; young branches brownish purple; leaves ovate to oblong-ovate, 4–7 cm. long, 2–3.5 cm. wide, acuminate, shallowly dentate, scattered stellate-pubescent on both sides, the petioles 4–6 mm. long, often slightly winged above; flowers white, solitary, rarely 2–3, the pedicels 5–12 mm. long, with 2 small bracts, sparingly stellate-pubescent; calyx-teeth lanceolate-deltoid, glabrous; petals about 20 mm. long; teeth of the filaments linear-lanceolate; capsules 4–5 mm. across.——Apr.–May. Rocks in mountains; Honshu (w. Kantō Distr.); rare.

18. KIRENGESHOMA Yatabe   KI-RENGE-SHŌMA ZOKU

Tall erect perennial herbs with stout short rhizomes; leaves opposite, petiolate, palmately lobed, exstipulate; cymes terminal and axillary; flowers large, yellow; calyx-tube subglobose, the teeth depressed; petals 5, convolute in bud; stamens 15, free; ovary subinferior, 3(–4)-locular; styles 3(–4), free, filiform; capsules broadly ovoid, loculicidally dehiscent, 3(–4)-valved; seeds many, flat, margined with a wing of unequal width.——One or two species, in Japan and Korea.

1. Kirengeshoma palmata Yatabe. KI-RENGE-SHŌMA. Stems glabrous, 80–120 cm. long; leaves membranous, orbicular-cordate, 10–20 cm. long and as wide, palmately lobed, thinly pilose with ventrally fixed hairs, toothed, the lobes deltoid, acuminate, the petioles long in the lower ones, sessile in the upper ones; inflorescence a ternately branched panicu-

late cyme; flowers large, erect, the pedicels deflexed in fruit; calyx-teeth undulate; petals erect, about 30 mm. long, narrowly obovate-oblong, subacute, deciduous; styles about 20 mm. long, glabrous, persistent; capsules glabrous, about 15 mm. long.——Aug. Woods in mountains; Shikoku, Kyushu; rare.

19. ITEA L.   ZUINA ZOKU

Evergreen or deciduous trees or shrubs; leaves alternate, simply toothed or entire, exstipulate, short-petiolate; inflorescence a raceme; flowers small, bisexual, white, 5-merous; calyx-tube short, the lobes persistent; petals deciduous or persistent, valvate in bud; stamens 5, alternate with the petals; ovary superior, 2-locular; styles 2, connate; capsules with a groove on each side; seeds flat.——About 10 species, in the temperate and warmer regions of e. Asia, one in N. America.

1. Itea japonica Oliv. *Reinia racemosa* Fr. & Sav.——ZUINA. Large deciduous shrub with glabrous yellow-green young branches; leaves membranous, vivid-green, oblong, ovate or ovate-oblong, 7–12 cm. long, 3–6 cm. wide, long-acuminate, serrulate, short-pilose on the nerves above, paler and often short-pilose beneath, the petioles 4–7 mm. long; inflorescence 7–20 cm. long, densely many-flowered, spreading-

puberulent on the axis, the pedicels 1–3 at the node, puberulent, 2–3 mm. long, bracteate; flowers about 3 mm. long; calyx-lobes narrowly deltoid, acute, about half as long as the petals; petals persistent, lanceolate, obtuse, 1-nerved; capsules slightly longer than the petals; styles short, the stigma capitate.——May. Honshu (s. Kinki Distr.), Shikoku, Kyushu.

20. RIBES L.   SUGURI ZOKU

Deciduous often armed shrubs; leaves alternate, petiolate, palmately lobed, usually exstipulate; flowers solitary to several in leaf-axils or in racemes, 5-merous or rarely 4-merous, bisexual or unisexual (plants sometimes dioecious); calyx rotate, the lobes slightly connate below; petals small; ovary inferior, 1-locular; styles 2, more or less connate at base; fruit a berry; seeds usually many.——About 150 species, in N. Hemisphere and S. America.

1A. Unarmed shrubs; pedicels jointed.
  2A. Flowers fascicled or solitary in axils.
    3A. Fruit with long glandular hairs; small epiphytic shrub. . . . . 1. *R. ambiguum*
    3B. Fruit glabrous; terrestrial shrub. . . . . 2. *R. fasciculatum*
  2B. Flowers in racemes.

1. Ribes ambiguum Maxim. *R. burejense* sensu Maxim., pro parte, non F. Schmidt; *R. cynosbati* sensu Thunb., non L. ——Yashabishaku. Sparsely branched small epiphytic shrub with rather stout stems and branches; leaves orbicular, 3–5 cm. in diameter, cordate, shallowly 3- to 5-lobulate, toothed, short-pubescent, sometimes glabrescent above, the petioles 1–3 cm. long, short-pubescent; flowers 1 or 2, the pedicels 5–10 mm. long, with 2 small deciduous bracteoles near the top; ovary with long brownish red glandular hairs; calyx-lobes oblong, the limb short; berry broadly ellipsoidal, 10–12 mm. long.——Apr.–May. Epiphytic on tree-trunks and branches in mountains; Honshu, Shikoku, Kyushu; rather rare.

Var. **glabrum** Ohwi. Ke-nashi-yashabishaku. Leaves and petioles glabrous.——Kyushu (Mount Gozendake in Chikugo Prov.).

2. Ribes fasciculatum Sieb. & Zucc. *R. japonicum* Carr., non Maxim.——Yabu-sanzashi. Erect dioecious shrub; leaves suborbicular, 3–4 cm. long, 3.5–5 cm. wide, truncate to slightly cordate, sparsely short-pilose especially beneath, 3- to 5-lobed, the lobes obtuse or subobtuse, irregularly toothed, the petioles 1–4 cm. long; flowers few, in axillary fascicles, the staminate cup-shaped, the pistillate ovoid, the pedicels short, bractless; calyx-limb extremely short, the lobes elliptic, recurved; petals minute; berry broadly ellipsoidal, red, 8–10 mm. long.——Mar.–Apr. Honshu (centr. distr. and westw.), Shikoku, Kyushu; rare.——Korea and China.

3. Ribes maximowiczianum Komar. *R. maximowiczii* Komar., non Batal.; *R. maximowiczianum* var. *umbrosum* Komar.; *R. tricuspe* Nakai; *R. alpinum* var. *mandshuricum* Maxim.; *R. alpinum* var. *japonicum* Maxim.; *R. distans* Jancz.; *R. maximowiczianum* var. *mandshuricum* (Maxim.) C. K. Schn.; *R. maximowiczianum* var. *japonicum* (Maxim.) C. K. Schn.; *R. distans* var. *japonicum* (Maxim.) Jancz.——Zari-gomi. Dioecious shrub with slender gray-brown branches; leaves deltoid-orbicular, 2–4 cm. in diameter or slightly narrower than broad, thinly pilose on both sides or nearly glabrous, often with a tuft of axillary hairs beneath, 3-lobed, the lobes acute to subacuminate, coarsely toothed, the lateral lobes often smaller and sometimes again 2-lobed, the petioles 7–10 mm. long; staminate inflorescence 7- to 10-flowered, the pistillate 2- to 6-flowered, 1–2.5 cm. long, the bracts lanceolate, 3–4 mm. long; flowers rotate, yellowish, 4–5 mm. across; calyx-limb extremely short, the lobes elliptic, glabrous; pedicels of the pistillate flowers about 1 mm. long, of the staminate 3–4

mm. long; fruit red, glabrous, ellipsoidal, about 7 mm. long. ——May–June. Honshu (centr. distr.).——Korea and Manchuria.

4. Ribes japonicum Maxim. Komagatake-suguri. Erect, spreading-puberulent shrub with rather stout branches; leaves membranous, orbicular, 7–15 cm. in diameter, 5-cleft, cordate, puberulent to nearly glabrous above, prominently puberulent and with sessile yellow-brown glandular dots beneath, the lobes short-acuminate, doubly toothed, the petioles 10–15 cm. long; racemes many-flowered, 10–15 cm. long, the pedicels 3–7 mm. long, the bracts linear, shorter than the pedicel; flowers rotate, 8–10 mm. across, yellow-green; calyx-limb short, the lobes broadly ovate, slightly recurved; petals small, obtriangulate; berry black, broadly ellipsoidal, 6–7 mm. long. ——May–July. Woods in mountains; Hokkaido, Honshu, Shikoku.

5. Ribes americanum Mill. *R. floridum* L'Hérit.—— Amerika-kuro-suguri. Resembles the preceding; leaves smaller, 3–8 cm. long and as wide, truncate or sometimes cordate, not prominently puberulous; racemes 4–6 cm. long; flowers more than 10, yellowish, campanulate, the bracts longer than the pedicels; ovary glabrous; calyx-limb nearly as long as the lobes; petals and stamens not exserted; berry glabrous, black.——May. A shrub of e. N. America rarely planted in our area.

6. Ribes sachalinense (F. Schmidt.) Nakai. *R. affine* var. *sachalinense* F. Schmidt; *R. laxiflorum* var. *japonicum* Jancz.; *R. laxiflorum* sensu auct. Japon., non Pursh——Toga-suguri. Procumbent sparingly branched shrub, the branches rather stout, puberulent while young; leaves membranous, orbicular, 4–10 cm. in diameter, deeply 5-cleft, loosely glandular-hairy on the nerves beneath; inflorescence 5–10 cm. long, loosely 7- to 10-flowered, short-glandular, the bracts linear, very much shorter than the pedicels; flowers rotate, 5–6 mm. across, yellow-green; ovary densely glandular-pubescent; calyx-limb extremely short, the lobes obovate-rounded, glabrous; berry glandular-hairy, broadly ellipsoidal, about 8 mm. long, dark red.——June–July. Hokkaido, Honshu (centr. distr.), Shikoku.——Sakhalin.

7. Ribes latifolium Jancz. *R. petraeum* var. *tomentosum* Maxim., pro parte——Ezo-suguri. Erect shrub, the young branchlets puberulent and often glandular-hairy; leaves orbicular, 5–10 cm. in diameter, or slightly broader than long, rather densely white-hairy to glabrous on upper side, cordate,

5-lobed, the lobes acute to obtuse, doubly toothed, the petioles 5–10 cm. long; racemes 5–8 cm. long, 10- to 20-flowered; flowers purple-red, short-campanulate, the bracts minute, ovate, the pedicels puberulent and often sparingly glandular-hairy; ovary glabrescent or glabrous; calyx-limb as long as the lobes, the lobes obovate, truncate-rounded at the tip, often puberulent on back; berry red, globose, about 7 mm. long.——June–July. Hokkaido.——s. Kuriles and Sakhalin.

8. Ribes triste Pall. Tokachi-suguri, Chishima-suguri. Procumbent shrub; leaves membranous, suborbicular, 5–10 cm. wide, cordate, glabrous on upper side, glabrate or slightly pilose beneath, 3- to 5-lobed, the lobes acute to obtuse, coarsely toothed, the petioles usually pilose; racemes to 3.5 cm. long, glandular-hairy; flowers purplish, sometimes greenish, rotate; calyx-lobes obliquely spreading, orbicular; berry dark red, 6–8 mm. across.——Hokkaido.——e. Asia, s. Kuriles, and Sakhalin.

9. Ribes sativum Syme. *R. rubrum* sensu auct. Japon., non L.——Aka-suguri.   Erect shrub, the branches some-what glandular-hairy when young; leaves suborbicular, 3–7 cm. in diameter, cordate, nearly glabrous on upper side, short-pilose beneath especially on the nerves, the lobes acute to obtuse, irregularly toothed, the petioles 3–6 cm. long, short-pilose; racemes nodding, rather many-flowered, nearly glabrous, the bracts ovate, minute; flowers rotate, the receptacle swollen, 5-angled; anther-locules distant; berry red.——Apr.–May. Cultivated in our area.——Europe.

10. Ribes horridum Rupr. *R. lacustre* Poir. var. *horridum* (Rupr.) Jancz.——Kuro-mi-no-hari-suguri.   Shrub, the branches elongate, yellowish, slightly lustrous, glabrous, densely yellowish prickly, more so on the nodes; leaves sub-orbicular, 3–6 cm. long and as wide, cordate, scattered prickly on both sides, glabrous, deeply 5-cleft, the lobes acute, often

again 3-lobed or shallowly incised, the petioles 3–6 cm. long, prickly; inflorescence 5- to 8-flowered, glandular-hairy, the bracts broadly lanceolate, about half as long as the pedicels; berry blackish red, long glandular-hairy.——Hokkaido (Tokachi Prov.).——e. Siberia, n. Korea, and Sakhalin.

11. Ribes sinanense F. Maekawa. *R. grossularioides* Maxim., non Steud.——Suguri.   Shrub with simple to 3-parted stout spines on the nodes, the internodes prickly; leaves suborbicular, 2–3.5 cm. in diameter, truncate to subcordate, slightly pilose to subglabrous on both sides, 3- to 5-cleft or -lobed, the lobes obtuse to subrounded, with few coarse obtuse teeth, the petioles 1.5–3 cm. long; pedicels slender, axillary, 1.5–3 cm. long, the bracts 2–3 mm. long, inserted at the middle of the pedicel; ovary glabrous to slightly pilose; calyx tubular-campanulate, the free portion about 8 mm. long, the limb twice as long as the lobes, the lobes linear-oblong, yellow-green, reflexed, rounded at the apex, glabrous inside, puberulent outside; petals reddish, erect; stamens as long as the style; anthers not tipped with a gland; berry red, glabrous, sub-globose, about 12 mm. long.——May–June. Mountains; Honshu (centr. distr.); rare.

12. Ribes grossularia L. Maru-suguri.   Allied to the preceding; spines simple to 3-parted; branches usually puberulent while young; leaves orbicular, 2–5 cm. in diameter, subcordate, usually puberulent on both sides, 3- to 5-cleft, the lobes obtuse, coarsely toothed; flowers 1–2, greenish; ovary puberulent and sometimes glandular hairy; calyx-limb about as long as the lobes, hairy inside, the lobes narrowly oblong, loosely puberulent outside, reflexed, rounded at apex; berry green, yellowish, sometimes reddish, globose to ellipsoidal.——Apr.–May. Cultivated for the edible fruit.——Europe, N. Africa, and Caucasus.

Fam. 102. PITTOSPORACEAE   Tobera Ka   Pittosporum Family

Shrubs or trees, sometimes scandent; leaves alternate, entire, toothed or incised, exstipulate; flowers bisexual, actinomorphic or slightly oblique, white, blue, yellow, or rarely reddish, solitary and terminal, in corymbs or in panicles, rarely in axillary fascicles; sepals 5, free or rarely connate at base, imbricate in bud; petals 5, imbricate in bud, usually with an erect to ascending claw, the limb spreading; stamens 5, alternate with the petals, free, the anthers versatile; style simple, the stigma minute to capitate; ovules many; fruit a capsule, rarely a berry.——About 10 genera, with about 200 species, mostly Australian.

1. PITTOSPORUM Banks & Soland., ex Gaertn.   Tobera Zoku

Usually evergreen often dioecious shrubs or trees; leaves entire or undulate-toothed, short-petiolate, alternate, often aggregated toward the tips of the branches, stipules absent; flowers in terminal cymes, panicles or umbels, sometimes solitary in the upper axils; sepals often connate at the base; petals deciduous, usually connivent thus forming a tube on the lower portion, the upper part spreading; anthers erect; ovary imperfectly 2- to 5-locular; fruit a globose, ovoid or obovoid capsule, loculicidally dehiscent, the valves woody or coriaceous; seeds rather plump, often viscid.——About 160 species, in Africa, Asia, Australia, and the Pacific Islands.

1A. Leaves coriaceous, rounded at apex; inflorescence a simple or branched umbel, short-pubescent. .... 1. *P. tobira*
1B. Leaves chartaceous, acuminate, sometimes the tip obtuse; inflorescence a few-flowered simple umbel, glabrous. ...... 2. *P. illicioides*

1. Pittosporum tobira (Thunb.) Ait. *Euonymus tobira* Thunb.——Tobera.   Large stout much-branched shrub; leaves narrowly obovate, 7–10 cm. long, 2–3 cm. wide, gla-brous, entire and slightly recurved on margin, short-petiolate, lustrous and dark green above; flowers white, changing to yellow, fragrant; sepals ciliate; petals spathulate, about 12 mm. long; capsules subglobose, about 12 mm. across, valves 3, thickened, nearly woody; seeds red.——Apr.–June. Sunny

slopes near the sea; Honshu (Kantō Distr. and westw.), Shi-koku, Kyushu; common, often planted in gardens and for hedges.——s. Korea, China, and Ryukyus.

2. Pittosporum illicioides Makino. *P. glabratum* sensu auct. Japon., non Lindl.——Koyasu-no-ki.   Glabrous green shrub, the branches rather slender, grayish brown, minutely lenticellate, terete; lower leaves alternate, approximate toward the top, deep green and slightly lustrous above, paler beneath,

6–10 cm. long, 2.5–4 cm. wide, gradually acuminate, some-times with an obtuse tip, petiolate, the margin barely undulate and cartilaginous; flowers 1–4, in the upper axils; petals about 8 mm. long; style about 2.5 mm. long, thickened toward the base, the stigma dilated; ovary 3- to 4-angled; capsules with thin-coriaceous valves.——May–June. Honshu (Harima Prov.); rare.

Fam. 103. HAMAMELIDACEAE MANSAKU KA Witch-hazel Family

Evergreen or deciduous shrubs or trees, with simple or stellate pubescence or glabrous; leaves alternate, rarely opposite, petiolate, simple or palmately lobed, entire or toothed, the stipules 2, deciduous or persistent, rarely absent; flowers in heads or spikes, rather small, unisexual or bisexual; perianth sometimes wholly reduced in the staminate flowers; calyx-tube adnate more or less to the ovary, the calyx-limb truncate or 5-lobed, the lobes valvate or imbricate in bud; petals 4, 5 or sometimes many or absent, linear to obovate, imbricate or valvate in bud; stamens 4 to many; ovary 2-locular, inferior, subinferior, or rarely superior; carpels 2; ovules 1 to many in each locule, anatropous, pendulous; styles subulate, persistent; capsules woody, bipartite.—— About 23 genera, with about 150 species, in the temperate to subtropical regions of both hemispheres, except Europe.

1A. Leaves palmately nerved; ovules several in each locule. . . . . 1. *Disanthus*
1B. Leaves penninerved; ovules solitary in each locule.
  2A. Petals present.
    3A. Leaves evergreen; flowers white, the petals linear. . . . . 2. *Loropetalum*
    3B. Leaves deciduous; flowers yellowish to brownish red.
      4A. Flowers 4-merous, in pairs on a short peduncle, the petals linear. . . . . 3. *Hamamelis*
      4B. Flowers 5-merous, in spikes, the petals spathulate to narrowly obovate. . . . . 4. *Corylopsis*
  2B. Petals absent; evergreen trees; ovary superior. . . . . 5. *Distylium*

1. DISANTHUS Maxim. MARUBA-NO-KI ZOKU

Deciduous glabrous shrub; leaves long-petiolate, palmately 5- to 7-nerved, entire, cordate or ovate-orbicular, obtuse to subacute, glaucescent beneath, the stipules rather large, caducous; flowers paired, bisexual, sessile on top of the peduncle; calyx-5-lobed, pilose, the lobes recurved; petals 5, linear-lanceolate, dark purple, caudate; stamens 5, very short; ovary superior; styles short; capsules obcordate, bifid, rather large; seeds several in each locule, black, shining.——One species in Japan and possibly also in China.

1. Disanthus cercidifolius Maxim. MARUBA-NO-KI, BENI-MANSAKU. Leaves 5–12 cm. long and equally wide, somewhat cordate, entire, smooth; stipules small in adult plants, large and leaflike in young vigorous plants; peduncles 3–10 mm. long; flowers about 12 mm. across; calyx hairy, the lobes short, recurved; capsules 15–17 mm. long and equally wide.——Oct.–Nov. Honshu (sw. part of centr. distr. and Aki Prov.), Shikoku; rare.——China (?).

2. LOROPETALUM R. Br. TOKIWA-MANSAKU ZOKU

Small evergreen tree, stellate-pubescent; leaves alternate, oblong or obliquely ovate, entire, penninerved, abruptly acute; stipules membranous, deciduous; flowers 6–8 in short pedunculate heads, white; calyx-lobes 4; petals 4, broadly linear; stamens 4, very short; ovary subinferior; styles 2; capsules broadly obovoid, 4-parted; seeds solitary in each locule, black, shining.——One species, in Japan, China, and n. India.

1. Loropetalum chinense Oliv. TOKIWA-MANSAKU. Small tree with stellate pubescence; branches dark brown, slender; leaves obliquely ovate or obliquely oblong, 2–4 cm. long, 1–2 cm. wide, deep green above, whitish beneath with raised veinlets, the petioles 3–5 mm. long; calyx-lobes ovate, 2–3 mm. long, obtuse; petals about 20 mm. long, about 2.5 mm. wide; anthers about 0.7 mm. long, the appendage of the connective nearly as long as the anther; capsules 6–7 mm. long. ——May. Honshu (Ise Prov.); rare.

3. HAMAMELIS L. MANSAKU ZOKU

Deciduous shrubs or trees with stellate pubescence; leaves short-petiolate, alternate, simple, oblique, deeply undulate-toothed; stipules deciduous; flowers bisexual, 4-merous, in axillary, short-pedunculate, few-flowered clusters; calyx 4-parted, adnate to the ovary at base, the lobes ovate, obtuse, persistent, stellate-pubescent; petals linear, yellow or reddish yellow; stamens 4, short, the staminodia alternate with the stamens; styles short; ovary subinferior or nearly superior; capsules ellipsoidal, 2-fid, the valves bifid again; seeds 2, black, shining, oblong.——Few species, in N. America and e. Asia.

1. Hamamelis japonica Sieb. & Zucc. *H. arborea* Ottol.; *H. zuccariniana* Ottol.——MANSAKU. Deciduous shrub or small tree; leaves rhombic-orbicular, broadly ovate or obovate, 4–12 cm. long, 3–8 cm. wide, slightly oblique at the base, deltoid on the upper half, subobtuse, stellate-pubescent while young, the lateral nerves 6 to 8 pairs, the petioles 5–12 mm. long; calyx-lobes elliptic, reflexed; petals about 2 cm. long, yellow; capsules with persistent calyx-lobes near the base.—— Mar.–Apr. Hokkaido, Honshu, Shikoku, Kyushu; very variable.

Var. **bitchuensis** (Makino) Ohwi. *H. bitchuensis* Makino ——Atetsu-mansaku. Leaves persistent stellate-pubescent. ——Honshu (Chūgoku Distr.).

Var. **obtusata** Matsum. *H. obtusata* (Matsum.) Makino ——Maruba-mansaku. Leaves subrounded on the upper half, rounded to subretuse at tip, the stellate hairs deciduous; petals yellow.——Mountains; Honshu (Japan Sea side).—— Forma **incarnata** (Makino) Ohwi. *H. incarnata* Makino

——Akabana-mansaku. Petals reddish.——Forma **flavo-purpurascens** (Makino) Rehd. *H. obtusata* forma *flavopurpurascens* Makino——Nishiki-mansaku. Petals yellow on upper part, reddish at base.

Hamamelis megalophylla Koidz. with larger leaves is doubtfully distinct.——Honshu (Kantō Distr.).

Hamamelis virginiana L., an American species, is occasionally cultivated; flowering in autumn.

4. CORYLOPSIS Sieb. & Zucc. Tosa-mizuki Zoku

Deciduous stellate-pubescent shrubs; leaves petiolate, simple, penninerved, toothed, broadly ovate, usually slightly oblique; stipules membranous, deciduous, rather large; flowers in axillary pendulous spikes, bisexual, yellow, sessile, precocious, bracteate, the bracts membranous, concave, the lower ones larger; calyx-lobes short, persistent, the tube adnate to the ovary; petals 5, clawed, obovate to spathulate; stamens 5, alternate with the staminodia; ovary subinferior, 2-locular; styles filiform, the stigma capitellate; ovules solitary in each locule; capsules subglobose or broadly obovoid, 2-valved, the valves bifid; seeds solitary in each locule, black, shining.——More than 10 species, in e. Asia and the Himalayas.

1A. Spikes 1- to 3-flowered; leaves 3–5 cm. long, 1.5–3 cm. wide. .... 1. *C. pauciflora*
1B. Spikes 5- to 10-flowered; leaves larger, 3–10 cm. long, 2–8 cm. wide.
  2A. Young branches, petioles, and inflorescence with yellowish hairs. .... 2. *C. spicata*
  2B. Young branches and other parts glabrous or nearly glabrous.
    3A. Leaves ovate-orbicular, short-dentate with prominently awn-pointed teeth; stamens about half as long as the petals.
                                                                 3. *C. glabrescens*
    3B. Leaves obovate, sometimes obovate-orbicular, with very short-awned teeth; stamens nearly as long as the petals. .... 4. *C. gotoana*

1. Corylopsis pauciflora Sieb. & Zucc. Hyūga-mizuki, Iyo-mizuki. Much-branched shrub, 1–2 m. high, with slender glabrous branches; leaves membranous, oblique, ovate, 3–5 cm. long, 1.5–3 cm. wide, acute, slightly cordate, glabrous on upper side, glaucescent and short stellate-hairy while young beneath, appressed-pilose on the under side of the nerves, undulate, short awned-toothed, the lateral nerves 5 to 7 pairs, the petioles 5–15 mm. long; flowers 1–3, pale yellow, about 15 mm. long; stamens slightly shorter than the petals.——Mar.–Apr. Mountains; Honshu (n. Kinki Distr. and Kaga, Echizen, and Mino Prov.); widely planted.——Formosa.

2. Corylopsis spicata Sieb. & Zucc. *C. kesakii* Sieb. & Zucc.——Tosa-mizuki. Shrub about 2 m. high, with rather stout branches, usually pilose when young; leaves ovate-orbicular to obovate-orbicular, abruptly acute, glabrous on upper side, glaucous and pilose beneath, short awn-toothed, the lateral nerves 6 or 7 pairs, the petioles 1–2.5 cm. long; spikes 7- to 10-flowered; flowers yellow, about 10 mm. long; stamens about

as long as the petals, the anthers dark purple.——Mar.–Apr. Mountains; Shikoku; rare; widely planted.

3. Corylopsis glabrescens Fr. & Sav. *C. kesakii* sensu auct. Japon., non Sieb. & Zucc.——Kirishima-mizuki. Large shrub; leaves glabrous on upper side, glaucous and sparingly stellate-pubescent beneath while young, abruptly acute, slightly cordate, the nerves appressed-pilose while young; spikes several-flowered; flowers yellow, about 15 mm. long; stamens half as long as the petals, the anthers purple.——Apr. Mountains; Kyushu (Kirishima Mts.).

4. Corylopsis gotoana Makino. Kōya-mizuki, Miyama-tosa-mizuki. Large shrub or small tree to 5 m. high; leaves with scattered, long, appressed hairs while young, glaucous and appressed-pilose on the nerves and stellate-pubescent beneath while young, usually becoming glabrate; flowers about 10 mm. long; petals obovate; stamens nearly as long as the petals, the anthers purplish.——Apr. Mountains; Honshu (centr. distr. and westw.), Shikoku, Kyushu; rather rare.

5. DISTYLIUM Sieb. & Zucc. Isu-no-ki Zoku

Evergreen stellate-pubescent trees; leaves coriaceous, short-petiolate, entire or few-toothed, penninerved; stipules membranous, deciduous; flowers polygamous (plants sometimes dioecious), apetalous, bracteate, in axillary spikes; calyx free from the ovary, 3- to 5-parted, the lobes imbricate in bud; stamens 2–8, the filaments short; ovary superior, usually densely stellate-pubescent; styles 2; capsules nearly woody, oblong to subglobose, cuspidate, 2-valved, the valves bifid; seeds oblong-cylindric, shining.—— Few species, in Japan, China, and the Himalayas.

1. Distylium racemosum Sieb. & Zucc. Isu-no-ki. Branchlets glabrate, yellowish stellate-hairy while young; leaves oblong to narrowly obovate, 3–7 cm. long, 1.5–3 cm. wide, acute to obtuse, sometimes undulate-toothed on the upper margin, stellate-pubescent while young; spikes axillary,

shorter than the leaves, stellate-pubescent; anthers red; capsules broadly ovoid, shallowly bifid, about 10 mm. long, yellow-brown tomentose.——Mar.–May. Honshu (s. Kantō Distr. and westw.), Shikoku, Kyushu; occasionally planted. ——Ryukyus. A phase with variegated leaves is known.

Fam. 104. ROSACEAE Bara Ka Rose Family

Herbs, shrubs or trees of various habit; leaves alternate, rarely opposite, simple or compound, usually stipulate; flowers actinomorphic; calyx superior or inferior, the lobes usually 5, imbricate in bud; petals usually as many as the calyx-lobes, rarely absent, perigynous, often orbicular with or without a short claw at base, imbricate in bud; stamens many, rarely definite, perigynous or inserted on the calyx or receptacle, inflexed or incurved in bud; receptacle (hypanthium) often adnate to the calyx-tube; carpels 1 to many, free or connate, sometimes adnate to the calyx-tube or perigynous receptacle, the ovules solitary or collaterally paired in each locule, anatropous; fruit various.——About 100 genera, with about 3,000 species, cosmopolitan, especially abundant in temperate regions.

1A. Fruit dehiscent; stipules sometimes absent.
  2A. Leaves simple, entire or lobed.
    3A. Stipules present; carpels solitary; inflorescence a corymbose panicle. .... 1. *Stephanandra*
    3B. Stipules absent; carpels 5; inflorescence an umbellate raceme or panicle. ....2. *Spiraea*
  2B. Leaves pinnate or ternately compound.
    4A. Shrubs; carpels connate at base, opposite the calyx-lobes; leaves pinnate; stipules present. .... 3. *Sorbaria*
    4B. Herbs; carpels free, alternate with the calyx-lobes; leaves ternately compound; stipules absent. .... 4. *Aruncus*
1B. Fruit indehiscent, dry or fleshy.
  5A. Ovary superior or rarely apparently inferior because of the tightly enclosing calyx-tube.
    6A. Carpels many, if 1 or 2, not drupelike; leaves usually compound.
      7A. Carpels not enclosed within a fleshy calyx-tube (hypanthium).
        8A. Carpels superior, not enveloped when ripe within the calyx-tube.
          9A. Filaments persistent.
            10A. Fruitlets (achenes) dry, not fleshy; plants unarmed.
              11A. Styles deciduous.
                12A. Shrubs with simple leaves.
                  13A. Flowers white, 4-merous, calyculate at base; leaves opposite. .... 5. *Rhodotypos*
                  13B. Flowers usually yellow, 5-merous, not calyculate; leaves alternate. .... 6. *Kerria*
                12B. Herbs, rarely shrubs; leaves compound.
                  14A. Carpels many.
                    15A. Receptacles accrescent after anthesis.
                      16A. Flowers yellow; receptacles not fleshy. .... 7. *Duchesnea*
                      16B. Flowers white; receptacles fleshy in fruit. .... 8. *Fragaria*
                    15B. Receptacles scarcely accrescent; herbs, rarely shrubs, with ternately or pinnately, rarely palmately compound leaves; flowers yellow, rarely white or blood-red. .... 9. *Potentilla*
                14B. Carpels few or rather few.
                  17A. Herbs; petals prominent; stamens many. .... 10. *Waldsteinia*
                  17B. Dwarf procumbent subshrubs; petals small; stamens 5. .... 11. *Sibbaldia*
            11B. Styles persistent, usually elongate after anthesis.
              18A. Calyx-lobes and petals 8 or 9; leaves simple, evergreen, obtusely toothed; styles plumose in fruit; dwarf shrubs.
                          12. *Dryas*
              18B. Calyx-lobes and petals 5; leaves pinnately compound; styles sometimes plumose in fruit; dwarf shrubs or herbs.
                          13. *Geum*
            10B. Fruitlets juicy in fruit; usually armed plants with prickles or bristles. .... 14. *Rubus*
          9B. Filaments deciduous after flowering; fruit an achene. .... 15. *Filipendula*
        8B. Carpels inferior, enveloped by a nonfleshy calyx-tube.
          19A. Petals absent.
            20A. Leaves pinnately compound; flowers not calyculate. .... 16. *Sanguisorba*
            20B. Leaves simple, palmately lobed; flowers calyculate at base. .... 17. *Alchemilla*
          19B. Petals present; calyx with hooked bristles; leaves pinnately compound. .... 18. *Agrimonia*
      7B. Carpels enclosed in a fleshy calyx-tube; usually armed shrubs. .... 19. *Rosa*
    6B. Carpel solitary; fruit a drupe; trees or large shrubs, usually deciduous, rarely evergreen; leaves simple, usually with a pair of glands near the base or on the upper portion of the petioles. .... 20. *Prunus*
  5B. Ovary inferior; carpels 2–5, connate and adnate to the receptacle; fruit a pome, sometimes berrylike.
    21A. Carpels bony; fruit drupelike. .... 21. *Crataegus*
    21B. Carpels with coriaceous or papyraceous walls at maturity; fruit a 1- to 5-loculed pome, each locule with 1 or more seeds.
      22A. Evergreen trees; leaves coriaceous, simple, toothed or entire.
        23A. Inflorescence a compound corymb or flat-topped panicle. .... 22. *Photinia*
        23B. Inflorescence an acute-topped panicle.
          24A. Calyx persistent; fruit rather large, yellow in maturity. .... 23. *Eriobotrya*
          24B. Calyx deciduous; fruit small, black. .... 24. *Rhaphiolepis*
      22B. Deciduous trees or shrubs; leaves simple or compound.
        25A. Flowers in umbels or fascicles, rarely in umbellike corymbs or solitary.
          26A. Carpels many-seeded; flowers short-pedicelled. .... 25. *Chaenomeles*
          26B. Carpels 1- or 2-seeded; flowers on rather elongate pedicels.
            27A. Styles connate at base; fruit usually without or with very few stone-cells in the flesh. .... 26. *Malus*
            27B. Styles free; fruit with abundant stone-cells. .... 27. *Pyrus*
        25B. Flowers in racemes or compound-corymbs.
         28A. Flowers in racemes. .... 28. *Amelanchier*
         28B. Flowers in compound corymbs.
          29A. Fruiting pedicels with prominent warts; leaves coriaceous, finely toothed; calyx persistent. .... 29. *Pourthiaea*
          29B. Fruiting pedicels without prominent warts; leaves pinnately compound or if simple the calyx deciduous after anthesis.
                        30. *Sorbus*

1. STEPHANANDRA Sieb. & Zucc. KOGOME-UTSUGI ZOKU

Deciduous shrubs; leaves alternate, petioled, simple, with irregular toothlike incisions, often slightly lobulate; stipules leaflike, persistent; inflorescence a terminal, corymbose panicle; flowers small, bisexual, ebracteolate, the pedicels bracteate at base; calyx persistent, cup-shaped, the teeth 5, ovate-deltoid, slightly imbricate in bud; petals 5, spathulate, small, white; stamens 10 to many, inserted on the throat of the calyx-tube, incurved; disc hairy; carpels solitary, hairy; style terminal, filiform; ovules pendulous, nearly apical; fruit a small hairy follicle enclosed in the calyx, 1- or 2-seeded; seeds rather small, subglobose, with abundant endosperm.——Few species, in the temperate regions of e. Asia.

1A. Stamens 10; leaves 3–6 cm. long, subpinnately cleft, with 4–6 pairs of lateral nerves; calyx-teeth elliptic, obtuse to rounded at apex.
1. S. incisa
1B. Stamens about 20; leaves 5–9 cm. long, shallowly lobed, with 6–8 pairs of lateral nerves; calyx-teeth ovate-deltoid, cuspidate.
2. S. tanakae

1. Stephanandra incisa (Thunb.) Zabel. *Spiraea incisa* Thunb.; *Spiraea flexuosa* Sieb. & Zucc.——KOGOME-UTSUGI. Much-branched shrub with slender terete nearly glabrous branchlets; leaves broadly deltoid-ovate, slightly pilose to nearly glabrous, caudately acuminate, whitish green beneath, pinnately cleft to lobed, incised and toothed, 3–6 cm. long, 2–3.5 cm. wide, the petioles 4–8 mm. long; stipules linear to broadly lanceolate, ascending, 1.5–7 mm. long; inflorescence paniculate, terminal on young branches, 2–6 cm. long, many-flowered; flowers creamy white, 4–5 mm. across; bracts caducous, broadly linear; petals slightly longer than the calyx; fruiting carpels globose, short-pubescent, about 2 mm. long; seeds shining, about 1.5 mm. across.——May–June. Thickets in low mountains; Hokkaido (Hidaka Prov.), Honshu, Shikoku, Kyushu; common.——Korea.

2. Stephanandra tanakae Fr. & Sav. KANA-UTSUGI. Resembles the preceding, the branches slightly stouter, flexuous, glabrous; leaves shallowly 3- to 5-lobed, with incised acute teeth, glabrous on upper side, slightly pubescent on nerves beneath, 5–9 cm. long, 3.5–6 cm. wide, caudate-acuminate at apex, the petioles 10–15 mm. long, slightly pubescent; stipules 5–12 mm. long; inflorescence glabrous, 4–10 cm. long, flowers slightly larger than in *S. incisa*.——May–June. Mountains; Honshu (Kinki, Tōkaidō, and Kantō Distr., and Ugo Prov.); rare.

2. SPIRAEA L. SHIMOTSUKE ZOKU

Shrubs or semishrubs; leaves alternate, usually simple, toothed; stipules often indistinct; flowers white to rose, small, in racemes, cymes or panicles; calyx persistent, with 4 or 5 teeth; petals small, usually as many as the sepals, sometimes many, inserted on the throat of the calyx-tube; stamens 20–60; disc thickened, adnate to the calyx-tube; carpels usually 5, inserted at the base of the calyx, free or connate at base; styles terminal; ovules many to few; follicles dehiscent on inner side, seeds pendulous, without endosperm.——About 100 species, in the temperate to cold regions of the N. Hemisphere.

1A. Inflorescence umbellate.
  2A. Inflorescence sessile on last year's branchlets, without leaves at base.
    3A. Leaves broadly lanceolate to elliptic or ovate.
      4A. Leaves broadly lanceolate to narrowly ovate-oblong, glabrous or nearly so; flowers single. .... 1. S. faurieana
      4B. Leaves ovate or broadly so, pilose beneath; flowers usually double. .... 2. S. prunifolia
    3B. Leaves linear-lanceolate. .... 3. S. thunbergii
  2B. Inflorescence peduncled or subsessile, terminal on the current year's leafy branchlets.
    5A. Calyx-lobes reflexed after anthesis; stamens longer than the petals; axis of inflorescence more or less distinct.
      6A. Leaves oblong to elliptic, entire or with few teeth toward the tip; branches more or less terete. .... 4. S. media var. sericea
      6B. Leaves ovate, with doubly incised-serrate margin on upper half; branches distinctly angled. .. 5. S. chamaedryfolia var. pilosa
    5B. Calyx-lobes erect or spreading, but not reflexed in fruit.
      7A. Buds flat, outer bud scales longer; leaves entire or with 2 or 3 teeth near the tip; inflorescence with a rather distinct axis.
6. S. nipponica
      7B. Buds terete, outer bud-scales shorter than the inner; leaves usually distinctly toothed, sometimes lobulate.
        8A. Leaves acute, rhombic-lanceolate, glabrous. .... 7. S. cantoniensis
        8B. Leaves obtuse to rounded at apex, flabellate-orbicular to rhombic-ovate, glabrous to pubescent.
          9A. Leaves glabrous or nearly so, obtuse to subrounded at apex. .... 8. S. blumei
          9B. Leaves pubescent beneath, obtuse to subacute at apex; nerves and veinlets impressed above, raised beneath. .. 9. S. nervosa
1B. Inflorescence a compound corymb or panicle.
  10A. Inflorescence a flat-topped to convex compound corymb.
    11A. Leaves obtuse; inflorescence convex at top. .... 10. S. betulifolia
    11B. Leaves acute to acuminate; inflorescence flat-topped.
      12A. Leaves thinly membranous, glabrous, concolorous; inflorescence on short branches of the current year, these rising usually from arcuate elongate branches of the preceding year. .... 11. S. miyabei
      12B. Leaves membranous to coriaceous, pubescent to subglabrate and pale green to glaucous beneath. .... 12. S. japonica
  10B. Inflorescence a panicle. .... 13. S. salicifolia

1. Spiraea faurieana C. K. Schn. *S. pruniflora* Honda ——EZO-NO-SHIJIMIBANA, KO-SHIJIMIBANA. Shrub; branches more or less angled, glabrous or short-pubescent while young; leaves broadly lanceolate, sometimes oblong, 2.5–4 cm. long, 6–12 mm. wide, acute, sometimes subobtuse, cuneate at base, serrulate except on the lower margin, nearly concolorous, glabrous or sparingly pubescent while young, penninerved, the petioles 1–2 mm. long; inflorescence 4- to 6-flowered, with few bracts at base; pedicels slender, 1–2 cm. long, glabrous; flowers white, single; calyx glabrous, the teeth deltoid, erect, short-pubescent within; fruiting carpels glabrous or sparingly pilose.——Apr.–May. Hokkaido, Honshu (n. distr.); rare.

2. **Spiraea prunifolia** Sieb. & Zucc. SHIJIMIBANA. Shrub; branchlets scarcely angled, short-pubescent; leaves ovate to broadly so, 2–3.5 cm. long, 1.5–2 cm. wide, obtuse to acute, serrulate, appressed-pubescent beneath, the petioles 3–7 mm. long, appressed-pubescent; inflorescence sessile, umbellate, 3- to 5-flowered, with few green bracts at base; pedicels 2–4 cm. long, thinly pubescent; flowers white, double, about 10 mm. across.——Apr.–May. A Chinese shrub, cultivated in gardens. The nomenclatorial type is based upon the double-flowered phase in cultivation.——Forma **simpliciflora** Nakai. HITOE-NO-SHIJIMIBANA. Flowers single.——The indigenous wild phase of Formosa, Korea, and centr. China.

3. **Spiraea thunbergii** Sieb. YUKI-YANAGI, KOGOMEBANA. Shrub; branches slender, elongate, arcuate above, slightly angled and short-pubescent while young; leaves linear-lanceolate, 2–4 cm. long, 3–6 mm. wide, acuminate to acute, acuminate at base, with upwardly directed acute teeth, glabrous except for thin pubescence on midrib while young, the petioles scarcely distinct; inflorescence sessile, 2- to 5-flowered, with few bracts at base; pedicels glabrous, 1–1.5 cm. long; flowers about 8 mm. across, single; calyx glabrous, the teeth erect, deltoid; petals broadly obovate, much longer than the stamens; fruiting carpels spreading, glabrous.——Mar.–Apr. A Chinese shrub, widely cultivated and extensively naturalized in Honshu (Kantō Distr. and westw.), Shikoku, Kyushu.

4. **Spiraea media** F. Schmidt var. **sericea** (Turcz.) Regel. *S. mombetsusensis* Franch.; *S. media* var. *mombetsusensis* (Franch.) Cardot; *S. sericea* Turcz.——EZO-SHIMOTSUKE. Shrub; branches slightly angled or terete, short-pubescent while young; leaves oblong to narrowly so, sometimes elliptic, 2–4 cm. long, 6–20 mm. wide, rounded to subacute at apex, entire or with few teeth on upper margin, silky-pubescent beneath and on margin while young, short-pubescent above only while young, often becoming glabrous on both sides, the petioles none or to 2 mm. long; inflorescence solitary on short young branchlets from the lateral buds on the previous year's branches, umbellately racemose with a short axis, many-flowered; pedicels 7–15 mm. long; flowers white, about 7 mm. across; calyx-teeth reflexed, rounded to obtuse; stamens longer than the petals; fruiting carpels short-pubescent, about 2 mm. long.——June–July. Hokkaido, Honshu (Mutsu Prov.).——Kuriles, Sakhalin, Ussuri, Manchuria, and Korea. The typical phase occurs from se. Europe to Siberia.

5. **Spiraea chamaedryfolia** L. var. **pilosa** (Nakai) Hara. *S. ulmifolia* var. *pilosa* Nakai; *S. ussuriensis* Pojark.——AIZU-SHIMOTSUKE. Shrub; young branches angled, glabrous or sparingly short-pubescent; leaves membranous, ovate to narrowly so, 3–6 cm. long, 1.5–3.5 cm. wide, acute, acutely duplicate-toothed except on lower margin, thinly puberulent with appressed short hairs above while young, more or less pubescent beneath, the petioles 3–8 mm. long, slightly pubescent; inflorescence a loosely pubescent umbellate raceme, with short axis; pedicels 10–15 mm. long; flowers many, white, about 10 mm. across; calyx-teeth reflexed, subacute; petals orbicular, shorter than the stamens; fruiting carpels connivent, about 3 mm. long, pubescent, the styles suberect.——May. Hokkaido, Honshu (centr. distr. and northw.).——Sakhalin, n. Korea,

Manchuria, Amur, Ussuri, and Siberia. The typical phase occurs from Europe to Siberia.

6. **Spiraea nipponica** Maxim. *S. ogawae* Nakai; *S. bracteata* Zabel, non Raf.——IWA-SHIMOTSUKE. Shrub; branches angled, glabrous; leaves chartaceous, narrowly oblong to obovate-orbicular, usually elliptic to broadly so, 1–2.5 cm. long, 7–15 mm. wide, rounded to very obtuse at apex, entire or with few obtuse teeth on upper margin, glabrous, glaucous beneath, the petioles 1.5–4 mm. long, glabrous; inflorescence terminal on short branchlets, glabrous or nearly so, usually with slightly developed axis, rather many-flowered; pedicels 1–1.5 cm. long; flowers white, 6–7 mm. across; calyx glabrous, the teeth deltoid, erect, acute, brown-pubescent within; petals orbicular, nearly as long as the stamens; fruiting carpels more or less brown-pubescent to nearly glabrous.——May–July. Mountains; Honshu (Kinki Distr. and eastw.).——Forma **rotundifolia** Makino, MARUBA-IWA-SHIMOTSUKE, has broader leaves.——Forma **oblanceolata** (Nakai) Ohwi. *S. nipponica* var. *oblanceolata* Nakai——NAGABA-IWA-SHIMOTSUKE, has narrower leaves.

Var. **tosaensis** (Yatabe) Makino. *S. tosaensis* Yatabe——TOSA-SHIMOTSUKE. Leaves broadly to narrowly oblanceolate, 1.5–3 cm. long, 3–8 mm. wide, entire or 2- or 3-toothed near the apex.——May. Mountains; Shikoku.

7. **Spiraea cantoniensis** Lour. KO-DEMARI. Glabrous shrub; branches rather slender, the young branchlets not angled; leaves lanceolate to narrowly rhombic-oblong, 2–5 cm. long, 6–20 mm. wide, cuneate to obtuse at base, loosely and irregularly incised-serrate, often glaucescent beneath, the nerves and veinlets raised beneath, the petioles 2–10 mm. long; inflorescence rather many-flowered, the axis sometimes slightly developed; pedicels 1–1.5 cm. long, sometimes with filiform bracteoles at base; flowers white, 7–10 mm. across, single; calyx-teeth deltoid, acute; disc short-pilose; fruiting carpels glabrous.——Apr.–May. Introduced Chinese shrub long cultivated in gardens.

8. **Spiraea blumei** G. Don. *S. kinashii* Koidz.; *S. amabilis* Koidz.; *S. ribisoidea* Koidz.; *S. hypoglauca* Koidz.; *S. tsushimensis* Nakai; *S. sikokualpina* var. *glabra* Koidz.——IWAGASA. Glabrous or nearly glabrous shrub, branches terete; leaves rhombic-ovate, broadly ovate, oblong or narrowly obovate, 1.5–3.5 cm. long, 1–3 cm. wide, obtuse to subrounded or subacute at apex, broadly cuneate to obtuse at base, irregularly toothed or incised except on the lower half, sometimes 3-lobed, the lateral nerves slightly impressed above, raised beneath, the petioles 3–10 mm. long; inflorescence umbellate, rather many-flowered; pedicels 10–15 mm. long, slender; flowers white, 6–8 mm. across; calyx-teeth erect, narrowly deltoid, acute; petals orbicular, as long as or slightly shorter than the stamens; fruiting carpels glabrous or slightly hairy on ventral side.——May. Mountains; Honshu (Kinki Distr. and westw.), Shikoku, Kyushu. Very variable.

Var. **pubescens** (Koidz.) Ohwi. *S. ribisoidea* var. *pubescens* Koidz.; *S. sikokualpina* Koidz.——IYO-NO-MITSUBA-IWA-GASA. Young branches and inflorescence short-pubescent.——Shikoku.

Var. **hayatae** (Koidz.) Ohwi. *S. hayatae* Koidz.; *S. miyajimensis* Makino——URAJIRO-IWAGASA. Leaves glabrous or slightly pubescent beneath; inflorescence, young branchlets, and fruit pubescent.——Honshu (Chūgoku Distr.).

9. **Spiraea nervosa** Fr. & Sav. *S. dasyantha* sensu auct.

Japon., non Bunge; *S. kiusiana* Nakai——IBUKI-SHIMOTSUKE. Shrub; branches not angled, short-pubescent while young; leaves ovate to broadly so or rhombic-oblong, 2–4 cm. long, 8–20 mm. wide, obtuse to subacute at apex, cuneate, obtuse, or rounded at base, irregularly incised-toothed, sometimes obsoletely 3-lobed, pubescent above while young, with pale yellowish appressed pubescence beneath, the nerves and veinlets impressed above, prominently raised beneath, the petioles 3–6 mm. long, pubescent; inflorescence solitary on short branchlets, short-pubescent, rather many-flowered; pedicels 10–15 mm. long; flowers white, 6–8 mm. across; calyx pubescent, the teeth spreading, deltoid, acute, thinly puberulent with short brownish soft hairs or nearly glabrous within; disc short-pubescent; fruiting carpels glabrous or hairy on ventral side.——Apr.–June. Honshu (Kinki Distr.), Shikoku, Kyushu.

Var. **angustifolia** (Yatabe) Ohwi. *S. dasyantha* var. *angustifolia* Yatabe; *S. chinensis* Maxim.; *S. yatabei* Nakai——HOSOBA-NO-IBUKI-SHIMOTSUKE, TŌ-SHIMOTSUKE. Fruiting carpels thinly hairy all over.——Honshu (Kinki Distr. and westw.), Shikoku, Kyushu.——China.

10. Spiraea betulifolia Pall. MARUBA-SHIMOTSUKE. Shrub; young branches slightly striate, glabrous or sparingly pubescent; leaves obovate to broadly obovate or orbicular-ovate, 15–50 mm. long, 10–40 mm. wide, rounded to obtuse at apex, usually broadly cuneate, sometimes rounded or truncate at base, toothed except near base, pale green to slightly glaucous beneath, the nerves pinnate, the petioles glabrous, 1–3 mm. long; inflorescence terminal, many-flowered, glabrous or with thin rather long pubescence while young; flowers white, 5–8 mm. across; calyx-teeth deltoid, reflexed in fruit; petals ovate-orbicular, shorter than the stamens; nectaries prominent; fruiting carpels glabrous or slightly pilose within.——July–Aug. High mountains; Hokkaido, Honshu (n. and centr. distr.); rather common.——e. Siberia, Sakhalin, and Kuriles.

Var. **aemiliana** (C. K. Schn.) Koidz. *S. aemiliana* C. K. Schn.; *S. betulifolia* subsp. *aemiliana* (C. K. Schn.) Hara; *S. stevenii* Rydb.; *S. beauverdiana* C. K. Schn.——EZO-NO-MARUBA-SHIMOTSUKE. Smaller in all parts; young branches usually more or less short-pubescent; leaves broad, rounded at base, 7–15 mm. long, 8–18 mm. wide, the veinlets impressed above; inflorescence smaller, usually short-pubescent; flowers 3–5 mm. across.——July–Aug. High mountains; Hokkaido.——Kuriles and Kamchatka.——Forma **glabra** (Hara) Ohwi. *S. betulifolia* var. *glabra* (Hara) Hara; *S. aemiliana* var. *glabra* Hara. Plant glabrous.

11. Spiraea miyabei Koidz. EZO-NO-SHIROBANA-SHIMO-TSUKE. Shrub; young branches angled, glabrous or nearly so; leaves concolorous, broadly lanceolate to narrowly ovate, 3–7 cm. long, 1–3 cm. wide, acuminate, cuneate at base, doubly toothed on margin with acute incised teeth, loosely white-pilose on midrib beneath near base and on margin when young, the petioles 4–6 mm. long, glabrous; inflorescence terminal on short young branchlets from last year's branches, many-flowered, short-pubescent; flowers white, about 8 mm. across; calyx glabrous, the teeth reflexed, deltoid, acute; petals orbicular, about half as long as the stamens; fruiting carpels with short spreading-hairs.——Apr. Hokkaido (sw. distr.), Honshu (n. distr.); rare.

12. Spiraea japonica L. f. *S. callosa* Thunb.; *S. albiflora* (Miq.) C. K. Schn.; *S. callosa* var. *albiflora* Miq.——SHIMOTSUKE. Shrub; branches usually not or scarcely angled, glabrous or pubescent; leaves lanceolate to ovate, sometimes broadly ovate, 1–8 cm. long, 0.8–4 cm. wide, acute to acuminate, rounded to narrowly cuneate at base, irregularly and loosely acute-toothed or incised-toothed, pale green to glaucous beneath, glabrous on both sides or pubescent beneath, the petioles 1–5 mm. long, glabrous to pubescent; inflorescence flat-topped, often lobed, rather large to small, sometimes short-pubescent, densely many-flowered; flowers pale pink, rarely white, 3–6 mm. across; calyx-teeth deltoid, erect, later obliquely spreading, acute; stamens longer than the petals; disc with nectary glands on the margin; fruit glabrous.——May–Aug. Mountains; Hokkaido, Honshu, Shikoku, Kyushu; rather common and very variable.——Korea, China to the Himalayas.

Var. **bullata** (Maxim.) Makino. *S. bullata* Maxim.——KO-SHIMOTSUKE. Small pubescent shrub; leaves small, broadly ovate, bullate above; flowers small, without glands.——Rarely cultivated in our area, not known as a wild plant, possibly of garden origin.

13. Spiraea salicifolia L. HOZAKI-SHIMOTSUKE. Shrub; young branches slightly angled, pubescent on upper portion; leaves rather thinly chartaceous, lanceolate to broadly lanceolate, 3–10 cm. long, 1–3 cm. wide, acuminate to acute, usually acutely duplicate-toothed, slightly pubescent on upper side while young, the petioles 1–3 mm. long; inflorescence terminal, densely many-flowered, 7–15 cm. long, short-pubescent; flowers pink, 5–7 mm. across; calyx pubescent, the teeth deltoid, acute, erect; stamens much longer than the petals; ovary glabrous.——July–Sept. Wet boggy places in mountains; Hokkaido, Honshu (n. and centr. distr.).——Distributed throughout the cold regions of the Old World.

3. SORBARIA A. Br. HOZAKI-NANA-KAMADO ZOKU

Deciduous shrubs; leaves large, alternate, petiolate, pinnate, the leaflets toothed, the stipules persistent; inflorescence a large terminal panicle; flowers small, white; calyx cup-shaped, the teeth 5, deltoid or rounded, short, reflexed; petals 5, suborbicular; stamens 20–50, inserted on the throat of the calyx-tube; carpels 5, opposite the calyx-lobes, connate at base; follicles ventrally dehiscent; seeds without endosperm.——More than 10 species, in the temperate regions of the Old World.

1. Sorbaria sorbifolia (L.) A. Braun var. **stellipila** Maxim. *S. stellipila* (Maxim.) C. K. Schn.——HOZAKI-NANA-KAMADO. Young branchlets usually stellate-pilose; stipules lanceolate to narrowly ovate, toothed; leaves 15–30 cm. long, minutely puberulent on petioles and axis, the pinnae 15–23, membranous, lanceolate or broadly so, 4–10 cm. long, 8–30 mm. wide, acuminate, acutely duplicate-toothed, more or less stellate-pilose beneath; panicles 10–20 cm. long, densely many-flowered, stellate-pilose, becoming glabrous; flowers white, 7–8 mm. across; stamens longer than the petals; styles about 2.5 mm. long; fruit 4–6 mm. long, pubescent.——July–Aug. Hokkaido, Honshu (centr. and n. distr., rare).——Forma **incerta** (C. K. Schn.) Kitag. *S. stellipila* var. *incerta* C. K. Schn.——EZO-HOZAKI-NANA-KAMADO. Stellate hairs on underside of leaves deciduous at anthesis. The typical phase occurs in nw. Asia.

4. ARUNCUS Kostel. YAMABUKI-SHŌMA ZOKU

Dioecious perennial herbs; rhizomes short-creeping, stout; leaves radical and cauline, alternate, ternately compound, petioled, toothed, exstipulate; inflorescence a terminal panicle of densely flowered, simple to again branched racemes, the bracts linear, often partly adnate to the pedicel; flowers small; calyx 5-toothed, the teeth valvate in bud; petals 5, small, spathulate, inserted on the throat of the calyx-tube; stamens about 20; carpels usually 3, rarely more, free, the ovules many, pendulous; follicles coriaceous, ventrally dehiscent, the seeds scobiform, without endosperm.——Few species in the cold temperate regions of the N. Hemisphere.

1. Aruncus dioicus (Walt.) Fern. var. **kamtschaticus** (Maxim.) Hara. *A. sylvester* var. *kamtschaticus* Maxim.; *A. kamtschaticus* (Maxim.) Rydb.; *A. sylvester* var. *tomentosus* Koidz.; *A. tomentosus* (Koidz.) Koidz.; *A. sylvester* var. *tenuifolius* Nakai; *A. sylvester* var. *americanus* Maxim. pro parte; *A. kyusianus* Koidz.; *A. dioicus* var. *tenuifolius* (Nakai) Hara ——YAMABUKI-SHŌMA. Glabrous to sparingly pubescent coarse perennial herb with rather stout branching rhizomes; stems 30–80 cm. long, with a few deciduous scales at base, terete, few-leaved; leaves few, 2- or 3-ternate, the leaflets thinly chartaceous, often lustrous, narrowly ovate to ovate-orbicular, 3–10 cm. long, 1–6 cm. wide, acuminate to caudate, incised and toothed, sometimes pinnately lobed; inflorescence 10–30 cm. long, short-pubescent; flowers many, white, short-pedicelled; petals broadly spathulate, small; fruit pendulous on a deflexed pedicel, the carpels lustrous, about 2.5 mm. long, the styles short, about 0.5 mm. long.——June–Aug. Hokkaido, Honshu, Shikoku, Kyushu; rather common and variable.——Kamchatka, Kuriles, Sakhalin, Ussuri, Manchuria, Korea, and China.

Var. **insularis** Hara. SHIMA-YAMABUKI-SHŌMA. Petals larger, elliptic, about 1.2–1.5 mm. long, firm; calyx-lobes larger; style longer, about 0.8 mm. long.——Izu Isls. (Oshima).

Var. **laciniatus** (Hara) Hara. *A. vulgaris* var. *laciniatus* Hara——KIREBA-YAMABUKI-SHŌMA. Leaflets deeply incised, ovate-lanceolate; petals in staminate flowers narrowly oblong, 2 mm. long.——Hokkaido (s. Hidaka Prov.).

Var. **astilboides** (Maxim.) Hara. *Spiraea aruncus* var. *astilboides* Maxim.; *A. astilboides* Maxim.; *A. sylvester* var. *astilboides* (Maxim.) Makino; *A. vulgaris* var. *astilboides* (Maxim.) Nemoto——MIYAMA-YAMABUKI-SHŌMA. Pedicels and follicles erect when mature.——High mountains; Honshu (Mount Hayachine in Rikuchu Prov.).

Var. **subrotundus** (Tatew.) Hara. *A. subrotundus* Tatew.; *A. sylvester* var. *subrotundus* (Tatew.) Ohwi——APOI-YAMABUKI-SHŌMA. Leaflets rather firm, ovate-orbicular to broadly ovate, rounded with a short cusp at apex, acutely toothed; fruit pendulous on a deflexed pedicel.——Hokkaido (Mount Apoi in Hidaka Prov.).

5. RHODOTYPOS Sieb. & Zucc. SHIRO-YAMABUKI ZOKU

Deciduous shrubs; leaves opposite, short-petioled, simple, penninerved, the stipules free, linear, hairy, membranous, pale green; flowers solitary, terminal on short branches, large, white, short-pedicelled; calyx persistent, the tube short, the segments 4, spreading, rather large, acutely toothed, bracts of the caliculus 4, small; petals 4, spreading, orbicular; stamens many; carpels 4; styles filiform, glabrous; ovules 2 in each carpel; achenes 1–4, black, plump.——A single species, in e. Asia.

1. Rhodotypos scandens (Thunb.) Makino. *Corchorus scandens* Thunb.; *Kerria tetrapetala* Sieb.; *R. tetrapetala* (Sieb.) Makino; *R. kerrioides* Sieb. & Zucc.——SHIRO-YAMABUKI. Shrub with spreading glabrous branches; leaves membranous, ovate, 4–6 cm. long, 2–4 cm. wide, acuminate, rounded at base, acutely double-toothed, more or less long-appressed yellowish pubescent beneath, the petioles 2–3 mm. long; flowers 3–4 cm. across, on pedicels 7–20 mm. long; calyx and pedicels pubescent, the lobes narrowly ovate, 1–1.5 cm. long, the bracts of the caliculus linear to lanceolate, 4–8 mm. long; achenes 7–8 mm. long, ellipsoidal, glabrous.——Apr.–May. Honshu (Chūgoku Distr.); often cultivated.——Korea and China.

6. KERRIA DC. YAMABUKI ZOKU

Deciduous branched shrubs; leaves short-petioled, simple, alternate, the stipules linear, pale brownish; flowers pedicellate, large, terminal and solitary on short branchlets, usually yellow; calyx persistent; petals 5, or many in a double-flowered cultivar; stamens many; carpels 5–8, glabrous, the ovules solitary in each locule, the styles filiform; achenes ovate-globose, rather plump.——A single species, in China and Japan.

1. Kerria japonica (Thunb.) DC. *Rubus japonicus* L.; *Corchorus japonicus* Thunb.——YAMABUKI. Much-branched shrub with slender, green, spreading, glabrous branches; leaves membranous, ovate or narrowly so, 3–7 cm. long, 2–3.5 cm. wide, long-acuminate, incised duplicate-toothed, vivid green, paler beneath, nearly glabrous; flowers 2.5–5 cm. across; calyx glabrous, the tube short, the segments broadly ovate to elliptic, spreading, denticulate; petals ovate-orbicular, spreading; achenes 4–5 mm. long.——Apr.–May. Common in mountains; Hokkaido, Honshu, Shikoku, Kyushu. Several cultivars are grown: Cv. **Plena**. *K. japonica* forma *plena* C. K. Schn. Flowers double.——Cv. **Stellata**. *K. japonica* forma *stellata* (Makino) Ohwi; *K. japonica* var. *stellata* Makino——KIKUZAKI-YAMABUKI. Flowers with more numerous petals than in Cv. Plena.——Cv. **Albescens**. *K. japonica* forma *albescens* (Makino) Ohwi; *K. japonica* var. *albescens* Makino——SHIROBANA-YAMABUKI. Flowers nearly white.

7. DUCHESNEA Smith   Hebi-ichigo Zoku

Stoloniferous perennial herbs; leaves alternate, 3(–5)-foliolate, toothed, petiolate, the stipules partially adnate to the base of the petioles; pedicels axillary, solitary; flowers yellow, bisexual; calyx persistent, the tube spreading, short, the segments 5, the bracts of the caliculus large, alternate with the calyx-segments, toothed; stamens 20–30; carpels many, free, inserted on a globose receptacle; achenes small, inserted on a cavity of the receptacle which is much accrescent and somewhat spongy in fruit.——Few species in temperate and warmer regions in e. and s. Asia.

1A. Achenes distinctly rugose-tubercled, not shining; fruiting receptacles pinkish white, not shining, 8–12 mm. across, with a whitish neck; plant smaller than the following. .... 1. *D. chrysantha*
1B. Achenes nearly smooth, glossy when fresh; fruiting receptacles bright red, glossy, 11–20 mm. across, with a red neck; plant larger than the preceding. .... 2. *D. indica*

1. Duchesnea chrysantha (Zoll. & Moritz.) Miq. *Fragaria chrysantha* Zoll. & Moritz.; *F. indica* var. *wallichii* Fr. & Sav., excl. syn.; *D. indica* auct. Japon., non Focke; *D. indica* var. *japonica* Kitam.; *D. formosana* Odashima——Hebi-ichi-go.   Rather soft green perennial herb with long, slender, leafy stolons; leaflets ovate to ovate-orbicular, 2–3.5 cm. long, 1–3 cm. wide, acute, doubly or incised-toothed, thinly pubescent with ascending hairs on both surfaces especially beneath, the lateral leaflets often shallowly 2-lobed, the petioles with short ascending hairs, the stipules narrowly ovate to broadly lanceolate; calyx-segments narrowly ovate, acute, 5–8 mm. long, the bracts of the caliculus slightly larger than the calyx-segments, spreading, obovate; petals 5–10 mm. long; fruit globose, about 10 mm. across, red, the achenes small, about 1.5 mm. long.——Apr.–June.   Sunny grassy places in lowlands, especially common around paddy fields; Hokkaido, Honshu, Shikoku, Kyushu.——Korea, Manchuria, China to Malaysia.

2. Duchesnea indica (Andr.) Focke. *Fragaria indica* Andrews; *D. fragiformis* J. E. Sm.; *Potentilla denticulosa* Ser.; *P. wallichiana* Ser.; *F. malayana* Roxb.; *F. roxburghii* Wight & Arn.; *D. indica* var. *major* Makino; *D. major* (Makino) Makino——Yabu-hebi-ichigo.   Closely resembles the preceding but slightly larger, deeper green; leaflets to 7 cm. long; receptacles dark-colored, glabrous; achenes nearly smooth.——Apr.–May.   Shady places in woods and on grassy slopes along streams and in ravines in low mountains; Hokkaido, Honshu, Shikoku, Kyushu; rather common.——China to India.

8. FRAGARIA L.   Oranda-ichigo Zoku

Stoloniferous perennial herbs more or less long-pubescent; rhizomes rather stout; leaves radical, alternate, 3-foliolate, petioled, the leaflets toothed, the stipules membranous, partially adnate to the base of the petiole; scapes radical, few-flowered; flowers white; calyx persistent, with a caliculus outside; petals 5–8, as many as the calyx-segments; stamens many; receptacle semiglobose to globose, much accrescent and globose to ellipsoidal and fleshy in fruit; carpels many, free, the style from the side of the carpel; achenes small, glabrous.——More than 10 species, in the temperate regions of the N. Hemisphere and in S. America.

1A. Scapes rather many-flowered; plants rather coarse; cultivated. .... 1. *F.* × *ananassa*
1B. Scapes 1- to 5-flowered; plants slender; indigenous.
  2A. Petals 5 (rarely 6), suborbicular; leaves green, paler beneath; anthers ovate.
    3A. Stamens longer than the gynoecium.
      4A. Pedicels with ascending hairs. .... 2. *F. nipponica*
      4B. Pedicels with spreading hairs. .... 3. *F. yezoensis*
    3B. Stamens nearly as long as the gynoecium. .... 4. *F. vesca*
  2B. Petals 7 or 8, obovate to narrowly so; leaves green and slightly glaucescent; anthers broadly elliptic. .... 5. *F. iinumae*

1. Fragaria × **ananassa** Duchesne. *F. chiloensis* var. *ananassa* (Duchesne) L. H. Bailey; *F. grandiflora* Ehrh.——Oranda-ichigo.   Rhizomes rather stout, with long stolons after anthesis; terminal leaflets broadly ovate or obovate, 5–8 cm. long, 4–6 cm. wide, dentate-serrate, whitish beneath with appressed pubescence especially on nerves beneath, the petioles with ascending to spreading hairs; flowers 5–15 on a scape, 2–2.5 cm. across, 5- to 6-merous; fruit globose to ovoid, 1.5–3 cm. across, the achenes in pits on the fleshy receptacle.——Apr.–May.   Hybrid of *F. chiloensis* × *F. virginiana*. Garden strawberry widely cultivated for the edible fruit.

2. Fragaria nipponica Makino. *F. vesca* sensu auct. Japon., non L.; *F. elatior* sensu auct. Japon., non Ehrh.; *F. yakusimensis* Masam.——Shirobana-no-hebi-ichigo.   Rhizomes rather thickened, short; terminal leaflets elliptic to broadly ovate, 2–5 cm. long, 1–3 cm. wide, subobtuse, loosely appressed-pubescent on upper side, pale green and prominently to distinctly appressed-pubescent especially on nerves beneath, the margin with ovate or subdeltoid teeth, the petioles with spreading pubescence; scapes 1- to 4-flowered, with spreading pubescence, as long as the leaves or slightly longer; flowers 1.5–2 cm. across, on ascending-pubescent pedicels; calyx-segments narrowly ovate, acuminate, appressed-pubescent outside; stamens 3–4 mm. long; receptacles loosely short-pilose in fruit, broadly ovoid to subglobose, about 10 mm. across, the achenes in pits on the receptacles, many, smooth or very obsoletely nerved.——May–July.   Mountains; Honshu (Kantō and centr. distr.), Kyushu (Yakushima).——Korea (Quelpaert Isl.).

3. Fragaria yezoensis Hara. *F. nipponica* var. *yezoensis* (Hara) Kitam.; *F. neglecta* sensu auct. Japon., non Lindem.——Ezo-kusa-ichigo.   Closely resembles the preceding; stoloniferous, the pubescence rather spreading; terminal leaflets rhombic-obovate, rounded at apex, dentate-serrate, with thinly silky pubescence on both sides, denser on nerves beneath and on margin, the petioles with spreading hairs; scapes to 15 cm. long, the pedicels with spreading pubescence; flowers 1.5–2 cm. across; calyx-segments ovate-lanceolate, acute; stamens about 3.5 mm. long; receptacles pilose; achenes with 1 or 2 obsolete nerves on ventral side.——Hokkaido (e. distr.).——s. Kuriles.

4. Fragaria vesca L. EZO-HEBI-ICHIGO.   Closely allied to No. 2, but slightly larger; terminal leaflets broadly rhombic-obovate; stamens shorter, 1.5–2.5 mm. long; receptacles glabrous or only short-pilose at base.——Grassy places in lowlands; Hokkaido; possibly not indigenous in our area.——Europe.

5. Fragaria iinumae Makino.   *Potentilla daisenensis* Honda——NŌGŌ-ICHIGO.   Rhizomes rather stout; terminal leaflets broadly obovate or cuneate-orbicular, 15–30 mm. long, 12–25 mm. wide, rounded at apex, broadly cuneate at base, glaucous to green above, prominently glaucous beneath, glabrous above, with appressed to ascending long soft pubescence beneath especially on the nerves, the petioles long-pubescent; scapes 1- to 3-flowered, with appressed to ascending pubescence; flowers 15–25 mm. across; stamens about 2 mm. long; receptacles glabrous, ovoid to narrowly ovoid in fruit, about 8 mm. across, the achenes in pits, smooth.——June–July. Alpine regions; Hokkaido, Honshu (centr. and n. distr. and Mount Daisen in Hōki Prov.).——Sakhalin.

9. POTENTILLA L.   KIJI-MUSHIRO ZOKU

Herbs or shrubs, glabrous or tomentose; leaves radical or alternate, rarely opposite on the flowering stems, digitately 3- to 7-foliolate or pinnate, the leaflets toothed or pinnately lobulate, the stipules adnate to the base of the petiole; flowers solitary in leaf-axils or in corymbs, yellow, white, rarely red; calyx persistent, the tube concave to urceolate, the lobes 5, alternate with the bracts of the caliculus, as many as the petals; stamens many; receptacles small, conical, sometimes pilose; carpels many, the style lateral or terminal, persistent or deciduous; achenes small, glabrous or pilose, sessile, 1-seeded.——More than 300 species, in temperate and cooler regions of the N. Hemisphere, rare in the mountains in the Tropics, a few in the S. Hemisphere.

1A. Receptacles not accrescent after anthesis.
  2A. Herbaceous or suffrutescent.
    3A. Flowers solitary in axils of much-elongate stolons, the pedicels 10–30 cm. long; leaves pinnate, the leaflets 13–19, densely white-woolly beneath; plants of wet places near the sea and in salt marshes. .... 1. *P. egedei* var. *groenlandica*
    3B. Flowers at least partially corymbose, the pedicels not more than 4 cm. long; plants of grassy, wooded or rocky places from lowlands to high mountains.
      4A. Leaflets 5, digitate. .... 2. *P. kleiniana*
      4B. Leaflets 3 or more, pinnate.
        5A. Stems decumbent and branching, very slender and elongate; flowers axillary, solitary, small. .... 3. *P. centigrana*
        5B. Stems erect or ascending only below; stolons if present not flower-bearing.
          6A. Stems scapose, less than 40 cm. long, with 1–3 small cauline leaves.
            7A. Leaflets glaucous beneath. .... 4. *P. dickinsii*
            7B. Leaflets green or white-villous beneath.
              8A. Leaflets 3; stolons absent; anthers ovate-globose to broadly ovoid.
                9A. Achenes long-pilose; leaves broadly cuneate, truncate, with 3 coarse teeth at apex. .... 5. *P. miyabei*
                9B. Achenes glabrous or nearly so; leaflets with few to several teeth on each side.
                  10A. Leaves densely white-tomentose beneath. .... 6. *P. nivea*
                  10B. Leaves green on both sides.
                    11A. Leaves rather densely long-pubescent, thick; flowers 3–4 cm. across. .... 7. *P. megalantha*
                    11B. Leaves loosely long-pilose, rather thin; flowers about 2 cm. across. .... 8. *P. matsumurae*
              8B. Leaflets 3–7, if 3, plant with stolons; anthers oblong to broadly lanceolate.
                12A. Leaves with dense white-woolly hairs beneath; plant without stolons. .... 9. *P. discolor*
                12B. Leaves green or pale green beneath, not white-woolly.
                  13A. Leaflets 5–7, rarely 9.
                    14A. Stolons absent.
                      15A. Leaflets 5–9, the lower gradually smaller. .... 10. *P. fragarioides* var. *major*
                      15B. Leaflets 3, sometimes with a pair of smaller ones on the upper part of the leaf-axis. .... 11. *P. togasii*
                    14B. Stolons elongate. .... 12. *P. stolonifera*
                13B. Leaflets 3, rarely 5.
                  16A. Leaflets usually rhombic-ovate, acute to acuminate, thinly membranous; flowers 1–7, 15–20 mm. across; rhizomes scarcely thickened, stolons much elongate, prominent. .... 13. *P. yokusaiana*
                  16B. Leaflets usually oblong, obtuse to subacute; flowers few to rather many, 1–1.5 cm. across; rhizomes short and thickened, often bearing rather short to slightly elongate stolons.
                    17A. Leaflets herbaceous; stipules of the bracts broadly obovate or flabellate, dilated at tip, prominently toothed on margin; anthers narrowly oblong, yellow. .... 14. *P. freyniana*
                    17B. Leaflets subcoriaceous, lustrous above; stipules of the bracts oblong to lanceolate, not dilated, with fewer smaller teeth near apex; anthers ovate-orbicular, pale yellow. .... 15. *P. riparia*
         6B. Stems erect or with ascending base, 30–100 cm. long, rather stout, leafy; flowers many.
           18A. Cauline leaves 3-foliolate, green; leaflets toothed.
             19A. Stipules adnate for half the length of the petioles; perennial; cauline leaves of 3 elliptic-ovate leaflets.
                          16. *P. cryptotaeniae*
             19B. Stipules adnate only at the base of the petioles; annual or biennial; cauline leaves of 3 (rarely 5) obovate to oblong, obtuse to subacute leaflets. .... 17. *P. norvegica*
           18B. Cauline leaves, at least the lower ones, pinnately compound; leaflets pinnately lobed to parted, with short white-woolly tomentum beneath.

1. Potentilla egedei Wormsk. var. **groenlandica** (Tratt.) Polunin. *P. pacifica* Howell; *P. anserina* sensu auct. Japon., non L.; *P. anserina* var. *orientalis* Card.; *P. anserina* var. *groenlandica* Tratt.——Ezo-tsuru-kimbai. Perennial herb with elongate slender stolons; radical leaves 10–40 cm. long; stipules scarious, adnate to the base of the petioles, the rachis with short accessory leaflets, the ordinary leaflets 13–19, oblong or narrowly so, coarsely toothed, 2–5 cm. long, 7–20 mm. wide, glabrous to long-pilose on upper side, with white-woolly hairs beneath, and long silky hairs on the nerves; leaves on the stolons very small; flowers 20–30 mm. across, pedicellate, solitary, axillary on stolons; calyx-segments ovate-deltoid, 4–6 mm. long, acute; bracts of caliculus entire, broadly lanceolate, slightly smaller than the calyx-segments; petals broadly obovate; anthers elliptic; styles about 2 mm. long; achenes smooth, rounded on back.——June–July. Wet saline places near the sea; Hokkaido, Honshu (n. distr.).——Korea, Ochotsk Sea region, Kuriles, Sakhalin, Kamchatka, and N. America.

2. Potentilla kleiniana Wight & Arn. *P. reptans* A. Gray; *Duchesnea sundaica* Miq.; *? P. savatieri* Card.——O-hebi-ichigo. Perennial herb with short rhizomes densely leafy at apex; stems 20–40 cm. long, erect, ascending above, few-leaved and branched, slightly ascending-pubescent; radical leaves long-petioled; stipules tightly adnate to the petiole, the free portion narrowly deltoid, acute, the cauline leaves with the stipules shortly adnate, the free portion ovate, often dentate, the leaflets 5 (or 1–3 on the upper cauline leaves), digitate, narrowly ovate to broadly oblanceolate, 1.5–5 cm. long, 8–20 mm. wide, obtuse, nearly glabrous above, paler and with appressed-pilose nerves beneath; inflorescence rather many-flowered, the pedicels 5–20 mm. long, with white ascending-pilose hairs; flowers about 8 mm. across; calyx-segments ovate, subacute; bracts of caliculus narrower; stamens broadly ovate; receptacle short-pilose at base; achenes broadly ovoid, 0.6 mm. long, slightly longitudinally rugose; styles short, nearly as long as the achenes.——May–June. Honshu, Shikoku, Kyushu; rather common.——Korea, China, India, and Malaysia.

3. Potentilla centigrana Maxim. *P. centigrana* var. *japonica* Maxim.; *P. centigrana* var. *mandshurica* Maxim.; *P. reptans* var. *trifoliata* A. Gray; *P. rosulifera* Lév.; *P. longepetiolata* Lév.——Hime-hebi-ichigo. Rather slender perennial herb; stems decumbent, later much elongate, very slender, branching and creeping, many-leaved, slightly spreading-pilose; stipules lanceolate to ovate, subacute; leaflets 3, thin, subsessile, obovate, elliptic or obovate-orbicular, 1–3 cm. long, 8–20 mm. wide, obtuse to rounded, nearly glabrous above, slightly glaucous and thinly appressed-pilose on nerves beneath; flowers appearing terminal, but soon axillary, solitary or in a terminal corymb in the upper ones, the pedicels slender, 1–4 cm. long; calyx-segments narrowly ovate, acute,

nearly as long as the bracts of the caliculus; petals about 3 mm. long, yellow; anthers elliptic; achenes broadly ovate, glabrous, with few irregular longitudinal lines; styles linear, about 0.7 mm. long.——Apr.–Aug. Wooded mountains; Hokkaido, Honshu, Shikoku, Kyushu.——Korea, n. China, and Ussuri.

4. Potentilla dickinsii Fr. & Sav. *P. ancistrifolia* var. *dickinsii* (Fr. & Sav.) Koidz.; *P. ancistrifolia* sensu auct. Japon., non Bunge——Iwa-kimbai. Short-pubescent perennial herb with stout short woody branching rhizomes; stems 10–20 cm. long, erect; radical leaves petioled; stipules lanceolate, acuminate; leaflets 3–5, subcoriaceous, the upper 3 larger, rhombic-ovate or elliptic, 2.5–5 cm. long, 1.5–3 cm. wide, acute or sometimes obtuse, with coarse acute deltoid teeth, green above, glaucous beneath; inflorescence few- to rather many-flowered; flowers about 1 cm. across, yellow; calyx-segments narrowly ovate, acuminate; bracts of caliculus lanceolate; anthers broadly ovate; receptacle with dense white hairs about 2 mm. long; achenes ovoid, brown, smooth, with few long curled hairs longer than the body; styles filiform, about 1.5 mm. long.——June–Aug. Rocks in mountains; Hokkaido, Honshu, Shikoku, Kyushu.

5. Potentilla miyabei Makino. Meakan-kimbai. Perennial herb with yellowish brown appressed strigose hairs; rhizomes creeping, short, branching, slightly woody; stems 3–10 cm. long, few-leaved, erect; leaves 3–5 cm. long; free segments of the stipules lanceolate, gradually acute; leaflets 3, subcoriaceous, broadly cuneate, 5–12 mm. long, 4–10 mm. wide, truncate and with 3 coarse teeth at apex, grayish green; flowers 1–5, about 15 mm. across; calyx-segments lanceolate, acute; bracts of caliculus similar to the calyx-segments but smaller; petals 5, 1.5 times as long as the calyx-segments, obovate-orbicular; receptacle with dense strigose hairs about 4 mm. long; achenes ovoid, brown, about 1.5 mm. long, smooth, with straight hairs about 4 mm. long at the base and with minutely appressed hairs on the upper portion; styles filiform, 3–3.5 mm. long.——July–Aug. Gravelly to sandy slopes, alpine; Hokkaido; rare.

6. Potentilla nivea L. *P. macrantha* Ledeb.; *P. matsuokana* Makino——Urajiro-kimbai. Perennial herb with stout short-creeping and branching rhizomes; stems 10–20 cm. long, 1- or 2-leaved; radical and cauline leaves beneath and the petioles with densely woolly white tomentum; stipules scarious, reddish brown, the free segments broadly lanceolate, acute, green and sometimes dentate in the cauline ones; leaflets 3, obovate-oblong to broadly ovate, 15–25 mm. long, 10–17 mm. wide, obtuse, slightly silky-pubescent and green above, with ovate marginal teeth; flowers 1–5, 1.5–2 cm. across; calyx-segments broadly lanceolate, acute; bracts of caliculus narrowly lanceolate or broadly linear; petals about 1.5 times as long as the calyx-segments, obovate-orbicular; receptacles short-pilose; achenes ovoid, about 1.5 mm. long, smooth, gla-

FIG. 8.—*Potentilla matsumurae* T. Wolf. 1, Habit; 2, leaf, upper side; 3, the same, lower side; 4, hairs on leaf margin; 5, 6, stipules; 7, upper cauline leaf; 8–11, bracts; 12, flower bud; 13, 14, flowers seen from above and below; 15, 16, caliculi; 17, hairs on the caliculus; 18, 19, sepals above and below; 20, petal; 21, 22, stamens; 23, pollen; 24, 25, carpels; 26, surface of the carpel; 27, fruit; 28, the same, perianth removed; 29, 30, achenes; 31, floral diagram.

brous.——July–Aug. Rocks and dry slopes in alpine regions; Hokkaido, Honshu (centr. distr.); rare.——Widespread in northern and alpine regions of the N. Hemisphere.

7. Potentilla megalantha Takeda. *P. fragiformis* subsp. *megalantha* (Takeda) Hult.; *P. fragiformis* sensu Maxim., non Willd.——CHISHIMA-KIMBAI. Densely long-silky perennial with short-creeping stout branching rhizomes; stems 10–30 cm. long, few-leaved; radical leaves tufted, 5–15 cm. long; stipules brown, membranous, the free segments broadly ovate, oblique, acute; leaflets 3, broadly obovate-cuneate, 3–4 cm. long and nearly as wide, with ovate obtuse coarse teeth, rather thick-herbaceous to coriaceous-herbaceous, densely pubescent especially beneath; flowers 3–7, 3–4 cm. across; calyx-segments narrowly ovate, subacute, 6–15 mm. long; bracts of caliculus ovate, obtuse, rather small; receptacle short-pilose; achenes broadly ovoid, about 1.5 mm. long, with obsolete striations, acutely keeled on the back; styles filiform, 3–3.5 mm. long. ——July–Aug. Rocky places or on rocks near the sea; Hokkaido.——Sakhalin, Kuriles, and Kamchatka.

8. Potentilla matsumurae T. Wolf. *P. gelida* sensu auct. Japon., non C. A. Mey.——MIYAMA-KIMBAI. FIG. 8. Perennial herb with more or less coarse pubescence; rhizomes stout, short, sparsely branching; stems 10–20 cm. long, few-leaved; stipules brown, subscarious, the segments broadly ovate, obtuse to subacute, green and ovate on the cauline leaves; leaflets 3, broadly obovate-cuneate, 1.5–3 cm. long and nearly as wide, slightly paler beneath, with few usually ovate acute teeth on each side except toward the base; flowers 1–10, about 20 mm. across; calyx-segments narrowly ovate, acute, 4–5 mm. long; bracts of caliculus oblong, obtuse; petals obovate-orbicular, about twice as long as the calyx-segments; receptacle short-pilose; achenes ovoid, smooth, 1.2–1.5 mm. long, nearly rounded on back; styles filiform, about 3 mm. long.——July–Aug. Dry alpine slopes; Hokkaido, Honshu (centr. and n. distr.); rather common and slightly variable.

Var. **lasiocarpa** Hara. KE-MIYAMA-KIMBAI. Carpels long-hairy.——Hokkaido and Honshu (n. distr.).

9. Potentilla discolor Bunge. TSUCHIGURI. Perennial herb with stout rhizomes; stems 15–40 cm. long, densely white-woolly while young, sparingly long-pilose, few-leaved; radical leaves tufted, 8–20 cm. long; stipules white-woolly, membranous, the segments short, lanceolate, acute, green and ovate with few teeth in the cauline ones; leaflets 3–7 (rarely 9), oblong to narrowly so, 2–4 cm. long, 8–15 mm. wide, obtuse to acute, green, with thin woolly pubescence above while young, densely white-woolly tomentose beneath, the nerves beneath with appressed long pilose hairs while young, the marginal teeth broadly ovate, subacute; inflorescence few- to many-flowered; flowers 12–15 mm. across; calyx-segments narrowly ovate, acute, 3–4 mm. long; bracts of caliculus slightly smaller; petals 5, about twice as long as the calyx-segments, broadly obovate; receptacle hairy; achenes ovoid, nearly smooth and glabrous, brown, about 1 mm. long, rounded on back; styles rather stout, about 1 mm. long.——Apr.–June. Honshu (Kinki Distr.), Kyushu.——Formosa, China, Manchuria, and Korea.

10. Potentilla fragarioides L. var. **major** Maxim. *P. fragarioides* var. *sprengeliana* (Regel) Maxim.; *P. sprengeliana* Lehm.; *P. rufescens* Fr. & Sav.——KIJI-MUSHIRO. Perennial more or less pubescent or pilose herb with short stout rhizomes; stems 5–30 cm. long, often ascending at base, with coarse long hairs; leaflets of the radical leaves 5–7 (rarely 3–9),

the lower pairs gradually smaller, the upper 3 nearly equal, larger, broadly ovate to elliptic, 1.5–5 cm. long, 1–3 cm. wide, obtuse to subrounded, the teeth broadly ovate or rather deltoid; stipules membranous, coarsely pilose outside, the segments broadly lanceolate, acute, green and ovate with few teeth in the cauline ones; flowers few to many, rather large, 15–20 mm. across; calyx-segments narrowly ovate to broadly lanceolate, subacute, 4–8 mm. long, the bracts of caliculus equal or slightly smaller; petals 1.5–2 times as long as the calyx-segments; receptacle hairy; achenes ovoid, finely rugulose, 1–1.2 mm. long, glabrous; styles rather stout, 1–1.5 mm. long.—— Mar.–Aug. Sunny slopes and waste grounds in lowlands and on mountains; Hokkaido, Honshu, Shikoku, Kyushu; very variable.——e. Asia. The typical phase occurs in Siberia, India, and China.

11. Potentilla togasii Ohwi. ECHIGO-KIJI-MUSHIRO. Thinly pubescent, green perennial; rhizomes stout, sometimes branched; radical leaves many, 15–35 cm. long, much shorter at anthesis, the leaflets usually 3, membranous, sometimes with a pair of much smaller ones on the petiole, the terminal rhombic-ovate, 4–10 cm. long, 2–6 cm. wide, toothed, acute or nearly so at both ends; flowers yellow, 15–20 mm. across; sepals and bracts of caliculus deltoid-lanceolate; petals 5, orbicular; receptacles white-pilose.——May–June. Honshu (Japan Sea side from Uzen to Etchu).

12. Potentilla stolonifera Lehm. *P. fragarioides* var. *flagellaris* Lehm.; *P. fragarioides* var. *stolonifera* (Lehm.) Maxim.; *P. fragariiformis* var. *japonica* A. Gray; *P. japonica* Bl.——TSURU-KIJI-MUSHIRO. Closely allied to No. 10; stolons elongate; stems rather fewer flowered.——Sunny slopes on mountains; Hokkaido, Honshu, Shikoku, Kyushu.——Kuriles, Sakhalin, and Kamchatka.

13. Potentilla yokusaiana Makino. *P. freyniana* var. *grandiflora* T. Wolf.——TSURU-KIMBAI. Rather soft vivid green perennial herb with loose pubescence and long slender loosely leaved stolons; rhizomes short, not or very slightly thickened; stems 10–20 cm. long; radical leaves 3-foliolate, petioled; stipules membranous, pale, the segments narrowly ovate, acute, green on the stolons; leaflets 3 or very rarely with an additional small pair on the upper part of the petiole, membranous, rhombic ovate to broadly so, acute to short-acuminate (acute to obtuse in the radical leaves), 1.5–4 cm. long, 1.2–3 cm. wide, rather deeply and acutely toothed; flowers few, 15–20 mm. across; calyx-segments narrowly ovate to broadly lanceolate, acuminate; bracts of caliculus nearly alike; petals 1.5 times as long as the calyx-segments; receptacle hairy on margin; achenes glabrous, the style about 2 mm. long, nearly filiform.——Apr.–June. Wooded mountains; Honshu, Shikoku, Kyushu.

14. Potentilla freyniana Bornm. *P. ternata* (Maxim.) Freyn, non C. Koch; *P. fragarioides* var. *ternata* Maxim.—— MITSUBA-TSUCHIGURI. Perennial short-pilose herb, with short slender stolons; rhizomes short and stout; stems 15–30 cm. long, slender; stipules of radical leaves pale green to brownish; leaflets 3 (rarely also with a pair of small accessory leaflets on the upper part of the petiole), oblong to ovate or obovate, rounded to subacute, toothed, rather thin, green above, often somewhat purplish beneath; flowers rather many, 10–15 mm. across; calyx-segments broadly lanceolate, acute; bracts of caliculus slightly smaller; petals 1.5 times as long as the calyx-segments; receptacle slightly pilose; achenes about 1 mm. long, pale, rugulose, glabrous; styles rather stout, about 1 mm. long.

——Apr.–June. Sunny slopes of hills; Hokkaido, Honshu, Shikoku, Kyushu; common.——Korea, Manchuria, and Amur.

15. Potentilla riparia Murata. TERI-HA-KIMBAI. Perennial herb with slender elongate stolons to 80 cm. long; rhizomes short, ascending, rather slender; leaves few, subcoriaceous, ternate, sparingly pilose, the terminal leaflet slightly larger than the lateral, elliptic to obovate-elliptic, 2–5 cm. long, 1–2 cm. wide, obtuse, cuneate at base; scapes slender, 10–25 cm. long, ascending, few-bracted and few-flowered; flowers pale yellow, 12–13 mm. across; bracts of caliculus shorter than the sepals, linear-oblanceolate; sepals 5, deltoid-lanceolate, 4–5 mm. long, acute; petals 5, obovate, slightly emarginate; receptacle pilose.——Apr.–June. Honshu (Kinki Distr. and westw.); rare.

16. Potentilla cryptotaeniae Maxim. *P. cryptotaeniae* var. *insularis* Kitag.——MITSUMOTO-SŌ. Perennial herb with spreading pilose hairs; stems 50–100 cm. long, slightly branched above; radical leaves withering before anthesis; lower cauline leaves long-petioled, the upper sessile; stipule-segments lanceolate to narrowly ovate, acute; leaflets 3, ovate or narrowly so, or ovate-oblong, 3–8 cm. long, 2–3.5 cm. wide, acute to short-acuminate, slightly appressed-pilose on upper side, paler beneath, the teeth ovate, acute, often duplicate; inflorescence branched, many-flowered; flowers about 10 mm. across; calyx-segments narrowly ovate, acuminate; bracts of caliculus broadly lanceolate, subobtuse to acute; petals yellow, broadly obovate, as long as the calyx-segments, about 4 mm. long; anthers small, ovate-orbicular; receptacles short-pilose; achenes many, about 0.8 mm. long, obsoletely rugulose; styles rather stout, about 1 mm. long.——June–Sept. Mountains; Hokkaido, Honshu, Kyushu.

17. Potentilla norvegica L. EZO-NO-MITSUMOTO-SŌ. Annual or biennial herb with long pilose hairs; stems 20–60 cm. long; leaves 5-foliolate in the radical ones, 3-foliolate in the cauline; stipules adnate to the petiole on lower half or only at the base; leaflets obovate to oblong, obtuse to acute, the teeth ovate to lanceolate; flowers many, densely arranged; calyx-segments narrowly ovate, acute; petals obovate, shorter than the calyx-segments; receptacles short-pilose; anthers broadly elliptic, small; achenes broadly ovoid, about 0.7 mm. long, rugulose, glabrous, keeled on back; styles rather stout, nearly as long as the achene.——Hokkaido, Honshu (Shinano Prov.).——Cold regions of the N. Hemisphere.

18. Potentilla chinensis Ser. KAWARA-SAIKO. Perennial herb with stout rhizomes; stems 30–70 cm. long, long-pubescent; leaves obovate to narrowly so; free portion of stipules linear-lanceolate, green and laciniate to deeply toothed in the cauline leaves; leaflets 15–29, the lower gradually smaller, the upper oblanceolate to narrowly oblong, 2–5 cm. long, 8–15 mm. wide, pinnately parted nearly to the midrib with acute lobes, nearly glabrous above, densely white-woolly-tomentose beneath; flowers many, 10–15 mm., sometimes to 20 mm. across; calyx-segments narrowly ovate, acute; bracts of caliculus lanceolate; anthers ovoid; receptacle hairy; achenes broadly ovoid, longitudinally rugose, about 1.3 mm. long, keeled on back; styles rather stout, about 1 mm. long.——June–Aug. Sandy sunny places, especially along rivers in lowlands; Honshu, Shikoku, Kyushu.——Korea, Manchuria, Amur, China, and Formosa.

19. Potentilla nipponica T. Wolf. *P. chinensis* var. *isomera* Fr. & Sav.; *P. isomera* (Fr. & Sav.) Koidz.——HIRO-HA-NO-KAWARA-SAIKO. Closely resembles the preceding, but the stems slightly shorter, often ascending at base; leaflets fewer, with shallower lobes; calyx-segments large, densely white-woolly; achenes smooth, glabrous.——June–Aug.——Sandy sunny places along rivers in lowlands; Hokkaido, Honshu (n. and centr. distr.).

20. Potentilla fruticosa L. *Dasiphora fruticosa* (L.) Rydb.——KINROBAI. Small branching shrub; leaflets elliptic to narrowly oblong, obtuse to acute, 1–2 cm. long, 3–10 mm. wide, long-pubescent on both sides at least while young, the uppermost lateral ones decurrent at base; flowers yellow, 2–2.5 cm. across, on white-hairy pedicels; calyx-segments ovate, 5–7 mm. long, with nerves raised in fruit; petals obovate; achenes ovoid, lustrous, loosely pubescent with long hairs.——June–Aug. Alpine regions; Hokkaido, Honshu (n. and centr. distr.).——Northern and alpine regions of the N. Hemisphere.

Var. **mandshurica** Maxim. *P. fruticosa* var. *leucantha* Makino——HAKUROBAI. Leaflets often mucronate; flowers white.——Alpine regions; Honshu (centr. and Kinki Distr.), Shikoku; rare.——e. Siberia and n. China.

21. Potentilla palustris (L.) Scop. *Comarum palustre* L.; *P. comarum* Nestl.——KUROBANA-RŌGE. Stems rather stout, creeping and branching below, 30–60 cm. long, with appressed silky pubescence above; leaflets 5–7 (rarely 3), nearly sessile, narrowly oblong, 4–7 cm. long, 12–30 mm. wide, green above, whitish beneath, with appressed silky pubescence on both sides or glabrate on upper side; flowers 15–20 mm. across; calyx-segments narrowly ovate, 7–15 mm. long, long-acuminate, pubescent; bracts of caliculus small; petals ovate, purplish, shorter than the calyx-segments; receptacles pilose.——June–Aug. Wet places and shallow water around ponds and along ditches in lowlands and on mountains; Hokkaido, Honshu (centr. and n. distr.).——Cold regions of the N. Hemisphere.

10. WALDSTEINIA Willd. KO-KIMBAI ZOKU

Perennial herb with creeping rhizomes; leaves radical or mostly so, petioled, 3-foliolate, toothed, incised or serrate, the stipules membranous; flowers solitary, scapose or few in corymbs; calyx persistent, the tube obconical, the segments 5; bracts of caliculus 5; petals 5, yellow; stamens many; carpels 2–6, free, the ovules solitary; styles nearly terminal, deciduous; achenes obliquely obovate.——Few species, in the N. Hemisphere.

1. Waldsteinia ternata (Steph.) Fritsch. *Dalibarda ternata* Steph.; *W. sibirica* Tratt.; *W. fragarioides* sensu auct. Japon., non Tratt.——KO-KIMBAI. Perennial herb with slender elongate creeping rhizomes; leaves radical, 3-foliolate, the leaflets slightly pilose on both sides, toothed, the terminal one broadly rhombic–obovate, 2–3 cm. long, 1.5–2.5 cm. wide, shallowly 3-lobed, the lateral ones oblique, the petioles 5–10 cm. long, slightly spreading-pubescent or glabrate; scapes 10–20 cm. long, nearly as long as the leaves, sometimes sparingly spreading-pubescent, 1- to 3-flowered; flowers yellow, about 2 cm. across; calyx-segments broadly lanceolate, subacute; petals spreading; achenes 3–4 mm. long, white-pilose.——May–June. Woods in mountains; Hokkaido, Honshu (centr. and n. distr.).——Sakhalin and Siberia.

11. SIBBALDIA L. TATEYAMA-KIMBAI ZOKU

Dwarf suffrutescent shrubs or herbs with creeping branched rhizomes; leaves radical and closely alternate, usually 3-foliolate, the stipules adnate to the petiole; inflorescence a scapose cyme; calyx persistent, the tube short, the segments 5, as many as and alternate with the bracts of the caliculus; petals 5, smaller than the calyx-segments, usually narrowly obovate, commonly yellow, rarely purplish; stamens 5; disc hairy; carpels 5–10, with a short hairy stipe at base; style ventral; ovules solitary; achenes 5–10, glabrous.——About 10 species, in northern and alpine regions in the N. Hemisphere.

1. Sibbaldia procumbens L. *Potentilla procumbens* (L.) Clairv.; *P. sibbaldii* Haller f.——TATEYAMA-KIMBAI. Dwarf suffrutescent herb with short creeping branches densely covered with stipules and 2–6 cm. long persistent bases of the old petioles; leaflets 3, nearly similar, cuneate, truncate and with 3(–5) large teeth at apex, usually 1–1.5 cm. long, 6–10 mm. wide; inflorescence rather densely about 10-flowered, the pedicels very short; calyx-segments narrowly ovate, 3.5–5 mm. long, acute; petals narrowly obovate; achenes ovoid, small, nearly smooth.——July–Aug. Dry alpine slopes; Honshu (centr. distr.); rare.——n. Korea, Sakhalin, n. Kuriles to Europe and N. America.

12. DRYAS L. CHŌNOSUKE-SŌ ZOKU

Dwarf shrubs with short, densely tufted or short-creeping branches; leaves evergreen, simple, petioled, elliptic to ovate, entire or pinnately lobulate, the stipules adnate to the petioles; pedicels erect, solitary, radical; flowers white, sometimes yellow, large; calyx persistent, cup-shaped or campanulate, not caliculate at base, the segments 8 or 9; petals 8 or 9, obovate; stamens many; carpels many; styles terminal, persistent and elongate after flowering, plumose in fruit; ovules solitary; achenes many, usually pubescent.——Few species, in alpine regions in the N. Hemisphere.

1. Dryas octopetala L. var. **asiatica** (Nakai) Nakai. *D. octopetala* sensu auct. Japon., non L.; *D. octopetala* forma *asiatica* Nakai; *D. tschonoskii* Juzep.; *D. nervosa* Juzep.——CHŌNOSUKE-SŌ. Decumbent, much-branched dwarf shrub; leaves subcoriaceous, ovate to broadly ovate-elliptic, 10–20 mm. long, 6–15 mm. wide, obtuse, shallowly and obtusely lobulate and slightly recurved on margin, green above with impressed midrib and nerves, densely white-tomentose beneath, the peti- ole 5–20 mm. long, white-pubescent; pedicels 3–10 cm. long, white-woolly; petals 10–15 mm. long, rounded at apex; achenes about 3 mm. long, pubescent; styles elongate, about 3 cm. long, plumose with white hairs.——July–Aug. Dry alpine slopes; Hokkaido, Honshu (centr. distr.); rare.——Kamchatka, Kuriles, Sakhalin, Ussuri, Ajan, and n. Korea. The typical phase occurs in the cold regions of the N. Hemisphere.

13. GEUM L. DAIKON-SŌ ZOKU

Perennial herbs or dwarf shrubs; leaves radical and often cauline, alternate, pinnate or pinnately parted, the leaflets all alike or the terminal larger, sometimes only a terminal one; stipules green, rather large or linear and small, more or less adnate on the petiole; flowers solitary or in corymbs, yellow or white, sometimes reddish purple; calyx with a short obconical tube, the segments persistent, alternate with the bracts of the caliculus; petals 5; stamens many, free; carpels many, on a slightly enlarged glabrous or hairy receptacle; styles terminal, filiform, sometimes jointed, persistent, sometimes deciduous only above the joint, sometimes pinnately hairy.——About 50 species, in the temperate to cold regions of the N. Hemisphere, a few in S. America and Africa.

1A. Procumbent dwarf glabrous evergreen shrub; leaves with nearly equal leaflets; flowers white, rather large; styles pinnate and elongate in fruit. . . . . 1. *G. pentapetalum*
1B. Erect coarsely pilose perennial herbs; leaves with the terminal leaflet much larger than the lateral; flowers yellow; styles only slightly elongate after anthesis and not pinnate.
  2A. Styles jointed, the upper portion deciduous after flowering.
    3A. Receptacle with yellowish brown bristly hairs 2–3 mm. long; pedicels velvety puberulent. . . . . .2. *G. japonicum*
    3B. Receptacle with very short whitish hairs less than 1 mm. long.
      4A. Plant hispid-hirsute; lateral leaflets of the radical leaves 2 to 5 pairs; styles without stipitate glands. . . . . 3. *G. aleppicum*
      4B. Plant setose with bristly hairs; lateral leaflets of the radical leaves very small, 1 or 2 pairs; styles with sparse stipitate glands at base. . . . . 4. *G. macrophyllum* var. *sachalinense*
  2B. Styles not jointed, persistent; leaves with a large orbicular-cordate terminal leaflet and very small or obsolete lateral leaflets.
                    5. *G. calthaefolium* var. *nipponicum*

1. Geum pentapetalum (L.) Makino. *Dryas penta-petala* L.; *Sieversia pentapetala* (L.) Greene; *G. anemonoides* Willd.; *S. dryadioides* Sieb. & Zucc.; *G. dryadioides* (Sieb. & Zucc.) Fr. & Sav.——CHINGURUMA, IWAGURUMA. Dwarf evergreen glabrous shrub with prostrate slender much-divided branches, about 5 mm. thick at base; leaves petioled, 3–6 cm. long, the leaflets 7–9, subcoriaceous, broadly oblanceolate to obovate, 6–15 mm. long, 4–8 mm. wide, acute, with irregular incisions, sometimes subtrilobed, lustrous and deep green above, the stipules linear; scapes about 10–20 cm. long, erect, with a 3- to 5-parted leaf, densely puberulent above; flowers solitary, white, 2–3 cm. across; calyx-segments broadly lanceolate, long-acute, slightly recurved after flowering, the bracts of the caliculus short, lanceolate; petals orbicular; achenes rather many, about 2 mm. long, fusiform, with white ascending hairs; styles elongate after anthesis, filiform, about 3 cm. long, long white spreading-plumose.——June–Aug. Wet slopes and sometimes sphagnum bogs in high mountains; Hokkaido, Honshu (n. and centr. distr.).——Kuriles, Sakhalin, Kamchatka, and Aleutians.

2. Geum japonicum Thunb. *G. iyoanum* Koidz.——
DAIKON-SŌ.   Perennial herb with short rhizomes; stems
simple or slightly branched, sometimes few together, sometimes solitary, short velvety-puberulent; leaves puberulent,
sometimes sparingly long-pilose, the lateral leaflets small, 1 or
2 pairs, sometimes obsolete, often with a pair of accessory
small leaflets between the larger ones, the terminal leaflet
orbicular to broadly ovate, 3–6 cm. long and as wide, rounded
to obtuse at apex, subcordate to broadly cuneate at base,
toothed, usually 3-lobed, the cauline leaves short-petioled, usually simple, deltoid-orbicular, often 3-lobed; flowers few, rarely
1, loosely arranged, long-pedicelled, about 1.5 cm. across, yellow; calyx-segments narrowly ovate-deltoid, acute, reflexed,
densely puberulent, sometimes with a few long spreading
hairs.——June–Sept. Woods and thickets in hills and low
mountains; Hokkaido, Honshu, Shikoku, Kyushu; common.
——China.

3. Geum aleppicum Jacq. *G. vidalii* Fr. & Sav.——
Ō-DAIKON-SŌ.   Coarse perennial herb 60–100 cm. high, with
spreading hispid hairs; radical leaves petioled, pinnate, the
lateral leaflets rather small, 2 to 5 pairs, alternate, with very
small accessory segments, the terminal leaflet rhombic-ovate to
orbicular, 5–10 cm. long, 3–10 cm. wide, acute to rounded at
apex, cuneate to subcordate at base, irregularly toothed, the
cauline leaves short-petioled, 3- to 5-foliolate, the stipules obovate, incised; flowers 3–10, long-pedicelled, 15–20 mm. across,
yellow.——June–Sept. Thickets and grassy places in lowlands
and low mountains; Hokkaido, Honshu; common.——e.
Europe, Asia Minor, Siberia, China, Korea, and N. America
(var.).

4. Geum macrophyllum Willd. var. **sachalinense**
(Koidz.) Hara. *G. fauriei* Lév.; *G. japonicum* var. *sachalinense* Koidz.; *G. aleppicum* var. *sachalinense* (Koidz.) Ohwi;
G. sachalinense (Koidz.) Makino, non Lév.——KARAFUTO-
DAIKON-SŌ.   Closely resembles the preceding, but with coarse
bristly setose hairs; lateral leaflets in the radical leaves much
smaller and fewer, often obsolete, the terminal leaflets orbicular, the cauline leaves simple, 3-lobed, the stipules usually
entire.——Thickets and woods in mountains; Hokkaido, Honshu (centr. and n. distr.).——Sakhalin.——The typical phase
occurs in Asia and N. America.

5. Geum calthaefolium Smith var. **nipponicum** (Bolle)
Ohwi. *G. calthaefolium* sensu auct. Japon., non Smith;
Parageum calthaefolium var. *nipponicum* (F. Bolle) Hara;
Acomastylis nipponicum F. Bolle; *A. calthaefolia* var. *nipponica* (F. Bolle) Hara——MIYAMA-DAIKON-SŌ.   Coarse perennial herb with yellowish spreading strigose hairs; rhizomes
stout, short, creeping and branching; radical leaves petioled,
the terminal leaflet orbicular, 5–12 cm. long and as wide, shallowly lobulate and irregularly toothed, cordate at base, lustrous
above, the lateral ones minute, few; scapes 10–30 cm. long,
erect, few-flowered; flowers 2–2.5 cm. across, yellow; calyx
and pedicels densely puberulent, the calyx-segments acute, obliquely spreading; achenes many, long hispid-hairy including
the lower portion of the slightly elongate style, 10–13 mm.
long in fruit.——July–Aug. Alpine slopes; Hokkaido, Honshu (n. and centr. distr.), Shikoku (Mount Ishizuchi).——
Kuriles. The typical phase occurs from the Aleutians and
Kamchatka to northern N. America.

14. RUBUS L.   KI-ICHIGO ZOKU

Usually armed shrubs, rarely herbs; leaves alternate, petioled, simple, palmately lobed or palmately to pinnately compound,
the stipules rarely lobed, adnate to the base of the petiole; inflorescence 1- to rather many-flowered, usually terminal on short
branches, in corymbs or sometimes panicles, the flowers usually bisexual, rarely unisexual; calyx without a caliculus, the tube
short, broad, the segments 5, persistent; petals 5, white or rose; stamens many; carpels many, sometimes few, inserted on a
raised receptacle, the ovary 1-locular, 2-ovuled; styles subterminal; fruitlets juicy or fleshy, aggregated in a globose to ovoid head,
the stones small.——A difficult genus with many species and numerous natural hybrids, chiefly occurring in N. Hemisphere.

1A. Leaves simple and toothed or palmately lobed to cleft.
  2A. Stipules and bracts laciniate; leaves usually evergreen; stems usually procumbent to ascending-arching.
    3A. Stipules persistent; flowers 1 or 2 on short lateral branchlets borne on stolons; calyx with dense spreading straight prickles.
                                                                                                             1. *R. pectinellus*
    3B. Stipules deciduous; flowers few, axillary, often in a terminal panicle; calyx not prickly.
      4A. Stems glabrous or puberulent only when young; leaves acuminate; calyx puberulent.
        5A. Leaves ovate; prickles rather stout, distinctly flattened and slightly recurved; flowers many, in a large terminal panicle.
                                                                                                             2. *R. lambertianus*
        5B. Leaves suborbicular or ovate-orbicular; prickles slender, very slightly flattened, not recurved; flowers few, axillary or rarely
            in short terminal panicles. .... 3. *R. hakonensis*
      4B. Stems prominently pubescent; leaves obtuse or acute; calyx long-pubescent.
        6A. Branches and inflorescence without gland-tipped hairs.
          7A. Stems thick, 3–6 mm. across, with slender straight or slightly recurved prickles; leaves broadly ovate to ovate-orbicular;
              flowers 2.5–3 cm. across; calyx-segments about 15 mm. long; bracts large. .... 4. *R. sieboldii*
          7B. Stems slender, less than 2.5 mm. across, scarcely prickly; leaves suborbicular; flowers about 1 cm. across; calyx-segments
              8–10 mm. long. .... 5. *R. buergeri*
        6B. Branches and inflorescence densely glandular. .... 6. *R. hatsushimae*
  2B. Stipules and bracts entire or with few teeth.
    8A. Low usually simple erect herbs; flowering stems rising from slender rhizomes; flowers unisexual (plants dioecious); stipules
        elliptic to ovate. .... 7. *R. chamaemorus*
    8B. Shrubs; flowering stems from the lateral bud on last year's aerial shoots; flowers bisexual.
      9A. Leaf blades peltate; stipules narrowly ovate to broadly lanceolate; fruit oblong-cylindric, consisting of many very densely ar-
          ranged fruitlets. .... 8. *R. peltatus*
      9B. Leaf blades basifixed; stipules narrower; fruit globose.
        10A. Leaves glaucous beneath; fruit enclosed in a semiglobose calyx; calyx rather thick, nearly glabrous outside, purplish, the
            segments densely puberulent inside; pedicels slender; bud-scales persistent. .... 9. *R. microphyllus*

10B. Leaves not glaucous, green or pale green beneath.
 11A. Flowers few, on short pedicels; bud-scales persistent. .... 10. *R. crataegifolius*
 11B. Flowers solitary, if more than 1 then borne together on slender pedicels; bud-scales usually deciduous.
  12A. Petals 4–6 mm. long; branches and other parts nearly always unarmed; leaves deeply palmately parted; pedicels long
   and slender. .... 11. *R. pseudoacer*
  12B. Petals 10–25 mm. long; plants usually prickly.
   13A. Calyx glabrous or nearly glabrous outside, the segments rather thick, densely white-villous inside and on margin;
    leaves undivided or shallowly 3-lobed. .... 12. *R. grayanus*
   13B. Calyx pubescent or the segments nearly glabrous on both sides, usually membranous.
    14A. Flowers 3–7, erect; young branchlets and pedicels with stipitate glands; leaves palmately cleft to lobed.
     13. *R. trifidus*
    14B. Flowers 1, rarely 2, nodding or pendulous.
     15A. Fertile branchlets and petioles not at all or scarcely prickly.
      16A. Leaves suborbicular, palmately 5-(3–7)-parted. .... 14. *R. chingii*
      16B. Leaves broadly to narrowly ovate, 3-lobed or undivided, very rarely 5-lobed, the outermost lobes very small.
       17A. Young branches and underside of leaves with short spreading pubescence. .... 15. *R. ribesoideus*
       17B. Young branches and underside of leaves with appressed to ascending pubescence. ...... 16. *R. ohsimensis*
     15B. Fertile branchlets and petioles with distinct prickles; flowers solitary on short pedicels.
      18A. Leaves beneath densely covered with appressed and spreading hairs; fertile branches with undivided leaves.
       17. *R. corchorifolius*
      18B. Leaves nearly glabrous or with only a few appressed hairs on nerves beneath; fertile branchlets with undivided
       or 3- to 5-cleft leaves. .... 18. *R. palmatus*
1B. Leaves 3- to many-foliolate.
 19A. Shrubs with woody branches; leaves ternate or pinnate, the leaflets 3 to many.
  20A. Flowers solitary or few, large; petals orbicular or ovate-orbicular, 10–20 mm. long.
   21A. Flowers deep reddish, nodding; petals ascending; plants unarmed. .... 19. *R. vernus*
   21B. Flowers white, with spreading petals.
    22A. Leaves ternate, the leaflets sessile or nearly so. .... 20. *R. nishimuranus*
    22B. Leaves pinnate, the leaflets 3 to many, if 3, the terminal one distinctly petioluled.
     23A. Carpophore short but distinct; petals ovate-orbicular; fruitlets small, many; calyx without prickles.
      24A. Plants usually glandular; branches terete.
       25A. Underside of leaves and calyx-tube with sessile discoid yellowish glands. .... 21. *R. minusculus*
       25B. Plants without sessile discoid glands; pedicels and other parts with stipitate glands (glandular hairs).
        26A. Glandular hairs 3–5 mm. long. .... 22. *R. sorbifolius*
        26B. Glandular hairs 0.2–2 mm. long.
         27A. Plants with glandular hairs only. .... 23. *R. croceacanthus*
         27B. Plants with glandular hairs and prominent eglandular pubescence. .... 24. *R. hirsutus*
      24B. Plants without glands; branches obtusely angled.
       28A. Fertile branches rising directly from the underground rhizomes; leaflets usually 5–7, rarely 3; flowers single; indig-
        enous. .... 25. *R. illecebrosus*
       28B. Fertile branches on last year's shoots; leaves 3- to 5-foliolate; flowers double; cultivated. ...... 26. *R. commersonii*
     23B. Carpophore absent; fruitlets rather large; petals obovate; calyx prickly. .... 27. *R. oldhamii*
  20B. Flowers few, small; petals obovate to elliptic, erect, connivent above or ascending, 5–8 mm. long.
   29A. Plants not glandular-hairy or with glandular hairs not more than 1.5 mm. long.
    30A. Prickles slender, scarcely broadened at base, not flattened; petals ascending.
     31A. Branches, petioles, and inflorescence glandular-pilose; leaves on fertile branchlets 3-foliolate; calyx usually prickly, the
      segments caudately short-acuminate. .... 28. *R. idaeus*
     31B. Branches and leaves usually eglandular, sometimes glandular-pilose only on pedicels; leaves on fertile branches 3- to 5-
      foliolate, slightly smaller; calyx usually not prickly, the segments often filiform-elongate. .... 29. *R. yabei*
    30B. Prickles rather stout and flattened laterally, often slightly recurved above; plants without glandular hairs; petals erect or
     connivent.
     32A. Leaves on fertile branchlets 5- to 7-foliolate; fruit red. .... 30. *R. coreanus*
     32B. Leaves on fertile branchlets 3-foliolate.
      33A. Leaflets acuminate or very acute; calyx not prickly.
       34A. Leaflets with short white tomentum beneath, the terminal one ovate-orbicular; fruit purplish black at maturity.
        31. *R. mesogaeus*
       34B. Leaflets later becoming green, the terminal one ovate to rhombic-elliptic. .... 32. *R. yoshinoi*
      33B. Leaflets obtuse to rounded; calyx more or less prickly. .... 33. *R. parvifolius*
   29B. Plant with long-stiped glandular hairs 2–5 mm. long, especially on the inflorescence, stems, and leaf-axis; calyx-segments
    connivent even in fruit. .... 34. *R. phoenicolasius*
 19B. Herbs with pedately 3- to 5-foliolate leaves.
  35A. Herbs, unarmed, with very slender filiform creeping stolons; leaflets obtuse; stipules brownish, rounded at apex. .. 35. *R. pedatus*
  35B. Herbs with woody base and long ascending to erect shoots, becoming decumbent; leaflets acuminate; stipules green.
   36A. Plant armed with prickles; calyx-segments erect in fruit. .... 36. *R. ikenoensis*
   36B. Plant unarmed; calyx-segments reflexed in fruit. .... 37. *R. pseudojaponicus*

1. Rubus pectinellus Maxim. Koba-no-fuyu-ichigo.
Small evergreen pubescent suffrutescent shrub with slender, scarcely flattened prickles; stems slender, slightly woody, creeping, slightly prickly, the erect fertile branches 10–20 cm. long, few-leaved; leaves orbicular-cordate, 3–5 cm. long, obtusely to subacutely toothed, sometimes obsoletely incised, the

petioles 4–7 cm. long, prickly; pedicels 1–5 cm. long, prickly; calyx ovoid, rather densely prickly, 15–20 mm. long, the lobes laciniate, short-pubescent inside, reflexed in fruit; petals narrowly obovate, about 10 mm. long; receptacles and carpels pilose, the stones 2.5 mm. long, less prominently rugose.——May–July. Woods; Honshu, Shikoku, Kyushu.

2. Rubus lambertianus Ser. Shimabara-ichigo. Branches erect, elongate, slightly scandent above, with rather stout, flattened prickles, glabrous to short-pubescent; leaves ovate or broadly so, 5–12 cm. long, 4–8 cm. wide, acuminate, cordate at base, asperous and strigose-pilose especially on nerves on upper side, with spreading hairs beneath, acutely serrulate, shallowly 3- to 5-lobulate, the lateral lobes short, sometimes obsolete, the petioles 2–4 cm. long, prickly and short-pubescent; corymbs in a narrow terminal panicle 15–20 cm. long, short-pubescent, the pedicels 3–10 mm. long; calyx with a short tube, the lobes narrowly ovate-deltoid, 5–7 mm. long, acuminate, ascending, with dense white tomentum on margin; petals 4–5 mm. long, obovate-spathulate; stones about 1.5 mm. long, obliquely rugose.——Oct.–Nov. Kyushu (Higo and Hizen Prov.).——Formosa and s. China.

3. Rubus hakonensis Fr. & Sav. *R. lambertianus* subsp. *hakonensis* (Fr. & Sav.) Focke——Miyama-fuyu-ichigo, Miyama-ichigo. Stems slender, woody, loosely prickly with slender, scarcely flattened, spreading prickles, glabrous to short-pubescent; leaves subcoriaceous, broadly ovate to ovate-orbicular, 5–8 cm. long, 4–7 cm. wide, short-acuminate, cordate at base, glabrate on both sides or short-pubescent beneath, especially on nerves, and on nerves above, 3- to 5-lobulate, the lateral lobules short or often obsolete, with short-awned teeth, the petioles 3–7 cm. long; inflorescence rather densely few-flowered, short-pubescent, the pedicels short; calyx pubescent, 7–8 mm. long, the tube short, the lobes nearly ovate, acuminate, sometimes 2- to 4-lobed, densely short-pubescent on margin; petals 4–5 mm. long, obovate; fruit red, the stones about 2 mm. long, obliquely rugose.——Aug.–Oct. Woods in mountains; Honshu (centr. distr. and westw.), Shikoku, Kyushu.——China (?).

4. Rubus sieboldii Bl. Hōroku-ichigo. Branches stout, elongate and subscandent above, with slender, spreading or slightly recurved scattered prickles, short-puberulent and with long spreading hairs; leaves coriaceous, broadly ovate to ovate-orbicular, 7–25 cm. long, 6–20 cm. wide, cordate at base, shallowly and obtusely lobulate and mucronate-toothed, with deciduous woolly hairs on upper side, with yellowish tomentum beneath, sparsely prickly on both sides, the petioles 2–7 cm. long, loosely prickly and densely puberulent; inflorescence axillary, densely few-flowered, the pedicels very short, the bracts large; calyx yellowish tomentose, the lobes broadly ovate to elliptic, sometimes obtusely toothed; petals about 15 mm. long and as wide; fruit red.——Mar.–June. Thickets and cut-over areas in low mountains and hills; Honshu (Izu Isls. and Kinki Distr.), Shikoku, Kyushu.——s. China and Ryukyus.

Rubus pseudosieboldii (Makino) Makino. *R. buergeri* var. *pseudosieboldii* Makino——Ō-fuyu-ichigo. An alleged hybrid of Nos. 4 and 5, with characters intermediate between the parents, especially the branches somewhat more slender than in No. 4.

5. Rubus buergeri Miq. Fuyu-ichigo. Shrub with slender creeping stems sparingly armed with short slender

slightly flattened prickles or often unarmed, densely short-pubescent; leaves chartaceous, deep green, nearly orbicular, 5–10 cm. long and as wide, cordate, rounded to subacute at apex, with short-awned small teeth, obsoletely 3- to 5-lobulate with rounded depressed lobules or often unlobed, short-pubescent on the nerves above, spreading-pilose beneath, the petiole 3–10 cm. long, densely pubescent; inflorescence axillary, densely few-flowered, often aggregated in a terminal panicle, the flowers short-pedicelled; calyx with long yellowish hairs, 8–10 mm. long; petals 7–8 mm. long; fruit red, the stones about 2.5 mm. long, pitted.——Aug.–Nov. Woods in low mountains; Honshu (Awa and Izu Prov. and westw.), Shikoku, Kyushu.——s. Korea, Formosa, and China.

6. Rubus hatsushimae Koidz. Tsukushi-aki-tsuru-ichigo. Leaves ovate, acute, cordate, indistinctly incised-lobed, acutely toothed, appressed-pubescent on upper side, densely white-pubescent beneath, yellowish hairy on nerves above, glandular-hairy on nerves beneath; calyx densely hairy and glandular-pilose outside, the segments lanceolate-filiform, 1- or 2-lacerate above; petals spathulate-obovate.——Reported from Kyushu (Shimabara in Hizen Prov.).

7. Rubus chamaemorus L. Horomui-ichigo. Unarmed dioecious herb, with branched, slender rhizomes; stems erect, 10–25 cm. long, simple, 2- or 3-leaved, slightly soft-pubescent, with few scalelike leaves at base; leaves deep green, reniform-orbicular, cordate, 4–7 cm. long and as wide, shallowly (3–)5-lobulate with depressed rounded lobes, nerves above palmately impressed, loosely appressed-pubescent to glabrate on upper side, spreading-pubescent beneath especially on nerves, the stipules membranous, 3–7 mm. long, obtuse; flowers unisexual, solitary, terminal; calyx-lobes narrowly ovate or elliptic, obtuse to rounded, soft-pubescent, sometimes with a few glandular hairs; petals white, 8–10 mm. long; stones about 3 mm. long, obsoletely striate on upper portion.——June–July. Sphagnum bogs; Hokkaido, Honshu (n. distr.).——Kuriles, Sakhalin, and n. Korea to Europe and N. America.

8. Rubus peltatus Maxim. Hasu-no-ha-ichigo. Shrub with rather stout glabrous often glaucous loosely prickly branches, the prickles spreading, flattened, slightly curved; leaves peltate, ovate-orbicular, cordate, 10–25 cm. long, 7–20 cm. wide, shallowly 5-lobed, the lobes acuminate, toothed, the nerves beneath with spreading hairs, the stipules 10–15 mm. long; receptacle densely short-pilose; flowers glabrous, solitary on short branchlets; calyx-segments narrowly ovate; fruit short-cylindric, about 4 cm. long; fruitlets very many, densely spreading-puberulent; stones prominently pitted.——May–June. Mountains; Honshu (Mino, s. Shinano, and Echizen Provs. and westw.), Shikoku, Kyushu; rather rare.

9. Rubus microphyllus L. f. *R. incisus* Thunb.; *R. pseudoincisus* O. Kuntze; *R. geifolius* O. Kuntze——Niga-ichigo. Shrub with rather slender branches usually purplish brown, with rather slender prickles, more or less glaucescent, the fertile branchlets scarcely elongate, 2- to 4-leaved toward base; leaves ovate to orbicular, 2–5 cm. long, 1.5–4 cm. wide, rounded to acute at apex, rounded to cordate at base, usually shallowly 3-lobed, toothed, distinctly glaucous beneath, the petioles sometimes slightly pubescent on the upper side; flowers 1(–3) on reddish purple pedicels 1–2.5 cm. long; calyx glabrous, 5–7 mm. long, reddish purple, the segments densely white-tomentose inside; petals white, 10–12 mm. long;

fruit enclosed in a persistent calyx, red; stones pitted.——Apr.–May. Low mountains; Honshu, Shikoku, Kyushu; common. ——China.

Var. **subcrataegifolius** (Lév. & Van't.) Ohwi. *R. crataegifolius* var. *subcrataegifolius* Lév. & Van't.; *R. koehneanus* Focke; *R. incisus* var. *koehneanus* (Focke) Koidz.——Mɪyama-nɪga-ichigo. Similar to the typical phase but larger, with few slender prickles, the fertile branchlets slightly elongate, to 7–15 cm. long; leaves 4–8 cm. long, acuminate, 3-lobed; flowers 1–3; calyx-segments caudately acuminate.——May–June. Mountains; Honshu; rather common.

10. Rubus crataegifolius Bunge. *R. wrightii* A. Gray; *R. morifolius* Sieb., non P. J. Muell.; *R. savatieri* L. H. Bailey ——Kuma-ichigo, Ezo-no-kuma-ichigo, Tachi-ichigo. Shrub with rather stout flattened prickles, the flowering branches few-leaved, slightly elongate, more or less pubescent; leaves broadly ovate to orbicular, 4–10 cm. long, 3.5–8(–10) cm. wide, acuminate to sharply acute, truncate to cordate at base, glabrous to loosely appressed-pubescent on upper side, with spreading hairs beneath especially on the nerves, 3(–5)-lobed or -cleft and irregularly toothed, the lateral lobes acute; inflorescence 2- to 6-flowered, rather densely short-pubescent, the pedicels short, 5–10 mm. long; calyx 8–10 mm. long, cup-shaped, short-pubescent outside, the lobes narrowly ovate, acuminate, densely white-tomentose on margin and inner side; petals white, obovate; fruit globose, red, the stones pitted, about 2 mm. long, the receptacle pilose.——Apr.–June. Waste grounds and clearings in mountains; Hokkaido, Honshu, Shikoku, Kyushu; common with variable pubescence.——China.

11. Rubus pseudoacer Makino. Mɪyama-momiji-ichigo. Scarcely armed small erect shrub, the fertile branchlets rather elongate, nearly glabrous; leaves membranous, orbicular, cordate, 6–10 cm. long and as wide, palmately 5- to 7-parted with acuminate, often 3-fid, incised and toothed segments, slightly pilose to glabrous on both sides; flowers few, loose, on slender glabrous pedicels 2–4 cm. long; calyx glabrous outside, the tube cup-shaped, the segments ovate, caudate, densely white-puberulent inside and on margin; petals broadly ovate, 4–6 mm. long; fruit globose, the fruitlets glabrous; stones pitted.——July–Aug. Mountains; Honshu (Kantō to Kinki Distr.), Shikoku; rare.

12. Rubus grayanus Maxim. Ryūkyū-ichigo, Shima-awa-ichigo. Glabrous unarmed or very sparingly prickly shrub with elongate branches; leaves ovate or elliptic, 6–10 cm. long, 4–6 cm. wide, obtuse to acute, rounded to shallowly cordate at base, usually undivided, with shallow teeth; flowers solitary on a short fertile branchlet, the pedicels rather slender, 1–1.5 cm. long; calyx with shallowly cup-shaped tube, the segments narrowly ovate-deltoid, acuminate, 6–10 mm. long, white-tomentose on margin and on inner side; fruit globose; stones pitted.——Kyushu (Yakushima and Tanegashima). ——Ryukyus and s. China.

Var. **chaetophorus** Koidz. Toge-ryūkyū-ichigo. Prickles distinct; leaves 3-lobed.——Kyushu (Oosumi Prov.).

13. Rubus trifidus Thunb. *R. hydrastifolius* A. Gray ——Kaji-ichigo. Stout bushy unarmed shrub to 1.5 m. high; branches elongate, terete, rather stout, pubescent and glandular-hairy; leaves orbicular, cordate to truncate at base, 5- to 7-cleft or shallowly parted with very acute and doubly toothed lobes, slightly lustrous and green on upper side; inflorescence few-flowered; flowers white, 3–4 cm. across; calyx

12–15 mm. long, the lobes narrowly ovate, subcaudate, white-tomentose on margin and on inner side; petals orbicular; fruit globose, yellow; stones pitted.——Apr.–May. Near seashores in warmer parts of Honshu (Pacific side, from Awa to Kii Prov., incl. Izu Isls.); rather common and much planted for the edible fruit.

Hybrids of *R. trifidus* with the second parent one of the following: 9, 15, 17, or 18: **R. × medius** O. Kuntze. Hɪme-kaji-ichigo; **R. × manazurensis** Hisauchi. Manazuru-kɪ-ichigo; **R. × laudabilis** Koidz. Toge-nashi-ichigo; **R. × koidzumii** Hisauchi. Ō-toge-nashi-ichigo; **R. × hisautii** Koidz. Yabu-awa-ichigo; **R. × kyusianus** Koidz., non Nakai; **R. × yenoshimanus** Koidz.; **R. × omogoensis** Koidz.; **R. × yakumontanus** Masam.; **R. × tanakae** O. Kuntze; **R. × ribifolius** Sieb. & Zucc.; others are reported.

14. Rubus chingii Hu. *R. officinalis* Koidz.——Gosho-ichigo. Very sparingly prickly shrub with rather stout branches, the flowering branches glabrous, short; leaves membranous, suborbicular, shallowly cordate at base, 5–10 cm. in diameter, deeply 5-cleft, with long-acuminate doubly toothed segments, appressed-pubescent on nerves on both sides or glabrous above, the stipules filiform; flowers solitary, on slender glabrous pedicels 2–3 cm. long; calyx with nearly flat tube, the segments ovate or oblong, 6–8 mm. long, rounded to obtuse and sometimes with a mucro at apex, short appressed-pubescent on both sides; petals white, about 15 mm. long; fruit globose, fruitlets with dense spreading hairs.——May. Mountains; Kyushu (n. distr.).——China.

15. Rubus ribesoideus Matsum. *R. trifidus* var. *tomentosus* Makino——Hachijō-ichigo, Birōdo-kaji-ichigo. Unarmed shrub, the branches rather stout, densely velvety-puberulent while young, the fertile branchlets short; leaves broadly deltoid-ovate, 5–7 cm. long, 4–6 cm. wide, acute to short-acuminate, truncate to cordate at base, 3- to 5-cleft or -lobed, coarsely and irregularly toothed, short-puberulent on both sides especially on nerves beneath; flowers solitary, the pedicels 1–2 cm. long, nodding above, densely velvety-puberulent; calyx-lobes ovate-oblong, mucronate, densely puberulent inside; petals ovate-orbicular, spreading, about 15 mm. in diameter; fruit globose, the fruitlets nearly glabrous.——Dec.–Apr. Near seashores in warmer regions; Honshu (Pacific side), Shikoku, Kyushu.

16. Rubus ohsimensis Koidz. Maruba-no-momiji-ichigo. Nearly unarmed shrub, the branches soft-pubescent while young, the flowering canes very short; leaves broadly ovate to ovate-orbicular, 3–7 cm. long, 3–6 cm. wide, acute, cordate at base, irregularly duplicate-toothed, 3- to 5-cleft, short-pubescent on nerves on both sides, the petioles pubescent; flowers solitary, the pedicels densely pubescent, 8–12 mm. long; calyx appressed-pubescent, the segments narrowly ovate, 6–8 mm. long, acute; petals ovate-orbicular, longer than the calyx-lobes; carpels slightly pubescent; fruit globose, the fruitlets glabrous; stones pitted.——Mar.–Apr. Near seashores; Honshu (Izu Isls.), Kyushu. Possibly a hybrid between Nos. 15 and 18.

17. Rubus corchorifolius L. f. *R. villosus* Thunb.; *R. oliveri* Miq.; *R. corchorifolius* var. *glaber* Matsum.——Birōdo-ichigo. Shrubs, the branches elongate, with rather stout flat prickles, densely pubescent when young with short soft spreading hairs, the flowering branchlets 2–5 cm. long, prickly; leaves membranous, elliptic-ovate or narrowly so, 4–10 cm.

long, 2–4 cm. wide, abruptly acuminate, slightly cordate at base, very rarely obsoletely 3-lobed, irregularly toothed, with ascending short hairs on nerves above, with short spreading hairs and appressed pubescence beneath; flowers solitary, white, the pedicels 7–12 mm. long, densely short-pubescent; calyx 7–10 mm. long, appressed-pubescent, the tube rather broad, shallowly cup-shaped, slightly depressed in the center, the lobes broadly lanceolate, appressed-pubescent also on the inner side; petals about 10 mm. long; fruit globose, the fruitlets densely short-pubescent, the receptacle short-pubescent.——Apr.–May. Honshu (Suruga Prov. and westw.), Kyushu.——s. Korea and China.

18. Rubus palmatus Thunb. *R. palmatoides* O. Kuntze; *R. edulis* Koidz.; *R. dulcis* Koidz.; *R. palmatus* var. *palmatus* O. Kuntze——NAGABA-MOMIJI-ICHIGO. Shrub, the branches with rather stout flat prickles, greenish, the flowering branchlets scarcely elongate, slightly pubescent while young; leaves ovate to narrowly so, 3–7 cm. long, 2.5–4 cm. wide, long-acute to acuminate, truncate to shallowly cordate at base, 3(–5)-lobed or -cleft, rarely undivided, with irregularly acute-toothed lobes, appressed-pubescent on nerves on both sides when young, the petioles often prickly; flowers solitary, white, nodding, the pedicels 5–10 mm. long, sparingly soft-pubescent; calyx 10–12 mm. long, the lobes membranous, more or less appressed-pubescent on both sides; petals broadly obovate, much longer than the calyx-segments; fruit globose, yellow, the fruitlets glabrous.——Mar.–May. Hills and low mountains; Honshu (Kinki Distr. and westw.), Shikoku, Kyushu.——Korea and China. Common and variable.

Var. **coptophyllus** (A. Gray) O. Kuntze. *R. coptophyllus* A. Gray; *R. palmatus* forma *coptophyllus* (A. Gray) O. Kuntze; *? R. fauriei* Lév. & Van't.; *? R. horiyoshitakae* Koidz. ——MOMIJI-ICHIGO. Leaves ovate to broadly so, palmately (3–)5-lobed to -lobulate.——Honshu (centr. and n. distr.).

Var. **kisoensis** (Nakai) Ohwi. *R. kisoensis* Nakai——KISO-KI-ICHIGO. Leaves scarcely divided, deeply cordate at base, the lateral nerves many.——Mountains; Honshu (centr. distr.); rare.

19. Rubus vernus Focke. *R. spectabilis* subsp. *vernus* Focke; *R. spectabilis* sensu auct. Japon., non Pursh——BENI-BANA-ICHIGO. Unarmed shrub with pale yellowish brown branches, soft-pubescent only while very young, the fertile branchlets rather short; leaves membranous, 3-foliolate, the leaflets loosely pubescent on both sides or only on the nerves beneath, the terminal leaflet broadly rhombic-obovate, acute, incised or doubly toothed, 3–7 cm. in diameter, the stipules narrowly lanceolate, the petioles pubescent; flowers solitary, nodding, deep pink, the pedicels 3–6 cm. long, pubescent and with a few stipitate glands; calyx 15–18 mm. long, pubescent and glandular-pubescent, the tube broad, the segments ovate, acute; petals ovate-orbicular, ascending to obliquely spreading, 15–20 mm. long; fruit globose, yellowish red, the fruitlets glabrous; stones about 4 mm. long, pitted.——June–July. Alpine regions; Honshu (centr. and n. distr.).

20. Rubus nishimuranus Koidz. *R. hachijoensis* Nakai; *R. toyorensis* Koidz.; *R. sacrosanctus* Hatus.——HACHIJŌ-KUSA-ICHIGO. Shrub with very sparingly prickly branches, the prickles slender, spreading, scarcely flattened, the fertile branchlets slightly elongate, 7–15 cm. long, pubescent and glandular-pilose; leaves 3-foliolate or sometimes the leaflets connate below in the upper ones, the terminal leaflets ovate-elliptic or narrowly ovate, 5–8 cm. long, 3–5 cm. wide, acuminate, doubly toothed, pubescent on both sides especially on the

nerves, scarcely to very shortly petioluled, the stipules lanceolate; flowers 1 or 2, the pedicels 3–6 cm. long, pubescent and glandular-pilose, often bracteate near the middle; calyx glandular-pilose and short-pubescent, 12–18 mm. long, the lobes rather thick, acuminate, densely puberulent inside; petals white, ovate-orbicular, 15–20 mm. in diameter; carpels nearly glabrous.——Apr.–May. Honshu (Kantō Distr. and Nagato Prov.), Kyushu (n. distr.).——Bonins.

21. Rubus minusculus Lév. & Van't. *R. succedaneus* Nakai & Koidz.; *R. rosaefolius* var. *tropicus* forma *minor* Makino——HIME-BARA-ICHIGO. Shrub with slender, slightly flattened, acuminate prickles, and with scattered yellowish sessile discoid glands on young branches, underside of leaves, leaf-axis, and calyx, the branches slightly prickly, short-pubescent while young, the flowering canes slightly elongate, sparingly prickly; leaflets 5–7, membranous, broadly lanceolate to narrowly ovate, 1–3 cm. long, acute to acuminate, short-pubescent on both sides especially on nerves beneath, doubly toothed, the petioles short-pubescent and with few prickles, the stipules lanceolate; flowers solitary, the pedicels 2–3 cm. long, short-pubescent; calyx 10–16 mm. long, short-pubescent, the tube nearly flat, the lobes narrowly ovate, caudately acuminate, short-puberulent inside; petals white, 12–15 mm. long, broadly obovate-orbicular; fruit globose, the receptacle pilose, the fruitlets glabrous; stones small, about 1.5 mm. long, pitted.——Apr.–May. Honshu (centr. distr. and westw.), Shikoku, Kyushu.

22. Rubus sorbifolius Maxim. *R. asper* sensu auct. Japon., non Wall. nec Presl; *R. rosaefolius* var. *sorbifolius* (Maxim.) Makino——KOJIKI-ICHIGO. Shrub with rather stout, flat prickles and long spreading glandular hairs, the branches stout, elongate, slightly prickly, glandular-pilose while young and sometimes also short-pubescent, the flowering branchlets elongate, 20–40 cm. long; leaflets (3–)5–7, membranous, narrowly ovate to broadly lanceolate, 4–8 cm. long, 1.5–4 cm. wide, acuminate, rounded to obtuse at base, doubly or irregularly acute-toothed, ascending-pubescent on upper side especially on the nerves, glandular-pilose and often slightly spreading-pubescent on nerves beneath, the petioles glandular-pubescent and sparingly prickly; inflorescence a terminal leafy corymbose soft-pubescent panicle with glandular hairs; calyx 10–15 mm. long, the lobes lanceolate-deltoid, caudate, densely white-puberulent inside and on margin, reflexed in fruit; petals obovate or oblong, about 10 mm. long; fruit oblong, the fruitlets very many, glabrous; stones about 1.5 mm. long, pitted.——May–June. Honshu (Tōkaidō and Kinki Distr.), Shikoku, Kyushu.——s. Korea, China, Formosa, (India ?).

23. Rubus croceacanthus Lév. *R. rosaefolius* sensu auct. Japon., non Smith; *R. rosaefolius* var. *tropicus* lusus *genuinus* Makino; *R. kinokuniensis* Koidz.; *R. isensis* Honda; *R. kiusianus* Nakai; *R. okinawensis* Koidz.——Ō-BARA-ICHIGO. Shrub, the branches with rather flat prickles and glandular-pilose, often also soft-pubescent, the flowering canes short, prickly; leaflets 1–5, broadly lanceolate to ovate, 2–7 cm. long, 1–3 cm. wide, acute or acuminate, rounded to obtuse at base, doubly toothed, appressed-pubescent on midrib above and on nerves beneath while young, the petioles glandular-pilose, prickly and slightly pubescent; flowers 1–3; calyx and pedicels glandular-pilose, the calyx lobes ovate-deltoid, caudate, white-tomentose inside and on the margin; petals nearly orbicular, about 15 mm. in diameter; receptacles hairy near base; fruitlets glabrous, aggregated in a broadly ellipsoidal head.——

Mar.–May. Honshu (s. Kantō Distr. and Kii Prov.), Shikoku, Kyushu.——s. Korea and Formosa.

24. Rubus hirsutus Thunb. *R. thunbergii* Sieb. & Zucc. ——KUSA-ICHIGO. Suffruticose or a small shrub with few slender prickles and dense spreading pubescence, often with short glandular hairs, the flowering branchlets rather short; leaflets membranous, 3 (sometimes 5 on vigorous shoots), ovate, 2.5–7 cm. long, 2–4 cm. wide, acute, rounded at base, doubly toothed, prominently hairy on both sides, the petioles soft-pubescent, sometimes with glandular hairs, the stipules lanceolate; flowers 1 or 2, white, erect, about 3 cm. across, the pedicels 3–6 cm. long, densely glandular-hairy; calyx-lobes lanceolate–deltoid, densely puberulent on both sides, glandular-pilose outside, caudately elongate above; petals obovate to ovate-orbicular, 15–20 mm. long; fruit red, globose, the fruit-lets many.——Apr.–May. Thickets and waste grounds in lowlands and low mountains; Honshu, Shikoku, Kyushu; rather common.——Korea and China.——Cv. **Harae**. *R. hirsutus* forma *harae* (Makino) Ohwi; *R. thunbergii* var. *harae* Makino——YAEZAKI-KUSA-ICHIGO. A double-flowered cultivar.——Forma **simplicifolius** (Makino) Ohwi. *R. thunbergii* var. *simplicifolius* Makino——MARUBA-KUSA-ICHIGO. Leaves simple, 3-cleft; gradually grading into the typical phase through forma **ohmatiensis** (Nakai) Ohwi. *R. ohmatiensis* Nakai——ŌMACHI-KI-ICHIGO. With simple and pinnately ternate leaves on the same plant.

25. Rubus illecebrosus Focke. *R. rosaefolius* var. *coronarius* f. *simpliciflorus* Makino; *R. commersonii* var. *simpliciflorus* (Makino) Makino; *R. commersonii* var. *illecebrosus* (Focke) Makino; *R. yakusimensis* Masam.——BARA-ICHIGO. Subshrub with long-creeping stolonlike rhizomes, the prickles flat, the flowering branches erect, 20–60 cm. long, obtusely angled, glabrous, slightly prickly; leaflets 3–7, lanceolate to broadly so, 3–8 cm. long, 1–2 cm. wide, acuminate, usually rounded at base, doubly acute-toothed, glabrous to thinly appressed-pubescent above, often appressed-pubescent on nerves beneath, the petioles glabrous, loosely prickly; flowers white, few to rather many, in a terminal corymb, the pedicels glabrous, sparingly prickly; calyx glabrous, the lobes narrowly ovate, caudate, white-tomentose on margin and inner side; petals ovate-orbicular, 15–20 mm. long; fruit subglobose or broadly ellipsoidal, red.——July–Aug. Thickets and waste grounds; Honshu (centr. distr.), Shikoku, Kyushu.

26. Rubus commersonii Poir. *R. eustephanos* var. *coronarius* Koidz.——TOKIN-IBARA. Nearly glabrous shrub with flat prickles, the flowering branchlets rather elongate, on last year's shoots, obtusely angled, sparingly prickly; leaflets 3–5, ovate-oblong to oblong, 2–5 cm. long, 1–3 cm. wide, doubly acute-toothed; flowers 1 or 2, white, 5–6 cm. across, double, the pedicels 5–10 cm. long, sparsely prickly; calyx-lobes broadly ovate, densely appressed-puberulent inside, thinly puberulent outside, abruptly mucronate or short-caudate.——May–June. Chinese species long cultivated in our area.

27. Rubus oldhamii Miq. *R. pungens* var. *oldhamii* (Miq.) Maxim.——SANAGI-ICHIGO. Shrub with rather slender slightly flattened prickles on the elongate, decumbent, slender branches and pedicels, the flowering branchlets short, sometimes much abbreviated, sparingly short-pubescent above; leaflets membranous, 5(–7), the lateral ones smaller, ovate, obtuse to acute, the terminal one rhombic to deltoid-ovate, 2–4 cm. long, 1–3 cm. wide, acuminate, acute to truncate at base, often 3-lobed, doubly incised-toothed, sparingly ap-

pressed-pubescent to nearly glabrous; flowers 1(–3); calyx 8–12 mm. long, short-pubescent outside and glandular hairy, with many slender scarcely flattened prickles, the lobes narrowly ovate-deltoid, long-acuminate, rather densely appressed-puberulent inside; petals white, obovate-spathulate, about 10 mm. long; carpels soft-puberulent, the receptacle short-pilose. ——May–June. Woods in mountains; Honshu, Shikoku, Kyushu; rather rare.

28. Rubus idaeus L. var. **aculeatissimus** Reg. & Til. *R. melanolasius* Focke; *R. idaeus* subsp. *sachalinensis* var. *matsumuranus* (Lév. & Van't.) Koidz.; *R. sachalinensis* Lév.; *R. idaeus* subsp. *inermis* Koidz.; *R. idaeus* subsp. *melanolasius* Focke——EZO-ICHIGO, KARAFUTO-ICHIGO. Shrub, the branches with slender spreading not flattened bristlelike prickles, glandular-pilose and short-pubescent, the flowering branchlets elongate, 15–40 cm. long; leaflets 3 (sometimes 5 on vigorous shoots), the lateral ones slightly smaller, the terminal rhombic-ovate to broadly ovate or elliptic, 4–7 cm. long, 2.5–5 cm. wide, acuminate to acute, obtuse to rounded at base, irregularly and coarsely acute-toothed, with dense short white-woolly hairs beneath, the petioles prickly and short-pubescent, sometimes also with glandular hairs; inflorescence several-flowered, prickly, short-pubescent and glandular-pilose, the flowers white; calyx-segments narrowly ovate-deltoid, densely white-tomentose inside; petals spathulate, 5–7 mm. long; fruit globose, red, the fruitlets densely woolly-pubescent.——June–July. Hokkaido.——e. Asia and N. America.——Forma **concolor** (Komar.) Ohwi. *R. melanolasius* var. *concolor* Komar.; *R. strigosus* var. *kanayamensis* (Lév. & Van't.) Koidz.; *R. kanayamensis* Lév. & Van't.——KANAYAMA-KI-ICHIGO. Leaves green, not white-tomentose beneath. Hokkaido.——Temperate to cold regions of e. Asia.

29. Rubus yabei Lév. & Van't. *R. idaeus* subsp. *nipponicus* Focke——MIYAMA-URAJIRO-ICHIGO. Closely resembles the preceding species, the plant more slender, usually less densely prickly, with glandular hairs usually only on pedicels and calyx; leaflets somewhat smaller, thinner in texture, often 5, sometimes 3; calyx usually without prickles, the lobes often filiform-elongate above.——June–July. Woods and thickets in high mountains; Honshu (n. and centr. distr.).——Forma **eglandulosus** (Ohwi, pro var.) Ohwi. NAEBA-KI-ICHIGO. Without glandular hairs.——Forma **marmoratus** (Lév. & Van't.) Ohwi. *R. marmoratus* Lév. & Van't.; *R. idaeus* subsp. *melanolasius* var. *marmoratus* (Lév. & Van't.) Hara; *R. yatsugatakensis* Koidz.——SHINANO-KI-ICHIGO. Without glandular hairs; leaves green and not woolly-tomentose beneath. Both forms occur in the high mountains of central Honshu.

30. Rubus coreanus Miq. *R. tokkura* Sieb.——TOKKURI-ICHIGO. Shrub, the branches with stout often curved flattened prickles, the flowering branchlets glabrous; leaflets rather thick, 5–7, glabrous on both sides or slightly pubescent on nerves beneath, with awn-tipped irregular teeth, the lateral leaflets ovate, acute to obtuse, the terminal leaflet slightly larger, rhombic-ovate to suborbicular, 4–6 cm. long, 3–6 cm. wide, very acute, often 3-lobed; corymbs terminal and axillary, rather densely short-pubescent; calyx thick, the lobes broadly lanceolate, caudate-acuminate, densely white-tomentose inside; petals shorter than the calyx-segments, obovate, about 5 mm. long, white, erect.——Rarely cultivated.——Korea and China.

Rubus hiraseanus Makino. Ō-TOKKURI-ICHIGO, is an alleged hybrid of No. 30 and No. 33.

31. Rubus mesogaeus Focke. *R. kinashii* Lév. & Van't.;

R. occidentalis sensu auct. Japon., non L.; *R. idaeus* var. *exsuccus* Fr. & Sav.——KURO-ICHIGO.   Shrub, the branches with rather stout slightly flattened prickles, the flowering branchlets slightly elongate, prickly and short-pubescent; leaflets 3, glabrous to slightly appressed-pubescent on upper side, with short white tomentum beneath, duplicately incised-serrate, abruptly acuminate, the lateral ones obliquely obovate, the terminal larger, ovate-orbicular, 6–12 cm. long, 4–8 cm. wide, the lateral nerves sparingly ascending-pilose beneath, the petioles prickly and pubescent; inflorescence 5- to 10-flowered, densely white-pubescent, the pedicels short; calyx-lobes broadly lanceolate, caudate-acuminate; petals erect, about half as long as the calyx-segments, 4–5 mm. long, broadly obovate; fruit globose, purplish black, the fruitlets nearly glabrous; stones prominently pitted.——June–July.   Hokkaido, Honshu, Shikoku, Kyushu.——China.

Var. **adenothrix** Momiyama. SHIMOKITA-ICHIGO.   Stems, petioles, and inflorescence with short glandular hairs.——Honshu (Mutsu Prov.).

32. Rubus yoshinoi Koidz.   KIBI-NAWASHIRO-ICHIGO. Shrub, the branches with rather flattened prickles, the flowering branchlets elongate, slightly prickly and pubescent; leaflets 3, acuminate to very acute, sparingly appressed-pubescent to glabrous and green above, with appressed-pubescence on nerves and whitish (but later becoming green) tomentose beneath, irregularly doubly acute-toothed, the terminal leaflet slightly larger, rhombic-oblong to ovate, 5–10 cm. long, 2.5–5 cm. wide, obtuse to nearly rounded at base, the petioles thinly pubescent and prickly; inflorescence terminal, few-flowered, the pedicels short; calyx thinly pubescent outside, the lobes narrowly ovate, abruptly caudate-acuminate, densely puberulent inside and on the margin; petals pinkish, obovate, 5–7 mm. long; carpels puberulent.——May–June. Honshu (Iwaki and Shinano Prov. and Chūgoku Distr.), Kyushu; rare.

33. Rubus parvifolius L.   *R. triphyllus* Thunb.; *R. chinensis* Thunb.——NAWASHIRO-ICHIGO.   Shrub, the branches with few rather slender short prickles flattened below, elongate and branching, the flowering branchlets elongate, pubescent, thinly prickly; leaflets 3, appressed-pubescent and often glabrate above, usually pubescent on nerves, with short white tomentum beneath, coarsely double-toothed, the lateral leaflets smaller, broadly obovate to cuneate-orbicular, acute to rounded, the terminal rhombic-orbicular to broadly obovate, 3–5 cm. long and as wide; inflorescence several-flowered, densely pubescent; calyx 5–8 mm. long, usually prickly, the lobes narrowly ovate, acuminate, with short white tomentum especially prominent on the inner side; petals pale pink, erect and connivent above, broadly obovate, 5–7 mm. long; fruit red, globose, the fruitlets puberulent.——May–July.   Waste places and roadsides in lowlands and low mountains; Hokkaido, Honshu, Shikoku, Kyushu; very common.

Var. **concolor** (Koidz.) Makino & Nemoto.   *R. triphyllus* var. *concolor* Koidz.——AO-NAWASHIRO-ICHIGO.   Leaves green, not tomentose. Rare.——Occurs with the typical phase.

Rubus adenochlamys Focke.   KARA-NAWASHIRO-ICHIGO. Resembling No. 33 but with glandular calyx and pedicels.—— Reported to occur in Honshu (Mount Ibuki in Oomi Prov.). ——China.

34. Rubus phoenicolasius Maxim.   EBIGARA-ICHIGO, URAJIRO-ICHIGO.   Rather stout shrub with slightly flattened spreading prickles and long dense glandular hairs on branches, inflorescence, and leaf-axis, the flowering branchlets elongate, prickly, glandular-pilose and pubescent; leaflets 3, green and glabrous to thinly pubescent on upper side, thinly prickly and glandular-pilose on nerves, with white-woolly tomentum beneath, doubly incised-dentate, the lateral leaflets broadly ovate, the terminal larger, ovate-orbicular, short-acuminate, subcordate at base, 5–8 cm. long and as wide; corymbs terminal and in the upper axils, in a large terminal panicle, usually not prickly, glandular-pilose and short-pubescent, the pedicels short; calyx 15–20 mm. long, the lobes lanceolate, gradually narrowed at tip, caudate, glandular-pilose and short-pubescent, erect, connivent above; petals obovate, 4–5 mm. long, pale pink; fruit globose, red, the fruitlets glabrous; stones about 2 mm. long, pitted.——June–July. Waste places and clearings in lowlands and mountains; Hokkaido, Honshu, Shikoku, Kyushu; common.——Korea and China.

35. Rubus pedatus Smith. KOGANE-ICHIGO.   Delicate slender herb, without prickles; stems filiform, much-elongate, creeping, rooting from the nodes, scattered-pilose while young; flowering branchlets in axils of radical leaves, 5–12 cm. long, erect, 1- or 2-leaved; leaves radical, 3-foliolate, green, membranous, thinly appressed-pilose on upper side, nearly glabrous beneath, the terminal leaflet slightly larger, rhombic-ovate, 1.5–3 cm. long, 1–2 cm. wide, obtuse to rounded at apex, cuneate at base, double or incised-toothed, the lateral leaflets often 2-cleft or divided, the petioles thinly short-pubescent, rarely also with short glandular hairs, the stipules elliptic to ovate, brown, scarious, broadly obtuse to rounded at apex, 2–4 mm. long; scapes 3–10 cm. long, short-pubescent, sometimes with glandular hairs; flowers solitary; calyx 8–12 mm. long, the lobes lanceolate, often toothed on upper margin, membranous; petals white, 6–7 mm. long; fruitlets few, yellowish red, glabrous; stones 3.5 mm. long, raised reticulate on the upper half.——June–July.   Damp coniferous woods in mountains; Hokkaido, Honshu (centr. and n. distr.).——N. America.

36. Rubus ikenoensis Lév. & Van't.   *R. japonicus* Maxim., non L.; *R. defensus* Focke; *R. tschonoskii* Prokhan. ——GOYŌ-ICHIGO, TOGE-GOYŌ-ICHIGO.   Perennial herb with woody base, the prickles very slender, spreading, bristlelike, the shoots radical, at first short, erect, afterward elongate, arcuate and decumbent, prickly, the flowering branchlets in the lower axils of the shoots, erect, rather elongate, thinly prickly or sparingly short-pubescent; leaflets 5, pedately arranged, narrowly rhombic-obovate, acuminate, cuneate at base, doubly incised-toothed, sometimes scattered appressed-pilose on upper side and on nerves beneath, green, the stipules 6–8 mm. long, lanceolate, acuminate; flowers 1 or 2, terminal, nodding, white; calyx 15–20 mm. long, short-pubescent and prickly, thinly mixed with glandular-hairs, the lobes elongate above; fruit red, the stones about 2.5 mm. long, irregularly and obliquely rugose.——May–July.   Woods in high mountains; Honshu (n. and centr. distr.).

Rubus kenoensis Koidz. KIREBA-KUMA-ICHIGO, is an alleged hybrid of No. 36 and *R. crataegifolius* (No. 10).

37. Rubus pseudojaponicus Koidz.   *R. triflorus* var. *japonicus* Maxim.——HIME-GOYŌ-ICHIGO, TOGE-NASHI-GOYŌ-ICHIGO.   Pubescent unarmed herb closely resembling the preceding species, the shoots more slender; stipules acute to subacute; calyx-lobes reflexed; petals narrowly ovate, white, about 10 mm. long; stones indistinctly rugose.——May–July. Woods in high mountains; Hokkaido, Honshu (n. and centr. distr.).

15. FILIPENDULA L.   SHIMOTSUKE-SŌ ZOKU

Perennial herbs; radical leaves pinnate, the terminal leaflets large, palmately lobed, toothed, the lateral leaflets small or sometimes obsolete, the cauline leaves often without lateral leaflets, the stipules membranous or foliaceous; inflorescence a cymose corymb; flowers many, small, bisexual or rarely unisexual (plants dioecious), white or pinkish; calyx-lobes 4 or 5, reflexed; petals 4 or 5, orbicular; stamens rather numerous, deciduous after anthesis, the filaments filiform; carpels 3–10, rarely to 15, free, usually short-stiped; ovules 2 in each carpel; style short, terminal; achenes or indehiscent follicles glabrous or ciliate, laterally flattened, stiped or sessile.——More than 10 species, in temperate to cooler regions of the N. Hemisphere.

1A. Achenes sessile, loosely contorted, semiorbicular, glabrous, rounded on outer margin at base; flowers white, unisexual; pedicels with crisped ascending eglandular hairs; lateral leaflets 3 or 4 pairs, very small. . . . . 1. *F. tsuguwoii*
1B. Achenes stiped, not contorted, broadly oblanceolate to oblong, ciliate or glabrous, equilateral; flowers white to pink, bisexual; pedicels glabrous or with spreading strict short hairs.
    2A. Pedicels short-hairy; flowers white; achenes long-ciliate; stipules with a large semiorbicular auricle on one side. . . 2. *F. kamtschatica*
    2B. Pedicels glabrous; flowers usually pink, rarely white; achenes ciliate or glabrous; stipules with or without a prominent auricle.
        3A. Stipules with a semirounded spreading foliaceous auricle on one side; lateral leaflets few-paired and very small or often obsolete.
                3. *F. auriculata*
        3B. Stipules not auricled; lateral leaflets at least in the lower leaves several-paired, much smaller than the terminal leaflet but prominent.
                4. *F. multijuga*

1. Filipendula tsuguwoi Ohwi. SHIKOKU-SHIMOTSUKE-sō. Dioecious herb; stems 50–70 cm. long, erect, 5-angled, few-leaved, glabrous except for short pubescence on upper part; radical and lower cauline leaves long-petioled, the terminal leaflets membranous, orbicular, deeply cordate, 8–13 cm. in diameter, palmately 5- to 7-cleft, the lobes ovate to narrowly so, long-acuminate, duplicate-toothed, green on both sides, thinly pilose on nerves or glabrous on upper side, with crisped yellowish brown pubescence near base beneath, the lateral leaflets 3 or 4 pairs, very unequal, the larger ones ovate, to 1–2 cm. long, the smaller ones 2–3 mm. long, the stipules of the cauline leaves semi-ovate, erect, acuminate, toothed, 1–1.5 cm. long; corymbs terminal and in the upper axils, the branches and pedicels white-pubescent; flowers unisexual, white; calyx-lobes elliptic, glabrous; petals 5, spreading, glabrous, 3.5–4 mm. long; stamens about 30, usually slightly longer than the petals in the staminate flowers, much shorter than the petals in the pistillate flowers; fruiting carpels 5–6, glabrous, sessile, loosely contorted, nearly orbicular-ovate, about 3 mm. long, the stigma small, hemispheric.——July. High mountains; Shikoku, Kyushu; rare.

2. Filipendula kamtschatica (Pall.) Maxim. *Spiraea kamtschatica* Pall.——ONI-SHIMOTSUKE. Stout perennial herb; stems erect, 1–2 m. long, hispid or hispid-pilose, angled; leaves large, the terminal leaflets orbicular, shallowly cordate, 15–25 cm. in diameter, palmately (3–)5-cleft, doubly incised-toothed, green, hispid, the lateral leaflets few, very small, often obsolete in the upper leaves, the stipules green, incised-toothed, with an obliquely spreading auricle at base on one side; inflorescence large, with short spreading straight hairs on pedicels and branches; flowers white or slightly pink, 6–8 mm. across; calyx-lobes reflexed; petals obovate-orbicular, nearly entire; stamens 1.5 times as long as the petals; carpels 5; achenes oblanceolate, long-ciliate.——June–Sept. Mountains; Hokkaido, Honshu (centr. and n. distr.); often gregarious along streams and ravines; variable as to indument. Sometimes distinguished are: Forma **pilosa** Koidz. with looser shorter hairs on stems and leaves, and forma **glabra** Koidz. with nearly glabrous stems and leaves.
3. Filipendula auriculata (Ohwi) Kitam. *F. purpurea*

var. *auriculata* Ohwi.——KOSHIJI-SHIMOTSUKE-sō. Rhizomes short and stout; stems erect, angled, 1–1.5 m. long, 10- to 13-leaved; radical leaves long-petioled, the terminal leaflets large, cordate, 5–15 cm. long, 10–20 cm. wide, 5- to 7-cleft, the lobes acuminate, serrulate, rarely obsoletely incised, nearly glabrous on both sides or thinly pilose on nerves beneath, the lateral leaflets usually 1 or 2 pairs or absent, ovate, 1–3 cm. long, toothed, the stipules mostly foliaceous, green, with semiorbicular, denticulate auricles; corymbs many-flowered, rather large, glabrous; flowers pink, 4–5 mm. across; achenes broadly oblanceolate, ciliate, stiped.——July–Aug. Low mountains; Honshu (Iwashiro, Echigo Prov.).

Filipendula purpurea Maxim. KYŌGANOKO. Leaves mostly without lateral leaflets, the stipules smaller, with a smaller often indistinct auricle on one side; flowers pink; achenes rarely ripening.——July–Aug. Long cultivated in gardens, and possibly a hybrid of *F. multijuga* and *F. auriculata*.——Forma **albiflora** (Makino) Ohwi. *F. purpurea* var. *albiflora* Makino——NATSU-YUKI-sō. Flowers white.—— Rarely cultivated in gardens.

4. Filipendula multijuga Maxim. *Spiraea palmata* Thunb., non L. f., nec Pall.——SHIMOTSUKE-SŌ. Nearly glabrous perennial herb; stems 30–100 cm. long; terminal leaflets orbicular, subcordate, 5–10 cm. in diameter, 5- to 7-cleft to -parted, toothed, the lateral ones of few pairs, unequal, ovate, 3–30 mm. long, toothed and often incised, often reduced and obscure in the upper leaves; stipules membranous, sometimes few-toothed, erect or ascending; inflorescence a terminal many-flowered corymb, glabrous; flowers usually rose, 4–5 mm. across; calyx-lobes 4 or 5, reflexed; petals ovate-orbicular, obsoletely toothed; carpels 4 or 5; achenes short-stiped, oblong, glabrous.——July–Aug. Mountains; Honshu (centr. distr. and westw.), Shikoku, Kyushu.

Var. **ciliata** Koidz. *F. ciliata* (Koidz.) Miyabe & Kudo ——AKABANA-SHIMOTSUKE. Achenes ciliate.——Honshu (centr. and Kantō Distr.).

Var. **yezoensis** Hara. *F. yezoensis* Hara——EZO-NO-SHI-MOTSUKE-sō. Calyx-lobes short-pilose inside; stipules smaller; petals nearly entire.——Hokkaido.

16. SANGUISORBA L.   WARE-MOKŌ ZOKU

Perennial herbs with stout creeping rhizomes; stems erect, leafy, ascending at base; leaves radical and cauline, alternate, imparipinnate, with dilated petioles, the stipules adnate at base, the leaflets oblong or elliptic, toothed, sessile or short-petioluled;

spikes densely many-flowered, terminal; flowers white, greenish or pink, small, bisexual, sessile, with small bracts and bracteoles at base; calyx-tube ovoid, enclosing an achene, constricted at apex, 4-angled or -winged, with 4 small petaloid lobes; petals absent; stamens 4, rarely to 12, deciduous; carpel single, 1-ovuled; style terminal, the stigma fimbriate or pubescent; achenes coriaceous.——More than a dozen species, in temperate and cooler regions of the N. Hemisphere.

1A. Spikes flowering first from the apex or all flowers opening at nearly the same time, rounded at the top; (rarely flowering first from the bottom, if so the spikes deep rose-purple and nodding).
  2A. Stamens 4; spikes 2–7 cm. long.
    3A. Spikes ellipsoidal to obovoid-oblong, erect, 1–2 cm. long; flowers deep red to blood-red; filaments scarcely longer than the calyx-lobes; leaflets usually petioluled. .... 1. *S. officinalis*
    3B. Spikes cylindric or short-cylindric, 2–7 cm. long, often nodding above; flowers white, greenish or suffused with red; filaments much longer than the calyx-lobes.
      4A. Anthers dark colored when dry; leaflets broadly linear to narrowly oblong, sessile or short-petioluled; leaves and stems without curled hairs; flowers white to reddish. .... 2. *S. tenuifolia*
      4B. Anthers pale yellowish brown when dried; leaflets oblong to ovate-orbicular or ovate-cordate.
        5A. Leaflets petioluled; spikes white, often partly suffused with red; stems and leaf-rachises without curled hairs. ... 3. *S. albiflora*
        5B. Leaflets nearly sessile; spikes pink; stems and leaf-rachises with reddish brown curled multicellular hairs. .... 4. *S. obtusa*
  2B. Stamens (4–) 6–12; spikes 4–10 cm. long, pinkish purple, nodding. .... 5. *S. hakusanensis*
1B. Spikes flowering from the bottom upward, lanceolate-cylindric, obtuse or subobtuse at top, often gradually narrowed above; stamens 4, much longer than the calyx-lobes; anthers pale yellowish brown when dry; flowers white or greenish, sometimes partially suffused with red. .... 6. *S. stipulata*

1. Sanguisorba officinalis L. WARE-MOKO. Glabrous perennial herb scarcely glaucescent throughout; rhizomes creeping, rather stout; stems erect, branching above, 30–100 cm. long, often reddish; radical leaves petioled, the leaflets 5–11, oblong to elliptic, sometimes ovate, 2.5–5 cm. long, 1–2.5 (–3.5) cm. wide, rounded at apex, cordate to rounded at base, deltoid-toothed, sessile or the petioles 6–30 mm. long, reddish, often with a small blade at the base, the cauline leaves smaller, short-petioled or sessile; spikes few, long-peduncled, 1–2 (2.5) cm. long, 6–8 mm. across, erect, blood-red; calyx-segments dark colored at base in fruit; anthers dark brown; stigma capitellate.——July–Oct. Meadows in lowlands and mountains; Hokkaido, Honshu, Kyushu; rather common.

Var. **carnea** (Fisch.) Regel. *S. carnea* Fisch.——EZO-WARE-MOKO. Flowers flesh-red.——Occurs with the typical phase.

Var. **pilosella** Ohwi. URAGE-WARE-MOKO. Leaves with spreading short hairs beneath.——Honshu (Shinano Prov.).

2. Sanguisorba tenuifolia Fisch. var. **alba** Trautv. & C. A. Mey. *S. yezoensis* Sieb., ex Miq.——NAGABO-NO-SHIRO-WARE-MOKO. Glabrous perennial with stout short-creeping rhizomes; stems 80–130 cm. long, erect, branching above; radical leaves petioled, the leaflets 11–15, broadly linear to narrowly oblong, 3–8 cm. long, 5–20 mm. wide, obtuse or acute, acutely toothed, usually sessile; spikes cylindric, erect or the longer ones nodding, 2–7 cm. long, 6–7 mm. across exclusive of the stamens, the axis short-pubescent; flowers greenish white or white, often partially reddish, about 3 mm. across; calyx-segments yellowish or brownish below in fruit; filaments white, 1.5–2.5 times as long as the calyx-lobes, slightly broadened above, the anthers dark purple.——Aug.–Oct. Wet meadows and wet places along streams in lowlands and mountains; Hokkaido, Honshu, Shikoku, Kyushu; very variable.

Var. **grandiflora** Maxim. *S. tenuifolia* var. *kurilensis* Kudo; *S. grandiflora* (Maxim.) Makino——CHISHIMA-WARE-MOKO. Stems shorter; leaflets broadly oblong to elliptic or broadly ovate; spikes thicker; filaments longer.——Northern and alpine.

Var. **parviflora** Maxim. *S. parviflora* (Maxim.) Takeda; *S. tenuifolia* var. *angustifolia* Miq.——KOBANA-NO-WARE-MOKO. Leaflets narrower; spikes narrowly cylindric.——Western part of our area.

Var. **purpurea** Trautv. & C. A. Mey. NAGABO-NO-AKA-WARE-MOKO. Flowers blood-red.——Honshu (centr. distr.), and Kyushu.——The typical phase and some other varieties occur in e. Siberia, Kamchatka, Kuriles, Sakhalin, Korea, and Manchuria.

3. Sanguisorba albiflora (Makino) Makino. *S. canadensis* var. *media* Maxim., pro parte; *S. obtusa* var. *albiflora* Makino; *S. obtusa* var. *contraria* Koidz.; *S. amoena* var. *contraria* (Koidz.) Koidz.——SHIROBANA-TO-UCHI-SO. Nearly glabrous perennial herb with stems 30–70 cm. long; radical leaves petioled, the leaflets 11–15, rather loosely arranged on the rachis, broadly to narrowly ovate or elliptic, 2.5–5 cm. long, 1.5–3 cm. wide, rounded at apex, cordate at base, glaucous beneath, toothed, the petiolules 5–20 mm. long, distinct, sometimes with a stipulelike small blade at base, the cauline leaves smaller, the petioles sometimes with white-appressed short hairs near the base; spikes 1–5 or sometimes more, erect or slightly nodding above, 3–6 cm. long, 9–12 mm. across exclusive of the stamens; flowers white, often partially reddish, the peduncles sometimes with short reddish brown hairs; stamens 6–8 mm. long, flat and dilated above; stigma rather large.——Aug.–Sept. Alpine meadows; Honshu (n. distr.); rather common.

4. Sanguisorba obtusa Maxim. NAMBU-TO-UCHI-SO. Resembles the preceding but not as tall, 30–50 cm. high; stems, leaf-rachises and sometimes also the midrib on the leaflets beneath near base with reddish brown curled multicellular hairs; leaflets 13–17, rather densely arranged on the rachis, thicker than in No. 3, nearly sessile, the veinlets more prominent beneath than in the preceding; spikes 4–7 cm. long, nodding, pale rose; stamens 8–10 mm. long; stigma smaller than in No. 3.——Aug.–Sept. Alpine meadows; Honshu (Mount Hayachine); rare.

5. Sanguisorba hakusanensis Makino. KARA-ITO-SO. Nearly glabrous or sparingly reddish brown hairy; stems 40–80 cm. long, erect, few-leaved, often branching above; radical leaves long-petioled, the leaflets 9–13, oblong or ovate-oblong, 3–6 cm. long, 1.5–3.5 cm. wide, rounded at apex, cordate to rounded at base, coarsely toothed, glaucescent beneath, the petiolules 3–7 mm. long, the cauline leaves smaller, often with soft appressed hairs beneath near base; spikes few, pendulous, long-cylindric, 4–10 cm. long, about 10 mm. across exclusive of the stamens; flowers deep rose-purple; stamens 7–10 mm.

long, the filaments flattened and broadened above, the anthers pale yellowish brown when dried, with a deep brown base.——Alpine meadows; Honshu (Japan Sea side of centr. distr.).——Korea (variety).

Var. **japonensis** (Makino) Ohwi. *S. canadensis* var. *japonensis* Makino; *S. japonensis* (Makino) Kudō.——Ezo-tō-uchi-sō. Differs from the typical phase in the centripetal inflorescence.——Rocks along valleys and ravines; Hokkaido; rare.

6. **Sanguisorba stipulata** Raf. *S. sitchensis* C. A. Mey.; *S. canadensis* var. *sitchensis* (C. A. Mey.) Koidz.; *S. sitchensis* var. *typica* Hara; *S. sitchensis* var. *pilosa* Hara; *S. sitchensis* var. *bracteosa* Hara; *S. kishinamii* Honda——Takane-tō-uchi-sō. Perennial herb usually nearly glabrous, with stout creeping rhizomes; stems 40–80 cm. long, often branching above; leaflets of radical leaves 11–13, rather thin, ovate-orbicular to oblong, 2–4 cm. long, 1.5–3 cm. wide, rounded at apex, cordate to rounded at base, glaucous beneath, acutely toothed, the petiolules 5–15 mm. long, the cauline leaves smaller, often with appressed soft short hairs near base; spikes few, erect, 3–8 cm. long, 7–10 mm. across exclusive of the stamens, very densely many-flowered, flowering first from the base, white, often slightly greenish; filaments 7–8 mm. long, slightly broadened above; stigma rather small.——Aug.–Sept. Alpine meadows; Hokkaido, Honshu (Kantō and northern part of the centr. distr.); rare.

Var. **riishirensis** (Makino) Hara. *S. riishirensis* Makino; *S. sitchensis* var. *riishirensis* (Makino) Kudo; *S. chlorantha* Nakai——Rishiri-tō-uchi-sō. With reddish brown curled hairs on stems and leaf-rachises.——Hokkaido.

17. ALCHEMILLA L. Hagoromogusa Zoku

Perennial or annual herbs usually with long silky pubescence; leaves alternate, orbicular-cordate, more or less palmately lobed, the stipules connate and adnate to the lower part of the petiole; flowers small, usually green, in dense corymbs or in heads without bracteoles; calyx with urceolate or obconical tube, the lobes 4 or 5, bracts of calicuus as many and aternate with them; petal absent; stamens 4; carpels 1–4, free, the style from the ventral side or from the base of the carpel, the ovules solitary; achenes 1–4, enclosed in a calyx-tube, membranous.——Many species, in temperate to cold regions and on high mountains of the N. Hemisphere.

1. **Alchemilla japonica** Nakai & Hara. *A. vulgaris* sensu auct. Japon, non L.——Hagoromogusa. Greenish, silky-pubescent perennial herb 10–30 cm. high, with stout creeping rhizomes; stems ascending from base; radical leaves long-petioled, orbicular, deeply cordate with a nearly closed sinus, 3–7 cm. in diameter, shallowly 7- to 9-lobed, toothed, soft-pubescent on both sides, the cauline leaves few, smaller, short-petioled, the stipules green, toothed; flowers densely subumbellate, about 3 mm. across; calyx hairy, the lobes 4, 1–1.2 mm. long; achenes solitary, ovoid, glabrous, 1.5 mm. long, subacute.——Alpine meadows; Hokkaido (Mount Yubari), Honshu (centr. distr.); very rare.

18. AGRIMONIA L. Kin-mizu-hiki Zoku

Usually hirsute perennial herbs with erect simple or branching stems; leaves alternate, imparipinnate, the leaflets toothed, often alternating with smaller ones, the stipules adnate to the lower part of the petioles; inflorescence an elongate terminal spike; flowers small, yellow, short-pedicelled, bearing a bract at the base of the pedicel and 2 bracteoles near the middle; calyx persistent, accrescent, the tube obconical, the lobes 5, connivent in fruit, bracts of the caliculus consisting of many hooked spines; petals 5, yellow; stamens 5–10; carpels 2, sessile; achenes 1 or 2, elliptic, coriaceous, enclosed in a slightly hardened calyx-tube.——More than 10 species, in the N. Hemisphere and S. America.

1. **Agrimonia pilosa** Ledeb. *A. eupatoria* sensu auct. Japon., non L.; *A. viscidula* Bunge; *A. dahurica* Willd. ex Ser.; *A. eupatoria* var. *pilosa* (Ledeb.) Nakai; *A. pilosa* var. *japonica* (Miq.) Nakai; *A. viscidula* var. *japonica* Miq.; *A. japonica* (Miq.) Koidz.——Kin-mizu-hiki. Coarsely hirsute perennial; rhizomes rather stout; stems erect, 30–80 cm. long, often branching above; leaves 5- to 7-foliolate, the leaflets gradually smaller in the lower ones, alternate with much smaller ones, the upper 3 uniform, larger, oblong, obovate or ovate-oblong, 3–6 cm. long, 1.5–3.5 cm. wide, acute to sub-obtuse, coarsely toothed, pale green and often with spreading short hairs on the nerves beneath; stipules green, coarsely toothed; inflorescence terminal, erect, 10–20 cm. long; flowers loosely arranged at anthesis, yellow; calyx-tube accrescent after anthesis, about 3 mm. long in fruit, sulcate, with a series of connivent hooked spines on upper edge; petals narrowly obovate, rounded at apex, 3–6 mm. long.——July–Oct. Meadows and roadsides in lowlands and mountains; Hokkaido, Honshu, Shikoku, Kyushu; rather common.——Korea, China, Siberia, and e. Europe.

Var. **nipponica** (Koidz.) Kitam. *A. nipponica* Koidz.——Hime-kin-mizu-hiki. Slender; leaflets 3–5, obtuse; flowers smaller.——Occurs with the typical phase.

19. ROSA L. Bara Zoku

Erect or scandent shrubs usually prickly on branches and often also on petioles and other parts, the prickles at base of petioles often larger than the others; leaves alternate, imparipinnate, the stipules erect, partly adnate to the lower portion of petiole, the leaflets toothed; flowers solitary or in corymbs or panicles, white, pinkish or yellow; calyx not caliculate, the tube globose to urceolate, becoming fleshy in fruit, the limb 5-lobed, with the disc inside of the calyx-tube; petals 5; stamens many; carpels many, sometimes few, inserted on the lower portion of the calyx-tube, sessile, free; style from the ventral side of the carpel, free or connate above; achenes enclosed in the fleshy calyx-tube, coriaceous.——About 100 species, in temperate to cold regions of the N. Hemisphere.

1A. Styles connate above, usually exserted from the calyx-tube.
  2A. Stipules pectinate; leaflets usually pubescent beneath. . . . . 1. *R. multiflora*
  2B. Stipules entire or toothed; leaflets usually glabrous, rarely pubescent on nerves beneath; styles usually hairy.
    3A. Leaflets 5–7, the terminal one often larger than the others, acute to caudately acuminate, ovate to ovate-elliptic or broadly lanceolate, 2–10 cm. long; stipules entire or nearly so.
      4A. Leaflets 5–10 cm. long; stipules entire; flowers 4–5 cm. across; calyx-segments 12–20 mm. long. . . . . 2. *R. sambucina*
      4B. Leaflets 1–4 cm. rarely to 5 cm. long; stipules often with few teeth; flowers 2–3 cm. across; calyx-segments 5–12 mm. long.
                                                                                         3. *R. luciae*
    3B. Leaflets 5–9, nearly uniform in size, elliptic to obovate-orbicular, obtuse to rounded or acute, lustrous and deep green above; stipules toothed. . . . . 4. *R. wichuraiana*
1B. Styles free.
  5A. Calyx-tube without prickles; leaflets 5–9.
    6A. Branches and prickles hairy; leaves rugose or minutely bullate above, glandular-dotted beneath. . . . . 5. *R. rugosa*
    6B. Branches and prickles nearly or entirely glabrous.
      7A. Branches with flat stipular prickles; leaflets glandular-dotted beneath; calyx-tube globose. . . . . 6. *R. davurica*
      7B. Branches without stipular prickles; leaflets not glandular-dotted; calyx-tube fusiform. . . . . 7. *R. acicularis*
  5B. Calyx-tube densely prickly; leaves 9- to 17-foliolate. . . . . 8. *R. hirtula*

1. Rosa multiflora Thunb. *R. polyantha* Sieb. & Zucc.; *R. wichurae* C. Koch; *R. multiflora* var. *thunbergiana* Thory ——No-IBARA. Erect branching shrub with prickles on the spreading rather stout usually glabrous branches and petioles; stipules green, membranous, short-pubescent and glandular-pilose, the free portion filiform at apex; leaflets 5–7, rather membranous, ovate, obovate or oblong, green, 2–4 cm. long, obtuse to acute, dull, nearly glabrous on upper side, paler and usually with short soft pubescence beneath, the petioles short-pubescent; panicles many-flowered, erect; flowers white, 2–3 cm. across, the pedicels often with short stipitate glands.—— May–June. Hokkaido, Honshu, Shikoku; common.

Var. **adenochaeta** (Koidz.) Ohwi. *R. adenochaeta* Koidz.; *R. polyantha* var. *adenochaeta* (Koidz.) Nakai——TSUKUSHI-IBARA. Leaflets large, 3–7 cm. long, subcoriaceous, less prominently pubescent beneath; inflorescence with prominent stipitate glands 1–2 mm. long; flowers rose, 3–4 cm. across. ——Kyushu (Higo, Satsuma, and Hyuga Prov.).

Rosa pulcherrima Koidz. is an alleged hybrid of No. 1 and No. 3.

2. Rosa sambucina Koidz. *R. moschata* sensu auct. Japon., non Mill.——YAMA-IBARA. Shrub; branches glabrous with recurved flat prickles; stipules membranous, entire, slightly glandular on margin, the free portion linear; leaflets 5, subcoriaceous, broadly lanceolate to narrowly ovate-oblong, (3–)5–10 cm. long, 1–2.5 cm. wide, long-acuminate, glabrous, pale beneath; flowers 5–20, in corymbs or flat-topped panicles, white, 4–5 cm. across, the pedicels 3–5 cm. long, loosely short-glandular-pilose; calyx-tube loosely glandular-pilose, the styles hairy.——May–June. Honshu (s. part of centr. distr. and western Kinki Distr.), Shikoku, Kyushu; rare.

3. Rosa luciae Fr. & Rochebr. *R. luciae* var. *oligantha* Fr. & Sav.; *R. oligantha* (Fr. & Sav.) Koidz.; *R. franchetii* Koidz.——Ō-FUJI-IBARA. Nearly glabrous shrub with elongate branching stems with rather flat prickles; stipules membranous, entire, slightly glandular on margin, the free portion short, filiform; leaflets 5, sometimes 7, rather thin, slightly lustrous on upper side, paler beneath, the terminal ones ovate to narrowly so, 2–4 cm. long, 1–2 cm. wide, acuminate to long-acuminate, the lateral ones acute to acuminate, the lowest pair often smaller; inflorescence a corymb, usually eglandular, the pedicels 6–10 mm. long; flowers few, 2–3 cm. across, usually white.——May–June. Thickets in low mountains; Honshu (Kantō and Tōkaidō Distr.).

Var. **onoei** (Makino) Momiyama. *R. onoei* Makino; *R. hakonensis* var. *onoei* (Makino) Koidz.; *R. sikokiana* Koidz.;

(?) *R. fujisanensis* var. *setifera* Koidz.; (?) *R. yakualpina* Nakai & Momiyama——NIOI-IBARA. Resembles the typical phase, the leaflets beneath and leaf-rachis pubescent.——Honshu (centr. distr. and westw.), Shikoku, Kyushu.

Var. **hakonensis** Fr. & Sav. *R. jasminoides* Koidz.; *R. oligantha* var. *jasminoides* (Koidz.) Koidz.——MORI-IBARA. Leaflets thin, glabrous, whitish beneath, the terminal ones ovate to broadly so, acute to abruptly acuminate; flowers 1 or 2.——May–June. Honshu (Kantō Distr. and westw.), Shikoku, Kyushu.

Var. **fujisanensis** Makino. *R. fujisanensis* (Makino) Makino——FUJI-IBARA. Leaflets (5–)7, paler beneath, glabrous or nearly so, the lateral ones slightly smaller, the terminal one subcoriaceous, ovate to elliptic-ovate, 1.5–2.5 cm. long, 1–1.5 cm. wide, acute or subacute; flowers few.——June–July. Honshu (Sagami, Suruga, and Kai Provs. and westw.), Shikoku.

Var. **paniculigera** (Makino) Momiyama. *R. paniculigera* Makino; *R. franchetii* var. *paniculigera* (Makino) Koidz.—— MIYAKO-IBARA. Stipules nearly entire, glandular on margin, membranous, the free portion broadly lanceolate, acuminate; leaflets 5–9, nearly equal, 1–2.5 cm. long, 7–15 mm. wide, slightly or scarcely lustrous above, glabrous, whitish beneath, the terminal ones elliptic to obovate-elliptic or oblong-ovate, obtuse to abruptly acuminate, the lateral ones obtuse to rounded; inflorescence usually panicled, glabrous or more often glandular-pilose, few- to many-flowered; flowers white, 2–3 cm. across.——May–June. Honshu (centr. distr. and westw.), Shikoku, Kyushu.

4. Rosa wichuraiana Crép. *R. luciae* sensu auct. Japon., pro parte; *R. tsusimensis* Nakai; *R. luciae* var. *yokoscensis* Fr. & Sav.; *R. yokoscensis* (Fr. & Sav.) Koidz.; *R. ampullicarpa* Koidz.; *R. wichuraiana* var. *ampullicarpa* (Koidz.) Honda——TERI-HA-NO-IBARA. Shrub with elongate, rather stout, prostrate, prickly branches; stipules green, toothed, loosely glandular-pilose, the free portion broadly lanceolate, acuminate; leaflets 5–9, coriaceous, nearly equal, obovate-orbicular, elliptic or broadly ovate, 1–2 cm. long, 8–15 mm. wide, obtuse or sometimes subacute, lustrous above, paler beneath, glabrous on both sides; inflorescence a short, glabrous, few- to many-flowered panicle; flowers white, about 3 cm. across.——May–June. Thickets in lowlands and low mountains, especially abundant near seashores; Honshu, Shikoku, Kyushu.——Ryukyus, Formosa, Korea, and China.

5. Rosa rugosa Thunb. *R. ferox* Lawrance; *R. kamtschatica* var. *ferox* (Lawrance) Géel; *R. rugosa* var. *thunbergiana*

C. A. Mey.; *R. rugosa* var. *ferox* (Lawrance) C. A. Mey.——
HAMA-NASU. Erect bushy shrub with stout densely short-pubescent branches with needlelike slender spines and stout flattened short-pubescent prickles; stipules broad, membranous, the free portion broadly ovate or deltoid; leaflets 7–9, nearly equal, oblong, elliptic or obovate, 3–5 cm. long, 2–3 cm. wide, obtuse to rounded, glabrous and minutely bullate or rugulose above, densely cinereous hairy and with sessile pale glandular dots; flowers 1–3, terminal, 6–10 cm. across, deep rose, the pedicels stout, erect, 1–3 cm. long, with slender prickles; calyx with a depressed-globose tube, the lobes 3–4 cm. long, appressed-pubescent and with slender prickles, sometimes with short stipitate glands; fruit subglobose, yellowish red, 2–2.5 cm. across.——June–Aug. Sandy shores; Hokkaido, Honshu on Pacific side south to n. Kantō and on Japan Sea side south to San'in Distr.).——Temperate and northern parts of e. Asia to the Kuriles, Kamchatka, and Sakhalin.

Rosa × iwara Sieb. *R. yesoensis* (Fr. & Sav.) Makino; *R. iwara* var. *yesoensis* Fr. & Sav.——KO-HAMA-NASU. Natural hybrid of No. 5 and No. 1, differing from No. 5 in the less prominently rugose upper leaf-surfaces, and with smaller flowers 3–4 cm. across.——Occurs rarely along seashores in Hokkaido and n. Honshu.

6. Rosa davurica Pall. *R. willdenowii* Spreng.; *R. cinnamomea* var. *davurica* (Pall.) C. A. Mey.; *R. marretii* Lév.; *R. rubrostipulata* Nakai——YAMA-HAMA-NASU, KARAFUTO-IBARA. Shrub with glabrous branches becoming purplish brown with age, the prickles flat, stipular; stipules membranous, brownish, entire, the free portion oblong or ovate, rounded, rarely acute or mucronate; leaflets 7–9, oblong, 2–3 cm. long, 1–1.5 cm. wide, subacute to obtuse, slightly appressed-pubescent to glabrous on upper side, whitish, glandular, and with appressed pubescence especially on midrib beneath; flowers 1–3, deep pink, 3–4 cm. across, with glabrous pedicels; calyx with a globose, glabrous tube, the lobes elongate, with white tomentum inside and on margin; fruit globose or ellipsoidal, about 12 mm. across, red.——June–July. Hokkaido, Honshu (Shinano Prov.); rare.——Sakhalin and Korea to Manchuria and e. Siberia.

7. Rosa acicularis Lindl. *R. fauriei* Lév., pro parte; *R. gmelinii* Bunge; *R. acicularis* var. *gmelinii* (Bunge) C. A. Mey.; *R. acicularis* var. *taquetii* Nakai——Ō-TAKANE-IBARA. Shrub with reddish brown, glabrous, elongate branches and needlelike slender prickles; stipules membranous, entire, glandular on margin, the free portion lanceolate to ovate, acute to acuminate; leaflets 5–7, membranous, oblong or ovate-oblong, 2–4 cm. long, acute to acuminate, vivid green, glabrous on upper side, slightly whitish and with appressed-pubescence on midrib beneath, the petioles sometimes glandular-pilose; flowers solitary, pink, 3–4 cm. across, the pedicels glandular-pilose, 3–5 cm. long; calyx with glabrous tube, the lobes elongate above, white-tomentose inside and on margin, sometimes glandular-pilose outside; fruit fusiform, 2–3 cm. long, 7–10 mm. across.——June–July. High mountains; Hokkaido, Honshu (n. and centr. distr.).——Siberia, Kamchatka, Sakhalin, Korea, and N. America (var.).

Var. **nipponensis** (Crép.) Koehne. *R. nipponensis* Crép.; *R. acicularis* var. *glauca* Fr. & Sav., non Regel——TAKANE-IBARA. Stipules broader, the free portion elliptic-ovate, often obtuse; leaflets 7–9, obtuse to rounded, 1–3 cm. long, 8–15 mm. wide, with minute acute teeth; pedicels with rather long glandular hairs; fruit obovoid.——June–July. High mountains; Honshu (centr. distr. and Oze), Shikoku.

8. Rosa hirtula (Regel) Nakai. *R. microphylla* var. *hirtula* Regel; *R. roxburghii* var. *hirtula* (Regel) Rehd. & Wils.——SANSHŌ-IBARA. Large, much-branched shrub with much-flattened stipular prickles; stipules entire, glandular on margin, the free portion linear-lanceolate; leaflets 9–19, ovate-oblong or oblong, 10–25 mm. long, 5–15 mm. wide, acute, serrulate with short-awned acute teeth, nearly glabrous on upper side, paler and prominently pubescent especially on midrib beneath, the petioles and rachis short-pubescent; flowers solitary, terminal, pale pinkish, 5–6 cm. across, the pedicels stout, short, with needlelike prickles; calyx with densely prickly tube, the lobes broadly elliptic, often with a small appendage near base; fruit globose, about 2 cm. across.——June. Honshu (Sagami, Kai, and Suruga Prov.).

20. PRUNUS L. SAKURA ZOKU

Deciduous or evergreen trees and shrubs; leaves alternate, simple, toothed, conduplicate or convolute in bud; flowers solitary, in fasciculate corymbs or in racemes, white or pink; calyx deciduous, rarely persistent, the tube (hypanthium) tubular to obconical, the lobes 5, imbricate in bud; petals 5, inserted on the throat of the calyx-tube; stamens 15 to many, inserted on the throat of the calyx-tube, the filaments free; carpels solitary, the style terminal, the ovules 2, collateral; drupe fleshy, the endocarp a smooth or rugose, 1-seeded stone.——About 200 species, in the temperate and cooler regions of the N. Hemisphere and in the Tropics of Asia and America.

1A. Inflorescence a fascicle, corymb, or few-flowered short raceme; calyx-lobes deciduous after anthesis.
  2A. Drupes shallowly sulcate; flowers sessile or nearly so (except No. 1).
    3A. Terminal bud not developed; lateral bud solitary; leaves convolute in bud.
      4A. Flowers pedicelled, usually in 3's; ovary and fruit glabrous. .... 1. *P. salicina*
      4B. Flowers nearly sessile; ovary and fruit usually hairy.
        5A. Leaves broadly ovate-orbicular, 4–7 cm. wide; petioles 20–35 mm. long; flowers very short-pedicelled. .... 2. *P. ansu*
        5B. Leaves obovate or elliptic, 3–5 cm. wide; petioles 1–2 cm. long; flowers nearly sessile. .... 3. *P. mume*
    3B. Terminal bud present; lateral buds in 3's, the central one bearing 1 or 2 sessile flowers; drupe usually hairy; leaves conduplicate in bud. .... 4. *P. persica*
  2B. Drupes not sulcate; flowers usually distinctly pedicelled.
    6A. Buds in 3's.
      7A. Leaves prominently pubescent; flowers sessile. .... 5. *P. tomentosa*
      7B. Leaves glabrous or loosely short-pubescent; flowers pedicelled.
        8A. Leaves ovate, sometimes ovate-lanceolate, acuminate, acutely duplicate-toothed; petioles 2–3 mm. long; fruit dark red-purple, with a depressed base. .... 6. *P. japonica*
        8B. Leaves lanceolate to ovate-oblong, acute, obtusely serrulate; petioles 4–6 mm. long; fruit not depressed at base, red.
                                                             7. *P. glandulosa*

6B. Buds solitary; leaves conduplicate in bud.
  9A. Calyx-lobes erect to spreading.
    10A. Glands present at or near the base of the leaves; flowers usually without a common peduncle; calyx, young branches, leaves beneath, and petioles usually pilose or pubescent.
      11A. Pedicels, petioles, and young branchlets with spreading hairs. .... 8. *P. apetala*
      11B. Pedicels, petioles, and young branchlets with appressed to ascending hairs.
        12A. Leaves with few pairs of lateral nerves, 3–5 cm. long or sometimes more, doubly incised-toothed, the teeth not ending in a gland. .... 9. *P. incisa*
        12B. Leaves with 7–14 pairs of lateral nerves, 5–12 cm. long, sparingly double-toothed, the teeth ending in a gland; calyx urceolate. .... 10. *P. pendula*
    10B. Glands present near top of the petioles; flowers often in a corymb with a distinct common peduncle; leaves rarely pilose or pubescent beneath; branchlets and calyx glabrous.
      13A. Calyx-tube campanulate-infundibuliform, 2–3 times as long as the lobes; flowers scarlet, not fully open, nodding; stones pitted. .... 11. *P. campanulata*
      13B. Calyx-tube tubular or campanulate-tubular, as long as to longer than the lobes; flowers pale pinkish, rarely white or greenish.
        14A. Leaves smaller, 4–8 cm. long, with acute incised double teeth, the petioles 7–15 mm. long. .... 12. *P. nipponica*
        14B. Leaves larger, 8–13 cm. long, with acute often awn-tipped simple or sparingly double teeth, the petioles 1.5–3 cm. long.
          15A. Leaves oblong to ovate-elliptic, with scarcely awn-tipped teeth, deep green above, whitish and not lustrous beneath.
            16A. Leaves usually oblong, rounded to acute at base, finely toothed; branchlets rather slender, dark grayish brown or dark brown; flowers in corymbs. .... 13. *P. jamasakura*
            16B. Leaves usually elliptic, rounded to shallowly cordate (at least some of them), more coarsely toothed; branchlets rather stout, dark brown; flowers usually in sessile umbels. .... 14. *P. sargentii*
          15B. Leaves obovate to elliptic, with rather coarse awn-tipped teeth, pale to vivid green above, pale or bright green and lustrous beneath.
            17A. Calyx-lobes entire or nearly so; bracts small, 4–6 mm. long; leaves thin, 4–6 cm. wide, with very short awn-tipped teeth; petioles and pedicels usually more or less spreading short-pubescent. .... 15. *P. verecunda*
            17B. Calyx-lobes serrulate with minute glandular-tipped teeth; bracts large, about 1 cm. long; leaves rather thick, 5–8 cm. wide, awn-teeth about 2 mm. long; petioles and pedicels glabrous. .... 16. *P. lannesiana*
  9B. Calyx-lobes reflexed; petals not retuse.
    18A. Involucral bracts (bud scales of flowering bud) persistent through anthesis; inflorescence a sessile umbel; leaves with short obtuse teeth. .... 17. *P. avium*
    18B. Involucral bracts or bud-scales early deciduous; inflorescence a short raceme with persistent green bracts; leaves doubly incised-toothed. .... 18. *P. maximowiczii*
1B. Inflorescence a many-flowered raceme.
  19A. Calyx deciduous after anthesis.
    20A. Leaves evergreen; racemes not leafy below.
      21A. Glands of leaves on upper portion of petioles; inflorescence solitary to 5-nate; leaves 10–20 cm. long, 4–7 cm. wide, usually acutely toothed. .... 19. *P. zippeliana*
      21B. Glands of leaves usually near the base of the blade; inflorescence solitary; leaves 6–10 cm. long, 2–3 cm. wide, the teeth sometimes obscure or sometimes large, spine-tipped. .... 20. *P. spinulosa*
    20B. Leaves deciduous; racemes with 1–3 leaves at base.
      22A. Leaves cordate, with rather awnlike mucronate teeth; glands of leaves at top of petioles. .... 21. *P. ssiori*
      22B. Leaves acute to obtusely rounded at base.
        23A. Glands of leaves at the top of the petioles; leaves with 10–13 pairs of lateral nerves; stamens shorter than the petals.
                                                       22. *P. padus*
        23B. Glands of leaves on lower margin of the blade; leaves with 7–8 pairs of lateral nerves; stamens longer than the petals.
                                                       23. *P. grayana*
  19B. Calyx persistent; leaves deciduous; racemes not leafy at base. .... 24. *P. buergeriana*

1. **Prunus salicina** Lindl. *P. triflora* Roxb.——Su-momo. Small tree with lustrous glabrous branches; leaves broadly oblanceolate to narrowly obovate, 5–10 cm. long, 2–4 cm. wide, abruptly acuminate, cuneate at base, subduplicately obtuse-serrulate, thinly pubescent beneath while young, becoming glabrous on both sides or slightly pilose in axils beneath, the petioles 1–2 cm. long, usually thinly pubescent and with 2–5 glands on the upper portion; flowers usually in 3's, white, 15–20 mm. across, the pedicels glabrous, 1–1.5 cm. long; drupe ovoid-globose, 5–7 cm. across, with a depression at base.——Apr. Chinese tree long cultivated in our area and sometimes naturalized.

2. **Prunus ansu** (Maxim.) Komar. *P. armeniaca* var. *ansu* Maxim.——Anzu. Small tree; leaves broadly ovate or ovate-orbicular, 6–8 cm. long, 4–7 cm. wide, caudately acuminate, rounded to subtruncate at base, glabrous, serrulate, the petioles 20–35 mm. long, glabrous; flowers pink to white, precocious; drupe red, short-pilose, nearly globose, the stones roughened.——Apr. Introduced from China and long cultivated in Japan.

3. **Prunus mume** Sieb. & Zucc. Ume. Small tree with green branchlets; leaves broadly ovate, 4–10 cm. long, long-acuminate, broadly cuneate at base, acutely serrulate, sparingly pilose or sometimes only on nerves beneath, the petioles short-pilose above; flowers 1 or 2, appearing before the leaves, white or pink; drupe subglobose, the stones rugose.——Feb.–Apr. Said to be indigenous in the mountainous parts of Kyushu. A much-planted Chinese tree with many cultivars grown.

4. **Prunus persica** (L.) Batsch. *Amygdalus persica* L.; *Persica vulgaris* Mill.; *Prunus persica* var. *vulgaris* Maxim.——Momo. Small tree with glabrous elongate branches; leaves lanceolate to broadly oblanceolate, 8–15 cm. long, 15–35 mm. wide, long-acuminate, serrulate, glabrescent, the petioles 4–10 mm. long, with glands on the upper portion; flowers pink to

white, on very short pedicels; calyx short-pubescent; drupe typically short-pilose, the stones very prominently rugose.——Apr.–May. A widely cultivated Chinese tree with many cultivars grown; said to be spontaneous in Kyushu.

5. Prunus tomentosa Thunb. *Cerasus tomentosa* (Thunb.) Wall.; *P. trichocarpa* Bunge; *C. trichocarpa* (Bunge) Wall.——YUSURA-UME. Shrub with dense pubescence on young branchlets, petioles and underside of leaves; leaves obovate, abruptly short-acuminate, 5–7 cm. long, 3–4 cm. wide, hairy on upper side, duplicately toothed, the petioles 2–5 mm. long; flowers 1 or 2, pale pink, small; ovary and calyx-lobes short-pubescent; drupe subglobose, red.——Apr. Cultivated for the flowers and the edible fruit.——China and Korea to the Himalayas.

6. Prunus japonica Thunb. *Cerasus japonica* (Thunb.) Loisel.——NIWA-UME. Shrub with slender elongate branches often puberulent while young; leaves ovate, rarely ovate-lanceolate, 3–7 cm. long, acuminate, rounded or sometimes cuneate at base, doubly acute-toothed, glabrous on both sides or short-pubescent on nerves beneath, the stipules longer than the petioles, laciniate, minutely toothed, the petioles 2–3 mm. long; flowers appearing with the leaves, 2 or 3, small, white or pale pink, single, the pedicels 5–10 mm. long; drupes subglobose.——Apr. Cultivated shrub of Chinese origin.

7. Prunus glandulosa Thunb. *Cerasus glandulosa* (Thunb.) Loisel.; *P. glandulosa* var. *glabra* Koehne; *P. japonica* var. *glandulosa* (Thunb.) Maxim.——HITOE-NO-NIWA-ZAKURA. Deciduous shrub with slender, elongate, glabrous or sometimes puberulent branches; leaves lanceolate to ovate-oblong, 3–8 cm. long, gradually tapering above to a suobtuse tip, glabrous on both sides or slightly puberulent on nerves beneath, obtusely serrulate, the petioles 4–6 mm. long; flowers 1 or 2, small, single, white or pink, the pedicels 1–2 cm. long; drupes red, about 1.2 cm. across.——Apr.–May. Cultivated.

8. Prunus apetala (Sieb. & Zucc.) Fr. & Sav. *Ceraseidos apetala* Sieb. & Zucc.; *P. ceraseidos* (Sieb. & Zucc.) Maxim.; *P. crassipes* Koidz.——CHŌJIZAKURA, MEJIROZAKURA, TANI-NOZOKI. Small tree with glabrescent or pilose branchlets; leaves obovate or cuneate-obovate, 5–8 cm. long, 3–5 cm. wide, caudately acuminate, spreading-pilose on both sides especially on nerves beneath, dull, with short, rather obtuse to acute coarse double teeth, the petioles 3–8 mm. long, spreading-pilose; flowers 1–3, in sessile umbels, nodding, small, pink to nearly white; calyx-tube slender, 7–10 mm. long, prominently pubescent, 3–4 times as long as the ovate lobes; petals 5–8 mm. long, retuse; style with a few spreading hairs on the lower part.——Apr. Low mountains; Honshu, Kyushu.

Var. **pilosa** (Koidz.) Wils. *P. matsumurana* Koehne; *P. ceraseidos* sensu Koidz., non Maxim.; *P. ceraseidos* var. *pilosa* Koidz.; *P. crenata* Koehne——OKU-CHŌJIZAKURA. Less prominently hairy on all parts; leaves acutely toothed, the petioles 6–10 mm. long; flowers larger; calyx-tube 2 or 3 times as long as the lobes; petals 8–12 mm. long; style usually glabrous.——Mountain form especially abundant on the Japan Sea side of Honshu (centr. distr. and northw.). Grades into the typical phase.

9. Prunus incisa Thunb. *Cerasus incisa* (Thunb.) Loisel.——MAMEZAKURA. Small tree, the young branchlets sometimes pubescent; leaves obovate or ovate, 3–5 cm. long, 2–3 cm. wide, short-acuminate, pubescent on upper side and on nerves beneath, doubly acute-toothed, the petioles 8–10 mm. long; flowers 1–3, in sessile umbels, white to rose, on

glabrous to thinly pubescent pedicels 1–2.5 cm. long; calyx-tube usually glabrous, 5–6 mm. long, about twice as long as the lobes; petals about 10 mm. long, retuse.——Apr.–June. Mountains; Honshu (centr. distr. and northw.).——Forma **yamadei** (Makino) Ohwi. *P. incisa* var. *yamadei* Makino——RYOKUGAKU-ZAKURA, MIDORI-ZAKURA. Young leaves, calyx, and pedicels green.

Var. **tomentosa** Koidz. *P. hisauchiana* Koidz. ex Hisauchi——YABUZAKURA. Leaves larger, broader; calyx-lobes clearly serrulate.——Low mountains; Honshu (w. Kantō Distr.).

Var. **kinkiensis** (Koidz.) Ohwi. *P. kinkiensis* Koidz.——KINKI-MAMEZAKURA. Differs from the typical phase in the longer calyx-tube and slightly larger caudate leaves.——Mountains; Honshu (w. part of centr. distr. and westw.).

Var. **bukosanensis** (Honda) Hara. *P. nipponica* var. *bukosanensis* Honda——BUKO-MAME-ZAKURA. Leaves larger, broadly obovate, 5–8 cm. long, 3–5 cm. wide, glabrescent on both sides.——Mountains; Honshu (Mount Buko in Musashi).

10. Prunus pendula Maxim. forma **ascendens** (Makino) Ohwi. *P. pendula* var. *ascendens* Makino; *P. itosakura* var. *ascendens* (Makino) Makino; *P. subhirtella* var. *ascendens* (Makino) Wils.; *P. aequinoctialis* Miyoshi; *P. subhirtella* var. *pendula* forma *ascendens* (Makino) Ohwi; *P. microlepis* Koehne; *Cerasus herincqiana* Lavall.; *P. herincqiana* (Lavall.) Spaeth——EDO-HIGAN, AZUMA-HIGAN. Large tree with longitudinally fissured bark, the branches elongate, rather slender, light grayish brown, ascending-pilose while young; leaves membranous, oblong to narrowly obovate, 6–12 cm. long, 3–4 cm. wide, caudately acuminate, with 10–14 pairs of obliquely spreading parallel lateral nerves, subdouble-toothed, glabrous to thinly ascending-pubescent on upper side, ascending- to appressed-pubescent beneath especially on nerves, the petioles pubescent; flowers 2–5, in sessile umbels, pink, the pedicels 1.5–3 cm. long, with whitish ascending hairs; calyx-tube slightly inflated at base, about 5 mm. long, more or less pubescent; petals about 12 mm. long, retuse; style with few spreading hairs on the lower part.——Apr. Mountains; Honshu, Shikoku, Kyushu; occasionally planted in parks and around temples.——Cv. **Pendula**. *P. pendula* Maxim.; *P. itosakura* Sieb. nom. subnud.; *P. herincqiana* sensu C. K. Schn. non Spaeth; *P. subhirtella* var. *pendula* (Maxim.) Y. Tanaka——ITOZAKURA, SHIDAREZAKURA. Branches pendulous. A much cultivated phase.

The following are alleged hybrids:

Prunus sieboldii (Carr.) Wittm. *Cerasus sieboldii* Carr.; *P. pseudocerasus* var. *sieboldii* (Carr.) Maxim.; *C. watereri* Lavall. ex Bean; *P. pseudocerasus* forma *watereri* (Lavall.) Koehne; *P. koidzumii* Makino; *P. fortis* Koidz.; *P. donarium* subsp. *fortis* (Koidz.) Koidz.——NADEN. Small tree with dark grayish brown rather stout branches more or less spreading-pubescent while young; leaves prominently spreading-pubescent, the glands small, at base of the blade, the pedicels pubescent; flowers semidouble.——Alleged hybrid of Nos. 8 and 16.

Prunus yedoensis Matsum. SOMEI-YOSHINO. Small tree with ascending-hairs on young branchlets, lower leaf surfaces, inflorescence, and calyx; leaves vivid green; flowers single, precocious.——Apr. Much planted on riverbanks and in parks. Alleged hybrid of Nos. 10 and 16.

Prunus subhirtella Miq. *P. miqueliana* Maxim., pro parte; *P. kohigan* Koidz.; *P. subhirtella* var. *oblongifolia* Miq.; *P. herincqiana* var. *ascendens* C. K. Schn., excl. syn.; *P. itosa-*

kura var. *subhirtella* (Miq.) Koidz.; *P. subhirtella* var. *glabra* Koidz.——Ko-HIGANZAKURA, CHIMOTO-HIGAN. Small tree with slender branches, ascending-pubescent while young; leaves obovate, 3–8 cm. long, duplicate-toothed, with 7–10 pairs of lateral nerves, the petioles pubescent; calyx-tube slightly inflated below, short-pubescent; style usually glabrous.——Mar.–Apr. Alleged hybrid of Nos. 9 and 10.——Cultivated only, unknown in the wild.

Var. **autumnalis** Makino. *P. autumnalis* (Makino) Koehne——JŪGATSU-ZAKURA. Differs in having double flowers, appearing in spring and in autumn.——Cultivated only, unknown in the wild.

11. Prunus campanulata Maxim. *P. cerasoides* var. *campanulata* (Maxim.) Koidz.——HI-KANZAKURA, KAN-HI-ZAKURA. Small tree with rather stout branches; leaves obovate, ovate or elliptic, 7–12 cm. long, 3.5–5 cm. wide, glabrous; flowers 1–3, scarlet, semi-expanding, nodding.——Feb.–Mar. Cultivated in warmer parts.——Ryukyus and Formosa.

12. Prunus nipponica Matsum. *P. iwagiensis* Koehne; *P. nikkoensis* Koehne; *P. nipponica* var. *iwagiensis* (Koehne) Koidz.——MINE-ZAKURA, TAKANEZAKURA. Shrub to small tree; leaves obovate to broadly so, caudately acuminate, 4–8 cm. long, 2–4.5 cm. wide, often thinly pilose beneath while young, sometimes appressed-pilose above, pale green beneath, incisely and doubly acute-toothed, the petioles 7–15 mm. long, glabrous; flowers 1–3, in sessile umbels or in short-peduncled umbellike corymbs, pink, the pedicels 2–3 cm. long, glabrous; calyx-tube glabrous, about 6 mm. long, about 1.5 times as long as the lobes; petals 10–12 mm. long, retuse.——May–July. High mountains; Hokkaido, Honshu (centr. and n. distr.).

Var. **kurilensis** (Miyabe) Wils. *P. ceraseidos* var. *kurilensis* Miyabe; *P. apetala* var. *iwozana* C. K. Schn.; *P. incisa* var. *kurilensis* (Miyabe) Koidz.; *P. kurilensis* (Miyabe) Miyabe ex Takeda; *P. nipponica* var. *pubescens* Koidz.——CHISHIMA-ZA-KURA. Petioles or pedicels or both pilose.——Hokkaido, Honshu (centr. and n. distr.).——Kuriles and Sakhalin.

13. Prunus jamasakura Sieb. ex Koidz. *P. donarium* Sieb., pro parte; *P. serrulata* sensu auct. Japon., non Lindl.; *P. pseudocerasus* sensu auct. Japon., non Lindl.; *P. pseudocerasus* var. *jamasakura* subvar. *glabra* Makino; *P. donarium* var. *glabra* (Makino) Koidz.; *P. jamasakura* var. *glabra* (Makino) Koidz.; *P. serrulata* var. *spontanea* (Maxim.) Makino; *P. mutabilis* Miyoshi——YAMAZAKURA. Deciduous tree with dark brown transversely striate bark and rather slender glabrous branchlets; leaves oblong, narrowly elliptic or ovate to obovate-oblong, 8–12 cm. long, 3–4.5(–5.5) cm. wide, deep green and usually glabrous on upper side, whitish and glabrous beneath, acutely serrulate, the petioles reddish, with a pair of sessile glands near the top; flowers few, in corymbs, 2.5–3 cm. across, pink, glabrous, the bracts obovate-cuneate, rounded, 3–6 mm. long; calyx-tube tubular, the lobes entire; drupe purplish black when mature.——Low mountains; Honshu (Kantō Distr. and westw.), Shikoku, Kyushu; common and variable.——Forma **pubescens** (Makino) Ohwi. *P. pseudocerasus* var. *jamasakura* subvar. *pubescens* Makino, pro parte——USUGE-YAMAZAKURA. Petioles or pedicels or both thinly puberulent.

Var. **chikusiensis** (Koidz.) Ohwi. *P. chikusiensis* Koidz.; *P. serrulata* var. *chikusiensis* (Koidz.) Hatus.——TSUKUSHI-ZAKURA. An insular phase with stout branches, thicker leaves, and larger bracts; flowers larger, nearly white.——Small islands around Kyushu.

14. Prunus sargentii Rehd. *P. pseudocerasus* var. *sachalinensis* F. Schmidt; *P. sachalinensis* (F. Schmidt) Koidz.; *P. jamasakura* var. *borealis* (Makino) Koidz.; *P. serrulata* var. *sachalinensis* (F. Schmidt) Makino; *P. serrulata* var. *borealis* Makino; *P. floribunda* Koehne——Ō-YAMAZAKURA, EZO-YAMA-ZAKURA, BENI-YAMAZAKURA. Resembles the preceding; bark dark chestnut-brown, the branchlets stout, glabrous, with viscid buds; leaves elliptic, ovate-elliptic or sometimes obovate-elliptic, 8–13 cm. long, 4–7 cm. wide, rounded or at least some of them subcordate at base, deep green and usually glabrous on upper side, glaucescent and glabrous beneath, with obliquely deltoid teeth, the petioles 1.5–3 cm. long, glabrous, reddish, with a pair of reddish sessile glands near the top; flowers 3–4 cm. across, in sessile umbels, glabrous, pink; calyx-lobes entire; drupe dark purple.——Apr.–May. Mountains; Hokkaido, Honshu (centr. and n. distr.).——Sakhalin and Korea.——Forma **pubescens** (Tatew.) Ohwi. *P. sargentii* var. *pubescens* Tatew.——KE-EZO-YAMAZAKURA. Pubescent on lower leaf surfaces, petioles or pedicels, or on all of them.

15. Prunus verecunda (Koidz.) Koehne. *P. jamasakura* var. *verecunda* Koidz.; *P. donarium* subsp. *verecunda* (Koidz.) Koidz.——KASUMIZAKURA. Resembling the preceding two species; bark grayish brown, the branchlets grayish to yellowish brown, glabrous; leaves obovate, obovate-elliptic or sometimes elliptic, rounded to cuneate, sometimes shallowly cordate at base, 8–12 cm. long, 4–6 cm. wide, vivid green and sometimes scattered-pilose on upper side, pale green to bright green and lustrous and often with spreading hairs beneath, with short-awned coarser marginal teeth than in Nos. 13 and 14, the petioles slightly reddish, with a pair of small, sessile glands on the upper part, usually more or less spreading-pilose; flowers pink to nearly white, 2.5–3 cm. across, the peduncles sometimes absent, the pedicels often spreading-pilose, the bracts 3–6 mm. long; calyx-lobes glabrous, usually entire.——Apr.–May. Mountains; Hokkaido, Honshu, Shikoku, Kyushu.——Korea.

16. Prunus lannesiana (Carr.) Wils. var. **speciosa** (Koidz.) Makino. *P. serrulata* var. *albida* subvar. *speciosa* (Koidz.) Makino; *P. donarium* subsp. *speciosa* (Koidz.) Koidz.; *P. speciosa* (Koidz.) Nakai——ŌSHIMA-ZAKURA. Resembles Nos. 13–15; branchlets stouter, grayish to pale brown, glabrous; leaves rather thick, obovate to obovate-elliptic, sometimes elliptic, 8–13 cm. long, 5–8 cm. wide, rounded to broadly cuneate at base, vivid green on upper side, pale green beneath, glabrous on both sides, with narrowly deltoid awn-tipped teeth about 2 mm. long, some of the teeth doubly toothed, the petioles nearly green or slightly reddish, with a pair of glands near the top; flowers in corymbs, white, sometimes pink, 3.5–4 cm. across, the pedicels rather stout, elongate, glabrous, the bracts flabellate-obovate, about 1 cm. long; calyx-lobes serrulate with gland-tipped minute teeth; fruit rather large, dark purple.——Mar.–Apr. Low mountains near the sea; Honshu (Awa, Izu Prov., and Izu Isls.).

Var. **lannesiana**. *Cerasus lannesiana* Carr.; *P. pseudocerasus* var. *hortensis* Maxim., pro parte maj.; *P. donarium* Sieb., pro parte; *P. serrulata* var. *lannesiana* (Carr.) Rehd.; *P. serrulata* var. *hortensis* (Maxim.) Makino——SATOZAKURA. Includes an assemblage of various cultivars with large single or double flowers.

17. Prunus avium L. SEIYŌ-MIZAKURA. Tree with pyramidal crown and stout branches; leaves ovate- to obovate-oblong, 8–15 cm. long, 5–7 cm. wide, abruptly short-acuminate, irregularly obtuse-toothed, spreading-pilose beneath especially on nerves, the petioles 3–5 cm. long; flowers 3–5, in umbels, white; calyx-tube contracted above.——Apr.–May. Europe. The cultivated edible cherry.

18. **Prunus maximowiczii** Rupr. *P. bracteata* Fr. & Sav. ——Miyamazakura. Small tree with rather prominent appressed hairs on young branchlets, inflorescence, petioles, and nerves of leaves beneath; leaves obovate or rhombic-elliptic, gradually narrowed above or short-caudate, 4–8 cm. long, 2.5–6 cm. wide, coarsely and acutely double-toothed, appressed-pilose to glabrate on upper side, the petioles 7–15 mm. long, the glands on or near base of leaf-margin; flowers 4–7, in loose corymbose racemes, white, small, the bracts ovate-orbicular, persistent, 5–8 mm. long, toothed, the pedicels 10–15 mm. long; calyx-tube obconical, somewhat longer than the denticulate ovate lobes, pilose; style pilose or glabrous at base.—— June. Mountains; Hokkaido, Honshu, Kyushu.——Sakhalin, Korea, Manchuria, and Ussuri.

19. **Prunus zippeliana** Miq. *P. macrophylla* Sieb. & Zucc., non Poir.——Bakuchi-no-ki. Evergreen tree with reddish brown exfoliating bark; leaves subcoriaceous, narrowly ovate-oblong or -elliptic, 10–20 cm. long, 4–7 cm. wide, gradually narrowed to the tip, glabrous or thinly soft-pubescent beneath, acutely toothed, the petioles 10–15 mm. long, with a gland on each side at the top; racemes short, 1- to 5-nate, densely many-flowered, nearly sessile, the bracts broadly deltoid, caducous, with pale brown pubescence; flowers white, 6–7 mm. across, the rachis and pedicels with short spreading hairs; calyx-tube broadly obconical, slightly hairy, the teeth shorter than the tube; petals reflexed, about 3 mm. long, nearly orbicular; ovary glabrous, the style 3–3.5 mm. long; drupe ovoid, about 18 mm. long, terminated by the persistent style base.——Sept.–Oct. Honshu (warmer part of Kantō Distr. and westw.), Shikoku, Kyushu.——Ryukyus and Formosa.

20. **Prunus spinulosa** Sieb. & Zucc. Rinboku. Small tree, the young branches often puberulent; leaves coriaceous, lustrous, narrowly oblong to narrowly obovate, 5–8 cm. long, 1.5–3 cm. wide, caudate-acuminate with an obtuse or acute tip, glabrous, more or less thickened on margin, remotely toothed or entire, the teeth ascending to obliquely spreading, spine-tipped, the petioles 5–10 mm. long, glabrous; inflorescence 2–8 cm. long, rather densely many-flowered, puberulent only while young, glabrescent, nearly sessile, the bracts caducous; flowers white, about 6 mm. across, the pedicels 3–7 mm. long; calyx-tube broadly obconical, longer than the loosely toothed ovate lobes; petals orbicular, about 3 mm. long, reflexed; drupes ellipsoidal or broadly ovoid, glabrous, about 8 mm. long.——Sept.–Oct. Warmer parts; Honshu (Kantō Distr. and westw.), Shikoku, Kyushu.——Ryukyus.

21. **Prunus ssiori** F. Schmidt. Shūrizakura, Miyama-inuzakura. Deciduous tree with glabrous branchlets; leaves oblong to ovate, 7–15 cm. long, 3–7 cm. wide, long-acuminate, usually cordate at base, glabrous except for axillary tufts of hairs beneath, serrulate with narrow, acute, simple or some-

times double small teeth, the glands near the top of the petiole; racemes 15–20 cm. long, rather densely many-flowered, glabrous, with few small leaves at base; flowers small, white, about 10 mm. across, the pedicels 7–10 mm. long; calyx with cup-shaped tube, the teeth short, semi-rounded, with minute glandular teeth; petals spreading, elongate-orbicular, 4–5 mm. long, nearly as long as the stamens, loosely toothed.——May–June. Woods in mountains; Hokkaido, Honshu (centr. distr. and northw.).——Sakhalin, Manchuria, Ussuri, and s. Kuriles.

22. **Prunus padus** L. *P. racemosa* Lam.; *Padus racemosa* (Lam.) C. K. Schn.——Ezo-no-uwa-mizuzakura. Deciduous tree; leaves obovate to elliptic or narrowly obovate, 5–9 cm. long, 3–6 cm. wide, abruptly acuminate, obtuse to rounded at base, acutely toothed, glabrous or with few axillary tufts of hairs beneath, the petioles 10–15 mm. long, glabrous, with glands at the top; racemes 10–12 cm. long, rather many-flowered, glabrous to puberulent; flowers white, 12–15 mm. across, the pedicels 8–12 mm. long; calyx with broadly obconical tube, the lobes ovate, glandular-toothed; petals 5–6 mm. long, orbicular, loosely toothed; stamens short.——May–July. Hokkaido.——Korea, Sakhalin, and widely distributed in temperate regions of the Old World.

23. **Prunus grayana** Maxim. *P. padus* var. *japonica* Miq.; *Padus grayana* (Maxim.) C. K. Schn.——Uwa-mizu-zakura. Tree with puberulent or glabrous young branches; leaves ovate to ovate-oblong, 6–10(–15) cm. long, 2.5–4.5(–5.5) cm. wide, caudate to abruptly acuminate, glabrous on both sides or pilose along the midrib beneath while young, acutely toothed, the petioles 6–8(–10) mm. long, glabrous or puberulent while young; racemes 12–20 cm. long inclusive of short few-leaved peduncles, rather dense, many-flowered, glabrous or puberulent on rachis; flowers white, 6–8 mm. across, the pedicels 3–10 mm. long; calyx with obconical tube, the teeth deltoid, short, pilose inside; petals suborbicular, about 3 mm. long; stamens longer than the petals; drupe ellipsoidal, acute, 6–7 mm. long, black.——Apr.–June. Hills and mountains; Hokkaido, Honshu, Shikoku, Kyushu.——Plants of the northern part of the range generally have larger leaves and more pilose calyx-teeth than those in the southern part of the distribution.

24. **Prunus buergeriana** Miq. *Laurocerasus buergeriana* (Miq.) C. K. Schn.——Inuzakura, Shirozakura. Tree, usually the branches puberulent when young; leaves chartaceous, somewhat lustrous, ovate-oblong to cuneate-oblong, 5–10 cm. long, 2–3.5 cm. wide, acuminate with an acute or sub-obtuse tip, with appressed or ascending minute teeth, glabrous on both sides except for pubescence along the midrib near the base beneath while young, the petioles 8–25 mm. long, sometimes puberulent on the upper side; racemes 5–10 cm. long, rather loosely flowered, puberulent; flowers white, about 5 mm. across; calyx with cup-shaped tube, the teeth short, broadly ovate, sometimes loosely toothed; petals broadly obovate, 2–3 mm. long, shorter than the stamens.——Apr.–June. Hills and mountains; Honshu, Shikoku, Kyushu.

21. CRATAEGUS L. Sanzashi Zoku

Shrubs or trees, branchlets often thornlike; leaves petiolate, alternate, simple to 3-lobed or pinnatifid, the stipules usually small, deciduous; cymes or corymbs terminal; flowers white or reddish; calyx urceolate or campanulate, the teeth 5; petals 5; stamens many; ovary inferior or semi-inferior, 1- to 5-locular; styles 1–5; ovules 2 in each locule, erect; drupes subglobose, red to black, sometimes yellow, small, the stones bony; seeds solitary in each locule, erect, flat.——A large complicated genus of the N. Hemisphere, especially abundant in e. N. America.

1A. Leaves cuneate to obovate-cuneate, broadest above the middle, cuneate at base, sometimes 3- to 5-lobed on upper margin.

        1. *C. cuneata*

1B. Leaves ovate to broadly so, broadest at or below the middle, truncate to broadly cuneate at base, pinnately cleft to incised.

  2A. Leaves pinnately cleft. .... 2. *C. oxyacantha*

  2B. Leaves incised or shallowly lobed.

    3A. Fruit red or yellowish red; leaves rather deeply lobed, villous beneath. .... 3. *C. maximowiczii*

    3B. Fruit black; leaves incised to shallowly lobed. .... 4. *C. chlorosarca*

1. Crataegus cuneata Sieb. & Zucc. *Mespilus cuneata* (Sieb. & Zucc.) C. Koch——SANZASHI. Spiny deciduous much-branched shrub with straight ascending spines 3–8 mm. long, the branches pubescent while young; leaves cuneate, 2.5–7 cm. long, 1–4 cm. wide, obtuse, slightly appressed-pubescent while young, often glabrate on upper side, shallowly to deeply 3-lobed, irregularly incised or toothed, sometimes entire, paler beneath, sessile or the petioles to 3 cm. long; inflorescence 2- to 6-flowered, terminal on short branches, sessile, pubescent; flowers about 18 mm. across; calyx-teeth spreading, ovate, subacute; petals 5, suborbicular, white; styles usually 5; fruit subglobose, red, 10–12 mm. across.——Apr.–May. Only cultivated.——China and Mongolia.——Forma **lutea** Matsum. KIMI-SANZASHI. Fruit yellow.

2. Crataegus oxyacantha L. SEIYŌ-SANZASHI. Tall deciduous shrub with elongate stout spines 7–20 mm. long, the young branchlets usually glabrous; leaves broadly ovate, 1.5–5 cm. long, 2–4 cm. wide, subobtuse, deep green, deeply (3–) 5-lobed, the lobes obtuse, slightly toothed on upper margin, glabrous or nearly so, the petioles 1–2 cm. long, glabrous or nearly so; inflorescence glabrous, several-flowered, terminal on short branches; flowers white, about 15 mm. across, single; styles 2–3; fruit red, subglobose, terminated by the persistent reflexed calyx-teeth.——Apr.–May. Cultivated in gardens. ——Europe.

Crataegus pinnatifida Bunge. *Crataegus oxyacantha* var. *pinnatifida* (Bunge) Regel——Ō-SANZASHI. Allied to the preceding but with few spines, pubescent inflorescence and the young branches with rather acute leaf-lobes, and the stones flat on ventral side.——Occasionally cultivated.——Korea, Manchuria, and e. Siberia.

3. Crataegus maximowiczii C. K. Schn. *C. sanguinea* var. *villosa* Maxim.; *Mespilus sanguinea* var. *villosa* (Maxim.) Asch. & Graebn.——ARAGE-SANZASHI, ŌBA-SANZASHI. Deciduous shrub, the young branches coarsely hairy, later glabrate, lustrous reddish brown, the spines 15–35 mm. long; leaves pubescent on both sides, 7–9 cm. long, 5–6 cm. wide, rather deeply 3-cleft in vigorous elongate shoots, the petioles hairy, 15–30 mm. long; inflorescence densely flowered, coarsely hairy along with the young fruits; flowers about 1.5 cm. across; styles 3–5, nearly terminal.——Hokkaido; rare.——e. Siberia, Manchuria, Sakhalin, and Korea.

4. Crataegus chlorosarca Maxim. *C. sanguinea* var. *schroederi* Regel; *Mespilus chlorosarca* (Maxim.) Asch. & Graebn.——KUROMI-SANZASHI. Small deciduous tree with rather spreading, stout spines, the young branches glabrous or slightly pubescent, becoming dark purple in age; leaves membranous, dull, ovate or broadly so, 5–10 cm. long, 3.5–5 cm. wide, acute, broadly cuneate at base, slightly pubescent, shallowly lobed to incised, irregularly and acutely toothed, the petioles 1–3 cm. long, loosely pubescent on the upper side; inflorescence rather many-flowered, glabrous or slightly pubescent; flowers white, 10–15 mm. across; calyx-teeth narrowly ovate-lanceolate, subacuminate, reflexed; stamens rather long; styles 5.——May–June. Hokkaido, Honshu (Sugadaira in Shinano).——Sakhalin.

Crataegus jozana C. K. Schn. may be only a variant of the preceding with denser pubescence.

22. PHOTINIA Lindl. KANAME-MOCHI ZOKU

Glabrous to pubescent shrubs or trees; leaves alternate, petioled, coriaceous, evergreen, simple, entire or toothed, the stipules sometimes foliaceous; flowers in terminal corymbs or panicles, small, white; calyx with campanulate or napiform tube adnate to the ovary or free above, the lobes 5, obtuse; petals 5, spreading; stamens rather numerous, inserted on the throat of the calyx-tube; ovary inferior or free on upper part, 2- to 5-locular; styles 2–5, free or connate, the stigmas dilated, truncate; ovules 2 in each locule, erect; drupe ovoid, with 1 or 2 seeds in each locule, erect, with a membranous or coriaceous testa.——About 40 species, in the temperate to warmer regions of e. Asia.

1. Photinia glabra (Thunb.) Maxim. *Crataegus glabra* Thunb.; *P. serrulata* Sieb. & Zucc., non Lindl.——KANAME-MOCHI. Small evergreen tree, glabrous except the inflorescence; leaves oblong to narrowly obovate, 7–12 cm. long, 2.5–4 cm. wide, acute to acuminate, acute at base, serrulate, lustrous, paler beneath, the petioles 10–17 mm. long, often slightly pubescent inside near base while very young; inflorescence many-flowered, 8–12 cm. across, the bracts erect, broadly lanceolate, caducous; flowers about 8 mm. across, white and slightly pinkish; calyx-teeth deltoid, acute, erect; petals white-puberulent inside at base; styles 2 or 3, white-puberulent on lower part; fruit subglobose, 4–5 mm. across, red.——May–June. Honshu (Tōkaidō Distr. and westw.), Shikoku, Kyushu.

23. ERIOBOTRYA Lindl. BIWA ZOKU

Evergreen trees; leaves short-petioled, simple, coriaceous, toothed, with straight nerves ending in the teeth; inflorescence a broad pyramidal panicle usually densely woolly-pubescent, the bracts deltoid-ovate, persistent; flowers rather small, white; calyx-teeth acute, persistent; petals 5, ovate to nearly orbicular, clawed; stamens about 20; ovary inferior with 2 ovules in each locule; styles 2–5, connate below; fruit an obovoid to globose pome, with persistent calyx-teeth at apex; the seeds few, large.——About 10 species, in e. Asia.

1. Eriobotrya japonica (Thunb.) Lindl. *Mespilus japonica* Thunb.; *Crataegus bibas* Lour.; *Photinia japonica* (Thunb.) Fr. & Sav.——Biwa. Woolly-pubescent coarsely branching tree with stout branches; leaves broadly oblanceolate to narrowly obovate, 15–25 cm. long, 3–5 cm. wide, acute, gradually narrowed at base, deep green, deciduous woolly-pubescent above, loosely toothed, sessile or the petioles about 1 cm. long, the stipules lanceolate-deltoid; inflorescence densely flowered, terminal, erect; fruit globose to pyriform, 3–4 cm. across, yellow.——Nov.–Dec.——Honshu (w. distr.), Shikoku, Kyushu.——China. Much planted for the edible fruit.

24. RHAPHIOLEPIS Lindl. Sharimbai Zoku

Evergreen glabrescent shrubs or trees; leaves alternate, petioled, coriaceous, entire or toothed, the stipules subulate; inflorescence a panicle or corymb, the bracts subulate, deciduous; flowers white or reddish; calyx with the tube adnate to the ovary at base, obconical or infundibuliform, the upper portion deciduous after anthesis, the lobes 5, subulate; petals 5, white to rosy clawed, oblong, acute to obtuse; stamens many, inserted on the throat of the calyx-tube; ovary inferior, 2-locular; styles 2, elongate, connate at base, obliquely thickened at apex, the stigma along the inner side at tip; ovules 2 in each locule, erect; berry 1- or 2-locular, 1- or 2-seeded, globose, purplish black or bluish black; seeds rather large, globose.——Few species, in Japan and China.

1. Rhaphiolepis umbellata (Thunb.) Makino. *Laurus umbellata* Thunb.; *Mespilus sieboldii* Bl.; *R. japonica* Sieb. & Zucc.; *R. ovata* Briot; *R. umbellata* forma *ovata* (Briot) C. K. Schn.——Sharimbai. Shrub with stout branches brownish pubescent while very young; leaves coriaceous, lustrous, ovate-oblong to obovate, 4–10 cm. long, 2–4 cm. wide, obtuse to subacute, acute at base, loosely and obtusely toothed to nearly entire, slightly recurved on margin, the veinlets slightly raised, the petioles 5–20 mm. long; inflorescence erect, more or less dense, 5–15 cm. long, densely brownish pubescent while young, becoming glabrate in fruit; calyx-tube infundibuliform, densely pubescent, the teeth deltoid to narrowly ovate, subacute; petals obovate, 10–13 mm. long, rounded and often toothed at apex; fruit globose, 7–10 mm. across, lustrous.—— Apr.–June. Thickets near seashores; Honshu, Shikoku, Kyushu. Commonly planted in hedges and parks; very variable. ——Korea (Quelpaert Isl.) and Ryukyus.——Cv. **Minor.** *R. minor* (Makino) Koidz.——Hime-sharimbai. Densely branched shrub; leaves ovate-oblong, 2–4 cm. long, 12–20 mm. wide, obtuse to subacute, entire or obtusely toothed; flowers rather small. A garden cultivar.

Var. **integerrima** (Hook. & Arn.) Rehd. *R. japonica* var. *integerrima* Hook. & Arn.; *R. mertensii* Sieb. & Zucc.; *R. umbellata* var. *mertensii* (Sieb. & Zucc.) Makino——Marubano-sharimbai. Leaves broader, broadly obovate to broadly elliptic, entire or with a few teeth.——Honshu (centr. and w. distr.), Shikoku, Kyushu.

Var. **liukiuensis** Koidz. *R. liukiuensis* (Koidz.) Nakai ——Hosoba-sharimbai. Leaves narrower, oblanceolate to narrowly oblong, 5–10 cm. long, 1–2 cm. wide, acute to obtuse, undulately and obtusely toothed or entire.——Rarely cultivated in our area; sometimes merges with the typical phase. ——Ryukyus, reported from Korea (Quelpaert Isl.).

25. CHAENOMELES Lindl. Boke Zoku

Deciduous shrubs or small trees, sometimes with spiny branches; leaves simple, short-petioled, toothed or crenate, the stipules often large in vigorous shoots; flowers solitary or fascicled, before or after the leaves, sometimes a few staminate only; calyx-lobes 5, entire or serrulate; petals typically 5, large; stamens 20 or more; ovary 5-locular; ovules many in each locule; styles 5, connate at base; pome rather large, with many brown seeds.——Few species, in e. Asia.

1A. Small shrub 30–100 cm. high, with long ascending-creeping base; branchlets scabrous with low verrucose tubercles; leaves obtusely toothed, obtuse or subacute at apex. .... 1. *C. japonica*
1B. Erect bushy shrub 0.5–3 m. high; branchlets smooth, glabrous or slightly pubescent while young; leaves densely and acutely toothed, acute to subacute. .... 2. *C. speciosa*

1. Chaenomeles japonica (Thunb.) Lindl. *Pyrus japonica* Thunb.; *Cydonia japonica* (Thunb.) Pers.; *Cydonia lagenaria* Loisel.; *Cydonia japonica* var. *lagenaria* (Loisel.) Makino; *Chaenomeles lagenaria* (Loisel.) Koidz.; *Chaenomeles japonica* var. *alpina* Maxim.; *Chaenomeles japonica* var. *pygmaea* Maxim.; *Pyrus maulei* Masters; *Chaenomeles japonica* var. *maulei* (Masters) Lavall.; *Chaenomeles maulei* (Masters) Lavall.; *Chaenomeles alpina* (Maxim.) Koehne——Kusa-boke, Shidomi, No-boke, Ko-boke. Small shrub 30–100 cm. high, creeping or ascending at base, the young branches with deciduous short yellowish hairs leaving verrucose scars after falling, the short branchlets often becoming spiny; leaves obovate to broadly so, 2–5 cm. long, 1–3.5 cm. wide, rounded to subacute, pale green and often lustrous beneath, glabrous on both sides, narrowed at base, the petiole 3–10 mm. long; flowers scarlet, about 2.5 cm. across, short-pediceled; calyx glabrous outside, the teeth semirounded, ciliate; styles usually glabrous; fruit subglobose, about 3 cm. across, yellowish.——Apr.–May. Low mountains; Honshu, Kyushu.——Forma **alba** (Nakai) Ohwi. *C. maulei* var. *alba* Nakai——Shirobana-kusa-boke. Flowers white. Rare.

2. Chaenomeles speciosa (Sweet) Nakai. *C. lagenaria* Koidz., excl. syn.; *Malus japonica* Andr.; *Cydonia japonica* Loisel., non Pers.; *Cydonia speciosa* Sweet; *Chaenomeles japonica* var. *genuina* Maxim.——Boke. Shrub with glabrous branches or with pale yellowish brown short hairs while young, the short branchlets often reduced to spines; leaves oblong or broadly lanceolate to ovate, 4–8 cm. long, 1.5–5 cm. wide, acute to subobtuse, narrowed below to a short petiole, serrulate, glabrous on both sides or more or less pubescent beneath along the midrib; flowers short-pedicelled, 2.5–3.5 cm. across, scarlet or white to reddish; calyx glabrous outside, the teeth semirounded to depressed-rounded, ciliate; styles glabrous or hairy at base.——A much-planted Chinese shrub, with many cultivars of garden origin.

26. MALUS Mill.　Ringo Zoku

Deciduous shrubs or trees, often spiny; leaves alternate, toothed, petioled, simple or sometimes lobed, convolute or conduplicate in bud, stipulate; inflorescence a corymb; flowers white or pale pink; calyx-teeth 5; petals 5, suborbicular; stamens 15–50, the anthers yellow; ovary inferior, 3- to 5-locular; styles 3–5, connate at base; fruit a pome usually without stone cells, sometimes with persistent calyx-teeth at apex, rarely punctulate on outer pericarp.——About 30 species, in the temperate regions of the N. Hemisphere.

1A. Leaves always undivided, convolute in bud.
 2A. Calyx-lobes ovate, as long as to shorter than the tube.
  3A. Calyx-lobes subobtuse; styles 4 or 5; fruit pyriform.
   4A. Flowers rose-colored, nodding, on slender pedicels 3–4 cm. long; leaves ovate to narrowly so; calyx-lobes erect. .. 1. *M. halliana*
   4B. Flowers white to pinkish, on pedicels 2–2.5 cm. long; leaves obovate; calyx-lobes spreading. .... 2. *M. spontanea*
  3B. Calyx-lobes acute to subacuminate; styles 3 or 4; fruit ellipsoidal. .... 3. *M. hupehensis*
 2B. Calyx-lobes narrowly lanceolate, longer than the calyx-tube, acuminate. .... 4. *M. baccata* var. *mandshurica*
1B. Leaves conduplicate in bud, usually lobed in those of vigorous shoots.
 5A. Calyx-limb deciduous after anthesis; fruit 5–7 mm. in diameter. .... 5. *M. sieboldii*
 5B. Calyx-limb persistent after anthesis; fruit 2–3 cm. in diameter. .... 6. *M. tschonoskii*

1.　Malus halliana Koehne. *Pyrus halliana* (Koehne) Voss——Kaidō.　Shrub, the branches thinly pubescent while young; leaves ovate to narrowly so or elliptic, 3–6 cm. long, 1–3 cm. wide, acute, acute to broadly cuneate at base, obtusely serrulate, nearly glabrous, the petioles about 1 cm. long, thinly pubescent; flowers deep pink, rather large, semi-expanded, nodding on long slender nearly glabrous purplish pedicels 3–6 cm. long; calyx nearly glabrous, the teeth narrowly ovate, acute to obtuse, erect, pubescent inside; styles 4 or 5, white-villous at base.——Apr.–May.　Introduced from China, with the following cultivars grown in our area:

Cv. **Parkmannii.** *Pyrus malus* var. *parkmannii* Temple; *P. parkmannii* Hort.; *P. halliana* var. *parkmannii* (Temple) L. H. Bailey; *M. floribunda* var. *parkmannii* (Temple) Koidz.——Yae-kaidō.　Flowers double.

Cv. **Pendula.** *M. floribunda* forma *pendula* Koidz.——Shidare-kaidō.　Branches pendulous.

2.　Malus spontanea (Makino) Makino. *M. floribunda* var. *spontanea* Makino; *M. halliana* var. *spontanea* (Makino) Koidz.——No-kaidō.　Small tree; leaves obovate or narrowly so or elliptic, 3–5 cm. long, 1.5–3 cm. wide, acute to subacute, cuneate to acute at base, glabrous although soft pubescent on upper side while young, with incurved minute teeth, the petioles 1–3 cm. long, slender, thinly pubescent; flowers white to pinkish, rather large, on slender glabrous pedicels 2–2.5 cm. long; calyx glabrous, the teeth ovate to narrowly so, subobtuse, spreading, white-pubescent inside; petals 5, spreading, broadly ovate; style 4, white-villous at base; fruit pyriform, 6–8 mm. across, red, rounded at apex.——May.　Mountains; Kyushu (Mount Kirishima).

3.　Malus hupehensis (Pampan.) Rehd. *Pyrus hupehensis* Pampan.; *M. theifera* Rehd.; *P. theifera* (Rehd.) L. H. Bailey——Tsukushi-kaidō, Cha-kaidō.　Small tree; leaves coriaceous, usually ovate, subobtuse to obtuse, toothed, densely white-woolly on the midrib on both surfaces and on the petioles while young, glabrate on both sides or the pubescence persistent on the nerves beneath; flowers white, on glabrous pedicels; calyx glabrous outside, the lobes deltoid-ovate, acute, villous inside; styles 4, villous at base; fruit rounded at both ends, on pedicels 2.5–3.5 cm. long.——Reported to be spontaneous in Kyushu (Higo Prov.).——China and n. India.

4.　Malus baccata Borkh. var. **mandshurica** (Maxim.) C. K. Schn. *M. cerasifera* Spach; *Pyrus baccata* var. *mandshurica* Maxim.; *M. mandshurica* (Maxim.) Komar.——Ezo-no-ko-ringo, Hiro-ha-zumi.　Small tree; leaves broadly elliptic-ovate, sometimes oblong, 4–10 cm. long, 2.5–5 cm. wide, abruptly acute, obtuse to broadly cuneate at base, rather loosely and irregularly serrulate, with yellowish brown soft pubescence while young, later glabrate or thinly pubescent on both sides, the petioles pubescent, 1.5–3.5 cm. long; flowers white, about 3.5 cm. across, the pedicels 3–4 cm. long, pubescent to nearly glabrous; calyx pubescent to glabrescent, the lobes 5, narrowly lanceolate, long-acuminate, white-pubescent inside; styles 4 or 5, white-villous inside; fruit dark red, globose, about 8 mm. across.——Reported to be spontaneous in Hokkaido, Honshu (centr. and n. distr.).——Manchuria, n. China, Ussuri, Korea, Sakhalin, and s. Kuriles. The typical phase occurs in Siberia and e. Asia.

5.　Malus sieboldii (Regel) Rehd. *Pyrus sieboldii* Regel; *P. toringo* Miq.; *M. sargentii* Rehd.; *P. sargentii* (Rehd.) Bean; *M. sieboldii* var. *koringo* forma *sargentii* (Rehd.) Koidz.——Zumi.　Large shrub or small tree, the young branches usually pubescent; leaves of fertile branches oblong, ovate or narrowly so, 3–8 cm. long, those of sterile shoots often broadly ovate, to 8 cm. long, 3- to 5-cleft, irregularly serrulate, acute or abruptly so, pubescent on both sides at least while young, the petioles 1.5–3 cm. long, rather slender, pubescent; flowers white, suffused with pink in bud, 2.5–3 cm. across, on slender, spreading, pubescent to glabrous pedicels 2–3.5 cm. long; calyx glabrous to pubescent, the lobes narrowly lanceolate, long-acuminate; styles 3–4, white-pubescent at base; fruit globose, 5–7 mm. across, red, rounded at both ends.——May–June.　Mountains; Hokkaido, Honshu, Shikoku, Kyushu; common and variable.——Korea.

Var. **zumi** (Matsum.) Asami. *M. zumi* (Matsum.) Rehd.; *Pyrus zumi* Matsum.; *M. baccata* var. *mandshurica* forma *zumi* (Matsum.) Matsum.——Ō-zumi.　Leaves nearly always undivided, those on the fertile branches nearly or partially not toothed; styles usually 4–5.——Honshu; rare.

6.　Malus tschonoskii (Maxim.) C. K. Schn. *Pyrus tschonoskii* Maxim.; *Eriolobus tschonoskii* (Maxim.) Rehd.; *Cormus tschonoskii* (Maxim.) Koidz.; *Macromeles tschonoskii* (Maxim.) Koidz.; *Docyniopsis tschonoskii* (Maxim.) Koidz.——Ō-urajiro-no-ki, Zumi-no-ki.　Tree, the young branches densely woolly-pubescent; leaves thinly chartaceous, ovate to broadly so, 5–13 cm. long, 4–8 cm. wide, abruptly acute to short-acuminate, rounded to subcordate at base, deciduous woolly-tomentose on both sides while young, becoming glabrous above, irregularly incised-toothed, and lobulate to incised, the petioles 2–4 cm. long, densely white-pubescent; flowers white, about 3 cm. across, the pedicels rather stout, 2–2.5 cm. long, densely pubescent; calyx-lobes deltoid-ovate; fruit globose, 2–3 cm. across, yellowish green, slightly reddish purple, densely woolly, pubescent only while young.——May. Honshu; rare.

27. PYRUS L.   NASHI ZOKU

Large often spiny shrubs or trees; leaves alternate, stipulate, toothed or entire, undivided or lobed, folded in bud, petiolate, often with pale brown woolly hairs while young; inflorescence an umbellate corymb usually with a short axis; flowers white, rarely pink, opening before or with the leaves; calyx-lobes spreading or reflexed; petals short-clawed, ovate-orbicular; stamens 20–30; styles 2–5, free; ovules 2 in each locule; fruit a pome, with cartilaginous walls; seeds nearly black.——About 20 species, in temperate regions of Eurasia and N. Africa.

1A. Calyx-lobes persistent after anthesis.
  2A. Leaves aristately acute-toothed. .... 1. *P. ussuriensis*
  2B. Leaves obtusely serrulate. .... 2. *P. communis*
1B. Calyx-lobes deciduous after anthesis.
  3A. Styles 2, rarely 3; leaves obtusely to subacutely crenate-toothed; ovary 2- or 3-locular. .... 3. *P. calleryana* var. *dimorphophylla*
  3B. Styles, rarely 4; leaves aristately toothed; ovary 5-locular. .... 4. *P. pyrifolia*

1. Pyrus ussuriensis Maxim. var. **aromatica** (Nakai & Kikuchi) Rehd. *P. aromatica* Nakai & Kikuchi——IWATE-YAMA-NASHI.   Tree; leaves ovate to broadly so, 6–11 cm. long, 4–6 cm. wide, the pedicels stout, 1.5–4.5 cm. long; fruit globose to ovoid-globose, 2–3.3 cm. across, 5-locular, obtuse at base or with depressed ends, green, the lenticels brown, minute, very numerous, the persistent calyx-lobes linear-lanceolate to lanceolate.——Apr. Honshu (n. distr.).

Var. **hondoensis** (Nakai & Kikuchi) Rehd. *P. hondoensis* Nakai & Kikuchi——AO-NASHI.   Leaves ovate, 5–7.5 cm. long, 3–4 cm. wide, caudately acuminate, serrulate; pedicels 0.8–1.5(–2) cm. long; fruit globose, slightly depressed at base, green, 2.5–3 cm. across, with minute brown lenticels, the calyx-lobes lanceolate, spreading.——Honshu (centr. distr.). ——The typical phase occurs in Korea, Manchuria, and Ussuri.

2. Pyrus communis L. SEIYŌ-NASHI.   Tree; leaves ovate-orbicular to elliptic, 4–10 cm. long, undulately toothed, glabrous or woolly-pubescent while young; pedicels 1.5–3 cm. long; fruit pyriform to subglobose.——Apr. European tree cultivated for the fruit.

3. Pyrus calleryana Decne. var. **dimorphophylla** (Makino) Koidz. *P. dimorphophylla* Makino——MAME-NASHI, INU-NASHI.   Small tree, the young branches densely woolly-pubescent, soon glabrate; leaves dimorphic, pubescent only while young, ovate to ovate-orbicular, sometimes narrowly ovate, 2–9 cm. long, 1.5–5.5 cm. wide, often 3-lobed in those on young vigorous shoots, with incurved teeth, the pedicels soft-pubescent, 2–3.5 cm. long; calyx-lobes deltoid to ovate-lanceolate, deciduous; styles 2, rarely 3; fruit 7–14 mm. across, rounded at base, pale brownish yellow, lenticellate.—— Apr. Honshu (centr. distr.); rare. The typical phase occurs in Korea and China.

4. Pyrus pyrifolia (Burm. f.) Nakai. *Ficus pyrifolia* Burm. f.; *P. serotina* Rehd.; *P. montana* Nakai; *P. kleinhofiana* Koidz.——YAMA-NASHI.   Small tree, the branches often woolly-pubescent while young; leaves chartaceous, ovate or narrowly so, 7–12 cm. long, 4–6 cm. wide, acuminate, rounded at base, glabrate on both sides, the pedicels 3–5 cm. long; calyx-lobes lanceolate; ovary (4–)5-locular; styles 5, rarely 4; fruit globose, 2–3 cm. across, lenticellate, brown.——Apr. Naturalized (?) in low mountains and around villages in mountains; Honshu, Shikoku, Kyushu.——China.

Var. **culta** (Makino) Nakai. *P. sinensis* var. *culta* Makino; *P. serotina* var. *culta* (Makino) Rehd.——NASHI.   An assemblage of numerous improved cultivars yielding edible fruit, 5–10 cm. across, the leaves broadly ovate to ovate-orbicular.

28. AMELANCHIER Medic.   ZAI-FURI-BOKU ZOKU

Deciduous trees or shrubs; leaves alternate, petioled, simple, toothed, conduplicate in bud, the stipules small, deciduous; inflorescence a terminal raceme; flowers white; calyx-tube campanulate, the teeth small, persistent; petals 5, obovate to lanceolate; stamens 10–20; ovary inferior, the ovules 2 in each locule; styles 2–5, free or connate at base, each separated by a false wall from the back of the locule; pome berrylike, bluish black.——About 25 species, mainly in N. America, a few in Europe and Asia.

1. Amelanchier asiatica (Sieb. & Zucc.) Endl. *Aronia asiatica* Sieb. & Zucc.; *Amelanchier canadensis* var. *japonica* Miq.; *Pyrus taquetii* Lév.; *P. vaniotii* Lév.——ZAI-FURI-BOKU. Tree, the branches white-pubescent while young, glabrate, slender; leaves obovate or broadly so, 4–7 cm. long, 2.5–4 cm. wide, abruptly acuminate, obtuse to rounded at base, with minute low teeth, thinly appressed-pubescent on upper side while young, more densely so beneath, becoming glabrous on both sides, the petioles 1–1.5 cm. long, white-pubescent to nearly glabrous; inflorescence a corymbose raceme, white-woolly, about 10-flowered, the bracts linear; petals 15–18 mm. long, 4–5 mm. wide, obtuse; ovary densely hairy at tip; styles 5, connate on lower half; fruit globose, 4–6 mm. across, terminated by the recurved calyx-lobes.——Apr.–May. Hills and mountains; Honshu, Shikoku, Kyushu; common.——Korea.

29. POURTHIAEA Decne.   KAMA-TSUKA ZOKU

Deciduous shrubs or trees; leaves alternate, stipulate, simple, short-petioled, toothed, the stipules deciduous; inflorescence a terminal, many-flowered corymb or cyme; flowers small, white, bisexual; calyx persistent; petals 5, stamens rather many; styles 2–4, connate at base; pome globose to ovoid, small, the seeds 2 in each locule.——Few species, in temperate to warmer regions of e. Asia.

1. Pourthiaea villosa (Thunb.) Decne. *Crataegus villosa* Thunb.; *Photinia villosa* (Thunb.) DC.; *Photinia laevis* var. *villosa* (Thunb.) Koidz.; *Pourthiaea laevis* var. *villosa* (Thunb.) Koidz.; *Photinia villosa* var. *typica* C. K. Schn.; *Pourthiaea villosa* var. *typica* (C. K. Schn.) Nakai——WATAGE-KAMA-TSUKA, Ō-KAMA-TSUKA. Large shrub; young branches and petioles white-pubescent; leaves chartaceous, broadly obovate or narrowly so, 4–12 cm. long, 2–6 cm. wide, abruptly acuminate, cuneate at base, pale green beneath, acutely serrulate, pubescent on both sides while young, often glabrous later, the petioles 2–8 mm. long; inflorescence with dense white woolly pubescence at anthesis, glabrous in fruit, 4–8 cm. across; flowers 7–10 mm. across; calyx-lobes deltoid, obtuse, white-pubescent; petals often softly puberulent on the claw; styles usually 3, villous at base; fruit obovoid, 8–10 mm. long, red, terminated by the persistent calyx-lobes.——Apr.–May. Low mountains and hills; Hokkaido, Honshu, Shikoku, Kyushu; very common and variable.——Korea and China.

Var. **zollingeri** (Decne.) Nakai. *Pourthiaea zollingeri* Decne.; *Photinia villosa* var. *zollingeri* (Decne.) C. K. Schn. ——KE-KAMA-TSUKA. Slightly pubescent; intermediate between the typical phase and the following:

Var. **laevis** (Thunb.) Stapf. *Myrtus laevis* Thunb.; *Crataegus laevis* (Thunb.) Thunb.; *Photinia laevis* (Thunb.) DC.; *Photinia villosa* var. *laevis* (Thunb.) Dipp.; *Pourthiaea laevis* (Thunb.) Koidz.——KAMA-TSUKA. With rather smaller leaves, glabrous except when young; inflorescence glabrous or nearly so, rather small, fewer flowered.——Occurs with the typical phase.

30. SORBUS L. NANA-KAMADO ZOKU

Deciduous trees or shrubs; leaves alternate, stipulate, simple to imparipinnate, toothed; inflorescence a compound terminal corymb, the bracts caducous; flowers rather small, white, sometimes pinkish; calyx with an obconical tube, the teeth 5; petals 5, ovate, orbicular, often pubescent inside near the base; stamens 15–20, the filaments subulate; ovary inferior or semisuperior; styles free or connate at base; ovules 2 in each locule, ascending; pome 2- to 5-locular, small, with 1 or 2 seeds in each locule, these ellipsoidal to ovoid.——About 80 species, in temperate and cold regions of the N. Hemisphere.

1A. Leaves pinnate; calyx-limb persistent.
  2A. Branches stout; leaflets usually acuminate, 7–15, usually the median ones largest; stipules rather small, deciduous; inflorescence usually large.
    3A. Leaflets 9–11, chartaceous, slightly lustrous; inflorescence nodding; styles 5. . . . . 1. *S. sambucifolia*
    3B. Leaflets 9–15, membranous to herbaceous, scarcely lustrous.
      4A. Leaflets entire on lower half, vivid green, slightly glaucescent; styles 5. . . . . 2. *S. matsumurana*
      4B. Leaflets toothed near to the base, deep green, scarcely glaucescent; styles 3–4. . . . . 3. *S. commixta*
  2B. Branches rather slender; leaflets rounded to acute at apex, 5–11, usually the upper ones largest; stipules often rather large, toothed; inflorescence small. . . . . 4. *S. gracilis*
1B. Leaves simple; calyx-limb deciduous after anthesis.
  5A. Inflorescence and leaves glabrous or nearly so, rarely with appressed pubescence on leaves beneath. . . . . 5. *S. alnifolia*
  5B. Inflorescence and leaves beneath densely white-woolly. . . . . 6. *S. japonica*

1. Sorbus sambucifolia (Cham. & Schltdl.) Roem. *Pyrus sambucifolia* Cham. & Schltdl.——TAKANE-NANA-KAMADO. Branching shrub about 2 m. high, the young branches thinly soft-pubescent, those of the 2d year's growth grayish brown; leaves 9- or 11-foliolate, the petioles 2–3 cm. long, the leaflets narrowly ovate to broadly lanceolate, 4–6 cm. long, 1–3 cm. wide, acuminate, deep green and lustrous, slightly paler beneath, thinly white-pubescent while young, glabrescent or quite glabrous on both sides, acutely toothed, the lateral nerves slightly raised beneath; inflorescence 5–7 cm. long, often slightly more than 10-flowered, nodding, especially in fruit, glabrous or thinly pubescent; flowers white to pinkish, 10–12 mm. across; calyx glabrous, the teeth erect in fruit, ciliate; petals ascending, nearly as long as the stamens; fruit elongate-globose, red.——June–Aug. High mountains; Hokkaido, Honshu (centr. and n. distr.); rather common.——Kuriles, Sakhalin, Kamchatka, and Ochotsk Sea region.

Var. **pseudogracilis** C. K. Schn. *S. pseudogracilis* (C. K. Schn.) Koehne——MIYAMA-NANA-KAMADO. Low dwarf shrub with scarcely nodding inflorescence; calyx-teeth usually slightly incurved above; fruit slightly smaller.——Alpine regions of Hokkaido and Honshu (n. and centr. distr.).

2. Sorbus matsumurana (Makino) Koehne. *Pyrus matsumurana* Makino; *S. sitchensis* sensu auct. Japon., non Roem. ——URAJIRO-NANA-KAMADO. Small tree glabrous except for the bud-scales, nodes of leaf-rachises and top of the ovary; leaflets 9–13, thickly membranous, oblong to narrowly so, 3.7–7 cm. long, 1–2 cm. wide, obtuse to abruptly acuminate with a mucro at apex, vivid green and slightly glaucescent, whitish beneath, with ascending to incurved acuminate teeth on upper half, the lateral nerves not raised; inflorescence erect, large, many-flowered, 5–8 cm. across; flowers white, about 1 cm. across; calyx-teeth incurved in fruit, brown-ciliate; petals spreading, ovate-orbicular, short-clawed; fruit ellipsoidal to globose, red.——June–Aug. Mountains to alpine zone; Hokkaido, Honshu (centr. and n. distr.).——Forma **pseudogracilis** (Koidz.) Ohwi. *S. matsumurana* var. *pseudogracilis* Koidz.——NANKIN–NANA-KAMADO-MODOKI. With a large toothed stipule subtending the inflorescence.

3. Sorbus commixta Hedl. *S. aucuparia* var. *japonica* Maxim.; *Pyrus americana* var. *microcarpa* Miq.; *S. japonica* (Maxim.) Koehne, non Hedl.; *Pyrus americana* var. *micrantha* C. Koch, non Torr. & A. Gray; *S. aucuparia* sensu auct. Japon., non L.; ? *S. parviflora* Hedl.; ? *S. micrantha* (Fr. & Sav.) Koidz.; ? *P. micrantha* Fr. & Sav., non Dum. Cours.; *S. chionophylla* Nakai; *S. pruinosa* Koehne; *S. heterodonta* Koehne ——NANA-KAMADO. Tree with glabrous or thinly pubescent young branchlets; leaves pinnate, 15–20 cm. long, the petioles 3–5 cm. long, glabrous or nearly so, the leaflets 9–15, lanceolate to narrowly ovate-oblong, 3–7 cm. long, 1–2.5 cm. wide, long-acuminate to acuminate, glabrous or nearly so, pale to whitish and sometimes slightly pubescent on the midrib beneath, with acuminate teeth, the lateral nerves slender; inflorescence 5–10 cm. across, densely many-flowered, glabrous or loosely brown-pubescent at anthesis, becoming glabrous in fruit; flowers white, 6–8 mm. across; calyx glabrous, the teeth erect or only slightly incurved in fruit; petals spreading, nearly orbicular, short-clawed, often thinly puberulent inside at base, as long as the stamens; fruit red, 4–5 mm. across, globose.——May–July. Mountains; Hokkaido, Honshu, Shikoku, Kyushu.

Var. **rufoferruginea** C. K. Schn. *S. rufoferruginea* (C. K. Schn.) C. K. Schn.; *S. nikkoensis* Koidz.; *S. diabolica* Koidz.; *S. rufoferruginea* var. *ferrugineocosta* Koidz.; *S. wilfordii* Koehne——SABIBA-NANA-KAMADO. Inflorescence, the calyx-tube at anthesis, and the leaves beneath along the midrib with long, soft, brown hairs.——Mountains; Honshu, Shikoku, Kyushu.

Var. **sachalinensis** Koidz. *S. yesoensis* Nakai; *Pyrus micrantha* var. *macrophylla* Card.; *S. macrophylla* (Card.) Koidz.——Ō-NANA-KAMADO, EZO-NANA-KAMADO. Leaflets larger, to 8–9 cm. long.——Hokkaido.——s. Kuriles and Sakhalin.

Sorbus × rikuchuensis Makino. *Crataegus rikuchuensis* (Makino) Makino——RIKUCHŪ-NANA-KAMADO. Leaves pinnately divided on lower half, pinnately parted to lobed on upper half, 4–8 cm. wide, glabrous on upper side, irregularly toothed, pubescent beneath, brown-pubescent on nerves beneath, 6–12 cm. long, the leaflets and lobes chartaceous, spreading.——Honshu (Rikuchu Prov.); rare; a hybrid involving *S. commixta* and an unknown second parent.

Hybrids of uncertain origin occurring in Honshu: **Sorbus × uzenensis** (Koidz.) Koidz.: × *Crataegosorbus uzenensis* (Koidz.) Koidz.; × *Ariosorbus uzenensis* (Koidz.) Koidz.——DEWA-NO-HAGOROMO-NANA-KAMADO.——**Sorbus × tangoensis** Koidz.: × *Crataegosorbus tangoensis* Koidz.; × *Ariosorbus tangoensis* (Koidz.) Koidz.——YOSANO-HAGOROMO-NANA-KAMADO.

4. **Sorbus gracilis** (Sieb. & Zucc.) C. Koch. *Pyrus gracilis* Sieb. & Zucc.; *S. schwerinii* C. K. Schn.; *S. pachyphylla* Koidz.; *S. gracilis* var. *yoshinoi* Koidz.; *S. viminalis* Koidz.——NANKIN-NANA-KAMADO. Shrub with the slender branches soft-pubescent while young; leaves pinnate, the stipules persistent, those under the inflorescence larger, flabellate, toothed, 1–2 cm. long and as wide, the petioles 1–3 cm. long, loosely soft-pubescent, often becoming glabrous, the leaflets 7–9(–11), the upper ones larger, oblong to narrowly obovate-oblong, 2–5.5 cm. long, 1–2.5 cm. wide, usually obtuse, sometimes rounded or acute, toothed on upper half, glabrous or nearly so above, glaucous to white and often with yellowish brown pubescence beneath, mixed with white hairs on midrib, the rachis usually pubescent; inflorescence 5–8 cm. long, rather many-flowered, often nodding, glabrous to slightly soft-pubescent; flowers creamy white, 5–8 mm. across; calyx-teeth deltoid, inflexed in fruit; petals as long as the stamens, 4–5 mm. long, reflexed, nearly orbicular, short-clawed; styles 2–4,

slightly white-pubescent near the base or glabrescent; fruit red, ellipsoidal or globose, 7–10 mm. long.——Apr.–June. Mountains; Honshu (Kantō Distr. and westw.), Shikoku, Kyushu.

5. **Sorbus alnifolia** (Sieb. & Zucc.) C. Koch. *Crataegus alnifolia* Sieb. & Zucc.; *Pyrus miyabei* Sarg.; *Micromeles alnifolia* (Sieb. & Zucc.) Koehne; *S. miyabei* (Sarg.) Mayr; *Aria tiliaefolia* Decne.——AZUKI-NASHI. Tree with purplish brown branches glabrous or thinly white-pubescent while young; leaves simple, chartaceous, ovate to broadly so, elliptic or obovate, 5–10 cm. long, 3.5–7 cm. wide, abruptly acuminate, thinly pubescent above while young, pale green and thinly appressed-pubescent beneath especially on the nerves while young, doubly toothed, the lateral nerves 8- to 10-paired, ascending, nearly straight, the petioles pubescent on the upper side while young, 10–20 mm. long, usually reddish; inflorescences terminal on short branches, compound-corymbose, rather many-flowered, glabrous to sparsely white-pubescent; flowers white, 13–16 mm. across; calyx-teeth deltoid, obtuse, spreading, deciduous, white-pubescent inside; petals 5, orbicular-ovate, slightly longer than the stamens; stamens about 20; styles 2, slightly pubescent at base; fruit oblong, red, slightly lenticellate, 7–8 mm. long.——May–June. Mountains; Hokkaido, Honshu, Shikoku, Kyushu; common.

Var. **lobulata** (Koidz.) Rehd. *Micromeles alnifolia* var. *lobulata* Koidz.——FUGIRE-AZUKI-NASHI. Leaves shallowly lobulate.——Honshu and Kyushu.——Korea.

Var. **submollis** Rehd. OKUSHIMO-AZUKI-NASHI. Leaves densely soft-pubescent beneath.——Hokkaido, Honshu.——Korea, Manchuria, China, and Ussuri.

6. **Sorbus japonica** (Decne.) Hedl. *Pyrus lanata* Miq., non D. Don; *Aria japonica* Decne.; *Micromeles japonica* (Decne.) Koehne; *Sorbus koehnei* Zabel——URAJIRO-NO-KI. Tree, with dense white pubescence on young branches, inflorescence, and leaves beneath; leaves submembranous, ovate-orbicular to broadly obovate, 6–12 cm. long, 4–9 cm. wide, shallowly lobed, toothed, white-woolly on upper side while young, densely white-woolly beneath, the lateral nerves 8 to 11 pairs, ascending, nearly straight, the petioles white-woolly, 1–2 cm. long; inflorescence a compound corymb, terminal on short branches, many-flowered, the bracts scarious, caducous, brown; flowers white, about 10 mm. across; calyx white-woolly, the teeth lanceolate-deltoid, acuminate; petals subrecurved, nearly orbicular to elliptic, slightly longer than the stamens; styles 2; fruit obovoid-globose or obovoid, 8–12 mm. long, slightly lenticellate, red.——May–June. Hills and mountains; Honshu, Shikoku, Kyushu.

Var. **calocarpa** Rehd. KIMINO-URAJIRO-NO-KI. Fruit 15 mm. long, orange-yellow, not lenticellate.——Honshu (Yumoto in Nikko).

Fam. 105. **LEGUMINOSAE** MAME KA Pea Family

Herbs, shrubs, trees, twining or scandent vines; leaves alternate, usually compound, stipulate and often stipellate; flowers usually bisexual; sepals 5, connate at base into a tube; petals 5, rarely fewer, papilionaceous or nearly actinomorphic; stamens 10, sometimes fewer or numerous, usually diadelphous, sometimes monadelphous, or free; ovary 1-locular, 1-carpellate, superior, the ovules 1 to many, the placenta parietal; fruit a legume, dehiscent, indehiscent, or rarely fleshy, sometimes jointed; seeds without endosperm.——About 550 genera, with about 13,000 species, cosmopolitan, especially abundant in the tropics.

I. Subfamily **MIMOSOIDEAE:** Usually trees; flowers actinomorphic, small; calyx-lobes or sepals and petals valvate in bud; stamens 5 to numerous, exserted.

1A. Stamens 10, free, slightly exserted; anthers with a deciduous, stiped gland at apex; tendril-bearing, evergreen, eglandular-leaved vines with spicate inflorescence. . . . . 1. *Entada*

1B. Stamens numerous, exserted, the filaments connate at base; anthers not glandular; deciduous trees without tendrils; petioles with glands; flowers in heads. . . . . 2. *Albizia*

II. Subfamily **CAESALPINOIDEAE:** Trees or herbs; flowers zygomorphic, the corolla not or only slightly papilionaceous; sepals and petals 5, imbricate in bud, the upper petal enclosed by the others; stamens 10 or sometimes fewer, free.

1A. Leaves pinnately compound.
  2A. Trees or shrubs, usually prickly or spiny; at least some of the leaves bipinnate; anthers versatile.
    3A. Flowers small, green or greenish yellow, in spikes; trees. . . . . 3. *Gleditsia*
    3B. Flowers rather large, yellow to orange; shrubs or vines. . . . . 4. *Caesalpinia*
  2B. Unarmed herbs (in ours); leaves simply pinnate (in ours); anthers basifixed. . . . . 5. *Cassia*
1B. Leaves bifid or composed of 2 leaflets; tendrils often present. . . . . 6. *Bauhinia*

III. Subfamily **PAPILIONATAE:** Trees, herbs, or vines; flowers distinctly papilionaceous; petals imbricate in bud, the uppermost petal (standard) on the outside; stamens 5 or 10, monadelphous, diadelphous or free.

1A. Stamens 10, free or nearly so; trees or shrubs, rarely herbs.
  2A. Leaves pinnate; leaflets many.
    3A. Legumes moniliform, not flattened. . . . . 7. *Sophora*
    3B. Legumes flattened.
      4A. Leaflets opposite or nearly so; winter-buds naked in the axils of leaves. . . . . 8. *Maackia*
      4B. Leaflets alternate; winter-buds entirely covered with the base of the petiole. . . . . 9. *Cladrastis*
  2B. Leaflets 3. . . . . 10. *Thermopsis*
1B. Stamens with filaments connate at base, monadelphous or diadelphous.
  5A. Legumes fleshy, indehiscent.
    6A. Legumes rather flattened; trees, leaflets 3–7. . . . . 11. *Pongamia*
    6B. Legumes not flattened; evergreen shrubs, leaflets 3(–7). . . . . 12. *Euchresta*
  5B. Legumes membranous to coriaceous, not fleshy.
    7A. Legumes jointed.
      8A. Filaments wholly connate into a complete tube; anthers dimorphic; leaflets 2 or 4. . . . . 13. *Zornia*
      8B. Filaments nearest the standard free; anthers all alike; leaflets 1 or 3, sometimes 5, or many.
        9A. Leaflets 1, 3, or 5.
          10A. Legumes 1-seeded, flat, indehiscent, not jointed.
            11A. Leaflets pinnately arranged; shrubs or perennial herbs. . . . . 14. *Lespedeza*
            11B. Leaflets digitately arranged; annuals. . . . . 15. *Kummerowia*
          10B. Legumes more than 1-seeded, jointed. . . . . 16. *Desmodium*
        9B. Leaflets many.
          12A. Leaflets even in number.
            13A. Legumes coiled, included within the calyx. . . . . 17. *Smithia*
            13B. Legumes elongate, straight, much exserted from the calyx. . . . . 18. *Aeschynomene*
          12B. Leaflets odd in number; legumes exserted. . . . . 19. *Hedysarum*
    7B. Legumes not jointed.
      14A. Leaves even-pinnate, with the rachis ending in a tendril or mucro.
        15A. Tube of the filaments obliquely truncate at apex; style slender, with a tuft of hairs at apex. . . . . 20. *Vicia*
        15B. Tube of the filaments rectangularly truncate at apex; style flattened, hairy on inner side. . . . . 21. *Lathyrus*
      14B. Leaves odd-pinnate or trifoliate, rarely digitate, without tendrils.
        16A. Twining or trailing herbs, with 3 (rarely 5 or 7) leaflets.
          17A. Leaves glandular-dotted beneath; racemes with axis not thickened on nodes; styles glabrous.
            18A. Legumes 3- to 8-seeded. . . . . 22. *Dunbaria*
            18B. Legumes 1- or 2-seeded. . . . . 23. *Rhynchosia*
          17B. Leaves not glandular-dotted.
           19A. Style bearded.
             20A. Keel dextrorsely semicontorted, spurred on one side. . . . . 24. *Azukia*
            20B. Keel not contorted or spurred. . . . . 25. *Vigna*
           19B. Styles glabrous, not bearded.
            21A. Leaflets 3–5; keel coiled. . . . . 26. *Apios*
            21B. Leaflets 3; keel not coiled.
              22A. Flowers yellowish; calyx nearly truncate at apex. . . . . 27. *Dumasia*
              22B. Flowers purple, red, or white; calyx usually with evident teeth.
                23A. Keel conspicuously larger than the standard and wing. . . . . 28. *Mucuna*
                23B. Keel, standard, and wings nearly equal.
                  24A. Stamens monadelphous; flowers and legumes of one form.
                    25A. Upper 2 calyx-teeth broad, larger than the lower; legumes broad, with a longitudinal rib along the outer suture on each side. . . . . 29. *Canavalia*
                    25B. Upper 2 calyx-teeth shorter and smaller than the lower; legumes narrower and smaller, hispid, without a rib on each side.

1. ENTADA Adans. MODAMA ZOKU

Tendril-bearing woody vines; leaves bipinnate, the uppermost pair of leaflets reduced to tendrils; racemes solitary or in pairs; flowers polygamous, yellowish, sessile, small; calyx toothed; petals equal, narrow, slightly curved; stamens 10, free, the anthers terminated by a gland; legumes large, flat, woody, jointed, rather many-seeded.——About 20 species, widespread in the Tropics.

1. Entada phaseoloides (L.) Merr. *E. scandens* Benth.; *Lens phaseoloides* L.——MODAMA. Evergreen, woody vine; leaves petioled, the rachis ending in a tendril, the pinnae 4, the leaflets lustrous, coriaceous, 2–4 on each pinna, oblong to obovate, 3–8 cm. long, subacute to obtuse; racemes pedunculate, usually aggregated in a terminal panicle; flowers sessile, yellowish green, about 6 mm. long; legumes 30–60 cm. long, 7–10 cm. wide, curved, constricted between the seeds which are 5 cm. in diameter.——Evergreen forests near seashores; Kyushu (Yakushima); rare.——Ryukyus, Formosa, and Tropics of the New and Old Worlds.

2. ALBIZIA Durazz. NEMU-NO-KI ZOKU

Trees or shrubs without spines; leaves bipinnate, often with cup-shaped glands on petioles or on the rachis, the stipules various; inflorescence of globose to narrow heads, usually in panicles; flowers 5-merous, bisexual or polygamous; calyx tubular, 5-toothed at apex; corolla funnel-shaped, the segments valvate in bud; stamens numerous, exserted, the filaments connate at base into a tube; legumes broadly linear, flattened.——About 50 species, widespread in the Tropics.

1A. Leaflets 36–58 on each pinna, narrowly semiovate or broadly and obliquely lanceolate, 10–15 mm. long, 2.5–4 mm. wide; pinnae 10–25 on a leaf. . . . . 1. *A. julibrissin*

1B. Leaflets 30–40 on each pinna, semiovate, 15–20 mm. long, 5–7 mm. wide; pinnae 8–10 on a leaf. . . . . 2. *A. glabrior*

1. Albizia julibrissin Durazz. *Mimosa julibrissin* (Durazz.) Scop.; *M. speciosa* Thunb., non Jacq.; *Acacia nemu* Willd.; *Albizia nemu* (Willd.) Benth.——NEMU-NO-KI. Deciduous tree with spreading, nearly glabrous branches; leaves petiolate, 20–30 cm. long inclusive of the petiole, the petioles with a cup-shaped gland on the upper side, the rachis and rachilla puberulent on upper side, the leaflets herbaceous, short-pubescent beneath and on margin, abruptly acute and mucronate, truncate at base, whitish beneath; panicles pubescent when young, the axis 2–4 cm. long, with more than 10 spreading branches each 3–4 cm. long and the heads about 20-flowered; flowers sessile, 7–9 mm. long, pinkish; stamens exserted, 35–40 mm. long; legumes glabrous, short-stiped, 10–13 cm. long, 17–18 mm. wide, 10- to 15-seeded.——June–July. Woods and riverbanks in lowlands; Honshu, Shikoku, Kyushu; very common.——s. Asia, as far west as Iran.

2. Albizia glabrior (Koidz.) Ohwi. *A. mollis* var. *glabrior* Koidz.——HIRO-HA-NEMU. Similar to the preceding, but the leaflets larger and fewer; branchlets loosely pubescent while young; leaves 15–25 cm. long, bipinnate, the petiole 4–6 cm. long, with a sessile ovoid gland on the upper side below the middle, the pinnae 4 or 5 pairs, 10–15 cm. long, short-petioled, the leaflets 15–20 pairs on each pinna, semiovate, entire, 15–20 mm. long, 5–7 mm. wide, asymmetrical at base, acute on inner side and truncate on outer side, loosely short-pubescent, whitish beneath; inflorescences of few umbellate undivided peduncles on the end of branchlets; flowers numerous, sessile, in terminal heads; corolla pinkish, about 8 mm. long; stamens about 25, about 3 cm. long.——Kyushu (Amakusa Isl.); very rare.——Tokara Isls.

3. GLEDITSIA L.   SAIKACHI ZOKU

Deciduous trees with polygamous flowers, some of the branchlets reduced to simple or branched spines; leaves pinnate to bipinnate, the leaflets sometimes undulately incised; flowers small, greenish, in racemes or spikes without bracteoles; petals and calyx-lobes 3–5; stamens twice as many as the calyx-lobes; legumes elongate, or nearly ovate, flattened, coriaceous or slightly fleshy.——About 12 species, in the warmer parts of Asia, N. America, and tropical Africa.

1. Gleditsia japonica Miq. *G. horrida* (Thunb.) Makino, non Willd.; *Fagara horrida* Thunb.; *Caesalpinoides japonicum* (Miq.) O. Kuntze——SAIKACHI.   Tree with nearly glabrous branches and branched spines; leaves 10–25 cm. long, simply or twice pinnate, short-pubescent on the rachis, the leaflets 12–24 on each pinna, oblique, narrowly ovate, 15–35 mm. long, 8–15 mm. wide, obtuse, entire or undulate-serrate, upper margin slightly dilated at base, nearly or quite glabrous, with raised veinlets beneath, green and slightly lustrous above, yellowish green beneath; spikes 6–8 cm. long, densely flowered, from the short leafy shoots; flowers yellowish, 3–4 mm. across, nearly sessile; calyx slightly pubescent, the lobes oblong, slightly shorter than the tube; petals yellowish, elliptic, puberulous; stamens short-exserted; legumes 20–25 cm. long, 3–3.5 cm. wide, pendulous, slightly arcuate, on a stipe about 2.5 cm. long.——May–June. Honshu (centr. distr. and westw.), Shikoku, Kyushu.

4. CAESALPINIA L.   JAKETSU-IBARA ZOKU

Trees or shrubs, often prickly, rarely scandent; leaves bipinnate, the leaflets herbaceous to coriaceous; flowers yellow or reddish, in axillary or terminal racemes often aggregated in a panicle, the bracts small, caducous, the bracteoles absent; calyx widely open, 5-lobed, the lobes imbricate in bud, the lowest one often larger; petals rounded, spreading, imbricate in bud; legumes ovate to lanceolate, flattened.——About 40 species, widely distributed in the warmer parts of the world.

1A. Deciduous shrub; leaflets 10–20 on each pinna, obovate or oblong, 15–20 mm. long, rounded at apex; flowers in racemes; pedicels 3–4 cm. long. .... 1. *C. japonica*
1B. Evergreen woody climber; leaflets 4–10 on each pinna, ovate or oblong, 3–5 cm. long, acute to obtuse at apex; flowers in panicles; pedicels 6–8 mm. long. .... 2. *C. nuga*

1. Caesalpinia japonica Sieb. & Zucc. *C. sepiaria* var. *japonica* (Sieb. & Zucc.) Makino; *C. crista* sensu Thunb., non L.; *C. sepiaria* sensu auct. Japon., non Roxb.——JAKETSU-IBARA.   Somewhat scandent shrub, 1–2 m. high; branches much elongate, with prominent retrorse prickles, short-pubescent while young; leaves 20–40 cm. long, prickly on rachis and petioles, the pinnae 6–16, the leaflets vivid green on upper side, whitish beneath, sometimes slightly pubescent, the lateral nerves not prominent; racemes terminal, 20–30 cm. long, pedunculate, the bracts narrowly ovate, caducous, 3–4 mm. long, densely brownish pubescent, the pedicels spreading, slender; flowers 25–30 mm. across, expanded, yellow; stamens red, densely villous on the lower half; legumes oblong, slightly arcuate, sessile, flat, glabrous, about 10 cm. long, 3 cm. wide, with a thick winglike outer suture.——Apr.–June. Woods and grassy places in lowlands and hills; Honshu (Kantō Distr. and westw.), Shikoku, Kyushu; rather common.——Ryukyus and China (?).

2. Caesalpinia nuga Ait. *C. chinensis* Roxb.; *C. paniculata* Desf.——NANTEN-KAZURA.   Glabrous prickly climber; leaves 20–30 cm. long, the leaflets 4–6 on each pinna, deep green and lustrous on upper side, glaucous beneath, with slightly raised veinlets on both sides; flowers 15–20 mm. across; legumes about 5 cm. long.——Kyushu (Yakushima).——India, Malay Peninsula, Formosa, China, and Ryukyus.

5. CASSIA L.   KAWARA-KETSUMEI ZOKU

Herbs, shrubs, or trees; leaves even-pinnate, stipulate, the petiole or rachis with a cuplike sessile or stiped gland; flowers yellow, sometimes white or reddish, in racemes or panicles, often fasciculate in leaf-axils, bracteate and bracteolate; calyx-segments 5, imbricate; petals 5, spreading, imbricate; stamens 10 or fewer, the anthers opening by pores or longitudinally dehiscent on the upper part; legumes various, terete to flat, 2-valved, dehiscent.——About 450 species, widely distributed in tropical and warm-temperate regions.

1. Cassia nomame (Sieb.) Honda. *C. mimosoides* var. *nomame* (Sieb.) Makino; *C. mimosoides* sensu auct. Japon., non L.; *Soja nomame* Sieb.——KAWARA-KETSUMEI.   Annual herb 30–60 cm. high; stems terete, sometimes branched, whitish puberulent; stipules subulate or subulate-lanceolate, long-acuminate, few-ribbed at base, persistent, appressed, 5–7 mm. long; leaves 3–7 cm. long, the petioles short, with a solitary sessile gland just below the insertion of the lowest leaflets, the leaflets 30–70, linear-oblong, 8–12 mm. long, 2–3 mm. wide, scattered short-pilose on margin; flowers supra-axillary, 1 or 2 together, 6–7 mm. across, yellow, the bract similar to the stipules but smaller, the bracteoles on the top of the pedicels, narrowly lanceolate, about 2 mm. long, acuminate; calyx-segments lanceolate or broadly so, 5–6 mm. long, acuminate, short-puberulent and scarious on the margins; petals obovate, slightly longer than the calyx; fertile stamens 4 with a reduced minute 5th stamen usually also present, the anthers truncate, smooth, glabrous, about 2.5 mm. long, the filaments less than 1 mm. long; style short; legumes flat, broadly linear, 3–4 cm. long, 5–6 mm. wide, short-pilose, sessile, dehiscent, 2-valved. ——Sandy riversides and waste grounds; Honshu, Shikoku, Kyushu; very common.——Korea, Manchuria, and China.

6. BAUHINIA L.   HAKAMA-KAZURA ZOKU

Trees, shrubs, or vines with tendrils; leaves simple, entire or 2-lobed, sometimes composed of 2-leaflets, the stipules usually small and deciduous; flowers white or reddish, in simple or paniculately compound racemes, the bracts and bracteoles small; calyx closed in bud or 5-toothed; petals 5; stamens 10 or partly reduced; legumes flattened, elliptic to broadly linear, membranous to ligneous, dehiscent or indehiscent.——About 200 species, widely distributed in the Tropics.

1. Bauhinia japonica Maxim. *Lasiobema japonicum* (Maxim.) de Wit.——HAKAMA-KAZURA. Woody vine; branches sulcate, with red-brown short appressed hairs while young; stipules ovate, obtuse, deciduous; petioles rather elongate; leaves somewhat coriaceous, usually rounded-cordate, bifid, 6–10 cm. long, densely reddish brown pubescent when young; tendrils flat, simple, or usually paired; racemes terminal, 8–15 cm. long, many-flowered, densely reddish brown pubescent, the bracts subulate, about 2 mm. long, the pedicels ascending, 15–20 mm. long, obsoletely bracteolate near the middle; calyx 5-lobed, about 6 mm. long; petals rather thick, 6–7 mm. long, pale green inside, densely reddish brown pubescent outside, with a claw about as long as the nearly orbicular limb; fertile stamens 3, exserted; ovary villous; legumes coriaceous, oblong or ovate, 5–8 cm. long, 25–30 mm. wide, 1- to 3-seeded, dehiscent, nearly sessile.——July. Woods near seashores; Honshu (Kii Prov.), Shikoku, Kyushu; rare.——Ryukyus.

7. SOPHORA L.   KURARA ZOKU

Perennial herbs or trees; leaves odd-pinnate, the stipules subulate or absent; flowers pale yellow, white, or purplish, in terminal racemes or panicles, the bracts small or absent, the bracteoles none; calyx campanulate to cup-shaped, the teeth short; legumes moniliform, not flattened, sometimes 4-angled, fleshy to ligneous, usually indehiscent.——About 20 species, in warmer parts of the world.

1A. Perennial herb with racemose inflorescence. .... 1. *S. flavescens*
1B. Trees or shrubs.
  2A. Shrub with racemose inflorescence. .... 2. *S. franchetiana*
  2B. Tree with paniculate inflorescence. .... 3. *S. japonica*

1. Sophora flavescens Ait. *S. angustifolia* Sieb. & Zucc.; *S. flavescens* var. *angustifolia* (Sieb. & Zucc.) Kitag.; *S. galegoides* Pall.; *S. kronei* Hance; *S. sororia* Hance——KURARA. Perennial herb, usually glabrous or sometimes with sparse, yellowish, short, appressed pubescence; stems erect, 80–150 cm. long, terete, slightly ligneous near the base, branching above; stipules linear or filiform, 5–8 mm. long; leaves petioled, 15–25 cm. long, the leaflets 15–40, narrowly oblong to broadly lanceolate, 2–4 cm. long, 7–15 mm. wide, obtuse to rather acute, slightly appressed-pubescent on both sides or on the paler underside only; racemes erect, terminal, 10–20 cm. long, many-flowered; flowers pale greenish yellow, rarely purplish, 15–18 mm. long; calyx loosely appressed-pubescent, oblique, 7–8 mm. long, the teeth depressed; legumes coriaceous, linear, 7–8 cm. long, 7–8 mm. across, obtusely 4-angled, constricted between the seeds, short-stiped, appressed-pilose while young.——June–July. Grassy places in lowlands and waste grounds; Honshu, Shikoku, Kyushu; common and somewhat variable, our plants generally with smaller leaflets than in the continental Asian phase.——Siberia, China, and Korea.

2. Sophora franchetiana Dunn. TSUKUSHI-MURE-SU-ZUME, CHIKUSHI-MURE-SUZUME. Much-branched shrub, densely brown appressed-pubescent; leaves usually with 11 leaflets, the leaflets elliptic, 2.5–3 cm. long, acute, chartaceous, nearly glabrous on upper side; racemes about 5 cm. long; flowers white; ovary pubescent; legumes 2.5–3 cm. long, glabrous.——Kyushu; very rare.——China.

3. Sophora japonica L. *Styphnolobium japonicum* (L.) Schott——ENJU. Deciduous tree; leaves 15–25 cm. long, the leaflets 9–15, ovate to narrowly so, 2.5 cm. long, 15–25 mm. wide, acute to subobtuse, sometimes with a minute stipel at base, deep green, glaucescent and appressed-pubescent beneath; panicles 15–30 cm. long, broadly pyramidal, with short appressed pubescence; flowers 12–15 mm. long, white to yellowish; calyx 3–4 mm. long, pubescent, the teeth depressed 4-angled, short-villous on the margins; legumes stiped, loosely appressed-pubescent, glabrate, 5–8 cm. long, rather fleshy.——July–Aug. A Chinese tree planted in our area.

8. MAACKIA Rupr. & Maxim.   INU-ENJU ZOKU

Deciduous trees or shrubs with exposed, scaled buds; leaves odd-pinnate, petioled, exstipellate, the leaflets entire; flowers pedicelled, white, in terminal panicles or compound racemes, the bracts narrowly lanceolate, small, caducous, the bracteoles obsolete; calyx rather inflated, with depressed teeth; standard reflexed; legumes flattened, elliptic to broadly linear, 2-valved, dehiscent, rather thin, winged at least on the outer suture, 1- to 5-seeded.——Few species in the temperate regions of e. Asia.

1A. Legumes oblong to narrowly ovate, 2–4 cm. long; leaflets 11–15, 2–4 cm. long; flowers about 7 mm. long; standard nearly entire at
  apex. .... 1. *M. tashiroi*
1B. Legumes broadly linear, 4–9 cm. long; leaflets 7–13, 3–7 cm. long; flowers 7–12 mm. long; standard prominently retuse at apex.
                                                  2. *M. amurensis*

1. **Maackia tashiroi** (Yatabe) Makino. *Cladrastis tashiroi* Yatabe——SHIMA-ENJU. Deciduous shrub with spreading branches; branchlets slightly pubescent; leaves petioled, 15–20 cm. long, the leaflets 11–15, somewhat coriaceous, oblong or elliptic, 2–4 cm. long, subacute at both ends, short appressed-pubescent beneath or on both surfaces, glaucescent beneath; inflorescence of 1–4 racemes, 5–12 cm. long, rather densely many-flowered, densely brown-puberulent, the bracts lanceolate, small; flowers short-pedicelled, pale yellow; calyx yellowish brown villous, the teeth deltoid, short; legumes loosely short-pubescent, 10–12 mm. wide, 1- to 3-seeded, with a narrow wing on the outer suture.——July–Aug. Wet places near seashores; Kyushu.——Ryukyus and Formosa.

2. **Maackia amurensis** Rupr. & Maxim. var. **buergeri** (Maxim.) C. K. Schn. *Cladrastis amurensis* var. *buergeri* Maxim.; *C. amurensis* var. *floribunda* Maxim. ex Fr. & Sav.; *C. amurensis* var. *vidalii* Fr. & Sav.; *M. buergeri* (Maxim.) Tatew.——INU-ENJU. Small deciduous tree; leaves petioled, 20–30 cm. long, the leaflets 7–11, ovate, 4–7 cm. long, acute to subobtuse, appressed-pubescent beneath; inflorescence of 3–7 racemes arranged in a panicle, densely many-flowered, yellowish or grayish brown villous, the bracts linear, small; flowers pedicelled, 10–12 mm. long; calyx 5–6 mm. long, villous; legumes broadly linear, flat, 6–9 cm. long, 8–10 mm. wide, appressed-pubescent while young, 3- to 6-seeded, with a narrow wing on the outer suture.——July–Sept. Hokkaido, Honshu.——A variable widespread species occurring in China, Formosa, Manchuria, Amur, Ussuri, and Korea.

Var. **pubescens** (Koidz.) Ohwi. *M. floribunda* Takeda; *M. floribunda* var. *pubescens* Koidz.——HANEMI-NO-INU-ENJU. Leaflets usually 9–13, rather small, narrowly ovate to oblong, 3–5 cm. long; flowers 7–10 mm. long; legumes 4–8 cm. long, 8–12 mm. wide, appressed-pubescent, the wing on the suture slightly broader, 1–1.5 mm. wide.——Honshu, Shikoku, Kyushu.

9. CLADRASTIS Raf.   FUJIKI ZOKU

Trees with buds covered by the base of the petioles; leaves odd-pinnate, leaflets alternate on the rachis, exstipellate; flowers white, in terminal panicles, the bracts and bracteoles obsolete or absent; calyx-teeth short, broad; legumes linear, flat, thin, short-stiped, wingless or winged on the margins.——Few species, in e. Asia and N. America.

1A. Leaflets with 8 to 13 pairs of lateral nerves, pale green and with raised veinlets beneath; legumes winged on both margins; winterbuds enveloped with a grayish white thin membrane. .... 1. *C. platycarpa*
1B. Leaflets with 13 to 15 pairs of lateral nerves, glaucous and without raised veinlets beneath; legumes wingless; winter-buds naked.
    2. *C. sikokiana*

1. **Cladrastis platycarpa** (Maxim.) Makino. *Sophora platycarpa* Maxim.; *Platyosprion platycarpum* (Maxim.) Maxim.——FUJIKI. Deciduous tree; leaves 20–30 cm. long, the petioles short, slightly short-pubescent, the leaflets 8–13, ovate-oblong, 5–10 cm. long, 2–4 cm. wide, acute to acuminate, slightly dilated on the upper margin at base, puberulous on the midrib above, slightly appressed-pubescent beneath; panicles 15–25 cm. long, the separate branches in racemes; flowers about 15 mm. long, the pedicels and calyx appressed brown-pubescent, the bracteoles minute on the middle portion of the pedicel; calyx 4–5 mm. long, with depressed semirounded teeth; legumes much compressed, short-stiped, broadly linear, 7–8 cm. long, 15–17 mm. wide inclusive of the narrow wings, 2- to 3-seeded, glabrous or nearly so.——June–July. Mountains; Honshu (centr. distr. and westw.), Shikoku.

2. **Cladrastis sikokiana** (Makino) Makino. *Sophora sikokiana* Makino——YUKU-NO-KI, MIYAMA-FUJIKI. Resembles the preceding but brownish woolly-pubescent while young; leaves slightly thinner, glaucous beneath, the lateral nerves regularly arranged; flowers 15–18 mm. long, white, in terminal panicles; calyx 5–6 mm. long, short-pubescent; legumes nearly wingless, slightly thickened on the margins.——June–Aug. Honshu (Kantō Distr. and westw.), Shikoku, Kyushu; rare.

10. THERMOPSIS R. Br.   SENDAI-HAGI ZOKU

Erect perennial herbs with creeping rhizomes; stipules prominent, free, leaflike, the lower ones sheathlike; leaves digitately 3-foliolate; flowers rather large, yellow, rarely purple, in terminal racemes, the bracts distinct, prominent, the bracteoles absent, the pedicels solitary; calyx narrowly campanulate with short equal teeth; standard orbicular, the keel nearly as long as the wing; stamens free; legumes linear, flat, straight or sometimes arcuate, subsessile, rather many-seeded.——About 20 species, in temperate or cooler regions of Siberia, n. India, e. Asia, and N. America.

1. **Thermopsis lupinoides** (L.) Link. *T. fabacea* (Pall.) DC.; *Sophora fabacea* Pall.; *S. lupinoides* L.——SENDAI-HAGI. Stems erect, usually not branched, 40–80 cm. long, soft-pubescent while young, soon glabrate; stipules ovate, 3–4 cm. long, usually obtuse; leaflets ovate, obovate, or elliptic, equal, 5–7 cm. long, 3–5 cm. wide, obtuse, soft-pubescent while young; racemes rather elongate, the bracts narrowly ovate, deciduous, longer than the pedicels; flowers 23–25 mm. long; calyx 10–12 mm. long, with narrowly ovate teeth about half as long as the tube; legumes straight, flat, 8–10 cm. long, 6–7 mm. wide, pubescent while young, 12- to 15-seeded.——May–Aug. Sand dunes near the sea; Hokkaido, Honshu, Shikoku, Kyushu; common northw.——Korea, Ussuri, Sakhalin, Kuriles, regions around the Ochotsk Sea, and N. America.

11. PONGAMIA Vent.   KURO-YONA ZOKU

Trees, rarely shrubs; leaves petioled, odd-pinnate, the stipules caducous, the leaflets 3–7, opposite, coriaceous, exstipellate; inflorescence an axillary raceme; flowers in 2's, 3's, or 4's, white, the bracts caducous, the bracteoles minute or absent; calyx campanulate, nearly truncate with obsolete teeth; standard broad, minutely pilose externally; stamens monadelphous, the upper filament free, short, the anthers versatile, oblong, slightly hairy at base; legumes elliptic, 1-seeded, slightly flattened, hard-fleshy, indehiscent.——A few species, in tropical Asia and Australia.

1. Pongamia pinnata (L.) Merr. *P. glabra* Vent.; *Cytisus pinnatus* L.——KURO-YONA. Tree with glabrous branches; leaves 15–30 cm. long, the leaflets 5–7, opposite, ovate or broadly so, 5–10 cm. long, 4–8 cm. wide, abruptly acuminate with an obtuse tip, glabrous; racemes peduncled, to 20 cm. long, many-flowered, short-pilose; flowers 15–18 mm. long, pedicelled; legumes glabrous, 5–6 cm. long, about 2.5 cm. wide, rather thick.——Near seashores; Kyushu (Yakushima); rare.——Ryukyus, Formosa, and the Tropics of Asia and Australia.

12. EUCHRESTA A. Benn. MIYAMA-TOBERA ZOKU

Evergreen shrubs; leaves odd-pinnate, the leaflets 3–7, entire, exstipellate; flowers in terminal or axillary racemes, white, the bracts minute, the bracteoles absent; calyx broadly campanulate with depressed teeth; anthers versatile; legumes fleshy, stiped, ellipsoidal, lustrous, indehiscent, 1- to few-seeded.——Few species in se. Asia.

1. Euchresta japonica Benth. MIYAMA-TOBERA. FIG. 9. Small evergreen, erect or decumbent shrub 30–80 cm. high, nearly simple or sometimes sparsely branched, with thick branches, short-pilose; leaves petioled, the leaflets 3, elliptic or obovate, 5–8 cm. long, 3–5 cm. wide, obtuse, glabrous, deep green and lustrous above, grayish short-pilose beneath, the lateral nerves not prominent; racemes terminal, peduncled, rather many-flowered, short-pilose, the bracts minute; flowers 10–13 mm. long; calyx 2.5–3 mm. long, with undulate teeth; legumes ellipsoidal, 18–20 mm. long inclusive of the 2–3 mm. long stipe, about 10 mm. wide, dark bluish purple.——July. Woods in mountains; Honshu (Kantō Distr. and westw.), Shikoku, Kyushu; rare.

13. ZORNIA Gmel. SUNAJI-MAME ZOKU

Herbs, sometimes suffrutescent; leaves palmately 2- or 4-foliolate, often with pellucid dots, exstipellate, the stipules linear, sometimes foliaceous or peltate; flowers sessile, solitary or in few-flowered racemes, in 2's between a pair of broad, peltate, flat, bracts, without bracteoles; calyx usually thin, 2-fid, the upper lobes 2-toothed, the lower 3-lobed; petals small; anthers of two forms, the larger basifixed and alternate with the smaller versatile ones; legumes flat, the joints smooth or spinulose.——More than a dozen species in America, one or two in Asia and Africa.

1. Zornia cantoniensis Mohlenbrock. *Z. diphylla* (L.) Pers.; *Hedysarum diphyllum* L.; *Z. angustifolia* J. E. Smith ——SUNAJI-MAME. Slender but firm, tufted perennial herb 20–50 cm. high; leaflets 2, lanceolate, 1.5–2 cm. long, 2–3 mm. wide, sparsely pubescent with ventrally attached hairs; bracts narrowly ovate; flowers yellow, 8–10 mm. long; legumes elliptic, 3–4 mm. long, puberulous, spinulose on both faces, joints flat.——June–Aug. Near seashores; Shikoku (Tosa Prov.); rare.——A variable species widely distributed in the Tropics of both hemispheres.

14. LESPEDEZA Michx. HAGI ZOKU

Perennial herbs or shrubs, usually pilose; leaves pinnately 3-foliolate, the leaflets entire, exstipellate, the stipules free, usually subulate, often deciduous; flowers purple or yellowish, in axillary racemes, often cleistogamous, the bracts small, the bracteoles 2; calyx-teeth equal or unequal, the upper 2 often connate; legumes ovate to orbiculate, flat, indehiscent, 1-seeded.——About 60 species, in the temperate regions of N. America and e. Asia.

1A. Flowers all alike, chasmogamic, rather large, 12–20 mm. long, rose-purple, rarely white; keel incurved at apex; legumes short-stiped within the calyx.
  2A. Winter-buds flattened, the scales distichous; horizontally spreading shrubs much branched in upper part; racemes sessile or subsessile.
    3A. Calyx-teeth ovate or broadly so, acute to subacute, not spine-pointed. .... 1. *L. buergeri*
    3B. Calyx-teeth gradually acuminate and spine-tipped. .... 2. *L. maximowiczii*
  2B. Winter-buds terete, the scales spirally arranged; erect or arching shrubs often with ascending branches; racemes, when elongate, distinctly peduncled.
    4A. Calyx-teeth spine-tipped; racemes short, sessile, usually not longer than the leaves. .... 3. *L. cyrtobotrya*
    4B. Calyx-teeth rounded to acuminate, not spine-tipped; racemes sometimes elongate.
      5A. Calyx-teeth rounded to acute, often obsoletely 3-nerved, usually not longer than the tube.
        6A. Leaflets membranous to herbaceous, at least when young with appressed to ascending hairs 0.3–0.7 mm. long; calyx-teeth oblong to lanceolate, obtuse to acute, often with darker nerves. .... 4. *L. bicolor*
        6B. Leaflets somewhat thicker, with sparse appressed short hairs about 0.2 mm. long; calyx-teeth usually obtuse to rounded, the nerves of the same color, indistinct. .... 5. *L. homoloba*
      5B. Calyx-teeth acute to acuminate, 1-nerved, the lowest longer than the tube. .... 6. *L. thunbergii*
1B. Flowers of two forms, the chasmogamic usually less than 12 mm. long, with the keel erect or scarcely incurved at apex, the cleistogamic apetalous or nearly so; legumes sessile or nearly so.
  7A. Racemes of chasmogamic flowers peduncled, usually longer than the leaves.
    8A. Densely pubescent with yellowish brown soft hairs; leaflets 3–4 cm. long, 12–20 mm. wide. .... 7. *L. tomentosa*
    8B. Glabrescent or loosely white-pubescent; leaflets 1–2.5 cm. long, 5–10 mm. wide.
      9A. Peduncles capillary, 3- to 5-flowered, glabrescent, spreading; calyx of chasmogamic flowers 4–5 mm. long. .... 8. *L. virgata*
      9B. Peduncles thicker, 5- to 15-flowered, appressed-pilose, ascending; calyx of chasmogamic flowers 5–6 mm. long.
                                  9. *L. davurica*
  7B. Racemes of chasmogamic flowers sessile or nearly so, shorter than the leaves.
    10A. Leaflets oblanceolate, linear, rarely obovate.

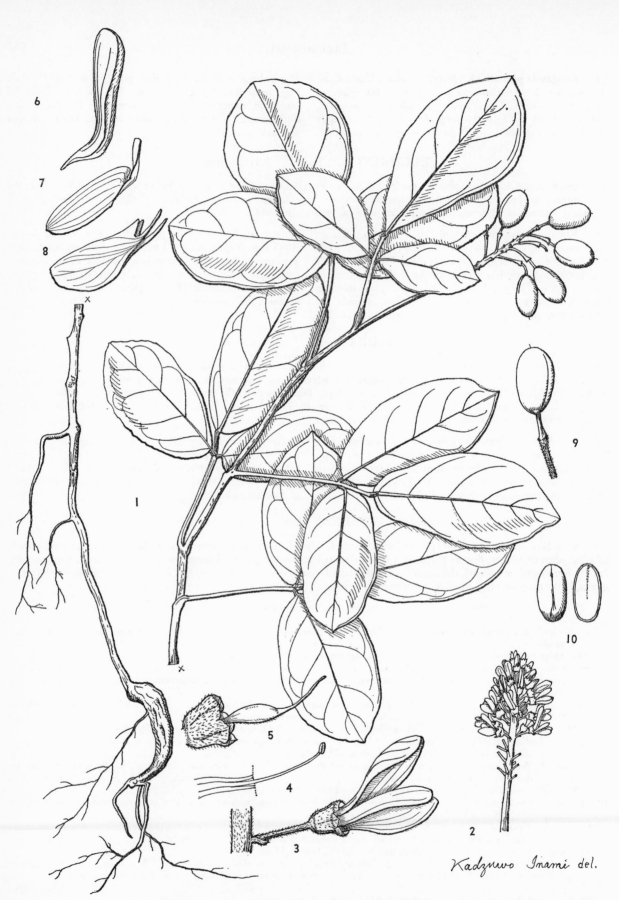

Fig. 9—*Euchresta japonica* Benth. 1, Habit; 2, raceme; 3, flower; 4, stamen; 5, pistil; 6, vexillum; 7, wing; 8, keel; 9, legume; 10, seeds.

Kadzuwo Inami del.

11A. Calyx-teeth of the cleistogamous flowers indistinctly 1-nerved; legumes of cleistogamous flowers 1.5–2 times as long as the calyx. .... 10. *L. cuneata*
11B. Calyx-teeth of the cleistogamous flowers clearly 3-ribbed; legumes of the cleistogamous flowers as long as to slightly shorter than the calyx. .... 11. *L. juncea*
10B. Leaflets obovate to obovate-orbicular; long prostrate or trailing slender herb with pubescent branches and leaves. .... 12. *L. pilosa*

1. Lespedeza buergeri Miq. *L. oldhamii* Miq.; *L. buergeri* var. *oldhamii* (Miq.) Maxim.——KI-HAGI. Shrub 40–80 cm. high; the branches horizontally spreading, slender, appressed-pilose; leaves petioled, the leaflets ovate to oblong, 2–4 cm. long, 1–3 cm. wide, usually acute, deep green, glabrous or glabrescent on upper side, appressed-pilose beneath; racemes to 7 cm. long, subsessile; flowers 10–12 mm. long, white with purple blotch; calyx-teeth equal, ovate, obtuse to subacute, as long as or shorter than the tube, obsoletely 1-nerved, the upper 2 connate, the bracts ovate, obtuse, few-nerved, the bracteoles broadly ovate, many-nerved, usually ciliolate, usually slightly longer than the calyx-tube; legumes oblong, loosely appressed-pilose, about 12 mm. long, about 4 mm. wide.——July–Sept. Rocky places along rivers in mountains; Honshu, Shikoku, Kyushu; common.——China.——Forma **albiflora** Honda. SHIROBANA-KI-HAGI. Flowers white; rare.

Var. **kinashii** Ohwi. TACHIGE-KI-HAGI. Hairs of branches and racemes spreading.——Occurs with the typical phase; rare.

2. Lespedeza maximowiczii C. K. Schn. *L. buergeri* var. *praecox* Nakai; *L. praecox* (Nakai) Nakai——CHŌSEN-KI-HAGI. Very closely allied to the preceding species; calyx-teeth broadly lanceolate, spine-tipped, longer than the tube.——July–Aug. Kyushu (Tsushima).——Korea.

3. Lespedeza cyrtobotrya Miq. MARUBA-HAGI, MIYAMA-HAGI. Erect or ascending herb, woody at base, 70–120 cm. high; leaves petioled, the leaflets usually elliptic or obovate, 2–3 cm. long, 15–20 mm. wide, obtuse to rounded, often retuse, glabrous above, short-pilose beneath; racemes axillary, sessile or nearly so, short, usually not exceeding the leaves, the bracts narrowly ovate, acute, 1-nerved, the bracteoles similar, rather narrow; flowers 10–15 mm. long, rose-purple; calyx-teeth longer than the tube, 1-nerved, spine-tipped; legumes nearly orbicular, beaked, white appressed-pilose, 6–7 mm. long, about 5 mm. wide.——Aug.–Oct. Meadows and grassy places in hills and in low mountains; Honshu, Shikoku, Kyushu; common.——Korea and China.

Var. **kawachiana** (Nakai) Ohwi. *L. kawachiana* Nakai ——KAWACHI-HAGI. Hairs of branches and racemes spreading.——Occurs with the typical phase; rare.

4. Lespedeza bicolor Turcz. EZO-YAMA-HAGI. Erect to ascending herb, often woody at base, 60–180 cm. high; stipules subulate, 3-nerved at base, the leaflets membranous to herbaceous, elliptic, 1.5–4 cm. long, 1–2.5 cm. wide, sparsely pubescent above only while young, loosely appressed-pilose to glabrate beneath, the hairs 0.5–0.7 mm. long; racemes longer to shorter than the leaves; flowers rose-purple, 12–17 mm. long, the bracts and bracteoles narrowly ovate, minute; legumes nearly orbicular, 5–7 mm. long and as wide, loosely white appressed-pilose.——July–Sept. Grassy places and thickets in lowlands and in mountains; Hokkaido, Honshu, Shikoku, Kyushu; common and variable.——Korea, Ussuri, Manchuria. and n. China.——Forma **acutifolia** Matsum. *L. bicolor* var. *intermedia* forma *acutifolia* Matsum.; *L. spicata* Nakai & F. Maekawa; *L. melanantha* Nakai; *L. bicolor* var. *japonica* Nakai; *L. floribunda* sensu auct. Japon., non Bunge; *L. bicolor* forma *microphylla* Miq.——YAMA-HAGI. With thicker, mostly acute leaflets and scattered shorter hairs be-

neath, about 0.3 mm. long.——Very common in our area and freely passing into the typical phase.——Forma **alba** (Bean) Ohwi. *L. bicolor* var. *alba* Bean. Flowers white.

5. Lespedeza homoloba Nakai. *L. nikkoensis* Nakai; *L. retusa* Nakai; *L. rotundiloba* Nakai; *L. sendaica* Nakai—— TSUKUSHI-HAGI, NIKKO-SHIRA-HAGI. Shrub with elongate branches, minutely appressed-pilose while young; leaflets rather thick, ovate to elliptic, 2–5 cm. long, 1–3 cm. wide, glabrous on upper side from the first, sparsely appressed-pilose beneath; racemes longer or shorter than the leaves; flowers 10–15 mm. long, rose-purple, the bracts and bracteoles narrowly ovate, minute, acute; legumes broadly obovate, about 8 mm. long, 5–6 mm. wide, appressed-pilose.——Aug.–Oct. Grassy places and thickets; Honshu, Shikoku, Kyushu; common.

6. Lespedeza thunbergii (DC.) Nakai. *Desmodium thunbergii* DC.; *D. penduliflorum* Oudem.; *L. penduliflora* (Oudem.) Nakai; *L. sieboldii* Miq.; *L. bicolor* var. *sieboldii* (Miq.) Maxim.; *L. grandis* Koidz.——MIYAGINO-HAGI. Stems sometimes woody at base, 100–180 cm. high, erect to arching above, short appressed-pilose while young; stipules small, subulate; leaflets narrowly oblong, sometimes elliptic or ovate, 3–5 cm. long, 1–2.5 cm. wide, usually acute, glabrous on upper side, with short appressed hairs beneath; racemes usually longer than the leaves, the bracts and bracteoles broadly lanceolate or narrowly ovate, small, acute; flowers 15–18 mm. long, rose-purple; legumes narrowly obovate to oblong, 10–13 mm. long, 4–5 mm. wide, appressed-pilose.—— June–Sept. Widely cultivated in our area; spontaneous in Honshu (Hokuriku and Chūgoku Distr.).——Reported to be spontaneous in China; very variable.——Forma **angustifolia** (Nakai) Ohwi. *L. intermedia* Nakai, non Britt.; *L. intermedia* var. *angustifolia* Nakai; *L. japonica* var. *intermedia* Nakai, var. *angustifolia* (Nakai) Nakai, var. *gracilis* Nakai, var. *retusa* Nakai, and var. *spicata* Nakai——BITCHŪ-YAMA-HAGI. Leaflet pilose above; flowers rose-purple.——Spontaneous in Honshu, Shikoku, Kyushu.——Korea and Manchuria. ——Cv. **Albiflora**. *Desmodium racemosum* var. *albiflorum* Sieb. ex Miq.; *L. bicolor* var. *intermedia* forma *albiflora* Matsum.; *L. japonica* L. H. Bailey; *L. sieboldii* var. *albiflora* (Sieb. ex Miq.) C. K. Schn.; *L. intermedia* var. *alba* Nakai ——SHIROBANA-HAGI. Leaflets thinly appressed-pilose on upper side; flowers white.——Long cultivated as an ornamental.——Cv. **Versicolor**. *L. penduliflora* forma *versicolor* (Nakai) Ohwi; *L. japonica* var. *versicolor* Nakai—— SOMEWAKE-HAGI. With white and rose-purple flowers on the same plant. Cultivated.

Var. **obtusifolia** (Nakai) Ohwi. *L. bicolor* var. *sieboldii* forma *sericea* Matsum.; *L. patens* Nakai; *L. patens* var. *obtusifolia* Nakai——KE-HAGI. Branches and sometimes racemes with spreading hairs; leaves glabrous on upper side. ——Aug.–Sept. Honshu (Uzen, Harima Prov. and elsewhere); often cultivated.——Forma **pilosella** Ohwi. UWAGE-KI-HAGI. Leaves minutely pilose above.

7. Lespedeza tomentosa (Thunb.) Sieb. ex Maxim. *L. villosa* (Willd.) Pers.; *Hedysarum tomentosum* Thunb.; *H. villosum* Willd.; *L. macrophylla* Bunge; *L. hirta* Miq., non Ellis——INU-HAGI. Erect stiff perennial herb, 80–150 cm.

high, densely yellowish brown ascending-pubescent; stipules linear, recurved, the leaflets oblong, obtuse to rounded, 3–6 cm. long, 10–30 mm. wide; chasmogamic flowers whitish, 8–10 mm. long, densely arranged on elongate, peduncled racemes, the bracts lanceolate, obtuse, small, the bracteoles broadly linear, about 2 mm. long, acuminate; calyx deeply 5-lobed, 5–7 mm. long, the lobes lanceolate, spine-tipped; legumes ovate, ascending-pilose; cleistogamic flowers rather many, fasciculate in the leaf-axils; calyx 4–5 mm. long; legumes 3–4 mm. long.——July–Sept. Grassy slopes in hills; Honshu, Kyushu.—— e. Siberia, Manchuria, China, India, Formosa, and Ryukyus.

8. **Lespedeza virgata** (Thunb.) DC. *Hedysarum virgatum* Thunb.; *L. swinhoei* Hance——MAKIE-HAGI. Slender stiff perennial with scattered, appressed, stiff hairs; stems 40–60 cm. long, erect, purplish, sparsely pilose, branching above or nearly simple; stipules subulate; leaflets oblong to elliptic, 7–20 mm. long, 5–10 mm. wide, rather thick and firm, glabrous on upper side, deep green, paler and sparsely pilose beneath; chasmogamic flowers few, 4–5 mm. long, on long peduncles; cleistogamic flowers few, axillary; calyx 2–2.5 mm. long; legumes equal, broadly ovate, about 4 mm. long, 3 mm. wide, slightly pilose.——Aug.–Sept. Honshu (Kantō Distr. and westw.), Shikoku, Kyushu.——Korea and Manchuria.

Var. **macrovirgata** (Kitag.) Kitag. *L. macrovirgata* Kitag. ——Ō-MAKIE-HAGI. Hairs at least partly spreading.——Honshu (Tamba Prov.).

9. **Lespedeza davurica** (Laxm.) Schindl. *Trifolium davuricum* Laxm.; *Hedysarum trichocarpum* Steph. ex Willd.; *L. trichocarpa* (Steph.) Pers.; *L. medicaginoides* Bunge; *L. fauriei* Lév.——ŌBA-MEDO-HAGI. Erect stiff perennial; stems 20–80 cm. long, short white appressed-pilose; stipules subulate; leaflets oblong or narrowly so, 15–30 mm. long, 5–10 mm. wide, nearly glabrous above, sparsely pilose beneath; chasmogamic flowers about 8 mm. long, in dense racemes as long as to shorter than the leaves, the peduncles rather stout, axillary, the bracts minute, the bracteoles narrowly lanceolate to subulate, 2–3 mm. long; calyx 5–6 mm. long, with narrowly lanceolate, 3-ribbed, spine-tipped lobes; cleistogamic flowers few, fasciculate in axils; calyx 4–5 mm. long; legumes broadly ovate, appressed-pilose, 3–3.5 mm. long.——Sept. Honshu (Hiratsuka in Sagami Prov.); possibly not indigenous.——e. Siberia, Manchuria, Korea, China, and Formosa.

10. **Lespedeza cuneata** (Dum. Cours.) G. Don. *Hedysarum sericeum* Thunb., non Mill.; *Anthyllis cuneata* Dum. Cours.; *L. argyrea* Sieb. & Zucc.; *Indigofera chinensis* Vogel; *L. sericea* (Thunb.) Miq., non Benth.; *L. juncea* var. *sericea* (Thunb.) Forbes & Hemsl.; *L. juncea* var. *subsericea* Miq.—— MEDO-HAGI. Tufted or solitary, stiffly erect perennial, copiously branched above, 60–100 cm. high, sparsely white appressed-pilose except the upper side of leaves; stems densely leafy; stipules lanceolate, 3-nerved; leaflets oblanceolate or cuneate, rounded to retuse, 7–25 mm. long, 2–4 mm. wide, with slightly incurved margins; chasmogamic flowers 6–7 mm. long, white with a purple blotch on the center of the standard, few, fasciculate in the leaf-axils; calyx 3–4 mm. long with narrowly lanceolate lobes; legumes ovate, as long as the calyx;

cleistogamic flowers few, fasciculate in leaf-axils; calyx 1.5–2 mm. long; legumes nearly orbicular, about 3 mm. long, longer than the calyx.——Aug.–Oct. Waste ground and grassy places in lowlands; Hokkaido, Honshu, Shikoku, Kyushu; common.——Korea, Manchuria, China, Ryukyus, Formosa, to India and Australia.

Var. **serpens** (Nakai) Ohwi. *L. serpens* Nakai; *L. sericea* var. *latifolia* Maxim., pro parte; *L. prostrata* Nakai, non Pursh ——HAI-MEDO-HAGI. Stems prostrate, branching, the branchlets thicker, spreading-pilose; leaflets thicker and broader; flowers with a purplish standard, the blotch at the center larger.——Aug.–Oct. Common in grassy places and abandoned lawns, especially abundant near the sea; Honshu, Kyushu.——China.

11. **Lespedeza juncea** (L. f.) Pers. *Hedysarum junceum* L. f.; *Trifolium cytisoides* Pall.; *T. hedysaroides* Pall.; *L. cytisoides* (Pall.) Nakai; *L. hedysaroides* (Pall.) Kitag.—— KARA-MEDO-HAGI, INU-MEDO-HAGI. Erect stiff perennial; stems slender, sparsely short white appressed-pilose; stipules subulate, small; leaflets oblanceolate to narrowly obovate, 10–15 mm. long, 2.5–6 mm. wide, rounded to retuse, pilose beneath; chasmogamic flowers 7–8 mm. long, yellowish white, with a purple blotch on the standard, several in fascicles in leaf-axils; calyx 4–5 mm. long with lanceolate, bristly-acuminate lobes; cleistogamic flowers several in leaf-axils; calyx 3–4 mm. long, the lobes lanceolate, 3-ribbed; legumes broadly ovate, pilose, about 3 mm. long.——Sept.–Oct. Honshu (Kantō Distr. and northw.); rare.——Korea, Manchuria, n. China, and e. Siberia.

12. **Lespedeza pilosa** (Thunb.) Sieb. & Zucc. *Hedysarum pilosum* Thunb.; *Desmodium pilosum* (Thunb.) DC.; *L. pilosa* var. *latifolia* Koidz.; *L. pilosa* var. *erecta* Hatus.; *L. sylvestris* Hatus.——NEKO-HAGI. Prostrate to trailing, sparsely branched perennial; branches elongate, slender, loosely leaved, 60–100 cm. long, with spreading whitish pubescence; stipules lanceolate, recurved, acuminate; leaflets membranous, broadly obovate, 1–2 cm. long, 8–15 mm. wide, retuse to rounded, green, pubescent beneath, sparsely pubescent to glabrate on upper side; racemes very short, sessile; chasmogamic flowers yellowish to white, few, 7–8 mm. long; calyxlobes nearly linear with a slightly broadened base; cleistogamic flowers 1–3 in upper leaf-axils; calyx about 1.5 mm. long; legumes broadly ovate, 3–4 mm. long, with long ascending pubescence.——July–Sept. Thickets and grassy places in lowlands and hills; Honshu, Shikoku, Kyushu; rather common. ——Korea and China.

L. × intermixta Makino. *L. juncea* var. *latifolia* Maxim., pro parte; *L. latissima* (Matsum.) Nakai; *L. juncea* var. *sericea* forma *latissima* Matsum.——TSURU-MEDO-HAGI. Hybrid of No. 10 and No. 12. Differs from *L. pilosa* in the scarcely prostrate stems and narrowly ovate to oblong leaflets with appressed to ascending pubescence. From *L. juncea* it differs in the scarcely erect stems, longer hairs, and the larger broader leaflets; legumes unknown.——Reported from Honshu.—— Ryukyus and Korea.

15. KUMMEROWIA Schindl. YAHAZU-SŌ ZOKU

Rather short-pilose annuals; leaves digitately 3-foliolate, the stipules soft-membranous, the leaflets subentire, exstipellate; flowers of 2 forms, chasmogamic and cleistogamic, both small, axillary, pedicelled, with a single bracteolelike bract subtending the calyx, the bracteoles persistent, ovate; calyx 5-lobed, the lobes pinnate-nerved; legumes flat, sessile, 1-seeded.——Two species, in e. Asia.

1A. Stems with downwardly curved, short, white hairs; leaves appressed-pilose on midrib and margin, nearly all alike, the leaflets obovate-oblong to narrowly oblong, obtuse or subacute; bracts and bracteoles 5- to 7-nerved; calyx loosely short-pilose; legumes slightly longer than the calyx, abruptly acute. . . . . 1. *K. striata*

1B. Stems with upwardly appressed short hairs; leaflets with obliquely spreading hairs on the midrib beneath and on margin; lower leaves spreading, the leaflets obovate, usually retuse, the upper leaves appressed to the stems or suberect; bracts and bracteoles 1- to 3-nerved; calyx glabrous; legumes twice as long as the calyx or longer, rounded at apex. . . . . 2. *K. stipulacea*

1. Kummerowia striata (Thunb.) Schindl. *Hedysarum striatum* Thunb.; *Lespedeza striata* (Thunb.) Hook. & Arn.; *Microlespedeza striata* (Thunb.) Makino; *M. makinoi* C. Tanaka——YAHAZU-SŌ. Much-branched, erect annual with slender branches; stipules narrowly ovate, few-nerved, 5–8 mm. long, erect, persistent, pale brown; leaves with very short petioles, the leaflets thin, 10–15 mm. long, 5–8 mm. wide; chasmogamic flowers about 5 mm. long, reddish purple, the bracteoles as long as the calyx-tube, ovate, subobtuse; calyx 3–3.5 mm. long, 5-lobed, the lobes ovate, subacute; calyx of cleistogamic flowers similar to those of the chasmogamic; legumes flat, nearly orbicular, about 3.5 mm. long, with short appressed scattered hairs.——Waste ground and roadsides;

Hokkaido, Honshu, Shikoku, Kyushu; common.——Manchuria, China, Korea, Ryukyus, and Formosa; naturalized in N. America.

2. Kummerowia stipulacea (Maxim.) Makino. *Lespedeza stipulacea* Maxim.; *Microlespedeza stipulacea* (Maxim.) Makino; *L. striata* var. *stipulacea* (Maxim.) Deb.; *L. striata* forma *adpressa* Matsum.; *L. striata* Maxim. pro parte, non Hook. & Arn.——MARUBA-YAHAZU-SŌ. Resembles the preceding; stems more profusely branched; leaves of 2 forms, the lower spreading, the upper nearly erect or appressed, rather densely arranged.——Aug.–Oct. Waste grounds and roadsides; Honshu, Kyushu; common.——Ussuri, n. China, Korea, and Manchuria.

16. DESMODIUM Desv. NUSUBITO-HAGI ZOKU

Herbs or shrubs, rarely small trees; leaves usually with free, linear or subulate stipules, the leaflets 1–7, pinnate, stipellate; flowers reddish purple or white, usually small, on loose, simple or compound racemes, the bracts various, the bracteoles sometimes present; calyx 5-toothed; filaments often connate into a tube; legumes sessile or long-stiped, flat, usually indehiscent, the joints smooth, glabrous or hairy.——About 180 species, in the warmer regions of the world.

1A. Calyx deeply 5-lobed, the segments or lobes much longer than wide; legumes sessile, shallowly constricted between the joints; flowers 3–5 mm. long; leaflets 3.

  2A. Flowers bracteolate; legumes with rusty pubescence; joints narrowly oblong, about 13 mm. long; erect shrub, horizontally spreading on upper part; terminal leaflets 7–12 cm. long. . . . . 1. *D. caudatum*

  2B. Flowers without bracteoles; legumes with pale minute pubescence; joints broadly elliptic, 3–4 mm. long; decumbent or prostrate, herbaceous or suffrutescent; terminal leaflets less than 3 cm. long.

    3A. Leaflets 5–15 mm. long; racemes 5- to 10-flowered. . . . . 2. *D. microphyllum*

    3B. Leaflets 2.5–3 cm. long; racemes many-flowered. . . . . 3. *D. heterocarpon*

1B. Calyx shallowly lobed with depressed deltoid teeth shorter than wide; legumes stiped, prominently constricted on lower side between the semi-obovate joints; flowers 3–10 mm. long; leaflets 3–7.

  4A. Leaflets 7, sometimes 5; flowers 8–10 mm. long; joints of legumes 12–15 mm. long. . . . . 4. *D. oldhamii*

  4B. Leaflets always 3.

    5A. Principal lateral nerves of leaflets not reaching margin; stipules deltoid-lanceolate, 2–4 mm. wide at base, semipersistent; flowers 5–7 mm. long.

      6A. Leaflets ovate; terminal leaflets only slightly larger than the lateral; filament-tube 5–6 mm. long; filament-tube, 10–18 mm. long, much longer than the stipe of legume.

        7A. Leaflets with raised, puberulous veinlets beneath; joints of legume 8–13 mm. long. . . . . 5. *D. laxum*

        7B. Leaflets nearly glabrous, without raised veinlets beneath; joints of legume 13–16 mm. long. . . . . 6. *D. leptopus*

      6B. Leaflets oblong-ovate to broadly lanceolate; terminal leaflets usually much larger than the lateral; filament-tube about 4 mm. long, only slightly shorter than the stipe of legume. . . . . 7. *D. laterale*

    5B. Principal lateral nerves of leaflets reaching margin; stipules subulate, usually not exceeding 1 mm. wide at base, commonly early deciduous; flowers 3–4 mm. long.

      8A. Leaflets broadest below the middle, ovate to oblong-ovate.

        9A. Stipe of legume short, 1–3 mm. long; leaves sparse on the stem. . . . . 8. *D. oxyphyllum*

        9B. Stipe of legume 5–7 mm. long; leaves approximate below middle of stem. . . . . 9. *D. fallax*

      8B. Leaflets broadest at or above the middle, broadly obovate; leaves sparse on stem. . . . . 10. *D. podocarpum*

1. Desmodium caudatum (Thunb.) DC. *Hedysarum caudatum* Thunb.; *H. laburnifolium* Poir.; *Catenaria laburnifolia* (Poir.) Benth.; *D. laburnifolium* (Poir.) DC.; *C. caudata* (Thunb.) Schindl.——MISO-NAOSHI. Short-pilose shrub with ascending to spreading branches and branchlets; stipules subulate; petioles 3–5 cm. long, slightly flattened and without groove on upper side; leaflets broadly lanceolate, 7–12 cm. long, acute, the lateral ones slightly smaller, deep green on upper side, slightly glaucous and with slightly raised veinlets beneath; racemes terminal and axillary, in leafy panicles, the bracteoles lanceolate, subtending the calyx, 0.5–0.8 mm. long; flowers 5–6 mm. long; legumes 5–7 cm. long, flat, the joints 4–6, slightly contracted at each end, narrowly oblong, 8–13

mm. long, 7–8 mm. wide, with rusty-brown uncinate hairs. ——Aug.–Sept. Woods; Honshu (Suruga Prov. and Chūgoku Distr.), Shikoku, Kyushu.——s. Korea, Ryukyus, Formosa, China to India, and Malaysia.

2. Desmodium microphyllum (Thunb.) DC. *Hedysarum microphyllum* Thunb.; *D. parvifolium* DC.——HIME-NO-HAGI. Decumbent, herbaceous or suffrutescent, with slender, elongate, loosely branched stems; stipules thinly membranous, lanceolate; leaflets elliptic, 5–10 mm. long, the lateral ones slightly smaller, loosely appressed-pilose beneath; racemes very loosely few-flowered, 1–5 cm. long, the axis flexuous, slender, the pedicels slender, 5–8 mm. long, with scattered spreading glandular hairs, the bracts similar to the stipules, deciduous; flowers 4–5 mm. long, reddish purple; legumes to 15 mm. long, sessile, the joints 3–4, minutely puberulous, elliptic, 3–4 mm. long, about 3 mm. wide.——Aug.–Oct. Honshu (centr. distr. and westw.), Shikoku, Kyushu; rare. ——Ryukyus, Formosa, China, India to the Philippines.

3. **Desmodium heterocarpon** (L.) DC. *Hedysarum heterocarpon* L.; *D. buergeri* Miq.; *D. heterocarpon* var. *buergeri* (Miq.) Hosokawa——SHIBA-HAGI. Diffuse suffrutescent shrub with procumbent or ascending slender branches, sparsely appressed-pilose while young; stipules lanceolate, 8–12 mm. long, long-acuminate, many-nerved; leaflets rounded to retuse, the terminal obovate, 2.5–3 cm. long, 1.5–2 cm. wide, the lateral obovate to oblong, 1.5–2.5 cm. long, 1–1.5 cm. wide, pilose while young; racemes 6–8 cm. long, with spreading hairs, the bracts narrowly ovate, caducous; flowers reddish purple, 4–5 mm. long; calyx rather deeply lobed; legumes about 15 mm. long, erect, sessile, the joints 4–5, about 3 mm. long and as wide, flat, with minute uncinate whitish hairs. ——Sept.–Oct. Grassy slopes and roadsides in lowlands; Honshu (centr. distr. and westw.), Shikoku, Kyushu; locally common.——Bonins, Korea (Quelpaert Isl.), Ryukyus, Formosa, China to India and Malaysia.

Var. **patulepilosum** (Ohwi) Ohwi. *D. buergeri* var. *patulepilosum* Ohwi. TACHIGE-SHIBA-HAGI. Hairs spreading on stems and branches.——Honshu (Kii Prov.).

4. **Desmodium oldhamii** Oliv. FUJI-KANZŌ. Puberulous and loosely hirsute, simple, erect perennial 50–120 cm. high; stipules linear-lanceolate, few-nerved; leaflets 5 or 7, ovate to oblong, 8–16 cm. long, 3–6 cm. wide, abruptly acuminate, the terminal slightly larger; racemes 1–5, terminal, elongate, sometimes to 30 cm. long, the bracts deciduous, the pedicels 7–12 mm. long; filament-tubes about 7 mm. long; legumes on a stipe 8–10 mm. long, the joints 1 or 2, broadly semi-obovate, 10–12 mm. long, with short uncinate pubescence on both surfaces.——Aug.–Sept. Thickets and woods; Honshu, Shikoku, Kyushu.——Korea, Manchuria, and China.

5. **Desmodium laxum** DC. *D. podocarpum* var. *laxum* (DC.) Baker; *D. austrojaponense* Ohwi; *D. tashiroi* sensu auct. Japon., pro parte, non Matsum.; *? D. laxum* var. *kiusianum* Matsum.; *D. gardneri* sensu auct. Japon., pro parte, non Benth. ——ŌBA-NUSUBITO-HAGI, SAIKOKU-NUSUBITO-HAGI. Perennial herb slightly woody at base, 60–100 cm. high; stems erect, usually simple, slightly pilose; stipules lanceolate, 3- to 5-nerved, 3–5 mm. long, deciduous; leaves chartaceous, deep green on upper side, slightly paler beneath, with appressed hairs especially on midrib on upper side, the terminal leaflets ovate to broadly so, 6–10 cm. long, 3–5 cm. wide, acuminate, the lateral leaflets slightly smaller, rounded-truncate at base on outer margin; racemes loosely flowered, 10–30 cm. long, short-pilose, the bracts subulate, minute; flowers about 7 mm. long, reddish purple; calyx-lobes depressed; legumes with stipe 10–15 mm. long, the joints 2 or 3, semi-obovate, 8–13 mm. long, minutely pubescent.——Aug.–Oct. Woods; Honshu (Tōtōmi Prov.), Shikoku, Kyushu; rare.——China and Formosa to India and Malaysia.

6. **Desmodium leptopus** A. Gray. *D. gardneri* Benth.;

D. tashiroi Matsum.——TOKIWA-YABU-HAGI. Closely allied to the preceding; leaflets glabrescent, with scarcely raised veinlets beneath; flowers slightly larger; legumes larger and the joints 13–16 mm. long.——Aug.–Oct. Kyushu (Yakushima and Tanegashima); rare.——Ryukyus, Formosa, China to India.

7. **Desmodium laterale** Schindl. *D. laxum* sensu auct. Japon., non DC.; *D. polycarpon* sensu auct. Japon., pro parte, non DC.——RYŪKYŪ-NUSUBITO-HAGI. Closely allied to No. 5 but smaller; leaves pale green and loosely short-pilose on the more prominently raised veinlets beneath, the terminal leaflets 5–10 cm. long, 1.5–3 cm. wide, the lateral ½–¾ as long as the terminal, scarcely or much dilated on outer margin at base.——Kyushu (Osumi Prov. and southw.).—— Ryukyus, Formosa, and China.

8. **Desmodium oxyphyllum** DC. *Hedysarum racemosum* Thunb., non Aubl.; *D. racemosum* (Thunb.) DC.; *D. thunbergii* DC. pro parte?; *D. podocarpum* var. *typicum*, var. *japonicum*, var. *polyphyllum*, var. *parvifolium* all of Maxim.; *D. japonicum* Miq., pro parte; *D. oxyphyllum* var. *japonicum* Matsum.——NUSUBITO-HAGI. Perennial herb, 60–120 cm. high, ligneous at base, thinly pilose, remotely leafy throughout; stipules linear to lanceolate, 3–7 mm. long, long-acuminate; leaflets 3, thin, acute, paler beneath, the terminal ovate to broadly so, 4–8 cm. long, 2.5–4 cm. wide, broadly cuneate at base, the lateral slightly smaller; racemes loosely flowered, to 30 cm. long, often slightly branched at base, the bracts deciduous; flowers 3–4 mm. long; calyx-teeth short; legumes on a stipe 2–8 mm. long, the joints 1–3, broadly semi-obovate, 5–7 mm. long, minutely pubescent on both surfaces.——July–Oct. Woods and thickets in lowlands; Hokkaido, Honshu, Shikoku, Kyushu; common.——Korea and China to N. India.

9. **Desmodium fallax** Schindl. *D. japonicum* Miq., pro parte; *D. japonicum* var. *dilatatum* Miq.; *D. podocarpum* var. *mandschuricum* Maxim.; *D. fallax* var. *mandschuricum* (Maxim.) Nakai; *D. fallax* var. *dilatatum* Nakai; *D. racemosum* var. *mandschuricum* (Maxim.) Ohwi; *D. racemosum* var. *dilatatum* (Nakai) Ohwi; *D. podocarpum* sensu auct. Japon., non DC.; *D. podocarpum* var. *membranaceum* Matsum.——YABU-HAGI. Closely resembles the preceding; leaves thinner, closely approximate near the middle of the stem, the leaflets narrowly ovate to ovate, less pubescent, prominently paler beneath.——Sept.–Oct. Woods and thickets in lowlands; Hokkaido, Honshu, Shikoku, Kyushu; common.—— Korea, Manchuria, Ussuri, and China.

10. **Desmodium podocarpum** DC. *D. japonicum* Miq. pro parte; *D. podocarpum* var. *japonicum* (Miq.) Maxim., pro parte; *D. podocarpum* var. *indicum* Maxim., pro parte; *D. podocarpum* var. *latifolium* Maxim. ex Matsum.; *D. oxyphyllum* var. *villosum* Matsum.; *D. racemosum* var. *villosum* (Matsum.) Ohwi; *D. maximowiczii* Makino——MARUBA-NUSUBITO-HAGI. Closely resembles No. 8; stems leafy throughout; leaves densely pubescent, the terminal leaflets broadly obovate.——Woods and thickets; Honshu, Shikoku, Kyushu; common.——Korea, Formosa, China to India.

17. SMITHIA Ait. SHIBA-NEMU ZOKU

Herbs with odd- or even-pinnate leaves, the leaflets small, exstipellate; stipules membranous, persistent, usually attached above the base and peltate; flowers yellowish or purplish, in axillary racemes, the bracts and bracteoles membranous or rather leaflike, persistent; calyx deeply 2-lobed, the upper lobes 2-toothed, the lower 3-toothed; legumes enclosed in the calyx, the joints few, strongly or slightly flattened, sometimes tuberculate.——Tropics of Asia and Africa.

1. **Smithia japonica** Maxim.——SHIBA-NEMU. Branched annual 30–50 cm. high; leaves ovate, 2–3 cm. long, the rachis ending in a short point, loosely hispid, the stipules pale brown, nerved, oblanceolate, peltate and attached near the center, the leaflets 8–14, membranous, oblanceolate, 5–12 mm. long, 2–4 mm. wide, rounded at apex, spine-tipped, loosely hispid on under side and margins; flowers bluish, on a short-peduncled subcapitate raceme, the bracteoles oblong, 4–5 mm. long, green, hispid-ciliate; calyx-lobes 5–6 mm. long, hispid-ciliate, many-nerved, the nerves connate at apex, the upper lobes ovate; legumes 7-jointed, the joints nearly orbicular, 1–1.5 mm. across, loosely papillate on both sides.——Sept.–Oct. Honshu (Suwō and Kii Prov.), Shikoku, Kyushu; rare.—— Ryukyus.

18. AESCHYNOMENE L.   KUSA-NEMU ZOKU

Herbs or suffrutescent shrubs; stipules ovate to subulate; leaves sometimes odd-pinnate, the leaflets many, exstipellate; flowers yellow, in axillary racemes, the bracts similar to the stipules, the bracteoles appressed to the calyx; calyx 2-lipped, the upper lobe 2-fid, the lower 3-fid; flowers with orbicular standard, the keel navicular; stamens diadelphous in 2 sets of 5; legumes stiped, the joints many, flattened, smooth or tubercled on the center, usually indehiscent.——About 70 species, in the Tropics and subtropics.

1. **Aeschynomene indica** L. KUSA-NEMU.   Green to somewhat glaucous, somewhat fleshy annual branched above, with scattered spreading hairs on stems, leaf-rachises, and peduncles, the hairs callose at base; stems erect, terete, green; leaves 5–10 cm. long, short-petioled, the stipules ovate to lanceolate, membranous, acuminate, attached slightly above the base, 7–12 mm. long, the leaflets 40–60, linear-oblong, 10–15 mm. long, 2–3.5 mm. wide, rounded at apex, glaucous beneath; racemes 2- or 3-flowered, loose, the bracts similar to the stipules but smaller, the bracteoles at the base of the calyx narrowly oblong, green; flowers about 1 cm. long, pale yellow; calyx membranous, about 5 mm. long, 2-lobed nearly to the base; legumes usually glabrous, linear, flattened, 3–5 cm. long, 5 mm. wide, the stipe 7–8 mm. long, the joints 6–8, obsoletely rugulose in the center on both sides.——July–Oct. Wet places near rivers and ponds in lowlands; Hokkaido, Honshu, Shikoku, Kyushu; common.——Tropics of Asia, Africa, and Australia.

19. HEDYSARUM L.   IWA-ŌGI ZOKU

Perennial herbs or suffrutescent shrubs; leaves odd-pinnate, the stipules membranous, the leaflets many, exstipellate, usually oblong; flowers reddish purple or pale yellow, in racemes, the bracts persistent or deciduous, the bracteoles linear, small; calyx-teeth equal or slightly unequal; keel nearly straight, obliquely truncate, unappendaged, longer than the wings; stamens diadelphous (9 and 1); legumes flat, several-jointed, the joints smooth or prickly, indehiscent.——About 70 species, mostly in cooler regions of the world.

1A. Flowers pale yellow; lateral teeth of calyx deltoid, ⅓–½ as long as the tube, the lower tooth shorter than the calyx-tube.
1. *H. vicioides*

1B. Flowers reddish purple; lateral teeth of calyx lanceolate, nearly as long as the tube, the lower tooth slightly longer than the calyx-tube.
2. *H. hedysaroides*

1. **Hedysarum vicioides** Turcz. *H. ussuriense* Schischk. & Komar; *H. esculentum* Ledeb.; *H. alpinum* var. *japonicum* B. Fedtsch.; *H. iwawogi* Hara——IWA-ŌGI, IWA-WŌGI, TATE-YAMA-ŌGI.   Tufted, more or less pubescent perennial with thick erect caudex; stems 25–80 cm. long, nearly simple or sparsely branched above; leaves petioled, 8–15 cm. long, the lower ones reduced to sheathing stipules; stipules membranous, deciduous, brown, many-nerved, on the ventral margin connate to form a narrowly lanceolate blade 12–20 mm. long, the leaflets 11–25, herbaceous to chartaceous, narrowly ovate or broadly lanceolate, 1.5–3 cm. long, 5–10 mm. wide, obtuse to rounded and obscurely mucronate at apex, with closely spaced slender lateral nerves, nearly glabrous above, usually pubescent and obscurely glandular-dotted beneath; racemes axillary, 3–8 cm. long, one-sided, many-flowered, the peduncles to 15 cm. long, the bracts caducous, narrowly to broadly lanceolate, 3–8 mm. long, 1–3.5 mm. wide, the pedicels 3–7 mm. long, white-pubescent, the bracteoles persistent, linear, shorter than the calyx; flowers 15–20 mm. long, pale yellow; keel longer than the standard and wings; legumes 3–4 cm. long, 3- or 4-jointed, the joints elliptic, flat, entire or obscurely undulate on margin, with reticulate venation on the faces, glabrous or minutely puberulous.——June–Aug. Sandy to gravelly slopes in high mountains, or in grassy places in lowlands in northern distr.; Hokkaido, Honshu (centr. distr. and northw.).——n. Korea to e. Siberia.

2. **Hedysarum hedysaroides** (L.) Schinz & Thell. *Astragalus hedysaroides* L.; *H. obscurum* L.; *H. sibiricum* Poir.; *H. neglectum* Ledeb.——CHISHIMA-GENGE, KARAFUTO-GENGE. Closely resembling the preceding, more prominently pubescent; leaflets slightly broader and fewer; flowers reddish purple; calyx-teeth longer; keel only slightly longer than the standard and wings; legumes glabrous to minutely pubescent. ——July–Aug. Hokkaido (Rebun Isl.).——Cooler regions and high mountains of n Eurasia.

20. VICIA L.   SORA-MAME ZOKU

Annuals or perennials usually with tendrils; leaves even-pinnate, terminated by a tendril or a mucro, exstipellate, the stipules usually semisagittate; flowers reddish purple, rarely yellowish, fasciculate in axils or in axillary racemes, the bracts small, the bracteoles absent; calyx often oblique, 5-lobed or 5-toothed, the lowest lobe or tooth sometimes longer; wings adnate to the middle of the keel; stamens diadelphous (9 and 1), the tube oblique at apex; style filiform, hairy all around or on back at apex, sometimes bearded under the stigma; legumes flattened, 2-valved, dehiscent, 1-locular, 2- to several-seeded, the seeds globular. ——About 150 species, in temperate northern regions of the N. Hemisphere and S. America.

1A. Flowers solitary or fasciculate in leaf-axils, usually without peduncles; style bearded below the stigma on outer side; stipules (in
　　 ours) with an obsolete depressed gland on the center.
　2A. Annual herbs, rarely perennial; calyx-teeth elongate, nearly equal, almost as long as the tube.
　　3A. Leaflets of upper leaves obovate or oblong; flowers about 2 cm. long; legumes 5–8 cm. long. 1. *V. sativa*
　　3B. Leaflets of upper leaves linear to oblong; flowers about 15 mm. long; legumes 3–5 cm. long. . . . . 2. *V. angustifolia*
　2B. Perennial herb; calyx-teeth short, unequal, ⅓–⅔ as long as the tube. . . . . 3. *V. sepium*
1B. Flowers in pedunculate racemes; styles short-pubescent all around; stipules not glandular.
　4A. Racemes few-flowered; flowers small, 3–5 mm. long.
　　5A. Legumes glabrous, 3- to 6-seeded; racemes 1- to 3-flowered. . . . . 4. *V. tetrasperma*
　　5B. Legumes short-pilose, 2-seeded; racemes 3- to 7-flowered. . . . . 5. *V. hirsuta*
　4B. Racemes many-flowered; flowers larger, 10–20 mm. long.
　　6A. Tendrils well developed, elongate; scandent or prostrate herbs.
　　　7A. Leaflets 12–28; scandent herbs.
　　　　8A. Annuals; limb of the standard distinctly shorter than the claw.
　　　　　9A. Plant soft-pubescent; flowers 15–17 mm. long; calyx-teeth subulate, the lower ones longer than the tube. . . . . 6. *V. villosa*
　　　　　9B. Plant loosely pubescent to glabrous, slender, but stiffer; flowers 10–15 mm. long; calyx-teeth shorter than the tube.
　　7. *V. dasycarpa* var. *glabrescens*
　　　　8B. Perennials; limb of the standard as long as the claw.
　　　　　10A. Plant dark reddish brown when dried; leaflets rather coriaceous, lanceolate to oblong, the lateral nerves forming an acute
　　　　　　　 angle with the midrib; stipules firm and dentate. . . . . 8. *V. amoena*
　　　　　10B. Plant green or yellowish green when dried; leaflets membranous to chartaceous; stipules rather soft, entire or bifid.
　　　　　　11A. Leaflets oblong to elliptic, chartaceous, the lateral nerves forming an angle of more than 40 degrees with the midrib.
　　　　　　　12A. Leaflets soft-pubescent, pale green; calyx-teeth distinct, nearly deltoid. 9. *V. japonica*
　　　　　　　12B. Leaflets nearly glabrous, yellowish green; calyx-teeth depressed. . . . . 10. *V. amurensis*
　　　　　　11B. Leaflets lanceolate to broadly linear, membranous, the lateral nerves ascending, forming an acute angle with the midrib.
　　11. *V. cracca*
　　　7B. Leaflets 4–10, fulvous-green when dried, chartaceous, with raised reticulate veinlets beneath; decumbent or prostrate perennial
　　　　 herb; tendrils absent in lower leaves. . . . . 12. *V. pseudo-orobus*
　　6B. Tendrils absent or nearly so; erect or ascending herbs.
　　　13A. Bracts not well developed or early deciduous.
　　　　14A. Leaflets 2, acute to obtuse, sometimes gradually acute, rather thick. . . . . 13. *V. unijuga*
　　　　14B. Leaflets more than 2.
　　　　　15A. Leaflets 4–6, elliptic or oblong, broadest near the middle, rather thick; tendrils sometimes slightly developed.
　　　14. *V. nipponica*
　　　　　15B. Leaflets 8–12, ovate to narrowly so or broadly lanceolate, broadest below the middle, thin; tendrils not well developed.
　　15. *V. venosa* var. *cuspidata*
　　　13B. Bracts ovate to broadly lanceolate, persistent, 3–5 mm. long; leaflets thin, long-acuminate.
　　　　16A. Leaflets 4–8. . . . . 16. *V. fauriei*
　　　　16B. Leaflets 2. . . . . 17. *V. bifolia*

1. Vicia sativa L. Ō-YAHAZU-ENDŌ, ZADOWIKKEN. Similar to the following, but stouter; climbing or trailing, sparsely pubescent annual or biennial to more than 1 m. high; leaflets 12–14, obovate to oblong, 10–20 mm. long, truncate to retuse and mucronate at apex; flowers usually 2 cm. or more long; calyx-teeth nearly equal; legumes 5–8 cm. long.——Cultivated for green manure.——w. Asia and Europe and widely cultivated elsewhere.

2. Vicia angustifolia L. var. **segetalis** (Thuill.) C. Koch. *V. segetalis* Thuill.; *V. sativa* var. *segetalis* (Thuill.) Ser.; *V. angustifolia* forma *segetalis* (Thuill.) Thuill. ex H. Boiss.; *V. sativa* sensu auct. Japon., non L.——YAHAZU-ENDŌ. Scandent, slightly pubescent annual; stems 60–150 cm. long; stipules bifid; leaflets 12–14, membranous, oblanceolate to broadly linear, 2–3 cm. long, 4–6 mm. wide, truncate to retuse and mucronate at apex, the lateral nerves forming an acute angle with the midrib; flowers reddish purple, 12–18 mm. long; calyx 8–12 mm. long, the teeth less than half as long as the tube, broadly linear, acuminate; legumes flat.——Common throughout Japan.——Forma **normalis** (Makino) Ohwi. *V. sativa* var. *normalis* Makino——TSURU-NASHI-YAHAZU-ENDŌ. Plant without tendrils. Rare in Japan.

Var. **minor** (Bertol.) Ohwi. *V. maculata* var. *minor* Bertol.; *V. angustifolia* var. *cordata* Boiss.; *V. angustifolia* sensu auct. Japon., non L.——HOSOBA-YAHAZU-ENDŌ. Of more slender habit; leaflets linear-lanceolate to nearly linear, 15–25 mm.

long, 2–5 mm. wide, rounded to truncate at apex, the lower ones usually obovate, 5–10 mm. long, retuse; flowers slightly smaller.——More rare in our area than the var. *segetalis*.——Apr.–June. Grassy places and waste ground in lowlands; Honshu, Shikoku, Kyushu; common. The typical phase occurs widely in temperate regions of the world.

3. Vicia sepium L. KARASU-NO-ENDŌ. Scandent perennial herb, with slender stolons; stems 60–120 cm. long, slightly pubescent; stipules dentate or bifid; leaflets 12–14, membranous, narrowly ovate, 15–30 mm. long, 10–15 mm. wide, obtuse to emarginate, indistinctly ciliate, the lateral nerves ascending and forming a rather acute angle with the midrib; flowers 1 or 2, 15–17 mm. long, violet-purple; calyx 7–8 mm. long, slightly pubescent, the lower tooth slightly elongate, lanceolate, 1/3–2/3 as long as the tube; legumes 3–4 cm. long, 8–9 mm. wide, glabrous.——May–July. Naturalized in Hokkaido and Honshu (Mount Ibuki in Oomi Prov.).——Europe.

4. Vicia tetrasperma (L.) Schreb. *Ervum tetraspermum* L.——KASUMAGUSA. Slender, slightly pubescent biennial; stems 30–60 cm. long; leaves nearly sessile, the stipules semi-sagittate, the leaflets 8–12, thin, linear-oblong, 12–17 mm. long, 2–4 mm. wide; peduncles axillary, 2–3 cm. long, slender, the bracts minute; flowers purple or bluish purple, about 5 mm. long; legumes oblong to elliptic, glabrous, 8–10 mm. long, about 4 mm. wide, 3- to 6-seeded.——Apr.–May. Waste grounds and grassy places in lowlands; Honshu, Shikoku, Kyushu; common.——Ryukyus, Formosa, Korea, China to Europe.

5. Vicia hirsuta (L.) S. F. Gray. *Ervum hirsutum* L.; *V. coreana* Lév.; *V. taquetii* Lév.——SUZUME-NO-ENDŌ. Loosely pubescent, scandent annual or biennial; stems 30–60 cm. long; leaves nearly sessile, the stipules bifid or dentate, the leaflets 12–14, thin, lanceolate, rounded to truncate at apex, 10–17 mm. long, 2–3 mm. wide; peduncles slender, 2–3 cm. long, loosely flowered; flowers pale bluish purple, 3–4 mm. long; legumes oblong, pubescent, about 8 mm. long, 3 mm. wide, 2-seeded.——Apr.–May. Waste grounds; Honshu, Shikoku, Kyushu; common.——Eurasia and N. Africa.

6. Vicia villosa Roth. BIRŌDO-KUSA-FUJI. Prominently pubescent, scandent annual or biennial; stems 80–120 cm. long; leaves very short-petioled, 8–12 cm. long, the stipules obliquely or semisagittate, the leaflets 16–24, oblong or elliptic to ovate, 15–20 mm. long, 7–10 mm. wide, soft; racemes 8–12 cm. long inclusive of the peduncle, densely many-flowered, one-sided; flowers about 15 mm. long, purple; calyx-teeth unequal, the lower elongate, linear, slightly longer than the tube; legumes flat, glabrous, narrowly oblong.——May–July. A European species, cultivated for green manure.

7. Vicia dasycarpa Tenore var. **glabrescens** (C. Koch) G. Beck. *V. villosa* var. *glabrescens* C. Koch——NAYO-KUSA-FUJI. Resembles the preceding but a much more slender, less pubescent, scandent herb with rather stiffer and more slender stems and shorter calyx-teeth.——June–July. A European plant sometimes grown with the preceding.

8. Vicia amoena Fisch. *V. rapunculus* Deb.; *V. amoena* var. *oblongifolia* Regel; *V. amoena* var. *lanata* Fr. & Sav.; *V. amoena* var. *sachalinensis* F. Schmidt ex Nakai——TSURU-FUJI-BAKAMA. Thinly pubescent or glabrescent, a scandent perennial herb; stems 80–180 cm. long; leaves very short-petioled, 8–15 cm. long, the stipules coarsely dentate, rather large, the leaflets 10–16, oblong to lanceolate, 15–30 mm. long, 4–10 mm. wide, rounded to subacute; racemes 4–8 cm. long, densely many-flowered, one-sided, on peduncles 2–7 cm. long; flowers purple, 12–15 mm. long, the bracts minute; calyx-teeth unequal, the lower tooth broadly linear, shorter than to nearly as long as the tube; legumes narrowly oblong, flat, 2–2.5 cm. long, about 5 mm. wide; glabrous.——Aug.–Oct. Thickets in lowlands and low mountains; Hokkaido, Honshu, Kyushu; rather common.——Siberia, China, Korea, Manchuria, Sakhalin, and s. Kuriles.

9. Vicia japonica A. Gray. *V. japonica* var. *comosa* H. Boiss.; *V. pallida* var. *japonica* (A. Gray) Maxim. ex Matsum. ——HIROHA-KUSA-FUJI. Allied to the preceding but slightly smaller and with more slender stems; stipules small, bifid; leaflets 10–14, elliptic to oblong, 10–20 mm. long, 6–10 mm. wide, rounded to retuse; racemes 2–3 cm. long, rather many-flowered; flowers pale bluish purple; calyx-teeth short; legumes 2–2.5 cm. long, about 6 mm. wide, several-seeded.——Apr.–Aug. Thickets and grassy places, seashores to the mountains; Hokkaido, Honshu; rather common.——Korea, Sakhalin, and s. Kuriles.

10. Vicia amurensis Oettingen. *V. pallida* Turcz., non Hook. & Arn.; *V. japonica* sensu Komar., non A. Gray; *V. vaniotii* Lév.——NOHARA-KUSA-FUJI. Intermediate between the two preceding; scandent perennial herb; stems 80–150 cm. long, short-pubescent, glabrate; stipules dentate or bifid; leaflets 15–30 mm. long, 8–12 mm. wide; racemes 2–5 cm. long.

——July–Sept. Hills and low mountains; Honshu (Kai and Shinano Prov.), Kyushu; rather rare.——Korea, Manchuria, Amur, and Ussuri.

11. Vicia cracca L. *V. cracca* var. *canescens* Maxim. ex Fr. & Sav.; *V. cracca* var. *japonica* Miq.; *V. oiana* Honda—— KUSA-FUJI. Rather soft, somewhat pubescent, vivid or gray-green perennial; stems scandent, 80–150 cm. long; leaves nearly sessile, 8–15 cm. long, the stipules bifid, lanceolate, the leaflets 18–24, lanceolate to broadly linear, 15–30 mm. long, 2–6 mm. wide; racemes densely many-flowered, 6–15 cm. long inclusive of the peduncle; flowers bluish purple, one-sided, 10–12 mm. long; lowest calyx-tooth shorter than the tube; legumes narrowly oblong, 2–3 cm. long, 6–7 mm. wide, glabrous, several-seeded.——May–July. Thickets and grassy places; Hokkaido, Honshu, Shikoku, Kyushu; common on hills. Widely distributed in temperate and cooler regions.

12. Vicia pseudo-orobus Fisch. & Mey. *V. pseudo-orobus* var. *tanakae* (Fr. & Sav.) Makino——ŌBA-KUSA-FUJI. Slightly pubescent to glabrate, prostrate to procumbent perennial; stems 80–150 cm. long; leaves very short-petioled, the stipules semi-ovate, dentate or 2-fid, the leaflets 4–10, chartaceous, often subalternate on axis, ovate, 3–5 cm. long, 15–30 mm. wide, acute to obtuse, yellowish brown when dry, the tendrils sometimes simple; racemes 4–6 cm. long, rarely sparsely branched at base, many-flowered, long-peduncled; flowers bluish purple, 13–15 mm. long; calyx-teeth depressed, deltoid, the lowest ones 0.5–0.8 mm. long; legumes short-stiped, narrowly oblong, flat, 25–30 mm. long, 6–7 mm. wide, glabrous, several-seeded, the stipe 2–3 mm. long, included within the calyx.——Aug.–Oct. Hokkaido, Honshu, Shikoku, Kyushu.——n. China and e. Siberia.

13. Vicia unijuga A. Br. *Orobus lathyroides* L.; *Ervum unijugum* (A. Br.) Alef.; *Lathyrus messerschmidtii* Fr. & Sav.; *V. austrohigoensis* Honda——NANTEN-HAGI. Perennial, slightly pubescent while young; stems erect to ascending, angular, 60–100 cm. long, branched above; leaves short-petioled, the stipules dentate or bifid, the leaflets 2, coriaceous to chartaceous, ovate to broadly lanceolate, 4–7(–10) cm. long, (8–)15–40 mm. wide, gradually acute, sometimes subobtuse; racemes 2–4 cm. long, one-sided, the peduncles variable in length, to 6 cm. long, or absent, sometimes branched at base; flowers bluish purple, 12–15 mm. long, the bracts inconspicuous; calyx-teeth short; legumes broadly lanceolate, 2.5–3 cm. long, 5–6 mm. wide, nearly sessile, glabrous.——June–Oct. Thickets and grassy places in lowlands and low mountains; Hokkaido, Honshu, Shikoku, Kyushu; very common and variable.——e. Siberia, China, Korea, and Sakhalin.——Forma **angustifolia** Makino. FUJIGAE-SŌ. Phase with narrow linear leaflets.

14. Vicia nipponica Matsum. *V. venosa* var. *capitata* Fr. & Sav.; *V. capitata* (Fr. & Sav.) Koidz., non Nakai—— YOTSUBA-HAGI. Perennial herb, short-pubescent while young, the stems rather stiff, angular, ascending, 30–80 cm. long; leaves petioled, the stipules semi-ovate, dentate, rather large, the leaflets 4–6, firmly chartaceous, elliptic, 2.5–5 cm. long, 15–30 mm. wide, with prominently raised veinlets beneath, the tendrils simple, short, often reduced to a spinelike point; racemes densely flowered, 1–3 cm. long, often branched at base, on peduncles about 5 cm. long; flowers bluish purple, 10–12 mm. long; calyx-teeth depressed-deltoid; legumes as in the preceding species.——July–Oct. Hokkaido, Honshu, Shikoku, Kyushu.——Korea and China.

15. Vicia venosa (Willd.) Maxim. var. **cuspidata** Maxim. *V. sexajuga* Nakai; *V. deflexa* Nakai; *V. senanensis* Nakai; *V. subcuspidata* (Nakai) Nakai; *V. venosa* var. *subcuspidata* Nakai——EBIRA-FUJI. Erect, slightly puberulous perennial; stems angular and sparsely branched above, 80–100 cm. long; leaves short-petioled, the stipules semi-ovate, dentate, or narrow and bifid, the leaflets 8–12, rather thin, lanceolate or narrowly ovate, 2.5–6 cm. long, 8–10 mm. wide, long-acuminate, the tendrils reduced to spines; racemes 2–4 cm. long, rather short-peduncled, one-sided, several to many-flowered; flowers 12–15 mm. long, bluish purple, the bracts inconspicuous; calyx-teeth short.——May–Sept. Woods; Honshu (centr. distr. and westw.), Kyushu (rare).——Forma **minor** (Nakai) Ohwi. *V. venosa* var. *minor* Nakai. With smaller ovate leaflets 10–25 mm. long, 8–15 mm. wide, acute.——Honshu (Chūgoku Distr.) and Kyushu. Intermediates often occur between the typical phase and var. *cuspidata*.——The typical phase occurs in e. Siberia, Manchuria, and Korea.

16. Vicia fauriei Franch. *V. venosa* var. *fauriei* (Franch.) Okuyama; *V. fauriei* var. *subsessilis* Nakai——TSU-GARU-FUJI. Erect, slightly pubescent perennial herb; stems angular, 60–80 cm. long, often branched above; leaves very short-petioled, the stipules obliquely ovate or semi-ovate, dentate, the leaflets 4–8, thinly chartaceous, ovate to broadly lanceolate, 4–10 cm. long, 2–3 cm. wide, long-acuminate or very acute, with reticulate veinlets especially prominent beneath; racemes densely flowered, one-sided, 2–3 cm. long, often branched at base, the peduncles absent or to 6 cm. long, the bracts lanceolate to ovate, green, 5–8 mm. long, acuminate, often serrate; flowers bluish purple, 12–15 mm. long; calyx-teeth short; legumes glabrous.——July–Aug. Hokkaido, Honshu (Echigo Distr. and northw.).——China.

17. Vicia bifolia Nakai. *V. unijuga* var. *bracteata* Fr. & Sav.; *V. bracteata* (Fr. & Sav.) Koidz.; *V. fauriei* var. *unijuga* Matsum.——MIYAMA-TANI-WATASHI. Resembling *V. unijuga*, but the bracts narrowly ovate to lanceolate, acuminate, sometimes toothed, 3–8 mm. long; leaves thinner, the leaflets acuminate.——June–Aug. Mountains; Honshu (centr. distr.); rather rare.——Korea.

21. LATHYRUS L.　RENRI-SŌ ZOKU

Erect or scandent herbs; leaves even-pinnate, the stipules often leaflike, sagittate, the rachis often ending in a tendril, the leaflets 2 to few, entire, rarely reduced; flowers usually showy, reddish purple, sometimes yellowish, in axillary racemes, the bracts minute, the bracteoles absent; wings nearly or completely free from the keel; stamens diadelphous (9 and 1), or sometimes monadelphous, the tube of the filaments scarcely oblique at apex; legumes flat, 2-valved, dehiscent, several to many-seeded.——About 100 species, in the temperate and cooler regions of the N. Hemisphere and S. America.

1A. Stems wingless, angular or ribbed.
  2A. Leaflets 2; flowers clear yellow. .... 1. *L. pratensis*
  2B. Leaflets 4–8, rarely 10.
    3A. Flowers many, yellow, changing to brown; calyx-teeth depressed-deltoid, much shorter than the tube; legumes 8–10 cm. long.
                                                        2. *L. davidii*
    3B. Flowers 3–6, purple; calyx-teeth (at least the lowest) as long as to longer than the tube; legumes 5–6 cm. long. .. 3. *L. maritimus*
1B. Stems winged.
  4A. Stems erect; tendrils simple; leaflets rather coriaceous, with raised veinlets on both sides. .... 4. *L. quinquenervius*
  4B. Stems often decumbent; tendrils usually branched; leaflets rather membranous, with raised veinlets only on the under side.
                                                         5. *L. palustris* var. *pilosus*

1. Lathyrus pratensis L. KIBANA-NO-RENRI-SŌ. Loosely pubescent to glabrous, rather slender perennial herb 50–100 cm. high; stems angular; stipules narrowly ovate, broadly lanceolate or semisagittate, acuminate, 15–30 mm. long; leaflets 2, oblong, subacute, 15–35 mm. long, 5–10 mm. wide, with palmate, nearly parallel nerves, the tendrils branched; racemes long-peduncled, 5- to 10-flowered, the bracts minute; flowers clear yellow, 15–20 mm. long; calyx-teeth rather equal, elongate-deltoid, as long as or slightly shorter than the tube; legumes broadly linear, 3–4 cm. long, about 6 mm. wide, glabrous.——Sparingly naturalized; Honshu (Mount Ibuki in Oomi Prov.).——Europe and nw. Asia.

2. Lathyrus davidii Hance. ITACHI-SASAGE. Nearly glabrous perennial herb; stems 80–120 cm. long, striate or obtusely angled in the upper part; leaves short-petioled, the stipules semisagittate, sometimes dentate, 15–30 mm. long, the leaflets 4–8, membranous, ovate or elliptic, obtuse, 3–8 cm. long, 2–4 cm. wide, nearly penninerved, glaucous beneath, the tendrils 2- or 3-branched; racemes many-flowered, 5–15 cm. long, on peduncles 5–10 cm. long; flowers yellow, changing to brown, 15–18 mm. long, suddenly curved upward near the middle; calyx-teeth depressed-triangular; legumes linear, flat, 8–10 cm. long, 5–6 mm. wide, glabrous, many-seeded.——

July–Aug. Woods and thickets in mountains; Hokkaido, Honshu (centr. distr. and northw.), Kyushu.——n. China, Manchuria, Ussuri, and Korea.

3. Lathyrus maritimus (L.) Bigel. *Pisum maritimum* L.; *L. japonicus* Willd.——HAMA-ENDŌ. Glabrous to soft-pubescent, glaucous perennial; stems 20–60 cm. long, angular, ascending from long-creeping, slender rhizomes; leaves short-petioled, the stipules obliquely ovate or semisagittate, 15–25 mm. long, very acute, the leaflets 8–12, ovate to oblong or elliptic, 15–30 mm. long, 10–20 mm. wide, acute to obtuse, nearly penninerved, the tendrils simple or 2- or 3-branched; racemes long-peduncled; flowers 25–30 mm. long, purple, rarely white; calyx-teeth lanceolate, the lower as long as to slightly longer than the tube; legumes narrowly oblong, about 5 cm. long, 1 cm. wide, glabrous, several-seeded.——Apr.–July. Sand dunes near seashores; Hokkaido, Honshu, Shikoku, Kyushu; common.——Throughout the N. Hemisphere.

4. Lathyrus quinquenervius (Miq.) Litv. *Vicia quinquenervia* Miq.; *L. palustris* var. *sericea* Franch.——RENRI-SŌ. Perennial herb short-pubescent while young; stems erect, 40–80 cm. long, 3-angled, the wings 1–2 mm. wide; leaves petioled, the stipules lanceolate, bifid, caudately long-acuminate, 10–25 mm. long, 1–3 mm. wide, the leaflets 2–6, linear to linear-lanceolate, 6–10 cm. long, 4–7 mm. wide, with nearly parallel nerves, the tendrils nearly always simple; racemes on peduncles 10–15 cm. long, 5- to 8-flowered; flowers 15–20 mm. long, purple, the lowest calyx-tooth lanceolate, nearly as long as the tube.——May–June. Wet grassy places and thickets along rivers in lowlands; Honshu, Kyushu.——Korea, Manchuria, China, and e. Siberia.

5. Lathyrus palustris L. var. **pilosus** (Cham.) Ledeb. *L. pilosus* Cham.; *L. palustris* subsp. *pilosus* var. *macranthus* (C. T. White) Fern.; *L. myrtifolius* subsp. *macranthus* C. T. White; *L. ugoensis* Matsum.; *L. miyabei* Matsum.; *L. pilosus* var. *miyabei* (Matsum.) Hara; *L. palustris* subsp. *pilosus* var. *macranthus* forma *miyabei* (Matsum.) Hara——Ezo-no-renri-

sō. Perennial herb pubescent while young; stems slender, ascending or decumbent at base, 40–80 cm. long, 3-angled, usually with a wing 0.5–2 mm. wide on 2 angles or nearly wingless; leaves short-petioled, the stipules lanceolate to narrowly ovate, bifid, the leaflets rather membranous, lanceolate to narrowly oblong, 3–5 cm. long, 8–15 mm. wide, pale green to glaucous and usually pubescent beneath, glabrous above, the tendrils 2- or 3-branched; flowers very similar to those of the preceding; legumes linear-oblong, flat, 4–5 cm. long, 7–8 mm. wide, glabrous.——May–Aug. Wet grassy places in lowlands; Hokkaido, Honshu, Kyushu (Tsushima).——Sakhalin, Kuriles, Korea, and widely distributed in the N. Hemisphere.

22. DUNBARIA Wight & Arn.　No-azuki Zoku

Scandent or twining herbs with scattered sessile glands especially on leaves beneath; stipules deciduous; leaves pinnately 3-foliolate, exstipellate; flowers yellowish, in axillary racemes, solitary or in pairs on nodes, the bracts caducous, the bracteoles absent, sometimes minute; calyx-teeth unequal, elongate, the upper 2 connate; legumes broadly linear, flat, 2-valved, dehiscent. ——Few species in tropical Asia.

1. Dunbaria villosa (Thunb.) Makino. *Glycine villosa* Thunb.; *D. subrhombea* (Miq.) Hemsl.; *Atylosia subrhombea* Miq.——No-azuki, Hime-kuzu. Prominently soft-pubescent, twining perennial; stems slender, with spreading or reflexed pubescence; stipules narrowly ovate-triangular, acute, about 2 mm. long, recurved, pubescent; leaflets rather densely short ascending-pubescent, with reddish brown, sessile, discoid glands beneath, the terminal leaflet rather larger than the lateral ones, depressed-rhombic, abruptly acute with an obtuse tip, 2–3 cm. long and as wide, stipels minute; racemes axillary, short-peduncled, loosely 3- to 8-flowered; flowers solitary

on each node of the raceme, 15–18 mm. across, yellow, the pedicels 6–8 mm. long; calyx about 10 mm. long, densely glandular-dotted and short-puberulent, sometimes loosely pilose, the lowest tooth longest, lanceolate, slightly longer than the tube; standard orbicular, short-clawed, with an obtuse tubercle on each side at base of limb, the keel semicircular and curved to the right, not spurred; legumes flat, broadly linear, 4.5–5 cm. long, about 8 mm. wide, short-pubescent, 6- or 7-seeded.——Aug.–Sept. Honshu (Kantō Distr. and westw.), Shikoku, Kyushu.——centr. China.

23. RHYNCHOSIA Lour.　Tankiri-mame Zoku

Twining to procumbent herb with scattered, yellowish brown, sessile, discoid glands; leaves pinnately 3-foliolate, stipellate, the stipules ovate or lanceolate; racemes axillary, the yellow flowers solitary or in pairs at each node, the bracts caducous; calyx 5-lobed; legumes flat, elliptic, 2-valved, dehiscent, 2-seeded.——About 100 species in Tropics.

1A. Terminal leaflet obovate, rounded at apex; lowest calyx-lobe longer than the tube. .... 1. *R. volubilis*
1B. Terminal leaflet ovate, short-acuminate at apex; all calyx-lobes shorter than the tube. .... 2. *R. acuminatifolia*

1. Rhynchosia volubilis Lour. Tankiri-mame. Twining grayish green perennial; stems rather slender, densely recurved short-pubescent; leaves petioled, the stipules narrowly ovate, acuminate, 4–5 mm. long, brown, several-nerved, the leaflets herbaceous, with sessile, yellowish brown, discoid glands on under side, dull yellow-villous, the terminal leaflet obovate or broadly so, 3–5 cm. long, 2.5–4 cm. wide, the stipels filiform, 2–3 mm. long; racemes nearly sessile, 10- to 20-flowered, 2–4 cm. long, the bracts elliptic, mucronate, caducous, brownish, 4–6 mm. long; flowers yellow, 8–10 mm. across; calyx densely puberulent, glandular, 6–8 mm. long, the lower 3 lobes lanceolate, acuminate; legumes elliptic, flat, slightly inflated above the seeds, slightly constricted between the seeds,

12–13 mm. long, about 8 mm. wide, puberulent, rarely glabrous except the margin, reddish when mature.——July–Sept. Thickets in hills and lowlands; Honshu (Kantō Distr. and westw.), Shikoku, Kyushu; rather common.——Korea, China, Ryukyus, Formosa, and Philippines.

2. Rhynchosia acuminatifolia Makino. *R. volubilis* var. *acuminata* Maxim.——Ōba-tankiri-mame, Tokiri-mame. Closely resembles the preceding, but vivid green and more slender, less densely pubescent; leaves thinner, the terminal leaflets ovate to narrowly so, 4–8 cm. long, 3–5 cm. wide; bracts 2.5–3.5 mm. long; calyx 4–5 mm. long, the lobes triangular; legumes puberulent, 15–17 mm. long, nearly sessile. ——Aug.–Sept. Honshu (centr. distr. and westw.), Kyushu.

24. AZUKIA Takah.　Azuki Zoku

Twining, rarely erect, more or less pilose annuals or perennials; leaves pinnately 3-foliolate, the stipules peltate, the stipels linear, basifixed, the leaflets entire or 2- or 3-lobed; racemes axillary, the bracts similar to the bracteoles, caducous, the bracteoles membranous to herbaceous, the nodes of the raceme prominently thickened; flowers yellow, or slightly greenish; calyx-teeth short; standard orbicular, short-clawed, the wings oblong, the left one embracing the semicircular curved keel, the right one spreading by the support of the spur on the keel, the keel horizontal and dextrorsely curved, with a horn-shaped spur on the right side; stamens diadelphous (9 and 1); style flattened, abruptly geniculate at the middle, fimbriate on one side below the stigma, extended beyond the discoid stigma into a subulate point; legumes linear or cylindric, sometimes rather flat, many-seeded.——More than 10 species, in e. and s. Asia.

1A. Bracteoles longer than the calyx-tube, 4–8 mm. long; hilum of the seeds flat.
　　2A. Terminal leaflets ovate to broadly so, broadest below the middle, abruptly acuminate; upper racemes very short-peduncled or subsessile; bracteoles 6–8 mm. long, about twice as long as the calyx-tube. .... 1. *A. angularis*
　　2B. Terminal leaflets narrowly ovate, broadest above the base, gradually acuminate; peduncles always distinct; bracteoles slightly longer than the calyx-tube, 4–5 mm. long. .... 2. *A. reflexopilosa*
1B. Bracteoles shorter than the calyx-tube, 2–2.5 mm. long; hilum of the seeds sulcate; very slender plants with the terminal leaflets 2–5 cm. long. .... 3. *A. nakashimae*

1. Azukia angularis (Willd.) Ohwi. *Dolichos angularis* Willd.; *Phaseolus angularis* (Willd.) W. F. Wight——AZUKI. Erect or slightly twining annual; cotyledons hypogeal; primary leaves simple, long-petioled, cordate; stems 30–80 cm. long, sparsely branched above, sometimes slightly twining above; ordinary leaves petioled, spreading-pilose especially on under side, the terminal leaflets broadly ovate, usually 3-lobed, 6–10 cm. long, 5–8 cm. wide, the stipels lanceolate, acuminate; racemes short-peduncled or subsessile, sometimes long-peduncled, the bracteoles narrowly ovate, few-nerved, acute; flowers 15–18 mm. across; legumes pendulous, linear, 6–10 cm. long, nearly glabrous, slightly inflated above the ellipsoidal seeds, the hilum flat.——Aug.–Sept. Widely cultivated in our area.

Var. **nipponensis** (Ohwi) Ohwi. *Phaseolus nipponensis* Ohwi; *P. angularis* var. *nipponensis* (Ohwi) Ohwi——YABU-TSURU-AZUKI. Twining and more slender; terminal leaflets narrowly ovate; racemes pilose; legumes blackish, 4–5 cm. long, the valves thinner.——Aug.–Sept. Thickets and grassy places in lowlands; Honshu, Shikoku, Kyushu; rather common.

2. Azukia reflexopilosa (Hayata) Ohwi. *Vigna reflexopilosa* Hayata; *Phaseolus reflexopilosus* (Hayata) Ohwi——Ō-YABU-TSURU-AZUKI. Coarsely pilose twining annual, allied to the preceding variety, but slightly larger, with longer peduncles, and elongate racemes and smaller bracteoles; legumes pendulous, 5–6 cm. long, 4–5 mm. wide, nearly straight, dark brown, the seeds black with paler blotches, truncate at both ends, 3.5 mm. long, the hilum flat.——Kyushu (Hyuga Prov.); rare.——Ryukyus and Formosa.

3. Azukia nakashimae (Ohwi) Ohwi. *Phaseolus nakashimae* Ohwi; *P. minimus* sensu auct. Japon., non Roxb.——

HIME-TSURU-AZUKI. Slender twining annual; stems elongate, retrorsely short-pilose while young; stipules small, 2–3 mm. long, the stipels acuminate, the terminal leaflets ovate, sometimes lanceolate, 2–5 cm. long, loosely appressed-pilose; racemes 2- to 5-flowered, long-peduncled, the bracteoles lanceolate, 1-nerved, acute, slightly shorter than the calyx-tube; flowers about 10 mm. across; legumes linear, 3–4 cm. long, smooth, the seeds ellipsoidal, dark brown.——July–Sept. Kyushu; rare.——Korea and Manchuria. The following are cultivated in our area.

Azukia umbellata (Thunb.) Ohwi. *Dolichos umbellatus* Thunb.; *Phaseolus calcaratus* Roxb.——TSURU-AZUKI, KANI-ME. Stems strongly twining, slender, soft-pubescent; cotyledons hypogeal; primary leaves lanceolate, long-petiolate; terminal leaflet of the ordinary leaves narrowly ovate; bracteoles linear, 1-ribbed, acuminate, about twice as long as the calyx; racemes elongate, long-peduncled; legumes narrow, slightly incurved, scarcely inflated over the seeds; seeds cylindric-oblong, the hilum sulcate.——Introduced from India.

Azukia radiata (L.) Ohwi. *Phaseolus radiatus* L.; *P. aureus* Roxb.; *Rudua aurea* (Roxb.) F. Maekawa——BUNDŌ, YAE-NARI. Stems erect or slightly twining, yellowish brown hispid; cotyledons epigeal; primary leaves broadly lanceolate, short-petioled; terminal leaflet of ordinary leaves orbicular-ovate, hispid, dark brown when dried, the stipels caudate, the bracteoles deciduous, broadly ovate, obtuse, several-ribbed, shorter than the calyx-tube; legumes spreading, straight, slightly flattened, hispid to scabrous, the seeds rather small, ellipsoidal, green or dark brown, the hilum sulcate.——Introduced from India.

25. VIGNA Savi　　SASAGE ZOKU

Twining or trailing, sometimes erect herbs; leaves pinnately 3-foliolate, the stipules rather small, sometimes peltate; racemes axillary, with thickened nodes, the bracteoles small, deciduous; flowers yellow to reddish purple, sometimes white, in pairs; legumes linear, rather cylindric, often inflated at the seeds, dehiscent, 2-valved, the seeds several to many, nearly globose to reniform, with a short hilum.——About 60 species, widespread in warmer and tropical regions.

1A. Keel distinctly beaked and incurved; leaflets membranous, coarsely pilose, acuminate; flowers large, 25–30 mm. long and as wide, bluish purple, becoming reddish brown; calyx-teeth narrowly lanceolate, acuminate, as long as the tube.
　　　　　　　　　　　　　　　　　　　　　　　　　　　　　　　　　　1. *V. vexillata* var. *tsusimensis*
1B. Keel not beaked; leaflets rather thick, chartaceous or herbaceous, often glabrescent, acute to obtuse; flowers smaller, 15–18 mm. long and as wide, yellow; calyx-teeth strongly depressed-triangular, much shorter than the tube. .... 2. *V. marina*

1. Vigna vexillata (L.) Benth. var. **tsusimensis** Matsum. AKA-SASAGE. Twining perennial, the stems rather slender, spreading to reflexed brown-pilose; stipules broadly lanceolate, few-nerved, 3–4 mm. long, acuminate, attached slightly above the base; leaflets coarsely pilose on both sides, the terminal 6–10 cm. long, 3–4 cm. wide, the stipels small, linear; racemes very short, umbellike, 2- to 4-flowered, on a long pedun-

cle 10–20 cm. long, the bracteoles lanceolate, caducous, the pedicels very short; calyx short-hispid, 8–10 mm. long, the teeth rather equal; keel gradually curved; legumes densely brown-hispidulous.——Sept.–Oct. Kyushu; rare. A variable widespread plant of tropical regions.

2. Vigna marina (Burm.) Merr. *Phaseolus marinus* Burm.; *Dolichos luteus* Sw.; *V. lutea* (Sw.) A. Gray; *V.*

retusa Walp.; *D. repens* L.; *V. repens* (L.) O. Kuntze, non Baker——HAMA-AZUKI. Trailing perennial; stems slender, sparsely appressed retrorse-pilose while young, soon glabrous; stipules narrowly ovate, 2–4 mm. long, rather acute, basifixed, obscurely few-nerved; leaflets sparsely pilose only while young, ovate or nearly obovate, 3–5 cm. long, 2.5–3.5 cm. broad, the stipels narrowly ovate, acute, about 1 mm. long; racemes 1–2 cm. long, few-flowered, peduncled, the bracteoles minute, ovate, acute, about 0.8 mm. long; calyx 2.5–3 mm. long, nearly glabrous, the teeth depressed-triangular, minutely ciliolate; legumes broadly linear, spreading, 3–5 cm. long, nearly or quite glabrous, 5- or 6-seeded, slightly inflated over the seeds. ——Apr.–Nov. Gravelly slopes near seashores; Kyushu (Yakushima); rare.——Ryukyus, Formosa, China, and tropical regions.

26. APIOS Moench. HODO-IMO ZOKU

Twining perennials; leaves petioled, pinnately compound, the stipules small, subulate, the leaflets 3–7, entire, the stipels linear or obsolete; flowers greenish yellow or purplish brown, in pairs or in 3's or 4's at each node of the axillary racemes, the bracteoles small, caducous; calyx campanulate, sometimes 2-lipped, the teeth obsolete; standard very broad, reflexed, the keel elongate, spirally coiled; stamens diadelphous; legumes broadly linear, 2-valved, dehiscent, straight or slightly curved, rather thick, many-seeded.——Few species in N. America, e. Asia, and India.

1. Apios fortunei Maxim. HODO-IMO. Tuberous-rooted twining perennial; stems slender, 100–150 cm. long, with scattered retrorsely appressed short hairs, soon glabrous; leaves petioled, the stipules broadly linear, 3–4 mm. long, 1-ribbed, deciduous, the leaflets 3 or 5, sparsely appressed-hispid on upper side, paler and nearly glabrous beneath, the terminal ovate, acuminate with an obtuse tip, 5–8 cm. long, 2.5–4 cm. wide; racemes 8–15 cm. long, peduncled, with thickened nodes, the pedicels jointed at apex, the bracts broadly linear, 1.5–2 mm. long, the bracteoles lanceolate, acute, about 1 mm. long, caducous; flowers pale greenish yellow, recurved, rather small; calyx with undulate obsolete teeth; legumes about 5 cm. long.——July–Aug. Sunny thickets and forest-borders; Hokkaido, Honshu, Shikoku, Kyushu. ——China.

27. DUMASIA DC. NO-SASAGE ZOKU

Twining herbs; leaves pinnately 3-foliolate, stipulate and stipellate; flowers pale yellow, solitary or in pairs at the nodes of axillary racemes, sometimes cleistogamous, the bracts narrow; calyx tubulose, oblique at truncate mouth, with 5 mucronate teeth; legumes sessile, broadly linear, 2-valved, dehiscent, inflated over the seeds.——Few species, in e. Asia, India, and N. America.

1. Dumasia truncata Sieb. & Zucc. NO-SASAGE. Cotyledons hypogeal; rhizomes slender, creeping, elongate; stems slender, nearly glabrous or retrorsely appressed-pubescent while young; stipules broadly linear, 3-nerved, 3–4 mm. long, acute, the leaflets thin, glabrous above, whitish and loosely appressed short-pilose beneath, the terminal leaflet narrowly ovate, 6–10 cm. long, 3–4 cm. wide, obtuse to subacute, the stipels subulate, about 1 mm. long; racemes peduncled, rather many-flowered, 2–5 cm. long, sometimes solitary or in pairs in axils; flowers 15–20 mm. long; calyx tubular, glabrescent, obliquely truncate at apex; legumes 3- to 5-seeded, oblanceolate, terete, 4–5 cm. long, 7–8 mm. across, glabrous, inflated over the seeds.——Aug.–Sept. Thickets and forest borders; Honshu, Kyushu.

28. MUCUNA Adans. TOBI-KAZURA ZOKU

Scandent, rarely erect, woody plants often with brownish stinging hairs; stipules present; leaflets petioled, pinnately 3-foliolate, usually stipellate; flowers in racemes, showy, purple or greenish yellow, pedicelled, usually in pairs, the racemes often cauline or on defoliated older branches, the bracts minute or caducous; calyx-tube short, the upper 2 teeth connate, the lowest longer than the others; standard short, the keel equal to or longer than the wing, beaked, incurved and cartilaginous above; stamens diadelphous, the anthers of 2 forms, the glabrous and basifixed ones alternate with the bearded versatile ones; ovary sessile; style filiform, the stigma terminal; legumes thick, ovate to linear, costate to winged.——About 30 species, widespread in the Tropics.

1A. Standard glabrous on margin; leaflets with prominent, finely reticulate elevated veinlets on both surfaces, appressed whitish pilose while young; lateral leaflets semicordate on the outer side at base; legumes about 40 cm. long. . . . . 1. *M. sempervirens*
1B. Standard ciliate; leaflets with less prominent barely elevated reticulate veinlets beneath, with appressed rusty strigose hairs especially beneath while young; lateral leaflets semirounded to truncate, rarely shallowly semicordate on the outer margin at base; legumes about 20 cm. long, 5-seeded, scarcely constricted. . . . . 2. *M. irukanda*

1. Mucuna sempervirens Hemsl. *M. japonica* Nakai ——AIRA-TOBI-KAZURA. Evergreen climber with woody stem, the branchlets glabrous, but with scattered pale yellow appressed hairs while very young; leaves petioled, the leaflets 3, deep green, lustrous, glabrous, but with scattered white appressed-hairs while young, with raised finely reticulate veinlets with a free veinlet within the network, the terminal leaflet oblong to elliptic, 7–15 cm. long, 4–8 cm. wide, acuminate to acute with a mucro at apex, obtuse at base; racemes 3- to 20-flowered, on peduncles 3.5–6.5 cm. long, fuscous-velvety, with sparse stinging hairs; calyx densely fuscous-velvety and with sparse stinging hairs; petals dark purple, the keel 6–8 cm. long, incurved and acutely acuminate; legumes about 40 cm. long, 7-seeded, velvety-puberulent, smooth on both surfaces, inflated on the seeds, both margins slightly thickened, not winged; seeds reniform, about 3 cm. long, flat on both surfaces.——May. Kyushu (Higo Prov.); very rare.——China.

2. Mucuna irukanda Ohwi. Iru-kanda, Ujiru-kanda. Woody climber; branches and leaves with rusty-brown retrorsely appressed stinging hairs while young; leaves petiolate, the leaflets chartaceous, shortly caudate-acuminate, the terminal oblong, 8–11 cm. long, 4–5 cm. wide, obtuse at base, with very fine barely raised veinlets; racemes densely many-flow-ered, about 15 cm. long, pendulous, subsessile to short-peduncled, rusty-brown velvety and with scattered brown stinging hairs, the pedicels 1–1.5 cm. long, usually in 3's at the nodes; flowers purple, with pale green standard, the wings dark purple, the keel 5–5.5 cm. long, cartilaginous, acute and white at apex, the remainder pale purple, paler toward base; legumes about 20 cm. long, 3 cm. wide, very shortly pilosulous, angled on each side near the margin, not winged, about 5-seeded.——May. Kyushu (Mageshima near Tanegashima); very rare.——Ryukyus.

29. CANAVALIA Adans. Nata-mame Zoku

Twining, rarely erect herbs; leaves pinnately 3-foliolate, the stipules small; flowers rather large, reddish or bluish purple or white, in axillary racemes, the nodes thickened, the bracts small, the bracteoles small and caducous; calyx campanulate, often 2-lipped, the teeth depressed; stamens diadelphous (9 and 1); legumes oblong, flat or rather inflated, 2-valved, dehiscent, ribbed or winged on outer edge.——About 30 species, widespread in the Tropics and subtropics.

1. Canavalia lineata (Thunb.) DC. *Dolichos lineatus* Thunb.——Hama-nata-mame. Twining or prostrate perennial herb; stems much elongate, glabrate; leaves long-petiolate; stipules ovate, 4–5 mm. long, herbaceous, deciduous, the scar of stipule thickened and later becoming a gland; leaflets subcoriaceous to rigidly chartaceous, often with scattered appressed hairs, nearly glabrous and yellowish green beneath, the terminal leaflets rounded to broadly obovate, 5–10 cm. long, 5–8 cm. wide; racemes 10- or more-flowered, long-peduncled, the nodes thickened and with a sessile gland at each; flowers in 2's or 3's, reddish purple, 25–30 mm. long, the bracts and bracteoles ovate, subobtuse, caducous, 2–3 mm. long; calyx about 1 cm. long, the teeth semirounded, the upper 2 larger, slightly connate; legumes puberulous at first, soon glabrous, oblong, rather flat, 5–6 cm. long, about 2.5 cm. wide, with 2 ridges along the outer suture.——July–Sept. Sandy beaches near seashores; Honshu (Tōkaidō Distr. and westw.), Shikoku, Kyushu; rather common.——Ryukyus, Formosa, and China.

30. PUERARIA DC. Kuzu Zoku

Twining or trailing herbs; leaves pinnately 3-foliolate, the stipules leaflike, sometimes peltate, the leaflets stipellate, entire or 2- or 3-lobed; flowers reddish purple, rarely white, in peduncled, erect, sometimes branched racemes, the bracts caducous, the bracteoles minute, the nodes of the racemes thickened; calyx deeply 5-lobed, the 2 upper lobes more or less connate, the standard usually with a callose spur on each side near the base, about as long as the wings and keel; stamens monadelphous; style beardless; legumes elongate, flat, linear, 2-valved, many-seeded.——More than a dozen species, in the Tropics of Asia.

1. Pueraria lobata (Willd.) Ohwi. *Pachyrrhizus thunbergianus* Sieb. & Zucc. nom. superfl.; *Dolichos lobatus* Willd.; *Pueraria thunbergiana* (Sieb. & Zucc.) Benth.; *D. trilobus* Houtt., non L.; *Pueraria triloba* (Houtt.) Makino——Kuzu. Perennial coarse herbs with woody base; stems much elongate, twining or prostrate, whitish puberulent, with coarse spreading or reflexed brown hispid-hairs; stipules lanceolate, subacute, medifixed, 15–20 mm. long, green; leaflets green, loosely appressed-hirsute on upper side, densely whitish pubescent beneath, the terminal lobe rhombic-orbicular, 10–15 cm. long and as wide, abruptly acuminate, some-times 3-lobed, the lateral lobes often bifid; racemes densely many-flowered, nearly sessile or short-peduncled, 10–20 cm. long; flowers reddish purple, rarely almost white, 18–20 mm. long, the bracts linear, 8–10 mm. long, 0.2–0.3 mm. wide, long-pilose, caducous, the bracteoles caducous, narrowly ovate or broadly lanceolate, acute; lowest calyx-lobe 1.5–2 times as long as the tube; legumes flat, densely dark brown spreading-hispid, linear, 6–8 cm. long, 8–10 mm. wide.——July–Sept. Thickets and thin woods; Hokkaido, Honshu, Shikoku, Kyushu; very common.——Korea and China.

31. GLYCINE L. Daizu Zoku

Twining to trailing, rarely erect herbs, usually hairy on stems and leaves; leaves with small stipules, pinnately 3-foliolate, stipellate; flowers small, purplish to white, on short axillary racemes, the bracts and bracteoles subulate, small, free; petals nearly equal in length; legumes flat, 2-valved, dehiscent, slightly inflated over the seeds.——About 10 species, in tropical and warmer regions of Africa, Asia, and Australia.

1A. Twining, usually slender, annual vine; bracteoles 1–2 mm. long; flowers 5–8 mm. long; pods 20–25 mm. long, 3–5 mm. wide, dark brown at maturity; seeds of an oblong type, 3–5 mm. long, 2–3.5 mm. wide, dark brown; indigenous. .... 1. *G. soja*
1B. Bushy, usually coarse and erect, annual herb; bracteoles 2.5–3.25 mm. long; flowers 6–7 mm. long; pods 25–75 mm. long, 8–15 mm. wide, yellowish brown at maturity; seeds of a globose or ovoid type, 6–11 × 5–8 mm., white to reddish black; cultivated.
2. *G. max*

1. Glycine soja Sieb. & Zucc. *G. ussuriensis* Regel & Maack; *G. javanica* Thunb. pro parte, non L.; *G. formosana* Hosokawa——Tsuru-mame, No-mame.   Twining annual; stems slender, elongate with retrorsely appressed to ascending brownish hispid hairs; leaves petioled, the stipules broadly lanceolate, slightly nerved, 2–3 mm. long, the leaflets 3, rarely digitately 5–7, entire, narrowly ovate to linear, 3–8 cm. long, 8–25 mm. wide, slightly paler beneath, loosely appressed-pilose, the stipels lanceolate, acuminate, 1.5–2 mm. long, 3-nerved; racemes axillary, few-flowered, sometimes very short; flowers pale purple, 5–8 mm. long; calyx-teeth nearly as long as the tube; legumes flat, linear-oblong, 2–2.5 cm. long, 4–5 mm. wide, densely pale brown pilose, 2- or 3-seeded.——Aug.-Sept. Thickets in lowlands; Honshu, Shikoku, Kyushu; common.——Korea, China, Manchuria, e. Siberia, and Formosa.

2. Glycine max (L.) Merr. *Phaseolus max* L.; *Dolichos soja* L.; *Soja hispida* Moench; *S. japonica* Savi; *S. viridis* Savi; *S. angustifolia* Miq.; *G. hispida* (Moench) Maxim.; *S. max* (L.) Piper; *G. gracilis* Skvortz.——Daizu.   Bushy, generally rather coarse, annual herb; leaves and legumes larger than in the preceding; plant with long spreading hirsute to pilose pale hairs.——A cultigen allegedly derived from *G. soja* now very common in cultivation throughout our area; known as soybean in English speaking countries.

32. GALACTIA P. Br.   Hagi-kazura Zoku

Twining or trailing, rarely erect herbs or suffrutescent subshrubs; leaves with small deciduous stipules, pinnately 3-foliolate, stipellate; flowers reddish purple or white, in axillary racemes, the bracts minute, setaceous, the bracteoles obsolete or minute; calyx-lobes acuminate, the upper 2 wholly connate, the lateral ones short, the lowest usually longer; standard ovate to orbicular, the wing adherent to the keel, which is nearly as long as the wing or longer, not beaked; stamens diadelphous or the upper one connate to the middle; ovary subsessile; style filiform, the stigma terminal; legumes linear, straight or incurved, flat, 2-valved, dehiscent.——About 50 species, in the warmer parts of the world especially in America.

1. Galactia tashiroi Maxim. *Galactia anisopoda* Ohwi ——Yonakuni-hagi-kazura, Yaeyama-hagi-kazura.   Diffuse, procumbent perennial; stems slender, rather firm, sparsely branched, short spreading whitish pubescent; leaves petioled, the stipules ovate to oblong-ovate, acute, 1–1.5 mm. long, appressed, caducous, the stipels obsolete, the rachis 2–5 mm. long, the leaflets subcoriaceous, broadly obovate or elliptic, 2–2.5 cm. long, 13–30 mm. wide, rounded at both ends or retuse at apex, with slightly recurved margins, grayish pubescent; racemes few-flowered, 2–4 cm. long, loose, slender, spreading-pubescent, the bracts ovate, caducous, the pedicels 1.5–3 mm. long, the bracteoles at the base of the calyx, ovate, acute, about 1 mm. long; calyx 4–5 mm. long, 5-lobed, puberulous, the upper pair wholly connate, the lateral lanceolate, acute; petals 13–15 mm. long, pale reddish purple; legumes flat, short-beaked at apex, 2–3 cm. long, 6–7 mm. wide, appressed-puberulous, 3- to 4-seeded, the seeds ellipsoidal, slightly compressed, about 5 mm. long.——Aug. Sunny gravelly slopes near seashores; Kyushu (Tanegashima); very rare.——Ryukyus. The typical phase has the leaflets glabrous on the upper side.

33. AMPHICARPAEA Ell.   Yabu-mame Zoku

More or less pubescent, twining to trailing herb; leaves pinnately 3-foliolate, stipulate and stipellate; flowers of 2 forms, the chasmogamic reddish purple or white, in short axillary racemes, the cleistogamic borne on filiform creeping branches near the ground, without petals or nearly so, the bracts nerved, similar to the stipules, persistent, the bracteoles absent to minute, subulate; calyx tubular, oblique, the teeth 4 or 5, rather unequal; stamens diadelphous; legumes of the chasmogamic flowers broadly linear or linear-oblong, flat, 2-valved, dehiscent, those of the cleistogamic fleshy, subterranean, obovoid to ellipsoidal, indehiscent, usually with 1 large seed.——About 15 species, tropical and temperate America, Japan, and Himalaya.

1. Amphicarpaea edgeworthii Benth. var. **japonica** Oliv. *Falcata japonica* (Oliv.) Komar.; *F. comosa* var. *japonica* (Oliv.) Makino——Yabu-mame.   Slender annual, thinly retrorse-pilose on stems, petioles, and racemes; stems slender, twining; stipules narrowly ovate, rather obtuse, 3–4 mm. long, nerved, persistent, the terminal leaflets ovate or broadly so, 3–6 cm. long, 2.5–4 cm. wide, obtuse to acute, appressed-pilose on upper side, paler or whitish with ascending pubescence beneath, the midrib beneath with spreading hairs; racemes axillary, short-peduncled, several-flowered; flowers pale purple, 15–20 mm. long; calyx with ascending pubescence, the teeth shorter than the tube; legumes flat, slightly curved, 2.5–3 cm. long, 7–8 mm. wide, glabrous but with appressed hairs on both margins.——Sept.–Nov. Woods and shaded places in lowlands; Honshu (Kantō Distr. and westw.), Shikoku, Kyushu; common.

Var. **trisperma** (Miq.) Ohwi. *Shuteria trisperma* Miq.; *A. trisperma* (Miq.) Bak. ex Jacks.——Usuba-yabu-mame. Plant uniformly appressed-pilose, with thinner leaves prominently whitish beneath; terminal leaflets slightly attenuate; flowers deeper purple.——Frequent in Hokkaido and n. Honshu, rare in Kyushu.——The typical phase and other variants occur from Korea, Manchuria, and Ussuri to the Himalayas.

34. INDIGOFERA L.   Koma-tsunagi Zoku

Herbs or shrubs with medifixed hairs; leaves odd-pinnate, rarely 1- or 3-foliolate, the stipules usually linear, the leaflets entire, often stipellate; flowers reddish purple, sometimes white, in axillary racemes, the bracts minute, the bracteoles absent; calyx small, obliquely toothed; standard broad, the keel laterally gibbous or pouched; stamens 10, monadelphous, the anthers with the connective extended at apex; legumes terete or angular, globose to linear, septate.——About 350 species, chiefly in the Tropics.

1. Indigofera pseudotinctoria Matsum. KOMA-TSU-NAGI. Appressed-pilose suffrutescent subshrub 30–50 cm. high, with many slender branches; leaves 4–8 cm. long; stipules minute, subulate, deciduous; leaflets oblong, 8–25 mm. long, 5–12 mm. wide, loosely appressed-pilose on both sides, glaucescent beneath, the stipels minute; racemes short-peduncled, 4–10 cm. long inclusive of the peduncle, densely many-flowered; flowers about 4 mm. long, pale reddish to nearly white; calyx white-puberulous, about 2 mm. long; legumes narrowly cylindric, straight, slightly pendulous, 25–30 mm. long, 2.5–3 mm. wide, short appressed-pilosulous, the seeds several, ellipsoidal, greenish yellow.——July–Sept. Grassy slopes and riverbanks in lowlands; Honshu (Kantō Distr. and westw.), Shikoku, Kyushu; very common.——centr. China.

2. Indigofera decora Lindl. *Hedysarum incarnatum* Willd., nom. superfl.; *I. incarnata* (Willd.) Nakai; *H. incanum* Thunb.——NIWA-FUJI. Subshrub with elongate, tufted, reddish, slender branches; leaves 7–20 cm. long inclusive of the short petiole; stipules linear, caducous; leaflets 9–13, thin, narrowly oblong, 2.5–4 cm. long, 1–1.5 cm. wide, acute, vivid green and nearly glabrous on upper side, glaucous beneath, the stipels minute; racemes axillary, 10–20 cm. long, many-flowered, on erect peduncles 3–10 cm. long; flowers 15–18 mm. long, reddish purple, rarely white; calyx-teeth deltoid, minutely pilose; legumes cylindric, linear, 3–4 cm. long, glabrous, straight.——May–June. River banks and old stone walls; Honshu, Kyushu; rather common.——centr. China.

3. Indigofera kirilowii Maxim. *I. koreana* Ohwi——CHŌSEN-NIWA-FUJI. Small shrub much-branched from the base; branches slender, glabrous; leaves 8–15 cm. long, short-petiolate; leaflets 9–13, thinly herbaceous, broadly ovate to elliptic, 1.5–2.5 cm. long, with scattered appressed medifixed hairs, sometimes nearly glabrous, glaucous beneath, mucronate; stipules and stipels short, subulate, caducous; racemes 5–10 cm. long, densely many-flowered, erect, the peduncles usually shorter than the raceme; flowers rose-colored, 12–15 mm. long, short-pedicelled; standard erect, oblong or elliptic, sometimes paler in color than the other petals, the wing and keels rather acute.——May–June. Kyushu (n. distr.); rare.——Korea, Manchuria, and n. China.

35. WISTERIA Nutt. FUJI ZOKU

Twining woody vines; leaves odd-pinnate, the stipules linear, usually caducous, the leaflets entire, often stipellate; racemes terminal on short shoots, elongate, many-flowered, pendulous; flowers showy, purple or white, on long pedicels, the bracts caducous, the bracteoles absent; calyx campanulate, the lobes unequal, triangular, the upper 2 short, the others longer; standard rounded, with 2 callosities at the base, the keel falcate, the wing auricled at the base; stamens diadelphous; legumes large and elongate, thickened, broadly linear, stiped, 2-valved, dehiscent, the seeds large.——Few species, in e. Asia and N. America.

1. Wisteria floribunda (Willd.) DC. *Glycine floribunda* Willd.; *Dolichos japonicus* Spreng.; *Kraunhia floribunda* (Willd.) Taub.; *Milletia floribunda* (Willd.) Matsum.——FUJI. Dextrorsely twining woody vine more or less pubescent while young and soon glabrous; leaflets 11–19; racemes flowering from the base; flowers purplish, rather smaller than in the following species, the bracts small.——May–July. Thickets and woods in hills and mountains; Honshu, Shikoku, Kyushu; common. Many cultivars are grown in gardens.

2. Wisteria brachybotrys Sieb. & Zucc. *Kraunhia brachybotrys* (Sieb. & Zucc.) Greene; *K. sinensis* var. *brachybotrys* (Sieb. & Zucc.) Makino; *K. floribunda* var. *brachybotrys* (Sieb. & Zucc.) Makino——YAMA-FUJI. Sinistrorsely twining woody vine; stipules linear, about 8 mm. long, caducous; leaflets 9–13, narrowly ovate, 4–6 cm. long, 15–30 mm. wide, pubescent on both sides when young, persisting on lower side, the stipels obscure; racemes 10–20 cm. long, the bracts caducous, ovate, acuminately caudate, densely brown-pubescent outside; flowers purple, opening nearly at the same time; ovary villous.——May–July. Mountains and hills; Honshu (w. distr.), Shikoku, Kyushu; rather rare.——Cv. **Alba.** *Kraunhia sinensis* var. *brachybotrys* forma *albiflora* Makino; *W. venusta* Rehd. & Wils.; *W. brachybotrys* var. *alba* W. Mill.——SHIRA-FUJI. Flowers white. Cultivated.

36. MILLETIA Wight & Arn. NATSU-FUJI ZOKU

Twining or trailing shrubs; leaves usually evergreen, odd-pinnate, the stipules small, acicular or subulate, the leaflets entire, stipellate; racemes axillary or in a terminal panicle; flowers greenish white or purple, the bracts and bracteoles usually subulate; calyx broadly campanulate, truncate or with depressed teeth; ovary many-ovuled; legumes broadly linear, slightly compressed, coriaceous, 2-valved, tardily dehiscent.——About 100 species, in tropical Asia.

1. Milletia japonica (Sieb. & Zucc.) A. Gray. *Wisteria japonica* Sieb. & Zucc.; *Kraunhia japonica* (Sieb. & Zucc.) Taub.——NATSU-FUJI, DOYŌ-FUJI. Dextrorsely twining deciduous shrub loosely short-pubescent but soon glabrous; stipules subulate, persistent, the leaflets 11–17, ovate to narrowly so, 1.5–4 cm. long, 1–2 cm. wide, gradually narrowed to the obtuse apex, rounded at base, slightly paler beneath, rather veiny on both surfaces, nearly glabrous, deep green above; racemes axillary, 10–20 cm. long inclusive of the short peduncle, rather densely many-flowered, nodding; flowers greenish white, pedicelled, about 15 mm. long; calyx about 4 mm. long, loosely appressed-puberulous, the teeth depressed-deltoid, the bracts and bracteoles acicular, less than 1 mm. long, persistent; ovary smooth, glabrous; legumes about 10 cm.

long, 8 mm. wide.——July–Aug. Thickets in lowlands; Honshu (centr. distr. and westw.), Shikoku, Kyushu; rather common.——Cv. **Microphylla**. *M. japonica* var. *microphylla*

Makino——HIME-FUJI, MEKURA-FUJI. A non-flowering cultivar, slightly pilose, with rather densely leafy branches and narrower leaflets than in the typical phase.

37. ROBINIA L. HARI-ENJU ZOKU

Shrubs or trees, often spiny; leaves odd-pinnate, the stipules spinelike or subulate, the leaflets entire, often stipellate; racemes axillary, pendulous; flowers showy, white or reddish purple, the bracts membranous, caducous; calyx-teeth short; standard reflexed, scarcely longer than the wing and keel; stamens diadelphous; legumes broadly linear, flat, narrowly winged on upper margin, 2-valved, dehiscent, several-seeded.——About 20 species, in N. and Centr. America.

1. Robinia pseudoacacia L. *Pseudacacia odorata* Moench——HARI-ENJU, NISE-AKASHIYA. Slightly pubescent to glabrate usually spiny tree; stipules often spiny, the leaflets 9–19, membranous, oblong, 2–5 cm. long, 1–2.5 cm. wide, rounded at apex, obtuse at base, glaucescent beneath; racemes pendulous, 10–15 cm. long; flowers showy, white, fragrant, 18–20 mm. long; lowest calyx-teeth lanceolate, the others triangular; legumes glabrous, about 8 cm. long, 12–15 mm. wide.——May–June. Cultivated and sometimes naturalized.——Introduced from the United States.

38. LOTUS L. MIYAKOGUSA ZOKU

Herbs or shrubs; leaves odd-pinnate, the stipules minute or absent, the leaflets 5–7, entire, the upper 3 usually at the top of the rachis, the lower 2 basal and simulating the stipules; flowers yellow or white, sometimes reddish, in axillary umbellike racemes, usually without bracts and bracteoles; calyx 5-toothed; standard ovate to rounded, the keel incurved; stamens diadelphous; ovary sessile, the style glabrous, the ovules many; legumes sessile, usually linear, compressed or somewhat terete, the seeds several.——About 150 species, widespread in the temperate and warmer regions.

1. Lotus corniculatus L. var. **japonicus** Regel. *L. japonicus* (Regel) K. Larsen——MIYAKOGUSA. Perennial, glabrous, prostrate herb with slender stems; leaflets obovate, 7–10 mm. long, slightly glaucescent, rounded with a mucro at apex, the lowest pair subacute, the rachis 2–10 mm. long; racemes umbellike, long-peduncled, with a ternate involucral bract smaller than but similar to the leaves; flowers 1–3, short-pedicelled, about 15 mm. long, yellow; calyx 6–7 mm. long, the lobes linear-lanceolate, longer than the tube.——Apr.–June. Fields and waste grounds; Hokkaido, Honshu, Shikoku, Kyushu; common.——Forma **versicolor** (Makino) Makino. *L. corniculatus* var. *versicolor* Makino——NISHIKI-MIYAKOGUSA. Flowers yellow at first, changing to reddish.——The typical phase occurs in Europe and Asia.

39. ASTRAGALUS L. GENGE ZOKU

Herbs or shrubs; leaves odd-pinnate, stipules often adnate to the base of the petiole, sometimes connate at the back, the leaflets entire, exstipellate; racemes on axillary or sometimes scapelike radical peduncles; flowers purple or yellowish, the bracts frequently small, membranous, the bracteoles minute or absent; calyx short-tubular, the teeth nearly equal; standard narrow, equaling to exceeding the wing and obtuse keel; stamens diadelphous; ovules many; legumes usually membranous, turgid, sometimes apparently 2-locular by intrusion of the suture.——About 1,500 species, in all temperate and cooler regions, except Australia.

1A. Flowers purple, rarely white; legumes apparently 2-locular.
  2A. Flowers in umbellike racemes; legumes broadly linear; bracts minute; standard as long as the keel; slender biennial herb of cultivated fields with elongate peduncles longer than the leaves. . . . . 1. *A. sinicus*
  2B. Flowers in more or less elongate racemes; legumes oblong; bracts prominent; standard longer than the other petals.
    3A. Stipules connate on the inner margin; flowers 10–20, 12–15 mm. long; bracts 3–5 mm. long, less than 1 mm. wide, 1-nerved; legumes sessile, not inflated. . . . . 2. *A. adsurgens*
    3B. Stipules free; flowers 5–10, rather large, 2–2.5 cm. long; bracts 7–10 mm. long, 2.5–3 mm. wide, few-nerved; legumes stiped, much inflated.
      4A. Flowers about 2.5 cm. long; standard obovate. . . . . 3. *A. yamamotoi*
      4B. Flowers about 2 cm. long; standard oblong. . . . . 4. *A. japonicus*
1B. Flowers yellow or greenish yellow.
  5A. Legumes linear or lanceolate, 2–5 cm. long, 3.5–6 mm. wide, apparently 2-locular; calyx-teeth linear, the lowest about 3 mm. long, about half to nearly as long as the tube.
    6A. Leaflets 13–19; racemes 2–3 cm. long; bracts very small, 1–2 mm. long, much shorter than the calyx; legumes broadly linear, erect, 3.5–4.5 cm. long. . . . . 5. *A. reflexistipulus*
    6B. Leaflets 25–31; racemes very short, very densely flowered, about 1 cm. long; bracts lanceolate, about as long as the calyx; legumes lanceolate, spreading, gradually curved upward, 2–3.5 cm. long. . . . . 6. *A. sikokianus*
  5B. Legumes elliptic or oblong, usually 1-locular; calyx-teeth short, usually triangular, much shorter than the calyx-tube.
    7A. Legumes with a broad groove on back, reniform in cross section, about 13 mm. long, nearly sessile; flowers 12–14 mm. long; leaflets 10–15 mm. long, 4–6 mm. wide. . . . . 7. *A. shiroumensis*
    7B. Legumes inflated, without a groove on back, oblong or lanceolate in cross section, about 3–5 cm. long, stipitate; flowers 15–20 mm. long; leaflets 15–30 mm. long, 5–20 mm. wide.
      8A. Bracts linear, 3–5 mm. long, about 1 mm. wide; leaflets 17–21; legumes much-inflated. . . . . 8. *A. shinanensis*
      8B. Bracts narrowly oblong, 5–10 mm. long, 2–4 mm. wide; leaflets 9–11; legumes only slightly inflated. . . . . 9. *A. secundus*

1. **Astragalus sinicus** L. *A. lotoides* Lam.; *Hedysarum japonicum* Basiner——Genge, Renge-sō. Biennial, soft, slightly pubescent herb; stems slender, ascending from a procumbent, branched base, 10–25 cm. long; leaves with petioles 2–5 cm. long, the stipules free, ovate, 3–6 mm. long, acute, the leaflets 9–11, thinly membranous, obovate to elliptic, 8–15 mm. long, rounded to emarginate; racemes umbellike, the peduncles 10–20 cm. long; flowers 7–10, about 12 mm. long, reddish purple, rarely white, the pedicels 1–2 mm. long; calyx about 4 mm. long, loosely white-pubescent, the teeth lanceolate, shorter than the tube; legumes erect or ascending in one direction, subsessile, blackish, 2–2.5 cm. long, about 4 mm. wide, glabrous, gradually beaked.——Apr.–June. Very commonly cultivated for green manure in rice-fields, occasionally naturalized in our area.——Formosa and China.

2. **Astragalus adsurgens** Pall. *A. fujisanensis* Miyabe & Tatew.; *A. laxmannii* subsp. *fujisanensis* (Miyabe & Tatew.) Kitag.——Murasaki-momenzuru. Stems ascending from a procumbent base, 10–30 cm. long, tufted, the caudex thickened; leaves 5–15 cm. long, short-petioled, the stipules broadly lanceolate, acute, connate on lower half on dorsal margins, the leaflets 17–21, oblong to narrowly so, 7–20 mm. long, 3–8 mm. wide, subobtuse at apex, slightly glaucescent, glabrous on upper side, loosely appressed-pilose beneath; racemes on somewhat longer peduncles, short-cylindric, 3–6 cm. long, densely many-flowered; flowers ascending, purple, rarely white, 12–15 mm. long, rather narrow; calyx 5–6 mm. long, with appressed-white and black pubescence, the teeth linear, about half as long as the tube; standard oblong, longer than the wings and keel; legumes loosely white and black appressed-puberulous, ascending, sessile, oblong, 3-angled, 2-locular, abruptly beaked.——July–Aug. Gravelly or sandy slopes in mountains; Hokkaido (Mount Oohira in Shiribeshi), Honshu (centr. distr. and northw.); rare.——Korea, Manchuria, Mongolia, and e. Siberia.

3. **Astragalus yamamotoi** Miyabe & Tatew. Kariba-ōgi. Stems rather stout, erect, thinly white-pubescent, 30–50 cm. long; leaves 8–10 cm. long, short-petioled, the stipules broadly lanceolate, free, obtuse, 6–10 mm. long, thinly membranous, the leaflets 9–13, oblong to narrowly ovate, 2–4 cm. long, 7–15 mm. wide, slightly retuse, slightly paler beneath, glabrescent; racemes 1.5–3 cm. long, 6- or 7-flowered, the bracts membranous; calyx 7–8 mm. long, glabrescent, the teeth elongate-triangular, 1–1.5 mm. long, slightly black-pubescent; standard narrowly obovate, keel nearly as long as the oblong wing, shorter than the standard; legumes ellipsoidal, glabrous, stipitate.——July. Hokkaido (Mount Kariba in Shiribeshi); rare.

4. **Astragalus japonicus** H. Boiss. *A. kurilensis* Matsum.——Ezo-momenzuru, Chishima-momenzuru. Stems rather stout, erect, 25–30 cm. long, thinly puberulous or glabrescent; stipules lanceolate or linear-lanceolate, acute to subobtuse, membranous, recurved above, pilose, the leaflets 7–15, oblong or elliptic, 2–3 cm. long, 7–13 mm. wide, retuse to rounded, glabrous above, pubescent beneath; racemes loosely 4- to 6-flowered, on peduncles about 10 cm. long, the bracts puberulous, twice as long as the pedicel; calyx obliquely campanulate, about 6 mm. long, white-pubescent, sparsely mixed with black hairs, the teeth very short; corolla about 2 cm. long, the standard oblong, emarginate, longer than the wing and the much shorter keel; legumes oblong, inflated,

stipitate, glabrous, pendulous, shallowly grooved on the ventral side, about 2 cm. long.——July. Grassy slopes in rocky places; Hokkaido (Mount Shari and Shiretoko Peninsula); very rare. ——Kuriles (Etorofu and Urupp).

5. **Astragalus reflexistipulus** Miq. *A. glycyphyllos* var. *reflexistipulus* (Miq.) Makino——Momenzuru. Slightly white-pubescent perennial with thick caudex; stems procumbent, elongate, ascending above, 30–50 cm. long; leaves 15–30 cm. long, the petioles 1–2 cm. long, the stipules free, broadly lanceolate, 5–8 mm. long, the leaflets 13–19, thinly membranous, oblong or ovate, 2–5 cm. long, 1–2.5 cm. wide, rounded to obtuse, pale green and short appressed-pubescent beneath, glabrous above; racemes 2–3 cm. long, 8- to 15-flowered on peduncles 3–10 cm. long; flowers about 13 mm. long, the bracts linear, 1–2 mm. long; calyx-teeth linear, the lower nearly as long as the tube; standard narrowly ovate, longer than the wings and keel; legumes subsessile, curved upward and erect, puberulous, gradually attenuate at both ends, many-seeded.——June–Aug. Hokkaido, Honshu.

6. **Astragalus sikokianus** Nakai. Naruto-ōgi. Slightly pubescent with whitish medifixed hairs; stems terete, elongate, ascending above; leaves 20–25 cm. long, short-petioled, the stipules free, membranous, lanceolate-triangular, acuminate, 8–10 mm. long, the leaflets membranous, oblong, 15–25 mm. long, 7–10 mm. wide, rounded and retuse, nearly glabrous above; racemes about 1 cm. long, densely 10- to 20-flowered, on peduncles 2–4 cm. long, solitary in upper axils, the bracts lanceolate, nearly as long as the calyx; calyx loosely white-pubescent, the teeth 1/2–2/3 as long as the tube, linear; corolla slender, yellowish white, about 13 mm. long, the keel rather short; legumes somewhat coriaceous, minutely appressed-pilose, abruptly acuminate, rounded and sessile at base, rather coriaceous.——Near seashores; Shikoku (Awa); rare.

7. **Astragalus shiroumensis** Makino. *A. arakawensis* Takeda——Shirouma-ōgi. Stems tufted, ascending from a stout caudex, 15–30(–50) cm. long, with loose white and blackish brown pubescence; leaves 4–6 cm. long, petioled, the stipules nearly free, narrowly ovate, 3–5 mm. long, obtuse to acute, the leaflets 11–13, narrowly oblong, obtuse to rounded, glabrous on upper side, thinly white-pilose and with dark brown hairs on the midrib beneath; racemes rather densely 5- to 10(–20)-flowered, on peduncles 5–8 cm. long, exceeding the leaves; flowers short-pedicelled; calyx about 4 mm. long, dark brown pubescent, the teeth short, triangular; legumes minutely appressed-puberulous with dark brown hairs, sometimes glabrescent, subsessile, pendulous.——July–Aug. Grassy and gravelly alpine slopes; central mountains of Honshu; very rare.

8. **Astragalus shinanensis** Ohwi. *A. frigidus* sensu auct. Japon., non A. Gray; *A. membranaceous* var. *obtusus* Makino——Tai-tsuri-ōgi. Stems tufted, erect from a stout caudex, 40–70 cm. long, sparsely branched, sparsely white-pubescent; leaves short-petioled, the stipules broadly lanceolate to linear, 5–10 mm. long, free, the leaflets narrowly ovate or oblong, 15–20 mm. long, 5–8 mm. wide, obtuse, glabrous on upper side, pubescent beneath; racemes 1.5–2.5 cm. long, densely 5- to 10-flowered, on peduncles 4–8 cm. long; flowers yellowish white, about 20 mm. long; calyx tubulose, 8–10 mm. long, brownish pubescent, the teeth short, triangular; legumes semirounded, pendulous, inflated, 3–4 cm. long inclusive of the stipe about 1 cm. long, about 15 mm. wide, with scattered

minute, appressed hairs.——Hokkaido (lowlands and high mountains), Honshu (high mountains of the centr. distr.); rare.

9. **Astragalus secundus** DC. *Phaca parviflora* Turcz.; *P. frigida* var. *parviflora* (Turcz.) Ledeb.; *A. frigidus* subsp. *parviflorus* (Turcz.) Hult.——RISHIRI-ŌGI. Slightly white-pubescent perennial; stems 20–30 cm. long, erect, simple; leaves 6–10 cm. long, on petioles to 2 cm. long, the stipules 10–20 mm. long, obtuse, membranous, green, the leaflets rather thin, 7–11, oblong to narrowly ovate, obtuse, 2–3 cm. long, 1–1.5 cm. wide, obtuse, glabrous on upper side; racemes densely 5- to 10-flowered, 1–2 cm. long, on peduncles 5–10 cm. long; flowers about 15 mm. long, pale yellow; calyx about 7 mm. long, tubulose, brownish pubescent especially on upper side, the teeth very short, triangular; legumes oblong, puberulent, slightly inflated.——July–Aug. Grassy alpine slopes; Hokkaido, Honshu (centr. distr.); rare.——Kuriles, Kamchatka, and e. Siberia.

40. OXYTROPIS DC. OYAMA-NO-ENDŌ ZOKU

Dwarf shrubs or small tufted usually acaulescent herbs; leaves odd-pinnate, the stipules adnate to the base of the petiole or free, the stipels none; racemes elongate or sometimes umbellike, on an axillary, usually radical scape or peduncle; flowers reddish purple, purple, or yellowish white, the bracts membranous, small or minute, the bracteoles absent; calyx tubulose, the teeth rather equal; standard erect or spreading, the keel pointed at apex, longer or shorter than the wing; ovary sessile or stipitate, many-ovuled; legumes rather inflated, sometimes with an intruding suture.——About 250 species, in temperate to frigid regions of the N. Hemisphere, especially abundant in Siberia, China, and centr. Asia.

1. **Oxytropis kudoana** Miyabe & Tatew. HIDAKA-GENGE. Acaulescent with appressed white-pubescence on petioles and lower part of peduncles; stipules softly scarious, rusty brown, 1- or 2-nerved, elliptic, rounded at apex, glabrescent, ciliate, the leaflets 9–15, obtuse, about 1 cm. long, 2–3 mm. wide; peduncles blackish pubescent on upper part, 2-flowered, the bracts narrowly ovate; flowers bluish purple; calyx blackish pubescent, the teeth lanceolate, about 3 mm. long, the tube about 6 mm. long; ovary with white hairs on the sides and black hairs on the ventral margin; legumes about 3 cm. long, about 1 cm. across, with a short included stipe.——July. Alpine regions; Hokkaido (Hidaka mountain range); rare.

2. **Oxytropis japonica** Maxim. OYAMA-NO-ENDŌ. Short-creeping, branched, dwarf suffrutescent subshrub; stems about 10 cm. long, densely covered at base with old marcescent leaves; leaves mostly radical, petioled, 3–8 cm. long, the stipules about 5 mm. long, narrowly ovate, obtuse, loosely pubescent, 1-nerved, connate below on ventral margins, adnate on the lower half to the base of the petiole, the leaflets 9–15, broadly lanceolate or narrowly ovate, 5–10 mm. long, 2–4 mm. wide, nearly acute, with slightly involute margins, appressed-pubescent on both sides; peduncles or scapes 1–5 cm. long, with dense ascending white hairs; flowers 1 or 2 (3), purple, 17–20 mm. long, the pedicels short, the bracts broadly lanceolate, membranous, 2–3 mm. long; calyx 10–12 mm. long, the teeth 3–4 mm. long, linear; legumes broadly lanceolate, 3–3.5 cm. long, 7–10 mm. wide, beaked at apex, loosely brownish short-pubescent.——July–Aug. Sandy or gravelly alpine slopes; Honshu (centr. distr.).

3. **Oxytropis yezoensis** Nakai. *O. japonica* var. *sericea* Koidz.——EZO-OYAMA-NO-ENDŌ. Closely allied to the preceding, differing in the ascending to spreading white-woolly pubescence on leaves and peduncles.——July–Aug. Sandy or gravelly alpine slopes; Hokkaido.

4. **Oxytropis megalantha** H. Boiss. REBUN-SŌ. Tufted perennial suffrutescent subshrub, with long dense ascending, white or grayish hairs; stems tufted, thick, short, ascending to short-creeping, from a stout often branched caudex; leaves radical, 5–8 cm. long, the stipules 15–20 mm. long, pubescent on the outside, the leaflets 8 to 11 pairs, rather thick, narrowly ovate-oblong to broadly ovate-lanceolate, 2–3 cm. long, 5–8 mm. wide, acute, with involute margins, less densely pubescent on upper side; peduncles or scapes about 20 cm. long, erect, about 7-flowered; flowers bluish purple, about 20 mm. long; calyx-tube about half as long as the corolla, with long white silky pubescence outside.——Hokkaido (Rebun Isl.); rare.

5. **Oxytropis hidakamontana** Miyabe & Tatew. HI-DAKA-MIYAMA-NO-ENDŌ. Acaulescent, with spreading, long white pubescence on the petioles and peduncles; stipules 1-nerved, lanceolate, acuminate to very acute, pubescent on the outside, the leaflets 11–17, oblong, 8–12 mm. long, 2–4 mm. wide, acute to subobtuse, densely pubescent on upper side, glabrescent beneath; racemes about 5-flowered, the bracts linear; flowers bluish purple; calyx with white and blackish hairs, the teeth lanceolate or linear-lanceolate, about 3 mm. long, the tube about 12 mm. long; ovary subsessile, appressed white-pubescent.——July. Hokkaido (Hidaka mountain range); rare.

6. **Oxytropis shokanbetsuensis** Miyabe & Tatew. MA-SHIKE-GENGE. Tufted, dwarf suffrutescent subshrub with scattered ascending white hairs; stems 10–15 cm. long, branched,

short-creeping on the ground; leaves petioled, 6–10 cm. long, the stipules 12–15 mm. long, lanceolate, loosely white-pubescent, 1-nerved, connate on the front at base, adnate to the base of the petiole on lower half on dorsal side, the leaflets 19–26, broadly lanceolate to narrowly ovate, 10–15 mm. long, 2.5–4 mm. wide, acute, with slightly involute margins, glabrous on upper side, appressed-pubescent beneath; peduncles or scapes 5–10 cm. long, about 5-flowered, nearly umbellate, the bracts broadly linear, about 10 mm. long, 1–1.5 mm. wide, the pedicels short; flowers 2–2.3 cm. long, reddish purple; calyx about 12 mm. long, with white and brown hairs, the teeth short, 1.5–2 mm. long; standard distinctly longer than the other petals, truncately bifid at apex.——Hokkaido (Mount Shokanbetsu in Teshio); rare.

7. Oxytropis rishiriensis Matsum. *O. campestris* sensu auct. Japon., non DC.——RISHIRI-GENGE. Short-creeping, lanate, suffrutescent subshrub; stems 10–15 cm. long; leaflets oblong, acute, the stipules adnate to the base of the petioles, ovate, acuminate, 1-nerved, pubescent; racemes rather many-flowered, the bracts lanceolate, much longer than the pedicel; flowers about 2 cm. long; calyx tubulose with short teeth; legumes sessile, acute, nearly glabrous, ascending.——Aug. Hokkaido (Mount Rishiri and Mount Yubari); rare.

41. TRIFOLIUM L.   SHAJIKU-SŌ ZOKU

Annual or perennial herbs; leaves digitately 3- to 7-foliolate, the stipules adnate to the petiole, the leaflets dentate; flowers red, purplish, white, rarely yellow, in axillary peduncled umbels, racemes, or heads, the bracts small or absent, sometimes forming a toothed involucre; calyx-teeth subequal; petals persistent, the wing longer than the keel, the claw adnate to the staminal tube, the keel exposing the stamens; stamens diadelphous; legumes small, oblong, terete, cylindric or obovate and flat, small, usually membranous, indehiscent, few-seeded.——About 300 species, in temperate regions of the N. Hemisphere, few in the S. Hemisphere.

1A. Flowers nearly sessile, bractless, 10–12 mm. long, reddish or rarely white; erect or ascending herb with yellowish spreading hairs, 3-foliolate leaves and sessile terminal heads. .... 1. *T. pratense*
1B. Flowers more or less distinctly pedicelled, with small bracts, white or reddish; heads peduncled.
  2A. Leaflets 4–7, rarely 3; petioles adnate to the stipules throughout their length; flowers deep red, rarely white, 15–18 mm. long; calyx 10 mm. long; stems erect or ascending from a thick caudex. .... 2. *T. lupinaster*
  2B. Leaflets 3; petioles adnate to the stipules only on the lower part; flowers 7–8 mm. long; calyx 3–3.5 mm. long.
    3A. Stems erect or ascending; petioles very unequal in length, the lower ones to 30 cm., the upper 1 to 2 cm. long; flowers reddish, rarely white. .... 3. *T. hybridum*
    3B. Stems long-creeping; petioles all elongate, nearly equal in length, usually 8–20 cm. long; flowers usually white. .... 4. *T. repens*

1. Trifolium pratense L. MURASAKI-TSUME-KUSA, AKA-TSUME-KUSA. More or less pubescent herb; stems branched, 30–60 cm. long; stipules membranous, appressed, caudate with a long setaceous tip, green, the leaflets ovate, elliptic, or oblong, 3–5 cm. long, often with a lunate band; heads rather large, globose to ovoid, subtended by a pair of leaves with much-dilated stipules at base; calyx-teeth subulate, unequal, not exceeding the corolla.——May–June. Naturalized in our area.——Europe and N. Africa.

2. Trifolium lupinaster L. SHAJIKU-SŌ. Perennial herb with a thickened caudex; stems firm, tufted, erect or ascending, 20–40 cm. long, usually simple, more or less soft-pubescent, several-leaved; petioles short, the stipules membranous, pale, acuminate; leaflets equal, oblanceolate, rarely narrowly obovate, 2–4 cm. long, 5–10 mm. wide; heads in upper axils, 10- to 20-flowered, the peduncles pubescent, 0.5–3 cm. long; flowers deep red; calyx-lobes subulate, the lowest longer.——June–Sept. Grassy slopes and rocky cliffs on volcanoes and near seashores; Hokkaido, Honshu (centr. distr. and northw.).——e. Europe to Asia.

3. Trifolium hybridum L. TACHI-ORANDA-GENGE. Nearly glabrous perennial; stems 30–50 cm. long, often branched, flexuous; stipules oblong, acuminate; leaflets ovate, 2–3.5 cm. long, 15–23 mm. wide, the petioles elongate; heads long-peduncled, axillary, globose, many-flowered; flowers pale reddish or white; calyx white, campanulate, the teeth subulate, nearly equal, green; legumes 2-seeded.——May–July. Naturalized in our area.——Europe, N. Africa, and w. Asia.

4. Trifolium repens L. SHIRO-TSUME-KUSA, ORANDA-GENGE. Glabrescent perennial with long-creeping entwined and matted branches; leaves radical, long-petioled, the stipules lanceolate, acuminate, membranous, greenish, the leaflets 3, obovate to obovate-orbicular, 15–25 mm. long, 10–25 mm. wide, emarginate and rounded at apex, green with a white band below the middle; peduncles or scapes longer than the leaves; heads many-flowered, globose, the pedicels recurved in fruit; flowers white with a greenish hue, sometimes reddish; legumes 4- to 6-seeded.——Apr.–July. Naturalized and very common in our area.——Europe, n. Africa, and w. Asia.

42. MEDICAGO L.   UMAGOYASHI ZOKU

Herbs; leaves pinnately 3-foliolate, with lateral nerves ending in the teeth, the stipules adnate to the petiole; flowers small, yellow or violet, the bracts small or absent; calyx-teeth 5, nearly equal; keel obtuse, shorter than the wings, spreading and free from the stamens, exposing the anthers; anthers of 1 form; style subulate, glabrous, the stigma nearly capitate; legumes spirally coiled, curved or rarely falcate, often spiny, 1- to several-seeded.——About 50 species, in Europe, n. Africa, and w. Asia.

1A. Perennial herb; flowers violet, about 7 mm. long; legumes soft-pubescent, smooth. .... 1. *M. sativa*
1B. Annuals or biennials; flowers yellowish, about 3 mm. long.
  2A. Legumes smooth, not spiny, about 2.5 mm. long, 1-seeded; flowers rather many. .... 2. *M. lupulina*
  2B. Legumes spiny, several-seeded; flowers few.
    3A. Glabrescent; stipules with laciniate margins, stems glabrous. .... 3. *M. hispida*
    3B. Pubescent; stipules entire or with few teeth; stems pubescent. .... 4. *M. minima*

1. Medicago sativa L. MURASAKI-UMAGOYASHI. Perennial slightly pubescent herb; stems erect, branched, 50–90 cm. long; stipules linear, dilated at base, nearly entire; leaflets elliptic to narrowly obovate, 2–3 cm. long, 6–10 mm. wide, cuneate at base, denticulate; racemes short and dense, axillary, peduncled, the bracts subulate; flowers violet; calyx 5–6 mm. long, the lobes linear, slightly longer than the tube; legumes 2- or 3-times coiled, 4–5 mm. across, slightly pilose, several-seeded.——Of e. Mediterranean origin, naturalized in our area.

2. Medicago lupulina L. KOMETSUBU-UMAGOYASHI. Short-pubescent annual or biennial 30–60 cm. high, with a branched procumbent base; stipules semiovate, somewhat toothed, the leaflets broadly obovate, 7–17 mm. long, 6–15 mm. wide, denticulate on the upper half; heads ovoid, on peduncles longer than the petioles, the bracteoles minute and subulate, the pedicels very short; calyx 1.5 mm. long; legumes semicircular, smooth, nerved, black when mature.——European introduction, naturalized in our area.

3. Medicago hispida Gaertn. *M. denticulata* Willd. ——UMAGOYASHI. Nearly glabrous annual, with procumbent stems; stipules laciniate, the leaflets broadly obovate, 1–2 cm. long, 7–15 mm. wide, retuse to rounded, cuneate at base, denticulate on upper half; heads rather few-flowered, axillary, the bracts minute, linear; calyx about 2 mm. long, the teeth subulate, incurved; legumes 2- to 3-times spirally coiled, 5–6 mm. across, nerved, with hooked spines on margins.—— Widely naturalized in our area.——Europe and n. Africa.

4. Medicago minima (L.) L. *M. polymorpha* var. *minima* L.——KO-UMAGOYASHI. Pubescent annual 10–20 cm. high, with firm, prostrate stems; stipules semiovate, entire or slightly toothed; leaflets obovate-cuneate, 7–10 mm. long, rounded at apex, denticulate on upper half; heads 2- to 5-flowered; calyx 2–2.5 mm. long; legumes 4- to 5-times spirally coiled, about 4 mm. across, spiny on the nerves and on margins.——Naturalized in our area.——Europe, Near East, and n. Africa.

43. MELILOTUS Juss.　SHINAGAWA-HAGI ZOKU

Annual or biennial herbs; leaves pinnately 3-foliolate, the stipules adnate to the petioles, subulate, the lateral nerves of the leaflets ending in the teeth; flowers small, yellow to white, rarely purple, in axillary racemes, the bracts small, the bracteoles absent; calyx-teeth nearly equal; petals deciduous, the standard usually longer than the keel and wings, the keel shorter than the wings, obtuse, not adnate to the stamens; stamens diadelphous; legumes short, ovate to nearly globose, few-seeded.—— About 20 species, in warmer and temperate regions of Eurasia.

1. Melilotus suaveolens Ledeb. SHINAGAWA-HAGI. Short-pubescent or nearly glabrous annual or biennial; stems erect, 50–90 cm. long, branched above; leaflets narrowly elliptic to broadly lanceolate, 1.5–3 cm. long, obtuse at apex, cuneate at base, toothed, glabrous; racemes axillary, 3–5 cm. long, rather densely flowered, elongate after anthesis, the peduncles 2–4 cm. long, the bracts linear-subulate, about 1 mm. long, slightly longer than the pedicels; flowers pale yellow, 3–4 mm. long; calyx thinly puberulent, 1.5 mm. long, teeth subulate; standard slightly longer than the wings and keel; legumes elliptic, about 3 mm. long, nearly smooth.——Waste ground and fields in lowlands; Honshu, Shikoku, Kyushu; rather common.——Korea, China, Indochina, and Siberia.

Melilotus indica (L.) All. KO-SHINAGAWA-HAGI. Flowers smaller, about 2 mm. long, with triangular-lanceolate calyx-teeth.——Naturalized in our area.

44. CROTALARIA L.　TANUKI-MAME ZOKU

Herbs or shrubs; leaves simple or digitately 3-(–7) foliolate, stipules free; flowers purplish or yellow, in terminal racemes or rarely solitary, rather large; calyx 5-lobed; standard large, the keel beaked; stamens monadelphous, the anthers of 2 forms, the larger basifixed ones alternate with the smaller versatile ones; legumes ellipsoidal, inflated, many-seeded.——About 350 species, in tropical and warmer regions.

1. Crotalaria sessiliflora L. *C. eriantha* Sieb. & Zucc. ——TANUKI-MAME. Erect annual, usually branched, long-appressed brown-villous except on upper side of leaves; stems 20–70 cm. long, terete; leaves 1-foliolate, nearly sessile, broadly linear to lanceolate, 4–10 cm. long, 3–10 mm. wide, acute to obtuse, the stipules linear, 3–5 mm. long; flowers blue, in terminal spikes, the bracts linear, 5–8 mm. long; calyx rusty-brown villous, accrescent after anthesis, ovoid, 10–15 mm. long; petals about 10 mm. long, slightly shorter than the calyx, the standard rounded-obovate, clawed, the limb with deeper colored striations on lower half near the middle, the wing horizontal, the keel white with a blue beak, convex on back below the middle, brownish villous on margin, 7–8 mm. long; style straight, with short, appressed hairs on one side; legumes enclosed in the calyx, oblong, 10–12 mm. long, glabrous, dehiscent.——July–Sept. Grassy places in lowlands; Honshu, Shikoku, Kyushu.——Korea, Manchuria, China, Ryukyus, Formosa to Malaysia, and India.

Fam. 106.　GERANIACEAE　FŪRO-SŌ KA　Geranium Family

Mostly herbs, sometimes shrubs; leaves alternate or opposite, stipulate, dentate or lobed, rarely entire; peduncles axillary, 1- to many-flowered; flowers rather large, bisexual, actinomorphic or more or less zygomorphic; calyx usually persistent, of 4 or 5 usually imbricate sepals; petals 4 or 5, usually imbricate; stamens 2 or 3 times as many as the sepals, the anthers versatile, 2-locular; carpels adnate to the axis, with a long beak at apex; ovules solitary or sometimes 2 in each locule, pendulous; fruit 3- to 5-lobed, each lobe with 1 or 2 seeds, dehiscing suddenly and spirally coiled.——About 11 genera, with about 650 species, chiefly in temperate regions.

1. GERANIUM L. Fūro-sō Zoku

Herbs with palmately or ternately divided or lobed leaves; peduncles axillary, 1- or 2-flowered; flowers actinomorphic, 5-merous; sepals and petals 5, the glands 5, alternate with the petals; stamens 10, perfect; ovary 5-lobed, the ovules 2 in each locule; seeds solitary in each carpel, the endosperm scanty or absent.——About 250 species, in the temperate regions and in high mountains of the Tropics.

1A. Flowers 2–3 cm. across; free portion of the styles 3–6 mm. long.
  2A. Column of the style distinctly longer than the free portion.
    3A. Stems, petioles, and pedicels with long spreading hairs; leaves rather shallowly lobed. .... 1. *G. eriostemon* var. *reinii*
    3B. Stems, petioles, and pedicels with recurved hairs; leaves finely and deeply lobed. .... 2. *G. erianthum*
  2B. Column of style short, nearly as long as to ⅓ as long as the free portion.
    4A. Pedicels erect in fruit. .... 3. *G. soboliferum*
    4B. Pedicels reflexed to spreading in fruit.
      5A. Stipules rather leaflike, small; petals usually pale rose with deeper colored striations.
        6A. Free portion of the styles 5–6 mm. long. .... 4. *G. krameri*
        6B. Free portion of the styles about 3 mm. long. .... 5. *G. yoshinoi*
      5B. Stipules rather broad, usually connate; petals rose-purple, rarely white, without deeper colored striations.
        7A. Petals white-pilose at base, on nerves above, and on margins; stipules connate or free; leaves deeply lobed, often with retrorse, appressed short hairs on stems and petioles. .... 6. *G. yesoense*
        7B. Petals white-pilose only on margin near base; stipules usually connate, broad; leaves less finely lobed, often spreading-hirsute or pilose on stems and petioles. .... 7. *G. shikokianum*
1B. Flowers about 1.5 cm. across or less; free portion of styles 1–3 mm. long.
  8A. Perennials.
    9A. Thickened root solitary; flowers usually solitary, pedunculate. .... 8. *G. sibiricum* var. *glabrius*
    9B. Thickened roots few; flowers usually 2 on each peduncle.
      10A. Leaves deeply lobed.
        11A. Leaves deeply 5-lobed, the lobes obovate, obtuse. .... 9. *G. thunbergii*
        11B. Leaves deeply 3-lobed except the lower ones, the lobes ovate, acute. .... 10. *G. wilfordii*
      10B. Leaves 3-sected. .... 11. *G. tripartitum*
  8B. Annual or biennial; leaves 3- or 5-parted, finely lobed. .... 12. *G. robertianum*

1. Geranium eriostemon Fisch. var. **reinii** (Fr. & Sav.) Maxim. *G. reinii* Fr. & Sav.; *G. japonicum* Fr. & Sav.——Gunnai-fūro. Rhizomes short, thick, with thickened roots; stems erect, 30–50 cm. long, with spreading somewhat reflexed hairs; radical and lower cauline leaves long-petiolate, 8–12 cm. across, deeply cordate at base, deeply 5- to 7-lobed, the lobes rhombic-obovate, acute, irregularly incised and loosely serrate; stipules free or connate, brown, membranous; flowers terminal, the pedicels erect, with long, spreading, gland-tipped hairs; sepals oblong, with spreading hairs; petals rose-purple, 12–15 mm. long; filaments long-hairy on lower half; free portion of styles 3–4 mm. long.——June–Aug. Mountains; Hokkaido, Honshu (centr. and northw.); rather rare.——Forma **onoei** (Fr. & Sav.) Hara. *G. onoei* Fr. & Sav.; *G. eriostemon* var. *onoei* (Fr. & Sav.) Nakai——Takane-gunnai-fūro. Stems lower; flowers deeper colored; leaves pilose chiefly on nerves beneath.——Alpine.
The typical phase occurs in Korea, China, and e. Siberia.

2. Geranium erianthum DC. *G. subumbelliforme* R. Knuth——Chishima-fūro. Perennial with thick, short rhizomes; stems erect, 30–40 cm. long, retrorsely short-pubescent; radical and lower cauline leaves depressed-orbicular, deeply 5-lobed nearly to the base, 6–10 cm. wide, deeply cordate, pubescent on both sides, the lobes narrowly rhombic-obovate, acute to acuminate, irregularly incised and coarsely toothed; flowers rather numerous, terminal, 2.5–3 cm. across, bluish purple, on short pedicels; sepals densely pubescent with long spreading, glandular hairs; filaments loosely long-pilose.——July–Aug. Grassy slopes near seashores; Hokkaido, Honshu (n. distr. and alpine region of Mount Hayachine in Rikuchu). ——e. Siberia, Sakhalin, Kuriles, Alaska, and N. America.

3. Geranium soboliferum Komar. *G. hakusanense* Matsum.; *G. soboliferum* var. *hakusanense* (Matsum.) Makino

——Asama-fūro. Stems 60–80 cm. long, retrorsely short-pilose; radical and lower cauline leaves long-petiolate, 7–10 cm. wide, 5- to 7-parted nearly to the base, the lobes rhombic or rhombic-cuneate, deeply lobed on upper half, acute, appressed-pilose above and on the nerves and nervules on under side; flowers rose-purple, about 3 cm. across, the pedicels retrorsely appressed-pilose; petals loosely white-pilose near base; free portion of styles about 6 mm. long.——Aug.–Sept. Grassy slopes in mountains; Honshu (centr. distr.); rare.——Korea and Manchuria.
Var. **kiusianum** (Koidz.) Hara. *G. kiusianum* Koidz.——Tsukushi-fūro. Leaves short-pilose beneath.——Wet places in mountains; Kyushu.

4. Geranium krameri Fr. & Sav. *G. sieboldii* Maxim.; *G. japonicum* sensu auct. Japon., non Fr. & Sav.——Tachi-fūro. Stems 60–80 cm. long, often decumbent at base, with retrorsely spreading hairs; leaves 7–10 cm. wide, rounded-cordate, deeply 5-parted, the lobes rhombic, bifid or 3-fid, incised and coarsely serrate, subacute, pilose especially on the nerves beneath, the stipules free, rather leaflike, narrow; peduncles 2-flowered, retrorsely spreading-hairy; flowers 2.5–3 cm. across, rose; petals with deeper colored striations, densely pilose at base; free portion of styles 5–6 mm. long.——Aug.–Sept. Honshu, Kyushu.——Korea, Manchuria, and Amur.——Forma **adpressipilosum** (Hara) Hara. *G. japonicum* var. *adpressipilosum* Hara——Fushige-tachi-fūro. Stems and petioles with retrorsely appressed hairs.——Honshu (centr. distr.), Kyushu.

5. Geranium yoshinoi Makino. Bitchū-fūro. Stems 40–70 cm. long, retrorsely appressed-pilose; radical leaves long-petiolate, the lower cauline 5-angled, rounded, depressed-cordate, 5–8 cm. wide, 5-parted to below the middle, the lobes rhombic-cuneate, 3-fid, with 2 or 3 coarse teeth, acute to subobtuse, appressed short-pilose above and on nerves beneath; stipules usually connate, small, rather leaflike; peduncles and pedicels retrorsely appressed short-pilose; flowers about 2 cm. across, rose; petals with deeper colored striations, loosely long-hairy near base on upper side; free portion of style 3–3.5 mm. long.——Honshu (Chūgoku Distr.).

6. Geranium yesoense Fr. & Sav. *G. dahuricum* sensu auct. Japon., non DC.——EZO-FŪRO. Rhizomes short, with thickened roots; stems 30–80 cm. long, from a decumbent base, retrorsely pilose; lower cauline leaves petiolate, 5-angled, orbicular, somewhat depressed, cordate at base, 5–10 cm. wide, deeply 5-parted nearly to the base, the lobes rhombic-obovate, 3-fid again and with a few deep incisions, obtuse to subacute, spreading-pilose; stipules free or connate, brown, membranous; flowers rose-purple, 2.5–3 cm. across; sepals with long spreading hairs; petals densely white hairy near base inside; free portion of style about 5 mm. long.——July–Aug. Hokkaido, Honshu (Mount Ibuki in Oomi, and eastw.); rather variable.

Var. **nipponicum** Nakai. HAKUSAN-FŪRO, AKANUMA-FŪRO. Less pubescent with shorter, appressed hairs on stems and petioles; sepals usually without long hairs.——Rather common in high mountains of Honshu (centr. and n. distr.).

Var. **pseudopalustre** Nakai. *G. miyabei* Nakai——HAMA-FŪRO. Similar to the preceding variety but the leaves shallowly incised, the lobes broader; sepals often sparsely long spreading-hairy.——Near seashores of Hokkaido and n. distr. of Honshu.——Kuriles.

7. Geranium shikokianum Matsum. IYO-FŪRO, SHI-KOKU-FŪRO. Stems 30–70 cm. long, coarsely pilose with recurved spreading hairs; lower cauline leaves petiolate, 5-angled, rounded, slightly depressed, 5–10 cm. wide, deeply 5-lobed to below the middle, the lobes rhombic-obovate, irregularly incised, subacute to acute, appressed short-pilose on upper side, with spreading hairs beneath especially on nerves; stipules broad, connate, scarious, brown; flowers rose-purple, 2.5–3 cm. across, the pedicels with short retrorsely curved hairs on upper part; sepals loosely long-pilose; free portion of styles 5–6 mm. long.——July–Sept. Honshu (Suruga Prov. and Chūgoku Distr.), Shikoku, Kyushu; rather variable.

Var. **kaimontanum** (Honda) Honda & Hara. *G. kaimontanum* Honda——KAI-FŪRO. Less pubescent; lobes of leaves acute; sepals without long hairs.——Honshu (Kai Prov.).

Var. **yamatense** Hara. YAMATO-FŪRO. Leaves deeply lobed and incised.——Mountains; Honshu (Yamato Prov.).

Var. **yoshiianum** (Koidz.) Hara. *G. yoshiianum* Koidz. YAKUSHIMA-FŪRO. Dwarf phase, 15–30 cm. high; leaves more densely pilose, 1–3 cm. across; pedicels nearly glabrous; flowers 2–2.5 cm. across.——High mountains of Yakushima in Kyushu.

8. Geranium sibiricum L. var. **glabrius** (Hara) Ohwi. *G. sibiricum* forma *glabrius* Hara——ICHIGE-FŪRO. Stems prostrate or long-decumbent, ascending, 30–80 cm. long, with scattered, retrorse, appressed hairs; cauline leaves rather thin, petiolate, 5-angled, orbicular, 4–7 cm. wide, cordate at base, 5-lobed to below the middle, the lobes narrowly rhombic, acute, 3-fid, irregularly incised and few-toothed, appressed-pilose on upper side, ascending-pilose beneath especially on nerves; stipules free, narrow; flowers solitary, rarely paired, pale rose to white, about 8 mm. across; sepals short-pilose; free portion of styles about 1 mm. long.——July–Sept. Grassy places in lowlands; Hokkaido, Honshu (n. distr.).——A variable plant occurring widely in Korea, China, Sakhalin, Kuriles, and Siberia.

9. Geranium thunbergii Sieb. & Zucc. *G. nepalense* var. *thunbergii* (Sieb. & Zucc.) Kudo——GEN-NO-SHŌKO, FŪRO-SŌ. Decumbent or prostrate at base, then ascending to erect, 30–50 cm. high, softly pubescent; leaves petiolate, depressed-orbicular, 3- to 5-lobed to below the middle, the lobes obovate, obtuse, shallowly 3-fid, irregularly few-toothed, appressed-pilose above, with curved ascending hairs on nerves beneath, the stipules free, narrow; flowers in pairs, 1–1.2 cm. across, pale rose or rose-purple, short-pubescent, with long spreading glandular hairs on sepals and pedicels; free portion of styles about 2 mm. long.——July–Oct. Grassy places in lowlands; Hokkaido, Honshu, Shikoku, Kyushu.——Kuriles, Ryukyus, Formosa, Korea, and China.

10. Geranium wilfordii Maxim. *G. iinumae* Nakai; *G. krameri* var. *iinumae* (Nakai) Nakai; *G. iinumae* var. *asiaticum* Koidz.——MITSUBA-FŪRO, FUSHIDAKA-FŪRO. Stems 40–80 cm. long, decumbent, with short, retrorsely appressed hairs; cauline leaves petiolate, 3- to 5-lobed to below the middle, shallowly cordate, 5–8 cm. wide, the lobes nearly rhombic, acute to acuminate, irregularly incised, appressed-pilose on the upper side and on the nerves beneath; stipules free, narrow; flowers in pairs, 1–1.5 cm. across; sepals short-pilose; petals pale rose with deeper colored nerves; free portion of styles about 3 mm. long.——July–Oct. Lowlands and hills; Hokkaido, Honshu, Shikoku, Kyushu.——Korea.

Var. **hastatum** (Nakai) Hara. *G. hastatum* Nakai, non Andr.——HOKAGATA-FŪRO. Lobes of upper leaves narrower, ovate-lanceolate to narrowly ovate, gradually narrowed at tip, with nearly equal teeth, the lateral lobes spreading; sepals loosely long-pilose.——Near Nikko in Honshu.——Korea, Manchuria, Amur, and China.

Var. **yezoense** Hiyama. EZO-MITSUBA-FŪRO. Leaves with appressed short pilose hairs beneath.——Hokkaido.

11. Geranium tripartitum R. Knuth. *G. robertianum* var. *glabrum* Fr. & Sav.; *G. wilfordii* Maxim., pro parte——KO-FŪRO. Stems 30–50 cm. long, decumbent, loosely and retrorsely pilose; leaves petiolate, 3-sected, loosely pilose on upper side and on nerves beneath, the segments 2–4 cm. long, rhombic-ovate, obtuse to subacute, the lateral segments obliquely ovate, often 2-fid, coarsely few-toothed, the petioles with retrorsely appressed and recurved short hairs, the stipules free and narrow; flowers in pairs, pale rose, 1.2–1.5 cm. across, the pedicels retrorsely short-pilose; sepals loosely long-pilose; free portion of styles about 1.5 mm. long.——Aug.–Sept. Honshu, Shikoku, Kyushu.——Korea (Quelpaert Isl.).

12. Geranium robertianum L. HIME-FŪRO, SHIO-YAKI-FŪRO. Rather soft, erect annual with spreading long soft hairs, becoming glandular in the upper ones; stem-leaves petiolate, 5-angled, 3(-5)-sected, soft, 3–6 cm. wide, the median segments broadly ovate, obtuse to acute, rather pinnately dissected, narrowed to a petiolelike base, the lobes finely dissected and incised; stipules free, small, membranous; peduncles 2-flowered, with long spreading glandular hairs; flowers pale rose, 8–12 mm. across; petals 8–10 mm. long, narrowed to the base; free portion of style about 0.8 mm. long.——May–Aug. Mountains; Honshu (Mount Ibuki in Oomi), Shikoku.——Formosa, China, Himalayas, Europe, N. Africa, and N. America.

Fam. 107. **OXALIDACEAE** Katabami Ka   Sorrel Family

Herbs, rarely shrubs, sometimes with bulbous rhizomes, the juice acid; leaves usually ternate, rarely pinnate or simple, the leaflets usually obcordate; stipules formed by the slightly swollen bases of the petioles one on each side; flowers bisexual, in umbels or dichotomous cymes, sometimes solitary, actinomorphic, sometimes cleistogamous; sepals 5; petals 5, white, rose-purple or yellow, contorted in bud; stamens 10–15; ovary 5-locular, 5-lobed, the styles free or connate, the ovules 1 to several in each locule; capsules globose to cylindric, loculicidally dehiscent; embryo erect in a fleshy endosperm.——About 5 genera, with more than 300 species, in the Tropics and subtropics, especially abundant in S. Africa and S. America.

1. **OXALIS** L. Katabami

Characters of the family.

1A. Acaulescent; leaves and peduncles radical; flowers solitary, white or rose.
  2A. Scapes few-flowered; naturalized plant growing in waste grounds and cultivated fields. .... 1. *O. martiana*
  2B. Scapes 1-flowered; indigenous plants growing in woods.
    3A. Leaflets obcordate or obtriangular, with obtuse or rounded angles; capsules ovoid-globose or oblong, 4–12 mm. long, with 1 or 2 seeds in each locule.
      4A. Plant smaller than the following, with slender rhizomes and the leaflets with more strongly rounded margins; petals 10–15 mm. long; capsules ovoid-globose, 4–6 mm. long. .... 2. *O. acetosella*
      4B. Plant larger than the preceding with stouter, thicker rhizomes and the margins of the leaflets less rounded; petals 15–18 mm. long in both chasmogamic and cleistogamic flowers; capsules oblong, 1–1.2 cm. long, or in cleistogamic flowers ovoid and 5–6 mm. long. .... 3. *O. griffithii*
    3B. Leaflets obtriangulate, with rather acute angles; capsules cylindric, about 2 cm. long, with 4 or 5 seeds on each locule; plant larger. 4. *O. obtriangulata*
1B. Caulescent; flowers yellow, 1 to several, in an axillary, nearly umbellate inflorescence.
  5A. Stems tufted, from a tap root, sometimes long-creeping; stipules small, but distinct, auriclelike; inflorescence subumbellate, the flowers 1–8; pedicels downwardly flexuous; seeds 1–1.5 mm. across. .... 5. *O. corniculata*
  5B. Stems distant, erect from slender elongate creeping rhizomes; stipules indistinct; inflorescence cymose, the flowers 1–2(–3); pedicels spreading or erect; seeds 1.5–2 mm. across. .... 6. *O. fontana*

1. Oxalis martiana Zucc. Murasaki-katabami. Weedy herb with a tap-root and numerous brownish bulbils; leaves all radical, thinly pubescent, petiolate, the leaflets 3, thin, obcordate, 2–4 cm. wide, broadly cuneate at base; scapes 10–20 cm. long, slender, thinly pubescent; flowers 5–10, umbellate on the scape, 5-merous, nodding in bud, the pedicels slender, 1–3 cm. long; sepals with a pair of brownish glands; petals 12–15 mm. long, obovate-spathulate, rose-purple.——Naturalized; Honshu, Shikoku, Kyushu.——Introduced from S. America.

2. Oxalis acetosella L. Ko-miyama-katabami. Thinly pubescent to nearly glabrous, delicate perennial with slender, elongate, few-scaled rhizomes; leaves all radical, the petioles 3–10 cm. long, jointed 2–3 mm. above the base, the stipules broadly ovate, the leaflets sessile, obcordate, 4–20 mm. long, 7–30 mm. wide, retuse, with rounded angles, often glabrous on upper side; scapes 5–8 cm. long, with a pair of small membranous bracts at or slightly above the middle; sepals narrowly ovate, obtuse, shorter than the stamens; petals white or pale rose-purple, 9–14 mm. long, narrowly obovate; capsules ovoid-globose, 4–6 mm. long.——May–June. Coniferous woods; Hokkaido, Honshu (centr. distr. and northw.).—— Europe and Asia.

3. Oxalis griffithii Hook. f. & Edgew. *O. acetosella* subsp. *griffithii* (Hook. f. & Edgew.) Hara; *O. acetosella* var. *vegeta* Tatew.; *O. japonica* sensu auct. Japon., non Fr. & Sav.; *O. acetosella* var. *japonica* Makino, excl. basionym.——Mi-yama-katabami. Larger; leaflets obcordate to depressed-obtriangular, 1–2.5 cm. long, 2–4 cm. wide; petals 15–18 mm. long; capsules oblong, 1–1.2 cm. long, or in cleistogamic flowers ovoid and 5–6 mm. long.——Apr.–May. Mountain woods; Honshu, Shikoku, Kyushu; rather common.——Formosa, China, and n. India.

4. Oxalis obtriangulata Maxim. Ō-yama-katabami. Similar to the preceding but larger; rhizomes thickened above with densely imbricate scales; leaves long-petioled, the leaflets obtriangular, about 2.5 cm. long, 4 cm. wide, truncate and retuse at apex; scapes bracteate on upper part; sepals oblong, longer than the filaments; capsules lanceolate-cylindric, acuminate, about 2 cm. long.——Mountain woods; Honshu, Shikoku.——Korea, Manchuria, and Ussuri.

5. Oxalis corniculata L. Katabami. Thinly pubescent perennial herb with acid juice; stems long-creeping from a thickened tap-root; leaves alternate, long-petioled, the leaflets depressed-obcordate, 1–2.5 cm. wide, retuse; stipules minute but distinct; peduncles erect, with linear or lanceolate bracts at apex, (1–)2- to 8-flowered; flowers yellow, about 8 mm. across; capsules cylindric, 1.5–2.5 cm. long, many-seeded; seeds broadly ovate, lenticular, transversely rugulose.——May–Sept. Waste grounds and roadsides in lowlands and hills; Hokkaido, Honshu, Shikoku, Kyushu; very common; variable. ——Cosmopolitan in all temperate and warm areas.

6. Oxalis fontana Bunge. *O. stricta* L. pro parte; *O. europaea* Jord.——Ezo-tachi-katabami. Closely related to the preceding species; rhizomes elongate, very slender, loosely and minutely scaly; stems 20–40 cm. long, usually simple; leaves long-petioled; stipules indistinct; inflorescence cymose; flowers 1 or 2, rarely 3, pedicellate, the bracts small; seeds rather large, 1.5–2 mm. across.——July–Oct. Thickets and thin woods; Hokkaido, Honshu (centr. distr. and northw.). ——Temperate Asia and N. America, introduced in Europe.

Fam. 108. **LINACEAE** Ama Ka   Flax Family

Herbs or shrubs; leaves alternate or opposite, simple, entire, the stipules absent or glandlike; flowers bisexual, actinomorphic; sepals 4 or 5, free or partly connate, imbricate; petals contorted, often clawed; stamens as many as the petals and alternate with them, sometimes with intervening staminodes, the filaments connate at base; ovary superior, 3- to 5-locular, the locules often divided, with 2 ovules in each locule, the styles 3–4, free, filiform; capsules loculicidally dehiscent; seeds flat, the endosperm copious or absent, the embryo erect, the cotyledons flat.——About 6 genera, with about 150 species, chiefly in temperate regions.

1. LINUM L. Ama Zoku

Herbs, sometimes woody, glabrous or rarely hairy; stipules absent or glandlike; leaves alternate or opposite, narrow, 1- to many-nerved; flowers yellow, blue, rose or white, in axillary or terminal racemes or loose racemelike cymes; sepals 5, entire, persistent; petals 5, deciduous; stamens 5, connate at base, usually alternating with small staminodes; glands 5, small, at the base of the filaments and opposite the petals; ovary 5-locular, with 2 ovules in each locule, the styles 5; capsules 5-valved; seeds with scanty endosperm, the embryo straight.——About 90 species, chiefly in the temperate and warmer regions.

1A. Sepals elliptic or broadly ovate, abruptly acute, 3–3.5 mm. long, with a few projecting black glands on margin; flowers about 1 cm. across; capsules 3.5–4 mm. across. . . . . 1. *L. stelleroides*
1B. Sepals ovate-oblong, acuminate, 5–7 mm. long, eglandular; flowers 1.5–2 cm. across; capsules about 7 mm. across. . . 2. *L. usitatissimum*

1. Linum stelleroides Planch. Matsuba-ninjin. Glabrous annual; stems terete, 40–60 cm. long, branched above; leaves broadly linear, 1–3 cm. long, 2–3 mm. wide, obtuse and mucronate or acute, 3-nerved, entire, slightly narrowed at base, sessile; flowers rose-colored, in loose racemes on branchlets, the pedicels 1–1.5 cm. long, erect; sepals elliptic or broadly ovate, 3–3.5 mm. long, abruptly acute, obsoletely 3-nerved at base, green with a few black projecting glands on the rather membranous margins; petals obovate; stigma terminal, capitellate; capsules globose, 3.5–4 mm. across; seeds oblong, flat, brown, lustrous, about 2 mm. long.——Aug.–Sept. Grassy slopes in mountains; Hokkaido, Honshu, Shikoku, Kyushu;

rather rare.——China, Korea, Manchuria, and e. Siberia.

2. Linum usitatissimum L. Ama. Glabrous annual; stems terete, 30–100 cm. long, often branching; leaves flat, broadly linear, 2–3.5 cm. long, 2–4 mm. wide, acute to acuminate, gradually and slightly narrowed at base, glaucous, sessile; flowers white or blue, the pedicels 2–4 cm. long; sepals ovate-oblong, 5–7 mm. long, entire, eglandular, acuminate, 3-nerved; petals obovate, about 10 mm. long; stigma decurrent along the inner side of styles; capsules globose, about 7 mm. across.——June–Aug. Widely cultivated and sometimes sparsely naturalized in our area.——Centr. Asia.

Fam. 109. ZYGOPHYLLACEAE Hamabishi Ka Caltrop Family

Shrubs or herbs woody at the base; branches often jointed at nodes; leaves opposite or alternate, pinnate or bifoliolate, the stipules persistent, often spinelike; flowers bisexual, actinomorphic or zygomorphic, white, red, or blue, rarely yellow; sepals 5, rarely 4, imbricate, rarely valvate; petals 4 or 5, rarely absent, imbricate or contorted, rarely valvate; disc usually present; stamens as many or 2 or 3 times as many as the petals, free, often with scales on inner side; ovary superior, usually 4- or 5-locular, rarely to 12-locular; style simple; ovules usually 2 in each locule, sometimes many; fruit various, but not a berry; seeds usually with endosperm.——About 21 genera, with about 160 species, in sandy areas and deserts of the Tropics and warmer regions.

1. TRIBULUS L. Hamabishi Zoku

Herbs with usually long-creeping branches; leaves even-pinnate, stipulate, opposite or one of each of the pairs reduced, thus appearing alternate; flowers solitary, pedunculate, axillary, white to yellow; sepals 5, often persistent, imbricate; petals 5, imbricate; disc annular, 10-lobed; stamens 10, at the base of the disc, the outer 5 slightly longer, opposite the petals, the inner with a gland outside at base; ovary sessile, 5- to 12-locular, the style short, the stigmas 5–12; ovules 1–5 in each locule; fruit 5-angled, indehiscent, separating into 5–12 horny or bony nutlets at maturity, with wings, spines, or tubercles on back; seeds without endosperm, the testa membranous.——More than a dozen species, in the Tropics and subtropics, especially abundant in Africa.

1. Tribulus terrestris L. Hamabishi. Annual or biennial, coarsely hirsute and puberulous on stems, leaf-rachis, and peduncles; stems branched, elongate, long-creeping, to 1 m. long, striate; leaves 1–6 cm. long including the short-petiole, the stipules free, lanceolate, about 3 mm. long, the leaflets of 4–8 pairs, oblong, 8–15 mm. long, 3–4 mm. wide, subobtuse, oblique at base, entire, white-pubescent especially on under side; peduncles 1–2 cm. long; sepals ovate-oblong, 4–5 mm.

long, gradually acute, deciduous, densely pilose on back; petals yellow, slightly exceeding the sepals; ovary densely pubescent; fruit about 1 cm. across, separating into 5 woody nutlets, each with firm tubercles and 2 thick spines on back. ——July–Oct. Sandy seashores; Honshu (Kantō Distr. and Wakasa Prov. and westw.), Shikoku, Kyushu.——Widespread in all tropical regions.

Fam. 110. RUTACEAE Mikan Ka Citrus Family

Shrubs or trees, sometimes herbs, with pellucid glands; leaves simple or compound, exstipulate; flowers bisexual, sometimes unisexual; sepals 4 or 5, imbricate; petals imbricate, free; stamens as many or twice as many as the petals, sometimes more numerous; disc usually present on the inner side of the stamens; ovary superior, the carpels 4 or 5, connate to free; ovules often 2 in each locule; fruit a berry, drupe, or capsule; seeds sometimes with endosperm, the embryo straight or curved.——About 100 genera, with about 1,000 species in warm temperate and tropical regions, especially abundant in Australia and S. Africa.

1A. Ovary 2- to 5-parted to the base or nearly so; style and stigma often connate; fruit a capsule or follicle or consisting of nutlets.
  2A. Ligneous; carpels usually with 2 ovules.
    3A. Inflorescence axillary or terminal; leaves usually pinnate.
      4A. Leaves alternate. .... 1. *Zanthoxylum*
      4B. Leaves opposite. .... 2. *Evodia*
    3B. Inflorescence supra-axillary; leaves simple, alternate. .... 3. *Orixa*
  2B. Herbaceous; carpels usually with 6–8 ovules. .... 4. *Boenninghausenia*
1B. Ovary connate or only slightly 2- to 5-lobed; style simple, terminal; fruit an indehiscent drupe or berry.
  5A. Fruit a drupe, with 2–5 stones.
    6A. Trees; leaves deciduous, pinnately compound, opposite. .... 5. *Phellodendron*
    6B. Shrubs; leaves evergreen, simple, alternate. .... 6. *Skimmia*
  5B. Fruit a berry.
    7A. Stamens 10, free; ovules solitary in each locule; styles very short, not jointed with the ovary, persistent. .... 7. *Glycosmis*
    7B. Stamens 20–30, often connate; ovules many in each locule; styles jointed with the ovary, deciduous. .... 8. *Citrus*

1. ZANTHOXYLUM L.   Sanshō Zoku

Shrubs or trees sometimes prickly, sometimes dioecious; leaves alternate, odd-pinnate, rarely 1- or 3-foliolate, the leaflets entire or obtusely toothed; inflorescence often paniculate-cymose; flowers small, green or greenish white, polygamous or unisexual; calyx 3- to 5-lobed, small, imbricate; petals 3–5 or absent; staminate flowers without a disc, the stamens 3–5; pistillate flowers with a small disc, the stamens absent or scalelike; carpels 1–5, each 1-locular, free; styles free or connate, the stigma capitate; ovules 2 in each locule, parallel; fruitlets 1–5, dehiscent, 1-seeded; seeds ellipsoidal, black, lustrous, the endosperm fleshy. ——About 150 species, chiefly in warmer regions.

1A. Flowers apetalous; branchlets usually with a large prickle on each side of the leaf-base; stamens alternate with the calyx-lobes.
  2A. Evergreen shrub with prickles much flattened laterally; petioles winged; leaflets 3–7, 3–8 cm. long. .... 1. *Z. planispinum*
  2B. Deciduous shrub with prickles thickened at base; petioles wingless; leaflets 11–19, 1–3.5 cm. long. .... 2. *Z. piperitum*
1B. Flowers with petals; branchlets irregularly prickly; stamens alternate with the petals.
  3A. Shrubs with slender branches; leaflets 1.5–5 cm. long, nearly destitute of pellucid glands except on the margins, not glaucous.
                                                           3. *Z. schinifolium*
  3B. Trees with rather thick branchlets; leaflets (3.5–)5–12 cm. long, with pellucid glands, usually glaucescent beneath.
    4A. Leaves 15–30 cm. long; leaflets 3.5–8 cm. long; branches of inflorescence glabrous or loosely puberulous. .... 4. *Z. fauriei*
    4B. Leaves 30–80 cm. long; leaflets 5–15 cm. long; branches of inflorescence rather densely puberulous on one side.
                                                            5. *Z. ailanthoides*

1. Zanthoxylum planispinum Sieb. & Zucc. *Z. alatum* Roxb. var. *planispinum* (Sieb. & Zucc.) Rehd. & Wils.; *Z. alatum* var. *subtrifoliolatum* Franch.——FUYUZANSHŌ. Dioecious shrub; prickles on branches at base of each leaf 2, spreading, 1.2–1.5 cm. long, prominently flattened and dilated at base; petioles 1–3 cm. long, winged; leaves pinnately 3- to 5(–7)-foliolate, the leaflets sessile, broadly lanceolate to narrowly oblong, 3–7 cm. long, 1–2.5 cm. wide, acute to acuminate, often emarginate, glabrous or spreading-pilose on the midrib beneath, with pellucid glands on margins, entire or obtusely toothed; inflorescence a nearly sessile axillary raceme often branched at base, 1–3 cm. long; flowers pale yellow.—— May. Low mountains; Honshu (Kantō and westw.), Shikoku, Kyushu.——Ryukyus, Formosa, China, and Korea.

2. Zanthoxylum piperitum (L.) DC. *Fagara piperita* L.——SANSHŌ. Dioecious shrub; prickles on branches at base of leaves 2, 5–8 mm. long; leaves puberulous while young, 5–15 cm. long, short-petiolate, the leaflets 11–19, sessile, broadly lanceolate to ovate, 1–3.5 cm. long, 6–12 mm. wide, emarginate, cuneate to obtuse at base, obtusely toothed with pellucid glands in the sinus; inflorescence terminal on short branches, pedunculate, many-flowered; flowers greenish yellow, small.——Apr.–May. Hills and mountains; Hokkaido, Honshu, Shikoku, Kyushu; rather common and frequently cultivated.——Korea, Manchuria, and China.——Forma **inerme** (Makino) Makino. *Z. piperitum* var. *inerme* Makino——ASAKURAZANSHŌ. Unarmed. The seeds and young leaves are used as a spice and the plant is often planted for this purpose.——Forma **brevispinosa** Makino. YAMA-ASA-KURAZANSHŌ. Prickles short. Intermediate between the typi-

cal phase and the preceding form.——Forma **ovalifoliolatum** (Nakai) Makino. *Z. ovalifoliolatum* Nakai——RYUJIN-ZAN-SHŌ. Leaflets 3–5, ovate.——Rarely cultivated.

3. Zanthoxylum schinifolium Sieb. & Zucc. *Z. mantchuricum* Benn.; *Fagara schinifolia* (Sieb. & Zucc.) Engl.; *F. mantchurica* (Benn.) Honda——INUZANSHŌ. Deciduous dioecious shrub with loosely prickly branches; leaves 8–20 cm. long, short-petiolate, the leaflets 13–23, sessile, broadly lanceolate to narrowly ovate, 1.5–4(–5) cm. long, 6–15 mm. wide, emarginate, sometimes minutely pubescent, obtusely toothed with a pellucid gland in each sinus; inflorescence a dense many-flowered cyme, terminal on branchlets; calyx-lobes minute; petals oblong, about 2 mm. long, greenish.——Aug. Low mountains; Honshu, Shikoku, Kyushu; common and variable.——Korea, China, and Manchuria.

4. Zanthoxylum fauriei (Nakai) Ohwi. *Fagara fauriei* Nakai; *F. shikokiana* Makino; ? *Z. ailanthoides* × *Z. schinifolium*——KO-KARASUZANSHŌ. Dioecious tree; branches thick, reddish when young, often glaucescent, loosely or densely and irregularly prickly; leaves glabrous, 15–30 cm. long, the petioles about 3 cm. long, the leaflets subsessile, broadly lanceolate, 3.5–8 cm. long, 1–1.5(–2) cm. wide, gradually narrowed to an emarginate apex, pellucid-dotted, minutely toothed with a pellucid gland in each sinus; inflorescence a dense many-flowered terminal cyme; calyx-lobes minute; petals oblong, about 2 mm. long.——Aug. Honshu (Tōkaidō and Kinki Distr.), Shikoku, Kyushu; rare.——s. Korea.

5. Zanthoxylum ailanthoides Sieb. & Zucc. *Fagara ailanthoides* (Sieb. & Zucc.) Engl.——KARASUZANSHŌ. Glabrous dioecious tree with thick, prickly branches; leaves 30–80 cm. long, petiolate, the leaflets 19–23, subsessile, broadly lanceolate, acuminate, emarginate, rounded at base, obtusely serrulate with a pellucid gland in each sinus; inflorescence a dense many-flowered terminal cyme; calyx-lobes minute, semi-rounded; petals greenish yellow, about 2.5 mm. long.——July-Aug. Mountains; Honshu, Shikoku, Kyushu; rather common.——s. Korea, China, Ryukyus, and Formosa.

2. EVODIA Forst.   Goshuyu Zoku

Unarmed trees or shrubs; leaves opposite, odd-pinnate, ternate or simple, entire, pellucid-dotted; cymes axillary, often compound in a terminal panicle; flowers unisexual, small; sepals 4 or 5, imbricate; petals 4 or 5, valvate or slightly imbricate; disc urceolate, with 4 or 5 undulate teeth; stamens 4 or 5, inserted on the base of the disc; ovary 4- or 5-parted, 4- or 5-locular; style cylindric, simple, the stigma 4-lobed; ovules 2 in each locule; fruitlets 4 or 5, coriaceous, dehiscent, 2-valved, 1(–2)-seeded; seeds ellipsoidal, the embryo straight, embedded in a fleshy endosperm.——About 50 species, in e. and s. Asia, Australia, and Polynesia.

1. **Evodia glauca** Miq.   *E. meliaefolia* sensu auct. Japon., non Benth.——Hama-sendan, Shima-kuroki, Urajiro-goshu-yu.   Dioecious tree with appressed-puberulous, rather thick branches; leaves 15–30 cm. long, on petioles 4–6 cm. long, the leaflets 11–19, petiolulate, obliquely ovate or narrowly ovate, 4–7 cm. long, appressed-puberulous on upperside while young, glaucous beneath, caudate, with an obtuse tip, entire, rounded on outer margin and acute on inner margin at base; cymes terminal, densely many-flowered, puberulous; flowers small; sepals 5, broadly ovate, ciliolate; petals 5, pubescent inside; filaments in staminate flowers hairy at base on outer side; ovary in pistillate flowers deeply 4- or 5-lobed, the style short and puberulous, the stigma rather large, with 4 or 5 undulate teeth; fruitlets densely white-puberulous inside; seeds ellipsoidal, about 3 mm. long.——Honshu (Nagato Prov.), Shikoku, Kyushu.——Ryukyus, Formosa, and China.

3. ORIXA Thunb.   Ko-kusagi Zoku

Deciduous, unarmed, dioecious shrub; leaves alternate, short-petiolate; flowers pale green, small, 4-merous, solitary or in racemes; sepals ovate, connate at base; petals elliptic, imbricate; disc 4-lobed in staminate flowers; stamens 4; ovary in pistillate flowers deeply 4-lobed, with a short common style and 4-lobed stigma; fruitlets 4, each 1-seeded and dehiscent, 2-valved; seeds black, globose.——One species in e. Asia.

1. **Orixa japonica** Thunb.   *Othera orixa* (Thunb.) Lam.; *Celastrus orixa* (Thunb.) Sieb. & Zucc.; *Evodia rami-flora* A. Gray; *C. japonicus* (Thunb.) K. Koch; *C. dilatatus* Thunb.——Ko-kusagi.   Erect foetid shrub with grayish branches, slightly puberulous while young; leaves alternate, two together on one side, the petiole 5–10 mm. long, the blade obovate, rhombic-ovate or elliptic, 5–12 cm. long, 3–7 cm. wide, abruptly acute with an obtuse tip, puberulous especially on nerves while young, entire or with a few irregular sinuses on margin; inflorescence axillary on the branches of the preceding year, the staminate inflorescences in few-flowered racemes, 2–3 cm. long, the bracts elliptic, membranous, deciduous, 3–5 mm. long, the pedicels 3–6 mm. long, the pistillate flowers solitary, the pedicels 1–2 cm. long; petals spreading, about 3 mm. long, greenish; fruitlets obliquely ellipsoidal, 8–10 mm. long, transversely nerved; seeds about 4 mm. long.——Apr.–May. Woods and thickets in low mountains; Honshu, Shikoku, Kyushu; common.——s. Korea and China.

4. BOENNINGHAUSENIA Reichenb.   Matsu-kaze-sō Zoku

Slender, thin perennials with pellucid glands; stems terete, branched; leaves alternate, ternately 2- to 3-pinnate, the leaflets entire; panicles compound, leafy at base; flowers white, bisexual; sepals small, 4, connate at base, persistent; petals 4, imbricate; disc urceolate; stamens 6–8, inserted on the base of the disc; ovary stipitate, 4-lobed, 4-locular, the style short, connate, deciduous; ovules 6–8 in each locule; fruitlets free, 4, dehiscent ventrally, the seeds few in each fruitlet, black, tubercled, reniform, the embryo fleshy.——Two species, India and e. Asia.

1. **Boenninghausenia japonica** Nakai.   *Ruta japonica* Sieb. ex Miq. in syn.; *B. albiflora* sensu auct. Japon., non Reichenb.; *B. albiflora* var. *japonica* (Nakai) S. Suzuki——Matsu-kaze-sō.   Glaucous, glabrous, flaccid perennial; stems erect, 40–80 cm. long, branched above; leaves deltoid, the petioles 5–10 cm. long, the leaflets obovate or elliptic, 1–2.5 cm. long, 7–20 mm. wide, rounded to emarginate, cuneate, acute or obtuse at base, pellucid-dotted, entire, white beneath; panicles terminal, large, the bracts narrowly obovate, 4–6 mm. long, simple; flowers white, on delicate pedicels; sepals oblong, about 1 mm. long; petals ascending-spreading, oblong, about 4 mm. long; stamens exserted; fruit on a stipe about 2 mm. long, the fruitlets ovoid, ascending, about 3 mm. long, the seeds dark brown, reniform, about 1.5 mm. long, tubercled.——Aug.–Oct. Woods in mountains; Honshu (Kantō Distr. and westw.), Shikoku, Kyushu; common.

5. PHELLODENDRON Rupr.   Kihada Zoku

Unarmed, deciduous, dioecious trees; leaves opposite, odd-pinnate, the leaflets petiolulate, ovate, with pellucid glands, nearly entire; cymes terminal, peduncled; flowers small, yellowish green; sepals 5–8, ovate, small; petals as many as and much longer than the sepals, oblong; stamens 5–6, exserted, reduced to staminodes in the pistillate flowers; ovary short-stipitate, 5-locular, the style short, the stigma 5-lobed; drupe black, nearly globose, the drupelets 5, each with a single seed, the endosperm scanty.——Few species, in e. Asia.

1. **Phellodendron amurense** Rupr. KIHADA. Tree with rather thick branches; leaves 20–40 cm. long, petiolate, the leaflets 5–13, membranous, narrowly ovate or ovate-oblong, 5–10 cm. long, 3–5 cm. wide, caudate-acuminate, rounded at base, usually ciliate while young or slightly pubescent beneath near base, deep green, glaucous beneath; cymes rather large, 5–7 cm. across, puberulous; flowers on short pedicels; sepals ovate-triangular, about 1 mm. long; petals narrowly oblong, pubescent inside, about 4 mm. long; fruit nearly globose, about 1 cm. across.——July. Woods in mountains; Hokkaido, Honshu, Shikoku, Kyushu.——Korea, n. China, Manchuria, Ussuri, and Amur. Variable.

Var. **sachalinense** F. Schmidt. *P. sachalinense* (F. Schmidt) Sarg.——HIRO-HA-NO-KIHADA. Bark thin; leaflets slightly broader, nearly destitute of marginal hairs; cymes nearly glabrous.——Hokkaido, Honshu.

Var. **japonicum** (Maxim.) Ohwi. *P. japonicum* Maxim.; *P. nikkomontanum* Makino——ŌBA-NO-KIHADA. Leaflets 9–15, rather smaller, spreading-pubescent beneath especially on nerves; cymes puberulous.——Honshu, Shikoku, Kyushu.

Var. **lavallei** (Dode) Sprague. *P. lavallei* Dode——MI-YAMA-KIHADA. Similar to the preceding variety, but with thicker, corky bark; leaflets broadly cuneate at base.——Hokkaido, Honshu.

6. SKIMMIA Thunb.   MIYAMA-SHIKIMI ZOKU

Glabrous, unarmed, evergreen shrubs; leaves simple, entire, alternate, short-petiolate, with pellucid glands; cymes terminal; flowers bisexual or polygamous, white, small, 4- or 5-merous; sepals imbricate; petals oblong, valvate or slightly imbricate, longer than the sepals; sterile carpels connate at base; pistillate flowers with 4–5 staminodes; ovary of 2–5 completely connate carpels, each with a single ovule; style short, the stigma 2- to 5-lobed; drupe obovoid or globose, with 2–4 stones, the endosperm fleshy. ——About 10 species, in e. Asia and the Himalayas.

1. **Skimmia japonica** Thunb. *S. oblata* Moore; *S. fragrantissima* Moore——MIYAMA-SHIKIMI. Dioecious shrub to 1.5 m. high, glabrous except for the puberulous inflorescence, the branches ascending, grayish; leaves often falsely verticillate, narrowly oblong to narrowly obovate, 6–12 cm. long inclusive of the short petioles, 2.5–3.5 cm. wide, abruptly acute and ending in an obtuse tip, gradually narrowed to the base, entire, lustrous on upper side, yellowish beneath; cymes 4–8 cm. across; flowers 4-merous, small, 5–6 mm. across, on puberulous pedicels; sepals and bracts minute, broadly triangular, minutely ciliate; petals narrowly oblong, 4–5 mm. long, with pellucid glands; stamens as long as the petals; fruit globose, about 7 mm. across, red, the style deciduous, 0.8 mm. long, with a somewhat capitate 4-fid stigma.——Apr. Woods in mountains; Honshu (Kantō Distr. and westw.), Shikoku, Kyushu; rather common.——Ryukyus, Formosa, and ? China. ——Forma **rugosa** (Yatabe) Ohwi. *S. japonica* var. *rugosa* Yatabe; *S. rugosa* (Yatabe) Makino——UCHI-DASHI-MIYAMA-SHIKIMI. Nerves of leaves impressed on upper side.

Var. **intermedia** Komar. *S. repens* var. *intermedia* (Komar.) Nakai; *S. japonica* var. *repens* forma *rugata* Ohwi—— KARAFUTO-SHIKIMI, UCHIKOMI-MIYAMA-SHIKIMI. Low shrub with ascending branches often creeping at base; leaves with impressed nerves on upper side.——A rare variant first found in Sakhalin; also occurs in Hokkaido and Honshu (centr. & n. distr.).——Forma **repens** (Nakai) Hara. *S. repens* Nakai; *S. japonica* var. *repens* (Nakai) Ohwi——MIYAMA-SHIKIMI. The leaves flat.——Hokkaido, Honshu, Shikoku, Kyushu; common in mountains.——Sakhalin and s. Kuriles.

7. GLYCOSMIS Correa   HANA-SHIMBŌGI ZOKU

Unarmed trees or shrubs; leaflets 1–7, alternate, entire or serrate; panicles axillary, rarely terminal; flowers small, bisexual; calyx 5-parted, the segments broad, imbricate; petals 5, free, oblong to rounded, imbricate; stamens 10, free, the filaments subulate, dilated at base, the anthers small, often with a gland at the apex or on the back; disc stipelike or cup-shaped; ovary 2- to 5-locular, the style very short, the ovules solitary in each locule; berry juicy, 1- to 3-seeded, the seeds oblong, the testa membranous, the endosperm absent.——About 10 species, in se. Asia and Australia.

1. **Glycosmis pentaphylla** (Retz.) Correa. *Limonia pentaphylla* Retz.; *G. citrifolia* (Willd.) Lindl.; *L. citrifolia* Willd.; *L. parviflora* Sims——HANA-SHIMBŌGI, GEKKITSU-MO-DOKI. Large evergreen shrub with minute brown hairs while young; petioles 1–2 cm. long, the leaf-rachis short, the leaflets 1–3, alternate, sessile, narrowly oblong, 7–12 cm. long, 2–3.5 cm. wide, obtuse to acute; panicles axillary, densely flowered; flowers small, short-pedicelled, 5-merous; petals about 3 mm. long; stamens 10, as long as the petals, glabrous; ovary glabrous.——Kyushu (southern distr., possibly not indigenous).——s. China, India, Indochina, and Malaysia.

8. CITRUS L.   MIKAN ZOKU

Usually armed, evergreen shrubs or trees; leaves alternate, simple, entire or obtusely toothed, coriaceous, lustrous, with pellucid glands, the petioles often winged; flowers bisexual, axillary, solitary, fascicled or in short panicles, white, fragrant; calyx cup-shaped or urceolate, 3- to 5-lobed; petals 4–8, narrowly oblong, imbricate; stamens 15 or more, in a few fascicles; disc large, cup-shaped or annulate; ovary 8- to 15-locular; style long, deciduous, the stigma capitate, the ovules several in each locule; berry large, globose, with several segments filled with juicy pulp composed of stalked pulp-vesicles; seeds 1–8 in each locule, without endosperm.——Many species in India and se. Asia.

1. Citrus tachibana (Makino) T. Tanaka. *C. auran-tium* var. *tachibana* Makino——TACHIBANA. Large glabrous shrub; branches 3-angled, with scattered spines 2–20 mm. long; leaves chartaceous, narrowly ovate, 3–8 cm. long, 1.5–3.5 cm. wide, acute with an emarginate tip, obtuse to broadly cuneate at base, entire, the petioles 5–12 mm. long, narrowly winged; flowers solitary or in pairs, terminal and axillary, the pedicels about 5 mm. long; calyx cup-shaped, the lobes depressed 3-angular; petals 5, ascending, white, narrowly obovate-oblong, about 12 mm. long, acute; stamens about 20; fruit depressed-globose, yellow when mature, 2–2.5 cm. across.——June. Honshu (Izu, Kii, and Nagato Prov.), Shikoku, Kyushu.——Korea (Quelpaert Isl.), Ryukyus, and Formosa.

Fam. 111. **SIMAROUBACEAE** NIGAKI KA Quassia Family

Trees or shrubs usually with bitter bark; leaves commonly alternate, pinnate, rarely simple, without pellucid glands; flowers bisexual or unisexual, actinomorphic, usually small, in panicles or spikes; sepals 3–5, more or less connate at base; petals 3–5 or absent; stamens twice as many as the sepals; ovary superior, surrounded by a prominent disc; carpels 2–5, free at base; style or stigma variously connate to the base; ovules 1 to several in each locule; fruit a drupe, rarely a berry or samara, the seeds with or without a scanty endosperm.——About 28 genera, with about 150 species, chiefly in the Tropics.

1. PICRASMA Bl. NIGAKI ZOKU

Deciduous trees; leaves alternate, odd-pinnate, fasciculate toward the top of the branchlets; flowers polygamous, small, in pedunculate, axillary cymes; sepals 4 or 5, small, ovate, persistent; petals as many as the sepals, oblong, nearly valvate; stamens as many as the petals and longer, inserted at the base of a 4- or 5-lobed disc; carpels 2–5, free, with a connate common style at the apex; fruit of 1–5 globose-obovoid druplets, the seeds without endosperm.——About 8 species, in the Tropics and subtropics.

1. Picrasma quassioides (D. Don) Benn. *Simaba quassioides* D. Don; *Rhus ailanthoides* Bunge; *P. ailanthoides* (Bunge) Planch.; *P. japonica* A. Gray——NIGAKI. Glabrescent tree; leaves 20–30 cm. long, the petioles 2–4 cm. long, the stipules caducous, small, lanceolate, the leaflets subsessile, oblique, narrowly ovate to broadly lanceolate or oblong, 4–10 cm. long, 1.5–3 cm. wide, long-acuminate with an obtuse tip, acute to broadly cuneate at base, obtusely toothed, glabrous or nearly so; cymes axillary, on peduncles 4–10 cm. long, puberulous, loosely many-flowered, 4–7 cm. across; flowers yellowish green; sepals ovate; petals obovate, persistent, spreading, 3–4 mm. long; drupelets obovoid-globose, 6–7 mm. long.——May-June. Woods in lowlands and hills; Hokkaido, Honshu, Shikoku, Kyushu; common.——Ryukyus, Formosa, Korea, China, and India.

Fam. 112. **MELIACEAE** SENDAN KA Mahogany Family

Trees or shrubs, rarely herbs; leaves usually alternate, pinnate, rarely simple, exstipulate; flowers bisexual, rarely unisexual, actinomorphic, commonly in axillary cymes; sepals 4–6, usually connate at base, imbricate, as many as the petals; stamens usually twice as many as the petals, commonly connate at base into a tube; ovary superior, 2- to 5-locular, with a disc at base, with 1 or 2 ovules in each locule; fruit a berry, drupe or capsule; seeds sometimes winged, the endosperm present or absent.——About 50 genera, with about 800 species, chiefly in the Tropics.

1. MELIA L. SENDAN ZOKU

Deciduous trees or shrubs; leaves alternate, twice-pinnate, the leaflets entire or toothed; flowers bisexual, in axillary panicles; calyx 5- or 6-parted, the segments small; petals 5 or 6, free, imbricate; stamens 10–12, the filaments connate into a tube, the anthers attached on the inner side above and alternate with the lobes at the summit of the tube; ovary on a short disc, 5- to 8-locular, terminated by a cylindrical style, the ovules 2 in each locule; fruit a drupe, with fleshy exocarp.——About 10 species, in s. Asia and Australia.

1. Melia azedarach L. *M. azedarach* var. *japonica* (G. Don) Makino; *M. japonica* G. Don——SENDAN. Tree with fine stellate pubescence when young; branches thick; leaves alternate, petiolate, large, bipinnate, tripinnate, or rarely pinnate, the leaflets petioluled, ovate or ovate-elliptic, 3–7 cm. long, 1.2–2.5 cm. wide, chartaceous, abruptly acuminate or caudate with an obtuse point, oblique at base, green on upper side, yellowish beneath; panicles axillary, 10–15 cm. long (inclusive of the peduncles 5–10 cm. long), 5–7 cm. wide, loosely many-flowered; flowers pale purple or rarely white, on short pedicels; sepals ovate-oblong, 2–2.5 mm. long; petals oblong-oblanceolate, 1–1.2 cm. long, spreading; stamens about 7 mm. long, the tube about 2 mm. across; drupes ellipsoidal, yellowish brown when mature, about 17 mm. long.——May-June. Thickets in lowlands and hills; Shikoku, Kyushu; widely planted in our area for ornamental purposes.——Bonins, Ryukyus, Formosa, e. and w. Asia.

Fam. 113. **POLYGALACEAE** HIME-HAGI KA Milkwort Family

Herbs or shrubs, sometimes small trees; leaves alternate, simple, exstipulate; flowers bisexual, zygomorphic; sepals 5, free, imbricate, the inner 2 larger, sometimes petallike; petals 3–5, the outer 2 free or connate with the lowest one, the upper 2 free, scalelike or absent; stamens 8 or 4 or 5, monadelphous, connate to above the middle, rarely free, often adnate to the petals;

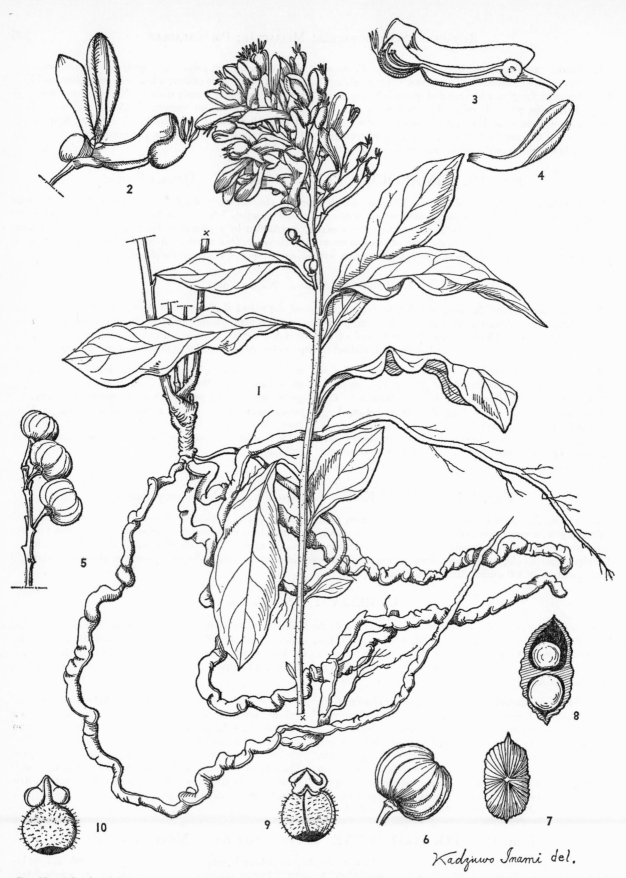

Kadjuwo Imami del.

Fɪɢ. 10.—*Polygala reinii* Fr. & Sav. 1, Habit; 2, flower; 3, longitudinal section of same, the lateral sepals removed; 4, lateral sepal; 5, fruiting raceme; 6, capsule; 7, the same, seen from the apex; 8, fruit transverse section; 9, 10, seeds.

anthers opening by terminal pores; ovary superior, usually 2-locular, the styles simple, the ovule usually solitary in each locule; fruit a capsule or drupe, the seeds often pilose and strophiolate, the endosperm present, the embryo straight.——About 10 genera, with about 700 species, in tropical to temperate regions.

1A. Sepals unequal, 2 larger and winglike; stamens 8; flowers relatively large. .... 1. *Polygala*
1B. Sepals slightly unequal; stamens 4; flowers small or minute. .... 2. *Salomonia*

1. POLYGALA L. Hime-hagi Zoku

Herbs, shrubs, or rarely small trees; leaves usually exstipulate; flowers usually in racemes, yellow or purple; sepals 5, unequal; petals 3, unequal, adnate to each other and to the stamen-tube, the lowest clawed; stamens 8 or very rarely fewer; ovary 2-locular, with one ovule in each locule; capsules loculicidally dehiscent, 2-seeded, the seeds ovoid or globose.——About 450 species, chiefly in temperate regions, the woody species chiefly in the warmer regions of the world.

1A. Sepals persistent; small perennial. .... 1. *P. japonica*
1B. Sepals deciduous after anthesis.
  2A. Annual with leaves 1–3 cm. long; flowers about 2 mm. long. .... 2. *P. tatarinowii*
  2B. Perennial with leaves 8–12 cm. long; flowers about 2 cm. long. .... 3. *P. reinii*

1. Polygala japonica Houtt. *P. sibirica* var. *japonica* (Houtt.) T. Itō ex T. Itō & Matsum.; *P. hondoensis* Nakai——Hime-hagi. Small perennial; roots elongate; stems 10–30 cm. long, slender, green, prostrate or decumbent from a short caudex, ascending above, tufted, with scattered minute curved hairs; leaves chartaceous, very short petiolate, elliptic to ovate or broadly lanceolate, 1–2.5(–3) cm. long, 3–12 mm. wide, acute to subobtuse at both ends, vivid green, slightly puberulous or glabrate, with a raised midrib beneath, the reticulated veinlets rather prominent above; racemes loosely few-flowered, 1–3 cm. long, the peduncles short, supra-axillary; flowers purple; lateral sepals ovate, obtuse, rather petallike, 5–7 mm. long, accrescent after anthesis, to 10 mm. long, becoming green in fruit, the remaining sepals lanceolate, 3–4 mm. long; petals connate, spreading above, 6–7 mm. long, the lowest one with a fimbriate appendage at apex; capsules flat, winged on the margins, rounded-cordate, retuse, 7–8 mm. wide, membranous, glabrous; seeds ellipsoidal, brown, about 2.5 mm. long, white-pilose, strophiolate on upper margin.——Apr.–July. Grassy slopes in lowlands and hills; Hokkaido, Honshu, Shikoku, Kyushu; very common.——Ryukyus, Formosa, China, Korea, Manchuria, and Ussuri.

2. Polygala tatarinowii Regel. *P. triphylla* Hamilt., non Burm.; *P. sieboldiana* Miq.——Hina-no-kinchaku. Nearly glabrous, erect, delicate, green annual; stems 4–12 cm. long, angular, often branched at base; leaves several, orbicular-ovate to elliptic, 1–3 cm. long, acute to obtuse, abruptly narrowed to a winged petiolelike short base, entire, ciliolate; racemes terminal, elongate after anthesis, to 8 cm. long, the

peduncles about 1 cm. long; flowers many, small, pale purple, about 2 mm. long, one-sided, densely disposed on the axis, the pedicels 1–1.5 mm. long; capsules rounded, flat, about 3 mm. wide, wingless; seeds ellipsoidal, 1–1.2 mm. long, puberulous, with a small strophiole at apex.——July–Oct. Honshu, Shikoku, Kyushu; rather rare.——Korea, n. China, e. Siberia, and India.

3. Polygala reinii Fr. & Sav. Kaki-no-ha-gusa. Fig. 10. Soft, puberulous to glabrous green perennial with thick roots and rather thick woody rhizomes; stems 20–35 cm. long, erect, usually simple, green, terete, rather thick; leaves thin, few, mostly toward the top of the stems, obovate-oblong to oblong, 8–17 cm. long, 3–7 cm. wide, abruptly acute to abruptly acuminate, acute or gradually narrowed at base, short-petiolate, thin, soft, green, slightly paler beneath; racemes solitary, terminal, about 10-flowered, 2–5 cm. long, erect; flowers short-pedicelled, yellow, about 2 cm. long, the 2 lateral sepals larger, broadly oblanceolate, obtuse, as long as the petals, the 2 smaller sepals oblong, obtuse to rounded, ciliate, the upper one rather longer, bowed out on the back, about 7 mm. long; petals 3, connate, about 2 cm. long, the lowest with a fimbriate appendage at apex; fruit depressed-orbicular, glabrous, about 10 mm. wide, longitudinally ribbed; seeds about 4 mm. long, reddish brown.——Woods in hills and low mountains; Honshu (Tōkaidō and Kinki Distr.); rather common.——Forma **angustifolia** (Makino) Ohwi. *P. reinii* var. *angustifolia* Makino——Nagaba-kaki-no-ha-gusa. Leaves narrow, lanceolate.

2. SALOMONIA Lour. Hina-no-kanzashi Zoku

Small herbs with alternate often much-reduced scalelike leaves; flowers minute, in terminal spikes; sepals nearly equal or the inner 2 larger; petals 3, connate into a corolla split above, unappendaged; stamens 4 (–5), the anthers opening by pores; ovary 2-locular; capsules membranous, flat, transversely oblong to obcordate, often ciliate, loculicidally dehiscent, the seeds globose, without endosperm, the strophioles minute or absent.——About 10 species, in the Tropics of Asia and Australia.

1. Salomonia oblongifolia DC. *S. stricta* Sieb. & Zucc. ——Hina-no-kanzashi. Glabrous, delicate annual; stems 8–25 cm. long, slender, striate, simple or with ascending branches; leaves nearly sessile, oblong or elliptic, 3–8 mm. long or the upper lanceolate ones to 12 mm. long, 1.5–4 mm. wide, mucronate, entire or with few spinules on the upper margin; spikes many-flowered, at first short and dense, later

elongating, 2–6 cm. long; flowers sessile, rose-purple, about 2 mm. long; sepals lanceolate, less than 1 mm. long; lateral petals shorter than the lower; capsules less than 2 mm. wide, glabrous, reniform, flat, with a few spines on the margins, the seeds about 0.8 mm. long, ellipsoidal, dull.——Aug.–Sept. Wet places in lowlands; Honshu, Shikoku, Kyushu; rare.—— Formosa, s. Korea, India, Malaysia, and Australia.

Fam. 114. EUPHORBIACEAE　Tōdaigusa Ka　Spurge Family

Herbs or shrubs, sometimes trees, usually with milky juice; leaves alternate, rarely opposite or verticillate, simple or compound, usually stipulate; flowers unisexual; sepals imbricate or valvate, sometimes absent; petals sometimes absent; stamens 1 to many, free or connate, the anthers 2 (–4)-locular; ovary usually 3-locular, the styles free or connate at base, the ovules 1 or 2 in each locule, pendulous; fruit a capsule or drupe, the seeds often with a distinct hilum, the endosperm fleshy, the embryo straight.——About 280 genera, with about 8,000 species, chiefly in the Tropics.

1A. Fruit a drupe; ovary incompletely 2- to 4-locular, with 2 ovules in each locule; evergreen trees or shrubs. .... 1. *Daphniphyllum*
1B. Fruit a capsule, rarely a drupe; ovary completely 3- to many-locular or 1-locular in *Antidesma*.
  2A. Inflorescence various, but not a cyathium.
    3A. Ovules 2 in each locule.
      4A. Inflorescence an axillary fascicle; ovary 3- to many-locular.
        5A. Styles slender; herbs or shrubs.
          6A. Staminate flowers without a reduced pistil.
            7A. Flowers with a disc. .... 2. *Phyllanthus*
            7B. Flowers without a disc. .... 3. *Glochidion*
          6B. Staminate flowers with a reduced pistil. .... 4. *Securinega*
        5B. Styles dilated; evergreen tree. .... 5. *Putranjiva*
      4B. Inflorescence a spike or raceme; ovary 1-locular; fruit a 1-seeded, indehiscent drupe. .... 6. *Antidesma*
    3B. Ovules solitary in each locule.
      8A. Sepals of staminate flowers valvate; stamens usually many.
        9A. Herbs.
          10A. Leaves alternate; anthers cylindric-oblong, usually curved. .... 7. *Acalypha*
          10B. Leaves opposite; anthers nearly globose. .... 8. *Mercurialis*
        9B. Trees; anthers oblong. .... 9. *Mallotus*
      8B. Sepals of staminate flowers imbricate; stamens 2 or 3; trees. .... 10. *Sapium*
  2B. Inflorescence a cyathium, consisting of much-reduced unisexual flowers surrounded by a cuplike involucre. .... 11. *Euphorbia*

1. DAPHNIPHYLLUM Bl.　Yuzuri-ha Zoku

Glabrous, evergreen, dioecious trees or shrubs; leaves alternate, entire, petiolate, exstipulate; flowers small, apetalous, in axillary racemes; sepals 3–8, minute; stamens 5–18, the filaments short, the anthers conspicuous; ovary incompletely 2- to 4-locular, with 2 ovules in each locule, the styles 2–4, short, reflexed; fruit an ellipsoidal drupe; endosperm abundant, the embryo minute, straight, at the top of the seed.——About 25 species, in e. and se. Asia.

1A. Sepals fully evident; staminodes in pistillate flowers distinct, sometimes with reduced anthers; veinlets of leaves rather scattered, scarcely raised beneath. .... 1. *D. macropodum*
1B. Sepals very small, less evident but distinct; staminodes in pistillate flowers scarcely developed; veinlets of leaves rather dense, often reticulate, raised beneath. .... 2. *D. teijsmannii*

1. Daphniphyllum macropodum Miq.　Yuzuri-ha. Evergreen, glabrous tree with thick green branchlets; leaves coriaceous, fascicled at the tips of branchlets, narrowly oblong or narrowly obovate-oblong, 15–20 cm. long, 4–7 cm. wide, abruptly acute, acute to obtuse at base, entire, deep green and lustrous on upper side, glaucous beneath, the lateral nerves 16–19 pairs, the petioles reddish, 4–6 cm. long; racemes axillary, 4–8 cm. long; flowers yellowish green, pedicellate, the staminate with about 10 stamens; ovary in the pistillate flowers ellipsoidal to broadly ovoid, about 2 mm. long, surrounded at the base by a series of staminodes, the styles 2(–4); drupe ellipsoidal, dark blue, about 1 cm. long.——May–June. Woods; Honshu (centr. distr. and westw.), Shikoku, Kyushu. ——s. Korea.——Cv. **Variegatum.** *D. macropodum* forma *variegatum* (Bean) Rehd.; *Tetranthera lhuysii* Carr.; *D. macropodum* var. *variegatum* Bean; *D. macropodum* var. *lhuysii* (Carr.) Nakai; *D. macropodum* forma *lhuysii* (Carr.) Ohwi——Fuiri-yuzuri-ha. Leaves variegated.——Cultivated. ——Forma **viridipes** (Nakai) Ohwi. *D. macropodum* var. *viridipes* Nakai——Aojiku-yuzuri-ha, Inu-yuzuri-ha. Petioles green, not reddish.

Var. **humile** (Maxim.) Rosenth. *D. humile* Maxim.; *D. jezoense* Bean——Ezo-yuzuri-ha. Shrub, branched at base; leaves 10–15 cm. long, thinner, with 12–15 pairs of lateral nerves.——Woods; Hokkaido, Honshu (Japan Sea side of centr. distr. and northw.).——Korea.

2. Daphniphyllum teijsmannii Zoll. *D. roxburghii* sensu auct. Japon., non H. Baill.; *D. buergeri* Muell.-Arg.; *D. glaucescens* sensu auct. Japon., non Bl.——Hime-yuzuri-ha. Evergreen, glabrous tree with thick branchlets; leaves coriaceous, oblong or narrowly so, 6–12 cm. long, abruptly acute to rounded at base, glaucous beneath, with 8–10 pairs of lateral nerves, the petioles 1.5–3 cm. long; racemes 4–8 cm. long, several-flowered; flowers pedicellate, the staminate with about 8-stamens; ovary in pistillate flowers broadly ovoid; fruit ellipsoidal, 8–10 mm. long.——May. Honshu (centr. distr. and westw.), Shikoku, Kyushu.——Ryukyus and Formosa.

Var. **hisautii** Hurusawa.　Suruga-hime-yuzuri-ha. Leaves thinner, chartaceous, with petioles 3–5 cm. long; fruiting racemes pendulous. With the typical phase.

2. PHYLLANTHUS L.　Ko-mikan-sō Zoku

Dioecious or monoecious herbs or shrubs; leaves usually small, entire, often distichous, although arranged like leaflets of a pinnate leaf; flowers unisexual, with a well-developed disc; sepals 4–6, in two cycles, sometimes petallike; stamens in staminate flowers usually 3, sometimes 2 to 5, the filaments free or connate; ovary in pistillate flowers 3- to several-locular, the styles sometimes bifid; fruit a capsule, sometimes fleshy; seeds 3-angled in cross section, rounded on back.——About 500 species, chiefly in the Tropics and subtropics, few in temperate regions.

1A. Monoecious annuals.
  2A. Capsules nearly sessile, transversely rugose; stems without ordinary leaves except at base, these on the branches, obovate-oblong or
       narrowly oblong, rounded, mucronate. .... 1. *P. urinaria*
  2B. Capsules short-pedicellate, smooth; stems and branchlets leafy; leaves broadly lanceolate or narrowly oblong, rather obtuse.
                                                       2. *P. matsumurae*
1B. Dioecious shrub; leaves elliptic or ovate, obtuse to subobtuse; fruit pedicellate, berrylike, smooth. .... 3. *P. flexuosus*

1. Phyllanthus urinaria L.

1. Phyllanthus urinaria L. *P. lepidocarpus* Sieb. & Zucc.——Ko-mikan-sō. Glabrous, monoecious annual often reddish in sunny places; stems erect, 5–30 cm. long, branched, with few ordinary leaves at bases of stems, and scale-leaves on stems subtending the branches, the branchlets ascending-spreading, slightly flattened, narrowly winged, 5–12 cm. long; leaves on branches obovate-oblong or narrowly oblong, 7–17 mm. long, 3–7 mm. wide, rounded, mucronate, obliquely rounded and subsessile at base, entire, minutely scabrous on margins, whitish beneath; flowers subsessile, about 4 mm. across; sepals 6, oblong; stamens 3, the anthers splitting transversely, the filaments connate; capsules depressed-globose, about 2.5 mm. across, pendulous, transversely rugose with raised, red-brown lines, the styles free, minute, bifid; seeds transversely rugose, about 1.2 mm. across.——July–Oct. Waste places and fields; Honshu, Shikoku, Kyushu; common.—— Widely scattered in the Tropics and subtropics.

2. Phyllanthus matsumurae Hayata. Hime-mikan-sō. Glabrous, monoecious annual; stems slender, sparsely branched at base, the branches narrowly winged or striate on one side; leaves broadly lanceolate to narrowly oblong, sometimes elliptic, 7–20 mm. long, 3–6 mm. wide, subacute to obtuse, rarely rounded at apex, obtuse and subsessile at base, entire, smooth, whitish beneath; flowers pale green, small; sepals of pistillate flowers 6, minute, oblong, reflexed in fruit; capsules pendulous, short pedicellate, depressed-globose, pale yellowish green, smooth, about 2.5 mm. across, the styles 3, free, short bifid; seeds yellowish brown, about 1.2 mm. long, irregularly and longitudinally striate with dark brown minute dots. ——Aug.–Oct. Thickets and grassy, and waste places in lowlands; Honshu, Shikoku, Kyushu; common but not abundant.

3. Phyllanthus flexuosus (Sieb. & Zucc.) Muell.-Arg. *Cicca flexuosa* Sieb. & Zucc.; *Glochidion flexuosus* (Sieb. & Zucc.) Fr. & Sav.; *Hemicicca japonica* H. Baill.; *P. japonicus* (H. Baill.) Muell.-Arg.; *H. flexuosa* (Sieb. & Zucc.) Hurusawa——Koban-no-ki. Glabrous, dioecious, low shrub; branchlets 8–15 cm. long, deciduous; leaves alternate, elliptic or ovate, 2–4.5 cm. long, 1–2.5 cm. wide, obtuse to broadly so, rounded at base, entire, smooth, vivid green to glaucous on upper side, glaucous or whitish beneath, the petioles 2–3 mm. long; flowers about 3.5 mm. across, the staminate short-pedicellate; sepals 4, rounded, dark reddish; disc annular; stamens 2, rarely 3, the anthers ellipsoidal, longitudinally dehiscent; pistillate flowers pale green, the pedicels 1–1.5 cm. long, gradually thickened on upper part; sepals 3, deciduous, elliptic; fruit depressed globose, about 6 mm. across, becoming berrylike when mature; styles connate and short-cylindric, deeply 3-lobed, the stigma entire; seeds about 3 mm. long.——May. Thickets in mountains; Honshu (Kinki Distr. and westw.), Shikoku, Kyushu.——China.

3. GLOCHIDION Forst.   Kanko-no-ki Zoku

Shrubs, trees or herbs; leaves alternate, often regularly arranged on branchlets in one plane; flowers unisexual, usually 5- to 6-merous, without a disc; sepals of staminate flowers 3–8, more or less connate at base, the connective exserted at tip; ovary in pistillate flowers 3- to several-locular, the styles usually connate in a short, thick column; fruit a capsule; seeds not arillate.—— About 150 species, in tropical Asia, the Pacific Islands, and Malaysia.

1A. Branches, underside of leaves especially on nerves, and flowers distinctly pubescent. .... 1. *G. triandrum*
1B. Branches and other parts glabrous.
  2A. Leaves obovate to cuneate-obovate, 2.5–8 cm. long, 1.2–3 cm. wide, cuneate at base. .... 2. *G. obovatum*
  2B. Leaves ovate-oblong to ovate, 7–18 cm. long, 4–8 cm. wide, rounded at base. .... 3. *G. hongkongense*

1. Glochidion triandrum (Blanco) C. B. Robinson. *Kirganelia triandra* Blanco; *G. bicolor* Hayata, excl. syn.; *G. hypoleucum* Hayata, non Boerl.; *G. hayatae* Croizat & Hara ——Urajiro-kanko-no-ki. Shrub with slender grayish brown branches with grayish hairs; leaves chartaceous, broadly oblanceolate to obovate, 5–8 cm. long, 1.5–2.5 mm. wide, glabrous or nearly so, green above, paler or becoming purplish brown when dry, with spreading to ascending curved whitish hairs especially on nerves beneath, the nerves and veinlets slightly raised beneath, the petioles 1–2 mm. long, pubescent; flowers in few-flowered axillary fascicles, small, the pedicels to 5 mm. long; sepals 6, about 1.5 mm. long, narrowly cuneate-obovate, rounded; stigmas about 6, connate; ovary white-pubescent.——Kyushu (Hizen Prov.); rare.——Formosa and China to Philippines.

2. Glochidion obovatum Sieb. & Zucc. *Phyllanthus obovatus* (Sieb. & Zucc.) Muell.-Arg.——Kanko-no-ki. Densely branched, dioecious, glabrous shrub with branchlets often becoming spinelike; leaves obovate to broadly cuneate, 2.5–8 cm. long, 1.2–3 cm. wide, rounded, obtuse or abruptly acute with obtuse tip, long-cuneate at base, whitish and becoming brownish when dry beneath, the petioles 1–3 mm. long; flowers several, fasciculate, axillary, short-pedicellate; sepals 6, about 1.5 mm. long; capsules depressed-globose, 6–8 mm. across, 4- or 5-locular; styles short, cylindric.——July–Oct. Honshu (Kii and Ise Prov. and Chūgoku Distr.), Shikoku, Kyushu.——Ryukyus.

3. Glochidion hongkongense Muell.-Arg. *Phyllanthus hongkongensis* (Muell.-Arg.) Muell.-Arg.——Kakiba-kanko-no-ki. Glabrous shrub with thick elongate branches; leaves obovate-, ovate-, or elliptic-oblong, sometimes ovate, 7–18 cm. long, 4–6 cm. wide, acute with an obtuse tip, obliquely rounded at base, green above, pale green or purplish brown beneath when dry, the petioles 4–6 mm. long; flowers in fascicles on short slightly supra-axillary peduncles, the pedicels short; sepals of pistillate flowers 5 or 6; capsules depressed-globose, 8–12 mm. across, about 6-locular, the style obtusely angled, cylindric.——Sept.–Oct. Thickets in lowlands; Kyushu (Yakushima and Tanegashima).——Ryukyus, Formosa, China, and Indochina.

4. SECURINEGA Juss.   HITOTSUBA-HAGI ZOKU

Shrubs with alternate, entire, rather small leaves; flowers in axillary fascicles, unisexual, the staminate small, subsessile, the sepals 5, the stamens 5 or fewer, free, the anthers longitudinally dehiscent, the abortive pistils small, these sometimes 2- or 3-fid at the apex; pistillate flowers on distinct pedicels; ovary 3-locular, the styles free, the ovules 2 in each locule; capsules sometimes slightly fleshy, 3-valved, the valves bifid.——More than 10 species, in temperate and warmer regions.

1. Securinega suffruticosa (Pall.) Rehd. var. **japonica** (Miq.) Hurusawa. *S. japonica* Miq.; *Phyllanthus fluggeoides* (Muell.-Arg.) Muell.-Arg.; *Flueggea japonica* (Miq.) Pax ——HITOTSUBA-HAGI. Dioecious, branching, glabrous shrub with slender green branchlets; leaves oblong or elliptic, sometimes obovate, 2–4 cm. long, 1.2–2.5 cm. wide, subacute to very obtuse, acute to broadly cuneate at base, entire, green on upper side, whitish beneath, the stipules less than 1 mm. long, deciduous; staminate flowers many, axillary, about 3 mm. across, greenish, on short pedicels 2–3 mm. long; sepals and stamens 5; pistillate flowers few or solitary, axillary, the pedi-cels 5–10 mm. long; capsules depressed-globose, about 4 mm. across, green, the styles 3, bifid, the lobes short, flat, broadly linear, retuse to obtuse; seeds slightly lustrous.——July. Thickets and grassy slopes in lowlands; Honshu, Shikoku, Kyushu; common.

Var. **amamiensis** Hurusawa. *Phyllanthus trigonocladus* Ohwi——AMAMI-HITOTSUBA-HAGI. Branchlets obtusely but distinctly trigonous, slightly thicker; leaves thicker; capsules depressed.——s. Kyushu (Mageshima near Tanegashima).——Ryukyus, Formosa. The Typical phase with larger, thinner leaves occurs in Korea, China, and e. Siberia.

5. PUTRANJIVA Wall.   TSUGE-MODOKI ZOKU

Trees with thick, entire, alternate leaves; flowers axillary, unisexual; sepals 3–6, slightly connate at base, the staminate numerous, on very short pedicels, the stamens 2 or 3 (–6), the filaments free or connate at base, the anthers erect, with distinct, parallel, longitudinally dehiscent locules, the ovary wholly reduced; pistillate flowers ovoid, with 2- or 3-locular ovaries; styles short, spreading, lobed or dentate; ovules 2 in each locule; drupe ovoid or globose, with a single ovoid seed, the endosperm fleshy.——Several species in s. Asia.

1. Putranjiva matsumurae Koidz. *P. roxburghii* sensu auct. Japon., non Wall.; *Drypetes matsumurae* (Koidz.) Kaneh.; *Liodendron matsumurae* (Koidz.) H. Keng—— TSUGE-MODOKI. Branches yellowish gray, lenticellate, the branchlets puberulous while young, slightly flattened and with raised striations; leaves alternate, in one plane, coriaceous, oblong or ovate, 5–10 cm. long, 2.5–4 cm. wide, short-caudate, obtusely tipped, oblique at base, obtusely and obscurely toothed, soon becoming glabrous, the veinlets reticulate, raised, the petioles 6–10 mm. long; inflorescence axillary; staminate flowers numerous, in racemes or spikes about 1 cm. long, the stamens 3; pistillate flowers 1 to few, on pedicels 6–12 mm. long; sepals 3 or 4, oblong to elliptic, obtuse, about 1.5 mm. long; drupes narrowly ovoid-oblong, about 2 cm. long, 1 cm. across, rather densely puberulous, the persistent styles 3 or 4, entire, grooved above.——Reported to occur in Kyushu (Osumi Prov.).——Ryukyus.

6. ANTIDESMA L.   YAMA-HIHATSU ZOKU

Dioecious trees or shrubs with alternate leaves; inflorescence a raceme or spike, solitary in axils or compound in terminal panicles; flowers and bracts small; sepals 3–5, small; stamens in staminate flowers 2–5, opposite the sepals, the ovary reduced; ovary of the pistillate flowers usually 1-locular, the styles 3, very short, often bifid at apex, the ovules 2 in each locule; drupe small, often oblique, 1-seeded; endosperm fleshy.——About 100 species, in the Tropics of Australia, Asia, and Africa.

1. Antidesma japonicum Sieb. & Zucc. YAMA-HIHATSU. Glabrous much-branched shrub with slender branches loosely puberulent while young; leaves broadly oblanceolate to narrowly obovate, 5–8(–10) cm. long, 2–4 cm. wide, acuminate, slightly oblique and acute to obtuse at base, entire, sometimes with scattered minute hairs on the midrib beneath, the lateral nerves 5–7 pairs, arcuate above, the petioles 3–5 mm. long; racemes nearly sessile, 4–6 cm. long, at first puberulent; staminate flowers nearly sessile, about 2 mm. across; sepals 3- to 5-lobed, the segments deltoid; stamens 2–5; pistillate flowers on spreading pedicels 2–4 mm. long; drupe obliquely ellipsoidal, red, about 5 mm. long, the persistent styles 3–4, short, spreading.——Honshu (Kii Prov.), Shikoku, Kyushu. ——Ryukyus and Formosa.

7. ACALYPHA L.   ENOKIGUSA ZOKU

Herbs or trees; leaves alternate, serrate, 3-, 5- or penni-nerved; flowers small, unisexual, in spikes, these sometimes compounded into panicles, the pistillate on the lower part of the spikes, the bracts often accrescent after anthesis; staminate flowers with the calyx connate in bud, 3-lobed at anthesis, the stamens many, usually 8; calyx of pistillate flowers of 3 or 4 sepals; ovary 3-locular; styles free, slender, often lacerate; ovules solitary in each locule; capsules small, the 3 valves bifid; seeds globose, with fleshy endosperm.——About 300 species, widespread in the tropics and in warmer regions.

1. Acalypha australis L. *A. virgata* sensu Thunb., non L.; *A. pauciflora* Hoffm.; *A. chinensis* Roxb.——ENOKIGUSA. Annual with short ascending hairs; stems erect, branched, 30–50 cm. long; leaves membranous, ovate to narrowly so, sometimes broadly lanceolate, 3–8 cm. long, 1.5–3.5 cm. wide, acute, rounded at base, coarsely toothed, sparsely appressed-pilose on upper side, pilose especially on nerves beneath, 3(–5)-nerved; inflorescence spicate, short-pedunculate, 1–5 in leaf-axils, 1–2 cm. long; pistillate flowers at base of the spike and surrounded by a cordate, obtusely toothed sessile bract 1–2 cm. long; staminate flowers small, naked, nearly sessile; capsules about 3 mm. across, pilose, nearly sessile; seeds broadly ovoid, dark brown, smooth, 1.5 mm. long.——Aug.–Nov. Waste places and cultivated fields in lowlands; Hokkaido, Honshu, Shikoku, Kyushu; common.——Ussuri, Korea, Manchuria, China, and Philippines. ——Forma **velutina** (Honda) Ohwi. *A. australis* var. *velutina* Honda——BIRŌDO-ENOKI-GUSA. With longer spreading hairs on all parts.

8. MERCURIALIS L.　YAMA-AI ZOKU

Pilose to glabrous herbs with usually serrate, penninerved, opposite leaves; inflorescence an axillary spike or raceme; pistillate flowers few, subsessile, in leaf-axils or at the base of staminate inflorescence; staminate flowers with a 3-lobed membranous calyx, stamens many, free; pistillate flowers with 3 sepals and 2 scales at base; ovary 2-locular; style nearly free, entire; capsules dehiscent, 2-valved; seeds ovoid, smooth or tubercled, with fleshy endosperm.——Few species, in Europe and e. Asia.

1. Mercurialis leiocarpa Sieb. & Zucc. YAMA-AI. Nearly glabrous perennial with long-creeping slender rhizomes; stems erect, simple, with a few pairs of leaves, the lower 2 or 3 nodes naked; leaves rather unequal in the pair, thin, ovate-oblong or oblong, 7–12 cm. long, 2.5–5 cm. wide, acute, rounded at base, obtusely toothed, with scattered coarse pubescence above and on nerves beneath, the petioles 1.5–3 cm. long, the stipules free, membranous, rather persistent, re-curved, lanceolate, 3–4 mm. long; spikes axillary, peduncled, few together, the staminate in the lower axils, loosely flowered with many stamens, the pistillate in upper axils, few-flowered; capsules 2-valved, green, the valves subglobose, about 4 mm. across, bifid, with few callose tubercles on the surface. ——Apr.–July. Woods in low mountains; Honshu, Shikoku, Kyushu; rather common.——Korea, Formosa, China, and India.

9. MALLOTUS Lour.　AKAME-GASHIWA ZOKU

Trees or shrubs often with lepidote or stellate hairs throughout; leaves alternate or opposite, entire or dentate, sometimes 3-lobed, flat, petiolate, penninerved or 3- to 7-nerved, sometimes with 2 sessile glands at base; spikes simple or compounded into panicles; staminate flowers small, fasciculate on the nodes of the rachis, the calyx 3- or 4-lobed, the stamens many, free; pistillate flowers few, solitary in the axil of a bract, the calyx 3- to 5-lobed; ovary 3-locular; styles free or connate at base; ovules solitary in each locule; capsules globose, hairy, tubercled, or glandular-lepidote, 3-valved, dehiscent, the valves bifid; seeds ovoid or globose, with fleshy endosperm.——About 80 species, in the Tropics of the Old World.

1. Mallotus japonicus (Thunb.) Muell.-Arg. *Croton japonicum* Thunb.——AKAME-GASHIWA. Deciduous tree, with dense reddish stellate scales while young, grayish afterward, the branches rather thick; leaves rounded-obovate to broadly ovate, 10–20 cm. long, 6–15 cm. wide, acute, rounded at base, shallowly 3-fid, sometimes entire or undulate, 3-nerved, with parallel veinlets, pale yellowish green beneath with yellow sessile glands, 2 glands just above the base, the petioles rather long; inflorescence a panicle, 8–20 cm. long, the branches elongate in the staminate; staminate flowers with calyx about 3 mm. long and numerous stamens 6–7 mm. long; pistillate flowers short-pedicellate on shorter branches of the panicles; capsules trigonous-globose, about 7 mm. across, densely lepidote and with yellowish brown glandular dots, and sparsely armed with soft spinelike protuberances, densely stellate-pilose while young, 3-valved, the valves bifid; seeds dark brown, about 4 mm. long, obscurely tuberculate.—— June. Thickets and thin woods in lowlands; Honshu, Shikoku, Kyushu; common.——Ryukyus, Formosa, China, and Korea.

10. SAPIUM R. Br.　SHIRAKI ZOKU

Usually glabrous trees or shrubs; leaves alternate, petiolate, flat, penninerved, usually 2-glandular at base of the blades or on the upper part of the petioles; spikes or racemes terminal, simple or rarely in panicles; staminate flowers many, small, the calyx 2- or 3-lobed, toothed, membranous, the stamens 2 or 3, free; pistillate flowers 1 to few on the lower part of the otherwise staminate inflorescence, the calyx 2- or 3-lobed; ovary 3-locular; styles sometimes connate at base, entire, reflexed; ovules solitary in each locule; fruit fleshy or berrylike, sometimes a capsule; seeds globose, arillate or not, with fleshy endosperm.——About 100 species, in the Tropics of the Old World.

1. Sapium japonicum (Sieb. & Zucc.) Pax & Hoffm. *Stillingia japonica* Sieb. & Zucc.; *Excoecaria japonica* (Sieb. & Zucc.) Muell.-Arg.; *Triadica japonica* (Sieb. & Zucc.) H. Baill.; *Shirakia japonica* (Sieb. & Zucc.) Hurusawa——SHIRAKI. Glabrous small tree with pale grayish branches; leaves broadly ovate, 7–15 cm. long, 5–10 cm. wide, abruptly acuminate to acute, rounded and with a pair of glands at base, vivid to deep green, entire, glaucous beneath, the petioles 2–3 cm. long, the stipules lanceolate, scarious, deciduous, 7–12 mm. long; racemes short-pedunculate, 5–10 cm. long; staminate flowers numerous, short-pedicellate; pistillate flowers on the lower part of the spikes or racemes, the pedicels 1–2 cm. long, the calyx membranous, 3-parted, about 3 mm. long; capsules depressed-globose, about 8 mm. across, yellow with brown blotches.——May–July. Thin woods in lowlands and mountains; Honshu, Shikoku, Kyushu; rather common.——China, Korea, and Ryukyus.

11. EUPHORBIA L.   TŌDAIGUSA ZOKU

Herbs or shrubs of various habit, with milky juice; leaves alternate or the upper opposite or verticillate, entire or lobed; inflorescence a cyathium composed of a cup-shaped glandular or lobed involucre that surrounds a few much-reduced staminate flowers and a central pistillate flower; staminate flowers with a single anther supported by a jointed filament and a linear or subulate sometimes obsolete bracteole; pistillate flowers with a single ovary on an accrescent stipe and sometimes with 3 scale-like perianth segments; ovary 3-locular; styles 3, free or connate at base, entire or bifid above; ovules solitary in each locule; capsules 3-valved, dehiscent, the valves bifid, the central column persistent; seeds with endosperm.——About 1,600 species, chiefly in warmer regions; sometimes split into several smaller genera.

1A. Glands of involucre with a petallike appendage; leaves opposite, small, the sides unequal; stipules small; seeds without appendage.
  2A. Perennial, glabrous, with thick caudex; stems tufted; leaves entire. .... 1. *E. atoto*
  2B. Annual, pubescent; stems single, simple or branched from base; leaves toothed.
    3A. Erect or ascending herb.
      4A. Capsules glabrous.
        5A. Leaves ovate-oblong, 4–12 mm. wide; cyathium on upper part of the branches. .... 2. *E. maculata*
        5B. Leaves linear, 4–6 mm. wide; cyathium axillary, with whitish glands. .... 3. *E. vachellii*
      4B. Capsules pilose. .... 4. *E. hirta*
    3B. Prostrate herbs, much-branched from the base.
      6A. Capsules glabrous. .... 5. *E. pseudochamaesyce*
      6B. Capsules pilose. .... 6. *E. supina*
1B. Glands of involucre without petallike appendage; leaves alternate except beneath inflorescence, the sides equal; stipules absent; seeds with an appendage or caruncle.
  7A. Annuals or biennials without rhizomes.
    8A. Glands of involucre entire, transversely elliptic; seeds reticulate. .... 7. *E. helioscopia*
    8B. Glands of involucre lunate; seeds foveolate. .... 8. *E. peplus*
  7B. Perennials with rhizomes.
    9A. Glands of involucre entire.
      10A. Capsules smooth, glabrous; staminate flowers without bracteoles; rhizomes long-creeping. .... 9. *E. ebracteolata*
      10B. Capsules verrucose, sometimes long-pilose; staminate flowers with bracteoles at base; rhizomes short.
        11A. Involucres with 4 glands.
          12A. Capsules glabrous, verrucose.
            13A. Leaves acute to obtuse; glands flat; capsules about 3.5 mm. across. .... 10. *E. pekinensis*
            13B. Leaves rounded to very obtuse; glands depressed; capsules about 6 mm. across. .... 11. *E. jolkinii*
          12B. Capsules pilose to subglabrous, obsoletely verrucose. .... 12. *E. togakusensis*
        11B. Involucres with 5 glands; capsules verrucose, glabrous. .... 13. *E. adenochlora*
    9B. Glands of involucre retuse or with projections on each side.
      14A. Glands with short obtuse projections on both sides.
        15A. Stems thick, 40–60 cm. long; leaves 5–8 cm. long, green. .... 14. *E. sendaica*
        15B. Stems slender, 10–30 cm. long; leaves 3–4 cm. long, the upper ones yellowish. .... 15. *E. esula*
      14B. Glands with long acute projections on both sides. .... 16. *E. sieboldiana*

1. **Euphorbia atoto** Forst. f. HAMA-TAIGEKI, SUNA-TAI-GEKI.   Glabrous perennial with thick roots; stems many from the stout caudex, much branched, with thickened nodes; leaves thick, ovate-oblong or narrowly ovate, 2–3 cm. long, 1–1.5 cm. wide, obtuse to mucronate, obliquely rounded to shallowly cordate at base, entire; cyathia in terminal and dense axillary cymes, the glands transversely oblong; capsules about 3 mm. across, smooth, glabrous; seeds smooth, tetrahedral-ellipsoidal, about 1.3 mm. long.——Sept.–Oct. Grassy slopes near seashores; Kyushu (Yakushima and Tanegashima).——Ryukyus, Formosa, s. China, India, Malaysia, and Pacific Isls.

2. **Euphorbia maculata** L. *E. hypericifolia* sensu auct. Japon., non L.; *Chamaesyce maculata* (L.) Small; *C. hyssopifolia* sensu auct. Japon., non Small——Ō-NISHIKI-SŌ.   Erect annual; stems 40–60 cm. long, terete, branched and short-puberulent, on one side in upper part; leaves rather thin, ovate-oblong or narrowly oblong, 1–3 cm. long, 3(–5)-nerved, obtuse, obliquely rounded at base, toothed, sparsely long-pilose beneath; cyathia few on the upper part of the branchlets, the glands transversely elliptic; capsules smooth, glabrous, about 1.7 mm. across, trigonous; seeds tetrahedral-ovoid, 1–1.2 mm. long, dark brown, with 2 or 3 transverse wavy lines.——Aug.–Sept. Naturalized in lowlands; Honshu.——N. America; naturalized in Europe.

3. **Euphorbia vachellii** Hook. & Arn.   *E. serrulata* Reinw., non Thuill.; *Chamaesyce vachellii* (Hook. & Arn.) Hara——MIYAKOJIMA-NISHIKI-SŌ, AWAYUKI-NISHIKI-SŌ.   Glabrous annual; stems erect, branched from base, 30–50 cm. long; leaves broadly linear or linear-lanceolate, 2–4 cm. long, 4–6 mm. wide, obtuse to subacute, obliquely rounded at base, toothed, the midrib prominently raised beneath, impressed above, the lateral nerves pinnate, inconspicuous; cymes axillary, dense, the cyathia glabrous, the glands white, semirounded; capsules glabrous, smooth, slightly less than 2 mm. across, obtusely angled, the styles short, erect; seeds obovoid, about 1 mm. long, obtusely 3-ridged, with about 3 transverse wavy lines.——July–Oct. Waste grounds and cultivated fields in lowlands; Kyushu (Tanegashima).——Ryukyus, Formosa, s. China, and Malaysia.

4. **Euphorbia hirta** L.   *Chamaesyce hirta* (L.) Millsp.; *E. pilulifera* sensu auct., non L.; *E. pilulifera* var. *hirta* (L.) Thell. SHIMA-NISHIKI-SŌ, TAIWAN-NISHIKI-SŌ.   Ascending or erect annual; stems often branched from base, 20–40 cm. long, puberulent and pilose with multicellular spreading hairs; leaves narrowly ovate-triangular, oblong-ovate or narrowly ovate, 2.5–4 cm. long, 8–15 mm. wide, subacute, obliquely cuneate to rounded at base, obtusely serrulate, 3- or 4-nerved, loosely short-pilose on upper side, yellowish, puberulent and

pilose beneath; cymes dense, axillary, the cyathia many, the glands transversely elliptic; capsules trigonous, 1.3 mm. across, short-puberulent; seeds 0.7–0.8 mm. long, ovoid, 3-ridged, with transverse wavy lines on the face.——Aug.–Oct. Waste places and cultivated fields in lowlands; Honshu (Kinki Distr. and westw.), Shikoku, Kyushu.——Ryukyus, Formosa, China, and the Tropics and subtropics generally.

5. Euphorbia pseudochamaesyce Fisch., Mey., & Lallem. *E. humifusa* sensu auct., non Willd.; *Chamaesyce humifusa* sensu auct. Japon., non Prokhan.; *C. pseudochamaesyce* (Fisch., Mey., & Lallem.) Komar.; *E. thymifolia* sensu Thunb., non L.——Nishiki-sō. Procumbent, diffuse, branched, delicate annual 10–30 cm. long, usually slightly white-pubescent; leaves obliquely oblong or obovate-oblong, 7–15 mm. long, 3–7 mm. wide, rounded at apex, obliquely rounded at base, serrulate; cymes terminal or axillary, often leafy, the cyathia few, the glands transversely elliptic; capsules glabrous, about 1.3 mm. across, trigonous, smooth; seeds grayish brown, ovoid, 0.7 mm. long, smooth.——July–Nov. Waste grounds and fields in lowlands; Honshu, Shikoku, Kyushu; common.—— Ryukyus, China, Korea, Manchuria, and temperate Asia.

6. Euphorbia supina Raf. *Chamaesyce supina* (Raf.) Moldenke; *E. littoralis* Raf.; *E. maculata* sensu auct., non L. ——Ko-nishiki-sō. Procumbent, diffuse, pubescent annual 10–30 cm. long, much branched from the base; leaves oblong, elliptic or obovate-oblong, 7–10 mm. long, 2.5–4.5 mm. wide, nearly rounded to obtuse, obliquely rounded at base, serrulate, usually with a reddish brown spot in the center; cymes axillary, dense, the cyathia few, short-pilose; capsules puberulous, exserted, about 1.8 mm. across, obtusely angled; seeds tetrahedral-ellipsoidal, about 0.6 mm. long, with 3 ridges and a few transverse wavy lines on the face.——July–Nov. Waste grounds and cultivated fields in lowlands; Hokkaido, Honshu, Shikoku, Kyushu; very common.——N. America.

7. Euphorbia helioscopia L. *Tithymalus helioscopius* (L.) Hill; *Galarhoeus helioscopius* (L.) Haw.——Tōdaigusa. Green biennial; stems sparsely branched at base, sparsely pubescent in upper part; leaves spathulate-obovate to obovate-cuneate, 1–3 cm. long, 6–20 mm. wide, rounded at apex, obtusely serrulate, green, the upper leaves similar, relatively large, the involucral leaves obovate or broadly so, relatively small, green; cyathia many, dense, the glands transversely elliptic, about 1 mm. long, yellowish green; capsules smooth, about 3 mm. across; seeds obovoid, brown, with raised reticulations, about 1.8 mm. across.——Apr. Waste grounds and riverbanks, sometimes cultivated fields in lowlands; Honshu, Shikoku, Kyushu; common.——Warmer regions of Eurasia.

8. Euphorbia peplus L. Chabo-taigeki. Slender, glabrous biennial with fibrous roots; stems 10–20 cm. long, sometimes branched at base; leaves rather loose, thin, broadly ovate to elliptic, 1–2 cm. long, 7–12 mm. wide, obtuse, narrowed below to a petiolelike base, entire, green, the midrib slender, the upper (verticillate) leaves similar and slightly larger, the involucral leaves broadly ovate, 1–1.2 cm. long; cyathia pale yellow, the gland lunate, about 1 mm. wide; capsules about 2 mm. across, smooth, 3-angled; seeds about 1.5 mm. long, impressed-punctate.——May. Naturalized; Kyushu (Nagasaki).——s. Europe and the Mediterranean region.

9. Euphorbia ebracteolata Hayata. *Galarhoeus ebracteolatus* (Hayata) Hara; *Tithymalus ebracteolatus* (Hayata) Hara——Marumi-nourushi. Stems erect, thick, terete, 40–

50 cm. long, the upper portions sparsely white-pubescent; leaves rather numerous, soft, broadly oblanceolate or obovate-oblong, 5–10 cm. long, 1.5–2.5 cm. wide, obtuse to rounded, gradually narrowed toward the base, entire, the midrib thick and slightly raised, sparsely white-pubescent beneath, the lateral nerves slender, the verticillate leaves similar but slightly shorter and broader, often dilated at base, the involucral leaves deltoid to ovate-deltoid, 2–3 cm. long, 1–2 cm. wide, obtuse, usually truncate at base; inflorescence rather loose, the glands of the involucres rounded-cordate, dark brown, about 2.5 mm. wide, the filaments much exserted; capsules smooth; styles slender about 3 mm. long, retuse, connate nearly to the middle.——Apr. Mountains; Hokkaido, Honshu, and (?) Kyushu.

10. Euphorbia pekinensis Rupr. *Galarhoeus pekinensis* (Rupr.) Hara; *Tithymalus pekinensis* (Rupr.) Hara; *E. lasiocaula* Boiss.; *E. onoei* Fr. & Sav.; *E. pekinensis* var. *japonensis* Makino; *E. pekinensis* var. *onoei* (Fr. & Sav.) Makino; *G. lasiocaulus* (Boiss.) Hurusawa; *E. watanabei* Makino; *G. watanabei* (Makino) Hara; *G. lasiocaulus* var. *watanabei* (Makino) Hurusawa——Taka-tōdai, Ibuki-taigeki, Fuji-taigeki. Stems erect from thick caudex, terete, often branched at base, 20–80 cm. long, usually white-pubescent with long curved hairs; leaves many, lanceolate to oblong, 2.5–8 cm. long, 6–12 mm. wide, obtuse or subacute, narrowed gradually toward the base, minutely toothed or nearly entire, deep green, glabrous, the involucral leaves broadly ovate, triangular-ovate, or ovate-orbicular, 5–12 mm. long; inflorescence loose or rather dense, the glands of the involucres oblong, dark brown or dark purple, about 1.5 mm. across; capsules about 3.5 mm. across, with 3 or 6 longitudinal series of tubercles; seeds broadly ellipsoidal, smooth, about 1.8 mm. long.——June–July. Grassy places in lowlands and mountains; Honshu, Shikoku, Kyushu; common and variable.——Korea, China, and Manchuria.

11. Euphorbia jolkinii Boiss. *Galarhoeus jolkinii* (Boiss.) Hara; *Tithymalus jolkinii* (Boiss.) Hara——Iwa-taigeki. Glabrous perennial; stems erect, thick, terete, 40–80 cm. long, often branched at base; leaves numerous, rather dense, oblanceolate to linear-oblanceolate, 4–7 cm. long, 8–12 mm. wide, rounded to very obtuse at apex, entire and gradually narrowed toward the base, green, the midrib prominently raised beneath, the verticillate leaves similar, the involucral leaves oblong or elliptic, 1–2 cm. long, 6–12 mm. wide, yellowish; glands of involucres transversely flabellate-elliptic, yellowish, depressed; capsules densely tuberculate, about 6 mm. across; seeds subglobose, about 3 mm. long, smooth.——Apr.–May. Rocky places near seashores; Honshu (Awa Prov. and westw.), Shikoku, Kyushu.——s. Korea, Ryukyus, and Formosa.

12. Euphorbia togakusensis Hayata. *Galarhoeus togakusensis* (Hayata) Hara; *Tithymalus togakusensis* (Hayata) Hara——Hakusan-taigeki, Miyama-nourushi. Stems erect, few, from a short-creeping rhizome, 40–80 cm. long, glabrous; leaves rather many, narrowly oblong or oblong-lanceolate, 5–7 cm. long, 1–2 cm. wide, obtuse or subacute, entire, glabrous, the midrib rather thick, slightly raised beneath, the verticillate leaves yellowish, rounded-ovate, 1–2 cm. long; inflorescence rather loose, the glands of the involucres transversely reniform-elliptic, brown, about 2 mm. wide; capsules about 5 mm. across, obsoletely verrucose, pilose to subglabrous; seeds subglobose, about 2.5 mm. long, smooth.——June–July. High mountains; Honshu (centr. and n. distr.).

13. Euphorbia adenochlora Morr. & Decne. *Galarhoeus adenochlorus* (Morr. & Decne.) Hara; *Tithymalus adenochlorus* (Morr. & Decne.) Hara; *E. palustris* sensu auct. Japon., non L.; *E. rochebrunii* Fr. & Sav.; *E. japonica* Boiss. ——Nourushi. Stems thick, terete, erect, 40–50 cm. long, often with loosely scattered long white hairs on upper part; leaves soft, broadly oblanceolate to oblong, 5–8 cm. long, 1–2 cm. wide, obtuse to rounded, entire, the midrib thick and slightly raised beneath, the lateral nerves slender, the verticillate leaves similar but shorter, the involucral leaves rounded to broadly ovate, 7–20 mm. long, rounded; inflorescence rather dense, yellowish, the glands of involucres reniform, about 2 mm. wide, yellowish; capsules about 6 mm. across, verrucose; seeds subglobose, smooth, about 3 mm. long.——Apr.–May. Grassy places near rivers in lowlands; Hokkaido, Honshu, Kyushu; common.

14. Euphorbia sendaica Makino. *Galarhoeus sendaicus* (Makino) Hara; *Tithymalus sendaicus* (Makino) Hara—— Sendai-taigeki. Rhizomes slender, long-creeping; stems erect, about 40 cm. long, glabrous; leaves rather loose, lanceolate or oblong-lanceolate, 5–8 cm. long, 1–2 cm. wide, obtuse, glabrous, the midrib rather thick and slightly raised beneath, the verticillate leaves rather small, the involucral leaves orbicular- or cordate-triangular, 6–15 mm. long, truncate at base; inflorescence loose, the glands of the involucres lunate-reniform, both ends extended and obtuse, about 2.5 mm. wide; capsules smooth.——May. Honshu (Kantō Distr. and northw.); rare.

15. Euphorbia esula L. *Galarhoeus esula* (L.) Rydb.; *E. nakaiana* Lév.; *G. nakaii* Hurusawa; *Tithymalus nakaianus* (Lév.) Hara.——Hagiku-sō. Roots rather thick, elongate; stems glabrous, erect, 15–30 cm. long, terete, rather slender; leaves rather numerous, densely arranged, oblanceolate to linear-oblanceolate, sometimes spathulate, 2–3 cm. long, 3–5 (–7) mm. wide, rounded or very obtuse, glabrous, entire, the upper leaves of sterile stems fasciculate, the verticillate leaves oblong, elliptic or ovate, rather short, broad, the involucral leaves yellow, cordate to reniform, 5–10 mm. long, 7–15 mm. wide, truncate at base; inflorescence rather dense, the glands of the involucres reniform, both ends slightly extended and obtuse; capsules 3–3.5 mm. across, smooth; styles short; seeds 1.7 mm. long, ovoid.——Apr.–May. Near seashores; Honshu (Tōkaidō), Kyushu.——Korea, Manchuria, Amur, Siberia, and Europe.

16. Euphorbia sieboldiana Morr. & Decne. *E. guilielmii* A. Gray; *E. taquetii* Lév. & Van't.; *Galarhoeus sieboldianus* (Morr. & Decne.) Hara; *E. tsukamotoi* Honda; *Tithymalus sieboldianus* (Morr. & Decne.) Hara; *E. tsukamotoi* Honda; *T. sieboldianus* (Morr. & Decne.) Hara——Natsu-tōdai. Glabrous perennial with slender, creeping rhizomes; stems erect, 20–40 cm. long; leaves rather loosely arranged, soft, narrowly oblong to broadly oblanceolate, 3–6 cm. long, 7–20 mm. wide, obtuse to very obtuse, gradually narrowed toward base, entire, the midrib slender and slightly raised beneath, the lateral nerves very slender, the verticillate leaves similar but slightly larger, the involucral leaves green, deltoid-ovate to ovate, sometimes deltoid, 1–4 cm. long, 8–25 mm. wide, obtuse to acute; inflorescence loose, the glands of involucres reniform, with an outward pointing caudate projection at each end, dark brown or dark purple, about 3 mm. wide; capsules about 3 mm. across, smooth; seeds broadly ovoid, about 2.3 mm. long, smooth.——Apr.–July. Grassy places and thickets in lowlands and mountains; Hokkaido, Honshu, Shikoku, Kyushu; very variable and rather common.——s. Sakhalin, s. Kuriles, Korea, and Manchuria.

Fam. 115. CALLITRICHACEAE   Awagoke Ka   Water Starwort Family

Soft, delicate herbs of wet places or aquatic; leaves opposite, linear to obovate, entire, without stipules; flowers unisexual, small, axillary, solitary or one of each sex in pairs; calyx and petals none; stamen 1 with 2 bracteoles, the filament long and slender, the anthers 2-locular; ovary sessile, 4-lobed, 4-locular; styles 2, elongate, free; ovules solitary in each locule, pendulous from the top of the locule; fruit small, 4-lobed, the lobes with acute or winged margins; seeds with a membranous testa, the endosperm fleshy, the embryo cylindric, terete, erect.——A single genus.

1. CALLITRICHE L.   Awagoke Zoku

Characters of the family. About 30 species, in wet places or aquatic; of worldwide distribution.

1A. Leaves obovate, small; fruit cordate-orbicular, retuse; styles very short, rather persistent, reflexed, not exceeding the sinus of the fruit; filaments much shorter than the fruit, slightly longer than the style. .... 1. *C. japonica*
1B. Leaves often dimorphic, spathulate-obovate or obovate in the aerial, linear in immersed ones; fruit elliptic, scarcely retuse; styles slender and erect, as long as the fruit; filaments slightly longer than the fruit. .... 2. *C. verna*

1. Callitriche japonica Engelm. Awagoke. Glabrous annual; stems procumbent and diffuse, branched from the base, 1–4 cm. long, slender, densely leafy; leaves obovate or ovate-orbicular, 3–6 mm. long, 2–3 mm. wide, rounded to obtuse, abruptly narrowed to a petiolelike base, entire, slenderly 3-nerved; filament about 0.3 mm. long, the anthers yellowish, minute; fruit brown, orbicular-cordate, retuse, less than 1 mm. long, wing-margined, the styles 2, slender, reflexed.——May–June. Wet places in lowlands; Hokkaido, Honshu, Shikoku, Kyushu.——Korea, Ryukyus, and Formosa.

2. Callitriche verna L. *C. fallax* Petrov.; *C. palustris* L., pro parte; *C. stagnalis* sensu auct. Japon., non Scop.; *C. elegans* Petrov.——Mizu-hakobe. Annual, often growing in stagnant water; stems slender, often much elongate and loosely leaved; immersed leaves linear, 7–15 mm. long, 1–1.5 mm. wide, truncate to emarginate, 1-nerved, the floating or aerial leaves spathulate-obovate or obovate, sometimes oblong, 6–12 mm. long, 3–5 mm. wide, rounded to emarginate, cuneate or gradually narrowed to the base, 3-nerved; filament very slender, 2–3 mm. long, the anthers cordate, yellowish; fruit elliptic to obovate-elliptic, less than 1 mm. long, scarcely emarginate, very weakly winged on margins, the styles erect, very slender, deciduous, as long as or slightly longer than the fruit; bracteoles white, 1/2–2/3 as long as the fruit.——May–Nov. Ditches and paddy fields in lowlands; Hokkaido, Honshu, Shikoku, Kyushu; common.——Korea, Kuriles, Sakhalin, Kamchatka, New Guinea, and widely distributed in the N. Hemisphere.

Fam. 116. **BUXACEAE**   Tsuge Ka   Boxwood Family

Evergreen shrubs or small trees; leaves opposite or alternate, exstipulate; flowers unisexual, small, apetalous, without a disc; calyx 4-lobed or in pistillate flowers sometimes to 12-lobed, sometimes absent; stamens 4, opposite the calyx-lobes, or many, the anthers basifixed; ovary superior, 3-locular, rarely 2- or 4-locular, the ovules 2 in each locule, collateral, rarely solitary, pendulous; fruit a capsule or fleshy and indehiscent; seeds with endosperm.——About 6 genera, with about 50 species in the Tropics, a few in temperate regions.

1A. Leaves alternate, dentate, usually 3-nerved; procumbent subshrub with spicate inflorescence; pistillate flowers below. .. 1. *Pachysandra*
1B. Leaves opposite, entire, penninerved; small to large shrubs with fasciculate flowers; pistillate flowers terminal. .... 2. *Buxus*

1. **PACHYSANDRA** Michx.   Fukki-sō Zoku

Procumbent or ascending suffrutescent subshrubs; stems rather thick; leaves thick, alternate, 3-nerved, dentate; flowers unisexual, white, in erect spikes; pistillate flowers below the staminate; staminate flowers of 4 sepals and stamens, the filaments thick, longer than the sepals; pistillate flowers with 4 or more sepals; ovary 2- or 3-locular, the ovules 1 or 2 in each locule; styles 2 or 3, spreading; fruit a drupe or composed of 2- or 3-horned dehiscent capsules.——About 5 species, in e. Asia and N. America.

1. **Pachysandra terminalis** Sieb. & Zucc.   Fukki-sō. Stems elongate, sparsely branched, creeping and ascending, 20–30 cm. long, green, at first puberulent; leaves fasciculate and falsely verticillate, deep green, coriaceous, thick, lustrous, rhombic-obovate, 5–10 cm. long inclusive of the petiole, 2–4 cm. wide, cuneate at base, with rather acute coarse teeth on upper margin, 3-nerved, puberulous on nerves above; spikes terminal, erect, sessile, 2–4 cm. long, often becoming lateral by the elongation of the lateral shoots; bracts and sepals broadly ovate, ciliolate, 2.5–3.5 mm. long; stamens usually 4, sometimes 3, about 8 mm. long; pistillate flowers short-pedicelled; fruit a glabrous ovoid drupe, white when mature, about 1.5 cm. long, the stones ovoid, about 5 mm. long, the styles 2, persistent, thick, suberect, with minute tubercles or papillae inside.——Apr.–May. Woods in lowlands and low mountains; Hokkaido, Honshu, Shikoku, Kyushu; frequently cultivated in gardens.——China.

2. **BUXUS** L.   Tsuge Zoku

Evergreen shrubs or trees with scaly buds; leaves opposite, short-petiolate, entire, coriaceous, slenderly penninerved, usually glabrous, lustrous; inflorescence fasciculate, axillary and terminal on the branches with a small apetalous pistillate flower in the center surrounded by several staminate ones; pistillate flowers with 6 sepals, apetalous; ovary 3-locular; styles 3; capsules globose or obovoid, 3-valved, loculicidally dehiscent, each valve terminated by the 2 hornlike persistent lobes of the styles; seeds 2 in each locule, lustrous, oblong.——About 30 species, in Eurasia and Centr. America, abundant in se. Asia.

1. **Buxus microphylla** Sieb. & Zucc.   *B. sempervirens* var. *microphylla* (Sieb. & Zucc.) Makino; *B. japonica* var. *microphylla* (Sieb. & Zucc.) Muell.-Arg.——Hime-tsuge. Erect shrub much branched from the base, to 60 cm. high, the branches slender; leaves narrowly obovate or narrowly oblong, 1.2–2 cm. long, 4–7 mm. wide, cuneate at base, slightly thinner than in the following varieties.——Mar.–Apr. Cultivated only, not known in the wild state.

Var. **japonica** (Muell.-Arg.) Rehd. & Wils. *B. japonica* Muell.-Arg.; *B. sempervirens* var. *suffruticosa* sensu Sieb., non L.; *B. sempervirens* var. *japonica* (Muell.-Arg.) Makino—— Tsuge, Asama-tsuge.   Erect shrub or small tree about 1–3 m. high, with a main trunk and grayish, glabrous branches, the branchlets 4-angled; leaves coriaceous, opposite, obovate, broadly obovate or oblong, 15–25 (–30) mm. long, 7–13 (–15) mm. wide, rounded, obtuse or retuse, acute at base, nearly sessile, entire, glabrous, lustrous, green above, paler beneath, the midrib raised on both surfaces, the lateral nerves faint, ascending, more distinct on the upper side, simple or sometimes forked; flowers pale yellow, several in axillary sessile fascicles; sepals 4, elliptic, membranous, small, about 2.5 mm. long, obtuse; stamens 4, 6–7 mm. long, the anthers pale yellow, about 1 mm. long, oblong; capsules ellipsoidal, about 1 cm. long, 3-fid; styles rather thick, persistent, ascending, each bifid when the capsules split; valves of capsule with two hornlike appendages at apex formed from the split styles; seeds black, oblong, 3-angled, smooth.——Mar.–Apr. Mountains; Honshu (Kantō Distr. and westw.), Shikoku, Kyushu; local and rare; sometimes cultivated.

Var. **riparia** (Makino) Makino. *B. sempervirens* var. *riparia* Makino; *B. riparia* (Makino) Makino——Ko-tsuge. Stems often procumbent; branches rather slender; leaves small, oblong or obovate-oblong, 1–2 cm. long, 5–8 mm. wide.—— Rocks along rivers in mountains; Honshu (Kii Prov. and Chūgoku Distr.).

Fam. 117. **EMPETRACEAE**   Gankō-ran Ka   Crowberry Family

Shrubs; leaves evergreen, alternate, small, densely covering the branchlets, with a pulvinus at the base, exstipulate; flowers unisexual or bisexual, small, in axillary or terminal heads; sepals 4–6, subpetaloid, nearly arranged in 2 cycles; petals absent; stamens 2–4, free, the anthers small, 2-locular, longitudinally dehiscent; disc absent; ovary sessile, globose; style short, variously lobed or divided; ovules solitary in each locule, campylotropous; fruit subglobose, drupelike, juicy, the stones 2 or more, each with a single seed; seeds with abundant endosperm, the embryo straight, the cotyledons small.——Three genera, with about seven species, in boreal and alpine regions of the N. Hemisphere and high mountain areas of S. America.

1. EMPETRUM L.   Gankō-ran Zoku

Small dioecious or monoecious evergreen shrubs with long procumbent stems and branches with globose winter-buds; leaves small, linear-oblong or linear, obtuse, thick, revolute on margins, densely covering the branchlets; flowers axillary, with scaly bracts at base; sepals 3, small, rather petaloid; stamens 3, with slender filaments; styles very short; stigma 6- to 9-toothed; fruit a berrylike drupe, with 6–9 stones.——Arctic, northern, and alpine regions of the N. Hemisphere and in the alpine regions of S. America; about 5 species.

1. Empetrum nigrum L. var. **japonicum** K. Koch. *E. nigrum* sensu auct. Japon., non L.; *E. nigrum* var. *asiaticum* Nakai; *E. kurilense* Vass.; *E. albidum* Vass.——Gankō-ran. Procumbent shrub about 10–20 cm. high, very densely branched, densely white-puberulent while young; leaves thick, coriaceous, lustrous, broadly linear, 5–6 mm. long, 0.7–0.8 mm. wide, obtuse to subrounded, spreading, slightly deflexed when old, the margins revolute and covering the under surface, both ridges slightly scabrous; flowers axillary, small, sessile, bisex-

ual or polygamous; filaments very slender, the anthers broadly ovate; fruit globose, sweet to taste, dark purple or blackish, about 6 mm. across, the stones several, semiglobose, about 1.5 mm. long, with a thick rounded back.——June–July. Forming broad, dense mats in alpine regions; Hokkaido, Honshu (centr. distr. and northw.); common. The typical phase and some other variants occur in the arctic, northern, and alpine regions of the N. Hemisphere.

Fam. 118. CORIARIACEAE   Doku-utsugi Ka   Coriaria Family

Shrubs with angular branches and scaly winter-buds; leaves opposite or verticillate, exstipulate; flowers bisexual or unisexual, small, greenish, solitary or in racemes, axillary; sepals 5, imbricate; petals small, keeled on inner side; stamens 10, free, or half of them adnate to the keel of the petals, the anthers large, longitudinally splitting; carpels 5–10, free, each 1-locular, the styles free, elongate, the ovules solitary in each locule, pendulous from the top of the locule, anatropous; fruitlets covered with the accrescent, juicy, persistent petals; seeds flat, the endosperm scanty, with a straight embryo in the center.——A single genus, in the temperate regions of e. Asia, New Zealand, the Mediterranean region, and Central and South America.

1. CORIARIA L.   Doku-utsugi Zoku

The characters of the family; 8 or 9 species.

1. Coriaria japonica A. Gray. Doku-utsugi. Glabrous, deciduous shrub; branches rather thick, elongate, the branchlets 4-angled; leaves nearly sessile, opposite, in 2 rows on branchlets (appearing as though the branchlets were pinnate leaves), ovate, or ovate-oblong, 6–8 cm. long, 2–3.5 cm. wide, gradually acuminate, rounded at base, 3-nerved; racemes pistillate and staminate, arising from lateral buds of the preceding year's growth, without subtending normal leaves; staminate racemes 2–3 cm. long, the bracts and 1 or 2 pairs of scalelike leaves obovate, membranous, 3–5 mm. long, the

pedicels about half as long as the bracts; sepals about 3.5 mm. long, membranous; stamens 8–10; pistillate racemes 8–15 cm. long, including the peduncles, the scalelike leaves rounded, sometimes green and somewhat leaflike, the bracts greenish, recurved, the pedicels 7–10 mm. long, spreading; petals accrescent after anthesis and enclosing the fruit, 5–6 mm. long, juicy, becoming dark reddish purple; fruitlets obliquely ovate, rather thick, 4 mm. long, with few longitudinal striations on the back.——May. Thickets in mountains and hills; Hokkaido, Honshu (Kinki Distr. and eastw.); locally abundant.

Fam. 119. ANACARDIACEAE   Urushi Ka   Sumac Family

Trees or shrubs; leaves alternate or rarely opposite, usually odd-pinnate, without stipules; flowers bisexual or unisexual, actinomorphic; sepals variously lobed; petals 3–7 or absent, free, rarely connate; disc present; stamens usually twice as many as the petals, rarely as many or numerous, the anthers 2-locular, longitudinally dehiscent; ovary superior, 1-locular, rarely 2- to 5-locular; styles 1–3; ovules solitary in each locule, pendulous; fruit usually a drupe; seeds without endosperm or endosperm very scanty, the cotyledons fleshy.——About 70 genera, with about 600 species, chiefly in the Tropics.

1A. Styles 3; ovary 1-locular, 1-ovuled; fruit 1-seeded. . . . . 1. *Rhus*
1B. Styles 5; ovary 5-locular, 5-ovuled; fruit 3- to 5-seeded. . . . . 2. *Choerospondias*

1. RHUS L.   Urushi Zoku

Deciduous or evergreen, dioecious or polygamous, trees or shrubs, sometimes scandent, with naked winter-buds; leaves alternate, ternate or odd-pinnate, rarely simple; flowers small, in axillary or terminal panicles; sepals 4–6, small, connate at base, persistent, imbricate; petals 4–6, imbricate; disc annular; stamens 4–6, at the base of the disc, free; ovary sessile; styles 3, free or connate, the stigma capitate; ovules solitary; fruit a small, slightly depressed drupe, the stones coriaceous or horny; cotyledons flat.——About 150 species, in temperate and tropical regions.

1A. Scandent woody plants with 3-foliolate leaves; inflorescence axillary. . . . . 1. *R. ambigua*
1B. Erect trees with odd-pinnate leaves, the leaflets more than 3.

2A. Panicles axillary; leaf rachis wingless; fruit whitish.
  3A. Fruit glabrous.
    4A. Branchlets and leaves glabrous. . . . . 2. *R. succedanea*
    4B. Branches and under side of leaflets on nerves pubescent.
      5A. Leaflets to 16 cm. long, 7 cm. wide; lateral nerves rather distant, more than 5 mm. apart. . . . . 3. *R. verniciflua*
      5B. Leaflets to 12 cm. long, 4 cm. wide; lateral nerves close together, about 5 mm. or less apart. . . . . 4. *R. sylvestris*
  3B. Fruit hispid; leaflets 4–10 cm. long, pubescent beneath. . . . . 5. *R. trichocarpa*
2B. Panicles terminal; leaf rachis winged; fruit red when mature, hairy. . . . . 6. *R. javanica*

1. Rhus ambigua Lavall. ex Dipp. *R. orientalis* (Greene) C. K. Schn.; *R. toxicodendron* var. *radicans* sensu auct. Japon., non A. Gray; *R. toxicodendron* var. *hispida* Engl.; *Toxicodendron orientale* Greene——Tsuta-urushi. Woody dioecious vine with aerial climbing roots; branchlets with short, appressed brown puberulence; leaves with petioles 3–6 cm. long, the leaflets 3, elliptic or ovate, 5–15 cm. long, 3–8(–10) cm. wide, abruptly acuminate, obtuse to subacute at base, the terminal one short-petioluled, the lateral ones nearly sessile, entire in the adult, often appressed-pubescent on nerves, with persistent axillary tufts of hairs beneath, glabrous on upper side; inflorescence axillary, 3–5 cm. long; flowers short-pedicellate, the petals pale yellow, about 3 mm. long, oblong, recurved above; fruit pale yellow, obliquely depressed-globose, 5–6 mm. long, glabrous or sparsely spiny-hispid.——May–June. Woods in mountains; Hokkaido, Honshu, Shikoku, Kyushu; common; poisonous.——Sakhalin, s. Kuriles, China, and Formosa.——Forma **rishiriensis** (Nakai) Hara. *R. rishiriensis* Nakai——Rishiri-tsuta-urushi. Stems to 2 m. long, scarcely scandent.——Reported from Hokkaido.

2. Rhus succedanea L. *R. succedanea* var. *japonica* Engl.; *Toxicodendron succedaneum* (L.) O. Kuntze——Haze, Ryūkyū-haze, Haze-no-ki. Deciduous, dioecious tree, glabrous except for yellowish brown pubescence on the margins and inner sides of the bud-scales; branches rather thick; leaves odd-pinnate, 25–35 cm. long inclusive of the petioles 6–8 cm. long, the leaflets 9–15, broadly lanceolate or narrowly oblong, 5–9 cm. long, 18–30 mm. wide, long-acuminate, acutish on inner side, semirounded on outer side at base, entire, sometimes glaucous beneath, the lateral nerves close together, 4 mm. or less apart, the rachis terete, wingless; panicles axillary, peduncled, 10–20 cm. long inclusive of the peduncles; flowers small, yellowish green; fruit depressed-globose, 8–10 mm. across, slightly flattened, lustrous, glabrous, smooth, pale yellow.——May–June. Honshu, Shikoku, Kyushu; often planted for yielding wax.——Ryukyus, Formosa, China, Malaysia, and India.

3. Rhus verniciflua Stokes. *R. vernicifera* DC.; *R. kaempferi* Sweet——Urushi. Deciduous, dioecious tree; branches thick, with numerous, prominently raised lenticels, with yellowish brown hairs while young; leaves 10–30 cm. long, on petioles 4–10 cm. long, the leaflets 9–11, ovate-oblong, sometimes ovate or oblong, 7–15 cm. long, 3–6 cm. wide, abruptly acuminate, subobtuse at base, slightly oblique in the lateral ones, entire, often pilose on upper side, spreading-pubescent at least on nerves beneath; panicles 15–30 cm. long,

with spreading yellowish brown hairs; flowers yellowish green; fruit slightly flattened, 6–8 mm. long, pale yellow, smooth, lustrous.——May. Cultivated in the warmer parts of our area for the production of lacquer.——China and India. Poisonous.

4. Rhus sylvestris Sieb. & Zucc. *Toxicodendron sylvestre* (Sieb. & Zucc.) O. Kuntze——Yama-haze, Haze. Deciduous tree; branches thick, densely yellowish brown pubescent; leaves 25–40 cm. long inclusive of the petiole, the leaflets 9–13, yellowish brown pubescent, broadly lanceolate to narrowly oblong, 7–12 cm. long, 2–4 cm. wide, short- or abruptly acuminate, obliquely rounded to broadly cuneate at base, entire, the lateral nerves numerous, rather densely parallel, slightly arcuate above; inflorescence a panicle 8–15 cm. long, spreading-pilose; fruit yellow-brown, rather flat, compressed-globose, 8–10 mm. across, glabrous.——May–June. Low mountains; Honshu (Tōkaidō Distr. and westw.), Shikoku, Kyushu.——Ryukyus, Formosa, China, and Korea.

5. Rhus trichocarpa Miq. *Toxicodendron trichocarpum* (Miq.) O. Kuntze——Yama-urushi. Deciduous, dioecious tree; branches thick, yellowish brown pubescent; leaves 25–45 cm. long inclusive of the petiole, the leaflets 13–17, ovate, obovate-elliptic or elliptic, 5–10 cm. long, 3–5 cm. wide, abruptly acute, slightly oblique at base, entire or coarsely toothed, yellowish brown pubescent beneath; panicles axillary, 15–30 cm. long inclusive of the peduncle, yellowish brown pubescent; flowers yellowish green, small, short-pedicellate; filaments slender, longer than the petals; fruit obliquely compressed-globose, 5–6 mm. across, yellowish, short-hispid, the exocarp easily peeled off and exposing the whitish mesocarp.——May–June. Hills and mountains; Hokkaido, Honshu, Shikoku, Kyushu; common.——s. Kuriles, Korea, and China.

6. Rhus javanica L. *R. chinensis* Mill.; *R. semialata* Murr.; *R. semialata* var. *osbeckii* DC.; *R. osbeckii* (DC.) Decne. ex Steud.——Nurude, Fushi-no-ki. Deciduous, dioecious tree; branches rather thick, densely yellowish brown pubescent when young; leaves 25–40 cm. long inclusive of the petiole, the rachis winged, the leaflets 7–13, oblong to ovate-oblong, 5–12 cm. long, 2.5–6 cm. wide, abruptly acuminate, cuneate to rounded at base, coarsely toothed, glabrate on upper side, brownish pubescent beneath; panicles terminal, broad, 15–30 cm. long, brownish pubescent; flowers yellowish, small; filaments nearly as long as the petals; fruit subglobose, reddish yellow, densely yellowish brown puberulent, about 4 mm. across.——Aug.–Sept. Hills, lowlands, and mountains; Hokkaido, Honshu, Shikoku, Kyushu; common.——Ryukyus, Formosa, Korea, Manchuria, and China to Malaysia.

2. CHOEROSPONDIAS Burtt & A. W. Hill   Chanchin-modoki Zoku

Polygamodioecious trees; leaves odd-pinnate, the leaflets 7–15, entire, the lateral nerves spreading; staminate flowers in panicles, the pistillate solitary or few; calyx cup-shaped, obtusely 5-lobed; petals 5, free, imbricate; stamens 10, the filaments connate and adnate to the 10-lobed disc at the base; ovary superior, ovoid, 5-locular; styles 5, free just below the apex, the stigma capitate; ovules solitary in each locule, pendulous; fruit ovoid, drupe-like, the stone surrounding the seeds horny, 5-locular, each with a membranous pore at apex.——A single species.

1. Choerospondias axillaris (Roxb.) Burtt & A. W. Hill var. **japonica** (Ohwi) Ohwi. *Poupartia fordii* var. *japonica* Ohwi; *P. fordii* sensu auct. Japon., non Hemsl.; *C. axillaris* sensu auct. Japon., non Burtt & A. W. Hill; *P. axillaris* sensu auct. Japon., non King & Prain——CHANCHIN-MODOKI. Deciduous apparently dioecious tree, with glabrous rather thick branches; leaves fasciculate at the ends of the branches, the leaflets 7–9, prominently petioluled, rather membranous, ovate to ovate-oblong, 4–8 cm. long, 1.5–3 cm. wide, acuminate, rounded to abruptly acute at base, entire, sparsely furfuraceous-puberulous on upper side while young, soon glabrous, slightly glaucescent and with sparse axillary tufts of hairs beneath; flowers chocolate-brown, the staminate many, in panicles, about 3 mm. across, the petals oblong, obtuse, 3 mm. long, recurved above, the stamens 10; pistillate flowers solitary in leaf-axils, about 8 mm. across, the peduncle with 2 caducous, membranous, narrow, ciliolate bracts at or above the middle, the petals about 7 mm. long, the staminodes 10, glabrous; fruits ellipsoidal, about 2.5 cm. long, edible.——May. Kyushu; rare. ——The typical phase occurs in China and India.

Fam. 120. AQUIFOLIACEAE   MOCHI-NO-KI KA   Holly Family

Trees or shrubs; leaves deciduous or evergreen, alternate, simple, exstipulate; flowers actinomorphic, bisexual or unisexual, in cymes, fascicles, or solitary; calyx-lobes imbricate; petals 4 or 5, free or connate at base, imbricate; stamens as many as the petals or rarely more, free, the anthers 2-locular, splitting longitudinally; disc absent; ovary with 3 or more locules, the style simple, sometimes absent, the ovules 1 or 2 in each locule, pendulous; fruit a drupe, with 3 or more 1-seeded stones; seeds with abundant endosperm, the embryo small, straight.——Three genera, with about 300 species, in the N. Hemisphere, a few in the Tropics, especially in America.

1. ILEX L.   MOCHI-NO-KI ZOKU

Dioecious or polygamous trees or shrubs; leaves alternate, entire or toothed; peduncles axillary, few-flowered, sometimes branched; flowers whitish, small, usually unisexual; calyx-lobes 4 or 5, small; corolla rotate, the petals 4 or 5; stamens as many as the petals; ovary 4 or 5 (–8)-locular; style short, thick or none; drupe globose, with 4–8 horny or crustaceous stones (pyrenes). ——About 300 species in temperate regions and in tropical and subtropical America.

1A. Leaves membranous, deciduous.
  2A. Pyrenes sulcate on back; leaves and flowers often fasciculate at ends of short spurlike branches. .... 1. *I. macropoda*
  2B. Pyrenes not sulcate, smooth; short spurs absent; leaves alternate on the elongate branches.
    3A. Fruit pendulous on slender peduncles 2–3.5 cm. long. .... 2. *I. geniculata*
    3B. Fruit on short peduncles, not pendulous.
      4A. Leaves rounded at base, 8–13 cm. long exclusive of the 2–3 cm. long petioles. .... 3. *I. micrococca*
      4B. Leaves cuneate to acute at base, the petioles 2–13 mm. long.
        5A. Leaves broadly oblanceolate, 4–13 cm. long, prominently cuneate at base; petioles 8–12 mm. long. .... 4. *I. nipponica*
        5B. Leaves elliptic to oblong, 2.5–8 cm. long, acute or subcuneate at base; petioles 4–8 mm. long. .... 5. *I. serrata*
1B. Leaves coriaceous, usually lustrous, evergreen.
  6A. Peduncles in axils of the current year's branches, 1- to few-flowered.
    7A. Leaves serrate.
      8A. Leaves 8–15 cm. long, obtusely appressed-serrate. .... 6. *I. chinensis*
      8B. Leaves usually not exceeding 4 cm. long.
        9A. Leaves without glandular dots beneath; fruit red when mature, the peduncles 1.5–4 cm. long. .... 7. *I. sugerokii*
        9B. Leaves glandular spotted beneath; fruit black when mature, on peduncles less than 1.5 cm. long. .... 8. *I. crenata*
    7B. Leaves entire.
      10A. Peduncles 3–5 cm. long, with 1–3 ripe fruits. .... 9. *I. pedunculosa*
      10B. Peduncles 1–2 cm. long, with 3–8 ripe fruits. .... 10. *I. rotunda*
  6B. Peduncles in axils of the preceding year's branches, short or very short.
    11A. Leaves with impressed veinlets on upper side; stems low, creeping, sparsely branched, the branches angular.
      12A. Leaves 2–4 cm. long. .... 11. *I. rugosa*
      12B. Leaves 5–7 cm. long. .... 12. *I.* × *makinoi*
    11B. Leaves flat or with scarcely impressed or raised veinlets.
      13A. Leaves entire or nearly so.
        14A. Shrub, long-creeping at base, with ascending branches; leaves 8–15 cm. long. .... 13. *I. leucoclada*
        14B. Trees; leaves 2.5–6 cm. long.
          15A. Fruit about 1 cm. across; leaves 5–8 cm. long. .... 14. *I. integra*
          15B. Fruit about 3 mm. across; leaves 3–6 cm. long. .... 15. *I. goshiensis*
      13B. Leaves toothed, sometimes obsoletely so.
        16A. Leaves 10–20 cm. long, thick-coriaceous, with distinct, mucronate teeth. .... 16. *I. latifolia*
        16B. Leaves 3–8 cm. long, thinner, the teeth obscure or confined to the upper half.
          17A. Pedicels 1–3 mm. long; petioles 4–8 mm. long. .... 17. *I. buergeri*
          17B. Pedicels 10–15 mm. long; petioles 10–16 mm. long. .... 18. *I. liukiuensis*

1. Ilex macropoda Miq. *I. dubia* var. *macropoda* (Miq.) Loesen.; *I. monticola* var. *macropoda* (Miq.) Rehd.; *I. montana* var. *macropoda* (Miq.) Fern.; *I. macropoda* var. *stenophylla* Koidz.——AO-HADA. Glabrous tree with grayish branches; leaves membranous, ovate or broadly so, 4–7 cm. long, 2.5–4 cm. wide, abruptly or shortly acuminate, acute to rounded at base, mucronate, serrate, sparsely puberulous or nearly glabrous on upper side, with spreading pubescence especially on nerves beneath, pale green and lustrous beneath, the petioles 1–2 cm. long; flowers on short spurs, greenish white, about 4 mm. across, the pedicels slender, 6–12 mm. long; calyx 4- to 5-lobed, ciliolate; petals elliptic, about 2.5 mm. long; stamens 4 or 5; fruit globose-ellipsoidal, about 7 mm. long, red, the pyrenes shallowly grooved.——Mountains; Hokkaido, Honshu, Shikoku, Kyushu; common.——Korea and China.——Forma **pseudomacropoda** (Loesen.) Hara. *I. dubia* var. *pseudomacropoda* Loesen.; *I. macropoda* var. *pseudomacropoda* (Loesen.) Nakai——Ke-nashi-ao-hada. Leaves glabrous beneath. Occurs with the typical phase.

2. **Ilex geniculata** Maxim. Fūrin-ume-modoki. Glabrous deciduous tree, with grayish brown slender branches; leaves membranous, elliptic, ovate, or oblong, 4–10 cm. long, 2.5–5 cm. wide, acuminate or abruptly so, rounded at base, with short obtuse and mucronate teeth on margin, sparsely puberulous on upper side, paler and often spreading-pilose, especially on the nerves beneath, the petioles 7–10 mm. long; flowers white, 1–3 on slender peduncles 2.5–3.5 cm. long, the pedicels 1–2 cm. long; calyx-lobes semirounded, ciliolate; petals about 2 mm. long; stamens shorter than the petals; fruit pendulous, globose, about 4 mm. across, the pyrenes smooth. ——June–Aug. Mountains; Honshu, Shikoku, Kyushu.

Var. **glabra** Okuyama. Okuno-fūrin-ume-modoki. Leaves glabrous beneath.——Honshu (n. distr.).

3. **Ilex micrococca** Maxim. Tama-mizuki. Glabrous, deciduous tree with rather stout grayish brown branchlets; leaves chartaceous, ovate-oblong or oblong, 8–13 cm. long, 3–4.5 cm. wide, abruptly acuminate and somewhat obtuse at tip, rounded to subacute at base, shallowly mucronate-toothed, yellowish green beneath, the costa impressed above, the veinlets slightly raised beneath, the petioles 2–3 cm. long, reddish; cymes axillary, rather densely many-flowered, the peduncle 1–1.5 cm. long; pistillate flowers about 3.5 mm. across; petals and reduced stamens 6–8; drupe globose, short pedicellate, red, about 3 mm. across.——May. Mountains; Honshu (Tōtōmi, Mino, and westw.), Shikoku, Kyushu.

4. **Ilex nipponica** Makino. *I. spathulata* Koidz.; *I. nemotoi* sensu Ohwi, non Makino; *I. serrata* var. *nipponica* (Makino) Ohwi——Miyama-ume-modoki, Hosoba-ume-modoki. Deciduous shrub with nearly glabrous branches; leaves rather thick-membranous, broadly oblanceolate to obovate, 4–13 cm. long, acuminate to rounded at apex, cuneate at base, with short mucronulate scattered teeth, slightly puberulous on both surfaces while young, the petioles 8–12 mm. long; pistillate flowers white, 3–7, clustered, about 0.7 mm. long, about 5 mm. across, on a short peduncle, the pedicels 4–8 mm. long; calyx-lobes triangular to semirounded, obtuse, ciliolate; fruit globose, red, about 6 mm. across.——May–June. Wet places in mountains; Honshu (Hokuriku Distr. and northw.).

5. **Ilex serrata** Thunb. *I. serrata* var. *sieboldii* (Miq.) Rehd.; *I. sieboldii* Miq.; *I. subtilis* Miq.; *I. serrata* var. *subtilis* (Miq.) Yatabe; *I. serrata* forma *subtilis* (Miq.) Ohwi; *I. nemotoi* Makino——Ume-modoki. Much-branched dioecious shrub with slender grayish brown branches puberulous while young; leaves oblong to elliptic or obovate-elliptic, 4–8 cm. long, abruptly acuminate to acute, acute to subcuneate at base, with minute mucronate teeth, puberulous on both surfaces, paler beneath, the petioles 4–8 mm. long; flowers pale purple to white, about 3.5 mm. across, the pistillate 1–7, the staminate 7–15, fasciculate on very short axillary peduncles, the pedicels 2–4

mm. long; fruit red, globose, about 5 mm. long.——June. Wet places in mountains; Honshu, Shikoku, Kyushu; common; variable.

Var. **argutidens** (Miq.) Rehd. *I. argutidens* Miq.——Inu-ume-modoki. Leaves and branches glabrous. With the typical phase.

Frequently cultivated are:——Cv. **Xanthocarpa.** *I. serrata* forma *xanthocarpa* (Rehd.) Rehd.; *I. serrata* var. *xanthocarpa* Rehd.——Kimi-no-ume-modoki. Fruit yellow.——Cv. **Leucocarpa.** *I. serrata* forma *leucocarpa* Beissn.——Shiro-ume-modoki. Fruit white.

6. **Ilex chinensis** Sims. *I. purpurea* Hassk.; *I. oldhamii* Miq.; *I. purpurea* var. *oldhamii* (Miq.) Loes.——Nanami-no-ki, Naname-no-ki. Nearly glabrous dioecious evergreen tree dark brown when dried, the branches dark gray; leaves rather coriaceous, oblong or narrowly so, 7–12 cm. long, 2.5–5 cm. wide, caudately elongate with an acute to subobtuse tip, with scattered short teeth, the petioles 1–1.5 cm. long; inflorescence solitary in axils of the current year's shoots, the peduncles 5–10 mm. long, rather thick and flat, the pedicels 2–7, 4–6 mm. long; flowers about 5 mm. across; calyx-lobes semirounded to deltoid, ciliolate; petals slightly reflexed; stamens 4; fruit globose, red, about 6 mm. across.——June. Mountains and hills; Honshu (Suruga Prov. and westw.), Shikoku, Kyushu.——China.

7. **Ilex sugerokii*** Maxim. var. **brevipedunculata** (Maxim.) S. Y. Hu. *I. sugeroki* forma *brevipedunculata* Maxim.; *I. sugeroki* subsp. *brevipedunculata* (Maxim.) Makino——Akami-no-inu-tsuge. Evergreen shrub with pale brown branches and often glaucous branchlets, spreading-puberulent while young; leaves coriaceous, many, ovate, oblong, or oblong-obovate, 2–3.5 cm. long, 1–2 cm. wide, acute with an obtuse tip, acute to obtuse at base, with short obtuse teeth on upper margin, lustrous, glabrous, deep green above, paler beneath, the midrib slightly raised on both surfaces, the lateral nerves obsolete, the petioles 2–4 mm. long, reddish; inflorescence glabrous, axillary on lower part of the branchlets, the pedicels 1–1.5 cm. long; pistillate flowers solitary, the staminate 1–3; stamens 4–5; fruit globose, red, about 7 mm. across.——July. High mountains; Hokkaido, Honshu (centr. distr. and northw.); locally abundant.

Var. **longipedunculata** (Maxim.) Makino. *I. sugeroki* forma *longipedunculata* Maxim.; *I. sugeroki* subsp. *longipedunculata* (Maxim.) Makino; *I. sugeroki* var. *longipedicellata* Koidz.——Ushi-kaba, Kuro-soyogo, Aburagi. Small tree; leaves elliptic, oblong, or ovate, 3–4 cm. long, 1.5–2.5 cm. wide; pedicels 2–3 cm. long.——July. Honshu (Mikawa and Shinano Prov. and westw.).

8. **Ilex crenata** Thunb. *Celastrus adenophylla* Miq.; *I. crenata* var. *typica* forma *genuina* Loes.——Inu-tsuge. Evergreen shrub or small tree with densely and repeatedly ramulose grayish branches, the branchlets minutely puberulous; leaves elliptic, oblong, or narrowly obovate, 1.5–3 cm. long, 6–20 mm. wide, acute or obtuse, acute or cuneate at base, with appressed teeth, glabrous, lustrous and deep green above, paler and with scattered sessile pellucid glands beneath, the midrib raised beneath, the lateral nerves obsolete, the petioles 1–5 mm. long, usually puberulous; staminate flowers

---

* Spelling "sugerokii" is based upon the recommendation in the International Code of Botanical Nomenclature in the use of "ii." The specific name is based on the personal name, Sukeroku Mizutani or Sugeroku Mizutani.

few, fasciculate, sometimes on very short peduncles, the pedicels 3–5 mm. long, the pistillate solitary; petals white, about 2 mm. long; stamens 4; fruit globose, black, about 6 mm. across.——June–July. Thickets and wet places in lowlands and mountains; Hokkaido, Honshu, Shikoku, Kyushu; common and very variable; often planted as a hedge, with a few cultivars in gardens.

Var. **fukasawana** Makino. TSUKUSHI-INU-TSUGE. Branches prominently angled; leaves chartaceous, oblong, 2.5–4 cm. long.——Kyushu.——(?) Amami-Ōshima.

Var. **mutchagara** (Makino) Ohwi. *I. mutchagara* Makino ——SHIMA-INU-TSUGE. Larger shrub with prominently angled branches; leaves thinner, broadly lanceolate to obovate-oblong, 2.5–5 cm. long, 1–1.5 cm. wide, usually obtuse, cuneate at base; pedicels of fruit 7–10 mm. long.—— Kyushu (s. distr. and Yakushima).——Ryukyus and Formosa.

Var. **paludosa** (Nakai) Hara. *I. radicans* Nakai; *I. crenata* subsp. *radicans* Tatew.; *I. crenata* var. *radicans* (Nakai) Ohwi; *I. radicans* var. *paludosa* Nakai——HAI-INU-TSUGE, YACHI-INU-TSUGE. Similar to the typical phase but with low creeping stems and branches.——Wet, swampy places; Hokkaido, Honshu (Japan Sea side).——s. Kuriles and s. Sakhalin.

9. Ilex pedunculosa Miq. *I. pedunculosa* forma *genuina* Loesen.; *I. fujisanensis* Sakata——SOYOGO. Glabrous evergreen shrub or small tree with grayish branches; leaves rather coriaceous, ovate-elliptic, 4–8 cm. long, 2.5–3 cm. wide, acuminate to abruptly cuspidate, entire, deep green and lustrous on upper side, dark-brown when dried, the midrib raised beneath, the lateral nerves obsolete, the petioles reddish, 1–2 cm. long; flowers white, about 4 mm. across, the peduncles 2–5 cm. long, the pedicels 1–5, umbellate, 1–2 cm. long; fruit red, globose, nodding, about 7 mm. across.——June. Mountains; Honshu (Shinano and Kai Prov. and westw.), Shikoku, Kyushu.——Forma **aurantiaca** (Koidz.) Ohwi. *I. pedunculosa* var. *aurantiaca* Koidz.——KIMI-SOYOGO. Fruit yellow.—— Forma **variegata** (Nakai) Ohwi. *I. pedunculosa* var. *variegata* Nakai——FUIRI-SOYOGO. Leaves variegated.

Var. **senjoensis** (Hayashi) Hara. *I. senjoensis* Hayashi—— TAKANE-SOYOGO. Low shrub with long creeping sparsely branching stems and 6-merous flowers.——High mountains of Honshu (Mount Senjo in Shinano Prov.).

10. Ilex rotunda Thunb. KUROGANE-MOCHI. Glabrous evergreen tree with dark brown branches; leaves coriaceous, elliptic to broadly elliptic, 5–8 cm. long, 3–4 cm. wide, entire, acute or acute with an obtuse tip, obtuse at base, brownish when dried, the midrib raised beneath, narrowly grooved on upper side, the petioles 1.5–2.5 cm. long; inflorescence few-flowered, the peduncles 7–10 mm. long, the pedicels shorter than the peduncles; flowers about 4 mm. across; petals elliptic, about 3 mm. long, reflexed; stamens 4 or 5, slightly longer than the petals; fruit globose or broadly ellipsoidal, red, 5–8 mm. long.——June. Honshu (Kantō Distr. and westw.), Shikoku, Kyushu; frequently cultivated in gardens.——Ryukyus, Formosa, China, and Korea.

11. Ilex rugosa F. Schmidt. TSURU-TSUGE. Dwarf, long-creeping, glabrous shrub, sparsely branched, angled, punctulate, green; leaves densely disposed, coriaceous, broadly lanceolate to ovate-oblong, 2–3.5 cm. long, 5–15 mm. wide, obtuse to acute with an obtuse tip, obtuse at base, obtusely serrate, lustrous, deep-green, paler beneath, dark brown when dry, the midrib, lateral nerves, and veinlets impressed on upper side, slightly raised beneath, the petioles 2–5 mm. long; flowers

white, about 5 mm. across, 1 to few, fasciculate, axillary, the pedicels 3–7 mm. long; fruit red, globose, about 5 mm. across.——June. Coniferous woods; Hokkaido, Honshu (Yamato Prov. and eastw.), Shikoku, Kyushu.——Sakhalin and s. Kuriles.

Var. **vegeta** Hara. MARUBA-TSURU-TSUGE. Larger leaves to 4 cm. long, 2.3 cm. wide.——Western (Kinki Distr. and westw.).

Var. **stenophylla** (Koidz.) Sugimoto. *I. stenophylla* Koidz. ——HOSOBA-TSURU-TSUGE. Leaves usually narrower, usually lanceolate to broadly linear.——Honshu, Shikoku, Kyushu.

12. Ilex × makinoi Hara. *I. rugosa* var. *fauriei* Loesen.; *I. fauriei* (Loes.) Makino, non Gand.——Ō-TSURU-TSUGE. Alleged hybrid of *I. leucoclada* × *I. rugosa*, differing from *I. rugosa* in the larger parts, scarcely punctulate branches, narrowly oblong, subacute leaves 5–7 cm. long, becoming grayish to brownish green when dried, and by the less prominently impressed nerves and veinlets.——Hokkaido, Honshu (n. distr.).

13. Ilex leucoclada (Maxim.) Makino. *I. integra* var. *leucoclada* Maxim.——HIME-MOCHI. Small evergreen, glabrous shrub with thick, elongate, terete, grayish branches ascending from a long creeping base; leaves somewhat coriaceous, narrowly oblong, broadly oblanceolate or narrowly obovate-oblong, 8–15 cm. long, 2–4 cm. wide, acute with an obtuse tip, entire or crenate-serrate on upper margin, flat and deep green on upper side, paler beneath, brownish when dried, the midrib raised beneath, the lateral nerves and veinlets slender, obsolete to only slightly raised on both surfaces, the petioles 1–1.5 cm. long, often glaucous; flowers white, about 7 mm. across, few, axillary, the peduncle nearly obsolete, the pedicels 1.2–2 cm. long in fruit; petals about 4 mm. long; fruit globose, red, about 10 mm. across.——May–June. Woods in mountains; Hokkaido (sw. distr.), Honshu (Japan Sea side as far west as Chūgoku Distr.).

14. Ilex integra Thunb. *Othera japonica* Thunb.; *I. othera* (Thunb.) Spreng.; *I. asiatica* Spreng.; *I. integra* var. *typica* Loes.——MOCHI-NO-KI. Small, glabrous, evergreen tree with thick, terete branches; leaves coriaceous, 5–8 cm. long, 2–3.5 cm. wide, abruptly acute, with an obtuse tip, acute to cuneate at base, entire, deep-green, lustrous, pale yellowish green with the midrib prominently raised beneath, the lateral nerves sometimes slightly distinct beneath, the petioles 7–12 mm. long; flowers yellowish green, about 8 mm. across, the pistillate 1 or 2, axillary, the staminate few, the pedicels 5–10 mm. long; petals spreading, oblong, 3.5–4 mm. long; stamens 4; fruit globose, red, about 1 cm. across.——Apr. Thickets and woods on low mountains especially near the sea; Honshu, Shikoku, Kyushu.——Ryukyus, Formosa, and China.—— Forma **ellipsoidea** (Y. Okamoto) Ohwi. *I. ellipsoidea* Y. Okamoto——INU-MOCHI. A rare spontaneous phase with ellipsoidal fruit.——Cv. **Xanthocarpa**. *I. integra* forma *xanthocarpa* (Matsum. & Nakai) Ohwi; *I. integra* var. *xanthocarpa* Matsum. & Nakai——KIMI-NO-MOCHI-NO-KI. Fruit yellow; cultivated.

15. Ilex goshiensis Hayata. *I. hanceana* sensu auct. Japon., non Maxim.; *I. hanceana* forma *goshiensis* (Hayata) Yamamoto; *I. hayataiana* Loes.——TSUGE-MOCHI. Small tree, the branches slender, much-branched, the branchlets, petioles, inflorescence, and calyx puberulous while young; leaves coriaceous, elliptic to obovate or oblong, 3–6 cm. long, 1.5–2.5 cm. wide, acute or obtuse with an obtuse tip, entire, grayish green when dried, flat and scarcely lustrous on upper side, the petioles 4–7 mm. long, often reddish, winged in upper part; inflorescence fasciculate, 2- to 7-flowered, the peduncles nearly obsolete, the pedicels 2–3 mm. long; calyx-lobes short, semirounded, puberulous and ciliolate; fruit globose, red, about 3.5 mm. across.——Honshu (Ise and Kii Prov.), Shikoku, Kyushu.——Ryukyus and Formosa.

16. Ilex latifolia Thunb. *I. macrophylla* Bl.; *I. tarajo* Hort. ex Goepp.——Tara-yō.   Evergreen glabrous tree, the branches thick, terete, the branchlets green, densely elevate-punctulate; leaves firmly coriaceous, oblong to narrowly so, 12–17 cm. long, 4–8 cm. wide, acute to acuminate with an obtuse tip, rounded to obtuse at base, callosely mucronate-serrate, deep green and lustrous above, yellowish green beneath, the midrib impressed above, raised beneath, the lateral nerves slightly distinct above, the petioles thick, 2–2.5 cm. long, green; flowers many, fasciculate, axillary, yellowish green, the pedicels 7–12 mm. long; stamens 4, about 3 mm. long, as long as to slightly longer than the recurved petals; fruits globose, red, about 8 mm. across.——May. Honshu (s. Kinki Distr. and westw.), Shikoku, Kyushu; frequently cultivated around temples.

17. Ilex buergeri Miq. *I. subpuberula* Miq.; *I. buergeri* var. *subpuberula* (Miq.) Loes.——Shii-mochi, Hizen-mochi. Small evergreen tree with rather slender, much ramulose and slightly angled branches, puberulous when young; leaves rather thinly coriaceous, oblong to narrowly so, or ovate-

oblong, 4–8 cm. long, 1.5–3 cm. wide, caudately acuminate to abruptly acute with an obtuse tip, acute or broadly cuneate at base, remotely obtuse-toothed sometimes on upper half, greenish yellow when dried, lustrous on upper side, the midrib prominently impressed above, raised beneath, the lateral nerves indistinct, the petioles rather slender, 4–8 mm. long; flowers about 6 mm. across, few to rather numerous, axillary, the peduncles absent, the pedicels 1–3 mm. long; petals recurved, 2.5–3 mm. long; stamens 4 or 5, erect, longer than the petals; fruits few, globose, red, 5–6 mm. across.——Apr.–May. Honshu (Chūgoku Distr.), Kyushu; rare.

Ilex × kiusiana Hatusima, Narihira-mochi, is a hybrid allegedly of *I. buergeri × I. integra*. Shrub about 4 m. high, the leaves 6–9 cm. long, the inflorescence peduncled, the pedicels 6–8 mm. long.——Reported from n. Kyushu (Fukuoka).

18. Ilex liukiuensis Loes. *I. mertensii* sensu auct. Japon., non Maxim.——Ryūkyū-mochi.   Glabrous tree; leaves coriaceous, obovate to oblong-elliptic, 3–7.5 cm. long, 2–3.5 cm. wide, obtuse to rounded, sometimes retuse or broadly short-acuminate, acute at base, nearly entire or remotely crenate on the recurved and thickened margin, dark brown when dry, the midrib raised beneath, the lateral nerves reticulate beneath, the petioles 1–1.6 cm. long; pistillate flowers 2–4 in axillary fascicles, the pedicels 9–14 mm. long; calyx-lobes 4–5, rounded, ciliate; fruit globose, 6 mm. across.——Reported from Kyushu (Yakushima and Tanegashima).——Ryukyus.

Fam. 121. **CELASTRACEAE**   Nishikigi Ka   Spindle-Tree Family

Trees, shrubs, or sometimes scandent vines; leaves opposite or alternate, simple, the stipules small, deciduous or absent; flowers usually in cymose inflorescences or fasciculate, usually bisexual, actinomorphic, small; calyx 4- or 5-lobed, imbricate; petals 4 or 5, imbricate, rarely valvate; stamens 4 or 5, alternate with the petals, usually on a flat disc or under the margin of the disc; ovary superior, sometimes adnate to the disc, 1- to 5-locular; style short, 3- to 5-lobed, the ovules usually 2 on the inner angle of each locule; fruit various, forming a loculicidal dehiscent capsule or an indehiscent dry fruit, drupe or berry; seeds usually with abundant fleshy endosperm. the embryo straight, the cotyledons flat, leaflike.——About 45 genera, with about 500 species, in temperate and tropical regions.

1A. Leaves alternate; scandent, deciduous shrubs.
  2A. Fruit a 3-winged samara; inflorescence a terminal panicle; seeds without an aril. . . . . 1. *Tripterygium*
  2B. Fruit a dehiscent capsule; inflorescence a terminal or axillary cyme, raceme, or panicle; seeds with an aril. . . . . 2. *Celastrus*
1B. Leaves opposite; erect or scandent trees or shrubs; fruit a capsule.
  3A. Flowers usually unisexual; petals usually connate at base, ascending and spreading above; capsules 1-locular and 1-seeded, dehiscent into 2 valves; leaves entire. . . . . 3. *Microtropis*
  3B. Flowers bisexual; petals free, horizontally spreading; capsules 3- to 5-locular with 1 or 2 seeds in each locule; leaves usually serrate.
                                                                 4. *Euonymus*

1. **TRIPTERYGIUM** Hook. f.   Kurozuru Zoku

Scandent shrubs; leaves alternate, large, petiolate, deciduous, serrate, the stipules linear, caducous; flowers polygamous, small, whitish, in rather large terminal panicles; calyx 5-lobed; petals 5; stamens 5 on margin of cup-shaped disc; ovary superior, triquetrous, incompletely 3-locular, with a short style at apex; ovules 2 in each locule, erect; fruit a 3-winged samara with an erect, linear, exarillate seed, the embryo small.——Few species in E. Asia.

1A. Panicles puberulous; leaves 5–15 cm. long, rounded at base; flowers 5–6 mm. across; fruit cordate at base. . . . . 1. *T. regelii*
1B. Panicles glabrous; leaves 5–10 cm. long, usually broadly cuneate at base; flowers 4–5 mm. across; fruit truncate to rounded at base.
                                                                2. *T. doianum*

1. Tripterygium regelii Sprague & Takeda. *T. wilfordii* sensu auct. Japon., non Hook. f.; *T. wilfordii* var. *regelii* (Sprague & Takeda) Makino——Kurozuru.   Scandent shrub; branches reddish brown, prominently warty, slightly angled, glabrous; leaves ovate to broadly so, or elliptic, 5–15 cm. long, 4–10 cm. wide, abruptly acuminate to acute, rounded

at base, obtusely serrate, vivid green, glabrous on upper side, sometimes papillose-pilose on nerves beneath, the petioles 1.5–3 cm. long; panicles 10–25 cm. long, the branches and branchlets papillose-puberulous; flowers greenish white, 5–6 mm. across; calyx-lobes obtusely deltoid; petals elliptic, nearly as long as the stamens; fruit broadly lanceolate, with 3 elliptic or cordate wings, 12–18 mm. long and as wide inclusive of the wing, retuse to cordate at both ends, pale green, glabrous.——July–Aug. Mountains; Honshu (Japan Sea side from centr. distr. northw., and Kii Prov.), Shikoku, Kyushu.——Korea and Manchuria.

2. **Tripterygium doianum** Ohwi. *T. regelii* var. *doianum* (Ohwi) Masam.——Koba-no-kurozuru. Similar to but slightly smaller than the preceding, glabrous in all parts; branches less prominently warty, rather slender; leaves ovate to broadly so, or elliptic, 5–10 cm. long, 3–5 cm. wide, acuminate, broadly cuneate at base, serrulate, the petioles 1.5–2 cm. long; panicles 6–10 cm. long, 4–5 cm. wide, the pedicels slender; flowers 4–5 mm. across, greenish white; calyx-lobes semi-rounded; stamens as long as the petals; fruit with 3 broad wings, elliptic to ovate-orbicular inclusive of the wings, 8–12 mm. long, 6–19 mm. wide, retuse, rounded to truncate at base, pale green.——July–Aug. Kyushu (s. distr., incl. Yakushima).

2. CELASTRUS L. Tsuru-ume-modoki Zoku

Deciduous or evergreen, dioecious or polygamous, usually scandent shrubs; leaves petiolate, alternate, serrate, the stipules small; flowers small, greenish or white, 5-merous, in axillary cymes or terminal panicles; calyx 5-lobed; petals oblong-ovate; disc entire or undulate ovary superior; style rather short, the stigma 3-lobed; fruit a dehiscent capsule, 3-valved, few-seeded; seeds covered with a fleshy orange or red aril.——About 30 species, in e. and s. Asia, Australia, and America.

1A. Stipules deciduous; leaves 5–12 cm. long.
  2A. Branches and inflorescence glabrous; leaves glabrous or papillose-scabrous beneath. . . . . 1. *C. orbiculatus*
  2B. Branches, inflorescence, and under side of leaves short-pilose, at least when young. . . . . 2. *C. stephanotiifolius*
1B. Stipules becoming spiny, persistent; leaves 2–5 cm. long. . . . . 3. *C. flagellaris*

1. **Celastrus orbiculatus** Thunb. *C. articulatus* Thunb.; *C. tatarinowii* Rupr.; *C. crispulus* Regel——Tsuru-ume-modoki. More or less scandent, deciduous, dioecious shrubs with glabrous, terete, grayish brown branches; leaves orbicular, obovate, or elliptic, 5–10 cm. long, 3–8 cm. wide, rounded and apiculate to obtuse at apex, broadly cuneate to abruptly acute at base, glabrous, with appressed obtuse teeth, paler beneath, the petioles 1–2 cm. long; inflorescence cymose, axillary, 1- to many-flowered on a short peduncle, cymose, glabrous, the pedicels slender, 3–5 mm. long; flowers pale green, 5-merous; petals narrowly oblong, about 4 mm. long; stamens half as long as the petals; fruit globose, pale greenish yellow, glabrous, about 8 mm. across, 3-valved, dehiscent, exposing the seeds, these covered with an orange-red aril; style persistent, 2–3 mm. long.——May–June. Thickets and grassy slopes in lowlands and mountains; Hokkaido, Honshu, Shikoku, Kyushu; common and often planted as a hedge.——s. Kuriles, Korea, Manchuria, Sakhalin, China, and Ussuri.——Forma **aureoarillatus** (Honda) Ohwi. *C. orbiculatus* var. *aureoarillatus* Honda——Kimino-tsuru-ume-modoki. Aril yellow.

Var. **punctatus** (Thunb.) Rehd. *C. punctatus* Thunb.; *C. kiusianus* Fr. & Sav.; *C. articulatus* var. *punctatus* (Thunb.) Makino; *C. articulatus* var. *humilis* Matsum.——Teri-ha-tsuru-ume-modoki. Leaves smaller, thicker, elliptic, more or less lustrous, 3–5 cm. long, the petioles shorter, 5–10 mm. long.——A southern phase, abundant usually near seashores, although transitional with the typical phase.

Var. **papillosus** (Nakai) Ohwi. *C. articulatus* var. *papillosus* Nakai; *C. lancifolius* Nakai; *C. strigillosus* Nakai——

Inu-tsuru-ume-modoki. Leaves papillose beneath on nerves and veinlets.——Honshu (centr. distr. and northw.), Hokkaido.

2. **Celastrus stephanotiifolius** (Makino) Makino. *C. articulatus* var. *pubescens* Makino and var. *stephanotiifolius* Makino; *C. insularis* sensu Ohwi, non Koidz.——Ō-tsuru-ume-modoki, Shitaki-tsuru-ume-modoki. Closely resembles the preceding; branches brownish, lenticellate, the young parts pilosulous; leaves elliptic to broadly so, sometimes orbicular or ovate-orbicular, 6–10 cm. long, 5–8 cm. wide, abruptly cuspidate to acute, rounded to acute at base, obtusely serrate, glabrous on upper side, paler and with short yellowish curved hairs on nerves and veinlets beneath, the petioles 2–3 cm. long; inflorescence few-flowered, the short peduncle and pedicels pilosulous; fruit globose, 5–6 mm. across.——May–June. Honshu (Kantō Distr. and westw.), Kyushu.——s. Korea.

3. **Celastrus flagellaris** Rupr. *C. ciliidens* Miq.——Iwa-umezuri. Scandent shrub; branches slender, with reddish brown exfoliating bark, the young parts minutely papillose; stipules becoming spiny; leaves broadly elliptic to ovate, 2–5 cm. long, 1.8–4 cm. wide, abruptly acuminate to acute, acute or broadly cuneate at base, with small spine-pointed teeth, sparsely scabrous on upper side, the veinlets distinct, the petioles 1–3 cm. long, slender; flowers 1–3, in axillary sessile cymes, the pedicels 2–6 mm. long; fruit globose, about 6 mm. across, the aril orange-red.——June. Mountains; Honshu (Kantō Distr. and westw.), Kyushu; rather rare.——Korea, Manchuria, Amur, and n. China.

3. MICROTROPIS Wall. Moku-reishi Zoku

Evergreen trees or shrubs; leaves opposite, petiolate, exstipulate, entire, coriaceous; flowers usually unisexual, small, whitish, in sessile or peduncled axillary cymes or fascicles; calyx of 5 imbricate, persistent sepals; petals 5 or none, imbricate, usually connate at base, rounded, rather thick, flat; disc absent or annular or of separate segments, sometimes adnate to the petals; ovary free, ovoid, 2- or 3-loculed, the style thick, the stigma small, 2- to 4-lobed, the ovules 2 in each locule; capsules ellipsoidal, coriaceous, 1-locular, 2-valved, 1-seeded; seed erect, stiped, oblong, the testa smooth, brown or red, sometimes fleshy outside, the endosperm copious, fleshy, the cotyledons leaflike.——About 70 species, in mountains of tropical Asia and e. Asia.

1. **Microtropis japonica** (Fr. & Sav.) H. Hallier. *Elaeodendron japonicum* Fr. & Sav.; *Cassine japonica* (Fr. & Sav.) O. Kuntze; *Otherodendron japonicum* (Fr. & Sav.) Makino ——Moku-reishi. A large, glabrous, evergreen, dioecious shrub with brownish gray branches; leaves coriaceous, elliptic or ovate, sometimes ovate-oblong, 5–10 cm. long, 2.5–4.5 cm.

wide, obtuse or obtusely acute, sometimes emarginate, abruptly acuminate at base, entire, with slightly recurved margins, deep green and lustrous above, the midrib rather slender, the lateral nerves slender, both slightly raised, the petioles 7–12 mm. long; flowers 5-merous, sessile, few, on a short peduncle 7–12 mm. long, pale green, about 5 mm. across, with 2 small bracteoles just beneath the calyx; calyx-lobes rounded, minutely toothed; petals orbicular, about 3 mm. long; disc cup-shaped; filaments short, borne on the margin of the disc, as long as the pale yellow anthers; capsules ellipsoidal, 1.5–2 cm. long, smooth, dehiscent and exposing a large red seed.——Mar.–Apr. Woods and thickets near seashores; Honshu (Izu, Awa, Sagami Prov.), Kyushu.——Ryukyus and Formosa.

4. EUONYMUS L.   Nishikigi Zoku

Evergreen or deciduous, erect or scandent shrubs or trees; leaves opposite, rarely alternate, entire or serrate; flowers small, bisexual or a few unisexual, in axillary cymes or solitary; calyx-lobes 4 or 5; petals attached below the disc, usually entire; stamens as many as the petals and calyx-lobes, opposite the latter, the filaments sometimes absent; disc discoid, 4- or 5-lobed; ovary below the disc and adnate to it, 3- to 5-locular; style short; ovules 2, rarely few in each locule; capsules 3- to 5-valved, sometimes winged, rarely spinose, with 1 or 2 seeds in each locule, the seeds entirely or partially enclosed in an aril, the endosperm fleshy.——About 150 species, in temperate to warmer regions of Asia, Europe, N. and Centr. America.

1A. Fruit divided nearly to the base into 1–4 lobes.
  2A. Deciduous, the leaves thin, membranous. .... 1. *E. alatus*
  2B. Evergreen, the leaves deep green, chartaceous to thinly coriaceous. .... 2. *E. lutchuensis*
1B. Fruit entire, not lobed, sometimes winged or sulcate.
  3A. Evergreen.
    4A. Fruits globose, not winged nor angular; branches terete.
      5A. Erect tree with smooth branches. .... 3. *E. japonicus*
      5B. Woody climber with obsoletely punctulate branches. .... 4. *E. fortunei*
    4B. Fruits angled, rather large.
      6A. Branches 4-angled. .... 5. *E. chibae*
      6B. Branches terete. .... 6. *E. tanakae*
  3B. Deciduous (apparently evergreen, leaves 7–12 cm. long in *E. lanceolatus*).
    7A. Filaments longer than the anthers; petioles 8–20 mm. long; peduncles short. .... 7. *E. sieboldianus*
    7B. Filaments very short; petioles 3–10 mm. long; peduncles elongate.
      8A. Branches warty; winter-buds short; inflorescence just above the leaves; fruit 4-angled, narrowed toward the base.
                                                                8. *E. oligospermus*
      8B. Branches nearly smooth; winter-buds elongate; inflorescence under the leaves; fruit rounded at base.
        9A. Flowers 4-merous.
          10A. Fruit with 4 broad wings; leaves obovate, 6–12 cm. long, 3–7 cm. wide; young branchlets pale green. .. 9. *E. macropterus*
          10B. Fruit globose, without wings or angles; leaves narrowly ovate, 6–8 cm. long, 2–3 cm. wide; young branchlets often whitish with waxy excrescence. .... 10. *E. yakushimensis*
        9B. Flowers 5-merous.
          11A. Ovary and fruit 5-locular.
            12A. Calyx-teeth denticulate; flowers purplish; fruit globose, wingless.
              13A. Stems erect; leaves ovate, thinly chartaceous, 3–8 cm. long. .... 11. *E. melananthus*
              13B. Stems creeping; leaves broadly oblanceolate, somewhat coriaceous, 7–12 cm. long. .... 12. *E. lanceolatus*
            12B. Calyx-teeth entire; flowers white to pale purple.
              14A. Fruit globose, wingless. .... 13. *E. oxyphyllus*
              14B. Fruit with 5 short deltoid wings. .... 14. *E. planipes*
        11B. Ovary and fruit 3- or rarely 4-locular; fruit with 3 or 4 rather broad, falcately deltoid wings near base; leaves rugulose on upper side in life; flowers purple, with 5 calyx-teeth and petals. .... 15. *E. tricarpus*

1. Euonymus alatus (Thunb.) Sieb. *Celastrus alatus* Thunb.; *E. alatus* var. *typicus* Regel; *E. striatus* var. *alatus* (Thunb.) Makino——Nishikigi.   Horizontally spreading, much-branched, deciduous shrub, the branches green, glabrous, with 4 corky wings; leaves short-petiolate, membranous, obovate to broadly oblanceolate, 1.5–7 cm. long, 1–4 cm. wide, acuminate to acute at both ends, glabrous, finely serrulate; cymes 1- to few-flowered, on slender peduncles about 2 cm. long, the pedicels short; flowers 4-merous, pale green, about 6 mm. across; calyx-teeth depressed-rounded, with minute teeth on margin; petals orbicular; fruiting carpels usually 1 or 2, narrowly obovoid, about 8 mm. long, pale red, 1-seeded; seeds ellipsoidal, 4–6 mm. long, enclosed in an orange-vermilion aril. ——May–June. Thickets and woods in lowlands and mountains; Hokkaido, Honshu, Shikoku, Kyushu; common and very variable.——Korea, Sakhalin, s. Kuriles, Ussuri, Manchuria and China.——Forma **subtriflorus** (Bl.) Ohwi. *E.* *subtriflorus* Bl.; *Celastrus striatus* Thunb.; *E. alatus* var. *subtriflorus* (Bl.) Fr. & Sav.——Ko-mayumi.   Branches wingless; leaves glabrous.——Forma **pilosus** (Loes. & Rehd.) Ohwi. *E. alatus* var. *pilosus* Loes. & Rehd.; *E. thunbergianus* Bl.——Ke-nishikigi.   Branches winged; leaves pilose on under side.——Forma **apterus** (Regel) Rehd. *E. alatus* var. *apterus* Regel; *E. alatus* var. *pubescens* Maxim., pro parte.——Ke-ko-mayumi.   Branches wingless; leaves pilose on under side.

2. Euonymus lutchuensis T. Ito.   Ryūkyū-mayumi.   Glabrous small evergreen tree with slender, green, 4-angled branchlets; leaves thinly coriaceous, broadly lanceolate, 3–7 cm. long, 1–3 cm. wide, long-acuminate, obtuse to acute at base, remotely and obtusely serrate, the lateral nerves obsolete, the petioles slender, 2–3 mm. long; inflorescence 2.5–5 cm. long, 1- to 3-flowered, the peduncles very slender; flowers 4-merous, about 6 mm. across, pale green; calyx-teeth depressed-rounded, minutely denticulate on margin; petals orbicular; carpels 1 or 2, rarely 3, oblong, about 8 mm. long; seeds enclosed in a yellowish aril.——Kyushu (s. distr.).——Ryukyus.

3. Euonymus japonicus Thunb. *E. japonicus* var. *typicus* Regel——MASAKI. Large, glabrous, evergreen shrub or small erect tree with green, smooth, rather thick terete branches; leaves thick-coriaceous, elliptic or obovate, sometimes oblong, 3–7(–10) cm. long, 2–4(–6) cm. wide. acute to obtuse, cuneate to acute at base, obtusely serrate, lustrous, the lateral nerves not prominent, the petioles 8–15(–20) mm. long, green; inflorescence 4–7 cm. long, densely rather many-flowered, on thick, rather flattened peduncles 2–5 cm. long; flowers 4-merous, pale green, about 5 mm. across; calyx-teeth semirounded, entire; petals elliptic; filaments longer than the anthers; fruit globose, about 7 mm. across, the aril orange-red.——June–July. Thickets and woods, especially common on slopes near the sea; Hokkaido, Honshu, Shikoku, Kyushu; very common and variable and much-planted.——Ryukyus, Korea, and China.

Var. **longifolius** Nakai. *E. yoshinagae* Makino——NAGABA-MASAKI. Leaves narrower, oblanceolate, 7–10 cm. long, 1.5–2 cm. wide.

Var. **macrophyllus** Regel. *E. japonicus* var. *obovatus* Nakai——ŌBA-MASAKI. Leaves much broader.

4. Euonymus fortunei (Turcz.) Hand.-Mazz. var. **radicans** (Sieb.) Rehd. *Masakia radicans* (Sieb.) Nakai; *Euonymous radicans* Sieb. ex Miq.; *E. japonicus* var. *radicans* Miq.; *E. repens* Carr.——TSURU-MASAKI. Evergreen glabrous climber, often creeping on the ground, with green, terete, obscurely elevated-punctate branchlets; leaves somewhat coriaceous, elliptic to oblong, 1–4 cm. long, obtuse to subacute, acute to cuneate at base, obtusely serrate, the petioles 5–10 mm. long; inflorescence 2–5 cm. long, few-flowered, on rather slender peduncles; flowers pale green; fruit 5–6 mm. across, globose.——June–July. Woods and thickets on low mountains; Hokkaido, Honshu, Shikoku, Kyushu; common and variable, with cultivars in gardens.——Korea, China, and Ryukyus.

Var. **villosus** (Nakai) Hara. *E. radicans* var. *villosus* Nakai——KE-TSURU-MASAKI. Leaves pubescent.

5. Euonymus chibae Makino. HIZEN-MAYUMI. Evergreen glabrous tree with 4-angled branches; leaves thinly coriaceous, elliptic or oblong, 6–8 cm. long, 2–5 cm. wide, abruptly acuminate to abruptly acute, abruptly acute at base, loosely and appressed obtuse-serrate, lustrous, the lateral nerves very weak, the petioles 1–1.5 cm. long; inflorescence few-flowered, on flat peduncles 2–4 cm. long; fruit obovoid-globose, 4-angled, 15–18 mm. long, 1–1.5 cm. across, pale brown when dry, with few seeds in each locule, the aril reddish brown.——Kyushu (Hizen and Satsuma Prov.).——Ryukyus and s. Korea.

6. Euonymus tanakae Maxim. *Genitia tanakae* (Maxim.) Nakai——KOKUTENGI. Evergreen glabrous small tree or large shrub with rather thick branches; leaves often falsely verticillate, somewhat coriaceous, narrowly obovate-oblong to obovate, 7–15 cm. long, 2.5–4.5 cm. wide, acute, abruptly acute, or rounded, cuneate to acute at base, minutely toothed, paler beneath, the nerves very weak, the petioles 1–2.5 cm. long; inflorescence 5–10 cm. long, rather densely few- to many-flowered, the peduncles 4–7 cm. long; flowers 4-merous, about 10 mm. across, greenish white; calyx-teeth very short, depressed; petals thick, orbicular, about 5 mm. long; filaments

longer than the anthers, inserted on the disc; fruit globose, with 4 prominent ridges, 1–1.5 cm. long and as wide; seeds few in each cell.——June. Near seashores; Kyushu; rare.——Ryukyus and Formosa.

7. Euonymus sieboldianus Bl. *E. vidalii* Fr. & Sav.; *E. yedoensis* Koehne; *E. hians* Koehne; *E. semiexsertus* Koehne; *E. dorsicostatus* Nakai; *E. maackii* var. *stenophyllus* Koidz.; *E. maackii* sensu auct. Japon., non Rupr.; *E. hamiltonianus* sensu auct. Japon., non Wall.——MAYUMI. Glabrous deciduous shrub with terete branches; leaves oblong, sometimes elliptic or ovate-oblong, 5–12 cm. long, 2–6 cm. wide, acuminate to abruptly acute, rounded to abruptly acute at base, minutely serrulate, green and scarcely lustrous on upper side, the lateral nerves slightly evident beneath, the petioles 8–20 mm. long; inflorescence 3–6 cm. long, loosely several-flowered with spreading pedicels 2–4 cm. long; flowers pale green, about 8 mm. across; calyx-lobes rounded, entire; petals oblong, about 4 mm. long, densely and minutely papillose inside; filaments 3 or 4 times as long as the dark purple anthers; fruit obtriangular-cordate, with 4-ridges, cuneately narrowed toward the base, 8–10 mm. long and as wide at apex.——May–June. Mountains and hills; Hokkaido, Honshu, Shikoku, Kyushu; common and very variable.——Korea, Manchuria, China, Sakhalin, and s. Kuriles.

Var. **nikoensis** (Nakai) Ohwi. *E. nikoensis* Nakai; *Euonymus hamiltonianus* var. *nikoensis* (Nakai) Blakel.——YUMOTO-MAYUMI. Leaves puberulous with papillose hairs especially on the nerves beneath, the fruit, and on the petioles.

8. Euonymus oligospermus Ohwi. *E. pauciflorus* sensu auct. Japon., non Maxim.; *E. pauciflorus* subsp. *oligospermus* (Ohwi) Okuyama; *E. pauciflorus* var. *japonicus* Koidz.——ITO-MAYUMI, ANDON-MAYUMI. Deciduous shrub with rather slender, glabrous, densely warty, dark brown branches; leaves membranous, oblong to ovate-oblong, 6–10 cm. long, 2.5–4 cm. wide, caudately short-acuminate, acute to subrounded at base, obtusely and minutely toothed, glabrous and dull on upper side, pale green and pubescent especially on nerves beneath, the petioles 4–7 mm. long; inflorescence 4–5 cm. long in fruit, the peduncles slender, bearing 1 or 2 fruits; calyx-teeth rounded, nearly entire; fruit 4-angled, about 10 mm. long, dilated and winged toward apex, obtriangular on the sides, deeply depressed.——Mountains; Honshu (Hinoemata near Ose in Iwashiro); very rare.

9. Euonymus macropterus Rupr. *E. ussuriensis* Maxim.——HIRO-HA-TSURIBANA. Glabrous, deciduous shrub with rather thick, terete branches; leaves obovate to broadly so, 6–12 cm. long, 3–7 cm. wide, abruptly acuminate or abruptly acute, broadly cuneate to rounded at base, obtusely toothed, pale green and the lateral nerves rather prominent beneath, the petioles 4–8 mm. long; inflorescence 7–12 cm. long including the long peduncles, umbellately branched and rather many-flowered; flowers 4-merous, pale green, about 6 mm. across; calyx-teeth depressed-rounded, entire; petals orbicular; fruit depressed, 4-ridged, about 1 cm. long, 2–2.5 cm. wide with 4 horizontally spreading wings broadest about the middle, the aril orange-red, the seeds reddish brown.——June–July. Woods and thickets in mountains; Hokkaido, Honshu (centr. distr. and eastw.), Shikoku.——s. Kuriles, Sakhalin, Korea, Manchuria, Ussuri, and Amur.

10. Euonymus yakushimensis Makino. AO-TSURIBANA. Glabrous deciduous shrub, the branches green, often whitish waxy, terete; leaves membranous, broadly lanceolate to ovate-oblong, 5–13 cm. long, 2–4 cm. wide, long-acuminate, acute to subrounded at base, minutely and obtusely toothed, pale green and the lateral nerves slightly raised beneath, the petioles 3–5 mm. long; inflorescence 5–10 cm. long, few-flowered on slender peduncles; flowers 4-merous, purplish; calyx-teeth depressed-rounded, entire; petals orbicular, about 5 mm. long; filaments very short; fruit pendulous, globose, smooth, 8–10 mm. across, the aril red; seeds yellow-brown.——Mountains; Kyushu (Yakushima); very rare.

11. Euonymus melananthus Fr. & Sav. SAWADATSU, AO-JIKU-MAYUMI. Rather small, deciduous, glabrous shrub with green, slenderly 4-striate branches; leaves chartaceous to membranous, ovate, sometimes broadly lanceolate, 3–8 cm. long, 1.5–3.5 cm. wide, abruptly acuminate and subacute, rounded to broadly cuneate at base, mucronate-serrulate, the nerves raised beneath, the petioles 2–3 mm. long; inflorescence 2–3 cm. long, 1- to 3-flowered, the peduncles slender and pendulous; flowers 5-merous, purplish, about 7 mm. across; calyx-teeth semirounded, minutely toothed on margins; petals orbicular; fruit globose, about 1 cm. across, smooth.——June–July. Woods in mountains; Honshu, Kyushu.

12. Euonymus lanceolatus Yatabe. MURASAKI-MAYUMI. Small, creeping, sparsely branched, glabrous shrub, the branches elongate, green, terete, obscurely 4-striate; leaves rather thick, broadly oblanceolate to broadly lanceolate, rarely nearly oblong, 7–12 cm. long, 1–3(–5) cm. wide, long-acuminate, acute to obtuse at base, obsoletely serrulate, deep green on upper side, the nerves not prominent, the petioles 4–7 mm. long; inflorescence 3 or 4 cm. long, loosely few-flowered on slender peduncles; flowers 5-merous, purplish, about 8 mm. across; calyx-teeth semirounded, minutely toothed on margin; petals orbicular; fruit globose, about 8 mm. across, smooth. ——July. Woods in mountains; Honshu.

13. Euonymus oxyphyllus Miq. *E. nipponicus* Maxim.; *E. yesoensis* Koidz.; *E. oxyphyllus* var. *magnus* Honda; *E. oxyphyllus* var. *yesoensis* (Koidz.) Blakel. and var. *nipponicus* (Maxim.) Blakel.; *Turibana oxyphylla* (Miq.) Nakai; *T. yesoensis* (Koidz.) Nakai; *T. nipponica* (Maxim.) Nakai; *Kalonymus yesoensis* (Koidz.) Prokhan.——TSURIBANA. Large deciduous glabrous shrub with terete branches; leaves

membranous, ovate or obovate, sometimes elliptic or narrowly obovate-oblong, 5–10 cm. long, 2–5 cm. wide, acuminate, acute to rounded at base, obtusely serrulate, with raised nerves beneath, the petioles 3–6 mm. long; inflorescence 7–15 cm. long, loosely flowered, the peduncles slender; flowers more than 10, umbellate on the peduncles, 5-merous, white or pale purple, about 7 mm. across; calyx-teeth much depressed, entire; petals orbicular; fruit globose, 10–12 mm. across, smooth. ——May–June. Thickets and woods in low mountains; Hokkaido, Honshu, Shikoku, Kyushu; common and variable.—— s. Kuriles, Korea, and centr. China.

14. Euonymus planipes (Koehne) Koehne. *E. latifolius* var. *planipes* Koehne; *E. oxyphyllus* var. *kuenbergii* Honda; *E. robustus* Nakai; *Turibana planipes* (Koehne) Nakai ——Ō-TSURIBANA. Large deciduous glabrous shrub with terete, green, rather thick branches; leaves obovate, elliptic, or oblong, 7–13 cm. long, 4–6 cm. wide, abruptly acuminate to abruptly acute, broadly cuneate to abruptly acute at base, obtusely serrulate, the nerves especially raised on the under side, the petioles 5–12 mm. long, inflorescence 7–15 cm. long, loosely umbellate, the peduncles slender; flowers more than 10, white to pale green, about 8 mm. across; calyx-teeth short, entire; petals orbicular; fruit globose, with 5 narrow wings about the middle.——May–June. Hokkaido, Honshu (centr. distr. and northw.).——Korea, Manchuria, Ussuri, and s. Kuriles.

15. Euonymus tricarpus Koidz. *E. sachalinensis* sensu auct. Japon., non Maxim.; *E. sachalinensis* var. *tricarpus* (Koidz.) Kudo——KURO-TSURIBANA. Glabrous, deciduous shrub with terete, green branches later becoming dark reddish brown; leaves membranous, rugulose in life, elliptic or obovate, 6–10(–12) cm. long, 3–4(–5) cm. wide, abruptly acute, rounded to acute at base, obtusely serrulate, the petioles 8–12 mm. long; inflorescence 5–7 cm. long, 2- or 3-flowered on pendulous, slender peduncles; flowers dark purple, about 8 mm. across; calyx-teeth 5, broadly ovate; petals 5, orbicular; fruit cordately globose, 1–1.2 cm. long, 1.2–1.5 cm. across, truncate at base, rounded at apex, with 3 falcately recurved obtuse wings toward the base.——July. Thickets in high mountains; Hokkaido, Honshu (centr. distr. and northw.). ——Sakhalin.

Fam. 122. **STAPHYLEACEAE** MITSUBA-UTSUGI KA Bladdernut Family

Trees or shrubs; leaves opposite, 3-foliolate or odd-pinnate, the stipules paired; flowers bisexual or unisexual, actinomorphic; sepals imbricate; petals inserted on the disc or on its margin, imbricate; stamens 5, free, alternate with the petals; carpels 2 or 3, connate at base, 2- or 3-locular, forming a lobed ovary; styles free or connate; ovules many, in 1 or 2 series on the ventral suture; fruit an inflated, bladderlike capsule dehiscent on the upper part, or a berry; seeds few, the endosperm scanty, the embryo straight, the cotyledons flat.——About 5 genera, with about 30 species, chiefly in the temperate regions of the N. Hemisphere.

1A. Deciduous trees; fruit dehiscent.
  2A. Fruit an inflated lobed capsule; lateral leaflets sessile; seeds without an aril. . . . . 1. *Staphylea*
  2B. Fruit a follicle; lateral leaflets petiolulate; seeds with a thin fleshy aril. . . . . 2. *Euscaphis*
1B. Evergreen trees; fruits fleshy and indehiscent; seeds without an aril. . . . . 3. *Turpinia*

1. **STAPHYLEA** L. MITSUBA-UTSUGI ZOKU

Shrubs or trees; leaves opposite, the stipules deciduous, the leaflets 3 (to 7 in exotic species), stipellate; flowers white, bisexual, in terminal panicles; sepals 5, deciduous; petals 5, erect; disc connate with the base of the sepals, free on the margin; stamens 5, erect; carpels 2 or 3, connate at base; styles free or connate at apex; ovules many, in 2 series on ventral sutures of the carpels, ascending, anatropous; capsules inflated, membranous, 2- or 3-lobed in upper part, 2- or 3-locular, with a few seeds in each locule, dehiscent on the inner side; seeds globose, exarillate, the endosperm fleshy.——About 8 species, in the temperate regions of the N. Hemisphere.

1. **Staphylea bumalda** (Thunb.) DC. *Bumalda trifolia* Thunb.——MITSUBA-UTSUGI. Shrub with spreading to ascending branches, the branches terete, grayish brown, glabrous; leaves trifoliolate, on petioles 2–3 cm. long, the lateral leaflets sessile, ovate or ovate-elliptic, 3–7 cm. long, 1.5–3 cm. wide, acuminate, acute or acuminate at base, mucronate-serrate, the terminal one attenuate on a petiolelike base, paler and glabrous (var. *glabra* Nakai) or pubescent on midrib beneath; inflorescence a racemelike panicle, terminal on young branch-lets, 5–8 cm. long; flowers white, the pedicels 8–12 mm. long; calyx-segments linear-oblong or oblanceolate, obtuse, erect, 7–8 mm. long, greenish white; petals simulating the sepals and as long, white; capsules slightly inflated, short-stiped, shallowly and broadly 2-fid, 2–2.5 cm. wide, the stigma at first connate; seeds lustrous, pale yellow, obovoid, about 5 mm. long.—— May–June. Woods in lowlands and hills; Hokkaido, Honshu, Shikoku, Kyushu; common.——Korea, Manchuria, and China.

2. EUSCAPHIS Sieb. & Zucc.　GONZUI ZOKU

Small trees; leaves opposite, odd-pinnate, stipulate, the leaflets serrate, stipellate; flowers 5-merous, bisexual, small, in terminal panicles; ovary 2- or 3-locular, with an annular disc at base, the stigma connate; fruit of 1–3 coriaceous dehiscent follicles, each 1- to 3-seeded; seeds enclosed in a thin aril.——A single species in e. Asia.

1. **Euscaphis japonica** (Thunb.) Kanitz. *Sambucus japonica* Thunb.; *E. staphyleoides* Sieb. & Zucc.; *Hebokia japonica* (Thunb.) Raf.——GONZUI. Small deciduous tree with glabrous, rather thick branches; leaves opposite, the petioles 4–6 cm. long, the axis nearly terete, 15–30 cm. long, the leaflets 7–11, rather thick, narrowly ovate or broadly lanceolate, 4–8 cm. long, 2–4 cm. wide, gradually acuminate, rounded to broadly cuneate at base, appressed-serrate, deep green and lustrous on upper side, often white-pubescent on midrib beneath near base, the petiolules 2–8 mm. long; inflorescence a long peduncled broadly pyramidal panicle 8–15 cm. across, many-flowered; flowers pale green, 4–5 mm. across; sepals and petals elliptic, persistent; follicles 1–3, ellipsoidal, arcuately spreading, acute, 1.5–2 cm. long, longitudinally ribbed, the seeds nearly globose, black, shining, about 5 mm. across.—— May(–June). Thickets and thin woods in lowlands and low mountains; Honshu (Kantō Distr. and westw.), Shikoku, Kyushu; very common.——Korea, Ryukyus, Formosa, and China.

3. TURPINIA Vent.　SHŌBEN-NO-KI ZOKU

Glabrous trees or shrubs with terete branches; leaves opposite, exstipulate, odd-pinnate, sometimes trifoliolate or unifoliolate, the leaflets serrate, coriaceous; panicles terminal and axillary; flowers small, white, bisexual; sepals and petals 5, imbricate; disc convex, toothed; stamens 5, free; ovary 3-locular and -lobed, the styles 3, often connate, the ovules ascending, anatropous; fruit ellipsoidal or nearly globose, fleshy or leathery, 3-locular; seeds angular, flat.——More than 10 species, in e. Asia, Malaysia, India, and America.

1. **Turpinia ternata** Nakai. *T. pomifera* sensu auct. Japon., non DC.——SHŌBEN-NO-KI. Evergreen glabrous tree with terete, red-brown branches; leaves with petioles 3–5 cm. long, the leaflets 3 or sometimes 1 or 2, thick-coriaceous, deep green, abruptly acuminate with an obtuse tip, appressed serrate, oblong or narrowly so, 7–12 cm. long, 2.5–5 cm. wide, the midrib raised beneath, the lateral leaflets slightly smaller, on petiolules 5–10 mm. long, the terminal leaflets on petiolules 1–3 cm. long; inflorescence pedunculate, terminal on young branchlets, 10–20 cm. long, many-flowered, puberulous; flowers pale green, about 5 mm. across; sepals elliptic; petals obovate, white, about 3.5 mm. long, slightly longer than the sepals; fruit fleshy, subglobose, 7–10 mm. across, becoming orange-red when mature; seeds few, grayish brown, with small elevated spots, 5–6 mm. long.——May–June. Shikoku, Kyushu.—— Ryukyus and Formosa.

Fam. 123. ICACINACEAE　KUROTAKI-KAZURA KA　Icacina Family

Trees or shrubs, sometimes scandent; leaves usually alternate, simple, exstipulate; flowers 4- or 5-merous, bisexual or unisexual, actinomorphic; calyx small, 4- to 5-lobed, the lobes imbricate, rarely valvate; petals 4 or 5, imbricate, rarely valvate, free or connate, rarely absent; stamens as many as the petals and alternate with them; disc sometimes developed; ovary 1-locular, rarely 3- to 5-locular, the ovules usually 2, pendulous from the top of the locule; style very short; fruit a 1-locular, 1-seeded drupe, rarely winged; seeds usually with endosperm, the embryo usually small and straight.——About 38 genera, with about 200 species, in the warmer parts of the world.

1. HOSIEA Hemsl. & Wils.　KUROTAKI-KAZURA ZOKU

Deciduous, sometimes dioecious woody climber; leaves membranous, alternate, long-petiolate, ovate-cordate; inflorescence a loose cyme or sometimes a compound axillary panicle, often reduced to a few flowers, flowers polygamous or unisexual; sepals 5, small, much shorter than the petals; corolla of 5 united petals, subrotate, the lobes caudate; stamens 5, alternate with the petals, the filaments shorter than the petals; disc of 5 fleshy and rounded scales; ovary ovoid, 1-locular, the ovules 2, pendulous; style short, simple, columnar, with a 5-lobed stigma; fruit fleshy, slightly flattened, ellipsoidal, 1-seeded; endosperm scanty, the cotyledons elliptic.——Two species, in e. Asia.

1. **Hosiea japonica** (Makino) Makino. *Natsiatum japonicum* Makino; *N. sinense* sensu auct. Japon., non Oliv.——KUROTAKI-KAZURA. Dioecious woody climber with grayish brown, loosely lenticellate, elongate branches, yellowish retrorse-pubescent; leaves thin, ovate-cordate, 8–18 cm. long, 5–12 cm. wide, acuminate with a mucro or short awn at apex, cordate to nearly truncate at base, irregularly and coarsely dentate with short-aristate teeth, pale yellow appressed-pubescent, the lateral nerves and chief veinlets ending in the teeth, the petioles 4–8 cm. long; flowers in axillary cymes, on slender pedicels 1–2.5 cm. long, about 8 mm. across, pendulous, pale yellow; calyx-lobes ovate, about 1 mm. long; corolla with a short tube at base, the lobes broadly lanceolate, recurved from the middle, caudate, papillose inside; filaments erect, about 3 mm. long, slightly longer than the corolla-tube; fruit rather flat, pendulous, about 1.5 cm. long, elliptic, red, the stone solitary, with raised reticulations on both sides.——May–June. Woods in mountains; Honshu (Chūgoku Distr.), Shikoku, Kyushu; rare.

Fam. 124. **ACERACEAE** KAEDE KA Maple Family

Trees and shrubs, monoecious or sometimes dioecious; leaves simple or compound, opposite, exstipulate; flowers bisexual or unisexual, small, actinomorphic, in terminal or axillary racemes, fascicles, or panicles; sepals 4 or 5, imbricate, rarely connate; petals 4 or 5, imbricate or absent; disc usually flat, rarely absent; stamens 4–10; ovary superior, 2-locular, 2-lobed; style simple, the stigma 2-fid; ovules 2 in each locule; fruit a pair of 1-seeded, broadly winged samaras, separating at maturity; seeds without endosperm, the testa thin.——Two genera, with about 100 species, chiefly in the temperate regions of the N. Hemisphere.

1. **ACER** L. KAEDE ZOKU

Deciduous or rarely evergreen trees or shrubs; leaves simple with palmately or rarely pinnately arranged nerves, or 3- to 7-foliolate flowers 5- or rarely 4-merous, in racemes, fascicles, panicles, or corymbs; disc usually annular and prominent, rarely absent; stamens 4–10, usually 8; styles or stigmas 2.——Nearly 100 species, in Eurasia, n. Africa, and N. America.

1A. Leaves simple
  2A. Inflorescence terminal on young leafy branches.
    3A. Inflorescence a corymb or panicle, sometimes racemelike with short branches on lower part.
      4A. Inflorescence corymbose or paniculate with short axis.
        5A. Leaves distinctly toothed.
          6A. Leaves palmately lobed.
            7A. Petioles and peduncles pubescent at least while young.
              8A. Flowers yellowish with scabrous anthers; petioles ½ to nearly as long as the leaves; leaves finely toothed.
                      1. *A. sieboldianum*
              8B. Flowers purplish with smooth anthers; petioles ¼–½ as long as the leaves; leaves incised-toothed. 2. *A. japonicum*
            7B. Petioles and peduncles glabrous from the first.
              9A. Leaves thinly chartaceous, glabrous or nearly so; ovary glabrous or thinly pubescent.
                10A. Leaves 9- to 13-lobed, cordate-rounded, slightly depressed. . . . . 3. *A. shirasawanum*
                10B. Leaves 5- to 7 (–9)-lobed, cordate-rounded. . . . . 4. *A. palmatum*
              9B. Leaves nearly membranous, usually loosely white-pilose on nerves beneath. 5. *A. tenuifolium*
          6B. Leaves undivided or shallowly 3-lobed with short lateral lobes. . . . . 6. *A. ginnala*
        5B. Leaves entire or with few coarse teeth.
          11A. Lobes of leaf acuminate with a short awned tip, glabrous or pubescent; ovary and fruit glabrous. . . . . 7. *A. mono*
          11B. Lobes of leaf elongate with an obtuse tip, pubescent on both sides, ovary and fruit pubescent. . . . . 8. *A. miyabei*
      4B. Inflorescence a racemelike panicle with elongate axis.
        12A. Leaves undivided, orbicular-cordate or broadly ovate, with appressed short teeth. . . . . 9. *A. distylum*
        12B. Leaves palmately lobed, sharply toothed.
          13A. Leaves irregularly and coarsely toothed; samara 1.5–2 cm. long, short-pubescent. . . . . 10. *A. ukurunduense*
          13B. Leaves finely and sharply toothed; samara 3–4 cm. long, with rusty brown, dusty puberulence. . . . . . . 11. *A. nipponicum*
    3B. Inflorescence a raceme with elongate axis.
      14A. Leaves undivided, penninerved with about 20 pairs of closely parallel lateral nerves. . . . . 12. *A. carpinifolium*
      14B. Leaves lobed, if undivided, with less than 10 pairs of lateral nerves.
        15A. Leaves of fertile branches ovate or ovate-triangular, with or without a small lateral lobe on each side.
          16A. Leaves 3–6 cm. long; petioles 1–3 cm. long. . . . . 13. *A. crataegifolium*
          16B. Leaves 8–12 cm. long; petioles 3–6 cm. long.
            17A. Pedicels 5–10 mm. long; leaves glabrous or nearly so beneath, ovate or ovate-triangular, if 3-lobed, the lateral lobes
              near the base. . . . . 14. *A. capillipes*
            17B. Pedicels 2–4 cm. long; leaves rusty brown-pubescent when young, 3-lobed with lateral lobes on the upper part.
                      15. *A. rufinerve*
        15B. Leaves of fertile branches palmately 5- to 9-lobed.
          18A. Leaves softly white-pubescent beneath; flowers 4-merous. . . . . 16. *A. argutum*
          18B. Leaves glabrous or rusty brown-pubescent on nerves or on axils beneath.
             19A. Racemes about 10-flowered; pedicels 8–15 mm. long; petals and stamens about 4 mm. long. . . . . . . 17. *A. tschonoskii*
             19B. Racemes about 20-flowered; pedicels 4–6 mm. long; petals and stamens 1.5–2 mm. long. . . . . 18. *A. micranthum*
  2B. Inflorescence from lateral leafless buds precocious.

20A.   Inflorescence corymbose, with a short axis; leaves and petioles pubescent; lobes of leaves acuminate with an obtuse tip.
19. *A. diabolicum*

20B.   Inflorescence a fascicle, without an axis and peduncle; leaves and petioles pubescent only while young; lobes of leaves acuminate.
20. *A. pycnanthum*

1B.   Leaves compound, 3(–5)-foliolate.
21A.   Flowers fasciculate, terminal on the branches; young branches, underside of leaves, and inflorescence coarsely pilose; petioles 2–3 cm. long. ....   21. *A. nikoense*
21B.   Flowers in slender racemes; leaves glabrous to pubescent; petioles 5–8 cm. long.
22A.   Leaflets 3, all alike; ovary glabrous or loosely pubescent; flowers 4-merous. ....   22. *A. cissifolium*
22B.   Leaflets 3–5, the terminal one larger; ovary pubescent; flowers 5-merous. ....   23. *A. negundo*

1.   Acer sieboldianum Miq.   *A. japonicum* var. *sieboldianum* (Miq.) Fr. & Sav.; *A. sieboldianum* var. *typicum* Maxim. ——Ko-hauchiwa-kaede, Itaya-meigetsu.   Tree with the young branches, inflorescence, and petioles white-pubescent; leaves chartaceous, rounded, 6–8 cm. wide, palmately 7- to 11-lobed, sometimes to the middle, shallowly cordate to truncate at base, white-pubescent while young, later nearly glabrous, the lobes narrowly ovate to broadly lanceolate, acuminate, sharply toothed, the petioles 3–4 cm. long; inflorescence a many-flowered compound pendulous corymb, the peduncles 3–5 cm. long; flowers small, pale yellow, 5-merous; sepals about 3 mm. long, pubescent on the outside; ovary white-pubescent; samaras nearly glabrous, the wings ascending, 1.5–2 cm. long inclusive of the nutlets.——May. Woods and thickets in mountains; Hokkaido, Honshu, Shikoku, Kyushu; common. Many cultivars are grown in gardens.

Var. **microphyllum** Maxim.   Hime-uchiwa-kaede.   Leaves smaller. Occurs with the typical phase.

2.   Acer japonicum Thunb.   *A. monocarpon* Nakai——Hauchiwa-kaede.   Tree with glabrous branches; leaves orbicular-cordate, 7–12 cm. wide, palmately 9–(7–11)-lobed, shallowly so, cordate at base, duplicate- or incised-toothed, with long scattered white hairs while young, soon glabrate except for an axillary tuft beneath, the lobes narrowly ovate, acuminate, the petioles 2–3.5 cm. long, pubescent while young; inflorescence corymbose, few-flowered, pendulous, white-pubescent while young; flowers 10–15 mm. across, 5-merous; sepals elliptic or oblong, dark reddish purple; samaras pendulous, pubescent while young, often glabrate, the wings ascending, 2–2.5 cm. long inclusive of the nutlets.——May. Woods and thickets on mountains; Hokkaido, Honshu; rather common and frequently planted; many cultivars are grown in gardens. ——Cv. **Aconitifolium.**   *A. japonicum* forma *aconitifolium* (Meehan) Rehd.   *A. japonicum* var. *aconitifolium* Meehan; *A. circumlobatum* var. *heyhachii* (Matsum.) Makino; *A. heyhachii* Matsum.——Mai-kujaku.   Leaves dissected almost to the base with deeply incised, rhombic-oblanceolate lobes.

Var. **insulare** (Pax) Ohwi.   *A. circumlobatum* Maxim.; *A. circumlobatum* var. *insulare* Pax; *A. japonicum* var. *circumlobatum* (Maxim.) Koidz.——Ō-meigetsu, Shinano-ha-uchiwa.   Wings of samaras horizontally spreading.——Honshu.

Var. **kobakoense** (Nakai) Hara.   *A. kobakoense* Nakai ——Kobako-hauchiwa.   Leaf-lobes simply and coarsely toothed.——Hokkaido.

3.   Acer shirasawanum Koidz.   Ō-itaya-meigetsu. Tree with glabrous branches; leaves chartaceous, orbicular, 6–10 cm. wide, palmately 9- to 11-lobed, shallowly so, somewhat cordate at base, rather prominently duplicate-serrate, white-pubescent on both sides while young, soon glabrous except for an axillary tuft beneath, the lobes ovate, acuminate, the petioles 3–8 cm. long, rather slender, glabrous; inflorescence a more than 10-flowered compound corymb, the peduncles 4–6 cm. long, glabrous; flowers pale yellow, 6–8 mm. across; sepals broadly lanceolate, about 4 mm. long; ovary with a few long hairs; samaras glabrous, the wings ascending, about 2 cm. long including the nutlets.——May–June. Mountains; Honshu, Shikoku.

4.   Acer palmatum Thunb.   *A. polymorphum* Sieb. & Zucc., non Spach; *A. polymorphum* var. *palmatum* (Thunb.) K. Koch; *A. formosum* Carr.; *A. palmatum* var. *thunbergii* Pax; *A. eupalmatum* Koidz.; *A. palmatum* var. *quinquelobum* K. Koch; *A. palmatum* forma *genuina* Miq.——Iroha-momiji, Takao-kaede, Momiji.   Tree, with the young buds loosely yellowish brown pubescent, soon glabrous; leaves chartaceous, suborbicular, 4–7 cm. wide, cordate-truncate or subcordate at base, glabrous, palmately 5- to 7-lobed often to the middle, the lobes lanceolate to broadly so, acuminate, rather prominently duplicate-serrate, the petioles glabrous, slender, 3–5 cm. long, reddish; inflorescence a glabrous, pendulous, more than 10-flowered compound corymb, the slender peduncles 3–4 cm. long; flowers dark red, 4–6 mm. across; sepals oblanceolate, about 3 mm. long, loosely brown-pubescent at apex; ovary glabrous or with few rusty brown soft hairs; samaras glabrous, ascending or spreading, the wing about 1.5 cm. long.——Apr.–May. Woods and thickets in lowlands and mountains; Honshu, Shikoku, Kyushu; very common and variable, with numerous cultivars grown in gardens.——Korea.

Var. **amoenum** (Carr.) Ohwi.   *A. amoenum* Carr.; *A. sanguineum* var. *amoenum* (Carr.) Koidz.; *A. palmatum* subsp. *amoenum* (Carr.) Hara; *A. palmatum* subsp. *septemlobum* Koidz., excl. syn.; *A. euseptemlobum* Koidz.——Ō-momiji.   Leaves larger, 6–10(–12) cm. wide, palmately 7(–9)-lobed, sometimes to the middle, the lobes rather regularly serrulate; wings of samara 2–2.5 cm. long inclusive of the nutlets. ——Apr.–May.   Mountains; Hokkaido, Honshu, Shikoku, Kyushu; very variable and frequently planted in gardens.—— Forma **latilobum** (Koidz.) Ohwi.   *A. palmatum* var. *latilobatum* Koidz.——Hiro-ha-momiji.   Lobes of the leaves broader, deltoid or deltoid-ovate.——Western part of Honshu and Shikoku.

Var. **matsumurae** (Koidz.) Makino.   *A. palmatum* subsp. *matsumurae* Koidz.; *A. matsumurae* (Koidz.) Koidz.; *A. ornatum* var. *matsumurae* (Koidz.) Koidz.——Yama-momiji. Leaves 5–8(–10) cm. wide, 7(5–9)-lobed to the middle, the lobes narrowly ovate to broadly lanceolate, prominently duplicate- or incised-serrate; wings of samaras 15–25 mm. long inclusive of the nutlets.——May. Mountains; Hokkaido, Honshu, Shikoku, Kyushu; common and variable; many cultivars of this variety are grown in gardens.

5.   Acer tenuifolium (Koidz.) Koidz.   *A. shirasawanum* var. *tenuifolium* Koidz.; *A. dissectum* var. *tenuifolium* (Koidz.) Koidz.——Hina-uchiwa-kaede.   Tree with glabrous branches; leaves membranous-chartaceous, thin, ovate-orbicular, 4–7 cm. wide, cordate at base, (5–)7- to 9-lobed, sometimes to the middle, sparsely white-pubescent, soon becoming glabrous except in the axils, or with very sparse persistent pubescence on nerves beneath, the lobes lanceolate, rhombic-lanceolate, or narrowly rhombic-ovate, acuminate, incised and doubly serrate acute-toothed, the petioles slender, 2–5 cm. long, glabrous or nearly so; inflorescence corymbose, pendulous, more than 10-flowered, on slender peduncles 1.5–4 cm. long; flowers pale yellow with a red-purple hue, about 5 mm. across; sepals oblong, about 3 mm. long; ovary rather densely white-pubescent; samaras loosely pubescent or glabrous, the wings ascending, about 2 cm. long including the nutlets.——May–June. Mountains; Honshu (Kantō Distr. and westw.), Shikoku, Kyushu.

6. Acer ginnala Maxim. *A. tataricum* var. *ginnala* (Maxim.) Maxim.; *A. tataricum* var. *aidzuense* Franch.; *A. aidzuense* (Fr.) Nakai——KARA-KOGI-KAEDE. Shrub or small tree with glabrous branches; leaves thick-chartaceous, elliptic-ovate, triangular-ovate or ovate, 3–6 cm. wide, often shallowly 3-fid, 5–10 cm. long, caudately acuminate with a subobtuse tip, glabrous on upper side, thinly pubescent on nerves beneath, the lateral lobes short, ascending, acute, incised and doubly serrate, the petioles 2–5 cm. long, glabrous; inflorescence loosely branched, compound-corymbose, rather many-flowered, the peduncles 2–3 cm. long, often with scattered sessile glands; flowers greenish yellow, small; sepals about 2.5 mm. long; samaras glabrous or nearly so, erect or slightly ascending, the wings 25–35 mm. long.——May. Wet places; Hokkaido, Honshu, Shikoku, Kyushu.——Korea, Manchuria, and e. Siberia.

7. Acer mono Maxim. *A. pictum* Thunb. 1784, non Thunb. 1783; *A. pictum* var. *mono* (Maxim.) Maxim.; *A. laetum* var. *parviflorum* Regel; *A. pictum* var. *parviflorum* (Regel) C. K. Schn.——ITAYA-KAEDE. Tree, usually with glabrous branches; leaves depressed-orbicular, 7–15 cm. wide, deeply to rather shallowly 5- to 7-lobed, cordate to somewhat truncate at base, usually glabrous on upper side, short-pubescent to glabrous except for the axillary tufts of hairs beneath, the lobes deltoid to lanceolate, entire or with few coarse teeth, acuminate and awn-tipped, the petioles 4–12 cm. long, glabrous to short-pubescent; inflorescence on short peduncles, corymbose, 4–6 cm. across; flowers yellowish green, small, 5–7 mm. across; sepals oblong; petals broadly oblanceolate; samaras glabrous or short-pubescent while young, nearly erect to ascending, 2–3 cm. long inclusive of the wing.——Apr.–May. Mountains; Hokkaido, Honshu, Shikoku, Kyushu; rather common and very variable.——Korea, Manchuria, Amur, Sakhalin, and China.

Var. **mayrii** (Schwerin) Koidz. ex Nemoto. *A. mayrii* Schwerin; *A. pictum* var. *mayrii* (Schwerin) Henry——AKA-ITAYA, BENI-ITAYA. Bark smooth; leaves reddish while young, depressed reniform-orbicular, shallowly 5-lobed, glabrous, the lobes short, broad; flowers larger; samaras often erect.——Hokkaido, Honshu (n. distr.).

8. Acer miyabei Maxim. *A. shibatae* Nakai; *A. miyabei* var. *shibatae* (Nakai) Hara——KUROBI-ITAYA. Tree, the branches grayish white, sometimes short-pubescent while young; leaves depressed, 5-angled, 8–15 cm. wide, 5-lobed, often beyond the middle, cordate at base, short-pubescent on both sides especially on nerves, the lobes 5-angled, caudate-acuminate with an obtuse tip, usually with a small obtuse lobule on each side near the middle, the petioles 4–15 cm. long, pubescent with spreading, short hairs; inflorescence paniculately corymbose on short peduncles with 10 or more flowers; flowers greenish yellow, about 8 mm. across; sepals oblong,

about 4 mm. long; petals as long as the sepals, linear-oblanceolate; stamens slightly exserted; samaras usually yellowish pubescent, horizontally spreading, 2–3 cm. long.——May. Hokkaido, Honshu (centr. distr. and northw.); rare.

9. Acer distylum Sieb. & Zucc. HITOTSUBA-KAEDE, MARUBA-KAEDE. Tree, the branches with pale brown crisped appressed hairs; leaves simple, ovate-cordate, 10–17 cm. long, 6–12 cm. wide, abruptly caudate-acuminate with an obtuse tip, deeply cordate at base, glabrous on upper side, rusty-pubescent beneath while young, appressed and obtusely toothed, the petioles 3–5 cm. long, appressed-pubescent while young; inflorescence pendulous, racemose, with short branches on lower part, 7–10 cm. long inclusive of the peduncles 3–5 cm. long, pale brown appressed-pubescent, the pedicels to 1 cm. long; flowers yellowish, 3–4 mm. across; sepals oblong, appressed-pubescent externally; petals as long as the sepals, oblanceolate; stamens longer than the sepals; samaras nearly erect to ascending, pale brown-villous while young, about 3 cm. long.——June. Mountains; Honshu (Kinki Distr. and eastw.), Kyushu.

10. Acer ukurunduense Trautv. & Mey. *A. dedyle* Maxim.; *A. spicatum* var. *ukurunduense* (Trautv. & Mey.) Maxim.; *A. caudatum* var. *ukurunduense* (Trautv. & Mey.) Rehd.——OGARABANA, HOZAKI-KAEDE. Tree, the branches yellowish puberulent; leaves orbicular and 5-angled, 8–13 cm. long, 8–15 cm. wide, cordate at base, often whitish and with dull yellowish pubescence beneath, palmately 5(–7)-lobed on outer third, the lobes ovate-triangular, acuminate, sharply serrate and incised, the petioles 6–12 cm. long; inflorescence an erect raceme, 10–15 cm. long inclusive of the peduncles 3–5 cm. long, densely many-flowered, with pale brown soft hairs, the pedicels about 10 mm. long; flowers yellowish; sepals broadly lanceolate, about 1 mm. long; petals narrowly oblanceolate, about 2 mm. long; samaras loosely short-pubescent or glabrous, ascending, 1.5–2 cm. long.——July–Aug. High mountains; Hokkaido, Honshu (centr. distr. and northw.), Shikoku.——Sakhalin, Korea, Manchuria, e. Siberia, and Kuriles.——Forma **pilosum** Nakai. USUGE-OGARABANA. Leaves less prominently pubescent beneath.——Occurs with the species.

11. Acer nipponicum Hara. *A. parviflorum* Fr. & Sav., non Ehrh.; *A. pennsylvanicum* subsp. *parviflorum* (Fr. & Sav.) Wessmael; *A. brevilobum* Hesse——TETSU-KAEDE. Tree, the young branches rusty-brown soft pubescent; leaves large, rather thin, cordate, 5-angled, depressed, 10–15 cm. long, 12–20 cm. wide, glabrous or nearly so on upper side, often thinly rusty brown curled-pubescent, persistently so on the nerves beneath, shallowly 5-lobed, the lobes broadly deltoid, often shallowly 3-fid, sharply double-serrate, acuminate, the petioles 8–17 cm. long, with persistent sparse brown pubescence on upper part; inflorescence erect, rather densely many-flowered, 10–20 cm. long, on a short peduncle 1–1.5 cm. long, short-branched in lower part, densely rusty-brown pubescent, the pedicels spreading, 4–7 mm. long; flowers 3–4 mm. across; sepals ovate, 1 mm. long; petals small; stamens 3 mm. long; samaras ascending to obliquely spreading, rusty brown pubescent, 3.5–4 cm. long.——June–July. Mountains; Honshu, Shikoku, Kyushu; rare.

12. Acer carpinifolium Sieb. & Zucc. CHIDORI-NO-KI. Dioecious tree with glabrous branches; leaves thin, ovate-oblong, 8–15 cm. long, 3.5–7 cm. wide, acuminate, cordate or shallowly so at base, regularly and sharply double-serrate, short-pubescent, often with long persistent appressed hairs on nerves beneath, the 18–23 pairs of lateral nerves ascending-

parallel, the petioles 1–1.5 cm. long; inflorescence nearly glabrous, racemose, pendulous, loosely few to rather many-flowered, 5–8 cm. long, on short peduncles, the pedicels 1–2.5 cm.
long in fruit; flowers pale green, 1–1.2 cm. across; sepals 4,
narrowly elliptic, long-pubescent, 5–6 mm. long; petals in
pistillate flowers narrowly ovate, about as long as the sepals,
often absent in staminate flowers; stamens 5–6; samaras soon
glabrous, ascending, 2.5–3 cm. long inclusive of the wing.——
Valleys and ravines in mountains; Hokkaido, Honshu, Shikoku, Kyushu; common.

13. Acer crataegifolium Sieb. & Zucc. Uri-kaede,
Me-uri-no-ki.   Large dioecious shrub with glabrous branches;
leaves ovate or broadly so, undivided or shallowly 3-fid near
the base, 4–7 cm. long, 3–4(–7) cm. wide, long-acuminate,
shallowly cordate at base, rather irregularly toothed, short
rusty brown pubescent while young, paler beneath, the petioles glabrous, 1–2 cm. long; inflorescence a glabrous raceme
3–5 cm. long, about 10-flowered, the peduncles 1–1.5 cm. long,
the pedicels slender, 4–10 mm. long; flowers yellowish green,
about 8 mm. across; sepals narrowly oblong, ascending, about
4 mm. long; petals spathulate-obovate, slightly longer than the
sepals; stamens about 8, about one-half as long as the sepals;
samaras glabrous, horizontally spreading to ascending, about
2 cm. long.——May. Woods and thickets in low mountains;
Honshu, Shikoku, Kyushu; common.

14. Acer capillipes Maxim. Hoso-e-kaede.   Tree, glabrous except the branches and leaves which are rusty-brown
crisped-pubescent while young; leaves thick-chartaceous, ovate
to broadly so, 3-angled, undivided or shallowly 3-fid and 5-
angled, 8–13 cm. long, 4–6(–10) cm. wide, long-acuminate,
cordate or rounded at base, irregularly and acutely serrulate,
often glaucescent beneath, the lateral lobes if present small, the
petioles 3–5 cm. long; inflorescence racemose, densely many-
flowered, about 10 cm. long, on short peduncles, the pedicels
4–8 mm. long, ascending; flowers pale green, 7–8 mm. across;
sepals 5, narrowly oblong, 3–4 mm. long; petals and stamens
equaling the sepals; samaras glabrous, nearly horizontally
spreading, about 15 mm. long.——May–June. Mountains;
Honshu (Kantō Distr. and westw.), Shikoku, Kyushu.

Var. **morifolium** (Koidz.) Hatus.   *A. morifolium* Koidz.;
A. insulare sensu Ohwi, pro parte, non Makino——Yaku-
shima-ogarabana.   Leaves broadly ovate, without axillary
tufts of hairs on the under surface.——Kyushu (Yakushima).

15. Acer rufinerve Sieb. & Zucc. Uri-hada-kaede. Tree
with greenish glabrous branches; leaves rather thick, flabellately 5-angled, 8–15 cm. long and as wide, shallowly 3-fid on
upper part, with the central lobes broadly deltoid, acuminate,
the lateral lobes acuminate, ascending, smaller, often with an
accessory smaller obsolete lobe on the outer side near the base,
sharply double-serrulate, the blade shallowly cordate to
rounded at base, with rusty brown long pubescence on the
nerves beneath while young; inflorescence a rather densely
many-flowered raceme, short-pedunculate, 5–10 cm. long, the
pedicels rather erect, 2–5 mm. long, rusty-brown pubescent;
flowers pale green, 8–10 mm. across; sepals 4, oblong, as long
as to slightly shorter than the petals; stamens about 8; samaras
ascending, 2.5–3 cm. long.——Woods and thickets in mountains; Honshu, Shikoku, Kyushu; common.

16. Acer argutum Maxim. Asa-no-ha-kaede.   Small
dioecious tree with white-puberulent branchlets; leaves orbicular, thin-membranous, palmately 5(–7)-lobed, cordate at base,
5–10 cm. long and as wide, white-puberulent at first beneath,

persistent on the nerves, the lobes broadly ovate, caudately
acuminate, rather regularly double-toothed, the petioles 3–10
cm. long, slender; inflorescence a few-flowered raceme, 3–6
cm. long, thinly white-puberulent, the pedicels ascending, glabrous, about 3 mm. long; petals short; stamens about 4, longer
than the sepals; samaras glabrous, nearly horizontally spreading, 2–2.5 cm. long.——May–June. Mountains; Honshu
(Kantō Distr. to s. Kinki Distr.), Shikoku.

17. Acer tschonoskii Maxim. Mine-kaede.   Shrub
or small tree with glabrous branches; leaves 5-angled, nearly
orbicular, 5–9 cm. long, 5–10 cm. wide, cordate to nearly
rounded at base, while young rusty-brown pubescent beneath,
on nerves near base, and on petioles, soon glabrous, palmately
5-lobed on outer half, the lobes rhomboid to rhombic-ovate,
long-acuminate, often with a short lobe on each side near base,
incised and doubly serrate, the petiole glabrous, 2–5 cm. long,
reddish; inflorescence loosely few-flowered, short-pedunculate,
2–5 cm. long, slightly rusty-brown pubescent on the nodes, the
pedicels ascending, 8–15 mm. long; flowers 8–10 mm. across,
pale yellow; sepals and petals nearly alike, spreading, linear-
oblanceolate, about 4 mm. long; stamens about 8, as long as
the petals; samaras glabrous, ascending to obliquely spreading,
2.5–3 cm. long.——June–July. Coniferous forests to the lower
alpine regions; Hokkaido, Honshu (centr. distr. and northw.).
——s. Kuriles.

18. Acer micranthum Sieb. & Zucc. Ko-mine-kaede.
Small tree or shrub with glabrous branches; leaves nearly
orbicular, 5–8 cm. long and as wide, palmately 5-lobed, beyond
the middle, cordate at the base, with rusty-brown curled
puberulence on nerves beneath while young, the lobes rhombic-
ovate to ovate, caudately long-acuminate, incised and doubly
serrate, the petioles 3–5 cm. long, glabrous; inflorescence a
rather dense, more than 10-flowered raceme, 4–7 cm. long, on
short peduncles, brownish pubescent on the nodes, the pedicels
4–7 mm. long, spreading to ascending; flowers small, about 4
mm. across, pale green; sepals oblong, about 1 mm. long;
petals oblong, about 2 mm. long; stamens 8, as long as the
petals; samaras glabrous, nearly horizontally spreading, 1.5–2
cm. long.——Mountains; Honshu, Shikoku, Kyushu.

19. Acer diabolicum Bl.   *A. purpurascens* Fr. & Sav.;
A. diabolicum var. *purpurascens* (Fr. & Sav.) Rehd.——Kaji-
kaede, Oni-momiji.   Dioecious tree, the branches grayish,
lenticellate, white-pilosulous when young; leaves orbicular,
depressed-cordate, 7–15 cm. long, 8–16 cm. wide, 5-lobed,
sometimes to the middle, glabrous on upper side, pilosulous
beneath, truncate to shallowly cordate at base, the terminal
lobe ovate, 5-angled, shallowly 3-fid on upper part, abruptly
acuminate with an obtuse tip, with few obtuse coarse teeth,
the outermost pair of lobes very short and small; inflorescence
a loose few-flowered, sessile corymb, 2–5 cm. long, glabrate,
the pedicels 1–2 cm. long, nearly erect; flowers 5-merous, pale
green; sepals and petals similar, oblong, suberect, 5–6 mm.
long; stamens 8; samaras appressed-pilosulous, erect or slightly
ascending, 3–4 cm. long.——Apr.–May. Mountains; Honshu,
Shikoku, Kyushu.

20. Acer pycnanthum K. Koch.   *A. rubrum* sensu auct.
Japon., non L.; *A. rubrum* var. *pycnanthum* (K. Koch) Makino——Hana-no-ki, Hana-kaede.   Tree, the branches glabrous, the nodes sparsely rusty-brown pubescent while young;
leaves ovate to orbicular, 3-fid, 4–7 cm. long, 3–6(–8) cm. wide,
rounded to shallowly cordate at base, glabrous, deep green on
upper side, glaucous beneath, the lobes 3-angled or ovate-tri-

angular, acuminate, irregularly serrate, the petioles glabrous, 3–6 cm. long; inflorescence fasciculate, few-flowered, from lateral winter-buds developing before the leaves, the pedicels elongate, glabrous, to 5 cm. long, pendulous; sepals of pistillate flowers 5, broadly lanceolate to narrowly ovate, about 3 mm. long, reddish; petals smaller; samaras glabrous, about 2 cm. long, nearly erect.——Apr. Wet places in mountains; Honshu (w. part of Tōkaidō and Shinano); rare; sometimes planted.

21. Acer nikoense Maxim. *A. maximowiczianum* Miq. ex Koidz.; *Negundo nikoense* (Maxim.) Nichols., non Miq. ——MEGUSURI-NO-KI. Tree, coarsely grayish spreading-hirsute on young branches, under side of leaves, pedicels, and petioles; leaves 3-foliolate, the petioles 2–3 cm. long, the lateral leaflets sessile, narrowly ovate to narrowly elliptic, 5–12 cm. long, 2–6 cm. wide, acute with an obtuse tip, slightly oblique at base, remotely obtuse-serrate except near base or subentire, glabrous on upper side when mature, glaucescent beneath, the 10–15 pairs of lateral nerves parallel, the terminal leaflet on a petiolule 5–10 mm. long; inflorescence about 3-flowered in a terminal fascicle; the pedicels 1–1.5 cm. long; sepals 5, as long as the petals; stamens about 10; samaras large, yellowish villous, ascending to obliquely spreading, 4–5 cm. long.—— May. Mountains; Honshu, Shikoku, Kyushu.

22. Acer cissifolium (Sieb. & Zucc.) K. Koch. *Negundo cissifolium* Sieb. & Zucc.; *N. nikoense* Miq.——MITSUDE-KAEDE.

Dioecious tree, white-puberulent on young branches; leaves 3-foliolate, the leaflets nearly all alike, on petiolules 5–10 mm. long, rather thinly membranous, ovate-elliptic or oblong, 5–8 cm. long, 2–3.5 cm. wide, caudately long-acuminate, cuneate at base, coarsely acute-serrate on upper half, thinly white appressed-puberulent especially on the nerves beneath, and with axillary tufts of white hairs, the lateral nerves 10–12 pairs; inflorescence racemose, pendulous, loosely white-puberulent, the racemes 4–15 cm. long, on peduncles 2–4 cm. long, the pedicels spreading to ascending, 4–6 mm. long; flowers small, about 5 mm. across, 4-merous; sepals about 1 mm. long; petals about 2.5 mm. long; stamens 4, slightly shorter than the petals, the filaments broadened at base; samaras glabrous or short-puberulent while young, nearly erect to ascending, 2.5–3 cm. long. ——Apr.–May. Mountains; Hokkaido, Honshu, Shikoku, Kyushu.

23. Acer negundo L. *Negundo aceroides* Moench; *A. fauriei* Lév. & Van't.——TONERIKOBA-NO-KAEDE. Dioecious tree with glabrous branches; leaves pinnately 3- to 5(–7)-foliolate, on petioles 5–8 cm. long, the leaflets ovate-elliptic, 5–10 cm. long, acuminate, coarsely serrate, glabrous or slightly pubescent beneath; staminate flowers in corymbs, the pistillate in racemes; ovary pubescent; samaras glabrous, ascending, 2.5–3.5 cm. long.——Apr. Widely cultivated in our area.——N. America.

Fam. 125. **HIPPOCASTANACEAE**   TOCHI-NO-KI KA   Horsechestnut Family

Trees and shrubs with palmately or pinnately compound, exstipulate, opposite leaves; flowers polygamous, zygomorphic, in terminal panicles; sepals 4 or 5, free or connate, imbricate in bud; petals 4 or 5, unequal, clawed; stamens 5–9, free; disc on outer side of the stamens; ovary superior, 3-locular, with 2 ovules in each locule; style and stigma simple; fruit usually 1-locular, 3-valved, dehiscent, usually 1-seeded; seeds large, with a large hilum, the endosperm absent.——Two genera, with about 25 species, in the temperate regions of the N. Hemisphere and S. America.

1. **AESCULUS** L.   TOCHI-NO-KI ZOKU

Deciduous trees or shrubs, with large sometimes resinous winter-buds; leaves long-petiolate, palmately 5- to 9-foliolate, serrate; flowers many, in erect terminal panicles; calyx campanulate or tubular, with 4 or 5 teeth; petals long-clawed at base.——About 24 species, in Europe, e. Asia, India, and N. America.

1. Aesculus turbinata Bl. *A. dissimilis* Bl.——TOCHI-NO-KI. Tree, with rusty-brown, long, curled, soft hairs while young, soon glabrate, the buds ovoid, resinous; leaves 5- to 7-foliolate, the leaflets narrowly obovate-oblong, the median largest, 20–35 cm. long, about 12 cm. wide, abruptly acuminate, gradually narrowed from above the middle to the sessile base, obtusely toothed, pubescent on nerves and in axils beneath, the lateral nerves about 20 pairs; inflorescence erect, narrow, 15–25 cm. long, 6–10 cm. across, with short spreading pubescence; flowers pinkish, rather dense, on one side of the panicle

branches, numerous, about 1.5 cm. across, the pedicels 3–5 mm. long; calyx-teeth 5, unequal, depressed-rounded; petals 4, recurved, on a short claw, pubescent; stamens exserted, 15–17 mm. long, arcuate; fruit obovoid-globose, about 3 cm. across, 3-valved, dehiscent; seeds chestnut-brown, lustrous, the hilum occupying about half of the surface of the seed.——May–June. Mountains, especially in ravines; Hokkaido, Honshu, Shikoku, Kyushu; common.——Forma **pubescens** (Rehd.) Ohwi. *A. turbinata* var. *pubescens* Rehd.——URAGE-TOCHI-NO-KI. Leaves soft-pubescent beneath.

Fam. 126. **SAPINDACEAE**   MUKUROJI KA   Soapberry Family

Trees, rarely herbs; leaves alternate or opposite, usually odd-pinnate, sometimes bipinnate or ternate, exstipulate; flowers unisexual or polygamous, actinomorphic or zygomorphic, small, usually in panicles; sepals 4 or 5, imbricate in bud, rarely valvate; petals 4 or 5, rarely absent, usually hairy or scaled near the base within; disc prominent, on outer side of the stamens; stamens usually 8 or 10 in 2 cycles, rarely 6 or 7, more or less connate below; ovary superior, usually 3-locular, deeply 3-lobed, the ovules usually solitary in each locule; style simple; fruit a capsule, berry, drupe, nut, or a samara; seeds without endosperm.——About 120 genera, with about 1,000 species, chiefly in the Tropics, a few in temperate regions.

1A. Leaflets entire; flowers small, actinomorphic; fruit berrylike. . . . . 1. *Sapindus*
1B. Leaflets toothed or pinnately lobed; flowers zygomorphic, rather large; fruit a capsule. . . . . 2. *Koelreuteria*

1. SAPINDUS L.   Mukuroji Zoku

Trees; leaves alternate, deciduous or evergreen, leaflets entire; flowers small, actinomorphic, unisexual, in large axillary or terminal panicles; sepals and petals 4 or 5; petals with 1 or 2 scales on the claw; disc annular; stamens 8–10; carpels 3, only one maturing into a drupe; seeds solitary, usually globose and arillate.——About 15 species, chiefly in the Tropics.

1. Sapindus mukorossi Gaertn. *S. abruptus* Lour.—— Mukuroji.   Large glabrous tree; leaves petiolate, odd-pinnate, the leaflets 8–12, rather coriaceous, ovate or lanceolate, 7–15 cm. long inclusive of the petiolules, 3–4.5 cm. wide, gradually acuminate with an obtuse tip, obliquely acute at base, entire, prominently reticulate-veined; panicles terminal, 20–30 cm. long, puberulous; flowers 4–5 mm. across, yellowish green; calyx-lobes and petals 4 or 5, the outer side of the latter and the lower half of the stamens pubescent; fruit globose, about 2 cm. across, glabrous, yellowish brown, lustrous, with the abortive carpels persistent on one side at base.——June. Honshu (centr. distr. and westw.), Shikoku, Kyushu.——Ryukyus, Formosa, China to India.

2. KOELREUTERIA Laxm.   Mokugenji Zoku

Deciduous trees; leaves alternate, odd-pinnate, sometimes bipinnate, the leaflets toothed; flowers yellow, rather large, zygomorphic, in large terminal panicles; calyx unequally 5-lobed; petals 4, lanceolate, turned upward, clawed, the limb with 2 upwardly turned appendages at the cordate base; disc undulate-toothed on upper margin; stamens 8; style trifid; fruit a bladderlike inflated capsule, loculicidally dehiscent; seeds globose, black, hard.——Few species in e. Asia.

1. Koelreuteria paniculata Laxm. *Sapindus chinensis* Murray; *K. paullinioides* L'Hérit.——Mokugenji.   Young branches often puberulous; leaves 25–35 cm. long, the petioles 3–8 cm. long, puberulent beneath while young, the leaflets sessile, herbaceous, ovate, 4–10 cm. long, 3–5 cm. wide, acuminate to acute, coarsely and irregularly toothed or incised, often pinnately lobed; inflorescence paniculate, 25–35 cm. long, the racemose branches elongate, spreading-puberulous; flowers about 1 cm. across, yellow, short-pedicelled; calyx-lobes oblong, about 2 mm. long; petals obtuse, about 8 mm. long, reddish at base; stamens slightly shorter than the petals, erect, the filaments loosely pubescent on lower half; capsules chartaceous, 3-angled, 3-valvate, dehiscent, ovoid, acute, rounded at base, glabrous, 4–5 cm. long, 2.5–3 cm. across; seeds about 7 mm. across.——July. Often naturalized near seashores and sometimes planted around temples of Honshu.——Korea and China.

Fam. 127. SABIACEAE   Awabuki Ka   Sabia Family

Erect or scandent woody plants; leaves alternate, simple or pinnate, exstipulate; flowers bisexual or unisexual, in cymes or panicles; sepals 4 or 5, sometimes 3, often connate at base, imbricate, unequal; petals 5, rarely 4, imbricate, the inner 2 often small and scalelike; stamens 5, opposite the petals, perfect or the outer 3 reduced to staminodes; ovary superior, usually with a disc at base, 2(–3)-locular, the ovules 2 or 1 in each locule; fruit indehiscent, 1(–2)-locular; seeds solitary, the endosperm absent. ——Four genera, with about 70 species, in tropical Asia and America.

1A. Stamens all perfect; flowers solitary or in axillary cymes; scandent shrubs with simple entire leaves. . . . . 1. *Sabia*
1B. Inner 2 stamens perfect; flowers in terminal panicles; erect trees with simple or pinnate, toothed or entire leaves. . . . . 2. *Meliosma*

1. SABIA Colebr.   Ao-kazura Zoku

Evergreen or deciduous climbers; leaves simple, entire; flowers bisexual, rarely polygamous, small, in axillary cymes or solitary; calyx deeply 4- or 5-lobed, the lobes imbricate; petals 5 or 4, nearly opposite the calyx-lobes, but longer; stamens 5 or 4, the anthers globose ovary with 5 acutely margined discs at base; style slender; fruit 1-seeded and entire or 2-seeded and deeply 2-lobed, the exocarp slightly fleshy, the mesocarp rather woody, reticulate.—About 20 species, in e. and s. Asia.

1. Sabia japonica Maxim. Ao-kazura.   Deciduous, scandent shrub, the branches deep green, the young branchlets puberulent; leaves ovate-elliptic or ovate-oblong, 5–7 cm. long, 2–4 cm. wide, acute with an obtuse tip, acute to rounded at base, deep green on upper side, whitish and with finely reticulate veinlets beneath, thinly spreading-pubescent on the midrib beneath, the petiole bases persistent, spinelike; flowers yellow, precocious, 5-merous, solitary in axils, peduncled, cup-shaped, nodding, 5–6 mm. across; peduncles slender, 1.5–2.5 cm. long in fruit; calyx minute; petals obovate, about 3.5 mm. long; stamens erect, slightly shorter than the petals; ovary 1; style gradually thickened at base, about 3 mm. long; fruit with 1 or 2 carpels, blue, about 6 mm. across, slightly flattened, with raised reticulate veinlets when dry.——Mar. Shikoku, Kyushu.——China.

2. MELIOSMA Bl.   Awabuki Zoku

Evergreen or deciduous trees or shrubs with naked buds; leaves simple or pinnate; flowers bisexual, rarely polygamous, in terminal or axillary panicles, relatively small; sepals 5 or 4; petals 5, unequal, the outer 3 rounded, imbricate, the inner 2 minute and often bifid or scalelike, sometimes adnate with the perfect stamens; stamens 5, the outer reduced to cuplike staminodes, the inner perfect, the anthers surrounded by a cup formed by the apex of the filaments; style short; fruit a small drupe, with a solitary 1-seeded stone.——About 50 species, in e. and s. Asia and tropical America.

1A. Leaves simple.
  2A. Leaves evergreen, coriaceous, entire or remotely serrate near tip, with prominently raised reticulated veinlets beneath, the lateral nerves 10–14 pairs, densely rusty-brown pubescent while young except the upper side; panicles erect. .... 1. *M. rigida*
  2B. Leaves deciduous, membranous to chartaceous, distinctly toothed, with slender, not prominently raised reticulate veinlets.
    3A. Leaves chartaceous, 10–25 cm. long, the lateral nerves 20–27 pairs, ending in short-awned teeth; panicles erect, broadly pyramidal, densely flowered, larger. .... 2. *M. myriantha*
    3B. Leaves membranous, 8–15 cm. long, the lateral nerves 7–10 pairs, ending in undulate or obtusely 3-angled teeth; panicles nodding, small, narrow. .... 3. *M. tenuis*
1B. Leaves pinnate.
  4A. Fruit red at maturity. .... 4. *M. oldhamii*
  4B. Fruit black-purple at maturity. .... 5. *M. hachijoensis*

1. Meliosma rigida Sieb. & Zucc. YAMA-BIWA. Small evergreen tree, the branchlets, under side of leaves, petioles, and inflorescence densely rusty- to yellow-brown soft woolly-pubescent; leaves simple, obovate-oblong, 12–20 cm. long, 4–6 cm. wide, shortly to abruptly acuminate, gradually cuneate from above the middle to the base, subentire or with few coarse short awned teeth, the lateral nerves 10–14 pairs, ending in the teeth, slightly curved upward, the veinlets prominently raised-reticulate beneath, the petioles 2–4 cm. long; panicles pyramidal, leafy below, 15–20 cm. long and as wide; flowers sessile, about 4 mm. across; sepals suborbicular, the 3 outer petals rounded-cordate, about 2.5 mm. across; fruit globose, black.——June. Woods and thickets in low mountains; Honshu (Tōkaidō and s. Kinki Distr.), Shikoku, Kyushu.——Ryukyus, Formosa, and China.

2. Meliosma myriantha Sieb. & Zucc. AWABUKI. Deciduous tree, the young branches with dull yellow short hairs, also the under side of the leaves especially the lateral nerves, the lenticels elliptic; leaves simple, chartaceous, narrowly cuneate-obovate to oblong, 10–25 cm. long, 4–8 cm. wide, short-acuminate, cuneate to obtuse at base, with short-awned teeth, glabrous on upper side, paler beneath, the lateral nerves 20–27 pairs, parallel, slightly arcuate at tip, the secondary nerves perpendicular to the lateral nerves, slender, the petioles 1–1.5 cm. long; inflorescence broadly pyramidal, 15–25 cm. long and as wide, densely branched, spreading; flowers very numerous, yellowish green, about 3 mm. across, short-pedicelled; sepals nearly orbicular; petals 1.5 mm. long, orbicular; perfect stamen solitary; fruit obliquely globose, with a persistent style-base, about 4 mm. across, red.——June. Thickets in low mountains; Honshu, Shikoku, Kyushu.——Korea.

3. Meliosma tenuis Maxim. *M. tenuiflora* Miq.——MI-YAMA-HAHASO. Large shrub, the branches slender, dark brown, with scattered pale brown ascending hairs while young; leaves simple, membranous, obovate, 8–15 cm. long, abruptly and broadly caudate and short-awned, acute and often decurrent to the petiole at base, coarsely deltoid- or undulate-toothed, with scattered multicellular hairs on upper side, pale green and with ascending pubescence on nerves and with axillary tufts of hairs beneath, the lateral nerves 7–10 pairs, ending in awns of the teeth, the petioles 1–1.5 cm. long; inflorescence narrowly pyramidal, nodding, 10–15 cm. long, 5–10 cm. across, acute at apex, short-pubescent, the rachis slender, more or less flexuous; flowers short-pedicelled, about 4 mm. across, yellow; sepals obtusely deltoid; petals orbicular, 2.5–3 mm. long; perfect stamens 2; fruit obliquely globose, about 4 mm. across, dark purple.——May–June. Thickets and woods in mountains; Honshu, Shikoku, Kyushu; rather common.

4. Meliosma oldhamii Miq. *M. rhoifolia* Maxim.; *Fraxinus fauriei* Lév.——FUSHI-NO-HA-AWABUKI, RYŪKYŪ-AWABUKI. Tree, with robust branches, golden-brown pubescent while young; leaves pinnate, the leaflets 9–15, chartaceous to membranous, very short-petiolate, ovate to ovate-lanceolate, 5–10 cm. long, 2–3.5 cm. wide, long-acuminate, cuneate to acute at base, with remote short-awned teeth, loosely pubescent on both sides, more densely so on the nerves beneath, the secondary nerves ending in the teeth; panicles large, the branches elongate, composed of numerous short few-flowered racemes; flowers short-pedicelled; petals 3 times as long as the sepals, orbicular; ovary villous; fruit red at maturity.——June. Honshu (Suwo Prov.), Kyushu (Tsushima); rare.——Ryukyus, Formosa, China, and s. Korea.

5. Meliosma hachijoensis Nakai. SAKU-NO-KI. Tree, pubescent with yellowish brown hairs; branches stout; leaves pinnate, 25–40 cm. long, densely pubescent, the petioles 4–6 cm. long; leaflets 13–19, thinly chartaceous, ovate to narrowly oblong, 7–12 cm. long, 2.5–5 cm. wide, acuminate, obliquely cuneate to obliquely rounded at base, with appressed acute teeth or entire, short-pilose on nerves above and near margin, paler beneath; panicles large, terminal and axillary, pubescent; flowers small, short-pedicelled; sepals 5, minutely ciliate, broadly ovate to elliptic, 0.3–1 mm. long; ovary pubescent; fruit subglobose, about 6 mm. long, glabrescent, black-purple when mature.——Honshu (Hachijo Isl.).

Fam. 128. BALSAMINACEAE   TSURI-FUNE-SŌ KA   Balsam Family

Soft, juicy herbs; leaves alternate or opposite, simple, penninerved, exstipulate; flowers zygomorphic, bisexual, solitary, racemose, or nearly umbellate; sepals 3, rarely 5, imbricate, unequal, the lowest with a spur; stamens 5, the filaments short, connate at the tip; ovary superior, 5-locular; stigmas 1–5, sessile; ovules many; fruit a juicy 5-valved capsule which suddenly explodes at maturity, casting the seeds for a considerable distance; seeds without endosperm, the embryo straight.——Two genera, with about 400 species, chiefly in tropical Asia and Africa, a few in temperate regions.

1. IMPATIENS L.   TSURI-FUNE-SŌ ZOKU

Soft juicy annual herbs, sometimes somewhat woody at the base; leaves alternate, opposite, or radical, serrate, often with glands near base of the petioles; peduncles axillary, solitary or in fascicles; flowers on slender pedicels, zygomorphic, usually large and showy, purple, yellow, or white; sepals 3, rarely 5, often colored, the lateral 2 reduced, the lower one largest, spurred; petals 3, the lateral ones bifid; capsules loculicidally dehiscent.——About 400 species, chiefly in tropical Asia and Africa, a few in e. Asia, Europe, and N. America.

1A. Flowers pale yellow; plant glabrous; leaves oblong to ovate, rounded to acute at base, obtusely toothed; pedicels very slender.
1. *I. noli-tangere*
1B. Flowers reddish purple, rarely white; plant sparsely hirsute on upper side of leaves and on peduncles; leaves rhombic-ovate to
-lanceolate, cuneate at base, those near the inflorescence rounded to cordate at base, rather densely serrate; pedicels rather thick.
2A. Spur circinnate; peduncles ascending to suberect, widely spaced above and not subtending the leaves; leaves cuneate at base.
2. *I. textori*
2B. Spur curved forward; peduncles spreading, subtending the leaves; leaves gradually narrowed to the base, those near the inflorescence
rounded, cordate, or obtuse at base. . . . . 3 . *I. hypophylla*

1. **Impatiens noli-tangere** L. Kɪ-ᴛsᴜʀɪ-ꜰᴜɴᴇ.   Glaucous and glabrous, the stems terete, much-branched, 30–60 cm. long; leaves with slender petioles progressively shorter toward the summit, the blades oblong to ovate, 4–8 cm. long, 2.5–4 cm. wide, obtuse to subacute, acute at base or the upper ones rounded to shallowly cordate at base, coarsely and obtusely serrate, mucronate-tipped, the basal teeth filiform; inflorescence nodding, on a slender peduncle, 1- to 5-flowered, in corymbose racemes, the bracts linear, small; flowers pale yellow, with reddish brown blotches inside, about 2 cm. across, with a gently and downwardly curved spur at base, often cleistogamous late in the season.——Wet shaded places in mountains; Hokkaido, Honshu, Shikoku, Kyushu; rather common.——Europe, Siberia, e. Asia, and w. N. America.

2. **Impatiens textori** Miq. *I. japonica* Fr. & Sav.——Tsᴜʀɪ-ꜰᴜɴᴇ-sō.   Stems much-branched, 40–80 cm. long, usually reddish; leaf-blades rhombic-ovate to narrowly rhomboidal, 6–15 cm. long, 3–7 cm. wide, acuminate, cuneate at base, vivid green, short awn-toothed, the upper leaves smaller, nearly sessile and rarely obtuse at base; peduncles ascending, elongate, straight, with fleshy spreading reddish hairs, the bracts lanceolate to deltoid-ovate, spreading, small, the pedicels 1–2 cm. long; flowers purplish red, with purple spots, about 3 cm. across, the spur prominently circinnate; fruit oblanceolate, 1–2 cm. long.——July–Sept. Wet shaded places on low mountains; Hokkaido, Honshu, Shikoku, Kyushu; rather common.——Korea.

3. **Impatiens hypophylla** Makino. Hᴀɢᴀᴋᴜʀᴇ-ᴛsᴜʀɪ-ꜰᴜɴᴇ.   Stems branched; leaves rhombic-elliptic to narrowly rhomboidal, 8–20 cm. long, 3–8 cm. wide, acuminate, gradually narrowed to the petiole, obtusely deltoid-toothed and mucronate-tipped, with scattered white multicellular curled hairs on upper side, and on nerves beneath, the upper leaves short-petiolate, rounded, subcordate to truncate at base; inflorescence racemosely few-flowered, on long peduncles placed just beneath the leaves, usually with whitish multicellular curled hairs on the lower part, arcuate above, the pedicels to 1.5 cm. long, slender, the bracts broadly lanceolate to deltoid-ovate; flowers purplish red, about 2 cm. across, with deeper colored spots, the spur forwardly curved; fruit linear-lanceolate.——July–Oct. Honshu (Kinki Distr.), Shikoku, Kyushu.

Var. **microhypophylla** (Nakai) Hara. *I. microhypophylla* Nakai.——Eɴsʜū-ᴛsᴜʀɪ-ꜰᴜɴᴇ-sō.   Flowers smaller, about 1 cm. across, without purple spots, the pedicels less than 1 cm. long.——Honshu (Tōtōmi Prov.).

Fam. 129. RHAMNACEAE   Kᴜʀᴏ-ᴜᴍᴇ-ᴍᴏᴅᴏᴋɪ Kᴀ   Buckthorn Family

Trees, often spiny, sometimes scandent, or herbs; leaves alternate, rarely opposite, simple, usually stipulate; flowers actinomorphic, bisexual or unisexual, green or white, in cymes, fascicles, or panicles; flowers 5- or 4-merous; calyx-lobes valvate; petals sometimes absent; stamens opposite the petals; receptacle cup-shaped, the disc distinct; ovary 2- or 3-locular, rarely 1- or 4-locular, superior or inferior, with 1 or sometimes 2 ovules in each locule; styles 2–4, connate at base; fruit a drupe or capsule, sometimes winged; endosperm scanty or absent.——About 45 genera, with about 550 species, widespread in temperate and warmer regions.

1A. Stipules often spinelike; leaves 3-nerved.
2A. Fruit dry, winged. . . . . 1. *Paliurus*
2B. Fruit a fleshy drupe. . . . . 2. *Zizyphus*
1B. Stipules not spinelike; branchlets often abbreviated and spinelike; leaves penninerved or 3-nerved.
3A. Leaves penninerved; peduncles and pedicels not fleshy-thickened.
4A. Leaves entire or nearly so. . . . . 3. *Berchemia*
4B. Leaves serrulate.
5A. Flowers nearly sessile, in spikes or panicles; leaves partly opposite; branches partly abbreviated and spinelike. . . . . 4. *Sageretia*
5B. Flowers pedicelled, usually in axillary fascicles or cymes.
6A. Ovary incompletely 2-locular; flowers bisexual; fruit with 1 or rarely 2 stones. . . . . 5. *Rhamnella*
6B. Ovary 2- to 4-locular; fruit with 2–4 stones; flowers bi- or unisexual. . . . . 6. *Rhamnus*
3B. Leaves 3-nerved; peduncles thickened and fleshy in fruit. . . . . 7. *Hovenia*

1. PALIURUS Mill.   Hᴀᴍᴀ-ɴᴀᴛsᴜᴍᴇ Zᴏᴋᴜ

Deciduous trees or shrubs, often procumbent or creeping; leaves alternate, in 2 rows, 3-nerved, entire or serrulate; flowers bisexual, 5-merous, small, in axillary or terminal cymes; stipules often spinose; ovary 2- or 3-locular, adnate to the disc; fruit 2- or 3-locular, woody, depressed, with a broad transverse wing above, with 1 seed in each locule.——Few species, in s. Europe and e. Asia.

1. **Paliurus ramosissimus** (Lour.) Poir. *Aubletia ramosissima* Lour.; *P. aubletia* (Lour.) Schult.——Hᴀᴍᴀ-ɴᴀᴛsᴜᴍᴇ. Deciduous much-branched shrub with grayish brown branches, often spiny in young individuals, the branchlets densely brownish-puberulent; leaves thinly coriaceous, broadly ovate, elliptic, or ovate-orbicular, 4–6 cm. long inclusive of the densely soft-

puberulous petioles, obtuse to rounded at both ends, obtusely serrulate, pale green beneath, soft-pubescent especially on nerves beneath when young, 3-nerved, the secondary nerves slender and ascending; cymes axillary, on the upper part of the branchlets, short-peduncled, densely brownish pubescent; flowers few, about 5 mm. across, pale green; calyx-lobes deltoid; petals broadly ovate, shorter than the sepals, less than 1 mm. long; fruit suborbicular, with a 3-lobed, dentate, broad wing at the margin of the broadly truncate apex, soft brownish puberulent.——Aug.–Sept. Near seashores; Honshu (Tōkaidō Distr. and westw.), Shikoku, Kyushu.——Ryukyus, Formosa, and China.

2. **ZIZYPHUS** Mill. Natsume Zoku

Deciduous or evergreen shrubs or trees; leaves alternate, short-petiolate, entire or serrulate, 3- to 5-nerved from the base, the stipules usually spinelike; flowers bisexual, 5-merous, usually yellow, small, in axillary cymes; ovary 2- to 4-locular; styles simple, 2-parted at tip; fruit a fleshy drupe, nearly globose to oblong.——About 40 species, in tropical and warmer regions of both hemispheres.

1. **Zizyphus jujuba** Mill. *Rhamnus zizyphus* L.; *Z. sativa* Gaertn.; *Z. vulgaris* Lam.; *Z. vulgaris* var. *spinosus* Bunge ——Sanebuto-natsume. Small deciduous tree or shrub, glabrous except sometimes thinly pubescent on the tips of the young branchlets and on the nerves of the leaves beneath, the branchlets often fascicled and partly deciduous in autumn; leaves ovate, 2.5–4 cm. long inclusive of the short petioles, 1–2.5 cm. wide, obtuse to subacute, rounded at base, obtusely serrulate, lustrous, the stipules often spinelike; cymes small, about 10-flowered, axillary and terminal; flowers pale green, 5–6 mm. across; calyx-lobes acute; petals small and recurved at tip; drupe ellipsoidal to globose, 1.5–2.5 cm. in diameter, glabrous, lustrous, smooth, dark reddish brown.——Apr.–May. Spontaneous in se. Europe, s. Asia; often cultivated and semi-naturalized in our area.

Var. **inermis** (Bunge) Rehd. *Z. vulgaris* var. *inermis* Bunge.——Natsume. Spineless cultivated phase.

3. **BERCHEMIA** Neck. Kuma-yanagi Zoku

Scandent or suberect, deciduous shrubs; leaves alternate, petiolate, entire or nearly so, penninerved, with arcuate parallel lateral nerves, the stipules not spinelike, slender, connate; flowers bisexual, 5-merous, greenish yellow, small, in terminal panicles or racemes, or sometimes in axillary cymes; calyx-lobes deltoid, erect; petals small; ovary 2-locular; style bifid; fruit an ellipsoidal or short-cylindric smooth drupe with a 2-locular stone.——About 10 species, in Asia, e. Africa, and N. America.

1A. Scandent or bushy shrubs with smooth greenish branches; stipules connate almost to the apex.
  2A. Racemes simple or with few branches at base.
    3A. Inflorescence 1–2.5 cm. long, simple; leaves 2.5–4 cm. long. .... 1. *B. pauciflora*
    3B. Inflorescence 5–8 cm. long, simple or with few branches at base; leaves 5–10 cm. long. .... 2. *B. longeracemosa*
  2B. Racemes many, spreading, in a terminal panicle. .... 3. *B. racemosa*
1B. Erect tree with a main stem; branches reddish brown, with scattered raised lenticels; stipules free. .... 4. *B. berchemiaefolia*

1. **Berchemia pauciflora** Maxim. Miyama-kuma-yanagi. Slightly scandent shrub elongate and much-branched, glabrous, green, becoming yellowish brown with age; leaves membranous, ovate or elliptic, 2.5–4 cm. long, 1.5–2.5 cm. wide, very obtuse at both ends or rounded at base, entire, glabrous and deep green on upper side, often yellowish and nearly glabrous except for yellowish brown thin pubescence on the nerves and with axillary tufts of hairs beneath, the petioles 5–10 mm. long, sparsely short-pubescent near point of attachment; racemes 1–2.5 cm. long, about 10-flowered, glabrous, the pedicels 2–3 mm. long; flowers about 3 mm. across; calyx-lobes deltoid, rather obtuse; fruit narrowly oblong, 7–8 mm. long.——Aug. Mountains; Honshu (Kantō Distr., Kai and s. Shinano Prov.); rare.

2. **Berchemia longeracemosa** Okuyama. Ho-naga-kuma-yanagi. Shrub with green glabrous branches; leaves membranous, ovate to broadly so, sometimes elliptic, 5–10 cm. long, 3.5–6 cm. wide, very obtuse or subrounded at both ends, entire, glabrous on upper side, yellowish and with axillary tufts of hairs beneath, the lateral nerves 7–11 pairs, the petioles 8–20 mm. long, glabrous; racemes terminal, simple or often with a few branches at base, densely many-flowered, glabrous, 5–8 cm. long, erect, the pedicels 3–4 mm. long; calyx-lobes deltoid, rather acute.——July–Aug. Mountains bordering Japan Sea side of Honshu from Uzen to Tajima Prov.

3. **Berchemia racemosa** Sieb. & Zucc. Kuma-yanagi. Subscandent, glabrous shrub; leaves ovate or oblong-ovate, 4–6 cm. long, 2.5–3 cm. wide, acute to subobtuse, rounded at base, glabrous and deep green on upper side, whitish or yellowish and glabrous except for axillary tufts or hairs beneath, the lateral nerves 7–8 pairs, the petioles 1–1.5 cm. long, reddish, glabrous; racemes many, spreading, many-flowered, glabrous, in terminal panicles 10-25 cm. long, the pedicels 2–3 mm. long; flowers greenish; calyx-lobes deltoid, acuminate; fruit oblong, 5–7 mm. long, reddish, black at maturity.——Aug. Thickets in low mountains; Hokkaido, Honshu, Shikoku, Kyushu; common.

Var. **magna** Makino. *B. magna* (Makino) Koidz.——Ō-kuma-yanagi. Leaves large, 5–10 cm. long; calyx-lobes shorter, acute.——Mountains; Shikoku, Kyushu; rather rare. ——Forma **pubescens** (Ohwi) Nemoto. *B. magna* var. *pubescens* Ohwi; *B. fagifolia* Koidz.——Ke-kuma-yanagi. Leaves with scattered pubescence beneath.——Honshu (Kinki Distr.), Shikoku, Kyushu.

4. **Berchemia berchemiaefolia** (Makino) Koidz. *Rhamnella berchemiaefolia* Makino; *Chaydaia berchemiaefolia* (Makino) Koidz.; *Berchemiella berchemiaefolia* (Makino) Nakai——Yokogura-no-ki. Small tree with glabrous, reddish brown branches and scattered lenticels; leaves membranous, oblong or ovate-oblong, 7–12 cm. long, 3–5 cm. wide, acuminate, acute to rounded and sometimes slightly oblique at base, entire or obsoletely undulate, glabrous on upper side, glaucous, thinly

pubescent near axils of nerves beneath, the lateral nerves 7–9 pairs, the petioles 6–10 mm. long, glabrous; cymes terminal in upper axils, glabrous, few-flowered, short-peduncled, the pedicels 2–4 mm. long; flowers 3–3.5 mm. across; calyx-lobes deltoid, rather acute; fruit narrowly oblong, 7–8 mm. long.——June. Mountains; Honshu (Rikuchu, Shinano, Echigo Prov. and westw.), Shikoku, Kyushu; rare.——s. Korea and China.

4. SAGERETIA Brongn.   KUROIGE ZOKU

Evergreen or deciduous shrubs sometimes scandent, the branchlets often reduced to spines; leaves coriaceous, nearly opposite, small, penninerved, entire or serrate, the stipules small, deciduous; spikes axillary, often in terminal panicles; flowers bisexual, 5-merous, small, whitish; petals inflexed at apex, the disc cup-shaped, 5-lobed; ovary 2- or 3-locular; style short, 2- or 3-lobed; fruit a small, subglobose drupe with 2 or 3 stones.——About 10 species, in e. and s. Asia, and N. America.

1. **Sageretia theezans** (L.) Brongn. *Rhamnus theezans* L.——KUROIGE.   Deciduous or semi-evergreen shrub with grayish, ramulose branches, densely short gray-pubescent, the short spurs often reduced to spines; leaves coriaceous, lustrous, broadly ovate, 2–3 cm. long, 1–1.5 cm. wide, obtuse with a mucro at apex, rounded at base, soon glabrous on upper side, with deciduous grayish lanate hairs beneath, appressed-serrulate, the veinlets rather prominent on both sides, the petioles 2–4 mm. long, short-pubescent; spikes in terminal panicles and on upper part of the branchlets, densely pubescent; flowers sessile, about 3.5 mm. across, in few-flowered fascicles, yellowish; calyx-lobes deltoid, pubescent outside; drupe purplish black, subglobose, about 5 mm. across.——Oct.   Shikoku, Kyushu (Fukuejima).——Ryukyus, Formosa, China, and India.

5. RHAMNELLA Miq.   NEKO-NO-CHICHI ZOKU

Deciduous trees or shrubs; leaves alternate, serrulate, penninerved, petiolate, the stipules not spinelike, narrow; flowers short-pedicelled, bisexual, 5-merous, greenish, small, in axillary cymes; petals small; stamens 5, perigynous; ovary incompletely 2-locular, semisuperior; style bifid; drupe oblong, black, the stone solitary, 1- or rarely 2-seeded.——Few species, in e. Asia.

1. **Rhamnella franguloides** (Maxim.) Weberb. *R. japonica* Miq.; *Microrhamnus franguloides* Maxim.——NEKO-NO-CHICHI.   Deciduous tree with brown branches, the young branchlets, inflorescence and petioles with thinly scattered spreading short hairs; leaves obovate-oblong, sometimes oblong, 6–12 cm. long, 2.5–4 cm. wide, caudate-acuminate, rounded to obtuse at base, serrulate, glabrous or with scattered spreading hairs on midrib beneath, the petioles 3–8 mm. long; cymes axillary, subsessile to short-peduncled, few-flowered, the pedicels 2–4 mm. long; flowers about 3.5 mm. across; calyx-lobes deltoid, with 3 weak nerves; petals small; stamens 5; fruit narrowly oblong, 8–10 mm. long, yellowish at first, black at maturity.——May–June.   Mountains; Honshu (Kinki Distr. and westw.), Shikoku, Kyushu.——s. Korea.

6. RHAMNUS L.   KURO-UME-MODOKI ZOKU

Deciduous or evergreen shrubs or trees, sometimes the short spurs spinelike; leaves alternate or nearly opposite, penninerved, serrate or rarely entire, stipulate; flowers bisexual, polygamous, or unisexual, small, yellowish green or whitish, 4- or 5-merous, sometimes apetalous, in axillary fascicles, umbels, or racemes; ovary 2- to 4-locular; style usually simple; fruit a subglobose drupe with 2–4, 1-seeded stones or rarely 1, the stones opening on the inner side.——More than 100 species, chiefly in the N. Hemisphere, a few in S. America and Africa.

1A. Winter-buds with scales; leaves often subopposite; flowers unisexual, the plants dioecious; seeds grooved on back.
  2A. Leaves subopposite or at least partly so; stipules small, linear; erect much branched shrubs.
    3A. Pairs of lateral nerves few, forming an acute angle with the midrib, prominently incurved at the tip; pedicels not more than 1.5 cm. long; short-spurs often becoming spiny.
      4A. Leaves small, 2–8 cm. long, 1–3 cm. wide, glabrous to pubescent.
        5A. Leaves broadly to narrowly obovate, 2–8 cm. long.
          6A. Branches grayish; leaves glabrous or puberulous. . . . . 1. *R. japonica*
          6B. Branches purplish brown; leaves puberulent on upper side when young, sparsely pubescent beneath especially on nerves.
                                                     2. *R. yoshinoi*
        5B. Leaves elliptic to narrowly oblong, or sometimes obovate in smaller ones, 5–10 cm. long. . . . . . . 3. *R. davurica* var. *nipponica*
      4B. Leaves larger, 10–20 cm. long, 3–8 cm. wide, membranous, much longer than the petioles, yellowish pubescent on nerves beneath. . . . . 4. *R. utilis*
    3B. Pairs of lateral nerves 17–23, ascending, parallel, curved at tip; petioles very short; pedicels slender, 2–4 cm. long; branches not spiny. . . . . 5. *R. costata*
  2B. Leaves always alternate; stipules rather broad; small unarmed shrub with creeping branches. . . . . 6. *R. ishidae*
1B. Winter-buds naked, without scales; leaves always alternate; flowers bisexual; seeds without a groove on back. . . . . 7. *R. crenata*

1. **Rhamnus japonica** Maxim. *R. japonica* var. *genuina* Maxim.——KURO-UME-MODOKI.   Much branched, glabrous, deciduous shrub with short spurs often reduced to spines; leaves broadly to narrowly obovate, rarely orbicular-obovate, 2–8 cm. long, 1–4 cm. wide, rounded to obtuse in smaller ones, abruptly acuminate with an obtuse tip in larger ones, long-cuneate at base, obtusely appressed-serrulate, glabrous or sparingly short-puberulous on upper side, pale green and glabrous or thinly puberulent on nerves beneath, the lateral nerves 3 or 4 pairs, forming an acute angle with the midrib; flowers yellowish green, about 4 mm. across, fasciculate in axils of lower part of branches, the pedicels 5–15 mm. long; calyx-

lobes elongate-deltoid, acute; drupe globose-obovoid, 6–8 mm. across, glabrous, black, the stones 2 or 3, sometimes 1.——Apr. –May. Hokkaido, Honshu, Shikoku, Kyushu; common.—— The leaves are very variable in shape and in pubescence.

2. Rhamnus yoshinoi Makino. *R. schneideri* Lév. & Van't.——Ao-ume-modoki, Kibino-kuro-ume-modoki. Erect, divaricately branched dioecious shrub with branchlets often reduced to spines, the branches and branchlets purplish brown; leaves alternate or fasciculate on short spurs, obovate or obovate-elliptic, 3–8 cm. long, 1–4 cm. wide, abruptly acuminate, cuneate at base, serrulate, appressed-puberulent on upper side when young, sparingly pubescent beneath especially on nerves, the petioles 5–20 mm. long, usually pubescent; pedicels 1–1.5 cm. long, fasciculate; flowers yellowish green, 5–6 mm. long; calyx-lobes 4, erect, acuminate; stamens 4, about 2/3 as long as the calyx-lobes; fruit obovoid-globose, glabrous, about 7 mm. across, with 2 or 3 stones.——May. Honshu (Chūgoku Distr.), Kyushu.——Korea and Manchuria.

3. Rhamnus davurica Pall. var. **nipponica** Makino. *R. nipponica* (Makino) Grubov.——Kuro-tsubara. Large, glabrous or nearly glabrous, deciduous, dioecious shrub; leaves narrowly oblong or often broadly obovate in the smaller ones, 5–10 cm. long, 2–3 cm. wide, acuminate with an obtuse tip or sometimes rounded at apex, cuneate at base in smaller ones, obtusely serrulate, the lateral nerves 4- or 5-pairs, suberect, forming an acute angle with the midrib, incurved at tip, the petioles 1–2 cm. long; flowers small, few in the lower axils of the branchlets, glabrous, yellowish green, 4–5 mm. across, the pedicels about 1 cm. long; calyx-lobes elongate-deltoid, acute; fruit obovate-globose, 6–8 mm. across, black, usually with 2 stones.——May. Honshu (centr. and n. distr.).——The typical phase with broader leaves occurs in Dahuria, n. China, Manchuria, and Korea.

4. Rhamnus utilis Decne. *R. sieboldiana* Makino—— Shiiboruto-no-ki, Siebold-no-ki. Small deciduous tree with glabrous branches, the short-spurs reduced to spines; leaves membranous, obovate-elliptic, 5–20 cm. long, 2.5–7 cm. wide, abruptly acute with an obtuse tip, acute to cuneate at base, obtusely serrulate, sometimes subentire, with scattered yellowish hairs beneath especially on nerves, the petioles 1–2 cm. long; flowers about 3.5 mm. across, the pedicels about 1 cm. long; fruit obovoid-globose.——May. Kyushu (near Nagasaki); rare.——China.

5. Rhamnus costata Maxim. Kuro-kamba. Deciduous unarmed shrub, black when dried, with fulvous glabrous branches; leaves thinly membranous, obovate-oblong to obovate, slightly oblique, 8–15 cm. long, 4–8 cm. wide, abruptly acute to acuminate, broadly short-cuneate or obtuse at base, obtusely serrulate, densely yellowish-brown pubescent beneath especially on nerves, the lateral nerves 17–23 pairs, ascending, rather closely parallel, scarcely to slightly incurved at tip, the petioles 3–6 mm. long; flowers few, in fascicles on the lower part of young branchlets, yellowish green, about 5 mm. across, the pedicels slender, 2–4 cm. long; calyx-lobes deltoid-lanceolate, acute, 2.5–3 mm. long; fruit globose-obovoid, 7–8 mm. across, black, with 2 stones.——May–June. Honshu, Shikoku, Kyushu; rather rare in mountains.——Forma **nambuana** (Honda) Hara. *R. costata* var. *nambuana* Honda——Nambukuro-kamba. Glabrescent. Occurs with the typical phase.

6. Rhamnus ishidae Miyabe & Kudo. Miyama-hanmodoki. Small, deciduous, unarmed, dioecious shrub with creeping grayish brown branches and rather thick, green, thinly and minutely hairy branchlets; leaves elliptic, broadly ovate, or obovate-elliptic, 4–8 cm. long, 3–5 cm. wide, obtuse to rather abruptly acute, rounded to subcordate at base, obtusely serrulate except near the base, glabrate on upper side, sometimes pubescent especially on nerves beneath, the lateral nerves 8–12 pairs, ascending, nearly parallel and incurved at tip, the petioles 5–6 mm. long; pistillate flowers green, solitary in the axils of the branchlets, the pedicels 5–8 mm. long, glabrous; calyx-lobes 5, ovate-deltoid, acuminate, 3-nerved; fruit globose, 7–8 mm. across, red, becoming black at maturity, with 3 stones.——Alpine regions; Hokkaido (Mount Yubari and Mount Apoi); rare.

7. Rhamnus crenata Sieb. & Zucc. *Frangula crenata* (Sieb. & Zucc.) Miq.——Iso-no-ki. Deciduous shrub with glabrous branches sparsely soft-puberulent when young; leaves obovate to oblong, 6–12 cm. long, 2.5–5.5 cm. wide, abruptly acuminate, rounded to obtuse at base, serrulate, pale green with raised veinlets and nerves beneath, the lateral nerves 6–10 pairs, ascending, inwardly curving near the margin, the petioles 8–15 mm. long, sparsely short-pubescent; cymes axillary, pilose, few-flowered, on short peduncles to 7 mm. long or sessile, the pedicels 5–10 mm. long; flowers about 5 mm. across; calyx-lobes narrowly deltoid, acute; fruit globose-obovoid, reddish purple at first, black at maturity, about 6 mm. across, with 3 stones.——June–July. Wet places in mountains and hills; Honshu, Shikoku, Kyushu.——Korea and China.

Var. **yakushimensis** Makino. *R. crenata* var. *acuminatifolia* (Hayata) Hatusima; *R. acuminatifolia* Hayata——Hosoba-iso-no-ki. Leaves narrow.——Kyushu.——Formosa.

7. HOVENIA Thunb. Kempo-nashi Zoku

Large deciduous trees without spines; leaves alternate, long-petiolate, 3-nerved from the base, ovate-orbicular, the stipules free, linear-lanceolate, caducous; flowers bisexual, 5-merous, purplish, small, in axillary and terminal cymes, the peduncles becoming fleshy in fruit; petals convolute and enclosing the stamens; disc hairy; style 3-lobed; fruit indehiscent, 3-locular, with a coriaceous exocarp, 3-seeded; seeds flat, lustrous, brown.——Few species in e. Asia.

1A. Fruit glabrous; leaves deep brown when dry, usually glabrous, with obtuse, obliquely triangular teeth; inflorescence glabrous. .... 1. *H. dulcis*

1B. Fruit brown-pubescent; leaves dark reddish brown when dry, usually pubescent beneath, with appressed partly obsolete teeth; inflorescence pubescent. .... 2. *H. tomentella*

1. Hovenia dulcis Thunb. *H. dulcis* var. *glabra* Makino ——Kempo-nashi. Tall, almost glabrous tree with lenticellate branches; leaves rather membranous, broadly ovate, 8–15 cm. long, 6–12 cm. wide, short-acuminate, rounded to shallowly cordate at base, with obliquely triangular obtuse teeth, green and glabrous on upper side, pale green and glabrous to thinly pubescent on nerves beneath, becoming deep brown when dry, 3-nerved from the base, the lateral nerves close to the margin from near the base; the petioles 2.5–6 cm. long, glabrous; cymes terminal and in the upper axils, glabrous, 3–5 cm. across, many-flowered, the pedicels short, the branches thickened in fruit; flowers pale green, about 7 mm. across; calyx-lobes about 2.5 mm. long, acute; disc sparsely pilose; fruit globose, about 7 mm. across, yellowish brown, glabrous, lustrous.——June–July. Hokkaido (Okushiri Isl.), Honshu, Shikoku, Kyushu.——Korea and China.

2. **Hovenia tomentella** (Makino) Nakai. *H. dulcis* var. *tomentella* Makino——KE-KEMPO-NASHI. Resembles the preceding; leaves rather thick, broadly ovate to cordate-ovate, with short, obtuse, partly obsolete teeth, glabrous and becoming dark reddish brown on upper side when dry, pale green and with red-brown short hairs beneath, sometimes only on the nerves; cymes and calyx-lobes densely reddish brown pubescent; fruit densely reddish brown long-pubescent.——June–July. Honshu, Shikoku.

Fam. 130. **VITACEAE** Budō Ka   Grape Family

Mostly woody vines with tendrils, rarely herbs or erect woody plants; leaves alternate, simple or palmately compound, stipulate; flowers bisexual or unisexual, actinomorphic, small, in cymes or panicles, usually opposite a leaf; calyx-lobes 4 or 5, rarely 3, 6, or 7, small or indistinct; petals as many as the sepals, valvate in bud; stamens opposite the petals; disc annular or lobed; ovary superior, 2(–6)-locular, the ovules 2 or 1 in each locule; style single, simple, or absent, the stigma capitate, peltate, or lobed; fruit a berry; seeds with endosperm and a small embryo, the cotyledons flat.——About 10 genera, with about 500 species, especially abundant in the Tropics.

1A. Petals connate at apex and not expanding, falling as a calyptra at anthesis; inflorescence a panicle; leaves usually simple; bark without lenticels, soon longitudinally fissured, the pith brown. . . . . 1. *Vitis*
1B. Petals free and expanded at anthesis; inflorescence usually a cyme; bark lenticellate, not fissured, the pith white.
  2A. Flowers 5-merous; inflorescence opposite a leaf; leaves simple or rarely compound.
    3A. Tendrils twining, without adhesive holdfasts; disc of the flowers cup-shaped, not adnate to the ovary. . . . . 2. *Ampelopsis*
    3B. Tendrils with adhesive holdfasts; disc adnate to the ovary. . . . . 3. *Parthenocissus*
  2B. Flowers 4-merous; inflorescence axillary; leaves palmate or digitate. . . . . 4. *Cayratia*

1. VITIS L.   Budō Zoku

Deciduous, rarely evergreen, tendril-bearing vines; leaves usually simple and lobed, rarely palmately compound, dentate; flowers bisexual and staminate, 5-merous, in panicles; calyx-lobes small or obscure; petals connate at apex, deciduous as a whole at anthesis; disc of 5 nectary glands; ovary 2-locular, with 2 ovules in each locule; style pyramidal, short; fruit a berry, the seeds 2–4, usually pyriform, abruptly rostrate at base, 2-grooved on the ventral side.——About 60 species, in temperate regions of the N. Hemisphere.

1A. Leaves deeply cordate at base; inflorescence relatively large, 10–25 cm. long.
  2A. Leaves 10–25 cm. wide; petioles 5–25 cm. long; branches rather thick.
    3A. Branches with spreading spiny protuberances. . . . . 1. *V. kiusiana*
    3B. Branches smooth.
      4A. Leaves 8–30 cm. long and as wide, reddish brown lanate beneath. . . . . 2. *V. coignetiae*
      4B. Leaves 7–15 cm. long and as wide, thinly brownish lanate beneath. . . . . 3. *V. amurensis*
  2B. Leaves 4–8 cm. wide, densely reddish brown lanate beneath; petioles 3–6 cm. long; branches rather slender. . . . . . . 4. *V. thunbergii*
1B. Leaves truncate to shallowly cordate at base with a wide open sinus; inflorescence relatively small, 4–12 cm. long.
  5A. Leaves distinctly 3-lobed, densely reddish brown lanate beneath, acute to obtuse, cordate at base. . . . . 4. *V. thunbergii*
  5B. Leaves undivided or obsoletely 3-lobed, at least the upper ones deltoid, acuminate, truncate to shallowly open-cordate at base.
    6A. Leaves rather thick, the teeth undulate and short. . . . . 5. *V. saccharifera*
    6B. Leaves thin, teeth deltoid. . . . . 6. *V. flexuosa*

1. **Vitis kiusiana** Momiyama.   KUMAGAWA-BUDŌ. Branches thick, terete, yellowish brown lanate when young, with small blackish spreading spinelike protuberances; leaves chartaceous, undulate-serrulate, sparsely yellowish brown arachnoid at first and soon glabrous on upper side, densely brown-lanate beneath, the lower oblong-ovate or narrowly ovate, slightly 3-lobed or angled, sometimes 5-angled-ovate, to 17 cm. long, 13 cm. wide, acute to acuminate, open-cordate at base, the petioles 9–13 cm. long, the upper leaves acute to subobtuse, with shorter petioles about 5 cm. long; staminate panicles longer than the leaves, narrow, to 25 cm. long, including the 6 cm. long peduncle, to 7 cm. across, with a tendril on the peduncle.——July. Kyushu (Higo Prov.); rare.

2. **Vitis coignetiae** Pulliat. *V. labrusca* var. *typica* α *grandifolia* Regel; *V. kaempferi* sensu Rehd., non K. Koch; *V. amurensis* var. *coignetiae* (Pulliat) Nakai——YAMA-BUDŌ. Branches terete, smooth, brown-araneous while young; leaves cordate-orbicular and somewhat 5-angled, 8–30 cm. long and as wide, deeply cordate to slightly open-cordate at base, with depressed-deltoid, mucronate teeth, shallowly 3-lobed, the median lobe depressed-deltoid, acute to subacute, more or less sparingly arachnoid when young, densely reddish brown lanate beneath; inflorescence to 20 cm. long, including the 3–6 cm. long peduncle, about 8 cm. wide, with tendrils; fruit globose, about 8 mm. across, purplish black when mature; seeds about 5 mm. long, broadly obovate, dark brown.——May–June. Thickets in mountains; Hokkaido, Honshu, Shikoku; common.——Korea and Sakhalin.

Var. **glabrescens** Nakai. *V. kaempferi* var. *glabrescens* (Nakai) Nakai; *V. coignetiae* forma *glabrescens* (Nakai) Nakai——TAKESHIMA-YAMA-BUDŌ. Leaves sparsely lanate, becoming green beneath.——Hokkaido.——Korea.

3. **Vitus amurensis** Rupr. var. **shiragae** (Makino) Ohwi. *V. shiragae* Makino——SHIRAGA-BUDŌ. More slender than

No. 2; branches rather thick, smooth, obsoletely angled, pale brown araneous while young; leaves membranous, orbicular-cordate or more or less angled, deeply cordate with a V-shaped sinus at base, 7–15 cm. long and as wide, shallowly deltoid-toothed, thinly arachnoid on upper side while young, dull green and pale lanate beneath, often shallowly 3-lobed, the median lobe slightly larger, deltoid-acuminate to acute, the petioles 5–11 cm. long, at first thinly lanate, later only the spreading bases of the hairs persistent; inflorescence 10–12 cm. long including the short peduncle, 3–5 cm. across, with tendrils.——June. Honshu (Chūgoku Distr.); rare.——Korea, Manchuria, and Amur (species).

4. **Vitis thunbergii** Sieb. & Zucc. *V. ficifolia* var. *thunbergii* (Sieb. & Zucc.) Lavall.; *V. thunbergii* var. *typica* Makino; *V. ficifolia* var. *pentaloba* Nakai; *V. ficifolia* var. *lobata* (Regel) Nakai; *V. labrusca* var. *typica* β *lobata* Regel; *V. kaempferi* K. Koch——EBIZURU. Branches slender, striate and obsoletely angled, more or less reddish brown arachnoid; leaves membranous, orbicular-cordate, 5-angled, 4–8 cm. long and as wide, rather open-cordate to deeply cordate at base, with 3(–5) lobes, these ovate to broadly so or deltoid, acute to obtuse, densely reddish brown lanate beneath, the teeth short or almost obsolete, short-mucronate; panicles 6–12 cm. long, including the peduncles, 2–4 cm. across, sometimes with tendrils; fruit globose, 5–6 mm. across, black; seeds 3.5 mm. long, dark reddish brown.——June–July. Thickets in hills and mountains; Hokkaido, Honshu, Shikoku, Kyushu; common.——China, Korea, and Formosa.

Var. **sinuata** (Regel) Rehd. *V. labrusca* var. *typica* δ *sinuata* Regel; *V. ficifolia* var. *sinuata* (Regel) Hara——KI-KUBA-EBIZURU. Leaves deeply lobed, the lobes coarsely sinuate on margin.——More frequent in the western part of our area.

5. **Vitis saccharifera** Makino. *V. flexuosa* var. *japonica* Makino, in syn.——AMAZURU, OTOKO-BUDŌ. Branches slender, obsoletely striate, the nodes often slightly thickened;

leaves cordate, cordate-orbicular, or deltoid-ovate, acute to acuminate, truncate or shallowly open-cordate at base, undulately short-toothed, glabrescent and deep green on upper side, reddish brown arachnoid, especially on nerves while young and with axillary tufts of hairs beneath, the upper leaves longer; panicles with or without tendrils, 4–6 cm. long, including the peduncles, 2–3 cm. across; fruit small, black when mature.——May–June. Honshu (Tōkaidō Distr. and westw.), Shikoku, Kyushu.

Var. **yokogurana** (Makino) Ohwi. *V. yokogurana* Makino——YOKOGURA-BUDŌ. Arachnoid hairs sparse on leaves beneath, persistent at maturity.

6. **Vitis flexuosa** Thunb. *V. flexuosa* forma *typica* Planch.——SANKAKU-ZURU, GYŌJA-NO-MIZU. Branches slender, slightly striate, glabrous; lower leaves depressed-deltoid to cordate-deltoid, slightly shorter to as long as wide, acute to acuminate, the upper leaves elongate-deltoid, rather thin, 4–10 cm. long, 4–8 cm. wide, truncate to shallowly open-cordate at base, green, with short deltoid teeth, glabrous on upper side, the under side with the persistent bases of pale brown arachnoid hairs and with axillary tufts of hairs; inflorescence without tendrils, 4–7 cm. long, including the peduncles, 2–3 cm. across; fruit globose, bluish black, about 7 mm. across; seeds obovate-orbicular, about 5 mm. long.——May–June. Mountains and hills; Honshu, Shikoku, Kyushu; common.——Korea, China, and Amami-Oshima.

Var. **tsukubana** Makino. *V. tsukubana* (Makino) F. Maekawa——USUGE-SANKAKU-ZURU. Young branches sparingly arachnoid; leaves glabrous above, thinly brown arachnoid beneath.——Honshu (Kantō and Kinki Distr.); rare.

Var. **rufotomentosa** Makino. *V. rufotomentosa* (Makino) Makino; *V. gilvotomentosa* Makino & F. Maekawa——KE-SANKAKU-ZURU. Leaves on upper side with only the base of the hairs persistent, arachnoid beneath.——Honshu (Kinki Distr., Wakasa Prov.), Shikoku, Kyushu.

2. AMPELOPSIS Michx.   NO-BUDŌ ZOKU

Woody or herbaceous climbing deciduous vines with tendrils; leaves alternate, simple, sometimes compound; cymes peduncled, dichotomously forked, opposite a leaf or terminal; flowers small, bisexual, green, 5-merous or rarely 4-merous; calyx-lobes small; petals small, spreading, free; stamens as many as and shorter than the petals; ovary 2-locular; disc cup-shaped, **entire** or undulate-toothed, adnate to the ovary; style slender; fruit a 1- to 4-seeded berry.——About 20 species, in e. Asia, centr. Asia, and N. America.

1A. Leaves simple, digitately lobed. .... 1. *A. brevipedunculata*
1B. Leaves ternately pinnate. .... 2. *A. leeoides*

1. **Ampelopsis brevipedunculata** (Maxim.) Trautv. *Cissus brevipedunculata* Maxim.; *C. humulifolia* var. *brevipedunculata* (Maxim.) Regel; *A. brevipedunculata* var. *maximowiczii* (Regel) Rehd.; *Vitis heterophylla* Thunb.; *A. heterophylla* (Thunb.) Sieb. & Zucc., non Bl.; *A. regeliana* Carr.; *A. heterophylla* var. *humulifolia* Hook f., excl. syn.; *A. citrulloides* Lebas; *A. brevipedunculata* forma *citrulloides* (Lebas) Rehd.——NO-BUDŌ. Herbaceous vine, woody at base; branches terete, glabrous or with pale brown multicellular spreading hairs, the nodes slightly thickened and jointed; leaves cordate, 3(–5)-lobed, 4–12 cm. long, 4–10 cm. wide, paler and thinly pilose beneath, with obtuse or acute-mucronate teeth, the median lobes ovate-deltoid, the lateral lobes smaller; inflorescence many-flowered, 3–6 cm. across; flowers about 3 mm. across; petals narrowly ovate-triangular, about

2.5 mm. long, deciduous; disc erect, entire; fruit globose, glabrous, 6–8 mm. across, purplish to blue in maturity; seeds 2, rarely 1 or 3, about 4 mm. long, orbicular with an extended tip, rounded at back.——Thickets in hills; Hokkaido, Honshu, Shikoku, Kyushu; very common and variable.——Korea, China, Manchuria, Ussuri, and s. Kuriles.

2. **Ampelopsis leeoides** (Maxim.) Planch. *Vitis leeoides* Maxim.——UDO-KAZURA. Deciduous woody vine with lenticellate, glabrous branches; leaves petiolate, ternately 2- or 3-pinnate, 12–30 cm. long, glabrous or nearly so, the leaflets petiolulate, ovate or oblong, sometimes narrowly oblong, 4–7 cm. long, acuminate, acute at base, with few appressed teeth, the lowest leaflets ternately parted or lobed; the tendrils 2-fid; inflorescence puberulous, large, apparently terminal and pedunculate; flowers short-pedicelled, greenish yellow, about 2

mm. across; disc 5-toothed; fruit globose, red.——July–Aug. Thickets and woods; Honshu (Kii, Yamato, and Chūgoku Distr.), Shikoku, Kyushu.——Ryukyus and Formosa.

Ampelopsis japonica (Thunb.) Makino. *Paullinia japonica* Thunb.; *A. serjaniaefolia* Bunge——Kagamigusa, Byaku-ren. Leaves palmate, with ternate-pinnate leaflets on a jointed, winged rachis; flowers pale yellow, in pedunculate cymes.——A climber from China, rarely cultivated for medicinal purposes.

3. PARTHENOCISSUS Planch.　Tsuta Zoku

Deciduous, rarely evergreen vines, the branchlets with white pith, the tendrils ending in a discoid sucker; leaves simple, ternate, or palmately compound, long-petiolate; cymes compound, opposite a leaf or crowded into a terminal panicle; flowers bisexual, rarely polygamous, small; petals 5, rarely 4, spreading; disc indistinct; style short, thick; ovary 2-locular, the ovules 2 in each locule; berry bluish black, 1- to 4-seeded.——About 10 species, in e. Asia, Himalayas, and N. America.

1. Parthenocissus tricuspidata (Sieb. & Zucc.) Planch. *Ampelopsis tricuspidata* Sieb. & Zucc.; *Cissus thunbergii* Sieb. & Zucc.; *Vitis inconstans* Miq.; *V. taquetii* Lév.; *Psedera tricuspidata* (Sieb. & Zucc.) Rehd.; *Psedera thunbergii* (Sieb. & Zucc.) Nakai.——Tsuta, Natsuzuta. Woody vine, glabrous except on nerves of leaves, especially beneath and on upper part of petioles; suckers of tendrils orbicular; leaves polymorphic, usually thick and lustrous, those on the short spurs 3-lobed, cordate at base, with rounded outer edges, the lobes deltoid and acuminate, 5–12 cm. long and as wide, with loose, obtuse and mucronate coarse teeth, leaves on the long shoots smaller and thinner, broadly ovate, sometimes trifoliolate; inflorescence short-pedunculate, usually on a short spur, 3–6 cm. long; fruit purplish black, globose, 5–7 mm. across; seeds 1–3, obovate-rounded, 4–5 mm. long.——June–July. Thickets and woods in hills and mountains; Hokkaido, Honshu, Shikoku, Kyushu; common and widely planted in hedges and for covering walls, with several cultivars grown in gardens.——Korea and China.

The North American **Parthenocissus quinquefolia** Planch., Amerika-zuta, with 5-foliolate leaves is sometimes cultivated in our area.

4. CAYRATIA Juss.　Yabu-karashi Zoku

Woody or herbaceous vines with usually divided tendrils; leaves alternate, ternate or pedately 5- to 9-foliolate, the leaflets petioluled; inflorescence axillary, cymose or umbellate; flowers small, 4-merous, bisexual; calyx-lobes indistinct; petals spreading; disc 4-lobed, small, thin, adnate to the ovary; style filiform or linear; berry 2- to 4-seeded, the seeds 1- or 2-grooved on the ventral side.——More than 10 species, in e. and s. Asia.

1A. Branchlets puberulous; inflorescence and petals with short granular hairs; terminal leaflets narrowly ovate or oblong; seeds with broad, transverse ridges on each side on back, the central part of the back with a narrow scarcely impressed longitudinal band.

　　　　　　　　　　　　　　　　　　　　　　　　　　　　　　　　　　　　　　1. *C. japonica*

1B. Branches and inflorescence glabrous; terminal leaflets broadly lanceolate; seeds with narrow, transverse, much raised ridges on each side on back, the central part with a longitudinal impressed band broader than the sides. .... 2. *C. corniculata*

1. Cayratia japonica (Thunb.) Gagnep. *Vitis japonica* Thunb.; *Cissus japonicus* (Thunb.) Willd.; *Columella japonica* (Thunb.) Merr.——Yabu-karashi. Herbaceous vine with much-elongate rhizomes; stems much elongate, branched, striate, the young parts furfuraceous, short-puberulent, purplish, flat, the tendrils usually bifid; stipules ovate-deltoid, scarious-margined, ciliate; leaves petiolate, pedately 5-foliolate, the terminal leaflets narrowly ovate or oblong, 4–8 cm. long, 2–4.5 cm. wide, short-acuminate, mucronate, acute at base, with shallow undulate, spreadingly mucronate teeth, puberulous on the midrib on both sides, the petiolules 1–3 cm. long, the lateral leaflets smaller; inflorescence with short granular hairs; flowers short-pedicellate, about 5 mm. across; petals 4, green, deflexed, deltoid-ovate, about 3 mm. long, obtuse with an inflexed tip, granular outside; disc yellowish or reddish; fruit globose, black; seeds on the back with broad transverse raised ridges on each side and with a narrow scarcely impressed longitudinal band.——July–Aug. Thickets and hedges in lowlands and hills; very common.——Ryukyus, China, India, and Malaysia.

Var. **dentata** (Makino) Honda. *Cissus japonica* var. *dentata* Makino——Hiiragi-yabu-karashi. Leaflets incisely dentate.——Kyushu.

2. Cayratia corniculata (Benth.) Gagnep. *Vitis corniculata* Benth.; *Cissus corniculata* (Benth.) Planch.; *C. yoshimurae* Makino; *Cayratia yoshimurae* (Makino) Honda——Akami-no-budō. Resembles the preceding species; stems slender, glabrous; leaflets thinner, green, the nerves on upper side minutely puberulous, nearly glabrous beneath, with undulate, spreading, mucronate shallow teeth, the terminal leaflets larger, broadly lanceolate, 4–9 cm. long, 1.5–3.5 cm. wide, acuminate, obtuse to acute at base; inflorescence glabrous, peduncled; fruit globose, 7–12 cm. across, purplish red at first, becoming black at maturity; seeds 1–4, broadly obovate, about 7 mm. long, with a narrow transverse raised ridge on each side, the central part with a longitudinal impressed band broader than the sides.——June. Kyushu (s. distr.); rare.——Ryukyus and China.

Fam. 131. ELAEOCARPACEAE　Horuto-no-ki Ka　Elaeocarpus Family

Trees with simple, alternate, rarely opposite, stipulate leaves; flowers usually bisexual, actinomorphic, 4- or 5-merous; sepals valvate, free or connate at base; petals free, rarely connate at base, often deeply laciniate, valvate or induplicate, rarely imbricate, not contorted, rarely absent; receptacle sometimes elevated and stipelike; stamens free, the anthers 2-locular, usually opening by pores, sometimes longitudinally dehiscent; ovary superior, 2- to many-locular, rarely 1-locular, the ovules usually many in each locule, the style simple; fruit a capsule or drupe; seeds often arillate.——About 7 genera, with about 120 species, in tropical and warm-temperate regions, especially in the S. Hemisphere.

1. ELAEOCARPUS L.   HORUTO-NO-KI ZOKU

Trees with alternate, rarely opposite, entire or toothed leaves; flowers in axillary racemes, actinomorphic, bisexual or polygamous; sepals usually valvate; petals laciniate or rarely entire, inserted on the base of the thickened receptacle, induplicate; stamens many, rarely 8–12; anthers linear, opening by transverse valves; ovary 2- to 5-locular, the ovules 2 to many in each locule; drupe with one usually bony tuberculate stone; seeds pendulous, one in each locule, the embryo fleshy.——About 60 species, in s. Asia, Malaysia, Australia, the Pacific Isls., and Madagascar.

1A. Petioles about half as long as the blade, 2.5–5 cm. long, thickened at apex; leaves narrowly elliptic or ovate-elliptic, rounded to obtuse at base; petals appressed-pubescent on back, deeply 2- to 5-toothed. .... 1. *E. japonicus*
1B. Petioles much shorter than the blade, 5–10 mm. long, not thickened at apex; leaves oblanceolate to oblong-lanceolate, acute and slightly decurrent on the petiole; petals glabrous on back, laciniate into filiform lobes. .... 2. *E. sylvestris* var. *ellipticus*

1. **Elaeocarpus japonicus** Sieb. & Zucc. *E. kobanmochi* Koidz.; *E. dioicus* Turcz.——KOBAN-MOCHI.   Evergreen tree with rather thick, terete branches, loosely white-silky while young; leaves long-petiolate, narrowly elliptic or ovate-elliptic, 6–10 cm. long, 3–5 cm. wide, abruptly acuminate with an obtuse tip, rounded to obtuse at base, obtusely appressed-toothed, thinly white-silky on both sides while young, glaucescent and loosely black-spotted beneath; racemes 4–6 cm. long, one-sided, rather many-flowered, the pedicels 4–6 mm. long; flowers about 5 mm. long; sepals and pedicels short white appressed-pubescent; petals cuneate-oblong, as long as the sepals, appressed-pubescent on both sides, deeply and obtusely 2- to 5-toothed; fruit subglobose, about 10 mm. long, grayish green. ——May–June.   Honshu (Kinki Distr. and westw.), Shikoku, Kyushu.——Ryukyus, Formosa, and China.

2. **Elaeocarpus sylvestris** (Lour.) Poir. var. **ellipticus** (Thunb.) Hara.   *E. decipiens* Hemsl.; *Prunus elliptica* Thunb.; *E. ellipticus* (Thunb.) Makino, non Smith; *E. japonicus* Turcz.; *E. makinoi* Kaneh.; *E. zollingeri* K. Koch——HORUTO-NO-KI, MOGASHI.   Evergreen tree with fulvous appressed-pubescent young branches and inflorescence; leaves short-petiolate, oblanceolate or oblong-lanceolate, 6–12 cm. long, 2–3.5 cm. wide, acute with an obtuse point or subobtuse, acute and slightly decurrent to the petiole, shallowly obtuse-toothed, glabrous; racemes 4–7 cm. long, rather many-flowered, short white appressed-pubescent, the pedicels 3–8 mm. long; petals obovate-cuneate, white, shorter than the appressed-pubescent sepals, deeply laciniate with filiform lobes, puberulous within at base; fruit narrowly ovoid, obtuse, 1.5–1.8 cm. long.——July–Aug.   Honshu (Awa and westw.), Shikoku, Kyushu.——Ryukyus and Formosa. The typical phase occurs in China and Indochina.

Fam. 132. **TILIACEAE**   SHINA-NO-KI KA   Linden Family

Trees or herbs, often with stellate or fasciculate hairs; leaves alternate, rarely opposite, stipulate; flowers bisexual, actinomorphic; sepals 5, rarely 3 or 4, usually valvate; petals as many as the sepals, contorted, sometimes imbricate, rarely absent; stamens 10 to many, the filaments free or connate at base, the anthers 4-locular; ovary 2- to 10-locular, with 1 to several ovules in each locule; style single, the stigma radiate; fruit a capsule or indehiscent nut or drupe, rarely a berry; seeds with endosperm, the embryo straight.——About 40 genera, with about 400 species, widely distributed in tropical and temperate regions.

1A. Herbs; inflorescence without a subtending leaflike bract.
  2A. Fruit globose and indehiscent, or separated into fruitlets, spinulose; receptacle more or less elongate; stamens usually many.
            1. *Triumfetta*
  2B. Fruit a 3-lobed, loculicidally dehiscent, elongate capsule, glabrous or stellate-pilose; receptacle not elongate; stamens 10–15.
            2. *Corchoropsis*
1B. Trees; inflorescence with a subtending leaflike bract adnate to the peduncle. .... 3. *Tilia*

1. TRIUMFETTA L.   RASEN-SŌ ZOKU

Stellate-pilose herbs or shrubs; leaves simple, toothed, 3- to 5-lobed or undivided; flowers yellow, few in dense axillary or fasciculate cymes; sepals 5, free, with a spinose to hornlike protuberance at apex; petals 5, thickened with a gland or depression at base, inserted on the base of the receptacle, sometimes absent; stamens many, free; ovary 2- to 5-locular, with 2 ovules in each locule; style filiform, the stigma 2- to 5-toothed; fruit dry, small, subglobose, spinulose, indehiscent or separating into fruitlets; seeds 1 or rarely 2 in each locule, with endosperm.——About 40 species, in the tropics, abundant in tropical America, Australia, and S. Africa.

1. **Triumfetta japonica** Makino.   *T. trichoclada* sensu auct. Japon., non Link; *T. annua* sensu auct. Japon., non L. ——RASEN-SŌ.   Coarse erect annual, 60–120 cm. high with spreading branches; leaves ovate to elliptic, 5–10 cm. long, 3–10 cm. wide, often obsoletely 3-lobed, the upper ones shorter and narrowly ovate, acuminate, cordate to rounded at base, deltoid-toothed, loosely hispid on both sides especially on nerves, 3- to 5-nerved, the petioles 0.5–10 cm. long, the stipules reflexed, linear, acuminate; cymes short, supra-axillary, few-flowered; flowers yellow, about 5 mm. across; sepals 5, broadly

linear, slightly pilose, with a hairy protuberance outside just below the apex; petals slightly shorter than the sepals; fruit globose, 6–7 mm. across, densely uncinate-spinulose, smooth or spreading-pilosulous at base.——Aug.–Oct. Honshu (Kantō Distr. and westw.), Shikoku, Kyushu.——Korea and Ryukyus.

2. CORCHOROPSIS Sieb. & Zucc. KARASU-NO-GOMA ZOKU

Stellate-pilose annuals; leaves petiolate, simple, dentate, 3- or 5-nerved, with short lateral nerves, the stipules linear, deciduous; flowers solitary, axillary, yellow, rather small, the pedicels with 3 filiform bracteoles on upper part; sepals 5, lanceolate, persistent; petals 5, slightly longer than the sepals, obovate; perfect stamens 10–15, free or connate at base, the staminodes 5, linear, longer than the stamens, slightly thickened on upper part; ovary 3-locular, with many ovules in each locule; style simple, subclavate, the stigma truncate and 3-toothed; capsules elongate, terete, loculicidally dehiscent into 3 valves.——Few species, in e. Asia.

1. Corchoropsis tomentosa (Thunb.) Makino. *Corchorus tomentosus* Thunb.; *Corchoropsis crenata* Sieb. & Zucc. ——KARASU-NO-GOMA. Erect, branched annual with stellate pubescence; leaves ovate, 4–8 cm. long, 2–4.5 cm. wide, sub-acuminate, rounded to subtruncate at base, obtusely toothed, densely stellate-pubescent on both sides, the petioles 0.5–5 cm. long, stellate-pubescent; peduncles axillary, slender, 1.5–3 cm. long, the bracteoles linear, ternately subverticillate, erect, stellate–pubescent; sepals recurved, linear-lanceolate, 6–8 mm. long, acuminate, stellate-pilose on the outside; petals obovate, 7–10 mm. long, staminodes linear, slightly thickened on upper part, 7–9 mm. long, granular-tuberculate; capsules 3–4 cm. long, about 3 mm. across, stellate-pilose; seeds ovoid, less than 3 mm. long, with obscure transverse ridges.——Aug.–Sept. Sunny grassy places and waste grounds in lowlands and low mountains; Honshu, Shikoku, Kyushu; rather common.——Korea and China.

3. TILIA L. SHINA-NO-KI ZOKU

Deciduous trees with simple or stellate hairs; leaves alternate, petiolate, usually obliquely cordate at base, toothed; inflorescence cymose, the peduncle adnate on a leaflike axillary or terminal bract; flowers bisexual, rather small, white or yellowish; sepals and petals 5, the latter with or without a petaloid scale at base inside; stamens all perfect, many, free or connate at base, on the receptacle; ovary 5-locular, with 2 ovules in each locule; style simple, the stigma 5-toothed; fruit globose, nutlike, indehiscent, 1- or 2-seeded, the seeds with endosperm.——About 30 species, in temperate regions of the N. Hemisphere.

1A. Petioles 8–15 mm. long; leaves obliquely and broadly ovate to oblong-ovate, 5–8 cm. long, 2.5–5 cm. wide; inflorescence furfuraceous or glabrescent. .... 1. *T. kiusiana*
1B. Petioles 2–7 cm. long; leaves obliquely 3-angled, broadly ovate to obliquely cordate-orbicular, 5–17 cm. long, 3–15 cm. wide.
  2A. Bract of the inflorescence adnate from base of the peduncle; branchlets and under surface of leaves usually densely pilose.
    3A. Leaves without axillary tufts of hairs on underside.
      4A. Staminodes slightly shorter than the sepals; fruits densely short-pilose; Chinese species planted around temples.
        2. *T. miqueliana*
      4B. Staminodes much longer than the sepals; fruits villous with both long red-brown hairs and short pubescence; indigenous.
        4. *T. rufovillosa*
    3B. Leaves with axillary tufts of hairs on under side; staminodes about as long as the sepals; fruits densely yellowish brown-hairy.
      3. *T. maximowicziana*
  2B. Bract adnate from about the middle part of the peduncle; leaves 5–8 cm. long, glabrous or nearly so except for axillary tufts of hairs beneath; young branches usually glabrous. .... 5. *T. japonica*

1. Tilia kiusiana Makino & Shiras. HERA-NO-KI. Young branchlets puberulous; leaves oblique, broadly ovate, ovate-elliptic, sometimes oblong, 5–8 cm. long, 2.5–5 cm. wide, abruptly caudate-acuminate, broadly and obtusely cuneate on one side at base, truncate to shallowly open-cordate or rounded on other side, serrate, puberulous on both sides, paler and with yellowish brown axillary hairs beneath, the petioles 8–15 mm. long; inflorescence furfuraceous or glabrescent; sepals about 4 mm. long; petals lanceolate, 5–6 mm. long; staminodes linear-oblong, as long as the sepals; fruit subglobose, 4–5 mm. long, densely brownish puberulous, without ribs or ridges.—— July. Honshu (Yamato, and w. Chūgoku Distr.), Shikoku, Kyushu.

2. Tilia miqueliana Maxim. *T. kinashii* Lév. & Van't.; *T. franchetiana* C. K. Schn.——BODAIJU. Branchlets densely grayish stellate-puberulent; leaves oblique, 3-angled, ovate to cordate-orbicular, 5–10 cm. long, 4–8 cm. wide, narrowly deltoid-acuminate, broadly cuneate to subrounded on one side at base, cordate on the other side, usually acutely serrate, glabrous on upper side, with grayish stellate hairs and without axillary tufts of hairs beneath, the lateral nerves 7–10 pairs, nearly parallel and straight, the petioles 1.5–7 cm. long, densely stellate-puberulous; sepals about 5 mm. long; petals slightly longer than the sepals, cuneate, rounded, ciliolate at tip; staminodes slightly shorter than the sepals; fruit subglobose, 7–8 mm. across, 5-ribbed at base, densely grayish brown puberulous.——Planted around temples.——China.

3. Tilia maximowicziana Shiras. *T. miyabei* Jack; *T. maximowicziana* var. *shirasawae* Engl.——ŌBA-BODAIJU. Branches rather thick, densely yellowish brown stellate-puberulous; leaves cordate-orbicular, 10–15 cm. long, 8–12 cm. wide, abruptly acuminate, obliquely cordate, or subtruncate on one side at base, deltoid-toothed, pale yellowish brown stellate-pubescent on both surfaces, densely so and with axillary tufts of hairs beneath, the tertiary nerves rather prominently raised beneath, the petioles 4–7 cm. long, densely pubescent; inflorescence inserted near base of the bract; sepals 7–8 mm. long; petals oblong-cuneate, slightly longer than the sepals; staminodes as long as the sepals, oblanceolate; fruit subglobose, 5-angled, 8–10 mm. long, grayish puberulous.——June–July. Hokkaido, Honshu (centr. distr. and northw.).

Var. **yesoana** (Nakai) Tatew. *T. miyabei* var. *yesoana*

Nakai——Moiwa-bodaiju. Less pubescent; leaves green beneath.——This may be only a juvenile form.

Tilia × noziricola Hisauchi. Nojiri-bodaiju. Hybrid of *T. maximowicziana* × *T. japonica*.

4. Tilia rufovillosa Hatus. ? *T. mankichiana* Makino ——Tsukushi-bodaiju. Resembles the preceding species; young branches densely grayish pubescent; leaves rather thin, oblique, cordate or ovate-cordate, 10–15 cm. long, 8–12 cm. wide, short-acuminate, acutely serrate, loosely stellate-pubescent only while young, densely grayish pubescent on nerves beneath, without axillary hairs, the lateral nerves 7–9 pairs, the petioles 3–5 cm. long, densely pubescent only while young; inflorescence 5- to 9-flowered, loosely stellate-pilose; sepals 4–5 mm. long, densely stellate-pilose; petals oblong-oblanceolate, about 8 mm. long; staminodes spathulate, about 7 mm. long; fruit subglobose, 6–8 mm. in diameter, obscurely 5-ribbed, with long dense reddish brown simple hairs and short grayish stellate hairs.——July. Honshu (Jakuchisan in Suwo), Kyushu (mountains of n. distr.).

5. Tilia japonica (Miq.) Simonk. *T. cordata* var. *japonica* Miq.; *T. ulmifolia* var. *japonica* (Miq.) Sarg. ex Mayr. ——Shina-no-ki. Branchlets glabrous; leaves oblique, cordate-orbicular, rarely broadly ovate-orbicular, 5–8(–10) cm. long and as wide, abruptly acuminate, obliquely cordate or rounded on one side at the base, rather irregularly and acutely serrate, glabrous on upper side, glabrescent and with pale brown axillary tufts of hairs beneath, the lateral nerves 4–6 pairs, the tertiary nerves very slender, the petioles 3–5 cm. long, rather slender; sepals 3–4 mm. long, densely appressed-puberulent outside, sparsely so inside; petals broadly oblanceolate, about 5 mm. long; stamens as long as the petals; staminodes shorter than the petals; fruit subglobose, about 5 mm. long, densely grayish brown puberulous.——July–Aug. Mountains; Hokkaido, Honshu, Shikoku, Kyushu.

Var. **leiocarpa** Nakai. Kenashi-shina-no-ki. Fruit glabrous.——Shikoku.

Fam. 133. MALVACEAE   Aoi Ka   Mallow Family

Herbs or woody plants usually with stellate hairs; leaves stipulate, alternate, simple, palmately or rarely pinnately nerved, often palmately lobed; flowers bisexual, actinomorphic; sepals 5, often connate at base, valvate, often with a caliculus; petals 5, contorted, usually connate with the stamens at base; stamens numerous, rarely 5, the filaments adnate into a tube, the anthers 1-locular; ovary superior, 2- to many-locular; styles and stigmas as many as the carpels; ovules 1 to many in each locule; fruit a capsule or composed of separate carpels, very rarely fleshy; seeds with scanty endosperm.——About 80 genera, with about 1,500 species, widely distributed in tropical and temperate regions.

1A. Fruit a capsule; flowers large; ovules 3 to many in each locule. .... 1. *Hibiscus*
1B. Fruit of few to many separate carpels; flowers usually small.
  2A. Bracteoles adnate to the calyx; carpels 5; style-branches 10; ovules solitary in each locule. .... 2. *Urena*
  2B. Bracteoles absent or distant from the calyx; carpels 5 to many, as many as the style-branches.
    3A. Ovules and seeds solitary in each locule. .... 3. *Sida*
    3B. Ovules and seeds few in each locule. .... 4. *Abutilon*

1. HIBISCUS L.   Fuyō Zoku

Herbs, shrubs, or small trees; leaves deciduous or evergreen, alternate, palmately nerved and lobed; flowers usually axillary, solitary, large; calyx campanulate, 5-lobed or -parted, usually caliculate at base; petals 5; staminal tube 5-toothed at apex, the anthers numerous; ovary 5-locular, with 3 to many ovules in each locule; style 5-lobed, the stigma capitate; capsules loculicidally dehiscent, 5-lobed.——About 200 species, chiefly in the Tropics.

1A. Flowers yellowish; woody plants; stipules large, narrowly ovate or oblong, obtuse; leaves undivided, densely pubescent beneath; bracteoles forming a partially connate cup-shaped caliculus.
  2A. Leaves 3.5–6 cm. long and as wide, depressed-serrulate, rounded to slightly cordate at base; stipules about 1 cm. long.
                                                                    1. *H. hamabo*
  2B. Leaves 8–15 cm. long and as wide, obsoletely undulate-dentate or subentire, deeply cordate at base; stipules 1.5–3 cm. long.
                                                                     2. *H. tiliaceus*
1B. Flowers reddish, yellowish or white; herbs or woody plants; stipules linear, acuminate; leaves lobed; bracteoles nearly free, not connate.
  3A. Herbs with nearly 3-parted and pinnately incised leaves; flowers pale yellow; calyx membranous, thinly pubescent with simple long hairs on nerves. .... 3. *H. trionum*
  3B. Woody plants with 3- to 5-lobed leaves; flowers reddish or white; calyx herbaceous, densely stellate-pubescent with indistinct nerves.
    4A. Leaves ovate, 3-lobed, subcuneate, sometimes obtuse at base. .... 4. *H. syriacus*
    4B. Leaves cordate-orbicular, angular, cordate at base. .... 5. *H. mutabilis*

1. Hibiscus hamabo Sieb. & Zucc. *H. tiliaceus* var. *hamabo* (Sieb. & Zucc.) Maxim.——Hamabō. Bushy shrub 1–2 m. high, with yellowish gray stellate pubescence on branchlets, undersides of leaves, outer side of stipules, bracteoles, and calyx; leaves deciduous, rather thick, depressed-orbicular to -obovate, or obsoletely 4-angled, 3–6 cm. long, 3–7 cm. wide, abruptly acute, rounded to subcordate at base, obtusely serrulate, 5- or sometimes 7-nerved from the base, thinly stellate-puberulent on upper side, the petioles 8–20 mm. long, the stipules about 1 cm. long, narrowly ovate-triangular, deciduous; flowers solitary, axillary, pedicellate, about 5 cm. across, pale yellow with a dark red center, the pedicel about 1 cm. long; bracteoles connate to middle, about one-half as long as the calyx, the free portion lanceolate; calyx-lobes deltoid-ovate, acute; petals 4–5 cm. long, obovate, rounded.——July–Aug. Near seashores; Honshu (Sagami Prov. and westw.), Shikoku, Kyushu.——Ryukyus and Korea (Quelpaert Isl.).

2. Hibiscus tiliaceus L. *H. tortuosus* Roxb.——YAMA-ASA, Ō-HAMABŌ. Small tree with rather thick branches, minutely puberulous and punctulate when young; leaves rather thick, cordate-orbicular, 8–15 cm. long, 10–20 cm. wide, abruptly acute, deeply cordate at base, entire or with obsolete undulate teeth, with minute yellowish gray stellate pubescence, 7- to 9-nerved, the petioles 1–10 cm. long, the stipules narrowly ovate-triangular, obtuse, 1.5–3 cm. long; flowers axillary or subterminal on the upper part of the branchlets, pale yellow, with a dark red center, 8–10 cm. across, the pedicels 3–5 cm. long, often with 2 stipulelike bracts on the upper part, the caliculus cup-shaped, with deltoid marginal teeth; calyx-lobes lanceolate, gradually acute, about 15 mm. long; petals broadly obovate, rounded; capsules globose, mucronate, about 1.5 cm. across, yellowish brown long-pilose; seeds many, dark brown, reniform and thick, about 4 mm. across, with scattered raised minute spots.——July–Aug. Near seashores; Kyushu (Yaku-shima and Tanegashima).——Ryukyus, Formosa, s. China, India, Malaysia, and Australia.

3. Hibiscus trionum L. *H. ternatus* Cav.——GINSEN-KA. Annual herb 30–60 cm. high; stems with scattered coarse spreading hairs and with short hairs; leaves petiolate, membranous, 3-parted nearly to the cuneate base, thinly hirsute, the terminal segment large, 3–10 cm. long, narrowly ovate or broadly lanceolate, obtuse, pinnately lobed and incised; flowers solitary, axillary, pale yellow, 4–5 cm. across, the bracteoles linear; calyx 1.5–2 cm. long, coarsely spreading-hirsute, membranous, with green nerves, the lobes deltoid; capsules globose, enclosed within the calyx, brown-hairy, about 1 cm. across. ——Aug.–Sept. Sometimes naturalized in our area; common in warmer regions of the Old World.

4. Hibiscus syriacus L. *H. chinensis* DC.——MUKUGE, HACHISU. Deciduous shrub with grayish branches, stellate-puberulent when young; leaves ovate, rhombic-ovate, or broadly ovate, 3-lobed, 4–10 cm. long, 2.5–5 cm. wide, sub-cuneate at base, 3- to 5-nerved, with scattered simple and stellate hairs, the terminal lobes narrowly deltoid, rather large, acuminate and sometimes with an obtuse tip, the petiole 7–20 mm. long, densely puberulous on upper side; flowers axillary and terminal, 5–6 cm. across, rose-purple, sometimes white (often double in cultivars), the pedicels about 1.5 cm. long, as long as the calyx, the bracteoles linear, nearly as long as the calyx; calyx stellate-puberulent, the lobes deltoid; fruit ovoid-globose, densely stellate-puberulent; seeds reniform, about 4–5 mm. across.——Aug.–Oct. Widely planted for hedges and in gardens, several cultivars are grown.——Introduced from China.

5. Hibiscus mutabilis L. FUYŌ. Deciduous shrub densely gray stellate-pubescent on branchlets, under side of leaves, peduncles, and calyx; leaves long-petiolate, 5-angled-orbicular, 10–20 cm. long and as wide, cordate, shallowly 5- or sometimes 3-lobed, with remote and shallow undulate teeth, stellate-pubescent and with minute spots on upper side, the lobes deltoid, acuminate; flowers solitary, pink, 10–13 cm. across, the peduncles to 12 cm. long, usually with an obscure joint above the middle; calyx 5-lobed, nearly to the middle, sometimes glandular-pilose, the lobes narrowly deltoid, the bracteoles longer than the calyx-tube; capsules globose, about 2.5 cm. across, long spreading-pilose, nerved; seeds many, reniform, about 2 mm. across, white-hirsute on the back.—— Aug.–Oct. Thickets; Kyushu; often cultivated for the large showy flowers, and sometimes naturalized in warmer regions in our area.——Ryukyus and China.

2. URENA L. BONTEN-KA ZOKU

Coarse herbs or shrubs; leaves angled or palmately lobed, with glands at the base of the midrib beneath; flowers rather small, sessile or short-pedunculate, rose-colored or yellow, the bracteoles forming a 5-lobed connate caliculus, adnate to the calyx-tube; calyx 5-lobed or -toothed; staminal tube truncate or 5-toothed, the anthers sessile or nearly so; ovary 5-locular, the ovules solitary in each locule; style-branches 10, the stigmas capitellate; fruitlets indehiscent, glochidiate-hairy, the seeds ascending.—— Few species, in Asia, Africa, and tropical America.

1. Urena lobata L. var. **tomentosa** (Bl.) Walp. *U. tomentosa* Bl.; *U. lobata* var. *scabriuscula* sensu auct. Japon., non A. Gray——ŌBA-BONTEN-KA, Ō-BONTEN-KA. Shrub, densely grayish stellate-puberulent, the branches elongate; leaves orbicular to broadly ovate and angled, 4–5 cm. long, 4–8 cm. wide, shallowly 3- to 5-lobed or undivided, rounded, cordate or broadly cuneate at base, the lobes deltoid, acute to obtuse, toothed, the petioles 3–10 cm. long; flowers axillary, solitary, on very short pedicels, rose-colored, the caliculus deeply 5-lobed, the lobes lanceolate, longer than the calyx; calyx about 5 mm. long, the lobes broadly lanceolate; fruitlets obovoid, 5–6 mm. long, stellate-hairy and glochidiate-pilose, the hairs about 1.5 mm. long; seeds reniform, about 4 mm. long, grayish brown.——Aug. Thickets and waste grounds; Kyushu (s. distr. incl. Yakushima and Tanegashima). The typical phase, widely distributed in the Tropics, is very variable.

Var. **sinuata** (L.) Gagnep. *U. sinuata* L.——BONTEN-KA. Leaves smaller and less densely puberulous, deeply 5-lobed, the lobes usually ovate, narrowed at the base; bracteoles as long as the calyx; petals glabrous.——Shikoku, Kyushu (s. distr. incl. Yakushima and Tanegashima).——Widely distributed in the Tropics of Asia.

3. SIDA L. KIN-GOJI-KA ZOKU

Herbs or shrubs; leaves simple or lobed; flowers yellow or white, sessile or peduncled, without bracteoles, solitary or in glomerate, racemose, capitate, or spicate inflorescences; calyx 5-toothed or -lobed; staminal tube shorter than the free part of the filaments; ovary 5- to many-locular, the ovules solitary in each locule; style-branches as many as the carpels, filiform or somewhat clavate, the stigmas capitate or truncate; fruitlets separating from the axis, rostrate or erostrate, indehiscent or 2-valvate and dehiscent, the seeds pendulous or horizontal.——About 130 species, abundant in America, occurring also in Africa and Asia.

1. **Sida rhombifolia** L. KIN-GOJI-KA. Stellate-puberulent perennial herb with terete branches; leaves narrowly ovate or rhomboid-ovate, rarely oblong, 1.5–5 cm. long, 1–2 cm. wide, obtuse, obtusely toothed, green on upper side, paler and rather densely stellate-pubescent beneath, the petioles 2–6 mm. long, densely puberulent; flowers axillary, yellow, about 1.5 cm. across, the pedicels 1–4 cm. long, jointed above the middle; calyx 5–7 mm. long, infundibuliform, 5-lobed, 10-ribbed, the lobes deltoid and acute; petals obovate; fruitlets 10, stellate-hairy on back, about 3 mm. long, 2-lobed with an awn at apex.——Aug. Kyushu (Yakushima and Tanegashima).——Widely distributed in the Tropics.

4. ABUTILON Gaertn. ICHIBI ZOKU

Herbs and shrubs with stellate-hairs; leaves angled or palmately lobed, sometimes cordate-orbicular; inflorescence axillary or terminal, without bracteoles; calyx often subtubulose, the lobes 5, valvate; petals 5, connate and adnate at base to the staminal tube; filaments free above; carpels 5 to many, the styles as many as the carpels, the ovules 2 to many in each carpel; fruitlets separating from the axis, sometimes awned, seeds 2 or more in each carpel, reniform.——About 100 species, chiefly in the Tropics.

1. **Abutilon theophrasti** Medik. *Sida abutilon* L.; *A. avicennae* Gaertn.——ICHIBI. Velvety annual; stems 50–150 cm. long; leaves cordate-orbicular, 5–12 cm. long and wide, abruptly acuminate, deeply cordate, with obtuse shallow teeth, palmately 7- to 9-nerved; flowers axillary and terminal, short-peduncled, yellow, about 2 cm. across; calyx-lobes longer than the tube, ovate, mucronate; fruit semiglobose, 15–20 mm. across, the fruitlets about 15 mm. long, dehiscent, membranous, hairy, with 2 ascending awns; seeds obliquely reniform, brown, puberulent, about 3.5 mm. across.——Aug.–Sept. Cultivated and sometimes naturalized.——s. Europe, Asia, America, and Australia.

Fam. 134. STERCULIACEAE AOGIRI KA Sterculia Family

Trees, rarely herbs, often with stellate hairs; leaves alternate, rarely subopposite, simple, or palmately compound, usually stipulate; inflorescence various; flowers bisexual or unisexual, actinomorphic; sepals 3–5, more or less connate, valvate; petals 5 or absent, often adnate to the staminal tube, convolute; stamens often connate into a tube at base, with as many staminodes; ovary, with 2–5 (–12) united carpels; styles usually 4 or 5, distinct or united; ovules usually 2 on inner angle of each locule; fruit dry, rarely a berry, indehiscent or dehiscent; seeds with fleshy endosperm or none.——About 50 genera, with about 750 species, chiefly in the tropics.

1A. Trees; flowers unisexual, apetalous; follicles dehiscent early before maturity; styles connate, with a dilated stigma; seeds several in each carpel. . . . . 1. *Firmiana*
1B. Herbs (in ours); flowers bisexual, petaliferous; capsules globose, separating into 5 fruitlets; styles free or connate at base; seeds solitary in each carpel. . . . . 2. *Melochia*

1. FIRMIANA Marsili AOGIRI ZOKU

Deciduous trees with alternate palmately lobed leaves; flowers in large terminal panicles, 5-merous, unisexual; sepals yellowish; petals absent; androgynophore well developed; carpels 5, free at base, connate at apex; style solitary; follicles stellately spreading, early dehiscent and leaflike; seeds globose, marginal on the dehiscent carpels.——About 10 species, chiefly in s. Asia, one in Africa.

1. **Firmiana simplex** (L.) W. F. Wight. *Hibiscus simplex* L.; *Sterculia platanifolia* L. f.; *S. tomentosa* Thunb.; *F. platanifolia* (L. f.) Schott & Endl.——AOGIRI. Monoecious tree, the branches green, the branchlets, young leaves on upper side, and petioles brown stellate-hairy while young; leaves large, long-petiolate, palmately 3- to 5-lobed; panicles large, many-flowered, yellowish brown villous; flowers yellowish; sepals 5, reflexed, linear-oblong, about 1.5 cm. long; filaments of the staminate flowers connate into a tube about 12 mm. long; androgynophore glabrous, about 5 mm. long; carpels early dehiscent after anthesis, stellately spreading, narrowed to a petiolelike base, exposing several marginal seeds.——July. Very common as a street tree, sometimes naturalized in the warmer parts of our area.——Ryukyus, Formosa, China, and Indochina.

2. MELOCHIA L. NOJI-AOI ZOKU

Woody or herbaceous, with simple and stellate hairs; leaves ovate or cordate, serrate; flowers usually small, 5-merous, in axillary or terminal subcapitate inflorescences; calyx 5-lobed or -toothed; petals spathulate or oblong; stamens opposite the petals, connate at base, the staminodes absent or toothlike; ovary sessile or short-stiped, 5-locular, with 2 ovules in each locule; styles 5, free or connate at base; capsules loculicidally dehiscent, 5-valved, the seeds solitary in each locule, obovoid, with scanty endosperm.——About 60 species, chiefly in the Tropics.

1. **Melochia corchorifolia** L. *M. concatenata* L.——NOJI-AOI. Loosely stellate-pilose branched annual 30–60 cm. high; leaves broadly ovate, or in the upper ones narrowly 3-angled ovate, 2.5–6 cm. long, 1.5–5 cm. wide, acute to subobtuse, the larger ones often shallowly and obsoletely 3-fid, irregularly toothed, coarsely hirsute on midrib on upper side and on nerves beneath; flowers in capitate inflorescences in upper axils and terminal, the bracteoles 4, subulate-linear,

ciliate; calyx cup-shaped, shorter than the bracteoles, 5-toothed; petals rose-colored, twice as long as the calyx and nearly as long as the bracteoles, obovate, about 7 mm. long; capsules depressed-globose, about 5 mm. across, coarsely hirsute, the seeds grayish brown, with black spots, about 2.5 mm. long.——Aug.–Oct. Shikoku, Kyushu.——Widely distributed as a tropical weed.

Fam. 135. ACTINIDIACEAE   MATA-TABI KA   Actinidia Family

Trees or shrubs, often scandent; leaves alternate, petiolate, simple, often membranous, toothed, without stipules; inflorescence small, in axillary cymes or fascicles; flowers unisexual, bisexual, or polygamous (plants rarely dioecious); sepals 5, imbricate, free; petals 5, imbricate or subcontorted; stamens many, free or adnate to the petals, the anthers versatile; ovary superior, (3–) 5- or many-locular, with many ovules in each locule; styles 5 to many, free or connate at the top; fruit a berry or capsule; seeds many with copious endosperm.——Four genera and about 280 species, in the Tropics, few in temperate regions.

1. ACTINIDIA Lindl.   MATA-TABI ZOKU

Deciduous scandent polygamodioecious shrubs; leaves alternate, long-petiolate, simple, toothed or rarely entire; flowers solitary or in axillary cymes, white, rarely reddish; sepals 5, imbricate; petals 5 or rarely 4, contorted; stamens many, the anthers versatile; ovary superior, many-locular; styles many, linear, radiate; fruit a many-seeded berry; seeds small, with endosperm.——About 25 species, in e. and s. Asia, and India.

1A. Leaves thick, green and lustrous; anthers dark-purple; fruit ellipsoidal.
  2A. Leaves green beneath; inflorescence more or less brown pubescent.
    3A. Inflorescence and sepals densely rusty-brown villous; ovary villous. . . . . 1. *A. rufa*
    3B. Inflorescence pale brown pubescent; sepals ciliate; ovary glabrous. . . . . 2. *A. arguta*
  2B. Leaves glaucescent beneath; inflorescence and sepals nearly glabrous. . . . . 3. *A. hypoleuca*
1B. Leaves membranous, vivid green, scarcely lustrous, those near the inflorescence white on upper half on upper side; anthers pale yellow; fruit narrowed toward apex.
  4A. Leaves rounded to acute at base; flowers usually 1, rarely 2 or 3, in axils of the middle portion of young branchlets, about 2.5 cm. in diameter; pith solid. . . . . 4. *A. polygama*
  4B. Leaves at least some of them cordate; flowers 3, rarely 1 or 2, in axils of lower portion of young branchlets; pith lamellate.
      5. *A. kolomikta*

1. **Actinidia rufa** (Sieb. & Zucc.) Planch. *Trochostigma rufa* Sieb. & Zucc.; *A. callosa* var. *rufa* (Sieb. & Zucc.) Makino; *A. rufa* var. *typica* Dunn; *A. arguta* var. (?) *rufa* (Sieb. & Zucc.) Maxim.; *A. kiusiana* Koidz.——SHIMA-SARU-NASHI, NASHI-KAZURA.   Branchlets lenticellate, elongate, reddish brown-woolly while young, soon becoming glabrous; leaves elliptic to broadly so, or ovate-elliptic, 6–13 cm. long, 4–8 cm. wide, acuminate or abruptly so, rounded to shallowly cordate at base, loosely serrate with shallow callose-pointed teeth, pale green, glabrous on upper side, rusty-brown pubescent on nerves beneath while young, persisting as axillary tuft of hairs, the petioles 3–6 cm. long; inflorescence on middle part of the branchlets, rusty-brown villous, 1- to 8-flowered, shorter than the petioles; flowers 1–1.5 cm. across; fruit broadly ellipsoidal, 2–3 cm. long, with light brown spots.——May–June. Honshu (Kii Prov., Chūgoku Distr.), Shikoku, Kyushu.——Ryukyus and s. Korea.

2. **Actinidia arguta** (Sieb. & Zucc.) Planch. ex Miq. *Trochostigma arguta* Sieb. & Zucc.; *A. rufa* var. *arguta* (Sieb. & Zucc.) Dunn; *A. callosa* var. *arguta* (Sieb. & Zucc.) Makino——SARU-NASHI.   Woody scandent shrub, branchlets nearly glabrous, smooth, the pith lamellate; leaves thin, elliptic or broadly so, sometimes broadly ovate, 6–10 cm. long, 3.5–7 cm. wide, abruptly short-acuminate, rounded to shallowly cordate at base, mucronate-serrulate, glabrous on upper side, rusty-brown pubescent on nerves beneath while young, persisting in the nerve axils, the petioles rather slender, 2–8 cm. long; inflorescence axillary on upper part of branchlets, 1- to 10-flowered, rusty-brown pubescent; flowers 1–1.3 cm. across, white; sepals ciliate, ovate-rounded; ovary glabrous; fruit broadly ellipsoidal, 2–2.5 cm. long.——May–July. Hokkaido, Honshu, Shikoku, Kyushu.——s. Kuriles, Sakhalin, Ussuri, Manchuria, and Korea.

Var. **platyphylla** (A. Gray) Nakai. *A. platyphylla* A. Gray; *A. japonica* Nakai——KOKUWA.   Under side of leaves callose-hairy.

3. **Actinidia hypoleuca** Nakai. URAJIRO-MATA-TABI. Branchlets glabrous, nearly smooth; leaves oblong, 6–10 cm. long, 3–6 cm. wide, abruptly short-acuminate, rounded to abruptly acute at base, mucronate-serrulate, whitish and with rusty-brown axillary tufts of hairs on the nerves beneath, the petioles 2–6 cm. long; inflorescence on upper part of branchlets, 1- to 10-flowered, puberulent; flowers white, 1–1.5 cm. across; sepals broadly ovate, glabrescent; ovary glabrous; fruit 2–3 cm. long, broadly ellipsoidal, with light colored spots. ——May–June. Honshu (Awa and westw.), Shikoku, Kyushu.

4. **Actinidia polygama** (Sieb. & Zucc.) Maxim. *Trochostigma polygama* Sieb. & Zucc.; *T. volubilis* Sieb. & Zucc.; *A. polygama* var. *latifolia* Miq.; *A. volubilis* (Sieb. & Zucc.) Miq.; *A. repanda* Honda, excl. basionym.——MATA-TABI. Branches rather slender, the pith solid, the branchlets brownish pubescent while young, sometimes thinly callose-hairy; leaves thin, broadly ovate to elliptic, sometimes oblong, 6–15 cm. long, 3.5–8 cm. wide, acuminate to abruptly so, rounded to shallowly cordate at base, mucronate-serrulate, usually with scattered callose hairs on upper side and on nerves beneath, with axillary tufts of pale brown hairs beneath, the petioles 3–6 cm. long; leaves near inflorescence white on upper half; inflorescence 1- to 3-flowered, axillary, pale brown-pubescent, shorter than the petioles; flowers white, pendulous, 2–2.5 cm. across; sepals ovate, membranous, white-puberulent near mar-

gin; ovary glabrous; fruit oblong, rostrate, 2–2.5 cm. long.——June–July. Thickets and woods in mountains; Hokkaido, Honshu, Shikoku, Kyushu; rather common.——Sakhalin, s. Kuriles, Korea, Manchuria, and Ussuri.

5. Actinidia kolomikta (Rupr. & Maxim.) Maxim. *Prunus kolomikta* Rupr. & Maxim.; *Kolomikta mandshurica* Regel; *Trochostigma kolomikta* (Rupr. & Maxim.) Rupr.——MIYAMA-MATA-TABI. Branches slender; pith lamellate; branchlets brownish pubescent while young; leaves thinly membranous, obovate, elliptic or broadly ovate, 7–12 cm. long, 4–8 cm.

wide, mucronate-serrulate, sometimes loosely callose-hairy on upper side, thinly brownish pubescent on both sides, especially on nerves, with axillary tufts of white hairs beneath, the petioles slender, 4–7 cm. long; inflorescence 3-, sometimes 1- or 2-flowered, axillary, loosely pubescent; flowers pendulous, 1–1.5 cm. across; sepals ovate, white-puberulous near margin; ovary glabrous; fruit narrowly oblong, 1.5–2 cm. long, narrowed toward tip.——June–July. Mountains; Hokkaido, Honshu (centr. distr. and northw.).——s. Kuriles, Sakhalin, Amur, and Manchuria.

Fam. 136. **THEACEAE**   Tsubaki Ka   Tea Family

Trees and shrubs; leaves alternate, simple, deciduous or evergreen, stipules absent; flowers usually solitary, sometimes in panicles, racemes, or in cymes, usually large, actinomorphic, bisexual, rarely unisexual; bracts usually just beneath the calyx, paired; sepals 5, free or connate at base, imbricate; petals 5, free or connate at base, imbricate or contorted; stamens many, in several series, rarely definite, free or connate; ovary superior, sessile, 3- to 5-locular; styles sometimes connate; ovules 2 or more, rarely 1 in each locule; fruit dehiscent or indehiscent; seeds with scanty endosperm.——About 33 genera, with about 480 species, abundant in the Tropics and subtropics, few in temperate regions.

1A. Fruit a dehiscent capsule; anthers versatile.
  2A. Evergreen; capsules with an axis; seeds wingless; ovules pendulous.
    3A. Sepals persistent; bracteoles 2; inflorescence apparently axillary; flowers pedicellate. .... 1. *Thea*
    3B. Sepals deciduous; bracteoles many, deciduous; inflorescence apparently terminal or subterminal; flowers sessile. ...... 2. *Camellia*
  2B. Deciduous; capsules without an axis; seeds winged; ovules ascending. .... 3. *Stewartia*
1B. Fruit fleshy or nearly so, usually indehiscent; anthers basifixed.
  4A. Flowers bisexual; leaves usually entire (in ours).
    5A. Anthers glabrous; ovules pendulous from the top of the ovary. .... 4. *Ternstroemia*
    5B. Anthers bearded; ovules median. .... 5. *Cleyera*
  4B. Flowers unisexual, the plants dioecious; leaves serrate. .... 6. *Eurya*

1. **THEA** L.   CHA-NO-KI ZOKU

Evergreen shrubs or small trees; leaves coriaceous, serrate; inflorescence apparently axillary, 1- to 3-flowered; flowers bibracteolate, bisexual, pedicellate, deflexed, actinomorphic; sepals 5, persistent; petals 5, imbricate, connate at base; stamens many, glabrous, connate at base into a short tube, the anthers versatile; ovary superior, usually 3-locular; styles 3, connate at base; ovules few in each locule, pendulous; capsules thick, rather ligneous, loculicidally dehiscent, 3-valved; seeds 1 or 2 in each locule, wingless, without endosperm, the embryo straight, the cotyledons thick.——About 10 species, in s. and e. Asia.

1. Thea sinensis L. *Camellia sinensis* (L.) O. Kuntze; *T. bohea* L.; *T. viridis* L.; *T. sinensis* var. *bohea* (L.) K. Koch——CHA-NO-KI. Much-branched evergreen shrub, the young branchlets thinly ascending-pilose; leaves lanceolate-oblong or oblong, 4–10(–15) cm. long, 2–3(–5) cm. wide, obtuse or subobtuse, acute at base, short-petiolate, serrulate, glabrous on both sides except for a few appressed hairs sometimes on the underside when young; flowers 1–3, in axillary or subterminal cymes, deflexed, 2–2.5 cm. across, white, the pedicels 5–15 mm.

long; capsules depressed-globose, about 2 cm. across, with 3 shallow indistinct grooves.——Oct.–Dec. Widely cultivated in warmer regions and allegedly spontaneous in Kyushu.——China.——Forma **rosea** (Makino) Ohwi. *T. sinensis* var. *rosea* Makino; *Camellia sinensis* forma *rosea* (Makino) Kitam.——BENIBANA-CHA. Flowers rose.——Forma **macrophylla** (Sieb.) Ohwi. *T. sinensis* var. *macrophylla* Sieb. ex Miq.; *T. macrophylla* (Sieb.) Makino——TŌ-CHA. Leaves larger.——Cultivated.

2. **CAMELLIA** L.   Tsubaki Zoku

Evergreen shrubs or trees; leaves coriaceous, short-petiolate, serrate; inflorescence apparently terminal or subterminal; flowers bisexual, solitary, large, white to red, sessile, the perules (bracteoles and sepals) deciduous; petals 5 (–7) (often numerous in cultivars), slightly connate at base; stamens many, the outer filaments connate into a tube, the anthers versatile; ovary 3-locular (in ours), with 4–6 ovules in each locule; styles slender, connate at base to free or nearly so; fruit a woody loculicidally dehiscent capsule, the axis persistent; seeds few, globose or angled, the embryo straight, the cotyledons thick.——About 80–90 species, in e. and s. Asia.

1A. Leaves 6–12 cm. long, 3–7 cm. wide; petioles 7–12 mm. long; petals connate at base; filaments connate into a tube; ovary glabrous.
  2A. Petioles glabrous; involucre of perules (bracteoles and sepals) 2–3 cm. long; petals often nearly erect to ascending-spreading, rounded to slightly retuse or emarginate; outer filaments united to form a tube for ½–⅔ their length; filaments white or creamy yellow.
                                          1. *C. japonica*
  2B. Petioles thinly pilose while young; involucre of perules 1.5–1.7, rarely 2 cm. long; petals wide-spreading, distinctly retuse to shallowly bifid; outer filaments united for only 2–3 mm. above base; filaments spreading bright yellow. .... 2. *C. rusticana*
1B. Leaves 3–5(–7) cm. long, 1–3 cm. wide; petioles 2–5 mm. long; petals nearly free, spreading; filaments only slightly connate at base; ovary villous. .... 3. *C. sasanqua*

1. **Camellia japonica** L. *Thea japonica* (L.) H. Baill.; *C. japonica* var. *spontanea* (Makino) Makino; *T. japonica* var. *spontanea* Makino——YABU-TSUBAKI. Evergreen tree or shrub with glabrous branches; leaves elliptic or oblong, 6–12 cm. long, 3–7 cm. wide, abruptly acuminate, serrulate, glabrous on both surfaces, deep green and lustrous, the petioles 8–15 mm. long; flowers solitary and subterminal, sessile, red, the bracteoles many, small, rounded, white-puberulous on outer side; sepals 5, rather unequal, ovate-orbicular, 1–2 cm. long; petals 5, 3–5 cm. long, often nearly erect to ascending-spreading, rounded to slightly retuse or emarginate, connate and adnate to the stamens at base; stamens numerous, the filaments white or creamy yellow, the outer ones united to form a tube for 1/2–2/3 their length, the anthers yellow; ovary glabrous; style trifid; capsules globose, 4–5 cm. across, with a thick pericarp; seeds large, dark-brown.——Feb.–Apr. Near seashores; Honshu, Shikoku, Kyushu; common.——Korea and Ryukyus.

Var. **hortensis** (Makino) Makino. *T. japonica* var. *hortensis* Makino——TSUBAKI. Includes cultivars with single to double, red, white, or blotched flowers.

Var. **macrocarpa** Masam. YAKUSHIMA-TSUBAKI, RINGO-TSUBAKI. Fruit larger, about 5 cm. across, the pericarp very thick, about 1.5 cm. thick; seeds rather small, darker brown.——Mountains of Kyushu (Yakushima) and Shikoku, extending to Amami-Oshima.

Camellia reticulata Lindl. TŌ-TSUBAKI. Leaves regularly serrulate, strongly veined; flowers large, semi-double, rose-colored; ovary hairy.——Native of China, but sometimes cultivated in our area.

2. **Camellia rusticana** Honda. *C. japonica* subsp. *rusticana* (Honda) Kitam.——YUKI-TSUBAKI, SARUIWA-TSUBAKI. Small tree or shrub with ascending or suberect glabrous branches; leaves thick, oblong to elliptic, sometimes obovate-oblong, 6–12 cm. long, 3.5–6 cm. wide, abruptly to gradually acuminate, acutely ascending-serrulate, glabrous, deep green and lustrous on upper side, green beneath, the petioles 5–8 mm. long; flowers subterminal on the branchlets, solitary, sessile, reddish, the bracteoles rather numerous, externally pilosulous; sepals orbicular, about 1 cm. long; petals 5, wide-spreading, distinctly retuse to shallowly bifid, 2.5–4 cm. long, about 2 cm. wide, shortly connate at base; outer filaments yellow, obliquely spreading, shortly united above base, the anthers deep yellow; styles 3- or 4-fid; fruit globose, about 2.5 cm. in diameter, with rather thick pericarp.——Apr.–June. Thickets and thin woods on mountain slopes; Honshu (Japan Sea side of centr. distr. and northw.).

3. **Camellia sasanqua** Thunb. *Thea sasanqua* (Thunb.) Noisette; *Thea miyagii* Koidz.; *T. sasanqua* var. *thunbergii* Pierre——SAZANKA. Small evergreen tree with slender branches; branchlets terete, ascending- or spreading-pilose; leaves thinly coriaceous, broadly oblanceolate to narrowly obovate or narrowly oblong, 3–7 cm. long, 1–3 cm. wide, acute with an obtuse tip or sometimes emarginate, obtusely serrulate, with ascending short hairs on the midrib on both sides, the lateral nerves and veinlets slender, only slightly raised on both sides, the petioles 2–5 mm. long, slightly pilose on the upper side; flowers subterminal, sessile, white (also pink to rose in cultivars), 4–7 cm. across; sepals depressed-orbicular, about 1 cm. long; petals obovate or oblong, retuse or shallowly bifid, widely spreading; stamens numerous, the filaments slightly connate at base; ovary densely white-pilose; capsules obovoid-globose, loosely long-pilose, 1.5–2 cm. across, with a rather thick pericarp.——Oct.–Dec. Thickets and grassy slopes in mountains; Kyushu. Many cultivars are grown in gardens.——Ryukyus.

3. STEWARTIA L. NATSU-TSUBAKI ZOKU

Deciduous shrubs or trees; leaves membranous; flowers bisexual, solitary, rather large, sessile or short-pedicellate, axillary, white or reddish, bibracteolate; sepals 5 or 6, rather unequal, connate at base; petals as many as the sepals, imbricate, connate at base; stamens usually adnate to the petals and connate at base, the anthers versatile; ovary 5-locular; styles 5, free or connate; ovules 2 in each locule, anatropous, ascending; capsules ovoid, woody, loculicidally dehiscent; seeds lenticular, entire or narrowly wing-margined, the endosperm scanty. About 8 species, in e. Asia and N. America.

1A. Leaves (4–)6–12 cm. long, appressed-pilose beneath, acute to abruptly short-acuminate; pedicels 1–6 cm. long; bracteoles always shorter than the sepals; flowers 6–7 cm. across; sepals depressed-orbicular; filaments glabrous; ovary and capsules pilose.
                                                      1. *S. pseudocamellia*
1B. Leaves 3–7(–8) cm. long, glabrous or pilose only on nerves beneath, acuminate to acute; pedicels less than 1 cm. long; bracteoles as long as or longer than the sepals; flowers 1.5–4 cm. across; sepals lanceolate to ovate; filaments white-hairy at base.
   2A. Ovary and capsules glabrous; branches grayish or dark brown; flowers 3.5–4 cm. across; sepals 12–17 mm. long. ...... 2. *S. serrata*
   2B. Ovary and capsules white-pilose; branches reddish brown; flowers 1.5–2 cm. across; sepals 4–7 mm. long. .... 3. *S. monadelpha*

1. **Stewartia pseudocamellia** Maxim. *S. grandiflora* Carr.——NATSU-TSUBAKI. Tree with grayish brown branches, the young branches usually with appressed white hairs on lower portion while young; leaves membranous, obovate or elliptic, (4–)6–12 cm. long, 3–5 cm. wide, acute to short-acuminate, acute to subcuneate at base, mucronate-serrate, green, glabrous on upper side, pilose and with axillary tufts of hairs beneath, the lateral nerves 6 or 7 pairs, the petioles 3–15 mm. long; pedicels axillary, 1–6 cm. long, glabrescent, the bracteoles 2, cordate-orbicular, 4–6 mm. long, much shorter than the sepals; flowers 6–7 cm. across; sepals depressed-orbicular, 8–12 mm. long, with dense long white silky hairs, tubercled on margin; petals white, 3–4 cm. long, obovate, irregularly denticulate, densely white-silky on back; filaments glabrous; ovary pilose; capsules 5(–6)-angled, conical, ovoid, gradually acuminate, about 1.5 cm. across, white-pilose.——July. Mountains; Honshu, Shikoku, Kyushu.

2. **Stewartia serrata** Maxim. HIKOSAN-HIME-SHARA. Tree with grayish to dark brown branchlets, densely spotted while young, glabrous or thinly long-pilose; leaves rather thick-membranous or chartaceous, ovate-elliptic or elliptic, 3–7(–8) cm. long, 1.5–3.5(–4) cm. wide, acuminate, acute at base, mucronate-serrate, glabrous or pilose on midrib on upper side, pale green with axillary tufts of hairs and loosely appressed-pilose on midrib beneath, the lateral nerves and midrib often reddish; flowers in upper axils, white, 3.5–4 cm. across, the pedicels 2–4 mm. long, the bracteoles leaflike, narrowly ovate, usually longer than the sepals; sepals broadly lanceolate, 12–17 mm. long, serrate; petals obovate, 2.5–3 cm. long, obtusely serrulate, white-silky on back; ovary glabrous; capsules ob-

ovoid, 5-angled, glabrous, about 1.5 cm. across, terminated by a portion of the persistent style; seeds planconvex, brown, obovate, about 8 mm. long, narrowly wing-margined.——Honshu (Tōkaidō and s. Kinki Distr.), Shikoku, Kyushu.—— Forma **epitricha** (Nakai) Ohwi. *S. epitricha* Nakai——ICHI-BUSA-HIME-SHARA. Leaves sparingly callose-pilose on upper side.——Kyushu (Mount Ichibusa in Higo).——Forma **sericea** (Nakai) Hara. *S. serrata* var. *sericea* Nakai——TŌGOKU-HIME-SHARA. Leaves pilose, especially on midrib beneath.——Honshu (Tōkaidō Distr.).

3. **Stewartia monadelpha** Sieb. & Zucc. HIME-SHARA. Shrub with slender, reddish brown branches, the branchlets glabrous or sparsely ascending-pilose; leaves chartaceous, ovate-elliptic or elliptic to oblong, 4–8 cm. long, 2–3 cm. wide, acumi-nate, acute at base, short mucronate-toothed, pilose on both sides or only on midrib on upper side, with axillary tufts of hairs beneath, the petioles slender, 2–12 mm. long, often reddish; flowers axillary, white, 1.5–2 cm. across, the pedicels 5–10 mm. long, the bracteoles longer than the sepals; sepals broadly lanceolate to ovate, 4–7 mm. long, entire, ciliate; petals obovate, 12–15 mm. long, obtusely toothed, white-silky on back; ovary white-pilose; capsules 5-angled, ovoid-pyramidal, narrowed above to a beak about 1 cm. across, white-pilose; seeds planoconvex, dark brown, smooth, obovate, about 5 mm. long, narrowly winged.——July–Aug. Mountains; Honshu (Kantō and s. Kinki Distr.), Shikoku, Kyushu.——Korea (Quelpaert Isl.).

4. TERNSTROEMIA Mutis   MOKKOKU ZOKU

Evergreen trees or shrubs; leaves coriaceous, entire or with undulate obtuse teeth; flowers bisexual (sometimes functionally unisexual, in ours), solitary or fasciculate, axillary, deflexed, bibracteolate; sepals 5, imbricate, persistent; petals 5, imbricate, connate at base, deciduous; stamens numerous, adnate to the base of the petals, the anthers glabrous, basifixed; ovary 2- or 3-locular; styles connate or nearly absent, the stigma 2- or 3-lobed or nearly entire; ovules 2, rarely 3–6 in each locule, pendulous; fruit fleshy, indehiscent or ultimately dehiscent; seeds rather large, the endosperm fleshy, sometimes absent, the embryo incurved.——About 100 species, in the Tropics of Asia and America, few in Africa.

1. **Ternstroemia gymnanthera** (Wight & Arn.) Sprague. *Cleyera gymnanthera* Wight & Arn.; *C. japonica* Thunb., pro parte; *Taonabo japonica* (Thunb.) Szysz.——MOKKOKU. Glabrous evergreen tree with rather thick, rusty-gray branches; leaves alternate, often closely approximate near the top of branchlets, narrowly cuneate-obovate, 4–7 cm. long, 1.5–2.5 cm. wide, rounded to very obtuse, cuneate to acuminate at base, entire, slightly recurved on margin, deep green and the midrib impressed on upper side, slightly paler and the midrib raised beneath, the lateral nerves and veinlets obscure, the petioles reddish, 5–10 mm. long; flowers axillary, deflexed, white, the pedicels simple, 1–2 cm. long, curved downward; sepals 5, orbicular, 3.5–4 mm. long, longer than the ovate acute bracteoles, irregularly denticulate; petals obovate-cuneate, about 8 mm. long; stamens many, the anthers longer than the filaments; fruit globose, about 10–12 mm. across, later dehiscent; seeds red.——July. Near seashores; Honshu (Tōkaidō and westw.), Shikoku, Kyushu.——s. Korea, Ryukyus, Formosa, China, India, and Borneo.——Forma **subserrata** (Makino) Ohwi. *T. japonica* var. *subserrata* Makino——HIME-MOKKOKU. Leaves obtusely serrate on upper margin.

5. CLEYERA Thunb.   SAKAKI ZOKU

Evergreen trees or shrubs; leaves alternate, coriaceous, petiolate, entire; flowers bisexual, solitary or fasciculate, usually axillary, relatively small, actinomorphic, pedicellate, the bracteoles minute or absent; sepals 5, imbricate; petals 5, imbricate, free or connate at base; stamens many, inserted on the base of the petals, the anthers bearded, basifixed, erect; ovary 2- or 3-locular; styles often elongate, shortly 2- or 3-fid; ovules centrally attached; fruit indehiscent, more or less fleshy, the endosperm fleshy. ——Few species, in e. and s. Asia, also Central and S. America.

1. **Cleyera japonica** Thunb., pro parte, emend. Sieb. & Zucc. *C. ochnacea* DC.; *Eurya ochnacea* (DC.) Szysz.; *Tristylium ochnaceum* (DC.) Merr.; *Sakakia ochnacea* (DC.) Nakai——SAKAKI. Evergreen glabrous tree with green branches; leaves alternate, coriaceous, narrowly oblong or ovate-oblong, 7–10 cm. long, 2–4 cm. wide, acute with an obtuse tip, acute at base, entire, deep green and lustrous on upper side, paler and the midrib raised beneath, the lateral nerves rather obscure, the petioles 5–10 mm. long; flowers 1–3, axillary, white, becoming yellowish with age, the pedicels 1–1.5 cm. long, the bracteoles minute, ovate-orbicular, caducous; sepals ovate-orbicular, about 3 mm. long, ciliolate; petals rather thick, narrowly oblong, 8–10 mm. long; anthers ovate, cuspidate, white, much shorter than the filaments; fruit globose, black, indehiscent.——(June)–July. Honshu (centr. distr. and westw.), Shikoku, Kyushu.——Korea, Ryukyus, Formosa, and China.

6. EURYA Thunb.   HI-SAKAKI ZOKU

Evergreen dioecious shrubs, usually with serrate leaves; flowers small, unisexual, sessile or short-pedicellate, solitary or in axillary fascicles, the bracteoles persistent; sepals 5, imbricate; petals 5, imbricate, connate at base; stamens rather many, rarely 5, adnate to the base of petals, glabrous, the anthers basifixed; styles 3, sometimes 2 or up to 5, nearly free to connate; ovules many in the center of each locule; fruit a berry, the seeds with fleshy endosperm.——About 100 species, chiefly in s. and e. Asia, a few in Central America.

1A. Branchlets glabrous or thinly ascending-pilose; leaves elliptic to oblong-lanceolate, acute with an obtuse tip or very obtuse at apex. .... 1. *E. japonica*
1B. Branchlets densely pilose; leaves narrowly obovate, rounded, often emarginate at apex. .... 2. *E. emarginata*

1. Eurya japonica Thunb.　*E. japonica* var. *uniflora* Bl.; *E. japonica* var. *hortensis* Bl.; *E. japonica* var. *montana* Bl.; *E. japonica* var. *pusilla* Bl.; *E. japonica* var. *multiflora* Miq.——HI-SAKAKI.　Evergreen shrub or small tree with glabrous or thinly ascending-pilose branches; leaves coriaceous, elliptic to oblong-lanceolate, 3–8 cm. long, 1–3 cm. wide, acute with an obtuse tip, acute at base, with erect obtuse teeth, glabrous, deep green and lustrous, pale green or yellowish green beneath, the midrib impressed on upper side, raised beneath, the petioles short; flowers 5–6 mm. across, glabrous, greenish yellow, short-pedicellate; berry globose, purplish black, 4–5 mm. across; seeds about 1.5 mm. long.——Mar.–Apr.　Thickets and thin woods in low mountains; Honshu, Shikoku, Kyushu; very common and variable.——Ryukyus, Formosa, s. Korea, China, India, and Malaysia.

Var. **australis** Hatus.　*E. pubicalyx* Ohwi.　KE-HI-SAKAKI. Branchlets, back of sepals, and pedicels pilose.——Honshu (Tō-kaidō Distr.), Kyushu (Yakushima).

Var. **yakushimensis** Makino.　*E. yakushimensis* (Makino) Makino——HIME-HI-SAKAKI.　Leaves elliptic-lanceolate to broadly lanceolate, 1–3 cm. long, 7–12 mm. wide.——Kyushu (Yakushima).

2. Eurya emarginata (Thunb.) Makino.　*Ilex emarginata* Thunb.——HAMA-HI-SAKAKI.　Densely branched shrub; branchlets densely yellowish brown ascending-pilose; leaves thick-coriaceous, narrowly obovate, 2–3.5 cm. long, 1–1.2 cm. wide, rounded to very obtuse, often emarginate, cuneate at base, glabrous, slightly recurved on margin, obtusely undulate-toothed, deep green and with impressed nerves on upper side, pale green and usually the midrib raised beneath, the petioles very short, sometimes obsolete; flowers axillary, the pedicels 1–3 in a group, 2–3 mm. long; sepals 5, depressed-rounded, 1.5–2 mm. long, glabrous, with a hyaline margin; petals pale yellow-green; berry globose, about 5 mm. across, purple-black, glabrous.——Dec.–Apr.　Near seashores; Honshu (Mikawa Prov. westw.), Shikoku, Kyushu.——Amami-oshima.

Fam. 137. GUTTIFERAE　OTOGIRI-SŌ KA　Garcinia Family

Trees, shrubs, or herbs, sometimes dioecious, with pellucid or black dots; leaves opposite or verticillate, rarely alternate, simple, the stipules absent; flowers bisexual, polygamous or unisexual, solitary or in cymes, actinomorphic; sepals 4 or 5, imbricate; petals 4 or 5, imbricate or contorted; stamens usually many, rarely few, connate in fascicles; ovary superior, 3- to 5-locular or 1-locular; styles as many as the carpels, free or connate; ovules usually many, on parietal placentas; fruit a capsule or drupe, sometimes berrylike; seeds without endosperm.——More than 45 genera, with about 650 species, chiefly in the Tropics, a few genera in temperate regions.

1A.　Stamens 9, in fascicles of 3's, the filaments connate from the base to the middle; glands 3, entire, between the fascicles of stamens; styles free; flowers rose-colored. .... 1. *Triadenum*
1B.　Stamens usually many, in fascicles of 3's or 5's, the filaments connate at base; glands absent; styles free or connate to the top; flowers yellow. .... 2. *Hypericum*

1. TRIADENUM Raf.　MIZU-OTOGIRI ZOKU

Erect glabrous perennials; leaves opposite, sessile or subsessile and clasping, entire, with pellucid dots; racemes short, few-flowered, terminal and axillary, the bracteoles 2, small, subtending the flowers; sepals subequal, with pellucid striations; petals 5, actinomorphic, imbricate or contorted, rose-colored; stamens 9, in fascicles of 3's, the filaments connate from the base to the middle, each fascicle alternate with the glands; glands 3, entire, between the fascicles of stamens; anthers with a pellucid gland at apex; ovary apparently 3-locular, with pellucid striations, the placentas much raised; ovules many, anatropous, horizontal; styles 3, free; capsules dehiscent, 3-valved; seeds small, short-cylindric, with raised reticulations.——Two species, one in e. Asia, the other in N. America.

1. Triadenum japonicum (Bl.) Makino.　*Elodea japonica* Bl.; *E. crassifolia* Bl.; *Hypericum virginicum* auct. Japon., non L.; *E. virginica* var. *asiatica* Maxim.; *H. crassifolium* (Bl.) Nakai——MIZU-OTOGIRI.　Rather soft, glabrous perennial with slender elongate rhizomes; stems erect, rarely branched, terete, 30–70 cm. long; leaves spreading, narrowly oblong or oblong-lanceolate, 4–8 cm. long, 1–2.5 cm. wide, obtuse or rounded at apex, rounded and sessile at base; inflorescence short-pedunculate, much shorter than the leaves, 1- to 3-flowered, the pedicels very short; flowers about 1 cm. across, rose-colored; sepals oblong, 3–4 mm. long; petals narrowly obovate, 6–7 mm. long; capsules ovoid, 8–10 mm. long, acute, with pellucid striations.——Aug.–Sept.　Wet places; Hokkaido, Honshu, Shikoku, Kyushu.——Korea, Manchuria, Ussuri, and Amur.

2. HYPERICUM L.　OTOGIRI-SŌ ZOKU

Herbs or rarely shrubs, leaves opposite, usually sessile, membranous, entire, usually with pellucid and black dots; flowers yellow, solitary or in panicles or cymes; sepals usually 5, as many as the petals; stamens usually many, in fascicles of 3's or 5's, the filaments usually connate at base, sometimes free; ovary 1-locular, with 3–5 parietal placentas, sometimes incompletely 3- to 5-locular; styles free or connate; ovules many, rarely few or 2 in each carpel; capsules septicidally dehiscent; seeds wingless.—— About 300 species, in temperate and subtropical regions of the N. Hemisphere, few in the S. Hemisphere.

1A.　Fascicles of stamens in 5's; flowers 4–6 cm. across. .... 1. *H. ascyron*
1B.　Fascicles of stamens in 3's; flowers 5–25 mm. across.
　2A.　Stems 4-ridged; leaves small, 3–10 mm. long; mostly small annuals with flowers 5–8 mm. across; stamens few, the anthers without a gland at tip; capsules without glands.
　　3A.　Stems 20–50 cm. long, erect; leaves ovate to broadly so, ascending except the lower ones; upper leaves lanceolate to linear, often bractlike; sepals somewhat truncate at apex. .... 2. *H. japonicum*

1. Hypericum ascyron L. *Roscyna japonica* Bl.——To-MOE-sō. Glabrous perennial; stems erect, branched, 4-angled, 50–80 cm. long; leaves lanceolate, 5–10 cm. long, 1–2 cm. wide, acute, sessile and slightly clasping, or abruptly narrowed and sometimes subsagittate at base, entire, pellucid-dotted, penninerved; flowers few, terminal, yellow to reddish, 4–6 cm. across; sepals ovate, about 1 cm. long, obtuse to acute, many-nerved; petals broadly falcate-ovate, 2.5–3.5 cm. long, without black dots; anthers without black dots; styles including the stigmas 6–8 mm. long, connate on lower half, the stigma punctate; capsules ovoid, 12–18 mm. long; seeds minutely reticulate, slightly more than 1 mm. long, with a raised line on one side.——July–Aug. Grassy places in mountains; Hokkaido, Honshu.——e. Siberia, Korea, China, and N. America.

Var. **longistylum** Maxim. *H. sagittifolium* Koidz.——Kōrai-tomoe-sō. Styles about 1 cm. long, 5-lobed on upper one-third.——Mountains of Kyushu.——Altai, Dahuria, Manchuria, Mongolia, and Korea.

2. Hypericum japonicum Thunb. *H. mutilum* sensu auct. Japon., non L.; *Sarothra japonica* (Thunb.) Y. Kimura ——Hime-otogiri. Glabrous annual or perennial; stems 20–50 cm. long, branched in upper part, 4-angled; leaves ascending to somewhat spreading, ovate to broadly so, 5–12(–15) mm. long, 3–8 mm. wide, subobtuse, semiclasping at base, with minute pellucid dots, those of the inflorescence narrow and bractlike; flowers 6–8 mm. across, the pedicels usually longer than the flowers; sepals 3–5 mm. long, broadly lanceolate; petals shorter than the sepals, few-nerved, without dots; styles 0.5–1 mm. long; stigmas capitellate; capsules ovoid, shorter than the sepals, without glands; seeds oblong, about 0.3 mm. long, with minute reticulations.——June–Aug. Wet places in lowlands; Honshu (Kinki Distr. and westw.), Shikoku, Kyushu.——Ryukyus, Formosa, s. Korea, China, India, and Australia.

3. Hypericum laxum (Bl.) Koidz. *H. yabei* Lév. & Van't.; *H. dominii* Lév.; *H. thunbergii* Fr. & Sav.; *H. japonicum* forma *yabei* (Lév. & Van't.) Makino; *Sarothra laxa* (Bl.) Y. Kimura; *Brathys laxa* Bl.——Koke-otogiri. Delicate annual or perennial; stems 5–20(–35) cm. long, much branched in upper part, 4-angled; leaves elliptic or ovate, 5–10 mm. long, 3–8 mm. wide, rounded at apex, rounded to shallowly cordate at base, sessile, semiclasping, spreading, the uppermost ones narrowly ovate, leaflike; flowers in cymes, 5–7 mm. across; sepals oblong, 3–4 mm. long, obtuse; petals 2.5–3 mm. long, 3-nerved, without dots; stamens 8–10; styles about 0.5 mm. long; capsules ovoid, 3 mm. long; seeds about 0.3 mm. long.——June–Sept. Wet places in lowlands; Hokkaido (sw. distr.), Honshu, Shikoku, Kyushu; common.——Korea and Ryukyus.

4. Hypericum sampsonii Hance. *H. electrocarpum* Maxim.——Tsuki-nuki-otogiri. Glabrous, erect perennial; stems terete, often branched in upper part, 50–80 cm. long; leaves membranous, narrowly ovate-oblong, 5–8 cm. long, 15–35 mm. wide, broadest at the base, obtuse, the leaf pairs connate at base, spreading, with pellucid and black dots; flowers few, terminal, 7–8 mm. across, yellow, the bracts linear, 1–2.5 mm. long; sepals oblong, 3–6 mm. long, obtuse; petals slightly longer than the sepals, broadly obovate, 5–6.5 mm. long; stamens many, slightly shorter than the petals, with a black dot at apex of the anther; styles about 1.5 mm. long, the stigmas capitellate; capsules ovoid, about 6 mm. long, with raised pellucid linear glands; seeds about 1 mm. long, with few longitudinal striations and minute cross-lines.——July. Honshu, Shikoku, Kyushu; rare.——From Formosa and China to India.

5. Hypericum pseudopetiolatum Keller. *H. kiusianum* Koidz.; *H. muraianum* Makino; *H. kosiense* Koidz.; *H. pentholodes* Koidz.——Sawa-otogiri. Loosely tufted perennial; stems 10–40 cm. long, terete, slender, often branched, usually ascending from base; leaves thin, rather soft, narrowly oblong to oblanceolate, 2–4 cm. long, 6–12 mm. wide, obtuse, narrowed to a petiolelike base, with pellucid and/or black dots, the upper leaves smaller; flowers about 10 mm. across; sepals narrowly oblong, 3.5–5.5 mm. long, obtuse, slightly unequal, often with raised black dots; petals 4–6 mm. long; styles 1–2 mm. long; capsules ovoid-ellipsoid or ovoid, 5–6 mm. long; seeds about 0.8 mm. long, minutely reticulate. ——July–Aug. Wet places in mountains; Hokkaido (sw. part), Honshu, Shikoku, Kyushu; rather common.

6. **Hypericum oliganthum** Fr. & Sav. *H. erectum* var. *obtusifolium* Bl.; *H. obtusifolium* (Bl.) Makino; *H. makinoi* Lév.; *H. conjunctum* Y. Kimura——Aze-otogiri. Glabrous perennial; stems terete, slender, somewhat tufted, often branched, ascending to creeping at base; leaves thin, narrowly oblong to ovate-oblong, 1.5–3.5 cm. long, 8–15 mm. wide, obtuse at both ends, with pellucid and/or black dots, sometimes black-dotted only on margin; inflorescence loosely few-flowered, the bracts small, leaflike; flowers 10–12 mm. across; sepals narrowly oblong, 5–8 mm. long, obtuse; petals 7–10 mm. long; styles 1–2 mm. long; capsules ovoid, 6–7 mm. long; seeds about 0.8 mm. long, minutely reticulate.——Wet places in lowlands; Honshu (Kantō Distr. and westw.), Shikoku, Kyushu; rather common.——s. Korea.

7. **Hypericum kinashianum** Koidz. *H. umbrosum* Y. Kimura; *H. conjunctum* var. *longistylum* Y. Kimura——Mi-yako-otogiri. Glabrous perennial; stems erect, terete, often branched above, 20–60 cm. high; leaves rather thin, broadly linear, lanceolate, or linear-oblong, 2–5 cm. long, 3–8 mm. wide, obtuse or subobtuse at both ends, black-dotted, the bracts linear, 2–4 mm. long; flowers 8–12 mm. across; sepals linear-oblong, 3–4.5 mm. long, obtuse; black-dotted and -striated; petals 4–7 mm. long, longer than the sepals; styles 2–3 mm. long; capsules ovoid, 5–6 mm. long; seeds minutely reticulate, about 0.8 mm. long.——Aug. Mountains; Honshu, Shikoku, Kyushu.

8. **Hypericum hakonense** Fr. & Sav. Ko-otogiri. Slender glabrous perennial; stems tufted, 15–30 cm. long, terete, ascending from base, often branched; leaves thin, linear-oblong or narrowly oblanceolate, 15–25 mm. long, 2–5 mm. wide, obtuse, narrowed to a petiolelike base, usually only with pellucid dots; inflorescence loosely about 10-flowered; flowers about 15 mm. across, the bracts linear to broadly so; sepals lanceolate or linear-oblong, about 3.5 mm. long, rather unequal, acute to obtuse; petals 7–8 mm. long; styles 2–3 mm. long; capsules ovoid-conical, 5–6 mm. long; seeds about 0.8 mm. long, minutely reticulate.——Aug.–Sept. Mountains; Honshu (centr. and Kantō Distr.).

9. **Hypericum sikokumontanum** Makino. Takane-otogiri. Loosely tufted, glabrous perennial, rather wide-spreading at the decumbent base; stems 10–25 cm. long, terete, sometimes branched; leaves linear-oblong, about 15 mm. long, 2–4 mm. wide, obtuse or subobtuse, obtuse to acute and semiclasping at base, sometimes with pellucid dots and with scattered black dots on the often slightly recurved margin; flowers few, 18–20 mm. across; sepals lanceolate, 5–7 mm. long, obtuse; petals 10–12 mm. long; styles slender, 3.5–5 mm. long; capsules about 6 mm. long; seeds about 0.8 mm. long, minutely reticulate.——July–Aug. High mountains; Shikoku.

Var. **hyugamontanum** (Y. Kimura) Ohwi. *H. hyuga-montanum* Y. Kimura——Kumoi-otogiri. Flowers slightly smaller; petals 8–10 mm. long; sepals 3–5 mm. long, often with protruding black dots on margin.——High mountains; Kyushu.

10. **Hypericum perforatum** L. var. **angustifolium** DC. *H. foliosissimum* Koidz.——Kogome-otogiri. Glabrous erect perennial; stems 30–80 cm. long, obsoletely striate, much-branched, often with short, leafy shoots in the leaf axils; leaves linear to linear-oblong, 1–2.5 cm. long, 1.5–3 mm. wide, obtuse at both ends, pellucid-dotted, often recurved on margin; flowers few, 15–18 mm. across; sepals lanceolate, 3.5–4.5 mm. long, acute; petals 10–12 mm. long; ovary with linear and elliptic glands; styles 4–5 mm. long, the stigma punctate.

——June–July. Naturalized; Honshu (near Tokyo and in Shima Prov.).——Europe.

11. **Hypericum yezoense** Maxim. *H. procumbens* Keller; *H. mororanense* Keller; *H. oliganthemum* Keller; *H. porphyrandrum* Lév. & Van't.——Ezo-otogiri. Glabrous perennial; stems somewhat tufted, usually 10–30 cm. long, erect, simple or branched, with 2 raised lines; leaves oblong to ovate-oblong, 8–25 mm. long, 3–8 mm. wide, obtuse, semiclasping at base, densely pellucid-dotted, the margin slightly recurved and black-dotted; flowers few, 15–20 mm. across, the bracts leaflike, ovate or oblong-ovate; sepals lanceolate, 4–6 mm. long, acute; petals 8–11 mm. long; styles 4–6 mm. long; capsules broadly ovoid, 5–8 mm. long; seeds about 1 mm. long.——July–Aug. Hokkaido, Honshu (northern distr.); rather rare.——s. Kuriles and Sakhalin.

Var. **momoseanum** Makino. Sei-taka-otogiri-sō. Stems 25–65 cm. long, solitary or few together; sepals abruptly acuminate.——Honshu (centr. distr.).

12. **Hypericum tosaense** Makino. Tosa-otogiri. Glabrous perennial; stems erect, 25–50 cm. long, with 2 raised lines; leaves narrowly oblong, obovate or ovate-oblong, 5–20 mm. long, 2–8 mm. wide, obtuse at both ends, densely pellucid-dotted, the margins slightly recurved and black-dotted; flowers few, 15–18 mm. across, the bracts small, linear or linear-lanceolate; sepals ovate, 2.5–3 mm. long, acute; petals about 1 cm. long; styles 3.5–4.5 mm. long, the stigma punctate; capsules ovoid-conical, about 7 mm. long; seeds about 1 mm. long, minutely reticulate.——Aug.–Sept. Shikoku.

13. **Hypericum erectum** Thunb. *H. erectum* var. *wichurae* Keller and var. *debile* Keller; *H. otaruense* Keller; *H. dielsii* Lév. & Van't.; *H. erectum* subsp. *vaniotii* Lév.; *H. matsumurae* Lév.; *H. amabile* Koidz.; *H. takeuchianum* Koidz.——Otogiri-sō. Glabrous perennial; stems solitary or few together, 20–60 cm. long, simple or slightly branched, terete; leaves broadly lanceolate to narrowly ovate, sometimes ovate-elliptic, 2–6 cm. long, 7–30 mm. wide, obtuse or subacute, rounded and semiclasping at base, black-dotted; inflorescence subpaniculately few- to rather many-flowered, the bracts leaflike, small; flowers 15–20 mm. across; sepals narrowly oblong, acute to suboutuse; petals 8–10 mm. long; style 3.5–4 mm. long; capsules 5–11 mm. long; seeds minutely reticulate, about 1 mm. long.——July–Aug. Grassy places and thin woods in hills and mountains; Hokkaido, Honshu, Shikoku, Kyushu; variable and very common.

14. **Hypericum kamtschaticum** Ledeb. *H. mutiloides* Keller; *H. paradoxum* Keller; *H. paramushirense* Kudo——Iwa-otogiri. Glabrous perennial; stems tufted, 10–30 cm. long, erect or ascending from base, usually simple, terete; leaves elliptic, oblong, or ovate-oblong, 2–5 cm. long, 1–3 cm. wide, obtuse to rounded at both ends, semiclasping at base, with black or rarely with pellucid dots, sometimes without dots, margins usually black-dotted, the bracts leaflike, ovate; flowers 15–25 mm. across; sepals oblong or lanceolate, obtuse or subacute, 6–8 mm. long; petals 8–15 mm. long; styles 4.5–6 mm. long; capsules 6–12 mm. long; seeds minutely reticulate, about 1 mm. long.——July–Aug. Alpine regions; Hokkaido, Honshu (n. and centr. distr.).——Kamchatka and Kuriles.

Var. **senanense** (Maxim.) Y. Kimura. *H. senanense* Maxim.——Shinano-otogiri. Leaves black-dotted only on margin, with pellucid dots on the surface; flowers 15–18 mm. rarely to 20 mm. across.——July–Aug. Alpine regions; Honshu (centr. and n. distr.).

Fam. 138. ELATINACEAE Mizo-hakobe Ka Waterwort Family

Herbs or subshrubs; leaves simple, opposite or verticillate, stipulate; flowers small, bisexual, actinomorphic, axillary, solitary or in cymes; sepals 3–5, free, imbricate; petals 3–5, persistent, imbricate; stamens as many or twice as many as the petals, free; ovary superior, 3- to 5-locular, the placenta axile; styles 3–5, free; ovules many; capsules septicidally dehiscent; seeds without endosperm.——Two genera, with about forty species, in tropical and temperate regions.

1. ELATINE L. Mizo-hakobe Zoku

Delicate herbs growing in muddy places or aquatic, glabrous; leaves opposite or verticillate, small; flowers minute, usually axillary, solitary, 3- or 4-, sometimes 5- or 2-merous; sepals membranous, without ribs, obtuse, connate at base; petals obtuse; ovary globose; capsules globose, membranous, the septa winglike; seeds narrow, often curved, minutely reticulate.——About 20 species, in temperate and tropical regions.

1. Elatine triandra Schk. *E. orientalis* Makino; *E. triandra* var. *orientalis* (Makino) Makino——Mizo-hakobe. Small soft annual; stems terete, decumbent, branched, 3–10 cm. long; leaves broadly lanceolate to narrowly ovate, 5–10 mm. long, 2–3 mm. wide, obtuse, narrowed to a short petiole-like base, lateral nerves delicate, in 2 or 3 pairs; flowers solitary, nearly sessile, about 1 mm. across, rose-colored; sepals 3, ovate, very obtuse, connate at base, 1 of them often smaller than the others; petals 3, elliptic, very obtuse, slightly longer than the sepals; stamens 3; styles 3, free, short, erect; capsules membranous, depressed-globose, about 2 mm. across; seeds many, oblong-cylindric, about 0.5 mm. long, slightly curved, rounded at both ends, with minute hexagonal reticulations.——June–Aug. Wet places and paddy fields in lowlands; Honshu, Shikoku, Kyushu.——The N. Hemisphere and S. America.

Fam. 139. VIOLACEAE Sumire Ka Violet Family

Herbs, rarely shrubs; leaves alternate, rarely opposite, simple, the stipules small or leaflike; flowers solitary or in panicles, zygomorphic or actinomorphic, bisexual, rarely polygamous, sometimes cleistogamous; sepals 5, persistent, imbricate, often produced at base as an appendage; petals 5, imbricate or contorted, usually unequal, the lower ones often largest and spurred; stamens 5, the anthers erect, introrse, the lower 2 often spurred; ovary superior, sessile, 1-locular, with 3 (or 2, 4, or 5) parietal placentae; style connate or rarely divided; ovules usually many, anatropous; fruit an abruptly dehiscent capsule or berry; seeds often appendaged, the endosperm fleshy.——About 16 genera, with about 800 species, in temperate and tropical regions.

1. VIOLA L. Sumire Zoku

Herbs or rarely shrubs; stipules persistent, often leaflike; leaves basal or alternate; peduncles axillary, usually solitary; flowers zygomorphic, some of them often cleistogamous; sepals nearly equal; petals spreading, the lower one spurred at base; styles solitary, simple, variously dilated or curved; stigma terminal or placed in front; capsules abruptly dehiscent, 3-valvate; seeds ovoid or ovoid-globose, the testa crustaceous, usually lustrous.——More than 400 species, chiefly in temperate regions, mostly in the N. Hemisphere, few in the S. Hemisphere.

1A. Flowers violet, purple or white, never yellow.
  2A. Plants stemless.
    3A. Leaves compound or deeply incised. .... 1. *V. dissecta*
    3B. Leaves simple, serrate.
      4A. Capsules and ovaries pilose; leaves rounded-cordate to -reniform.
        5A. Sepals obtuse; spur saccate and short; rhizomes thick, densely noded; leaves green on upper surface, the petioles and peduncles with spreading white hairs.
          6A. Stolons absent or very short; leaves ovate-cordate, subacute with an obtuse tip. .... 2. *V. collina*
          6B. Stolons well developed; leaves cordate-orbicular, usually rounded at the tip. .... 3. *V. hondoensis*
        5B. Sepals subacute; spur long; rhizomes slender and short; leaves white-variegated on upper surface, the petioles and peduncles short-pubescent. .... 21. *V. variegata* var. *nipponica*
      4B. Capsules and ovaries glabrous, or if puberulous, the leaves much longer than wide.
        7A. Rhizomes thick, densely noded; stolons absent.
          8A. Rhizomes thick and short, erect; bracts 8–13 mm. long; flowers white; capsules ellipsoidal. .... 4. *V. keiskei*
          8B. Rhizomes thick and rather long-creeping; bracts 3–8 mm. long.
            9A. Leaves depressed-cordate or orbicular-cordate, short-obtuse; spur 5–7 mm. long; capsules produced exclusively from the cleistogamous flowers. .... 40. *V. mirabilis* var. *subglabra*
            9B. Leaves narrowly deltoid-ovate or deltoid-cordate, acute to acuminate; flowers before the leaves, the spur saccate, 2.5–5 mm. long.
              10A. Petioles usually persistent on rhizomes over winter; flowers rather small, 8–13 mm. long. .... 5. *V. yazawana*
              10B. Petioles usually not persistent on rhizomes over winter; flowers larger, 15–20 mm. long.
                11A. Stipules pale green, 5–10 mm. long; adult leaves cordate; flowers pale rose-purple. .... 6. *V. rossii*
                11B. Stipules dark brown, 10–18 mm. long; flowers pale purple.
                  12A. Adult leaves cordate-rounded. .... 7. *V. vaginata*
                  12B. Adult leaves broadly deltoid-lanceolate to -ovate. .... 8. *V. bissetii*
        7B. Rhizomes slender or almost absent, sometimes rather thick and short; stolons often present.

13A.   Leaves truncate to shallowly cordate at base; petioles rather broad, often winged.
   14A.   Petioles distinctly winged on upper portion.
      15A.   Spur 3–4 mm. long, terete; plants of wet places, with white roots; petioles usually longer than the blades; flowers white with purple striations. . . . . 9. *V. patrinii*
      15B.   Spur 5–7 mm. long; plants of grassy fields, with orange-brown roots; petioles as long as to shorter than the blades; flowers dark-purple, or rarely white with purple striations. . . . . 10. *V. mandshurica*
   14B.   Petioles very narrowly winged or wingless.
      16A.   Flowers rose-purple.
         17A.   Flowers deep rose-purple; petals 10–13 mm. long; leaves rather thick, 1–3 cm. long, 0.8–2.5 cm. wide.
                                                         13. *V. phalacrocarpa*
         17B.   Flowers pale rose-purple; petals 15–20 mm. long; leaves thin, 3–6 cm. long, 2–4 cm. wide. . . . . . . 16. *V. hirtipes*
      16B.   Flowers purple.
         18A.   Leaves rather thin, ovate to broadly so; roots slender; flowers pale purple to violet. . . . . 14. *V. japonica*
         18B.   Leaves rather thick, lanceolate, deltoid-lanceolate to -ovate; roots rather thick.
            19A.   Flowers purple; petals 10–14 mm. long, the lateral ones glabrous; leaves 3–6(–8) cm. long.   11. *V. yedoensis*
            19B.   Flowers deep purple, small; petals 8–10 mm. long, the lateral ones bearded inside; leaves 1.5–4 cm. long.
                                                         12. *V. minor*
13B.   Leaves deeply cordate at base; petioles slender and wingless.
   20A.   Spur saccate, very short; flowers white with purple striations on the lip; petals often of different lengths.
      21A.   Rhizomes slender, long-creeping, distinctly noded.
         22A.   Leaves orbicular-reniform, rounded and sometimes with an abrupt short tip; rhizomes with persistent petiole bases of preceding year. . . . . 17. *V. blandaeformis*
         22B.   Leaves broadly ovate-orbicular to ovate, short-acuminate; rhizomes without persistent petiole bases of preceding year.
                                                     22. *V. shikokiana*
      21B.   Rhizomes short.
         23A.   Leaves ovate-elliptic to ovate; sepals recurved toward tip; plant often white-pubescent. .. 20. *V. maximowicziana*
         23B.   Leaves ovate, deltoid, or broadly deltoid-lanceolate; sepals erect; plant glabrous or puberulous.
            24A.   Leaves ovate to ovate-cordate, depressed-toothed. . . . . 18. *V. pumilio*
            24B.   Leaves deltoid-cordate to narrowly ovate, usually broadly deltoid-ovate, the teeth slightly dilated at tip.
                                                    19. *V. boissieuana*
   20B.   Spur elongate, cylindric; flowers purple to white; petals nearly all equal in length.
      25A.   Leaves at flowering time ovate or reniform-cordate, nearly as long as wide.
         26A.   Leaves broadly ovate-elliptic to cordate-orbicular, obtuse to rounded. . . . . 21. *V. variegata* var. *nipponica*
         26B.   Leaves broadly ovate to rounded-cordate, acute.
            27A.   Flowers pale purple; leaves rather thick, broadly ovate, purplish beneath, the teeth short, undulate, with a notch at each end. . . . . 24. *V. tokubuchiana*
            27B.   Flowers pale violet to whitish; leaves thin, ovate to ovate-cordate, sometimes purplish beneath, the outer margin of teeth flat on lower half, semirounded on upper half.
               28A.   Leaves ovate; flowers white with purple striations. . . . . 15. *V. yezoensis*
               28B.   Leaves ovate-cordate; flowers pale violet. . . . . 23. *V. selkirkii*
      25B.   Leaves at flowering time broadly lanceolate to ovate, usually distinctly longer than wide.
         29A.   Leaves deltoid-ovate to narrowly so, usually sparingly pilose, especially those of summer, the teeth undulate, semi-rounded on upper half; flowers rose-purple. . . . . 25. *V. takedana*
         29B.   Leaves narrowly deltoid-ovate to broadly lanceolate, glabrous or sparsely short-hairy on upper surface, the teeth flat, indistinctly notched at both ends.
            30A.   Flowers deep rose-purple; leaves narrowly deltoid-ovate. . . . . 26. *V. violacea*
            30B.   Flowers rose-purple; leaves broadly deltoid-lanceolate. . . . . 27. *V. makinoi*
2B.   Plants with distinct aerial stems or leafy stolons.
   31A.   Radical leaves absent at anthesis; stems erect.
      32A.   Leaves much longer than wide; stipules toothed.
         33A.   Leaves deltoid-lanceolate, broadest at the truncate base. . . . . 28. *V. raddeana*
         33B.   Leaves broadly lanceolate, broadest below the middle, cuneate at base. 29. *V. thibaudieri*
      32B.   Leaves ovate to ovate-cordate; stipules pinnately incised. . . . . 30. *V. acuminata*
   31B.   Radical leaves present at anthesis; stems usually ascending or decumbent at base.
      34A.   Leaves ovate, broadly cuneate to subtruncate at base, broadly decurrent on the petiole; stoloniferous; coarsely pilose or glabrescent. . . . . 31. *V. diffusa* var. *glabella*
      34B.   Leaves cordate, reniform-cordate, or ovate-cordate.
         35A.   Capsules subglobose, pilose. . . . . 3. *V. hondoensis*
         35B.   Capsules ellipsoidal or oblong, glabrous.
            36A.   Stipules pinnately lobed to distinctly incised; leaves brown-spotted when dry.
               37A.   Flowers usually cauline only; spur rather thick and slightly flattened.
                   38A.   Stipules divided for less than half their length; flowers pale violet, the lateral petals bearded inside. 33. *V. sacchalinensis*
                   38B.   Stipules divided for more than half their length; petals glabrous.
                      39A.   Plants to 40 cm. high, in woods or in wet places; rhizomes herbaceous, creeping; flowers pale violet.
                                                     34. *V. kusanoana*
                    39B.   Plants 5–10 cm. high, in dunes along seacoast; rhizomes woody, erect to ascending; flowers deep violet.
                                                    35. *V. senamiensis*
              37B.   Flowers radical and cauline; spur rather slender and long.

40A. Spur 10–15 mm. long, as long as or longer than the lip. . . . . 32. *V. rostrata* var. *japonica*
40B. Spur 4–5 mm. long, straight, much shorter than the lip.
  41A. Flowers pale violet, not fragrant; peduncles usually glabrous.
    42A. Rhizomes woody, elongate; leaves orbicular-cordate to deltoid-cordate, thick, lustrous above, with raised veinlets above when dry. . . . . 36. *V. faurieana*
    42B. Rhizomes soft and very short, whitish; leaves herbaceous.
      43A. Leaves cordate to reniform. . . . . 37. *V. grypoceras*
      43B. Leaves cordate to reniform in the radical, ovate to narrowly deltoid-ovate in the upper cauline.
          39. *V. ovato-oblonga*
  41B. Flowers purple, with a white center, fragrant; peduncles usually puberulous. 38. *V. obtusa*
36B. Stipules entire or with few indistinct teeth; leaves green when dry.
  44A. Leaves 4–7 cm. wide; stems suberect; rhizomes creeping, thick, densely noded; flowers violet, 15–20 mm. across.
    45A. Chasmogamic flowers always radical and usually sterile; fertile cleistogamous flowers cauline.
          40. *V. mirabilis* var. *subglabra*
    45B. Chasmogamic and cleistogamic flowers cauline. . . . . 41. *V. langsdorffii*
  44B. Leaves smaller, 2–3.5 cm. wide; stems decumbent or creeping; rhizomes very short or absent; flowers white with purple striations on the lip, 6–8 mm. across. . . . . 42. *V. verecunda*
1B. Flowers yellow.
  46A. Petals glabrous; leaves reniform or reniform-cordate, leaf-blades 1–2 cm. long.
    47A. Leaves thin, soft, usually sparingly pilose at least in part; wet or moist places in high mountains. . . . . 43. *V. biflora*
    47B. Leaves thick, firm, usually quite glabrous; gravelly places in alpine regions. . . . . 44. *V. crassa*
  46B. Lateral petals bearded inside; leaves ovate-cordate to depressed-cordate, leaf-blades 2.5–8 cm. long.
    48A. Leaves reniform, 2–3 cm. long, rounded to broadly retuse, incised-toothed; styles glabrous. . . . . 45. *V. alliariaefolia*
    48B. Leaves cordate or ovate, 3–8 cm. long, with a more or less extended apex; styles bearded on each side at apex.
      49A. Rhizomes creeping, often elongate; appendage of sepals semirounded.
        50A. Stems terete, glabrous; stem-leaves cordate to narrowly ovate, acuminate. . . . . 46. *V. brevistipulata*
        50B. Stems 4-angular, densely puberulous; stem-leaves thicker in texture, cordate with an obtuse or acute apex. . . 47. *V. yubariana*
      49B. Rhizomes short, usually erect; appendage of sepals ovate. . . . . 48. *V. orientalis*

1. Viola dissecta Ledeb. var. **sieboldiana** (Maxim.) Nakai. *V. pinnata* L. var. *sieboldiana* Maxim.; *V. sieboldiana* (Maxim.) Makino; *V. chaerophylloides* var. *sieboldiana* (Maxim.) Makino; *V. albida* var. *chaerophylloides* forma *sieboldiana* (Maxim.) F. Maekawa; *V. chaerophylloides* sensu Ohwi, pro parte, non W. Becker.——HIGO-SUMIRE. Rhizomes short; leaf-blades green, 1.5–2.5 cm. long and as wide (vernal), to 3 cm. long, 3–4 cm. wide (summer), palmately 5-nate, the lobes linear with few incisions, pinnately parted or divided, glabrous or short-pilose on upper surface and on margin, rarely also on nerves beneath, the petioles 4–12 cm. long, 2 or 3 times as long as the blades, the stipules broadly linear, adnate to the petiole; peduncles 5—12 cm. long; flowers slightly fragrant, white with purple striations; sepals lanceolate, 5–8 mm. long, subobtuse, the appendage ovate, or 4-angled and bifid, acutely few-toothed; petals 10–12 mm. long, the lateral ones slightly bearded inside at base, the spur short-cylindric, 4–5 mm. long; capsules ellipsoidal, glabrous, 6–10 mm. long.——Apr.–May. Mountains; Honshu, Shikoku, Kyushu; rather common.

Var. **chaerophylloides** (Regel) Makino forma **eizanensis** (Makino) E. Itô. *V. dissecta* var. *eizanensis* Makino; *V. chaerophylloides* var. *eizanensis* (Makino) Ohwi——EIZAN-SUMIRE. Vernal leaves pedately dissected, the summer leaves ternate, the segments broadly lanceolate, to 10 cm. long, 4 cm. wide, incised, flowers purplish.——Honshu, Shikoku, Kyushu.—— Forma **simplicifolia** (Makino) Ohwi. *V. chaerophylloides* forma *simplicifolia* Makino; *V. eizanensis* var. *simplicifolia* (Makino) Makino; *V. eizanensis* forma *simplicifolia* (Makino) F. Maekawa——HITOTSUBE-EZO-SUMIRE. Leaves simple. Sometimes occurs with the dissected-leaved phase.——Forma **chaerophylloides.** *V. pinnata* L. var. *chaerophylloides* Regel; *V. chaerophylloides* (Regel) W. Becker; *V. sieboldiana* var. *chaerophylloides* (Regel) Nakai; *V. albida* Palib. var. *chaerophylloides* (Regel) F. Maekawa——NANZAN-SUMIRE. Rather small; leaves finely dissected.——Honshu, Shikoku,

Kyushu.——Korea, Manchuria. The typical phase occurs in e. Siberia.

2. Viola collina Bess. *V. hirta* var. *collina* (Bess.) Regel; *V. teshioensis* Miyabe & Tatew.——MARUBA-KE-SUMIRE. Densely spreading-pilose on leaves, petioles and peduncles; rhizomes thick, creeping, densely noded; stolons nearly absent; leaf-blades ovate-cordate or subcordate, 2–3.5 cm. long, 2–3 cm. wide (vernal), to 6 cm. long and as wide in summer, subacute with an obtuse tip, short-toothed, the petioles at anthesis 3–10 cm. long, at fruiting time to 20 cm. long, sometimes slightly wing-margined; peduncles 4–6 cm. long, much shorter than the petioles in fruit, loosely pilose; flowers pale purple; sepals narrowly oblong or narrowly ovate, 5–6 mm. long, obtuse, ciliolate, appendages semirounded, short; petals 10–12 mm. long, the lateral ones sparingly bearded, the spur 3–4 mm. long, ellipsoidal; capsules subglobose, 6–8 mm. long, densely pubescent.——Apr.–May. Dry woods in mountains; Hokkaido, Honshu (n. & c. distr.).——Korea, Sakhalin, Manchuria, Siberia, e. Europe, and Caucasus.

3. Viola hondoensis W. Becker & H. Boiss. *V. nipponica* Maxim., nom. prov.; *V. hirta* var. *nipponica* (Maxim.) Fr. & Sav.——AOI-SUMIRE. Conspicuously spreading white-pubescent; rhizomes thick, short; stolons slender, bearing flowers; leaf-blades cordate-orbicular, 1–3 cm. long, 1.5–4 cm. wide, to 6 cm. long, 8 cm. wide in fruit, rounded or nearly so at the tip, deeply cordate, undulately toothed, the petioles at anthesis 2–8 cm. long, at fruiting time to 25 cm. long; peduncles 4–7 cm. long, from the leaf rosettes and stolons, usually pilose; flowers pale violet or nearly white; sepals narrowly oblong, 5–7 mm. long, obtuse, ciliolate, the appendage deltoid or trapezoid, acute to obtuse; petals 10–13 mm. long, the lateral ones slightly bearded inside, the lip purple-striate, the spur oblong, short, 3–4.5 mm. long; capsules chiefly from the cleistogamous flowers, globose, densely white-pubescent, 5–6 mm. long.——Mar.–Apr. Moist woods; Hokkaido, Honshu, Shikoku, Kyushu.

4. Viola keiskei Miq. *V. keiskei* var. *typica* Makino; *V. okuboi* var. *glabra* Makino; *V. keiskei* var. *glabra* (Makino) W. Becker——MARUBA-SUMIRE.　Glabrous or glabrescent; rhizomes short, rather thick, erect, densely noded; leaf-blades cordate, sometimes ovate-cordate, 2–3 cm. long, 2–4 cm. wide at anthesis, usually rounded or very obtuse, undulately toothed, 5–6 cm. long and as wide at fruiting time, the petioles 2–10 cm. long at anthesis, to 20 cm. long at fruiting time; peduncles glabrous, 5–10 cm. long; flowers white; sepals narrowly oblong, 6–8 mm. long, obtuse to subobtuse, the appendage 4-angled, with few teeth; petals 10–14 mm. long, the lateral ones glabrous or only slightly bearded, the lip purple-striate, the spur 6–7 mm. long; capsules ovoid-ellipsoid, 7–8 mm. long, the valves glabrous, with an obtuse keel.——Apr.–May. Mountains; Honshu, Shikoku, Kyushu; rather common.—— Korea.——Forma **okuboi** (Makino) F. Maekawa. *V. keiskei* var. *okuboi* Makino; *V. okuboi* (Makino) Makino; *V. keiskei* var. *typica* W. Becker——KE-MARUBA-SUMIRE.　Plant pilose. Occurs with the typical phase and equally as common.

5. Viola yazawana Makino.　HIME-SUMIRE-SAISHIN. Sparingly short-pilose on leaves and sometimes on upper part of petioles; rhizomes rather thick, long-creeping, firm, densely noded, with a few persistent petiole bases of previous year; leaves few, appearing after anthesis, blades broadly ovate-cordate, 3–4 cm. long and as wide, acuminate, sometimes with an obtuse tip, deeply cordate at base, with scattered undulate teeth, the petioles slender, 8–12 cm. long, the stipules adnate at base to the petiole, the free portion lanceolate, 3–4 mm. long; peduncles 7–8 cm. long; flowers white; sepals narrowly ovate to broadly lanceolate, 5–8 mm. long, acute, the appendage semirounded, entire; petals 8–13 mm. long, sometimes to 15 mm. long, glabrous, the lip purple-striate, the spur saccate, 2.5–3 mm. long and as wide; capsules 7–8 mm. long, glabrous, purple-spotted.——June. Mountains; Honshu (w. Kantō and e. centr. distr.); rare.

6. Viola rossii Hemsl. *V. matsumurae* Makino——AKE-BONO-SUMIRE.　Sparingly pilose on leaves especially on nerves beneath, rarely glabrescent; rhizomes rather thick, creeping, densely noded; leaves appearing after anthesis, rather thick, blades 4–7 cm. long, 4–8 cm. wide, abruptly acuminate with an obtuse tip, cordate, undulately toothed, the petioles 10–25 cm. long; stipules free, thinly membranous, lanceolate, 7–10 mm. long, pale green; peduncles 10–15 cm. long; flowers pale rose-purple; sepals oblong, 7–8 mm. long, obtuse, the appendage obtusely 4-angled, usually entire; petals 15–20 mm. long, rounded to retuse at apex, the lateral ones glabrous or only slightly bearded, the spur thick and short, saccate, 3–4 mm. long; capsules glabrous, indistinctly brown-spotted, 1–1.5 cm. long.——Apr.–May. Mountains; Honshu, Shikoku, Kyushu. ——Korea, China, and Manchuria.

7. Viola vaginata Maxim. *V. franchetii* H. Boiss.—— SUMIRE-SAISHIN.　Sparingly white-pilose on leaves beneath and rarely on upper part of petioles; rhizomes creeping, thick, densely noded; leaves appearing after anthesis, rather thick, blades 5–14 cm. long and as wide, abruptly acuminate, cordate, undulately toothed, glabrous on upper side, the petioles usually shorter than the peduncles at anthesis, 15–25 cm. long at fruiting time; stipules brown, scarious, free, 12–18 mm. long, lanceolate; peduncles few, 10–15 cm. long; flowers relatively large, pale purple; sepals broadly lanceolate, obtuse, 6–8 mm. long, the appendage ovate or 4-angled, toothed; petals 15–20 mm. long, rounded to emarginate, glabrous, the lip purple-

striate, the spur globose, saccate, 4–5 mm. long; capsules 12–15 mm. long, usually from cleistogamous flowers.——Apr.–June. Mountains; Hokkaido, Honshu (Japan Sea side, as far west as Kinki Distr.). A phase with white flowers is known.

8. Viola bissetii Maxim. *V. vaginata* var. *angustifolia* Yatabe——NAGABA-NO-SUMIRE-SAISHIN.　Glabrous or slightly pilose on underside of leaves especially near base; rhizomes rather thick, creeping, densely noded; leaves appearing after anthesis, few, rather thick, blades narrowly deltoid-ovate to broadly deltoid-lanceolate, 5–10 cm. long, 3–5 cm. rarely up to 8 cm. wide, gradually acuminate to acute, deeply cordate, the petioles at first shorter than the peduncles, to 7–15 cm. long at fruiting time; stipules brownish, free, deltoid-lanceolate, (5–)8–12 mm. long; peduncles 5–12 cm. long; flowers relatively large, pale purple; sepals broadly lanceolate, 6–8 mm. long, obtuse, the appendage 4-angled ovate, toothed; petals 15–18 mm. long, glabrous, the lip purple-striate toward base, the spur globose, saccate, 4–5 mm. long.——Apr. Mountains; Honshu (Kantō Distr. and westw.), Shikoku, Kyushu.—— Forma **albiflora** Nakai has white flowers.

9. Viola patrinii DC. *V. primulifolia* L. pro parte—— SHIROBANA-SUMIRE.　Glabrous or sometimes spreading-pilose on nerves of leaves and lower part of petioles; rhizomes short, with white, divided roots; leaves suberect, blades deltoid-lanceolate or broadly oblong-lanceolate, 2.5–7 cm. long, 1–2 cm. wide, obtuse to subacute, truncate at base, sparsely or obsoletely toothed, the petioles conspicuously winged in upper part, especially in those developed after anthesis, 4–10 cm. long, longer than the blades; peduncles 7–15 cm. long; flowers white, with purplish striations; sepals lanceolate, 4–7 mm. long, acute, the appendage semirounded, sometimes with few obtuse teeth; petals 10–13 mm. long, the lateral ones slightly bearded inside, the lip purple-striate, the spur terete, 3–4 mm. long, ellipsoidal; capsules about 8 mm. long.——Apr.–May. Wet places in lowlands; Hokkaido, Honshu, Shikoku, Kyushu.——Korea, Manchuria, Kuriles, Sakhalin, and Siberia.

10. Viola mandshurica W. Becker. *V. patrinii* var. *chinensis* sensu auct. Japon., non Ging.; *V. patrinii* var. *macrantha* Maxim.——SUMIRE.　Glabrous, sometimes pilose; rhizomes short, with orange-brown divided roots; leaf-blades rather thick, deltoid-lanceolate to broadly oblong-lanceolate, 3–8 cm. long, 1–2.5 cm. wide, obtuse, truncate to broadly cuneate at base, the summer leaves often narrowly ovate-deltoid, sometimes shallowly cordate, the teeth undulate, the upper leaves remote and indistinct, the petioles 3–15 cm. long, as long as to shorter than the blades, winged in upper part; peduncles 5–20 cm. long; flowers deep purple, rarely white with purple striations; sepals lanceolate, 5–8 mm. long, acute, the appendage semirounded, usually entire; petals 12–17 mm. long, the lateral ones bearded inside, rarely glabrous, the spur short-cylindric, 5–7 mm. long.——Apr.–May. Grassy places and fields in lowlands and hills; Hokkaido, Honshu, Shikoku, Kyushu; very common.——s. Kuriles, Korea, Manchuria, China, and Formosa.

Var. **boninensis** (Nakai) Mizushima. *V. boninensis* Nakai ——ATSUBA-SUMIRE.　Maritime phase with thicker lustrous leaves, often elongate-deltoid in spring.——Near seashores.

11. Viola yedoensis Makino. *V. philippica* subsp. *munda* W. Becker, pro parte——NOJI-SUMIRE.　Rather conspicuously short spreading-pilose; rhizomes short; leaf-blades rather thick, broadly deltoid-lanceolate, 3–6 cm. long, 1–2 cm. wide at anthesis, sometimes to 8 cm. long, and 3 cm. wide in fruit, subobtuse, truncate to broadly cuneate, or slightly cordate at base, with scattered obtuse undulate teeth, the petioles usually shorter than the blades, 2–5 cm. long or sometimes to 15 cm. long, slightly winged in upper part; peduncles as long as or slightly shorter than the leaves; flowers purple; sepals broadly lanceolate, 5–7 mm. long, acute or subobtuse, glabrous, the appendage rounded, entire to obtusely toothed; petals 10–14 mm. long, glabrous, the spur slender, cylindric, 5–7 mm. long.——Apr. Grassy fields in lowlands; Honshu, Shikoku, Kyushu.——Korea, Manchuria, and China.

12. Viola minor (Makino) Makino. *V. patrinii* var. *minor* Makino——HIME-SUMIRE. Usually glabrous, rarely puberulous; rhizomes short; leaf-blades deep green and slightly lustrous on upper side, broadly deltoid-lanceolate at anthesis, somewhat narrowly deltoid in fruit, 1.5–4 cm. long, 1–1.5 cm. wide, obtuse to subacute, truncate to slightly cordate at base, undulately toothed or sometimes entire toward the tip, the petioles 2–4 cm. long at anthesis, to 8 cm. long at fruiting time, wingless or slightly winged in upper part; peduncles 4–10 cm. long, usually longer than the leaves; flowers deep-purple, relatively small; sepals broadly lanceolate, acute, 3–4 mm. long, the appendage semirounded; petals 8–10 mm. long, the lateral ones bearded inside, the spur slender, 3–4 mm. long.——Apr. Lowlands; Honshu, Shikoku, Kyushu.

13. Viola phalacrocarpa Maxim. *V. prionantha* var. *latifolia* Miq. pro parte; *V. conilii* Fr. & Sav.; *V. pseudoprionantha* W. Becker; *V. nipponica* Makino; *V. phalacrocarpoides* Makino; *V. reinii* W. Becker; *V. ishidoyana* Nakai——AKANE-SUMIRE, OKA-SUMIRE. Glabrous to densely short-pubescent; rhizomes short, somewhat tufted; leaf-blades ovate to narrowly so, 1–3 cm. long, 0.8–2.5 cm. wide at anthesis, to 8 cm. long in fruit, obtuse to subacute, shallowly cordate to truncate-rounded at base, obtusely undulate-toothed, the petioles 3–10 cm. long at anthesis, to 20 cm. long at fruiting time, narrowly winged in upper part; peduncles 5–10 cm. long; flowers deep rose-purple; sepals broadly lanceolate, 5–7 mm. long, acute, the appendage deltoid, acute, or 4-angled, with acute teeth; petals 10–13 mm. long, the lateral ones bearded inside, the spur slender, 6–8 mm. long.——Apr. Lowlands; Hokkaido, Honshu, Shikoku, Kyushu; common.——Korea, Manchuria, and China.

14. Viola japonica Langsd. *V. prionantha* var. *latifolia* Miq., pro parte; *V. oedemansii* W. Becker; *V. lactiflora* sensu auct. Japon., non Nakai; *V. metajaponica* Nakai——KO-SU-MIRE. Mostly glabrous or sparingly pilose on upper leaves and petioles; rhizomes short; blades of vernal leaves ovate to broadly so or narrowly deltoid-ovate, 2–5 cm. long, 1.5–3.5 cm. wide, obtuse or subobtuse, cordate to slightly so at base, obtusely toothed, the summer leaves often elongate-deltoid, to 8 cm. long, the petioles 2–8 cm. long, narrowly winged in upper part; peduncles 6–12 cm. long, usually glabrous; flowers purple; sepals broadly lanceolate, acute, 5–7 mm. long, the appendage elliptic, often obtusely toothed; petals 1–1.5 cm. long, rounded at apex, the lateral ones bearded inside or glabrous, the spur cylindric, 6–8 mm. long.——(Mar.)–Apr.——Lowlands; Honshu, Shikoku, Kyushu; common.——s. Korea and Ryukyus.

15. Viola yezoensis Maxim. *V. pycnophylla* Fr. & Sav.; *V. yatabei* Makino; *V. biacuta* W. Becker——HIKAGE-SUMIRE. Spreading whitish pilose on leaves and peduncles; rhizomes slender, short, the roots slender, often with adventitious buds; leaf-blades thin, sometimes obscurely white-variegated along the nerves, ovate, 3–6 cm. long, 2.5–4 cm. wide in anthesis, to 10 cm. long and 8 cm. wide at fruiting time, subobtuse, cordate at base, obtusely toothed, the petioles 5–10 cm. long at anthesis, to 20 cm. long at fruiting time, narrowly winged in upper portion; peduncles 7–12 cm. long; flowers white, with purple striations, relatively large; sepals broadly lanceolate, 8–10 mm. long, subacute, ciliate, the appendage ovate to broadly lanceolate, somewhat elongate, acute, entire or with few lobes; petals 15–20 mm. long, the lateral ones bearded inside, the lip purple-striate, the spur 7–8 mm. long, cylindric.——Apr.–May. Woods in mountains; Hokkaido, Honshu, Shikoku, Kyushu.

16. Viola hirtipes S. Moore. *V. miyabei* Makino; *V. hirtipes* var. *typica* and var. *miyabei* (Makino) Nakai——SAKURA-SUMIRE. Rather long-spreading soft white-pubescent, rarely subglabrous; rhizomes short; leaf blades ovate to narrowly so, rarely deltoid-ovate, 3–6 cm. long, 2–4 cm. wide, obtuse or subobtuse, cordate at base, undulately obtuse-toothed, the petioles 5–15 cm. long, slightly winged near the apex; peduncles 7–12 cm. long; flowers pale rose-purple; sepals lanceolate, 7–8 mm. long, subacute, the appendage semirounded, entire, glabrous; petals 15–20 mm. long, retuse to rounded at apex, the lateral ones bearded inside, the spur cylindric, 7–8 mm. long.——May. Mountains; Hokkaido, Honshu, Shikoku, Kyushu.——Korea, Manchuria, and Ussuri.——Forma **rhodovenia** (Nakai) Hiyama ex F. Maekawa. *V. hirtipes* var. *rhodovenia* Nakai——CHISHIO-SUMIRE. Leaves with nerves on upper side purple-brown.

17. Viola blandaeformis Nakai. *V. blanda* var. *violascens* Nakai; *V. blanda* sensu auct. Japon., pro parte, non Willd.——USUBA-SUMIRE. Glabrous, slender; rhizomes rather thick, short, creeping, densely noded, with persistent petiole bases of preceding year; stolons present; leaf-blades membranous, orbicular-reniform, 1–2 cm. long, 1.5–3 cm. wide, rounded and sometimes with an abrupt short tip, deeply cordate, obsoletely toothed, the petioles 2–5 cm. long; peduncles 4–6 cm. long; flowers white; sepals broadly lanceolate, 3–4 mm. long, acute, the appendage semirounded, entire; petals 8–10 mm. long, glabrous, the lip with purple striations, the spur about 2 mm. long, saccate.——July. Wet places in coniferous forests; Hokkaido, Honshu (centr. distr. and westw.).——Kuriles and Kamchatka.

Var. **pilosa** Hara. *V. blanda* sensu auct. Japon., non Willd.; *V. hultenii* W. Becker——KE-USUBA-SUMIRE. Leaves short-pilose.——Hokkaido, Honshu (centr. distr. and northw.).

18. Viola pumilio W. Becker. FUMOTO-SUMIRE. Small, glabrous or sparingly puberulous on upper side of leaves; rhizomes very short; leaf-blades ovate to ovate-cordate, 1–2(–2.5) cm. long, 1–1.8 cm. wide, obtuse, cordate at base, often white-variegated on upper side, often purplish beneath, depressed-toothed, the petioles 2–5 cm. long; peduncles 4–7 cm. long; flowers white, small; sepals broadly lanceolate, 3–4 mm. long, acute, the appendage semirounded, entire; petals 7–8(–10) mm. long, the lateral ones usually bearded inside, the lip short, narrow, the spur about 2.5 mm. long, saccate; capsules 5–7 mm. long, pale green.——Apr.–May. Sunny hills; Honshu, Shikoku, Kyushu.

Viola sieboldii Maxim., with glabrous petals, possibly is a form of *V. pumilio*.

19. Viola boissieuana Makino. *V. pseudoselkirkii* Nakai——HIME-MIYAMA-SUMIRE. Very slender, glabrous or with short hairs on upper side of leaves; rhizomes short; leaf-blades thin, deltoid-cordate to narrowly ovate, usually broadly deltoid-ovate, 0.8–2 cm. long, 1–2 cm. wide, acute to obtuse, deeply cordate with an obtuse sinus, often white-variegated above, the teeth few, slightly dilated on upper part of outer margin, the petioles 2–7 cm. long; peduncles 5–10 cm. long; flowers white, small; sepals lanceolate, acute, 3–4.5 mm. long, the appendage entire, obtuse; petals 7–10 mm. long, the lateral ones usually bearded inside, the lip rather short, purple-striate, the spur 2–3 mm. long, saccate, rounded.——May. Woods; Honshu (w. Kantō Distr. and westw.), Shikoku, Kyushu.

Var. **iwagawae** (Makino) Ohwi. *V. iwagawae* Makino ——YAKUSHIMA-SUMIRE.   Plants smaller; leaves nearly deltoid, 5–10 mm. long and as wide, truncate-cordate at base; capsules 3–5 mm. long.——Kyushu (Yakushima).

20. Viola maximowicziana Makino. *V. selkirkii* var. *major* Yatabe——KO-MIYAMA-SUMIRE.   Often white-pubescent; rhizomes short, the roots often with adventitious buds; leaf-blades spreading, ovate-elliptic to ovate, 2–3.5 cm. long, 1.5–3 cm. wide, acute or subacute, cordate at base, undulately toothed, often white-variegated on upper side, sometimes purplish beneath, the petioles 2–5 cm. long, 1 or 2 times as long as the blades, the stipules lanceolate; peduncles 5–10 cm. long, the bracts 3–6 mm. long, usually on upper part; flowers white; sepals broadly lanceolate, about 3 mm. long, acute, usually pilose, recurved toward tip, the appendage short, entire; petals rather narrow, 8–10 mm. long, the lateral ones glabrous to slightly bearded inside, the lip short, purple-striate, the spur somewhat globose, 2–3 mm. long.——May. Woods; Honshu (Kantō Distr. and westw.), Shikoku, Kyushu.

21. Viola variegata Fisch. var. **nipponica** Makino. *V. umemurae* Makino——GENJI-SUMIRE, IYO-SUMIRE.   Short-pubescent especially on leaves, peduncles, sepals, and capsules; rhizomes slender, short; leaf-blades rather thick-membranous, broadly ovate-elliptic to cordate–orbicular, 2.5–5 cm. long and as wide, obtuse to rounded, cordate at base, often white-variegated on upper surface, purplish beneath, obtusely toothed, the petioles 2–5 cm. long at anthesis, to 15 cm. long at fruiting time; peduncles 6–10 cm. long, shorter than the leaves in fruit, the bracts linear, 3–10 mm. long, near the middle on the peduncle; sepals lanceolate, subacute, 3–7 mm. long, the appendage semirounded or somewhat squarrose, entire, rounded to retuse at the tip; petals pale purple, 8–11 mm. long, the lateral ones bearded inside, the spur cylindric, 7–8 mm. long; capsules puberulent, 4–5 mm. long.——Apr. Honshu (Kantō and centr. distr., Mimasaka Prov.), Shikoku (Iyo); rare.——The typical phase occurs in Korea, Manchuria, and Siberia.

22. Viola shikokiana Makino. *V. epipsila* Ledeb. var. *acuminata* Nakai——SHIKOKU-SUMIRE.   Rhizomes slender, creeping, remotely nodded; leaf-blades few, membranous, ovate-orbicular to broadly ovate, 2–4 cm. long and as wide, short-acuminate, cordate at base, obtusely undulate-toothed, sparingly short-pilose on upper side and on nerves beneath, the petioles 4–15 cm. long; peduncles 5–10 cm. long, longer than the leaves in anthesis; flowers white, 5–6 mm. long, somewhat precocious; sepals lanceolate or narrowly so, subobtuse, the appendage short, deltoid or squarrose; petals narrowly oblong, 8–15 mm. long, the lateral ones glabrous or slightly bearded, the lip slightly shorter than the other petals, purple-striate, the spur saccate, globose, 2–3 mm. long; capsules glabrous.——

Apr.–May. Mountains; Honshu (w. Kantō and Tōkaidō Distr.), Shikoku, Kyushu.

23. Viola selkirkii Pursh. *V. kamtschatica* Ging.; *V. selkirkii* var. *glabra* Miq.——MIYAMA-SUMIRE.   Plant slender, usually loosely pilose on leaves and peduncles; rhizomes short, slender; leaf-blades ovate-cordate, 2–3 cm. long and as wide at anthesis, to 5 cm. long and as wide at fruiting time, abruptly acute or short-acuminate, deeply cordate, undulately toothed, the petioles 3–10 cm. long, 1.5–3 times as long as the blades; peduncles 5–8 cm. long, about as long as the leaves; flowers pale violet; sepals lanceolate, subacuminate, 5–7 mm. long, the appendage ovate-deltoid, subacute or with few subacute small teeth at apex, sometimes glabrous; petals 12–15 mm. long, glabrous, the spur short-cylindric, 6–8 mm. long. ——May. Mountains; Hokkaido, Honshu (centr. distr. and northw.).——Widely distributed in the cooler temperate regions of the N. Hemisphere.

24. Viola tokubuchiana Makino. *V. nikkoensis* Nakai ——FUJI-SUMIRE.   Sparingly short pilose on leaf-blades, sometimes on the peduncles and petioles; rhizomes often elongate; leaf-blades broadly ovate, 1.5–2 cm. long, 1–1.5 cm. wide, subacute, cordate, white-variegated, often purplish beneath, the summer leaves to 4 cm. long and as wide, with short undulate teeth notched at each end, the petioles 2–8 cm. long, 1 to 2 times as long as the blades; stipules linear; peduncles 5–8 cm. long, the bracts linear, 3–8 mm. long; flowers pale purple; sepals lanceolate, acute, 4–5 mm. long, the appendage short; petals 8–12 mm. long, the lateral ones bearded inside, the spur 4–6 mm. long, cylindric; capsules glabrous, 6–7 mm. long.——Apr. Honshu (Kantō and centr. distr.).

25. Viola takedana Makino. *V. scabrida* Nakai; *V. tokubuchiana* var. *takedana* (Makino) F. Maekawa——HINA-SUMIRE.   Glabrous or sparsely short-pilose on leaf-blades on upper side and on nerves beneath, especially in summer leaves; rhizomes short; leaf-blades rather thick, deltoid-ovate to narrowly so, sometimes broadly ovate, 3–6 cm. long, 2–4.5 cm. wide, acuminate to gradually acute with a minute obtuse tip, deeply cordate, undulately toothed, nerves on upper side often white, the petioles 3–10 cm. long, 1.5 to 2 times as long as the blades; stipules broadly linear; peduncles 5–8 cm. long, the bracts filiform-linear, 5–7 mm. long; flowers rose-purple; sepals broadly lanceolate, acute, 6–7 mm. long, the appendage ovate, subobtuse, often few-toothed; petals 12–15 mm. long, the lateral ones glabrous or slightly bearded inside, the spur short-cylindric, 6–7 mm. long; capsules 6–8 mm. long.——Apr.–May. Woods in mountains; Honshu.

26. Viola violacea Makino. SHIHAI-SUMIRE.   Usually glabrous or rarely short-pubescent on upper side of leaves; rhizomes short; leaf-blades deltoid-ovate to narrowly so, sometimes ovate, 2–4 cm. long, 1–2 cm. wide at anthesis, to 5 cm. long, 3 cm. wide at fruiting time, nerves on upper side white, purplish beneath, obtusely short-toothed, the petioles 2–10 cm. long, 1 to 2 times as long as the blades; peduncles 5–8 cm. long; flowers deep rose-purple; sepals broadly lanceolate, 3–4 mm. long, subacute, the appendage rounded, usually entire; petals 8–12 mm. long, glabrous, the spur rather slender, cylindric, 5–6 mm. long, curved slightly upward; capsules 6–7 mm. long.——Apr. Honshu (centr. distr. and westw.), Shikoku, Kyushu.——Korea.

27. Viola makinoi H. Boiss. *V. obtusosagitta* Koidz.;

V. violacea var. *makinoi* (H. Boiss.) Hiyama——MAKINO-SU-MIRE.   Usually glabrous; rhizomes short; leaf-blades broadly deltoid-lanceolate, 2–4 cm. long, 1–1.5 cm. wide at anthesis, to 5 cm. long and 3 cm. wide at fruiting time, acuminate with an obtuse tip, sinuately cordate at base, sometimes obsoletely white-variegated on upper side, with scattered, obscure teeth, the petioles 4–10 cm. long, 1.5 to 3 times as long as the blades; peduncles 4–10 cm. long; flowers rose-purple; sepals broadly lanceolate, 3–4 mm. long, acute, the appendage short, rounded, entire; petals 8–10 mm. long, glabrous, the spur short-cylindric, 6–8 mm. long; capsules usually from short-peduncled cleistogamic flowers.——Apr.–May. Mountains; Honshu, Shikoku, Kyushu.

28. Viola raddeana Regel. *V. raddeana* var. *japonica* Makino; *V. deltoidea* Yatabe——TACHI-SUMIRE.   Glabrous or sometimes sparsely papillose and short pilose on nerves of leaves beneath and on bracts; rhizomes short; stems 30–50 cm. long, erect, tufted; cauline leaves deltoid-lanceolate, 4–8 cm. long, 1–2(–3) cm. wide, gradually narrowed to a subobtuse tip, broadest at the truncate to very slightly cordate base, obscurely obtuse-toothed; stipules rather large, green, linear-lanceolate to lanceolate, 3–6 cm. long, acutely few-toothed, the petioles 2–5 cm. long, winged in upper part; peduncles 5–10 cm. long, axillary on upper part of stem; flowers pale-purple; sepals lanceolate, 5–6 mm. long, very acute at the tip, the appendage short, entire; petals 8–10 mm. long, the lateral ones slightly bearded inside, the lip purple-striate, shorter than the other petals, the spur semiglobose, 1.5–2 mm. long.——May. Wet places in lowlands; Hokkaido, Honshu (Kantō Distr.), Kyushu.——Korea, Manchuria, and Amur.

29. Viola thibaudieri Fr. & Sav. TADE-SUMIRE.   Glabrous; rhizomes short; stems tufted, 25–35 cm. long, green; lowermost leaves reduced, the principal ones broadly lanceolate, 7–10 cm. long, 15–25 mm. wide, gradually acuminate, broadest below the middle, cuneately narrowed in lower third, with scattered obtuse teeth and the margins papillose, subsessile or short-petioled; stipules brown, linear-lanceolate, 1.5–2 cm. long, long-acuminate with few linear teeth, the lower ones strongly connate along the dorsal margins; peduncles 5–6 cm. long, slender; sepals narrowly lanceolate, 5–6 mm. long, acuminate, the appendage short, obtuse or retuse; petals 12–13 mm. long, rather narrow, the lateral ones slightly white-bearded inside at base, the lip short, purple-striate, the spur ellipsoidal, short, about 3 mm. long; capsules usually on short-peduncled cleistogamic flowers.——May. Mountains; Honshu (Shinano Prov.); very rare.——Korea.

30. Viola acuminata Ledeb. *V. micrantha* Turcz., non Presl; *V. laciniosa* A. Gray——EZO-NO-TACHI-TSUBO-SUMIRE. Plant whitish puberulent; rhizomes short; stems erect, tufted, 20–40 cm. long; lower leaves reduced, the cauline cordate or ovate-cordate, 2.5–4 cm. long, 3–5 cm. wide, the uppermost ones sometimes ovate-deltoid to cordate, longer than broad, abruptly acuminate with an obtuse tip, with short depressed teeth, the petioles or some of them longer than the blades, those uppermost shorter than the blades, wingless; stipules oblong, pinnately incised, 15–25 mm. long; peduncles 5–10 cm. long, with a pair of linear bracts on upper half; flowers pale violet; sepals narrowly lanceolate, 7–10 mm. long, acuminate, often short white-pubescent, the appendage semirounded to retuse; petals 8–13 mm. long, the lateral ones white-bearded inside at base, the spur saccate and globose, 3–4 mm. long.

——May–June. Hokkaido, Honshu.——Korea, China, Manchuria, s. Kuriles, Sakhalin, and e. Siberia.——Forma **glaberrima** (Hara, pro var.) Ohwi. (?) *V. micrantha* subsp. *shikokuensis* W. Becker; (?) *V. acuminata* forma *shikokuensis* (W. Becker) F. Maekawa——KENASHI-EZO-NO-TACHI-TSUBO-SUMIRE. Glabrous except the lateral petals.——Honshu and reported from Shikoku.

31. Viola diffusa Ging. var. **glabella** H. Boiss. *V. kiusiana* Makino——TSUKUSHI-SUMIRE, HAI-SUMIRE.   Coarsely white-pilose or glabrescent; rhizomes short, leafy; stoloniferous; leaf-blades spreading, ovate, 2–5 cm. long, 1.5–3 cm. wide, obtuse, broadly cuneate to subtruncate at base, broadly decurrent on the petiole, obtusely toothed, the petioles 1 or 2 times as long as the blades, winged; stipules narrowly lanceolate, with linear teeth; peduncles radical, from the stolons, 3–10 cm. long; flowers white, slightly purple-tinged, small; sepals narrowly ovate or broadly lanceolate, 3–4 mm. long, gradually acute, the appendage rounded, short; petals 6–8 mm. long, glabrous, the lip purple-striate, much shorter than the other petals, the spur semiglobose, very short, only slightly exserted between the sepals; capsules not exserted from the sepals.——Apr. Kyushu (Satsuma, Higo, and Hizen Prov.). ——The typical phase occurs in China and Formosa.

32. Viola rostrata Muhl. var. **japonica** (W. Becker & H. Boiss.) Ohwi. *V. rostrata* subsp. *japonica* W. Becker & H. Boiss.; *V. rostrata* sensu auct. Japon., non Muhl.——NAGA-HASHI-SUMIRE.   Glabrous; rhizomes short; stems 2–4 cm. long at anthesis; leaves radical and cauline, the radical ones rather thick, orbicular-cordate, 2–4 cm. long and as wide, abruptly acute with an obtuse tip, obtusely short-toothed, the midrib raised above; upper leaves usually ovate-deltoid, acuminate with an obtuse tip, shallowly cordate, short-petioled or subsessile, the petioles 2–5 cm. long; stipules red-brown, narrowly ovate-oblong, about 1 cm. long, deeply laciniate-lobed; upper stipules linear-lanceolate, few-toothed; peduncles near base or radical, longer than the radical leaves; flowers pale violet; sepals broadly lanceolate, 5–6 mm. long, acute, the appendage semirounded, entire; petals 12–14 mm. long, glabrous, the spur cylindric, ascending, 10–15 mm. long; capsules from short-peduncled cleistogamic flowers on upper part of stems.——Apr.–May. Shaded places in low mountains; Honshu (Japan Sea side of northern distr. as far west as Inaba Prov.).——The typical phase occurs in N. America.

33. Viola sacchalinensis H. Boiss. *V. komarovii* W. Becker; *V. mutsuensis* W. Becker; *V. sylvestriformis* W. Becker; *V. canina* var. *kamtschatica* Ging.——AINU-TACHI-TSUBO-SUMIRE.   Rhizomes rather short; stems few, ascending, 2–5 cm. long at anthesis, 10–25 cm. long at fruiting time; radical leaf-blades cordate-orbicular or reniform-cordate, 2–3 cm. long and as wide at anthesis, 3–4 cm. long and as wide at fruiting time, subacute or obtuse, vivid green, flat, soft, obscurely undulate-toothed, the petioles 2 to 3 times as long as the blades; cauline leaves few, short-petiolate; stipules pinnately lobed; peduncles axillary to the stem-leaves; flowers pale violet; sepals narrowly lanceolate, 4–6 mm. long, acuminate, the appendage semirounded, entire; petals 10–15 mm. long, the lateral ones bearded inside, the spur narrowly oblong, 4–6 mm. long; capsules from short-peduncled cleistogamic flowers on upper part of stems.——May–June. Hokkaido, Honshu (n. distr.).——s. Kuriles, Sakhalin, Kamchatka, n. Korea, Manchuria, and Siberia.

34. Viola kusanoana Makino. Ō-TACHI-TSUBO-SUMIRE. Usually glabrous; rhizomes densely noded, short-creeping; stems erect, ascending at base, 5–10 cm. long at anthesis, to 40 cm. long at fruiting time; radical leaf-blades cordate-rounded or depressed, 3–5 cm. long and as wide at anthesis, to 6 cm. long and as wide at fruiting time, densely brown-spotted, subacute to obtuse, vivid green, membranous, cordate at base, obtusely undulate-toothed, the petioles 1.5 to 2.5 times as long as the leaves; cauline leaves short-petiolate; stipules deeply laciniate-lobed; peduncles axillary to the stem-leaves; flowers pale violet, rather large; sepals lanceolate, 5–8 mm. long, acute, the appendage semirounded, obtusely toothed or entire; petals glabrous, 15–18 mm. long, the spur short-cylindric, 6–8 mm. long; capsules chiefly from short-peduncled cleistogamic flowers on upper part of stems.——Apr.–May. Wet grassy places sometimes in thin woods; Hokkaido, Honshu, Kyushu.——Forma **pubescens** (Nakai) Mizushima. *V. kusanoana* var. *pubescens* Nakai——KE-Ō-TACHI-TSUBO-SUMIRE. Plant short-pubescent. Occurs with the species, but especially abundant in northern districts.

35. Viola senamiensis Nakai. *V. grayi* Fr. & Sav., pro parte, nom., excl. syn.; *V. grayi* var. *glabra* W. Becker; *V. grypoceras* var. *grayi* (Fr. & Sav.) Nakai——ISO-SUMIRE, SE-NAMI-SUMIRE. Usually glabrous; rhizomes thick, elongate, often branched, rather ligneous, ascending or erect; stems tufted, 5–10 cm. long; radical leaves thick, depressed-orbicular, 1–3 cm. long, 1.5–3.5 cm. wide, very obtuse, cordate, obtusely toothed, rather long-petioled; stipules deeply pinnately lobed; peduncles usually cauline; flowers deep-violet; sepals lanceolate to broadly so, 6–7 mm. long, acute, the appendage semi-rounded, subentire; petals 13–15 mm. long, glabrous, the spur rather short, white, about 5 mm. long; capsules glabrous. ——May. Sand-dunes near seashores; Hokkaido, Honshu (Japan Sea side as far south as Echigo Prov., also Rikuchu and Mutsu Prov.).

36. Viola faurieana W. Becker. TERIHA-TACHI-TSUBO-SUMIRE. Glabrous; rhizomes densely noded; stems ascending, few-leaved, very short at anthesis, to 20 cm. long at fruiting time; radical leaf-blades orbicular-cordate to somewhat orbicular-deltoid, 2–3 cm. long and as wide, rounded to obtuse, shallowly cordate or subtruncate at base, or sometimes deltoid-cordate to broadly deltoid-ovate, to 4 cm. long, 3.5 cm. wide, obtusely toothed, nerves and veinlets slender, raised on upper side when dry, the petioles 2 to 2.5 times as long as the blades; stipules deeply linear-lobed; cauline leaves rather few, short-petioled; peduncles radical and cauline, with a pair of bracts on upper half; flowers pale violet; sepals broadly lanceolate, 3.5–5 mm. long, subacute, the appendage semirounded, entire; petals 8–12 mm. long, glabrous, the spur 4–5 mm. long. ——May. Woods; Honshu (Japan Sea side from Mutsu to Echigo).

37. Viola grypoceras A. Gray. *V. canina* var. *japonica* Ging; *V. sylvestris* var. *japonica* (Ging.) Makino; *V. sylvestris* var. *grypoceras* (A. Gray) Maxim.; *V. reichenbachiana* sensu Fr. & Sav., non Jord.; *V. longipedunculata* Fr. & Sav.; *V. pruniflora* Nakai——TACHI-TSUBO-SUMIRE. Usually glabrous; rhizomes short, densely noded; stems few, ascending or decumbent, short at anthesis, 10–30 cm. long at fruiting time; radical leaf-blades cordate to depressed-cordate, 1.5–2.5 cm. long, 2–3 cm. wide, obtuse to very obtuse, cordate, shallowly toothed, the petioles 3–7 cm. long, 2 to 4 times as long as the blades; cauline leaves simulating the radical; stipules deeply

laciniate-lobed; peduncles 6–10 cm. long, radical and cauline, with a pair of linear-bracts on upper half; flowers pale violet, inodorous; sepals lanceolate, 5–7 mm. long, acute, the appendage semirounded, with a rounded entire tip; petals 12–15 mm. long, glabrous, the spur short-cylindric, 6–8 mm. long.——Apr.–May. Thin woods in lowlands and mountains; Hokkaido, Honshu, Shikoku, Kyushu; common and very variable.——Forma **pubescens** (Nakai) Mizushima. *V. grypoceras* var. *pubescens* Nakai——KE-TACHI-TSUBO-SUMIRE. Plant short-pubescent.——Common especially in northern areas.

Var. **rhizomata** (Nakai) Ohwi. *V. rhizomata* Nakai—— TSURU-TACHI-TSUBO-SUMIRE. Stems decumbent, over-wintering, new buds appearing from the apex; leaves smaller, depressedly deltoid-orbicular, subtruncate at base.——High mountains on Japan Sea side of Honshu (from Tajima to Echigo).

Var. **exilis** (Miq.) Nakai. *V. sylvestris* var. *exilis* Miq.; *V. coreana* H. Boiss.——KO-TACHI-TSUBO-SUMIRE. Smaller plant with decumbent stems and depressedly deltoid leaves truncate at base.——Honshu (western distr.), Shikoku and Kyushu.

Var. **imberbis** (A. Gray) Ohwi. *V. sylvestris* var. *imberbis* A. Gray; *V. krugiana* W. Becker——ISO-TACHI-TSUBO-SUMIRE. Stems tufted; leaves thicker, rather lustrous.——Maritime form near seashores; Honshu, Shikoku, Kyushu.

Var. **hichitoana** (Nakai) F. Maekawa. *V. hichitoana* Nakai; *V. kusanoana* var. *hichitoana* (Nakai) Ohwi——SHICHITŌ-SUMIRE. Plant larger with larger leaves.——Izu-shichitō Islands.

38. Viola obtusa (Makino) Makino. *V. ovato-oblonga* var. *obtusa* Makino——NIOI-TACHI-TSUBO-SUMIRE. Usually papillose-puberulous throughout; stems ascending or erect, usually very short at anthesis, 10–30 cm. long at fruiting time; radical leaf-blades cordate or deltoid-cordate, 1–2 cm. long and as wide at anthesis, to 3 cm. long and as wide at fruiting time, obtuse to rather rounded at apex, cordate, obtusely short-toothed, the petioles 1 to 3 times as long as the blades; cauline leaves at fruiting time narrowly ovate-deltoid, 2.5–4 cm. long, 1.5–2.5 cm. wide, shallowly cordate; stipules pinnately linear-laciniate; peduncles radical and cauline; flowers rose-purple with a white center, fragrant; sepals lanceolate, 6–8 mm. long, acute, the appendage semirounded, obtusely toothed; petals 12–15 mm. long, glabrous, the spur 6–7 mm. long; capsules glabrous, usually from short-peduncled cleistogamic flowers on upper part of stems.——Apr.–May. Thickets and sunny slopes; Honshu, Kyushu.——Forma **nuda** (Ohwi) F. Maekawa. *V. obtusa* var. *nuda* Ohwi——KENASHI-NIOI-TACHI-TSUBO-SUMIRE. Glabrous.——Grows with the typical phase.

39. Viola ovato-oblonga (Miq.) Makino. *V. sylvestris* forma *ovato-oblonga* Miq.——NAGABA-TACHI-TSUBO-SUMIRE. Glabrous; rhizomes short, densely noded; stems ascending or erect, 20–40 cm. long in fruit; radical leaf-blades cordate-orbicular to cordate, 2–2.5 cm. long, 2–3 cm. wide, very obtuse to subrounded at apex, cordate, undulate-toothed, the petioles 1.5–4 times as long as the blades; stipules pinnately laciniate-lobed; lower cauline leaves narrowly ovate-deltoid, subacute to obtuse, shallowly cordate, the upper ones short-petioled, lanceolate to narrowly deltoid-ovate, acuminate with an obtuse tip, shallowly cordate, obsoletely toothed; stipules narrowly lanceolate, pinnately incised; peduncles radical and cauline; flowers pale violet; sepals lanceolate, 5–7 mm. long, acute, the appendage rounded, short, entire; petals 12–15 mm. long, glabrous, the spur 7–8 mm. long, short-cylindric; capsules chiefly from short-peduncled cleistogamic flowers on upper part of stems.——Apr.–May. Sunny slopes and thickets in hills and low mountains; Honshu (Kinki Distr. and westw.), Shikoku, Kyushu.——Forma **pubescens** (Nakai) F. Maekawa. *V. ovato-oblonga* var. *pubescens* Nakai——KE-NA-GABA-TACHI-TSUBO-SUMIRE. Plant pubescent.——Grows with the typical phase.——s. Korea.

40. Viola mirabilis L. var. **subglabra** Ledeb. *V. brachysepala* Maxim.; *V. mirabilis* var. *glaberrima* W. Becker——IBUKI-SUMIRE. Rhizomes creeping, rather short; stems erect, 15–30 cm. long, with 2 subsessile cauline leaves at the top and sometimes with a third long-petiolate one; leaf-blades membranous, depressed-cordate or orbicular-cordate, 2–3(–4.5) cm. long, 2.5–4(–6) cm. wide, very shortly obtuse, cordate, with flat obtuse teeth, the petioles 7–15 cm. long; stipules thinly membranous, broadly lanceolate, acute to obtuse, glabrous or loosely glandular-ciliate, the cauline 5–8 mm. long, acute, usually short-ciliate, nearly free; peduncles radical; flowers pale violet, chasmogamic, rather large, usually sterile; sepals narrowly ovate or broadly lanceolate, 6–8 mm. long, subacute, the appendage depressed-rounded, short, entire; petals 13–15 mm. long, the lateral ones bearded inside at base, the spur 5–7 mm. long, oblong; capsules 8–10 mm. long, from short-peduncled cleistogamic flowers, terminal on the stems.——Apr. Honshu.——e. Siberia, Korea, Manchuria. The typical phase occurs in Europe, Caucasus, and Siberia.

41. Viola langsdorffii Fisch. *V. kamtschadalorum* W. Becker & Hult.; *V. langsdorffii* var. *caulescens* Ging.; *V. mirabilis, β langsdorffii, γ canescens, δ hispidula,* and var. *kusnetzoffii α glabra* Regel; *V. langsdorffii* subsp. *sachalinensis* W. Becker; *V. kurilensis* Nakai; *V. sapporoensis* Franch.——ŌBA-TACHI-TSUBO-SUMIRE. Glabrous or slightly pilose on leaves beneath; rhizomes creeping; stems erect, 20–30(–40) cm. long, 3- to 4-leaved; leaf-blades cordate or orbicular-cordate, 3–7 cm. long, 4–8 cm. wide, slightly extended and obtuse, cordate, undulate-toothed, the petioles 7–15 cm. long; stipules broadly lanceolate, acuminate, slightly connate at base, nearly entire, the cauline membranous, greenish, semi-ovate, 1–2 cm. long, acute, entire or with few fine teeth; peduncles from upper part of stems; flowers pale violet, rather large; sepals broadly lanceolate, 7–8 mm. long, acute, the appendage short, sometimes with few obsolete teeth; petals 15–20 mm. long, the lateral ones bearded inside at base, the spur short, saccate, 3–4 mm. long; capsules glabrous, 12–15 mm. long.——May–July. Wet grassy places; Hokkaido, Honshu (Oze in Kotsuke).——Sakhalin, Kuriles, and Kamchatka.

42. Viola verecunda A. Gray. *V. alata* subsp. *verecunda* (A. Gray) W. Becker; *V. japonica* Fr. & Sav., non Langsd.; incl. var. *typica* Fr. & Sav., var. *subaequiloba* Fr. & Sav., var. *decumbens* Fr. & Sav., and var. *pusilla* Fr. & Sav.——TSUBO-SUMIRE. Glabrous; rhizomes short; stems ascending or decumbent, 5–20(–30) cm. long; radical leaf-blades reniform-cordate or deltoid-reniform, 1.5–2.5 cm. long, 2–3.5 cm. wide, obtuse or deltoid-rounded, deeply open-cordate, obtusely undulate-toothed, the petioles 2 to 4 times as long as the blades; cauline leaves depressed-cordate, deltoid-cordate or deeply open-cordate, subacute to obtuse, the upper cauline leaves short; stipules membranous, green, lanceolate or linear-oblong, 7–20 mm. long, entire or with few obsolete teeth; peduncles cauline; flowers white; sepals broadly lanceolate, 4–5 mm. long, subacute, the appendage short, rounded, entire; petals 8–10 mm.

long, the lateral ones slightly bearded inside at base, the lip shorter than the other petals, purple-striate, the spur short, saccate, 2–3 mm. long.——Apr.–May. Wet grassy places in lowlands and hills; Hokkaido, Honshu, Shikoku, Kyushu; very common and variable.——Korea, China, Manchuria, Sakhalin, s. Kuriles, Ussuri, Amur, and Formosa.

Var. **fibrillosa** (W. Becker) Ohwi. *V. fibrillosa* W. Becker——MIYAMA-TSUBO-SUMIRE. Small alpine phase with short stems, smaller leaves and stipules.——High mountains in Honshu (Japan Sea side).

Var. **yakusimana** (Nakai) Ohwi. *V. yakusimana* Nakai——KOKE-SUMIRE. An extremely dwarf phase with leaves 3–7 mm. wide and petals 4–5 mm. long.——High mountains; Kyushu (Yakushima).

Var. **semilunaris** Maxim. *V. semilunaris* (Maxim.) W. Becker——AGI-SUMIRE. Leaves depressed, broadly cordate with an open sinus at base.——Occurs with the typical phase.

Var. **excisa** (Hance) Maxim. *V. excisa* Hance——HIME-AGI-SUMIRE. Resembles var. *semilunaris* but smaller and with long-creeping stems.

43. Viola biflora L. KIBANA-NO-KOMA-NO-TSUME. Rhizomes short, creeping, closely noded; stems ascending, 5–20 cm. long, loosely 3- or 4-leaved; leaf-blades reniform-cordate, 1–2 cm. long, 1.5–3.5 cm. wide, rounded at apex, cordate, undulate-toothed, sparingly pilose on upper side and on nerves beneath, the petioles 2–10 cm. long, 2–6 times as long as the blades; upper cauline leaves short-petioled; stipules green, narrowly ovate or broadly lanceolate, 3–5 mm. long, suboptuse to subacute, entire, loosely ciliolate; peduncles cauline, 2–5 cm. long, with a pair of minute bracts; flowers yellow; sepals lanceolate, suboptuse, 3–5 mm. long, the appendage short, rounded, entire; petals 7–10 mm. long, glabrous, the lip with dark brown striations, the spur short, semirounded, 1.5–2 mm. long.——June–July. Wet places in high mountains; Hokkaido, Honshu (centr. distr. and northw.), Shikoku, Kyushu (Yakushima).——Cooler regions of the N. Hemisphere.

44. Viola crassa Makino. *V. biflora* var. *crassifolia* Makino——TAKANE-SUMIRE. Glabrous or nearly so, deep-green, suffused with red-brown; rhizomes short, creeping, densely noded; stems 5–12 cm. long, 3- or 4-leaved; leaf-blades thick, lustrous, reniform-cordate or sometimes depressed-cordate, 1–2 cm. long, 2–4.5 cm. wide, rounded at apex, undulate-toothed, the petioles 3 to 4 times as long as the blades; stipules ovate, 3–4 mm. long, suboptuse; peduncles cauline, 3–5 cm. long, with a pair of small bracts on upper half; flowers deep yellow; sepals narrowly oblong or broadly lanceolate, 4–5 mm. long, obtuse, the appendage semirounded, entire; petals 10–12 mm. long, glabrous, the lip with dark-brown striations, the spur short, semiglobose, about 1 mm. long, slightly protruding between the sepals.——July–Aug. Gravelly places in alpine regions; Hokkaido, Honshu (centr. and n. distr.); rare.——Kuriles and Kamchatka.

45. Viola alliariaefolia Nakai. *V. glabella* var. *renifolia* Koidz.; *V. brevistipulata* var. *renifolia* (Koidz.) F. Maekawa——JIN'YŌ-KI-SUMIRE. Loosely soft-pubescent on nerves of leaves beneath and on margins; stems erect, 10–20 cm. long, 3- or 4-leaved; radical leaf-blades solitary or none, reniform, 2–3 cm. long, 3–6 cm. wide, rounded to broadly retuse, cordate, obtusely toothed or irregularly incised-toothed, the petioles 2 to 4 times as long as the blades, the cauline leaves simulating the basal, the upper ones usually smaller and approximate, on very short petioles; stipules ovate, 2–3 mm.

long, acute with obsolete glandular teeth or subentire; peduncles 2–5 cm. long, axillary in the upper stem leaves, with short bracts; flowers yellow; sepals lanceolate, 4–5 mm. long, obtuse, the appendage short, entire; petals 10–12 mm. long, the lateral ones slightly bearded inside at base, the spur short, about 2 mm. long; styles glabrous.——July.   Alpine regions; Hokkaido.

46. Viola brevistipulata (Fr. & Sav.) W. Becker. *V. pubescens* var. *brevistipulata* Fr. & Sav.; *V. glabella* sensu auct. Japon., non Nutt.; *V. kishidae* Nakai; *V. flaviflora* Nakai; *V. brevistipulata* var. *acuminata* Nakai; *V. brevistipulata* var. *minor* Nakai; *V. lasiostipes* Nakai——ŌBA-KI-SUMIRE.   Glabrous or slightly pubescent with rather thick, creeping rhizomes; stems erect, 15–30 cm. long, naked at base, 3- or 4-leaved on upper part; radical leaves 1 or 2, or absent, blades 5–8(–10) cm. long and as wide, abruptly acute, sometimes with an obtuse tip, cordate, obtusely undulate-toothed, the petioles 2 or 3 times as long as the blades; cauline leaves unequally petioled; stipules broadly ovate, 4–7 mm. long, acute, entire, ascending-spreading; peduncles 1–3, 3–7 cm. long, axillary in the stem-leaves, with a pair of minute bracts; flowers yellow; sepals narrowly lanceolate, 6–8 mm. long, gradually narrowed to an obtuse tip, the appendage very short, semirounded, entire; petals 12–15 mm. long, obovate, the lateral ones bearded inside, the spur very short, slightly protruding between the sepals.——June–July.   High mountains; Hokkaido, Honshu (Mount Daisen in Hōki Prov., n. Kinki, and centr. distr. northw.); rather common.

Var. **laciniata** W. Becker. *V. laciniata* (W. Becker) Koidz.——FUGIRE-SUMIRE.   Leaves laciniately cut.——High mountains; Hokkaido.

Var. **hidakana** (Nakai) S. Watanabe. *V. hidakana* Nakai——EZO-KI-SUMIRE.   Leaves thicker, rather firm, lustrous on both sides, often purplish beneath.——July. Alpine; Hokkaido.

47. Viola yubariana Nakai. *V. glabella* var. *crassifolia* Koidz.——SHISOBA-KI-SUMIRE.   Short-pilose on stems and petioles; stems 4–5 cm. long; radical leaves thick, orbicular, 2.5–4 cm. wide, abruptly acuminate with an obtuse tip, cordate with closed or overlapping edges, undulate-toothed, purplish beneath; cauline leaves 3–4, sessile except the lowest; stipules orbicular or broadly ovate, nearly entire; peduncles axillary to the stem-leaves, 1–2(–3.5) cm. long, with a pair of small bracts; sepals 4–5 mm. long, lanceolate or broadly so, acute or obtuse, the appendage semirounded; petals about 10 mm. long, yellow, with reddish nerves, the spur very short; capsules 6–7 mm. long.——Hokkaido (alpine zone of Mount Yubari); rare.

48. Viola orientalis (Maxim.) W. Becker. *V. uniflora* var. *orientalis* Maxim.; *V. xanthopetala* Nakai; *V. uniflora* sensu auct. Japon., non L.; *V. uniflora* var. *glabricapsula* Makino——KI-SUMIRE.   Puberulous or nearly glabrous except the leaves; rhizomes usually erect, with rather numerous, horizontally spreading, thick roots; stems erect, 10–15 cm. long, 3(–4)-leaved, naked at base; radical leaf-blades cordate, 2.5–4 cm. long and as wide, acute, sometimes with an obtuse tip, undulate-toothed, the petioles tinged red-brown, 3–5 times as long as the leaves; cauline leaves 3, broadly ovate to cordate, sometimes rounded at base, the lower ones short-petioled, usually rather distant, the upper ones nearly opposite, subsessile; stipules broadly ovate, 2–3 mm. long, acute, subentire; peduncles 2–4 cm. long, with a pair of small bracts; flowers yellow; sepals lanceolate, 6–8 mm. long, gradually narrowed to an obtuse tip, the appendage ovate, entire, short; petals 12–15 mm. long, the lateral ones sometimes bearded inside, the spur very short.——Apr.–May. Mountains; Honshu (Suruga, Bingo), Kyushu (n. distr.).——Korea, Manchuria, and Ussuri.

Fam. 140. FLACOURTIACEAE   IIGIRI KA   Flacourtia Family

Trees or shrubs; leaves simple, alternate; stipules caducous; flowers usually bisexual but often unisexual (the plants monoecious or dioecious); sepals imbricate or valvate, sometimes free, often similar to the petals; petals large, small, or absent, sometimes with a scale inside, imbricate; stamens numerous, rarely few, free; ovary superior, 1-locular, the placentae parietal; ovules 2 or more; styles or stigmas as many as the placentae; fruit usually a berry or drupe, indehiscent; endosperm fleshy.——About 70 genera, with about 500 species, chiefly in the Tropics.

1A.   Leaves coriaceous, 4–8 cm. long, acute at base, penninerved, the petioles 2–5 mm. long, glandless; inflorescence axillary. . . 1. *Xylosma*
1B.   Leaves herbaceous, 10–20 cm. long, cordate, palmately nerved, the petioles elongate, with glands at top; inflorescence terminal. . . 2. *Idesia*

1. XYLOSMA G. Forst.   KUSUDO-IGE ZOKU

Dioecious or polygamous trees or shrubs usually with axillary spines; leaves coriaceous, toothed, short-petioled; stipules absent; flowers small, in axillary panicles, with small bracts; sepals 4–5(–7), slightly connate at base, imbricate, usually ciliate; petals absent; stamens many; disc glandlike or annular and surrounding the stamens; ovary 1-locular, free, the placentae 2–3(–6); styles short, alternate with the placentae, the stigmas slightly thickened and lobed; fruit a berry; seeds usually few, rather flat, usually obovate.——About 60 species, in the Tropics.

1. Xylosma japonicum (Walp.) A. Gray. *Apactis japonica* Thunb.; *Hisingera racemosa* Sieb. & Zucc., non Presl; *Flacourtia japonica* Walp.; *H. japonica* Sieb. & Zucc.; *X. racemosa* (Sieb. & Zucc.) Miq.; *Myroxylon japonicum* (Thunb.) Makino; *X. congestum* Merr.; *X. apactis* Koidz.——KUSUDO-IGE.   Evergreen dioecious shrub with spinelike short spurs in young individuals; branches red-brown, short-pubescent when young; leaves ovate to oblong-ovate, 4–8 cm. long, 3–4(–5) cm. wide, acuminate sometimes with an obtuse tip, acute at base, rounded when young, toothed, glabrous, the petioles 2–5 mm. long; flowers in very short axillary racemes, yellowish white, about 2.5 mm. across, short-pedicelled; sepals nearly round, short-pubescent on outer side; stamens much longer than the sepals; berry globose, black, about 5 mm. across; style very short, the stigma thickened; seeds 2–3, ovoid, about 4 mm. in diameter, rounded on back, brown with black striations.——Sept. Near seashores; Honshu (Kinki Distr. and westw.), Shikoku, Kyushu.——Ryukyus, Formosa, and China.

2. IDESIA Maxim.   Iigiri Zoku

Deciduous dioecious tree; branches thick, gray-brown, glabrous; leaves large, ovate-cordate or deltoid-cordate, 10–20 cm. long, 8–20 cm. wide, acuminate, shallowly cordate or truncately rounded at base, sparsely toothed, palmately 5- to 7-nerved, green, glaucous and white-hairy beneath chiefly on the midrib near base, the petioles reddish, about as long as the blades, with two sessile oblong glands at the top, young leaves with 1 or 2 additional pairs of petiolar glands; inflorescence paniculate, large, 20–30 cm. long, terminal and axillary, often somewhat subracemose, yellowish puberulent; flowers yellowish, the pedicels 1–1.5 cm. long, the staminate 13–16 mm. across, the pistillate about 8 mm. across; sepals often with obscure teeth, nearly as long as the stamens; petals absent; stamens numerous, on a small disc, the filaments hairy; staminodia of pistillate flowers small; ovary glabrous, globose; placentae 5 (3–6), parietal; ovules many; styles about 2 mm. long, deciduous, the stigmas clavate-globose; berry 8–10 mm. across, orange-red, with numerous small seeds.——One species in e. Asia.

1. Idesia polycarpa Maxim. *Polycarpa maximowiczii* Linden ex Carr.; *Cathayeia polycarpa* (Maxim.) Ohwi——Iigiri.   Branches thick, gray-brown, glabrous; leaves long-petiolate, ovate-cordate or deltoid-cordate, 10–20 cm. long, 8–20 cm. wide, acuminate, shallowly cordate or truncately rounded at base, rather loosely toothed, green on upper side, glaucous and white-hairy chiefly on the midrib at base of leaf beneath, petioles reddish, about as long as the leaves, with 2 oblong sessile glands at apex, often with 1 to 2 more pairs of accessory glands in young individuals; inflorescence pendulous, 20–30 cm. long, yellowish puberulent, the pedicels 1–1.5 cm. long; staminate flowers 13–16 mm. across, the sepals often with obscure teeth, nearly as long as the stamens; pistillate flowers about 8 mm. across; ovary glabrous, globose; styles about 2 mm. long, deciduous, the stigma clavate-globose; fruit globose, 8–10 mm. across, orange-red.——Apr. Hills and mountains; Honshu, Shikoku, Kyushu; often planted.——Ryukyus, Formosa, China, and Korea.

Fam. 141. STACHYURACEAE   Kibushi Ka   Stachyurus Family

Deciduous or evergreen trees or shrubs; leaves alternate, simple, toothed, petioled; stipules small, caducous; flowers bisexual or sometimes unisexual, actinomorphic, subsessile, in axillary pendulous racemes; sepals and petals 4, imbricate; stamens 8, free; ovary superior, incompletely 4-locular by the intrusion of the parietal placentae; styles short, the stigma 4-lobed; ovules many; fruit a berry with a coriaceous pericarp; seeds many, soft arillate, with endosperm, the embryo straight.——One genus with few species, in e. Asia and the Himalayas.

1. STACHYURUS Sieb. & Zucc.   Kibushi Zoku

1. Stachyurus praecox Sieb. & Zucc. Kibushi.   Large deciduous shrub; branches brown to red-brown, glabrous, lustrous; leaves membranous to herbaceous, elliptic-ovate, oblong, or narrowly ovate, 7–12 cm. long, 3–6 cm. wide, long-acuminate, rounded at base, toothed, glabrous or thinly pubescent especially on nerves beneath, the petioles 1–2.5(–3) cm. long; racemes precocious, rather densely many-flowered, subsessile, 4–10 cm. long, pendulous, glabrous, the bracts deltoid, small, about 2 mm. long; flowers subsessile, about 7 mm. long, campanulate, pale yellow to greenish yellow; petals about twice as long as the sepals, cuneate-obovate; fruit ellipsoidal to nearly globose, yellowish at maturity, 8–12 mm. long.——Mar.–Apr. Thickets and thin woods in mountains; Hokkaido (sw. distr.), Honshu, Shikoku, Kyushu; common.

Var. **matsuzakii** (Nakai) Makino. *S. lancifolius* Koidz.; *S. matsuzakii* Nakai; *S. ovalifolius* Nakai——Namban-kibushi, Enoshima-kibushi, Hachijō-kibushi.   Maritime form with thicker branches, larger and longer leaves 10–15 cm. long, and larger fruit 10–15 mm. long.——Thickets and thin woods near seashores; Honshu (Kantō Distr. and westw.), Shikoku, Kyushu.——Ryukyus.

Fam. 142. THYMELAEACEAE   Jinchōge Ka   Mezereum Family

Plants woody, rarely herbaceous; leaves opposite or alternate, simple, usually small; stipules absent; flowers usually actinomorphic, usually in terminal heads, spikes, or racemes, bisexual or unisexual (plants then mostly dioecious); calyx tubular, sometimes corollalike, the lobes 4 or 5, imbricate; petals absent or represented by 4–12 scales, inserted on the tube or throat of calyx; stamens 2 to many, usually as many as the calyx-lobes and opposite them or twice as many, the disc scalelike, annular or cup-shaped, sometimes absent; ovary superior, 1- to 2-locular; style lateral in 1-locular ovaries, the stigma capitate; ovules solitary in each locule, pendulous, anatropous; fruit indehiscent; seeds with or without endosperm, the embryo straight.——About 40 genera, with about 450 species, widely distributed in both hemispheres.

1A. Inflorescence in terminal or axillary heads or in very short racemes, usually bracteate; disc absent or annular, sometimes lobed on one side; leaves usually alternate.
2A. Style short or absent, the stigma capitate. . . . . 1. *Daphne*
2B. Style cylindric, the stigma elongate, linear; inflorescence nodding. . . . . 2. *Edgeworthia*

1B. Inflorescence in terminal racemes or spikes, sometimes in panicles, bractless; disc of one or more scales; leaves usually opposite, sometimes alternate. .... 3. *Wikstroemia*

1. DAPHNE L. JINCHŌGE ZOKU

Evergreen or deciduous shrubs; inflorescence capitate or short-racemose, sometimes paniculate; flowers bisexual or unisexual (plants sometimes dioecious); calyx-lobes 4, the tube cylindric, often dilated at base, without an appendage in the throat; stamens 8, inserted on the calyx-tube, the filaments very short, the anthers oblong, the connective indistinct; disc absent or annular, sometimes lobed on one side; ovary 1-locular, sessile or nearly so; style often absent, the stigma capitate; berry globose or oblong, sometimes surrounded by the persistent calyx-tube, sometimes naked; testa crustaceous, the endosperm scanty or absent; cotyledons fleshy.——About 50 species, in Europe and Asia.

1A. Leaves coriaceous, lustrous, evergreen, deep green; branchlets dark purple-brown; flowers white.
  2A. Calyx-tube short-pilose externally; inflorescence terminal on last year's branchlets. .... 1. *D. kiusiana*
  2B. Calyx-tube glabrous externally; inflorescence on current year's branchlets. .... 2. *D. miyabeana*
1B. Leaves membranous, appearing in autumn, deciduous in summer, green or glaucous; branchlets gray-brown; flowers yellowish.
  3A. Leaves oblanceolate, 1–2 cm. wide, rather acute to obtuse; flowers greenish yellow. .... 3. *D. pseudomezereum*
  3B. Leaves narrowly cuneate-obovate, 2–3 cm. wide, rounded to retuse with a minute mucro at apex; flowers yellow.
                                                                   4. *D. kamtschatica* var. *jezoensis*

1. Daphne kiusiana Miq. *D. odora* var. *kiusiana* (Miq.) Keissler; *D. cannabina* sensu auct. Japon., non Wall.——KOSHŌ-NO-KI. Evergreen dioecious shrub, glabrous except the inflorescence; leaves alternate, rather softly coriaceous, lustrous, oblanceolate, 7–14 cm. long, 1.2–3.5 cm. wide, acute to acuminate with an obtuse tip, gradually narrowed to the base, entire, short-petiolate; flowers white, in dense terminal heads on last year's branchlets, the pedicels very short, finely white-puberulous; calyx tube 7–8 mm. long, short-pilose externally, the lobes spreading, about half as long as the tube; stamens 8, in 2 series, the anthers of the upper series half-exserted from the throat; berry red, ellipsoid-globose, about 8 mm. long; seeds ellipsoidal, about 6 mm. long.——Apr. Honshu (s. Kantō Distr. and westw.), Shikoku, Kyushu.——Ryukyus. Closely related to *D. odora*.

Daphne odora Thunb. *D. japonica* Thunb.——JINCHŌGE. Chinese shrub commonly cultivated in our area, slightly larger than *D. kiusiana* in all respects; flowers larger, white or rose-purple outside, glabrous; plants cultivated in this country are staminate.

2. Daphne miyabeana Makino. KARASU-SHIKIMI. Small evergreen shrub often puberulous on young branchlets; leaves thinly coriaceous, oblanceolate, 5–10 cm. long, 1–2.5 cm. wide, acute to acuminate with an obtuse or subacute tip, dark green above, entire, glabrous, usually lustrous, gradually narrowed to a short petiole; heads terminal on current year's branchlets, puberulous; flowers white, on very short pedicels; calyx-tube 5–6 mm. long, glabrous externally, the lobes spreading, ovate, 3-nerved; stamens 8, in 2 series, the anthers of upper series semi-exserted from the throat; ovary glabrous; berry ellipsoidal, about 8 mm. across, vermilion.——Rocks

in high mountains; Hokkaido, Honshu (centr. distr. and northw.); rare.

3. Daphne pseudomezereum A. Gray. ONI-SHIBARI, NATSU-BŌZU. Glabrous, dioecious shrub with thick gray-brown branches; leaves oblanceolate, 5–10 cm. long, 8–20 mm. wide, obtuse, acute, or rounded at apex, slightly glaucous, gradually narrowed to a short petiole; inflorescence a few-flowered, sessile, axillary head, nearly terminal on the branchlets; flowers yellowish green, the pedicels very short; calyx-lobes narrowly ovate, about half as long as the tube, the tube 6–8 mm. long; stamens 8, in 2 series, the anthers of upper series semi-exserted from the throat; berry red, ellipsoidal, about 8 mm. long, usually 2-seeded.——Mar.–Apr. Mountains; Honshu, Shikoku, Kyushu.

4. Daphne kamtschatica Maxim. var. **jezoensis** (Maxim.) Ohwi. *D. jezoensis* Maxim.——NANIWAZU. Glabrous, dioecious shrub with thick, terete, pale yellow-brown branchlets; leaves rather thick, cuneate-obovate to narrowly ovate, 4–8 cm. long, 2–3 cm. wide, rounded to obtuse, sometimes retuse with a minute mucro at apex, cuneately narrowed to a short petiole, vivid green or slightly glaucous on upper side, glaucous beneath; inflorescence axillary, sessile, usually more than 10-flowered, glabrous; flowers yellow, on very short pedicels, the staminate slightly larger; calyx-lobes ovate, those of the upper series 4–5 mm. long, slightly shorter than the tube, the calyx-tube 5–8 mm. long, glabrous, greenish, with 8 obtuse longitudinal ridges; stamens 8, in 2 series, the anthers bright yellow, semi-exserted from the throat, the pollen orange-yellow; ovary glabrous.——Apr.–May. Hokkaido, Honshu (Japan Sea side as far west as Noto Prov.).——The typical phase occurs in Sakhalin, Kuriles, Ussuri, and Kamchatka.

2. EDGEWORTHIA Meisn. MITSUMATA ZOKU

Deciduous shrubs with membranous, alternate, entire leaves; inflorescence a dense axillary head on branches of the preceding year; flowers bracteate, bisexual; calyx-lobes 4, the tube cylindric, white-pubescent externally, without appendage in throat; stamens 8, in 2 series on the calyx-tube, the filaments very short, the anthers oblong; disc annular, very short, slightly lobed; ovary sessile, 1-locular, hairy; stigma elongate, linear, papillose; fruit surrounded by the persistent calyx-tube, scarcely fleshy.——Two species, one in India, another in e. Asia.

1. Edgeworthia papyrifera Sieb. & Zucc. *E. chrysantha* sensu auct. Japon., non Lindl.; *E. tomentosa* Nakai, excl. syn.——MITSUMATA. Ternately branched shrub with rather

thick yellow-brown branches appressed-pubescent while young; leaves membranous, lanceolate to broadly so, 8–15 cm. long, 2–4 cm. wide, acute to acuminate, acuminate at base, appressed-pubescent especially on underside, whitish beneath, the petioles 5–8 mm. long, appressed-pubescent; inflorescence densely white-villous, many-flowered, globose, capitate, on nodding peduncles, subtended by few narrowly ovate caducous bracts while young; calyx-lobes elliptic, yellow inside, about 5 mm. long, the tube sessile, 12–14 mm. long; anthers of the upper stamens slightly exserted from the throat.——Mar.–Apr. Introduced from China, now much cultivated in our area for making paper.

3. WIKSTROEMIA Endl.　GAMPI ZOKU

Shrubs or rarely trees; leaves evergreen or deciduous, opposite or alternate; inflorescence terminal, spicate, racemose, or capitate, sometimes in compound terminal panicles, bractless, bisexual; calyx-lobes 4, spreading, the tube elongate, without appendages on the throat; stamens 8, in 2 series on the calyx-tube, the upper 4 often slightly exserted from the throat, the filaments very short, the anthers oblong, the connective indistinct; disc with 4 or 2, very rarely with single scales; ovary sessile or nearly so, 1-locular, usually pubescent; style very short, the stigma large, globose; fruit a berry or rather dry, the endosperm scanty or absent; cotyledons fleshy.——About 50 species, in e. Asia, Malaysia, and the Pacific Islands.

1A. Inflorescence mostly capitate; disc-scales fleshy, somewhat globose-thickened at apex; leaves terminal at tip of branches.
　2A. Leaves somewhat coriaceous, 2–4 cm. long, 1.2–2.2 cm. wide; ovary and fruit short-stiped. .... 1. *W. kudoi*
　2B. Leaves thin, about 15 cm. long, 6 cm. wide; ovary and fruit long-stiped. .... 2. *W. capitellata*
1B. Inflorescence usually a raceme; disc-scales thin; leaves scattered on the branches.
　3A. Glabrous shrubs with opposite leaves.
　　4A. Inflorescence 2- to 4-flowered, without an axis; flowers white. .... 3. *W. albiflora*
　　4B. Inflorescence about 10-flowered, with an elongate axis; flowers yellow. .... 4. *W. trichotoma*
　3B. More or less pubescent shrubs, with alternate leaves.
　　5A. Inflorescence in simple solitary heads on branchlets; leaves ovate, silky on both sides. .... 5. *W. sikokiana*
　　5B. Inflorescence in terminal, usually branched racemes; leaves glabrous to slightly pubescent.
　　　6A. Leaves oblong, glabrous or sparsely pubescent. .... 6. *W. ganpi*
　　　6B. Leaves ovate, more or less appressed-pubescent. .... 7. *W. pauciflora*

1. **Wikstroemia kudoi** Makino. *Diplomorpha kudoi* (Makino) Masam.; *Daphnimorpha kudoi* (Makino) Nakai ——SHAKUNAN-GAMPI.　Branched shrub to 1.5 m. high, with thick, terete, densely leaf-scarred branches; leaves rather coriaceous, alternate, sessile, becoming dense toward the top of branchlets, obovate, 2–4 cm. long, 1.5–2.2 cm. wide, cuneate toward base, obtuse to subacute, glabrous, paler beneath, entire, sessile; heads few, 2- or 3-flowered, terminal, 2–3 cm. long, glabrous; flowers nearly sessile; calyx-lobes 4, ovate-lanceolate, about 5 mm. long, the tube about 8 mm. long, glabrous; upper 4 stamens inserted at the middle on the tube; disc-scales linear, erect; ovary appressed-pilose; fruit about 5 mm. long, enclosed in the calyx-tube, very short stipitate.—— June. High mountains; Kyushu (Mount Miyanouradake in Yakushima); rare.

2. **Wikstroemia capitellata** Hara. *Diplomorpha capitellata* Hara; *Daphnimorpha capitellata* (Hara) Nakai—— TSUCHIBI-NO-KI.　Deciduous shrub to 2 m. high; branches thick, with numerous large leaf-scars; leaves thin, alternate, dense at the top of branches, obovate to lanceolate-obovate, to 15 cm. long and 6 cm. wide, rounded to obtuse, cuneately narrowed at base, entire, glabrous, whitish beneath, the petioles 2–5 mm. long; inflorescence capitate, terminal, glabrous, 5–10 mm. long, sometimes with a few branches; flowers 1 to many, subsessile; calyx 10–14 mm. long, glabrous, the tube on back slightly inflated below the middle, the lobes about 4 mm. long; stamens inserted at the middle of the tube; fruit ovoid, about 5 mm. long, acute, long-stiped.——Wet woods in mountains; Kyushu (Hyuga Prov.); very rare.

3. **Wikstroemia albiflora** Yatabe. *W. gynopoda* Maxim.; *Diplomorpha albiflora* (Yatabe) Nakai——MIYAMA-GAMPI.　Glabrous shrub with slender, red-brown branches and branchlets; leaves membranous, opposite, ovate or elliptic, 1.5–3 cm. long, 1–2 cm. wide, acute to obtuse at both ends, entire, whitish beneath, the petioles 1–2 mm. long; peduncles terminal or subterminal, 0.8 mm. long, slender, simple; inflorescence more or less capitate, 2- to 4-flowered; flowers white, on short pedicels 1–2 mm. long; calyx-lobes elliptic, 2.5–3 mm. long, the tube about 10 mm. long; stamens of the upper series on the throat, the lower 4 median; fruit narrowly ovoid, 4–5 mm. long, distinctly stiped, with few long white hairs or glabrate.——May. Mountains; Honshu (Mount Odaigahara in Yamato), Shikoku, Kyushu; rare.

4. **Wikstroemia trichotoma** (Thunb.) Makino. *Queria trichotoma* Thunb.; *Passerina japonica* Sieb. & Zucc.; *W. japonica* (Sieb. & Zucc.) Miq.; *W. ellipsocarpa* Maxim.; *Diplomorpha ellipsocarpa* (Maxim.) Nakai; *D. trichotoma* (Thunb.) Nakai——KI-GAMPI.　Glabrous much-branched shrub with slender reddish branchlets; leaves membranous, ovate or oblong ovate, 2.5–4.5 cm. long, 1–3 cm. wide, subacute, usually rounded at base, entire, whitish beneath, the petioles short, about 2 mm. long; racemes few- to about 10-flowered, on an axis 2–15 mm. long, the pedicels very short; peduncles slender, 1–2 cm. long, often branched, terminal and in upper axils; calyx yellow, the tube 6–7 mm. long, the lobes elliptic, about 2 mm. long; anthers of the 4 upper stamens slightly exserted from the throat, the lower ones on upper part of the tube; fruit ovoid, about 5 mm. long, short-stiped, glabrous.——Aug.– Sept. Grassy slopes and thickets in hills; Honshu (Kinki Distr. and westw.), Shikoku, Kyushu; rather common.——Korea.

5. **Wikstroemia sikokiana** Fr. & Sav. *Diplomorpha sikokiana* (Fr. & Sav.) Nakai——GAMPI.　Silky pubescent shrub; leaves membranous, more or less alternate, ovate, 3–5 cm. long, 1.5–2 (2.5) cm. wide, densely silky on both sides, especially so beneath, the petioles 2–3 mm. long; inflorescence capitate, few-flowered, terminal, the peduncles terminal, simple, 5–10 mm. long; calyx yellow, white-silky, subsessile, the tube 8–10 mm. long, the lobes ovate, about 3 mm. long; anthers of upper stamens half-exserted from the throat, the lower series above the middle of the tube; ovary white-pubescent; fruit ovoid-fusiform, about 6 mm. long, sparsely white-

hairy, short-stiped.——May–June. Honshu (Tōkaidō Distr. and westw.), Shikoku, Kyushu.

6. Wikstroemia ganpi (Sieb. & Zucc.) Maxim. *Passerina ganpi* Sieb. & Zucc.; *W. canescens* var. *ganpi* (Sieb. & Zucc.) Miq.; *Diplomorpha ganpi* (Sieb. & Zucc.) Nakai——Ko-GAMPI. Shrub with white-appressed pubescence on branchlets, inflorescence, and petioles; leaves membranous, oblong or rarely ovate-oblong, 2–4 cm. long, 8–20 mm. wide, obtuse to subacute, sparsely pubescent or nearly glabrous, with rather conspicuous raised nerves beneath, the petioles 1–2 mm. long; inflorescence a raceme, densely few-flowered, very short, terminal on the branchlets and in upper axils, the peduncles short; calyx lobes narrowly ovate, about 2.5 mm. long, the tube white-pubescent, pale rose, 7–8 mm long; anthers of upper stamens on the throat, the lower series on upper half of the tube; fruit ovoid-fusiform, about 4 mm. long, loosely white-pilose.——Aug. Honshu (Kantō Distr. and westw.), Shikoku, Kyushu.

7. Wikstroemia pauciflora Fr. & Sav. *W. ganpi* var. *pauciflora* (Fr. & Sav.) Maxim.; *Diplomorpha pauciflora* (Fr. & Sav.) Nakai; *W. franchetii* Koidz.——SAKURA-GAMPI, MI-YAMA-KO-GAMPI. Shrub with loosely pubescent branches and leaves, the branchlets much branched; leaves membranous, nearly alternate, ovate, 2–3 cm. long, 1–2 cm. wide, acute, subacute to rounded at base, slightly paler on underside, appressed-pubescent especially beneath, slenderly nerved, the petioles about 2 mm. long; inflorescence very short, few-flowered, terminal and axillary in the upper axils, peduncled, often slightly branched, rather densely white-pubescent; calyx-lobes oblong, 1.5 mm. long, the tube about 6 mm. long, appressed-pubescent; anthers of the upper 4 stamens two-thirds exserted from the throat, the lower 4 on the upper part of the tube; fruit narrowly ovoid, about 3 mm. long, loosely appressed-pilose, stipitate.——July–Aug. Honshu (Idzu, Sagami, and Suruga Prov.).

Var. **yakusimensis** Makino. *W. yakusimensis* (Makino) Nakai; *Diplomorpha yakusimensis* (Makino) Masam.——SHIMA-SAKURA-GAMPI. Plant slightly larger; leaves 3–7 cm. long, 1.5–3 cm. wide, often acuminate; inflorescence more branched; flowers more numerous; anthers of lower stamens nearer the throat.——Aug. Kyushu (Bungo, Hyuga, and Osumi Prov. in Yakushima.).

Fam. 143. ELAEAGNACEAE   GUMI KA   Oleaster Family

Woody plants with scalelike or stellate hairs; leaves alternate, rarely opposite, entire; flowers bisexual or unisexual (plants dioecious or polygamodioecious), solitary, fasciculate, or in racemes; calyx tubular, constricted above the ovary, the lobes 2 to 4, valvate, rarely depressed and truncate; stamens on the calyx-tube, 4, and alternate with calyx-lobes, or 8, the filaments free, the anthers 2-locular; staminodes absent; ovary sessile, 1-locular; style terminal, linear, the stigma on one side; ovules solitary, basal, erect, anatropous; fruit enclosed in the thickened fleshy or juicy calyx-tube, berrylike or drupaceous; seeds erect, the endosperm scanty or absent, the embryo straight.——Three genera, with about 50 species, in N. America, e. Asia, Europe, and Malaysia.

1. ELAEAGNUS L.   GUMI ZOKU

Deciduous or evergreen trees or shrubs often with spines (reduced branches); leaves alternate, short-petioled; flowers axillary, solitary or in fascicles, without common peduncles, bisexual or the plants polygamous; calyx-lobes 4, the tube (receptacle) cylindric or campanulate, somewhat constricted near the base, the lower part adnate to the ovary; stamens 4, inserted on the tube, the filaments very short; fruit drupelike, the stones ellipsoidal, longitudinally sulcate.——About 45 species, s. Europe, abundant in e. Asia and N. America.

1A. Evergreen trees or shrubs with coriaceous leaves, flowering in autumn, fruiting in spring.
  2A. Petioles 1–2.5 cm. long; leaves orbicular-ovate to broadly elliptic-ovate, 4–6 cm. wide, abruptly acuminate, densely white-scaly beneath. . . . . 1. *E. macrophylla*
  2B. Petioles 7–12 mm. long; leaves oblong or ovate-elliptic, less than 3.5 cm. wide.
    3A. Leaves oblong to ovate-elliptic, often caudate, brown-scaly beneath. . . . . 2. *E. glabra*
    3B. Leaves oblong to narrowly so, acute to obtuse, prominently undulate on margin, with white and brown scales beneath; branchlets often spinelike. . . . . 3. *E. pungens*
1B. Deciduous shrubs or small trees with membranous leaves, flowering in spring, fruiting in autumn.
  4A. Underside of leaves scaly, without stellate hairs.
    5A. Branchlets white-scaly; flowers 1–7 in axils, the pedicels 3–12 mm. long; calyx-lobes small, one-half to two-thirds as long as the tube, the tube gradually narrowed to the ovary. . . . . 4. *E. umbellata*
    5B. Branchlets brown-scaly; flowers 1–3 in axils, the pedicels 1–5 cm. long; calyx-lobes broadly ovate, the tube constricted above the ovary.
      6A. Leaves abruptly acuminate with an obtuse tip, acute to rounded at base; flowering late May to July. . . . . 5. *E. montana*
      6B. Leaves, especially at flowering time, obtuse to acute at both ends; flowering April to early May. . . . . 6. *E. multiflora*
  4B. Underside of leaves with scales and stellate hairs; young branches with few stellate hairs.
    7A. Flowers usually solitary in axils, the pedicels elongate and nodding, 1.5–5 cm. long in fruit.
      8A. Branchlets and petioles with red-brown scales; leaves oblong, obovate, oblanceolate, sometimes ovate-elliptic; fruit obovoid, with a narrow stipelike base. . . . . 7. *E. murakamiana*
      8B. Branchlets and petioles with pale red-brown or fulvous scales; leaves lanceolate to ovate or obvate-elliptic, often caudate; fruit ellipsoidal to oblong.
        9A. Calyx-tube about 7 mm. long; fruit broadly ellipsoidal, 6–7 mm. long. 8. *E. matsunoana*
        9B. Calyx-tube about 4–5 mm. long; fruit oblong-fusiform, about 12 mm. long. . . . . 9. *E. takeshitae*
    7B. Flowers 1–3 in axils, the pedicels straight, about 1 cm. long in fruit. . . . . 10. *E. yoshinoi*

1. **Elaeagnus macrophylla** Thunb. *E. macrophylla* var. *typica* C. K. Schn.——MARUBA-GUMI. Slightly scandent evergreen shrub with white and pale brown scales while young; leaves coriaceous, orbicular-ovate to broadly elliptic-ovate, sometimes orbicular, 5–7 cm. long, 4–6 cm. wide, acute to abruptly acuminate with an obtuse tip or rounded at both ends, loosely white-scaly on upper surface while young, densely white-scaly beneath, the petioles 1–2.5 cm. long; flowers few in axils, the pedicels 5–8 (–10) mm. long, densely white-scaly, sometimes also with reddish brown scales; calyx campanulate, the tube abruptly narrowed at base, 4–5 mm. long, the lobes broadly ovate, slightly shorter than the tube; fruit oblong, white-scaly, 1.5–2 cm. long.——Oct.–Nov. Thickets near seashores; Honshu (Kantō Distr. and westw.), Shikoku, Kyushu. ——s. Korea and Ryukyus. Hybrids are:

Elaeagnus × maritima Koidz. *E. hisauchii* Makino; *E. liukiuensis* Rehd.——AKABA-GUMI, ŌBA-TSURU-GUMI. Hybrid of *E. glabra × E. macrophylla*.

Elaeagnus × submacrophylla Serv. *E. nikaii* Nakai—— Ō-NAWASHIRO-GUMI. Hybrid allegedly of *E. macrophylla × E. pungens*.

2. **Elaeagnus glabra** Thunb. TSURU-GUMI. Scandent evergreen shrub with spreading, rather slender, red-brown scaly branches; leaves rather coriaceous, oblong to ovate-elliptic, 4–8 cm. long, 2.5–3.5 cm. wide, often caudate with an obtuse tip, rounded to acute at base, often stellate-scaly on upper surface while young, becoming glabrate, densely red-brown scaly beneath, the petioles 7–10 mm. long; flowers few, axillary, fasciculate, the pedicels red-brown scaly, 4–7 mm. long; calyx-tube narrow, 4–5 mm. long, about twice as long as the lobes; fruit oblong, 12–18 mm. long, red-brown scaly.——Oct.–Nov. Thickets in hills; Honshu (Kantō Distr. and westw.), Shikoku, Kyushu; common.

Elaeagnus × reflexa Morren & Decne. *E. glabropungens* Maxim. ex Nakai; *E. hypoargentea* Hatusima——MARUBA-TSURU-GUMI, URAGIN-TSURU-GUMI. Hybrid of *E. glabra × E. pungens*.

3. **Elaeagnus pungens** Thunb. NAWASHIRO-GUMI. Evergreen shrub; branchlets densely brown-scaly, frequently reduced to spines; leaves coriaceous, oblong to narrowly so, 5–8 cm. long, 2.5–3.5 cm. wide, obtuse to acute with an obtuse tip, rounded at base, obsoletely toothed with prominently undulate margins, glabrous and lustrous on upper side, densely whitish and usually also brown-scaly beneath; petioles 10–12 (–15) mm. long, brown-scaly; flowers few, fasciculate in axils, the pedicels 5–8 mm. long, brown-scaly; calyx-tube 6–7 mm. long, rather thick, abruptly narrowed at base, the lobes ovate-rounded, about half as long as the tube; fruit oblong, about 1.5 cm. long, brown-scaly.——Oct.–Nov. Thickets in lowlands especially near the sea; Honshu (centr. distr. and westw.), Shikoku, Kyushu; common.

4. **Elaeagnus umbellata** Thunb. *E. crispa* Thunb.; *E. umbellata* var. *typica* C. K. Schn.——AKI-GUMI. Deciduous shrub with densely white-scaly branchlets becoming grayish with age; leaves membranous, oblong-lanceolate, sometimes narrowly oblong or broadly oblanceolate, 4–8 cm. long, 1–2(–2.5) cm. wide, subobtuse to acute with an obtuse tip, acute to gradually so at base, thinly white-scaly on upper surface while young, densely and persistently so beneath, the petioles 5–10 mm. long, white-scaly; flowers 1–7, fasciculate in axils, densely white-scaly, the pedicels 3–6(–8) mm. long at anthesis to 12 mm. long in fruit; calyx-tube slender, gradually narrowed at base, 5–7 mm. long, the lobes narrowly ovate, about one-half to two-thirds as long as the tube, acute to acuminate; fruit subglobose to broadly ellipsoidal, 6–8 mm. long.——May. Thickets and thin woods in lowlands and hills; Hokkaido (w. distr.), Honshu, Shikoku, Kyushu. Common and variable.

Var. **coreana** (Lév.) Lév. *E. coreana* Lév.; *E. umbellata* var. *borealis* Ohwi——KARA-AKI-GUMI, MICHI-NO-KU-AKI-GUMI. Leaves sparsely stellate-pilose on upper side.——Honshu (n. distr.).——Korea.

Var. **rotundifolia** Makino. *E. fragrans* Nakai——MARUBA-AKI-GUMI. Branchlets thicker; leaves obovate to ovate, to 5 cm. wide.——Abundant near seashores; Honshu, Shikoku, Kyushu.

5. **Elaeagnus montana** Makino. MAME-GUMI. Deciduous shrub with densely brown-scaly young branchlets; leaves chartaceous to somewhat membranous, elliptic, ovate to oblong, 3–8.5 cm. long, 1.5–4 cm. wide, rather abruptly acuminate with an obtuse tip (especially at flowering time), acute to rounded at base, loosely white-scaly or glabrous on upper side, densely white- and red-brown scaly beneath, the petioles 5–8 mm. long; flowers 1–3, usually solitary, axillary, the pedicels about 1 cm. long at anthesis, 1–3 cm. long in fruit, thickened in upper part, whitish scaly; calyx-tube abruptly constricted above the ovary, the lobes large, broadly ovate, 3–4 mm. long, acute or abruptly so; fruit ovoid-ellipsoid, about 1 cm. long. ——May–July. Mountains; Honshu, Shikoku, Kyushu.

Var. **ovata** (Maxim.) Araki. *E. longipes* var. *ovata* Maxim.; *E. tsukubana* Makino; *E. nikoensis* Nakai; *E. multiflora* var. *tsukubana* (Makino) Ohwi——TSUKUBA-GUMI, NIKKŌ-NATSU-GUMI. Branchlets red-brown scaly; leaves green, with fulvous stellate hairs on upper surface while young, densely white-scaly and loosely brown stellate-hairy beneath; calyx-tube red-brown scaly.——Honshu (Iwaki to Shinano Prov.).

6. **Elaeagnus multiflora** Thunb. *E. longipes* A. Gray; *E. multiflora* var. *crispa* (Maxim.) Serv.; *E. longipes* var. *crispa* Maxim., excl. syn.——NATSU-GUMI. Deciduous shrub or small tree with red-brown scales on young branchlets; leaves elliptic, ovate to obovate-oblong, 3–10 cm. long, 2–5 cm. wide, obtuse to acute at both ends (especially at flowering time), loosely stellate-pilose while young, densely white- and brown-scaly beneath; flowers 1 or 2, usually solitary, axillary, densely white- and sparsely brown-scaly; corolla-tube constricted at base, nearly as long to 1.5 times as long as the broadly ovate abruptly acute lobes; fruit oblong, about 1.5 cm. long, on a slender pendulous pedicel 1.5–5 cm. long.——Apr.–May. Thickets and thin woods in lowlands and mountains; Hokkaido, Honshu, Shikoku, Kyushu; common and variable; often cultivated.

Var. **hortensis** (Maxim.) Serv. *E. edulis* Carr.; *E. longipes* var. *hortensis* Maxim.——TŌ-GUMI. Leaves elliptic, loosely stellate-pilose on upper side while young; fruit larger.——Occurs with the typical phase.

Var. **angustifolia** (Nakai) Makino & Nemoto. *E. longipes* var. *angustifolia* Nakai——HOSOBA-NATSU-GUMI. Leaves narrow, loosely white-scaly on upper side.

Var. **jucundicocca** (Koidz.) Ohwi. *E. jucundicocca* Koidz.; *E. isensis* Makino——SAI-GUMI, ISUZU-GUMI. Leaves narrow, with white-stellate scalelike hairs on upper side while young; pedicels 4–5 cm. long.——Honshu (Kinki Distr.).

7. **Elaeagnus murakamiana** Makino. ARIMA-GUMI. Deciduous shrub, the branchlets densely red-brown scaly and

loosely stellate-pilose while young; leaves membranous, oblong, obovate, oblanceolate or sometimes ovate-elliptic, 4–8 cm. long, 1–4 cm. wide, obtuse to abruptly acuminate with an obtuse tip, with stellate hairs on upper side while young, densely white-scaly and with fulvous-stellate hairs especially on nerves near base beneath, the petioles 3–5 mm. long; flowers 1 to 3 in axils, the pedicels 2.5–3 cm. long in fruit, with white and brown scales and pale stellate pubescence; calyx-tube 6–8 mm. long, abruptly constricted above the ovary, with white-stellate hairs, the lobes broadly ovate, 3–4 mm. long; fruit brown-scaly, obovoid, 7–8 mm. long, abruptly narrowed to a stipelike base.——Apr.–May. Honshu (Tōkai and Kinki Distr.), Shikoku.

8. Elaeagnus matsunoana Makino. HAKONE-GUMI. Deciduous shrub; branchlets gray-brown, densely fulvous-scaly, with fulvous-stellate hairs while young; leaves membranous, broadly lanceolate to narrowly ovate, 4–8 cm. long, 1.5–3.5 cm. wide, caudate-acuminate to acute with an obtuse tip, obtuse, rounded to subacute at base, stellate-pilose on upper side, densely white-scaly and loosely stellate-pilose beneath, the petioles 3–5 mm. long; flowers solitary, with pale yellow scales and stellate hairs; calyx-tube rather broad, abruptly constricted above the ovary, about 7 mm. long, the lobes rounded-ovate, 4–5 mm. long, acute; fruit broadly ellipsoidal, small, 6–7 mm. long, the pedicels 2.5–4 cm. long, thinly stellate-pilose at first, later only with scales.——May. Honshu (Musashi, Kai and to Suruga Prov.); rare.

9. Elaeagnus takeshitae Makino. KATSURAGI-GUMI. Much-branched, deciduous shrub with slender branches, the branchlets gray-brown, with dense red-brown stellate scales; leaves membranous, ovate-elliptic to ovate-oblong, 4–6 cm. long, 1.5–2.5 cm. wide, caudate-acuminate, rather obtuse at the tip, obtuse to acute at base, conspicuously stellate-hairy on both sides, loosely brown-scaly beneath, the petioles 2–4 mm. long, brown-scaly and stellate-pilose; flowers solitary; calyx-tube slender, 4–5 mm. long, constricted above the ovary, the lobes ovate, acute, slightly shorter than the tube; fruit oblong-fusiform, about 12 mm. long, the pedicels 2.5–4 cm. long, at first fulvous-pilose and -scaly, later only scaly.——May. Honshu (Kawachi, Izumi, and Kii Prov.); rare.

10. Elaeagnus yoshinoi Makino. NATSU–ASADORI. Large deciduous shrub; branches dark brown, the young branchlets densely stellate-pilose; leaves membranous, obovate-elliptic, oblong, to ovate-elliptic, 5–12 cm. long, 3–6 cm. wide, acute to short-acuminate with an obtuse tip, rounded to obtuse or broadly cuneate at base, with brown scales while young, later densely stellate-hairy on both surfaces, especially so beneath, the petioles 5–8 mm. long, with fulvous-stellate hairs; flowers 1 to 3, fasciculate in axils; calyx-tube slender, 8–10 mm. long, slightly narrowed at base, white stellate-pilose, the lobes broadly ovate, acute, about half as long as the tube; fruit obovoid-oblong, about 1 cm. long, the pedicels erect, straight, about 1 cm. long, densely fulvous-stellate hairy.——May. Honshu (Chūgoku Distr.).

Fam. 144. LYTHRACEAE   MISO-HAGI KA   Loosestrife Family

Herbs or woody plants; leaves opposite or verticillate, rarely alternate, stipules absent or very small; flowers usually actinomorphic, bisexual, solitary or paniculate; calyx tubular, the lobes valvate, often with an appendage between the lobes; petals on the throat of calyx-tube, usually wrinkled in bud, sometimes absent; stamens 4 or 8, sometimes fewer, inserted on the tube; anthers 2-locular; ovary superior, perfectly or imperfectly 2- to 6-locular, rarely 1-locular; style simple, solitary; ovules many, the placentation typically axile; capsules longitudinally to transversely or sometimes irregularly dehiscent; seeds many, without endosperm, the embryo straight.——About 22 genera, with about 500 species, chiefly in the Tropics.

1A. Trees or shrubs; placentae continuous on the style; stamens many, with long filaments; capsules loculicidally dehiscent; seeds winged.
   1. *Lagerstroemia*
1B. Herbs; placentae interrupted, not continuous on the style; stamens 2–12; capsules septicidally or irregularly dehiscent; seeds unwinged.
   2A. Capsules not horizontally striate.
      3A. Capsules irregularly dehiscent. . . . . 2. *Ammannia*
      3B. Capsules septicidally dehiscent. . . . . 3. *Lythrum*
   2B. Capsules transversely striate, septicidally dehiscent. . . . . 4. *Rotala*

1. LAGERSTROEMIA L.   SARU-SUBERI ZOKU

Evergreen or deciduous trees or shrubs; leaves opposite, simple; stipules small, caducous; racemes terminal or axillary, often compound and paniculate, the bracts and bracteoles caducous, the pedicels jointed; flowers 5- to 8-merous; calyx nearly hemispherical or obconical, sometimes ridged or winged, the appendage absent or short; petals often clawed; stamens many, inserted on the calyx-tube, the longer ones opposite the lobes; ovary 3- to 6-locular; ovules many; style slender; capsules ellipsoidal, woody, loculicidally dehiscent, 3- to 6-valved; seeds winged. ——About 30 species, in se. Asia, Malaysia, and Australia.

1A. Branches and inflorescence puberulous; leaves 3.5–8 cm. long, with 3–5 pairs of lateral nerves, the petioles 2–3 mm. long.
   1. *L. subcostata*
1B. Branches and inflorescence glabrous; leaves 8–10 cm. long, with 8–13 pairs of lateral nerves, the petioles 8–10 mm. long. . . 2. *L. fauriei*

1. Lagerstroemia subcostata Koehne. *L. subcostata* var. *hirtella* Koehne——SHIMA-SARU-SUBERI.   Deciduous, much-branched tree, branches gray-brown, terete, 4-striate when young, spreading-puberulent; leaves ovate, elliptic, or obovate, 3.5–8 cm. long, 2–3 cm. wide, acute to short-acuminate, subacute at base, glabrous on upper side, with 3–5 pairs of lateral nerves, white spreading-pubescent at least on the axils of nerves beneath, the petioles 2–3 mm. long; panicles terminal, 5–10 cm. across, obtuse, densely many-flowered, puberulent, the pedicels 5–8 mm. long, jointed at base, the subtending bracts small; calyx-tube hemispherical-obconical, about 3 mm. long, glabrescent, 10- to 12-ribbed, the lobes 5–6, ovate, acute; limb of the petals white, ovate, wrinkled, 5–6 mm. long, slightly longer than the linear claw; capsules ellipsoidal, 8–10 mm. long;

seeds obliquely oblong, inclusive of the wing, about 6 mm. long.——July–Aug. Kyushu (Yakushima and Tanegashima). ——Ryukyus and Formosa.

Lagerstroemia indica L. SARU-SUBERI. Branchlets 4-winged; leaves obtuse to rounded at apex; flowers large; calyx-tube inconspicuously nerved. Introduced from China. Much planted as an ornamental in gardens.

Lagerstroemia × amabilis Makino. KO-SARU-SUBERI. Hybrid of *L. indica × L. subcostata*. Sometimes cultivated.

2. **Lagerstroemia fauriei** Koehne. YAKUSHIMA-SARU-SU-BERI. Deciduous tree with glabrous somewhat terete branches; leaves coriaceous, oblong, sometimes ovate, 8–10 cm.

long, about 5 cm. wide, slightly obtuse, subrounded and abruptly petioled at base, glabrous except for axillary tufts of hairs beneath, with 8–13 pairs of lateral nerves, the veinlets prominently reticulate beneath, the petioles 8–10 mm. long; panicles 5–10 cm. long, slightly narrower than the length, rather densely flowered, glabrous, the branches 4-angled, the pedicels 3–7 mm. long; flowers 6-merous; calyx about 4.5 mm. long and as wide, 12-nerved, the lobes deltoid, without appendages; limb of the petals 4 mm. long, about as long as the claw, 4-angled, orbicular, cordate; stamens 30–36; ovary 6-locular; style about 8–9 mm. long.——Kyushu (Yakushima).

2. AMMANNIA L. HIME-MISO-HAGI ZOKU

Annual herbs growing in wet places or aquatic; stems and branches often 4-angled; leaves opposite, sessile, 1-nerved; inflorescence cymose, sessile or pedunculate, the bracteoles small, membranous, hyaline or white; flowers small, 4(–6)-merous; calyx-tube campanulate or urceolate, often becoming globose or obovoid in fruit; stamens 2–8, inserted on the tube; ovary incompletely 2- to 4-locular or 1-locular; capsules globose or ellipsoidal, often slightly exserted from the calyx-tube, thinly membranous, irregularly dehiscent; seeds numerous, minute.——About 20 species, mostly in the Tropics, also warm-temperate N. America and e. Asia.

1A. Flowers petaliferous; leaves more or less auricled at base.
  2A. Capsules exserted from the calyx, 2–2.5 mm. across .... 1. *A. multiflora*
  2B. Capsules not exserted from the calyx, 3.5–4 mm. across. .... 2. *A. coccinea*
1B. Flowers apetalous; leaves narrowed at the base; capsules 1–1.5 mm. across, exserted from the calyx. .... 3. *A. baccifera*

1. **Ammannia multiflora** Roxb. *A. parviflora* DC.; *A. multiflora* var. *parviflora* (DC.) Koehne; *A. japonica* Miq.—— HIME-MISO-HAGI. Annual, 20–30 cm. high; stems erect, 4-angled, short-branched; leaves spreading, broadly linear to lanceolate-oblong, 2.5–5 cm. long, 3–12 mm. wide, or 7–20 mm. long, 1–4 mm. wide in those of the branches, acute to subobtuse, slightly narrowed and rounded at the clasping and auriculate base; flowers few, in dense sessile axillary cymes, the pedicels about as long as the flowers; flowers about 1.5 mm. across; calyx obconical, obsoletely 4-angled, the lobes deltoid; petals minute, whitish; stamens 4; capsules globose, 2–2.5 mm. across, red-brown, lustrous, half-exserted from the calyx; styles distinct, slightly less than 1 mm. long, the stigmas rather thickened; seeds minute, brown.——Sept.–Nov. Wet places in lowlands, especially in paddy fields; Honshu, Shikoku, Kyushu. ——Tropics and subtropics of Africa, Asia, and Australia.

2. **Ammannia coccinea** Rottb. HOSOBA-HIME-MISO-HAGI. Glaucescent glabrous annual; stems simple or branched at base, quadrangular, 30–50 cm. long; leaves broadly linear to lanceolate, 3–5 cm. long, 3–6 mm. wide, acute, auriculate and sessile

at base, entire; flowers few, in dense axillary fascicles, the pedicels less than 1 mm. long; calyx about 2.5 mm. long, short-tubulose at anthesis, hemispherical, 3–4 mm. long in fruit, whitish with green nerves, teeth short, green; petals minute, purplish; stamens slightly exserted; capsules globose, not exserted from the calyx, 3.5–4 mm. across; styles about 1.5 mm. long.——Kyushu (Hizen Prov.); introduced.——N. and S. America, Philippines, Micronesia, and Ryukyus.

3. **Ammannia baccifera** L. *A. indica* Lam.; *A. discolor* Nakai——SHIMA-MISO-HAGI, NAGATO-MISO-HAGI. Erect glabrous much-branched annual; stems 10–50 cm. long, 4-angled; leaves membranous, oblanceolate to linear, 5–50 mm. long, 1–10 mm. wide, narrowed at both ends, entire; flowers few, in dense axillary fascicles, the pedicels 0.2–1.5 mm. long, glabrous; calyx 1–2 mm. long, the lobes 0.5–0.7 mm. long; petals absent; stamens 4, the anthers subglobose; styles 0.3–0.5 mm. long; capsules globose, 1–1.5 mm. across, slightly exserted from the calyx.——Paddy fields; Honshu (Nagato Prov.); introduced.——Africa, se. Asia, Europe, and Australia.

3. LYTHRUM L. MISO-HAGI ZOKU

Annuals, perennials, or rarely half-shrubs; leaves opposite, verticillate, or rarely alternate, entire; inflorescence cymose, often in compound spikes or umbels, or flowers solitary in axils, actinomorphic or slightly zygomorphic, 4- to 6-merous, dimorphic or trimorphic; calyx-tube tubular or rarely campanulate, appendage distinct, rarely obscure; petals prominent or absent; stamens 4–12, the ventral ones inserted higher than the dorsal; ovary sessile or nearly so; style distinct, rarely short; capsules septicidally dehiscent, 2-valved, the valves sometimes 2-fid; seeds 8 to many, small.——About 25 species, in the Tropics and temperate regions.

1A. Plant scabrous with papillose hairs; leaves rounded or subcordate at base, semiclasping; appendage of calyx-lobes 1.5–2 mm. long, erect.
                                                     1. *L. salicaria*
1B. Plant glabrescent; leaves acute to broadly cuneate at base, sessile but not clasping; appendage of calyx-lobes 0.5–0.7 mm. long, spreading.
                                                     2. *L. anceps*

1. **Lythrum salicaria** L. *L. salicaria* var. *intermedium* Koehne; *L. salicaria* var. *vulgare* DC.——EZO-MISO-HAGI. Rhizomatous perennial, usually scabrous on stems, leaves be-

neath, and calyx; stems erect, slightly branched, 4-striate, 50–100 cm. long; leaves opposite, lanceolate to broadly so, 4–6 cm. long, 8–15 mm. wide, subobtuse or acute, rounded or subcordate and semiclasping at base; spikes erect, terminal, 20–35 cm. long, the bracts ascending, broadly lanceolate to deltoid-ovate, 5–12 mm. long; flowers rose-purple, 1–3 in axils, cymose; peduncles and pedicels very short; calyx-tube 5–8 mm. long, 12-ribbed, with 6 deltoid teeth, the appendage spinelike, erect, 1.5–2 mm. long; petals cuneate-oblong, 7–8 mm. long.——Aug. Wet places in lowlands and hills; Hokkaido, Honshu, Shikoku, Kyushu; rather common.——Europe, N. Africa, Afghanistan, n. India, China, and e. Asia.

2. Lythrum anceps (Koehne) Makino. *L. virgatum* sensu Miq. non L.; *L. salicaria* var. *anceps* Koehne; *L. salicaria* subsp. *anceps* (Koehne) Hara——MISO-HAGI. Resembling the preceding but more slender and nearly glabrous; leaves usually acute, acute to broadly cuneate at base, sessile or sub-sessile, not clasping; floral bracts usually spreading, broadly lanceolate to ovate-oblong; appendages of calyx-tube short, 0.5–0.7 mm. long, spreading.——Wet places in lowlands and hills; Honshu, Shikoku, Kyushu.——Korea.

4. ROTALA L.　KIKASHIGUSA ZOKU

Glabrous annuals or perennials; leaves opposite or verticillate, sessile; flowers small, 3- to 6-merous, usually sessile, solitary in axils or in terminal spikes or panicles, 2-bracteolate at base; calyx nearly globose, campanulate, or urceolate, 3- to 6-lobed, often with a spinelike appendage between the lobes, with annular nectary glands at base; petals persistent, deciduous, or absent; stamens 1–6, opposite the calyx-lobes; ovary incompletely 2- to 4-locular; ovules few to many; style absent or conspicuous; capsules septicidally dehiscent, cartilaginous, the pericarp with minute transverse striae; seeds minute.——About 40 species, in the Tropics and warm temperate regions.

1A. Leaves verticillate, shortly truncate or minutely bimucronate at apex.
  2A. Flowers apetalous; stamens 2 or 3; leaves 3- or 4-verticillate. .... 1. *R. mexicana*
  2B. Flowers petaliferous; stamens 4; leaves 5- to 12-verticillate. .... 2. *R. hippuris*
1B. Leaves opposite, obtuse, rounded, or subacute at apex.
  3A. Flowers or fascicles of flowers axillary, white to reddish.
    4A. Leaves broadly linear to broadly lanceolate-oblong, 6–25 mm. long, obtuse or subacute. .... 3. *R. leptopetala* var. *littorea*
    4B. Leaves obovate-oblong, 5–10 mm. long, rounded at apex.
      5A. Leaves with translucent cartilaginous margins; styles 0.5–0.7 mm. long, slender, distinct, with a capitate stigma; stamens 4.
                                                    4. *R. indica*
      5B. Leaves without translucent cartilaginous margins; styles shorter, with a capitellate stigma; stamens 2. ...... 5. *R. elatinomorpha*
  3B. Flowers in a terminal spike, red-purple. .... 6. *R. rotundifolia*

1. Rotala mexicana Cham. & Schltdl. *R. apetala* F. Muell.; *Hypobrichia spruceana* Benth.; *R. mexicana* var. *typica* Koehne, and var. *spruceana* (Benth.) Koehne——MIZU-MATSUBA. Glabrous reddish annual 3–10 cm. high, usually creeping and branching at base, erect in upper part, rather soft; leaves 3- to 4(–5)-verticillate, narrowly lanceolate to broadly linear, 6–10 mm. long, 1.5–2 mm. wide, truncate and often bimucronate, narrowed to a sessile base; flowers solitary in axils, sessile, about 0.8 mm. long, reddish, usually 5-merous, the bracteoles thinly membranous, linear, as long as the calyx; calyx semiglobose in fruit, the lobes deltoid; stamens 2 or 3; capsules globose, about twice as long as the calyx; style absent.——Aug.–Oct. Paddy fields and wet places in lowlands; Honshu, Shikoku, Kyushu; common.——The Tropics and warm-temperate regions of the world.

2. Rotala hippuris Makino. MIZU-SUGINA. Usually glabrous perennial with slender creeping rhizomes; stems erect, terete, branched at base; leaves 5- to 12-verticillate, the submerged ones filiform, 2–3 cm. long, 2-lobulate at apex, the aerial ones linear, 5–10 mm. long, 0.7–1 mm. wide, obtusely truncate to obscurely bimucronate at apex, scarcely narrowed at base; flowers solitary in axils, sessile, the bracteoles linear, shorter than the calyx-tube; calyx-tube about 0.7 mm. long, globose in fruit, the lobes 4, deltoid, short; petals obovate, about twice as long as the calyx-lobes, retuse; stamens 4; capsules globose, about 1.5 mm. across; styles short and slender, the stigmas capitellate.——Sept.–Oct. Wet places in lowlands; Honshu (Owari and Ise Prov.); very rare.

3. Rotala leptopetala (Bl.) Koehne var. **littorea** (Miq.) Koehne. *Ammannia littorea* Miq.——MIZU-KIKASHIGUSA. Erect, branched annual often ascending from base; stems terete, 10–30 cm. long; leaves broadly linear to broadly lanceolate-oblong, 6–25 mm. long, 2–5 mm. wide, obtuse to subacute,

obtuse to subrounded at the sessile base; leaves on upper branches much smaller and narrower; flowers solitary in axils, sessile, often reddish, about 0.7 mm. long; calyx-tube cup-shaped in fruit, the lobes usually 4, very small, acuminate; petals oblanceolate, as long as the calyx-tube, persistent; capsules globose, about 2 mm. across, reddish, much exserted from the calyx-tube.——Aug.–Nov. Paddy fields and wet places in lowlands; Honshu, Shikoku, Kyushu; common.——The typical phase occurs in e. Asia to Malaysia, and s. Asia to Afghanistan.

4. Rotala indica (Willd.) Koehne. *Peplis indica* Willd.; *Ammannia peploides* Spreng.; *Ameletia uliginosa* Miq.——KIKASHIGUSA. Glabrous annual; stems creeping and branched at base, ascending to erect and short-branched in upper part; leaves opposite, flat, obovate-oblong or obovate, sometimes nearly oblong, 5–10 mm. long, 3–5 mm. wide, rounded at apex, narrowed to the sessile base, with translucent cartilaginous margins; leaves of short upper branches smaller; flowers solitary in axils, sessile, the bracteoles linear, as long as the calyx; calyx tubular-campanulate, about 1.5 mm. long, the lobes 4, narrowly deltoid, acuminate, spine-tipped; petals small; stamens 4; capsules ellipsoidal, not exserted, included within the calyx; styles slender, 0.5–0.7 mm. long, the stigmas capitate.——Aug.–Oct. Paddy fields and wet places in lowlands; Hokkaido (sw. distr.), Honshu, Shikoku, Kyushu; common.——e. Asia, Malaysia, and India.

5. Rotala elatinomorpha Makino. HIME-KIKASHIGUSA. Glabrous annual; stems creeping, the branches erect 4–7 cm. long, loosely leafy; leaves thin, opposite, obovate-oblong, 3–10 mm. long, 1.5–4 mm. wide, rounded at apex, narrowed to the sessile base, without translucent cartilaginous margins; flowers sessile, solitary in axils, about 1.5 mm. long, the bracteoles linear, acuminate, as long as the calyx-tube; calyx-tube short-cylindric, 4-angled, without appendages, the lobes 4, narrowly

Plate 12

Pittosporum tobira (Thunb.) Ait. (Pittosporaceae). Cape Ashizuri, Tosa Prov., s. Shikoku. (Photo J. Ohwi.)

Rhus javanica L. (Anacardiaceae). Narashino near Tokyo, Honshu. (Photo J. Ohwi.)

Plate 13

Kandelia candel (L.) Druce (Rhizophoraceae). Kumano, Tanegashima Island, Kyushu. (Photo S. Ouchiyama.)

Angelica ursina (Rupr.) Maxim. (Umbelliferae). Rubeshibe, Kitami Prov., Hokkaido. (Photo M. Tatewaki.)

Plate 14

Helwingia japonica (Thunb.) F. G. Dietr. (Cornaceae). Fuji-Yamonaka, Honshu. (Photo Lindquist and Nitzelius.)

Rhododendron japonicum (A. Gray) Sur. (Ericaceae). Red flowered phase; Mount Hakkoda, Honshu. (Photo J. L. Creech.)

deltoid, acuminate, about one-third as long as the calyx-tube; petals less than half as long as the calyx-lobes, obovate-oblong, obtuse; stamens 2; styles very short, with a capitellate stigma; capsules globose or ellipsoidal, about 1 mm. across.——Sept.–Oct. Wet places in lowlands; Honshu (Shimōsa and westw.), Shikoku; rare.

6. Rotala rotundifolia (Roxb.) Koehne. *Ammannia rotundifolia* Roxb.——HOZAKI-KIKASHIGUSA. Glabrous perennial; rhizomes slender, creeping; stems simple or sparsely branched, 10–20 cm. long, erect, with ascending base, loosely leafy, purplish red; leaves membranous, opposite, elliptic to obovate-elliptic, sometimes nearly orbicular, 5–8 mm. long, 3.5–6 mm. wide, rounded at apex, narrowed to the sessile base; inflorescence erect, 3–6 cm. long, small-leafy; flowers many in a terminal spike, red-purple, about 2.5 mm. across, the bracteoles lanceolate, slightly longer than the calyx-tube; calyx-tube obconical, about 1 mm. long, the lobes 4, deltoid, acute, slightly shorter than the tube; petals 4, obovate, 1.5 mm. long, red-purple; stamens not exserted from calyx-lobes; capsules dehiscent, 4- or rarely 3-valved; styles short, the stigma capitate.——May. Kyushu (Chikuzen, Chikugo, Higo Prov.).——Ryukyus, Formosa, China, and India.

Fam. 145. **RHIZOPHORACEAE** HIRUGI KA Mangrove Family

Trees or shrubs of tidal marshes; branches inflated and jointed at the nodes; leaves usually coriaceous and simple, opposite, stipulate, rarely alternate and exstipulate; inflorescence axillary; flowers bisexual; calyx-tube free or adnate to the ovary, the lobes 3–14, persistent, valvate; petals small, sometimes laciniate, convolute or involute; stamens inserted on the disc, as many as the petals or more, often in pairs, opposite the petals; anthers 4- or many-locular; ovary usually inferior, 2- to 6-locular, rarely 1-locular; ovules inserted on inner angle near top of the locule in pairs or rarely more; styles free or connate; fruit usually indehiscent, 1-seeded; seeds sometimes with endosperm, often germinating on the plant and with a well-developed much-elongating hypocotyl; cotyledons cylindric or connate.——About 17 genera, with about 60 species, the chief component of tropical mangrove forests.

1. **KANDELIA** Wight & Arn. ME-HIRUGI ZOKU

Evergreen small tree with terete branches; leaves thick-coriaceous, opposite, simple, petiolate, elliptic, obtuse, entire, the stipules intrapetiolar, caducous; inflorescence a dichotomous cyme, the peduncles axillary; flowers few, white, rather large; calyx with a cuplike caliculus, the tube short, adnate to the ovary at base, lobes 5–6, linear-lanceolate, coriaceous, valvate; petals 5–6, on a fleshy disc, bifid, the lobes filiform-laciniate; stamens many, the filaments capillary; ovary semi-inferior, 1-locular, conical; style filiform, the stigma 3-lobed; ovules 6, in pairs on a central axis; fruit ovoid, coriaceous; seeds 1, pendulous; germinating on the plant, with the hypocotyl much-elongating before dropping to the ground.——A single species, in se. Asia.

1. Kandelia candel (L.) Druce. *Rhizophora candel* L.; *K. rheedii* Wight & Arn.——ME-HIRUGI, RYŪKYŪ-KŌGAI. Plant glabrous, the branchlets rather thick, with slightly inflated nodes; leaves thick, deep green, lustrous, oblong to obovate-oblong, sometimes elliptic, 5–10 cm. long, 3–5 cm. wide, rounded and with a mucro at apex, acute at base, the petioles 9–10 mm. long, narrowly winged; inflorescence shorter than the leaves, short-peduncled, about 10-flowered, the bracteoles about 3 mm. long, sepallike; calyx-tube short, the lobes linear-oblong, reflexed, about 16 mm. long; petals narrow, laciniate, about 1 cm. long; fruit ovoid, about 3 cm. long, surrounded by the persistent reflexed calyx-lobes.——Aug. Muddy places near the mouth of rivers; Kyushu (s. distr. including Yakushima and Tanegashima).——India, Malaysia, s. China, Formosa, and Ryukyus.

Fam. 146. **ALANGIACEAE** URI-NO-KI KA Alangium Family

Shrubs or trees; leaves simple, alternate, exstipulate; flowers white or pale yellow, actinomorphic, bisexual, in axillary cymes, pedicels jointed; calyx small, truncate, or the limb 4- to 10-toothed; petals 4–10, valvate, usually linear, revolute, sometimes connate at base; stamens as many as the petals and alternate with them, or 2 to 4 times as many as the petals, pubescent inside, the anthers 2-locular, linear, longitudinally dehiscent; ovary inferior, 1- or 2-locular; style simple, clavate, or 2- or 3-fid; ovules solitary, pendulous; drupe 1-seeded, crowned by the calyx-teeth and a disc; seeds with endosperm.——A single genus with about 20 species, in e. Asia, Malaysia, Australia, and Africa.

1. **ALANGIUM** Lam. URI-NO-KI ZOKU

Characters of the family.

1A. Leaves cordate-orbicular, 4-angled, palmately 3- to 5(–7)-lobed, usually short-pubescent especially beneath; petals 6, 3–3.5 cm. long; stamens about 3 cm. long, the anthers about ½ their length, the connective glabrous. .... 1. *A. platanifolium*
1B. Leaves obliquely obovate to oblong, sometimes ovate, entire, obliquely truncate at base, nearly glabrous except for axillary tufts beneath; petals 7, about 2 cm. long; stamens about 18 mm. long, the anthers about ⅓ their length, the connective densely appressed yellowish pubescent on inner side at base. .... 2. *A. premnifolium*

1. Alangium platanifolium (Sieb. & Zucc.) Harms var. **platanifolium.** *Marlea platanifolia* Sieb. & Zucc.; *A. platanifolium* (Sieb. & Zucc.) Harms; *A. platanifolium* var. *genuinum* Wanger.; *M. platanifolia* var. *typica* Makino——MOMIJI-URI-NO-KI. Leaves deeply 3- to 5(–7)-cleft, the lobes ovate to narrowly so, caudately long-acuminate.——Honshu (western distr.), Shikoku, Kyushu; rare.——Korea.

Var. **trilobum** (Miq.) Ohwi. *Marlea macrophylla* Sieb. &

Zucc.; *M. platanifolia* var. *triloba* Miq.; *M. platanifolia* var. *macrophylla* (Sieb. & Zucc.) Makino; *A. platanifolium* var. *macrophyllum* (Sieb. & Zucc.) Wanger.; *M. macrophylla* var. *trilobata* Nakai——URI-NO-KI. Large shrub; branches usually short-pilose when young, gray-brown; leaves thinly membranous, cordate-orbicular, 7–20 cm. long and as wide, 4-angled, slightly cordate at base, usually slightly short-pilose and green above, paler and usually prominently short-pubescent beneath, shallowly 3- to 5(–7)-lobed, the lobes deltoid, caudately long-acuminate, the petioles 3–10 cm. long, short-pubescent; inflorescence loosely few-flowered; limb of calyx about 1 mm. long, minutely toothed; petals white, revolute at anthesis, linear, 3–3.5 cm. long, about 2.5 mm. wide; stamens 12, the filaments short-pilose on back at base, the anthers glabrous; drupe ellipsoidal, 7–8 mm. long, blue, glabrous.——June. Woods in mountains; Hokkaido, Honshu, Shikoku, Kyushu; rather common.——Korea and Manchuria.

2. **Alangium premnifolium** Ohwi. *Marlea premnifolia* (Ohwi) Honda; *A. begoniaefolium* sensu auct. Japon., non Roxb.; *A. chinense* var. *nipponicum* Masam.——SHIMA-URI-NO-KI. Leaves obliquely obovate to oblong, sometimes ovate, 10–15 cm. long, 5–10 cm. wide, abruptly obtuse, obliquely truncate at base, entire, with axillary tufts of hairs beneath, the petioles 2–4 cm. long; inflorescence short, 2- to 5-flowered; calyx glabrous, the limb minutely toothed; petals 7, broadly linear, about 2 cm. long, pale yellow-pubescent inside; stamens about 18 mm. long, the anthers about 1/3 as long, the connective densely appressed yellowish pubescent on inner side at base; style as long as the stamens, glabrous; stigma 2-fid; drupe ellipsoidal, 8–12 mm. long.——Kyushu (Sata in Oosumi, Yakushima, and Tanegashima).——Ryukyus.

Fam. 147. **MYRTACEAE** Futo-momo Ka Myrtle Family

Trees or shrubs; leaves simple, usually entire, opposite, rarely alternate, glandular-dotted, stipules usually absent; flowers usually actinomorphic, bisexual, or polygamous by abortion; calyx-tube more or less adnate to the ovary, the lobes 3 or more, imbricate, valvate, or irregularly lobed; petals 4 or 5, rarely absent, on the margin of the disc; stamens numerous, on the margin of the disc, incurved in bud, sometimes erect, the filaments often connate at base; ovary inferior, 1- to many-locular; ovules many, on axile, rarely on parietal placentae; fruit loculicidally dehiscent or indehiscent; seeds nearly destitute of endosperm; embryo straight or coiled.——About 70 genera, with about 3,000 species, mostly tropical.

1. **SYZYGIUM** Gaertn. ADEKU ZOKU

Trees or shrubs; leaves coriaceous, sometimes membranous, opposite, penninerved; inflorescence a terminal or sometimes axillary cyme, usually ternately branched; flowers 4- sometimes 5-merous, rather small; calyx obconical, obovoid, or clavate, not much longer than the ovary, the margin of the limb truncate, undulate, or short-lobed; petals usually not spreading, calyptrately deciduous at antheis, or absent; stamens numerous, prominently incurved in bud; anther-locules parallel, longitudinally dehiscent; ovary usually 2-locular; ovules many in each locule; style simple, filiform; fruit a few-seeded berry; testa loosely adnate to the pericarp; cotyledons 2.——Many species in the Tropics of the Old World.

1. **Syzygium buxifolium** Hook. & Arn. *Eugenia microphylla* Abel; *S. microphyllum* (Abel) Masam., non Gamble; *E. sinensis* Hemsl.——ADEKU. Evergreen, glabrous, much-branched small tree; branches 4-angled, slender, dark brown; leaves rather thick, elliptic, broadly ovate or broadly obovate, 2–4 cm. long, 1–2.5 cm. wide, rounded to very obtuse, abruptly acute at base, the midrib impressed on upper side, raised beneath, the nerves pinnate, parallel, ascending; petioles 2–4 mm. long; flowers rather numerous, about 3 mm. across, bracteolate at base; calyx narrowly campanulate, about 3 mm. long, the teeth very depressed and short; corolla deciduous at anthesis; fruit globose, black, about 7 mm. across.——Kyushu (s. distr.). ——Ryukyus, Formosa, s. China, Indochina, and Bonins.

Jambosa jambos (L.) Millsp. *Eugenia jambos* L.—— FUTO-MOMO. Branches thick; leaves lanceolate 10–20 cm. long; flowers about 4 cm. across.——Naturalized in Kyushu (Yakushima and Tanegashima).——India; much cultivated in tropical countries.

Fam. 148. **MELASTOMATACEAE** No-botan Ka Melastoma Family

Herbs or woody plants, rarely scandent; leaves simple, opposite or verticillate, 3- to 9-nerved from the base or rarely penninerved; stipules absent; flowers bisexual, usually showy, actinomorphic; calyx short-tubular, adnate to the ovary or sometimes connected by the septa, or free, the lobes imbricate or valvate in bud; petals imbricate, free; stamens as many or twice as many as the petals, the filaments free, incurved in bud, the anthers 2-locular, basifixed, usually with 1 or 2 pores, the connective often produced as an appendage at base; ovary 2- to many-locular; style simple; ovules many, on axile, basal, or parietal placentae; fruit a capsule or berry; seeds usually small, endosperm absent.——About 150 genera, with about 4,000 species, chiefly in the Tropics and subtropics.

1A. Flowers small, 7–17 mm. across, in axillary fascicles or terminal panicles; calyx-lobes persistent; seeds straight.
  2A. Flowers in axillary fascicles; stamens 4, homomorphic, the connective without an appendage. .... 1. *Blastus*
  2B. Flowers in terminal cymes or panicles; stamens 8, dimorphic, the larger ones with a bifid appendage on the connective in front at base. .... 2. *Bredia*
1B. Flowers larger, 3–8 cm. across, in terminal heads or cymes; calyx-lobes deciduous; seeds curved.
  3A. Anthers homomorphic, the connective scarcely extended at base, without an appendage; fruit a dry, regularly dehiscent capsule; herbs (in ours). .... 3. *Osbeckia*
  3B. Anthers dimorphic, the larger ones with 2 tubercles or spurs in front at base; fruit a drupe, indehiscent or irregularly dehiscent; shrubs. .... 4. *Melastoma*

1. BLASTUS Lour. MIYAMA-HASHI-KAMBOKU ZOKU

Glabrous or short-pilose shrubs with slender, terete branches; leaves membranous, opposite, oblong, long-acuminate, 3- to 5-nerved, petiolate; flowers small, fasciculate in axils, short-pedicelled; calyx-tube short-pubescent, ellipsoidal, the limb truncate with 4 minute teeth on margin; petals 4, ovate, convolute; stamens 4, homomorphic, the filaments filiform, the anthers linear-lanceolate, the locules distinct and divergent at base, the connective without an appendage; ovary with 4 septa and joined with the calyx-tube, 4-locular, short-pilose at apex; capsules obovoid-globose, with 4 obscure grooves, enclosed in the calyx-tube; seeds falcate.——Few species, in se. Asia, and India.

1. Blastus cochinchinensis Lour. *Anplectrum parviflorum* Benth.; *B. parviflorus* (Benth.) Triana——MIYAMA-HASHI-KAMBOKU. Shrub, with yellow, sessile, mealy glands throughout; branches slender; leaves membranous, oblong or oblong-lanceolate, 8–15 cm. long, 2.5–5 cm. wide, long-caudate, acute at base, entire, often purplish beneath, the 5 nerves slender, raised beneath, the outermost pair very slender not reaching the apex, the veinlets transversely parallel, very slender, the petioles slender, 2–4 cm. long, grooved above; flowers white, 1–6 in axillary fascicles, 7–8 mm. across, the pedicels 3–5 mm. long; calyx-tube tubular-campanulate, 2–3 mm. long; petals 4, deltoid-ovate, about 3 mm. long, deflexed, short-clawed at base; anthers about 4 mm. long, the locules with a callose base; capsules urceolate-globose, about 3 mm. long.——July–Aug. Dense woods in evergreen forests; Kyushu (Yakushima); rare. ——Ryukyus, Formosa, s. China, Indochina, and e. India.

2. BREDIA Bl. HASHI-KAMBOKU ZOKU

Glabrous or puberulous shrubs; leaves petioled, ovate to oblong, often oblique, acute, serrulate, 5-nerved; flowers in terminal cymes or panicles, pedicelled, rather small, rose-purple; calyx pilose, the tube obconical and obtusely 4-angled, the limb short, the lobes 4, persistent, deltoid, mucronate below the apex; petals 4, obovate; stamens 8, dimorphic, the anthers narrowly lanceolate, incurved, the larger with a bifid appendage on the connective in front at base, the smaller with a globose appendage of connective in front at base, with a short protuberance on back; ovary 4-locular, adnate to the calyx-tube, with 8 septa and with a membranous crown at apex; style ascending, the stigma punctate; capsules semiglobose with a depressed apex surrounded by a crown, enclosed in the obconical calyx-tube; seeds obovoid, small.——Few species, in se. Asia.

1. Bredia hirsuta Bl. HASHI-KAMBOKU. Shrub with spreading callose hairs and short puberulence; branches terete; leaves ovate to narrowly so, or ovate-oblong, 4–10 cm. long, 2–5 cm. wide, more or less unequal, sometimes slightly oblique, acute to rather acuminate, slightly cordate at base, callose-denticulate, 5- to 7-nerved, green on upper side, whitish beneath, the petioles 1–7 cm. long; inflorescence 8–12 cm. long, ascending, rather many-flowered, puberulous, the pedicels 5–12 mm. long, with 2 minute bracteoles at base; flowers rose with a purple hue, about 1.5 cm. across; calyx-tube about 4 mm. long, narrowed to the pedicel, the lobes small, deltoid-lanceolate, about 1 mm. long, obtuse, spreading; petals obovate-rhombic, about 8 mm. long, mucronate; anthers 8, the larger ones 4–5 mm. long, rose, the smaller ones about 2.5 mm. long; capsules obconical, about 7 mm. long, the crown nearly entire.——July–Oct. Kyushu (Kagoshima and Yakushima). ——Ryukyus.

3. OSBECKIA L. HIME-NO-BOTAN ZOKU

Herbs or shrubs, strigose, the branches usually 4-angled; leaves rather firm, 3- to 7-nerved; flowers terminal, solitary or in heads or panicles, rose or rose-purple, usually bracteate at base; calyx with pinnate scales or stellately fascicled hairs, the tube urceolate or ovoid, the lobes 4–5, linear or lanceolate, deciduous; petals 4–5, obovate, deciduous; stamens 8 or 10, homomorphic, the anthers often beaked, the connective inflated but not appendaged at base, with 2 tubercles in front, rarely with a short spur at base on back; ovary 4- to 5-locular, more or less adnate to calyx-tube at base, the apical portion free, with a tuft of hairs; capsules enclosed in the calyx-tube, 4- to 5-valved; seeds small, curved.——About 50 species, India, Malaysia, e. Asia, and Australia.

1. Osbeckia chinensis L. *O. japonica* Naud.——HIME-NO-BOTAN. Perennial 30–50 cm. high; stems erect, firm, rather slender, 4-angled, strigose on the angles, sparsely branched; leaves broadly lanceolate, 3–6 cm. long, 8–15 mm. wide, acute, subrounded at base, spreading, strigose-hairy, entire, 5-nerved, the petioles 1–2 mm. long; flowers few, sessile, in terminal heads, the bracts ovate, obtuse, ciliate, slightly shorter than the calyx-tube; calyx-tube urceolate, glabrous, 5–6 mm. long, the lobes 4, membranous, ovate-oblong, subobtuse, as long as the tube, with a short callose tubercle terminated by a tuft of long straight hairs between the lobes; petals broadly obovate, 12–15 mm. long, ciliate near tip; anthers about 4 mm. long, short-beaked, the connective slightly thickened at base; capsules enclosed in the calyx-tube, subglobose, 4-valved; seeds minutely tubercled.——Aug.–Oct. Grassy places in lowlands; Honshu (Kii Prov.), Shikoku, Kyushu.——Ryukyus, Formosa, China, and Malaysia.

4. MELASTOMA L. NO-BOTAN ZOKU

Strigose shrubs; leaves coriaceous, opposite, elliptic to lanceolate, entire, 3- to 7-nerved, petioled; flowers terminal, large, solitary, fasciculate, or in panicles, rose-purple, rarely white, with 2 bracts at base; calyx-tube ovoid or urceolate, the lobes 4–7, usually 5, ovate or lanceolate, deciduous, often with a tooth between the lobes; petals 4–7, usually 5; stamens 8–14, usually 10, dimorphic, the anthers slender and slightly incurved, the larger ones usually violet-purple, the connective much extended at base, bifid or emarginate, the smaller ones yellow, the connective not extended, with 2 tubercles in front; ovary ovoid, free or with 5(4–7) septa, hairy at the mucronate apex; fruit a coriaceous or fleshy, indehiscent or irregularly dehiscent berry.——About 80 species, tropical Asia, Pacific Islands, and Australia.

1. Melastoma candidum D. Don. *M. macrocarpum* D. Don; *M. septemnervium* Lour., non Jacq.; *M. nobotan* Bl.; *M. candidum* var. *nobotan* (Bl.) Makino——No-botan. Shrub, the branches rather thick, obscurely 4-angled, densely appressed-strigose and with flattened whitish scalelike hairs; leaves somewhat coriaceous, oblong to ovate-elliptic, 5–12 cm. long, 2.5–6 cm. wide, acute, obtuse to rounded at base, entire, appressed-strigose on both sides, 5- to 7-nerved, the veinlets slender, parallel, the petioles 1–2 cm. long; flowers 3–7, terminal, 6–8 cm. across, rose-purple, the bracts lanceolate to narrowly ovate, caducous, slightly shorter than the calyx-tube; calyx-tube urceolate-campanulate, 8–10 mm. long, the lobes 5, narrowly deltoid, acuminate, as long as the tube; anthers dimorphic, the larger ones about 1 cm. long, with the elongate portion of connective as long as the anther and shallowly bifid at the tip, the smaller ones 8–9 mm. long, the connective without an elongate portion; berry coriaceous, the upper half free, exserted, with flattened hairs.——July–Aug. Reported to occur in Kyushu (Yakushima); rare.——Ryukyus, Formosa, China, and Indochina.

## Fam. 149. ONAGRACEAE	Akabana Ka	Evening Primrose Family

Herbs, or rarely shrubs, sometimes aquatic; leaves simple, opposite or alternate; stipules deciduous or absent; flowers bisexual, actinomorphic, usually solitary in axils or in spikes; calyx adnate to the ovary, the lobes 4 or 5, sometimes 2, valvate; petals as many as the calyx-lobes, convolute or imbricate, rarely absent; stamens as many or twice as many as the calyx-lobes; anthers 2-locular, longitudinally dehiscent; ovary inferior or half-inferior, 2- to 6-locular, rarely the locules incomplete; ovules 1 to many, on axile placentae; style simple; fruit a capsule or nutlike; seeds 1 to many, endosperm absent, embryo straight.—— About 37 genera, with about 600 species, in both hemispheres, especially abundant in the New World.

1A.	Ovary half-inferior, 1-locular, 1-seeded; fruit relatively large, with 2–4 spines, indehiscent; aquatic herbs with rhomboidal coarsely toothed leaves. .... 1. *Trapa*
1B.	Ovary inferior, usually 2- to 6-locular, with 1 to many small seeds.
  2A.	Flowers dimerous; fruit indehiscent, obovoid, uncinate-pilose, 1- or 2-locular, each locule 1-seeded. .... 2. *Circaea*
  2B.	Flowers tetramerous; fruit a many-seeded capsule.
    3A.	Capsules loculicidally dehiscent; seeds comose; petals usually rose, white, or red-purple. .... 3. *Epilobium*
    3B.	Capsules septicidally dehiscent or irregularly ruptured; seeds not comose; petals yellow or absent. .... 4. *Ludwigia*

### 1. TRAPA L.	Hishi Zoku

Aquatic herbs; leaves floating, rhombic, coarsely toothed, the petioles often spongy and partially inflated; flowers axillary, solitary, short-pedicelled; calyx-tube short, subtending the ovary, the limb 4-lobed, the lobes persistent, becoming at least partly spine-like; petals 4, inserted on the disc, superior to the ovary; stamens 4; ovary half-inferior, 1-locular, adnate at base to the calyx-tube, the free portion conical, terminated by the persistent style, the stigma capitate; ovules solitary, pendulous; fruit bony, indehiscent, obconical, with 2–4 large spines, 1-seeded; seeds rather large, testa membranous, cotyledons unequal.——Few species, in tropical and temperate regions of the Old World.

1A.	Plant prominently pubescent on petioles, pedicels, calyx, and underside of leaves; leaves 3–8 cm. wide; fruit 3–4 cm. long, inclusive of the usually spreading spines. .... 1. *T. japonica*
1B.	Plant glabrous except leaves beneath; leaves 1–2 cm. wide; fruit 2–3 cm. across, inclusive of the usually ascending spines; plant smaller than the preceding. .... 2. *T. incisa*

1. Trapa japonica Flerov. *T. bispinosa* sensu auct. Japon., non Roxb.; *T. natans* var. *bispinosa* Makino, excl. syn.; *T. bispinosa* var. *iwasakii* Nakano, and var. *iinumai* Nakano ——Hishi. Annual; stems slender, much elongate, with filiform pinnately divided roots at each node; leaves rosulate, ovate-rhombic or broadly rhombic, 2.5–5 cm. long, 3–8 cm. wide, obtuse to subacute, broadly cuneate or subtruncate at base, entire on lower half, irregularly coarse-toothed on upper margin, glabrous and lustrous on upper side, prominently pubescent beneath especially on nerves, the petioles 10–20 cm. long, pubescent, the inflated portion lanceolate, 1–5 cm. long; pedicels short, pubescent, 2–4 cm. long, deflexed in fruit; flowers white to pale rose, about 1 cm. across; calyx soft-pubescent, the dorsal and ventral calyx-lobes deciduous; fruit bony, rather flat, obtriangular, 3–4 cm. long inclusive of the spines, with a hornlike projection at the center and with a solitary spine at each end, the spines retrorsely scabrous toward tip.——July– Oct. Lakes and ponds; Hokkaido, Honshu, Shikoku, Kyushu; common.——Korea, Manchuria, and China.

Var. **rubeola** (Makino) Ohwi. *T. natans* var. *rubeola* Makino——Mebishi. Petioles reddish; fruit 4-spined.——Honshu.

2. Trapa incisa Sieb. & Zucc. *T. bispinosa* var. *incisa* (Sieb. & Zucc.) Fr. & Sav.; *T. natans* var. *incisa* (Sieb. & Zucc.) Makino——Himebishi. Resembling the preceding species but smaller; stems very slender; leaves deltoid-rhombic or depressed-rhombic, 1–2 cm. long and as wide, acute or subacute, broadly cuneate and entire on lower half, with few, prominent deltoid teeth on upper margin, glabrous on upper side, glabrous or loosely pubescent on nerves near base beneath, the petioles 1–8 cm. long, slender, usually glabrous except at apex; flowers white or pale rose, 6–8 mm. across; fruit slightly flattened, obtriangular, 2–3 cm. long inclusive of the spines, with a hornlike projection at the center, with an ascending or obliquely spreading spine at each end, dorsal and ventral calyx-lobes becoming spinose and descending in fruit.——July–Oct. Lakes and ponds; Honshu, Shikoku, Kyushu.

2. CIRCAEA L.   Mizu-tama-sō Zoku

Perennial herbs, glabrous or short-pubescent; leaves membranous, petioled, opposite, ovate, toothed or subentire; flowers dimerous, in terminal or axillary racemes, small, white or rose, the bracts absent or small; calyx-tube ovoid, adnate to and constricted above the ovary, the lobes 2; petals 2, obcordate or obovate, inserted on the margin of the ovary-disc; stamens 2, alternate with the petals; ovary 1- or 2-locular; style filiform, the stigma capitate; ovules usually solitary and ascending on the inner side of each locule, rarely paired and collateral; fruit obovoid, indehiscent, uncinate-pilose, with 1 seed in each locule.——About 10 species in the temperate to northern parts of the N. Hemisphere.

1A. Fruit 2-locular, often sulcate, 2-seeded, obovoid-clavate to globose; petals much shorter than the calyx-lobes; plants 20–60 cm. high, glabrous to pubescent.
  2A. Plant green throughout; stems and often the axis of inflorescence glandular-puberulent, and loosely long spreading pubescent; leaves ovate-cordate, puberulent; fruit globose, sulcate; pedicels shorter to as long as the fruit. . . . . 1. *C. cordata*
  2B. Stems and inflorescence often reddish, glabrous or puberulent; leaves ovate to ovate-oblong, cuneate to rounded at base; fruit obovoid; pedicels as long as to longer than the fruit.
    3A. Axis of inflorescence rather thick, usually short-pubescent, rarely glabrous; fruit broadly obovoid, sulcate; pedicels as long as to 1.5 times as long as the fruit; leaves mostly cuneate, rarely rounded at base; petals deeply bifid.
      4A. Stems recurved-puberulent; leaves slightly puberulent; axis of inflorescence glabrous or loosely puberulent. . . . . . . 2. *C. mollis*
      4B. Stems glabrous or slightly puberulent; leaves nearly glabrous except the puberulent margins; axis of inflorescence rather densely short glandular-puberulent. . . . . 3. *C. quadrisulcata*
    3B. Axis of inflorescence slender, glabrous; fruit narrowly obovoid or obovoid-clavate, not sulcate; pedicels 3 to 5 times as long as the fruit; leaves rounded at base; petals emarginate. . . . . 4. *C. erubescens*
1B. Fruit 1-locular, not sulcate, 1-seeded, clavate; petals nearly as long as the calyx-lobes; plant 5–10 cm. high, rarely 20 cm. high, glabrous, reddish; leaves deltoid-ovate, 1–3 cm. long, cordate. . . . . 5. *C. alpina* var. *caulescens*

1. **Circaea cordata** Royle. *C. cardiophylla* Makino——Ushitaki-sō.   Green perennial 40–60 cm. high, with creeping rhizomes; stems branched, glandular-puberulent and loosely long spreading-pubescent; leaves ovate-cordate to broadly ovate, 7–12 cm. long, 5–8 cm. wide, abruptly acuminate, shallowly cordate to rounded at base, subentire or indistinctly mucronate-toothed, puberulent, the petioles 3–10 cm. long, with spreading short and long hairs; inflorescence 7–15 cm. long, rather densely spreading glandular-puberulent, the axis rather thick; petals with 2 deep narrow lobes, about 1/3–1/2 as long as the calyx-lobes; fruit nearly globose, about 3 mm. across, sulcate, densely yellow-brown uncinate-hairy and loosely puberulent, the pedicels shorter than to as long as the fruit.——Moist woods; Hokkaido, Honshu, Shikoku.——Korea, Manchuria, Ussuri, China, Formosa, and n. India.

2. **Circaea mollis** Sieb. & Zucc. *C. quadrisulcata* Fr. & Sav., excl. syn.; *C. coreana* Lév.; *C. lutetiana* var. *taquetii* Lév. ——Mizu-tama-sō.   Rhizomatous and stoloniferous; stems 20–50 cm. long, recurved-puberulent; leaves narrowly ovate to narrowly ovate-oblong, 5–10 cm. long, 2–3 cm. wide, ascending-puberulent, gradually acuminate, cuneate to acute, rarely rounded at base, loosely mucronate-toothed, the petioles 1–4 cm. long, puberulous; inflorescence elongate after anthesis, to 15 cm. long, the axis rather thick, glabrous or loosely puberulent; calyx-lobes broadly ovate-elliptic; petals obovate, moderately or deeply bifid, 1/2–2/3 as long as the calyx-lobes; fruit broadly obovoid, 3–4 mm. across, sulcate, uncinate-pilose, the pedicels 1–1.5 times as long as the fruit.——Aug.–Sept. Woods in mountains; Hokkaido, Honshu, Shikoku, Kyushu.——Korea, China, and Indochina.
Var. **ovata** (Honda) Hara. *C. quadrisulcata* var. *ovata* Honda——Hiro-ha-mizu-tama-sō.   Leaves ovate, rounded to slightly cordate at base, long-petioled.——Honshu (Shimotsuke and Iwaki Prov.); rare.

3. **Circaea quadrisulcata** (Maxim.) Fr. & Sav. *C. lutetiana* sensu auct. Japon., non L.; *C. lutetiana* forma *quadrisulcata* Maxim.; *C. mollis* var. *maximowiczii* Lév.; *C. maximowiczii* (Lév.) Hara——Yama-tani-tade.   Stems 30–40 cm. long, glabrous or slightly puberulent; leaves narrowly ovate to narrowly ovate-oblong, 4–14 cm. long, 2–5 cm. wide, acuminate, rounded to shallowly cordate to cuneate at base, obscurely mucronate-toothed, glabrous except the puberulent margins, the petioles shorter than the leaves; inflorescence elongate after anthesis, the axis rather densely short-glandular-puberulent; petals about 2/3 as long as the calyx-lobes, bifid to the middle; fruit broadly obovoid, about 3 mm. across, sulcate, uncinate-pilose, the pedicels as long as to 1.5 times as long as the fruit. ——Moist grassy places; Hokkaido, Honshu (centr. distr.). ——Sakhalin, Korea, Manchuria, n. China, e. Siberia, and Altai.

4. **Circaea erubescens** Fr. & Sav. *C. lutetiana* var. *intermedia* Lév.; *C. lutetiana* subsp. *erubescens* (Fr. & Sav.) Lév.; *C. kawakamii* Hayata——Tani-tade.   Slenderly stoloniferous, nearly glabrous rubescent herb, thinly puberulent while very young; stems 20–50 cm. long; leaves ovate or narrowly so, 3–7 cm. long, 1.5–3.5 cm. wide, acuminate, rounded at base, sparsely short-toothed, the petioles 2–5 cm. long, reddish; inflorescence glabrous, the axis slender, 6–10 cm. long; calyx usually reddish; petals obovate, emarginate, about 1/2 as long as the calyx-lobes; fruit narrowly ovoid to obovoid-clavate, 2–2.5 mm. across, not sulcate, short, uncinate-pilose, the pedicels slender, 3–5 times as long as the fruit.——July–Sept. Moist woods; Hokkaido, Honshu, Shikoku, Kyushu.——Formosa and China.

5. **Circaea alpina** L. var. **caulescens** Komar. *C. caulescens* (Komar.) Nakai——Miyama-tani-tade.   Delicate, reddish, glabrous, stoloniferous herb, 5–10 cm. high, rarely 20 cm. high; stems glabrous or loosely puberulent; leaves deltoid-ovate to ovate, 1–3 cm. long, 7–20 mm. wide, acute, open-cordate to rounded-truncate at base, thinly puberulent on upper surface or only toward margin, with few acute teeth, the petioles 1–2 cm. long, reddish; inflorescence glabrous, 3–5 cm. long; calyx-lobes reddish; petals nearly as long as the calyx-lobes; fruit clavate, about 1.5 mm. across, not sulcate, 1-locular, 1-seeded, loosely soft uncinate-pilose, the pedicels slender, 1.5–2.5 times as long as the fruit.——July–Aug. Moist woods in mountains; Hokkaido, Honshu, Shikoku, Kyushu; rather common.——Widely distributed in the temperate and northern parts of the N. Hemisphere.

3. EPILOBIUM L. AKABANA ZOKU

Herbs or half-shrubs; leaves opposite or some of them alternate, entire or toothed; flowers solitary in axils or in terminal racemes or spikes, white, rose, or red-purple; calyx-tube adnate to ovary, linear, 4-angled or terete, the limb 4-lobed, deciduous; petals 4, obovate or obcordate; stamens 8; ovary 4-locular; styles prominent, the stigmas clavate or 4-lobed; ovules many, ascending, in 2 series on inner angle of the locule; capsules elongate, linear, 4-angled, loculicidally dehiscent, 4-valved, the axis persistent; seeds many, ascending, comose.——About 160 species, especially in the temperate and northern parts of the N. Hemisphere, abundant also in New Zealand.

(Treatment chiefly that of Dr. Hiroshi Hara.)

1A. Petals large, about 1.5 cm. long, entire, rose-purple; stamens and styles declined, the stigma deeply 4-lobed; plant 1–1.5 m. high, robust, with alternate leaves and a terminal racemose inflorescence. . . . . 1. *E. angustifolium*
1B. Petals smaller, 3–12 mm. long, emarginate, white to pale rose; stamens and styles erect; plants smaller and more slender, with more or less opposite leaves and axillary flowers, the inflorescence seldom racemose.
  2A. Stigma 4-lobed; petals 7–12 mm. long.
   3A. Plant large, about 1 m. high; stems long spreading-pilose; pedicels, ovary, and calyx densely short-glandular and long-pubescent; flowers 12–15 mm. long. . . . . 2. *E. hirsutum* var. *villosum*
   3B. Plant smaller, 15–60 cm. high; stems and flowers with curved puberulence; flowers 7–11 mm. long. . . . . 3. *E. montanum*
  2B. Stigma simple, capitate to clavate; petals mostly shorter.
   4A. Leaves nearly entire or obscurely serrulate; ovary densely white appressed-puberulent.
    5A. Seeds oblanceolate, 1.5–2.7 mm. long; flowers 5–9 mm. long; calyx 4–5 mm. long; leaves linear to ovate-lanceolate, obtuse to subacute; plant with stolons. . . . . 4. *E. palustre* var. *lavandulaefolium*
    5B. Seeds narrowly oblong, 1–1.5 mm. long, obtuse at both ends; flowers 3.5–4.5 mm. long; calyx 2.5–3.5 mm. long; leaves narrowly oblong or oblong-lanceolate, short-acute to obtuse; plant without stolons. 5. *E. fastigiatoramosum*
   4B. Leaves distinctly serrulate; ovary short-hairy.
    6A. Median leaves linear or narrowly lanceolate, 1–5 mm. wide.
     7A. Rhizomes with a terminal scaly subterranean winter bud; leaves with 1–2(–4) small teeth on each side; calyx-lobes minutely mucronate; pedicels 1–3.5 cm. long; stems slender. . . . . 6. *E. fauriei*
     7B. Rhizomes with subsessile innovation-shoots on the neck; leaves finely toothed; calyx-lobes prominently mucronate; pedicels 5–12 mm. long; stems usually stouter. . . . . 7. *E. formosanum*
    6B. Median leaves broader, oblong-lanceolate to broadly ovate, (3–)5–35 mm. wide.
     8A. Stigma capitate.
      9A. Stems densely long curved-puberulent along the ridges; leaves puberulent on nerves and on margin; ovary sparsely glandular-puberulent or the hairs not glandular. . . . . 8. *E. amurense*
      9B. Stems curved-puberulent or glabrous; ovary appressed curved-pubescent or glabrous. . . . . 9. *E. cephalostigma*
     8B. Stigma clavate.
      10A. Plants stout, (15–)35–90 cm. high; leaves usually rounded at base, prominently toothed; ovary densely glandular- or curved-puberulent.
       11A. Plant stoloniferous; stems without ridges; leaves 2–10 cm. long, 0.5–3 cm. wide; ovary usually prominently glandular-puberulent; seeds 1.5–1.8 mm. long, the coma often red-brown. . . . . 10. *E. pyrricholophum*
       11B. Plant with subsessile innovation shoots at base; stems ridged; leaves 4–9 cm. long, 1.5–2.5 cm. wide; ovary densely curved-puberulent and sparingly glandular-puberulent; seeds 0.8–1.4 mm. long, the coma whitish.
           11. *E. glandulosum* var. *asiaticum*
      10B. Plants usually smaller, 5–30 cm. high; leaves narrowed at base, often obsoletely toothed; ovary glabrous or with glandular or curved puberulence.
       12A. Seeds without papillae; flowers 3.5–4.5 mm. long; pedicels thinly glandular-puberulent. . . . . 12. *E. shiroumense*
       12B. Seeds minutely papillose; flowers 4–6 mm. long.
        13A. Pedicels thinly glandular-puberulent, 3–20 mm. long in fruit. . . . . 13. *E. foucaudianum*
        13B. Pedicels thinly curved-puberulent, 22–35 mm. long in fruit. . . . . 14. *E. dielsii*

1. Epilobium angustifolium L. *Chamaenerion angusti-folium* (L.) Scop.; *E. spicatum* Lam.——YANAGI-RAN.   Rhizomatous perennial, 1–1.5 m. high; stems stout, erect, usually simple, terete, glabrous or sparsely curved-puberulent; leaves numerous, alternate, lanceolate, 8–15 cm. long, 1–3 cm. wide, acuminate, narrowed toward the subsessile base, subentire, puberulent on the midrib and glaucous beneath; racemes terminal, elongate, many-flowered, the bracts linear, the pedicels 8–30 mm. long; flowers 2–3 cm. across; ovary densely curved-puberulent; calyx deeply 4-cleft to just above the ovary, the lobes oblanceolate, about 1.5 cm. long; petals rose-purple, obovate-orbicular, about 1.5 cm. long, entire, short-clawed at base; stamens and styles declined, the stigmas deeply 4-lobed; capsules 8–10 cm. long.——July–Aug. Gregarious in grassy places; Hokkaido, Honshu (centr. distr. and northw.); rather

common.——Widely distributed in the temperate and northern parts of the N. Hemisphere.

2. Epilobium hirsutum L. var. **villosum** Hausskn. Ō-AKABANA.   With creeping rhizomes and stout stolons; stems about 1 m. long, erect, rather stout, much branched, terete, densely long spreading-pilose and short glandular-puberulent; leaves linear-oblong to linear-lanceolate, 3–10 cm. long, 5–18 mm. wide, acute, slightly narrowed at base, sessile, clasping, acutely toothed, long-pubescent on both sides; flowers rose, 12–15 mm. long, mucronate in bud, the pedicels short, densely short-glandular and long spreading-pubescent; calyx 9–11 mm. long, the lobes 7–9 mm. long; petals broadly obovate, bifid, 10–12 mm. long; stigmas 4-fid; capsules 5–8 cm. long, the pedicels 4–12 mm. long; seeds oblong-obovate, 1–1.2 mm. long, rounded at apex, very densely papillose.——July–Aug. Honshu (Sado and Iwashiro Prov.); rare.——The species is widely distributed in Eurasia and N. Africa.

3. Epilobium montanum L. *E. parviflorum* var. *menthoides* sensu auct. Japon., non Hausskn.——EZO-AKABANA. Rhizomes short, with sessile innovation shoots on the neck; stems 15–60 cm. long, terete, without ridges, densely curved-puberulent, especially on upper portion; leaves ovate or oblong-ovate, the lower ones obtuse, the upper ones acute, rounded to subcordate, very short-petioled, irregularly serrulate, thinly puberulous; median leaves 2–5 cm. long, 10–23 mm. wide; flowers 7–11 mm. long, pale rose, subobtuse at apex in bud, the pedicels densely appressed white-puberulent and short glandular-pilose; calyx 5–7.5 mm. long; stigma 4-fid; capsules to 7 cm. long, curved-puberulent, the pedicels 8–15 mm. long; seeds oblong, 1–1.3 mm. long, rounded at apex, obtuse at base, densely papillose, the coma whitish.——July–Aug. Wet places in mountains; Hokkaido, Honshu (centr. distr. and northw.).——Siberia, Himalayas, Caucasus, Asia Minor, and Europe.

4. Epilobium palustre L. var. **lavandulaefolium** Lecoq & Lamotte. *E. dahuricum* sensu auct. Japon., non Fisch.; *E. lineare* sensu auct. Japon., non Muhl.——HOSOBA-AKABANA, YANAGI-AKABANA. Plant with elongate rhizomes and stolons; stolons with minute opposite scalelike leaves and a small ellipsoidal winter-bud at the tip; stems erect, 10–40 (–60) cm. long, usually without ridges, with minute curved puberulence on upper portion; leaves linear to ovate-lanceolate, 2–6 cm. long, 2–15 mm. wide, obtuse to subacute, narrowly cuneate or obtuse at base, obscurely toothed, minutely curved-puberulent on upper side and curved-puberulent on nerves beneath and on margin, sessile or nearly so; flowers rose to white, 5–9 mm. long, the pedicels short, densely curved-puberulent; ovary densely white appressed-puberulent and with a few glandular hairs; calyx 4–5 mm. long; petals 5–8 mm. long; stigmas clavate, much shorter than the style; capsules 5.5–8 cm. long, with curved puberulence especially on the angles, the pedicels 1–2.5 cm. long; seeds oblanceolate, 1.5–2.7 mm. long, obtuse, narrowly cuneate at base, with small papillae on the surface, the coma nearly white.——July–Aug. Wet places in lowlands; Hokkaido, Honshu (centr. distr. and northw.).——The species is widely distributed in Europe, Asia, and N. America.

5. Epilobium fastigiatoramosum Nakai. *E. palustre* var. *mandjuricum* Hausskn.——EDA-UCHI-AKABANA. Sometimes with elongate rhizomes and with stolons; stems much branched or simple, 20–40 cm. long, without ridges, densely leafy, with curved puberulence in upper portion; leaves narrowly oblong to oblong-lanceolate, 1.5–4.5 cm. long, 3–10 mm. wide, short-acute to obtuse, cuneate to subrounded at base, broadest about the middle, subentire, sparsely puberulent on upper side and on nerves beneath, the petioles less than 1 mm. long; flowers 3.5–4.5 mm. long, the buds minutely mucronate; ovary densely white-puberulent and with few glandular hairs; calyx 2.5–3.5 mm. long; stigmas clavate, shorter than the style; capsules 3–6 cm. long, the pedicels 5–15 mm. long; seeds narrowly oblong, 1–1.5 mm. long, obtuse at both ends, densely papillose.——Hokkaido; rare.——Korea, Manchuria, China, Ussuri, Amur, and Dahuria.

6. Epilobium fauriei Lév. *E. kitadakense* Koidz.——HIME-AKABANA. Often gregarious; rhizomes slender, usually long-ascending, the terminal, ovoid, sessile, winter-bud with fleshy scalelike leaves; stems nearly erect, slender, densely leafy, 3–20(–40) cm. long, without ridges, sometimes with 2 series of minute hairs, curved white-puberulent especially on upper portion; leaves linear, 1–2.5(–3) cm. long, 1–3(–5) mm. wide, subobtuse, cuneately narrowed to the very short petiole, 1- or 2(–4)-toothed on each side, often with axillary bulbils resembling winter-buds in late autumn; flowers white to rose, 4–6(–8) mm. long, the pedicels and ovary white-appressed curved-puberulent; calyx 3–4(–5) mm. long, the lobes minutely mucronate; stigmas clavate, much shorter than the styles; capsules 2–3.5(–4.5) cm. long, the pedicels 1–3.5 cm. long; seeds oblanceolate-oblong, 0.8–1.3 mm. long, abruptly narrowed at the tip, cuneate with a mucro at base, densely papillose, the coma whitish.——July–Sept. Wet gravelly places in mountains; Hokkaido, Honshu (centr. distr. and northw., and Mount Daisen in Hōki Prov.).——Kuriles.

7. Epilobium formosanum Masam. *E. cephalostigma* var. *linearifolium* Hisauchi; *E. sohayakiense* Koidz.——TODAI-AKABANA, SAIYŌ-AKABANA. Often gregarious, bearing innovation-shoots on the neck of the rhizomes; stems 20–35 cm. long, rather stout, without ridges, densely leafy, with curved puberulence; leaves linear to lanceolate, 1–3(–4) cm. long, 1.5–5 mm. wide, subacute, cuneately narrowed at base, minutely toothed, the petioles 1–3(–4) mm. long; flowers 3.5–5 mm. long, rose to white, distinctly mucronate at apex in bud, the pedicels and ovary rather densely puberulent; calyx 3–4 mm. long, the lobes distinctly mucronate; stigmas thick-clavate, much shorter than the styles; capsules 3–4 cm. long, the pedicels 5–12(–18) mm. long; seeds oblong, 0.8 mm. long, rounded at apex, obtuse at base, densely papillose, the coma white.——High mountains; Honshu (s. Shinano and westw.), Shikoku.——Formosa.

8. Epilobium amurense Hausskn. *E. origanifolium* var. *pubescens* Maxim.; *E. gansuense* Lév.; *E. montanum* sensu auct. Japon., non L.; *E. miyabei* Lév.; *E. shikotanense* Takeda; *E. ovale* Takeda——KEGON-AKABANA. Rhizomes with innovation-shoots on the neck, the short stolons with small scalelike leaves and an innovation-shoot at the tip; stems 6–40(–70) cm. long, densely rather long curved-puberulent along the 2 ridges, otherwise glabrous; leaves lanceolate-oblong to ovate, 1.5–6.7 cm. long, 6–30 mm. wide in the median ones, acute, narrowed to the very short petiole, irregularly incurved-toothed, puberulent on nerves and on margin; flowers usually rose, distinctly mucronate at apex in bud, glandular-puberulent or curved-puberulent; calyx 3.5–6 mm. long; petals 4–7(–8) mm. long; stigmas capitate; capsules 2.5–5(–6.5) cm. long, glabrous or sparingly puberulent, the pedicels 4–10(–22) mm. long; seeds linear-oblong, 0.8–1.2 mm. long, rounded at apex, slightly narrowed toward the obtuse base, densely and minutely papillose, the coma whitish.——July–Aug. Streams in mountains; Hokkaido, Honshu, Shikoku.——Sakhalin, Kuriles, Manchuria, Amur, and Formosa.

9. Epilobium cephalostigma Hausskn. *E. affine* sensu auct. Japon., non Bong.; *E. calycinum* Hausskn.; *E. consimile* sensu auct. Japon., non Hausskn.; *E. coreanum* Lév.; *E. consimile* var. *japonicum* Nakai; *E. sugaharae* Koidz.——IWA-AKABANA. Rhizomes short, with subsessile innovation-shoots on the neck; stems 15–60 cm. long, minutely puberulent on upper part and on branches, only on ridges in lower portion; leaves lanceolate or oblong-lanceolate, sometimes ovate-lanceolate, 1.5–9 cm. long, 5–33 mm. wide in the median ones, acute, cuneately narrowed to the very short petiole, irregularly serrate, the upper leaves much smaller; flowers 4.5–8 mm. long, rose to white, distinctly mucronate at apex in bud, the pedicels short, curved-puberulent; ovary puberulent; calyx 4–6(–7) mm. long; stigmas capitate, the styles slender, much longer than the stigmas; capsules 3.5–8 cm. long, the pedicels 4–15 mm. long; seeds lanceolate-oblong, 0.8–1.2 mm. long, rounded at apex, slightly narrowed toward the obtuse base, minutely papillose, the coma white.——July–Aug. Moist woods in mountains; Hokkaido, Honshu, Kyushu.——Sakhalin, s. Kuriles, Manchuria, and China.

Var. **nudicarpum** (Komar.) Hara. *E. nudicarpum* Komar.——Ke-nashi-iwa-akabana. Glabrous except margins of leaves.——Mountains; Hokkaido, Honshu (centr. distr.).——Korea and Manchuria.

10. Epilobium pyrricholophum Fr. & Sav. var. **pyrricholophum.** *E. tetragonum* var. *japonica* Miq.; *E. japonicum* Hausskn.; *E. oligodontum* Hausskn.; *E. pyrricholophum* var. *japonicum* (Miq.) Hara; *E. quadrangulum* Lév.; *E. chrysocoma* Lév.; *E. arcuatum* Lév.; *E. nakaianum* Lév.; *E. axillare* Franch.——Akabana. Stolons slender, loosely leafy; stems terete, 15–90 cm. long, more or less curved-puberulent, densely glandular-puberulent on upper part; leaves ovate to ovate-lanceolate, 2–10 cm. long, 0.5–3 cm. wide in the median ones, acute with an obtuse tip, rounded to subcordate at base, somewhat irregularly toothed; flowers 5–10 mm. long, rose; ovary densely glandular-puberulent, sometimes also sparingly curved-puberulent; calyx 4–6.5 mm. long; stigmas clavate, slightly shorter than the styles; capsules 3–8 cm. long, glandular-puberulent, the pedicels 7–15 mm. long; seeds lanceolate-oblong, 1.5–1.8 mm. long, rounded at apex, narrowed to the obtuse base, densely papillose, the coma usually red-brown.——July–Sept. Wet places in mountains and lowlands; Hokkaido (Oshima Prov.), Honshu, Shikoku, Kyushu; common.——Korea and China.

Var. **curvatopilosum** Hara. *E. rouyanum* Lév.; *E. hakkodense* Lév.; *E. makinoense* Lév.; *E. prostratum* Lév.; *E. kiusianum* Nakai; *E. myokoense* Koidz.——Mutsu-akabana. Stems puberulent and with glandular hairs on the uppermost part; leaves mostly curved-puberulent; pedicels and ovary curved- and glandular-puberulent.——Hokkaido, Honshu, Kyushu.——s. Kuriles.

11. Epilobium glandulosum Lehm. var. **asiaticum** Hara. *E. maximowiczii* Hausskn.; *E. glandulosum* forma *brevifolia* Hausskn.; *E. glandulosum* sensu auct. Japon., non Lehm.; *E. punctatum* Lév.; *E. minutiflorum* sensu auct. Japon., non Hausskn.——Karafuto-akabana, Azuma-akabana. Rhizomes short, with subsessile innovation-shoots on the neck; stems erect, 35–80 cm. long, usually branched, with slender longitudinal ridges, glabrous or curved-puberulent on the ridges at base, curved- and glandular-puberulent on upper part; median leaves oblong-lanceolate, 4–9 cm. long, 1.5–2.5 cm. wide, acute, rounded or subcordate at base, irregularly toothed, glabrescent, sessile or very short-petioled; flowers small, usually 3.5–5 mm. long, rarely to 7 mm. long, minutely mucronate at apex in bud, the pedicels very short, curved- and glandular-puberulent; ovary densely curved- and sparingly glandular-

puberulent; calyx 3–4 mm. long; stigmas clavate; capsules 4–6.5 cm. long, the pedicels 3–6 mm. long; seeds ovate-oblong, 0.8–1.4 mm. long, subrounded at apex, subacute at base, densely papillose, the coma whitish.——Wet places; Hokkaido, Honshu (centr. distr. and northw.).——Sakhalin and s. Kuriles. ——The species occurs in Korea, Ussuri, Kamchatka, Alaska, and N. America.

12. Epilobium shiroumense Matsum. & Nakai. Shirouma-akabana. With rhizomes and small-leaved stolons; stems often gregarious, 5–30 cm. long, usually simple, curved-puberulent on the obscure ridges, glandular-puberulent on upper part; leaves oblong or ovate-lanceolate, 12–30 mm. long, 3–11 mm. wide, subobtuse in the lower ones, subacute in the upper ones, cuneate and obtuse at base, minutely and obsoletely toothed, nearly glabrous, the petioles 1–4 mm. long; flowers small, rose; calyx 3–3.5 mm. long, minutely mucronate; stigmas clavate, about ½ as long as the styles; capsules 2–5 cm. long, glabrescent, the pedicels 15–30 mm. long, glandular-puberulent; seeds oblong-oblanceolate, 1.2–1.4 mm. long, narrowed at both ends, obtuse, acute at base, without papillae, the coma white.——July–Aug. Wet places in mountains; Hokkaido, Honshu (centr. distr.).

13. Epilobium foucaudianum Lév. *E. sertulatum* sensu auct. Japon., non Hausskn.; *E. lucens* Lév.; *E. hornemannii* sensu auct. Japon., non Reichenb.; *E. hornemannii* var. *foucaudianum* (Lév.) Hara.——Miyama-akabana. Plants with short rhizomes and small-leaved stolons; stems often gregarious, 5–22 cm. long, usually simple, curved-puberulent on the obscure ridges, glandular-puberulent on upper part; median leaves thin, oblong-lanceolate to oblong-ovate or ovate, 10–40 mm. long, 3–15 mm. wide, obtuse or rounded at apex, narrowed at base, often with obsolete short teeth, glandular-puberulent, the petioles 1–5 mm. long; upper leaves acute, rather prominently toothed; flowers 4–6 mm. long, rose, the pedicels glandular; calyx 3–4.5(–5) mm. long; stigmas clavate, ½ as long or slightly shorter than the style; capsules 2–6 cm. long, the pedicels 3–20(–25) mm. long; seeds oblanceolate, 1.1–1.3 mm. long, slightly narrowed at apex, subacute at base, minutely papillose.——July–Aug. High mountains; Hokkaido, Honshu (centr. distr. and northw.).——Sakhalin.

14. Epilobium dielsii Lév. *E. nakaharanum* Nakai——Ashiboso-akabana. With short rhizomes and small-leaved stolons; stems often gregarious, 3–10(–15) cm. long, curved-puberulent on the ridges; median leaves lanceolate to ovate-lanceolate, or oblong, 10–22 mm. long, 3.5–7 mm. wide, subobtuse, obscurely serrulate, glabrous except the puberulent midrib, the petioles 1–4 mm. long; upper leaves subobtuse; flowers few, 4–5 mm. long, rose, the pedicels puberulent; ovary glabrate or sometimes sparingly puberulent and glandular; calyx 3–4 mm. long; stigmas clavate, shorter than the styles; capsules 2.5–5 cm. long, the pedicels 22–35 mm. long, sparsely curved-puberulent; seeds oblanceolate, 1–1.2 mm. long, minutely papillose.——July–Aug. High mountains; Hokkaido, Honshu (centr. distr. and northw.).——s. Kuriles.

4. LUDWIGIA L. Mizu-yuki-no-shita Zoku

Herbs or subshrubs, often aquatic; leaves usually membranous, alternate, rarely opposite, entire, rarely toothed; flowers solitary, usually axillary, sessile or pedicelled, 2-bracteolate on pedicels or on the lower part of ovary; calyx-tube cylindrical or obconical, scarcely extended beyond the ovary, the lobes 3–6, persistent or nearly so; petals 3–6 or absent; stamens usually twice as many as the calyx-lobes; ovary 4- or 5-locular; styles simple, the stigmas 3- to 6-grooved or -lobed; ovules many, in many series on the inner angle of the placentae; capsules linear to obovoid, terete, angled, or winged, at the apex with a disc and persistent calyx-lobes, dehiscent through a rupture of the pericarp; seeds many, small.——About 40 species, in tropical and warm-temperate regions of the world, especially abundant in America.

1A. Pedicels longer than the fruit; bracteoles on the lower half of ovary, deciduous; petioles with 2 stipulelike glands at base; flowers large, 2–2.5 cm. across, bright yellow; capsules cylindric-clavate. . . . . 1. *L. stipulacea*
1B. Pedicels absent or very short; bracteoles very small at the base of flower; petioles without stipulelike glands at base; flowers smaller.
  2A. Leaves 3–8 cm. long; flowers with petals; capsules cylindrical, 1.5–5 cm. long.
    3A. Stamens 8; calyx-lobes 6–8 mm. long; petals 6–10 mm. long; capsules narrowly cylindric, about 4 mm. across.
                                                                    2. *L. octovalvis* var. *sessiliflora*
    3B. Stamens 4; calyx-lobes 2–4 mm. long; petals minute; capsules linear-cylindric, 1.5–2 mm. across.
      4A. Seeds in one longitudinal series in each locule; petals minute. . . . . 3. *L. epilobioides*
      4B. Seeds in 2 series in each locule; petals about 4 mm. long. . . . . 4. *L. greatrexii*
  2B. Leaves 1–2.5 cm. long; flowers apetalous; capsules ellipsoidal to nearly globose, 4–5 mm. long. . . . . 5. *L. ovalis*

1. Ludwigia stipulacea (Ohwi) Ohwi. *Jussiaea stipulacea* Ohwi; *J. repens* sensu auct. Japon., non L.; *Ludwigia adscendens* var. *stipulacea* (Ohwi) Hara——Mizu-kimbai. Nearly glabrous rhizomatous perennial somewhat viscid in upper part in life; stems terete, ascending or creeping at base, erect above, 25–30 cm. long; leaves alternate, oblanceolate, oblong-lanceolate to narrowly oblong, 4–7 cm. long, 1.3–2.5 cm. wide, obtuse to subacute or sometimes rounded at apex, gradually narrowed at base, entire, the petioles 1–1.5 cm. long, with 2 stipulelike glands at base, the lateral nerves rather many, parallel, ascending; flowers axillary, solitary, 2–2.5 cm. across, bright yellow, the bracteoles on lower part of ovary, broadly ovate, about 2 mm. long; calyx-lobes 4, sometimes 5, lanceolate, 8–10 mm. long, acuminate; petals broadly obovate, about 13 mm. long, emarginate; stamens twice as many as the calyx-lobes, in 2 series, the disc densely white-villous; capsules cylindric-clavate, 2–2.5 cm. long, narrowed to the base, the pedicels 3–4 cm. long; seeds obliquely 4-angled, about 2 mm. across.——July–Sept. Ponds and ditches in lowlands; Hokkaido, Honshu, Shikoku, Kyushu.

2. Ludwigia octovalvis (Jacq.) Raven var. **sessiliflora** (Micheli) Raven. *Jussiaea octonervia* var. *sessiliflora* Micheli; *J. suffruticosa* L.; *J. villosa* Lam.; *L. pubescens* var. *villosa* (Lam.) Hara——Kidachi-kimbai. Coarsely hirsute, suffruticose herb; stems erect, branched, 40–100 cm. long; leaves lanceolate, 4–8 cm. long, 5–12 mm. wide, subobtuse, short-petiolate; flowers solitary in axils, pale yellow, subsessile, the bracteoles minute or absent; ovary and calyx more or less hirsute while young; calyx-lobes ovate, 6–8 mm. long; petals nearly orbicular, 6–10 mm. long, the disc with 4 tufts of hairs; capsules narrowly cylindric, 3–5 cm. long, about 4 mm. wide, terminated by the persistent calyx-lobes, narrowed to the base; seeds many, rounded, emarginate.——Aug.–Oct. Wet places; Kyushu (Yakushima).——Widely distributed in the Tropics.

3. Ludwigia epilobioides Maxim. *Nematopyxis japonica* Miq.; *Jussiaea fauriei* Lév.; *J. japonica* Lév.; *J. philippiana* Lév.; *J. prostrata* sensu auct. Japon., non Lév.; *L. prostrata* sensu auct. Japon., non Roxb.——Chōji-tade. Annual, scattered-puberulent while young; stems erect or ascending, branched, slightly ridged, reddish, 30–60 cm. long; leaves alternate, lanceolate or oblong-lanceolate, 3–12 cm. long, 1–3 cm. wide, gradually acute, acuminate at base, entire, with 7–15 pairs of lateral nerves, the petioles 5–15 mm. long; flowers sessile; ovary minutely appressed-pilose; calyx-lobes ovate, 2–4 mm. long, acute; petals minute; stamens 4; capsules linear-cylindric, somewhat obtusely angled, 1.5–3 cm. long, 1.5–2 mm. wide; seeds fusiform, about 1 mm. long, in one longitudinal series in each locule, enveloped in a spongy pericarp, with brown longitudinal striations.——Aug.–Oct. Wet places in lowlands especially in paddy fields; Honshu, Shikoku, Kyushu; common. ——Widely distributed throughout the Tropics.

4. Ludwigia greatrexii Hara. *Jussiaea greatrexii* (Hara) Hara——Usuge-chōji-tade. Much resembling the preceding, but with somewhat larger flowers, white-villous on the disc, and smaller seeds arranged in 2 longitudinal series in each locule, and the young shoots more pubescent; stems short-pubescent while young; leaves puberulent on both sides especially while young, 7–8 cm. long, 1–2 cm. wide; flowers subsessile, densely appressed-puberulent, the bracteoles minute or absent; calyx-lobes usually 5, 3–4 mm. long; petals obovate, 4 mm. long; stamens usually 5, the disc white-villous; capsules usually 5-locular, the endocarp rather thick, not corky, not closely adherent to the seeds; seeds oblong, slightly curved, 0.8 mm. long, with an obscure ridge.——Honshu (Kadzusa, and Suruga Prov.), Kyushu; rare.

5. Ludwigia ovalis Miq. Mizu-yuki-no-shita. Glabrous perennial; stems long-creeping at base, 20–40 cm. long, terete; leaves alternate, broadly ovate, ovate-orbicular to elliptic-orbicular, 1–2.5 cm. long, 1–2 cm. wide, rounded to very obtuse, abruptly narrowed at base, entire, the lateral nerves of 4–6 pairs, the petiole to 7 mm. long; flowers solitary, sessile, apetalous, pale green-yellow, the bracteoles small, at base of the ovary; calyx-lobes 4, deltoid, about 2 mm. long, acute; stamens 4; capsules ellipsoidal or nearly globose, 4–5 mm. long; seeds ovoid, about 0.8 mm. long, red-brown, lustrous, with an appendage on one side.——July–Oct. Wet places; Honshu, Shikoku, Kyushu.——Formosa.

Fam. 150. **HALORAGACEAE** Ari-no-tō-gusa Ka   Water-milfoil Family

Herbs or shrubs, sometimes aquatic; leaves alternate, opposite, or verticillate, the immersed ones often finely dissected, stipules absent; flowers bisexual or unisexual, solitary, in corymbs or panicles, often very small; calyx-tube adnate to the ovary, the lobes obscure or 2–4, slightly imbricate or valvate; stamens 2–8, the anthers basifixed, longitudinally split; ovary inferior, 1- to 4-locular; styles 2–4; ovules pendulous, as many as the styles; fruit a small nut or drupe, often winged, indehiscent; seeds with endosperm.——About 7 genera, with about 100 species, in the Tropics and temperate regions.

1A. Terrestrial herbs or rarely subshrubs ligneous at base; leaves opposite; flowers solitary or more often racemose; petals 4; stamens usually 8. . . . . 1. *Haloragis*
1B. Aquatic herbs; leaves verticillate (in ours); flowers axillary or in a terminal spike; petals 2–4 or absent; stamens 2–8. . . . 2. *Myriophyllum*

1. HALORAGIS Forst. Ari-no-tō-gusa Zoku

Glabrous or strigose perennials or subshrubs; leaves small, usually coriaceous, entire or serrate, sometimes pinnately lobed, often with cartilaginous margins, the lower ones often opposite, the upper alternate; bracteoles 2 or absent; flowers small, short-pedicelled, axillary or in racemes, usually deflexed or nodding, unisexual or bisexual; calyx-tube with 4 or 8 ribs, the lobes usually 4, short, sometimes peltate; petals usually 4, membranous to coriaceous, valvate; stamens usually 8, with short filaments; ovary 2- to 4-locular or 1-locular; styles 2–4 or absent, the stigmas usually pinnate; ovules solitary in each locule; fruit an indehiscent small drupe, striate, winged, or spined.——About 60 species, chiefly in Australia, few in N. America, e. and s. Asia, and the Pacific Islands.

1A. Leaves broadly linear, 1–1.5 mm. wide, often with a broadly linear to linear-lanceolate lobe or tooth on each side above the middle; inflorescence simple. .... 1. *H. walkeri*
1B. Leaves ovate to broadly so, 5–10 mm. wide, few-toothed; inflorescence branched. .... 2. *H. micrantha*

1. Haloragis walkeri Ohwi. Hosōba-ari-no-tō-gusa. Rather firm glabrous perennial herb, much branched from a somewhat ligneous scarcely thickened ascending base; stems slender, weakly 4-striate, simple or rarely sparsely branched, erect, leafy; leaves rather firm, cauline, opposite, sometimes alternate in upper ones, broadly linear, 1–2 cm. long, 1–1.5 mm. wide, gradually narrowed at base, sessile, scabrid on margin, simple and entire or more often 3-parted above the middle, the segments linear-lanceolate to broadly linear, acuminate with an obtuse tip; median leaves larger, 5–7 mm. long, the lateral ones obliquely spreading to ascending, 2–5 mm. long or sometimes reduced to an obtuse tooth, the midrib slightly raised beneath; spikes terminal, bracteate, erect, loosely flowered; flowers solitary or in pairs, the short pedicels minutely bibracteolate; calyx-teeth minute, ovate-deltoid, papillose-scabrid; petals 4, navicular, 1.5 mm. long, obtuse and subcucullate at apex, pilose on back; stamens 8.——July. Wet places; Kyushu (Mageshima near Tanegashima).

2. Haloragis micrantha (Thunb.) R. Br. *Gonocarpus micranthus* Thunb.——Ari-no-tō-gusa. Glabrous perennial; stems erect from a creeping much-branched base, 10–30 cm. long, slender, 4-angled; leaves usually opposite, the upper ones alternate, ovate to broadly so, 7–15 mm. long, 5–10 mm. wide, acute or subacute, rounded at base, somewhat cartilaginous and translucent on margin, few-toothed, the petioles 0.5–1 mm. long; racemes 3–10 cm. long, loosely many-flowered, usually paniculately compound, the bracts minute, the pedicels slender, 0.3–0.5 mm. long, curved downward; flowers small, deflexed, brownish yellow; calyx-tube broadly obovoid, about 1 mm. long in fruit, with 8 raised ribs, the lobes 4, ovate, about ½ as long as the calyx-tube, erect, subacute, slightly cordate at base; petals oblong, inflexed and mitrate at apex, 2–2.5 times as long as the calyx-lobes, deciduous.——July–Sept. Grassy places in mountains; Hokkaido, Honshu, Shikoku, Kyushu.——Ryukyus, Formosa, China, Malaysia, and Australia.

2. MYRIOPHYLLUM L. Fusa-mo Zoku

Soft, glabrous, often aquatic herbs; leaves opposite, alternate, or verticillate, linear to ovate, entire or toothed, sometimes pinnately lobed, the immersed ones often pinnately dissected into filiform segments; flowers unisexual, small, solitary in axils or in a terminal spike, sessile, the staminate uppermost, the pistillate below; calyx-tube of staminate flowers very short, the lobes 4 or rarely 2; petals 2–4; stamens 2–8, filiform; calyx-tube of pistillate flowers 4-grooved, the lobes obscure or 4, linear, minute; petals minute or absent; ovary 4- or rarely 2-locular; styles 4, short, pinnate, usually reflexed; ovules solitary in each locule; drupe 4-grooved or separating into 4 nutlike fruits, the pericarp coriaceous or slightly fleshy, sometimes tuberculate on back.——About 40 species, in wet places or aquatic.

1A. Flowers axillary. .... 1. *M. verticillatum*
1B. Flowers in spikes.
  2A. Aerial leaves pinnately lobed; plants monoecious.
    3A. Aerial leaves green; petals and calyx-tube smooth. .... 2. *M. spicatum*
    3B. Aerial leaves whitish on upper side; petals of staminate flowers and calyx-tube of pistillate flowers minutely papillose. 3. *M. oguraense*
  2B. Aerial leaves usually linear; plants dioecious. .... 4. *M. ussuriense*

1. Myriophyllum verticillatum L. Fusa-mo. Aquatic perennial herb; stems often much elongate, sparsely branched, with cylindric winter buds in axils; leaves verticillate in 4's, sessile, the immersed ones 3–4 cm. long, pinnately dissected into capillary segments, but the aerial ones linear-oblong, 1–1.5 cm. long, 3–5 mm. wide, pinnately cleft into narrowly linear ascending lobes; flowers solitary in axils of aerial leaves, the staminate terminal; petals 4, narrowly oblong, about 3 mm. long, obtuse; stamens 8, the anthers 1.5–2 mm. long; pistillate flowers below the staminate; calyx-tube urceolate, with 4 obtuse angles; fruit ovoid-globose, about 2.5 mm. long, 4-grooved, crowned by the persistent calyx-lobes and remnant of the styles.——May–July. Ponds and ditches in lowlands; Hokkaido (sw. distr.), Honshu, Shikoku, Kyushu; common.

——Widely distributed in temperate regions of the world.

2. Myriophyllum spicatum L. Hozaki-no-fusa-mo. Aquatic monoecious perennial herb; stems elongate, often 1.5 m. long, sparsely branched; leaves verticillate in 4's, sessile, 2–3.5 cm. long, pinnately cleft or dissected, the segments filiform, entire, the aerial leaves green; inflorescence aerial, spicate, 3–8 cm. long, with 4 flowers at each node, the bracts narrowly oblong, spreading, 1–1.5 mm. long, entire, obtuse; staminate flowers on upper part of the spike, the petals 4, about 2.5 mm. long, obtuse, deciduous at anthesis; stamens 8, the anthers 1.5 mm. long; calyx-tube in the pistillate flowers nearly campanulate, 4-grooved, about 1 mm. long, the lobes minute; fruit globose-ovoid, about 2.5 mm. long, separating into 4 nutlike fruits, smooth on back.——May–Oct. Ponds and ditches; Hokkaido, Honshu, Shikoku, Kyushu; common.——Widely distributed in warmer and temperate regions of the world.

Var. **muricatum** Maxim. TOGE-HOZAKI-NO-FUSA-MO. Margins of fruitlets on back with callose tubercles.

3. **Myriophyllum oguraense** Miki. OGURA-NO-FUSA-MO. Resembles the preceding species; winter-buds elongate, 6–8 cm. long, about 3 mm. across; leaves pinnately cleft, verticillate in 4's, the immersed ones 5 cm. long, the segments about 2.5 cm. long, the aerial ones about 1 cm. long, the lobes ascending, linear, whitish on upper side in life; pistillate flowers in lower leaf axils, the calyx-tube minutely papillose; staminate flowers in upper leaf axils, the petals 4, minutely papillose; stamens 8.——Nov. Ponds and ditches; Honshu (Kinki Distr.), Shikoku.

4. **Myriophyllum ussuriense** (Regel) Maxim. *M. verticillatum* var. *ussuriense* Regel——TACHI-MO. Winter-buds 6–8 mm. long, to 3 mm. wide; stems 5–20 cm. long, much elongate in the immersed ones; leaves verticillate in 3's or 4's, pinnately parted, broadly lanceolate, 5–10 mm. long, the segments linear, short, entire, the upper leaves linear, entire; flowers unisexual (plants dioecious); petals of the staminate flowers 4, obovate-oblong, about 2 mm. long, obtuse; stamens 8, the anthers about 1 mm. long; calyx-tube of pistillate flowers globose-urceolate, with short tubercles.——June–Sept. Wet places and ponds; Hokkaido, Honshu, Shikoku, Kyushu.——Ussuri, Amur, Manchuria, Korea, and Formosa.

Fam. 151. HIPPURIDACEAE   SUGINA-MO KA   Marestail Family

Aquatic or marsh herbs; stems erect; leaves simple, verticillate, entire; flowers small, solitary in axils, sessile, bisexual or unisexual; calyx-tube subglobose, terete, entire; petals absent; stamens solitary, superior, the filaments rather short, the anthers basifixed, longitudinally split on inner side; ovary 1-locular; style solitary, papillose the full length; ovules solitary, pendulous from the top of the locule; drupe coriaceous, the stone crustaceous, 1-seeded; seeds short-cylindric, the testa membranous, endosperm scanty, embryo cylindric.——A single genus, with about 3 species, in cooler regions of the N. Hemisphere and S. America.

1. HIPPURIS L.   SUGINA-MO ZOKU

Characters of the family.

1. **Hippuris vulgaris** L. SUGINA-MO. Mostly glabrous aquatic herb with long-creeping rhizomes; stems erect, simple, 20–50 cm. long; leaves verticillate in 6's to 12's, entire, the aerial ones broadly linear to narrowly oblong, 1–1.5 cm. long, about 2 mm. wide, obtuse, spreading, the immersed ones 2–6 cm. long, 2–3 mm. wide, 1-nerved; flowers axillary in the upper aerial leaves, solitary, sessile; calyx about 1 mm. long, the limb entire; anthers broadly ovate, embracing the style on innerside at first; style linear, slender; fruit narrowly oblong, about 2 mm. long, smooth.——July–Aug. Wet places and shallow ponds; Hokkaido, Honshu (Osenuma in Kotsuke and northw.).——Widely distributed in cooler parts of the N. Hemisphere.

Fam. 152. THELIGONACEAE   YAMATOGUSA KA   Theligonum Family

Herbs; leaves somewhat fleshy, ovate, the lower opposite, the upper alternate, entire; stipules membranous; flowers unisexual (plants monoecious), axillary, nearly sessile; perianth of staminate flowers with 2–5 sepals; stamens 10–30, the anthers linear, on long capillary filaments; perianth of the pistillate flowers small, connate, 3- or 4-toothed; ovary inferior; carpels 1; ovules with single integument; stigma 1, lateral; fruit a small achene.——One genus, with about four species, in the Mediterranean region and e. Asia.

1. THELIGONUM L.   YAMATOGUSA ZOKU

Characters of the family.

1. **Theligonum japonicum** Okubo & Makino. *Cynocrambe japonica* (Okubo & Makino) Makino——YAMATOGUSA. Perennial; stems loosely branched and creeping at base, terete, 15–30 cm. long, with a longitudinal line of recurved short hairs; leaves ovate or narrowly so, 1–3 cm. long, 8–20 mm. wide, short-acuminate to obtuse, entire, loosely short-pubescent, the petioles 5–10 mm. long; stipules slightly connate at base, membranous, 2–3 mm. long; staminate flowers 1 or 2 on each node, subsessile, 8–10 mm. long just before anthesis, the perianth-segments 3, membranous, valvate, narrowly oblong, obtuse, strongly circinnate at anthesis, about 10-nervulose; stamens 20–25, free, the anthers pendulous on capillary filaments, 4–5 mm. long, linear; pistillate flowers small, axillary, sessile, with a bracteole at base, the calyx obovoid, coarsely pilose; stigma from the small bilobed body attached laterally to the calyx, minutely papillose; achenes narrowly obovate or obovate-elliptic, 3–3.5 mm. long, flat, coarsely pilose on both sides.——Apr.–June. Woods in mountains; Honshu (Kantō and Kinki Distr.), Shikoku, Kyushu.

Fam. 153. ARALIACEAE   UKOGI KA   Ginseng Family

Evergreen or deciduous trees, shrubs, or herbs, rarely scandent; stems often spiny or prickly; leaves alternate, rarely opposite or falsely verticillate, simple, or palmately to pinnately compound; stipules absent or at base of petioles; inflorescence in umbels or heads; flowers bisexual or polygamous (plants sometimes dioecious), actinomorphic, small; calyx superior, small, entire or dentate; petals 3 or more, usually 5, valvate or slightly imbricate in bud, usually free; stamens free, usually as many as the petals and alternate with them, the disc superior to the ovary, often inserted at the base of the style; ovary 1- to many-locular; styles free or connate; ovules solitary and pendulous from the top of the locule, anatropous; fruit a berry or drupe; seeds with copious endosperm and a small embryo.——About 60 genera, with about 900 species, chiefly in the Tropics.

1A. Leaves pinnately compound. . . . . 1. *Aralia*
1B. Leaves simple or palmately compound.
  2A. Herbs; leaves palmately compound, falsely verticillate on top of a naked simple stem; petals more or less imbricate in bud.
                                                                   2. *Panax*
  2B. Woody plants; petals valvate in bud.
    3A. Evergreen woody vine, climbing or scrambling, with adventitious roots on the stems; leaves simple, entire or 2- or 3-lobed.
                                                                  3. *Hedera*
    3B. Trees or shrubs, not vines.
      4A. Leaves evergreen.
        5A. Leaves palmately compound. . . . . 4. *Schefflera*
        5B. Leaves simple, palmately lobed or partly undivided.
          6A. Shrubs; leaves 7- to 9-lobed; inflorescence a large terminal panicle. . . . . 5. *Fatsia*
          6B. Trees; leaves entire and undivided, or 3-lobed in young individuals; inflorescence simple or slightly branched.
                                                  6. *Dendropanax*
      4B. Leaves deciduous.
        7A. Leaves palmately compound into leaflets.
          8A. Stones of the fruit laterally flattened; leaves (in ours) 5-foliolate. . . . . 7. *Acanthopanax*
          8B. Stones of the fruit dorsally flattened; leaves 3-foliolate. . . . . 8. *Evodiopanax*
        7B. Leaves simple, or often palmately lobed.
          9A. Shrub, densely aculeolate-prickly on stems and leaves; styles free; leaves with spines; calyx-teeth obscure. . . . 9. *Oplopanax*
          9B. Tree, with large flattened prickles on stems and branches; styles mostly connate; leaves unarmed; calyx-teeth minute.
                                                    10. *Kalopanax*

1. ARALIA L.   TARA-NO-KI ZOKU

Shrubs or herbs, sometimes prickly; leaves pinnately compound or ternate-pinnate, toothed; stipules adnate to base of the petioles; umbels usually compound, the bracts small; flowers usually polygamous; calyx-limb distinct, truncate or minutely 5-toothed; petals 5, ovate, slightly imbricate in bud; stamens 5; ovary usually 5-locular; styles as many as the locules of ovary, free or connate toward the base; fruit a berrylike drupe, the stones 2–5, ovate or oblong, flat, rounded.——About 30 species, in e. and s. Asia, Malaysia, Australia, and N. America.

1A. Deciduous shrub, often densely prickly on stems and leaves; inflorescence a compound raceme, the main axis not developed. . . 1. *A. elata*
1B. Herbs, without prickles; inflorescence simple or sparsely branched, the umbels terminal, the axis simple or elongate.
  2A. Plant green, loosely pilose throughout, except the flowers; flowers pale green. . . . . 2. *A. cordata*
  2B. Plant purplish, mostly glabrous throughout, except for scattered appressed hairs on leaves; flowers purplish. . . . . 3. *A. glabra*

1. Aralia elata (Miq.) Seem. *Dimorphanthus elatus* Miq.; *A. canescens* Sieb. & Zucc.; *D. mandshuricus* Rupr. & Maxim.; *A. mandshuricus* (Rupr. & Maxim.) Maxim.; *A. spinosa* var. *canescens* (Sieb. & Zucc.) Fr. & Sav. and var. *glabrescens* Fr. & Sav.——TARA-NO-KI. Deciduous erect large shrub with simple or sparsely branched thick prickly stems; leaves alternate, very large, terminal and densely rosulate on the branches, deltoid in outline, 50–100 cm. long, twice pinnate, prickly, the leaflets 5–9 on each pinna, subsessile, ovate, elliptic, or narrowly ovate, 5–12 cm. long, 2–7 cm. wide, acuminate, irregularly toothed, loosely pubescent on both sides, densely so on nerves and whitish beneath; inflorescence a terminal compound raceme, very large, 30–50 cm. long, densely brownish curved-pubescent, the main axis very short or obsolete; flowers white, about 3 mm. across; fruit globose, black, about 3 mm. across, the stones with obscure granular striations; styles 5, free, filiform, slightly recurved, about 1.5 mm. long.——Aug. Thickets and thin woods in lowlands and hills; Hokkaido, Honshu, Shikoku, Kyushu; common.——Korea, Sakhalin, Amur, Ussuri, Manchuria, and s. Kuriles.

Var. **subinermis** Ohwi. *A. spinosa* var. *canescens* Fr. & Sav.; *A. elata* var. *canescens* (Fr. & Sav.) Nakai——MEDARA. Plant less prickly, densely brownish or yellowish pubescent. ——Occurs with the typical phase.

2. Aralia cordata Thunb. *A. edulis* Sieb. & Zucc.; *A. nutans* Fr. & Sav.; *Dimorphanthus edulis* (Sieb. & Zucc.) Miq. ——UDO. Large stout perennial herb 1–1.5 m. high, loosely short-pilose throughout except the flowers; leaves alternate, few, large, deltoid in outline, 50–100 cm. long, twice pinnate, the petioles long, with a small stipule on each side at base, the leaflets 5–7 on each pinna, ovate to oblong, or ovate-elliptic, 5–30 cm. long, 3–20 cm. wide, acuminate, toothed, sessile or short-petioluled; inflorescence a loosely branched terminal raceme, axillary on upper part of stem; flowers pale green, about 3 mm. across; fruit ovoid-globose, about 2 mm. long, glabrous, purple-black, the stones smooth; styles about 1 mm. long, spreading, slightly recurved, connate toward base to form a column.——Aug. Thickets and thin woods especially along streams and ravines; Hokkaido, Honshu, Shikoku, Kyushu; frequently cultivated and the shoots eaten as a vegetable.—— Sakhalin, Manchuria, Korea, and China.

Var. **sachalinensis** (Regel) Nakai. *A. racemosa* var. *sachalinensis* Regel——KARAFUTO-UDO. Calyx-tube pubescent.—— Hokkaido and Honshu.

3. Aralia glabra Matsum. MIYAMA-UDO. Nearly glabrous purplish herb 80–100 cm. high, with thick rhizomes; stems rather stout, few-leaved; leaves petioled, twice ternate-pinnate, the leaflets membranous, usually petioluled, ovate-cordate to ovate or ovate-elliptic, 3–10 cm. long, 3–6 cm. wide, acuminate-caudate, toothed, with appressed callose hairs on upper side, nerves beneath often short-pilose; inflorescence racemose-umbellate, branched, terminal and from upper axils, peduncled, glabrous or nearly so, often purplish, umbels with very short bracts, the pedicels very slender, 1–2 cm. long; flowers about 3 mm. across; petals purplish green, minutely papillose on upper margin, subglobose, about 3 mm. long, purple-black, glabrous; styles about 1 mm. long, connate toward base, free and subrecurved at tip.——June–Aug. Dense woods in mountains; Honshu (Kantō and centr. distr.); rather rare.

2. PANAX L.   Ninjin Zoku

Perennial polygamodioecious herbs with thick rhizomes or roots; stems simple, with few membranous scales at base; leaves 3–5, falsely verticillate and terminal on the naked stems, petioled, palmately compound, the leaflets petioluled; inflorescence umbellate, the umbels solitary or few, terminal, on long peduncles, the pedicels jointed below the fertile flowers, not jointed in the staminate flowers; calyx-teeth 5; petals 5, slightly imbricate in bud; stamens 5, the disc flat; ovary 3- or 4-locular; styles 3 or 4, free in fertile flowers, connate in the staminate; fruit with a rather fleshy pericarp, stones 2–4.——Few species, in e. Asia, India, and N. America.

1. **Panax japonicus** C. A. Mey. *P. schinseng* var. *japonica* Nees; *P. quinquefolium* var. *subsessilis* Miq.; *P. repens* Maxim.; *Aralia repens* (Maxim.) Makino; *A. quinquefolia* var. *repens* (Maxim.) Burkill; *P. ginseng* var. *japonicum* Makino ——Tochiba-ninjin.   Perennial 50–80 cm. high with rather thick, nodose, creeping rhizomes; stems erect, simple, glabrous; leaves 3–5 in terminal false-verticils, palmately 3- to 7-, usually 5-foliolate, the petioles 5–10 cm. long, the leaflets membranous, green, irregularly toothed, the lateral ones gradually smaller, the median oblanceolate to obovate or obovate-oblong, 10–30 cm. long inclusive of the petiolule, 2–7 cm. wide, long-acuminate to caudate; umbels solitary or sometimes 2–4, many-flowered, the bracts minute, smooth, the pedicels 1–2 cm. long, slender, slightly scabrous; flowers yellowish green, about 3 mm. across; fruit globose, 4–5 mm. long, red, sometimes black above; styles 2(–4), free, recurved, about 1.5 mm. long.—— June–Aug. Woods in mountains; Hokkaido, Honshu, Shikoku, Kyushu; rather rare.

Panax schinseng Nees. Ninjin, Otane-ninjin.   Closely resembles the preceding species, but differs chiefly in the very short indistinct rhizomes and stout, fusiform, rarely forked main roots. Often cultivated in our area as a Chinese drug plant.——Korea and Manchuria.

3. HEDERA L.   Kizuta Zoku

Evergreen woody climbers or scramblers with adventitious hold-fast roots on the stems; leaves simple or lobed, often with stellate hairs or scales; stipules absent; umbels forming a paniclelike inflorescence, the bracts minute or absent; flowers bisexual; calyx-limb obsolete and entire or 5-toothed; petals 5; stamens 5; ovary 5- or 4-locular, subinferior; styles 5, connate the whole length, the stigmas obsoletely 5-lobed; fruit a berrylike subglobose drupe, with 4 or 5 stones; seeds solitary in each stone, the embryo ruminate.——Few species, in Europe, N. Africa, and Asia.

1. **Hedera rhombea** (Miq.) Bean. *H. helix* Thunb., non L.; *H. helix* forma *rhombea* Miq.; *H. japonica* W. Paul, non Jungh.; *H. japonica* Tobler; *H. tobleri* Nakai——Kizuta, Fuyuzuta.   Branches gray-brown, the scales stellately 15- to 20-lobed, yellowish; leaves on flowering branches coriaceous, broadly lanceolate to ovate-orbicular, usually rhombic-ovate, 3–6 cm. long, 2–4 cm. wide, subobtuse, acute to rounded at base, deep green and lustrous on upper side, paler beneath, simple, entire or 2- or 3-lobed, the petioles 2–5 cm. long; juvenile leaves on nonflowering vegetative shoots subdeltoid, palmately 3(–5)-lobed, truncate to cordate at base; flowers green-yellow, 4–5 mm. across, the pedicels 1–1.5 cm. long; fruit black, globose, 6–7 mm. across, with an orbicular terminal disc about 4 mm. across and short persistent style.——Oct.–Nov. Thickets and woods especially near the sea; Honshu, Shikoku, Kyushu; common.——Forma **pedunculata** (Nakai) Hatusima. *H. pedunculata* Nakai——Nagabo-kizuta.   Peduncles and pedicels longer, the latter 2–4 cm. long.——Kyushu.——Ryukyus.

4. SCHEFFLERA Forst.   Fuka-no-ki Zoku

Evergreen shrubs or trees, sometimes epiphytic, without prickles or spines; leaves mostly palmately compound, petioled, the leaflets coriaceous, entire or with irregular incisions; stipules prominent, usually adnate to base of petioles; inflorescence racemose or paniculate, the umbels rarely subcapitate, the bracts woolly, the pedicels not jointed beneath the flower; calyx-limb entire or toothed; petals 5–8, sometimes more, valvate; stamens as many as the petals; styles connate to form a column, or very short; fruit subglobose; seeds flattened.——Many species in the Old World Tropics.

1. **Scheffiera octophylla** (Lour.) Harms. *Aralia octophylla* Lour.; *Agalma octophyllum* (Lour.) Seem.; *Heptapleurum octophyllum* (Lour.) Hance; *A. lutchuense* Nakai ——Fuka-no-ki.   Evergreen tree; branchlets thick, spreading, with large V-shaped leaf-scars, densely brownish stellate-puberulent while young; leaves 7- to 10-foliolate, the petioles 15–30 cm. long, the leaflets narrowly oblong or narrowly obovate-oblong, unequal, 7–20 cm. long, 3–7 cm. wide, entire or in young plants often irregularly pinnately incised, abruptly acute, acute to obtuse at base, the petiolules 2–5 cm. long; stipules semirounded, adnate on back to base of petioles; inflorescence terminal, rather much branched, about 20 cm. long,

the bracts persistent; flowers greenish, 4–5 mm. across, the pedicels 4–6 mm. long; calyx-teeth short; fruit subglobose, about 5 mm. across, glabrous, the disc less than 1.5 mm. across; column of styles about 1 mm. long, the stigmas disclike, with a slightly depressed center; seeds 4–6.——Nov.–Jan. Woods and thickets; Kyushu (s. distr. incl. Yakushima and Tanegashima).——Ryukyus, Formosa, China, and Indochina.

5. FATSIA Decne. & Planch.   YATSUDE ZOKU

Evergreen shrubs without spines or prickles; leaves large, palmately 7- to 9-lobed, the petioles dilated at base; stipules absent; inflorescence in large terminal panicles, the lateral umbels staminate, the terminal ones bisexual, the bracts caducous, the bracteoles membranous, caducous, narrow; flowers polygamous; petals 4–6, valvate in bud; stamens as many as the petals; ovary 2- to 6-locular; styles of the bisexual flowers free, with a capitate stigma, of the staminate short, connate; fruit globose, fleshy, the stones sulcate or rugose on back, seeds flattened.——Few species, in e. Asia.

1. Fatsia japonica (Thunb.) Decne. & Planch. *Aralia japonica* Thunb.; *A. sieboldii* K. Koch——YATSUDE. Shrub, unbranched or sparsely branched, with stout branches about 1.5 cm. thick, and large lunate leaf-scars; stems rusty-brown woolly while young; leaves thick-coriaceous and lustrous above, soon glabrate, nearly orbicular, 20–40 cm. across, cordate to cordate-truncate at base, long-petioled, palmately 7- to 9-lobed, the lobes narrowly ovate-elliptic, acuminate, toothed; panicles terminal, 20–40 cm. long, broadly pyramidal, with caducous white bracts in bud; umbels many-flowered, the bracteoles caducous, scarious, brown; flowers white, about 5 mm. across, 5-merous, the disc convex, about 3 mm. across in fruit; calyx-teeth indistinct; fruit subglobose, black, about 5 mm. across; styles 5, slender, about 1.5 mm. long.——Oct.–Nov. Woods near the sea; Honshu (Pacific side from Iwaki Prov. through Tōkaidō to Kii Prov.), Shikoku, Kyushu; rather common and frequently cultivated, with several cultivars grown in gardens. ——Ryukyus and s. Korea.

6. DENDROPANAX Decne. & Planch.   KAKURE-MINO ZOKU

Evergreen glabrous trees or shrubs; leaves simple, entire or 3- to 5-lobed; stipules small, intrapetiolar, adnate to the base of petioles on back, sometimes absent; umbels simple or branched, the bracts small or absent, the pedicels not jointed; flowers bisexual or rarely polygamous; calyx-limb distinct, entire or 5-toothed; petals rather thick, valvate in bud; stamens 5; ovary 5- to 8-locular; styles connate at least to middle or nearly the whole length, fruit globose or nearly so, sulcate when dried, the pericarp fleshy or juicy, the stones flat or subtriangular.——About 30 species, in e. Asia, the Malay Peninsula, Centr. and S. America.

1. Dendropanax trifidus (Thunb.) Makino. *Acer trifidum* Thunb.; *Hedera japonica* Jungh.; *Fatsia ? mitsde* (Sieb.) de Vriese; *Aralia mitsde* Sieb.; *Textoria japonica* Miq.; *D. japonicus* (Jungh.) Seem.; *Gilibertia japonica* (Jungh.) Harms; *G. trifida* (Thunb.) Makino; *T. trifida* (Thunb.) Nakai—— KAKURE-MINO. Small glabrous evergreen tree with gray-brown branches and rather thick green branchlets; leaves of inflorescence broadly rhombic-ovate to ovate, 7–12 cm. long, 3–8(–12) cm. wide, broadly cuneate at base, with 3 nerves rising from the base, entire or sometimes partly 2- or 3-lobed, often deeply 3- to 5-cleft and broadly truncate at base in young plants, with a small obsolete gland beneath among the reticulate veinlets, the petioles 2–10 cm. long; stipules adnate to the swollen base of petioles; umbels usually solitary and terminal, subglobose, with a flat disc 3–4 mm. across at apex of peduncle, the pedicels 6–15 mm. long; flowers yellowish green, 4–5 mm. across; calyx-teeth undulate; fruit black, broadly ellipsoidal, about 8 mm. long, with a terminal disc about 2 mm. across; styles 5, connate and forming a short column; seeds with 3 obtuse ridges on back.——Honshu (Kantō Distr. and westw.), Shikoku, Kyushu.

7. ACANTHOPANAX Miq.   UKOGI ZOKU

Deciduous shrubs or trees, glabrous or with crisped pubescence, the branches often prickly; leaves palmately compound, the leaflets toothed; stipules often long-ciliate; umbels simple or paniculately compound; flowers polygamous or bisexual; calyx-teeth minute or obsolete; petals usually 5, valvate in bud; stamens as many as the petals; ovary 2- to 5-locular; styles free or connate to the top; fruit globose or slightly flattened, the pericarp juicy, the stones 2–5; seeds flattened laterally.——About 30 species, in e. and s. Asia.

1A. Styles 2, free or connate in lower half.
  2A. Umbels from the short spurs, with the leaves rosulate at base.
    3A. Leaves glabrous beneath.
      4A. Leaflets herbaceous, 2–4 cm. long, rounded at apex, long-cuneate at base, the lateral nerves and veinlets prominent and slightly raised beneath. . . . . 1. *A. japonicus*
      4B. Leaflets membranous, 2–7 cm. long, acute to obtuse, acute to acuminate at base, the nerves not raised. . . . . 2. *A. spinosus*
    3B. Leaves short-pubescent beneath, the nerves not raised. . . . . 3. *A. nikaianus*
  2B. Umbels terminal, on elongate loosely leaved shoots; branches with slender prickles. . . . . 4. *A. trichodon*
1B. Styles 2–7, connate nearly the full length, the free portion very short.
  5A. Styles 5–7.
    6A. Leaflets sessile, 3–7 cm. long; umbels solitary and terminal, on short spurs with rosulate leaves at base of peduncle; prickles thick; styles 5- to 7-lobed at apex; flowering May and June. . . . . 5. *A. sieboldianus*

6B. Leaflets petioluled, 5–15 cm. long; umbels 1–3, terminal, on long loosely leaved shoots; prickles slender; styles minutely 5-lobed at apex; flowering August and September.

7A. Inflorescence usually with 3 umbels; plant nearly glabrous throughout. . . . . 7. *A. hypoleucus*

7B. Inflorescence usually with a solitary umbel; petiolules and nerves of leaflets pubescent beneath. . . . . 8. *A. senticosus*

5B. Styles 2.

8A. Shrub; styles 2–2.5 mm. long; inflorescence densely gray-brown crisped-pubescent, the axis with several to 10 or more umbels on long, loosely leaved shoots; branches with thickened prickles; leaflets 5–12 cm. long. . . . . 6. *A. divaricatus*

8B. Tree; styles 0.3–0.5 mm. long; inflorescence glabrous, with abbreviated axis, with few lateral branches and numerous umbels; branches unarmed; terminal leaflets 10–15 cm. long. . . . . 9. *A. sciadophylloides*

1. **Acanthopanax japonicus** Fr. & Sav. *A. nipponicus* Makino, saltem pro parte; *A. commixta* Nakai, pro parte——OKA-UKOGI. Shrub with gray-brown, glabrous, prickly branches; leaves glabrous, the petioles 2–8 cm. long, the leaflets 5, herbaceous, nearly equal in size, obovate to broadly cuneate-oblanceolate, 2–4 cm. long, 1–2.5 cm. wide, rounded, long-cuneate at base, sessile, margins toward tip with mucronulate, incurved, obtusely acute teeth or sometimes rather irregularly incised, with a thin membrane and tuft of hairs in axil of nerves beneath; umbels many-flowered, glabrous, the peduncles glabrous, 4–6 cm. long, the pedicels 4–6 mm. long; fruit black, slightly flattened, globose, 5–6 mm. across; styles 2.5 mm. long, free or connate on lower half.——May. Honshu, Kyushu.

2. **Acanthopanax spinosus** (L. f.) Miq. *Panax spinosum* L. f.; *Aralia pentaphylla* Thunb.——YAMA-UKOGI, ONI-UKOGI. Shrub usually with prickly branches; petioles 3–7 cm. long, glabrous, with a tuft of crisped hairs at apex; leaflets 5, membranous, slightly unequal in length, narrowly obovate to obovate-cuneate, 2–7 cm. long, 1–3.5 cm. wide, acute to obtuse, acute to acuminate toward the petiolelike base, glabrous, with obscure incurved teeth, a thin membrane, and a tuft of hairs in the axil of nerves beneath; umbels many-flowered, the peduncles 2–5 cm. long, the pedicels 8–12 mm. long; fruit black, rather flattened, globose, 5–6 mm. across; styles slender, 2–2.5 mm. long, free or connate only at base.——May. Honshu; common.

3. **Acanthopanax nikaianus** Koidz. ex Nakai. *A. nipponicus* Nakai——URAGE-UKOGI. Shrub usually with prickly branches; leaves 5-foliolate, the petioles 3–7(–10) cm. long, the leaflets membranous, rhombic- or broadly obovate, 1.5–5 cm. long, 1–2.5 cm. wide, acute, acuminate to cuneate at base, doubly serrate to somewhat incised, sparsely strigose on upper side, callose-pilose beneath; umbels glabrous, rather many-flowered, the peduncles 2–3(–4) cm. long, the pedicels 5–8 mm. long; fruit globose, about 5 mm. long; styles slender, about 2 mm. long, nearly free.——May. Honshu (Kinki Distr. and westw.), Kyushu.

4. **Acanthopanax trichodon** Fr. & Sav. MIYAMA-UKOGI. Prickly shrub with grayish white branches; petioles 2–5 cm. long, purplish red; leaflets thin, rhombic- or cuneate-obovate, 2–6 cm. long, 1–2 cm. wide, acuminate with an acute tip, somewhat cuneate at base, loosely spinulose on margin and nerves on upper side; umbels terminal on new shoots, purplish red, glabrous, the peduncles 2–4 cm. long, slender; flowers many, the pedicels 1–2 cm. long, slender; fruit slightly flattened, globose, about 6 mm. across, purple-black; styles about 1 mm. long, free, somewhat appressed to the fruit.——May–June. Mountains; Honshu (Kantō, Tōkaidō, and Kinki Distr.); rare.

5. **Acanthopanax sieboldianus** Makino. *Aralia pentaphylla* Sieb. & Zucc., non Thunb.——UKOGI, HIME-UKOGI. Shrub with prickly, grayish, glabrous branches; petioles 3–10 cm. long; leaflets (3–)5, rather unequal, the terminal ones rather thick, deep green, oblanceolate to obovate-oblong, 3–7 cm. long, 1–2.5 cm. wide, acuminate at base, short-toothed, glabrous; umbels terminal, solitary, on short spurs, glabrous, the peduncles 5–10 cm. long, rather thick, the pedicels 1–2 cm. long; fruit globose, 6–7 mm. across, not flattened, black, with 5–7 stones; styles 5–7, about 1.2 mm. long, connate on lower half, the upper free portion recurved.——May–June. Cultivated in our area for medicinal and culinary uses.——China.

6. **Acanthopanax divaricatus** (Sieb. & Zucc.) Seem. *Panax divaricatus* Sieb. & Zucc.; *Kalopanax divaricatus* (Sieb. & Zucc.) Miq.; *A. asperatus* Fr. & Sav.——KE-YAMA-UKOGI, ONI-UKOGI. Shrub; branches with broadened prickles, densely gray-brown crisped-pubescent when young; leaves 5-foliolate, the petioles 3–8 cm. long, shorter than to as long as the leaflets, densely crisped-pubescent, often short-prickly, the leaflets ovate or obovate-oblong, 5–12 cm. long, 2–6 cm. wide, acuminate to acute, acuminate at base, doubly toothed, puberulent on nerves on upper side, paler green and densely gray-brown crisped-pubescent beneath especially on nerves, the lateral nerves pinnate, the petiolules 0.8 mm. long; inflorescence paniculately racemose, terminal, with several to 10 or more umbels, on long loosely leaved new shoots, densely gray-brown crisped-pubescent; flowers in the terminal umbels bisexual, the others staminate, the pedicels 3–10 mm. long; fruit globose, about 6 mm. across, black, pubescent near the persistent calyx-teeth; styles connate the full length, slightly bifid at apex, slender, 2–2.5 mm. long.——Aug.–Sept. Mountains; Hokkaido, Honshu, Shikoku, Kyushu; rather rare.——China.

7. **Acanthopanax hypoleucus** Makino. *A. fauriei* Harms; *A. higoensis* Hatus.——URAJIRO-UKOGI. Shrub; branches rusty-brown, with needlelike prickles or unarmed, glabrous; petioles 3–12 cm. long, purplish; leaflets thinly chartaceous, obovate-oblong, 7–15 cm. long, 2.5–5 cm. wide, acuminate, acute to acuminate at base, toothed, green on upper side, glaucous beneath, glabrous, the petiolules 5–10 mm. long; inflorescence on new shoots, glabrous, sparsely branched; umbels 2 or 3, rather long-peduncled, the pedicels many, rather slender, 1–2 cm. long; fruit globose, 6–8 mm. across, purple-black; style wholly connate, 1–1.2 mm. long, obsoletely 5-lobed; stones 5.——Aug. Mountains; Honshu (Shinano, Kai, Musashi Prov.); rare.

8. **Acanthopanax senticosus** (Rupr. & Maxim.) Harms. *Hedera senticosa* Rupr. & Maxim.; *Eleutherococcus senticosus* (Rupr. & Maxim.) Maxim.; *A. eleutherococcus* (Maxim.) Makino——EZO-UKOGI. Shrub, branches gray-brown, with acicular prickles especially around the base of petioles; petioles 3–8 cm. long, loosely crisped-pilose or glabrous; leaflets membranous, obovate-oblong or narrowly obovate, 6–10 cm. long, 2–4 cm. wide, abruptly acuminate, acute at base, acutely double-toothed, loosely brownish crisped-pilose beneath especially on the nerves, the petiolules rather densely pilose, 3–10 mm. long; inflorescence glabrous, terminal on the new shoots, with a single umbel, the peduncles 2.5–4 cm. long, the pedicels 1–1.8 cm.

long; fruit ellipsoidal, glabrous, about 6 mm. long, with persistent calyx-limb and styles on top; styles short, wholly connate, 1–1.8 mm. long, the stigmas small, peltate, obscurely 5-lobed; stones 5.——Aug. Hokkaido.——Sakhalin, Korea, Manchuria, Amur, and n. China.

9. **Acanthopanax sciadophylloides** Fr. & Sav. *Kalopanax sciadophylloides* (Fr. & Sav.) Harms——Koshi-abura. Unarmed mostly glabrous tree with light brown crisped hairs on young growth; branchlets gray-brown, the lenticels elliptic; petioles 7–30 cm. long, with a slightly adnate stipule at base; leaves 5-foliolate, the leaflets unequal, the lateral ones smaller, the terminal ones obovate-oblong to obovate or elliptic, 10–15 cm. long, 4–7 cm. wide, acuminate, cuneate or acuminate at base, aristate-toothed, paler to slightly glaucous with persistent axillary tufts of crisped hairs beneath, the petiolules 1–2 cm. long; inflorescence terminal on the new shoots, glabrous, with an abbreviated main axis, the branches slender, 7–12 cm. long, spreading, sparsely branched, with 1–5 umbels on the upper half, the bracts small, deltoid, the pedicels 4–7 mm. long; flowers yellow-green, small; calyx-teeth depressed-deltoid; petals recurved, 1.5 mm. long, about half as long as the stamens; fruit globose, slightly flattened, 4–5 mm. across, black-purple; styles 0.3–0.5 mm. long, connate and slightly bifid at apex; stones flat, with 2 shallow grooves on each side.——Aug. Hills; Hokkaido, Honshu, Shikoku, Kyushu; common.

8. EVODIOPANAX Nakai　Taka-no-tsume Zoku

Deciduous trees, unarmed, glabrous; leaves petiolate, alternate, exstipulate; leaflets 3, serrulate, the petioles swollen at base; inflorescence a corymbose-panicle, on short spurs, glabrous; umbels several; flowers small, the bracts and bracteoles minute or absent; calyx-limb absent; petals 5, valvate in bud; stamens 5; styles 2, connate toward base; ovary 2-locular; fruit a berrylike drupe, black; stones 2, rather flattened dorsi-ventrally, flat on innerside, convex on back.——Two species, in Japan and China.

1. **Evodiopanax innovans** (Sieb. & Zucc.) Nakai. *Panax innovans* Sieb. & Zucc.; *Acanthopanax innovans* (Sieb. & Zucc.) Fr. & Sav.——Taka-no-tsume, Imo-no-ki.　Small tree with gray-brown branches; leaflets thinly membranous, 3, sometimes 1 or 2, the lateral ones slightly oblique at base, the terminal elliptic or ovate-elliptic, 5–15 cm. long, 2.5–6 cm. wide, acuminate at both ends, sessile, with minute teeth ending in a short awn, glabrous, except a tuft of hairs in axil of nerves beneath, the petioles rather slender, 5–12 cm. long; inflorescence terminal on short spurs, with several pedunculate umbels on upper part, usually with a single branch at base, 7–12 cm. long, the pedicels many, 5–12 mm. long; flowers yellow-green; fruit globose, 5–6 mm. across, black; styles rather slender, about 1 mm. long, arcuate toward tip.——May. Hills; Hokkaido, Honshu, Shikoku, Kyushu; common.

9. OPLOPANAX Miq.　Haribuki Zoku

Deciduous shrubs unbranched or sparsely branched, densely armed with slender spreading prickles; leaves long-petiolate, often peltate, palmately lobed, toothed, densely prickly, exstipulate; inflorescence a terminal panicle or racemelike panicle with an elongated main axis, rather much branched; flowers bisexual and staminate, greenish; calyx-teeth indistinct; petals 5, valvate in bud; stamens 5; styles 2, free or connate toward base; ovary 2-locular; drupe berrylike, red, slightly flattened; stones 2, laterally flattened.——Three species, in e. Asia, and N. America.

1. **Oplopanax japonicus** (Nakai) Nakai. *O. horridus* var. *japonicus* (Nakai) Hara and var. *brevipes* Hara; *Echinopanax japonicus* Nakai——Haribuki.　Shrub about 1 m. tall; stems nearly simple, stout, about 1–1.5 cm. thick throughout, yellowish brown, the prickles 5–10 mm. long; petioles armed, with firm crisped hairs; leaves membranous, orbicular to orbicular-cordate, 20–40 cm. long and nearly as wide, palmately 7- to 9-lobed, loosely armed on nerves on both sides, with long firm crisped hairs especially on nerves beneath, the lobes 3-parted, long-acuminate, irregularly toothed, setulose on margin; inflorescence solitary, terminal, racemelike, or with short branches at base; umbels many, 10–20 cm. long, 3–5 cm. across, densely crisped-pilose and with prickles, the bracts membranous, lanceolate, caducous, the pedicels 3–10 mm. long; fruit globose-obovoid, 6–7 mm. long, red at maturity; styles slender, 1.5–2 mm. long.——June–July. Coniferous woods; Hokkaido, Honshu (n. and centr. distr. including Mount Omine in Yamato), Shikoku.

10. KALOPANAX Miq.　Harigiri Zoku

Deciduous tree with thick grayish terete branches and thick prickles, mostly on the young growth; leaves alternate, often densely aggregated near top of branches, palmately 5- to 9-lobed, the petioles 10–25 cm. long, the lobes 10–30 cm. long and as wide, acuminate, truncate to cordate at base, minutely toothed, usually brownish crisped-pubescent beneath, especially on nerves, and with axillary tufts of hairs beneath, soon becoming glabrous; stipules adnate on innerside to swollen base of petioles; inflorescence terminal on the new shoots, with an abbreviated main axis, composed of several umbellate-pedunculate racemes, the branches 8–15 cm. long, spreading, the bracts 1–2 cm. long, caducous; flowers small, yellow-green, bisexual, the pedicels 7–10 mm. long; calyx-limb 5-toothed; petals 5, valvate in bud, the disc convex; stamens 5; styles 2, 1.5–2 mm. long, connate almost to the top, with 2 short stigma-lobes on innerside; fruit a drupe, globose, 4–5 mm. across, blue-black, with a juicy pericarp; stones 2, rarely 3, prominently ribbed on back, laterally 2-grooved, flat inside.——A single species, in e. Asia.

1. **Kalopanax septemlobus** (Thunb.) Koidz. *Acer pictum* Thunb.; *A. septemlobum* Thunb.; *Panax ricinifolium* Sieb. & Zucc.; *K. ricinifolius* (Sieb. & Zucc.) Miq.; *Acantho-* *panax ricinifolius* (Sieb. & Zucc.) Seem.; *K. pictus* (Thunb.) Koidz. ex Rehd.——Harigiri, Sen-no-ki.　Tree, branches thick, grayish, armed with broad-based erect or slightly re-

curved prickles; leaves 10–30 cm. long and as wide, 5- to 9-lobed, the lobes acuminate, minutely toothed, soon glabrous on upper side, usually pale-brown curled-pubescent, especially on nerves beneath and with tufts of axillary hairs on the nerves on lower side, truncate to cordate at base, the petioles 10–25 cm. long; inflorescence terminal on the current year's shoots, glabrous or nearly so, the main axis very short and obsolete, the branches few to more than 10, 8–15 cm. long, spreading, with few umbels at the apex, the bracts 1–2 cm. long, caducous, the pedicels 7–10 mm. long; fruit globose, 4–5

mm. across; styles connate, 1.5–2 mm. long, rather slender, bifid at apex.——July–Aug. Mountains; Hokkaido, Honshu, Shikoku, Kyushu; common.——s. Kuriles, Sakhalin, Korea, Manchuria, China, and Ussuri.

Var. **lutchuensis** (Nakai) Ohwi. *K. autumnalis* Koidz.; *K. ricinifolium* var. *lutchuense* Nakai; *K. sakaguchiana* Koidz.; *K. pictus* var. *lutchuensis* (Nakai) Nemoto——MIYAKO–DARA, RYŪKYŪ–HARIGIRI. Southern glabrous phase.——Kyushu.——Ryukyus.

Fam. 154. **UMBELLIFERAE**   SERI KA (SANKEI KA)   Carrot Family

Herbs, rarely more or less woody; stems usually fistulose, rarely solid; leaves alternate, usually prominently divided and compound, the upper petioles frequently embracing the young inflorescence with a dilated sheathlike base; flowers bisexual or polygamous, in simple or compound umbels, sometimes capitate; calyx adnate to the ovary, usually 5-toothed; petals 5, valvate or slightly imbricate in bud, free, deciduous, often incurved at apex; stamens 5, alternate with the petals, the filaments incurved in bud; ovary inferior, bilocular; styles 2, often slightly accrescent, usually with a swollen base (stylopodium); ovules solitary in each locule, pendulous; fruit dry, bilocular, separating into the 2 carpels, these usually suspended from the top of a slender prolongation of the fruit axis called a carpophore, with 5 primary ribs and sometimes 4 intermediate or secondary ones, vittae (oil-tubes) often present; seeds with abundant endosperm and a small embryo.——About 200 genera, with about 3,000 species, cosmopolitan.

1A. Umbels simple or irregularly compound.
  2A. Leaves simple, orbicular-cordate, usually palmately lobed or incised; carpels compressed laterally, without vittae.
    3A. Carpels 5-ribbed, without transverse veinlets; leaves with a pair of free stipules at base of petiole; involucral bracts small, or wanting. . . . . 1. *Hydrocotyle*
    3B. Carpels 7- to 9-ribbed, with transverse veinlets forming areoles; leaves without stipules; involucre conspicuous. . . . . . . 2. *Centella*
  2B. Leaves 3- to 5-parted or -divided, the segments or leaflets lobed; carpels essentially terete in cross section, with vittae.
    4A. Carpels densely prickly or tuberculate; ovulate flowers sessile with prominent calyx-teeth. 4. *Sanicula*
    4B. Carpels smooth; flowers more or less pedicellate, without evident calyx teeth. . . . . 8. *Cryptotaenia*
1B. Umbels regularly compound.
  5A. Fruit more or less compressed; carpels essentially terete in cross section.
    6A. Fruit densely prickly. . . . . 7. *Torilis*
    6B. Fruit smooth, rarely loosely setose or tuberculate.
      7A. Carpels setose, caudate at base and decurrent on the pedicel. . . . . 6. *Osmorhiza*
      7B. Carpels not caudate or decurrent on the pedicel.
        8A. Flowers yellow; leaves simple, entire. . . . . 3. *Bupleurum*
        8B. Flowers white or purplish; leaves lobed or dentate, usually compound.
          9A. Carpels 5–8 mm. long, with indistinct ribs. . . . . 5. *Anthriscus*
          9B. Carpels 2–5 mm. long, with distinct ribs.
            10A. Umbellets small, 1- to 3-flowered, with very short, unequal pedicels; umbels regularly compound. .. 9. *Pternopetalum*
            10B. Umbellets many-flowered (at least more than 10-flowered).
              11A. Stylopodium prominent, conical to nearly subglobose, as long or longer than wide.
                12A. Calyx-teeth prominent, deltoid, acute.
                  13A. Carpels with slender, filiform ribs and numerous vittae; woodland plants. . . . . 10. *Spuriopimpinella*
                  13B. Carpels with thick, acute ribs and a solitary vitta at each interval; plants of wet, grassy places.
                                                23. *Pterygopleurum*
                12B. Calyx-teeth obscure or obsolete.
                  14A. Carpels subterete in cross section, with thick ribs. . . . . 11. *Oenanthe*
                  14B. Carpels 5-angled, scarcely compressed, with very slender ribs.
                    15A. Vittae numerous; styles short. . . . . 12. *Pimpinella*
                    15B. Vittae lacking; styles elongate. . . . . 13. *Aegopodium*
              11B. Stylopodium depressed, wider than long.
                16A. Carpels with slender, filiform ribs much narrower than the intervals; woodland plants. . . . . 16. *Chamaele*
                16B. Carpels with thick, more or less corky ribs often broader than the intervals; plants of wet places.
                  17A. Styles very short, from a slender minute stylopodium; carpels with vittae solitary in the intervals; leaves simply pinnate. . . . . 20. *Apodicarpum*
                17B. Styles rather short, from a broad, prominent but short stylopodium.
                  18A. Leaves 2–4 times ternately pinnate; vittae solitary at the intervals. . . . . 18. *Cicuta*
                  18B. Leaves simply pinnate; vittae numerous. . . . . 19. *Sium*
  5B. Fruit terete or compressed dorsally in cross section; carpels semiorbicular to strongly flattened dorsally, sometimes with winged lateral ribs.
    19A. Carpels semiorbicular or lunate in cross section, the lateral ribs unwinged or all ribs with equal narrow wings.
      20A. Stylopodium conical, as long or longer than wide; carpels with thickened ribs.
        21A. Small herbs, 10–30 cm. high (in ours); pericarp layers persistently adnate; carpel ribs obtuse. . . . . 14. *Cnidium*
        21B. Rather stout herbs, 60–100 cm. high; pericarp layers separating; carpel ribs acute, almost winged. . . . . . 15. *Pleurospermum*

20B. Stylopodium depressed.
  22A. Carpels pubescent.
    23A. Carpels puberulent, with filiform ribs. . . . . 17. *Seseli*
    23B. Carpels pubescent, with prominent acute ribs much wider than the intervals. . . . . 21. *Glehnia*
  22B. Carpels glabrous, with slender, acute, noncorky ribs. . . . . 22. *Tilingia*
19B. Carpels strongly flattened dorsally, the lateral ribs often more prominently winged than the dorsal ribs.
  24A. Lateral ribs on carpels equaling or only very slightly broader than the dorsal ribs.
    25A. Ribs solid and subequal; calyx teeth minute or obsolete; petals at tip incurved, retuse, or bifid. . . . . 24. *Ligusticum*
    25B. Ribs hollow, the lateral ones membranous, winged; calyx teeth obsolete; petals only slightly inflexed and entire at apex.
                                                          25. *Coelopleurum*
  24B. Lateral ribs on carpels much broader than the dorsal ribs.
    26A. Vittae extending the full length of the carpel; petals all equal.
      27A. Dorsal ribs on fruit winged.
        28A. Fruit-wings thick; calyx-teeth evident. . . . . 26. *Dystaenia*
        28B. Fruit-wings thin; calyx-teeth obsolete. . . . . 27. *Conioselinum*
      27B. Dorsal ribs on fruit scarcely winged.
        29A. Pericarp single-layered; wings on fruit thinly membranous; calyx-teeth usually evident. . . . . 29. *Ostericum*
        29B. Pericarp multilayered; wings thicker; calyx-teeth usually minute, indistinct.
          30A. Lateral wings on fruit membranous, those of the 2 carpels not closely appressed. . . . . 28. *Angelica*
          30B. Lateral wings thick, those of the 2 carpels closely appressed. . . . . 30. *Peucedanum*
    26B. Vittae clavate, not reaching the base of the mericarp; petals unequal, the marginal ones much larger. . . . . 31. *Heracleum*

1. HYDROCOTYLE L.   Chidomegusa Zoku

Annual or perennial herbs with creeping or shortly erect stems usually rooting from the nodes; leaves petiolate, with membranous stipules, herbaceous, usually simple, palmately lobed or incised, sometimes peltate; umbels capitate, small, opposite the leaves or axillary, solitary or in terminal spikes, the involucre usually of small deciduous bracts, or wanting; flowers bisexual, white to pale green; calyx-teeth obsolete; petals entire, acute, usually valvate in bud; disc flat; styles filiform; fruit compressed laterally, the commissure narrow, the carpels compressed laterally, the ribs slender, the vittae obsolete or nearly so.——About 100 species, in the Tropics and warm-temperate regions.

1A. Leaves usually 3–6 cm. in diameter; umbels solitary to several in fascicles, sessile to long-pedunculate; fruits 15–40, 1–1.2 mm. long.
                                          1. *H. javanica*
1B. Leaves 0.7–2.5 cm. in diameter; umbels solitary, pedunculate; fruits 2–10, rarely as many as 15.
  2A. Umbels cauline; stems or branches ascending to erect; leaves 2–2.5(–3.5) cm. in diameter; stipules prominent, brown, scarious; pedicels short but evident; fruits about 1.5 mm. long.
    3A. Leaves cordate, with margins often overlapping at base; lobes much-depressed, rounded; upper leaves often smaller and much shorter than the inflorescence. . . . . 2. *H. ramiflora*
    3B. Leaves cordate with an open sinus; lobes depressed-deltoid, obtuse; upper leaves usually longer or only slightly shorter than the inflorescence. . . . . 3. *H. maritima*
  2B. Umbels radical; stems creeping; leaves 5–20 mm. in diameter; stipules small and very thin; pedicels nearly lacking.
    4A. Petioles glabrous or with a few long hairs at apex.
      5A. Leaves 0.8–1.5 cm. in diameter, often evergreen, lobed on upper 1/3–1/4, the basal margins approximate; fruits few to more than 10, about 1 mm. long. . . . . 4. *H. sibthorpioides*
      5B. Leaves 0.5–2 cm. in diameter, deciduous in winter, lobed on upper half, the basal margins not approximate; fruits few, about 1.3 mm. long. . . . . 5. *H. yabei*
  4B. Petioles thinly pubescent with crisped, white, recurved, short hairs; fruits about 0.6 mm. long. . . . . 6. *H. dichondroides*

1. Hydrocotyle javanica Thunb. *H. nepalensis* Hook.; *H. polycephala* Wight & Arn.; *H. javanica* var. *polycephala* (Wight & Arn.) Masam. and var. *laxa* Masam.——ŌBA-CHI-DOME. Perennial; stems long-creeping, with few-leaved ascending branches 5–25 cm. long, crisped-puberulent, sparingly branched in upper part; leaf-blades reniform-orbicular, 3–6 cm. in diameter, obtuse, deeply cordate, with callose hairs on upper side, shallowly 5- to 9-lobed, the lobes depressed-deltoid, obtusely double-dentate, the petioles of the radical leaves 7–15 cm. long, of the cauline 5–10 mm. long, the stipules membranous, rounded, about 3 mm. wide; umbels rather many-flowered, usually borne only on the branches; fruit short-pedicellate, the styles short.——July–Oct. Woods in low mountains; Honshu (Kantō Distr. and westw.), Shikoku, Kyushu.——Tropics of Asia.

2. Hydrocotyle ramiflora Maxim. *H. wilfordii* Maxim.——Ō-CHIDOME. Perennial, glabrous except pubescent on the upper side of leaves and petioles; stems long-creeping, slender, with erect or ascending branches 10–15 cm. long, few-leaved; leaf-blades depressed-orbicular with overlapping margins, about 2–2.5(–3.5) cm. in diameter, the lobes much depressed-orbicular with overlapping basal margins, rounded at apex, shallowly toothed, the petioles 4–8 cm. long, the upper ones very short, the stipules membranous, 2–2.5 mm. wide, often brownish with darker spots, toothed; peduncles 2–5 cm. long, much longer than the leaves, the pedicels short; fruits more than 10, the styles slender.——June–Sept. Lawns and sunny places in lowlands and low mountains; Hokkaido, Honshu, Shikoku, Kyushu.——Korea.

3. Hydrocotyle maritima Honda. *H. wilfordii* sensu auct. Japon., non Maxim.; *H. ramiflora* var. *maritima* (Honda) Hiroe——NO-CHIDOME. Resembles the preceding; branches 7–15 cm. long; leaf-blades orbicular-cordate, 2–3(–3.5) cm. in diameter, somewhat 5-angled, with an open sinus at base, palmately 5-lobed on the upper 1/2–2/3, glabrous or loosely long-pilose on upper side, usually long-pilose beneath, the lobes broadly obovate-cuneate, incised and few-toothed; peduncles cauline, shorter (or the upper slightly longer) than the leaves,

1–3 cm. long, the pedicels short; fruits more than 10, the styles rather long and slender.——June–Sept. Lawns and roadsides in lowlands and hills; Honshu, Shikoku, Kyushu; common. ——Korea, China, and Bonins.

4. Hydrocotyle sibthorpioides Lam. *H. rotundifolia* Roxb.——CHIDOME. Glabrous evergreen perennial; stems slender, creeping, subterranean; leaves all radical, the blades orbicular-reniform, with overlapping basal margins, 1–1.5 cm. in diameter, the lobes shallow and very obtuse, 3- to 5-toothed, the petioles 1–3 cm. long, the stipules thin, membranous, soon withering, about 1.5 mm. wide; peduncles radical, shorter or slightly longer than the leaves, 5–12 mm. long; fruits few to 10, rarely more, the styles short, slender.——June–Oct. Shaded places in the lowlands; Honshu, Shikoku, Kyushu; common. ——Widely distributed in tropical and warm-temperate Asia.

5. Hydrocotyle yabei Makino. *H. japonica* Makino; *H. yabei* var. *japonica* (Makino) Hiroe——HIME-CHIDOME. Perennial, in autumn producing short-cylindric winter-buds at the end of slender, elongate, creeping stems, glabrous or with a few long hairs at the summit of the petiole; stems filiform, long-creeping, loosely leafy; leaves radical, deciduous in autumn, the blades reniform-orbicular, with closed to open basal margins, 5(–7)-lobed, 5–20 mm. in diameter, the lobes broadly cuneate, with 3–5 obtuse marginal teeth, the petioles 7–20 mm. long, the stipules thin, membranous, about 1.2 mm. wide; peduncles radical, shorter or slightly longer than the leaves, filiform, the pedicels obsolete or nearly so; fruits 2–4(–10), the styles slender.——June–Oct. Woods in hills and mountains; Honshu, Shikoku, Kyushu.

6. Hydrocotyle dichondroides Makino. *H. delicata* auct. Japon., non Elmer; *H. sibthorpioides* var. *dichondroides* (Makino) Hiroe——KE-CHIDOME. Delicate evergreen herb with very slender, filiform, creeping stems often thinly white, recurved-puberulent; leaf-blades thin, membranous, reniform-orbicular, with an open sinus at base, 3–10 mm. in diameter, glabrous or short-puberulent on the nerves above, shallowly (on upper 1/5–1/3) 5- to 7-lobed, the lobes truncate-rounded at apex, with 3–5 obscure teeth, the petioles 5–15 mm. long, sparsely white-puberulent, the stipules about 1 mm. wide; peduncles 5–20 mm. long, very slender, the pedicels obsolete or very short; fruit very small, about 0.8 mm. long, usually pale green, the styles short, slender.——July–Sept. Kyushu (s. distr.).——Ryukyus, Formosa.

2. CENTELLA L. TSUBO-KUSA ZOKU

Perennial herbs, rarely suffrutescent, usually with long-creeping stems rooting at the nodes; stipules absent; leaves long-petiolate, the blades usually reniform-cordate, dentate, palmately veined; umbels solitary or in fascicles, axillary, rather few-flowered, rarely 1-flowered, the involucral bracts few, usually thinly membranous; calyx-teeth obsolete; disc flat or slightly convex; petals flat, obtuse, slightly imbricate in bud; fruit flattened laterally, the carpels rather firm, with 7–9 nerves connected by transverse veinlets, the vittae absent, the styles filiform.——About 20 species, especially abundant in S. Africa.

1. Centella asiatica (L.) Urban. *Hydrocotyle asiatica* L.; *H. repanda* Pers.——TSUBO-KUSA. Perennial herb, usually woolly-pubescent in young parts; stems long-creeping, rooting at nodes, the scalelike leaves in false pairs; ordinary leaves long-petiolate, fascicled in the axils of scale-leaves, the blades reniform-rounded, 2.5–5 cm. wide, deeply cordate with an open sinus at base, with depressed teeth, the petioles 4–10 (–20) cm. long; umbels axillary, on peduncles 2–8 mm. long, 2- to 5-flowered, the pedicels very short or obsolete; involucral bracts 2, ovate, membranous, persistent; fruit about 3 mm. long, somewhat flattened, depressed ovate-globose, the carpels with raised reticulations, loosely pubescent when young, soon glabrate.——May–Aug. Old stone walls and rocky, sunny places in lowlands and hills especially near seashores; Honshu (s. Kantō Distr., Echigo Prov. and westw.), Shikoku, Kyushu; common.——s. Korea, Ryukyus, Formosa, China, and the warmer regions of the Old World.

3. BUPLEURUM L. MISHIMA-SAIKO ZOKU

Usually perennial, sometimes annual herbs or (rarely) shrubs, glabrous and often glaucous throughout; leaves entire, narrow, sometimes ensiform, often clasping at base; umbels usually compound, subdivided into umbellets, the involucre and involucel usually of leaflike bracts and bractlets, these sometimes small, or lacking; flowers usually pedicellate; calyx-teeth obsolete; petals yellow, incurved at apex; styles short; stylopodium flat; fruit compressed laterally, sulcate on the commissure, the carpels 5-angled in cross section, the ribs thick or slender, the vittae solitary at the intervals or sometimes nearly obsolete, or dispersed as oil glands, the carpophore bifid; seeds terete or angled in cross section, the face concave.——About 90 species, widely distributed in warm-temperate regions of the Old World.

1A. Leaves laterally flattened, rather firm, green, with prominent parallel nerves, narrowed at base, broadly linear to narrowly lanceolate, not clasping; involucel of small bractlets; stems 40–70 cm. long. .... 1. *B. falcatum*
1B. Leaves flattened dorsiventrally, thin, glaucous at least beneath, broadly lanceolate to oblanceolate, the upper ones rounded or auriculate at base, clasping.
    2A. Stems 5–10 (–15) cm. long; umbellets 3–5, the rays shorter or slightly longer than the involucre, the involucel longer than the fruit, with broadly elliptic or ovate, abruptly aristate-mucronate bractlets. .... 2. *B. triradiatum*
    2B. Stems 20–100 cm. long; umbellets numerous, the rays much longer than the involucre, the involucel longer or sometimes slightly shorter than the fruit, the bractlets acuminate.
        3A. Rhizomes thick and short; stolons absent; stems 20–100 cm. long, branched; bractlets of involucel usually lanceolate to narrowly oblong, rarely elliptic. .... 3. *B. longiradiatum*
        3B. Rhizomes slender; stolons filiform; stems 20–40 cm. long, usually simple; bractlets of involucel ovate to obovate-elliptic, or obovate-oblong, usually longer than the pedicels. .... 4. *B. nipponicum*

1. **Bupleurum falcatum** L. *B. scorzoneraefolium* Willd.; *B. falcatum* var. *scorzoneraefolium* (Willd.) Ledeb.; *B. chinense* DC.; *B. falcatum* var. *komarowii* Koso-Pol.; *B. stenophyllum* (Nakai) Kitag.; *B. scorzoneraefolium* var. *stenophyllum* Nakai——MISHIMA-SAIKO. Rhizomes short, thick, with thickened roots; stems erect, 40–70 cm. long, branched above; radical leaves 10–30 cm. long, including the narrow petiolelike base, the cauline leaves rather thick and firm, broadly linear to narrowly lanceolate, 4–10 cm. long, 5–15 mm. wide, acuminate and with a short awn at tip, with several prominent parallel nerves, green on both sides, gradually narrowed to a sessile base; umbels numerous, with 2–7 rays, the involucre 0–15 mm. long, with linear-lanceolate bracts much shorter than the rays; umbellets 5- to 10-flowered, the involucels of broadly linear to oblong bractlets 2.5–4 mm. long, shorter than the fruiting pedicels; fruit ellipsoidal, 2.5–3.5 mm. long.——Aug.–Oct. Grassy slopes in hills; Honshu, Shikoku, Kyushu; common.——Siberia, Mongolia, China, Korea, Manchuria, the Caucasus, and Europe.

2. **Bupleurum triradiatum** Adams. *B. triradiatum* var. *alpinum* Rupr.; *B. ranunculoides* var. *triradiatum* forma *alpinum* Rupr. ex Wolff——REBUN-SAIKO. Rhizomes short, thick, branched; stems tufted, 5–10(–15) cm. long, few-leaved; radical leaves oblanceolate to broadly so, 5–10 cm. long, acute, gradually narrowed to the base, glaucous, the cauline leaves broadly oblanceolate to narrowly oblong or narrowly ovate, 1–3 cm. long, 5–10 mm. wide, rounded and clasping at base; umbels usually solitary, sometimes 2, the involucre with 3(–5) broadly elliptic or suborbicular bracts 1–2 cm. long, the rays 3(–5), 1–1.5 cm. long; umbellets 3–5, rather many-flowered, the involucel with 5(–6) broadly elliptic to broadly ovate bractlets, abruptly aristate-mucronate, longer than the fruit, the pedicels short; fruit oblong.——July–Aug. High mountains; Hokkaido; rare.——Sakhalin, Kamchatka, and e. Siberia.

3. **Bupleurum longiradiatum** Turcz. var. **breviradiatum** F. Schmidt. *B. sachalinense* F. Schmidt; *B. longiradiatum* var. *sachalinense* (F. Schmidt) H. Boiss.——HOTARU-SAIKO. Rhizomes short, with thick roots; stolons absent; stems 20–100 cm. long, branched with many umbels on upper part; radical leaves long-petiolate, broadly lanceolate to narrowly oblong, usually membranous, glaucous beneath, the lower cauline similar but subsessile, the upper ones 5–15 cm. long, 2–3.5 cm. wide, acute to acuminate with a minute awn at apex, sessile and clasping at the rounded-auriculate base; umbels rather numerous, 10- to 15-flowered, the involucral bracts 2 or 3, small, the involucel with 3–5 lanceolate to narrowly oblong bractlets 2–3 mm. long, the pedicels 4–5 mm. long, longer than the involucel; carpels with prominent ribs.——Aug.–Oct. Mountains; Hokkaido, Honshu, Shikoku, and Kyushu.——s. Kuriles and Sakhalin.

Var. **longiradiatum**. *B. longiradiatum* var. *genuinum* Wolff; *B. yokoyamae* Miyabe & Tatew.——Ō-HOTARU-SAIKO. Stems taller; pedicels more slender, 6–7(–15) mm. long, 5–7 times as long as the flowers and 2–3 times as long as the fruit.——Hokkaido, Honshu, and Kyushu; rare.——China, Manchuria, Korea, and Amur.

Var. **shikotanense** (Hiroe) Ohwi. *B. shikotanense* Hiroe; *B. aureum* sensu auct. Japon., non Fisch.——KOGANE-SAIKO. Stems 20–30 cm. long; leaves rather thick; umbels few, the pedicels 2–3 mm. long, shorter than to as long as the fruit; bractlets of the involucel often slightly longer than the fruit, 4–5 mm. long, oblong.——Hokkaido (Kitami and Nemuro Prov.).——s. Kuriles.

4. **Bupleurum nipponicum** Koso-Pol. *B. multinerve* forma *minus* auct. Japon., non DC.——HAKUSAN-SAIKO. Rhizomes slender; stems erect, 20–40(–50) cm. long, simple, slender, few-leaved, sparingly branched; lower cauline leaves oblanceolate or narrowly so, narrowed and clasping at base, the median and upper ones lanceolate to broadly so, 3–10(–12) cm. long, 1–2 cm. wide, subacuminate, clasping at the rounded-auriculate base, the uppermost leaves deltoid-ovate, short; umbels terminal, 1–5, the rays 5–8, the involucral bracts 1(–3), broadly ovate, the involucel bractlets 5 or 6, ovate, obovate-elliptic or obovate-oblong, abruptly acuminate, 6–8 mm. long, the pedicels 2–3.5 mm. long in fruit, slender; fruit 3–3.5 mm. long, oblong, the carpels with slender and prominent ribs.——July–Aug. High mountains; Honshu (centr. distr., and northw.).

Var. **yesoense** (Nakai) Hara. *B. yesoense* Nakai——HOSOBA-NO-KOGANE-SAIKO. Leaves acuminate; bractlets narrowly ovate, acuminate.——Hokkaido (Mountains of Hidaka Prov.).

4. SANICULA L.   UMA-NO-MITSUBA ZOKU

Glabrous, perennial or rarely annual herbs; leaves ternately compound, palmately or rarely pinnately lobed; umbels simple or irregularly compound, the bracts prominent or small; flowers unisexual or polygamous, the fertile ones in the central part of the umbel 1 to many, sessile, the staminate pedicellate, usually marginal; calyx-lobes prominent, free in the fertile flowers, often connate in the staminate; petals imbricate in bud, emarginate and incurved at tip; disc flat, the styles usually filiform; fruit globose or ellipsoidal, the carpels orbicular in cross section, aculeolate, uncinate, or tuberculate, without a carpophore, the vittae 5 or numerous, often obscure; seeds plane or concave on the face.——About 40 species, widely dispersed over the world, except in Australia and the Arctic.

1A. Fruit densely uncinate-prickly; radical leaves 6–20 cm. wide; plants stout.
   2A. Stems much-branched, with few petiolate, alternate leaves, and with many umbels in upper part. .... 1. *S. chinensis*
   2B. Stems with a single pair or rarely 2 pairs of sessile leaves on upper part, naked or with a single petiolate leaf near base; umbels solitary or few.
     3A. Involucral bracts small, 1–2 mm. long, much shorter than the staminate flowers and fruits; petiolate cauline leaves sometimes present; umbels with 1–5 rays, the peduncles 5–30 cm. long; flowers white or purplish. .... 2. *S. kaiensis*
     3B. Involucral bracts foliaceous, 10–20 mm. long, much longer than the flowers and fruits; petiolate cauline leaves absent; umbels 1–5, the peduncles 3–6 cm. long; flowers dark purple. .... 3. *S. rubriflora*
1B. Fruit densely armed with straight prickles or tuberculate; radical leaves 3–5 (–7) cm. wide; plants rather slender or delicate.
   4A. Umbels 1–3, on peduncles 1–3 cm. long, many-flowered; involucral bracts 4–10 mm. long, spreading; fertile flowers 1–4; fruit densely tuberculate, the upper tubercles rather fleshy-spiny. .... 4. *S. tuberculata*
   4B. Umbels 3–5, on peduncles 4–7 mm. long, few-flowered; involucral bracts minute, 2–3 mm. long, aristate-acicular; fertile flowers solitary; fruit densely prickly with small callose spines. .... 5. *S. lamelligera*

1. **Sanicula chinensis** Bunge. *S. europaea* sensu auct. Japon., non L.; *S. europaea* var. *elata* Makino, excl. syn.; *S. elata* var. *japonica* Koidz.; *S. elata* var. *chinensis* (Bunge) Makino——UMA-NO-MITSUBA. Rhizomes short and thick; stems 30–60 cm. long, much branched above; radical leaves with petioles 10–20 cm. long, the blades 5-angled, reniform-cordate, 5–10 cm. wide, glabrous, trisect, the lateral segments 2-parted, the lobes and segments rhombic-obovate or rhombic, acute, sometimes shallowly 2- or 3-lobed and incised, dentate with awn tipped teeth, the cauline leaves few, shorter petiolate, the upper sessile and smaller with the lateral lobes often undivided; leaves of the inflorescence 3-parted, often bracteate, small; peduncles 3–20 mm. long, few-flowered; staminate flowers very short-pedicellate, the involucral bracts 2–3 mm. long, longer than the staminate flowers; fruits 2–4, usually 3, globose-ovoid, sessile, 5–6 mm. long, including the persistent acicular calyx-teeth 1.5–2 mm. long, the uncinate prickles firm, spreading, about 1.5 mm. long.——July–Sept. Woods in low mountains; Hokkaido, Honshu, Shikoku, Kyushu; common. ——s. Kuriles, s. Sakhalin, Korea, Manchuria, and China.

2. **Sanicula kaiensis** Makino & Hisauchi. YAMANASHI-UMA-NO-MITSUBA. Rhizomes short and thick; stems elongate after anthesis, 15–60 cm. long, simple, with a pair of sessile terminal leaves, sometimes with a petiolate leaf at the middle bearing an axillary branch; radical leaves with petioles 10–30 cm. long, the blades rather thin, reniform-cordate, 4–10 cm. wide, glabrous, trisect, the lateral segments 2-parted, the segments and lobes cuneate-obovate, shallowly 3-fid, incised and acutely dentate, acute, the median cauline leaves similar to the radical, with shorter petioles, the terminal leaves sessile or nearly so, with the lateral segments sometimes undivided, and the segments narrower, cuneate-oblong; umbels 1–5, unequal, from the axils of the upper leaves, 5- to 10-flowered, the peduncles 1–5, unequal, ascending, 5–30 cm. long, the involucral bracts lanceolate, spreading, usually 2–3 mm. long, the pedicels of the staminate flowers 2–3 mm. long; fruit ovoid-ellipsoidal, about 4 mm. long including the persistent calyx-lobes, the uncinate prickles rather slender, 3 mm. long, tuberculate at base, often very short in the lower ones.——May–June, fruiting July–Sept. Mountains; Honshu (Kai and Shinano Prov.); very rare.

3. **Sanicula rubriflora** F. Schmidt. KUROBANA-UMA-NO-MITSUBA. Rhizomes short and thick; stems with a pair of terminal leaves, elongate after anthesis, naked, 20–50 cm. long; radical leaves with petioles 20–40 cm. long, the blades thin, smooth, reniform-orbicular, 6–20 cm. wide, trisect, the lateral segments 2-parted, all segments and lobes obovate-cuneate, obtuse, shallowly and acutely 3- to 5-fid or incised, acutely den-

tate; cauline leaves a single pair at summit of stem, sessile, trisect, the segments slightly narrower, the lateral segments undivided; umbels 1–5, axillary in the paired leaves, densely many-flowered, the peduncles 3–6 cm. long, the involucral bracts foliaceous, linear-oblanceolate, spreading, green, 10–20 mm. long; staminate flowers dark purple, the pedicels 1–2 mm. long, 1–2 times as long as the calyx-lobes; fruits sessile, 1–3 in each umbel, about 4 mm. long including the calyx lobes, the uncinate prickles firm, ascending, about 2 mm. long.——May–June, fruiting June–July. Honshu (Mount Himegamiyama in Rikuchu, Azusayama in Shinano, and elsewhere); rare.——Amur, Manchuria, and Korea.

4. **Sanicula tuberculata** Maxim. FUKIYA-MITSUBA. Rhizomes short and thick; stems 8–20 cm. long, slender, usually with 2 pairs of terminal leaves; radical leaves with petioles 5–15 cm. long, the blades 5-angled, reniform-orbicular, glabrous, 3–5(–7) cm. wide, trisect, the lateral segments 2-parted, the segments and lobes cuneate-obovate, shallowly 2- or 3-fid, acute, mucronate-dentate, paler beneath; cauline leaves subsessile, slightly smaller than the radical, with the lateral segments usually undivided; umbels 1–3, in the axils of the terminal cauline leaves, the peduncles 1–3 cm. long; involucral bracts linear-lanceolate, 4–10 mm. long, spreading; staminate flowers about 10, the pedicels 2–3 mm. long, 2 or 3 times as long as the lanceolate calyx-lobes; fruits 1–4 in each umbel, about 3.5 mm. long including the calyx-lobes, depressed-globose, densely tuberculate, the spines or prickles short and thick, erect, the upper tubercles 0.5–0.7 mm. long, rather fleshy-spiny. ——May, fruiting in June. Honshu (w. Tōkaidō Distr. and westw.), Shikoku, Kyushu; rare.——Korea.

5. **Sanicula lamelligera** Hance. *S. satsumana* Maxim.; *S. wakayamensis* Masam.; *S. lamelligera* var. *wakayamensis* (Masam.) Murata——HIME-UMA-NO-MITSUBA. Rhizomes short and thick; stems erect, 6–15 cm. long, slender, subscapose; radical leaves with petioles 5–10 cm. long, the blades 5-angled, cordate-orbicular, 2–3.5 cm. wide, trisect, deep green on upper side, paler beneath, the segments rhombic, 2- or 3-lobed, subobtuse, with awn-tipped incurved teeth, glabrous, often narrowed to a short petiolelike base; cauline leaves 1 or 2, small, bracteate, often 3-lobed; umbels 3–5, few-flowered, the peduncles 1–3 cm. long, the involucral bracts aristate-acicular, 2–3 mm. long; pedicels of the staminate flowers 2–2.5 mm. long, 3–4 times as long as the acicular calyx-lobes; fruit ovoid, 2–2.5 mm. long including the calyx-lobes, the prickles or spines callose, ascending, slender, firm, straight, about 0.5 mm. long. ——May. Honshu (Kii Prov.), Kyushu (s. distr.); rare.——Ryukyus, Formosa, and China.

5. ANTHRISCUS Hoffm. SHAKU ZOKU

Annuals or perennials, usually pubescent on the leaves and upper part of the stems; leaves pinnately or ternately compound, the segments pinnately divided, lobed and toothed; umbels numerous, the involucral bracts 1 or 2 or none, the bractlets of the involucel entire, numerous; flowers white, often polygamous; calyx-teeth minute or obsolete; petals often unequal, with an incurved tip; stylopodium usually conical, entire; fruit slightly compressed laterally, with a sulcate commissure, lanceolate or oblong, the carpels terete or subterete with the ribs distinct only in the upper part, glabrous or minutely spiny, the vittae very slender, solitary in the intervals or obscure, the carpophore sometimes bifid; seeds subterete or dorsally compressed, grooved on the face.——About 13 species in Eurasia.

1. **Anthriscus sylvestris** (L.) Hoffm. *A. nemorosa* (Bieb.) Spreng.; *Chaerophyllum nemorosum* Bieb.; *A. sylvestris* var. *aemula* Woron.; *A. sylvestris* subsp. *nemorosa* (Bieb.) Koso-Pol.——SHAKU, KOJAKU. Perennial; taproot thick-

ened; stems hollow, simple, 80–140 cm. long, erect, branched above, glabrous or with reflexed white hairs at base; radical leaves long-petiolate, the blades green, herbaceous, 20–50 cm. long, deltoid, 2–3 times ternate and pinnately parted, the ultimate lobes acute or acuminate, toothed, with spreading, white, setulose hairs at least on the nerves beneath, the cauline leaves similar and gradually reduced upward, the upper petioles reduced to a white-pilose sheath; umbels numerous, glabrous, the involucre absent, the umbellets 5–12, glabrous, on rays 3–4 cm. long, rather many-flowered, the bractlets of the involucels 4–8, recurved, pale green, membranous, oblong to lanceolate, abruptly caudately acuminate, 3–8 mm. long, ciliate; pedicels smooth, short-setulose near the apex; fruit lanceolate, dark gray

and slightly lustrous at maturity, 5–8 mm. long, glabrous or with short, ascending, callose tubercles, as long as to half as long as the pedicels.——May–June., fruiting June–July. Hokkaido, Honshu, Shikoku, Kyushu; common in mountains.—— e. Europe, Caucasus, Siberia, Kamchatka, Kuriles, Sakhalin, Korea, Manchuria, and China.

Var. **hirtifructus** (Ohwi) Hara. *A. nemorosa* var. *hirtifructus* Ohwi——KEJAKU. Rather densely pilose, the fruit yellowish setulose.——Occurs with the typical phase.

6. OSMORHIZA Raf.　　YABU-NINJIN ZOKU

More or less pilose perennials; leaves triternate, the lobes or segments pinnately lobed or toothed; umbels few to rather numerous, the involucre and involucel minute or absent; flowers white; calyx-teeth obsolete; petals obovate or oblong, retuse, incurved at apex; stylopodium conical, entire; fruit narrowly oblanceolate or broadly linear, gradually narrowed at base, decurrent on the pedicels, the carpels 5-angled, nearly terete in cross section, the ribs evident, setose, the vittae numerous, weak or obscure, the carpophore slender, bifid, the seeds flattened dorsally, grooved on the face.——Few species in India, e. Asia, N. and S. America.

1. **Osmorhiza aristata** (Thunb.) Makino & Yabe. *Chaerophyllum aristatum* Thunb.; *O. japonica* Sieb. & Zucc. ——YABU-NINJIN. White-pilose perennial with thick ascending rhizomes; stems erect, sparingly branched above, 40–60 cm. long, solid, few-leaved; radical leaves long-petiolate, the blades deltoid, 2 or 3 times ternately pinnate, 10–20 cm. long, the primary pinnae deltoid-ovate, acute, the ultimate segments thin, ovate or oblong, toothed, green, the cauline leaves sessile, shortly sheathing at base; umbels 2 or 3 on long peduncles, the involucre of lanceolate, acuminate, recurved, deciduous bracts, the rays 3–6, 5–10 cm. long, smooth, spreading-ascending, the umbellets 5- to 10-flowered, the involucel bractlets 5 or 6, lanceolate, recurved, acuminate, membranous, 5–8 mm. long; fruit linear-oblanceolate, 18–20 mm. long, about 1.5 mm. wide,

appressed-setose, narrowed gradually toward base, the pedicels about as long as the fruit, loosely setose on upper part, the styles slender, 2.5–3 mm. long, including the conical stylopodium.——Apr.–May. Woods in hills and low mountains; Hokkaido, Honshu, Shikoku, Kyushu; very common.—— Korea, China, and Formosa to India.

Var. **montana** Makino. *O. amurensis* F. Schmidt; *O. montana* (Makino) Makino——MIYAMA-YABU-NINJIN, ONAGA-YABU-NINJIN. Less pubescent, the primary leaf segments acuminate, the ultimate segments or lobes smaller, narrower. ——May–June, fruiting June–Aug. Woods in mountains; Hokkaido, Honshu (centr. distr. and northw.).——e. Siberia, Korea, Sakhalin, and s. Kuriles.

7. TORILIS Spreng.　　YABUJIRAMI ZOKU

Coarsely hirsute annuals or perennials; leaves 2 or 3 times ternately pinnate, with small ultimate segments; umbels numerous, usually opposite the leaves, the bracts and bractlets few, linear; flowers white; calyx-teeth deltoid-lanceolate, firm, persistent, acute; petals obovate, emarginate at the incurved or slightly bifid apex; stylopodium conical, the styles short; fruit ovoid, with a groove on the commissure, the carpels densely prickly with scaberulous prickles, the vittae solitary in the intervals, 2 on the commissure; seeds deeply grooved on the face.——About 20 species, chiefly in the Mediterranean region, a few in S. Africa and e. Asia.

1A. Fruits 4–10, ovoid, 2.5–3 mm. long, in a dense umbellet; fruit-prickles short, gradually upward-curved. ....1. *T. japonica*
1B. Fruits 3–6, oblong, 4.5–6 mm. long, in a loose umbellet; fruit-prickles rather long, spreading. .... 2. *T. scabra*

1. **Torilis japonica** (Houtt.) DC. *Caucalis japonica* Houtt.; *T. anthriscus* Gmel., non Gaertn.; *T. anthriscus* var. *japonica* (DC.) H. Boiss.——YABUJIRAMI. Biennial with short, appressed, setose hairs throughout; stems erect, 30–70 cm. long, retrorsely hirsute; lower cauline leaves petiolate, green, ovate-deltoid, 5–10 cm. long, acute, twice ternately pinnate, appressed-hirsute, the ultimate segments narrowly ovate, pinnatifid, the upper cauline leaves rather small, sessile, acuminate; umbels few, pedunculate, the involucral bracts 4–8, linear, about half as long as the rays, suberect, the involucel bractlets linear, appressed to the pedicels, rather unequal, as long as to shorter than the fruit, the pedicels 2–4 mm. long, ascending; fruits 4–10, crowded, ovoid, 2.5–3(–3.5) mm. long, densely covered with short, scabrous, slightly uncinate prickles. ——May–July. Grassy places and thickets in lowlands and hills; Hokkaido, Honshu, Shikoku, Kyushu; common.——

Europe and Caucasus, Africa, n. and e. Asia, naturalized in s. Asia.

2. **Torilis scabra** (Thunb.) DC. *Chaerophyllum scabrum* Thunb.; *Caucalis scabra* (Thunb.) Makino——O-YABU-JIRAMI. Resembling the preceding species; stems 30–70 cm. long, branched above; lower cauline leaves 5–10 cm. long, the ultimate segments pinnatifid; umbels pedunculate, the rays 2–3, rarely 4, 2–3 cm. long, the bracts of the involucre 1 or none, short, filiform, the involucel bractlets linear, ascending, as long as the fruiting pedicel; fruits 3–6, oblong, 4.5–6 mm. long, about as long as the pedicels, appressed-pilose between the spreading, scabrous, slightly uncinate prickles.—— May–July. Grassy places and thickets in lowlands and hills; Honshu, Shikoku, Kyushu; common.——Ryukyus, Formosa, China, and s. Korea.

8. CRYPTOTAENIA DC.   Mitsuba Zoku

Glabrous perennials; leaves radical and cauline, trisect, the leaflets broad, incised-dentate; umbels irregularly compound, the peduncles, rays, and pedicels very unequal, the bracts and bractlets small or none; calyx-teeth obsolete; petals small, incurved at apex; stylopodium conical, entire; fruit oblong, compressed, laterally grooved on the commissure, the carpels teretely 5-angled in cross section, the commissure plane or nearly so, the ribs subequal and evident, the vittae beneath and between the ribs, the carpophore bifid.——Few species in e. Asia, Africa, and N. America.

1. Cryptotaenia japonica Hassk. *C. canadensis* sensu auct. Japon., non DC.; *C. canadensis* var. *japonica* (Hassk.) Makino——Mitsuba. Glabrous perennial with short rhizomes and rather thick roots; stems erect, slightly branched above, 30–60 cm. long; radical and lower cauline leaves long-petiolate, the blades cordate-deltoid, 5–15 cm. wide, 3-foliolate, the leaflets sessile, the terminal one broadly rhomboid-ovate to broadly ovate, 3–8 cm. long, 2–6 cm. wide, acuminate, abruptly narrowed to the acuminate base, incised- or irregularly dentate, rarely deeply lobed, the lateral leaflets at base with a slightly dilated outer margin, the leaves of the inflorescence very short, often undivided, linear to lanceolate; umbellets 1- to 4-flow-ered, erect, narrow, the pedicels very unequal, 3–15 mm. long, erect, the involucel bractlets linear, short, the flowers white; fruit 3–4 mm. long, glabrous, the carpels oblong-cylindric, narrowed at apex, the styles erect, the stylopodium elongate-conical.——June–July, fruiting in Aug. Woods in hills and mountains; Hokkaido, Honshu, Shikoku, Kyushu; common and frequently cultivated as a vegetable.——Korea, China, Ryukyus, and s. Kuriles.——Forma **atropurpurea** (Makino) Ohwi. *C. japonica* var. *atropurpurea* Makino——Murasaki-mitsuba. Dark purple throughout. Rarely cultivated as an ornamental.

9. PTERNOPETALUM Franch.   Iwa-sentō-sō Zoku

Small or slender, glabrous to pilose perennials; stems erect or ascending, simple or sparingly branched above; leaves ternately decompound, the upper with often narrow, elongate segments; umbels compound, numerous, on the long rays, the pedicels 2 or 3, very unequal, short, erect, the involucel small, the flowers white, bisexual; calyx-teeth sometimes obsolete, often acute; petals emarginate, incurved at apex; stylopodium conical, the styles erect, slender; fruit oblong or ovoid, compressed laterally, usually with a groove on each side, glabrous, the carpels 5-angled, subterete in cross section, the ribs usually slender, the vittae 1–3 in the intervals; seeds flat on the face, the carpophore bifid.——About 10 species, in India, China, and Japan.

1. Pternopetalum tanakae (Fr. & Sav.) Hand.-Mazz. *Carum tanakae* Fr. & Sav.; *Cryptotaeniopsis tanakae* (Fr. & Sav.) H. Boiss.——Iwa-sentō-sō. Glabrous, delicate or slender perennial; rhizomes creeping and short stolons also developed; stems 10–25 cm. long, erect, simple, 1- or 2-leaved; radical leaves petiolate, ovate-deltoid in outline, 3–4 cm. wide, 2 or 3 times ternately pinnate, the ultimate segments short-petiolulate, usually ovate, 4–6 mm. long, deeply 3-parted, usually few-lobed, minutely scaberulous on the margin, the upper cauline petiolate, 1 or 2 times divided, the ultimate segments linear, 1–3 cm. long, about 2 mm. wide, acute; umbels 1 or sometimes 2, glabrous, the involucre lacking, the rays 10–20, slender, 2.5–3 cm. long, the umbellets 2- or sometimes 3-flow-ered, the involucel of small linear bractlets, the pedicels unequal, to 2 mm. long, erect; fruit glabrous, 2–2.5 mm. long, the styles short, the calyx-teeth very small.——May–June. Damp mountain woods; Honshu (Kantō Distr. and westw.), Shikoku, Kyushu.——s. Korea.

10. SPURIOPIMPINELLA Kitag.   Ka-no-tsume-sō Zoku

Rather slender perennials; stems erect, hollow, loosely few-leaved; leaves radical and cauline, petiolate, usually ternately decompound, membranous, the segments pinnately lobed and dentate; umbels few or solitary, the involucral bracts few or none, the rays slender, the umbellets with many bractlets; flowers bisexual; calyx-teeth prominent; petals white, equal, incurved at apex; stylopodium conical, the styles filiform; fruit glabrous, compressed laterally, ovoid or ellipsoidal, the carpels nearly orbicular in cross section, the ribs equal, filiform, the vittae many and slender, the seeds adnate to the pericarp.——About 10 species, in e. Asia and e. Siberia.

1A. Fruit longer than wide; leaves with short, ascending, setose hairs on nerves on both surfaces and on margin; upper leaves usually once ternate, with narrowly ovate to broadly lanceolate leaflets; petiolules and upper part of petioles with callose hairs.
                                                                                                      1. *S. calycina*
1B. Fruit as wide as long; leaves membranous, coarsely pilose on both surfaces and on the margin; upper leaves usually twice ternate, with ovate leaflets; petiolules and petioles subglabrous or pilose. .... 2. *S. nikoensis*

1. Spuriopimpinella calycina (Maxim.) Kitag. *Pimpinella calycina* Maxim.——Ka-no-tsume-sō. Stems 50–80 cm. long, branched above, glabrous; radical and lower cauline leaves petiolate, twice ternate, often simply ternate in the upper ones, the leaflets sessile or nearly so, membranous, broadly ovate or obovate, 3–8 cm. long, 2–4 cm. wide, acute to abruptly acuminate, dentate, with short, ascending, setose hairs on nerves on both surfaces and on the margin, the leaf-lets of the upper leaves narrowly ovate to broadly lanceolate, 8–12 cm. long, 2–4 cm. wide, long-acuminate; umbels solitary or few, the rays 2–4 cm. long, slender, the involucral bracts few, linear, 2–8 mm. long, the umbellets few- to more than 10-flowered, the involucel bractlets few, linear, 2–3 mm. long, the pedicels slender, minutely tuberculate on the inner side, 6–10 mm. long; fruit ovoid to narrowly so, 4–5 mm. long, the ribs slender, the calyx-teeth deltoid, erect, the styles slender and recurved, scarcely 2 mm. long.——Aug., fruiting Sept.–Oct. Damp mountain woods; Hokkaido, Honshu, Shikoku, Kyushu.

2. Spuriopimpinella nikoensis (Yabe ex Hisauchi) Kitag. *Pimpinella nikoensis* Yabe ex Hisauchi; *P. koreana* (Yabe) Nakai——HIKAGE-MITSUBA. Perennial; rhizomes short; stems 50–80 cm. long, slightly branched above; radical leaves long-petiolate, the cauline petiolate or sessile, 10–15 cm. long and wide, biternately compound, the leaflets usually sessile, thinly membranous, green, ovate to obovate, 1.5–4 cm. long, 1–3 cm. wide, usually abruptly caudately acuminate, dentate, sparsely pilose on both surfaces, short-ciliate; umbels few, the involucral bracts lacking, or 1–3, linear, 5–20 mm. long, the rays about 10, 15–35 mm. long, minutely tuberculate on the inner side, the umbellets more than 10-flowered, the bracts of the involucel 2–5 mm. long, linear, the pedicels slender, 8–12 mm. long, minutely tuberculate on the inner side; calyx-teeth elongate-deltoid, erect; fruit ovoid-globose, 3–4 mm. long, only a single carpel often maturing, the styles about 1 mm. long, recurved.——Aug., fruiting Sept.–Oct. Mountain woods; Honshu (Kantō Distr. and westw.), Shikoku, Kyushu. ——Forma **dissecta** (Nakai) Ohwi. *Pimpinella nikoensis* var. *dissecta* Nakai——HAGOROMO-HIKAGE-MITSUBA. Leaves finely dissected.

11. OENANTHE L.   SERI ZOKU

Glabrous perennials growing in wet places; leaves pinnate or rarely bladeless; umbels compound, the bracts and bractlets numerous, rarely absent; flowers white; calyx-teeth acute; petals emarginate, incurved at apex; stylopodium conical, narrow; fruit ovoid or oblong, terete, with a broad, flat commissure, the carpels semi-orbicular in cross section, the ribs corky, rather thick, the lateral ones broader than the subequal dorsal ribs, the vittae solitary in the intervals, the carpophore absent; seeds dorsally compressed, with a flat face.——About 30 species, in the N. Hemisphere, S. Africa, and Australia.

1. Oenanthe javanica (Bl.) DC. *Sium javanicum* Bl.; *Falcaria javanica* (Bl.) DC.; *O. stolonifera* DC.; *O. linearis* DC.; *O. benghalensis* DC.; *S. laciniatum* Bl.; *O. lacinata* (Bl.) Zoll.; *Dasyloma subbipinnatum* Miq.; *D. japonicum* Miq.; *O. stoloniferum* var. *japonica* (Miq.) Maxim.——SERI. Glabrous perennial with slender roots; stems erect, 20–40 cm. long, angled, striate, from a long-creeping, branched base; leaves petiolate, 7–15 cm. long, deltoid or deltoid-ovate, once or twice pinnately compound, the ultimate segments ovate or narrowly so, 1–3 cm. long, 7–15 mm. wide, acute or subacuminate, irregularly dentate, sometimes deeply lobed; umbels pedunculate, opposite the leaves, the rays 5–15, usually minutely tuberculate on the angles, 1–2.5 cm. long, the umbellets 10- to 25-flowered, the involucel bractlets few, linear, shorter than the fruit, the pedicels 2–5 mm. long; flowers white; fruit ellipsoidal, about 2.5 mm. long, the carpels with corky ribs, the calyx-teeth persistent, the styles slender, suberect, about 1.5 mm. long.——July–Aug. Ditches, ponds, paddy fields, and wet places in lowlands; Hokkaido, Honshu, Shikoku, Kyushu; common.——Ryukyus, Formosa, China, Malaysia, India, and Queensland (Australia).

12. PIMPINELLA L.   MITSUBAGUSA ZOKU

Perennials or rarely annuals; leaves pinnately or sometimes ternately compound, rarely simple, dentate; umbels compound, the bracts and bractlets few or lacking; calyx-teeth obsolete; petals usually white, with an entire or emarginate incurved apex; stylopodium conical, depressed or cushionlike, entire; fruit ovoid or ellipsoidal, compressed laterally, with a concave commissure, the carpels 5-angled, subterete in cross section, the ribs equal, slender, filiform, the vittae numerous, the carpophore bifid; seeds compressed dorsally, with a nearly plane face.——About 100 species in the N. Hemisphere and S. Africa.

1A. Fruit broadly ovoid, 1.5 mm. long, short-pubescent; cauline leaves usually 3-foliolate or simple, the leaflets irregularly acute-dentate or incised. .... 1. *P. diversifolia*
1B. Fruit narrowly ovoid, 2.5–3 mm. long, glabrous; cauline leaves 5–9 foliolate; the leaflets often ternately cleft to divided, coarsely and acute-dentate. .... 2. *P. thellungiana* var. *gustavohegiana*

1. Pimpinella diversifolia DC. *P. sinica* Hance; *Platyraphe japonica* Miq.——MITSUBAGUSA. Stems erect, 80–100 cm. long, terete, branched, with short whitish hairs; lower cauline leaves simple or 3-foliolate, petiolate, sparsely short-pubescent, the leaflets herbaceous, orbicular-cordate, obtuse or rounded, 4–7 cm. long and as wide, cordate to truncate at base, irregularly acute-dentate or incised, the upper leaves sessile, the leaflets ovate or broadly so, acute, deeply lobed, the terminal one slightly larger than the others, 3–7 cm. long, 2–4 cm. wide, the lateral leaflets sessile; umbels rather numerous, short-pubescent, the involucral bracts solitary or none, the rays about 10, 1.5–2 cm. long, the umbellets about 10-flowered, the involucel of few, linear, small bractlets, the pedicels about 5 mm. long; flowers white; fruit broadly ovoid, short-pubescent, 1.5 mm. long, the styles short, about 0.5 mm. long.——Aug., fruiting Sept. Kyushu.——China and n. India.

2. Pimpinella thellungiana H. Wolff var. **gustavo-**hegiana (Koidz.) Kitam. *P. magna* sensu auct. Japon., non L.; *P. gustavohegiana* Koidz.; *P. thellungiana* sensu auct. Japon., non H. Wolff——TSUKUSHI-BŌFŪ. Stems terete, erect, branching, 30–80 cm. long, densely white-puberulent; radical and lower cauline leaves petiolate, 8–20 cm. long, oblong-ovate, the leaflets 5–9, sessile, broadly ovate, obtuse to rounded, 2–3.5 cm. long, 1.5–3 cm. wide, often ternately cleft or divided, coarsely acute-dentate, loosely puberulent on upper side, densely so beneath, the upper leaflets 3 or 5, sessile; umbels few, glabrous, pedunculate, the rays 10–15, 2.5–3 cm. long, nearly smooth, the involucre and involucel lacking; flowers crowded, rather numerous, the pedicels 5–8 mm. long, suberect, somewhat flattened; fruit narrowly ovoid, 2.5–3 mm. long, glabrous, the styles spreading obliquely, slender, about 1.2 mm. long.——Aug. Mountains; Kyushu (Bungo).—— The typical phase occurs from Dahuria to China and Formosa.

13. AEGOPODIUM L. Ezo-bōfū Zoku

Perennials with rhizomes; leaves once or twice ternately compound, the ultimate segments broad, dentate; umbels 1 to many, the involucral bracts and bractlets few or lacking; flowers white; calyx-teeth obscure; petals emarginate, incurved at apex; stylopodium thick, nearly conical, entire; fruit ovoid or ovoid-ellipsoidal, glabrous, slightly compressed laterally, the carpels 5-angled in cross section, very slightly flattened dorsally, the ribs slender, filiform, equal, the vittae absent, the carpophore bifid; seeds flat on the face.——Few species in Eurasia.

1. **Aegopodium alpestre** Ledeb. Ezo-bōfū. Glabrous perennial except the leaf-margins and inner side of the rays; rhizomes short and slender; creeping stolons with thickened nodes; stems rather slender, erect, 30–50 cm. long, sparsely branched above, terete, hollow, few-leaved; radical and lower cauline leaves long-petiolate, deltoid, 2–3 times ternately compound, the ultimate segments thinly membranous, ovate to narrowly so, 1–2.5 cm. long, 8–20 mm. wide, acuminate to acute, irregularly dentate, often 3-lobed, the upper leaves sessile, small; umbels 1–3, on long peduncles, the involucre and involucel lacking, the rays 8–12, 2–3 cm. long, the umbellets more than 10-flowered, the pedicels 5–8 mm. long; fruit ovoid-oblong, 3–3.5 mm. long, the stylopodium subglobose, the styles slender, about 1 mm. long.——June–July, fruiting Aug.–Sept. Damp mountain woods; Hokkaido, Honshu (Kantō Distr. and northw.).——s. Kuriles, Sakhalin, n. Korea, China, Manchuria, e. Siberia, and Altai.

14. CNIDIUM Cuss. Hamazeri Zoku

Perennials or biennials; leaves petiolate, long-sheathed at base, 1–3 times pinnately compound; umbels rather large, with many rays, without an involucre, the involucel present; flowers white; calyx-teeth obsolete; petals obovate with an incurved apex; stylopodium elongate-conical with the styles spreading at apex; fruit ovoid-globose, very slightly compressed laterally, the carpels subterete in cross section, the ribs thick, obtuse and unwinged, usually corky, the vittae solitary in the intervals, 2 or 3 on the commissure; seeds 5-angled in cross section with a nearly flat face.——More than 10 species, in Eurasia.

1. **Cnidium japonicum** Miq. Hamazeri. Glabrous biennial with simple, thickened roots; stems few, erect or decumbent at base, 10–30 cm. long, striate, branched above; radical and cauline leaves petiolate, ovate-oblong, 3–6 cm. long, pinnately compound, the leaflets 5–7, sessile or the lowest rarely short-petiolulate, ovate-deltoid, 7–12 mm. long, deeply pinnately lobed, lustrous, smooth on the margin; umbels pedunculate, glabrous, the rays few, stout, 5–10 mm. long, the bracts and bractlets few, linear, 2–5 mm. long, the pedicels 2–4 mm. long, the umbellets about 10; flowers white, crowded; fruit ovoid, about 3 mm. long, the carpels with equal, thick, corky, rather acute ribs, the calyx-teeth obsolete, the stylopodium conical, the styles 0.3–0.5 mm. long.——Aug.–Oct. Sandy places near seashores; Hokkaido, Honshu, Shikoku, Kyushu; common.——Korea, China, Ryukyus, and Bonins.

15. PLEUROSPERMUM Hoffm. Ō-kasa-mochi Zoku

Perennials or biennials; stems often stout; leaves ternately pinnate, the ultimate segments incised or dentate, sometimes pinnately lobed; umbels compound, large, the rays numerous, the bracts and bractlets numerous, membranous or foliaceous, sometimes pinnately lobed; flowers white, rather large; calyx-teeth minute; petals obovate or cuneate with an incurved apex; stylopodium flat or convex, entire; fruit ovoid or globose, nearly orbicular in cross section, with a broad commissure, the carpels convex and with subequal, acute, narrowly winged ribs, the vittae usually solitary (rarely 2) in the intervals, the outer pericarp loosely connected with the inner, the carpophore bifid; seeds compressed dorsally, with a deep groove on the commissural side.——About 15 species in Eurasia, especially the Himalayas.

1. **Pleurospermum camtschaticum** Hoffm. *P. uralense* Hoffm.; *P. austriacum* sensu auct. Japon., non Hoffm.; *P. austriacum* subsp. *uralense* (Hoffm.) Somm.——Ō-kasa-mochi. Stout, glabrescent perennial; stems stout, erect, hollow, usually simple, short-branched above, leafy; rhizomes simple, short and thick; radical and lower cauline leaves petiolate, broadly ovate-deltoid, twice ternately pinnate, 20–40 cm. long, the ultimate segments narrowly ovate, acuminate, sessile, membranous, glabrous except for the papillose nerves and margins, pinnatifid, acutely incised-dentate, 4–15 cm. long; umbels few, the terminal one largest, semiglobose, with numerous, densely papillose-puberulent rays 7–15 cm. long, the bracts of involucre and bractlets of involucel numerous, foliaceous, green, scarious-margined, the pedicels 1.5–3 cm. long; flowers rather large; fruit ovoid, 6–7 mm. long, the styles about 2 mm. long, the stylopodium conical, the ribs scabrous.——July–Aug., fruiting Sept.–Oct. Mountains; Hokkaido, Honshu (centr. and n. distr.).——Kuriles, Sakhalin, Korea, Manchuria, Kamchatka, and Siberia.

16. CHAMAELE Miq. Sentō-sō Zoku

Small perennials; stems scapelike, leafless; radical leaves 2 or 3 times ternately pinnate, the leaflets small, incised to lobed; umbels with few rays, the involucre and involucel lacking, the umbellets few-flowered, the central one often sessile; flowers small, white; calyx-teeth obscure; petals shortly incurved at apex; styles short, sometimes recurved; stylopodium flat, entire; fruit oblong or ovoid-oblong, compressed laterally, glabrous, grooved on the commissure, the carpels nearly 5-angled in cross section, with slender, filiform, equal ribs, the vittae absent; seeds nearly orbicular in cross section, the carpophore bifid.——A single species, in Japan.

1. Chamaele decumbens (Thunb.) Makino.

1. Chamaele decumbens (Thunb.) Makino. *Sium decumbens* Thunb.; *C. tenera* Miq.; *Aegopodium tenerum* (Miq.) Yabe——Sentō-sō. Delicate, glabrous perennial with rather thick, short rhizomes; leaves radical and petiolate, the petioles dilated and stipulelike in those at base, the blade ovate-deltoid, 3–6 cm. long, 2 or 3 times ternately pinnate, the ultimate segments membranous, ovate or deltoid, rather acute, 5–10 mm. long, deeply lobed; peduncles 10–25 cm. long, the involucre and involucel lacking, the rays 3–5, 3–5 cm. long, the central one often very short, the umbellets 7- to 10-flowered, the pedicels unequal, 1–5 mm. long, papillose on the inner side; flowers white, small; fruit 2.5–3.5 mm. long, the styles spreading-recurved, 0.5–1 mm. long.——Apr.–May. Woods; Hokkaido, Honshu, Shikoku, Kyushu; rather common and variable in shape of the leaflets.——Forma **japonica** (Yabe) Ohwi. *C. japonica* Makino ex Yabe in syn.; *C. decumbens* var. *japonica* Makino; *Aegopodium tenerum* var. *japonicum* Yabe——Miyama-sentō-sō. Ultimate segments slender, linear.——Forma **dilatata** Satake & Okuyama. Ibuki-sentō-sō. Leaves once or sometimes twice compound, larger, the ultimate segments broader.

17. SESELI L. Ibuki-bōfū Zoku

Usually pilose perennials; stems often angled; leaves ternately pinnate or pinnately compound, the ultimate segments usually narrow; umbels compound, the involucral bracts many or absent, the bractlets of the involucel numerous; flowers white; calyx-teeth evident; petals rather broad, emarginate at the incurved apex; styles usually slender; stylopodium flat or convex, with undulate teeth on the margin; fruit ovoid or oblong, short-pubescent, nearly orbicular in cross section, the carpels compressed dorsally, the ribs equal or nearly so, rather prominent, obtuse, the vittae usually solitary in the intervals and rarely also under the ribs, the carpophore bifid; seeds semi-orbicular in cross section, with a nearly flat face.——About 10 species, in Eurasia.

1. Seseli libanotis (L.) K. Koch var. **japonica** H. Boiss. *S. libanotis* sensu auct. Japon., non K. Koch; *S. libanotis* subsp. *japonica* (H. Boiss.) Hara; *S. ugoensis* Koidz.; *Libanotis ugoensis* (Koidz.) Kitag.; *S. libanotis* subsp. *japonica* forma *ugoensis* (Koidz.) Hara——Ibuki-bōfū. Pubescent perennial, the thick rhizome at the apex sparsely fibrous; stems 40–80 cm. long, branched; radical and lower cauline leaves petiolate, 2 or 3 times pinnately compound, ovate, the ultimate segments ovate, deeply cleft, 2–3 cm. long, the lobes lanceolate, finely ciliate, the upper leaves smaller, sessile, less divided; umbels pedunculate, papillose-pilose on the peduncles at tip, the rays more than 10, 2–3 cm. long, papillose-puberulent; involucre absent, the involucel bractlets short, few, appressed to the short pedicels; fruit ovoid, about 2 mm. long, papillose-puberulent, the calyx-teeth elongate-deltoid, the styles about 1 mm. long, the stylopodium convex.——Aug.–Sept. Hokkaido, Honshu (Yamato Prov. and northw.).——s. Kuriles. The typical phase occurs in Europe and w. Asia.

18. CICUTA L. Dokuzeri Zoku

Tall, glabrous perennials usually of wet places; leaves pinnate; umbels compound, many-rayed; bracts of the involucre few or absent; involucel bractlets few, small; flowers white; calyx-teeth small, acute; petals rather broad, incurved and emarginate at apex; stylopodium rather thick, flat, entire; fruit ovoid to depressed ovoid-globose, slightly compressed laterally, slightly grooved on the commissure, the carpels slightly flattened dorsally, suborbicular in cross section, the ribs thick, obtuse, broad, rather corky, the pericarp membranous, the vittae solitary in the intervals, the carpophore bifid; seeds suborbicular or flattened dorsally in cross section, convex on the face.——Few species in wet places in the N. Hemisphere.

1. Cicuta virosa L. *C. nipponica* Fr. & Sav.; *C. virosa* var. *nipponica* (Fr. & Sav.) Makino; *C. virosa* var. *tenuifolia* Koch; *C. virosa* var. *latisecta* Koso-Pol.——Dokuzeri. Large, glabrous perennial with thick, green, jointed rhizomes; stems much branched, 60–100 cm. long; radical and lower cauline leaves long-petiolate, deltoid-ovate in outline, 30–50 cm. long, twice pinnately compound, the ultimate segments linear-lanceolate to broadly lanceolate, 3–8 cm. long, 7–20 mm. wide, acute, acutely dentate, the upper leaves subsessile, smaller and less divided; umbels glabrous, rather numerous, the involucral bracts absent, the rays about 20, 3–7 cm. long, spreading, the umbellets more than 10-flowered, the bractlets of the involucel few, linear-filiform, 3–7 mm. long; pedicels slender, 8–10 mm. long; fruit ovoid-globose, about 2.5 mm. long, green, with thick yellow ribs, the calyx-teeth deltoid, minute, the styles slender, about 1 mm. long.——June–Aug. Wet places; Hokkaido, Honshu, Shikoku, Kyushu.——Europe, Siberia, China, Manchuria, Korea, and perhaps Alaska.

19. SIUM L. Mukago-ninjin Zoku

Glabrous herbs of wet places, with thickened spongy roots; stems elongate, leafy; leaves usually simply pinnate, the leaflets dentate; umbels compound, the bracts and bractlets numerous, usually prominent, green; calyx-teeth acute; petals emarginate and incurved at apex, white; stylopodium thickened, depressed-conical, entire; fruit ovoid or oblong, flattened or laterally grooved on each side on the commissure, the carpels nearly 5-angled in cross section, the ribs thick, obtuse, equal, often corky, the vittae numerous, the carpophore obscure, or undivided; seeds nearly orbicular in cross section, the seed-face plane.——About 10 species, in the N. Hemisphere and Africa.

1A. Rays 5–10, rather stout; involucral bracts few, often deflexed; pedicels 10–15, rather stout, 3–6 mm. long; fruit ovoid-globose or obovoid.
  2A. Leaflets 7–11 (very rarely 5), the terminal distinctly petiolulate; axillary bulbils none; fruit obovoid. . . . . 1. *S. suave*
  2B. Leaflets 3–5, the terminal sessile or short-petiolulate; axillary bulbils present; fruit ovoid-globose. . . . . 2. *S. sisarum*
1B. Rays 2–5, very slender; involucral bracts 1 or none; pedicels 2–7 (–10), very slender, 7–10 mm. long; leaflets 3–5, the terminal sessile or nearly so; fruit ovoid. . . . . 3. *S. serra*

1. Sium suave Walt. var. **nipponicum** (Maxim.) Hara. *S. nipponicum* Maxim.——SAWAZERI, NUMAZERI. Rhizomes short-creeping, with thick spongy white roots; stems 60–100 cm. long, branched above, glabrous; radical and lower cauline leaves long-petiolate, 20–40 cm. long, simply pinnate, the leaflets 7–9 (rarely 11), sessile except the terminal one, narrowly ovate, sometimes broadly lanceolate, 3–10 cm. long, 1–3 cm. wide, acuminate, rarely obtuse, acutely dentate, the upper cauline leaves with fewer and smaller, narrower, lanceolate leaflets; umbels glabrous, the rays 7–12, the bracts and bractlets broadly linear, deflexed, 2–6 mm. long, the pedicels 10–20, 3–7 mm. long; fruit obovoid, 2.5–3 mm. long, the calyx-teeth deltoid, minute, the styles short, slender.——July–Sept. Hokkaido, Honshu, Shikoku, Kyushu.

Var. **ovatum** (Yatabe) Hara. *S. ovatum* Yatabe——HIRO-HA-NUMAZERI. Leaflets ovate.

Var. **suave** *S. cicutaefolium* Schrank; *S. lineare* Michx.; *Apium cicutaefolium* (Schrank) Benth. & Hook. f. ex Forbes & Hemsl.——TŌ-NUMAZERI, HOSOBA-NUMAZERI. Leaflets 7–17, narrow, 5–15 cm. long, linear to lanceolate; calyx-teeth obsolete.——Hokkaido, Honshu.——Korea, China, Siberia to e. Europe and N. America.

2. Sium sisarum L. *S. ninsi* L.——MUKAGO-NINJIN. Rhizomes short, with thickened, whitish, spongy roots; stems 30–80 cm. long, flexuous and branched above; lower cauline leaves petiolate, 5- or sometimes 3-foliolate, the leaflets sessile or nearly so, lanceolate to linear, rarely narrowly oblong, 2–5 cm. long, 5–10 mm. wide, obtuse to acute, with short-awned, acute teeth, scaberulous on the margin, the upper leaves smaller and usually 3-foliolate, with small purplish bulbils in the axils; bracts and bractlets lanceolate, scarious-margined, 2–4 mm. long, the rays about 10, about 2 cm. long, the pedicels 10–20, 3–5 mm. long; fruit ovoid-globose, about 2 mm. long, the calyx-teeth minute, the stylopodium thick, the styles about 0.5 mm. long, nearly appressed to the fruit.——Aug.-Sept., fruiting Oct. Hokkaido, Honshu, Shikoku, Kyushu.——Korea and China.

3. Sium serra (Fr. & Sav.) Kitag. *Pimpinella serra* Fr. & Sav.——TANI-MITSUBA. Glabrous perennial with thickened, spongy, white, fusiform roots; stems erect, branched above, 60–80 cm. long; cauline leaves petiolate, ternately 3-foliolate or pinnately 5-foliolate, the leaflets sessile, subequal, ovate to broadly lanceolate, thinly membranous, long-acuminate, minutely serrulate, the upper leaves small, short-petiolate; umbels few, delicate, the involucral-bracts solitary, filiform or none, the rays 2–5, very slender, 1–2 cm. long, the involucel bractlets few, filiform, 1–2 mm. long, the pedicels 2–7 (rarely –10), 7–10 mm. long, very slender; fruit ovoid, about 2 mm. long, glabrous, the calyx-teeth minute, the styles about 0.5 mm. long, slender and spreading.——Aug. Hokkaido, Honshu (centr. and n. distr.).

20. APODICARPUM Makino   EKISAIZERI ZOKU

Small, glabrous perennial, of wet places; leaves simply pinnate; umbels compound, the bracts and bractlets few, small; calyx-teeth lacking; petals white, ovate-orbicular, acute, slightly incurved at apex; styles very short; stylopodium small, flat; fruit ellipsoidal, compressed laterally, with a groove on each side of the commissure, the carpels connate, suborbicular in cross section, the ribs rather thick, obtuse, equal, the vittae solitary in the intervals, 2 in the commissure, the carpophore lacking; seeds 5-angled, suborbicular in cross section, slightly inflated on the ventral side.——A single species, in Japan.

1. Apodicarpum ikenoi Makino. *Apium ikenoi* (Makino) Drude——EKISAIZERI. Perennial, with short rhizomes; stems erect, sparsely branched, 5–20 cm. long; radical and lower cauline leaves petiolate, simply pinnate, ovate-oblong in outline, 4–7 cm. long, the leaflets sessile except the terminal one, equal, 7–9, narrowly to broadly ovate, 1–2 cm. long, 7–15 mm. wide, acute to subobtuse, irregularly dentate, the upper leaves 3- to 5-foliolate, smaller, sessile; umbels few, long-pedunculate, the lower ones opposite the leaves, the involucral bracts few, linear-lanceolate, 3–8 mm. long, rather unequal, the involucel bractlets few, linear, 2–4 mm. long, spreading, the rays 3–5, unequal, glabrous, about 3 cm. long, the pedicels rather stout, unequal, 2–5 mm. long; fruit ellipsoidal or obovoid-ellipsoidal, 2.5–3 mm. long, the styles very slender, nearly erect.——May, fruiting through June. Wet muddy places along rivers in the lowlands; Honshu (Musashi and Owari Prov.); rare.

21. GLEHNIA F. Schmidt   HAMA-BŌFŪ ZOKU

Pubescent perennials with thick, elongate rhizomes; leaves twice ternately pinnate with elliptic, dentate, often 3-lobed leaflets; umbels few, terminal and from the upper axils, densely pubescent, without an involucre, the bracts of the involucels few, linear; calyx-teeth evident, linear-deltoid, acute, 1-nerved; petals white, obovate, emarginate, with an incurved tip; styles elongate; stylopodium disc-shaped, prominent, convex; fruit ellipsoidal, densely pubescent, nearly orbicular in cross section, the carpels with thick, rather winglike prominent ribs, the lateral ribs slightly broader than the dorsal, the vittae numerous, the carpophore bifid, the pericarp corky; seeds semi-orbicular in cross section, with a plane face.——One species, in e. Asia and w. N. America.

1. Glehnia littoralis F. Schmidt. *Phellopterus littoralis* Benth. & Hook. f.——HAMA-BŌFŪ. Plant densely long white-pubescent; rhizomes often elongate, rather thick; stems short, usually simple, 5–30 cm. long, few-leaved; radical and lower cauline leaves long-petiolate, spreading, usually deltoid or ovate deltoid, 10–20 cm. long, once or twice pinnately ternate, the leaflets often 3-lobed, rather thick, elliptic, obovate-elliptic or ovate-orbicular, 2–5 cm. long, 1–3 cm. wide, obtuse to rounded at apex, often partly suffused with red, often glabrous except on the nerves, irregularly denticulate, often callose on the margin; umbels 1, sometimes to 3, densely long white-villous, the rays more than 10, 4–6 cm. long, spreading, crowded, the umbellets densely 20- to 40-flowered, the bracts of the involucel linear, not exceeding the flowers; calyx-teeth 0.5–1 mm. long; styles 1.5–2 mm. long; fruit about 4 mm. long, villous.——June–July. Sandy beaches along the sea; Hokkaido, Honshu, Shikoku, Kyushu; common.——s. Kuriles, Sakhalin, Ussuri, Ochotsk Sea region, Manchuria, China, Korea, Ryukyus, Formosa, and the Pacific Coast of N. America.

22. TILINGIA Regel   SHIRANE-NINJIN ZOKU

Glabrous perennials with thick rhizomes; stems rather slender, few-leaved, often scapose; leaves twice to several times ternately compound, toothed or dissected into linear segments; umbels simple or compound, the involucral bracts 1 to few, the bractlets of the involucel few small and linear; calyx-teeth deltoid, acute, small but evident; petals white, emarginate and incurved at apex; styles rather short, the stylopodium conical, entire; fruit oblong or ovoid-oblong, slightly compressed laterally, the carpels 5-angled, suborbicular in cross section, with slender filiform ribs, the vittae solitary in the intervals, 2–4 on the commissure; seeds compressed, with the flat side on the commissure.——Few species in e. Siberia, Manchuria, Korea, and Japan.

1A. Stems 50–80 cm. long, leafy; leaves twice or once ternate, the leaflets broadly ovate to broadly lanceolate, thin, 2–8 cm. long, acuminate; umbels nearly glabrous, few. .... 1. *T. holopetala*
1B. Stems 7–20 cm. long, few-leaved; leaves finely ternately dissected, the ultimate segments linear to broadly ovate, rather thick, 5–20 mm. long; rays minutely puberulent on the inner side.
  2A. Ultimate leaf-segments broadly ovate to narrowly oblong, few-toothed to -cleft; styles nearly as long as or slightly longer than the stylopodium. .... 2. *T. ajanensis*
  2B. Ultimate leaf-segments linear-filiform, entire; styles 2.5–3.5 times as long as the stylopodium. .... 3. *T. tachiroei*

1. Tilingia holopetala (Maxim.) Kitag. *Carum holopetalum* Maxim.; *Ligusticum holopetalum* (Maxim.) Hiroe & Constance——IBUKIZERI. Rhizomes ligneous, shortly branched; stems 50–80 cm. long, leafy, branched; radical and lower cauline leaves long-petiolate, 1- or 2-ternate, the leaflets thin, broadly ovate to broadly lanceolate, irregularly dentate and usually pinnately incised or lobed, 2–8 cm. long, acuminate, membranous, with microscopically raised marginal cells, the upper leaves smaller, sessile, ultimately reduced to bracts; umbels few, pedunculate, nearly glabrous, the involucral bracts 1 or 2, linear, the rays 8–10, 3–4 cm. long, the peduncle and rays at apex with winglike ridges, the pedicels 8–12 mm. long, the involucral bractlets few, linear, 4–8 mm. long; calyx-teeth evident; styles short, about as long as the stylopodium; fruit ovoid-oblong, 3.5–4 mm. long.——Aug., fruiting Sept.–Oct. Grassy places in mountains; Hokkaido, Honshu (centr. and n. distr.).

2. Tilingia ajanensis Regel. *Conioselinum kamtschaticum* var. *alpinum* Rupr.; *Selinum tilingia* (Regel) Maxim.; *Cnidium ajanense* (Regel) Drude; *Cnidium tilingia* (Regel) Takeda; *Ligusticum ajanense* (Regel) Koso-Pol.——SHIRANE-NINJIN. Small perennial, glabrous except for papillae on the umbels beneath and on the inner side of the rays; rhizomes short, rather thick; stems 7–20 cm. long, rarely to 35 cm. long, leafless, or 1- to 3-leaved; radical and lower cauline leaves petiolate, 2–4 times ternately compound, the ultimate segments broadly ovate to narrowly oblong, thick-membranous or herbaceous, usually with finely dissected linear to narrowly

oblong lobes, 5–20 mm. long, lustrous and deep green on upper side, glabrous, the upper leaves very small, often reduced to sheaths; umbels 1–3, the rays 5–10, 1.5–2.5 cm. long, the involucral bracts 1, 2 or few, the umbellets 20- to 30-flowered, the bractlets of the involucel few, linear, the pedicels 3–6 mm. long, nearly as long as the involucel; flowers white, sometimes purplish; calyx-teeth minute; styles nearly as long as or slightly longer than the stylopodium; fruit ovoid, 2.5–3 mm. long; vittae 4 on the commissure.——July–Aug., fruiting Sept.–Oct. Alpine slopes; Hokkaido, Honshu (centr. and n. distr.).——Sakhalin, Kuriles, Kamchatka, Ochotsk, and e. Siberia.

3. Tilingia tachiroei (Fr. & Sav.) Kitag. *Seseli tachiroei* Fr. & Sav.; *Cnidium tachiroei* (Fr. & Sav.) Makino; *Ligusticum tachiroei* (Fr. & Sav.) Hiroe & Constance——MIYAMA-UIKYŌ. Resembles No. 2; glabrous except for papillae on the umbels below and on the inner side of rays; rhizomes short, thick; stems 10–20(–35) cm. long, 1- to 4-leaved; radical and lower cauline leaves 3–10 cm. long, long-petiolate, deltoid-ovate, several times finely dissected into linear-filiform entire segments, these 5–15 mm. long, 0.5–0.75 mm. wide, acute, entire, smooth, the upper leaves with fewer segments or reduced to sheaths; umbels 1–4; fruit ovoid-oblong, about 3 mm. long; styles slender, about 1.2 mm. long, 2.5–3.5 times as long as the stylopodium, the calyx-teeth evident; vittae 2 on the commissure.——Aug., fruiting Sept.–Oct. High mountains; Hokkaido, Honshu (centr. and n. distr.), Shikoku (Mount Tsurugi).——Korea.

23. PTERYGOPLEURUM Kitag.   SHIMURA-NINJIN ZOKU

Glabrous perennial with short rhizomes and fleshy white roots; stems erect, fistulose, branched above, angular; leaves once or twice ternately compound, the ultimate segments linear, entire, with thickened margins; umbels compound, the bracts and bractlets few, entire, narrow; flowers white; calyx-teeth deltoid, acute; petals equal, with an incurved and shallowly bifid apex; styles short, becoming recurved, the stylopodium depressed-conical with undulate margin; fruit glabrous, slightly compressed laterally, oblong, the carpels semi-orbicular in cross section, with thick, acute ribs, the vittae solitary in the intervals, 2 on the commissure, the carpophore bifid; seeds compressed and angular, with a flat face.——One species, in Japan.

1. Pterygopleurum neurophyllum (Maxim.) Kitag. *Edosmia neurophyllum* Maxim.; *Carum neurophyllum* (Maxim.) Fr. & Sav.; *Sium neurophyllum* (Maxim.) Hara ——SHIMURA-NINJIN. Stems angled, erect, loosely branched above, 80–120 cm. long; leaves rather firm, 2 or 3 times ternately compound, 10–20 cm. long, deltoid-orbicular in outline, the ultimate segments linear, 4–10 cm. long, 3–5 mm. wide, entire, acuminate, narrowed at base, scaberulous on the margins, the upper leaves smaller and simpler, ternate, sessile, directly attached by a narrow sheath; umbels compound, the

peduncles and rays with winglike ridges at the apex, the rays 8–10, rather firm, smooth, 2.5–4 cm. long, the involucral bracts 7–12 mm. long, appressed to the outer rays, bracts of the involucel appressed to the pedicels, 2–5 mm. long, the pedicels 6–10 mm. long; fruit 3.5–4 mm. long, the styles about 0.5 mm. long, appressed to and nearly as long as the stylopodium.—— Aug.–Sept. Wet places along rivers in lowlands and plateaus; Honshu (Musashi Prov.), Kyushu (Tano in Bungo); rare. ——Korea.

24. LIGUSTICUM L.   MARUBA-TŌKI ZOKU

Glabrous perennials; leaves ternately or pinnately compound; umbels compound, involucral bracts numerous, few, or absent, bractlets of the involucel numerous, usually entire; flowers white; calyx-teeth small or obsolete; petals with an incurved emarginate or bifid apex; stylopodium thick, sometimes conical, often undulate on the margin; fruit ovoid or oblong, nearly terete to dorsally compressed in cross section, the commissure broad, the carpels compressed dorsally with a flat commissural side, the ribs prominent, acute or winglike, the lateral ribs often broader than the dorsal, the vittae numerous, slender, sometimes obsolete, the carpophore bifid; seeds compressed dorsally, the seed-face flat or concave, free from the pericarp when mature. ——About 50 species, in the N. Hemisphere.

1A. Leaves twice ternate; rays 15–20, 2–3.5 cm. long; fruit narrowly oblong, 8–11 mm. long. . . . . 1. *L. hultenii*
1B. Leaves simply ternate; rays 5–8, 1–2 cm. long; fruit broadly ovate, about 2 mm. long. . . . . 2. *L. tsusimense*

1. Ligusticum hultenii Fern. *Angelica hultenii* (Fern.) Hiroe; *L. scothicum* sensu auct. Japon., non L.——MARUBA-TŌKI.   Rhizomes thick; stems glabrous, erect, sparsely branched above, terete, striate; leaves twice ternate, broadly deltoid-orbicular in outline, 10–25 cm. long and as wide, the ultimate segments broadly rhombic-ovate, often 3-lobed, acute to obtuse, 3–5 cm. long and as wide, dentate, glabrous, the upper leaves small, sessile, the uppermost reduced to short sheaths; umbels few, glabrous, the peduncles with winged ridges minutely serrulate in the upper portion, the rays 15–20, 2–3.5 cm. long, the involucral bracts few, membranous, linear, 8–10 mm. long, the involucel bractlets few, linear, shorter than the pedicels, the pedicels 5–8 mm. long; fruit narrowly oblong, 8–11 mm. long, the carpels flat, semi-orbicular in cross section, the ribs subequal, prominent; calyx-teeth minute; styles short, ascending, about 0.5 mm. long, the stylopodium globose-conical.——July–Aug., fruiting Sept.–Oct. Near the seashore; Hokkaido, Honshu (n. distr.).——Korea, Ussuri, Sakhalin, Kuriles, Ochotsk, Kamchatka, and Alaska.

2. Ligusticum tsusimense Yabe. *Ostericum tsusimense* (Yabe) Kitag.; *Pimpinella tsusimensis* (Yabe) Hiroe & Constance——TSUSHIMA-NODAKE, TSUSHIMA-TŌKI.   Rhizomes short, few; roots rather thick; stems erect, branched above, slender, faintly striate, 25–40 cm. long; radical and lower cauline leaves long-petiolate, deltoid or 5-angled, ternate; leaflets thickish, sometimes 2- or 3-lobed, whitish beneath, coarsely toothed, rhombic-ovate or the lateral obliquely ovate, 3–5 cm. long, 15–25 mm. wide, acute, cuneate at base, sometimes pilose above, scabrous on the margins, the veinlets slender to obsolete, the upper leaves reduced to oblanceolate sheaths; umbels long-pedunculate, few, the rays 5–8, 1–2 cm. long, subequal, minutely papillose on the inner side, the bractlets of the involucel few, linear-filiform, nearly as long as the pedicels, often incurved, the pedicels about 10, 2–5 mm. long; fruit broadly ovoid, about 2 mm. long, nearly round in cross section, the carpels with distinct, filiform, slender ribs.——Aug., fruiting in Oct.; Kyushu (Tsushima).

25. COELOPLEURUM Ledeb.   EZO-NO-SHISHI-UDO ZOKU

Tall, stout perennials; leaves large, ternately compound, the leaflets dentate; umbels compound; flowers white; calyx-teeth obsolete; petals oblong, entire, incurved at apex; stylopodium flat, often undulate on the margin; fruit ovoid or ovoid-oblong, slightly compressed dorsally, with a broad commissure, the carpels flat, often becoming corky on back, the dorsal ribs thick, keel-like, the lateral winged, the vittae numerous, the carpophore bifid; seeds flattened dorsally, nearly plane on face.——Few species, in Siberia and w. N. America.

1A. Stems often more than 1 m. long; rays numerous, densely pubescent; pedicels 4–7 mm. long; fruit often pubescent while young.
  1. *C. lucidum*
1B. Stems usually 40–70 cm. long; rays 20–30, densely puberulent; pedicels 1–1.5 cm. long, slender, smooth; fruit glabrous.
  2. *C. multisectum*

1. Coelopleurum lucidum (L.) Fern. var. **gmelinii** (DC.) Hara. *Archangelica gmelinii* DC.; *Angelica gmelinii* (DC.) Wormsk. ex DC. in syn.; *C. gmelinii* (DC.) Ledeb.——EZO-NO-SHISHI-UDO. Stout, 1–1.5 m. high, glabrous except short-pubescent on the upper part; leaves twice ternate, the ultimate segments ovate to rhombic-ovate, usually cuneate at base, often 3-lobed, mostly regularly dentate, pilose on the nerves, the upper leaves reduced to inflated sheaths; involucral bract solitary or absent, densely short yellowish brown pubescent on upper part, the rays numerous, densely pubescent, unequal, the involucel-bractlets many, broadly linear, nearly as long as the pedicels, the pedicels 4–7 mm. long; calyx-teeth absent; fruit oblong, 6–8 mm. long, about as long as or longer than the pedicels, often pubescent when young, with nearly equal, thick, corky ribs, the lateral ribs slightly broader than the dorsal, the vittae solitary in the intervals, the ribs 2 on the commissure; seed drupelike, with a slightly convex face.——Hokkaido.——Kuriles, Sakhalin, Ochotsk, and Kamchatka.

Var. **trichocarpum** (Hara) Hara. *Angelica trichocarpa* Hara; *C. trichocarpum* (Hara) Kitag.——EZO-YAMA-ZENKO.

Involucre of several bracts; leaflets smaller, 1.5–3.5 cm. long, irregularly acutely dentate.——Aug. Mountains; Hokkaido.

2. Coelopleurum multisectum (Maxim.) Kitag. *Angelica multisecta* Maxim.——MIYAMA-ZENKO.   Roots thick; stems erect, 40–70 cm. long, sparsely branched, stout, striate, glabrous, except pubescent below the umbels; leaves 3 to 5 times ternate, deltoid in outline, 10–15 cm. long, glabrous, the leaflets rather thin, oblong, ovate or rhombic-ovate, 1–3 cm. long, 7–20 mm. wide, acuminate, irregularly dentate, the upper leaves small, reduced to inflated sheaths; umbels few, long-pedunculate, the rays 20–30, spreading, densely puberulent, 4–6 cm. long, bractlets of the involucel as long as to shorter than the pedicels, the pedicels 20–40, smooth, slender, 1–1.5 cm. long; fruit oblong, glabrous, somewhat compressed dorsally, 4.5–5 mm. long, the ribs prominent, rather corky, thickened, the lateral ones slightly broader than the dorsal, the vittae solitary in the intervals, 2–4 on the commissure, the styles slender, about 1.5 mm. long, the stylopodium conical, the calyx-teeth obsolete, the carpophore bifid.——July–Aug., fruiting in Sept.——Alpine; Honshu (centr. distr.).

26. DYSTAENIA Kitag.   SERI-MODOKI ZOKU

Stems rather stout; radical and cauline leaves petiolate, 2 or 3 times ternately pinnate, the ultimate segments toothed and incised, often 3-lobed; umbels compound, the involucral bracts lacking, the involucel of many, narrow bractlets; flowers white; calyx-teeth evident, acute; petals equal, deeply emarginate at the incurved apex; stylopodium depressed-conical, undulate on the margin, the styles slender; fruit broadly ellipsoidal, compressed dorsally, with a broad commissure, the dorsal ribs thin, winged, the lateral more broadly winged, the vittae irregular, weak, numerous between and beneath the ribs, seeds semiorbicular, with a flat face, the carpophore bipartite.——Two species, in Japan and Korea.

1. Dystaenia ibukiensis (Yabe) Kitag. *Ligusticum ibukiense* Yabe; *Angelica ibukiensis* Makino ex Yabe——SERI-MODOKI, TANI-SERI-MODOKI. Perennial herb, white-papillose beneath the stem-nodes and in the inflorescence; stems 30–70 cm. long, striate, branched above; leaves twice ternately pinnate, deltoid in outline, 10–25 cm. long, the ultimate segments narrowly rhombic-ovate to deltoid-ovate or orbicular, 2–4 cm. long, 1–2 cm. wide, pinnately lobed or incised and dentate, acuminate, scaberulous on the nerves and margins, the veinlets prominent; umbels few, the rays 10–20, 2–4 cm. long, the involucel of filiform-linear, scabrous bractlets slightly longer than the pedicels, the pedicels 5–10 mm. long; fruit 5–6 mm. long, the styles reflexed, 1.5–2 mm. long.——July–Aug., fruiting Sept.–Oct.; Honshu (centr. and n. distr.).

27. CONIOSELINUM Fisch.   MIYAMA-SENKYŪ ZOKU

Perennials with fistulose stems; leaves twice ternately pinnate, the ultimate segments pinnatifid; umbels compound, the rays numerous, the involucral bracts few or absent, the bractlets of the involucel narrow; flowers white; calyx-teeth usually obsolete; petals obovate with an incurved and emarginate apex; stylopodium conical; fruit oblong, compressed dorsally, the carpels flat, the ribs membranous, winglike, the lateral much broader than the dorsal, the vittae 1 to few in the intervals, 4–8 on the commissure, the carpophore bifid; seed with a plane face.——Species few, in the N. Hemisphere.

1A. Primary leaflets acute to short-acuminate, firmly membranous; fruit oblong, the dorsal wings broad. .... 1. *C. kamtschaticum*
1B. Primary leaflets long-acuminate, thinly membranous; fruit ellipsoidal with narrowly winged dorsal ribs. .... 2. *C. filicinum*

1. Conioselinum kamtschaticum Rupr. *C. gmelinii* sensu Hult., non Coult. & Rose——KARAFUTO-NINJIN. Perennial, glabrous except the inflorescence; stems striate, simple or sparingly branched above; rhizomes short, thick, with thickened roots; radical and lower cauline leaves long-petiolate, triternate, deltoid in outline, 10–20 cm. long, the primary leaflets acute to short-acuminate, the ultimate segments narrowly to broadly ovate, pinnatifid, acute, incised-dentate, 15–35 mm. long, 1–2 cm. wide, scaberulous on the margins, the veinlets distinct, the upper leaves small, reduced to basal sheaths; umbels few, the involucral bracts few, linear, spreading, shorter than the rays or lacking, the rays 10–20, 2–3 cm. long, papillose on the inner side and on the upper part of the peduncle, the involucel of a few, linear bractlets as long as or slightly shorter than the flowers, the pedicels 20–30, 6–8 mm. long; fruit oblong, about 4 mm. long, the dorsal wings broad, the styles slender, appressed to the stylopodium, about 1 mm. long; vittae 2 or 3 in the intervals, 6 on the commissure.——Aug., fruiting in Oct. Grassy places near the sea; Hokkaido.——Sakhalin, Kuriles, Kamchatka and e. Siberia to Alaska.

2. Conioselinum filicinum (H. Wolff) Hara. *C. univittatum* sensu auct. Japon., non Turcz.; *C. nipponicum* Hara; *Peucedanum filicinum* H. Wolff——MIYAMA-SENKYŪ. Rhizomes stout; stems striate, glabrous except the inflorescence, sparsely branched above, 40–80 cm. long; radical and lower cauline leaves thinly membranous, long-petiolate, ternately tripinnate, deltoid in outline, 15–25 cm. long, the primary leaflets long-acuminate, the ultimate segments narrowly ovate or oblong, acute to acuminate, pinnatifid, 2–5 cm. long, 1–3 cm. wide, incised-dentate, scaberulous on the margins, the veinlets slender; umbels few, the rays 15–25, 4–6 cm. long, papillose-puberulent on the inner side and on the upper part of the peduncle, the involucral bracts 2 or 3, linear, 1–2 cm. long or lacking, the bractlets of the involucel few, filiform-linear, as long as to slightly longer than the flowers, the pedicels 7–10 mm. long, the styles about 1 mm. long, slender, recurved; fruit elliptic, 4–5 mm. long, the dorsal wings narrow; vittae solitary in the intervals, 2 on the commissure.——Aug., fruiting in Oct. Alpine regions; Hokkaido, Honshu (centr. and n. distr.).——s. Kuriles.

28. ANGELICA L.   SHISHI-UDO ZOKU

Usually stout perennial herbs with robust, leafy, fistulose or rarely solid stems; leaves ternately or pinnate-ternate compound, the ultimate segments often large, dentate; umbels compound, few to numerous, with numerous rays, the bracts and bractlets small or lacking; flowers white to dark purple; calyx-teeth minute or obsolete; petals obovate to oblanceolate, narrowed at base, usually with an incurved and emarginate apex; stylopodium flat, entire; fruit ovate, dorsally much compressed, with a broad commissure, the carpels lunate or planoconvex in cross section, the lateral ribs membranous, winged, the dorsal usually prominent but unwinged, the vittae 1–3 in the intervals, 2 to several on the commissure, the carpophore bifid; seeds compressed with a flat or slightly concave face.——More than 50 species, in temperate regions of the N. Hemisphere.

1A. Leaves simply pinnate; leaflets lanceolate with regular minute teeth, the lower often 2- or 3-lobed; sheaths of upper leaves somewhat inflated; carpels semiorbicular in cross section, with very narrowly winged lateral ribs. .... 1. *A. cartilaginomarginata*
1B. Leaves usually ternately or pinnate-ternate compound; carpels strongly compressed dorsally, planoconvex in cross section, with prominently winged lateral ribs.
  2A. Small or medium-sized herbs; leaf-segments 0.5–7 cm., rarely to 10 cm. long.
    3A. Leaflets thick or thickish, not herbaceous.
      4A. Ultimate leaf-segments dissected into linear to lanceolate lobes 2–5 mm. wide; upper leaves reduced to much-inflated sheaths.

5A. Rays 10–15; ultimate leaf-segments 5–10 mm. long, ovate to rhombic-ovate, dissected into lanceolate lobes. . . 2. *A. longeradiata*
5B. Rays 20–40; ultimate leaf-segments 2–6 cm. long, dissected into linear to lanceolate, acuminate lobes. . . . . 3. *A. ubatakensis*
4B. Ultimate leaf-segments ovate or broadly so, dentate, rarely incised or lobed, if the ultimate lobes linear-lanceolate, then the sheaths of the upper leaves not or scarcely inflated.
  6A. Leaves smooth, the ultimate segments gradually narrowed toward apex; rays 20–40, rarely fewer.
    7A. Vittae numerous; involucel of few bractlets. . . . . 4. *A. acutiloba*
    7B. Vittae solitary in the intervals, 2 on the commissure; involucel of 1 or 2 bractlets or lacking. . . . . 5. *A. shikokiana*
  6B. Leaf margins smooth or microscopically roughened, the leaflets or ultimate leaf-segments abruptly acuminate, acute or obtuse; rays 5–30.
    8A. Rays 20–30; leaflets deeply 3-cleft and incised, the teeth coarse and few; stems puberulent. . . . . 6. *A. saxicola*
    8B. Rays 10–20, rarely to 25; leaflets sometimes deeply cleft, the teeth numerous.
      9A. Leaflets and their lobes acute; terminal leaflets sessile, winged and decurrent at base; petals incurved but not emarginate.
7. *A. decursiva*
      9B. Leaflets and their lobes abruptly acuminate; terminal leaflets usually petiolulate.
        10A. Flowers dark purple, rather large; stems and petioles glabrous. . . . . 8. *A. gigas*
        10B. Flowers white, small; stems and petioles sparsely pubescent. . . . . 9. *A. hakonensis*
3B. Leaflets thin and herbaceous.
  11A. Sheaths of upper leaves not inflated; rays 10–15, very unequal; styles very short. . . . . 10. *A. inaequalis*
  11B. Sheaths of upper leaves inflated; rays 20–60.
    12A. Ovary glabrous; leaflets ovate to narrowly so; rays 20–40; vittae 2 on the commissure. . . . . 11. *A. polymorpha*
    12B. Ovary puberulent; leaflets broadly lanceolate to narrowly ovate; rays 40–60; vittae 4 on the commissure. . . 12. *A. genuflexa*
2B. Large, stout herbs with thick, erect stems; leaf-segments 5–15 cm. long, sometimes longer, often with a decurrent, winged base; upper leaves reduced to inflated sheaths.
  13A. Involucel of few linear to lanceolate bractlets.
    14A. Seashore plants with milky juice; leaflets thick, dark green and lustrous on upper surface.
      15A. Leaflets short-pubescent on nerves above, equally and minutely toothed; fruit elliptic; juice pale milky-white. . . 14. *A. japonica*
      15B. Leaflets glabrous above, acutely and slightly irregularly toothed; fruit oblong; juice distinctly yellow. . . . . 15. *A. keiskei*
    14B. Inland plants with colorless juice; leaflets relatively thin and nearly dull above.
      16A. Ultimate segments of the leaves lanceolate to narrowly ovate-oblong, acuminate, narrowed at base, prominently confluent and decurrent; bractlets lanceolate, several-nerved; fruit elliptic. . . . . 13. *A. dahurica*
      16B. Ultimate segments of the leaves ovate, elliptic or broadly rhombic-ovate, acute, usually rounded and not decurrent at base; bractlets linear; fruit oblong. . . . . 16. *A. edulis*
  13B. Involucel lacking or sometimes of 1 or 2 bractlets.
    17A. Leaflets usually glabrous, usually somewhat whitish beneath.
      18A. Rays of umbels up to 30 cm. long; ultimate segments of the leaves usually prominently decurrent and confluent. . . 17. *A. ursina*
      18B. Rays of umbels up to 12 cm. long; ultimate segments of the leaves usually distinct or only the terminal confluent or decurrent. . . . . 18. *A. anomala*
    17B. Leaflets pubescent, paler beneath; rays of umbels 3–16 cm. long. . . . . 19. *A. pubescens*

1. Angelica cartilaginomarginata (Makino) Nakai. *Peucedanum cartilaginomarginatum* Makino ex Yabe; *P. makinoi* Nakai; *A. cartilaginomarginata* var. *matsumurae* (H. Boiss.) Kitag.; *Sium matsumurae* H. Boiss.; *A. confusa* Nakai; *A. crucifolia* Komar.——HIME-NODAKE. Rhizomes short, with a few thickened roots; stems erect, branched above, rather slender, glabrous, 50–80 cm. long; radical and lower cauline leaves petiolate, simply pinnate, 10–25 cm. long, the leaflets or segments 5–9, sometimes 3, lanceolate, the lower 2- or 3-lobed, sessile or short-petiolulate, often 3- to 5-fid near the base, upper ones decurrent to the base, the terminal lobes often 3-fid, rather firm, 4–8 cm. long, 7–20 mm. wide, acute, with minute marginal teeth, often loosely papillate-pilosulous on the nerves on upper side, paler and prominently veined beneath, slightly cartilaginous on the margin, the upper leaves reduced to slender much-inflated sheaths; umbels few, on long slightly papillate peduncles, the rays 8–12, unequal, 2–5 cm. long, white-papillate on the inner side, the pedicels 3–7 mm. long; fruit broadly ovate-elliptic, suborbicular in cross section, glabrous, about 2 mm. long, with very narrowly winged lateral ribs, the calyx-teeth obsolete, the stylopodium short and small.—— Aug.–Oct., fruiting Sept.–Oct.; Honshu (Chūgoku Distr.), Shikoku, Kyushu.——Korea and Manchuria.

2. Angelica longeradiata (Maxim.) Kitag. *Selinum longeradiatum* Maxim.; *Cnidium longeradiatum* (Maxim.) Yabe; *C. yakushimense* Masam. & Ohwi; *A. longeradiata* var. *yakushimensis* (Masam. & Ohwi) Kitag.——TSUKUSHI-ZERI. Rhizomes thick; stems often scapiform or to 4-leaved, 5–25 cm.

long, slightly puberulent in upper part; radical and lower cauline leaves with petioles 2–8 cm. long, the blades ternately to pinnately decompound, deltoid-ovate, obtuse, glabrous, the leaflets rhombic-ovate to ovate, 5–10 mm. long, 3–8 mm. wide, deeply 2- to 5-lobed or -parted, the ultimate segments lanceolate, acute, entire, often papillose on the nerves on upper side, the upper leaves reduced, their sheaths elliptic, inflated; umbels 1–5, long-pedunculate, the rays 10–15, rather unequal, 2–5 cm. long, papillate-puberulent on the inner side and on the upper part of the peduncle and the inner side of the pedicels, the umbellets 20–30, the pedicels 2–5 mm. long, as long as or slightly shorter than the involucel, bractlets of the involucel few, linear; fruit compressed, elliptic, 2.5–3 mm. long, the dorsal ribs slender, short, rather prominent toward base, the lateral ribs rather thick, winged, the styles slender, nearly as long as the stylopodium.——Aug.–Sept., fruiting Oct.; Honshu (Mount Hiruzen in Mimasaka), Kyushu.

3. Angelica ubatakensis (Makino) Kitag. *Peucedanum ubatakense* Makino.——UBATAKE-NINJIN. Rhizomes thick, with thick elongate roots; stems to 40 cm. long, often scapiform or to 4-leaved, short-pubescent especially in upper part; radical and lower cauline leaves petiolate, deltoid-orbicular, 5–25 cm. long, ternately to pinnately decompound, the leaflets and ultimate segments dissected into linear and lanceolate lobes, 2–6 cm. long, acuminate, scabrous on the nerves on upper side, the upper leaves reduced, their sheaths elliptic, inflated; umbels few, long-pedunculate, the rays rather unequal, 20–40, papillose-puberulent on the inner side, 3–6 cm. long, the involucel bractlets few, longer or shorter than the pedicels, linear, the calyx-teeth minute; fruit glabrous, elliptic, about 3 mm. long, the lateral ribs winged, the vittae solitary in the intervals, 2 on the commissure.——Aug. Mountains; Shikoku, Kyushu; rare.

4. Angelica acutiloba (Sieb. & Zucc.) Kitag. *Ligusticum acutilobum* Sieb. & Zucc.; *Sium triternatum* Miq.; *A. ibukicola* Makino; *L. ibukicola* Makino——Tōki. Rhizomes short and thick, with few thick roots; stems glabrous, striate, 30–70 cm. long, solid; radical and lower cauline leaves long-petiolate, once or twice ternately pinnate, deltoid in outline, 10–25 cm. long, glabrous, the leaflets 5–10 cm. long, rather thin, 3-cleft, cuneate, truncate, or rounded at base, the ultimate segments lanceolate, often shallowly 3-lobed, 10–25 mm. wide, long-acuminate, doubly or simply and acutely dentate, the reticulate veinlets fine but prominent; umbels few, papillate, the rays 30–40, 3–8 cm. long, papillate, the involucel of a few, filiform-linear bractlets, the pedicels slender, 7–18 mm. long; fruit oblong, 4–5 mm. long, slightly compressed, narrowed toward base, the carpels with slender ribs, the lateral ribs slightly winged, the vittae 3 or 4 in the intervals, 4 on the commissure.——Aug., fruiting Sept.–Oct. Mountains; Honshu (Kantō distr.); cultivated for medical purposes.

Var. **iwatensis** (Kitag.) Hikino. *A. iwatensis* Kitag.; *Ligusticum japonicum* Maxim.——Miyama-tōki, Nambu-tōki, Iwate-tōki. Ultimate segments of the leaves narrowly ovate, usually rounded to shallowly cordate, rarely broadly cuneate at base, 1–2.5 cm. wide; carpels 4–6 mm. long, 2–2.5 mm. wide, the vittae 3–6 in the intervals, 8–10 on the commissure.—— Honshu; sometimes cultivated for medicinal purposes.

Var. **lanceolata** (Tatew.) Ohwi. *Ligusticum linearilobum* var. *lanceolatum* Tatew.; *A. acutiloba* var. *lineariloba* (Koidz.) Hikino; *L. linearilobum* Koidz.; *A. stenoloba* Kitag.——Hosoba-tōki. Leaves more deeply cut, the leaflets broadly linear, 3–8 mm. wide.——High mountains; Hokkaido (Mount Yubari and Apoi); rare.

5. Angelica shikokiana Makino. Inu-tōki. Rhizomes thick; stems 50–80 cm. long, glabrous, striate; radical and lower cauline leaves long-petiolate, once or twice ternately compound, deltoid or ovate-deltoid, 20–30 cm. long, glabrous, the terminal leaflets often deeply 3-cleft, the ultimate segments broadly lanceolate to narrowly ovate, long-acuminate, usually rounded at base, 4–8 cm. long, 1.5–3 cm. wide, usually shallowly dentate, rather thin, the reticulate veinlets prominent, the upper leaves reduced, and the sheaths rather prominently inflated, narrowly oblong to broadly oblanceolate; umbels few, minutely papillate-puberulent on the inner side of the rays and pedicels, the rays 15–30, 3–6 cm. long, the bractlets of the involucel 1 or 2 or lacking, the pedicels 5–10 mm. long; fruit narrowly oblong, about 5 mm. long, the carpels strongly compressed, emarginate at base, the lateral ribs broad, the dorsal ones slender, the styles short but slender; vittae solitary in the intervals, 2 on the commissures.——Aug. Mountains; Shikoku, Kyushu.

Var. **tenuisecta** Makino. *A. tenuisecta* (Makino) Makino ——Kawa-zenko. Leaves finely dissected, the leaflets 3–5 cm. long, 8–15 mm. wide, paler beneath.——Mountains of Honshu (Kii Prov.).

6. Angelica saxicola Makino ex Yabe. *Peucedanum saxicola* Makino ex Makino & Nemoto——Ishizuchi-bōfū. Rhizomes short, with thickened elongate roots; stems 20–80 cm. long, cylindrical, striate, puberulent; radical and lower cauline leaves long-petiolate, once or twice ternately com-

pound, 5–25 cm. long, the leaflets deltoid to deltoid-orbicular, 2–5 cm. long, 1–2 cm. wide, deeply 3-cleft and incised, rhombic-ovate to narrowly ovate, acuminate, with a few coarse acute teeth, scaberulous only on the nerves on upper side, glaucous beneath, the upper leaves reduced, their sheaths obovately inflated; umbels few, densely short-pubescent, the rays 20–30, 3–6 cm. long, the umbellets 25- to 40-flowered, the involucel of a few, linear bractlets often longer than the flowers, the pedicels 5–10 mm. long; calyx-teeth minute but evident, deltoid; ovary glabrous, the styles slender, recurved, 2 or 3 times as long as the stylopodium; fruit oblong.——July–Aug. High mountains; Shikoku (Mount Ishizuchi in Iyo); rare.

7. Angelica decursiva (Miq.) Fr. & Sav. *Porphyroscias decursiva* Miq.; *Peucedanum porphyroscias* (Miq.) Makino; *Peucedanum decursivum* (Miq.) Maxim.; *Panax fallax* Miq. ——Nodake. Usually purplish and nearly glabrous perennial herb; rhizomes short, with thickened roots; stems 80–150 cm. long, cylindrical, prominently striate, with white pith; radical and lower cauline leaves large, long-petiolate, ternately pinnate, broadly deltoid-ovate, usually pubescent on the nodes and the upper part of the petiolules, the leaflets 3(–5), usually deeply 3- to 5-cleft to -parted, the segments rather firm, usually decurrent on a winglike sessile base, narrowly ovate to oblong, 5–10 cm. long, 2–4 cm. wide, acute, cuneate to rounded at base, rather regularly and acutely serrate, sometimes pinnately incised, scaberulous on the nerves on upper side and on the margin; upper leaves much reduced, with an obovate much-inflated sheath; umbels pedunculate, the peduncles and upper side of the rays papillate-puberulent, the rays 10–20, 3–6 cm. long, the umbellets 20- to 30-flowered, the involucel with few linear bractlets, the pedicels often nearly glabrous, 5–10 mm. long; flowers dark purplish; fruit broadly elliptic, about 5 mm. long, flat, glabrous, the carpels ribbed, the lateral ribs rather thick, winged, the vittae 1–4 in the intervals, 4–6 on the commissure.——Sept.–Oct., fruiting Oct.–Nov. Grassy places in woods in lowlands and hills; Honshu (Kantō Distr. and westw.), Shikoku, Kyushu; common.——Korea to e Siberia and China.——Forma **angustiloba** (Makino) Ohwi. *Peucedanum decursivum* var. *angustilobum* Makino——Hosoba-nodake. Leaflets narrow.——Forma **albiflora** (Maxim.) Nakai. *Peucedanum decursivum* var. *albiflorum* Maxim.—— Shirobana-nodake. Flowers white.

8. Angelica gigas Nakai. Oni-nodake. Resembles the preceding species; rhizomes thickened; stems sometimes fistulose, 1–2 m. long, glabrous; radical and lower cauline leaves long-petiolate, once (sometimes twice) ternate, the leaflets 3-parted, sometimes again 2- or 3-cleft, the segments oblong or ovate, 6–12 cm. long, rounded to obtuse at base, abruptly acuminate, doubly and acutely serrate, the terminal segment rhombic-ovate, cuneate at base, the upper leaves modified to elliptic, much-inflated purplish sheaths; umbels papillate-pubescent with crowded purplish brown hairs, the involucral bract solitary or lacking, the rays 15–20, the umbellets rather densely 25- to 40-flowered, the involucel of few linear-lanceolate bractlets; flowers rather large, dark purple; ovary minutely punctate; fruit about 8 mm. long, 5 mm. wide, the vittae solitary in the intervals.——Aug. Mountains; Shikoku, Kyushu.——Korea and Manchuria.

9. Angelica hakonensis Maxim. Iwa-ninjin. Rhizomes short and thick, with thick elongate roots; stems 60–120 cm. long, rather stout, branched above, terete, sparsely whitish pubescent beneath the umbels; radical and lower cauline leaves petiolate, 15–30 cm. long, deltoid-ovate in outline, twice or thrice ternately pinnate, the leaflets narrowly ovate to broadly rhombic-ovate, acuminate, cuneately narrowed toward base, often 2- or 3-cleft, 3–6 cm. long, deeply and acutely serrate, scaberulous on the nerves on both surfaces and on the margin, the upper leaves modified to elliptic inflated sheaths; umbels few, the rays 10–25, 3–5 cm. long, rather stout, with short, dense, sordid yellow hairs, the bractlets of the involucel few to several, linear, longer than the pedicels, the pedicels 20–30, rather stout, 2–8 mm. long, puberulent on the inner side; flowers small, white; fruit about 6 mm. long, elliptic, glabrous, the carpels prominently ribbed, the lateral ribs winged, the vittae solitary in the intervals, 4 on the commissure, the styles short, about as long as the stylopodium.——Aug.–Sept. Mountains; Honshu (Kantō Distr. and Suruga Prov.).

Var. **nikoensis** (Yabe) Hara. *A. nikoensis* Yabe——No-DAKE-MODOKI. Leaves once to twice compound, the leaflets larger, 5–10 cm. long, 3–4 cm. wide.——Honshu (Kantō Distr.).

10. Angelica inaequalis Maxim. HANABIZERI. Glabrous herb with short rhizomes and few, rather thick roots; stems fistulose, cylindric, stout or slender, sparsely branched above, 60–80 cm. long; radical and lower cauline leaves long-petiolate, twice or thrice ternately compound, herbaceous, the leaflets often again 2- or 3-cleft to -parted, the ultimate segments thin, narrowly ovate, rhombic-ovate or broadly ovate, 3–7 cm. long, 2–3.5 cm. wide, acuminate, cuneate to rounded at base, irregularly dentate, incised and often pinnately lobed, paler beneath, minutely papillate on the nerves on upper side and on margin; sheaths of the upper leaves narrow, green, with a much-reduced blade; umbels few, the long peduncles, inner side of the rays, and pedicels minutely papillate at apex, the rays 10–15, very unequal, 1–8 cm. long, the involucel with a few, linear-filiform bractlets as long as the pedicels, the pedicels 20–30, unequal, 3–12 mm. long; fruit oblong, glabrous, emarginate at base, the dorsal ribs filiform, the lateral ones broadly winged, the vittae 3 or 4 in the intervals, 6 on the commissure, the calyx-teeth lax, sometimes evident, the styles very short, shorter than the stylopodium.——Aug.–Sept., fruiting Oct. Honshu (Kantō Distr. and westw.), Shikoku, Kyushu.

11. Angelica polymorpha Maxim. SHIRANE-SENKYŪ, SUZUKA-ZERI. Rhizomes short, with few, thickened roots; stems erect, glabrous, branched above, 80–150 cm. long; radical and lower cauline leaves long-petiolate, 3 or 4 times ternately pinnate, deltoid or deltoid-ovate in outline, 20–30 cm. long, herbaceous, glabrous or nearly so, the leaflets ovate to narrowly so, 3–6 cm. long, acuminate to acute, often again 3-cleft, pinnately incised or toothed, the sheaths of the much-reduced upper leaves oblong to broadly lanceolate; umbels few to rather numerous, papillate on the upper portion of the peduncles and the inner side of the rays and pedicels, the rays 20–40, rather unequal, 4–6 cm. long, the umbellets 20- to 40-flowered, the pedicels 5–12(–15) mm. long, the involucel with a few linear bractlets nearly as long as the pedicels; ovary glabrous; fruit elliptic, emarginate at both ends, flat, glabrous, 4–5 mm. long, the carpels ribbed, the lateral ribs rather broadly winged, the styles 1–1.5 mm. long, the vittae solitary in the intervals, 2 on the commissure.——Sept.–Oct., fruiting Oct.–Nov. Mountains; Honshu, Shikoku, Kyushu.——Korea and Manchuria.

12. Angelica genuflexa Nutt. *A. refracta* F. Schmidt; *A. yabeana* Makino; *A. refracta* var. *yabeana* (Makino) Koidz.;

A. caudata H. Boiss.——ŌBA-SENKYŪ, EZO-ŌBA-SENKYŪ. Rhizomes thick; stems fistulose, rather soft, glabrous, branched above, 80–150 cm. long; radical and lower cauline leaves petiolate, once or twice ternately pinnate, rather large, the petiolules of the primary leaflets divaricate, often reflexed, the leaflets herbaceous, broadly lanceolate to narrowly ovate, 3–10 cm. long, 1–2.5(–3) cm. wide, long-acuminate, narrowed toward base, irregularly and acutely dentate, papillate on the nerves on both surfaces and on the margin; umbels rather numerous, the peduncles densely papillate-puberulent at summit and on the inner side of the rays and pedicels, the rays 40–60, the involucel with a few, linear-filiform bractlets shorter than to as long as the pedicels, the pedicels 30–50, slender, 10–18 mm. long, the ovary puberulent; fruit broadly lanceolate, emarginate at both ends, 4–5 mm. long, glabrous, the lateral ribs broadly winged, the vittae solitary in the intervals, 4 on the commissure, the styles slender, 1.5–2 mm. long.——July–Sept., fruiting Aug.–Oct. Along mountain streams and ravines; Hokkaido, Honshu (centr. and n. distr.).——Sakhalin, Kuriles, Kamchatka, and w. N. America.

13. Angelica dahurica (Fisch.) Benth. & Hook. f. *Callisace dahurica* Fisch.——Ō-SHISHI-UDO. Gigantic perennial herb; stems erect, branched above, 1–2 m. long, very stout, glabrous, puberulent on upper part; radical and lower cauline leaves petiolate, large, twice or thrice ternately pinnate, the terminal leaflets decurrent at base, deeply 3-cleft to -parted, the ultimate segments rather thick, narrowly lanceolate-oblong, 5–10 cm. long, 2–5 cm. wide, acuminate, coarsely serrate, deep green and often scaberulous on margins and on the nerves on upper side, whitish and puberulent especially on the nerves beneath, sheaths of the much-reduced upper leaves ovate or oblong, inflated; umbels few, the peduncles, rays, and pedicels rather densely papillate-puberulent, the rays 20–40, 4–6 cm. long, the involucel bractlets few, broadly lanceolate, caudately acuminate, as long as the pedicels, the pedicels 4–10 mm. long; fruit elliptic, glabrous, flat, slightly emarginate at base, 8–9 mm. long, the dorsal ribs slender, filiform, the lateral ribs winged, the vittae 1(–2) in the intervals, 2–4 in the commissure.——July–Aug. Mountains; Honshu.——Korea, Manchuria, and e. Siberia.

14. Angelica japonica A. Gray. *A. kiusiana* Maxim. ——HAMA-UDO, ONI-UDO. Gigantic perennial herb; stems stout, thick, 50–100 cm. long, often dark purplish, puberulent in upper part; radical and lower cauline leaves petiolate, large, broadly ovate-deltoid, once or twice (rarely thrice) ternately pinnate, glabrous, the ultimate leaflets rather thick, oblong to ovate-elliptic, 7–10 cm. long, 2–4 cm. wide, acute, smooth margined, obliquely rounded at base, lustrous, paler and with finely reticulate veinlets beneath, equally and finely toothed, the terminal leaflets often again 3-fid, decurrent toward base, sheaths of the much-reduced upper leaves elliptic, inflated; umbels rather numerous, puberulent, the rays 30–40, puberulent on the inner side, 4–7 cm. long; bractlets of the involucel several, broadly lanceolate, caudately acuminate, shorter to slightly longer than the pedicels, the pedicels 20–40, 5–12 mm. long; fruit elliptic, flat, glabrous, 6–7 mm. long, the dorsal ribs filiform, the lateral ones rather thick, winged, the vittae solitary in the intervals, 4 on the commissure.——Apr.–June. Near seashores; Honshu (Kantō Distr. and westw.), Shikoku, Kyushu.——Ryukyus.

15. Angelica keiskei (Miq.) Koidz. *Archangelica keiskei* Miq.; *Angelica utilis* Makino ex Yabe——ASHITABA. Rhizomes thick and short, with few elongate roots; stems stout, branched above, 80–120 cm. long, glabrous; radical and lower cauline leaves long-petiolate, usually evergreen, twice or thrice ternately pinnate, obtusely deltoid in outline, 20–60 cm. long, glabrous, the leaflets rather thick, usually deeply 2- or 3-cleft to -parted, usually rounded at base, deep green and lustrous, the segments usually ovate, 5–10 cm. long, 3–6 cm. wide, acute, toothed, with impressed veinlets on upper side, the sheaths on the upper part elliptic, inflated; umbels rather numerous, the peduncles in upper part, the inner side of the rays and pedicels puberulent, the rays 10–20, the pedicels 20–40; bractlets of the involucel several, linear or broadly so, slightly broadened at base, caudately elongate, slightly longer or shorter than the pedicels; fruit oblong, glabrous, 6–8 mm. long, flat, the dorsal ribs short and rather thick, the lateral ones winged, the vittae solitary in the intervals, 4 on the commissure, the styles short, as long as the stylopodium.——May–Oct. Near seashores; Honshu (Kantō Distr. and Izu Prov.).

16. Angelica edulis Miyabe ex Yabe. AMA-NYŪ, MARU-BA-AMA-NYŪ.   Gigantic perennial herb; stems 1–3 m. long, purplish, glabrous; leaves once or twice ternately compound, glabrous except on the nodes, the leaflets herbaceous, cordate at base, slightly lustrous, paler beneath, 10–20 cm. long and as wide, again 2- or 3-cleft, the segments ovate to broadly rhombic-ovate, 3–10 cm. wide, abruptly acuminate to acute, irregularly dentate, often again 2- or 3-lobed, the terminal leaflets broadly ovate to deltoid-orbicular, 3-lobed to -parted, abruptly acuminate, the sheaths of the upper much-reduced leaves narrowly obovate to oblong; umbels often rather numerous, the peduncles in upper part, the inner side of the rays, and pedicels papillate-puberulent, the rays 30–50, unequal, 3–12 cm. long, the involucel with linear bractlets shorter than the pedicels, the pedicels 30–60, slender, 5–15 mm. long, spreading; fruit oblong, emarginate at base, glabrous, the dorsal ribs filiform, the lateral ones winged, the vittae solitary in the intervals, 2 on the commissure, the styles slender, as long as the stylopodium.——Aug. Mountain woods; Hokkaido, Honshu (centr. and n. distr.).——? Sakhalin.

17. Angelica ursina (Rupr.) Maxim. *Angelophyllum ursinum* Rupr.; *Angelica japonica* A. Gray, pro parte——EZO-NYŪ.   Gigantic perennial herb; stems 1–3 m. long, very stout, glabrous, slightly pubescent beneath the umbels; leaves large, twice or thrice ternately pinnate, the leaflets large, again pinnatifid to pinnately parted, the ultimate segments narrowly ovate-oblong, acuminate, often decurrent at base, irregularly dentate, glabrous or sparsely pubescent on the nerves and paler beneath, the terminal leaflets obovate-rhombic, cleft, acuminate, the upper leaves much reduced, with an ovoid, much-inflated sheath; umbels rather numerous, large, the involucral bract solitary or lacking, the rays 60–100 or more, 3–30 cm. long, papillate-puberulent on the inner side as in the pedicels, the umbellets 30- to 40-flowered, the pedicels 1–3 cm. long, the involucel obsolete; fruit broadly elliptic, about 7 mm. long, the carpels with filiform dorsal ribs and winged lateral ribs, the

vittae 1(–2) in the intervals, 4 on the commissure.——Hokkaido, Honshu (n. distr.).——Sakhalin, Kuriles, and Kamchatka to e. Siberia.

18. Angelica anomala Lallem. *Peucedanum angelicae-folium* Turcz.; *A. montana* var. *angustifolia* Ledeb.; *A. sachalinensis* Maxim.; *Angelophyllum dauricum* Rupr.; *Angelica refracta* var. *glaucaphylla* Koidz.; *A. pubescens* var. *glabra* Yabe; *A. glabra* (Yabe) Makino——EZO-NO-YOROIGUSA, EZO-NO-YOROIGUSA.   Gigantic perennial herb, glabrous except the inflorescence; stems 1–2 m. long, fistulose, often purplish, branched above; leaves twice or thrice ternately pinnate, large, deltoid in outline, the leaflets rather thick, often again 2- or 3-fid to -parted, whitish beneath, the ultimate segments narrowly oblong to narrowly ovate, acuminate, often decurrent toward base, deep green, scaberulous on the nerves on upper side, glabrous or sparingly pubescent on the under side, acutely dentate, smooth and somewhat callose on the margin, the upper leaves much reduced, their sheaths obovate and inflated, often puberulent; umbels rather numerous, rather densely puberulent on the upper portion of the peduncles, rays, and pedicels, the rays 30–70, 3–12 cm. long, the involucel lacking, the pedicels 40–60, 5–15 mm. long, slender; fruit obovate-elliptic, 6–7 mm. long, emarginate at base, the carpels with low dorsal ribs and rather narrowly winged lateral ribs, the vittae solitary in the intervals, 4 on the commissure, the styles slender, about 1 mm. long.——Aug.   Hokkaido, Honshu (centr. and n. distr.).——e. Siberia, Korea, Manchuria, n. China, and Sakhalin.

19. Angelica pubescens Maxim. *A. polyclada* Franch.; *A. myriostachys* Koidz.; *A. schishiudo* Koidz.——SHISHI-UDO, UDO-TARASHI, TAKAO-KYŌKATSU.   Stems stout, green, terete, 1–2 m. long, sparingly pubescent; leaves large, about thrice ternately pinnate, the leaflets ovate to elliptic, 5–10 cm. long, sometimes oblong, abruptly acuminate, often decurrent at base, acutely dentate, pale green beneath, the upper leaves much reduced, with obovate, inflated sheaths; umbels rather numerous, the bracts and bractlets lacking, the rays numerous, 3–16 cm. long, puberulent, the pedicels 25–40, slender, 1–1.5 cm. long, puberulent; fruit oblong, 7–8 mm. long, emarginate at both ends, the dorsal ribs slender and filiform, the lateral ones broadly winged, the vittae 1–3 in the intervals, 2–6 on the commissure, the styles short, shorter than the stylopodium.——Aug.   Hills and low mountains; Honshu, Shikoku, Kyushu; rather variable.

Var. **matsumurae** (Yabe) Ohwi.   *A. matsumurae* Yabe——MIYAMA-SHISHI-UDO.   Larger, glabrescent mountain phase with the ultimate leaf-segments broadly lanceolate to ovate-oblong, acuminate to long-acuminate, rather regularly dentate.——Higher mountains of Honshu.——Forma **muratae** Ohwi.   *A. pubescens* forma *glabra* (Koidz.) Murata, non Yabe; *A. matsumurae* var. *glabra* Koidz.——KENASHI-MIYAMA-SHISHI-UDO.   Glabrous phase.

29. OSTERICUM Hoffm.   YAMAZERI ZOKU

Perennials with glabrous, striate or sulcate stems; leaves ternately compound, the segments toothed; umbels compound, the involucre of few bracts, the involucel of several bractlets, the flowers white, the calyx-teeth 5, persistent, the petals obovate, incurved and emarginate at apex, the styles slender, the stylopodium depressed-conical; fruit elliptic, flattened dorsally, the carpels flat, the dorsal ribs filiform, the lateral ones extended into a thin broad wing, the pericarp thin, consisting of one layer of convex cells, free from the seed at least on the marginal side, the vittae 1–3 in the intervals, 2–8 on the commissure, the carpophore 2-fid, the seed often stonelike, flat.——About 10 species, e. Asia to e. Europe.

1A. Stems 50–80 cm. long, leafy, rather stout and branched above; leaves rather large, 10–30 cm. long; leaflets thin and herbaceous, ovate, coarsely dentate, 3–6 cm. long; rays 4–7. .... 1. *O. sieboldii*
1B. Stems 15–30 cm. long, slender; leaves smaller; leaflets 1–3 cm. long, rather thick, dissected into broadly linear segments 2–3 mm. wide; rays 7–15. .... 2. *O. florenti*

1. Ostericum sieboldii (Miq.) Nakai. *Peucedanum sieboldii* Miq.; *Angelica miqueliana* Maxim.; *O. miquelianum* (Maxim.) Kitag.——Yamazeri. Glabrous perennial herb with short thick rhizomes and few thickened roots; stems erect, branched, rather stout, 50–80 cm. long, leafy; leaves long-petiolate, deltoid in outline, 10–30(–40) cm. long, 2 or 3 times ternately pinnate, the leaflets thin, sessile or short-petiolulate, ovate to broadly so, 3–6 cm. long, glabrous or nearly so except on the margin, coarsely dentate, acute to acuminate, cuneate to rounded at base, rarely 2- or 3-fid, the upper leaves reduced, their sheaths broadly oblanceolate to narrowly oblong; umbels several, the peduncles beneath the umbels and on the inner side of the rays and pedicels very slightly papillate-puberulent, the rays 4–7, 1.5–3 cm. long, the involucral bracts few, lanceolate or lacking, the bractlets of the involucel several, narrowly lanceolate, thinly membranous on the margin, as long as the pedicels, the pedicels 15–20, slender, 5–10(–12) mm. long; fruit elliptic, glabrous, emarginate at both ends, 3.5–4 mm. long, the dorsal ribs slender, the lateral ribs extended into a broad wing, the calyx-teeth small, erect, the stylopodium minute, as broad as the style.——July–Oct. Mountain woods; Honshu, Shikoku, Kyushu; common.——Korea, Manchuria, and n. China.

2. Ostericum florentii (Fr. & Sav.) Kitag. *Angelica florentii* Fr. & Sav.——Miyama-ninjin. Glabrous perennial with short slightly thickened rhizomes and slender stolons; stems erect, 15–30 cm. long, slender, few-leaved; radical and lower cauline leaves petiolate, deltoid- to broadly ovate in outline, 2 or 3 times pinnate or ternately pinnate, the leaflets rather thick, ovate to broadly so, 1–3 cm. long, dissected into broadly linear segments, 2–3 mm. wide, entire or sometimes 3-fid, acute, scaberulous on the margin, the upper leaves reduced, with an oblong or narrowly oblong inflated sheath; umbels, inner side of rays, and pedicels slightly papillate-puberulent, the rays 7–15, ascending, 1.5–3 cm. long, the involucral bracts 1–4, membranous, dilated at base, shorter than the rays, linear, the bractlets of the involucel about 10, linear-filiform, shorter to slightly longer than the flowers, the pedicels 20–30, 5–10 mm. long; fruit broadly ovate-elliptic, flat, glabrous, 5–6 mm. long, emarginate at base, the dorsal ribs slender and rather prominent, the lateral ones extended into a broad wing, the calyx-teeth minute, the styles slightly longer than the stylopodium.——Aug.–Sept. Mountains; Honshu (Kantō and centr. distr.).

30. PEUCEDANUM L.   Kawara-bōfū Zoku

Perennials, frequently pilose; leaves ternately or pinnately compound, the ultimate segments often dentate; umbels compound, the rays numerous, the involucre and involucel of few small bracts and bractlets or these lacking; flowers white, sometimes yellow; calyx-teeth minute or obsolete; petals obovate or cuneate, incurved and retuse at apex, the stylopodium usually small, undulate on margin; fruit elliptic, strongly compressed, the carpels slightly convex on the back, the dorsal ribs filiform to rather thick, the lateral ones rather thick and extended into a wing, the wings of the two carpels appressed, the vittae 1(–3) in the intervals; seeds flattened, slightly convex on the back, the seed-face flat or slightly concave.——About 160 species, in temperate regions of the N. Hemisphere of the Old World, and Africa.

1A. Stout maritime plants, slightly glaucous, bluish green; leaves 20–60 cm. long, twice or thrice ternately pinnate, rather thick; leaflets obovate-cuneate, with coarse teeth on margin near tip; fruit puberulent. .... 1. *P. japonicum*
1B. Plants of lowlands or mountains, often rather large, green; leaves 5–20 cm. long, once or twice ternate or ternately pinnate, rather thin; leaflets ovate, ovate-orbicular or rhombic-ovate, usually dentate at least on the upper half; fruit glabrous.
  2A. Cauline leaves several, ternately pinnate; leaflets broadly ovate to deltoid, pinnatifid and incised, the upper sheaths narrow, not inflated; umbels few to numerous, the involucel of 4–12 filiform-linear bractlets, shorter than to as long as the pedicels; fruit broadly ovate-elliptic, about 3.5 mm. long; styles about 0.5 mm. long. .... 2. *P. terebinthaceum*
  2B. Cauline leaves few, ternately compound; leaflet elliptic or obovate, incised or coarsely dentate, often 2- or 3-lobed, the upper sheaths inflated; umbels 1–3, the involucel absent; fruit oblong, 8–10 mm. long; styles about 2 mm. long. .... 3. *P. multivittatum*

1. Peucedanum japonicum Thunb. Botan-bōfū. Stout bluish green, glaucous herb, glabrous in lower part, short-pubescent in the upper part, with thick roots and short rhizomes; stems branched, 60–100 cm. long, solid, fibrous at base; leaves rather thick, long-petiolate, 20–60 cm. long, twice or thrice ternately pinnate, the leaflets obovate-cuneate, 3–6 cm. long, obtuse with a minute mucro at apex, often 3-lobed, irregularly deeply dentate, with coarse teeth on upper margin, the veinlets reticulate with free veinlets in each net, the upper leaves reduced, with a membranous, uninflated sheath; umbels numerous, on rather short peduncles, the rays 10–20, somewhat flattened, ascending, 2–3.5 cm. long, densely puberulent especially on the inner side, the involucre absent, the involucel of few to about 10 lanceolate bractlets as long as to shorter than the pedicels, the pedicels 20–30, slightly longer than the fruit; fruit elliptic, puberulent, 4–5 mm. long, the dorsal ribs slender and filiform, the lateral narrowly winged, the vittae 3–4 in the intervals, 8 on the commissure, the styles short and slender, as long as the stylopodium.——July–Sept. Sandy places near seashores; Honshu (Kantō Distr., Kaga Prov. and westw.), Shikoku, Kyushu; rather common.——China, Formosa, and Philippines.

2. Peucedanum terebinthaceum (Fisch.) Fisch. ex Turcz. *Selinum terebinthaceum* Fisch. ex Trevir.; *P. deltoideum* Makino; *P. terebinthaceum* var. *deltoideum* (Makino) Makino——Kawara-bōfū, Yama-ninjin. Rhizomes short, fibrous at the apex, with slightly branched, rather thick roots; stems erect, branched, leafy, glabrous, 30–80 cm. long; leaves petiolate, 5–10 cm. long, acute, broadly ovate in outline, ternately pinnate, glabrous, the leaflets broadly ovate to deltoid,

3–5 cm. long, acute with a mucro at apex, decurrent to a narrowly winged base, somewhat pinnately cleft and again incised, with a few acute teeth, the upper leaves reduced, their sheaths narrowly oblanceolate, not inflated; umbels few to numerous, the rays 10–15, 1–2.5 cm. long, slightly papillate-puberulent on the inner side and on the pedicels, the involucral bracts 1(–2) or lacking, the involucel of 4–12 filiform-linear bractlets, slightly shorter than to as long as the flowers, the pedicels 20–30, 5–10 mm. long; fruit broadly ovate-elliptic, 3–4 mm. long, glabrous, the dorsal ribs low and slender, filiform, the lateral ones narrowly winged, the styles very slender, about 0.5 mm. long, the vittae solitary in the intervals, 2 on the commissure.——Aug.–Sept. Hokkaido, Honshu, Shikoku, Kyushu.——e. Siberia, Manchuria, and Korea.

3. Peucedanum multivittatum Maxim. HAKUSAN-BŌFŪ. Glabrous perennial herb with thick, short nonfibrous rhizomes with thickened roots; stems erect, simple or sparingly branched, fistulose, few-leaved, often pilose on the nodes; radi-

cal and lower cauline leaves long-petiolate, deltoid to broadly ovate in outline, 5–8 cm. long, once to sometimes twice ternately compound, the leaflets narrowly ovate to obovate-orbicular, 3–5 cm. long, obtuse to acuminate, often 2- or 3-lobed, incised or coarsely dentate, glabrous, the upper leaves reduced, often with narrow leaflets, their sheaths obovate or elliptic and inflated; umbels 1–3, the peduncles, inner side of rays, and pedicels brownish short-pubescent, the rays 8–15, 3–5 cm. long, the involucre and involucel absent, the pedicels 10–20, 1–1.5 cm. long; fruit oblong, glabrous, 8–10 mm. long, the dorsal ribs low, filiform, the lateral ones narrowly winged, the vittae numerous, slender, the calyx-teeth broadly lanceolate to narrowly ovate, minute, the styles slender, 1.5–2 mm. long.——Aug., fruiting Oct. Alpine slopes; Hokkaido, Honshu (centr. distr. and northw.).——Forma **linearilobum** (Tatew.) Ohwi. *P. multivittatum* var. *linearilobum* Tatew.——EZO-NO-HAKUSAN-BŌFŪ. Leaflets dissected and cut into broadly linear lobes.

31. HERACLEUM L.   HANA-UDO ZOKU

Pubescent perennials; leaves pinnately or ternately compound, the leaflets often large, dentate; umbels usually large, the involucre lacking or of few bracts, the involucel of numerous bractlets, the calyx-teeth usually inconspicuous, the petals cuneate or rhombic, white, rarely yellow, usually unequal, inflexed and bifid to emarginate at apex, the stylopodium conical or minute; fruit orbicular, obovate or elliptic, much flattened, glabrous or pubescent, the carpels flat with very slender dorsal ribs and broadly winged lateral ones, the vittae usually solitary in the intervals, clavate, shorter than the carpels, the carpophore 2-parted; seeds with a flat face.——About 60 species, in temperate, regions of the N. Hemisphere.

1A. Leaves 3-foliolate; stems 1–2 m. long; high mountains. . . . . 1. *H. dulce*
1B. Leaves 3- to 5-foliolate; stems 70–100 cm. long; usually in low mountains and in lowlands. . . . . 2. *H. moellendorffii*

1. Heracleum dulce Fisch. *H. lanatum* sensu auct. Japon., non Michx.; *H. lanatum* subsp. *asiaticum* Hiroe; *H. lanatum* var. *asiaticum* (Hiroe) Hara——Ō-HANA-UDO. Stems stout, fistulose, 1–2 m. long, pubescent or nearly glabrous, branched above; radical and lower cauline leaves long-petiolate, large, pinnately 3-foliolate, the leaflets broadly ovate-cordate to broadly ovate, 20–40 cm. long, paler beneath, glabrous or usually pubescent, cordate to rounded at base, 2- or 3-cleft, the lobes acuminate and shallowly 2- or 3-lobed, incised, and acutely dentate, the upper leaves reduced, with inflated sheaths; umbels rather numerous, large, the rays 20–30, pubescent on the inner side, 8–12 cm. long, the involucre absent, the involucel of several narrowly lanceolate, filiform-attenuate bractlets shorter than the pedicels, the pedicels slender, unequal, 5–25 mm. long; flowers rather large, white, the petals unequal; fruit strongly flattened, broadly obovate, glabrescent, 7–8 mm. long, the dorsal ribs filiform, the lateral ones broadly winged, the vittae linear-oblanceolate, solitary in the intervals, 2 on the commissure, the styles 1.5–2.5 mm. long, the stylopodium conical.——June–Aug. Grassy high mountain slopes; Hokkaido, Honshu; rather common.——Alaska, Aleutians, Kamchatka, Kuriles, and Sakhalin.

2. Heracleum moellendorffii Hance. *H. lanatum* sensu auct. Japon., pro parte, non Michx.; *H. inperpastum* Koidz.; *H. nipponicum* Kitag.; *H. lanatum* subsp. *moellendorffii*

(Hance) Hara——HANA-UDO. Resembles the preceding, the stems shorter, 70–100 cm. long, usually loosely pubescent; radical and lower cauline leaves petiolate, large, pinnately 3- to 5-foliolate, short-pubescent on the under side and on the petioles, the terminal leaflets rounded-cordate, deeply 3-fid, the lateral ones broadly ovate or deltoid-ovate, 2- or 3-fid, both 7–20 cm. long, the ultimate segments or lobes abruptly acuminate, incised and acutely dentate; umbels large, the rays 20–30, 7–10 cm. long, pubescent on the inner side, the involucre lacking or of few bracts, the involucel of linear-filiform bractlets, lanceolate at base, shorter than the pedicels, the pedicels 25–30, unequal, to 2 cm. long; fruit strongly flattened, nearly glabrous, the carpels nearly the same as in the preceding species.——May–June. Hills, low mountains and lowlands; Honshu (Kantō Distr. and westw.), Shikoku, Kyushu.

Var. **tsurugisanense** (Honda) Ohwi. *H. tsurugisanense* Honda; *H. barbatum* subsp. *moellendorffii* var. *tsurugisanense* (Honda) Hiroe——TSURUGI-HANA-UDO. A smaller mountain phase.

Var. **akasimontanum** (Koidz.) Ohwi. *H. lanatum* var. *akasimontanum* Hara; *H. akasimontanum* Koidz.——HOSOBA-HANA-UDO. A mountain phase with much-dissected leaflets and broadly linear ultimate segments.——Honshu (Akaishi Mts.).

Fam. 155. CORNACEAE   MIZU-KI KA   Dogwood Family

Trees or shrubs, sometimes herbs; leaves opposite or alternate, simple; stipules absent or pinnatifid; flowers small, in panicles or racemes, often capitate, perfect (or unisexual and the plants dioecious), actinomorphic; calyx-tube adnate to the ovary, the limb 4- or 5-lobed or nearly entire; petals 4 or 5, rarely absent, imbricate or valvate in bud; stamens as many as the petals, alternate with them, the anthers short, 2-locular, longitudinally dehiscent; disc nearly flat; ovary inferior, 1- to 4-locular; styles simple or lobed; ovules pendulous, solitary in each locule, anatropous, with 1 integument; drupe 1- to 4-stoned, the endosperm copious.——About 15 genera, with more than 100 species, chiefly in the N. Hemisphere.

1. HELWINGIA Willd. HANA-IKADA ZOKU

Glabrous deciduous dioecious shrubs; leaves simple, alternate, toothed, the stipules free, filiform; flowers few, small, fasciculate, borne on the midrib on the upper side of leaves, the pedicels short; calyx-lobes obsolete; petals 3–5, commonly 4, valvate in bud; stamens 3–5, the filaments rather thick; disc flat in the staminate, depressed-conical in the pistillate flowers; ovary 3- or 4-locular; styles connate, the stigmas 3–5; drupe berrylike, 1- to 4-stoned.——Three species, in e. Asia and the Himalayas.

1. Helwingia japonica (Thunb.) F. G. Dietr. *Osyris japonica* Thunb.; *H. rusciflora* Willd.——HANA-IKADA. Glabrous shrub about 2 m. high, with green branchlets; leaves ovate or elliptic, 3–10 cm. long, 2–6 cm. wide, acuminate, short awn-toothed, the petioles 2–4 cm. long; flowers pale green, 4–5 mm. wide, the staminate few, the pistillate 1–3, the pedicels 3–4 mm. long; petals 4(3–5), ovate-deltoid; fruit black, globose, about 7 mm. across; stones 2–4, oblong, 6–7 mm. long, with raised reticulations on surface.——May. Woods and thickets in hills and low mountains; Hokkaido (sw. distr.), Honshu, Shikoku, Kyushu; common.——Ryukyus.

2. AUCUBA Thunb. AO-KI ZOKU

Evergreen dioecious shrubs with rather thick branchlets; leaves opposite, simple, petiolate, toothed; stipules absent; flowers small, in terminal panicles; staminate flowers small, the calyx 4-toothed, the petals 4, valvate, ovate or lanceolate, the stamens 4, with short filaments; disc fleshy; pistillate flowers; small, the calyx-tube ovoid, with 4 teeth on the limb, the petals 4, disc fleshy, the style thick, short, with an entire stigma, the ovary 1-locular, the ovules pendulous; drupe ellipsoidal, red, rarely white or yellow; stones oblong, the endosperm fleshy.——Few species, in e. Asia and the Himalayas.

1. Aucuba japonica Thunb. *Eubasis dichotoma* Salisb.——AO-KI. Nearly glabrous evergreen shrub to 3 m. high, with thick terete green dichotomously forked branches; leaves oblong to ovate-oblong, rarely broadly lanceolate or elliptic, 8–20 cm. long, 2–10 cm. wide, black when dried, acute to acuminate with an obtuse tip, subacute at base, obtusely toothed, lustrous and deep green on upper side, the petioles 2–3 cm. long; flowers about 7 mm. across, the petals short-acuminate, reflexed, brown to dark purple, rarely green inside, the staminate in panicles 7–10 cm. long, the pistillate in pani- cles 1–2 cm. long; peduncles and pedicels short-pubescent; fruit ellipsoidal to ovoid-ellipsoidal, red, glabrous, lustrous, 15–20 mm. long.——Mar.–May. Woods in lowlands and mountains; Honshu (Kantō Distr. and westw.), Shikoku, Kyushu; common. Variable and with many cultivars grown in gardens.

Var. **borealis** Miyabe & Kudo. HIME-AO-KI. Plant smaller, usually minutely pubescent on petioles, under side of leaves, and on young branchlets.——Hokkaido and Japan Sea side of Honshu.

3. CORNUS L. MIZU-KI ZOKU

Deciduous, rarely evergreen, trees and shrubs or rarely evergreen herbs; leaves simple, petiolate, entire, alternate or opposite; inflorescence terminal in corymbs or umbels; involucres prominent or absent; flowers bisexual, small, usually white; calyx-tube urceolate or campanulate; petals 4, oblong, valvate in bud, deciduous; ovary 2-locular; styles columnar, simple, the stigma capitate, clavate, or truncate; ovules solitary in each locule; drupe globose or ellipsoidal, the stones solitary, 2-locular, 1-seeded in each locule, the testa membranous, the endosperm fleshy.——About 40 species, in temperate regions of the N. Hemisphere.

1. Cornus controversa Hemsl. *C. macrophylla* sensu auct. Japon., non Wall.; *C. ignorata* sensu Fr. & Sav., non K. Koch——MIZU-KI. Deciduous tree with grayish bark longitudinally fissured when old; branchlets glabrous, green, sometimes reddish; leaves alternate, broadly ovate to elliptic, 5–12 cm. long, 3–8 cm. wide, abruptly short acuminate, rounded to cuneate at base and slightly decurrent, green above, white be- neath, short appressed-hairy on upper side when young, persistent beneath, the lateral nerves of 5–8 pairs, the petioles reddish, glabrescent, 1.5–6 cm. long; inflorescence many-flowered, cymose, minutely puberulent; flowers white; calyx-tube with appressed white hairs; petals narrowly oblong, 4–5 mm. long; filaments about 5 mm. long, the anthers versatile; styles about half as long as the filaments, the stigma capitate-truncate; fruit globose, black, 6–7 mm. across, the stones pitted at tip.——May–June. Mountains; Hokkaido, Honshu, Shikoku, Kyushu; common.——Korea and China.

2. **Cornus brachypoda** C. A. Mey. *C. crispula* Hance; *C. macrophylla* sensu Rehd., pro parte, non Wall.——KUMANO-MIZU-KI. Resembles the preceding; branches glabrous or nearly so, becoming brownish when dry; leaves opposite, ovate to narrowly so or ovate-oblong, acuminate, often abruptly so, rounded to obtuse at base, whitish beneath, the lateral nerves of 6–8 pairs, the petioles 1.5–3.5 cm. long; inflorescence cymose, pedunculate; petals 4–5 mm. long, as long as to slightly longer than the filaments, anthers about 1.5 mm. long; styles shorter than the filaments, the stigma subglobose, truncate; fruit black, globose, about 5 mm. across, the stones unpitted.——June–July. Mountains; Honshu, Shikoku, Kyushu.——China.

3. **Cornus officinalis** Sieb. & Zucc. *C. mascula* var. *japonica* Sieb. ex Miq.; *Macrocarpium officinale* (Sieb. & Zucc.) Nakai——SANSHUYU, HARU-KOGANEBANA, AKI-SANGO. Deciduous tree; branches glabrous, terete, glaucous, with exfoliating bark; leaves ovate to narrowly so or narrowly elliptic, 4–10 cm. long, 2.5–5(–6) cm. wide, entire, acuminate with an obtuse tip, rounded at base, loosely appressed-pilose above while young, the hairs persistent beneath, also with axillary tufts of yellow-brown hairs on under side, the lateral nerves of 6 or 7 pairs, the petioles 6–10 mm. long; inflorescence 20- to 30-flowered, very short-pedunculate; involucral bracts 4, yellow-green, early deciduous, 6–8 mm. long, elliptic, acute; flowers appearing before the leaves, yellow, 4–5 mm. across, the pedicels about 10 mm. long, loosely appressed-pilose; fruit oblong, about 15 mm. long, red; stone smooth.——Apr. Cultivated in our area.——Korea and China.

4. **Cornus kousa** Buerg. ex Hance. *Benthamia japonica* Sieb. & Zucc.; *C. japonica* sensu G. Don; *Benthamidia japonica* (Sieb. & Zucc.) Hara; *Benthamia viridis* Nakai; *Benthamia kousa* (Buerg. ex Hance) Nakai; *Cynoxylon japonica* (Sieb. & Zucc.) Nakai; *C. kousa* (Buerg. ex Hance) Nakai; *Dendrobenthamia japonica* (Sieb. & Zucc.) Hutchins.——YAMABŌSHI. Deciduous tree with glabrescent rather slender branches; leaves elliptic to ovate-orbicular, 6–12 cm. long, 3.5–7 cm. wide, abruptly acuminate, abruptly acute to rounded at base, loosely appressed-hairy, green on upper side, paler and with axillary tufts of brown hairs beneath, the lateral nerves of 4 or 5 pairs, the petioles 5–10 mm. long; inflorescence terminal, the peduncles 5–10 cm. long, erect, loosely appressed-pilose while young; involucral bracts 4, spreading, longer than the inflorescence, persistent, narrowly ovate, 3–6 cm. long, acuminate, usually white, with few parallel nerves; flowers 20–30, in a dense globose head, yellow-green; fruit globose, red, fleshy, the syncarp 1–1.5 cm. across; stones ellipsoidal, smooth.——June–July. Mountains; Honshu, Shikoku, Kyushu; common.——Korea and China.

Cornus florida L. AMERIKA-YAMABŌSHI, of e. United States, is sometimes planted in parks and gardens.

5. **Cornus canadensis** L. *Chamaepericlymenum canadense* (L.) Asch. & Graebn.——GOZEN-TACHIBANA. Small evergreen herb with slender, long-creeping rhizomes; stems erect, slender, 5–15 cm. long, 4-angled, glabrous to sparsely appressed-pilose, usually unbranched; flowering stems with a pair of terminal leaves and with few smaller axillary leaves, thus appearing 6-leaved at the top of stem; leaves subsessile, narrowly obovate to rhombic-elliptic, 3–6 cm. long, 1–2.5 cm. wide, acute at both ends or acuminate at the base, appressed-pilose on both sides or only on upper surface, sometimes nearly glabrous on both sides, the lateral nerves rising near the base, of 2 or 3 pairs; peduncles erect, 1.5–3 cm. long; involucral bracts 4, broadly ovate, 7–10 mm. long, acute, white, persistent, 5- to 7-nerved; flowers 10–25, short-pedicelled; petals reflexed, about 1.5 mm. long, lanceolate-deltoid; petals usually white; ovary densely appressed-hairy; fruit globose, red, 5–6 mm. across, appressed-pilose; stones oblong, shallowly grooved.——June–July. Coniferous woods; Hokkaido, Honshu (centr. distr. and northw.), Shikoku (alpine); common.——Amur, Sakhalin, Kuriles, Kamchatka, Korea, and N. America.

6. **Cornus suecica** L. *Chamaepericlymenum suecicum* (L.) Asch. & Graebn.——EZO-GOZEN-TACHIBANA. Resembles the preceding; stems erect, 5–20 cm. long, simple or sparingly branched, 4-angled, glabrous to loosely appressed-pilose, with 4 or 5 distinct pairs of opposite leaves; leaves sessile, ovate-elliptic to oblong or sometimes broadly ovate, 1.5–3 cm. long, 1–2 cm. wide, subrounded to subacute, rounded to subacute at base, sparingly appressed-pilose on upper side, the lateral nerves of 2(–3)-pairs, rising near the base; inflorescence terminal and solitary, 10- to 20-flowered, the peduncles 1–2 cm. long; involucral bracts 4, broadly elliptic to ovate-orbicular, sometimes rhombic, 6–8 mm. long, subobtuse, white, 5- to 7-nerved; flowers short-pedicelled; petals purplish; ovary usually sparingly appressed-hairy; fruit globose, red, 5–6 mm. across, glabrous; stones slightly grooved.——July. Coniferous woods in lowlands; Hokkaido (e. distr.); rare.——Europe, Siberia, Korea, Manchuria, Sakhalin, Kuriles, Kamchatka, and N. America.

2. Metachlamydeae   Gamopetalae

Fam. 156. **DIAPENSIACEAE**   IWA-UME KA   Diapensia Family

Small shrubs or herbs; leaves simple, small and densely arranged on branchlets or loosely arranged and rather small; flowers bisexual, actinomorphic, solitary or in racemes or sometimes headlike, white, rose, or purple; calyx 5-lobed, persistent, the lobes imbricate in bud; corolla 5-lobed, the lobes imbricate; stamens adnate to and as many as the corolla-lobes and alternate with them, the scalelike or spathulate staminodia, when present, opposite the corolla-lobes; anthers 1- or 2-locular, usually longitudinally dehiscent, the disc absent; ovary superior, 3-locular, the ovules few to many, on axile placentae; style solitary, the stigma 3-lobed; capsule loculicidally dehiscent; seeds small, with fleshy endosperm; embryo cylindric.——About 6 genera, with 10 species, in Europe, Asia, N. America.

1A. Staminodia absent; densely matted subshrubs; leaves very small, 7–15 mm. long, entire, obovate-spathulate; flowers solitary and terminal. .... 1. *Diapensia*
1B. Staminodia alternate with the stamens, scalelike or spathulate-linear; typically not densely matted; leaves flattish, rather large, 2–6 cm. long, long-petiolate, toothed, ovate to depressed-orbicular; flowers in racemes, solitary or few. .... 2. *Shortia*

1. DIAPENSIA L.   Iwa-ume Zoku

Glabrous evergreen densely matted subshrubs with densely leafy branchlets; leaves coriaceous, alternate, spathulate, obtuse, entire; flowers solitary, terminal, erect, pedunculate, white or pink, sometimes purplish; calyx 2- or 3-bracteate at the base, the segments broadly ovate, obtuse; corolla campanulate-infundibuliform, 5-lobed; stamens 5, inserted on the throat of the corolla-tube, the filaments short, the anther-locules divergent, vertically dehiscent; staminodes absent; style filiform, the stigma obtusely and shortly 3-lobed; ovules numerous; capsules subglobose, 3-valved, the seeds numerous, nearly tetrahedral, small.——Few species, in the Himalayas, China, and the cooler regions of the N. Hemisphere.

1. Diapensia lapponica L. var. **obovata** F. Schmidt. *D. obovata* (F. Schmidt) Nakai; *D. lapponica* subsp. *obovata* (F. Schmidt) Hult.; *D. lapponica* sensu auct. Japon., non L. ——Iwa-ume.   Dwarf evergreen densely matted subshrub, 1–3 cm. high, glabrous; leaves thick, obovate-spathulate, 7–15 mm. long, 3–5 mm. wide, rounded to retuse, entire, long-decurrent and petiolelike at base, subclasping, lustrous above, yellow-green beneath, irregularly rugose on upper side, with 1 or 2 short impressed lateral nerves when dried, the midrib strongly impressed above; peduncles 1–2 cm. long, solitary, terminal on the branchlets, nearly naked; bracts 2 or 3, elliptic, shorter than the calyx-segments, the lowermost bracts often near the center of the peduncle; calyx-segments oblong, obtuse to rounded, 5–6 mm. long, green, erect; corolla white, 1–1.2 cm. across, the tube broad, the lobes elliptic-orbicular, ascending, entire, rounded at apex; styles rather persistent, narrowly cylindric, about 5 mm. long, truncate; capsules globose, erect, about 3 mm. across, surrounded by the persistent calyx.——June–July. Dry slopes and rocky cliffs in alpine regions; Hokkaido, Honshu (centr. and n. distr.); common. ——Sakhalin, Kuriles, Kamchatka, Alaska, and e. N. America.

2. SHORTIA Torr. & Gray   Iwa-uchiwa Zoku

Evergreen glabrous perennials with creeping branched stems; leaves radical, petiolate, ovate to depressed-orbicular, lustrous, undulately toothed; flowers in racemes, solitary or few, scapose, one-sided, bracteate, rose-colored or white; calyx-segments 5, obtuse, nerved, persistent, imbricate in bud; corolla-lobes 5, campanulate to infundibuliform, obtusely toothed to laciniate; anthers dorsally attached, the staminodes linear to scalelike or clavate; style columnar, obsoletely 3-lobed; ovules many; capsules globose, erect, dehiscent, 3-valved, surrounded by the persistent calyx; seeds many, small.——Few species, in e. Asia and N. America.

1A. Flowers solitary on the scape; corolla broadly campanulate, glabrous inside, the lobes deeply toothed; staminodes papillose, scalelike; calyx-segments glabrous, slenderly many-nerved. .... 1. *S. uniflora*
1B. Flowers few on the scape or peduncle; corolla narrowly infundibuliform, with white mammillate-hairs inside, the lobes laciniate; staminodes linear, inserted on the tube below the middle; calyx-segments ciliate while young, 4- to 7-nerved. .. 2. *S. soldanelloides*

1. Shortia uniflora (Maxim.) Maxim. *Schizocodon uniflorus* Maxim.——Iwa-uchiwa.   Stems slender, creeping, rather firm, sparingly branched; leaves few, approximate near the top of the stems and branches, long-petiolate, firm, slightly lustrous, depressed-orbicular to cordate-orbicular, 2.5–7 cm. long and as wide, rounded to retuse, cordate to rounded at the base, undulately toothed, with slightly raised nerves; scapes erect, to 15 cm. long, with few bud-scales at base, the bracts about 3, acute or subacute, broadly lanceolate, usually closely contiguous or the lowest widely separated from the calyx, slightly shorter than the calyx-segments; calyx segments green, oblong, glabrous, slenderly many-nerved, 8–12 mm. long, thick-papery; corolla pale rose-colored, 2.5–3 cm. across, glabrous, the lobes deeply and obtusely toothed; capsules ovoid-globose, 4–5 mm. across, mucronate, the valves somewhat coriaceous; seeds dark brown, oblong, about 0.7 mm. long, finely veined.——Apr.–May. Woods in mountains; Honshu (Kinki Distr. and eastw.).

2. Shortia soldanelloides (Sieb. & Zucc.) Makino. *Schizocodon soldanelloides* Sieb. & Zucc.; *Schizocodon soldanelloides* var. *genuinus* Makino; *Shortia soldanelloides* var. *genuina* (Makino) Makino——Iwa-kagami.   Glabrous perennial with creeping, loosely branched, slender stems and branches; leaves few, coriaceous, disposed mostly at end of stem, long-petiolate, orbicular to ovate-cordate, 3–6 cm. long, 3–5 cm. wide, deep green and lustrous on upper side, short-toothed; scapes 10–30 cm. long, with few bud-scales at base; racemes one-sided, 3- to 10-flowered, the bracts lanceolate or broadly linear, shorter than the calyx, the pedicels shorter than the flowers, erect after anthesis; calyx segments narrowly oblong, 5–10 mm. long, loosely ciliate near tip when young, very ob-

tuse, 4- to 7-nerved, becoming coriaceous-chartaceous in fruit; corolla rose-colored, narrowly infundibuliform, 1–1.5 cm. across, white-hairy at base inside, the lobes ascending, laciniate; capsules globose, 3–4 mm. across, the style persistent, 7–10 mm. long.——Apr.–June. Woods and thickets in mountains and sunny places in alpine regions; Hokkaido (w. distr.), Honshu, Shikoku, Kyushu.——Forma **alpina** (Maxim.) Makino. *Schizocodon soldanelloides* forma *alpinus* Maxim.——Ko-iwa-kagami.   Dwarf alpine phase.
Var. **minima** (Makino) Masam. *Shortia soldanelloides* forma *minima* Makino; *S. yakusimensis* Masam.——Hime-ko-iwa-kagami.   Very dwarf phase with few-toothed or entire leaves. Flowers mostly solitary.——High mountains of Yakushima.
Var. **magna** Makino. *Schizocodon soldanelloides* var. *magnus* (Makino) Hara; *Shortia magna* (Makino) Makino; *Schizocodon magnus* (Makino) Honda——Ō-iwa-kagami.   Larger montane phase with leaves 8–12 cm. long and as wide.—— Rather abundant in the western region.
Var. **ilicifolia** (Maxim.) Makino. *Schizocodon soldanelloides* var. *ilicifolius* (Maxim.) Makino; *S. ilicifolius* Maxim.; *Shortia ilicifolia* (Maxim.) Takeda——Hime-iwa-kagami. Leaves smaller, ovate-orbicular, with few, coarse, deltoid teeth nearly confined to the upper half.——Woods in mountains; Honshu, Kyushu; rather rare.
Var. **intercedens** Ohwi. *Schizocodon ilicifolia* var. *intercedens* (Ohwi) Yamazaki——Yama-iwa-kagami.   Leaves broadly ovate, abruptly narrowed at tip, whitish beneath, with 5–15 coarse teeth on each side; flowers white.——Woods in mountains; Honshu (Tōkaidō); rare.

Fam. 157. **CLETHRACEAE**   Ryōbu Ka   White Alder Family

Trees and shrubs with simple alternate exstipulate leaves; flowers bisexual, in terminal racemes or panicles; calyx 5-parted, the segments imbricate in bud, persistent; corolla lobes 5, free, imbricate in bud; stamens 10–12, free, the anthers reflexed in bud, sagittate, opening by pores; disc absent; ovary superior, 3-locular; style solitary, 3-lobed; ovules many, the placentae axile; capsules globose, loculicidally dehiscent, 3-valved, the axis persistent; seeds many, flat or 3-angled, sometimes winged; endosperm fleshy, the embryo cylindric.——Two genera, with about 30 species, in e. Asia, N. America, and Madeira.

1. **CLETHRA** L.   Ryōbu Zoku

Evergreen or deciduous trees or shrubs, usually stellate-pubescent; leaves short-petiolate, usually toothed; flowers white or pink.——About 30 species, in N. America, e. Asia, and Madeira.

1. Clethra barbinervis Sieb. & Zucc.   *C. kawadana* Yanagita; *C. barbinervis* var. *kawadana* (Yanagita) Hara; *C. repens* Nakai——Ryōbu.   Deciduous tree with smooth, yellow-brown bark, the young branchlets glabrous or minutely stellate-pubescent; leaves herbaceous to chartaceous, alternate, often approximate toward the top of branchlets, broadly oblanceolate to narrowly cuneate-obovate, abruptly acuminate to acute, acutely toothed, green and glabrous or sparingly stellate-pubescent on upper side, paler beneath and with loosely appressed hairs, especially on the nerves, the axillary hairs tufted, the lateral nerves 8 to 15 pairs; inflorescence 8–15 cm. long, more or less paniculate, many-flowered, densely stellate-pubescent; flowers white, pedicelled, 6–8 mm. across; calyx-segments ovate-elliptic, pilose; petals elliptic, rounded at apex, irregularly minutely toothed; fruit depressed, erect, exceeding the calyx, long-hairy, 4–5 mm. across; seeds elliptic, flat, about 1 mm. long, narrowly margined, with raised reticulations on the surface.——July–Sept. Hills and mountains; Hokkaido, Honshu, Shikoku, Kyushu; common and very variable.——Korea (Quelpaert Isl.).

Fam. 158. **PYROLACEAE**   Ichi-yaku-sō Ka   Shinleaf Family

Small shrubs or herbs sometimes saprophytic; leaves simple, alternate, rarely verticillate or opposite, sometimes scalelike and without chlorophyll, without stipules; flowers bisexual, actinomorphic, solitary or in terminal racemes or corymbs; calyx 5-parted, rarely absent; petals 5, sometimes 4; stamens 10, sometimes 8, the anthers opening by pores; ovary 5- or 4-locular; superior, many-ovuled; style and stigma simple; fruit a 5- or sometimes 4-valved capsule, rarely a berry; seeds minute, numerous, the testa transparent, the endosperm copious.——About 11 genera, with about 50 species, chiefly in the temperate regions of the N. Hemisphere.

1A. Plants green; anthers opening by pores, reflexed in bud.
  2A. Leaves radical, ovate, elliptic or depressed-orbicular, long-petiolate; capsules dehiscent from the base; herbs.
    3A. Flowers in racemes, not fully expanded; valves of capsules with webby pubescence on margin. . . . . 1. *Pyrola*
    3B. Flowers solitary, fully expanded; valves of capsules without webby pubescence on margin. . . . . 2. *Moneses*
  2B. Leaves cauline, lanceolate or narrowly oblong, short-petiolate; capsules dehiscent from the apex; flowers in corymbs or umbels, not fully expanded; subshrubs. . . . . 3. *Chimaphila*
1B. Plants saprophytic, without chlorophyll, with scalelike leaves; anthers vertically dehiscent, erect in bud.
  4A. Fruit a capsule; ovary 5-locular, with axile placentae. . . . . 4. *Monotropa*
  4B. Fruit a berry; ovary 1-locular, with parietal placentae. . . . . 5. *Monotropastrum*

1. **PYROLA** L.   Ichi-yaku-sō Zoku

Glabrous evergreen perennial herbs, often stoloniferous; leaves usually radical, long-petiolate, entire or toothed; flowers in simple racemes on nearly naked scapes, bracteate, nodding, white, reddish or purplish, rarely yellowish; calyx persistent, deeply 5-parted; petals 5, not fully expanded, connivent at tip, deciduous; stamens 10, erect, the anthers reflexed in bud, 4-locular, opening by pores; ovary globose, 5-locular; style simple, erect or descending, the stigma depressed, obsoletely 5-lobed; ovules numerous, on the spongy placentae of the persistent axis, disc absent, or 10-toothed; capsules depressed-globose, 5-locular, dehiscent; seeds many, small, the testa produced at both ends, the endosperm fleshy.——More than 20 species, in the temperate regions of the N. Hemisphere.

1A. Inflorescence glabrous; flowers in loose racemes, 1–1.5 cm. across; pores of anthers short-tubulose; disc absent; pollen in tetrads.
  2A. Style descending, 6–10 mm. long.
    3A. Leaves reniform-orbicular, with overlapping basal margins; scapes usually completely naked except at base; calyx-segments suborbicular, very obtuse, much shorter than wide. . . . . 1. *P. renifolia*
    3B. Leaves elliptic to depressed-orbicular, scarcely cordate; scapes usually with few scalelike leaves; calyx-segments deltoid to lanceolate, acuminate to acute, longer or as long as wide.
      4A. Calyx-segments deltoid or ovate-deltoid, as long as to only slightly longer than wide; slender plants with few-flowered scapes.
        5A. Leaves depressed-orbicular, slightly shorter than wide; bracts lanceolate, only slightly longer than to nearly as long as the pedicel. . . . . 2. *P. nephrophylla*
        5B. Leaves elliptic, slightly longer than wide; bracts broadly linear, slightly shorter than to slightly longer than the pedicel. 3. *P. alpina*
      4B. Calyx-segments broadly lanceolate, 2½–4 times as long as wide; stouter plants sometimes with 10-flowered scapes.

6A. Flowers white, bracts deltoid-lanceolate, gradually narrowed at tip, very acute. . . . . 4. *P. japonica*

6B. Flowers deep rose-colored; bracts oblong-lanceolate, abruptly acute. . . . . 5. *P. incarnata*

2B. Style straight, not curved, 3–4 mm. long; calyx-segments elliptic, subacute; leaves elliptic; bracts broadly lanceolate, acute.

6. *P. faurieana*

1B. Inflorescence minutely papillose; leaves ovate, subacute to obtuse, subverticillate on lower creeping part of stems; flowers in a one-sided rather dense raceme, about 5 mm. across, greenish; pores of anthers not tubulose; disc 10-lobed; pollen grains single.

7. *P. secunda*

1. **Pyrola renifolia** Maxim. *P. sodanellifolia* H. Andres. ——JIN-YŌ-ICHI-YAKU-SŌ. Rhizomes very slender, long-creeping; leaves 1–3, approximate, dull, deep green but paler along the nerves above, pale green beneath, reniform-orbicular or subreniform, 1–3 cm. long, 1.5–4 cm. wide, obsoletely toothed, rounded at apex, the petioles 2–5 cm. long; scapes 10–20 cm. long, rarely with a minute scale; flowers 2–4, 1–1.2 cm. across, white to greenish, the bracts linear-lanceolate, 1–2 mm. long, much shorter than the pedicel; calyx-segments suborbicular or depressed-deltoid, obtuse to rounded at the tip; style exserted; capsules 5–6 mm. across.——Damp coniferous woods in mountains; Hokkaido, Honshu (centr. and n. distr.).——s. Kuriles, Sakhalin, Korea, Manchuria, and Amur.

2. **Pyrola nephrophylla** (H. Andres) H. Andres. *P. media* sensu auct. Japon., non Sw.; *P. rotundifolia* 3. *nephrophylla* H. Andres——MARUBA-NO-ICHI-YAKU-SŌ. Rhizomes slender, long-creeping; leaves 2–5, approximate, depressed-orbicular, 1–2.5 cm. long, 1.5–3.5 cm. wide, rounded at apex, rounded-truncate or slightly cordate at base, obsoletely toothed, the petioles 2–5 cm. long; scapes 10–20 cm. long, 1- to 3-scaled, the scales broadly lanceolate, membranous, 7–12 mm. long, acuminate; flowers 4–7, rarely to 10, white, 1–1.2 cm. across, the bracts lanceolate, membranous, acuminate, 4–7 mm. long, nearly as long as the pedicels; calyx-segments deltoid, acuminate or very acute; styles exserted; capsules 4–6 mm. across. ——Damp woods in mountains; Hokkaido, Honshu, Shikoku, Kyushu.——s. Kuriles.

3. **Pyrola alpina** H. Andres. *P. elliptica* var. *minor* Maxim.; *P. maximowicziana* Makino——KOBA-NO-ICHI-YAKU-SŌ. Rhizomes slender, long-creeping; leaves 4–8, approximate or nearly so, deep green, elliptic, obsoletely toothed, rounded to abruptly mucronate at apex, rounded to subtruncate, rarely somewhat acute at base, the petioles 1–3 cm. long; scapes 10–20 cm. long, erect, the scales linear-lanceolate, acuminate, 3–7 mm. long, sometimes absent; flowers 3–5, rarely to 7, white, 12–15 mm. across, the bracts broadly linear, gradually acuminate, 3–5 mm. long; calyx-segments deltoid, acuminate; styles exserted; capsules 4–5 mm. across.——July–Aug. Woods in mountains; Hokkaido, Honshu.

4. **Pyrola japonica** Klenze ex Alef. *P. asarifolia* var. *japonica* Miq.; *P. rotundifolia* var. *albiflora* Maxim.; *P. elliptica* var. *intermedia* H. Boiss.; *P. denticulata* Koidz.; *P. incarnata* var. *japonica* (Klenze ex Alef.) Koidz.——ICHI-YAKU-SŌ. Rhizomes creeping; leaves 1–5, rarely to 8, approximate, elliptic, 4–7 cm. long, 2.5–4.5 cm. wide, rounded to obtuse at apex, rounded to subacute at base, deep green with paler nerves on upper side, rather loosely and obsoletely toothed, the petioles 3–8 cm. long; scapes 15–30 cm. long, angled, 5- to 12-flowered, often with 1 or sometimes 2 scales near the middle, the bracts linear-lanceolate, gradually narrowed from the base, long-acuminate, 5–8 mm. long, with a raised costa on the back, longer than to as long as the pedicels; flowers white, 12–15 mm. across; calyx-segments broadly lanceolate to narrowly ovate, 2.5–3 times as long as wide, acute; styles exserted; capsules 7–8 mm. across.——June–July. Woods in foothills; Hokkaido, Honshu, Shikoku, Kyushu.——Korea, Manchuria,

and Formosa.——Forma **subaphylla** (Maxim.) Ohwi. *P. subaphylla* Maxim.; *P. rotundifolia* var. *incarnata* forma *subaphylla* (Maxim.) Makino; *P. japonica* var. *subaphylla* (Maxim.) H. Andres——HITOTSUBA-ICHI-YAKU-SŌ. Leaves mostly reduced to scales. Occurs rarely with the typical phase.

5. **Pyrola incarnata** Fisch. *P. rotundifolia* var. *purpurea* Bunge; *P. rotundifolia* var. *incarnata* DC.; *P. asarifolia* var. *incarnata* (Fisch.) Fern.; *P. tokugoensis* Nakai——BENIBANA-ICHI-YAKU-SŌ. Rhizomes creeping, the plants forming an extensive loose mat; leaves 3–5, approximate, elliptic or ovate-elliptic, rounded to very obtuse, rounded or scarcely cordate, rarely subacute at base, lustrous above, deep green, obsoletely short-toothed, the petioles 3–5 cm. long; scapes 10–25 cm. long, angled, the scales 1–3, broadly lanceolate to narrowly oblong, 7–10 mm. long; flowers 7–15, deep rose-colored, 12–15 mm. across, the bracts broadly lanceolate, membranous, acuminate, 5–8 mm. long, longer than the pedicels; calyx-segments narrowly ovate to broadly lanceolate, acuminate, 3–4 times as long as wide; styles exserted; capsules 7–8 mm. across.——June–Aug. Thickets and shaded grassy places in mountains; Hokkaido, Honshu (centr. and n. distr.); locally abundant. ——Kuriles, Kamchatka, e. Siberia, Korea, Manchuria, n. China, and N. America.

6. **Pyrola faurieana** H. Andres. *P. minor* sensu auct. Japon., non L.——KARAFUTO-ICHI-YAKU-SŌ. Rhizomes creeping, slender; leaves 3–7, approximate, ovate-elliptic, ovate-orbicular, or broadly elliptic, 2.5–3.5 cm. long, 2–3 cm. wide, rounded and mucronulate at apex, rounded or shallowly cordate at base and abruptly decurrent, obsoletely short-toothed, the petioles 2–4 cm. long; scapes 10–20 cm. long, angular, naked except for a few basal scales; racemes rather densely 7- to 17-flowered; flowers white, sometimes pinkish, about 6 mm. across, the bracts lanceolate, membranous, 4–6 mm. long; calyx-segments elliptic or broadly ovate, subacute, about 1.5 times as long as wide; styles thick, scarcely curved; anthers short.——July–Aug. Hokkaido, Honshu (mountains in n. distr.); rare.——s. Kuriles and Sakhalin.

7. **Pyrola secunda** L. *Orthilia parviflora* Raf.; *Ramischia secunda* (L.) Garcke; *O. secunda* (L.) House——KO-ICHI-YAKU-SŌ. Rhizomes creeping, with few whorls of false leaf verticels in 3's or 4's; leaves thinly coriaceous to herbaceous, ovate to broadly so or sometimes elliptic, 1.5–3 cm. long, 1–2 (–2.5) cm. wide, acute to subobtuse, rounded at base, the petioles 1–1.5 cm. long; scapes slender, 7–12 cm. long, loosely and minutely papillose, the scales 1–3, narrowly ovate or broadly lanceolate, 2–5 mm. long, rather densely 8- to 15-flowered; flowers greenish white, one-sided, about 5 mm. across, the bracts membranous, broadly oblanceolate or broadly lanceolate, 3–5 mm. long, abruptly acute, the margins somewhat hyaline, the pedicels slender, rather prominently papillose; calyx-segments depressed-orbicular, minutely toothed, about as long as wide; styles long-exserted; capsules about 4 mm. across.——July–Aug. Damp coniferous woods; Hokkaido, Honshu (centr. and n. distr.).——Widely distributed in boreal areas of the Northern Hemisphere.

2. MONESES Salisb.   ICHIGE-ICHI-YAKU-SŌ ZOKU

Small evergreen glabrous stoloniferous perennial; leaves radical, petiolate, somewhat coriaceous; flowers solitary, terminal on scapes, white or pink; calyx persistent; stamens 10 or 8, the filaments slender, glabrous, the anthers erect, reflexed in bud, 2-locular, with 2 short tubular pores at apex, disc 10- or 8-lobed; ovary 5-locular, superior, globose; style erect, columnar, slightly thickened at apex; ovules on thickened placentae on the inner angle of the locule; capsules subglobose, 5-ridged, 5-locular, 5-valved, the valves septate; seeds numerous, small, with a netlike loose testa at both ends.——One species of wide distribution in the N. Hemisphere.

1. Moneses uniflora (L.) A. Gray. *Pyrola uniflora* L.; *M. grandiflora* S. F. Gray——ICHIGE-ICHI-YAKU-SŌ. Leaves 2–4 at the apex of the rhizomes, ovate-orbicular or nearly orbicular, 1–1.5 cm. long, rounded to very obtuse, subrounded to broadly cuneate at base, minutely toothed, shortly decurrent on the petioles, deep green and slightly lustrous, paler beneath, the petioles 6–8 mm. long; scapes 5–10 cm. long, loosely 1- to 2-scaled in upper part, the scales membranous, oblong to obovate, 2–3 mm. long, obtuse, ciliolate; flowers white, about 2 cm. across, nodding; petals 5 or 4, orbicular, sessile, spreading; calyx-segments 5(sometimes 4), ovate-elliptic, about 2.5 mm. long, rounded at apex, ciliolate; style 4–5 mm. long, the stigma 5-lobed; capsules erect, about as long as the style, depressed obovate-globose.——July. Woods; Hokkaido; very rare.——Sakhalin, Korea, China, Siberia to Europe, and N. America.

3. CHIMAPHILA Pursh   UMEGASA-SŌ ZOKU

Herbs or subshrubs; leaves petiolate, evergreen, falsely verticillate, lanceolate to narrowly ovate, coriaceous, lustrous, toothed; flowers solitary or in terminal corymbs, white or purplish, bracteolate; calyx 5-lobed, persistent, the segments obtuse; petals 5, orbicular, concave, spreading to reflexed, sessile, deciduous; stamens 10, the filaments short, dilated and ciliate on the lower half, the anthers 2-locular, opening by the terminal pores; style short, obconical, the stigma rounded, entire or with 5 undulate teeth; ovules very many; capsules subglobose, 5-locular; seeds small, the testa loosely reticulate at both ends.——Few species, in Europe, N. America, and Asia.

1A. Leaves broadly lanceolate, broadest below the middle, toothed nearly to the base; flowers 1 or sometimes 2; bracts broadly lanceolate; calyx-segments narrowly ovate to broadly lanceolate, 2–4 times as long as wide, nearly as long as the corolla. ...... 1. *C. japonica*
1B. Leaves oblanceolate, broadest below the apex, long-cuneate below, toothed only on upper half; flowers 3–9; bracts broadly linear; calyx-segments ovate-orbicular, about as long as wide, ¼ as long as the corolla. .... 2. *C. umbellata*

1. Chimaphila japonica Miq. *C. astyla* Maxim.—— UMEGASA-SŌ. Subshrub; stems simple or scarcely branched, erect, terete, 5–10 cm. long, slender, few-leaved, glabrous; leaves opposite or in 3's, alternating with the scalelike leaves, coriaceous, rather lustrous, broadly lanceolate, 2–3.5 cm. long, 6–10 mm. wide, acute or rarely obtuse, mucronate, rounded to subacute at base, deep green with paler midrib above, glabrous, few-toothed, the petioles 6–8 mm. long; peduncles erect, with 1 or 2 bracts on the upper portion, papillose; flowers 1 or sometimes 2, nodding, white, about 1 cm. across, not fully open; calyx-segments membranous, 6–7 mm. long, irregularly toothed; corolla-lobes obovate-orbicular, 7–8 mm. long; capsules depressed-globose, about 5 mm. across, the stigma sessile.——June–July. Dry woods in hills and low elevations in the mountains; Hokkaido, Honshu, Shikoku, Kyushu.——Sakhalin, s. Kuriles, Korea, Manchuria, and China.

2. Chimaphila umbellata (L.) Barton. *Pyrola umbellata* L.; *C. corymbosa* Pursh——Ō-UMEGASA-SŌ. Low subshrub; stems 5–12 cm. long, rather slender, erect, simple or sparsely branched, glabrous; leaves verticillate, rather many, thick-coriaceous, cuneate-oblanceolate, 3–5 cm. long, 6–10 mm. wide, obtuse to subacute, lustrous, glabrous, few-toothed above the middle, the nerves impressed above, the petioles 3–6 mm. long; peduncles 5–10 cm. long, papillose, 3- to 9-flowered, the pedicels erect, 1.5–2.5 cm. long, the bracts broadly linear, 3–4 mm. long; flowers white, sometimes roseate, about 8 mm. across, not fully open; calyx-segments ovate-orbicular, about 2 mm. long, irregularly toothed; corolla-lobes suborbicular; capsules depressed globose, about 6 mm. across; stigma nearly sessile.——July. Dry woods especially near the sea; Hokkaido, Honshu (Hitachi Prov. and northw.); rare.——Sakhalin to Europe, and N. America.

4. MONOTROPA L.   SHAKUJŌ-SŌ ZOKU

Saprophytic, fleshy, usually white or reddish, black when dry; stems simple, terete, with alternate scalelike leaves; flowers solitary or in a terminal raceme, nodding; calyx-lobes 4 or 5, deciduous; petals 4 or 5, longer than the sepals, deciduous, oblong, often saccate at base; stamens 8–10, sometimes 12, erect, filiform-linear, the anthers short; disc adnate to the base of ovary, 8–10 or rarely 12-toothed; capsules 4- or 5-locular, erect, loculicidally dehiscent, the valves septate within, the placentae remaining on the persistent axis; style columnar, erect, the stigma infundibuliform; seeds small, very many.——Few species, in the temperate regions of the N. Hemisphere.

1A. Flowers terminal and solitary; style thick and short. .... 1. *M. uniflora*
1B. Flowers in a terminal raceme; style longer than the ovary. .... 2. *M. hypopitys*

1. Monotropa uniflora L. *M. uniflora* var. *nipponica* Makino; *M. morisoniana* Michx.; *Hypopitys uniflora* (L.) Crantz——GINRYŌ-SŌ-MODOKI, YŪREI-SŌ-MODOKI, AKI-NO-GIN-RYŌ-SŌ. White, fleshy, saprophytic herb, becoming black when dry; stems erect, simple, terete, 10–30 cm. long, with leaflike scales; scales suberect, narrowly oblong or narrowly ovate-oblong, 1–2 cm. long, 5–8 mm. wide, obtuse to rounded, rather fleshy, glabrous or only slightly pubescent under the flower; flowers terminal, solitary, tubular-campanulate, nodding, about 2 cm. long; sepals deciduous, scalelike; petals 5, fleshy, cuneate or obovate-oblong, erect, irregularly denticulate in upper part, usually long-pubescent inside, gibbous, deciduous; capsules ellipsoid-globose, about 12 mm. long, erect; style thick, 3–4 mm. long, the stigma glabrous.——Aug.–Oct. Damp woods in hills and mountains; Honshu, Shikoku, Kyushu.——N. America, e. Asia, and India.

2. Monotropa hypopitys L. *Hypopitys multiflora* Scop.; *H. monotropa* Crantz; *M. hypopitys* var. *japonica* Fr. & Sav.——SHAKUJŌ-SŌ. Pale yellow-brown, rather fleshy saprophytic herb becoming black when dry, usually pubescent on stems and inflorescence; scales 20–30, suberect, ovate-oblong or broadly lanceolate, 1–1.5 cm. long, 5–7 mm. wide, usually acuminate, the upper ones often irregularly toothed, rather fleshy; flowers few, tubular-campanulate, 1–1.5 cm. long, in a nodding raceme; petals cuneate-oblong, deciduous after anthesis, rather fleshy, saccate at base, rounded to obtuse and irregularly toothed at apex; capsules ellipsoid-globose, 5–7 mm. long; style rather thick, 3–5 mm. long.——July. Woods in mountains; Hokkaido, Honshu, Shikoku, Kyushu.——Europe, Siberia, Sakhalin, s. Kuriles, China, Formosa, Korea, Manchuria, and N. America.

5. MONOTROPASTRUM H. Andres   GINRYŌ-SŌ ZOKU

Usually white, saprophytic, fleshy herbs; stems simple, scaly; flowers tubular-campanulate, nodding, solitary; calyx of separate, scalelike lobes which simulate the upper scalelike leaves; petals 3–5, fleshy, erect, saccate at base, slightly longer than the sepals; stamens about 10, erect, the filaments linear, usually pubescent, the anther-locules horizontally divergent, tranversely split; disc 10-lobed, adnate to the base of the ovary; ovary ovoid, 1-locular, with parietal placentation; style very short, thick, the stigma infundibuliform, indigo-blue on the margin, depressed in the center; berry nodding, ovoid-globose; seeds very many, small.——Few species, in e. Asia and India.

1. Monotropastrum globosum H. Andres. *Monotropa uniflora* sensu auct. Japon., pro parte; *Monotropa uniflora* var. *tripetala* Makino; *Monotropastrum globosum* var. *tripetalum* (Makino) Honda——GINRYŌ-SŌ, YŪREI-TAKE, MARUMI-NO-GINRYŌ-SŌ. Usually white, rarely somewhat roseate, fleshy herb becoming black when dry; stems erect, 10–20 cm. long, terete, often long-pubescent in upper part; scales 10–20, suberect, alternate, oblong, narrowly or broadly ovate, 7–20 mm. long, 5–10 mm. wide, rounded to obtuse, sometimes denticulate; flowers terminal, 15–25 mm. long; sepals 1–3, oblong; petals 3–5, cuneate-oblong, 1.5–2 cm. long, 5–10 mm. wide, rounded, saccate at base, white pubescent inside; filaments white-pubescent, the anthers papillose; style short, the stigma dilated, 3–5 mm. across; berry erect, ovoid-globose, abruptly narrowed to the very short thick style; seeds ovoid, small.——Apr.–Aug. Damp woods in hills and mountains; Hokkaido, Honshu, Shikoku, Kyushu.——Sakhalin, s. Kuriles, Ryukyus, and Korea.

Fam. 159. ERICACEAE   TSUTSUJI KA   Heath Family

Shrubs or small trees with alternate, rarely opposite or verticillate, deciduous or evergreen, simple, entire or toothed exstipulate leaves; flowers bisexual, actinomorphic or slightly zygomorphic, solitary, in racemes or panicles, or sometimes umbellate; calyx persistent, usually 4- or 5-lobed; corolla usually gamopetalous, rarely of separate petals, infundibuliform, campanulate, or urecolate, the lobes 4 or 5 or rarely fewer or many, imbricate or contorted in bud; stamens as many or twice as many as the corolla-lobes, inserted at the base of disc; anthers 2-locular, often appendaged, opening by pores or very rarely longitudinally dehiscent; ovary superior, sometimes inferior, 2- to 5-locular, the ovules many; style and stigma single and solitary; fruit a capsule or berry; seeds small, numerous, with endosperm; embryo small, in the center of the seed.——About 70 genera, with more than 1,500 species, chiefly in the temperate and cooler regions of both hemispheres and in mountains of the Tropics.

1A. Ovary superior.
  2A. Fruit a septicidal capsule; seed-coat loose, prominently striate, the seeds often winged; anthers without appendage.
    3A. Petals free.
      4A. Petals 3; stamens 6; ovary 3-locular; leaves deciduous, glandless; flowers in racemes or panicles. . . . . 1. *Tripetaleia*
      4B. Petals 5; stamens 5–10; ovary 5-locular; leaves evergreen, with sessile discoid glands; flowers in umbellate corymbs. . . 2. *Ledum*
    3B. Petals united.
      5A. Leaves ovate to lanceolate, flat.
        6A. Flowers tubular-campanulate or urceolate, 4-merous, nodding; stamens not exserted. . . . . 3. *Menziesia*
        6B. Flowers infundibuliform, rotate or sometimes tubular-campanulate, mostly 5-merous; stamens usually exserted.
                                            4. *Rhododendron*
      5B. Leaves usually linear, with prominently recurved or revolute margins.
        7A. Leaves alternate.
          8A. Flowers solitary and terminal, the corolla rotate, deeply 4-parted; stamens 8; capsules 4-valved. . . . . 5. *Bryanthus*
          8B. Flowers umbellate and terminal, the corolla urceolate or campanulate, shallowly 5-lobed; stamens 10; capsules with 5 bilobed valves. . . . . 6. *Phyllodoce*
        7B. Leaves opposite; flowers subumbellate; corolla infundibuliform-campanulate; anthers longitudinally split. . . . . 7. *Loiseleuria*
  2B. Fruit a loculicidal capsule or rarely berrylike.
    9A. Fruit a capsule, rarely berrylike.
      10A. Leaves opposite or ternate.
        11A. Leaves opposite, decussate, scalelike, closely appressed to the branches and branchlets; flowers axillary. . . . . 8. *Cassiope*
        11B. Leaves ternate or partly opposite, small, flat, spreading, rather loosely arranged on the stems and branches; flowers in a terminal verticil of 3 or in short racemes. . . . . 9. *Arcterica*

1. **TRIPETALEIA** Sieb. & Zucc.   Ho-tsutsuji Zoku

Deciduous shrubs with entire, short-petiolate, alternate leaves; flowers in terminal panicles or racemes, white, sometimes roseate, bracteate and bracteolate; calyx small and cuplike with the margin toothed or of free segments; petals 3, free, reflexed; stamens 6, the filaments often dilated, the anthers longitudinally split above, without an appendage; ovary 3-locular, the ovules many in each locule; style slender, more or less curved upward, the stigma obsoletely 3-lobed; capsules septicidally dehiscent, often short-stipitate; seeds ovoid or ellipsoidal.——Two species, in Japan.

1A. Sepals free, 4–5 mm. long; capsules nearly sessile, 4–5 mm. across; inflorescence a few-flowered raceme; branches nearly terete, obtusely striate; leaves rounded to obtuse; bracts foliaceous, 5–10 mm. long. .... 1. *T. bracteata*

1B. Sepals united, the calyx cup-shaped, 1–2 mm. long and wide, with 5 undulate-deltoid teeth; capsules distinctly stipitate, about 3 mm. across; inflorescence a rather many-flowered panicle; branches acutely angled; leaves usually acute, the tip obtuse; bracts 2–4 mm. long. .... 2. *T. paniculata*

1. Tripetaleia bracteata Maxim. *Elliottia bracteata* (Maxim.) Benth. & Hook. f.; *Botryostege bracteata* (Maxim.) Stapf——Miyama-ho-tsutsuji, Hako-tsutsuji. Much-branched shrub with red-brown, nearly terete, obtusely striate branchlets; leaves obovate, 2.5–6 cm. long, 12–20 mm. wide, rounded to very obtuse and mucronate, subcuneate at base, glabrous or slightly ciliate while young, the petioles 1–2 mm. long; racemes 3- to 8-flowered, 6–10 cm. long, inclusive of the peduncles, puberulous, the bracts elliptic or obovate, green and foliaceous, 5–10 mm. long, rounded to very obtuse; flowers greenish white; sepals acute or obtuse, slightly ciliolate; petals oblong, recurved, 8–10 mm. long.——July–Aug. High mountains; Hokkaido, Honshu (centr. and n. distr.); common northw.

2. Tripetaleia paniculata Sieb. & Zucc. *T. paniculata* var. *angustifolia* Maxim.; *Gautiera triquetra* Sieb. & Zucc. pro parte; *Elliottia paniculata* (Sieb. & Zucc.) Benth. & Hook. f.; *T. yakushimensis* Nakai; *T. paniculata* var. *yakushimensis* (Nakai) Kitam.; *T. paniculata* var. *latifolia* Maxim.——Ho-tsutsuji. Erect branched shrub with red-brown, 3-angled, glabrous branchlets; leaves obovate, to narrowly ovate-elliptic, 3–8 cm. long, 1.5–3 cm. wide, acute to obtuse, acute at base, entire, spreading-puberulent beneath, the midrib puberulent on upper and lower side, the petioles 1–2 mm. long; panicles 7–15 cm. long inclusive of the peduncle, rather densely many-flowered, puberulent, the bracts subulate-linear, 2–4 mm. long; flowers small, white to roseate; calyx cup-shaped, about 2 mm. across; petals 6–8 mm. long; fruit about 3 mm. across, the stipe short, about 1 mm. long.——Aug.–Sept. Mountains; Hokkaido, Honshu, Shikoku, Kyushu; rather common.

2. **LEDUM** L.   Iso-tsutsuji Zoku

Small decumbent or ascending glandular shrubs; leaves evergreen, alternate, short-petiolate, usually rusty-pubescent, rather coriaceous, linear to lanceolate, entire with recurved or revolute margins; flowers many, white, in terminal sessile corymbs, the pedicels bracteate; calyx small, 5-toothed; petals 5, imbricate in bud, free; stamens 5 to 10, exserted, the filaments filiform, the anthers small; disc small, annular, 8- to 10-lobed; ovary 5-locular, densely glandular-dotted; styles filiform, the stigma 5-lobed; capsules ellipsoidal; seeds many.——Few species, in the cooler parts of the N. Hemisphere.

1. Ledum palustre L. var. **diversipilosum** Nakai. *L. palustre* var. *dilatatum* sensu auct. Japon., non Wahlenb.; *L. palustre* var. *nipponicum* Nakai, and var. *yesoense* Nakai; *L. palustre* var. *groenlandicum* sensu Nakai, non Rosenvinge; *L. hypoleucum* Komar.——Iso-tsutsuji. Rusty-tomentose, erect, aromatic shrub, 30–50 cm. high; leaves coriaceous, lanceolate, 3–5 cm. long, 5–12 mm. wide, obtuse to acute, sometimes mucronate, deep green, entire, with yellow sessile glands especially while young, usually glaucous and white-papillose beneath, recurved on margin, the petioles 1–5 mm. long; corymbs densely many-flowered, terminal and sessile, densely puberulent, often glandular-dotted, the pedicels 1–3 cm. long, the bracts ovate, acute, caducous; flowers white, 8–10 mm. across; petals elliptic, glabrous; capsules oblong, 5-valved, obtuse at both ends, 4–5 mm. long, glandular-dotted, nodding; styles nearly as long as to slightly shorter than the capsules. ——June–July. Wet boggy slopes and high moors; Hokkaido, Honshu (n. distr.); gregarious.——Sakhalin, s. Kuriles, Korea, and e. Siberia.

Var. **decumbens** Ait. *L. palustre* var. *angustifolium* Hook.; *L. decumbens* (Ait.) Lodd. ex Steud.; *L. palustre* subsp. *decumbens* (Ait.) Hult.——HIME-ISO-TSUTSUJI. Low decumbent shrub with broadly linear, obtuse, revolute leaves 1–2 cm. long, 2–3 mm. wide, the petioles 1–2 mm. long, the pedicels slightly shorter; flowers smaller.——July. High moors in alpine regions; Hokkaido (Mount Daisetsu).——Kuriles, Korea, Siberia, and N. America.

3. MENZIESIA J. E. Smith YŌRAKU-TSUTSUJI ZOKU

Glabrous or pilose deciduous shrubs; leaves alternate, petiolate, elliptic, entire, membranous; flowers rather small, in terminal umbels or abbreviated racemes, nodding, white, greenish, or rose-purple, the pedicels with a caducous bract at base; calyx 4- or 5-parted; corolla urceolate, campanulate, or tubular-globose, 4- to 5-lobed, the lobes imbricate in bud; stamens as many or twice as many as the corolla-lobes, not exserted, the anthers unappendaged, opening by pores; style slender; capsules oblong to globose, erect, septicidally dehiscent, 4- to 5-valved; seeds minute, slightly attenuate at both ends.——About 10 species, in N. America and Japan.

1A. Leaves pale green beneath; corolla obliquely urceolate, 5–7 mm. long; stamens 5, as many as the corolla-lobes. ...... 1. *M. pentandra*
1B. Leaves usually glaucous beneath; corolla tubular-campanulate, scarcely urceolate, 10–17 mm. long; stamens 8 or 10, twice as many as the corolla-lobes.
  2A. Ovary glabrous; leaves glabrous on margin; corolla 4-lobed, glandular-pilose outside. .... 2. *M. goyozanensis*
  2B. Ovary hairy; leaves ciliate at least while young; corolla 4- to 5-lobed, glabrous outside.
    3A. Ovary glandular-hairy; corolla-lobes 5, not ciliate. .... 3. *M. ciliicalyx*
    3B. Ovary with eglandular hairs; corolla-lobes 4, glandular-ciliolate. .... 4. *M. purpurea*

1. **Menziesia pentandra** Maxim. KO-YŌRAKU-TSUTSUJI. Much-branched pilose pubescent shrub; leaves flat, narrowly elliptic, 2.5–5 cm. long, 1–2.5 cm. wide, acute at both ends, membranous, coarsely pilose with ascending hairs on upper side and on midrib beneath, green above, paler beneath; flowers 3–6, umbellate, yellowish, 5–7 mm. long, the pedicels 1.5–3 cm. long, obliquely spreading, slightly nodding, becoming erect in fruit, puberulous and glandular-pilose; calyx 2–3 mm. across, 5-parted, glandular-ciliate; corolla obliquely urceolate, puberulent inside at tip and on lobes; stamens 5, glabrous, the anthers broadly lanceolate, about 1.5 mm. long; ovary puberulent and glandular-pilose; style about 4 mm. long; capsules ovoid-globose, 3–4 mm. in diameter, 5-valved.——May–June. Mountains; Hokkaido, Honshu, Shikoku, Kyushu.——Sakhalin and s. Kuriles.

2. **Menziesia goyozanensis** M. Kikuchi. GOYŌZAN-YŌRAKU. Much-branched shrub with glabrous branchlets; leaves oblong, 2.5–4 cm. long, 1–2 cm. wide, glabrous on margin, long appressed-hairy toward the margin on upper side, glaucous and nearly glabrous beneath except for the appressed hairy costa; pedicels 3–6 together, about 1 cm. long, nodding, glandular-pilose; calyx small, 4-fid, lobes broadly ovate; corolla tubular, 13–15 mm. long, 2–3 mm. across, glandular-pilose outside, short-pubescent inside, pale yellow, rose-colored outside; stamens 8, filaments pilose on lower half; ovary glabrous.——June. Honshu (Mount Gōyōzan in Rikuchu Prov.).

3. **Menziesia ciliicalyx** (Miq.) Maxim. (?) *Andromeda ciliicalyx* Miq.; *M. ciliicalyx* var. *tubiflora* Koidz. ex Nakai; *M. tubiflora* Koidz.——USUGI-YŌRAKU. Shrub to 1 m. high, erect, branched, the inner bud-scales glandular-ciliate, the young branchlets glabrous or sparsely puberulous; leaves membranous, ovate-oblong or obovate, sometimes oblong, 2.5–5 cm. long, 1.5–2.5 cm. wide, subacute to subobtuse, loosely pilose especially toward the margin, ciliate at first, the hairs later falling and leaving minute obsolete teeth, glaucous and with appressed coarse hairs on the midrib beneath, the petioles 2–4 mm. long; flowers 3–8, umbellate, tubular-campanulate, slightly oblique, pale yellow or pale greenish yellow, 13–17 mm. long, nodding, the pedicels usually 2–3 cm. long, glandular-pilose; corolla white-pubescent inside; anthers slender, about 4 mm. long; ovary with few glandular hairs; capsules about 4 mm. long, 5-valved; style 10–14 mm. long.——May–June. Mountains and hills; Honshu (Totomi and Echizen Prov. and westw.); rather common.

Var. **lasiostipes** Ohwi. KENAGA-USUGI-YŌRAKU. Hairs of pedicels and of calyx margin longer, partly eglandular.——Mountains; Honshu (Mino, Mikawa, and s. Shinano).

Var. **multiflora** (Maxim.) Makino. *M. multiflora* Maxim. ——URAJIRO-YŌRAKU. Differs from the typical phase in the elongate, linear-lanceolate to lanceolate, ciliate calyx-segments 4–6 mm. long; anthers 2–3 mm. long; styles 7–10 mm. long; corolla subcampanulate, usually purplish, the umbels with a short axis 1–6 mm. long.——Mountains and hills; Hokkaido, Honshu, Shikoku; variable and rather common northw.

Var. **bicolor** Makino. *M. ciliicalyx* Maxim., pro parte; *M. multiflora* var. *bicolor* Makino——TSURIGANE-TSUTSUJI. Umbels usually without the axis; calyx-lobes subrounded to elliptic, 1–2 mm. long; anthers about 2 mm. long.——May–June. Mountains and hills; Hokkaido, Honshu, Shikoku.

Var. **purpurea** Makino. *M. lasiophylla* Nakai; *M. multiflora* var. *purpurea* (Makino) Ohwi——MURASAKI-TSURIGANE-TSUTSUJI. Bud-scales puberulent on the margins; leaves coarsely long-pilose on upper side; pedicels with long glandular and eglandular hairs; corolla purple; calyx-lobes short, 1–2 mm. long.——June. Mountains; Honshu (Hakone).——Forma **glabrescens** (Nakai) Ohwi. *M. lasiophylla* var. *glabrescens* Nakai——FUJI-TSURIGANE-TSUTSUJI. Leaves glabrous on upper side.——Honshu (Kai and Suruga Prov.).

4. Menziesia purpurea Maxim. YŌRAKU-TSUTSUJI. Shrub with glabrous branchlets; leaves ovate or obovate-elliptic, 2.5–4.5 cm. long, 1.5–2 cm. wide, obtuse, rarely retuse, entire, sparsely ciliate, minutely glandular-hairy on upper side, glaucous and sometimes sparsely glandular-puberulent beneath, the midrib beneath loosely appressed-hairy, the petioles 3–6 mm. long; flowers 4–8, umbellate, campanulate, 12–15 mm. long, rose-purple, usually 4-merous, the pedicels loosely glandular-pilose, 1.5–2.5 cm. long; calyx-lobes glandular-ciliate; capsules 3–4 mm. long, short white-hairy while young.——Mountains; Kyushu.

4. RHODODENDRON L. TSUTSUJI ZOKU

Deciduous or evergreen shrubs or sometimes small trees; leaves alternate, petiolate, entire; flowers pedicellate, usually in terminal umbellike racemes or trusses; calyx 4- to 5(–8)-parted; corolla rotate, campanulate, or infundibuliform, rarely tubular, usually slightly oblique or regular, 4- to 5(–8)-lobed; stamens as many as or twice as many as the corolla-lobes, the anthers usually opening by terminal pores; ovules many; style slender, the stigma capitate; capsules ovoid or oblong, septicidally dehiscent, 4- to 8-valved; seeds minute, the testa loose and protruding at each end.——More than 600 species, in the cooler and temperate regions of the N. Hemisphere and in high mountains of the Tropics (one in Australia).

1A. Leaves and ovary with sessile discoid glandular scales.
  2A. Leaves 2–8 cm. long.
    3A. Flowers pale greenish yellow; pedicels 1–1.5 cm. long. .... 1. *R. keiskei*
    3B. Flowers rose-purple; pedicels 5–10 mm. long.
      4A. Leaves obtuse to subacute, 2–4 cm. long, partly evergreen; flowers 2.5–3.5 cm. across. .... 2. *R. dauricum*
      4B. Leaves acute to acuminate, 3–7 cm. long, deciduous; flowers 4–5 cm. across. .... 3. *R. mucronulatum*
  2B. Leaves 1–2 cm. long; flowers rose-purple. .... 4. *R. parvifolium*
1B. Leaves and ovary without sessile glandular scales.
  5A. Leaves evergreen, thick-coriaceous, glabrous or lanate beneath, not strigose.
    6A. Bud-scales deciduous.
      7A. Leaves rounded to shallowly cordate at base, usually glabrous beneath. .... 5. *R. brachycarpum*
      7B. Leaves acute at base, usually densely rusty-tomentose beneath. .... 6. *R. metternichii*
    6B. Bud-scales persistent.
      8A. Leaves thickly brown-tomentose beneath, linear-oblanceolate, 10–15 cm. long; flowers pale rose; erect shrub. .. 7. *R. makinoi*
      8B. Leaves glabrous beneath, obovate-oblong, 3–5 cm. long; flowers pale greenish yellow; prostrate shrub. .... 8. *R. aureum*
  5B. Leaves deciduous, usually membranous, rarely partially evergreen, often strigose.
    9A. Flowers in umbels from nonleaf-forming buds.
      10A. Flowers solitary from the lateral buds; stamens of unequal length, the 2 upper ones much shorter, densely white-barbate, the 3 lower ones glabrous, longer; petioles glandular-pilose. .... 9. *R. semibarbatum*
      10B. Flowers 1 to several from a terminal bud.
        11A. Flowers campanulate, nodding or pendulous; leaves obovate or oblong. .... 10. *R. nipponicum*
        11B. Flowers infundibuliform, ascending or horizontal.
          12A. Flowers and young vegetative shoots terminal.
            13A. Branchlets strigose; leaves sometimes semi-evergreen, alternate.
              14A. Corolla white, pilose inside, tubular, the lobes spreading, small.
                15A. Corolla tubular-campanulate, short-pubescent outside; anthers longitudinally dehiscent; ovary 3-locular.
                      11. *R. tsusiophyllum*
                15B. Corolla short-tubular to infundibuliform-campanulate, glabrous outside; anthers opening by pores; ovary 4- to 5-locular. .... 12. *R. tschonoskii*
              14B. Corolla usually colored, glabrous inside, infundibuliform, large, 1.5–6 cm. across.
                16A. Bud-scales not viscid; calyx and pedicels without glandular hairs.
                17A. Leaves glabrescent, coriaceous and often evergreen; flowers rose-colored, glabrous. .... 13. *R. tashiroi*
                17B. Leaves strigose or long-hairy, deciduous or sometimes partially evergreen.
                  18A. Leaves glabrous beneath except on midrib; flowers solitary, terminal; corolla about 1.5 cm. across.
                      14. *R. serpyllifolium*
                  18B. Leaves pilose beneath; corolla 2–6 cm. across.
                    19A. Flowers with the leaves in June and July.
                    20A. Leaves lanceolate, acute; stamens 5; anthers dark purple. .... 15. *R. indicum*
                    20B. Leaves elliptic (summer leaves obovate); stamens 9 or fewer; anthers light in color, not dark purple.
                      16. *R. tamurae*
                  19B. Flowers with or before the leaves in April and May.
                    21A. Stamens 6–9; flowers rose-purple; leaves deciduous. .... 17. *R. komiyamae*
                    21B. Stamens 5(–8).
                      22A. Leaves usually lanceolate, acute, semi-evergreen; flowers rose-purple, 2–3 cm. across. .. 18. *R. tosaense*
                      22B. Leaves ovate, elliptic, or obovate.
                        23A. Flowers 4–5 cm. across, red; branches erect or ascending. .... 19. *R. kaempferi*
                        23B. Flowers 2–3 cm. across, rose-purple; branches spreading. .... 20. *R. kiusianum*
           16B. Bud-scales viscid, at least on the inner side; calyx and pedicels with glandular hairs.
              24A. Young branches and leaves appressed-strigose.
                25A. Calyx-lobes appressed-hairy, subacute; leaves not coriaceous, mostly deciduous; flowers rose-purple, fragrant.
                    21. *R. yedoense* var. *poukhanense*
                25B. Calyx-lobes glandular-ciliate, obtuse; leaves coriaceous, deep green and lustrous above, evergreen; flowers red, inodorous. .... 22. *R. scabrum*

24B. Young branches with appressed nonglandular hairs and spreading, somewhat glandular-tipped hairs.
    26A. Stamens 5; ovary glandular-pilose. . . . . 23. *R. macrosepalum*
    26B. Stamens 10; ovary strigose. . . . . 24. *R. ripense*
13B. Branchlets glabrous or pilose; leaves fully deciduous, mostly fasciculate at ends of branchlets.
  27A. Leaves in 3's, terminal on the branchlets.
    28A. Ovary short glandular-pubescent, sometimes eglandular on the upper part.
      29A. Stamens 5; leaves glandular-dotted, glabrous or sparingly pubescent on upper side. . . . . 25. *R. dilatatum*
      29B. Stamens 10; leaves while young with long straight hairs.
        30A. Leaves broadly rhombic; petioles glandular-pilose; young branchlets glabrous. . . . . 26. *R. decandrum*
        30B. Leaves ovate, acute; petioles with long eglandular hairs; young branchlets hairy. . . . . 27. *R. viscistylum*
    28B. Ovary densely hirsute, the hairs eglandular.
      31A. Flowers usually after the leaves, red to deep rose; stamens nearly equal in length.
        32A. Leaves lustreless; petioles with loose coarse hairs, the pedicels with coarse hairs only on the lower portion; flowers red, before or with the leaves, April and May. . . . . 28. *R. weyrichii*
        32B. Leaves lustrous; pedicels, petioles, and capsules appressed woolly-pubescent; flowers after the leaves, May to July.
          33A. Flowers deep rose, May and June; pedicels 5–8 mm. long. . . . . 29. *R. sanctum*
          33B. Flowers red, July; pedicels 6–13 mm. long. . . . . 30. *R. amagianum*
      31B. Flowers usually before the leaves, rose-purple; stamens unequal in length.
        34A. Petioles glabrous at least on the lower part.
          35A. Pedicels and calyx densely pilose; leaves glabrous beneath. . . . . 31. *R. mayebarae*
          35B. Pedicels glabrous on upper part or loosely pilose; leaves with white appressed-hairs beneath while young.
            36A. Midrib of leaves beneath glabrous or short pilose; capsules thickly cylindric, 4–5 mm. across.
                32. *R. kiyosumense*
            36B. Midrib of leaves beneath densely barbate near the base; capsules slenderly cylindric, 2–3 mm. across.
                33. *R. nudipes*
        34B. Petioles pilose throughout.
          37A. Petioles, pedicels, and young branchlets appressed-pubescent; midrib of leaves beneath glabrous or glabrescent; capsules ovoid, 7–10 mm. in diameter. . . . . 34. *R. reticulatum*
          37B. Petioles and basal part of leaf midrib beneath densely long-pubescent.
            38A. Style with short glandular hairs on lower half; capsules short-cylindric. . . . . 35. *R. wadanum*
            38B. Style glabrous; capsules ovoid-cylindric. . . . . 36. *R. lagopus*
  27B. Leaves in 5's, verticillate, terminal on the branchlets; flowers white. 37. *R. quinquefolium*
12B. Flowers from a terminal bud, the vegetative shoots lateral.
  39A. Leaves in 5's, terminal on the branchlets, elliptic, 2.5–4.5 cm. long; branchlets quite glabrous; flowers 1 or 2, deep rose-colored; ovary glabrous. . . . . 38. *R. pentaphyllum*
  39B. Leaves alternate, the upper ones more or less approximate, oblanceolate to obovate, 5–10 cm. long; branchlets coarsely pilose while young; flowers 2–8; ovary densely pilose.
    40A. Flowers rose-purple; stamens 10. . . . . 39. *R. albrechtii*
    40B. Flowers orange-yellow, sometimes vermilion, or clear yellow; stamens 5. . . . . 40. *R. japonicum*
9B. Flowers in loose erect racemes on leafy shoots of current growth; bracts partially persistent. . . . . 41. *R. camtschaticum*

---

1. Rhododendron keiskei Miq. *R. keiskei* var. *cordifolium* Masam.——HIKAGE-TSUTSUJI. Evergreen shrub to abou 1 m. high, with yellow discoid sessile scales; leaves somewhat coriaceous, lanceolate to broadly so, or rarely narrowly oblong, 4–8 cm. long, 1.2–2 cm. wide, abruptly acute to shallowly cordate at base, mostly glabrous, often puberulous on the upper side near base, sparsely scaly beneath, entire, the margin slightly recurved, the petioles 2–6 mm. long; flowers 2–5, nearly umbellate, pale greenish yellow, the pedicels 1–1.5 cm. long; calyx small, with 5 undulate teeth; corolla 5-lobed, infundibuliform, 2.5–3 cm. across, with discoid sessile scales outside; stamens 10, the filaments slightly pilose near base; style glabrous; capsules cylindric, 1–1.2 cm. long, 2.5–3 mm. across. ——Apr.–May. Rocky wooded slopes in mountains; Honshu (Kantō Distr. and westw.), Shikoku, Kyushu.

Var. **hypoglaucum** Suto & Suzuki. URAJIRO-HIKAGE-TSUTSUJI. Leaves obovate-oblong, usually slightly glaucous beneath, with sessile discoid scales on both surfaces while young, persistent only beneath; corolla puberulent outside, scarcely scaly.——May. Honshu (Mount Ozaku in Shimotsuke); rare.

2. Rhododendron dauricum L. EZO-MURASAKI-TSUTSUJI, TOKIWA-GENKAI. Partly evergreen branched shrub, the branchlets with sessile discoid scales; leaves rather coriaceous, oblong to ovate-oblong, 2.5–4 cm. long, 1–1.5 cm. wide, obtuse to subacute, acute at base, with dense sessile discoid brownish scales beneath, the petioles 2–5 mm. long; flowers rose-purple, solitary, the pedicels short, covered by the bud scales at anthesis, 6–8 mm. long in fruit; calyx small, with 5 undulate teeth; corolla broadly infundibuliform, about 3 cm. across, scattered puberulent outside; stamens 10, the filaments white-pubescent at the base; styles glabrous; capsules cylindric, scaly, about 1 cm. long, 2.5–3 mm. across.——May. Hokkaido.——Korea, n. China, and e. Siberia.

3. Rhododendron mucronulatum Turcz. *R. dauricum* var. *mucronulatum* (Turcz.) Maxim.——KARA-MURASAKI-TSUTSUJI. Deciduous shrub 1–2 m. high, with sessile discoid scales throughout and often coarse hairy; leaves somewhat coriaceous, narrowly elliptic to broadly lanceolate, 3–7 cm. long, 1.2–2.5 cm. wide, pale green beneath, acute to acuminate at both ends, sometimes glabrous, the petioles 3–10 mm. long; flowers solitary, before the leaves, broadly infundibuliform, 4–5 cm. across, rose-purple, the pedicels 5–10 mm. long; calyx small, 5-toothed; corolla loosely pubescent outside; stamens 10, the filaments white-pubescent at the base; style glabrous; capsules cylindric, 1–1.5 cm. long, 3–4 mm. across.——Apr. Honshu (Chūgoku Distr.), Kyushu (n. distr.).——Korea, China, and Ussuri.

4. Rhododendron parvifolium Adams. *R. palustre* Turcz. ex DC.——Sakai-tsutsuji. Mostly erect, much-branched evergreen shrub to 1 m. high, densely clothed with sessile discoid scales, the branchlets slender, puberulous; leaves somewhat coriaceous, narrowly elliptic or lanceolate-oblong, 1–2 cm. long, obtuse to subacute, abruptly acute at base, glabrous, sometimes brownish beneath, the midrib impressed on upper side, the petioles 1–3 mm. long; flowers 2–5, 1.5–2 cm. across, rose-purple, the pedicels 3–5 mm. long in fruit; calyx-lobes broadly ovate, 1–1.5 mm. long, obtuse; corolla glabrous, infundibuliform; stamens 10; style glabrous; capsules ovoid-oblong, 5–6 mm. long, about 3 mm. across.——Wet places in lowlands; Hokkaido (near Nemuro); rare.——Sakhalin, Korea, and e. Siberia.

5. Rhododendron brachycarpum D. Don. *R. fauriei* Franch.; *R. brachycarpum* var. *lutescens* Koidz.; *R. fauriei* var. *lutescens* (Koidz.) Takeda; *R. brachycarpum* var. *rosae-florum* Miyoshi——Hakusan-shakunage. Evergreen shrub 1–3 m. high, the young branchlets sometimes soft brown-pubescent; leaves somewhat coriaceous, narrowly oblong, 6–15 cm. long, 2–5 cm. wide, obtuse to rounded, mucronate, rounded to shallowly cordate at base, entire, slightly recurved on margins, glabrous, deep green on upper side, paler beneath, the petioles often reddish, 1–1.5 cm. long; inflorescence 5- to 15-flowered, on a very short axis, the pedicels brown-pubescent, the bracts and bud-scales deciduous; calyx very small; corolla white or pinkish, pale green spotted on inside of upper lobe, glabrous, infundibuliform, 3–4 cm. across; stamens 10, the filaments loosely white-puberulent on lower half; ovary rusty-pubescent; style 1–1.5 cm. long, glabrous, ascending; capsules short-cylindric, becoming glabrous at maturity, 1–1.7 cm. long, 5–6 mm. in diameter.——June–July. Coniferous woods in mountains; Hokkaido, Honshu (centr. and n. distr.).—— Forma **nemotoanum** Makino. *R. brachycarpum* var. *nemoto-anum* (Makino) Nakai——Nemoto-shakunage. Stamens petaloid.——Honshu.

Var. **roseum** Koidz. *R. brachycarpum* var. *roseum* (Makino) Nakai; *R. fauriei* var. *rufescens* Nakai——Shirobana-shakunage. Leaves beneath with dense, short, pale brown hairs.——Occurs with the typical phase.

6. Rhododendron metternichii Sieb. & Zucc. *Hymen-anthes japonica* Bl.; *R. metternichii* var. *heptamerum* Maxim. ——Tsukushi-shakunage. Large evergreen shrub to 4 m. high, with thick terete branches rusty-tomentose while young; leaves coriaceous, oblanceolate, 8–20 cm. long, 2.5–5 cm. wide, acute at both ends, deep green, lustrous, entire, with slightly recurved margins, densely rusty-tomentose beneath, the petioles 1–2 cm. long; inflorescence many-flowered, sessile, the bracts and bud-scales deciduous, the pedicels 2–5 cm. long, rusty-pubescent while young; corolla infundibuliform, 4–5 cm. across, 7-lobed, rose-colored, glabrous; stamens 14, the filaments white granular-pubescent on the lower half; ovary densely rusty-villous; styles glabrous, 3–3.5 cm. long; capsules ovoid-cylindric, 15–25 mm. long, 7–8 mm. in diameter, with persistent brown hairs.——May–June. Dense woods in mountains; Honshu (w. Shinano and westw.), Shikoku, Kyushu.

Var. **hondoense** Nakai. Hon-shakunage. Tomentum on the under side of leaves, appressed, thinner and paler. Occurs with the typical phase.

Var. **pentamerum** Maxim. *R. degronianum* Carr.; *R. pentamerum* (Maxim.) Matsum. & Nakai——Shakunage, Azuma-shakunage. Shrub to 2 m. high; leaves oblanceolate to oblong-lanceolate, 8–15 cm. long, obtuse, acute at base, with pale brown tomentum beneath; flowers 5-merous, 3–4 cm. across; styles glabrous, 2.5–3 cm. long; capsules short-cylindric, 1.5–2 cm. long, 5–7 mm. in diameter.——May–June. Woods in high mountains; Honshu (centr. and n. distr.).

Var. **yakushimanum** (Nakai) Ohwi. *R. yakushimanum* Nakai——Yakushima-shakunage. Tomentum more dense on leaves beneath, on the pedicels, and on the ovary; flowers 5-merous.——High mountains; Kyushu (Yakushima).

7. Rhododendron makinoi Tagg. *R. metternichii* var. *pentamerum* forma *angustifolium* Makino; *R. stenophyllum* Makino, non Hook. f.; *R. metternichii* var. *angustifolium* (Makino) Bean——Hosoba-shakunage. Erect evergreen shrub with thick, brown-pubescent branchlets, the bud-scales remaining on the branches for several years; leaves coriaceous, lustrous, deep green, linear-oblanceolate, 10–15 cm. long, 1–2 cm. wide, acute at both ends, with slightly impressed veinlets on upper side, thickly brown-tomentose beneath, the petioles 1–2 cm. long; inflorescence rather many-flowered, the bud-scales persistent at base, the pedicels brown-villous; calyx very small; corolla pale rose-colored, 3–4 cm. across, infundibuliform, 5-lobed; stamens 10; ovary brown-villous; style glabrous, about 3 cm. long; capsules oblong, brown-pubescent.——Mountains; Honshu (Tōtōmi and Mikawa Prov.).

8. Rhododendron aureum Georgi. *R. chrysanthum* Pall.; *R. officinale* Salisb.——Kibana-shakunage. Prostrate evergreen shrub, the branches thick, minutely puberulent while young or nearly glabrous, the bud-scales persistent; leaves thick-coriaceous, glabrous, deep green, slightly paler beneath, obovate-oblong, obovate, or oblong, 3–4(–6.5) cm. long, 1.5–2 (–2.5) cm. wide, rounded to very obtuse, slightly cuneate and acute to obtuse at base, veinlets impressed above, the petioles 1–1.5 cm. long; inflorescence 3- to 10-flowered, sessile, terminal, the pedicels 2.5–5 cm. long, brown-pubescent; calyx very small; corolla 2.5–3.5 cm. across, pale greenish yellow, glabrous; stamens 10, puberulous toward the base of filaments; ovary brown-pubescent; style 1.5–2 cm. long, glabrous, longer than the stamens; capsules narrowly oblong, 1–1.5 cm. long.—— July–Aug. Alpine slopes; Hokkaido, Honshu (centr. and n. distr.); rare.——n. Korea, Kuriles, Sakhalin, Kamchatka, and e. Siberia.——Forma **senanense** (Yabe) Hara. *R. chrysanthum* forma *senanense* Yabe——Yae-kibana-shakunage. The stamens petaloid.——Honshu (centr. and n. distr.).

Rhododendron × nikomontanum Nakai. Nikkō-kibana-shakunage. Hybrid of *R. aureum* × *R. brachycarpum*.

9. Rhododendron semibarbatum Maxim. *Azaleastrum semibarbatum* (Maxim.) Makino——Baika-tsutsuji. Deciduous shrub about 2 m. high, the young branches slender, erect or ascending, white-puberulous and glandular-pilose; leaves membranous, fasciculate and terminal on the branchlets, elliptic to narrowly so, 3–6 cm. long, 1.5–2.5 cm. wide, acute, rounded to obtuse, or sometimes subacute at base, serrulate, green on upper side, often glaucous beneath, the midrib white-puberulent on both sides and loosely glandular-pilose beneath, the petioles 5–10 mm. long, glandular-pilose; flowers about 1.5 cm. across, white, suffused rose-purple, the pedicels white-puberulous and glandular-pilose, 5–10 mm. long; calyx with 5 short lobes, glandular-pilose; corolla short-tubular, 5-lobed; stamens unequal in length, the 2 upper filaments shorter and densely white-pubescent throughout, the 3 lower stamens with longer glabrous exserted filaments; styles glabrous; capsules globose, glandular-pilose, about 5 mm. across.——June–July. Thickets in mountains; Honshu, Shikoku, Kyushu.

10. Rhododendron nipponicum Matsum. *Azalea nipponica* (Matsum.) Copel.——ŌBA-TSUTSUJI. Deciduous shrub 1–2 m. high, with red-brown, peeling, lustrous bark, the branches thick, glandular-pilose while young; leaves alternate, membranous, obovate or oblong, 5–12 cm. long, 3–6 cm. wide, cuneate and sessile or nearly so at base, densely spreading ciliate, coarsely pilose and often somewhat glandular-hairy; flowers few, white, in terminal umbels, the pedicels 1.5–2 cm. long, glandular-pilose; calyx-lobes 5, ovate, obtuse; corolla 5-lobed, tubular-campanulate, 1.5–2 cm. long, nodding, short-pubescent inside at base; stamens 10, scarcely exserted; style 1–1.5 cm. long, rarely glandular; capsules oblong-ovoid, oblique, glandular-pilose, about 1 cm. long, 5–6 mm. wide. ——July–Aug. Mountains; Honshu (n. and Hokuriku distr.); rather rare.

11. Rhododendron tsusiophyllum Sugimoto. *R. tanakae* (Maxim.) Ohwi, non Hayata; *Tsusiophyllum tanakae* Maxim.——HAKONE-KOME-TSUTSUJI. Prostrate or decumbent, much-branched shrub to 0.5 m. high, the branches densely brown-strigose; leaves rather thick, nearly sessile, elliptic to narrowly so, sometimes obovate, 7–10 mm. long, 4–6 mm. wide, acute at both ends, mucronulate, brown-strigose on upper side, on margins and midrib beneath; flowers 1–3, white, umbellate, the pedicels densely brown-strigose, 1–3 mm. long; calyx minute; corolla shallowly 4- or 5-lobed, tubular-campanulate, 7–10 mm. long, white-puberulent externally; stamens 5, included, the filaments pilose, the anthers longitudinally dehiscent; style glabrous; ovary 3-locular, densely strigose; capsules broadly ovoid.——June–July. Mountains; Honshu (Sagami, Izu, Musashi, and Suruga Prov.); rare.

12. Rhododendron tschonoskii Maxim. *R. tschonoskii* var. *typicum* forma *pentamerum* Makino and forma *tetramerum* Makino——KOME-TSUTSUJI. Densely branched deciduous shrub, the branches, leaves, pedicels and capsules strigose; leaves ovate, oblong, or elliptic, 7–25 mm. long, 4–12 mm. wide, acute to subacute at both ends, sessile or subsessile; flowers 2–5, white, terminal, the pedicels 3–8 mm. long; corolla short-tubular to infundibuliform-campanulate, 6–8 mm. long, 5–10 mm. across, glabrous, but puberulent inside at the base, the lobes 4 or 5, sometimes shorter than the tube; stamens 4 or 5, slightly exserted, the filaments pubescent at base; style 2–10 mm. long; ovary 4- or 5-locular; capsules ovoid-conical, obtuse, 4–5 mm. long, 3–4 mm. across.——July. Rocky places in mountains; Hokkaido, Honshu, Shikoku, Kyushu.——Korea (Mount Chiisan).

Var. **trinerve** (Franch.) Makino. *R. trinerve* Franch.——Ō-KOME-TSUTSUJI. Slightly larger in all parts; leaves oblong to broadly oblanceolate, 1.5–4.5 cm. long, 7–15 mm. wide, with 1 (rarely 2) pair of prominent lateral nerves; corolla 7–10 mm. long, 7–12 mm. across; stamens longer than the style; styles 7–10 mm. long, glabrous or brownish strigose near base; capsules ovoid, 4–5 mm. in diameter, brownish strigose.——July–Aug. Honshu (n. and Hokuriku distr.).

13. Rhododendron tashiroi Maxim. SAKURA-TSUTSUJI. Evergreen or partially deciduous erect shrub, the branches, petioles, pedicels, and ovary strigose; leaves rather thick, in 3's (rarely 2's), terminal on the branches, elliptic, oblong, obovate, or sometimes ovate, deep green and lustrous on upper side, acute at both ends, long appressed-pubescent on upper side while young, the hairs somewhat persistent beneath, the petioles 5–10 mm. long; flowers 2 or 3, rose-colored, terminal, in umbels, the pedicels 1–2 cm. long; calyx small, obsoletely

5-lobed; corolla 5-lobed, infundibuliform, 3.5–4 cm. across, glabrous, spotted inside at tip; stamens 10, glabrous; styles glabrous, exserted, about 2.5 cm. long; capsules ovoid-oblong, oblique, 10–12 mm. long.——Thickets and woods in hills and mountains; Kyushu (s. distr.).——Ryukyus and Formosa.

14. Rhododendron serpyllifolium (A. Gray) Miq. *Azalea serpyllifolia* A. Gray——UNZEN-TSUTSUJI. Much-branched deciduous small shrub, the branches, leaves, and capsules purple-brown strigose; leaves alternate, mostly clustered toward the tip of the branchlets, broadly lanceolate, 8–15 (–20) mm. long, 3–6 mm. wide, obtuse to acute, narrowed at base, glabrous beneath except the strigose midrib, strigose on the upper surface, the petioles to 2 mm. long; flowers solitary, red to nearly white, terminal, the pedicels 2–3 mm. long; calyx 5-toothed; corolla about 1.5 cm. across, infundibuliform, glabrous; stamens 5, the filaments granular-pilose in the lower part; styles glabrous, 1.5–2 cm. long, long-exserted; capsules short-pedicelled, ovoid, about 4 mm. in diameter.——Apr.–May. Mountains; Honshu (Kantō Distr. and westw.), Shikoku, Kyushu.

15. Rhododendron indicum (L.) Sweet. *Azalea indica* L.; *A. indica* var. *lateritia* Lindl.; *R. lateritium* (Lindl.) Planch.; *R. indicum* var. *lateritium* (Lindl.) DC.——SATSUKI-TSUTSUJI, SATSUKI. Semi-evergreen or deciduous, red-brown strigose, much-branched shrub to 1 m. high; leaves often clustered toward the tip of the branchlets, rather thick, lanceolate to broadly so, 2–3.5 cm. long, 5–8(–10) mm. wide, acute, mucronate, acuminate at base, the petioles 1–3 mm. long; flowers 1 or rarely 2, usually red, following the new leaves, the pedicels about 1 cm. long; calyx-lobes 5, ovate-orbicular; corolla 2.5–3.5 cm. across, broadly infundibuliform, glabrous, 5-lobed, spotted inside on upper lobes; stamens 5, the filaments granular-pilose on lower half, the anthers dark purple; styles glabrous, 3–5 cm. long; capsules ovoid, hairy, 7–8 mm. long.——June–Aug. Rocks in ravines; Honshu (w. Kantō Distr. and westw.), Kyushu. Widely cultivated with numerous cultivars grown in gardens; also extensively used as a parent of hybrids.

16. Rhododendron tamurae (Makino) Masam. *R. indicum* var. *tamurae* Makino; *R. simsii* var. *tamurae* (Makino) Kaneh. & Hatus.; *R. eriocarpum* Nakai, excl. basionym.—— MARUBA-SATSUKI. Resembles the preceding; plant reddish brown-strigose; vernal leaves mostly elliptic, the summer ones obovate, 2–4 cm. long, 8–15 mm. wide, obtuse to rounded or sometimes acute; petioles 2–8 mm. long; flowers 1–2, red, terminal; pedicels about 1 cm. long; calyx-lobes 5, broadly ovate, 2–3 mm. long; corolla broadly infundibuliform, 3–4 cm. across, 5-lobed; stamens 9 or fewer, the filaments short-granular, the antheers usually light-colored; styles glabrous; capsules 7–10 mm. long, densely long-hairy, narrowly ovoid.—— June–July. Thickets and thin woods in hills and low elevations in the mountains; Kyushu (Southern islands including Yakushima and Tanegashima).

17. Rhododendron komiyamae Makino. ASHI-TAKA-TSUTSUJI. Brown-strigose deciduous shrub or small tree to 10 m. high; leaves alternate, rather thin, often disposed toward the top of the branchlets, broadly lanceolate to narrowly ovate-oblong, 2–3 cm. long, 8–12 mm. wide, acute at both ends, short-petiolate; summer leaves smaller, deciduous the following spring; flowers 1–3, rose-purple, terminal, in umbels, the pedicels 5–10 mm. long; calyx-lobes ovate, about 2 mm. long; corolla infundibuliform, 2–3 cm. across, 5-lobed, glabrous;

stamens 6–9, the filaments loosely granular-pilose on lower half, the anthers light colored; styles glabrous, about 2 cm. long; capsules elongate, conical, about 8 mm. long, 3 mm. across.—— May–June. Mountains; Honshu (Suruga).

18. Rhododendron tosaense Makino. *R. miyazawae* Nakai & Hara.——Fuji-tsutsuji, Men-tsutsuji. Brown-strigose, semi-evergreen, much-branched shrub 1–2 m. high; leaves rather thin, dimorphic, the vernal ones lanceolate or narrowly oblong, 1.5–3 cm. long, 5–8 mm. wide, acute at both ends, short-petiolate, the summer leaves persistent, oblanceolate or broadly linear, 1–3 cm. long, 2–4 mm. wide, abruptly acute; flowers 2–3, rose-purple, in umbels, the pedicels 3–7 mm. long; calyx-lobes 5, ovate-orbicular, 1–1.5 mm. long; corolla infundibuliform, 2–3 cm. across, 5-lobed, glabrous; stamens 5, the filaments loosely white granular-pilose; style glabrous, 2–2.5 cm. long, longer than the stamens; capsules narrowly ovoid, 5–7 mm. long.——Apr.–May. Honshu (Kinki Distr. and westw.), Shikoku, Kyushu.

19. Rhododendron kaempferi Planch. *R. indicum* var. *kaempferi* (Planch.) Maxim.; *R. obtusum* var. *kaempferi* (Planch.) Wils.——Yama-tsutsuji. Brown-strigose, much-branched shrub 1–3 m. high; leaves alternate, rather thin, the vernal ones elliptic, or sometimes oblong to broadly ovate, 3–5 cm. long, 1–2.5(–3) cm. wide, the summer leaves smaller, narrowly obovate, acute to obtuse at both ends, the petioles 1–4 mm. long; flowers 2–3, brick-red, rarely white or purple, terminal, in umbels, the pedicels 5–20 mm. long; calyx-lobes 5, small, usually elliptic, 2–4 mm. long; corolla infundibuliform, 3–4 cm. across, 5-lobed; stamens 5, the filaments sparsely granular-pilose on lower half; styles glabrous; capsules conical. ——Apr.–May. Sunny slopes in hills and mountains; Hokkaido, Honshu, Shikoku, Kyushu. Very common and variable.

Rhododendron obtusum (Lindl.) Planch. *Azalea obtusa* Lindl.——Kirishima-tsutsuji. Widely cultivated and probably of hybrid origin involving *R. kaempferi* and several other species.

20. Rhododendron kiusianum Makino. *R. indicum* var. *amoenum* a *japonicum* Maxim.; *R. amoenum* var. *japonicum* (Maxim.) Zabel; *R. kaempferi* var. *japonicum* (Maxim.) Rehd.; *R. obtusum* forma *japonicum* (Maxim.) Wils.; (?) *R. sataense* Nakai; *R. indicum* var. *japonicum* (Maxim.) Makino——Miyama-kirishima. Brown-strigose, low, much-branched, often prostrate or decumbent shrub about 1 m. high; vernal leaves oblong to elliptic, 8–30 mm. long, acute at both ends, very short-petiolate, the summer leaves smaller, more or less persistent, narrowly obovate-spathulate; flowers 2–3, rose-purple, terminal, in umbels; calyx-lobes 5, short, broadly ovate, 2–3 mm. long; corolla infundibuliform, 2–3 cm. across, glabrous, 5-lobed; stamens 5, the filaments loosely granular-pilose on lower half or glabrous, the anthers often purplish; capsules ovoid, about 7 mm. in diameter, the pedicels 5–8 mm. long.——May. High mountains; Kyushu.

21. Rhododendron yedoense Maxim. var. **poukhanense** (Lév.) Nakai. *R. poukhanense* Lév.; *R. coreanum* Rehd.——Chōsen-yama-tsutsuji. Brown-strigose shrub 1–2 m. high; leaves narrowly oblong or broadly oblanceolate, 3–8 cm. long, 1–3 cm. wide, acute at both ends, loosely strigose on upper side, densely so beneath especially on the nerves, the petioles 1–5 mm. long; flowers 2–3, pale rose-purple, terminal, in umbels, the pedicels 7–10 mm. long; calyx 5-lobed, the lobes narrowly ovate, subobtuse to subacute, 4–8 mm. long;

corolla 5–6 cm. across, infundibuliform, 5-lobed, with deeply colored spots inside on the upper lobes; stamens (7–)10, the filaments glabrous to papillose-pilose on the lower half, the anthers purplish; styles glabrous or with appressed hairs near base; capsules ovoid, 8–10 mm. in diameter, long-pilose.—— Apr.–May. Kyushu (Tsushima).——Korea.

22. Rhododendron scabrum G. Don. *R. sublanceolatum* Miq.; *Azalea sublanceolata* (Miq.) O. Kuntze; *R. indicum* var. *sublanceolatum* (Miq.) Makino——Kerama-tsutsuji, Tō-tsutsuji. Brown-strigose, evergreen branched shrub; leaves rather thick, narrowly elliptic or narrowly ovate-elliptic, 4–10 cm. long, 1.5–3 cm. wide, deep green and lustrous on upper side, acute at both ends, the petioles 5–10 mm. long; flowers 1–4, vermilion, terminal, in umbels, the pedicels 1–2 cm. long; calyx-lobes 5, elliptic, 3–6 mm. long, long glandular-ciliate, obtuse to rounded; corolla infundibuliform, 6–7 cm. across, 5-lobed; stamens 10, the filaments granular-pilose at base; capsules hairy, about 1 cm. long.——Mar.–Apr. Frequently planted in southern Kyushu.——Ryukyus.

23. Rhododendron macrosepalum Maxim. *R. lineari-folium* var. *macrosepalum* (Maxim.) Makino——Mochi-tsutsuji. Mostly deciduous shrubs about 1 m. high; vernal leaves narrowly oblong to elliptic or ovate, 4–7 cm. long, 1–3 cm. wide, dull, acute at both ends, with spreading glandless and glandular hairs especially on under side and on the petioles, the petioles 3–7 mm. long; summer leaves sometimes persistent over winter, smaller and oblanceolate, obtuse, the petioles 3–7 mm. long; flowers 1–5, pale rose-purple, terminal, the pedicels 1.5–3 cm. long, glandular-hairy; calyx-lobes green, linear-lanceolate, acute, 2–3 cm. long, viscid; corolla about 5 cm. across, infundibuliform; stamens 5, the filaments granular-pilose; styles glabrous; capsules ovoid, 1–1.2 cm. in diameter, long glandular-pilose.——May. Sunny hillsides; Honshu (Izu, Kai, and westw. to e. Chūgoku Distr.), Shikoku; common and with numerous cultivars grown in gardens.

Rhododendron × tectum Koidz. Miyako-tsutsuji. Hybrid of *R. kaempferi × R. macrosepalum*.

24. Rhododendron ripense Makino. *R. mucronatum* var. *ripense* (Makino) Wils.——Kishi-tsutsuji. Brown-sometimes glandular-hairy shrub about 1 m. high; vernal leaves lanceolate or oblanceolate, 3–5 cm. long, 8–15 mm. wide, acute at both ends, the summer leaves usually persistent over winter, smaller, oblanceolate, usually subobtuse; flowers 1–3, rose-purple, terminal, in umbels, the pedicels 1–1.5 cm. long; calyx-lobes 5, green, lanceolate, 1–2 cm. long, gradually acute, short glandular-ciliate; corolla infundibuliform, 5-lobed; stamens 10, the filaments granular-pilose; styles glabrous; capsules ovoid, about 8 mm. in diameter, strigose.——Apr.–May. Honshu (Chūgoku Distr.), Shikoku, Kyushu (n. distr.).

25. Rhododendron dilatatum Miq. *R. reticulatum* forma *pentandrum* Wils.——Mitsuba-tsutsuji. Deciduous nearly glabrous shrub 1–2 m. high, with gray-brown branches; leaves in 3's, terminal on the branchlets, broadly ovate-rhombic, 4–7 cm. long, 3–5 cm. wide, very acute, broadly cuneate to subrounded at base, often minutely glandular-dotted on both sides, glabrous or sparsely hairy on the upper side and on the upper portion of petioles, entire, the petioles 5–15 mm. long, often puberulent; flowers 1–3, purplish, terminal, before the leaves, the pedicels 5–20 mm. long, rather densely short glandular-hairy; calyx-teeth very short; corolla 3–4 cm. across, 5-lobed; stamens 5, the filaments glabrous; styles glabrous; ovary densely glandular-puberulous; capsules ovoid-cylindric, oblique, 1–1.2 cm. long, about 4 mm. across, short-glandular.——Apr. Honshu (Kantō, Tōkaidō, and Kinki Distr.).

26. Rhododendron decandrum Makino. *R. dilatatum* var. *decandrum* Makino——Tosa-no-mitsuba-tsutsuji. Much-branched shrub resembling the preceding species; leaves broadly rhombic, more prominently brown-hairy, glabrescent with age, often short glandular-puberulent on upper side, the pedicels about 1 cm. long; calyx-teeth small, glandular-pilose; stamens 10, the anthers purplish; ovary densely short glandular-pubescent; capsules about 1 cm. long, short glandular-hairy.——Mar.–Apr. Honshu (Kinki Distr.), Shikoku, Kyushu. Possibly only a geographical variant of the preceding.

27. Rhododendron viscistylum Nakai. Takakuma-mitsuba-tsutsuji. Much-branched shrub 2–3 m. high, brown-pilose while young, glabrescent; leaves in 3's, terminal on the branchlets, ovate or oblong-ovate, 1.5–4.5 cm. long, 7–13 mm. wide, very acute, broadly cuneate at base, whitish and with obscure sessile glands beneath, the petioles 3–5 mm. long; flowers 1–2, purple, the pedicels 5–6 mm. long, glandular-puberulent; calyx-teeth short; stamens 10, unequal, the filaments glabrous; ovary ovoid, glandular.——June. Mountains; Kyushu (southern distr.).

28. Rhododendron weyrichii Maxim. *R. shikokianum* Makino; *R. shikokianum* var. *nudistemon* Koidz.——On-tsutsuji, Tsukushi-aka-tsutsuji. Rather large shrub to 5m. high, rusty-brown pubescent while young, glabrescent, the branches purplish; leaves in 3's, terminal on the branches, broadly ovate or ovate-orbicular, sometimes broadly rhombic, 3.5–8 cm. long, 2.5–6 cm. wide, acute, broadly cuneate to rounded at base, the petioles 5–10 mm. long; flowers 1–3, red, terminal, before or with the leaves; corolla 4–5 cm. across; stamens usually 10; styles glabrous or hairy near base; capsules shortly ovoid-cylindric, about 1 cm. long, often oblique at base.——Apr.–May. Honshu (s. Kinki Distr.), Shikoku, Kyushu.——Korea (Quelpaert Isl.).

29. Rhododendron sanctum Nakai. Jingū-tsutsuji. Large deciduous rusty-brown pubescent shrub to 5 m. high; leaves usually in 3's, terminal on the branchlets, broadly rhombic to broadly ovate-rhombic, 4–8 cm. long, 3–6 cm. wide, slightly lustrous and long rusty-brown hairy on upper side, paler and the midrib long-hairy beneath, abruptly acute at both ends or obtuse at base, the petioles 5–10 mm. long; flowers 3–4, deep rose, in umbels, terminal, the pedicels 5–8 mm. long; calyx-teeth small, obsolete; corolla glabrous; stamens 10, the anthers purple-brown; styles glabrous; capsules ovoid-cylindric, 1–1.5 cm. long, oblique at base, densely brown-pubescent.——May–June. Mountains; Honshu (Tōkaidō to s. Kinki Distr.).

30. Rhododendron amagianum Makino. *Azalea amagiana* Makino——Amagi-tsutsuji. Large deciduous shrub to 5 m. high, the branches rather slender, densely white-pubescent while young; leaves in 3's, terminal on the branchlets, broadly ovate-rhombic, 5–8 cm. long and as wide, abruptly acute, obtuse to acute at base, with scattered long brown hairs on upper side, rusty-brown appressed-pubescent beneath, especially dense on the midrib, the petioles 5–10 mm. long; flowers 2–4, red, the pedicels 6–13 mm. long, densely appressed rusty-brown pubescent; calyx-teeth minute; corolla 5–6 cm. across; stamens 10; styles white-pubescent at base; capsules narrowly ovoid-cylindric, 1.5–2 cm. long, densely brown-pilose.——July. Mountains; Honshu (Izu Prov.).

31. Rhododendron mayebarae Nakai & Hara. Nango-ku-mitsuba-tsutsuji. Deciduous shrub, the branchlets essentially glabrous; leaves rhombic-ovate, glabrous, long-hairy on upper side and on margin only while young, the nerves beneath brown-pubescent while young; flowers before the leaves, purple, the pedicels 1–1.5 cm. long, densely coarse brown-hairy; calyx-teeth obscure; corolla infundibuliform, about 3.5 cm. across, 5-lobed; stamens usually 10, unequal, the filaments glabrous; ovary ovoid, densely coarse brown-hairy; styles glabrous.——Apr.–May. Kyushu.

32. Rhododendron kiyosumense Makino. *Azalea kiyosumensis* Makino; *R. shimidzuanum* Honda ex Makino & Nemoto——Kiyosumi-mitsuba-tsutsuji. Deciduous shrub; branches whorled, slender, glabrous; leaves broadly rhombic, 3–6 cm. long, 2–5 cm. wide, abruptly short-acuminate, rounded to broadly cuneate at base, loosely brown-pilose while young; flowers purple, before or with the leaves, the pedicels 1–1.2 cm. long, loosely appressed brown-hairy especially on lower part; calyx-teeth minute, ciliate; corolla 3–3.5 cm. across, 5-lobed; stamens 10; style glabrous; capsules short-cylindric, oblique, about 10 mm. long, 3–5 mm. wide, brown-strigose.——Apr.–May. Honshu (s. Kantō and Tōkaidō to s. Kinki Distr.).

33. Rhododendron nudipes Nakai. *R. nagasakianum* Nakai; *R. yakumontanum* Masam.——Saikoku-mitsuba-tsutsuji. Large deciduous shrub; branches glabrous; leaves broadly rhombic, 4–8 cm. long, 3–6 cm. wide, abruptly acute, broadly cuneate at base, long brown-hairy on upper side and below while young soon becoming glabrous, the midrib beneath and upper portion of petioles persistently pale-brown villous; petioles 4–8 mm. long; flowers rose-purple, solitary or in pairs, the pedicels short, 6–10 mm. long, loosely brown-pubescent; calyx minute; corolla 3–4 cm. across; stamens 10; style glabrous; ovary densely pale-brown coarse hairy; capsules about 1 cm. long, oblique, narrowly ovoid, coarsely hairy.——May. Mountains; Honshu (Echigo, n. Shinano, and westw.), Kyushu.

34. Rhododendron reticulatum D. Don. *R. rhombicum* Miq.——Koba-no-mitsuba-tsutsuji. Deciduous branched shrub; leaves broadly ovate to ovate-rhombic, 4–7 cm. long, 2.5–4 cm. wide, abruptly acuminate to acute, sometimes rounded at base, long brown-hairy on upper side while young, pale green and pubescent beneath especially on midrib, glabrescent, the petioles 7–12 mm. long; flowers rose-purple, the pedicels 8–12 mm. long, with coarse ascending hairs; calyx-teeth depressed, minute, often obscurely glandular-dotted on margin; corolla 3–4 cm. across; stamens 10, glabrous; styles glabrous; capsules ovoid, 7–10 mm. in diameter, oblique, with ascending coarse brown hairs.——Apr.–May. Mountains; Honshu (Tōtōmi, Shinano, and westw.), Shikoku, Kyushu.——Forma **albiflorum** (Makino) Makino. *R. rhombicum* var. *albiflorum* Makino; *R. sakawanum* Makino——Shirobana-koba-no-mitsuba-tsutsuji. Flowers white.

Var. **ciliatum** Nakai. Arage-mitsuba-tsutsuji. Hairs more persistent on branches and leaves.——Occurs with the typical phase.

35. Rhododendron wadanum Makino. *R. glandulistylum* Komatsu——Tōgoku-mitsuba-tsutsuji. Deciduous brown-hairy shrub 2–4 m. high, with slender nearly glabrous branchlets; leaves broadly ovate-rhombic, 3–6 cm. long, 2–5 cm. wide, acute, mucronate, broadly cuneate at base, with long deciduous brown hairs on upper side, whitish and pale brown-pubescent beneath, the petioles short, 2–5 mm. long, rather dense brown-pubescent; flowers rose-purple, the pedicels 8–12 mm. long, with glandless and glandular hairs; corolla 3–4 cm. across; stamens 10, glabrous; style glandular on lower half; ovary densely pale brown-hairy; capsules short-cylindric, oblique, 1–1.2 cm. long, brown-hairy.——May. Mountains; Honshu (centr. and n. distr.).

36. Rhododendron lagopus Nakai. *R. wadanum* var. *lagopus* (Nakai) Hara——DAISEN-MITSUBA-TSUTSUJI. Deciduous rather high shrub, with glabrous branches; leaves in 3's, terminal, broadly rhombic, 3–6.5 cm. long, 2–5 cm. wide, acute, cuneate at base, pubescent only while young, the hairs persistent on midrib beneath; flowers solitary, rose-purple, the pedicels about 1 cm. long; calyx minute; corolla about 4 cm. across; stamens 10, glabrous; styles glabrous; ovary densely white-villous; capsules shortly ovoid-cylindric, 1–1.3 cm. long, oblique at base, brown-pilose.——Mountains; Honshu (centr. distr. and westw.), Shikoku.

37. Rhododendron quinquefolium Biss. & Moore. GOYŌ-TSUTSUJI, SHIRO-YASHIO. Large deciduous shrub 4–6 m. high, with much-branched slender glabrous branches; leaves membranous, in 5's, terminal, obovate-elliptic or rhombic-elliptic, obtuse, mucronate, usually reddish, ciliate or ciliolate, cuneate below, short white-pilose especially on the midrib above and below, the petioles 1–3 mm. long; flowers white, the pedicels glabrous to brown-pilose, sometimes loosely glandular-pilose, 1.5–2.5 cm. long; calyx-lobes lanceolate to deltoid, sometimes white-pilose, 1.5–3 mm. long; corolla 3–4 cm. across; stamens 10, the filaments white-pilose on lower half; styles glabrous or soft-pilose near base; ovary glabrous, sometimes scattered soft-pilose; capsules ovoid- to oblong-cylindric, 1–1.5 cm. long, 3.5–5 mm. across, with minute external warts. ——May–June. Mountains; Honshu, Shikoku.

38. Rhododendron pentaphyllum Maxim. *R. quinquefolium* var. *roseum* Rehd.——AKEBONO-TSUTSUJI. Branched shrub 3–4 m. high, with glabrous branchlets; leaves rather firm, elliptic, 2.5–4.5 cm. long, 1.7–2.5 cm. wide, acute at both ends, mucronulate, ciliate, loosely callose hairy on upper side, glabrous or with loose glandular hairs on midrib and near base beneath, green on both sides, the petioles 2–5 mm. long, loosely long-hispid on margin; flowers deep rose-colored, the pedicels 1–1.5 cm. long, glabrous; corolla about 5 cm. across; stamens 10, glabrous; styles and ovary glabrous; capsules narrowly oblong, 12–15(–20) mm. long, 5–7 mm. across.——Apr.–May. Mountains; Honshu (Kii Prov. and westw.), Shikoku, Kyushu.

Var. **nikoense** Komatsu. *R. nikoense* (Komatsu) Nakai; *R. pentaphyllum* var. *villosum* Koidz.——AKA-YASHIO, AKAGI-TSUTSUJI. Pedicels glandular; five of the filaments pilose on lower portion.——Honshu (Iwaki Prov. and westw.), Shikoku, Kyushu.

39. Rhododendron albrechtii Maxim. MURASAKI-YASHIO, MIYAMA-TSUTSUJI. Deciduous shrub 1–2 m. high, with rather slender, glandular-pilose branchlets; leaves relatively thin, mostly clustered near the ends of the branchlets, obovate to broadly lanceolate, 5–10 cm. long, acute to obtuse or rounded at the tip, with short thickened hairs on upper side and on margin, rarely glaucescent, the nerves beneath loosely glandular-pilose while young, the midrib spreading-pilose; flowers 3–6, terminal, deep rose to rose-purple, 3–4 cm. across, the pedicels 1–2 cm. long, densely long glandular-hairy; calyx minute, 5-toothed; corolla white-pubescent on the upper side within; stamens 10, usually densely white-pubescent near the base, styles glabrous; capsules ovoid, 8–12 mm. in diameter. ——May–June. Mountains; Hokkaido, Honshu (Echizen and northw.).

40. Rhododendron japonicum (A. Gray) Suring. *Azalea japonica* A. Gray; *R. japonicum* var. *glaucophyllum* Hara; *A. mollis* var. *glabrior* Regel; *R. glabrius* (Regel) Nakai—— RENGE-TSUTSUJI. Deciduous shrub 1–2 m. high, with rather slender, loosely pilose branchlets; leaves membranous, oblanceolate, 5–10 cm. long, 1.5–3 cm. wide, obtuse, sometimes acute or rounded and mucronate, narrowed gradually toward the base, vivid-green above, sparingly ciliate, often glaucous and minutely puberulent on the nerves beneath, loosely pilose on both sides, the petioles 3–7 mm. long; flowers 2–8, terminal, with the leaves, the pedicels 1.5–3 cm. long, often minutely white-puberulent and long-spreading glandular-hairy; corolla vermilion, orange to clear golden-yellow, 5–6 cm. across, puberulent outside; stamens 5, the filaments white-pilose near the base; styles glabrous; capsules oblong-cylindric, 2–2.5 cm. long, short white- and long brown-hairy.——Apr.–June. Mountains; Hokkaido (sw. distr.), Honshu, Shikoku, Kyushu; locally common and gregarious.

41. Rhododendron camtschaticum Pall. *Therorhodion camtschaticum* (Pall.) Small; *R. camtschaticum* var. *pallasianum* Komar.——EZO-TSUTSUJI. Deciduous shrub 10–30 cm. high, the young branches loosely coarse-hairy; leaves thin-chartaceous, broadly to oblong-obovate, 2–4 cm. long, 1–2 cm. wide, rounded and mucronate, cuneate, long-ciliate, loosely coarse-pilose on both sides or only on the under side; inflorescence racemose, 1- to 3-flowered, the bracts rather leaflike, green, elliptic to obovate, glandular-pilose, the pedicels 1.5–3 cm. long; calyx-lobes narrowly elliptic to broadly lanceolate, 3-nerved, rather large; corolla red, slightly tinged with purple, 2.5–3.5 cm. across; stamens 10, the filaments densely pilose at the base; style pilose at base; ovary villous; capsules narrowly ovoid.——July–Aug. Alpine regions; Hokkaido, Honshu (Mount Iwate in Rikuchu Prov.).——Sakhalin, Ochotsk Sea area, Kuriles, Kamchatka, Alaska, and Aleutians.

Var. **barbatum** (Nakai) Tatew. *Therorhodion camtschaticum* var. *barbatum* Nakai; *R. camtschaticum* subsp. *intercedens* Hult.——ARAGE-EZO-TSUTSUJI. Very glandular-hairy.—— Hokkaido.

5. BRYANTHUS Gmel. CHISHIMA-TSUGAZAKURA ZOKU

Prostrate evergreen much-branched shrub; leaves small, dense, alternate, linear; inflorescence racemose, pedunculate; flowers 4-merous, rose-colored, pedicellate; calyx 4-sected; corolla 4-merous, nearly free to the base, half-open; stamens 8, the filaments filiform, glabrous, the anthers not appendaged, opening by pores; style columnar, the stigma 4-lobed; ovary 4-locular, subglobose; capsules depressed-globose, coriaceous, septicidally dehiscent, 4-valved; seeds many, ovoid.——One species, in Japan and Kamchatka.

1. Bryanthus gmelinii D. Don. *Andromeda bryantha* L.; *Menziesia bryantha* (L.) Sw.; *B. musciformis* Nakai—— CHISHIMA-TSUGAZAKURA. Plant 2–3 cm. high, the branches minutely pubescent; leaves coriaceous, lustrous, 3–4 mm. long, nearly 1 mm. wide, lenticular in cross section, obtuse, the margin often obscurely glandular-puberulent, the midrib impressed on both sides, white-puberulent beneath while young; inflorescence 2–4 cm. long, white-puberulent, densely few-flowered; flowers 5–6 mm. across; calyx-segments ovate, obtuse, about 1 mm. long; corolla lobes oblong, about 3 mm. long; capsules about 4 mm. across.——July–Aug. Dry gravelly and rocky alpine slopes; Hokkaido, Honshu (centr. and n. distr.); rare.——Kuriles and Kamchatka.

5. PHYLLODOCE Salisb. TSUGAZAKURA ZOKU

Small evergreen shrub; leaves many, alternate, coriaceous, linear, subobtuse, minutely toothed, the margins strongly revolute; flowers few, axillary, nearly terminal, 2-bracteate; calyx 5-lobed, acuminate; corolla urceolate or campanulate, with 5 short lobes; stamens 10, included, the filaments filiform, the anthers opening by obovate pores; styles erect, the stigmas 5-lobed; capsules ovoid to depressed-globose, 5-locular, septicidally dehiscent, 5-valved, the valves bifid at apex; seeds many, ovoid.——Few species, in alpine and Arctic regions of the N. Hemisphere.

1A. Corolla campanulate; calyx-lobes ovate, minutely ciliate, glabrous on the back; capsules depressed-globose. . . . . 1. *P. nipponica*
1B. Corolla urceolate.
  2A. Calyx-lobes ovate, glabrous on back, minutely ciliate; corolla pink. . . . . 2. *P.* × *alpina*
  2B. Calyx-lobes narrowly ovate to broadly lanceolate, glandular-pilose on back especially toward the base.
    3A. Flowers pale yellowish green; calyx-lobes green, broadly lanceolate; anthers yellow, the filaments glabrous. . . . . 3. *P. aleutica*
    3B. Flowers purplish; calyx-lobes purplish, narrowly ovate; anthers purplish, the filaments ciliate at base. . . . . 4. *P. caerulea*

1. Phyllodoce nipponica Makino. *P. taxifolia* sensu auct. Japon., non Salisb.; *P. amabilis* Stapf; *P. empetriformis* var. *amabilis* (Stapf) Rehd.; *P. nipponica* var. *amabilis* (Stapf) Stoker——TSUGAZAKURA. Branched, suberect, densely leafy small shrub 7–20 cm. high, the branches angular, white-puberulent in the grooves; leaves spreading, strongly revolute and flattened, sessile, broadly linear, 5–8 mm. long, about 1.5 mm. wide, abruptly obtuse, with minute gland-tipped teeth while young, the midrib impressed on upper side, white-puberulent beneath; flowers nodding, rose to white, the pedicels 2–2.5 cm. long, glandular-pilose and white-puberulent; calyx-lobes ovate, glabrous, minutely ciliolate, 1.5–2 mm. long; corolla 6–7 mm. long, campanulate.——June–Aug. Rocky places, alpine; Honshu (centr. distr. and westw.), Shikoku.

Var. **oblongo-ovata** (Tatew.) Toyokuni. *P. tsugaefolia* var. *oblongo-ovata* Tatew.; *P. nipponica* var. *tsugaefolia* (Nakai) Ohwi; *P. tsugaefolia* Nakai; *P. nipponica* subsp. *tsugaefolia* (Nakai) Toyokuni——NAGABA-TSUGAZAKURA. Plant larger; leaves 8–12 mm. long, about 2 mm. across, the pedicels 2.5–3.5 cm. long.——July–Aug. Alpine; Hokkaido, Honshu (n. distr.).

2. Phyllodoce × alpina Koidz. *P. hybrida* Nakai—— KO-TSUGAZAKURA. Small evergreen shrub, the branches puberulent in the grooves; leaves coriaceous, lustrous, linear, 5–7 mm. long, slightly less than 1.5 mm. wide, minutely toothed, abruptly subobtuse, glabrous, the midrib white-puberulent beneath while young; flowers nodding, pink, the pedicels 2–5, erect, 2–3 cm. long, spreading glandular-hairy and white-puberulent; calyx-lobes ovate, 2.5–3 mm. long, subacute, glabrous, ciliolate; corolla urceolate, 7–8 mm. long.——July–Aug. Alpine; Honshu (centr. distr.); very rare.——Hybrid of Nos. 1 and 3.

3. Phyllodoce aleutica (Spreng.) A. Heller. *Menziesia aleutica* Spreng.; *P. pallasiana* D. Don; *P. taxifolia* var. *aleutica* (Spreng.) Herd.——AO-NO-TSUGAZAKURA. Small evergreen ascending or decumbent shrub 10–30 cm. high, forming dense mats; branches densely leafy, puberulent; leaves linear, 8–14 mm. long, about 1.5 mm. wide, flattened, abruptly obtuse, sessile, glabrous, grooved on upper side, minutely toothed, the teeth glandular while young, the midrib beneath elevated and white-puberulent; flowers 4–7, nodding, the pedicels 2.5–4 cm. long, puberulent and glandular-pilose; calyx-lobes broadly lanceolate, 4–5 mm. long, densely glandular-pilose especially toward the base on back; corolla urceolate, 7–8 mm. long, glabrous, pale yellowish green.——July–Aug. Wet alpine slopes; Hokkaido, Honshu (centr. and n. distr.); rather common especially on Japan Sea side.——Sakhalin, Kuriles, Kamchatka, and Alaska.

4. Phyllodoce caerulea (L.) Bab. *Andromeda caerulea* L.; *A. taxifolia* Pall.; *P. taxifolia* (Pall.) Salisb.——EZO-NO-TSUGAZAKURA. Evergreen decumbent much-branched shrub 10–25 cm. high, with puberulent densely leafy branchlets; leaves linear, 7–12 mm. long, about 1.5 mm. wide, abruptly obtuse, minutely toothed, glabrous, grooved on upper side, the midrib raised and white-puberulent beneath; flowers 2–5, nodding, the pedicels 2–2.5 cm. long, glandular-pilose and loosely puberulent; calyx-lobes narrowly ovate, purplish, about 4 mm. long, densely glandular-pilose, especially toward the base on back; corolla urceolate, 7–8 mm. long, purplish, loosely glandular-puberulent externally, or glabrous.——July–Aug. Alpine; Hokkaido, Honshu (n. distr.).——Northern or Arctic regions of the N. Hemisphere.

7. LOISELEURIA Desv. MINEZUŌ ZOKU

Depressed or prostrate evergreen shrub with decumbent densely leafy branches; leaves small, coriaceous, thick, opposite, narrow, lustrous, with revolute margins; flowers small, white to rose, 4- to 5-merous, erect, in groups of 1 to 5, terminal on the branchlets, pedicelled; calyx 4- to 5-parted, coriaceous; corolla broadly campanulate, the lobes 4 or 5, rounded, imbricate in bud;

stamens 4 or 5, included, the anthers longitudinally split; ovary subglobose, 2- to 4-locular; style short, the stigma capitate; ovules many in each locule; capsules subglobose, septicidal, erect, the valves 2–4, ovate; seeds many, ovoid.——One species, in Arctic and alpine regions of the N. Hemisphere.

1. Loiseleuria procumbens (L.) Desv. *Azalea procumbens* L.; *Chamaeledon procumbens* (L.) Link——Minezuō. Branches glabrous; leaves spreading, narrowly oblong to broadly lanceolate, 6–12 mm. long, 2.5–3 mm. wide, obtuse, acute at base, entire, glabrous and with a groove above, deep green, densely white-puberulent except on the raised midrib, the petioles very short, ascending, white-puberulent on upper side; flowers 4–5 mm. across, pale rose, glabrous, the pedicels erect, glabrous, 4–10 mm. long; calyx-lobes broadly lanceolate, obtuse, about 2 mm. long; corolla 4- to 5-lobed, the lobes deltoid-ovate; capsules ovate-globose, acute, 3–4 mm. long, the persistent style slightly shorter than the capsule.——July. Dry alpine slopes and rocky cliffs; Hokkaido, Honshu (centr. and n. distr.); rather common.——Kuriles, Sakhalin, and widely distributed in alpine and Arctic regions of the N. Hemisphere.

8. CASSIOPE D. Don IWA-HIGE ZOKU

Dwarf evergreen shrubs, the branches often 4-angled, densely scaly-leaved; leaves small, decussate, triangular, scalelike or linear, flat or grooved in the center; flowers solitary, axillary, white or pink, nodding, bracteate, pedicellate; calyx-lobes 4 or 5, free, imbricate in bud; corolla campanulate, the short lobes often recurved; stamens 4–10(–12), included, the anthers with 2 awn-like terminal dorsal appendages opening by pores; ovary 4- or 5-locular, the ovules many; capsules globose, 4- or 5-valved, loculicidally dehiscent, erect; seeds often elongate at both ends.——Few species, in Arctic and alpine regions of the N. Hemisphere.

1. Cassiope lycopodioides (Pall.) D. Don. *Andromeda lycopodioides* Pall.——Iwa-hige. Decumbent much-branched evergreen shrub; branches 4-angled, very slender, densely leafy; leaves scalelike, deep green, appressed to the branches and branchlets, ovate, 1.5–2 mm. long, obtuse, with inflexed hyaline margins, convex on the back; flowers solitary, axillary, 5-merous, nodding or pendulous, pink, the pedicels 1.5–3 cm. long; calyx-lobes ovate, about 2 mm. long, obtuse, with hyaline margins; corolla about 8 mm. long, urceolate-campanulate; capsules globose, about 3 mm. in diameter, erect.——July. Alpine rock-cliffs and crevices; Hokkaido, Honshu (centr. and n. distr.).——Kuriles, Sakhalin, Kamchatka, and n. Pacific regions.

9. ARCTERICA Cov. KOMEBA-TSUGAZAKURA ZOKU

Dwarf evergreen shrub; leaves thick-coriaceous, verticillate in 3's, small; flowers in short terminal racemes, short-pediceled, bracteate and 2-bracteolate; calyx 5-sected, the segments narrowly ovate, obtuse; corolla urceolate, small, nodding, with 5 small lobes imbricate in bud; stamens 10, the anthers opening by pores, with 2 awnlike appendages at base on back; ovary 5-locular; capsules erect, loculicidal; style columnar, the stigma 5-lobed; seeds minute, fusiform-ovoid.——One species.

1. Arcterica nana (Maxim.) Makino. *Andromeda nana* Maxim.; *Cassiope oxycoccoides* A. Gray; *Pieris nana* (Maxim.) Makino; *Arcterica oxycoccoides* (A. Gray) Cov.——Komeba-tsugazakura. Erect shrub, 5–10 cm. high with sparsely branched erect branches 1–2 mm. thick, from subterranean creeping stems or stolons; leaves short-petiolate, spreading, ovate-oblong to oblong, 5–10 mm. long, 3–5 mm. wide, obtuse with a mucronate gland at apex, flat, entire, slightly recurved on margins, deep green and lustrous with the midrib im-pressed above, paler and glabrous beneath; flowers nodding, often in 3's, the bracts persistent, coriaceous, narrowly ovate, 3–4 mm. long, the bracteoles and calyx-segments slightly smaller and obtuse; corolla 4–5 mm. long; capsules subglobose, about 3.5 mm. across, minutely puberulent, slightly longer than the persistent style, depressed at apex.——July. Dry alpine slopes and rocky cliffs; Hokkaido, Honshu (centr. and n. distr.); rather rare.——Kuriles and Kamchatka.

10. HARRIMANELLA Cov. JI-MUKADE ZOKU

Dwarf evergreen slender procumbent shrubs with densely leafy branches; leaves alternate, entire, linear, thick, glabrous or sometimes ciliate; flowers solitary and terminal, one-sided, bractless or the pedicels bracteate; calyx-lobes 5, free, elliptic, imbricate in bud; corolla broadly campanulate, deeply 5-lobed, the lobes imbricate in bud; stamens 10, with 2 dorsal awnlike terminal appendages on the apical part of the anther; styles short, gradually thickened at base; ovary 5-locular; capsules globose, erect, loculicidally dehiscent.——Two species, in Arctic and subarctic regions of the N. Hemisphere.

1. Harrimanella stelleriana (Pall.) Cov. *Andromeda stelleriana* Pall.; *Cassiope stelleriana* (Pall.) DC.——Ji-mukade. Branches wiry and elongate, prostrate, glabrous or slightly puberulent; leaves thick, spreading, narrowly obovate-oblong or broadly oblanceolate, 2–3 mm. long, about 1 mm. wide, very obtuse, upper side flat, lustrous, deep green, glabrous, sessile; flowers solitary and terminal, ebracteate, the pedicels 3–5 mm. long, puberulent; calyx lobes obovate-oblong, about 2.5 mm. long, coriaceous, deciduous in fruit, rounded, reddish, glabrous; corolla 4–5 mm. long; capsules globose, about 3.5 mm. in diameter.——July–Aug. Dry rocky slopes, alpine; Hokkaido, Honshu (centr. distr.); rare.——Kuriles, Kamchatka, and N. America.

11. CHAMAEDAPHNE Moench YACHI-TSUTSUJI ZOKU

Rather low evergreen shrub; leaves coriaceous, flat, alternate, short-petiolate; flowers small pedicelled, in bracteate racemes in the upper axils of branches; calyx-lobes 5, free, persistent, imbricate in bud, 2-bracteate at base; corolla tubular, 5-toothed, the teeth recurved; stamens 10, included, the anthers opening by short cylindric pores; ovary 5-locular, globose; style columnar; ovules many; capsules depressed-globose, loculicidal, 5-valved; seeds in 2 series in each locule, obtusely angled.——One species.

1. **Chamaedaphne calyculata** (L.) Moench. *Andromeda calyculata* L.; *Cassandra calyculata* (L.) D. Don—— YACHI-TSUTSUJI, HOROMUI-TSUTSUJI. Shrub, loosely branched, 30–100 cm. high with erect to ascending branches; leaves narrowly oblong to obovate-oblong, 1–5 cm. long, obtuse to rounded with a mucro at apex, obtuse, subrounded, or acute at base, very short-petiolate, entire or with few obscure teeth near tip, with sessile discoid glands on both surfaces; inflorescence 4–12 cm. long, with discoid sessile glands, the flowers on one side, on short pedicels; bracts leaflike, the bracteoles smaller than the calyx-segments, subtending the calyx; calyx-segments lanceolate; corolla white, 6–7 mm. long; capsules about 4 mm. across.——May–June. Bogs; Hokkaido.——In cooler regions of the N. Hemisphere.

12. EPIGAEA L. IWA-NASHI ZOKU

Small evergreen trailing or creeping subshrubs; branches coarsely hirsute; leaves alternate, short-petioled, subcoriaceous; flowers in terminal or axillary racemes or fascicles, white to rose-colored, subsessile, with persistent bracts and bracteoles; calyx-lobes 5; corolla infundibuliform or tubular-campanulate, 5-lobed; stamens 10, included, the anthers 2-locular, without appendages, longitudinally dehiscent or opening by pores; ovary 5-locular, hirsute, many-ovuled; capsules depressed-globose, loculicidally dehiscent, the placentae thickened, fleshy.——Two species, one in Japan and another in e. N. America.

1. **Epigaea asiatica** Maxim. *Parapyrola trichocarpa* Miq.——IWA-NASHI. Prostrate shrub with ascending branches 10–25 cm. long, with reddish brown gland-tipped hairs; stems slender, woody at base; leaves oblong to ovate-oblong, 4–10 cm. long, 2–4 cm. wide, abruptly cuspidate, cordate, ciliate, deep green and lustrous on upper side, the petioles 5–20 mm. long; racemes terminal, 1–2 cm. long, nodding, the bracts and bracteoles 8–10 mm. long, chartaceous, acuminate, ascending, the pedicels short, short-pilose; calyx-lobes chartaceous, narrowly ovate, acuminate, 7–8 mm. long, glabrous, ciliolate; corolla tubular-campanulate, pale roseate, about 10 mm. long; styles minutely puberulous; capsules about 1 cm. in diameter, short-papillate-hairy.——Apr.–May. Open woods in mountains; Hokkaido (sw. distr.), Honshu. The fruit is edible.

13. LEUCOTHOË D. Don IWA-NANTEN ZOKU

Deciduous or evergreen shrubs; leaves alternate, petiolate, oblong to lanceolate, usually toothed; flowers in racemes, white bracteate and bracteolate; calyx-lobes 5, acute, persistent; corolla ovoid, urceolate, or tubular, with 5 small lobes; stamens 10, the anthers narrowed above, usually ending in 2–4 awns; ovary 5-locular, the ovules many; capsules depressed-globose, loculicidally dehiscent, 5-valved, membranous; seeds small, many.——About 35 species, in N. and S. America, Madagascar, Himalayas, and e. Asia.

1A. Leaves evergreen, toothed; inflorescence on last year's growth; corolla tubular, 1.5–2 cm. long; anthers with 4 awns. . . . . 1. *L. keiskei*
1B. Leaves deciduous, entire; inflorescence on the new growth; corolla urceolate, about 4 mm. long; anthers awnless. . . . . . . 2. *L. grayana*

1. **Leucothoë keiskei** Miq. *Paraleucothoë keiskei* (Miq.) Honda——IWA-NANTEN. Glabrous evergreen shrub with elongate, often declined branches; leaves thick, lustrous, deep green, narrowly ovate to broadly lanceolate, 5–8 cm. long, 1.5–3 cm. wide, caudate, shallowly toothed, the petioles 5–8 mm. long; inflorescence racemose, terminal and axillary, nodding or drooping, 3–5 cm. long; flowers bracteate and bracteolate, the pedicels 1–1.5 cm. long; calyx-lobes broadly ovate, about 2.5 mm. long, subacute, ciliolate; corolla white, 1.5–2 cm. long; filaments pilose; capsules erect, 7–8 mm. in diameter, the pedicels curved upward in fruit; styles 1–1.5 cm. long.——July–Aug. Shaded wet rocky cliffs in mountains; Honshu (Kantō, Tōkaidō, and Kinki Distr.).

2. **Leucothoë grayana** Maxim. var. **oblongifolia** (Miq.) Ohwi. *L. chlorantha* var. *oblongifolia* Miq.; *Eubotryoides grayana* var. *oblongifolia* (Miq.) Hara——HANAHIRI-NO-KI. Deciduous erect glabrous shrub 30–100 cm. high; leaves elliptic to obovate or ovate-oblong, rarely broadly lanceolate, 3–8 cm. long, 1.5–5 cm. wide, acute to obtuse, ciliate, glabrous or loosely pilose on both sides, sometimes glaucescent beneath, the petioles to 3 mm. long; inflorescence a many-flowered, one-sided, leafy, ascending raceme 5–15 cm. long, the pedicels 3–10 mm. long; flowers nodding, pale green; calyx-segments ovate, with few glandular hairs on margin; corolla urceolate, 4–5 mm. long; filaments pilose; ovary glabrous to pilose; capsules depressed-globose, erect, 4–5 mm. in diameter.——July–Aug. Sunny slopes in mountains; Hokkaido, Honshu (centr. and n. distr.).

Var. **grayana**. *L. grayana* var. *typica* H. Boiss.; *Eubotryoides grayana* (Maxim.) Hara——HIRO-HA-HANAHIRI-NO-KI. Leaves larger and broader.——Hokkaido, Honshu (n. distr.).

14. PIERIS D. Don ASEBI ZOKU

Evergreen glabrous or puberulent shrubs or small trees; leaves alternate, coriaceous, petiolate, toothed or rarely entire; inflorescence a terminal axillary panicle or raceme, the bracts single, the bracteoles 2 or absent; calyx-lobes 5, valvate in bud, persistent; corolla ovoid or urceolate, 5-toothed; stamens 10, the anthers longitudinally dehiscent on upper half, with 2 reflexed awnlike dorsal appendages at base; ovary 5-locular; capsules loculicidally dehiscent, the valves not thickened on margin; seeds small, many.——About 8 species, in e. Asia, Himalayas, and N. America.

1. **Pieris japonica** (Thunb.) D. Don. *Andromeda japonica* Thunb.——AsEBI. Shrub 2–4 m. high; leaves thick, broadly oblanceolate, 3–8 cm. long, 1–2 cm. wide, lustrous, deep green, serrulate, glabrous, the petiole short; inflorescence pendulous, densely many-flowered, minutely puberulent; flowers white, nodding, short-pedicellate; calyx-segments broadly lanceolate; corolla 6–8 mm. long; filaments short-pilose; capsules depressed-globose, erect, 5–6 mm. in diameter. ——Apr.–May. Sunny hills; Honshu, Shikoku, Kyushu; rather common.

15. LYONIA Nutt. NEJIKI ZOKU

Evergreen or deciduous shrubs or small trees, the branches sometimes angled; leaves alternate, petiolate, entire or serrulate, sometimes scaly-pubescent; inflorescence terminal; flowers in axillary fascicles, racemes, or panicles, the bracts single, the bracteoles 2 or absent; calyx 4- to 8-lobed, the lobes valvate in bud, persistent; corolla ovoid, urceolate, or tubular-campanulate, short-toothed; stamens 8–16, the filaments with or without 2 short appendages near the top, the anthers obtuse, without appendages, opening by pores; capsules subglobose or ovoid, loculicidally dehiscent, the valves thickened and ligneous on margin; seeds small.——About 30 species, Himalayas, e. Asia, and N. America.

1. **Lyonia ovalifolia** (Wall.) Drude var. **elliptica** (Sieb. & Zucc.) Hand.-Mazz. *Andromeda elliptica* Sieb. & Zucc.; *Pieris ovalifolia* sensu auct. Japon., non D. Don; *P. ovalifolia* var. *elliptica* (Sieb. & Zucc.) Rehd. & Wils.; *P. elliptica* (Sieb. & Zucc.) K. Koch; *L. elliptica* (Sieb. & Zucc.) Okuyama; *L. neziki* Nakai & Hara——NEJIKI. Small deciduous tree or large shrub with smooth yellow-brown bark and terete branches; leaves chartaceous, broadly ovate, ovate-elliptic, or narrowly ovate-oblong, 6–10 cm. long, 2–6 cm. wide, abruptly acuminate, rounded to subcordate at base, white-pilose beneath especially near base; racemes 3–6 cm. long, few-leaved toward the base, from axillary buds of the last year's shoots, the bracts and bracteoles caducous, the pedicels short; calyx-lobes ovate, acute; corolla white, tubular-urceolate, 8–10 mm. long, loosely puberulent; ovary sparsely pilose; capsules 3–4 mm. in diameter, glabrous.——June. Hills and low elevations in the mountains; Honshu, Shikoku, Kyushu; common. ——Formosa and China.

16. ANDROMEDA L. HIME-SHAKUNAGE ZOKU

Small bog shrubs; stems erect, simple or sparsely branched; leaves coriaceous, alternate, short-petiolate, broadly linear to lanceolate, entire and revolute, glaucous beneath; flowers small, white to rose-colored, pendulous, few in terminal sessile umbels, the pedicels bracteate and 2-bracteolate; calyx deeply 5-fid, valvate, the lobes ovate, acute; corolla urceolate, 5-toothed; stamens 10, the anthers with 2 reflexed awns at apex; ovary 5-locular, many-ovuled; capsules subglobose, loculicidally dehiscent; seeds small, obtuse.——Two species, in bogs of the N. Hemisphere.

1. **Andromeda polifolia** L. HIME-SHAKUNAGE. Small glabrous, glaucous shrub, 10–30 cm. high, the ascending stems sparingly branched; leaves broadly linear to oblong-lanceolate, 1.5–3.5 cm. long, 3–7 mm. wide, acute, very short-petiolate, with recurved margins, glaucous beneath; corolla 5–6 mm. long, the pedicels erect, 1–2 cm. long, slender, usually reddish; capsules 3–4 mm. across, ovate-globose.——May–July. Bogs; Hokkaido, Honshu (centr. and n. distr.).——Widely distributed in the cooler regions of the N. Hemisphere.

17. ENKIANTHUS Lour. DŌDAN-TSUTSUJI ZOKU

Usually deciduous or sometimes evergreen shrubs; leaves oblanceolate to ovate-elliptic, serrulate or entire; flowers small, pendulous, in umbels or corymbose-racemes, white to red; calyx 5-fid, the segments persistent; corolla urceolate to campanulate, 5-toothed, the teeth entire or laciniate, the tube rounded or with 5 gibbosities at base; stamens 10, the anthers ventrally dehiscent with 2 awnlike dorsal appendages on back near apex; ovary 5-locular; ovules few in each locule; capsules ellipsoidal, loculicidally dehiscent; seeds few, oblong, rather large, often winged on the margins.——More than 10 species, in e. Asia and the Himalayas.

1A. Flowers in umbels, before or with the leaves; corolla with 5 gibbosities at base, the lobes entire. .... 1. *E. perulatus*
1B. Flowers in racemes after the leaves.
  2A. Corolla urceolate; capsules pendulous; seeds wingless.
    3A. Racemes white-pilose. .... 2. *E. subsessilis*
    3B. Racemes glabrous. .... 3. *E. nudipes*
  2B. Corolla campanulate; capsules erect; seeds winged.
    4A. Corolla-lobes entire. .... 4. *E. campanulatus*
    4B. Corolla-lobes irregularly laciniate. .... 5. *E. cernuus*

1. **Enkianthus perulatus** (Miq.) C. K. Schn. *Andromeda perulata* Miq.; *E. japonicus* Hook. f.; *E. perulatus* var. *japonicus* (Hook. f.) Nakai——DŌDAN-TSUTSUJI. Much-branched deciduous shrub with glabrous branchlets; leaves obovate or elliptic-ovate, 2–4 cm. long, 1–1.5 cm. wide, abruptly acute, cuneate at base, serrulate, glabrous, the midrib puberulous on upper side, white-puberulent beneath, the petioles 5–8 mm. long; calyx-lobes linear-lanceolate, acuminate, about 2 mm. long, glabrous; corolla pale green, ovoid-urceolate, about 8 mm. long, the pedicels 1–2 cm. long, glabrous, recurved in flower, erect in fruit; capsules narrowly oblong, erect, about 8 mm. long.——Apr. Honshu (Izu, Tōkaidō Distr.), Shikoku, Kyushu. Frequently cultivated.

2. **Enkianthus subsessilis** (Miq.) Makino. *Andromeda subsessilis* Miq.; *A. nikoensis* Maxim.; *E. nikoensis* (Maxim.) Makino; *Meisteria subsessilis* (Miq.) Nakai; *Tritomodon sub-*

sessilis (Miq.) F. Maekawa——ABURA-TSUTSUJI. Much-branched erect shrub 1–3 m. high; leaves obovate or elliptic, 2–3 cm. long, 1–1.5 cm. wide, acute, acuminate at base, serrulate, white-pilose on midrib on upper side, loosely brownish pilose beneath, the petioles 1–3 mm. long, short-pilose; racemes pendulous, 5- to 10-flowered, the axis 2–3 cm. long, with short spreading white pubescence, the bracts minute, the pedicels slender, 1–2.5 cm. long, glabrous except at base; calyx-lobes ovate, about 2 mm. long, acuminate, short-awned, white-ciliolate; corolla white, urceolate, about 5 mm. long; capsules glabrous, ellipsoidal, about 4 mm. long, pendulous; styles 3–3.5 mm. long; seeds narrowly oblong, about 3 mm. long, not winged.——May–July. Mountains; Honshu (centr. and n. distr.).

3. Enkianthus nudipes (Honda) Ohwi. *Meisteria nudipes* Honda; *Tritomodon nudipes* (Honda) Honda——Ko-ABURA-TSUTSUJI. Closely allied to the preceding; leaves elliptic or obovate-elliptic, 2–3 cm. long, 1–1.5 cm. wide, acute, acuminate to long-acuminate at base, minutely serrate, glabrous on upper side, except short white-hairy on the midrib when young, pale brown appressed-pubescent beneath, the petioles 1–2 mm. long, glabrescent; racemes glabrous, pendulous, 5- to 10-flowered; calyx-lobes very acute; capsules globose-ellipsoidal, 3.5 mm. long; styles 2.5–3 mm. long.——July. Mountains; Honshu (Mikawa and Mino Prov., westw. to s. Kinki).

4. Enkianthus campanulatus (Miq.) Nichols. *Andromeda campanulata* Miq.; *Meisteria campanulata* (Miq.) Nakai; *Tritomodon campanulatus* (Miq.) F. Maekawa——SARASA-DŌDAN, FŪRIN-TSUTSUJI. Deciduous shrub 4–5 m. high; leaves obovate-elliptic or obovate, 3–7 cm. long, 1.5–3.5 cm. wide, subacute, serrulate, loosely pilose to glabrous on upper side, sparsely tufted brown-hairy on the nerve axils beneath, the petioles 5–12 mm. long; racemes 5- to 15-flowered, pendulous, with scatered pale brown crisp hairs, the pedicels 1–2 cm. long; calyx-lobes broadly lanceolate, acuminate, about 3 mm. long; corolla campanulate, 8–12 mm. long, pale yellow with dark red striations, the limb shallowly 5-toothed on upper quarter; style included, 5–6 mm. long; capsules elliptic, 5–7 mm. long.——June. Mountains; Hokkaido, Honshu (Kinki Distr. and eastw.).——Forma **albiflorus** (Makino) Makino. *E. campanulatus* var. *albiflorus* Makino; *Meisteria campanulata* var. *albiflora* (Makino) Nakai; *Tritomodon campanulatus* forma *albiflorus* (Makino) Hara——SHIROBANA-FŪRIN-TSUTSUJI. Flowers white.

Var. **longilobus** (Nakai) Makino. *E. longilobus* (Nakai) Ohwi; *Meisteria longiloba* Nakai; *Tritomodon longilobus* (Nakai) Honda——TSUKUSHI-DŌDAN. Leaves elliptic to broadly so; corolla 5-toothed on upper 1/3 to 1/2; styles 4–5 mm. long.——Mountains; Kyushu.

Var. **palibinii** (Craib) Bean. *E. rubicundus* Matsum. & Nakai; *E. ferrugineus* Craib; *E. palibinii* Craib; *Meisteria rubicunda* (Matsum. & Nakai) Nakai; *E. campanulatus* var. *rubicundus* (Matsum. & Nakai) Makino——BENI-SARASA-DŌDAN. Leaves obovate to broadly so or oblong; corolla 5-toothed on upper 1/3; styles 3–4 mm. long; capsules about 4 mm. long.——June–July. Mountains; Honshu (n. Kantō and Echigo).

Var. **sikokianus** Palib. *E. sikokianus* (Palib.) Ohwi; *Meisteria sikokiana* (Palib.) Nakai; *Tritomodon sikokianus* (Palib.) Honda——KAINAN-SARASA-DŌDAN. Leaves obovate to broadly so; calyx-lobes narrowly ovate; corolla broadly campanulate, rose-colored, 5-toothed; styles slightly exserted, 4–5 mm. long.——May–June. Honshu (s. Kinki Distr.), Shikoku.

5. Enkianthus cernuus (Sieb. & Zucc.) Makino. *Meisteria cernua* Sieb. & Zucc.; *Tritomodon japonicus* Turcz.; *E. meisteria* Maxim.; *Andromeda cernua* (Sieb. & Zucc.) Miq.; *E. nipponicus* Palib.; *Tritomodon cernuus* (Sieb. & Zucc.) Honda——SHIRO-DŌDAN. Much-branched deciduous shrub; leaves obovate, 2–5 cm. long, 1–2 cm. wide, subacute, acute to acuminate at base, serrulate, pilose on upper side, brown-pubescent on nerves beneath especially dense near base, the petioles 3–7 mm. long; racemes pendulous, 5- to 12-flowered, loosely brown-pubescent, the pedicels 7–15 mm. long; calyx-lobes ovate, about 2 mm. long, abruptly long-acuminate; corolla broadly campanulate, 6–8 mm. long, white, the limb 5-toothed, the teeth laciniate; styles sometimes very slightly exserted, 4.5–5.5 mm. long; capsules 5–6 mm. long.——May–June. Mountains; Honshu (w. distr.), Shikoku, Kyushu.——Forma **rubens** (Maxim.) Ohwi. *Andromeda cernua* var. *rubens* Maxim.; *Tritomodon cernuus* var. *rubens* (Maxim.) Honda——BENI-DŌDAN. Flowers rubescent.

Var. **matsudae** (Komatsu) Makino. *E. matsudae* Komatsu; *Meisteria matsudae* (Komatsu) Nakai; *Tritomodon matsudae* (Komatsu) F. Maekawa——CHICHIBU-DŌDAN. Leaves broadly lanceolate to narrowly ovate, rather coarsely serrulate, slightly brown-pilose especially on midrib beneath; calyx-lobes narrowly ovate to broadly lanceolate; corolla broadly campanulate, deep red; style slightly exserted.——Apr.–May. Honshu (from Kantō to Kinki Distr.).

18. GAULTHERIA L.   SHIRATAMA-NO-KI ZOKU

Small evergreen shrubs, occasionally epiphytic, often decumbent, frequently coarsely hirsute; leaves coriaceous, alternate, toothed; flowers small, in axillary racemes, rarely solitary by reduction, white or reddish, bracteate and bracteolate, the braceoles alternate or opposite, sometimes connate, deciduous or persistent; calyx 5-lobed, usually accrescent after anthesis and surrounding the capsule; corolla urceolate or campanulate, the limb 5-toothed; stamens 10, the anthers 2-locular, the locules tubulose at tip and usually 1- or 2-awned, rarely unappendaged; capsules covered by the fleshy calyx, loculicidally dehiscent, 5-valved; seeds small, many.——More than 100 species, in e. and s. Asia, N. and S. America, and Australia.

1A. Flowers in racemes; fruit white when mature; anthers 4-awned; plant nearly glabrous. .... 1. *G. miqueliana*
1B. Flowers solitary; fruit red when mature; anthers unappendaged; stems, inflorescence, and calyx spreading red-brown coarse hairy.
　　　2. *G. adenothrix*

1. Gaultheria miqueliana Takeda. *G. pyroloides* Hook. & Thoms. ex Miq., pro parte——SHIRATAMA-NO-KI, SHIRO-MONO. Small, erect, deep green sparingly branched ever-green shrub, with creeping subterranean stems; leaves coriaceous, lustrous, pale green beneath, oblong to obovate-oblong, obtuse, subacute at the base, toothed, loosely puberulent on the midrib on upper side, the veinlets reticulate, impressed on upper side, slightly raised beneath; inflorescence 2–6 cm. long, few-flowered, puberulent, the pedicels recurved, 5–7 mm. long, the bracts broadly ovate, obtuse, the bracteoles broadly lanceolate; calyx-lobes deltoid, obtuse, accrescent after anthesis, becoming white and fleshy when mature, glabrous; corolla globose, white, about 6 mm. across; styles about 3 mm. long; fruit globose, about 6 mm. across.——June–July. Coniferous woods and subalpine thickets; Hokkaido, Honshu (centr. and n. distr. and Mount Daisen in Hoki Prov.).——Sakhalin, Kuriles, and Aleutians.

2. **Gaultheria adenothrix** (Miq.) Maxim. *Andromeda adenothrix* Miq.; *Diplycosia adenothrix* (Miq.) Nakai——

Aka-mono.　Brownish red hairy small shrub with subterranean creeping stems; aerial stems 10–30 cm. long, sparingly branched, red-brown and glandular-tipped hairy; leaves coriaceous, lustrous, broadly ovate, 1.5–3 cm. long, 1–2 cm. wide, acute to short-acuminate, undulate-toothed, glabrate, the veinlets prominent, the petioles 1–2 mm. long; flowers solitary, on nodding peduncles 2–4 cm. long, the bracts and bracteoles several, ovate, acute; calyx-lobes ovate, long-hirsute, red and fleshy when mature, surrounding the capsule; corolla campanulate, white, 7–8 mm. long; capsules about 8 mm. across.——May–July. Coniferous woods and subalpine thickets; Hokkaido, Honshu, Shikoku.

19. ARCTOUS Niedenzu　　Urashima-tsutsuji Zoku

Dwarf creeping subshrubs; leaves mostly fascicled at top of branchlets, alternate, petiolate, toothed, persisting over winter although not green; flowers few, pedicellate, in terminal sessile umbels, nodding; calyx 4- or 5-lobed; corolla urceolate, with 4 or 5 short lobes; stamens 8 or 10, the anthers 2-locular, opening by pores, with 2 protuberances on back; ovary 4- or 5-locular, 1-ovuled in each locule; drupe globose, black or red, juicy, 4- or 5-stoned.——About 3 species, in arctic and alpine regions of the N. Hemisphere.

1. **Arctous alpinus** (L.) Niedenzu var. **japonicus** (Nakai) Ohwi. *A. alpinus* sensu auct. Japon., non Niedenzu; *Arctous japonicus* Nakai; *Arctostaphylos alpina* var. *japonica* (Nakai) Hult.; *Arctous ruber* Nakai excl. syn.——Urashima-tsutsuji.　Stems creeping, the plants matted, 2–5 cm. high; leaves thickish, mostly terminal on the branches, obovate to broadly so, 3–7 cm. long, 1–2.5 cm. wide, obtuse to rounded, with prominently impressed veinlets on upper side, serrulate, puberulent at base of the midrib on upper side, pale green and glabrous beneath, the petioles winged; flowers white; corolla about 4 mm. long, pilose inside; drupe glabrous, black, 7–8 mm. in diameter.——June–July. Dry alpine slopes; Hokkaido, Honshu (centr. and n. distr.); rather rare.——Kuriles, Sakhalin. The typical phase occurs in Siberia, Europe, and N. America.

20. CHIOGENES Salisb.　　Harigane-kazura Zoku

Slender evergreen creeping subshrubs; leaves small, alternate, very short-petiolate; flowers small, axillary, solitary, nodding, the pedicels bibracteolate; calyx-tube adnate to the lower half of the ovary, 4-lobed; corolla short-campanulate, deeply 4-fid; stamens 8, the filaments short and broad, the anthers short, 2-locular, each locule barely 2-pointed and opening by a large median chink; ovary 4-locular, many-ovuled; berry globose or obovoid; seeds minute, rather flattened.——Two species, one in Japan, another in e. N. America.

1. **Chiogenes japonica** A. Gray. *C. hispidula* var. *japonica* (A. Gray) Makino; *Gaultheria japonica* (A. Gray) Sleumer——Harigane-kazura.　Prostrate; stems very slender, much branched, hardly more than 1 mm. in diameter; leaves elliptic or obovate-elliptic, 5–10 mm. long, 3–6 mm. wide, obtuse, undulate-margined, glabrous on upper side, scattered puberulent beneath, the petioles less than 1 mm. long; flowers white, nodding, the pedicels glabrous, about 2 mm. long, bibracteolate; fruit obovoid, 5–7 mm. in diameter, white, deflexed.——Coniferous woods; Honshu (centr. and n. distr.); very rare.

21. VACCINIUM L.　　Su-no-ki Zoku

Evergreen or deciduous shrubs or subshrubs; leaves usually alternate, simple, entire or toothed, sessile or petiolate, linear-lanceolate to suborbicular; flowers solitary or in axillary or terminal leafy racemes, bracteate and bracteolate; calyx-teeth 4 or 5, sometimes obsolete; corolla tubular, urceolate, or campanulate, the limb 4- or 5-toothed or parted nearly to the base, the teeth recurved; stamens 8 or 10, the anthers sometimes spurred on back, opening by pores; ovary inferior, 4- to 10-locular; berry red, purple-black, or white, several- to many-seeded, the calyx-lobes persistent.——About 150 species, in cooler and Arctic regions of the N. Hemisphere and at high elevations in the Tropics.

1A. Corolla urceolate, campanulate, or tubular, the lobes short, not recurved.
　2A. Evergreen shrubs.
　　3A. Large shrub 1–3 m. high; leaves 2.5–6 cm. long, acuminate; racemes 2–5 cm. long; berry purple-black. .... 1. *V. bracteatum*
　　3B. Small shrub 5–20 cm. high; leaves 1–2.5 cm. long, rounded to very obtuse; racemes less than 2 cm. long; berry red.
　　　　　　　　　　　　　　　　　　　　　　　　　　　　　　　　　　　　　　　2. *V. vitis-idaea*
　2B. Deciduous shrubs.
　　4A. Ovary 4- or 5-locular.
　　　5A. Anthers awnless; dwarf subshrub 2–5 cm. high; corolla campanulate; fruit red. .... 3. *V. praestans*
　　　5B. Anthers awned; shrubs more than 5 cm. high; corolla urceolate.

6A. Branches angled; flowers solitary in axils of new shoots.
  7A. Leaves membranous, vivid green in dried specimens; fruit red. . . . . 4. *V. yatabei*
  7B. Leaves chartaceous, brownish in dried specimens, sometimes glaucescent.
    8A. Leaves entire, elliptic to broadly ovate, the veinlets slender and only slightly raised beneath. . . . . 5. *V. ovalifolium*
    8B. Leaves serrulate, broadly ovate to ovate-orbicular or sometimes elliptic, the veinlets prominently raised and reticulate
      beneath. . . . . 6. *V. shikokianum*
  6B. Branches terete; flowers 1–3(–4) in short terminal corymbs from shoots of previous season; leaves obovate or obovate-orbicular,
    rarely elliptic, entire; fruit purple-black. . . . . 7. *V. uliginosum*
4B. Ovary 10-locular.
  9A. Inflorescence few-flowered, on previous year's branchlets; small shrubs.
    10A. Branches slightly flattened, glabrous; leaves lanceolate to narrowly ovate, nearly glabrous; fruit purple-black.
                                                              8. *V. yakushimense*
    10B. Branches terete, more or less puberulent; leaves broadly lanceolate to broadly ovate, puberulent.
      11A. Calyx-tube and young fruit 5-angled; fruit red, the receptacle rather broad. . . . . 9. *V. hirtum*
      11B. Calyx-tube and young fruit terete, not angled; fruit black, the receptacle rather narrow. . . . . 10. *V. smallii*
  9B. Inflorescence many-flowered, racemose, terminal on the new branchlets; large shrubs.
    12A. Inflorescence and young branchlets with stipitate glands; leaves chartaceous, callose-hairy toward the margin, the hairs gland-
      tipped only while young; bracts rather small, thinly chartaceous. . . . . 11. *V. oldhamii*
    12B. Inflorescence and young branchlets short-pubescent or glabrescent, not glandular-pilose; leaves rather coriaceous, scarcely
      pilose above; bracts usually large, chartaceous.
      13A. Plant mostly pilose. . . . . 12. *V. ciliatum*
      13B. Plant mostly glabrous. . . . . 13. *V. sieboldii*
1B. Corolla deeply 4-parted, the lobes recurved.
  14A. Erect shrub; leaves deciduous, usually toothed. . . . . 14. *V. japonicum*
  14B. Usually creeping subshrubs; leaves evergreen, entire.
    15A. Leaves 5–12 mm. long; peduncles puberulent. . . . . 15. *V. oxycoccum*
    15B. Leaves 3–6 mm. long; peduncles usually glabrous. . . . . 16. *V. microcarpum*

1. **Vaccinium bracteatum** Thunb. *V. buergeri* Miq.; *V. donianum* var. *ellipticum* Miq.; *V. idzuroei* Fr. & Sav.; *V. taquetii* Lév.; *Pieris fauriei* Lév.; *P. coreana* Lév.——SHA-SHAMBO. Much-branched large evergreen shrub or small tree, 1–3 m. high, the branches rather slender, gray-brown or gray, glabrous or punctate-puberulent while very young; leaves coriaceous, lustrous, elliptic-ovate or narrowly elliptic, 2.5–6 cm. long, 1–2.5 cm. wide, obsoletely serrulate, acuminate, with 2 glands near the base, glabrous or puberulent on the midrib on upper side, the petioles 2–4(–6) mm. long; racemes 2–5 cm. long, puberulent, more than 10-flowered, the bracts lanceolate, the flowers nodding; corolla white, tubular-ovoid, 5–7 mm. long, puberulent or glabrous; fruit globose, about 5 mm. in diameter, purple-black, glaucescent, the calyx-teeth persistent.——July. Low elevations in the mountains; Honshu (s. Kantō Distr. and westw.), Shikoku, Kyushu.——s. Korea, Ryukyus, and China.

Var. **lanceolatum** Nakai. NAGABA-SHASHAMBO. Leaves broadly oblanceolate, 4–8 cm. long, 1–2 cm. wide.——Occurs with the typical phase.

2. **Vaccinium vitis-idaea** L. *V. vitis-idaea* var. *minus* Lodd.; *V. jesoense* Miq.——KOKE-MOMO. Dwarf evergreen subshrub, 5–20 cm. high, simple to sparingly branched, with long-creeping subterranean stems; leaves coriaceous, 1–2.5 cm. long, 5–13 mm. wide, obovate-elliptic to obovate, rounded to very obtuse, lustrous, deep green on upper side, paler with dark spots beneath, glabrous on both sides, slightly recurved on margin, the petioles 0.5–1.5 mm. long; flowers nodding, in short terminal racemes; corolla campanulate, 6–7 mm. long, 4-lobed, white to reddish; berries globose, red, 4–7 mm. in diameter.——June–July. Coniferous woods and alpine slopes; Hokkaido, Honshu, Shikoku, Kyushu; common northward, rare in Shikoku and Kyushu.——Korea, Kuriles, Sakhalin to Europe, and N. America.

3. **Vaccinium praestans** Lamb. IWA-TSUTSUJI. Dwarf deciduous scarcely branched glabrous subshrub, 2–5 cm. high, with long-creeping subterranean branches; leaves chartaceous to submembranous, broadly ovate, broadly elliptic, or obovate-orbicular, 3–6 cm. long, 3–5 cm. wide, rounded to acute, serrulate, often loosely pilose on margin and on nerves beneath; flowers 1–3, terminal, puberulent, the bracts and bracteoles herbaceous, the pedicels rather short; corolla campanulate, 5–6 mm. long, white to reddish; berries globose, about 1 cm. in diameter, red, the receptacles 4–5 mm. across.——June–July. Coniferous woods; Hokkaido, Honshu (centr. and n. distr.); rare.——Kuriles, Sakhalin, Kamchatka, Amur, and Ussuri.

4. **Vaccinium yatabei** Makino. *V. myrtillus* sensu auct. Japon., non L.; *V. myrtillus* var. *yatabei* (Makino) Matsum. & Komatsu——HIME-USU-NO-KI, AO-JIKU-SU-NO-KI. Small glabrous branched deciduous shrub 10–20 cm. high, the stems 2–4 mm. across at base, the branches green becoming gray-brown in age, acutely 4-angled; leaves thinly membranous, ovate to broadly so, 1–2 cm. long, 7–13 mm. wide, acute, mucronate, vivid green when dried, serrulate, sparingly puberulent on the nerves, very short-petioled; flowers with the leaves, solitary, the pedicels glabrous, short; calyx-teeth short; corolla greenish white, urceolate, 4–5 mm. long; berry globose, red.——May–June. Coniferous woods; Honshu (centr. distr. and Kantō); rare.

5. **Vaccinium ovalifolium** J. E. Smith. *V. axillare* Nakai; (?) *V. yesoense* Koidz.; *V. ovalifolium* var. *membranaceum* H. Boiss.——KURO-USUGO. Glabrous much-branched deciduous shrub 30–100 cm. high; stems 5–15 mm. thick near base, the branches yellow-brown, acutely 4-angled; leaves subchartaceous, elliptic to broadly ovate, 2–4 cm. long, 1.5–2.5 cm. wide, obtuse, acute to obtuse at base, glabrous, entire, slightly glaucescent beneath, weakly nerved on both sides, subsessile; flowers solitary, nearly terminal, nodding, the pedicels ebracteate, 7–12 mm. long, glabrous; calyx-limb short, obsoletely toothed; corolla urceolate, about 5 mm. long, pale green; berries blue-black, globose, bloomy, 8–10 mm. in diameter, sweet and edible.——June–July. Alpine; Hokkaido, Honshu (centr. and n. distr.); rather common.——e. Asia and N. America.——Forma **obovoideum** (Takeda) Hara. *V.*

axillare var. *obovoideum* Takeda——NAGAMI-KURO-USUGO. Berries obovoid.——Forma **platyanthum** (Nakai) Hara. *V. axillare* var. *platyanthum* Nakai.——MIYAMA-KURO-USUGO. Corolla depressed-urceolate.

Var. **coriaceum** H. Boiss. *V. axillare* var. *coriaceum* (H. Boiss.) Hara; *V. chamissonis* sensu auct. Japon., non Bong.——EZO-KURO-USUGO. Leaves broadly ovate, serrulate, especially on lower half.——Occurs with the typical phase.—— Forma **angustifolium** (Nakai) Ohwi. NAGABA-KURO-USUGO. Narrow-leaved phase.

6. Vaccinium shikokianum Nakai. *V. ovalifolium* var. *coriaceum* H. Boiss. pro parte——MARUBA-USUGO, SHIKOKU-USUGO. Closely resembles the preceding; branches gray-brown, glabrous, acutely angled; leaves subcoriaceous, broadly ovate to ovate-orbicular, or sometimes broadly elliptic, 1.5–2.5 cm. long, 1–2.5 cm. wide, rounded to obtuse, sometimes shallowly cordate at base, serrulate, glabrous, glaucous and with prominently raised veinlets beneath, subsessile; flowers solitary, opening with the leaves, pale greenish, nodding, the pedicels 5–7 mm. long, naked; calyx-limb short; corolla urceolate, about 5 mm. long; awns of the anthers slightly exceeding the anther length; berries blue, globose, about 8 mm. in diameter.——Alpine; Honshu (n. distr. and Japan Sea side of centr. distr.); rare.

7. Vaccinium uliginosum L. KURO-MAME-NO-KI. Branched deciduous shrub 30–70 cm. high; stems to 1–2 cm. across, the branches terete, glabrous, brownish; leaves thick-chartaceous, obovate to obovate-orbicular, rarely elliptic, 1.5–2.5 cm. long, 1–2 cm. wide, rounded to very obtuse, acute to subcuneate at base, entire, whitish and with raised veinlets beneath, the petioles 1–2 mm. long; flowers few, glabrous, the pedicels 5–7 mm. long; calyx-teeth deltoid; corolla 5–7 mm. long, urceolate, greenish to pinkish; berries globose, 6–7 mm. in diameter, blue-black, bloomy.——June–July. Alpine; Hokkaido, Honshu (centr. and n. distr.).——Mountainous regions of the N. Hemisphere.

8. Vaccinium yakushimense Makino. AKU-SHIBA-MO-DOKI. Nearly glabrous branched deciduous shrub, branches green, slightly flattened, with a deep groove on each side; leaves lanceolate to narrowly ovate, 3–6 cm. long, 1–1.5 cm. wide, long-acuminate, acute to subrounded at base, serrulate, glabrous except the midrib puberulent on the upper side; flowers 1 or 2 near the top of last year's branchlets, the pedicels glabrous, about 3 mm. long; corolla small, urceolate; berries obovoid-globose, purple-black, about 9 mm. in diameter. ——High mountains; Kyushu (Yakushima); rare.

9. Vaccinium hirtum Thunb. *V. buergeri* Miq., pro parte; *V. buergeri* var. *minor* Miq.; *V. pterocarpum* Nakai; *V. kiusianum* Koidz.; *V. hirtum* var. *kiusianum* (Koidz.) Hara; *V. motosukeanum* Koidz.; *V. usunoki* Nakai; *V. buergeri* var. *usunoki* (Nakai) Koidz.——USU-NO-KI, KAKUMI-SU-NO-KI. Much-branched erect shrub to 1 m. high, the young branchlets green, rather slender, curved-puberulent, obscurely 2-grooved; leaves slightly bitter in taste, membranous, narrowly to broadly ovate, sometimes broadly lanceolate, 2–5 cm. long, 1–2.5 cm. wide, acuminate, usually rounded at base, serrulate, vivid green and sometimes reddish, usually rounded at base, puberulent especially the midrib on upper side, white-pubescent beneath near base; flowers 1–3, on shoots of last year, greenish to pinkish; calyx 5-toothed; berries obovoid-globose or subglobose, red.——May. Thickets and pine woods in hills and low elevations in the mountains; Hokkaido, Honshu, Shikoku, Kyushu; rather common and very variable.——Forma **lasiocarpum** (Koidz.) Ohwi. *V. hirtum* var. *lasiocarpum* Koidz.; *V. lasiocarpum* (Koidz.) Nakai——KE-USU-NO-KI. Prominently white-pubescent throughout.

10. Vaccinium smallii A. Gray. *V. hirtum* var. *smallii* (A. Gray) Maxim.——ŌBA-SU-NO-KI. Much-branched erect shrub about 1 m. high, the branches terete, the branchlets slender, often shallowly 2-grooved, usually curved-puberulent; leaves membranous, slightly acid in taste, oblong, elliptic, or broadly ovate, 3–8 cm. long, 1.5–3 cm. wide, acuminate to acute, acute to obtuse at base, serrulate, puberulent on the midrib on upper side and somewhat so on the midrib beneath, the petioles 1–1.5 mm. long; flowers 1–3, on the shoots of last year, nodding, pedicellate, greenish white to pinkish; calyx-tube not angled, the teeth oval to obtuse-deltoid, commonly with few minute stipitate glands on margin; corolla campanulate, 6–7 mm. long; berries globose, 7–8 mm. in diameter, not angled, purple-black, the receptacle narrow.——Apr.–May. Thickets and sunny places in mountains; Hokkaido, Honshu (centr. and n. distr.); rather common.——s. Kuriles and Sakhalin.

Var. **glabrum** Koidz. *V. politum* Koidz.; *V. hirtum* var. *versicolor* Koidz.; *V. versicolor* (Koidz.) Nakai; *V. versicolor* var. *glabrum* (Koidz.) Hara; *V. versicolor* var. *minus* (Nakai) Hara; *V. kansaiense* Koidz.——SU-NO-KI. Leaves rather small, 2–4 cm. long, 1–2 cm. wide, usually less hairy. ——Thickets and pine woods in hills and low elevations in the mountains; Honshu, Shikoku.

11. Vaccinium oldhamii Miq. NATSU-HAZE. Erect shrub 1–3 m. high, the branches dark brown, the branchlets gray-brown, short curved-puberulent and stipitate-glandular while young, terete, rather slender; leaves elliptic, oblong or ovate, 3–8 cm. long, 2–4 cm. wide, acute to short-acuminate, acute or rarely rounded at base, serrulate with glandular-tipped teeth, puberulent on midrib and often scattered coarse hairy on upper side, pale green and coarsely pilose on midrib beneath, with 2 sessile glands on the margin near base, the petioles 1–2.5 mm. long, puberulent; racemes 3–6 cm. long, terminal on young new shoots, rather densely many-flowered, curved-puberulent and stipitate glandular, the bracts lanceolate to narrowly ovate, 4–7 mm. long, acuminate, somewhat persistent in fruit, the pedicels 1–3 mm. long; calyx-teeth depressed-deltoid; corolla campanulate, 4–5 mm. long, pale yellow-brown and reddish, glabrous, 5-toothed; berries black, bloomy, globose, 6–7 mm. in diameter, the calyx-teeth persistent.——May–June. Thickets and woods at low elevations in the mountains; Hokkaido, Honshu, Shikoku, Kyushu; rather common.——s. Korea and China.

12. Vaccinium ciliatum Thunb. *V. santanense* Koidz. ——ARAGE-NATSU-HAZE. Large erect shrub, the branches dark brown or gray brown, terete, short-puberulent while young; leaves elliptic, obovate-elliptic, or broadly ovate, 3–5 (–8) cm. long, 2–3(–4.5) cm. wide, acute to subobtuse, acute or rarely subrounded at base, entire, puberulent on midrib on upper side, spreading coarse-hairy beneath, the petioles 1–3 mm. long; racemes elongate, 3–7 cm. long, terminal on new shoots, rather densely many-flowered, the bracts narrowly ovate to lanceolate, the pedicels 3–4 mm. long, puberulent; berries globose, black, bloomy, 5–7 mm. in diameter.——Honshu (Tango, Suwō, and Nagato Prov.), Kyushu.

13. Vaccinium sieboldii Miq. *V. donianum* var. *elliptica* Miq.; *V. longeracemosum* Fr. & Sav.; *V. nagurae* Makino;

FIG. 11.—*Vaccinium sieboldii* Miq. 1, Flowering branch; 2, fruiting branch; 3, berry; 4, flower; 5, 6, stamens; 7, style; 8, flower.

Kadzuwo Inami del.

V. ciliatum var. *sieboldii* (Miq.) Okuyama——Nagabo-natsu-haze. Fig. 11. Large shrub, the branches usually gray-brown, the branchlets terete, with loose whitish curved hairs while young; leaves elliptic, ovate-elliptic, or obovate, 4–8 cm. long, 2–4 cm. wide, acuminate to acute, acute at base, entire, deep green and loosely puberulent on midrib on upper side, glabrous or the midrib scattered coarse-hairy beneath with 2 sessile glands on margin near base, the petioles 2–4 mm. long, usually glabrous; racemes 5–10 cm. long, rather densely many-flowered, glabrous or scattered white-pilose, the pedicels 4–8 mm. long, the bracts lanceolate to obovate, 5–10 mm. long; corolla tubular-campanulate, 4–5 mm. long; berries globose, bloomy, 5–6 mm. in diameter.——Honshu (Mikawa and Kii Prov.), reported to occur in Kyushu (Satsuma).

14. Vaccinium japonicum Miq. *Oxycoccus japonicus* (Miq.) Makino; *Hugeria japonica* (Miq.) Nakai; *Oxycoccoides japonicus* (Miq.) Nakai; *V. japonicum* var. *fauriei* Sleumer; *Hugeria incisa* F. Maekawa——Aku-shiba. Deciduous sparingly branched shrub 30–60 cm. high, the branches green, the branchlets terete or often slightly flattened, with 2 shallow grooves, usually glabrous or sometimes puberulent; leaves ovate to broadly lanceolate, 2–6 cm. long, 1–3 cm. wide, acuminate, rounded to subcordate at base, vivid green and puberulent on midrib on upper side, often whitish or sometimes pale green and glabrous beneath, serrulate, subsessile, the teeth glandular while young; flowers solitary in leaf axils, the peduncles slender, 1–2 cm. long, glabrous, linear-bracteolate near the base; calyx-teeth short-deltoid; corolla 7–10 mm. long, white or pink, lobed nearly to the base, the lobes lanceolate, strongly recurved; stamens 8, the anthers red-brown, yellow at apex, linear, the filaments white-pilose; berries red, about 7 mm. in diameter; seeds impressed-punctate.——June–July. Woods in mountains and hills; Hokkaido, Honshu, Shikoku, Kyushu.——s. Korea.

Var. **ciliare** Matsum. *Oxycoccus japonicus* var. *ciliaris* (Matsum.) Makino & Nemoto; *Hugeria japonica* var. *ciliaris* (Matsum.) Nakai——Ke-aku-shiba. Leaves usually smaller and the young branchlets and peduncles glandular-pilose.——Honshu (w. distr.) and Shikoku.

15. Vaccinium oxycoccum L. *Oxycoccus vulgaris* Hill; *O. quadripetalus* Gilib.; *O. palustris* Pers.——Tsuru-koke-momo. Stems very slender, long-creeping, sparsely branched, puberulent while young, becoming dark red-brown; leaves coriaceous, oblong to narrowly ovate, obtuse to subacute, rounded at base, glabrous, lustrous and deep green, paler and glaucous beneath, entire, the margins slightly recurved, the midrib impressed on upper side, slightly raised beneath, the petioles about 1 mm. long; flowers nodding, the peduncles 3–4 cm. long, deflexed at tip, with 2 small linear bracteoles near the middle; corolla 7–10 mm. long, rose-colored, the segments broadly lanceolate, obtuse, strongly recurved; berries red, globose, about 1 cm. in diameter; seeds longitudinally rugose.——June–July. Sphagnum bogs; Hokkaido, Honshu (centr. and n. distr.).——Kuriles, Sakhalin, and widely distributed in the cooler regions of the N. Hemisphere.

16. Vaccinium microcarpum (Turcz.) Schmalh. *Oxycoccus microcarpus* Turcz.; *O. palustris* var. *pusillus* Dunal; *O. pusillus* (Dunal) Nakai——Hime-tsuru-koke-momo. Resembles the preceding but much slenderer; stems about 0.5 mm. across, puberulent while young, becoming dark-brown; leaves coriaceous, oblong-ovate, 3–6 mm. long, 2–2.5 mm. wide, obtuse to subacute, deep green and lustrous on upper side, glabrous, entire, paler and glaucous beneath, the margins slightly recurved, the petioles less than 1 mm. long; flowers nodding, rose-colored, the peduncles about 2 cm. long, slender, nearly glabrous, deflexed at tip, the bracteoles 2, minute, on lower half of the peduncles; corolla-segments broadly lanceolate, strongly recurved, obtuse, 6–7 mm. long; anthers yellow; styles slender, longer than the stamens; berries red, nodding, globose, 6–7 mm. in diameter.——Sphagnum bogs; Hokkaido, Honshu (Shinano and northw.); rare.——Kuriles, Sakhalin, N. Korea, and widely distributed in the N. Hemisphere.

Fam. 160. **MYRSINACEAE** Yabu-kōji Ka   Myrsine Family

Mostly small evergreen shrubs or small trees, sometimes suffrutescent undershrubs; leaves simple, usually alternate, rarely opposite, entire or toothed, without stipules; inflorescences axillary or in terminal panicles, racemes, or fascicles, often on special articulating lateral branches; flowers bisexual, sometimes unisexual, 4- to 5(–6)-merous, white or pink; calyx persistent, 4- to 5(–6)-lobed; corolla gamopetalous, campanulate, cup-shaped, or rotate; stamens inserted on corolla-tube or at the base, always opposite the lobes, the anthers longitudinally splitting or opening by pores; ovary superior or semi-inferior (in *Maesa* only), 1-locular, the ovules 1 to many, imbedded in a basal or free central placenta; style simple, the stigma entire, lobed, or fringed; fruit a drupe or berry; seeds 1 or many (in *Maesa*), the endosperm fleshy or horny.——About 25 genera, with about 1,000 species, chiefly in tropical and subtropical regions.

1A. Ovary and fruit semi-inferior; flowers bibracteolate; seeds many. .... 1. *Maesa*
1B. Ovary and fruit superior; flowers without bracteoles; seeds solitary, globose.
  2A. Corolla-lobes dextrorsely contorted; flowers bisexual; inflorescence paniculate, corymbose, or subumbellate. .... 2. *Ardisia*
  2B. Corolla-lobes valvate or imbricate in bud, not contorted; flowers fascicled in axils, usually unisexual, rarely bisexual. .... 3. *Myrsine*

1. **MAESA** Forsk.   Izu-senryō Zoku

Erect or decumbent, evergreen shrubs or small trees; leaves toothed or entire, glabrous or pubescent, usually glandular within; inflorescence terminal or axillary, racemose or paniculate; flowers small, bisexual or unisexual, 5-merous, with 2 bracteoles at base or on the pedicels; calyx-lobes persistent on sides or near apex of semi-inferior ovary; corolla tubular, cup-shaped or campanulate, the lobes glandular-striate within; stamens inserted on the corolla-tube, the filaments distinct, nearly as long as the ovoid to reniform anthers; ovary with a globose placenta, many-ovuled; style columnar, slender, the stigma entire or 3- to 5-lobed; fruit a globose to ovoid, fleshy berry; seeds small, many, angular.——About 100 species, chiefly in the Tropics of the Old World, a few in Australia and Africa.

1A. Corolla tubular-campanulate, the tube 2 or 3 times longer than the ascending to erect lobes; bracteoles broadly ovate to reniform, rounded to very obtuse. .... 1. *M. japonica*
1B. Corolla broadly campanulate, the tube nearly as long as the spreading lobes; bracteoles lanceolate, small, acute. .... 2. *M. tenera*

1. Maesa japonica (Thunb.) Moritzi. *Doraena japonica* Thunb.; *M. doraena* (Thunb.) Bl.; *M. coriacea* Champ.——Izu-senryō. Dioecious shrub about 1 m. high, with much-elongate arcuate-declining branches; branchlets glabrous, green, loosely lenticellate; leaves oblong to elliptic, sometimes broadly lanceolate or broadly elliptic, 5–17 cm. long, 2–5 cm. wide, acuminate, rounded to acute at base, loosely short-toothed or subentire, deep green and slightly lustrous on upper side, glabrous on both sides except for loose brownish granular short hairs while very young, the lateral nerves 5–8 pairs, weak, ending in the teeth, the petioles 1–1.5 cm. long; racemes axillary, sometimes compound in the staminate ones, 1–2 cm. long; bracteoles broadly ovate to reniform, rounded or obtuse, glandular-striate; flowers about 5 mm. long; calyx-teeth deltoid; corolla yellowish white, tubular-campanulate, the lobes about 1/3 as long as the tube; fruit globose, about 5 mm. across, milky white, brownish striate.——Apr.–June.

Woods in low mountains; Honshu (s. Kantō Distr. and westw.), Shikoku, Kyushu.——Ryukyus, Formosa, China, and Indochina.

2. Maesa tenera Mez. *M. formosana* Mez; *M. sinensis* sensu auct. Formos., non DC.——Shima-Izu-senryō, Karaubagane. Glabrous branched shrub; leaves elliptic or oblong, 8–13 cm. long, 3–5 cm. wide, acute at both ends, coarsely undulate-toothed or subentire, the petioles 1–1.5 cm. long; inflorescence axillary, a simple or rarely a compound raceme, 1–5 cm. long; bracteoles lanceolate, acute; calyx-teeth ovate; corolla 2.5–3 mm. long, white or slightly purplish, the tube about as long as the spreading rounded lobes; fruit globose, 3.5–4 mm. across, obsoletely striate, the upper 1/4–1/3 exserted beyond the calyx-lobes.——Mar.–Apr. Woods and thickets in hills and low mountains; Kyushu (Osumi, Satsuma Prov., incl. Koshiki Isl.).——Ryukyus, Formosa, and China.

2. ARDISIA Sw. YABU-KŌJI ZOKU

Trees or shrubs or creeping suffrutescent undershrubs; leaves usually alternate, sometimes opposite or subverticillate in the suffrutescent species, entire, undulate or sometimes toothed, usually with marginal glands; inflorescence a terminal or axillary panicle, corymb, or umbellate corymb; flowers small, bisexual or polygamous, 4- or 5-merous, white or pink, usually glandular; sepals almost free or more or less united at base, dextrorsely contorted, sometimes imbricate or valvate; corolla gamopetalous, the lobes dextrorsely contorted; stamens inserted at base of corolla-tube, the filaments short, the anthers erect, dehiscing by longitudinal slits or rarely by pores; ovary superior, 1-locular, the placenta globose, bearing 3–12 ovules in 1–3 series; fruit a berrylike globose drupe, with a solitary seed, the persistent style simple, slender, the stigma apiculate, punctate.——About 250 species, chiefly in the Tropics of Asia and America, few in Australia.

1A. Small shrubs with long slender, more or less creeping stems, the aerial stems ascending, scarcely branched, 5–30 cm. long, 2–4 mm. in diameter; leaves verticillately arranged; inflorescence lateral, leafless.
2A. Creeping stems without leaves; stem-leaves rather coriaceous, 6–13 cm. long, many-toothed.
3A. Stems and peduncles with very short and minute glandular hairs; peduncles 1–1.5 cm. long; calyx-lobes broadly ovate, abruptly acute. .... 1. *A. japonica*
3B. Stems and peduncles with very short and minute glandular hairs as well as spreading multicellular hairs about 0.5 mm. long; peduncles 1.5–2.5 cm. long; calyx-lobes ovate, acuminate. .... 2. *A. montana*
2B. Creeping stems leafy; stems slender; leaves chartaceous, 2–6 cm. long, with less than 10 coarse teeth on each side; plant with long multicellular hairs. .... 3. *A. pusilla*
1B. Small shrubs or trees without creeping stems, the erect stems usually branched; leaves alternate; inflorescence lateral or subterminal, usually on special lateral branches, these sometimes leafless.
4A. Leaves with distinctive albuminous marginal glands, toothed or entire; fruit red; small sparsely branched shrubs.
5A. Leaves oblong, acute to abruptly acuminate with an obtuse tip; stems and inflorescence glabrous; flowering branches few, on the upper part of the stems, spreading; inflorescence terminal, sessile. .... 4. *A. crenata*
5B. Leaves lanceolate or broadly so, gradually narrowed to an obtuse tip, green; upper part of stem and the inflorescence more or less minutely granular-puberulent; inflorescence axillary and long-peduncled. .... 5. *A. crispa*
4B. Leaves without distinctive marginal glands, entire or slightly undulate; fruit dark purplish; large much-branched shrubs.
6A. Branches and peduncles rather thick; calyx-segments broadly ovate, subobtuse; leaves pale grayish brown when dried, the midrib not prominent. .... 6. *A. sieboldii*
6B. Branches and peduncles rather slender; calyx-segments deltoid, short, subacute; leaves darkened when dried, dark purplish brown beneath, the midrib rather slender. .... 7. *A. quinquegona*

1. Ardisia japonica (Thunb.) Bl. *Bladhia japonica* Thunb.——Yabu-kōji. Usually a low unbranched suffrutescent shrub with minute hairs on upper part of stems, inflorescence, and young petioles; creeping stems elongate, sparingly branched, leafless; stems ascending, 10–30 cm. long, 2–3 mm. in diameter, terete, with few whorls of verticillate leaves; leaves oblong or narrowly so, 6–13 cm. long, 2–5 cm. wide, acute at both ends, short mucronate-toothed, lustrous, deep green, glabrate, the petioles 7–13 mm. long; inflorescence in the axils of leaves or scalelike leaves, the peduncles slender,

spreading, 1–1.5 cm. long, the pedicels 2–5, subumbellate, 7–10 mm. long; flowers white, nodding, 6–8 mm. across, rotate; calyx-segments broadly ovate, abruptly acute, ciliolate; fruit globose, red, 5–6 mm. across.——July–Aug. Woods in hills and low mountains; Hokkaido (Okushiri Isl.), Honshu, Shikoku, Kyushu.——Korea, China, and Formosa.

Var. **angusta** (Nakai) Makino & Nemoto. *Bladhia japonica* var. *angusta* Nakai——Hosoba-yabu-kōji. Leaves lanceolate, 2–5 cm. long, 6–20 mm. wide.——Honshu (Izu-oshima), Kyushu (Yakushima).

2. Ardisia montana (Miq.) Sieb. ex Fr. & Sav. *A. japonica* var. *montana* Miq.; *Bladhia montana* (Miq.) Nakai ——Ō-TSURU-KŌJI. Resembles the preceding; creeping stems elongate, the ascending stems simple, 10–30 cm. long, 2–3 mm. in diameter, terete, with spreading pale brown multicellular hairs on the upper part; leaves subcoriaceous, in a few whorls on upper part of stems, usually in 3's or in 4's, obovate-oblong to oblong, 5–13 cm. long, 2–4 mm. wide, acute, rather coarsely and acutely toothed, with spreading hairs on midrib beneath while young, the petioles 7–15 mm. long; peduncles usually from axils of scalelike leaves between the whorls of ordinary leaves, 1.5–2.5 cm. long, slender, the pedicels 7–12 mm. long; flowers 2–6, white, 6–8 mm. across; calyx-segments ovate or narrowly so, short-acuminate, ciliolate; fruit red, 5–6 mm. across, globose.——May. Woods; Honshu (Awa Prov., Izu-shichito Isl. and westw.), Kyushu. Intermediate between *A. japonica* and *A. pusilla.*

3. Ardisia pusilla A. DC. *Bladhia villosa* Thunb.; *A. villosa* (Thunb.) Mez, non Roxb.——TSURU-KŌJI. Creeping stems long and loosely branched, with whorls of verticillate leaves, the ends or branches becoming aerial shoots, densely hirsute with long, soft, spreading, reddish brown, multicellular hairs; leaves chartaceous, in verticils of 3's to 5's, ovate to ovate-oblong, 2–6 cm. long, 1.5–3 cm. wide, acute at base, with a few coarse subacute dentate teeth on each side, long-hirsute on both sides, the lateral nerves rather prominent beneath, ending in teeth, the petioles 5–10 mm. long; inflorescence from the axils of scalelike leaves, 2–3 cm. long, slender, long-pilose and minutely granular-puberulent; flowers 2–4, the pedicels 7–12 mm. long; calyx-segments lanceolate, acuminate; corolla white, 6–7 mm. across; fruit globose, red, 5–6 mm. across.——June–Aug. Woods; Honshu (Izu Isls. and Awa Prov. westw.), Shikoku, Kyushu.——s. Korea, Ryukyus, Formosa, China, and Philippines.——Forma **liukiuensis** (Nakai) Ohwi. *Bladhia villosa* var. *liukiuensis* Nakai; *A. villosa* var. *liukiuensis* (Nakai) Makino & Nemoto——RYŪKYŪ-TSURU-KŌJI. Leaves with many prominent teeth.——Occurs in the warmer parts of our area.

4. Ardisia crenata Sims. *A. lentiginosa* Ker-Gawl.; *A. crispa* DC., excl. syn.; *Bladhia crenata* (Sims) Hara; *B. lentiginosa* (Ker-Gawl.) Nakai——MANRYŌ. Glabrous shrub 30–60 cm. high; stems erect, grayish brown, simple except for a few spreading special flowering and articulating branches in upper part; leaves alternate, thick, oblong, 7–12 cm. long, 2–4 cm. wide, green on upper side, reddish brown when dry, acute to short-acuminate with an obtuse tip, gradually narrowed and acuminate at base, undulately toothed with an albuminous gland between the teeth, with pellucid dots and dark brown inner glands, the lateral nerves weak and not prominent, the petioles 5–10 mm. long; inflorescence terminal on the special lateral flowering branches, sessile, or the branches leafless and becoming pedunclelike, sometimes corymbosely branched, the pedicels 1–1.5 cm. long; calyx-segments ovate, subacute or subobtuse, not ciliate, spreading; corolla about 8 mm. across, white; fruit globose, red, about 6 mm. across.——July. Woods in low mountains; Honshu (s. Kantō Distr. and westw.), Shikoku, Kyushu.——Ryukyus,

Formosa, Korea, China to India.——Forma **taquetii** (Lév.) Ohwi. *A. crispa* var. *taquetii* Lév.; *A. taquetii* Lév. ex Nakai; *Bladhia crenata* var. *taquetii* (Lév.) Hara——Ō-MI-MANRYŌ. A robust phase.

5. Ardisia crispa (Thunb.) DC. *Bladhia crispa* Thunb.; *A. hortorum* Maxim.; *A. punctata* sensu auct. Japon., non Lindl.——KARA-TACHIBANA. Small shrub, minutely brown granular-puberulent on young stems and inflorescence; stems simple, erect, 20–40 cm. long, terete; leaves alternate, lanceolate to broadly so, 8–18 cm. long, 1.5–3.5 cm. wide, gradually narrowed to an obtuse tip, acute at base, undulately toothed, with a resinous marginal gland in sinus, vivid green, slightly paler beneath, smooth, glabrous, the petioles 8–10 mm. long; inflorescence about 10-flowered, umbellate on ends of special articulate flowering pedunclelike branches 4–7 cm. long, these usually bearing a few leaves or leaflike bracts, the pedicels about 10 mm. long; flowers white, 7–8 mm. across; calyx-lobes narrowly oblong, obtuse, not ciliate, recurved above; fruit globose, red, 6–7 mm. across.——July. Woods in hills and low mountains; Honshu (Awa Prov. and westw.), Shikoku, Kyushu.——Formosa, Ryukyus, and China.

6. Ardisia sieboldii Miq. *Bladhia sieboldii* (Miq.) Nakai——MOKU-TACHIBANA. Large erect loosely branched shrub, the branches grayish brown or reddish gray, rather thick, terete; leaves rather thick, broadly oblanceolate to narrowly obovate-oblong, 7–12 cm. long, 2.5–4 cm. wide, rounded with an abrupt obtuse tip, acuminate to acute at base, entire, lustreless and grayish brown when dried, with scattered obscure pellucid dots, with pale brown scurfy pilose hairs beneath while young, the midrib rather thick, impressed on upper side, raised beneath, the lateral nerves obliquely spreading, many, weak, the petioles 5–8 mm. long; inflorescence subterminal on special articulating lateral pedunclelike rather thick branches 2.5–4.5 cm. long, the pedicels 7–10 mm. long; calyx-segments broadly ovate, obtuse, pale brown-scurfy, ciliolate; corolla about 7 mm. across; fruit globose, dark purple, 7–8 mm. across. ——Shikoku, Kyushu.——Ryukyus, Formosa, and China.

7. Ardisia quinquegona Bl. *Bladhia quinquegona* (Bl.) Nakai; *A. pentagona* DC.; *B. pseudoquinquegona* Masam.——SHISHI-AKUCHI, MIYAMA-AKUCHI. Large branched shrub, the branches grayish brown, the branchlets rather slender, terete, brown-scurfy while young, purplish brown when dried; leaves alternate, rather thin, elliptic or oblanceolate to narrowly obovate, 6–10 cm. long, 2–3.5 cm. wide, long-acuminate to abruptly acute with an acute to obtuse tip, acuminate at base, entire, dark brown when dried, slightly purplish beneath, brown-scurfy when very young, with obscure glands within, the midrib slender, raised beneath, impressed on the upper side, the lateral nerves many, parallel, obliquely spreading, not prominent; inflorescence axillary, umbellate, sometimes branched, on slender peduncles 3–4 cm. long; calyx-segments deltoid or narrowly so, short, rather acute; corolla about 6 mm. across; fruit obtusely 5-angled or globose or depressed-globose, with obsolete longitudinal striations when dried, 4–6 mm. across.——Kyushu (Yakushima and Tanegashima).——Ryukyus, Formosa, s. China, and Indochina.

3. MYRSINE L.   TAIMIN-TACHIBANA ZOKU

Evergreen, sometimes deciduous shrubs or trees; leaves alternate, entire or toothed; inflorescence at nodes or in axils of leaves, fasciculate or umbellate; flowers bisexual or unisexual, small, 4- or 5(–6)-merous; calyx-segments imbricate or valvate;

corolla deeply parted; stamens with short filaments, the anthers splitting longitudinally; ovary ovoid, 1-locular; style short, the stigma punctate, flat, lobed or ligulate; fruit a globose berry, with crustaceous endocarp; seed solitary, with entire or ruminate endosperm.——Several species, in tropical and subtropical regions from Africa to e. Asia.

1A. Prostrate stoloniferous shrub with slender branches; fruit red. . . . . 1. *M. stolonifera*
1B. Large erect shrub with thick branches; fruit purplish black. . . . . 2. *M. seguinii*

1. Myrsine stolonifera (Koidz.) Walker. *Anamtia stolonifera* Koidz.; *M. marginata* sensu auct. Japon., non Hook. & Arn.; *Rapanea stolonifera* (Koidz.) Nakai; *A. mezii* Masam.——Tsuru-manryō, Tsuru-akami-no-ki. Glabrous shrub, sparingly branched, the branches prostrate or decumbent, slender, 1–1.5 m. long, elongate, terete, reddish brown, becoming dark brown in age; leaves alternate, somewhat coriaceous, oblong, broadly lanceolate to narrowly obovate, 3–8 cm. long, 1–3 cm. wide, acute at both ends, entire or with few mucronate teeth near tip, deep green, paler beneath, with impressed obsolete glands, the midrib slightly raised, the petioles 5–10 mm. long; flowers few, fasciculate in leaf-axils, the pedicels 3–4 mm. long, with a small bract at base; calyx-segments narrowly ovate-oblong, about 1 mm. long; corolla-lobes oblanceolate, obtuse, 2–2.5 mm. long; style rather thickened at base, the stigma punctate; fruit red, globose, about 5 mm. across.——Nov.–Dec. Thickets in hills; Honshu (Yamato and Suwō Prov.), Kyushu (Yakushima); very rare.——Formosa and China.

2. Myrsine seguinii Lév. *Rapanea neriifolia* Mez; *M. neriifolia* Sieb. & Zucc., non Casar.; *M. capitellata* sensu auct. Japon., non Wall.; *M. capitellata* var. *neriifolia* Kanitz; *Athruphyllum seguinii* (Lév.) Nakai; *A. neriifolia* (Mez) Hara——Taimin-tachibana. Large erect dioecious shrub or small tree, the branches rather thick, erect, the young branchlets dark purple; leaves evergreen, coriaceous, oblanceolate or lanceolate, 5–12(–19) cm. long, 1–2.5(–4) cm. wide, obtuse, gradually narrowed at base, entire, glabrous, deep green and lustrous on upper side, the midrib raised beneath, the lateral nerves slender, not prominent, the petioles 2–13 mm. long; flowers few, fasciculate in axils of the last year's leaves, the pedicels 1–3 mm. long, with scalelike bracts at base; calyx-segments ovate, subobtuse, ciliolate; corolla greenish white, the lobes ovate, spreading-recurved; style thick and short, the stigma ligulate; fruit globose, 6–7 mm. across, purplish black. ——Apr. Honshu (Awa Prov. and westw.), Shikoku, Kyushu.——Ryukyus, Formosa, China, and Indochina.

Fam. 161. **PRIMULACEAE**   Sakura-sō Ka   Primrose Family

Herbs often with creeping rhizomes; leaves simple, exstipulate, radical or cauline, alternate, opposite, or verticillate, rarely pinnately lobed; flowers in racemes, or panicles, or in simple or superposed verticils, bisexual, usually actinomorphic; calyx 4- to 9-, usually 5-lobed, usually persistent; corolla rotate, hypocrateriform, or infundibuliform, sometimes subcampanulate, rarely absent, imbricate in bud, the limb 4- to 9-lobed; stamens as many as the corolla-lobes and opposite them, inserted on the tube, staminodia alternate with the stamens, rarely present; ovary superior, 1-locular; ovules many, the placentation free-central; seeds small, many, the endosperm copious.——About 30 genera, with about 500 species, chiefly in the N. Hemisphere, few in the Tropics and the S. Hemisphere.

1A. Corolla-lobes contorted, or the corolla absent.
  2A. Capsules circumscissile, the top falling off as a lid; prostrate annual; flowers scarlet or blue. 1. *Anagallis*
  2B. Capsules longitudinally dehiscent.
    3A. Corolla absent; stamens inserted at the base of ovary; rather fleshy saline herbs with small opposite leaves and axillary flowers.
                                                                                                2. *Glaux*
    3B. Corolla present; stamens inserted on the corolla-tube or at the base of corolla.
      4A. Corolla 7 (5–9)-parted; testa loosely inclosing the seeds. . . . . 3. *Trientalis*
      4B. Corolla 5(–6)-lobed; testa closely adherent to the seed. . . . . 4. *Lysimachia*
1B. Corolla-lobes imbricate in bud.
  5A. Ovary superior.
    6A. Stamens inserted at the base of the corolla-tube; leaves all radical; inflorescence umbellate, scapose. . . . . 5. *Cortusa*
    6B. Stamens inserted on the corolla-tube.
      7A. Leaves radical and sometimes rosulate; inflorescence umbellate or verticillate, scapose.
        8A. Flowers large, the corolla-tube longer than the calyx. . . . . 6. *Primula*
        8B. Flowers small, the corolla-tube shorter than to as long as the calyx. . . . . 7. *Androsace*
      7B. Leaves cauline, alternate; inflorescence a leafy raceme. . . . . 8. *Stimpsonia*
  5B. Ovary subinferior; leaves cauline, alternate; flowers small, in racemes. . . . . 9. *Samolus*

1. ANAGALLIS L.   Ruri-hakobe Zoku

Usually glabrous erect or creeping annuals or perennials; leaves cauline, opposite, ternate, or sometimes all or the upper ones alternate, entire, sessile or short-petiolate; flowers rather small, axillary, pedicellate, blue-purple to red, bisexual; calyx 5-parted, the segments lanceolate or broadly linear, spreading; corolla 5-parted, rotate, the lobes contorted; stamens inserted at the base of corolla-lobes, the filaments filiform or subulate, hairy, the anthers obtuse; ovary glabrous; style filiform; capsules globose, circumscissile; seeds small, many, conical, flat on back.——About 24 species, in Europe, Africa, Asia, and temperate S. America.

1. Anagallis arvensis L. *A. femina* Mill.; *A. coerulea* Schreb.——Ruri-hakobe. Glabrous whitish or glaucous annual or biennial; stems much branched, 4-angled, 10–30 cm. long; leaves opposite, ovate to narrowly lanceolate, 1–2.5 cm. long, 5–15 mm. wide, spreading, entire, acute to subobtuse, rounded at base, sessile; flowers solitary, axillary, the pedicels

2–3 cm. long; calyx-segments linear-lanceolate, long-acuminate, 4–6 mm. long, 1-ribbed; corolla blue-purple to reddish, 1–1.3 cm. across, the lobes obovate-orbicular, often ciliolate; capsules globose, about 4 mm. in diameter; seeds dark brown, about 1 mm. across, densely warty.——Mar.–May. Honshu (Izu Islands and westw.), Shikoku, Kyushu.——Ryukyus, Formosa, and temperate and tropical regions generally.

2. GLAUX L.   UMI-MIDORI ZOKU

Rhizomatous rather fleshy glabrous perennial; stems terete, erect or decumbent at base; leaves opposite or rarely somewhat ternate, small, entire; flowers axillary, subsessile, rather small, white to rose; calyx broadly campanulate, 5-lobed, the lobes oblong, obtuse, persistent, imbricate in bud; corolla absent; stamens perigynous, alternate with the calyx-segments, the filaments nearly filiform, the anthers cordate, dorsifixed; ovary ovoid, with a filiform style at apex; ovules semi-anatropous; capsules 5-valved, with the persistent style base at apex, covered on lower half with the calyx; seeds few, flat on the dorsal side, convex and with a hilum on the ventral side, the testa brown.——One species, along seashores and estuaries and in wet alkaline soils in the N. Hemisphere.

1. **Glaux maritima** L. var. **obtusifolia** Fern.   *G. maritima* sensu auct. Japon., non L.——UMI-MIDORI, SHIO-MATSUBA. Plant glabrous, lustrous, deep green and slightly glaucous; stems erect or ascending at base, simple or slightly branched, 5–20 cm. long; leaves rather dense on the stems, spreading, broadly lanceolate to obovate-oblong, 6–15 mm. long, 3–6 mm. wide, rounded to obtuse, obtuse at base, sessile, with impressed dots beneath, the midrib and nerves indistinct; flowers 6–7 mm. across; capsules globose, glabrous, 3–4 mm. across.—— May–Aug. Wet places by the sea near high tide mark; Hokkaido, Honshu (centr. and n. distr.).——Kuriles, Sakhalin, and generally in e. Asia to N . America. The typical phase is widespread in Eurasia and N. America.

3. TRIENTALIS L.   TSUMA-TORI-SŌ ZOKU

Delicate glabrous perennials with slender creeping rhizomes; stems simple slender; leaves in a single whorl near top of the stem; flowers erect, white; calyx deeply 5- to 9-fid, the segments linear-lanceolate, green, spreading, persistent; corolla rotate, deeply 7(5–9)-lobed, the lobes narrowly ovate, entire, contorted in bud; stamens 5–9, inserted on the throat of the corolla-tube, the filaments filiform, the anthers linear, obtuse; ovary globose; style slender; ovules many; seeds few, the testa loose, coriaceous, finely reticulate.——Two species, in the cooler regions of the N. Hemisphere.

1. **Trientalis europaea** L. var. **europaea**.   *T. europaea* var. *eurasiatica* R. Knuth——TSUMA-TORI-SŌ.   Stems erect, terete, 7–25 cm. long, with a few scalelike or small leaves at base; leaves 5–10 in a whorl near the top, spreading, broadly oblanceolate, lanceolate, or elliptic, sometimes ovate or obovate-oblong, 2–7 cm. long, 1–2.5 cm. wide, acute to subobtuse, acuminate at base, entire, subsessile, the lateral nerves incurved on upper part, each confluent with the upper one; peduncles slender, 2–3 cm. long; flowers 1.5–2 cm. across; calyx-segments 4–7 mm. long, acuminate; corolla-lobes acute; fruit globose, 2.5–3 mm. in diameter.——June–July. Coniferous woods; Hokkaido, Honshu (centr. and n. distr.), Shikoku.——Kuriles, Sakhalin, Korea, to Eurasia, Alaska, and e. N. America.—— Forma **ramosa** Iljinsky ex Hegi. *T. ramosa* Koidz.—— Ō-TSUMA-TORI-SŌ.   Rare phase with branched stems.

Var. **arctica** (Fisch.) Ledeb. *T. arctica* Fisch.——KO-TSUMA-TORI-SŌ.   Upper cauline leaves gradually larger, the apical leaves smaller, obovate-cuneate, obtuse to subrounded. ——Coniferous woods; Hokkaido, Honshu (centr. and n. distr.); rare.——Sakhalin, Kuriles, e. Siberia, and N. America.

4. LYSIMACHIA L.   OKA-TORA-NO-O ZOKU

Erect or prostrate, glabrous to pubescent, sometimes glandular-dotted herbs; leaves opposite, alternate, or verticillate, entire; flowers axillary or terminal, solitary or in racemes, corymbs, or panicles, white, yellow, or red, rarely blue-purple, (4–)5- to 6-merous; calyx-lobes persistent; corolla infundibuliform, hypocrateriform, or broadly campanulate, (4–)5- to 6-lobed, the tube often very short, the lobes contorted in bud; stamens inserted on the tube, sometimes with staminodia in an alternate whorl, the anthers obtuse; ovary globose or ovoid; style simple, filiform; capsules longitudinally dehiscent or indehiscent; seeds few to many, with a thin testa.——About 110 species, chiefly in the temperate regions of the N. Hemisphere, few in the Tropics, Australia, N. America, and Africa.

1A. Flowers yellow.
  2A. Leaves opposite, rarely verticillate.
    3A. Flowers in racemes or panicles; stems erect; leaves narrow, subsessile.
      4A. Flowers in axillary spikelike racemes; style and stamens long-exserted; corolla-lobes linear-oblanceolate. .... 1. *L. thyrsiflora*
      4B. Flowers in terminal panicles; style and stamens shorter than the narrowly ovate corolla-lobes. .... 2. *L. vulgaris* var. *davurica*
    3B. Flowers solitary and axillary or rarely few together and terminal; stems decumbent or prostrate; leaves distinctly petiolate.
      5A. Leaves usually longer than wide, without brown striations and dots; calyx-segments linear-lanceolate.
        6A. Leaves ovate to broadly so, acute; corolla 5–7 mm. long; stems ascending at first, then creeping; calyx-segments acuminate.
          3. *L. japonica*
        6B. Leaves broadly ovate to broadly ovate-elliptic, rounded to very obtuse; corolla about 15 mm. long; stems creeping from the
           first; calyx-segments obtuse. .... 4. *L. tashiroi*
      5B. Leaves shorter than to slightly longer than wide; calyx-segments oblanceolate to obovate-spathulate.
        7A. Leaves with dark brown glandular striations and dots; corolla without striations. .... 5. *L. tanakae*

1. Lysimachia thyrsiflora L. *Naumburgia guttata* Moench; *N. thyrsiflora* (L.) Reichenb.; *L. kamtschatica* Gand. ——YANAGI-TORA-NO-O. Perennial with long-creeping rhizomes; stems erect, terete, simple or sometimes sparsely branched, 30–60 cm. long, with reduced scalelike leaves at base; leaves opposite, oblanceolate or broadly so, 4–10 cm. long, 1.5–2.5 cm. wide, obtuse or subobtuse, entire, sessile, glabrous and green, paler and long brownish tomentose beneath, black-spotted; racemes spikelike, densely many-flowered, axillary midway on the stems, the peduncles 2–3 cm. long, erect; corolla yellow, the lobes broadly linear-oblanceolate, 4–5 mm. long, obtuse, with few black spots on upper part; capsules globose, black-spotted, about 2.5 mm. in diameter, the persistent style about 4 mm. long.——May–July. Bogs and wet places; Hokkaido, Honshu (Hitachi, Kotsuke, Shimotsuke, Echizen, and northw.).——Kuriles, Sakhalin, Korea, and cooler parts of the N. Hemisphere.

2. Lysimachia vulgaris L. var. **davurica** (Ledeb.) R. Knuth. *L. davurica* Ledeb.; *L. vulgaris* var. *typica* R. Knuth, pro parte, and auct. Japon.——KUSA-REDAMA. Perennial herb, rhizomatous and stoloniferous; stems terete, erect, 40–80 cm. long, simple or sparsely branched, minutely glandular-puberulent on upper part; leaves opposite, rarely verticillate in 3's or 4's, lanceolate to narrowly ovate, 4–12 cm. long, 1–4 cm. wide, acuminate, acute to subobtuse at base, sessile, entire, deep green, often whitish beneath, black-spotted; flowers in a many-flowered terminal leafy panicle or in whorls in the upper axils, the pedicels 7–12 mm. long, bracteate; flowers yellow, 12–15 mm. across; calyx-segments narrowly ovate-deltoid, acuminate, with a line of black glands near margin; corolla-lobes narrowly ovate, granular-papillose inside; capsules globose, about 4 mm. in diameter; style persistent, 5–6 mm. long.——July–Aug. Wet places; Hokkaido, Honshu, Kyushu (mountains).——Temperate regions of e. Asia.

3. Lysimachia japonica Thunb. KO-NASUBI. Pubescent perennial; stems soon creeping, leafy; leaves opposite, ovate or broadly so, 1–2.5(–3) cm. long, 7–20 mm. wide, with pellucid dots, acute, entire, rounded at base, the petioles 5–10 mm. long; flowers solitary, axillary, yellow, 5–7 mm. across, erect in anthesis, soon deflexed, the pedicels 3–8(–12) mm. long; calyx-segments linear-lanceolate, acuminate, as long as the corolla or slightly accrescent after anthesis; corolla-lobes ovate-deltoid, 5–7 mm. long; capsules globose, 4–5 mm. across, long-pubescent at tip; seeds black, ellipsoidal, 1-ridged, densely warty, about 1 mm. long.——May–June. Thickets and roadsides in lowlands and mountains; Hokkaido, Honshu, Shikoku, Kyushu; common.——Ryukyus, Formosa, China, and Malaysia.

4. Lysimachia tashiroi Makino. ONI-KO-NASUBI. Long-creeping, brownish pubescent perennial; stems loosely branched, 5–10 cm. long; leaves opposite, broadly ovate to broadly ovate-elliptic, 2–4 cm. long, 1.5–3 cm. wide, rounded to very obtuse, entire, pubescent, the petioles 7–12 mm. long; flowers 1–4 toward the top of stems, yellow, about 15 mm. across, the pedicels 15–20 mm. long; calyx-segments linear-lanceolate, 7–10 mm. long, obtuse, long-pubescent; corolla-lobes ovate-oblong or oblong, about 15 mm. long, much longer than the calyx-lobes; ovary short-pubescent; style about 6.5 mm. long.——July. Kyushu (Buzen Prov.); rare.

5. Lysimachia tanakae Maxim. MIYAMA-KO-NASUBI. Long-creeping herb, brownish hairy on the stems, petioles, and flowers; leaves reniform-orbicular or ovate-cordate, 1–2 cm. in diameter, pubescent while young, rounded to retuse, mucronate, entire, black-dotted and striate, the petioles 5–15 mm. long; calyx-segments oblanceolate, obtuse, 5–6 mm. long; corolla yellow, 8–10 mm. across, about 1.5 times as long as the calyx, the lobes ovate-oblong, the pedicels 2–3.5 cm. long.——June–Aug. Honshu (Kii Prov.), Shikoku, Kyushu; rare.

6. Lysimachia ohsumiensis Hara. HETSUKA-KO-NASUBI. Stems long-creeping, often rooting at the nodes, pubescent with soft-spreading long hairs; leaves ovate-cordate, 7–13 mm. long, 6–10 mm. wide, acutish, rounded to slightly cordate at base, pubescent, petioled; pedicels axillary, slender, solitary, 15–30 mm. long; calyx-segments obovate-spathulate, 3–4 mm. long; short-acuminate; corolla about 1 cm. wide, yellow, glandular-striate; anthers oblong, about 1.2 mm. long.——Kyushu (Ohsumi Prov.).

7. Lysimachia sikokiana Miq. MOROKOSHI-SŌ, YAMA-KUNEMBO. Fragrant when dry; rhizomes short; stems erect or ascending at base, angled, 30–80 cm. long, simple or sparingly short-branched, minutely glandular-hairy; leaves alternate, ovate to broadly lanceolate, acuminate at both ends, entire, glabrescent, deep green, pale green beneath, the petioles 1–2.5 cm. long; flowers yellow, 1–1.2 cm. across, the peduncles slender, 4–8 cm. long, axillary, solitary, ebracteate; calyx-segments ovate-deltoid, acuminate, minutely glandular-ciliolate; corolla 2 or 3 times as long as the calyx, the lobes narrowly oblong; anthers about 4 mm. long, erect; capsules globose, whitish, about 6 mm. in diameter, deflexed, the valves crustaceous.——Woods at low elevations in the mountains; Honshu (Awa, Izu Prov. and westw.), Shikoku, Kyushu.——Ryukyus and Formosa.

8. Lysimachia leucantha Miq. *L. candida* var. *samolina* R. Knuth; *L. candida* var. *leucantha* (Miq.) Makino——SAWA-TORA-NO-O, MIZU-TORA-NO-O. Glabrous, erect perennial, 40–60 cm. high, usually unbranched; stems angled; leaves narrowly oblanceolate to broadly linear, 2–4.5 cm. long, 3–5 mm. wide, obtuse, entire, black-dotted, sessile or subsessile; inflorescence racemose, terminal, many-flowered, erect, the bracts linear, the pedicels 12–20 mm. long, 2 or 3 times as long as the bracts; calyx-segments lanceolate, acuminate, entire; corolla white, about 5 mm. across, the lobes obovate, obtuse; capsules small.——Apr. Wet places, near ponds and rivers in lowlands; Honshu, Kyushu; rare.——Korea.

9. Lysimachia acroadenia Maxim. *L. keiskeana* Miq.; *L. decurrens* sensu auct. Japon., saltem pro parte, non G. Forst.; *L. decurrens* var. *acroadenia* (Maxim.) Makino——Mɪ-ʏᴀᴍᴀ-ᴛᴀɢᴏʙō, Gɪɴʀᴇɪ-ᴋᴀ. Erect perennial herb 30–60 cm. high, with short rhizomes; stems slightly branched, angled, green, minutely glandular-hairy in upper part; leaves broadly lanceolate to narrowly ovate, usually narrowly ovate-oblong, 5–10 cm. long, 1–3 cm. wide, acuminate at both ends, glabrous, entire, green, paler and with minute red-brown spots beneath, the petioles 1–2 cm. long; racemes to 30 cm. long, terminal, loosely 10- to 30-flowered, the bracts linear, filiform, the pedicels 1–1.5 cm. long, obliquely spreading in fruit; flowers white to pinkish, partially open, 5–6 mm. long; calyx-segments lanceolate, acuminate; corolla-lobes suberect, cuneate, rounded; capsules globose, 4–5 mm. in diameter, the valves somewhat crustaceous.——June–July. Wet places in mountain woods; Honshu, Shikoku, Kyushu; not abundant.——Korea (Quelpaert Isl.).

10. Lysimachia fortunei Maxim. *L. fortunei* forma *suboppositifolia* Honda——Nᴜᴍᴀ-ᴛᴏʀᴀ-ɴᴏ-ᴏ. Rhizomatous and stoloniferous; stems 40–70 cm. long, simple or sparingly branched, reddish above the base; leaves oblanceolate to oblong-oblanceolate, 4–7 cm. long, 1–1.5 cm. wide, obtuse, mucronate, entire, sessile, pellucid-dotted, pale green beneath; racemes erect, rather densely many-flowered, 10–20 cm. long, glabrous or minutely glandular-puberulent, the pedicels obliquely spreading, 3–4 mm. long, the bracts linear, as long as the pedicels; flowers white, 5–6 mm. across; calyx-segments ovate-oblong, obtuse, black-dotted on back; corolla-lobes spreading, obovate, rounded; capsules globose, 2–2.5 mm. in diameter.——July–Aug. Wet margins of rivers and ponds in lowlands; Honshu, Shikoku, Kyushu.——Korea, China, Formosa, and Indochina.

11. Lysimachia barystachys Bunge. Nᴏᴊɪ-ᴛᴏʀᴀ-ɴᴏ-ᴏ. Perennial 70–100 cm. high, reddish above the base; stems terete, simple or sparingly branched, scattered brownish pubescent; leaves many, lanceolate to oblanceolate, 6–10 cm. long, 8–15 mm. wide, obtuse to subacute, entire, with pale inner glands, thinly brownish pubescent on margins, the lower and

often the upper subsessile; racemes terminal, to 30 cm. long, one-sided, densely many-flowered, the pedicels ascending, 4–7 mm. long, the bracts linear, as long as to slightly longer than the pedicels; calyx-segments narrowly ovate-oblong, obtuse; corolla white, 7–12 mm. across, the lobes narrowly oblong, about 4 times as long as the calyx; capsules globose, about 2.5 mm. across.——June–July. Sunny hills and low elevations in the mountains; Honshu (centr. and s. distr.); common.——Korea, n. China, and Manchuria.

12. Lysimachia clethroides Duby. *L. sororia* Miq.; *L. clethroides* var. *sororia* (Miq.) R. Knuth——Oᴋᴀ-ᴛᴏʀᴀ-ɴᴏ-ᴏ. Perennial herb with long creeping rhizomes; stems terete, reddish above the base, 50–100 cm. high, rather stout, usually simple, erect, glabrous or loosely pubescent on upper part; leaves broadly lanceolate to narrowly ovate, 2–5 cm. wide, acuminate, entire, often short-pilose on upper side, pale green and usually glabrous beneath, with pale inner glands, sessile or subsessile; racemes densely many flowered, erect, 10–20 cm. long in anthesis, to 40 cm. long in fruit, the pedicels obliquely spreading, 6–10 mm. long; calyx-segments narrowly ovate-oblong, obtuse; corolla white, 8–12 mm. across, the lobes narrowly oblong, spreading; capsules ovoid-globose, about 2.5 mm. in diameter.——June–July. Sunny hills and low elevations in the mountains; Hokkaido, Honshu, Shikoku, Kyushu; very common.——Korea, Manchuria, China, and Indochina.

Lysimachia pilophora (Honda) Honda. *L. fortunei* var. *pilophora* Honda.——Iɴᴜ-ɴᴜᴍᴀ-ᴛᴏʀᴀ-ɴᴏ-ᴏ. Alleged hybrid of *L. clethroides* and *L. fortunei.*——Honshu (Kantō Distr.).

13. Lysimachia mauritiana Lam. *L. lineariloba* Hook. & Arn.; *L. lubinioides* Sieb. & Zucc.; *L. nebeliana* Gilg——Hᴀᴍᴀ-ʙᴏssᴜ. Glabrous slightly fleshy biennial; stems few together or solitary, often reddish, erect, 10–40 cm. long, stout, terete, often short-branched in upper part; leaves obovate-oblong to obovate or spathulate-oblanceolate, 2–5 cm. long, 1–2 cm. wide, obtuse to rounded, entire, lustrous, with black inner glands, sessile or subsessile; racemes erect, terminal, often compound-paniculate, very densely flowered, 4–12 cm. long in fruit, the pedicels spreading, 1–2 cm. long, about as long as the leaflike bracts; flowers white to pinkish, 1–1.2 cm. across; calyx-segments broadly lanceolate, obtuse, often with a few black glandular dots on back; corolla-lobes narrowly cuneate-oblong, obliquely spreading; capsules globose, 4–6 mm. in diameter.——May–July. Near the sea; Hokkaido, Honshu, Shikoku, Kyushu; common.——Korea, Ryukyus, Formosa, China, Pacific Islands, and India.

5. CORTUSA L. Sᴀᴋᴜʀᴀ-sō-ᴍᴏᴅᴏᴋɪ Zᴏᴋᴜ

Pubescent perennial with short-creeping rhizomes; leaves radical, long-petiolate, orbicular-cordate, palmately 7- to 9-lobed, dentate and incised; stems scapose, the pedicels slender, with a small bract at base; flowers umbellate, usually rose-purple; calyx broadly campanulate, deeply 5-fid, the lobes persistent; corolla infundibuliform-campanulate, short-tubular, appendaged on the throat, the lobes 5, spreading, obtuse, imbricate in bud; stamens 5, inserted at the base of the corolla, the filaments very short, connate by a membrane, the anthers acuminate; ovary superior, ovoid; style simple, filiform; ovules many; capsules ovoid-globose, 5-valved; seeds many.——Few species, in Europe, n. Asia, Himalayas, and e. Asia.

1. Cortusa matthioli L. var. **jozana** (Miyabe & Tatew.) Hara. *C. jozana* Miyabe & Tatew.; *C. coreana* var. *jozana* (Miyabe & Tatew.) Hara——Eᴢᴏ-ɴᴏ-sᴀᴋᴜʀᴀ-sō-ᴍᴏᴅᴏᴋɪ. Radical leaves reniform-orbicular, 6–10 cm. long, 8–10 cm. wide, shallowly and palmately lobed, loosely long-pubescent on up-

per side, on nerves beneath, and on petioles, the lobes 9–13, ovate to narrowly so, usually again shallowly 3-lobed and loosely toothed, the petioles 8–12 cm. long; scapes 20–35 cm. long, long-pubescent especially on the lower half, longer than the leaves, erect; inflorescence umbellate, terminal, the pedicels minutely glandular-hairy, the bracts oblanceolate, few-toothed; calyx-lobes lanceolate, acuminate, 3–4 mm. long, nearly as long as the tube, glabrous; corolla rose-purple, 15–17 mm. across, the lobes 5, narrowly ovate, obtuse, 5–7 mm. long; anthers 5–6 mm. long, shorter than the style.——June. Mountains; Hokkaido (Jozankei).

Var. **congesta** (Miyabe & Tatew.) Ohwi. *C. sachalinensis* var. *congesta* Miyabe & Tatew.; *C. coreana* var. *congesta* (Miyabe & Tatew.) Hara——REBUN-SAKURA-SŌ-MODOKI.

Leaves more densely pubescent and deeply cordate; flowers rather dense.——Hokkaido (Rishiri and Rebun Isls.).

Var. **yezoensis** (Miyabe & Tatew.) Hara. *C. sachalinensis* var. *yezoensis* Miyabe & Tatew.; *C. coreana* var. *yezoensis* (Miyabe & Tatew.) Hara——SAKURA-SŌ-MODOKI. Leaves rather densely pubescent and the teeth larger and more acute.——Mountains; Hokkaido (Oshima and Ishikari Prov.). The species in the broad sense occurs from Sakhalin, Korea, and n. China to Europe.

6. PRIMULA L. SAKURA-SŌ ZOKU

Perennials with short rhizomes; leaves usually radical, rarely densely arranged on short cauduculi, obovate-spathulate or ovate-orbicular and subsessile or reniform-orbicular and long-petiolate, entire or palmately lobed, usually toothed; flowers in terminal umbels or superimposed on the scapes, rarely solitary, white to lavender or rose-purple, rarely yellow, the bracts small, without bracteoles; calyx tubular to infundibuliform, 5-lobed; corolla salverform or campanulate, often with appendages on the throat, the limb 5-lobed, imbricate in bud, entire or again 2-lobed; stamens inserted on the tube or throat, included, the filaments very short, the anthers obtuse; ovary globose or ovoid; style simple, slender; capsules globose to cylindric, 5-valved; seeds many, peltately attached.——More than 200 species, in Europe, Asia, Malaysia, especially abundant in w. China and the Himalayas, few in N. and S. America.

1A. Leaves glabrous, sometimes farinose, decurrent below to the petiole or sessile.
  2A. Small herbs with leaves less than 10 cm. long; scapes 6–20 cm. long; flowers in single, terminal umbels.
    3A. Leaves with few coarse teeth on upper margin, rather fleshy, not farinose, quite glabrous and green, without raised nerves beneath; capsules elliptic or ovoid-globose, shorter or slightly longer than the calyx; bud-scales with incurved margins; plants of wet slopes and bogs in high mountains.
      4A. Flowers rather large, fully expanding, usually rose-purple, the tube 7–8 mm. long; corolla-lobes obtriangular or obovate-deltoid, rather deeply bifid, the lobes ascending. . . . . 1. *P. cuneifolia*
      4B. Flowers smaller, not fully expanding, white with a yellow eye, the tube about 5 mm. long; corolla-lobes elliptic or obovate, shallowly bifid, the lobes not divergent. . . . . 2. *P. nipponica*
    3B. Leaves with small obtuse teeth on upper margin, yellow- or white-farinose or sometimes glabrous, with raised midrib and nerves beneath; capsules short-cylindric, usually longer than the calyx; bud-scales with recurved margins; plants not of bogs.
      5A. Leaves not farinose, broadly ovate, abruptly narrowed to a petiolelike base. . . . . 3. *P. macrocarpa*
      5B. Leaves white- to yellow-farinose beneath.
        6A. Mealiness prominent, yellow. . . . . 4. *P. modesta*
        6B. Mealiness thin, white. . . . . 5. *P. yuparensis*
  2B. Rather large herb with leaves 15–40 cm. long; scapes 40–80 cm. long; flowers in superimposed umbels. . . . . 6. *P. japonica*
1B. Leaves and scapes pubescent, not farinose, ovate to reniform-orbicular, long-petiolate, incised or lobulate and toothed.
  7A. Leaves ovate to ovate-elliptic; calyx broadly infundibuliform, much longer than the depressed-conical capsule. . . . . 7. *P. sieboldii*
  7B. Leaves ovate-orbicular to reniform; calyx tubular to tubular-campanulate, shorter to longer than the oblong to cylindric capsule.
    8A. Rhizomes rather long-creeping, the scales persistent. . . . . 8. *P. hidakana*
    8B. Rhizomes tufted or short-creeping, the scales membranous, soon withering and not persistent.
      9A. Scapes 15–40 cm. long, much longer than the leaves; leaves deeply incised to lobed.
        10A. Leaves incised and minutely toothed; flowers usually rose-purple, salverform; corolla-lobes spreading, the tube 12–14 mm. long. . . . . 9. *P. jesoana*
        10B. Leaves deeply incised or lobed, scarcely toothed; flowers white, infundibuliform; corolla-lobes obliquely spreading, the tube 6–8 mm. long. . . . . 10. *P. takedana*
      9B. Scapes 5–10(–15) cm. long, scarcely longer than the leaves; leaves scarcely lobed.
        11A. Calyx 10–15 mm. long, the lobes prominently long-pubescent. . . . . 11. *P. kisoana*
        11B. Calyx 5–8 mm. long, the lobes glabrescent or glabrous.
          12A. Leaves usually without multicellular long hairs above; corolla-tube 12–20 mm. long. . . . . 12. *P. tosaensis*
          12B. Leaves pubescent above, at least while young, with long whitish hairs; corolla-tube 8–10 mm. long. . . . . . . 13. *R. reinii*

1. Primula cuneifolia Ledeb. *P. cuneifolia* var. *dubyi* Pax——EZO-KOZAKURA. Glabrous, rather fleshy, vivid green perennial with short rhizomes; leaves 7–10, radical, obovate-cuneate, 2–4 cm. long, 1–1.5 cm. wide, rounded, cuneately narrowed to a petiolelike base, flat, without distinct nerves, with a few coarse teeth on upper margin; scapes 5–15 cm. long, umbellately 3- to 6-flowered, the bracts linear, entire, small, the pedicels 5–10 mm. long, minutely glandular-dotted; flowers rose-purple, about 2 cm. across, fully expanding; calyx 4–6 mm. long, 5-cleft, the teeth lanceolate, obsoletely mucronate; corolla-tube 7-8 mm. long, shorter than the limb; capsules broadly ovoid-ellipsoidal, about 5 mm. long, as long as to slightly shorter than the calyx.——July–Aug. Wet alpine slopes; Hok-

kaido.——Kuriles, Sakhalin, Ochotsk Sea region, Aleutians, and Alaska.

Var. **hakusanensis** (Franch.) Makino. *P. hakusanensis* Franch.; *P. cuneifolia* sensu auct. Japon., non Ledeb.; *P. cuneifolia* var. *tanigawaensis* Tatew.——HAKUSAN-KOZAKURA. Larger; leaves 3–8 cm. long, 1–2.5 cm. wide, the teeth more numerous, unequal, narrower; scapes 5–20 cm. long, 1- to 10-flowered, the pedicels 1–2 cm. long.——Wet alpine slopes; Honshu (centr. and n. distr.).

Var. **heterodonta** (Franch.) Makino. *P. heterodonta* Franch.; *P. hakusanensis* var. *heterodonta* (Franch.) Takeda——MICHI-NO-KU-KOZAKURA. Plant larger and the leaves broader.——Mount Iwaki in Honshu.

2. **Primula nipponica** Yatabe. HINAZAKURA. Resembles *P. cuneifolia* but smaller; radical leaves 5–10, rather fleshy, cuneate or obovate-cuneate, 2–4 cm. long, 5–15 mm. wide, very obtuse, with few coarse teeth toward the tip, flat, without distinct nerves; scapes 7–15 cm. long, 1- to 8-flowered, the bracts small, linear, the pedicels 1–2 cm. long, loosely granular-dotted; flowers not fully expanding, 12–15 mm. across; calyx narrowly cup-shaped, about 4 mm. long, 5-toothed to the middle; corolla white, the tube about 5 mm. long, yellow on the throat, the lobes obliquely spreading; capsules broadly ovoid-ellipsoid, slightly longer than to as long as the calyx.——Wet alpine slopes; Honshu (Uzen, Rikuchu, and northw.).

3. **Primula macrocarpa** Maxim. *P. farinosa* var. *mistassinica* sensu auct. Japon., non Pax; *P. hayaschinei* Petitm.——HIME-KOZAKURA. Small, glabrous perennial with short rhizomes; leaves few, not farinose, broadly ovate or oval, 1–3(–5) cm. long inclusive of the petiolelike base, irregularly toothed on the upper margin, nerves very slightly impressed on upper side; scapes 5–10 cm. long; flowers 1–4, terminal, the bracts linear, the pedicels erect, 7–20 mm. long; calyx tubular, 4–6 mm. long, 5-toothed to about the middle; corolla white, about 10 mm. across, the tube as long as the calyx; capsules cylindric, 6–8 mm. long.——June. Alpine; Honshu (Mount Hayachine in Rikuchu); rare.

4. **Primula modesta** Biss. & Moore. *P. farinosa* var. *luteofarinosa* forma *japonica* Fr. & Sav.; *P. farinosa* var. *modesta* (Biss. & Moore) Makino; *P. modesta* var. *shikokumontana* Miyabe & Tatew.——YUKIWARI-SŌ. Rhizomes short, often covered with the persistent marcescent leaves of the previous season; leaves broadly cuneate-oblanceolate, 3–6(–10) cm. long, 1–1.5 cm. wide, obtuse or subobtuse, gradually long-narrowed to a petiolelike base, glabrous and not farinose on upper side, densely yellow-farinose beneath, short-toothed on the slightly recurved margin; scapes 7–15 cm. long, 3- to 10-flowered; flowers small, terminal, the pedicels about 1 cm. long in flower, 1.5–3(–4) cm. long in fruit, loosely farinose; calyx about 5 mm. long, short-tubular, 5-toothed to the middle; corolla about 15 mm. across, rose-colored, the tube about 6 mm. long; capsules short-cylindric, 5–8 mm. long.——May–June. Mountains; Hokkaido, Honshu (centr. and n. distr.), Shikoku, Kyushu.

Var. **fauriei** (Franch.) Takeda. *P. fauriei* Franch.; *P. farinosa* var. *fauriei* (Franch.) Miyabe; *P. fauriei* var. *samanimontana* Tatew.; *P. modesta* var. *samanimontana* (Tatew.) Nakai——YUKIWARI-KOZAKURA. Smaller, the leaves broadly ovate, obsoletely toothed, prominently recurved on the margin, abruptly narrowed and long-decurrent on the petiolelike base; pedicels 1–2 cm. long.——June–July. Mountains and rocky cliffs near the sea; Hokkaido, Honshu (n. distr.).——s. Kuriles.

Var. **matsumurae** (Petitm.) Takeda. *P. matsumurae* Petitm.——REBUN-KOZAKURA. Rather stout; leaves oblanceolate, obscurely toothed, gradually narrowed to the base; scapes stouter, about 10-flowered; capsules slightly longer.——Hokkaido (Rebun Isl. and Teshio Prov.).

5. **Primula yuparensis** Takeda. YŪBARI-KOZAKUBA. Resembles somewhat the preceding species; rhizomes short; leaves few, broadly lanceolate to spathulate-cuneate, 1.5–3 cm. long, 1–1.5 cm. wide, obtuse to very obtuse, minutely toothed, decurrent on a petiolelike base, slightly recurved toward the tip, slightly white-farinose beneath; scapes 4–6 cm. long, 2- to

3-flowered at the top, the bracts narrow, small, slightly inflated at base on back, the pedicels about 1 cm. or sometimes to 2 cm. long; calyx about 7 mm. long, slightly white-farinose, short-tubular, 5-toothed to the middle; corolla pale rose-purple, about 1.5 cm. across, the tube twice as long as the calyx; capsules slightly longer than the calyx.——July–Aug. Alpine regions; Hokkaido (Mount Yubari); rare.

Primula sorachiana Miyabe & Tatew. SORACHI-KOZAKURA. Flowers 3–10(–17) on the scape, the pedicels slightly declined; calyx 4–5 mm. long, the tube half as long as the lobes; corolla 10–13 mm. across; capsules nearly as long as the calyx.——Reported from Kanayama, at the foot of Mount Yubari in Hokkaido. This may be only a variant of *P. yuparensis*.

6. **Primula japonica** A. Gray. KURIN-SŌ. Rhizomes stout and branched, loosely tufted; leaves few, glabrous, obovate-oblong to oblong-oblanceolate, 15–40 cm. long, 5–13 cm. wide, very obtuse, gradually narrowed below to a petiolelike often reddish base, with short deltoid teeth, the midrib thick, white, usually reddish on under side; scapes terete, 40–80 cm. long, glabrous, the flowers whorled at intervals, the bracts linear, the pedicels obliquely spreading, ascending in fruit, 2–3 cm. long; calyx narrowly cup-shaped, 6–8 mm. long, slightly accrescent after anthesis, 7–10 mm. long in fruit, 5-toothed; corolla rose-purple, rarely white, 2–2.5 cm. across, the tube narrow, 15–17 mm. long, the lobes spreading; capsules globose, as long as to slightly shorter than the calyx, 7–8 mm. long.——June–July. Wet places along streams in mountains; Hokkaido, Honshu, Shikoku, Kyushu; frequently cultivated.——Formosa.

7. **Primula sieboldii** E. Morren. *P. cortusoides* sensu auct. Japon., non L.; *P. cortusoides* var. *patens* Turcz.; *P. cortusoides* var. *sieboldii* (E. Morren) Nichols.; *P. gracilis* Stein.——SAKURA-SŌ. White-pubescent perennial; rhizomes short-cheeping; leaves ovate to ovate-elliptic, 4–10 cm. long, 3–6 cm. wide, obtuse, cordate at base, shallowly incised and irregularly toothed, more or less wrinkled on the upper surface especially while young, the petioles 1–4 times as long as the blades; scapes 15–40 cm. long, umbellately 7- to 20-flowered at the top, the bracts narrowly lanceolate, the pedicels 2–3 cm. long in fruit, minutely puberulent; calyx 8–10(–12) mm. long, tubular in anthesis, infundibuliform in fruit, 5-cleft; corolla rose-colored, sometimes white, rarely rose-purple in cultivated races, 2–3 cm. across, the tube 10–13 mm. long; capsules depressed conical-globose, about 5 mm. across.——Apr.–May. Wet grassy places in lowlands along rivers; Hokkaido (s. distr.), Honshu, Kyushu. Much cultivated.——Korea, Manchuria, and e. Siberia.

8. **Primula hidakana** Miyabe & Kudo. HIDAKA-IWAZAKURA. Rhizomes creeping, covered with brown marcescent scales and persistent leaf bases; leaves 1–3, orbicular to deeply reniform-orbicular, 1.5–4.5 cm. long, 2.5–5.5 cm. wide, loosely pilose on upper side, or glabrous, pilose on the nerves beneath, shallowly 7-lobed, the lobes depressed-deltoid, irregularly toothed, the petioles 3–17 cm. long; scapes 5–12 cm. long, longer than the leaves, glabrous, 1- to 2-flowered, the bracts lanceolate, ciliate, small, the pedicels 1–2 cm. long; calyx scattered short-pilose, nearly tubular, 6–7 mm. long, 5-cleft, the lobes subobtuse, ciliate; corolla rose-colored, about 2.5 cm. across, the tube about 1 cm. long, the lobes minutely ciliolate, the throat yellow; capsules oblong-cylindric, 11–13 mm. long, about twice as long as the calyx.——May. Rocky cliffs and ravines, alpine; Hokkaido (Hidaka Prov.); rare.

Var. **kamuiana** (Miyabe & Tatew.) Hara. *P. kamuiana* Miyabe & Tatew.——KAMUI-KOZAKURA. Petioles, underside of leaves, and scapes more densely pubescent; pedicels and calyx pubescent; corolla-lobes ciliolate.——Alpine; Hokkaido (Ishikari Prov.).

9. Primula jesoana Miq. *P. hondonensis* Nakai & Kitag.; *P. jesoana* var. *glabra* Takeda——Ō-SAKURA-SŌ. Rhizomes short-creeping; leaves membranous, cordate to reniform, 4–8 cm. long, 6–12 cm. wide, shallowly 7- to 9-lobed, glabrous on upper side, thinly puberulent beneath, the lobes deltoid or again lobulate, mucronate, the petioles glabrous; scapes 20–40 cm. long, 1.5–2.5 times as long as the leaves, glandular-puberulent in upper part; flowers several, in one or sometimes 2 whorls, the bracts broadly linear, small, obtuse, the pedicels 1–2 cm. long; calyx nearly tubular, 7–8 mm. long, 5-cleft, the lobes lanceolate, obtuse; corolla 1.5–2.5 cm. across, rose-purple, the tube 12–14 mm. long; capsules 7–12 mm. long, ovoid-oblong. ——July–Aug. High mountains; Hokkaido (w. distr.), Honshu (centr. and n. distr.); rare.

Var. **pubescens** (Takeda) Takeda & Hara. *P. jesoana* forma *pubescens* Takeda; *P. yesomontana* Nakai & Kitag.; *P. yezomontana* var. *nudiuscula* Nakai & Kitag.——EZO-Ō-SAKURA-SŌ. Leaves shallowly lobed, long-pubescent on the nerves beneath and on lower part of scapes and petioles.——Hokkaido.

10. Primula takedana Tatew. TESHIO-KOZAKURA. Leaves reniform-orbicular, 3.5–4 cm. long, 4.5–5(6) cm. wide, short-pilose on upper side, long-pubescent beneath, deeply incised, the lobes usually 3- to 5-lobulate, without or with few teeth, the petioles 6–12 cm. long, long-pubescent; scapes about 15 cm. long, longer than the leaves, long-pubescent at the base, puberulent in upper part, the pedicels longer than the calyx; calyx 5–7 mm. long, 5-cleft; corolla white, infundibuliform, the tube 6–8 mm. long, the lobes slightly shorter than the tube, ascending; capsules short-cylindric, about 12 mm. long, longer than the calyx.——May–June. High mountains; Hokkaido (Teshio Prov.).

11. Primula kisoana Miq. *P. shikokiana* (Makino) Nakai; *P. kisoana* var. *shikokiana* Makino——KAKKO-SŌ, KISO-KOZAKURA. Rhizomes short; leaves reniform-orbicular, 5–10 cm. in diameter, rather prominently wrinkled, slightly recurved on margin, vivid to deep green, short-pubescent on upper side, long-pubescent beneath, shallowly incised, the lobes obtuse, mucronate, the nerves strongly impressed on upper side, the petioles densely long-pubescent; scapes 10–15 cm. long, rather stout, densely long-pubescent, umbellately 2- to 8-flowered at the top, the bracts lanceolate, pubescent; flowers usually rose-purple; calyx 10–12 mm. long, densely crisp-pubescent, 5-cleft, the lobes linear-lanceolate; corolla 2–3 cm. across,

the tube 15–20 mm. long; capsules shorter than the calyx, ovoid-globose, about 5 mm. long.——May. Honshu (Kantō and south centr. distr.), Shikoku; rare.

12. Primula tosaensis Yatabe. IWAZAKURA, TOSAZAKURA. FIG. 12. Rhizomes short; leaves orbicular to ovate-cordate, 4–7 cm. in diameter, rounded, glabrate on upper side, loosely long-pubescent on the nerves beneath, shallowly incised and toothed, the petioles once to twice as long as the leaves, long-pubescent; scapes 10–15 cm. long, long-pubescent on the lower part, glabrous above, the bracts broadly linear, the pedicels 2–5, usually umbellate, 1–2.5 cm. long; flowers rose-purple, 2.5–3 cm. across; calyx 6–8 mm. long, 5-cleft, the lobes gradually narrowed, obtuse, callose-pointed; corolla-tube 15–20 mm. long, narrow; capsules long-cylindric, 1.5–2.5 cm. long, erect, 2 or 3 times as long as the calyx.——Apr.–May. Wet shaded rocks in mountains; Honshu (s. Kinki Distr. and Mino Prov.), Shikoku, Kyushu; rare.

Var. **brachycarpa** (Hara) Ohwi. *P. tosaensis* forma *brachycarpa* Hara; *P. senanensis* Koidz.; *P. reinii* var. *brachycarpa* (Hara) Ohwi; *P. reinii* var. *ovatifolia* Ohwi——SHINA-NO-KOZAKURA. Leaves orbicular or ovate-cordate; corolla tube about 15 mm. long; capsules oblong-cylindric to oblong, 5–13 mm. long.——Wet rocky cliffs in mountains; Honshu (south centr. distr. and Kantō); rare.

Var. **rhodotricha** (Nakai & F. Maekawa) Ohwi. *P. rhodotricha* Nakai & F. Maekawa——CHICHIBU-IWAZAKURA. Leaves orbicular-cordate, obsoletely incised; capsules about 10 mm. long.——Rocky wet shady cliffs in mountains; Honshu (south centr. distr. and Kantō); rare.

13. Primula reinii Fr. & Sav. *P. okamotoi* Koidz.; *P. hakonensis* Nakai——KO-IWAZAKURA. Rhizomes slender, short; leaves few, reniform to orbicular-cordate, 1–2(–6) cm. long, 1–3(–7) cm. wide, the margins recurved, shallowly 7- to 9-lobulate, pubescent above at least while young, pale green and long-pubescent especially on the nerves beneath, the petioles 3–10 cm. long, soft-pubescent; scapes 5–10 cm. long, slender, long-pubescent at base, 1- to 5-flowered at the apex, the pedicels 1–2 cm. long in fruit, the bracts lanceolate; flowers rose-purple, salverform, 2–3 cm. across; calyx 5–8 mm. long, 5-cleft; corolla-tube 8–10 mm. long; capsules as long as to slightly longer than the calyx, oblong to short-cylindric, 5–13 mm. long.——May. Wet shaded rocky cliffs in mountains; Honshu (Kantō and centr. distr. westw. to Mount Omine in Yamato); rare.

Var. **kitadakensis** (Hara) Ohwi. *P. kitadakensis* Hara; *P. hisauchii* Miyabe & Tatew.——KUMOI-KOZAKURA. Leaves and petioles less densely pubescent, the blades with acute lobes and teeth.——Honshu (Kantō and centr. distr.).

7. ANDROSACE L. TOCHINAI-SŌ ZOKU

Delicate, low annuals or perennials; stems or cauliculi absent or short and much branched, densely leafy; flowers solitary or more commonly umbellate on scapes, small, white or pink; calyx 5-lobed or -cleft, persistent, the lobes erect, incurved or spreading in fruit; corolla salverform or infundibuliform, the tube short, not exserted beyond the calyx, constricted on the throat, the limb 5-lobed; stamens 5, included, the filaments very short, the anthers obtuse; style usually short; ovules few to many, semi-anatropous; capsules globose or ovoid, dehiscent, 5-valved; seeds flattened dorsally, the testa wrinkled.——About 100 species, mostly alpine or montane, few in the warmer regions of the N. Hemisphere.

1A. Annual or biennial; stemless; leaves radical, depressed-orbicular to nearly round, toothed, truncate at base and abruptly petiolate; pedicels 2–4 cm. long; calyx cleft nearly to the base, the lobes spreading, ovate. .... 1. *A. umbellata*
1B. Perennial; stems short and much branched, leaves cauline, verticillate, broadly oblanceolate, entire, gradually narrowed to a short petiolelike base; pedicels 2–3.5 mm. long; calyx cleft half way, the lobes erect, short-deltoid. .... 2. *A. lehmanniana*

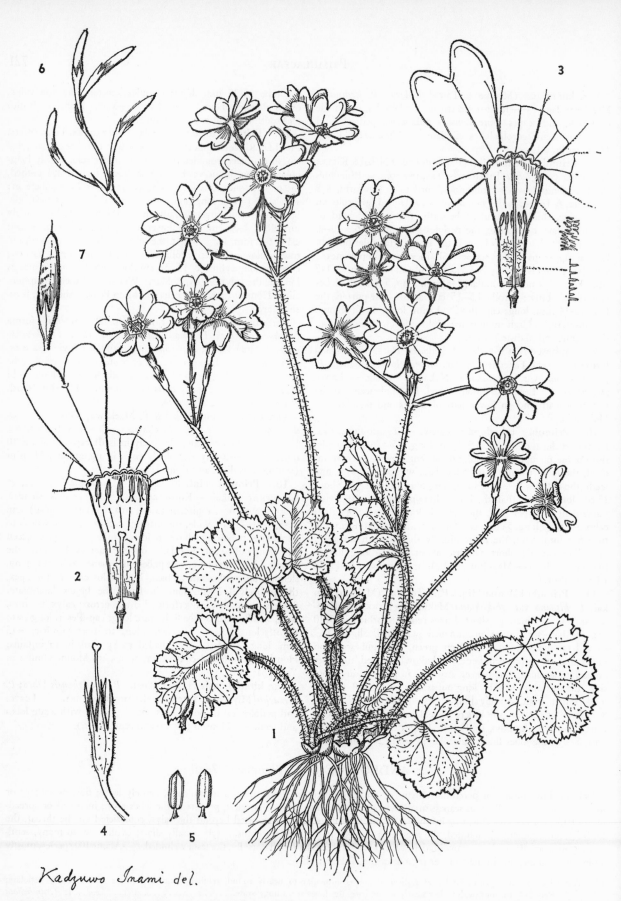

Kadzuwo Inami del.

Fig. 12.—*Primula tosaensis* Yatabe. 1, Habit; 2, flower with short style; 3, flower with long style; 4, calyx; 5, stamens; 6, portion of inflorescence in fruit; 7, capsule.

1. **Androsace umbellata** (Lour.) Merr. *Drosera umbellata* Lour.; *A. saxifragaefolia* Bunge; *A. patens* Wright ex A. Gray——RYŪKYŪ-KOZAKURA. Annual or biennial, with short spreading pubescence; leaves radical, 10–30, spreading, subrounded or sometimes ovate-rounded, 5–15 mm. in diameter, flat, deltoid-toothed, truncate and abruptly petiolate, the petioles 1–2 cm. long; scapes 1–25, 5–10 cm. long, the bracts ovate to lanceolate, 4–7 mm. long, the pedicels 4–10, spreading, slender, 2–3.5 cm. long; calyx accrescent after anthesis, spreading, 5-parted nearly to the base, the lobes green, ovate, 2–3 mm. long in anthesis, 4–5 mm. long in fruit, acute; corolla white, 4–5 mm. across, 5-lobed; capsules ovoid-globose, about 4 mm. in diameter, 5-valved, the valves white, membranous. ——Mar.–Apr. Lowlands; Honshu (Chūgoku Distr.), Shikoku, Kyushu.——Korea, Formosa, and se. Asia.

2. **Androsace lehmanniana** Spreng. *A. villosa* var. *latifolia* Ledeb.; *A. chamaejasme* var. *capitata* R. Knuth, and var. *arctica* R. Knuth; *A. chamaejasme* var. *paramushirensis* Kudo; *A. chamaejasme* subsp. *lehmanniana* (Spreng.) Hult.——TOCHINAI-SŌ, CHISHIMA-KOZAKURA. Small perennial, loosely white-pubescent; stems short, slender, much-branched, with several dense whorls of leaves near top; leaves thick, pale yellow-green, broadly oblanceolate to narrowly obovate, 5–12 mm. long, 2–5 mm. wide, very obtuse, narrowed to a short petiole-like base, entire, loosely long-pubescent, without a distinct midrib; scapes 3–4 cm. long, white-pubescent; flowers 2–4, umbellate, white, 5–6 mm. across, the bracts narrowly oblong, obtuse, 2–3(–4) mm. long, the pedicels 3–5 mm. long, slightly glandular; calyx cup-shaped, 5-lobed on upper half, the lobes ovate, obtuse, erect; corolla salverform, with a yellow center, the limb spreading, the lobes broadly elliptic; capsules enveloped by the calyx.——July. Alpine; Hokkaido, Honshu (Mount Hayachine in Rikuchu); rare.——Kuriles, Sakhalin, and n. Korea to Alaska, n. China, Siberia, and Turkestan.

8. STIMPSONIA C. Wright ex A. Gray   HOZAKIZAKURA ZOKU

Glandular-pilose small annual with erect or ascending, rather simple leafy stems; leaves alternate; inflorescence a loose terminal raceme; flowers small, solitary in axils of bractlike leaves; calyx 5-parted, the lobes linear-oblong, spreading in fruit; corolla hypocrateriform, the tube slightly longer than the calyx, not constricted, hairy on the throat, the lobes 5, imbricate in bud; stamens 5, on the tube, the filaments short-filiform, the anthers not exserted, obtuse; ovary globose; ovules many, semi-anatropous; capsules globose, 5-valved to the base; seeds small, many.——One species in e. Asia.

1. **Stimpsonia chamaedryoides** C. Wright. *Primula veronicoides* Petitm.——HOZAKIZAKURA. Plant 3–12 cm. high, with multicellular spreading glandular hairs; stems simple, erect; radical and lower few cauline leaves elliptic to broadly ovate, 1–2 cm. long, 7–12 mm. wide, rounded at base, obtusely toothed, the petioles about as long or shorter than the blades; upper leaves sessile, ovate-orbicular, almost duplicate-toothed, the upper bracteal leaves lanceolate, 2–5 mm. long, entire; flowers 2–10, 4–5 mm. across, white, short-pedicellate; calyx 2–3 mm. long; corolla-tube about 2.5 mm. long, about as long as the lobes; capsules about 2.5 mm. across.——Apr. Kyushu (Yakushima); rare.——Ryukyus, Formosa, and China.

9. SAMOLUS L.   HAI-HAMA-BOSSU ZOKU

Glabrous herbs, rarely suffrutescent; leaves radical and cauline, alternate, linear to obovate, entire; flowers small, white, in terminal racemes or corymbs, the pedicels often bracteate near the middle; calyx-tube adnate to the lower part of ovary, the limb persistent, 5-lobed; corolla subcampanulate, with a short tube, the limb 5-lobed; stamens 5, inserted on the tube, alternate with the staminodes, the filaments short, the anthers obtuse to acuminate; ovary globose, the style short; ovules many, semi-anatropous; capsules ovoid-globose, 5-valved; seeds minute.——About 10 species, chiefly in the S. Hemisphere, few in the N. Hemisphere.

1. **Samolus parviflorus** Raf. *S. floribundus* H. B. K.; *S. valerandi* var. *floribundus* (H. B. K.) Reichenb.; *S. valerandi* sensu auct. Japon., non L.——HAI-HAMA-BOSSU. Glabrous perennial 10–30 cm. high, ascending to decumbent, with slender short rhizomes; radical leaves obovate or broadly elliptic, 2–6 cm. long, 1–2 cm. wide, entire, rounded to very obtuse, gradually narrowed to a petiolelike base, minutely brown-spotted, the cauline leaves abruptly narrowed to a short petiole-like base, 1–3 cm. long; racemes often leafy at base, loosely 10- to 20-flowered, 4–10 cm. long, the pedicels ascending-spreading, slender, 1–2 cm. long, the bracts minute, near the middle of the pedicel; flowers white, about 1.5 mm. across; calyx accrescent after anthesis, the teeth deltoid, minute; capsules globose, about 2.5 mm. across.——June–Aug. Hokkaido, Honshu.——N. America.

Fam. 162. PLUMBAGINACEAE   ISO-MATSU KA   Leadwort Family

Herbs or shrubs, sometimes scrambling or scandent, exstipulate; flowers bisexual, actinomorphic, often in one-sided racemes or umbels, the bracts often sheathing, scarious; calyx frequently ribbed, often scarious between the lobes; corolla-lobes imbricate, usually small, persistent; stamens 5, opposite the corolla-lobes and more or less inserted on the tube, the anthers 2-locular, longitudinally dehiscent; ovary superior, 1-locular; styles 5, free or connate; ovules solitary, pendulous; fruit a utricle, sometimes circumscissile; seeds with or without endosperm; embryo large, erect.——About 10 genera, chiefly in the Mediterranean region and centr. Asia, a few in N. America.

1. LIMONIUM Mill.   ISO-MATSU ZOKU

Perennials or rarely annuals, acaulescent or with short, thick, branched stems; leaves rosulate, dense, alternate, flat, entire or pinnately divided; scapes or peduncles often branched; inflorescence a cyme, corymb, or panicle, the bracts scalelike, with a bonelike midrib and broad scarious margins; calyx usually infundibuliform or tubular, 5- or 10-ribbed, scarious, plicate at base;

petals distinct or united only at base; stamens inserted at base of petals; styles usually free; fruit enclosed in the calyx, indehiscent or variously dehiscent, the endosperm scanty or copious.——About 120 species, chiefly of deserts, seashores, and high mountains of the N. Hemisphere, few in Australia, and S. Africa.

1A. Herbaceous biennial; leaves radical, 8–15 cm. long, 1.5–3 cm. wide; scapes 20–50 cm. long; corolla yellow. .... 1. *L. tetragonum*
1B. Shrubby perennial; leaves fascicled and terminal on the branches, 2–4 cm. long, 4–7 mm. wide; scapes 7–15 cm. long; corolla pinkish.
                                                                                                                     2. *L. wrightii*

1. Limonium tetragonum (Thunb.) Bullock. *Statice tetragona* Thunb.; *S. japonica* Sieb. & Zucc.; *L. japonicum* (Sieb. & Zucc.) O. Kuntze——HAMA-SAJI.   Glabrous biennial; leaves radical, rather thick and firm, rosulate, oblong-spathulate, 8–15 cm. long, 1.5–3 cm. wide, obtuse, mucronate, entire, 3-nerved with very weak veinlets; scapes few, erect, 20–50 cm. long, much branched from the base, the bracts narrowly deltoid, acute, 2–3 mm. long; inflorescence spicate, 2–4 cm. long, the bracts scarious-margined, rounded, cuspidate, about 4 mm. long; flowers few, axillary, the bracteoles 2, small, scarious; calyx 5–6 mm. long, pink in upper part, with short appressed white hairs on back, 5-ribbed, the ribs cuspidate; corolla yellow, slightly longer than the calyx; fruit fusiform, 2.5 mm. long.——Sept.–Nov. Sandy places near the sea; Hon-

shu (Rikuchu Prov. and southw.), Shikoku, Kyushu.——Korea and Manchuria.
    2. Limonium wrightii (Hance) O. Kuntze. ISO-MATSU, MURASAKI-ISO-MATSU.   Glabrous shrubby perennial 5–10 cm. high, much branched; leaves fascicled and terminal on the branches, thick-coriaceous, oblanceolate, 2–4 cm. long, 4–7 mm. wide, rounded at apex, clasping at the base, entire, the nerves and midrib obscure; scapes terminal on the branches, 7–15 cm. long; inflorescence spicate, 1–2 cm. long, one-sided, the bracts scarious-margined, rounded at apex, about 5 mm. long; corolla pinkish; calyx about 5 mm. long, with 5 appressed-pilose ribs. ——Aug.–Oct. Rocks along the sea; Honshu (Izu Isls.), Kyushu (Yakushima). A yellow–flowered variety, var. *luteum* (Hara) Hara occurs in the Bonins, Ryukyus, and Formosa.

Fam. 163. **EBENACEAE**   KAKI-NO-KI KA   Ebony Family

Usually dioecious shrubs or trees; leaves alternate, simple, entire, exstipulate; calyx 3- to 6-lobed, persistent, often accrescent after anthesis; corolla 3- to 6-lobed; stamens inserted at base of corolla, 2–4 times as many or rarely as many as the corolla-lobes and alternate with them (staminodia usually present in pistillate flowers), the filaments free or connate into pairs, the anthers 2-locular, introrse, longitudinally dehiscent; ovary superior, with 3 or more locules; styles often free; ovules 1–2 in each locule, pendulous and anatropous; fruit a berry; seeds with a thin pericarp, the embryo straight, about half as long as the endosperm.——About 6 genera, with about 300 species, in the Tropics and subtropics of both hemispheres.

1. DIOSPYROS L.   KAKI-NO-KI ZOKU

Dioecious shrubs or trees; leaves alternate, petiolate; inflorescence axillary, produced on the new growth, sometimes 1-flowered; flowers polygamous; calyx 3- to 7-lobed, hairy, accrescent after anthesis; corolla urceolate, tubular, campanulate or rotate, more or less pubescent externally, 3- to 7-lobed, the lobes contorted in bud; fertile stamens 4 to many, commonly 16, inserted at the base of the tube, the filaments shorter than the anthers, free or variously connate; staminodia in pistillate flowers much-reduced; ovary globose or conical-globose, the locules either twice as many as the styles and 1-ovuled, or as many as the styles and 2-ovuled; styles 1–4, often connate at base, rarely obscure; fruit globose or ovoid, fleshy; seeds oblong, lustrous, somewhat laterally compressed.——About 200 species, mainly tropical, a few temperate.

1A. Leaves evergreen; young branchlets grayish puberulent; calyx-lobes not reflexed. .... 1. *D. morrisiana*
1B. Leaves deciduous; young branchlets glabrous or brownish pilose; calyx-lobes reflexed.
    2A. Young branchlets glabrous or glabrescent; leaves lustrous on upper side, often whitish beneath; corolla glabrous outside; fruit 1.5–2 cm. across.
        3A. Young branches short-pubescent; petioles 8–12 mm. long. .... 2. *D. lotus*
        3B. Young branches glabrous; petioles 1–2 cm. long. .... 3. *D. japonica*
    2B. Young branchlets brownish pilose; leaves scarcely lustrous on upper side, pale green beneath; corolla pubescent especially on outside of lobes; fruit 4–8 cm. across. .... 4. *D. kaki*

1. Diospyros morrisiana Hance. *D. nipponica* Nakai ——TOKIWAGAKI.   Evergreen small tree or large shrub; branches lenticellate, the branchlets grayish puberulent; leaves coriaceous, narrowly elliptic to oblong, 6–9 cm. long, 2–3.5 cm. wide, short-acuminate with an obtuse tip, acuminate at base, glabrous, deep green on upper side, brownish appressed-hairy beneath while very young, entire; flowers short-pedicelled, pale yellow; calyx-lobes deltoid, brownish pilose; corolla campanulate, 7–8 mm. long, glabrous outside; ovary glabrous; fruit globose, 1.5–2 cm. across.——June. Honshu (Izu Prov. and westw. through Tōkaidō and Kii Prov. to Chūgoku Distr.), Shikoku, Kyushu.——Ryukyus, Formosa, and s. China.
    2. Diospyros lotus L. MAMEGAKI.   Deciduous tree; branches grayish, the branchlets short-appressed yellow-gray

puberulent while very young; leaves elliptic, oblong, or ovate-elliptic, 6–12 cm. long, 5–7 cm. wide, abruptly acute, entire, soft-pubescent while very young, glaucous and grayish white-pubescent especially on nerves beneath, the petioles 8–12 mm. long; flowers pale yellow, axillary; calyx-lobes short-pilose outside while young, regularly pilose inside; corolla campanulate; fruit globose, about 1.5 cm. across.——June. Frequently cultivated in our area.——Korea and China to Asia Minor.
    3. Diospyros japonica Sieb. & Zucc. *D. lotus* var. *glabra* (DC.) Makino; *D. kaki* var. *glabra* A. DC.; *D. lotus* var. *japonica* (Sieb. & Zucc.) Franch.——SHINANOGAKI. Branches glabrous; leaves glabrous or nearly so, the petioles 1–2 cm. long.——June. Honshu (Izu Prov. and westw.), Shikoku, Kyushu.——Ryukyus and China.

4. Diospyros kaki Thunb. *D. kaki* var. *domestica* Makino——KAKI-NO-KI. Deciduous tree, the branches lenticellate, gray-brown or gray-white, young branchlets prominently brownish pilose; leaves broadly elliptic, obovate or ovate-elliptic, 7–17 cm. long, 4–10 cm. wide, abruptly acute, rounded to broadly cuneate or acute at base, entire, puberulent on midrib on upper side, usually brownish pilose beneath, the petioles 1–1.5 cm. long; flowers pale yellow; corolla campanu-late, pubescent especially on the outside of lobes; fruit globose, ovoid, 4–8 cm. in diameter.——June. Widely cultivated in Honshu, Shikoku, Kyushu.

Var. **sylvestris** Makino. YAMAGAKI. Wild phase, more densely pubescent; leaves rather small and narrower; ovary pilose.——Honshu (w. distr.), Shikoku, Kyushu.——s. Korea and China.

Fam. 164. **SYMPLOCACEAE** HAI-NO-KI KA Symplocos Family

Evergreen or deciduous shrubs or trees; leaves simple, alternate, exstipulate; flowers in axillary or terminal spikes, racemes, or panicles, or sometimes solitary, actinomorphic, bisexual or rarely polygamous; calyx-lobes 5, valvate in bud; corolla 3- to 11-parted or -divided; stamens inserted on the corolla, 4 to many, free or variously connate, in 1–4 series, the anthers 2-locular, longitudinally dehiscent; ovary inferior or subinferior, 2- to 5-locular; ovules 2–4 in each locule, pendulous; style single, slender; fruit a drupe or berry, crowned with the persistent calyx-lobes; seeds solitary in each locule, endosperm copious; embryo straight or curved.——One genus with about 300 species, in Asia, Australia, and America, sometimes divided into several smaller genera.

1. **SYMPLOCOS** Jacq. HAI-NO-KI ZOKU

Characters of the family. The Japanese species are sometimes placed into the segregate genera *Palura* Hamilt. ex Miers, and *Dicalyx* Lour. (*Bobua* DC.).

1A. Deciduous trees and shrubs; inflorescence paniculate; flowers polygamous "*Palura*."
  2A. Leaves broadly obovate, 3–5 cm. wide, caudate, coarsely serrate; fruit bluish black. .... 1. *S. coreana*
  2B. Leaves obovate to oblong, usually narrower, the teeth smaller and incurved.
    3A. Leaves obovate-cuneate, sometimes oblong or elliptic, abruptly acute to short-acuminate; fruit bluish, rarely white.
                                                 2. *S. chinensis* var. *leucocarpa*
    3B. Leaves oblong to elliptic, acuminate to acute at both ends, or ovate and rounded at the base; fruit black. .... 3. *S. paniculata*
1B. Evergreen trees and shrubs; inflorescence spicate or racemose; flowers bisexual.
  4A. Inflorescence racemose; pedicels 8–15 mm. long, as long as to longer than the flowers. .... 4. *S. myrtacea*
  4B. Inflorescence spicate or capitate; pedicels absent or shorter than the flowers.
    5A. Flowers sessile; inflorescence spicate, headlike or elongate; bracts and bracteoles persistent.
      6A. Spikes elongate, 1–7 cm. long; fruit 5–6 mm. long.
        7A. Branchlets yellow-brown pilose; leaves chartaceous, rather thin, 4–6 (–8) cm. long, long-caudate, slightly lustrous beneath; petioles 2–4 mm. long; spikes 1–3 cm. long, simple, yellow-brown pilose; corolla-lobes about 2 mm. long; fruit broadly ellipsoidal. .... 5. *S. lancifolia*
        7B. Branchlets glabrous or brownish appressed-pilose while young; leaves coriaceous, whitish and dull beneath; petioles 7–12 mm. long; spikes 3–7 cm. long, often branched at base, brown-pilose; corolla-lobes about 3 mm. long; fruit urceolate.
        8A. Leaves 10–15 cm. long, with slender, scarcely raised lateral nerves beneath; branches slender; fruit about 5 mm. long.
                                               6. *S. theophrastaefolia*
        8B. Leaves 15–23 cm. long, with raised lateral nerves beneath; branches thick; fruit about 6 mm. long. ...... 7. *S. kotoensis*
      6B. Spikes very short and nearly headlike; fruit 1–2 cm. long.
        9A. Young branchlets, petioles, and midrib of leaves beneath reddish; leaves oblong-lanceolate to broadly oblanceolate, glaucous beneath, deep green on upper side. .... 8. *S. glauca*
        9B. Young branchlets, petioles, and midrib of leaves beneath yellow-green; leaves elliptic to narrowly oblong, pale green beneath, vivid- to yellow-green on upper side.
          10A. Leaves 10–15 cm. long, the midrib not raised on upper side; branches often slightly angled; fruit 2–2.2 cm. long.
                                               9. *S. tanakae*
          10B. Leaves 4–7 cm. long, with a raised midrib on upper side; branches distinctly trigonous while young; fruit 1–1.5 cm. long.
                                              10. *S. lucida*
    5B. Flowers short-pedicelled; bracts and bracteoles caducous; young branchlets glabrous, trigonous, gray-brown; leaves narrowly elliptic, 5–8 cm. long, coriaceous, with an impressed midrib on upper side; fruit ovoid-oblong, 7–8 mm. long. .. 11. *S. prunifolia*

1. Symplocos coreana (Lév.) Ohwi. *S. crataegoides* forma *major* Fr. & Sav.; *Cotoneaster coreanus* Lév.; *S. argutidens* Nakai; *Palura argutidens* (Nakai) Nakai; *Palura coreana* (Lév.) Nakai——TANNA-SAWA-FUTAGI. Large deciduous shrub, with exfoliating bark; branches gray-brown, glabrous; leaves broadly obovate, 5–8 cm. long, 3–5 cm. wide, caudate, broadly cuneate at base, coarsely serrate, pale green and pilose especially on nerves beneath, the petioles 3–7 mm. long; panicles 4–7 cm. long, pilose or glabrous; flowers white, 6–7 mm. across, on short slender pedicels; calyx-teeth membranous, ovate-rounded; stamens slightly longer than the corolla, more or less fasciculate in 5 groups; fruit obliquely ovoid, bluish black, 6–7 mm. in diameter.——June. Honshu (Kantō Distr. and westw.), Shikoku, Kyushu.——Korea (Quelpaert Isl.).

2. Symplocos chinensis (Lour.) Druce var. **leucocarpa** (Nakai) Ohwi. *S. paniculata* var. *leucocarpa* Nakai; *Palura paniculata* var. *leucocarpa* (Nakai) Nakai; *S. crataegoides* var. *leucocarpa* (Nakai) Makino & Nemoto; *Palura chinensis* var. *leucocarpa* (Nakai) Hara.——SHIROMI-NO-SAWA-FUTAGI. Large deciduous shrub, with longitudinally fissured bark; branches gray-brown, short curved-hairy while young, glabrescent; leaves obovate-cuneate, sometimes oblong or elliptic, 4–7(–10) cm. long, abruptly acute to short-acuminate, sometimes rounded at apex, with short incurved teeth, scattered-pilose on both surfaces and on the nerves beneath, the petiole 3–8 mm. long; panicles 3–6 cm. long, pilose, the bracts membranous, linear, small, caducous; flowers white, 7–8 mm. across; calyx-teeth ovate-rounded, membranous; stamens slightly longer than the corolla; fruit obliquely ovoid, 6–7 mm. long, white.——May–June. Honshu; rare.——Forma **pilosa** (Nakai) Ohwi. *S. crataegoides* sensu auct. Japon., non Hamilt.; *Palura paniculata* var. *pilosa* Nakai; *P. ciliata* Nakai; *P. pilosa* Nakai ex Honda; *P. chinensis* var. *pilosa* (Nakai) Nakai——SAWA-FUTAGI, RURIMI-NO-USHI-KOROSHI, NISHIGORI. Fruit blue.——Common in mountains and hills; Hokkaido, Honshu, Shikoku, Kyushu.——Korea, Manchuria, and China. The typical phase accurs in Formosa, s. China, Indochina, Philippines, and the Himalayas.

3. **Symplocos paniculata** (Thunb.) Miq. *Prunus paniculata* Thunb.; *Palura paniculata* (Thunb.) Nakai——KUROMI-NO-NISHIGORI, SHIRO-SAWA-FUTAGI, NISHIGORI. Large deciduous shrub with exfoliating bark, branches glabrous; leaves oblong or elliptic, rarely somewhat ovate, 5–10 cm. long, 2.5–4 cm. wide, acuminate to acute at both ends, incurved-toothed, the petioles 5–10 mm. long; inflorescence 3–7 cm. long, nearly glabrous; flowers white, about 8 mm. across, short-pedicellate; calyx-teeth ovate, rounded, membranous; fruit ovoid-globose, black, 6–7 mm. in diameter.——May–June. Honshu (centr. distr. westw.).

Var. **pubescens** (Nakai) Ohwi. *Palura paniculata* var. *pubescens* Nakai; *P. chinensis* var. *pubescens* (Nakai) Nakai; *S. tanakana* Nakai; *P. tanakana* (Nakai) Nakai——KUROMI-NO-SAWA-FUTAGI. Bark transversely fissured; young branchlets and underside of leaves pilose.——Honshu (w. distr.), Shikoku, Kyushu.——s. Korea.

4. **Symplocos myrtacea** Sieb. & Zucc. *Bobua myrtacea* (Sieb. & Zucc.) Miers; *Dicalyx myrtacea* (Sieb. & Zucc.) Hara ——HAI-NO-KI. Small glabrous evergreen tree; branches dark brown, slender, the young branchlets terete, pale green to vivid green when dried; leaves thinly coriaceous, ovate or narrowly so, 4–7 cm. long, 1.5–2.5 cm. wide, caudate, abruptly mucronate, rounded to abruptly acute at base, obtuse-toothed, lustrous, pale green beneath, the petioles slender, 8–15 mm. long; racemes 3- to 6-flowered, loose, nearly corymbose, the pedicels 8–15 mm. long, slender, the bracteoles lanceolate, membranous, long-hairy on back, slightly longer than the calyx, caducous; calyx-limb short, the teeth ovate-deltoid, small, less than 1 mm. long; corolla about 12 mm. across; stamens nearly as long as the corolla; fruit narrowly ovoid, 7–8 mm. in diameter.——May. Honshu (sw. Kinki Distr. and westw.), Shikoku, Kyushu.

5. **Symplocos lancifolia** Sieb. & Zucc. *S. leptostachys* Sieb. & Zucc.; *S. lancifolia* var. *leptostachys* (Sieb. & Zucc.) Miq.; *Bobua lancifolia* (Sieb. & Zucc.) Miers; *B. leptostachys* (Sieb. & Zucc.) Miers; *Dicalyx lancifolia* (Sieb. & Zucc.) Hara ——SHIROBAI. Small evergreen tree with slender, terete, gray-brown branches, the branchlets yellow-brown pilose; leaves narrowly ovate to broadly lanceolate, 4–6(–8) cm. long, 1.5–2.5(–3) cm. wide, caudate, mucronate, acute or sometimes rounded at base, deep green and lustrous with the midrib puberulent on upper side, slightly paler and appressed-hairy beneath while young, undulate-toothed, the petioles 2–4 mm. long; inflorescence spicate, 1–3 cm. long, densely yellow-brown pilose; flowers 10–15, sessile, about 4 mm. across, the bracts and bracteoles broadly deltoid, as long as the calyx-tube; calyx-teeth elliptic, 0.7 mm. long; fruit broadly obovoid-ellipsoidal,

5–6 mm. long, black.——Sept.–Oct. Honshu (Kinki Distr. and westw.), Shikoku, Kyushu.——Ryukyus.

6. **Symplocos theophrastaefolia** Sieb. & Zucc. *S. spicata* sensu auct. Japon., non Roxb.; *Bobua theophrastaefolia* (Sieb. & Zucc.) Miers; *Dicalyx theophrastaefolia* (Sieb. & Zucc.) Hara ——KANZABURŌ-NO-KI. Small evergreen tree with terete, rather slender, brownish branches, the branchlets brown, appressed-pilose while young; leaves coriaceous, narrowly elliptic or narrowly oblong, sometimes broadly lanceolate, 10–15 cm. long, 3–5 cm. wide, caudate to acuminate, acute to acuminate at base, undulate-toothed, pale green or sometimes very slightly glaucous beneath, glabrous on both surfaces, the petioles 7–12 mm. long; spikes often branched at base, 3–5 cm. long, densely brown-pilose; flowers sessile, 7–8 mm. across, the bracts and bracteoles broadly deltoid, as long as the calyx-tube, densely brown-pilose; calyx-teeth subrounded, ciliolate; fruit depressed-urceolate, about 5 mm. long.——Aug.–Sept. Honshu (Suruga Prov. and westw.), Shikoku, Kyushu.——Ryukyus, Formosa, and China.

7. **Symplocos kotoensis** Hayata. *S. lithocarpoides* Nakai; *Bobua lithocarpoides* (Nakai) Nakai; *B. kotoensis* (Hayata) Yamamoto; *Dicalyx kotoensis* (Hayata) Hara——AOBA-NO-KI. Resembles the preceding, the branches thick, terete, appressed brown-pubescent while young; leaves vivid green, coriaceous, elliptic to narrowly oblong, sometimes broadly lanceolate, 15–23 cm. long, 3.5–8 cm. wide, short-caudate to short-acute, acute at base, glabrous and lustrous on upper side, scattered-pilose, pale green and somewhat glaucous beneath, entire to obsoletely undulate-toothed near tip, midrib and lateral nerves raised beneath, the petioles 7–12 mm. long, stout; spikes often branched at base, 4–7(–10) cm. long, sparsely woolly in anthesis, the bracts and bracteoles densely brown-pubescent; fruit black, ovoid-urceolate, about 6 mm. long, the calyx persistent.——Aug.-Sept. Kyushu (Yakushima and Tanegashima).——Ryukyus and Formosa.

8. **Symplocos glauca** (Thunb.) Koidz. *Laurus glauca* Thunb.; *S. neriifolia* Sieb. & Zucc.; *Bobua neriifolia* (Sieb. & Zucc.) Miers; *Myrsine thunbergii* T. Tanaka; *B. glauca* (Thunb.) Nakai; *Dicalyx glauca* (Thunb.) Migo——MIMITSUBAI. Small evergreen tree with stout terete red-brown branches, thinly brown-pubescent while young; leaves thick, oblong-lanceolate to broadly oblanceolate, 8–15 cm. long, 2–3.5 cm. wide, acute to acuminate, acute at base, entire or with few short teeth on upper margin, deep green and lustrous, glaucous and the midrib reddish beneath, the petioles 1–1.5 cm. long; inflorescence very short, axillary, densely brown-woolly; flowers sessile, about 6 mm. across; anthers purplish; fruit oblong-ovoid, 12–15 mm. in diameter.——July–Aug. Honshu (Awa Prov. and westw.), Shikoku, Kyushu.——Ryukyus, Formosa, China, and Indochina.

9. **Symplocos tanakae** Matsum. *Bobua tanakae* (Matsum.) Masam.; *Dicalyx tanakae* (Matsum.) Hara——HIRO-HA-NO-MIMITSUBAI. Small glabrous evergreen tree, yellow-green when dry, the branchlets rather stout, slightly angled; leaves narrowly oblong, 10–15 cm. long, 2–4 cm. wide, acute with an obtuse tip, acuminate at base, lustrous above, undulate-toothed on upper margin, the midrib slightly impressed or flat on upper side, the midrib raised beneath; flowers 8–10 mm. across, sessile, in axillary headlike spikes, the bracts and bracteoles depressed-cordate, glabrous, ciliolate; calyx-teeth broadly elliptic, about 3 mm. long, obsoletely ciliolate; fruit ellipsoid-globose, 2–2.2 cm. in diameter, black.——Dec.

Shikoku, Kyushu (Hiuga Prov., Yakushima and Tanegashima).——Ryukyus.

10. Symplocos lucida Sieb. & Zucc. *Laurus lucida* Thunb.; *Hopea lucida* (Thunb.) Thunb.; *S. japonica* DC.; *S. japonica* var. *lucida* (Sieb. & Zucc.) K. Koch; *Bobua japonica* (DC.) Miers; *B. lucida* (Sieb. & Zucc.) Kaneh. & Sasaki, non Miers——Kuro-ki. Resembles the preceding but smaller, more densely branched, the young branches trigonous, rather stout, yellow-green when dry; leaves coriaceous, elliptic or oblong, 4–7 cm. long, 2–3.5 cm. wide, acute or obtuse, acute at base, scarcely undulate-toothed near tip, the midrib slightly raised on both sides, the petioles 7–15 mm. long, rather thick; flowers sessile, in dense axillary heads, gray-brown, hairy, the axis very short, the bracts and bracteoles depressed-cordate; calyx-teeth cordate, ciliolate; corolla about 8 mm. across; fruit oblong, black, about 15 mm. long.——Mar.–Apr. Honshu (Izu and westw.), Shikoku, Kyushu; rather common.——Ryukyus, Formosa, and Korea (Quelpaert Isl.).

11. Symplocos prunifolia Sieb. & Zucc. *Bobua prunifolia* (Sieb. & Zucc.) Miers; *Dicalyx prunifolia* (Sieb. & Zucc.) Hara——Kurobai. Small, much-branched evergreen, glabrous tree, with branches terete, grayish to dark brown, rather thick; leaves coriaceous, narrowly elliptic, narrowly ovate-elliptic, or oblong, 5–8 cm. long, 2–3 cm. wide, caudate, acute at base, undulate-toothed, yellow-green beneath, the midrib flat or slightly impressed on upper side, raised beneath, the petioles usually purplish brown, 1–1.5 cm. long; racemes simple, 10- to 30-flowered, rather dense, 3–7 cm. long, brownish puberulent, the pedicels 1.5–3 mm. long, the bracts orbicular, 3 mm. long, the bracteoles ovate, membranous, shorter than the calyx; calyx-teeth depressed-orbicular, ciliolate; fruit narrowly ovate, black, 6–7 mm. long.——Apr.–May. Honshu (Kazusa and Awa Prov. and westw.), Shikoku, Kyushu.——s. Korea and Ryukyus.——Forma **uiae** (Makino) Ohwi. *S. prunifolia* var. *uiae* Makino; *Bobua prunifolia* var. *uiae* (Makino) Nakai; *B. uiae* (Makino) Nakai; *Dicalyx prunifolia* var. *uiae* (Makino) Hara——Maruba-kurobai. Leaves elliptic to orbicular, 3–5 cm. long, 2.5–4.5 mm. wide, rounded at both ends.——Kii Prov. in Honshu.

Fam. 165. STYRACACEAE Ego-no-ki Ka Styrax Family

Shrubs or trees usually with stellate hairs; leaves alternate, simple, exstipulate; flowers actinomorphic, bisexual, in terminal, axillary, sometimes compound racemes; calyx-tube more or less adnate to the ovary, the teeth valvate in bud; corolla 4- to 7-lobed, valvate or imbricate in bud; stamens as many as and alternate with the corolla-lobes or twice as many, usually inserted on the tube, the anthers 2-locular, longitudinally dehiscent; ovary superior or inferior, 3- to 5-locular; ovules 1 to many in each locule, anatropous; style slender, 3- to 5-lobed; fruit a drupe or capsule, the calyx persistent; seeds sometimes winged, the endosperm copious.——About 8 genera, with about 100 species, chiefly in the warmer and tropical regions of the N. Hemisphere, few in Africa.

1A. Ovary superior; flowers in terminal or axillary racemes. .... 1. *Styrax*
1B. Ovary inferior; flowers in paniculate one-sided racemes. .... 2. *Pterostyrax*

1. STYRAX L. Ego-no-ki Zoku

Shrubs or trees with stellate hairs, very rarely glabrous; leaves evergreen or deciduous, simple, entire or toothed; flowers usually white, in terminal or axillary racemes, usually pendulous; calyx campanulate, adnate at base to ovary or free, the limb entire or minutely 5-toothed; corolla actinomorphic, 5-lobed, the lobes valvate or imbricate in bud; stamens 10, inserted at the base or on the tube of the corolla, the filaments free or connate at base; ovary superior, 3-valved, the ovules few; styles slender; fruit 1- or 2-seeded, irregularly dehiscent, the pericarp dry or fleshy.——About 100 species, in temperate regions of the N. Hemisphere.

1A. Flowers few, on long slender pedicels; buds not covered by the base of the petioles; leaves ovate to narrowly oblong, 4–8 cm. long, 2–4 cm. wide, simple-hairy to glabrous beneath. .... 1. *S. japonica*
1B. Flowers few to many, short-pedicelled; buds enclosed within the inflated base of the petioles; leaves obovate- to ovate- or rhombic-orbicular, 1.5 times as long as wide or less, stellate-pubescent beneath.
    2A. Leaves 10–20 cm. long; flowers rather many, in long racemes; pedicels longer than the calyx; corolla deeply 5-cleft, the tube slightly longer than the calyx and much shorter than the lobes. .... 2. *S. obassia*
    2B. Leaves 5–8 cm. long; flowers few, in short racemes; pedicels 0–3 mm. long, much shorter than the calyx; corolla shallowly 5-lobed, the tube longer than the lobes and calyx. .... 3. *S. shiraiana*

1. Styrax japonica Sieb. & Zucc. *Cyrta japonica* (Sieb. & Zucc.) Miers——Ego-no-ki. Much-branched small tree; branchlets minutely stellate-pubescent while young; leaves membranous, ovate to narrowly oblong, 4–8 cm. long, 2–4 cm. wide, acuminate to acute, acute at base, undulately toothed, usually glabrous and vivid green on upper side, pale green with tufts of axillary hairs beneath, the lateral nerves of 3–5 pairs, the petioles 3–7 mm. long; flowers 1–4, pendulous, in axillary terminal or corymbose racemes, the pedicels 2–3 cm. long; calyx 4–5 mm. long; corolla about 2.5 cm. across, the tube not exceeding the calyx in length, the lobes narrowly ovate, densely white stellate-puberulent externally; fruit ovoid or ellipsoidal, 1–1.2 cm. in diameter, mucronate.——May–June. Thickets and thin woods in mountains and hills; Hokkaido, Honshu, Shikoku, Kyushu; very common.——Ryukyus(?), Korea, China, Formosa, and Philippines.

Var. **jippei-kawamurae** (Yanagita) Hara. *S. jippei-kawamurae* Yanagita——Ōba-ego-no-ki. Larger plant with leaves 5–10 cm. long, 3–5 cm. wide; flowers slightly larger.——Honshu (Izu Isls.).

2. Styrax obassia Sieb. & Zucc. Haku-unboku. Small tree with horizontally spreading stellate-pubescent branches while young, becoming dark brown and the outer layer exfoliating; leaves membranous, ovate to obovate-orbicular, 10–

20 cm. long, 8–20 cm. wide, abruptly acute, usually rounded at base, mucronate-toothed, vivid green and usually glabrous above, densely white stellate-hairy beneath, lateral nerves of 7–8 pairs, the petioles 5–25(–30) mm. long; racemes 10–20 cm. long, rather many-flowered, the flowers stellate-pubescent; calyx-limb mucronate; corolla about 2 cm. long, deeply 5-cleft; fruit white, stellate-pubescent, ovoid, mucronate.——May–June. Hokkaido, Honshu, Shikoku, Kyushu; rather common.—— Korea, Manchuria, and China.

3. **Styrax shiraiana** Makino. *Strigilia shiraiana* (Makino) Nakai——Ko-HAKUUNBOKU. Large shrub, the young branches stellate-pubescent, becoming gray-brown or purple-brown, the outer layer of bark often exfoliating; leaves broadly rhombic to broadly rhombic-obovate, 5–8 cm. long, 4–7 cm. wide, rounded to obtuse, broadly cuneate at base, abruptly mucronate, deep green on upper side, irregularly large-toothed near the tip, pale green and stellate-hairy while young beneath, the petioles 7–15 mm. long, stellate-hairy; racemes 3–6 cm. long, few-flowered, densely stellate-hairy, the bracts and bracteoles persistent, linear, shorter than the calyx, the pedicels 0–3 mm. long; calyx-teeth narrowly deltoid; corolla 15–20 mm. long; filaments connate at base into a short tube; fruit ellipsoidal, 1–1.2 cm. long, stellate-hairy, mucronate.——June. Mountains; Honshu (Kantō Distr. and westw.), Shikoku, Kyushu; rare.

2. PTEROSTYRAX Sieb. & Zucc.   ASAGARA ZOKU

Small deciduous stellate-hairy trees or shrubs; leaves alternate, simple, rather large, membranous, toothed, petiolate; flowers many, in one-sided paniculate racemes, white, bisexual, actinomorphic; calyx narrowly obconic or campanulate, 5-toothed; petals 5, scarcely connate at base, imbricate in bud; stamens 10, the filaments nearly free to connate and forming a tube; ovary inferior or sometimes subinferior, 3- to 5-locular; ovules 4, in 2 series in each locule; styles slender, slightly longer than the stamens; fruit a dry angular or winged 1- to 2-seeded drupe.——About 4 species, in e. Asia.

1A. Leaves 7–13 cm. long, pale green beneath, without long hairs, with 5–7 pairs of lateral nerves; fruit 5-angled, densely gray-white, stellate-hairy. .... 1. *P. corymbosa*
1B. Leaves grayish 10–20 cm. long, long white-hairy along the nerves beneath, especially in the axils, with (6–) 8–12 pairs of lateral nerves; fruit 10-angled, hispid and minutely stellate-hairy. .... 2. *P. hispida*

1. **Pterostyrax corymbosa** Sieb. & Zucc. *Halesia corymbosa* (Sieb. & Zucc.) Nichols.——ASAGARA. Small much-branched tree with gray-brown branches, loosely appressed stellate-hairy while young; leaves elliptic, or sometimes broadly ovate or broadly obovate, rarely oblong, 7–13 cm. long, 4–8 cm. wide, abruptly acute, rounded or sometimes broadly cuneate at base, minutely toothed or subentire, loosely appressed-stellate hairy on both sides, the petioles 1–3 cm. long; panicles nodding, 8–12 cm. long, white stellate-hairy; flowers white, sub-sessile, pendulous; calyx-teeth elongate-deltoid; corolla white, 8–10 mm. long, semi-expanding, the petals broadly oblanceolate; stamens tubular-connate, the longer ones alternate with the shorter; fruit obovoid, 8–12 mm. long, 5-winged, stellate-hairy and villous, the style base persistent.——May–June. Mountains; Honshu (Kinki Distr. and westw.), Shikoku, Kyushu.——China.

2. **Pterostyrax hispida** Sieb. & Zucc. *P. micrantha* Sieb. & Zucc.; *Halesia hispida* (Sieb. & Zucc.) Mast.; *Decavenia hispida* (Sieb. & Zucc.) Koidz.; *D. micrantha* (Sieb. & Zucc.) Koidz.; *D. japonica* Koidz.——ŌBA-ASAGARA. Small tree with pale brown glabrous branches; leaves oblong, obovate-oblong, or narrowly elliptic, 10–20(–25) cm. long, 5–8(–10) cm. wide, abruptly acute to short-acuminate, acute to subrounded at base, mucronate-toothed, green and glabrous on upper side, except for scattered stellate hairs on the nerves, usually whitish and puberulent or nearly glabrous beneath, the nerves white-pubescent, the petioles 7–25 mm. long; panicles nodding, 10–20 cm. long, villous and white stellate-hairy; flowers nearly sessile, pendulous; calyx-teeth deltoid; corolla semi-expanding, 6–8 mm. long, the lobes oblanceolate, 6–8 mm. long; fruit narrowly obovoid, 7–8 mm. long, 10-ribbed, rather densely long-spreading yellow-hispid, the style base persistent.——June. Mountains; Honshu, Shikoku, Kyushu.

Fam. 166. **OLEACEAE**   MOKUSEI KA   Olive Family

Evergreen or deciduous trees, shrubs, or rarely vines; leaves opposite, very rarely alternate, simple or pinnate; stipules absent; flowers bisexual or unisexual, actinomorphic; calyx lobed or toothed, rarely reduced; corolla usually 4-lobed, or the petals 4 and free or rarely absent, imbricate or rarely induplicate in bud; stamens 2, rarely 4, disc absent; ovary superior, 2-locular; ovules 2 in each locule, pendulous or ascending, the placentation axile; fruit a capsule or berry or drupelike; seeds usually with endosperm. ——About 20 genera, with more than 400 species, chiefly in temperate and tropical regions.

1A. Fruit a berry or drupelike; leaves simple.
  2A. Corolla-lobes much longer than the tube. .... 1. *Chionanthus*
  2B. Corolla-lobes as long as to shorter than the tube.
    3A. Flowers in terminal panicles; corolla-lobes valvate in bud. .... 2. *Ligustrum*
    3B. Flowers in axillary fascicles or in cymes; corolla-lobes imbricate in bud. .... 3. *Osmanthus*
1B. Fruit a dry capsule or samara.
  4A. Fruit a capsule; leaves simple or rarely pinnate.
    5A. Flowers yellow; corolla-lobes imbricate in bud; branches hollow or with lamellate pith; leaves toothed or entire, simple or ternate.
                                                      4. *Forsythia*
    5B. Flowers white, rose-purple, or lilac, in terminal and axillary much-branched thyrsoid panicles; corolla-lobes valvate in bud; branches with solid pith; leaves usually simple, entire, rarely pinnatifid. .... 5. *Syringa*
  4B. Fruit a samara; leaves pinnately compound. .... 6. *Fraxinus*

1. CHIONANTHUS L. Hitotsuba-tago Zoku

Shrubs or trees, glabrous or pubescent; leaves opposite, entire, sometimes toothed; flowers in trichotomous panicles; calyx minute, 4-lobed; corolla white, the lobes linear, much longer than the tube; stamens 2, on the tube, the filaments short, the anthers exserted or included, the connective mucronate; ovary 2-locular; ovules 2 in each locule, pendulous; style short, the stigma thick, retuse or subbifid; drupe ellipsoidal, with a firm endocarp; seeds solitary, rarely 2 or 3, the endosperm fleshy.——Two or three species, in e. Asia and N. America.

1. **Chionanthus retusus** Lindl. & Paxt. *C. chinensis* Maxim.; *C. serrulatus* Hayata; *C. retusus* var. *serrulatus* (Hayata) Koidz.——Hitotsuba-tago. Tree with gray-brown branches, puberulent while young; leaves oblong, elliptic, or ovate, 5–10(–15) cm. long, 2.5–6 cm. wide, obtuse or acute, entire or doubly toothed in the young leaves, green and glabrous on upper side, puberulent on the impressed midribs, brownish pubescent on lower part of the midrib beneath, the petioles 1.5–3 cm. long, puberulent; panicles terminal, 7–12 cm. long, glabrous, the pedicels 7–10 mm. long; calyx deeply 4-lobed, 2–3 mm. long, the lobes lanceolate, acuminate; corolla white, the tube 1–1.5 times as long as the calyx, the lobes linear-oblanceolate, 1.5–2 cm. long, 2–4 mm. wide; fruit ellipsoidal, black, about 1 cm. long.——May. Honshu (centr. distr. incl. Kantō), Kyushu (Tsushima); rare.——Korea, China, and Formosa.

2. LIGUSTRUM L. Ibota-no-ki Zoku

Evergreen or deciduous shrubs or trees; leaves opposite, entire; flowers white, small, in terminal, much-branched or sometimes reduced panicles; calyx cup-shaped or short-tubular, truncate or shortly 4-toothed; corolla infundibuliform or tubular, with ascending or spreading lobes, valvate in bud; stamens 2, on the tube, the filaments short, the anthers often exserted; ovary 2-locular, the ovules 2 in each locule, pendulous; fruit drupelike, somewhat juicy, the endocarp membranous or chartaceous; seeds 1–3, with fleshy endosperm.——About 50 species, in temperate to warmer regions of Eurasia, also in Malaysia and Australia.

1A. Corolla-tube as long as the lobes or nearly so; anthers oblong, rounded at both ends, 1.2–2 mm. long, shorter than the filaments; styles exserted from the tube.
  2A. Evergreen; leaves coriaceous, lustrous.
    3A. Leaves elliptic or ovate-elliptic, 5–8 (–10) cm. long; corolla-lobes slightly shorter than the tube; fruit usually ellipsoidal.
        1. *L. japonicum*
    3B. Leaves ovate to broadly so, 6–10 (–12) cm. long; corolla-lobes half as long as the tube; fruit ellipsoid-globose. .... 2. *L. lucidum*
  2B. Deciduous; leaves herbaceous, not lustrous, puberulent beneath while young or glabrous on both sides, narrowly obovate to broadly oblanceolate. .... 3. *L. salicinum*
1B. Corolla-tube usually 2–3 times or rarely only slightly longer than the lobes; anthers broadly lanceolate, narrowed to the tip, 2–4 mm. long, longer than the filaments; style not exserted from the tube.
  4A. Anthers as long as to longer than the corolla-lobes.
    5A. Panicles much branched; leaves half-evergreen, somewhat coriaceous, 4–10 cm. long, 2–5 cm. wide. .... 4. *L. ovalifolium*
    5B. Panicles reduced, racemose; leaves deciduous, herbaceous to membranous, 2–5 (–8) cm. long, 1–2 (–3.5) cm. wide.
        5. *L. tschonoskii*
  4B. Anthers about half as long as the corolla-lobes.
    6A. Leaves oblong, usually obtuse; racemes 2–3 cm. long, elongate, several- to rather many-flowered, densely pilosulous.
        6. *L. obtusifolium*
    6B. Leaves rhombic-ovate or elliptic, mostly acute; racemes 1–1.5 cm. long, headlike, 4- to 8-flowered, sparsely pilose, peduncled.
        7. *L. ibota*

1. **Ligustrum japonicum** Thunb. *Ligustridium japonicum* (Thunb.) Spach; *Ligustrum kellerianum* Vis.; *Ligustrum taquetii* Lév.——Nezumi-mochi, Tama-tsubaki. Large evergreen glabrous shrub; branches gray-brown or grayish; leaves thick-coriaceous, elliptic or broadly ovate-elliptic, rarely ovate, 5–8(–10) cm. long, 2.5–4.5 cm. wide, entire, acute at both ends or sometimes subrounded at base, dark green and lustrous on upper side, yellow-green and obscurely punctulate beneath, the midrib impressed on upper side, often raised and reddish brown beneath, the petioles 5–12 mm. long; panicles pyramidal, 5–12 cm. long and as wide; corolla 5–6 mm. long, the tube as long as to slightly longer than the lobes; anthers oblong, 1.5–2 mm. long, filaments nearly as long as the corolla-lobes; styles exserted, 4–5 mm. long; fruit purple-black, ellipsoidal, 8–10 mm. long.——June. Woods and thickets in lowlands and hills; Honshu (centr. distr. and westw.), Shikoku, Kyushu. A few cultivars of this plant are grown in gardens.

Var. **pubescens** Koidz. *L. rotundifolium* var. *pubescens* (Koidz.) Hatus.——Ke-nezumi-mochi. Inflorescence and young branches puberulent.——Occurs with the typical phase.

Var. **crassifolium** Hisauchi. *L. iwaki* Hisauchi, in syn.——Iwaki. Branchlets puberulent; leaves thicker, elliptic to suborbicular, 1–5 cm. long, 1–4 cm. wide, obtuse to very obtuse.——Cultivated in our area, possibly spontaneous in the Ryukyus.

Var. **rotundifolium** Bl. *L. coriaceum* Carr.; *L. lucidum* var. *coriaceum* (Carr.) Decne.; *L. japonicum* var. *coriaceum* Bl. ex Lavall.——Fukuro-mochi. Shrub with puberulent branches; leaves ovate-elliptic to nearly orbicular, rather short-petioled, densely arranged on the branchlets; panicles narrow, very dense and interrupted, the axis puberulent, 3–6 cm. long, with very short branches; corolla-tube rather elongate; filaments short, the anthers shorter than the corolla-lobes.——Cultivated.

2. **Ligustrum lucidum** Ait. Tō-nezumi-mochi. Resembles the preceding, but larger, sometimes a small tree; leaves coriaceous, ovate to broadly so, 6–10(–12) cm. long, 3–5 cm. wide, gradually narrowed to the acute tip, rounded to subacute

at base, lustrous, pale green and obscurely punctulate beneath, the petioles brownish, 1–2 cm. long; panicles pyramidal, 10–18 cm. long and as wide; corolla 3–4 mm. long, the tube slightly shorter than the lobes; anthers oblong, 1.2–1.5 mm. long, the filaments elongate; styles 2.5–3 mm. long; fruit ellipsoid-globose, purple-black, 8–10 mm. in diameter.——June–July. Chinese species widely cultivated in our area.

3. Ligustrum salicinum Nakai. *L. mayebaranum* Koidz. ——YANAGI-IBOTA, HANA-IBOTA.   Small deciduous tree, the branchlets sparsely puberulent; leaves narrowly obovate to broadly oblanceolate, sometimes oblong, 6–10(–12) cm. long, 2–3.5(–6.5) cm. wide, usually gradually narrowed at both ends, deep green and glabrous above, paler and puberulent while young beneath, the petioles 8–15 mm. long; panicles large, elongate, 15–20 cm. long, 10–15 cm. wide, puberulent, sometimes nearly glabrous; corolla-lobes as long as to slightly shorter than the tube, the filaments longer than the lobes; anthers oblong, about 1.5 mm. long, half as long as the filaments; styles exserted from the tube, about 4 mm. long; fruit oblong, purple-black, 8–10 mm. long.——May. Honshu (Kinki Distr. and westw.), Kyushu; rare.——s. Korea.

4. Ligustrum ovalifolium Hassk. *L. japonicum* var. *ovalifolium* (Hassk.) Bl.; *L. medium* Fr. & Sav.; *L. tsushimense* Nakai; *L. hisauchii* Makino——ŌBA-IBOTA, OKA-IBOTA. Semi-evergreen glabrous shrub; leaves thick, lustrous, elliptic, obovate, or ovate, 4–10 cm. long, 2–5 cm. wide, acute or sometimes obtuse at both ends, deep green and the midrib impressed on upper side, pale green and the midrib raised beneath, the petioles 3–10 mm. long; panicles pyramidal-deltoid, 5–10 cm. long and as wide; corolla-lobes about half as long as the tube, slightly longer than the filaments; anthers broadly lanceolate, about 2.5 mm. long, longer than the filaments; fruit nearly globose, purple-black, about 8 mm. in diameter.——June–July. Thickets near the sea; Honshu, Shikoku, Kyushu; rather common.

Var. **heterophyllum** (Bl.) Nakai. *L. ciliatum* var. *heterophyllum* Bl.; *L. medium* var. *pubescens* Koidz.; *L. hisauchii* var. *pubescens* Makino; *L. macrocarpum* var. *pubescens* (Makino) Nakai; *L. medium* var. *psilorhachis* Nakai——KE-ŌBA-IBOTA, YABU-IBOTA, KE-OKA-IBOTA.   Panicles and young branches puberulent.——Honshu (Kantō Distr.).——Korea.

Var. **pacificum** (Nakai) Mizushima. *L. pacificum* Nakai ——HACHIJŌ-IBOTA.   Corolla-tube as long as to slightly longer than the lobes.——Thickets near the sea; Honshu (Izu Isls.).

5. Ligustrum tschonoskii Decne. *L. acuminatum* Koehne; *L. yesoense* Nakai; *L. ciliatum* var. *tschonoskii* (Decne.) Mansf.; *L. rufum* Nakai; *L. yuhkianum* Koidz.—— MIYAMA-IBOTA.   Much-branched deciduous shrub with gray branches and slender gray-pilosulous branchlets; leaves membranous, broadly lanceolate to ovate, or sometimes rhombic-ovate, 2–5(–8) cm. long, 1–2(–3.5) cm. wide, usually acute, sometimes acuminate or obtuse at both ends, usually appressed-puberulent above, pale green and pilosulous especially on the nerves beneath, the petioles 2–5 mm. long, usually pilosulous; panicles narrowly pyramidal, subracemose, slender, 3–5(–7) cm. long, 1.5–2.5 cm. wide, rather densely puberulent; corolla 6–7 mm. long, the tube rather slender, 1.5–2 times as long as the lobes; filaments shorter than the corolla-lobes, the anthers lanceolate, 2.5–3.5 mm. long; styles 2–3 mm. long; fruit globose, purple-black, about 8 mm. in diameter.——June–July. Thickets and thin woods in mountains; Hokkaido, Honshu, Shikoku, Kyushu; common and variable.

Var. **glabrescens** Koidz. *L. acuminatum* var. *glabrum* Koidz.; *L. yesoense* var. *glabrum* (Koidz.) Nakai——EZO-IBOTA.   Glabrescent; leaves pilosulous only on the nerves beneath.——Occurs with the typical phase.

Var. **macrocarpum** (Koehne) Mansf. *L. macrocarpum* Koehne; *L. acuminatum* var. *macrocarpum* (Koehne) C. K. Schn.——ŌMI-IBOTA.   Leaves larger and fruit to 12 mm. long.——Occurs with the typical phase.

Var. **kiyozumianum** (Nakai) Ohwi. *L. kiyozumianum* Nakai.——KIYOZUMI-IBOTA.   Leaves broadly ovate, obovate, or broadly elliptic, and slightly pilosulous; panicles pilosulous, narrow, 3–7 cm. long, 2–4 cm. wide.——Honshu (se. Kantō Distr.).

Var. **epile** Ohwi. *L. kiyozumianum* var. *glabrescens* Nakai ——KE-NASHI-KIYOZUMI-IBOTA.   Glabrous variant of the preceding variety, and occurs with it.

6. Ligustrum obtusifolium Sieb. & Zucc. *L. ibota* Sieb., nom. seminud., non Sieb. & Zucc.; *L. ibota* var. *angustifolium* Bl.; *L. ibota* var. *obovatum* Bl.; *L. ciliatum* var. *spathulatum* Bl.; *L. ibota* var. *obtusifolium* (Sieb. & Zucc.) Koidz.—— IBOTA-NO-KI.   Much-branched deciduous shrub with slender grayish pilosulous branches while young; leaves membranous, oblong-lanceolate, narrowly oblong, or broadly oblanceolate, 2–5(–7) cm. long, 7–20(–25) mm. wide, obtuse or sometimes subacute at both ends, the petioles 1–2 mm. long; panicles racemose, 2–3 cm. long, densely flowered, densely pilosulous; calyx usually pilosulous; corolla 7–9 mm. long, the tube 1.5–2.5 times as long as the lobes; filaments very short, the anthers broadly lanceolate, 2–2.5 mm. long, about half as long as the corolla-lobes; styles 3–4(–4.5) mm. long; fruit nearly globose, purple-black, 5–6 mm. in diameter.——June. Thickets and thin woods in lowlands and hills; Hokkaido, Honshu, Shikoku, Kyushu; very common and variable.——Ryukyus and Korea.

Var. **leiocalyx** (Nakai) Hara. *L. tschonoskii* var. *leiocarpa* Nakai——SETTSU-IBOTA.   Calyx glabrous.——Occurs with the typical phase.

Var. **regelianum** (Koehne) Rehd. *L. regelianum* Koehne; *L. ibota* var. *regelianum* (Koehne) Rehd.——ONI-IBOTA. Plant prominently pilose.——Occurs with the typical phase.

Var. **velutinum** (Bl.) Hara. *L. ibota* var. *velutinum* Bl. ——BIRŌDO-IBOTA.   Leaves prominently pilosulous beneath. ——Occurs with the typical phase.

7. Ligustrum ibota Sieb. ex Sieb. & Zucc. *L. ciliatum* Sieb. ex Bl.; *L. ibota* var. *ciliatum* (Sieb.) Decne.; *L. ibota* forma *glabrum* Nakai——SAIKOKU-IBOTA.   Deciduous shrub with slender, gray-brown spreading branches, puberulent on young growth; leaves membranous, ovate, elliptic, or oblong, 1–4 cm. long, 7–25 mm. wide, acute or sometimes obtuse, rounded to acute at base, glabrous to thinly appressed-pilosulous above, pale green and pilose at least on the midrib beneath, the petioles 1–2 mm. long; panicles very much reduced, racemose, few-flowered, headlike, the axis 3–8 mm. long, thinly pilosulous; calyx glabrous; corolla 7–8 mm. long, the tube 3–4 times as long as the lobes; filaments very short, the anthers about 2.5 mm. long; styles slender, 3–3.5 mm. long; fruit globose, about 7 mm. across, purple-black.——May–June. Mountains; Kyushu.

Var. **microphyllum** Nakai. *L. ciliatum* var. *microphyllum* Nakai——KOBA-NO-IBOTA.   A small-leaved phase, occurring with the species.

3. OSMANTHUS Lour. MOKUSEI ZOKU

Glabrous evergreen dioecious trees or large shrubs; leaves coriaceous, opposite, simple, penninerved, petiolate, entire or toothed; flowers small, unisexual or polygamous, axillary, fasciculate, short-racemose; calyx short, 4-toothed; corolla short-tubular, 4-lobed, imbricate in bud; stamens 2, rarely 4, the filaments short, the anthers ovoid, subextrorse; ovary 2-locular, superior; styles short, the stigma entire; ovules 2 in each locule; drupe ellipsoidal or globose, the endocarp thick; seeds solitary, with a thin testa, endosperm fleshy.——More than 10 species in e. Asia, one in se. U.S.A.

1A. Leaves entire or serrulate, the midrib impressed above, the lateral nerves distinctly raised beneath; wild and cultivated species.
  2A. Leaves narrowly oblong or broadly oblong-lanceolate; flowers yellow to orange. .... 1. *O. aurantiacus* var. *thunbergii*
  2B. Leaves oblong or narrowly ovate-oblong; flowers white. .... 2. *O. fragrans*
1B. Leaves entire or coarsely spiny-toothed, the midrib flat above, the lateral nerves slightly raised beneath; indigenous species.
  3A. Leaves 3–5(–7) cm. long, with few large teeth or entire. .... 3. *O. heterophyllus*
  3B. Leaves 7–15 cm. long, usually entire, or sometimes with large teeth in young specimens.
    4A. Leaves relatively thin to subcoriaceous, long-acuminate. .... 4. *O. insularis*
    4B. Leaves very thickly coriaceous, abruptly acute. .... 5. *O. rigidus*

1. Osmanthus aurantiacus (Makino) Nakai var. **thunbergii** (Makino) Honda. *Osmanthus intermedius* Nakai; *Olea fragrans* Thunb.; *Osmanthus fragrans* var. *thunbergii* Makino ——USUGI-MOKUSEI. Large glabrous evergreen shrub with pale gray-brown branches; leaves narrowly oblong or broadly oblong-lanceolate, 7–12 cm. long, 2.5–4 cm. wide, acuminate, acute at base, deep green above, serrulate especially toward the tip or subentire, the petioles 7–15 mm. long; flowers about 5 mm. across, pale yellow, the pedicels 7–10 mm. long.—— Sept. Kyushu (s. distr.); planted chiefly in the western parts of our area.——China and India.

Var. **auranthiacus.** *O. fragrans* var. *aurantiacus* Makino; *O. aurantiacus* (Makino) Nakai; *O. fragrans* forma *aurantiacus* (Makino) P. S. Green——KIN-MOKUSEI. Flowers orange-yellow. Commonly cultivated in our area.

2. Osmanthus fragrans Lour. *O. asiaticus* Nakai; *O. fragrans* var. *latifolius* Makino; *O. latifolius* (Makino) Koidz. ——GIN-MOKUSEI. Large glabrous shrub with grayish branches; leaves coriaceous, oblong, sometimes narrowly so or narrowly ovate-oblong, 8–13 cm. long, 3–5 cm. wide, abruptly acuminate, acute to nearly rounded at base, deep green, serrulate or entire, the petioles 7–13 mm. long; flowers about 5 mm. across, white, the pedicels 7–10 mm. long.——Sept. Thought to be spontaneous in mountains of Kyushu; widely cultivated in our area.——China.

3. Osmanthus heterophyllus (G. Don) P. S. Green. *Ilex aquifolium* nec. Thunb., non L.; *I. heterophylla* G. Don; *Olea ilicifolia* Hassk.; *Olea aquifolium* Sieb. & Zucc.; *Osmanthus aquifolium* Sieb.; *I. odora* Sieb. ex Miq.; *Olea aquifolium* var. *subintegra* (Miq.) Miq.; *Osmanthus heterophyllus* Hort. ex Koch; *Olea aquifolium* var. *ilicifolia* Hort. ex Dipp.; *Osmanthus ilicifolius* (Hassk.) Hort. ex Carr.; *Osmanthus diversifolius* Hort. ex Mouill.; *O. aquifolium* f. *subangulatus* Makino; *O. integrifolius* Hayata; *O. ilicifolius* var. *subangulatus* (Makino) Makino; *O. ilicifolius undulatifolius* Makino; *O. ilicifolius* f. *subangulatus* (Makino) Makino & Nemoto——HIIRAGI. Much-branched large shrub or tree; branches pale grayish brown, branchlets minutely papillose, puberulent while young; leaves coriaceous, dimorphic, elliptic, ovate, oblong, or obovate-elliptic, 3–5(–7) cm. long, 2–3 cm. wide, acuminate, lustrous and dark green above, pale green or yellowish green beneath, glabrous, coarsely spiny-toothed in young and vigorous individuals, entire and obtuse in adult trees, the petioles 7–12 mm. long; flowers white, 4–5 mm. across, the pedicels 5–12 mm. long; calyx-lobes ovate-deltoid, entire; fruit elliptic, purple-black, 12–15 mm. long.——Oct.– Dec. Honshu (Kantō Distr. and westw.), Shikoku, Kyushu. ——Formosa. Several cultivars are grown in gardens.

Osmanthus × fortunei Carr. *Olea japonica* Sieb., nom. nud.; *Osmanthus fortunei* var. *cordifolius* Carr.; *O. fortunei* var. *ovatus* Hort. ex Carr.; *O. ilicifolius* var. *latifolius* Hort. ex Mouill.; *O. latifolius* Hort. ex Mouill. pro syn.; *O. aquifolium* var. *ilicifolius latifolius* Hort. ex Nichols.; *O. japonicus* Sieb. ex Makino; *O. aquifolium* var. *japonicus* (Sieb. ex Makino) Makino——HIIRAGI-MOKUSEI. Leaves rather large, semi-lustrous, elliptic, 6–10 cm. long, 3–5 cm. wide, entire or coarsely spine-toothed; calyx-lobes denticulate.——Hybrid of *O. fragrans* × *O. heterophyllus*. Much cultivated.

4. Osmanthus insularis Koidz. *O. zentaroanus* Makino; *O. hachijoensis* Nakai——NATAORE-NO-KI, SATSUMA-MOKUSEI. Glabrous evergreen tree with grayish branches and often slightly flattened branchlets; leaves relatively thin to subcoriaceous, narrowly oblong or ovate-oblong, 7–10(–12) cm. long, 2.5–4(–5) cm. wide, acuminate, acute to acuminate at base, entire, or with 3–10 coarse spiny teeth on each side in young individuals, the petioles 1.5–2.5 cm. long, dilated in upper part; flowers white, the pedicels 7–10 mm. long; calyx-lobes deltoid, minutely toothed; fruit ellipsoidal, black, 15–20 mm. long.—— Nov.–Dec. Honshu (Wakasa Prov. and westw. and Izu Isls.), Shikoku, Kyushu.——Ryukyus, Bonins, s. Korea (Hamilton Isl.).

5. Osmanthus rigidus Nakai. Ō-MOKUSEI. Glabrous evergreen large shrub with gray-white branches and thick, lenticellate branchlets; leaves very thick-coriaceous, oblong, 8–15 cm. long, 3–6 cm. wide, acute at both ends, entire, the petioles reddish, 2–3 cm. long, dilated in upper part; flowers white, 6–7 mm. across, the pedicels about 1 cm. long, recurved in fruit; calyx-lobes deltoid, entire.——Oct.–Nov. Kyushu (s. distr.); possibly not spontaneous.——Tokara Isls. (according to S. Hatusima).

4. FORSYTHIA Vahl RENGYŌ ZOKU

Deciduous shrubs; branches hollow or with lamellate pith; leaves simple or 3-foliolate, opposite, entire or toothed; flowers heterostylous, precocious, short-pedicelled, yellow; calyx-tube short, the lobes 4; corolla 4-lobed, with a short broad tube, the limb spreading, the lobes much longer than the tube, imbricate in bud; stamens 2, inserted at base of corolla, the filaments short, the anthers scarcely exserted; ovary 2-locular, with 4- to 10-ovules in each locule; capsules 2-valved; seeds winged, endosperm wanting.——About 6 species, 1 in Albania, the others in e. Asia.

1A. Internodes hollow; leaves sometimes ternately compound; calyx-lobes oblong, (5–) 6–7 mm. long, as long as the corolla-tube. .... 1. *F. suspensa*

1B. Internodes with lamellate pith; leaves undivided; calyx-lobes elliptic, 2.5–3.5 mm. long, shorter than the corolla-tube.
   2A. Leaves glabrous; branches rather distinctly 4-angled. .... 2. *F. viridissima*
   2B. Leaves with persistent pubescence beneath; branches nearly terete, with 2 shallow obsolete grooves. .... 3. *F. japonica*

1. Forsythia suspensa (Thunb.) Vahl. *Ligustrum suspensum* Thunb.; *Syringa suspensa* Thunb.; *Lilac perpensa* Lam.; *Rangium suspensum* (Thunb.) Ohwi; *F. sieboldii* Dipp.; *F. suspensa* var. *sieboldii* (Dipp.) Zabel——RENGYŌ. Glabrous, somewhat sprawly shrub; internodes hollow, the nodes with solid pith; leaves simple or ternately compound, ovate to broadly so, or elliptic-ovate, 4–8 cm. long, 3–5 cm. wide, acute, rounded to broadly cuneate at base, acutely toothed except at base, deep green, pale yellow-green beneath, the petioles 1–1.5 cm. long; flowers solitary from a lateral bud, the pedicels 1–1.5 cm. long, usually longer than the bud-scales; calyx-lobes oblong, 6–7 mm. long, ciliate; corolla about 2.5 cm. across, the lobes obovate-elliptic, about 1.5 cm. long, 8–10 mm. wide; fruit ovoid, about 1.5 cm. long, with scattered warty lenticels.——Mar.–Apr. Chinese shrub much cultivated in our area.

2. Forsythia viridissima Lindl. *Rangium viridissimum* (Lindl.) Ohwi——SHINA-RENGYŌ. Glabrous deciduous shrub with 4-angled, gray-brown branches, the pith lamellate; leaves rather thick, oblong to lanceolate, rarely obovate-oblong, 6–10 cm. long, 1–3 cm. wide, broadest at or above the middle, acute, cuneate at base, short-toothed on the upper half, deep green, pale green beneath, the petioles 7–10 mm. long; pedicels 5–10 mm. long, as long as to slightly longer than the bud-scales; calyx-lobes elliptic, ciliate, 2.5–3 mm. long, erect, half as long as the corolla-tube; corolla 2–2.5 cm. across, the lobes narrowly oblong, about 15 mm. long, about 7 mm. wide, slightly recurved on margin; fruit broadly ovoid, about 1.5 cm. long, nearly rounded at base.——Apr. Chinese shrub rarely cultivated in our area.

Forsythia koreana Nakai. *F. viridissima* var. *koreana* (Nakai) Rehd.——CHŌSEN-RENGYŌ. Leaves usually broadest on lower half; calyx slightly shorter than the corolla-tube, the lobes obliquely ascending.——Rarely cultivated.——Korea.

3. Forsythia japonica Makino. *Rangium japonicum* (Makino) Ohwi——YAMATO-RENGYŌ. Deciduous shrub; branches grayish yellow-brown, nearly terete, with 2 shallow obscure grooves or indistinctly 4-striate, glabrous; leaves membranous, ovate to broadly so, 7–12(–15) cm. long, 4–6(–8) cm. wide, short-acuminate, usually toothed, glabrous, pale green and more or less white-hairy, especially on the nerves beneath, the petioles 6–8 mm. long, white-pubescent; pedicels 2–3 mm. long, slightly shorter than the bud-scales; calyx-lobes elliptic, 2.5–3 mm. long, ciliate; corolla about 2.5 cm. across, the lobes narrowly oblong, about 12 mm. long, 5–6 mm. wide; fruit broadly ovoid, about 1 cm. long.——Apr. Mountains; Honshu (Chūgoku Distr.); rare.

5. SYRINGA L. HASHIDOI ZOKU

Shrubs or trees, glabrous or pubescent; branches with solid pith; leaves opposite, petiolate, entire or rarely pinnatifid; inflorescence a paniculate thyrse; flowers white, rose-purple or lilac, rather small; calyx campanulate, small, irregularly toothed; corolla tubular, the lobes 4, longer or shorter than the tube, valvate in bud; stamens 2, inserted on the throat or on the tube of corolla, the filaments short or rather long; ovary 2-locular; stigma bifid; ovules 2 in each locule; capsules narrowly oblong, terete or flattened, 2-valved; seeds flat, winged on back, the endosperm fleshy.——About 30 species in Eurasia.

1. Syringa reticulata (Bl.) Hara. *Ligustrum reticulatum* Bl.; *Ligustrina amurensis* var. *japonica* Maxim.; *S. amurensis* var. *japonica* (Maxim.) Fr. & Sav.; *S. japonica* Maxim. ex Decne.——HASHIDOI. Deciduous small tree with glabrous, gray-brown branches; leaves ovate to broadly so, or ovate-orbicular, 6–8(–12) cm. long, 3.5–6(–9) cm. wide, abruptly acuminate or acute, rounded to shallowly cordate or sometimes abruptly acute at base, entire, deep green and glabrous above, pale green and short, white-pubescent beneath especially on the midrib, the veinlets reticulate beneath; panicles large and rather broad, 15–25 cm. long, 10–15 cm. wide, densely many-flowered, glabrous or puberulent; flowers white, fragrant, about 5 mm. across, short-pedicellate; calyx-teeth depressed; corolla subinfundibuliform, the lobes ovate, about 2.5 mm. long, longer than the tube; filaments slightly longer than the corolla-lobes, the anthers ellipsoidal, about 1.5 mm. long; capsules narrowly oblong, 15–20 mm. long, smooth to loosely lenticellate; seeds flat, wing-margined.——June–July. Mountains; Hokkaido, Honshu, Shikoku, Kyushu.

Syringa vulgaris L. *Lilac vulgaris* Lam. MURASAKI-HA-SHIDOI. Often cultivated.——Introduced from se. Europe.

6. FRAXINUS L. TONERIKO ZOKU

Deciduous, rarely evergreen, often dioecious trees; leaves opposite, pinnately compound, very rarely simple, petiolate, toothed or entire; flowers small, in terminal or axillary panicles or racemes, polygamous or unisexual; calyx minute, 4-lobed; petals 2, 4, or absent, free or united at base; stamens 2; ovary 2-locular, with 2-ovules in each locule; samara winged, 1-seeded; seeds narrowly oblong.——About 70 species, widely distributed in the N. Hemisphere, also in Mexico and Java.

1A. Panicles from lateral buds, without leaves at base; leaflets sessile.
   2A. Leaf-rachis and inflorescence glabrous; samara about 4 cm. long, 8–15 mm. wide. .... 1. *F. spaethiana*
   2B. Leaf-rachis and inflorescence red-brown woolly; samara about 2.5–3.5 cm. long, 7–8 mm. wide. .... 2. *F. mandshurica* var. *japonica*
1B. Panicles on young shoots, with few pairs of leaves at base.
   3A. Petals absent; calyx small, cup-shaped, with small teeth, persistent; leaflets usually petiolate, often with multicellular crisped hairs at base.
      4A. Leaflets abruptly acute to short-acuminate. .... 3. *F. japonica*
      4B. Leaflets long-acuminate to caudate. .... 4. *F. longicuspis*

3B. Petals present; calyx minute or nearly absent, rarely with few small narrow deciduous teeth; leaflets usually sessile, with unicellular, spreading hairs at base or glabrous.
    5A. Bud-scales closely appressed.
        6A. Young branches, petioles, and inflorescence with very minute spreading hairs and minute often stipitate glands. 5. *F. sieboldiana*
        6B. Young branches, petioles, and inflorescence glabrous or pilose with spreading stiff hairs. 6. *F. lanuginosa*
    5B. Outer 2 bud-scales spreading at tip; young branches, petioles, and inflorescence glabrous. . . . . 7. *F. apertisquamifera*

1. **Fraxinus spaethiana** Lingelsh. *F. mandshurica* var. *shioji* Kudo; *F. commemoralis* Koidz.; *F. verecunda* Koidz.——SHIOJI. Tall deciduous tree with stout, glabrous, grayish yellow-brown branchlets; leaves 25–30 cm. long, 7- to 9-foliolate, mostly glabrous or occasionally scattered-hairy on the leaflets beneath, the petioles shallowly and rather narrowly grooved on upper side, dilated at base and somewhat clasping; terminal leaflets petiolate, narrowly cuneate at base, 8–15 cm. long, 3–7 cm. wide, the lateral ones gradually smaller, broadly oblanceolate, rarely narrowly oblong, sessile, slightly oblique, minutely toothed, acuminate to abruptly acute; panicles 10–15 cm. long, glabrous; flowers apetalous; calyx cup-shaped, irregularly toothed; samaras pendulous, broadly lanceolate, 4 cm. long, 8–15 mm. wide, usually narrowed at both ends, the pedicels 5–10 mm. long.——Mountains; Honshu (Kantō Distr. and westw.), Shikoku, Kyushu.

Var. **nipponica** (Koidz.) Hara. *F. nipponica* Koidz.——YAMA-SHIOJI, KAI-SHIJI-NO-KI. Petioles and leaf-axis puberulent.——Honshu (Kai and Harima Prov. and elsewhere).

2. **Fraxinus mandshurica** Rupr. var. **japonica** Maxim. *F. excelsissima* Koidz.; *F. mandshurica* sensu auct. Japon., non Rupr.——YACHI-DAMO. Large deciduous tree; branches 4-angled; leaves 7- to 11-foliolate, the leaflets sessile except the terminal one, narrowly oblong, 5–15 cm. long, 3–4 cm. wide, acuminate, toothed, obliquely cuneate at base, rusty-brown woolly at the attachment, spreading-pilosulous along midrib and nerves beneath, usually glabrous above; calyx minute, obconical, deciduous and not persistent in fruit; corolla absent; samaras broadly oblanceolate, obtuse, narrowly cuneate toward the base, 2.5–3.5 cm. long, 7–8 mm. wide.——Apr.–May. Wet places in mountains; Hokkaido, Honshu (centr. and n. distr.).——The typical phase with the base of leaflets less densely woolly occurs in Sakhalin, Korea, Manchuria, and northern e. Asia.

3. **Fraxinus japonica** Bl. *F. kantoensis* Koidz.; *F. nakaiana* Koidz.——TONERIKO. Deciduous tree with thick, slightly flattened branchlets; leaves 20–35 cm. long inclusive of the petioles, 5- or 7(–9)-foliolate, the leaflets narrowly to broadly ovate or broadly lanceolate, 5–15 cm. long, 3–6 cm. wide, spreading white-pilose along the midrib beneath, abruptly acute to short-acuminate, toothed, obliquely rounded to cuneate at base, petiolate; calyx small, cuplike, irregularly toothed, persistent; corolla absent; samaras oblanceolate, 3–4 cm. long, 5–7 mm. wide, usually obtuse, narrowly long-cuneate at base.——Wet places in lowlands and mountains; Honshu (centr. and n. distr.); often planted on borders of paddy fields.——Forma **intermedia** (Nakai) Hara. *F. intermedia* Nakai; *F. koshiensis* Koidz.——NAGAMI-NO-TONERIKO. Samaras narrower than in the typical phase.

Var. **stenocarpa** (Koidz.) Ohwi. *F. stenocarpa* Koidz.; *F. spaethiana* sensu auct. Japon., non Lingelsh.——DEWA-NO-TONERIKO. Leaflets oblong to broadly oblanceolate; samaras narrowly lanceolate.——Honshu (Dewa Prov.).

4. **Fraxinus longicuspis** Sieb. & Zucc. *F. yamatensis* Na-kai; *F. borealis* Nakai——YAMOTO-AO-DAMO. Deciduous tree; branches gray-brown, pale brown crisped-hairy while young; leaves 15–25 cm. long inclusive of the petioles, 5- or 7-foliolate, the rachis and petioles crisped-hairy while young, usually slightly reddish, the leaflets broadly lanceolate, broadly oblanceolate or sometimes oblong-lanceolate, 5–10 cm. long, 2–3 cm. wide, long-acuminate to caudate, short-toothed, narrowed toward the base, petiolate; calyx cup-shaped, minutely toothed, persistent; corolla absent; samaras oblanceolate, 3–4 cm. long, about 6 mm. wide, retuse to obtuse.——Mountains; Honshu, Shikoku, Kyushu.

Var. **pilosella** (Honda) Hara. *F. borealis* var. *pilosella* Honda——URAGE-Ō-TONERIKO. Leaflets pubescent beneath.——Occurs with the typical phase.

Var. **latifolia** Nakai. (?) *F. pubinervis* Bl.; (?) *F. obovata* Bl.; (?) *F. satsumana* Koidz.——HIRO-HA-AO-DAMO, TSUKUSHI-TONERIKO. Leaflets broader, ovate to narrowly so.——Occurs with the typical phase.

5. **Fraxinus sieboldiana** Bl. *F. sieboldiana* var. *angustata* Bl.; *F. angustata* (Bl.) Hatus.; *F. tobana* Honda; *F. longicuspis* var. *sieboldiana* (Bl.) Lingelsh.——MARUBA-AO-DAMO. Deciduous tree; buds closed, the scales gray-puberulent to nearly glabrous outside; branchlets slender, gray-brown, spreading-puberulent while young; leaves 5-(–7)-foliolate, 10–15(–20) cm. long inclusive of the petioles, the petioles and rachis puberulent while young, the leaflets narrowly to broadly ovate, sometimes broadly lanceolate, 5–10(–13) cm. long, 1.5–3.5 cm. wide, long-acuminate, glabrous or with spreading white hairs along the midrib beneath, toothed or nearly entire, subsessile; calyx minutely toothed; petals 4, 6–7 mm. long, free, linear-oblanceolate, white; samaras oblanceolate to linear-oblanceolate, 2.5–3 cm. long, about 5 mm. wide.——Apr.–May. Mountains and hills; Honshu, Shikoku, Kyushu; common.——Korea.

6. **Fraxinus lanuginosa** Koidz. *F. sieboldiana* var. *pubescens* Koidz.——AO-DAMO, KOBA-NO-TONERIKO. Closely related to the preceding; young branches, inflorescence, and petioles glabrous or pilose with spreading stiff hairs; buds smooth or pilose with spreading stiff unicellular (rarely bicellular) hairs; leaflets generally distinctly serrate and sometimes larger.——Mountains; Hokkaido, Honshu, Shikoku, Kyushu.

7. **Fraxinus apertisquamifera** Hara. *F. sambucina* Koidz., excl. syn.——MIYAMA-AO-DAMO. Deciduous small tree with rather slender, glabrous branchlets; leaves rather thin, 10–20 cm. long inclusive of the petioles, (5-)7-foliolate, glabrous or nearly so, the leaflets broadly lanceolate to narrowly ovate or oblong, 5–12 cm. long, 2–4 cm. wide, long-acuminate, sessile except the terminal one, acutely toothed, obliquely acute to rounded at base, hairy along the midrib beneath; panicles glabrous; calyx minute or absent; petals 4, free, very narrow, white; samaras narrowly oblanceolate, obtuse to retuse, 2.5–3 cm. long, about 5 mm. wide.——Mountains; Honshu.

Fam. 167. **LOGANIACEAE** Fuji-utsugi Ka Logania Family

Herbs, shrubs, or trees; leaves opposite, rarely alternate or verticillate, entire or toothed, simple, stipulate; inflorescence cymose to thyrsiform, many-flowered, rarely reduced to 1–3 flowers; flowers bisexual, actinomorphic, 4- to 5-merous; calyx-lobes valvate, rarely imbricate in bud; corolla tubular, 4- to 10-lobed, the lobes contorted, imbricate, or valvate in bud; stamens on the tube, as many as the lobes, and alternate with them, rarely solitary; ovary superior, (1–)2- to 4-locular; ovules 1 to many in each locule, the placentation axile or parietal in a unilocular ovary; styles solitary; fruit a capsule, berry, or drupe; embryo straight, the endosperm fleshy or bony.——About 30 genera, with about 800 species, of wide distribution in the Tropics and subtropics.

1A. Fruit a many-seeded capsule; erect herbs or woody plants.
  2A. Herbs; corolla-lobes valvate in bud. .... 1. *Mitrasacme*
  2B. Shrubs; corolla-lobes imbricate in bud. .... 2. *Buddleia*
1B. Fruit a few-seeded berry; scandent evergreen shrubs. .... 3. *Gardneria*

1. **MITRASACME** Labill. Ai-nae Zoku

Small or delicate herbaceous annuals or perennials; leaves opposite, small, entire, the petioles joined by a very short sheath from the adjacent one; flowers small, in terminal and axillary fascicles or umbels, 1 to many, white or yellowish; calyx campanulate, 4-lobed; corolla campanulate or cup-shaped, 4-lobed, valvate in bud; stamens 4, inserted on the tube, the anthers usually not exserted from the tube; ovary 2-locular; ovules many; styles 2, connate while young, later becoming bifid to the base; capsules truncate to bifid at the apex; seeds small.——About 30 species, chiefly in Australia, few in the Pacific Islands and tropical Asia.

1A. Stems leafy only on lower part, papillose-hairy toward the base; leaves ovate to oblong, 7–15 mm. long. .... 1. *M. pygmaea*
1B. Stems leafy throughout, nearly glabrous; leaves lanceolate to broadly linear, 3–8 mm. long. .... 2. *M. alsinoides* var. *indica*

1. Mitrasacme pygmaea R. Br. *M. nudicaulis* Reinw.; *M. chinensis* Griseb.; *M. polymorpha* sensu auct. Japon., non R. Br.——Ai-nae. Delicate annual, 5–15 cm. high; stems branched at base or simple, terete, with few pairs of leaves at base, scapelike in upper part; leaves ovate or oblong, 7–15 mm. long, 3–6 mm. wide, acute or obtuse, obsoletely 3-nerved, often short-pilosulous on both sides, always so on the margin; flowers 3–5, verticillate, erect, white, about 3 mm. across, the pedicels 1–4 cm. long; capsules globose, bifid, about 3 mm. long.——Aug.–Oct. Sunny places in lowlands; Honshu, Shikoku, Kyushu.——Korea, Ryukyus, Formosa, China, India, Malaysia, and Australia.

2. Mitrasacme alsinoides R. Br. var. **indica** (Wight) Hara. *M. alsinoides* sensu auct. Japon., non R. Br.; *M. indica* Wight——Hime-nae. Delicate annual, 5–15 cm. high; stems often branched at base, leafy to the apex, loosely granular-puberulent on upper part; leaves of few pairs, spreading, lanceolate to broadly linear, 3–8 mm. long, 1–2 mm. wide, acute to acuminate, mucronate, obsoletely 1-costate, glabrous; flowers white, about 2.5 mm. across, the pedicels 7–20 mm. long; capsules globose, bifid; style about 2.5 mm. long.——Aug.–Oct. Wet grassy places in lowlands; Honshu, Shikoku, Kyushu.——Ryukyus, Formosa, China, India, Malaysia, and Australia.

2. **BUDDLEIA** L. Fuji-utsugi Zoku

Shrubs or trees, rarely herbs, usually pubescent; leaves opposite, rarely alternate or verticillate, entire or toothed; inflorescence spiciform, paniculate, or thyrsoid, densely flowered; flowers small, white, yellowish, or blue-purple, 4-merous; calyx campanulate, 4-toothed; corolla tubular, the lobes 4, spreading, imbricate in bud; stamens inserted on tube or throat, the anthers subsessile; ovary 2-locular, many-ovuled; capsules septicidally dehiscent, the valves sometimes bifid; seeds many, often winged.——About 100 species, in tropical and subtropical regions of Asia, America, and S. Africa.

1A. Branches 4-winged; anthers below the middle of the tube. .... 1. *B. japonica*
1B. Branches nearly terete, scarcely angled; anthers above the middle of the tube. .... 2. *B. venenifera*

1. Buddleia japonica Hemsl. *B. insignis* sensu auct. Japon., non Carr.——Fuji-utsugi. Deciduous, branched shrub 50–150 cm. high, the branches 4-winged, loosely pale brown stellate-hairy while young; leaves broadly lanceolate to narrowly ovate, 8–20 cm. long, 2–5 cm. wide, long-acuminate, loosely mucronate-toothed or entire, green and glabrate above, pale green and usually stellate-hairy beneath, the petioles 2–3 mm. long; panicles narrow, 10–25 cm. long, one-sided, many-flowered; corolla pale purple, 15–20 mm. long, pale brown-tomentose externally, the tube gently curved; fruit ovoid, 6–8 mm. long, with persistent corolla and calyx.——July–Oct. Sunny slopes and thickets in mountains; Honshu, Shikoku.

2. Buddleia venenifera Makino. *B. curviflora* sensu auct. Japon., saltem pro parte, non Hook. & Arn.; *B. curviflora* var. *venenifera* (Makino) Makino——Urajiro-fuji-utsugi. Densely stellate-tomentose shrub, except the leaves on upper side; branches terete; leaves narrowly ovate, sometimes broadly lanceolate, 7–15 cm. long, 3–6 cm. wide, caudate, rounded to subacute at base, entire or obsoletely undulate-toothed, very thinly pubescent on upper side while young, the petioles 5–15 mm. long; panicles 10–30 cm. long, one-sided and densely many-flowered, declined in upper part, spiciform; corolla pale purple, densely stellate-hairy externally, the tube gently curved, about 15 mm. long; anthers above the middle of the tube; capsules ovoid, 5–6 mm. long.——July–Nov. Sunny slopes and waste grounds in hills and low elevations in the mountains; Shikoku, Kyushu.——Forma **calvescens** Ohwi. *B. venenifera* forma *kofujii* Ohwi——Ko-fuji-utsugi. Leaves green on both surfaces.

3. GARDNERIA Wall. HŌRAI-KAZURA ZOKU

Glabrous, scandent, evergreen shrubs; leaves opposite, entire; flowers axillary, solitary or in trichotomous cymes or panicles, rather small; calyx 4- to 5-parted; corolla nearly rotate, deeply 4- or 5-lobed, thick, valvate in bud; stamens 4 or 5, inserted on the throat, the anthers erect, subsessile, free and approximate, or slightly connate; ovary 2-locular, the ovules 1 to few in each locule; berry with fleshy pulp; seeds rather flat.——About 6 species, in India, China, Formosa, Japan, and Korea.

1A. Flowers in groups of 3–10; corolla-lobes glabrous inside, spreading; anthers glabrous, 2.5–3 mm. long, about half as long as the corolla-lobes; leaves narrowly oblong to broadly oblanceolate, 7–12 cm. long. .... 1. *G. multiflora*
1B. Flowers 1(–3); corolla-lobes puberulous inside, recurved; anthers with a dense tuft of hairs on back at base, 5–6 mm. long, slightly shorter than the corolla-lobes; leaves ovate to ovate-oblong, 5–10 cm. long. .... 2. *G. nutans*

1. Gardneria multiflora Makino. *Pseudogardneria multiflora* (Makino) Pampan.——CHITOSE-KAZURA. Glabrous, evergreen, scandent shrub, the branches terete, with a pulvinuslike leaf-scar on each side of the node; leaves narrowly oblong to broadly oblanceolate, rarely lanceolate, 7–12 cm. long, 1.5–3.5 cm. wide, acuminate, subobtuse at base, deep green and lustrous, entire, the petioles 7-10 mm. long; inflorescence trichotomously 3- to 10-flowered, the peduncle 1–2 cm. long, the bracts small; calyx-lobes semirounded, ciliolate, about 0.6 mm. long; corolla about 12 mm. across, yellow, 5-parted nearly to the base, the lobes glabrous inside, linear-oblong, spreading, obtuse; anthers glabrous, narrowly oblong, 2.5–3 mm. long, about half as long as the corolla-lobes. ——June. Honshu (Chūgoku Distr.).——China.

2. Gardneria nutans Sieb. & Zucc. *Pseudogardneria nutans* (Sieb. & Zucc.) Racib.——HŌRAI-KAZURA. Glabrous, evergreen, scandent shrub, the branches nearly terete, elongate, with a raised pulvinus like leaf-scar on each side; leaves rather thick, ovate or ovate-oblong, 5–10 cm. long, 2–4 cm. wide, abruptly acuminate to acute, abruptly acute at base, deep green and lustrous above, entire, the petioles 7–12 mm. long; inflorescence usually 1-flowered, rarely 3-flowered; flowers pendulous, white, the peduncles 1–2 cm. long; calyx-lobes semirounded, ciliolate, less than 1 mm. long; corolla 5-parted nearly to the base, the lobes deltoid-lanceolate, puberulous inside; anthers erect, 5–6 mm. long, with a dense tuft of hairs on back; fruit globose, red, about 1 cm. across, 3- or 4-seeded.——July. Honshu (Awa Prov. and westw.), Shikoku, Kyushu; rather rare.

Gardneria insularis Nakai, EISHŪ-KAZURA, is reported to occur in Kyushu (Tsushima and Buzen Prov.).——s. Korea.

Fam. 168. GENTIANACEAE RINDŌ KA Gentian Family

Perennials or annuals, sometimes aquatic; leaves opposite or alternate, exstipulate; flowers bisexual, rarely polygamous, actinomorphic, usually showy; calyx tubular or of separate sepals, the segments imbricate in bud; corolla 4- to 12-lobed, the lobes contorted, rarely imbricate or induplicate in bud; stamens inserted on the tube, as many as and alternate with the lobes; ovary superior, usually 1(–2)-locular, the placentae parietal; fruit a capsule, rarely a berry; seeds usually many, the endosperm copious.——About 60 genera, with about 800 species, chiefly in temperate to subarctic, and sometimes subtropical regions.

1A. Leaves opposite or rarely verticillate, simple, entire; corolla-lobes contorted in bud; terrestrial.
  2A. Style filiform, usually deciduous; anthers spirally contorted. .... 1. *Centaurium*
  2B. Style short and thick, persistent, sometimes absent; anthers erect, not spirally contorted.
    3A. Corolla without glands or glandular-grooves at base of lobes.
      4A. Scandent, slender; leaves petiolate; flowers not erect.
        5A. Fruit berrylike; flowers 5-merous. .... 2. *Tripterospermum*
        5B. Fruit a capsule; flowers 4-merous. .... 3. *Pterygocalyx*
      4B. Erect, not scandent; leaves sessile or nearly so; flowers erect. .... 4. *Gentiana*
    3B. Corolla with large glands or glandular-grooves at base of lobes.
      6A. Corolla not spurred at base, rotate or rarely campanulate, the lobes dextrorsely contorted. .... 5. *Swertia*
      6B. Corolla spurred at base, campanulate, the lobes sinistrorsely contorted. .... 6. *Halenia*
1B. Leaves alternate, sometimes trifoliolate, entire or toothed; corolla-lobes induplicate in bud; aquatic or marsh herbs.
  7A. Leaves 3-foliolate or reniform, not floating; capsules dehiscent; marsh plants.
    8A. Leaves 3-foliolate; flowers in racemes; capsules 2-fid. .... 7. *Menyanthes*
    8B. Leaves simple, reniform; flowers in cymes; capsules 4-fid. .... 8. *Fauria*
  7B. Leaves ovate to orbicular, deeply cordate, floating; capsules indehiscent; aquatic. .... 9. *Nymphoides*

1. CENTAURIUM Hill SHIMA-SEMBURI ZOKU

Erect, annual or perennial herbs; leaves opposite, sometimes clasping; cymes often subspicate; flowers axillary, rose, yellowish, or white, rather small; calyx tubular, 4- or 5-lobed; corolla salverform or rotate, the lobes 4 or 5, spreading, contorted in bud; stamens inserted on the tube, the anthers spirally contorted; ovary 1-locular; style filiform, the stigma 2, lamellate; capsules bifid; seeds many, small.——About 30 species, in temperate to subtropical regions of the N. Hemisphere and Australia.

1. Centaurium japonicum (Maxim.) Druce. *Erythraea japonica* Maxim.; *E. australis* sensu auct. Japon., non R. Br.; *E. spicata* sensu auct. Japon., non Pers.——SHIMA-SEMBURI. Glabrous, glaucescent annual 10–40 cm. high; stems simple or branched, leafy throughout, erect, angled; leaves oblong, elliptic, or obovate, 1–2.5 cm. long, 5–10 mm. wide, obtuse, spreading, sessile, entire; flowers sessile, axillary; calyx 6–8 mm. long, 5-lobed nearly to the base; corolla-tube as long as the calyx, the lobes narrowly oblong, 3–4 mm. long; anthers half as long as the corolla-lobes; capsules oblong-cylindric, 8–10 mm. long, longer than the calyx, sessile.——July–Aug. Kyushu (Yakushima).——Ryukyus and Formosa.

2. TRIPTEROSPERMUM Bl.　Tsuru-rindō Zoku

Scandent or trailing glabrous perennials; leaves opposite, usually 3-nerved, petiolate; flowers 1–3, blue-purple or white, terminal and axillary, 5-merous; calyx tubular, 5-winged or -angled, the lobes narrow; corolla tubular or tubular-campanulate, broadened toward tip, the lobes contorted, often plaited between the lobes; stamens inserted on the tube, the filaments filiform; ovary stipitate, 1-locular; style slender, the stigmas 2, lamellate; berry globose to ellipsoidal; seeds many, small, 3-winged.——Few species, e. Asia, India, and Malaysia.

1. Tripterospermum japonicum (Sieb. & Zucc.) Maxim. *Convolvulus trinervis* Thunb.; *Crawfurdia japonica* Sieb. & Zucc.; *Crawfurdia trinervis* (Thunb.) Makino, non Dietr. nec Hassk.; *Gentiana trinervis* (Thunb.) Marq.——Tsuru-rindō. Scandent or trailing perennial, 40–80 cm. high, with short, slender, creeping rhizomes; stems slender, simple or sparingly branched, purplish; leaves deltoid-ovate to -lanceolate, 4–8 cm. long, 12–35 mm. wide, gradually acuminate, shallowly cordate to rounded at base, 3-nerved, deep green above, pale green and often purplish beneath, the petioles 5–15 mm. long; flowers pale blue-purple; calyx-tube 6–8 mm. long, narrowly 5-winged, the lobes linear, as long as to slightly longer than the tube; corolla about 3 cm. long, gradually narrowed at base, the lobes narrowly deltoid, plaited; fruit ellipsoid-globose, red-purple, about 8 mm. across, slightly exserted from the corolla, the stipe 1.5–2 cm. long; stigma linear-lanceolate.——Aug.–Oct. Woods in mountains; Hokkaido, Honshu, Shikoku, Kyushu.——s. Kuriles, Sakhalin, Korea, Formosa, and China.

3. PTERYGOCALYX Maxim.　Hosoba-no-tsuru-rindō Zoku

Glabrous, scandent or trailing perennials; leaves opposite, usually 3-nerved, petiolate; flowers blue-purple or white, solitary, terminal and axillary, 4-merous, sometimes 5-merous; calyx tubular; corolla tubular, 4- to 5-lobed, the lobes contorted; stamens inserted on the corolla-tube; ovary 1-locular, stipitate; style short, the stigmas 2, lamellate; capsules 2-valved, dehiscent, many-seeded; seeds small, winged on margin.——Few species, in e. Asia, India, and China.

1. Pterygocalyx volubilis Maxim. *Crawfurdia ptery-gocalyx* Hemsl.; *C. volubilis* (Maxim.) Makino——Hosoba-no-tsuru-rindō.　Resembles *Tripterospermum japonicum* in habit; leaves broadly lanceolate to linear-lanceolate, 2–4 cm. long, 5–10 mm. wide, long-acuminate, acute at base, subtri-nerved, entire, petiolate; flowers short-pedicellate; calyx-tube 1.5–2 cm. long, 5-winged and often deeply fissured on one side, the lobes 4, broadly linear, 3–5 mm. long, laterally flattened and winglike; corolla 3–3.5 cm. long, shallowly 4-cleft, the lobes not plaited; style slender, 2–3 mm. long; capsules narrowly oblong, about 10 mm. long, the stipe slightly shorter than the capsule; seeds minute, thin-winged on margin.——Sept.–Oct.　Mountains; Hokkaido, Honshu, Shikoku, Kyushu; rare.——Korea, Manchuria, n. China, and Ussuri.

4. GENTIANA L.　Rindō Zoku

Erect perennials or sometimes annuals; leaves opposite, rarely verticillate, sessile, entire; flowers 4- to 7-merous, axillary or terminal, sessile or pedunculate, erect, usually large, blue-purple, blue, white, or rarely yellow; calyx tubular, rarely fissured on one side; corolla tubular, salverform or infundibuliform, sometimes narrowly campanulate, the lobes contorted in bud, rarely ciliate; stamens inserted on the corolla-tube, usually included; ovary 1-locular, often stipitate at base; style short, the stigmas 2, lamellate; capsules sessile or stipitate, 2-valved; seeds many.——About 500 species, widely distributed in temperate zones, and high mountains of the Tropics.

1A. Annuals or biennials without rhizomes, erect.
  2A. Corolla without plaits between the lobes; capsules sessile or short-stipitate.
    3A. Corolla-lobes without a fringe at junction with throat; seeds brown, ellipsoidal, densely spinulose-tuberculate.
      4A. Corolla-lobes fimbriate; plant 5–40 cm. high; flowers 2.5–3.5 cm. long, the peduncles (1–)5–12 cm. long. .... 1. *G. yabei*
      4B. Corolla-lobes entire; plant 5–10 cm high; flowers about 15 mm. long, the peduncles 7–15 mm. long. .... 2. *G. contorta*
    3B. Corolla-lobes with a fringe at junction with throat; seeds pale brown, globose, smooth.
      5A. Calyx-lobes ovate-cordate, obtuse, auriculate, as long as broad. .... 3. *G. auriculata*
      5B. Calyx-lobes linear-lanceolate to oblong-lanceolate, not auriculate, acute to acuminate, much longer than broad.
        6A. Crown at the base of the corolla-lobes, laciniate into capillary filaments.
          7A. Calyx-lobes nearly as long as the tube; cauline leaves minutely papillose on margin; flowers 16–22 mm. long; calyx about half as long as the corolla. .... 4. *G. takedae*
          7B. Calyx-lobes 3-6 times as long as the tube; cauline leaves smooth on margin; flowers 2.5–3 cm. long; calyx half to nearly as long as the corolla. .... 5. *G. yuparensis*
        6B. Crown at the base of the corolla-lobes, finely laciniate into slender, stouter, straight filaments ending in an obtuse tip; stems simple or branched from the base; flowers 15–17 mm. long; calyx 5–6 mm. long, 1/3–1/2 as long as the corolla-tube, divided nearly to the base. .... 6. *G. secta*
  2B. Corolla plaited between the lobes; capsules long-stipitate.
    8A. Calyx-lobes ovate or obovate, gradually spreading or slightly reflexed; leaves rosulate at base; cauline leaves smaller and ascending, spreading in those above, ovate or narrowly so, short awn-tipped. .... 7. *G. squarrosa*
    8B. Calyx-lobes lanceolate or narrowly deltoid, erect.
      9A. Rosulate leaves present; cauline leaves appressed to the stem.
        10A. Leaves smooth, acute, not awn-tipped; peduncles 1–3 cm. long; corolla 2–2.5 times as long as the calyx. .. 8. *G. thunbergii*
        10B. Leaves papillose on margin in the lower ones, awn-tipped; flowers sessile or short-pedunculate; corolla 1.5 times as long as the calyx. .... 9. *G. aquatica*

9B.   Rosulate leaves absent; cauline leaves gradually spreading. ....10. *G. zollingeri*
1B.   Perennial herbs with rhizomes or creeping stems.
  11A.   Leaves opposite; flowering stems single or few together.
    12A.   Rhizomes absent or obscure; stems creeping; leaves 7–15 mm. long, 3–6 mm. wide; seeds ellipsoid-fusiform, smooth.
      13A.   Plaits of corolla deltoid, incurved and covering the throat; corolla-tube narrow, not dilated. .... 11. *G. jamesii*
      13B.   Plaits of corolla spreading at anthesis; corolla-tube gradually dilated in upper part. .... 12. *G. nipponica*
    12B.   Rhizomes short or elongate, sometimes stout; creeping stems absent; leaves longer (except in No. 13) and wider.
      14A.   Flowers short-pedunculate; rhizomes with short innovation shoots with leaves rosulate at apex; seeds rather flat, winged.
        15A.   Rosette and cauline leaves about the same size, 8–17 mm. long; flowers pale blue, about 2 cm. long; capsules long-stipitate.
          13. *G. glauca*
        15B.   Rosette leaves 8–15 cm. long, the cauline leaves few, 2–5 cm. long; flowers pale yellow, blue-green spotted, 3.5–5 cm. long;
          capsules short-stipitate. .... 14. *G. algida*
      14B.   Flowers sessile; rhizomes without rosulate leaves; seeds lanceolate, with a short tail at each end.
        16A.   Stems with few pairs of leaves; leaves ovate to rhombic-oblong or obovate, subpetiolate; calyx-lobes ovate or ovate-cordate,
          spreading. .... 15. *G. sikokiana*
        16B.   Stems with few to many pairs of leaves; leaves lanceolate to broadly so, sessile; calyx-lobes erect, linear to linear-lanceolate.
          17A.   Leaves green, papillose on margin in the upper ones; stems usually red-brown. .... 16. *G. scabra* var. *buergeri*
          17B.   Leaves glaucescent, nearly smooth; stems scarcely red-brown.
            18A.   Flowers 22–30 mm. long; corolla-lobes not fully spreading. .... 17. *G. makinoi*
            18B.   Flowers 4–5 cm. long; corolla-lobes spreading. .... 18. *G. triflora*
  11B.   Leaves verticillate in 4's, linear-lanceolate, 1–2 cm. long, 2–3 mm. wide; stems 7–30 cm. long; flowers 6- to 8-merous.
          19. *G. yakushimensis*

1.   Gentiana yabei Takeda & Hara. *G. detonsa* var. *albiflora* Yabe; *Gentianella yabei* Hara, in syn.——SHIROUMA-RINDŌ.   Smooth biennial; stems erect, few-flowered, 5–40 cm. long, simple or sparingly branched, with few pairs of leaves; lower leaves obovate-spathulate, obtuse to rounded, the median ones largest, ovate-oblong or broadly lanceolate, 2–7 cm. long, 1–2.5 cm. wide, subacute, sessile, ascending, the upper leaves smaller and narrower; flowers 4- to 5-merous, 2.5–3.5 cm. long, the peduncles (1–)5–12 cm. long, the bracts absent; calyx-tube 1–1.5 cm. long, 4- or 5-angled, the lobes erect, deltoid-lanceolate, unequal, usually shorter than the tube, acuminate; corolla 25–35 mm. long, pale blue, the lobes short, obtuse, fimbriate near base; capsules shorter than the corolla, short-stipitate; seeds minutely spinulose.——Aug.–Sept. Alpine slopes; Honshu (Mount Shirouma in Shinano); rare.

2.   Gentiana contorta Royle. *Gentianella contorta* (Royle) H. Smith; *Gentianopsis contorta* (Royle) Ma; *G. yamatsutae* Kitag.——HIRO-HA-HIGE-RINDŌ, CHICHIBU-RINDŌ. Smooth, erect, biennial herb; stems simple to sparingly branched, 5–10 cm. long, obsoletely 6-striate, loosely few-leaved; median cauline leaves ovate- to obovate-elliptic, 1–1.5 cm. long, 8–10 mm. wide, rounded to very obtuse, abruptly narrowed at base, sessile, flat, spreading, entire, 1-nerved, the lower leaves smaller, the upper ones slightly smaller and obtuse at base; flowers solitary, about 15 mm. long, the peduncles 7–15 mm. long; calyx 13–15 mm. long, tubular, slightly narrowed at base, 4-ribbed, the lobes 4, deltoid-lanceolate, 3–4 mm. long, acuminate, slightly unequal, keeled, erect; corolla tubular, about 18 mm. long, the lobes 4, about 4 mm. long, oblong, obtuse, entire.——Sept. Rocks in mountains; Honshu (Mount Jumonjitoge in Chichibu); rare.——Manchuria, China, and Himalayas.

3.   Gentiana auriculata Pall. *G. fauriei* Lév. & Van't.——CHISHIMA-RINDŌ.   Smooth, glabrous biennial; stems erect, 4-angled, 5–20 cm. long, sparingly branched or simple; basal leaves small, oblanceolate-spathulate; cauline leaves narrowly ovate to broadly lanceolate, 1–3 cm. long, 5–15 mm. wide, subacute, sessile, obliquely spreading; flowers few, blue, subsessile or pedunculate, 2.5–3 cm. long; calyx-tube 7–10 mm. long, narrowly 5-winged, the lobes ovate-cordate, obtuse, auriculate, 5–7 mm. across, minutely papillose on the margin; co-rolla about twice as long as the calyx, the lobes 5, nearly half as long as the tube, their crown deltoid-ovate, finely laciniate; capsules sessile, as long as the corolla; seeds globose, smooth. ——Aug. Hokkaido.——Sakhalin, Kuriles, northern regions of e. Asia to the Aleutian Isls.

4.   Gentiana takedae Kitag. *G. amarella* var. *uliginosa* sensu auct. Japon., non Griseb.; *G. amarella* sensu auct. Japon., non L.——ONOE-RINDŌ.   Glabrous annual or biennial; stems erect, 5–25 cm. long, simple or sparsely branched, somewhat 4-angled, smooth, few-flowered; basal leaves small, oblanceolate, narrowed at base; cauline leaves broadly lanceolate or narrowly oblong, 1–3 cm. long, 5–10 mm. wide, obtuse, minutely papillose on margin, sessile; flowers pale blue-purple, 16–22 mm. long, 4- to 5-merous, the peduncles 0–2 cm. long; calyx-tube 4–6 mm. long, the lobes unequal, lanceolate, acute, nearly as long as the tube; corolla about twice as long as the calyx, the lobes about 1/4–1/3 as long as the tube; their crown about 2/3 as long as the corolla-lobes, whitish, finely laciniate; capsules as long as the corolla, sessile; seeds globose, pale brown, smooth.——Aug.–Sept. Alpine; Honshu (mountains of Shinano Prov.); rare.

5.   Gentiana yuparensis Takeda. *G. yezoalpina* Koidz.; *G. yuparensis* var. *yezoalpina* (Koidz.) Kudo——YŪBARI-RINDŌ, EZO-ONOE-RINDŌ.   Glabrous biennial; stems 10–20 cm. long, 4-angled; basal leaves obovate-spathulate; cauline leaves oblong to narrowly ovate, 1.5–3 cm. long, 7–12 mm. wide, obtuse, sessile, smooth; flowers 1–10, 2.5–3 cm. long, rose-purple, 5-merous, the peduncles 5–10 mm. long; calyx with hairlike papillae on the angles and on margins of the lobes, the tube 2–4 mm. long, the lobes very unequal, 3–6 times as long as the tube, lanceolate, 8–15 mm. long, acuminate or acute, often recurved; corolla 1½–2 times as long as the calyx-lobes, the lobes 1/3 as long as the tube; crown of the corolla-lobes about 2/3 as long as the corolla-lobes, deeply cut into capillary filaments, whitish.——Aug. Alpine; Hokkaido.

6.   Gentiana secta (Satake) Ohwi. *G. takedae* var. *secta* Satake; *Gentianella secta* (Satake) Satake; *Gentianella pulmonaria* subsp. *secta* (Satake) Toyokuni——SAMPUKU-RINDŌ. Glabrous biennial; stems 5–20 cm. long, somewhat 4-angled, simple or branched; cauline leaves narrowly oblong or broadly lanceolate in the upper ones, 1–2.2 cm. long, 4–8 mm. wide, obtuse, sessile, spreading; flowers 1–20, 15–17 mm. long, 5-merous, pale blue, the peduncles 5–15 mm. long; calyx 5–6 mm. long, the tube 0.5–1 mm. long, the lobes unequal, erect, lanceolate, 4–5 mm. long, subobtuse, minutely papillose on margin; corolla 2.5–3 times as long as the calyx, the lobes about 1/3 as long as the tube, ovate, obtuse; crown of the corolla-lobes whitish, erect, about half as long as the primary corolla-lobes, bifid and finely laciniate; capsules sessile, scarcely exserted beyond the persistent corolla; seeds creamy yellow, globose, smooth.——Aug.–Sept. Alpine; Honshu (Akaishi Mts.); rare.

7. Gentiana squarrosa Ledeb. *G. aquatica* sensu Thunb., non L.——Koke-rindō. Small biennial, 2–10 cm. high, usually branched from the base and somewhat tufted, minutely papillose; basal leaves rosulate, rhombic-ovate to narrowly so, 1–4 cm. long, 5–12 mm. wide, spreading, rather thick, acuminate, awn-pointed; cauline leaves ovate to narrowly so, 5–10 mm. long, 2–5 mm. wide, awn-tipped, sessile and short-sheathing at base; flowers few to rather numerous, short-pedunculate; calyx-tube 4–6 mm. long, the lobes ovate or obovate, gradually spreading or slightly reflexed, awn-pointed, about half as long as the tube, 2–4 mm. long; corolla 12–15 mm. long, pale blue, about twice as long as the calyx, the plaits of the corolla slightly smaller than the lobes, sometimes 2-fid; capsules long-stipitate, exserted; seeds broadly fusiform, nearly smooth.——Mar.–June. Sunny fields in lowlands; Honshu, Kyushu; rather common.——Korea, Formosa, n. India, China, and Siberia.

8. Gentiana thunbergii (G. Don) Griseb. *Ericala thunbergii* G. Don; *G. japonica* Maxim., non Roem. & Schult.——Haru-rindō. Glabrous biennial, 5–15 cm. high, usually branched from the base; basal leaves rosulate, spreading, ovate to narrowly so, or rhombic-ovate, 1–3 cm. long, 7–22 mm. wide, acute, entire, rather narrowly translucent on margin, the cauline leaves ovate-lanceolate, 5–12 mm. long, 2–8 mm. wide, ascending, short-sheathed at the base; flowers blue, 5-merous, 2.5–3.5 cm. long, few, the peduncles 1–3 cm. long; calyx-tube 8–10 mm. long, 5-angled, the lobes erect, deltoid-lanceolate, acuminate, about 1/3 as long as the tube; corolla 2–2.5 times as long as the calyx, the plaits subrounded, toothed; capsules long-stipitate, exserted; seeds fusiform, indistinctly reticulate.——Mar.–May. Sunny places in lowlands and mountains; Honshu, Shikoku, Kyushu; rather common.——Korea, Manchuria, and China.

Var. **minor** Maxim. *G. minor* (Maxim.) Nakai——Tate-yama-rindō. Stems few or solitary; flowers 1.5–2 cm. long, pale blue; seeds slightly larger.——Wet boggy places in mountains; Hokkaido, Honshu (centr. and n. distr.).

9. Gentiana aquatica L. *G. humilis* Stev.; *G. pseudohumilis* Makino——Hina-rindō. Glabrous, dwarf biennial, 3–5 cm. high; basal leaves rosulate, obovate, elliptic, or nearly orbicular, 4–8 mm. long, 2–4 mm. wide, acute, recurved awn-tipped, papillose on margin; cauline leaves spathulate to broadly oblanceolate, few, 4–7 mm. long, ascending, short-sheathed at base; flowers pale blue, 1 to few, 10–16 mm. long, 5-merous, sessile or short-pedunculate; calyx-tube 8–12 mm. long, 5-angled, the lobes erect, deltoid-ovate or broadly lanceolate, acuminate, 1/6–1/3 as long as the tube; corolla about 1½ times as long as the calyx; capsules short, long-stipitate, exserted; seeds fusiform, longitudinally lined on one side.——June. Alpine; Honshu (Mount Yatsugatake); very rare.——Caucasus and Siberia.

Var. **laeviuscula** (Ohwi) Ohwi. *G. pseudohumilis* var. *laeviuscula* Ohwi.——Ko-hina-rindō. Radical leaves broader, 8–15 mm. long, 6–12 mm. wide; flowers smaller.——Alpine; Honshu (Mts. Akaishi in Shinano and Nikko in Shimotsuke); rare.

10. Gentiana zollingeri Fawc. *G. aomorensis* Lév.——Fude-rindō. Biennial, 5–10 cm. high; basal leaves small, few; cauline leaves relatively thick, ovate to broadly so, 5–15 mm. long, 4–10 mm. wide, short-sheathed at base, abruptly cuspidate, often red-purple beneath, the margins whitish and thickish; flowers relatively few, subsessile or very short-pedunculate; calyx-tube 5–7 mm. long, the lobes broadly lanceolate, about half as long as the tube, acuminate, awn-tipped; corolla 18–25 mm. long, 2–2½ times as long as the calyx, blue; capsules short, stipitate, exserted; seeds very minute, fusiform, rather lustrous.——Apr.–May. Sunny slopes in lowlands and mountains; Hokkaido, Honshu, Shikoku, Kyushu; rather common.——Korea, Sakhalin, s. Kuriles, Manchuria, and China.

11. Gentiana jamesii Hemsl. *G. nipponica* var. *kawakamii* Makino; *G. kawakamii* (Makino) Makino——Rishiri-rindō, Kumoma-rindō. Glabrous, erect perennial, 5–12 cm. high; stems 4-angled, often red-purple, creeping at base; cauline leaves rather thick, broadly lanceolate to oblong, 7–15 mm. long, 3–6 mm. wide, obtuse, sessile, whitish on margin; flowers few, dark blue-purple, sessile, 2.5–3 cm. long; calyx-tube 6–8 mm. long, the lobes ovate, obtuse, spreading, 1/4–1/3 as long as the tube; corolla tube narrow, the lobes 1/4–1/3 as long as the tube, the plaits deltoid, deeply toothed, incurved and closing the throat; capsules long-stipitate, slightly exserted; seeds fusiform, smooth.——July–Sept. Alpine; Hokkaido.——Korea, Kuriles.

Var. **robusta** (Hara) Ohwi. *G. nipponica* var. *robusta* Hara——Iide-rindō. Somewhat larger; leaves 3–7 mm. wide; flowers 22–30 mm. long; calyx-lobes erect or ascending; corolla-tube slightly thicker.——Wet places, alpine; Honshu (Mount Iide in Iwashiro Prov.).

12. Gentiana nipponica Maxim. *G. makinoi* Lév. & Van't., non Kusn.——Miyama-rindō. Resembles the preceding species but smaller, 5–10 cm. high; stems obsoletely 4-angled; leaves broadly lanceolate to narrowly ovate-oblong, 5–12 mm. long, 3–5 mm. wide, obtuse, with a translucent margin, rather thick and deep green, short-sheathed at base; flowers 1–4, blue, 15–22 mm. long, sessile; calyx-lobes narrowly ovate, acute and recurved, about 1/3 as long as the tube; corolla 2–2½ times as long as the calyx, the tube gradually broadened in upper part, the plaits spreading in anthesis.——Aug.–Sept. Wet places, alpine; Hokkaido, Honshu (centr. and n. distr.).

13. Gentiana glauca Pall. *G. glauca* var. *major* Ledeb.——Yokoyama-rindō. Glabrous, erect perennial, 5–10 cm. high, short-creeping at the base and densely covered with old leaves; leaves rather thick, elliptic or oblong, 8–17 mm. long, 6–12 mm. wide, obtuse, slightly glaucescent, sessile; flowers 1 to few, pale blue, about 2 cm. long, the peduncles 0–3 mm. long; calyx-tube 4–5 mm. long, the lobes half as long as the tube, deltoid-lanceolate, subacute; plaits of corolla entire, about 1/4 as long as the corolla-lobes, obtuse, deltoid; capsules long-stipitate, exserted, broadly lanceolate; seeds reticulate, with irregular small wings.——Aug.–Sept. Alpine; Hokkaido (Ishikari Prov.); rare.——Kuriles, Kamchatka, Ochotsk Sea region, and western N. America.

14. Gentiana algida Pall. *G. frigida* var. *algida* (Pall.) Froel.; *G. algida* var. *sibirica* Kusn.——TŌYAKU-RINDŌ. Yellowish green, glabrous perennial, 10–25 cm. high, with short rhizomes; stems scapose, few-leaved, angled; rosette leaves rather thick, linear-oblanceolate to broadly linear, 8–15 cm. long, 5–10 mm. wide, subobtuse, gradually narrowed and sheathed at the base, the cauline leaves few, lanceolate, 2–5 cm. long, 5–10 mm. wide, short-sheathed at base; flowers 3.5–5 cm. long, few, pale yellow with blue-green spots, the peduncles 0–10(–20) mm. long; calyx half as long as the corolla, the lobes unequal, about half as long as the tube, deltoid-lanceolate, suberect, acute; plaits of corolla short, few-toothed; capsules short-stipitate, scarcely exserted; seeds reticulate, with 3–4 narrow wings.——Alpine slopes; Honshu (centr. and n. distr.).——Korea, Siberia, and N. America.——Forma **igarashii** (Miyabe & Kudo) Toyokuni. *G. algida* var. *igarashii* Miyabe & Kudo; *G. igarashii* Miyabe & Kudo——KUMOI-RINDŌ. Plant low, about 10 cm. high; flowers 1–3, larger, to 5 cm. long.——Alpine; Hokkaido.

15. Gentiana sikokiana Maxim. ASAMA-RINDŌ. Glabrous, erect perennial, 7–20 cm. high, with slender, short rhizomes; leaves spathulate, ovate, rhombic-ovate, or obovate, 3–8 cm. long, 1.5–4 cm. wide, acute or subacuminate, green above, pale green beneath, 3-nerved, slightly undulate-margined; flowers 3–10, nearly terminal, 4–5 cm. long, blue, with green spots, 5-merous; calyx-tube 6–10 mm. long, the lobes broadly ovate or ovate-cordate, spreading, rather unequal; plaits of corolla nearly entire; capsules lanceolate, stipitate, not exserted; seeds lanceolate, reticulate, narrowly winged.——Oct.–Nov. Woods in mountains; Honshu (s. Kinki Prov.), Shikoku, Kyushu.

16. Gentiana scabra Bunge var. **buergeri** (Miq.) Maxim. *G. buergeri* Miq.; *G. scabra* var. *intermedia* Kusn.; *G. scabra* var. *buergeri* subvar. *saxatilis* Honda; *G. saxatilis* (Honda) Honda; *G. scabra* var. *orientalis* Hara; *G. subpetiolata* Honda——RINDŌ. Rhizomes rather short; stems 20–100 cm. long, with 10–20 pairs of leaves, 4-striate, sometimes decumbent or ascending, commonly red-brown tinged; leaves lanceolate to narrowly deltoid-ovate, 4–8 cm. long, 1–3 cm. wide, gradually acuminate, 3-nerved, green above, paler beneath, usually sessile, papillose on margin in the upper ones, the bracts narrowly lanceolate, rather small; flowers sessile, 4.5–6 cm. long, few to rather numerous, blue; calyx-tube 12–18 mm. long, the lobes rather unequal, linear-lanceolate; plaits of corolla deltoid, often toothed; capsules stipitate, not exserted; seeds broadly lanceolate, short-caudate at both ends.——Sept.–Nov. Thickets, grassy places, and wet meadows in lowlands and low elevations in the mountains; Honshu, Shikoku, Kyushu; common. The typical phase occurs in Korea, Manchuria, and n. China.——Forma **stenophylla** (Hara) Toyokuni. *G. scabra* var. *buergeri* forma *angustifolia* Kusn., non *G. scabra* var. *bungeana* forma *angustifolia* Kusn.; *G. scabra* var. *stenophylla* Hara——HOSOBA-RINDŌ. Narrow-leaved phase, the leaves linear to linear-lanceolate. Often growing in wet places. Known hybrids are: **Gentiana** × **brevidens** Fr. & Sav. (*G. makinoi* × *G. scabra* var.); and **Gentiana** × **iseana** Makino. ISE-RINDŌ (*G. scabra* var. × *G. sikokiana*).

17. Gentiana makinoi Kusn. OYAMA-RINDŌ. Glabrous, slightly glaucous erect perennial, 30–60 cm. high, with thick stout rhizomes; stems terete, with 10–20 pairs of leaves; lower leaves reduced to short sheaths, the upper leaves broadly lanceolate to narrowly ovate, 3–6 cm. long, 8–25 mm. wide, gradually narrowed to the tip, obtuse, broadly cuneate to rounded at base, glaucescent, entire, 3-nerved, sessile; bracts leaflike; flowers 1–7, on upper part of stem, sessile, 22–30 mm. long, dark blue; calyx-tube 6–8 mm. long, subtruncate at apex, the lobes rather unequal, very short to short-linear; plaits of corolla truncate; capsules stipitate, broadly lanceolate, as long as the corolla; seeds minutely reticulate, caudate at both ends.——Aug.–Sept. Wet slopes and bogs in high mountains; Honshu (centr. distr.).

18. Gentiana triflora Pall. var. **japonica** (Kusn.) Hara. *G. rigescens* var. *japonica* Kusn.; *G. axillariflora* Lév. & Van't.; *G. naitoana* Lév. & Faurie; *G. jesoana* Nakai; *G. axillariflora* var. *naitoana* (Lév. & Faurie) Koidz.——EZO-RINDŌ. Glabrous, slightly glaucescent, erect perennial, 30–80 cm. high, with thick rhizomes; stems pale green to very slightly brownish red; lower few pairs of leaves reduced to short sheaths; upper leaves lanceolate to broadly so, 6–10 cm. long, 1–2.5 cm. wide, 3-nerved, gradually narrowed to the tip, entire, sessile, glaucescent beneath; bracts linear-oblanceolate, shorter than the calyx or sometimes leaflike and longer than the calyx; flowers 4–5 cm. long, dark blue, sessile; calyx-tube subtruncate at the apex, 12–15 mm. long, the lobes unequal, erect, sometimes toothlike or leaflike; plaits of corolla short, nearly truncate; capsules stipitate, not exserted from the corolla; seeds minutely reticulate, caudate at both ends.——Aug.–Sept. Lowlands and mountains; Hokkaido, Honshu (n. distr. to n. Kinki Distr. along the Japan Sea side).——s. Kuriles and Sakhalin. The typical phase occurs in Sakhalin, Korea, and e. Siberia.

Var. **montana** (Hara) Hara. *G. axillariflora* var. *montana* Hara——EZO-OYAMA-RINDŌ. Stems low, few-flowered; flowers 3.5–4.5 cm. long.——Alpine; Hokkaido, Honshu (n. distr.).——s. Kuriles and Sakhalin.

Var. **horomuiensis** (Kudo) Hara. *G. horomuiensis* Kudo; *G. axillariflora* var. *horomuiensis* (Kudo) Hara——HOROMUI-RINDŌ. Leaves linear-lanceolate.——Boggy places; Hokkaido.

19. Gentiana yakushimensis Makino. *Kudoa yakushimensis* (Makino) Masam.——YAKUSHIMA-RINDŌ. Glabrous, tufted, erect perennial, 7–30 cm. high, with stout elongate rhizomes; stems few, simple, striate, densely many-leaved; leaves verticellate in 4's, thick and lustrous, linear-lanceolate to linear-oblong, 1–2 cm. long, 2–3 mm. wide, obtuse, short-sheathed at base, deep green, whitish beneath; flowers solitary, terminal, sessile, 3.5–4 cm. long, blue-purple; calyx about 1 cm. long, subtruncate at apex, the lobes 6–8, lanceolate, obtuse, shorter than the tube; corolla-lobes 6–8, the plaits bifid, linear; capsules short-stipitate, included; seeds angled, about 3.5 mm. long including the filiform tail at one end.——Aug. Alpine. Kyushu (Yakushima); rare.

5. SWERTIA L. SEMBURI ZOKU

Simple or branched perennials and annuals; leaves opposite, radical and cauline; flowers blue or purple, sometimes yellow, in cymes, panicles, or corymbs, 4- to 5-merous; calyx 4- to 5-parted, the lobes linear or lanceolate, 1- to 3-nerved; corolla rotate, rarely campanulate, the lobes dextrorsely contorted, glandular-grooved at the base, the grooves naked or covered with fleshy hairs; stamens inserted on the base of the corolla; ovary 1-locular; style short or absent, the stigmas 2, lamellate; capsules septicidally dehiscent, 2-valved; seeds many, often wing-margined.——About 80 species, in Europe, Asia, Africa, and America.

1A. Stigma sessile, decurrent on the ovary; nerves of corolla-lobes yellowish red, also when dried; small annual or biennial.
1. *S. carinthiaca*

1B. Stigma sessile or stalked, not decurrent on the ovary; nerves of corolla-lobes not yellowish red; small or rather large annuals or perennials.

  2A. Annuals or biennials without rhizomes; seeds not winged.

    3A. Radical leaves not well developed; cauline leaves linear to broadly lanceolate; seeds smooth or nearly so.

      4A. Glandular-grooves of corolla-lobes solitary; flowers small, 4-merous. . . . . 2. *S. tetrapetala*

      4B. Glandular-grooves of corolla-lobes in pairs; flowers rather large, usually 5-merous.

        5A. Calyx-lobes lanceolate to broadly so, gradually and slightly narrowed at base; leaves usually oblanceolate, 3–10 mm. wide; plant not bitter to taste. . . . . 3. *S. diluta* var. *tosaensis*

        5B. Calyx-lobes linear or broadly so, not narrowed at base; leaves linear to lanceolate, 1–5(–8) mm. wide; plants bitter to taste.

          6A. Stems, margin of calyx-lobes, and upper portion of pedicels minutely punctate; hairs around the glandular grooves undulately striate under magnification. . . . . 4. *S. pseudochinensis*

          6B. Plant usually smooth; hairs around the glandular-grooves smooth under magnification. . . . . 5. *S. japonica*

    3B. Radical leaves large; cauline leaves ovate to oblong, or lanceolate; seeds muricate.

      7A. Radical leaves persistent through anthesis, the cauline leaves much smaller; flowers nodding; glandular-grooves solitary.
6. *S. tashiroi*

      7B. Radical leaves withering before anthesis, the cauline leaves broader and prominent; flowers erect; glandular-grooves in pairs.

        8A. Glandular-grooves naked; flowers in panicles; calyx-lobes broadly oblanceolate, about 1/3 as long as the corolla; seeds less than 1 mm. long; style absent. . . . . 7. *S. bimaculata*

        8B. Glandular-grooves hairy on margin; flowers fasciculate in leaf-axils; calyx-lobes linear, as long as the corolla; seeds 3–3.5 mm. long; style distinct. . . . . 8. *S. swertopsis*

  2B. Rhizomatous perennial; glandular-grooves of corolla in pairs on each lobe; seeds wing-margined. . . . . 9. *S. perennis*

1. **Swertia carinthiaca** Wulfen. *Lomatogonium carinthiacum* (Wulfen) Reichenb.; *Pleurogyna carinthiaca* G. Don; *Gentiana stelleriana* Cham. & Schltdl.; *Lomatogonium carinthiacum* var. *stellerianum* (Cham. & Schltdl.) Fern.——HIME-SEMBURI.   Glabrous annual or biennial 2–10 cm. high; stems simple or branched; lower leaves obovate-spathulate, small, the median and upper ones oblong to ovate, 1–1.5 cm. long; flowers pale blue, pedicelled; calyx deeply 4- to 5-parted, the segments unequal, oblong, elliptic, or ovate, subacute, 3-nerved, 4–7 mm. long; corolla deeply 4- to 5-lobed, the tube very short, glabrous, the lobes oblong, about 1 cm. long, yellow-red nerved, with 2 small pocketlike glands at base; capsules as long as or slightly longer than the corolla; stigma decurrent on the ovary; seeds fusiform or oblong, smooth, about 0.6 mm. long.——Aug.–Sept. Alpine; Honshu (centr. distr.); very rare.——Widely distributed in the alpine regions of the N. Hemisphere.

2. **Swertia tetrapetala** Pall. *S. pallasii* G. Don; *Ophelia papillosa* Fr. & Sav.; *O. yesoensis* Fr. & Sav.; *S. yesoensis* (Fr. & Sav.) Matsum.; *S. bissetii* S. Moore & Burkill——CHISHIMA-SEMBURI.   Glabrous erect annual or biennial; stems simple or sparingly branched, 4-angled; basal leaves broadly oblanceolate-spathulate, the median and upper leaves spreading, deltoid-lanceolate or lanceolate, 2–3.5 cm. long, 7–15 mm. wide, gradually narrowed to the tip, acute to obtuse, sessile; flowers in a rather dense terminal panicle, 8–12 mm. across, blue, dark purple-spotted, pedicelled, 4-merous; calyx-segments lanceolate, about 1/3 as long as the corolla, acute; corolla-lobes elliptic or ovate-oblong, glandular-grooved near the center, with thick flexuous hairs; capsules slightly longer than to as long as the corolla; seeds smooth, subovoid, slightly flattened, brown.——Aug.–Sept. Sandy grassy slopes near the sea; Hokkaido; rare.——Kuriles, Sakhalin, Ochotsk Sea region, and Kamchatka.

Var. **yezoalpina** (Hara) Hara. *S. yezoalpina* Hara; *S. chrysantha* var. *yezoalpina* (Hara) Satake; *S. micrantha* Takeda; *S. chrysantha* Honda & Tatew.——TAKANE-SEMBURI. Inflorescence rather loosely flowered; flowers pale blue, about 7 mm. across, the glandular-grooves usually slightly broader, naked or with few hairs on margin; capsules longer than the corolla.——Aug.–Sept. Alpine slopes; Hokkaido (Hidaka

Prov.), Honshu (centr. and n. distr.); rare.——Korea and ? e. Siberia.

3. **Swertia diluta** (Turcz.) Benth. & Hook. f. var. **tosaensis** (Makino) Hara. *S. tosaensis* Makino——INU-SEMBURI.   Glabrous annual or biennial; stems often branched from the base, 4-angled; radical leaves small, broadly oblanceolate to obovate, the cauline leaves oblanceolate, oblong-lanceolate, or rarely lanceolate, 2–5 cm. long, 3–10 mm. wide, obtuse; inflorescence a narrow panicle; flowers white, 5-merous, pedicelled; calyx-segments lanceolate to broadly so, obtuse, about half as long to as long as the corolla; corolla-lobes 8–12 mm. long, the glandular-grooves in pairs, lanceolate, with long flexuous hairs on margin; capsules slightly longer than the corolla, narrowly ovoid; seeds nearly globose, almost smooth.——Honshu, Shikoku, Kyushu.——Korea and China.

4. **Swertia pseudochinensis** Hara. *S. chinensis* sensu auct. Japon.; *S. chinensis* forma *violacea* Makino——MURASAKI-SEMBURI.   Annual or biennial; stems often branched, dark purple, minutely papillose; radical leaves oblanceolate, small, the cauline leaves lanceolate or linear-lanceolate, 2–4 cm. long, 3–5(–8) mm. wide, acute; flowers few to many, pedicelled, pale blue-purple, 5-merous; calyx-segments broadly linear to linear-lanceolate, half as long to nearly as long as the corolla; corolla-lobes 1–1.5 cm. long, with dark-colored nerves, the glandular-grooves in pairs, each surrounded by fleshy, flexuous hairs; capsules nearly as long as the corolla, broadly lanceolate; seeds nearly globose, smooth.——Aug.–Oct. Thickets in lowlands and hills; Honshu, Shikoku, Kyushu.——Korea, n. China, Manchuria, and Amur.

5. **Swertia japonica** (Roem. & Schult.) Makino. *Gentiana japonica* Roem. & Schult.; *Ophelia japonica* (Schult.) Griseb.——SEMBURI.   Glabrous annual or biennial; stems simple or branched from the base, 4-angled, purplish; radical leaves small, oblanceolate, the cauline linear or broadly so, rarely linear-oblanceolate, 1.5–3.5 cm. long, 1–3 mm. wide, acute or subobtuse, the margins revolute; flowers in dense terminal panicles, 5-merous; calyx-segments linear to linear-lanceolate, 1/2–2/3 as long as the corolla; corolla-lobes broadly lanceolate, 12–17 mm. long, white, with purple nerves, the glandular-grooves in pairs, with long fleshy hairs on margin; capsules slightly longer than the corolla, lanceolate; seeds

nearly smooth, subglobose.——Aug.-Oct. Sunny slopes and thickets in lowlands and hills; Hokkaido, Honshu, Shikoku, Kyushu; rather common.——Korea, and China.

6. Swertia tashiroi (Maxim.) Makino. *Ophelia tashiroi* Maxim.; *S. tashiroi* var. *cruciata* F. Maekawa——HETSUKA-RINDŌ. Rather stout glabrous erect biennial 30–60 cm. high; stems terete, subscapose; leaves rosulate, oblong or obovate-spathulate, 8–30 cm. long, 3–8 cm. wide, acute to short-acuminate, entire, rather fleshy; cauline leaves ovate to narrowly lanceolate, 2–5 cm. long, 5–20 mm. wide, the upper ones becoming bracteate; panicles terminal, few-flowered, the bracts small, 1–10 mm. long, the pedicels deflexed, thickened above; calyx-segments 4–5, narrowly deltoid, acuminate, 2–4 mm. long, about as long as the tube; corolla-lobes 4–5, broadly oblanceolate, 12–15 mm. long, acuminate, greenish, with an orbicular gland about 3 mm. across slightly above the center, purple-brown spotted on the upper half; capsules as long as the corolla, 2-valved, sessile; seeds small, oblong-cylindric, short-tubercled.——Dec.-Jan. Kyushu (s. distr. incl. Yakushima and Tanegashima).——Ryukyus (Amami-Oshima).

7. Swertia bimaculata (Sieb. & Zucc.) Hook. f. & Thoms. *Ophelia bimaculata* Sieb. & Zucc.——AKEBONO-SŌ. Glabrous erect annual or biennial 50–80 cm. high; stems often much branched, with 4 elevated striations; rosulate leaves oblong, large, long-attenuate to a petiolelike base, with several parallel nerves; lower cauline leaves ovate to broadly lanceolate, 5–12 cm. long, 1–6 cm. wide, acuminate, cuneate or acuminate at base; flowers erect, 5-merous, in rather loose panicles, the pedicels 1–5 cm. long; calyx-segments broadly oblanceolate, obtuse, much longer than the tube, about half as long as the corolla; corolla-lobes broadly cuneate-oblanceolate, 10–13 mm. long, creamy white with green spots on upper half and a pair of orbicular, brownish glands slightly above the center, about 1.5 mm. across; capsules slightly longer than the corolla; seeds small, dark brown, with minute warty papillae. ——Sept.-Oct. Wet grassy places in lowlands and mountains; Hokkaido, Honshu, Shikoku, Kyushu.

8. Swertia swertopsis Makino. *Swertopsis umbellata* Makino; *Swertia umbellata* (Makino) Makino, non Ruiz & Pav.——SHINONOME-SŌ. Glabrous erect annual or biennial; stems simple or sparingly branched, 4-striate; cauline leaves ovate to narrowly so, or oblong, acute to acuminate, 5(–7)-nerved, acutely narrowed to a petiolelike base; flowers white, fasciculate, 5-merous, the pedicels 5–10 mm. long; calyx-segments linear, as long as the corolla, keeled on back; corolla subcampanulate, 8–10 mm. long, the lobes ovate, with purple spots inside, the glands paired, surrounded by filiform hairs; capsules less than 1 cm. long, narrowly ovoid; seeds rather large, oblong, 3–3.5 mm. long, densely warty.——Aug. Mountains; Honshu (Izu Prov. and westw.), Shikoku, Kyushu; rare.

9. Swertia perennis L. var. **cuspidata** Maxim. *S. obtusa* Ledeb. var. *cuspidata* (Maxim.) Hara; *S. cuspidata* (Maxim.) Kitag.——MIYAMA-AKEBONO-SŌ. Glabrous rhizomatous erect perennial 12–30 cm. high; stems simple, solitary, striate; leaves alternate or opposite, the radical few, broadly ovate or elliptic, 3–8 cm. long, 2.5–4 cm. wide, obtuse to acute, 5- to 7-nerved, with a long petiolelike base 1–3 times as long as the leaf-blade, the cauline leaves small; inflorescence few-flowered, corymbose, often slightly branched, the bracts broadly lanceolate, the pedicels 1–3 cm. long; flowers 5-merous; calyx-lobes unequal, broadly linear, 1/2–2/3 as long as the corolla; corolla-lobes lanceolate, 17–20 mm. long, dark blue, with dark purple nerves, the glands paired, small, near the base of the corolla-lobes, surrounded by hairs about 2 mm. long; capsules 2/3 as long as the corolla; seeds wing-margined.——Aug.-Sept. Alpine; Hokkaido, Honshu (centr. and n. distr.).

Var. **stenopetala** (Regel & Tiling) Maxim. *S. obtusa* var. *stenopetala* Regel & Tiling——EZO-MIYAMA-AKEBONO-SŌ. Flowers smaller, about 12 mm. long; corolla-lobes less prominently narrowed toward tip.——Hokkaido.——Sakhalin, Kuriles, e. Siberia, and Alaska.

6. HALENIA Borkh.   HANA-IKARI ZOKU

Annuals, biennials, or perennials; leaves opposite, entire; cymes axillary and terminal, few-flowered or sometimes compound and paniculate; flowers yellowish, sometimes purplish, 4-merous; calyx deeply 4-parted, the lobes lanceolate; corolla campanulate, rarely rotate, 4-lobed, the lobes spurred near base, ovate or oblong, sinistrorsely contorted in bud; stamens inserted near the base of corolla; ovary 1(–2)-locular; style short or absent; stigmas 2, lamellate; capsules 2-valved; seeds smooth, globose or ellipsoidal.——About 20 species, in e. Europe, Asia, N. and S. America.

1. Halenia corniculata (L.) Cornaz. *Swertia corniculata* L.; *H. sibirica* Borkh.; *H. japonica* Gand.; *H. deltoides* Gand.——HANA-IKARI. Glabrous, erect annual or biennial, 10–60 cm. high; stems erect, often branched, 4-angled; leaves oblong to narrowly ovate, 3–10 cm. long, 1.5–4 cm. wide, acuminate, narrowed to a petiolelike base, 3(–5)-nerved, minutely papillose on nerves beneath and on margin; flowers 4-merous, short-pedicelled, pale yellow, greenish after anthesis; calyx divided nearly to the base, the segments linear, minutely papillose, 1/2–2/3 as long as the corolla; corolla narrowly ovoid, 6–10 mm. long, the spur 3–7 mm. long, rarely absent; capsules lanceolate, as long as the corolla, sessile; seeds ellipsoidal, smooth, terete, about 1 mm. long.——Sept. Sunny slopes in mountains; Hokkaido, Honshu, Shikoku, Kyushu. ——e. Europe, Siberia, Korea, Manchuria, Kuriles, Sakhalin, and Kamchatka.

7. MENYANTHES L.   MITSU-GASHIWA ZOKU

Glabrous, rather soft, slightly thickened perennial herb growing in wet places and in shallow water, with thick, creeping-rhizomes; leaves radical, long-petiolate, alternate, 3-foliolate, the petioles sheathed at base; flowers white or bluish, 5-merous, in a terminal simple raceme on radical scapes, with membranous bracts; calyx deeply parted; corolla infundibuliform, 5-lobed, the lobes induplicate in bud, barbed on inner side; stamens inserted on the corolla-tube; ovary 1-locular, with 5 glands at base; style slender, with 2 lamellate stigmas; capsules globose, 2-lobed; seeds few, orbicular, flat, lustrous, salmon-red, rather large.—— One species throughout the cooler regions of the N. Hemisphere.

1. Menyanthes trifoliata L. *M. palustris* S. F. Gray—— MIZU-GASHIWA, MITSU-GASHIWA. Rhizomes 7–10 mm. across, few-leaved in upper part; leaflets rather thick, oblong, ovate-elliptic, or rhombic-elliptic, sessile, 4–8 cm. long, 2–5 cm. wide, obtuse, obtusely toothed, or nearly entire, sessile; scapes terete, 20–40 cm. long, the bracts broadly ovate, 3–8 mm. long, the pedicels 1–2.5 cm. long; flowers 1–1.5 cm. across; calyx short, the lobes elliptic; capsules 5–7 mm. across; seeds 2.5–3 mm. across.——Apr.–Aug. Wet boggy places and margins of lakes and ponds; Hokkaido, Honshu, Kyushu; rather common.

8. FAURIA Franch.   IWA-ICHŌ ZOKU

Glabrous, rather fleshy scapose perennial of wet slopes, with stout creeping leafy rhizomes; leaves radical, rather thick and soft, obtusely denticulate, petiolate, sheathing at base; inflorescence terminal, a dichotomous or trichotomous 10- to 30-flowered cyme, the bracts herbaceous; flowers 5-merous; calyx-tube adnate to the ovary; corolla infundibuliform, deeply 5-lobed, the lobes induplicate in bud, with longitudinal ridges inside; stamens inserted on the corolla-tube; ovary semi-inferior, 1-locular, with 5 glands at base; style with a peltate stigma; capsules cylindric, 2-valved, the valves 2-lobed; seeds ellipsoidal, rather flat, gray-white, lustrous, rather numerous.——One species, in Japan and w. N. America.

1. Fauria crista-galli (Menz, ex Hook.) Makino. *Menyanthes crista-galli* Menzies ex Hook.; *Villarsia crista-galli* (Menzies) Griseb.; *Fauria japonica* Franch.; *Nephrophyllidium crista-galli* (Menzies ex Hook.) Gilg——IWA-ICHŌ. Leaves long-petiolate, reniform, 3–6 cm. long, 4–10 cm. wide, somewhat retuse, cordate, obtusely denticulate, pedately 7- to 15-nerved, the petioles 2–5 times as long as the blades; scapes 15–40 cm. long; flowers pedicelled, white, about 12 mm. across, the bracts broadly lanceolate, obtuse; calyx-lobes deltoid-ovate, obtuse, about 8 mm. long; capsules cylindric, 1–1.5 cm. long; seeds about 2.5 mm. long.——July–Aug. Wet alpine slopes; Hokkaido, Honshu (n. and centr. distr. on Japan Sea side). ——s. Kuriles and w. N. America.

9. NYMPHOIDES Hill   ASAZA ZOKU

Floating aquatic perennial herbs; leaves alternate, petiolate, ovate to orbicular, deeply cordate, sometimes peltate, entire or undulate; flowers 1 to many, in fascicles, yellow or white, 5-merous, pedicellate; calyx-lobes oblong or lanceolate; corolla rotate, the lobes induplicate in bud, hairy on margin or on inner side; stamens inserted at base of corolla, the filaments short; ovary 1-locular, with 5 small glands at base; style short or elongate, the stigmas 2(–3), lamellate; capsules ovoid or oblong, indehiscent or irregularly ruptured; seeds hairy or glabrous, sometimes tubercled or winged.——About 20 species, chiefly in the Tropics, especially abundant in s. Asia.

1A. Flowers yellow, large; calyx 8–15 mm. long; seeds flat, with long tubercles on margin; leaves often with the lower edges slightly united, subpeltate; petioles sheathed at base. . . . . 1. *N. peltata*
1B. Flowers white, small; calyx 3–6 mm. long; seeds plump, smooth, lustrous; leaves alternate, not united on lower margin; petioles auriculate at base.
  2A. Leaves large, 7–20 cm. across; pedicels 3–7(–10) cm. long; flowers about 1.5 cm. across; corolla with long hairs inside; capsules not longer than the calyx. . . . . 2. *N. indica*
  2B. Leaves small, 2–6 cm. across; pedicels 1–3 cm. long; flowers about 8 mm. across; corolla-lobes ciliate; capsules slightly longer than the calyx. . . . . 3. *N. coreana*

1. Nymphoides peltata (Gmel.) O. Kuntze. *Menyanthes nymphoides* L.; *N. flava* Hill; *Limnanthemum peltatum* Gmel.; *L. nymphoides* (L.) Hoffmanns. & Link; *L. nymphoides* var. *japonica* Miq.; *N. nymphoides* (L.) Britt. & Brown——ASAZA. Rhizomes creeping; stems elongate, remotely leaved; leaves floating, rather thick and soft, ovate to orbicular, 5–10 cm. across, deeply cordate with overlapping lobes or sometimes slightly peltate, undulately toothed, the petioles elongate, dilated and sheathing at base, the pedicels 2–12 cm. long; calyx 8–15 mm. long, the lobes broadly lanceolate, obtuse, rather thick; corolla yellow, 3–4 cm. across, long-ciliate; capsules narrowly ovoid, slightly longer than the calyx, narrowed at apex; style about 3 mm. long; seeds flat, obovate, about 3 mm. long, tubercled on margin, minutely scabrous on both sides.——June–Aug. Ponds and shallow lakes in lowlands; Honshu, Shikoku, Kyushu; common.——Korea, Formosa, China, and generally in the temperate regions of the N. Hemisphere.

2. Nymphoides indica (L.) O. Kuntze. *Menyanthes indica* L.; *Limnanthemum indicum* (L.) Griseb.——GAGA-BUTA. Stems slender, remotely 1- to 3-leaved; leaves orbicular-cordate or ovate-orbicular, 7–20 cm. across, subentire, deeply lobed, the petioles decurrent on the stem, 1–2 cm. long, auriculate, clasping; flowers few, fasciculate, white, about 1.5 cm. across, the pedicels 3–10 cm. long; calyx-lobes broadly lanceolate, subobtuse, 4–6 mm. long; corolla-lobes long-hairy inside; capsules oblong, 4–5 mm. long, not exserted beyond the calyx; style about 2 mm. long; seeds gray-brown, lustrous, nearly globose, smooth, about 0.8 mm. across.——July–Sept. Ponds and shallow water in lowlands; Honshu, Shikoku, Kyushu; common.——Korea, Formosa, China, se. Asia to Australia, and Africa.

3. Nymphoides coreana (Lév.) Hara. *Limnanthemum coreanum* Lév.; *L. parvifolium* sensu Miki, non Griseb.——HIME-SHIRO-ASAZA. Stems slender, elongate, 1- to 2-leaved; leaves ovate-cordate or orbicular-cordate, 2–6 cm. across, deeply 2-lobed, entire, the petioles decurrent on the stem, unequal in length, 1–10 cm. long, auriculate; flowers few, white, about 8 mm. across, fasciculate, the pedicels 1–3 cm. long; calyx-lobes broadly lanceolate, subacute, 3–4 mm. long, membranous; corolla-lobes ciliate; capsules ellipsoidal, 4–5 mm. long, slightly longer than the calyx; style a little less than 1 mm. long.——Aug. Ponds and shallow water; Honshu, Shikoku, Kyushu. ——Korea and Ryukyus.

Fam. 169. APOCYNACEAE Kyōchiku-tō Ka Dogbane Family

Usually evergreen woody plants, sometimes herbs, often scandent, with milky juice; leaves opposite, rarely verticillate or alternate, simple, entire, without stipules; flowers bisexual, actinomorphic; calyx often glandular inside, 4- to 5-lobed, imbricate in bud; corolla salverform or infundibuliform, the lobes contorted in bud, very rarely valvate; stamens 4 or 5, inserted on the tube, the filaments usually free, the anthers sagittate, free or connivent around the stigma, the pollen granular; ovary superior, 1-locular, with 2 parietal placentae, or of 2 free carpels; style solitary; ovules more than 2 in each carpel; fruit a follicle, berry, or drupe; seeds usually with endosperm, often winged or comose.——About 300 genera, with about 1,100 species, warm-temperate to subtropical and tropical.

1A. Herbs, sometimes somewhat woody near the base.
  2A. Anthers free from the stigma, not appendaged at base; corolla salverform, villous on the tube inside, the lobes sinistrorsely contorted in bud; styles filiform; seeds not comose. .... 1. *Amsonia*
  2B. Anthers adherent to the stigma, appendaged at base; corolla campanulate, glabrous, the lobes dextrorsely contorted in bud; styles very short or absent; seeds comose. .... 2. *Apocynum*
1B. Shrubs or woody vines.
  3A. Scandent; corolla 1–3 cm. wide, the throat naked; leaves opposite.
    4A. Calyx with 5 or 10 glands inside; stamens inserted on the upper part of the corolla-tube; seeds without a beak; inflorescence corymbose, few-flowered. .... 3. *Trachelospermum*
    4B. Calyx without glands or with small glands inside; stamens inserted near the base of the corolla-tube; seeds beaked; inflorescence paniculate, many-flowered. .... 4. *Anodendron*
  3B. Shrubs or small trees; corolla 4–5 cm. wide, the throat appendaged; leaves in 3's or 4's at the nodes. .... 5. *Nerium*

1. AMSONIA Chōji-sō Zoku

Glabrous or hairy erect herbs or subshrubs; leaves alternate, membranous; flowers 5-merous, usually blue, in terminal corymbs or panicles; calyx 5-parted, without glands, the segments narrow; corolla salverform, the tube cylindric, narrow, slightly inflated in upper part, villous within, without scales on the throat, the lobes narrow, sinistrorsely contorted; anthers obtuse, unappendaged on the tube, free; carpels 2, follicular; style solitary, filiform, the stigma thick, with a reflexed membranous appendage; ovules many, in 2 series; follicles 2, cylindric; seeds many, oblong, obliquely truncate at both ends, not comose.——Few species, in N. America and Japan.

1. **Amsonia elliptica** (Thunb.) Roem. & Schult. *Tabernaemontana elliptica* Thunb.——Chōji-sō. Glabrous perennial; stems erect, 40–80 cm. high, terete, slightly branched in upper part; leaves alternate or opposite, lanceolate, 6–10 cm. long, 1–2 cm. wide, acuminate, subsessile, deep green; inflorescence cymose, many-flowered, terminal; flowers blue, about 13 mm. across; calyx-lobes 1–2 mm. long; corolla-tube about 8 mm. long, the lobes narrowly oblong, nearly as long as the tube; follicles obliquely spreading, 5–6 cm. long, glabrous, terete; seeds narrowly oblong, 7–10 mm. long, obliquely truncate at both ends, brown, irregularly rugose.——May–June. Grassy places along rivers in lowlands; Hokkaido, Honshu, Kyushu.——Korea and China.

2. APOCYNUM L. Bashi-kurumon Zoku

Usually glaucescent perennial herbs or subshrubs; leaves alternate or opposite, with weak, pinnate nerves; flowers small, in dense to loose cymes; calyx small, deeply 5-lobed, without glands inside, the lobes acute; corolla campanulate, 5-lobed, with 5 short appendages inside, glabrous, dextrorsely contorted; stamens alternate with the appendages of the corolla, inserted on the lower portion of corolla-tube, the filaments very short, the anthers convergent, slightly adherent to stigma, the anther-locules appendaged at base; carpels 2; stigma sessile; follicles elongate; seeds comose, small.——Few species, in s. Europe, Asia, and N. America.

1. **Apocynum basikurumon** Hara. *A. venetum* sensu auct. Japon., non L.; *A. venetum* var. *basikurumon* Hara, in syn.; *Trachomitum venetum* var. *basikurumon* (Hara) Hara ——Bashi-kurumon. Glabrous except the flowers; rhizomes thick; stems 40–80 cm. long, branched, reddish; leaves of the main stems alternate, those of the branches opposite, oblong-ovate to oblong, 2–5 cm. long, 7–15 mm. wide, rounded to obtuse, mucronate, broadly cuneate at base and short-petiolate, callose on the margin; inflorescence a terminal panicle, the bracts thinly membranous, narrow, caducous; calyx deeply 5-parted, slightly puberulent, the lobes about 2 mm. long; corolla pale rose, 6–7.5 mm. long, papillose-puberulous on both sides, 5-lobed.——July. Near seashores; Hokkaido (sw. distr.), Honshu (Japan Sea side from Mutsu to Echigo Prov.).

3. TRACHELOSPERMUM Lem. Teika-kazura Zoku

Scandent evergreen, woody or rarely herbaceous, glabrous or sometimes pilose; leaves opposite, penninerved; cymes loosely flowered, terminal and axillary; flowers white, often changing to yellow; calyx small, 5-parted, with 5 or 10 glands inside; corolla salverform, without scales on the throat, the lobes 5, spreading, falcate-obovate; stamens 5, on the tube within, the anthers acuminate, partially exserted from the throat, connate and adnate to the stigma; carpels 2, with a single elongate style; ovules many; follicles elongate, cylindric; seeds comose.——More than 10 species, chiefly in the warmer and subtropical regions of e. Asia, few in N. America.

1A. Stamens inserted on the middle of the corolla tube or slightly above on the corolla-tube; anthers wholly included within the tube; calyx rather large, the lobes broader above the middle, longer than the narrow part of the corolla-tube; corolla puberulent on the throat. .... 1. *T. jasminoides* var. *pubescens*
1B. Stamens inserted on the upper part of the tube; anthers slightly exserted from the throat; calyx small, the lobes broadest at the middle or on the lower half; corolla usually glabrous on the throat.
  2A. Corolla-tube 7–8 mm. long; calyx-lobes broadly lanceolate, 2–3 mm. long, usually glabrous. .... 2. *T. asiaticum*
  2B. Corolla-tube 6–7 mm. long; calyx-lobes ovate-orbicular, not more than 1 mm. long, white-ciliolate. .... 3. *T. liukiuense*

1. Trachelospermum jasminoides (Lindl.) Lem. var. **pubescens** Makino. *T. asiaticum* var. *pubescens* (Makino) Nakai——KE-TEIKA-KAZURA.   Scandent, woody evergreen, the inflorescence, young branches, and underside of leaves pubescent; leaves elliptic to narrowly so, 4–8 cm. long, 2–4.5 cm. wide, obtuse or acute, short-petiolate; flowers white, changing to yellow, 2–2.5 cm. across; calyx thin, loosely pilose, 5–6 mm. long, the lobes obtuse and spreading; corolla-tube 7–8 mm. long, the narrow part as long as the broadened upper portion or somewhat shorter, the lobes obliquely obtriangular; anthers included within the tube; style short, not exceeding the calyx; follicles 10–15 cm. long.——May. Honshu (centr. and w. distr.), Shikoku, Kyushu.——Korea, and China.
2. Trachelospermum asiaticum (Sieb. & Zucc.) Nakai. *Malouetia asiatica* Sieb. & Zucc.; *Parechites thunbergii* A. Gray; *Rhynchospermum japonicum* Sieb. ex Lavall.; *T. divaricatum* Kanitz; *T. crocostomum* Stapf; *T. majus* Nakai——TEIKA-KAZURA.   Scandent, glabrous or pubescent on young branches; leaves elliptic or narrowly so, sometimes broadly oblanceolate, 3–6(–8) cm. long, 1.5–3 cm. wide, acute with an obtuse tip, acute and short-petiolate at base, lustrous above; flowers white,

changing to yellow, 2–2.5 cm. across, fragrant; calyx 3–4 mm. long, about half as long as the narrow portion of the corolla-tube, the lobes narrowly ovate to broadly lanceolate, obtuse to acute, often white-puberulent, with an ascendingly recurved tip; corolla-tube 7–8 mm. long, the narrow portion about twice as long as the broadened portion, the lobes narrowly falcate-obtriangular; style elongate, about twice as long as the calyx; follicles rather slender, elongate.——May–June.   Thickets in hills and low mountains; Honshu, Shikoku, Kyushu; common; variable.——Korea.
3. Trachelospermum liukiuense Hatus. *T. divaricatum* var. *brevisepalum* C. K. Schn., pro parte——RYŪKYŪ-TEIKA-KAZURA, OKINAWA-TEIKA-KAZURA.   Nearly glabrous except the young branches; leaves obovate-oblong or elliptic, 4–6 cm. long, 1.5–3 cm. wide, acute with an obtuse tip, acute and petiolate at base; flowers white, about 1.5 cm. across; calyx about 2 mm. long, much shorter than the corolla-tube, the lobes ovate-orbicular or broadly ovate, obtuse, not more than 1 mm. long, white-ciliolate, suberect; corolla-tube 6–7 mm. long, the narrow portion 2–3 times as long as wide; style elongate, about 3 times as long as the calyx.——Kyushu (s. distr.).——Ryukyus.

4. ANODENDRON A. DC.   SAKAKI-KAZURA ZOKU

Scandent, evergreen, woody scrambler; leaves opposite, penninerved; inflorescence paniculate; flowers small, the bracts and bracteoles persistent; calyx 5-lobed, small, without glands or with 5 small glands inside; corolla salverform, 5-lobed, the tube inflated at base, the throat narrow, unappendaged; stamens inserted near base of corolla-tube, the filaments very short, the anthers connate and adnate to the stigma, short-appendaged at base; carpels 2, free, with a solitary very short style; ovules many; follicles divaricate, narrowed to the tip; seeds flat, beaked, comose.——Few species, in India, China, and Japan.

1. Anodendron affine (Hook. & Arn.) Druce. *Holarrhaena affinis* Hook. & Arn.; (?) *Aganosma laevis* Champ.; *Anodendron laeve* (Champ.) Maxim.; *Anodendron suishaense* Hayata——SAKAKI-KAZURA.   Glabrous woody climber; leaves coriaceous, oblanceolate to oblanceolate-oblong, 6–8(–10) cm. long, 1–2.5 cm. wide, acute with an obtuse tip, acute at base, entire, dark green above, pale green beneath, the petioles 5–10 mm. long; inflorescence depressed-pyramidal, short-pedunculate, the bracts and bracteoles narrowly deltoid, small; flowers

pale yellow; calyx about 3 mm. long, deeply 5-parted; corolla 8–10 mm. across, the tube 4–5 mm. long, about as long as the lobes, the lobes narrowly falcate-oblong, white papillose-hairy inside; follicles deltoid-lanceolate, broadened at base, horizontally spreading, 8–10 cm. long, 18 mm. across at base; seeds many, long comose.——May–June.   Honshu (warmer parts of Awa and Sagami Prov. and westw.), Shikoku, Kyushu.——Ryukyus, Formosa, China, and India.

5. NERIUM L.   KYŌCHIKU-TŌ ZOKU

Evergreen trees or large shrubs; leaves in 3's, rarely opposite or in 4's; inflorescence cymose, flowers short-pedicelled, 5-merous; calyx 5-parted, with numerous glands inside at base; corolla salverform or infundibuliform, large, 4–5 cm. across, the upper portion of the throat with many long hairlike scaly appendages; stamens inserted on the upper portion of the tube, the anthers connate, adnate to the stigma, with a long filiform appendage at the apex, callose-appendaged at the base; carpels free, with a long common style; follicles long and firm; seeds villous, comose.——Few species, in subtropical Asia to the Mediterranean.

1. Nerium indicum Mill. *N. odorum* Soland. ex Ait. ——KYŌCHIKU-TŌ.   Large shrub with green, rather stout, puberulous branches; leaves narrowly lanceolate, 7–15 cm. long, 8–20 mm. wide, acuminate at both ends; flowers 4–5 cm.

across, usually red, rose, or rarely yellowish, often double in cultivars; follicles linear, erect; seeds comose and long pale brown-villous.——Aug.   Indian plant long cultivated in our area.

Fam. 170. ASCLEPIADACEAE   GAGA-IMO KA   Milkweed Family

Perennial herbs or shrubs, often scandent, sometimes succulent; leaves opposite, verticillate, rarely alternate, simple, exstipulate; flowers cymose, bisexual, actinomorphic; calyx-lobes imbricate or open in bud; corolla 5-lobed, the lobes contorted or valvate in bud, with 5, sometimes 1 or many appendages (corona) adnate to the corolla-tube or to the stamens; stamens 5, the filaments free or coherent in a tube, the anthers 2-locular, adnate to the stigma and forming a gynostegium, the pollen coherent into waxy or granular masses (pollinia); ovary superior; carpels 2, with a marginal placenta; styles free, the stigma conical, sometimes discoid; fruit a follicle; seeds usually comose, the endosperm scanty.——About 220 genera, with about 2,000 species, in the Tropics and warmer regions, especially abundant in Africa.

1A. Pollinia pendulous on caudicles.
  2A. Corona short and depressed, much shorter than the gynostegium, the lobes alternate with the stamens; stigma long-beaked; cymes racemelike. .... 1. *Metaplexis*
  2B. Corona as long as to longer than the gynostegium; cymes usually umbellike. .... 2. *Cynanchum*
1B. Pollinia erect on caudicles.
  3A. Pollinia small, discoid. .... 3. *Tylophora*
  3B. Pollinia large, usually pyriform.
    4A. Leaves very thick, fleshy; inflorescence umbellate; corolla rotate. .... 4. *Hoya*
    4B. Leaves membranous to coriaceous; corolla-tube prominent.
      5A. Corolla campanulate. .... 5. *Marsdenia*
      5B. Corolla salverform. .... 6. *Stephanotis*

1. METAPLEXIS R. Br. GAGA-IMO ZOKU

Scandent herbs or subshrubs; leaves opposite, petiolate, cordate; cymes racemelike; flowers rather small; calyx 5-parted, with 5 glands inside at the base; corolla rotate-campanulate, deeply 5-lobed, the lobes narrow, contorted in bud, long-hairy inside; corona annular, the lobes 5, alternate with the stamens and inserted on the stamen-tube; stamens coherent into a short tube, the anthers with an incurved cordate membrane at the apex, the pollinia solitary in each anther-locule, ovoid-ellipsoidal, pendulous on the apex of a short caudicle; stigma entire or elongate to a long bifid beak; follicles rather thick; seeds comose.——Few species, in e. Siberia and e. Asia.

1. Metaplexis japonica (Thunb.) Makino. *Pergularia japonica* Thunb.; *M. stauntonii* Schult.; *Urostelma chinense* Bunge; *M. chinensis* Decne.——GAGA-IMO. Sparsely white-pubescent scandent perennial herb, woody at base; leaves ovate-cordate, 5–10 cm. long, 3–6 cm. wide, acute, abruptly cuspidate, cordate, glaucous beneath, the petiole shorter than the blade; racemes 2–5 cm. long, pedunculate, the bracts linear, shorter than the pedicel; calyx 4–5 mm. long, 5-parted nearly to the base, the segments deltoid-lanceolate, membranous, as long as to slightly shorter than the pedicels, acuminate; corolla pale purple or white, about 1 cm. across, broadly campanulate, deeply 5-lobed, the lobes lanceolate, recurved at tip, villous inside; follicles puberulent or glabrous, 8–10 cm. long, about 2 cm. across, broadly lanceolate; seeds flat, obovate, 6–8 mm. long, narrowly winged, with short, radiating striations on both surfaces.——Aug. Sunny slopes and thickets in hills and low mountains; Hokkaido, Honshu, Shikoku, Kyushu.——s. Kuriles, Korea, China, and Manchuria.

2. CYNANCHUM L. KAMOMEZURU ZOKU

Erect or scandent herbs, rarely subshrubs; leaves usually opposite, petiolate or sessile; cymes umbellike, sometimes racemelike, often branched, usually pedunculate; flowers small, white, greenish, or purplish; calyx 5-parted; corolla rotate or campanulate-rotate, deeply 5-lobed, the lobes contorted, rarely valvate in bud, the corona inserted at base of the stamen-tube, truncate to 5-parted, as long as to longer than the gynostegium, often with secondary inner lobes inside; anthers with a membranous appendage at apex, the pollinia solitary in each anther-locule, pendulous or sometimes horizontal on the caudicle; follicles thick or slender; seeds comose.——More than 100 species, in warm-temperate and subtropical regions.

1A. Leaves ovate-cordate; rhizomes long-creeping; corona-lobes highly adnate on the ventral side to the gynostegium; membranous appendage of the anthers relatively large and prominent.
  2A. Corolla-lobes slightly reflexed; corona-lobes oblong or narrowly obovate, with an appendage inside much longer than the gynostegium.
    1. *C. caudatum*
  2B. Corolla-lobes obliquely spreading; corona-lobes obovate-orbicular, slightly shorter than the gynostegium, without an appendage inside.
    2. *C. wilfordii*
1B. Leaves linear to narrowly deltoid-ovate, rarely cordate; rhizomes short; corona-lobes fleshy, as long as the gynostegium or much shorter; membranous appendage of anthers small, white at apex.
  3A. Plant erect; leaves linear to narrowly lanceolate, acuminate at both ends, the petioles 1–3 mm. long; flowers yellow-brown.
    3. *C. paniculatum*
  3B. Plants erect, with broad leaves, or more or less scandent; flowers white, purple-brown, or sometimes yellowish.
    4A. Leaves sessile, or short-petiolate and clasping at base.
      5A. Leaves glaucescent, glabrous, sessile. .... 4. *C. amplexicaule*
      5B. Leaves yellow-green, crisped-hairy on both surfaces, the petioles 3–4 mm. long. .... 5. *C. krameri*
    4B. Leaves usually distinctly petiolate, not clasping at base.
      6A. Leaves large, 3–15 cm. wide, of few pairs, more or less approximate toward the top of the stems or the stems more elongate and becoming scandent with remote pairs of smaller leaves.
        7A. Larger leaves ovate-orbicular to broadly elliptic, 7–15 cm. wide, rounded at base.
          8A. Corolla-lobes thinly white-pubescent inside; stems often elongate and scandent; follicles horizontally spreading to slightly deflexed. .... 7. *C. grandifolium*
          8B. Corolla-lobes glabrous on both sides; stems erect; follicles ascending. .... 8. *C. magnificum*

7B. Larger leaves oblong to narrowly so, 3–5 cm. wide; corolla-lobes glabrous on both sides. .... 9. *C. katoi*
6B. Leaves smaller and usually narrower, 1–7 cm. wide, of many pairs.
  9A. Stems erect from base, barely scandent at the tip.
    10A. Corolla-lobes thinly pubescent outside, glabrous inside; leaves densely soft-pubescent. .... 10. *C. atratum*
    10B. Corolla-lobes glabrous outside; leaves glabrous or slightly pubescent on nerves.
      11A. Leaves oblong, ovate, or elliptic, with few pairs of lateral nerves prominently arcuate at tip; corolla-lobes glabrous inside.
        12A. Leaves obtuse, mucronate. .... 11. *C. japonicum*
        12B. Leaves acuminate.
          13A. Inflorescence usually pedunculate, terminal; pedicels 1–3 cm. long; corolla about 15 mm. across. 12. *C. ascyrifolium*
          13B. Inflorescence sessile, axillary; pedicels 7–10 mm. long; corolla about 7 mm. across. .... 13. *C. inamoenum*
      11B. Leaves lanceolate, with 8–15 pairs of lateral nerves; corolla-lobes glabrous or puberulent inside. ...... 6. *C. multinerve*
  9B. Stems scandent from the base or on upper half.
    14A. Stems erect on lower half; lower leaves ovate-oblong to broadly lanceolate; flowers few.
      15A. Corolla-lobes glabrous on both surfaces; inflorescence nearly sessile; corona half as long as the gynostegium.
                                                                                                    14. *C. nipponicum*
      15B. Corolla-lobes puberulent, especially on inner side; inflorescence short-pedunculate; corona nearly as long as the
              gynostegium. .... 15. *C. doianum*
    14B. Stems scandent from near the base, much elongate; lower leaves linear to broadly ovate-cordate; flowers rather numerous.
      16A. Corolla yellow-green, the lobes white-puberulent inside; corona-lobes yellowish, deltoid, obtuse, about as long as the
              gynostegium; leaves 5–10 mm. wide, the petioles 3–8 mm. long. .... 16. *C. ambiguum*
      16B. Corolla yellowish to purple-brown, the lobes nearly glabrous.
        17A. Corona-lobes as long as the gynostegium, broadly ovate, yellowish; leaves narrowly ovate-deltoid, rather promi-
                nently puberulent beneath. .... 17. *C. austrokiusianum*
        17B. Corona-lobes much shorter than the gynostegium, usually purplish or brownish. .... 18. *C. sublanceolatum*

1. Cynanchum caudatum (Miq.) Maxim. *Endotropis caudata* Miq.; *C. ikema* (Sieb.) Ohwi——Ikema. Stems branched, scandent, minutely crisped-pubescent while young, glabrescent; leaves membranous, ovate-cordate, abruptly short-caudate, vivid green above, whitish beneath, the petiole shorter than the blade; cymes racemelike, many-flowered, the peduncles 2–2.5 times as long as the petioles, the pedicels slender, 1–2 cm. long, puberulent on inner side; flowers white; calyx-segments ovate, obtuse; corolla-lobes slightly reflexed, with revolute margins, narrowly oblong, 4–5 mm. long, obtuse, puberulent inside; corona yellow, the lobes highly adnate to the gynostegium for more than half the length of the corolla; follicles 8–10 cm. long, 8–10 mm. wide, usually solitary; seeds obcuneate-ovate, 7–8 mm. long, flat, winged, rounded and toothed on basal margin.——July–Aug. Hokkaido, Honshu, Shikoku, Kyushu.——s. Kuriles and China.

2. Cynanchum wilfordii (Maxim.) Hemsl. *Cynoctonum wilfordii* Maxim.; *Vincetoxicum wilfordii* (Maxim.) Fr. & Sav.——Ko-ikema. Much-branched, nearly glabrous scandent herb, puberulent while very young; leaves thick-membranous, broadly ovate, 5–10 cm. long, 4–8 cm. wide, shortly acuminate to abruptly acute, deeply cordate, the petiole shorter than the blade; flowers umbellate, the peduncles as long as to shorter than the petioles, 1–4 cm. long, the pedicels 5–8 mm. long; flowers pale yellow-green; calyx-lobes broadly lanceolate, acute; corolla-lobes broadly ovate, about 3 mm. long, rather thick, obliquely spreading, incurved on margin, puberulent inside; follicles lanceolate, about 8 cm. long, about 1 cm. wide.——July–Aug. Grassy hillsides; Honshu, Shikoku, Kyushu.——Korea and Manchuria.

3. Cynanchum paniculatum (Bunge) Kitag. *Asclepias paniculata* Bunge; *Pycnostelma chinense* Bunge ex Decne.; *P. paniculatum* (Bunge) K. Schum.; *Vincetoxicum pycnostelma* Kitag.——Suzu-saiko. Erect, glabrous perennial; stems nearly simple, 40–100 cm. long, loosely leaved; leaves rather thick, ascending, narrowly lanceolate to linear, 6–12 cm. long, 4–15 mm. wide, acuminate at both ends, deep green above, whitish beneath, slightly recurved and scabridulous on margin, the petioles 1–3 mm. long; peduncles axillary and terminal, slightly branched; cymes rather loosely flowered, the

pedicels slender; calyx-segments broadly lanceolate; corolla yellow-brown, the lobes narrowly ovate, 7–8 mm. long, glabrous; corona-lobes ovate, obtuse, erect, slightly shorter than the gynostegium; follicles narrowly lanceolate, 6–8 cm. long; seeds narrowly ovate, 4–5 mm. long, entire, narrowly winged, smooth.——July–Aug. Grassy slopes in hills and mountains; Hokkaido, Honshu, Shikoku, Kyushu; rather common.——Korea, Manchuria, China, and Dahuria.

4. Cynanchum amplexicaule (Sieb. & Zucc.) Hemsl. *Vincetoxicum amplexicaule* Sieb. & Zucc.; *V. brandtii* Fr. & Sav.; *C. brandtii* (Fr. & Sav.) Matsum.——Rokuon-sō. Erect, glabrous perennial; stems erect, sometimes branched, 40–100 cm. long; leaves rather firm, elliptic, ovate-oblong, or oblong, 4–8 cm. long, 2–4 cm. wide, rounded to subacute, mucronate, subauriculate and clasping at base, sessile, glaucous above, duller beneath; inflorescence as long as the leaves, pedunculate, sparsely branched, densely flowered, the pedicels rather short, 3–7 mm. long; calyx-segments narrowly ovate, sometimes ciliolate; corolla yellowish, the lobes narrowly ovate, about 3 mm. long; corona-lobes subrounded, short; follicles narrowly lanceolate, about 5 cm. long, 6–7 mm. wide.——July–Aug. Shikoku, Kyushu.——Korea, Manchuria, Mongolia, and China.——Forma **castaneum** (Makino) Ohwi. *C. amplexicaule* var. *castaneum* Makino——Kurobana-rokuon-sō. Flowers purplish brown.

5. Cynanchum krameri (Fr. & Sav.) Matsum. *Vincetoxicum krameri* Fr. & Sav.; *V. amplexicaule* var. *krameri* (Fr. & Sav.) Maxim.——Maruba-no-funabara-sō. Resembles the preceding; stems crisped pubescent; leaves obovate, rounded to emarginate, cordate-truncate at base, shortly crisped pubescent on both sides, the petioles 3–4 mm. long, the upper ones sessile; cymes umbellate, few-flowered, the peduncles not exceeding the leaves; calyx-segments densely pubescent, deltoid-lanceolate; corolla-lobes short-pilose on both surfaces, the pedicels as long as the flowers; corona as long as the gynostegium, the lobes ovate-orbicular, rather fleshy, pale brown.——Reported from Honshu (Nagato Prov.) and Kyushu (Hirato Isl.).

6. Cynanchum multinerve (Fr. & Sav.) Matsum. *Vincetoxicum multinerve* Fr. & Sav.——Hosoba-no-rokuon-sō.

Stems erect, about 30 cm. long, short-pubescent; leaves lanceolate, 9–10 cm. long, 2–3.5 cm. wide, acuminate, obtuse at base, short-pilose on both surfaces especially on nerves beneath, with 8–15 pairs of lateral nerves, the petioles 4–7 mm. long; cymes short-pedunculate, few-flowered, short-pubescent; flowers about 1.5 cm. across; calyx-segments glabrous to puberulent, lanceolate; corolla whitish, glabrous on both surfaces, the lobes oblong, rounded at apex; corona as long as the gynostegium, brown, the lobes ovate-orbicular.——Reported to occur in Honshu.

Var. **kiyohikoanum** (Honda) Ohwi. *C. kiyohikoanum* Honda——AKI-NO-KUSA-TACHIBANA. Flowers dark purple; corolla-lobes puberulent inside.——Honshu (Aki Prov.).

7. **Cynanchum grandifolium** Hemsl. *Vincetoxicum macrophyllum* Sieb. & Zucc.; *C. macrophyllum* (Sieb. & Zucc.) Matsum., non Thunb.; *C. kiusianum* Nakai; *C. lasiocarpum* Koidz.——TSUKUSHI-GASHIWA, Ō-TACHI-GASHIWA. Stems 50–100 cm. long, scandent, puberulent or nearly glabrous, with the largest leaves below the middle, those above remote and smaller; larger leaves thinly membranous, sparsely puberulent, ovate-orbicular to ovate or broadly elliptic, 12–25 cm. long, 10–15 cm. wide, abruptly acuminate, rounded to subacute at base, the petioles 3–6 cm. long; inflorescence axillary and terminal, usually short-pedunculate, much branched, rather large, puberulent, the pedicels 1–2 cm. long; calyx-segments lanceolate, acuminate; corolla purple-brown or dark purple, 8–10 mm. across, the lobes lanceolate, obtuse, white crisped-pubescent inside; follicles spreading to slightly deflexed, linear-lanceolate, usually puberulent, 5–8 cm. long, about 4 mm. wide; seeds broadly lanceolate, about 12 mm. long, flat, wingless, smooth.——Mountains; Honshu (w. distr.), Shikoku, Kyushu.

Var. **nikoense** (Maxim.) Ohwi. *Vincetoxicum macrophyllum* var. *nikoense* Maxim.; *C. nikoense* (Maxim.) Makino; *C. macrophyllum* var. *nikoense* (Maxim.) Matsum.——TSURU-GASHIWA. Inflorescence short, rather few-flowered; follicles more puberulent.——Woods in mountains; Honshu, Shikoku.

8. **Cynanchum magnificum** Nakai. *C. nikoense* Makino, excl. syn.——TACHI-GASHIWA. Closely resembles the preceding species; stems 30–60 cm. long, erect, usually loosely puberulent, with few pairs of larger leaves, close together near the top; larger leaves ovate-orbicular to rhombic-elliptic, or broadly elliptic, 10–15 cm. long, 7–13 cm. wide, abruptly acute or shortly acuminate, the petioles 2–5 cm. long; inflorescence near the top of the stems, rather densely flowered, small, puberulent, the pedicels 1–2 cm. long; corolla-lobes lanceolate, glabrous on both sides, obtuse, about 5 mm. long; corona small, the lobes semirounded; follicles narrowly lanceolate, 5–7 cm. long, 5–6 mm. wide, nearly glabrous, ascending to suberect.——May–June. Woods in mountains; Honshu, Shikoku.

9. **Cynanchum katoi** Ohwi. KUSANAGI-OGOKE. FIG. 13. Rather slender, thinly puberulent perennial; stems 30–100 cm. long, usually becoming somewhat scandent; largest leaves thinly membranous, of few pairs close together below the middle of the stem, narrowly elliptic or oblong, 7–15 cm. long, 3–5 cm. wide, acuminate, acute to acuminate at base, the petioles 7–12 mm. long, the upper smaller leaves of few pairs; inflorescence loosely rather many-flowered, rather large, branched, glabrous or nearly so, the pedicels 5–18 mm. long, filiform; calyx-segments lanceolate; corolla purplish, 7–9 mm. across, the lobes oblong, glabrous on both sides; corona-lobes ovate-deltoid, obtuse, small; follicles lanceolate, 4–5 cm. long, 7–8 mm. wide;

seeds oblong, 6–7 mm. long, obsoletely dark-spotted.——June. Honshu (Tōkaidō and e. Kinki Distr.); rather rare.——Forma **albescens** (Hara) Ohwi. *C. katoi* var. *albescens* Hara——SHIROBANA-KUSANAGI-OGOKE. Flowers white.

10. **Cynanchum atratum** Bunge. *Vincetoxicum atratum* (Bunge) Morr. & Decne. ex Decne.——FUNABARA-SŌ, ROKUON-SŌ. White-pubescent, especially on the inflorescence, upper part of stems, and lower side of leaves; stems erect, simple, 40–80 cm. long, rather thick; leaves elliptic, broadly ovate or ovate-orbicular, 6–10(–15) cm. long, 3–7(–10) cm. wide, acute to abruptly so, or sometimes rounded and mucronate, the petioles 8–12 mm. long; inflorescence umbellate, in the upper axils, densely flowered, sessile or very short-pedunculate, the pedicels shorter than the flowers; calyx-segments lanceolate, slightly shorter than the corolla-tube; corolla 5-lobed, thinly puberulent outside, dark purple and glabrous inside, the lobes ovate-oblong, 6–8 mm. long, obtuse; corona-lobes elliptic, rounded, nearly as long as the gynostegium; follicles broadly lanceolate, densely puberulent, 7–8 cm. long, about 1.5 cm. wide.——May–June. Mountains; Hokkaido, Honshu, Shikoku, Kyushu.——Korea, Manchuria, and China. ——Forma **viridescens** Hara. AO-FUNABARA-SŌ. Flowers green.

Cynanchum × sakaianum Honda. MIYAMA-FUNABARA-SŌ. Hybrid of *C. ascyrifolium* × *C. atratum*.

11. **Cynanchum japonicum** Morr. & Decne. *Vincetoxicum japonicum* Morr. & Decne., ex Decne.; *Vincetoxicum japonicum* var. *verum* Maxim.; *V. japonicum* var. *grayanum* Maxim.; *C. grayanum* (Maxim.) Koidz.——IYO-KAZURA, SUZUME-NO-OGOKE. White crisped-hairy on the stems, the nerves of both leaf surfaces, and the inflorescence; stems simple or sparingly branched, often somewhat scandent, 30–80 cm. long; leaves elliptic, elliptic-ovate, or obovate, 3–10 cm. long, 2–7 cm. wide, rounded to obtuse and often mucronate, rounded at base, the petioles 3–10 mm. long; inflorescence umbellate, sometimes sparingly branched, pedunculate or subsessile, rather densely flowered, the pedicels 5–10 mm. long; calyx-segments broadly lanceolate; corolla about 8 mm. across, deeply 5-lobed, creamy, the tube shorter than the calyx, the lobes narrowly ovate, 4–5 mm. long, glabrous on both sides; corona-lobes obovoid-orbicular, nearly as long as the gynostegium; follicles broadly lanceolate, 5–6 cm. long, about 1.5 cm. across, glabrous; seeds broadly ovate, 8–10 mm. long, narrowly winged.——May–July. Thickets and grassy slopes near seashores; Honshu, Shikoku, Kyushu; rather common.——Korea and China.

Var. **albiflorum** (Fr. & Sav.) Hara. *Vincetoxicum purpurascens* var. *albiflorum* (Fr. & Sav.) Hara; *C. purpurascens* var. *albiflorum* (Fr. & Sav.) Matsum.; *C. albiflorum* (Fr. & Sav.) Koidz.——SHIROBANA-KUSA-TACHIBANA, BIRŌDO-IYO-KAZURA. Leaves thinner, abruptly acute, obtuse to rounded at base.——Occurs with the typical phase.——Forma **puncticulatum** (Koidz.) Ohwi. *C. puncticulatum* Koidz.; *C. japonicum* var. *puncticulatum* (Koidz.) Hara; *Vincetoxicum rubellum* Fr. & Sav.; *C. rubellum* (Fr. & Sav.) Matsum.——KUROBANA-IYO-KAZURA. Flowers purple.

Cynanchum purpurascens Morr. & Decne. *Vincetoxicum japonicum* var. *purpurascens* (Morr. & Decne.) Maxim., partim.——MURASAKI-SUZUME-NO-OGOKE. Leaves narrower, acuminate to cuspidate; flowers on rather long pedicels; calyx loosely pilose.——Described from a cultivated plant. Allied to *C. japonicum*.

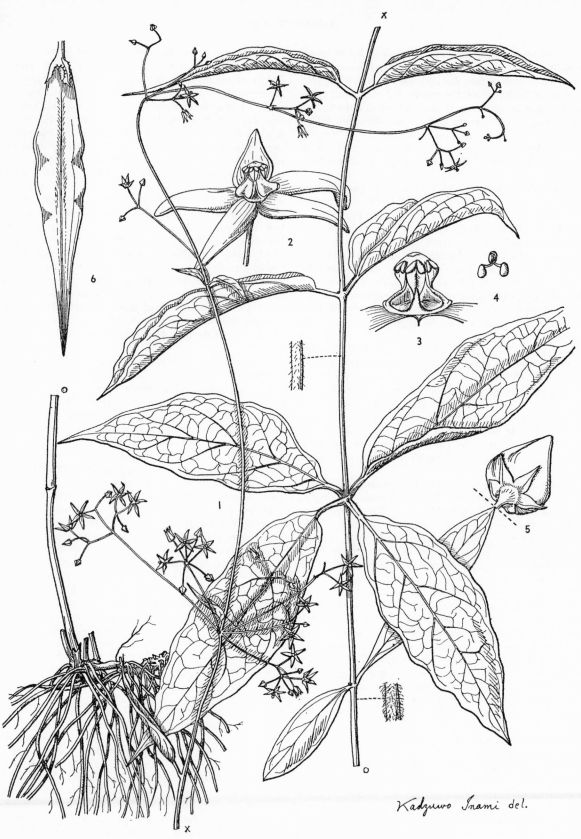

Kadzuwo Inami del.

Fig. 13.—*Cynanchum katoi* Ohwi. 1, Habit; 2, flower; 3, gynostegium; 4, pollinia; 5, flower-bud; 6, follicle.

12. Cynanchum ascyrifolium (Fr. & Sav.) Matsum. *Vincetoxicum acuminatum* Decne.; *V. ascyrifolium* Fr. & Sav.; *C. acuminatifolium* Hemsl.; *C. acuminatum* (Decne.) Matsum., non Thunb.——KUSA-TACHIBANA. Sparsely crisped-hairy on the inflorescence and nerves of leaves beneath; stems erect, 30–60 cm. long, with several pairs of leaves; leaves membranous to firmly so, green above, paler beneath, oblong, elliptic, or ovate, 8–15 cm. long, 4–8 cm. wide, acuminate to short-acuminate, the petioles 1–2 cm. long; inflorescence terminal and in upper axils, pedunculate, few-branched, the pedicels 1–3 cm. long; calyx-segments broadly lanceolate; corolla white, about 15 mm. across, the lobes narrowly ovate, 8–10 mm. long, glabrous on both surfaces; corona-lobes ovate-deltoid, slightly shorter than the gynostegium; follicles broadly lanceolate, glabrous, 4–6 cm. long, about 8 mm. wide; seeds broadly ovate, about 7 mm. long, entire, scarcely winged.——June. Honshu (Kantō Distr. and westw.), Shikoku.——Korea and Manchuria.

13. Cynanchum inamoenum (Maxim.) Loes. *Vincetoxicum inamoenum* Maxim.——EZO-NO-KUSA-TACHIBANA. Short-pubescent; stems erect, 30–50 cm. long; leaves membranous, of few pairs, elliptic or narrowly so, to ovate, 6–10 cm. long, 3–6 cm. wide, short-acuminate or acute, rounded to broadly cuneate and obtuse at base, the petioles 7–10 mm. long; inflorescence few-flowered, umbellate, sessile or very short-pedunculate; flowers small, yellowish, about 7 mm. across; corolla deeply 5-lobed, glabrous; corona-lobes nearly as long as the gynostegium, broadly deltoid; follicles obliquely spreading, lanceolate, glabrous, 4–5 cm. long, about 5 mm. wide; seeds broadly ovate, winged, about 5 mm. long.——July. Hokkaido.——Sakhalin, Korea, Manchuria, and n. China.

14. Cynanchum nipponicum Matsum. Ō-AO-KAMOME-ZURU. Somewhat crisped-hairy; stems slender, erect on lower half, much elongate and scandent, 40–100 cm. long; leaves of the lower erect portion larger, membranous, broadly lanceolate to narrowly oblong-ovate, 5–12 cm. long, 1–2(–3) cm. wide, gradually acuminate, rounded to shallowly cordate-truncate at base, the petioles 2–4 mm. long; inflorescence in upper axils; flowers yellow, 7–8 mm. across, few, in umbels, the peduncles nearly absent, the pedicels 3–4 mm. long, usually shorter than the flowers; corolla-lobes narrowly ovate, glabrous on both surfaces, obtuse; corona-lobes depressed-deltoid, about half as long as the gynostegium; follicles broadly lanceolate, glabrous, 4–5 cm. long, 5–7 mm. wide; seeds ovate-elliptic, about 5 mm. long, winged.——July–Sept. Mountains; Honshu (centr. and n. distr.).——Korea.——Forma **abukumense** (Koidz.) Hara. *C. abukumense* Koidz.——NAGABA-KURO-KA-MOMEZURU. Flowers purple.

Var. **glabrum** (Nakai) Hara. *C. glabrum* Nakai——TACHI-KAMOMEZURU, KUROBANA-KAMOMEZURU. Leaves thicker; corona-lobes ovate-deltoid, slightly shorter than the gynostegium.——Honshu (Kinki Distr. and westw.), Shikoku, Kyushu.——Korea.

15. Cynanchum doianum Koidz. SATSUMA-BYAKUZEN. Crisped-hairy perennial; stems scandent, loosely leaved; leaves membranous, narrowly oblong to oblong-lanceolate, to 10 cm. long, 4.5 cm. wide, puberulent on both surfaces, the petioles to 5 mm. long; inflorescence pedunculate, umbellate; flowers about 11 mm. across, yellowish; corolla-lobes lanceolate, puberulent on both sides especially on the inner surface, obtuse; corona-lobes as long as the gynostegium.——June–Aug. Kyushu (s. distr.).

16. Cynanchum ambiguum (Maxim.) Matsum. *Vincetoxicum ambiguum* Maxim.——AO-KAMOMEZURU. Sparsely crisped-puberulent perennial herb; stems long-scandent and branched; leaves rather thick-membranous, broadly- to linear-lanceolate, 3–10 cm. long, 5–10 mm. wide, acuminate, rounded to truncate at base, the petioles 3–8 mm. long; inflorescence umbellike, rather few-flowered, on axillary peduncles about 1 cm. long, the pedicels 5–10 mm. long; flowers yellow-green, 7–8 mm. across; corolla-lobes deltoid-lanceolate, 3–4 mm. long, obtuse, glabrous outside, short white-puberulent inside; corona-lobes deltoid, yellowish, obtuse, nearly as long as the gynostegium; follicles narrowly lanceolate, glabrous, 4–5 cm. long, 5–6 mm. wide.——Aug.–Oct. Honshu (Yamato Prov.), Shikoku, Kyushu.

17. Cynanchum austrokiusianum Koidz. NANGOKU-KAMOMEZURU. Perennial with short crisped hairs, especially on the leaves beneath; stems scandent, branched, slender; leaves membranous, narrowly ovate-deltoid or broadly ovate, 5–9 cm. long, 2–4 cm. wide, gradually acuminate, shallowly cordate at base, the petioles 1–3 cm. long; inflorescence umbellike, the peduncles 0–2 cm. long, the pedicels slender, 1–1.5 cm. long; flowers white, 10–15 mm. across; corolla-lobes lanceolate, gradually narrowed to the tip, obtuse, 7–8 mm. long; corona-lobes broadly ovate, obtuse, yellowish, as long as the gynostegium; follicles broadly lanceolate, 5–6 cm. long, 7–8 mm. wide, glabrous; seeds broadly ovate, 6–7 mm. long, narrowly winged, with dark brown spots.——Aug.–Sept. Kyushu (s. distr.).

18. Cynanchum sublanceolatum (Miq.) Matsum. var. **sublanceolatum.** *Tylophora sublanceolata* Miq.; *C. sublanceolatum* (Miq.) Matsum.; *Vincetoxicum sublanceolatum* var. *dickinsii* Fr. & Sav.; *C. macranthum* var. *dickinsii* (Fr. & Sav.) Ohwi; *C. dickinsii* (Fr. & Sav.) Nakai——KOBA-NO-KA-MOMEZURU. Perennial with more or less curved hairs; stems slender, elongate and branched, scandent, loosely leafy; lower leaves lanceolate to broadly lanceolate, 4–5 cm. long, 1–2 cm. wide, gradually acuminate, shallowly cordate to sometimes rounded at base, the petioles 7–15 mm. long; upper leaves 1–5 cm. long, 5–15 mm. wide, usually rounded at base, the petioles 3–10 mm. long; inflorescence umbellike, simple or sometimes few-branched, nearly sessile or short-pedunculate, the pedicels 7–10 mm. long; flowers 6–8 mm. across, dark purple; corolla-lobes lanceolate, gradually narrowed and obtuse, minutely puberulent inside at base; corona-lobes about half as long as the gynostegium, ovate-deltoid; follicles usually solitary, lanceolate, 5–7 cm. long, 6–7 mm. wide; seeds broadly ovate, very narrowly winged, about 6 mm. long.——July–Sept. Thickets in hills and low mountains; Honshu (Kantō Distr. westw. to Kinki).

Var. **albiflorum** (Fr. & Sav.) Hara. *Tylophora japonica* var. *albiflora* Fr. & Sav.; *Vincetoxicum sublanceolatum* var. *albida* Fr. & Sav.; *C. sublanceolatum* var. *albidum* (Fr. & Sav.) Matsum.; *C. albidum* (Fr. & Sav.) Koidz.——AZUMA-KAMO-MEZURU. Flowers white.——Occurs with the typical phase.

Var. **macranthum** (Maxim.) Matsum. *Vincetoxicum sublanceolatum* var. *macranthum* Maxim.; *C. franchetii* Nakai; *Tylophora macrantha* (Maxim.) Makino; *C. matsudanum* Koidz.——SHIROBANA-KAMOMEZURU, ŌBANA-KAMOMEZURU. Slightly larger in every respect; flowers 10–20 mm. across, light yellow.——Occurs with the typical phase.

Var. **auriculatum** (Fr. & Sav.) Matsum. *Vincetoxicum sublanceolatum* var. *auriculatum* Fr. & Sav.; *C. harunamontanum* Honda; *C. retropilum* Nakai——JŌSHŪ-KAMOMEZURU.

Petioles in lower leaves 8–15 mm. long; flowers dark purple; corona-lobes rather depressed.——Aug. Honshu (centr. distr. and Kantō).

3. TYLOPHORA R. Br.   Ō-KAMOMEZURU ZOKU

Herbs, sometimes subshrubs, usually scandent, rarely suberect; leaves opposite, entire; cymes branched or rarely umbellike, sometimes racemelike; flowers small; calyx deeply 5-lobed or parted, the lobes acute; corolla rotate, deeply 5-lobed, the lobes contorted or nearly valvate in bud; corona-lobes 5, fleshy, forming a connate annulus at base, adnate to the stamens; stamens inserted on the corolla-tube or on the throat, the filaments short, connate, the anthers erect, short, with an incurved membrane at apex, the pollinia solitary in each locule, discoid, globose to ovoid, erect or laterally attached to a slender horizontally spreading caudicle; follicles lanceolate, acuminate, smooth; seeds comose.——Many species in se. and e. Asia, the Pacific Islands, and Australia, few in Africa. In habit scarcely distinguishable in our area from *Cynanchum* except by the more strongly scandent branches and generally looser inflorescence.

1A. Leaves membranous, more or less cordate at base.
  2A. Corona-lobes much shorter than the gynostegium, erect. . . . . 1. *T. nikoensis*
  2B. Corona-lobes at least half as long as the gynostegium, radially spreading.
    3A. Leaves rather small, 3–7 cm. long, acute; inflorescence diffuse and much branched; corolla-lobes glabrous inside. . . 2. *T. floribunda*
    3B. Leaves rather larger, 7–12 cm. long, acuminate; inflorescence small, often umbellike; corolla-lobes puberulent inside, at least on
      upper part. . . . . 3. *T. aristolochioides*
1B. Leaves coriaceous or subcoriaceous, obtuse or rounded at base.
  4A. Leaves elliptic or ovate-elliptic, prominently short-pubescent beneath; inflorescence rather umbellike, slightly branched.
                                          4. *T. tanakae*
  4B. Leaves lanceolate, glabrous except on midrib above; inflorescence much branched. . . . . 5. *T. japonica*

1. Tylophora nikoensis (Fr. & Sav.) Matsum. *Vincetoxicum nikoense* Fr. & Sav.——KO-KAMOMEZURU. Scandent nearly glabrous or thinly puberulent perennial herb; stems slender, elongate, branched; lower leaves membranous, narrowly deltoid-ovate or narrowly ovate, 4–7 cm. long, 1.2–2.5 cm. wide, smaller in the upper ones, acuminate, shallowly cordate with an open sinus at base, glabrate to thinly puberulent, the petioles 5–15 mm. long; inflorescence branched, loosely many-flowered, longer than to nearly as long as the leaves, nearly glabrous, the pedicels 5–10 mm. long; calyx-lobes small, ovate-deltoid; corolla dark purple, 4–5 mm. across, the lobes spreading, narrowly ovate; corona-lobes shorter than the gynostegium; follicles lanceolate, glabrous, 4–5 cm. long, 4–5 mm. wide; seeds ovate, 4–5 mm. long.——July–Aug. Thickets in lowlands and mountains; Honshu, Shikoku, Kyushu.

2. Tylophora floribunda Miq. *Vincetoxicum floribundum* (Miq.) Fr. & Sav.; *T. sikokiana* Matsum.——TOSA-KAMOMEZURU. Scandent herb with short rhizomes, glabrous or nearly so, crisped-hairy on nerves of the upper leaf surface; stems slender, elongate, much branched; lower leaves narrowly oblong-ovate to broadly lanceolate, 3–7 cm. long, 1–3 cm. wide, abruptly acuminate, cordate at base, puberulent on upper side, the petioles 1–2 cm. long; inflorescence diffuse and much branched, slender, loosely many-flowered, usually longer than the leaves, glabrous, the pedicels 4–7 mm. long; calyx-segments ovate-deltoid, small; corolla dark purplish, 3–4 mm. across, the lobes glabrous, narrowly ovate; corona-lobes depressed-globose, spreading, about half as long as the gynostegium; follicles horizontally spreading, narrowly lanceolate, about 5 cm. long, glabrous.——Aug. Honshu (Chūgoku Distr.), Shikoku, Kyushu.——Korea and China.

3. Tylophora aristolochioides Miq. *Vincetoxicum aristolochioides* (Miq.) Fr. & Sav.; *T. yoshinagae* Makino——Ō-KAMOMEZURU. Much-branched, slender, scandent perennial, nearly glabrous except for fine crispate hairs on nerves of upper leaf surface while young; larger leaves membranous,

deltoid-ovate to broadly deltoid-lanceolate, 7–12 cm. long, 2–5 cm. wide, long-acuminate, shallowly cordate at base, often obscurely auriculate, puberulent on upper side, the petioles 1.5–3 cm. long; upper leaves smaller; inflorescence small, slightly branched, umbellike, usually shorter to slightly longer than the petiole, glabrous, the pedicels 3–10 mm. long; calyx-segments deltoid, small; corolla dark purplish, 5–7 mm. across, the lobes narrowly ovate, obtuse, white-pubescent inside; corona-lobes radially spreading, obovate, obtuse, shorter than the gynostegium; follicles horizontally spreading, narrowly lanceolate, 4–6 cm. long, 4–5 mm. wide, glabrous; seeds narrowly ovate, narrowly margined, about 6 mm. long.——July. Mountains; Hokkaido, Honshu, Shikoku, Kyushu; rather common.

4. Tylophora tanakae Maxim. *Vincetoxicum tanakae* (Maxim.) Fr. & Sav.——TSURU-MŌRINKA. Scandent much-branched slender perennial herb, rather densely yellowish short-crispate and soft pubescent on stems and lower leaf surfaces; leaves subcoriaceous, broadly ovate-elliptic, elliptic, or ovate, 4–7 cm. long, 2–4.5 cm. wide, rounded to obtuse, rounded at base, glabrate and lustrous above, paler beneath, the petioles 5–12 mm. long; inflorescence pedunculate or subsessile, many-flowered, rather umbellike, slightly branched, nearly as long as the leaves, the pedicels slender, nearly glabrous, 7–10 mm. long; calyx-segments small, ovate; corolla 6–7 mm. across, the lobes narrowly deltoid-ovate, glabrous; corona-lobes erect, ovate-deltoid, about half as long as the gynostegium; follicles horizontally spreading, lanceolate, often loosely pubescent, sometimes glabrous, 5–6 cm. long, 8–10 mm. wide; seeds broadly ovate, narrowly winged, about 6 mm. long.——Aug. Honshu (Izu Prov.), Kyushu.——Ryukyus.

5. Tylophora japonica Miq. *Vincetoxicum sieboldii* Fr. & Sav.; *T. liukiuensis* Matsum.; *Henrya augustiniana* forma *liukiuensis* (Matsum.) Makino; *H. liukiuensis* (Matsum.) Koidz.; *Cynanchum sieboldii* (Fr. & Sav.) Ohwi——TOKIWA-KAMOMEZURU. Nearly glabrous scandent herb with firm slender striate branches; leaves thick and rather firm, lanceolate to broadly so, 5–8 cm. long, 1–2(–2.5) cm. wide, gradually acuminate, rarely acute, sometimes obtuse at base, glabrous except on midrib above, deep green and slightly lustrous above, pale green beneath, the petioles 5–12 mm. long; inflorescence much branched, as long as to slightly longer than the leaves, the pedicels 3–5 mm. long; calyx-segments ovate, slightly pilose; corolla 7–8 mm. across, purplish, the lobes ovate; corona-lobes orbicular, about half as long as the gynostegium; follicles narrowly lanceolate, about 9 cm. long.——June–July. Warmer regions; Shikoku, Kyushu.——Ryukyus.

4. HOYA R. Br. SAKURA-RAN ZOKU

Scandent herbs or subshrubs, often creeping on rocks or tree-trunks; leaves opposite, fleshy, coriaceous; inflorescence in umbellate cymes, sessile or pedunculate, axillary; flowers relatively large, rarely small; calyx small, deeply 5-lobed; corolla rotate, deeply 5-lobed, the lobes spreading or reflexed, valvate in bud; corona-lobes 5, fleshy, adnate to the stamen-tube, radially spreading; stamens inserted at the base of the corolla, the filaments short, connate, the anthers adhering around the stigma, with an erect or incurved membrane at apex, pollinia solitary in each locule, pyriform, erect; fruit a follicle.——About 100 species, in the warmer regions of e. and s. Asia, Malaysia, and Australia.

1. **Hoya carnosa** (L. f.) R. Br. *Asclepias carnosa* L. f.; *H. roundifola* Sieb.; *H. motoskei* Teijsm. & Binn.; *H. carnosa* var. *japonica* Sieb. ex Maxim.——SAKURA-RAN. Branches rather thick, creeping, glabrous; leaves subcoriaceous, elliptic, 5–10 cm. long, 3–5 cm. wide, rounded to obtuse, mucronate, obtuse at base, fleshy, lustrous, pilose beneath, petiolate; inflorescence umbellate, densely flowered, the peduncles axillary, thick, short; flowers 12–15 mm. across, pale rose, on rather slender elongate pedicels; corolla-lobes deltoid-ovate, obtuse, densely short-papillose, hairy on inside; corona-lobes short, narrowly deltoid, horizontally spreading, shining; follicles linear, 10–14 cm. long, 6–7 mm. across; seeds oblanceolate, 4–5 mm. long, slightly contracted at apex, comose.——June–Oct. Kyushu.——Ryukyus, Formosa, and tropical Asia.

5. MARSDENIA R. Br. KIJO-RAN ZOKU

Shrubs or subshrubs, usually scandent, rarely suberect; leaves opposite; cymes umbellike or somewhat branched, terminal and axillary, pedunculate; calyx 5-parted, the lobes obtuse; corolla campanulate or infundibuliform, rarely rotate, 5-lobed, the throat constricted, sometimes hairy inside, the lobes contorted in bud, obtuse; corona-lobes 5, erect; anthers with a small membrane at apex, the pollinia erect, pyriform; follicles often fleshy; seeds comose.——About 70 species, tropical and subtropical.

1. **Marsdenia tomentosa** Morr. & Decne. *M. acuta* T. Tanaka, excl. syn.——KIJO-RAN. Woody evergreen climber with rather thick elongate branches; leaves petiolate, orbicular, 7–12 cm. in diameter, abruptly acute, truncate-rounded or shallowly cordate at base; inflorescence axillary, short-pedunculate, shorter or as long as the petioles, 3–6 cm. long, densely pilose; flowers white, about 5 mm. across; corolla campanulate, hairy on the throat; follicles 13–15 cm. long, rather thick.——Aug.–Sept. Warmer regions; Honshu (Kantō Distr. and westw.), Shikoku, Kyushu.——Ryukyus and s. Korea.

6. STEPHANOTIS Thouars SHITAKI-SŌ ZOKU

Glabrous scandent shrubs; leaves opposite, coriaceous; cymes umbellike, axillary, short-pedunculate; flowers rather large, white; calyx 5-parted, the segments somewhat foliaceous, lanceolate; corolla salverform or subinfundibuliform, the tube long, the lobes contorted in bud; corona-lobes 5, erect, sometimes obsolete, inserted at the back of the stamens; anthers erect, with a small membrane at apex, the pollinia solitary in each locule, erect on short caudicles; follicles rather thick; seeds comose.——About 15 species, in e. Asia, Malaysia, and Madagascar.

1. **Stephanotis lutchuensis** Koidz. var. **japonica** (Makino) Hatus. *S. japonica* Makino——SHITAKI-SŌ. Much-elongate woody climber with yellowish pubescent young branches; leaves thick-membranous, oblong to broadly ovate, 6–12 cm. long, 3–6 cm. wide, acute to abruptly acuminate, rounded to shallowly cordate at base, with sparse persistent crispate hairs on nerves on both surfaces, the petioles 1.5–2.5 (–3) cm. long; peduncles axillary, shorter than to as long as the petioles, 2- or 3-flowered at the top, the pedicels longer than the peduncle, shorter than the flowers; flowers white, fragrant; calyx-segments 1/2–3/4 as long as the corolla-tube, rounded to subacute; corolla-tube 12–14 mm. long, hairy inside, the lobes lanceolate, obliquely spreading, subobtuse, about 20 mm. long.——June. Honshu (Awa Prov. and westw.), Shikoku, Kyushu.

Var. **lutchuensis**. *S. lutchuensis* Koidz. OKINAWA-SHITAKI-ZURU. Resembles the preceding but smaller; leaves oblong or ovate, 6–10 cm. long, 3–6 cm. wide, usually rounded at base, the petioles 1.5–2 cm. long; peduncles axillary, shorter than the petioles, the pedicels 2–5, about 1.5 cm. long, nearly as long as the flowers; flowers white, fragrant; corolla-tube 7–8 mm. long, hairy inside, to twice as long as the calyx, the lobes broadly lanceolate, slightly reflexed, subobtuse, about 10 mm. long.——June–July. Kyushu (s. distr.).——Ryukyus.

Fam. 171. **CONVOLVULACEAE** HIRUGAO KA Morning-glory Family

Herbs or often scandent woody plants, often with milky juice; leaves simple, alternate, exstipulate; flowers bisexual, actinomorphic, the bracts often involucrelike; calyx-segments usually free, imbricate in bud, persistent; corolla usually infundibuliform, the lobes contorted in bud; stamens inserted on the lower part of the corolla and alternate with the lobes, the anthers 2-locular, longitudinally dehiscent; ovary often surrounded by a disc at base, 1- to 4-locular; ovules 1 or 2 in each locule, erect; style erect, terminal; fruit dehiscent and capsular or indehiscent and fleshy; seeds often hairy, with scanty endosperm; cotyledons infolded or wrinkled.——About 50 genera, with more than 1,200 species, chiefly tropical and subtropical, abundant in Asia and America.

1A. Plants green, with chlorophyll, nonparasitic; stamens without scales at base; leaves well developed.
 2A. Ovary 2-lobed; styles 2; small creeping herbs; corolla broadly campanulate; flowers solitary in axils, small. . . . . 1. *Dichondra*
 2B. Ovary unlobed; style 1, sometimes very short.
  3A. Scandent herbs, sometimes slightly woody at base; leaves orbicular to ovate-cordate; flowers 2 cm. or more long; style elongate; fruit a capsule.
   4A. Bracts small, not covering the calyx; ovary 2- to 4-locular. . . . . 2. *Ipomoea*
   4B. Bracts large, covering the calyx; ovary 1-locular. . . . . 3. *Calystegia*
  3B. Scandent evergreen shrubs; leaves elliptic to oblong; flowers 10 mm. long; style nearly absent; fruit a 1-seeded berry. . . 4. *Erycibe*
1B. Plants not green, without chlorophyll, parasitic; stamens with small scales at base; leaves reduced to scales. . . . . 5. *Cuscuta*

1. DICHONDRA Forst. AOIGOKE ZOKU

Delicate creeping herbs; leaves cordate-orbicular or reniform, small; flowers axillary, solitary, small; calyx-segments nearly equal, often spathulate; corolla broadly campanulate, deeply 5-lobed, the lobes induplicate in bud; filaments filiform, the anthers small; ovary 2-lobed, the lobes usually 2-locular and 2-ovuled; styles 2, filiform, the stigma capitate; capsules membranous, usually densely pubescent, 2-parted, erect, each lobe 1(–2)-seeded; seeds subglobose, smooth; cotyledons oblong-linear.——Few species in Centr. and S. America, one widely distributed in tropical and subtropical regions.

1. **Dichondra repens** Forst. *Sibthorpia evolvulacea* L. f.; *Dichondra micrantha* Urb.; *Dicrondra evolvulacea* (L. f.) Britt.——AOIGOKE. Thinly yellowish pubescent or nearly glabrous; stems very slender, creeping, matted; leaves loosely arranged, reniform- to orbicular-cordate, 5–10 mm. long, 8–20 mm. wide, rounded to retuse, deeply cordate, the petioles slender, 1–4 cm. long; flowers sessile or on short delicate peduncles shorter than the petioles, yellow, about 3 mm. across; calyx-segments or sepals obovate-oblong; corolla-lobes oblong, slightly longer than the calyx; lobes of capsules globose, about as long as the calyx, loosely long-hairy.——Apr.–Aug. Around dwellings and along roadsides in lowlands; Honshu (warmer parts of w. distr.), Shikoku, Kyushu.——s. Korea, Ryukyus, Formosa, and the Tropics generally.

2. IPOMOEA L. SATSUMA-IMO ZOKU

Herbs, rarely shrubs, usually twining; leaves alternate, entire or palmately lobed; inflorescence cymose, often capitate, sometimes in terminal panicles, the bracts persistent or caducous, the peduncles axillary, 1- to many-flowered; flowers purple, red, or white, rarely yellow; calyx-segments sometimes unequal; corolla infundibuliform or campanulate, the limb 5-plicate in bud, entire, rarely deeply 5-lobed; stamens not exserted, the filaments filiform or dilated below, sometimes hairy; ovary 2- or 4-locular and 4-ovuled, or 3-locular and 6-ovuled; style filiform, the stigma globose or 2-lobed; capsules globose; seeds glabrous or pilose; cotyledons broad, often bifid.——About 400 species, in warmer areas of the world.

1A. Calyx-segments elliptic to oblong, obtuse, mucronate; ovary 2-locular, 4-ovuled; leaves orbicular, obovate-orbicular to elliptic, retuse to bilobed, glabrous. . . . . 1. *I. pes-caprae*
1B. Calyx-segments lanceolate, gradually narrowed to tip; ovary 3-locular, 6-ovuled; leaves cordate, abruptly acuminate, scattered-pilose.
                                2. *I. indica*

1. **Ipomoea pes-caprae** (L.) Sweet. *Convolvulus pes-caprae* L.; *I. biloba* Forsk.; *I. pes-caprae* subsp. *brasiliensis* (L.) Oostst.; *C. brasiliensis* L.——GUMBAI-HIRUGAO. Glabrous creeping perennial, the branches terete; leaves rather thick, long-petiolate, orbicular, obovate-orbicular to elliptic, sometimes orbicular-cuneate, 3–8 cm. long, 4–10 cm. wide, retuse to shallowly 2-lobed, cuneate to truncate-rounded at base, the midrib rather strong, the lateral nerves nearly parallel, ascending; peduncles longer than the petioles; inflorescence cymose, 1–2 (–6)-flowered, the pedicels 1–4 cm. long, flattened in fruit, the bracts small, caducous; calyx-segments membranous, broadly elliptic to oblong, 6–8 mm. long, rounded to very obtuse, mucronate; corolla rose-purple, infundibuliform, 5–6 cm. across; ovary 2-locular, 4-ovuled; capsules about 2 cm. across, depressed-globose; seeds semiglobose, brown-villous, about 7 mm. across.——Aug.–Nov. Sandy places along seashores; Shikoku, Kyushu.——Pantropic.

2. **Ipomoea indica** (Burm.) Merr. *Convolvulus indicus* Burm.; *I. congesta* R. Br.; *Pharbitis congesta* (R. Br.) Hara; *P. insularis* Choisy; *I. insularis* (Choisy) Steud.; *P. indica* Hagiwara, excl. syn.——NO-ASAGAO. More or less pilose scandent annual; stems slender, branched; leaves membranous, cordate, 5–10 cm. long, 4–8 cm. wide, entire, abruptly acuminate, long-petiolate; peduncles 2–5 cm. long, shorter than the petioles, 1- to 3-flowered, the pedicels shorter than the calyx, the bracts linear-lanceolate; calyx-segments lanceolate, gradually narrowed to tip, green, about 2 cm. long; corolla blue-purple, 7–8 cm. long, 6–7 cm. across; ovary 3-locular and 6-ovuled; style about half as long as the corolla.——June–Dec. Thickets in lowlands and hills; Honshu (Izu Isls., Kii), Shikoku, Kyushu.——se. Asia and Australia.

3. CALYSTEGIA R. Br. HIRUGAO ZOKU

Creeping or scandent herbs, glabrous or nearly so; leaves alternate, entire or palmately lobed; peduncles axillary, 1-flowered; bracts 2, large, persistent, covering the calyx; flowers rather large, rose, white, or purplish; calyx-segments mostly equal or the inner ones shorter; corolla campanulate or infundibuliform, the limb 5-plicate, entire or 5-angled, sometimes slightly 5-lobed; stamens not exserted, the disc annular; ovary 1-locular or partially 2-locular, 4-ovuled; style filiform, the stigmas 2, flat, ovate or oblong; capsules globose; seeds glabrous; cotyledons broad, often 2-lobed.——Few species, in tropical and temperate regions.

1A. Leaves reniform-cordate, shorter than broad, rounded; flowers about 5 cm. across. .... 1. *C. soldanella*
1B. Leaves sagittate, acute to obtuse.
  2A. Auricles or lateral lobes of leaves spreading, usually acute; flowers 3–3.5 cm. long; peduncles with narrow flexuous wings on upper part. .... 2. *C. hederacea*
  2B. Auricles or lateral lobes of leaves spreading to deflexed; flowers 5–6 cm. long; peduncles not winged.
    3A. Main portion of leaves narrowly ovate to oblong, obtuse, mucronate; bracts obtuse to retuse. .... 3. *C. japonica*
    3B. Main portion of leaves deltoid-ovate, usually acute; bracts obtuse to subacute. .... 4. *C. sepium*

1. Calystegia soldanella (L.) Roem. & Schult. *Convolvulus soldanella* L.——HAMA-HIRUGAO. Glabrous, rhizomatous perennial; stems creeping and branched; leaves long-petiolate, reniform-cordate, 2–3 cm. long, 3–5 cm. wide, retuse to rounded, sometimes undulate on margin; peduncles usually longer than the leaves, terete; bracts broadly ovate or broadly ovate-deltoid, 1–1.3 cm. long, obtuse, mucronate; corolla pale rose, the limb obsoletely 5-angled, 4–5 cm. across.——May–June. Sandy seashores; Hokkaido, Honshu, Shikoku, Kyushu; common.——Seashores of Eurasia and the Pacific Islands.

2. Calystegia hederacea Wall. *Convolvulus japonicus* Thunb.; *Calystegia japonica* (Thunb.) Miq. non Choisy; *C. sepium* var. *japonica* (Thunb.) Makino——KO-HIRUGAO. Nearly glabrous twining perennial herb with slender rhizomes; stems slender; leaves long-petiolate, sagittate, the lateral lobes (auricles) spreading or slightly deflexed, usually acute and bilobed, the main (central) lobe lanceolate-deltoid, 4–6 cm. long, 3–6 cm. wide, gradually narrowed to tip, obtuse; flowers solitary, axillary, rose, the bracts deltoid-ovate, acute, rounded at base, 1–2 cm. long, longer than the calyx.——June–Aug. Sunny grassy places and thickets in lowlands and hills; Honshu, Shikoku, Kyushu; common.——se. Asia.

3. Calystegia japonica Choisy. *C. sepium* var. *japonica*
Makino, pro parte; *C. subvolubilis* sensu auct. Japon., non Don——HIRUGAO. Slightly pubescent twining perennial with slender creeping rhizomes; stems elongate; leaves petiolate, sagittate or hastate, the lateral lobes (auricles) deflexed, small, ovate, obtuse, rarely with a tooth on inner margin, the main lobe broadly lanceolate to narrowly ovate-deltoid, obtuse to subacute, 5–10 cm. long, 2–7 cm. wide, including the lateral lobes; flowers solitary, long-pedunculate, rose, shorter than the leaves, the bracts ovate, slightly cordate, 2–2.5 cm. long.——July–Aug. Thickets and sunny grassy slopes in lowlands; Hokkaido, Honshu, Shikoku, Kyushu.——Korea and China.

4. Calystegia sepium (L.) R. Br. var. **americanum** (Sims) Matsuda. *Convolvulus sepium* var. *americanus* Sims; *Calystegia sepium* var. *communis* (Tryon) Hara; *C. sepium* sensu auct. Japon., non R. Br.——HIRO-HA-HIRUGAO. Rhizomes creeping; stems twining; leaves long-petiolate, subhastate, deltoid-ovate to deltoid, 4–8 cm. long, 3–7 cm. wide, with an obtusely angled lateral lobe on each side near base, acute to abruptly acuminate, mucronate, cordate; flowers solitary, long-pedunculate, the bracts ovate, shallowly cordate, 2–3 cm. long; corolla rose-colored.——Hokkaido, Honshu. Temperate regions of the N. Hemisphere.

4. ERYCIBE Roxb. HORUTO-KAZURA ZOKU

Scandent evergreen shrubs; leaves coriaceous; inflorescence cymose, often racemelike or paniclelike, often brown-pubescent; flowers small, yellow; calyx-segments orbicular, nearly equal; corolla campanulate, short-tubular, the lobes 2-fid, induplicate in bud; stamens included, the filaments short, the anthers acuminate; ovary globose, 1-locular, 4-ovuled; stigmas nearly sessile, large, subglobose; berry fleshy, ovoid, 1-seeded; cotyledons often 2-parted.——More than 50 species, in India and se. Asia.

1. Erycibe henryi Prain. *E. obtusifolia* sensu auct. Japon., non Benth.; *E. acutifolia* Hayata——HORUTO-KAZURA. Subscandent evergreen shrub; branches gray-yellow, glabrous; leaves elliptic to oblong, 5–10(–15) cm. long, 3–5(–7) cm. wide, glabrous, coriaceous, penninerved, with rather prominent veinlets beneath, abruptly acute with an obtuse tip, acute at base, the petioles 1.5–2 cm. long; inflorescence paniculate, many-flowered, axillary, with short gray-yellow hairs, the pedicels nearly as long as the calyx; calyx globose, the segments about 3 mm. long, yellow-villous; corolla about 10 mm. long, 5-lobed, densely yellow-pubescent along the ribs on back.——Kyushu (s. distr.).——Ryukyus, Formosa, and China.

5. CUSCUTA L. NE-NASHI-KAZURA ZOKU

Plants without chlorophyll, parasitic, twining; leaves reduced to scales; flowers in axillary racemes, small, white or rose, sessile or pedicelled, ebracteate; calyx-segments (4–)5, equal, free or connate at base; corolla campanulate, ovoid or globose, the lobes 4–5, short, imbricate in bud; scales on inner side of corolla lobed or laciniate, as many as the lobes and alternate with the stamens; stamens slightly exserted, inserted on the tube or throat, the filaments short-filiform or almost absent; ovary completely or incompletely 2-locular, 4-ovuled; styles 2, often united; capsules globose or ovoid, dry or slightly fleshy, irregularly or transversely dehiscent; seeds glabrous; cotyledons inconspicuous.——More than 100 species, in temperate and warmer regions.

1A. Styles connate; capsules ellipsoid-ovoid, circumscissile; valves dry and herbaceous; flowers spicate. .... 1. *C. japonica*
1B. Styles free, slender; fruit depressed-globose, irregularly rupturing or circumscissile; valves thinly membranous; flowers mostly fasciculate.
  2A. Stigma oblong, about as thick as the style. .... 2. *C. europaea*
  2B. Stigma capitellate.
    3A. Corolla shorter than the capsule and covering the lower half, with 2-lobed small scales within. .... 3. *C. australis*
    3B. Corolla much longer than the capsule and completely covering it, with rather prominent laciniate scales. .... 4. *C. chinensis*

1. Cuscuta japonica Choisy. *C. systyla* Maxim.; *C. japonica* var. *thyrsoidea* Engelm.——NE-NASHI-KAZURA. Rather fleshy, glabrous annual; stems terete, much-branched, pale
yellow with red striations and spots; leaves scalelike, deltoid, about 2 mm. long; flowers sessile or very short-pedicelled, many to rather numerous, in short dense axillary spikes, pale

yellow; calyx-segments orbicular to elliptic, rounded at apex, about 1 mm. long, erect; corolla campanulate, rather fleshy, 3.5–4 mm. long, 5-lobed, the lobes oblong; anthers narrowly oblong; capsules ellipsoid-ovoid, about 4 mm. long, enveloped by the persistent corolla at first, naked when ripe, circumscissile; styles connate, about 0.5 mm. long, the stigmas 2, short, oblong; seeds 2.5–3 mm. across.——Aug.–Oct. Lowlands and low mountains; Hokkaido, Honshu, Shikoku, Kyushu; common.——Ryukyus, China, Korea, Manchuria, and Amur.

2. Cuscuta europaea L. *C. major* Gilib.——KUSHIRO-NE-NASHI-KAZURA. Glabrous; scalelike leaves membranous, deltoid; flowers sessile, rather numerous or few, densely fascicled; calyx-segments membranous, elliptic, 1–1.5 mm. long, rounded at apex; corolla membranous, broadly campanulate, about twice as long as the calyx, 5-lobed, the tube broad, the scales inconspicuous, the lobes deltoid, obtuse; capsules membranous, depressed-globose, about 3 mm. across, enveloped by the persistent corolla at first, often naked when ripe, circumscissile; styles slender, the stigma oblong, as thick as the style. ——July–Sept. Hokkaido.——Sakhalin, Siberia, Europe, and Africa.

3. Cuscuta australis R. Br. *C. millettii* Hook & Arn.; *C. obtusifolia* var. *australis* (R. Br.) Engelm.; *C. sojagena* Makino; *C. hygrophilae* Pears.; *C. hawakamii* Hayata——MAME-DAOSHI. Very slender glabrous annual; leaves scalelike,

ovate, thinly membranous; flowers few, densely fascicled, sessile; calyx-segments thinly membranous, depressed-deltoid, obtuse, 1/3 to 1/2 as long as the corolla; corolla about 2 mm. long, thinly membranous, the tube with minute bifid scales nearly as long as the broadly ovate, obtuse lobes; stamens as long as the corolla-lobes, the anthers minute, ovate-deltoid; capsules about 3 mm. across, thinly membranous, irregularly dehiscent; styles slender, about 1 mm. long, the stigma punctate; seeds about 0.3 mm. across.——July–Oct. Hokkaido, Honshu, Shikoku, Kyushu.——Ryukyus, Formosa, China, se. Asia, and Australia.

4. Cuscuta chinensis Lam. *Grammica aphylla* Lour.; *C. maritima* Makino——HAMA-NE-NASHI-KAZURA. Very slender, glabrous; scalelike leaves deltoid-ovate, membranous; flowers few, rather densely fascicled, short-pedicelled; calyx-segments deltoid, about 2 mm. long, united at base; corolla about 2.5 mm. long, 5-lobed to the middle, the lobes ovate, subacute, thinly membranous, the scales of the tube inside laciniate; stamens about half as long as the lobes, the anthers broadly ovate; capsules depressed-globose, about 2 mm. across, about half as long as and enveloped by the corolla; styles slender, about 1 mm. long, the stigma capitellate; seeds about 1.5 mm. across.——June–Oct. Honshu (near seashores of centr. and w. distr.), Shikoku, Kyushu.——Ryukyus, Formosa, Korea, China, Malaysia to Australia, and Africa.

Fam. 172. POLEMONIACEAE   HANA-SHINOBU KA   Polemonium Family

Herbs or rarely shrubs; leaves alternate, sometimes opposite, entire or pinnate; inflorescence in terminal and axillary corymbs or panicles; flowers bisexual, actinomorphic, rarely bilabiate; calyx-segments 5; corolla gamopetalous, the lobes 5, contorted in bud; stamens 5, inserted on the tube and alternate with the lobes, the anthers 2-locular, longitudinally split; ovary superior, inserted on the disc, (2–)3-locular, sessile; ovules 1 or more, on inner angles of the locule; style solitary, the stigmas (2–)3; fruit a loculicidally dehiscent or rarely indehiscent capsule; seeds usually with copious endosperm.——About 12 genera, with about 300 species, chiefly in N. and S. America, few in the temperate regions of the Old World.

1. POLEMONIUM L.   HANA-SHINOBU ZOKU

Perennials, sometimes glandular pilose; leaves alternate, pinnatisect; calyx campanulate, often more or less accrescent after anthesis; corolla infundibuliform, broadly campanulate or subrotate, 5-lobed; filaments elongate, declined, hairy at base; ovary ovoid, 3-locular; ovules 2–12 in each locule; capsules ovoid, 3-valved; seeds sometimes narrowly winged, often viscid.——About 20 species, chiefly in N. America, few in Eurasia.

1A. Flowers densely arranged, 11–15 mm. long, short-pedicelled; seeds about 2 mm. long. . . . . 1. *P. kiushianum*
1B. Flowers usually loosely arranged, 18–25 mm. long, long-pedicelled; seeds 3–4 mm. long.
2A. Disc 5-lobed; flowers 2–2.5 cm. long; corolla-lobes emarginate. . . . . 2. *P. yezoense*
2B. Disc entire, undulate on margin; flowers 18–20 mm. long; corolla-lobes rounded to acute. . . . . 3. *P. acutiflorum*

1. Polemonium kiushianum Kitam. *P. caeruleum* sensu auct. Japon., saltem pro parte, non L.——HANA-SHINOBU. Rhizomes short-creeping; stems 70–100 cm. long, erect, striate; lower leaves 17–25 cm. long, petiolate, pinnatisect, the leaflets 10 to 12 pairs, broadly lanceolate, sessile, 2–3 cm. long, acuminate; upper leaves gradually smaller, sessile, the segments fewer, glabrous; inflorescence glandular-pilose; flowers short-pedicelled, blue; calyx 5–7 mm. long, the segments as long as to slightly longer than the tube, acuminate; corolla-lobes obtuse, short-pilose on back and margin; capsules globose, 4–5 mm. long; seeds nearly fusiform, angled, nearly wingless.—— June–Aug. Mountains; Kyushu.

2. Polemonium yezoense (Miyabe & Kudo) Kitam. *P. caeruleum* subsp. *vulgare* var. *yezoense* Miyabe & Kudo—— EZO-NO-HANA-SHINOBU. Stems 35–80 cm. long, obsoletely ribbed; lower leaves 12–16 cm. long, petiolate, pinnatisect, the

leaflets 9 to 12 pairs, lanceolate, sessile, acuminate to acute, glabrous; upper leaves gradually smaller and sessile, with fewer segments; inflorescence a loose corymb, glandular-pilose; flowers rather long-pedicelled, 22–25 mm. long; calyx deeply 5-lobed, 10–12 mm. long, the lobes narrowly lanceolate, acute; corolla blue, the lobes elliptic, emarginate, loosely ciliate.—— May–Aug. Hokkaido.

3. Polemonium acutiflorum Willd. var. **nipponicum** (Kitam.) Ohwi. *P. caeruleum* sensu auct Japon., non L.; *P. nipponicum* Kitam.——MIYAMA-HANA-SHINOBU. Stems 40–80 cm. long, ridged; lower leaves petiolate, 10–12 cm. long, pinnatisect, the leaflets 8 to 9 pairs, broadly lanceolate, 2–2.5 cm. long, gradually acute; upper leaves gradually smaller and sessile, with fewer leaflets; inflorescence loosely corymbose, glandular-pilose; flowers rather long-pedicelled, about 2 cm. long, blue; calyx 10–12 mm. long, deeply 5-lobed, the lobes lanceolate, acute; corolla-lobes subacute or obtuse, sparingly ciliate; capsules globose, about 5.5 mm. long; seeds 3–4 mm. long.——July–Aug. Alpine slopes; Honshu (centr. distr.); rare.

Var. **laxiflorum** (Regel) Ohwi. *P. caeruleum* var. *vulgare* lus. *laxiflorum* Regel; *P. caeruleum* subsp. *vulgare* var. *laxi-*

florum (Regel) Miyabe & Kudo; *P. caeruleum* subsp. *vulgare* var. *racemosum* Miyabe & Kudo; *P. laxiflorum* (Regel) Kitam. ——KARAFUTO-HANA-SHINOBU. Calyx 7.5–9.5 mm. long, rather deeply 5-lobed, the lobes subobtuse to acute.——June–July. Hokkaido.——The typical phase occurs in Sakhalin, Kuriles, and the cold regions of the N. Hemisphere.

Fam. 173. BORAGINACEAE   MURASAKI KA   Borage Family

Usually strigose-hirsute herbs, shrubs, or trees; leaves alternate, very rarely opposite, simple, exstipulate; flowers in scorpioid cymes, actinomorphic, rarely oblique, usually bisexual; calyx-segments imbricate or rarely valvate in bud; corolla-lobes contorted or imbricate in bud; stamens inserted on the corolla-tube, as many as the lobes and alternate with them; anthers 2-locular, longitudinally split; disc sometimes absent; ovary superior, 2-locular, becoming falsely 4-locular at maturity; style terminal or gynobasic; ovules 4, 2 in each locule, erect or spreading; fruit a drupe or composed of 4 nutlets; seeds with or without endosperm. ——About 90 genera, with about 1,500 species, nearly cosmopolitan.

1A. Fruit a drupe, undivided, with 1–4 nutlets; style terminal.
  2A. Trees or large shrubs; style without annulus near apex. . . . . 1. *Ehretia*
  2B. Herbs or shrubs; style with an annulus below the bifid stigma. . . . . 2. *Messerschmidia*
1B. Fruit of 4 nutlets; style gynobasic.
  3A. Nutlets attached near apex.
    4A. Nutlets flat or gently convex on back. . . . . 3. *Cynoglossum*
    4B. Nutlets excavated on back. . . . . 4. *Omphalodes*
  3B. Nutlets attached near center or base.
    5A. Nutlets prickly or spinulose.
      6A. Perennials; inflorescence bractless or bracteate at base; nutlets with a small attachment at the gynobase, sometimes without prickles; rosulate radical leaves present in anthesis. . . . . 5. *Eritrichium*
      6B. Annuals or biennials; inflorescence bracteate; rosulate radical leaves not present in anthesis.
        7A. Pedicels and mature calyx erect or ascending; nutlets as long as the gynobase, attached nearly the whole length along the keeled ventral side. . . . . 6. *Lappula*
        7B. Pedicels and mature calyx reflexed; nutlets twice as long as the gynobase, attached on the upper half of ventral face.
            7. *Hackelia*
    5B. Nutlets neither prickly nor spinulose, smooth, wrinkled or short-tubercled.
      8A. Nutlets short-tubercled, attached on ventral side to the convex gynobase, the attachment with a ringlike raised margin.
            8. *Bothriospermum*
      8B. Nutlets smooth or wrinkled, attached at the base to the flat gynobase.
        9A. Corolla-lobes imbricate in bud.
          10A. Corolla infundibuliform-campanulate or tubular-campanulate, the tube prominent.
            11A. Nutlets bony and lustrous, not compressed.
              12A. Nutlets ovoid, not hooked at apex.
                13A. Corolla-throat decorated inside with 5 well-developed vertical lines of hairs. . . . . 9. *Buglossoides*
                13B. Corolla-throat lacking very well developed vertical lines of hairs. . . . . 10. *Lithospermum*
              12B. Nutlets elongate and hooked at apex. . . . . 11. *Ancistrocarya*
            11B. Nutlets sometimes fleshy, usually flattened dorsally. . . . . 12. *Mertensia*
          10B. Corolla rotate, 5-scaled on the throat; nutlets tetrahedral or slightly flattened dorsally. . . . . 13. *Trigonotis*
        9B. Corolla-lobes contorted in bud; nutlets slightly flattened. . . . . 14. *Myosotis*

1. EHRETIA L.   CHISHA-NO-KI ZOKU

Smooth or scabrous shrubs or trees; leaves alternate, entire or toothed; flowers small, usually white, in cymes, corymbs or panicles; calyx small, 5-parted, the segments linear or broader, imbricate or not overlapping in bud; corolla rotate or tubular, the lobes 5, imbricate in bud, spreading; stamens 5, inserted on the tube, the anthers usually exserted; ovary 2-locular, with an imperfect septum in each locule or 4-locular; style solitary, terminal, the stigma bifid; drupe small, 2- or 4-stoned.——About 50 species, chiefly in the warmer regions of the Old World.

1A. Young branches and upper side of leaves scabrous; leaves 5–12 cm. wide; inflorescence corymbose-paniculate, rather flat-topped; corolla 8–10 mm. long; fruit 1–1.5 cm. across. . . . . 1. *E. dicksonii* var. *japonica*
1B. Young branches and leaves glabrous; leaves 3–7 cm. wide; inflorescence paniculate, pyramidal; corolla 4–5 mm. long; fruit 4–5 mm. across. . . . . 2. *E. ovalifolia*

1. Ehretia dicksonii Hance var. **japonica** Nakai. *E. macrophylla* sensu auct. Japon., non Wall.; *E. dicksonii* var. *liukiuensis* Nakai; *E. dicksonii* var. *velutina* Nakai——MARU-BA-CHISHA-NO-KI.   Deciduous tree with rather thick scabrous branches while young; leaves broadly elliptic to broadly obovate, 5–17 cm. long, 5–12 cm. wide, sometimes nearly orbicular, prominently scabrous and appressed-strigose on upper side, densely short-hairy beneath, irregularly deltoid-serrate, the petioles 3–4 cm. long; panicles terminal, short-pilose, nearly flat-topped; calyx about 4 mm. long, deeply 5-fid, the lobes broadly lanceolate or narrowly oblong, obtuse to subacute, short-pilose; corolla 8–10 mm. long, white; fruit globose, 1–1.5

cm. across, yellow when mature.——May–June. Warmer parts; Honshu (Awa Prov. and westw.), Shikoku, Kyushu. ——Ryukyus, Formosa, and China.

2. Ehretia ovalifolia Hassk. *E. serrata* var. *obovata* Lindl.; *E. acuminata* var. *obovata* (Lindl.) Johnston; *Cordia thyrsiflora* Sieb. & Zucc.; *E. thyrsiflora* (Sieb. & Zucc.) Nakai; *E. acuminata* sensu auct. Japon., non R. Br.——CHISHA-NO-KI. Smooth, nearly glabrous deciduous tree with rather slender branches; leaves obovate to obovate-oblong, 5–12(–18) cm.

long, 3–7(–10) cm. wide, abruptly acute, with short appressed teeth, the petioles 1.5–2.5 cm. long; inflorescence a large densely many-flowered pyramidal panicle, glabrous to slightly short-pilose; calyx about 2 mm. long, deeply 5-lobed, the lobes broadly elliptic, rounded at apex, rarely ciliolate; corolla small, 4–5 mm. long, white; fruit 4–5 mm. across.——June. Honshu (Chūgoku Distr.), Shikoku, Kyushu.——Ryukyus, Formosa, China, and Korea (Quelpaert Isl.).

2. MESSERSCHMIDIA L. SUNABIKI-SŌ ZOKU

Herbs or shrubs; leaves alternate, entire; flowers small, usually in branched terminal cymes; calyx 5-parted; corolla-tube cylindric, unappendaged in the throat, the lobes 5, induplicate in bud; stamens 5, included and inserted on the dilated portion of the corolla-tube; ovary 4-locular, undivided; style terminal, simple, with an annulus below the stigma; stigma usually bifid; ovules pendulous; fruit a drupe, the exocarp corky, stones 4, free.——Three species, in Europe, Asia, Australia, and S. America.

1A. Rhizomatous perennial herb; inflorescence corymbose; flowers pedicelled; fruit hairy, 5–7 mm. wide; leaves 7–30 mm. wide, appressed grayish strigose on both sides. . . . . 1. *M. sibirica*
1B. Large shrub 1–5 m. high; inflorescence branches scorpioid; flowers sessile; fruit glabrous, about 4 mm. across; leaves 3–6 cm. wide, densely whitish silky on both surfaces. . . . . 2. *M. argentea*

1. Messerschmidia sibirica L. *Tournefortia sibirica* L.; *M. arguzia* L.; *Heliotropium japonicum* A. Gray; *M. sibirica* var. *latifolia* (DC.) Hara; *Tournefortia arguzia* var. *latifolia* DC.——SUNABIKI-SŌ. Perennial herb, long appressed-strigose, with creeping rhizomes; stems 30–50 cm. long, rather fleshy, sometimes branched; leaves oblanceolate or oblong-lanceolate, 4–10 cm. long, 7–30 mm. wide, obtuse, sessile; inflorescence corymbose, few-flowered, terminal or often appearing lateral, pedunculate; flowers white, short-pedicellate; calyx-segments lanceolate, 3–5 mm. long; corolla-tube slender, 6–7 mm. long, longer than the limb, densely appressed-pubescent, the limb shallowly 5-lobed, with a yellow center; fruit broadly ellipsoidal, 8–10 mm. long, 5–7 mm. wide, obtusely 4-angled, pubescent, depressed at apex, with a short style.——May–Aug. Sand-dunes along the sea; Hokkaido, Honshu, Shikoku, Kyushu; rather common.——Korea, s. Siberia, and w. Europe.

2. Messerschmidia argentea (L. f.) Johnston. *Tournefortia argentea* L. f.——MONPA-NO-KI, HAMA-MURASAKI-NO-KI. Densely leafy shrub, 1–5 m. high, with thick, terete branches; leaves fleshy, narrowly obovate to obovate-spathulate, 10–20 cm. long, 3–6(–9) cm. wide, rounded to obtuse or retuse, appressed whitish silky-hairy, the lateral nerves strongly arcuate, ascending; inflorescence terminal, rather large, pedunculate, branched in upper part, silvery gray hairy, the branches scorpioid, many flowered; flowers small, white, sessile; calyx-segments as long as the corolla-tube; corolla-tube 1.5–2 mm. long, the limb about 4 mm. across, deeply 5-lobed, the lobes elliptic, slightly appressed-pilose on back near base; anthers semi-exserted; fruit globose, about 4 mm. across, glabrous.——Aug.–Nov. Seashores; Kyushu (Tanegashima); very rare. ——Ryukyus, Formosa, se. Asia to Australia, and Madagascar.

3. CYNOGLOSSUM L. Ō-RURI-SŌ ZOKU

Usually tall biennial or perennial herbs, coarsely gray-white strigose; radical leaves often long-petiolate; racemes usually elongate, sometimes in a loose panicle, secund, bractless or rarely bracteate, often dichotomously forked; flowers pedicelled or nearly sessile, blue to white; calyx 5-parted; corolla small, the tube short, with obtuse or reniform appendages on throat, the lobes 5, imbricate in bud; stamens not exserted; ovules horizontal; nutlets 4, prickly, depressed, apically attached to the gynobase, flat, gently convex on back.——About 70 species, in temperate regions and mountains of the subtropics.

1A. Lower part of stems with coarse spreading hairs about 2 mm. long; leaves coarsely hispid; racemes rather loosely flowered, the branches ascending; nutlets about 3 mm. long excluding the prickles. . . . . 1. *C. asperrimum*
1B. Stems with hairs less than 1 mm. long; leaves with short coarse or appressed hairs; racemes rather densely flowered, the branches obliquely spreading; nutlets 2–2.5 mm. long excluding the prickles.
2A. Leaves 4–7 cm. long, relatively thick; hairs on leaves and stems longer. . . . . 2. *C. lanceolatum*
2B. Leaves 10–15 cm. long, relatively thin; hairs shorter. . . . . 3. *C. zeylanicum*

1. Cynoglossum asperrimum Nakai. *Paracynoglossum asperrimum* (Nakai) Popov; *C. furcatum* sensu auct. Japon., non Wall.——ONI-RURI-SŌ. Coarsely strigose biennial; stems 40–80 cm. long, much branched in upper part, rather stout, hairs of lower half spreading, about 2 mm. long; lower leaves petiolate, the median ones sessile, oblong-lanceolate, 10–20 cm. long, 2–3 cm. wide, acuminate at both ends, coarsely strigose on upper side and on nerves beneath; leaves on branches 3–10 cm. long, 7–20 mm. wide; racemes 15–20 cm. long, loosely flowered; flowers pale blue, about 3 mm. across; calyx-segments reflexed in fruit.——June–Aug. Riverbanks and valleys, some-

times in clearings in mountains; Hokkaido, Honshu, Shikoku, Kyushu.——s. Kuriles and Korea.

Var. **tosaense** (Nakai) Hara. *C. tosaense* Nakai——TOSA-RURI-SŌ. Leaves less densely hairy on upper side.——Occurs with the typical phase.

Var. **yesoense** Nakai. EZO-RURI-SŌ. Inflorescence more prominently hairy.——Occurs with the typical phase.

2. Cynoglossum lanceolatum Forsk. *C. formosanum* Nakai; *C. micranthum* sensu auct. Japon., non Desf.——TAIWAN-RURI-SŌ, SHIMA-RURI-SŌ. Biennial, scabrous, densely short grayish strigose, the hairs retrorse on lower part of stems, spreading on the lower leaves beneath to ascending or up-wardly appressed; stems 30–50 cm. long, branched in upper part; lower leaves petiolate, the median sessile or short-petio-late, narrowly oblong to oblong-lanceolate, 4–7 cm. long, 1–2 cm. wide, acute, acuminate at base, upper leaves 1.5–3 cm. long; racemes densely flowered, 10–20 cm. long; flowers about 3 mm. across, white or bluish; calyx-segments obliquely spread-ing in fruit.——July–Aug. Kyushu.——Ryukyus, Formosa, and widely spread in all warmer regions.

3. Cynoglossum zeylanicum (Vahl) Thunb. *C. fur-catum* sensu auct. Japon., non Wall.; *Anchusa zeylanica* Vahl ——USUBA-RURI-SŌ. Biennial; hairs on lower part of stems and leaves beneath retrorse, otherwise the hairs short and

upwardly appressed; stems 40–70 cm. long, branched; lower leaves long-petiolate, the median ones sessile or short-petiolate, narrowly oblong or broadly lanceolate, 10–15 cm. long, 2–4 cm. wide, acute to acuminate, gradually narrowed at base, rather densely appressed-pilose above, sparsely long-pubescent beneath especially on nerves, the upper leaves sessile, 3–10 cm. long, obtuse at base; racemes 10–20 cm. long, rather densely flowered; flowers bluish, sometimes white, about 4 mm. across, the pedicels short; calyx-segments spreading in fruit. July–Aug. Kyushu.——s. Korea, Ryukyus, Formosa, and s. Asia.

Var. **villosulum** (Nakai) Ohwi. *C. villosulum* Nakai—— Ō-RURI-SŌ. Hairs obliquely spreading to retrorse.——Honshu and (?) Shikoku.

4. OMPHALODES Moench RURI-SŌ ZOKU

Annuals or perennials, glabrous or hirsute; radical leaves petiolate or narrowed to a petiolelike base; cauline leaves few, alternate; inflorescence a simple or forked raceme sometimes bracteate at base, loosely flowered; flowers white or blue, pedicelled; calyx 5-parted or lobed, slightly accrescent in fruit; corolla rotate, gibbous or scaly on the throat, the lobes 5, imbricate in bud; stamens included; nutlets 4, depressed, excavated on back, attached to a flat or convex gynobase with a broad attachment on upper portion of ventral face, the dorsal margin extended to form a smooth ringlike or glochidiate crown; seeds nearly horizontal.—— About 20 species, in Europe, N. Africa, and Asia.

1A. Stems 1(–3), erect; inflorescence usually leafless, dichotomously forked. .... 1. *O. krameri*
1B. Stems few, ascending from a short-decumbent base; inflorescence with few leaflike bracts at base, simple, not forked.
   2A. Leaves with long and short hairs; rhizomes rather stout, short. .... 2. *O. japonica*
   2B. Leaves with long hairs of one sort; rhizomes slender and proliferating from the axil of the lowest bract. .... 3. *O. prolifera*

1. Omphalodes krameri Fr. & Sav. RURI-SŌ. Perennial with coarse spreading white hairs; rhizomes short, rather thick; stems solitary or few, erect, simple, 25–40 cm. long; leaves broadly oblanceolate, 7–15 cm. long, 1.5–3.5 cm. wide, acute, the radical leaves long-petiolate, the cauline sessile or narrowed to a short petiolelike base; inflorescence pedunculate, dichotomously forked, usually leafless, loosely flowered, erect; flowers blue, 1–1.5 cm. across, the pedicels 1–1.5 cm. long; calyx-segments accrescent after anthesis, narrowly ovate, 5–8 mm. long; nutlets orbicular, about 3.5 mm. across, slightly depressed, excavated on the dorsal face, glochidiate on the margin of the erect crown.——Apr.–June. Woods in mountains; Hokkaido, Honshu.——Forma **alba** (T. Ito) Hara. *Omphalodes krameri* var. *alba* T. Ito.——SHIROBANA-RURI-SŌ. Flowers white.

Var. **laevisperma** (Nakai) Ohwi. *O. laevisperma* Nakai ——ECHIGO-RURI-SŌ. Nutlets smooth, not spiny.——Honshu (Echigo Prov.).

2. Omphalodes japonica (Thunb.) Maxim. *Cynoglossum japonicum* Maxim.——YAMA-RURI-SŌ. Coarsely spreading-hirsute perennial with short, stout rhizomes; stems scapiform, few, ascending from a decumbent base, 12–20 cm. long, sometimes branched near the base; radical leaves oblanceolate or broadly so, 7–15 cm. long, 1–2.5 cm. wide, acute, gradually narrowed to a petiolelike base, with short and long hairs on

both sides, the cauline leaves few, smaller, sessile, 1–6 cm. long, becoming bracteate in the uppermost; racemes simple, not branched, loosely several-flowered, with few bracts at base, the pedicels deflexed after anthesis, 8–15 mm. long; calyx-segments narrowly ovate, accrescent after anthesis, 5–8 mm. long; corolla about 1 cm. across, blue; nutlets smooth, about 3 mm. long. ——Apr.–May. Woods in mountains; Honshu (Iwashiro Prov. and southw.), Shikoku, Kyushu.——Forma **albiflora** S. Okamoto. SHIROBANA-YAMA-RURI-SŌ. Flowers white.

Var. **echinosperma** Kitam. TOGE-YAMA-RURI-SŌ. Nutlets short-glochidiate on margin of the crown.——Shikoku; rare.

3. Omphalodes prolifera Ohwi. HAI-RURI-SŌ. FIG. 14. Perennial herb uniformly hirsute with long spreading hairs; rhizomes slender; radical leaves 4 or 5, oblanceolate, 12–25 cm. long, 2.5–4 cm. wide, abruptly acuminate, gradually attenuate at base; scapes 1 to 3, decumbent after anthesis; cauline leaves 2 or 3, remote, sessile, oblong-lanceolate, the upper rounded at base, 4–8 cm. long; racemes terminal, simple, loosely 4- to 8-flowered, the bracts 1 to 3, similar to the cauline leaves, the lowest bract always proliferous in the axil in fruit; corolla rose-colored, the limb nearly 1 cm. across, 5-parted, the lobes orbicular; nutlets smooth, about 2 mm. long.——May. Honshu (Owari Prov.); rare.

5. ERITRICHIUM Schrad. MIYAMA-MURASAKI ZOKU

Tufted hispid or silky perennials; leaves alternate, usually narrow; racemes simple or branched, bractless or rarely bracteate at base, the flowers very rarely axillary; calyx deeply 5-fid, the segments accrescent after anthesis; corolla blue or white, small, the tube short, the throat usually 5-scaled, the lobes 5, imbricate in bud, spreading, obtuse; stamens 5, inserted on the tube, included, the filaments short, the anthers obtuse; gynobase convex to columnar; nutlets usually 4, erect, areolate on the inner face, smooth, rugose, or tubercled, rarely glochidiate on the margin; seeds straight, rarely curved.——About 10 species, in Eurasia.

Kadzuwo Inami. del.

FIG. 14.—*Omphalodes prolifera* Ohwi. 1, Habit; 2, plant in fruit; 3, flower; 4, the same, side view; 5, calyx; 6, the same, expanded with the pistil showing; 7, corolla cut open; 8, stamen and appendage on throat of corolla; 9, fruit; 10, fruitlet.

1. **Eritrichium nipponicum** Makino. *Hackelia nipponica* (Makino) Brand; *E. yesoense* Nakai——MIYAMA-MURA-SAKI. Dwarf grayish strigose perennial with a thick perpendicular caudex; stems 10–20 cm. long, tufted, ascending from a decumbent base; radical leaves rosulate, closely imbricate, rather many, linear-lanceolate, 3–6 cm. long, 4–6 mm. wide, obtuse or subacute, sessile; cauline leaves about 10, lanceolate, 1–2.5(–3.5) cm. long, 3–5 mm. wide; racemes few, terminal, several-flowered, bracteate in the lower flowers, the bracts broadly linear, 3–7 mm. long, the pedicels erect, 7–12 mm. long; calyx-segments narrowly oblong; corolla blue, about 8 mm. across, the tube shorter than the calyx; nutlets ascending, about 1.5 mm. long, puberulent on the back, with a series of marginal glochidiate bristles.——July–Aug. Gravelly or sandy alpine slopes; Hokkaido, Honshu (centr. distr.); rare. ——Sakhalin.

6. LAPPULA Moench NO-MURASAKI ZOKU

Coarsely hirsute annuals or biennials; leaves alternate, narrow; inflorescence a bracteate raceme; flowers small, blue or white, on short erect pedicels; calyx-segments 5, erect; corolla short tubular, 5-scaled on the throat; gynobase long and narrow; nutlets tetrahedral or flattened, with a narrow areole occupying nearly the whole length of the prominently keeled ventral face, usually glochidiate on the margin; style usually longer than the nutlets.——About 15 species, in temperate regions.

1. **Lappula echinata** Gilib. *Myosotis lappula* L.; *Echinospermum lappula* (L.) Lehm.——NO-MURASAKI. Coarsely hirsute annual; stems erect, 20–30(–50) cm. long; leaves oblanceolate to broadly linear, 3–5 cm. long, 3–5 mm. wide; racemes few or more than 10, suberect, about 10 cm. long, loosely flowered, the pedicels 2–3 mm. long, shorter than the bracts, erect; flowers about 3 mm. across, pale blue; nutlets tetrahedral, with callose bumps or tubercles on back, glochidiate on margin.——July. Naturalized in our area.——Europe to Manchuria, n. China and Korea, naturalized in N. America.

7. HACKELIA Opiz OKA-MURASAKI ZOKU

Annuals, biennials, or perennials, grayish pilose or subglabrous; stems often tall and much branched; leaves alternate, narrow; flowers small, pedicelled, bracteate or bractless, the pedicels reflexed after anthesis; calyx deeply 5-fid, the segments ovate or narrowly so; corolla-tube short, with 5 scales on the throat, the lobes 5, imbricate in bud, spreading; stamens 5, inserted on the tube, included, the filaments very short, the anthers obtuse; gynobase broadly conical; nutlets erect with a large ovate or deltoid areole near the center of the ventral side, keeled only on the upper half of the ventral side, glochidiate on the margin; style not exceeding the nutlets.——About 40 species, in temperate regions of Eurasia and N. America.

1A. Calyx-segments 1–3 mm. long, reflexed after anthesis; nutlets only glochidiate. .... 1. *H. deflexa*
1B. Calyx-segments 1.3–1.5 mm. long, spreading; nutlets glochidiate and tuberculate. .... 2. *H. matsudairae*

1. **Hackelia deflexa** (Wahlenb.) Opiz. *Myosotis deflexa* Wahlenb.; *Echinospermum deflexum* (Wahlenb.) Lehm.; *Lappula deflexa* (Wahlenb.) Garcke——OKA-MURASAKI. Annual or biennial; stems 20–60 cm. long, much branched; leaves linear-lanceolate, 2–5(–9) cm. long, 3–10(–20) mm. wide, obtuse or acute, petiolate only in the lower ones, loosely pilose on both surfaces; racemes 10–15 cm. long, the pedicels bracteate, 2–5 mm. long, as long as the bracts, deflexed after anthesis; calyx-segments lanceolate, 1–3 mm. long, reflexed after anthesis; corolla 3(–5) mm. across, pale blue to white, yellow on the throat; nutlets tetrahedral, smooth, glochidiate.——June–Aug. Reported to be spontaneous in Hokkaido (Ishikari Prov.).——Sakhalin and Korea to Europe; naturalized in N. America.
2. **Hackelia matsudairae** (Makino) Ohwi. *Echi-nospermum matsudairae* Makino; *Lappula matsudairae* (Makino) Druce——IWA-MURASAKI. About 70 cm. high, slender, erect, branched and appressed-hairy in upper part, with spreading hairs at base; leaves linear to oblong, coarsely hirsute, the lower ones petiolate; racemes 3–10 cm. long, bracteate, loosely flowered, appressed-hirsute, the pedicels 4–5 mm. long, spreading or deflexed, as long as or in the upper ones shorter than the lanceolate bract; calyx-segments spreading, oblong, 1.3–1.5 mm. long, spreading, shorter than the nutlets; corolla small; nutlets about 2 mm. long, ovate-deltoid on back, with small callose tubercles, glochidiate; style short.——July–Aug. Reported from Honshu (Mount Togakushi in Shinano); very rare.

8. BOTHRIOSPERMUM Bunge HANA-IBANA ZOKU

Decumbent to erect, rather small, hirsute or strigose annuals or biennials; leaves alternate, lanceolate to ovate; flowers small, bluish or white, axillary or in loose bracteate racemes; calyx deeply 5-fid, the lobes lanceolate, slightly accrescent after anthesis; corolla-tube short, 5-scaled on the throat, the lobes 5, imbricate, spreading; stamens 5, inserted on the tube, the anthers obtuse; gynobase short, oblong; nutlets erect, rounded on back, granular-tuberculate, with an ovate or oblong subdepressed areole with entire or dentate margin on the ventral side.——Few species in Asia.

1. **Bothriospermum tenellum** (Hornem.) Fisch. & Mey. *Anchusa tenella* Hornem.; *Cynoglossum diffusum* Roxb.; *B. asperugoides* Sieb. & Zucc.; *B. tenellum* var. *asperugoides* (Sieb. & Zucc.) Maxim.; *B. perenne* Miq.——HANA-IBANA. Short-strigose, ascending to decumbent annual or biennial; stems 5–30 cm. long, tufted; leaves oblong or elliptic, 2–3 cm. long, 1–2 cm. wide, rounded and mucronate or obtuse, petiolate; bracteal leaves 5–15(–25) mm. long; pedicels supra-axillary, much shorter than the bracteal leaves and deflexed after anthesis; flowers pale blue, 2–3 mm. across; nutlets ellipsoidal, about 1.5 mm. long, about 1 mm. across, densely callose-tuberculate, the areole smooth-margined.——Mar.–Dec. Sunny places in lowlands and hills; Hokkaido, Honshu, Shikoku, Kyushu; common.——Ryukyus, Formosa, China, Korea, Manchuria, and se. Asia.

9. BUGLOSSOIDES Moench   INU-MURASAKI ZOKU

Annuals or perennials; leaves alternate; racemes one-sided, simple to twice or thrice branched, usually elongate, prominently bracteate; calyx 5-fid, the lobes narrow; corolla blue or white, infundibuliform or salverform, sometimes incurved, plaited, usually appressed-hairy outside, with 5 longitudinal lines of hairs inside, the lobes spreading to ascending, imbricate in bud; stamens included; gynobase flat or depressed-pyramidal; nutlets 1–4, erect to strongly divergent, smooth or rough, rounded or angulate, the attachment basal or oblique.——About 10 species, in Europe, n. Africa, and Asia.

1A. Flowers 15–18 mm. across, blue or rarely white; nutlets smooth; leaves oblong, 1–2 cm. wide; perennial with innovation shoots at base. . . . . 1. *B. zollingeri*
1B. Flowers 4–5 mm. across, white; nutlets rugose; leaves lanceolate to broadly linear, 1–3(–7) mm. wide; biennial without innovation shoots at base. . . . . 2. *B. arvense*

1. Buglossoides zollingeri (DC.) Johnston. *Lithospermum zollingeri* DC.; *L. japonicum* A. Gray; *L. confertiflorum* Miq.; *Plagiobotrys zollingeri* (DC.) Johnston——HOTARU-KAZURA.   Perennial; stems slender, erect, with elongate innovation shoots at the base; leaves oblong to narrowly so, rarely broadly oblanceolate, 2.5–6 cm. long, 1–2 cm. wide, acute, sessile, with callose-based strigose hairs on upper surface, the midrib distinct; flowers in upper axils, 15–18 mm. across; calyx-segments linear-lanceolate, acuminate; corolla blue or rarely white; nutlets smooth, white, 2.5–3 mm. long.——Apr.–May.   Sunny grassy slopes; Hokkaido, Honshu, Shikoku, Kyushu.——Korea, China, and Formosa.

2. Buglossoides arvense (L.) Johnston. *Lithospermum arvense* L.——INU-MURASAKI.   Rather grayish appressed-strigose biennial; stems erect, 20–40 cm. long, nearly simple or branched; leaves cauline, rather thick, narrowly lanceolate to broadly linear, 1–3 cm. long, 1–3(–7) mm. wide, obtuse, ascending, 1-nerved, the margin often slightly recurved; flowers white, 4–5 mm. across, 6–7 mm. long; nutlets gray, ovoid, obtuse, rugose.——Apr.–June.   Sunny grassy places in lowlands and low mountains; Honshu, Shikoku, Kyushu.—— The temperate regions of Europe and Asia.

10. LITHOSPERMUM L.   MURASAKI ZOKU

Usually perennials from a strong, frequently dye-stained taproot; stems simple or somewhat branched, hispid, villous, or strigose, erect, or spreading; leaves cauline, usually many; racemes scorpioid, simple or paired, few to many-flowered, the bracts usually many; calyx 5-fid, the lobes linear-cuneate to lanceolate; corolla yellow, orange, or white, or rarely bluish, salverform, infundibuliform, or tubular, the lobes imbricate in bud, often with 5 appendages on the throat; gynobase flat to broadly pyramidal; nutlets erect, usually ovoid or ellipsoidal, convex on back, smooth or rarely rugose, the areole large, basal, flat, convex, or excavated.——About 44 species, in Eurasia, Asia, and America.

1. Lithospermum erythrorhizon Sieb. & Zucc. *L. murasaki* Sieb., nom. seminud.; *L. officinale* sensu auct. Japon., non L.; *L. officinale* var. *japonica* Miq.; *L. officinale* var. *erythrorhizon* (Sieb. & Zucc.) Maxim.——MURASAKI.   Erect perennial with coarse hirsute stems; roots thick, purple-stained when dried; stems few, erect, 40–70 cm. long, many-leaved; leaves ascending, lanceolate to oblong-lanceolate, gradually narrowed to an obtuse tip, sessile, with few parallel nerves; inflorescence spikelike, with leaflike bracts; calyx-segments broadly linear, obtuse, longer than the corolla-tube; corolla white, 6–7 mm. long, about 4 mm. across, with shallow transverse appendages on the throat; nutlets gray-white, shining, smooth, about 3 mm. long, ovoid.——June–July.   Grassy slopes in mountains and hills; Hokkaido, Honshu, Shikoku, Kyushu. ——Korea, Manchuria, China, and Amur.

11. ANCISTROCARYA Maxim.   SAWA-RURI-SŌ ZOKU

Tall perennial herb with short erect strigose hairs; leaves alternate, oblong, fasciculate near the middle; racemes several or in pairs, loosely flowered, bractless or bracteate only in lower part; calyx deeply 5-fid, the lobes linear; corolla with a rather short tube, ascending-puberulent, scaleless on the throat, the lobes 5, imbricate in bud, spreading; stamens 5, included, the anthers obtuse; style filiform, the stigma capitate; nutlets 1–2, bony, lustrous, subulate, with a hooked beak, the gynobase pyramidal, with a large, basal flabelliform areole.——One species, in Japan.

1. Ancistrocarya japonica Maxim. SAWA-RURI-SŌ.   Rhizomes short; stems 50–80 cm. long, short appressed-strigose, with about 10 leaves near the middle; leaves chartaceous, oblong to obovate-oblong, 10–20 cm. long, 3–7 cm. wide, short-acuminate, gradually narrowed to the base, sessile, deep green and scabrous above with spotted callosities, pale green and appressed-puberulent beneath, lateral nerves 3–5 pairs, fused to the upper ones at tip; racemes 6–12 cm. long, short appressed-strigose, the bracts linear, absent or present only in a few lower flowers; flowers erect, blue-purple, on very short pedicels; calyx-segments linear-lanceolate to linear; corolla 10–13 mm. long, tubular-campanulate, the lobes semirounded; nutlets gray-white, gradually narrowed from the rounded broad base to a subulate hooked beak, 8–10 mm. long.——May–June. Woods in mountains; Honshu (Kantō Distr. and westw.), Shikoku, Kyushu.——Forma **albiflora** (Honda, pro var.) Hara. SHIROBANA-SAWA-RURI-SŌ.   Flowers white.

12. MERTENSIA Roth HAMA-BENKEI-SŌ ZOKU

Smooth or strigose, sometimes rather fleshy perennials; leaves alternate, often spotted; racemes terminal, sometimes loosely few-flowered and cymelike or in a compound panicle; flowers blue to purplish, rarely white, pedicelled, bractless or bracteate only in the lower ones; calyx 5-lobed, corolla infundibuliform, with a broad, naked, plaited, or scaled throat, the limb shallowly 5-lobed; stamens slightly exserted; gynobase convex or pyramidal; nutlets 4, erect, sometimes fleshy, smooth or rugose, keeled on the ventral face, sometimes with a winged margin, the areole elliptic and slightly excavated on inner angle of ventral face.——About 20 species, in cooler regions of the N. Hemisphere.

1A. Seashore species; stems procumbent; inflorescence glabrous; nutlets fleshy, not wing-margined. .... 1. *M. asiatica*
1B. Alpine species; stems erect; inflorescence short appressed-strigose; nutlets not fleshy, broadly wing-margined.
2. *M. pterocarpa* var. *yezoensis*

1. **Mertensia asiatica** (Takeda) Macbr. *M. maritima* subsp. *asiatica* Takeda; *M. maritima* sensu auct. Japon., non S. F. Gray——HAMA-BENKEI-SŌ. Stems tufted, fleshy, procumbent, to 1 m. long, many-leaved; radical and lower leaves long-petiolate, oblong, elliptic, obovate to broadly so, 3–8 cm. long, 2–6 cm. wide, slightly fleshy, glabrous on both sides or loosely short callose-based hairy, without distinct nerves, the upper leaves subsessile; inflorescence glabrous; flowers pedicelled, bracteate, blue, 8–12 mm. long; calyx-segments glabrous, acute, as long as the corolla-tube; nutlets fleshy.—— Sandy seashores; Hokkaido, Honshu (n. distr.).——Sakhalin, Kuriles, Korea, and Aleutians.

2. **Mertensia pterocarpa** (Turcz.) Tatew. & Ohwi var.

yezoensis Tatew. & Ohwi. *M. rivularis* var. *japonica* Takeda, pro parte; *M. yezoensis* Tatew. & Ohwi, in syn.——Ezo-RURI-SŌ. Rhizomatous perennial; stems erect, 20–40 cm. long, simple or short-branched in upper part; radical leaves ovate-cordate, long-petiolate, the cauline leaves ovate, 3–5 cm. long, 2–3 cm. wide, shortly acuminate, abruptly narrowed to rounded and sessile or subpetiolate at base, 7- to 9-nerved, with appressed callose-based hairs above, glabrescent beneath; inflorescence pedunculate, short appressed-strigose; flowers 10–12 mm. long, blue; calyx lanceolate, pilose, acute, slightly shorter than the corolla-tube; nutlets flat, broadly wing-margined.——Aug. Alpine; Hokkaido; rare. The typical phase occurs in the s. Kuriles.

13. TRIGONOTIS Stev. TABIRAKO ZOKU

Annuals, biennials, or often slender or sometimes delicate perennials, pilose to nearly glabrous; leaves alternate, entire, simple; racemes simple or branched, bractless or bracteate at base, rarely with the flowers axillary in the leaflike bracts; flowers pedicelled or subsessile, blue, rose, or white; calyx deeply 5-lobed, slightly accrescent after anthesis; corolla rotate, 5-scaled on the throat, the lobes 5, imbricate in bud; stamens included; gynobase flat; nutlets 4, erect, smooth, sometimes puberulent, tetrahedral and slightly flattened dorsally.——About 30 species, in s. and e. Asia, Malaysia, and e. Siberia.

1A. Pedicels longer than the calyx, ascending, spreading, or reflexed after anthesis.
  2A. Flowers 6–10 mm. across; perennial with 5- to 15-flowered racemes.
    3A. Flowers supra-axillary on the leaflike bracts.
      4A. Plant appressed-strigose. .... 1. *T. nakaii*
      4B. Plant spreading-strigose on lower part of stems, petioles, and underside of leaves. .... 2. *T. radicans*
    3B. Flowers bractless or bracteate only in the lower ones.
      5A. Flowering stems suberect; stolons absent; inflorescence 8- to 15-flowered, commonly branched or sometimes simple.
3. *T. guilielmii*
      5B. Flowering stems decumbent; stolons well developed; inflorescence 3- to 10-flowered, simple and unbranched. .... 4. *T. icumae*
  2B. Flowers about 2 mm. across; biennial with many-flowered racemes. .... 5. *T. peduncularis*
1B. Pedicels shorter than the calyx, never reflexed, appressed to the axis; flowers about 3 mm. across. .... 6. *T. brevipes*

1. **Trigonotis nakaii** Hara. *T. radicans* sensu auct. Japon., non Maxim.——CHŌSEN-KAMEBA-SŌ. Somewhat tufted perennial, 10–15 cm. high, loosely appressed-strigose; stems erect in flower, decumbent in fruit; leaves ovate, 1.5–4 cm. long, acute, rounded to shallowly cordate at base, the radical ones long-petiolate, the upper short-petiolate; flowers supra-axillary on the leaflike bracts, the pedicels 1–2 cm. long, ascending in anthesis, nodding or pendulous in fruit, calyx-segments accrescent after anthesis, 2.5–7 mm. long, lanceolate, acute; corolla 7–10 mm. across, nearly white, the tube about 2 mm. long, short-pilose at base inside, the lobes orbicular, 4–5 mm. across; stamens inserted above the middle of the tube, the anthers about 0.8 mm. long, obtuse; nutlets puberulent, dark-brown, about 2 mm. long on back.——Mountains; Kyushu (Higo Prov.); rare.——Korea and Manchuria.

2. **Trigonotis radicans** (Turcz.) Stev. *Myosotis radicans* Turcz.; *Omphalodes aquatica* Brand; *O. sericea* Maxim.; *T. sericea* Ohwi——KE-RURI-SŌ. Somewhat tufted perennial,

spreading-strigose on lower part of stems, petioles, and underside of leaves, appressed-hairy on other parts; stems much elongate and decumbent; leaves ovate, 2.5–5 cm. long, 1–3 cm. wide, rounded and mucronate or subacute, rounded to shallowly cordate at base, the petiole long in the radical ones, much shorter in those above; flowers supra-axillary, the pedicels 1–2 cm. long; calyx-segments 2–5 mm. long, narrowly oblong; corolla 8–10 mm. across; stamens inserted on the middle of the tube, the anthers obtuse, about 0.8 mm. long.——Mountains; Kyushu (Higo Prov.); rare.——Korea and e. Siberia.

3. **Trigonotis guilielmii** A. Gray, ex Guerke. *Eritrichium guilielmii* A. Gray——TACHI-KAMEBA-SŌ. Loosely appressed-strigose perennial herb with creeping rhizomes; stems erect, 20–40 cm. long, simple; leaves ovate or broadly so, 3–7 cm. long, acute, the radical ones long-petiolate, shallowly cordate, the cauline short-petiolate, truncate to broadly cuneate at base; racemes 2- to 4-branched or sometimes solitary, 8- to 15-flowered, bractless, the pedicels ascending, 1–1.5 cm. long;

corolla white or pale blue, 7–10 mm. across, the tube 1.5 mm. long, smooth, glabrous; stamens inserted above the middle of the tube, the anthers about 0.5 mm. long; nutlets dark brown, very short-stipitate, puberulent, about 2 mm. long on the back.——May–June. Wet places in mountains; Hokkaido, Honshu.

4. Trigonotis icumae (Maxim.) Makino. *Omphalodes icumae* Maxim.——Tsuru-kameba-sō.   Perennial, short appressed-strigillose, stoloniferous; stems 7–20 cm. long, soon decumbent; leaves ovate, 3–5 cm. long, 1.5–2.5 cm. wide, acute, truncate to shallowly cordate or rounded at base, the lower ones long-petiolate; inflorescence soon becoming lateral, unbranched, loosely 3- to 10-flowered, bractless, the pedicels 1–1.5 cm. long, ascending; corolla about 10 mm. across, pale blue, the tube about 2 mm. long; stamens inserted above the middle of the tube, the anthers about 0.5 mm. long.——May–June. Honshu (centr. and n. distr.); rare.

5. Trigonotis peduncularis (Trevir.) Benth. *Myosotis peduncularis* Trevir.; *Eritrichium japonicum* Miq.——Tabi-rako, Kiurigusa.   Appressed-pilosulous biennial; stems branched at base, somewhat tufted, 10–30 cm. long, erect, slender; leaves oblong or ovate, 1–3 cm. long, 6–20 mm. wide, obtuse, mucronate, subcuneate at base, the lower ones petiolate, the upper ones sessile; racemes rather loosely many-flowered, 5–25 cm. long, often leafy at the base, the pedicels ascending, 3–9 mm. long, spreading in fruit, clavately thickened at tip; flowers pale blue, about 2 mm. across; stamens inserted on the middle of the corolla-tube, the anthers about 0.3 mm. long; nutlets short-stipitate, glabrous or puberulent, about 1 mm. long on the back.——Apr.–May. Grassy places and cultivated fields in lowlands; Hokkaido, Honshu, Shikoku, Kyushu; very common.——Ryukyus, Korea, and widely distributed in the temperate regions of Asia.

6. Trigonotis brevipes (Maxim.) Maxim. *Eritrichium brevipes* Maxim.——Mizu-tabirako.   Rhizomatous and stoloniferous, appressed-pilose perennial; stems somewhat tufted, 10–40 cm. long, erect, simple or slightly branched; leaves oblong or ovate, 1.5–3 cm. long, 1–2 cm. wide, obtuse and mucronate, or rounded at both ends, petiolate; racemes 1–5, sessile, 5–10 cm. long, densely many-flowered, the pedicels 1–2 mm. long or subsessile, erect, appressed to the axis; calyx-segments 0.5–1 mm. long; corolla pale blue, 2.5–3 mm. across, the tube less than 1 mm. long; stamens inserted on the middle of the tube, the anthers about 0.3 mm. long; nutlets dark-brown, smooth, lustrous, tetrahedral, without appendages, about 1 mm. long on back.——Wet places along streams and ravines in mountains; Honshu, Shikoku, Kyushu; rather common.—— ? China.

Var. **coronata** (Ohwi) Ohwi. *Trigonotis coronata* Ohwi ——Koshiji-tabirako.   Nutlets appendaged.——May–July. Wet places along streams in mountains; Honshu (especially on Japan Sea side from Kinki to Hokuriku Distr.).

14. MYOSOTIS L.   Wasurenagusa Zoku

Usually slender annuals or perennials, nearly glabrous to densely pilose; leaves alternate, narrow; racemes simple or sometimes branched, bractless or rarely with few bracts at base; flowers pedicelled or subsessile, blue, rose, or white, rarely yellowish; calyx shallowly 5-lobed, the lobes slightly accrescent after anthesis; corolla-tube short, 5-scaled or tubercled on the throat, the lobes 5, spreading, contorted in bud; stamens included; nutlets tetrahedral, ovoid, erect, lustrous, smooth, usually black.—— About 40 species, in the temperate regions of the N. Hemisphere.

1A. Hairs on calyx short, straight, closely appressed. .... 1. *M. scorpioides*
1B. Hairs on calyx hooked, at least some of them.
  2A. Corolla blue, rarely white; fruiting pedicels longer than the calyx.
    3A. Limb of corolla flat, 6–8 mm. across. .... 2. *M. sylvatica*
    3B. Limb of corolla spreading-ascending, about 3 mm. across. .... 3. *M. arvensis*
  2B. Corolla yellow, changing to blue before withering, 2–3 mm. across; calyx longer than the pedicel. .... 4. *M. discolor*

1. Myosotis scorpioides L. *M. palustris* Hill——Wasurenagusa.   Resembles *M. sylvatica,* but with creeping or decumbent stems; calyx shallowly lobed, loosely appressed-pilose. ——Wet places in Honshu (Shinano Prov.) and Hokkaido (Nemuro Prov.).——Introduced from Europe and naturalized.

2. Myosotis sylvatica (Ehrh.) Hoffm. *M. scorpioides* var. *sylvatica* Ehrh.; *M. intermedia* sensu auct. Japon., non Link——Ezo-murasaki, Miyama-wasurena-sō.   Perennial with coarse spreading hairs; rhizomes short-creeping; stems 20–40 cm. long, slender, loosely branched in upper part; radical leaves spathulate, rounded at apex; cauline leaves oblanceolate, 2–6 cm. long, 7–12 mm. wide, obtuse, the middle and upper ones sessile and semiclasping; racemes 10–25 cm. long, bractless or bracteate only at base; calyx shorter than the pedicels, with spreading hooked hairs; corolla blue, 6–8 mm. across; nutlets ovoid, dark brown, smooth, lustrous, 1.5 mm. long, slightly flattened, with a ridge on the ventral side.——May–July. Woods in mountains; Hokkaido, Honshu (centr. distr.); rare.——Eurasia and Africa.

3. Myosotis arvensis (L.) Hill.   Nohara-murasaki. Resembles *M. sylvatica,* but with smaller flowers about 3 mm. across.——Honshu (Shirakawa in Iwaki Prov.).——Introduced from Europe and naturalized.

4. Myosotis discolor Pers. *M. versicolor* (Pers.) J. E. Smith——Hama-wasurenagusa.   Small annual or biennial with flowers 2–3 mm. across, yellow at the beginning, changing to blue.——Sandy places; Honshu (Echigo and elsewhere).——Introduced from Europe and now naturalized.

Fam. 174. **VERBENACEAE**   Kuma-tsuzura Ka   Verbena Family

Herbs, shrubs, or trees, commonly with 4-striate or angled branches; leaves opposite or sometimes verticillate, simple or compound; stipules absent; flowers bisexual, zygomorphic; calyx 4- to 5-lobed or toothed, persistent; corolla gamopetalous, tubular, 4- to 5-lobed, the lobes imbricate in bud; stamens 4, rarely 2 or 5, inserted on the corolla, the anthers 2-locular, the locules often divergent, longitudinally split; ovary superior, 2- to 8-locular; style simple, terminal; ovules 1 or 2 in each locule, erect or rarely pendulous; fruit a drupe or berry; seeds with a straight embryo, the endosperm scanty.——About 90 genera, with about 2,600 species, mostly in warmer regions.

1A. Inflorescence a spike or raceme.
  2A. Calyx tubular, 5-toothed; nutlets 4. .... 1. *Verbena*
  2B. Calyx short, bilabiate or 2- to 4-lobed; nuts 2. .... 2. *Lippia*
1B. Inflorescence a loose or dense cyme.
  3A. Fruit a drupe; woody (in ours).
    4A. Flowers nearly actinomorphic; stamens equal; leaves usually with sessile discoid glands. .... 3. *Callicarpa*
    4B. Flowers zygomorphic; stamens didymous, more or less unequal.
      5A. Drupe with a single 4-locular stone; corolla-tube short.
        6A. Corolla 4-lobed, the lower lip slightly larger than the upper; fruit fleshy. .... 4. *Premna*
        6B. Corolla 5-lobed, the lower lip much larger than the upper; fruit dry. .... 5. *Vitex*
      5B. Drupe with 4, 1-loculard stones; corolla-tube much elongate. .... 6. *Clerodendrum*
  3B. Fruit a capsule; herbaceous (in ours). .... 7. *Caryopteris*

1. VERBENA L.   Kuma-tsuzura Zoku

Herbs or subshrubs, glabrous or with simple hairs; leaves opposite, rarely alternate or verticillate, usually toothed or divided; spikes terminal; flowers small, sessile, solitary in the axils of bracts; calyx tubular, 5-ribbed, 5-toothed; corolla-tube sometimes dilated in upper part, the limb spreading, subbilabiate, the lobes 5, nearly equal, the 2 posterior lobes outermost in bud, the anterior and median innermost; stamens 4, didymous, included; ovary 4-parted, the ovules 4; style short, unequally 2-fid; fruit enclosed in the calyx, dry, with 4 nutlets.——About 120 species, chiefly in the New World, few in the Old World.

1A. Leaves ovate, pinnately lobed to cleft, petiolate; spikes linear, much elongate after anthesis, to 30 cm. long, becoming loose in fruit.
                                                                1. *V. officinalis*
1B. Leaves lanceolate, merely toothed on margin, sessile; spikes ellipsoidal to oblong-cylindric, 1–2 cm. long, densely flowered.
                                                                2. *V. bonariensis*

1. Verbena officinalis L. *V. spuria* L.——Kuma-tsuzura. Loosely short-hirsute perennial; stems 4-angled, usually branched, 30–80 cm. long; leaves cauline, ovate, 3–10 cm. long, 2–5 cm. wide, pinnately lobed to cleft, the lobes again incised, abruptly narrowed to a winged petiole; spikes much elongate after anthesis, to 30 cm. long; flowers pale purple, about 4 mm. across, the bracts lanceolate, nearly as long as the calyx; calyx about 2 mm. long; nutlets about 1.5 mm. long, with few longitudinal ribs on back.——June–Nov. Weed in waste grounds and along roadsides in lowlands; Honshu, Shikoku, Kyushu.——A nearly cosmopolitan weed.

2. Verbena bonariensis L. Sanjaku-bābena. Stems to more than 1 m. high, much branched, 4-angled; leaves lanceolate, 7–12 cm. long, 1–3 cm. wide, rounded and clasping at base, scarious; spikes many, in terminal corymbs, densely many-flowered; flowers small, blue-purple, about 1 cm. long, exserted from the bracts.——Sometimes cultivated and naturalized in the western part of Honshu and Kyushu.——Introduced from S. America.

Verbena × hybrida Voss. Bijozakura. Flower heads 1.5–2 cm. across; calyx slender-tubular, about 12 mm. long, glandular-pilose.——Of garden origin and often cultivated.

2. LIPPIA L.   Iwadare-sō Zoku

Shrubs or herbs, glabrous or with simple hairs; leaves opposite, sometimes verticillate, toothed or entire; inflorescence spicate; flowers small, sessile, solitary in axils of bracts; calyx small, sometimes bilabiate or 2-winged, 2- or 4-lobed; corolla tubular, the limb ascending and spreading, 5-lobed; stamens 4, anther-locules parallel; ovary 2-locular; ovules solitary in each locule; stigma rather thick; nutlets 2.——About 100 species, chiefly in the New World.

1. Lippia nodiflora (L.) L. C. Rich. *Verbena nodiflora* L.; *Phyla chinensis* Lour.; *P. nodiflora* (L.) Greene; *L. nodiflora* var. *sarmentosa* Schauer——Iwadare-sō. Perennial; stems long-creeping or decumbent, much branched; leaves rather fleshy, obovate, 2–4 cm. long, 8–18 mm. wide, rounded to obtuse, long-cuneate at base, coarsely toothed toward tip, subsessile, 1-nerved; spikes axillary, unbranched, long-pedunculate, ellipsoidal to short-cylindric, 8–20 mm. long, 6–8 mm. across, very densely many-flowered, the bracts flabellate, about 2.5 mm. long, cuspidate; calyx short, flattened, with 2 narrow wings; corolla about 2 mm. across, slightly exserted between the bracts, subbilabiate, rose-purple; fruit broadly obovate, about 2 mm. long; nutlets slightly corky.——July–Oct. Sandy seashores; Honshu (Kantō Distr. and westw.), Shikoku, Kyushu; common.——Ryukyus, Formosa, China, and widely distributed in the warmer regions of the world.

3. CALLICARPA L.   Murasaki-shikibu Zoku

Shrubs or trees, villous, stellate-pubescent, or glabrate; leaves opposite, toothed, usually with sessile discoid glands; cymes axillary, many-flowered, sessile or pedunculate; flowers nearly regular, small, white, rose, or purplish; calyx short-campanulate, truncate or 4-lobed; corolla tubular, the limb 4-lobed, the lobes nearly equal, imbricate in bud; stamens 4, equal, the anther-locules parallel; ovary incompletely 2-locular, with 2 ovules in each locule; stigma 2-fid, broad and short; drupe globose, juicy, 4- or few-stoned.——About 40 species, in the warmer parts of the N. Hemisphere, except S. America and Africa, abundant in se. Asia.

1A. Plants glabrous or thinly pubescent; calyx glabrous, with very short teeth.
  2A. Leaves caudate, glandular-dotted on both sides. .... 1. *C. shikokiana*
  2B. Leaves acuminate to acute glandular-dotted on underside only.

3A. Cymes supra-axillary; anthers broadly ellipsoidal. . . . . 2. *C. dichotoma*
3B. Cymes axillary.
    4A. Corolla 1 mm. long, not glandular-dotted; branches slightly 4-angled; leaves with 12 to 14 pairs of nerves. 3. *C. takakumensis*
    4B. Corolla 3–5 mm. long, glandular-dotted; branches terete; leaves with 5 to 9 pairs of nerves. . . . . 4. *C. japonica*
1B. Plants densely soft-pubescent to villous; calyx pubescent, 4-fid.
    5A. Leaves 5–10 cm. long, rounded to obtuse at base; branches and leaves with whitish stellate hairs less than 1 mm. long; calyx-lobes lanceolate; flowers 4–5 mm. long, about 10 in a cyme; anthers 1.5–2 mm. long. . . . . 5. *C. mollis*
    5B. Leaves 15–30 cm. long, gradually narrowed at base; branches and petioles with pinnately branched hairs 1.5–3 mm. long; calyx-lobes linear; flowers about 1.5 mm. long, very many in a cyme; anthers about 0.7 mm. long. . . . . 6. *C. kochiana*

1. Callicarpa shikokiana Makino. *C. yakusimensis* Koidz.——TOSA-MURASAKI, YAKUSHIMA-KO-MURASAKI. Deciduous shrub, scattered short-pilose; branches slender, terete; leaves membranous, 5–12 cm. long, 1.5–3 cm. wide, caudate, cuneate at base, coarsely and obtuse toothed, glandular-dotted on both sides, loosely pilose, especially on nerves on both sides; petioles 5–10 mm. long; cymes axillary, short-pedunculate, rather many-flowered; corolla puberulent, scarcely glandular, about 3 mm. long, 4-lobed to the middle; anthers broadly ellipsoidal, about 0.6 mm. long; fruit purple, about 2 mm. across.——July–Sept. Warmer districts; Shikoku, Kyushu.

2. Callicarpa dichotoma (Lour.) K. Koch. *Porphyra dichotoma* Lour.; *C. purpurea* Juss.; *C. gracilis* Sieb. & Zucc.; *C. japonica* var. *dichotoma* (Lour.) Bakhuizen——KO-MURASAKI, KO-SHIKIBU. Deciduous shrub with slender purplish branches, minutely stellate-hairy while young; leaves rather thick, obovate-oblong or obovate, 3–6 cm. long, 1.5–3 cm. wide, abruptly acuminate, cuneate to acute at base, loosely puberulent above and on nerves beneath, coarsely serrate near tip, the petioles 1–4 mm. long; cymes supra-axillary, 10- to 20-flowered, the peduncles 1–1.5 cm. long; corolla pale purple, glabrous, about 3 mm. long; anthers broadly ellipsoidal, about 0.6 mm. long; fruit about 3 mm. across, purple.——July–Aug. Honshu (Rikuchu Prov. and westw.), Shikoku, Kyushu.——Korea, Ryukyus, Formosa, and China.

3. Callicarpa takakumensis Hatus. TAKAKUMA-MURASAKI. Shrub with slightly 4-angled glabrous branches; leaves membranous, oblong-lanceolate, 10–18 cm. long, 2.5–3.7 cm. wide, acuminate at both ends, obsoletely serrulate, grayish puberulent on nerves above, densely discoid-glandular beneath, the lateral nerves 12–14 pairs, the petioles about 1 cm. long; cymes axillary, densely many-flowered, about 3 cm. across, the peduncles about 1.3 cm. long; calyx glabrous; corolla about 1 mm. long, puberulent outside; stamens about twice as long as the corolla; fruit about 2 mm. across.——Aug. Kyushu (Mount Takakuma in Osumi); rare.

4. Callicarpa japonica Thunb. *C. murasaki* Sieb.; *C. japonica* forma *angustifolia* Miq.——MURASAKI-SHIKIBU. Deciduous shrub; branches terete, minutely stellate-puberulent while young, glabrescent; leaves membranous to thinly chartaceous, obovate, ovate or oblong, 6–12 cm. long, 2.5–4.5 cm. wide, acuminate to acute at both ends, serrulate, nearly glabrous to loosely puberulent on both surfaces, yellowish glandular-dotted below, the lateral nerves 5–9 pairs, the petioles 2–10 mm. long; cymes rather many-flowered, axillary, short-pe-

dunculate, scattered stellate-puberulent; corolla pale purple, 3–5 mm. long, usually puberulent and glandular-dotted; anthers oblong, about 2 mm. long; fruit purple, about 3 mm. across.——June–Aug. Hokkaido (s. distr.), Honshu, Shikoku, Kyushu; common and variable.——Ryukyus, Formosa, China, and Manchuria.

Var. **luxurians** Rehd. *C. australis* Koidz.; *C. japonica* subsp. *luxurians* (Rehd.) Masam. & Yanagih.; *Premna staminea* Maxim.——Ō-MURASAKI-SHIKIBU. Leaves thicker, larger, rather lustrous above, 10–15(–20) cm. long, 4–6(–7) cm. wide, serrulate; cymes larger with thicker branches.——July–Aug. Lowlands near the sea; Honshu, Shikoku, Kyushu; rather common.

5. Callicarpa mollis Sieb. & Zucc. *C. zollingeriana* Schauer——YABU-MURASAKI. Deciduous shrub, stellate-hairy on young branches; leaves thinly membranous, ovate to broadly so or ovate-elliptic, 5–10(–12) cm. long, 2.5–5 cm. wide, acuminate, rounded to obtuse at base, serrate, densely pubescent above with soft simple hairs, densely stellate-pubescent beneath, minutely glandular-dotted on both surfaces, the petioles 3–5 mm. long; cymes often more than 10-flowered or less, densely stellate-pubescent; calyx 5-parted, the lobes lanceolate, densely stellate-pubescent or with variously branched grayish hairs; corolla 4–5 mm. long, the tube nearly as long as the calyx; fruit purple, 3–4 mm. across.——June–July. Honshu (Rikuchu Prov. and southw.), Shikoku, Kyushu.——Korea.

Callicarpa × shirasawana Makino. INU-MURASAKI-SHIKIBU. Hybrid of *C. japonica* × *C. mollis*.——Honshu.

6. Callicarpa kochiana Makino. *C. tomentosa* sensu auct. Japon., non Willd.; *C. longiloba* Merr.——BIRŌDO-MURASAKI, ONI-YABU-MURASAKI. Deciduous shrub with pinnately branched yellow-brown hairs; branches thick, terete, villous; leaves chartaceous, narrowly oblong, broadly lanceolate, or ovate-oblong, 15–30 cm. long, 4–8 cm. wide, long-acuminate, gradually narrowed at base, serrulate, with gray-yellow stellate and pinnately branched hairs beneath, obsoletely glandular-dotted, the lateral nerves 8–12 pairs, the petioles 2–3.5 cm. long; cymes densely many-flowered, short-pedunculate; calyx 5-parted, the lobes broadly linear; corolla about 1.5 mm. long; anthers oblong, about 0.5 mm. long; fruit about 2 mm. across, enclosed by the calyx, white.——Aug. Honshu (s. Kinki Distr.), Shikoku, Kyushu (s. distr.).——Formosa, China, and Indochina.

4. PREMNA L.   HAMA-KUSAGI ZOKU

Shrubs or trees; leaves opposite, entire or toothed; cymes axillary, in terminal corymbs, panicles, or racemes; flowers small, zygomorphic, white, yellowish, or purplish, often polygamous; calyx small, campanulate, truncate or toothed; corolla-tube rather short, often hairy inside the throat, the limb 4-lobed, the lobes somewhat unequal, the dorsal ones larger, retuse or lip-like; stamens didymous, rarely exserted; ovary 2-locular, with 2 ovules in each locule; style bifid; fruit a small globose fleshy drupe, with a persistent calyx at base, the stone solitary, 4-locular, sometimes 2- or 3-locular.——About 40 species, chiefly in warmer regions.

1. Premna japonica Miq. *P. microphylla* sensu auct. Japon., non Turcz.; (?) *P. glabra* A. Gray ex Maxim.; *P. luxurians* Nakai; *P. japonica* var. *luxurians* (Nakai) Sugimoto——HAMA-KUSAGI. Deciduous much-branched small tree or large shrub, furfuraceous while very young; branches gray-brown, terete, the nodes with thickened pulvinilike scars; leaves membranous, ovate to broadly so, or elliptic, 5–12 cm. long, 2.5–7 cm. wide, acuminate, obtuse at the tip, coarsely few-toothed on upper half or entire, the petioles 1–3 cm. long; panicles pedunculate, 6–10 cm. long, loosely branched, few-flowered, the bracts linear, truncate; flowers pale greenish yellow, 8–10 mm. long, tubular-campanulate; calyx irregularly toothed; corolla glandular outside, the lobes short; fruit obovoid-globose, dark purple, 3–3.5 mm. across.——May–June. Near seashores; Honshu (s. Kinki, Chūgoku Distr.), Shikoku, Kyushu.——Ryukyus and Formosa.

5. VITEX L.   HAMAGŌ ZOKU

Glabrous or hairy shrubs or trees; leaves opposite, simple or usually palmately compound, entire or toothed; cymes axillary or terminal; flowers zygomorphic, white, purplish, or yellowish, the bracts small; calyx (3-)5-toothed; corolla-limb subbilabiate, 5-lobed, the 2 inner lobes often the smallest, the outermost ones largest, sometimes retuse; stamens 4, usually exserted, didymous; ovary imperfectly 2-locular, later becoming 4-locular, with 2 ovules in each locule; drupe globose, dry or slightly juicy, surrounded usually by the accrescent calyx, the stone bony, 4-locular.——About 100 species, chiefly in the Tropics, few in temperate regions of Europe and e. Asia.

1. Vitex rotundifolia L. f. *V. ovata* Thunb.; *V. trifolia* var. *simplicifolia* Cham.; *V. trifolia* var. *unifoliata* Schauer; *V. trifolia* var. *ovata* (Thunb.) Makino——HAMAGŌ. Procumbent or ascending shrub densely gray-white puberulent; branches 4-angled; leaves herbaceous, broadly ovate to broadly elliptic, 2–5 cm. long, 1.5–3 cm. wide, obtuse to rounded, abruptly acute at base, entire, green and thinly puberulent above, densely grayish puberulent beneath; panicles terminal, densely flowered, 4–7 cm. long, with very short branches; corolla purplish, about 13 mm. long; style about 15 mm. long; fruit globose, 5–7 mm. across, enclosed on lower half by the persistent calyx.——July–Sept. Sandy places by the sea; Honshu, Shikoku, Kyushu.——Korea, Bonins, Ryukyus, Formosa to se. Asia, Pacific Islands, and Australia.

6. CLERODENDRUM L.   KUSAGI ZOKU

Shrubs or trees; leaves opposite, rarely ternate, entire, sometimes obsoletely toothed or shallowly lobed, cymes in upper axils, sometimes forming a terminal corymb or panicle; flowers zygomorphic, rather large, white, blue, or red; calyx campanulate, infundibuliform or tubular, 5-lobed, often accrescent after anthesis, green or variously colored; corolla-tube much-elongate, the limb 5-lobed, spreading or reflexed, the lobes nearly equal; stamens 4, long-exserted, involute in bud; ovary incompletely 4-locular, the ovules 1 in each locule; style elongate, bifid; drupe surrounded at base by the persistent calyx; stones 4.——About 100 species, in the Tropics and subtropics.

1A. Leaves narrowly deltoid to deltoid-ovate, 8–15 cm. long, 3-nerved from the base; corymbs rather many-flowered, much-branched, terminal; calyx acutely angled, deeply 5-lobed, about 15 mm. long, about half as long as the corolla-tube. ...... 1. *C. trichotomum*
1B. Leaves elliptic, ovate to broadly so, 3–7 cm. long, obscurely penninerved; corymbs usually 3-flowered, axillary; calyx not angled, with 5 depressed teeth, 3–4 mm. long, shorter than the corolla-tube. .... 2. *C. neriifolium*

1. Clerodendrum trichotomum Thunb. *Siphonanthus trichotomus* (Thunb.) Nakai; *C. serotinum* Carr.——KUSAGI. Large shrub or small tree with rather thick terete branches; leaves membranous, deltoid-ovate, narrowly deltoid, or broadly ovate, 8–15 cm. long, 5–10 cm. wide, gradually acuminate, somewhat rounded at base, obsoletely toothed or subentire, 3-nerved from the base, sometimes thinly pubescent above, pubescent beneath especially on nerves, obsoletely glandular-dotted, long-petiolate; cymes in large terminal corymbs; calyx deeply 5-lobed, about 15 mm. long, reddish, the lobes ovate, acuminate, spreading in fruit; corolla-tube 2–2.5 cm. long, the limb 2.5–3 cm. across, white; fruit globose, dark blue, lustrous, 6–7 mm. across, subtended by the persistent spreading calyx.——Aug.–Sept. Hokkaido, Honshu, Shikoku, Kyushu; common.——Korea, Manchuria, China, Ryukyus, Formosa, and Philippines.——Forma **ferrugineum** (Nakai) Ohwi. *C. trichotomum* var. *ferrugineum* Nakai——BIRŌDO-KUSAGI. Plant densely pubescent.——Occurs with the typical phase.

Var. **esculentum** Makino. SHŌRO-KUSAGI. Leaves cordate, long-acuminate; inflorescence shorter; calyx-lobes narrower.——Occurs with the typical phase.

Var. **yakusimense** (Nakai) Ohwi. *Siphonanthus yakusimensis* Nakai; *C. yakusimense* (Nakai) Nakai——AMA-KUSAGI. Plant less pubescent and with fewer flowers and thicker lustrous leaves.——Occurs with the typical phase.

2. Clerodendrum neriifolium Wall. ex Schaur. *Volkameria neriifolia* Roxb.; *C. inerme* sensu auct. Japon., non Gaertn.——IBOTA-KUSAGI, GASHANGI. Slightly scandent shrub; branches grayish brown, gray-white puberulent while young; leaves chartaceous, elliptic, ovate to broadly so, 3–7 cm. long, 1.5–3.5 cm. wide, acute at both ends, sometimes obtuse at apex, entire, lustrous, puberulent while very young, the petioles 5–10 mm. long; cymes long-pedunculate, axillary, 3-flowered; calyx 3–4 mm. long at anthesis, becoming obconical, 8–10 mm. long in fruit; corolla-tube 2–2.5 cm. long, the limb about 15 mm. across, pinkish; fruit obovoid, 6–8 mm. across, covered on lower half by the persistent calyx.——Wet banks along rivers; Kyushu (Tanegashima); rare.——Ryukyus, Formosa, China, Burma, Malaysia to Australia.

Clerodendrum japonicum (Thunb.) Sweet. *Volkameria japonica* Thunb.; *C. squamatum* Vahl.——HIGIRI. Malayan shrub often cultivated in the warmer parts of our area as an ornamental.

7. CARYOPTERIS Bunge KARIGANE-SŌ ZOKU

Herbs or shrubs, often puberulent; leaves opposite, dentate or entire; cymes many-flowered, axillary, sometimes in a narrow terminal compound panicle; flowers zygomorphic, blue, rose, or rarely white; calyx campanulate, 5-lobed, more or less accrescent in fruit; corolla tubular, the limb 5-lobed, the anterior one longer, placed innermost in the bud, concave, sometimes fimbriate, the others equal, oblong or obovate, flat; stamens 4, didymous, involute in bud, exserted at anthesis; ovary imperfectly 4-locular, with 1 ovule in each locule; style long, bifid; fruit dry, shorter than the persistent calyx, separating into 4 nutlets.——About 10 species, in e. Asia from Mongolia to the Himalayas.

1A. Plant green, 40–100 cm. high; leaves 8–15 cm. long; corolla-lobes entire. .... 1. *C. divaricata*
1B. Plant cinereous, 30-60 cm. high; leaves 3–6 cm. long; anterior lobe of corolla larger, fimbriate. .... 2. *C. incana*

1. **Caryopteris divaricata** Maxim. *Clerodendron divaricatum* Sieb. & Zucc., non Jack——KARIGANE-SŌ. Ill-scented thinly-pubescent perennial herb, 40–100 cm. high; stems 4-angled, green, branched; leaves membranous, ovate to broadly so, 8–15 cm. long, 4–8 cm. wide, green, obtusely toothed, short-acuminate, shallowly cordate to rounded at base, long-petiolate; panicles long-pedunculate, loosely few-flowered, the peduncles and pedicels often scattered glandular-pilose; calyx obconical, 2–3 mm. long in anthesis, 5–6 mm. long in fruit, the teeth deltoid; corolla blue-purple, the tube 8–10 mm. long, the limb oblique, the largest lobe reflexed, slightly longer than the tube; stamens 3–3.5 cm. long; style 3–3.5 cm. long; nutlets obovoid, 4–4.5 mm. long, net-veined, glandular-dotted.—— Aug.–Sept. Woods in mountains; Hokkaido, Honshu, Shikoku, Kyushu.——Korea and China.

2. **Caryopteris incana** (Thunb.) Miq. *Nepeta incana* Thunb.; *Barbula sinensis* Lour.; *N. japonica* Willd.; *C. mastacanthus* Schauer; *C. ovata* Miq.——DANGIKU. Perennial herb, 30–60 cm. high, woody at the base, cinereous, incurved short-pubescent; stems nearly terete, erect, simple or branched; leaves ovate, 3–6 cm. long, 1.5–3 cm. wide, subacute, broadly cuneate to rounded at base, with few coarse teeth on each side, densely short grayish pubescent beneath, the petioles 5–15 mm. long; cymes short-pedunculate, densely many-flowered; calyx 2–3 mm. long at anthesis, 5–6 mm. long in fruit, deeply lobed; corolla about 7 mm. long, blue-purple, the anterior lobe larger, fimbriate; mature carpels winged, about 2 mm. long, deciduous with the seeds.——Sept.–Oct. Kyushu (w. distr. including Tsushima).——Korea, China, and Formosa.——Forma **candida** (C. K. Schn.) Hara. *C. incana* var. *candida* C. K. Schn.——SHIROBANA-DANGIKU. A white-flowered phase, often cultivated.

Fam. 175. **LABIATAE** SHISO KA   Mint Family

Herbs, rarely shrubs, usually aromatic; stems and branches commonly 4-angled; leaves opposite or verticillate, simple; stipules absent; inflorescence composed of dichasial or circinnate cymes forming a simulated whorl (verticillaster); flowers bisexual, usually bilabiate; calyx 5-toothed, regular or bilabiate; corolla tubular, the limb 4- to 5-lobed, often bilabiate; stamens 4 or 2, inserted on the tube, the anthers 2-locular, often with divergent locules, usually longitudinally split; ovary superior, deeply 4-lobed; style solitary, gynobasic, usually bifid; ovules 4, erect; fruit of 4 achenelike nutlets, the endosperm scanty or absent.—— About 160 genera, with about 3,000 species, cosmopolitan, especially abundant in the Mediterranean basin and w. Asia.

1A. Ovary shallowly 4-lobed; nutlets with a large areole on inner face near base.
  2A. Stamens 2; corolla nearly actinomorphic, 4-lobed, the lower (outer) lobe rather larger than the others. .... 1. *Amethystea*
  2B. Stamens 4; corolla zygomorphic, the upper (inner) lobe very small.
    3A. Upper lip entire or shallowly 2-lobed, the midlobe of lower lip bifid or emarginate. .... 2. *Ajuga*
    3B. Upper lip deeply 2-fid, the midlobe of the lower lip much larger than the others, sometimes bifid. .... 3. *Teucrium*
1B. Ovary deeply 4-lobed; nutlets with a small areole at base.
  4A. Calyx gibbous on upper (inner) side; upper lip of the corolla incurved, the tube often geniculate and gibbous at base.
                                                                                4. *Scutellaria*
  4B. Calyx not gibbous.
    5A. Upper lip of corolla galeate or much concave on back.
      6A. Calyx 13- to 15-nerved; upper (inner) pair of stamens longer than the others.
        7A. Anthers not approximate, separate or distant, anther-locules nearly parallel.
          8A. Flowers in dense spikes; stamens exserted. .... 5. *Agastache*
          8B. Flowers axillary or in a loose raceme; stamens not exserted. .... 6. *Meehania*
        7B. Anthers approximate, anther-locules strongly divergent.
          9A. Calyx-teeth oblique at the throat, with subequal narrow teeth.
            10A. Stems erect; cymes dense, interrupted, in terminal spikes; anther-locules divergent. .... 7. *Nepeta*
            10B. Stems procumbent or creeping; flowers in the leaf axils; anther-locules rectangular, each pair crossing. ..... 8. *Glechoma*
          9B. Calyx-teeth equal, the upper teeth much broader than the others. .... 9. *Dracocephalum*
      6B. Calyx 5- to 10-nerved; upper (inner) pair of stamens shorter than the others.
        11A. Calyx distinctly bilabiate, 10-nerved, closed in fruit, the upper lip flat, erect, toothed, the lower 2 teeth incurved at tip.
                                                                                         10. *Prunella*
        11B. Calyx not distinctly bilabiate, open in fruit, with nearly equal teeth.
          12A. Calyx much inflated in fruit, membranous-papery, campanulate; corolla gradually inflated in upper part, with a concave upper lip; nutlets flattened dorsally, winged on upper margin. .... 11. *Chelonopsis*
          12B. Calyx scarcely accrescent after anthesis, nor inflated; nutlets wingless.
            13A. Nutlets obovate, rounded at apex, obtusely 2-angled, unequally lenticular.
              14A. Anther-locules transversely split, ciliate on the inner side of the locule. .... 12. *Galeopsis*
              14B. Anther-locules longitudinally split, not ciliate. .... 13. *Stachys*

13B. Nutlets cuneate or obovate-cuneate at the base, truncate or obliquely truncate at apex, acutely 3-angled.
    15A. Anther-locules parallel; upper lip of corolla suberect, densely villous on back; anthers glabrous. . . . . 14. *Leonurus*
    15B. Anther-locules divergent; upper lip of corolla arcuately incurved, galeate, glabrous or soft-pubescent on back; anthers barbate or glabrous. . . . . .15. *Lamium*
5B. Upper lip of corolla not concave on back.
  16A. Fertile stamens 2.
    17A. Anthers of fertile stamens 1-locular or apparently 2-locular; connective elongate at base and jointed with the filament; corolla clearly bilabiate. . . . . 16. *Salvia*
    17B. Anthers of fertile stamens 2-locular, the locules approximate or confluent at tip; connective not elongate at base; corolla nearly actinomorphic.
      18A. Nutlets subglobose, without angles or ridges, rounded at apex. . . . . 17. *Mosla*
      18B. Nutlets broadly cuneate, triangular, obliquely truncate and obsoletely undulate-toothed at apex. . . . . 18. *Lycopus*
  16B. Fertile stamens 4.
    19A. Stamens erect, spreading, or ascending under the upper lip of corolla; corolla actinomorphic or bilabiate, the upper lip 2-lobed, the lower lip 3-lobed.
      20A. Stamens, at least the lower pair, ascending or arcuate beneath the upper lip of corolla, not exserted; anthers approximate; nutlets globose. . . . . 19. *Clinopodium*
      20B. Stamens erect or spreading, not ascending beneath the upper lip.
        21A. Calyx bilabiate; nutlets globose, obtuse.
          22A. Small shrubs with small leaves; corolla distinctly bilabiate; nutlets rounded, rather flattened, smooth. . . 20. *Thymus*
          22B. Herbs; corolla indistinctly bilabiate.
            23A. Annuals; spikes densely many-flowered; nutlets reticulate-veined. . . . . 21. *Perilla*
            23B. Perennials; spikes loosely flowered; nutlets veinless. . . . . 22. *Perillula*
        21B. Calyx actinomorphic; nutlets not globose.
          24A. Inflorescence not one-sided.
            25A. Anther-locules distinct; flowers or some of them in spaced verticils. . . . . 23. *Mentha*
            25B. Anther-locules confluent at tip, becoming 1-locular; flowers in dense terminal spikes; stamens long-exserted.
              26A. Corolla bilabiate, the tube longer than the calyx; bracts caducous; filaments glabrous. . . . . . 24. *Leucosceptrum*
              26B. Corolla nearly actinomorphic, the tube usually shorter than the calyx; bracts persistent; filaments often long-pilose. . . . . 25. *Dysophylla*
          24B. Inflorescence a one-sided spike.
            27A. Calyx deeply 5-fid, the lobes lanceolate. . . . . 26. *Keiskea*
            27B. Calyx 5-toothed, the teeth deltoid. . . . . 27. *Elsholtzia*
    19B. Stamens declined along inner side of the lower navicular lip of corolla; corolla bilabiate, the upper lip 4-lobed, the lower lip undivided. . . . . 28. *Plectranthus*

1. AMETHYSTEA L. Ruri-hakka Zoku

Erect nearly glabrous annual; leaves 3- to 5-parted and incised, the floral leaves gradually reduced; verticils loosely arranged, rather many-flowered, pedunculate, forming a narrow terminal panicle; flowers small, blue, pedicelled; calyx globose-campanulate, 10-nerved, the teeth 5, equal; corolla-tube shorter than the calyx, without a ring inside, the limb 4-lobed, the anterior lobes slightly larger; fertile stamens 2, placed in front, ascending, exserted between the upper corolla-lobes; anthers 2-locular, the locules spreading, slightly confluent; nutlets obovoid, the areole on ventral side.——One species.

1. **Amethystea caerulea** L. Ruri-hakka. Stems erect, branched, 40–80 cm. long; leaves ovate to narrowly so, 3–6 cm. long, 2–4 cm. wide, 3(–5)-parted, broadly cuneate at base, the median segment lanceolate, incised, acuminate with an obtuse tip, the lateral segments smaller, the petioles 1–2 cm. long; upper leaves smaller and petiolate, the lateral lobes often re- duced; calyx 2.5–3 mm. long, slightly accrescent after anthesis, 5-lobed, the lobes linear-lanceolate; corolla slightly longer than the calyx; nutlets rounded at apex, 1.5 mm. long, net-veined, the areole obovate, covering lower 2/3 of the body on ventral side.——Sept.–Oct. Honshu, Kyushu.——Korea, Manchuria, China to Turkey.

2. AJUGA L. Kiran-sō Zoku

Mostly perennials, sometimes stoloniferous; leaves coarsely toothed or incised, rarely entire; verticils 2- to many-flowered, dense, often forming a terminal spike; flowers subsessile; calyx ovate or globose-campanulate, 10- to many-nerved, equally 5-toothed or -lobed; corolla tubular, widened at the tip, the limb bilabiate with the upper lip short, entire or sometimes bifid, the midlobe of the lower lip larger, emarginate or bilobed, the lateral lobes smaller; stamens 4; nutlets obovate, with raised reticulations, the areole on ventral face.——About 50 species, in temperate regions and high mountains in the Tropics.

1A. Corolla-tube about 2 cm. long; leaves cordate or partially truncate or rounded at base, prominently incised.
  2A. Stems 30–50 cm. long; stolons absent; leaves acutely incised, 6–10 cm. long. . . . . 1. *A. incisa*
  2B. Stems 10–20 cm. long; stolons present; leaves with few obtuse incisions, 2–5 cm. long. . . . . 2. *A. japonica*
1B. Corolla-tube 6–12 mm. long; leaves cuneate to gradually narrowed toward base.
  3A. Verticils forming a distinct terminal spike.
    4A. Leaves 5–8 cm. long, 3–6 cm. wide; stems 30–40 cm. long. . . . . 3. *A. ciliata* var. *villosior*
    4B. Leaves 2–4(–6) cm. long, 7–25 mm. wide; stems 10–25 cm. long.

5A.   Plant gray-green, prominently white-pubescent perennial herb without stolons; upper leaves abruptly reduced to bracts.
                                                          4. *A. nipponensis*

5B.   Plant vivid-green, hirsute perennial herb with long stolons; upper leaves gradually reduced to bracts. . . . . 5. *A. shikotanensis*
 3B.   Verticils not forming a distinct terminal spike.
   6A. Stems erect; lower leaves reduced to scales.
     7A. Corolla white with purple striations, 10–11 mm. long at back; stems erect. . . . . 6. *A. yezoensis*
     7B. Corolla purple, 13–16 mm. long at back; stems ascending. . . . . 7. *A. makinoi*
   6B. Stems decumbent; lower leaves not scalelike.
     8A. Stems decumbent, tufted, 5–10 (–15) cm. long, stolons absent. . . . . 8. *A. decumbens*
     8B. Plant stoloniferous, 1–5 cm. long. . . . . 9. *A. pygmaea*

1. Ajuga incisa Maxim. H<small>IIRAGI</small>-sō.   Thinly pubescent or nearly glabrous short-rhizomatous perennial; stems erect, rather stout, 30–50 cm. long; leaves membranous, ovate to broadly so, 6–10 cm. long, 4–6 cm. wide, acute, rounded to shallowly cordate at base, acutely incised and toothed, the petioles 3–5 cm. long; verticils often loosely arranged in the lower ones, becoming capitate and spicate above, the bracts broadly lanceolate, slightly longer than the calyx; calyx 5-fid, glabrescent, the lobes lanceolate; corolla dark blue-purple, the tube about 20 mm. long, ampliate above, the limb rather small, the upper lip short, broad, 2-lobed; nutlets obovate-oblong, 2.5 mm. long.——May–June. Damp woods along streams; Honshu (Kantō and centr. distr.).

2. Ajuga japonica Miq. *A. japonica* var. *grossedentata* Fr. & Sav.; *A. grossedentata* Fr. & Sav.——Ō<small>GI</small>-<small>KAZURA</small>. Plant loosely pubescent, with long multicellular crisped hairs; stolons from lower nodes after anthesis; rhizomes short; flowering stems 8–20 cm. long, erect, few-leaved, with few scalelike leaves at base, obtusely angled; leaves spreading, broadly ovate to ovate-cordate, 2–5 cm. long, 1.5–3.5 cm. wide, obtuse, sometimes acute, usually with 2 or 3 short incisions on each side, the petioles 2–5 cm. long; leaves on stolons rather small, loosely arranged, the petioles 1–2 cm. long; verticils few-flowered, the lower loosely arranged, the upper few pairs forming a short headlike spike, the bracts prominent, rather leaflike; corolla pale purple, the tube about 2 cm. long, ampliate above, the upper lip about 4 mm. long, shallowly 2-lobed, the lobes broadly elliptic, rounded at apex; nutlets about 2 mm. long.——Apr.–May. Damp woods in mountains; Honshu, Shikoku, Kyushu; rare.

3. Ajuga ciliata Bunge var. **villosior** A. Gray ex Nakai. *A. ciliata* sensu auct. Japon., non Bunge——K<small>AI</small>-<small>JINDŌ</small>. Prominently pubescent with multicellular long whitish hairs; rhizomes short-creeping; stems erect, often slightly purplish red, 30–40 cm. long, sometimes slightly branched, rather stout, 4-angled; lower leaves spathulate or scalelike, small, deciduous, the middle one ovate to broadly so, 3–8 cm. long, 2–4.5 cm. wide, subacute, broadly cuneate at base, with few irregular coarse teeth, the petioles 5–10 mm. long, winged in upper part; bracts narrowly ovate, obtuse, often purplish red, longer than the calyx; verticils dense, forming a terminal cylindrical spike; flowers blue-purple, the tube 10–12 mm. long, the upper lip 1–1.5 mm. long, 2-lobed; nutlets about 2 mm. long.——May–June. Hokkaido, Honshu, Kyushu.

4. Ajuga nipponensis Makino. *A. decumbens* var. *typica* Fr. & Sav.; *A. genevensis* var. *pallescens* Maxim., pro parte——J<small>ŪNI</small>-<small>HITOE</small>.   Grayish soft pubescent; rhizomes short; stems suberect, 10–25 cm. long, with few pairs of scalelike leaves at base; leaves 2 to 4 pairs, the median ones obovate-spathulate, broadly oblanceolate, or oblong, 3–5 cm. long, 1.5–2(–3) cm. wide, obtuse to rounded, with few coarsely undulate teeth, the winged petiole 1.5–3 cm. long; spikes terminal, solitary, 4–8 cm. long, the lowest verticil usually distant, the bracts

broadly oblanceolate; flowers pale blue, many, the upper lip ovate, about 1.5 mm. long, shortly 2-fid, about 2/5 as long as the lower lip, the lower lip nearly as long as the tube, retuse and mucronate; nutlets about 1.5 mm. long.——Apr.–May. Low mountains; Honshu, Shikoku.

5. Ajuga shikotanensis Miyabe & Tatew. *A. genevensis* var. *pallescens* Maxim.; *A. reptans* var. *japonica* Makino; *A. glabrescens* Makino, excl. syn.; *A. pallescens* (Maxim.) Nakai, non Price & Metcalf——T<small>SURU</small>-<small>KAKO</small>-<small>SŌ</small>. Plant stoloniferous; rhizomes short; stems usually solitary, simple, erect, 10–25 cm. long, rather prominently long-hirsute; basal leaves rosulate; cauline leaves few, narrowly obovate to spathulate-oblong, 2–4 cm. long, 1–2 cm. wide, obtuse, cuneate at base, sessile or subsessile, loosely hirsute, with few undulate teeth, the upper leaves gradually reduced to bracts; leaves of stolons spathulate, petiolate; spikes with 5–10 rather loose verticils; flowers pale purple; corolla-tube 6–7 mm. long, the upper lip scarcely 1 mm. long, bilobed, the lower lip erect, scarcely shorter than the tube, spreading, the midlobe cuneate-obovate, notched at the apex; nutlets about 1.5 mm. long.——May–June. Sunny hills; Honshu; rather rare.——s. Kuriles.

6. Ajuga yesoensis Maxim. N<small>ISHIKIGOROMO</small>, K<small>IMMON</small>-sō. Sparingly crisped-pilose; rhizomes short; stems erect, 8–15 cm. long, rather slender, with 2 or 3 pairs of scales at the base; leaves 3 or 4 pairs; the median ones oblong to broadly ovate, 2–4 cm. long, 1–2 cm. wide, obtuse, cuneate to abruptly narrowed at base, coarsely and obtusely few-toothed, green and often purplish along the nerves above, pale purplish beneath, the petioles winged, 1–2 cm. long, the upper leaves rather small and short-petiolate; verticils rather loose, few-flowered; upper lip of corolla erect, 2.5–3 mm. long, deeply 2-lobed, the lower lip 1/3–1/2 as long as the tube, the lateral lobes broadly lanceolate, slightly narrower than the median; nutlets 1.2–1.5 mm. long.——Apr.–May. Hokkaido, Honshu, Shikoku; rather common.

Var. **tsukubana** Nakai. T<small>SUKUBA</small>-<small>KIMMON</small>-sō. Upper lip semirounded, subentire, much shorter than the lower lip, about 1 mm. long.——Honshu, Kyushu.

7. Ajuga makinoi Nakai. T<small>ACHI</small>-<small>KIRAN</small>-sō. Perennial herb, long-pubescent on stems and leaves; stems several together, 5–20 cm. long, ascending, decumbent in fruit, with several pairs of scalelike oblanceolate leaves 1–2 cm. long at base; cauline leaves oblong or oblong-spathulate, acute and long-decurrent on the petiole at base, 3–5 cm. long, 1–1.5 cm. wide in anthesis, 8–10 cm. long, 2–3.5 cm. wide in summer leaves, with few obtuse lobes on margin; flowers several in leaf-axils, subsessile; closely related to *A. yesoensis* but larger; flowers blue-purple, the upper lip erect, 2–3 mm. long, deeply 2-lobed.——Apr.–June. Honshu (Tōkaidō, W. Musashi Prov. and s. Shinano Prov.).

8. Ajuga decumbens Thunb. *A. decumbens* var. *sinuata* Fr. & Sav.; *A. decumbens* var. *glabrescens* Fr. & Sav.; *A. devestita* Lév. & Van't.——K<small>IRANE</small>-sō, J<small>IGOKU</small>-<small>NO</small>-<small>KAMA</small>-<small>NO</small>-<small>FUTA</small>.

Prominently whitish crisped-pubescent; rhizomes short; stems 5–10(–15) cm. long, decumbent or ascending, purplish, with few pairs of leaves; rosulate leaves few, broadly oblanceolate, obtuse, 4–6 cm. long, 1–2 cm. wide, dark green, coarsely undulate-toothed, often purplish; cauline leaves oblong to ovate, narrowed to a short petiolike base; cymes axillary, few-flowered; flowers blue-purple, with darker striations; corolla-tube about 8 mm. long, the upper lip semirounded, about 2 mm. long, retuse or shallowly bilobed; nutlets about 2 mm. long.——Mar.–May. Thin woods and hedges in hills and low mountains; Honshu, Shikoku, Kyushu; common.——Korea, Manchuria, and China.

Hybrids: **Ajuga** × **bastarda** Makino. KIRAN-NISHIKI-

GOROMO. (*A. decumbens* × *A. yesoensis*); **Ajuga** × **mixta** Makino. JŪNI-KIRAN-SŌ. (*A. decumbens* × *A. nipponensis*).

9. **Ajuga pygmaea** A. Gray. HIME-KIRAN-SŌ. Plant stoloniferous, loosely white crisped-pubescent; stems 1–5 cm. long; radical leaves rosulate, broadly oblanceolate to narrowly obovate, 2–4 cm. long, 8–10 mm. wide, obtuse, with few coarsely sinuate teeth, petiolate; flowering stems 2–5 cm. long, radical or from the tip of the stolons, the cauline leaves few, small; verticils few, usually shorter than the leaves, approximate; corolla-tube slender, 7–8 mm. long, the upper lip about 3 mm. long, deeply 2-lobed; nutlets about 1.5 mm. long.——Mar.–Apr. Kyushu.——Ryukyus and Formosa.

3. TEUCRIUM L. NIGA-KUSA ZOKU

Annuals or biennials, sometimes perennials or shrubs; leaves entire to pinnatilobed; verticils 2- or rarely many-flowered, often forming a terminal spike, the bracts foliaceous or bractlike; calyx tubular or campanulate, 10-nerved, 5-toothed, the teeth equal or the inner ones broader; corolla-tube without an annulus inside, the limb 2-lipped, the upper lip minute, deeply 2-fid, the segments connate to the lateral lobes of the lower lip, the lower lip 3-lobed, the lateral lobes small, erect, the median lobe much larger than the others, sometimes bifid, concave on back; stamens 4, didymous, exserted from the fissure between the lobes of the upper lip, the anterior pair longer; anther-locules divergent and confluent at tip; nutlets obovoid, with raised reticulations, the areole large, oblique or lateral, often occupying more than half the length of the body.——About 100 species, mostly in warmer regions of the N. Hemisphere.

1A. Calyx uniformly glandular-hairy; bracts broadly lanceolate; flowers in dense spikes; leaves broadly cuneate at base; upper portion of stems and axis of spikes usually with recurved hairs. .... 1. *T. viscidum* var. *miquelianum*
1B. Calyx glabrescent or minutely glandular-puberulent on upper half, or uniformly long-pilose throughout, rarely sparsely glandular-puberulent.
   2A. Stems 10–25 cm. long, with long spreading hairs at least on the upper nodes; bracts broadly lanceolate; calyx glandular-hairy.
      3A. Calyx-teeth incurved in fruit, the upper 3 teeth obtuse in fruit; plants usually densely spreading-pilose; leaves ovate to broadly so, 2.5–4 cm. long. .... 2. *T. veronicoides*
      3B. Calyx-teeth erect or only slightly incurved above, acute to acuminate in fruit; plants glabrous except the long hairs on petioles and nodes; leaves narrowly ovate, 3–5 cm. long. .... 3. *T. teinense*
   2B. Stems 30–70 cm. long, usually recurved-puberulent; bracts broadly linear; calyx nearly glabrous or loosely puberulent on upper part, not glandular-puberulent, the teeth erect or slightly recurved, acute to acuminate; leaves 5–10 cm. long. ...... 4. *T. japonicum*

1. **Teucrium viscidum** Bl. var. **miquelianum** (Maxim.) Hara. *T. stoloniferum* var. *miquelianum* Maxim.; *T. viscidum* sensu auct. Japon., non Bl.; *T. stoloniferum* sensu auct. Japon., pro parte, non Roxb.; *T. miquelianum* (Maxim.) Kudo——TSURU-NIGA-KUSA. Perennial, 25–40 cm. high, the stems, inflorescence, petioles, and underside of leaves recurved-puberulent; leaves thin, narrowly to broadly ovate, 4–8(–10) cm. long, 2–4 cm. wide, very acute, usually broadly cuneate at base, irregularly toothed, sometimes thinly puberulent above, the petioles 1.5–3 cm. long; spikes 3–5 cm. long, one-sided, the bracts broadly lanceolate, nearly as long as the calyx; flowers 8–10 mm. long, short-pedicellate; calyx broadly ovoid in fruit, 3–3.5 mm. across, uniformly glandular-puberulent, the upper calyx teeth deltoid, obtuse, the lower 2 narrowly deltoid, acute; nutlets nearly orbicular, inflated-lenticular, 1.2 mm. in diameter, the scar orbicular.——July–Sept. Wet places in lowlands and hills; Honshu, Shikoku, Kyushu. The typical phase occurs in the Ryukyus and Formosa, and Malaysia to India.

2. **Teucrium veronicoides** Maxim. EZO-NIGA-KUSA, HIME-NIGA-KUSA. Perennial stoloniferous herb densely spreading-pilose; stems 20–30 cm. long, erect, sometimes branched; leaves ovate or broadly so, 2.5–4 cm. long, 1.5–2.5 cm. wide, subacute, broadly cuneate to shallowly cordate at base, toothed, the petioles 1–2 cm. long; spikes 4–8 cm. long, one-sided, rather loosely flowered, the bracts rather shorter than the calyx, broadly lanceolate, 1-toothed; flowers about

8 mm. long, pedicellate; calyx descending in fruit, often brownish and lustrous, 3–3.5 mm. long, with loose long glandular hairs on upper part, the upper 3 teeth deltoid, obtuse, the lower 2 narrowly deltoid, very acute; corolla glabrous, the lobes rather narrow; nutlets orbicular, about 1.5 mm. in diameter, with obsolete reticulations.——Aug. Woods in mountains; Hokkaido, Honshu; rather rare.——Korea.

Var. **brachytrichum** Ohwi. INU-NIGA-KUSA. Plant with short spreading hairs and the leaves coarsely toothed.——Honshu (Uzen Prov.).

3. **Teucrium teinense** Kudo. TEINE-NIGA-KUSA. Stoloniferous perennial, nearly glabrous, long-hairy on the nodes; stems 20–30 cm. long, erect; leaves narrowly ovate to ovate-oblong, 3–5 cm. long, 1.5–2.5 cm. wide, acuminate to very acute, broadly cuneate to rounded at base, irregularly toothed, long-hirsute on both sides or only on the upper side, the petioles 1–1.5 cm. long; spikes short; bracts broadly lanceolate; calyx in fruit membranous, about 4 mm. long, very sparsely glandular-puberulent, the teeth narrowly deltoid, acute to acuminate, the innermost tooth sometimes deltoid and obtuse; nutlets about 1.5 mm. long.——Aug. Hokkaido, Honshu (Mount Gassan in Uzen).

4. **Teucrium japonicum** Houtt. *T. nepetoides* Lév.; *T. brevispicum* Nakai——NIGA-KUSA. Glabrescent perennial herb; stems erect, 30–70 cm. long, branched, rather stout, 4-angled, often loosely recurved-puberulent; leaves narrowly ovate-oblong or broadly lanceolate, 5–10 cm. long, 2–3.5 cm.

wide, very acute, rounded to subtruncate at base, irregularly toothed, usually loosely short-pilose on the nerves beneath; spikes 3–10 cm. long, sometimes branched at base, rather densely flowered; bracts broadly linear; calyx in fruit 4–5 mm. long, nearly glabrous or loosely puberulent on upper part, usually not glandular-pilose, the teeth narrowly deltoid, or the 2 anterior ones lanceolate, acuminate to acute; nutlets about 1.5 mm. long.——July. Hokkaido, Honshu, Shikoku, Kyushu.——Korea.

4. SCUTELLARIA L.   Tatsu-nami-sō Zoku

Perennials or annuals, rarely subshrubs, erect or decumbent; leaves often toothed; flowers paired in the verticils, blue or rose-colored, sometimes yellow; calyx campanulate, bilabiate, the lips entire, short, broad, closed in fruit; corolla-tube long, often geniculate and gibbous, erect, the limb bilabiate, the upper lip erect, galeate, sometimes retuse, the lower lip spreading, the lateral lobes connate to the upper lip or to the midlobe of the lower lip; anthers approximate in pairs, the lower ones more or less reduced and 1-locular, the upper ones 2-locular; nutlets orbicular, often slightly compressed, usually short-tubercled.——About 20 species, widely distributed in temperate regions and mountains of the Tropics.

1A. Flowers 6–8 mm. long; corolla-tube erect from the base or slightly geniculate.
  2A. Flowers axillary, solitary, not forming a spike.
    3A. Nearly glabrous to minutely puberulent; leaves narrowly deltoid-ovate or broadly deltoid-lanceolate; petioles 1–3 mm. long.
                                                                                   1. S. dependens
    3B. Spreading-hairy on leaves and calyx; leaves cordate; petioles 1–2 cm. long. .... 2. S. guilielmii
  2B. Flowers in a terminal spike; leaves broadly ovate-deltoid, with a few deep teeth on each side. .... 3. S. shikokiana
1B. Flowers 1.5–3 cm. long; corolla-tube erect from the prominently geniculate base (except gently curved upward in S. pekinensis).
  4A. Flowers in terminal spikes.
    5A. Flowers 15–22 mm. long.
      6A. Stems with upwardly curved hairs or subglabrous.
        7A. Lower lip of corolla about twice as long as the upper lip, straight. .... 4. S. pekinensis
        7B. Lower lip of corolla about as long as the upper lip.
          8A. Leaves 8–20 mm. wide, ovate to ovate-cordate; spikes few-flowered. .... 5. S. amabilis
          8B. Leaves 10–15 mm. wide, ovate-deltoid; spikes many-flowered. .... 6. S. laeteviolacea
      6B. Stems with spreading or downwardly curved hairs.
        9A. Stems densely spreading-pubescent.
          10A. Internodes shorter than the larger leaves. .... 7. S. abbreviata
          10B. Internodes longer than the leaves.
            11A. Leaves flat, thinly long-hairy. .... 8. S. kurokawae
            11B. Leaves minutely uneven on the surface, densely short-hairy. .... 9. S. indica
        9B. Stems with downwardly curved hairs.
          12A. Stems and leaves puberulent; leaves rounded to broadly cuneate at base. .... 10. S. muramatsui
          12B. Stems and leaves long-pubescent; leaves cordate at base.
            13A. Leaves tending to be approximate on upper part of stem; spikes short, 0.5–3 cm. long. .... 11. S. brachyspica
            13B. Leaves tending to be approximate on lower part of stem; spikes often elongate, 2–8 cm. long. .... 12. S. maekawae
    5B. Flowers 27–32 mm. long. .... 13. S. iyoensis
  4B. Flowers in axils of cauline leaves. .... 14. S. strigillosa

1. Scutellaria dependens Maxim. *S. oldhamii* Miq.——Hime-namiki. Nearly glabrous to minutely puberulent, green, slender, perennial; stems erect, sometimes branched, 10–30 cm. long, acutely 4-angled, puberulent on the angles; leaves narrowly ovate-deltoid or deltoid-lanceolate, 1–2 cm. long, 6–10 mm. wide, obtuse, rounded to shallowly cordate at base, slightly scabrous on margin and on nerves above, with 1 or 2 teeth near base, entire on upper half, the petioles 1–3 mm. long; calyx 1.5 mm. long in anthesis, 2.5–3 mm. long in fruit; corolla white, the lower lip twice as long as the upper, about 1.5 mm. long; nutlets about 0.7 mm. long, densely granular-mammillate. June–Aug.——Hokkaido, Honshu, Shikoku (?), Kyushu.——Korea, Manchuria, and e. Siberia.

2. Scutellaria guilielmii A. Gray. *S. hederacea* sensu auct. Japon., non Kunth & Bouché; *S. tanakae* Fr. & Sav.; *S. suzukiana* Honda——Ko-namiki. Slender, green, perennial; stems erect, branched, 20–40 cm. long, angled, glabrous or loosely spreading-pubescent; cauline leaves reniform-cordate or cordate, 1–2 cm. wide, rounded at apex, few-toothed, pubescent on both sides, petiolate; floral leaves ovate or narrowly so, 1–2 cm. long, 5–15 mm. wide, obtuse, nearly sessile; calyx 2.5 mm. long in anthesis, 5–6 mm. long in fruit, spreading-pubescent; corolla white, 7–8 mm. long, the upper lip about half as long as the erect lower lip; nutlets wing-margined, about 2 mm. across inclusive of the wing, toothed, the lower surface with long linear-filiform white tubercles, the upper surface loosely papillose.——May. Honshu (Awa Prov. and westw.), Shikoku, Kyushu.——Ryukyus.

3. Scutellaria shikokiana Makino. Miyama-namiki. Green, glabrescent perennial with slender branched rhizomes; stems 5–15 cm. long, angled, glabrous to spreading glandular-hairy often on the upper part; leaves membranous, ovate-deltoid, 2–3 cm. long, 1.5–2.5 cm. wide, subobtuse, truncate or very broadly cuneate at base, often scattered hairy, few-toothed, the petioles 1.5–2.5 cm. long; spikes loosely flowered, 1–5 cm. long, nearly glabrous; calyx 2 mm. long in anthesis, 4 mm. long in fruit, with long spreading hairs; corolla white, greenish yellow when dried, 7–8 mm. long, the lower lip descending, 2.5 mm. long, about twice as long as the suberect upper lip; nutlets about 0.5 mm. long, densely mammillate.——July–Aug. Honshu (Kantō Distr. and westw.), Shikoku, Kyushu.——Forma **pubicaulis** Ohwi. Ke-miyama-namiki. Plant rather prominently puberulent on upper part of stems.——Kyushu.

4. Scutellaria pekinensis Maxim. var. **ussuriensis** (Regel) Hand.-Mazz. *S. japonica* var. *ussuriensis* Regel; *S.*

indica var. *ussuriensis* (Regel) Komar.; *S. dentata* Lév.; *S. ussuriensis* (Regel) Kudo; *S. transitra* var. *ussuriensis* (Regel) Hara.——Ezo-TATSU-NAMI-SŌ. Nearly glabrous perennial; stems erect, slender, slightly branched, few-leaved, often with upwardly curved white hairs on the nodes; leaves thinly membranous, broadly deltoid-ovate, 2–4 cm. long, 1.5–2.5 cm. wide, subobtuse, shallowly cordate, toothed; spikes 3–6 cm. long, loosely flowered; calyx 2–2.5 mm. long in anthesis, about 5 mm. long in fruit; corolla-tube gently curved upward from the base, 15–16 mm. long, pale purple, the lower lip about twice as long as the upper lip, straight, ascending; nutlets about 0.7 mm. long, densely mammillate, gray-fulvous.——July–Aug. Woods in mountains; Hokkaido, Honshu.——Korea, Manchuria, and e. Siberia.

Var. **transitra** (Makino) Hara. *S. transitra* Makino; *S. ussuriensis* var. *transitra* (Makino) Nakai; *S. ussuriensis* var. *tomentosa* Koidz.——YAMA-TATSUNAMI-SŌ. Plant with upwardly curved white hairs throughout.——Hokkaido, Honshu, Shikoku, Kyushu.——Korea. The typical phase occurs in n. China and Manchuria.

5. Scutellaria amabilis Hara. YAMAJI-NO-TATSU-NAMI-SŌ. Rhizomes slender, creeping; stems 15–25 cm. long, ascending-puberulent especially on the angles; leaves broadly ovate to ovate-cordate, 1–2 cm. long, 8–20 mm. wide, truncate to shallowly cordate at base, coarsely obtuse-toothed, puberulent on both sides, the petioles to 2 cm. long; spikes short, few-flowered; calyx 2.5–3 mm. long, glandular-pilose; corolla 2–2.5 cm. long, blue-purple, the upper lip about 4 mm. long, galeate, the lower lip spreading, as long as the upper.——May–June. Honshu.

6. Scutellaria laeteviolacea Koidz. *S. indica* var. *japonica* forma *humilis* Makino——SHISOBA-TATSU-NAMI-SŌ. Rhizomes short; stems erect, 5–15 cm. long, ascending-puberulent; leaves few, ovate-deltoid, 1–2 cm. long, 1–1.5(–2) cm. wide, obtuse, truncate at base, coarsely obtuse-toothed, with ascending hairs on both sides, purplish beneath; spikes usually short, 1–3 cm. long, many-flowered, ascending-puberulent; calyx 2.5 mm. long in flower, about 5 mm. long in fruit; corolla 17–20 mm. long, purplish, the lower lip spreading, the upper lip arcuate, galeate, as long as the lower; nutlets about 1 mm. long, densely mammillate.——May–June. Honshu, Shikoku, Kyushu.

Var. **discolor** (Hara) Hara. *S. kiusiana* Hara, incl. var. *discolor* Hara——TSUKUSHI-TATSU-NAMI-SŌ. Plants taller, 20–30 cm. high; leaves larger, 2–4 cm. long; spikes more or less elongate.——Honshu. (w. distr.), Kyushu.

Var. **yakusimensis** (Masam.) Hara. *S. indica* var. *yakusimensis* Masam.——YAKUSHIMA-NAMIKI. Plant smaller, often with decumbent stems and rather numerous leaves.——Kyushu (Yakushima).

7. Scutellaria abbreviata Hara. *S. japonica* var. *ussuriensis* forma *humilis* Matsum. & Kudo.——TŌGOKU-SHISOBA-TATSU-NAMI-SŌ. Resembles the preceding; stems 5–30 cm. long, with spreading white hairs; leaves reniform to ovate-cordate, 2–3(–5) cm. long and as wide; spikes 2–5 cm. long, spreading-hairy; calyx 3 mm. long in anthesis, 6 mm. long in fruit; corolla purple, 20–22 mm. long, puberulent.——June–July. Honshu.

8. Scutellaria kurokawae Hara. IGA-TATSU-NAMI-SŌ. Stems 10–30 cm. long, prominently spreading-hairy; leaves ovate to broadly so, or ovate-cordate, 1.2–2.5 cm. long, 1.2–2 cm. wide, obtuse, truncate to shallowly cordate at base, obtuse-toothed, thinly long-hairy, often purplish beneath, the petioles 1–2.5 cm. long, spreading-hairy; spikes loosely flowered; calyx 3 mm. long, often with glandular hairs; corolla pale purple, about 2 cm. long.——June. Honshu (centr. distr.).

9. Scutellaria indica L. TATSU-NAMI-SŌ. Rhizomes slender, creeping; stems erect, 20–40 cm. long, prominently white-pilose; leaves loosely few-paired, deltoid-cordate to broadly ovate, obtuse, cordate at base, obtusely toothed, spreading-hairy, often prominently pubescent on both sides, the petioles 5–20 mm. long, the upper leaves sometimes orbicular, 1–2.5 cm. in diameter; spikes 3–8 cm. long, with prominent spreading hairs; calyx about 3 mm. long in anthesis, 6–7 mm. long in fruit; corolla 18–22 mm. long, geniculate at the base, becoming erect above, pale purple; nutlets about 1 mm. long, densely mammillate.——May–June. Sunny hills; Honshu, Shikoku, Kyushu; common.——Korea, China, Indochina, Formosa, and Ryukyus.

Var. **parvifolia** (Makino) Makino. *S. indica* var. *japonica* forma *parvifolia* Makino; *S. parvifolia* (Makino) Koidz.——KOBA-NO-TATSU-NAMI, BIRŌDO-TATSU-NAMI. Stems shorter, 5–20 cm. long, long-decumbent at base; leaves small, usually about 1 cm. long and as wide, more sparsely toothed.——Apr.–June. Honshu (centr. and w. distr.), Shikoku, Kyushu.

Var. **tsusimensis** (Hara) Ohwi. *S. tsusimensis* Hara——ATSUBA-TATSU-NAMI-SŌ. Plant rather large, 10–30 cm. high; leaves 2–4 cm. long and as wide, shallowly cordate to truncate at base, with impressed nerves above.——Honshu (w. distr.), Kyushu.——Korea.

10. Scutellaria muramatsui Hara. DEWA-NO-TATSU-NAMI-SŌ. Stems ascending, 10–30 cm. long, very short recurved-puberulent; leaves deltoid-ovate to ovate, sometimes ovate-cordate in the lower ones, 2–3.5 cm. long, 1–2.5 cm. wide, obtuse, rounded to broadly cuneate at base, scattered undulate-toothed, loosely pilose above, puberulent on the nerves beneath; spikes 3–5 cm. long, the axis with simple spreading and glandular hairs; calyx 1.5–2 mm. long; corolla 15–18 mm. long, purple.——May–June. Honshu (n. and Hokuriku Distr.).

11. Scutellaria brachyspica Nakai & Hara. *S. indica* var. *brachyspica* Nakai ex Hara, in syn.——OKA-TATSU-NAMI-SŌ. Stems erect, 10–50 cm. long, loosely leafy, densely recurved-hairy; leaves deltoid-ovate to ovate, 1.5–5 cm. long, 1–4 cm. wide, obtuse, truncate to subcordate at base, coarsely obtuse-toothed, spreading-pilose on both sides, the petioles to 2 cm. long, densely recurved- to spreading-hairy; spikes short, 0.5–3(–5) cm. long, densely flowered, the axis with spreading and glandular hairs; corolla 15–25 mm. long; nutlets about 1.5 mm. across, densely mammillate.——May–June. Honshu, Shikoku; rather common.

12. Scutellaria maekawae Hara. *S. indica* var. *intermedia* Nakai ex Hara, in syn.——HONAGA-TATSU-NAMI-SŌ. Stems 7–20 cm. long, recurved-hairy on the angles; leaves ovate to ovate-cordate, 1–3 cm. long, 1.5–2.5 cm. wide, very obtuse, obtuse-toothed, purplish beneath, pilose toward margin above and on nerves beneath; spikes 2–8 cm. long, the axis recurved-hairy; calyx 2.5 mm. long in anthesis, about 5 mm. long in fruit; corolla 2–2.2 cm. long, purple; nutlets 1.5 mm. long, densely papillose.——June. Honshu (centr. and w. distr.).

Var. **pubescens** Hara. SHIMAJI-TATSU-NAMI-SŌ. Plant more densely hairy.——Occurs with the typical phase.

13. Scutellaria iyoensis Nakai. HANA-TATSU-NAMI-SŌ. Stems erect, 10–40 cm. long, 4-angled, ascending-puberulent; leaves broadly lanceolate to narrowly ovate, 3–6 cm. long, 1–2.5 cm. wide, acute, cuneate to subtruncate at base, usually glandular-dotted on both sides, coarsely deltoid-toothed, ciliolate, the petioles 1–5 mm. long; spikes 2–5 cm. long, loosely flowered, the pedicels appressed-pilose; calyx 3–3.5 mm. long; corolla 27–32 mm. long, blue-purple.——May–June. Honshu (Chūgoku Distr.), Shikoku.

14. Scutellaria strigillosa Hemsl. *S. cordifolia* var. *pubescens* Miq.; *S. scordiifolia* var. *hirta* F. Schmidt; *S. galericulata* forma *hirta* (F. Schmidt) Koidz.; *S. scordiifolia* var. *sachalinensis* Matsum. & Kudo; *S. scordiifolia* var. *nipponica* Matsum. & Kudo; *S. schmidtii* Kudo——NAMIKI-SO. Rhizomes slender, long-creeping; stems 10–40 cm. long, 4-angled, pubescent; leaves elliptic to oblong, 1.5–3.5 cm. long, 1–1.5 cm. wide, truncate to rounded at base, more or less pubescent on both sides, sometimes densely so, the petioles 1–4 mm. long, with few minute teeth on both sides; flowers solitary in axils of upper cauline leaves; calyx about 3 mm. long in anthesis, about 5 mm. long in fruit; corolla 2–2.2 cm. long, erect from the arcuate base, blue-purple; nutlets about 1.8 mm. long, semirounded, densely mammillate.——July–Aug. Sandy places along seashores; Hokkaido, Honshu, Shikoku, Kyushu (Tsushima); very variable.——Sakhalin, Kuriles, Korea, and Manchuria.

Var. **yezoensis** (Kudo) Kitam. *S. yezoensis* Kudo—— EZO-NAMIKI-SŌ, Ō-NAMIKI-SŌ. Stems softer, glabrescent, upwardly puberulent only on the angles; leaves ovate or narrowly ovate, obtuse to subacute at apex.——Wet places; Hokkaido, Honshu (n. distr.).——Kuriles, Sakhalin, and Korea.

5. AGASTACHE Clayt. ex Gronov. KAWA-MIDORI ZOKU

Erect herbs; leaves toothed; verticils many-flowered, often forming a dense terminal spike, the bracteoles linear; flowers blue or purple; calyx tubular, 5-toothed, 15-nerved, oblique at the mouth, the upper teeth rather longer than the others; corolla-tube as long as to slightly longer than the calyx, broadened at tip, the limb bilabiate, the upper lip suberect, emarginate to bifid, the lower lip somewhat spreading, 3-lobed, the midlobe broader, undulately toothed; stamens 4, slightly exserted, the anthers 2-locular, the locules parallel; style bifid; nutlets ovoid, smooth.——Few species, in e. Asia and N. America.

1. Agastache rugosa (Fisch. & Mey.) O. Kuntze. *Lophanthus rugosus* Fisch. & Mey.; *Cedronella japonica* Hassk.; *L. formosanus* Hayata——KAWA-MIDORI. Perennial herb; stems erect, 40–100 cm. long, 4-angled, branched in upper part, white-puberulent; leaves thinly membranous, ovate-cordate or ovate, 5–10 cm. long, 3–7 cm. wide, acuminate, cordate to rounded at base, toothed, nearly glabrous above, often whitish beneath, the petioles 1–4 cm. long; spikes 5–15 cm. long, about 2 cm. across, densely many-flowered, often interrupted below, bracteate; calyx 5–6 mm. long, the teeth narrowly deltoid, acute, about 1/3 as long as the calyx; corolla purple, 8–10 mm. long, the limb short; stamens long-exserted; nutlets obovate-elliptic, compressed-trigonous, 1.8 mm. long, rounded and without a ridge at apex, puberulent above.——Aug.–Oct. Grassy places in mountains especially along streams and valleys; Hokkaido, Honshu, Shikoku, Kyushu.——Korea, Manchuria, e. Siberia, China, and Formosa.

6. MEEHANIA Britt. RASHŌMON-KAZURA ZOKU

Soft green sometimes stoloniferous perennials; leaves opposite, petiolate, toothed; verticils few, few-flowered, distant, the bracts leaflike or small, the bracteoles small, subulate; calyx campanulate or tubular-campanulate, 15-nerved, bilabiate, the upper lip bifid, the lower ones 3-fid, the teeth narrowly deltoid, acute; corolla-tube gradually broadened to the tip, the limb bilabiate, the upper lip bifid, concave on back, the lower lip 3-lobed; stamens 4, parallel, not exserted; style 2-fid at apex; nutlets ovate, smooth.——Few species, in e. Asia and N. America.

1A. Stolons elongate, leafy; verticils forming a terminal spike; plant with long-spreading whitish hairs. .... 1. *M. urticifolia*
1B. Stolons absent; verticils not forming a distinct spike; plant without long-spreading hairs. .... 2. *M. montis-koyae*

1. Meehania urticifolia (Miq.) Makino. *Dracocephalum urticifolium* Miq.; *Cedronella urticifolia* (Miq.) Maxim.; *Glechoma urticifolia* (Miq.) Makino——RASHŌMON-KAZURA. More or less long-pubescent stoloniferous green herb; flowering stems erect, 15–30 cm. long, with few pairs of leaves; cauline leaves thinly membranous, deltoid-cordate to ovate-cordate, 2–5 cm. long, 2–3.5 cm. wide, acute, coarsely obtuse-toothed, the petioles 2–5 cm. long; leaves of the stolons to 10 cm. long; flowers 3–12, in loose terminal one-sided spikes, the bracts broadly ovate, 7–15 mm. long, the pedicels and sometimes also the floral axis minutely puberulent; calyx about 1 cm. long; corolla purple-blue, 4–5 cm. long, the upper lip suberect, 8–10 mm. long, the lower lip with a larger, obcordate, retuse median lobe and with spreading, dark-spotted lateral lobes; nutlets narrowly obovate, about 3 mm. long, loosely puberulent.——Apr.–May. Damp woods in mountains; Honshu, Shikoku, Kyushu.——Manchuria and Korea.

2. Meehania montis-koyae Ohwi. *M. urticifolia* var. *montis-koyae* (Ohwi) Ohwi——OCHI-FUJI. Perennial herb without stolons; stems 20–25 cm. long, slender, minutely puberulent, erect or ascending, loosely few-leaved; leaves thinly membranous, deltoid-cordate, 2–4 cm. long and as wide, obtuse, sparingly pilose, undulate-toothed, the petioles spreading, 1.5–3 times as long as the blades; flowers few, solitary in axils, short-pedicelled; calyx tubular-campanulate, about 1 cm. long, glandular-puberulent; corolla about 4 cm. long, sparingly pubescent externally, the tube gradually broadened in upper part, the limb bilabiate, the upper lip bifid, the lower lip 3-fid, the median lobe larger.——May. Woods in mountains; Honshu (Kii Prov. and Chūgoku Distr.); rare.

7. NEPETA L.   Inu-hakka Zoku

Erect herbs; leaves toothed; verticils dense, forming a terminal spike; calyx tubular, straight or slightly curved, 15-nerved, oblique at the mouth and with 5 equal teeth; corolla-tube broadened in upper part, narrowed at base, the limb bilabiate, the upper lip erect, slightly concave on back, bifid, the lower lip 3-lobed, the midlobe large, bilobed, spreading; stamens 4, the upper pair ascending, the lower pair suberect, the anther-locules divergent; style bifid, the lobes equal; nutlets elliptic, smooth. ——About 150 species, in the temperate regions of the Old World.

1A. Leaves ovate to broadly lanceolate, usually 6–12 cm. long, sessile or the petioles 3–5 mm. long; flowers purple, 22–30 mm. long; calyx 8–10 mm. long, the teeth acuminate with an obtuse tip. . . . . 1. *N. subsessilis*
1B. Leaves deltoid-ovate, 3–6 cm. long, shallowly cordate, broadest above the base, the petioles 1–3 cm. long; flowers pale purple, 8–10 mm. long; calyx 5–7 mm. long, the teeth spine-tipped. . . . . ——————2. *N. cataria*

1. Nepeta subsessilis Maxim. *Glechoma subsessilis* (Maxim.) O. Kuntze; *N. fauriei* Lév.——Misogawa-sō. Tall perennial herb; stems rather stout, erect, often branched, 50–100 cm. long, puberulent; leaves submembranous, broadly ovate to broadly lanceolate, usually narrowly ovate, 6–12 cm. long, 2.5–8 cm. wide, acuminate, rounded at base, obtusely toothed, puberulent on both sides, more densely so on nerves beneath, subsessile or the petioles 3–5 mm. long; cymes many-flowered, sometimes short-pedunculate; calyx 8–10 mm. long, with raised nerves, 5-fid, the teeth linear-lanceolate, acuminate, the 2 dorsal teeth connate on lower 1/3; corolla purple, 22–30 mm. long; nutlets obovate-oblong, compressed-trigonous, about 3 mm. long, loosely white-pilose on upper side, rounded at apex.——July–Aug. Wet shaded slopes in mountains; Honshu, Shikoku.

Var. **yesoensis** Fr. & Sav. Ezo-misogawa-sō. Petioles about 1 cm. long; corolla about 30 mm. long, rarely white. ——Hokkaido and Honshu (centr. and n. distr.).——Forma **albiflora** Tatew. Shirobana-ezo-misogawa-sō. Flowers white.

2. Nepeta cataria L. *N. minor* Mill.——Inu-hakka, Chikuma-hakka. Prominently white-puberulent perennial herb 50–100 cm. high; stems branched; leaves deltoid-ovate, 3–6 cm. long, 2–3.5 cm. wide, acute, shallowly cordate at base, coarsely toothed or incised, the petioles 1–3 cm. long; spikes often paniculately branched, dense, 2–4 cm. long; calyx 5–7 mm. long, spreading-puberulent, distinctly ribbed, the teeth spine-tipped; corolla pale purple, 8–10 mm. long; nutlets elliptic, 1.5 mm. long, slightly compressed, black-brown, smooth, slightly inflated on ventral face, the areole transversely broadened, white.——July–Aug. Naturalized in Honshu (Shinano Prov. and northw.).——Korea, China, to w. Asia and Europe.

8. GLECHOMA L.   Kakidōshi Zoku

Creeping or ascending perennial herbs; leaves toothed; verticils few-flowered, distant; flowers blue or purple, pedicelled; calyx tubular or campanulate, 15-nerved, slightly oblique at mouth, bilabiate, the upper lip 3-lobed, connate at base, the lower lip bifid with narrow lobes; corolla-tube ampliate in upper part, the limb bilabiate, the upper lip rather flat, erect, retuse, the lower lip spreading, 3-lobed, the midlobe large, retuse; stamens 4, parallel, ascending under the upper lip of corolla; style 2-fid, the lobes equal; nutlets smooth.——Few species, in the N. Hemisphere.

1. Glechoma hederacea L. var. **grandis** (A. Gray) Kudō. *G. hederacea* subsp. *grandis* (A. Gray) Hara; *Nepeta glechoma* var. *grandis* A. Gray; *G. grandis* (A. Gray) Kudo, in syn.; *G. hederacea* var. *minor* Honda——Kakidōshi. Perennial creeping herb; stems tufted, erect, 5–20 cm. long; creeping stems 50 cm. or more long, with spreading hairs; leaves reniform to cordate, 1.5–2.5 cm. long, 2–3 cm. wide, coarsely obtuse-toothed, slightly pubescent on both sides, long-petiolate; flowers 1–3 in leaf-axils, the pedicels short, puberulent; calyx 7–9 mm. long, spreading-pilose, the teeth short spine-tipped; corolla rose-purple, 15–25 mm. long, the upper lip retuse, the lower lip obliquely spreading, dark-spotted, twice as long as the upper; nutlets ellipsoidal, 1.8 mm. long, rounded, slightly compressed, keeled on the ventral face, obscurely glandular-dotted, nearly smooth.——Apr.–May. Hedges and thickets in lowlands and low mountains; Hokkaido, Honshu, Shikoku, Kyushu; common.——Temperate regions of e. Asia, Formosa, and Korea.

9. DRACOCEPHALUM L.   Musha-rindō Zoku

Annuals, biennials, or perennials, sometimes subshrubs, with erect or ascending stems; leaves entire or toothed, sometimes palmately pinnatifid; cymes many-flowered, axillary, often in terminal spikes; flowers purple, blue, or rarely white; calyx tubular, 15-nerved, bilabiate, the upper lip 3-lobed, more or less connate at base, the lower lip 2- to 4-lobed; corolla-tube ampliate in upper part, the limb bilabiate, the upper lip erect, entire, concave on back, the lower lip spreading, 3-lobed, the midlobe large, shallowly lobed at apex; stamens 4, didymous, the anther-locules 2 or reduced to 1; nutlets smooth.——About 40 species, in Eurasia.

1. Dracocephalum argunense Fisch. ex Link. *D. ruyschiana* var. *speciosum* Ledeb.; *D. ruyschiana* var. *japonicum* A. Gray; *D. ruyschiana* var. *argunense* (Fisch.) Nakai; *D. japonicum* (A. Gray) Kudo——Musha-rindō. Rhizomes short; stems 15–40 cm. long, simple or branched, rather prominently whitish recurved-hairy, usually with a tuft of leaves from the abbreviated lateral branches in axils; cauline leaves rather thick, linear or narrowly oblong-lanceolate, 2–5 cm. long, 2–5 mm. wide, obtuse, acute at base, entire, recurved on margin, glabrous and lustrous above, loosely ascending-puberulent on the raised midrib beneath, the petioles 1–3(–7) mm. long; spikes 2–5 cm. long, densely flowered, the bracts narrowly ovate, shorter than the calyx except the lower ones; calyx 12–15 mm. long, rather unequally 5-lobed to near the middle, narrow, usually spreading-pubescent, with raised nerves, the lobes short spine-tipped; corolla 3–3.5 cm. long, blue-purple, pubescent externally; anthers bearded; nutlets ovate-orbicular, compressed-trigonous, black, about 2.5 mm. long, rounded on both ends, the areole transversely oblong, white.——June–July. Sunny dry slopes in mountains; Hokkaido, Honshu (Kinki Distr. and northw.).——Korea, n. China, Manchuria, and e. Siberia.

10. PRUNELLA L.　　Utsubogusa Zoku

Perennial herbs with opposite, entire to pinnately lobed leaves; verticils many, few-flowered, in dense terminal spikes; flowers purplish or white; calyx tubulose, about 10-nerved, bilabiate, the upper lip flat, truncate, with 3 short teeth, the lower lip bifid, the lobes lanceolate; corolla-tube with a ring of hairs inside near base, the limb bilabiate, the upper lip galeate, the lower lip 3-lobed, with a larger midlobe concave on back, and reflexed lateral lobes; stamens 4, didymous, the filaments often toothed at top; style bifid, the lobes equal; nutlets oblong, smooth.——Few species, in temperate regions and in mountains of the tropics.

1A. Petioles 10–30 mm. long; stems usually ascending or short-decumbent at base; flowers about 2 cm. long, blue-purple; filaments with a tooth on upper part. . . . . 1. P. vulgaris
1B. Petioles 0–5(–10) mm. long; stems usually erect from the base; flowers about 3 cm. long, rose-purple; filaments without teeth.
　　2. P. prunelliformis

1. Prunella vulgaris L. var. **lilacina** (Nakai) Nakai. *P. vulgaris* forma *lilacina* Nakai; *P. vulgaris* sensu auct. Japon., pro parte, non L.; *P. asiatica* Nakai; *P. vulgaris* subsp. *asiatica* (Nakai) Hara——Utsubogusa. Stoloniferous; stems, leaves and inflorescence coarsely white-pilose; stems ascending or short-decumbent at base, erect, 20–30 cm. long, with few pairs of leaves, sometimes branched; leaves broadly lanceolate to ovate, usually ovate-oblong, 3–6 cm. long, 2–4 cm. wide, subobtuse, cuneate to rounded at base, with few mucronate teeth, the petioles 1–3 cm. long, the upper leaves usually sessile; spikes densely many-flowered, 3–8 cm. long, about 2 cm. across, the bracts depressed-cordate, long-ciliate, broadly cuspidate; calyx 7–10 mm. long, the teeth short spine-tipped; flowers blue-purple, about 2 cm. long; corolla toothed on the midlobe of lower lip; filaments with a tooth on upper part; nutlets about 1.6 mm. long, yellow-brown, lustrous, obtusely compressed-trigonous, rounded at apex, with few pale longitudinal lines on margin and on both faces, the areole small. ——June–Aug. Sunny places in lowlands and mountains; Hokkaido, Honshu, Shikoku, Kyushu; common.

Var. **aleutica** Fern. *P. japonica* Makino; *P. vulgaris* var. *japonica* (Makino) Kudo——Miyama-utsubogusa. Stems suberect from the base, estoloniferous, innovation shoots short.

——Hokkaido, Honshu (n. and centr. distr.).——Kuriles, Korea, cooler regions of e. Asia, and the Aleutians. Grades into the typical phase which occurs in Europe.

2. Prunella prunelliformis (Maxim.) Makino. *Dracocephalum prunelliforme* Maxim.; *Prunellopsis prunelliformis* (Maxim.) Kudo——Tateyama-utsubogusa. Scattered coarse-pilose; rhizomes short; stems usually tufted, erect, 25–50 cm. long, with 5–10 pairs of leaves; leaves narrowly to broadly ovate, 3–8 cm. long, 1.5–4 cm. wide, subacute to obtuse, rounded or sometimes broadly cuneate at base, sessile or short-petiolate; spikes 1–5 cm. long, about 2.5 cm. across, densely many-flowered, the bracts depressed deltoid-ovate, long-ciliate; calyx 1–1.5 cm. long, glabrous inside, the teeth spine-tipped; corolla rose-purple, about 3 cm. long, glabrous to scattered long-pubescent outside, papillose on lower half of the tube inside, with a ring of hairs about 1.5 mm. above the base, midlobe of lower lip prominently toothed on margin; anther-locules pubescent on margin; nutlets lenticular, about 2 mm. long, rounded at apex, yellow-brown, rather lustrous, with few longitudinal striations on both faces and on margin, the areole slightly convex.——Aug. Alpine slopes; Honshu (centr. and n. distr.).

11. CHELONOPSIS Miq.　　Jakō-sō Zoku

Tall erect herbs sometimes slightly woody at base; leaves coarsely toothed; verticils loosely 2- to 10-flowered, distant; flowers relatively large; calyx papyraceous, campanulate, 10-nerved, much inflated in fruit, bilabiate, the upper lip 3-toothed, the lower lip 2-toothed; corolla-tube ampliate, the limb subbilabiate, the upper lip retuse, the lower lip 3-lobed, ascending, the midlobe larger than the others; stamens 4, didymous, the anther-locules bearded in front; style with 2 equal lobes; nutlets flattened dorsally, veiny, winged on upper margin.——Few species, in Japan and China.

1A. Peduncles 2–12 mm. long, nearly as long as the petiole. . . . . 1. C. moschata
1B. Peduncles 3–4 cm. long, much longer than the petiole. . . . . 2. C. longipes

1. Chelonopsis moschata Miq. *C. moschata* var. *subglabra* Miq.; *C. subglabra* (Miq.) Koidz.; *C. yagiharana* var. *jesoensis* Koidz.; *C. yagiharana* Hisauchi & Matsuno; *C. subglabra* var. *jesoensis* (Koidz.) Hara; *C. moschata* var. *jesoensis* (Koidz.) Miyabe & Tatew.; *C. moschata* var. *lasiocalyx* Hayata ——Jakō-sō. Rhizomatous; stems tufted, erect, 60–100 cm. long, declined or ascending in upper part, simple, loosely pilose or nearly glabrous; leaves membranous, narrowly obovate or broadly oblanceolate, 10–20 cm. long, 3–10 cm. wide, long-acuminate, usually auriculate at base, scattered-pilose above, loosely spreading-pilose on nerves beneath, the petioles 5–12 mm. long; flowers 1–3, in axillary cymes; calyx 1–1.5 cm. long at anthesis, ovoid-globose, to 18 mm. long in fruit, the lower lip at anthesis shorter than the upper; corolla rose-colored, tubular, 4–4.5 cm. long, with a short upper lip; nutlets 7–8 mm. long, elliptic, glabrous, flattened.——Aug.–Sept. Wet places along streams and valleys; Hokkaido, Honshu, Shikoku.

2. Chelonopsis longipes Makino. *C. moschata* var. *longipes* Makino——Tani-jakō-sō. Stems 50–100 cm. long, usually purplish in upper part, often loosely spreading-hairy; leaves broadly lanceolate to narrowly obovate-oblong, long-acuminate, auriculate at base, the nerves pilose on both sides, the petioles 5–10 mm. long; peduncles axillary, 3–4 cm. long, puberulent, 1- to 3-flowered, the bracts filiform; calyx ap-pressed-puberulent, 7–8 mm. long at anthesis, 15–18 mm. long and much inflated in fruit, the lower lip shorter, with short teeth; corolla 3.5–4 cm. long, rose-purple; nutlets about 1 cm. long.——Sept.–Oct.   Low mountains; Honshu (s. Kantō Distr. and westw.), Shikoku, Kyushu.

12. GALEOPSIS L.   Chishima-odoriko-sō Zoku

Herbs; leaves opposite, toothed; cymes axillary, sometimes forming a terminal spike, 6- to many-flowered; calyx tubular-campanulate, 10-nerved, the teeth 5, short spine-tipped; corolla-tube erect, broadened on the throat, without or with an im-perfect ring of hairs inside at base, the limb bilabiate, the upper lip ovate, arcuate, the lower lip spreading, 3-lobed, the lateral lobes ovate, the midlobe obovate, with 2 conical tubercles near base; stamens 4, exserted, the anther-locules transversely split; nutlets lenticular, obovate, obtusely margined.——About 7 species, in Eurasia, naturalized in N. America.

1. Galeopsis bifida Boenn.   *G. tetrahit* var. *parviflora* Benth.; *G. tetrahit* var. *bifida* Syme; *G. tetrahit* sensu auct. Japon., non L.; *G. bifida* var. *emarginata* Nakai——Chishima-odoriko-sō, Itachiji-sō.   Erect prominently hispid annual; stems branched, 25–50 cm. long; leaves rather thin, ovate or narrowly so, sometimes ovate-oblong, 4–8 cm. long, 2–3.5 cm. wide, acuminate with an obtuse tip, usually broadly cuneate, obtusely to acutely toothed, the petioles 1–2 cm. long; cymes dense, few, subsessile, the bracteoles linear, spinelike, nearly as long as the calyx or calyx-tube; calyx 7–8 mm. long, long-hispid, loosely glandular-pilose on lower half, the teeth nearly as long as the tube, obliquely spreading, prominently spine-tipped; corolla about 10 mm. long, rather slender; nutlets about 2.5 mm. long, smooth, obovate-orbicular.——July–Aug. Hokkaido, Honshu (Nikko, Kamikochi in Shinano).——Sa-khalin, Kuriles, Korea, Siberia, and Europe.

13. STACHYS L.   Inu-goma Zoku

Annuals, biennials, perennials, or small shrubs; leaves toothed; verticils 2- to many-flowered, or forming a terminal spike; flowers subsessile, rose-purple, white, or yellow; calyx 5- or 10-nerved, 5-toothed; corolla-tube usually with a ring of hairs inside, the limb bilabiate, the upper lip erect, dilated, or galeate, the lower lip spreading, 3-lobed, the middle lobe larger; stamens 4, didymous, the anther locules longitudinally split; nutlets ovate or oblong, truncate at apex.——About 200 species, nearly cos-mopolitan, except Australia.

1. Stachys japonica Miq.   *S. baicalensis* var. *japonica* (Miq.) Komar.; *S. aspera* var. *japonica* (Miq.) Maxim.; *S. aspera* var. *chinensis* forma *glabrata* Nakai; *S. japonica* forma *glabra* Kudo; *S. riederi* var. *japonica* (Miq.) Hara——Ke-nashi-inu-goma, Chōsen-inu-goma.   Stoloniferous, smooth and nearly glabrous except white-hairy on nodes of stems; stems erect, 30–60 cm. long; leaves lanceolate, deltoid-lanceolate or linear-lance-olate, 4–8 cm. long, 1–2.5 cm. wide, subacute, truncate to rounded at base, toothed, smooth, the petioles 5–15 mm. long; floral bracts sessile, 7–15 mm. long; verticils several-flowered, subsessile; flowers pinkish, 12–15 mm. long; calyx 6–8 mm. long, usually loosely puberulent, the teeth spreading, short spine-tipped, narrowly deltoid, slightly shorter than the tube; nutlets suborbicular, 1–7 mm. long, unequally biconvex, ob-tuse on margin, rounded at apex.——July–Aug. Wet places in lowlands; Hokkaido, Honshu, Kyushu; rather common. ——Sakhalin and s. Kuriles.

Var. **intermedia** (Kudo) Ohwi.   *S. japonica* forma *inter-media* Kudo.——Inu-goma.   Stems retrorsely scabrous, the leaves with the midrib often scabrous beneath.——Hokkaido, Honshu, Shikoku, Kyushu; common.

Var. **villosa** (Kudo) Ohwi.   *S. japonica* forma *villosa* Kudo——Ezo-inu-goma.   Leaves and stem-angles coarsely hispid.——Hokkaido, Honshu.

S. affinis Bunge.   *S. sieboldii* Miq.; *S. tubifera* Naud.—— Chorogi.   Cultivated plant of Chinese origin yielding edible tuberlike rhizomes.

14. LEONURUS L.   Me-hajiki Zoku

Erect herbs; leaves opposite, coarsely toothed and incised, or in the lower ones palmately parted; verticils many-flowered, loose; calyx obconical or tubular-campanulate, nearly equally 5-toothed, 5-nerved, the teeth ascending, spine-tipped; corolla-tube short, the limb bilabiate, the upper lip nearly erect, oblong, entire, white-villous on back, the lower lip spreading, 3-lobed; stamens 4, didymous, the anthers glabrous; nutlets smooth, compressed-trigonous, truncate.——About 10 species, in Europe and Asia.

1A. Leaves palmately lobed to parted; flowers 12–15 mm. long; calyx 6–7 mm. long, densely white-puberulent, gray-green or whitish green. .... 1. *L. sibiricus*
1B. Leaves coarsely toothed or incised; flowers 25–32 mm. long; calyx 15–18 mm. long, loosely hirsute, green to yellowish green.
2. *L. macranthus*

1. Leonurus sibiricus L.   *L. japonicus* Houtt.; *Stachys artemisia* Lour.——Me-hajiki.   Gray-green biennial with white short-appressed pilose hairs; stems 50–100 cm. long, erect, branched; radical leaves ovate-cordate, obtusely toothed and incised, long-petiolate, withering before anthesis; cauline leaves 5–10 cm. long, pinnatisect, cuneate at base, long-peti-oled, the upper leaves gradually smaller, often entire, lanceo-late or linear; flowers few, dense, 12–15 mm. long, the bracte-oles usually shorter than the calyx, spinelike; calyx-teeth short spine-tipped, ascending in fruit; corolla rose-colored; nutlets black, subcuneate, acutely 3-angled, slightly compressed, truncate at apex, glabrous.——July–Sept. Honshu, Shikoku, Kyushu.——Ryukyus, Formosa, Korea, and China.

2. Leonurus macranthus Maxim. *L. japonicus* Miq., non Houtt.——Kise-wata. Pubescent perennial; stems stout, firm, erect, simple or slightly branched; leaves chartaceous, ovate or narrowly so, 6–10 cm. long, 3–6 cm. wide, acute to acuminate, broadly cuneate to truncate at base, incised or with large, deep, acute teeth, the petioles 1–5 cm. long, the upper leaves gradually smaller, broadly lanceolate to narrowly ovate, often entire, the petioles about 1 cm. long; flowers few, in interrupted verticils, subsessile, 25–32 mm. long, pink; calyx-teeth prominently spine-tipped, the lower lip of corolla shorter than the upper; nutlets about 2.5 mm. long, cuneate-obovate, acutely 3-angled, truncate at apex, black.——Aug.–Sept. Grassy places and thickets in hills and mountains; Hokkaido, Honshu, Shikoku, Kyushu.——Korea, Manchuria, and China.

15. LAMIUM L.　Odoriko-sō Zoku

Annuals or perennials; leaves usually cordate, toothed; verticils densely many-flowered, distant or approximate in upper ones, the bracteoles short, linear or lanceolate, sometimes spinelike; flowers purple, white, rarely yellow, sometimes dimorphic; calyx tubular or obconical-campanulate, 5-nerved, 5-toothed; corolla-tube usually exserted, widened toward the throat, the limb bilabiate, the upper lip erect and concave on back or galeate, the lower lip spreading, 3-lobed, the median lobe larger, retuse, usually narrowed at base; stamens 4, didymous, the anther-locules divergent, barbate or glabrous; nutlets trigonous, smooth or minutely tubercled, usually with acute margin.——About 40 species, in temperate regions of Eurasia and Africa.

1A. Annuals or biennials; flowers less than 2 cm. long, often cleistogamous; corolla-tube usually without a ring of hairs inside.
  2A. Floral leaves sessile, rather loosely arranged throughout, green; calyx with incurved or erect teeth. .... 1. *L. amplexicaule*
  2B. Floral leaves short-petiolate, approximate toward top of stems, often somewhat purplish; calyx with obliquely spreading teeth.
                                                                            2. *L. purpureum*
1B. Perennials; flowers 15–40 mm. long; corolla-tube with a ring of hairs inside.
  3A. Flowers 15–22 mm. long; corolla-tube erect; calyx-teeth deltoid-lanceolate, spine-tipped; anther-locules glabrous.
    4A. Leaves not approximate at any point, loosely arranged.
      5A. Plant 10–25 cm. high, tuberous, stoloniferous; calyx-teeth nearly equal. .... 3. *L. chinense* var. *tuberiferum*
      5B. Plants 40–70 cm. high, rhizomatous; outer 2 calyx-teeth longer and connate at base. .... 4. *L. ambiguum*
    4B. Leaves of 2 or 3 pairs, crowded toward the top of short stems. .... 5. *L. humile*
  3B. Flowers 3.5–4 cm. long; corolla-tube slightly curved upward near base; calyx-teeth linear-filiform, not spine-tipped; anther-locules bearded. .... 6. *L. album* var. *barbatum*

1. Lamium amplexicaule L. Hotoke-no-za. Winter annual; stems 10–30 cm. long, sometimes much-branched near base, reddish, sparsely reflexed-puberulent; leaves mostly basal, cordate or reniform, 1–2 cm. long and as wide, loosely ascending-pilose, sparsely round-toothed, the floral leaves suborbicular or flabellate, 1–2.5 cm. wide, sessile; flowers 17–20 mm. long, sometimes cleistogamous, in dense sessile verticils, red; calyx about 5 mm. long with incurved or erect teeth; nutlets about 2 mm. long, trigonous, rounded at apex, narrowed below, rounded on back, usually white-spotted.——May–June. Roadsides, cultivated fields, and waste grounds in lowlands; Honshu, Shikoku, Kyushu; common.——Ryukyus, Formosa, China, Korea to Europe, N. Africa, and widely adventive elsewhere.

2. Lamium purpureum L. Hime-odoriko-sō. Winter annual with decumbent branched base; stems erect, 10–25 cm. long, rather soft and stout, slightly pubescent; leaves long-petiolate, orbicular-cordate, 1.5–3 cm. long and as wide, obtusely round-toothed, long-pubescent, the floral leaves slightly larger, ovate-orbicular or depressed, often tinged purplish, much crowded toward the top of stems, short-petiolate; flowers dense in sessile verticils, about 1 cm. long; calyx-teeth ciliate, obliquely spreading; nutlets broadly obovate, 1.5 mm. long, acutely 3-angled, obliquely truncate at apex, rounded on back.——Apr.–May. Naturalized in our area.——Europe and Asia Minor and widely adventive in N. America.

3. Lamium chinense Benth. var. **tuberiferum** (Makino) Murata. *Leonurus tuberiferus* Makino; *Matsumurella tuberifera* (Makino) Makino; *Lamium tuberiferum* (Makino) Ohwi——Hime-kise-wata. Stoloniferous perennial herb with tubers; stems 10–25 cm. long, simple or branched at base, recurved-pubescent in lower part; leaves deltoid to rhombic-ovate, 1–3.5 cm. long, 1–2.5 cm. wide, obtuse, broadly cuneate to subtruncate at base, coarsely obtuse-toothed, coarsely pubescent on both sides; flowers 1–3 in axils, pale purple, 15–20 mm. long; calyx pubescent, 6–7 mm. long, 5-toothed, the 2 outer lobes slightly broader; lower lip of corolla about 1 cm. long, spreading, the lateral lobes somewhat reflexed.——Apr.–May. Kyushu (s. distr.).——Ryukyus. The typical phase occurs in Formosa and China.

4. Lamium ambiguum (Makino) Ohwi. *Leonurus ambiguus* Makino; *Loxocalyx ambiguus* (Makino) Makino——Manekigusa, Yama-kise-wata. Rhizomes rather slender, creeping; stems erect, 40–70 cm. long, often branched, recurved white-pubescent on the angles; lower leaves cordate, the petioles 2–7 cm. long, the median deltoid-ovate or deltoid-orbicular, 3–6 cm. long and as wide, obtuse to acuminate with an obtuse tip, truncate or very broadly cuneate at base, coarsely obtuse-toothed or incised, shallowly cordate in the lower ones, pubescent on both surfaces especially on nerves beneath; floral bracts leaflike or broadly ovate and smaller, the petioles 1–2 cm. long; flowers 1–3 in axils, dark red-purple, 18–20 mm. long; calyx loosely puberulent, about 8 mm. long in back, 11 mm. long in front, oblique at the mouth, the teeth narrowly deltoid, spine-tipped, ascending; lower lip of corolla 7–8 mm. long, spreading; nutlets obovate-cuneate, 2.5–3 mm. long, 3-angled, rounded at apex.——Sept. Woods in mountains; Honshu (Sagami Prov. and westw.), Shikoku, Kyushu.

5. Lamium humile (Miq.) Maxim. *Ajuga humilis* Miq.; *Loxocalyx humilis* (Miq.) Makino; *Ajugoides humilis* (Miq.) Makino——Yama-jiō. Rhizomes slender; stolons filiform, branched and creeping; stems 5–10 cm. long, ascending from the base, rather densely white-pubescent, simple, or with 2 or 3 pairs of approximate leaves toward the top: leaves obovate to obovate-orbicular or obovate-oblong, rounded at apex, cuneate to acute at base, coarsely toothed, short-ascending pubescent above and on nerves beneath, the petioles 1–5 mm. long; flowers 1–3 in axils, 15–18 mm. long; calyx 7–8 mm. long, 5-toothed, prominently hairy; flowers pinkish; lower lip of corolla 5–6 mm. long, spreading; nutlets about 2 mm. long, acutely angled, glabrous, obliquely truncate at apex.——Aug. Woods in mountains; Honshu (Izu Prov. and westw.), Shikoku, Kyushu.

6. Lamium album L. var. **barbatum** (Sieb. & Zucc.) Fr. & Sav. *L. barbatum* Sieb. & Zucc.; *L. petiolatum* sensu auct. Japon., non Royle; *L. album* var. *petiolatum* Nakai, excl. syn. ——Odoriko-sō. Stems tufted, rather stout, erect, short-ascending at the base, nearly glabrous or long-pilose on the nodes; leaves triangular to ovate-cordate or broadly ovate, 5–10 cm. long, 3–8 cm. wide, obtuse to short caudate, shallowly cordate to rounded at base, loosely pubescent above and on nerves beneath, petiolate; verticils loose; flowers dense, white to pinkish, 3–4 cm. long; calyx 13–18 mm. long, 5-toothed, the teeth obliquely spreading, linear-filiform, ciliate, gradually broadened at base; corolla-tube slightly curved upward near base, the upper lip of corolla arcuate, white-pubescent externally, the lower lip reflexed, the midlobe large, the lateral ones with a linear appendage; anther locules bearded; nutlets cuneate-obovate, about 3 mm. long, acutely 3-angled, truncate at apex.——Apr.–June. Thickets and thin woods especially along streams in lowlands to high mountains; Hokkaido, Honshu, Shikoku, Kyushu; common.——s. Kuriles, Sakhalin, Korea, China, and Manchuria. The typical phase occurs from Eurasia to N. Africa.

16. SALVIA L.　Akigiri Zoku

Herbs or shrubs; leaves entire to pinnately parted, verticils 2- to many-flowered, often forming a terminal spike, raceme, or a compound terminal panicle, rarely loosely arranged; flowers sessile or short-pedicelled, red, purple, white, or yellow, the bracteoles usually minute; calyx bilabiate, the upper lip entire or 3-toothed, the lower lip 2-lobed, the throat naked; corolla bilabiate, the tube usually with a ring of hairs inside near base, the upper lip erect, often galeate, the lower lip spreading, 3-lobed, the midlobe larger; stamens 2, the filaments short, the arcuate connective jointed, with a fertile anther-locule at one end and a rudimentary locule opposite; nutlets smooth.——More than 500 species, widespread in temperate and warmer regions.

1A. Corolla 5–12 mm. long; calyx 3–7 mm. long.
 2A. Biennial herb; leaves ovate-oblong or broadly lanceolate, acute at base; spikes usually in a compound panicle; corolla 4–5 mm. long; fertile anther-locules about 0.5 mm. long, not exserted from the upper lip of corolla; nutlets loosely warty. .... 1. *S. plebeia*
 2B. Perennial herbs; leaves when simple broadly ovate and rounded to shallowly cordate at base; spikes simple or slightly compound; corolla 7–12 mm. long; fertile anther-locules 1–2 mm. long, often exserted from under the upper lip of corolla; nutlets smooth.
  3A. Calyx not pilose inside. .... 2. *S. omerocalyx*
  3B. Calyx white-pilose inside near middle.
   4A. Corolla-tube sparsely pilose inside; flowering March to July. .... 3. *S. ranzaniana*
   4B. Corolla-tube with a distinct annular-pilose ring inside; flowering July to November.
    5A. Corolla 11–13 mm. long; corolla-tube much longer than the calyx, with a ring of hairs near the base inside; stamens and styles slightly exserted. .... 4. *S. japonica*
    5B. Corolla 6–10 mm. long; corolla-tube slightly longer than the calyx, with a ring of hairs near the middle; stamens and styles much exserted.
     6A. Corolla 6–8 mm. long; anther-connectives 3–5 mm. long; leaves relatively thick with raised nerves beneath; stems usually white spreading-pilose. .... 5. *S. isensis*
     6B. Corolla 8–10 mm. long; anther-connectives 7–8 mm. long; leaves nearly membranous; stems pilose on the nodes.
      6. *S. lutescens*
1B. Corolla 3–4 cm. long; calyx 7–12 mm. long.
 7A. Stems and inflorescence not glandular-hairy; leaves hastate.
  8A. Flowers yellow; corolla with a ring of hairs near the base of the tube inside, the hairs smooth and obtuse at the tip under magnification. .... 7. *S. nipponica*
  8B. Flowers purple; corolla uniformly scattered-hairy inside tube, the hairs roughened and acute at the tip under magnification.
   8. *S. glabrescens*
 7B. Stems and inflorescence glandular-hairy; leaves rounded-cordate. .... 9. *S. koyamae*

1. Salvia plebeia R. Br. *Ocimum virgatum* Thunb.; *S. brachiata* Roxb.; *S. minutiflora* Bunge——Mizo-kōju, Yukimi-sō. Erect biennial; stems erect, usually branched in upper part, recurved-puberulent on the angles; radical leaves rosulate, long-petiolate, larger than the cauline, withering away before anthesis; cauline leaves ovate-oblong to broadly lanceolate, 3–6 cm. long, 1–2 cm. wide, obtuse, acute to cuneate at base, obtusely toothed, thinly puberulent on both surfaces, the petioles 1–3 cm. long; spikes terminal, 8–10 cm. long, rather prominently spreading-puberulent, simple or more often branched at base; calyx 2.5–3 mm. long at anthesis, to 4 mm. long in fruit, puberulent and glandular-dotted; corolla pale purple, 4–5 mm. long; nutlets loosely warty.——May–June. Wet places, especially around paddy fields; Honshu, Shikoku, Kyushu.——Korea, China, Formosa, India to Malaysia, and Australia.

2. Salvia omerocalyx Hayata. Tajima-tamura-sō. Erect perennial herb; stems 20–40 cm. long, leafy toward base; leaves long-petiolate, glabrous; leaflets 3–5, the terminal ones membranous, oblong or ovate-oblong, 3–3.5 cm. long, obtuse, rounded to shallowly cordate at base, crenate-toothed, the petiolules 3–5 mm. long, the lateral leaflets much smaller, 1–2 cm. long; spikes 5–15 cm. long, thinly pubescent, the lower verticils loosely arranged, the upper ones dense, 4- to 6-flowered, the floral bracts small; calyx tubular, 8–10 mm. long, bilabiate, sparingly glandular-pilose; corolla 1 cm. long, deep blue, the tube slightly longer than the calyx, the limb spreading; stamens long-exserted.——Honshu (Tajima and Tango Prov.).

Var. **prostrata** Satake. Hai-tamura-sō. Stems prostrate, sparingly leafy on lower parts.——Honshu (Echizen and Wakasa Prov.).

3. **Salvia ranzaniana** Makino. *S. rosulata* Nakai; *S. japonica* var. *pumila* Fr. & Sav.; *S. chinensis* var. *pumila* (Fr. & Sav.) Makino; *S. vernalis* Kudo——Haru-no-tamura-sō. Small perennial, puberulent especially on petioles and spikes; stems erect, simple, 10–20 cm. long, scapelike, with long spreading hairs below the nodes and with small leaves near the middle; radical and lower cauline leaves ovate or narrowly so, 3–6 cm. long, twice or sometimes once pinnate, sparingly pilose, long-petiolate, the terminal leaflets larger, rhombic-orbicular, 1–1.5 cm. long, petioluled, the lateral ones sessile or petioluled, 2 or 3 pairs, ovate or broadly so, 7–12 mm. long, obtuse, few-toothed; spikes 3–6 cm. long, the bracts shorter than the pedicels; calyx 5–8 mm. long; corolla white, about 8 mm. long, glabrescent outside; nutlets elliptic, 1.5 mm. long. ——Mar.–July. Honshu (centr. and w. distr.), Shikoku, Kyushu.——Ryukyus (?) and Formosa (var.).

4. **Salvia japonica** Thunb. *S. chinensis* Benth.; *S. diversifolia* Miq.; *S. japonica* var. *integrifolia* Fr. & Sav.; *S. japonica* var. *ternata* Fr. & Sav.; *S. japonica* var. *bipinnata* Fr. & Sav.; *S. fushimiana* Koidz.; *S. tsushimensis* Nakai; *S. japonica* var. *chinensis* (Benth.) Peter-Stibal——Aki-no-tamura-sō. Fig. 15. Perennial; stems 20–80 cm. long, sometimes branched, glabrous or puberulent, loosely leafy; leaves simple, or often once- or twice-pinnate, ternate, long-petiolate, the leaflets broadly ovate to rhombic, 2–5 cm. long, glabrous or sparingly pilose above; spikes terminal, often branched at base, prominently puberulent, 10–25 cm. long; corolla pale purple, rarely white (forma **albiflora** Hiyama), 1–1.3 cm. long, sparingly pubescent outside, with a ring of hairs near the base; stamens slightly exserted; nutlets 1.5–2 mm. long.——July–Nov. Woods and thickets in hills and mountains; Honshu, Shikoku, Kyushu.——Formosa, China, and Korea.——Forma **longipes** (Nakai) Murata. *S. longipes* (Nakai) Satake; *Polakiastrum longipes* Nakai——Inu-tamura-sō. An abnormal phase with longer pedicels, green flowers, and sterile stiped ovary.——Honshu; rare.

5. **Salvia isensis** Nakai ex Hara. Shimaji-tamura-sō. Perennial; stems erect, 10–60 cm. long, densely long spreading-hirsute or sometimes nearly glabrate; leaves approximate toward the base, long-petiolate, pinnate, the leaflets rhombic-ovate to oblong, 1.5–2.5 cm. long, obtuse, cuneate to subcordate at base, crenate-toothed, the lateral leaflets of 2 pairs; spikes elongate, the verticils loosely 2- to 8-flowered, the bracts ovate-lanceolate; flowers purplish, 6–8 mm. long; calyx about 6 mm. long; corolla-tube nearly as long as the calyx, with a ring of hairs slightly below the middle inside; stamens long-exserted, the connectives 3–5 mm. long; nutlets elliptic, about 2 mm. long.——July–Nov. Honshu (Ise, Owari, and Mikawa Prov.).

6. **Salvia lutescens** (Koidz.) Koidz. var. **intermedia** (Makino) Murata. *S. japonica* var. *intermedia* Makino; *S. omerocalyx* var. *intermedia* (Makino) F. Maekawa; *S. japonica* sensu auct. Japon., pro parte, non Thunb.; *S. lutescens* forma *lobatocrenata* (Makino) G. Nakai; *S. japonica* var. *intermedia* forma *lobatocrenata* Makino——Natsu-no-tamura-sō. Perennial; stems 40–80 cm. long, erect, usually with long spreading hairs on nodes; leaves once or twice pinnate or ternate, long-petiolate, the leaflets narrowly ovate to broadly so, the terminal ones slightly larger, 2–7 cm. long, the bipinnate ones 1–2.5 cm. long; spikes terminal, 10–25 cm. long, often branched at base, with long spreading hairs; corolla deep purple, about 1 cm. long, long-pubescent externally; nutlets 1.5–2 mm. long.——July–Aug. Honshu (centr. and w. distr.).

Var. **lutescens**. Usugi-natsu-no-tamura-sō. Flowers pale yellow.——Honshu (Suzuka mts. in e. Kinki Distr.).

Var. **stolonifera** G. Nakai. Dando-tamura-sō. Resembles var. *intermedia* but stolons developed after anthesis.——Honshu (Tōkaidō, from Izu Prov. to Mikawa Prov.).

Var. **crenata** (Makino) Murata. *S. japonica* var. *intermedia* forma *crenata* Makino; *S. japonica* var. *crenata* Makino; *S. omerocalyx* var. *crenata* (Makino) F. Maekawa——Miyama-tamura-sō, Ke-natsu-no-tamura-sō. Mountain phase with pale purple flowers, and the anther connectives usually glabrous.——Honshu (centr. and n. distr.)

7. **Salvia nipponica** Miq. *S. nipponica* forma *argutidens* Makino; *S. polakioides* Honda; *S. mayebarae* Honda; *S. nipponica* forma *lutea* Makino——Kibana-akigiri. Perennial, sometimes with thickened narrowly fusiform roots; stems 20–40 cm. long, usually soft-pubescent, decumbent or ascending at base, becoming erect; leaves hastate, 5–10 cm. long, 4–7 cm. wide, abruptly acuminate, cordate at base, pilose, the long petioles pilose; spikes 10–20 cm. long, spreading-pilose, the bracts ovate, longer than the pedicels; calyx with scattered long-spreading hairs; corolla pale yellow, 2.5–3.5 cm. long, bilabiate and bifid almost to the middle; nutlets obovate-orbicular, obtusely 3-angled, slightly flattened, smooth, about 2 mm. long.——Aug.–Oct. Woods and thickets in mountains; Honshu, Shikoku, Kyushu; rather common.

Var. **trisecta** (Matsum.) Honda. *S. trisecta* Matsum.——Mitsuba-kotoji-sō. Leaves 3-parted.——Shikoku and Kyushu.

8. **Salvia glabrescens** Makino. *S. nipponica* var. *glabrescens* Fr. & Sav.; *S. nipponica* var. *purpurea* Makino; *S. nipponica* subsp. *robusta* Koidz.; *S. robusta* (Koidz.) Makino——Akigiri. Closely allied to *S. nipponica*, the stems somewhat taller; leaves less prominently pilose, usually sagittate, rachis usually with minute crispate glandular-tipped hairs; corolla purple, the tube sparingly scaberulous on inner side.——Honshu (centr. and n. distr.).

9. **Salvia koyamae** Makino. Shinano-akigiri. Resembles the preceding, the plant slightly taller; leaves 8–15 cm. long, prominently pubescent on both sides; rachis and calyx densely long glandular-pubescent; flowers pale yellow.——Honshu (Matsubara-mura in Shinano Prov.); rare.

17. MOSLA Hamilt. Inu-kōju Zoku

Mostly pubescent annuals; leaves lanceolate-linear to ovate-orbicular, toothed; verticils 2-flowered, distant or in terminal spikes, the bracts small or ovate and slightly longer than the calyx; calyx campanulate, accrescent after anthesis, subequally 5-lobed or bilabiate, the lobes linear-lanceolate, the upper lip of 3 more or less connate lobes, the tube in fruit often slightly inflated in front; corolla subbilabiate, the tube with or without a ring of hairs inside, the lower lip 3-lobed; fertile stamens 2, erect, staminodia 2; nutlets subglobose, reticulate or smooth.——About 10 species, in e. Asia and in s. Asia from India to Malaysia.

1A. Bracts lanceolate or linear-lanceolate, shorter than to as long as the pedicels; calyx bilabiate in fruit, the pedicels 2–4 mm. long.
  2A. Stems and axis of spikes rather prominently puberulent. .... 1. *M. punctulata*
  2B. Stems and axis of spikes glabrous, the nodes sparsely pubescent, with long whitish hairs. .... 2. *M. dianthera*

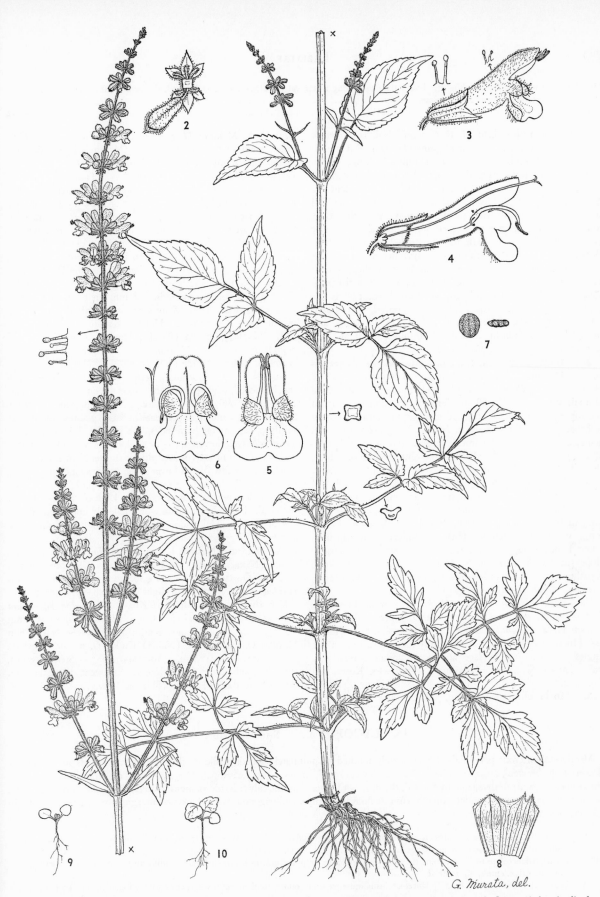

Fig. 15.—*Salvia japonica* Thunb. 1, Habit showing a cross section of stem and petiole; 2, verticil, top view; 3, flower; 4, longitudinal section of a flower; 5, front view of flower late in anthesis; 6, the same, a few days later; 7, pollen; 8, calyx cut open; 9, 10, seedlings.

1B.   Bracts ovate or broadly so, slightly shorter than to as long as the calyx; calyx subequally 5-lobed, the pedicels 0.5–1 mm. long.
    3A.   Leaves broadly linear to lanceolate. .... 3. *M. chinensis*
    3B.   Leaves ovate. .... 4. *M. japonica*

1. Mosla punctulata (J. F. Gmelin) Nakai. *Ocimum punctatum* Thunb., non L. f.; *O. punctulatum* J. F. Gmelin; *O. scabrum* Thunb.; *M. punctata* (Thunb.) Maxim.; *Orthodon scaber* (Thunb.) Hand.-Mazz.; *Orthodon punctulatum* (J. F. Gmelin) Ohwi——INU-KŌJU.   Erect puberulent annual; stems usually reddish, 20–60 cm. long; leaves ovate, 2–4 cm. long, 1–2.5 cm. wide, acute at apex, abruptly acute, rounded or cuneate at base, toothed, puberulent above, loosely puberulent on nerves beneath, the petioles 1–2 cm. long; spikes elongate, 4–10 cm. long, rather densely flowered, the bracts 2.5–3 mm. long, lanceolate, nearly as long as the pedicels; calyx 2–3 mm. long in anthesis, about 4 mm. long in fruit, the teeth acute; corolla rose-colored, 3–4 mm. long; nutlets obovate, nearly 1 mm. across, slightly compressed, with raised reticulations.——Sept.–Oct. Grassy slopes in lowlands and hills; Hokkaido, Honshu, Shikoku, Kyushu.——Ryukyus, Formosa, China, Korea, and Manchuria.

2. Mosla dianthera (Hamilt.) Maxim.   *Lycopus dianthera* Hamilt. ex Roxb.; *M. grosseserrata* Maxim.; *Orthodon grosseserratum* (Maxim.) Kudo——HIME-JISO.   Branched annual; stems erect, 20–50 cm. long, recurved-puberulent on the angles, white-hairy on the nodes; leaves membranous, ovate to broadly so, or rhombic-ovate, 2–4 cm. long, 1–2.5 cm. wide, acute, coarsely toothed, glabrous or scattered appressed-pilose above, glabrous or scattered long-pubescent beneath near base, the petioles 1–3 cm. long; spikes 3–7 cm. long, rather loosely flowered, the axis white-pubescent on nodes, the pedicels 2–4 mm. long, thinly puberulent, nearly as long as the linear-lanceolate bracts; calyx 2–3 mm. long at anthesis, about 5 mm. long in fruit, teeth of upper lip broad; corolla about 4 mm. long, white or sometimes pinkish; nutlets ovate-orbicular, about 1 mm. long, obscurely reticulate.——Sept.–Oct. Grassy slopes in lowlands and hills; Hokkaido, Honshu, Shikoku, Kyushu; rather common.——Korea, Manchuria, China, India, Formosa, and Malaysia.

Var. **nana** (Hara) Ohwi.   *Orthodon hirtum* Hara; *O. hirtum* forma *nanum* (Hara) Hara; *O. grosseserratum* var. *nanum* Hara; *O. tenuicaule* Koidz.; *O. mayebaranum* Honda——HIKAGE-HIME-JISO.   Stems thinly long-pubescent on the angles; leaves on both surfaces and axis of spikes loosely long-pubescent.——Sept.–Oct. Honshu, Kyushu.——s. Korea and Formosa.

3. Mosla chinensis Maxim.   *M. japonica* Maxim. var.
angustifolia Makino; *M. coreana* Lév.; *M. angustifolia* (Makino) Makino; *Orthodon chinense* (Maxim.) Kudo; *O. japonicum* var. *angustifolium* (Makino) Kudo; *O. angustifolium* (Makino) Masam.; *O. coreanum* (Lév.) Honda——HOSOBA-YAMA-JISO.   Annual; stems 10–30 cm. long, often branched, purplish, recurved-puberulent; leaves broadly linear to lanceolate, 1.5–3 cm. long, 2–6 mm. wide, subobtuse, acuminate at base, obscurely short-toothed, puberulent on both sides, the petioles 5–10 mm. long; spikes 1–3 cm. long, densely flowered, the bracts broadly ovate, 5–7 mm. long, cuspidate, appressed to and as long as or slightly shorter than the calyx, much longer than the pedicels; flowers about 4 mm. long; calyx about 3 mm. long at anthesis, 7–8 mm. in fruit, puberulent; nutlets orbicular, about 1.5 mm. across, slightly flattened, with obscure pits.——Sept.–Oct. Honshu, Kyushu.——China.

4. Mosla japonica (Benth.) Maxim.   *Orthodon japonicum* Benth. ex Oliv.; *Micromeria* (?) *perforata* Miq.; *Mosla perforata* (Miq.) Koidz.; *O. perforatum* (Miq.) Ohwi——YAMA-JISO.   Erect annual; stems 10–40 cm. long, purplish red, branched, white-puberulent, especially on the nodes; leaves ovate to narrowly so, 1–3 cm. long, 7–17 mm. wide, subacute, few-toothed, acute to subcuneate at base, glabrescent or very thinly pubescent on both sides, the petioles 3–10 mm. long; spikes densely flowered, 1–4 cm. long, short-pubescent, the bracts ovate to broadly so, cuspidate, shorter than the calyx in fruit except a few of the lower ones; calyx about 3 mm. long in anthesis, 7–8 mm. long in fruit; corolla pink, about 3 mm. long; nutlets nearly orbicular, about 1.3 mm. across, slightly flattened, obsoletely reticulate with shallowly impressed areoles.——Sept.–Oct. Hokkaido, Honshu, Shikoku, Kyushu.——s. Korea.

Var. **hadae** (Nakai) Kitam.   *Orthodon japonicum* var. *hadae* (Nakai) Ohwi; *M. hadae* Nakai; *O. hadae* (Nakai) Kudo——Ō-YAMA-JISO.   Long spreading-pubescent on stems and leaves; corolla white to slightly purplish.——Honshu.

Var. **thymolifera** (Makino) Kitam.   *Orthodon japonicum* var. *thymoliferum* (Makino) Ohwi; *M. thymolifera* Makino; *O. thymoliferum* (Makino) Kudo; *M. leucantha* Nakai, non Hayata; *O. leucanthum* (Nakai) Kudo; *O. nakaii* Okuyama——SHIROBANA-YAMA-JISO, AO-YAMA-JISO.   Stems green, spreading-pubescent; leaves loosely pubescent above and on nerves beneath.——Honshu (Shimosa Prov.).

18. LYCOPUS L.   SHIRO-NE ZOKU

Mostly stoloniferous perennials; leaves acutely toothed to pinnately lobed, sessile or short-petiolate; verticils loose, densely flowered, the bracteoles small or short, subulate; flowers small, subsessile; calyx campanulate, equally 4- or 5-lobed, naked on the throat; corolla campanulate, equally 5-lobed, the tube not longer than the calyx; fertile stamens 2, erect, the anther-locules parallel, the staminodia 2 or none; nutlets smooth, cuneate-obovate, compressed-trigonous with thickened margins, truncate.——About 10 species, in e. Asia and N. America.

1A.   Calyx-teeth ovate, obtuse or subacute, not spine-tipped; stems puberulent; leaves membranous, ovate; nutlets longer than the calyx, with 4 obsolete obtuse teeth near apex. .... 1. *L. uniflorus*
1B.   Calyx-teeth lanceolate to linear, spine-tipped; stems glabrous or pubescent on the nodes, rarely uniformly scaberulous; nutlets shorter than the calyx, entire, not toothed.
    2A.   Leaves rhombic-ovate to broadly lanceolate, subobtuse, coarsely obtuse-toothed, narrowed below to a winged petiolelike base.
                                                                                                      2. *L. ramosissimus*
    2B.   Leaves broadly linear to broadly lanceolate, at least the upper ones acuminate or acute and coarsely acute-toothed.

3A. Leaves 8–15 cm. long, 1.5–4 cm. wide, at least the upper ones acuminate at base, sessile or subsessile; stems usually rather stout, to 1 m. high. .... 3. *L. lucidus*
3B. Leaves 4–8 cm. long, 5–15 mm. wide, slightly narrowed toward the base, somewhat auriculate, very short-petiolate; stems 30–70 cm. long, rather slender. .... 4. *L. maackianus*

1. Lycopus uniflorus Michx. *L. parviflorus* Maxim.; *L. virginicus* var. *parviflorus* (Maxim.) Makino; *L. coreanus* Lév. ——EZO-SHIRO-NE. Perennial; stolons terminated by a fusiform tuber; stems erect, 20–40 cm. long, somewhat branched, soft-puberulent, purplish; leaves membranous, rhombic-ovate or narrowly ovate, obtuse or acute with an obtuse tip, cuneate or acute at base, scattered obtuse-toothed, glabrate or thinly puberulent and glandular-dotted on both sides, short-petiolate; flowers few in each verticil, dense, white, about 2 mm. long; calyx membranous, 1 mm. long at anthesis, to 1.5 mm. long in fruit, 5-fid, the teeth ovate; nutlets broadly cuneate at base, thickened on margin, compressed-trigonous, smooth, truncate, with 4 obsolete undulate teeth on the broad anterior apex.—— Aug.–Sept. Wet places; Hokkaido, Honshu, Shikoku, Kyushu.——Sakhalin, Kuriles, Korea, Manchuria, e. Siberia, and N. America.

2. Lycopus ramosissimus (Makino) Makino var. **japonicus** (Matsum. & Kudo.) Kitam. *L. coreanus* auct. Japon., non Lév.; *L. maackianus* auct. Japon., pro parte, non Makino; *L. japonicus* Matsum. & Kudo; *L. coreanus* var. *ramosissimus* (Makino) Nakai; *L. maackianus* var. *ramosissimus* Makino ——HIME-SARUDAHIKO, SARUDAHIKO. Stoloniferous perennial; stems mostly erect, 10–80 cm. long, usually branched, often slightly scabrous, white-pilose at the nodes; leaves narrowly rhombic-ovate to broadly lanceolate, 2–4 cm. long, 1–2 cm. wide, subobtuse, narrowed to a winged petiolelike base, coarsely obtuse-toothed; flowers in loose verticils; calyx about 3 mm. long, 5-toothed, the teeth narrowly deltoid, spine-tipped; corolla white, about 3 mm. long; nutlets 1.5 mm. long, broadly cuneate at base, truncate and very slightly undulate at apex. ——Aug.–Nov. Wet places; Hokkaido, Honshu, Shikoku, Kyushu.——Korea, Manchuria, and China.

Var. **ramosissimus.** *L. maackianus* var. *ramosissimus* Makino; *L. coreanus* var. *ramosissimus* (Makino) Nakai—— HIME-SARUDAHIKO. Stems low, 10–30 cm. long, much-branched; leaves smaller.——Honshu, Shikoku, Kyushu.—— Korea.

3. Lycopus lucidus Turcz. SHIRO-NE. Stoloniferous; stems somewhat thickened at the base, erect, simple, to 1 m. long, stout, glabrous or nearly so except for white hairs at nodes; leaves rather broadly lanceolate to narrowly elliptic, 8–15 cm. long, 1.5–4 cm. wide, acuminate at both ends, lustrous above, glabrous on both surfaces or thinly pilose on the nerves beneath, sessile or subsessile, the upper leaves slightly smaller; flowers white, about 5 mm. long; calyx 4–5 mm. long, 5-toothed, the teeth linear-lanceolate, spine-tipped; nutlets about 2 mm. long, entire and rounded-truncate at apex, broadly cuneate at base.——Aug.–Oct. Wet places; Hokkaido, Honshu, Shikoku, Kyushu.——e. Asia and N. America.

Var. **hirtus** Regel. *L. europaeus* var. *parvifolia* Miq.; *L. lucidus* var. *formosanus* Hayata——KE-SHIRO-NE. Stems and underside of leaves long-pilose.——Hokkaido and Honshu. ——Temperate e. Asia.

4. Lycopus maackianus (Maxim.) Makino. *L. lucidus* forma *angustifolia* Miq.; *L. lucidus* var. *maackianus* Maxim.; *L. angustus* Makino——HIME-SHIRO-NE. Stoloniferous; stems erect, 30–70 cm. long, rather slender, usually simple, sometimes branched, glabrous except for thin white hairs on the nodes; lower and median leaves rather thick, broadly lanceolate, 4–8 cm. long, 5–15 mm. wide, acuminate, rounded, obtuse, or auriculately rounded at base, deeply acute-toothed, glabrous, the petioles very short, about 1 mm. long, the upper leaves narrowly lanceolate to broadly linear, 1.5–3 cm. long, 1.5–5 mm. wide, sometimes entire; flowers few, about 5 mm. long, white; calyx about 4 mm. long, 4-lobed, the teeth deltoid-lanceolate, spine-tipped; nutlets about 1.5 mm. long, obovate-cuneate, truncate-rounded at apex.——Aug.–Oct. Wet places; Hokkaido, Honshu, Shikoku, Kyushu.——Korea, Manchuria, and e. Siberia.

19. CLINOPODIUM L. TŌBANA ZOKU

Perennial herbs; leaves usually toothed, the upper ones reduced to bracts; verticils loose or in a terminal spike, densely or loosely many-flowered, the pedicels bracteolate; calyx tubular, 13-nerved, bilabiate, the upper lip often spreading, 3-toothed, the lower lip bifid, the throat often long-pilose within; corolla erect or slightly curved, often inflated on the throat, the limb bilabiate, the upper lip entire or retuse, the lower lip 3-lobed; fertile stamens 4, didymous, the lower pair longer; nutlets globose, smooth, not angled, small.——About 50 species, in temperate regions.

1A. Bracteoles longer than or rarely as long as the pedicels, long spreading-hispid; verticils not forming a terminal spike; leaves and stems long-hispid.
  2A. Corolla 15–25 mm. long; bracteoles 3–6 mm. long. .... 1. *C. macranthum*
  2B. Corolla 8–12 mm. long; bracteoles 5–8 mm. long. .... 2. *C. chinense*
1B. Bracteoles 1–3 mm. long, shorter than to as long as the pedicels, usually not long-pilose; verticils forming a terminal spike; leaves and stems glabrous to finely pilose, rarely long-pubescent.
  3A. Leaves with distinct sessile discoid glands beneath; calyx with long spreading hairs. .... 3. *C. micranthum*
  3B. Leaves without distinct sessile glands beneath; calyx glabrous or puberulent. .... 4. *C. gracile*

1. Clinopodium macranthum (Makino) Hara. *C. chinense* var. *macranthum* Makino; *Satureia chinensis* var. *macrantha* (Makino) Matsum. & Kudo; *S. macrantha* (Makino) Kudo, non C. A. Mey.——MIYAMA-KURUMABANA. Rhizomes short-creeping; stems 10–40 cm. long, with recurved hairs especially on the angles; leaves broadly ovate or ovate-orbicular, sometimes narrowly ovate, 3–5 cm. long, 2–3.5 cm. wide, acute to obtuse, rounded to shallowly cordate at base, toothed, long-pilose on both surfaces, the petioles 1–5(–7) mm. long; floral leaves bractlike, the bracteoles 3–6 mm. long, linear, with long spreading hairs; calyx 6–8 mm. long, prominently spreading-hairy and long glandular-pilose; corolla 15–25 mm. long, pale rose-purple, puberulous externally; nutlets nearly orbicular, slightly flattened, pale brown, smooth.——Aug. High mountains; Honshu (n. and Hokuriku Distr.).

2. Clinopodium chinense (Benth.) O. Kuntze var. **parviflorum** (Kudo) Hara. *Satureia chinensis* sensu auct. Japon., non Briq.; *S. chinensis* var. *parviflora* Kudo; *S. chinensis* var. *megalantha* Kudo——KURUMABANA. Stems erect from a decumbent base, 15–40 cm. long, white-pilose, often short-branched in upper part; leaves ovate, 2–4 cm. long, 1–2.5 cm. wide, acute, rounded or often broadly cuneate at base, toothed in the upper ones, more or less whitish pilose on both sides, penninerved, the petioles 2–20 mm. long; verticils densely many-flowered, the bracteoles linear, 5–8 mm. long, with long spreading hairs; calyx often reddish, 6–8 mm. long, long spreading-pilose; corolla 8–12 mm. long, pale rose-purple, puberulent externally; nutlets orbicular, slightly compressed, about 1 mm. across.——Aug.–Sept. Woods and thickets in hills and low mountains; Hokkaido, Honshu, Shikoku, Kyushu; common and very variable.——s. Kuriles and Korea.

Var. **shibetchense** (Lév.) Koidz. *Calamintha umbrosa* var. *shibetchensis* Lév.; *Clinopodium japonicum* Makino; *Satureia makinoi* Kudo, pro parte——YAMA-KURUMABANA, Ō-MIYAMA-KURUMABANA. Plant green, not reddish; flowers white; calyx short glandular-hairy.——Hokkaido, Honshu, Shikoku, Kyushu.——Korea.

Var. **glabrescens** (Nakai) Ohwi. *C. japonicum* Makino, pro parte; *Satureia ussuriensis* var. *glabrescens* Nakai——AO-MIYAMA-TŌBANA, Ō-MIYAMA-TŌBANA. Plant green, less pilose; flowers white; bracteoles shorter; calyx shorter, with short-spreading and glandular hairs.——Honshu, Kyushu.——Korea.

Var. **chinense.** *Calamintha chinensis* Benth.; *C. clinopodium* var. *chinensis* (Benth.) Miq.; *Satureia chinensis* (Benth.) Briq.——OKINAWA-KURUMABANA. Much more densely hairy.——Kyushu.——Ryukyus, Formosa, and China.

3. Clinopodium micranthum (Regel) Hara. *Hedeoma micrantha* Regel; *Calamintha umbrosa* forma *robustior* Miq.; *C. umbrosa* var. *japonica* Fr. & Sav.; *Clinopodium fauriei* var. *japonicum* (Fr. & Sav.) Hara; *C. omuranum* Honda——INU-TŌBANA. Stems somewhat tufted, 20–50 cm. long, rather slender, green, pilose; leaves thin, ovate to narrowly so, 2–5 cm. long, 1–2.5 cm. wide, subacute, broadly cuneate at base, the petioles 5–20 mm. long, toothed, usually sparsely pilose, with distinct sessile discoid glands beneath; verticils dense, prominently pilose, mostly forming a terminal spike up to 5 cm. long; calyx 4–5 mm. long, green, with long spreading and glandular hairs; corolla white, often pink, 5–6 mm. long; nutlets orbicular, slightly flattened, less than 1 mm. across, smooth.——Aug.

–Oct. Woods in low mountains; Hokkaido, Honshu, Shikoku, Kyushu; rather common.

Var. **yakusimense** (Masam.) Hara. *Satureia yakusimensis* Masam.; *S. ussuriensis* var. *yakusimensis* (Masam.) Masam. YAKUSHIMA-TŌBANA. Small plant with decumbent stems; leaves 1–2 cm. long.——Kyushu (Yakushima).

4. Clinopodium gracile (Benth.) O. Kuntze. *Calamintha* (?) *gracilis* Benth.; *Clinopodium confine* (Hance) O. Kuntze; *Calamintha confinis* Hance; *Satureia gracilis* (Benth.) Bailey; *S. confinis* (Hance) Kudo——TŌBANA. Stems slender, 10–30 cm. long, tufted, decumbent at base, puberulent; leaves ovate to broadly so, 1–3 cm. long, 0.8–2 cm. wide, subobtuse, rounded to broadly cuneate at base, toothed, nearly glabrous on both sides or loosely pilose above and on nerves beneath, the petioles 5–15 mm. long; verticils prominently puberulent, the upper ones forming a loose terminal spike 1–4 cm. long; calyx 3.5–4 mm. long, puberulent on nerves; corolla pale rose, 5–6 mm. long; nutlets obovate-orbicular, less than 1 mm. long.——Apr.–Aug. Woods in lowlands and low mountains; Honshu, Shikoku, Kyushu; common and very variable. ——Ryukyus, Formosa, China, s. Korea, Malaysia, and India.

Var. **multicaule** (Maxim.) Ohwi. *Calamintha multicaulis* Maxim.; *Clinopodium multicaule* (Maxim.) O. Kuntze; *Satureia multicaulis* (Maxim.) Matsum. & Kudo——YAMA-TŌBANA. Somewhat taller, usually green, with rather longer hairs on stems and petioles; leaves ovate or broadly so, rather acutely toothed, the petioles 5–10 mm. long; verticils looser, forming a short spike 1–2 cm. long; calyx green, 5–6(–7) mm. long; corolla white, 7–9 mm. long.——June–Oct. Woods in mountains; Honshu (centr. and w. distr.), Shikoku, Kyushu; common.——Korea.

Var. **minimum** (Hara) Ohwi. *Clinopodium minimum* Hara——KOKE-TŌBANA. Small plant; leaves 4–7 mm. long; calyx 3–4 mm. long; corolla 5–7 mm. long.——Kyushu (Yakushima).

Var. **latifolium** (Hara) Ohwi. *C. multicaule* var. *latifolium* Hara.——HIRO-HA-YAMA-TŌBANA. Stems to 40 cm. long; leaves larger; calyx with longer hairs.——Honshu (Iwashiro Prov., Kantō and centr. distr.).

Var. **sachalinense** (F. Schmidt) Ohwi. *Calamintha umbrosa* var. *sachalinensis* F. Schmidt; *Satureia sachalinensis* (F. Schmidt) Kudo; *Clinopodium sachalinensis* (F. Schmidt) Koidz.; *S. ussuriensis* Kudo, pro parte——MIYAMA-TŌBANA. Stems taller, less densely pilose; leaves larger, to 6 cm. long; spikes 2–7 cm. long; calyx 3.5–4 mm. long, puberulent on lower half; corolla 5–6 mm. long.——Hokkaido, Honshu (n. distr.).——s. Kuriles and Sakhalin.

20. THYMUS L.   IBUKI-JAKŌ-SŌ ZOKU

Shrubs or subshrubs; leaves small, entire or nearly so; verticils usually few-flowered, remote, or forming a short spike, the bracts small; calyx 10- to 13-nerved, bilabiate, with a dense tuft of hairs on throat inside, the upper lip broad, 3-lobed, the lower lip with 2 narrow ciliate lobes; corolla prominently bilabiate, without a ring of hairs inside, the limb slightly bilabiate, the upper lip erect, retuse, the lower lip spreading, 3-lobed; stamens 4, of equal length or the lower ones longer, the anthers 2-locular; nutlets somewhat flattened, smooth.——About 35 species, Europe to e. Asia, and Africa.

1. Thymus quinquecostatus Celak. *T. serpyllum* sensu auct. Japon., non L.; *T. serpyllum* var. *ibukiensis* Kudo; *T. serpyllum* var. *przewalskii* Komar.; *T. przewalskii* (Komar.) Nakai; *T. przewalskii* var. *laxa* Nakai; *T. quinquecostatus* var. *ibukiensis* (Kudo) Hara; *T. quinquecostatus* var. *japonicus*

Hara——IBUKI-JAKŌ-SŌ. Aromatic dwarf prostrate shrub with ascending, more or less white-pilose branchlets; leaves narrowly to broadly ovate, 5–10 mm. long, 3–8 mm. wide, obtuse, entire, glandular-dotted on both surfaces, long white-ciliate at base, the lateral nerves arcuate, 2 or 3 pairs, rather abruptly short-petiolate; flowers in a short terminal spike; upper lip of calyx toothed; corolla rose-purple, 7–8 mm. long, about 5 mm. across; nutlets orbicular, slightly flattened, about 1 mm. across.——June–July. Rocky sunny places in high mountains, or sometimes near the sea; Hokkaido, Honshu, Kyushu.——Korea, n. China, Manchuria, Mongolia, and Siberia.

21. PERILLA L. Shiso Zoku

Erect annual; leaves opposite, toothed; verticils 2-flowered, in dense or loose terminal spikes; floral leaves bractlike; flowers small; calyx campanulate, 5-toothed, slightly accrescent in fruit, bilabiate, inflated in front at base, the upper lip 3-toothed, the lower lip 2-fid; corolla-tube short, the limb 5-lobed, the lower lobes slightly larger; stamens 4, nearly equal, erect or ascending; nutlets globose, reticulate.——Few species, in e. Asia and India.

1. Perilla frutescens (L.) Britt. var. **japonica** (Hassk.) Hara. *P. ocimoides* var. *japonica* Hassk.——E-goma. Erect branched annual; stems 20–70 cm. long, long-pubescent; leaves green, broadly ovate, 7–12 cm. long, 5–8 cm. wide, abruptly acute or acuminate, rounded to broadly cuneate at base, toothed, long-pubescent especially on nerves, long-petiolate; spikes 4–8 cm. long, rather one-sided, densely many-flowered, the axis prominently pubescent, the bracts deltoid-ovate, longer than the pedicels; calyx 3–4 mm. long in anthesis, 8–12 mm. long in fruit, long-pubescent; corolla whitish, 4–5 mm. long; nutlets globose, slightly compressed, about 2 mm. across, reticulate.——Aug.–Oct. Widely cultivated for oil-yielding seeds; naturalized in our area.——se. Asia.

Var. **citriodora** (Makino) Ohwi. *P. ocimoides* var. *typica* forma *citriodora* Makino; *P. citriodora* (Makino) Nakai—— Remon-e-goma. Stems rather densely short soft-pubescent; leaves loosely pubescent beneath; spikes elongate after anthesis, to 10–18 cm. long, rather loosely flowered and the hairs shorter; calyx 6–8 mm. long in fruit; nutlets 1.2–1.5 mm. across.——Hills and mountains; Honshu, Shikoku, Kyushu.

Var. **hirtella** (Nakai) Makino & Nemoto. *P. hirtella* Nakai——Tora-no-o-jiso. Spikes slender; calyx smaller, 5–6 mm. long in fruit.——Honshu. May be a hybrid of var. *citriodora* × var. *crispa*.

Var. **crispa** (Thunb.) Deane. *Ocimum crispum* Thunb.; *Dentidia nankinensis* Lour.; *P. ocimoides* var. *crispa* (Thunb.) Benth.; *P. frutescens* var. *nankinensis* (Lour.) Britt.—— Chirimen-jiso. Stems soft-pubescent; leaves green or deep purple, long-pubescent beneath especially on nerves, thinly pilose above, the margins wrinkled; spikes 5–15 cm. long, rather long-pubescent; calyx 7–10 mm. long in fruit; nutlets about 1.5 mm. across.——Widely cultivated.——Forma **viridis** (Makino) Makino. *P. ocimoides* vars. *crispa, viridis, discolor,* and *purpurea* of Makino.——Shiso. Plants with margins of the leaves flat.

22. PERILLULA Maxim. Suzu-kōju Zoku

Stoloniferous perennials; leaves opposite, small, toothed; verticils 2- to 6-flowered, in loose, simple or subcompound racemes; floral leaves small, linear, bractlike; calyx obliquely campanulate-obconical, 10-nerved, bilabiate, accrescent, the upper lip 3-lobed, the lower one 2-lobed; corolla small, campanulate-infundibuliform, the tube short and broad, without a ring of hairs inside, the limb obsoletely bilabiate, the upper lip retuse, the lower one shallowly 3-lobed; stamens 4, slightly didymous, the anthers with parallel locules; nutlets elliptic, small.——A single species, in Japan.

1. Perillula reptans Maxim. Suzu-kōju. Stems 20–40 cm. long, tuberlike and decumbent at base, then erect, 4-angled, green, with scattered white recurved hairs; leaves membranous, ovate or rhombic-ovate, 2–4 cm. long, 1–2.2 cm. wide, acute at apex, broadly cuneate at base, coarsely toothed, thinly pilose above and on nerves beneath, the petioles 1–2 cm. long; racemes terminal, short-pedunculate, 4–10 cm. long, loosely flowered, recurved white-puberulent, the bracts linear, 1–3(–5) mm. long, the pedicels spreading, 5–7 mm. long; calyx 2.5 mm. long at anthesis, 6–7 mm. long and deflexed in fruit, lobes of lower lip obscurely spine-tipped, longer than the upper lip, the upper lip with 3 short teeth; corolla white, 5–6 mm. long, about 4 mm. across; nutlets narrowly elliptic, slightly compressed, 1.7 mm. long.——Aug.–Dec. Honshu (w. Tōkaidō and Kinki Distr.), Shikoku, Kyushu.——Ryukyus.

23. MENTHA L. Hakka Zoku

Aromatic perennial herbs; leaves usually toothed; verticils densely many-flowered, globose, remote or spicate, the bracteoles minute; flowers pedicellate, small, white to rose; calyx campanulate or tubular, 5-toothed, often with long hairs on throat; corolla campanulate, the limb equally 4-lobed; stamens 4, equal, erect, the anther-locules distinct, parallel; nutlets ovate, small, smooth or reticulate.——About 20 species, in temperate and warmer parts of the N. Hemisphere.

1A. Stems, leaves, and calyx pubescent; leaves distinctly toothed; calyx-teeth acuminate; verticils loose, many.
                                                 1. *M. arvensis* var. *piperascens*
1B. Stems, leaves, and calyx nearly glabrous, the nodes of stems sometimes pubescent; leaves entire or nearly so; calyx-teeth obtuse; verticils 2–6, forming a short 10- to 20-flowered umbel-like spike. . . . . 2. *M. japonica*

1. Mentha arvensis L. var. **piperascens** Malinv. *M. arvensis* var. *sachalinensis* Briq.; *M. sachalinensis* (Briq.) Kudo; *M. haplocalyx* sensu auct. Japon., non Briq.; *M. haplocalyx* var. *sachalinensis* (Briq.) Briq. ex Kudo; *M. sachalinensis* forma *pilosa* Hara——Hakka. Perennial; stems 20–40 cm. long, scattered-pubescent; leaves ovate to narrowly so, oblong, 2–5 cm. long, 1–2.5 cm. wide, acute at both ends, acutely toothed, glandular-dotted on both sides, the petioles

3–10 mm. long; verticils many flowered; cymes sessile or rarely short-pedunculate, the pedicels slightly shorter than the calyx, glabrous or puberulous; calyx 2.5–3 mm. long, the teeth spreading-ciliate, narrowly deltoid, long-acuminate; corolla white or pale rose, 4–5 mm. long, the tube as long as the calyx; nutlets elliptic, slightly compressed, 0.7 mm. long, very broadly cuneate and 3-angled at base.——Aug.–Oct. Hokkaido, Honshu, Shikoku, Kyushu.——Korea, Sakhalin, and e. Asia. The typical phase occurs in temperate regions of the N. Hemisphere.

2. Mentha japonica (Miq.) Makino. *Micromeria japonica* Miq.; *Satureia japonica* (Miq.) Matsum. & Kudo; *Mi-*

cromeria yezoensis Miyabe & Tatew.——HIME-HAKKA. Nearly glabrous perennial; stems 20–40 cm. long, erect, sometimes branched, slender, green, the internodes short, puberulent; leaves many, ovate-oblong, obtuse, acute at base, the lateral nerves few, obscure, the petioles to 1.5 mm. long; verticils 2–6, forming a short spike, the pedicels longer than the calyx; calyx about 2.5 mm. long, glabrous, glandular-dotted, the teeth deltoid-ovate, erect, obtuse; corolla pale purple, 3.5 mm. long, the tube slightly longer than the calyx; nutlets about 0.8 mm. long, elliptic, slightly flattened, very broadly cuneate and mucronate at the base.——Aug.–Oct. Wet places in lowlands to mountains; Hokkaido, Honshu.

24. LEUCOSCEPTRUM Smith   TENNIN-SŌ ZOKU

Subshrubs or herbs, woody at base; leaves toothed, broadly lanceolate to broadly elliptic; bracts scalelike, broad, depressed-orbicular, clasping, caducous; spike elongate, terminal; verticils few-flowered, short, the pedicels short; calyx tubular, 5-toothed; corolla bilabiate, the tube longer than the calyx, slightly inflated, the limb slightly bilabiate, the upper lip 2-lobed, the lower 3-lobed; stamens 4, nearly equal, long-exserted, the filaments glabrous or pubescent below; nutlets cuneate, truncate at apex, acutely angled, smooth.——Few species, Japan to the Himalayas.

1A. Subshrub; bracts orbicular, short-cuspidate; leaves usually oblong to elliptic. .... 1. *L. stellipilum*
1B. Herb with a slightly woody base; bracts long-cuspidate; leaves usually lanceolate to narrowly oblong. .... 2. *L. japonicum*

1. Leucosceptrum stellipilum (Miq.) Kitam. & Murata. *Comanthosphace stellipila* (Miq.) S. Moore; *Elsholtzia stellipila* Miq.; *Pogostemon stellipila* (Miq.) Benth. & Hook. f.; *C. tajimensis* Makino——MIKAERI-SŌ, ITOKAKE-SŌ. Subshrub with slightly branched slender woody base; stems 40–100 cm. long, smooth, lustrous, green and prominently stellate-pilose while young, pale yellow-brown the second year; leaves membranous, oblong to broadly elliptic, 10–20 cm. long, 6–12 cm. wide, rounded or short-cuneate at base, toothed, stellate-pubescent above and especially on the nerves beneath, the petioles 1–5 cm. long; spikes erect, densely flowered, 10–18 cm. long, sessile, stellate-puberulent, the bracts densely imbricate, short-cuspidate, 6–8 mm. long, 8–10 mm. wide, caducous; corolla 8–10 mm. long, rose-colored.——Sept.–Oct. Woods in mountains; Honshu (centr. and w. distr.).

Var. **tosaense** (Makino) Kitam. & Murata. *Comanthosphace stellipila* var. *tosaensis* (Makino) Makino——Ō-MARUBA-NO-TENNIN-SŌ. Leaves stellate-pubescent while very young. Reported from Shikoku and Kyushu.

2. Leucosceptrum japonicum (Miq.) Kitam. & Murata. *Elsholtzia japonica* Miq.; *E. sublanceolata* Miq.; *E. barbinervis* Miq.; *Comanthosphace japonica* (Miq.) S. Moore; *C. sublanceolata* (Miq.) S. Moore; *C. barbinervis* (Miq.) S. Moore; *C. stellipila* var. *japonica* (Miq.) Matsum. & Kudo; *C. hakonensis* Koidz.; *C. stellipila* var. *barbinervis* (Miq.) Ohwi; *C. stellipila* var. *sublanceolata* (Miq.) Ohwi; *C. stellipila* var. *japonica* forma *sublanceolata* (Miq.) Matsum. & Kudo and forma *barbinervis* (Miq.) Matsum. & Kudo; *C. japonica* var. *barbinervis* (Miq.) Makino——TENNIN-SŌ. Closely allied to the preceding; herb with a short woody base; stems erect, slightly 4-angled, glabrescent, slightly stellate-pilose on upper part; leaves lanceolate to oblong, 3–7 cm. wide, cuneate at base, acuminate, slightly stellate-pilose only while young; spike terminal, erect, the bracts scalelike, closely imbricate, caducous, orbicular, conspicuously cuspidate; corolla pale yellow.——Sept.–Oct. Woods; Hokkaido, Honshu, Shikoku.

25. DYSOPHYLLA Bl.   MIZU-TORA-NO-O ZOKU

Perennials or annuals; leaves verticillate or opposite; verticils many-flowered, densely spicate, the bracts persistent; calyx ovate-campanulate, equally 5-toothed, without hairs on throat; corolla nearly actinomorphic, the tube usually shorter than the calyx, the limb spreading, nearly equally 4-lobed, the upper lobe entire or retuse; stamens 4, long-exserted, rather unequal, erect or slightly recurved, the filaments often long-pilose near the middle, the anthers subglobose, the anther-locules connate at tip; nutlets ovate or oblong, smooth.——About 20 species, in se. Asia and Australia.

1A. Perennial with slender rhizomes; spikes (excluding filaments) 7–10 mm. across; flowers 7–8 mm. long, inclusive of the stamens; filaments with a dense tuft of long crisped hairs near middle; stems usually simple, sometimes branched at base; leaves usually verticillate in 4's. .... 1. *D. yatabeana*
1B. Annual without rhizomes; spikes 4–5 mm. across; flowers about 3 mm. long, inclusive of the stamens; filaments loosely short-pilose on lower half; stems usually branched in upper half; leaves verticillate in 3's to 6's. .... 2. *D. verticillata*

1. Dysophylla yatabeana Makino. *D. linearis* var. *yatabeana* (Makino) Kudo——MIZU-TORA-NO-O, MURASAKI-MIZU-TORA-NO-O. Rhizomes slender, creeping; stems erect, 30–50 cm. long, many-leaved, rather soft, glabrous or slightly pubescent on the nodes, usually simple; inflorescence terminal, spicate; leaves usually verticillate in 4's, linear or broadly linear,

3–7 cm. long, 2–5 mm. wide, obtuse, narrowed at both ends, entire, glabrous on both sides or loosely short-pubescent on midrib above, nearly sessile; spikes 2–8 cm. long, densely very many-flowered; flowers nearly sessile, rose-colored, 7–8 mm. long inclusive of the long exserted stamens, the bracts lanceolate, as long as the calyx, pubescent; calyx about 3 mm. long, pubescent; filaments with a dense tuft of long crisped hairs near middle; nutlets ovate, biconvex, about 0.7 mm. long, obtusely angled, narrowed below and subtruncate at base.——Aug.–Oct. Wet places; Honshu, Shikoku, Kyushu.——Korea.

2. **Dysophylla verticillata** (Roxb.) Benth. *Mentha verticillata* Roxb.; *D. japonica* Miq.——Mizu-neko-no-o. Annual; stems 10–50 cm. long, sometimes short-creeping at base, usually branched in upper half, glabrous or short-pubescent

on the nodes; leaves verticillate in 3's to 6's(–10's), linear, 2–5 cm. long, 2–4 mm. wide, glabrous; spikes 2–5 cm. long, densely very many-flowered; flowers white, about 3 mm. long inclusive of the stamens; calyx about 1.5 mm. long, the teeth obliquely ascending in fruit; filaments loosely short-pilose on lower half; nutlets about 0.7 mm. long, pale brown, ovate, narrowed at base.——Aug.–Oct. Wet places; Honshu, Shikoku, Kyushu.——Korea, Ryukyus, Formosa, China, and se. Asia.

26. KEISKEA Miq. Shimobashira Zoku

Perennials; leaves opposite, toothed; verticils 2-flowered, forming one-sided spikes, the bracts lanceolate, small, persistent; flowers pedicelled; calyx campanulate, deeply 5-fid, with lanceolate lobes, hairy on the throat; corolla-tube broadened in upper part, with a ring of hairs inside, the limb slightly bilabiate, the lower lip 3-lobed, the midlobe slightly larger; stamens 4, didymous, exserted, the upper pair shorter, the filaments glabrous; nutlets ovate or oblong, smooth.——Few species, in e. Asia.

1. **Keiskea japonica** Miq. *K. japonica* var. *lancifolia* Nakai; *K. japonica* var. *hondoensis* Nakai——Shimobashira. Perennial herb; stems about 60 cm. long, slightly pilose or glabrous, often branched in upper part; leaves thinly chartaceous, broadly lanceolate to narrowly ovate, 6–15(–20) cm. long, 2–5.5 cm. wide, long-acuminate, abruptly to long-acuminate, rarely subobtuse at base, acutely toothed except near the base, puberulent on midrib above, glabrous or thinly pilose on the

nerves beneath, the petioles 5–30 mm. long; spikes 5–12 cm. long, short-pilose, one-sided, the bracts broadly linear; calyx 3 mm. long at anthesis, 5–6 mm. long in fruit, nearly as long as the pedicels, exceeding the bracts; corolla white, about 7 mm. long, shallowly lobed; nutlets usually solitary, globose, dark brown, 1.5–2 mm. across, with darker reticulations.——Sept.–Oct. Mountains; Honshu (Kantō Distr. and westw.), Shikoku, Kyushu.

27. ELSHOLTZIA Willd. Naginata-kōju Zoku

Herbs or shrubs; leaves opposite, usually toothed; verticils many-flowered, in dense one-sided spikes; floral bracts persistent, broadly ovate or obovate-orbicular, dense; calyx equally 5-toothed, without long hairs on the throat, inflated and accrescent in fruit; corolla small, the tube nearly as long as the calyx, the limb 4-lobed, subbilabiate, the upper lip erect, retuse, the lower lip spreading; stamens 4, exserted, ascending or spreading, the lower pair longer; nutlets ovate to oblong.——About 20 species, in temperate regions of Eurasia, N. Africa, and Malaysia.

1A. Bracts depressed-orbicular, nearly glabrous on back, short-ciliate, the awns 1–2 mm. long. . . . . 1. *E. ciliata*
1B. Bracts narrowly flabellate, short-pubescent on back, long-ciliate, the awns 2–3 mm. long. . . . . 2. *E. nipponica*

1. **Elsholtzia ciliata** (Thunb.) Hylander. *Sideritis ciliata* Thunb.; *Mentha patrinii* Lepech.; *E. patrinii* (Lepech.) Garcke; *E. cristata* Willd.; *E. interrupta* Ohwi——Naginata-kōju. Erect branched annual, 30–60 cm. long; stems obtusely angled, loosely soft-pubescent; leaves thinly membranous, broadly to narrowly ovate, 6–10 cm. long, 2.5–6 cm. wide, short-acuminate with an obtuse tip, cuneate to broadly cuneate at base, toothed, thinly pubescent above and on nerves beneath, the petioles winged in upper part; spikes terminal and axillary, 5–10 cm. long, 7–8 mm. across, densely many-flowered, one-sided, the bracts depressed-orbicular, short-ciliate, rounded at base, nearly glabrous on back, as long as to slightly longer than the calyx, often purplish, the awns 1–2 mm. long; flowers pale rose-colored, about 5 mm. long; calyx 3–3.5 mm. long in fruit, the teeth spine-tipped; nutlets narrowly obovate, about 1 mm. long, slightly flattened, broadly rounded at apex, cuneate at base.——Sept.–Nov. Hokkaido, Honshu, Shikoku,

Kyushu; common.——s. Kuriles, Korea, Formosa, China, temperate Asia to w. Europe.

2. **Elsholtzia nipponica** Ohwi. *E. oldhamii* var. *nipponica* Ohwi; *E. argyi* var. *nipponica* (Ohwi) Murata——Futobo-naginata-kōju. Resembles the preceding; stems 30–80 cm. long, obtusely angled, short-pubescent; leaves ovate or broadly so, 2.5–6 cm. long, 1.5–4 cm. wide, acuminate or acute, broadly cuneate at base, toothed, glabrous or thinly pubescent on nerves beneath, thinly pilose above, petiolate; spikes 2–5 cm. long, about 10 mm. across, densely many-flowered, one-sided, the bracts long-ciliate, narrowly flabellate, short-pubescent on back, the awns 2–3 mm. long; corolla rose-colored, 4–5 mm. long; nutlets obovate, slightly compressed, about 1 mm. long. ——Sept.–Oct. Sunny slopes along streams and valleys in mountains; Honshu (Kantō Distr. and westw.), Kyushu. The typical phase occurs in China.

28. PLECTRANTHUS L'Hérit. Yama-hakka Zoku

Perennial herbs or sometimes shrubs; leaves toothed; verticils several- to many-flowered, in loose simple or branched racemes; calyx campanulate in anthesis, equally 5-toothed or bilabiate, the upper lip 3-toothed, the lower lip 2-fid, accrescent in fruit; corolla bilabiate, the tube elongate, gibbous, the upper lip 3- or 4-lobed, the lower lip entire, often navicular; stamens 4, didymous, declined; nutlets ovate or oblong, smooth or granular, sometimes short-pilose.—About 150 species, in Asia, Australia, and Africa.

1A. Nutlets white-pilosulous; corolla dark purple. . . . . 1. *P. trichocarpus*
1B. Nutlets glabrous; corolla pale blue, blue-purple or white.
  2A. Corolla 5–12 mm. long, the tube 1–3 times as long as the limb.
    3A. Calyx nearly equally 5-lobed, with fine ascending hairs; corolla 5–10 mm. long, with small dark spots on upper lip inside.
      4A. Flowers 5–7 mm. long, pale blue; stamens or style exserted; leaves 6–15 cm. long. . . . . 2. *P. japonicus*
      4B. Flowers 8–10 mm. long, blue-purple; stamens and styles included; leaves 3–4 cm. long. . . . . 3. *P. inflexus*
    3B. Calyx bilabiate, with spreading hairs or pilose; corolla 6–12 (–15) mm. long, without dark spots, the lower lip bifid, usually longer than the upper.
      5A. Leaves broadly lanceolate to ovate, acuminate.
        6A. Calyx-teeth acute; bracts ovate to ovate-cordate. . . . . 4. *P. umbrosus*
        6B. Calyx-teeth mucronate; bracts broadly lanceolate to ovate. . . . . 5. *P. shikokianus*
      5B. Leaves ovate-orbicular to ovate, caudate. . . . . 6. *P. kameba*
  2B. Corolla 15–20 mm. long, the tube 4–6 times as long as the limb.
    7A. Cymes short, the pedicels puberulent, less than 1 cm. long; upper calyx-teeth short, obtuse. . . . . 7. *P. longitubus*
    7B. Cymes effuse, the pedicels slender, glabrous, usually 1–2.5 cm. long; upper calyx-teeth lanceolate, acute. . . . . 8. *P. effusus*

1. Plectranthus trichocarpus Maxim. *P. inconspicuus* Maxim., non Miq.; *Isodon trichocarpus* (Maxim.) Kudo; *Amethystanthus trichocarpus* (Maxim.) Nakai——KUROBANA-HIKI-OKOSHI. Rhizomatous; stems tufted, 50–150 cm. long, branched in upper part, rarely minutely recurved-puberulent on the angles; leaves broadly deltoid-ovate to broadly lanceolate, 6–15 cm. long, 2.5–7 cm. wide, acuminate, broadly cuneate and gradually decurrent at base to a winged petiole, toothed, usually thinly pilose above, appressed-puberulent on nerves beneath; verticils loose, distant, axillary, on long slender peduncles, those in the upper part forming a loose terminal raceme, the pedicels slender; calyx puberulent, 3–3.5 mm. long in fruit, the teeth deltoid; corolla dark purple, 5–6 mm. long, the tube as long as the lower lip and calyx; nutlets obovoid, scarcely flattened, about 1.5 mm. long, white-pilosulous.——Aug.–Sept. Along streams and in clearings in mountains; Hokkaido, Honshu (Tango Prov. and eastw.).

2. Plectranthus japonicus (Burm.) Koidz. *Scutellaria* (?) *japonica* Burm.; *Ocimum rugosum* Thunb.; *P. maximowiczii* Miq.; *P. buergeri* Miq.; *P. glaucocalyx* var. *japonicus* (Burm.) Maxim.; *Isodon japonicus* (Burm.) Hara; *I. glaucocalyx* var. *japonicus* (Maxim.) Kudo; *Amethystanthus japonicus* (Burm.) Nakai——HIKI-OKOSHI. Stems 50–100 cm. long, with recurved short hairs on the angles; leaves ovate to narrowly so, 6–12(–15) cm. long, 3.5–7 cm. wide, short-acuminate or acute, truncate to cuneate-truncate at base, paler beneath and pilose at least on nerves; racemes loose, terminal and in upper axils, elongate, the pedicels short; calyx 3–4 mm. long, gray-puberulent, the teeth deltoid; corolla pale purple, 5–7 mm. long, the tube slightly inflated at base on back, about twice as long as the calyx, slightly longer or nearly as long as the lower lip; nutlets ellipsoidal, slightly flattened, obsoletely reticulate, loosely granular above.——Sept.–Oct. Honshu, Shikoku, Kyushu.——Korea.

3. Plectranthus inflexus (Thunb.) Vahl ex Benth. *Ocimum inflexum* Thunb.; *Isodon inflexus* (Thunb.) Kudo; *Amethystanthus inflexus* (Thunb.) Nakai; *I. inflexus* var. *macrophyllus* (Maxim.) Kudo; *I. inflexus* var. *transticus* Kudo ——YAMA-HAKKA. Stems 40–80 cm. long, branched, with recurved short hairs especially on the angles; leaves broadly deltoid-ovate or broadly ovate, 3–6 cm. long, 2–4 cm. wide, acute to subobtuse, broadly cuneate and decurrent on the winged petiole, coarsely toothed, thinly pilose above and on nerves beneath; racemes loose, terminal and axillary, narrow, elongate, the pedicels and peduncles of the cymes short; calyx 5–6 mm. long in fruit, ascending-puberulent, the teeth narrowly deltoid; corolla blue-purple, rarely white, 8–10 mm. long, the tube rather broad, as long as the lower lip, about

twice as long as the calyx; nutlets slightly compressed, orbicular, 1.2 mm. long, with darker reticulations, smooth.——Sept.–Oct. Thin woods and thickets in hills and low mountains; Hokkaido, Honshu, Shikoku, Kyushu; common and variable.

4 Plectranthus umbrosus (Maxim.) Makino. *P. inflexus* var. *umbrosus* Maxim.; *Isodon umbrosus* (Maxim.) Hara; *Amethystanthus umbrosus* (Maxim.) Nakai; *I. inflexus* var. *umbrosus* (Maxim.) Kudo——INU-YAMA-HAKKA. Stems 60–80 cm. long, recurved-hairy on the angles; leaves broadly lanceolate or narrowly ovate-oblong, 5–10 cm. long, 1–3.5 cm. wide, long-acuminate, cuneate and long decurrent at base, acutely-toothed, thinly pilose on both surfaces especially on nerves beneath, petiolate; racemes loose, terminal and axillary, 10–25 cm. long, the bracts ovate to ovate-cordate, the pedicels rather long; calyx spreading-puberulent and glandular-dotted, about 4 mm. long in fruit, the teeth narrowly deltoid, acute; corolla blue-purple, 8–10 mm. long, the tube rather broad, gibbous, about twice as long as the calyx and slightly longer than the lower lip; nutlets orbicular, 1.2 mm. long, slightly flattened, obsoletely reticulate, smooth.——Sept.–Oct. Mountains; Honshu (Fuji and Hakone regions).

5. Plectranthus shikokianus (Makino) Makino. *Isodon shikokianus* (Makino) Hara; *I. excisus* var. *shikokianus* (Makino) Kudo; *I. lanceus* (Nakai) Kudo; *P. lanceus* Nakai; *P. axillariflorus* Honda; *Amethystanthus lanceus* (Nakai) Nakai; *A. manabeanus* Honda; *I. manabeanus* (Honda) Hara ——TAKA-KUMA-HIKI-OKOSHI. Stems 50–80 cm. long, with recurved short hairs, especially on the angles; leaves broadly lanceolate or narrowly ovate, 5–10 cm. long, 1.5–2.5 cm. wide, long-acuminate, cuneate at base, acutely toothed, thinly pilose above, loosely puberulent beneath or nearly glabrous on both surfaces, petiolate; racemes 5–15 cm. long, rather narrow; bracts broadly lanceolate to ovate; calyx 3.5–5 mm. long in fruit, loosely puberulent, the teeth narrowly deltoid, mucronate; corolla about 10 mm. long, rather thick, the tube nearly as long as the lower lip, gibbous, 2 or 3 times as long as the calyx; nutlets broadly elliptic, about 1.2 mm. long, rather smooth, slightly flattened.——Sept.–Oct. Woods and thickets in mountains; Honshu (centr. and s. distr.), Shikoku, Kyushu.

6. Plectranthus kameba (Okuyama) Ohwi. *Isodon kameba* Okuyama; *Plectranthus excisus* sensu auct. Japon., non Maxim.; *I. excisus* Kudo, excl. basionym; *Amethystanthus excisus* Nakai, excl. basionym——KAMEBA-HIKI-OKOSHI. Tufted perennial; stems 50–100 cm. long, with recurved short hairs on the nodes; leaves ovate-orbicular or ovate, 5–8 cm. long and as wide, truncate to incised at apex, with a long-caudate tip, 2–5 cm. long, cuneate-truncate at the winged-petiolate base, acutely toothed, thinly puberulent on the nerves of both sides, sparingly pilose above, the upper leaves often without appendage at apex, few-toothed; racemes 5–20 cm. long, rather slender, the pedicels short, spreading-puberulent; calyx about 7 mm. long in fruit, spreading-puberulent, the teeth narrowly deltoid, acute; corolla blue-purple, 8–10 mm. long, the tube rather broad, gibbous, twice as long as the calyx, 1.5 times as long as the lower lip; nutlets orbicular, about 2 mm. long, slightly flattened, smooth.——Aug.-Sept. Woods in mountains; Honshu (Kantō and eastern centr. distr.).

Var. **excisinflexus** (Nakai) Ohwi. *Amethystanthus excisinflexus* Nakai——TAIRIN-YAMA-HAKKA. Corolla larger, 10–13 mm. long.——Honshu (Hokuriku and n. distr.).

Var. **hakusanensis** (Kudo) Ohwi. *Isodon excisus* var. *hakusanensis* Kudo; *Amethystanthus excisus* var. *hakusanensis* (Kudo) Nakai——HAKUSAN-KAMEBA-HIKI-OKOSHI. Appendage of leaves broader and toothed or incised.——Honshu (high mountains of w. Hokuriku Distr.).

Var. **latifolius** (Okuyama) Ohwi. *Isodon umbrosus* var. *latifolia* Okuyama——KŌSHIN-YAMA-HAKKA. Leaves larger, 8–15 cm. long, 4–7 cm. wide, without a distinct caudate appendage at apex.——Honshu. (s. centr. distr.).

7. **Plectranthus longitubus** (Miq. *Isodon longitubus* (Miq.) Kudo; *Amethystanthus longitubus* (Miq.) Nakai; *P. longitubus* var. *contractus* Maxim.; *I. longitubus* var. *cantractus* (Maxim.) Kudo——AKI-CHŌJI. Stems 70–100 cm. long, recurved-puberulent on the angles; leaves narrowly ovate, 7–12 cm. long, 2.5–4 cm. wide, acuminate, cuneate at the base, toothed, puberulent on nerves beneath, petiolate; racemes 10–20 cm. long, one-sided, the pedicels rather long, the cymes sessile or short-pedunculate; calyx puberulent, about 6 mm. long in fruit, the teeth deltoid-ovate, short, very short-mucronate; corolla 15–20 mm. long, blue-purple, the tube rather broad, gibbous, 4–6 times as long as the limb; nutlets orbicular, about 2 mm. across, slightly flattened, smooth.——Aug.-Oct. Mountains; Honshu (w. Tōkaidō Distr. and westw.), Shikoku, Kyushu.

8. **Plectranthus effusus** (Maxim.) Ohwi. *P. longitubus* var. *effusus* Maxim.; *Isodon longitubus* var. *effusus* (Maxim.) Kudo; *I. effusus* (Maxim.) Hara; *Amethystanthus effusus* (Maxim.) Honda——SEKIYA-NO-AKI-CHŌJI. Closely allied to No. 7; cymes effuse, with slender, elongate, horizontally spreading peduncles; upper calyx-teeth lanceolate, acute.——Woods in mountains; Honshu (e. Tōkaidō and Kantō Distr.).

Fam. 176. SOLANACEAE NASU KA Nightshade Family

Herbs or woody plants with exstipulate, simple, alternate leaves; flowers bisexual, usually actinomorphic; calyx 4- to 6-lobed, persistent, often accrescent after anthesis; corolla gamopetalous, rotate to tubular, usually 5-lobed, induplicate or valvate in bud; stamens inserted on the tube, alternate with the corolla-lobes, the anthers 2-locular, longitudinally dehiscent or opening by pores; ovary 2-locular, the locules often divided by false septa, the ovules many, on axile placentae; style terminal; fruit a capsule or berry; seeds with copious endosperm.——About 75 genera, with about 2,000 species, chiefly in the Tropics.

1A. Shrubs. . . . . 1. *Lycium*
1B. Herbaceous or sometimes woody at the base.
  2A. Fruit a capsule; corolla and calyx campanulate. . . . . 2. *Scopolia*
  2B. Fruit a berry; corolla rotate or campanulate.
    3A. Anthers longitudinally split, distinct.
      4A. Calyx accrescent after anthesis, enclosing the fruit.
        5A. Calyx bladdery, loosely enclosing the fruit. . . . . 3. *Physalis*
        5B. Calyx accrescent and tightly enclosing the fruit. . . . . 4. *Physaliastrum*
      4B. Calyx scarcely accrescent after anthesis, not enclosing the fruit. . . . . 5. *Tubocapsicum*
    3B. Anthers opening by pores, surrounding the style, free or connate. . . . . 6. *Solanum*

1. LYCIUM L. KUKO ZOKU

Mostly spiny shrubs or small trees; leaves entire, slender and terete or flat, often fasciculate; peduncles usually solitary in axils, rarely fasciculate; flowers white, rose-purple, or yellowish; calyx campanulate, truncate, irregularly 3- to 5-toothed, slightly accrescent after anthesis; corolla tubular, infundibuliform, campanulate or urceolate, 5-lobed, the throat often inflated; stamens often exserted, the anthers short; ovary 2-locular, the ovules often few; berry globose or ovoid.——About 100 species, chiefly in temperate and warmer regions, especially abundant in the S. Hemisphere.

1. **Lycium chinense** Mill. *L. barbatum* var. *chinense* Ait.——KUKO. Glabrous deciduous spiny shrub; leaves elliptic or narrowly oblong, 2–4(–6) cm. long, 1–2(–3) cm. wide, entire, short-petiolate; flowers pale purple or white, 1–1.5 cm. long, with darker striations near base, the pedicels slender, nearly as long as the flowers; stamens exserted, the filaments densely bearded near base; fruit narrowly ovoid-ellipsoid, about 1.5–2 cm. long, red.——Aug.-Nov. Thickets and riverbanks in lowlands; Honshu, Shikoku, Kyushu; common.——Korea, Manchuria, China, Ryukyus, and Formosa.

2. SCOPOLIA Jacq. HASHIRI-DOKORO ZOKU

Erect, fleshy, glabrous herbs; stems rather stout, slightly branched; leaves membranous, entire; flowers pendulous, brown-purple or greenish, solitary, on slender axillary pedicels; calyx membranous, broadly campanulate, truncate or shallowly 5-lobed at apex, accrescent and covering the fruit; corolla campanulate, 5-angled or shallowly 5-lobed, with 2 or 3 lobes larger in bud and covering the others, the limb plicate; stamens not exserted, the anthers ovoid, longitudinally split; ovary conical, 2-loculed; ovules many; style filiform; capsules circumscissile, enclosed in the calyx, subglobose; seeds nearly reniform, papillose.——Few species, in Japan, Himalayas, and Europe.

1. Scopolia japonica Maxim. HASHIRI-DOKORO. Glabrous stout rhizomatous perennial; stems 30–60 cm. long, loosely branched in upper part; leaves petioled, narrowly oblong, 10–20 cm. long, 3–7 cm. wide, abruptly narrowed and acute at both ends, entire or with 1 or 2 teeth in the lower ones, petiolate; flowers 3–5 cm. long; calyx unequally 5-lobed; corolla about 2 cm. long, purplish yellow, 5-lobed; fruit globose, about 1 cm. across, enclosed in the accrescent calyx; seeds reniform, about 2.5 mm. across, reticulate with depressed dots. ——Apr.–May. Mountains; Honshu, Shikoku, Kyushu.

3. PHYSALIS L.   HŌZUKI ZOKU

Annuals or perennials often slightly woody at base, with simple or stellate hairs; leaves entire or rarely pinnatilobed; flowers small, usually solitary in axils, pedicelled, blue-purple, yellowish, or whitish, often purplish toward base; calyx membranous, campanulate or conical, shallowly 5-lobed or cleft, becoming bladdery in fruit, 5-angled or prominently 10-ribbed, the lobes connivent at tip; corolla nearly rotate or very broadly campanulate, the limb plicate in bud, 5-angled or shallowly 5-lobed; style slender, the stigma bifid; ovules many; berry globose, enclosed in the inflated calyx; seeds flat.——About 50 species, chiefly in N. and S. America, few in temperate regions of the Old World.

1A. Rhizomes long-creeping; flowers 1.5–2 cm. across. . . . . 1. *P. alkekengi* var. *franchetii*
1B. Rhizomes absent or nearly so.
  2A. Cultivated; calyx loosely short-hairy, the teeth linear-lanceolate in fruit. . . . . 2. *P. angulata*
  2B. Spontaneous; calyx loosely pilose on the prominent winglike ribs, the hairs multicellular, broadened at base, the teeth ovate-deltoid
    in fruit. . . . . 3. *P. chamaesarachoides*

1. Physalis alkekengi L. var. **franchetii** (Mast.) Makino. *P. franchetii* Mast.; *P. bunyardii* Hort.; *P. franchetii* var. *bunyardii* Makino——HŌZUKI. Nearly glabrous perennial with long creeping rhizomes; stems 40–80 cm. long; leaves broadly ovate, 5–12 cm. long, 3.5–9 cm. wide, acute to obtuse, rounded to broadly cuneate at base, scattered incised-toothed and short-pilose on margin, petiolate; flowers solitary, the pedicels 3–4 cm. long; calyx short-tubular, 5–6 mm. long at anthesis, subtruncate at the base, deeply 5-lobed, the lobes narrowly deltoid, obtuse, short-ciliate; corolla creamy, subrotate; fruit globose, red, about 1.5 cm. across, enclosed by the finely veined bladdery red calyx about 4–5 cm. long.——June–Aug. Hokkaido, Honshu, Shikoku, Kyushu.——Korea, n. China. The typical phase occurs in Europe and w. Asia.——Forma **monstrosa** Miq. YŌRAKU-HŌZUKI. An abnormal sterile phase; spike pendulous, many-bracteate but without flowers. Cultivated.

2. Physalis angulata L. *P. ciliata* Sieb. & Zucc.; *P. minima* sensu auct. Japon., non L.——SENNARI-HŌZUKI. Slightly pubescent annual; stems 30–40 cm. long, the branches obliquely spreading; leaves broadly ovate, 3–7 cm. long, 2–5 cm. wide, short-acute, rounded at base, nearly entire or with few obscure teeth, petiolate; flowers solitary in axils, pale yellow, the pedicels about 1 cm. long, pubescent; calyx 4–5 mm. long at anthesis, ovoid, about 2.5 cm. long in fruit, short-pubescent especially on the angles, green; corolla about 8 mm. long; fruit globose, about 1 cm. across.——Aug.–Oct. A tropical American plant cultivated in our area and often naturalized.

3. Physalis chamaesarachoides Makino. *Physaliastrum chamaesarachoides* (Makino) Makino——YAMA-HŌZUKI. Plant sparsely puberulent; stems 30–60 cm. long, branched, soft; leaves thinly membranous, ovate or elliptic-ovate, 6–10 cm. long, 3–5 cm. wide, acuminate or abruptly short-acuminate, rounded to cuneate at base, green, usually with few irregular coarse teeth, the petioles winged in upper part; flowers 1 or 2 in leaf-axils, pendulous, white, 7–8 mm. long, the pedicels glabrous, 1–2 cm. long; calyx 3–4 mm. long at anthesis, ovoid-globose, 12–15 mm. long in fruit, loosely pilose on the prominent winglike ribs; fruit globose, about 1 cm. across, loosely enclosed by the persistent, membranous, 10-ribbed calyx.——Aug.–Oct. Woods; Honshu (Kanto Distr. and westw.), Shikoku, Kyushu.

4. PHYSALIASTRUM Makino   IGA-HŌZUKI ZOKU

Perennial branched herbs; leaves ovate, often oblique at base, petiolate; flowers axillary, solitary or fascicled, small, pendulous; calyx short-campanulate, globose or ellipsoidal, 5-lobed, accrescent, tightly enclosing the fruit; corolla campanulate, 5-lobed, the lobes short, induplicate in bud, with a gland on the inside toward the base; style slender, the stigma bilobed; ovules many; berry pendulous, slightly fleshy, globose or ellipsoidal, enclosed by the accrescent calyx; seeds flat, rugose, depressed-punctate.——Few species, in Japan and China.

1A. Leaves ovate to broadly so, obtuse or acute, with a long winged petiole; calyx long-pubescent at anthesis, the hairs becoming soft-
    spiny in fruit; corolla about 10 mm. long and as wide; fruit globose, as long as the calyx; filaments glabrous. . . . . 1. *P. japonicum*
1B. Leaves oblong or ovate-elliptic, acuminate, with a short petiole; calyx short-pubescent at anthesis, rarely with short spinelike hairs;
    corolla 1.5–2 cm. long and as wide; fruit ellipsoidal, rarely globose, shorter than the calyx; filaments hairy. . . . . 2. *P. savatieri*

1. Physaliastrum japonicum (Fr. & Sav.) Honda. *Chamaesaracha japonica* Fr. & Sav., pro parte; *C. echinata* Yatabe; *P. echinatum* (Yatabe) Makino; *P. japonicum* (Fr. & Sav.) Kitam.——IGA-HŌZUKI. Perennial herb with short rhizomes; stems branched, 50–70 cm. long, often with scattered long-spreading hairs; leaves membranous, ovate to broadly so, 4–8 cm. long, 3–5 cm. wide, obtuse to acute, abruptly narrowed at base to a rather long winged petiole, entire, ciliate; flowers 2 or 3, rarely solitary in the axils, pendulous, small, pale yellowish, the pedicels 2–3 cm. long, sometimes long pubescent, thickened at apex; calyx long-pubescent in flower, the teeth depressed-deltoid, undulate, the hairs becoming soft spiny in fruit; corolla puberulent externally; filaments glabrous; fruit globose, 8–10 mm. across, enclosed in the persistent calyx, white at maturity.——July–Oct. Hokkaido, Honshu, Shikoku, Kyushu.——Korea, Manchuria, and China.

2. Physaliastrum savatieri (Makino) Makino. *Chamaesaracha japonica* Makino; *C. savatieri* Makino; *C. watanabei* Yatabe; *P. japonicum* Honda, pro parte——Ao-HŌZUKI. Perennial herb with short rhizomes; stems branched, soft, thinly soft-pubescent; leaves membranous, oblong, sometimes ovate-elliptic, 6–12 cm. long, 2.5–4.5 cm. wide, acuminate at both ends, loosely short-pubescent, ciliate, short-petiolate; flowers 1 or sometimes 2 in axils, greenish, the pedicels 2–3 cm. long in fruit, thickened at apex; calyx often thinly short-pubescent, the teeth deltoid, the hairs becoming soft spiny in fruit; corolla puberulent outside, densely pubescent inside at base with branched hairs; filaments soft spreading-pubescent; fruit ellipsoidal, 1–1.3 cm. long, enclosed by the green persistent urceolate calyx longer than the fruit.——June–July. Woods in mountains; Honshu, Shikoku.——Forma **kimurae** (Makino) Ohwi. *P. kimurae* Makino——TAKAO-HŌZUKI. Fruit globose.——Honshu (Mount Takao in Musashi Prov.).

5. TUBOCAPSICUM Makino   HADAKA-HŌZUKI ZOKU

Perennial rhizomatous herbs; stems erect, branched; leaves nearly entire; flowers fasciculate in axils, pendulous, yellowish; calyx short, truncate, nearly entire, slightly accrescent and not enclosing the fruit; corolla broadly campanulate, 5-lobed, valvate in bud; anthers ovate, with parallel locules; style elongate, the stigma dilated and capitate; ovules many; berry juicy, small, globose, red; seeds flat, depressed-punctate.——Few species, in e. Asia and India.

1A. Leaves thinly membranous, acuminate; pedicels somewhat thickened at apex; corolla with the lobes deltoid, acute. .. 1. *T. anomalum*
1B. Leaves rather thick, obtuse; pedicels much thickened at apex; corolla with the lobes oblong, obtuse. .... 2. *T. obtusum*

1. Tubocapsicum anomalum (Fr. & Sav.) Makino. *Capsicum anomalum* Fr. & Sav.; *C. cordiforme* var. *truncata* Miq.; *Solanum anodontum* Lév. & Van't.——HADAKA-HŌZUKI. Nearly glabrous perennial herb with obliquely spreading branches; leaves oblong or elliptic, 8–18 cm. long, 4–10 cm. wide, acuminate or abruptly acuminate at both ends, entire or obsoletely and remotely undulate-toothed, the petioles short, winged; flowers 2–5 in axils, sometimes solitary, yellowish, the pedicels slender, pendulous, 15–25 mm. long, thickened at the apex in fruit; calyx shallow, truncate, glabrous; corolla about 8 mm. across, 5-lobed, the lobes recurved above, lanceolate-deltoid, acute with a minute obtuse tip; fruit globose, 7–10 mm. across, red, naked.——Aug.–Oct. Woods in hills and mountains; Honshu, Shikoku, Kyushu.——Ryukyus, Formosa, Philippines, and India.

2. Tubocapsicum obtusum Kitam. *T. anomalum* var. *obtusum* Makino——MARUBA-HADAKA-HŌZUKI. Resembles the preceding; leaves thicker, ovate or narrowly so, 8–12 cm. long, 4.5–5.5 cm. wide, obtuse, entire, gradually narrowed to a short petiole; flowers 2–5 in axils, the pedicels 2–2.5 cm. long in fruit, about 7 mm. long in anthesis; calyx truncate, about 5 mm. across; corolla with the lobes oblong, obtuse; fruit globose, 8–12 mm. across, red.——Woods and thickets near seashores; Honshu (Kii Prov.), Shikoku, Kyushu.

6. SOLANUM L.   NASU ZOKU

Herbs or woody plants, sometimes scandent; stems sometimes prickly, glabrous or stellate-pubescent, often viscid; leaves alternate, entire or pinnatifid; cymes regularly branched or one-sided, axillary or in terminal panicles or corymbs; flowers yellow, white, or purplish; calyx campanulate or spreading, 5- to 10-toothed, rarely accrescent in fruit; corolla rotate, the limb plicate in bud, angled or shallowly lobed; filaments inserted on the throat, very short, the anthers connivent around the style, longitudinally splitting or opening by pores; ovary 2(–4)-loculed, many-ovuled; berry juicy, sometimes becoming dry, globose to oblong; seeds flat.——About 1,500 species, in the Tropics and warmer regions, especially abundant in S. America.

1A. Flowers axillary, fasciculate; calyx 10-cleft, densely yellowish pubescent, the lobes linear. .... 1. *S. biflorum*
1B. Flowers in supra-axillary or extra-axillary pedunculate cymes; calyx undulately 5-lobed or -toothed.
  2A. Flowers in simple umbellate cymes.
    3A. Cymes with a short axis; leaves rather thick, broadly ovate, acute or obtuse, deep green; corolla and fruit 6–7 mm. across.
              2. *S. nigrum*
    3B. Cymes without axis; leaves thinly membranous, ovate, acuminate, vivid green; corolla and fruit narrower. .. 3. *S. photeinocarpum*
  2B. Flowers in dichotomous elongate cymes.
    4A. Anthers lanceolate, 4–5 mm. long, narrowed at tip; berry ellipsoidal; leaves entire. .... 4. *S. megacarpum*
    4B. Anthers oblong or narrowly so, 2.5–3.5 mm. long, not narrowed at the tip; fruit usually globose.
      5A. Plants glabrous or nearly so; calyx-teeth short, distinctly shorter than wide.
        6A. Anther-locules obliquely truncate at apex; leaves often incised or deeply 3- to 5-lobed, long-acuminate, rounded at base when entire, cordate when lobed. .... 5. *S. japonense*
        6B. Anther-locules subtruncate at apex; leaves entire, acute to acuminate, rounded to broadly cuneate at base.
              6. *S. maximowiczii*
    5B. Plant densely glandular-tipped hairy; calyx-teeth depressed-deltoid to broadly ovate; anthers nearly truncate at apex; leaves pinnately cleft, rarely entire. .... 7. *S. lyratum*

1. Solanum biflorum Lour. *Lycianthes biflora* (Lour.) Bitter; *S. decemdentatum* Roxb.——MEJIRO-HŌZUKI. Perennial erect herb, 60–100 cm. high, ligneous at base, the branches pubescent while young, becoming glabrous and gray-yellow; leaves oblong, ovate, or narrowly so, 6–13 cm. long, 3–7 cm. wide, acute, rounded or abruptly narrowed at base, entire or slightly undulate-toothed, pubescent above and on nerves beneath or on both surfaces, the petiole winged in upper part; flowers 1 or 2 in axils, the pedicels 5–10 mm. long, prominently yellow-brown spreading-pubescent; calyx 10-cleft, 6–10 mm. across at anthesis, 1–2 cm. across in fruit, densely yellowish pubescent, the teeth broadly linear; corolla about 1.5 cm.

across, 5-cleft, the teeth narrowly ovate-deltoid; anthers 3–4 mm. long, narrowed at tip, the locules opening by an oblique apical pore; berries globose, red.——May–Oct. Warm regions; Honshu (s. centr. distr.), Shikoku, Kyushu.——Ryukyus, Formosa, China, India to Malaysia, and Hawaii.

2. Solanum nigrum L. INU-HŌZUKI. Nearly glabrous or scattered short-pubescent erect annual; stems 30–60 cm. long, angled, branched; leaves rather thick, broadly ovate, 6–10 cm. long, 4–6 cm. wide, acute or obtuse, rounded to broadly cuneate at base, entire or undulate-toothed, deep green, the petioles short, winged in upper part; cymes umbellate, supra-axillary, the axis 3–8 mm. long, the peduncles 1–3 cm. long, the pedicels 7–12 mm. long; flowers 6–7 mm. across, white; calyx-teeth depressed-deltoid or ovate-rounded; anthers narrowly oblong; berries globose, black, 6–7 mm. across.——Aug.–Oct. Waste grounds and roadsides; Hokkaido, Honshu, Shikoku, Kyushu; common.——Cosmopolitan.

3. Solanum photeinocarpum Nakamura & Odashima. *S. nigrum* var. *pauciflorum* Liou——TERIMI-NO-INU-HŌZUKI. Closely resembles the preceding; leaves thinner, acuminate, vivid green; inflorescence umbellate, without an axis; flowers fewer, smaller; corolla and fruit narrower; seeds smaller.——Honshu, Shikoku, Kyushu.——Ryukyus, Formosa, and s. China.

4. Solanum megacarpum Koidz. *S. dulcamara* var. *macrocarpum* Maxim.; *S. macrocarpum* (Maxim.) Kudo; *S. nipponense* var. *macrocarpum* (Maxim.) Makino & Nemoto ——Ō-MARUBA-NO-HOROSHI, MARUBA-NO-HOROSHI. Somewhat scandent, nearly glabrous perennial; leaves thinly membranous, ovate to narrowly so, 5–8 cm. long, 2.5–4.5 cm. wide, acute, rounded to shallowly cordate at base, entire, short-ciliolate, petiolate; cymes supra-axillary, loosely rather many-flowered, the peduncles 2–4 cm. long, the pedicels slender; calyx-teeth depressed; corolla pale purple, 1–1.2 cm. long, 5-cleft, the teeth reflexed, lanceolate; anthers lanceolate, 4–5 mm. long, opening by small pores; berries ellipsoidal, 12–15 mm. long, red.——July–Sept. Hokkaido, Honshu (centr. and n. distr.).——Sakhalin and s. Kuriles.

5. Solanum japonense Nakai. *S. dulcamara* var. *heterophyllum* Makino; *S. heterophyllum* (Makino) Nakai; *S. nipponense* Makino; *S. gracilescens* Nakai, 1930, non Nakai ex Makino, 1926——YAMA-HOROSHI, HOSOBA-NO-HOROSHI. Nearly glabrous scandent perennial herb with slender, elongate branches; leaves thinly membranous, deltoid-lanceolate to narrowly deltoid-ovate, 4–8 cm. long, 1–3.5 cm. wide, gradually long-acuminate, rounded at base, entire or undulate-toothed, sometimes incised to deeply 3- to 5-lobed and cordate, often thinly pilose on margin and on nerves above; peduncles 1–3 cm. long, loosely few-flowered; corolla 6–7 mm. long, pale purple, 5-cleft, the lobes lanceolate, reflexed; anthers narrowly oblong, 2.5–3 mm. long, the locules obliquely truncate at apex; berries globose, red, 6–7 mm. across.——July–Oct. Woods and thickets in mountains; Hokkaido, Honshu, Shikoku, Kyushu. ——Korea, Manchuria, and China.——Forma **xanthocarpum** (Makino) Hara. *S. nipponense* forma *xanthocarpum* Makino ——KIMI-NO-YAMA-HOROSHI. Fruit yellow.

Var. **takaoyamense** (Makino) Hara. *S. takaoyamense* Makino; *S. nipponense* var. *takaoyamense* (Makino) Makino—— TAKAO-HOROSHI. Leaves toothed; fruit ellipsoidal.——Honshu.

6. Solanum maximowiczii Koidz. *S. dulcamara* var. *ovatum* sensu auct. Japon., non Dunal; *S. gracilescens* Nakai ex Makino——MARUBA-NO-HOROSHI. Nearly glabrous scandent perennial herb with elongate, slender branches; leaves narrowly oblong or narrowly ovate, 5–10 cm. long, 1.5–4 cm. wide, acute to acuminate, rounded to broadly cuneate at base, entire, ciliolate, the petiole winged; cymes rather large, loosely flowered, the peduncles 2–4 cm. long; corolla pale purple, 5–6 mm. across, 5-cleft, the lobes lanceolate, reflexed; anthers 2–2.5 mm. long, narrowly oblong, subtruncate at apex; berries globose, red, 7–10 mm. across.——Aug.–Oct. Honshu, Shikoku, Kyushu.——Ryukyus.

7. Solanum lyratum Thunb. *S. dulcamara* var. *pubescens* Bl.; *S. dulcamara* var. *lyratum* (Thunb.) Sieb. & Zucc. ——HIYODORI-JŌGO. Scandent, densely glandular-pubescent perennial herb; leaves ovate, 3–8 cm. long, 2–4 cm. wide, acuminate or acute, sometimes obtuse, cordate at base, usually with 1 or 2 pairs of lobes near base, rarely entire; cymes loosely flowered, the peduncles 1–4 cm. long; calyx-teeth depressed-deltoid to broadly ovate; corolla white, 7–8 mm. long, 5-cleft, the lobes lanceolate, reflexed; anthers narrowly oblong, about 3 mm. long, nearly truncate at apex; fruit red, about 8 mm. across.——Aug.–Oct. Thickets in hills and low mountains; Hokkaido, Honshu, Shikoku, Kyushu.——Korea, China, Ryukyus, Formosa, and Indochina.——Forma **xanthocarpum** (Makino) Hara. *S. lyratum* var. *xanthocarpum* Makino—— KIMI-NO-HIYODORI-JŌGO. Fruit yellow. Rare.

Fam. 177. SCROPHULARIACEAE   GOMA-NO-HA-GUSA KA   Figwort Family

Herbs or shrubs, rarely trees; leaves alternate or opposite, sometimes verticillate; stipules absent; flowers bisexual, usually zygomorphic; calyx-lobes imbricate or valvate in bud; corolla gamopetalous, the limb 4- to 5(–8)-lobed, usually more or less bilabiate, the lobes imbricate in bud; stamens usually 4, didymous, or 2, inserted on the tube and alternate with the lobes, the 5th one absent or present as a staminode, rarely fertile; ovary superior, sessile, entire, usually completely 2-locular; style terminal; ovules many, on axile placentae; fruit a capsule; seeds many, with fleshy endosperm.——About 180 genera, with about 3,000 species, essentially cosmopolitan, especially abundant in temperate regions.

1A. Upper lip of corolla (the inner lobes) covering the others in bud.
  2A. Corolla spurred; capsules opening by pores; inflorescence a raceme or spike or the flowers solitary in axils. .... 1. *Linaria*
  2B. Corolla not spurred; capsules commonly dehiscent by valves.
    3A. Inflorescence compound.
      4A. Tree; leaves 20–40 cm. wide. .... 2. *Paulownia*
      4B. Herbs or subshrubs; leaves smaller.
        5A. Staminodes absent or scalelike; calyx 5-lobed. .... 3. *Scrophularia*
        5B. Staminodes filiform or spathulate; calyx 5-cleft. .... 4. *Pentstemon*
    3B. Inflorescence simple.
      6A. Anterior pair of stamens inserted on the corolla-tube, sometimes absent.

7A. Fertile stamens 4.
  8A. Corolla nearly actinomorphic, rotate; anther-locules confluent; acaulescent. .... 5. *Limosella*
  8B. Corolla bilabiate, with a distinct tube; caulescent.
    9A. Flowers solitary in axils.
      10A. Calyx tubular, prominently ridged or winged, the teeth very short. .... 6. *Mimulus*
      10B. Calyx-tube not winged or prominently ribbed, the lobes lanceolate to ovate or linear. .... 7. *Limnophila*
    9B. Flowers in short racemes; calyx-tube short but distinct. .... 8. *Mazus*
7B. Fertile stamens 2.
  11A. Calyx tubular, 5-lobed, 5-angled; corolla nearly rotate. .... 9. *Microcarpaea*
  11B. Calyx 5-cleft to -parted; corolla bilabiate, with a distinct tube.
    12A. Leaves chiefly cauline, uniformly flat or linear.
      13A. Calyx subtended by two bracteoles, the teeth imbricate in bud; anthers glabrous. .... 10. *Gratiola*
      13B. Calyx not bracteolate, the teeth valvate in bud; anthers pilose. .... 11. *Deinostema*
    12B. Leaves of two forms, the radical subelongate, the cauline scalelike. .... 12. *Dopatrium*
  6B. Anterior pair of stamens inserted on the corolla-throat.
    14A. Leaves palmately veined, entire or obscurely toothed; anther-locules emarginate at base; seeds smooth. ...... 13. *Lindernia*
    14B. Leaves pinnately veined, toothed; anther-locules not emarginate at base; seeds with several rows of small pits.
      15A. Flowers small, usually 4–10 mm. long; calyx wingless; capsules usually longer than the calyx. .... 14. *Vandellia*
      15B. Flowers larger, 1.5–3 cm. long; calyx broadly 5-winged; capsules not exceeding the calyx. .... 15. *Torenia*
1B. Outer (anterior) or lateral lobes of corolla enveloping the others in bud.
  16A. Corolla not bilabiate, the lobes flat.
    17A. Stamens 4 or 5.
      18A. Stems long-creeping below ground; leaves radical, alternate, long-petiolate, pinnately cleft; flowers radical; pedicels spirally coiled after anthesis. .... 16. *Ellisiophyllum*
      18B. Stems erect; leaves cauline, opposite, or alternate in the upper portion, sessile, entire or toothed; flowers cauline, nearly sessile, the upper ones forming a terminal spike. .... 17. *Centranthera*
    17B. Stamens 2.
      19A. Stems leafy; stamens inserted on corolla-tube, long-exserted; fruit a dehiscent capsule.
        20A. Corolla with lobes longer than the tube; capsules obtuse to retuse at apex; leaves opposite. .... 18. *Veronica*
        20B. Corolla with lobes much shorter than the tube; capsules acute at apex; leaves alternate or verticillate. .... 19. *Veronicastrum*
      19B. Stems scapelike, with few pairs of small leaves; stamens inserted on the corolla-throat, not exserted; fruit consisting of 2 nutlets. .... 20. *Lagotis*
  16B. Corolla clearly bilabiate, with a galeate or erect upper lip and spreading lower lip.
    21A. Calyx 2-bracteolate.
      22A Stems with many imbricate scalelike leaves on lower part; capsules dehiscent only on one side; calyx 4-lobed. .. 21. *Monochasma*
      22B. Stems without scalelike leaves; capsules dehiscent into 2 valves; calyx 5-lobed. .... 22. *Siphonostegia*
    21B. Calyx without bracteoles.
      23A. Ovules 2 in each locule; capsules 1- to 4-seeded. .... 23. *Melampyrum*
      23B. Ovules several to many in each locule; capsules several to many-seeded.
        24A. Annuals or biennials; upper lip of corolla recurved on margin.
          25A. Leaves toothed or lobed; flowers white or yellowish; capsules symmetrical; seeds longitudinally striate. ... 24. *Euphrasia*
          25B. Leaves pinnately dissected; flowers usually reddish; capsules somewhat asymmetrical; seeds reticulate.

                                                      25. *Phtheirospermum*
        24B. Perennials or rarely biennials; upper lip of corolla not recurved; leaves usually pinnately lobed to dissected; flowers white, yellowish, or reddish; capsules oblique, usually dehiscent on one side. .... 26. *Pedicularis*

1. LINARIA Mill. Un-ran Zoku

Herbs, rarely subshrubs; lower leaves usually opposite or verticillate in 3's or 4's, the upper or sometimes all leaves alternate, entire, toothed, or lobed; flowers solitary in axils or the upper sometimes forming a spike or raceme, white, yellow, or rose-purple, the pedicels without bracteoles; calyx 5-cleft, the lobes imbricate in bud; corolla-tube spurred, the upper lip erect, bifid, the lower lip spreading, 3-lobed, with protuberances or teeth closing the throat; stamens didymous, the anther-locules parallel, not connate; capsules ovoid-globose, usually opening by pores near the top; seeds sometimes winged.——About 100 species, in temperate regions of the N. Hemisphere, chiefly in the Old World.

1. Linaria japonica Miq. *L. geminiflora* F. Schmidt; *L. japonica* var. *geminiflora* (F. Schmidt) Nakai——Un-ran. Glaucous, glabrous perennial; stems terete, erect, more or less branched, 15–40 cm. long, from an ascending base; leaves opposite or verticillate in 3's or 4's, alternate in the upper ones, elliptic to elliptic-lanceolate, 1.5–3 cm. long, 5–15 mm. wide, obtuse to subacute, sessile, obscurely 3-nerved; racemes short, the pedicels short; flowers pale yellow, 15–18 mm. long, the spur 5–10 mm. long; capsules 6–8 mm. long; seeds winged, about 3 mm. long.——Aug.–Oct. Sandy places along seashores; Hokkaido, Honshu, Shikoku; common.——Kuriles, Sakhalin, Korea, Ussuri, and Manchuria.

2. PAULOWNIA Sieb. & Zucc. Kiri Zoku

Trees; leaves opposite, large, petiolate, broadly ovate or angled-cordate, 20–40 cm. wide, soft-pubescent; panicles terminal, many-flowered, with opposite branches; flowers purplish, large; calyx 5-parted, the segments obtuse; corolla-tube elongate, inflated in upper part, the lobes 5, obliquely spreading, rounded, nearly equal; stamens 4, didymous, not exserted, the anther-locules divergent, not connate; staminodia absent; ovules many; capsules coriaceous, rather large, broadly ovoid, acuminate; seeds small, numerous, with membranous wings.——Few species, in warmer regions of e. Asia, one long cultivated in our area.

1. Paulownia tomentosa (Thunb.) Steud. *Bignonia tomentosa* Thunb.; *P. imperialis* Sieb. & Zucc.——KIRI. Deciduous tree with glandular-pubescent branches and leaves; leaves large, long-petiolate, entire or shallowly 3- to 5-lobed; flowers before the leaves; calyx yellow-brown villous; corolla 5-6 cm. long, pale purple, yellow-striate inside; capsules 3-4 cm. long.——May-June. Much cultivated in our area and reportedly wild in Kyushu.——(?) Korea (Utsuryo Island).

3. SCROPHULARIA L.   GOMA-NO-HA-GUSA ZOKU

Usually foetid herbs or subshrubs; leaves opposite, or alternate in the upper ones, toothed or entire, sometimes pinnatifid; cymes often forming a terminal panicle; flowers small, green-purple, yellow-green, or sometimes brown; calyx 5-cleft, the lobes obtuse, rounded; corolla-tube ellipsoidal, gibbous, the lobes 5, small, flat, the 2 posterior lobes often longer than the others, the anterior one spreading; stamens didynamous, the anther-locules parallel, connate; capsules ovoid, septicidal; seeds ovoid, rugose, wingless.——More than 200 species, chiefly in the N. Hemisphere, abundant in the Mediterranean region.

1A. Stems winged; leaves slightly fleshy; petioles winged, semiclasping; capsules 7-10 mm. long, 5-7 mm. wide. .... 1. *S. grayana*
1B. Stems 4-angled, wingless; leaves membranous or chartaceous.
  2A. Roots more or less thickened; rhizomes not prominent; capsules ovoid-conical; leaves chartaceous.
    3A. Calyx-lobes broadly ovate; panicles terminal, very narrow and spikelike, densely flowered; flowers yellow-green, 6-7 mm. long; capsules about 5 mm. long, about 3 mm. across. .... 2. *S. buergeriana*
    3B. Calyx-lobes narrowly ovate to lanceolate; panicles broader and looser; flowers dark purple, slightly larger; capsules 5-8 mm. long, 4-6 mm. across.
      4A. Calyx-lobes narrowly ovate, acute to subobtuse; leaves serrulate. .... 3. *S. kakudensis*
      4B. Calyx-lobes lanceolate, acuminate; leaves minutely serrulate. .... 4. *S. toyamae*
  2B. Roots slender, fibrous; rhizomes short, thick and prominent; capsules ovoid-globose to globose; leaves membranous.
    5A. Cymes axillary, 1- to 3-flowered, shorter than the leaves, not forming a terminal panicle; flowers 9-11 mm. long; capsules ovoid-globose. .... 5. *S. musashiensis*
    5B. Cymes forming a loose terminal panicle; flowers 7-9 mm. long; capsules globose, abruptly acute. .... 6. *S. duplicatoserrata*

1. Scrophularia grayana Maxim. ex Komar. *S. alata* A. Gray, non Gilib.——EZO-HINA-NO-USU-TSUBO. Stems rather stout, 4-winged, usually pubescent on the nodes and upper parts; leaves slightly fleshy, narrowly ovate to deltoid-ovate, 8-15 cm. long, 4-8(-10) cm. wide, acute, rounded to broadly cuneate at base, regularly toothed, nearly glabrous on both sides, the petioles narrowly winged, semiclasping; cymes rather many-flowered, pedunculate, forming a terminal panicle, longer than the bracts, the branches usually loosely glandular-pilose; pedicels 1-2 cm. long; calyx-lobes ovate-orbicular, rounded at apex; corolla 12-15 mm. long, yellow-green, tinged pale brownish purple; capsules 7-10 mm. long, 5-7 mm. wide, ovoid-conical; seeds narrowly ovate, dark brown, with few obsolete longitudinal grooves.——May-Aug. Near seashores; Hokkaido, Honshu (Rikuchu and Noto Prov. and northw.).——s. Kuriles and Sakhalin.

2. Scrophularia buergeriana Miq. *S. buergeriana* var. *distantiflora* Miq.; *S. oldhamii* Oliv.; *S. oldhamii* var. *distantiflora* (Miq.) Fr. & Sav.——GOMA-NO-HA-GUSA. Stems erect, 80-150 cm. high, 4-angled, glabrous, nearly simple; leaves narrowly deltoid or deltoid-oblong, 5-10 cm. long, 2.5-5 cm. wide, acuminate, truncate to broadly cuneate at base, the petioles wingless or very slightly winged; cymes few-flowered, forming a spikelike rather dense panicle, 1-2 cm. long inclusive of the peduncle, minutely glandular-pilose, the bracts small; calyx-lobes broadly ovate, subacute or subobtuse; corolla yellow-green, 6-7 mm. long; capsules ovoid, about 5 mm. long, about 3 mm. across; seeds rather small.——July-Aug. Grassy places along rivers in lowlands; Honshu, Kyushu.——Korea, Manchuria, and China.

3. Scrophularia kakudensis Franch. Ō-HINA-NO-USU-TSUBO, HINA-NO-USU-TSUBO. Stems 4-angled, erect, often much branched, thinly pubescent or glabrous at base; leaves narrowly deltoid to deltoid-oblong, acute, truncate to rounded at base, serrulate, thinly pubescent especially beneath, petiolate; cymes several-flowered, pedunculate, forming a terminal rather loose panicle, the pedicels 7-15 mm. long, glandular-pilose; calyx-lobes narrowly ovate or ovate-deltoid, acute to subobtuse; corolla dark purple, 8-10 mm. long; capsules ovoid.——Aug.-Oct. Mountains; Hokkaido, Honshu, Shikoku, Kyushu.——Korea.

4. Scrophularia toyamae Hatus. ex Yamazaki. TSUSHIMA-HINA-NO-USU-TSUBO. Stems erect, simple, glabrous, 4-angled; leaves ovate, 5-7 cm. long, 2.5-4 cm. wide, subacute, shallowly cordate at base, very minutely serrulate, thinly short-pilose on both surfaces; cymes forming a narrow glandular-pilose terminal panicle, the pedicels 5-15 mm. long; calyx-lobes lanceolate, acuminate; corolla purple, about 8 mm. long; capsules ovoid-conical, acute, 5-8 mm. long.——July-Aug. Reported from Kyushu (Tsushima).

5. Scrophularia musashiensis Bonati. SATSUKI-HINA-NO-USU-TSUBO. Stems 40-80 cm. long, 4-angled, usually simple, often loosely pubescent; leaves thinly membranous, ovate-oblong or ovate, 7-15 cm. long, 3-8 cm. wide, subacute, rounded or cuneate at base, doubly toothed or incised, the petioles narrowly winged; cymes axillary, 1- to 3-flowered, loose, long-pedunculate, loosely glandular-pilose; flowers 9-11 mm. long, the pedicels slender, 2-3(-4) cm. long; calyx-lobes ovate, obtuse; capsules ovoid-globose, 7-10 mm. long.——(Apr.-)May. Honshu (w. Kantō Distr. and Shinano Prov.).

6. Scrophularia duplicatoserrata (Miq.) Makino. *S. alata* var. *duplicatoserrata* Miq.; *S. musashiensis* var. *surugensis* Honda; *S. duplicatoserrata* var. *surugensis* (Honda) Hara——HINA-NO-USU-TSUBO, YAMA-HINA-NO-USU-TSUBO. Stems 40-60 cm. long, 4-angled, glabrous or thinly pubescent at base; leaves thinly membranous, ovate to oblong-ovate, acute to acuminate, rounded to broadly cuneate at base, glabrous on both sides or rarely thinly pubescent above and on nerves beneath, doubly toothed, petiolate; cymes forming a loose terminal panicle, scattered glandular-pilose, the pedicels slender, 1(-3) cm. long; calyx-lobes ovate or broadly so, obtuse to subacute; capsules 6-7 mm. long, globose, abruptly acute.——Honshu (Tōkaidō Distr. and westw.), Shikoku, Kyushu.

4. PENTSTEMON Mitchell   IWABUKURO ZOKU

Perennial herbs or subshrubs; leaves opposite, the lower petiolate, the upper reduced to bracts; inflorescence a terminal, often leafy, bracteate panicle, rarely a raceme; flowers showy, rose-purple, purplish, whitish, rarely pale yellow; calyx 5-parted or cleft, the lobes imbricate in bud; corolla-tube usually elongate, sometimes inflated on one side, the limb bilabiate; stamens 4, didymous, not exserted; staminode filiform; capsules septicidal, the valves entire.——About 150 species, in N. America and Mexico, one in e. Asia.

1. **Pentstemon frutescens** Lamb. IWABUKURO. Rhizomes ascending and branched; stems erect, somewhat tufted, 10–20 cm. long, few-leaved, with 2 longitudinal lines of hairs; lower few leaves scalelike, the upper ones rather thick and fleshy, ovate-oblong or broadly lanceolate, 4–7 cm. long, 1.5–3 cm. wide, acute, narrowed or obtuse at base, mucronate-toothed, short-pilose on margin, sessile; flowers few, the pedicels 5–10 mm. long, as long as or slightly longer than the lanceolate bracts, rather densely glandular-tipped pubescent; calyx 12–15 mm. long, 5-parted, the lobes narrowly lanceolate; corolla about 2.5 cm. long, tubular, pale purple, loosely pubescent externally; capsules narrowly ovoid, as long as the calyx; seeds many, flat, winged, about 2 mm. long.——Aug. Gravelly and sandy alpine slopes; Hokkaido, Honshu (n. distr.). ——Kuriles, Sakhalin, e. Siberia, Kamchatka, and Aleutians.

5. LIMOSELLA L.   KITAMI-SŌ ZOKU

Dwarf glabrous tufted acaulescent often stoloniferous herbs; leaves tufted, radical, alternate, linear to oblong or ovate; pedicels radical, axillary, bractless; flowers white or rose-colored, minute; calyx campanulate, 5-toothed; corolla rotate, actinomorphic, 5-lobed, the lobes nearly equal, spreading, ovate-rounded, imbricate in bud; stamens 4, inserted on the tube, the anther-locules confluent; ovary 2-locular at base, without the septum at tip; style short, the stigma subcapitate; capsules indehiscent or tardily dehiscent, the valves membranous; seeds many, small, ovate.——Few species, chiefly S. African, 1 or 2 nearly cosmopolitan.

1. **Limosella aquatica** L. KITAMI-SŌ. Delicate glabrous annual; leaves radical, tufted, broadly linear to narrowly oblong, 4–10 mm. long, 1–3 mm. wide, 2–5 cm. long including the petiole; stolons short, nodeless, with a tuft of leaves at the top; pedicels 4–20 mm. long, ascending, appressed to the ground in fruit; flowers white to pinkish, about 2.5 mm. across; calyx thinly membranous, shorter than the fruit; corolla lobes broadly ovate; capsules 2-lobed, ellipsoidal, about 3 mm. long; seeds elliptic, about 0.3 mm. long, somewhat reticulate. ——May–Oct. Muddy or sandy places around springs and along rivers; Hokkaido, Honshu, Kyushu; very rare.—— Cooler regions of the N. Hemisphere.

6. MIMULUS L.   MIZO-HŌZUKI ZOKU

Erect or decumbent, glabrous to pubescent herbs; leaves opposite, entire or toothed; flowers yellow, rose, or purplish, solitary, in axils or the upper ones sometimes in terminal racemes, the pedicels bractless; calyx tubular, 5-ribbed, 5-toothed, the ribs often winglike; corolla-tube sometimes inflated on one side, the upper lip 2-lobed, the lower lip 3-lobed, with 2 tubercles or ridges inside the throat; stamens 4, didymous; style with 2 lamellate stigmas; capsules ellipsoidal to linear, enclosed in the calyx, loculicidal. ——About 60 species, in temperate regions, except Europe and the Mediterranean area, especially abundant in w. N. America, also in S. America.

1A. Calyx-teeth 5, mucrolike; corolla 1.5–2 cm. long; leaves petiolate except the upper ones, abruptly acute at base.
　　　　　　　　　　　　　　　　　　　　　　　　　　　　　　　　1. M. nepalensis var. japonicus
1B. Calyx with 5 short deltoid lobes; corolla 2.5–3 cm. long; leaves sessile, rounded at base. .... 2. M. sessilifolius

1. **Mimulus nepalensis** Benth. var. **japonicus** Miq. ex Maxim. *M. nepalensis* forma *japonica* Miq.; (?) *Torenia inflata* Miq.; *M. tenellus* var. *japonicus* (Miq.) Hand.-Mazz. ——MIZO-HŌZUKI. Glabrous perennial; stems 10–30 cm. long, 4-angled; leaves ovate to broadly so, 1.5–4 cm. long, 1–2.5 cm. wide, subacute or obtuse, abruptly acute to broadly cuneate or sometimes rounded at base, loosely few-toothed, short-petiolate or sessile in the upper ones; pedicels longer or slightly shorter than the leaves, 1.5–3 cm. long; calyx 8–10 mm. long at anthesis, 10–15 mm. in fruit, with 5 narrow wings, the teeth mucrolike, less than 1 mm. long; corolla pale yellow, 1.5–2 cm. long, with 2 lines of hairs inside near tip; capsules oblong, 8–10 mm. long, enclosed in the persistent slightly inflated calyx; seeds minute, smooth.—— June–Aug. Wet places and along streams in mountains; Hokkaido, Honshu, Shikoku, Kyushu; common.——Korea.

2. **Mimulus sessilifolius** Maxim. ŌBA-MIZO-HŌZUKI. Glabrous perennial; stems erect, simple, 4-angled, 20–30 cm. long; basal leaves scalelike; cauline leaves broadly to narrowly ovate, 3–6 cm. long, 1.5–3.5 cm. wide, short-acuminate or acute, rounded at base, sessile, sometimes obtuse in the lower ones, coarsely acute-toothed, short-pilose on margin, the lateral nerves of 2 or 3 pairs arising near the base, the pedicels 1–3 cm. long, usually half as long as the leaves; calyx 8–13 mm. long, 5-angled, the lobes deltoid, acuminate; corolla pale yellow, yellow-brown spotted at base inside, 2.5–3 cm. long, with 2 rows of long papillose hairs; capsules not exserted from the persistent calyx.——July–Aug. Wet grassy places in high mountains; Hokkaido, Honshu (centr. and n. distr.).——Sakhalin and s. Kuriles.

7. LIMNOPHILA R. Br.　SHISO-KUSA ZOKU

Glabrous or pubescent soft mostly marsh herbs; aerial leaves opposite or verticillate, dentate or incised, the aquatic leaves sometimes dissected into filiform segments; flowers solitary in axils, pedicelled, the upper ones often forming a terminal raceme; bracteoles small, linear, inserted at the base of calyx; calyx 5-cleft or -lobed, the lobes lanceolate to ovate, or linear, imbricate in bud, equal or the posterior ones larger; corolla with the upper lip entire or 2-lobed, the lower lip spreading, 3-lobed; stamens 4, didymous; stigmas 2, lamellate; capsules broadly ovoid to oblong, usually obtuse, 4-valved, loculicidal or septicidal; seeds many, small, reticulate.——About 30 species, in Asia, Africa, and Australia.

1A. Leaves verticillate in 5's to 8's, pinnately dissected or parted; flowers rose-purple, rarely white, sessile or the pedicels shorter than the calyx.
　2A. Stems and lower portion of calyx pubescent; calyx lobes linear-lanceolate, distinctly longer than the capsule; pedicels absent or
　　　nearly so. .... 1. L. sessiliflora
　2B. Stems, calyx, and pedicels glabrous; calyx lobes narrowly deltoid, as long as to slightly shorter than the capsule, shorter than the
　　　calyx-tube; pedicels as long as or shorter than the calyx. .... 2. L. indica
1B. Leaves opposite, narrowly oblong, obtusely toothed; flowers white, the pedicels longer than to as long as the calyx. .... 3. L. aromatica

1. Limnophila sessiliflora Bl. *Hottonia sessiliflora* Vahl; *Ambulia sessiliflora* (Vahl) Baill.——KIKU-MO. Perennial, terrestrial or sometimes aquatic; aerial stems terete, reddish, pubescent, the basal stems creeping and branched, 10–30 cm. long or much longer in the aquatic phase; aerial leaves verticillate in 5's to 8's, spreading, the pinnatifid segments linear or narrowly lanceolate, 1–2 cm. long, 3–7 mm. wide, acute, glabrate; aquatic leaves 2 or 3 times pinnately dissected into filiform segments; flowers solitary in axils, sessile or nearly so; calyx 5–7 mm. long, 5-cleft, loosely pubescent at base, the lobes lanceolate; corolla rose-purple, 6–10 mm. long, tubular; capsules ovoid-globose, about 4 mm. long; seeds oblong, terete, about 0.6 mm. long.——Paddy fields, shallow ponds, and ditches in lowlands; Honshu, Shikoku, Kyushu; common. ——Korea, Ryukyus, Formosa, China to India.

2. Limnophila indica (L.) Druce. *Hottonia indica* L.; *Gratiola trifida* Willd.; *L. gratioloides* R. Br.; *Ambulia trichophylla* Komar.; *A. indica* (L.) W. F. Wight; *A. stipitata* Hayata; *L. stipitata* (Hayata) Makino & Nemoto——KO-KIKU-MO, TAIWAN-KIKU-MO.　　Resembles the preceding; stems glabrous; leaves sessile, verticillate in 5's or 6's, glabrous, pinnately lobed in the upper half, 1–1.5 cm. long, 3–5 mm. wide, gradually narrowed at base; aquatic leaves finely dissected into linear segments; flowers solitary in axils, the pedicels glabrous,

shorter than or as long as the calyx; calyx greenish, 3–3.5 mm. long, membranous, as long as the capsules, glabrous, 5-lobed, the lobes deltoid, acute or with an obtuse tip; capsules about 3 mm. across, broadly ellipsoidal; seeds broadly ellipsoidal, about 0.3 mm. long, dark brown.——Sept. Honshu; rare.——Manchuria, Korea, Formosa, China, India, Malaysia, and Australia.

3. Limnophila aromatica (Lam.) Merr. *Ambulia aromatica* Lam.; *Gratiola aromatica* Pers.; *L. punctata* Bl.; *L. gratissima* Bl.; *A. gratissima* (Bl.) Nakai——SHISO-KUSA. Nearly glabrous annual; stems terete, short-decumbent and branched at base; leaves rarely verticillate in 3's, narrowly oblong to lanceolate, sometimes oblanceolate, 1.5–3 cm. long, 3–7 (–10) mm. wide, acute to obtuse at both ends, obtusely short-toothed, glabrous, glandular-dotted beneath, sessile; pedicels 7–15 mm. long; bracteoles short-linear, much shorter than the calyx; calyx 5–7 mm. long, 5-cleft, the lobes narrowly lanceolate; corolla about 10 mm. long, creamy white; capsules broadly ovoid, slightly shorter than the calyx; seeds reniform, with 2 ridges on the back, about 0.3 mm. long.——Sept.–Oct. Wet places in lowlands; Honshu, Shikoku, Kyushu; common. ——s. Korea, Ryukyus, Formosa, China, India, Malaysia, and Australia.

8. MAZUS Lour.　SAGIGOKE ZOKU

Small perennial often stoloniferous herbs; lower leaves and those of stolons opposite, the upper leaves alternate, toothed; racemes terminal, the bracts small, the bracteoles absent or on the pedicel; flowers pale purple or white; calyx broadly campanulate, short-tubular, 5-cleft; corolla-tube shorter or slightly exserted from the calyx, the upper lip erect, 2-lobed, the lower lip 3-lobed, spreading or ascending, with 2 elevated papillose lines on the throat; stamens 4, didymous; stigma bifid with lamellate lobes; capsules globose, loculicidal; seeds many, small, ovoid.——About 10 species, in e. Asia, India, Malaysia, and Australia.

1A. Stolons much-elongate; flowers pale purple, rarely white, 1.5–2 cm. long. .... 1. M. miquelii
1B. Stolons absent.
　2A. Flowers 1.3–2 cm. long; leaves mostly basal. .... 2. M. fauriei
　2B. Flowers 1–1.2 cm. long; leaves also cauline. .... 3. M. japonicus

1. Mazus miquelii Makino. *Lindernia japonica* Thunb. pro parte; *M. rugosus* var. (?) *stoloniferus* Maxim.; *M. rugosus* var. *macranthus* Fr. & Sav.; *M. rugosus* var. *rotundifolius* Fr. & Sav.; *M. rotundifolius* (Fr. & Sav.) Koidz.; *M. rotundus* Furumi; *M. stoloniferus* (Maxim.) Makino; *M. englerianus* Bonati——MURASAKI-SAGIGOKE.　　Thinly pubescent stoloniferous perennial; stems 7–15 cm. long, few-leaved at base; leaves obovate, elliptic, or broadly ovate, 4–7 cm. long inclusive of the petiole, 1–1.5 cm. wide, very obtuse, undulately obtuse-

toothed, the petioles winged in upper part; leaves of stolons short-petiolate, sometimes ovate-orbicular, 1.5–2.5 cm. long; racemes loosely few-flowered, puberulent, the pedicels longer than the calyx; calyx 7–10 mm. long, 5-lobed; corolla slightly flattened dorsiventrally, the lobes of the bilobed upper lip narrowly oblong, the lower lip inside with 2 raised lines of yellow-brown clavate tubercles; capsules subglobose, about 4 mm. long.——Apr.–May. Wet places, especially abundant around paddy fields in lowlands; Honshu, Shikoku, Kyushu.——

Forma **albiflorus** (Makino) Makino. *M. japonicus* var. *albiflorus* Makino——SHIROBANA-SAGIGOKE, SAGIGOKE. White-flowered phase.

Var. **contractus** Makino. JAKAGO-SŌ. Leaves strongly incurved and bullate, very numerous on the stems.——Possibly of garden origin.

2. **Mazus fauriei** Bonati. *M. japonicus* var. *tenuiracemus* Hayata; *M. taihokuensis* Masam.; *M. tenuiracemus* (Hayata) Hatus.——SEITAKA-SAGIGOKE. Somewhat pubescent perennial; leaves opposite, nearly radical, obovate-spathulate, 3–4 cm. long inclusive of the winged petiole, 1–1.5 cm. wide, rounded at apex, obtusely toothed; stems leafy near base, 8–15 cm. long, loosely 5- to 10-flowered, scattered-puberulent, the pedicels longer than the calyx; calyx 5–7 mm. long, 5-cleft; corolla 13–20 mm. long, pale purple, the upper lip divided into 2 narrowly oblong lobes, the lower lip with 2 yellowish raised bands inside; capsules globose, 3–4 mm. across; seeds minute.——Apr. Kyushu (s. distr.).——Ryukyus and Formosa.

3. **Mazus japonicus** (Thunb.) O. Kuntze. *Lindernia japonica* Thunb., pro parte; *M. rugosus* Lour.; *Tittmannia obovata* Bunge——TOKIWA-HAZE. Somewhat pubescent perennial; stems 5–15 cm. long, leafy mostly on lower half or toward the base; basal leaves opposite, obovate, elliptic-cuneate or oblong-spathulate, 2–6 cm. long inclusive of the petiole, 8–15 mm. wide, rounded at apex, coarsely obtuse-toothed; cauline leaves subsessile or short-petiolate; racemes loosely few-flowered, the pedicels longer than the calyx, puberulent; calyx 5–10 mm. long, 5-fid; corolla white to purplish, 1–1.2 cm. long, the upper lip about half as long as the lower, the lower lip with 2 raised yellow clavate-pilose bands inside; fruit globose, 3–4 mm. across.——Apr.–Oct. Waste grounds, cultivated fields, and riverbanks in lowlands; Hokkaido, Honshu, Shikoku, Kyushu; very common.——Korea, Ryukyus, China, and India.

9. **MICROCARPAEA** R. Br. SUZUME-NO-HAKOBE ZOKU

Minute, creeping, much-branched annual; leaves opposite, narrow, entire, sessile; flowers solitary in axils, sessile, very minute, without bracteoles; calyx tubular, 5-ribbed, shortly 5-lobed; corolla-tube short, the limb 5-lobed, spreading, imbricate in bud, the 2 posterior (upper) lobes connate at base; stamens 2, inserted on tube, the anther-locules confluent; ovules many; capsules ovoid, 2-grooved, loculicidal; seeds ovoid.——One species, in e. Asia, India, Malaysia, and Australia.

1. **Microcarpaea minima** (Koenig) Merr. *Paederota minima* Koenig; *M. muscosa* R. Br.——SUZUME-NO-HAKOBE. Plant very delicate, glabrous except the calyx-lobes; stems much branched, appressed to the ground, many-leaved; leaves spreading, broadly linear to broadly lanceolate, 3–5 mm. long, 1(–2) mm. wide, obtuse, shorter than the internodes of branches, semiclasping at base, entire, sessile; flowers solitary in axils, sessile; calyx 2.5–3 mm. long, tubular-campanulate, 5-ribbed, 5-lobed, the lobes narrowly deltoid, acute, sparsely ciliate, slightly spreading at tip; corolla minute, pink; capsules oblong, nearly as long as the calyx-tube, enclosed in the calyx; seeds fusiform-ovoid, obtuse, yellow-brown, nearly smooth, about 0.3 mm. long.——July–Oct. Paddy fields and wet riverbanks in lowlands; Honshu, Shikoku, Kyushu.——Korea, Ryukyus, Formosa, China, India, Malaysia, and Australia.

10. **GRATIOLA** L. Ō-ABU-NO-ME ZOKU

Erect or decumbent, glabrous or pilose herbs; leaves opposite, entire or toothed; pedicels solitary in axils, usually bibracteolate; calyx deeply 5-cleft to -parted, slightly unequal; corolla-tube bilabiate, the upper lip entire or shortly 2-lobed, the lower lip 3-lobed, the lobes rounded; stamens 2, the anther-locules distinct, not confluent, the staminodia 2 or absent; capsules ovoid, often acute, 4-valved, loculicidal or septicidal, the valves incurved on margin, free from the placentae; seeds many, small, striate, or reticulate.——About 25 species, chiefly in temperate and warmer regions.

1. **Gratiola japonica** Miq. *G. micrantha* Fr. & Sav.——Ō-ABU-NO-ME. Glabrous annual; stems rather fleshy, erect, 10–25 cm. long, usually branched at base; leaves spreading, lanceolate, subobtuse, entire, weakly few-nerved from the base, sessile; flowers solitary in axils, sessile, the bracteoles simulating the calyx-lobes; calyx 3–4 mm. long, 5-parted nearly to the base, the segments lanceolate, subobtuse; corolla white, about 6 mm. long, the limb rather short, bilabiate; capsules globose, 4–5 mm. across, slightly longer than the calyx; seeds lanceolate, about 0.7 mm. long.——May–July. Wet places especially in paddy fields; Honshu, Kyushu.——Korea, Amur, and Ussuri.

2. **Gratiola virginiana** L. *G. fluviatilis* Koidz.——KA-MIGAMO-SŌ. Annual, about 15 cm. high; stems somewhat fleshy, puberulent, sparsely branched at base; leaves ovate, 2.5–5.5 cm. long, 1.5–2.8 cm. wide, very obtuse, broadly cuneate at base, 5-nerved, irregularly obtuse-toothed, loosely pilose on margin; flowers solitary, white, about 8 mm. long, in axils, the pedicels pilose, 3–10 mm. long, the bracteoles broadly linear, 2.5 mm. long; calyx 5-parted to the base, the segments linear-lanceolate, pilose, slightly accrescent after anthesis, to 7 mm. long in fruit; corolla bilabiate; stamens 2, the filaments densely pilose, staminodes absent; capsules cordate, about 5 mm. long.——Oct. Found once in Honshu near Kyoto.——N. America.

11. **DEINOSTEMA** Yamazaki SAWA-TŌGARASHI ZOKU

Annual herbs with fibrous roots; stems erect; leaves opposite, sessile, entire, obscurely veined; chasmogamic flowers long-pedicelled in the axils of upper leaves, solitary; calyx campanulate, 5-parted, ebracteolate; corolla tubulose-campanulate, the limb dilated, bilabiate; 2 posterior stamens perfect, included, the filaments short, adnate on the corolla-tube; anther locules short-pilose, the 2 anterior anthers reduced to staminodes; ovary glabrous, ovate-globose; capsules ovate-orbicular, shorter than the calyx; seeds many, oblong.——Two species in e. Asia.

1. Deinostema adenocaulum (Maxim.) Yamazaki. *Gratiola adenocaula* Maxim.; *G. violacea* var. *adenocaula* (Maxim.) Makino——Maruba-no-sawa-tōgarashi. Annual, thinly pilose with spreading glandular hairs on upper portion; stems erect, more or less branched at base; leaves ovate or elliptic, 5–8 mm. long, 3–6 mm. wide, acute to subobtuse, rounded and weakly few-nerved from the base, entire, sessile; pedicels slender, 1–2 cm. long in chasmogamic flowers, sessile or less than 10 mm. long in the cleistogamic; calyx 3–4 mm. long, 5-parted nearly to the base, the segments linear-lanceolate; corolla pale purple, about twice as long as the calyx; capsules ellipsoidal, 1.5–3 mm. long, shorter than the calyx; seeds oblong-lanceolate, about 0.5 mm. long, finely reticulate.——July–Sept. Wet places and paddy fields; Honshu, Shikoku, Kyushu.——Korea (Quelpaert Isl.).

2. Deinostema violaceum (Maxim.) Yamazaki. *Gratiola violacea* Maxim.; *G. axillaris* Nakai; *G. saginoides* var. *violacea* (Maxim.) Matsum.——Sawa-tōgarashi. Glabrous, thinly glandular-pilose on pedicels; stems erect, often branched at base, terete, 3–10 cm. long, sometimes decumbent at base; leaves spreading, linear-lanceolate or broadly linear, 7–10 mm. long, about 2 mm. wide, long-acuminate, entire; pedicels 1–1.5 cm. long or absent in cleistogamic flowers; calyx 4–6 mm. long, 5-parted nearly to the base, the segments linear-lanceolate; corolla purple, bilabiate, about 1½ times as long as the calyx, the limb nearly as long as the tube; cleistogamic flowers often fasciculate in axils; capsules oblong, about 3 mm. long; seeds narrowly ovoid-oblong, about 0.5 mm. long, reticulate.——Aug.–Oct. Wet places; Honshu, Shikoku, Kyushu.——Korea, Manchuria, and Ryukyus.

12. DOPATRIUM Hamilt. ex Benth.　　Abu-no-me Zoku

Glabrous herbs; leaves opposite, small, approximate toward the base of stems, the cauline leaves scalelike, loosely arranged; bracteoles absent; flowers purplish, solitary in upper axils; calyx small, deeply 5-fid, the segments imbricate in bud; corolla-tube narrow, broadened at the throat, the upper lip short, 2-lobed, the lower lip larger, spreading, 3-lobed; stamens 2, the anther-locules parallel, not confluent, the staminodia inserted on the tube; capsules globose or ovoid, scarcely grooved, loculicidal, the valves entire or 2-fid; seeds small.——Few species, in the Tropics of Africa, Asia, and Australia.

1. Dopatrium junceum (Roxb.) Hamilt. *Gratiola juncea* Roxb.——Abu-no-me. Glabrous, fleshy annual; stems erect, 10–30 cm. long, usually tufted, branched at base; basal leaves fleshy, rosulate, lanceolate or linear-oblong, 1.5–2.5 cm. long, 3–5 mm. wide, obtuse, with few obscure parallel nerves, sessile, the cauline leaves gradually smaller, narrowly oblong, 1–3 mm. long, erect or appressed, very obtuse; pedicels axillary in the upper nodes, solitary, 0–1 cm. long; calyx about 1.5 mm. long, deeply 5-cleft, the lobes obtuse; corolla purple, 5–6 mm. long; capsules globose, much longer than the calyx, 2.5–3 mm. long; seeds ovoid, reticulate, about 0.2 mm. in diameter.——Aug.–Oct. Wet places in lowlands; Honshu, Shikoku, Kyushu.——Ryukyus, Formosa, China, and India.

13. LINDERNIA All.　　Azena Zoku

Herbs; leaves opposite, usually entire, palmately veined; flowers small, solitary in axils, without bracteoles, sessile or pedicelled, sometimes forming a terminal elongate or umbellate raceme; calyx tubular-campanulate, 5-parted or -toothed; corolla tubular or ampliate above, the upper lip erect, 2-lobed or retuse, the lower lip larger, 3-lobed; stamens 4 or 2, the upper 2 fertile, the lower fertile or reduced to staminodia; capsules linear to globose, often longer than the calyx, septicidally dehiscent.——Many species, in warmer and temperate regions of the world.

1. Lindernia pyxidaria L. *Vandellia pyxidaria* (L.) Maxim.——Azena. Glabrous annual; stems branched below, erect or ascending, 7–15 cm. long; leaves oblong to narrowly so, 1.5–3 cm. long, 5–12 mm. wide, obtuse at both ends, entire, with 5(–3) parallel nerves, slightly lustrous above, sessile; flowers solitary in axils, the pedicels 2–2.5 cm. long, mostly longer or as long as the leaves; calyx 3–4 mm. long, the segments slightly shorter than the capsule; corolla rose-purple, about 6 mm. long; capsules ellipsoidal, 3.5–4.5 mm.

long; style 1–1.5 mm. long; seeds smooth.——Aug.–Oct. Wet places in the lowlands; Hokkaido, Honshu, Shikoku, and Kyushu.——Korea, China, Formosa, and the warmer regions of Eurasia.

2. Lindernia dubia (L.) Pennell. *Gratiola dubia* L.; *Ilysanthes dubia* (L.) Barnhart——Amerika-azena. Leaves toothed, as long as to longer than the pedicels; capsules oblong; style 2.5–3.5 mm. long; seeds weakly lineolate.——Naturalized in Honshu (Kinki Distr.).——N. America.

14. VANDELLIA L.　　Aze-tōgarashi Zoku

Small annual or short-lived perennial herbs; leaves opposite, often toothed, pinnately veined; flowers small, solitary in axils, sessile or pedicelled, rarely in terminal racemes, ebracteolate; calyx 5-parted to -toothed; corolla-tube tubulose or slightly ampliate above, the limb bilabiate; stamens 4, all perfect, or the 2 anterior reduced, the posterior fixed to the throat of corolla, anther-locules divaricate; capsules linear to ellipsoidal, usually longer than the calyx, septicidally dehiscent; seeds with several rows of small pits.——Many species, in warmer areas of the world.

1A. Capsules broadly linear, 2 or 3 times as long as the calyx.
  2A. Fertile stamens 2; leaves of upper portion of racemes abruptly smaller, bractlike; margin of calyx-segments narrowly scarious;
    corolla rose-purple. . . . . 1. *V. anagallis* var. *verbenaefolia*
  2B. Fertile stamens 4; leaves of upper portion of racemes well developed; calyx-segments not scarious; corolla creamy-white.
                                                                            2. *V. angustifolia*
1B. Capsules ellipsoidal or globose, scarcely longer than the calyx.
  3A. Calyx shallowly 5-lobed, the lobes narrowly deltoid; corolla purple; capsules oblong. . . . . 3. *V. crustacea*
  3B. Calyx 5-lobed nearly to the base, the lobes linear; corolla white; capsules narrowly fusiform-ovoid. . . . . 4. *V. setulosa*

1. Vandellia anagallis (Burm.) Yamazaki var. **verbenaefolia** (Colsm.) Yamazaki. *Lindernia verbenaefolia* (Colsm.) Pennell; *Gratiola verbenaefolia* Colsm.; *Ruellia serrata* Thunb.; *V. pachypoda* Fr. & Sav.; *Bonnaya veronicaefolia* var. *verbenaefolia* (Colsm.) Hook. f.; *L. pachypoda* (Fr. & Sav.) Matsum.; *L. anagallis* var. *verbenaefolia* (Colsm.) Hara; *Ilysanthes serrata* (Thunb.) Makino——SUZUME-NO-TŌGARASHI. Glabrous annual; stems branched at base, 8–20 cm. long; leaves oblanceolate to broadly so, 2–4 cm. long, 5–10 mm. wide, obtuse or subacute, gradually narrowed to the base, sessile, obtusely toothed; racemes loose, terminal, leafy at base, abruptly linear-bracteate in upper part, the pedicels thick and spreading, nearly as long as the capsules; calyx 5–6 mm. long, the segments linear, scarious on margin; corolla rose-purple, about 1 cm. long; fertile stamens 2, the anthers without appendages; capsules linear, 2–2.5 times as long as the calyx.——Aug.–Oct. Wet places in lowlands; Honshu, Shikoku, Kyushu.——Ryukyus, Formosa, India and Malaysia.

2. Vandellia angustifolia Benth. *Lindernia angustifolia* (Benth.) Wettst.; *V. cymulosa* Miq.; *L. cymulosa* (Miq.) Matsum.——AZE-TŌGARASHI. Glabrous annual; stems branched at base, 8–20 cm. long; leaves lanceolate or oblong-lanceolate, sometimes narrowly lanceolate, 1–3(–4) cm. long, 3–5(–8) mm. wide, acute or subobtuse, narrowed at base, obsoletely short-toothed, obsoletely penninerved; flowers solitary in upper axils, the pedicels spreading, nearly as long as the capsules; calyx 4–5 mm. long, the segments linear; corolla creamy white, about 1 cm. long; lower anther-locule of the fertile stamens short spine-tipped; capsules linear, 3 or 4 times as long as the calyx.——Aug.–Oct. Wet places in lowlands; Honshu, Shikoku, Kyushu.——Korea, China, India, and Malaysia.

3. Vandellia crustacea (L.) Benth. *Lindernia crustacea* (L.) F. Muell.; *Capraria crustacea* L.; *Torenia crustacea* (L.) Cham. & Schltdl.——URI-KUSA. Thinly puberulent annual; stems 7–15 cm. long, ascending from a branched base; leaves ovate or narrowly deltoid-ovate, sometimes oblong, 1–2 cm. long, 6–12 mm. wide, subobtuse to obtuse, rounded at base, obtusely few-toothed, short-petiolate; flowers solitary in axils, the uppermost flowers forming a terminal cyme, the pedicels slender, ascending, 1–2.5 cm. long; calyx 4–5 mm. long, with 5 narrowly deltoid-acute shallow lobes; corolla purple, about 1 cm. long; capsules oblong, slightly shorter than the calyx; seeds elliptic, obsoletely impressed-punctate.——Aug.–Oct. Wet cultivated fields and riverbanks in lowlands; Hokkaido, Honshu, Shikoku, Kyushu; common.——s. Korea, Ryukyus, Formosa, China, India, and Malaysia.

4. Vandellia setulosa (Maxim.) Yamazaki. *Torenia setulosa* Maxim.; *Lindernia setulosa* (Maxim.) Tuyama——SHISOBA-URI-KUSA. Thinly pilosulous annual; stems ascending, branched, 10–20 cm. long, glabrous or thinly coarse-hairy on upper portion; leaves broadly ovate, 1–1.5 cm. long, 8–12 mm. wide, obtuse, rounded and short-petiolate at base, obtusely toothed, thinly pilosulous especially on upperside; pedicels 1–1.5 cm. long, longer than the leaves; calyx 5–7 mm. long, the segments linear, cleft nearly to the base, incurved and appressed to the fruit, sparsely coarse-pilosulous; corolla white, 7–8 mm. long; capsules narrowly fusiform-ovoid, acute, 4–5 mm. in diameter; seeds minute, slightly angled.——Aug.–Oct. Honshu (Kii and Ise Prov.), Shikoku, Kyushu.——Ryukyus (Shichito).

15. TORENIA L. TSURU-URI-KUSA ZOKU

Glabrous or pilose herbs; leaves opposite, entire or toothed; racemes short or elongate, few-flowered, the pedicels not bracteolate; calyx tubular, usually with 5 winglike ribs, 3- to 5-toothed or bilabiate; corolla tubular, sometimes ampliate in upper part, the upper lip erect, retuse or 2-lobed, the lower lip larger, 3-lobed; stamens 4, toothed or with a filiform appendage at base, the lower pair inserted on the throat; capsules oblong, enclosed in the calyx, septicidal; seeds many, rugose, impressed-areolate.——About 30 species, in the Tropics and warmer regions of Asia and Africa.

1. Torenia glabra Osb. *T. kiusiana* Ohwi——KOBANA-TSURU-URI-KUSA. Short-pubescent perennial; stems decumbent, 15–20 cm. long, branched, slender; leaves ovate to deltoid-ovate, 2–2.5 cm. long, 1–1.5 cm. wide, acute, rounded to truncate at base, appressed-toothed, the petioles 5–10 mm. long; flowers solitary in axils, pedicelled; calyx 13–17 mm. long, 5-toothed, 5-winged; corolla about 15 mm. long; capsules about 13 mm. long, short-cylindric.——Kyushu (s. distr.).——Ryukyus (Amami-oshima), China.

16. ELLISIOPHYLLUM Maxim. KIKUGARA-KUSA ZOKU

Delicate perennial herb; rhizomes long-creeping, loosely leaved; leaves alternate, long-petiolate, pinnately parted, the segments coarsely obtuse-dentate; pedicels axillary, radical, elongate, 1-flowered, bracteolate, coiled after anthesis; flowers small, white; calyx 5-cleft, the lobes 3-nerved; corolla infundibuliform, 5-lobed; stamens 4, inserted on the throat, equal, slightly exserted, the anther-locules distinct; style elongate, the stigma bifid; ovary 2-locular, each locule few-ovuled; capsules 2-locular, globose; seeds orbicular, convex on back, attached at the center on ventral face, puberulent, viscid when moistened.——One species, in e. Asia and India.

1. Ellisiophyllum pinnatum (Wall.) Makino. *Ourisia pinnata* Wall. ex Benth.; *Hornemannia pinnata* (Wall.) Benth.; *E. reptans* Maxim.; *Moseleya pinnata* (Wall.) Hemsley ——KIKUGARA-KUSA, HOROGIKU.   Plant pubescent; stems slender; leaves thin, soft, membranous, distant, broadly ovate-deltoid to broadly ovate, 2.5–6 cm. long, 2–5 cm. wide, long-petiolate, deeply pinnatiparted with few coarse rounded and toothed segments; flowers solitary in axils, 8–10 mm. across, white, the pedicels radical, 3–6 cm. long; calyx 5–7 mm. long, 5-lobed; fruit globose, 4–5 mm. across; seeds few, about 1.5 mm. across.——May–June. Woods in mountains; Honshu (s. Kinki and Chūgoku Distr.), Shikoku.——Formosa, China, and India.

17. CENTRANTHERA R. Br.   GOMA-KUSA ZOKU

Rather firm usually strigose and scabrous erect herbs; leaves opposite or alternate in upper ones, obtuse, entire or few-toothed; flowers rose or yellowish, solitary in axils, the upper ones forming a terminal spike, the pedicels very short, 2-bracteolate; calyx spathelike, split on one side, acute, entire or with 3–5 short lobes; corolla-lobes 5, subequal, spreading, the tube elongate, ampliate above, the limb obsoletely bilabiate; stamens 4, didymous, not exserted, in pairs; capsules loculicidal; seeds minute, very many.——Several species, in Asia, Malaysia, and Australia.

1. Centranthera cochinchinensis (Lour.) Merr. var. **lutea** (Hara) Hara.   *C. hispida* sensu auct. Japon., non R. Br.; *C. brunoniana* sensu auct. Japon., non M.L.; *Razumovia cochinchinensis* var. *lutea* Hara——GOMA-KUSA.   Strigose and scabrous erect annual; stems often branched, terete, 20–40 cm. long; leaves opposite, the upper ones alternate, broadly linear to linear-lanceolate, 2–5 cm. long, 3–7 mm. wide, nearly entire, sessile; flowers nearly sessile, yellowish, 15–20 mm. long, the bracteoles narrowly lanceolate, shorter than and appressed to the calyx; calyx shallowly split in front, acute, 7–10 mm. long at anthesis, 10–15 mm. long in fruit, longer than the capsule; seeds truncate at apex, cuneate at base, obliquely lineolate.——Aug.–Oct. Wet grassy places in lowlands; Honshu, Shikoku, Kyushu. The typical phase occurs in Korea, Ryukyus, Formosa, China, and Indochina.

18. VERONICA L.   KUWAGATA-SŌ ZOKU

Herbs or shrubs; leaves at least sometimes opposite or verticillate, simple; flowers solitary in axils or in terminal or axillary racemes, without bracteoles; calyx 4- or 5-parted, the outer 2 segments sometimes connate; corolla rotate, the limb spreading, 4- or 5-lobed; stamens 2, inserted on the tube, exserted; style entire; ovules many or rarely 2 in each locule; capsules flat or inflated, 2-grooved, loculicidal or septicidal; seeds few to many, ovate or orbicular, attached at the ventral face, convex on back, smooth or rugose.——About 300 species, chiefly in the temperate regions of the N. Hemisphere, Australia, and New Zealand.

1A. Flowers in terminal spikes or racemes, or solitary and axillary.
  2A. Perennials; flowers in terminal spikes or racemes; plants of woods or grassy places.
    3A. Herbs 30–150 cm. high, rather many-leaved; inflorescence a dense spike (racemose in No. 3).
      4A. Calyx and leaves beneath white-tomentose. .... 1. *V. ornata*
      4B. Calyx green, glabrous to pubescent; leaves usually green on both surfaces.
        5A. Leaves narrowed at base.
          6A. Leaves slightly dilated and clasping at base; inflorescence and young branches with long-spreading multicellular hairs.
                2. *V. sieboldiana*
          6B. Leaves not dilated and clasping at base; inflorescence and stems puberulent or with upwardly directed hairs.
            7A. Lowest lateral nerves of leaves ⅓–⅔ the length of the blades. .... 3. *V. linariaefolia*
            7B. Lowest lateral nerves shorter. .... 4. *V. rotunda* var. *subintegra*
        5B. Leaves rounded to truncate-cordate at base.
          8A. Leaves sessile, or with petioles 3–5 mm. long, ovate, rounded at base. .... 5. *V. subsessilis*
          8B. Leaves, at least the median ones, with petioles 1–3 cm. long.
            9A. Leaves with scattered multicellular hairs on both sides, slightly scabrous above. .... 6. *V. kiusiana*
            9B. Leaves short-pubescent or thinly tomentose on both sides or sometimes glabrous, not scabrous.
              10A. Leaves with many prominent acute teeth. .... 7. *V. subincanovelutina*
              10B. Leaves with rather loose appressed teeth. .... 8. *V. denkichiana*
      3B. Herbs less than 30 cm. high, with few pairs of leaves; inflorescence racemose, loosely few- to rather many-flowered.
        11A. Leaves distinctly petiolate, ovate to broadly lanceolate, pinnately incised to parted. .... 9. *V. schmidtiana*
        11B. Leaves sessile or subsessile, ovate or elliptic, toothed or subentire.
          12A. Plants short rhizomatous; leaves distinctly toothed, ovate; inflorescence few-flowered, about 1 cm. long.
            13A. Flowers 7–8 mm. across; styles 3–6 mm. long; plant rather prominently pubescent; leaves 1.5–2.5 cm. long.
                10. *V. stelleri* var. *longistyla*
            13B. Flowers 4–5 mm. across; styles 1–2 mm. long; plant sparingly pubescent; leaves 1–2 (2.5) cm. long. ... 11. *V. nipponica*
          12B. Plants from a long creeping base; leaves subentire, elliptic, sometimes ovate; inflorescence rather many-flowered, 3–15 cm. long.
            14A. Inflorescence curved-puberulent; leaves 5–10 mm. long. .... 12. *V. serpyllifolia*
            14B. Inflorescence subglandular-puberulent; leaves 7–15 mm. long. .... 13. *V. tenella*

1. Veronica ornata Monjus. *V. incana* sensu auct. Japon., non L.——Tōtei-ran. Erect rhizomatous perennial, white-tomentose on stems, lower side of leaves, and inflorescence; stems terete, 40–60 cm. long, usually simple; leaves opposite, lanceolate or oblanceolate, 5–10 cm. long, 1.5–2 cm. wide, subobtuse to acute, short-toothed toward the tip, the midrib raised beneath; spikes usually simple, solitary, erect, 8–12 cm. long, densely many-flowered, the pedicels 2–3 mm. long; calyx 3–4 mm. long, 4-cleft; corolla blue-purple, 7–8 mm. across; style 6–7 mm. long.——Aug.–Oct. Pine woods near seashores; Honshu (n. Kinki Distr. on Japan Sea side); rare.

2. Veronica sieboldiana Miq. *V. sieboldiana* var. *integrifolia* Maxim. ex Matsum.——Hama-tora-no-o. Perennial with short rhizomes; stems erect, 20–30 cm. long, long spreading-pubescent while young; leaves on innovation shoots long-petiolate, ovate, broadly cuneate at base, toothed, soft-pubescent, the cauline leaves rather coriaceous, oblong, 3–5 cm. long, 1.5–2.5 cm. wide, lustrous above, obtuse, slightly dilated and clasping at base, obscurely short-toothed, lustrous above, the margins slightly recurved; racemes usually solitary, 7–12 cm. long, erect, densely many-flowered, long-pubescent, the pedicels 4–7 mm. long, longer than to as long as the linear bracts; calyx narrowly ovate, ciliate; corolla blue-purple, 6–8 mm. across; capsules deltoid-obcordate, 4–5 mm. long, longer than the calyx; style 5–6 mm. long.——Sept.–Oct. Seashores; Kyushu (Isls. of western coast).——Ryukyus (Kerama Isl.).

3. Veronica linariaefolia Pall. ex Link. *V. paniculata* var. *angustifolia* Benth.; *V. spuria* var. *angustifolia* Benth. ex Makino——Hosoba-hime-tora-no-o. Rhizomes short; stems erect, 40–70 cm. long, often branched in upper part; leaves rather thick, oblanceolate to linear-oblanceolate, 4–8 cm. long, 5–8 mm. wide, acute, loosely few-toothed near tip, with ascending hairs especially on midrib beneath; spikes erect, terminal, densely many-flowered, 10–30 cm. long; calyx about 2 mm. long, 4-cleft, the lobes obtuse, ciliate; corolla about 6 mm. across, blue-purple; capsules slightly longer than the calyx, depressed-orbicular, inflated, retuse; style 6–7 mm. long; seeds

minute.——Aug.–Oct. Honshu (s. Kinki Distr. and westw.), Shikoku, Kyushu.——Korea, Formosa, and temperate e. Asia.

Var. **dilatata** (Nakai & Kitag.) Nakai & Kitag. *V. angustifolia* var. *dilatata* Nakai & Kitag.——Ō-hosoba-tora-no-o. Leaves broadly lanceolate to obovate-oblong.——Occurs with the typical phase.

4. Veronica rotunda Nakai var. **subintegra** (Nakai) Yamazaki. *V. spuria* var. *subintegra* Nakai; *V. komarovii* Monjus.; *V. spuria* var. *paniculata* sensu auct. Japon., non Maxim.——Hime-tora-no-o, Yama-tora-no-o. Rhizomes short; stems erect, 40–80 cm. long, terete, with scattered upwardly curved short soft hairs; leaves opposite, narrowly ovate to broadly lanceolate, 5–10 cm. long, 1.5–2.5 cm. wide, acuminate, acute to acuminate at base, irregularly acute-toothed, nearly glabrous, the nerves beneath ascending-puberulent, sessile or short-petioled; spikes 1 or few, erect, 10–20 cm. long, short-pubescent; calyx 4-cleft, the lobes acute with an obtuse tip; corolla blue-purple, about 8 mm. across; capsules depressed-orbicular or broadly obovate, longer than the calyx. ——Aug.–Oct. Mountains; Honshu, Shikoku, Kyushu; rare. ——Korea, Manchuria, and e. Siberia.

5. Veronica subsessilis (Miq.) Carr. *V. longifolia* var. *subsessilis* Miq.——Ruri-tora-no-o. Stems 40–80 cm. long, with scattered, ascending, soft, short hairs; leaves narrowly ovate or deltoid-ovate, 5–10 cm. long, 2.5–5 cm. wide, acute to very acute, rounded at base, finely toothed, soft-pubescent especially on nerves beneath, sessile or the petioles 3–5 mm. long; spikes 1–4, 10–20 cm. long, pubescent; calyx about 3 mm. long, 4-cleft, the lobes narrowly ovate, prominently ciliate, acute; corolla 8–10 mm. across.——July–Aug. Honshu (Mount Ibuki in Oomi); also planted in gardens.

6. Veronica kiusiana Furumi. Hiro-ha-tora-no-o. Stems erect, 50–70 cm. long, soft-hairy while young; leaves deltoid-ovate or narrowly deltoid, 5–8 cm. long, 3–6 cm. wide, subacute, truncate to shallowly cordate at base, mucronate-toothed, scabrous above, the petioles 1–2.5 cm. long; spikes 10–25 cm. long, simple, erect, densely many-flowered, short-pubescent, the pedicels 3–4 mm. long, longer than the calyx and bracts; calyx about 3 mm. long, 4-cleft, the lobes ovate, acute, ciliate; corolla blue-purple, 6–9 mm. across; style 7–10 mm. long.——Aug.–Sept. Mountains; Honshu (centr. distr.), Kyushu.——Korea.

7. Veronica subincanovelutina Koidz.   *V. longifolia* var. *grayi* F. Schmidt; *V. grayi* Miyabe & Kudo, non Armstr.; *V. longifolia* var. *villosa* Furumi; *V. miyabei* Nakai——Birōdo-tora-no-o, Ezo-ruri-tora-no-o.   Rhizomes short; stems erect, 50–80 cm. long, terete, densely white-pubescent in the upper portion; leaves deltoid-lanceolate to deltoid-ovate, 6–12 cm. long, 1.5–4 cm. wide, acuminate to acute, truncate-cordate to truncate at base, serrulate, loosely white-pubescent or glabrous above, densely white-pubescent beneath, the petioles 1–2 cm. long; spikes few, many-flowered, sometimes solitary, 6–12 cm. long, densely white-pubescent; calyx 3–4 mm. long, 4-cleft, the lobes broadly lanceolate, subacute, with minute capitate hairs on margin, usually loosely pubescent on back; corolla about 7 mm. across, blue-purple; capsules obcordate, about 4 mm. long; seeds elliptic, flat, about 1 mm. long.——July–Sept. Hills and mountains on Japan Sea side; Hokkaido, Honshu (centr. and n. distr.).——Korea.

Var. **glabrescens** (Hara) Yamazaki.   *V. miyabei* var. *glabrescens* Hara; *V. miyabei* var. *hondoensis* Hara——Yama-ruri-tora-no-o.   Leaves nearly glabrous, the nerves beneath with recurved hairs.——Occurs with the typical phase.

8. Veronica denkichiana Honda.   *V. holophylla* var. *maritima* Nakai——Echigo-tora-no-o.   Stems erect, terete, 40–80 cm. long, scattered short-pilose or glabrous; leaves herbaceous, broadly lanceolate to deltoid-lanceolate or sometimes deltoid-ovate, 5–10 cm. long, 1.5–3.5 cm. wide, acute or acuminate, rounded to shallowly cordate or cuneate at base, short-toothed, glabrous, the midrib beneath loosely incurved-puberulent, the petioles 1–3 cm. long; spikes few, sometimes solitary, 8–12 cm. long, densely many-flowered, incurved-puberulent, the pedicels shorter to nearly as long as the calyx; calyx about 3 mm. long, 4-cleft, the lobes obtuse, ciliate with incurved short hairs; corolla blue-purple, about 8 mm. across; capsules broadly obovate, 4–5 mm. long; seeds elliptic, flat, about 1 mm. long.——Aug.–Oct. Near seashores; Honshu (Hokuriku and n. distr., along the Japan Sea).

9. Veronica schmidtiana Regel.   *V. yezoalpina* (Koidz.) Takeda; *V. senanensis* var. *yezoalpina* Koidz.——Kikuba-kuwagata.   Rhizomes rather short; stems 10–25 cm. long, simple, leafy, short-pubescent; rosulate and lower leaves slightly fleshy, long-petiolate, broadly lanceolate to ovate, sometimes ovate-deltoid, 2–4 cm. long, 1–2 cm. wide, obtuse, truncate to broadly cuneate at base, glabrous to thinly pubescent, pinnately incised or parted, the lobes obtuse, 2- to 4-toothed; racemes loosely 10- to 30-flowered, 5–10 cm. long, soft-pubescent; calyx 4–6 mm. long, 4-parted; corolla pale blue-purple, with deeper striations, 10–12 mm. across; capsules obovate, glabrous, retuse, slightly longer to as long as the calyx.——June–Aug. Alpine slopes; Hokkaido.——Sakhalin and s. Kuriles.

Var. **bandaiana** Makino.   *V. bandaiana* (Makino) Takeda; *V. schmidtiana* var. *daisenensis* (Makino) Ohwi; *V. senanensis* Maxim.; *V. senanensis* var. *daisenensis* (Makino) Hara; *V. daisenensis* Makino; *V. schmidtiana* var. *senanensis* (Maxim.) Ohwi——Bandai-kuwagata, Miyama-kuwagata, Daisen-kuwagata.   Leaf-lobes shallow, subacute.——Alpine regions; Honshu (n. and centr. distr. and Mount Daisen in Hoki Prov.).

10. Veronica stelleri Pall. var. **longistyla** Kitag.   *V. yesoensis* Nakai——Ezo-hime-kuwagata.   Rhizomes short-creeping; stems somewhat tufted, slender, erect, 7–15 cm. long, white-pubescent, with 5–8 pairs of leaves; leaves ovate to broadly so, 1.5–2.5 cm. long, 8–15 mm. wide, subobtuse, rounded at base, few-toothed, slightly pubescent, sessile; racemes loosely few-flowered, long-pubescent, the upper bracts small, oblanceolate, shorter than the pedicels, the lower bracts leaflike; calyx 4-parted, 4–6 mm. long, the segments oblanceolate, obtuse; corolla pale blue-purple, 7–8 mm. across; style slender, 3–6 mm. long, longer than the calyx.——July–Aug. Alpine slopes; Hokkaido.——Sakhalin, s. Kuriles, and n. Korea. The typical phase occurs in Kamchatka, Kuriles, Aleutians, and Alaska.

11. Veronica nipponica Makino.   *V. stelleri* sensu auct. Japon., non Pall.——Hime-kuwagata.   Rhizomes slender, short, branched; stems 7–12 cm. long, simple, with 4–6 pairs of leaves, short-pubescent; leaves ovate or narrowly so, sometimes oblong, 1–2(–2.5) cm. long, 6–10 mm. wide, obtuse, rounded to obtuse at base, slightly pubescent, short-toothed, sessile or nearly so; racemes few-flowered, usually short-pubescent, the lower bracts leaflike; calyx 4-parted, 4–6 mm. long, the segments oblanceolate, obtuse; corolla pale blue-purple, 4–5 mm. across; style 1(–2) mm. long, shorter than the calyx; capsules flat, broadly obovate, retuse, loosely short-pilose; seeds elliptic, flat, about 1 mm. long.——July–Aug. Alpine slopes in mountains of Japan Sea side; Honshu (centr. and n. distr.).

Var. **shinanoalpina** Hara.   *V. stelleri* sensu auct. Japon., pro parte, non Pall.——Shinano-hime-kuwagata.   Capsules entire, obtuse, not retuse.——Alpine slopes in mountains; Pacific side of Honshu (centr. and n. distr.).

12. Veronica serpyllifolia L.   Ko-tengu-kuwagata. Delicate creeping perennial herb; stems to 10 cm. long, puberulent; leaves elliptic, 5–10 mm. long, 4–7 mm. wide, obtuse to rounded, obscurely few-toothed, nearly glabrous, sessile or very short-petiolate; racemes terminal, 2–5 cm. long, loosely many-flowered, the bracts lanceolate, longer than the pedicels; calyx about 3 mm. long, as long as the pedicels, the lobes oblong, obtuse, nearly as long as the capsules; corolla about 3 mm. across, pale blue-purple; capsules retuse, about 3 mm. long, about 4 mm. across; style about 2 mm. long, slender; seeds elliptic, flat, about 1 mm. long.——July. Naturalized in Honshu (mountains of Kozuke Prov.).——Europe.

13. Veronica tenella All.   *V. humifusa* Dicks.; *V. serpyllifolia* var. *humifusa* (Dicks.) Vahl; *V. serpyllifolia* sensu auct. Japon., non L.——Tengu-kuwagata.   Resembles the preceding but slightly larger and the stems longer creeping, loosely puberulent, 10–20 cm. long; leaves ovate to broadly so, 7–15 mm. long, 6–13 mm. wide, obtuse, rounded to very obtuse at base, nearly glabrous, obscurely toothed, sessile or very short-petiolate; racemes terminal, 5–10 cm. long, loosely 10- to 20-flowered, often glandular-puberulent, the bracts lanceolate, shorter than the ascending pedicels; calyx lobes 4, about 3 mm. long, oblong, obtuse, nearly as long as the capsules; corolla pale blue, 5-7 mm. across; capsuels 4–5 mm. across; style about 3 mm. long.——June–Aug. Woods and grassy places in coniferous woods; Hokkaido, Honshu (centr. and n. distr.).——Korea, Sakhalin, Kuriles, Siberia, Europe, and N. America.

14. Veronica peregrina L.   Mushi-kusa.   Glabrous to glandular-puberulent slightly fleshy biennial herb; stems tufted, branched from base, 5–20 cm. long; leaves linear to lanceolate, 1.5–2 cm. long, 3–5 mm. wide, obtuse to subacute, few-toothed or subentire, sessile; racemes loosely flowered, leafy bracteate, the pedicels about 1 mm. long; calyx 3.5–4.5 mm. long, the segments narrowly lanceolate, subobtuse; corolla pale rose, 2–3 mm. across; style about 0.3 mm. long; capsules depressed-orbicular, retuse, shorter than to as long as the calyx; seeds minute.——Apr.–May. Wet places especially common along riverbanks in lowlands.——Honshu, Shikoku, Kyushu.——Eurasia and N. America.

Var. **xalapensis** (H.B.K.) St. John & Warren. *V. peregrina* var. *pubescens* Honda; *V. xalapensis* H.B.K.——KE-MUSHI-KUSA. Stems and capsules puberulent.——Occurs with the typical phase.

15. Veronica arvensis L. TACHI-INU-NO-FUGURI. Annual or biennial short-pubescent herb; stems terete, slender, erect, 10–30 cm. long, often branched at base; leaves opposite, ovate to broadly so, or ovate-orbicular, 1–2 cm. long, 7–15 mm. wide, obtuse, rounded and sessile at base, with few obtuse teeth, nerved from the base; floral leaves alternate, gradually smaller above, broadly lanceolate, obtuse, longer or sometimes in the upper ones slightly shorter than the calyx, the pedicels less than 1 mm. long; calyx segments 4, broadly lanceolate, 4–6 mm. long, longer than to as long as the capsules; corolla blue, about 4 mm. across; capsules obcordate, deeply notched at apex, about 4 mm. across.——Apr.–Oct. Naturalized in our area.——Europe, Africa, and Asia.

16. Veronica persica Poir. *V. tournefortii* Gmel., non Vill.; *V. buxbaumii* Tenore, non F. W. Schmidt——Ō-INU-NO-FUGURI. Soft-pubescent annual or biennial; stems 10–30 cm. long, loosely leafy, from a creeping or ascending branched base; leaves opposite, deltoid or broadly ovate-deltoid, 1–2 cm. long and as wide, rounded at the base, coarsely toothed, short-petiolate; upper leaves alternate, broadly ovate, subsessile; flowers solitary in axils, the pedicels 1.5–4 cm. long, slender, longer than the leaves; calyx 6–10 mm. long, the segments narrowly ovate, obtuse, slightly longer than and appressed to the capsule; corolla 7–10 mm. across, blue, with darker striations; capsules depressed-obcordate, 5 mm. long, about 1 cm. across, shallowly and broadly 2-lobed, with prominent raised reticulations; style about 3 mm. long; seeds elliptic, about 1.5 mm. long, finely rugose, rather thick, concave on the ventral face.——Apr.–May. Widely naturalized in our area.——Eurasia and Africa.

17. Veronica hederaefolia L. FURASABA-SŌ. Pubescent annual or biennial; stems 5–15 cm. long, ascending to erect from a decumbent branched base; lower leaves opposite, petiolate, the upper ones alternate, subsessile, depressed-cordate or depressed-orbicular, 7–10 mm. long, 8–12 mm. wide, shallowly 5 (3 or 7)-lobed, the lobes entire, the terminal lobe larger; pedicels longer than to as long as the leaves; calyx 4–5 mm. long, the segments membranous, ovate or ovate-deltoid, prominently ciliate; corolla 2–2.5 mm. across, pale blue; capsules depressed-globose, with 4 shallow grooves, glabrous; style nearly 1 mm. long; seeds elliptic, 2.5 mm. long, black-brown, rather thick, convex and transversely rugose on the back, deeply concave on the ventral face.——Mar.–May. Sparingly naturalized in our area.——Eurasia and Africa.

18. Veronica didyma Tenore var. **lilacina** Yamazaki. *V. caninotesticulata* Makino; *V. agrestis* auct. Japon., non L.; *V. polita* auct. Japon., non Fries——INU-NO-FUGURI. Rather short pubescent annual or biennial; stems 5–15 cm. long, erect from a short-decumbent or ascending branched base; leaves opposite in lower ones, alternate in the inflorescence, ovate-orbicular or

depressed-cordate, 6–10 mm. long and as wide, rounded at the base, 5- to 7-lobed, short-petiolate or sessile in the upper ones, the pedicels nearly as long as the leaves; calyx herbaceous, 3–6 mm. long, the segments ovate, few-nerved, obtuse; corolla 3–4 mm. across, rose-purple; capsules reniform, inflated, about 6 mm. across, prominently puberulent; style about 1 mm. long; seeds ovate, about 1.2 mm. long, obsoletely rugose.——Mar.–May. Sunny places and waste grounds in lowlands; Hokkaido, Honshu, Shikoku, Kyushu.——Ryukyus, Formosa, China, and Korea.

19. Veronica onoei Fr. & Sav. MARUBA-KUWAGATA, GUMBAIZURU. Stems rather slender, long-creeping, leafy with long spreading hairs; leaves broadly elliptic, broadly ovate-orbicular, or broadly ovate, 1.5–3 cm. long, 1–2.5 cm. wide, rounded to obtuse, rounded at base, serrulate, glabrous, sparingly ciliate, the petioles 2–5 mm. long; racemes rather densely many-flowered, from leaf-axils, minutely glandular-hairy, the peduncles 3–6 cm. long; calyx 3–4 mm. long, the segments oblanceolate, obtuse; corolla blue, about 8 mm. across; capsules obovate, flat, retuse, loosely glandular-pilose, 5–6 mm. long.——July. Mountains; Honshu (ne. Shinano and Kozuke Prov.).

20. Veronica cana Wall. var. **miqueliana** (Nakai) Ohwi. *V. miqueliana* Nakai; *V. cana* sensu auct. Japon., non Wall. ex Benth.——KUWAGATA-SŌ. Short-pubescent perennial; stems 15–30 cm. long, decumbent at the base; leaves narrowly to broadly ovate, 3–6(–8) cm. long, 2–3.5 cm. wide, acute, toothed, rounded, sometimes shallowly cordate or broadly cuneate at base, the petioles 5–15 mm. long; racemes 3–8 cm. long, pedunculate or subsessile, loosely 3- to 10-flowered, terminal and lateral, short-pubescent, the pedicels suberect, nearly as long as the calyx and the broadly linear bract; calyx 4–6 mm. long, the segments oblanceolate, usually ciliate, acute; corolla pale rose, deeply striate, 8–13 mm. across; capsules depressed-deltoid, flat, loosely ciliolate, nearly as long as the calyx, 5–6 mm. long, 7–10 mm. wide, with a shallow notch at apex; style 3–4 mm. long.——May–June. Woods in mountains; Honshu, Shikoku, Kyushu.

Var. **takedana** Makino. *V. miqueliana* var. *takedana* (Makino) Nemoto; *V. takedana* (Makino) Nakai.——KO-KUWAGATA. Plant dwarf, 5–15 cm. high; leaves 1–2.5 cm. long; racemes 1- to 3-flowered, not exceeding the leaves.——Honshu (centr. and w. distr.), Shikoku, Kyushu.——Formosa.

Var. **decumbens** Makino. *V. japonensis* Makino——YAMA-KUWAGATA, TSURU-KUWAGATA. Stems 5–15 cm. long, from a long-creeping rooting base, pubescent; leaves usually broadly ovate, sometimes ovate, short-toothed; calyx-segments not ciliate; capsules subrhombic-deltoid.——July. Woods in high mountains; Honshu (centr. distr.).

Var. **occidentalis** (Murata) Ohwi. *V. japonensis* var. *occidentalis* Murata. NISHINO-YAMA-KUWAGATA. Similar to the preceding variety; stems appressed-pubescent; calyx-segments acuminate.——Honshu (w. distr.).——The typical phase occurs in the Himalayas and sw. China.

21. Veronica laxa Benth. *V. melissaefolia* auct. Japon., non Poir.; *V. thunbergii* A. Gray——HIYOKU-SŌ. Perennial, rhizomatous, creeping herb, prominently white-pubescent; stems erect, 30–50 cm. long, simple; leaves about 10 pairs, sessile, membranous, ovate, 2.5–4 cm. long, 1.5–3 cm. wide, obtuse, toothed, rounded at base; racemes in the upper leaf axils loosely 10- to 30-flowered, 10–20(–30) cm. long, short-pedunculate, the pedicels as long as to slightly shorter than the calyx and the linear-oblanceolate bract; calyx 3–5 mm. long, the segments broadly oblanceolate, obtuse, longer than the capsule; corolla about 6–8 mm. across, pale rose-purple; capsules depressed-obcordate, 4–5 mm. long, 5–6 mm. wide, flat, broadly and shallowly emarginate, ciliate; style 2.5–3 mm. long.——May–July. Grassy places in mountain valleys; Hokkaido, Honshu, Shikoku, Kyushu.——China and Himalayas.

22. Veronica javanica Bl. *V. murorum* Maxim.; *V. cana* var. *glabrior* Miq.; *V. murorum* var. *glabrior* (Miq.) Maxim.; *V. glabrior* (Miq.) Koidz.——Hama-kuwagata. Annual or biennial; stems erect, 15–35 cm. long, branched in lower part, with a line of short hairs; leaves membranous, opposite, deltoid-ovate or deltoid-cordate, 2–3 cm. long, 1–2.5 cm. wide, subobtuse, shallowly cordate at base, obtusely toothed, loosely pubescent on both sides, the petioles 5–10 mm. long; racemes on the upper part of stems, 3–6 cm. long, loosely few-flowered, incurved-puberulent, the pedicels 1–2 mm. long, suberect, the bracts broadly linear, as long as to slightly longer than the calyx; calyx 3–5 mm. long, the segments oblanceolate, obtuse, longer than the capsule; corolla 2–2.5 mm. across; capsules depressed-cordate, 3 mm. long, 4 mm. across, retuse, loosely ciliate; seeds about 0.5 mm. long.——Apr.–May. Lowlands near the sea; Honshu (Miyake Isl. in Izu Prov.), Shikoku, Kyushu.——Ryukyus, Formosa, China, Malaysia to India and Africa, and S. America.

23. Veronica undulata Wall. ex Roxb.——Kawajisa. Glabrous slightly fleshy annual or biennial; stems terete, erect, 30–60 cm. long; leaves lanceolate or oblong-lanceolate, 4–7 cm. long, 8–15 mm. wide, subacute, rounded to shallowly cordate at base, undulately serrulate, sessile and semiclasping; racemes rather loosely many-flowered, 5–12 cm. long, on short axillary peduncles, the pedicels spreading, 4–6 mm. long, the bracts broadly linear, nearly as long as the pedicels; calyx segments 4, narrowly oblong, 3–4 mm. long, obtuse, slightly longer than to as long as the capsule; corolla about 4 mm. across, pale blue-purple; capsules nearly orbicular, about 3 mm. across; style about 1.5 mm. long; seeds about 0.3 mm. long.——May–June. Wet places in lowlands; Honshu, Shikoku, Kyushu.——Ryukyus, Formosa, China, Korea to Europe, and N. America.

24. Veronica americana (Raf.) Schwein. *V. beccabunga* var. *americana* Raf.——Ezo-no-kawajisa. Glabrous, slightly fleshy creeping perennial; stems terete, 30–50 cm. long; leaves opposite, broadly lanceolate or narrowly ovate, 4–6 cm. long, 1–2.5 cm. wide, rounded at base, short-toothed, the petioles 2–4 mm. long; racemes 5–10 cm. long, axillary, pedunculate, loosely 10- to 20-flowered, the pedicels spreading, 7–12 mm. long, the bracts shorter than the pedicels; calyx 3–4 mm. long, the segments acute; corolla blue-purple, about 6 mm. across; capsules inflated, orbicular, as long as to slightly shorter than the calyx, 3–4 mm. in diameter.——July–Aug. Wet places; Hokkaido.——Sakhalin, Kuriles, e. Siberia, and N. America.

19. VERONICASTRUM Heister ex Fabricius　　Kugai-sō Zoku

Frequently declined, rarely erect perennials or subshrubs; leaves alternate or verticillate, simple, toothed; flowers in dense elongate spikes, ebracteolate, subsessile; calyx 5-parted, the segments acuminate; corolla tubular, blue-purple, the lobes 4, nearly equal, or the upper lobe slightly larger; stamens 2, inserted on the tube, exserted; style slender; ovules many; capsules ovoid-conical, loculicidal or septicidal; seeds many, small.——About 10 species in the temperate regions of the northern hemisphere, especially abundant in the Himalayas, China, Formosa, Philippines, and Japan; one in e. United States.

1A. Leaves alternate; stems declined; spikes axillary, 1–3 cm. long.
　　2A. Leaves and stems rather densely pale-brown pubescent; spikes 1–1.5 cm. long. .... 1. *V. villosulum*
　　2B. Leaves and stems glabrous or very sparsely puberulent; spikes 2–3 cm. long.
　　　　3A. Stems, petioles, and leaves scattered-puberulent. .... 2. *V. tagawae*
　　　　3B. Plant glabrous throughout. .... 3. *V. axillare*
1B. Leaves verticillate; stems erect; spikes terminal, 20–35 cm. long. .... 4. *V. sibiricum*

1. Veronicastrum villosulum (Miq.) Yamazaki. *Botryopleuron villosulum* (Miq.) Makino; *Calorhabdos villosula* (Miq.) Makino; *Paederota villosula* Miq.; *C. axillaris* var. *villosula* (Miq.) Benth. & Hook. f. ex Makino in syn.——Suzu-kake-sō. Plant densely pubescent; rhizomes short; stems terete, leafy, much elongate, declined in upper part, slender; leaves rather thick, ovate, 5–13 cm. long, 3–5 cm. wide, long-acuminate, broadly cuneate at base, with obliquely deltoid acute teeth, short-petiolate; spikes nearly globose, 1–1.5 cm. long, densely many-flowered, sessile, the bracts lanceolate, about 6 mm. long, acuminate; calyx-segments linear-lanceolate, 2.5–4 mm. long; corolla deep violet-purple, the tube 6–7 mm. long, spreading-pilose inside except at base, the limb about 5 mm. across, the lobes acute; anthers about 1.5 mm. long; ovary broadly ovoid; style more than 10 mm. long.——Aug.–Oct. Thickets; Honshu (Mino Prov.); very rare.

2. Veronicastrum tagawae (Ohwi) Yamazaki. *Botryopleuron tagawae* Ohwi; *Calorhabdos tagawae* (Ohwi) Masam.——Kinokuni-suzu-kake. Perennial herb 1–2 m. high; stems terete, arcuate and declined, loosely recurved-puberulent; leaves ovate or oblong, 8–15 cm. long, 3–5 cm. wide, acuminate, broadly cuneate at base, loosely recurved-puberulent beneath and on nerves, with appressed short teeth on margin, short-petiolate; spikes nearly sessile, 2–3 cm. long, densely many-flowered, the bracts lanceolate, acuminate, about 2 mm. long, glabrous; calyx-segments slightly unequal, linear-lanceolate, as long as the bracts, acuminate; corolla blue-purple, about 6 mm. long, the tube pilose inside, the lobes 4, deltoid, acute; anthers about 1 mm. long; style persistent, about 8 mm. long; capsules broadly ovoid, about 2.5 mm. in diameter.——Aug.–Sept. Honshu (Kii Prov.); rare.

3. Veronicastrum axillare (Sieb. & Zucc.) Yamazaki. *Botryopleuron axillare* (Sieb. & Zucc.) Hemsl.; *Calorhabdos axillaris* (Sieb. & Zucc.) Benth. & Hook. f. ex S. Moore; *Paederota axillaris* Sieb. & Zucc.; *P. bracteata* Sieb. & Zucc. ——Tora-no-o-suzu-kake. Glabrous perennial 1–2 m. high; stems terete, slender, often declined in upper part; leaves herbaceous, oblong or ovate, 5–10 cm. long, 2.5–5 cm. wide, acuminate, rounded to broadly cuneate at base, with obliquely deltoid teeth, short-petiolate; spikes axillary, 1.5–3 cm. long, densely many-flowered; flowers blue-purple, nearly sessile, the bracts linear-lanceolate, acuminate, slightly longer than the calyx, smooth or sparingly ciliate; calyx-segments 5, linear-lanceolate, 3–4 mm. long, long-acuminate, glabrous; corolla about 5 mm. long, the tube about 3 mm. long, the lobes acuminate; anthers about 0.8 mm. long; style about 5 mm. long; capsules ovoid-globose, slightly flattened.——Honshu (Tōkaidō Distr. and westw.), Shikoku, Kyushu.——China.

4. Veronicastrum sibiricum (L.) Pennell. *Veronica sibirica* L.; *Leptandra sibirica* (L.) Nutt.; *Veronica virginica* (L.) var. *sibirica* (L.) Nakai; *Veronica virginica* (L.) var. *japonica* (Raf.) Nakai; *Eustachys japonica* Raf.; *Veronica japonica* (Raf.) Steud.; *Veronica virginica* (L.) var. *zuccarinii* Koidz,; *Veronicastrum sibiricum* (L.) var. *yesoense* Hara—— KUGAI-SŌ, KUKAI-SŌ. Short-pubescent to nearly glabrous peren-nial with short rhizomes; stems terete, erect, 80–150 cm. long, simple or sparingly branched in upper part; inflorescence terminal; spikes erect, 20–35 cm. long, very densely many-flowered; leaves verticillate, in 4's to 6's, broadly lanceolate to oblong, 10–15 cm. long, 2–5 cm. wide, acuminate, acute at base, acutely toothed; calyx deeply 5-cleft, the lobes broadly lanceolate, slightly unequal; corolla blue-purple; style and stamens exserted; capsules ovate-conical, 4–5 mm. long, acute, with a groove on each side; seeds about 4 mm. long, oblong.——July–Sept. Grassy places in mountains and lowlands; Hokkaido, Honshu, Shikoku, Kyushu.——Korea, n. China, Manchuria, Sakhalin, and Siberia.

20. LAGOTIS Gaertn.   URUPPU-SŌ ZOKU

Rather fleshy perennials; radical leaves obovate to oblong, entire, incised or undulate-toothed; stems scapelike, erect or de-cumbent at base, few-leaved; spikes terminal, densely many-flowered, globose to cylindric; flowers in spikes, blue-purple, sessile, solitary in axils of bracts; calyx membranous, split in front, sometimes 2- or 3-lobed or -toothed; corolla-tube narrow, broadened on the throat, the limb bilabiate, spreading, the upper lip entire or 2-lobed, the lower lip 3-lobed; stamens 2, in-serted on the throat, shorter than the lobes; ovary 2-locular, 2-ovuled; fruit consisting of 2 nutlets, enclosed in the calyx.—— About 10 species, in the cooler regions of N. Asia, alpine regions of central Asia, and N. America.

1A. Filaments shorter than to as long as the anthers; stamens about half as long as the upper lip of corolla. .... 1. *L. glauca*
1B. Filaments longer than the anthers, inserted at the base on each side of upper lip of corolla, nearly as long as to slightly shorter than the upper lip; stigma entire. .... 2. *L. stelleri* var. *yesoensis*

1. Lagotis glauca Gaertn. *Gymnandra gmelinii* Cham. & Schltdl.——URUPPU-SŌ. Glabrous fleshy short-rhizomatous perennial; radical leaves long-petiolate, deltoid-orbicular to broadly ovate, 4–10 cm. long and as wide, obtuse to rounded, shallowly cordate to rounded at base, deep green, undulately obtuse-toothed, lustrous above, nerveless, long-petiolate; stems scapelike, 10–25 cm. long, rather stout, with few small orbicu-lar-ovate or broadly cuneate leaves 1.5–3 cm. long, arising from a tuft of radical leaves; spikes solitary, terminal, 3–10 cm. long, very densely many-flowered, erect; flowers pale blue-purple, sessile, the bracts narrowly elliptic, longer than the calyx, ob-tuse, with broad hyaline margins; calyx long-pubescent on margin, with 2 broad teeth at apex; corolla 8–10 mm. long. ——Aug. Sunny gravelly or sandy places in alpine regions; Hokkaido (Rebun Isl.), Honshu (alpine regions of centr. distr.); rare.——Kuriles, Sakhalin, Kamchatka, Ochotsk Sea.

Var. **takedana** (Miyabe & Tatew.) Kitam. *Lagotis takedana* Miyabe & Tatew.——YŪBARI-SŌ. Differs from the typi-cal phase in the smaller parts; leaves narrower; stems 10–20 cm. long.——Hokkaido (Mount Yubari); rare.

2. Lagotis stelleri (Cham. & Schltdl.) Rupr. var. **yeso-ensis** Miyabe & Tatew. *L. yesoensis* Tatew.——HOSOBA-URUP-PU-SŌ. Glabrous rhizomatous perennial; stems scapelike, 10–15 cm. long, few-leaved in upper part; radical leaves narrowly ovate or broadly oblong-lanceolate, 4–8 cm. long, 2–4 cm. wide, obtuse, rounded to abruptly narrowed at base, rather fleshy, obtuse-toothed, long-petiolate; cauline leaves ovate or broadly so, 1–2.5 cm. long, obtusely toothed, sessile; spike cylindric, solitary and terminal, 4–6 cm. long, 1.5–2 cm. across, the bracts narrowly ovate, longer than the calyx; flowers dense; calyx membranous, truncate or rounded at apex, pubescent on margin; corolla about 8 mm. long, the lower lip shorter than the upper; stigma entire.——Aug. Alpine regions; Hok-kaido (Mount Daisetsu).

Further study may reveal that these two species are con-specific.

21. MONOCHASMA Maxim.   KUCHINASHI-GUSA ZOKU

Small herbs; stems tufted, leafy; lower leaves scalelike, the upper ones larger, usually opposite, linear-lanceolate, entire; flowers solitary in axils, short-pedicelled, bracteoles 2, inserted at the base of calyx; calyx tubular, 4-lobed, enclosing the capsule, ac-crescent in fruit; corolla tubular, the throat slightly ampliate, the upper lip arcuate, galeate, 2-fid, with recurved margins, the lower lip longer than the upper, 3-lobed, the midlobe longer than the lateral ones and 2-grooved on the throat; stamens inclosed, didymous, inserted on the tube, the filaments filiform; ovary imperfectly 2-locular, many-ovuled; capsules ovoid, acu-minate, loculicidal, enclosed in the calyx; seeds small, many, flat.——Few species, in e. Asia.

1A. Plant white-woolly only at base; stems with elongate internodes; leaves only opposite; corolla about 10 mm. long, shorter than the calyx-lobes. .... 1. *M. japonicum*
1B. Plant prominently white-woolly throughout except on the inflorescence; stems with short internodes; leaves opposite, ternate, or sometimes nearly alternate; corolla 20–25 mm. long, longer than the calyx-lobes. .... 2. *M. savatieri*

1. Monochasma japonicum (Maxim.) Makino. *M. she-areri* var. *japonicum* Maxim. ex Fr. & Sav.——KUCHINASHI-GUSA, KAGARIBI-SŌ. Biennial herb; stems tufted, 10–30 cm. long, nearly terete, reddish, loosely white-woolly at base, nearly glabrous or puberulent in upper part; basal leaves broadly sub-ulate, 1–5 mm. long, entire, sessile, the upper leaves linear, to 4 cm. long, 3(–4) mm. wide, acute, entire, sessile; flowers solitary in axils, the pedicels 3–10(–20) mm. long, puberulent, the bracteoles 7–15 mm. long, nearly leaflike; calyx-tube with 10 raised ribs, the lobes broadly linear, leaflike, 1–2 cm. long, about 2 mm. wide; corolla pale rose-purple, about 10 mm. long; capsules glabrous, narrowly ovoid, acute or short-beaked, about 8 mm. long, enclosed in the calyx; seeds ovoid, many, about 1 mm. long.——May–June. Woods in lowlands and low mountains; Honshu, Shikoku, Kyushu.

2. Monochasma savatieri Franch., ex Maxim.　Usu-YUKI-KUCHINASHI-GUSA.　White-woolly biennial (?); stems tufted, 15–30 cm. long; basal leaves 3–5 mm. long, scalelike;

upper leaves rather many, ascending to obliquely spreading, narrowly lanceolate, 1–2(–3) cm. long, gradually acute, entire; inflorescence green, glandular-pubescent, the bracts leaflike; flowers several, the pedicels 3–5 mm. long, the bracteoles leaflike, 1–1.5 cm. long; calyx-tube with 10 raised ribs, the lobes nearly as long as the tube; corolla 20–25 mm. long, much longer than the calyx-lobes, the tube gradually ampliate, white, with a rosy hue on the upper corolla lobe, the lower lip about half as long as the tube.——Apr.–May. Kyushu (Hondo-mura in Amakusa Isl.).——China.

22. SIPHONOSTEGIA Benth.　Hiki-yomogi Zoku

Usually pilose or pubescent annuals or perennials; stems erect, sometimes branched; lower leaves opposite, the upper ones alternate, entire or pinnately parted; inflorescence spicate; flowers solitary in axils of bracts, sessile or nearly so, 2-bracteolate at the base of calyx; calyx tubular, 10-ribbed, 5-lobed, the linear segments nearly as long as the tube; corolla tubular, yellowish or purplish, the limb bilabiate, the upper lip erect, galeate, entire, the lower lip 3-lobed, with 2 longitudinal crests inside toward the base; stamens 4, didymous, inserted on the tube, ascending under the upper lip; capsules narrowly oblong, subacute, loculicidal; seeds very many, minute.——Few species, in the warmer parts of Asia.

1A. Scabrous annual, without glandular hairs; leaf-lobes slightly broadened at base; calyx-lobes equal, $\frac{1}{4}$–$\frac{1}{2}$ as long as the tube, suberect; corolla clear yellow, the lower lip glabrous inside, the upper lip dark brown on lower margin, acute, with a minute tooth at the tip. . . . . 1. *S. chinensis*
1B. Glandular-pubescent annual; leaf-lobes often prominently broadened at base; calyx-lobes unequal, one slightly smaller than the others, $\frac{1}{2}$–$\frac{2}{3}$ as long as the tube, ascending; corolla grayish yellow, the lower lip villous inside, the upper lip red-brown on upper side, truncate, entire at tip. . . . . 2. *S. laeta*

1. Siphonostegia chinensis Benth.　Hiki-yomogi. Scabrous annual; stems erect, 30–60 cm. long, simple or branched; leaves deltoid, 2–3.5 cm. long, pinnately parted, the segments linear-lanceolate, acute, with 1–3 teeth, the petioles winged, the upper leaves smaller, 3-parted; flowers solitary in axils of the rather leafy spike, very short-pedicelled; calyx narrowly tubular, 12–15 mm. long, 2.5–4 mm. across, prominently ribbed, the bracteoles short; calyx-lobes lanceolate; corolla about 2.5 cm. long, clear yellow, ampliate near the throat, the upper lip loosely long-pubescent on back, the lower lip glabrous inside with 2 raised crests; capsules enclosed in the calyx, as long as the calyx-tube; seeds ovoid, about 0.5 mm. in diameter.——Aug. Meadows and grassy slopes in hills and low mountains; Hokkaido, Honshu, Shikoku, Kyushu.——China, Korea, Manchuria, Formosa, and s. Kuriles.

2. Siphonostegia laeta S. Moore.　*S. laeta* var. *japonica* Matsum.; *S. japonica* (Matsum.) Matsum. ex Furumi——Ō-HIKI-YOMOGI.　Glandular-pubescent annual; stems 30–60 cm. long, much branched in upper part; leaves deltoid, 2.5–4 cm. long, acuminate, pinnately parted, the segments acutely toothed, the petioles winged, the upper leaves ovate, incised or entire; flowers very short-pedicelled; calyx-tube membranous, 12–17 mm. long, ribbed, the lobes lanceolate, unequal; corolla about 2.5 cm. long, grayish yellow, loosely short-pubescent externally, the upper lip red-brown on upper side, truncate, entire at the tip, the lower lip with 2 raised crests, villous inside; anthers obtuse; capsules broadly lanceolate, glabrous, slightly shorter than the calyx-tube.——Aug.–Sept. Honshu (Kantō Distr. and westw.), Shikoku.——China.

23. MELAMPYRUM L.　Mamako-na Zoku

Glabrous or pubescent annuals; leaves opposite, entire or the upper ones incised; flowers yellowish or rose-purple, subsessile, ebracteolate, solitary in axils, often forming a terminal spike, the bracts usually colored; calyx tubular, 5-toothed, the upper teeth often larger than the others; corolla tubular, the limb bilabiate, the upper lip erect, flat, short, galeate, obtuse and often bidentate, the lower lip 3-lobed; stamens 4, didymous, inserted on underside of the upper lip; ovary 2-locular, with 2 ovules in each locule; capsules flat, ovate, oblique, loculicidal; seeds 4 or fewer, strophiolate.——About 40 species, in temperate regions of the N. Hemisphere.

1A. Corolla white on each side of throat within; bracts acute or acuminate.
　　2A. Bracts green; leaves oblong-lanceolate or narrowly ovate, short-acute to acuminate. . . . . 1. *M. roseum*
　　2B. Bracts rose-purple; leaves linear-lanceolate to lanceolate, long-acuminate. . . . . 2. *M. setaceum*
1B. Corolla orange-yellow on each side of throat within; bracts obtuse. . . . . 3. *M. laxum*

1. Melampyrum roseum Maxim.　*M. henryanum* Soó; *M. jedoense* Miq.; *M. roseum* var. *typicum* Fr. & Sav.——Tsushima-mamako-na.　Stems 4-angled, erect, 30–50 cm. long, with ascending branches, puberulent; median leaves narrowly ovate to oblong-lanceolate, 5–7 cm. long, 1.5–2.5 cm. wide, acuminate, rounded at base, entire, loosely puberulent on

both sides, the petioles 7–10 mm. long; spikes elongate, the bracts green, smaller than the median leaves, with few soft spines on margins; flowers 15–20 mm. long; calyx 3–4 mm. long, glabrous or sparsely puberulent, the lobes acute; corolla red, granular-puberulent externally, hairy inside; capsules ovate, acuminate, about 8 mm. long, densely puberulent toward the top; seeds ellipsoidal, black, about 3 mm. long, strophiolate at base.——Aug.–Oct. Honshu (centr. distr. and westw.), Kyushu.——Korea, Manchuria, and China.

Var. **ovalifolium** Nakai ex Beauverd. *M. ovalifolium* Nakai; *M. roseum* subsp. *ovalifolium* (Nakai) Beauverd; *M. aristatum* (Beauverd) Soó——MARUBA-MAMAKO-NA. Spikes usually congested, the bracts long-acuminate, with rather numerous spreading soft spines on margin; calyx glabrous or pubescent, the teeth usually acuminate.——Kyushu (Iki and Tsushima).——Korea.

Var. **japonicum** Fr. & Sav. *M. ciliare* Miq.; *M. jedoense* var. *luxurians* Miq.; *M. roseum* var. *ciliare* (Miq.) Nakai; *M. roseum* subsp. *japonicum* Nakai ex Matsum., incl. vars. *genuinum, salicifolium,* and *resupinatum* of Nakai; *M. roseum* var. *sendaiense* Beauverd.——MAMAKO-NA. Similar to var. *ovalifolium* but densely long-pubescent on axis of spikes and calyx.——Hokkaido, Honshu, Shikoku, Kyushu; common. ——Korea.

2. **Melampyrum setaceum** (Maxim.) Nakai. *M. roseum* var. *setaceum* Maxim. ex Palib.; *M. roseum* var. *alpinum* Kitam.; *M. kawasakianum* Kitam.——HOSOBA-MAMA-KO-NA. Stems slightly angled, erect, 30–60 cm. long, usually much branched, terete, loosely puberulent; median leaves linear-lanceolate to lanceolate, 2–4 cm. long, 3–4 mm. wide, long-acuminate, thinly short-pubescent, short-petiolate; bracts linear to broadly deltoid-lanceolate, prominently setose on margin, the upper ones rose-purple; calyx 3–4.5 mm. long, puberulent to glabrous, the lobes acuminate, slightly awn-tipped; corolla red, 15–18 mm. long; capsules 8–9 mm. long, puberulent.—— Aug.–Oct. Honshu (Kinki Distr. and westw.), Shikoku, Kyu-hu.——Korea and Manchuria.

Var. **latifolium** Nakai. *M. nakaianum* var. *latifolium* (Nakai) Tuyama——Ō-HOSOBA-MAMAKO-NA. Broad-leaved phase with leaves 3–4.5 cm. long, 6–8 mm. wide.——Kyushu.

3. **Melampyrum laxum** Miq. var. **nikkoense** Beauverd. *M. australe* forma *edentatum* Tuyama; *M. laxum* forma *nikkoense* (Beauverd) Soó——MIYAMA–MAMAKO-NA. Stems 20–50 cm. long, puberulent, with ascending branches; median leaves broadly lanceolate to narrowly ovate, acuminate, rounded to abruptly acute at base, petiolate; bracts loose, small, obtuse, entire; calyx puberulent or nearly glabrous, about 3 mm. long, the teeth thick, obtuse, short; corolla red, rarely white, 13–18 mm. long; capsules ovate, acuminate, about 1 cm. long, densely puberulent.——Aug.–Sept. Woods in mountains; Hokkaido, Honshu, Kyushu.

Var. **laxum.** *M. australe* Nakai ex Tuyama; *M. roseum* var. *brevidens* Kitam.; *M. laxum* var. *brevidens* (Kitam.) Hara; *M. laxum* forma *australe* Nakai, pro parte——SHIKOKU-MAMAKO-NA, UBA-MAMAKO-NA. Bracts often hastate, more or less setose-dentate on lower margin; corolla 13–18 mm. long, red.——Honshu (centr. distr. and westw.), Shikoku, Kyushu.

Var. **arcuatum** (Nakai) Soó. *M. arcuatum* Nakai—— TAKANE-MAMAKO-NA. Stems 15–20 cm. long; leaves ovate-elliptic to narrowly ovate, 2–4 cm. long, 7–15 mm. wide; bracts loose, broadly lanceolate, small; calyx-lobes rounded; corolla yellowish, 10–12 mm. long.——July–Aug. Damp woods in high mountains; Honshu (Kai, s. Shinano, and w. Musashi Prov.).

Var. **yakusimense** (Tuyama) Kitam. *M. yakusimense* Tuyama——YAKUSHIMA-MAMAKO-NA. Resembles var. *arcuatum,* but with broader, ovate or oblong bracts and red corollas.——Kyushu (Yakushima).

24. EUPHRASIA L. KOGOMEGUSA ZOKU

Branched or simple erect annuals or perennials; leaves small, opposite, cuneate, sessile, toothed or incised, or pinnately lobed; flowers sessile or short-pedicelled, in axils of leaflike bracts, white, yellowish, or rarely purplish, ebracteolate; calyx tubular or campanulate, 4-cleft, the lobes obtuse or acute; corolla-tube ampliate, the limb bilabiate, the upper lip erect, shallowly 2-lobed or bidentate, the lower lip spreading, 3-lobed, the lobes obtuse or retuse; stamens 4, didymous; capsules oblong or obovate, flattened, loculicidal; seeds longitudinally striate.——About 100 species, in cooler regions of the N. Hemisphere, Australia, S. America, and in mountains of Malaysia.

1A. Leaves narrowly obovate to narrowly oblong, more than twice as long as wide.
  2A. Corolla large, 7–10 mm. long on back.
    3A. Calyx less than half as long as the corolla. . . . . 1. *E. insignis*
    3B. Calyx half as long as the corolla or longer.
      4A. Bracts with acute teeth. . . . . 2. *E. pubigera*
      4B. Bracts short awn-toothed. . . . . 3. *E. hachijoensis*
  2B. Corolla small, 5–6 mm. long on back. . . . . 4. *E. microphylla*
B. Leaves orbicular, broadly obovate, or rhombic-elliptic, less than twice as long as wide.
  5A. Corolla large, 7–10 mm. long.
    6A. Plants 2–10 cm. high, usually simple, rarely sparingly branched.
      7A. Hairs not curled or glandular. . . . . 5. *E. yabeana*
      7B. Hairs curled and glandular on upper part. . . . . 6. *E. matsumurae*
    6B. Plants 10–30 cm. high, usually much branched.
      8A. Leaves scarcely scabrous; corolla 8–10 mm. long; lower lip of corolla much longer than the upper. . . . . 7. *E. iinumae*
      8B. Leaves prominently scabrous; corolla 7–8 mm. long; lower lip of corolla slightly longer than the upper. . . . . . . 8. *E. multifolia*
  5B. Corolla small, 4–5(–6) mm. long; lower lip nearly as long as to slightly longer than the upper lip; teeth of leaves and calyx-lobes acuminate, short awn-tipped. . . . . 9. *E. maximowiczii*

1. **Euphrasia insignis** Wettst. MIYAMA-KOGOMEGUSA. Stems 7–15 cm. long, simple or branched, white-puberulent; median leaves narrowly obovate, 6–12 mm. long, 3–6(–7) mm. wide, cuneate at base, with 2 or 3 acute or subacute teeth on each side, sessile; leaves of the inflorescence rhombic-obovate or obovate, scabrous on upper side, with 2 or 3 very acute or mucronate teeth; calyx 4–6 mm. long, 4-lobed, the lobes linear-lanceolate, very acuminate; corolla 8–10 mm. long, the lobes

of lower lip shallowly 2-lobed; capsules obovate, retuse, appressed-pilose on margin; seeds 4–6, about 1.5 mm. long.——Aug.–Sept. High mountains; Honshu (centr. distr.).

Var. **togakusiensis** (Y. Kimura) Y. Kimura. *E. koidzumii* var. *togakusiensis* Y. Kimura——TOGAKUSHI-KOGOMEGUSA. Teeth of leaves acute, those of bracts awn-tipped; calyx-lobes short awn-tipped.——Honshu (Mount Togakushi in Shinano Prov.).

Var. **japonica** (Wettst.) Ohwi. *E. japonica* Wettst.——HOSOBA-KOGOMEGUSA. Calyx-lobes acute or subacute, sometimes very acute, not cuspidate.——Honshu (Hokuriku Distr.).

2. **Euphrasia pubigera** Koidz. MATSURA-KOGOMEGUSA. Stems 6–12 cm. long, white-pilose; median leaves ovate, acute, cuneate toward the base, glabrous, with 2–3(–4) acute teeth on each side, the bracts leaflike, scabrous on margin; calyx about 4 mm. long, the lobes broadly linear, as long as the tube, scabrous on margin; corolla prominently puberulent externally, about 7.5 mm. long, the lower lip deeply 3-lobed, the lobes retuse, ciliolate; style scattered pilose at the tip.——July. Honshu (Mount Odaigahara in Yamato).

3. **Euphrasia hachijoensis** Nakai. HACHIJŌ-KOGOMEGUSA. Stems 8–15 cm. long, white-puberulent; median leaves narrowly ovate to narrowly obovate, 6–10 mm. long, 3–5 mm. wide, cuneate at base, sparingly scabrous on margin, with 3(–4) acute teeth on each side, the bracts quite numerous, similar to the median leaves but usually narrowly obovate, smooth, short awn-toothed; calyx about 5 mm. long, nearly smooth, 4-lobed, the lobes linear-lanceolate, long-acuminate and awn-tipped; corolla 7–8 mm. long, the lower lip distinctly longer than the upper, the lobes retuse; capsules narrowly oblong-obovate, slightly shorter than the calyx, about 6-seeded, loosely ciliolate.——Aug. Honshu (Mount Hachijofuji on Hachijo Isl.).

4. **Euphrasia microphylla** Koidz. NAYONAYO-KOGOMEGUSA. Stems 5–10 cm. long, very slender, white-puberulent, simple or branched; median leaves narrowly obovate, 4–8 mm. long, 2–3(–4) mm. wide, obtuse, long-cuneate toward the base, 1- or 2-toothed on each side; bracts obovate, 3–5 mm. long, 2–3 mm. wide, 1- or 2-toothed on each side; calyx deltoid, 2.5–3.5 mm. long, shallowly toothed, the lobes obtuse; corolla 5–6 mm. long, the lower lip slightly longer than the upper, the lobes shallowly retuse; capsules obovate, nearly as long as the calyx, retuse.——Aug. Alpine regions; Shikoku.

5. **Euphrasia yabeana** Nakai. HINA-KOGOMEGUSA. Stems 2–10 cm. long, simple or sparingly branched, white-puberulent, few-leaved; leaves broadly obovate or broadly ovate, 3–7 mm. long, 2–5 mm. wide, obtuse, acute to cuneate at base, nearly glabrous, 1- or 2-toothed on each side, the bracts simulating the leaves but sometimes flabellate; calyx 3–4 mm. long, nearly glabrous, shallowly lobed, the lobes narrowly ovate-deltoid, subacute; capsules obovate, about as long as the calyx, loosely pilose on margin; seeds 4–6.——Aug. Alpine regions; Honshu (sw. part of n. and centr. distr.).

6. **Euphrasia matsumurae** Nakai. KOBA-NO-KOGOMEGUSA. Stems 3–10 cm. long, simple or sparsely branched, white-puberulent at base, glandular crispate-pubescent on the upper portion; leaves of few pairs, nearly glabrous, 1- to 3-toothed on each side, the bracts flabellate or obovate, 5–8 mm. long, 4–6 mm. wide, rounded or obtuse, with 2–4 teeth on each side; calyx 3.5–5 mm. long, shallowly lobed, the lobes deltoid, obtuse; corolla 8–10 mm. long, the lower lip shallowly 2-lobed;

capsules broadly obovate, slightly shorter than the calyx, loosely long-pilose on the margin, about 8-seeded.——Aug. Alpine regions; Honshu (Kantō and centr. distr.).

7. **Euphrasia iinumae** Takeda. KOGOMEGUSA. Stems 10–20 cm. long, branched, white-puberulent; median leaves obovate, rhombic or rhombic-elliptic, 7–10 mm. long, 5–7 mm. wide, obtuse, cuneate at base, glabrous, with 2–4 teeth on each side, the bracts simulating leaves but broader and shorter, the teeth obtuse or subobtuse; calyx 4–5 mm. long, shallowly 4-lobed, the lobes broadly lanceolate, acute or subacute; corolla 8–10 mm. long, the lower lip deeply retuse, much longer than the upper; capsules obovate-oblong, slightly shorter than the calyx, loosely long-pilose on margin, 4- to 6-seeded.——Mountains; Honshu (Mount Ibuki in Oomi Prov.).

Var. **makinoi** (Takeda) Ohwi. *E. makinoi* Takeda——TOSA-NO-KOGOMEGUSA. Leaves broader, the bracts cuspidate or short awn-toothed; calyx-lobes acuminate and sometimes minutely awn-tipped.——Honshu (Suwo Prov.) and Shikoku.

Var. **kiusiana** (Y. Kimura) Ohwi. *E. kiusiana* Y. Kimura——KYŪSHŪ-KOGOMEGUSA. Leaves narrower; teeth of the bracts and calyx-lobes awn-tipped.——Mountains of Honshu (Chūgoku Distr.) and Kyushu.

Var. **idzuensis** (Takeda) Ohwi. *E. idzuensis* Takeda——IZU-KOGOMEGUSA. Bracts with the teeth minutely cuspidate; calyx-lobes usually minutely awn-tipped.——Honshu (Izu Prov.).

8. **Euphrasia multifolia** Wettst. TSUKUSHI-KOGOMEGUSA. Stems 20–30 cm. long, white-puberulent, branched above the middle; leaves rather numerous, the median ones ovate to broadly so, or elliptic, 7–11 mm. long, 4–6 mm. wide, obtuse, obtuse to rounded at base, scabrous on upper surface near margin, with 3 or 4 teeth on each side, the bracts simulating the leaves but somewhat smaller, prominently scabrous on upper surface near margin, sometimes cuspidate-toothed; calyx deltoid-lanceolate, acuminate, prominently scabrous; corolla prominently puberulent, (6–)7–8 mm. long, the lower lip slightly longer than the upper, the lobes retuse; capsules obovate-oblong, ciliate, slightly shorter than the calyx.——Sept.–Oct. Honshu (Shinano Prov. and Chūgoku Distr.), Kyushu.

Var. **kirishimana** Y. Kimura. KUMOI-KOGOMEGUSA. Teeth of the leaves more acute.——s. Kyushu.

9. **Euphrasia maximowiczii** Wettst. *? E. officinalis* var. *nervosa* Miq.——TACHI-KOGOMEGUSA. Stems erect, 20–40 cm. long, white-puberulent, branched in upper part; median leaves broadly ovate or nearly ovate-orbicular, 6–12 mm. long, 5–10 mm. wide, obtuse, rounded at base, glabrous or thinly pilose on the nerves beneath, scabrous near the margin, the teeth obtuse in the upper leaves, acuminate and often awn-pointed in the lower, the bracts small, simulating the leaves, scabrous, the teeth acuminate, awn-tipped; calyx 3.5–5 mm. long, shallowly lobed, the lobes deltoid-lanceolate, acuminate, usually awn-tipped, scabrous; corolla 4–5(–6) mm. long, the lower lip 3-lobed, slightly longer to as long as the upper lip, the lobes retuse; capsules obovate-oblong, slightly shorter than the calyx, loosely ciliate, about 12-seeded.——Aug.–Oct. Mountains; Hokkaido, Honshu, Shikoku, Kyushu.——Korea and Manchuria.

Var. **yezoensis** Hara. *E. yezoensis* Hara; *E. maximowiczii* Wettst. pro parte——EZO-KOGOMEGUSA. Copiously coarse whitish hairy; bracts with acuminate but not awn-tipped teeth.——Hokkaido, Honshu. (n. distr.)——s. Kuriles.

25. PHTHEIROSPERMUM Bunge Ko-shiogama Zoku

Glandular-pubescent annuals or biennials; leaves opposite, pinnately dissected; flowers solitary in upper axils, short-pedicelled, ebracteolate, usually reddish; calyx campanulate, 5-lobed, the upper lobe smaller; corolla-tube rather broad, ampliate, the limb bilabiate, the upper lip erect, short, broadly 2-lobed, the lower lip elongate, spreading, 3-lobed, with 2 raised points inside at base, open on throat; stamens 4, didymous, inserted below the upper lip; style spathulate, shortly 2-fid; capsules flat, beaked, loculicidal, somewhat asymmetrical; seeds many, ovoid, reticulate.——Few species, in Japan, China, and the Himalayas.

1. Phtheirospermum japonicum (Thunb.) Kanitz.

Gerardia japonica Thunb.; *P. chinense* Bunge——Ko-shio-gama. Glandular-hairy annual herb; stems simple or usually branched, erect, 30–60 cm. long; leaves deltoid-ovate, 3–5 cm. long, 2–3.5 cm. wide, acute, pinnately divided, serrulate, petiolate, the lateral segments usually obtuse, the upper leaves gradually smaller; calyx obliquely 5-lobed, 5–7 mm. long, slightly accrescent after anthesis, the lobes green, oblong, toothed; corolla pale rose-purple, about 2 cm. long; capsules narrowly ovate, 8–12 mm. long, 4–6 mm. wide, oblique, much longer than and exserted from the calyx, glandular-pubescent; seeds many, elliptic, about 1 mm. long, with winglike longitudinal and horizontal markings.——Sept.–Oct. Meadows and thickets in lowlands and low mountains; Hokkaido, Honshu, Shikoku, Kyushu.——Formosa, China, Korea, and Manchuria.

26. PEDICULARIS L. Shiogama Zoku

Perennials or rarely biennials; leaves opposite, alternate, or verticillate, toothed or pinnatilobed, sometimes pinnately compound; floral bracts small, sessile; inflorescence spicate; flowers solitary in the bracts, white, yellowish, or reddish; calyx tubular or campanulate, often split in front or back, 2- to 5-toothed; corolla-tubular, ampliate, the limb bilabiate, the lower lip spreading, 3-lobed or undivided, the upper lip or helmet erect, galeate, entire or with a tooth on each side, or beaked at apex; stamens 4, didymus, inserted inside the upper lip; capsules flat, ovate or lanceolate, oblique, usually beaked, loculicidal; seeds many, often reticulate or longitudinally striate.——About 500 species, in temperate regions of the N. Hemisphere and S. America, especially abundant in sw. China, centr. Asia, and the Himalayas.

1A. Helmet (upper lip of corolla) developed as an incurved beak.
  2A. Leaves alternate or opposite; calyx deeply split in front, oblique, subentire or with 2 or 3 minute teeth on back at apex; lower lip of corolla only barely lobed, longer than the helmet.
    3A. Leaves opposite, pinnately divided; capsules deltoid-lanceolate, 3 to 4 times as long as the calyx. .... 1. *P. keiskei*
    3B. Leaves alternate, rarely opposite, pinnately incised or doubly toothed; capsules narrowly deltoid-ovate or narrowly ovate, 1.5–2 times as long as the calyx.
      4A. Flowers pale yellow; helmet abruptly long and slenderly beaked. .... 2. *P. yezoensis*
      4B. Flowers rose-purple, rarely white; helmet gradually somewhat short and stoutly beaked. .... 3. *P. resupinata*
  2B. Leaves verticillate in 4's (3's to 6's); calyx not split on one side, toothed at apex; lower lip of corolla with 3 nearly equal lobes, shorter than to as long as the helmet; corolla rose-purple. .... 4. *P. chamissonis* var. *japonica*
1B. Helmet obtuse or rounded, entire or with a minute tooth on each side at apex, not forming a beak.
  5A. Leaves verticillate in 4's (3's to 6's); flowers rose-purple, rarely white.
    6A. Helmet 6–7 mm. long, about ½ as long as the corolla-tube.
      7A. Stems 5–15 cm. long; spikes densely flowered, 2–3(–5) cm. long; flowers about 15 mm. long; capsules nearly horizontally spreading. .... 5. *P. verticillata*
      7B. Stems 20–40 cm. long; spikes rather loosely flowered, 10–20 cm. long; flowers about 20 mm. long; capsules sometimes slightly deflexed. .... 6. *P. refracta*
    6B. Helmet about 2.5 mm. long, about ¼–⅓ as long as the corolla-tube. .... 7. *P. spicata*
  5B. Leaves opposite or alternate.
    8A. Cauline leaves alternate; stems 5–20 cm. long, rarely to 30 cm. long.
      9A. Stems 5–15 cm. long, rarely to 20 cm. long.
        10A. Flowers rose-purple or deep red.
          11A. Radical leaves bipinnately parted, the segments acute. .... 8. *P. apodochila*
          11B. Radical leaves pinnately parted, the lobes toothed or shallowly incised, obtuse. .... 9. *P. koidzumiana*
        10B. Flowers yellowish, the helmet brown-yellow on upper part. .... 10. *P. oederi*
      9B. Stems 20–30 cm. long; flowers many, yellow. .... 11. *P. venusta* var. *schmidtii*
    8B. Cauline leaves opposite; stems stout, usually 30–60 cm. long.
      12A. Stems unbranched; spikes terminal, solitary; flowers pale rose. ....12. *P. nipponica*
      12B. Stems branched above; spikes (1–) 3–5; flowers red-purple.
        13A. Calyx 5-toothed or split on one side, truncate; lower lip of corolla ascending. .... 13. *P. gloriosa*
        13B. Calyx with 2 lateral lobes; lower lip of corolla spreading or slightly recurved. ....14. *P. ochiaiana*

1. Pedicularis keiskei Fr. & Sav. Seriba-shiogama.

Nearly glabrous slender perennial; stems slender, 30–50 cm. long, sometimes somewhat branched in upper part; radical leaves absent at anthesis, the median ones short-petioled, thinly membranous, narrowly ovate in outline, 4–7 cm. long, 2–4 cm. wide, pinnately divided, the segments broadly lanceolate and toothed, acute, the upper leaf-segments adnate at base to the rachis; flowers sparse, the bracts small, simulating the median leaves; calyx membranous, about 4 mm. long, deeply split in front, obtuse, with 2 or 3 minute teeth at apex on back side; corolla pale yellow, about 2 cm. long, the beak nearly as long as the basal part, the lower lip ascending, minutely 3-lobulate; capsules about 12 mm. long, 3 mm. across near base, gradually narrowed to tip, deltoid-lanceolate in outline, acuminate;

seeds fusiform, smooth, longitudinally striate, subtruncate, with a short rounded point at one end.——Aug.–Sept. Damp woods in high mountains; Honshu (southern half of centr. distr.).

2. Pedicularis yezoensis Maxim. *P. resupinata* var. *glabricalyx* Miq.; *P. glabricalyx* (Miq.) Koidz.——Ezo-shiogama.   Glabrous or nearly glabrous perennial; stems 30–60 cm. long, nearly erect, in small tufts; cauline leaves alternate, broadly deltoid-lanceolate or rarely narrowly ovate, 3–5 cm. long, 1–2 cm. wide, acute, truncate-rounded at base, pinnately double-toothed, short-petiolate; floral bracts leaflike, ovate-deltoid; flowers pale yellow; calyx 4–6 mm. long, glabrous or puberulent, deeply split anteriorly, with 2 or 3 minute teeth at apex; corolla about 2 cm. long, the beak slender, falcate, as long as to slightly longer than basal part, the lower lip ascending, shortly 3-lobulate; capsules oblique, narrowly ovate, about 1 cm. long, abruptly mucronate at apex; seeds fusiform, about 2.5 mm. long, obtuse at both ends, smooth.——Aug.–Sept. High mountains and alpine regions; Hokkaido, Honshu (centr. and n. distr.).——Sakhalin.——Forma **pubescens** (Hara) Ohwi. *P. yezoensis* var. *pubescens* Hara——Birōdo-Ezo-shiogama.   Short-pubescent throughout, especially on the calyx.——Hokkaido.——Sakhalin.

3. Pedicularis resupinata L. *P. leveilleana* Bonati; *P. vaniotiana* Bonati——Shiogama-giku.   Nearly glabrous to short-pubescent perennial; stems ascending at base, becoming erect, 40–100 cm. long, purplish red, loosely leaved; leaves alternate or sometimes subopposite, broadly lanceolate to broadly deltoid-lanceolate, sometimes narrowly oblong-ovate, 5–8 cm. long, 1.5–2.5(–3) cm. wide, acute, rounded to truncate-rounded at base, penninerved, obtusely incised and serrulate, short-petiolate; floral bracts leaflike, the upper ones often ovate-rhombic; calyx 5–10 mm. long, deeply split in front, rounded-truncate and obtusely 2- or 3-toothed at apex on backside, soft-pubescent or sometimes nearly glabrous; corolla about 2 cm. long, red-purple, with a darker helmet, the helmet gradually curved and short-beaked, the lower lip ascending; capsules subovate, 8–12 mm. long, acute; seeds fusiform, about 2.5 mm. long, obscurely striate, smooth, obtuse at both ends.——Aug.–Oct. Meadows in hills and mountains; Hokkaido, Honshu, Shikoku, Kyushu; rather common.——Kuriles, Sakhalin, Kamchatka, Korea, e. Siberia, and China.

Var. **caespitosa** Koidz. Tomoe-shiogama.   Flowers in short densely congested spikes, paler in color, the helmet somewhat deeper colored.——Alpine regions; Honshu (centr. distr.).

Var. **microphylla** Honda. Mikawa-shiogama.   Leaves smaller, narrower, rather densely arranged on usually simple or sparsely branched stems; beak shorter.——Wet boggy places; Honshu (Mikawa Prov.).

4. Pedicularis chamissonis Stev. var. **japonica** (Miq.) Maxim. *P. japonica* Miq.; *P. fauriei* Bonati; *P. japonica* var. *maximowiczii* Nakai, pro parte.——Yotsuba-shiogama. Stems 20–60 cm. long, in small tufts, erect, with sparse lines of short hairs or nearly glabrous; radical leaves withering before anthesis, the cauline ones in 4's (3's to 6's), narrowly ovate or broadly deltoid-lanceolate, 3–5 cm. long, 1.5–3 cm. wide, acute, pinnately divided, nearly glabrous, short-petiolate, the segments 7 to 12 pairs, lanceolate, pinnately incised and toothed; spikes usually solitary, loosely flowered, 5–25 cm. long, the bracts linear, longer than the calyx, toothed or incised, glabrous or pubescent; calyx 5–7 mm. long, 5-toothed;

corolla reddish, about 1.5 cm. long, the helmet dark red, abruptly long-beaked, the lower lip spreading, deeply 3-lobed; capsules nearly elliptic, 8–10 mm. long, 4–5 mm. wide, mucronate; seeds fusiform, about 2.5 mm. long, longitudinally striate and cross-lined.——July–Sept. Alpine regions; Hokkaido, Honshu (centr. and n. distr.)——Sakhalin and s. Kuriles.——The typical phase occurs in n. Kuriles, Kamchatka, and the Aleutians.

5. Pedicularis verticillata L. *P. amoena* sensu auct. Japon., non Adams——Takane-shiogama.   Monocarpic densely pubescent herb; stems 5–15 cm. long, simple or somewhat branched at the base; leaves cauline, verticillate in 4's (2's to 6's), narrowly oblong or ovate-oblong, 2–3 cm. long, 5–10 mm. wide, obtuse, petiolate, pinnately parted to divided, the pinnae 5 to 7 pairs, oblong, toothed; spikes terminal, solitary, head-like or very short, densely flowered, 2–5 cm. long, the bracts longer than the calyx, deltoid, pinnately or palmately lobed or cleft, cuneate at base; calyx membranous, 3–5 mm. long at anthesis, to 8 mm. long in fruit, split to the middle in front, with 10 darker colored nerves, the 5 deltoid teeth minute; corolla red, about 1.5 cm. long, the helmet declined, rounded at apex, entire, the lower lip ascending, 3-lobed; capsules oblique to nearly horizontally spreading, about 15 mm. long; seeds about 3 mm. long, reticulate, with elliptic areolae.——July–Aug. Alpine regions; Hokkaido, Honshu (centr. distr.).——n. Kuriles and Korea to Europe, and N. America.

6. Pedicularis refracta (Maxim.) Maxim. *P. verticillata* var. *refracta* Maxim.——Tsukushi-shiogama.   Plant rather prominently whitish crispate-pubescent; stems erect, tufted or solitary, 20–40 cm. long, simple or branched at base; leaves verticillate in 4's, narrowly oblong to narrowly ovate, the basal ones long-petiolate, the cauline short-petiolate, 3–5 cm. long, 1–1.5 cm. wide, obtuse, pinnately parted or divided, the segments of several pairs, oblong, deeply toothed; spikes 10–20 cm. long in fruit, the bracts as long as the calyx or longer, broadly lanceolate, pinnately lobed; calyx 5–8 mm. long, shallowly split in front, with 10 cross-lined ribs, deltoid-toothed on the back at apex; corolla about 2 cm. long, red; anthers of lower and upper pairs contiguous; capsules horizontally spreading or slightly deflexed, broadly deltoid-lanceolate, about 1.5 cm. long.——May. Meadows in mountains; Kyushu (centr. distr.).

7. Pedicularis spicata Pall. Hozaki-shiogama.   Monocarpic herb, prominently whitish pubescent throughout, especially on the spikes; stems simple or somewhat tufted, often short-branched in upper part; radical leaves withering before anthesis, the median cauline ones short-petiolate, usually verticillate in 4's, lanceolate or deltoid-lanceolate, 3–5 cm. long, about 1 cm. wide, subobtuse, pinnately cleft to parted, the segments serrulate; spikes rather densely flowered, the bracts leaflike, shallowly lobed, shorter than to as long as the flowers, longer than the calyx, the upper bracts broadly ovate, smaller; calyx 3–4 mm. long, undulate-toothed; corolla red, 1–1.5 cm. long, the helmet short, declined, rounded and entire at apex, the lower lip obliquely spreading, 3-lobed; capsules suberect, 6–8 mm. long, obliquely ovate.——July–Sept. Naturalized in Hokkaido (Tokachi Prov.).——Korea, Manchuria, e. Siberia, and n. China.

8. Pedicularis apodochila Maxim. *P. rubens* var. *japonica* Maxim.——Miyama-shiogama.   Stems 5–15 cm. long, rather stout, leafless or 1- or 2-leaved; basal leaves rather many, rosulate, narrowly ovate to ovate-oblong, 4–8 cm. long,

2–3 cm. wide, abruptly acute, long-petiolate, pinnately divided, the segments ovate-oblong, acute, pinnately parted; cauline leaves simulating the basal, alternate, the uppermost smaller; spikes solitary, terminal, erect, densely 10- to 20-flowered, subcapitate or somewhat elongate in fruit, to 4 cm. long, prominently pubescent, the lower bracts leaflike, usually longer than to as long as the flowers, the upper bracts ternate-pinnate, long-attenuate at base, as long as or slightly longer than the calyx; calyx 8–12 mm. long, tubular, membranous, the teeth oblanceolate, 3–5 mm. long; corolla red, 2–2.5 cm. long, ascending, the helmet rounded at apex, with a minute tooth on each side immediately below the apex, the lower lip spreading, shorter than the helmet, deeply 3-lobed; anthers of upper and lower pairs contiguous; capsules 12–18 mm. long, suberect, acute, mucronate; seeds about 3 mm. long, with longitudinal rows of depressed, minute, 4-angled impressed dots.——July–Aug. Alpine regions; Hokkaido, Honshu (centr. and n. distr.).

9. Pedicularis koidzumiana Tatew. & Ohwi. BENI-SHIO-GAMA, RISHIRI-SHIOGAMA.   Rhizomes thick; stems 5–10 cm. long, stout, simple, few-leaved or leafless; radical leaves many, long-petiolate, oblong-ovate, 1.5–3 cm. long, 1–1.5 cm. wide, subobtuse, pinnately divided, the segments of 7 to 10 pairs, oblong, obtuse, toothed; cauline leaves absent or few, similar to the radical; spikes solitary, densely few-flowered, subcapitate, 1–2 cm. long, unbranched, the lower bracts leaflike, much longer than the calyx, the upper ones shorter than the calyx; calyx short-tubular, about 1 cm. long, the teeth oblong-spathulate, few-toothed, obtuse; corolla red, about 3 cm. long, the helmet rounded at apex, with a shallow minute tooth on each side immediately below the apex, the lower lip spreading, shorter than the helmet, 3-lobed; capsules about 1.5 cm. long, erect, mucronate; seeds 2.5–3 mm. long, reticulate.——Hokkaido (Rishiri Isl.).——Sakhalin.

10. Pedicularis oederi Vahl. *P. versicolor* Wahlenb. ——KIBANA-SHIOGAMA.   Plant long-pubescent, becoming nearly glabrous; rhizomes stout, with few membranous scale-like leaves on the neck; stems 7–15 cm. long, rather stout, few-leaved; radical leaves few, petiolate, oblanceolate or oblong-lanceolate, 3–5 cm. long, 8–15 mm. wide, obtuse, pinnately divided, the segments 10 to 18 pairs, ovate-oblong, obtuse, toothed; cauline leaves similar but sometimes smaller; spikes solitary, 3–5 cm. long, simple, densely 10- to 20-flowered, the lower bracts leaflike, the upper ones reduced to a small pinnately lobed blade, with a dilated petiole; calyx 1–1.5 cm. long, tubular, membranous, usually loosely long-pubescent, the teeth 2–4 mm. long, narrowly deltoid, with a small toothed blade at apex; corolla erect, about 2.5 cm. long, yellowish, rather narrow, the helmet obtuse, brownish on upper half, not toothed, entire, the lower lip much shorter than the helmet, spreading, 3-lobed; capsules about 1.5 cm. long, cuspidate; seeds minutely impressed-punctulate, reticulate.——July–Aug.   Alpine regions; Hokkaido.——Kuriles, Sakhalin, and northern regions of the N. Hemisphere.

11. Pedicularis venusta Schangin var. **schmidtii** T. Itō. NEMURO-SHIOGAMA.   Perennial with stout rhizomes; stems (15–)20–30(–50) cm. long, prominently whitish pubescent, few-leaved, simple; radical leaves pinnately divided, long-petiolate, the cauline leaves alternate, similar to the radical, broadly lanceolate or narrowly ovate-oblong, 7–12 cm. long, 3–4 cm. wide, subacute, short-petiolate, pinnately parted, the segments broadly lanceolate to narrowly ovate, 10 to 15 pairs, pinnately cleft or lobed, white-pubescent beneath, mucronately acute-

toothed; spikes solitary, simple, densely many-flowered, 5–12 cm. long, prominently whitish pubescent, the lower bracts rather leaflike, longer than the flowers, pinnately lobed, the lobes incised, the upper bracts slightly shorter than the calyx, palmately lobed, cuneate at base; calyx 7–8 mm. long, erect, deltoid, toothed; corolla pale yellow, about 2.5 cm. long, the helmet shallowly bilobed, minutely toothed, the lower lip much shorter than the helmet, spreading, 3-lobed; capsules cuspidate, obliquely spreading, 1–1.2 cm. long; seeds reticulate.——July. Rocky slopes near the sea; Hokkaido.——s. Kuriles, Sakhalin, and Ochotsk Sea region.

12. Pedicularis nipponica Makino.   *P. praeclara* Franch. ——ONI-SHIOGAMA.   Perennial, prominently white-pubescent on spikes especially while young; stems 30–60 cm. long, rather stout, simple, with few pairs of opposite leaves on lower part; radical and lower cauline leaves ovate to narrowly so, abruptly acute, narrowed toward the base, long-petiolate, pinnately parted or cleft, the segments incised or pinnately lobed, acutely serrulate, the upper leaves gradually smaller, becoming bracteate; spikes rather loose, 5–25 cm. long, the bracts broadly obovate, 1–1.5 cm. long, doubly toothed; calyx 8–12 mm. long, toothed; corolla pale rose, suberect, about 3.5 cm. long, the helmet erect, subobtuse, villous on upper margin, the lower lip ascending, as long as the helmet, shallowly 3-lobed. ——Aug.–Oct.   Wet places in mountains; Honshu (Hokuriku and n. distr.).

13. Pedicularis gloriosa Biss. & Moore.   *P. sceptrum-carolinum* forma *japonica* Miq.——HANKAI-SHIOGAMA.   Long-pubescent on lower part of stems and leaves, glabrate in upper part; stems 50–80 cm. long, often sparsely branched; radical and lower cauline leaves opposite, ovate or broadly so, 10–30 cm. long, 8–15 cm. wide, acute, long-petiolate, pinnately parted or divided, the segments oblong-ovate, 5 to 8 pairs, acute, pinnately cleft or lobed, incised and mucronate-toothed; spikes few or rarely solitary, 2–5 cm. long, about 10-flowered, glabrous or sparsely long-pubescent, the bracts cordate or cordate-orbicular, short-toothed; calyx 7–8 mm. long, with obscure minute teeth, about 3 mm. long; corolla reddish, about 3 cm. long, obliquely spreading, the helmet obtuse, entire, white-villous on the margin, the lower lip slightly shorter than the helmet, obliquely spreading, shallowly 3-lobed; capsules ovate-orbicular, 12–15 mm. long, abruptly narrowed at apex to a reflexed beak.——Aug.–Oct.   Woods in mountains; Honshu (Kantō and e. Tōkaidō Distr.).

Var. **iwatensis** (Ohwi) Ohwi.   *P. iwatensis* Ohwi—— IWATE-SHIOGAMA.——Stems angled on lower portion; calyx-lobes oblong with few short teeth; bracts ovate-deltoid, usually long-pubescent on upper margin.——Honshu (n. distr.).

14. Pedicularis ochiaiana Makino.   *P. gloriosa* var. *ochiaiana* (Makino) Masam.——YAKUSHIMA-SHIOGAMA.   Perennial, long-pubescent at base, glabrous in upper part; stems 20–50 cm. long, loosely few-leaved on the lower half; radical leaves long-petiolate, the cauline opposite, broadly deltoid-ovate, 7–15 cm. long, 6–12 cm. wide, acute, petiolate, pinnately divided or parted, the segments 5 to 8 pairs, broadly lanceolate, pinnately cleft or lobed, irregularly toothed, the lower segments often petiolulate; upper leaves abruptly smaller; spikes 1 or few, 3–5 cm. long, glabrous, about 10-flowered, the lower bracts rather leaflike, shorter than the flowers, the upper bracts slightly shorter to slightly longer than the calyx, broadly deltoid-ovate, 3-lobed or pinnately lobed, broadly cuneate at base, with a short, broad dilated petiole; calyx 6–7 mm. long, with 2 lateral rounded lobes, obsoletely toothed; corolla obliquely spreading, about 3 cm. long, the helmet gently incurved, obtuse, entire, villous on margin, the lower lip spreading from the base, deeply 3-lobed, the lobes subreflexed; capsules oblique, ovate, to 15 mm. long, with a short reflexed beak; seeds about 4 mm. long, reticulate.——Aug.–Sept. Mountains; Kyushu (Yakushima); rare.

Fam. 178.   **PEDALIACEAE**   Goma Ka   Sesame Family

Annuals or perennials; leaves opposite or the upper alternate, simple, not stipulate; flowers bisexual, zygomorphic; sepals 4 or 5, often connate, sometimes split on one side; corolla often oblique, the limb 5-lobed, the lobes imbricate in bud; fertile stamens 4, rarely 2, alternate with the corolla-lobes, the anthers in pairs, contiguous, longitudinally split; ovary superior or semi-inferior, 1-locular with 2 intruded parietal placentae, or 2- to 4-locular; ovules 1 to many; fruit a capsule or nut, or drupelike; seeds without endosperm.——About 12 genera, with about 60 species in tropical and warmer regions, especially abundant in Africa.

1.   TRAPELLA F. W. Oliv.   Hishi-modoki Zoku

Aquatic herb with elongate slender stems; leaves opposite, the submersed ones narrowed at base, the floating deltoid-orbicular or reniform-cordate, toothed; flowers solitary in axils, long-pedicellate; calyx-tube adnate to the ovary, with 5 hornlike appendages near the apex, these elongate after anthesis, spreading, antennalike; corolla tubular, ampliate, 5-lobed, the lobes imbricate in bud; stamens inserted on the tube, the upper pair sterile, elongate; ovary semi-inferior, 2-locular, the 1 fertile locule with 2 ovules; stigma dilated, shortly bilobed; fruit indehiscent, narrow, 1-seeded; seeds 4-ridged, narrow.——One species.

1. Trapella sinensis F. W. Oliv. *Trapa antennifera* Lév.; *Trapella sinensis* var. *antennifera* (Lév.) Hara; *Trapella antennifera* (Lév.) Glueck; *Trapella sinensis* var. *infundibularis* Glueck——Hishi-modoki.   Plant glabrous; steems slender, elongate; submersed leaves lanceolate to broadly so, loosely toothed, the floating 2–3.5 cm. long, 2.5–4 cm. wide, very obtuse or rounded at apex, undulately toothed, cordate at base, 3-nerved, the petioles 1–3 cm. long; flowers pedicellate, 2–2.5 cm. long, 1.5–2 cm. across, pale rose, sometimes cleistogamous; calyx-teeth about 2 mm. long; fruit cylindric, narrowed at base, often winged, 12–20 mm. long, 3–4 mm. across, with 3–5 spreading spines longer than the fruit, these often hooked.—— July–Sept. Ponds and shallow lakes in lowlands; Honshu, Kyushu.——Korea and China.

Fam. 179.   **OROBANCHACEAE**   Hama-utsubo Ka   Broom-rape Family

Root parasites without chlorophyll; scapes usually simple, often with scalelike leaves toward the base; flowers bracteate, usually in dense spikes, bisexual, zygomorphic; calyx 4- or 5-toothed or variously lobed; corolla often incurved, the limb oblique or bilabiate, the lobes 5, imbricate in bud, the 2 dorsal (inner) ones innermost in bud; fertile stamens 4, didymous, inserted on lower part of tube, alternate with the lobes, the median or 5th one absent or reduced to a staminode, the anthers in pairs, longitudinally split; ovary superior, 1-locular, the placentation parietal; ovules many; capsules usually enclosed in the calyx, 2-lobed; seeds many, minute, with fleshy endosperm.——About 15 genera, with about 150 species. Cosmopolitan.

1A. Flowers on very long ebracteolate basal pedicels; calyx spathelike, split in front nearly to the base, entire; anthers with one well-developed locule. . . . . 1. *Aeginetia*
1B. Flowers sessile or short-pedicellate, in terminal spikes; calyx of separate or variously connate sepals; anthers with 2 well-developed locules.
  2A. Sepals 4 or 5, connate.
    3A. Calyx united, 4- or 5-toothed.
      4A. Calyx cup-shaped, 5-toothed; flowers sessile, without a disc. . . . . 2. *Boschniakia*
      4B. Calyx campanulate, 4-toothed; flowers short-pedicellate, with a disc at base of ovary in front. . . . . 3. *Lathraea*
    3B. Calyx irregularly 2-fid, the 1- or 2-toothed segments free or connate at base in front. . . . . 4. *Orobanche*
  2B. Sepals 2(or 3), free, sometimes much reduced. . . . . 5. *Phacellanthus*

1.   AEGINETIA L.   Namban-giseru Zoku

Stems very short, simple, few-scaled, narrow; flowers few, large, without bracteoles, very long-pedicelled; calyx spathelike, split in front nearly to the base, acute or obtuse, usually entire; corolla-tube broad and elongate, incurved, the limb spreading, slightly bilabiate, the lobes 5, broad, the dorsal (inner) 2 lobes connate at base; stamens not exserted, the anthers with one well-developed locule; placentae 2, free, broad, irregularly lobed; stigma large, peltate; capsules many-seeded; seeds small, impressed-punctulate.——Few species, in s. Asia.

1A. Calyx acute, 1.5–3 cm. long, pale rose-purple striate; corolla-lobes entire. . . . . 1. *A. indica*
1B. Calyx obtuse, (2.5–)3–4(–5) cm. long, pale rose-purple; corolla-lobes minutely toothed. . . . . 2. *A. sinensis*

1. Aeginetia indica L. var. **gracilis** Nakai. *A. indica* sensu auct. Japon., non L.; *A. japonica* Sieb. & Zucc.——Namban-giseru.   Reddish glabrous saprophytic slightly fleshy herb; stems short, branched, scarcely emergent above ground, scales (bracts) few, narrowly deltoid, 5–10 mm. long; flowers horizontal, pale rose-purple, 3–5 cm. long, the pedicels erect, 20–30 cm. long, naked, 1-flowered; calyx spathelike, 2–3 cm. long, obtusely keeled on the back; corolla tubular, the lobes depressed-rounded, spreading, about 7 mm. long; capsules ovoid-globose, 1–1.5 cm. long.——July–Sept. Grassy places in lowlands and low mountains; Hokkaido, Honshu, Shikoku, Kyushu; common.——Ryukyus.

Var. **sekimotoana** (Makino) Makino. *A. sekimotoana* Makino——HIME-NAMBAN-GISERU. Plants 10–20 cm. high; scales narrow, 4–6 mm. long; calyx 15–20 mm. long; corolla 2–3 cm. long.——Mountains; Honshu (Kantō Distr.). The typical phase occurs from India to Malaysia.

2. **Aeginetia sinensis** G. Beck. *A. japonica* sensu auct. Japon., non Sieb. & Zucc.; *A. indica* forma *japonica* Bakhuiz.,

excl. syn. Sieb.——Ō-NAMBAN-GISERU. Plant slightly larger than the preceding; stems and pedicels stouter; pedicels 20–40 cm. long; calyx 2.5–4 cm. long, pale rose-purple, obtuse; corolla 4–6 cm. long, the lobes spreading, depressed-rounded, 8–10 mm. long, 1–1.5 cm. wide, obsoletely serrulate.——July–Sept. Honshu, Shikoku, Kyushu.——Ryukyus, and China.

2. BOSCHNIAKIA C. A. Mey.   ONIKU ZOKU

Stems stout, simple, scaly, terouslike, fissured, and naked at base; spikes stout, many-flowered, solitary, simple; flowers sessile, ebracteolate, solitary in axils of bracts; calyx cup-shaped, with 5 unequal teeth; corolla-tube recurved, the limb bilabiate, the upper lip erect, entire or shallowly bilobed, concave, the lower lip with 3 short lobes; stamens subexserted, the anther locules parallel, equal, not spurred or tuberculate; placentae 2, each deeply 2-lobed; stigma dilated, shallowly 2-fid; seeds minute, sub-globose, reticulate.——Three or four species, in Asia and N. America.

1. **Boschniakia rossica** (Cham. & Schltdl.) Fedtsch. & Flerov. *Orobanche rossica* Cham. & Schltdl.; *B. glabra* C. A. Mey.——ONIKU. Root parasite on *Alnus maximowiczii*; stems stout, yellow-brown, terete, erect, 15–30 cm. long, densely scaly, especially in lower part; scales rather thick, narrowly deltoid, 7–10 mm. long, erect, obtuse, the upper ones ascending or obliquely spreading, with rather membranous margins, glabrous; spikes about half as long as the whole stem, densely

many-flowered, the bracts deltoid or narrowly so, often pubescent on margin at base; calyx truncate, with 3 teeth in front, undulate on posterior margin; corolla erect, recurved, about 15 mm. long, lip loosely ciliate, the upper rounded, the lower small, 3-lobed.——July–Aug. Alpine areas among bushes; Hokkaido, Honshu (centr. and n. distr.).——Sakhalin, Kuriles, Kamchatka, Korea, e. Siberia, and nw. America.

3. LATHRAEA L.   YAMA-UTSUBO ZOKU

Whitish, yellowish, bluish, or sometimes rose-colored root parasites; rhizomes branched, scaly; stems simple, with scalelike leaves; racemes or spikes densely or loosely flowered; flowers short-pedicelled, ebracteolate, with a disc at the base of ovary; calyx campanulate, 4-toothed, the lobes broad, valvate in bud; corolla-tube suberect, the limb bilabiate, the upper lip retuse, sometimes galeate, the lower lip shorter than the upper, truncate or shallowly 3-lobed; stamens shorter than to nearly as long as the upper lip; ovary with 2 bifid placentae; capsules 2-valved; seeds many, small, globose, rugulose.——Few species, in Eurasia.

1. **Lathraea japonica** Miq. *L. nakaharae* Makino——YAMA-UTSUBO. Plants whitish, glabrous or glabrescent, with short branched creeping rhizomes, densely fleshy-scaly; scapes erect, 10–30 cm. long, loosely few-scaled; spikes densely many-flowered, 5–13 cm. long, the bracts membranous, ascending, broadly lanceolate or narrowly ovate, 3–7 mm. long, longer than the pedicel, obtuse; calyx 5–8 mm. long, the teeth deltoid;

corolla tubular, 12–15 mm. long, the upper lip broad, retuse; anthers with a short awn at one end.——Apr.–June. Woods in mountains; Honshu, Shikoku, Kyushu.

Var. **miqueliana** (Fr. & Sav.) Ohwi. *Clandestina japonica* Miq.; *L. miqueliana* Fr. & Sav.——KE-YAMA-UTSUBO. Plant prominently pubescent.

4. OROBANCHE L.   HAMA-UTSUBO ZOKU

Variously colored parasitic herbs; scapes usually more or less fleshy, scaly, rarely branched; flowers solitary in axils of bracts, sessile or nearly so, bracteolate or ebracteolate, forming a terminal spike; calyx irregularly 2-fid or divided into 2 lateral segments, these 2-lobed or entire, rarely with a well-developed dorsal segment, the lobes membranous, usually acuminate; corolla-tube usually ampliate, the limb more or less bilabiate, the upper lip erect, entire or 2-lobed, the lower lip spreading, 3-lobed, sometimes with tubercles or plaits between the lobes; stamens not exserted, the anthers usually mucronate at base; placentae 4; seeds minute, globose.——About 100 species, in the N. Hemisphere, chiefly in the Old World, few in N. America.

1A. Corolla pale purple, about 2 cm. long, white-woolly externally. .... 1. *O. coerulescens*
1B. Corolla pale yellow with purple striations, about 15 mm. long, short glandular-pubescent externally. .... 2. *O. minor*

1. **Orobanche coerulescens** Steph. ex Willd. *O. nipponica* Makino; *O. japonensis* Makino; *O. akiana* Honda——HAMA-UTSUBO. Plants pale bluish purple, white-pubescent; stems rather stout, 10–30 cm. long inclusive of the spikes, erect, thickened at base; scalelike leaves at base of stem looser, narrowed and membranous above, lanceolate or narrowly ovate, 1–1.5 cm. long, loosely white-pubescent; spikes rather densely many flowered, the bracts narrowly deltoid; flowers about 2 cm. long; calyx membranous, about half as long as the corolla; upper lip of corolla broad, retuse, the lower lip spreading, 3-lobed, with undulate margins; capsules narrowly ellipsoidal, about 1 cm. long.——May–July. Parasitic on *Artemisia*, along

rivers and near seashores; Hokkaido, Honshu, Shikoku, Kyushu.——Formosa, Korea, Ryukyus, China, Siberia, and e. Europe.

2. **Orobanche minor** Sutton. YASE-UTSUBO. Stems simple from a thickened base, 10–40 cm. long; leaves scalelike; flowers loosely racemose on upper part of stem, solitary in axils of caudate-lanceolate bracts; calyx deeply bifid, the segments 2-lobed, lobes unequal; corolla bilabiate, pale yellow, purple-striate, about 15 mm. long, the margin of lobes minutely toothed.——Naturalized in Honshu; root parasite on *Trifolium* spp.——Introduced from Europe.

5. PHACELLANTHUS Sieb. & Zucc.   KIYOSUMI-UTSUBO ZOKU

Glabrous saprophytes without chlorophyll; stems short, simple; leaves scalelike; flowers solitary in axils of bracts, very short-pedicelled, in short dense spikes; bracteoles absent; calyx of 2(or 3) separate, sometimes much-reduced sepals; corolla ascending, tubular, slightly ampliate, deeply 2-lobed, the upper lip broad, erect, entire, the lower lip short, ascending, 3-lobed, glabrous; stamens usually 4, not exserted, the anther locules parallel, without appendages; ovary 2- to 4-lobed, the placentae 4(–10); stigma thick, densely papillose; capsules ovoid.——A single species, in e. Asia.

1. **Phacellanthus tubiflorus** Sieb. & Zucc. *P. continen-talis* Komar.; *Tienmuia triandra* Hu——KIYOSUMI-UTSUBO. Plant white or pale yellowish, glabrous; stems rather thick, 5–10 cm. long; scales ovate or elliptic, 4–8 mm. long, obtuse or rounded at apex, more or less membranous on margins, erect or suberect, the bracts oblong or narrowly ovate, sometimes spathulate-oblong, 1–1.5(–2) cm. long, membranous or nearly so, fragile; flowers about 10, in a terminal solitary short head-like spike, white, changing to yellow in age, ascending to erect, very short-pedicellate; sepals usually 2, linear or filiform, 1–2 cm. long, lateral, free, sometimes partly or wholly absent; corolla (2–)2.5–3 cm. long, the tube rather narrow, the upper lip 7–8 mm. long, flabellate-cuneate, sometimes retuse; capsules about 1 cm. long, ellipsoid-ovoid.——Apr.–July. Hokkaido, Honshu, Shikoku, Kyushu.——Korea, China, Manchuria, Ussuri, and Sakhalin.

Fam. 180. GESNERIACEAE   IWA-TABAKO KA   Gloxinia Family

Herbs sometimes shrubs, rarely trees; leaves radical or opposite, equal or paired and unequal, sometimes one of the pair completely reduced and apparently alternate; flowers bisexual, usually zygomorphic, sometimes nearly actinomorphic, often showy; calyx usually tubular, free or adnate to the ovary, the lobes valvate, rarely imbricate in bud; corolla sometimes more or less bilabiate or nearly regular, the limb usually oblique, the lobes imbricate with the dorsal lobe innermost in bud; stamens 2 or 4, sometimes 5, the anthers usually paired, contiguous or connate, 2-locular, longitudinally split, the disk ringlike or only on one side, sometimes divided; ovary superior or inferior, 1-locular, with 2 parietal or intruded placentae; ovules many; fruit a capsule, rarely a berry; seeds small, many.——About 100 genera, with about 1,500 species, chiefly in the Tropics of both hemispheres.

1A. Fruit a capsule.
  2A. Corolla long-tubular, zygomorphic; fertile stamens 2; capsules linear.
    3A. Small shrubs with evergreen, ternate leaves; seeds caudate at each end. .... 1. *Lysionotus*
    3B. Herbs with radical leaves (in ours); seeds with a minute obtuse point on each end. .... 2. *Opithandra*
  2B. Corolla rotate, actinomorphic; fertile stamens 5; capsules lanceolate; leaves radical, large. .... 3. *Conandron*
1B. Fruit berrylike; fertile stamens 4; subshrubs or herbs with silky hairs; flowers small. .... 4. *Rhynchotechum*

1. LYSIONOTUS D. Don   SHISHIN-RAN ZOKU

Shrubs or subshrubs, often epiphytic; leaves coriaceous or membranous, cauline, ternate, evergreeen, entire or toothed; cymes terminal and in upper axils, sometimes reduced to a single flower, the bracts small, caducous; calyx 5-parted, the segments narrow, the 2 anterior ones connate; corolla purplish or violet, the tube elongate, straight, ampliate, the limb broadly bilabiate, the upper lip 2-lobed, the lower one 3-lobed; fertile stamens 2, anterior, inserted above the middle on the tube, not exserted, the filaments with a tooth at the apex longer than the anthers, the disc annular; style long, the stigmas 2, lamellate, short and broad; placentae connate on the center with the ovules borne on the incurved margins; capsules elongate-linear, 4-valvate; seeds linear, with a capillary appendage at each end.——Few species, in e. Asia and the Himalayas.

1. **Lysionotus pauciflorus** Maxim. SHISHIN-RAN. Small epiphytic glabrous shrub, 5–20 cm. high; stems short, usually simple, 2–3 mm. thick, pale grayish brown; leaves coriaceous, ternate, oblong-lanceolate, 3–6 cm. long, evergreen, 8–15 mm. wide, subobtuse, acute at base, scattered mucronate-toothed, with an impressed midrib above, the lateral nerves obscure, slightly recurved on margin, the petioles 2–3 mm. long; flowers solitary in axils, the pedicels about 1 cm. long, with a subtending caducous bract; sepals linear-lanceolate, about 3 mm. long; corolla about 3 cm. long, pale rose-colored; capsules pendulous, 4–8 cm. long; seeds minute, fusiform, with a tail nearly as long as the body on each end.——Epiphytic on large trees in mountains; Honshu (Izu, Yamato, Oki Prov. and westw.), Shikoku, Kyushu; rare.——Ryukyus (Amami-oshima).

2. OPITHANDRA B. L. Burtt   IWAGIRI-SŌ ZOKU

Acaulescent herb; leaves radical, petiolate, usually toothed; peduncles axillary, scapelike; flowers umbellate, pedicellate; calyx 5-parted; corolla purple-violet, the tube elongate, the limb bilabiate, the upper lip 2-lobed, the lower 3-lobed; fertile stamens 2, extrorse, not exserted, the filaments straight; anther-locules parallel.——One species, in Japan.

1. **Opithandra primuloides** (Miq.) B. L. Burtt. *Chirita primuloides* (Miq.) Ohwi; *Boea primuloides* Miq.; *Didymocarpus primuloides* (Miq.) Maxim.; *Oreocharis primuloides* (Miq.) Benth. & Hook. f. ex C. B. Clarke——IWAGIRI-SŌ. Densely pubescent perennial; rhizomes short; leaves rather fleshy, radical, ovate, elliptic to orbicular, 4–10 cm. long, 3–7 cm. wide, obtuse, cordate, grayish green, coarsely depressed-deltoid and short mucronate-toothed, long-petiolate; inflorescence umbellate, about 10-flowered, the peduncles elongate, 10–20 cm. long, basal, in axils of the radical leaves, leafless, the bracts broadly linear, 2–4 mm. long, the pedicels 2–3 cm. long, slender; flowers horizontal; calyx 5-parted, 4–5 mm. long, the

segments linear-lanceolate; corolla about 2 cm. long, pale purple-violet, with deeper-colored striations, soft-pubescent externally, the upper lip rather short, the midlobe of the lower lip erect, slightly spreading above, the tube narrow, slightly recurved at base; upper 2 stamens fertile; capsules linear, 2.5–4.5 cm. long, glabrous, the style very short, the stigmas 2, persistent; seeds minute, fusiform, smooth.——May–June. Rocks in mountains; Honshu (Kinki Distr. and westw.), Shikoku, Kyushu.

3. CONANDRON Sieb. æ Zucc. IWA-TABAKO ZOKU

Rhizomes short, creeping, densely dark brown long-hairy; leaves radical, few, large, toothed, rugose while young; scapes with few small bracts; cymes terminal; flowers pedicelled, purple, actinomorphic; calyx 5-toothed, the segments narrow, imbricate in bud; corolla rotate, deeply 5-lobed, the lobes narrowly deltoid, subacute, imbricate in bud; stamens 5, all fertile, inserted at base of corolla; filaments short, the anthers erect, oblong, connate into a tube around the style, the anther-locules parallel; disc absent; placentae 2-lobed, prominently intruded and raised within the ovary; ovary imperfectly 2-locular; ovules marginal on the placentae; style linear, the stigma capitellate; capsules lanceolate, rather short, 2-valved; seeds minute, fusiform, smooth.——Two species, in Japan and Indochina.

1. Conandron ramondioides Sieb. & Zucc. IWA-TABAKO. Soft green perennial; scapes slender, 10–30 cm. long, 10- to 40-flowered, the bracts broadly linear; leaves pendulous, solitary, elliptic-ovate, 10–30 cm. long, 5–15 cm. wide, abruptly short-acuminate, lustrous, irregularly mucronate-toothed, wing-petiolate; calyx-segments 4–6 mm. long; corolla about 1.5 cm. across, purple, rarely white, the lobes narrowly deltoid, obtuse, slightly contracted at base, the pedicels 8–15 mm. long, much longer than the bracts; anthers shorter than the narrow apical appendages; capsules broadly lanceolate, 1.5–2 times as long as the calyx, acute.——Aug. Wet rocky cliffs in mountains; Honshu, Shikoku, Kyushu; rather common.——Ryukyus and Formosa.——Forma **pilosum** (Makino) Ohwi. *C. ramondioides* var. *pilosum* Makino ——KE-IWA-TABAKO. Rather short-pilose on scapes, cymes, calyx and sometimes also on nerves of leaves beneath.——Honshu (centr. distr.).

4. RHYNCHOTECHUM Bl. YAMA-BIWA-SŌ ZOKU

Long silky-pubescent perennials or subshrubs; stems erect, low, scarcely branched; leaves alternate or opposite; cymes simple or slightly branched, axillary, pedunculate; flowers small, white; calyx 5-toothed, the segments linear; corolla subrotate or rotate-campanulate, deeply 5-lobed, the lobes equal, broad; stamens 4, inserted at base of corolla, short; staminodes small or absent; disc annular; stigma entire, rather thick; ovules evenly distributed over the placentae except on the inner face at the center; fruit berrylike, small, ovoid to subglobose; seeds minute, ovoid.——About 12 species, in India, Philippines, Formosa, Ryukus, Japan, and China.

1. Rhynchotechum discolor (Maxim.) Burtt var. **austrokiushiuense** (Ohwi) Ohwi. *Isanthera discolor* sensu auct. Japon., non Maxim.; *I. discolor* var. *austrokiushiuensis* Ohwi——TAMAZAKI-YAMA-BIWA-SŌ. Stems 30–40 cm. long, nearly always simple, soft, terete, gray, densely yellowish brown long-woolly; leaves herbaceous, alternate, broadly oblanceolate or obovate-oblong, 10–25 cm. long, 3–8 cm. wide, acute, acuminate at base, mucronate-toothed, rather fleshy, pale gleen above, paler beneath, with yellowish brown long-woolly hairs beneath while young, these persistent on the nerves, densely punctulate beneath, with 8–15 pairs of parallel ascending arcuate lateral nerves, the petioles 2–7 cm. long; peduncles 1.5–3 cm. long; cymes solitary, 20- to 30-flowered, dense and in headlike glomerules, the bracts lanceolate; flowers about 7 mm. across, nearly sessile; calyx 6–8 mm. long, the segments broadly linear; corolla-lobes rounded; filaments longer than the anthers; fruit oblong, 6–7 mm. long.——Aug.-Oct. Woods in low mountains; Kyushu (Osumi, Yakushima and Tanegashima).

Fam. 181. LENTIBULARIACEAE TANUKI-MO KA Bladderwort Family

Herbs, often aquatic or in wet places, insectivorous; leaves alternate or basal and rosulate, often reduced to scales or bladders; flowers bisexual, zygomorphic; calyx 2- to 5-parted, the segments not overlapping or slightly imbricate in bud; corolla spurred at base, the limb bilabiate, the lobes 5, imbricate in bud; stamens 2, inserted at the base of corolla, staminodes 2, the anthers 1- or 2-locular, longitudinally split; disc absent; ovary superior, 1-locular; stigma often sessile; ovules many, on basal placentae; fruit a capsule, 4- or 2-valved or irregularly dehiscent; seeds many, without endosperm.——Four or five genera, with more than 250 species, of wide distribution.

1A. Calyx 2-lipped; corolla personate, the upper lip erect; anthers parallel, nearly 2-locular; leaves finely dissected, with bladders; aquatic or growing in wet muddy places. .... 1. *Utricularia*
1B. Calyx 5-lobed; corolla merely bilabiate with an open throat, the limb spreading; anthers transverse, 1-locular; leaves rosulate, entire; wet rocky cliffs or in bogs. .... 2. *Pinguicula*

1. UTRICULARIA L. TANUKI-MO ZOKU

Soft delicate herbs, aquatic or in wet muddy places; stems usually elongate, leafy; leaves finely dissected into capillary segments, floating or immersed, with small bladders at intervals, or aerial and simple, spathulate, small; scapes simple or slightly branched, erect, rarely elongate and scandent, often scaly, 1- to many-flowered; flowers on a terminal raceme, the bracts small, the pedicels solitary, often 2-bracteolate; calyx 2-toothed or -parted, often slightly accrescent after anthesis; corolla often incurved, spurred, the upper lip erect, entire or 2-lobed, the lower lip usually larger, spreading, 2- to 3-lobed or again lobulate; filaments often thickened, incurved; stigma-lobes lamellate, sometimes reduced to a single lobe; capsules irregularly dehiscent or 2-valved; seeds ovoid, compressed above or lenticular, reticulate, striate, or rugulose, sometimes spinulose.——About 250 species, in wet muddy soil, or in ponds and lakes. Cosmopolitan.

1A. Plants terrestrial, of wet muddy places; aerial leaves linear or spathulate, simple.
  2A. Bracts and scalelike leaves on scapes oblanceolate, dorsifixed; calyx minutely mammillate externally; spur incurved; plant becoming blackish when dry. .... 1. *U. racemosa*
  2B. Bracts and scalelike leaves on scapes lanceolate or ovate-elliptic, basifixed; calyx smooth; spur outwardly curved; plants not becoming blackish when dry.
    3A. Scapes 1–3 cm. long; flowers 2–3 mm. across, nearly sessile; capsules about 1 mm. long, nearly as long as the calyx.
                                                             2. *U. nipponica*
    3B. Scapes 5–20 cm. long; flowers 4–5 mm. across, pedicelled; capsules 3–4 mm. long, shorter than the calyx.
      4A. Flowers pale purple, rarely white; leaves spathulate. .... 3. *U. yakusimensis*
      4B. Flowers yellow; leaves linear. .... 4. *U. bifida*
1B. Plants aquatic; leaves all submerged, finely dissected; aerial leaves absent.
  5A. Leaves of two kinds, the floating ones trichotomously or dichotomously dissected, 5–10(–15) mm. long, the rachis not elongate, the immersed leaves less dissected, bladder-bearing; pedicels not reflexed in fruit; plants growing in mud.
    6A. Floating leaves capillary, with 3–5 segments, nearly bladeless in the immersed ones, each leaf uniformly with 1–3 bladders.
                                                              5. *U. exoleta*
    6B. Floating leaves dissected into 5–10, flat, linear lobes without bladders, the immersed ones with reduced blades and with well-developed bladders.
      7A. Floating leaves with rather narrow entire segments. .... 6. *U. multispinosa*
      7B. Floating leaves with rather broad segments loosely and minutely toothed on upper half. .... 7. *U. intermedia*
  5B. Leaves all alike, 2–5 cm. long, with an elongate rachis, pinnately much-dissected into narrow segments; plants free floating.
    8A. Leaves always with well-developed bladders; scapes without cleistogamous flowers at base.
      9A. Pedicels 6–10(–12) mm. long, deflexed and thickened above after anthesis; scapes without scalelike leaves; spur and upper lip of corolla minutely puberulent. .... 8. *U. pilosa*
      9B. Pedicels 1.5–2.5 cm. long, arcuate after anthesis, not thickened; scapes usually with few scalelike leaves; corolla glabrous.
        10A. Floating stems thicker than the scape, with slender leafless respiratory branches; flowers about 15 mm. across, sterile.
                                                 9. *U. japonica*
        10B. Floating stems more slender than the scape, without respiratory branches; flowers about 10 mm. across, fertile; capsules globose.
          11A. Spur longer than the lower lip, about 8 mm. long; teeth and spinules of leaf-segments often obscure.
                                               10. *U. siakujiiensis*
          11B. Spur shorter than the lower lip; teeth and spinules of leaf-segments distinct. .... 11. *U. tenuicaulis*
    8B. Leaves usually bladderless, occasionally with small bladders; scapes with solitary single cleistogamic flowers at base and on adjacent nodes of floating stems. .... 12. *U. dimorphantha*

1. Utricularia racemosa Wall. ex Walp. *U. caerulea* Wight, non L.; *Calpidisca takenakae* Nakai——HOZAKI-NO-MIMIKAKIGUSA.   Delicate herb of wet muddy places, dimorphic; subterranean stems capillary, with 1 to few remote aerial leaves, these simple, minute, narrowly oblanceolate to narrowly obovate-spathulate; scapes slender, 10–30 cm. long; scales few, loose, appressed to the scape, oblanceolate, 2–3.5 mm. long, long-acuminate at both ends, dorsifixed; flowers 4–10, rather loose, the bracts simulating the scales, the bracteoles linear, 1–1.5 mm. long, as long as the pedicels; calyx broadly ellipsoidal, about 2.5 mm. long, densely mammillate externally; corolla purple, about 4 mm. across, the spur about twice as long as the lower lip, incurved; capsules globose, 2.5–3 mm. across, enclosed in and as long as the persistent calyx; style about 0.5 mm. long; seeds wingless.——June–Sept. Wet muddy places in lowlands and hills; Hokkaido, Honshu, Shikoku, Kyushu; rather common.——Korea, China, Formosa, and India.

2. Utricularia nipponica Makino. HIME-MIMIKAKI-GUSA.   Subterranean stems capillary, loosely bladder-bearing; aerial leaves distant, linear, obtuse; scapes 1–3 cm. long, very delicate, erect, 1- to 3-flowered; scalelike leaves 2 or 3, loose, narrowly deltoid, basifixed, acuminate, about 0.5 mm. long, the bracts basifixed, rather membranous, small, the pedicels as

long as to slightly longer than the bracts; calyx smooth, about 1 mm. long, enclosing the capsule in fruit, the lobes not accrescent after anthesis, rounded and nearly all alike; corolla purplish, 2–3 mm. across, the spur curved forward, longer than the lower lip, obtuse; capsules broadly ellipsoidal, about 1 mm. long, as long as the calyx; seeds minute, smooth, ovoid, wingless.——Aug.–Oct. Wet muddy places in lowlands; Honshu (Tōkaidō and Ise Prov.); rare.

3. Utricularia yakusimensis Masam. *U. yakusimensis* f. *albida* (Makino) Hara; *U. affinis* sensu auct. Japon., non Wight——MURASAKI-MIMIKAKIGUSA.   Glabrous slender herb; subterranean stems capillary, with bladders; scapes erect, very slender, 7–15 cm. long; aerial leaves small, oblanceolate or narrowly so, very obtuse; scalelike leaves few, ovate to narrowly so, membranous, obtuse or acute, 0.5–0.8 mm. long, basifixed; flowers 1–3(–4), loose, the bracts similar to the scalelike leaves, slightly longer, 1–1.5 mm. long, acute, the bracteoles 2, linear, shorter than the bracts, the pedicels erect or suberect, 2–3 mm. long in anthesis, 3–6 mm. long in fruit, winged; calyx-segments membranous, broadly ovate, obtuse, 2–3 mm. long at anthesis, accrescent and 4–5 mm. long in fruit, erect, entire, enclosing the capsule; corolla 3.5–4 mm. across, pale purple, rarely white, the spur 2–3 mm. long, directed downward, obtuse; capsules broadly ellipsoidal or globose, about 3 mm. long, much shorter than the calyx; seeds with raised oblique striations.——Aug.–Sept. Wet muddy places in lowlands and hills; Hokkaido, Honshu, Shikoku, Kyushu.

4. Utricularia bifida L. MIMIKAKIGUSA. Glabrous very slender herb; subterranean stems capillary, with bladders; aerial leaves small, broadly linear to oblanceolate; scapes very slender, erect, 7–15 cm. long; scalelike leaves membranous, few, loose, ovate or narrowly so, obtuse or acute, about 1 mm. long, basifixed; flowers 2–7(–10), loose, the bracts ovate, acute, the bracteoles 2, linear, about half as long as the bracts, the pedicels 2–3 mm. long, arcuate-spreading, 3–7 mm. long in fruit, winged; calyx-segments membranous, broadly ovate, nearly as long as the pedicels, obtuse; corolla 3.5–4 mm. long, yellow, the spur directed downward, about 3 mm. long, acute; capsules globose, about 3.5 mm. long; seeds with raised oblique striations.——Aug.–Oct. Wet muddy places in lowlands and hills; Honshu, Shikoku, Kyushu.——Formosa, China, India, Malaysia, and Australia.

5. Utricularia exoleta R. Br. *U. nagurae* Makino——MIKAWA-TANUKI–MO, ITO-TANUKI–MO. Glabrous, very delicate sedentary aquatic herb, in shallow water; stems circinnate; leaves 5–10 mm. long, 1- to 3-times dichotomously forked, the segments capillary, very narrow, minutely toothed; bladders absent or 1–3 on a leaf, very small; scapes 5–8(–12) cm. long; scalelike leaves 1–3, distant, membranous, ovate, basifixed, 0.5–1 mm. long; flowers 1–3, the bracts broadly ovate, retuse, the pedicels slender, slightly curved, 3–6 mm. long in fruit, terete; calyx membranous, about 2 mm. long, the segments orbicular or broadly obovate, rounded at apex, scarcely accrescent after anthesis, enclosing the lower half of capsule; corolla yellow, 5–6 mm. across, the spur obtuse, curved forward, as long as to slightly longer than the lower lip; capsules depressed-globose, about 3 mm. across; seeds elliptic, about 1.2 mm. across inclusive of the wing, lenticular, the winged-margin undulate-toothed.——Aug.–Nov. Shallow water in paddy fields and ponds in lowlands; Honshu (Tōkaidō and Kinki Distr.), Kyushu; rare.——Formosa, India, Africa, and Australia.

6. Utricularia multispinosa (Miki) Miki. *U. minor* var. *multispinosa* Miki; *U. minor* sensu auct. Japon., non L.——HIME-TANUKI–MO. Delicate sedentary aquatic herb in shallow water; leaves rather sparse, 3- to 4-times dichotomously forked, 8–13 mm. long, with a very short rachis, the segments linear-filiform, slightly flattened, entire, with a minute apical spine in the summer leaves, with 2 or 3 spines in the autumn leaves; bladders few, loose; branches below ground short leaved, with several bladders; scapes about 10 cm. long, slender; scalelike leaves few, scattered, ovate-deltoid, small; flowers loose, 4–8, the bracts similar to the scalelike leaves, the pedicels 4–5 mm. long, obliquely spreading; calyx-segments erect, broadly ovate, 2–2.5 mm. long, rounded at apex, about twice as long as the capsule; corolla pale yellow, about 8 mm. across, the spur short-conical; capsules globose; seeds winged, orbicular.——Shallow ponds; Hokkaido, Honshu.——Kuriles and Korea (?).

7. Utricularia intermedia Heyne. (?) *U. ochroleuca* sensu auct. Japon., non Hartm.——KO-TANUKI–MO. Rather delicate sedentary aquatic herb; leaves rather dense, 4–8 mm. long, 2- to 3-times ternately or dichotomously dissected into linear, flat, acute segments; summer leaves with few spinule-tipped minute teeth on margin and at tip of segments; leaves

absent in subterranean branches, only bladders present; scapes 5–15 cm. long, 2- to 5-flowered; scalelike leaves 1 or 2, small, the bracts ovate, the pedicels 7–10 mm. long, ascending; calyx-segments about 3 mm. long; corolla 12–15 mm. across, pale yellow, the spur cylindric, parallel with the lower lip and as long as the latter.——June–Sept. Shallow ponds especially in high moors; Hokkaido, Honshu.——Kuriles, Korea, and widely spread in the temperate and northern regions of the N. Hemisphere.

8. Utricularia pilosa (Makino) Makino. *U. vulgaris* var. *pilosa* Makino——NO-TANUKI–MO. Floating aquatic herb; stems rather thick, elongate; leaves 3–4 cm. long, much dissected into fine capillary loosely toothed or smooth segments, spine-tipped; bladders prominent, 2–2.5 mm. long; scapes thicker than the aquatic stems, 8–20 cm. long, without scalelike leaves; flowers 4–10, loosely arranged, the bracts thinly membranous, ovate, about 2 mm. long, acute, the pedicels deflexed after anthesis, thickened in upper part, 6–10(–12) mm. long; calyx-segments ovate-elliptic, 2.5–3 mm. long, rounded at apex, accrescent after anthesis, to 4–5 mm. long in fruit, spreading; corolla 6–7 mm. across, pale yellow, the spur obliquely descended, obtuse, slightly shorter than the lower lip, loosely puberulent; capsules globose, 4–5 mm. across; style persistent, as long as the capsule, broadened at base; seeds obtusely 5- or 6-angled.——Aug.–Oct. Shallow ponds and ditches in lowlands; Honshu, Shikoku.——s. Korea.

9. Utricularia japonica Makino. *U. vulgaris* sensu auct. Japon., non L.——TANUKI-MO. Rather stout floating aquatic herb; respiratory branches filiform, nearly leafless; leaves rather densely arranged, 3–6 cm. long, pinnately divided, the segments loosely spine-toothed; bladders few to rather numerous per leaf; scapes 20–30 cm. long, the scalelike leaves few, thinly membranous, broadly ovate, 2–3 mm. long, obtuse; flowers 4–7, sterile, the bracts simulating the scalelike leaves, the pedicels 1.5–2.5 cm. long, obliquely spreading, arcuate after anthesis; calyx segments 2, equal, elliptic, 3–4 mm. long, rounded at apex; corolla pale yellow, about 15 mm. across, the spur obliquely descending, about 6 mm. long, obtuse, shorter than the lower lip.——July–Sept. Ponds and ditches in lowlands and hills; Hokkaido, Honshu, Shikoku, Kyushu; common.——Sakhalin, s. Kuriles, and Manchuria.

10. Utricularia siakujiiensis Nakajima. SHAKUJII-TANUKI–MO. Resembles the preceding; plant free-floating; respiratory branches absent; leaves pinnately dissected, the segments slender, obscurely toothed or nearly entire, spines very short; bladders few; scapes 10–20 cm. long, rather thick, the scalelike leaves few, membranous, narrowly ovate, about 2 mm. long, obtuse; flowers 5–10, the bracts broadly ovate, 3–3.5 mm. long, obtuse, clasping at base, the pedicels ascending, about 2 cm. long, slender; upper calyx-segment slightly larger than the lower, ovate, about 3 mm. long, obtuse to rounded at apex; corolla about 10 mm. across, yellow, the spur about 8 mm. long, longer than and descending under the lower lip, straight, obtuse; capsules globose.——Aug.–Sept. Ponds; Honshu (near Tokyo).

11. Utricularia tenuicaulis Miki. ITO-TANUKI–MO. Resembles the preceding; plants free-floating, slender, respiratory branches absent; leaves pinnately dissected into finely spinule-toothed segments; bladders few; scapes 10–15 cm. long, with few ovate, membranous scalelike leaves; flowers 3–7, the bracts broadly ovate, membranous, rounded at apex, about 3 mm.

long, the pedicels 1.5–2 cm. long, ascending, deflexed in fruit, sometimes slightly thickened above; calyx-segments membranous, broadly ovate, 3–4 mm. long, obtuse, the upper ones slightly longer or nearly as long; corolla about 1 cm. across, pale yellow, the spur shorter than the lower lip; capsules globose, about 4 mm. across, with a beaklike short style at tip; seeds angled, truncate at the ends.——Aug.–Sept. Ponds; Honshu, Shikoku, Kyushu.

12. Utricularia dimorphantha Makino. Fusa-tanuki-mo.   Floating stems rather slender; respiratory branches absent; leaves pinnately dissected, the segments very slender, 3–4 cm. long, the teeth and spines not prominent; bladders absent or rarely 1 or 2, small; scapes 7–15 cm. long, rather slender, with a solitary cleistogamic flower at base and in axils of floating stems near the base of scape; pedicels of cleistogamic flowers short, 3–5 mm. long; scalelike leaves on scapes absent; flowers 3–10, the bracts membranous, elliptic to broadly ovate, 2–2.5 mm. long, subacute, the pedicels 4–5 mm. long, ascending in fruit, spreading or deflexed; calyx about 2 mm. long, the segments equal, orbicular, sometimes retuse, scarcely accrescent in fruit; corolla pale yellow, about 1 cm. across, the spur about 3 mm. long, shorter than the lower lip, straight, ovate-oblong, curved forward; capsules globose, about 3 mm. across, nearly as long as the calyx, with a very short style at apex; seeds angled, with convex faces.——July–Sept. Ponds; Honshu (centr. and Kinki Distr.).

2. PINGUICULA L.   Mushi-tori-sumire Zoku

Perennial acaulescent herbs of wet rocky cliffs and sphagnum moors; leaves rosulate, radical, entire, somewhat fleshy with glandular-tipped hairs and dots above, more or less involute on margin; scapes 1- or 2-flowered, leafless, ebracteate; flowers terminal, purple or violet, sometimes yellowish; calyx bilabiate, the upper lip 3-lobed, the lower lip retuse or 2-parted; corolla bilabiate, the lobes spreading, entire or retuse, the upper lip shorter than lower, the throat open; filaments slightly curved, the anthers transverse, the locules confluent; upper stigma lobe larger, often fringed, the lower lobe small or absent; capsules 2- or 4-valved; seeds oblong, rugose.——About 30 species, in the temperate and cooler regions of the N. Hemisphere, sub-Antarctic, and high mountains of S.America.

1A. Scapes 1-flowered; flowers 12–18 mm. long; capsules ovoid-conical, 4–6 mm. long. .... 1. *P. vulgaris* var. *macroceras*
1B. Scapes often branched at base, 1- or 2(–3)-flowered; flowers 5–7 mm. long; capsules depressed-obovate, 2.5–3.5 mm. long.
    2. *P. ramosa*

1. Pinguicula vulgaris L. var. **macroceras** (Pall. ex Link) Herd. *P. macroceras* Pall. ex Link——Mushi-tori-sumire.   Rather fleshy, pale green, fragile perennial; leaves few, spreading, rosulate, short-petiolate, narrowly ovate to ovate-oblong, 3–5 cm. long, 1–2 cm. wide, obtuse, more or less incurved on margin, glandular-hairy on upper side, becoming soft-pubescent toward base; scapes 1-flowered, 5–15 cm. long, loosely short glandular-puberulent in upper part; calyx 5-parted, 2–4 mm. long, the segments oblong, obtuse; corolla violet, the upper lip about half as long as the lower, subrounded at apex, the spur linear, 7–10 mm. long, obtuse, abruptly thickened toward the base; capsules ovoid-conical; seeds pale brown, obscurely reticulate, obovate-fusiform, about 0.8 mm. long.——July–Aug.   Wet rocky cliffs, rarely on wet mountain slopes; Hokkaido, Honshu (centr. and n. distr.), Shikoku.——Kuriles and widely distributed in mountain areas adjacent to the northern Pacific area. The typical phase is widely distributed in mountainous areas of the N. Hemisphere.

2. Pinguicula ramosa Miyoshi. *P. villosa* var. *ramosa* (Miyoshi) Tamura——Kōshin-sō.   Delicate perennial; leaves few, spreading, rosulate, pale green, ovate or elliptic, 7–15 mm. long, 5–8 mm. wide, obtuse or rounded at apex, incurved on margin, short-petiolate; scapes slender, 3–8 cm. long, often branched at base or near the middle, glandular-puberulent; flowers pale violet, 5–7 mm. long; calyx 2 mm. long, 5-fid, the lobes subequal, obtuse; upper lip of corolla short, the spur linear, 5–7 mm. long, obtuse, abruptly broadened toward base; capsules depressed-obovoid, 2.5–3.5 mm. long, often somewhat retuse.——June–July.   Wet shaded rocky cliffs in mountains; Honshu (n. Kantō Distr.); very rare.

Fam. 182. ACANTHACEAE   Kitsune-no-mago Ka   Acanthus Family

Herbs or shrubs, rarely scandent; leaves opposite, often with internal cystoliths; stipules absent; flowers bisexual, zygomorphic, often with a prominent bract; calyx-segments 4 or 5, imbricate or valvate, rarely reduced to an annulus; corolla bilabiate, often with a suppressed upper lip, the lobes imbricate or contorted in bud; stamens 4, didymous or 2, inserted on the corolla-tube, alternate with the lobes, the filaments free or more or less connate in pairs, the anthers 1- or 2-locular, longitudinally split; ovary superior, sessile, 2-locular, the placentae axile; style solitary, simple; ovules 2 or more in each locule; fruit a capsule, explosively dehiscent; endosperm usually absent.——About 240 genera, with about 2,200 species, mostly in the Tropics, few in temperate regions.

1A. Corolla-lobes nearly equal, scarcely bilabiate.
　2A. Corolla-lobes contorted in bud, the tube usually longer than the lobes. .... 1. *Strobilanthes*
　2B. Corolla-lobes imbricate in bud, the tube often not exceeding the lobes in length. .... 2. *Codonacanthus*
1B. Corolla distinctly bilabiate.
　3A. Seeds more than 4 in the capsule; corolla-lobes contorted in bud. .... 3. *Hygrophila*
　3B. Seeds 4 or fewer in the capsule; corolla-lobes imbricate in bud.
　　4A. Bracts large and leaflike, slightly connate at base, enclosing 1 to several flowers; anther-locules 2, both alike, parallel, not mucronate. .... 4. *Dicliptera*
　　4B. Bracts small, the flowers spicate (in ours); anther-locules 2, not alike, one of them not mucronate and above the other, the one below mucronate or spurred at base. .... 5. *Justicia*

1. STROBILANTHES Bl.   Ise-hanabi Zoku

Erect or decumbent herbs or shrubs; leaves opposite, entire or toothed, sometimes the pairs unequal; flowers blue-purple or white, relatively large, solitary in axils of bracts, sessile or short-pedicelled, forming spikes or compound panicles; bracts sometimes leaflike, often small, the bracteoles narrow, small or absent; calyx 5-parted or toothed, the segments linear, equal or unequal; corolla-tube ampliate, the limb spreading, the lobes 5, contorted in bud, the upper 2 sometimes more or less connate at base; stamens usually 4, didymous, the upper pair sometimes reduced to staminodes, the anther-locules alike and parallel, obtuse; stigma with one of the lobes reduced; ovules 2(–3) in each locule; capsules 4-seeded, the retinaculum acute.——About 200 species, chiefly in India, also in Malaysia to China and Japan.

1A. Subshrub; leaves rather thick, lustrous and minutely reticulate above, lanceolate or broadly so, 8–15 mm. wide; corolla 1.5–2 cm. long. .... 1. *S. japonica*
1B. Herbs, sometimes woody at base; leaves membranous, broadly ovate or obovate-oblong, 3–6 cm. wide; corolla 2–5.5 cm. long.
  2A. Leaves obovate or ovate-oblong, 8–20 cm. long, 3–6 cm. wide, gradually narrowed at base; young parts brown appressed-puberulent; corolla 4.5–5.5 cm. long. .... 2. *S. cusia*
  2B. Leaves broadly ovate or broadly deltoid-ovate, 4–10 cm. long, 3–6 cm. wide, abruptly narrowed at base to a winged petiole; young parts long white-pubescent; corolla 2–3 cm. long. .... 3. *S. oligantha*

1. Strobilanthes japonica (Thunb.) Miq. *Ruellia japonica* Thunb.; *Rostellularia japonica* (Thunb.) Kanitz——Ise-hanabi. Subshrub; stems tufted, branched, terete, 30–60 cm. long, the nodes puberulent when young, thickened with age; leaves rather thick, lanceolate to broadly so, 3–5 cm. long, 8–15 mm. wide, acute with an obtuse tip, acuminate at base, deep green, lustrous, glabrous above with transversely aligned linear cystoliths, paler and short appressed-pilose on nerves beneath, loosely and undulate short-toothed or nearly entire, the petioles 2–5 mm. long; flowers sessiile, more than 10, in a sessile short spike, the bracts small, leaflike, longer than the calyx; calyx about 6 mm. long; corolla pale purple, 1.5–2 cm. long; style loosely pubescent.——Sept.–Oct. Chinese plant long cultivated in our gardens, sometimes naturalized in warmer areas.——Ryukyus, Formosa, China, Indochina, and India.

2. Strobilanthes cusia (Nees) O. Kuntze. *Goldfussia cusia* Nees; *S. flaccidifolius* Nees——Ryūkyū-ai. Perennial herb dark green when dried; stems 30–70 cm. long, rather thick, slightly branched, obtusely angled, brown-puberulent while young; leaves obovate or ovate-oblong, 8–20 cm. long, inclusive of the 1–2 cm long petiole, 3–6 cm. wide, acute with an obtuse tip, gradually narrowed at the base, mucronately or undulately toothed, glabrous and obsoletely lineolate with the cystoliths above, loosely brownish puberulent beneath on nerves while young, the lateral nerves 5 or 6 pairs; flowers sessile in a loose terminal spike, the bracts leaflike, rather small, caducous; calyx 1–1.4 cm. long, one segment often larger than the others; corolla 4.5–5.5 cm. long, pale purple.——Kyushu (s. distr.).——Ryukyus, Formosa, China, Indochina, and India.

3. Strobilanthes oligantha Miq. Suzumushibana, Suzumushi-sō. Perennial herb with short rhizomes; stems 30–60 cm. long, obtusely angled, loosely branched, with long white pubescence on upper part; leaves broadly ovate or broadly deltoid-ovate, 4–10 cm. long, 3–6 cm. wide, acute with an obtuse tip, abruptly narrowed at base to the winged petiole, obtusely toothed, loosely long-pubescent and obsoletely lineolate with short linear cystoliths above, long-pubescent beneath especially on nerves, the petioles 1–5 cm. long; flowers few, sessile, in a dense subcapitate short spike, the bracts small, leaflike, longer than the calyx, long-pubescent; calyx 6–10 mm. long, long-pubescent; corolla pale purple, 2–3 cm. long, often loosely pubescent externally; seeds broadly elliptic, about 3 mm. long, brown-puberulent.——Aug.–Oct. Woods in mountains; Honshu (Kinki Distr. and westw.), Shikoku, Kyushu.

2. CODONACANTHUS Nees   Arimori-sō Zoku

Small erect glabrous herbs; leaves entire; flowers rather small, white or pale purple, short-pedicelled, loosely arranged on elongate, one-sided racemes, the bracts and bracteoles minute, subtending the pedicels; calyx nearly 5-parted, the lobes short, narrow, nearly equal; corolla broadly campanulate, short-tubular, 5-lobed, the lobes flat, imbricate in bud, the upper 2 lobes slightly narrower, innermost in bud, the lower median lobe outermost; fertile stamens 2, the anthers 2-locular, broadly divergent, versatile; stigma capitate; ovules 2 in each locule; capsules oblong; seeds 4 or fewer, orbicular, smooth, lustrous.——Two species, India to Japan and se. Asia.

1. Codonacanthus pauciflorus (Nees) Nees. *Asystasia pauciflora* Nees; *C. acuminatus* Nees——Arimori-sō. Small perennial herb; stems erect, 30–60 cm. long, slightly branched, often decumbent at base, with a line of short hairs; leaves oblong, 4–8 cm. long, 1–3 cm. wide, acute at both ends, entire or slightly undulate, deep green above, glabrous except loosely puberulent on nerves beneath while young, short-petiolate; racemes terminal, simple or slightly branched at base, peduncled, 8–15 cm. long, loosely about 10-flowered, puberulent, the bracts and bracteoles linear-lanceolate, 1–2 mm. long, the pedicels suberect, short, slender, 2 or 3 times as long as the bract; calyx nearly as long as the pedicels, the segments broadly linear, erect; corolla white, 8–10 mm. long and as wide, the lobes ovate, rounded at apex, longer than the tube; capsules oblanceolate, narrowed to a stipelike base, 12–15 mm. long.——Oct. Woods in lowlands; Kyushu (Yakushima and Tanegashima).——Ryukyus, Formosa, China, and India.

3. HYGROPHILA R. Br.   Ogi-no-tsume Zoku

Erect or decumbent herbs sometimes with axillary spines; leaves entire; flowers fasciculate in axils, nearly sessile, rarely solitary, the bracteoles linear; calyx-segments 5, narrow, slightly unequal, connate at base; corolla-tube slightly ampliate, the limb bilabiate, the lobes contorted in bud, the upper lip erect, 2-toothed or shallowly 2-lobed, concave, the lower lip nearly spreading, strongly 3-lobed, with 2 plaits inside at base; stamens 4, fertile, didymous, the anther-locules parallel, equal, obtuse; disc obscure; stigma-lobes linear, the upper lobes much reduced; ovules 4 to many, rarely 2 in each locule; capsules many, more than 4-seeded; seeds obliquely ovate or orbicular.——About 30 species, widely distributed in the Tropics.

1. Hygrophila lancea (Thunb.) Miq. *Justicia lancea* Thunb.——OGI-NO-TSUME.   Stems short–creeping then erect, 30–60 cm long, branched, obtusely angled, loosely puberulent in upper part; leaves lanceolate or linear-lanceolate, 5–10 cm. long, 5–15 mm. wide, subobtuse, narrowed below to a short petiole, entire or slightly undulate, often loosely pilose on nerves beneath; flowers fasciculate in leaf-axils, sessile, the bracts and bracteoles shorter than the calyx; calyx 6–8 mm. long, the segments broadly linear, connate on lower half, erect, usually loosely ciliate; corolla pale purple, 1–1.2 cm. long; capsules 1–1.2 cm. long, oblong-lanceolate, acute, about 15-seeded.——Sept.–Oct.   Honshu (Tōtōmi and westw.), Shikoku, Kyushu.——Ryukyus and Formosa.

4. DICLIPTERA Juss.   HAGURO-SŌ ZOKU

Herbs; leaves entire; flowers reddish or blue-purple, sessile, the bracts paired, leaflike, ovate or lanceolate, slightly connate at base, subtending 1 to many flowers; calyx scarious, 5-parted; corolla-tube slender, the limb bilabiate, the upper lip innermost in bud, entire or retuse, erect at anthesis, concave, the lower lip often broader than the upper one, spreading, rather flat, entire or shortly 3-lobed; stamens 2, inserted on the throat, the anthers 2-locular, alike, the locules parallel or of different lengths, usually not mucronate; ovules 2 in each locule; capsules rather flat, ovate or orbicular; seeds 4 or fewer, smooth or minutely tubercled.——More than 60 species, in the Tropics and subtropics.

1. Dicliptera japonica (Thunb.) Makino var. **subrotunda** Matsuda. *D. japonica* var. *elliptica* Matsuda——HAGURO-SŌ.   Thinly pilose or glabrous perennial; stems 20–50 cm. long, often branched; leaves membranous, narrowly ovate-oblong, rarely broadly lanceolate, 5–10 cm. long, 1–3 cm. wide, acuminate with an obtuse tip, the bracts broadly ovate or elliptic, 1–2.5 cm. long, obtuse, usually rounded at base, free, ciliolate on margin, the petioles 5–16 mm. long; calyx 2.5–3 mm. long, the segments narrow, acuminate; corolla pale rose-colored, 2.5–3 cm. long, bilabiate nearly to the middle, the lips narrowly oblong, equal, minutely 3-lobed; style 2–2.5 cm. long; capsules enclosed by a pair of leaflike persistent bracts, oblanceolate, 1–1.2 cm. long, loosely short-pilose; seeds elliptic, about 2.5 mm. long, minutely tuberculate.——July–Oct.   Honshu (Kantō Distr. and westw.), Shikoku, Kyushu.

Var. **japonica.** *Dianthera japonica* Thunb.; *Justicia japonica* (Thunb.) Vahl; *Dicliptera japonica* (Thunb.) Makino; *D. japonica* var. *ciliata* (Matsuda) Ohwi; *D. japonica* var. *elliptica* forma *ciliata* Matsuda——FUCHIGE-HAGURO-SŌ.   Bracts long-ciliate.——Kyushu (s. distr.).——China and Korea.

5. JUSTICIA L.   KITSUNE-NO-MAGO ZOKU

Herbs, rarely shrubs; leaves entire; flowers rather small, white or rose-colored, solitary in axils of bracts, forming a terminal spike (in ours), cyme, or panicle, the bracts and bracteoles various in shape; calyx 5-parted, the segments equal, narrow; corolla-limb bilabiate, the upper lip innermost in bud, the lower lip shallowly 3-lobed, spreading; fertile stamens 2, inserted on the throat, the anthers 2-locular, not alike, one of them not mucronate and above the other, short-spurred; ovules 2 in each locule; seeds 4, or fewer, orbicular, smooth or tubercled.——About 250 species, mostly in the Tropics.

1. Justicia procumbens L. var. **leucantha** Honda. *J. japonica* Thunb.; *J. procumbens* sensu auct. Japon., non L. ——KITSUNE-NO-MAGO.   Short-pubescent annual; stems 10–40 cm. long, from a much-branched decumbent base; leaves membranous, ovate, 2–4 cm. long, 1–2 cm. wide, subentire or loosely undulate-toothed, the petioles 5–15 mm. long; spikes 2–5 cm. long, densely flowered, the bracts, bracteoles, and calyx-segments nearly alike, narrowly lanceolate, 5–7 mm. long, pubescent on an elevated midrib, ciliate on a scarious margin; corolla slightly longer than the calyx, 7–8 mm. long, slightly puberulent, white or pale rose-colored, the lower lip longer than the upper, ovate, trilobulate, with darker reddish dots; one locule of the anther with an obtuse spur nearly as long as the locule; capsules nearly as long as the calyx; seeds minutely rugose.——Aug.–Oct.   Cultivated fields and waste grounds in lowlands; Honshu, Shikoku, Kyushu; very common.——Korea, Manchuria, China, Ryukyus, and Formosa. The typical phase occurs in e. Asia, India, Malaysia, and Australia.

Fam. 183.   MYOPORACEAE   HAMA-JINCHŌ KA   Myoporum Family

Shrubs or rarely trees sometimes with scalelike or pinnate hairs; leaves alternate, rarely opposite, simple; stipules absent; flowers axillary, solitary or fascicled, bisexual, zygomorphic; calyx 5-parted, imbricate or not overlapping in bud; corolla gamopetalous, usually 5-lobed, the lobes imbricate in bud; stamens 4 or as many as the corolla-lobes and alternate with them, inserted on the corolla-tube, the anthers longitudinally dehiscent; ovary superior, usually 2-locular, 2(–8) ovules in each locule, rarely many locular; style simple; ovules pendulous; fruit a drupe; seeds with a straight or slightly curved embryo, nearly destitute of endosperm.——About 5 genera, with about 110 species, mainly in Australia and the Pacific Islands, few in e. Asia.

1. MYOPORUM Banks and Soland.   HAMA-JINCHŌ ZOKU

Shrubs with glabrous or viscid parts; leaves usually alternate, simple; flowers small, white or purple; calyx 5-parted, not accrescent in fruit; corolla subcampanulate or infundibuliform, the lobes 5(–6), equal or the lower slightly larger; ovary 2- to 10-locular, the locules nearly equal, around the central axis, 1-ovuled in each locule, or rarely 2-locular, with 2 ovules in each locule; drupe ovoid or subglobose, indehiscent, 2- to 10-seeded; seeds in one horizontal series around the axis, rarely longitudinally arranged, pendulous from the inner angle of the locule.——About 25 species, in Australia, the Pacific Islands, Malaysia, and se. Asia.

1. Myoporum bontioides (Sieb. & Zucc.) A. Gray. *Pentacoelium bontioides* Sieb. & Zucc.; *Polycoelium bontioides* (Sieb. & Zucc.) A. DC.; *Polycoelium chinense* A. DC.—— HAMA-JINCHŌ. Evergreen shrub, sparingly furfuraceous while very young, the branches rather stout; leaves rather coriaceous, narrowly oblong or broadly oblanceolate, 5–10 cm. long, 1.5–3 cm. wide, entire, acute with an obtuse tip, acuminate at base, with a raised midrib beneath, lateral nerves obscure, the petioles about 1 cm. long; few in leaf-axils, purple, one-sided, 2.5–3 cm. long, about 2 cm. across, the pedicels 1–1.5 cm. long and thickened in upper part; calyx 5-parted nearly to the base, the segments narrowly deltoid-ovate, 4–6 mm. long, acuminate; fruit about 1 cm. across, with a long curved style while young, globose when mature; seeds few in one series around the axis.——Oct.–Nov. Wet places near the sea; Kyushu (western coast and Tanegashima).——Ryukyus, Formosa, and China.

Fam. 184. PHRYMACEAE   HAE-DOKU-SŌ KA   Phryma Family

Erect perennial; leaves opposite, toothed; spikes elongate, narrow, terminal or on upper part of the branches; flowers small, solitary in the axils of bracts, erect in bud, spreading at anthesis, deflexed in fruit; calyx tubular, 5-toothed, the 3 dorsal teeth becoming spinose and hooked in fruit, the 2 ventral teeth unchanged; corolla tubular, the limb bilabiate, the upper lip erect, shallowly 2-lobed, the lower one longer, spreading, 3-lobed; stamens 4, didymous; ovary oblique, 1-locular, 1-ovuled; style terminal, shallowly 2-fid; ovules erect, orthotropous, basal; fruit enclosed in the persistent calyx, the pericarp membranous, the testa adnate to the pericarp; endosperm absent.——One genus, with a single species.

1. PHRYMA L.   HAE-DOKU-SŌ ZOKU

Characters of the family.

1. Phryma leptostachya L. var. **asiatica** Hara. *P. leptostachya* sensu auct. Japon., non L.; *P. humilis* Koidz.; *P. nana* Koidz.; *P. leptostachya* var. *humilis* (Koidz.) Hara; *P. leptostachya* var. *nana* (Koidz.) Hara——HAE-DOKU-SŌ. Roots rather thick; stems obtusely angled, erect, retrorsely puberulent, often branched in upper part; leaves thinly membranous, ovate or broadly deltoid-ovate, 7–10 cm. long, 4–7 cm. wide, short-acuminate or abruptly acute with an obtuse tip, broadly cuneate to truncate at base, more or less pilose on both sides especially on nerves, toothed, petiolate; spikes 10–20 cm. long, loosely flowered, peduncled, often branched at base, puberulent on the axis; calyx 5-ribbed, lanceolate, about 3 mm. long in anthesis, 5–6 mm. long in fruit, nearly glabrous, the hooked spines about 1.5 mm. long; corolla pale rose or white, about 5 mm. long.——June–Aug. Woods in hills and low elevations in mountains; Hokkaido, Honshu, Shikoku, Kyushu; common and slightly variable.——Forma **oblongifolia** (Koidz.) Ohwi. *P. leptostachya* var. *oblongifolia* (Koidz.) Honda; *P. oblongifolia* Koidz.——NAGABA-HAE-DOKU-SŌ. Leaves ovate-oblong, mostly on the upper part of stems.—— Often occurs with var. *asiatica*.——Korea, China, Himalayas, and e. Siberia. The typical phase occurs in N. America.

Fam. 185. PLANTAGINACEAE   ŌBA-KO KA   Plantain Family

Herbs with radical, alternate or opposite simple leaves often dilated on the petiole; flowers bisexual, actinomorphic, in spikes; calyx herbaceous, 4-parted or -divided, sometimes the outer 2 more or less connate; corolla gamopetalous, scarious, 3- or 4-lobed, the lobes imbricate in bud; stamens 1 to 4, inserted on the tube and alternate with the lobes or inserted at the base of the ovary; anthers 2-locular, longitudinally split; ovary superior, 1- to 4-locular; style simple; ovules 1 or more, basal or attached to the axis; fruit a circumscissile capsule or bony nut; seeds peltately attached, the embryo straight, rarely curved, in the center of the fleshy endosperm.——Three genera, with about 300 species. Cosmopolitan.

1. PLANTAGO L.   ŌBA-KO ZOKU

Annuals or perennials with radical leaves; flowers bisexual, rarely polygamous or unisexual, small; stamens long-exserted, usually on much-elongate capillary filaments; ovary 2-locular or falsely 4-locular, each locule with 1 to many ovules; capsules membranous, usually circumscissile; seeds peltately attached on the ventral face, angled or planoconvex, sometimes grooved on the ventral face.——About 200 species, of wide distribution.

1A. Flowers bisexual; perennials; corolla-lobes spreading, not closing the top of the capsule.
  2A. Seeds flat on the ventral face; spikes densely to loosely flowered, elongate; leaves oblong to ovate-orbicular, spreading to ascending.
    3A. Seeds 4–12; spikes densely more than 20-flowered.
      4A. Seeds glabrous or rarely soft-pubescent; leaves usually ovate, long-petiolate.
        5A. Seeds 4–6(–10), 1.5–1.8 mm. long, black-brown; valves conic-subglobose or conic-ovoid. .... 1. *P. asiatica*
        5B. Seeds 8–12, 0.8–1.2 mm. long, dark brown; valves subglobose. .... 2. *P. major* var. *japonica*
      4B. Seeds usually long-pubescent; leaves oblong, short-petiolate. .... 3. *P. camtschatica*
    3B. Seed solitary; spikes loosely 10- to 20-flowered. .... 4. *P. hakusanensis*
  2B. Seeds 2, grooved on the ventral face, very convex on back; spikes very densely many-flowered, narrowly ovate to short-cylindric; leaves lanceolate, erect. .... 5. *P. lanceolata*
1B. Flowers dioecious or polygamous; annual or biennial; corolla-lobes erect, enclosing upper part of the capsule. ...... 6. *P. virginica*

1. **Plantago asiatica** L. *P. major* var. *asiatica* (L.) Decne.; *P. asiatica* var. *densiuscula* Pilger; *P. asiatica* var. *lobulata* Pilger; *P. yezoensis* Pilger; *P. asiatica* var. *yesoensis* (Pilger) Hara; *P. mohnikei* Miq.; *P. major* forma *contorta* Makino; *P. contorta* (Makino) Ikeno——ŌBA-KO. Perennial; leaves thinly herbaceous, radical, rosulate, ovate or broadly so, rarely elliptic, 5–15 cm. long, 3–8 cm. wide, obtuse, with few parallel nerves, abruptly long-petiolate, glabrous or nearly so, the petioles dilated; scapes few to many, 10–50 cm. long, erect; spikes solitary, densely many-flowered, usually glabrous, the bracts narrowly ovate, shorter than the calyx, obtusely keeled on back, slightly dilated at base; flowers sessile; calyx-segments obovate-elliptic, about 2 mm. long, rounded at apex, white-scarious, green, with a stout rib on back; stamens long-exserted, the anthers cordate-orbicular, mucronate; capsules ovoid-oblong, nearly twice as long as the calyx; seeds 4–6, rarely 8–10, black-brown, elliptic, planoconvex.——Apr.–Oct. Woods and waste grounds in lowlands and mountains; Hokkaido, Honshu, Shikoku, Kyushu; very common and variable. ——s. Korea, Kuriles, Sakhalin, Ryukyus, Formosa, China, e. Siberia, and Malaysia.

Var. **yakusimensis** (Masam.) Ohwi. *P. yakusimensis* Masam.——YAKUSHIMA-ŌBA-KO. Plant smaller; leaves soft-pubescent.——Honshu (Izu Isls.), Kyushu (Yakushima).

2. **Plantago major** L. var. **japonica** (Fr. & Sav.) Miyabe. *P. japonica* Fr. & Sav.——TŌ-ŌBA-KO. Usually glabrous perennial; leaves rather thick-herbaceous, ovate or ovate-elliptic, 10–30 cm. long, 5–15 cm. wide, few-nerved, suboptuse, long-petiolate; scapes erect, few, 50–100 cm. long, rather stout; spikes erect, usually glabrous, brownish when dry, the bracts narrowly ovate, obtuse, obtusely keeled on back, rarely slightly gibbous; calyx-segments elliptic, about 2 mm. long, rounded at apex, more or less white-scarious on margin; capsules ellipsoidal, about twice as long as the calyx; seeds about 10, small, dark brown, plano-convex, 0.8–1.2 mm. long.——July–Aug. Wet places in lowlands near the sea; Honshu, Shikoku, Kyushu.——Forma **yezomaritima** (Koidz.) Ohwi. *P. yezomaritima* Koidz.; *P. japonica* var. *yezomaritima* (Koidz.) Hara ——TERIHA-ŌBA-KO. Leaves thicker.——Wet places near the sea; Hokkaido.

Plantago togashii Miyabe & Tatew. ISO-ŌBA-KO. A smaller plant; leaves to 10 cm. long; scapes 10–20 cm. long. ——Resembles forma *yezomaritima* of the preceding species and possibly not distinct from it. Reported from Hokkaido (Okushiri Isl.).

3. **Plantago camtschatica** Cham. *P. villifera* Franch.; *P. depressa* subsp. *camtschatica* (Link) Pilger——EZO-ŌBA-KO. White-pubescent, rarely glabrescent perennial with thick rhizomes; leaves oblong to narrowly so, 5–10 cm. long, 2.5–4 cm. wide, acute, short-petiolate and slightly dilated at base, few-nerved; scapes erect, 15–30 cm. long; spike solitary, 3–10 cm. long, densely many-flowered, the bracts glabrous, elliptic or ovate, shorter than the calyx; calyx glabrous, ellipsoidal, 2–2.5 mm. long, the segments very obtuse, white-scarious on margin, with a broad green midrib; capsules narrowly ovoid or ovoid-oblong, 1.5–2 times as long as the calyx; seeds 4, elliptic or

ovate, dark brown, 1.2–1.5 mm. long.——May–Aug. Along sandy seashores; Hokkaido, Honshu, Kyushu.——Forma **glabra** (Makino & Honda) Ohwi. *P. camtschatica* var. *glabra* Makino & Honda——KE-NASHI-EZO-ŌBA-KO. Plant glabrescent.——Korea, Sakhalin, Kuriles, and Ochotsk Sea region.

Plantago depressa Willd. MUJINA-ŌBA-KO. Closely allied to the preceding species, but with smaller flowers and anthers. Sparingly naturalized in our area.——n. China, Korea, Manchuria, and e. Siberia.

4. **Plantago hakusanensis** Koidz. *P. mohnikei* sensu auct. Japon., non Miq.——HAKUSAN-ŌBA-KO. Rhizomes short and stout; leaves slightly fleshy, soft, elliptic to oblong, 3–5(–8) cm. long, 2–3(–4) cm. wide, few-nerved, glabrous or loosely pubescent, narrowed to a rather short petiolelike base; scapes 7–12 cm. long, slender, glabrous, often ascending from base; spikes simple, loosely 10- to 20-flowered, the bracts broadly ovate, broadly thin-scarious on margin, as long as to shorter than the calyx; calyx-segments elliptic, 2–2.5 mm. long, with a broad green back, rounded at apex; anthers cordate-orbicular, rather large; capsules 1.5–2 times as long as the calyx, ovoid-ellipsoid; seeds solitary, oblong, black-brown, about 2 mm. long, slightly depressed on the ventral side.——July–Aug. Wet alpine slopes on Japan Sea side; Honshu (centr. and n. distr.).

5. **Plantago lanceolata** L. HERA-ŌBA-KO. More or less white-pubescent perennial with thick rhizomes; leaves erect, lanceolate, 10–20 cm. long, 1.5–3 cm. wide, acute or gradually so, few-nerved, gradually narrowed to a long petiole; scapes erect, 40–60 cm. long; inflorescence densely many-flowered, narrowly ovoid in anthesis, slightly elongate and short-cylindric in fruit, 3–5 cm. long, the bracts nearly orbicular, mucronate, as long as to slightly longer than the calyx, thinly scarious, translucent, with a narrow midrib; calyx-segments thinly scarious, translucent, about 2.5 mm. long, rounded at apex, with a narrow green loosely pilose midrib; style very long; anthers ovoid, 2–2.5 mm. long, short-mucronate, cordate at base; capsules ovoid-globose; seeds 2, black-brown, oblong, 1.7–2 mm. long, with a broad groove on the ventral face.——June–Aug. Widely naturalized; introduced from Europe.

6. **Plantago virginica** L. TACHI-ŌBA-KO, TSUBOMI-ŌBA-KO. Annual or biennial with long soft whitish pubescence; leaves narrowly obovate to broadly oblanceolate, 4–5 cm. long, 1.5–3 cm. wide, few-nerved, gray-green, gradually narrowed to a rather long petiole; scapes 10–30 cm. long, rather slender, striate; spikes 10–20 cm. long, densely many-flowered, pale yellow to brownish white, the bracts narrowly ovate, shorter than the calyx, pubescent on the prominently raised midrib; calyx-segments oblong, about 2 mm. long, pubescent especially on the midrib; corolla-lobes narrowly ovate, as long as the calyx-segments, erect and covering the upper part of capsule, 1-nerved; capsules scarcely longer than the calyx-segments, broadly ovoid; seeds 2, pale red-brown, about 1.8 mm. long, with a groove on the ventral face.——May–June. Recently naturalized in our area; introduced from N. America.

Fam. 186. **RUBIACEAE**   AKANE KA   Madder Family

Herbs, shrubs, or trees; leaves opposite or rarely verticillate, entire or toothed, simple; stipules free or connate, sometimes leaflike, interpetiolar or intrapetiolar; flowers usually bisexual, actinomorphic, rarely slightly zygomorphic, solitary or variously arranged; calyx-tube adnate to the ovary; corolla usually tubular, the lobes 4–10, contorted or imbricate in bud, sometimes valvate; stamens as many as and alternate with the corolla-lobes; ovary 2- to several-locular, the placentae axile or apparently basal, or the ovary rarely 1-locular with parietal placentae; ovules 1 to many in each locule; fruit a capsule, berry, or drupe; seeds usually with endosperm.——About 350 genera, with about 4,500 species, nearly cosmopolitan, especially abundant in the Tropics.

1A. Ovary many-ovuled.
  2A. Herbs or subshrubs.
    3A. Inflorescence a helicoid cyme (cinicinnus); capsules laterally flattened, obcordate. .... 1. *Ophiorrhiza*
    3B. Inflorescence in cymes or panicles; capsules not flattened, usually globose or hemispherical. .... 2. *Hedyotis*
  2B. Woody plants.
    4A. Flowers in heads; fruit dry.
      5A. Bracteoles between the flowers scalelike; unarmed erect trees or large shrubs. .... 3. *Adina*
      5B. Bracteoles between the flowers absent or minute; scandent shrubs or small trees with hooked spines at the nodes. .. 4. *Uncaria*
    4B. Flowers solitary or in corymbose cymes; fruit a berry, often fleshy.
      6A. Inflorescence terminal.
        7A. Corolla-lobes valvate in bud; one calyx-segment of some outer flowers becoming enlarged and petaloid. ...... 5. *Mussaenda*
        7B. Corolla-lobes contorted in bud; calyx-segments equal, not petaloid. .... 6. *Tarenna*
      6B. Inflorescence or flowers axillary.
        8A. Ovary 2-locular; flowers usually in cymes or fascicles. .... 7. *Randia*
        8B. Ovary 1-locular; flowers solitary in leaf-axils. .... 8. *Gardenia*
1B. Ovary with 1 ovule in each locule.
  9A. Woody plants.
    10A. Flowers umbellate. .... 9. *Morinda*
    10B. Flowers solitary or fascicled, sometimes in cymes.
      11A. Fruit fleshy.
        12A. Flowers in cymes. .... 10. *Psychotria*
        12B. Flowers solitary or fascicled.
          13A. Flowers 5-merous; spines absent. .... 11. *Lasianthus*
          13B. Flowers 4-merous; spines present. .... 12. *Damnacanthus*
      11B. Fruit dry, capsular. .... 13. *Leptodermis*
  9B. Herbaceous plants.
    14A. Stipules small, between the leaves.
      15A. Scandent herbs with much-elongate stems. .... 14. *Paederia*
      15B. Erect or decumbent herbs.
        16A. Stems ascending to erect; leaves not evergreen; fruit dry. .... 15. *Pseudopyxis*
        16B. Stems long-creeping on the ground; leaves evergreen; fruit a berry. .... 16. *Mitchella*
    14B. Stipules relatively large, leaflike, with the aspect of verticillate leaves.
      17A. Corolla rotate.
        18A. Fruit fleshy; flowers usually 5-merous. .... 17. *Rubia*
        18B. Fruit dry; flowers usually 3- or 4-merous. .... 18. *Galium*
      17B. Corolla infundibuliform, sometimes nearly campanulate. .... 19. *Asperula*

1. OPHIORRHIZA L.   SATSUMA-INAMORI ZOKU

Erect or decumbent perennial herbs or subshrubs; leaves opposite, often unlike; stipules interpetiolar, solitary or 2; flowers in helicoid cymes, short-pedicelled, usually small, white, reddish, or greenish, often bracteolate; calyx-tube short, the lobes 5, small, persistent; corolla infundibuliform or tubular, the lobes 5, valvate in bud; stamens 5, inserted at base of the throat, the disc annular; ovary 2-locular; style long, the stigma 2-fid; ovules many; capsules rather flat, obcordate; seeds many, small, angled, the endosperm fleshy.——About 100 species, abundant in tropical Asia and the Pacific Islands, few in Australia and Japan.

1A. Inflorescence with linear bracts; flowers 1.5–2 cm. long, pedicelled; corolla-lobes white-pubescent inside. .... 1. *O. japonica*
1B. Inflorescence without bracts; flowers about 5 mm. long, nearly sessile; corolla-lobes densely yellow-puberulous inside. .... 2. *O. pumila*

1. Ophiorrhiza japonica Bl. SATSUMA-INAMORI, KI-DACHI-INAMORI-SŌ. Long-creeping perennial herb with brownish short crisped hairs; stems slender, erect, 10–20 cm. long, simple; leaves narrowly ovate to narrowly oblong-ovate, 3–5 cm. long, 1.5–2 cm. wide, often reddish when dried, acute or subacuminate, acute at base, nearly glabrous above, pilose on the nerves beneath, the lateral nerves 7 or 8 pairs, the petioles 1–3 cm. long; flowers few to more than 10, white, in a rather dense cyme, the peduncles 2–5 cm. long, the bracts linear, 3–7 mm. long, puberulent; corolla tubular; fruit depressed-obtriangular, flattened, about 4 mm. long, about 1 cm. across at the apex; seeds minute, elliptic, punctulate.——Nov.–Apr. Woods

in low mountains; Honshu (Awa and westw.), Shikoku, Kyushu.——Ryukyus, Formosa, and China.
    Var. **tashiroi** (Maxim.) Ohwi. *O. tashiroi* Maxim.——NAGABA-INAMORI-SŌ. Leaves narrowly lanceolate, 5–12 cm. long, 1–2.5 cm. wide.——Kyushu (Yakushima).
    2. Ophiorrhiza pumila Champ. *O. inflata* Maxim.; *O. pumila* var. *inflata* (Maxim.) Masam.——CHABO-INAMORI, YAEYAMA-INAMORI-SŌ. Stems slightly branched and decumbent at base, slender, 10–20 cm. long, with short yellowish puberulence especially toward the top; leaves broadly oblanceolate to narrowly oblong or rarely narrowly ovate, acute, puberulent on nerves beneath, the petioles 5–20 mm. long; cymes

terminal, densely yellow-puberulent, the peduncles to 10 cm. long or sometimes absent; flowers about 10, subsessile, the bracts and bracteoles absent; calyx-teeth minute; corolla about 5 mm. long; fruit flat, depressed, obtriangulate, about 3 mm. long, 6–8 mm. wide; seeds minute, puncticulate, obtusely angled.——Kyushu (Yakushima); rather rare.——Ryukyus, Formosa, and China.

2. HEDYOTIS L. Futaba-mugura Zoku

Herbs or shrubs, rarely scandent; leaves opposite or rarely verticillate, linear to ovate-orbicular; stipules often adnate to the petiole and forming a short sheath; flowers usually white, small, in cymes or panicles; calyx-tube short, the lobes 4, short, persistent; corolla rotate, infundibuliform, or hypocrateriform, sometimes campanulate, the lobes valvate in bud; ovary 2-locular; style slender, the stigma shortly 2-lobed or entire; ovules many; fruit a capsule; seeds small, angled, globose, or planoconvex. ——More than 200 species, chiefly in the tropics and the warmer regions, abundant in s. Asia.

1A. Leaves linear to broadly so.
  2A. Stipules short-toothed; leaves herbaceous; calyx-tube depressed-globose, with 4 obtuse shallow sinuses, the teeth separate in fruit, obliquely recurved. .... 1. *H. diffusa*
  2B. Stipules spinulose-toothed; leaves mostly coriaceous; calyx-tube ovoid-globose, nearly sessile, the teeth contiguous at base, erect.
                                                            2. *H. tenelliflora*
1B. Leaves oblanceolate, narrowly oblong to elliptic, or broadly ovate.
  3A. Flowers (1–)3, pedicelled, terminal and in upper axils; leaves rather fleshy, oblanceolate to oblong. .... 3. *H. biflora* var. *parvifolia*
  3B. Flowers more than 3, fascicled in leaf-axils, very short-pedicelled; leaves herbaceous or membranous, narrowly oblong to broadly ovate.
    4A. Plant slightly white-pubescent or nearly glabrous, black when dry; lateral nerves of leaves 4(–5) pairs. .... 4. *H. lindleyana*
    4B. Plant prominently pubescent with yellowish long hairs, yellow-green when dried; lateral nerves of leaves 3 pairs.
                                                  5. *H. chrysotricha*

1. Hedyotis diffusa Willd. *H. diffusa* var. *longipes* Nakai; *Oldenlandia diffusa* (Willd.) Roxb.; *O. angustifolia* var. *pedicellata* Miq.——Futaba-mugura. Annual, glabrous except the slightly scabrous leaf margins; stems branched and ascending from the base, 10–20 cm. long; leaves herbaceous, linear, 1–3.5 cm. long, 1.5–3 mm. wide, 1-nerved, with a minute awn at apex, slightly narrowed to a nearly sessile base; flowers axillary, white, about 2 mm. across, sessile or on very short pedicels to 3 mm. long; calyx-teeth acuminate, spreading, about 1.5 mm. long; corolla 4-lobed, the lobes as long as the tube; calyx-tube globose in fruit, about 5 mm. across; seeds minute, many, angled.——Sept.–Oct. Waste grounds, cultivated fields, and grassy slopes in lowlands; Honshu, Shikoku, Kyushu.——Korea, Ryukyus, Formosa, China, and tropical Asia.

2. Hedyotis tenelliflora Bl. *H. angustifolia* Cham. & Schltdl.; *Oldenlandia tenelliflora* (Bl.) O. Kuntze——Ke-nioi-gusa. Perennial, rather firm, glabrous, slightly scabrous on leaf-margins; stems tufted, erect, slender, 10–30 cm. long, sparsely branched, 4-angled; leaves broadly linear, 2–3 cm. long, 2–3 mm. wide, short-awned at apex, obtuse and subsessile at base, 1-nerved, the margins recurved; stipules short, the linear lobes 2 or 3, 2–5 mm. long; flowers axillary, nearly sessile; calyx-teeth erect, acuminate, 2–2.5 mm. long; corolla white, about 2 mm. long; capsules ovoid-globose, about 3 mm. long, adnate to the calyx.——July–Sept. Lowlands; Kyushu (Yakushima).——Ryukyus, Formosa, s. China, Malaysia, and India.

3. Hedyotis biflora (L.) Lam. var. **parvifolia** Hook. & Arn. *Oldenlandia crassifolia* DC.——Sonare-mugura. Rather fleshy glabrous perennial; stems tufted, branched, decumbent or ascending at base, 5–20 cm. long; leaves oblong, obovate, or broadly spathulate-oblanceolate, 1–2.5 cm. long, 7–12 mm. wide, rounded or very obtuse, narrowed to a petiole-like base, 1-nerved, slightly recurved on margin, lustrous above; stipules small, 2-toothed; flowers on upper part of branches, in 1–3's, the pedicels 3–10 mm. long; calyx-teeth short, broadly ovate-deltoid, erect, 1–1.5 mm. long; corolla white, 1.5–2 mm. long; calyx-tube in fruit depressed, obovoid-globose, 4–5 mm. across, subtruncate; seeds minute, rather many, ovoid.——Aug.–Nov. Rocks near seacoast; Honshu (Awa and westw.), Shikoku, Kyushu.——Ryukyus, Formosa, China, Philippines, and India.

4. Hedyotis lindleyana Hook. var. **hirsuta** (L. f.) Hara. *Oldenlandia hirsuta* L. f.; *H. hirsuta* (L. f.) Spreng., non Lam.; *O. japonica* Miq.; *H. japonica* (Miq.) Masam.——Ha-shikagusa. Annual, more or less whitish pubescent; stems decumbent, elongate, branched; leaves herbaceous, ovate, 2–4 cm. long, 1–2 cm. wide, acute, rounded to broadly cuneate at base, with 4(–5) pairs of ascending arcuate lateral nerves, the petioles 5–15 mm. long; stipules semirounded, short, with 2–4 filiform lobes; flowers few in leaf-axils, sessile or short-pedicelled; calyx-teeth ovate, about 2 mm. long, obtuse, spreading, recurved in fruit; corolla white, infundibuliform, with a narrow tube, 3–4 mm. long; calyx-tube depressed-globose in fruit, about 4 mm across; seeds small, many, ovoid, minutely impressed-punctulate.——Aug.–Oct. Woods and grassy places along streams in hills and low mountains; Honshu, Shikoku, Kyushu.——China and Malaysia.

Var. **glabra** (Honda) Hara. *Oldenlandia hirsuta* var. *glabra* Honda; *O. glabra* (Honda) Honda, non O. Kuntze; *O. hondae* Hara——Ō-hashikagusa. Plant larger, glabrate or slightly pubescent; leaves to 4.5 cm. long.——Aug.–Oct. Woods in mountains (n. Kantō and n. distr. to Echigo Prov.).

5. Hedyotis chrysotricha (Palib.) Merr. *Anotis chrysotricha* Palib.; *Oldenlandia kiusiana* Makino; *O. chrysotricha* (Palib.) Chun——Koban-mugura. Plant prominently yellowish long soft pubescent; stems creeping; leaves oblong or ovate, sometimes elliptic, 1–2 cm. long, 7–12 mm. wide, acute, sometimes minutely mucronate, rounded at base, with 3 pairs of lateral nerves arising from the lower part, the petioles 1–3 mm. long; stipules with 1 or 2 filiform lobes; flowers few in axils, the pedicels 3–6 mm. long; calyx-lobes broadly lanceolate, acuminate, 3–3.5 mm. long; corolla infundibuliform, 5–6 mm. long, 4-fid, the lobes minutely papillose inside.——June–Oct. Kyushu (w. distr.).——China.

3. ADINA Salisb.   TANI-WATARI-NO-KI ZOKU

Trees or shrubs; leaves opposite, petiolate; stipules deciduous; flower-heads solitary or in peduncled panicles or racemes, bracteate or ebracteate, the receptacles or shortened axis hairy; flowers many, small, with scalelike bracteoles at base; calyx connate, the tube angled, the lobes 5, sometimes deciduous; corolla tubular, elongate, the lobes 5, valvate in bud; stamens 5, inserted on the corolla-throat, the disc cup-shaped; ovary 2-locular; style elongate, the stigma clavate or capitate; ovules many; capsules septicidal, 2-valved, the valves inflexed, opening internally; seeds oblong, winged on each side.——Several species, in the Tropics of Asia and America.

1A. Leaves evergreen, broadly oblanceolate, acute at base, the petioles 4–10 mm. long; heads solitary, axillary, slenderly pedunculate.
           1. *A. pilulifera*
1B. Leaves deciduous, ovate, rounded to shallowly cordate at base, the petioles 3–5 cm. long; heads few in a terminal raceme, on thick and
           rather short peduncles. .... 2. *A. racemosa*

1. Adina pilulifera (Lam.) Franch. *Nauclea orientalis* L., pro parte; *Cephalanthus pilulifera* Lam.; *A. globiflora* Salisb.; *A. globiflora* var. *macrophylla* Nakai; *N. nipponica* Masam.——TANI-WATARI-NO-KI.   Large, slender, much-branched evergreen shrub or small tree, puberulent on young branchlets and inflorescence; leaves somewhat coriaceous, broadly oblanceolate or narrowly oblong, 5–10 cm. long, 2–3.5 cm. wide, acuminate with an obtuse tip, acute at base, with 7–9 pairs of slender lateral nerves, the petioles 4–10 mm. long; stipules 4, free, lanceolate, 3–6 mm. long, acuminate, caducous; peduncles 4–6 cm. long, slender, the heads about 12 mm. across; corolla pale yellow, tubular, about 5 mm. long, glabrous outside; capsules narrowly cuneate, about 2.5 mm. long, with broadly linear persistent calyx-lobes at apex.——Aug.-Oct. Kyushu (s. distr. and Amakusa Isl.).——China and Indochina.

2. Adina racemosa (Sieb. & Zucc.) Miq. *Nauclea racemosa* Sieb. & Zucc.——HETSUKA-NIGAKI, HA-NIGAKI.   Deciduous tree with rather stout glabrous branches; leaves rather coriaceous, ovate or oblong-ovate, 8–12 cm. long, 4–7 cm. wide, abruptly short-acuminate, rounded or shallowly cordate at base, with about 10 pairs of prominent lateral nerves beneath, lustrous and glabrous above, short-pubescent on nerves beneath while young, long-petiolate; stipules connate; inflorescence a several-headed raceme, the heads about 15 mm. across, the peduncles 1.5–2 times as long as the head; corolla about 7 mm. long, tubular, pale yellow, puberulent externally; fruit cuneate at base, about 3 mm. long, with minute calyx-teeth at apex.——July–Aug. Shikoku, Kyushu.——Ryukyus, Formosa, and China.

4. UNCARIA Schreb.   KAGI-KATSURA ZOKU

Scandent woody plants, with curved spiny short spurs; leaves opposite, stipulate; heads axillary, peduncled, usually solitary; flowers yellowish, usually pilose, sessile or pedicelled, the bracteoles absent or minute; calyx-tube fusiform, the lobes 5; corolla-tube elongate, narrow, the lobes valvate in bud; anthers 2-spinulose at base; disc obscure; ovary fusiform, 2-locular; style long and exserted, the stigma capitate; ovules many; capsules 2-locular, septicidal, 2-valved, bifid; seeds many, long-winged on each side.——More than 30 species, mainly in Malaysia and e. India, few in e. Asia, Africa, and S. America.

1. Uncaria rhynchophylla (Miq.) Miq. *Nauclea rhynchophylla* Miq.; *Ourouparia rhynchophylla* (Miq.) Matsum. KAGI-KATSURA, KAGI-KAZURA. FIG. 16.   Plant with horizontally spreading branches; leaves oblong or ovate, 5–12 cm. long, 3–7 cm. wide, abruptly acuminate, rounded or abruptly acute at base, with 4 or 5 pairs of lateral nerves, lustrous and green above, glaucous beneath, the petioles about 1 cm. long; stipules nearly free, 4 on a node, broadly linear, caducous; spines solitary in axils, thickened at base, gently recurved toward tip; heads many-flowered, about 2 cm. across, solitary, terminal and axillary, the peduncles 2–3 cm. long; flowers pale green, the tube glabrous outside, 7–8 mm. long; fruit fusiform, slightly pilose, about 4.5 mm. long, sessile, with persistent short calyx-teeth at apex; seeds minute, long wing-tipped.——July. Woods; Honshu (Awa Prov. and westw.), Shikoku, Kyushu.

5. MUSSAENDA L.   KONRON-KA ZOKU

Plants woody, rarely herbaceous, often slightly scandent; leaves opposite, sometimes ternate, rather large, stipulate; flowers small, yellowish, many, forming a terminal cyme, pedicelled, the bracts and bracteoles deciduous; calyx-lobes 5, the tube oblong or obconic, the outer calyx-segment of some outer flowers becoming large and petaloid, colored or white; corolla infundibuliform, the tube long, lobes 5, valvate in bud; ovary 2-locular; style elongate, the stigma 2-lobed; ovules many; fruit usually fleshy; seeds many, small.——About 50 species, in the tropical regions of Asia, Africa, and the Pacific Islands.

1A. Leaves narrowly oblong, 3–4 cm. wide, acuminate; stipules free; inflorescence slightly pubescent; bracts and calyx-lobes broadly
           linear. .... 1. *M. parviflora*
1B. Leaves broadly ovate, 6–12 cm. wide, cuspidate; stipules connate; inflorescence prominently puberulent; bracts and calyx-lobes lanceolate. .... 2. *M. shikokiana*

1. Mussaenda parviflora Miq. KONRON-KA.   Evergreen scandent shrub, sparingly appressed-pilose with coarse hairs; leaves oblong or narrowly so, 8–13 cm. long, 3–4(–5) cm. wide, acuminate, acute at base, with about 5 pairs of lateral nerves, the petioles about 1 cm. long; stipules 4–7 mm. long; inflorescence rather many-flowered; calyx-lobes 5–6 mm. long, as long as or shorter than the deciduous bract, the petaloid lobe broadly ovate, 3–4 cm. long, white, stipitate; corolla

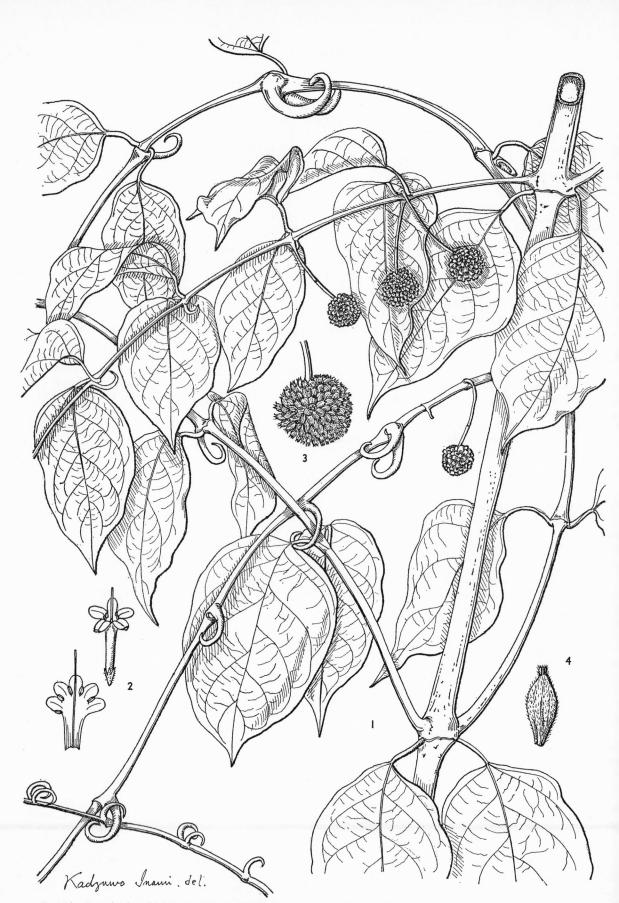

Kadzuwo Inami. del.

Fig. 16.—*Uncaria rhynchophylla* (Miq.) Miq. 1, Flowering branch; 2, flowers; 3, inflorescence at fruiting stage; 4, capsule.

12–15 mm. long, the lobes short, yellow-villous inside near base; fruit purple-black, ellipsoidal, 8–10 mm. long.——Kyushu (Yakushima).——Ryukyus and Formosa.

2. Mussaenda shikokiana Makino. Hiro-ha-konron-ka. Deciduous shrub, prominently short-pilose on inflorescence and young vegetative parts, the branches brownish; leaves thin, broadly ovate or broadly elliptic, abruptly acute to cuspidate, rounded and decurrent to a winged petiole or acuminate at base, with 7–10 pairs of lateral nerves, the peti-

oles 1–5 cm. long; stipules 2 at the node, ovate-deltoid, caducous, 8–10 mm. long; inflorescence rather many-flowered; calyx-lobes thin, lanceolate or broadly so, 7–10 mm. long, acuminate, as long as to slightly shorter than the bract, the petaloid lobe ovate-orbicular or broadly ovate, whitish, 3–5 cm. long; corolla 13–15 mm. long, densely white-villous externally, the lobes short, acuminate; fruit ellipsoidal, 8–10 mm. long.——May–July. Honshu (Kii Prov.), Shikoku, Kyushu.

6. TARENNA Gaertn. Gyokushin-ka Zoku

Trees or shrubs; leaves opposite, petiolate, stipulate; flowers in terminal cymes, the pedicels usually ebracteolate; calyx 5(4–6)-toothed; corolla hypocrateriform or infundibuliform, the lobes contorted in bud; filaments short or nearly absent; ovary 2-locular, few- to many-ovuled in each locule; style elongate, usually exserted, the stigma fusiform or slender; berry coriaceous or fleshy, 2-locular, with 1 to many seeds in each locule; seeds with fleshy or cartilaginous endosperm.——More than 40 species, mainly in the Tropics of Asia, Africa, and the Pacific Islands, few in Australia.

1. Tarenna gyokushinkwa Ohwi. *Webera corymbosa* sensu auct. Japon., non Willd.; *Chomelia corymbosa* sensu auct. Japon., non K. Schum.; *T. zeylanica* sensu auct. Japon., non Gaertn.——Gyokushin-ka. Large evergreen shrub, blackish when dry; young branchlets, underside of leaves on nerves, petioles, and inflorescence short appressed-pilose; branches of 2d year gray-brown; leaves chartaceous or thinly herbaceous, oblong, 10–15 cm. long, 4–7 cm. wide, short-acuminate or acute at both ends, deep green above, the peti-

oles 1–2 cm. long; stipules 4, connate into 2 ovate-deltoid lobes about 5 mm. long; inflorescence terminal, many-flowered, with small deltoid bracts, the pedicels 1–1.5 cm. long; flowers white; calyx short-toothed; corolla hypocrateriform, the tube about 3 mm. long, the lobes 5, spreading, oblanceolate, about 1 cm. long; anthers linear, elongate, on very short filaments; fruit broadly ellipsoidal, about 1 cm. long, 4-seeded.——June–Aug. Kyushu.——Ryukyus and Formosa.

7. RANDIA L. Misao-no-ki Zoku

Trees or shrubs, sometimes scandent, sometimes with spines; leaves usually coriaceous, opposite, often reduced and scalelike in those subtending the inflorescence branches; stipules often connate at base; flowers large, axillary, solitary or in cymes or fascicles, very rarely terminal, white or yellowish, rarely reddish; calyx-limb truncate or toothed, the lobes rarely leaflike; corolla hypocrateriform, infundibuliform or campanulate, the lobes 4 or 5, contorted in bud; ovary 2(–4)-locular; style rather thick, with a clavate or bifid stigma; berry 2-locular, many-seeded.——More than 150 species, in the Tropics, abundant in Asia and Africa.

1. Randia cochinchinensis (Lour.) Merr. *Aidia cochinchinensis* Lour.; *Stylocoryne racemosa* Cav.; *Webera densiflora* Wall.; *R. densiflora* (Wall.) Benth.; *R. racemosa* (Cav.) F. Vill.; *R. nipponensis* Makino——Misao-no-ki. Small evergreen unarmed glabrous tree; leaves rather coriaceous, lustrous, oblong to broadly lanceolate, 7–12(–15) cm. long, 2.5–5 cm. wide, acuminate, acute at base, with 4–6 pairs of lateral nerves, young leaves beneath with a few hairs in axillary tufts, the petioles about 1 cm. long; stipules 2, deltoid, 5–7

mm. long, long-cuspidate, deciduous; inflorescence a rather many-flowered nearly sessile cyme, the bracts narrowly deltoid, small, minutely ciliolate, the pedicels short; calyx small; corolla yellow, the tube short, the lobes 4, reflexed, narrowly oblong, 7–8 mm. long, as long as the anthers; stigma elongate, entire; fruit globose, 6–7 mm. across, black when mature.——May–June. Honshu (Kii Prov.), Shikoku, Kyushu.——Ryukyus, Formosa, China, Malaysia, India, and Australia.

8. GARDENIA L. Kuchi-nashi Zoku

Rarely spiny shrubs or trees; leaves opposite, rarely ternate; flowers axillary, solitary, white or yellow; calyx-limb tubular, bractlike, or lobed, usually persistent; corolla hypocrateriform, campanulate, or infundibuliform, the tube longer than the calyx, the limb 5- to 9-fid, the lobes contorted in bud; ovary 1-locular or falsely 2-locular; style thick, more so above, entire or bifid; ovules many; fruit sessile, many seeded.——About 100 species, of wide distribution in the Tropics and subtropics.

1. Gardenia jasminoides Ellis. *G. florida* L.——Korin-kuchi-nashi. Branched evergreen shrub, thinly furfuraceous on young branchlets; leaves broadly lanceolate, narrowly oblong, or broadly rhombic-oblanceolate, 6–12 cm. long, 1.5–4 cm. wide, attenuate, acute with an obtuse tip, deep green above, glabrous, short-petiolate; stipules deciduous, bractlike, connate at base; flowers solitary, short-pedicelled; calyx-lobes broadly linear, 1–2 cm. long, laterally compressed, as long as to longer than the tube, persistent; corolla white,

hypocrateriform, the tube about 3 cm. long, the lobes 6 or 7, spreading, broadly oblanceolate, 2.5–3 cm. long; anthers linear; fruit obovoid or oblong, 1.5–2.5 cm. in diameter.——June–July. Kyushu; variable.——Forma **grandiflora** (Lour.) Makino. *G. grandiflora* Lour.; *G. florida* var. *grandiflora* (Lour.) Fr. & Sav.——Kuchi-nashi. Flowers larger, with broader corolla-lobes.——Spontaneous in Honshu (Suruga Prov. and westw.), Shikoku, Kyushu.——Ryukyus, Formosa, and China. Much planted in our gardens.

Var. **radicans** (Thunb.) Makino. *G. radicans* Thunb.; *G. florida* var. *radicans* (Thunb.) Matsum.——Ko-kuchi-nashi.   Much-branched shrub with ascending branches, de-cumbent at base; leaves oblanceolate, 4–6(–8) cm. long, 1–2 cm. wide; flowers small, single or double. Known only in cultivation.

9. MORINDA L.   Yaeyama-aoki Zoku

Trees or shrubs, sometimes scandent; leaves opposite or ternate, usually membranous or slightly fleshy; stipules 2, forming a sheath at the base of the petioles; heads pedunculate, axillary or terminal, often umbellate; flowers white, rarely unisexual; calyx-limb short, persistent, truncate or obsoletely toothed; corolla infundibuliform or hypocrateriform, the tube short, the lobes 5(4–7); ovary 2 (or imperfectly 4)-locular; fruit connate and forming a headlike, juicy, many-stoned syncarp; stones free or few-connate, 1-seeded.——About 60 species, in the Tropics, mainly in Asia and the Pacific Islands.

1. Morinda umbellata L. *Stigmanthus cymosus* Lour. ——Hanagasa-no-ki.   Nearly glabrous scandent or twining shrub; leaves chartaceous, oblong or narrowly obovate, 7–12 cm. long inclusive of a petiole about 1 cm. long, 2.5–4.5 cm. wide, acute at both ends, with tufts of axillary hairs beneath; stipules connate and sheathlike, 3–4 mm. long, membranous, truncate; heads umbellate, puberulent, terminal, about 10, the simple peduncles 1–2.5 cm. long, few-flowered; calyx shallowly cup-shaped, connate in fruit; corolla about 4 mm. long, hypocrateriform, short-tubular, the lobes 4, white-villous inside at base; syncarp irregularly globose, about 8 mm. across, black when mature.——July–Aug. Kyushu (Yakushima and Tane-gashima).——Ryukyus, Formosa, China to Malaysia, and India.

10. PSYCHOTRIA L.   Bochōji Zoku

Trees or shrubs, sometimes scandent; leaves opposite, rarely verticillate in 3's or 4's; stipules interpetiolar, often forming a connate sheath; flowers in a terminal or rarely axillary thyrse, cyme, or panicle, sometimes fascicled in leaf-axils, or capitate, small, white, reddish, or yellow, bracteate or ebracteate, the bracts sometimes involucrelike; calyx-tube short; corolla hypocrateri-form, subrotate to infundibuliform, the lobes 5(4–6), valvate in bud; ovary 2-locular; style short, 2-lobed; ovules solitary in each locule, anatropous, drupe small, with 2 stones, the stones often grooved.——More than 500 species, in the Tropics, especially abundant in America.

1A. Erect shrub or small tree; leaves 12–20 cm. long; fruit red. .... 1. *P. rubra*
1B. Scandent, with aerial roots; leaves 2–4.5 cm. long; fruit white. .... 2. *P. serpens*

1. Psychotria rubra (Lour.) Poir. *Antherura rubra* Lour.; *P. elliptica* Ker-Gawl.; *P. reevesii* Wall.; *Aucubaephyllum lioukiense* Ahlburg——Bochōji, Ryūkyū-aoki.   Large glabrous shrub slightly tinged with red when dry; branchlets terete, rather thick, green; leaves thick, broadly oblanceolate, 12–20 cm. long, 2.5–5 cm. wide, acuminate with a minute obtuse tip, gradually narrowed at base, with 10–13 pairs of weak lateral nerves, the petioles 1–2 cm. long; cymes terminal, sessile, 4–7 cm. across, many-flowered; flowers small, white, about 5 mm. long, pedicelled; calyx nearly truncate at apex; corolla infundibuliform, with white hairs on throat inside, the lobes slightly less than half as long as the corolla; fruit globose, red, about 6 mm. long; seeds with longitudinal grooves on back.——Aug. Woods in low mountains and hills; Kyushu (Yakushima).——Ryukyus, Formosa, China, and Indochina.

2. Psychotria serpens L. *P. scandens* Hook. & Arn.——Shiratama-kazura.   Scandent woody plant, climbing by aid of aerial roots; leaves thick, elliptic or ovate-elliptic, 2–4.5 cm. long, 1–2.2 cm. wide, obtuse or subacute, acute at base, the lateral nerves obscure, the petioles 2–4 mm. long; stipules 2 at each node, ovate, caducous; inflorescence a terminal cyme, rather many-flowered, the bracts small; flowers white, 4–5 mm. long, on short pedicels; fruit globose or ellipsoidal, 4–5 mm. long, white at maturity.——On tree trunks in low mountains and hills; Honshu (Kii Prov.), Shikoku, Kyushu.——Ryukyus, Formosa, China, and Indochina.

11. LASIANTHUS Jack   Ruri-mi-no-ki Zoku

Shrubs; leaves opposite, short-petioled, usually entire, with prominent transverse veinlets; stipules interpetiolar, usually cadu-cous; flowers small, 5-merous, fasciculate in leaf-axils, bracteate, white, greenish, or yellowish; calyx-tube and limb short, teeth or lobes persistent; corolla infundibuliform or hypocrateriform, the lobes 4–9, valvate in bud; ovary 4- to 9-locular; style 4- to 9-lobed; ovules solitary in each locule, erect, anatropous; drupe 4- to 9-stoned.——More than 100 species, in tropical regions of Asia, few in e. Asia, Australia, Cuba, and w. Africa.

1A. Calyx-teeth shorter than the limb of calyx; corolla about 1 cm. long.
2A. Petioles and underside of leaves thinly appressed-pilose or nearly glabrous. .... 1. *L. japonicus*
2B. Petioles and underside of leaves especially on nerves with spreading hairs. .... 2. *L. satsumensis*
1B. Calyx-teeth longer than to as long as the limb of calyx; corolla 6–7 mm. long.
3A. Petioles and nerves on underside of leaves appressed-pubescent; young branches slightly appressed-pubescent or glabrous.
3. *L. tashiroi*
3B. Petioles and nerves on underside of leaves brownish yellow spreading-pubescent; young branches densely brownish hairy.
4. *L. tawadae*

1. **Lasianthus japonicus** Miq. *Mephitidia japonica* (Miq.) Nakai——RURI-MI-NO-KI. Evergreen shrub with slender, elongate glabrous or slightly appressed-pilose branches; leaves chartaceous, narrowly oblong or broadly oblanceolate, 8–15 cm. long, 2–4 cm. wide, caudate-acuminate, acute to obtuse at base, glabrous above, nearly glabrous to sparingly appressed-pilose beneath, the lateral nerves spreading and arcuate above, the petioles about 1 cm. long; flowers few in leaf-axils, nearly sessile, white; calyx short, the limb about 1 mm. long, short-toothed; fruit globose, dark blue, about 4 mm. across.——Woods in low mountains; Honshu (Izu Prov. and westw.), Shikoku, Kyushu.——Ryukyus, Formosa, and s. China.

2. **Lasianthus satsumensis** Matsum. *Mephitidia satsumensis* (Matsum.) Nakai; *L. japonicus* var. *satsumensis* (Matsum.) Makino——SATSUMA-RURI-MI-NO-KI. Closely resembles the preceding species; branches slightly pilose with ascending brownish hairs; leaves 8–10 cm. long, caudate-acuminate, nearly glabrous or thinly pilose on nerves above, with spreading brownish hairs beneath especially on nerves, the petioles spreading-pilose, about 1 cm. long; flowers few in leaf-axils, nearly sessile; calyx short, the limb about 1 mm. long, with depressed teeth; fruit globose, about 4 mm. across, dark blue.——Woods in low mountains; Honshu (Kii Prov.), Shikoku, Kyushu.

3. **Lasianthus tashiroi** Matsum. *Mephitidia tashiroi* (Matsum.) Nakai——RYŪKYŪ-RURI-MI-NO-KI, MIYAMA-RURI-MI-NO-KI. Resembles the preceding; branches with scattered appressed short pubescence while young, or glabrous; leaves firmly chartaceous, oblong or ovate-oblong, sometimes broadly oblanceolate, 6–12 cm. long, 2–4 cm. wide, caudate-acuminate, subacute at base, glabrous and lustrous on upper side, thinly appressed-hairy on nerves beneath, the petioles about 1 cm. long; flowers white, 6–7 mm. long, few in leaf-axils, nearly sessile; calyx-teeth longer than to as long as the limb of calyx, suberect, narrowly deltoid, subacute; corolla 6–7 mm. long, glabrous, the lobes 5, ovate, about half as long as the tube, white-pubescent inside; anthers about half as long as the corolla-lobes; fruit globose, about 4 mm. across.——Kyushu (Yakushima).——Ryukyus, Formosa, Hainan Island, and Philippines.

4. **Lasianthus tawadae** Ohwi. *L. tashiroi* var. *pubescens* Matsum.; *Mephitidia tashiroi* var. *pubescens* (Matsum.) Nakai——KE-HADA-RURI-MI-NO-KI, KE-MIYAMA-RURI-MI-NO-KI. Closely resembles the preceding; young branches densely ascending yellow-brown pubescent; leaves chartaceous, oblong to narrowly so, or obovate-oblong, 6–10 cm. long, 2–4 cm. wide, caudate-acuminate, glabrous above, densely spreading yellow-brown pubescent beneath, especially on nerves, the lateral nerves obliquely spreading, arcuate, the petioles yellow-brown pubescent, 5–8 mm. long; flowers few, in leaf-axils, sessile, densely brownish yellow hairy; calyx short, the lobes or teeth linear-deltoid, 2.5–3 mm. long, acute; fruit globose, about 4 mm. across.——Kyushu (Yakushima).——Ryukyus.

12. DAMNACANTHUS Gaertn. f.　ARIDŌSHI ZOKU

Evergreen much-branched frequently spinose shrubs; leaves opposite, coriaceous, ovate, subsessile; flowers rather small, 4-merous, white, 1 or 2 in leaf-axils, short-pedicellate, with minute bracteoles at base; calyx small, the limb 4- or 5-toothed, the teeth persistent; corolla infundibuliform, the lobes 4(–5), valvate in bud; ovary 2- to 4-locular; style slender, filiform, the stigma 2- to 4-lobed; ovules solitary in each locule; drupe 1- to 4-stoned.——Few species, in e. Asia and Bengal.

1A. Spines 1–3 mm. long or nearly absent, not longer than the petioles; roots moniliform; leaves 8–13 cm. long. .... 1. *D. macrophyllus*
1B. Spines 1–2 cm. long; root not moniliform; leaves 1–5 cm., rarely to 6.5 cm. long.
  2A. Spines about 1 cm. long, ¼–½ as long as the leaves. .... 2. *D. major*
  2B. Spines 1–2 cm. long, as long as to half as long as the leaves. .... 3. *D. indicus*

1. **Damnacanthus macrophyllus** Sieb. ex Miq. *D. major* var. *submitis* Maxim. ex Regel; *D. major* var. *macrophyllus* (Sieb.) Maxim.; *D. indicus* var. *macrophyllus* (Sieb.) Makino——ŌBA-JUZUNE-NO-KI, ŌBA-NO-ARIDŌSHI. Roots moniliform; plant glabrous or minutely puberulous on young parts, the branches grayish; leaves rather thin, lustrous, elliptic, ovate-elliptic, or oblong, 3–8 cm. long, 1.5–4 cm. wide, short-acuminate, rounded at base, glabrous, spines absent or 1–3 mm. long; flowers 1 or 2 in leaf-axils, the pedicels 2–3 mm. long; calyx-teeth short, acuminate, about 1 mm. long; corolla white; fruit globose, red, 5–6 mm. across.——Shikoku, Kyushu.

Var. **giganteus** (Makino) Koidz. *D. indicus* var. *giganteus* Makino; *D. giganteus* (Makino) Nakai.——NABAGA-JUZUNE-NO-KI. Leaves broadly lanceolate or narrowly oblong, 5–13 cm. long, long-acuminate, acute at base.——Honshu (Mikawa Prov. and westw.), Shikoku, Kyushu.

Var. **parvispinis** (Koidz.) Takah.——*D. major* var. *parvispinis* Koidz.; *D. moniliformis* Koidz.; *D. minutispinis* Koidz.; *D. major* var. *minutispinis* Koidz.; *D. major* var. *moniliformis* (Koidz.) Hara——MARUBA-JUZUNE-NO-KI. Leaves smaller, broadly ovate-rotundate, rounded at base.——Honshu (Kinki Distr. and westw.), Kyushu.

2. **Damnacanthus major** Sieb. & Zucc. *D. indicus* var. *major* (Sieb. & Zucc.) Makino; *D. indicus* var. *latifolius* Nakai——JUZUNE-NO-KI. Much-branched shrub, sometimes with thickened roots, the branches minutely puberulent, grayish; leaves somewhat coriaceous, lustrous, short-petiolate, ovate or broadly so, 1.5–4 cm. long, 1–2 cm. wide, acute to short-acuminate, rounded or very obtuse at base, glabrescent, the spines about 1 cm. long; flowers 1 or 2 in leaf-axils, short-pedicelled; calyx short, the teeth narrowly deltoid, 1.5–2 mm. long; corolla white, 15–18 mm. long, the lobes short, about 4 mm. long; fruit red, globose, about 5 mm. across.——May. Woods; Honshu (ne. Kantō Distr. and westw.), Shikoku, Kyushu; variable.

Var. **parvifolius** Koidz. KOBA-NO-NISE-JUZUNE-NO-KI. Leaves smaller, 1–3.5 cm. long.——Honshu (Izu Prov. and westw.), Kyushu.——Ryukyus and s. Korea.

Var. **lancifolius** (Makino) Ohwi. *D. indicus* var. *lancifolius* Makino; *D. lancifolius* (Makino) Koidz.——HOSOBA-NISE-JUZUNE-NO-KI. Leaves broadly lanceolate, 3–6.5 cm.

long, gradually narrowed at both ends.——Honshu (Suruga, Echizen, and westw. to Kinki Distr.).

3. Damnacanthus indicus Gaertn. f. Aridōshi. Much-branched shrub, crisped puberulent on branches, petioles, and sometimes on leaves of midrib beneath; branches slender, horizontally spreading; leaves coriaceous, lustrous, broadly ovate or broadly elliptic, 1–2.5 cm. long, 7–20 mm. wide, abruptly acute and short-awned, rounded at base, short-petioled, the spines 1–2 cm. long, as long as the leaves; flowers 1 or 2 in leaf-axils, short-pedicelled, white, about 15 mm. long; calyx short, about 1.5 mm. long, the teeth narrowly deltoid; corolla tubular, slightly ampliate, the lobes 4, spreading, very short; fruit globose, 5–6 mm. across, red.——May. Honshu (Kantō Distr. and westw.), Shikoku, Kyushu.——Ryukyus, s. Korea, China, ne. India, and Thailand.

Var. **microphyllus** Makino. Hime-aridōshi. Plants much branched; leaves small, about 1 cm. long; occurs with the typical phase.

Var. **ovatus** Koidz. Tamagoba-aridōshi. Spines prominent; leaves ovate.——Honshu (centr. and w. distr.), Shikoku.

13. LEPTODERMIS Wall. Shichōge Zoku

Slender much-branched shrubs; leaves short-petioled, opposite, lanceolate, with broad persistent stipules; flowers dimorphic, white or purple, subtended by 2 connate bracteoles, in sessile terminal cymes or heads on branches and on short spurs; calyx-limb 5-toothed; corolla narrowly infundibuliform, the lobes (4 or)5, valvate in bud; ovary 5-locular; style slender, 5-lobed; ovules solitary in each locule, anatropous; capsules 5-valved, oblong, ribbed, 1-seeded in each valve; seeds linear.——About 30 species, in e. Asia, e. Bengal, and the Himalayas.

1. Leptodermis pulchella Yatabe. Shichōge. Foetid-smelling shrub, while young with dark gray puberulent branches; leaves narrowly ovate or broadly lanceolate, 1.5–4 cm. long, subacute, acute at base, glabrous or thinly puberulent on midrib above, the petioles 1–3 mm. long, the stipules 2, depressed-deltoid, short spine-tipped; flowers few, sessile, the bracteoles connate at base, membranous, 3–4 mm. long, about as long as and surrounding the calyx in anthesis; calyx-teeth elliptic, about 2 mm. long, erect, ciliolate; corolla purple, 15–18 mm. long, puberulent externally; fruiting calyx oblong, 7–8 mm. long.——July. Honshu (s. Kinki Distr.), Shikoku.

14. PAEDERIA L. Hekuso-kazura Zoku

Scandent, foetid herbs or shrubs; leaves opposite, petiolate; stipules deciduous; flowers rather small, in axillary and terminal cymes, bracteolate; calyx-limb 4- or 5-toothed, persistent; corolla tubular-campulate or infundibuliform, granular-pilose, the tube rather broad, the lobes 4 or 5, induplicate in bud; ovary 2-locular; styles 2, filiform, often connate at base; ovules solitary in each locule, erect, anatropous; fruit globose, the pericarp crustaceous; stones 1-seeded, flattened.——About 20 species, in the Tropics of Asia and S. America.

1. Paederia scandens (Lour.) Merr. var. **mairei** (Lév.) Hara. *P. foetida* sensu auct. Japon., non L.; *P. tomentosa* sensu auct. Japon., non Bl.; *P. tomentosa* var. *mairei* Lév.——Hekuso-kazura, Yaitobana. Scandent, somewhat pubescent perennial herb often woody at base; leaves petiolate, membranous or thinly herbaceous, ovate to narrowly so, or the upper ones broadly lanceolate, 4–10 cm. long, 1–7 cm. wide, short-acuminate, cordate or sometimes rounded at base; stipules 2, ovate, green, obliquely ascending; cymes rather many-flowered; flowers sessile in the terminal ones, short-pedicelled in the lateral ones; calyx-teeth short, obliquely spreading; corolla-tube white, red-purple inside, about 10 mm. long, rather broad, the limb spreading, 7–8 mm. across, the lobes depressed, white, undulate on margin; fruit yellow, globose, about 5 mm. across.

——Aug.–Sept. Thickets and thin woods in lowlands and low elevations in the mountains; Hokkaido, Honshu, Shikoku, Kyushu; very common and variable.——Korea, China to Philippines.

Var. **longituba** (Nakai) Hara. *P. longituba* Nakai——Tsutsu-naga-yaitobana. Corolla-tube narrow.——Honshu, Kyushu.

Var. **maritima** (Koidz.) Hara. *P. chinensis* var. *maritima* Koidz.——Hama-saotome-kazura. Glabrescent maritime phase with thicker lustrous leaves.——Thickets near seashores; Honshu, Shikoku, Kyushu.——China.

Var. **velutina** (Nakai) Nakai. *P. chinensis* var. *velutina* Nakai——Birōdo-yaitobana. Leaves densely villous.——Often found with the typical phase.

15. PSEUDOPYXIS Miq. Inamori-sō Zoku

Small erect herbs with creeping rhizomes and multicellular pubescence; leaves membranous, opposite, petiolate, ovate; stipules intrapetiolar; flowers white or pale purple, 1 or 2, terminal and in upper axils, the pedicels short, 2-bracteolate; calyx short, the limb 5-parted, the segments broadly lanceolate at anthesis, accrescent, net-veined in fruit; corolla narrowly infundibuliform, lobes 5, induplicate in bud; ovary 4- or 5-locular; style slender, 4- or 5-lobed; ovules solitary in each locule, ascending from the inner angle, anatropous; fruit small, surrounded by stellately spreading calyx-lobes, circumscissile; seeds obovate, longitudinally grooved.——Two species, in Japan.

1A. Stems 5–10 cm. long, uniformly pubescent; upper 2 pairs of leaves frequently approximate and falsely verticillate; pedicels about 1 mm. long; corolla pale purple, about 2.5 cm. long, the lobes half as long as the tube. .... 1. *P. depressa*
1B. Stems 15–30 cm. long, with 2 series of short hairs; leaves in pairs, the upper leaves not as above; pedicels 4–6 mm. long; corolla white, 8–10 mm. long, the lobes nearly as long as the tube. .... 2. *P. heterophylla*

1. Pseudopyxis depressa Miq. *P. longituba* Fr. & Sav.——INAMORI-SŌ. Pubescent perennial herb, with slender creeping rhizomes; stems 5–10 cm. long, with 2 or 3 pairs of leaves; leaves petiolate, ovate or deltoid-ovate, 3–6 cm. long, acute or subacute, truncate to rounded, sometimes shallowly cordate, the lower leaves sometimes 1–2 cm. long; upper 2 pairs of leaves frequently approximate and falsely verticillate; calyx long-pubescent, the lobes narrowly deltoid in fruit, about 5 mm. long, acuminate and spine-tipped; corolla glabrous or thinly puberulent, pale purple, about 2.5 cm. long, the lobes half as long as the tube, ovate, about 7 mm. long; seeds about 2.5 mm. long, with raised longitudinal lines.——May–June. Woods in mountains; Honshu (centr. and s. distr.), Shikoku, Kyushu.

2. Pseudopyxis heterophylla (Miq.) Maxim. *Oldenlandia heterophylla* Miq.; *Lysimachia quadriflora* Petitm.; *P. quadriflora* (Petitm.) Koidz.——SHIROBANA-INAMORI-SŌ. Loosely pubescent perennial herb with slender woody rhizomes; stems 15–30 cm. long, with 2 series of short hairs; leaves 4–6 pairs, ovate or deltoid-ovate, 2–6 cm. long, 1–2.5 cm. wide, acute, broadly cuneate or sometimes nearly rounded at base, the petioles 1–1.5 cm. long, the uppermost leaves sessile; calyx-lobes obliquely spreading in fruit, ovate or broadly elliptic, the lobes abruptly acute, about 5 mm. long; corolla white, 8–10 mm. long, glabrous; seeds broadly elliptic, about 2.5 mm. long, with raised slender longitudinal lines and horizontal short striae.——July–Aug. Honshu (centr. distr.).

16. MITCHELLA L.   TSURU-ARIDŌSHI ZOKU

Small creeping evergreen herbs; leaves opposite, short-petiolate, small, with minute stipules; flowers small, in pairs, axillary and terminal, white, bractless, on a short common peduncle; calyx short, the limb 3- to 6-toothed, persistent; corolla infundibuliform, hairy on the throat inside, the limb 3- to 6-lobed, valvate in bud; ovary 4-locular; style slender, 4-fid; ovules 4, erect, anatropous; fruit 4-stoned, the stones bony, each 1-seeded.——Two species, one in e. Asia and the other in eastern N. America.

1. Mitchella undulata Sieb. & Zucc. *M. repens* var. *undulata* (Sieb. & Zucc.) Makino——TSURU-ARIDŌSHI. Glabrous evergreen deep green herb; stems slender, leafy, elongate, much branched, creeping; leaves short-petioled, deltoid-ovate, ovate or broadly so, 1–1.5 cm. long, 7–12 mm. wide, subacute to obtuse, rounded at base, slightly undulate on margin; flowers sessile, in pairs on a short common peduncle; ovary connate; corolla white, about 15 mm. long, the tube slender, 1.5–2 times as long as the lobes; fruit globose, red, about 8 mm. across.——June. Woods in mountains; Hokkaido, Honshu, Shikoku, Kyushu; rather common.——s. Korea.

17. RUBIA L.   AKANE ZOKU

Herbs, sometimes woody at base; leaves apparently in verticils of 4–8, very rarely opposite, stipulate; flowers small, usually 5-merous, in axillary and terminal cymes, the pedicels jointed with the calyx; calyx-teeth obscure; corolla rotate or subcampanulate, the lobes 4 or 5, valvate in bud; filaments short; ovary 2-locular or 1-locular by reduction, each locule with an erect amphitropous ovule; styles 2, free or connate, short, the stigma capitate; fruit fleshy, (1–)2-locular, (1–)2-seeded.——About 35 species, in temperate to warmer regions of the N. Hemisphere, S. Africa, and S. America.

1A. Leaves broadly lanceolate, gradually narrowed toward base, sessile or nearly so. .... 1. *R. jesoensis*
1B. Leaves ovate to lanceolate, rounded to cordate at base, distinctly petiolate.
  2A. Stems erect, not retrorsely aculeolate. .... 2. *R. chinensis*
  2B. Stems usually scandent, retrorsely aculeolate.
    3A. Leaves 4 in each verticil. .... 3. *R. akane*
    3B. Leaves, at least on the main stems, 6(–12) in each verticil, but on the branches 4–6.
      4A. Leaves in 4's to 6's on main stems, cordate to ovate-cordate. .... 4. *R. hexaphylla*
      4B. Leaves in 6's to 10's on main stems, ovate-oblong, shallowly cordate to rounded at base. .... 5. *R. cordifolia* var. *pratensis*

1. Rubia jesoensis (Miq.) Miyabe & Miyake. *Galium jesoense* Miq.; *R. tatarica* var. *grandis* F. Schmidt; *R. grandis* (F. Schmidt) Komar.——AKANE-MUGURA, Ō-AKANE. Rhizomes slender, creeping; stems 40–80 cm. long, erect, usually simple, 4-angled, retrorsely aculeolate on angles; leaves in verticils of 4's, lanceolate, 5–8 cm. long, 6–15 mm. wide, acuminate at both ends, retrorsely aculeolate on margin and on nerves beneath, spreading pilosulous beneath, sessile or nearly so; cymes axillary, rather many-flowered, slightly shorter than to as long as the leaves, peduncled, 15–25 mm. across; flowers small, whitish; ovary smooth; corolla rotate, 3–4 mm. across; fruit globose.——June–July. Hokkaido, Honshu (n. and Hokuriku distr.).——s. Kuriles, Sakhalin, Korea, and Ussuri.

2. Rubia chinensis Regel & Maack var. **glabrescens** (Nakai) Kitag. (?) *R. mitis* Miq.; *R. mitis* forma *glabrescens* Nakai; *R. pedicellata* Nakai——Ō-KINUTA-SŌ. Smooth perennial herb with slender creeping rhizomes; stems simple, nearly erect; leaves thinly membranous, ovate or ovate-oblong, 6–10 cm. long, 2.5–5 cm. wide, acuminate, rounded to shallowly cordate at base, glabrous or nearly so, 5- to 7-nerved, the petioles 1–2 cm. long; cymes sometimes paniculate, terminal and in upper axils; flowers greenish white, on slender pedicels; ovary glabrous; corolla 4- or 5-lobed, rotate, 3–4 mm. across; fruit globose, black when mature.——May–July. Woods in mountains; Hokkaido, Honshu, Shikoku, Kyushu.

Var. **chinensis.** MANSEN-Ō-KINUTA-SŌ. Plant prominently pilose.——Honshu (centr. distr.).——Korea, Manchuria, N. China, and Amur regions.

3. Rubia akane Nakai. *R. cordifolia* sensu auct. Japon., non L.; *R. cordata* var. *mungista* Miq., excl. syn.; *R. cordata* var. *laxa* Nakai——AKANE. Perennial herb with much-branched yellowish rhizomes; stems much branched, elongate, 4-angled, retrorsely aculeolate on the angles; leaves in verticils of 4's, petiolate, ovate to ovate-cordate, 3–7 cm. long, 1–3 cm.

wide, acuminate, cordate at base, with 5 nerves from the base, retrorsely aculeolate on petioles, on margins, and on nerves of leaves beneath; inflorescence subpaniculate, axillary and terminal, often compound, the pedicels short; flowers pale yellow, 3.5–4 mm. across; ovary smooth; corolla rotate, 4- or 5-lobed; fruit black.——Aug.–Oct. Thickets and hedges in lowlands and low elevations in the mountains; Honshu, Shikoku, Kyushu; rather common.——Korea, China, and Formosa.

4. Rubia hexaphylla (Makino) Makino. *R. cordifolia* var. *hexaphylla* Makino——Ō-AKANE. Resembles the preceding but larger; stem-leaves in verticils of 4's to 6's, cordate or ovate-cordate, 5–13 cm. long, 3–7 cm. wide, long-petiolate, (5 or)7-nerved, the leaves on branches smaller, broadly deltoid-ovate, rounded at base; inflorescence rather few-flowered, the pedicels 4–7 mm. long; ovary glabrous, smooth; corolla cam-

panulate-rotate, 5-lobed, about 3 mm. across; fruit black.—— July–Sept. Mountains; Honshu, Kyushu.——Korea.

5. Rubia cordifolia L. var. **pratensis** Maxim. *R. pratensis* (Maxim.) Nakai; *R. cordifolia* var. *lancifolia* sensu auct. Japon., non Regel——KURUMABA-AKANE, TSUSHIMA-AKANE. Resembles No. 3; leaves rather thick, on main stems in verticils of 6's to 10's, on the branches in 4's to 6's, ovate-oblong, 2–4 cm. long, 1–2 cm. wide, cuspidate or abruptly acute, rounded or slightly cordate at base, petiolate, usually 5-nerved, prominently retrorsely aculeolate on margin and on nerves beneath; inflorescence paniculate, axillary and terminal; flowers pedicelled, about 3.5 mm. across; ovary glabrous; corolla 5-lobed, rotate; fruit black.——Oct. Honshu (Chūgoku Distr.), Kyushu (n. distr.).——Korea, Manchuria, Ussuri, and Amur.

18. GALIUM L. YAE-MUGURA ZOKU

Annuals, biennials, or perennials, often retrorsely scabrous; stems erect or scandent; leaves whorled, (3–)4 to many in a whorl; flowers small, usually 4-merous, in axillary cymes or terminal panicles, white, pale yellow, or greenish, bractless, the pedicels jointed with the calyx; calyx-teeth obsolete; corolla rotate, (3–)4-lobed, the lobes valvate in bud; stamens (3–)4, the filaments short; ovary 2-locular; styles 2, short; ovules solitary in each locule; fruit usually dry, globose, separating at maturity into 2 seedlike indehiscent 1-seeded carpels, smooth or scabrous, often prickly.——About 300 species, nearly worldwide, except Australia.

1A. Leaves on the main stems in verticils of 6's to 10's.
  2A. Stems erect; leaves linear, short-pubescent, especially at the nodes; flowers greenish white to pale yellow. .... 1. *G. verum*
  2B. Stems weak and decumbent or creeping, retrorsely aculeolate on the angles; flowers white to pale green.
    3A. Leaves not aculeolate on margin.
      4A. Cymes nearly all terminal.
        5A. Leaves narrowly oblong, rounded to obtuse, mucronate, deep green when dry. .... 2. *G. trifloriforme*
        5B. Leaves narrowly ovate or broadly lanceolate, acute, mucronate, black-green when dry. .... 3. *G. japonicum*
      4B. Cymes axillary and terminal; leaves narrowly ovate-oblong. .... 4. *G. triflorum*
    3B. Leaves retrorsely aculeolate on margin.
      6A. Straggling annual or biennial; leaves awn-tipped; inflorescence axillary and terminal; flowers pale green. ...... 5. *G. spurium*
      6B. Perennial, very slender, simple or sparsely branched, with slender creeping rhizomes; leaves short-mucronate; inflorescence nearly terminal; flowers white. .... 6. *G. dahuricum*
1B. Leaves on main stems in verticils of 4's, sometimes in 5's or 6's.
  7A. Leaves retrorsely aculeolate on margin, in verticils of 5's to 6's on main stems.
    8A. Leaves rounded to very obtuse; corolla 3- or 4-lobed; fruitlets subglobose, glabrous. ..... 7. *G. trifidum* var. *brevipedunculatum*
    8B. Leaves rounded or obtuse, mucronate-tipped; corolla 4-lobed; fruitlets ellipsoidal, usually with hooked prickles.
                                                     8. *G. pseudoasprellum*
  7B. Leaves not retrorsely aculeolate, in verticils of 4's(–5's) on main stems.
    9A. Leaves 1-nerved or obscurely so.
      10A. Leaves nearly glabrous or loosely appressed-pilose on margin and on upper surface near margin.
        11A. Stems 5–20 cm. long, usually simple; leaves in a verticil often unequal in size; inflorescence nearly terminal.
          12A. Leaves 1–4 cm. long inclusive of the petiole; fruitlets with hooked prickles. .... 9. *G. paradoxum*
          12B. Leaves 0.6–1.2 cm. long inclusive of the petiole; fruitlets puberulent. .... 10. *G. yakusimense*
        11B. Stems 30–80 cm. long, often branched; leaves in a verticil equal or nearly equal in size; inflorescence axillary and terminal.
          13A. Pedicels much longer than the fruit; bracts 2(–3); fruitlets glabrous. .... 11. *G. niewerthii*
          13B. Pedicels shorter than the fruit; bracts solitary; fruitlets rather densely pilosulous. .... 12. *G. kikumugura*
      10B. Leaves with obliquely spreading long hairs on midrib beneath and on margin.
        14A. Pedicels 1–3 mm. long; leaves yellow-green or pale green when dry.
          15A. Leaves ovate-oblong, elliptic, or ovate. .... 13. *G. trachyspermum*
          15B. Leaves oblong-linear or linear-lanceolate. .... 14. *G. gracilens*
        14B. Pedicels (3–)5–10 mm. long; leaves vivid green when dry. .... 15. *G. pogonanthum*
    9B. Leaves distinctly 3-nerved.
      16A. Leaves thinly membranous, rounded and mucronulate-tipped, rarely acute; fruitlets densely prickly. .. 16. *G. kamtschaticum*
      16B. Leaves thicker, gradually acute with an obtuse tip, or acute to acuminate; fruitlets glabrous or short-hairy.
        17A. Inflorescence a large terminal many-flowered panicle.
          18A. Leaves lanceolate to ovate-oblong, acuminate or acute with an obtuse tip; inflorescence densely flowered, the branches short; pedicels very short. .... 17. *G. boreale* var. *kamtschaticum*
          18B. Leaves ovate to narrowly so, acuminate; inflorescence loosely flowered, with rather slender branches; pedicels slender.
                                                          18. *G. kinuta*
        17B. Inflorescence loosely cymose, terminal, 3- to 7-flowered. .... 19. *G. nakaii*

1. **Galium verum** L. var. **asiaticum** Nakai. *G. verum* var. *luteum* Nakai, excl. syn.; *G. verum* var. *japonalpinum* Nakai——Kibana-kawara-matsuba. Perennial herb; stems erect, slightly branched in upper part, short-pubescent especially at the nodes; leaves in verticils of 8's to 10's(–12's), linear, 2–3 cm. long, 1.5–3 mm. wide, short spine-tipped, pubescent beneath; inflorescence a terminal and axillary densely many-flowered panicle; flowers greenish white to pale yellow, about 2.5 mm. across; fruitlets ellipsoidal, glabrous.——July–Oct. Grassy slopes in mountains; Hokkaido, Honshu, Kyushu; very variable in flower color and in the indument.——Forma **luteolum** Makino. Usuiro-kawara-matsuba. With pale yellow flowers.——Forma **nikkoense** (Nakai) Ohwi. *G. verum* var. *lacteum* sensu auct. Japon., non Maxim.; *G. verum* var. *nikkoense* Nakai——Kawara-matsuba. With pale flowers, and the leaves glabrous on upper surface.

Var. **trachycarpum** DC. *G. verum* var. *ruthenicum* Nakai, excl. syn.——Ezo-kawara-matsuba. Ovary densely pilose; flowers pale yellow.——Grassy places in mountains; Hokkaido, Honshu; variable.——Forma **album** Nakai. Chōsen-kawara-matsuba. Flowers white.——Forma **intermedium** Nakai. Usuki-kawara-matsuba. Flowers creamy white.——Forma **tomentosum** Nakai. Ezo-no-ke-kawara-matsuba. Prominently pubescent throughout.——The species occurs widely from e. Asia to e. Europe.

2. **Galium trifloriforme** Komar. *G. japonicum* Makino, pro parte; *G. nipponicum* Makino, pro parte——Oku-kuruma-mugura, Chōsen-kuruma-mugura. Perennial herb, deep green when dry; stems ascending, 20–50 cm. long, 4-angled, retrorsely scabrous on the angles; leaves usually in verticils of 6's, narrowly oblong, 2.5–4 cm. long, 7–10 mm. wide, rounded to obtuse, mucronate, acute at base, the midrib usually smooth; cymes loosely few-flowered, axillary and terminal; flowers white, pedicelled; fruitlets with long hooked spines, densely so.——July–Aug. Damp woods in mountains; Hokkaido, Honshu, Shikoku, Kyushu.——Kuriles, Sakhalin, Korea, Manchuria, and China.

3. **Galium japonicum** Makino. *G. japonicum* Makino, pro parte; *G. nipponicum* Makino, pro parte; *G. trifloriforme* var. *nipponicum* (Makino) Nakai——Kuruma-mugura. Resembles the preceding, blackish when dry; stems usually smooth; leaves narrowly ovate or broadly lanceolate, acute, mucronate, smooth except the margin and the upper surface near the margin; fruitlets with long prickles.——May–July. Damp woods in mountains; Hokkaido, Honshu, Shikoku, Kyushu.——Korea.

4. **Galium triflorum** Michx. Yatsugatake-mugura. Perennial; stems decumbent, 20–40 cm. long, retrorsely prickly on the angles; leaves in verticils of 6's, narrowly ovate-oblong, 1–3 cm. long, abruptly cuspidate, with retrorsely directed prickles on midrib beneath; inflorescence axillary and terminal, several-flowered; flowers white, small, pedicelled; fruitlets relatively long-prickled.——July. Damp coniferous woods; Honshu; very rare.——Kuriles, Sakhalin, and northern regions of the N. Hemisphere.

5. **Galium spurium** L. var. **echinospermon** (Wallr.) Hayek. *G. vaillantii* DC.; *G. strigosum* Thunb.; *G. agreste* var. *echinospermon* Wallr.; *G. spurium* var. *vaillantii* (DC.) Gaud.; *G. aparine* var. *echinospermon* (Wallr.) Farwell——Yae-mugura. Straggling annual or biennial; stems elongate, much branched, retrorsely scabrous; leaves in verticils of 6's to 8's, narrowly oblanceolate or broadly linear, 1–3 cm. long, 1.5–4 mm. wide, obtuse and short awn-tipped, gradually narrowed at base, retrorsely scabrous on margin and on midrib beneath; inflorescence few flowered, axillary and terminal; flowers small, pale green, on spreading pedicels; fruit with hooked prickles.——May–July. Grassy slopes, waste grounds and thickets in lowlands and low elevations in the mountains; Hokkaido, Honshu, Shikoku, Kyushu; very common.——Sakhalin, Korea, China, Ryukyus, to Europe and Africa.

Var. **spurium**. *G. agreste* var. *leiospermon* Wallr.; *G. aparine* var. *spurium* (L.) Wimmer——Toge-nashi-yae-mugura. Fruitlets not prickly.——The typical phase is doubtfully naturalized in Hokkaido.——Eurasia.

6. **Galium dahuricum** Turcz. var. **tokyoense** (Makino) Cufodontis. *G. tokyoense* Makino; *G. asprellum* var. *tokyoense* (Makino) Nakai——Hana-mugura. Very slender simple or sparsely branched perennial with slender creeping rhizomes; stems 30–60 cm. long, retrorsely scabrous on the angles; leaves usually in verticils of 6's, oblanceolate to narrowly so, 2–3 cm. long, 4–7 mm. wide, rounded, slightly retuse, or obtuse, short-mucronate, retrorsely scabrous on margin and on the midrib beneath; inflorescence rather many-flowered; flowers white, pedicelled; fruit glabrous.——June–July. Wet grassy places along rivers in lowlands; Honshu (centr. and n. distr.).——Korea and Manchuria.

Var. **dahuricum**. *G. asprellum* var. *dahuricum* Maxim.——Ezo-mugura. Leaves acuminate, short awn-tipped.——Reported from Hokkaido.——Korea and the temperate regions of e. Asia.

7. **Galium trifidum** L. var. **brevipedunculatum** Regel. *G. trifidum* var. *pacificum* Wieg.; *G. trifidum* sensu auct. Japon., non L.——Hosoba-no-yotsuba-mugura. Rather delicate perennial, black when dry; stems 15–40 cm. long, often weakly scabrous-prickly; leaves in verticils of 4's or sometimes in 5's, narrowly oblong to oblanceolate, sometimes slightly unequal in the same whorl, rounded to very obtuse at apex, narrowed at base, with sparse retrorse prickles on margins and on midrib beneath; cymes few-flowered, terminal; flowers small, white, 3- or 4-lobed, the pedicels spreading; fruitlets subglobose, glabrous.——May–July. Wet boggy places in lowlands; Hokkaido, Honshu, Shikoku, Kyushu.——e. Asia and N. America. The typical phase occurs in Eurasia.

8. **Galium pseudoasprellum** Makino. *G. asprellum* var. *lasiocarpum* Makino——Ōba-no-yae-mugura. Green perennial; stems much elongate and branched, ascending, with retrorse prickles; leaves in verticils of 5's or 6's on the main branches, the uppermost ones in 4's or 5's, broadly oblanceolate to narrowly obovate-oblong, or oblanceolate in the upper ones, 1.5–3 cm. long, 5–10 mm. wide, obtuse to rounded, mucronate-tipped, sometimes acuminate in the upper ones, retrorse-prickly on the margins and on the midrib beneath; inflorescence axillary and terminal, loosely flowered, with slender branches and pedicels; corolla 4-lobed; fruitlets ellipsoidal, usually with hooked prickles.——Aug.–Oct. Mountains; Hokkaido, Honshu, Shikoku, Kyushu.——Korea, Manchuria, and China.

9. **Galium paradoxum** Maxim. *G. stellariifolium* Fr. & Sav.——Miyama-mugura. Rather soft perennial, deep green when dried; stems 10–25 cm. long, smooth, with 3 to 5 whorls of leaves; leaves in verticils of 4's, one of the pairs smaller than the other, ovate to broadly so, 1–4 cm. long, 1–2 cm. wide, acute, rounded or acute at base, loosely strigose on margin and on upper side toward the margin, the petioles 4–12 mm. long; cymes pedunculate, few-flowered; flowers white; fruit long-

pedicellate, with hooked prickles.——July–Aug. Damp coniferous woods in mountains; Hokkaido, Honshu, Shikoku, Kyushu.——Korea, China, Manchuria, and e. Siberia.

10. Galium yakusimense Masam. Yakushima-mugura. Delicate perennial; stems 5–10 cm. long, smooth, with few whorls of leaves; leaves in verticils of 4's, rather unequal in a whorl, narrowly obovate or obovate-oblong, 6–10(–12) mm. long, 2–3 mm. wide, acute, minutely awn-tipped, narrowed at base, subsessile, deep green, loosely strigose on margins and on midrib beneath; cymes few-flowered, the pedicels rather slender, 2–5 mm. long; fruit puberulent.——Aug.–Oct. Kyushu (Yakushima); rare.

11. Galium niewerthii Fr. & Sav. Yabu-mugura. Perennial; stems elongate, slender, smooth, 4-angled, 50–70 cm. long; leaves in verticils of 4's or 5's, narrowly obovate to oblong, 1–1.5(–2) cm. long, abruptly cuspidate, acute at base, loosely strigose on margin, on upper side toward margin, and on under side; inflorescence axillary, loosely few-flowered, the pedicels much longer than the fruit, the bracts 2(–3); fruitlets oblong, glabrous.——Aug. Honshu (Kantō Distr.).

12. Galium kikumugura Ohwi. *G. brachypodion* Maxim., non Jord.——Kiku-mugura. Perennial; stems much elongate, smooth, 30–50 cm. long; leaves membranous, vivid green when dry, narrowly ovate or oblong, 1–1.5(–2.5) cm. long, 5–8 mm. wide, rounded to obtuse, mucronate, thinly strigose on margin and on upper surface near the margin; cymes terminal and axillary, few-flowered, on rather long, slender peduncles, the pedicels unequal, 1–5 mm. long, shorter than the fruit, the bracts solitary; flowers white, sometimes bracteate at base; fruitlets oblong, rather densely pilosulous. ——May–June. Hokkaido, Honshu, Shikoku, Kyushu.

13. Galium trachyspermum A. Gray. *G. gracile* sensu auct. Japon., non Bunge; *G. bungei* var. *trachyspermum* (A. Gray) Cufodontis——Yotsuba-mugura. Delicate nearly erect perennial; stems slender, 4-angled, glabrous or nearly so, 30–50 cm. long; leaves verticillate in 4's, ovate-oblong, elliptic, or ovate, 1–1.5 cm. long, 3–6 mm. wide, obtuse or subacute, rounded at base, with spreading whitish hairs on margin and on lower surface; cymes axillary, densely about 10-flowered, the pedicels 1–3 mm. long; flowers small, pale yellowish green; fruit relatively small, densely warty.——May–June. Thickets in hills and low elevations in the mountains; Hokkaido, Honshu, Shikoku, Kyushu.——Korea and China.——Forma **hispidum** (Matsuda) Ohwi. *G. gracile* forma *hispidum* Matsuda——Ke-yotsuba-mugura. With prominent white hairs throughout.——Occurs with the typical phase.

Var. **nudicarpum** Honda. Ke-nashi-yotsuba-mugura. Fruitlets smooth or nearly so.——Occurs with the typical phase.

14. Galium gracilens (A. Gray) Makino. *G. trachyspermum* var. *gracilens* A. Gray——Hime-yotsuba-mugura. Resembles the preceding; stems 30–40 cm. long; leaves in verticils of 4's, oblong-linear or linear-lanceolate, 6–12 mm. long, 1.5–2.5 mm. wide, obtuse at both ends, with spreading white hairs on margin and on midrib; cymes rather small, loosely flowered, the pedicels slender; flowers relatively small, pale green; fruit rather small, densely warty.——May–June. Hills and mountains; Honshu, Shikoku, Kyushu.——Ryukyus, Formosa, China, and Korea.

15. Galium pogonanthum Fr. & Sav. *G. trachyspermum* var. *setuliflorum* A. Gray; *G. setuliflorum* (A. Gray) Makino——Yama-mugura. Resembles species Nos. 13 and 14, but slightly larger; stems 30–50 cm. long, slender, glabrous;

leaves in verticils of 4's, broadly linear or oblong-lanceolate, 1.5–2.5(–3) cm. long, 2–3.5 mm. wide, acute, narrowed to the base, slightly scabrous on margin and on midrib beneath; cymes terminal or nearly so, trichotomously branched, few-flowered, the pedicels 3–10 mm. long; flowers pale green, the corolla-lobes usually with few simple long hairs externally; fruit densely warty.——May–June. Woods and thickets in mountains; Honshu, Shikoku, Kyushu.——Korea.——Forma **trichopetalum** (Nakai) Ohwi. *G. trichopetalum* Nakai—— Oyama-mugura. Stems spreading-pilose.——Forma **nudiflorum** (Makino) Ohwi. *G. setuliflorum* var. *nudiflorum* Makino.——Ke-nashi-yama-mugura. Corolla glabrous.

16. Galium kamtschaticum Steller ex Roem. & Schult. *G. obovatum* sensu auct. Japon., non H.B.K.——Ezo-no-yotsuba-mugura. Small perennial 5–15 cm. high; stems smooth, glabrous, weak; leaves thinly membranous, in verticils of 4's, elliptic to broadly so, sometimes obovate-elliptic or oblong, 1–2.5 cm. long, 7–12 mm. wide, rounded and mucronulate-tipped, 3-nerved, rather long coarse-hairy on margins and upper surface; cymes nearly terminal, trichotomously branched, few- to about 10-flowered; flowers pale yellow-green, the pedicels spreading, rather thick; fruit densely hooked-prickly.—— July–Aug. Damp coniferous woods in mountains; Hokkaido, Honshu (centr. and n. distr.).——Forma **intermedium** Takeda. Ke-nashi-ezo-no-yotsuba-mugura. Plant short-hairy. ——Occurs with the typical phase.——Kuriles, Sakhalin, Korea, Ussuri, Kamchatka, and N. America.

Var. **acutifolium** Hara. *G. kamtschaticum* var. *oreganum* sensu auct. Japon., non Piper——Ōba-no-yotsuba-mugura. Plant slightly larger; stems 15–30 cm. long; leaves oblong, narrowly obovate, or elliptic, 2–3.5 cm. long, 1–2 cm. wide, obtuse and mucronate or subacute.——Damp coniferous woods; Hokkaido, Honshu, Shikoku.——Kuriles and Sakhalin.

17. Galium boreale L. var. **kamtschaticum** Maxim. *G. boreale* var. *ciliatum* Nakai; *G. boreale* var. *genuinum* sensu auct. Japon., non Gren. & Godr.——Ezo-kinuta-sō. Perennial; stems rather firm, erect, 20–30 cm. long, often branched, puberulent; leaves in verticils of 4's, lanceolate to ovate-oblong, 1.5–3 cm. long, 2–5 mm. wide, acute to acuminate with an obtuse tip, 3-nerved, puberulent above, on margin, and on the nerves beneath; inflorescence a terminal densely flowered panicle, the pedicels very short; flowers white; fruitlets soft-pubescent.——July–Aug. Hokkaido.——Sakhalin, Kuriles, and Kamchatka. The typical phase is widely spread throughout the cooler regions of the N. Hemisphere.

18. Galium kinuta Nakai & Hara. *G. boreale* var. *japonicum* Maxim.; *G. japonicum* (Maxim.) Makino & Nakai (1908), non Makino (1903)——Kinuta-sō. Perennial; stems erect, glabrous, purplish, simple, 40–50 cm. long, with several whorls of leaves; leaves chartaceous, in verticils of 4's, ovate to narrowly so, 3–5 cm. long, 1–2 cm. wide, gradually acuminate, 3(–5)-nerved, short-strigose on margin and on upper surface, especially on the nerves; inflorescence loosely flowered, in a terminal rather large panicle, with slender long branches and pedicels; flowers many, rather loosely arranged, white; fruitlets glabrous.——July–Aug. Mountains; Honshu, Shikoku; rather variable.——Forma **bracteatum** (Nakai) Hara. *G. japonicum* var. *bracteatum* Nakai——Otogiri-kinuta-sō. With broad bracts.——Forma **viridescens** (Matsum. & Nakai) Hara. *G. japonicum* var. *viridescens* Matsum. & Nakai——Ao-kinuta-sō. With green stems and inflorescence.

19. **Galium nakaii** Kudo ex Hara. *G. japonicum* var. *intermedium* Nakai——MIYAMA-KINUTA-SŌ. Perennial; stems erect, 15–25(–40) cm. long, glabrous, 4-angled; leaves in verticils of 4's, ovate to ovate-lanceolate, 2.5–5 cm. long, 1–2.5 cm. wide, usually acute, 3-nerved, strigose on margin and nerves; inflorescence terminal and in upper axils, 1–3 cm. long, much shorter than the leaves, 3- to 7-flowered; fruitlets strigose, the hairs incurved.——July–Aug. Mountains; Hokkaido, Honshu (n. distr.).

19. ASPERULA L. KURUMABA-SŌ ZOKU

Herbs; stems 4-angled; leaves in whorls, sometimes opposite in the upper ones; flowers small, white or reddish, the pedicels jointed with the calyx; calyx 4-toothed or absent; corolla infundibuliform, rarely subcampanulate, the lobes 4, spreading, valvate in bud stamens 4, on the throat, the filaments filiform; ovary 2-locular, with 1 ovule in each locule; styles 2, often connate at base; fruit dry or slightly fleshy, divided into 2 fruitlets; seeds solitary in each fruitlet, adnate to the pericarp.——About 80 species, in Asia, Europe, and Australia.

1A. Leaves in verticils of 6's to 10's, 2.5–4 cm. long; fruitlets with spreading long hooked prickles. . . . . 1. *A. odorata*
1B. Leaves in verticils of 4's, 8–15 mm. long; fruitlets glabrous. . . . . 2. *A. trifida*

1. **Asperula odorata** L. KURUMABA-SŌ. Resembles *Galium japonicum,* deep green when dry; stems erect, simple, with few whorls of leaves, 25–40 cm. long, 4-angled, glabrous; leaves in verticils of 6's to 10's, narrowly oblong or oblong-lanceolate, 2.5–4 cm. long, 5–10 mm. wide, obtuse and mucronate or acute, acute at base, strigose on margin and on midrib beneath; cymes terminal, trichotomously branched, few- to many-flowered; flowers white, infundibuliform, 4–5 mm. across; fruitlets with spreading long hooked prickles.——June–July. Woods in mountains; Hokkaido, Honshu.——Korea, s. Kuriles, Sakhalin, and Europe.

2. **Asperula trifida** Makino. *Galium shikokianum* Nakai ——USU-YUKI-MUGURA. Slender green perennial; stems 20–30 cm. long, erect, glabrous, 4-angled; leaves in verticils of 4's, ovate or narrowly so, 8–15 mm. long, 5–8 mm. wide, obtuse, sometimes minutely mucronate, strigose on margin and on nerves beneath; cymes rather loose, on upper portion of the stem, few- to many-flowered; flowers pedicelled, white, infundibuliform, about 2.5 mm. across; fruit glabrous.——June–Aug. Honshu (Tōkaidō Distr. and westw.), Shikoku, Kyushu.

Fam. 187. CAPRIFOLIACEAE SUI-KAZURA KA Honeysuckle Family

Shrubs or rarely herbs; leaves opposite, simple, rarely pinnately compound; stipules small or more often absent; flowers bisexual, actinomorphic or zygomorphic, usually in cymes; calyx adnate to the ovary, 5-lobed or -toothed; corolla gamopetalous, sometimes bilabiate, the lobes imbricate in bud; stamens inserted on the corolla-tube and alternate with the lobes, the anthers 2-locular, longitudinally split; ovary 2- to 5-locular; style solitary, simple, terminal; fruit a berry or capsule; seeds with copious endosperm.——About 18 genera, with about 300 species, mainly in the temperate regions of the N. Hemisphere.

1A. Corolla usually rotate, small, actinomorphic; style very short or absent.
  2A. Leaves pinnately compound; fruit berrylike, 3- to 5-seeded. . . . . 1. *Sambucus*
  2B. Leaves simple, rarely lobed or cleft; fruit a 1-seeded drupe. . . . . 2. *Viburnum*
1B. Corolla tubular or campanulate, relatively large, sometimes zygomorphic; style elongate.
  3A. Stamens 4.
    4A. Fruit narrowly oblong, the calyx-lobes persistent, slightly accrescent after anthesis; erect much-branched shrubs. . . . . . . 3. *Abelia*
    4B. Fruit ovoid to ovoid-globose, the calyx-lobes deciduous; creeping subshrub. . . . . 4. *Linnaea*
  3B. Stamens 5.
    5A. Herbs; ovules solitary in each locule; corolla gibbous. . . . . 5. *Triosteum*
    5B. Shrubs.
      6A. Fruit a 2-valved capsule; seeds many; corolla nearly actinomorphic. . . . . 6. *Weigela*
      6B. Fruit a berry; seeds few; corolla usually zygomorphic, rarely actinomorphic. . . . . 7. *Lonicera*

1. SAMBUCUS L. NIWATOKO ZOKU

Small trees or shrubs, sometimes merely giant herbs with stout stems; leaves opposite, pinnately compound, the leaflets toothed, the petioles sometimes with few discoid glands; flowers small, white, yellowish, or reddish, bracteolate, in large umbellate corymbs; calyx-teeth 3 to 5; corolla rotate or rotate-campanulate, 3- to 5-lobed, the lobes imbricate, rarely valvate in bud; stamens 5, the anther-locules splitting on outer margin; ovary 3- to 5-locular; style short, 3-lobed; ovules solitary in each locule; drupe berrylike, 3- to 5-seeded.——About 30 species, in the temperate regions and mountains of the Tropics except S. Africa.

1A. Large shrub without discoid glands on the paniculate inflorescence. . . . . 1. *S. sieboldiana*
1B. Large herb, rarely somewhat woody at base, with few large obconical glands on the corymbose inflorescence. . . . . 2. *S. chinensis*

1. **Sambucus sieboldiana** Bl. ex Graebn. *S. racemosa* var. *sieboldiana* Miq.——NIWATOKO. Large shrub with grayish yellow-brown lenticellate branchlets; leaves 12–30 cm. long, petiolate, the leaflets (5–)7, oblong or narrowly so, 4–12 cm. long, toothed; panicles terminal, pyramidal, pedunculate, with minute warty tubercles on the branches; flowers white, 3–4 mm. across; fruit red, rarely yellow or orange-yellow, globose or ellipsoidal, about 4 mm. across; stones obsoletely rugose.——Apr. Thickets and woods in hills and low elevations in the mountains; Honshu, Shikoku, Kyushu; very common and variable.——Korea.

Var. **miquelii** (Nakai) Hara. *S. buergeriana* Bl. ex Nakai; *S. racemosa* var. *miquelii* Nakai; *S. racemosa* var. *pubescens* Schwerin, non Dipp.; *S. latipinna* var. *miquelii* (Nakai) Nakai; *S. miquelii* Nakai ex Komar. & Alisova——Ezo-NIWATOKO. Warty tubercles of inflorescence elongate and hairlike.—— Woods and thickets in mountains; Hokkaido, Honshu (centr. distr.).——s. Kuriles, Sakhalin, Korea, and Ussuri.

2. **Sambucus chinensis** Lindl. (?) *S. thunbergii* G. Don; *S. thunbergiana* Bl.; *S. javanica* sensu auct. Japon., non Bl.; *Ebulis chinensis* (Lindl.) Nakai; *S. formosana* Nakai—— SOKUZU.　Rather stout green erect herb with creeping rhizomes; stems 1–1.5 m. long; leaves petiolate, rather large, the leaflets 5–7, broadly lanceolate to narrowly ovate, 5–17 cm. long, 2–6 cm. wide, toothed; inflorescence flat-topped, large, nearly glabrous, glands yellowish, nearly obconical; flowers white, 3–4 mm. long, on slender pedicels; fruit globose, about 4 mm. long, red.——Aug.–Sept.　Thickets and woods in hills and low elevations in the mountains; Honshu, Shikoku, Kyushu.——China, Ryukyus, Formosa, and Bonins.

2. VIBURNUM L.　　GAMAZUMI ZOKU

Shrubs or trees; leaves opposite, simple, or rarely palmately lobed, toothed; stipules usually small or absent; flowers in axillary or terminal subumbellate corymbs or panicles, white or pink, bisexual or sometimes the peripheral ones sterile with a large corolla; calyx-teeth small; corolla rotate or tubular, the lobes 5, imbricate in bud; stamens 5; ovary 1- to 3-locular; style short; ovules solitary in each locule; drupe fleshy or dry, 1-seeded.——About 150 species, in the temperate to warmer regions of the N. Hemisphere and S. America.

1A. Corolla rotate.
　2A. Inflorescence with large sterile flowers on the periphery.
　　3A. Leaves 3-lobed, 3-nerved from the base, with sessile discoid glands at the top of petioles; flowering branchlets usually with 2 pairs of leaves. . . . . 1. *V. opulus* var. *calvescens*
　　3B. Leaves toothed, not lobed, penninerved, glandless; flowering branchlets with one pair of leaves.
　　　4A. Leaves cordate; inflorescence without a common peduncle. . . . . 2. *V. furcatum*
　　　4B. Leaves rounded to cuneate at base; inflorescence with a common peduncle. . . . . 3. *V. plicatum*
　2B. Inflorescence without large sterile flowers on the periphery.
　　5A. Inflorescence a panicle with a short main axis. . . . . 4. *V. sieboldii*
　　5B. Inflorescence flat-topped, without an elongate main axis.
　　　6A. Inflorescence large, usually many-flowered, erect.
　　　　7A. Stamens shorter than the corolla; leaves subcoriaceous.
　　　　　8A. Evergreen shrub or small tree, glabrous. . . . . 5. *V. japonicum*
　　　　　8B. Deciduous shrub, stellate-pilose on young branchlets, under side of leaves and inflorescence. . . . . 6. *V. brachyandrum*
　　　　7B. Stamens longer than the corolla; leaves membranous or herbaceous.
　　　　　9A. Petioles 1–1.5 cm. long; stipules absent.
　　　　　　10A. Leaves coarsely pilose on both sides; young branches and inflorescence scabrous. . . . . 7. *V. dilatatum*
　　　　　　10B. Leaves usually glabrous above; young branches and inflorescence with scattered long-appressed hairs. . . . . 8. *V. wrightii*
　　　　　9B. Petioles 2–5 mm. long; stipules subulate, minute. . . . . 9. *V. erosum*
　　　6B. Inflorescence rather small and few-flowered, nodding or nearly pendulous. . . . . 10. *V. phlebotrichum*
1B. Corolla tubular or hypocrateriform, the tube longer than the lobes.
　11A. Evergreen shrubs; inflorescence a panicle; leaves coriaceous.
　　12A. Leaves 8–20 cm. long, the petioles 2.5–4 cm. long. . . . . 11. *V. awabuki*
　　12B. Leaves 5–8 cm. long, the petioles 5–10 mm. long. . . . . 12. *V. suspensum*
　11B. Deciduous shrubs; inflorescence a corymb; leaves membranous.
　　13A. Corolla-tube 1–1.3 cm. long, with spreading lobes about half as long as the tube. . . . . 13. *V. carlesii*
　　13B. Corolla-tube shorter with erect lobes about 1 mm. long. . . . . 14. *V. urceolatum*

1. **Viburnum opulus** L. var. **calvescens** (Rehd.) Hara. *V. sargentii* var. *calvescens* Rehd.; *V. sargentii* Koehne; *V. opulus* var. *sargentii* (Koehne) Takeda; *V. pubinerve* Bl. ex Nakai——KAMBOKU.　Deciduous shrub often white-pubescent on lower side of leaves and inflorescence; flowering branchlets usually with 2 pairs of leaves; leaves obovate-orbicular, 6–10 cm. long and as wide, truncate, shallowly cordate, or broadly cuneate at base, 3-lobed, incised, and few-toothed, 3-nerved from the base, the lobes more or less cuspidate or short-caudate, the petioles 3–4 cm. long, with a pair of sessile discoid glands at the top; inflorescence many-flowered, flat-topped, 6–12 cm. across, with a few large bractlike neutral flowers on the outer margin; anthers dark purple; fruit globose, 8–10 mm. across, red.——May–June.　Mountains; Hokkaido, Honshu, Shikoku, Kyushu.——Korea, Amur, Ussuri, Manchuria, China, s. Kuriles, and Sakhalin.

2. **Viburnum furcatum** Bl.　Ō-KAME-NO-KI, MUSHIKARI. Large shrub, yellowish furfuraceous while young, the branches dark purple; leaves ovate-orbicular or broadly ovate, 7–15 cm. long, 5–10 cm. wide, abruptly acute, cordate at base, toothed, with 7–10 pairs of parallel lateral nerves, the petioles reddish, 2–4 cm. long; inflorescence sessile, 7–12 cm. across, 2-leaved at the base, many-flowered, with few neutral flowers on margin; stamens about half as long as the corolla-lobes; fruit ellipsoidal, about 8 mm. long, becoming red, then black at maturity.—— May.　Mountains; Hokkaido, Honshu, Shikoku, Kyushu; rather common northward.——s. Kuriles, Sakhalin, and Korea (Quelpaert and Dagelet Isls.).

3. **Viburnum plicatum** Thunb. var. **tomentosum** (Thunb.) Miq. *V. tomentosum* Thunb.; *V. plicatum* forma *tomentosum* (Thunb.) Rehd.——YABU-DEMARI.　Large deciduous shrub pubescent while young, the branches graybrown; leaves ovate to broadly so, or obovate, 7–10 cm. long inclusive of the petioles, 4–7 cm. wide, abruptly acute, rounded to cuneate at base, with deltoid mucronate teeth except near base, pale green beneath, with 10–15 pairs of parallel lateral nerves, the petioles 1.5–2.5 cm. long; inflorescence rather large, terminal on short 2-leaved branchlets, peduncled, with large

sterile flowers on margin, the bracts and bracteoles small, deciduous; stamens about 1.5 times as long as the corolla-lobes, the anthers pale yellow; fruit broadly ellipsoidal, 5–6 mm. long, becoming red, then black at maturity.——May–June. Thickets in mountains; Honshu (Kantō Distr. and westw.), Shikoku, Kyushu.——Formosa and China.

Var. **glabrum** (Koidz.) Hara. *V. tomentosum* var. *glabrum* Koidz. ex Nakai; *V. amplissimum* Satake——KE-NASHI-YABU-DEMARI. Plant larger, less pubescent, soon almost glabrous; leaves to 15 cm. long.——Honshu (Japan Sea side of centr. and n. distr.).

4. **Viburnum sieboldii** Miq. GOMA-KI. Large deciduous foetid-smelling shrub with gray-brown branches, the branchlets, under side of leaves especially on nerves, and inflorescence with fascicled white hairs while young; leaves obovate-oblong, oblong or obovate, 8–15 cm. long, inclusive of petioles, 15–25 mm. long, 4–8 cm. wide, abruptly acute to rounded, cuneate or acute at base, toothed toward the tip, pale green beneath, with 9–12 pairs of parallel lateral nerves, the petioles 15–25 mm. long; inflorescence a panicle, 7–15 cm. across, on short 2- or 4-leaved branchlets; stamens shorter than the corolla-lobes; fruit ellipsoidal, about 1 cm. long, red.—— Apr.–May. Thickets in lowlands and low mountains; Honshu, Shikoku, Kyushu.

Var. **obovatifolium** (Yanagita) Sugimoto. *V. obovatifolium* Yanigita——MARUBA-GOMA-KI. Leaves obovate to broadly so, to 25 cm. long.——Japan Sea side of Honshu.

5. **Viburnum japonicum** (Thunb.) Spreng. *Cornus japonica* Thunb.; *V. buergeri* Miq.; *V. macrophyllum* Bl.—— HAKUSAN-BOKU. Large evergreen foetid-smelling shrub or small tree; branches dark purple-brown, glabrous, often glandular-dotted while young, as well as the inflorescence; leaves subcoriaceous, glabrous, lustrous, broadly ovate, broadly obovate, or ovate-orbicular, 7–20 cm. long, 5–17 cm. wide, abruptly acute, broadly cuneate or rounded at base, undulate-toothed toward the tip, minutely glandular-dotted beneath, with 5–8 pairs of parallel lateral nerves, the petioles 1.5–5 cm. long; inflorescence on short 2-leaved branchlets, short-peduncled, glabrous, 8–15 cm. across; stamens slightly shorter than the corolla-lobes; fruit ovoid, 6–8 mm. long, red.——Mar.–Apr. Honshu, Kyushu.——Formosa and Ryukyus.

Var. **fruticosum** Nakai. KO-HAKUSAN-BOKU. Smaller ascending shrub less than 2 m. high.——Honshu (Izu Peninsula and Izu Isls.).

6. **Viburnum brachyandrum** Nakai. *V. dilatatum* var. *brachyandrum* (Nakai) Makino——SHIMA-GAMAZUMI. Deciduous shrub; branches dark purple-brown, long-pubescent and stellate-pilose while young; leaves broadly obovate, obovate-orbicular, or rhombic-orbicular, 8–13 cm. long, abruptly acute, rounded to cuneate at base, deltoid-toothed, the petioles 1–1.5 cm. long, villous; inflorescence relatively large, short-peduncled, terminal on short 2-leaved branchlets, long-pubescent and stellate-pilose while young; stamens shorter than the corolla-lobes.——Apr. Honshu (Izu Isl.).

7. **Viburnum dilatatum** Thunb. *V. lantana* var. *japonicum* Fr. & Sav.——GAMAZUMI. Coarsely pilose deciduous shrub; young branches and inflorescence more or less scabrous; leaves obovate to broadly so, broadly ovate or obovate-orbicular, 3–12 cm. long, 2–8 cm. wide, abruptly acute, rounded to obtuse or sometimes cuneate at base, short mucronate-toothed, pale green and glandular-dotted beneath, coarsely pilose on both sides, the petioles (5–)10–15 mm. long; inflorescence 4–10 cm.

across, peduncled, on short 2-leaved branchlets; stamens longer than the corolla-lobes; fruit ovoid, about 5 mm. long, red.—— May–June. Thickets in hills and low elevations in the mountains; common and extremely variable.——Forma **xanthocarpum** Rehd. KIMINO-GAMAZUMI. Yellow-fruited.

Viburnum hizenense Hatus. *V. dilatatum* subsp. *hizenense* Hatus.; *V. dilatatum* var. *hizenense* (Hatus.) Ohwi. Alleged hybrid of *V. dilatatum* and *V. japonicum*. Plants nearly glabrous except the nerves of leaves beneath, the young branchlets sometimes short-pilose.——Kyushu (n. distr.).

8. **Viburnum wrightii** Miq. MIYAMA-GAMAZUMI. Deciduous shrub with purple-brown branches, these glabrous when young; inflorescence with scattered long-appressed hairs while young; leaves ovate-orbicular, broadly obovate, or broadly ovate, 7–12 cm. long, 4–9 cm. wide, abruptly acuminate, rounded to broadly cuneate at base, mucronate-toothed, usually glabrous above, usually minutely glandular-spotted and appressed-pubescent on nerves beneath, the petioles 1–1.5 cm. long, purplish red, long appressed-pubescent; inflorescence on short 2-leaved branchlets, glabrous or long-pubescent, sometimes with minute glandular dots; fruit ovoid, red, 6–7 mm. long.——May–June. Mountains; Hokkaido, Honshu, Shikoku, Kyushu.——Sakhalin, Korea, and China.

9. **Viburnum erosum** Thunb. *V. erosum* var. *punctata* Fr. & Sav.; *V. erosum* var. *laevis* Fr. & Sav.; *V. erosum* var. *furcipila* Fr. & Sav.; (?) *V. shikokianum* Koidz.——KOBA-NO-GAMAZUMI. Deciduous shrub with gray or purplish branches; young branches, leaves, and inflorescence with stellate and short coarse hairs; stipules subulate, minute; leaves narrowly ovate, elliptic, oblong or obovate, 3–10 cm. long, 2–5 cm. wide, acute, obtusely rounded or cuneate at base, toothed, the petioles pilose, 2–5 mm. long; inflorescence 4–8 cm. across, short-peduncled, on short 2-leaved branchlets; fruit red, broadly ovoid, 6–7 mm. long.——May. Sunny hills and low elevations in the mountains; Honshu (Kantō Distr. and westw.), Shikoku, Kyushu; very common and variable.——Forma **xanthocarpum** (Sugimoto) Hara. *V. erosum* var. *xanthocarpum* Sugimoto——KIMI-NO-KOBA-NO-GAMAZUMI. Yellow-Fruited. ——Korea and China.

Var. **taquetii** (Lév.) Rehd. *V. taquetii* Lév.——SAIKOKU-GAMAZUMI. Leaves irregularly incised.——Honshu (w. distr.), Shikoku, Kyushu.

10. **Viburnum phlebotrichum** Sieb. & Zucc. OTOKO-YŌZOME. Slender deciduous shrub with gray-brown branches; young branchlets, upper surface of leaves, and inflorescence glabrous or thinly long-pubescent; leaves nearly membranous, ovate, elliptic, or rhombic-ovate, 4–8 cm. long, 2.5–4 cm. wide, acute or acuminate, obtuse to subacute at base, coarsely mucronate-toothed, with appressed silky-pubescence on nerves, the petioles 3–7 mm. long, usually glabrous; inflorescence 3–6 cm. across, rather small, few-flowered, nodding or nearly pendulous on 2-leaved short branchlets; flowers white, often pinkish; stamens shorter than the corolla-lobes, suberect; fruit broadly ovoid, 7–8 mm. long, red.——May. Mountains; Honshu, Shikoku, Kyushu.

11. **Viburnum awabuki** K. Koch. *V. odoratissimum* sensu auct. Japon., non Ker-Gawl.; *V. odoratissimum* var. *awabuki* (K. Koch) K. Koch ex Ruempler——SANGO-JU. Large evergreen nearly glabrous shrub or small tree; branches rather stout, gray-brown; leaves coriaceous, narrowly oblong, 8–20 cm. long, 4–8 cm. wide, obtuse or acute at both ends, undulately short-toothed on upper half or nearly entire, lustrous

above, pale green beneath, with 6–8 pairs of upwardly directed lateral nerves, the petioles 2.5–4 cm. long, stout; inflorescence paniculate, rather large, on short 4-leaved branchlets, peduncled, rusty-brown pubescent on nodes when young; calyx-limb shallowly 5-lobed; corolla about 6 mm. long, the lobes about 2 mm. long, spreading; filaments short; fruit ellipsoidal, 7–8 mm. long, red.——June. Lowlands, especially near the sea; Honshu (s. Kantō Distr. and westw.), Shikoku, Kyushu; frequently planted for hedges.——s. Korea, Ryukyus, and Formosa.

12. Viburnum suspensum Lindl. *V. sandankwa* Hassk.——GOMO-JU, KO-URUME. Evergreen much-branched shrub with red-brown branches, minutely stellate-pilose; leaves thick, coriaceous, elliptic or obovate-elliptic, 5–8 cm. long, 2–4 cm. wide, rounded or very obtuse, obtuse to acute at base, with obtuse appressed teeth, lustrous, glabrous, with impressed nerves and veinlets above, pale green and nearly glabrous beneath except the axillary tufts of hairs, the lateral nerves of 3 or 4 pairs, the petioles 5–10 mm. long; inflorescence terminal, 2–4 cm. long, on the previous year's branches, peduncled; calyx-limb 5-lobed; corolla about 1 cm. long, the lobes 2–3 mm. long, spreading; fruit ellipsoidal, about 5 mm. long, red.——Mar.–Apr.——Sometimes cultivated in our area.——Ryukyus and Formosa.

13. Viburnum carlesii Hemsl. *Solenolantana carlesii* (Hemsl.) Nakai——Ō-CHŌJI-GAMAZUMI. Deciduous much-branched shrub; stems, leaves and inflorescence stellate-pilose and tufted furfuraceous-puberulent; leaves ovate-orbicular or broadly ovate, abruptly acute, 3–7 cm. long, 2.5–6 cm. wide, rounded to shallowly cordate at base, mucronate-toothed, with

4–6 pairs of lateral nerves, the petioles 3–5 mm. long; inflorescence cymose, short-peduncled, 2–5 cm. across, on short 2-leaved branchlets; calyx-limb 5-lobed; corolla white or pale rose, the tube 1–1.3 cm. long, the lobes about half as long as the tube; fruit ellipsoidal, about 8 mm. long, black.——Apr.–May. Kyushu (Tsushima Isl.).——Korea.

Var. **bitchiuense** (Makino) Nakai. *V. bitchiuense* Makino; *V. carlesii* var. *syringaeflora* Hutchins.; *V. burejaeticum* sensu auct. Japon., non Regel & Herder; *Solenolantana carlesii* var. *bitchiuensis* (Makino) Nakai——CHŌJI-GAMAZUMI. Leaves oblong or ovate, acute, the petioles 4–8 mm. long; flowers smaller.——Honshu (Chūgoku Distr.), Shikoku (Azukijima), Kyushu (Buzen Prov.).——Korea.

14. Viburnum urceolatum Sieb. & Zucc. *V. urceolatum* forma *brevifolium* Makino——YAMA-SHIGURE, MARUBA-MIYAMA-SHIGURE. Nearly glabrous deciduous shrub with purple-brown branches; leaves ovate to ovate-oblong, 6–12 cm. long, 2.5–5 cm. wide, acuminate to acute, usually rounded at base, obtusely serrulate, deep green and glabrous above, pale green and the nerves minutely stellate-puberulent beneath, with 4 or 5 pairs of lateral nerves, the petioles 1–2 cm. long, minutely stellate-puberulent; inflorescence 3–4 cm. across, on 4- rarely 2-leaved branchlets, puberulent, peduncled; corolla 3–3.5 mm. long, short-tubular, the lobes erect, about 1 mm. long; stamens exserted; fruit broadly ellipsoidal, about 6 mm. long, red.——June–July. Woods in mountains; Honshu (Kinki Distr. and westw.), Shikoku, Kyushu.

Var. **procumbens** Nakai. MIYAMA-SHIGURE. Stems long-creeping.——Damp coniferous woods; Honshu (centr. and n. distr.).

3. ABELIA R. Br.   TSUKUBANE-UTSUGI ZOKU

Erect or suberect shrubs; leaves opposite, petiolate, entire or toothed; stipules absent; inflorescence usually a 3- or rarely 1-flowered axillary or terminal cyme; flowers usually sessile, white, pink, reddish or yellowish, 2- to 4-bracteate at base; calyx-tube narrow, flat, the lobes 2–5, persistent; corolla tubular-campanulate or infundibuliform, the limb 5-lobed, nearly regular; stamens 4; ovary 3-locular, only one locule maturing; fruit coriaceous, 1-seeded, narrowly oblong, the calyx-lobes persistent, slightly accrescent after anthesis.——About 30 species, in e. Asia and the Himalayas, 2 or 3 in Mexico.

1A. Branches scarcely thickened at the nodes, not fissured; corolla campanulate-infundibuliform, the tube ampliate, much longer than the calyx lobes; fruit barely flattened.
  2A. Calyx-lobes 2 or 3; peduncles 2- to 7-flowered. .... 1. *A. serrata*
  2B. Calyx-lobes 4 or 5; peduncles 1- or 2-flowered.
    3A. Calyx-lobes 5. .... 2. *A. spathulata*
    3B. Calyx-lobes 4. .... 3. *A. tetrasepala*
1B. Branches much thickened at the nodes, with 6 longitudinal fissures; corolla hypocrateriform, the tube scarcely ampliate, as long as or shorter than the calyx-lobes; fruit flattened. .... 4. *A. integrifolia*

1. Abelia serrata Sieb. & Zucc. *Linnaea serrata* (Sieb. & Zucc.) Graebn.——KO-TSUKUBANE-UTSUGI. Deciduous shrub 1–1.5 m. high, with irregularly fissured bark, young branches often puberulent; leaves rhombic-ovate, 1–2.5 cm. long, 5–10 mm. wide, obtuse or acuminate with an obtuse tip, obtusely few-toothed, short-pilose on both sides, short-petiolate; flowers (1–)2–7, short-pedunculate, bracteate and bracteolate at base; calyx-lobes 2, sometimes 3, oblong or elliptic, often shallowly 2-lobed; corolla campanulate-infundibuliform, 1–1.5 cm. long, pale yellow.——May. Hills and mountains; Honshu (centr. distr. and westw.), Shikoku, Kyushu; variable.

Var. **buchwaldii** (Graebn.) Nakai. *Linnaea buchwaldii* Graebn.; *A. buchwaldii* (Graebn.) Rehd.——KIBANA-TSUKU-

BANE-UTSUGI. Leaves larger, 3–5 cm. long; corolla 2–3 cm. long.——Honshu (centr. and w. distr.), Shikoku, Kyushu.

Var. **gymnocarpa** (Graebn.) Nakai. *Linnaea gymnocarpa* Graebn.——HIRO-HA-TSUKUBANE-UTSUGI. Resembles the preceding variety, the plant glabrous.——Honshu (centr. distr.).

Var. **tomentosa** (Koidz.) Nakai. ONI-TSUKUBANE-UTSUGI. Plant pilose.——Shikoku.

Var. **integerrima** Nakai. MARUBA-TSUKUBANE-UTSUGI. Leaves and calyx-lobes entire.——Honshu (w. distr.).

2. Abelia spathulata Sieb. & Zucc. *A. spathulata* var. *elliptica* Miq.——TSUKUBANE-UTSUGI. Deciduous shrub with gray-brown bark; young branches red-brown, often puberulent; leaves narrowly rhombic-ovate to broadly so, 2.5–6 cm.

long, 1.5–3.5 cm. wide, abruptly acuminate with an obtuse tip, few-toothed, more or less pilose on both sides, short-petiolate; flowers (1 or) 2, sessile, short-pedunculate, bracteate and bracteolate at base; calyx-lobes 5, persistent, oblanceolate or oblong; corolla pale yellow, 2–3 cm. long.——May. Sunny thickets in mountains; Honshu, Shikoku, Kyushu; common and variable.

Var. **sanguinea** Makino. *A. sanguinea* Makino; *A. curviflora* Nakai; *A. sanguinea* var. *purpurascens* Honda——Beni-bana-no-tsukubane-utsugi. Plant slender; corolla reddish, 1.5–2 cm. long.——Honshu.

Var. **stenophylla** Honda. *A. spathulata* var. *macrophylla* Honda & var. *ionostachya* Honda——Ugo-tsukubane-utsugi. Usually larger in all parts; branchlets and leaves spreading-pilose.——Honshu (Japan Sea side of n. distr.).

3. **Abelia tetrasepala** (Koidz.) Hara & Kurosawa. *A. spathulata* var. *tetrasepala* Koidz.; *A. spathulata* var. *subtetrasepala* Makino——Ō-tsukubane-utsugi, Me-tsukubane-utsugi. Resembles No. 2, but the petioles and underside of

leaves spreading-pilose especially on midvein; calyx pilose, the segments 4; corollas large, 3–4 cm. long, with slender tube.——Honshu (Iwashiro and westw.), Shikoku, Kyushu.

4. **Abelia integrifolia** Koidz. *A. shikokiana* Makino; *Zabelia shikokiana* (Makino) Makino; *Z. integrifolia* (Koidz.) Makino.——Iwa-tsukubane-utsugi. Much-branched deciduous shrub; branches with 6 longitudinal fissures, the nodes much thickened, young branchlets red-brown, glabrous or with recurved coarse hairs; leaves chartaceous, obovate, or obovate-oblong, sometimes ovate-elliptic, 2–5 cm. long, 1.5–2.5 cm. wide, acute at both ends, entire or with few incisions, coarsely pilose on both sides especially on margin and on nerves beneath, short-petiolate; flowers 1 or 2, short-pedunculate, bracteate and bracteolate at base; calyx-lobes oblanceolate; corolla whitish, hypocrateriform, about 2 cm. long, the tube as long as or shorter than the calyx-lobes, the lobes spreading, 3–4 mm. long; fruit flattened, about 1.5 cm. long, lustrous.——May–June. Mountains; Honshu (Kantō Distr. and westw.), Shikoku, Kyushu; rather rare.

4. LINNAEA Gronov. ex L.   Rinne-sō Zoku

Subshrubs with slender creeping branches; leaves evergreen, small, opposite, nearly orbicular, short-petiolate, few-toothed; stipules absent; peduncles terminal on erect short branchlets, naked, with 2 pedicelled flowers at apex, the bracts and bracteoles small; flowers pale rose-colored; calyx-tube ovoid, the limb deciduous, 5-sected; corolla campanulate-infundibuliform, the lobes 5, nearly equal, imbricate in bud; stamens 4; ovary 3-locular, only 1 locule maturing; style slender; ovules many, pendulous; fruit subglobose, coriaceous, 1-seeded; seeds oblong, with fleshy endosperm, the embryo cylindric.——A single widely distributed species in cooler regions of the N. Hemisphere.

1. **Linnaea borealis** L. Rinne-sō. Branches about 1 mm. thick, brownish, puberulent when young; leaves ovate-orbicular to obovate-orbicular, 7–15 mm. long, and as wide, rounded at apex, few-toothed, usually short appressed-pilose and the nerves impressed above; peduncles 5–10 cm. long, short glandular-pilose; flowers nodding, the pedicels 6–15 mm.

long; calyx-segments lanceolate, 2–3 mm. long; corolla 8–10 mm. long, the bracteoles broadly ovate, as long as the ovary and surrounding it.——June–July. Damp coniferous woods in mountains; Hokkaido, Honshu (centr. and n. distr.); rather rare.——Cooler regions of the N. Hemisphere inclusive of the Kuriles and Sakhalin.

5. TRIOSTEUM L.   Tsuki-nuki-sō Zoku

Erect herbs; leaves opposite, sessile, often perfoliate; flowers usually axillary, sometimes in a terminal spike, yellow, white or purple, with 2 bracteoles at base; calyx-tube ovoid, the lobes 5, persistent; corolla tubular-campanulate, gibbous, the lobes oblique, unequal, imbricate in bud; stamens 5; ovary 3- to 5-locular; style elongate; ovules solitary in each locule; berry coriaceous or fleshy, 2(–5)-locular, 2- or 3-seeded.——Few species, in the Himalayas, e. Asia, and N. America.

1A. Leaves pinnatifid into 5 acuminate lobes, scarcely perfoliate at base; flowers in a terminal pedunculate spike, about 10 mm. long.

1. *T. pinnatifidum*

1B. Leaves entire or with an obtuse sinus on each side, perfoliate at base; flowers 1 or 2 in axils of upper leaves, about 2.5 cm. long.

2. *T. sinuatum*

1. **Triosteum pinnatifidum** Maxim. *T. murayamae* Ohwi——Hozaki-tsuki-nuki-sō. Densely pubescent herb with long spreading hairs throughout, and with short glandular hairs on upper part; stems erect, about 30 cm. long; leaves in few pairs, obovate-rhombic, 7–12 cm. long, nearly as wide, abruptly narrowed to broadly cuneate at base, scarcely perfoliate, pinnatifid, the lobes long-acuminate; spikes erect, solitary, 2–5 cm. long including the peduncle; flowers sessile; calyx-limb minute, 5-dentate; corolla tubular, obliquely spreading, about 10 mm. long, pale green, the lobes short, brownish inside; drupe nearly globose, about 8 mm. across, white when ripe, juicy and somewhat sweet, the pyrenes 3, ovate, biconvex. ——May. Honshu (Kai Prov.); very rare.——China.

2. **Triosteum sinuatum** Maxim. Tsuki-nuki-sō. Perennial glandular-pubescent herb; stems 60–80 cm. long, terete, green; leaves ovate or elliptic, 10–15 cm. long, 6–8 cm. wide, acute, narrowed below and usually perfoliate, spreading, entire or with an obtuse sinus on each side; flowers 1 or 2 in axils of upper leaves, sessile, erect; calyx-lobes lanceolate or narrowly oblong, obtuse, erect, 6–10 mm. long; corolla about 2.5 cm. long, yellow-green with purple-brown striations inside, soft-pubescent; fruit ovoid-globose, obtusely 3-angled, about 1 cm. long and as wide, pubescent, crowned with the persistent calyx-lobes.——May–June. Honshu (Shinano Prov.).——China to Amur.

6. **WEIGELA** Thunb.   Tani-utsugi Zoku

Deciduous shrubs; leaves opposite, petiolate, toothed; flowers in axillary and terminal often trichotomously divided corymbs; calyx-tube narrow, the lobes 5, deciduous or persistent; corolla campanulate-infundibuliform, nearly actinomorphic, white or reddish, sometimes yellowish, the tube ampliate, the limb nearly equally 5-lobed, the lobes imbricate in bud; stamens 5; style slender; ovary 2-locular, many-ovuled; capsules coriaceous or slightly woody, narrow, 2-valved, septicidally dehiscent, the axis and base of style persistent; seeds many.——Ten or more species in e. Asia.

1A. Flowers white or reddish; calyx-limb regular, 5-lobed or -divided.
  2A. Calyx-limb divided to the middle; seeds wingless.
    3A. Leaves glabrous on upper side. .... 1. *W. florida*
    3B. Leaves pilose on both surfaces. .... 2. *W. praecox*
  2B. Calyx-limb divided to the base; seeds narrowly winged.
    4A. Corolla glabrous outside.
      5A. Leaves glabrous beneath or with short coarse hairs on nerves.
        6A. Leaves 8–15 cm. long, 4–10 cm. wide, glabrous or nearly so; corolla-tube abruptly ampliate on upper half.
                                                                              3. *W. coraeënsis*
        6B. Leaves 6–10 cm. long, 3–6 cm. wide, short-pilose on nerves beneath; corolla-tube gradually ampliate on upper half.
                                                                        4. *W. decora*
      5B. Leaves densely whitish short-pubescent beneath. .... 5. *W. hortensis*
    4B. Corolla pilose outside.
      7A. Ovary, young branches, and underside of leaves ascending-pilose; flowers usually white at anthesis, becoming reddish with age; petioles 5–8 mm. long. .... 6. *W. japonica*
      7B. Ovary, young branchlets, and underside of leaves spreading-pilose; flowers usually reddish at anthesis; petioles 1–5 mm. long.
        8A. Plant sparsely pubescent; petioles 1–3 mm. long. .... 7. *W. sanguinea*
        8B. Plant densely pubescent; petioles 3–5 mm. long. .... 8. *W. floribunda*
1B. Flowers pale yellow; calyx-limb bilabiate.
  9A. Flowers pedunculate; calyx-limb persistent in fruit. .... 9. *W. middendorffiana*
  9B. Flowers sessile; calyx-limb deciduous in fruit. .... 10. *W. maximowiczii*

1. Weigela florida (Bunge) A. DC. *Calysphyrum floridum* Bunge; *Diervilla florida* (Bunge) Sieb. & Zucc.; *W. pauciflora* DC.; *W. rosea* Lindl.——Ō-beni-utsugi. Shrub with the young branches glabrous or with 2 longitudinal lines of short hairs; leaves ovate-oblong or obovate, 4–10 cm. long, 2–4 cm. wide, acuminate, toothed, glabrous above, densely pubescent on nerves beneath, the petioles 1–3 mm. long; flowers solitary in leaf-axils; calyx-tube glabrous or with scattered coarse hairs, the lobes 8–12 mm. long, lanceolate, acuminate, pilose; corolla 3–4 cm. long, soft-puberulent externally, reddish; style as long as the corolla; capsules 15–22 mm. long.——Kyushu.——Korea, n. China, and Manchuria.

2. Weigela praecox (Lemoine) L. H. Bailey. *Diervilla praecox* Lemoine——Birōdo-utsugi. Shrub with 2 longitudinal lines of hairs on young branches; leaves ovate or ovate-oblong, 4–7 cm. long, 2–3 cm. wide, prominently pubescent on both sides especially on nerves beneath, sessile or the petioles to 2 mm. long; flowers 1 or 2 in leaf-axils, sessile; ovary loosely pilose; calyx-segments lanceolate, 7–10 mm. long, unequally connate at base, pilose; corolla 3–4 cm. long, short-pubescent externally, infundibuliform-campanulate, abruptly ampliate, reddish; style included; capsules about 15 mm. long. ——May–June. Kyushu (reported from mountains of Chikuzen Prov.).——Korea and Manchuria.

3. Weigela coreënsis Thunb. *Diervilla grandiflora* Sieb. & Zucc.; *W. grandiflora* (Sieb. & Zucc.) K. Koch; *D. japonica* var. *grandiflora* Miq.——Hakone-utsugi. Nearly glabrous shrub with thick gray-brown branches; leaves thick, broadly elliptic or obovate-elliptic, 8–15 cm. long, 4–10 cm. wide, abruptly acuminate, abruptly acute at base, glabrous or nearly so, lustrous above, the petioles 1–1.5 cm. long; flowers 2–8 in axillary short-pedunculate cymes; ovary glabrous; calyx-lobes broadly linear, nearly glabrous; corolla 3–4 cm. long, the tube abruptly ampliate on upper half, white at first, becoming red with age (rarely not); style not exserted; capsules 2–3 cm.

long.——May–(June). Near seashores; Hokkaido, Honshu, Shikoku, Kyushu.

Var. **fragrans** (Ohwi) Hara. *W. fragrans* Ohwi——Nioi-utsugi. Flowers fragrant; corolla short, about 2 cm. long.——Honshu (Izu Isls.).

4. Weigela decora (Nakai) Nakai. *Diervilla decora* Nakai; *W. japonica* var. *decora* (Nakai) Okuyama; *W. versicolor* Sieb. & Zucc., pro parte; *D. floribunda* var. *versicolor* Rehd.; *D. nikoensis* Nakai; *W. hakonensis* Nakai——Nishiki-utsugi. Shrub with gray-brown branches, the young branchlets glabrous or with 2 longitudinal lines of hairs; leaves obovate-elliptic, 6–10 cm. long, 3–6 cm. wide, abruptly acuminate, acute at base, slightly lustrous and nearly glabrous above, short-pilose on nerves beneath, the petioles about 1 cm. long, nearly glabrous; flowers 2 or 3 in leaf axils, often short-pedunculate; ovary glabrous or scattered hairy; calyx-segments broadly linear, slightly pilose; corolla white, reddish with age, the tube gradually ampliate on upper half, essentially glabrous; style exserted.——May–June. Honshu, Shikoku, Kyushu.——Forma **unicolor** (Nakai) Hara. *W. decora* var. *unicolor* Nakai——Benibana-nishiki-utsugi. Flowers reddish from the first.

Var. **rosea** (Makino) Hara. *Diervilla fujisanensis* var. *rosea* Makino; *W. fujisanensis* var. *rosea* (Makino) Nakai——Fuji-beni-utsugi. Leaves longer, acute; calyx-lobes longer, more densely pilose; corolla reddish.——Honshu (centr. distr.).——Forma **fujisanensis** (Makino) Hara. *Diervilla fujisanensis* Makino; *W. fujisanensis* (Makino) Nakai——Fuji-beni-utsugi. Corolla white, becoming reddish with age.

5. Weigela hortensis (Sieb. & Zucc.) K. Koch. *Diervilla hortensis* Sieb. & Zucc.; *D. japonica* var. *hortensis* (Sieb. & Zucc.) Maxim.——Tani-utsugi. Shrub with gray-brown branches, glabrous or often loosely pilose while young; leaves ovate-elliptic, ovate-oblong, or obovate, sometimes ovate, 6–10 cm. long, 2.5–5 cm. wide, acuminate, acute at base, short-toothed, nearly glabrous above, densely whitish short-pubescent beneath, the petioles 4–8 mm. long; flowers 2 or 3 in upper leaf-axils, often short-pedunculate; ovary glabrous or scattered pilose; calyx-segments lanceolate-linear, 3–7 mm. long, slightly pilose; corolla reddish, 3–3.5 cm. long, glabrous or nearly so externally, gradually ampliate; style subexserted.——May–June. Hokkaido, Honshu.——Forma **albiflora** (Sieb. & Zucc.) Rehd. *Diervilla hortensis* var. *albiflora* Sieb. & Zucc.; *W. nivea* Carr.——SHIROBANA-UTSUGI. Flowers white.

6. **Weigela japonica** Thunb. *Diervilla japonica* (Thunb.) DC.; *D. versicolor* Sieb. & Zucc., pro parte; *D. floribunda* var. *versicolor* (Sieb. & Zucc.) Rehd.; *W. floribunda* var. *versicolor* (Sieb. & Zucc.) Rehd.——TSUKUSHI-YABU-UT-SUGI. Shrub with gray-brown branches often pilose while young; leaves oblong, ovate-oblong, or elliptic, 3–5 cm. wide, 6–10 cm. long, acuminate, minutely toothed, loosely pubescent above while young, rather prominently ascending-pubescent beneath especially on nerves, the petioles 5–8 mm. long; flowers 1 or 2 in upper leaf-axils, sometimes short-pedunculate; ovary ascending-pilose; calyx-segments broadly linear, 7–10 mm. long, pilose; corolla 3–3.5 cm. long, white, becoming reddish with age, short crisped pilose externally, the tube rather prominently ampliate; style exserted.——May. Honshu (Chūgoku Distr.), Shikoku, Kyushu.

7. **Weigela sanguinea** (Nakai) Nakai. *Diervilla sanguinea* Nakai——KE-UTSUGI, BIRŌDO-UTSUGI. Shrub, sparsely pubescent; leaves ovate-oblong to ovate-elliptic, 6–12 cm. long, 2.5–6 cm. wide, acuminate, pilose on upper surface while young, spreading-pilose beneath especially on nerves, the petioles 1–3 mm. long; flowers 1 or 2 in upper leaf-axils, sometimes short-pedunculate; ovary pilose; calyx-segments 7–10 mm. long, pilose; corolla deep reddish, 3–3.5 cm. long, pubescent outside; style exserted.——May–June. Mountains; Honshu (centr. distr.).——Forma **leucantha** (Nakai) Hara. *Diervilla sanguinea* var. *leucantha* Nakai——SHIROBANA-BIRŌDO-UTSUGI. Flowers white.

8. **Weigela floribunda** (Sieb. & Zucc.) K. Koch. *Diervilla floribunda* Sieb. & Zucc.; *D. versicolor* Sieb. & Zucc.; *W. japonica* var. *floribunda* (Sieb. & Zucc.) Hara, in syn.——YABU-UTSUGI. Shrub, densely pubescent; leaves elliptic, ovate-oblong, or ovate, 6–12 cm. long, 2.5–6 cm. wide, acuminate, toothed, loosely pilose above, densely whitish spreading-pilose especially on nerves beneath, the petioles 3–5 mm. long; flowers 1–3 in upper leaf-axils, subsessile or sessile; ovary pilose; corolla reddish, 3–3.5 cm. long, short-pubescent outside; style exserted.——May–June. Honshu (centr. and w. distr.), Shikoku. Variable.

Var. **nakaii** (Makino) Hara. *Diervilla sanguinea* var. *nakaii* Makino——CHISHIO-UTSUGI. Plants less densely pilose throughout; corolla white, becoming red with age.——Honshu (centr. distr.).——Forma **kariyosensis** (Nakai) Hara. *W. floribunda* var. *kariyosensis* Nakai——NISHIKI-BIRŌDO-UT-SUGI. Flowers white. Intermediate between *W. floribunda* and *W. sanguinea*.——Honshu (mountains in Kai and Musashi Prov.).

9. **Weigela middendorffiana** (Carr.) K. Koch. *Diervilla middendorffiana* Carr.; *Macrodiervilla middendorffiana* (Carr.) Nakai; *Calyptrostigma middendorffiana* (Carr.) Trautv. & Mey.——UKON-UTSUGI. Shrub with exfoliating bark, young branches with 2 longitudinal lines of hairs; leaves oblong or narrowly ovate, acute or abruptly acuminate, acutely toothed, short-pilose on nerves of upper surface, coarsely hairy beneath especially on nerves, subsessile; flowers 1 or 2 in upper leaf-axils and terminal on branchlets; calyx deeply bifid, the upper lobes broadly oblanceolate, 3-toothed, the lower lobes narrower, entire or 2-lobed; corolla 3.5–4 cm. long, pale yellow; capsules narrowly fusiform-oblong, 1.5–2 cm. long, longitudinally striate, glabrous, crowned with persistent calyx-lobes.——July. Thickets on alpine slopes; Hokkaido, Honshu (n. distr.).——s. Kuriles, Sakhalin, Ussuri, Amur, and Ochotsk Sea region.

10. **Weigela maximowiczii** (S. Moore) Rehd. *Diervilla middendorffiana* var. *maximowiczii* S. Moore; *D. maximowiczii* (S. Moore) Makino; *Calyptrostigma maximowiczii* (S. Moore) Makino; *D. sakaii* Makino & Honda; *Weigelastrum maximowiczii* (S. Moore) Nakai——KIBANA-UTSUGI. Deciduous shrub with gray-brown branches, with 2 longitudinal lines of spreading hairs; leaves ovate-oblong, narrowly obovate, or ovate, 4–8(–10) cm. long, 2–3 cm. wide, abruptly acuminate, toothed, loosely appressed-hairy on both sides, sessile; flowers sessile; ovary glabrous; calyx-lobes 1–1.5 cm. long, loosely pilose, the limb deciduous in fruit; corolla pale yellow, 3.5–4 cm. long; capsules 2–3 cm. long.——May–June. Mountains; Honshu (centr. and n. distr.).

7. LONICERA L. SUI-KAZURA ZOKU

Erect or scandent shrubs; leaves opposite, petiolate, sessile, or perfoliate, usually entire; flowers usually paired, sessile, on a common peduncle, usually with connate bracts and bracteoles at base; calyx-tube ovoid or subglobose, the limb short, 5-toothed, deciduous or persistent; corolla tubular-infundibuliform or campanulate, usually zygomorphic, rarely actinomorphic, the limb oblique or bilabiate, often nearly regular, the lobes imbricate in bud; stamens 5; ovary 2- or 3-locular; style elongate; ovules many; berry 1- to 3-locular, each locule few seeded; seeds ovoid or oblong, with a crustaceous testa.——About 180 species, mainly in cooler regions of the N. Hemisphere, few species in Centr. America, Malaysia, and N. Africa.

1A. Scandent shrubs with hollow stems; fruit black or blue-black.
  2A. Bracts leaflike, ovate; young branchlets, outer side of corolla, and young leaves more or less pubescent. .... 1. *L. japonica*
  2B. Bracts minute, linear or linear-lanceolate; plants glabrous or soft-puberulent.
    3A. Leaves glabrous, not glandular-dotted beneath. .... 2. *L. affinis*
    3B. Leaves soft-puberulent, with red-brown glandular dots beneath. .... 3. *L. hypoglauca*
1B. Erect or nearly erect shrubs, with pithy or rarely entirely hollow stems; fruit red.
  4A. Branches partially hollow; ovary and fruit usually paired, not connate.
    5A. Peduncles 2–4 mm. long, shorter than to as long as the petioles; calyx-limb cup-shaped, 2–3 mm. long, 5-lobed.
                                                      4. *L. maackii*
    5B. Peduncles 1–2.5 cm. long, much longer than the petioles; calyx-limb shorter.
      6A. Leaves 6–10 cm. long, very acuminate. .... 5. *L. chrysantha*
      6B. Leaves 2–4 cm. long, obtuse or acute with an obtuse tip.

7A.   Flowers distinctly bilabiate; leaves usually acute at base, glaucescent beneath. .... 6. *L. demissa*
7B.   Flowers nearly actinomorphic; leaves usually obtuse at base, pale green beneath. .... 7. *L. morrowii*
4B.   Branches solid with white or brown pith; ovary and fruit solitary, or if paired, usually more or less connate.
   8A.   Bracteoles bladderlike, enveloping the ovary, juicy in fruit, becoming blue-black and bloomy at maturity; corolla creamy-yellow.
       8. *L. caerulea*
   8B.   Bracteoles not enclosing the fruit, sometimes separate, scarcely accrescent, cup-shaped or bifid.
     9A.   Corolla nearly actinomorphic (somewhat bilabiate in No. 9).
       10A.   Flowers paired.
         11A.   Bracteoles absent or obsolete.
           12A.   Flowers white or pale yellow; bracts broadly ovate to broadly lanceolate; leaves 2–7 cm. wide.
              13A.   Leaves subcoriaceous, loosely setose especially on margin and on midrib, obtuse, mucronate-tipped. ..... 9. *L. harae*
              13B.   Leaves membranous, pubescent at least beneath, acute or nearly acuminate.
                 14A.   Stamens not exserted; bracts large, ovate to broadly so; flowers opening with the leaves; leaves obsoletely veined; peduncles 1–2 cm. long. .... 10. *L. strophiophora*
                 14B.   Stamens exserted; bracts broadly lanceolate; flowers opening before the leaves; leaves rather prominently veined; peduncles 3–8 mm. long. .... 11. *L. praeflorens*
           12B.   Flowers rose-purple; bracts linear; leaves usually less than 1.5 cm. wide.
              15A.   Peduncles 1.5–2.5 cm. long; leaves broadly lanceolate. .... 12. *L. linderifolia*
              15B.   Peduncles 5–10 mm. long; leaves ovate or oblong. .... 13. *L. konoi*
         11B.   Bracteoles prominent, nearly as long as the ovary. .... 14. *L. ramosissima*
       10B.   Flowers solitary, rarely paired. .... 15. *L. gracilipes*
     9B.   Corolla distinctly bilabiate.
       16A.   Leaves loosely long-ciliate, obsoletely veined above; petioles 8–10 mm. long; bud-scales deciduous.
         17A.   Peduncles 1–1.2 cm. long; leaves glabrous beneath; corolla-tube nearly as long as the limb. .... 16. *L. cerasina*
         17B.   Peduncles 1.5–4.5 cm. long; leaves usually pilose beneath on nerves.
           18A.   Bracteoles nearly as long as the ovary; peduncles 1.5–2 cm. long; stamens exserted. .... 17. *L. vidalii*
           18B.   Bracteoles minute or absent; peduncles 2.5–4.5 cm. long; stamens not exserted. .... 18. *L. alpigena* var. *glehnii*
       16B.   Leaves glabrous, rarely ciliate, prominently veined above; petioles 5(–7) mm. long or absent; bud-scales persistent.
         19A.   Leaves sessile or the petioles about 1 mm. long, obtuse or rounded; peduncles 6–12 mm. long; flowers deep red.
                19. *L. chamissoi*
         19B.   Leaves with petioles 3–5(–7) mm. long, acuminate or acute, rarely obtuse; peduncles longer.
           20A.   Peduncles 2.5 cm. long; leaves with slightly raised veinlets above; flowers white. .... 20. *L. tschonoskii*
           20B.   Peduncles 1–2 cm. long; leaves strongly veined above.
              21A.   Leaves glabrous, not ciliate, rarely thinly pilose on midrib beneath; flowers white. .... 21. *L. mochidzukiana*
              21B.   Leaves rarely loosely long-pilose beneath, long-ciliate; flowers deep red. .... 22. *L. sachalinensis*

1.   Lonicera japonica Thunb. *L. acuminata* var. *japonica* Miq.; *L. fauriei* Lév. & Van't.   Sui-kazura, Nindō. Scandent shrub; branches terete, hollow, glandular and prominently spreading-pubescent when young; leaves broadly lanceolate to ovate-elliptic, 3–7 cm. long, 1–3 cm. wide, acute or obtuse, mucronate-tipped, usually rounded at base, entire or the leaves of young plants often incised-sinuate, pubescent, pale green beneath, the petioles 3–8 mm. long, peduncles 3–10 mm. long, the bracts leaflike, ovate, 1–2 cm. long, the bracteoles free, about 1 mm. long, pilose, elliptic; calyx-tube glabrous, the lobes ovate, about 1 mm. long, ciliate; corolla 3–4 cm. long, soft-pubescent externally, white, changing to yellow with age, the tube narrow, the limb as long as the tube, bilabiate; fruit blue-black, 6–7 mm. across.——May–June. Thickets in hills and mountains; Hokkaido, Honshu, Shikoku, Kyushu; common.——Korea, Manchuria, and China.

Var. **repens** (Sieb.) Rehd. *L. flexuosa* Thunb.; *L. brachypoda* var. *repens* Sieb.; *L. japonica* var. *flexuosa* (Thunb.) Nichols.; *L. japonica* var. *chinensis* Bak.——Teri-ha-nindō. Less pubescent; corolla white to slightly purplish, the limb longer than the tube.——Honshu, Shikoku, Kyushu.——China.

2.   Lonicera affinis Hook. & Arn.   *L. buergeriana* Bl. ——Hama-nindō, Inu-nindō.   Scandent shrub; branches terete, glandular and white-puberulent while young; leaves broadly ovate or ovate-elliptic, 5–8 cm. long, 3–5 cm. wide, acute or short-acuminate, rounded at base, often glabrous on both sides, whitish beneath, mucronate, the petioles 5–10 mm. long, glandular and white-puberulent while young; flowers paired, terminal, the peduncles axillary, 3–7 mm. long, the bracts lanceolate-deltoid, spreading, about 2 mm. long, the bracteoles free, semirounded, slightly pilose; calyx glabrous, the lobes narrowly deltoid; corolla 4–6 cm. long, glabrous externally, white, changing to yellow with age, the tube about half as long as the lobes, narrow; fruit globose, about 7 mm. in diameter, blue-black, white-bloomy.——May–June. Honshu (Kii Prov.) Shikoku, Kyushu.——Ryukyus and China.

3.   Lonicera hypoglauca Miq.   *L. mollissima* Bl., ex Maxim.; *L. affinis* var. *pubescens* Maxim.; *L. affinis* var. *hypoglauca* (Miq.) Rehd.; *L. affinis* var. *mollissima* (Bl.) Makino ——Kidachi-nindō, Tō-nindō.   Scandent shrub; branches slender, elongate, terete, rather prominently soft-puberulent; leaves rather thin, ovate or oblong-ovate, 4–8(–10) cm. long, 1.5–5 cm. wide, abruptly acuminate at apex, rounded at base, glabrous above, glaucescent, puberulent and with red-brown glandular dots beneath; peduncles short, puberulent, the bracts deltoid-lanceolate, hairy, about 2 mm. long, the bracteoles small, free, semirounded; calyx-tube glabrous, the lobes narrowly deltoid, less than 1 mm. long; corolla bilabiate, 4–5 cm. long, slightly puberulent externally, white, changing to yellow with age.——May–June. Honshu (Tōkaidō, Chūgoku Distr.), Shikoku, Kyushu.——Ryukyus, Formosa, and China.

4.   Lonicera maackii (Rupr.) Maxim. *Xylosteum maackii* Rupr.——Hana-hyōtan-boku.   Shrub; young branches obtusely angled, puberulent; leaves membranous, ovate-oblong or obovate-elliptic, 5–8 cm. long, 2.5–3.5 cm. wide, acuminate, rounded or abruptly narrowed at base, slightly pubescent on both sides while young, the indument persistent on margin and on midrib on both sides, the petioles 3–5 mm. long; peduncles 2–4 mm. long, loosely glandular-dotted, the bracts membranous, linear, slightly longer than the ovary, the bracteoles connate, as long as the ovary, shallowly 2-lobed, ciliate; calyx-limb cup-shaped, 2–3 mm. long, irregularly 5-lobed, deciduous; corolla white, slightly pubescent externally, about 2 cm. long, the tube short; fruit black, globose, 3–4 mm. in diameter.——May. Mountains; Honshu (centr. and n. distr.); rare.——Korea, Manchuria, Amur, Ussuri, and n. China.

5. Lonicera chrysantha Turcz. *Xylosteum gibbiflorum* Rupr.; *L. xylosteum* var. *chrysantha* (Turcz.) Regel——NEMURO-BUSHIDAMA. Shrub; young branches terete, slightly pubescent and minutely glandular-dotted; leaves obovate, ovate-elliptic, or oblong-ovate, 6–10 cm. long, 3–6 cm. wide, very acuminate, acute to rounded at base, pubescent especially on nerves beneath, the petioles 3–7 mm. long, the peduncles 1–2 cm. long, with spreading hairs or glabrate, the bracts broadly linear, 2–4 times as long as the ovary, the bracteoles elliptic, free, about half as long as the ovary; calyx-limb very short, the 5 teeth short; corolla pale yellow, 1.2–1.5 cm. long, the tube short, gibbous; fruit red, about 6 mm. across.——June. Hokkaido.——s. Kuriles, Sakhalin, Korea, and e. Siberia.

6. Lonicera demissa Rehd. *L. ibotaeformis* Nakai——IBOTA-HYŌTAN-BOKU. Much-branched shrub; young branches, leaves and inflorescence ascendingly short-pilose; leaves thin, broadly oblanceolate to narrowly obovate, or sometimes narrowly oblong, 1.5–3.5 cm. long inclusive of the very short petiole, 7–12 mm. wide, acute with an obtuse tip, acute at base, glaucescent beneath; peduncles 8–15 mm. long, slender, the bracts membranous, linear, 2–3 times as long as the ovary, the bracteoles free, nearly orbicular, nearly as long as the ovary; calyx-tube minutely glandular-dotted, the limb with 5 short obtuse teeth; corolla bilabiate, pale yellow, 1–1.2 cm. long, short-pubescent externally, the tube short, gibbous; fruit red. ——June. Mountains; Honshu (southern part of centr. distr.).

7. Lonicera morrowii A. Gray. *Caprifolium morrowii* (A. Gray) O. Kuntze——KINGIN-BOKU. Prominently pubescent, much-branched shrub with terete branches; leaves oblong to elliptic, or ovate-elliptic, 2.5–4 cm. long, 1–2(–2.5) cm. wide, usually obtuse to sometimes subacute, rounded at base, pale green beneath; peduncles 1–1.5 cm. long, the bracts linear, somewhat leaflike, 2–4 times as long as the ovary, the bracteoles free, elliptic, slightly shorter than the ovary; calyx-limb deeply cleft, the segments broadly lanceolate, less than 1 mm. long; corolla about 1.3 cm. long, white, changing to yellow with age, pubescent externally, the tube short, slightly gibbous; fruit red, about 6 mm. across.——May. Hokkaido, Honshu, Shikoku.——Cv. **Xanthocarpa.** *L. morrowii* forma *xanthocarpa* (Nash) Hara; *L. morrowii* var. *xanthocarpa* Nash ——KIMI-NO-KINGIN-BOKU. Fruit yellow.——Cultivated.

8. Lonicera caerulea L. var. **edulis** Turcz. ex Herd. *L. caerulea* sensu auct. Japon., pro parte; *L. edulis* Turcz. ex Freyn; *L. caerulea* subsp. *edulis* (Turcz.) Hult.——KE-YO-NOMI. Much-branched shrub with exfoliating bark; the young branches prominently pubescent; leaves oblong, ovate-elliptic, or ovate, obtuse to subacute, or sometimes rounded at the tip, 2.5–4(–5) cm. long, 1–2 cm. wide, rounded to obtuse at base, prominently pubescent, short-petiolate; peduncles 2–10 mm. long, the bracts linear, 2–3 times as long as the calyx-tube, the bracteoles connate, bladderlike, completely enveloping the ovary, blue-black, bloomy, juicy when ripe; calyx-limb

short, with 5 short teeth; corolla creamy-white, infundibuliform, 1–1.3 cm. long, loosely pubescent externally, the lobes ascending more than half of the entire length; stamens usually slightly longer than the corolla-lobes; fruit sweet and edible, ellipsoidal, bloomy.——June–July. Hokkaido.——Kuriles, Sakhalin, e. Siberia, and Kamchatka.

Var. **emphyllocalyx** (Maxim.) Nakai. *L. emphyllocalyx* Maxim.; *L. caerulea* var. *altaica* sensu auct. Japon., non Sweet, incl. forma *emphyllocalyx* (Maxim.) Rehd.; *L. caerulea* var. *glabrescens* forma *longibracteata* C. K. Schn.; *L. caerulea* var. *longibracteata* (C.K. Schn.) Hara——KUROMI-NO-UGUISU-KAGURA. Plants less hairy, indument shorter, or nearly glabrate.——Hokkaido, Honshu (centr. distr.).

Var. **venulosa** (Maxim.) Rehd. *L. reticulata* Maxim., non Raf., non Champ.; *L. venulosa* Maxim.——MARUBA-YONOMI. Plants nearly glabrous; leaves prominently venulous beneath. ——The typical phase of the species occurs in w. Europe to Japan, and N. America.

9. Lonicera harae Makino. *L. harae* var. *tashiroi* Nakai——TSUSHIMA-HYŌTAN-BOKU, NOYAMA-HYŌTAN-BOKU. Shrub with grayish branches; young branchlets terete, with coarse hairs or nearly glabrous; leaves subcoriaceous, elliptic or ovate-elliptic, 3–6 cm. long, 2–4 cm. wide, obtuse to rounded, mucronate-tipped, usually rounded at base, short-petiolate, loosely setose especially on margin and midrib at least while young, becoming glabrous, nearly concolorous, with prominently raised veinlets on both surfaces; peduncles 3–8 mm. long, glabrous, the bracts lanceolate, 1.5–2.5 times as long as the calyx-tube; bracteoles absent; calyx-limb short, subtruncate; flowers with the leaves; corolla about 1 cm. long, somewhat bilabiate, yellow to white, the tube nearly as long as the limb, slightly gibbous.——Mar.–Apr. Shikoku, Kyushu.——Korea and Manchuria.

10. Lonicera strophiophora Franch. *L. pilosa* Maxim.; *L. amherstii* Dipp., pro parte——ARAGE-HYŌTAN-BOKU, ŌBA-HYŌTAN-BOKU. Shrub; branches gray-brown, slightly pilose while very young; leaves thinly membranous, broadly ovate to ovate-oblong, 5–10 cm. long, acuminate to acute, or sometimes obtuse, obtusely rounded at base, coarsely pilose on both surfaces especially on midrib beneath, often glaucous beneath, the petioles 3–7 mm. long, loosely pilose and sparingly glandular-pilose; flowers with the leaves, the peduncles deflexed, 1–2 cm. long, loosely glandular-pilose, the bracts membranous, ovate to broadly so, 1–2 cm. long, subacute, pale green, ciliate, weakly many-nerved, bracteoles absent; calyx-limb short, irregularly lobed; corolla pale yellow, narrowly infundibuliform, 2–2.5 cm. long, loosely pilose or glabrous externally; fruit red, globose.——Apr.–June. Mountains; Hokkaido, Honshu.

Var. **glabra** Nakai. DAISEN-HYŌTAN-BOKU. Ovary and lower half of style glabrous.——Honshu (centr. and westw.), Shikoku.

11. Lonicera praeflorens Batal. *L. kaiensis* Nakai——HAYAZAKI-HYŌTAN-BOKU. Shrub with gray-brown exfoliating bark; young branchlets glabrous or puberulent; leaves membranous, broadly ovate, 3–6 cm. long, 2–4 cm. wide, acute or abruptly so, obtusely rounded at base, pubescent, especially on underside, glaucescent and the veinlets rather prominently raised, the petioles 2–4 mm. long; flowers precocious, the peduncles glabrous, 3–8 mm. long, the bracts broadly lanceolate or narrowly ovate, 2–4 times as long as the ovary, ciliate, bracteoles absent; calyx-tube glabrous, the limb irregularly toothed, sparsely glandular-pilose on margin; corolla infundibuliform, nearly actinomorphic, glabrous externally, 12–15 mm. long, the tube obsoletely gibbous, as long as the limb; stamens exserted; fruit red, about 8 mm. across.——May. Mountains; Honshu (centr. distr.); rare.——Korea, Manchuria, and Ussuri.

12. Lonicera linderifolia Maxim. YABU-HYŌTAN-BOKU. Much-branched shrub with exfoliating bark; young branches puberulent; leaves membranous, broadly lanceolate or narrowly ovate-oblong, sometimes narrowly oblong, 2–4(–4.5) cm. long, 1–1.5 cm. wide, acute, sometimes with an obtuse tip, glaucescent and puberulent beneath, the petioles 2–3 mm. long; peduncles slender, 1.5–2.5 cm. long, puberulent; bracts linear, 3–5 mm. long, nearly glabrous, bracteoles absent; calyx glabrous, the calyx-limb short, nearly truncate; corolla purplish, about 10 mm. long.——Honshu (Mount Hayachine in Rikuchu Prov.); rare.

13. Lonicera konoi Makino. *L. tobitae* Nakai——KO-GOME-HYŌTAN-BOKU, KUMOI-HYŌTAN-BOKU. Much-branched small shrub with gray-brown branches and exfoliating bark; young branchlets slender, loosely puberulent; leaves rather thin, ovate, elliptic, or obovate, 1–2(–2.5) cm. long, 6–8 mm. wide, obtuse or subacute, subacute at base, glaucous, puberulent especially along the midrib beneath, the petioles 2–3 mm. long; peduncles short-pilose, 5–10(–13) mm. long; bracts linear, glabrous, 2 or 3 times as long as the ovary, bracteoles obsolete; calyx-limb short, entire, glabrous; corolla deep purple, glabrous, infundibuliform, nearly actinomorphic, about 5 mm. long, the lobes 1.5 mm. long, the tube somewhat gibbous; fruit red, about 4 mm. across, few-seeded.——May. Alpine; Honshu (Mount Yatsugatake and Akaishi mountain range in centr. distr.); rare.

14. Lonicera ramosissima Fr. & Sav. KO-UGUISU-KA-GURA. Much-branched shrub with slender branches; young branchlets short-pilose or nearly glabrous; leaves ovate, ovate-elliptic, or broadly ovate, 2–4 cm. long, 1–2.5 cm. wide, acute or obtuse, rounded or obtusely rounded at base, usually appressed-pubescent above, glaucescent and with incurved hairs beneath, the petioles 2–3 mm. long; peduncles slender, 1.5–2.5 cm. long, pubescent or glabrous; bracts lanceolate, 1.5–3 times as long as the ovary, the bracteoles connate, obcordate, forming a lobe about half as long as the ovary; calyx-limb very short; corolla infundibuliform, pale yellow, often sparingly pilose externally, 12–15 mm. long, the tube long, gibbous; style long-exserted; fruit red.——May–June. Mountains; Honshu (centr. distr.).

Var. **kinkiensis** (Koidz.) Ohwi. *L. kinkiensis* Koidz.——KINKI-HYŌTAN-BOKU. Leaves narrower, acuminate, acute or subobtuse at the tip with shorter hairs on the upper side; bracts nearly as long as the ovary, slender, the bracteoles somewhat larger.——Honshu (Settsu, Tanba, and Aki Prov.).

Var. **fudzimoriana** (Makino) Nakai. *L. fudzimoriana* Makino; *L. bukoensis* Nakai; *? L. ramosissima* var. *borealis* Koidz.——CHICHIBU-HYŌTAN-BOKU. Leaves broader, elliptic or oblong, rounded to cordate at base.——High mountain or northern phase; Honshu (Kantō Distr., Kai Prov. and n. distr.).

15. Lonicera gracilipes Miq. *Xylosteum phylomelae* Jacob-Makoy; *L. phylomelae* Carr.; *L. gracilipes* var. *genuina* Maxim.——YAMA-UGUISU-KAGURA. Branched shrub with slender, usually glabrous branches and branchlets; leaves narrowly to broadly ovate, ovate-elliptic, or rhombic-ovate, 2.5–6

cm. long, 1.5–5 cm. wide, acute with an obtuse tip, obtuse to acute at the base, whitish and sometimes brown-spotted beneath, usually coarsely pilose especially on midrib beneath, the petioles 2–5 mm. long; flowers solitary, rarely in pairs; peduncles slender, pendulous, 1.5–2.5 cm. long, usually glabrous; bracts solitary or rarely 2, narrowly lanceolate, 3–7 mm. long, bracteoles usually absent; calyx-limb shallowly cup-shaped, irregularly short-toothed; corolla pink, narrowly infundibuliform, nearly actinomorphic, 1.5–2 cm. long, the lobes short; style exserted; fruit ellipsoidal, red, about 1 cm. long.——Apr.–May. Hokkaido (s. distr.), Honshu, Shikoku, Kyushu; variable.——Forma **glabra** (Miq.) Hiyama. *L. gracilipes* var. *glabra* Miq.——UGUISU-KAGURA. Plant nearly glabrous.

Var. **glandulosa** Maxim. *L. tenuipes* Nakai; *L. tenuipes* var. *glandulosa* (Maxim.) Nakai; *L. glandulosa* (Maxim.) Koidz.; *L. macrocalyx* Nakai——MIYAMA-UGUISU-KAGURA. Plant coarsely pilose, the peduncles and ovaries glandular-pilose.——Higher elevations in the mountains.

16. Lonicera cerasina Maxim. *L. shikokiana* Makino USUBA-HYŌTAN-BOKU. Shrub with glabrous terete branchlets; leaves membranous, oblong to narrowly so, 7–12 cm. long, 2–4 cm. wide, long-acuminate with a minute obtuse tip, acute or subrounded at base, ascending-hairy on margin, sometimes scattered-pilose on upper surface, glabrous beneath, sparingly glandular-pilose, the petioles 7–10 mm. long; peduncles 1–1.2 cm. long, glabrous; bracts linear, 1–1.5 times as long as the ovary, the bracteoles nearly free, elliptic, slightly shorter than the ovary; calyx-limb very short; corolla bilabiate, yellowish white, about 1 cm. long, the tube nearly as long as the limb, gibbous; fruit red.——Apr.–May. Honshu (Kinki Distr. and westw.), Shikoku.

17. Lonicera vidalii Fr. & Sav. ONI-HYŌTAN-BOKU. Shrub; branches rather thick, terete, minutely glandular-dotted while young; leaves rather thin, elliptic, broadly ovate, or ovate-oblong, 6–12 cm. long, 2–6 cm. wide, abruptly acute to short-acuminate with an obtuse tip, abruptly acute to sometimes rounded at base, green above, paler beneath, short-pilose on nerves above, coarsely pilose beneath especially on the midrib, the petioles 1–1.5 cm. long; peduncles 1.5–2 cm. long, rather thick; bracts linear, nearly as long as to slightly longer than the ovary, the bracteoles orbicular, nearly as long as the ovary, connate at the base; calyx-limb very short; corolla bilabiate, about 12 mm. long, glabrous outside, pale yellow, gibbous; stamens slightly exserted; fruit red, about 7 mm. long.——May. Mountains; Honshu (Echigo, Shinano Prov. and Chūgoku Distr.).——s. Korea.

18. Lonicera alpigena L. var. **glehnii** (F. Schmidt) Nakai. *L. alpigena* sensu auct. Japon., non L.; *L. glehnii* F. Schmidt; *L. watanabeana* Makino; *L. watanabeana* var. *viridissima* Nakai——EZO-HYŌTAN-BOKU, SURUGA-HYŌTAN-BOKU, ŌBA-BUSHIDAMA. Resembles the preceding species; young branchlets obtusely angled, glabrous; leaves ovate-oblong or narrowly ovate, 6–10 cm. long, 2.5–5 cm. wide, acuminate, rounded to shallowly cordate at base, ciliate, puberulent on nerves above, spreading-hairy beneath especially on nerves, the petioles minutely glandular-dotted, 5–10 mm. long; peduncles 2.5–4.5 cm. long, sometimes minutely glandular-dotted; bracts linear, 3–5 times as long as the ovary, the bracteoles minute or absent; calyx-limb nearly absent; corolla bilabiate, yellowish, glabrous outside, 12–15 mm. long, deeply lobed, gibbous; stamens included; fruit red, about 7 mm. long.——Hokkaido, Honshu (n. distr. to high mountains of centr. distr.).——Sakhalin and s. Kuriles.

19. Lonicera chamissoi Bunge. Chishima-hyōtan-boku.   Glabrous small shrub with slightly glaucescent terete branchlets; leaves chartaceous to membranous, broadly ovate to elliptic, 2–4(–5) cm. long, 1.5–2.5 cm. wide, obtuse or rounded, rounded at base, with slightly raised veinlets beneath, sessile or the petioles up to 2 mm. long; peduncles 6–12 mm. long, the bracts and bracteoles broadly ovate or orbicular, less than 1 mm. long; calyx-limb short, minutely 5-toothed; corolla bilabiate, about 1 cm. long, deep red, glabrous outside, gibbous; fruit red.——June–July. Alpine; Hokkaido, Honshu (centr. distr. and northw.).——Kuriles, Sakhalin, Kamchatka, Amur, and Ochotsk Sea region.

20. Lonicera tschonoskii Maxim. *L. brandtii* Fr. & Sav.——Ō-hyōtan-boku.   Shrub; branches terete, obsoletely 4-striate, glabrous; leaves membranous or thinly chartaceous, narrowly obovate, oblong, or ovate-oblong, 7–13 cm. long, 3–5(–6) cm. wide, acuminate, acute at base, glabrous, often puberulent on midrib above, loosely pilose on margin and on midrib beneath while young, glaucous beneath, the petioles 7–13 cm. long; peduncles slender, 2–5 cm. long, glabrous; bracts lanceolate, about 3 mm. long, the bracteoles free, broadly ovate, about 2.5 mm. long; calyx-limb 5-cleft, deciduous; corolla bilabiate, about 15 mm. long, white, glabrous outside, gibbous; fruit red.——July. Alpine; Honshu (centr. distr.).

21. Lonicera mochidzukiana Makino. Nikko-hyō-tan-boku.   Nearly glabrous shrub; branches prominently 4-angled; leaves chartaceous, ovate-oblong to ovate, sometimes broadly lanceolate, 4–8 cm. long, 1.5–3 cm. wide, acute to acuminate, rounded at base, glabrous, rarely thinly pilose on midrib beneath, with the tertiary veinlets rather prominently raised on both sides, the petioles 3–5 mm. long; peduncles rather thick, 1–2 cm. long; bracts narrowly ovate or broadly lanceolate, 1–2 mm. long, the bracteoles orbicular, connate at base, about 1 mm. long, usually more or less glandular-dotted as well as the bracts; calyx-limb deeply 5-cleft, the lobes deciduous; corolla bilabiate, about 1 cm. long, white; fruit red.——June. Mountains; Honshu (Kantō Distr. and Kai Prov.).

Var. **nomurana** (Makino) Nakai. *L. nomurana* Makino ——Yama-hyōtan-boku.   Leaves broadly ovate to ovate-orbicular, 2.5–5 cm. long, 1.5–3 cm. wide, acute to obtuse.——Mountains; Honshu (Tōkaidō Distr., Kii Prov., and Chūgoku Distr.).

Var. **filiformis** Koidz. Akaishi-hyōtan-boku.   Leaves lustrous on upper surface; bracts filiform; segments of calyx-limb nearly as long as the ovary.——Alpine; Shikoku.

22. Lonicera sachalinensis (F. Schmidt) Nakai. *L. maximowiczii* var. *sachalinensis* F. Schmidt——Benibana-hyōtan-boku.   Shrub; branches glabrous, nearly terete, with 2 or 4 obtuse angles; leaves rather firmly chartaceous, ovate-oblong or ovate, sometimes elliptic, 4–8 cm. long, 2–3.5 cm. wide, acute to acuminate, sometimes subobtuse, rounded to acute at base, long-ciliate, glabrous or rarely loosely long-pilose beneath, with the tertiary veinlets prominent and raised, the petioles glabrous, 3–5 mm. long; peduncles 1.5–2 cm. long, glabrous; bracts broadly lanceolate to ovate, 1–3 mm. long, the bracteoles about 0.5 mm. long, connate, forming a broad lobe; calyx-limb deeply 5-cleft; corolla bilabiate, 8–10 mm. long, deep red; fruit red.——June–July. Hokkaido.——s. Kuriles, Sakhalin, n. Korea, Manchuria, and Ussuri.

Fam. 188. **ADOXACEAE** Rempuku-sō Ka   Moschatel Family

Family of a single genus and species, widely distributed in the N. Hemisphere.

1. **ADOXA** L.   Rempuku-sō Zoku

1. Adoxa moschatellina L. Rempuku-sō.   Delicate perennial herb 8–15 cm. high, with slender creeping rhizomes; radical leaves long-petiolate, 1- to 2-ternate, the leaflets ternately dissected and lobulate; cauline leaves a single pair, similar to the radical but short petiolate and less dissected, smaller; flowers sessile, small, greenish, 4–6 mm. across, capitate, 4-merous in the terminal one, 5- or 6-merous in the lateral ones, sessile; calyx 2- or 3-lobed; corolla rotate, 4- to 6-cleft, disc absent; stamens 4–6, bifid, the anthers 1-locular; ovary partially superior, 3- to 5-locular, the styles 3- to 5-lobed; ovules solitary in each locule, pendulous; drupe 1- to 5-stoned; seeds with endosperm, the embryo small.——Apr.–May.   Damp woods in mountains and hills; Hokkaido, Honshu (Kinki Distr. and northw.). The phase in e. Asia and India is without fragrance and is sometimes distinguished from the European phase as var. *inodora* Falconer ex C. B. Clarke.

Fam. 189. **VALERIANACEAE** Omina-eshi Ka   Valerian Family

Herbs often foetid when dry; leaves radical or opposite, often dissected, stipules absent; inflorescence usually determinate, cymose; flowers bisexual, sometimes unisexual (plants sometimes polygamodioecious), usually small, somewhat zygomorphic; calyx epigynous, plumose or sometimes coronate; corolla gamopetalous, tubular, infundibuliform or rotate, often gibbous or spurred at base, the lobes imbricate in bud; stamens 1–4 (rarely 5), on the corolla-tube, alternate with the lobes, the anthers 2-locular; ovary inferior, basically 3-locular, with 1 locule fertile; style solitary, slender; ovules solitary, pendulous; fruit a cypselate achene, the embryo straight, endosperm absent.——About 10 genera, with about 400 species, in the N. Hemisphere and S. America, few in Africa.

1A.   Stamens 4; sterile locules of the achenes often winglike; calyx-lobes small, coronate. .... 1. *Patrinia*
1B.   Stamens usually 3; sterile locules of the achenes usually inconspicuous; calyx-lobes much divided and plumose. .... 2. *Valeriana*

1. PATRINIA Juss.   OMINA-ESHI ZOKU

Erect perennial herbs; leaves once or twice pinnately lobed, sometimes incised; cymes much branched, paniclelike, the bracts narrow, free, sometimes with a large bracteole appressed to the fruit; flowers usually yellow, sometimes white; calyx-limb small, obtusely toothed, coronate; corolla-tube short, the limb rather regularly 5-lobed; stamens 4; style entire; sterile locules of the achenes often winglike.——About 15 species, in e. and centr. Asia.

1A. Corolla gibbous or spurred.
    2A. Leaves palmately lobed to cleft. .... 1. *P. triloba*
    2B. Leaves pinnately incised. .... 2. *P. gibbosa*
1B. Corolla not gibbous or spurred.
    3A. Stems 7–15 cm. long, naked or with 1 or 2 pairs of leaves; radical leaves many. .... 3. *P. sibirica*
    3B. Stems 50–100 cm. long, with more than 2 pairs of cauline leaves.
        4A. Achenes winged; flowers white; leaves coarsely pilose. .... 4. *P. villosa*
        4B. Achenes wingless; flowers yellow; leaves nearly glabrous or sparingly puberulent. .... 5. *P. scabiosaefolia*

1. **Patrinia triloba** Miq. var. **triloba.** *P. palmata* var. *gibbosa* Makino; *P. gibbiferum* Nakai; *P. triloba* var. *gibbosa* (Makino) Matsum.——KO-KINREI-KA, HAKUSAN-OMINA-ESHI. Spur shorter than in var. *palmata,* saccate.——Mountains; Honshu (Kinki Distr. and eastw.).

Var. **palmata** (Maxim.) Hara. *P. palmata* Maxim.—— KINREI-KA. Stoloniferous; rhizomes short-creeping; stems erect, nearly simple, 20–60 cm. long, glabrous at base, with 2 lines of short hairs in upper part; leaves petiolate, 5-angled, orbicular, 4–8 cm. long and as wide, cordate, palmately (3–)5-lobed to -cleft, the midlobe rhombic-ovate, caudate, irregularly incised and acutely toothed, nearly glabrous or coarsely pilose especially on nerves beneath; cymes terminal, many-flowered; corolla 7–8 mm. across, spurred at base, the spur 2.5–3 mm. long; stamens as long as the spreading corolla-lobes; achenes oblong, glabrous, about 4 mm. long, flat, with a thick broad band on the back, the wing twice as broad as the fruit.——July–Aug. Mountains; Honshu (centr. distr. and westw.), Kyushu.

Var. **kozushimensis** Honda. *P. kozushimensis* (Honda) Honda——SHIMA-KINREI-KA. Leaves thicker, nearly glabrous; calyx-teeth more prominent.——Maritime; Honshu (Izu Isls.).

Var. **takeuchiana** (Makino) Ohwi. *P. takeuchiana* Makino——Ō-KINREI-KA. Plant larger; leaves to 15 cm. long; corolla with a short-cylindric spur; stamens exserted.—— Honshu (Mount Aobayama in Tango).

2. **Patrinia gibbosa** Maxim. MARUBA-KINREI-KA. Rhizomes short-creeping; stems rather stout, erect, 50–70 cm. long, loosely short-pubescent or nearly glabrous; leaves broadly ovate or ovate-elliptic, 7–15 cm. long, acuminate to acute, rounded to broadly cuneate at base, pinnately incised and coarsely toothed, the petioles winged on upper part, the upper leaves subsessile, usually short-pilose on nerves beneath; cymes many-flowered, short-pubescent; corolla yellow, 4–5 mm. across, spurred at base; stamens exserted; achenes oblong, flat, broadly winged.——July–Aug. Mountains; Hokkaido, Honshu (n. distr. and Sado Isl.).

3. **Patrinia sibirica** (L.) Juss. *Valeriana sibirica* L.—— TAKANE-OMINA-ESHI, CHISHIMA-KINREI-KA. Rhizomes stout, elongate; stems 7–15 cm. long, naked or with 1 or 2 pairs of leaves; radical leaves many, rosulate, elliptic or obovate, 2–4 cm. long, 1–2 cm. wide, obtusely toothed, incised or usually pinnately cleft, glabrous or scattered pilose, the petioles winged on upper part; inflorescence rather many-flowered, the branches with prominent tuberclelike white hairs on one side; calyx-teeth broadly ovate, slightly accrescent, rather small; corolla yellow, about 4 mm. across; achenes elliptic, about 4 mm. long, the wing orbicular, 5–6 mm. across.——July–Aug. Alpine; Hokkaido.——Kuriles, Sakhalin, and e. Siberia.

4. **Patrinia villosa** (Thunb.) Juss. *Valeriana villosa* Thunb.; *Fedia villosa* Vahl; *P. villosa* var. *japonica* Lév.; *P. villosa* var. *sinensis* Lév.; *P. sinensis* (Lév.) Koidz.——OTOKO-ESHI. Rather stout short-rhizomatous perennial; stolons well developed; stems sometimes branched, erect, 50–100 cm. long, leafy, densely whitish retrorse-pilose, especially on lower portion, nearly glabrous in upper portion; leaves simple and ovate to pinnately divided, 3–15 cm. long, toothed, more or less coarsely white-pilose, petiolate or sessile in the upper ones; inflorescence many-flowered, branched, the branches with 2 pilose lines of spreading or slightly deflexed white hairs; corolla white, about 4 mm. across, spurless; achenes obovate, 2–3 mm. long, rounded on back, thick, the wing orbicular-cordate, 5–6 mm. across.——Aug.–Oct. Sunny hills and low elevations in the mountains; Hokkaido, Honshu, Shikoku, Kyushu; common.——Ryukyus, Korea, Manchuria, and China.

Patrinia × hybrida Makino. OTOKO-OMINA-ESHI. Hybrid of *P. scabiosaefolia* × *P. villosa.*——Reported to occur in Honshu and Kyushu.

5. **Patrinia scabiosaefolia** Fisch. *P. serratulaefolia* Fisch.; *P. hispida* Bunge; *P. parviflora* Sieb. & Zucc.—— OMINA-ESHI. Rhizomes rather thick, creeping; stems erect, 60–100 cm. long, rather stout, glabrescent in upper part, more or less coarsely hairy at base; leaves mostly cauline, subsessile or short-petiolate, pinnately divided, the lowest divisions smaller and stipulelike or auriculate, the upper divisions broadly lanceolate, coarsely toothed, nearly glabrous or sparingly puberulent on both sides, sparingly scabrous on margin; inflorescence rather flat-topped, many-flowered, the branches with short white tuberclelike hairs on one side; corolla yellow, 3–4 mm. across, the tube nearly as long as the lobes, loosely long-hairy inside, ampliate; achenes wingless, oblong, 3–4 mm. long, more or less flat, 1-nerved on the ventral side, the dorsal side with a swollen appendage at base.——Aug.–Oct. Sunny grassy places in hills and mountains; Hokkaido, Honshu, Shikoku, Kyushu; common.——Kuriles, Sakhalin, Ryukyus, Formosa, Korea, China, Manchuria, and e. Siberia.

2. VALERIANA L.   KANOKO-SŌ ZOKU

Glabrous to hairy perennials (in ours); leaves entire or toothed, usually pinnate to pinnatifid or rarely bipinnatifid, often decurrent; inflorescence determinate, either aggregate-dichasial and thyrsoid or the cymes compound, dense and more or less scorpioid, bracteate; flowers bisexual or unisexual (plants gynodioecious or polygamodioecious), white to pink; calyx initially involute, epigynous, later spreading, much divided and usually plumose; corolla infundibuliform, subcampanulate, or rotate, the tube gibbous or straight, the 5 lobes equal or subequal; stamens 3, rarely 4, the anthers included or exserted; achenes 3-nerved on the back, 1-nerved on face.——About 200 species, in the N. Hemisphere and S. America, few in the mountains of e. Africa, 1 in S. Africa.

1A. Stems rather stout, erect, 40–80 cm. long; leaves with deeply toothed segments, the lateral segments 2–5 cm. long; corolla 5–7 mm. long with a long tube; stamens and style long-exserted. . . . . 1. *V. fauriei*
1B. Stems slender, becoming decumbent, 20–40 cm. long; leaves with shallowly toothed segments, the lateral segments 1–2.5(–3) cm. long; corolla about 2 mm. long, with a short tube; stamens and style not exserted. . . . . 2. *V. flaccidissima*

1. Valeriana fauriei Briq. *V. officinalis* sensu auct. Japon., non L.; *V. officinalis* var. *angustifolia* Miq.; *V. officinalis* var. *latifolia* Miq.; *V. nipponica* Nakai——Kanoko-sō. Rhizomes short; stolons slender; stems erect, rather stout, 40–80 cm. long, with dense tuberclelike long white hairs on the nodes; radical leaves rather small, usually absent in anthesis; cauline leaves pinnate, the segments 5–7, oblong-lanceolate to oblanceolate, 2–5(–7) cm. long, 7–15(–25) mm. wide, coarsely obtuse-toothed, sessile, with ascending minute papillae on margin; inflorescence terminal, densely many-flowered, the bracts linear; corolla 5–7 mm. long, rosy, with a slender long tube, gibbous, the lobes about 1/5 the whole length; stamens long-exserted; achenes lanceolate, about 4 mm. long.——Mountains; Hokkaido, Honshu, Shikoku, Kyushu; rather rare.——Korea, Manchuria, Formosa, Sakhalin, and s. Kuriles.
2. Valeriana flaccidissima Maxim. *V. hardwickii* var.

leiocarpa Miq.——Tsuru-kanoko-sō. Rhizomes short; stolons very slender and elongate; stems soft, slender, decumbent and flaccid after anthesis, 20–40(–60) cm. long, short-pilose on the nodes; radical leaves and those of the stolons broadly ovate, 1.5–4 cm. long, simple or 3-foliolate; cauline leaves of 2 or 3 pairs, petiolate or the upper sessile, 3–7(–9)-foliolate, short-pilose on margin, the terminal leaflets ovate or narrowly so, 2–5 cm. long, the lateral ones mostly ovate, 1–2(–3) cm. long, smaller, the lowest pair often stipulelike or auricled; inflorescence densely flowered, the bracts linear, short; corolla white, sometimes pinkish, infundibuliform, about 2 mm. long, obsoletely gibbous; stamens and style not exserted; achenes 2–2.5 mm. long, broadly lanceolate.——Apr.–June. Wet woods and thickets along streams and ravines; Honshu, Shikoku, Kyushu.

Fam. 190. **DIPSACACEAE** Matsumushi-sō Ka   Teasel Family

Herbs, rarely subshrubs; leaves opposite or verticillate; stipules absent; flowers bisexual, zygomorphic, usually in dense heads, enveloped by an epicalyx; calyx cup-shaped, often lobed and divided into pappuslike segments; corolla gamopetalous, 4- or 5-lobed, the lobes imbricate in bud; stamens usually 4, rarely 2 or 3, inserted near base of corolla-tube, alternate with the lobes, the filaments free, the anthers 2-locular, longitudinally split; ovary 1-locular and 1-ovuled, inferior; style slender; seeds with a straight embryo.——About 9 genera, with about 160 species, indigenous of the Old World, abundant in the Mediterranean region, a few naturalized elsewhere.

1A. Plant spiny or prickly, especially on the stem; heads globose or ellipsoidal; bracts of the receptacle and involucre spine-tipped.
1. *Dipsacus*
1B. Plant not spiny or prickly; heads depressed, subglobose to ovoid-conical; bracts of receptacle and involucre not spine-tipped.
2. *Scabiosa*

1. **DIPSACUS** Nabe-na Zoku

Coarse spiny to prickly annuals, biennials, or perennials; leaves opposite, coarsely toothed, pinnately lobed, or divided; bracts of the involucre and of the receptacle firm, spine-tipped; flowers blue or whitish, in dense globose or ellipsoidal heads; involucel investing the ovary and achene 4-leaved; calyx-tube adnate to the ovary, the limb cup-shaped; corolla-limb nearly regularly 4-lobed; stamens 4; achenes more or less adnate to the involucel or free, 8-ribbed, usually crowned with the persistent calyx-limb.——About 12 species, in Europe, Asia, and n. Africa.

1. Dipsacus japonicus Miq. Nabe-na.   Course stout prickly biennial; stems 1 m. long or longer, erect, branched, longitudinally striate, with scattered weak spines or prickles; leaves opposite, the lower simple to pinnately compound, the segments oblong-ovate, rhombic-ovate, or ovate, 6–15 cm. long, acutely toothed, winged-petiolate, the upper leaves smaller, sessile, usually simple; heads pedunculate, strict, glo-

bose or broadly ellipsoidal, 2–3 cm. long, many-flowered, with linear bracts 5–20 mm. long at base, the bracts on receptacles cuneate, 5–8 mm. long, densely appressed-puberulent on back, short spine-tipped; corolla rose-purple, 5–6 mm. long; achenes about 6 mm. long, cuneate-oblong, puberulent on upper half.——Aug.–Oct. Mountains; Honshu, Shikoku, Kyushu.——Korea, China, and Manchuria.

2. **SCABIOSA** L.   Matsumushi-sō Zoku

Pubescent or sometimes nearly glabrous herbs, sometimes woody at the base; leaves entire or toothed to pinnate; heads depressed globose to ovoid-conical, many-flowered, usually pedunculate; involucral bracts 1- or 2-seriate; bracts on receptacles linear, small or absent; involucels adnate at base to the achene, the expanded portion cuplike, 2- to 4(–8)-ribbed, toothed on the margin; corolla-limb 4- or 5-lobed, usually blue, sometimes white; stamens 4 or 2; style slender; achenes adnate about half way to the involucel, crowned with the calyx-teeth.——About 60 species, in Europe, Africa, Asia, abundant in the Mediterranean region.

1. Scabiosa japonica Miq. *S. fischeri* var. *japonica* (Miq.) Nakai; *S. comosa* var. *japonica* (Miq.) Tatew.; *S. tschiliensis* var. *japonica* (Miq.) Hurusawa——Matsumushi-sō. Slightly pubescent biennial herb with a thick root; stems 30–80 cm. long, erect, branched; cauline leaves slightly thickened at base, pinnately compound, 5–10 cm. long, the lobes ovate, incised or again cleft into nearly linear segments; peduncles very long; heads at anthesis flat, 2.5–5 cm. across, globose in fruit, about 1.5 cm. across; corollas blue, the marginal rays elongate; fruit fusiform, about 3.5 mm. long, gla-brous, invested by an 8-ribbed pilose involucel.——Aug.–Oct. Sunny grassy slopes in mountains; Hokkaido, Honshu, Shikoku, Kyushu; common.

Var. **acutiloba** Hara. *S. jezoensis* Nakai——Ezo–matsumushi-sō. Lobes of leaves more acute at apex.——Northern phase; Hokkaido, Honshu (n. distr.).

Var. **alpina** Takeda. *S. fischeri* var. *longiseta* Hara——Takane–matsumushi-sō. Plant dwarf; flowers rather large; calyx rather long-spined.——Alpine; Honshu and Shikoku.

Fam. 191. CUCURBITACEAE   Uri Ka   Gourd Family

Tendril-bearing scandent (rarely otherwise) herbs, sometimes woody at base, often scabrid; flowers unisexual, actinomorphic (plants monoecious or dioecious); calyx of the staminate flowers short-tubular or obconical, the lobes imbricate or not overlapping in bud; corolla gamopetalous or polypetalous, the petals imbricate or induplicate-valvate in bud; stamens usually 3, free or variously connate, one of the anthers 1-locular, the others 2-locular, locules straight or flexuous; calyx-tube of the pistillate flowers adnate to the ovary; ovary inferior, with 3 parietal placentae often meeting in the center; ovules many or sometimes few; style simple or rarely 3, free, the stigma thickened; fruit a pepo, usually fleshy; seeds usually flat, without endosperm.—— About 100 genera, with about 860 species, chiefly in the Tropics and warmer temperate regions.

1A. Flowers rather large, more than 1 cm. across; fruit more than 2 cm. long.
  2A. Flowers less than 6 cm. across, white; corolla finely fimbriate on margin; fruit 5–10 cm. long. .... 1. *Trichosanthes*
  2B. Flowers 6–7 cm. across, yellow; corolla not fimbriate; fruit 4–5 cm. long. .... 2. *Thladiantha*
1B. Flowers small and less than 1 cm. across; fruit small, to 2 cm. long.
  3A. Leaves simple or often lobed; stamens free or connate only at base.
    4A. Stamens 5; corolla-lobes caudate; fruit circumscissile; seeds 2–4, rather large. .... 3. *Actinostemma*
    4B. Stamens 3; corolla-lobes acute; fruit not circumscissile.
      5A. Seeds horizontal within the fruit, many. .... 4. *Melothria*
      5B. Seeds perpendicular within the fruit, solitary in each locule. .... 5. *Schizopepon*
  3B. Leaves palmately compound, (3-)5-foliolate; stamens connate, forming a column; fruit black. .... 6. *Gynostemma*

1. TRICHOSANTHES L.   Karasu-uri Zoku

Annuals or perennials often with thickened roots; leaves 3- to 5-lobed or undivided; tendrils with 1–5 branches; flowers unisexual (plants often dioecious), the staminate solitary or in racemes, more than 1 cm. across, white; calyx-limb 5-lobed; corolla hypocrateriform, the limb nearly 5-parted, the segments fimbriate on margin; stamens 3, the filaments very short, the anther-locules S-shaped; pistillate flowers solitary; ovary oblong, 1-locular, the placentae 3–5, parietal, bifid; style short, the stigma capitate, entire or shallowly 3-lobed; ovules many, horizontal or slightly pendulous; fruit oblong or ovoid-globose, rarely elongate-cylindric, 5–10 cm. long, many-seeded.——About 50 species, in s. and e. Asia, and Australia.

1A. Stems soon glabrous; seeds flat, 1-locular, usually ellipsoidal, without a longitudinal band.
  2A. Calyx-segments toothed; leaves palmately 5- to 7-cleft, scabrous; fruiting peduncles 2–3 cm. long; seeds truncate. .... 1. *T. bracteata*
  2B. Calyx-segments entire.
    3A. Calyx-tube of pistillate flowers 2–2.5 cm. long; calyx-segments in both sexes 5–6 mm. long; fruiting peduncles 7–25 cm. long; fruit red. .... 2. *T. multiloba*
    3B. Calyx-tube of pistillate flowers 3.5–4 cm. long, calyx-segments in both sexes 10–12 mm. long; fruiting peduncles 2–3 cm. long; fruit yellow. .... 3. *T. kirilowii* var. *japonica*
1B. Stems puberulent; seeds slightly inflated, usually 3-locular, usually broader than long, with a longitudinal broad band on the face.
  4A. Calyx-segments 3–4 mm. long; tendrils with 1 or 2 branches; fruit rounded at apex. .... 4. *T. cucumeroides*
  4B. Calyx-segments 8–10 mm. long; tendrils with 2 or 3 branches; fruit beaked at apex. .... 5. *T. rostrata*

1. Trichosanthes bracteata (Lam.) Voigt. *Modecca* (?) *bracteata* Lam.; *T. shikokiana* Makino——Ō-karasu-uri. Roots thickened; stems glabrous; leaves petiolate, cordate or reniform-cordate, 8–15 cm. long and as wide, scabrous, loosely short-pubescent beneath especially on nerves while young, palmately 5- to 7-cleft, the terminal lobe ovate or ovate-deltoid; tendrils 3-lobed; fruiting peduncles 2–3 cm. long; staminate spikes elongate, to 3 cm. long; bracts large, green, 3–4 cm. long, obovate, finely lobulate on margin; calyx-segments narrowly ovate or broadly lanceolate, 12–15 mm. long, 5–7 mm. wide, toothed; fruit broadly ellipsoidal, about 7 cm. long, red; seeds gray, oblong, 12–13 mm. long, truncate.——Aug.–Sept. Shikoku, Kyushu.——Ryukyus, Formosa, China, India, and Malaysia.

2. Trichosanthes multiloba Miq. Momiji-karasu-uri. Roots thickened; stems and leaves soft brown-pubescent while young; leaves orbicular-cordate, short-pilose on both sides and slightly scabrous, palmately 5- to 9-cleft, the lobes broadly lanceolate or nearly obovate, acuminate; fruiting peduncles 7–25 cm. long; staminate spikes 10–25 cm. long; bracts broadly ovate or obovate, 1–1.5 cm. long, toothed, the calyx-segments broadly linear to narrowly lanceolate; calyx-tube of pistillate flowers 2–2.5 cm. long; fruit ovoid-globose, about 10 cm. long, red, with longitudinal orange striations; seeds broadly elliptic, 10–11 mm. long, dark brown.——Aug. Honshu (Kinki Distr. and westw.), Shikoku, Kyushu.

3. Trichosanthes kirilowii Maxim. var. **japonica** (Miq.) Kitam. *Gymnopetalum japonicum* Miq.; *T. quadricirra*

Miq.; *T. japonica* Regel——Kɪ-ᴋᴀʀᴀsᴜ-ᴜʀɪ. Roots thickened; stems slender, soft brown-pubescent while young; leaves ovate to orbicular, 5-angled, cordate, palmately 3- to 5(–7)-lobed or -cleft, short-pilose above; fruiting peduncles 2–3 cm. long; staminate spikes 10–20 cm. long; bracts broadly ovate to obovate, 1.5–2.5 cm. long, coarsely toothed; calyx-segments broadly linear, 8–12 mm. long; calyx-segments of pistillate flowers 3.5–4 cm. long; fruit ovate-globose, to 10 cm. long, yellow, the peduncles 2–3(–4.5) cm. long; seeds oblong to ovate-orbicular, 11–14 mm. long, dark brown.——Aug.–Sept. Hokkaido (Okushiri Isl.), Honshu, Shikoku, Kyushu.——Ryukyus. The typical phase with pale brown seeds, occurs in the temperate to warmer regions of e. Asia.

4. Trichosanthes cucumeroides (Ser.) Maxim. *Bryonia cucumeroides* Ser.; *Platygonia kaempferi* Naud.——Kᴀʀᴀsᴜ-ᴜʀɪ. Roots fasciculate, thickened; stems slender, white-puberulent; leaves membranous, the tendrils with 1 or 2 branches; ovate-cordate to cordate-reniform, usually 6–10 cm. long and as wide, usually 3- to 5-lobed or angled, sometimes 5-cleft, white-puberulent; staminate spikes 2–10 cm. long; bracts small, lanceolate, 2–3 mm. long; calyx-segments 3–4 mm. long; corolla-tube in staminate flowers 6–7 cm. long; fruit globose to oblong, red, 5–7 cm. long, the peduncles about 1 cm. long; seeds brown, about 1 cm. long, 7–9 mm. wide, with a raised longitudinal band 4–5 mm. wide.——Aug.–Sept. Honshu, Shikoku, Kyushu.——Ryukyus, Formosa, and China.

5. Trichosanthes rostrata Kitam. Kᴇ-ᴋᴀʀᴀsᴜ-ᴜʀɪ, Tᴇɴɢᴜ-ᴋᴀʀᴀsᴜ-ᴜʀɪ. Roots elongate, somewhat thickened; stems puberulent, 5-angled; tendrils with 2 or 3 branches; leaves thick-membranous, broadly ovate-cordate or reniform-cordate, 3- to 5-lobed, puberulent, with scattered multicellular hairs on nerves beneath; staminate spikes 6–12 cm. long; bracts cuneate or narrowly oblong, 1–1.5 cm. long, toothed on upper margin; calyx-segments 8–10 mm. long; corolla-tube about 7 cm. long; fruit 4.5–7 cm. long, beaked, red, the peduncles 1.5–2 cm. long; seeds dark brown, 7–8 mm. long, 6.5–7.5 mm. wide, with a longitudinal band 3.5–4 mm. wide.——Aug. Kyushu (Oosumi Prov.).——Ryukyus.

2. THLADIANTHA Bunge   Ō-sᴜᴢᴜᴍᴇ-ᴜʀɪ Zᴏᴋᴜ

Dioecious perennials with tendrils and tuberous roots; leaves ovate, rarely deeply 3-lobed to pedately compound into 3–7 leaflets; tendrils simple or bifid; flowers yellow, the staminate in racemes or solitary; calyx-tube short-campanulate to rotate, with 1–3 scalelike appendages on the throat, the lobes linear; corolla campanulate, with 5 large, reflexed lobes; stamens in pairs; calyx of pistillate flowers without scalelike appendages on the throat; staminodes 5, elongate; style cylindric, 3-lobed, the stigma dilated or 2-lobed; fruit oblong, indehiscent; seeds many, obovate, flat.——About 24 species, in centr. to e. Asia, and tropical Asia.

1. Thladiantha dubia Bunge. Ō-sᴜᴢᴜᴍᴇ-ᴜʀɪ. Perennial herb, densely long spreading-hairy with simple tendrils; leaves ovate-cordate, irregularly mucronate-dentate; flowers yellow, 6–7 cm. across, the staminate axillary, solitary, the pedicels densely pubescent; calyx short-campanulate; corolla-lobes oblong, reflexed; stamens in pairs; fruit oblong, 4–5 cm. long.——Naturalized; Honshu (Shinano and Iwaki Prov.).——Korea, Manchuria, and China.

3. ACTINOSTEMMA Griff.   Gᴏᴋɪᴢᴜʀᴜ Zᴏᴋᴜ

Scandent monoecious herbs sometimes woody at base; leaves petiolate, hastate, cordate, coarsely toothed; tendrils simple or bifid; flowers unisexual, small, in axillary panicles or racemes, the pedicels jointed in the middle; calyx rotate, 5-parted, the segments slender, glandular on margin; corolla rotate, 5-parted, the segments narrow, caudate; stamens 5, free, the anther-locules oblong; pistillate flowers on lower portion of inflorescence, the staminate on the upper portion; ovary 1-locular; style short, the stigmas 2, reniform; ovules 2 to 4, pendulous; fruit small, asperous, circumscissile; seeds 2–4, flat, scabrous.——About 6 species, in e. Asia and India.

1. Actinostemma lobatum (Maxim.) Maxim. *Mitrosicyos lobatus* Maxim.; *M. racemosus* Maxim.; *Karivia* (?) *longicirrha* Miq.; *A. racemosum* (Maxim.) Cogn.; *A. japonicum* Miq.——Gᴏᴋɪᴢᴜʀᴜ. Slender annual, loosely short-pubescent on stems, leaves, and inflorescence; leaves thinly membranous, petiolate, narrowly hastate, or deltoid-cordate, 5–10 cm. long, 2.5–7 cm. wide, acuminate, sometimes with an obtuse tip, sparsely short-toothed, sometimes shallowly 3- to 5-lobed; panicles racemiform, to 12 cm. long; flowers small, greenish yellow; calyx-segments linear-lanceolate, slightly shorter than the corolla-lobes, caudate; corolla-lobes narrowly ovate-oblong, 5–6 mm. long, caudate; pistillate flowers solitary, near the base of staminate inflorescence, the peduncles about 1 cm. long; fruit ovoid, soft-spiny, about 1 cm. in diameter, circumscissile; seeds 2, large.——Aug.–Dec. Thickets along rivers in lowlands; Honshu, Shikoku, Kyushu; common.

Actinostemma palmatum (Makino) Makino. *A. lobatum* var. *palmatum* Makino——Tsᴜᴛᴀʙᴀ-ɢᴏᴋɪᴢᴜʀᴜ. Obscurely known plant. Leaves orbicular to reniform, deeply 3-cleft, the lobes again 2- or 3-lobed.——Once collected near Chiba in Honshu.

4. MELOTHRIA L.   Sᴜᴢᴜᴍᴇ-ᴜʀɪ Zᴏᴋᴜ

Slender scandent monoecious or rarely dioecious herbs; leaves lobed or undivided; tendrils simple; flowers unisexual, small, yellowish or white, the staminate in racemes or corymbs, rarely solitary; calyx campanulate, shortly 5-toothed; corolla 5-parted nearly to the base; stamens 3, the anthers free or slightly connate, the locules straight; pistillate flowers solitary, usually on very slender peduncles; placentae 3; style short, the stigma linear or capitate, bifid; fruit juicy, small; seeds flat, horizontal, many.——About 85 species, of wide distribution in all warmer regions.

1A. Staminate flowers solitary or sometimes in terminal racemes; anthers nearly sessile; seeds without thickened margins; leaves deltoid-ovate to ovate-cordate; fruiting peduncles 1–2 cm. long. .... 1. *M. japonica*
1B. Staminate flowers few, umbellate; anthers with pilose filaments; seeds with somewhat thickened margins; leaves deltoid; fruiting peduncles 6–8 mm. long. .... 2. *M. perpusilla*

1. **Melothria japonica** (Thunb.) Maxim. *Bryonia japonica* Thunb.; *M. regelii* Naud.; *M. japonica* var. *major* Cogn. ——SUZUME-URI. Slender annual, sparingly pubescent when young; stems very slender; leaves thinly membranous, deltoid-ovate to ovate-cordate, 3–6 cm. long, 4–8 cm. wide, acute with an obtuse tip, cordate, coarsely short mucronate-toothed, often obsoletely 3-lobed, scabrous above; staminate flowers usually solitary in leaf-axils or sometimes in terminal racemes, white, 6–7 mm. across, on slender pedicels; calyx-segments linear, short; corolla-lobes deltoid-ovate, subobtuse; anthers nearly sessile; fruit globose, about 1 cm. across, green, grayish white at maturity, smooth, the peduncles 1–2 cm. long; seeds grayish, 5–6 mm. long.——Aug.–Sept. Wet places along rivers and ditches in lowlands; Honshu, Shikoku, Kyushu; common. ——Korea (Quelpaert Isl.).

2. **Melothria perpusilla** (Bl.) Cogn. var. **perpusilla**. HOSOGATA-SUZUME-URI. Leaves ovate-cordate, the basal sinus 5–15 mm. long; calyx-lobes reflexed.——Reported from Kyushu (Oosumi Prov.).——se. Asia and Malaysia.

Var. **deltifrons** Ohwi. SATSUMA-SUZUME-URI. Monoecious slender herb; stems loosely short-pubescent while young; leaves thinly membranous, deltoid, 4–7 cm. long and as wide, acuminate, mucronate, truncate at base, mucronate-dentate, nearly glabrous beneath, petiolate; staminate inflorescence few-flowered, umbellate, the long peduncles capillary, 2–3 cm. long, the flowers 4–5 mm. across, the pedicels 4–5 mm. long; pistillate flowers solitary; calyx-lobes minute, not reflexed; filaments pilose; fruit globose, 8–10 mm. across, the peduncles 6–8 mm. long; seeds pallid with somewhat thickened margins. ——Sept. Kyushu (Satsuma Prov.).

5. SCHIZOPEPON Maxim.   MIYAMA-NIGA-URI ZOKU

Slender scandent polygamodioecious annual herbs; leaves thinly membranous, green, petiolate, ovate- to orbicular-cordate, toothed; tendrils bifid; flowers small, white, the staminate in racemes; calyx rotate; corolla rotate, 5-parted; stamens 3, free, the anthers straight, on short filaments; perfect flowers solitary in axils, long-pedicelled; ovary 3-locular; style 3-lobed, the stigmas again 2-lobed; ovules solitary in each locule; fruit berrylike, ovoid, dehiscing with the 3 valves involute on margin; seeds ovate, flat, perpendicular, with rather thicker slightly undulate margins.——Few species, in e. Asia.

1. **Schizopepon bryoniaefolius** Maxim. *S. bryoniaefolius* var. *japonica* Cogn.; *S. bryoniaefolius* var. *paniculatus* Komar.——MIYAMA-NIGA-URI. Slender annual herb slightly short-pubescent or nearly glabrous, with slender, elongate branches; leaves ovate- to orbicular-cordate, 5–12 cm. long and as wide, acuminate, mucronate-dentate, often loosely pilose above, puberulent on the nerves beneath, shallowly 5- to 7-lobed; flowers about 5 mm. across; calyx-segments lanceolate; corolla-lobes narrowly ovate, subobtuse; fruit obliquely ovoid, about 1 cm. long, smooth, the peduncles slender, capillary, 1–10 cm. long; seeds 1–3, about 1 cm. long, brown, flat, roughened on both surfaces, truncate on margin.——Aug.–Sept. Damp coniferous woods in mountains; Hokkaido, Honshu, Kyushu.——s. Kuriles, Sakhalin, Korea, and e. Siberia.

6. GYNOSTEMMA Bl.   AMACHAZURU ZOKU

Scandent herbs sometimes slightly woody at base; leaves membranous, petiolate, narrowly ovate, palmately compound, (3-) 5-foliolate, toothed; tendrils once or twice branched; flowers small, unisexual, in axillary panicles, greenish or white; staminate calyx small, 5-lobed, the corolla rotate, 5-cleft, the segments lanceolate, involute in bud; stamens 5, the filaments connate, forming a column; ovary in pistillate flowers 2- or 3-locular; styles 2 or 3, connate at base, the stigma retuse or bifid; ovules 2 in each locule; berry small, black, globose, 1- to 3-seeded; seeds flat, with irregular warty tubercles.——Few species, in India, Malaysia, and e. Asia.

1. **Gynostemma pentaphyllum** (Thunb.) Makino. *Vitis pentaphylla* Thunb.; *Cissus pentaphylla* (Thunb.) Willd.; *G. pedata* Bl.——AMACHAZURU. Slender nearly glabrous perennial herb; rhizomes creeping; stems slender, loosely white-pubescent on the nodes; leaves membranous, palmately compound, petiolate; leaflets 5 (3 or 7), narrowly ovate-elliptic to narrowly ovate, the terminal one 4–8 cm. long, 2–3 cm. wide, acuminate or obtuse, puberulent on the nerves above, penninerved, the veins parallel; inflorescence a panicle often racemiform, 8–15 cm. long; flowers yellowish green; calyx-lobes small; corolla-segments lanceolate, caudate, about 2 mm. long; fruit blackish green, globose, 6–8 mm. across, with a transverse line on upper half; seeds about 4 mm. long.——Aug.–Sept. Thickets in lowlands; Hokkaido, Honshu, Shikoku, Kyushu; common.——Korea, China, India, and Malaysia.

Fam. 192. **CAMPANULACEAE**   KIKYŌ KA   Campanula Family

Herbs, rarely woody plants, usually with milky juice; leaves usually alternate, simple, exstipulate; flowers often large and showy, perfect, actinomorphic or zygomorphic; calyx adnate to the ovary, the lobes 5(3–10); corolla superior, gamopetalous, tubular or campanulate, often 1- or 2-labiate, the lobes valvate in bud; stamens as many as and alternate with the corolla-lobes, inserted on the base of corolla-tube or free, the anthers free or connate; ovary usually inferior, 2- to 10-locular, with axile placentae, the ovules usually many; fruit a capsule or berry, the seeds usually small, many, the embryo straight, the endosperm fleshy.—— About 60 genera, with about 1,500 species. Cosmopolitan.

1A. Flowers actinomorphic; anthers free.
  2A. Capsule dehiscent laterally, or, if indehiscent, inferior to the calyx.
    3A. Disc tubular or cup-shaped, surrounding the base of style. .... 1. *Adenophora*
    3B. Disc flat or absent.
      4A. Corolla campanulate, rarely infundibuliform.
        5A. Capsules dehiscent, 3- to 5-valved. .... 2. *Campanula*
        5B. Capsules scarcely dehiscent. .... 3. *Peracarpa*
      4B. Corolla divided nearly to the base into narrow segments. .... 4. *Phyteuma*
  2B. Capsule dehiscent apically, or, if indehiscent, superior to the calyx.
    6A. Calyx inferior; fruit a berry. .... 5. *Campanumoea*
    6B. Calyx superior; fruit a capsule.
      7A. Erect herbs; stigma-lobes narrow.
        8A. Plants 20–40 cm. high; flowers about 2.5 cm. across; carpels alternate with the calyx-lobes and stamens. .. 6. *Wahlenbergia*
        8B. Plants 40–100 cm. high; flowers 4–5 cm. across; carpels opposite the calyx-lobes and stamens. .... 7. *Platycodon*
      7B. Vinelike herbs; stigma-lobes broad. .... 8. *Codonopsis*
1B. Flowers zygomorphic; anthers united into a tube or ring around the style; capsule dehiscent apically. .... 9. *Lobelia*

1. ADENOPHORA Fisch.   Tsurigane-ninjin Zoku

Perennial herbs; stems from a thick caudex; leaves alternate or verticillate, entire or toothed; flowers pendulous, in loose terminal racemes or panicles; calyx-limb 5-parted; corolla campanulate, blue, 5-lobed; stamens free from the corolla, the filaments broadened and pilose at base; disc thick, surrounding the base of the style; ovary inferior, 3-locular; stigma lobes 3, linear; capsules inferior, with a rather broad top, crowned with the persistent calyx-segments, dehiscent between the ribs; seeds ovate, flat.——More than 50 species, in Eurasia.

1A. Stems weak, declined or nodding above; pedicels capillary, mostly longer than the flowers.
  2A. Disc tubular, about 2 mm. long, longer than wide. .... 1. *A. maximowicziana*
  2B. Disc cup-shaped, wider than long.
    3A. Leaves narrowly ovate to broadly lanceolate; corolla 15–25 mm. long; style not exserted. .... 2. *A. takedae*
    3B. Leaves ovate-cordate in the lower ones, lanceolate in the upper ones; corolla about 10 mm. long; style long-exserted.
           3. *A. hatsushimae*
1B. Stems erect or ascending; pedicels stouter, mostly shorter than the flowers.
  4A. Corolla campanulate; pedicels usually as long as or longer than the calyx.
    5A. Disc cup-shaped, 1–2 mm. long, wider than long.
      6A. Leaves alternate. .... 5. *A. nikoensis*
      6B. Leaves usually verticillate. .... 6. *A. pereskiaefolia* var. *heterotricha*
    5B. Disc tubular, longer than wide.
      7A. Calyx-segments linear, usually toothed. .... 4. *A. triphylla*
      7B. Calyx-segments lanceolate, entire.
        8A. Leaves usually verticillate, subsessile; stems erect, 60–100 cm. long; inflorescence paniculate with long spreading branches.
           7. *A. divaricata*
        8B. Leaves usually alternate, petiolelike at base; stems decumbent, 10–30 cm. long; inflorescence in racemes with short branches.
           8. *A. tashiroi*
  4B. Corolla infundibular-campanulate; pedicels usually as long as to shorter than the ovary.
    9A. Leaves sessile; lower half of filaments ovate-oblong or elliptic, abruptly narrowed at tip.
      10A. Plant with white spreading hairs; calyx-segments and bracts lanceolate, acuminate. .... 9. *A. stricta*
      10B. Plant nearly glabrous; calyx-segments and bracts ovate, obtuse. .... 10. *A. palustris*
    9B. Leaves petiolate, usually more or less cordate at base; lower half of filaments lanceolate, gradually narrowed at tip.
           11. *A. remotiflora*

1. Adenophora maximowicziana Makino. *A. verticillata* var. *marsupiiflora* sensu auct. Japon., non Trautv.——Hina-shajin. Plant glabrous, the main roots rather short, slender; stems slender, 40–60 cm. long, declined; radical leaves ovate or ovate-cordate, long-petiolate, the median leaves membranous, linear-lanceolate, 8–20 cm. long, 5–10 mm. wide, gradually acuminate, gradually narrowed to a short petiolelike base, with appressed teeth; upper leaves small, rarely lanceolate; inflorescence terminal, corymbose or subpaniculate, loosely few- to about 10-flowered, the bracteoles absent or minute, the pedicels (2–)3–6 cm. long; calyx-segments filiform-linear, 4–6 mm. long, entire; corolla campanulate, about 1 cm. long; style 18–20 mm. long; disc tubular, about 2 mm. long. ——Aug.–Oct. Calcareous mountains; Shikoku.

2. Adenophora takedae Makino. *Campanula rotundifolia* sensu auct. Japon., non L.——Iwa-shajin. Plant glabrous, with rather thick main roots; stems slender, declined,

30–70 cm. long; leaves membranous, narrowly ovate, lanceolate to linear, 7–15 cm. long, 1–3 cm. wide, incurved- to appressed-toothed, sometimes with scattered appressed hairs above; inflorescence a terminal open leafy raceme, few- to more than 10-flowered, the pedicels 2–5 cm. long, the bracteoles small, filiform-linear; calyx-segments slightly curved, spreading, filiform-linear, 5–8 mm. long, loosely glandular-toothed; corolla campanulate, 15–25 mm. long; style not exserted; disc cup-shaped, about 1 mm. long.——Sept.–Oct. Rocks along valleys and ravines in mountains; Honshu (Sagami, Suruga, Kai, and s. Shinano Prov.).

Var. **howozana** (Takeda) Sugimoto. *A. howozana* Takeda——Hōō-shajin. Stems 5–15 cm. long; leaves slightly thicker, obsoletely toothed; flowers smaller.——High mountains; Honshu (Kai and s. Shinano Prov.).

3. Adenophora hatsushimae Kitam. Tsukushi-iwa-shajin. Glabrous slender plant, 20–40 cm. high; radical and lower cauline leaves membranous, ovate-cordate, 3–4 cm. long, 1.8–2.8 mm. wide, acuminate, long-petiolate, incised-toothed, short-pilose above, glabrous beneath; median leaves smaller, elongate-ovate, about 2.5 cm. long, the petioles 1–2 cm. long; upper leaves small, lanceolate, bractlike; flowers in a terminal raceme, the pedicels filiform; calyx-segments linear, 3–5 mm. long, loosely toothed; corolla about 1 cm. long; style long-exserted, nearly 2 cm. long; disc about 0.8 mm. long.——Oct. Rocks in mountains; Kyushu (Shiiba in Hyuga Prov.).

4. Adenophora triphylla (Thunb.) A. DC. *Campanula triphylla* Thunb.; *C. tetraphylla* Thunb.; *A. tetraphylla* Fisch. ex Jacks.; *A. verticillata* var. *latifolia* Miq.; *A. verticillata* var. *angustifolia* Miq.; *A. verticillata* var. *triphylla* Miq.——SAIYŌ-SHAJIN. Plant nearly glabrous to white-pilose; main roots thick; stem terete, erect, 40–100 cm. long, sometimes branched; leaves usually verticillate in 4's, rarely alternate, usually oblong or ovate-elliptic, sometimes lanceolate or broadly linear, 4–8 cm. long, 5–40 mm. wide, obtuse to subacuminate at both ends, toothed; inflorescence usually a rather loose many-flowered erect panicle with elongate branches, the pedicels usually shorter than the flowers; calyx-segments 3–5 mm. long, filiform-linear, subentire, similar to the bracteoles; corolla urceolate–campanulate, slightly constricted above, 8–11 mm. long; style long-exserted; disc tubular, 2–3 mm. long.——Sept.–Oct. Kyushu.——Ryukyus, Formosa, and China. Extremely variable.

Var. **japonica** (Regel) Hara. *A. pereskiaefolia* var. *japonica* Regel; *A. verticillata* sensu auct. Japon., pro parte; *A. thunbergiana* Kudo; *A. triphylla* subsp. *aperticampanulata* Kitam.——TSURIGANE-NINJIN. Corolla campanulate, not constricted above, 13–22 mm. long; style slightly exserted.——Grassy places in lowlands and mountains; Hokkaido, Honshu, Shikoku, Kyushu; common.

Var. **hakusanensis** (Nakai) Kitam. *A. hakusanensis* Nakai; *A. thunbergiana* var. *hakusanensis* (Nakai) Hara.——HAKUSAN-SHAJIN. Stems 30–50 cm. long; panicle-branches short; flowers densely arranged.——Alpine regions; Hokkaido, Honshu.

Var. **puellaris** (Honda) Hara. *A. puellaris* Honda——OTOME-SHAJIN. Stems low; leaves linear, alternate, 1–3 mm. wide; flowers in racemes; corolla 1 cm. long, campanulate; style exserted.——Shikoku (Mount Higashi-akaishi in Iyo).

5. Adenophora nikoensis Fr. & Sav. *A. polymorpha* sensu auct. Japon., non Ledeb.; *A. coronopifolia* sensu auct. Japon., non Fisch.; *A. polymorpha* var. *coronopifolia* sensu auct. Japon., non Trautv.——HIME-SHAJIN. Plant glabrescent; main roots rather thick; stems erect, 20–40 cm. long; leaves usually alternate, linear-lanceolate to broadly lanceolate, rarely narrowly oblong, 3–7(–10) cm. long, 5–20 mm. wide, acute to long-acuminate, acute at base, toothed, sessile or subsessile; inflorescence a loose few-flowered raceme, rarely subpaniculate, the bracteoles absent or short, linear, the pedicels 5–10(–20) mm. long, rather slender; calyx-segments linear, 4–8 mm. long, usually toothed; corolla campanulate, 1.5–2.5 (–3) cm. long; style as long as to slightly longer than the corolla; disc cup-shaped, 1–2 mm. long.——Aug.–Sept. Alpine regions; Honshu (centr. and n. distr.); extremely variable.

Var. **petrophila** (Hara) Hara. *A. petrophila* Hara——MYŌGI-SHAJIN. Stems elongate, rather slender; leaves lance-olate, much-elongate, falcate.——Honshu (Kozuke and Musashi Prov.).

Var. **stenophylla** (Kitam.) Ohwi. *A. lamarckii* forma *multiloba* Takeda; *A. polymorpha* var. *lamarckii* forma *multiloba* (Takeda) Makino & Nemoto; *A. nikoensis* forma *nipponica* (Kitam.) Hara; *A. nikoensis* forma *stenophylla* (Kitam.) Hara; *A. nipponica* forma *multiloba* (Takeda) Kitam.; *A. nipponica* var. *stenophylla* Kitam.; *A. nipponica* Kitam.; *A. nikoensis* var. *multiloba* (Takeda) Ohwi, non Honda; *A. nikoensis* forma *multifida* Hara——MIYAMA-SHAJIN. With the habit of the typical phase; calyx-segments lanceolate, entire.——High mountains of centr. and n. Honshu, inclusive of Yamato Prov.

6. Adenophora pereskiaefolia (Fisch.) Fisch. var. **heterotricha** (Nakai) Hara. *A. moiwana* var. *heterotricha* Nakai; *A. moiwana* Nakai; *A. onoi* Kitam.; *A. ishiyamae* Miyabe & Tatew.; *A. latifolia* sensu auct. Japon., non Fisch.——MOIWA-SHAJIN. Pilose or nearly glabrous; main roots thick; stems usually simple and erect, 30–50 cm. long; radical leaves reniform or cordate-ovate, long-petiolate; cauline leaves in 3's or 4's, rarely alternate or opposite, lanceolate to ovate, 2–8 cm. long, 1.5–4 cm. wide, acute to short-acuminate, sessile or subsessile, toothed; inflorescence a rather densely flowered raceme, the branches usually not elongate, the pedicels rather short; calyx-segments lanceolate, entire, 3–6 mm. long; corolla campanulate, 15–20 mm. long; style slightly exserted; disc cup-shaped, 1–1.5 mm. long.——Aug.–Sept. Hokkaido, Honshu (alpine regions of n. distr.).——s. Kuriles.

7. Adenophora divaricata Fr. & Sav. *A. polymorpha* var. *divaricata* (Fr. & Sav.) Makino——FUKUSHIMA-SHAJIN. Plant nearly glabrous to slightly spreading-pilose; stems 60–100 cm. long, erect; cauline leaves 3–5, verticillate, alternate or opposite, ovate-elliptic or narrowly ovate, 5–10 cm. long, 2–4 cm. wide, acute to short-acuminate, toothed, subsessile; inflorescence a rather large terminal loose panicle with long spreading branches, the pedicels rather long, the bracteoles linear; calyx-segments lanceolate, 4–8 mm. long, entire; corolla campanulate, 15–20 mm. long; style as long as to slightly longer than the corolla; disc tubular, 2–3 mm. long.——Aug.–Sept. Mountains; Honshu (centr. and n. distr.).——Korea and Manchuria.

8. Adenophora tashiroi (Makino & Nakai) Makino & Nakai. *A. polymorpha* var. *tashiroi* Makino & Nakai——SHIMA-SHAJIN. Plant nearly glabrous; main roots thick; stems 10–30 cm. long, often decumbent; leaves rather thick, usually alternate, ovate or broadly so, 1.5–3 cm. long, 1–2 cm. wide, coarsely toothed, the lower and median leaves acute, gradually or abruptly narrowed to a petiolelike base, the upper leaves small, narrow, sessile; inflorescence a terminal few-flowered raceme, sometimes 1-flowered, the pedicels rather elongate; calyx-segments lanceolate, 4–6 mm. long, entire; corolla campanulate, 16–20 mm. long; style not exserted; disc tubular, about 2.5 mm. long.——Kyushu (Fukue Isl. in Hizen Prov.).——s. Korea (Quelpaert Isl.).

9. Adenophora stricta Miq. *A. polymorpha* var. *stricta* (Miq.) Makino——MARUBA-NO-NINJIN, TŌ-SHAJIN. Plant sparsely white spreading-pilose; main roots stout; stems erect, 60–100 cm. long; radical leaves reniform-cordate, long-petiolate; cauline leaves alternate, ovate or oblong, 3–7 cm. long, 1–3 cm. wide, acute, acute to obtuse at base, coarsely toothed, sessile; inflorescence rather densely racemose, often branched at base, the bracts and bracteoles lanceolate, acuminate, the pedicels 2–4 mm. long; calyx-segments lanceolate, 7–10 mm. long, the tube densely white-pilose; corolla campanulate, 1.5–2 cm. long; style not exserted.——Aug.–Sept. Honshu (n. distr.).

Var. **lancifolia** Honda. NAGABA-NO-NINJIN.   Leaves narrower; inflorescence glabrescent.——Honshu (Mutsu Prov.).

10. Adenophora palustris Komar.   YACHI-SHAJIN. Main roots thick; stems strict, purplish, 60–100 cm. long, glabrous; leaves rather thick, elliptic or oblong, 3–6 cm. long, 1.5–2.5 cm. wide, subacute at both ends, toothed, scattered-pilose on the nerves beneath, sessile, the upper leaves gradually smaller; inflorescence narrowly racemose, the pedicels 1–4 mm. long, the subtending bracteoles broadly ovate to broadly lanceolate, 3–4 mm. long, often toothed; calyx-segments ovate, 3–4 mm. long, obtuse, sometimes toothed, erect; corolla cam-

panulate, 1–1.5 cm. long; style scarcely exserted; disc tubular, about 2 mm. long.——Aug. Wet places; Honshu (Mikawa Prov. and Chūgoku Distr.); rare.

11. Adenophora remotiflora (Sieb. & Zucc.) Miq. *Campanula remotiflora* Sieb. & Zucc.——SOBA-NA.   Glabrous or rarely papillose-pilose; stems 40–100 cm. long, sometimes branched above; leaves ovate-cordate to broadly lanceolate, 5–20 cm. long, 3–8 cm. wide, acuminate, acute to cordate at base, acutely toothed, petiolate, the upper leaves small and sessile; inflorescence a large loose panicle, the pedicels short, the bracteoles minute; calyx-segments lanceolate, 5–8 mm. long; corolla campanulate, bluish, 2–3 cm. long; style usually not exserted.——Aug. Woods in mountains; Honshu, Shikoku, Kyushu.——Korea and Manchuria.

2. CAMPANULA L.   HOTARU-BUKURO ZOKU

Perennials or rarely annuals; leaves simple, usually alternate, entire or toothed; flowers solitary, axillary or terminal, sometimes forming a terminal panicle, usually blue, sometimes purple or white; calyx-tube adnate to the ovary, the segments 5, often with an appendage on the sinus; corolla campanulate, rarely infundibuliform or rotate, 5-lobed; stamens free from the corolla, the filaments broadened at base; disc indistinct; ovary inferior, 3- or 5-locular, with many ovules in each locule; stigma lobes 3 or 5, narrow; capsules usually flat at the top, crowned with the persistent calyx-segments, the valves dehiscent between the lateral ribs; seeds small.——About 250 species, in the N. Hemisphere, mostly in the Mediterranean region and w. Asia.

1A. Stems 40–80 cm. long, relatively stout and thick, many leaved; radical leaves not prominent at anthesis.
  2A. Flowers pendulous, tubular-campanulate, on short pedicels; bracts not clasping the base of the flowers. .... 1. *C. punctata*
  2B. Flowers erect, narrowly campanulate, gradually broadened, sessile; bracts clasping the base of the flowers.
                                                  2. *C. glomerata* var. *dahurica*
1B. Stems 5–15 cm. long, slender, few-leaved; radical leaves prominent at anthesis.
  3A. Leaves with minute undulate teeth; calyx-segments broadly deltoid-lanceolate, entire or obsoletely toothed. ...... 3. *C. chamissonis*
  3B. Leaves with spreading mucronate teeth; calyx-segments broadly linear or linear-lanceolate, with few mucronate teeth.
                                             4. *C. lasiocarpa*

1. Campanula punctata Lam.   HOTARU-BUKURO. Coarsely hirsute perennial, short-rhizomatous, sometimes stoloniferous; stems erect, usually simple, 40–80 cm. long; radical leaves ovate-cordate, long-petiolate; cauline leaves deltoid-ovate to broadly lanceolate, acuminate, usually with an obtuse tip, rounded to cuneate at base, irregulary obtuse-toothed, winged-petiolate or sessile; calyx-segments linear-lanceolate to deltoid-lanceolate, with a reflexed appendage in the sinus; corolla rose-purple, with dark spots, the lobes rather short, long-ciliate.——June. Grassy slopes in lowlands and low mountains; Hokkaido, Honshu, Shikoku, Kyushu; common.

Var. **hondoensis** (Kitam.) Ohwi.   *C. hondoensis* Kitam.; *C. punctata* subsp. *hondoensis* (Kitam.) Kitam.——HONDO-HOTARU-BUKURO.   Calyx-segments without a reflexed appendage in the sinuses; corolla deeper colored, slightly spotted; seeds narrowly winged.——Mountains; Honshu.

Var. **microdonta** (Koidz.) Ohwi.   *C. microdonta* Koidz.; *C. punctata* subsp. *microdonta* (Koidz.) Kitam.——SHIMA-HOTARU-BUKURO.   Less prominently hairy on all parts; flowers smaller, paler in color; calyx-segments with or without a reflexed appendage in the sinuses; seeds nearly wingless.—— Near seashores; Honshu (n. and centr. distr. on the Pacific side).

2. Campanula glomerata L. var. **dahurica** Fisch.   *C. cephalotes* Fisch. ex Nakai; *C. glomerata* var. *speciosa* (Hornem.) A. DC.; *C. speciosa* Hornem., non Gilib.; *C. glomerata* sensu auct. Japon., non L.——YATSUSHIRO-SŌ.   Perennial, short-hairy; rhizomes short-creeping; stems erect, 40–80 cm. long, simple; lower cauline leaves broadly lanceolate to narrowly ovate, 5–10 cm. long, 1–3 cm. wide, gradually acute,

rounded to cuneate at base, irregularly serrulate, winged-petiolate, the upper sessile and semiclasping; flowers about 10, erect, sessile, in a subcapitate terminal panicle; calyx-lobes linear or linear-lanceolate; corolla blue-purple, 2–2.5 (–3) cm. long, 5-lobed.——Aug.–Sept. Mountains; Kyushu.——Korea, Manchuria, and e. Siberia. The typical phase occurs in Eurasia.

3. Campanula chamissonis Fed.   *C. pilosa* var. *dasyantha* (Bieb.) Herd.; *C. dasyantha* Bieb.——CHISHIMA-GIKYŌ. Rhizomes slender, slightly thickened; stems 5–15 cm. long, few-leaved; radical leaves somewhat rosulate, oblanceolate to narrowly oblong, 2–4 cm. long, 0.6–1 cm. wide, obtuse, gradually narrowed at base to a long petiole, minutely undulate-toothed, lustrous above, glabrous or with scattered long hairs on margin and on midrib beneath; flowers solitary, terminal; calyx long-pubescent, the segments erect, broadly deltoid-lanceolate, 8–15 cm. long, acute; corolla blue, campanulate, 3–4 cm. long with long soft white hairs on margin and inside.——Aug. Gravelly and sandy alpine slopes; Hokkaido, Honshu (centr. and n. distr.).——Kuriles, Sakhalin, Aleutians, and Alaska.

4. Campanula lasiocarpa Cham.   *C. algida* Fisch.—— IWAGIKYŌ.   Rhizomes elongate, branched, slender; stems 5–12 cm. long, loosely long-pilose in upper part; radical leaves few, rosulate, oblanceolate or narrowly obovate, 1.5–3 cm. long, 4–7 mm. wide, rounded to subacute, petiolate, with a few scattered mucronate teeth, loosely long-pilose on margin of petiole; cauline leaves few, sessile, small; flowers solitary; calyx prominently spreading-pilose, the segments ascending, broadly linear, 8–10 mm. long, few-toothed; corolla blue, 2–2.5 cm. long, glabrous; capsules erect.——Aug. Gravelly and sandy alpine slopes; Hokkaido, Honshu (centr. and n. distr.). ——Sakhalin, Kuriles, Kamchatka, Aleutians, and Alaska.

3. PERACARPA Hook. f. & Thoms.   TANIGIKYŌ ZOKU

Delicate rather fleshy perennial herb with slender rhizomes; leaves alternate; flowers solitary, pedicellate, axillary or terminal, erect; calyx-tube adnate to the ovary, obconical, the limb 5-parted, the segments deltoid; corolla campanulate, 5-cleft; stamens free from the corolla, the filaments narrow, the anthers free, linear; ovary inferior, 3-locular, with many ovules in each locule; stigma lobes 3, narrow; fruit nodding, indehiscent, crowned with the calyx-segments, the pericarp membranous.——A single species, in e. Asia and the Himalayas.

1. Peracarpa carnosa (Wall.) Hook. f. & Thoms. var. **circaeoides** (F. Schmidt) Makino. *Campanula circaeoides* F. Schmidt; *P. circaeoides* (F. Schmidt) Feer; *P. carnosa* sensu auct. Japon., non Hook. f. & Thoms.——TANIGIKYŌ. Plant nearly glabrous; rhizomes slender, creeping, much branched; stems erect, soft, 5–15 cm. long, glabrous, few-leaved at top; leaves ovate-orbicular to broadly ovate, 8–25 mm. long, 6–20 mm. wide, obtuse, rounded at base, obtusely toothed, scattered short-hairy, the petioles 5–15 mm. long; corolla white or purplish, 4–8 mm. long, 5-parted, the segments broadly lanceolate;

fruit ellipsoidal, 5–6 mm. long; seeds lustrous, fusiform, about 1.5 mm. long.——May–Aug. Damp woods in mountains; Hokkaido, Honshu, Shikoku, Kyushu.——s. Kuriles, Sakhalin, Kamchatka, and Korea (Quelpaert Isl.).

Var. **pumila** Hara. TSUKUSHI-TANIGIKYŌ. Plant smaller; leaves orbicular; seeds 2–2.5 mm. long.

Var. **kiusiana** Hara. EDAUCHI-TANIGIKYŌ. Plant larger; leaves acute to subacute, short-toothed.——At lower elevations; Kyushu.——The typical phase occurs in northern India.

4. PHYTEUMA L.   SHIDE-SHAJIN ZOKU

Perennial herbs; leaves alternate, the radical prominent, the cauline sometimes few and small; flowers usually sessile, in terminal spikes or heads, sometimes short-pedicelled, the bracts small or involucrelike; calyx-tube adnate to the ovary, the limb 5-parted; corolla 5-parted nearly to the base, the segments linear, spreading or revolute, sometimes connate also at the tip; stamens free from the corolla, the filaments broadened at base, the anthers free; ovary inferior, 2- or 3-locular, many-ovuled, the stigma 2- or 3-lobed; capsules laterally dehiscent.——About 40 species, in Europe, Asia, and N. America, abundant in the Mediterranean region.

1. Phyteuma japonicum Miq. *Campanula japonica* (Miq.) Vatke; *Asyneuma japonicum* (Miq.) Briq. SHIDE-SHAJIN. Rhizomes creeping; stems erect, 50–100 cm. long, longitudinally striate, sparingly pilose; leaves membranous, ovate to narrowly so or oblong, 5–12 cm. long, 2.5–4 cm. wide, irregularly toothed, the lower petiolate, the median with a short winged petiolelike base, the upper gradually smaller and sessile; inflorescence spicate, terminal, often branched at base,

glabrous, the bracts linear, the pedicels 1–3 mm. long; calyx-segments linear, 4–6 mm. long, entire; corolla bluish-purple, the segments spreading and slightly recurved, linear, 1–1.2 cm. long, about 1 mm. wide; filaments as long as the linear anthers, ciliate on lower part; style 1–1.2 cm. long; stigma 3-lobed; capsules depressed-globose, 5–6 mm. across.——Aug. Mountains; Honshu, Kyushu.——Korea, Manchuria, Amur, and Ussuri.

5. CAMPANUMOEA Bl.   TSURUGIKYŌ ZOKU

Usually scandent perennial herbs with stout rhizomes; leaves cauline, opposite, petiolate, usually cordate; flowers terminal, axillary, or on short leafless branches, solitary; calyx-tube short, broad, adnate to the middle or to the upper portion of the ovary, the limb 4- to 6-parted; corolla broadly campanulate, inserted on the upper portion of the calyx-tube, the limb 4- to 6-lobed; stamens free from the corolla; ovary inferior, 4- to 6-locular, many-ovuled; stigma 4- to 6-lobed; berry subglobose, indehiscent; seeds small.——Few species, in India, Malaysia, and e. Asia.

1A. Ascending; leaves narrowly ovate to broadly lanceolate, acuminate, rounded to very obtuse at base; calyx-segments linear, remotely pinnately lobed, 1–2 mm. wide; free portion of corolla 6–8 mm. long; seeds flattened, smooth. . . . . 1. *C. lancifolia*

1B. Scandent; leaves ovate-cordate to cordate, obtuse to subobtuse; calyx-segments broadly lanceolate, 3–4 mm. wide; free portion of corolla about 12 mm. long; seeds scarcely flattened, minutely reticulate. . . . . 2. *C. maximowiczii*

1. Campanumoea lancifolia (Roxb.) Merr. *Campanula lancifolia* Roxb.; *Codonopsis truncata* Wall. ex A. DC.; *Campanumoea axillaris* Oliv.; *C. truncata* (Wall.) Diels——TANGEBU, TAIWAN-TSURUGIKYŌ, SHIMAGIKYŌ. Glaucous ascending herb with elongate 4-angled spreading branches; leaves opposite, narrowly ovate to broadly lanceolate, 7–15 cm. long, 2–6 cm. wide, acuminate, rounded to very obtuse at base, mucronate-toothed, the petioles 7–10 mm. long, the upper leaves smaller; flowers terminal and solitary in upper leaf axils, the pedicels 7–20 mm. long, the bracts 2, subtending the flower; calyx-tube wholly adnate to the ovary, flat, 3–4 mm. across at anthesis, to 1 cm. long in fruit, the segments 5–6, linear, spreading, 6–10 mm. long, remotely pinnately lobed; corolla-tube adnate to the ovary in lower half, the free portion 6–8 mm. long,

the lobes 5–6, narrowly deltoid, about 4 mm. long; anthers linear, about 2 mm. long; style short-pilose; berry subglobose, nearly 1 cm. across, with the remains of the corolla-tube at the top; seeds flattened, smooth.——Autumn. Kyushu (Tanegashima, according to S. Hatusima).——Ryukyus, Formosa, se. Asia, and Malaysia.

2. Campanumoea maximowiczii Honda. *C. japonica* Maxim., non Sieb. & Morr.; *C. javanica* var. *japonica* (Maxim.) Makino——TSURUGIKYŌ. Glaucous scandent perennial; stems slender, elongate, terete; leaves thinly membranous, opposite and alternate, cordate or ovate-cordate, 3–5 cm. long, 2.5–5 cm. wide, subobtuse, obsoletely undulate-toothed, glaucous, long-petiolate; flowers solitary, pendulous, adjacent to a leaf, the pedicels 1–2 cm. long; calyx adnate to the ovary at base, the segments broadly lanceolate, 8–10 mm. long, 3–4 mm. wide, entire, subacute; corolla adnate at base to the ovary, the free portion about 12 mm. long, purplish inside, the lobes 5, acute, narrowly ovate, recurved at tip; fruit depressed-globose, 1–1.2 cm. across, 5-locular; seeds scarcely flattened, minutely reticulate.——Aug.–Sept. Mountains; Honshu (Kantō Distr. and westw.), Shikoku, Kyushu.——Formosa.

6. WAHLENBERGIA Schrad. Hinagikyō Zoku

Annuals or perennials; leaves alternate, rarely opposite; inflorescence a centrifugal panicle or sometimes 1-flowered; flowers usually bluish purple, nodding; calyx-tube adnate to the ovary, the segments 5; corolla campanulate, infundibuliform, tubular or rotate, 5-lobed; stamens free from the corolla, the filaments often broader at base; ovary inferior or subsuperior, 2- to 5-locular, many-ovuled; stigma-lobes 2–5, narrow; capsules loculicidally dehiscent, the valves alternate with the calyx-segments; seeds small.——About 100 species, chiefly in the S. Hemisphere, few in tropical America and Eurasia.

1. **Wahlenbergia marginata** (Thunb.) A. DC. *Campanula marginata* Thunb.; *Lobelia campanuloides* Thunb.; *W. gracilis* sensu auct. Japon., non Schrad.——Hinagikyō. Slender perennial herb; stems erect, slender, 20–40 cm. long, striate, often branched and loosely spreading-pilose at base; radical and lower cauline leaves broadly spathulate or oblanceolate, 2–4 cm. long, 3–8 mm. wide, often undulate on the margin; upper leaves small, few; flowers few, terminal and in upper axils, long-pedicelled; calyx-segments erect, persistent, lanceolate, 2–3 mm. long; corolla blue, infundibular-campanulate, 5–8 mm. long, 5-cleft; capsules erect, obconical, 6–8 mm. long.——May–Aug. Grassy sunny places in lowlands and low mountains; Honshu (s. Kantō Distr., Etchū Prov. and westw.), Shikoku, Kyushu.——Ryukyus, Formosa, China, and Korea.

7. PLATYCODON A. DC. Kikyō Zoku

Erect perennial herbs with thickened roots; leaves alternate, sometimes opposite or verticillate, simple, toothed; flowers large, mostly toward the top, the upper ones opening first; calyx-tube adnate to the ovary, the lobes 5; corolla blue, sometimes lilac or white, broadly campanulate, 5-lobed; stamens free from the corolla, the filaments broadened at base; ovary inferior, many-ovuled; stigma-lobes 5, linear; capsules obovate, erect, conical at top, loculicidally dehiscent, the valves opposite the calyx-lobes; seeds oblong, flat.——A single species in e. Asia.

1. **Platycodon grandiflorum** (Jacq.) A. DC. *Campanula grandiflora* Jacq.; *C. glauca* Thunb.; *C. gentianoides* Lam.; *P. grandiflorum* var. *glaucum* (Thunb.) Sieb. & Zucc.; *P. chinense* Lindl. & Paxt.; *P. glaucum* (Thunb.) Nakai——Kikyō. Glabrous or short-pilose on leaves beneath; stems 40–100 cm. long, sometimes branched at top; leaves narrowly to broadly ovate, 4–7 cm. long, 1.5–4 cm. wide, acute, broadly cuneate to acute at base, glaucescent especially beneath, sessile or subsessile; flowers mostly terminal, pedicelled, slightly nodding; calyx-lobes 0.5–5 mm. long, deltoid-lanceolate; corolla 4–5 cm. across; capsules 5-lobed, the valves subcoriaceous, attenuate; seeds about 2 mm. long.——Aug.–Sept. Grassy slopes in hills and mountains; Hokkaido, Honshu, Shikoku, Kyushu.——Korea, N. China, Manchuria, and Ussuri. Frequently planted as an ornamental and for medicine.

8. CODONOPSIS Wall. ex Roxb. Tsuru-ninjin Zoku

Perennial scandent herbs with thick tuberous or fusiform roots; leaves opposite, alternate, or in false verticils of 4's, petiolate; flowers nodding, terminal on short branches, rather large, usually greenish to purplish, sometimes white or blue; calyx-tube adnate to the ovary, the lobes 5, rather broad; corolla campanulate or broadly tubular, slightly adnate to the ovary at base, 5-lobed; stamens free from the corolla; ovary inferior or subinferior, truncate to short-conical at the broad top, depressed, 3- to 5-locular, many-ovuled, the stigma-lobes 3–5; capsules loculicidally dehiscent at the top; seeds oblong, sometimes winged.——About 40 species, in India and e. Asia.

1A. Tubers narrowed toward the end; calyx-segments 2–2.5 cm. long, 6–10 mm. wide; corolla 2.7–3.5 cm. long, purple-spotted inside; fruit 2–2.5 cm. across; seeds dull, winged on one side. . . . . 1. *C. lanceolata*
1B. Tubers not narrowed toward the end; calyx-segments 1–1.5 cm. long, 4–6 mm. wide; corolla 2–2.5 cm. long, deep-purple inside on upper half, spotted on lower half; fruit 1–1.3 cm. across; seeds lustrous, wingless. . . . . 2. *C. ussuriensis*

1. **Codonopsis lanceolata** (Sieb. & Zucc.) Trautv. *Campanumoea lanceolata* Sieb. & Zucc.; *Glossocomia hortensis* Rupr.; *Campanumoea japonica* Sieb. & Morr.——Tsuru-ninjin. Plant usually glabrous, often pilose while very young; tubers obovoid-fusiform; main stems much elongate, branched, with remote alternate scalelike leaves; lateral branches short, with a whorl of leaves at the end; leaves thinly membranous, narrowly oblong to ovate, 3–10 cm. long, 1.5–4 cm. wide, acute at both ends, glabrous or nearly so, glaucous especially beneath, with obsolete undulate teeth or nearly entire; flowers solitary, pedicelled, terminal on the short lateral branches, pendulous; calyx-segments narrowly ovate or narrowly ovate-oblong, subacute, green, 2–2.5 cm. long, 6–10 mm. wide; corolla pale green outside, purple-spotted inside, 2.7–3.5 cm. long; fruit 2–2.5 cm. across; seeds dull, winged on one side.——Aug.–Oct. Woods in low mountains and hills; Hokkaido, Honshu, Shikoku, Kyushu.——Forma **emaculata** (Honda) Hara. *C. lanceolata* var. *emaculata* Honda——Midori-tsuru-ninjin. Corolla without spots inside.——Korea, Manchuria, China, and Ussuri.

2. **Codonopsis ussuriensis** (Rupr. & Maxim.) Hemsl. *Glossocomia ussuriensis* Rupr. & Maxim.; *G. lanceolata* var.

obtusa Regel; *G. lanceolata* var. *ussuriensis* (Rupr. & Maxim.) Regel; *C. lanceolata* var. *ussuriensis* (Rupr. & Maxim.) Trautv.——Baa-sobu.   Closely allied to the preceding species but the plant smaller; tubers shorter, not narrowed at the end; stems more slender, often spreading-pilose at base; leaves usually ovate, sometimes ovate-elliptic, 2–4.5 cm. long, 1.2–2.5 cm. wide, obtuse or subacute, usually prominently white-pilose

especially on the glaucescent lower surface, rarely nearly glabrous; calyx-segments 1–1.5 cm. long, 4–6 mm. wide; corolla 2–2.5 cm. long, deep purple inside on upper half, spotted on lower half; fruit 1–1.3 cm. across; seeds lustrous, wingless.—— Aug. Woods in mountains and lowlands; Hokkaido, Honshu, Shikoku, Kyushu.——Korea, Manchuria, Ussuri, and Amur.

9. LOBELIA L.   Mizo-kakushi Zoku

Shrubs or herbs; leaves alternate; flowers axillary, sometimes forming a terminal raceme, pedicelled, usually without bracteoles; calyx-tube adnate to the ovary, the segments 5, sometimes unequal; corolla oblique or incurved, the tube dorsally split to the base, the limb bilabiate or 1-labiate, the 2 upper lobes smaller, erect or spreading; stamens usually free from the corolla, forming a tube around the style, the anthers at least partially united with a tuft of hairs at the apex; ovary inferior or subsuperior, 2-locular, many-ovuled; stigma shallowly 2-lobed; capsules loculicidal, apically dehiscent, 2-valved.——About 200 species, nearly cosmopolitan.

1A. Stems rather stout, 50–100 cm. long, densely many-leaved; flowers 3–3.5 cm. long, in dense narrow terminal racemes.
                                                          1. *L. sessilifolia*
1B. Stems slender, 30–40 cm. long, rather loosely few-leaved; flowers 5–12 mm. long, axillary or in loose racemes.
  2A. Stems long-creeping, branched at base, ascending; flowers solitary in leaf-axils, the pedicels 1.5–3 cm. long, the corolla 1–1.2 cm. long; capsules obconical-clavate, 5–7 mm. long, deflexed after anthesis. . . . . 2. *L. chinensis*
  2B. Stems erect, rarely creeping; flowers in loose terminal racemes, the pedicels 1–1.5 cm. long, the corolla 5–6 mm. long; capsules obovoid-globose, 2.5–3 mm. long, not deflexed after anthesis. . . . . 3. *L. hancei*

1. Lobelia sessilifolia Lamb. *L. camtschatica* Pall. ex Spreng.; *Rapuntium kamtschaticum* Presl——Sawagikyō. Glabrous perennial; rhizomes short, thick, creeping; stems stout, terete, erect, simple, 50–100 cm. long, densely many-leaved; median leaves lanceolate, 4–7 cm. long, 5–15 mm. wide, gradually attenuate to a minute obtuse point, shallow-toothed, sessile, the upper leaves gradually smaller, becoming bracteate; flower spikes terminal, densely many-flowered, the pedicels 5–12 mm. long; corolla bilabiate, lobed to the middle, the lower lip 3-lobed, the lobes loosely long-pubescent on margin; capsules 8–10 mm. long; seeds flat, ovate, smooth, lustrous, about 1.5 mm. long.——Aug.–Sept. Wet places in lowlands and mountains; Hokkaido, Honshu, Shikoku, Kyushu.——Kuriles, Sakhalin, Formosa, Korea, Manchuria, and e. Siberia.

2. Lobelia chinensis Lour. *L. radicans* Thunb.; *Rapuntium chinense* (Lour.) Presl——Mizo-kakushi, Aze-mushiro. Glabrous perennial; stems slender, long-creeping, branched at base, the branches usually simple, ascending, 3–15 cm. long; leaves in 2 series, subsessile, lanceolate, 1–2(–2.5) cm. long, 2–4(–7) mm. wide, obtuse, with a few obsolete teeth on each side; flowers solitary or paired, axillary, erect, white to rose-purple, deflexed after anthesis, the pedicels 1.5–3 cm. long; corolla glabrous, 1–1.2 cm. long, 5-lobed, 2 upper and 3 lower; capsules obconical-clavate, 5–7 mm. long; seeds broadly ovate,

about 0.2 mm. long, smooth, punctulate, reddish brown.—— June–Oct. Wet places especially around paddy fields in lowlands; Hokkaido, Honshu, Shikoku, Kyushu; common.—— Ryukyus, Formosa, China, Korea, India, and Malaysia.

3. Lobelia hancei Hara. *L. chinensis* sensu auct. Japon., non Lour.; *L. trigona* sensu auct. Japon., non Roxb.; *L. chinensis* var. *cantonensis* E. Wimm. ex Danguy——Tachi-mizo-kakushi.   Glabrous (annual?) herb; stems erect, soft, rather slender, slightly branched or simple; leaves scattered, broadly lanceolate to narrowly ovate, 1–2 cm. long, 3–6 mm. wide, obtuse to acute, obsoletely mucronate-toothed, sessile; flowers few on upper part of stem, forming a loose terminal raceme, the bracts leaflike, the pedicels 1–1.5 cm. long, ebracteolate; corolla 5–6 mm. long, bluish purple, 5-lobed; capsules obovoid-globose, 2.5–3 mm. long, rounded at base; seeds light brown, smooth, ellipsoidal, obtusely 3-angled, about 0.4 mm. long.——Aug.–Oct. Kyushu.——Ryukyus, Formosa, China, and Indochina.

Lobelia inflata L. Plant spreading-pubescent; leaves oblong, 3–6 cm. long, obtusely toothed; racemes about 10-flowered, the pedicels 3–6 mm. long; calyx inflated after anthesis; corolla 7–10 mm. long, blue-purple or pinkish. Cultivated and often naturalized in our area.——Introduced from N. America.

Fam. 193. GOODENIACEAE   Kusa-tobera Ka   Goodenia Family

Herbs or subshrubs, rarely prickly; leaves alternate or opposite; stipules absent; flowers solitary or in panicles, bisexual, zygomorphic; calyx tubular, adnate to or rarely free from the ovary, 5-lobed; corolla gamopetalous, 2- or rarely 1-labiate, 5-lobed, the lobes valvate or often induplicate-valvate; stamens 5, free or barely inserted on the corolla-tube, alternate with the lobes, the anthers 2-locular, sometimes adherent around the style, the stigma covered with a cup-shaped or bilabiate indusium; ovary inferior, 1- or 2(–4)-locular; ovules solitary or few in each locule; seeds small, flat, the endosperm copious.——About 12 genera, with about 300 species, chiefly Australian, few widely distributed in the Tropics.

Plate 15

Hypochaeris crepidioides (Miyabe & Kudo) Tatew. & Kitam. (Compositae). Mount Apoi in Hidaka Prov., Hokkaido. (Photo M. Tatewaki.)

Viburnum plicatum var. *glabrum* (Koidz.) Hara (Caprifoliaceae). Koide, Echigo Prov., n. Honshu. (Photo J. Ohwi.)

Plate 16

Trichosanthes bracteata (Lam.) Voigt (Cucurbitaceae). Cape Ashizuri, Tosa Prov., s. Shikoku. (Photo J. Ohwi.)

Farfugium japonicum (L.) Kitam. (Compositae). Roadside near Irozaki in Idzu Prov., Honshu. (Photo Junpei Sato.)

1. SCAEVOLA L.   KUSA-TOBERA ZOKU

Shrubs or subshrubs; leaves alternate or rarely opposite, entire or toothed; flowers axillary, sometimes with a subtending pair of bracts, sometimes on dichotomous peduncles, the pedicels sometimes absent; calyx adnate to the ovary, the limb short, annular or 5-lobed; corolla oblique, the tube split to the base on back, the 5-lobes nearly equal or the 2 dorsal shorter; anthers free, distinct; ovary inferior or nearly so, 2-locular, with a single ovule in each locule, or 1-locular with 1 or 2 ovules; indusium of the stigma cup-shaped, the stigma truncate or bifid; fruit indehiscent, the exocarp fleshy or thinly membranous, the endocarp hard or crustaceous; seeds solitary.——About 100 species, mainly in Australia, few in the Pacific Islands, e. Asia, Africa, and tropical America.

1. **Scaevola sericea** Vahl. *S. frutescens* (Mill.) Krause.; *Lobelia frutescens* Mill. pro parte; *S. koenigii* Vahl; *S. koenigii* var. *glabra* Matsum.; *S. frutescens* var. *glabra* (Matsum.) Masam.——KUSA-TOBERA.   Shrub, glabrous or nearly so, with tufts of long hairs in the leaf-axils; branches thick, becoming slightly woody, densely leafy toward the top; leaves alternate, fleshy, narrowly obovate, 10–18 cm. long, 4–8 cm. wide, rounded, obsoletely toothed on upper margin, sessile or subsessile; inflorescence an axillary pedunculate cyme, about 10-flowered, the bracts narrowly lanceolate, opposite, with a tuft of long hairs at base, the pedicels 1–1.5 cm. long; corolla about 2–2.5 cm. long, white, soon becoming sordid yellow, densely long-pubescent on the tube inside; indusium of the stigma white-ciliate; drupe ellipsoidal, pale green, 7–8 mm. long.——June–Oct. Near seashores; Kyushu (Yakushima and Tanegashima).——Ryukyus, Formosa, Pacific Islands, Australia, and Madagascar.

Fam. 194.   COMPOSITAE   KIKU KA   Composite Family

(Contributed by Siro Kitamura.)

Herbs or shrubs, rarely trees; leaves alternate or opposite; stipules absent; heads subtended by involucral bracts, terminal, solitary or variously arranged, composed of 1 to many flowers; receptacle glabrous, setose, or paleaceous; flowers inserted on the receptacle, bisexual or unisexual (plants sometimes dioecious or monoecious); corollas tubular, actinomorphic, 4- or 5-lobed or ligulate and zygomorphic, 2- to 5-toothed; stamens 5, rarely 4, the filaments on the corolla-tube and alternate with the lobes, the anthers basifixed, connate and forming a tube around the style, usually appendaged at the apex, often caudate or sagittate at base; ovary inferior, 1-locular; style slender, usually bifid; ovule solitary, erect, anatropous; achenes 1-seeded, often crowned with a setose, chaffy, or plumose pappus (modified calyx-limb); endosperm absent.——About 1,000 genera, with about 20,000 species, cosmopolitan.

1A. Disc-corollas tubular; ligulate florets marginal or absent, pistillate or neutral; sap not milky. (TUBULIFLORAE).
  2A. Styles below the branches, slender, sometimes filiform, often flattened, entire or bifid, rarely capitate or rounded, mammillate to papillose, not hispid, not nodosely thickened (rarely penicillate in *Cacalia, Syneilesis,* and *Achillea*).
    3A. Anthers sagittate, often caudate.
      4A. Style-branches linear, subacute, puberulent; heads all discoid, the florets bisexual. .... 1. *Vernonia*
      4B. Style-branches obtuse.
        5A. Corolla of tubular florets shallowly 4- or 5-toothed.
          6A. Involucral bracts translucent or white, scarious, petallike.
            7A. Bisexual florets fertile. .... 2. *Gnaphalium*
            7B. Bisexual florets sterile.
              8A. Pappus-bristles of bisexual florets thickened and flat at apex; plants dioecious. .... 3. *Antennaria*
              8B. Pappus-bristles of bisexual florets not flattened; plants not dioecious.
                9A. Pappus-bristles connate into a ring at base. .... 4. *Leontopodium*
                9B. Pappus-bristles free at base. .... 5. *Anaphalis*
          6B. Involucral bracts green, herbaceous, coriaceous, or membranous, not scarious.
            10A. Bisexual florets fewer than the pistillate florets. .... 6. *Blumea*
            10B. Bisexual florets usually more than the pistillate florets.
              11A. Heads with ligulate florets. .... 7. *Inula*
              11B. Heads all discoid, without ligulate florets. .... 8. *Carpesium*
        5B. Corolla bilabiate or deeply 5-lobed.
          12A. Ligulate florets present in vernal heads; corollas distinctly bilabiate. .... 9. *Leibnitzia*
          12B. Ligulate florets none, heads all tubular; corollas of tubular florets deeply 5-cleft.
            13A. Pappus-bristles scabrous. .... 10. *Pertya*
            13B. Pappus-bristles plumose.
              14A. Heads few-flowered. .... 11. *Ainsliaea*
              14B. Heads 1-flowered. .... 12. *Diaspananthus*
    3B. Anthers not or very shortly caudate at base.
      15A. Heads unisexual.
        16A. Involucral bracts of staminate heads connate; pistillate heads with a single achene. .... 13. *Ambrosia*
        16B. Involucral bracts of staminate heads not connate; pistillate heads with 2 achenes. .... 14. *Xanthium*
      15B. Heads bisexual and fertile.

17A. Style-branches nearly terete, obtuse, short-papillose; leaves opposite or verticillate.
    18A. Anthers appendaged at apex; pappus-bristles setose. .... 15. *Eupatorium*
    18B. Anthers without appendage at apex; pappus-bristles clavate. .... 16. *Adenostemma*
17B. Style-branches flat, smooth, truncate or appendaged at apex.
    19A. Style-branches with a lanceolate or deltoid hairy appendage at apex; leaves alternate.
        20A. Ligulate corollas broad, petallike, spreading.
            21A. Ligulate corollas yellow. .... 17. *Solidago*
            21B. Ligulate corollas white or reddish.
                22A. Ligulate florets fewer than the tubular florets; heads more than 1 cm. wide.
                    23A. Pappus absent. .... 18. *Gymnaster*
                    23B. Pappus present.
                        24A. Pappus bristles less than 1 mm. long, lacerate. .... 19. *Kalimeris*
                        24B. Pappus more than 2 mm. long, at least of the tubular florets.
                            25A. Pappus dimorphic, of the tubular florets setose, of the ligulate florets very short and lacerate.
                                       20. *Heteropappus*
                        25B. Pappus all bristlelike.
                            26A. Involucral bracts narrow, in a few series, not definitely imbricate; ligulate florets in 2 or more series, narrow. .... 21. *Erigeron*
                            26B. Involucral bracts broadly imbricate; heads less than 1 cm. wide; ligulate florets in 1 or 2 series. 22. *Aster*
                22B. Ligulate florets more numerous than the tubular florets.
                    27A. Achenes not beaked. .... 23. *Myriactis*
                    27B. Achenes beaked.
                      28A. Stems erect, branched, many-leaved. .... 24. *Rhynchospermum*
                      28B. Stems scapiform, few-leaved. .... 25. *Lagenophora*
        20B. Ligulate corollas filiform, long or short, erect.
            29A. Pappus well-developed. .... 26. *Conyza*
            29B. Pappus wanting .... 27. *Dichrocephala*
    19B. Style-branches truncate or rarely with very short or narrow appendages.
        30A. Pappus-bristles elongate, well developed.
            31A. Plants usually dioecious; bisexual florets with a short bifid style; leaves all radical, those of the inflorescence scalelike. .... 28. *Petasites*
            31B. Plants hermaphroditic; leaves radical and cauline.
                32A. Annuals; leaves alternate; corollas all tubular and fertile. .... 29. *Erechtites*
                32B. Perennials; leaves opposite.
                    33A. Heads usually with ligulate florets. .... 30. *Arnica*
                  33B. Heads all discoid; leaves alternate.
                    34A. Style-branches appendaged, short-strigose at apex. .... 31. *Emilia*
                    34B. Style-branches truncate, penicillate, or obtuse at apex.
                      35A. Leaves very thick, involute in the early stages, the lower and radical with a short sheath at base; achenes densely hairy. .... 32. *Farfugium*
                    35B. Leaves membranous, revolute in the early stages.
                      36A. Radical and cauline leaves with a short sheath at base.
                        37A. Style-branches obtuse; heads with ligulate florets; involucral bracts 1-seriate but of two forms.
                                       33. *Ligularia*
                      37B. Style-branches truncate; heads without ligulate florets.
                        38A. Achenes beaked; involucres with caliculus at base; corollas yellow. .... 34. *Miricacalia*
                        38B. Achenes not beaked; involucres without caliculus (rarely caliculate); corollas whitish. ... 35. *Cacalia*
                  36B. Leaves not sheathed at base.
                      39A. Corollas yellow; ligulate corollas present or rarely absent. .... 36. *Senecio*
                    39B. Corollas whitish or reddish; ligulate corollas absent. .... 37. *Syneilesis*
    30B. Pappus absent or very short, crownlike, scalelike, or awnlike.
        40A. Involucral bracts scarious on margin.
            41A. Receptacles chaffy; leaves toothed or bipinnate. .... 38. *Achillea*
            41B. Receptacles naked.
                42A. Ligulate florets more numerous than the disc florets; annuals with small axillary heads. .... 39. *Centipeda*
                42B. Ligulate florets fewer than the disc florets.
                    43A. Heads erect; appendage of anthers oblong.
                      44A. Receptacle conical; achenes ribless on back, 3-ribbed in front. .... 40. *Matricaria*
                      44B. Receptacle convex, hemispheric, sometimes flat; achenes striate. .... 41. *Chrysanthemum*
                  43B. Heads nodding; appendage of anthers lanceolate; pistillate corollas tubular. .... 42. *Artemisia*
        40B. Involucral bracts herbaceous, green.
            45A. Tubular florets sterile; receptacle not chaffy. .... 43. *Adenocaulon*
            45B. Tubular florets fertile; receptacle chaffy.
                46A. Achenes of the ligulate florets 3-angled, or of the tubular florets laterally flattened.
                  47A. Involucral bracts spreading, glandular-pubescent; scales of receptacle enveloping or surrounding the achene.
                                   44. *Siegesbeckia*
                47B. Involucral bracts appressed; scales of receptacle folded.
                    48A. Involucral bracts 2-seriate; scales of receptacle very narrow; annuals. .... 45. *Eclipta*
                    48B. Involucral bracts 1-seriate; scales on receptacle convex or folded; perennials. .... 46. *Wedelia*
              46B. Achenes more or less dorsally flattened.

49A. Outer scales of receptacle longer than the inner; achenes without hooked awns. . . . . 47. *Synedrella*
49B. Outer scales of receptacle shorter than the inner; achenes with retrorsely hooked awns.
    50A. Ligulate florets sterile; style-branches with acute-deltoid hairy appendages; leaves opposite (at least in the lower). . . . . 48. *Bidens*
    50B. Ligulate florets, if present, pistillate and fertile; style-branches with obtuse hairy appendages; leaves alternate. . . . . 49. *Glossogyne*
2B. Styles below the branches nodosely thickened, often hispid.
  51A. Heads many-flowered, not densely aggregated.
    52A. Achenes villous, heads with finely dissected subtending bristlelike bracteal leaves. . . . . 50. *Atractylodes*
    52B. Achenes glabrous.
      53A. Achenes basally attached.
        54A. Filaments papillose-hairy; style-branches connate.
          55A. Pappus-bristles scabrous; stems winged (in ours). . . . . 51. *Carduus*
          55B. Pappus-bristles plumose.
            56A. Plants polygamous; rhizomes not creeping. . . . . 52. *Cirsium*
            56B. Plants dioecious; rhizomes long-creeping. . . . . 53. *Breea*
        54B. Filaments glabrous; style-branches free, reflexed afterward.
          57A. Achenes 15-ribbed; involucral bracts with a keeled appendage on back. 54. *Hemistepta*
          57B. Achenes obsoletely 4-angled or striate; involucral bracts without a keeled appendage on back. . . . . 55. *Saussurea*
      53B. Achenes obliquely attached.
        58A. Anthers caudate, the tails free. . . . . 56. *Serratula*
        58B. Anther-tails connate into a tube around the filaments. . . . . 57. *Synurus*
  51B. Heads 1-flowered, densely aggregated in globose compound heads. . . . . 58. *Echinops*
1B. Corollas all ligulate; sap milky. (LIGULIFLORAE).
  59A. Pappus absent. . . . . 59. *Lapsana*
  59B. Pappus bristlelike or plumose.
    60A. Pappus-bristles plumose.
      61A. Receptacle scaly (paleaceous). . . . . 60. *Hypochoeris*
      61B. Receptacle not scaly.
        62A. Pappus-bristles with intricate soft hairs. . . . . 61. *Scorzonera*
        62B. Pappus-bristles with distinct hairs not intricate. . . . . 62. *Picris*
    60B. Pappus bristlelike.
      63A. Achenes beaked, warty-tuberculate near base of beak; leaves radical. 63. *Taraxacum*
      63B. Achenes not beaked, or, if beaked, then not tuberculate.
        64A. Achenes truncate.
          65A. Achenes cylindric, not flattened; corolla-tube slightly shorter to as long as the limb. . . . . 64. *Hieracium*
          65B. Achenes slightly flattened.
            66A. Pappus-bristles few, 1-seriate; corolla-tube much shorter than the limb; heads erect in anthesis. . . . . 65. *Hololeion*
            66B. Pappus-bristles 2- or 3-seriate, firm; corolla-tube as long as to slightly shorter than the limb; heads nodding in anthesis. . . . . 66. *Prenanthes*
        64B. Achenes narrowed or beaked at tip.
          67A. Subshrubs with the main axis bearing only radical leaves; scapes always lateral; achenes slightly flattened, shortly narrowed at tip. . . . . 67. *Crepidiastrum*
          67B. Herbs with the main axis bearing only the inflorescence.
            68A. Achenes terete, scarcely flattened, ribbed; corolla-tube ¼–½ as long as the limb. . . . . 68. *Crepis*
            68B. Achenes more or less distinctly flattened.
              69A. Achenes long-beaked, with 10 equal wings or ribs. . . . . 69. *Ixeris*
              69B. Achenes unequally ribbed.
                70A. Achenes distinctly flattened, with a prominent beak; corolla-tube as long as to half as long as the limb.

70. *Lactuca*

              70B. Achenes slightly flattened, narrowed at tip, not beaked.
                71A. Pappus deciduous as a whole. . . . . 71. *Sonchus*
                71B. Pappus persistent or the bristles separately deciduous. . . . . 72. *Youngia*

1. VERNONIA Schreb.   YAMBARU-HIGOTAI ZOKU

Subshrubs or herbs, erect, rarely scandent; leaves alternate, hairy or glabrous; heads in corymbs, sometimes paniculate; involucres campanulate or subglobose, the bracts many-seriate, the outer series becoming smaller, sometimes spreading, acute, acuminate, awn-tipped, or obtuse; receptacles alveolate, denticulate or rarely with crisped hairs; florets all tubular, fertile; corollas purplish, 5-lobed; anthers obtuse or sagittate at base, the auricles connate; style-branches linear, subacute, puberulent; achenes angled or striate, callose at base, glabrous or hairy; pappus-bristles 1- or 2-seriate, the outer row short, the inner scaberulous, deciduous or persistent.——About 650 species, in warmer regions of America, Asia, and Africa.

1. **Veronia cinerea** Less. YAMBARU-HIGOTAI, MURA-SAKI-MUKASHI-YOMOGI.   Perennial; stems 40–100 cm. long, gray-pubescent, branched above; cauline leaves rhombic-ovate or ovate, 3.5–6.5 cm. long, 1.5–3 cm. wide, acute, cuneate at base, mucronate- or undulate-toothed, puberulent above, soft grayish pubescent beneath especially on nerves, the petioles 10–25 mm. long, the upper leaves gradually smaller, narrower, broadly lanceolate to linear; inflorescence corymbose-panicu-

late; heads many, rose-purple, short-pedunculate; florets about 20; involucres campanulate, 4–5 mm. long, pubescent, glandular-dotted; bracts 4-seriate, the outer linear, 1.5–2 mm. long, acuminate, the median linear, the inner linear-lanceolate, awn-tipped; achenes terete, about 2 mm. long, densely short-hispid, glandular-dotted; pappus-bristles whitish, 4–5 mm. long, 2-seriate.——Oct.–Nov. Kyushu.——Ryukyus, Formosa, and tropical regions of Asia.

2. GNAPHALIUM L.   Hahakogusa Zoku

Densely woolly herbs sometimes woody at base; leaves alternate, linear or lanceolate, entire, usually sessile, sometimes decurrent; heads small, many, usually densely arranged, unisexual; pistillate florets marginal, the corollas filiform, denticulate or shallowly 3- or 4-lobed; bisexual fertile florets central, the corollas tubular, 5-toothed; involucres subglobose or campanulate-globose, the bracts more or less scarious; receptacle flat, alveolate; anthers sagittate, caudate; style of bisexual florets with terete or slightly flattened branches; achenes oblong, small, not ribbed, papillose; pappus-bristles slender, usually scabrous, 1-seriate. ——Widely distributed, especially in temperate regions of both hemispheres and in mountains of the Tropics.

1A. Involucres pale yellow.
  2A. Leaves white-woolly on upper surface; style shorter than the corollas; biennial. .... 1. G. affine
  2B. Leaves green on upper surface; style longer than the corollas; annual. .... 2. G. hypoleucum
1B. Involucres dark brownish.
  3A. Perennial, often stoloniferous; involucres campanulate, about 5 mm. long; cauline leaves 2–2.5 cm. long. .... 3. G. japonicum
  3B. Annual; involucres subglobose, about 2 mm. long; cauline leaves 4–5 cm. long. .... 4. G. uliginosum

1. Gnaphalium affine D. Don. *G. multiceps* Wall. ex DC.; *G. luteo-album* var. *multiceps* (Wall.) Hook. f.; *G. confusum* DC.——Hahakogusa. White-woolly biennial; stems 15–40 cm. long; radical leaves small; cauline leaves spathulate or oblanceolate, 2–6 cm. long, 4–12 mm. wide, rounded, sometimes mucronate, entire, white-woolly on upper surface, decurrent; heads in dense corymbs; involucres globose-campanulate, about 3 mm. long, 3.5 mm. wide; bracts pale yellow, 3-seriate, ovate to oblong, obtuse; style shorter than the corolla; achenes 0.5 mm. long; pappus-bristles pale yellowish, about 2.2 mm. long.——Apr.–June. Waste grounds and cultivated fields in lowlands; Hokkaido, Honshu, Shikoku, Kyushu; common.——Korea, China, Ryukyus, Formosa to Indochina, Malaysia, and India.

2. Gnaphalium hypoleucum DC. *G. confertum* Benth. ——Aki-no-hahakogusa. Annual; stems white-woolly, 30–60 cm. long, branched above; cauline leaves rather numerous, linear, 4–5 cm. long, 2.5–7 mm. wide, acute, auriculate, semiclasping, green and short crisped-pilose on upper surface, densely white-woolly beneath; heads in corymbs; involucres globose-campanulate, about 4 mm. long, 6–7 mm. wide; bracts 5-seriate, scarious, yellowish, the outer short, white-pubescent on back; style longer than the corolla; achenes punctulate; pappus sordid-white.——Sept.–Oct. Honshu, Shikoku, Kyushu.——Korea, China, Formosa, Philippines to India.

3. Gnaphalium japonicum Thunb. Chichikogusa. Perennial; flowering stems tufted, 8–25 cm. long, simple, white-woolly, often stoloniferous; radical leaves rosulate, linear-oblanceolate, 2.5–10 cm. long, 4–7 mm. wide, green and thinly woolly above, densely white-woolly beneath, the cauline few, linear, 2–2.5 cm. long, 2–4 mm. wide, the uppermost subtending the inflorescence, lanceolate; involucres campanulate, about 5 mm. long, 4–5 mm. wide; bracts 3-seriate, obtuse, red-brown, the outer broadly elliptic, the inner narrowly oblong; achenes about 1 mm. long; pappus white, about 3 mm. long. ——May–Oct. Sunny places in lowlands and mountains; Honshu, Shikoku, Kyushu; common.——Korea, China, Ryukyus, and Formosa.

4. Gnaphalium uliginosum L. Hime-chichikogusa, Ezo-no-hahakogusa. Annual; stems often tufted, 15–35 cm. long, usually more or less branched, white-woolly, leafy to the top; radical leaves small, the lower cauline oblanceolate-linear, 4–5 cm. long, 3–5 mm. wide, white woolly-tomentose on both surfaces; involucres subglobose, about 2 mm. long, 5 mm. wide; bracts 3-seriate, dark brownish, the outer short, broadly ovate, obtuse, the inner oblong to lanceolate, acute; achenes about 0.1 mm. long, punctulate; pappus whitish, about 1.5 mm. long.——Aug.–Oct. Wet cultivated fields; Hokkaido, Honshu (n. distr.).——n. China, Korea, Kamchatka, Siberia to Europe, and N. America.

3. ANTENNARIA Gaertn.   Ezo-no-chichikogusa Zoku

White-woolly dioecious perennials; radical leaves rosulate, the cauline alternate; heads in corymbs, unisexual; pistillate involucres campanulate, the bracts many-seriate, whitish and petallike above, the outer woolly on back; receptacle glabrous; pistillate florets filiform; staminate corollas tubular, 5-toothed, obtuse; style of staminate florets capitate; anthers sagittate, the tails connate, acuminate; achenes oblong; pappus bristles many, connate at the base, falling together in a ring.——About 55 species, in the N. Hemisphere, abundant in N. America.

1. Antennaria dioica (L.) Gaertn. *Gnaphalium dioicum* L.——Ezo-no-chichikogusa. Rhizomes slender, creeping; leaves of the sterile innovation shoots rosulate, spathulate, 15–25 mm. long, 5–6 mm. wide; stems 6–25 cm. long, simple, white-woolly; radical leaves rosulate, spathulate, 18–25 mm. long, 5–8 mm. wide, rounded, mucronate, green or slightly white-woolly above, densely white-woolly beneath, the median leaves linear, erect, small, acute, 1–1.5 cm. long, about 3 mm. wide; heads few, densely arranged; involucres 7 mm. long in the staminate, 10–14 mm. long in the pistillate; bracts linear to broadly so, whitish on upper half, rounded; pappus white, about 4 mm. long.——June–Aug. Hokkaido (Kitami Prov.).——Kuriles, Kamchatka, Sakhalin, Mongolia, Siberia to Europe.

4. LEONTOPODIUM R. Br.   Usu-yuki-sō Zoku

Low tufted white-woolly perennial herbs, rarely woody at the base; leaves entire, alternate, lingulate to linear; heads heterogamous, often unisexual, in a dense headlike corymb surrounded by many spreading bracteate leaves; involucres subglobose or campanulate-globose, the bracts 3-seriate, nearly equal, erect, oblong, scarious and brown on margin, usually woolly on back; pistillate florets marginal, filiform, the corollas broadened at tip, 5-toothed; staminate florets central, tubular, the corollas slightly broadened at tip; styles entire; anthers sagittate, caudate; achenes oblong, flattened; pappus-bristles slender, 1-seriate, short-setulose or scabrous, slightly connate at base.——More than 30 species, in Eurasia and S. America.

1A. Cauline leaves lanceolate to narrowly oblong; stems rather many-leaved. . . . . 1. *L. japonicum*
1B. Cauline leaves linear to narrowly oblanceolate; stems few-leaved.
  2A. Cauline leaves or some of them sheathing at base.
    3A. Leaves of innovation shoots linear-lanceolate, 3–8 cm. long. . . . . 2. *L. hayachinense*
    3B. Leaves of innovation shoots linear to oblanceolate.
      4A. Flowering stems 6–15 cm. long; heads 4–10; achenes about 1.2 mm. long, pilose. . . . . 3. *L. fauriei*
      4B. Flowering stems 4–7 cm. long; heads 2–3; achenes about 1.5 mm. long, glabrous. . . . . 4. *L. shinanense*
  2B. Cauline leaves long-petiolate, not sheathing at base. . . . . 5. *L. discolor*

1. **Leontopodium japonicum** Miq. *Gnaphalium sieboldianum* Fr. & Sav.——Usu-yuki-sō. Rhizomes tufted; stems 25–55 cm. long, leafy to the top; median cauline leaves lanceolate to narrowly oblong, 4–6.5 cm. long, 5–15 mm. wide, acute to acuminate, abruptly narrowed at base, sessile, green and glabrous or thinly woolly above, whitish woolly beneath; bracteal leaves loosely arranged, smaller than the upper leaves, sordid-yellow tomentose on upper surface; involucres 4–5 mm. long, about 5 mm. across; bracts acute or acuminate; achenes about 1 mm. long, more or less papillose.——July–Oct. Hokkaido, Honshu.——China.——Forma **orogenes** (Hand.-Mazz.) Ohwi. *L. japonicum* var. *orogenes* Hand.-Mazz.——Yama-usu-yuki-sō. Stems 20–40 cm. long; leaves about 3 cm. long, 6–9 mm. wide; inflorescence branched; heads many.——Kyushu, Honshu.——Forma **perniveum** (Honda) Ohwi. *L. perniveum* Honda——Kawara-usu-yuki-sō. Leaves about 2 cm. long, densely tomentose.——Mountains; Honshu (Shinano Prov.).

Var. **shiroumense** Nakai ex Kitam. *L. japonicum* forma *shiroumense* (Nakai) Ohwi——Takane-usu-yuki-sō. Plants to 10 cm. high, with smaller leaves and fewer heads.——Alpine regions; Honshu.

Var. **spathulatum** (Kitam.) Murata. *L. spathulatum* Kitam.; *L. japonicum* forma *spathulatum* (Kitam.) Ohwi——Ko-usu-yuki-sō. Plants quite low, tufted and with smaller spathulate leaves.——Alpine regions; Honshu (Yamato Prov.) and Shikoku.

2. **Leontopodium hayachinense** (Takeda) Hara & Kitam. *L. alpinum* subsp. *campestre* var. *hayachinense* Takeda; *L. discolor* var. *hayachinense* (Takeda) Takeda & Beauverd——Hayachine-usu-yuki-sō. Stems 10–20 cm. long, loosely leaved; leaves of innovation shoots linear-lanceolate, 3–8 cm. long, 3–5 mm. wide, usually acute, green and woolly on upper surface, gray-woolly beneath; radical leaves of flowering stems and lower cauline leaves small, the median leaves erect, lanceolate, 3–5 cm. long, 4–6 mm. wide; bracteal leaves 5–15, linear-lanceolate to narrowly oblong, 7–30 mm. long, 1–6 mm. wide, acute to acuminate; heads 4–8; involucres globose, about 5 mm. long; bracts obtuse, woolly; achenes about 1.6 mm. long, papillose.——July–Aug. Alpine regions; Honshu (Mount Hayachine in Rikuchu).

3. **Leontopodium fauriei** (Beauverd) Hand.-Mazz. *L. alpinum* sensu auct. Japon., pro parte, non Cass.; *L. alpinum* subsp. *fauriei* Beauverd; *L. alpinum* var. *fauriei* (Beauverd) Beauverd——Miyama-usu-yuki-sō, Hina-usu-yuki-sō. Flowering stems 6–15 cm. long, woolly; leaves of innovation shoots linear-oblanceolate, 2.5–6 cm. long, acute, long-petiolate, slightly woolly on upper surface, densely so beneath; lower cauline leaves linear, 1.5–3 cm. long, 2–3 mm. wide, gradually narrowed at base, sheathing the stems, densely yellowish gray woolly; bracteal leaves 8–13, linear, to 22 mm. long, 2(–4) mm. wide; heads densely arranged, 4–10; involucres globose, about 4 mm. long; bracts acute, long-hairy on back, with brown scarious margins; achenes about 1.2 mm. long, 4-angled, pilose.——July–Aug. Alpine regions; Honshu (n. distr.).

Var. **angustifolium** Hara & Kitam. Hosoba-hana-usu-yuki-sō. Plant more slender, with narrower leaves 1–1.5 mm. wide.——Honshu (Mount Shibutsu and Tanigawa in n. Kantō Distr.).

4. **Leontopodium shinanense** Kitam. *L. alpinum* sensu auct. Japon., non Cass.——Hime-usu-yuki-sō, Koma-usu-yuki-sō. Flowering stems slender, 4–7 cm. long, woolly; leaves of innovation shoots oblanceolate, 8–20 mm. long, 2–4 mm. wide, usually acute, long-petiolate, loosely gray-woolly on upper surface, densely so beneath; radical and lower cauline leaves spathulate or oblanceolate, the median narrowly lingulate, 14–20 mm. long, 2.5–3 mm. wide, obtuse, sheathing the stems, densely gray-woolly on both sides; bracteal leaves 6–9, lanceolate, 7–14 mm. long, 1–3.5 mm. wide, acute; heads 2 or 3; involucres about 4 mm. long; bracts acute or acuminate, with brownish scarious margins; achenes about 1.5 mm. long, glabrous.——July–Aug. Alpine regions; Honshu (Mount Nishi-koma in Shinano).

5. **Leontopodium discolor** Beauverd. *L. japonicum* subsp. *sachalinense* Takeda; *L. sachalinense* (Takeda) Miyabe & Kudo——Ezo-usu-yuki-sō. Flowering stems 13–33 cm. long, woolly; leaves of the innovation shoots linear-lingulate, 3.5–8 cm. long, 5–7 mm. wide, acute, mucronate, long-petiolate, glabrous or thinly woolly and green above, whitish woolly beneath; median cauline leaves lanceolate-lingulate, 3–5.5 mm. long, (1.5–)5–8 mm. wide, long-narrowed and petiolelike at base, glabrous to loosely white-woolly on upper surface, densely whitish woolly beneath; bracteal leaves lanceolate to narrowly oblong, acute to acuminate at both ends; heads 5–22, densely arranged; involucres 3–4 mm. long; bracts acute to obtuse, brown, scarious on the margins, densely woolly on back; achenes about 1 mm. long, 4-angled, glabrous. ——July–Sept. Hokkaido.——Sakhalin.

5. ANAPHALIS DC.   Yama-haha-ko Zoku

Grayish woolly or villous, often stoloniferous, dioecious or polygamodioecious perennial herbs; leaves alternate, entire, linear to lanceolate, sometimes oblong, often decurrent; heads in dense or loose corymbs, rarely solitary, heterogamous or homogamous, unisexual or sometimes the pistillate with a few staminate flowers in the center; involucres globose-campanulate, the bracts 5- to 8-seriate, imbricate, often whitish on upper half, brownish on lower half, the outer short, woolly on back; receptacle alveolate or pilose; pistillate florets marginal, fertile, the bisexual central, few, sterile; staminate corollas filiform, 2- to 4-toothed; corollas of bisexual florets tubular, 5-toothed; style of the bisexual florets bifid; achenes oblong; pappus-bristles 1-seriate, slender, scaberulous, deciduous.——About 35 species, in temperate regions of the N. Hemisphere.

1A. Leaves distinctly decurrent.
  2A. Involucres 5–7 mm. long, the bracts 5-seriate; cauline leaves 4–6 cm. long. .... 1. *A. sinica*
  2B. Involucres 9–10 mm. long, the bracts 6- to 7-seriate; cauline leaves 6–10 cm. long. .... 2. *A. alpicola*
1B. Leaves not decurrent. .... 3. *A. margaritacea*

1. Anaphalis sinica Hance. *A. pterocaulon* (Fr. & Sav.) Maxim.; *Gnaphalium pterocaulon* Fr. & Sav.; *A. todaiensis* Honda——Yahazu-haha-ko.   Stems 20–35 cm. long, usually simple, densely grayish tomentose; cauline leaves oblanceolate, 4–6 cm. long, 1–1.5 cm. wide, obtuse, mucronate, sessile, decurrent on the winged stems, green and often viscid above, sparsely grayish tomentose while young above, densely so and persistent beneath; inflorescence a globose corymb, 3–7 cm. across; involucres campanulate to globose, 5–7 mm. long; bracts 5-seriate, those of the pistillate snow-white on upper half, brownish and pubescent on lower half, the inner bracts acute. ——Aug.–Sept. Sunny places in mountains; Honshu (Kantō Distr. and westw.), Shikoku, Kyushu.

Var. **viscosissima** (Honda) Kitam.   *A. viscosissima* Honda ——Kuriyama-haha-ko.   Hairs dense, glandular.——Honshu (Kuriyama in Shimotsuke Prov.).

Var. **morii** (Nakai) Ohwi.   *A. sinica* subsp. *morii* (Nakai) Kitam.; *A. morii* Nakai; *A. yakushimensis* Masam.——Tanna-yahazu-haha-ko.   Stems tufted, densely leafy, 5–20 cm. long; cauline leaves thicker, 1.5–2 cm. long, 3–7 mm. wide; heads fewer, sometimes solitary.——High mountains in Kyushu (Higo Prov. and Yakushima).——Quelpaert Isl., Korea, and China.

2. Anaphalis alpicola Makino.   *A. apoiensis* Nakai—— Takane-yahazu-haha-ko.   Stems tufted, 10–20 cm. long, grayish tomentose; leaves of sterile innovation shoots oblanceolate, 6–10 cm. long, 1–1.8 cm. wide, acute, gradually narrowed at base, grayish woolly tomentose on both sides; radical leaves smaller, the median lanceolate, 4–6 cm. long, 9–12 mm. wide, acuminate to obtuse, sessile, decurrent; inflorescence a dense corymb; involucres globose-campanulate, 9–10 mm. long; bracts 6- or 7-seriate, the outer short, broadly ovate, obtuse, white, brown-red on lower half.——Aug. Alpine slopes; Hokkaido, Honshu (centr. and n. distr.).

3. Anaphalis margaritacea (L.) Benth. & Hook. f. var. **margaritacea**. *Gnaphalium margaritaceum* L.; *Antennaria margaritacea* (L.) R. Br.; *Antennaria cinnamomea* DC.; *Anaphalis margaritacea* forma *latifolia* Kudo——Hiro-ha-yama-haha-ko.   Leaves 1.5–3 cm. wide.——Hokkaido.——Northern e. Asia and N. America.

Var. **angustior** (Miq.) Nakai.   *Antennaria cinnamomea* var. *angustior* Miq.; *Anaphalis margaritacea* subsp. *angustior* (Miq.) Kitam.——Yama-haha-ko.   Rhizomes elongate; stems rather firm, 30–70 cm. long, grayish white woolly, many-leaved; median and upper leaves thick, linear-lanceolate to lanceolate or narrowly oblong, 6–9 cm. long, 1–1.5 cm. wide, acuminate to obtuse, semiclasping, recurved on margin, 3-nerved, green and woolly above when young, long grayish to brownish villous beneath; inflorescence a corymb; heads many; involucres globose, about 5 mm. long; bracts 6-seriate, the outer ovate or oblong, brownish scarious, the median oblong, obtuse, white at tip, brown at base, the inner lanceolate.—— Aug.–Sept. Sunny slopes in mountains, rarely alpine; Hokkaido, Honshu (centr. and n. distr.); rather common.—— China.

Var. **angustifolia** (Fr. & Sav.) Hayata.   *Antennaria japonica* Schultz-Bip.; *Gnaphalium margaritaceum* var. *angustifolium* Fr. & Sav.; *Anaphalis japonica* Maxim.; *A. margaritacea* var. *japonica* (Maxim.) Makino; *A. margaritacea* subsp. *japonica* (Schultz-Bip.) Kitam.——Hosoba-no-yama-haha-ko. Median leaves linear, 3–6 cm. long, 3(2–6) mm. wide, green and woolly above; involucres globose.——Aug.–Oct. Mountains; Honshu (Kinki Distr. and westw.), Shikoku, Kyushu.

Var. **yedoensis** (Fr. & Sav.) Ohwi.   *A. margaritacea* subsp. *yedoensis* (Fr. & Sav.) Kitam.; *Gnaphalium yedoense* Fr. & Sav.; *A. yedoensis* (Fr. & Sav.) Maxim.——Kawara-haha-ko. Stems branched in upper part, many-leaved; leaves linear, 3–6 cm. long, about 1.5 mm. wide, 1-nerved, green and woolly above, densely woolly beneath; involucral bracts 3-seriate, yellowish brown on lower half, the outer very short.—— Sunny places along rivers; Hokkaido, Honshu, Shikoku, Kyushu; rather common.

6. BLUMEA DC.   Tsuru-haguma Zoku

Annual or perennial herbs often woody at base, villous or woolly throughout; leaves alternate, sessile or petiolate, mucronate-toothed to laciniate; inflorescence paniculate or in spikelike panicles; heads heterogamous; involucres campanulate-globose, the bracts imbricate or reflexed, 4- or 5-seriate; corolla of pistillate florets filiform, minutely 2- or 3-toothed; perfect florets tubular, 5-toothed; anthers sagittate, tails connate, caudate-acuminate; style-branches narrow, compressed or filiform, papillose on back; achenes cylindric, hirsute, 5- to 10-ribbed; pappus-bristles capillary, 1-seriate, often caducous, whitish or rubescent.——More than 100 species, in the Tropics of Asia, Africa, and Australia.

1. Blumea conspicua Hayata.   *B. fruticosa* Koidz.—— Tsuru-yabu-tabira-ko, Ō-kibana-mukashi-yomogi.   Stems stout, 1–2 m. long, erect, striate, puberulent, branched in upper part; lower cauline leaves chartaceous, many, large, obovate-oblong, 25–35 cm. long, 8–11 cm. wide, abruptly acute, gradually narrowed at base, winged-petiolate, mucronate-toothed, villous while young; upper leaves distant, smaller; heads many, in large panicles, the pedicels 1–3 cm. long; involucres subglobose, about 9 mm. long, 18–20 mm. wide; bracts 5-seriate, acute, gray-hirsute on back, the outer corolla 6 mm. long; achenes cylindric, 1.5 mm. long, hirsute, 10-ribbed; pappus sordid-rubescent, 6 mm. long.——Kyushu (Yakushima and Tanegashima).——Ryukyus.

7. INULA L.   OGURUMA ZOKU

Perennial glabrous or villous sometimes glandular-pilose herbs often woody at base; inflorescence a corymb or panicle; heads sometimes solitary; leaves alternate, entire; heads heterogamous; involucres globose to campanulate, the bracts many-seriate, often nearly equal; receptacle glabrous, alveolate; pistillate flowers marginal, 1- to many-seriate, fertile, the corollas ligulate, yellow, 3-toothed; perfect florets many, the corollas 5-toothed; anthers sagittate, tails setiform; style branches in perfect florets linear, rounded at apex, minutely papillose on back; achenes pilose or glabrous, many-ribbed; pappus-bristles many, mostly equal, setulose.——About 100 species, in Eurasia and Africa.

1A. Radical leaves rosulate. . . . . 1. *I. ciliaris*
1B. Radical leaves not rosulate.
  2A. Achenes glabrous; veinlets of leaves raised beneath. . . . . 2. *I. salicina* var. *asiatica*
  2B. Achenes pilose; veinlets of leaves not raised beneath. . . . . 3. *I. britannica*

1. Inula ciliaris (Miq.) Maxim. *Erigeron ciliaris* Miq.——MIZUGIKU. Stems 25–50 cm. long, subscapose, densely hairy or glabrescent at base, rarely slightly branched; radical leaves rosulate, spreading, spathulate, 4–10 cm. long, 8–15 mm. wide, obtuse or subacute, entire, hairy at first, becoming glabrate; cauline leaves becoming gradually smaller in upper part, few, obsoletely clasping at base; upper leaves ovate-lanceolate, obtuse, broadly clasping at base; uppermost leaf 8–10 mm. long; heads solitary or few; involucres subglobose, 8–10 mm. long, with many subtending bracteal leaves; bracts 4- or 5-seriate, equal, sparsely glandular, the outer narrowly oblong, densely hairy; ligulate florets yellow, glandular-dotted, 14–15 mm. long; achenes about 1.5 mm. long, 10-ribbed, loosely pilose; pappus whitish, about 4.5 mm. long, the bristles scaberulous.——June–Oct. Wet places; Honshu (Kinki Distr. and eastw.).

Var. **glandulosa** Kitam. OZE-MIZUGIKU. Leaves beneath densely glandular-dotted.——Honshu (Ose in Kotzuke).

2. Inula salicina L. var. **asiatica** Kitam. *I. involucrata* Miq., non Kalenic.; *I. kitamurana* Tatew. ex Kitam., in syn.——KASEN-SŌ. Stems slender, 60–80 cm. long, densely leaved, pilose, branched at the tip; radical leaves scalelike, becoming larger in the upper ones; median leaves chartaceous, lanceolate, 5–8 cm. long, 1–2 cm. wide, gradually narrowed at tip, acute, broadly clasping at base, loosely mucronate-toothed, ciliate, scabrous on upper surface, scattered pilose on the veinlets beneath; upper leaves becoming smaller and reduced to bracts below the involucres; involucres subglobose, about 1 cm. long, about 2 cm. wide; bracts about 4-seriate, the outer broadly lanceolate, acute, ciliolate, spreading or ascending, the inner linear, gradually narrowed at tip; ligulate corollas about 9 mm. long, 2 mm. wide, 3-toothed; achenes about 1.5 mm. long, glabrous, 10-ribbed; pappus about 8 mm. long.——July–Sept. Wet places; Hokkaido, Honshu, Shi-

koku, Kyushu.——Korea, Manchuria, and Siberia. The typical phase occurs in Europe and w. Siberia.

3. Inula britannica L. var. **chinensis** (Rupr.) Regel. *I. japonica* Thunb.; *I. britannica* var. *japonica* (Thunb.) Fr. & Sav.; *I. britannica* sensu auct. Japon., non L.; *I. chinensis* Rupr. ex Maxim.——OGURUMA. Stems 20–60 cm. long, appressed-pilose, sometimes glabrescent; radical and lower leaves smaller than the median; median leaves lanceolate to oblong, 5–10 cm. long, 1–3 cm. wide, subacute, abruptly narrowed at base, sessile or semiclasping, loosely mucronate-serrulate, appressed-pilose or nearly glabrous on both surfaces; upper leaves gradually smaller; heads few or solitary, sometimes with subtending bracteal leaves; involucres subglobose, 7–8 mm. long, 15–17 mm. across; bracts 5-seriate, nearly equal, the outer lanceolate, the inner narrow, scarious, ciliolate; ligulate florets 1-seriate, the corollas yellow, 16–19 mm. long; achenes about 1 mm. long, 10-ribbed, pilose; pappus about 5 mm. long, the bristles minutely scabrous.——June–Oct. Wet places in lowlands, especially along rivers; Hokkaido, Honshu, Shikoku, Kyushu; common.——Korea, Manchuria, and China.——Cv. **Plena.** YAE-OGURUMA. A double-flowered cultivar.

Var. **ramosa** Komar. EDA-UCHI-OGURUMA. Stems to 1 m. long, branched in upper part; heads many, small; leaves lanceolate or linear-lanceolate; involucral bracts nearly all alike.——Kyushu, Shikoku——Korea, Manchuria, and China.

Var. **linariaefolia** (Turcz.) Regel. *I. linariaefolia* Turcz.; *I. britannica* subsp. *linariaefolia* (Turcz.) Kitam.; *I. britannica* var. *maximowiczii* Regel——HOSOBA-OGURUMA. Stems 30–70 cm. long; median leaves linear to linear-lanceolate, 4–9 cm. long, 6–10 mm. wide, recurved on margin, glandular-dotted beneath; involucres 4–6 mm. long, 8–14 mm. across; bracts 4-seriate, glandular-dotted; ligulate corollas 8–10 mm. long, glandular-dotted on back; pappus about 3 mm. long.——Honshu, Kyushu.——Korea, Manchuria, China, and Siberia.

8. CARPESIUM L.   YABU-TABAKO ZOKU

Branched perennials; leaves alternate, entire or toothed; heads all discoid, heterogamous; involucres subglobose or depressed-globose; bracts 3- or 4-seriate, the outer herbaceous or with herbaceous appendages, the inner scarious, oblong, obtuse, sometimes acute; receptacle flat; pistillate florets marginal, many-seriate, the corollas 3- to 5-toothed; the bisexual florets fertile, many, the corollas 4- or 5-toothed, yellowish; anthers sagittate with setiform appendages; style of bisexual florets linear, rather flat, rounded at apex; achenes glabrous, many-grooved, with a short glandular beak, crowned by a cartilaginous ring; pappus absent.——Ten species or more, Eurasia.

1. Carpesium abrotanoides L. *C. thunbergianum* Sieb. & Zucc.; *C. abrotanoides* var. *thunbergianum* (Sieb. & Zucc.) Makino——Yabu-tabako. Roots fusiform, woody; stems 50–100 cm. long, stout, terete, leafy, pubescent toward the top, dichotomously forked; lower cauline leaves broadly elliptic to oblong, 20–28 cm. long, 8.5–15 cm. wide, obtuse to acute, narrowed at base to a broadly winged petiole, irregularly mucronate-toothed, glandular-dotted beneath, short-pubescent on both sides; upper leaves oblong, sessile, gradually smaller, acute; heads many, sessile, axillary, usually bractless, deflexed in anthesis; involucres campanulate-globose; bracts in 3 series, the outer shortest, ovate, acuminate, short-pubescent, the median and inner oblong, rounded at apex; achenes about 3.5 mm. long, with a beak about 0.7 mm. long.——Sept.–Nov. Woods; Hokkaido, Honshu, Shikoku, Kyushu.——Korea and China.

2. Carpesium glossophyllum Maxim. *C. hieracioides* Lév.——Saji-gankubi-sō. Rhizomes short, creeping; stems 25–50 cm. long, subscapose, densely spreading-pubescent, slightly branched in upper part; radical leaves rosulate, oblanceolate-lingulate, densely pubescent on both sides; cauline leaves few, scattered, smaller, oblong-lanceolate, obtuse, sometimes acute; upper leaves linear-lanceolate, few, very small; heads subglobose, terminal, nodding, long-pedunculate, with many usually linear subtending bracteal leaves; involucres cupuliform, 6–8 mm. long, 8–15 mm. wide; bracts 5-seriate, the outer shortest, reflexed, pubescent on back, the median and inner oblong, obtuse, ciliolate; achenes 3.5–4 mm. long. ——Aug.–Oct. Woods; Honshu, Shikoku, Kyushu.——Ryukyus and Korea (Quelpaert Isl.).

3. Carpesium rosulatum Miq. Hime-gankubi-sō. Rhizomes short; stems erect, slender, subscapose, 15–45 cm. long, densely pubescent, branched; radical leaves rosulate, spathulate-lanceolate, 6–15 cm. long, 1.2–3 cm. wide, obtuse to rounded, narrowed at base, sessile, sparsely mucronate-toothed, densely pubescent; cauline leaves smaller, narrowly oblanceolate, sessile, obtuse; heads depressed-globose, 1–30, terminal, on slender peduncles, nodding; involucres tubular-campanulate, about 6.5 mm. long, 5 mm. across, bractless or with few small obovate subtending bracteal leaves; bracts 3-seriate, the outer ovate, shortest, reflexed, the median narrowly oblong, obtuse, ciliolate, the inner linear; achenes about 3.5 mm. long.

——Aug.–Oct. Woods; Honshu (Kantō Distr. and westw.), Shikoku, Kyushu.——Korea (Quelpaert Isl.).

4. Carpesium macrocephalum Fr. & Sav. *C. eximium* C. Winkl.——Ō-gankubi-sō. Stems crisped-pubescent, with thickened branches below the heads; lower leaves large, broadly ovate, 30–40 cm. long, 10–13 cm. wide, acute, winged-petiolate, with irregular, coarse, double teeth, short-pubescent on both sides especially on nerves; median cauline leaves gradually smaller, obovate-oblong, acute, abruptly narrowed on lower half; upper leaves small, narrow, acuminate; heads 25–35 mm. across; involucres cupuliform, 8–10 mm. long, 23–30 mm. across; outer bracts similar to the bracteal leaves, the median oblong-linear, acute, densely pubescent, the inner linear-spathulate, 5.5–6 mm. long.——Aug.–Oct. Honshu (centr. distr. and northw.).——Korea and Manchuria.

5. Carpesium koidzumii Makino. *C. triste* var. *abrotanoides* Matsum. & Koidz.; *C. divaricatum* var. *abrotanoides* (Matsum. & Koidz.) Kitam., in syn.——Hosoba-gankubi-sō. Rhizomes thickened; stems erect, 70–100 cm. long, often reddish, pubescent, branched in upper half; radical and lower leaves ovate-oblong, about 18 cm. long, 3–5 cm. wide, acuminate, rounded at base, long winged-petiolate; median leaves narrowly oblong, 12–15 cm. long, 3–5 cm. wide, acute at base, short-petiolate, loosely mucronate-toothed, sparsely pubescent on both sides; upper leaves lanceolate, gradually smaller; leaves on branches lanceolate, 2–5 cm. long, acuminate at both ends; heads in racemes, short-pedunculate, nodding; involucres obovoid, about 4.5 mm. long, 5–6 mm. across; bracts 4-seriate, the outermost often caudate, the median and inner oblong, obtuse, ciliolate; achenes about 3 mm. long, glandular-tuberculate.——Aug.–Oct. Woods; Honshu, Shikoku, Kyushu.

6. Carpesium hosokawae Kitam. Banjin-gankubi-sō. Rhizomes short; stems 50–70 cm. long, densely pubescent, often purplish, branched in upper half; lower cauline leaves ovate-oblong, long-petiolate, 10–14 cm. long, 2.5–3.5 cm. wide, obtuse to acuminate, cuneate at base, on long wingless petioles; median leaves lanceolate, acuminate, cuneate at base; upper leaves linear-lanceolate, gradually smaller; heads 4–5 mm. across, solitary on the branches, nodding at anthesis, subtended by many involucral bracts longer than the heads; involucres campanulate-globose, about 4 mm. long, 4–5 mm. across; bracts 4-seriate, the outermost shortest, ovate, mucronate, pubescent, the median narrowly oblong, rounded at apex, denticulate, the inner narrow; achenes about 2.5 mm. long. ——Aug.–Oct. Woods; Honshu (Kinki Distr. and westw.), Kyushu.——Formosa.

7. Carpesium divaricatum Sieb. & Zucc. *C. divaricatum* var. *pygmaea* Miq.; *C. erythrolepis* Lév.——GANKUBI-SŌ. Rhizomes short-creeping; stems erect, 25–150 cm. long, densely pubescent; lower leaves ovate to ovate-oblong, 7–23 cm. long, obtuse to acute, rounded or sometimes shallowly cordate or truncate, rarely short-cuneate at base, irregularly mucronate-toothed, glandular-dotted beneath, the petioles shorter or as long as the blades, short-winged; median leaves oblong, acuminate, cuneate at base; upper leaves smaller, sessile; heads 6–8 mm. across, terminal and solitary on the branches or sub-racemose, nodding in anthesis, the subtending bracteal leaves 2–4, lanceolate, reflexed, 2–5 times as long as the head; involucres ovoid-globose, 5–6 mm. long; bracts 4-seriate, the outer broadly ovate, mucronate, the median oblong, rounded, the inner linear; achenes about 3.5 mm. long.——Aug.–Oct. Honshu, Shikoku, Kyushu.——Korea, Manchuria, China, Ryukyus, and Formosa.

8. Carpesium matsuei Tatew. & Kitam. *C. koidzumii* var. *matsuei* (Tatew. & Kitam.) Hara——NOPPORO-GANKUBI-SŌ. Rhizomes thick; stems 40–100 cm. long, pubescent, branched at the top; lower and median leaves membranous, ovate or broadly so, the blades 7.5–14 cm. long, 5–11 cm. wide, acute or short-acuminate, shallowly cordate-rounded to cuneate at base, entire or loosely mucronate-toothed, loosely pubescent on both sides, the petioles 2.5–11 cm. long; upper leaves gradually smaller, lanceolate to broadly so, cuneate at base; heads terminal on the branches, nodding in anthesis, peduncu-late, the subtending bracteal leaves 3–5, lanceolate, longer than the head; involucres subglobose, about 5 mm. long, 11 mm.

across; bracts 3-seriate, equal in length, the outer ovate-oblong, acute, the inner obtuse; achenes about 3.5 mm. long, slightly viscid.——Aug.–Oct. Hokkaido, Honshu (centr. and n. distr.).

9. Carpesium cernuum L. *C. glossophylloides* Nakai; *C. taquetii* Lév.——KO-YABU-TABAKO. Stems 50–100 cm. long, thickened, densely white-pubescent and crisped-puberu-lent at base, much branched; lower cauline leaves spathulate-oblong, 8–25 cm. long, 4–6 cm. wide, acute or obtuse, abruptly narrowed at base, winged-petiolate, doubly mucronate-toothed or undulate mucronate-toothed, rather densely white-pubes-cent on both sides; median leaves slightly smaller, oblong, obtuse, or sometimes acuminate; heads nodding in anthesis, the subtending bracteal leaves many, linear-lanceolate, 2–5 cm. long; involucres cupuliform, 7–8 mm. long; bracts scari-ous, the outer leaflike, broad, white-pubescent, the inner nar-rowly oblong, obtuse; achenes linear, 4.5–5 mm. long.——July–Sept. Hokkaido, Honshu, Shikoku, Kyushu.——Temperate regions, Europe to China, Ryukyus, Formosa, and Korea.

10. Carpesium triste Maxim. MIYAMA-YABU-TABAKO, GANKUBI-YABU-TABAKO. Stems 40–100 cm. long, densely spreading-pubescent especially at base, branched in upper part; lower cauline leaves ovate-oblong, 13–20 cm. long, 3–5 cm. wide, rounded at base, irregularly mucronate-toothed, densely pubescent on both sides, long winged-petiolate; median leaves narrower, long-acuminate; upper leaves gradually smaller, lanceolate or linear-lanceolate, acuminate at both ends; heads pedunculate, the subtending bracteal leaves many, linear-lanceolate, as long as or longer than the heads, reflexed; in-volucres campanulate; bracts mostly all alike, the outer oblong-lanceolate, acute, green, ciliate, the inner lanceolate, scarious; achenes about 3.5 mm. long.——Aug.–Oct. Hokkaido, Hon-shu, Shikoku, Kyushu.——China (?).

9. LEIBNITZIA Cass. SEMBON-YARI ZOKU

Perennials; stems scapiform, scaly-leaved; radical leaves many, pinnately lobed; heads dimorphic, the vernal ones ligulate, the autumnal cleistogamous and tubular; involucres tubular; bracts imbricate, few-seriate, linear; receptacle flat, alveolate; ligulate florets fertile or sterile; ligulate corollas bilabiate, the outer lip elongate, 3-toothed; tubular corollas slightly bilabiate; autumnal florets fertile; anthers sagittate, the tails connate, acuminate; achenes fusiform, more or less flattened, hairy; pappus-bristles many, smooth or scaberulous, persistent.——Five species, in Asia.

1. Leibnitzia anandria (L.) Nakai. *Tussilago anandria* L.; *Perdicium tomentosum* Thunb.; *Tussilago lyrata* Willd.; *Gerbera anandria* (L.) Schultz-Bip.——SEMBON-YARI, MURA-SAKI-TAMPOPO. Rhizomes short; stems 10–20 cm. long, sim-ple, cobwebby-lanate while young, loosely scaly; radical leaves broadly oblanceolate, 5–16 cm. long, 13–45 mm. wide, pin-nately cleft or lobed, sometimes ovate-cordate or deltoid-ovate by reduction of the lateral lobes, cobwebby-lanate especially beneath; heads about 15 mm. across; involucres 8–10 mm.

long; bracts broadly linear, obtuse; ligulate corollas 10–12 mm. long, white, red-purple outside; achenes 4.5–6 mm. long; pappus-bristles about as long as the achene, scaberulous.—— Vernal phase, Apr.–June.——Autumnal phase with taller stems to 60 cm. high; involucres about 15 mm. long; achenes about 6 mm. long, more or less hairy; pappus brown, 11 mm. long, the bristles nearly smooth.——Oct.–Nov.——Grassy hills and low mountains; Hokkaido, Honshu, Shikoku, Kyushu.—— Sakhalin, s. Kuriles, Manchuria, Siberia, China, and Formosa.

10. PERTYA Schultz-Bip. KŌYA-BŌKI ZOKU

Herbs or subshrubs; leaves alternate, toothed, sometimes divided, sessile or petiolate; heads solitary or sometimes racemose or paniculate; florets all tubular, bisexual, fertile; involucres campanulate; bracts many-seriate, imbricate, firmly membranous or rather coriaceous, rounded to acute; receptacle flat, rarely villous; corollas tubular, deeply 5-lobed; anthers sagittate, the tails connate, often laciniate; style shortly 2-fid, with obtuse reflexed lobes; achenes obovoid-oblong, narrowed at base, many-ribbed, glabrous or hairy; pappus-bristles many, unequal, scabrous.——About 16 species, in e. Asia and India.

1A. Stems woody and branched; rhizomes indistinct; heads solitary and terminal on the branchlets.
  2A. Leaves ovate, appressed-pilose; heads on elongate branchlets. . . . . 1. *P. scandens*
  2B. Leaves oblong, nearly glabrous; heads on abbreviated branchlets. . . . . 2. *P. glabrescens*
1B. Stems herbaceous, simple; rhizomes elongate; heads in racemes, spikes, or panicles.
  3A. Leaves long-petiolate.
    4A. Leaves 3-lobed; heads 1-flowered, in panicles; achenes short-pilose; receptacle glabrous. . . . . 3. *P. triloba*
    4B. Leaves dentate or irregularly incised; heads few-flowered, in spikes; achenes glabrous; receptacle setose. . . . . 4. *P. robusta*
  3B. Leaves sessile, obovate-oblong, 10–30 cm. long, cuneate at base. . . . . 5. *P. rigidula*

1. Pertya scandens (Thunb.) Schultz-Bip. *Erigeron scandens* Thunb.; *P. ovata* Maxim.——Kōya-bōki. Stems slender, woody, short-pilose, gray-brown, much branched; leaves on previous season's branches many, loosely arranged, ovate, 2–5 cm. long, 1.5–4.5 cm. wide, obtuse to acute, rounded at base, subsessile, loosely mucronate-toothed, 3-nerved, appressed-pilose; leaves of current season 3 or 4, fasciculate, deciduous; heads solitary, terminal; involucres 13–14 mm. long; bracts distinctly imbricate, acute, the outer ovate, the median ovate-oblong, the inner narrowly oblong; florets about 13, about 15 mm. long; achenes densely pilose, about 5.5 mm. long.——Sept.–Nov. Thin woods in hills and low mountains; Honshu, Shikoku, Kyushu.——China.

2. Pertya glabrescens Schultz-Bip. *P. scandens* sensu auct. Japon., non Schultz-Bip.; *P. glabrescens* var. *viridis* Nakai; *P. scandens* var. *schultziana* Franch.——Nagaba-kōya-bōki. Rhizomes thickened; stems slender, to 120 cm. long, nearly glabrous, much branched; leaves on previous season's branches ovate, 25–35 mm. long, 15–25 mm. wide, obtuse to acute, rounded at base, very short-petiolate, mucronate-toothed, glabrous on both sides, 3-nerved; leaves of current season 3–5, fasciculate on the nodes of the last year's branches, oblong, 3–6.5 cm. long, 1–1.7 cm. wide, acute at both ends, serrulate, lustrous, sparsely pilose; heads terminal on short spurs; florets 15–18 mm. long; achenes 6.5–7 mm. long, 10-ribbed, appressed-pubescent.——Aug.–Nov. Thin woods in hills and low mountains; Honshu, Shikoku, Kyushu.

3. Pertya triloba (Makino) Makino. *Macroclinidium trilobum* Makino; *P. fauriei* Franch.——Oyari-haguma. Stems erect, 45–85 cm. long, nearly glabrous; radical and lower cauline leaves scalelike; median leaves alternate, chartaceous, oblong, 10–13 cm. long, 7–13 cm. wide, mucronate, 3-lobed, long-petiolate, coarsely mucronate-toothed, glabrescent above, crisped-puberulent on nerves and veinlets beneath and on margin; upper leaves gradually smaller, becoming bractlike; heads many, paniculate, 1-flowered, subsessile; receptacles glabrous; involucres 14–17 mm. long, narrowly tubular; bracts ovate, the median oblong, the inner linear-oblong; achenes 8.5–9 mm. long, short-pilose.——Sept.–Oct. Woods; Honshu (n. distr. and n. Kantō Distr.).

4. Pertya robusta (Maxim.) Beauverd. *Macroclinidium robustum* Maxim.; *P. macroclinidium* Makino——Kashiwaba-haguma. Rhizomes creeping, knotty; stems erect, simple, 30–70 cm. long, thinly puberulent at top; leaves mostly near the median portion of the stem, ovate to ovate-oblong, 10–20 cm. long, 7–12 cm. wide, acute, rounded to cuneate at base, dentate or irregularly incised, puberulent on both sides, ciliate,

the petioles to 12 cm. long; leaves of the inflorescence sessile, small, ovate; heads in spikes, few-flowered, sometimes geminate; receptacles setose, the hairs 0.5–1 mm. long; involucres 17–27 mm. long, 8–18 mm. across; bracts many-seriate, rather coriaceous, the outer depressed-ovate, rounded at apex, the median ovate, purplish brown, the inner narrowly oblong; achenes 10–11 mm. long, glabrous; pappus 13–15 mm. long, the bristles minutely scabrous, unequal.——Sept.–Nov. Woods in mountains; Honshu, Shikoku, Kyushu.

Var. **kiushiana** Kitam. *Macroclinidium robustum* var. *kiushianum* (Kitam.) Honda——Tsukushi-kashiwaba-haguma. Heads short-pedunculate; leaves often broader; involucral bracts furfuraceous-puberulent on back.——Kyushu.

5. Pertya rigidula (Miq.) Makino. *Eupatorium rigidulum* Miq.; *Macroclinidium rigidulum* (Miq.) Makino; *M. verticillatum* Fr. & Sav.——Kurumaba-haguma. Rhizomes creeping, knotty; stems 60–90 cm. long, nearly glabrous; leaves coriaceous, mostly 7 or 8 in false verticils near the median portion of the stems, obovate-oblong, 10–30 cm. long, 3.5–11.5 cm. wide, mucronate, cuneate at base, mucronate-toothed, subtrinerved, sessile; leaves of the inflorescence small, broadly lanceolate; heads on slender peduncles, in panicles, bracteate; florets 7–9; involucres 18–20 mm. long, about 1 cm. across; bracts rounded at apex, the outer 1.5–2 mm. long, depressed-ovate; achenes 6–8 mm. long, coarsely pilose.——Oct.–Nov. Woods in mountains; Honshu (Kinki Distr. and eastw.).

The following interspecific hybrids are reported:

Pertya × hybrida Makino. *Macroclinidium hybridum* (Makino) Matsum.; *P. macrophylla* Nakai; *Macropertya hybrida* (Makino) Honda——Kakoma-haguma, Ōba-kōya-bōki. Annual; median leaves about 8 cm. long, 6 cm. wide, 3-nerved, the petioles about 1 cm. long; upper leaves small; heads solitary in axils of upper leaves, forming a terminal spike; involucres 17–19 mm. long.——Hybrid of *P. robusta* × *P. scandens*.

Pertya × suzukii Kitam. *Macroclinidium suzukii* Kitam., in syn.——Iwaki-haguma. Median leaves chartaceous, approximate, irregularly dentate; florets 3 or 4.——Honshu (Iwaki Prov.).——Hybrid of *P. triloba* × *P. rigidula*.

Pertya × koribana (Nakai) Makino & Nemoto. *Macroclinidium koribanum* Nakai; *P. triloba* var. *koribana* (Nakai) Makino——Sendai-haguma. Median leaves petiolate, alternate, elliptic, about 12 cm. long, about 6 cm. wide, acute, cuneate at base, loosely mucronate-toothed, the petioles 2–2.5 cm. long; heads 1-flowered, in panicles.——Honshu (near Sendai).——Hybrid of *P. triloba* × *P. robusta*.

11. AINSLIAEA DC.   Momiji-haguma Zoku

Perennial herbs; leaves petiolate, radical or on the median part of stems, entire to lobed; heads in racemes, spikes, or panicles, often nodding; involucres narrowly tubular; bracts imbricate; receptacles glabrous; florets few, bisexual, fertile or rarely sterile, rarely cleistogamous, the corolla lobes 5, linear, reflexed, the anthers sagittate, the tails connate, usually acuminate;

style-branches short, flat, rounded to truncate at apex, divergent; achenes oblanceolate, rather flattened, many-nerved, truncate at apex, narrowed at base, glabrous or coarsely pilose; pappus-bristles 1-seriate, plumose, sometimes absent.——About 20 species, in e. Asia, Philippines, and India.

1A. Leaves linear, 4.5–7 mm. wide. . . . . 1. *A. linearis*
1B. Leaves more than 1 cm. wide, sometimes lobed.
  2A. Leaves entire, ovate, rounded at apex, cordate at base, the petioles densely brown-lanate. . . . . 2. *A. fragrans* var. *integrifolia*
  2B. Leaves dentate or divided, the petioles not brown-lanate.
    3A. Leaves dentate or lobed.
      4A. Leaves 1–3 cm. long, 5-angled or shallowly 5-lobed. . . . . 3. *A. apiculata*
      4B. Leaves 6–16 cm. long.
        5A. Leaves reniform-cordate or orbicular, palmately 7- to 11-lobed, shorter than to as long as wide. . . . . 4. *A. acerifolia*
        5B. Leaves oblong, sagittate, hastate or ovate, longer than broad, irregularly dentate and shallowly incised. . . . . 5. *A. cordifolia*
    3B. Leaves palmately 2- to 5-parted, the segments 2- or 3-lobed. . . . . 6. *A. dissecta*

1. Ainsliaea linearis Makino. *A. faurieana* Beauverd ——HOSOBA-HAGUMA. Rhizomes short; stems tufted or solitary, 17–40 cm. long, purplish; leaves densely aggregated near the median part of stems, linear, 4–8 cm. long, 4.5–7 mm. wide, obtuse, mucronate-tipped, entire or undulate and mucronate-toothed, slightly recurved on margin, nearly glabrous, 1- or sometimes obscurely 3-nerved, gradually narrowed below to a winged petiole; inflorescence a bracteate raceme; heads rose-purple, 3-flowered, the peduncles 3–15 mm. long; involucres 7–9 mm. long; bracts obtuse, the outer ovate, the inner linear, about 0.5 mm. wide; achenes 3–4 mm. long, densely coarse-pilose; pappus brownish.——July–Nov. Rocks along rivers in mountains; Kyushu (Yakushima).

2. Ainsliaea fragrans Champ. var. **integrifolia** (Maxim.) Kitam. *A. integrifolia* (Maxim.) Makino; *A. cordifolia* var. *integrifolia* Maxim.——MARUBA-TEISHŌ-SŌ. Rhizomes short-creeping, slender; stems 45–60 cm. long, densely brown-pubescent, usually simple; leaves 4–5, in a false verticil on lower part of stems, ovate, 6–10 cm. long, 3.5–6 cm. wide, mucronate, deeply cordate, mucronate-toothed, densely brown-lanate on both sides, long-petiolate; heads spicate, nodding in anthesis, subsessile; involucres narrowly tubular, 12–17 mm. long; bracts glabrous, the outer obtuse, ovate, the inner lanceolate and acuminate, about 1 mm. wide; achenes about 5 mm. long, 1 mm. wide, appressed brown-pilose; pappus about 10 mm. long, brownish.——Nov.–Dec. Kyushu (s. distr.).——s. China.

3. Ainsliaea apiculata Schultz-Bip. *A. apiculata* var. *typica* Masam.; *A. apiculata* var. *rotundifolia* Masam.——KIKKŌ-HAGUMA. Rhizomes slender, creeping; stems 8–30 cm. long, rather prominently pubescent, simple or branched; leaves chartaceous, crowded on lower portion of stems, ovate, cordate, or reniform, 1–3 cm. long and as wide, long-pubescent on both sides, 5-angled or shallowly 5-lobed, the lobes obtuse, mucronulate, the terminal lobe often elongate, the petiole about twice as long as the blade; heads in racemes, short-pedunculate; florets often cleistogamous; involucres 10–15 mm. long; bracts 5-seriate, the outer about 1 mm. long, ovate, the inner linear; achenes about 4.5 mm. long, densely short-pilose; pappus about 7 mm. long.——Sept.–Nov. Woods in mountains and hills; Hokkaido, Honshu, Shikoku, Kyushu.——Korea.

Var. **acerifolia** Masam. *A. apiculata* var. *multiscapa* Masam.; *A. apiculata* var. *ovatifolia* Masam. and var. *scapifolia* Masam.; *A. liukiuensis* Beauverd——RYŪKYŪ-HAGUMA. Leaves rather deeply lobed or oblong with an elongate apical lobe.——Kyushu (Yakushima).

4. Ainsliaea acerifolia Schultz-Bip. MOMIJI-HAGUMA. Stems 35–80 cm. long, loosely long-pubescent, simple; leaves mostly near the median part of the stem, in a false whorl of 4–7, reniform-cordate or orbicular, 6–12 cm. long, 6.5–19 cm. wide, palmately 7- to 11-lobed or lobulate, acuminate, mucronate-toothed, loosely pubescent on both sides, the petioles 5–13 cm. long, wingless, pubescent; heads spicate, many, nodding in anthesis, the peduncles about 2 mm. long, minutely bracteolate; involucres 12–15 mm. long, tubular; bracts many-seriate, scarious, the outer broadly ovate, very short, the inner oblong, 2–3 mm. wide, obtuse; achenes about 9 mm. long, 2 mm. wide; pappus 10–11 mm. long, brownish to purplish.——Woods in mountains and hills; Honshu (Kinki Distr. and westw.), Shikoku, Kyushu.

Var. **subapoda** Nakai. *A. affinis* Miq.; *A. acerifolia* var. *affinis* (Miq.) Kitam.——OKU-MOMIJI-HAGUMA. Leaves shallowly lobed.——Aug.–Oct. Honshu.——Korea., Manchuria, and China.

5. Ainsliaea cordifolia Fr. & Sav. *A. maruoi* Makino ——TEISHO-SŌ. Rhizomes creeping, slender, knotty; stems 20–70 cm. long, pubescent, simple or rarely branched, with oblong buds at base; leaves 4–7 on lower part of stems, falsely verticillate, oblong, sagittate, hastate, or ovate, to 16 cm. long, 12.5 cm. wide, obtuse to acute, cordate, irregularly dentate and shallowly incised, often purple-maculate beneath, long-appressed-pubescent on both sides, the petioles to 12 cm. long; upper leaves 1.5–3 cm. long, deltoid, bractlike; heads in racemes, the peduncles 2–3 mm. long, prominently bracteolate; involucres 11–14 mm. long, the median and outer bracts narrowly oblong, obtuse; achenes glabrous; pappus 9–11 mm. long, brownish or purplish.——Sept.–Nov. Woods in mountains; Honshu (s. Kantō, Tōkaidō, s. Kinki Distr.), Shikoku.

6. Ainsliaea dissecta Fr. & Sav. ENSHŪ-HAGUMA. Rhizomes slender, creeping; stems 10–33 cm. long, glabrous, usually simple; leaves densely arranged on lower half of stems, often falsely verticillate, rounded, 2.3–6.5 cm. long, cordate, long-petiolate, palmately 2- to 5-parted to -divided, the segments 2- or 3-lobed, coarsely toothed, nearly glabrous to loosely puberulent on both sides, the petioles wingless, about twice as long as the blades; heads in racemes, the pedicels 3–4 mm. long, with many minute bracteoles; involucres about 10 mm. long; bracts many-seriate, obtuse, the outer ovate, the inner linear, about 1 mm. across, acute; achenes glabrous.——Sept.–Oct. Woods in mountains; Honshu (Tōkaidō Distr.).

12. DIASPANANTHUS Miq. KUSA-YATSUDE ZOKU

Perennial herb; leaves long-petiolate, radical, rosulate, palmately cleft; inflorescence paniculate; heads many, 1-flowered, the florets bisexual, fertile, the involucres narrow, tubular, the bracts many-seriate, imbricate, scarious; corolla equally 5-cleft, actinomorphic, dark purple or white, the lobes reflexed; anthers sagittate, the auricles contiguous, connate, acuminate; style shortly bifid, the branches rounded at apex, papillose on back; achenes narrowly oblong, somewhat flattened, truncate at apex, contracted at base, densely hirsute; pappus bristles setose, pinnate, in 1 series.——One species, in Japan.

1. Diaspananthus palmatus Miq. *D. uniflorus* (Schultz-Bip.) Kitam.; *Ainsliaea uniflora* Schultz-Bip.——KUSA-YATSUDE. Rhizomes short-creeping, with persistent bases of previous year's stems; stems 40–100 cm. long, with short brown pubescence on lower part; leaves approximate near base of stems, long-petiolate, spreading, orbicular, 6–14 cm. long, cordate, palmately cleft, the lobes usually 7, 2- or 3-lobulate, acute, mucronately toothed, loosely pubescent on both sides or only on under surface, the petioles as long as to longer than the blades; heads nodding, pedicelled; involucres narrowly tubular, about 11 mm. long, 2 mm. across; bracts obtuse, the outer ovate, about 0.5 mm. long, the inner linear, about 1 mm. wide; corolla dark purple; achenes about 7 mm. long, 1.5 mm. wide, densely hirsute.——Sept.–Nov. Woods in mountains; Honshu (Tōkaidō and s. Kinki Distr.), Shikoku, Kyushu.

13. AMBROSIA L. BUTA-KUSA ZOKU

Annuals or perennials; leaves opposite or alternate, usually dissected or compound; heads unisexual, the pistillate heads 1 or few together, below the staminate heads; involucres of staminate heads obconical, the bracts 7–12, connate; pistillate heads 1-flowered, ellipsoidal to obovoid, with 4–8 small tubercles at apex; staminate florets 5–20; anthers nearly free; achenes ovoid; pappus absent.——More than 20 species, in America and the Mediterranean region.

1. Ambrosia artemisiifolia L. var. **elatior** (L.) Descourt. *A. elatior* L.; *A. artemisiifolia* subsp. *diversifolia* Piper; *A. diversifolia* (Piper) Rydb.——BUTA-KUSA. Erect annual; stems 30–100 cm. long, branched, soft-pubescent; leaves opposite or alternate, 2 or 3 times pinnately dissected, green, loosely pubescent above, gray-green and soft-pubescent beneath, 3–11 cm. long; staminate heads in racemes, nodding, short-pedicelled; involucral bracts connate, pubescent and glandular; pistillate heads few, 4–5 mm. long in fruit.——Aug.–Sept. Widely naturalized in our area.——N. America.

14. XANTHIUM L. ONA-MOMI ZOKU

Large branched annuals; leaves alternate, coarsely dentate, petiolate; heads unisexual, the staminate densely aggregated, terminal, globose; involucral bracts spreading, 1-seriate, oblong-lanceolate; receptacle cylindric, scaly; corollas tubular, 5-toothed, the filaments and anthers free, obtuse at base; style clavate at apex; pistillate heads fasciculate in leaf-axils; outer series of involucral bracts oblong-lanceolate, spreading, small, the inner forming an ellipsoidal or ovoid, sharply prickly, 2-beaked utricle enclosing 2 florets; corollas absent; style-branches filiform, exserted; achenes 2 in each head, oblong, enclosed within the utricle.——More than 20 species, chiefly in America.

1. Xanthium strumarium L. *X. sibiricum* Patr. ex Widder; *X. japonicum* Widder; *X. strumarium* var. *japonicum* (Widder) Hara——ONA-MOMI. Short-pubescent; stems 20–100 cm. long, rather stout; leaves ovate-deltoid, 5–15 cm. long, acute, shallowly cordate to broadly cuneate at base, irregularly acute-toothed, often obsoletely 3-lobed, slightly scabrous on both sides, long-petiolate; utricles sessile, oblong, elliptic, or ovoid, 10–18 mm. long, 6–12 mm. wide, densely puberulent, 2-beaked, the beak 1.5–2 mm. long, the prickles 1.5–2 mm. long.——Aug.–Oct. Waste grounds; Hokkaido, Honshu, Shikoku, Kyushu.——Ryukyus, Formosa, China, Korea to Europe; naturalized in N. America.

15. EUPATORIUM L. FUJIBAKAMA ZOKU

Perennials or subshrubs; leaves opposite or verticillate, dentate or variously parted; heads many, small, short-pedicelled, in dense corymbs, homogamous; involucres tubular, the bracts loosely imbricate, few-seriate; receptacle flat, alveolate, glabrous; florets 5–15, tubular; corollas regular, 5-toothed; anthers obtuse, entire at base; style-branches elongate, obtuse or flat; achenes cylindric, truncate at apex, 5-angled, often glandular or pilose.——More than 600 species, mainly in N. and S. America, a few in e. Asia and Europe.

1A. Leaves obtuse to acuminate.
  2A. Leaves sessile, abruptly acute with an obtuse tip, 3-nerved, sometimes deeply 3-lobed. .... 1. *E. lindleyanum*
  2B. Leaves acute to acuminate.
    3A. Leaves short-petiolate, cuneate at base.
      4A. Leaves usually glandular-dotted beneath; stems slightly scabrous. .... 2. *E. chinense*
      4B. Leaves not glandular-dotted beneath; stems glabrous except near tip. .... 3. *E. fortunei*
    3B. Leaves rounded or subtruncate at base. .... 4. *E. variabile*
1B. Leaves 3-cleft, the segments incised and pinnately cleft, caudately long-acuminate. .... 5. *E. yakushimense*

1. Eupatorium lindleyanum DC.

1. Eupatorium lindleyanum DC. *E. kirilowii* Turcz.; *E. lindleyanum* var. *trifoliolatum* Makino——SAWA-HIYODORI. Rhizomes short; stems 40–70 cm. long, usually simple, densely crisped-puberulent especially toward the top; leaves opposite, the lower cauline small, the median rather thick, lanceolate or linear-lanceolate, 6–12 cm. long, 1–2 cm. wide, mucronulate, abruptly cuneate at base, subsessile, sometimes deeply 3-cleft, usually distinctly 3-nerved, with resinous glandular dots beneath, prominently crisped-pubescent on both sides, sparsely toothed; heads 4–5 mm. long, 5-flowered; involucral bracts about 10, 2-seriate; achenes about 2.5 mm. long; pappus white. ——Aug.–Oct. Wet grassy places in lowlands and low mountains; Hokkaido, Honshu, Kyushu.——Ryukyus, Formosa, Korea, Manchuria, China, and Philippines.

Var. **eglandulosum** Kitam. HOSHI-NASHI-SAWA-HIYODORI. Leaves pubescent or nearly glabrous, without glandular-dots beneath.——Honshu (centr. and w. distr.), Kyushu.

2. Eupatorium chinense L. var. simplicifolium

2. Eupatorium chinense L. var. **simplicifolium** (Makino) Kitam. *E. japonicum* Thunb.; *E. japonicum* var. *simplicifolium* Makino; *E. fortunei* var. *simplicifolium* (Makino) Nakai; *E. sachalinense* var. *oppositifolium* Koidz.—— HIYODORIBANA. Stems tall, to 2 m. long, crisped-pubescent; lower leaves small, the median large, opposite, ovate-oblong or elliptic, rather regularly acute-toothed, 10–18 cm. long, 3–8 cm. wide, short-acuminate, cuneate at base, glandular-dotted beneath, loosely crisped-pubescent on both sides, with 6–7 pairs of lateral nerves, the petioles 1–2 cm. long; inflorescence loosely corymbose; heads 5–6 mm. long, 5-flowered; involucral bracts 2-seriate, the outer very short, scarious, rounded at apex; achenes about 3 mm. long, glandular or pilose; pappus white. ——Aug.–Oct. Hokkaido, Honshu, Shikoku, Kyushu.—— Korea, Manchuria, China, and Philippines; very variable.—— Forma **tripartitum** (Makino) Hara. *E. japonicum* var. *tripartitum* Makino——MITSUBA-HIYODORIBANA. Leaves not glandular-dotted beneath.——Honshu, Kyushu.

Var. **angustatum** (Makino) Hara. *E. japonicum* var. *tripartitum* forma *angustatum* Makino; *E. laciniatum* Kitam.; *E. japonicum* var. *angustatum* (Makino) Kitam.——SAKEBA-HIYODORI. Leaves 3-sected, incised or pinnately lobed, without glandular dots beneath.——Honshu (Izu Prov., Tōkaidō Distr. and westw.), Kyushu.

Var. **dissectum** (Makino) Hara. *E. japonicum* var. *dissectum* Makino; *E. laciniatum* var. *dissectum* (Makino) Kitam. ——KIKUBA-HIYODORI. Resembles the preceding variety; leaves with glandular dots beneath.——Honshu (Kinki Distr. and westw.), Shikoku, Kyushu.

Var. **sachalinense** (F. Schmidt) Kitam. *E. japonicum* var. *sachalinense* F. Schmidt; *E. glehnii* F. Schmidt ex Trautv.; *E. sachalinense* (F. Schmidt) Makino; *E. hakonense* var. *inter-medium* Nakai; *E. japonicum* subsp. *sachalinense* (F. Schmidt) Kitam.——YOTSUBA-HIYODORI. Leaves oblong to lanceolate-oblong, acutely or doubly toothed, verticillate in 3's or 4's, with resinous glandular dots beneath.——Hokkaido, Honshu (centr. and n. distr.).——Sakhalin and s. Kuriles.

Var. **hakonense** (Nakai) Kitam. *E. hakonense* Nakai; *E. glehnii* var. *hakonense* (Nakai) Hara; *E. japonicum* subsp. *sachalinense* var. *hakonense* (Nakai) Kitam.——HOSOBA-YO-TSUBA-HIYODORI. Resembles the preceding variety; leaves lanceolate to linear-lanceolate; inflorescence often looser.—— Mountains; Honshu (centr. and w. distr.).

3. Eupatorium fortunei

3. Eupatorium fortunei Turcz. *E. stoechadosmum* Hance; *E. chinensis* var. *tripartitum* Miq.; *E. japonicum* var. *fortunei* (Turcz.) Pampan.——FUJIBAKAMA. Stems 1–1.5 m. long, tufted, thinly crisped-puberulent or glabrous at base; lower leaves small; median leaves opposite, usually deeply 3-cleft, rarely undivided, acutely toothed, short-petiolate, lustrous above, the terminal segments oblong or oblong-lanceolate, 8–13 cm. long, 3–4.5 cm. wide, short-acuminate to acute, cuneate to rounded at base, the lateral lobes lanceolate or oblong-lanceolate, acuminate, slightly pilose; upper leaves smaller, usually not divided, oblong; inflorescence a dense corymb; heads 7–8 mm. long, 5-flowered; involucral bracts about 10, 2- or 3-seriate, obtuse; achenes about 3 mm. long; pappus white.—— Aug.–Sept. Grassy slopes and river-beds; Honshu (Kantō Distr. westw.), Shikoku, Kyushu; sometimes cultivated.—— Korea and China.

4. Eupatorium variabile

4. Eupatorium variabile Makino. YAMA-HIYODORI. Stems 40–100 cm. long, glabrous at base, branched, crisped-puberulent at top; median leaves opposite, ovate, oblong-ovate, or oblong-lanceolate, 5–7 cm. long, 2.5–3.5 cm. wide, short-acuminate or short-acute, truncate, shallowly cordate or rounded at base, toothed, nearly glabrous on both sides, the petioles 2–3 cm. long; upper leaves slightly smaller; inflorescence dense, 3–4 cm. across; heads 4–5 mm. long, 5-flowered; involucral bracts 8–9, scarious, 2-seriate, the outer very short; achenes 2–3 mm. long, pilose at tip.——Sept.–Nov. Kyushu. ——Ryukyus.

5. Eupatorium yakushimense

5. Eupatorium yakushimense Masam. & Kitam. *E. variabile* Makino pro parte——YAKUSHIMA-HIYODORI. Stems about 1 m. long, slender, glabrous at base; median leaves opposite, 9–14 cm. long, once or twice 3-cleft, the median segments largest, lanceolate, very long-acuminate, the lateral segments smaller, acutely toothed, loosely pilose on upper side, glandular-dotted beneath, the petioles 2–2.5 cm. long; heads 4–6 mm. long, reddish, 5-flowered; involucral bracts about 10, scarious, acute to obtuse; achenes black, about 3 mm. long, 5-angled, glandular-dotted; pappus white.——July–Aug. Along rivers; Kyushu (Yakushima).

16. ADENOSTEMMA Forst. NUMA-DAIKON ZOKU

Perennial herbs with leafy stems; leaves opposite, often petiolate, toothed; heads small, in loose corymbs; involucres subglobose; bracts herbaceous, 2-seriate, equal, acute, free or connate; receptacle flat and alveolate, florets all tubular, bisexual, fertile; corollas 4- or 5-toothed; anthers obtuse, without appendage at apex, obtuse to subtruncate at base; style-branches elongate, flat, obtuse; achenes obtusely 3-angled, glandular-dotted or tuberculate, rounded at apex; pappus of 3 or 4 clavate gland-tipped bristles.——About 10 species, in warmer and tropical regions, especially in America.

1. Adenostemma lavenia

1. Adenostemma lavenia (L.) O. Kuntze. *Verbesina lavenia* L.; *A. viscosum* Forst.——NUMA-DAIKON. Stems erect, 30–100 cm. long, puberulent at top, the branches often spreading; lower leaves small; median leaves ovate or ovate-oblong, 4–20 cm. long, 3–12 cm. wide, rounded-cuneate at base, obtusely toothed, sparsely pilosulous, the petioles 1–6 cm.

long; heads about 4 mm. long, pedunculate; involucral bracts 2-seriate, equal, connate at base, narrowly oblong, rounded at apex, reflexed; styles long-exserted; achenes about 4 mm. long, tubercled or glandular-dotted.——Sept.–Nov. Wet places in lowlands and mountains; Honshu, Shikoku, Kyushu; common.——Korea, Ryukyus, Formosa, China, and the temperate to tropical parts of Asia and Australia.

17. SOLIDAGO L.   Aki-no-kirin-sō Zoku

Perennial herbs; leaves alternate, toothed; heads small, in spikes or panicles, heterogamous; involucres campanulate; bracts scarious, imbricate, 3- or 4-seriate; receptacle flat; ligulate florets 1-seriate; pistillate florets fertile, the corollas ligulate, yellow; tubular florets fertile, yellow, 5-toothed; anthers obtuse, entire at base; style-branches flat, with a lanceolate appendage at apex; achenes cylindric, 8- to 12-ribbed, truncate at apex, gradually narrowed at base, glabrous or pilose; pappus-bristles many, nearly equal, scabrous.——About 130 species, in the N. Hemisphere, especially in N. America.

1A. Stems densely leaved; leaves linear-lanceolate; achenes densely hirsute. ....1. S. yokusaiana
1B. Stems loosely leaved; leaves ovate to lanceolate; achenes glabrous to scattered-pilose. ....2. S. virga-aurea

1. **Solidago yokusaiana** Makino. *S. virga-aurea* var. *linearifolia* Franch. ex Makino; *S. virga-aurea* var. *angustifolia* Makino, non Gaudich.——Okinagusa, Aoyagi-sō. Rhizomes rather thick; stems tufted, 16–60 cm. long, densely leaved and densely short-pubescent at top; radical leaves lanceolate, long-petiolate; median cauline leaves many, linear-lanceolate, 4–7 cm. long, 2–5 mm. wide, acute, entire or loosely incurved-toothed, narrowed to a petiolelike base; upper leaves smaller and linear; heads subracemose, subtended by linear bracteate leaves; involucral bracts 3-seriate, 1-nerved, narrowly oblong, obtuse, the outer 1.5 mm. long, the inner 5–6 mm. long; ligulate corollas about 7 mm. long, 1.5 mm. wide; achenes densely hirsute.——Aug.–Oct. Rocks along rivers in low mountains; Honshu, Shikoku, Kyushu.——Ryukyus.

2. **Solidago virga-aurea** L. var. **leiocarpa** (Benth.) Miq. *S. decurrens* Lour.; *Amphirhapis leiocarpa* Benth.; *S. cantoniensis* Lour.; *S. virga-aurea* sensu auct. Japon., non L.; *S. virga-aurea* subsp. *leiocarpa* (Benth.) Hult.——Koganegiku, Miyama-aki-no-kirin-sō. Stems 15–70 cm. long, often branched at top, puberulent; cauline leaves many, on winged petioles, narrowly oblong to ovate, 4–10 cm. long, 1.5–4 cm. wide, acute, rounded to cuneate at base, acutely toothed, often puberulent on both surfaces, the petioles winged; upper leaves gradually narrower, acuminate; heads 12–15 mm. across, in panicles, bractless or with few small bracteal leaves; involucral bracts 3-seriate, 1-nerved, the outer ovate-lanceolate, 2–3 mm. long, the inner lanceolate, 6–6.5 mm. long, acute to acuminate; ligulate corollas about 8.5 mm. long, about 2 mm. wide, yellow, 1-seriate; achenes glabrous to scattered-pilose.——Aug.–Sept. Open grassy slopes or woods in mountains; Hokkaido, Honshu, Shikoku, Kyushu; very common.——Sakhalin, Kuriles, Korea, Manchuria, China, and Philippines.——Forma **paludosa** (Honda) Kitam. *S. virga-aurea* var. *paludosa* Honda; *S. japonica* var. *paludosa* (Honda) Honda——Kirigamine-aki-no-kirin-sō. Leaves linear-lanceolate, acuminate.——Wet boggy places; Honshu (centr. distr.).

Var. **minutissima** Makino. *S. virga-aurea* var. *yakushimensis* Nakai; *S. minutissima* (Makino) Kitam.; *S. yakushimensis* (Nakai) Masam.——Issun-kinka. Stems less than 10 cm. long; lower leaves ovate or oblong-lanceolate, 15–35 mm. long; achenes glabrous.——High mountains; Kyushu (Yakushima).

Var. **praeflorens** Nakai. *S. decurrens* var. *praeflorens* (Nakai) Kitam.; *S. hachijoensis* Nakai——Hachijō-aki-no-kirin-sō. Stems low, branched; leaves ovate, acutely toothed; achenes pilose at apex.——Honshu (Izu Isl.).——Bonins.

Var. **gigantea** Nakai. *S. mirabilis* Kitam.; *S. decurrens* var. *gigantea* (Nakai) Ohwi——Ō-aki-no-kirin-sō, Oku-kogane-giku. Stems 60–80 cm. long, stout; cauline leaves broadly ovate, ovate-oblong to suborbicular, 10–15 cm. long, 7–10 cm. wide, truncate-cordate at base; heads dense; achenes pilose on upper half.——Lowlands near seashore; Hokkaido (sw. distr.), Honshu (n. distr.).

Var. **asiatica** Nakai. *S. japonica* Kitam.; *S. japonica* var. *ovata* Honda; *S. virga-aurea* sensu auct. Japon., non L.——Aki-no-kirin-sō. Resembles var. *leiocarpa;* involucres narrowly campanulate, the bracts obtuse.——Aug.–Oct. Lowlands and low mountains, Honshu, Shikoku, Kyushu.——Korea.——The typical phase occurs in Eurasia and N. America.

18. GYMNASTER Kitam.   Miyama-yomena Zoku

Erect perennials; leaves alternate, glabrous or with short coarse hairs; heads ligulate, subglobose to campanulate; involucral bracts 2- or 3-seriate, nearly equal, imbricate; receptacle conical, alveolate; ligulate florets pistillate, 1- or 2-seriate, blue or white, fertile; tubular florets many, bisexual, fertile, yellow, the corollas 5-toothed; achenes flat, oblong, glabrous, ribbed; pappus absent.——Few species, in e. Asia.

1. **Gymnaster savatieri** (Makino) Kitam. *Boltonia savatieri* Makino; *Aster savatieri* Makino; *Asteromoea savatieri* Makino, in syn.——Miyama-yomena. Rhizomes short-creeping; stems 20–50 cm. long, branched at top, green; radical leaves oblong or ovate-oblong, 4–6 cm. long, 2.5–3 cm. wide, coarsely incised-toothed, deep green above, short-pubescent on both sides, winged-petiolate; cauline leaves scattered; heads long-pedunculate, in loose corymbs, 3.5–4 cm. across; involucral bracts lanceolate, 2-seriate; ligulate corollas blue or white, 17–20 mm. long, 3–3.5 mm. wide; achenes 3–4 mm. long.——May–July. Woods in mountains; Honshu, Shikoku, Kyushu.

Var. **pygmaeus** (Makino) Kitam. *Aster savatieri* var. *pygmaeus* Makino; *Asteromoea pygmaea* (Makino) Kitam.;

Aster yokusaianum Kitam. ex Hara; *G. pygmaeus* (Makino) Kitam.——SHINJUGIKU. Rhizomes long-creeping; stems about 10 cm. long; leaves ovate, 2–3 cm. long; heads solitary on long peduncles, smaller; involucres about 3 mm. long, 6–7 mm. across; ligulate corollas 7–9 mm. long; achenes about 2.5 mm. long.——Woods in mountains; Honshu (Kinki and Chūgoku Distr.), Shikoku.

19. KALIMERIS Cass. YOMENA ZOKU

Erect perennial herbs; leaves alternate, ovate, elliptic to linear, toothed, sometimes pinnatified; heads subglobose; involucral bracts equal or the outer shorter, few-seriate, herbaceous or with scarious margins; receptacle convex, alveolate; achenes flat, obovate or oblong, margined, 0- to 2-ribbed, sometimes pilose; pappus-bristles very short, 0.3–1 mm. long, connate at base in a ring; style of the bisexual florets with a deltoid or lanceolate appendage; anthers entire and obtuse at base.——Few species, in e. Asia.

1A. Involucral bracts 3-seriate, loosely imbricate; style-branches deltoid; leaves ovate or lanceolate.
    2A. Leaves short-puberulent beneath; heads about 2.5 cm. across; achenes about 2.5 mm. long; pappus bristles about 0.3 mm. long, few.
                                                1. *K. pinnatifida*
    2B. Leaves glabrous beneath; heads about 3–3.5 cm. across; achenes 3–3.5 mm. long; pappus bristles 0.5–1 mm. long, many, reddish.
                                                2. *K. incisa*
1B. Involucral bracts 2-seriate, equal; style-branches lanceolate; radical leaves cordate, acuminate to acute, long-petiolate. . 3. *K. miqueliana*

1. Kalimeris pinnatifida (Maxim.) Kitam. *Aster indicus* var. *pinnatifida* Maxim.; *Asteromoea pinnatifida* (Maxim.) Koidz.; *Aster pinnatifidus* (Maxim.) Makino; *Asteromoea indica* var. *pinnatifida* (Maxim.) Matsum.; *Aster iinumae* Kitam.——YŪGAGIKU. Rhizomes creeping; stems 40–150 cm. long, short-pilose; median leaves membranous, oblong to ovate-oblong, 7–8 cm. long, 3–4 cm. wide, sessile, pinnatifid, incised-toothed, green, the lobes linear, 3 or 4 pairs, obtuse, mucronate, short-puberulent on both sides especially toward the margin; leaves on branches short, linear-lanceolate; heads about 2.5 mm. across, in loose corymbs, long-pedicelled; outer involucral bracts linear, obtuse, the inner narrowly oblong, green on back, loosely short-puberulent; ligulate corollas 12–13 mm. long, 2–2.5 mm. wide, bluish or rosy; achenes about 2.5 cm. long; pappus bristles about 0.3 mm. long, few.——July–Oct. Hills and low mountains; Honshu.——Cv. **Hortensis**. *K. pinnatifida* forma *hortensis* (Makino) Ohwi; *Aster indicus* var. *pinnatifidus* forma *hortensis* Makino; *K. pinnatifida* var. *hortensis* (Makino) Kitam.; *Aster iinumae* forma *hortensis* (Makino) Hara——CHŌSEN-GIKU. With larger tubular florets.——Cultivated.

2. Kalimeris incisa (Fisch.) DC. *Aster incisus* Fisch.; *Asteromoea incisa* (Fisch.) Koidz.; *K. polycephala* Cass.; *Aster pinnatifidus* forma *robustus* Makino; *A. macrodon* Lév. & Van't.——CHŌSEN-YOMENA, Ō-YŪGAGIKU. Rhizomes long-creeping; stems 1–1.5 m. long, loosely short-pilose in upper part, branched in upper half; lower and median cauline leaves oblong-lanceolate to lanceolate, 8–10 cm. long, about 2.5 cm. wide, acuminate, sessile, loosely incised, lustrous above, short-pilose on margin, glabrous beneath; upper leaves linear-lanceolate, acuminate at both ends; heads 3–3.5 cm. across, on long pedicels; involucral bracts 3-striate, the outer slightly shorter, lanceolate, acute, the inner oblong, obtuse; ligulate corollas about 18 mm. long, 2.5 mm. wide; achenes 3–3.5 mm. long; pappus 0.5–1 mm. long, reddish.——July–Sept. Meadows and waste grounds in lowlands; Honshu (Chūgoku), Shikoku, Kyushu; common.——Korea, Manchuria, n. China, and Siberia.

Kalimeris yomena Kitam. *K. incisa* var. *yomena* Kitam.; *Aster yomena* (Kitam.) Honda; *A. indicus* sensu auct. Japon., non L.——YOMENA. Leaves dentate, lustrous; pappus about 0.5 mm. long.——Common in lowlands; Honshu, Shikoku, Kyushu.——s. Korea.

3. Kalimeris miqueliana (Hara) Kitam. *Biotia japonica* Miq.; *Boltonia japonica* (Miq.) Fr. & Sav.; *Aster japonicus* (Miq.) Fr. & Sav., non Less.; *Asteromoea japonica* (Miq.) Matsum.; *Aster miquelianus* Hara——ŌBA-YOMENA. Rhizomes creeping; stems 35–90 cm. long, slender, sometimes loosely pilose; radical leaves cordate, 4–9 cm. long, 3.5–6 cm. wide, acuminate to acute, coarsely incised-serrate, more or less pilose on both sides, long-petiolate; upper leaves smaller, ovate, shallowly cordate at base, short-petiolate; uppermost leaves lanceolate, bractlike; heads 22–25 mm. across, on slender pedicels; involucral bracts 2-seriate, equal, herbaceous, lanceolate, ciliolate; ligulate corollas 13–16 mm. long, about 2 mm. wide; style branches lanceolate; achenes 4–4.5 mm. long.——Aug.–Oct. Shikoku, Kyushu.

20. HETEROPAPPUS Less. HAMABE-NOGIKU ZOKU

Biennial erect or decumbent herbs; stems branched; leaves alternate, spathulate-oblong or ovate-oblong, toothed or entire, pilose or glabrous; heads in loose corymbs, ligulate; heads subglobose; involucral bracts herbaceous, 2-seriate, subequal, acuminate; receptacle slightly convex, alveolate; ligulate florets pistillate, blue-purple or white, fertile; tubular florets yellow, fertile, the corolla 5-toothed; anthers obtuse, entire at base; style-branches flat, with a deltoid-acute appendage at apex; achenes much flattened, obovate, densely pilose, rounded at apex, acute at base; pappus of the ligulate florets very short and crownlike, of the tubular florets setose, 1–4 mm. long, brownish red, scabrous.——e. Asia.

1. Heteropappus hispidus (Thunb.) Less. *Aster hispidus* Thunb.; *H. subserratus* Sieb. & Zucc.; *A. hispidus* var. *mesochaeta* Fr. & Sav.; *A. hispidus* var. *heterochaeta* Fr. & Sav.; *H. decipiens* Maxim.——ARENO-NOGIKU. Stems 0.3–1 m. long, densely leafy, coarsely hirsute, branched at tip; radical leaves oblanceolate, 7–13 cm. long, 1–1.5 cm. wide, loosely short-pubescent on both sides; cauline leaves oblanceolate to linear, 5–7 cm. long, 4–20 mm. wide, obtuse, narrowed at base, entire or loosely toothed, slightly recurved on margin, ciliolate; upper leaves smaller, linear, 8–15 mm. long, bractlike; heads 3–5 cm. across; involucres 7–8 mm. long, the bracts nearly equal, herbaceous, linear-lanceolate or the inner row rhombic-lanceolate, acuminate, puberulent; ligulate corollas 14–24 mm. long, 2.5–4 mm. wide; achenes 2.5–3 mm. long.——Sept.–Nov. Lowlands; Honshu (centr. distr. and westw.), Shikoku, Kyushu.——Korea, Manchuria, China, and Formosa.

Key to the varieties of H. hispidus

1A. Pappus of the tubular florets 1–2 mm. long; involucres about 1 cm. across. .... var. *leptocladus*
1B. Pappus of the tubular florets 3–4 mm. long; involucres 1.5 cm. across.
  2A. Plant glabrous; stems 1.5 m. long; leaves linear. .... var. *koidzumianus*
  2B. Plant more or less hirsute.
    3A. Stems decumbent; radical leaves rather thick, small, spathulate, petiolate; involucral bracts rather thick. .... var. *arenarius*
    3B. Stems erect, branched. .... var. *insularis*

Var. **leptocladus** (Makino) Kitam. *Aster leptocladus* Makino; *H. leptocladus* (Makino) Matsum.——Yanagi-no-giku. Stems slender, densely leafy, densely branched at top, nearly glabrous; median leaves linear, acute, glabrous; heads 2.5–3 cm. across; involucres about 1 cm. across; ligulate corollas about 13 mm. long; achenes obovate, about 3 mm. long, pubescent.——Oct. Lowlands; Shikoku.

Var. **koidzumianus** (Kitam.) Kitam. *H. koidzumianus* Kitam.; *Aster koidzumianus* (Kitam.) Nemoto——Buzen-no-giku. Plant glabrous; stems much branched, 1.5 m. long, with slender branches; cauline leaves linear-spathulate; upper leaves gradually smaller, linear; heads 3.5–4 cm. across; involucres 6.5–7.5 mm. long; achenes about 3 mm. long, hirsute; pappus of the tubular florets 3–4 mm. long, brownish red.——Rocks; Kyushu (Yabakei in Buzen).

Var. **arenarius** (Kitam.) Kitam. *H. arenarius* Kitam.;

Aster arenarius (Kitam.) Nemoto——Hamabe-nogiku. Stems branched from the base, decumbent, ascending, glabrescent; radical leaves spathulate, nearly glabrous to loosely hirsute, prominently ciliate, petiolate; heads about 3.5 cm. across; involucres about 7 mm. long.——July–Oct. Near seashores; Honshu (Japan Sea side from Etchu Prov. and westw.), Kyushu.

Var. **insularis** (Makino) Kitam. *Aster insularis* Makino; *H. insularis* (Makino) Matsum.——Sonare-nogiku. Plant glabrescent; roots fusiform; stems much branched from the base; leaves glabrous, ciliolate, lustrous; heads 4–4.5 cm. across; involucres 7–9 mm. long; bracts glabrous, lanceolate, acuminate; ligulate corollas about 18 mm. long, 3–3.5 mm. wide; achenes 4–5 mm. long, hirsute.——Aug.–Oct. Near seashores; Shikoku.

21. ERIGERON L.   Mukashi-yomogi Zoku

Annuals or perennials with erect, sometimes branched stems; leaves alternate, narrow, entire or toothed; heads on naked peduncles or scapes, ligulate or eligulate, subglobose or tubular-campanulate; involucral bracts equal, in a single series, membranous, not definitely imbricate, linear-lanceolate, acuminate, pilose on back; receptacle slightly convex, alveolate; ligulate florets pistillate, fertile, few-seriate, equal, very narrow, blue or white; tubular florets bisexual, the corollas tubular, yellow, fertile; anthers entire at base; style-branches obtuse; achenes flat, ribbed on margin, pilose; pappus-bristles 1-seriate, simple, slender, scabrous.——About 150 species, mostly in temperate regions, a few tropical.

1A. Limb of ligulate corolla longer than the tubular part, exserted from the pappus, usually prominent.
  2A. Heads 3–3.5 cm. across, solitary or few on short scapes.
    3A. Radical leaves long-petiolate, elliptic. .... 1. *E. miyabeanus*
    3B. Radical leaves spathulate-oblong. .... 2. *E. thunbergii*
  2B. Heads 2 cm. or less across, many on tall stems.
    4A. Ligulate corollas involute when dry, often becoming indistinct; pistillate florets with capillary corollas present; stems often purplish red. .... 3. *E. acris*
    4B. Ligulate corollas prominent, flat; capillary corollas absent; stems green.
      5A. Inflorescence nodding before anthesis; leaves vivid green, the cauline oblong to broadly oblanceolate, more or less clasping at base; ligulate corollas usually pinkish; May to June. .... 4. *E. philadelphicus*
      5B. Inflorescence erect; leaves pale green, the cauline ovate to ovate-lanceolate, not clasping at base; ligulate corollas white or purplish rose; July to October. .... 5. *E. annuus*
1B. Limb of ligulate corolla shorter than the tubular part, not exserted from the pappus, minute, in a few series, rarely absent.
  6A. Involucres 4–5 mm. long; receptacle 2–4 mm. across in fruit.
    7A. Involucres about 5 mm. long; cauline leaves linear. .... 6. *E. bonariensis*
    7B. Involucres about 4 mm. long; cauline leaves narrowly oblanceolate. .... 7. *E. sumatrensis*
  6B. Involucres 2.5–4 mm. long; receptacle 1.5–2.5 mm. across in fruit. .... 8. *E. canadensis*

1. **Erigeron miyabeanus** Tatew. & Kitam. *Aster miyabeanus* Tatew. & Kitam.——Miyama-nogiku. Plant densely pubescent with multicellular crisped-hairs; rhizomes 6–8 mm. in diameter, covered with old petioles, the innovation-shoots many, short; scapes about 15 cm. long, solitary; radical leaves elliptic, the inner 2.5–3.5 cm. long, about 2 cm. wide, rounded to obtuse at both ends, sometimes cuneately narrowed at base, 3-nerved, coarsely toothed on upper margin, long-petiolate; cauline leaves loosely arranged, small; heads about 3.5 cm. across, blue-purple, showy; involucres about 1 cm. long, subglobose, densely pubescent, the bracts 3-seriate, equal, broadly linear, acuminate, purplish; ligulate corollas about 13

mm. long, 2 mm. wide; achenes about 2.5 mm. long, hirsute; pappus sordid-white.——July. Hokkaido (Kitami Prov.).

2. Erigeron thunbergii A. Gray. *Aster dubius* (Thunb.) Onno ex Kitam.; *Inula dubia* Thunb.; *A. japonicus* Less.——AZUMAGIKU. Rhizomes short and slender; innovation shoots short; flowering scapes tufted or solitary, 10–37 cm. long, usually simple, densely pubescent; radical leaves spathulate-oblong, 4.5–8 cm. long, 1–2 cm. wide, sometimes few-toothed, ciliate, nearly glabrous on both surfaces; radical leaves at the base of scapes 1.8–3 cm. long, 5–9 mm. wide, spathulate, rounded at apex, entire, densely pubescent; cauline leaves few, loosely arranged, small, 1–3 cm. long, 2–5 mm. wide, densely pubescent; heads solitary, 3–3.5 cm. across; involucres about 8 mm. long, densely pubescent; bracts 3-seriate, equal, narrowly lanceolate, about 1 mm. wide, acuminate, 3-nerved; ligulate corollas 3-seriate, about 15 mm. long, 1.5 mm. wide, blue-purple; pappus brownish red, about 5 mm. long.——Apr.–June. Grassy slopes in lowlands and mountains; Honshu (n. and centr. distr.).

Var. **glabratus** A. Gray. *Aster consanguineus* Ledeb.; *E. dubius* var. *alpicola* Makino; *E. alpicola* (Makino) Makino; *E. dubius* var. *glabratus* (A. Gray) Makino; *E. thunbergii* subsp. *glabratus* (A. Gray) Hara——MIYAMA-AZUMAGIKU. Plant less densely shorter pubescent; pappus sordid white, 2.5–3.5 mm. long.——July–Aug. Alpine regions; Hokkaido, Honshu (centr. and n. distr.).——s. Kuriles, Sakhalin, Korea, Manchuria, Kamchatka, and Siberia.

Var. **angustifolius** (Tatew.) Hara. *E. glabratus* var. *angustifolius* Tatew.; *E. tsuneoi* Tatew.; *Aster tsuneoi* (Tatew.) Miyabe & Tatew.; *A. dubius* var. *angustifolius* (Tatew.) Hara ——APOI-AZUMAGIKU. Leaves narrower, the radical 3–4 cm. long, 2–4 mm. wide; ligulate corollas often white; pappus 2–2.5 mm. long.——Alpine slopes; Hokkaido (Mount Apoi).

Var. **heterotrichus** (Hara) Hara. *E. heterotrichus* Hara; *Aster heterotrichus* (Hara) Hara ex Kitam., in syn.——JŌSHŪ-AZUMAGIKU. Leaves of the sterile rosettes linear-spathulate, glabrous; cauline leaves 4 or 5, broadly linear, narrower; ligulate corollas 10–13 mm. long; style-branches slender; pappus sordid white, about 3.5 mm. long.——Alpine slopes; Honshu (Mount Tanigawa and Mount Shibutsu in Kotsuke).

3. Erigeron acris L. *E. kamtschaticus* var. *hirsutus* F. Schmidt; *E. acris* var. *hirsutus* (F. Schmidt) Miyabe & Miyake ——EZO-MUKASHI-YOMOGI. Perennial, sometimes biennial; rhizomes short; stems 15–55 cm. long, hirsute, often branched at top; radical leaves spathulate, long winged-petiolate; lower cauline leaves oblanceolate, 3.5–6.5 cm. long, 6–12 mm. wide, obtuse, mucronate, entire or mucronate-toothed, short-hirsute on both sides, winged-petiolate; median leaves linear-oblong, 3–6 cm. long, 4–16 mm. wide, usually entire, sessile; upper leaves linear, gradually smaller; heads long-pedunculate, about 17 mm. across, in racemes or loose corymbs; involucres tubular-campanulate, 6–9 mm. long; bracts linear-lanceolate, acuminate, hirsute; ligulate corollas rose-purple, 7–7.5 mm. long, about 0.2 mm. wide; achenes 2–2.5 mm. long, pilose; pappus 4–6 mm. long, whitish or reddish.——Aug. High mountains; Hokkaido, Honshu (centr. distr.).——Manchuria, China, Kuriles, Sakhalin, Korea, Siberia to Europe, and N. America.

Var. **kamtschaticus** (DC.) Herd. *E. kamtschaticus* DC.; *E. acris* var. *drobachensis* sensu Kitam., non Blytt——MUKASHI-YOMOGI, YANAGI-YOMOGI. Stems purplish, much branched at top; lower and median leaves 5–12 cm. long, 5–13

mm. wide; heads many, in panicles or corymbs; involucral bracts puberulent.——Mountains; Hokkaido, Honshu (centr. distr.).——Kamchatka, Sakhalin, and Kuriles.

Var. **linearifolius** (Koidz.) Kitam. *E. kamtschaticus* var. *linearifolius* Koidz.; *E. koidzumii* Honda.——HOSOBA-MUKASHI-YOMOGI. Stems purplish red, sometimes slightly pubescent at base; median leaves many, linear, 3–6 cm. long, about 2.5 mm. wide, usually acute, entire, glabrous or ciliolate; heads many; involucres about 5 mm. long, the bracts puberulent.——Honshu (n. and centr. distr. and Mount Daisen in Hoki Prov.), Shikoku.

Var. **amplifolius** Kitam. HIRO-HA-MUKASHI-YOMOGI. Stems short-hirsute; leaves larger, broadly oblanceolate in the lower ones; median leaves many, oblong-lanceolate, 6–10 cm. long, 17–27 mm. wide, usually nearly entire, short-hirsute on both sides; involucres about 6.5 mm. long, the bracts puberulent.——Hokkaido (Ishikari Prov.), Honshu (Mount Azuma in Uzen Prov.).

4. Erigeron philadelphicus L. HARU-JION. Erect pubescent biennial, 30–60 cm. high; radical leaves spathulate or oblong, about 8 cm. long, 2 cm. wide; median cauline leaves oblong to broadly oblanceolate, 5–6 cm. long, 1.5–2 cm. wide, clasping or narrowed and rounded at base; heads in dense corymbs, erect in fruit, about 2 cm. across; involucres pubescent; ligulate florets many; corollas narrow, linear, pinkish.——May–June. Naturalized in our area; common.——Introduced from N. America.

5. Erigeron annuus (L.) Pers. *Aster annuus* L.; *Stenactis annua* (L.) Cass.——HIME-JOON. Annual; stems 30–100 cm. long, coarsely hirsute, much branched at top; leaves membranous, coarsely acute-toothed, the lower cauline ovate to ovate-lanceolate, 4–15 cm. long, 1.5–3 cm. wide, pilose on both surfaces, narrowed to a winged petiole; upper leaves narrowly ovate to lanceolate, acute at both ends, acute-toothed, coarsely pilose on margin and on midrib beneath; heads erect, in corymbs, about 2 cm. across; involucres long-pilose; ligulate corollas white or purplish rose, linear, 7–8 mm. long, about 1 mm. wide, as long as to slightly longer than the involucre. ——July–Oct. Waste grounds and cultivated fields in lowlands; very common. Introduced from N. America.—— Forma **discoides** Mar.-Vict. & Rouss.——BŌZU-HIME-JOON. Heads without ligulate florets. Rare.

6. Erigeron bonariensis L. *E. linifolius* Willd.—— ARECHI-NOGIKU. Gray-green annual, 30–50 cm. high, puberulent and sparingly long-pilose; lower leaves oblanceolate, sometimes coarsely toothed; median and upper leaves linear; heads in racemes or racemose panicles; involucres gray-hirsute, about 5 mm. long; ligulate florets many, the corollas very short, white, shorter than the pappus.——A pantropic weed rather rare in our area.

7. Erigeron sumatrensis Retz. *E. flahaultianus* (Sennen) Thell.; *E. musashiensis* Makino——Ō-ARECHI-NOGIKU. Gray-green annual, coarsely long-hirsute and short-pilose; radical leaves oblanceolate, 6–10 cm. long, 1.5–3 cm. wide, coarsely toothed; cauline leaves narrowly oblanceolate, gradually acute; upper leaves gradually smaller, narrow, entire; inflorescence many-headed; involucres about 4 mm. long, pilose; ligulate corollas very short, inconspicuous.——Aug.–Oct. Waste grounds and roadsides; very common in lowlands.—— Pantropic.

8. Erigeron canadensis L. HIME-MUKASHI-YOMOGI.

Loosely hirsute annual, 0.5–1.5 m. high; lower leaves oblanceolate, dentate or entire, 7–10 cm. long, 1–1.5 cm. wide, the upper linear; heads many, small, in large terminal panicles; involucres loosely pilose, 2.5–4 mm. long; florets rather few; ligulate florets white, very small.——Aug.–Oct. Waste grounds, cultivated fields, and roadsides; very common.—— Introduced from N. America.

Var. **levis** Makino. KENASHI-HIME-MUKASHI-YOMOGI. Plant glabrescent.

22. ASTER L. SHION ZOKU

Perennials, rarely annuals; stems sometimes scapelike, sometimes leafy and branched; leaves alternate, entire to lobed; heads mostly on leafy branchlets, in loose corymbs or rarely solitary, ligulate; involucres tubular, campanulate, or subglobose, the receptacle slightly convex or flat, alveolate; involucral bracts imbricate; ligulate florets pistillate, fertile, 1- or 2- seriate, bluish or white; tubular florets many, bisexual, usually fertile, the corollas tubular, yellow, 5-toothed; anthers entire; style branches flat, deltoid or lanceolate; achenes more or less flat, ribbed on margin, with 0–4 ribs on the face, the pappus-bristles elongate, few or many, in 1 or 2 series, scabrous.——Many species in the N. Hemisphere, especially abundant in N. America, few in S. Africa and S. America.

1A. Perennials; pappus-bristles not elongate after anthesis.
  2A. Involucral bracts of two lengths, the outer ones short; pappus-bristles many.
    3A. Achenes not flattened.
      4A. Leaves ovate; achenes glabrous or finely hirsute.
        5A. Heads 3–4 cm. across; petioles broadly winged; style-branches lanceolate at apex. .... 1. *A. komonoensis*
        5B. Heads about 2 cm. across.
          6A. Leaves cuneate at base, gradually decurrent on the petiole. .... 2. *A. sekimotoi*
          6B. Leaves cordate at base, petioles narrowly winged. .... 3. *A. scaber*
      4B. Leaves lanceolate; achenes coarsely hirsute.
        7A. Leaves rugulose above, usually obtuse; peduncles long, leafless; pappus-bristles many. .... 4. *A. rugulosus*
        7B. Leaves smooth and lustrous above, acuminate; peduncles short, bracteate; pappus-bristles rather few. .... 5. *A. sohayakiensis*
    3B. Achenes flat.
      8A. Involucral bracts acuminate; pits on receptacle lacerate on margin.
        9A. Leaves oblong, short-petiolate; heads many; achenes glandular-dotted. .... 6. *A. glehnii*
        9B. Leaves cordate, long-petiolate; heads few; achenes hirsute. .... 7. *A. dimorphophyllus*
      8B. Involucral bracts obtuse, rarely acute.
        10A. Heads small, many, some of the tubular florets sterile. .... 8. *A. fastigiatus*
        10B. Heads relatively large, the tubular florets fertile.
          11A. Pits on receptacle entire; heads large, 3–4 cm. across; involucral bracts reddish.
            12A. Stems 48–85 cm. long; median leaves 6–10 cm. long, 1–2 cm. wide; involucral bracts rounded at apex; achenes about 2 mm. long; heads few. .... 9. *A. maackii*
            12B. Stems 1–2 m. long; median leaves usually 20–35 cm. long, 6–10 cm. wide; involucral bracts acuminate; achenes about 3 mm. long; heads many. .... 10. *A. tataricus*
          11B. Pits on receptacle lacerate on margin; heads smaller, 1.5–3 cm. across; involucral bracts not reddish.
            13A. Involucral bracts in 4 series; heads solitary or to 3 on each branch. .... 11. *A. tenuipes*
            13B. Involucral bracts in 2 or 3 series; heads usually more than 3 on each branch. .... 12. *A. ageratoides*
  2B. Involucral bracts nearly equal; pappus-bristles few.
    14A. Involucral bracts acuminate, scarious on margin; plant of waste grounds. .... 13. *A. kantoensis*
    14B. Involucral bracts acute to subobtuse, without scarious margins; plants of seashores.
      15A. Leaves densely puberulent, the median ones broadly spathulate; achenes oblong; involucral bracts acuminate.
                14. *A. spathulifolius*
      15B. Leaves scattered puberulent, the median ones ovate-spathulate; achenes obovate; involucral bracts obtuse to rounded.
                15. *A. pseudoasagrayi*
1B. Annual or biennial, or short-lived perennial; involucral bracts few; achenes flat, the inner ones often not fertile; pappus-bristles elongate after anthesis. .... 16. *A. tripolium*

1. Aster komonoensis Makino. KOMONOGIKU, TAMA-GIKU. Rhizomes short-creeping; scapes 15–50 cm. long, glabrous at base, loosely short-hairy and loosely branched in upper part; sterile branches short; rosette leaves ovate, 2.5–7 cm. long, 2.5–4 cm. wide, long-petiolate, the lower cauline leaves usually withering before anthesis, the median ones sparsely arranged, oblong, 3.5–7 cm. long, 1.5–2 cm. wide, acute, narrowed to a winged petiole or gradually narrowed to the clasping base, coarsely toothed on upper half, the upper leaves small; heads in loose corymbs, 3–4 cm. across; involucres 6–10 mm. long, globose; bracts 3-seriate, glabrous, the outer elliptic or orbicular, shorter than the inner; ligulate corollas 11–17 mm. long, 2–2.5 mm. wide; achenes about 3 mm. long, hirsute.——July–Sept. Rocks in mountains; Honshu (s. Kinki Distr.), Shikoku.

2. Aster sekimotoi Makino. NAGABA-SHIRAYAMA-GIKU. Stems erect, about 1.5 m. long, branched in upper part; lower cauline leaves firmly chartaceous, ovate-oblong, 14–17 cm. long, about 5 cm. wide, gradually acuminate, cuneate at base, gradually decurrent on the long petiole, loosely mucronate-toothed, short-pilose above and on nerves beneath, the upper leaves gradually smaller, lanceolate, acuminate at both ends; heads in loose corymbs, about 2 cm. across, on short-pilose peduncles; involucres depressed-globose; bracts 3-seriate, 3-nerved, glabrous, rounded at apex, the outer short; ligulate corollas about 12.5 mm. long, 2 mm. wide.——Oct. Honshu (Oosawa in Shimotsuke Prov.). This is a hybrid, allegedly of *A. rugulosus* × *A. scaber*.

3. Aster scaber Thunb. *Doellingera scabra* (Thunb.) Nees; *Biotia discolor* Maxim.; *B. corymbosa* var. *discolor*

(Maxim.) Regel——SHIRAYAMA-GIKU.   Rhizomes short and thick; stems 1–1.5 m. long, stout, glabrous, branched in upper part, the branches puberulent; radical leaves usually withering before anthesis, cordate, long-petiolate, the lower cauline leaves chartaceous, cordate, long-petiolate, 9–24 cm. long, 6–18 cm. wide; short-acuminate, cordate, singly or doubly toothed, deep green above, rather prominently scabrous on both sides, the median ones narrowly winged-petiolate, gradually becoming smaller, ovate-deltoid, acute, cordate to truncate at base; heads in loose corymbs, 18–24 mm. across, pedunculate; involucral bracts 3-seriate, oblong, 1.5–5 mm. long, rounded at apex, scarious on margin; ligulate corollas 11–15 mm. long, white, about 3 mm. wide.——Aug.–Oct. Woods and thickets in hills and low mountains; Hokkaido, Honshu, Kyushu; rather common.——Korea, Manchuria, and China.

4. Aster rugulosus Maxim. SAWA-SHIROGIKU.   Rhizomes long-creeping; stems erect, 50–60 cm. long, simple, glabrous, slenderly branched in upper part; radical leaves withering before anthesis, the lower cauline leaves rather thick, long-petiolate, linear-lanceolate, 7–17 cm. long, about 15 mm. wide, gradually narrowed to the tip, mucronate, gradually narrowed at base, loosely toothed or subentire, rugulose above and scabrous near the slightly recurved margin, glabrous beneath, the median leaves lanceolate, 7–10 cm. long, about 1 cm. wide, sessile or short-petiolate, the upper leaves few and small; heads few, terminal, about 27 mm. across; involucral bracts 3-seriate, the outer short, broadly ovate, rounded at apex, the inner oblong, about 4.5 mm. long, rounded at apex, reddish; ligulate corollas white, 1–1.3 cm. long, 2–2.5 mm. wide, turning reddish following anthesis; achenes setulose.——Aug.–Oct. Wet places; Honshu, Shikoku, Kyushu.

5. Aster sohayakiensis Koidz. *A. ohtanus* Makino——HOSOBA-NOGIKU, KISHŪ-GIKU.   Rhizomes creeping; stems 30–60 cm. long, glabrous, branched in upper part; radical and lower cauline leaves withering before anthesis or sometimes present at anthesis, lanceolate, 9–13.5 cm. long, about 1.5 cm. wide, acuminate, gradually narrowed at base, long-petiolate, pubescent toward the margin and sometimes also on the nerves beneath, otherwise glabrous, scattered mucronate-toothed, lustrous and with impressed veins above, with raised veins beneath, the upper leaves small, linear, sessile; heads on slender peduncles, about 18 mm. across; involucres campanulate, the outer bracts 1–2 mm. long, oblong, the inner 4–4.5 mm. long, rounded to subacute; ligulate corollas 8.5–11 mm. long, 2–2.5 mm. wide.——July–Nov. Along streams, especially on rocks in mountains; Honshu (s. Kinki Distr.).

6. Aster glehnii F. Schmidt. *A. korsakoviensis* Lév. & Van't.——EZO-GOMA-NA.   Rhizomes creeping, thick; stems 1–1.5 m. long, powdery-puberulent, branched in upper part; lower leaves withering before anthesis, the median ones chartaceous, many, densely arranged, oblong, remotely short-toothed, usually glandular-dotted beneath, densely puberulent on both sides, short-petiolate, the uppermost ones much-reduced, linear; heads in dense corymbs, many, about 15 mm. across; involucres tubular, 4–5 mm. long; bracts 2 or 3 seriate, green, equal, linear, the outer acute, densely puberulent, the inner acuminate; ligulate corollas white, about 1 cm. long, 1.5–2 mm. wide; achenes hairy, glandular-dotted.——Aug.–Oct. Grassy places; Hokkaido.——Sakhalin and s. Kuriles.

Var. **hondoensis** Kitam. *A. glehnii* sensu auct. Japon., non F. Schmidt——GOMA-NA.   Leaves loosely puberulent; involucres smaller, 3–3.5 mm. long.——Grassy slopes and woods in mountains; Honshu.

7. Aster dimorphophyllus Fr. & Sav. *A. dimorphophyllus* var. *indivisus* Makino; *A. dimorphophyllus* var. *divisus* Makino——TATEYAMA-GIKU.   Rhizomes slender, creeping; stems 30–55 cm. long, slender, glabrous, slightly branched in upper part; radical leaves withering before anthesis, the lower ones thin, long-petiolate, deltoid-ovate, 3–5.5 cm. long, 2.5–5 cm. wide, acute, cordate, undulate, incised or palmately lobed, sometimes 3-sected and again 2- or 3-lobed, sometimes puberulent on both sides; cauline leaves loosely arranged, the median ones slightly larger than the lower, deltoid-ovate, acuminate, truncate to shallowly cordate at base, petiolate, the upper leaves smaller, becoming linear in the uppermost ones; heads loosely corymbose, few, on long peduncles, 2.3–3 cm. across; involucres tubular, the bracts unequal, acute; ligulate corollas few, 9–15 mm. long, 2–3 mm. wide; achenes hirsute.——July–Oct. Woods in mountains; Honshu (Izu, Sagami, and Suruga Prov.).

8. Aster fastigiatus Fisch. *Turczaninowia fastigiata* (Fisch.) DC.; *A. micranthus* Lév. & Van't.; *A. micranthus* var. *achilleiformis* Lév.; *Calimeris japonica* Schultz-Bip.——HIME-SHION.   Rhizomes short, indistinct; stems 30–100 cm. long, erect, densely puberulent and branched in the upper part; lower leaves rather thick-chartaceous, linear-lanceolate or lanceolate, 5–12 cm. long, 4–15 mm. wide, gradually narrowed at both ends, subpetiolate at base, scattered mucronate-toothed, often slightly recurved on margin, short setose-pilose near upper margin, whitish and glandular-dotted, with minute appressed hairs beneath; cauline leaves rather numerous, gradually smaller in those above, linear to linear-lanceolate, the uppermost ones 2–3 mm. long, acute, sessile; heads many, in dense corymbs, slender-pedunculate, small, 7–9 mm. across, some of the tubular florets sterile; involucres tubular, about 4 mm. long, the bracts 4-seriate, oblanceolate, obtuse, densely puberulent; ligulate corollas white, 5–6.5 mm. long, about 1 mm. wide; achenes about 1.2 mm. long, puberulent, glandular-dotted.——Aug.–Oct. Waste grounds especially along rivers in lowlands; Honshu, Shikoku, Kyushu.——Korea, Manchuria, Dahuria, and China.

9. Aster maackii Regel. *A. koidzumanus* Makino; *A. horridifolius* Lév. & Van't.——HIGO-SHION.   Rhizomes creeping; stems erect, 45–85 cm. long, many-leaved, branched in upper part, minutely rough-puberulent; radical and lower cauline leaves rather thick-chartaceous, withering before anthesis, the median ones lanceolate, 6–10 cm. long, 1–2 cm. wide, acute to acuminate, gradually narrowed and sessile at base, sometimes subtrinerved, scattered mucronate-toothed, with short rough hairs on both surfaces, the upper leaves gradually smaller, narrowly oblong, 1.5–2 cm. long, obtuse, entire; heads about 4 cm. across, showy, in loose corymbs, few, the peduncles long, very densely short-setulose, with 2 or 3 bracts, 5–10 mm. long at base; involucres 8–9 mm. long, subglobose, rounded at apex, the bracts in 3 series, scarious, purplish at tip, the outer narrowly oblong, rounded at apex; ligulate corollas 19–20 mm. long, 2–2.5 mm. wide; achenes about 2 mm. long, densely coarse-hirsute.——Aug.–Oct. Wet places in mountains; Kyushu.——Korea and Manchuria.

10. Aster tataricus L. f. *A. trinervius* var. *longifolia* Fr. & Sav.; *A. nakaii* Lév. & Van't.; *A. tataricus* var. *minor* Makino; *A. tataricus* var. *nakaii* (Lév. & Van't.) Kitam.——SHION.   Rhizomes short; stems 1–2 m. long, erect, slightly branched in upper part, loosely setulose; radical leaves withering before anthesis, spathulate-oblong, obtuse, gradually narrowed to a long winged petiole, loosely setulose on both sides, the median leaves long-petiolate, 20–35 cm. long, 6–10 cm. wide, ovate or oblong, short-acuminate, rounded or gradually narrowed to a long petiole, acutely toothed, the upper leaves gradually smaller, narrowly oblong to lanceolate, acuminate at both ends, subsessile; uppermost leaves linear, 3–5 mm. long; heads 25–33 mm. across, long-pedunculate, many, in large corymbs, the peduncles densely setulose; involucres sub-globose, about 7 mm. long, the bracts 3-seriate, lanceolate, acuminate, short-pilose, scarious on margin; ligulate corollas blue-purple, 16–17 mm. long, 3–3.5 mm. wide; achenes pilose, about 3 mm. long.——Aug.-Oct. Wet places; Honshu (Chūgoku Distr.) Kyushu.——Korea, Manchuria, n. China, Mongolia, and Siberia.

11. Aster tenuipes Makino. KURUMAGIKU. Stems about 30 cm. long, short-pubescent, much branched; radical leaves rosulate, withering before anthesis, oblanceolate, about 7 cm. long, 2.5 mm. wide, acute, narrowed at base, coarsely toothed toward tip, puberulent above, the cauline leaves many, linear-lanceolate, about 5 cm. long, 6 mm. wide, acuminate, narrowed at base, sessile, obsoletely 3-nerved, glabrous on both sides, loosely toothed; heads solitary or to 3 on each branch, about 2 cm. across, the branches slender, densely leaved, the leaves linear, 6–15 mm. long, entire, resembling involucral bracts; involucres about 7 mm. long; bracts 4-seriate, the outer ovate, the inner narrowly oblong, obtuse, 1-nerved; ligulate corollas about 9 mm. long, 1.5 mm. wide; achenes densely hairy.——Aug. Honshu (s. Kinki Distr.); sometimes cultivated in gardens.

12. Aster ageratoides Turcz. *A. trinervius* Roxb., non Gilib.——NOYAMA-KINGIKU. Very polymorphic species occurring from e. Asia to India.

Key to the varieties of *A. ageratoides*

1A. Involucres about 1 cm. across; bracts viscid at tip; leaves oblong-ovate, rounded at base, sessile.
  2A. Stems 35–65 cm. long; heads in loose corymbs. .... var. *viscidulus*
  2B. Stems 20–30 cm. long; heads solitary. .... var. *alpinus*
1B. Involucres smaller; bracts not viscid.
  3A. Leaves abruptly narrowed at base. .... var. *sugimotoi*
  3B. Leaves cuneate at base.
    4A. Marginal florets with large tubular corollas. .... var. *tubulosus*
    4B. Marginal florets with ligulate corollas.
      5A. Stems and leaves densely pubescent; cauline leaves semiclasping. .... var. *semiamplexicaulis*
      5B. Stems and leaves scattered-pubescent.
        6A. Leaves broadly auriculate-clasping at base, lanceolate, long-acuminate. .... var. *yoshinaganus*
        6B. Leaves not auriculate-clasping at base.
          7A. Heads about 3 cm. across, white; involucres campanulate-globose, 7–8 mm. long. .... var. *robustus*
          7B. Heads about 2.5 cm. across; involucres 4–5 mm. long.
            8A. Leaves abruptly narrowed on lower half; involucres tubular, about 4 mm. long; ligulate corollas white, rarely blue-purple. .... var. *harae*
            8B. Leaves gradually narrowed at base; involucres subglobose; ligulate corollas usually blue-purple.
              9A. Leaves oblong to ovate. .... var. *ovatus*
              9B. Leaves linear to lanceolate.
                10A. Leaves linear-lanceolate to lanceolate, about 10 cm. long, 15 mm. wide. .... var. *angustifolius*
                10B. Leaves linear, 3–6 cm. long, 2–5 mm. wide.
                  11A. Stems 20–90 cm. long, short-pilose. .... var. *ripensis*
                  11B. Stems 30–40 cm. long, setulose. .... var. *microcephalus*

Var. **viscidulus** (Makino) Kitam. *A. trinervius* var. *viscidulus* Makino; *A. viscidulus* (Makino) Makino——HAKONE-GIKU, MIYAMA-KONGIKU. Rhizomes indistinct; stems tufted, 35–65 cm. long, densely pubescent; median leaves many, oblong, lanceolate, or ovate-oblong, 4–7 cm. long, 15–17 mm. wide, acute to acuminate, dilated and rounded at base, subtrinerved, very sparsely mucronate-toothed, densely short-pilose on both sides, sometimes glandular-dotted beneath, short-petiolate or sessile; heads 23–25 mm. across, in loose corymbs, long-pedunculate; involucres tubular-globose, about 5 mm. long; bracts 4-seriate, the outer very short, the median oblong, rounded at apex, scarious on margin, viscid above; ligulate corollas blue-purple, 12–13 mm. long, about 2.5 mm. wide.——July–Oct. Woods in mountains; Honshu (Kantō and e. Tōkaidō Distr.).

Var. **alpinus** Koidz. *A. viscidulus* var. *alpinus* (Koidz.) Kitam.——TAKANE-KONGIKU. Stems 20–30 cm. long; heads solitary.——Alpine; Honshu (Akaishi Mts.).

Var. **sugimotoi** (Kitam.) Kitam. *A. sugimotoi* Kitam.; *A. trinervius* var. *sugimotoi* (Kitam.) Kitam.; *A. ageratoides* subsp. *sugimotoi* (Kitam.) Kitam.——AKIHAGIKU, KIYOSUMI-GIKU. Stoloniferous; stems about 50 cm. long, setulose; median leaves approximate, ovate, 5.5–8 cm. long, 3–4 cm. wide, acuminate, setulose on both sides, with raised netted veinlets, petiolate; heads about 16 mm. across; involucres tubular, about 4 mm. long, the outer bracts small, the inner obtuse; ligulate corollas about 6 mm. long, 2 mm. wide.——Oct.–Nov. Woods in mountains; Honshu (s. Kantō and Tōkaidō Distr.).

Var. **tubulosus** (Makino) Ohwi. *A. trinervius* var. *congestus* forma *tubulosus* Makino; *A. tubulosus* (Makino) Kitam.——CHOKUZAKI-YOME-NA. Stems 40–60 cm. long; cauline leaves rather thick, oblong-lanceolate, 3-nerved, acuminate, lustrous, glabrous to loosely short-pilose; heads 2–2.3 cm. across; involucres about 7 mm. long, tubular, bilabiate, purplish; ligulate corollas 10–11 mm. long, tubular, bilabiate, purplish.——Cultivated in gardens.——Forma **albiflorus** Kitam. SHIROBANA-CHOKUZAKI-YOME-NA. Flowers white.

Var. **semiamplexicaulis** (Makino) Ohwi. *Calimeris amplexifolia* Sieb. & Zucc.; *Amphyrapis japonica* Miq.; *Aster trinervius* var. *semiamplexicaulis* Makino; *A. semiamplexicaulis* (Makino) Makino; *A. ageratoides* subsp. *amplexifolius* (Sieb. & Zucc.) Kitam.——INAKAGIKU, YAMASHIROGIKU. Densely white-pubescent; cauline leaves oblong-lanceolate, 6–12 cm.

long, 1.5–3 cm. wide, acuminate, abruptly narrowed at base, semiclasping; heads about 2 cm. across; ligulate corollas about 12 mm. long, whitish, sometimes slightly bluish.——Aug.-Oct. Woods in mountains; Honshu (Tōkaidō, Kinki Distr. and westw.), Shikoku, Kyushu.

Var. **yoshinaganus** (Kitam.) Ohwi. *A. ageratoides* subsp. *yoshinaganus* Kitam.; *A. yoshinaganus* Kitam.——SHIKOKU-KONGIKU. Stems slender, often short-pilose; median leaves linear-lanceolate, 12–13 cm. long, loosely short-hairy or glabrous; heads about 15 mm. across, on slender peduncles; involucres about 3 mm. long; ligulate corollas purpurascent. ——Aug.-Oct. Shikoku.

Var. **robustus** (Koidz.) Makino & Nemoto. *A. trinervius* var. *robustus* Koidz.; *A. ageratoides* subsp. *megalocephalus* Kitam.——Ō-YAMASHIROGIKU, ŌBANA-SHIRO-YOME-NA. Stems stout, nearly glabrous; median leaves oblong, 12–15 cm. long, 3–6 cm. wide, acuminate, mucronate, gradually narrowed at base, 3-nerved, loosely mucronate-toothed; involucres 7–8 mm. long, the bracts green, broadly linear, obtuse; ligulate corollas about 13 mm. long, white.——Oct.-Nov. Honshu (n. Kantō Distr.).

Var. **harae** (Makino) Kitam. *A. trinervius* var. *harae* Makino; *A. leiophyllus* var. *purpurascens* Honda——SAGAMI-GIKU. Stems about 70 cm. long, purpurascent; leaves linear-lanceolate or lanceolate, long-acuminate, short-petiolate at base, 3-nerved; ligulate corollas purpurascent or white; involucral bracts purplish brown near tip.——Sept.-Oct. Honshu (s. Kantō Distr. and Suruga Prov.).——Forma **leucanthus** Honda. *A. leiophyllus* Fr. & Sav.; *A. leiophyllus* var. *oligocephalus* Nakai——YAMASHIROGIKU, SHIRO-YOME-NA. Stems about 1 m. long; leaves oblong-lanceolate, abruptly contracted and narrowed to the base, coarsely acute-toothed, green, lustrous, slightly scabrous above; heads many, small; ligulate corollas white; style-branches slender, oblong-lanceolate.—— Commonest phase in Honshu, Shikoku, Kyushu.——Formosa. ——Forma **stenophyllus** (Kitam.) Ohwi. *A. ageratoides* var. *stenophyllus* Kitam.——HOSOBA-NO-SHIRO-YOME-NA. Leaves linear-lanceolate, about 15 cm. long, 1–1.3 cm. wide, long-acuminate, abruptly narrowed to the base.——Honshu (s. Kinki and Tōkaidō Distr.).——Forma **sawadanus** (Kitam.) Ohwi. *A. ageratoides* var. *sawadanus* Kitam.—— KINTOKI-SHIRO-YOME-NA. Stems short, densely pubescent; leaves pubescent; heads few; involucres small; ligulate corollas white; achenes densely hirsute.——Honshu (Sagami and Kai Prov.).——Forma **ovalifolius** (Kitam.) Ohwi. *A. ageratoides* var. *ovalifolius* Kitam.——TAMABA-SHIRO-YOME-NA. Cauline leaves ovate, 10–12 cm. long, 4.5–6 cm. wide, acute, rounded-cuneate at base, coarsely incised-toothed, sessile; heads dense; ray-flowers white; style-branches deltoid, acute. ——Honshu (centr. and n. distr.).——Forma **tenuifolius** (Kitam.) Ohwi. *A. ageratoides* var. *tenuifolius* Kitam.—— NAGABA-NO-SHIRO-YOME-NA. Leaves very thin, linear-lanceolate, 15–20 cm. long, 1.5–3 cm. wide, long-acuminate, sparsely incised-dentate; flowers white.——Honshu and Shikoku.

Var. **ovatus** (Fr. & Sav.) Nakai. *A. trinervius* var. *ovata* Fr. & Sav.; *A. trinervius* var. *congesta* Fr. & Sav.; *A. ageratoides* var. *adustus* sensu auct. Japon., non Maxim.; *A. ageratoides* subsp. *ovatus* (Fr. & Sav.) Kitam.——NO-KONGIKU. Rhizomes creeping; stems 50–100 cm. long, stout, often much branched, the branches densely setulose; cauline leaves oblong or ovate, 6–12 cm. long, 3–5 cm. wide, acute to obtuse, cuneate to rounded at base, setulose on both sides, the upper leaves

oblong to lanceolate; heads blue-purple or purple, about 2.5 cm. across; involucres herbaceous; bracts 3-seriate, 4.5–5 mm. long, greenish, broadened at tip and purplish, 1-nerved, short-pilose; achenes 1.5–3 mm. long, hirsute, the pappus 4–6 mm. long, whitish to reddish.——Aug.-Oct. Mountains and hills; Honshu, Shikoku, Kyushu; very common.——Forma **humilis** (Nakai) Ohwi. *A. ageratoides* var. *humilis* Nakai—— KOMACHIGIKU. Stems much branched, 15–20 cm. long; leaves thicker, about 4.5 cm. long, obtuse, cuneate-acuminate at base, sessile; heads many in pyramidal corymbs, the peduncles stouter.——Forma **vernalis** (Honda) Ohwi. *A. ageratoides* var. *vernalis* Honda——HARU-NO-KONGIKU. Stems short; leaves acute at both ends; spring flowering.——Honshu (Suruga Prov. and Izu Isl.).——Forma **littoricola** (Kitam.) Ohwi. *A. ageratoides* var. *littoricola* Kitam.——HAMA-KON-GIKU. Leaves broader, obtuse.——Near seashores.——Forma **yezoensis** (Kitam. & Hara) Ohwi. *A. ageratoides* var. *yezoensis* Kitam. & Hara——EZO-NO-KONGIKU. Median leaves abruptly narrowed on lower half.——Hokkaido.——Cv. **Hortensis**. *A. trinervius* forma *hortensis* Makino——KONGIKU. Heads larger, deep blue-purple. Cultivated.

Var. **angustifolius** Kitam. *A. ageratoides* subsp. *angustifolius* (Kitam.) Kitam.; *A. mayasanensis* Kitam.——HOSOBA-KONGIKU. Stems 50–100 cm. long; median leaves linear-lanceolate, 8–10 cm. long, 1–1.5 cm. wide, narrowed at both ends; heads few, 2–2.5 cm. wide, showy; involucres subglobose, about 5.5 mm. long; bracts 2-seriate, the outer about 3 mm. long, obtuse, green, the inner purplish, the pappus whitish.——Aug.-Nov. Ravines and valleys; Honshu (Kinki Distr. and westw.), Shikoku, Kyushu.

Var. **ripensis** (Makino) Ohwi. *A. microcephalus* var. *ripensis* Makino; *A. ageratoides* subsp. *ripensis* (Makino) Kitam.——TANIGAWA-KONGIKU. Stems 20–90 cm. long, short-pilose, much branched in upper part; median leaves rather firm, linear-lanceolate, 3–6 cm. long, 2–5(–10) mm. wide, 1-nerved, acuminate, gradually narrowed to a sessile base; heads long-pedunculate, blue-purple; ligulate corollas 11–13 mm. long.——Aug.-Nov. Ravines in mountains; Honshu (s. Kinki Distr.), Shikoku, Kyushu.

Var. **microcephalus** (Miq.) Ohwi. *A. ageratoides* subsp. *microcephalus* (Miq.) Kitam.; *A. microcephalus* (Miq.) Fr. & Sav.; *Calimeris microcephala* Miq.; *A. trinervius* var. *microcephalus* (Miq.) Makino——SEMBONGIKU, MISUGIKU. Rhizomes short-creeping; stems tufted, slender, erect, 30–40 cm. long, branched in upper part, setulose; median leaves many, approximate, linear, 3.5–6 cm. long, 3.5–5 mm. wide, acuminate, narrowed at base, mucronate, toothed, thinly setulose above, 1-nerved, nearly glabrous beneath; involucres subglobose, about 3.5 mm. long; bracts 3-seriate, linear, obtuse, 1-nerved, puberulent, green; ligulate corollas about 9 mm. long, about 1.2 mm. wide, pale blue.——Oct.-Nov. Ravines; Honshu.

13. Aster kantoensis Kitam. *A. altaicus* sensu auct. Japon., non Willd.; *A. hispidus* var. *isochaeta* Fr. & Sav., excl. syn.——KAWARA-NOGIKU. Stems about 50 cm. long, with short rigid hairs, densely leaved, much branched in upper part; radical and lower cauline leaves withering before anthesis, the median ones linear, 6–7 cm. long, 3–5 mm. wide, 1(–3)-nerved, obtuse, long-attenuate, petiolate, entire, short-setulose on margin and on underside, the upper leaves smaller, narrowly linear; heads 3.5–4 cm. across, in panicles, long-pedunculate; involucres 7–16 mm. long; bracts 2-seriate, broadly linear, equal, acuminate, scarious on margin, short-setulose on back; ligulate corollas about 17 mm. long, 3 mm. wide; pappus brownish red.——Oct.–Nov. Waste grounds along rivers; Honshu (sw. Kantō Distr. and Suruga Prov.).

14. **Aster spathulifolius** Maxim. *A. fauriei* Lév. & Van't.; *A. feddei* Lév. & Van't.; *A. oharae* Nakai; *Heteropappus oharae* (Nakai) Kitam.; *Erigeron oharae* (Nakai) Batsch; *E. feddei* (Lév. & Van't.) Batsch——DARUMA-GIKU. Stems slightly woody, about 25 cm. long, branched from base, densely hirsute; lower cauline leaves thick, rosulate while young, withering before anthesis, obovate or orbicular, 3–9 cm. long, 1.5–5.5 cm. wide, rounded at apex, cuneate at base, entire or obtusely toothed, densely puberulent on both sides, winged-petiolate or sessile, the upper leaves very short on the peduncles, spathulate, 12–16 mm. long, becoming bractlike below the heads; heads 35–40 mm. across, solitary and terminal on long peduncles; involucres subglobose, 8–9 mm. long; bracts 3-seriate, linear, equal, acuminate, pilose; ligulate corollas blue-purple, 13–16 mm. long, 2.5–3 mm. wide; achenes oblong, the pappus whitish.——July–Nov. Rocks near seashores; Honshu (Chūgoku Distr.), Kyushu.——Korea.

15. **Aster pseudoasagrayi** Makino. ISO-KANGIKU, KAN-YOME-NA. Perennial; stems radiately much branched, slender, many-leaved, incurved-setulose, ascending above; lower leaves withering before anthesis, the median leaves rather thick, spreading, ovate-spathulate, 3–10 cm. long, 5–9 mm. wide, rounded at apex, entire, incurved short-setulose on both sides, deep green, more or less lustrous above, petiolate, the uppermost leaves subtending the heads, very small, bractlike; heads many, showy, terminal, 3.5–4 cm. across; involucres subglobose, about 6 mm. long; bracts 3-seriate, the outer gradually smaller, rounded at apex, the median oblong, obtuse, loosely pilose; ligulate corollas purplish, about 14 mm. long, 2.5–3 mm. wide; achenes obovate, the pappus reddish purple.——Nov.–Dec. Rarely cultivated. Presumably indigenous, although it has not been found in the wild.

16. **Aster tripolium** L. *A. macrolephus* Lév. & Van't.; *A. tripolium* var. *integrifolium* Miyabe & Kudo——URAGIKU, HAMA-SHION. Annual, or sometimes a short-lived perennial, without distinct rhizomes; stems 25–55 cm. long, glabrous, erect, branched in upper part, reddish at base; lower and median leaves slightly fleshy, linear-lanceolate, 6.5–10 cm. long, 6–12 mm. wide, acuminate, semiclasping, entire or with short mucronate-teeth, glabrous, the upper leaves linear, bractlike; heads loosely or densely corymbose, 16–22 mm. across; involucres tubular, about 7 mm. long and as wide; bracts 3-seriate, membranous, loose, the outer lanceolate, 2.5–3 mm. long, obtuse, loosely ciliolate, the inner narrowly oblong, about 2 mm. wide, rounded at apex, purplish; ligulate corollas 12–16 mm. long, about 2 mm. wide, purplish; pappus about 5 mm. long at anthesis, 14–16 mm. long in fruit.——Aug.–Nov. Wet places near seashores; Hokkaido, Honshu, Shikoku, Kyushu.——Sakhalin, Korea, n. China, Manchuria, and e. Siberia.

23. MYRIACTIS Less.   HIME-KIKU-TABIRA-KO ZOKU

Simple or branched perennials; leaves alternate, toothed or lyrately pinnate; heads small, solitary, long-pedunculate; receptacle convex, alveolate; involucres subglobose, the bracts membranous, 2-seriate, equal; ligulate florets fertile, 3- to 5-seriate, the corollas white or reddish; tubular florets campanulate, bisexual, many, fertile, yellow, the anthers obtuse and entire at base; style-branches flat, acute, mammillate; achenes oblong, flat, nerved on margin or 1-ribbed on the face, narrowed at both ends, crowned by a ring, bearing glands at apex; pappus absent. ——Few species in mountains of s. Asia and Japan.

1. **Myriactis japonensis** Koidz. *Solenogyne japonensis* (Koidz.) Masam.——HIME-KIKU-TABIRA-KO. Rhizomes slender; stems scapose, 3–12 cm. long, densely white-woolly at base, puberulent above, simple or with 1 or 2 branches at base; radical leaves rosulate, oblanceolate-oblong, 2.5–4 cm. long, 7–14 mm. wide, pinnately cleft, the lateral lobes 1 to 4 pairs, spreading, unequal, obtuse, toothed, rather densely pubescent on both surfaces, the terminal lobe orbicular, the cauline leaves 2 or 3, small, narrow, entire or toothed; heads on long peduncles, solitary; involucres about 2.5 mm. long, 5 mm. across, the bracts 2-seriate, equal, oblong, obtuse, puberulent; ligulate corollas about 1 mm. long, 0.5 mm. wide; achenes about 2 mm. long.——July–Sept. High mountains; Kyushu (Yakushima).

24. RHYNCHOSPERMUM Reinw.   SHŪBUN-SŌ ZOKU

Erect branched perennial often with spreading branches; leaves lanceolate, loosely toothed; heads very small, in racemes or spikes; involucres campanulate, the bracts unequal, obtuse, with thinly membranous margins; ligulate florets fertile, in 2 series, the corolla rather thick, whitish; disc florets bisexual, fertile, the corolla tubular, 5-cleft; anthers obtuse at base; style-branches flat, deltoid at apex; achenes flat, narrowed to both ends, nerved on margin, nerveless on the face, beaked in those of the ligulate florets, beakless in the disc florets; pappus absent or of few slender bristles.——One species, in e. Asia, Malaysia, and India.

1. **Rhynchospermum verticillatum** Reinw. ex Bl. *R. formosanum* Yamamoto; *Leptocoma racemosa* Less.; *R. verticillatum* var. *subsessilis* Oliv. ex Miq.——SHŪBUN-SŌ. Rhizomes short; stems 50–100 cm. long, mealy-puberulent, usually divaricately branching in upper part; leaves membranous, the lower oblanceolate or oblong-oblanceolate, 7–15 cm. long, 17–32 mm. wide, acute, narrowed to the petiole at base, shallowly undulate-toothed on upper half, with short setulose hairs on both sides, the middle leaves many, densely arranged, lanceolate, acuminate at both ends, gradually smaller in the upper; heads terminal or axillary; involucres broadly campanulate, about 2.5 mm. long; bracts in 3 series, oblong; ligulate florets in 2 series, the corollas whitish, 1.2 mm. long, 0.5 mm. wide; achenes of ligulate florets beaked, flat, about 4 mm. long, those of the disc florets beakless.——Aug.–Nov. Woods in hills and low mountains; Honshu (Kantō Distr.), Shikoku, Kyushu.——s. Korea, Ryukyus, Formosa, Malaysia, and India.

25. LAGENOPHORA Cass. KOKE-SEMBONGIKU ZOKU

Small delicate perennials; stems scapiform; leaves obovate to oblanceolate, pinnately cleft to toothed; heads terminal, small; involucres subglobose, the bracts 2- to 4-seriate, unequal, scarious on margin; receptacles flat; ligulate florets pistillate, fertile, 2- or 3-seriate, the corolla strap-shaped or tubular, white; tubular florets few, bisexual, sterile, the anthers obtuse, entire at base; style-branches flat, acuminate, mammillate; achenes much flattened, nervelike on margin, not ribbed on the face, glabrous, short-beaked; pappus wanting.——About 10 species, in Australia, Asia, S. America, and Pacific Islands.

1. **Lagenophora stipitata** (Labill.) Druce var. **microcephala** (Benth.) Domin. *L. billardieri* var. *microcephala* Benth.; *L. crepidioides* Koidz.; *L. billardieri* sensu auct. Japon., non Cass.——KOKE-SEMBONGIKU. Rhizomes slender; scapes 3.5–12 cm. long, simple, slender, puberulent especially toward the top; radical leaves rosulate, obovate or broadly spathulate, 12–30 mm. long, 7–13 mm. wide, rounded at apex, cuneately narrowed to the petiole, shallowly undulate, mucronate-toothed, densely pubescent, the cauline leaves few, 1–1.5 mm. long, linear; heads scapose, solitary; involucres about 2.5 mm. long, glabrous; bracts 3- or 4-seriate, the outer lanceolate, about 1 mm. long, obtuse, the inner linear, obtuse, purplish; ligulate florets 3-seriate, the corolla about 2.5 mm. long, about 0.5 mm. wide; achenes about 2.5 mm. long, glandular-dotted.——July–Oct. Honshu (Aki Prov.), Kyushu.——Ryukyus and Formosa. The typical phase occurs in the Pacific Islands.

26. CONYZA L. IZU-HAHA-KO ZOKU

Herbs usually grayish hairy; leaves alternate, toothed, sometimes pinnately cleft; involucres campanulate, the bracts 3- or 4-seriate, scarious, linear-lanceolate, unequal, acuminate; receptacles subglobose, alveolate; ligulate florets pistillate, many-seriate, fertile, the corollas capillary, shorter than the style, lacerate at apex; tubular florets few, bisexual, fertile, the corollas 5-lobed; anthers obtuse, entire at base; style-branches flat, obtuse, mammillate; achenes very small, flat, narrowed toward the ends, with nervelike margins, nerveless on the face, hairy; pappus whitish or reddish, the bristles nearly equal, 1-seriate, slender, scabrous or short-pilose.——About 50 species, widely distributed in the Tropics and subtropics, few in temperate regions.

1. **Conyza japonica** (Thunb.) Less. *Erigeron japonicum* Thunb.——IZU-HAHA-KO, WATA-NA, YAMA-JIŌGIKU. Annual or biennial, grayish soft-villous; stems 25–55 cm. long, often branched in upper part; lower leaves oblong, 5–13 cm. long, 1.2–4 cm. wide, rounded at both ends, or cuneately narrowed at base, obtusely toothed, the median leaves clasping, the upper ones smaller, lanceolate; heads densely arranged, terminal; involucres about 5.5 mm. long, the bracts 3-seriate, narrow; ligulate florets many, capillary, indistinct; achenes flattened, about 1 mm. long, pale brown; pappus 4.5 mm. long, whitish to brownish red.——Apr.–June. Sunny places; Honshu (sw. Kantō Distr. and westw.), Shikoku, Kyushu.——Ryukyus, Formosa, China, India, and Malaysia.

27. DICHROCEPHALA L'Hérit. BUKURYŌ-SAI ZOKU

Branched puberulent annuals; leaves alternate, pinnately cleft; heads small, in racemes; involucres depressed-cupuliform, the bracts 1-seriate, ovate, acuminate, scarious; receptacles convex, globose, alveolate; ligulate florets 7- or 8-seriate, fertile, the corollas tubular, white; tubular florets bisexual, fertile, 5-lobed; anthers obtuse, entire at base; style-branches flat, deltoid at apex, acute; achenes broadly oblanceolate, flat, nervelike on margin, with 2 nerves on each face, glandular-dotted; pappus wanting; disc of ovary with 2 white bristles in fruit.——About 5 species, in s. Asia and Africa.

1. **Dichrocephala bicolor** (Roth) Schltdl. *Cotula bicolor* Roth; *Grangea latifolia* Lam.; *D. latifolia* (Lam.) DC.——BUKURYŌ-SAI, BUKURYŪ-SAI. Roots narrowly fusiform; stems erect, branched, 20–35 cm. long, densely puberulent; lower leaves thin, lyrate, 9–13 cm. long, 4–5 cm. wide, puberulent on both sides, pinnately cleft, the terminal lobes ovate, obtusely toothed, the lateral lobes 1 or 2 pairs, distant, obovate or oblong, spreading, the median and upper leaves gradually smaller, the uppermost narrowly oblong, 1–2 cm. long, undivided; heads in racemes, about 2.5 mm. across at anthesis, 4 mm. long in fruit, long-pedunculate; involucral bracts about 0.8 mm. long, ovate, acuminate; ligulate corollas 0.5 mm. long, glandular-dotted.——Apr.–Nov. Clay soils; Honshu (Hachijō Isl. in Izu Prov.), Shikoku, Kyushu.——Ryukyus, Formosa, s. China, Malaysia, India, and Africa.

28. PETASITES Hill FUKI ZOKU

Dioecious white-pubescent or white-woolly perennial herbs; rhizomatous; leaves radical, large, cordate or reniform; scapiform stems with alternate parallel-nerved scalelike leaves; heads sometimes with ligulate flowers, in a raceme or in a thyrse; involucres campanulate, with small scalelike leaves at base, the bracts equal, 1-seriate; receptacle flat; ligulate corollas filiform and truncate or very small; florets of the pistillate heads fertile; tubular florets in the nonfruiting heads with sterile bisexual flowers; pappus snow-white in pistillate florets, the bristles elongate, slender, many, minutely scabrous, not developed in the bisexual florets.——About 20 species, in the N. Hemisphere.

1. **Petasites japonicus** (Sieb. & Zucc.) Maxim. *Nardosmia japonica* Sieb. & Zucc.——FUKI. Scapes 5–45 cm. long; radical leaves after anthesis long-petiolate, reniform-orbicular, 15–30 cm. wide, short mucronate-toothed, with crisped pubescence above and cobwebby tomentum beneath while young; heads in corymbs; involucres 5–6 mm. long, slightly accrescent after anthesis in the fertile ones; achenes about 3.5 mm. long, glabrous; pappus about 12 mm. long.——

Apr.–May. Moist woods and thickets along streams and ravines in hills and mountains; Honshu, Shikoku, Kyushu; frequently cultivated as a vegetable.——Korea, China, and Ryukyus.

Var. **giganteus** (F. Schmidt) Nichols. *P. giganteus* F. Schmidt ex Trautv., non Fuss; *P. amplus* Kitam.; *P. japonicus* subsp. *giganteus* (F. Schmidt) Kitam.——AKITABUKI. Plant larger; leaves to 1.5 m. across, the petioles to 2 m. long.—— Hokkaido, Honshu (n. distr.).——Sakhalin to s. and centr. Kuriles.

29. ERECHTITES Raf.   TAKEDAGUSA ZOKU

Coarse erect annuals; leaves alternate, simple, often hairy; heads many-flowered, tubular, in paniculate corymbs; involucres cylindric, with subtending bractlets at base; involucral bracts 1-seriate, linear, acute; receptacle naked; florets all fertile, the marginal pistillate with a slender corolla; achenes oblong, narrowed toward the apex; pappus copious, with very fine, soft white hairlike bristles.——America and Australia.

1A. Stems coarsely pubescent; leaves irregularly short-toothed. .... 1. *E. hieracifolia*
1B. Stems nearly glabrous; leaves irregularly double-toothed. .... 2. *E. valerianaefolia*

1. Erechtites hieracifolia (L.) Raf. *Senecio hieracifolius* L.——DANDO-BOROGIKU. Annual, 50–100 cm. high; stems soft, striate, coarsely pubescent; leaves lanceolate to oblong, acute, irregularly short-toothed; heads rather many, yellowish to brownish green, about 15 mm. long; involucral bracts linear; florets greenish, with a very slender tubular corolla 1–1.7 cm. long; pappus white, about 14 mm. long.—— Nov.–Dec. Naturalized on deforested mountain slopes.—— Introduced from N. America.

2. Erechtites valerianaefolia DC. TAKEDAGUSA. Allied to the preceding but differing in the nearly glabrous stems, irregularly doubly toothed leaves, purplish heads, and pale rose-purple tips of the pappus-bristles.——Naturalized; introduced from S. America.

30. ARNICA L.   USAGIGIKU ZOKU

Perennial herbs with simple or branched stems; leaves opposite, toothed, the upper ones sometimes alternate; heads relatively large, usually ligulate, sometimes eligulate; involucres tubular or subglobose, without bracteate leaves or caliculus at base, the bracts herbaceous, 2-seriate, nearly equal, acute; receptacle hairy; ligulate florets yellow; tubular florets bisexual, the corolla 5-toothed; anthers obtuse at base; achenes cylindric, truncate, narrowed at base, hairy, striate; pappus-bristles 1-seriate, rather firm, short-pilose.——More than 100 species, mainly in N. America, few in the cooler regions of the Old World.

1A. Heads eligulate. .... 1. *A. mallotopus*
1B. Heads ligulate.
  2A. Heads solitary; stems and leaves prominently pubescent; cauline leaves 3- to 5-paired. .... 2. *A. unalascensis*
  2B. Heads to 5 in the inflorescence; stems and leaves nearly glabrous; cauline leaves 10- to 20-paired. .... 3. *A. sachalinensis*

1. Arnica mallotopus (Fr. & Sav.) Makino. *Mallotopus japonicus* Fr. & Sav.——CHŌJIGIKU, KUMAGIKU. Rhizomes creeping; stems tufted, 20–85 cm. long, nearly glabrous, densely crisped-pubescent in upper part, loosely branched; radical leaves scalelike, the lower cauline ones withering before anthesis, the median leaves thick, oblong-lanceolate, 7–12.5 cm. long, 12–28 mm. wide, acuminate, sessile and short-sheathing at base, irregulary mucronate-toothed, more or less scabrous on both sides or nearly glabrous, the upper leaves slightly smaller, lanceolate, 7–23 mm. long, the uppermost alternate, small; heads 5–20, in corymbs, 12–20 mm. across, nodding, on long slender densely white-villous peduncles; involucres tubular, about 1 cm. long, 1–1.5 cm. across, the bracts nearly glabrous, lanceolate, without ligulate florets; achenes 4–5.5 mm. long; pappus brown, 6–7 mm. long.——Aug.–Oct. Wet slopes in high mountains; Honshu, Shikoku.

2. Arnica unalascensis Less. EZO-USAGIGIKU. Rhizomes long-creeping; stems 12–35 cm. long, simple, leafy at base, naked on upper half, with a single flower at apex, prominently crisped-pubescent; radical leaves small, spathulate, often withering before anthesis, the lower cauline leaves chartaceous, large, usually opposite, spathulate, 6–12 cm. long, 2–3 cm. wide, obtuse, narrowed at base to a sheathless petiole, loosely toothed, puberulent, the median leaves somewhat smaller, ovate-lanceolate, obtuse, sessile; heads solitary, long-pedunculate, erect, 4–5.5 cm. across; involucres subglobose, 9–15 mm. long, the bracts 20–28, 2-seriate, lanceolate; ligulate corollas 15–24 mm. long, 5–7 mm. wide; tubular corollas glabrous; achenes 4–4.5 mm. long; pappus brownish, 5.5–7 mm. long. ——July–Aug. Alpine slopes; Hokkaido, Honshu (centr. and n. distr.).——Kuriles, Kamchatka, and Aleutians.

Var. **tschonoskyi** (Iljin) Kitam. & Hara. *A. tschonoskyi* Iljin; *A. unalascensis* sensu auct. Japon., non Less.——USAGIGIKU, KINGURUMA. Corollas slightly pilose.——Often growing with the typical phase.

3. Arnica sachalinensis (Regel) A. Gray. *A. chamissonis* sensu F. Schmidt non Less.; *A. chamissonis* var. *sachalinensis* Regel——Ō-USAGIGIKU, KARAFUTO-KINGURUMA. Rhizomes long-creeping; stems 36–56 cm. long, stout, glabrous, short-branched in upper part; leaves opposite, united at base into a short sheath, the radical and lower cauline leaves withering before anthesis, the median leaves thick-chartaceous, lanceolate, 9–13 cm. long, 13–28 mm. wide, acuminate, narrowed at base to a short sheath about 5 mm. long, mucronate-toothed; setose-ciliolate, glabrous on both sides, the upper leaves rather small, ovate-lanceolate; heads of the inflorescence about 5, solitary on the branches, to 6.5 cm. across, nodding in anthesis, erect in fruit, long-pedunculate, densely brownish crisped-hairy; involucres subglobose, 17–20 mm. long, 3–4 cm. across, the bracts about 12, 2-seriate, oblong-lanceolate, glabrous, short-ciliate; ligulate florets 1-seriate, the corolla about 3 cm. long, about 6 mm. wide, densely pilose on the tube; achenes about 7 mm. long; pappus about 1 cm. long, reddish brown.——June–Sept. Reported to occur in Hokkaido (Ishikari Prov.).—— Sakhalin and Ochotsk Sea region.

31. EMILIA Cass.   Usu-beni-nigana Zoku

Glaucous annuals; leaves alternate, approximate toward the base of the stems, often pinnatifid, clasping at base; heads long-pedunculate, nodding before anthesis, without ligulate florets; involucres tubular, bractless at base, the bracts 1-seriate, equal, elongate after anthesis; receptacle glabrous; flowers equal, bisexual, fertile, the corollas capillary, 5-lobed; anthers with a narrow appendage at apex, obtuse at base; style-branches appendaged, short-strigose at apex; achenes 5-angled, truncate at both ends; pappus-bristles white, many, slender, scabrous.——Tropics of Asia and Africa; naturalized in America.

1. **Emilia sonchifolia** (L.) DC. *Cacalia sonchifolia* L.; *Senecio sonchifolius* (L.) Moench——Usu-beni-nigana. Nearly glabrous, glaucous annual; stems 20–45 cm. long, branched in upper part; lower leaves approximate, 5–10 cm. long, 25–65 mm. wide, lyrately pinnatifid, loosely short-pubescent, usually purplish beneath, winged-petiolate, clasping at base, the median leaves loose, somewhat smaller, ovate-lanceolate, acute, sagittate at base, entire or irregularly serrulate, sessile; heads 2–5, terminal, 8 mm. long, accrescent, to 14 mm. long, in corymbs; involucres tubular, the bracts broadly linear, acute; corolla pale rose; achenes about 3 mm. long, setulose on the ribs; pappus about 8 mm. long.——Aug.–Oct. Waste grounds; Honshu (Kinki Distr. and westw.), Shikoku, Kyushu.——Ryukyus, Formosa, China, and the Tropics of Asia and Africa.

Emilia sagittata (Vahl) DC. *E. flammea* Cass.; *Cacalia sagittata* Vahl; *C. coccinea* Sims——Beni-nigana. Flowers red or orange-yellow.——Often cultivated in gardens.——India.

32. FARFUGIUM Lindl.   Tsuwabuki Zoku

Evergreen perennial herbs; radical leaves tufted, involute in bud, long-hairy while young, cordate, reniform, or flabellate, long-petiolate, the petioles with a dilated base; cauline leaves bractlike; heads in corymbs, ligulate; involucres tubular, with subtending bracteate leaves, the bracts 1-seriate, equal, free; receptacle toothed around the pits; ligulate flowers 1-seriate; tubular flowers bisexual, many, fertile; achenes terete, slightly narrowed at the ends, densely hairy; pappus-bristles persistent, bristle-like, many, short-pilose.——Few species, in e. Asia.

1A. Leaves reniform, rounded at apex, mucronate-toothed or nearly entire. .... 1. *F. japonicum*
1B. Leaves ovate-cordate, abruptly acute to short-acuminate, doubly acute-toothed. .... 2. *F. hiberniflorum*

1. **Farfugium japonicum** (L.) Kitam. *Tussilago japonica* L.; *Arnica tussilaginea* Burm.; *Senecio kaempferi* DC.; *Ligularia kaempferi* (DC.) Sieb. & Zucc.; *L. tussilaginea* (Burm.) Makino——Tsuwabuki. Grayish brown-woolly or nearly glabrous perennial with thick rhizomes; radical leaves thick, long-petiolate, reniform, 4–15 cm. long, 6–30 cm. wide, semi-evergreen, lustrous above, mucronate-toothed or nearly entire; scapes 30–75 cm. long, bracteate; heads loosely corymbose, 4–6 cm. across, yellow, pedunculate; involucres 12–15 mm. long; ligulate corollas 3–4 cm. long, about 6 mm. wide; achenes 5–6.5 mm. long; pappus 8–11 mm. long, dark brown.——Oct.–Dec. Near seashores; Honshu (Iwaki and Echizen Prov. and westw.), Shikoku, Kyushu; common and frequently cultivated in gardens.

Var. **giganteum** (Sieb. & Zucc.) Kitam. *Ligularia tussilaginea* var. *gigantea* (Sieb. & Zucc.) Makino; *L. gigantea* Sieb. & Zucc.; *Senecio sieboldii* Schultz-Bip.——Ō-tsuwabuki. Larger spontaneous phase.——Kyushu. Sometimes cultivated in gardens.

2. **Farfugium hiberniflorum** (Makino) Kitam. *Ligularia hiberniflora* Makino; *Senecio hiberniflorus* Makino, in syn.——Kan-tsuwabuki. Rhizomes stout; radical leaves somewhat fleshy, cordate or ovate-cordate, 5.5–20 cm. long, 7.5–20 cm. wide, abruptly acute to acuminate, cordate to subtruncate at base, irregularly double-toothed, gray-woolly beneath, long-petiolate; scapes branched in upper part, 25–42 cm. long, densely grayish white-woolly, loosely scaly-leaved; heads about 3 cm. across, in corymbs, yellow, pedunculate; involucres broadly campanulate, about 1 cm. long, the bracts purplish; ligulate corollas 17–18 mm. long; pappus snow-white, 6.5–7 mm. long.——Sept.–Dec. Thickets and woods in low mountains; Kyushu (Yakushima and Tanegashima).

33. LIGULARIA Cass.   Me-takara-kō Zoku

Perennial herbs; leaves usually cordate, reniform, or ovate-oblong, petiolate, the cauline leaves few, the uppermost bractlike, alternate, sheathed at base; heads relatively large, in corymbs or racemes, the peduncles often bracteate; involucres tubular or campanulate, the bracts equal, 1-seriate, of two forms, sometimes connate; ligulate florets pistillate, the corollas yellow; tubular florets bisexual, fertile, the corolla tubular; style branches obtuse; achenes terete, striate, beakless, glabrous; pappus elongate or short, scabrous or pilose.——Many species, in Eurasia.

1A. Radical leaves erect, glaucous, ovate-oblong or oblong.
  2A. Involucral bracts free. .... 1. *L. fauriei*
  2B. Involucral bracts connate. .... 2. *L. angusta*
1B. Radical leaves spreading or ascending, green, as long as broad or broader than long.
  3A. Ligulate florets absent or 1–3; involucres narrowly tubular; bracts 5. .... 3. *L. stenocephala*
  3B. Ligulate florets 5–10; involucres tubular-campanulate; bracts 7–13.
    4A. Involucres 11–12 mm. long, 5–14 mm. across, bracteate, the pedicels bracteate.
      5A. Involucres 8–10 mm. across; pappus as long as the corolla. .... 4. *L. sibirica*
      5B. Involucres 5–14 mm. across; pappus much shorter than the corolla. .... 5. *L. hodgsonii*
    4B. Involucres 16–25 mm. long, 16–18 mm. across, bractless, the pedicels bractless.
      6A. Leaves palmately parted, strongly revolute while young; pappus 6–7.5 mm. long. .... 6. *L. japonica*
      6B. Leaves merely toothed, revolute in bud; pappus about 12 mm. long. .... 7. *L. dentata*

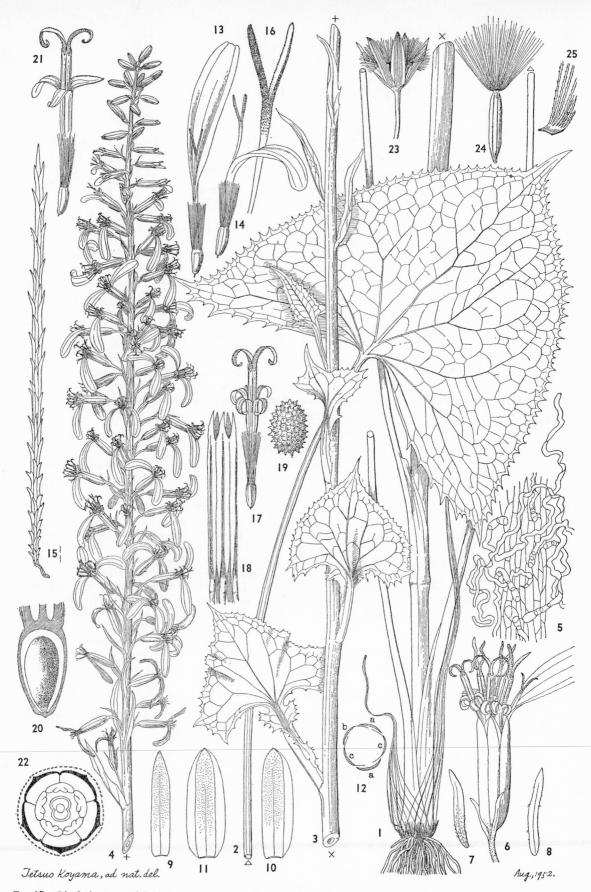

Tetsuo Koyama, ad nat. del.   Aug., 1952.

Fig. 17.—*Ligularia stenocephala* (Maxim.) Matsum. & Koidz. 1, 2, 3, Plant showing basal leaf and scape; 4, inflorescence; 5, hairs on axis of inflorescence; 6, flower head; 7, 8, bractlets; 9, 10, 11, involucral bracts; 12, arrangement of involucral bracts; 13, 14, ligulate florets; 15, pappus-bristle; 16, stigma branches; 17, tubular floret; 18, anthers; 19, pollen; 20, longitudinal section of ovary; 21, anomalous tubular floret; 22, floral diagram; 23, head in fruiting stage; 24, achene; 25, basal part of pappus. Specimen from Umagaeshi, Nikko, Honshu.

1. **Ligularia fauriei** (Franch.) Koidz. *Senecio fauriei* Franch.; *Senecillis fauriei* (Franch.) Kitam.——MICHI-NO-KU-YAMA-TABAKO. Rhizomes short; stems erect, 60–100 cm. long, glabrous; radical leaves chartaceous, ovate-oblong or oblong, 10–25 cm. long, 7–13.5 cm. wide, rounded, truncate at base or cuneately narrowed to long winged petioles, loosely mucronate-toothed, glaucous on both sides; cauline leaves 3–4, the lower ones larger, erect, about 20 cm. long, 7 cm. wide, sessile or with a winged petiole, the petiole sheathing, the median and upper leaves narrowly oblong, sessile, the upper-most bracteate, lanceolate or broadly so; racemes to 20 cm. or more long, centripetal, the pedicels about 7 mm. long, with a minute bract near the middle; heads many, nodding in anthesis; involucres tubular, about 6 mm. long, about 3 mm. across, glabrous, the bracts free, 5; ligulate florets 2 or 3, the corollas about 16 mm. long, about 3 mm. wide; pappus about 3 mm. long, rusty brown.——June. Honshu (Shimotsuke, Iwaki and northw.).

2. **Ligularia angusta** (Nakai) Kitam. *Cyathocephalum angustum* Nakai; *L. schmidtii* sensu auct. Japon., non Maxim. ——YAMA-TABAKO. Rhizomes short; stems slender, 1–1.3 m. long, glabrous, glaucous, usually simple; radical leaves chartaceous, erect, ovate-oblong, 17–30 cm. long, 13–19 cm. wide, rounded, truncate at base and cuneately narrowed to a partially winged petiole, undulate-toothed, glaucous on both sides; cauline leaves 3, the lower ones oblong, erect, 15–23 cm. long, about 10 cm. wide, clasping and somewhat sheathing at base, the median and uppermost ones smaller; racemes simple or rarely branched, about 30 cm. long, centripetal; heads nodding in anthesis, the pedicels 5–10 mm. long, with a small bract at the middle; involucres 6–8 mm. long, 3–3.5 mm. across, tubular, glaucous, connate on lower half, blackish at tip; ligulate florets 3(–5), the corollas about 17 mm. long, about 3 mm. wide; achenes 4.5–6 mm. long, the pappus about 2 mm. long, rusty brown.——May–June. Mountains; Honshu (Kantō and centr. distr.).

3. **Ligularia stenocephala** (Maxim.) Matsum. & Koidz. *Senecio stenocephalus* Maxim.; *S. cacaliaefolius* var. *stenocephalus* (Maxim.) Franch.; *S. stenocephalus* var. *comosus* Fr. & Sav.——ME-TAKARA-KŌ. FIG. 17. Rhizomes stout; stems slender, to 1 m. long, glabrous; radical leaves thinly chartaceous, cordate- or reniform-hastate, to 24 cm. long, 20 cm. wide, abruptly acuminate, deeply cordate, acute-toothed, short-pilose on nerves beneath, long-petiolate; cauline leaves 3, the lower ones larger, the petiole sheathing at base; racemes densely many-headed, densely pubescent, the bracteate leaves broadly lanceolate, small; heads centripetal, many, the pedicels 1–3.5 cm. long; involucres narrowly tubular, 10–12 mm. long, 2.5–3 mm. across, the bracts 5; ligulate florets absent or 1–3, the corollas 2–2.5 cm. long, 3–4 mm. wide.——June–Sept. Wet places in mountains; Honshu, Shikoku, Kyushu.——China and Formosa.

4. **Ligularia sibirica** Cass. var. **speciosa** DC. *Cineraria fischeri* Ledeb.; *C. speciosa* Schrad. & Link; *Hoppea speciosa* Reichenb.; *L. speciosa* (Schrad.) Fisch. & Mey.; *L. fischeri* (Ledeb.) Turcz.; *L. sibirica* sensu auct. Japon., non Cass.; *L. euodon* Miq.; *L. fischeri* var. *euodon* (Miq.) Kitam.—— Ō-TAKARA-KŌ. Rhizomes stout; stems 1–2 m. long, loosely cobwebby-hairy at base, short-pubescent in upper part; radical leaves chartaceous, large, reniform-cordate, to 32 cm. long, 40 cm. wide, regularly toothed, long-petiolate; cauline leaves 3, the lower ones adjacent to the radical, short-petiolate,

sheathing at base, the upper ones short-petiolate, small; inflorescence to 75 cm. long, racemose, rarely branched; heads many, centripetal, 4–5 cm. across, the pedicels 1–9 cm. long, with 1 bract; involucres tubular-campanulate, 10–12 mm. long, 8–10 mm. across, the bracts 8–9, oblong; ligulate florets 5–9, the corollas about 25 mm. long, 3–4 mm. wide; pappus 6–10 mm. long, brownish or purplish.——July–Oct. Wet places in mountains; Honshu, Shikoku, Kyushu.——Korea, Sakhalin, Manchuria, e. Siberia, and China.

Var. **kaialpina** (Kitam.) Kitam. *L. sibirica* subsp. *kaialpina* Kitam., in syn.——KAI-TAKARA-KŌ. Stems 30–50 cm., rarely to 70 cm. long; leaves smaller; heads 4–11, in a corymb, the pedicels slender, 1–7 cm. long; involucres tubular, about 11 mm. long, about 5 mm. wide, the bracts 7–8; ligulate florets usually 5, the corollas 25–28 mm. long; pappus 6.5 mm. long, brownish, much shorter than the florets.——Wet places in high mountains; Honshu (centr. distr.).

5. **Ligularia hodgsonii** Hook. f. *L. yezoensis* Franch.; *L. calthaefolia* sensu Matsum., non Maxim.——TŌGEBUKI, EZO-TAKARA-KŌ, ONI-O-TAKARA-KŌ. Rhizomes stout; stems 30–80 cm. long, cobwebby-hairy only on the inflorescence; radical leaves chartaceous, long-petiolate, reniform, 4.5–13 cm. long, 7.5–27 cm. wide, rounded to short-mucronate at apex, regularly mucronate-toothed, glabrous, long-petiolate; cauline leaves usually 2, the lower one long-petioled and long-sheathing at the dilated base, the upper one smaller, the uppermost bracts oblong; inflorescence a corymb; heads 5–9, 4–5 cm. across, the pedicels 1–8 cm. long, cobwebby-hairy and crisped-puberulent; involucres campanulate, 11–12 mm. long, 8–10 mm. across, with 2 small bracteal leaves at base, the bracts 8 or 9, oblong, acute; ligulate florets 7–12, the corollas about 27 mm. long, 5–8 mm. wide; achenes 6–7 mm. long, the pappus about 1 cm. long, reddish, as long as the corolla.——July–Aug. Mountains; Hokkaido, Honshu (n. distr.).——s. Kuriles and Sakhalin.

6. **Ligularia japonica** (Thunb.) Less. *Arnica japonica* Thunb.; *Senecio japonicus* (Thunb.) Schultz-Bip.; *Erythrochaete palmatifida* Sieb. & Zucc.——HANKAI-SŌ. Rhizomes stout; stems about 1 m. long, glaucescent, glabrous, purple-spotted; radical leaves chartaceous, long-petiolate, cordate-orbicular, to 30 cm. long and as wide, strongly revolute while young, the segments palmately cleft to parted, the lobes pinnately lobed or incised, densely pubescent beneath while young; cauline leaves 3, with broadly sheathing petioles, the upper leaves smaller; inflorescence a corymb, the pedicels 2.5–20 cm. long, rather stout, crisped-pubescent; heads 2–8, about 10 cm. across; involucres campanulate-tubular, 18–24 mm. long and as wide, the bracts 9–12, elliptic, acuminate; ligulate florets about 10, the corollas 5–6.5 cm. long, 8–10 mm. wide; achenes about 9 mm. long, the pappus 6–7.5 mm. long, rusty brown.——June–Aug. Mountains; Honshu (w. distr.), Shikoku, Kyushu.——Korea, China, and Formosa.

7. **Ligularia dentata** (A. Gray) Hara. *Erythrochaete dentata* A. Gray; *L. clivorum* Maxim.; *Senecio clivorum* (Maxim.) Maxim.; *L. japonica* var. *clivorum* (Maxim.) Makino——MARUBA-DAKE-BUKI, MARUBA-NO-CHŌRYŌ-SŌ. Rhizomes short; stems 40–100 cm. long, glabrous; radical leaves large, thinly chartaceous, long-petiolate, reniform-orbicular, about 30 cm. long, 38 cm. wide, deeply cordate, revolute in bud, rather regularly mucronate-toothed, loosely short-pubescent on nerves above; cauline leaves 2, the petioles dilated and clasping at base; heads few to many, in corymbs, 7–10 cm. across, the pedicels 2–9 cm. long, densely crisped-pubescent;

involucres tubular-campanulate, 16–20 mm. long, 16–28 mm. wide, densely crisped-pubescent, the bracts 9–13, free or slightly connate, oblong, short-cuspidate; ligulate florets about 10, the corollas 4–5 cm. long, 7–8 mm. wide; achenes about 9 mm. long, the pappus about 12 mm. long, reddish.——July–Aug. Meadows in mountains; Honshu.——China.

Ligularia × yoshizoeana (Makino) Kitam.   *Senecillis*

yoshizoeana (Makino) Kitam.; *L. japonica* var. *yoshizoeana* Makino——Dakebuki, Chōryō-sō.   Hybrid of *L. japonica* × *L. dentata;* lower cauline leaves reniform, about 30 cm. long, 13 cm. wide, palmately cleft, the lobes oblong, toothed at apex; heads about 7 cm. across; involucres campanulate-globose, 17 mm. long and as much across; pappus about 1 cm. long.——Rarely cultivated in gardens.

34. MIRICACALIA Kitam.   Ō-momijigasa Zoku

Large perennial herbs; leaves few, alternate, lower ones large, long-petiolate, palmately cleft, short-sheathed at base; heads in racemes, nodding, homogamous; involucral bracts many, 1-seriate; receptacles glabrous; style-branches penicillate at apex; anthers long-exserted, obtuse at base; achenes cylindric, slightly flattened, striate, beaked, narrowed and stiped at base, glabrous; pappus scabrous, brown, with persistent bristles as long as the achene.——One species, in Japan.

1. Miricacalia makineana (Yatabe) Kitam.   *Senecio makineanus* Yatabe; *Cacalia makineana* (Yatabe) Makino; *C. iinumae* Makino; *S. iinumae* Makino; *S. makinoi* C. Winkler ——Ō-momijigasa, Momijigasa, Tosa-no-momijigasa. Stems 55–80 cm. long, densely pubescent with multicellular short curled hairs; cauline leaves 3, scattered, with crisp short hairs on both sides, the lower long-petiolate, large, peltate, orbicular, 25–33 cm. wide, deeply cordate, palmately cleft, the lobes 9–12, oblong, acuminate, usually shallowly 3-lobed, mucronate-toothed, the median leaf smaller, short-petiolate; bracteal leaves ovate-oblong, 1–1.5 cm. long; heads many, the peduncles 1–4 cm. long; involucres tubular, thinly membranous, about 15 mm. long, 12–14 mm. across, densely crisped-puberulent; bracts of caliculus 6–7, ovate-oblong, 6–8 mm. long, reflexed in anthesis; bracts lanceolate; florets about 21; corollas yellow, about 11 mm. long; achenes 7–8 mm. long, the beak about 2 mm. long; pappus about 10 mm. long.——July–Sept. Woods in mountains; Honshu, Shikoku, Kyushu.

35. CACALIA L.   Kōmori-sō Zoku

Perennial herbs; leaves alternate, revolute in bud, the long petioles sometimes short-sheathed at base; inflorescence a corymb, raceme, or panicle; heads without ligulate florets; involucres usually narrowly tubular, sometimes with small bractlets at base, rarely caliculate, the bracts 1-seriate, free; receptacle usually flat, glabrous; tubular florets 1–20, usually few, whitish or pale yellow, bisexual, all fertile; style-branches elongate, nearly clavate and short-pilose or penicillate, reflexed in anthesis; anthers obtuse or sagittate, auricles connate, elongate or short, the filaments often globosely thickened above; achenes terete, truncate, slightly narrowed at tip, beakless, gradually narrowed at base, glabrous, many-ribbed; pappus-bristles many, snow-white, rarely reddish, minutely scabrous.——About 50 species, in Asia and America.

1A. Cauline leaves many; petioles not entirely clasping the stem.
  2A. Involucres not caliculate; pappus snow-white.
    3A. Involucral bracts 3; corolla deeply 5-fid; leaves reniform, toothed, not clasping at base. .... 1. *C. adenostyloide*s
    3B. Involucral bracts 5–8; corolla shallowly 5-fid.
      4A. Leaves palmately cleft.
        5A. Leaves with raised veinlets; involucres 5–6 mm. long; pappus about 5 mm. long. .... 2. *C. tebakoensis*
        5B. Leaves with scarcely raised veinlets; involucres 9–10 mm. long; pappus 6.5–8 mm. long. .... 3. *C. delphiniifolia*
      4B. Leaves not palmately cleft.
        6A. Leaves bearing bulbils in the axils.
          7A. Leaves reniform, the petioles clasping at base. .... 5. *C. auriculata*
          7B. Leaves cordate, the petioles auriculate, not clasping at base. .... 4. *C. farfaraefolia*
        6B. Leaves without bulbils in the axils.
          8A. Involucral bracts 5.
            9A. Petioles clasping at base; inflorescence paniculate. .... 5. *C. auriculata*
            9B. Petioles not clasping at base; inflorescence corymbose.
              10A. Leaves 5–18 cm. long, 9–27 cm. wide, mucronate-toothed. .... 6. *C. nikomontana*
              10B. Leaves 3.5–4.5 cm. long, 5–6.5 cm. wide, coarsely incised. .... 7. *C. shikokiana*
          8B. Involucral bracts more than 5.
           11A. Leaves reniform, 3–9 cm. long, 5–12 cm. wide, long-caudate; stems 22–42 cm. long; heads 1–7. ...... 8. *C. nipponica*
           11B. Leaves hastate to deltoid-reniform, 25–35 cm. long, 30–40 cm. wide, short-acuminate; stems 1–2 m. long; heads numerous. .... 9. *C. hastata*
  2B. Involucres caliculate; pappus reddish. .... 10. *C. kiusiana*
1B. Cauline leaves few; petioles entirely clasping the stem, forming a sheath at base.
  12A. Leaves reniform, toothed. .... 11. *C. amagiensis*
  12B. Leaves orbicular, palmately cleft.
    13A. Leaves peltate. .... 12. *C. peltifolia*
    13B. Leaves basifixed. .... 13. *C. yatabei*

1. Cacalia adenostyloides (Fr. & Sav.) Matsum.   *Senecio adenostyloides* Fr. & Sav.——Kani-kōmori. Stems 60–100 cm. long, rather slender, glabrous, naked at base; larger

leaves sometimes 2 or 3, membranous, reniform, 6–11 cm. long, 10–20 cm. wide, short-cuspidate, cordate, irregularly incised-toothed, glabrous on both sides, the petioles 3–13 cm.

long, wingless, the upper leaves gradually smaller, bractlike, oblong and short-petiolate or linear-lanceolate; heads many, nodding in anthesis, in a narrow panicle, the pedicels 2–5 mm. long, with short crisped-pubescence, the bractlets 1–3, linear; involucres narrow-tubular, 8–9 mm. long, about 1.5 mm. across, without a caliculus, the bracts 3, narrowly oblong, obtuse; tubular florets 3–5, whitish, 8–8.5 mm. long; achenes about 6 mm. long, linear, the pappus snow-white, about 6 mm. long.——Aug.–Sept. Woods in mountains; Honshu (centr. and n. distr. incl. Yamato Prov.), Shikoku.

Cacalia × koidzumiana Kitam. A hybrid of *C. adenostyloides* × *C. hastata* var. *farfaraefolia*.——Honshu (Shinano Prov.).

2. **Cacalia tebakoensis** (Makino) Makino. *C. delphiniifolia* var. *tebakoensis* Makino——TEBAKO-MOMIJIGASA. Stoloniferous; stems 25–85 cm. long, slender; median cauline leaves membranous, few, long-petiolate, palmately 5- to 7-cleft, 3.5–10 cm. long, 5–17 cm. wide, the segments oblong, acuminate, incised-serrate, loosely short-pubescent on both sides, with prominently raised veinlets, the upper leaves rather abruptly smaller, short-petiolate, 3- to 5-cleft; leaves of the inflorescence linear and small; heads many, in a panicle, the pedicels 1–5 mm. long, with 3 or 4 minute ovate bractlets; involucres tubular, about 6 mm. long, about 3 mm. across, the bracts 5, narrowly oblong, obtuse; tubular florets 5 or 6, the corollas 7–7.5 mm. long; achenes about 5 mm. long, linear, the pappus bristles many, whitish, about 5 mm. long.——Aug.–Oct. Woods in mountains; Honshu (Kantō, Tōkaidō Distr. and Yamato Prov.), Shikoku, Kyushu.

3. **Cacalia delphiniifolia** Sieb. & Zucc. *Senecio zuccarinii* Maxim.; *C. zuccarinii* (Maxim.) Hand.-Mazz.——MOMIJIGASA, MOMIJI-SŌ. Stems about 80 cm. long, the branches with short crisped pubescence; median cauline leaves long-petiolate, 15 cm. long, to 20 cm. wide, palmately 7-cleft, the segments thin, oblong, acuminate, irregularly incised-toothed, glabrous above, densely silky-hairy beneath while young, loosely so afterward, the petioles to 14 cm. long, wingless, the upper leaves gradually smaller, short-petiolate, 3- or 5-cleft; leaves of the inflorescence linear and small; heads many, in a panicle, ascending at anthesis, the pedicels 3–12 mm. long, the bractlets 1–3, ovate-lanceolate, 1–1.5 mm. long; involucres 8–9 mm. long, 3–4 mm. across, without a caliculus, mealy-puberulent, the bracts 5, narrowly oblong, obtuse; tubular florets 5, white to purplish, about 8.5 mm. long; achenes about 5 mm. long, the pappus snow-white, 6.5–8 mm. long.——Aug.–Oct. Woods in mountains; Hokkaido, Honshu, Shikoku, Kyushu.

× **Senecillicacalia telphusaefolia** Koidz. KANI-O-TAKARA-KŌ. A bigeneric hybrid of *Cacalia delphiniifolia* × *Ligularia sibirica* var. *speciosa*.——Honshu (Yamashiro Prov.).

4. **Cacalia farfaraefolia** Sieb. & Zucc. *C. bulbifera* sensu auct. Japon., non Maxim.——USUGE-TAMABUKI. Stems 50–140 cm. long, cobwebby only while young, paniculately branched above and densely crisped-puberulent; median cauline leaves few, large, 11–15 cm. long, 13–21 cm. wide, acute, cordate, coarsely toothed and incised-lobulate or mucronate-toothed near base, minutely toothed toward tip, puberulent above, loosely cobwebby or glabrous beneath, the petioles 7–15 cm. long, wingless, not auriculate at base, with small globose bulbils in axils, the upper leaves abruptly smaller, ovate to broadly so, sometimes oblong, bractlike, the uppermost one linear; heads many, in a narrow panicle, the short pedicels

with few bractlets; involucres 9–10 mm. long, 2–3 mm. wide, the bracts 5; tubular florets 5–6; achenes about 6.5 mm. long, the pappus 7 mm. long, snow-white.——Aug.–Oct. Woods in mountains; Honshu (Kantō Distr. and westw.), Shikoku, Kyushu.

Var. **bulbifera** (Maxim.) Kitam. *Senecio bulbifera* Maxim.; *C. bulbifera* (Maxim.) Maxim.——TAMABUKI. Leaves densely cobwebby-hairy beneath.——Woods in mountains; Hokkaido, Honshu (Kantō Distr. and northw.).

Var. **acerina** (Makino) Kitam. *C. lobatifolia* Maxim.; *C. bulbifera* var. *acerina* Makino; *C. bulbifera* var. *lobata* Makino——MIYAMA-KŌMORI-SŌ, MOMIJI-TAMABUKI. Stems slender, 25–60 cm. long; median cauline leaves 2–4, deltoid-cordate or deltoid-reniform, irregularly palmately cleft, the lobes entire, obtuse, mucronate.——Sept.–Oct. Mountains; Honshu (Tōkaidō, s. Kinki Distr.), Shikoku, Kyushu.

5. **Cacalia auriculata** DC. var. **kamtschatica** (Maxim.) Matsum. *Senecio dahuricus* var. *kamtschaticus* Maxim.; *C. kamtschatica* (Maxim.) Kudo——MIMI-KŌMORI. Stems erect, 60–120 cm. long, stout, short-branched in upper part, short crisped-pubescent; median cauline leaves 3 or 4, thinly chartaceous, usually reniform, 7–17 cm. long, 11–25 cm. wide, abruptly short-acute, cordate, irregularly incised-toothed, sometimes with 1 or 2 rather large acute lobes, glabrous on both sides or with short crisped hairs on the nerves beneath, the petioles 4.5–9 cm. long, auriculate at base, the upper leaves abruptly smaller, linear; heads many, in a narrow racemelike panicle, the pedicels short, with 3–4 bractlets about 2 mm. long; involucres not caliculate, glabrous, 8–10 mm. long, the bracts 5, narrowly oblong, obtuse; tubular florets 3–6, the corollas about 8.5 mm. long; achenes 4–5 mm. long, the pappus about 5 mm. long, snow-white.——July–Sept. Woods in mountains; Hokkaido, Honshu (n. distr.).——Manchuria, Ussuri, Kamchatka, Aleutians. The typical phase occurs in Manchuria, Dahuria, Ussuri, and e. Siberia.

Var. **bulbifera** Koidz. *C. bulbifera* Koidz.; *C. matsumurana* Kudo; *C. kamtschatica* var. *bulbifera* (Koidz.) Kitam.——KOMOCHI-MIMI-KŌMORI. Plants with bulbils in leaf-axils.——Hokkaido.

6. **Cacalia nikomontana** Matsum. *C. zigzag* Honda——Ō-KANI-KŌMORI. Stems 30–100 cm. long, more or less crisped-puberulent; median cauline leaves somewhat approximate, membranous, long-petiolate, reniform, 5–18 cm. long, 10–27 cm. wide, short-acuminate, shallowly cordate, often 5-angled, 5-nerved, mucronate-toothed, with short crisped hairs on nerves beneath, the upper leaves abruptly smaller, narrowly ovate; heads many, often in a corymb, the pedicels 5–10 mm. long, with many linear, small bractlets; involucres 8–10 mm. long, with 1–2 bractlets at base, the bracts 5, broadly linear, obtuse; tubular florets 5–6 mm. long, the corollas about 8 mm. long; achenes about 4.5 mm. long; pappus snow-white, 6–6.5 mm. long.——Aug.–Oct. Woods in mountains; Honshu.

Cacalia × **cuneata** (Honda) Kitam. *C. zigzag* var. *cuneata* Honda——HAKUSAN-KANI-KŌMORI. A hybrid of *C. hastata* var. *ramosa* × *C. nikomontana*.——Honshu (Mount Hakusan in Kaga Prov.).

7. **Cacalia shikokiana** Makino. *Senecio farfaraefolius* var. *humilis* Makino——HIME-KŌMORI-SŌ. Stems 25–35 cm. long, branched and with short crisped pubescence in upper part; median cauline leaves thinly chartaceous, long-petiolate, reniform, 3.5–4.5 cm. long, 5–6.5 cm. wide, acuminate, cordate

to truncate at base, 5-angled, coarsely incised, mucronate, loosely crisped–puberulent on both sides, the petioles wingless, not auricled, the upper leaves gradually smaller, the uppermost linear; heads rather many, dense, corymbose, the pedicels 7–11 mm. long, often bracteate; involucres 8–9 mm. long, glabrous, the bracts 5, narrowly oblong; tubular florets 7, the corollas about 7 mm. long; pappus snow-white, about 5 mm. long.——Oct. Mountains; Honshu (Kii Prov.), Shikoku.

8. **Cacalia nipponica** Miq. *Senecio farfaraefolius* var. *nipponicus* (Miq.) Maxim.; *C. farfaraefolia* var. *nipponica* (Miq.) Matsum.——Tsukushi-kōmori-sō. Stems 22–42 cm. long, slender, flexuous, sometimes with short crisped pubescence; median cauline leaves few, approximate, long-petiolate, depressed-reniform, 3–9 cm. long, 5–12 cm. wide, 5-cleft, abruptly caudate, truncate, rarely cordate, the terminal lobes lanceolate, large, the lateral ones of 2 pairs, spreading, deltoid-lanceolate, incised-toothed toward tip, crisped-puberulent on both sides or glabrous beneath, the petioles wingless, the upper leaves abruptly smaller; heads 1–7, on long bracteate pedicels; involucres 8–10 mm. long, glabrous, the bracts 7 or 8, broadly linear; tubular florets 12–14, the corollas 7–7.5 mm. long; achenes 4.5–6 mm. long, the pappus snow-white, 5–5.5 mm. long.——Aug.–Oct. Woods in mountains; Kyushu.

9. **Cacalia hastata** L. var. **orientalis** (Kitam.) Ohwi. *C. hastata* var. *glabra* sensu auct. Japon., non Ledeb.; *C. hastata* sensu auct. Japon., non L.; *C. hastata* subsp. *orientalis* Kitam.——Yobusuma-sō. Stems 1–2 m. long, stout, branched, with short crisped pubescence on upper portion; median cauline leaves chartaceous, hastate to deltoid-reniform, usually 25–35 cm. long, 30–40 cm. wide, short-acuminate, cordate, undivided or 2-lobed, puberulent on both sides or only on underside, the lobes acuminate, mucronate-toothed, the petioles 9–13 cm. long, usually broadly winged and auriculate-clasping, the upper leaves gradually smaller, hastate to oblong, becoming linear in the inflorescence, short-petiolate; heads numerous, 10–12 mm. long, about 5 mm. across, often in large terminal panicles, the bracts 5–8, obtuse; tubular florets 6–9, the corollas 8–9 mm. long; achenes 5–8 mm. long, the pappus about 7 mm. long.——Aug.–Oct. Woods in mountains; Hokkaido, Honshu (n. Kantō Distr. and northw.).——Korea, Sakhalin, s. Kuriles, Kamchatka, and Manchuria; variable.

Var. **tanakae** (Fr. & Sav.) Kitam. *C. aidzuensis* Koidz.; *C. aidzuensis* var. *yukii* Kitam.; *Senecio farfaraefolius* var. *tanakae* Fr. & Sav.——Inu-dōna. Median leaves deltoid-reniform, rounded and short-acuminate, more or less cordate, irregularly toothed, the petioles broadly winged; involucral bracts 6 to 8, 7–8 mm. long; tubular florets 8 or 9; achenes about 6 mm. long.——Honshu (n. Kantō Distr. and northw.).

Var. **chokaiensis** (Kudo) Kitam. *C. chokaiensis* Kudo ——Kobana-no-kōmori-sō. Stems slender; leaves deltoid-reniform, 5-angled, the petioles not auriculate at base; involucral bracts 8–10.——Honshu (n. distr.).

Var. **nantaica** (Komatsu) Kitam. *C. nantaica* Kamatsu ——Nikkō-kōmori. Stems 50–120 cm. long; leaves hastate, the petioles not clasping at base; involucres 10–12 mm. long, the bracts 7–9; tubular florets 10–15.——Honshu (Mount Nantai in Nikko).

Var. **ramosa** (Maxim.) Kitam. *Senecio farfaraefolius* var. *ramosa* Maxim.; *C. tschonoskii* Koidz.; *C. farfaraefolia* var. *ramosa* (Maxim.) Matsum.——Ōba-kōmori. Stems slender, to 1.5 m. long; leaves hastate, about 15 cm. long, about 22 cm.

wide, acuminate, truncate or shallowly cordate, sometimes 2-lobed on each side, the petioles usually not auricled at base; heads usually on pedicels 5–10 mm. long; involucres 7–8 mm. long, the bracts 5–7; tubular florets 9–10; achenes 4–6.5 mm. long.——Honshu (centr. distr. and Kinki).

Var. **farfaraefolia** (Maxim.) Ohwi. *Senecio farfaraefolius* Maxim. var. *farfaraefolius*; *C. farfaraefolia* (Maxim.) Matsum.; *C. nipponica* var. *farfaraefolia* (Maxim.) Koidz.; *C. yakushimensis* Masam.; *C. maximowicziana* Nakai & F. Maekawa—— Kōmori-sō. Stems slender, 30–70 cm. long; median cauline leaves long-petiolate, shallowly 5-lobed, 8–10 cm. long, 13–15 cm. wide, cordate to subtruncate at base or depressed-hastate, the terminal lobe largest, acuminate, irregularly denticulate, the lateral lobes smaller, often entire; heads rather small, loose; involucres 7.5–10 mm. long, the bracts 6 or 7; tubular florets 6–10; achenes 4–4.5 mm. long, the pappus about 5 mm. long, snow-white.——Mountains; Honshu (Kantō to Kinki Distr.).

Var. **alata** (F. Maekawa) Kitam. *C. maximowicziana* var. *alata* F. Maekawa; *C. crucifolia* F. Maekawa, in syn.——Oku-yama-kōmori. Petioles of median cauline leaves winged and auricled at base.——Mountains; Honshu (southern half of centr. distr.).

10. **Cacalia kiusiana** Makino. Momiji-kōmori. Stems 70–80 cm. long, short-branched in upper part, crisped-pubescent; median cauline leaves chartaceous, long-petiolate, reniform, 10–15 cm. long, 12.5–18 cm. wide, slightly cordate to rarely subtruncate at base, palmately 5-lobed, the lobes deltoid, acuminate, mucronate-toothed, green above, paler and densely soft-puberulent beneath, the upper leaves gradually smaller, the uppermost ones ovate; heads many in a racemelike panicle, the pedicels short, bractless or with small bracts; involucres about 9 mm. long, about 5 mm. across, the bracts 5, narrowly oblong, obtuse, glabrous, the bractlets about 7, ovate, about 1.5 mm. long at base; tubular florets 6 or 7, the corollas about 1 cm. long; achenes glabrous, the pappus about 8 mm. long, brownish red.——Aug.–Oct. Kyushu (s. distr.).

11. **Cacalia amagiensis** Kitam. *Miricacalia amagiensis* (Kitam.) Nakai——Izu-kani-kōmori. Rhizomes creeping; stems 45–60 cm. long, branched, short crisped-pubescent on the upper portion; leaves few, usually 2(–3), chartaceous-membranous, the lower one long-petiolate, reniform, 12–13 cm. long, about 22 cm. wide, rounded at apex, deeply cordate, loosely pilose above, net-veined and loosely silky along veins, lobulate, minutely mucronate-toothed, the petioles 11–12 cm. long, wingless, densely white-pubescent, short-sheathed at base, the median leaf slightly smaller, 7–8 cm. long, 10–12 cm. wide, the petioles 3–3.5 cm. long, the floral leaf ovate or narrower and bractlike; heads many, in racemelike panicles, the pedicels with 3 or 4 small bractlets; involucres 11 mm. long, 2–3 mm. wide, the bracts 5, broadly linear, acute; tubular florets 4 or 5, the corollas about 11 mm. long; pappus white, 7–9 mm. long. ——Sept.–Oct. Mountains; Honshu (Izu Prov.).

12. **Cacalia peltifolia** Makino. *Miricacalia peltifolia* (Makino) Nakai——Taimingasa. Rhizomes short-branched; stems 1–2 m. long, usually branched and crisped-puberulent in upper part; cauline leaves about 3, thin, the lower one largest, peltate, suborbicular, 35–55 cm. across, deeply palmately cleft, deep green, crisped-puberulent above, puberulent on nerves beneath, the lobes 9–14, the larger ones oblanceolate-oblong, 2- or 3-lobulate, irregularly toothed, the petioles 30–65 cm. long, wingless, short-sheathed at base, the median leaf

long-petiolate, slightly smaller, the uppermost very small, sometimes bractlike; heads in large panicles, densely crisped-puberulent, the pedicels with 3–4 small bractlets; involucres 9–11 mm. long, about 4 mm. across, yellow-green, crisped-puberulent, the bracts 5, narrowly oblong, obtuse, longitudinally striate; tubular florets 6, the corollas white, changing to brown, 10–11 mm. long; pappus about 8 mm. long, sordid-white.——Sept.–Oct. Woods along streams and ravines in mountains; Honshu (centr. distr. and westw.).

13. Cacalia yatabei Matsum. & Koidz. *C. palmata* Makino, excl. syn.; *Miricacalia yatabei* (Matsum. & Koidz.) Nakai——Yama-taimingasa, Taimingasa-modoki. Plant stoloniferous and with long-creeping rhizomes; stems 75–90 cm. long, short-branched, with short crisped-pubescence on upper portion; cauline leaves 3, thin, the lower largest, long-petiolate, 17–24 cm. long, 30–35 cm. wide, somewhat cordate, green and glabrous above, paler and short crisped-pubescent on nerves beneath, prominently veined on both sides, palmately pinnaticleft, the lobes oblong, 9 or 10 pairs, usually 2- or 3-lobulate, the lobules acuminate, loosely and irregularly mucronate-denticulate, the petioles 17–19 cm. long, wingless, short-sheathed at base, the upper leaves gradually smaller, shorter-petiolate; heads many, in terminal racemose panicles; involucres 5–10 mm. long, about 1.5 mm. across, sometimes with bractlets at base, the bracts 5, broadly linear, obtuse; tubular florets 5 or 6, the corollas 8.5–10 mm. long; achenes 5–6 mm. long; pappus about 9 mm. long, sordid-white.——Aug.–Oct. Woods in mountains; Honshu, Shikoku.

Var. **occidentalis** F. Maekawa ex Kitam. *Miricacalia maekawae* Nakai——Nishi-no-yama-taimingasa. Involucral bracts 3 or 4; tubular florets 2–4.——Woods in mountains; Honshu (centr. distr. and westw.), Shikoku, Kyushu.

36. SENECIO L.   Kion Zoku

Perennials, rarely annuals or biennials; leaves revolute in bud, cauline or radical, alternate, entire to pinnately parted, not sheathing at base; heads in corymbs or falsely so, homogamous or heterogamous; involucres tubular, sometimes subglobose, sometimes bracteate at base, the bracts 1-seriate, usually free; receptacle flat or convex; ray flowers pistillate, fertile, the corollas 1-seriate (rarely absent), yellow; tubular florets bisexual, fertile; style-branches often divergent, dilated and short-pilose at tip, sometimes rounded and mammillate at apex; anthers obtuse, sagittate or bifid at base; achenes terete, truncate at apex; pappus-bristles scabrous, slender, many.——About 1,200 species. Cosmopolitan.

1. Senecio vulgaris L. No-borogiku. Annual or biennial, slightly fleshy, sparingly cobwebby-hairy while young; stems branched, 10–40 cm. long, somewhat purplish; leaves 3–5 cm. long, 1–2.5 cm. wide, pinnately cleft to deeply undulate-sinuate, irregularly toothed, narrowed below and semiclasping at base, the lower ones petiolate; heads many to few; involucres cylindric, 6–8 mm. long, the caliculus very small, the bractlets lanceolate, 2–3 mm. long, blackish at the tip, the bracts linear, acute, blackish at tip; ligulate corollas absent; tubular corollas yellow; achenes terete, appressed-pilosulous.——Mar.–Oct. Sometimes flowering the whole year.——European weed, widely naturalized in our area.

2. Senecio scandens Hamilt. ex D. Don. *S. hibernus* Makino——Taikin-giku, Miyuki-giku. Stems elongate, branched, scandent, 2–5 m. long, densely pubescent while young, glabrous; cauline leaves elongate-deltoid, usually 7–10 cm. long, 3.5–4.5 cm. wide, acuminate, truncate to hastate at base, irregularly incised-toothed or subentire, often lobed in the lower ones, pubescent on both sides, the petioles 1–2 cm. long; inflorescence paniculately corymbose, terminal, the branches much spreading, densely pubescent; heads 13–14 mm. across, the pedicels 5–10 mm. long; involucres 5–6 mm. long, about 5 mm. across, tubular, the bracts 8, lanceolate, acute; ligulate corollas yellow, about 9 mm. long, about 2 mm. across; achenes about 3 mm. long, short-pilose, the pappus 5–6 mm. long, snow-white.——Nov.–Mar. Near seashores; Honshu (Kii Prov.), Shikoku.——Formosa, China, Philippines, and India.

3. Senecio nemorensis L. *S. fuchsii* Gmel.; *S. nemorensis* var. *fuchsii* (Gmel.) Koch; *S. nemorensis* var. *subinteger* Hara; *S. ovatus* var. *japonicus* Nakai——Kion, Higo-omina-eshi. Rhizomes short; stems 50–100 cm. long, leafy, branched in upper part; median leaves lanceolate to ovate-oblong, acuminate, narrowed to a short petiole or abruptly narrowed and semiclasping, irregularly serrulate, glabrous or with short crisped hairs on both sides; heads many, 17–25 mm. across, in corymbs, short-pedicellate; involucres 6–7 mm. long and as wide, caliculate at base, the bractlets linear, 2–8 mm. long, the bracts 9–12, narrowly oblong, deltoid at apex; ligulate corollas about 5, yellow, 13–19 mm. long, 1.5–3.5 mm. wide; achenes 3.5–4 mm. long, glabrous, the pappus 6–6.5 mm. long, whitish.——Aug.–Sept. High mountains; Hokkaido, Honshu, Shikoku, Kyushu.——Formosa, China, Korea, Sakhalin, s. Kuriles, Siberia to Europe.

4. Senecio cannabifolius Less. *S. palmatus* Pall. ex Ledeb.; *S. palmatus* var. *genuinus* Herd.; *S. palmatus* var. *davuricus* Herd.; *S. cannabifolius* var. *davuricus* (Herd.) Kitag.——HANGON-SŌ. Rhizomes creeping; stems erect, 1–2 m. long, often reddish; lower and median cauline leaves petiolate, 10–20 cm. long, 9–15 cm. wide, pinnately divided, more or less crisped-puberulent beneath, the terminal segments lanceolate, acuminate, incurved-serrulate, the lateral segments of 1 or 2 pairs, ascending, the petioles auriculate at base, the upper leaves 3-parted to -cleft, petiolate, gradually smaller, simpler, and lanceolate in the uppermost ones; heads many, about 2 cm. across, in large corymbs, the pedicels slender; involucres tubular, 5–6 mm. long and as wide, caliculate, the bractlets linear, 2–3 mm. long, few, the bracts lanceolate, acute; ligulate florets 5–7, the corollas yellow, 12–13 mm. long, 1.5–2 mm. wide; achenes about 3 mm. long, glabrous, the pappus twice as long as the achenes, yellowish white.——July–Sept. Mountains; Hokkaido, Honshu (centr. and n. distr.); rather common.——Sakhalin, Kuriles, Kamchatka, Korea, Manchuria, to e. Siberia, n. China, and Aleutians.

Var. **integrifolius** (Koidz.) Kitam. *S. palmatus* var. *integrifolius* Koidz.; *S. otophorus* Maxim.——MIMI-KION, HITO-TSUBA-HANGON-SŌ. Leaves undivided, simple.——Hokkaido, Honshu.——Manchuria.

5. Senecio argunensis Turcz. *S. jacobaea* var. *grandiflorus* Turcz. ex DC.——KŌRINGIKU. Rhizomes woody; stems erect, single or tufted, 65–100 cm. long, cobwebby-hairy, branched in upper part; median leaves ovate-oblong to oblong, 8–10 cm. long, 4–6 cm. wide, pinnately parted to cleft, the lobes about 6 pairs, ascending, narrow, incised, acute to obtuse, with a broad sinus between the lobes, sessile, deep green and glabrous above, loosely cobwebby-hairy, with raised nerves beneath, the upper leaves gradually smaller, pinnately lobed; heads many, 2–2.5 cm. across, in corymbs; involucres subglobose, about 6 mm. long, 10 mm. wide, caliculate, the bractlets many, 3–5 mm. long, acuminate, the bracts oblong; ligulate corollas deep yellow, about 15 mm. long, about 3 mm. wide; achenes 2–3 mm. long, glabrous, the pappus about 5.5 mm. long, whitish.——Aug.–Oct. Mountains; Kyushu (n. distr.); rare.——Korea, China, Manchuria, and Dahuria.

6. Senecio pseudoarnica Less. *Arnica maritima* (L.) Koidz., non L. f.——EZO-OGURUMA. Fleshy perennial herb with thick rhizomes; stems 30–50 cm. long, stout, usually simple, cobwebby-hairy, densely leafy; radical and lower cauline leaves smaller than the median ones, the median oblong or ovate-oblong, 12–15 cm. long, 4.5–5.5 cm. wide, obtuse, slightly narrowed at base and semiclasping, mucronate-toothed, cobwebby-hairy beneath, the uppermost ones slightly smaller, obtuse to acuminate; heads 1–30, 3.5–4.5 cm. across, in peduncled corymbs; involucres subglobose, 1–1.2 cm. long, with

many bracteal leaves 2–2.5 cm. long, the bracts narrowly oblong, acuminate; ligulate corollas yellow, 17–24 mm. long, 3–5 mm. wide; achenes grayish, about 5 mm. long, glabrous, the pappus about 1 cm. long, whitish, brownish at base.—— July–Sept. Near seashores; Hokkaido, Honshu (Mutsu Prov.).——Korea, Manchuria, Amur, Ussuri, Kamchatka, and N. America.

7. Senecio nikoensis Miq. SAWAGIKU, BOROGIKU. Plant stoloniferous, rhizomes short; stems erect, 40–100 cm. long, white-pubescent, often branched in upper part; median and lower cauline leaves thinly membranous, ovate-oblong, 5.5–14 cm. long, 2.5–8.5 cm. wide, soft-pubescent on both surfaces, deeply pinnatifid to pinnatiparted, petiolate, the segments spreading, 4–6 pairs, linear-oblong, toothed or incised; heads nearly umbellate, about 12 mm. across, on slender bractless pedicels; involucres about 5 mm. long, 7 mm. across, the bracts narrowly oblong, acute; ligulate corollas pale yellow, about 8 mm. long, 1 mm. wide; achenes about 1.5 mm. long, short-pilose; pappus snow-white, 6–7 mm. long.——June–Aug. Damp woods in mountains; Hokkaido, Honshu, Shikoku, Kyushu.

8. Senecio takedanus Kitam. *S. flammeus* var. *alpina* Takeda——TAKENE-KŌRIN-KA. Rhizomes short; stems simple, 20–40 cm. long, cobwebby and sometimes short-pubescent; lower and median cauline leaves spathulate-oblong, 5–10 cm. long, 1.5–3 cm. wide, obtuse, clasping at base, irregularly mucronate-toothed, cobwebby-hairy and often with short crisped hairs, the uppermost leaves gradually smaller, narrower, broadly clasping at base; heads usually 4–5, in umbels, 2–2.5 cm. across, the pedicels purplish puberulent, thickened at top, with small purplish linear bractlets; involucres cup-shaped, 7–10 mm. long, 1–1.6 cm. wide, dark purplish; ligulate corollas dark orange-red, about 1 cm. long, 2 mm. wide; achenes about 4 mm. long, hairy; pappus 6–7 mm. long, whitish.——Aug. Alpine slopes; Honshu (centr. distr.); rare.

9. Senecio flammeus Turcz. ex DC. *S. longeligulatus* Lév. & Van't.——TAKENE-KŌRINGIKU. Rhizomes slender, short; stems simple, 15–40 cm. long, cobwebby and puberulent; radical leaves withering before anthesis, oblong, cuneately narrowed at base, long-petiolate, the lower cauline leaves oblanceolate-oblong, 8–9 cm. long, 2.3–2.5 cm. wide, obtuse, narrowed to a winged semiclasping slightly decurrent petiole, irregularly mucronate-toothed in upper half, puberulent and cobwebby on both sides, the median leaves oblong, the upper ovate-lanceolate, gradually much reduced; heads 2–7, in false umbels, 3–3.2 cm. across, the pedicels with few bractlets; involucres cup-shaped, about 5 mm. long, 12 mm. across, dark purplish at tip, the bracts lanceolate, acute; ligulate corollas dark orange-red, incurved on margin, reflexed at anthesis, 13–22 mm. long, about 1.5 mm. wide; achenes 2.5–3 mm. long, hairy, the pappus whitish, about 5.5 mm. long.——Aug. Mountains; Kyushu (n. distr.).——Korea, Manchuria, n. China, and Dahuria.

Var. **glabrifolius** Cufodontis. *S. flammeus* sensu auct. Japon., non Turcz.; *S. flammeus* subsp. *glabrifolius* (Cufodontis) Kitam.; *S. flammeus* forma *glabrescens* Hara——KŌRIN-KA. Stems 30–60 cm. long; heads 3–13; involucres 5–8 mm. long; achenes about 3 mm. long, densely hairy, the pappus 6.5–8 mm. long.——July–Oct. Sunny places in mountains; Honshu.——Korea.

10. Senecio kawakamii Makino. MIYAMA-OGURUMA. Rhizomes short-decumbent; stems 17–30 cm. long, erect, subscapose, short-pubescent and cobwebby; lower leaves rosulate, approximate, oblong, 6–10 cm. long, about 2 cm. wide, obtuse, short-pubescent and cobwebby while young, irregularly mucronate-toothed, winged-petiolate, the upper cauline leaves gradually smaller, oblanceolate-oblong, semiclasping, the uppermost linear, bractlike; heads 3–7, in false umbels, 2–3 cm. across, the peduncles slender, minutely glandular-pubescent and cobwebby-hairy; involucres cup-shaped, about 5 mm. long, 1–1.2 cm. across, the bracts 12–15, glandular-ciliate, acute; ligulate corollas yellow, 9–12 mm. long, 3–3.5 mm. wide; achenes 3–3.5 mm. long, glabrous, the pappus accrescent, 6–7 mm. long, snow-white.——July–Aug. Alpine slopes; Hokkaido.——s. Kuriles and Sakhalin.

11. Senecio integrifolius (L.) Clairv. var. **spathulatus** (Miq.) Hara. *S. kirilowii* Turcz. ex DC.; *S. campestris* sensu auct. Japon., non DC.; *S. fauriei* Lév. & Van't.; *S. integrifolius* subsp. *kirilowii* (Turcz.) Kitag.; *S. integrifolius* subsp. *fauriei* (Lév. & Van't.) Kitam.; *S. aurantiacus* var. *spathulatus* Miq.; *S. aurantiacus* var. *elatior* Miq.; *S. aurantiacus* var. *foliosa* Miq.——OKA-OGURUMA. Rhizomes short-ascending; stems scapose, erect, 20–65 cm. long, densely cobwebby and slightly purplish in upper part; radical leaves rosulate, oblong, rarely obovate-oblong, 5–10 cm. long, 1.5–2.5 cm. wide, obtuse, minutely mucronate, gradually narrowed at base, usually sessile, rarely short-petiolate, irregularly mucronate-toothed, cobwebby on both surfaces, the cauline leaves few, the lower ones ascending, lanceolate, 7–11 cm. long, 1–1.5 cm. wide, obtuse, semiclasping at base, decurrent, the upper ones gradually smaller, narrowly lanceolate, acuminate, broadly clasping at

base; heads 3–9, in corymbs or false umbels, 3–4 cm. across, pedunculate; involucres about 8 mm. long, 11 mm. across, the bracts lanceolate, acuminate; ligulate corollas yellow, 12–16 mm. long, 2–3 mm. wide; achenes about 2.5 mm. long, densely pilose, the pappus about 11 mm. long, snow-white.——May–June. Sunny slopes in hills and low elevations in mountains; Honshu, Shikoku, Kyushu; common.——Korea, Manchuria, China, and Formosa. The typical phase occurs from Siberia to Europe.

12. Senecio pierotii Miq. *S. campestris* var. *subdentatus* Maxim.; *S. subdentatus* var. *pierotii* (Miq.) Cufodontis——SAWA-OGURUMA. Rhizomes short, ascending; stems erect, 50–80 cm. long, rather stout, simple, hollow, cobwebby; radical leaves present at anthesis, rather fleshy and thick, often long-petiolate, oblong or narrowly so, sometimes lanceolate, 12–25 cm. long, 1.5–7 cm. wide, obtuse, narrowed or cuneate at base, subentire or mucronately toothed, cobwebby while young; cauline leaves rather numerous, the lower ones oblong to lanceolate, semiclasping at base, the uppermost gradually smaller, acuminate, broadly clasping at base; heads 6–30, in false umbels or corymbs, 3.5–5 cm. wide, long-pedunculate; involucres cup-shaped, without a caliculus, 7–8.5 mm. long, 1.5–2 cm. wide, the bracts broadly lanceolate, acuminate; ligulate corollas 11–16 mm. long, about 2 mm. wide, yellow; achenes glabrous, the pappus snow-white, 9–13 mm. long.——Apr.–June. Wet places at low elevations in mountains; Honshu, Shikoku, Kyushu.——Ryukyus.

37. SYNEILESIS Maxim. YABUREGASA ZOKU

Rather stout perennial herbs; radical leaves peltate, palmately parted, cobwebby while young; cauline leaves alternate, few, the petioles entirely clasping the stem; heads erect in anthesis, eligulate, in corymbs or in panicles; involucres narrowly tubular, with 2–3 linear bractlets at base, the bracts usually 5, 1-seriate, free, rather thick; receptacle flat, glabrous; tubular florets bisexual, fertile, the corollas tubular, 5-lobed, whitish or reddish; style-branches elongate, obtuse to depressed-deltoid or penicillate, sometimes pilose; anthers short-sagittate, the anther-tails pollen-bearing, connate; achenes terete, striate; pappus-bristles many, nearly equal, scabrous; cotyledons 1, convolute, orbiculate, sublobed.——Few species, in e. Asia.

1A. Heads in flat-topped corymbs.
  2A. Leaf-segments narrow, 4–8 mm. wide. .... 1. *S. aconitifolia*
  2B. Leaf-segments broader, 1–1.7 cm. wide. .... 2. *S. tagawae*
1B. Heads in panicles; leaf-segments 2–4 cm. wide. .... 3. *S. palmata*

1. Syneilesis aconitifolia (Bunge) Maxim. *Cacalia aconitifolia* Bunge; *Senecio aconitifolius* (Bunge) Turcz. ex Forbes & Hemsl.——HOSOBA-YABUREGASA. Rhizomes short-creeping, 6–8 mm. across; stems 70–120 cm. long, brownish, glabrous, simple, with 2 main leaves; lower leaf peltate, 20–30 cm. wide, palmately parted or deeply cleft, long-petiolate, the segments 7–9, once to twice bifid, the lobes 4–8 mm. wide, acuminate, irregularly acute-toothed, revolute in bud, densely cobwebby while young, whitish beneath, the median leaf slightly smaller, petiolate, 12–24 cm. wide, 4- or 5-cleft, the upper leaves bractlike; heads many, 6–7 mm. across, in dense compound corymbs, the pedicels 6–16 mm. long, linear-bracteate; involucre 9–12 mm. long, purple-brown, the bracts 5, oblong, obtuse, glabrous; tubular florets 8–10, reddish; achenes about 5 mm. long, the pappus sordid white to reddish, 8–10 mm. long.——Aug. Honshu (Tanba Prov.).

2. Syneilesis tagawae Kitam. *S. aconitifolia* var. *tagawae* Kitam.——YABUREGASA-MODOKI. Rhizomes short; stems about 1 m. long, glabrous; inflorescence branched; prin-

cipal cauline leaves 2, distant, the lower one thick, long-petiolate, peltate, orbicular, about 25 cm. across, glabrous above, whitish and loosely pilose beneath, palmately parted, the main segments 6, usually twice bifid, the lobes lanceolate, 10–17 mm. wide, acuminate, the median leaf short-petiolate, palmately parted, about 14 cm. across, the segments 3 or 4, bifid, the uppermost leaves very small, lanceolate or narrowly so, 2–7 mm. long; heads in corymbs, 7–10 mm. long, pedicelled, with 3 or 4 small linear bractlets at base; involucres about 1 cm. long, the bracts 5; pappus sordid-white.——Aug. Shikoku (Tosa Prov.); rare.

3. Syneilesis palmata (Thunb.) Maxim. *Arnica palmata* Thunb.; *Senecio palmatus* (Thunb.) Less.; *S. krameri* Fr. & Sav.; *Cacalia palmata* (Thunb.) Makino; *C. thunbergii* Nakai——YABUREGASA. Rhizomes short; stems 70–120 cm. long, glaucous, silky while young; cauline leaves 2(–3), chartaceous, distant, the lower one long-petiolate, peltate, orbicular, 35–40(–50) cm. across, palmately 7- to 9-parted, the segments often twice bifid and the lobes 2–4 cm. wide, acuminate, acutely toothed, pubescent while young, often whitish and at first silky beneath, the median leaves slightly smaller, short-petiolate; heads many, 8–10 mm. across, in panicles, the pedicels puberulent, with small bractlets; involucre (7–)9–10 mm.

long, the bracts 5, glabrous; achenes 4.5–6 mm. long; pappus sordid-white, minutely scabrous.——July–Oct. Woods in low mountains; Honshu, Shikoku, Kyushu.——Korea.

38. ACHILLEA L.   Nokogiri-sō Zoku

Perennial herbs; leaves alternate, oblong-linear, toothed to twice pinnate, sometimes woolly; heads relatively small, usually in dense corymbs, rarely solitary, ligulate; involucres subglobose or globose-campanulate, the bracts 2- or 3-seriate, the outer ones short, thickened on back, scarious on margin, dark brownish; receptacle convex or conical, chaffy; style-branches truncate and penicillate at apex; anthers bifid at base, usually acute; achenes much-flattened, oblong, truncate at apex, glabrous; pappus absent.——About 100 species, in Europe, Asia, N. America and N. Africa, especially abundant in the Mediterranean region.

1A. Leaves serrulate, not clasping, the median limb (undivided portion) 3.5–10 mm. wide; involucral bracts and chaff densely hairy.
                                                                                                            1. A. ptarmica
1B. Leaves serrate, bipinnately cleft, clasping, the median limb 1–7 mm. wide; involucral bracts less densely pubescent; chaff glabrous or
       only sparingly hairy. .... 2. A. sibirica

1. **Achillea ptarmica** L. var. **macrocephala** (Rupr.) Masam. *A. macrocephala* Rupr.; *A. ptarmica* var. *speciosa* sensu Herd., excl. syn. and sensu auct. Japon.; *A. ptarmica* subsp. *macrocephala* (Rupr.) Heim.——Ezo-nokogiri-sō. Stems 10–85 cm. long, densely appressed-pubescent on upper portion; median cauline leaves oblong-linear or linear-lanceolate, 3–7 cm. long, 4–11 mm. wide, usually obtuse, sessile and semiclasping, regularly toothed, the leaf limb 3.5–10 mm. wide, the teeth small, incurved, acute, serrulate; heads usually many, in corymbs, rarely solitary; involucres semiglobose, about 5 mm. long, 9–11 mm. wide, densely silky-hairy; bracts 2-seriate, oblong, obtuse, the outer short; ligulate florets 2-seriate, 12–19, the corollas whitish, 6–7 mm. long, about 4 mm. wide; achenes about 2 mm. long, 1 mm. wide.——July–Oct. Hokkaido, Honshu (centr. distr. and northw.).——Kuriles, Sakhalin, Kamchatka, and e. Asia.

Var. **yezoensis** Kitam. Hosoba-Ezo-nokogiri-sō. Stems to 50 cm. long, often with prominent buds in the axils; leaves linear, 5–6.5 cm. long, 3–6 mm. wide, toothed, not clasping; heads slightly smaller; involucres 3–4 mm. long, 7.5–9 mm. wide; ligulate florets 6–8.——Hokkaido (Kamuikotan in Ishikari Prov.).

2. **Achillea sibirica** Ledeb. *A. mongolica* Fisch.; *A. sibirica* var. *typica* Herd.; *A. sibirica* subsp. *mongolica* (Fisch.) Heim.——Nokogiri-sō, Hagoromo-sō. Stems 50–100 cm. long, densely pubescent in upper part; cauline leaves 6–10 cm. long, 7–15 mm. wide, obtuse, sessile, the lobes semiclasping, pectinate, the lobules oblong-lanceolate, acute, acutely toothed, the median leaf limb 1–7 mm. wide, silky-hairy while young; heads many, 7–9 mm. across, in dense corymbs; involucres globose-campanulate, slightly pubescent, 5 mm. long and as wide, the bracts oblong, the outer shorter; ligulate florets 5–7, the corollas 3.5–4.5 mm. long, 2.5–3 mm. wide; achenes about 3 mm. long, about 1 mm. wide, glabrous.   Very variable.

Key to the varieties of *A. sibirica*

1A. Leaves bipinnately cleft to parted.
   2A. Heads less than 10 mm. across; ligulate corollas 3.5–4.5 mm. long. .... var. *sibirica*
   2B. Heads 12–14 mm. across; ligulate corollas about 6 mm. long.
       3A. Ligulate florets 8–12. .... var. *camtschatica*
       3B. Ligulate florets 6–8. .... var. *angustifolia*
1B. Leaves minutely toothed or incised-toothed.
   4A. Leaves minutely toothed; ligulate florets 9–11. .... var. *pulchra*
   4B. Leaves regularly incised-toothed; ligulate florets 5–6. .... var. *brevidens*

Var. **sibirica**. Nokogiri-sō, Hagoromo-sō. Heads less than 10 mm. across; ligulate corollas 3.5–4.5 mm. long.——July–Nov. Hokkaido, Honshu.——Sakhalin, s. Korea, Manchuria, China, e. Siberia, Kamchatka, Aleutians, and N. America.——Forma **discoidea** Regel. *A. ptarmicoides* Maxim.; *A. sibirica* var. *ptarmicoides* (Maxim.) Makino——Yama-nokogiri-sō. Ligulate corollas 3 mm. long.

Var. **camtschatica** (Heim.) Ohwi. *Ptarmica kamtschatica* Rupr. ex Komar.; *A. sibirica* subsp. *camtschatica* Heim.——Shumushu-nokogiri-sō.   Cauline leaves lanceolate or narrowly oblong, 3–9 cm. long, 5–22 mm. wide, pinnate-toothed; involucres densely pubescent; ligulate florets 8–12.——Hokkaido (Teshio and Kitami Prov.).——Sakhalin, Kuriles, and Kamchatka.

Var. **angustifolia** (Hara) Ohwi. *A. sibirica* subsp. *japonica* Heim.; *A. pulchra* var. *angustifolia* Hara——Horoman-nokogiri-sō.   Stems elongate and branched; leaves bipinnately parted or cleft; involucres less densely hairy; ligulate florets

6–8, the corollas about 6 mm. long, 4–5 mm. wide.——Hokkaido, Honshu (centr. and n. distr.).——Sakhalin and Kuriles.

Var. **pulchra** (Koidz.) Ohwi. *A. pulchra* Koidz.; *A. speciosa* var. *pulchra* (Koidz.) Nakai; *A. sibirica* subsp. *pulchra* (Koidz.) Kitam.——Akabana-nokogiri-sō.   Leaves 5–7 cm. long, 9–13 mm. wide, minutely toothed, the median leaf-limb about 7 mm. wide; involucres loosely hairy; ligulate florets 9–11, the corollas pink or white, 5–7 mm. long, 4.5–5 mm. wide.——Hokkaido.

Var. **brevidens** (Makino) Ohwi. *A. cartilaginea* Miq.; *A. ptarmicoides* forma *brevidens* Makino; *A. sibirica* var. *ptarmicoides* subvar. *brevidens* Makino; *A. sibirica* subsp. *subcartilaginea* Heim.——Aso-nokogiri-sō.   Leaves regularly incised-toothed, the median leaf limb 2–6.5 mm. wide; heads somewhat smaller; ligulate florets 5–6, the corollas 5–6 mm. long, 3.5–4 mm. wide, white or pink.——Mountains of Kyushu.

39. CENTIPEDA Lour.   TOKIN-SŌ ZOKU

Small annuals; stems decumbent, much branched, leafy; leaves alternate, spathulate, toothed; heads small, axillary, pedicelled or sessile; involucres small, the bracts 2-seriate, spreading, equal, oblong, with a narrow membranous margin; pistillate florets marginal, many-seriate, fertile, the corollas tubular, about 0.2 mm. long; bisexual florets central, fertile, the corollas about 0.5 mm. long, deeply 4-lobed; style-branches in bisexual florets short, obtuse, the anthers obtuse at base, without appendages; achenes 5-angled, hairy, narrowed at base.——Few species, in Madagascar, India, e. Asia, Australia, and S. America.

1. Centipeda minima (L.) A. Br. & Asch. *C. orbicularis* Lour.; *Artemisia minima* L.; *Myriogyne minuta* Less.; *C. minuta* (Less.) C. B. Clarke——TOKIN-SŌ.   Green, thinly cobwebby or glabrescent annual; stems slender, 5–20 cm. long, many-leaved, decumbent or procumbent, much branched and often rooting; leaves alike, nearly equal, spathulate, 7–20 mm. long, obtuse, with few teeth toward the tip, glandular-spotted beneath, sessile; heads depressed-globose, green, 3–4 mm. across, axillary, sessile or subsessile; involucral bracts oblong, equal; ligulate florets minute, the corollas green; achenes about 1.3 mm. long, minutely hirsute, 5-angled.——July–Nov. Waste grounds and roadsides; Hokkaido, Honshu, Shikoku, Kyushu; common.——Ryukyus, Formosa, e. Siberia, Korea, China, India, Malaysia, and Australia.

40. MATRICARIA L.   SHIKAGIKU ZOKU

Annuals or biennials, sometimes strongly scented; leaves alternate, 2 or 3 times pinnate, the ultimate segments linear; heads small or relatively large, pedicelled, solitary, ligulate or eligulate; involucres subglobose, the bracts 4-seriate; receptacle conical; ligulate florets marginal, 1-seriate, fertile, the corollas white; tubular florets many, bisexual, the corollas yellow, 4- or 5-lobed; anthers obtuse at base; style branches truncate and mammillate at apex; achenes oblong, truncate at apex, ribless on back, 3-ribbed in front, often with 2 oil-glands, the pappus crownlike or absent.——About 50 species, in the N. Hemisphere and S. Africa.

1A. Ligulate florets absent. . . . . 1. *M. matricarioides*
1B. Ligulate florets present. . . . . 2. *M. tetragonosperma*

1. Matricaria matricarioides (Less.) Porter. *Artemisia matricarioides* Less.; *Tanacetum matricarioides* Less.; *M. discoidea* DC.; *Santolina suaveolens* Pursh; *M. suaveolens* (Pursh) Buchen., non L.——KO-SHIKAGIKU, OROSHAGIKU. Slightly pubescent or nearly glabrous annual; stems 10–30 cm. long, often much branched; cauline leaves narrowly oblong or oblanceolate, 2–5 cm. long, 8–18 mm. wide, bipinnate, the pinnae many, the ultimate segments linear, acute, about 0.5 mm. wide, the basal ones clasping, the upper leaves gradually smaller; heads erect on stout pedicels; involucres 3–3.5 mm. long, 6–9 mm. across; bracts 4-seriate, oblong, the outer as long as the middle, longer than the innermost, rounded at apex; achenes glabrous, oblong, 1.5 mm. long, 3-ribbed on both sides, the pappus very short, cuplike, white.——July–Oct. Waste grounds especially near seashores; Hokkaido.——nw. America, e. Asia, and Europe.

2. Matricaria tetragonosperma (F. Schmidt) Hara & Kitam. *Chamaemelum tetragonospermum* F. Schmidt; *Tripleurospermum inodorum* var. *ambiguum* Reichenb. ex Herd.; *T. ambiguum* Fr. & Sav.; *M. ambigua* Miyabe——SHIKAGIKU. Short-pubescent to nearly glabrous annual; stems 15–50 cm. long, often branched; leaves subsessile, narrowly oblong, 5.5–18 cm. long, 2–5 cm. wide, thrice pinnate, the ultimate segments 0.3–0.6 mm. wide, acute, the upper leaves clasping, smaller; heads few, 3.5–4 cm. across; involucres 7–8 mm. long, the median bracts longest, narrowly oblong; ligulate corollas 15–18 mm. long, 3–4.5 mm. wide; achenes about 3 mm. long, nearly 4-angled, with a pair of black spots on both sides below the apex.——July–Aug. Sandy places near seashores; Hokkaido.——Sakhalin, Kuriles, Manchuria, and Siberia.

41. CHRYSANTHEMUM L.   KIKU ZOKU

Perennial or annual herbs, sometimes subshrubs; leaves alternate, pinnately divided or undivided; heads solitary on branchlets or in corymbs; involucres subglobose, the bracts many-seriate, imbricate, appressed, often scarious and brownish on margin; receptacle convex, hemisphaeric, sometimes flat; pistillate florets marginal, fertile, 1- to many-seriate, the corollas ligulate or tubular, 3-lobed; disc flowers fertile, rarely sterile, the corollas 5-lobed; anthers obtuse and entire at base, with an oblong appendage at apex; style-branches truncate at apex; achenes terete or 5- to 10-ribbed, sometimes slightly flattened, glabrous, the pappus scalelike.——About 200 species, in the N. Hemisphere, and S. Africa.

1A. Heads eligulate, in dense corymbs, the achenes not viscid when moistened, with a very short crownlike pappus; leaves bipinnate.
                                                    1. *C. vulgare*
1B. Heads ligulate or, if eligulate, the achenes without a pappus.
  2A. Plant shrubby; achenes with a very short crownlike pappus. . . . . 2. *C. nipponicum*
  2B. Plant herbaceous; achenes without a crownlike pappus.
    3A. Achenes not viscid when moistened.
      4A. Achenes 5-ribbed; ligulate florets fertile; leaves or lobes of leaves not linear.
        5A. Leaves pinnately lobed or cleft, or 3-cleft; achenes with the margins not raised. . . . . 3. *C. yezoense*
        5B. Leaves pinnately parted and again pinnately lobed or toothed; achenes with slightly raised edges at apex. . . . . 4. *C. weyrichii*
      4B. Achenes 10-ribbed; ligulate florets sterile; leaves linear or deeply 3- to 5-parted into linear segments. . . . . 5. *C. lineare*
    3B. Achenes viscid when moistened.

6A. Leaves densely woolly beneath.
　7A. Leaves silvery-woolly beneath.
　　8A. Leaves truncate at base; heads 4–5 cm. across .... 6. *C. ornatum*
　　8B. Leaves broadly cuneate at base; heads narrower.
　　　9A. Leaves with shallow pinnate lobes in those of upper part; near seashores.
　　　　10A. Involucres about 10 mm. across, the outer bracts linear. .... 7. *C. shiwogiku*
　　　　10B. Involucres about 6 mm. across, the outer bracts ovate. .... 8. *C. pacificum*
　　　9B. Leaves deeply pinnately cleft into narrow lobes; involucres 3–4 mm. across; alpine. .... 9. *C. rupestre*
　7B. Leaves grayish white-woolly beneath.
　　11A. Leaves usually cuneate at base; involucral bracts nearly equal.
　　　12A. Leaves obovate or oblanceolate, shallowly pinnatilobed, grayish woolly on upper side. .... 10. *C. yoshinaganthum*
　　　12B. Leaves ovate or broadly so, 3-cleft, green on upper side. .... 11. *C. makinoi*
　　11B. Leaves slightly cordate at base; outer involucral bracts shorter than the inner ones. .... 12. *C. japonense*
6B. Leaves slightly pubescent beneath.
　13A. Leaves pinnately cleft, cuneate at base; involucres 13 mm. across, the outer bracts linear. .... 13. *C. cuneifolium*
　13B. Leaves rarely deeply cleft, truncate to cordate at base.
　　14A. Heads solitary on the branchlets, 3–7 cm. across; outer involucral bracts linear. .... 14. *C. zawadskii*
　　14B. Heads many, in corymbs.
　　　15A. Heads 12–20 mm. across, in false umbels.
　　　　16A. Involucres about 5 mm. long, the outer bracts linear; stems erect from the base. .... 15. *C. boreale*
　　　　16B. Involucres 6.5 mm. long, the outer bracts oblong or ovate; stems decumbent. .... 16. *C. okiense*
　　　15B. Heads often more than 25 mm. across, in corymbs.
　　　　17A. Heads about 2.5 cm. across; outer involucral bracts ovate or oblong. .... 17. *C. indicum*
　　　　17B. Heads 3–4.5 cm. across; outer involucral bracts oblong or linear. .... 18. *C. aphrodite*

1. Chrysanthemum vulgare (L.) Bernh. *Tanacetum vulgare* var. *boreale* (Fisch.) Makino ex Makino & Nemoto; *T. boreale* Fisch. ex DC.; *T. vulgare* var. *boreale* (Fisch.) Trautv. & Mey.; *T. vulgare* sensu auct. Japon., non L.——Ezo-yomogigiku. Rhizomes long-creeping; stems 60–70 cm. long, slightly cobwebby-pubescent, branched in upper part; median cauline leaves oblong, 15–25 cm. long, 7–11 cm. wide, often clasping at base, slightly cobwebby-pubescent, bipinnate, the pinnae about 12-pairs, approximate, lanceolate, 12–20 mm. wide, acuminate, the pinnules 2–3 mm. wide, acute, callose-toothed, glandular-spotted on upper side, the upper leaves gradually smaller and less divided; heads in dense corymbs, short-pedicelled, about 1 cm. across; involucres depressed-globose, 4 mm. long; bracts 4-seriate, the outer ovate-oblong, obtuse, shorter than the median; ligulate florets 1-seriate, the tubular corollas about 2.5 mm. long, yellow; achenes about 2 mm. long, with a very short crownlike pappus.——July–Sept. Hokkaido.——Korea, Manchuria, Sakhalin, and e. Siberia.

2. Chrysanthemum nipponicum (Franch.) Matsum. *Leucanthemum nipponicum* Franch. ex Maxim.——Hama-giku. Stems to 1 m. long, branched, puberulent in upper part; cauline leaves approximate, spathulate, 2–9 cm. long, 1.3–2 cm. wide, obtuse, obtusely undulate-toothed, gradually narrowed and entire at base, fleshy, puberulent, lustrous, 1-nerved, sessile, the upper leaves gradually smaller, narrowly oblong; heads long-pedicelled, solitary, about 6 cm. across; involucres about 1 cm. long; bracts 4-seriate, puberulent, brownish on margin, the outer ovate-oblong, obtuse, the inner narrowly oblong; receptacle glabrous; ligulate corollas white, 22–28 mm. long, 5–6 mm. wide; achenes 3–4 mm. long.—— Near seashores; Honshu (Kantō Distr. and northw.).

3. Chrysanthemum yezoense Maekawa. *C. arcticum* sensu auct. Japon., non L.; *C. arcticum* subsp. *maekawanum* Kitam.——Ko-hamagiku. Plant stoloniferous and with long-creeping rhizomes; stems 10–50 cm. long, purplish, slightly pubescent; radical and lower cauline leaves long-petiolate, ovate-cuneate or broadly so, 1–4 cm. long and as wide, pinnately cleft or lobed, the lobes oblong, obtusely toothed on upper margin, fleshy and nearly glabrous, glandular-punctate, the upper leaves spathulate, sometimes with a pair of stipule-like appendages at base; heads solitary on branches, long-pedunculate, erect; outer involucral bracts linear, 1–1.8 mm. wide, rounded at apex, the median as long as the outer, oblong, about 3 mm. wide, with brown scarious margins, slightly reddish on upper half, the inner bracts slightly shorter; ligulate florets 1-seriate, the corollas white, about 2 cm. long, 4–6 mm. wide; achenes about 2 mm. long.——Sept.–Dec. Rocks near seashores; Hokkaido, Honshu (n. Kantō Distr. and northw.).

Chrysanthemum × miyatojimense Kitam. Miyato-jimagiku. Hybrid of *C. morifolium* × *C. yezoense*.—— Honshu (Rikuzen Prov.).

4. Chrysanthemum weyrichii (Maxim.) Miyabe. *Leucanthemum weyrichii* Maxim.; *C. littorale* Maekawa; *C. weyrichii* var. *littorale* (Maekawa) Kudo——Pireo-giku, Ezo-no-sonaregiku. Plant stoloniferous; stems 10–30 cm. long, simple or branched, purple, soft-pubescent; radical and lower cauline leaves long-petiolate, fleshy, suborbicular, palmately 5-cleft or parted, the segments oblong, pinnately lobed to parted, slightly pubescent or nearly glabrous, the upper leaves smaller, pinnately cleft or linear and undivided; heads solitary, 4–4.5 cm. across, long-pedunculate; outer and median involucral bracts equal in length, the outer linear, about 1 mm. wide, the median oblong, about 3 mm. wide, the inner somewhat shorter; ligulate florets 1-seriate, the corollas pink or white, 15–16 mm. long, 4–4.5 mm. wide; achenes about 2 mm. long. ——Aug.–Oct. Rocks near seashores; Hokkaido.——Sakhalin.

5. Chrysanthemum lineare Matsum. Mikoshi-giku, Hosoba-no-seitakagiku. Plants stoloniferous; stems 30–100 cm. long, often branched, puberulent; lower leaves 4.5–9 cm. long, often withering before anthesis, sessile, subentire to 3-lobed or -cleft, or rarely pinnately cleft, the lobes 1 or 2 pairs, erect, linear, often incised to the base; heads solitary, 3–6 cm. across; involucres 5–6 mm. long, slightly woolly; bracts equal in length, the outer narrowly linear, rounded at apex, the inner oblong, with broad scarious margins; ligulate florets 1-seriate, sterile, the corollas white, about 2 cm. long, 5 mm. wide; achenes with 10 prominent ribs.——Sept.–Nov. Wet places in mountains; Honshu, Kyushu.——Korea and Manchuria.

6. Chrysanthemum ornatum Hemsl. *C. sinense* var. *satsumense* Yatabe; *C. decaisneanum* var. *satsumense* (Yatabe) Makino——SATSUMA-GIKU. Rhizomes creeping, becoming woody; stems 25–50 cm. long, stout, branched, densely silvery-woolly; lower leaves thick, 4–6 cm. long, 4–5.5 cm. wide, pinnately lobed, truncate, long-petiolate, the lobes of 2 pairs, undulately toothed, green, with a white upper margin, with dorsifixed dense white-villous hairs beneath, the petioles 1.5–3.5 cm. long, the upper leaves gradually smaller; heads many, 4–5 cm. across; involucres about 6 mm. long; bracts 3-seriate, the outer linear or narrowly oblong, half as long as the inner, obtuse, the inner bracts oblong; ligulate florets 1-seriate, the corollas 18–22 mm. long, 4–5 mm. wide, white or somewhat pinkish; achenes about 1.5 mm. long.——Oct.–Dec. Near seashores; Kyushu (s. distr.).

7. Chrysanthemum shiwogiku Kitam. *C. decaisneanum* Matsum., excl. syn.; *C. decaisneanum* var. *discoideum* Makino; *C. ornatum* var. *discoideum* (Makino) Nakai——SHIOGIKU. Rhizomes creeping, woody; stems ascending, decumbent at base, 25–35 cm. long, densely silvery-hairy; lower leaves thick, obovate, 4–5 cm. long, 2.5–3 cm. wide, cuneate at base, the petioles 1–1.5 cm. long, pinnately lobed on upper half, the lobes broad, with 1 or 2 obtuse teeth or entire, green above, densely hairy with dorsifixed silvery hairs; heads in corymbs, on slender peduncles; involucres 5–6 mm. long; bracts 3-seriate, the outer linear or narrowly oblong, often densely hairy, the median and inner elliptic, broadly brownish scarious on margin; ligulate florets pistillate, the corollas often tubular, 3- or 4-lobed, sometimes slightly ligulate; achenes about 1.8 mm. long.——Nov.–Dec. Near seashores; Shikoku.

Var. **kinokuniense** Shimotomai & Kitam. KI-NO-KUNI-SHIOGIKU. Leaves narrower, often subspathulate, smaller, short-lobed near the top; involucres narrower.——Honshu (Kii and Shima Prov.).

Var. **ugoense** Hara & Mori. UGO-SHIOGIKU. Leaves broader, rhombic-obovate, 3–4 cm. wide, with deeper lobes; involucres slightly smaller, the outer bracts broader, ovate to ovate-oblong.——Rocky cliffs near seashores; Honshu (Ugo Prov.).

8. Chrysanthemum pacificum Nakai. *Pyrethrum marginatum* sensu Maxim., non Miq.; *C. marginatum* Miq. ex Matsum., non Korth.——ISOGIKU. Rhizomes long-creeping; stems arcuate-ascending, 30–40 cm. long, densely leafy at the top, densely silvery-tomentose above; cauline leaves thick, oblanceolate to obovate, 4–8 cm. long, 1.5–2.5 cm. wide, cuneate at the base, short-petiolate, pinnately lobed, green and nearly glabrous on upper side, glandular-dotted, white-margined, densely covered with silvery peltate hairs, the lobes often entire, obtuse, rarely 1- or 2-toothed; heads many, in corymbs; involucres depressed-globose, 4–5 mm. long, 5–6 mm. across; bracts 3-seriate, the outer short, ovate, white-hairy, the inner rhombic-elliptic; marginal corollas tubular, rarely ligulate, about 3 mm. long, 3- or 4-lobed, white or yellow; achenes about 1.5 mm. long.——Near seashores; Honshu (Kazusa, Sagami, Izu, and Suruga Prov. incl. Izu Isls.).

9. Chrysanthemum rupestre Matsum. & Koidz. *C. pallasianum* var. *japonicum* (Fr. & Sav.) Matsum.; *Pyrethrum pallasianum* var. *japonicum* Fr. & Sav.——IWA-INCHIN, IN-CHIN-YOMOGI. Rhizomes short-creeping; stems densely tufted, 10–18 cm. long, densely white-hairy, leafy to the top; cauline leaves obovate, sessile, 2–2.5 cm. long, 1–2 cm. wide,

pinnately parted or 3-parted, the segments usually 2 pairs, narrowly linear, obtuse, green, pubescent above while young, silvery-tomentose beneath; heads 3–4 mm. across, few to rather numerous, in terminal corymbs, eligulate; involucres depressed-globose, about 2.5 mm. long; bracts 3-seriate, the outer short, oblong or ovate, the inner oblong, broadly scarious on the margins; ligulate florets few or absent, the corollas small, tubular, shorter than the disc florets.——Aug.–Sept. Rocky cliffs in high mountains; Honshu (centr. and n. distr.); rare.

Chrysanthemum × konoanum Makino. TOGAKUSHI-GIKU. Hybrid of *C. makinoi* × *C. rupestre*.——Reported to occur in Honshu (Mount Togakushi in Shinano Prov.).

10. Chrysanthemum yoshinaganthum Makino ex Kitam. NAKAGAWA-NOGIKU. Rhizomes woody; stems ascending, decumbent at base, about 60 cm. long, much branched, with dorsifixed hairs; cauline leaves rather coriaceous, densely arranged, obovate or oblanceolate-cuneate, 5–6 cm. long, 1.5–2.5 cm. wide, deeply 3-cleft to -lobed, the lobes oblong, obtuse, obtusely toothed, green, short-pubescent above, 3-nerved, with grayish appressed dorsifixed tomentum beneath, the petioles 5–10 mm. long, the upper leaves smaller; heads 3–4 mm. across, solitary or in loose corymbs, on long rigid peduncles; involucres 9–10 mm. long; bracts 3-seriate, the outer linear, 6–10 mm. long, about 1 mm. wide, gray-pubescent, the median and inner 7–8 mm. long, oblong, rounded at apex; ligulate florets 1- or 2-seriate, the corollas whitish or pinkish, 18–20 mm. long, about 4 mm. wide; achenes about 1.5 mm. long.——Dec. Valleys and ravines; Shikoku.

11. Chrysanthemum makinoi Matsum. & Nakai. *Pyrethrum sinense* var. *japonicum* Maxim.; *C. sinense* var. *japonicum* (Maxim.) Maxim. ex Matsum.; *C. japonicola* Makino; *C. makinoi* var. *japonicum* (Maxim.) Nakai; *C. makinoi* var. *laciniatum* Nakai; *C. makinoi* var. *elatum* Nakai——RYŪNŌGIKU. Plants stoloniferous, becoming woody; stems ascending, 40–80 cm. long, often branched, the branches gray-tomentose; lower cauline leaves ovate or broadly so, cuneate rarely subtruncate at base, short-petiolate, green and short-pubescent on upper side, with grayish dorsifixed tomentum, 3-cleft or rarely-lobed, the lobes coarsely toothed, chartaceous; heads solitary, 2.5–5 cm. across, on slender peduncles; involucres about 7 mm. long; bracts equal in length, the outer narrowly linear, obtuse, gray-tomentose, the inner oblong, brownish and scarious on margin; ligulate florets white or pinkish, rarely yellow, the corollas 1–1.5 cm. long, 2–3 mm. wide, rarely tubular; achenes about 1.8 mm. long.——Oct.–Nov. Sunny hills and mountains; Honshu, Shikoku; common.

Var. **wakasaense** (Shimotomai) Kitam. *C. wakasaense* Shimotomai——WAKASA-HAMAGIKU. Stems stouter; leaves larger, usually 5–10 cm. long, 2.5–5.5 cm. wide; heads many, about 2.5 cm. across, on slender peduncles.——Seashores; Honshu (Echizen to Tajima Prov.).

12. Chrysanthemum japonense (Makino) Nakai. *C. sinense* var. *spontaneum* Makino; *C. sinense* var. *sinense* Makino; *C. morifolium* var. *genuinum* forma *japonense* Makino; *C. japonense* var. *debile* Kitam.——NOJIGIKU, SETO-NOJIGIKU. Stoloniferous; stems ascending, stout, decumbent at base, branched above, densely gray-tomentose; basal leaves withering before anthesis; lower cauline leaves broadly ovate, 3–5 cm. long, 2.5–4 cm. wide, usually shallowly cordate to truncate at base, the petioles 1.5–3 cm. long, pinnately cleft, the lateral lobes 2(–1)-pairs, elliptic, approximate, rounded at apex, obtusely toothed, the terminal lobes larger,

3-lobed, loosely pubescent above, with gray dorsifixed tomentum beneath, the upper leaves gradually smaller; heads rather numerous, sometimes loosely corymbose, 3–5.5 cm. across; involucres about 8 mm. long; bracts 3-seriate, the outer linear or linear-oblong, half as long as the inner or longer, rounded at apex, gray-tomentose, the inner oblong; ligulate florets 1(–2)-seriate, the corollas 1–2 cm. long, 3–5 mm. wide, white or pinkish, sometimes yellowish; achenes about 1.8 mm. long.——Oct.–Nov. Along the seacoast; Shikoku, Kyushu.

Var. **ashizuriense** Kitam. ASHIZURI-NOJIGIKU.   Leaves thicker, smaller, more densely tomentose beneath, usually 3-lobed; heads smaller, more numerous.——Along the seacoast; Shikoku.

13. Chrysanthemum cuneifolium Kitam. WAJIKI-GIKU.   Stems slender, much branched, elongate, becoming decumbent at base, with gray dorsifixed pubescence while young; median cauline leaves approximate, obovate-cuneate, about 6.5 cm. long, up to 3 cm. wide, long-tapering at base, usually 3-lobed, rarely 5-lobed, the lobes chartaceous, oblong, erect, acute to obtuse, mucronate-toothed, green and short-pubescent above, with dorsifixed short pubescence beneath, 3- to 5-nerved, the petioles about 2 cm. long; heads in loose corymbs, 23–25 mm. across, on slender peduncles, with few lanceolate bractlets; involucres about 7 mm. long; bracts nearly equal in length, rounded at apex, the outer linear, about 1 mm. wide, the median ovate-oblong, the inner elliptic; ligulate corollas 12.5–14 mm. long, about 2 mm. wide; achenes about 1.5 mm. long.——Nov. Shikoku (Wajiki in Awa).

14. Chrysanthemum zawadskii Herbich. *C. sibiricum* Fisch. ex Turcz.; *Leucanthemum sibiricum* var. *acutilobum* DC.; *C. sibiricum* var. *acutilobum* (DC.) Komar.; *C. zawadskii* var. *acutilobum* (DC.) Sealy; *C. zawadskii* var. *sibiricum* Sealy; *C. hakusanense* Makino——IWAGIKU.   Rhizomes often long-creeping, woody, simple or branched, purplish; radical and lower cauline leaves long-petiolate, broadly ovate, 1–3.5 cm. long, 1–4 cm. wide, bipinnate, the ultimate segments 1–2 mm. wide, acute or obtuse, glandular-dotted, loosely appressed-pubescent to nearly glabrous, lustrous above, the petioles 2–4.5 cm. long; upper leaves gradually smaller and sometimes clasping at base, the uppermost ones linear; heads solitary, 3–6 cm. across; involucres 6–7 mm. long; bracts 3-seriate, nearly equal in length, the outer narrowly linear, rounded, the inner oblong; ligulate florets 1-seriate, the corollas 1.5–3 cm. long, 3–6.5 mm. wide; achenes about 2 mm. long.——Aug.–Oct. Mountains; Honshu (Echizen, Kaga, and Yamato Prov.), Shikoku, Kyushu.——Korea, n. China, Manchuria, Siberia, and the Carpathian Mountains.

Var. **latilobum** (Maxim.) Kitam. *Leucanthemum sibiricum* DC.; *L. sibiricum* var. *latilobum* Maxim.; *C. sibiricum* var. *latilobum* (Maxim.) Komar.; *C. naktongense* Nakai; *C. erubescens* Stapf.; *C. rubellum* Sealy——CHŌSEN-NOGIKU.   Stems stout, to 1 m. long; leaves long-petiolate, ovate or broadly so, truncate to shallowly cordate at base; upper leaves often cuneate at the base, pinnati-cleft to -lobed, the lateral lobes usually 4, oblong, obtuse, toothed or shallowly lobulate; heads to 8 cm. across.——Aug.–Oct. Kyushu (Tsushima, Hirado Isl., and Isomadake in Satsuma).——Korea, Manchuria, and n. China.

Var. **campanulatum** (Makino) Kitam. *C. sibiricum* var. *campanulatum* Makino——OGURAGIKU.   Heads 3–5 cm.

across; ligulate florets many, the corollas broad, often connate; tubular florets few, the corollas to 2 cm. long; leaves broadly ovate, pinnately lobed.——Cultivated in our area.

15. Chrysanthemum boreale (Makino) Makino. *C. indicum* var. *boreale* Makino; *C. lavandulaefolium* sensu auct. Japon., non Fisch.; *Artemisia debilis* Kitam.; *C. seticuspe* var. *boreale* (Makino) Hand.-Mazz.——ABURAGIKU, AWA-KOGANE-GIKU.   Rhizomes short; stems tufted, erect, 1–1.5 m. long, branched in upper part, rather prominently white-pubescent; median cauline leaves oblong-ovate, 5–7 cm. long, 4–6 cm. wide, slightly cordate or truncate and abruptly narrowed at base, pinnately parted, the segments 2 pairs, oblong, obtuse, acutely incised-toothed, with a broad sinus at base, loosely pubescent above, with dorsifixed pubescence beneath, the petioles 1–2 cm. long; heads nearly corymbose, terminal, many, about 15 mm. across, nodding after anthesis; involucres about 4 mm. long; bracts 3- or 4-seriate, the outer linear or narrowly oblong, soft-pubescent, the inner elliptic; ligulate corollas 5–7 mm. long, 1.5–2 mm. wide, yellow; achenes about 1 mm. long.——Oct.–Nov. Honshu, Kyushu (Iki and Tsushima).——Korea, Manchuria, and n. China.

Chrysanthemum × leucanthum Makino. SHIROBANA-ABURAGIKU.   Hybrid of *C. boreale* × *C. makinoi*.

16. Chrysanthemum okiense Kitam. *C. boreale* var. *okiense* (Kitam.) Okuyama——OKI-NO-ABURAGIKU.   Rhizomes creeping; stems to 60 cm. long, often decumbent, branched above, gray-pubescent; median cauline leaves chartaceous, dense, ovate, obtuse, truncate or shallowly cordate at base, falsely stipulate, short-pubescent above, with soft dorsifixed pubescence beneath, deeply pinnaticleft, petiolate, the lateral lobes 2 pairs, oblong, obtuse, irregularly incised-toothed; uppermost leaves spathulate, pinnatilobed or linear and entire; heads many, in false umbels, 15–20 mm. across, the pedicels densely white-pubescent; involucres about 5 mm. long; bracts 3-seriate, the outer narrowly oblong or ovate, obtuse, loosely short-pubescent, the inner oblong; ligulate corollas about 8 mm. long, 2.5 mm. wide; achenes about 1.5 mm. long.——Nov. Honshu (Oki Isl.).

17. Chrysanthemum indicum L. *C. sabinii* Lindl.; *C. indicum* var. *coreanum* Lév.——ABURAGIKU, HAMA-KANGIKU, SHIMA-KAN-GIKU.   Plant stoloniferous and with creeping rhizomes; stems leafy, slightly decumbent with deciduous soft dorsifixed pubescence; median cauline leaves chartaceous, ovate-oblong, 3–5 cm. long, 2.5–4 cm. wide, slightly cordate to truncate and abruptly narrowed at base, falsely stipulate, the petioles 1–2 cm. long, pinnately cleft, the terminal lobe largest, obtuse, the lateral lobes 2 pairs, elliptic to oblong, toothed, with a narrow or rather broad sinus between the lobes, short-pubescent above, with soft dorsifixed pubescence beneath; heads about 2.5 cm. wide, in loose terminal corymbs, the pedicels slender; involucres 5–6 mm. long; bracts 4-seriate, the outer small, oblong or ovate, the inner shorter than the median; ligulate corollas yellow, 11–13 mm. long, 2.5–3 mm. wide; achenes about 1.8 mm. long.——Oct.–Dec. Honshu (Kinki Distr. and westw.), Shikoku, Kyushu.

Var. **albescens** Makino. *C. indicum* subsp. *albescens* (Makino) Kitam.——SHIROBANA-KANGIKU.   Ligulate corollas white.

Var. **iyoense** Kitam. IYO-ABURAGIKU.   Branches prominently soft-pubescent; leaves somewhat smaller, about 4 cm. long, 2.5 cm. wide, pinnately parted, the segments 2 pairs, oblong, pinnately cleft, green and pubescent above, rather densely pubescent with dorsifixed hairs.——Shikoku (Iyo Prov.).

Chrysanthemum × ogawae Kitam. HI-NO-MISAKIGIKU. Hybrid of *C. indicum × C. shiwogiku* var. *kinokuniense.*

18. Chrysanthemum aphrodite Kitam. SANINGIKU. Stoloniferous; stems about 1 m. long, woody at base, glabrous, decumbent, the branches white-pubescent; median and upper cauline leaves thin, broadly ovate, 5–7 cm. long, 4.5–5.5 cm. wide, cordate and long-decurrent at base, scabrous and green above, white-pubescent beneath, pinnately cleft, the lobes of 2 pairs, oblong, rounded, mucronate-toothed; heads 3–4 cm. across, in terminal corymbs; involucres 8 mm. long; bracts 3-seriate, the outer linear or oblong, rounded, the inner ovate; ligulate florets 14–24, the corollas 17–23 mm. long, 3–4 mm. wide, yellow or white, sometimes rose-colored; achenes about 2 mm. long.——Nov.–Dec. Honshu (Japan Sea side from Etchu Prov. westw.).

Chrysanthemum × shimotomai Makino. NIJIGAHAMA-GIKU. Hybrid of *C. aphrodite × C. pacificum.*

42. ARTEMISIA L. YOMOGI ZOKU

Entomophilous herbs or subshrubs, aromatic or bitter to taste, often silky or cobwebby; leaves alternate, entire, incised, or 1 to 3 times pinnate; heads small, usually nodding at anthesis, erect in fruit, in racemes, panicles, or solitary, heterogamous; disc florets without ligulate florets; involucres subglobose to ovoid; bracts few-seriate, imbricate, the outer small or nearly as long as the inner, scarious on margin; receptacle flat or globose, often hairy; marginal florets pistillate, 1-seriate, fertile, the central ones bisexual, fertile or sterile, the heads sometimes without marginal florets and hence homogamous; pistillate corollas with slender tubes, shortly 2- or 3-lobed, the central florets with 5-lobed actinomorphic corollas; anthers with lanceolate appendages; achenes obovoid, terete, or usually with 2 obsolete wings, glabrous or hairy, rounded and with a disc at apex, without a pappus.——About 200 species, in the N. Hemisphere.

1A. Central florets sterile.
  2A. Heads 4 mm. wide; cauline leaves not clasping at base; sterile shoots elongate, decumbent; radical leaves gray-woolly, pinnately parted. .... 1. *A. congesta*
  2B. Heads 1–3 mm. wide; cauline leaves clasping at base.
    3A. Heads 1–1.2 mm. across; leaf-segments filiform. .... 2. *A. scoparia*
    3B. Heads 1.5–3 mm. across; leaf-segments of some of them broader.
      4A. Subshrub; ultimate leaf-segments filiform. .... 3. *A. capillaris*
      4B. Herb; ultimate leaf-segments broader, not filiform. .... 4. *A. japonica*
1B. Central florets fertile.
  5A. Receptacles glabrous.
    6A. Annuals.
      7A. Ultimate leaf-segments acute; heads rounded at base.
        8A. Leaves thrice pinnate, the rachis entire; heads about 1.5 mm. across. .... 5. *A. annua*
        8B. Leaves bipinnate, the rachis pectinate; heads about 5–6 mm. wide. .... 6. *A. apiacea*
      7B. Ultimate leaf-segments obtuse; heads obconical, about 4 mm. wide. .... 7. *A. fukudo*
    6B. Perennials.
      9A. Style-branches acuminate; median cauline leaves obovate or spathulate, cuneate at base. .... 8. *A. keiskeana*
      9B. Style-branches truncate.
        10A. Leaves usually rather densely silky-pubescent, rarely glabrous.
          11A. Heads large, 8–14 mm. across.
            12A. Leaves spathulate, simple, incised-toothed. .... 12. *A. pedunculosa*
            12B. Leaves 1–3 times pinnate.
              13A. Cauline leaves 5–9 cm. long, 2–3 times pinnate.
                14A. Cauline leaves thrice pinnate, the ultimate segments about 1 mm. wide. .... 13. *A. sinanensis*
                14B. Cauline leaves bipinnate, the ultimate segments 1.5–2 mm. wide. .... 14. *A. arctica*
              13B. Cauline leaves 1.5–2 cm. long, palmately 3- to 5-sected. .... 10. *A. trifurcata*
          11B. Heads smaller, 4–6.5 mm. across.
            15A. Stems 7–16 cm. long; leaves palmately divided.
              16A. Heads in corymbose racemes, 4.5 mm. long, about 6.5 mm. wide. .... 10. *A. trifurcata*
              16B. Heads in dense corymbs, 4–4.5 mm. across. .... 9. *A. glomerata*
            15B. Stems 25–40 cm. long; leaves bipinnate or simple.
              17A. Leaves bipinnate. .... 11. *A. laciniata*
              17B. Leaves simple. .... 8. *A. keiskeana*
        10B. Leaves cobwebby, often densely so beneath, rarely glabrous.
          18A. Heads 8–10 mm. wide; leaves pinnately cleft, densely cobwebby and grayish on both sides. .... 15. *A. stelleriana*
          18B. Heads 1–6 mm. wide.
            19A. Plants without stolons; ultimate leaf segments relatively small, 1.5–4.5 mm. wide. .... 16. *A. iwayomogi*
            19B. Plants stoloniferous; ultimate leaf segments relatively large.
              20A. Heads 2 mm. long, about 1 mm. across. .... 17. *A. jeddei*
              20B. Heads more than 1 mm. across.
                21A. Leaves white-spotted above.
                  22A. Median leaves bipinnate, the ultimate segments linear. .... 18. *A. dubia*
                  22B. Median leaves 3-fid, the lateral lobes lanceolate to oblong. .... 19. *A. gilvescens*
                21B. Leaves not white-spotted above.

1. Artemisia congesta Kitam. ONI-OTOKO-YOMOGI. Rhizomes long-creeping, 4–7 mm. across; flowering stems 40–65 cm. long, branched above, gray-woolly while young; sterile shoots elongate, decumbent, with a tuft of rosulate leaves at apex; basal leaves rather thick, long-petiolate, densely gray-woolly while young, ovate or broadly so in outline, 5.5–8 cm. long, 6–7 cm. wide, pinnately parted, the segments oblong-cuneate, pinnately cleft or incised, the cauline leaves petiolate, 5–7 cm. long, 1–2 mm. wide, bipinnate, gray-woolly to glabrescent, the upper leaves short-petiolate, smaller, pinnate, the uppermost leaves becoming bracteate, linear, entire; heads many, in a spikelike panicle; involucres globose, 3.5 mm. long, 4 mm. wide; bracts 4-seriate, the outer smallest, ovate, obtuse, the inner elliptic; achenes 1.3 mm. long.——Sept. Near seashores; Hokkaido (Kojima in Oshima Prov.), Honshu (Bentenjima in Mutsu).

2. Artemisia scoparia Waldst. & Kitaib. *A. capillaris* var. *scoparia* (Waldst. & Kitaib.) Pampan. HAMA-YOMOGI. Roots fusiform, branched; flowering stems 60–90 cm. long, much branched above, sometimes with purplish sterile shoots at base provided with a tuft of rosulate leaves at the apex; cauline leaves 3–5 cm. long, 2–5 cm. wide, once to twice pinnate, broadly clasping at base, the segments filiform, elongate, the upper leaves gradually smaller; heads very many, small, 1.2–1.5 mm. long, 1–1.2 mm. wide, in large panicles, the pedicels short; involucres globose, glabrous; bracts 3-seriate, the outer shortest, broadly ovate or elliptic, obtuse, the median and inner elliptic.——Sept.–Nov. Waste grounds; Honshu. ——Korea, Manchuria, n. India, China, Siberia, and Europe.

3. Artemisia capillaris Thunb. *A. capillaris* var. *arbuscula* Miq.——KAWARA-YOMOGI. Subshrub; stems somewhat woody, 30–100 cm. long, much branched, silky-pubescent while young; sterile shoots with a tuft of rosulate leaves at the tip; upper leaves of the sterile shoots long-petiolate, broadly clasping at base, bipinnatisect, 1.5–3.8 cm. long, the ultimate segments fiiliform, 0.3–1(–2) mm. wide, usually densely silky, the median cauline leaves 1.5–9 cm. long, 1–7 cm. wide, broadly clasping at base, bipinnatisect, glabrous to densely silky, the upper leaves smaller, pinnatisect; heads very numerous, globose or ovoid, glabrous, 1.5–2 mm. long and as wide, in large panicles; involucral bracts 3- or 4-seriate; outer very small, ovate, obtuse, the inner elliptic, rounded at apex, keeled on back; achenes 0.8 mm. long.——Sept.–Oct. Along rivers and seashores; Honshu, Shikoku, Kyushu.——Korea, Manchuria, China, Formosa, and Philippines.

4. Artemisia japonica Thunb. *Chrysanthemum japonicum* Thunb.; *A. subintegra* Kitam.; *A. manshurica* Komar. ——OTOKO-YOMOGI. Herb; nonflowering stems elongate, with a tuft of leaves at the tip; leaves of the sterile shoots spathulate, 3.5–8 cm. long, 1–3 cm. wide, rounded, pinnately lobed and toothed, loosely silky-pubescent on both sides; flowering stems 40–100 cm. long, branched above, rarely cobwebby; median leaves rather fleshy, cuneate-spathulate, 4–8 cm. long, toothed, 3-fid, pinnately parted, or pinnately divided, the bracteate leaves linear, small; heads small, many, ovoid-globose or ellipsoidal, slightly lustrous, yellow-green, glabrous, in panicles; outer involucral bracts small, ovate, obtuse, the inner elliptic, rounded at apex; achenes 0.8 mm. long.——Aug.–Nov. Sunny hills and low elevations in the mountains; Hokkaido, Honshu, Shikoku, Kyushu; common.——Ryukyus, Formosa, Korea, Manchuria, China, and Philippines.

Var. **macrocephala** Pampan. *A. desertorum* sensu auct. Japon., non Spreng.; *A. japonica* var. *desertorum* sensu auct. Japon., non Maxim.; *A. littoricola* Kitam.——HAMA-OTOKO-YOMOGI. Heads larger, about 2 mm. wide; leaves pinnate or bipinnately cleft.——Near seashores; Hokkaido, Honshu (Ugo Prov.).——s. Kuriles and Sakhalin.

5. Artemisia annua L. KUSO-NINJIN. Annual; stems to 1.5 m. long, glabrous, much branched; median cauline leaves broadly clasping at base, ovate in outline, 4.5–7 cm. long, tripinnate, the pinnae oblong, spreading, the ultimate segments acute, about 0.4 mm. wide, mealy-puberulent above, glandular-dotted, the rachis entire, the upper leaves smaller; heads globose, small, about 1.5 mm. long, very numerous; involucres glabrous; bracts 2- or 3-seriate, the outer narrowly oblong, green, the median and inner elliptic, green on back; achenes 0.7 mm. long.——Aug.–Oct. Waste grounds and cultivated fields; Honshu, Shikoku.——Formosa, Korea, Manchuria, Siberia, China, India, w. Asia, and e. Europe; naturalized in w. Europe and N. America.

6. **Artemisia apiacea** Hance. *A. carvifolia* var. *apiacea* (Hance) Pampan.; *A. thunbergiana* Maxim.——KAWARA-NIN-JIN, NO-NINJIN. Biennial; stems 40–150 cm. long, glabrous, much branched; radical and lower cauline leaves withering before anthesis, oblong in outline, 9–15 cm. long, 3.5–5.5 cm. wide, bipinnate, the primary segments oblong, spreading, pectinate on upper part, the secondary segments 1.5–2 mm. wide, acutely incised, dentate, glabrous on both sides, the median leaves oblong, about 6 cm. long, bipinnatiparted, the primary segments loose, the ultimate segments incised-dentate, usually 0.5 mm. wide; heads many, in large panicles, nodding in anthesis, subglobose, 3.5–4 mm. long, 5–6 mm. wide, on slender pedicels; involucres globose; the bracts 3-seriate, the outer narrowly oblong, slightly shorter, the median and inner oblong, equal in length, green and punctate on back, yellow-scarious on margin; achenes about 1 mm. long.——Aug.-Sept. Along rivers, cultivated fields, and waste grounds; Honshu, Shikoku, Kyushu.——Korea, China, and Manchuria.

7. **Artemisia fukudo** Makino. FUKUDO, HAMA-YOMOGI. Biennial or a short-lived perennial; stems 30–50 cm. long, purplish, cobwebby while young, much branched from the base; radical leaves thick, rosulate, withering before anthesis, 14–21 cm. long inclusive of the petiole, flagellate, palmately 2- or 3-parted, the segments linear, about 2 mm. wide, rounded at apex, cobwebby while young, the lower cauline leaves withering before anthesis, long-petiolate, 9–12 cm. long, pinnately parted, the ultimate leaf segments linear, about 2 mm. wide, obtuse, cobwebby while young, the upper leaves 3-lobed or linear and entire, 1–3 cm. long; heads obconical, 3–5 mm. long, about 4 mm. wide, in panicles; involucral bracts green, the median ovate-oblong, obtuse; achenes 1.2–2 mm. long.——Sept.-Oct. Clay soils near seashores; Honshu (Kinki Distr. and westw.), Shikoku, Kyushu.——Korea.

8. **Artemisia keiskeana** Miq. INU-YOMOGI. Perennial; rhizomes stout; sterile shoots decumbent, with a tuft of rosulate leaves at the tip; median cauline leaves chartaceous, obovate or spathulate, 3–10 cm. long, 1.5–4.5 cm. wide, rounded at apex, cuneate, coarsely obtuse-toothed, green, short-pubescent above, glandular-dotted, pale green and silky beneath; flowering stems 30–80 cm. long, branched and often silky above; median leaves obovate to spathulate, 4.5–8.5 cm. long, 17–40 mm. wide, obtuse, coarsely incised-dentate, cuneate at base, sometimes short-pubescent above, rather densely silky beneath, the upper leaves gradually smaller, obovate, incised or entire; heads globose, 3–3.5 mm. long and as wide, in narrow panicles, the pedicels nodding in anthesis; involucres glabrous; bracts 3- or 4-seriate, the outer smaller, about 1 mm. long, broadly ovate, obtuse, the median elliptic, green on back; style branches acuminate; achenes about 2 mm. long.——Aug.-Oct. Mountains; Hokkaido, Honshu, Shikoku, Kyushu.——Korea, Manchuria, and n. China.

9. **Artemisia glomerata** Ledeb. *A. leontopodioides* Fisch. ex Bess.; *A. curilensis* Spreng. ex Bess.——HAHA-KO-YOMOGI. Rhizomes creeping, 1–3 mm. across; stems 7–15 cm. long, slender, densely white-silky, short-branched above; leaves of the sterile shoots radical, obovate or flabellate in outline, 1.2–4.6 cm. long, dilated and clasping at base, petioles 5–32 mm. long, bipalmately cleft, the segments linear, about 1 mm. wide, obtuse, densely silky on both sides, the median leaves spathulate or flabellate, 13–20 mm. long, 3- or 4-cleft or bipalmately cleft, densely silky, the petioles winged, the upper-most leaves often oblong, 2–2.5 mm. wide, entire; heads glo-

bose-campanulate, about 5 mm. long, 4–4.5 mm. wide, in dense corymbs, pedicelled; involucres densely silky, the bracts 3-seriate, equal in length, elliptic to obovate-oblong; achenes about 1.5 mm. long.——Aug. Rocky cliffs in alpine regions; Honshu (Mount Nishikoma and Kotadake).——Sakhalin, Kuriles, Siberia, Kamchatka, and Alaska.

10. **Artemisia trifurcata** Steph. ex Spreng. *A. hetero-phylla* Bess.; *A. trifurcata* var. *heterophylla* (Bess.) Kudo——NAGAE-HAHA-KO-YOMOGI. Rhizomes stout, slightly branched; stems tufted, 12–16 cm. long, gray-silky, branched above; rosulate leaves of the sterile shoots to 8 cm. long, densely gray-silky on both sides, bipalmately divided, long-petiolate, the median cauline leaves 1.5–2.5 cm. long, 6–12 mm. wide, palmately 3- to 5-sected, densely silky on both sides, the segments obtuse, about 1 mm. wide, the petioles 6–11 mm. long, the upper leaves linear, 2–2.5 mm. wide; heads in corymbose racemes, long-pedicellate, nodding in anthesis, about 4.5 mm. long, 6.5 mm. wide; involucres densely silky, subglobose, the bracts 2-seriate, equal in length, elliptic, rounded; achenes about 2 mm. long.——Aug. Hokkaido (Mount Daisetsu).

Var. **pedunculosa** (Koidz.) Kitam. *A. glomerata* var. *pedunculosa* Koidz.; *A. yezoensis* Tatew. & Kitam.——EZO-HAHA-KO-YOMOGI. Heads larger, about 8 mm. across, with 30 to 40 florets.——Kuriles, Kamchatka, Siberia, and N. America.

11. **Artemisia laciniata** Willd. *A. sacrorum* var. *major* forma *japonica* Pampan.; *A. laciniata* var. *glabriuscula* forma *dissecta* Pampan.——SHIKOTAN-YOMOGI. Rhizomes decumbent, stout, woody; stems 25–40 cm. long, cobwebby while young, branched above; rosulate leaves of the sterile shoots oblong in outline, 12–18 cm. long inclusive of the petioles, 5.5–6 cm. wide, bipinnate, green above, appressed-silky beneath, the primary segments about 8 pairs, elliptic, the secondary segments oblong, pinnaticleft, the ultimate segments acute, 0.5–0.7 mm. wide, the median leaves similar to the radical but smaller, short-petiolate, the uppermost leaves linear and small; heads globose, 2.5–4 mm. long, 4–6 mm. wide, pedicelled, in a narrow panicle; involucres glabrous; bracts 4-seriate, the outer 2.5 mm. long, broadly ovate, the median and inner broadly elliptic; achenes about 1.2 mm. long.——Aug.-Oct. Hokkaido (Nemuro Prov. and Rebun Isl.).——s. Kuriles, Sakhalin, Korea, Manchuria, China, India, Siberia, and Europe.

12. **Artemisia pedunculosa** Miq. MIYAMA-OTOKO-YO-MOGI. Rhizomes 2.5–6 mm. thick; plant stoloniferous; stems 15–35(–45) cm. long, silky-pubescent while young, usually purplish, short-branched above; rosulate leaves of the sterile shoots chartaceous, obovate-spathulate, 2.7–9 cm. long, 13–23 mm. wide, rounded at apex, incised-toothed, green above, silky-pubescent beneath, winged-petiolate, the radical leaves of flowering stems withering before anthesis, the median leaves sessile, oblanceolate-spathulate, 2.5–6 cm. long, 6–16 mm. wide, rounded at apex, gradually narrowed at base, obtusely incised-toothed; heads subglobose, about 6 mm. long, 8–10 mm. wide, nodding in anthesis, long-pedicelled, in simple or compound racemes; involucres glabrous, the bracts nearly equal in length, 3-seriate, ovate-elliptic to elliptic, yellowish; achenes about 2 mm. long, short-pilose while young.——Aug.-Sept. Alpine; Honshu (centr. distr.).

13. **Artemisia sinanensis** Yabe. TAKANE-YOMOGI. Rhizomes stout; stems 20–50 cm. long, rather stout, densely silky-pubescent while young, short-branched above; cauline leaves

of the sterile shoots similar, elliptic or obovate, 12–15 cm. long inclusive of the long petioles, silky, tripinnate, silky-pubescent, the ultimate segments linear, about 1 mm. wide, the median and upper cauline leaves smaller, short-petiolate or sessile, elliptic, bi- or tripinnate; heads 6–7 mm. long, 12–14 mm. wide, pedicelled, in narrow panicles; involucres glabrous, the bracts 3-seriate, equal in length, ovate to elliptic or oblong.——Aug. Alpine slopes; Honshu (centr. distr., Uzen and Iwashiro Prov.).

14. Artemisia arctica Less. *A. norvegica* sensu auct. Japon., non Fries; *A. arctica* var. *sachalinensis* F. Schmidt—— SAMANI-YOMOGI. Rhizomes stout, elongate; stems 20–50 cm. long, stout, yellow-brown villous while young, short-branched above; radical leaves persistent in anthesis, long-petiolate, ovate-elliptic, in outline, 3.5–4 cm. long, bipinnately parted, the primary segments 2 or 3 pairs, irregularly palmately cleft, the ultimate segments entire, acute or toothed, brownish white villous while young, the cauline leaves gradually smaller, with false stipules at base, the upper leaves oblong, once or twice pinnate, sessile and clasping, the ultimate segments 1.5–2 mm. wide; heads subglobose, about 5 mm. long, 10 mm. wide, in simple or compound racemes, involucres dark brown, glabrous, bracts 3-seriate, the outer ovate, 2.5–3 mm. long, obtuse, the inner elliptic.——July– Sept. Alpine; Hokkaido, Honshu (n. distr.).——Sakhalin, Kuriles, Siberia, and N. America.——Forma **villosa** (Koidz.) Kitam. *A. norvegica* var. *villosa* Koidz.; *A. arctica* var. *villosa* (Koidz.) Tatew.; *A. czekanowskiana* sensu auct. Japon., non Trautv.——SHIRO-SAMANI-YOMOGI. Plants densely villous.——Hokkaido.

15. Artemisia stelleriana Bess. *A. stelleriana* var. *vesiculosa* Fr. & Sav.; *A. stelleriana* var. *sachalinensis* Nakai—— SHIRO-YOMOGI. Plant stoloniferous; rhizomes elongate, branched, 2–5 mm. across; stems 20–65 cm. long, nodding in upper part before anthesis, densely white-villous; leaves of the sterile shoots rather thick, ovate to oblong in outline, 3–7 cm. long inclusive of the long petioles, pinnately cleft, the segments 2-cleft, oblong, obtuse, densely cobwebby and grayish on both sides, the median cauline leaves 3–9 cm. long, 2–5 cm. wide, pinnately parted, the segments 2 or 3 pairs, narrowly oblong, entire or 2-fid, obtuse, white-woolly above, densely white-tomentose beneath, the upper leaves gradually smaller, pinnately cleft or entire; heads globose, 7 mm. long, 8–10 mm. wide, in narrow panicles; involucres densely long-woolly, the bracts 4-seriate, nearly equal, oblong to narrowly so; achenes about 3 mm. long.——July–Oct. Sand dunes along the seacoast; Hokkaido, Honshu (n. Kantō Distr. and northw.).—— Korea, Sakhalin, Kuriles, Kamchatka, and Ochotsk Sea region; naturalized in Europe and N. America.

16. Artemisia iwayomogi Kitam. *A. sacrorum* var. *latiloba* sensu auct. Japon., non Ledeb.——IWA-YOMOGI. Rhizomes woody; stems tufted, 50–100 cm. long, slightly woody at base, branched above; median cauline leaves bipinnate, ovate in outline, acute, truncate at base, green, short-pubescent on upper surface, glandular-dotted beneath, cobwebby while young, the petioles 2–3 cm. long, the segments 6–10 pairs, spreading, approximate, oblong, obtuse, the ultimate segments pinnately lobed or incised-dentate, 1.5–4.5 mm. wide, the upper leaves gradually smaller, the uppermost oblanceolate; heads many, globose, 3–3.5 mm. long and as much across, in panicles; involucres nearly glabrous, the outer bracts 1 mm. long, oblong, the median ovate, the inner elliptic; achenes 1.5 mm.

long.——Sept.–Oct. Hokkaido.——s. Kuriles, Sakhalin, Korea and Manchuria.——Forma **laciniiformis** (Nakai) Kitam. *A. sacrorum* var. *laciniiformis* Nakai. Leaves much-dissected.

17. Artemisia feddei Lév. & Van't. *A. lavandulaefolia* var. *feddei* (Lév. & Van't.) Pampan.——HIME-YOMOGI. Plant stoloniferous; rhizomes elongate, creeping; stems 1–1.2 m. long, cobwebby, often purplish, branched above; median cauline leaves thin, 3–7 cm. long, 28–65 mm. wide, pinnately parted, the segments 2 or 3 pairs, linear, obtuse, thinly cobwebby above, densely white-villous beneath, the upper ones gradually smaller, the uppermost bracteate; heads numerous, small, sessile, about 2 mm. long, about 1 mm. wide; involucres loosely cobwebby; bracts 4-seriate, the outer short, broadly ovate, obtuse, the median suborbicular, the inner elliptic; achenes about 1.1 mm. long.——Grassy places; Honshu, Shikoku, Kyushu.——Korea, Manchuria, China, and Formosa.

18. Artemisia dubia Wall. *A. lavandulaefolia* DC. pro parte; *A. umbrosa* Turcz.; *A. vulgaris* var. *umbrosa* Bess.; *A. codonocephala* Diels; *A. araneosa* Kitam.——KESHŌ-YOMOGI. Rhizomes elongate; plant stoloniferous; stems stout, 1.5–2 m. long, prominently cobwebby and whitish, short-branched, the branches nodding toward the top while young; median cauline leaves with false stipules at base, 11–14 cm. long, about 8 cm. wide, bipinnate, densely white-punctate and silky-cobwebby above, densely grayish tomentose beneath, the petioles 2.5–3 cm. long, the primary segments usually 2 pairs, the ultimate segments linear, 5–7 mm. wide, gradually narrowed to tip, mucronate, entire, the upper leaves gradually smaller, pinnately cleft, 3-fid, or entire, short-petiolate or sessile, the lobes linear, acuminate, silky-cobwebby; heads globose to campanulate, 3–3.5 mm. long, about 3 mm. wide, in leafy narrowly pyramidal panicles 25–30 cm. long; involucres densely cobwebby; bracts 3- or 4-seriate, the outer ovate or ovate-acuminate, 1.8–2.5 mm. long, the inner oblong; achenes about 1 mm. long.——Kyushu.——Korea, Manchuria, n. China, Dahuria to India.

19. Artemisia gilvescens Miq. *A. vulgaris* var. *gilvescens* (Miq.) Nakai——WATA-YOMOGI. Stems 30 cm. high or higher, densely woolly, short-branched above, the branches nodding; median cauline leaves chartaceous, 6–7 cm. long, 2.8–4.5 cm. wide, sessile, cuneate at base, 3-fid, the terminal lobe oblong, acute, entire or with 1 tooth on each side, the lateral lobes oblong-lanceolate, smaller, acuminate, green, densely white-punctate and short-pubescent above, densely woolly beneath, the upper leaves gradually smaller, oblong, acute, cuneate at base, sessile, entire, slightly recurved; panicles narrow; heads tubular, about 4 mm. long, 2 mm. wide, densely long-woolly; involucral bracts 3-seriate; outer ovate, obtuse, rather short, the median and inner oblong, rounded at the tip; achenes 0.8 mm. long.——Honshu (Nagato Prov.), Shikoku.——China.

20. Artemisia monophylla Kitam. *A. vulgaris* var. *integrifolia* sensu auct. Japon., non L.; *A. viridissima* var. *japonica* Pampan.——HITOTSUBA-YOMOGI, YANAGI-YOMOGI. Rhizomes creeping; stems 70–100 cm. high, tufted, densely cobwebby while young, branched above; cauline leaves many, chartaceous, approximate, oblong-lanceolate, 6.5–14 cm. long, 2–4 cm. wide, gradually acuminate at both ends, mucronate, acutely toothed, green and slightly cobwebby above, long whitish tomentose beneath, short-petiolate, the upper leaves gradually smaller, lanceolate, acuminate; heads campanulate, subsessile, 3–4 mm. long, 2–3 mm. wide, cobwebby, in narrow panicles; achenes about 2 mm. long.——Aug.-Oct. High mountains; Honshu (centr. and n. distr. incl. high mountains of Mimasaka and Hōki Prov. in Chūgoku Distr.).

21. Artemisia unalaskensis Rydb. *A. vulgaris* var. *kamtschatica* Bess.; *A. vulgaris* var. *coarctata* sensu auct. Japon., non Bess.; *A. nipponica* (Nakai) Pampan.; *A. opulenta* Pampan.; *A. vulgaris* var. *nipponica* Nakai; *A. verlotorum* Hult., non Lamotte; *A. pampaninii* Kitam.——CHISHIMA-YOMOGI, EZO-Ō-YOMOGI. Plant stoloniferous; rhizomes elongate; stems 30–100 cm. long, short-pubescent toward the top; median cauline leaves thinly chartaceous, 10–15 cm. long, 7–10 cm. wide, pinnate to bipinnately parted or cleft, the petioles 13–30 mm. long, the primary segments 3 pairs, oblong, the secondary segments lanceolate, usually acuminate or acute, entire or toothed, green and thinly silky-pubescent on the upper side, densely grayish tomentose beneath, the upper leaves gradually smaller, short-petiolate or sessile, pinnately cleft, the uppermost leaves linear-lanceolate; heads campanulate or globose-campanulate, 4–4.5 mm. long, 3–5 mm. wide, in a narrowly pyramidal panicle; involucres slightly cobwebby and pale yellowish; bracts membranous, 4-seriate, nearly equal in length, the outer ovate-lanceolate, acuminate, the median oblong.——Aug.-Sept. Alpine regions; Honshu (centr. distr.).——Kuriles, Sakhalin, Kamchatka, and Aleutians.

22. Artemisia stolonifera (Maxim.) Komar. *A. vulgaris* var. *stolonifera* Maxim.; *A. vulgaris* var. *kiusiana* Makino; *A. vulgaris* var. *stolonifera* lusus *incana* Regel; *A. integrifolia* var. *stolonifera* (Maxim.) Pampan.——HIRO-HA-YAMA-YOMOGI, HIRO-HA-NO-HITOTSUBA-YOMOGI. Plants stoloniferous; rhizomes long-creeping; stems 50–100 cm. long, slender, cobwebby, glabrous, short-branched near the top; median cauline leaves chartaceous, many, ovate-oblong, rarely obovate-oblong in outline, 7–14 cm. long, 4.5–8 cm. wide, acute or sometimes obtuse, cuneate and falsely stipulate at base, winged-petiolate, pinnately lobed to cleft, the lobes oblong, 2 pairs, approximate, acute, rarely obtuse, toothed, cobwebby, becoming glabrous above, densely gray-woolly beneath, the upper leaves gradually becoming smaller; heads many, globose-campanulate, 4–4.5 mm. long, 3–4 mm. wide, usually aggregated on one side of the branched panicles; involucres cobwebby; bracts 3-seriate, the outer ovate or oblong, acute, the inner rounded at apex; achenes 1.8 mm. long.——Aug.-Oct. Mountains; Honshu (Chūgoku Distr.), Kyushu.——Korea and Manchuria.

23. Artemisia koidzumii Nakai. *A. ursorum* Hult.; *A. koidzumii* var. *arguta* Kitam.; *A. megalobotrys* Nakai; *A. nipponica* var. *megalobotrys* (Nakai) Pampan.——Ō-WATA-YOMOGI, HIRO-HA-URAJIRO-YOMOGI. Plants stoloniferous; rhizomes creeping; stems stout, rarely slender, 35–100 cm. high, long grayish woolly; median cauline leaves rather thick, obovate, 4.5–18 cm. long, 3–11 cm. wide, cuneate, with false stipules at base, sessile or winged-petiolate, pinnately cleft or lobed, the lobes 2(–3) pairs, oblong, obtuse or acute, few-toothed or entire, cobwebby on the upper surface while young, densely long grayish tomentose beneath, the upper leaves gradually smaller, oblong or lanceolate, acute; heads globose or globose-campanulate, about 5 mm. long, 4–6 mm. wide, villous; involucres densely long-pubescent; bracts 3-seriate, the outer ovate, subacute, the median broadly ovate, obtuse, the inner oblong, rounded.——Aug.-Oct. Near seashores; Hokkaido.

Var. **megalophylla** Kitam. *A. megalophylla* Kitam., no-

men provis.——ŌBA-YOMOGI. Larger and stouter in habit; leaves deeply pinnately cleft, the segments 3-lobed; heads cobwebby.——Honshu (Bentenjima in Mutsu).

Var. **tsuneoi** (Tatew. & Kitam.) Kitam. *A. tsuneoi* Tatew. & Kitam.——MASHŪ-YOMOGI. Heads smaller; leaves deeply pinnately cleft, 6.5–7 cm. long, the lobes 2 pairs, subentire.——Hokkaido (near Mashū Lake).

24. Artemisia momiyamae Kitam. *A. princeps* var. *momiyamae* (Kitam.) Hara——YUKI-YOMOGI. Plants stoloniferous; rhizomes creeping; stems about 1 m. long, much-branched, white-villous; lower cauline leaves rather thick, falsely stipulate, oblong, about 12 cm. long, 6 cm. wide, densely white cobwebby-tomentose on upper side, pinnately divided, the segments 3 pairs, oblong, pinnaticleft, the median leaves 7–9 cm. long, 4.5–5 cm. wide, deeply pinnatisect, the segments 2 pairs, spreading or sometimes obliquely so, linear-lanceolate, entire, acute to obtuse, cobwebby above, long white-villous beneath, the upper leaves 3-parted, gradually smaller, the lobes linear-lanceolate, acute; heads numerous, tubular or oblong-campanulate, about 4 mm. long, 1.5–2 mm. wide, densely cobwebby, in panicles; involucral bracts 3-seriate, the outer broadly ovate, the inner oblong; achenes about 1.5 mm. long.——Oct.-Nov. Near seashores; Honshu (Sagami Prov. and Izu Isls.).

25. Artemisia rubripes Nakai. *A. vulgaris* var. *parviflora* Maxim.; *A. nipponica* var. *rubripes* (Nakai) Pampan.; *A. venusta* Pampan.——YABU-YOMOGI. Plants stoloniferous; rhizomes stout, long-creeping; stems 1–2 m. long, nearly glabrous, much branched; median and lower cauline leaves chartaceous, large, 12–21 cm. long, 9–12 cm. wide, falsely stipulate, bipinnately divided, the segments many, narrow, 5–7 mm. wide, acuminate, loosely toothed, green, the midrib hairy on the upper side, densely gray-cobwebby beneath, the upper leaves gradually smaller, pinnate or bipinnately divided, the segments narrow, the uppermost leaves linear, acuminate; heads numerous, small, globose-campanulate, 2–2.5 mm. long, 1.5 mm. wide, in compound racemes or spikes; involucral bracts slightly cobwebby, 3-seriate, the outer oblong or ovate, obtuse, the inner oblong, rounded at apex.——Aug.-Sept. Kyushu.——Korea, Manchuria, and n. China.

26. Artemisia montana (Nakai) Pampan. *A. montana* var. *latiloba* Pampan.; *A. vulgaris* var. *indica* forma *montana* Nakai; *A. nipponica* var. *elata* Pampan.; *A. gigantea* Kitam.; *A. vulgaris* var. *yezoana* Kudo; *A. shikotanensis* Kitam.——Ō-YOMOGI, EZO-YOMOGI, YAMA-YOMOGI. Plants stoloniferous; rhizomes creeping; stems 1.5–2 m. long, stout; median leaves chartaceous, usually 15–19 cm. long, 6–12 cm. wide, cuneate at base, winged-petiolate, deeply pinnately cleft to parted, the segments oblong or oblong-lanceolate, acuminate, the lateral lobes 2 or 3 pairs, entire or often incised-dentate, loosely cobwebby or nearly glabrous above, grayish woolly beneath, the upper leaves gradually smaller, lanceolate, 3-fid or entire; panicles large, the heads many, 3.5 mm. long, 3 mm. wide; involucres cobwebby; bracts 3-seriate, the outer short, broadly ovate, the inner oblong, rounded at apex; achenes 1.5–2 mm. long.——Aug.-Oct. Woods in mountains; Hokkaido, Honshu (Kinki Distr. and eastw.).——Sakhalin, and s. Kuriles.

27. Artemisia princeps Pampan. *A. vulgaris* var. *maximowiczii* Nakai; *A. pleiocephala* var. *insularis* Pampan.; *A. montana* var. *nipponica* Pampan.; *A. parvula* Pampan.——YOMOGI, KAZUZAKI-YOMOGI. Plants stoloniferous; rhizomes creeping; stems 60–120 cm. long, tufted, crisped-pubescent, cobwebby toward the top, usually branched; median cauline leaves chartaceous, falsely stipulate at base, elliptic, 6–12 cm. long, 4–8 cm. wide, pinnately parted to cleft, the segments 2 to 4 pairs, oblong-lanceolate, acuminate to obtuse, toothed or deeply incised, rarely entire, green above, white-tomentose beneath, the upper leaves gradually smaller, pinnately cleft or 3-fid, the segments lanceolate, entire or rarely toothed, the bracteate leaves linear, small; inflorescence paniclelike, consisting of compound racemes; heads very many, dense, 2.5–3.5 mm. long, about 1.5 mm. wide; involucres oblong-campanulate, loosely cobwebby; bracts 4-seriate, oblong or the outer broadly ovate, the inner rounded; achenes about 1.5 mm. long.——Sept.–Oct. Waste grounds and thickets in lowlands and low elevations in the mountains; Honshu, Shikoku, Kyushu; very common.——Bonins and Korea.

28. Artemisia indica Willd. *A. dubia* sensu auct. Japon., non Wall.; *A. dubia* var. *grata* forma *asiatica* Pampan.; *A. vulgaris* var. *indica* Maxim.; *A. asiatica* Nakai; *A. dubia* var. *orientalis* Pampan.; *A. dubia* var. *septentrionalis* Pampan.; *A. princeps* var. *orientalis* (Pampan.) Hara——NISHI-YOMOGI, YOMOGI. Plant stoloniferous and with rhizomes; stems 50–120 cm. long, arachnoid or sometimes glabrous at base, branched above; median cauline leaves chartaceous, falsely stipulate, oblong or elliptic in outline, 7–12 cm. long, 3.3–10 cm. wide, pinnately parted, the segments 2 or 3 pairs, oblong or oblong-lanceolate, rarely lanceolate, obtuse, toothed or entire, pinnatilobed, green, slightly cobwebby to subglabrate above, densely white-woolly beneath, the upper leaves gradually smaller, pinnatiparted or 3-fid, the uppermost lanceolate, entire; panicles narrow; heads very many, dense, globose-campanulate, 3.5–4 mm. long, 2.5–3 mm. wide; involucres 3–4 mm. long, slightly cobwebby; bracts 3- or 4-seriate, the outer broadly ovate, obtuse, the inner oblong and rounded; achenes about 2 mm. long.——Sept.–Dec. Waste ground; Honshu (Chūgoku Distr.), Kyushu.——Ryukyus, Formosa, China, and India.

29. Artemisia schmidtiana Maxim. *A. sericea* sensu auct. Japon., non Weber; *A. sericea* var. *schmidtiana* (Maxim.) Kitam., in syn.——ASAGIRI-SŌ. Rhizomes creeping; stems tufted, 16–30 cm. long, white-villous, much branched above; median cauline leaves many, 3–4.5 cm. long, petiolate, bipalmately divided, the segments linear, about 1 mm. wide, obtuse, densely silvery-villous on both sides, the upper leaves gradually smaller, palmately or pinnately divided, the uppermost linear; heads many, 4–5 mm. wide, in panicles, the pedicels bracteate; involucres globose or depressed-globose, about 4 mm. long, 4–5 mm. wide, densely silky-pubescent; bracts 3-seriate, the outer slightly shorter, ovate or elliptic, the inner obovate, rounded at apex; receptacle densely white-hairy, the hairs about 1.2 mm. long; achenes about 1.5 mm. long.——Aug.–Oct. High mountains and near seashores; Hokkaido, Honshu (Japan Sea side of centr. and n. distr.).——s. Kuriles and Sakhalin.

30. Artemisia kitadakensis Hara & Kitam. *A. borealis* var. *wormskjoldii* sensu auct. Japon., non Bess.; *A. sericea* var. *kitadakensis* Kitam.——KITADAKE-YOMOGI. Rhizomes densely tufted, woody; flowering stems tufted, 20–30 cm. long, simple, those of the sterile shoots 10–20 cm. long; lower cauline leaves scalelike, 4–6 mm. long, sometimes 3-lobed, the median leaves 22–26 mm. long, about 2 cm. wide, the petioles about 1 mm. long, twice ternately divided, the segments linear, about 1 mm. wide, acute, white-villous on both sides, the upper leaves gradually smaller, petiolate, once or twice ternately divided, the bracteate leaves absent or simple and linear; heads nodding, about 8 mm. across, pedicelled, in simple racemes; involucres subglobose, densely silky-pubescent, about 5 mm. long, 8 mm. wide; bracts 4-seriate, nearly equal in length, the outer ovate, acute, the inner elliptic, rounded at apex, yellow-green, villous on back.——Aug. Alpine regions; Honshu (Akaishi and Shirane mt. ranges in Kai and Shinano Prov.); rare.

43. ADENOCAULON Hook.   NOBUKI ZOKU

Perennial herbs; stems erect, branched, with gland-tipped hairs in upper part; leaves membranous, alternate, petiolate, densely glandular-hairy beneath; heads in panicles, small, the disc florets of 2 kinds; involucres subglobose, the bracts herbaceous, 5–7, essentially 1-seriate, equal, reflexed after anthesis; receptacles convex, glabrous; marginal florets 7–11, pistillate, fertile, the corollas broadly campanulate, 4- or 5-lobed; central florets 7–18, sterile, the corollas tubular, 5-lobed, the anthers sagittate, with small acute appendage at apex; style obsoletely bifid in bisexual florets, short-bifid in the marginal pistillate florets; achenes clavate-obovate, with stipitate glands; pappus absent.——Few species, in America and Asia.

1. Adenocaulon himalaicum Edgew. *A. adhaerescens* Maxim.; *A. bicolor* var. *adhaerescens* (Maxim.) Makino——NOBUKI. Rhizomes creeping; stems 60–100 cm. long, cobwebby, with stipitate glands on upper portion; lower leaves thin, reniform to cordate, 7–13 cm. long, 11–22 cm. wide, undulately lobulate or toothed, green, nearly glabrous on the upper side, densely white-woolly beneath, the petioles winged, 10–20 cm. long, the median and upper leaves gradually smaller, the uppermost broadly lanceolate, becoming bracteate; heads white, pedicelled, with stipitate glands; involucres about 2.5 mm. long, 5 mm. wide, subglobose, the bracts 5–7, broadly ovate, obtuse; achenes spreading, 6–7 mm. long.——Aug.–Oct. Woods in mountains; Hokkaido, Honshu, Shikoku, Kyushu.——s. Kuriles, Korea, Manchuria, China to the Himalayas.

44. SIEGESBECKIA L.   ME-NAMOMI ZOKU

Erect branched annuals, more or less glandular-pilose; leaves ovate, toothed; heads small, in loose panicles, yellow, the florets of 2 kinds; involucral bracts herbaceous, 5, spreading, linear-spathulate, with stipitate glands on upper half; receptacles chaffy, the scales dilated and enveloping the florets, outer ring of scales coriaceous, densely glandular-hairy on the back, the innermost ones membranous; ligulate florets all pistillate, 1-seriate, fertile, the corollas 3-lobulate, spreading; tubular florets bisexual, fertile, the corollas 5-toothed; anthers sagittate; style-branches in bisexual florets short, slightly flattened, acute; achenes 4-angled, truncate, incurved at tip, narrowed at the base, glabrous, blackish; pappus absent.——About 10 species, widely distributed in warmer regions of the world.

1A. Branches forked in upper part; leaves chartaceous, irregularly lobulate. . . . . 1. *S. orientalis*
1B. Branches not forked; leaves thinly membranous, usually toothed.
2A. Stems and leaves uniformly soft-pubescent on both sides; pedicels not glandular; achenes about 2 mm. long. . . . . 2. *S. glabrescens*
2B. Stems and lower side of leaves densely white-pubescent; pedicels usually glandular-pilose; achenes 2.5–3.5 mm. long. . 3. *S. pubescens*

1. Siegesbeckia orientalis L. *S. orientalis* forma *angustifolia* Makino; *S. humilis* Koidz.——TSUKUSHI-ME-NA-MOMI. Annual herb; stems erect, to 20–70 cm. long, the branches ascending-spreading, forked in upper part, densely short-pubescent; median cauline leaves chartaceous, long-petiolate, ovate-oblong to deltoid-ovate, 5–14 cm. long, 3–12 cm. wide, short-acuminate to acute, truncate to cuneate at base, irregularly obtuse-toothed, 3-nerved, irregularly lobulate on lower margin, densely short-pubescent on both surfaces, glandular-dotted beneath, the upper leaves gradually smaller and narrower, short-petiolate, oblong, obtuse; heads many, 16–21 mm. wide, the long pedicels sometimes glandular-pilose; involucral bracts 5, spreading, glandular-pilose; ligulate corollas 2.2–2.5 mm. long; achenes glabrous.——Apr.–Oct. Honshu (s. Kantō Distr., Kii Prov. and Chūgoku Distr.), Shikoku, Kyushu.——Ryukyus, Formosa, China, Malaysia to India, Caucasus, Africa, and Australia.

2. Siegesbeckia glabrescens (Makino) Makino. *S. orientalis* forma *glabrescens* (Makino) Makino——KO-ME-NA-MOMI. Annual herb; stems erect, 35–100 cm. long, usually branched, short appressed-pilose; median cauline leaves ovate-deltoid, 5–13 cm. long, 3.5–11 cm. wide, short-acuminate, truncate and cuneate at base, 3-nerved, irregularly toothed, glandular-dotted beneath, densely short appressed-pubescent on both surfaces, winged-petiolate, the upper leaves gradually

smaller, oblong, subsessile, the uppermost linear; heads many, the pedicels 1–3 cm. long, densely short-pubescent; involucral bracts 5, spreading, nearly equal in length, spathulate, puberulent and densely glandular-pilose; ligulate corollas 1.5–2.5 mm. long, yellow, 3-toothed; achenes 4-angled, about 2 mm. long, glabrous.——Sept.–Oct. Waste grounds and roadsides; Hokkaido, Honshu, Shikoku, Kyushu.——Korea, Manchuria, China, Ryukyus, and Formosa.

3. Siegesbeckia pubescens (Makino) Makino. *S. orientalis* forma *pubescens* Makino; *S. orientalis* sensu auct. Japon., pro parte, non L.——ME-NAMOMI. Annual herb; stems 60–120 cm. long, branched, densely spreading white-pubescent, especially in upper part; median cauline leaves ovate to deltoid-ovate, 7.5–19 cm. long, 6.5–18 cm. wide, short-acuminate, truncate to rounded and narrowed at base, irregularly toothed, 3-nerved, short appressed-pubescent on both surfaces, the nerves beneath densely white-pubescent, the petioles winged, 6–12 cm. long; heads many, in loose terminal corymbs, the pedicels 15–35 mm. long, densely glandular-pilose; involucral bracts 5, linear, 10–12 mm. long, spreading, rounded at apex, glandular-pilose at base; ligulate corollas about 3.5 mm. long, shallowly 2- or 3-lobed; achenes 4-angled, 2.5–3.5 mm. long, glabrous.——Sept.–Oct. Waste grounds; Hokkaido, Honshu, Shikoku, Kyushu; common.——Korea, Manchuria, and China.

45. ECLIPTA L. TAKASABURŌ ZOKU

Branched strigose annual herb; leaves opposite, toothed, heads small, terminal on the stems and branches, pedicelled, heterogamous; involucres campanulate; receptacle convex, the bracts herbaceous, in 2 series, the inner short; outer chaff lanceolate, the inner linear and bristlelike; ligulate florets pistillate, nearly in 2 series, fertile, the corollas white; disc florets bisexual, fertile, the corollas tubular, white, 4-lobed; anthers very shortly bifid at base; style-branches obtuse and mammillate at apex; achenes of ligulate florets 3-angled, of the disc florets compressed, 4-angled, truncate and depressed at apex, with 1 to 3 minute teeth on margin, coarsely hairy, tuberculate on both surfaces.——One species, in the warmer parts of the world.

1. Eclipta prostrata (L.) L. *Verbesina prostrata* L.; *V. alba* L.; *E. alba* (L.) Hassk.; *E. thermalis* Bunge——TAKA-SABURŌ. Stems erect or ascending, branched, strigose-pilose, 10–60 cm. long; leaves lanceolate, 3–10 cm. long, 5–25 mm. wide, gradually acuminate, narrowed to the base, sessile or very short-petiolate, serrulate, subtrinerved, densely strigose-pilose on both surfaces; heads on pedicels 2–4.5 cm. long; involucres campanulate at anthesis, 5 mm. long, 6–7 mm.

across, to 11 mm. across in fruit, the bracts 5 or 6, green, oblong, acute; ligulate corollas white, 2.5–3 mm. long, 0.4 mm. wide, entire or bifid; achenes about 2.8 mm. long, trigonous in the ligulate florets, compressed-tetragonous in the disc ones, ribbed on margin.——July–Nov. Wet places, especially common in paddy fields in lowlands; Honshu, Shikoku, Kyushu.——Korea, China, Ryukyus, Formosa and in warmer regions of the world.

46. WEDELIA Jacq. HAMA-GURUMA ZOKU

Perennial herbs or subshrubs, scabrous or sometimes strigose; stems creeping or scandent, elongate; leaves opposite; heads terminal and in upper axils, peduncled, heterogamous; involucres subglobose, the bracts 1-seriate, green, coarsely strigose; receptacles convex, the scales convex or folded; ligulate florets fertile, the corollas spreading, yellow; tubular florets bisexual, fertile, the corollas tubular, 5-toothed; anthers appendaged at apex, entire at base; styles in bisexual florets acute, hairy on back; achenes of the disc florets oblong-cuneate or obovate, flattened laterally, trigonous in the ligulate ones; pappus crownlike, toothed or ringlike, with 1 or 2 small short deciduous spines at apex.——More than 50 species, in all warmer regions.

1A. Pappus vase- or cup-shaped; peduncles 6–12 cm. long; involucres 8–9 mm. long. . . . . 1. *W. chinensis*
1B. Pappus crownlike, with 1 or 2 deciduous short bristles; peduncles 1–7 cm. long; involucres 4–6 mm. long.
2A. Stems creeping; leaves acute to obtuse, cuneate at base, sparsely toothed.
3A. Leaves ovate, 3–12 cm. long, thick-chartaceous; heads usually in 3's; achenes about 3 mm. long. . . . . 2. *W. robusta*
3B. Leaves oblong, 1.5–4.5 cm. long, thick-coriaceous; heads usually solitary; achenes 3.5–4 mm. long. . . . . 3. *W. prostrata*
2B. Stems scandent; leaves 7–14 cm. long, acuminate, rounded at base, much-toothed. . . . . 4. *W. biflora*

1. **Wedelia chinensis** (Osbeck) Merr. *Solidago chinensis* Osbeck; *Verbesina calendulacea* L.; *W. calendulacea* (L.) Less., non Pers.——KUMANO-GIKU, HAMA-GURUMA. Stems prostrate, elongate, rooting from the nodes, appressed-pilose; leaves chartaceous, lanceolate, sometimes oblong, 2–7 cm. long, 6–12 mm. wide, acute, abruptly narrowed at base, mucronate-tipped, sessile or short-petiolate, appressed-pilose on both surfaces, mucronate-toothed; heads 2–2.5 cm. across, solitary on erect branches, the peduncles 6–12 cm. long; involucres subglobose, 8–9 mm. long, the bracts 5, nearly equal in length, 1-seriate, oblong, acute or sometimes obtuse, short appressed-pubescent; achenes obovate, 3.5 mm. long, 1.5–2 mm. wide, coarsely hairy at tip; pappus vase- or cup-shaped.——May–Sept. Seashores; Honshu (Izu and Kii Prov.), Shikoku, Kyushu.——Ryukyus, Formosa, China, Malaysia, and India.

2. **Wedelia robusta** (Makino) Kitam. *W. prostrata* var. *robusta* Makino——Ō-HAMA-GURUMA. Stems much elongate, creeping, rooting at the nodes, appressed-strigose; leaves thick-chartaceous to somewhat coriaceous, ovate, 3–12 cm. long, 1.5–6 cm. wide, acute, cuneate at base, loosely obtuse-toothed, appressed-strigose on both surfaces, the petioles 3–28 mm. long; heads in 3's or sometimes solitary, 2–2.5 cm. across; involucres subglobose, the bracts ovate, densely short-strigose on back; achenes about 3 mm. long, 2–2.5 mm. wide, cuneate at base, 3- or 4-angled, short-strigose at tip; pappus crownlike, at first strigose-hairy, with 1 or 2 bristles in fruit.——Apr.-Sept. Seashores; Honshu (Kii Prov.), Shikoku, Kyushu.

3. **Wedelia prostrata** (Hook. & Arn.) Hemsl. *Eclipta dentata* Lév. & Van't.; *Wollastonia prostrata* Hook. & Arn., non DC.; *Verbesina prostrata* Hook. & Arn.——HAMA-GURUMA, NEKO-NO-SHITA. Stems long-creeping, rooting at the nodes, coarsely hirsute; flowering branches ascending, leafy, densely strigose; cauline leaves thick-coriaceous, oblong, sometimes ovate or lanceolate, 1.5–4.5 cm. long, 4–14 mm. wide, usually obtuse, rarely acute, cuneate at base, loosely toothed, 3-nerved, coarsely strigose, short-petiolate; heads 16–22 mm. long, usually solitary, terminal; involucres subglobose, the bracts ovate, 2–3.5 mm. wide, coarsely strigose; ligulate florets 1-seriate, the corollas yellow, 8–11 mm. long, about 4 mm. wide; achenes 3.5–4 mm. long, 2 mm. wide, densely strigose, 3- or 4-angled.——July–Oct. Near seashores; Honshu (Kantō Distr. and westw.), Shikoku, Kyushu.——Korea, Ryukyus, Formosa, Bonins, China, and Indochina.

4. **Wedelia biflora** (L.) DC. ex Wight. *Wollastonia biflora* (L.) DC.; *Verbesina biflora* L.; *Acmella biflora* Spreng.——KIDACHI-HAMA-GURUMA, TOKIWA-HAMA-GURUMA. Stems elongate, branched, scandent, coarsely appressed-strigose; leaves thick-chartaceous, ovate, 7–14 cm. long, acuminate, rounded at base, much toothed, appressed-strigose on both surfaces, the petioles 12–23 mm. long; heads 3–6, terminal, 2–3 cm. wide, the pedicels slender; involucral bracts ovate-lanceolate or narrowly ovate, gradually narrowed to tip, densely appressed-strigose; ligulate corollas 9–13 mm. long, 3.5–4.5 mm. wide, yellow, 2- or 3-toothed; achenes 3–3.5 mm. long, 2–2.5 mm. wide, cuneate at base, 3-angled, coarsely strigose toward the tip.——May–Oct. Near seashores; Kyushu (Yakushima and Tanegashima).——Bonins, Ryukyus, Formosa, China, Malaysia, India to Australia, and Africa.

47. SYNEDRELLA Gaertn.　FUSHIZAKI-SŌ ZOKU

Strigose annual or biennial branched herbs; leaves opposite, toothed, on winged petioles; heads sessile or pedunculate, axillary, solitary or in a dichasium, heterogamous; involucres globose, the bracts and outer chaff oblong, the inner chaff slightly smaller and narrowly oblong, scarious; ligulate florets 2-seriate, fertile, the corollas bifid; tubular florets fertile, the corollas 4-lobed; anthers shortly bifid at base; achenes of the ligulate florets much flattened dorsally, winged, the wings lacerate and awned at apex; achenes of the tubular florets narrowly oblong, flat, 2-awned.——Two species, in the Tropics of Asia and America.

1. **Synedrella nodiflora** (L.) Gaertn. *Verbesina nodiflora* L.——FUSHIZAKI-SŌ. Annual herb; stems erect, 25–30 cm. long, with long internodes, short-pilose, dichotomously branched; median leaves ovate, 5–12 cm. long, 2–6 cm. wide, acute to obtuse, cuneate at base, short-toothed, strigose on both surfaces, 3-nerved, the petioles short, briefly connate at base to the opposite petiole; upper leaves smaller, 3.5–4 cm. long; inflorescence globose, in leaf-axils, the heads sessile or short-pedicelled; involucres tubular, about 7 mm. long, 5–7 mm. wide, the bracts and outer chaff oblong, obtuse; ligulate corollas yellow, 3.5 mm. long, about 1 mm. wide, bifid; achenes of the tubular florets 4.5 mm. long, coarsely strigose, the short-strigose awns 2–3.5 mm. long.——Honshu (Hachijo Isl.).——Ryukyus, Formosa, s. China, India, Indochina, Malaysia, and tropical America.

48. BIDENS L.　SENDANGUSA ZOKU

Erect branched annuals; lower leaves opposite, the upper sometimes alternate, toothed, sometimes 1 to 3 times ternate or pinnate; heads solitary or in corymboselike panicles, heterogamous; involucral bracts herbaceous, often spreading, 1-seriate, the outer chaff oblong, scarious; ligulate florets 1-seriate, sterile, rarely pistillate, sometimes absent, the corollas yellow or white, spreading; tubular florets bisexual, fertile, the corollas 4- or 5-toothed; achenes flattened dorsally or nearly 4-angled, cuneate or linear, beakless, with 2 to 4 usually retrorsely spinulose awns.——About 230 species, in temperate to warmer regions, especially abundant in America.

1A. Achenes obovate-cuneate, truncate.
  2A. Achenes 4-angled, 4-awned; tubular corollas 5-toothed. ....1. *B. cernua*
  2B. Achenes flat, 2- rarely 3- or 4-awned; tubular corollas 4-toothed.
    3A. Heads ligulate; median leaves pinnately compound, at least the terminal leaflets distinctly petiolate. .... 2. *B. frondosa*
    3B. Heads eligulate; median leaves pinnately parted.
      4A. Body of achenes 7–11 mm. long. .... 3. *B. tripartita*
      4B. Body of achenes 4.5–5.5 mm. long. .... 4. *B. radiata* var. *pinnatifida*

1B. Achenes linear, narrowed to the tip.
5A. Achenes 2-awned; leaves much dissected, the ultimate segments about 2 mm. wide; tubular corollas 4-toothed. .... 5. *B. parviflora*
5B. Achenes 3- or 4-awned; leaves at most 2 or 3 times pinnate; tubular corollas 5-toothed.
6A. Involucral bracts spathulate, broadened at tip; leaves usually ternate. .... 6. *B. pilosa*
6B. Involucral bracts lanceolate, not broadened at tip; leaves usually 2 or 3 times pinnate.
7A. Terminal leaf segments narrow, acuminate, few-toothed. .... 7. *B. bipinnata*
7B. Terminal leaf segments ovate, short-acuminate, rather prominently toothed. .... 8. *B. biternata*

1. **Bidens cernua** L. *B. cernua* var. *radiata* DC.—— YANAGI-TA-UKOGI. Stems 25–90 cm. long, branched, short-pilose or glabrate; leaves lanceolate, 8–17 cm. long, acuminate at both ends, mucronate-toothed, minutely scabrous on both surfaces; upper leaves slightly dilated, clasping at base; heads ligulate; involucral bracts 5–7, herbaceous, lanceolate, about 2 cm. long; ligulate corollas 12 mm. long, about 5 mm. wide, yellow; tubular corollas 5-toothed; achenes narrowly cuneate, 4-angled, retrorsely spinulose on the angles, the awns 4, 2 or 3 mm. long.——Sept. Wet places; Hokkaido, Honshu (n. distr.).——Korea, China, and widely distributed in the N. Hemisphere.

2. **Bidens frondosa** L. AMERIKA-SENDANGUSA. Stems 1–1.5 m. long, nearly glabrous; leaves and leaflets petiolate, the leaflets 3–5, lanceolate to lanceolate-ovate, 3–13 cm. long, coarsely toothed; heads ligulate; involucral bracts foliaceous, the inner bracts and chaff 5–9 mm. long; ligulate corollas short, yellow; tubular corollas 4-toothed; achenes cuneate, the outer ones 5–7 mm. long, the inner ones 7–10 mm. long, the awns 2.5–5 mm. long.——Sept.–Oct. North American weed widely naturalized in lowlands in our area.

3. **Bidens tripartita** L. *B. taquetii* Lév. & Van't.; *B. shimadae* Hayata; *B. tripartita* var. *cernuaefolia* Sherff.—— TA-UKOGI. Stems 20–150 cm. long, glabrous, branched; median leaves simple, oblong-lanceolate, or more commonly 3(–5)-parted, the terminal segments larger, oblong-lanceolate, toothed, gradually narrowed at both ends, the lateral segments spreading, the petioles more or less winged, 5–13 cm. long; heads eligulate, the involucral bracts oblanceolate, 1.5–4.5 cm. long in fruit, the outer chaff oblong, 9–11 mm. long; tubular corollas 4-toothed; achenes 7–11 mm. long, retrorsely spinulose on margin and ribs, the well-developed awns 2, 3–4 mm. long, with 1 or 2 additional imperfect ones at the tip. ——Aug.–Oct. Hokkaido, Honshu, Shikoku, Kyushu.—— Formosa, Ryukyus, China, Korea, Eurasia to N. Africa, and Australia.

Var. **repens** (D. Don) Sherff. *B. repens* D. Don; *B. minuscula* Lév. & Van't.——HAI-TA-UKOGI. Stems 1–40 cm. long; leaves or the segments unequally and coarsely toothed; achenes 6.5–7.5 mm. long, glabrous; awns 2, 2–3 mm. long, retrorsely spinulose or smooth.——Reported in Honshu (Oomi Prov.).——Korea, China, India, and Malaysia.

4. **Bidens radiata** Thuill. var. **pinnatifida** (Turcz.) Kitam. *B. tripartita* var. *pinnatifida* Turcz. ex DC.; *B. maximowicziana* Oett.——EZO-NO-TA-UKOGI, KIKUBA-TA-UKOGI. Stems 15–70 cm. long, nearly 4-angled, nearly glabrous; median leaves about 15 cm. long inclusive of the petiole, glabrous, pinnately parted, the segments 2 or 3 pairs, usually linear-lanceolate, acuminate, with incurved coarse teeth, the terminal segments larger, lanceolate, acuminate, lobulate; involucral bracts 12–14, 1-seriate, foliaceous, 2–3.5 cm. long, rarely to 7 cm. long, linear-lanceolate, pinnately lobulate, the outer chaff about 8 mm. long; ligulate corollas absent; tubular corollas 4-toothed; achenes flat, 4.5–5.5 mm. long, glabrous, retrorsely spinulose on margins, the awns 2, 2.5–3 mm. long.——Aug.–

Sept. Wet places; Hokkaido.——s. Kuriles, Sakhalin, Korea, Manchuria, and Dahuria. The typical phase occurs from Siberia to Europe.

5. **Bidens parviflora** Willd. HOSOBA-SENDANGUSA. Stems 20–70 cm. long, often short-hairy; median leaves 2 or 3 times pinnately compound, short-hairy above, with longer hairs on nerves beneath, the petioles 2–3 cm. long, the leaflets pinnately divided and again pinnatilobed to pinnatiparted, or toothed, the ultimate segments oblong, about 2 cm. wide, acute; heads eligulate; outer involucral bracts herbaceous, 4 or 5, lanceolate, 4 mm. long in anthesis, 6–9 mm. long in fruit, gradually narrowed at apex, the chaff 10–12 mm. long in fruit; tubular florets 6–12; achenes linear, 13–16 mm. long, about 1 mm. wide, tetragonous, compressed, short-strigose, the awns 2, 3–3.5 mm. long.——Sept. Honshu (centr. distr. and westw.), Kyushu.——Korea, Manchuria, China, and e. Siberia.

6. **Bidens pilosa** L. KO-SENDANGUSA. Stems 25–85 cm. long, loosely crisped-pilose in upper portion; median leaves 12–19 cm. long inclusive of the petiole, usually ternate, loosely short-hairy on both sides, the lateral segments ovate, acute, short-petiolate, the terminal segments ovate or narrowly so, short-acuminate; involucral bracts 7 or 8, spathulate, 3–4 mm. long in anthesis, 5 mm. long in fruit, the inner chaff to 9 mm. long, narrow; heads eligulate; tubular corollas 5-toothed; achenes linear, 7–13 mm. long, about 1 mm. wide, flat, 4-angled, short-strigose, the awns 3 or 4, 1.5–2.5 mm. long.—— Kyushu (Hyuga Prov.). Widespread in tropical regions.

Var. **minor** (Bl.) Sherff. *B. sundaica* var. *minor* Bl.; *B. pilosa* var. *radiata* Schultz-Bip.; *B. albiflora* Makino; *B. pilosa* var. *albiflora* Maxim.——SHIROBANA-SENDANGUSA. Heads with 5–7 sterile ligulate florets, the corollas 5–8 mm. long, 3.5–5 mm. wide, white.——Honshu and Kyushu.——Korea, China, Formosa, Ryukyus, and generally in the tropics.

7. **Bidens bipinnata** L. *B. pilosa* var. *bipinnata* (L.) Hook. f.——KOBA-NO-SENDANGUSA. Stems 25–85 cm. long, slightly hairy on upper part; median leaves 11–19 cm. long, slightly hairy on both sides, bipinnately parted, the petioles 3.5–5 cm. long, the lower segments often 2 or 3, or pinnately cleft, the terminal segments narrow, acuminate, few-toothed near base; involucral bracts 5–7, lanceolate, 2.5 mm. long at anthesis, to 5 mm. long in fruit, the outer chaff 3.5–4 mm. long in anthesis, 6–8 mm. long in fruit; ligulate florets 1–3, sterile, the corollas yellow, 5–6 mm. long, 2.5–3 mm. wide; achenes linear, 12–18 mm. long, about 1 mm. wide, flat, 3 or 4 angled, more or less short-strigose, the awns 3 or 4, 3–4 mm. long.——Aug.–Oct. Waste grounds; Honshu.——Korea, Manchuria, China, Malaysia, India, Australia, Europe, and America.

8. **Bidens biternata** (Lour.) Merr. & Sherff. *Coreopsis biternata* Lour.; *B. chinensis* Willd.; *B. pilosa* sensu auct. Japon., non L.; *B. robertianaefolia* Lév. & Van't.——SENDAN-GUSA. Stems 30–150 cm. long, subtetragonous, loosely crisped-pilose; median leaves 9–15 cm. long inclusive of the 5-cm.-long petiole, prominently soft-pubescent on both surfaces, once or twice pinnately parted, the terminal segments

ovate, short-acuminate, rather prominently toothed, the lateral segments ovate, sometimes pinnatifid in the lower ones; involucral bracts 8–10, linear, 3–6.5 mm. long, acute, the outer chaff 5–6 mm. long; ligulate florets absent or 1–5, sterile, the corollas yellow, 5.5 mm. long, 2.5–3 mm. wide; achenes linear, 9–19 mm. long, about 1 mm. wide, compressed, 4-angled, short-strigose, the awns 3 or 4, 3–4 mm. long.——Sept.–Nov. Wet places and roadsides; Honshu (Kantō Distr. and westw.),

Shikoku, Kyushu.——Korea, Manchuria, China, Formosa, India, Malaysia, Australia, Asia Minor, and Africa.

Var. **mayebarae** (Kitam.) Kitam. *B. mayebarae* Kitam. ——Maruba-ta-ukogi.　Lower stems 50–60 cm. long; leaves simple, ovate-oblong, 5–13 cm. long, 2.5–6.5 cm. wide, acute, truncate-rounded at base, serrulate, the petioles winged in upper part; achenes very densely setulose.——Near seashores; Kyushu (Higo Prov.).

49. GLOSSOGYNE Cass.　Seriba-no-sendangusa Zoku

Glabrous perennials with a thick caudex; stems erect, few-leaved, branched; radical leaves long-petiolate, pinnately parted, the cauline alternate; heads small, solitary, heterogamous; involucral bracts and outer chaff oblong, slightly connate and somewhat inflated at base, the inner chaff narrowed and flat; ligulate florets 1-seriate, fertile, the corollas spreading, 3-toothed, the tubular corollas 4-toothed; anthers obtuse at base; achenes glabrous, flattened dorsally, linear, truncate, 2-awned, the awns retrorsely spinulose.——About 6 species, in tropical Asia and Australia.

1. Glossogyne tenuifolia Cass.　Seriba-no-sendangusa. Perennial herb, woody at the base; stems 20–30 cm. long, somewhat tufted; radical leaves persistent, long-petiolate, 4.5–9 cm. long, pinnately parted, the segments 2 or 3 pairs, linear, about 2 mm. wide, the lower segments 8–20 mm. long, the median leaves few, petiolate, 3–4 cm. long, pinnately parted, the upper ones smaller, linear; involucral bracts and chaff about 3 mm. long; ligulate corollas about 3.5 mm. long, 2–2.5 mm. wide, 3-toothed; achenes 6.5–7 mm. long.——Honshu (Hachijo Isl. in Izu).——Bonins, Formosa to s. China, Malaysia, and Australia.

50. ATRACTYLODES DC.　Okera Zoku

Erect, sometimes branched, stiff herbs; leaves alternate, entire or pinnate, spinulose on margin; florets tubular, bisexual and pistillate, the corollas deeply 5-lobed; involucres campanulate or tubular, the subtending bracteal leaves finely dissected and bristlelike, the bracts many-seriate, obtuse, entire, the innermost row purplish at the tip; receptacle flat, bristly-setose; style thickened at the tip in bisexual flowers, the lobes short, deltoid, mammillate on back, reflexed in the pistillate flowers; achenes terete, truncate, villous; pappus bristles nearly equal in length, pinnate, connate at base into a ring.——Few species, in Japan and China.

1. Atractylodes japonica Koidz. ex Kitam.　*Atractylis ovata* sensu auct. Japon., non Thunb.; *Atractylis ovata* var. *ternata* Komar.; *Atractylodes lyrata* var. *ternata* (Komar.) Koidz.——Okera.　Perennial with elongate, knotty rhizomes 5–8 cm. long, 1.5–3 cm. wide; stems 30–100 cm. long, terete; radical leaves withering before anthesis, the cauline firmly chartaceous, long-petiolate, pinnately 3- to 5-sected, 8–11 cm. long, short-spinulose on margin, the segments elliptic or obovate-oblong, the terminal larger, glabrous and lustrous above, slightly cobwebby beneath, with prominent veinlets, the upper leaves small and often simple, subsessile; heads terminal, erect, 15–20 mm. across, the bracteal leaves 2-seriate, as long as the head, bipinnate, the segments bristlelike; involucres campanulate, 10–12 mm. long, 12–14 mm. wide, the bracts 7- or 8-seriate, the corollas whitish, 10–12 mm. long; pappus brownish, 8–9 mm. long.——Sept.–Oct. Hills and mountains; Honshu, Kyushu; common.——Korea and Manchuria.

51. CARDUUS L.　Hire-azami Zoku

Erect, branched herbs; leaves alternate, often long-decurrent and winged, sinuate-toothed or pinnately cleft, spinulose; heads red-purple or white; involucres campanulate to globose, the bracts many-seriate, narrow, spine-tipped, entire, erect or recurved; receptacles setose; florets all alike, fertile, the corollas 5-lobed; style thickened and mammillate with elongate slightly spreading connate branches; filaments papillose-hairy; achenes glabrous, oblong, slightly flattened, striate, truncate, abruptly narrowed at base; pappus-bristles many, firm, simple, minutely scabrous.——About 100 species, in Eurasia and n. Africa.

1. Carduus crispus L.　Hire-azami.　Biennial herb; stems 70–100 cm. long, branched, crisped-pubescent, with spine-toothed wings; lower cauline leaves oblong-lanceolate, 5–20 cm. long, obtuse to acute, decurrent on the winged base, sessile, pinnately cleft to parted, the segments obtuse, toothed, spine-margined, with cobwebby hairs beneath while young, the upper leaves smaller; heads many; involucres campanulate, about 20 mm. long, the bracts 7- or 8-seriate, the outer row gradually smaller, the median row linear-lanceolate, spine-tipped, spreading or recurved at tip; flowers red-purple or rarely white; achenes oblong, about 3 mm. long, 1.5 mm. wide; pappus about 15 mm. long.——May–July. Waste grounds, roadsides, and cultivated fields in lowlands; Honshu, Shikoku, Kyushu.——Europe, Caucasus, and e. Asia.

52. CIRSIUM Adans.　Azami Zoku

Biennial or perennial usually spiny herbs; leaves alternate, simple or divided; heads terminal, polygamous, peduncled or sessile and spicate, nodding or erect, purplish, reddish, or white; involucres globose, campanulate, or tubular, cobwebby-hairy, often viscid, the bracts many-seriate, erect or spreading, entire or spinulose; receptacles setose; florets equal, bisexual, the corollas 5-

lobed; style-branches connate except at tip, with a tuft of hairs at base of the branches; filaments papillose-hairy; achenes glabrous, obsoletely 4-angled; pappus-bristles many-seriate, unequal, the longer ones plumose, slightly thickened in upper part.——More than 150 species, in Europe, N. Africa, Asia, N. and Centr. America.

1A. Involucres depressed-globose, 6.5–8.5 cm. across, the bracts coriaceous, 5–9 mm. wide, spinulose-margined. .... 1. *C. purpuratum*
1B. Involucres globose, campanulate, or tubular, 1.5–2.5 cm. across, the bracts 1–4 mm. wide, entire or obsoletely spinulose.
  2A. Biennial; corolla capillary, the narrow part 1.5–2.5 times as long as the remainder. .... 2. *C. pendulum*
  2B. Perennial; corolla broader, the narrow part slightly longer than to shorter than the remainder.
    3A. Leaves decurrent.
      4A. Heads nodding.
        5A. Cauline leaves entire to pinnately cleft, the lobes of few pairs. .... 3. *C. kamtschaticum*
        5B. Cauline leaves pectinate, the lobes of 8–12 pairs.
          6A. Outer involucral bracts linear-lanceolate; leaves or some of them pinnately divided, the segments narrow.
              4. *C. pectinellum*
          6B. Outer involucral bracts oblong-lanceolate; leaves or some of them pinnately parted to cleft, the segments broad.
              5. *C. apoiense*

      4B. Heads erect.
        7A. Plants about 30 cm. high; involucres cobwebby. .... 6. *C. ugoense*
        7B. Plants 1.5–2 m. high; involucres not cobwebby. .... 7. *C. grayanum*
    3B. Leaves not decurrent.
      8A. Radical leaves not withered at anthesis, rosulate.
        9A. Achenes 4–4.5 mm. long, not narrowed at base. .... 8. *C. sieboldii*
        9B. Achenes 3–4 mm. long, narrowed at base.
          10A. Involucral bracts recurved or spreading.
            11A. Plants 0.5–2 m. high; heads 4–5.5 cm. across; bracteal leaves of peduncles 2–4.
              12A. Leaves glaucous, sparingly spinulose, not lustrous.
                13A. Radical leaves broadly elliptic; involucral bracts lanceolate. .... 9. *C. yezoense*
                13B. Radical leaves lanceolate-elliptic; involucral bracts narrowly linear. .... 10. *C. tenuisquamatum*
              12B. Leaves yellow-green, spinulose, lustrous. .... 11. *C. lucens*
            11B. Plants 15–50 cm. high; heads 3–4 cm. across; bracteal leaves of peduncles obsolete.
              14A. Radical leaves sessile, deeply pinnatifid; cauline leaves slightly clasping at base; involucres 15–20 mm. wide.
                12. *C. tashiroi*
              14B. Radical leaves long-petiolate, entire to pinnately lobed; cauline leaves not clasping at base; involucres 2.5–3 cm. wide.
                13. *C. hidaense*

          10B. Involucral bracts erect.
            15A. Involucral bracts mostly ovate; narrow portion of corolla longer than the remainder.
              16A. Stems and leaves cobwebby; spines of the bracteal leaves stout, 10–15 mm. long; heads subsessile; involucres 17–18 mm. long; corollas 18–22 mm. long. .... 14. *C. spinosum*
              16B. Stems and leaves densely short-pubescent; spines of the bracteal leaves 2–3 mm. long; heads on short peduncles; involucres 22–25 mm. long; corollas 27–28 mm. long. .... 15. *C. maritimum*
            15B. Involucral bracts mostly linear to lanceolate; narrow portion of corolla shorter than the remainder.
              17A. Heads erect, viscid. .... 16. *C. japonicum*
              17B. Heads nodding and viscid, not erect and viscid.
                18A. Heads erect, not viscid.
                  19A. Radical leaves elliptic to elliptic-lanceolate, lustrous and uniformly green on upper side; heads few.
                  17. *C. aomorense*
                  19B. Radical leaves oblong-lanceolate, lustreless, the nerves reddish on upper side; heads many. .... 18. *C. tanakae*
              18B. Heads nodding, viscid.
                20A. Outer and median involucral bracts ovate-lanceolate to oblong-lanceolate; leaves slightly cobwebby beneath, relatively thick.
                  21A. Narrower portion of corolla-tube half as long as the remainder, the lobes ⅓ to ½ as long as the broadest part of the corolla; leaf-segments of radical leaves slightly toothed, often densely cobwebby on both sides.
                  19. *C. nipponense*

                  21B. Narrow portion of corolla-tube about as long as the remainder, the lobes slightly shorter than the broadest part of the corolla; leaf-segments of radical leaves incised-toothed, loosely cobwebby beneath.
                  20. *C. chokaiense*

                20B. Outer and median involucral bracts lanceolate-linear to lanceolate; leaves glabrous beneath, relatively thin.
                21. *C. maruyamanum*

      8B. Radical leaves withering before anthesis.
        22A. Involucral bracts spinulose on margin. .... 22. *C. magofukui*
        22B. Involucral bracts entire.
           23A. Involucral bracts many-ribbed on back. .... 23. *C. dipsacolepis*
          23B. Involucral bracts 1-ribbed on back.
            24A. Limb of corolla 1.5–2 times as long as the broadest part of corolla-tube; leaves linear, 5–10 mm. wide. .. 24. *C. lineare*
            24B. Limb of corolla shorter to as long as the broadest part of corolla-tube.
              25A. Involucres narrow, tubular.
                26A. Cauline leaves clasping.
                  27A. Heads erect or ascending.
                    28A. Involucral bracts much recurved.

29A. Leaves and heads densely arranged. .... 25. *C. confertissimum*
29B. Leaves and heads not densely arranged.
 30A. Leaves auriculate-clasping. .... 26. *C. aidzuense*
 30B. Leaves merely clasping.
  31A. Leaves often falcate; heads loosely arranged. .... 27. *C. bitchuense*
  31B. Leaves not falcate; heads fasciculate-spicate. .... 28. *C. buergeri*
28B. Outer and median involucral bracts barely recurved.
 32A. Cauline leaves toothed to shallowly pinnatifid; involucral bracts 9-seriate, the outer ones ovate.
                                       29. *C. heiianum*
 32B. Cauline leaves deeply pinnatifid; involucral bracts 6-seriate, the outer ones lanceolate or ovate-lanceolate.
                                       30. *C. suzukaense*
27B. Heads nodding or pendulous.
 33A. Leaf-segments lanceolate, the spines 3–12 mm. long. .... 31. *C. gyojanum*
 33B. Leaf-segments oblong-lanceolate, the spines 2–4 mm. long.
  34A. Plants 1.5–2 m. high, leafy throughout; leaves deeply pinnatifid, not white-variegated; involucres about 2.5 cm. across. .... 32. *C. kagamontanum*
  34B. Plants 50–80 cm. high, subscapose, densely leafy at base; leaves much cleft and incised, obsoletely white-variegated on upper surface; involucres about 2 cm. across. .... 33. *C. longepedunculatum*
26B. Cauline leaves not clasping.
 35A. Heads on slender peduncles; leaves oblong or lanceolate, the segments spreading.
  36A. Involucres narrowly tubular, 1.4–2 cm. across, the bracts elongate, prominently recurved at tip. 34. *C. effusum*
  36B. Involucres campanulate, about 2.2 cm. across, the bracts relatively short, barely recurved at tip.
                                     35. *C. gratiosum*
 35B. Heads sessile; leaves elliptic or elliptic-lanceolate, the segments erect.
  37A. Involucral bracts elongate, acute, recurved.
   38A. Involucres 1.7–2 cm. wide; leaf-segments oblong, short-acuminate, the spines 1–2 mm. long.
                             36. *C. congestissimum*
   38B. Involucres 1.5–1.7 cm. wide; leaf-segments oblong-lanceolate, acuminate, the spines 3–10 mm. long.
                              37. *C. spicatum*
  37B. Involucral bracts short, acuminate, barely recurved.
   39A. Involucres 1–1.7 cm. across, green, viscid, scarcely cobwebby; leaves loosely cobwebby beneath. 38. *C. tenue*
   39B. Involucres 1.7–2 cm. across, purplish, scarcely viscid, densely cobwebby; leaves densely cobwebby beneath.
                             39. *C. microspicatum*
25B. Involucres campanulate or campanulate-globose.
 40A. Heads nodding at anthesis.
  41A. Leaves not clasping.
   42A. Involucres campanulate or tubular-campanulate, 2–3 cm. across.
    43A. Outer involucral bracts as long as the others.
     44A. Leaves elliptic-lanceolate. .... 40. *C. nipponicum*
     44B. Leaves linear-lanceolate.
      45A. Involucres 2–2.3 cm. across; corolla about 16 mm. long; leaf-segments acuminate.
                           41. *C. yakushimense*
      45B. Involucres about 2.5 cm. across; corolla 17–21 mm. long; leaf-segments short-acuminate.
                           42. *C. chikushiense*
    43B. Outer involucral bracts shorter than the others.
     46A. Involucral bracts not recurved at tip. .... 43. *C. indefensum*
     46B. Involucral bracts shortly recurved at tip. .... 44. *C. hanamakiense*
   42B. Involucres campanulate-globose to depressed-globose, 3.5–4.5 cm. across.
    47A. Leaf-segments acuminate, the spines 2–5 mm. long; heads nodding at anthesesis; involucres campanulate-globose, the outer bracts lanceolate. .... 45. *C. suffultum*
    47B. Leaf-segments acute, the spines 2–3 mm. long; heads erect at anthesis; involucres depressed-globose, the outer bracts lanceolate to linear-lanceolate. .... 46. *C. hachijoense*
  41B. Leaves clasping.
   48A. Leaves densely white-cobwebby beneath. .... 47. *C. norikurense*
   48B. Leaves green beneath.
    49A. Leaves narrowly oblong-lanceolate, the segments of about 8 pairs, prominently spiny. .... 48. *C. senjoense*
    49B. Leaves elliptic-lanceolate, the segments 3–6 pairs, usually subspinose.
     50A. Heads erect at anthesis, about 3 cm. across; involucral bracts linear-lanceolate, rather thin.
                             49. *C. amplexifolium*
     50B. Heads nodding at anthesis, 3–4 cm. across; involucral bracts lanceolate to ovate, rather thick.
      51A. Outer and median involucral bracts viscid on back. .... 50. *C. matsumurae*
      51B. Outer and median involucral bracts not viscid.
       52A. Leaf-segments 5 or 6 pairs, spreading; corolla about 15 mm. long. .... 51. *C. babanum*
       52B. Leaf-segments 3 or 4 pairs, erect; corolla about 17 mm. long. .... 52. *C. ganjuense*
 40B. Heads erect at anthesis.
  53A. Cauline leaves deeply pinnatifid, the segments prominently incised-toothed; corolla about 15 mm. long.
                             53. *C. nambuense*
  53B. Cauline leaves pinnately cleft to entire, the segments loosely lobulate; corolla 18–20 mm. long.
   54A. Cauline leaves dilated at base and broadly clasping. .... 54. *C. homolepis*
   54B. Cauline leaves slightly dilated and semiclasping. .... 55. *C. inundatum*

1. **Cirsium purpuratum** (Maxim.) Matsum. *Cnicus purpuratus* Maxim.——FUJI-AZAMI. Large stout perennial; stems tufted, 50–100 cm. long, densely leafy at base, ascending, branched, cobwebby; radical leaves large, narrowly oblong, 50–70 cm. long, 17–30 cm. wide, pinnately cleft, the lobes narrowly elliptic, 5 or 6 pairs, acute to acuminate, spinulose, cobwebby on both sides, glabrescent, the cauline leaves gradually smaller, sessile, partially clasping, sometimes auriculate at base; heads large, nodding, long-pedunculate; involucres depressed-globose, 6.5–8.5 cm. across, purplish, glabrous; bracts about 10-seriate, coriaceous, the outer and median acute, 5–9 mm. wide, densely spinulose-margined, reflexed; corollas rose-purple, 23–29 mm. long.——Aug.-Oct. Gravelly and sandy slopes in mountains; Honshu (Kantō and centr. distr.).——Forma **albiflorum** (Kitam.) Kitam. *C. purpuratum* var. *albiflorum* Kitam.——SHIROBANA-FUJI-AZAMI. Corollas white.

Cirsium × misawaense Nakai. MISAWA-AZAMI. Hybrid of *C. nipponicum* var. *incomptum* × *C. purpuratum*.——Honshu (centr. distr.).

2. **Cirsium pendulum** Fisch. *C. falcatum* Turcz.; *Cnicus pendulus* (Fisch.) Maxim.; *Cnicus hilgendorffii* Fr. & Sav.; *Cirsium hilgendorffii* (Fr. & Sav.) Makino——TAKA-AZAMI. Biennial; rhizomes short; roots narrowly fusiform at base; stems erect, 1–2 m. long, branched and cobwebby on upper portion; radical and lower cauline leaves large, short-petiolate, 40–50 cm. long, about 20 cm. wide, caudate, pinnately cleft, the lobes about 5 pairs, linear, glabrous, the median leaves narrowly elliptic, 15–25 cm. long, sessile; heads many, pendulous at anthesis; involucres ovoid, about 2 cm. long, 2.5–3.5 cm. wide, slightly cobwebby; bracts 8-seriate, the median and outer linear, often reflexed, with a dark midrib; corollas capillary, 17–22 mm. long, purplish.——Aug.-Oct. Grassy places in lowlands; Hokkaido, Honshu (Kantō Distr. and northw.).——Korea, Manchuria, Amur, Ussuri, and Dahuria.——Forma **albiflorum** (Makino) Kitam. *C. pendulum* var. *albiflorum* Makino——SHIROBANA-TAKA-AZAMI. Corollas white.

Cirsium × iburiense Kitam. IBURI-AZAMI. Hybrid of *C. aomorense × C. pendulum*.——Hokkaido (Iburi Prov.).

3. **Cirsium kamtschaticum** Ledeb. *C. kamtschaticum* var. *genuinum* Herd. and var. *weyrichii* Herd.; *Cnicus kamtschaticus* (Ledeb.) Maxim.; *Cnicus weyrichii* (Herd.) Maxim.; *Cnicus borealis* Kitam.——CHISHIMA-AZAMI, EZO-AZAMI. Rhizomes stout, woody; stems erect, usually 1–2 m. long, stout, simple or branched, hairy or glabrous, cobwebby in upper part; lower and median cauline leaves rather thin, elliptic to narrowly so, 17–35 cm. long, acute to acuminate, narrowed at base, sessile or short-petiolate, decurrent, spinulose, entire or pinnately cleft, the lobes usually of few pairs, nearly glabrous above, crisped-pubescent, cobwebby-hairy beneath while young; heads many, nodding in anthesis; involucres globose-campanulate, cobwebby, 17–20 mm. long, about 4 cm. wide; bracts about 7-seriate, the outer and median linear-lanceolate, spreading; corolla 15–17 mm. long, purplish. ——July–Sept. Hokkaido.——Sakhalin, Kuriles, Kamchatka, and Aleutians.

4. **Cirsium pectinellum** A. Gray. *Cnicus pectinellus* Maxim.; *Cirsium mamiyanum* Koidz.——EZO-NO-SAWA-AZAMI. Stems 50–150 cm. long, slender, often hairy, cobwebby in upper portion, usually branched; radical leaves elliptic, caudate, abruptly narrowed at base, pinnately divided, the long winged petiole 30–45 cm. long, the segments 10 or 11 pairs, linear-lanceolate, 3–8 cm. long, 1–1.5 cm. wide, often deflexed, pecti-

nate, glabrous on upper side, paler and pubescent on nerves beneath, the spines about 3 mm. long; cauline leaves elliptic to lanceolate, sessile or short-petiolate, decurrent, pectinate; heads pendulent on slender peduncles; involucres depressed-globose, about 1.6 cm. long, 3 cm. wide, slightly cobwebby; bracts 7-seriate, the outer and median linear-lanceolate, spreading; corollas purplish, 15–16 mm. long.——July–Aug. Wet places in lowlands; Hokkaido.——Sakhalin and Kuriles.

Var. **alpinum** Koidz. MIYAMA-SAWA-AZAMI. Stems shorter, 25–27 cm. long; leaf-segments broader.——Alpine; Hokkaido (centr. distr.).

5. **Cirsium apoiense** Nakai. APOI-AZAMI. Stems 50–80 cm. long, crisped-pubescent, slightly branched in upper portion; lower cauline leaves elliptic-lanceolate, acuminate, pinnately parted, the segments of 7 or 8 pairs, approximate, oblong, 17–22 cm. long, about 7.5 cm. wide, toothed, glabrous on both sides, spines 2–4 mm. long, winged-petiolate, decurrent, wings prominently spiny; heads simple, nodding at anthesis; involucres globose-campanulate, about 1.5 cm. long, 2–2.8 cm. wide, cobwebby; bracts 6-seriate, the outer oblong-lanceolate, acuminate, spine-tipped, the median row lanceolate; florets about 15 mm. long.——Aug. Hokkaido (Mount Apoi in Hidaka Prov.).

6. **Cirsium ugoense** Nakai. UGO-AZAMI. Stems erect, 30 cm. or more long, loosely pubescent or glabrous, sometimes cobwebby in upper portion; median cauline leaves closely approximate, elliptic to oblong-lanceolate, 15–20 cm. long, acute to acuminate, spine-margined, sessile, sometimes clasping and decurrent, pinnately parted to entire, the segments approximate, ovate to lanceolate, loosely pubescent on upper side, pubescent on nerves beneath, sometimes glabrous on both sides, the spines 2–3 mm. long; heads sessile or short-pedunculate; involucres slightly cobwebby, globose-campanulate, 1.5–2.2 cm. long, 2.5–3.5 cm. wide; bracts 6-seriate, erect, nearly equal, the outer sometimes slightly shorter, short spine-tipped; corollas purplish, 16–17 mm. long.——Aug. Alpine regions; Honshu (n. distr.).

7. **Cirsium grayanum** (Maxim.) Nakai. *Cnicus kamtschaticus* var. *grayanus* Maxim.; *Cirsium kamtschaticum* var. *grayanum* (Maxim.) Matsum.——MARUBA-HIRE-AZAMI. Rhizomes stout; stems 1.5–2 m. long, short-pubescent, branched in upper portion, loosely cobwebby on the peduncles; median and lower cauline leaves elliptic-lanceolate, 20–25 cm. long, acute to acuminate, narrowed at base, clasping and decurrent, entire to pinnately parted, spiny, the segments about 6 pairs, spreading, oblong-lanceolate, acute, toothed, glabrous on upper side, short-pubescent on nerves beneath, the spines 3–5 mm. long; heads few, approximate, erect at anthesis; involucres campanulate-globose, about 1.5 cm. long, 2.5–3 cm. wide; bracts 6-seriate, the outer shorter, linear-lanceolate, the outer and median spine-tipped, sometimes spinulose on margin; corolla purplish, 16–17 mm. long.——July–Sept. Wet places in lowlands; Hokkaido (s. distr.), Honshu (Mutsu Prov.).

8. **Cirsium sieboldii** Miq. *Cnicus sieboldii* (Miq.) Maxim.; *Cnicus reinii* Fr. & Sav.; *Cirsium reinii* (Fr. & Sav.) Matsum.; *Cirsium longipes* Nakai——MA-AZAMI, KISERU-AZAMI. Stems subscapose, loosely leaved, 50–100 cm. long, cobwebby or pubescent in upper portion; radical leaves rosulate, much larger than the cauline, ligulate, 15–55 cm. long, winged-petiolate, usually pinnatifid, the lobes spreading or erect, usually narrow and approximate, incised, glabrous on both sides, the spines 2–3 mm. long; cauline leaves few, gradually smaller, lanceolate or broadly so, pinnately lobed to in-

cised, sessile; heads solitary, terminal, nodding at anthesis, erect in fruit, long-pedunculate; involucres campanulate-globose, about 2.2 cm. long, 3–4 cm. wide; bracts sometimes cobwebby, 7-seriate, acute to obtuse, the outer very short, 3–4 mm. long, spine-tipped, appressed; corollas purplish, 17–19 mm. long; achenes 4–4.5 mm. long.——Sept.–Oct. Wet places along rivers and streams in lowlands; Honshu, Shikoku, Kyushu.

Var. **austrokiushianum** (Kitam.) Kitam. *C. austrokiushianum* Kitam.——SATSUMA-AZAMI. Leaves narrowly elliptic, often prominently spinose; heads larger, often with 1–3 linear bracts at base; involucres 25–27 mm. long, 4–5 cm. wide, the bracts loosely imbricate; corolla 20–23 mm. long.——Wet places along rivers in Kyushu (s. distr.).

Cirsium × pilosum Kitam. KE-MA-AZAMI. Hybrid of *C. sieboldii × C. tanakae.*——Honshu (Shirakawa in Iwaki Prov.).

9. **Cirsium yezoense** (Maxim.) Makino. *Cnicus yezoensis* Maxim.; *Cirsium maximowiczii* var. *riparium* Koidz.; *Cirsium riparium* (Koidz.) Koidz.; *Cirsium yoshizawae* Koidz.——SAWA-AZAMI. Stems 1–2(–3) m. long, cobwebby, branched, leafy; radical leaves thin, large, broadly elliptic, 50–60 cm. long, about 30 cm. wide, pinatifid, the lobes of about 5 pairs, lanceolate-elliptic, irregularly toothed, short spine-tipped, glaucous and mealy-puberulent above, glabrous beneath; cauline leaves gradually smaller, lanceolate to broadly elliptic, pinnatifid; heads solitary on the branches, nodding, usually subtended by few bracteal leaves; involucres depressed-globose, 2.5–2.7 cm. long, 4–4.5 cm. wide; bracts thin, flat, cobwebby, about 7-seriate, the outer lanceolate, obliquely spreading, the median nearly unarmed; corollas reddish, 18–21 mm. long; achenes 3–4 mm. long.——Sept.–Oct. Woods; Hokkaido (s. distr.), Honshu (Kinki Distr. and northw.).

10. **Cirsium tenuisquamatum** Kitam. SAMBE-SAWA-AZAMI. Stems 50–100 cm. long, branched, cobwebby; radical leaves rosulate, lanceolate-elliptic, 25–50 cm. long, 12–20 cm. wide, usually obtuse, narrowed below to a winged petiole, pinnately cleft, the lobes about 5 pairs, oblong, coarsely toothed, short-spinulose, deep green and slightly scaberulous above, glabrous beneath; cauline leaves gradually smaller, oblong-lanceolate, slightly clasping or not at all, pinnatifid to toothed, the spines short; heads solitary on the branches, nodding, subtended by 2–4 lanceolate, spinulose-toothed bracteal leaves; involucres depressed-globose, 2–2.2 cm. long, 3.5–4.5 cm. wide, cobwebby; bracts about 6-seriate, the outer narrowly linear, acuminate, 0.5–1 mm. wide, about half as long as the inner, the inner and median obliquely spreading, spine-tipped, mealy dotted, the inner recurved; corollas purplish, 21–22 mm. long.——Oct.–Nov. Woods along rivers; Honshu (Chūgoku Distr.).

11. **Cirsium lucens** Kitam. TERI-HA-AZAMI. Stems 1–1.5 m. long, densely leafy, cobwebby, branched; radical leaves thin, elliptic, 55–60 cm. long, 20–30 cm. wide, narrowed at base to a winged petiole, lustrous and nearly glabrous above, yellow-green, loosely cobwebby and short-hairy on nerves beneath, pinnately cleft, the lobes 4–7 pairs, oblong, coarsely toothed, the spines thick, 5–6 mm. long; upper leaves gradually smaller, lanceolate, pinnatifid or coarsely toothed, acute, partially clasping at base, nearly glabrous on both sides; heads solitary, nodding, bracteal leaves at base few or absent; involucres depressed-globose, cobwebby, about 1.8 cm. long, 3.5 cm. wide; bracts rather thin, flat, 6- or 7-seriate, obliquely

spreading at the tip, entire, spine-tipped, the outer linear-lanceolate, the median ovate-lanceolate; corollas pink, 19–20 mm. long.——Sept.–Oct. Along rivers; Kyushu.

12. **Cirsium tashiroi** Kitam. WATAMUKI-AZAMI. Stems scapose, slender, 30–50 cm. long, rarely to 1 m. long, hairy and cobwebby, usually slightly branched; radical leaves very thin, rosulate, lanceolate-elliptic, usually 15–30 cm. long, acute to acuminate, gradually narrowed at base, subsessile, deeply pinnatifid, the segments 7–9 pairs, approximate, ovate, irregularly 3- to 5-lobed, glabrous or pubescent on upper side, pubescent on nerves beneath, the spines 2–3(–5) mm. long; cauline leaves few, small, oblong or lanceolate, acuminate, slightly clasping, pinnately lobulate or incised; heads few, nodding, on long peduncles, the bracteal leaves 1–3, obsolete; involucres cobwebby, campanulate-tubular, 1.5–1.7 cm. long, 1.5–2.5 cm. wide; bracts thin, 6-seriate, erect or spreading, short spine-tipped, the outer linear-lanceolate; corollas purplish, 18–21 mm. long.——Sept.–Oct. Woods in mountains; Honshu (Kinki Distr. and Tōtōmi Prov.).

13. **Cirsium hidaense** Kitam. HIDA-AZAMI. Stems 15–50 cm. long, branched, hairy and cobwebby; radical leaves rosulate, lanceolate-elliptic, 15–37 cm. long, acute to obtuse, long-petiolate, entire to pinnately lobed, spinose, the lobes oblong or ovate, toothed or incised, sometimes hairy on both sides; cauline leaves gradually becoming smaller in the upper ones, acute to acuminate, sessile, entire to pinnately lobed, spinose; heads terminal, nodding at anthesis; involucres campanulate, cobwebby, 1.5–1.9 cm. long, 2.5–3 cm. wide; bracts thin, 7-seriate, the outer and median elongate, linear-lanceolate, long-acuminate, short spine-tipped; corollas purplish, 16–20 mm. long.——Sept.–Nov. Woods; Honshu (Mino and Hida Prov.).

14. **Cirsium spinosum** Kitam. OIRAN-AZAMI. Stems stout, branched, cobwebby; radical leaves rather thick, oblong, about 30 cm. long, long-petioled, pinnately parted, the segments spreading, ovate, approximate, toothed, spine-tipped, glabrous above, cobwebby beneath, rarely short-pubescent on nerves beneath; cauline leaves gradually smaller, elliptic or broadly ovate, auriculate-clasping, pinnately lobed to cleft, the lobes prominently spiny, cobwebby beneath; heads subsessile, congested, the spines of the bracteal leaves 10–15 mm. long; involucres campanulate, 1.7–1.8 cm. long, 2–2.5 cm. wide; bracts 5- or 6-seriate, erect, spine-tipped, nearly equal in length or the outer slightly shorter, the outer ovate, acuminate, ciliolate; corollas 18–20 mm. long.——Aug.–Oct. Near seashores; Kyushu (s. distr.).——Ryukyus.

15. **Cirsium maritimum** Makino. HAMA-AZAMI. Stems ascending, 15–60 cm. long, branched at base, hairy and cobwebby; radical leaves spreading, narrowly oblong, 15–35 cm. long, obtuse, narrowed at base, spiny, short–petiolate, pinnately parted, the segments rather thick, elliptic, variously lobed, deep green and lustrous above, densely pubescent on nerves beneath, the cauline leaves narrowly oblong, sessile, scarcely clasping at base; heads erect, short-pedunculate, in loose corymbs, usually with few rather large bracteal leaves at base; involucres campanulate, 2.2–2.5 cm. long, 2.2–2.8 cm. wide; bracts 6-seriate, short spine-tipped, the outer elongate, ovate-lanceolate, obliquely ascending in upper half; corollas purplish to white, 27–28 mm. long.——July–Sept. Near seashores; Honshu (Izu Prov. and westw.), Shikoku, Kyushu.

16. **Cirsium japonicum** DC. *Cnicus japonicus* (DC.) Maxim., and var. *intermedius* Maxim.; *Cirsium lacinulatum*

Nakai; *C. senile* Nakai; *C. maackii* var. *kiusianum* Nakai——No-AZAMI.   Plant 0.5–1 m., rarely to 2 m. high; stems often densely pubescent at base; radical leaves obovate-oblong, 15–30 cm. long, narrowed at base, loosely pubescent above, pubescent on nerves beneath, pinnately cleft to toothed, the lobes 5 or 6 pairs, oblong, short spine-tipped, the median cauline leaves oblong, broadly clasping at base, pinnately cleft; heads usually terminal, erect, viscid; involucres depressed-globose, 1.5–2 cm. long, 2.5–4 cm. wide; bracts 6- or 7-seriate, slightly cobwebby, linear, the outer very short, somewhat longer in the inner, acuminate, short spine-tipped, entire, viscid on back; corollas purplish or rose-colored, 18–23 mm. long.——June–Aug. (–Oct.). Lowlands and low elevations in the mountains; Honshu, Shikoku, Kyushu; very common and variable; sometimes cultivated for cut-flowers.

Var. **horridum** Nakai. TOGE-AZAMI.   Stems 15–20 cm. long, densely leafy; spines 5–7 mm. long.——Alpine; Shikoku.

Var. **ibukiense** Nakai. *C. ibukiense* (Nakai) Nakai——MIYAMA-KO-AZAMI.   Stems short, densely leafy; spines stout, 3–5 mm. long.

Var. **vestitum** Kitam. KESHŌ-AZAMI.   Leaves densely cobwebby and white beneath.——Honshu (Tanba and Settsu Prov. westw.), Shikoku.

Var. **ussuriense** (Regel) Kitam. *Cirsium littorale* var. *ussuriense* Regel; *Cirsium maackii* Maxim.; *Cnicus japonicus* var. *maackii* (Maxim.) Maxim.; *Cnicus maackii* var. *koraiensis* Nakai; *Cirsium maackii* var. *koraiense* (Nakai) Nakai; *Cirsium asperum* Nakai——KARA-NO-AZAMI.   Leaves thicker, darker when dried; heads slightly nodding to erect; involucres very viscid.——Kyushu (Tsushima).——Manchuria, Ussuri, and Korea.

Var. **diabolicum** (Kitam.) Kitam. *Cirsium diabolicum* Kitam.——ONI-ŌNO-AZAMI.   Stems densely long-pubescent, cobwebby above; leaves deeply pinnatifid, the lobes about 10 pairs; heads nodding in anthesis, very viscid.——Mountains of Japan Sea side of Honshu.

17. Cirsium aomorense Nakai. *Cnicus japonicus* var. *yezoensis* Maxim.; *Cirsium maximowiczii* Nakai; *C. yesoanum* Nakai; *C. japonicum* var. *yezoense* (Maxim.) Matsum.——ŌNO-AZAMI, AOMORI-AZAMI.   Stems about 50 cm. long, subscapose, often densely cobwebby; radical leaves much larger than the cauline, elliptic to elliptic-lanceolate, 30–60 cm. long, 11–25 cm. wide, narrowed at base, lustrous and glabrous above, slightly cobwebby to glabrous beneath, once to twice pinnatifid, the lobes incised; lower cauline leaves semiclasping; median leaves loosely arranged, small, clasping, narrowly oblong, pinnatifid, loosely cobwebby beneath; heads long-pedunculate, erect at anthesis; involucres depressed-globose, 1.8–2 cm. long, 3–4 cm. wide, cobwebby; bracts 6- or 7-seriate, linear, short spine-tipped, the outer often elongate; corollas 20–22 mm. long.——Aug.–Sept. Grassy places in lowlands; Hokkaido, Honshu (n. distr.).

Cirsium perplexissimum Kitam. MAYOWASE-AZAMI.   Hybrid of *C. aomorense* × *C. heiianum*.

18. Cirsium tanakae (Fr. & Sav.) Matsum. *Cnicus tanakae* Fr. & Sav.; *Cirsium autumnale* Kitam.——NOHARA-AZAMI.   Stems 50–100 cm. long, scapose, pubescent, glabrous, or cobwebby; radical leaves 25–40 cm. long, oblong-lanceolate, narrowed at base, glabrous or loosely cobwebby on upper side, with reddish nerves, loosely cobwebby beneath, often pubescent on nerves, pinnatifid, the lobes 8–12 pairs, incised; cauline leaves smaller, scattered, oblong or lanceolate,

pinnatifid, clasping; heads erect, sometimes with 1 to 3 bracteale leaves at base; involucres campanulate-globose, cobwebby, 1.6–1.8 cm. long, 2–2.5 cm. wide; bracts 6- or 7-seriate, linear-lanceolate, short spine-tipped, the outer short or slightly elongate; corollas purplish, 17–18 mm. long.——Aug.–Oct. Grassy places; Honshu (n. and centr. distr.).

Var. **nikkoense** (Nakai) Kitam. *Cirsium nikkoense* Nakai; *C. japonicum* subsp. *nikkoense* (Nakai) Nakai——NIKKŌ-AZAMI.   Leaves twice pinnately cleft, the lobes narrower and congested.——Honshu (Nikko in Shimotsuke).

Var. **niveum** (Kitam.) Kitam. *C. nikkoense* var. *niveum* Kitam.——OKINA-AZAMI.   Stems white-villous; leaves glabrous to whitish cobwebby beneath, the lobes narrower.——Mountains of Honshu (Shinano and Echigo Prov.).

19. Cirsium nipponense (Nakai) Koidz. *C. nipponense* var. *spinulosum* Kitam.; *C. japonicum* subsp. *yesoense* var. *nipponense* Nakai; *C. maximowiczii* var. *nipponense* (Nakai) Nakai; *C. maximowiczii* var. *glutinosum* Nakai, pro parte.——ONI-AZAMI, ONI-NO-AZAMI.   Stems 50–100 cm. long, stout, crisped-pubescent and cobwebby on upper portion; radical leaves large, oblong to elliptic-lanceolate, 35–65 cm. long, obtuse to short-acuminate, narrowed at base, long-petiolate, pinnatifid, the lobes 5–8 pairs, spreading, ovate or oblong, acute, slightly toothed, spinulose, cobwebby on both sides; cauline leaves gradually smaller, acuminate, broadly auriculate-clasping, pinnatifid, the lobes 5–8 pairs, oblong, hairy above, cobwebby beneath while young, the spines 5–8 mm. long; heads nodding at anthesis, often with bracteal leaves at base; involucres globose, viscid, 2–2.2 cm. long, 3–4.5 cm. wide, spine-tipped; bracts 6-seriate, the outer short, ovate, acuminate, often spinulose on margin; corollas purplish, 15–18 mm. long. ——June–Sept. Mountains; Honshu (n. and centr. distr.).

20. Cirsium chokaiense Kitam. *C. maximowiczii* var. *glutinosum* Nakai, pro parte.——CHŌKAI-AZAMI.   Stems 1–1.5 m. long, densely pubescent in upper portion; radical leaves rosulate, oblong, about 40 cm. long, acute, pinnatifid, the lobes about 6 pairs, oblong, acute, incised-toothed, densely crisped-pubescent above, loosely cobwebby beneath and pubescent on nerves; median cauline leaves oblong-lanceolate, 17–20 cm. long, long-acuminate, broadly clasping, lobed to cleft, the spines 5–20 mm. long; heads nodding at anthesis, with 1 or 2 bracteal leaves; involucres depressed-globose, prominently viscid, about 2.7 cm. long, 4–5 cm. wide, short-acuminate, the outer bracts short, ovate-lanceolate, the median oblong-lanceolate; corollas about 18 mm. long, purplish.——Aug. Alpine regions; Honshu (Mount Chōkai in Ugo Prov.).

21. Cirsium maruyamanum Kitam. MURA-KUMO-AZAMI.   Stems 50–90 cm. long, crisped-pubescent and cobwebby-hairy; radical leaves rosulate, narrowly to broadly oblong, 20–35 cm. long, pinnately cleft to parted, the lobes 8–9 pairs, spreading, often 3-lobulate, spinulose, glabrous on both sides; cauline leaves many, gradually smaller, oblong, acuminate, broadly clasping, pinnatifid, loosely pubescent to glabrous on upper side; heads nodding at anthesis, often bracteate; involucres depressed-globose, about 2 cm. long, about 3.5 cm. wide, loosely cobwebby; bracts about 5-seriate, viscid on the midrib, blackish, the outermost lanceolate to linear-lanceolate, about 1/5 as long as the inner, acuminate, very short spine-tipped; corollas purplish, about 20 mm. long.——May–June. Wet places in lowlands; Honshu (Chūgoku Distr.).

22. Cirsium magofukui Kitam. INABE-AZAMI.   Stems nearly glabrous, branched; cauline leaves elliptic to elliptic-lanceolate, 20–25 cm. long, acuminate, abruptly narrowed at base, sessile, slightly clasping, pinnatifid, the lobes about 5 pairs, oblong-lanceolate, acuminate, coarsely toothed, spinulose, nearly glabrous on both sides; heads solitary, long-pedunculate; involucres campanulate-globose, about 2.7 cm. long, about 4.5 cm. wide; bracts 6-seriate, long-acuminate, the outer lanceolate or ovate-lanceolate, spine-tipped, the median oblong-lanceolate, 3.5–4 mm. wide, strongly reflexed, loosely spinulose on margin; corollas purplish, about 21 mm. long.——Oct. Honshu (Ise Prov.).

23. Cirsium dipsacolepis (Maxim.) Matsum. *Cnicus dipsacolepis* Maxim.——MORI-AZAMI, YABU-AZAMI. Rhizomes short; roots fusiform, to 2 cm. across, unbranched; stems 50–100 cm. long, pubescent or glabrous; lower cauline leaves long-petiolate, lanceolate-oblong, sometimes broadly ovate or lanceolate, 15–20 cm. long, rarely to 40 cm. long, pinnately cleft to incised, the lobes chartaceous, spreading, coarsely incised, the lobules short spine-tipped, pubescent to glabrous on both sides; heads solitary on the branches, erect at anthesis, sometimes bracteate at base; involucres depressed-globose, 2–3 cm. long, 3–4 cm. wide, cobwebby; bracts 6- to 7-seriate, many-ribbed on back, the outer elongate, spreading, lanceolate, the median longer and broader than the inner, 2–3 mm. wide, obliquely spreading at tip, often ciliate; corollas purplish or rose-colored, 17–18 mm. long.——Sept.–Oct. Lowlands; Honshu, Shikoku, Kyushu.

Var. **calcicola** (Nakai) Kitam. *C. calcicola* Nakai——AKIYOSHI-AZAMI. Stems 30–60 cm. long; lower cauline leaves broadly lanceolate, 10–12 cm. long, irregularly incised-serrate, crisped-pubescent on both sides, sometimes cobwebby beneath, sessile; involucres about 17 mm. long, 22–35 mm. wide; bracts 5-seriate, the outer short, broadly lanceolate, acuminate, not recurved at tip; corollas 18–19 mm. long.——Honshu (Nagato Prov.).

24. Cirsium lineare (Thunb.) Schultz-Bip. *Carduus linearis* Thunb.; *Spanioptilon lineare* (Thunb.) Less.; *Cnicus linearis* (Thunb.) Benth.——YANAGI-AZAMI. Rhizomes short; roots fusiform; stems about 1 m. long, glabrous or pubescent, branched and cobwebby on upper portion; median cauline leaves many, linear, 6–20 cm. long, 5–10 mm. wide, gradually attenuate, narrowed at base, decurrent, entire or irregularly undulate-incised, sessile or short-petiolate, the lobules spinulose on margin, glabrous above, loosely cobwebby and distinctly 1-nerved beneath, the upper leaves linear, smaller; heads solitary on the branches, erect; involucres globose, cobwebby, about 1.5 cm. long, 2–2.5 cm. wide; bracts 6- to 7-seriate, 1-ribbed on back, the outer and median linear, appressed, spine-tipped; corollas purplish, about 17 mm. long.——Sept.–Oct. Lowlands; Honshu (Suwō Prov.), Shikoku, Kyushu.

Var. **discolor** Nakai. URAYUKI-YANAGI-AZAMI. Leaves whitish cobwebby.

25. Cirsium confertissimum Nakai. *C. confertissimum* var. *saxatile* Nakai——KO-IBUKI-AZAMI. Stems 50–100 cm. long, hairy, densely leafy, much branched above; radical and lower cauline leaves elliptic-oblanceolate or oblong, 25–45 cm. long, acute, short-petiolate, pinnatifid, the lobes 8 or 9 pairs, oblong, approximate, spreading, acute, coarsely incised at base, short spine-tipped, often white-variegated above, loosely pubescent to glabrous beneath; cauline leaves very many, 12–18 cm. long, caudate, broadly clasping, the lobes 4 or 5 pairs; heads many, erect; involucres tubular, 1.4–1.5 cm. long, 1.5–2 cm. wide, cobwebby and viscid; bracts 6-seriate, the outer short,

deltoid, acute, the median oblong, acute to obtuse, scarcely spine-tipped; corollas purple, 15–18 mm. long.——Sept.–Oct. Honshu (Mount Ibuki in Oomi).

Var. **herbicola** Nakai. IBUKI-AZAMI. Larger in all respects.——Woods; occurs with the typical phase.

26. Cirsium aidzuense Nakai. AIZU-HIME-AZAMI. Stems 1–1.5 m. long, prominently hairy; median cauline leaves oblong, 16–20 cm. long, 10–12 cm. wide, acuminate, narrowed at base, short-spined, puberulent above, slightly cobwebby beneath, the midrib hairy beneath, broadly auriculate-clasping, pinnately parted to divided, the segments 4 or 5 pairs, oblong, acute, coarsely toothed, short spined-tipped, the upper leaves smaller and narrower; inflorescence large; heads erect or ascending, on long slender peduncles; involucres tubular, 1.6–1.7 cm. long, 1.7–2 cm. wide, cobwebby; bracts 8-seriate, the outer short, lanceolate to ovate, about 2 mm. long, the median oblong, acute, recurved or straight; corollas purplish, 15–17 mm. long.——Aug.–Oct. Honshu (Iwashiro Prov.).

27. Cirsium bitchuense Nakai. BITCHŪ-AZAMI. Rhizomes stout; stems 1–2 m. long, pubescent; median cauline leaves often falcate, oblong-lanceolate or lanceolate, 20–30 cm. long, acuminate, narrowed at base, short-pubescent above, cobwebby and pubescent on the nerves beneath, broadly clasping, spinulose, entire to pinnately lobed, the lobes 4 or 5 pairs, acuminate, coarsely toothed, the spines 3–4 mm. long; heads small, rather many, loose, on long slender peduncles; involucres tubular, 1.6 cm. long, 1.6–1.9 cm. wide; bracts viscid, 9-seriate, the outer very short, linear to ovate, acuminate, short spine-tipped, the median row ovate-lanceolate, acute, appressed; corollas purplish, 17–18 mm. long.——Sept.–Oct. Woods; Honshu (Chūgoku Distr.).

Var. **manisanense** Kitam. MANISAN-AZAMI. Leaves larger, 25–40 cm. long, cobwebby; bracts scarcely or not viscid.——Honshu (Tajima and Inaba Prov.).

28. Cirsium buergeri Miq. *Cnicus buergeri* (Miq.) Maxim.——HIME-AZAMI, HIME-YAMA-AZAMI. Stems 1–2 m. long, pubescent; median cauline leaves oblong-lanceolate, 15–25 cm. long, long-acuminate, slightly narrowed at base, sometimes white-variegated, glabrous or crisped-pubescent above, pubescent on nerves beneath, clasping, pinnatifid, the lobes 4–6 pairs, lanceolate, acuminate, spines stout, 3–4(10) mm. long; heads small, many, in spikelike fascicles, usually erect; involucres cobwebby, tubular, 1.6–1.9 cm. long, 2–2.5 cm. wide; bracts 8-seriate, sometimes slightly viscid on back, the outer very short, ovate, acute, the median appressed, narrowly oblong, acute to obtuse, spine-tipped; corollas purplish, 16–18 mm. long.——Aug.–Nov. Mountains; Honshu (Kinki Distr. and westw.), Kyushu.

Cirsium × connexum Kitam. KASUGAI-AZAMI. Hybrid of *C. buergeri × C. lucens.*

29. Cirsium heiianum Koidz. *Cnicus buergeri* var. *albrechtii* Maxim.; *Cirsium albrechtii* (Maxim.) Kudo; *C. buergeri* var. *albrechtii* (Maxim.) Nakai; *C. ruderale* Nakai——EZO-YAMA-AZAMI. Stems 1.5–2 m. long, pubescent while young; median cauline leaves oblong or elliptic-lanceolate, acute, narrowed at base, sessile, usually clasping, toothed, spine-tipped, sometimes pinnately lobed, the lobes chartaceous, ovate, acute, glabrous above, puberulent or loosely cobwebby beneath, glabrescent; heads small, many, in spikelike fascicles or in corymbs; involucres campanulate-tubular, 1.3–1.8 cm. long, 1.5–2 cm. wide, cobwebby; bracts about 9-seriate, the outer very short, ovate, acute, the median oblong-lanceolate, re-

curved, short spine-tipped; corollas purplish, 16–17 mm. long.
——Aug.–Sept. Hokkaido, Honshu (n. distr.).

30. Cirsium suzukaense Kitam. Suzuka-azami. Stems
1–1.5 m. long, densely pubescent or cobwebby; lower cauline
leaves elliptic or elliptic-lanceolate, about 30 cm. long, caudate,
glabrous or short-pubescent on both sides, sessile, sometimes
clasping, deeply pinnatifid, the lobes 4 or 5 pairs, acuminate,
the spines 5–10 mm. long; heads sessile or short-pedunculate,
erect or ascending; involucres tubular, 1.6–1.9 cm. long, 2–2.3
cm. wide, cobwebby, viscid; bracts 6-seriate, short-recurved
and spine-tipped, the outer small, lanceolate or ovate-lanceolate, acuminate, the median oblong-lanceolate, acuminate;
corollas purplish, 14–20 mm. long.——Sept.–Oct. Honshu
(Kinki and Tōkaidō Distr.).

31. Cirsium gyojanum Kitam. Gyōja-azami. Stems
about 70 cm. long, pubescent; median cauline leaves elliptic-
lanceolate or oblong-lanceolate, caudate, narrowed at base,
sparingly pubescent on both sides, sessile, clasping, deeply-pin-
natifid, the lobes 4–7 pairs, lanceolate, acuminate, the spines 3–
12 mm. long; heads small, many, nodding in anthesis; involucres tubular, 1.4–1.7 cm. long, 1.2–1.8 cm. across, cobwebby,
prominently viscid, the bracts 7-seriate, the outer short, lanceolate or ovate-lanceolate, acute to obtuse, short spine-tipped, the
median appressed, oblong-lanceolate; corollas purplish, 14–16
mm. long.——Aug.–Sept. Mountains; Honshu (Mount Oo-
mine in Yamato).

32. Cirsium kagamontanum Nakai. *C. kagamontanum* var. *spinuliferum* Kitam.; *C. buergeri* var. *sparsum*
Nakai——Kaga-no-azami. Stems 1.5–2 m. long, glabrous,
leafy throughout; lower cauline leaves large, elliptic or elliptic-
lanceolate, 30–50 cm. long, acute, narrowed at base, glabrous
above, thinly silky beneath, clasping, deeply pinnatifid, the
lobes about 5 pairs, oblong-lanceolate, the spines 2–3 mm. long;
heads small, pendulous at anthesis, many, on long peduncles,
bractless; involucres campanulate, cobwebby, viscid, about 1.6
cm. long, 2.5 cm. wide; bracts 7- or 8-seriate, the outer short,
oblong or ovate, obtuse, short spine-tipped, the median narrowly oblong, appressed; corollas purplish, 16–17 mm. long.
——Aug.–Oct. Mountains; Honshu.

33. Cirsium longepedunculatum Kitam. *C. buergeri*
var. *glutinosum* Kitam., pro parte.——Nagae-no-azami, Ne-
bari-hime-azami. Stems 50–80 cm. long, subscapose, densely
leafy at base, loosely so above, glabrous, the branches cobwebby
toward the top; lower cauline leaves thin, large, subrosulate,
elliptic to elliptic-lanceolate, 30–40 cm. long, acuminate, abruptly narrowed at base, glabrous and usually white-variegated
above, loosely silky beneath, sessile and clasping or short-
petiolate, pinnatifid, the lobes about 4 pairs, again pinnatifid
or coarsely incised, the spines 3–4 mm. long; upper leaves
few, lanceolate to linear, small; heads small, nodding at anthesis, the peduncles slender, very long; involucres campanulate-
globose, 1.6–1.8 cm. long, about 2 cm. wide, prominently
viscid; bracts 7-seriate, the outer very short, lanceolate or
ovate, acute, spine-tipped, the median oblong, appressed;
corollas purplish, 15–18 mm. long.——Aug.–Oct. Mountains;
Honshu (n. Kinki Distr.).

34. Cirsium effusum (Maxim.) Matsum. *Cnicus effusus*
Maxim.——Hosoe-no-azami. Stems about 1 m. long, glabrous, densely leafy; branches short-pubescent and loosely cobwebby; median cauline leaves oblong to linear-lanceolate, 20–40
cm. long, long-acuminate, narrowed at base, loosely short-
pubescent above, short-pubescent on nerves beneath, sessile,

deeply pinnatifid, the lobes about 8 pairs, ovate, short-acuminate, trilobulate nearly to the base, the spines 5–10 mm. long;
upper leaves linear; heads in racemes, short-pedunculate; involucres narrowly tubular, 1.5–1.8 cm. long, 1.4–2 cm. wide,
scarcely cobwebby; bracts 6-seriate, recurved, the outer lanceolate or ovate, acuminate, long spine-tipped, the median oblong,
long-acuminate, often spinulose-margined; corollas purplish,
17–18 mm. long.——Sept.–Oct. Woods in mountains; Honshu (Sagami, Suruga, Kai, and Musashi Prov.).

35. Cirsium gratiosum Kitam. Hōki-azami. Stems
60 cm. or more long, cobwebby; lower cauline leaves oblong,
about 30 cm. long, acuminate, cuneate at base, nearly glabrous
on both sides, short-petiolate, the lobes 9 pairs, acuminate,
coarsely toothed, spines stout; upper leaves gradually smaller,
sessile; heads nodding, long-pedunculate, in paniculate corymbs; involucres campanulate, about 1.6 cm. long, 2.2 cm.
wide; bracts 7- or 8-seriate, the margins scarious, undulate,
shortly recurved, the outer short, ovate, acute, short spine-
tipped, the median oblong, acute; corollas purple, 16–17 mm.
long.——Aug. Honshu (Akaishi Mt. range in Shinano).

Var. **alpinum** (Nakai) Kitam. *C. effusum* var. *alpinum*
Nakai——Miyama-hosoe-no-azami. Involucral bracts elon-
gate, short spine-tipped.——Honshu (Mount Yatsugatake in
Shinano Prov.).

36. Cirsium congestissimum Kitam. Hittsuki-azami.
Stems 1–1.5 m. long, cobwebby above; radical leaves elliptic or
elliptic-lanceolate, short-acuminate, narrowed at base, pinnatifid, the lobes 5 or 6 pairs, oblong, irregularly incised, spines
1–2 mm. long, glabrous above, short-pubescent on midrib
beneath; median and lower cauline leaves elliptic or elliptic-
lanceolate, 25–40 cm. long, short-acuminate, narrowed at base;
upper leaves small, rounded at base, semiclasping; heads congested, many, short-pedunculate or sessile; involucres tubular,
1.5–1.6 cm. long, 1.7–2 cm. wide, recurved, spine-tipped, the
outer bracts lanceolate or ovate-lanceolate, the median oblong-
lanceolate, elongate; corollas purplish, 16–19 mm. long.——
Sept.–Oct. Slopes in mountains; Honshu (Kinki and Chūgoku Distr.).

37. Cirsium spicatum (Maxim.) Matsum. *Cnicus
spicatus* Maxim.——Yama-azami. Stems 1.5–2 m. long,
stout, glabrous or sometimes pubescent, densely leafy; median
cauline leaves chartaceous, elliptic to elliptic-lanceolate, 25–35
cm. long, long-acuminate, cuneate at base, glabrous on both
sides or short-pubescent beneath, sessile, pinnatifid, the lobes
4–6 pairs, oblong-lanceolate, long-acuminate, often 3-lobulate
nearly to the base, spines 3–10 mm. long; heads many, small,
in dense spikes, erect or ascending; involucres tubular, more or
less cobwebby, 1.4–1.5 cm. long, 1.5–1.7 cm. wide; bracts 7-
seriate, recurved, the outer short, lanceolate or ovate, acuminate, stout spine-tipped, the median narrowly oblong, sometimes spinulose-margined; corollas purplish, 17–18 mm. long.
——Mountains; Shikoku, Kyushu.

38. Cirsium tenue Kitam. Usuba-azami. Stems 1–
1.5 m. long, rather slender, glabrous at base; median and lower
cauline leaves thin, elliptic-lanceolate, acuminate, narrowed at
base, sometimes whitish above along the puberulent nerves,
loosely cobwebby beneath, sessile or short-petiolate, pinnatifid,
the lobes 5–7 pairs, oblong-lanceolate to lanceolate, acuminate,
sometimes coarsely incised near base, spines 5–10 mm. long;
heads spicate, sometimes short-pedunculate, ascending in anthesis; involucres tubular, scarcely cobwebby, prominently viscid,
1.6–1.7 cm. long, 1–1.7 cm. wide; bracts 7-seriate, the outer

short, lanceolate to ovate, acute to acuminate, the median oblong or narrowly so, usually acuminate, often spreading at apex; corollas purplish, 17–18 mm. long.——Sept.–Oct. Honshu (Bitchū Prov.).

Var. **ishizuchiense** Kitam. ISHIZUCHI-USUBA-AZAMI. Lobes of radical leaves about 8 pairs, long-acuminate; spines of cauline leaves large and stout, 6–15 mm. long; involucres long-acuminate and spreading.——Shikoku.

39. Cirsium microspicatum Nakai. AZUMA-YAMA-AZAMI. Stems 1.5–2 m. long, densely cobwebby in upper portion; lower and median cauline leaves thin, elliptic or elliptic-lanceolate, 23–60 cm. long, abruptly acuminate, narrowed at base, sometimes white-variegated and mealy-puberulent or glabrous above, densely cobwebby and puberulent beneath, short-petiolate or sessile, pinnately cleft to lobed, the lobes lanceolate, 4 or 5 pairs, narrowly acuminate, often irregularly 3-lobulate nearly to the base, spines 5–10 mm. long; heads spicate, erect or ascending, usually pedunculate, rarely sessile; involucres tubular-campanulate, densely cobwebby, 1.5–1.8 cm. long, 1.7–2 cm. wide, purplish; bracts 7-seriate, the outer short, ovate or lanceolate, acuminate, spine-tipped, the median narrowly oblong; corollas purplish, 18–20 mm. long.——Sept.–Oct. Mountains; Honshu (Kantō to Kinki Distr.).

Var. **kiotense** Kitam. OHARAME-AZAMI. Outer and median involucral bracts obtuse, very rarely acute, appressed, scarcely spine-tipped.——Honshu (Kinki and w. Hokuriku Distr.).

Var. **yechizenense** Kitam. ECHIZEN-AZAMI. Leaves usually elliptic, pinnatilobed, the lobes entire, spinulose, short spine-tipped; heads nodding; involucres campanulate, 12–17 mm. long, the bracts imbricate, obtuse or acute, scarcely spine-tipped.——Honshu (Echizen Prov.).

40. Cirsium nipponicum (Maxim.) Makino. *Cnicus nipponicus* Maxim.——NAMBU-AZAMI. Stems 1–2 m. long, cobwebby to glabrous; median cauline leaves thick, elliptic-lanceolate, 20–30 cm. long, acuminate, narrowed at base, entire or incised-toothed, glabrous above, crisped-pubescent or sometimes glabrous beneath, sessile or short winged-petiolate, sometimes pinnatifid, the spines 1–2 mm. long, the lobes 5 or 6 pairs, acuminate, lobulate nearly to the base; heads many, pedunculate, nodding in anthesis; involucres campanulate, sometimes cobwebby; bracts 7-seriate, the outer and median often lanceolate and elongate, long-recurved or spreading, sometimes short and spreading; corollas purplish, 16–20 mm. long.——Aug.–Oct. Honshu (n. and centr. distr.).——Forma **lanuginosum** (Nakai) Kitam. *C. pseudopendulum* Nakai, excl. syn.; *C. nipponicum* var. *lanuginosum* Nakai——URAGE-HIME-AZAMI. Leaves cobwebby beneath.

Var. **sendaicum** (Nakai) Kitam. *Cirsium sendaicum* Nakai; *C. matsushimense* Kitam.——MATSUSHIMA-AZAMI. Lower and median leaves pinnatiparted, the segments 5 pairs, glabrous above, short-pubescent beneath.——Honshu (n. distr.).

Var. **amplexicaule** (Nakai) Kitam. *C. sendaicum* var. *amplexicaule* Nakai——DAKIBA-NAMBU-AZAMI. Differs from var. *sendaicum* in having cauline leaves slightly clasping.——Honshu (Matsushima in Rikuzen).

Var. **comosum** (Fr. & Sav.) Kitam. *Cnicus comosus* Fr. & Sav.; *Cirsium comosum* (Fr. & Sav.) Matsum.——IGA-AZAMI. Heads congested or short-pedunculate; involucral bracts and leaves with thick stout spines.——Grassy places near seashore; Honshu (s. Kantō Distr.).

Var. **incomptum** (Maxim.) Kitam. *Cnicus suffultus* var. *incomptus* Maxim.; *Cirsium suffultum* var. *incomptum* (Maxim.) Matsum.; *Cnicus incomptus* (Maxim.) Fr. & Sav.; *Cirsium incomptum* (Maxim.) Nakai; *C. fauriei* Nakai; *C. nipponicum* var. *incomptum* (Maxim.) Kitam.; *C. tonense* Nakai; *C. spicatum* Matsum., pro parte——TAI-AZAMI. Stems 1–2 m. long; leaves pinnatifid, sometimes merely toothed; spines shorter; heads often rather long-pedunculate; involucral bracts with a shorter spine-tip.——Honshu (Kantō, Tō-kaidō, and e. Kinki Distr.).

Var. **sawadae** (Kitam.) Kitam. *C. comosum* var. *sawadae* Kitam.——HAKONE-AZAMI. Stems often hairy; leaves pinnately cleft, the lobes very narrow, long-acuminate; heads campanulate, about 25 mm. wide, ascending at anthesis, short-pedunculate or sessile.——Mountains; Honshu (Sagami and Suruga Prov.).

Var. **yatsugatakense** (Nakai) Kitam. *C. yatsugatakense* Nakai; *C. comosum* var. *yatsugatakense* (Nakai) Kitam.——YATSUGATAKE-AZAMI. Median leaves long-acuminate, short-pubescent on both sides, often cobwebby beneath, sometimes partially clasping, the lobes 4–6 pairs, spines 5–10 mm. long.——High mountains; Honshu (Shinano, Iwashiro Prov.).

Var. **yoshinoi** (Nakai) Kitam. *C. yoshinoi* Nakai; *C. inflatum* Kitam.——YOSHINO-AZAMI. Leaves entire to pinnately cleft, acuminate, not clasping; spines 2–4 mm. long; heads many, sometimes in racemes, nodding at anthesis; involucres about 1.6 cm. long, 1.5–2 cm. wide, the outer bracts short, recurved toward the tip, short spine-tipped, sometimes viscid.——Honshu (Kinki and Chūgoku Distr.).

Var. **shikokianum** Kitam. SHIKOKU-AZAMI. Differs from the preceding variety in the slightly larger heads with spreading marginal florets; involucral bracts slightly elongate, acutely recurved, short-spined.——Shikoku.

41. Cirsium yakushimense Masam. YAKUSHIMA-AZAMI. Stems 30–40 cm. long, densely leafy, cobwebby and often short-pubescent; lower cauline leaves lanceolate, 17–27 cm. long, narrowly long-acuminate, narrowed at base, usually short-pubescent above, glabrous or the nerves pubescent beneath, sessile or the petiole winged, deeply pinnatifid, the lobes 6–9 pairs, acuminate, spines stout, 5–10 mm. long; median leaves sometimes partially clasping; heads in spikes, erect or ascending at anthesis, bracteate; involucres campanulate, 1.6–1.7 cm. long, 2–2.3 cm. wide; bracts 6-seriate, slightly cobwebby, the outer ovate, acuminate, elongate, spine-tipped, the median oblong, acuminate; corollas purplish, about 16 mm. long.——Mountains; Kyushu (Yakushima).

42. Cirsium chikushiense Koidz. NOMA-AZAMI. Stems 70–100 cm. long, densely cobwebby; lower cauline leaves lanceolate, 22–33 cm. long, narrowly long-acuminate, short-petiolate or sessile, pinnatifid, the lobes 5 or 6 pairs, ovate, short-acuminate, irregularly 3-lobulate, loosely cobwebby on both sides, the spines very stout, 8–16 mm. long, the median leaves very long-acuminate; involucres campanulate-globose, slightly cobwebby, about 1.9 cm. long, 2.5 cm. wide; bracts 6-seriate, the outer elongate, linear-lanceolate, the median elongate, ovate or oblong-acuminate, erect or sometimes recurved; corollas purplish, 17–21 mm. long.——Oct.–Nov. Kyushu (Satsuma and Oosumi Prov.).

43. Cirsium indefensum Kitam. TOGE-NASHI-AZAMI. Stems about 2 m. long, densely white-villous while young; lower cauline leaves elliptic-lanceolate, about 40 cm. long, gradually narrowed at tip, cuneate at base, short-petiolate or sessile, pinnatifid, the lobes 5 or 6 pairs, ovate or oblong, acute, scarcely spiny, sometimes short-pubescent above, loosely cobwebby beneath; heads small, many, nodding at anthesis, on slender peduncles; involucres tubular, 1.8–2 cm. long, 2.5–3 cm. wide, cobwebby; bracts 8-seriate, appressed, scarcely spine-tipped, sometimes slightly viscid at tip, the outer oblong or ovate, about 2 mm. long, 1 mm. wide, acute, the median oblong, abruptly acute; corollas purplish, 19–20 mm. long.——Nov. Honshu (Chūgoku Distr.).

44. Cirsium hanamakiense Kitam. HANAMAKI-AZAMI. Stems 1.5–2 m. long, glabrous at base, short-pubescent on upper part; lower cauline leaves elliptic or elliptic-lanceolate, 35–45 cm. long, short-acuminate, cuneate at base, short-petiolate, deeply pinnatifid, the lobes 5 or 6 pairs, oblong to ovate, acuminate, glabrous on upper side, glabrous to short-silky beneath, the spines 3–7 mm. long; heads small, many, nodding at anthesis, on rather long peduncles; involucres campanulate, 1.5–1.7 cm. long, 2.4–2.9 cm. wide, loosely cobwebby; bracts 8-seriate, densely brown-pubescent on margin, the outer very short, lanceolate to ovate-lanceolate, the median oblong, acuminate, scarcely spiny, recurved; corollas purplish, 15–16 mm. long. ——Sept.–Oct. Along ravines and valleys; Honshu (n. distr.).

45. Cirsium suffultum (Maxim.) Matsum. *Cnicus suffultus* Maxim.; *C. suffultus* var. *pexus* Maxim.; *C. pexus* (Maxim.) Fr. & Sav.; *Cirsium pexum* (Maxim.) Nakai; *C. kiushianum* Nakai; *C. ogatae* Koidz.——TSUKUSHI-AZAMI, TSUKUSHI-YAMA-AZAMI. Stems about 1 m. long, loosely cobwebby; radical and lower cauline leaves elliptic or elliptic-lanceolate, 30–40 cm. long, short-acuminate, petiolate, pinnaticleft, the lobes 7 or 8 pairs, oblong-lanceolate, acuminate, irregularly pinnatilobed, white-variegated, glabrous to loosely pubescent above and below, especially on nerves beneath, the spines 2–5 mm. long; median leaves 25–35 cm. long, with stout spines; heads nodding at anthesis, bracteate at base; involucres campanulate-globose, cobwebby, about 2 cm. long, 2.5–4.5 cm. wide; bracts 6-seriate, the outer lanceolate, acuminate, erect or spreading, the median ovate- to oblong-lanceolate, elongate, acuminate; corollas purplish, 22–24 mm. long.——Sept.–Nov. Shikoku, Kyushu; common.

46. Cirsium hachijoense Nakai. HACHIJŌ-AZAMI. Stems 0.5–1.5 m. long, cobwebby; lower cauline leaves elliptic-lanceolate, 25–60 cm. long, short-acuminate to acute, petiolate, pinnatifid, the lobes ovate to oblong, 5 or 6 pairs, acute, coarsely incised-toothed, glabrous on both sides or short-pubescent beneath, the spines 2–3 mm. long; heads erect at anthesis, usually bracteate; involucres depressed-globose, 1.8–2 cm. long, 3–4 cm. wide, cobwebby; bracts 6-seriate, the outer and median linear-lanceolate to lanceolate, elongate, spreading or slightly recurved on upper half; corollas purplish, 16–17 mm. long.—— Honshu (Hachijō and Ooshima in Izu).

47. Cirsium norikurense Nakai. NORIKURA-AZAMI. Stems 1–1.5 m. long, short-pubescent; median cauline leaves oblong-lanceolate, 24–28 cm. long, acuminate, abruptly narrowed at base, sessile, sometimes clasping, pinnately lobed, the lobes 5 or 6 pairs, ovate, acute, glabrous on upper side, densely white-cobwebby beneath, the spines 2–3 mm. long; heads solitary on the branches, nodding at anthesis, sometimes bracteate; involucres campanulate-globose, about 1.7 cm. long, 2.5–3.5 cm. wide, cobwebby; bracts 5- or 6-seriate, much-elongate toward the tip, much recurved, the outer linear-lanceolate or ovate-acuminate, scarcely spine-tipped, the median oblong-lanceolate; corollas purplish, about 16 mm. long.——Aug.–Sept. Woods

and grassy places along streams in high mountains; Honshu (centr. distr.).

Var. **integrifolium** Kitam. *C. kurobense* Honda——YUKI-AZAMI, MARUBA-NORIKURA-AZAMI. Leaves undivided, spine-toothed only on margin.——Honshu (centr. distr.).

48. Cirsium senjoense Kitam. SENJŌ-AZAMI. Stems 70–100 cm. long, cobwebby and short-pubescent; lower cauline leaves narrowly oblong-lanceolate, 16–30 cm. long, acuminate, sessile, clasping, deeply pinnatifid, the lobes 8 pairs, ovate-acuminate, 3-lobulate, crisped-pubescent on upper side, pubescent and cobwebby beneath, the spines 3–7 mm. long; heads 1–3 on the branches, nodding at anthesis; involucres globose-campanulate, 1.5–1.7 cm. long, 1.5–2.5 cm. wide, cobwebby; bracts 5-seriate, usually viscid, scarcely spine-tipped, the outer lanceolate, elongate-acuminate, spreading on upper half; corolla purplish, about 16 mm. long.——Aug. Mountains; Honshu (Akaishi and Shirane Mt. ranges).

Var. **kurosawae** Kitam. KUROSAWA-AZAMI. Leaves not clasping. Occurs with the typical phase.

49. Cirsium amplexifolium (Nakai) Kitam. *C. nipponicum* var. *amplexifolium* Nakai; *C. nipponicum* var. *purpureum* Koidz.; *C. sendaicum* Nakai, pro parte; *C. tobae* Nakai——DAKIBA-HIME-AZAMI. Stems 1.5–2 m. long, glaucescent, glabrous; cauline leaves oblong or elliptic-lanceolate, 18–28 cm. long, acuminate, abruptly narrowed at base, broadly auriculate-clasping, spinulose-toothed or pinnately cleft, the lobes 5–7 pairs, oblong, acute to short-acuminate, coarsely toothed at base, scattered short-pubescent to glabrous on both sides, the spines 3–4 mm. long; heads solitary on the branches, long-pedunculate, erect at anthesis, about 3 cm. across, often bracteate; involucres campanulate, about 1.6 cm. long, about 2 cm. wide, nearly glabrous; bracts 6-seriate, the outer and median linear-lanceolate, acuminate, scarcely spine-tipped, elongate, recurved in upper half, very rarely spinulose on margin; corollas purplish, 17–20 mm. long.——July–Sept. Honshu (n. distr.).

Var. **muraii** (Kitam.) Kitam. *C. muraii* Kitam.——KINKA-AZAMI. Leaf-lobes ovate-linear, reflexed, long-acuminate, the spines 5–15 mm. long; inflorescence with larger leaves.—— Honshu (Kinkasan in Rikuchu).

50. Cirsium matsumurae Nakai. *C. hokkokuense* Kitam.——HAKUSAN-AZAMI. Stems about 2 m. long, densely leaved, glabrous; lower cauline leaves elliptic or elliptic-lanceolate, about 30 cm. long, short-acuminate, cuneate at base, broadly auriculate-clasping, deeply pinnatifid, the lobes about 5 pairs, spreading, oblong, obtuse to acute, coarsely spine-toothed, glabrous above, loosely silky beneath; heads few, 3–4 cm. across, usually on long slender short-pubescent peduncles, nodding at anthesis; involucres globose-campanulate, sometimes cobwebby, 1.5 cm. long, 2–2.5 cm. wide; bracts 6-seriate, scarcely spine-tipped, viscid on back, obtuse to acute, the outer short, lanceolate to ovate, spreading, oblong; corollas purplish, 16–18 mm. long.——Aug.–Oct. High mountains; Honshu (nw. part of centr. distr.).

Var. **pubescens** Kitam. KE-HAKUSAN-AZAMI. Stems densely pubescent.

Var. **dubium** Kitam. HOKKOKU-AZAMI. Leaves entire to pinnatifid, clasping or not clasping; heads smaller; involucres shorter, the outer bracts short-recurved and densely cobwebby. ——Honshu (nw. part of centr. distr.).

51. Cirsium babanum Koidz. DAI-NICHI-AZAMI. Stems pubescent; median cauline leaves narrowly elliptic-oblanceolate, short-acuminate, semiclasping, pinnatifid, the lobes 5 or 6 pairs, distant, ovate-acuminate, coarsely incised, pubescent on both sides, the spines 3–8 mm. long, the upper leaves clasping; heads solitary, long-pedunculate, nodding at anthesis; involucres globose-campanulate, about 2 cm. long, about 4 cm. wide, the bracts 6-seriate, the outer gradually narrowed, 3–3.5 cm. long, longer than the inner; corollas purplish, about 15 mm. long.——Aug. High mountains; Honshu (Hida Mt. range).

Var. **otayae** (Kitam.) Kitam. *C. otayae* Kitam.; *C. fauriei* Nakai, pro parte——TATEYAMA-AZAMI. Stems 40–100 cm. long; median cauline leaves broader, broadly clasping at base, the lobes approximate, pubescent on both sides, the spines 1–2 mm. long; involucres about 15 mm. long, 2.5–3 cm. wide, the outer and median bracts spreading to recurved on upper half; corolla-limb shorter than the broader part of the tube.—— Aug. High mountains; Honshu (Hida Mt. range and Mount Hakusan in Kaga).

52. Cirsium ganjuense Kitam. *C. amplexifolium* sensu Nakai, non Kitam.——GANJU-AZAMI. Stems cobwebby, glabrate; cauline leaves elliptic-lanceolate to oblong, acute to acuminate, broadly clasping at base, spine-tipped, entire to pinnately lobed, the lobes 3 or 4 pairs, oblong, acute, glabrous on both sides or slightly cobwebby beneath; heads 1 or 2 on the branches; involucres campanulate-globose, about 2 cm. long, 2.5–3.5 cm. wide, more or less cobwebby; bracts rather thick, outer and median spreading on upper half; corollas purplish, about 17 mm. long.——Aug.–Sept. Alpine regions; Honshu (Mount Iwate and Mount Hayachine in Rikuchu Prov.).

53. Cirsium nambuense Nakai. NAMBU-TAKANE-AZAMI. Stems about 50 cm. long, often densely short-pubescent, cobwebby on upper part; median and lower cauline leaves oblong-lanceolate, 20–35 cm. long, acuminate, clasping, deeply pinnatifid, the lobes 8–11 pairs, usually ovate, lobulate, short-pubescent above and on nerves beneath, the spines 2–5 mm. long; heads 2–4, terminal, erect at anthesis, often bracteate; involucres globose-campanulate, 1.2–2 cm. long, 2.5–3 cm. wide, slightly cobwebby; bracts 5-seriate, lanceolate-linear, equal in length, outer and median elongate, erect; corollas purplish, about 15 mm. long.——Aug.–Sept. Alpine regions; Honshu (n. distr.).

54. Cirsium homolepsis Nakai. *C. inundatum* subsp. *homolepis* (Nakai) Kitam.——OZENUMA-AZAMI. Stems 50–100 cm. long, glabrous at base, cobwebby on the branches; radical leaves large, to 65 cm. long; median cauline leaves oblong-lanceolate, 15–30 cm. long, acute to acuminate, dilated at base, broadly clasping, pinnately cleft, the lobes narrow, acuminate, lobulate, short-spined, glabrous on both sides; heads erect at anthesis; involucres campanulate, 1.8–2.8 cm. long, 2.5–3.5 cm. wide; bracts 7-seriate, usually equal in length, loosely cobwebby, the outer linear; corollas purplish, 19–20 mm. long.——Aug.–Sept. Wet places; Honshu (Ozegahara Moor).

55. Cirsium inundatum Makino. TACHI-AZAMI. Stems 1–1.5 m. long, glabrous, the branches short-pubescent and cobwebby; median cauline leaves oblong or oblong-lanceolate, 15–25 cm. long, acute to acuminate, abruptly narrowed at base, slightly dilated, semiclasping or winged-petiolate, spine-toothed to pinnatifid, the lobes about 4 pairs, acute to acuminate, incised, short-pubescent or glabrous on both sides, the spines 2–6 mm. long; heads many, terminal, erect at anthesis, usually bractless; involucres globose-campanulate, about 1.9 cm. long, 2.5–3 cm. wide, slightly cobwebby, the outer bracts linear-lanceolate, elongate, erect, sometimes ascending, ciliolate on margin; corollas purplish, 18–20 mm. long.——Aug.–Oct. Wet places; Honshu (n. and centr. distr.).

Var. **alpicola** (Nakai) Ohwi. *C. alpicolum* Nakai; *C. greatrexii* Miyabe & Kudo; *C. inundatum* subsp. *alpicola* (Nakai) Kitam.——MINE-AZAMI. Leaves broader, elliptic-lanceolate, 17–30 cm. long, pinnatifid, semiclasping; outer and median involucral bracts ascending to spreading; corollas 17–20 mm. long.——High mountains; Hokkaido (sw. distr.), Honshu (n. distr.).

53. BREEA Less.   ARECHI-AZAMI ZOKU

Erect, dioecious, perennial herbs with long-creeping rhizomes; leaves alternate, loosely cobwebby, toothed or pinnatifid, spinulose; heads terminal, many-flowered, homogamous; involucres campanulate, the bracts narrow, many-seriate; receptacle setose; staminate corollas capillary, 5-lobed, the anthers sagittate, awnlike or lacerate; pistillate corollas capillary, long-tubular; pappus-bristles plumose, connate at base; style thickened above, the branches obtuse; achenes glabrous, oblong, 4-angled.—— Mediterranean region, Europe, Siberia, to e. Asia.

1A. Stems 50–100 cm. long; involucres of the pistillate flowers 15–20 mm. long. . . . . 1. *B. setosum*
1B. Stems 25–50 cm. long; involucres of the pistillate flowers 23 mm. long. . . . . 2. *B. segetum*

1. Breea setosum (Bieb.) Kitam. *Cephalonoplos setosum* (Bieb.) Kitam.; *Cirsium setosum* Bieb.; *C. setosum* var. *subulatum* Ledeb.; *C. arvense* sensu auct. Japon., non L.——EZO-NO-KITSUNE-AZAMI. Slightly cobwebby perennial; stems 50–100 cm. long; lower and median cauline leaves oblong-lanceolate, 10–20 cm. long, obtuse, spinulose, gradually narrowed at base, sessile, flat, coarsely incised-toothed; heads rather numerous, erect, the staminate smaller, with involucres about 13 mm. long, the pistillate involucres 15–20 mm. long, the bracts 8-seriate, the outer short, the median lanceolate; achenes about 2.5 mm. long.——Aug.–Oct. Waste grounds and roadsides in lowlands; Hokkaido, Honshu (n. distr.).——Korea, Manchuria, China, Siberia, and e. Europe.

2. Breea segetum (Bunge) Kitam. *Cephalonoplos segetum* (Bunge) Kitam.; *Cirsium segetum* Bunge; *Cnicus segetum* (Bunge) Maxim.——ARECHI-AZAMI. Stems 25–50 cm. long; lower and median cauline leaves oblong-lanceolate, 7–10 cm. long, obtuse, entire to spinulose-toothed, narrowed at base, sessile, rounded at base in upper ones; heads few, long-pedunculate; involucres of the staminate heads about 18 mm. long, in the pistillate 23 mm. long, the bracts 8-seriate, short-acuminate and dark colored at the tip, the outer very short, oblong-lanceolate, the median lanceolate; corolla 17–20 mm. long in the staminate flowers, about 26 mm. long in the pistillate, purplish.——May–Aug. Kyushu (Tsushima).——Korea, Manchuria, and n. China.

54. HEMISTEPTA Bunge KITSUNE-AZAMI ZOKU

Biennial herbs with erect branched stems; leaves alternate, lyrate, pinnatifid, densely whitish tomentose beneath, spineless; heads many, homogamous; involucres ovoid, the bracts many-seriate, imbricate, the outer and median with a keeled appendage on back; receptacle nearly flat, setose; florets bisexual, fertile, capillary, purplish, the narrower portion of tube very long; anthers sagittate-connate, the tails laciniate; style branches rounded-truncate and granular-tuberculate at the tip, tuberculate on back, becoming recurved; filaments glabrous; achenes glabrous, oblong, 15-ribbed; pappus-bristles 2-seriate, the outer very short, few, persistent, rather broad, truncate at apex, the inner plumose, elongate, deciduous.——Few species, in India, China, Australia, and e. Asia.

1. Hemistepta lyrata Bunge. *H. carthamoides* (Hamilt.) O. Kuntze; *Cnicus multicaulis* DC.; *Serratula carthamoides* Hamilt. ex Roxb., non Poir.; *Aplotaxis carthamoides* DC.; *Cirsium lyratum* Bunge; *Saussurea affinis* Spreng.; *Serratula tinctoria* Sieb., non L.—KITSUNE-AZAMI. Stems 60–80 cm. long, often branched; leaves broadly oblanceolate, 10–20 cm. long, pinnately parted, the terminal segments largest, deltoid, the lateral segments 7 or 8 pairs, rather narrow, green above, white-tomentose beneath; heads usually many, erect; involucres globose, 12–14 mm. long, the bracts 8-seriate, with a cristate appendage on back at tip, the outer short, ovate-deltoid, acute; corollas purplish.——May–June. Waste grounds, roadsides and cultivated fields in lowlands; Honshu, Kyushu; common.——Ryukyus, Formosa, China, Korea, Manchuria, India, and Australia.

55. SAUSSUREA DC. TŌ-HIREN ZOKU

Biennials or perennials; leaves toothed or pinnatifid, ending in a callose point; heads usually small, in corymbs or racemes, sometimes solitary, homogamous; involucres globose to campanulate or tubular, the bracts many-seriate, imbricate, not spiny, sometimes with a scarious apical appendage; receptacle glabrous or setose; florets usually rose-purple, sometimes white, bisexual, fertile; corollas tubular, the limb 5-lobed; anthers sagittate, the filaments glabrous; achenes glabrous, obsoletely 4-angled; pappus-bristles 2-seriate, the outer short, scabrous to plumose, deciduous, the inner plumose, persistent, connate at base.——About 150 species, abundant in the mountains of Asia, few in Europe and N. America.

19A. Outer involucral bracts ovate or lanceolate, long-acuminate or cuspidate.
 20A. Involucres 5–8 mm. across.
  21A. Heads few; leaves sagittate at base. . . . . 18. *S. sagitta*
  21B. Heads many; leaves hastate to cordate or cuneate at base. . . . . 19. *S. spinulifera*
 20B. Involucres 8–12 mm. across.
  22A. Outer involucral bracts recurved.
   23A. Leaves not decurrent, usually consistently undulate to sinuate. . . . . 20. *S. yoshinagae*
   23B. Leaves decurrent, mucronate-toothed to undulately lobulate. . . . . 21. *S. nipponica*
  22B. Outer involucral bracts not recurved.
   24A. Leaves not decurrent.
    25A. Lower cauline leaves oblong, not sinuate, hastate. . . . . 22. *S. amabilis*
    25B. Lower cauline leaves ovate, sinuate, cordate. . . . . 23. *S. sinuatoides*
   24B. Leaves decurrent. . . . . 24. *S. pennata*
19B. Outer involucral bracts ovate-deltoid, acute to obtuse.
 26A. Involucral bracts obtuse; leaves oblong, glandular-dotted, pinnately parted to toothed, sometimes entire. . . 25. *S. maximowiczii*
 26B. Involucral bracts acute to cuspidate.
  27A. Inflorescence a loose corymb; leaves not decurrent. . . . . 26. *S. insularis*
  27B. Inflorescence racemose; leaves decurrent. . . . . 27. *S. tanakae*

1. Saussurea pulchella Fisch. ex DC. *Serratula pulchella* (Fisch.) Sims; *Theodorea pulchella* (Fisch.) Cass.; *Saussurea japonica* sensu auct., pro parte, non DC.; *S. japonica* var. *pulchella* (Fisch.) Koidz.; *S. koraiensis* Nakai; *S. pulchella* vars. *subintegra, pinnatifida, ovata,* and *alata* of Regel——HIME-HIGOTAI. Biennial; stems 30–150 cm. long, puberulent, winged or wingless; radical and lower cauline leaves petiolate, oblong to narrowly so or elliptic, 12–18 cm. long, pinnatiparted, the segments 6–10 pairs, lanceolate, puberulent on both sides, densely glandular-dotted beneath; heads many; involucres broadly campanulate, 11–13 mm. long, 10–14 mm. wide; bracts 6- or 7-seriate, with a lobulate scarious apical appendage, the outer ovate, the median narrowly oblong, the inner linear; corollas purplish, 11–13 mm. long.——Aug.–Oct. Grassy slopes in mountains; Hokkaido, Honshu, Shikoku, Kyushu.——Korea, Sakhalin, Manchuria, and e. Siberia.

2. Saussurea japonica (Thunb.) DC. *Serratula japonica* Thunb.; *S. amara* var. *integra* DC.; *S. scabra* Less.; *S. taquetii* Lév. & Van't.; *S. pulchella* var. *japonica* (Thunb.) Herd.——HINA-HIGOTAI. Biennial herb; stems 50–150 cm. long, striate, loosely glandular-puberulent; radical leaves long-petiolate, oblong, 20–35 cm. long, gradually narrowed at both ends, usually pinnatifid, sometimes undivided, the terminal lobes oblong-lanceolate, obtuse, the lateral lobes 7 or 8 pairs, narrowly oblong, usually pinnately lobulate, glandular-puberulent on both sides; cauline leaves gradually reduced; heads many; involucres tubular, 8.5–12.5 mm. long, 5–8 mm. wide, cobwebby, the bracts 6-seriate, rounded, with a scarious apical appendage; corollas purplish, 10–14 mm. long.——Oct.–Nov. Grassy slopes in lowlands and mountains; Kyushu.——Korea, Manchuria, China, and Formosa.

3. Saussurea chionophylla Takeda. YUKIBA-HIGOTAI. Small perennial herb, 4–10 cm. high; stems simple, densely white-cobwebby; radical and lower cauline leaves coriaceous, persistent at anthesis, spreading, petiolate, ovate to broadly so, 4–8.5 cm. long, cordate to truncate at base, mucronate-toothed, cobwebby above while young, densely white-woolly beneath; median and upper leaves few, truncate to cuneate at base; heads in corymbs, in groups of 5–11; involucres campanulate, 12–14 mm. long, 10–12 mm. wide, cobwebby; bracts 5-seriate, the outer ovate, about half as long as the inner, mucronate, the inner linear, acuminate; receptacles naked; corollas purplish, 11–12 mm. long; achenes 5.5–6.5 mm. long.——Aug. Alpine; Hokkaido (Mount Yubari).

4. Saussurea tobitae Kitam. SHINANO-TŌ-HIREN.

Stems 60–70 cm. long, nearly terete, purplish, mealy-puberulent; lower cauline leaves chartaceous, deltoid to elongate-deltoid, 10–13 cm. long, caudate, truncate to subcordate at base, toothed, loosely pubescent on both sides, the petioles narrowly winged; median and upper leaves many, gradually smaller, the uppermost leaves oblong to linear-lanceolate; heads rather many, in loose corymbs, very short pedicellate; involucres globose, about 11 mm. wide, more or less cobwebby; bracts 4-seriate, the outer ovate, mucronate-acuminate, sometimes shortly recurved, black-tinged, the inner linear, acuminate; corollas about 11 mm. long.——Aug. Honshu (Kirigamine in Shinano Prov.). Possibly a hybrid of *S. tanakae* and *S. triptera*.

5. Saussurea franchetii Koidz. MIYAMA-KITA-AZAMI. Stems 50–70 cm. long, narrowly winged, densely brownish puberulent in upper part; lower and median cauline leaves cordate, 6–11 cm. long, mucronate-acuminate, cordate, abruptly mucronate-toothed, puberulent above, hairy along the nerves beneath, decurrent, petiolate; upper leaves gradually reduced, broadly ovate to lanceolate; heads in groups of 5–8, corymbose; involucres globose, about 13 mm. long and as wide; bracts 5-seriate, cobwebby, the outer and median herbaceous, as long as the inner, oblong to narrowly so, acuminate, the median and inner white-woolly on back, spreading above, the inner linear, blackish purple, appressed-puberulent; corollas purplish, about 11 mm. long.——Aug. Alpine regions; Honshu (n. distr.).

6. Saussurea brachycephala Franch. IWATE-HIGOTAI. Stems 19–70 cm. long, narrowly winged, puberulent, the branches densely brown-puberulent on upper portion; radical and lower cauline leaves broadly to narrowly ovate, mucronate-acuminate, rarely obtuse, hastate or cordate at base, denticulate, brown-puberulent on both sides, usually long winged-petiolate; median and upper leaves gradually reduced, decurrent; heads few, in corymbs, short-pedicelled; involucres campanulate, about 16 mm. long and as wide, cobwebby, densely brown-puberulent; bracts 5-seriate, the outer slightly more than half the length of the inner, ovate-lanceolate, acute, erect, about 4 mm. wide at base, the median lanceolate, acuminate, as long as the inner; corollas purplish, 11–12 mm. long. ——Aug.–Sept. Alpine; Honshu (n. distr.).

7. Saussurea nikoensis Fr. & Sav. *S. tanakae* var. *phyllolepis* Maxim.; *S. tanakae* var. *intermedia* Matsum. & Koidz.; *S. grandifolia* var. *nikoensis* (Fr. & Sav.) Koidz.; *S. sikokiana* var. *intermedia* (Matsum. & Koidz.) Nakai.——SHIRANE-

AZAMI, AKI-NO-YAHAZU-AZAMI. Stems 35–65 cm. long, narrowly winged, densely brownish crisped-puberulent; radical leaves ovate or oblong, 5–12 cm. long, mucronate-acuminate, deeply cordate to subhastate at base, mucronate-toothed, brown-puberulent on both sides especially toward the margin on upper side and veins beneath, the petioles narrowly winged, the median and upper leaves gradually smaller, ovate to lanceolate, cordate to cuneate at base; heads in groups of 2–8, in loose corymbs, pedicelled; involucres globose-campanulate, 15–16 mm. long, 13–15 mm. wide, densely brownish crisped-puberulent, dark purple on upper portion; bracts 5-seriate, the outer ovate-lanceolate, 2–4 mm. wide at base, as long as the median and inner ones, acuminate, often recurved, the inner oblong-linear, acuminate; corollas purplish, about 10 mm. long.——Aug.–Sept. High mountains; Honshu (Iwaki Prov., n. Kantō Distr. and Shinano Prov.).

Var. **sessiliflora** (Koidz.) Kitam. *S. sessiliflora* (Koidz.) Kitam.; *S. tanakae* var. *sessiliflora* Koidz.; *S. sikokiana* var. *sessiliflora* (Koidz.) Nakai——KURO-TŌ-HIREN. Heads usually 2 or 3, rather large, sessile.——Honshu (n. and centr. distr.).

Var. **involucrata** (Matsum. & Koidz.) Kitam. *S. involucrata* Matsum. & Koidz.; *S. grandifolia* var. *nikoensis* subvar. *involucrata* (Matsum. & Koidz.) Koidz.——NIKKŌ-TŌ-HIREN. Involucral bracts very large, recurved.——Honshu (Nikko in Shimotsuke).

8. **Saussurea gracilis** Maxim. *S. bicolor* Lév. & Van't. HOKUCHI-AZAMI. Stems slender, cobwebby while young, the branches flat; radical leaves rosulate, long-petiolate, elongate-deltoid, rarely deltoid, 6–11 cm. long, acuminate, hastate to cordate or saggitate at base, mucronate-toothed, glabrous and pale green above, villous or white-tomentose beneath, the upper leaves gradually smaller, lanceolate to linear, short-petiolate or sessile; heads in loose corymbs, rarely solitary, the pedicels 5–30 mm. long; involucres tubular, 13–16 mm. long, 8–14 mm. wide, slightly cobwebby, purplish; bracts 8–11 seriate, 7-nerved on back, the outer ovate-lanceolate to ovate, mucronate-acute, the median oblong or oblong-lanceolate, the inner linear; corollas purplish, about 12 mm. long.——Aug.–Oct. Mountains; Honshu (w. Tōkaidō Distr. and westw.), Shikoku, Kyushu.——Korea.

9. **Saussurea scaposa** Fr. & Sav. *S. reinii* Franch.——KIRISHIMA-HIGOTAI. Stems 22–50 cm. long, slender, leafy; radical leaves narrowly oblong to broadly lanceolate, acuminate to acute, narrowed below to a winged petiole or truncate at base, usually pinnately lobed, rarely toothed, the lobes spreading, obtuse or acute, crisped-pubescent above, glabrous beneath, the median leaves gradually smaller, broadly linear to broadly lanceolate, slightly clasping and short-decurrent; heads 3 or 4 in a group; involucres narrowly tubular, 9–11 mm. long, about 6 mm. wide, cobwebby; bracts 5-seriate, loosely arranged, the outer short, about 2/5 as long as the inner, ovate-lanceolate or ovate, mucronate, the median oblong, the inner linear, acute, purplish; corollas purplish, 9–10 mm. long.——Aug.–Oct. Grassy slopes in mountains; Shikoku, Kyushu.

10. **Saussurea modesta** Kitam. NEKO-YAMA-HIGOTAI. Stems about 50 cm. long, angled, glabrous, the branches much-flattened; radical leaves linear-lanceolate, about 15 cm. long, 18 mm. wide, acuminate, gradually narrowed at base to a winged petiole; lower cauline leaves slightly clasping and short-decurrent, loosely toothed, glabrous on both sides, the upper gradually reduced, linear; heads in groups of 2 or 3,

sometimes loosely corymbose; involucres tubular, cobwebby, purplish at tip, 10–12 mm. long, 6–8 mm. wide; bracts 5-seriate, acute, the outer about half as long as the inner, ovate to ovate-lanceolate, mucronate, the median oblong, mucronate, the inner linear; corollas purplish, about 9 mm. long.——Oct. Mountains; Honshu (Chūgoku Distr.).

11. **Saussurea kirigaminensis** Kitam. KIRIGAMINE-TŌ-HIREN. Stems erect, 35–70 cm. long, nearly simple, narrowly winged, glabrous to pubescent; radical and lower cauline leaves linear-oblanceolate, 15–30 cm. long, short-acuminate, gradually long-attenuate at base to a winged petiole, sparsely mucronate-toothed, or undulately mucronate-toothed, loosely hairy above, puberulent on midrib beneath, the upper leaves gradually reduced, linear-lanceolate, slightly clasping to more or less decurrent; heads in groups of 5–9, densely arranged; involucres tubular, 9–11 mm. long, about 6 mm. wide, purplish, cobwebby; bracts loose, 5-seriate, the outer half as long as the inner, ovate-lanceolate, acuminate, the median mucronate-acute, the inner linear, acute; corollas purplish, about 9 mm. long.——Honshu (Mount Kirigamine in Shinano Prov.).

12. **Saussurea ussuriensis** Maxim. *S. ussuriensis* vars. *genuina, incisa,* and *pinnatifida* of Maxim.; *S. grandifolioides* Nakai——KIKU-AZAMI. Stems 30–120 cm. long, very leafy, wingless, glabrous; radical leaves ovate to oblong-ovate, 7–18 cm. long, acuminate-mucronate, cordate, truncate, or subhastate at base, puberulent on both sides or glabrous, long-petiolate, usually pinnately lobulate to cleft, the lobes 3 to 7 pairs, obovate, toothed or lobulate, the upper cauline gradually reduced, ovate to lanceolate, truncate to cuneate at base; heads many, in dense corymbs, very short pedicellate; involucres tubular, 12–13 mm. long, 4–7 mm. across, cobwebby; bracts 5- to 7-seriate, the outer very short, oblong or ovate, about 2 mm. long, cuspidate, the inner linear, obtuse; corollas purplish, 11–13 mm. long.——Sept.–Oct. Mountains; Honshu, Kyushu.——Ussuri, Manchuria, Korea, and n. China.

Var. **nivea** Kitam. USU-YUKI-KIKU-AZAMI. Leaves white-cobwebby beneath.——Honshu (Bingo Prov.).——Korea.

13. **Saussurea triptera** Maxim. YAHAZU-HIGOTAI. Stems 30–55 cm. long, narrowly to broadly winged, prominently puberulent; radical leaves ovate or ovate-oblong, 5.5–12.5 cm. long, acuminate-mucronate, cordate or sometimes truncate at base, mucronate-toothed to undulate-lobulate, crisped-pubescent on both sides, the lobes dilated above, rounded at apex, mucronate-toothed, the upper cauline leaves broadly ovate to lanceolate, short winged-petiolate or sessile, decurrent, gradually smaller toward tip; heads 7–24, in corymbs, short-pedicellate; involucres tubular, about 11 mm. long, 5 mm. across, cobwebby; bracts 5-seriate, the outer about half as long as the inner, broadly ovate, mucronate, the median oblong, mucronate, the inner linear, acuminate; corollas purplish, about 10 mm. long.——Aug.–Sept. Mountains; Honshu (s. part of centr. distr.).

Var. **major** (Takeda) Kitam. *S. kaimontana* forma *major* Takeda; *S. triptera* forma *major* (Takeda) Kitam.——MIYAMA-HIGOTAI. Heads 2–7, loosely arranged, sometimes long-pedicellate; involucres 12–13 mm. long, 5–10 mm. across, the bracts somewhat larger, and the mucronate tip often longer.

Var. **minor** (Takeda) Kitam. *S. kaimontana* forma *minor* Takeda; *S. kinbuensis* Nakai; *S. triptera* forma *minor* (Takeda) Ohwi——TAKANE-HIGOTAI. Stems 10–20 cm. long, prominently winged; heads 3–5.——Alpine.

Var. **kaialpina** (Nakai) Kitam.   *S. triptera* forma *kaialpina* (Nakai) Ohwi;   *S. kaialpina* Nakai——Shirane-higotai. Stems low; heads solitary; involucres about 15 mm. long, 12–15 mm. across, the bracts long-mucronate.——Alpine.

Var. **hisauchii** (Nakai) Kitam.   *S. hisauchii* Nakai; *S. triptera* forma *hisauchii* (Nakai) Ohwi——Tanzawa-higotai. Stems 40–60 cm. long, taller, wingless or scarcely winged; heads in loose corymbs or subspicate, sometimes long-pedicellate; involucres 13–15 mm. long, 8–10 mm. wide, the bracts 6- or 7-seriate; corollas 11–12 mm. long.——Montane; Honshu (centr. distr.).

14. Saussurea riederi Herd. var. **yezoensis** Maxim.   *S. riederi* sensu auct. Japon., non Herd.; *S. nambuana* Koidz.; *S. riederi* subsp. *yezoensis* (Maxim.) Kitam.——Nagaba-kita-azami. Stems leafy, narrowly winged, crisped-puberulent in upper portion; radical leaves long, broadly ovate or elongate-deltoid, 5–8.5 cm. long, cuspidate, mucronate-toothed, truncate to cordate, loosely puberulent on both sides, long-petiolate, the cauline becoming gradually smaller toward tip, clasping; heads in dense corymbs, very short-pedicellate; involucres tubular, 8–12 mm. long, cobwebby, dark-purplish on upper side; bracts 4- to 6-seriate, the outer lanceolate or oblong, acute to acuminate, the inner broadly linear, acuminate; corollas purplish, about 10 mm. long.——Aug.–Sept. Alpine regions; Hokkaido, Honshu (n. distr.).——s. Kuriles.

Var. **kudoana** (Tatew. & Kitam.) Kitam.   *S. kudoana* Tatew. & Kitam.; *S. tokubuchii* Kudo ex Kitam.; *S. riederi* subsp. *kudoana* Kitam., in syn.——Hidaka-tō-hiren. Cauline leaves smaller, lanceolate; involucral bracts acute.——Hokkaido (Mount Apoi in Hidaka Prov.).

Var. **elongata** Kitam.   *S. yezoensis* Franch. excl. syn.; *S. riederi* var. *yezoensis* forma *elongata* (Kitam.) Ohwi——Ezo-tō-hiren. Stems 30–60 cm. long; heads many; involucres about 12 mm. long; corollas 11–12 mm. long.——Hokkaido and Honshu (n. distr.).

Var. **insularis** Tatew. & Kitam.   *S. riederi* subsp. *yezoensis* var. *insularis* Tatew. & Kitam.——Rebun-tō-hiren. Stems 5–13 cm. long; involucres about 12 mm. long; corollas about 13 mm. long.——Hokkaido (Rishiri and Rebun Isls.).

Var. **japonica** Koidz.   *S. riederi* forma *japonica* (Koidz.) Ohwi——Oku-kita-azami. Outer and median involucral bracts long-caudate.——Honshu (Mount Chokaisan in Ugo Prov.).

Var. **daisetsuensis** (Nakai) Kitam.   *S. riederi* forma *daisetsuensis* (Nakai) Ohwi; *S. yezoensis* var. *daisetsuensis* Nakai——Daisetsu-higotai. Stems stout, many-leaved, about 1 m. long; leaves widely decurrent on the stem; heads many, dense, with elongate involucral bracts.——Hokkaido (Mount Daisetsu). The typical phase occurs in the Kuriles and Kamchatka.

15. Saussurea acuminata Turcz. ex DC. var. **sachalinensis** (F. Schmidt) Herd.   *S. sachalinensis* F. Schmidt; *S. acuminata* subsp. *sachalinensis* (F. Schmidt) Kitam.——Kara-futo-azami. Stems 50–100 cm. long, densely leafy, winged, crisped-pubescent or glabrous; radical leaves ovate-oblong, 15–17 cm. long, short-acuminate, usually truncate at base, mucronate callose-toothed, scabrous toward the margin, loosely cobwebby on both sides, short crisped-hairy on the veinlets beneath, long-petiolate, the cauline acuminate, truncate or gradually narrowed at base, short winged-petiolate and decurrent; inflorescence densely corymbose; involucres narrowly cylindric, 11–12 mm. long, 5–6 mm. wide, the bracts 4-seriate,

the outer ovate, short-mucronate, green, longer than the inner, the inner linear, scarious, often purplish at apex; corollas purplish, 13–14 mm. long.——July–Sept. Hokkaido.——Sakhalin. The typical phase occurs in Manchuria and e. Siberia.

16. Saussurea yanagisawae Takeda.   Usu-yuki-tō-hiren.   Stems 6–30 cm. long, crisped-puberulent and cobwebby, narrowly winged or sometimes wingless; radical and lower cauline leaves petiolate, lanceolate or oblong-lanceolate, 4–7.5 cm. long, acuminate, cuneate to truncate at base, irregularly toothed, prominently crisped-puberulent above and along the veinlets beneath, cobwebby on both sides while young, the median and upper leaves gradually smaller, lanceolate; heads 4–8, in dense corymbs; involucres campanulate, 11–12 mm. long and as wide, purplish at the tip and cobwebby; bracts 4-seriate, the outer as long as the inner, broadly lanceolate, long-acuminate, sometimes spreading, the innermost lanceolate, acute; corollas purplish, about 11 mm. long.——Aug.–Sept. Alpine regions; Hokkaido. Very variable.——Forma **nivea** (Koidz.) Ohwi.   *S. tilesii* var. *nivea* Koidz.——Yukiba-tō-hiren.   Leaves white-tomentose beneath.——Forma **imperialis** (Koidz.) Ohwi.   *S. imperialis* Koidz.; *S. tilesii* var. *imperialis* (Koidz.) Koidz.——Takane-kita-azami.   Leaves rather loosely cobwebby on both sides; lower leaves ovate, 9–10 cm. long, about 5 cm. wide, rather irregularly coarse-toothed.——Forma **vestita** (Kitam.) Ohwi.   *S. yanagisawae* var. *vestita* Kitam.——Yukiba-takane-kita-azami.   Stems about 20 cm. long; lower leaves ovate, 8–9 cm. long, 5–5.5 cm. wide, acuminate, truncate to cordate at base, cobwebby above, densely white-woolly beneath.——Forma **elegans** (Koidz.) Ohwi.   *S. tilesii* var. *elegans* Koidz.; *S. yanagisawae* var. *elegans* (Koidz.) Nakai——Ō-takane-kita-azami.   Stems 18–27 cm. long; radical and lower cauline leaves about 9 cm. long, 4 cm. wide, rounded to cuneate at base, loosely cobwebby on both sides, irregularly coarse-toothed.——Forma **angustifolia** (Nakai) Ohwi.   *S. yezoensis* var. *angustifolia* Nakai; *S. yanagisawae* var. *angustifolia* (Nakai) Kitam.——Hosoba-ezo-higotai.   Stems about 22 cm. long; cauline leaves linear-lanceolate, about 12 cm. long, about 1.2 cm. wide, cuneate at base, petiolate, not decurrent, loosely mucronate-toothed, nearly glabrous on both sides.

17. Saussurea fauriei Franch.   Fōrii-azami.   Stems 1.5–2 m. long, stout, densely crisped-puberulent, winged, densely leafy, much branched in upper portion; cauline leaves on short winged petioles, ovate, ovate-oblong, or elliptic-lanceolate, 14–19 cm. long, short-acuminate, cordate to truncate or cuneate at base, mucronately callose-toothed, glabrous above, densely gray-pubescent and whitish beneath, decurrent at base on the winged stem, the upper leaves linear-lanceolate, acuminate; heads in very dense corymbs; involucres narrowly tubular, green, 10–12 mm. long, 4–5 mm. wide; bracts 5-seriate, slightly brownish toward the tip, cobwebby on margin, the outer short, obtuse to rounded, the median narrowly oblong, the inner linear; corollas 1 cm. long.——July–Sept. Rocks near the sea; Hokkaido.——s. Kuriles.

18. Saussurea sagitta Franch.   *S. sagitta* var. *calvescens* Franch.; *S. sagitta* var. *blepharolepis* Franch.; *S. pseudosagitta* Honda——Yahazu-tō-hiren.   Stems 30–45 cm. long, glabrous; median cauline leaves deltoid–hastate, petiolate, 6–8 cm. long, acuminate, sagittate, mucronately callose-toothed, ciliolate, glabrous above, sometimes short crisped-puberulent on the veins beneath, the upper leaves becoming smaller, oblong-lanceolate to lanceolate, long-acuminate, truncate to cuneate at base, short-petiolate; heads few or often solitary, on slender pedicels; involucres tubular, 9–10 mm. long, 6–8 mm. wide, cobwebby; bracts 6-seriate, sometimes purplish toward the margin, the outer short, acuminate, ovate, the median narrowly oblong, rounded, the inner linear, acute; corollas pale purple, about 10 mm. long.——Aug. Alpine regions; Honshu (centr. and n. distr.).

19. Saussurea spinulifera Franch. TOGE-KIKU-AZAMI. Stems 40–70 cm. long, puberulent, winged, branched on upper portion; radical and lower cauline leaves long-petiolate, ovate or deltoid-ovate, acuminate, mucronate, hastate to cordate or cuneate at base, mealy-puberulent above, short-pubescent on the veinlets, pinnately lobed at petiolar end, the median and upper leaves gradually smaller, short-petiolate, broadly ovate to lanceolate, hastate to cuneate at base; heads many, loosely arranged, long-pedicellate; involucres tubular, 12–14 mm. long, 5–7 mm. wide, cobwebby, purplish; bracts 4-seriate, imbricate, the outer ovate to narrowly so, about ¼ as long as the inner, the median oblong, mucronately callose-tipped, the inner linear, acuminate; corollas 11.5 mm. long.——Sept.–Oct. Woods in mountains; Honshu (Sagami, Suruga, and Kai Prov.).

20. Saussurea yoshinagae Kitam. *S. nipponica* subsp. *yoshinagae* (Kitam.) Kitam.——TOSA-TŌ-HIREN. Stems 55–65 cm. long, densely crisped-puberulent; radical leaves chartaceous, rosulate, ovate, oblong to narrowly so, 7–12 cm. long, 3.5–7 cm. wide, acute, mucronate, cordate or truncate at base, winged-petiolate, sometimes sinuately lobulate at the petiolar end, mucronate-toothed, mealy-puberulent on both sides, the cauline leaves gradually becoming smaller, short-petiolate to sessile, ovate to lanceolate; heads few, rarely racemose; involucres campanulate, about 15 mm. long, 15–17 mm. wide; bracts 6-seriate, the outer oblong-lanceolate, 3–3.5 mm. long, acute, the median lanceolate, acute, green, often spreading, brown-pubescent on back, the inner linear, acuminate, purplish; corollas purplish, 11–13 mm. long.——Sept.–Oct. Shikoku.

21. Saussurea nipponica Miq. *S. yamatensis* Honda ——ŌDAI-TŌ-HIREN. Stems 50–100 cm. long, slender, mealy-puberulent in upper portion, narrowly winged; radical and lower cauline leaves long-petiolate, narrowly to broadly ovate, 12–18 cm. long, acuminate, mucronate, cordate at base, obsoletely mucronate-toothed, mealy-puberulent above and on veinlets beneath, the median leaves shallowly cordate to cuneate at base, narrowly decurrent, the upper leaves much reduced; heads 7–15, campanulate, short-pedicellate; involucres tubular, 10–14 mm. long, 7–14 mm. wide; bracts 5-seriate, densely gray-puberulent on back, the outer about ¼ as long as the inner, lanceolate to ovate, acute, shortly recurved at tip, the inner linear, acuminate; corollas purplish, 10–13 mm. long. ——Aug.–Oct. Woods in mountains; Honshu (Kinki Distr.), Shikoku. Very variable.

Var. **kiushiana** (Franch.) Ohwi. *S. kiushiana* Franch.; *S. higomontana* Honda; *S. yakushimensis* Masam.; *S. nipponica* subsp. *kiushiana* (Franch.) Kitam.——TSUKUSHI-TŌ-HIREN. Stems stout; leaves broader, thicker, broadly ovate to subreniform, usually broadly decurrent on the stem; involucres 13–17 mm. long.——Kyushu.

Var. **kurosawae** (Kitam.) Ohwi. *S. kurosawae* Kitam.; *S. nipponica* subsp. *kurosawae* (Kitam.) Kitam.——ABE-TŌ-HIREN. Stems broadly winged; leaves thinner; involucres about 10 mm. long; outer involucral bracts broadly ovate, the median elliptic.——Honshu (Suruga Prov.).

Var. **muramatsui** (Kitam.) Ohwi. *S. muramatsui* Kitam.; *S. nipponica* subsp. *muramatsui* (Kitam.) Kitam.——TOGA-HIGOTAI. Lower leaves cordate, the petioles often broadly winged; heads erect, densely corymbose, terminal on the branches; outer and median involucral bracts linear-lanceolate, prominently recurved.——Honshu (n. distr., Japan Sea side).

Var. **hokurokuensis** (Kitam.) Ohwi. *S. nipponica* subsp. *hokurokuensis* Kitam.——HOKUROKU-TŌ-HIREN. Involucres glabrous on upper portion, the bracts rather shorter; petioles and stems prominently winged.——Honshu (Hokuriku or Japan Sea side).

Var. **robusta** (Makino) Ohwi. *S. nipponica* subsp. *sikokiana* (Makino) Kitam.; *S. sikokiana* Makino; *S. tanakae* var. *robusta* Makino; *S. nipponica* var. *sikokiana* (Makino) Ohwi ——Ō-TŌ-HIREN. Stems stout, with wings often to 2 cm. wide; petioles often winged; involucres 15–16 mm. long, the bracts larger, the outer broadly ovate, the median elliptic, long recurved-tipped.——High mountains; Shikoku.

Var. **sendaica** (Franch.) Ohwi. *S. nipponica* subsp. *sendaica* (Franch.) Kitam.; *S. tanakae* var. *sendaica* Franch.; *S. nikoensis* subsp. *sendaica* (Franch.) Kitam.; *S. sugimurae* Honda; *S. tanakae* var. *crinita* Franch.——NAMBU-TŌ-HIREN. Radical leaves broadly ovate or deltoid-ovate, 10–16 cm. long, loosely pubescent on upper side; involucres 15–18 mm. long, cobwebby, the bracts 7- or 8-seriate, recurved on upper half, the outer shorter than the median.——Honshu (Kantō and n. distr.).

Var. **savatieri** (Franch.) Ohwi. *S. nipponica* subsp. *savatieri* Kitam.; *S. triptera* var. *savatieri* Franch.; *S. nipponica* var. *nikoensis* Nakai; *S. sinuatoides* var. *serrata* Nakai; *S. tsukubensis* Nakai——ASAMA-HIGOTAI. Stems slender; involucral bracts narrower, long-acuminate.——Honshu (Kantō Distr. and ne. Shinano Prov.).

Var. **glabrescens** (Nakai) Kitam. *S. sawadae* Kitam.; *S. sinuatoides* var. *glabrescens* Nakai——KINTOKI-HIGOTAI. Median leaves ovate, undulately lobulate or lobed; heads 1–3, terminal; involucres cobwebby.——Honshu (Hakone).

22. Saussurea amabilis Kitam. *S. obvallata* Nakai, non Wall. ex C. B. Clarke——KŌSHŪ-HIGOTAI. Stems 40–60 cm. long, wingless, sparingly branched; radical leaves long-petiolate, oblong, 8–25 cm. long, acute, hastate at base, mucronate-toothed, pale bluish green beneath, short appressed-puberulent on both sides, the median and upper leaves gradually becoming smaller, short-petiolate or sessile, oblong-lanceolate to lanceolate, gradually acuminate, hastate to cuneate at base; heads 2 or 3, the short peduncles glandular and puberulent, often bracteate; involucres campanulate, about 17 mm. long, 18–20 mm. wide; bracts 7-seriate, the outer and median somewhat shorter than the inner, lanceolate, about 3 mm. wide, acuminate, cobwebby, glandular and short-puberulent on back, long recurved on upper half, the inner linear, acuminate; corollas purplish, 12.5–14 mm. long.——Aug.–Oct. Rocky cliffs in mountains; Honshu (w. Kantō Distr. and Kai Prov.).—— Forma **pinnatiloba** Kitam. KIREBA-KŌSHIŪ-HIGOTAI. Leaves pinnately lobed on lower half.

23. Saussurea sinuatoides Nakai. *S. sinuata* forma *japonica* Nakai——TAKAO-HIGOTAI. Stems slender, 35–60 cm. long, hairy; radical and lower cauline leaves long-petiolate, ovate to broadly so, 7–11 cm. long, mucronate-acuminate, cordate at base, deeply sinuate, irregularly mucronate-toothed, puberulent on both sides, the cauline gradually becoming smaller above, short-petiolate or sessile, cordate to cuneate at base; heads in racemes or loose corymbs, pedicelled; involucres campanulate, 17–18 mm. long, densely cobwebby; bracts 7-seriate, acuminate, the outer about half as long as the inner, lanceolate, recurved on upper half, the inner membranous; corollas purplish, 13–14 mm. long.——Sept.–Oct. Woods in mountains; Honshu (Musashi and Sagami Prov.).

24. Saussurea pennata Koidz. MIYAMA-TŌ-HIREN. Stems slender, 30–40 cm. long, glabrous; lower and median cauline leaves chartaceous, sometimes elongate-deltoid, 7–9 cm. long, long-acuminate, sagittate or rarely truncate at base, decurrent, mucronate-toothed, ciliolate, puberulent above and on veinlets beneath, the upper leaves gradually smaller, short-petiolate or sessile, ovate-oblong to lanceolate, truncate to cuneate at base; heads few, in loose corymbs, with slender pedicels; involucres tubular, 12–13 mm. long, 8–10 mm. wide, the bracts 6-seriate, loosely cobwebby, the outer lanceolate or ovate-lanceolate, acuminate, recurved in upper half, nearly glabrous, the inner membranous, linear, acute; corollas purplish, 12 mm. long.——July–Sept. Honshu (Yamato Prov.), Shikoku.

25. Saussurea maximowiczii Herd. MIYAKO-AZAMI. Stems 50–150 cm. long, wingless, loosely short-pubescent, glandular-dotted; radical and lower cauline leaves long-petiolate, large, oblong, 11–30 cm. long, acute or cuspidate, cuneate at base, pinnately parted, the lateral segments 4 to 6 pairs, oblanceolate, mucronate, entire to irregularly sinuate-incised, the terminal segments short-pubescent on both sides, often loosely glandular-dotted beneath, the cauline leaves gradually smaller toward tip, oblong to lanceolate, entire to pinnatiparted, short-petiolate or sessile, semiclasping; heads many, in loose corymbs, on slender pedicels; involucres narrowly tubular, 10–14 mm. long, about 6 mm. wide, cobwebby; bracts 8-seriate, the inner gradually longer, broadly linear, obtuse, the outer very short, ovate, the median oblong, ending in a callose tip; corollas purplish, 11–13 mm. long.——Aug.–Oct. Grassy places in mountains; Honshu, Kyushu.——Korea, Manchuria, and Amur.

26. Saussurea insularis Kitam. SHIMA-TŌ-HIREN. Stems about 50 cm. long, cobwebby; lower cauline leaves chartaceous, long-petiolate, deltoid-ovate, 9–10 cm. long, 7.5–10 cm. wide, subcaudately acuminate, sagittate-cordate at base, irregularly mucronate-dentate, loosely pubescent above, cobwebby beneath while young, the upper leaves gradually becoming smaller, densely whitish cobwebby; heads in loose corymbs, short-pedicelled; involucres campanulate, about 11 mm. long and as wide, cobwebby; bracts 7-seriate, the outer deltoid, about 2 mm. long, obtuse, mucronate, appressed, the inner linear, acute, densely cobwebby; corollas about 11 mm. long.——Sept. Mountains; Kyushu (Tsushima).

27. Saussurea tanakae Fr. & Sav. *S. tanakae* var. *pycnolepis* Franch.——SEITAKA-TŌ-HIREN, AKI-NO-YAHAZU-AZAMI. Stems 70–100 cm. long, puberulent; lower cauline leaves chartaceous, cordate to cordate-deltoid, 8–15 cm. long, cuspidate, cordate to truncate at base, toothed, puberulent on both sides, the petioles sometimes winged, decurrent, the upper leaves gradually becoming smaller, ovate or oblong, acute, rounded at base, entire, sessile; heads in racemes or rarely subcorymbose; involucres campanulate, about 17 mm. long, 12–15 mm. wide, densely silky-woolly; bracts 9-seriate, imbricate, the outer short, broadly ovate, obtuse, mucronate, much shorter than the inner, the median oblong, obtuse, the inner linear, acute, membranous; corollas purplish, about 13 mm. long.——Sept.–Oct. Grassy places in mountains; Honshu.——Korea.

Hybrids are: **Saussurea × mirabilis** Kitam. KŌSHŪ-TŌ-HIREN. (*S. kirigaminensis* × *S. tanakae*); **Saussurea × rara** Kitam. KAI-TŌ-HIREN. (*S. amabilis* × *S. tanakae*); **Saussurea × karuizawensis** Hara (*S. tanakae* × *S. ussuriensis*); **Saussurea × satowii** Kitam. ONGATA-HIGOTAI. (*S. sinuatoides* × *S. tanakae*).

56. SERRATULA L.   TAMURA-SŌ ZOKU

Erect, often branched perennial herbs; leaves alternate, pinnately parted, toothed, not spiny; involucres campanulate, the bracts many-seriate, the outer gradually shorter, broad, acuminate, short spine-tipped; receptacle densely setose; florets of outer row sterile, filiform, the corollas deeply 3- to 5-lobed, the central florets fertile, the corollas tubular, 5-lobed; filaments mammillate; anthers sagittate; style branches slender, with a tuft of hairs at base; achenes glabrous, terete, truncate at apex, gradually narrowed at base, obliquely attached; pappus-bristles many, unequal in length, scaberulous.——Many species in the N. Hemisphere.

1. Serratula coronata L. var. **insularis** (Iljin) Kitam. *S. insularis* Iljin; *S. koreana* Iljin, pro parte; *S. coronata* sensu auct. Japon., non L.——TAMURA-SŌ. Rhizomes creeping, woody; stems 30–140 cm. long, striate, usually pubescent; lower and radical leaves long-petiolate, ovate-oblong, pinnately parted, the segments 6 or 7 pairs, oblong, acute, narrowly cuneate at base, irregularly toothed, white-puberulent on both sides, the upper leaves smaller, short-petiolate or sessile; heads erect, long-pedunculate; involucres campanulate, 20–27 mm. long, yellow-green to purplish, slightly cobwebby; bracts 7-seriate, the outer lanceolate or broadly so, the median and inner acuminate, spine-tipped, the inner narrowly lanceolate, scarious; corollas pale rose-purple; achenes about 6 mm. long, 1.5 mm. wide; pappus 11–14 mm. long.——Sept.–Oct. Mountains; Honshu, Shikoku, Kyushu.——Korea. The typical phase occurs in Korea, Manchuria, and Siberia.

57. SYNURUS Iljin   YAMA-BOKUCHI ZOKU

Large branched perennials; leaves alternate, ovate or ovate-elliptic, petiolate, incised-toothed to pinnately cleft, mucronate-toothed; heads large, nodding in anthesis; involucres globose or globose-campanulate, large, the bracts many-seriate, acuminate, entire, the outer gradually smaller and recurved, the inner erect; receptacle densely setose; florets dark purple or yellowish to whitish; style 2-fid, the branches erect, obtuse, with a tuft of hairs at base; anther appendages connate around the filaments; achenes oblong, glabrous, truncate at apex, many-ribbed; pappus-bristles unequal in length, slightly scabrous.——e. Asia, Siberia, and Mongolia.

1A. Narrow part of corolla-tube slightly shorter than to as long as the other part; leaves deltoid, hastate at base; involucral bracts 2 to 3
   mm. wide. .... 1. *S. excelsus*
1B. Narrow part of corolla-tube much shorter than the other part.
   2A. Involucral bracts needle-like. .... 2. *S. palmatopinnatifidus*
   2B. Involucral bracts linear to lanceolate. .... 3. *S. pungens*

1. **Synurus excelsus** (Makino) Kitam. *Serratula excelsa* (Makino) Makino; *S. atriplicifolia* var. *excelsa* Makino——HABA-YAMA-BOKUCHI. Rhizomes stout; stems stout, 1–2 m. long, purplish, cobwebby, branched in upper portion; lower cauline leaves long-petiolate, deltoid, 10–20 cm. long, acute, mucronate, hastate at base, irregularly incised, mucronate-toothed, green above, densely white-woolly beneath, the upper leaves gradually smaller, ovate-deltoid, truncate to cuneate at base, short-petiolate to sessile; heads rather congested, on stout peduncles; involucres globose, transversely cobwebby, about 3 cm. long, 3–5 cm. wide; bracts many-seriate, the outer gradually smaller, linear-lanceolate, the median 2–3 mm. wide; florets dark-purple; achenes about 6 mm. long, 3 mm. wide, toothed; pappus about 20 mm. long, brownish.——Sept.–Oct. Honshu, Shikoku, Kyushu.

2. **Synurus palmatopinnatifidus** (Makino) Kitam. *Serratula deltoides* var. *palmatopinnatifida* Makino; *S. palmatopinnatifida* (Makino) Kitam.——KIKUBA-YAMA-BOKUCHI. Rhizomatous; stems 70–100 cm. long, cobwebby, purplish; lower cauline leaves long-petiolate, ovate or ovate-oblong, 15–23 cm. long, acute, mucronate, usually deeply cordate at base, palmately pinnatifid, the segments oblong or lanceolate, acute, irregularly mucronate-toothed or denticulate, green above, white-tomentose beneath, the upper leaves smaller, short-petiolate or sessile, truncate to cuneately rounded at base; involucres globose-campanulate, about 26 mm. long, 3–4 cm. wide, loosely cobwebby; bracts many-seriate, needlelike, spine-tipped, the outer recurved, the inner erect, 1.5 mm. wide; achenes about 6 mm. long, about 2.5 mm. wide, glabrous, toothed; pappus brownish, about 20 mm. long, rather unequal in length.——Oct. Mountains; Honshu (Kinki Distr. and westw.), Shikoku, Kyushu.——Korea.

Var. **indivisus** Kitam. *Serratula deltoides* Makino, excl. syn.; *S. atriplicifolia* sensu Makino, non Benth.——YAMA-BOKUCHI. Leaves undivided.——Honshu, Shikoku, Kyushu.——Korea.

3. **Synurus pungens** (Fr. & Sav.) Kitam. *Rhaponticum pungens* Fr. & Sav.; *Serratula pungens* Fr. & Sav.; *S. atriplicifolia* var. *pungens* (Fr. & Sav.) Makino——Ō-YAMA-BOKUCHI. Rhizomes ascending to creeping; stems 1–1.5 m. long, stout, purplish, slightly cobwebby; lower cauline leaves long-petiolate, ovate-oblong or ovate, sometimes broadly ovate, 15–35 cm. long, acute, mucronate, deeply cordate at base, incised mucronate-toothed, white or sometimes reddish woolly-tomentose beneath, the upper leaves smaller, short-petiolate or sessile, ovate, cordate to rounded at base; heads nodding; involucres campanulate-globose, about 3 cm. long, 3.5–4.5 cm. wide, dark purple; bracts many-seriate, acuminate, spine-tipped, the outer gradually smaller, the median linear-lanceolate, 1.5–2 mm. wide, the inner erect, linear; corollas purplish; achenes about 6 mm. long, 3 mm. wide; pappus-bristles brownish, unequal in length, about 16 mm. long.——Sept.–Oct. Hokkaido (sw. distr.), Honshu (centr. and n. distr.).

Var. **giganteus** Kitam. ONI-YAMA-BOKUCHI. Heads larger; pappus bristles to 21 mm. long.——Honshu (centr. distr.).

58. ECHINOPS L.   HIGOTAI ZOKU

Large stout perennial herbs; leaves alternate, pinnately lobed to divided, spiny; heads 1-flowered, sessile, densely congested in compound globose heads; involucral bracts many-seriate, the outer needlelike, the median spathulate, awned, the inner narrowly rhomboid or oblong, short-awned; florets all fertile; corollas tubular, deeply 5-lobed, broader part of the tube short; style-branches relatively thick, erect, mammillate on back, thickened and papillose-puberulent below the middle; achenes densely hairy; pappus crownlike, the bristles many, short, scalelike.——s. Europe and S. Africa, few in Asia and tropical Africa.

1. **Echinops setifer** Iljin. *E. sphaerocephalus* sensu auct. Japon., non L.; *E. dahuricus* sensu auct. Japon., non Fisch.——HIGOTAI. Stems stout, about 1 m. long, densely white-villous, slightly branched in upper part, deeply striate at base; radical leaves long-petiolate, about 20 cm. long, pinnately parted, the segments acuminate, short spine-margined, slightly cobwebby above, white-woolly beneath, decurrent, the cauline leaves oblong, 15–25 cm. long, acuminate, sessile, pinnately lobed; heads sessile, very many, densely congested in a compound globose head, about 5 cm. across; involucres 14–21 mm. long, 4–5 mm. wide, the bracts 16–18; florets blue; achenes terete, densely brownish hairy.——Aug.–Sept. Low mountains; Honshu (Mino, Bingo Prov.), Shikoku, Kyushu.——Korea.

59. LAPSANA L.   YABU-TABIRA-KO ZOKU

Annual or biennial herbs; leaves alternate, coarsely toothed to pinnate; heads small, in loose panicles, the peduncles often long; involucres cylindric-campanulate; outer involucral bracts few, small, the inner 1-seriate, nearly alike; receptacle flat, glabrous; flowers all ligulate, the corollas truncate, 5-toothed, yellow; achenes oblong, often slightly flattened, many-ribbed, rounded at apex, narrowed at base, glabrous; pappus absent.——About 10 species, in temperate regions of the Old World.

1A. Involucres ovoid to ovoid-globose; achenes oblong, about 2.5 mm. long, without appendages at apex, distinctly shorter than the involucres. .... 1. *L. humilis*
1B. Involucres ellipsoidal or oblong; achenes oblong-lanceolate, 4–4.5 mm. long, with 2 short hornlike appendages at apex, nearly as long as the involucres. .... 2. *L. apogonoides*

1. Lapsana humilis (Thunb.) Makino. *Prenanthes humilis* Thunb.; *L. parviflora* A. Gray——Yabu-tabira-ko. Sparingly pubescent biennial; radical leaves tufted, petiolate, 5–15 cm. long, 1.5–3 cm. wide, lyrately pinnatiparted, the terminal segments largest, rounded at apex; scapiform stems ascending, 10–30 cm. long, few-leaved, branched, several-headed; heads small, nodding in fruit; involucres about 4.5 mm. long, ovoid to ovoid-globose, the inner bracts 7 or 8, broadly lanceolate; achenes oblong, about 2.5 mm. long, red-brown, slightly flattened, ribbed on both sides.——Riverbanks, paddy fields, cultivated fields, and thickets in lowlands; Hokkaido, Honshu, Shikoku, Kyushu; common.

× **Lapsoyoungia musashiensis** (Hiyama) Hiyama. *Lapsana musashiensis* Hiyama——Oni-yabu-tabira-ko. A bigeneric hybrid of *Lapsana humilis* × *Youngia japonica*.——Honshu (Kantō Distr.).

2. Lapsana apogonoides Maxim. Ko-oni-tabira-ko. Resembles *L. humilis;* biennial sparingly pilose herb; radical leaves tufted, 4–10 cm. long, 1–2 cm. wide, petiolate, lyrately pinnatiparted, usually obtuse, mucronate, the terminal segments large; scapiform stems ascending, 10–25 cm. long, few-leaved, the several heads in loose corymbs; heads nodding in fruit; involucres about 4.5 mm. long, as long as the achenes, ellipsoidal or oblong; inner involucral bracts oblong-lanceolate; achenes pale yellow-brown, flattened, 3-angled, minutely mammillate, oblong-lanceolate, 4–4.5 mm. long, ribbed, with 2 short hornlike appendages at apex.——Apr.–June. Riverbanks, cultivated fields, and thickets in lowlands; Honshu, Shikoku, Kyushu; common.——Korea and China.

60. HYPOCHOERIS L.　Ōgon-sō Zoku

Perennials, rarely annuals or biennials; radical leaves tufted, entire to pinnatifid; stems scapiform and bracteate or slightly branched and leafy; heads relatively large, on long peduncles, yellow; involucres oblong-cylindric or campanulate, the bracts many-seriate, imbricate, the outer gradually smaller; receptacle flat, scaly (paleaceous); corolla ligulate, truncate and toothed at apex; achenes oblong or linear, 10-ribbed, sometimes flattened, truncate and toothed at apex, slightly narrowed at base; pappus-bristles 1-seriate, plumose.——About 50 species, chiefly in temperate regions, especially abundant in S. America.

1. Hypochoeris crepidioides (Miyabe & Kudo) Tatew. & Kitam. *Picris crepidioides* Miyabe & Kudo; *Achyrophorus crepidioides* (Miyabe & Kudo) Kitag.——Ezo-kōzori-na. Perennial with sparse coarse hairs; stems 15–40 cm. long, simple, few-leaved at base; radical leaves few, oblong-lanceolate to oblong-spathulate, 8–13 cm. long, 2–3 cm. wide, obtuse, irregularly toothed, the lower cauline leaves similar to the radical, the upper oblong, sessile, rather small, clasping; heads solitary, about 5 mm. across, yellow; involucral bracts many, blackish, the outer ovate-oblong, elongate-linear, about 7 mm. long, 1.5–2.5 mm. wide, with coarse black and short white-woolly

hairs, the innermost membranous, 2.2 cm. long, 1.5 mm. wide; achenes linear-lanceolate, 1–1.3 cm. long, beaked, rugose; pappus sordid-white, plumose, the bristles about 1 cm. long.——July. Alpine regions; Hokkaido (Mount Apoi).

Hypochoeris ciliata (Thunb.) Makino. *Arnica ciliata* Thunb.; *H. grandiflora* Ledeb.——Ōgon-sō. Perennial with few-leaved stems; leaves hispid on midrib beneath and on margin; involucral bracts glabrous, 4–6 mm. wide at base; achenes beakless. Rarely cultivated in our area.——Korea and e. Asia.

61. SCORZONERA L.　Futanami-sō Zoku

Perennial, rarely annual or biennial herbs often with woolly hairs; leaves alternate, entire to pinnate; heads relatively large, long-pedunculate; involucres tubular or campanulate, the bracts many-seriate, imbricate, the outer gradually smaller; receptacle flat or alveolate, rarely hairy; corollas ligulate, yellow, truncate, 5-toothed; achenes linear, terete, or angled, rarely winged, narrowed at tip; pappus-bristles many-seriate, unequal in length, toothed or plumose.——About 100 species, in Europe, N. Africa, and Asia.

1. Scorzonera rebunensis Tatew. & Kitam. *S. radiata* sensu auct. Japon., non Fisch.; *S. radiata* forma *humilis* Koidz.——Futanami-sō. Perennial herb with stout rhizomes, densely covered with brownish marcescent leaves at base; scapiform stems simple, 10–20 cm. long, with few scalelike leaves, short-pubescent above; radical leaves tufted, glabrous, oblanceolate or spathulate, 3-nerved, acute, as long as to

shorter than the stems; heads solitary, relatively large, erect; involucral bracts narrowly scarious-margined, glabrous, rather fleshy; corollas ligulate, yellow, the tube puberulent; achenes glabrous; pappus white, the bristles elongate, some scabrous and some short-plumose.——July–Aug. Hokkaido (Rebun Isl.).

62. PICRIS L.　Kōzori-na Zoku

Erect branched herbs usually hispid or coarsely hirsute on stems and leaves; leaves radical and cauline; heads pedunculate; involucres ovoid-urceolate or campanulate, the inner bracts 1-seriate, nearly equal, the outer many; receptacles flat, short-pilose; corollas ligulate, yellow, truncate, 5-toothed; achenes linear or oblong, sometimes angled, 5- to 10-ribbed, rather abruptly narrowed at apex, sometimes beaked; pappus usually in a single series, plumose, sometimes also with an outer series of smaller unbranched bristles.——About 40 species, in Europe, Asia, and N. Africa.

1. Picris hieracioides L. var. **glabrescens** (Regel) Ohwi. *P. japonica* Thunb.; *P. japonica* var. *glabrescens* Regel; *P. flexuosa* Thunb.; *P. hieracioides* var. *japonica* (Thunb.) Regel; *P. hieracioides* sensu auct. Japon., non L.;

P. hieracioides subsp. *japonica* (Thunb.) Krylov——Kōzori-na. Biennial brownish red hispid herb; stems terete, 30–80 cm. long, leafy, often branched; lower and radical leaves oblanceolate, 6–15 cm. long, 1–4 cm. wide, mucronate or acutely toothed, subobtuse, gradually narrowed to a petiolelike base, the median cauline leaves sessile, the upper ones linear-lanceolate, sometimes semiclasping; heads 2–2.5 cm. across, yellow; involucres green, often slightly black-tinged, the inner bracts linear-lanceolate, 8–12 mm. long, about 1.5 mm. wide, hispid and short-pubescent on back, the outer very short; achenes red-brown, narrowly fusiform, about 4 mm. long, longitudinally ribbed, nearly beakless; pappus sordid-white, 7–8 mm. long.——May–Oct. Grassy places in lowlands and mountains; Hokkaido, Honshu, Shikoku, Kyushu; common.——Sakhalin.

Var. **jessoensis** (Tatew.) Ohwi. *P. jessoensis* Tatew.; *P. hieracioides* subsp. *jessoensis* (Tatew.) Kitam.; *P. japonica* var. *jessoensis* (Tatew.) Ohwi——Hosoba-kōzori-na. Plant less hispid; radical leaves rather prominent, the cauline narrow, fewer; involucral bracts blackish green; corollas deep yellow. ——Serpentine rocks in mountains; Hokkaido.

Var. **akaishiensis** Kitam. *P. japonica* var. *akaishiensis* (Kitam.) Ohwi——Akaishi-kōzori-na. Leaves linear-lanceolate; heads smaller; involucral bracts less hispid.—— Honshu (centr. distr.).

Var. **mayebarae** (Kitam.) Ohwi. *P. hieracioides* subsp. *mayebarae* Kitam.; *P. japonica* var. *mayebarae* (Kitam.) Ohwi.——Higo-kōzori-na. Plants moderately hispid; involucral bracts blackish green; corollas deep yellow.——Flowering in late autumn; Kyushu (Higo Prov.).

Var. **alpina** Koidz. *P. kamtschatica* Ledeb; *P. hieracioides* subsp. *kamtschatica* (Ledeb.) Hult.——Kanchi-kōzori-na. Plant short, dark green, very hispid; involucres dark green, the bracts prominently hispid; corollas deep yellow.——Hokkaido and alpine regions of Honshu (centr. and n. distr.).—— Kuriles, Kamchatka, and Aleutians. The typical phase occurs in Eurasia; introduced in N. America.

63. TARAXACUM Wiggers Tampopo Zoku

Acaulescent perennial herbs; leaves radical, entire to pinnate; heads on a scape; involucres campanulate or oblong, the bracts herbaceous, the innermost 1-seriate, elongate, the outer short, many-seriate, often reflexed at tip; receptacle flat, glabrous; corolla ligulate, truncate, 5-toothed; achenes oblong or narrowly so, 4- or 5-angled or slightly compressed, 10-ribbed, narrowed at base, beaked, often warty-tuberculate near base of beak; pappus-bristles simple, many, slender, unequal in length.——Many species, in temperate to Arctic regions, especially abundant in the N. Hemisphere.

1A. Outer involucral bracts reflexed from the base; introduced. .... 1. *T. officinale*
1B. Outer involucral bracts erect or ascending, not reflexed; indigenous.
  2A. Involucres dark green; corollas orange-yellow; alpine or boreal plants with dense usually brown scales on neck of caudex.
    3A. Involucres small, 12–15 mm. long at anthesis.
      4A. Leaves pinnately divided, the sinus reaching the midrib, the terminal segments usually acuminate; outer involucral bracts oblong, not corniculate. .... 2. *T. yuparense*
      4B. Leaves pinnately cleft to parted, the sinus reaching part-way to the midrib, the terminal segments usually obtuse, rarely acute; outer involucral bracts ovate, corniculate. .... 3. *T. trigonolobum*
    3B. Involucres large, 15–21 mm. long at anthesis.
      5A. Outer involucral bracts very short, about ⅓ as long as the inner; leaves lingulate, toothed or pinnately cleft to parted.
                                                        4. *T. yatsugatakense*
      5B. Outer involucral bracts half or more than half as long as the inner.
        6A. Outer involucral bracts about half as long as the inner, with broad membranous margins. .... 5. *T. platypecidum*
        6B. Outer involucral bracts more than half as long as the inner.
          7A. Involucres pruinose, 15–18 mm. long, the outer bracts oblong to oblong-lanceolate, not corniculate, obsoletely membranous on margin. .... 6. *T. alpicola*
          7B. Involucres not pruinose, 17–19 mm. long, the outer bracts oblong to ovate-oblong, usually corniculate, distinctly membranous on margin. .... 7. *T. shikotanense*
  2B. Involucres bright to pale green; corollas yellow or white; lowland and mountain plants with few brown scales on neck of caudex.
    8A. Corollas yellow; leaves and bracts green.
      9A. Involucres 13–14 mm. long at anthesis.
        10A. Outer involucral bracts less than half as long as the inner, oblong-lanceolate to ovate-lanceolate, corniculate or not.
                                                    8. *T. japonicum*
        10B. Outer involucral bracts about half to more than half as long as the inner.
          11A. Outer involucral bracts about half as long as the inner.
            12A. Outer involucral bracts ovate to ovate-oblong, corniculate. .... 9. *T. kiushianum*
            12B. Outer involucral bracts oblong, usually not corniculate, rarely so. .... 10. *T. kuzakaiense*
          11B. Outer involucral bracts more than half as long as the inner. .... 11. *T. variabile*
      9B. Involucres 15–20 mm. long at anthesis.
        13A. Outer involucral bracts less than half as long as the inner, broadly ovate or ovate-oblong, obtuse; leaves pectinate.
                                                     12. *T. pectinatum*
        13B. Outer involucral bracts about half to more than half as long as the inner.
          14A. Outer involucral bracts more than half as long as the inner.
            15A. Outer involucral bracts blackish on back, ovate-lanceolate; achenes about 5 mm. long, 1 mm. wide.
                                            13. *T. maruyamanum*
            15B. Outer involucral bracts green.
              16A. Outer involucral bracts oblong, relatively short-corniculate. .... 14. *T. sendaicum*
              16B. Outer involucral bracts oblong-lanceolate, long-corniculate. .... 15. *T. longeappendiculatum*
         14B. Outer involucral bracts about half as long as the inner.
           17A. Outer involucral bracts oblong or lanceolate.
            18A. Outer involucral bracts small-corniculate or not so. .... 16. *T. elatum*
            18B. Outer involucral bracts prominently corniculate. .... 17. *T. ceratolepis*

1.   Taraxacum officinale Weber. *Leontodon taraxacum* L.——Seiyō-tampopo.   Rather robust soft perennial; leaves variously pinnatifid to deeply toothed, usually 20–30 cm. long, 2.5–5 cm. wide; heads rather large, 4–5 cm. across, clear yellow, the outer involucral bracts usually narrowly lanceolate, about half as long as the inner, not appendaged just below the apex; achenes about 4 mm. long, more or less muricate.——Mar.–May.   Naturalized around towns and cities.——Introduced from Europe.

2.   Taraxacum yuparense H. Koidz.   *T. officinale* var. *lividum* subvar. *dissectissimum* Koidz.——Takane-tampopo, Yūbari-tampopo.   Leaves pinnately divided, the sinus reaching to the midrib, the segments acute to acuminate, acutely toothed on upper margin, the petioles slender; scapes 20–45 cm. long; involucral bracts not corniculate, the outer oblong, short-acuminate, appressed, scarious on margin, the innermost lanceolate to linear-lanceolate, acute; achenes prominently 5- or 6-ribbed, with rather large spinelike tubercles.——Alpine regions; Hokkaido (Ishikari and Hidaka Prov.).

3.   Taraxacum trigonolobum Dahlst.   *T. chamissonis* Greene, pro parte; *T. yesoalpinum* Nakai; *T. livens* H. Koidz.; *T. crassicollum* H. Koidz.——Kumoma-tampopo.   Roots thickened; leaves rather thick, oblanceolate or spathulate-oblanceolate, pinnately cleft to parted, the sinus reaching partway to the midrib, the terminal segments usually obtuse; heads about 4 cm. across; outer involucral bracts appressed to slightly spreading, ovate or ovate-orbicular, small-corniculate, about half as long as the inner, the inner linear; achenes 4.5–5 mm. long, with spinelike tubercles.——Alpine regions; Hokkaido (Ishikari Prov.).——Kuriles, Kamchatka, Aleutians, and Alaska.

4.   Taraxacum yatsugatakense H. Koidz.   Yatsugatake-tampopo.   Plant about 25 cm. high; leaves suberect, oblanceolate, lingulate, toothed or pinnately cleft to parted; scapes white woolly-pubescent on upper part; heads 2.5–3 cm. across; involucres rounded at base, the bracts not corniculate, the outer appressed, ovate or ovate-rounded, 1/3 as long as the inner, nearly glabrous on margin, the inner linear-lanceolate, glabrous.——Alpine regions; Honshu (Mount Yatsugatake in Shinano and in Akaishi Mt. range).

5.   Taraxacum platypecidum Diels.   *T. otagirianum* Koidz.; *T. imbricatum* H. Koidz.; *T. pruinosum* H. Koidz.; *T. saxatile* H. Koidz.; *T. sugawarae* H. Koidz.; *T. koidzumii* Nemoto; *T. albomarginatum* Kitam.; *T. multisectum* Kitag. ——Odasamu-tampopo.   Roots thickened; leaves oblanceolate, 10–20 cm. long, 1–5 cm. wide, irregularly pinnatifid, the petioles slightly purplish; heads about 4 cm. across; outer involucral bracts ovate to broadly so, about half as long as the inner, sometimes small-corniculate, broadly membranous on margin, white-pubescent on upper surface; achenes pale yellow-brown, about 3.5 mm. long, 1 mm. wide, with spreading spinelike tubercles toward the tip.——Alpine regions; Hokkaido (Shiribeshi Prov.).——Sakhalin, Korea, Manchuria, and China.

6.   Taraxacum alpicola Kitam.   *T. officinale* var. *lividum* sensu auct. Japon., non K. Koch; *T. japonense* Nakai; *T. kurohimense* H. Koidz.; *T. togakushiense* H. Koidz.——Miyama-tampopo, Tateyama-tampopo.   Roots thickened; leaves lingulate, 15–30 cm. long, 4–6 cm. wide, obtuse or rounded, petiolelike at base, usually incised; scapes longer than the leaves at anthesis, loosely pubescent, villous under the head; heads about 4 cm. across; involucres pruinose, 15–18 mm. long, the outer bracts erect, lanceolate-oblong or oblong, not corniculate, the inner 14–15 mm. long, linear-lanceolate; achenes oblong-linear, 3.5–4 mm. long, slightly flattened, pale yellow-brown, with spinelike tubercles on upper part, smooth at base.——July–Aug.   Alpine regions; Honshu (centr. and n. distr.).

Var. **shiroumense** (H. Koidz.) Kitam.   *T. shiroumense* H. Koidz.——Shirouma-tampopo.   Involucral bracts usually corniculate.——Honshu (centr. distr.).

7.   Taraxacum shikotanense Kitam.   Shikotan-tampopo, Nemuro-tampopo.   Roots thick; leaves oblanceolate, obtuse, gradually narrowed and petiolelike at base, pinnately lobed to cleft; heads large, about 5 cm. across; involucres 17–19 mm. long, the outer bracts erect, oblong to ovate-oblong, narrowed at tip and corniculate, dark purplish, distinctly membranous on margin; achenes brown to pale yellow-brown, oblong, about 4 mm. long, 1.3 mm. wide, narrowed at tip, with dense spinelike tubercles on upper part, gradually less so toward the base.——July.   Near seashores; Hokkaido.——Kuriles.

8.   Taraxacum japonicum Koidz.   *T. kansaiense* Koidz.; *T. bitchuense* H. Koidz.; *T. liukiuense* H. Koidz.; *T. okinawense* H. Koidz.——Kansai-tampopo, Kansei-tampopo. Plant about 20 cm. high; leaves lyrate, pinnately cleft, obtuse, acute or rounded, nearly glabrous; scapes few, slender, longer than the leaves at anthesis; heads 2–3 cm. across; outer involucral bracts 2-seriate, oblong, lanceolate or ovate-lanceolate, obtuse, less than half as long as the inner, the inner linear-lanceolate or ovate-lanceolate, obtuse, 12–14 mm. long, corniculate or sometimes not so; achenes about 5 mm. long, pale yellow-brown with spinelike tubercles on upper part.——Honshu (Kinki Distr. and westw.), Shikoku, Kyushu.——Ryukyus.

9.   Taraxacum kiushianum H. Koidz.   *T. shikokianum* Kitam.; *T. imaizumii* H. Koidz.; *T. japonicum* var. *macrolepis* H. Koidz.——Tsukushi-tampopo.   Leaves broadly oblanceolate, 10–30 cm. long, acute to obtuse, pinnatiparted; scapes many, longer than the leaves at anthesis, loosely pubescent, villous toward the tip; heads 3–4 cm. across; outer involucral bracts appressed, broadly ovate to ovate-oblong, corniculate, the inner 1–1.4 cm. long, blackish on back; achenes slightly flattened, brownish, 4–5 mm. long, about 1 mm. wide.——Shikoku, Kyushu (n. distr.).

10.   Taraxacum kuzakaiense Kitam.   Kuzakai-tampopo.   Leaves linear-oblanceolate, loosely pubescent, 5–11 cm. long, 7–20 mm. wide, obtuse to acute, irregularly pinnatisect;

scapes shorter than to as long as the leaves at anthesis, densely cobwebby toward the tip, villous under the head; involucres small, the outer bracts appressed, oblong, acuminate, pubescent toward the tip, usually not corniculate or sometimes so, the inner lanceolate, glabrous; achenes brownish, narrowly oblong, about 3 mm. long, 1 mm. wide, with spinelike tubercles on upper part.——Honshu (Rikuchu Prov.).

11. Taraxacum variabile Kitam. *T. platycarpum* var. *variabile* (Kitam.) H. Koidz.——KITSUNE-TAMPOPO. Leaves spreading, lingulate-lanceolate, 7–10 cm. long, 1.5–2 cm. wide, obtuse, pubescent, pinnately cleft; scapes commonly shorter than the leaves, prominently hairy; heads small, 3–3.5 cm. across; involucres small, 12–14 mm. long, the outer bracts erect, ovate-oblong, gradually acute to acuminate, usually not corniculate or sometimes so, more than half as long as the inner, reddish at tip, the inner bracts linear-lanceolate, scarious on margin, blackish, with a small hornlike tubercle at apex; achenes oblong, 4–4.5 mm. long, 1 mm. wide, angled, with spinelike tubercles on upper part.——Honshu (Sagami Prov.).

12. Taraxacum pectinatum Kitam. *T. nemotoi* H. Koidz.; *T. numajirii* H. Koidz.——KUSHIBA-TAMPOPO. Leaves many, linear-lanceolate, 10–30 cm. long, 3–4 cm. wide, acute, pinnately parted, pectinate, the segments linear, the terminal one smaller; scapes many, shorter than to as long as the leaves, loosely cobwebby, villous under the head; heads 3.5–4 cm. across; involucres 15–20 mm. long, about 2 cm. wide, the outer bracts appressed, corniculate or rarely not so, white-ciliate, broadly ovate or ovate-oblong, obtuse, less than half as long as the inner, the inner bracts linear-lanceolate, distinctly scarious on margin, blackish; achenes pale yellow-brown or brownish, oblong, 4.5 mm. long, 1.3 mm. wide, somewhat warty, with a spinelike apical appendage.——Honshu (Etchu Prov., Kinki Distr. and westw.), Shikoku.

13. Taraxacum maruyamanum Kitam. OKI-TAMPOPO. Roots thickened; leaves many, oblanceolate, 18–28 cm. long, 3–4.5 cm. wide, glabrous on both sides, pinnately cleft, the terminal segment small, obtuse; scapes many, as long as to longer than the leaves, glabrescent, except pubescent at apex; heads 3.5–4 cm. across; involucres 16–18 mm. long at anthesis, 2–2.2 cm. long in fruit, the outer bracts 2-seriate, appressed, ovate-lanceolate, blackish on back, acuminate, corniculate or not so, the inner bracts oblong-lanceolate or lanceolate, green or with broad scarious margins, blackish toward the tip; achenes brownish, narrowly oblong, about 5 mm. long, 1 mm. wide, with a spinelike apical appendage.——Honshu (Oki Isl.).

14. Taraxacum sendaicum Kitam. *T. platycarpum* var. *sendaicum* (Kitam.) H. Koidz.——SENDAI-TAMPOPO. Leaves few, lanceolate-lingulate, about 15 cm. long, 2–2.5 cm. wide, acute, pubescent while young, usually pinnately lobed to cleft, rarely coarsely toothed; scapes many, longer than to as long as the leaves at anthesis, pubescent, villous under the head; heads about 3 cm. across; involucres about 13 mm. long, the outer bracts oblong, acute, relatively short-corniculate, the inner lanceolate, blackish at apex, small-corniculate.——Honshu (Rikuzen Prov.).

15. Taraxacum longeappendiculatum Nakai. *T. micranthum* Kitam.; *T. tokaiense* Kitam.; *T. candidum* H. Koidz.; *T. nakaii* H. Koidz.; *T. longeappendiculatum* var. *tokaiense* (Kitam.) Kitam.——HIRO-HA-TAMPOPO, TŌKAI-TAMPOPO. Leaves oblanceolate or spathulate-oblanceolate, toothed or pinnately lobed, pubescent while young, rounded, mucro-

nate; scapes loosely pubescent, villous under the head; heads about 3 cm. across; involucres 15–16 mm. long, the outer bracts erect, oblong to lanceolate, long-corniculate, the inner narrowly lanceolate; achenes oblong, slightly flattened, 3–4 mm. long, 1.2–1.4 mm. wide, pale yellow-brown and somewhat greenish, about 1.3 mm. wide with spinelike tubercles toward the tip, sparingly tubercled at base.——Honshu (Awa Prov., Tōkaidō Distr., s. Kinki and Chūgoku Distr.).——Forma **alboflavescens** Kitam. Flowers pale yellow.——Honshu (Mino Prov.).

16. Taraxacum elatum Kitam. *T. japonicum* Koidz., pro parte; *T. kisoense* H. Koidz.; *T. minoense* H. Koidz.; *T. yetchuense* H. Koidz.——SEITAKA-TAMPOPO. Leaves few, ascending, linear-oblanceolate, 25–35 cm. long, 5–8 cm. wide, deeply pinnatifid, the terminal segments small, acute, mucronate; scapes longer than to as long as the leaves, cobwebby while young, villous under the head; heads 3.5–4 cm. wide; involucres about 2 cm. long, the outer bracts appressed or erect, oblong or lanceolate, dark purplish toward the tip, corniculate or not so, the inner linear-lanceolate; achenes gray to brown, narrowly oblong, about 4 mm. long, with spinelike tubercles on upper part, warty-tubercled or nearly smooth on lower part.——Honshu (w. part of centr. distr., Oomi and Ise Prov.).

Var. **ibukiense** Kitam. IBUKI-TAMPOPO. Involucral bracts dark green, narrowly white-membranous on margin.——Honshu (Mount Ibuki in Oomi Prov.).

17. Taraxacum ceratolepis Kitam. KENSAKI-TAMPOPO. Leaves obovate-oblong, about 30 cm. long, 5–7.5 cm. wide, obtuse, pinnately cleft; scapes many, somewhat shorter than the leaves at anthesis, much elongate after anthesis; heads about 5 cm. across; involucres about 21 mm. long, the outer bracts somewhat spreading, oblong, prominently corniculate, the inner lanceolate, corniculate; achenes dark brown, oblong, 4–5 mm. long, slightly flattened, with spinelike tubercles at apex, slightly warty-tubercled toward the base.——Honshu (Tanba and Izumo Prov.).

18. Taraxacum platycarpum Dahlst. *T. officinale* var. *platycarpum* (Dahlst.) Nakai; *T. denticorne* H. Koidz.; *T. hitachiense* H. Koidz.; *T. luteopapposum* H. Koidz.; *T. tsurumachii* Kitam.; *T. platycarpum* var. *ecorniculatum* H. Koidz.——KANTŌ-TAMPOPO. Leaves oblanceolate, obtuse, rarely acute, pinnately incised or parted, 20–30 cm. long, 2.5–5 cm. wide; scapes as long as to shorter than the leaves at anthesis, villous under the head; heads 3.5–4.5 cm. across; involucres 15–18 mm. long, the outer bracts narrowly ovate to broadly lanceolate, erect, prominently corniculate, green, the inner linear-lanceolate; achenes oblong or fusiform, 4.5–5 mm. long, 1–1.2 mm. wide, yellow-brown with spreading spinelike tubercles.——Honshu (Kantō Distr., Kai and Suruga Prov.).

19. Taraxacum hondoense Nakai ex H. Koidz. *T. albociliatum* H. Koidz.; *T. brachyphyllum* H. Koidz.; *T. denticulatomarginatum* H. Koidz.; *T. filicinum* H. Koidz.; *T. fujisanense* H. Koidz.; *T. inugawense* H. Koidz.; *T. miyabei* H. Koidz.; *T. napifolium* H. Koidz.; *T. neorhodobasis* H. Koidz.; *T. patulum* H. Koidz.; *T. rhodobasis* H. Koidz.; *T. robustius* H. Koidz.; *T. sonchifolium* H. Koidz.; *T. tobae* H. Koidz.; *T. towadense* H. Koidz.; *T. venustum* H. Koidz.; *T. yuhkii* H. Koidz.; *T. tenuifolium* H. Koidz., non K. Koch ——EZO-TAMPOPO. Leaves rather numerous, oblanceolate, lanceolate, or lingulate, subacute, sometimes obtuse, toothed to pinnately parted, to 35 cm. long, 2–8 cm. wide; scapes villous under the head; heads 2–4 cm. wide; involucres to 2.5 cm.

long, the outer bracts erect or appressed, many, ovate to broadly so, obtuse, corniculate or not so, green, the inner lanceolate or linear-lanceolate, blackish on back; achenes slightly flattened, obovate-oblong, 4–5 mm. long, brown, with spinelike tubercles.——Hokkaido, Honshu (centr. and n. distr.).——Forma **alboflavescens** H. Koidz. Usujiro-tam-popo. Corollas pale yellow.——Forma **rubicundum** (Koidz.) H. Koidz. Beni-tampopo. Corollas reddish.

20. Taraxacum arakii Kitam. *T. albofimbriatum* H. Koidz.; *T. iyoense* H. Koidz.; *T. arakii* var. *ecorniculatum* H. Koidz.——Yamazato-tampopo. Leaves ascending, linear-lanceolate, to 30 cm. long, 4–5 cm. wide, obtuse, pinnately cleft; scapes many, shorter than to as long as the leaves at anthesis, cobwebby while young, villous under the head; involucres about 18 mm. long, the outer bracts appressed, oblong-ovate or ovate, caudate, corniculate or not so, scarious, reddish at the tip, densely woolly on margin, green on midrib, the inner linear-lanceolate, green, not corniculate, dark reddish at the tip; achenes pale yellow-brown or brownish, oblong-fusiform, about 5 mm. long, with spinelike tubercles on upper part and warty-tubercles on lower part.——Honshu (Kinki and Chūgoku Distr.), Shikoku.

21. Taraxacum albidum Dahlst. *T. officinale* var. *albiflorum* Makino; *T. albiflorum* (Makino) Koidz.; *T. hideoi* Nakai ex H. Koidz.——Shirobana-tampopo. Leaves lanceolate to oblanceolate, 20–30 cm. long, 3–5 cm. wide, obtuse or subacute, pinnately parted to cleft; scapes usually longer, rarely shorter than the leaves at anthesis, glabrous, white-pubescent under the head; heads 3.5–4.5 cm. across; involucres 17–20 mm. long at anthesis, the outer bracts narrowly ovate-oblong, ascending, prominently corniculate; achenes narrowly obovoid-oblong, pale yellow-brown, with spinelike tubercles on upper part and warty tubercles below except at the base, slightly flattened, about 4 mm. long.——Honshu (Kantō, Tōkaidō, Kinki, and Chūgoku Distr.), Shikoku, Kyushu.——Forma **sulphureum** (H. Koidz.) Kitam. *T. albidum* var. *sulphureum* H. Koidz.——Kibana-shiro-tampopo. Corollas pale yellow.——Honshu (Chūgoku Distr.), Kyushu.

22. Taraxacum shinanense H. Koidz. *T. denudatum* H. Koidz.; *T. nambuense* H. Koidz.; *T. shinanense* var. *nambuense* (H. Koidz.) H. Koidz.——Usugi-tampopo. Leaves tufted, spathulate-oblanceolate to linear-oblanceolate, pinnately lobed or parted, 10–30 cm. long, 2–8 cm. wide, obtuse, mucronate; scapes as long as to shorter than the leaves at anthesis; heads 3–4.5 cm. across; involucres 1.7–2 cm. long, the outer bracts appressed or slightly spreading, ovate or ovate-oblanceolate, small corniculate or not so, the inner narrowly lanceolate, usually not corniculate; achenes 5–6 mm. long, obovoid-oblong, with loose spinelike tubercles toward the tip, with warty tubercles toward the base.——Honshu (Shinano, Hitachi Prov. and n. distr.).

64. HIERACIUM L.   Miyama-kōzori-na Zoku

Perennial herbs; stems leafy; leaves alternate, cauline or radical, entire or toothed; heads solitary, or in loose panicles or corymbs, yellow to orange-yellow, rarely red; involucres tubular-campanulate, usually with blackish hairs, the bracts narrow, herbaceous, the inner nearly equal; receptacle flat, glabrous or hairy; corollas all lingulate, the tube slightly shorter to as long as the limb, truncate, 5-toothed at apex; achenes cylindric, smooth, 10- or 15-ribbed and terete, or 4- or 5-ribbed and angled, truncate at apex; pappus-bristles many, 1- or 2-seriate, firm, simple, persistent.——About 400 species, especially abundant in Europe, few in N. and S. America, Asia, and S. Africa.

1A. Stems 10–30 cm. long, few-leaved; cauline leaves, or some of them, oblanceolate, obute to subobtuse, glandular-hairy and with long spreading coarse hairs, mucronate-toothed; achenes about 2 mm. long. .... 1. *H. japonicum*
1B. Stems 40–80 cm. long, many-leaved; cauline leaves, or some of them, lanceolate to linear-lanceolate, acuminate, scabrous on margin and often on underside, entire or with few short teeth; achenes 3–3.5 mm. long. .... 2. *H. umbellatum* var. *japonicum*

1. Hieracium japonicum Fr. & Sav. Miyama-kōzori-na. Perennial rhizomatous herb; stems 10–30(–40) cm. long, few-leaved; radical leaves membranous, oblanceolate, 5–10(–15) cm. long, 1–2.5 cm. wide, obtuse to subobtuse, mucronate-toothed, glandular-hairy and with long-spreading coarse brownish hairs; lower cauline leaves similar to the radical, sessile, clasping at base, the upper leaves small; heads few, deep yellow, 1.5–2 cm. across; pedicels, upper part of stems, and back of involucral bracts with long-spreading coarse brown hairs and dense red-brown short hairs; involucres blackish, 6–8 mm. long; achenes about 2 mm. long; pappus pale brown.——July–Aug. Alpine grassy slopes; Honshu (centr. and n. distr.).

2. Hieracium umbellatum L. var. **japonicum** Hara.

H. umbellatum var. *serotinum* sensu auct. Japon., non DC.——Yanagi-tampopo. Perennial; stems erect, simple, rarely branched, firm, terete, 40–80 cm. long; cauline leaves many, the median rather thick, linear-lanceolate to lanceolate, 4–12 cm. long, 5–12 mm. wide, acuminate, sessile, scabrous on margin and often on underside, entire or with few short teeth; heads few, in terminal subumbellate corymbs, yellow, 2.5–3.5 cm. wide, the pedicels white-puberulent; involucres nearly glabrous, greenish black, 1–1.3 cm. long; achenes red-brown, 3–3.5 mm. long, brownish.——Aug.–Sept. Grassy slopes in mountains; Hokkaido, Honshu, Shikoku.——s. Kuriles, Sakhalin, and Korea. The typical phase occurs in Europe, Asia, N. Africa, and N. America.

65. HOLOLEION Kitam.   Sui-ran Zoku

Nearly glabrous stoloniferous perennials; leaves narrow, alternate; heads erect at anthesis; involucres terete, cylindric, the bracts herbaceous, the inner nearly equal, the outer gradually becoming shorter; receptacle flat, alveolate; corollas all lingulate, yellow, truncate, 5-toothed, the tube much shorter than the limb; achenes linear-oblong, 6–9 mm. long, 4-angled, truncate; pappus-bristles 1-seriate, nearly equal, persistent or deciduous, minutely scabrous.——About 3 species, in e. Asia.

1A. Outer involucral bracts lanceolate to narrowly deltoid, acute; achenes 7.5–9 mm. long; peduncles 3.5–10 cm. long. .... 1. *H. krameri*
1B. Outer involucral bracts ovate-lanceolate, obtuse; achenes 5.5–6 mm. long; peduncles 1.5–3(–6) cm. long. .... 2. *H. maximowiczii*

1. **Hololeion krameri** (Fr. & Sav.) Kitam. *Hieracium krameri* Fr. & Sav.——SUI-RAN. Nearly glabrous perennial with soft white ascending rhizomes and subterranean slender naked stolons; stems simple or branched, 50–100 cm. long, erect; radical leaves few, linear-lanceolate, 15–50 cm. long, 1–3 cm. wide, acuminate, loosely toothed or mucronate-toothed, glaucous beneath; cauline leaves similar, narrow, elongate, sessile, the upper filiform-linear, small; peduncles 3.5–10 cm. long; heads loosely arranged, 3–3.5 cm. across, yellow, 10–30-flowered; involucres about 12 mm. long, the outer bracts lanceolate to narrowly deltoid, acute; achenes linear, 4-angled, 7.5–9 mm. long, about 1 mm. wide; pappus about 6 mm. long, pale brown.——Oct.–Nov. Wet boggy places; Honshu (Tōkaidō, Kinki Distr. and westw.), Shikoku, Kyushu.

2. **Hololeion maximowiczii** Kitam. *Hieracium hololeion* Maxim.; *H. sparsum* subsp. *hololeion* (Maxim.) Zahn——CHŌSEN-SUI-RAN, ITO-SUI-RAN. Rhizomes with innovation-shoots; stems 50–100 cm long; radical leaves linear-lanceolate, acute with a mucro at apex; lower cauline leaves 15–40 cm. long, 5–30 mm. wide, entire, the median smaller, sessile; heads in loose corymbs, yellow, about 3 cm. across; involucres about 13 mm. long, the outer bracts ovate-lanceolate, obtuse; achenes 4-angled, 5.5–6 mm. long, about 1 mm. across; pappus about 7 mm. long, pale yellow.——Sept.–Oct. Wet places; Kyushu.——Korea, Manchuria, and Amur.

66. PRENANTHES L. FUKUŌ-SŌ ZOKU

Erect or sometimes scandent herbs; leaves alternate, sagittate-cordate, coarsely pinnatifid, the upper sessile and auriculate-clasping; heads usually in loose slender panicles, nodding at anthesis; involucres tubular, the inner bracts 1- or 2-seriate, equal, the outer few, small; receptacle flat, glabrous; corollas truncate, 5-toothed, the tube as long as to slightly shorter than the limb; achenes narrowly oblong, slightly flattened, 3- to 5-angled or terete, 10-ribbed, subtruncate, crowned with the persistent base of pappus; pappus-bristles 2- or 3-seriate, simple, deciduous.——About 30 species, in Europe, Asia, and N. America.

1A. Stems leafy only near base; leaves usually palmately 3- to 7-cleft to -parted, cordate; stems 40–60 cm. long, usually with long-spreading coarse glandular hairs; corollas blue-white; outer involucral bracts about ⅓ as long as the inner; achenes 3.5–4.5 mm. long. ... 1. *P. acerifolia*
1B. Stems leafy throughout; leaves often pinnately parted, sagittate; stems 80–150 cm. long, usually without glandular hairs; corollas pale yellow; outer involucral bracts about half as long as the inner; achenes 7–9 mm. long. ... 2. *P. tanakae*

1. **Prenanthes acerifolia** (Maxim.) Matsum. *Nabalus acerifolius* Maxim.——FUKUŌ-SŌ. Stoloniferous; stems 40–60 cm. long, leafy only near base, simple, long coarse glandular-hairy; radical leaves long-petiolate, ovate-cordate, about 6–10 cm. long and as wide, coarsely toothed and incised, glandular-hairy on underside, usually palmately 3–7-cleft to -parted, the petioles distinctly winged, the wings in lower cauline leaves broadly clasping the stems; panicles large, the heads about 1.5 cm. across; involucres gray-green, 1–1.2 cm. long, with long coarse hairs on back, the outer bracts ovate to narrowly oblong, about 1/3 as long as the inner; corollas blue-white; achenes 3.5–4.5 mm. long; pappus about 7 mm. long.——Aug.–Sept. Mountains; Honshu, Shikoku, Kyushu.——Forma **heterophylla** Matsum. & Koidz.——MARUBA-FUKUŌ-SŌ. Leaves undivided, rounded at base.——Honshu.——Forma **nipponica** (Fr. & Sav.) Matsum. & Koidz. *Nabalus nippon-icus* (Fr. & Sav.) Makino——FUKUŌ-NIGA-NA. Plants sparsely coarse-hirsute; involucral bracts smooth.——Honshu.

2. **Prenanthes tanakae** (Fr. & Sav.) Koidz. *Nabalus tanakae* Fr. & Sav. ex Tanaka & Ono; *P. ochroleuca* var. *tanakae* (Fr. & Sav.) Koidz.——Ō-NIGA-NA. Stoloniferous; stems 80–150 cm. long, leafy throughout, usually simple, smooth or sparsely coarse-hairy; median cauline leaves ovate-deltoid, sagittate, long winged-petiolate, shallowly incised or pinnately cleft to parted, the segments retrorse, 8–17 cm. long, 6–12 cm. wide, often with spreading coarse hairs on nerves beneath, the upper leaves narrowly lanceolate, smaller, pinnately few-cleft or entire, acuminate, sessile; panicles large, the heads 3.5–4 cm. across; corollas pale yellow; involucres nearly glabrous, the outer bracts lanceolate, about half as long as the inner; achenes 7–9 mm. long, linear; pappus as long as the achene.——Sept.–Oct. Honshu (Kinki Distr. and eastw.).

67. CREPIDIASTRUM Nakai AZE-TŌ-NA ZOKU

Glabrous glaucous suffrutescent herbs with stout simple or branched stems; leaves rosulate or cauline, alternate, entire or pinnately incised; flowering branches lateral, leafy; heads rather small, in corymbs, yellow or white; involucres narrow, the inner bracts equal, 1-seriate, the outer small, few; corollas all ligulate, truncate, 5-toothed; achenes slightly flattened, with 10 longitudinal ribs; pappus bristles 1-seriate, slightly scabrous, deciduous.——About 10 species, in e. Asia.

1A. Leaves spathulate, 1–2 cm. wide, rounded at apex, shallowly toothed. ... 1. *C. keiskeanum*
1B. Leaves not spathulate, 2–7 cm. wide.
  2A. Radical leaves ovate to narrowly so, 2–4 cm. wide, usually obtuse, the petioles 2–3 mm. wide; cauline leaves obtuse, 1–2 cm. wide, rather thin. ... 2. *C. lanceolatum*
  2B. Radical leaves obovate or obovate-elliptic, 4.5–7 cm. wide, rounded at apex, the petioles 5–7 mm. wide; cauline leaves rounded or sometimes subacute, 2–4 cm. wide, rather thick. ... 3. *C. platyphyllum*

1. **Crepidiastrum keiskeanum** (Maxim.) Nakai. *Crepis keiskeana* Maxim.; *Lactuca keiskeana* (Maxim.) Makino; *Ixeris keiskeana* (Maxim.) Stebbins——AZE-TŌ-NA. Caudex stout, short, rhizomelike, with a tuft of rosulate leaves at apex; branches solitary to several, lateral, decumbent; radical leaves spathulate, 3–10 cm. long, 1–2 cm. wide, gradually rounded at tip, shallowly toothed; heads many, rather densely arranged, about 1.5 cm. across, yellow; involucres about 1 cm. long.——Sept.–Dec. Rocky places near the sea; Honshu (Izu Prov. and westw.), Shikoku, Kyushu.——Forma **pinnatilobum** Hisauchi. SOTETSUBA-AZE-TŌ-NA. Leaves pinnately lobed to cleft.

2. **Crepidiastrum lanceolatum** (Houtt.) Nakai. *Pre-*

nanthes lanceolata Houtt.; *P. integra* Thunb.; *Youngia lanceolata* (Houtt.) DC.; *Crepis tanegana* Miq.; *Lactuca lanceolata* (Houtt.) Makino; *Crepidiastrum integrum* (Thunb.) T. Tanaka; *Ixeris lanceolata* (Houtt.) Stebbins; *Crepis integra* (Thunb.) Miq.——HOSOBA-WADAN.   Caudex stout, short, rhizomelike, with a tuft of rosulate leaves at apex; inflorescence branches lateral, decumbent; radical leaves ovate to narrowly so, rarely elliptic, 8–18 cm. long, 2–4 cm. wide, obtuse, rarely rounded at apex, narrowed at base, entire, the petioles 2–3 mm. wide, the cauline leaves obtuse, 1–2 cm. wide, clasping at base; flowering branches 10–40 cm. long; heads many, yellow, about 1.5 cm. across; involucres 8–9 mm. long.——Oct.–Jan. Rocky places near the sea; Honshu (w. distr.), Shikoku, Kyushu.——s. Korea, Ryukyus, and Formosa——Forma **pinnatilobum** (Maxim.) Nakai. *Crepis integra* var. *pinnatiloba* Maxim.; *Lactuca quercus* Lév. & Van't.; *Crepidiasterum quercus* (Lév. & Van't.) Nakai——HAMA-NAREN, SOTETSU-NA.   Leaves pinnately lobed.——Occurs with the typical phase.

3. Crepidiastrum platyphyllum (Fr. & Sav.) Kitam. *Crepis integra* var. *platyphylla* Fr. & Sav.; *Lactuca platyphylla* (Fr. & Sav.) Makino; *Crepidiastrum lanceolatum* var. *latifolium* Nakai; *Ixeris lanceolata* subsp. *platyphylla* (Fr. & Sav.) Stebbins; *C. lanceolatum* var. (?) *platyphyllum* (Fr. & Sav.) Tuyama——WADAN.   Caudex stout, short, erect, rhizomelike, with a tuft of rosulate leaves at apex; inflorescence branches axillary, decumbent; scapes 20–40 cm. long; radical leaves rather thick, obovate or obovate-elliptic, 8–18 cm. long, 4.5–7 cm. wide, rounded at apex, narrowed at base, entire, the petioles 5–7 mm. wide; cauline leaves broadly obovate-elliptic, 2–4 cm. wide, usually rounded to subacute, the uppermost subtending the inflorescence, clasping; heads many, very congested, yellow, about 1 cm. across; involucres 6–7 mm. long.——Sept.–Nov. Seashores; Honshu (s. Kantō and Tōkaidō Distr.).——s. Korea.

68. CREPIS L.   FUTA-MATA-TAMPOPO ZOKU

Glabrous to coarsely hirsute perennials, sometimes annuals or biennials; stems leafy or scapiform; leaves radical and/or cauline, alternate; heads peduncled, solitary or in panicles; involucres tubular, campanulate, or inflated at base in fruit, the inner bracts nearly equal, 1-seriate, the outer usually much shorter, few-seriate; receptacle flat or depressed, glabrous or hairy; corollas all lingulate, truncate, 5-toothed; achenes many-ribbed, terete, narrowed above or beaked; pappus-bristles many, slender, simple.—— About 200 species, in Eurasia, Africa, and N. America.

1A. Radical leaves pinnatifid or sinuately shallow-lobed to runcinate; stems pubescent toward summit; heads 1 or 2. .. 1. *C. hokkaidoensis*
1B. Radical leaves spathulate, remotely repand-denticulate, mucronate-toothed; stems glabrous; heads 3 to 8. .... 2. *C. gymnopus*

1. Crepis hokkaidoensis Babcock. *C. burejensis* sensu auct. Japon., non F. Schmidt; *C. miyabei* Tatew. & Kitam.—— FUTA-MATA-TAMPOPO.   Rhizomes vertical or oblique; stems 5–20 cm. long, erect, rather stout, simple or rarely branched, with brown multicellular hairs and short white puberulence; radical leaves pinnatifid or sinuately shallow-lobed to runcinate, 5–15 cm. long, 1–3 cm. wide, acute, acuminate, or obtuse, gradually narrowed to a narrowly winged petiole, slightly pubescent on both sides; cauline leaves few, sessile, semiclasping; upper leaves linear and small; heads 1 or 2 on long peduncles, yellow; involucres 1–1.5 cm. long, with brown, often glandular, long hairs; achenes black-brown, about 1 cm. long, 1 mm. across, gradually narrowed toward apex, about 20-ribbed, puberulent; pappus 7.5–8 mm. long, white.——July–Aug. Alpine slopes; Hokkaido.——s. Kuriles and Sakhalin.

2. Crepis gymnopus Koidz. *Youngia gymnopus* (Koidz.) Hara——EZO-TAKANE-NIGA-NA.   Rhizomes vertical or oblique; stems scapiform, 20–45 cm. long, rather slender, loosely woolly pubescent while young, becoming glabrous; leaves all radical, spathulate, 8–15 cm. long, 1.2–3 cm. wide, rounded at apex, remotely repand-denticulate, mucronate-toothed, glabrous or loosely short hairy on upper side; scalelike leaves of stems absent or single; bracts linear-lanceolate; heads 3 to 8, the pedicels 1.5–6 cm. long, erect, yellow; involucres 1–1.2 cm. long, narrow; achenes brown, 4.5–5 mm. long, about 0.6 mm. across, 20-ribbed, smooth; pappus snow-white, about 5 mm. long.——Alpine regions; Hokkaido.

69. IXERIS Cass.   NIGA-NA ZOKU

Low, usually glaucous annuals, biennials, or perennials; leaves radical and cauline, alternate; heads few to many, in corymbs; involucres tubular, the bracts few-seriate, glabrous, herbaceous, the inner much larger than the outer, the outer ones often calyculate; receptacles flat, naked; corollas lingulate, truncate, 5-toothed, yellow, white, or purplish achenes somewhat flattened, narrowly oblong, equally 10-ribbed, long-beaked; pappus-bristles many, slender, scaberulous, nearly equal.——About 20 species, in e. Asia.

1A. Stolons much elongate; stems scapose or scapiform.
  2A. Leaves palmately 3- to 5-cleft, -parted or -divided; outer involucral bracts rather prominent, about 2/5–1/2 as long as the inner; beak of achenes about 1 mm. long; sandy beaches. .... 1. *I. repens*
  2B. Leaves spathulate, entire or shallowly toothed; outer involucral bracts few, short, about ¼–⅓ as long as the inner; beak of achenes longer.
    3A. Involucres 12–14 mm. long; beak of achene 2.5 mm. long. .... 2. *I. japonica*
    3B. Involucres about 8–10 mm. long; beak of achene 3 mm. long. .... 3. *I. stolonifera*
1B. Stolons absent or short; stems usually distinctly leafy (except in nos. 9 and 10).
  4A. Heads nodding at maturity, in false umbels; cauline leaves sagittate; achenes deeply sulcate, acutely winged; biennial.
                                      4. *I. polycephala*
  4B. Heads erect at maturity, in loose corymbs; cauline leaves auriculate; achenes shallowly grooved, obtusely winged; perennials.

5A. Beak 1/5–1/3 as long as the body of achene.
  6A. Involucres 4–5 mm. long at anthesis; achenes about 2.5 mm. long. . . . . 5. *I. makinoana*
  6B. Involucres 6–9 mm. long at anthesis; achenes 3–5 mm. long.
    7A. Radical leaves broadly linear to oblong-lanceolate, rarely nearly oblong, acute to acuminate; heads in loose corymbs; involucres 7–9 mm. long. . . . . 6. *I. dentata*
    7B. Radical leaves orbicular to broadly ovate, rounded; heads in dense corymbs; involucres about 6 mm. long. . . 7. *I. nipponica*
5B. Beak nearly as long as the body of achene.
  8A. Involucres about 5 mm. long; pappus brown. . . . . 8. *I. laevigata* var. *lanceolata*
  8B. Involucres 9–11 mm. long; pappus white.
    9A. Radical leaves many, broadly linear, usually entire; outer involucral bracts somewhat unequal, often ascending at tip.
      9. *I. tamagawaensis*
    9B. Radical leaves rather few, spathulate, mucronate-toothed or incised; outer involucral bracts nearly equal, appressed.
      10. *I. chinensis* var. *strigosa*

1. Ixeris repens (L.) A. Gray. *Prenanthes repens* L.; *Lactuca repens* (L.) Benth.——HAMA-NIGA-NA. Stems rather slender, much elongate below ground, remotely leafy; leaves rather thick, long-petiolate, 3- to 5-angled, cordate, palmately 3- to 5-cleft, -parted, or -divided, 3–5 cm. long and as wide, the segments broadly elliptic, rounded at apex, 2- to 3-lobed and obsoletely toothed; scapes axillary from radical leaves, 3–15 cm. long, 2- to 5-headed, the lowest bracts sometimes subfoliaceous; heads about 3 cm. across, rather many-flowered; involucres 11–14 mm. long, the outer involucral bracts rather prominent, about 2/5–1/2 as long as the inner, the inner bracts 6–8; achenes about 5 mm. long; pappus white, 5–6 mm. long.——May–July. Sandy beaches; Hokkaido, Honshu, Shikoku, Kyushu; common.——Formosa, Ryukyus, China, Korea, Manchuria, and Kamchatka.

2. Ixeris japonica (Burm.) Nakai. *Lapsana japonica* Burm.; *Prenanthes debilis* Thunb.; *I. debilis* (Thunb.) A. Gray; *Lactuca debilis* (Thunb.) Benth.; *I. japonica* formae *integra* and *sinuata* of Nakai——JI-SHIBARI, Ō-JI-SHIBARI, TSURU-NIGA-NA. Perennial herb with slender, creeping, branched, leafy stems; leaves oblanceolate to spathulate-elliptic, 6–20 cm. long inclusive of the petioles, 1.5–3 cm. wide, obtuse to rounded, entire or loosely toothed on the lower half, long-petiolate; scapes 10–30 cm. long, sometimes with a single leaf; heads 1–5, 2.5–3 cm. across, rather many-flowered, pedicelled, yellow; involucres 12–14 mm. long, the inner bracts about 8; achenes about 4 mm. long, narrowly fusiform, deeply sulcate, acutely winged, the beak about 2.5 mm. long; pappus white, about 7 mm. long.——May. Cultivated fields and waste grounds in lowlands; Hokkaido, Honshu, Shikoku, Kyushu; common.——Ryukyus and Korea.

3. Ixeris stolonifera A. Gray. *Lactuca stolonifera* (A. Gray) Benth.——IWA-NIGA-NA, JI-SHIBARI, HAI-JI-SHIBARI. Prostrate, very slender, branched perennial; stems remotely leafy; leaves thin, ovate-orbicular, broadly ovate to broadly elliptic, 2–8 cm. long inclusive of the petioles, 2–8 cm. wide, rounded at both ends, entire or loosely toothed on margin; scapes slender, 8–15 cm. long; heads 1–3, 2–2.5 cm. across, yellow; involucres 8–10 mm. long, the inner bracts usually 9 or 10; achenes narrowly fusiform, about 3 mm. long, rather narrowly winged, the beak as long as the body of achene; pappus white, about 5 mm. long.——May–June. Cultivated fields, waste grounds, and open mountain slopes; Hokkaido, Honshu, Shikoku, Kyushu; common.——Forma **sinuata** (Makino) Ohwi. *Lactuca stolonifera* var. *sinuata* Makino——KIKUBA-JI-SHIBARI. Leaves pinnately incised.——Forma **capillaris** (Nakai) Ohwi. *Lactuca capillaris* (Nakai) Makino & Nemoto; *I. stolonifera* subsp. *capillaris* (Nakai) Kitam.——MIYAMA-IWA-NIGA-NA. Slender alpine phase.

Ixeris × nikoensis Nakai. Alleged hybrid of *I. stolonifera* × *I. tamagawaensis*.

4. Ixeris polycephala Cass. *Lactuca polycephala* (Cass.) Benth. & Hook. f.; *L. matsumurae* Makino; *I. matsumurae* (Makino) Nakai; *L. biauriculata* Lév. & Van't.——NO-NIGA-NA. Glaucous biennial; stems 15–40 cm. long, erect, often branched at base; radical leaves linear-lanceolate, 10–25 cm. long, 7–15 mm. wide, acuminate, narrowed to a petiolelike base, entire; cauline leaves few, 7–15 cm. long, sessile, sagittate; inflorescence umbelliform; heads about 8 mm. across, pedicelled, yellow, nodding at maturity; involucres 7–8 mm. long in fruit, the inner bracts 8–9, narrowly lanceolate; achenes fusiform, obtuse at base, 3–3.5 mm. long, deeply sulcate, acutely winged, the beak about 1.5 mm. long; pappus snow-white, about 5 mm. long.——May. Wet places around paddy fields; Honshu, Shikoku, Kyushu; common.——Ryukyus, Formosa, Korea, China, India to the Caucasus.——Forma **dissecta** (Makino) Ohwi. *Lactuca matsumurae* var. *dissecta* Makino——KIKUBA-NO-NIGA-NA. Leaves pinnately dissected.

Ixeris × sekimotoi Kitam. NO-JI-SHIBARI. Hybrid of *I. japonica* × *I. polycephala*.

5. Ixeris makinoana (Kitam.) Kitam. *Lactuca thunbergii* var. *angustifolia* Makino; *L. dentata* var. *angustifolia* (Makino) Makino; *L. makinoana* Kitam.; *I. dentata* var. *angustifolia* (Makino) Nakai——HOSOBA-NIGA-NA. Perennial herb; stems erect, 30–40 cm. long, branched at the top, loosely few-leaved; radical leaves linear or linear-oblanceolate, 6–13 cm. long, 3–7 mm. wide, acuminate at both ends, entire; cauline leaves simulating the radical, sessile, suberect; heads many, loosely arranged, yellow, 6–7 mm. across; involucres 5–6 mm. long in fruit, the inner bracts about 5, broadly linear; achenes about 2.5 mm. long, lanceolate, rather narrowly ribbed, the beak very short; pappus 2.5–3 mm. long, brown.——May–June. Honshu (Kantō Distr. and westw.), Shikoku, Kyushu.

6. Ixeris dentata (Thunb.) Nakai. *Prenanthes dentata* Thunb.; *I. thunbergii* A. Gray; *Lactuca thunbergii* (A. Gray) Maxim.; *L. dentata* (Thunb.) Robins.——NIGA-NA. Perennial herb with short rhizomes, rarely with short stolons; stems erect, 20–50 cm. long, rather slender, few-leaved; radical leaves broadly linear to oblong-lanceolate, rarely oblong, long-petiolate, the blades 3–10 cm. long, 5–30 mm. wide, acute to acuminate, rarely rounded and mucronate, entire or remotely mucronate-toothed, rarely shallowly incised; cauline leaves rather short, sessile, auriculate-clasping; heads in loose corymbs, about 15 mm. across, yellow or rarely white, usually 5-flowered; involucres 7–9 mm. long, the inner bracts 5–6; achenes 3–3.5 mm. long, beak less than 1 mm. long; pappus brown, about 4 mm. long.——May–July. Sunny places in lowlands and low mountains; Hokkaido, Honshu, Shikoku, Kyushu; very common.——s. Kuriles, Ryukyus, and Korea.

Var. **albiflora** (Makino) Nakai. *Lactuca thunbergii* var. *albiflora* Makino; *I. dentata* var. *amplifolia* forma *leucantha* Hara——SHIROBANA-HANA-NIGA-NA. Plant somewhat larger;

leaves slightly broader; florets 7–8, the corollas white.——
Mountains; Hokkaido, Honshu.——Forma **amplifolia**
(Kitam.) Hiyama. *I. dentata* var. *amplifolia* Kitam.——
HANA-NIGA-NA.　　Corollas yellow; more common than the
white-flowered phase.

Var. **alpicola** (Takeda) Ohwi. *Lactuca thunbergii* lusus
alpicola Takeda; *L. dentata* var. *flaviflora* subvar. *alpicola*
(Takeda) Makino; *L. dentata* var. *alpicola* (Takeda) Makino;
I. alpicola (Takeda) Nakai——TAKANE-NIGA-NA.　　Stems low;
leaves narrower, shorter; heads fewer, about 1 cm. long, the
florets 6–9, yellow; achenes slightly longer.——Rocky alpine
slopes; Hokkaido, Honshu, Shikoku, Kyushu (Yakushima).

Var. **kimurana** (Kitam.) Ohwi. *I. kimurana* Kitam.; *I.
dentata* subsp. *kimurana* (Kitam.) Kitam.——KUMOMA-NIGA-
NA.　　Alpine phase intermediate between var. *alpicola* and
var. *albiflora* forma *amplifolia*.——Forma **albescens** Kitam.
SHIROBANA-KUMOMA-NIGA-NA.　　Flowers white.——Alpine.

Var. **stolonifera** (Kitam.) Nemoto. *I. albiflora* A. Gray;
Lactuca albiflora (A. Gray) Maxim.; *L. dentata* var. *stolonifera* Kitam.——HAI-NIGA-NA.　　Stems creeping; florets 5 in
a head.——Honshu, Shikoku, Kyushu.

Ixeris musashiensis Makino & Hisauchi. TAKASAGO-NIGA-
NA.　　Alleged hybrid of *I. dentata* and an undetermined second parent.

7. Ixeris nipponica Nakai. *Lactuca nipponica* Nakai ex
Makino & Nemoto——ISO-NIGA-NA.　　Perennial (?) with
short rhizomes; stems 10–20 cm. long; radical leaves orbicular
or broadly ovate, sometimes obovate, rounded at apex, abruptly petiolate, mucronate-toothed, the blades 3–5 cm. long,
about 3 cm. wide; cauline leaves few, broadly ovate or orbicular-cordate, 3–8 cm. long, about 3 cm. wide, auriculate-clasping; heads about 1 cm. across, in dense interrupted corymbs,
yellow; involucres about 6 mm. long, the inner bracts 6–8.
——June. Sandy beaches; Honshu (Echigo Prov.).

8. Ixeris laevigata (Bl.) Schultz-Bip. var. **lanceolata**
(Makino) Kitam. *Lactuca oldhamii* Maxim.; *L. dentata* var.
lanceolata Makino; *L. stenophylla* Makino; *I. oldhamii*
(Maxim.) Kitam.——ATSUBA-NIGA-NA.　　Perennial herb;

flowering stems 20–40 cm. long, few-leaved; radical leaves oblong to lanceolate, blades 6–12 cm. long, 1–2 cm. wide, acuminate, mucronate-toothed, acute at base, petiolate; cauline
leaves gradually narrowed below to a petiolelike base; inflorescence
many-headed; heads 8–10 mm. across, about 10-flowered; involucres about 5 mm. long, the bracts about 8; achenes narrowly lanceolate, 4–5 mm. long inclusive of the beak; pappus
about 3 mm. long, brown.——Kyushu (Oosumi and Satsuma
Prov.).——Ryukyus and Formosa. The typical phase occurs
in the Philippines, s. China, Indochina, and Malaysia.

9. Ixeris tamagawaensis (Makino) Kitam. *Lactuca
tamagawaensis* Makino; *L. versicolor* var. *arenicola* Makino;
I. chinensis subsp. *arenicola* (Makino) Kitam.——KAWARA-
NIGA-NA.　　Perennial herb with tufted rhizomes; stems erect,
15–30 cm. long, naked or 1-leaved; radical leaves many,
broadly linear, 8–15 cm. long, 3–5 mm. wide, acute to acuminate, usually entire or slightly mucronate-toothed, narrowed to
a short petiolelike base; heads few to rather many, pedicelled,
15–20 mm. across, pale yellow, 25- to 30-flowered; involucres
9–10 mm. long, the outer bracts somewhat unequal, the inner
about 10; achenes lanceolate, about 6 mm. long inclusive of the
beak nearly as long as the body; pappus white, 5–6 mm. long.
——June–Aug. Waste grounds along rivers; Honshu, Kyushu.——Korea and Formosa.

10. Ixeris chinensis (Thunb.) Nakai var. **strigosa** (Lév.
& Van't.) Ohwi. *Prenanthes chinensis* Thunb., pro parte;
Lactuca strigosa Lév. & Van't.——TAKASAGO-SŌ.　　Perennial
herb; flowering stems erect, 20–40 cm. long, naked or 1-leaved;
radical leaves rather few, spathulate, 7–15 cm. long, 5–15 mm.
wide, acute to acuminate, narrowed to a rather broad petiolelike base, sparsely mucronate-toothed or incised; cauline leaves
clasping but scarcely auriculate; heads rather many, white to
purplish, about 20-flowered, 2–2.5 cm. across; involucres 9–11
mm. long, the outer bracts nearly equal, the inner about 8;
achenes narrowly lanceolate, about 6 mm. long inclusive of
the beak slightly shorter than the body; pappus white, about
6 mm. long.——Apr.–June. Honshu, Shikoku, Kyushu.——
Korea. The typical phase occurs in China and e. Siberia.

70. LACTUCA L.　AKI-NO-NOGESHI ZOKU

Glabrous or sometimes slightly hairy annual, biennial, or perennial herbs; stems usually tall; leaves radical and cauline, alternate, entire or coarsely toothed, sometimes pinnately cleft; cauline leaves usually auriculate-clasping; heads in panicles, sessile or
pedicelled; involucres tubular, the bracts few-seriate, the outer smaller; receptacle flat, glabrous; corollas ligulate, truncate, 5-
toothed, the tube half to as long as the limb; achenes ovate or lanceolate, distinctly flattened, narrowed at both ends, few-ribbed
on both sides, the beak with a ring at apex; pappus-bristles many-seriate, simple, persistent or individually deciduous.——About
70 species, in temperate and warmer regions of the N. Hemisphere.

1A. Achenes barely flattened, narrowly oblong or narrowly fusiform; perennial herbs with purplish corollas.
  2A. Heads about 1 cm. across; inner involucral bracts 7 or 8, the outer bracts much shorter than the inner, caliculelike; leaves pinnately
    divided; inflorescence pyramidal, paniculate. . . . . 1. *L. sororia*
  2B. Heads 3–3.5 cm. across; inner involucral bracts about 15, the outer bracts unequal; leaves entire or incised; inflorescence flat-topped,
    corymbiform. . . . . 2. *L. sibirica*
1B. Achenes distinctly flattened, elliptic; annual or biennial herbs with yellowish corollas.
  3A. Petioles scarcely or not clasping.
    4A. Achenes elliptic or obovate-elliptic, about 5 mm. long, 2.5 mm. wide, 1-ribbed on each side; involucral bracts about 2.5 mm.
      wide. . . . . 3. *L. indica*
    4B. Achenes obovate-oblong, about 4 mm. long, 1.5 mm. wide, 3-ribbed on each side; involucral bracts about 1.5 mm. wide.
      . . . . 4. *L. raddeana*
  3B. Petioles distinctly clasping, except the uppermost ones. . . . . 5. *L. triangulata*

1. Lactuca sororia Miq.　*L. polypodiifolia* Franch.;
Mycelis sororia (Miq.) Nakai; *M. sororia* var. *nudipes* Migo;
L. sororia var. *nudipes* (Migo) Kitam.; *L. sororia* var. *glabra*

Kitam.——MURASAKI-NIGA-NA.　　Perennial short rhizomatous
herb, sometimes with short stolons; stems erect, usually simple, hollow, 70–120 cm. long, glabrous; cauline leaves petiolate,

10–20 cm. long, 5–10 cm. wide, pinnately divided, the terminal segments usually deltoid, acuminate, mucronate-toothed, the petioles wingless, upper leaves membranous, gradually smaller, lanceolate, broadly winged and petiolelike at base, nearly glabrous; panicle 15–30 cm. long, glabrous; heads about 1 cm. across, rather few-flowered; involucres 1–1.2 cm. long, the inner bracts 7–8, the outer ones much shorter than the inner; achenes lanceolate, nearly beakless, 3–3.5 mm. long, weakly ribbed; pappus white, about 6 mm. long.——June–Aug. Honshu, Shikoku, Kyushu.

2. **Lactuca sibirica** (L.) Benth. *Sonchus sibiricus* L.; *Mulgedium sibiricum* (L.) Less.——EZO-MURASAKI-NIGA-NA. Glabrous perennial herb; stems erect, usually simple, 60–100 cm. long; leaves lanceolate, 7–12 cm. long, 1–2.5 cm. wide, acuminate, mucronate-toothed, sometimes incised, slightly glaucous beneath, sessile; inflorescence a corymb; heads few to rather many, 3–3.5 cm. across, blue-purple; involucres 12–15 mm. long, the inner bracts about 15, the outer ones unequal; achenes lanceolate, scarcely beaked, ribbed, smooth; pappus white, about 1 cm. long.——July–Aug. Hokkaido.——Kuriles, Sakhalin, n. Korea, Siberia, and Europe.

3. **Lactuca indica** L. var. **laciniata** (O. Kuntze) Hara. *Prenanthes laciniata* Houtt.; *P. squarrosa* Thunb.; *L. laciniata* (Houtt.) Makino; *L. squarrosa* var. *laciniata* O. Kuntze——AKI-NO-NOGESHI. Large glabrous annual or biennial; stems erect, stout, usually simple or branched toward the top, 80–150 cm. long; leaves lanceolate or narrowly oblong in outline, 10–30 cm. long, acuminate, usually pinnately incised or parted, glaucous beneath; upper leaves usually entire, small; inflorescence a narrow cylindrical panicle 20–40 cm. long; heads about 2 cm. across, pale yellow; involucres 12–15 mm. long, the inner bracts about 8; achenes about 5 mm. long inclusive of a very short beak, 2.5 mm. wide, 1-ribbed on each side; pappus white, 7–8 mm. long.——Sept.–Nov. Grassy places in lowlands; Hokkaido, Honshu, Shikoku, Kyushu; common.——Korea, Ryukyus, and Formosa.——Forma **indivisa** (Makino)

Hara. *L. laciniata* forma *indivisa* Makino——HOSOBA-AKI-NO-NOGESHI. Leaves undivided, narrowly to broadly lanceolate. ——Occurs with the typical phase.

Var. **dracoglossa** (Makino) Kitam. *L. dracoglossa* Makino——RYŪZETSU-SAI. Leaves large, entire.——Cultivated.

4. **Lactuca raddeana** Maxim. var. **raddeana**. *L. alliariaefolia* Lév. & Van't.——CHŌSEN-YAMA-NIGA-NA. Achenes 2–3 mm. long, 4- or 5-ribbed on both sides; florets 9–14.—— Kyushu.——e. Asia.

Var. **elata** (Hemsl.) Kitam. *L. raddeana* sensu auct. Japon., non Maxim.; *L. elata* Hemsl.——YAMA-NIGA-NA. Annual or biennial herb; stems erect, terete, usually simple, 60–80 cm. long, sometimes with multicellular crisped hairs; leaves membranous, ovate or deltoid-ovate, glaucescent beneath; lower leaves often pinnately divided, with a long wingless petiole; median leaves truncate, with a broadly winged petiole, scarcely clasping; upper leaves smaller, broadly rhombic-lanceolate, gradually narrowed at base, sometimes sparingly hairy on upper surface; inflorescence a cylindrical panicle, 20–40 cm. long; heads narrow, about 1 cm. across; florets 8–10, pale yellow; involucres about 1 cm. long, the inner bracts about 5; achenes obovate-oblong, about 4 mm. long inclusive of the very short beak, 1.5 mm. wide, 3-ribbed on each side; pappus white, about 6 mm. long.——Aug.–Sept. Mountains; Hokkaido, Honshu, Shikoku, Kyushu.——China and Indochina.

Lactuca × aogashimaensis Kitam. YAMA-AKI-NO-NOGESHI. Hybrid of *L. indica* × *L. raddeana*.

5. **Lactuca triangulata** Maxim. MIYAMA-AKI-NO-NOGESHI. Glabrous biennial herb; stems erect, up to 1 m. long, terete, simple or sometimes branched, rather loosely leafy; leaves thin, deltoid or deltoid-cordate, acuminate, irregularly mucronate-toothed, whitish beneath, the petioles winged, clasping; upper leaves smaller and narrower, the petioles not clasping; inflorescence rather many-headed.——Aug. Hokkaido, Honshu (mountains of centr. distr.).——Korea, Manchuria, and Sakhalin (var.).

71. SONCHUS L. HACHIJŌ-NA ZOKU

Annual, biennial, or perennial herbs sometimes woody at base; leaves alternate, usually auriculate-clasping, often with callose hairs or spinose on margin; involucres ovoid or campanulate, the bracts many-seriate; receptacle flat, glabrous; corollas lingulate, yellow, truncate, 5-toothed; achenes slightly flattened, 10- to 20-ribbed, slightly narrowed at both ends, beakless, glabrous; pappus-bristles many-seriate, very slender, snow-white, connate at base into a ring, deciduous as a whole.——About 45 species, indigenous of the Old World, few widely distributed and naturalizing.

1A. Perennial herbs with long rhizomes; heads 4–5 cm. across. .... 1. *S. brachyotis*
1B. Annual or biennial herbs without rhizomes; heads about 2 cm. across.
  2A. Cauline leaves irregularly spine-toothed; achenes scaberulous, longitudinally striate. .... 2. *S. oleraceus*
  2B. Cauline leaves rather densely spine-toothed; achenes nearly smooth, 3-ribbed on each side. .... 3. *S. asper*

1. **Sonchus brachyotis** DC. *S. shzucinianus* Turcz. ex Trautv.; *S. arvensis* var. *uliginosus* sensu auct. Japon., non Trautv.——HACHIJŌ-NA. Slightly glaucous long-creeping rhizomatous perennial herb; rhizomes with short white woolly hairs on upper part while young; stems erect, 30–80 cm. long, usually simple; cauline leaves rather numerous, oblong-lanceolate to narrowly oblong, 10–20 cm. long, 2–5 cm. wide, obtuse, glaucous beneath, obtusely rounded and auriculate-clasping at base, often incised or pinnately lobed; heads few, 4–5 cm. across, subumbellate, yellow; involucres 1.5–2 cm. long, the outer bracts ovate-deltoid; achenes about 3 mm. long, slightly scabrous, with few longitudinal ribs on both sides; pappus about 12 mm. long.——Aug.–Oct. Sunny grassy places and cultivated fields; Hokkaido, Honshu, Shikoku, Kyushu.——

Kuriles, Sakhalin, Korea, China, Siberia, and Altai Mts.

2. **Sonchus oleraceus** L. NOGESHI, HARU-NO-NOGESHI, KESHI-AZAMI. Nearly glabrous annual or biennial; stems stout, 50–100 cm. long, often branched, hollow; lower leaves oblong-lanceolate, 15–50 cm. long, 5–8 cm. wide, winged-petiolate, pinnately parted to divided, irregularly spine-toothed; median leaves sessile, auriculate-clasping; heads about 2 cm. across, umbelliform, yellow, the peduncles glandular-hairy; involucres 1.2–1.5 cm. long; achenes narrowly obovate, 2.5–3 mm. long, with scaberulous ribs; pappus about 6 mm. long.——May–Aug. Waste grounds and roadsides in lowlands; Hokkaido, Honshu, Shikoku, Kyushu; common.—— Sakhalin, Korea, Formosa, Ryukyus, Europe, and N. America.

3. **Sonchus asper** (L.) Hill. *S. oleraceus* var. *asper* L.

——Oni-nogeshi.   Nearly glabrous annual or biennial herb; stems branched, 50–100 cm. long, hollow; leaves ovate-oblong, large, often incised or pinnately parted, rather closely spine-toothed; upper leaves dilated at base, auriculate; heads about 2 cm. across, umbelliform, yellow; involucres 1.2–1.3 cm. long; achenes narrowly obovate, flattened, smooth or nearly so, with 3 ribs or nerves on each side; pappus 7–8 mm. long.——May–Oct.   European weed, naturalized in our area.

72. YOUNGIA Cass.   Oni-tabira-ko Zoku

Pubescent or glabrous annuals, biennials, or perennials; stems usually much branched; leaves alternate, usually pinnately divided, often prominently dilated at the base and clasping; heads small, in corymbs or panicles, pedicelled; involucres tubular-campanulate, the inner bracts 1-seriate; corollas ligulate, yellow, truncate, 5-toothed; achenes oblong-linear, slightly flattened, many-ribbed, beakless or short-beaked; pappus-bristles 1-seriate, many, persistent or the bristles separately deciduous.——About 50 species, in e. Asia, India, and Australia.

1A. Plant pubescent; inflorescence a corymbose-panicle; achenes beakless; pappus-bristles persistent. .... 1. Y. japonica
1B. Plant glabrous; inflorescence a terminal or axillary corymb; achenes short-beaked; pappus-bristles separately deciduous from the basal ring.
  2A. Leaves, at least the upper ones, auriculate-clasping, mucronate-toothed to pinnately divided.
    3A. Florets 10–13; leaves usually entire, rarely pinnately divided, the upper with large rounded auricles; inner involucral bracts 7 or 8.
                                                      2. Y. denticulata
    3B. Florets 5 or 6; leaves pinnately divided, slightly auriculate-clasping; inner involucral bracts about 5. ...... 3. Y. chelidoniifolia
  2B. Leaves not clasping, subentire or with obscure mucronate teeth; florets 5 or 6; inner involucral bracts about 5. ...... 4. Y. yoshinoi

1. **Youngia japonica** (L.) DC. *Prenanthes japonica* L.; *P. multiflora* Thunb.; *P. lyrata* Thunb.; *Y. thunbergiana* DC.; *Crepis japonica* (L.) Benth.; *Y. japonica* subsp. *elstonii* (Hochreut.) Babcock & Stebbins; *Crepis japonica* var. *elstonii* Hochreut.——Oni-tabira-ko.   Pubescent biennial; stems erect, often branched at base, 20–80 cm. long, naked or with 1–3 leaves; radical leaves tufted, oblanceolate, acute, narrowed to a petiolelike base, pinnately parted, loosely mucronate-toothed, the terminal segment largest, deltoid-ovate; heads 7–8 mm. across, many, in terminal corymbose panicles, yellow; involucres 5–6 mm. long, the inner bracts about 8, the outer much smaller, ovate; achenes about 1.8 mm. long, gradually narrowed at the tip, brown, with slender scaberulous ribs; pappus white, about 3 mm. long, persistent.——May–June. Waste grounds, cultivated fields, and roadsides; Hokkaido, Honshu, Shikoku, Kyushu; very common.——Ryukyus, Formosa, China, India, Polynesia, and Australia.

2. **Youngia denticulata** (Houtt.) Kitam. *Paraixeris denticulata* (Houtt.) Nakai; *Prenanthes denticulata* Houtt.; *P. hastata* Thunb.; *Y. hastata* DC.; *Y. chrysantha* Maxim.; *Lactuca denticulata* (Houtt.) Maxim.; *Ixeris denticulata* (Houtt.) Nakai——Yakushi-sō.   Slightly glaucescent annual or biennial; stems 30–70 cm. long, much branched, often purplish red; cauline leaves membranous, spathulate, rarely pinnately divided, 5–10 cm. long, 2–5 cm. wide, rather abruptly winged-petiolate, mucronate-toothed, glaucous beneath; upper leaves lingulate, sessile, auriculate-clasping; corymbs terminal and axillary, few-headed; heads about 1.5 cm. across, yellow, nodding in fruit, the florets 10–13; involucres dark green, the inner bracts 7 or 8; achenes lanceolate, 2.5–3 mm. long inclusive of the short beak, punctate-scaberulous.——Sept.–Nov. Sunny slopes and roadsides in mountains; Hokkaido, Honshu,

Shikoku, Kyushu; common.——Korea, Manchuria, China, and Indochina.——Forma **pinnatipartita** (Makino) Kitam. *Lactuca denticulata* var. *typica* lusus *pinnatipartita* Makino ——Hana-yakushi-sō.   Leaves pinnatiparted.

3. **Youngia chelidoniifolia** (Makino) Kitam. *Paraixeris chelidoniifolia* (Makino) Nakai; *Lactuca chelidoniifolia* Makino; *Ixeris chelidoniifolia* (Makino) Stebbins——Kusa-no-ōba-nogiku.   Rather slender annual or biennial; stems 20–50 cm. long, branched from base or sometimes simple; leaves thin, petiolate, pinnately divided, the segments 3–6 pairs, ovate, 1–2 cm. long, with few incisions or coarsely toothed, the petioles and rachis wingless, slightly auriculate-clasping; upper leaves becoming gradually smaller; heads about 10 mm. across, few, in a corymb, the florets 5 or 6; involucres narrowly tubular, about 6 mm. long, the inner bracts about 5, dark green; achenes linear-lanceolate, about 4 mm. long inclusive of the short beak, punctulate, scaberulous at the tip; pappus as long as the achene.——Sept.–Oct. Honshu (Shimotsuke and Yamato Prov.), Shikoku, Kyushu (Mount Ichifusa in Higo).——Korea and Manchuria.

4. **Youngia yoshinoi** (Makino) Kitam. *Paraixeris yoshinoi* (Makino) Nakai; *Lactuca denticulata* var. *yoshinoi* Makino; *L. yoshinoi* (Makino) Makino & Nakai; *Ixeris yoshinoi* (Makino) Kitam.——Iwa-yakushi-sō, Nagaba-yakushi-sō.   Much-branched annual or biennial herb 30–40 cm. high; leaves oblong-lanceolate or broadly lanceolate, 7–15 cm. long, 1–3 cm. wide, narrowed to a petiolelike base, subentire or loosely mucronate-toothed; heads about 1.2 cm. across, yellow, the florets 5 or 6; involucres 6 or 7 mm. across, the inner bracts about 5, dark green; achenes about 4 mm. long inclusive of the short beak, minutely scabrous; pappus about 4 mm. long.——Oct.   Honshu (Bitchu Prov.).

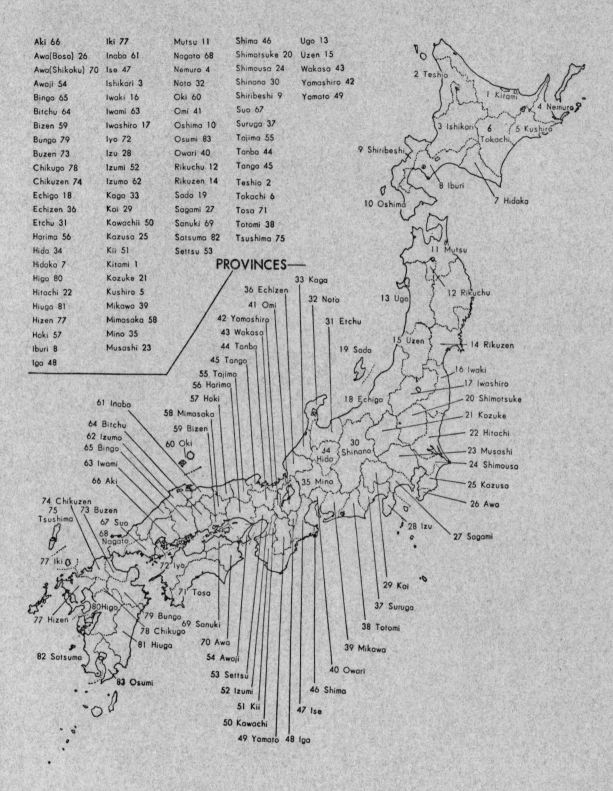

Aki 66
Awa(Boso) 26
Awa(Shikoku) 70
Awaji 54
Bingo 65
Bitchu 64
Bizen 59
Bungo 79
Buzen 73
Chikugo 78
Chikuzen 74
Echigo 18
Echizen 36
Etchu 31
Harima 56
Hida 34
Hidaka 7
Higo 80
Hitachi 22
Hiuga 81
Hizen 77
Hoki 57
Iburi 8
Iga 48

Iki 77
Inaba 61
Ise 47
Ishikari 3
Iwaki 16
Iwami 63
Iwashiro 17
Iyo 72
Izu 28
Izumi 52
Izumo 62
Kaga 33
Kai 29
Kawachii 50
Kazusa 25
Kii 51
Kitami 1
Kozuke 21
Kushiro 5
Mikawa 39
Mimasaka 58
Mino 35
Musashi 23

Mutsu 11
Nagato 68
Nemuro 4
Noto 32
Oki 60
Omi 41
Oshima 10
Osumi 83
Owari 40
Rikuchu 12
Rikuzen 14
Sado 19
Sagami 27
Sanuki 69
Satsuma 82
Settsu 53

Shima 46
Shimotsuke 20
Shimousa 24
Shinano 30
Shiribeshi 9
Suo 67
Suruga 37
Tajima 55
Tanba 44
Tango 45
Teshio 2
Tokachi 6
Tosa 71
Totomi 38
Tsushima 75

Ugo 13
Uzen 15
Wakasa 43
Yamashiro 42
Yamato 49

PROVINCES—

2 Teshio
1 Kitami
4 Nemuro
3 Ishikari
6 Tokachi
5 Kushiro
9 Shiribeshi
8 Iburi
7 Hidaka
10 Oshima
11 Mutsu
13 Ugo
12 Rikuchu
15 Uzen
14 Rikuzen
16 Iwaki
17 Iwashiro
20 Shimotsuke
21 Kozuke
22 Hitachi
23 Musashi
24 Shimousa
25 Kazusa
26 Awa
27 Sagami
28 Izu
18 Echigo
19 Sada
33 Kaga
36 Echizen
41 Omi
32 Noto
31 Etchu
42 Yamashiro
43 Wakasa
44 Tanba
45 Tango
55 Tajima
56 Harima
57 Hoki
61 Inaba
64 Bitchu
62 Izumo
65 Bingo
63 Iwami
66 Aki
58 Mimasaka
59 Bizen
60 Oki
74 Chikuzen
75 Tsushima
73 Buzen
67 Suo
68 Nagato
77 Iki
72 Iyo
71 Tosa
80 Higo
79 Bungo
78 Chikugo
77 Hizen
81 Hiuga
82 Satsuma
83 Osumi
69 Sanuki
70 Awa
54 Awaji
53 Settsu
52 Izumi
51 Kii
50 Kawachi
49 Yamato
48 Iga
46 Shima
47 Ise
40 Owari
39 Mikawa
38 Totomi
37 Suruga
29 Kai
34 Hida
30 Shinano
35 Mino

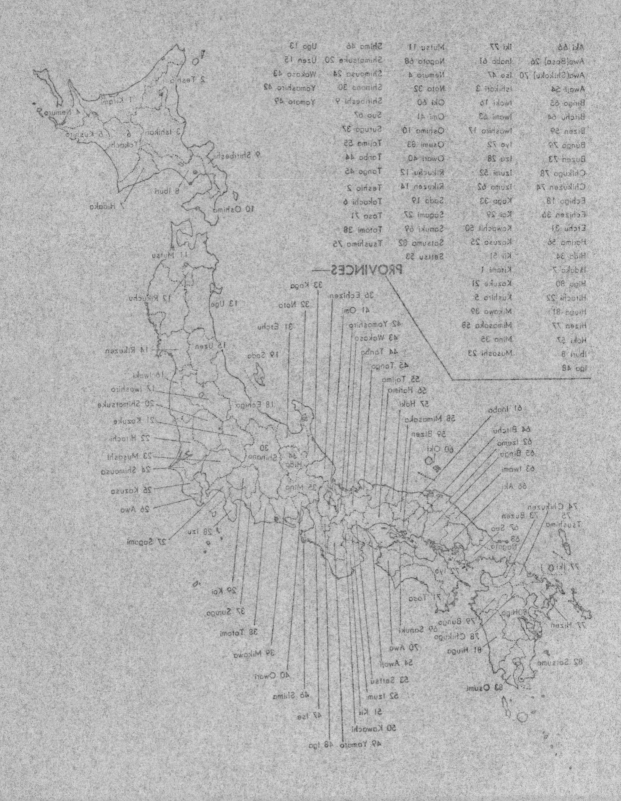

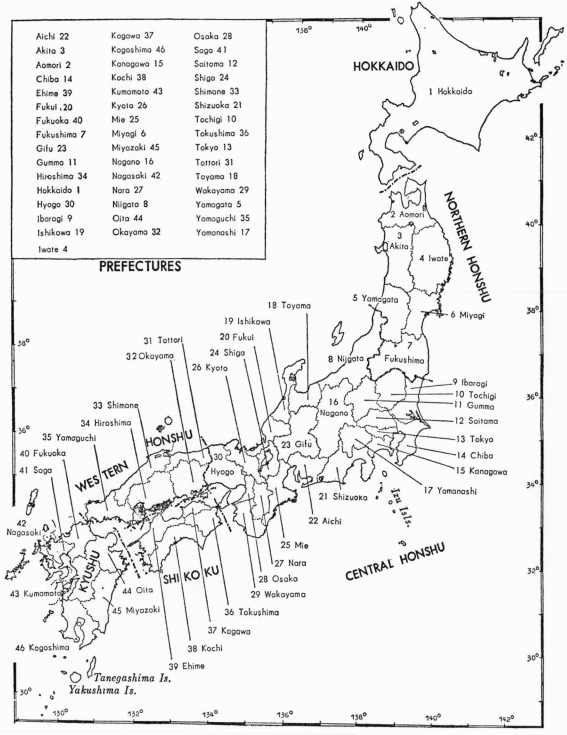

Aichi 22	Kagawa 37	Osaka 28
Akita 3	Kagoshima 46	Saga 41
Aomori 2	Kanagawa 15	Saitama 12
Chiba 14	Kochi 38	Shiga 24
Ehime 39	Kumamoto 43	Shimane 33
Fukui 20	Kyoto 26	Shizuoka 21
Fukuoka 40	Mie 25	Tochigi 10
Fukushima 7	Miyagi 6	Tokushima 36
Gifu 23	Miyazaki 45	Tokyo 13
Gumma 11	Nagano 16	Tottori 31
Hiroshima 34	Nagasaki 42	Toyama 18
Hokkaido 1	Nara 27	Wakayama 29
Hyogo 30	Niigata 8	Yamagata 5
Ibaragi 9	Oita 44	Yamaguchi 35
Ishikawa 19	Okayama 32	Yamanashi 17
Iwate 4		

PREFECTURES

Drawn by T. Koyama

Abel: Clarke Abel, 1780–1826, United Kingdom.
Abrams: Leroy Abrams, 1874–1956, U.S.A. (flora of Pacific States, U.S.A.).
Abrom.: Johannes Abromeit, 1857–1946, Germany.
Adams: Johannes Michael Friedrich Adams, 1780–1838, U.S.S.R.
Adans.: Michel Adanson, 1727–1806, France.
Aellen: Paul Aellen, 1896—, Switzerland.
Agardh: Jacob Georg Agardh, 1813–1901, Sweden.
Ahlburg: Hermann Ahlburg, ?–1878, Germany (*Aucubaephyllum*).
Airy-Shaw: Herbert Kenneth Airy-Shaw, 1902—, United Kingdom (Coniferae).
Ait.: William Aiton, 1731–1793, United Kingdom.
Ait. f.: William Townsend Aiton, 1766–1849, United Kingdom.
Akahori: Akira Akahori, 赤堀　昭, Japan.
Akasawa: Toshiyuki Akasawa,赤澤時之, 1915—, Japan (Burmanniaceae).
Akiyama: Shigeo Akiyama, 秋山茂雄, 1906—, Japan (*Carex*).
Alef.: Friedrich Georg Christoph Alefeld, 1820–1872, Germany.
Alisova: Evgeniia Nikolaevna Alisova-Klobukova, 1899—, U.S.S.R.
All.: Carlo Allioni, 1725–1804, Italy.
Alston: Arthur Hugh Garfit Alston, 1902–1958, United Kingdom (*Selaginella, Eugenia*).
Ames: Oakes Ames, 1874–1950, U.S.A. (Orchidaceae).
Anderss.: Nils Johan Andersson, 1821–1880, Sweden (*Salix*, Gramineae).
Andr.: Henry C. Andrews, 1796–1828, United Kingdom.
André: Édouard François André, 1840–1911, France.
H. Andres: Heinrich Andres, 1883–1957, Germany (Pyrolaceae).
Andrz.: Antoni Lukianovich Andrzeiovski, 1784–1868, U.S.S.R.
Ångstr.: Johan Ångstrom, 1813–1879, Sweden.
Ant.: Franz Antoine, 1815–1886, Austria (Coniferae).
Araki: Yeiichi Araki, 荒木英一, 1904–1955, Japan (*Hosta*).
Arcang.: Giovanni Arcangeli, 1840–1921, Italy.
Arduino: Luigi Arduino, 1759–1834, Italy.
Armstr.: John B. Armstrong, fl. 1850, New Zealand.
Arn.: George Arnold Walker-Arnott, 1799–1868, United Kingdom.
Asai: Yasuhiro Asai, 淺井康宏, 1933—, Japan (*Epipactis*).
Asami: Yoshichi Asami, 淺見與七, 1894—, Japan (*Malus*).
Asch.: Paul Friedrich August Ascherson, 1834–1913, Germany.
Aubl.: Jean Baptiste Christophe Fusée Aublet, 1720–1778, France (French Guiana).
Auct.: Auctorum, of Authors.
Avé-Lall.: Julius Léopold Eduard Avé-Lallemant, 1803–1867, Germany.
Bab.: Charles Cardale Babington, 1808–1895, United Kingdom.
Babcock: Ernest Brown Babcock, 1877–1955, U.S.A. (Compositae).
Backer: Cornelis Andries Backer, 1874–1963, Netherlands (Java).
F. M. Bailey: Frederick Manson Bailey, 1827–1915, United Kingdom (Australia).
L. H. Bailey: Liberty Hyde Bailey, 1858–1955, U.S.A. (*Carex, Rubus,* cultivated plants).
Baill.: Henri Ernest Baillon, 1827–1895, France.
Bak.: John Gilbert Baker, 1834–1920, United Kingdom (ferns, Liliaceae, Amaryllidaceae).
Bak. f.: Edmund Gilbert Baker, 1864–1949, United Kingdom.
Bakhuizen: Reinier Cornelis Bakhuizen van den Brink, 1881–1945, Netherlands (Java).
Balansa: Benedict Balansa, 1825–1891, France (Indochina).
Balbis: Giovanni Battista Balbis, 1765–1831, Italy.
Banks: Sir Joseph Banks, 1743–1820, United Kingdom.
Barnhart: John Hendley Barnhart, 1871–1949, U.S.A. (Lentibulariaceae).
Baroni: Eugenio Baroni, 1865–1943, Italy.
W. Barton: William Paul Crillon Barton, 1786–1856, U.S.A.
Basiner: Theodor Friedrich Julius Basiner, 1817–1862, (*Hedysarum*).
Batal.: Alexander Batalin, 1847–1898, U.S.S.R.

BATEM.: James Bateman, 1811–1897, United Kingdom.
BATSCH: August Johann Georg Karl Batsch, 1761–1802, Germany (horticulture).
BAUMG.: Johann Christian Gottlob Baumgarten, 1765–1845, Germany.
BEAL: William James Beal, 1833–1924, U.S.A. (Gramineae).
BEAN: William Jackson Bean, 1863–1947, United Kingdom.
BEAUVERD: Gustave Beauverd, 1867–1942, Switzerland (Compositae).
P. BEAUV.: Ambroise Marie François Joseph Palisot de Beauvois, 1752–1820, France (Gramineae).
BECC.: Odoardo Beccari, 1843–1920, Italy (Palmae).
BECHERER: Alfred Becherer, 1897–, Switzerland.
G. BECK: Günther Beck von Mannagetta, 1856–1931, Austria.
W. BECKER: Wilhelm Becker, 1874–1920, Germany (*Viola*).
BECKM.: John Beckmann, 1739–1811, Germany.
BEDD.: Richard Henry Beddome, 1830–1911, United Kingdom.
BEETLE: Allan Ackerman Beetle, 1913–, U.S.A. (Cyperaceae).
BEISSN.: Ludwig Beissner, 1853–1927, Germany (Coniferae).
BELLARDI: Carol Antonio Ludovico Bellardi, 1741–1826, Italy.
A. BENN.: Alfred William Bennett, 1833–1902, United Kingdom (*Potamogeton*).
J. BENN.: John Joseph Bennett, 1801–1876, United Kingdom.
BENTH.: George Bentham, 1800–1884, United Kingdom.
BERCHTOLD: Friedrich, Graf von Berchtold, 1781–1876, Austria.
BERG.: Peter Jonas Bergius, 1730–1790, Sweden.
BERGER: Alwin Berger, 1871–1931, Germany (*Sedum*).
BERNH.: Johann Jacob Bernhardi, 1774–1850, Germany.
BERTERO: Carlo Giuseppe Bertero, 1789–1831, Italy.
BERTHEL.: Sabin Berthelot, 1794–1880, France.
BERTOL.: Antonio Bertoloni, 1775–1869, Italy.
BESS.: Wilibald Swibert Joseph Gottlieb von Besser, 1784–1842, Austria; Poland (flora of Russia).
BEURL.: Pehr Johan Beurling, 1800–1866, Sweden (flora of Scandinavia).
BIEB.: Baron Friedrich August Marschall von Bieberstein, 1768–1826, Germany (flora of Russia).
BIGEL.: Jacob Bigelow, 1787–1879, U.S.A. (flora of Boston.).
BINN.: Simon Binnendijk, 1821–1883, Netherlands (Java).
BISS. & MOORE: James Bisset, 1843–1911, and Spencer Moore, United Kingdom.
BITTER: Friedrich August Georg Bitter, 1873–1927, Germany.
BIV.: Baron Antonio Bivona-Bernardi, 1778–1834, Sicily.
BL.: Karl Ludwig von Blume, 1796–1862, Netherlands (Malaysia).
BLAKE: Sidney Fay Blake, 1892–1959, U.S.A. (Compositae).
S. T. BLAKE: Stanley Thatcher Blake, 1910–, Australia (Gramineae).
BLAKEL.: Ralph Antony Blakelock, 1915–1963, United Kingdom.
BLANCO: Francisco Manuel Blanco, 1780–1845, Spain (Philippines).
BLUFF & FINGERH.: Mathias Joseph Bluff, 1805–1837, and Karl Anton Fingerhut, flourished 1821–1833, Germany.
BLYTT: Matthias Numsen Blytt, 1789–1862, Norway; Sweden.
BOBROV: Eugenii Grigorievich Bobrov, 1902–, U.S.S.R.
BOECKLR.: Johann Otto Boeckeler, 1803–1899, Germany (Cyperaceae).
BOEHMER: Georg Rudolph Boehmer, 1723–1803, Germany (*Boehmeria*).
BOEHMER: Louis B. Boehmer (& Co.), 1882–1908, (nurseryman in Japan).
BOENN.: Clemens Maria Friedrich von Boenninghausen, 1785–1864, Germany.
BOERL.: Jacob Gijsbert Boerlage, 1849–1900, Netherlands.
BOISS.: Edmond Pierre Boissier, 1810–1885, Switzerland (flora of Orient).
H. BOISS.: Henri de Boissieu, 1875–1912, Switzerland (Saxifragaceae, Cruciferae).
BOIVIN: Bernard Robert Boivin, 1916–, Canada.
BOLLE: Friedrich Bolle (contemporary), Germany.
BONATI: Gustave Bonati, 1873–1927, France (Scrophulariaceae).
BONG.: August Heinrich Gustav Bongard, 1786–1839, U.S.S.R. (Sitka, Bonin).
E. BONN.: Edmond Bonnet, 1848–1922, France (*Eragrostis*).
BONPL.: Aimé Jacques Alexandré Bonpland, 1773–1858, France (S. America).
BOOTT: Francis Boott, 1792–1863, United Kingdom; U.S.A. (*Carex*).
BORB.: Vincze von Borbás, 1844–1905, Hungary.
BOREAU: Alexandré Boreau, 1803–1875, France.
A. BORISS.: Antonina Georgievna Borisova (Borissova), 1903–, U.S.S.R.
BORKH.: Moritz Balthasar Borkhausen, 1760–1806, Germany.
BORNM.: Joseph Friedrich Nicolaus Bornmueller, 1862–1948, Germany.

BORY & CHAUB.: Baron de Jean Baptiste Geneviève Marcelin Bory de St. Vincent, 1778–1846, and Louis Anastase Chaubard, 1785–1854, France.

v. D. BOSCH: Roelof Benjamin van den Bosch, 1810–1862, Netherlands.

BOTSCH.: V. P. Botschantzev, Contemporary, U.S.S.R.

BOUCHÉ: Carl David Bouché, 1809–1881, Germany.

BOULENGER: George Albert Boulenger, 1858–1937, Belgium (*Rosa*).

BOWLES: Edward Augustus Bowles, 1865–1954, United Kingdom.

N.E. BR.: Nicholas Edward Brown, 1849–1934, United Kingdom.

R. BR.: Robert Brown, 1773–1858, United Kingdom.

P. BR.: Patrick Browne, 1720–1790, Ireland.

BRAND: August Brand, 1863–1931, Germany (Polemoniaceae, Boraginaceae, Symplocaceae).

A. BRAUN: Alexander Carl Heinrich Braun, 1805–1877, Germany.

BREIT.: Wilhelm Breitenbach, 1856–.

BRIOT: Pierre Louis Briot, 1804–1888, France.

BRIQ.: John Isaac Briquet, 1870–1931, Switzerland.

BRITT.: Nathaniel Lord Britton, 1859–1934, U.S.A.

BRITT. & BROWN: Nathaniel Lord Britton and Addison Brown.

BRITTEN: James Britten, 1846–1924, United Kingdom.

BRONGN.: Adolphe Théodore Brongniart, 1801–1876, France.

BROT.: Felix da Silva Avellar Brotero, 1744–1828, Portugal.

BROUSS.: Pierre Marie Auguste Broussonet, 1761–1801, France.

BROWN: Addison Brown, 1830–1913, U.S.A.

BSP.: N. L. Britton, Emerson Ellick Sterns, 1846–1906, and Justus Ferdinand Poggenburg, 1840–1893, U.S.A.

BUBANI: Pietro Bubani, 1806–1888, Italy.

L. v. BUCH: Christian Leopold von Buch, 1774–1854, Germany.

BUCH.-HAM.: Francis Buchanan, or Lord Hamilton, 1762–1828, United Kingdom.

BUCHEN.: Franz Georg Philipp Buchenau, 1831–1906, Germany (Juncaceae).

BUC'HOZ: Pierre Joseph Buc'hoz, 1731–1807, France.

BUEK: Heinrich Nikolaus Buek II, 1796–1879, Germany.

BUERG.: Heinrich Buerger (Bürger), 1804(?1806)–1858, Germany; Netherlands (collector in Japan).

BULLOCK: Arthur Allman Bullock, 1906–, United Kingdom.

BUNGE: Alexander Andreievich von Bunge, 1803–1890, U.S.S.R.

BUREAU: Edouard Bureau, 1830–1918, France.

BURGSD.: Friedrich August Ludwig von Burgsdorf, 1747–1802, Germany.

BURKILL: Isaac Henry Burkill, 1870–, United Kingdom.

BURM.: Johannes Burman, 1706–1779, Netherlands (Ceylon, India).

BURM. F.: Nicolaus Laurens Burman, 1733–1793, Netherlands (India).

BURRET: Carl Ewald Max Burret, 1883–, Germany (Palmae, Tiliaceae).

BURTT: Brian Lawrence Burtt, 1913–, United Kingdom.

E. BUSCH: Elizaveta Aleksandrovna Bush (Busch), 1886–1960, U.S.S.R.

N. A. BUSCH: Nikolai Adolfovich Bush (Busch), 1869–1941, U.S.S.R.

BUSE: L. H. Buse, 1819–1888, Netherlands (Gramineae of Malaysia).

BYHOUWER: Jan Tijs Pieter Bijhouwer (Byhouwer), 1898–1938, Netherlands.

CALLIER: Alfons Callier, 1866–1927, Germany (dendrology).

A. CAMUS: Aimée Camus, 1879–, France (Gramineae).

E. G. CAMUS: Edmond Gustave Camus, 1852–1915, France (Cyperaceae).

CARD.: Jules Cardot, 1860–1934, France (Rosaceae).

CARR.: Elie-Abel Carrière, 1816–1896, France (dendrology).

CARRUTH.: William Carruthers, 1830–1922, United Kingdom (dendrology).

CASAR.: Giovanni Casaretto, 1812–1879, Italy.

CASP.: Johann Xavier Robert Caspary, 1818–1887, Germany (*Hydrilla,* Nymphaeaceae).

CASS.: Comte de Alexandré Henri Gabriel de Cassini, 1781–1832, France (Compositae).

CAV.: Antonio José Cavanilles, 1745–1804, Spain.

ČELAK.: Ladislav Josef Čelakovsky, 1834–1902, Czechoslovakia.

CHAIX: Abbé Dominique Chaix, 1730–1799, France.

CHAM.: Adelbert Ludwig von Chamisso, 1781–1838, Germany (poet-naturalist).

CHAMP.: Lt. Col. John George Champion, 1815–1854, United Kingdom (Hongkong).

CHANDLER: Marjorie Elizabeth Jane Chandler, 1897–, United Kingdom.

CHANG: Chiao-Chien Chang, 張 兆 騫, 1900–, China (Compositae).

CHASE: Mary Agnes Chase, 1869–1963, U.S.A. (Gramineae).

CH'EN: Feng-Huei Ch'en, 陳 封 懷, 1901–, China (Compositae).

CHENG: Wan-Chün Cheng, 鄭 萬 鈞, 1903—, China (dendrology).
CHEVAL.: Auguste J. B. Chevalier, 1873–1956, France.
CHIEN: Sung-shu Chien (Ch'ung-shu Ch'ieng), 錢 崇 澍, 1883—, China.
CHING: Ren-ch'ang (R. C. Ching), 秦 仁 昌, 1898—, China (ferns).
CHODAT: Robert Hippolyte Chodat, 1865–1934, Switzerland.
CHOISY: Jacques Denys Choisy, 1799–1859, Switzerland (Convolvulaceae).
C. CHR.: Carl Friedrik Albert Christensen, 1872–1942, Denmark (ferns).
CHRIST: Konrad Hermann Heinrich Christ, 1833–1933, Switzerland (ferns).
CHUN: Woon-Young Chun, 陳 煥 鏞, 1895—, China (dendrology).
CHURCH: George Lyle Church, 1903—, U.S.A. (Gramineae).
CLAIRV.: Joseph Philippe de Clairville, 1742–1830, Switzerland.
C. B. CLARKE: Charles Baron Clarke, 1832–1906, United Kingdom (Cyperaceae, Compositae, Commelinaceae).
CLAUSEN: Robert Theodore Clausen, 1911—, U.S.A.
CLAYT.: John Clayton, 1685–1773, Virginia.
CLEYER: André Cleyer, ?–1697 or 1698, Germany (resident of Japan 1682–86; *Cleyera*).
COGN.: Célestin Alfred Cogniaux, 1841–1916, Belgium (Cucurbitaceae, Melastomataceae).
COHEN-STUART: Combert Pieter Cohen-Stuart, 1889—, Netherlands (Theaceae).
COLEBR.: Henry Thomas Colebrooke, 1765–1837, United Kingdom (India).
COLSM.: Johannes Colsmann, 1771–1830, Denmark.
COMM.: Philibert Commerson, 1727–1773, France.
CONSTANCE: Lincoln Constance, 1906—, U.S.A. (Umbelliferae).
T. COOKE: Theodore Cooke, 1836–1910, United Kingdom.
COPEL.: Edwin Bingham Copeland, 1873–1964, U.S.A. (ferns).
CORNAZ: Edouard Cornaz, 1825–1911, France.
CORNER: Edred John Henry Corner, 1906—, United Kingdom (Malay Penins.; *Ficus*).
CORREA: José Francisco Correa da Serra, 1751–1823, Portugal.
COSS.: Ernest Saint-Charles Cosson, 1819–1889, France.
COSTE: Abbé Hyppolyte Jacques Coste, 1858–1924, France.
COULT.: John Merle Coulter, 1851–1928, U.S.A.
COURT.: Richard Joseph Courtois, 1806–1836, Belgium.
COV.: Frederick Vernon Coville, 1867–1937, U.S.A. (Ericaceae, Juncaceae).
COWAN: John Macqueen Cowan, 1892–1960, United Kingdom.
CRAIB: William Grant Craib, 1882–1933, United Kingdom.
CRANTZ: Heinrich Johann Nepomuk von Crantz, 1722–1797, Austria.
CRÉP.: François Crépin, 1830–1903, Belgium.
CROIZAT: Léon Camille Marius Croizat, 1894—, U.S.A.; Venezuela (Euphorbiacae).
CROOM: Hardy Bryon Croom, 1797–1887, U.S.A.
CUFODONTIS: Georg Cufodontis, 1896—, Austria.
CUMING: Hugh Cuming, 1791–1865, United Kingdom (Philippines).
CURT.: William Curtis, 1746–1799, United Kingdom (Botanical Magazine).
CYR.: Domenico Cyrillo, 1739–1799, Italy.
DAHLST.: Gustav Adolph Hugh Dahlstedt, 1856–1934, Sweden (*Taraxacum*).
DALLA-TORRE: Karl Wilhelm von Dalla-Torre, 1850–1928, Germany.
DALLA-TORRE & SARNTH.: K. W. von Dalla-Torre, and Ludwig von Sarnthein.
DALLIM.: William Dallimore, 1871–1959, United Kingdom (horticulture, dendrology).
DANDY: James Edgar Dandy, 1903—, United Kingdom.
DANGUY: Paul Danguy, 1862–1942, France.
DANSER: Benedictus Hubertus Danser, 1891–1943, Netherlands (Loranthaceae).
DAVEAU: Jules Alexandré Daveau, 1852–1929, France.
DAVID: Armand David, 1826–1900, France (collector in China).
DEANE: Walter Deane, 1848–1930, U.S.A.
DC.: Augustin Pyramus de Candolle, 1778–1841, Switzerland.
A. DC.: Alphonse Louis Pierre Pyramus de Candolle, 1806–1893, Switzerland.
C. DC.: Casimir de Candolle, 1836–1918, Switzerland.
DEB.: Jean Odon Debeaux, 1826–1910, France (flora of China).
DE BRUYN: Ary Johannes de Bruyn (Bruijn), 1811–1895, Netherlands.
DECNE.: Joseph Decaisne, 1807–1882, France.
DEGL.: Jean Vincent Yves Degland, 1773–1841, France.
DEL.: Alire Raffeneau Delile, 1778–1850, France.
DELAVAY: Abbé Jean Marie Delavay, 1834–1895, France (collector in China).
DESCOURT.: Michel Étienne Descourtilz, 1775–1836, France.
DESF.: Réné Louiche Desfontaines, 1750–1833, France.

DESR.: Louis Auguste Joseph Desrousseaux, 1753–1838, France.

DESV.: Augustin Nicaise Desvaux, 1784–1856, France.

DE VRIESE: Willem Hendrik de Vriese, 1806–1862, Netherlands (Malaysia).

DEWEY: Rev. Chester Dewey, 1784–1867, U.S.A. (*Carex*).

DIELS: Friedrich Ludwig Emil Diels, 1874–1945, Germany.

DIETR.: Friedrich Gottlieb Dietrich, 1768–1850, Germany.

A. DIETR.: Albert Dietrich, 1795–1856, Germany.

D. N. F. DIETR.: David Nathanael Friedrich Dietrich, 1800–1888, Germany.

F. G. DIETR.: Friedrick Gottlieb Dietrich, 1768–1850, Germany.

DIPP.: Ludwig Dippel, 1827–1914, Germany (dendrology).

DODE: Louis Albert Dode, 1875–1943, France (dendrology).

DOELL: Johann Christoph Doell, 1808–1885, Germany.

DOI: Tohei Doi, 土 井 藤平, 1882–1946, Japan.

DOMIN: Karel Domin, 1882–1954, Czechoslovakia (*Koeleria*).

D. DON: David Don, 1799–1841, United Kingdom.

G. DON: George Don, 1798–1856, United Kingdom.

DONN: James Donn, 1758–1813, United Kingdom.

DOUGL.: David Douglas, 1798–1834, United Kingdom.

DREJER: Solomon Thomas Nicolai Drejer, 1813–1842, Denmark (*Carex*).

DROBOV: Vassilii Petrovich Drobov, 1885–1956, U.S.S.R. (Gramineae).

DRUCE: George Claridge Druce, 1850–1932, United Kingdom.

DRUDE: Carl Georg Oscar Drude, 1852–1933, Germany.

DRYAND.: Jonas Carlsson Dryander, 1748–1810, Sweden; United Kingdom.

DUBOIS: François Noel Alexandré Dubois, 1752–1824, France.

DUBY: Jean Étienne Duby, 1798–1885, Switzerland.

DUCHART.: Pierre Étienne Simon Duchartre, 1811–1894, France.

DUCHESNE: Antoine Nicolas Duchesne, 1747–1827, France.

DULAC: Abbé Joseph Dulac, France.

DUM. COURS.: Georges Louis Marie Dumont de Courset, 1746–1824, France.

DUMORT.: Count Barthélemy Charles Joseph Dumortier, 1797–1887, Belgium.

DUNAL: Michel Felix Dunal, 1789–1856, France.

DUNN: Stephen Troyte Dunn, 1868–1938, United Kingdom.

DURAND & SCHINZ: Théophile Alexis Durand, 1855–1912, Belgium, and Hans Schinz, 1858–1941, Switzerland.

DURAZZ.: Antonio Durazzini, Italy.

D'URV.: Jules Sebastian Cesar Dumont D'Urville, 1790–1842, France.

DYER: William Turner Thiselton-Dyer, 1843–1928, United Kingdom.

DYKES: William Rickatson Dykes, 1877–1925, United Kingdom (*Iris, Tulipa*).

EATON: Daniel Cady Eaton, 1834–1895, U.S.A. (ferns).

A. A. EATON: Alvah Augustus Eaton, 1865–1908, U.S.A. (ferns).

EBERM.: Carl Heinrich Ebermaier, 1802–1870, Germany.

EDGEW.: Michael Pakenham Edgeworth, 1812–1881, United Kingdom (Bengal).

EHRENB.: Christian Gotfried Ehrenberg, 1795–1876, Germany.

EHRH.: Friedrich Ehrhart, 1742–1795, Germany.

ELL.: Stephen Elliott, 1771–1830, U.S.A.

ELLIS: John Ellis, 1710–1776, Ireland.

ELMER: Adolph Daniel Edward Elmer, 1870–1942, U.S.A. (Philippines).

ELWES: Henry John Elwes, 1846–1922, United Kingdom (*Lilium,* Coniferae).

ENDL.: Stephan Friedrich Ladislaus Endlicher, 1804–1849, Austria.

ENGELM.: George Engelmannn, 1809–1884, Germany; U.S.A.

ENGL.: Heinrich Gustav Adolph Engler, 1844–1930, Germany.

EXELL: Arthur Wallis Exell, 1901–, United Kingdom.

FABR.: Philipp Konrad Fabricius, 1714–1774, Germany.

FARW.: Oliver Atkins Farwell, 1867–1944, U.S.A.

FASSETT: Norman Carter Fassett, 1900–1954, U.S.A.

FAURIE: Urban Faurie, 1847–1914, France (collector in Japan).

FAWC.: William Fawcett, 1851–1926, United Kingdom.

FED.: Andrei Aleksandrovich Fedorov, 1908–, U.S.S.R.

FEDDE: Friedrich Karl Georg Fedde, 1873–1942, Germany (Papaveraceae, Berberidaceae).

B. FEDTSCH.: Boris Aleksevitch Fedtschenko, 1872–1947, U.S.S.R.

O. FEDTSCH.: Olga Alexandrovna Fedtschenko, 1845–1921, U.S.S.R.

FÉE: Antoine Laurent Apollinare Fée, 1789–1874, Germany (Ferns).

FEER: Heinrich Feer, 1857–1892, Switzerland (Campanulaceae).

Fenzi: Eduard Otto Fenzi, 1808–1879, Austria.

Fenzl: Eduard Fenzl, 1808–1879, Austria (Caryophyllaceae).

Fern.: Merritt Lyndon Fernald, 1873–1950, U.S.A.

Fieber: Franz Xavier Fieber, 1807–1812, Hungary.

Finet: Achille Eugene Finet, 1863–1913, France (Orchidaceae).

Fiori: Adriano Fiori, 1865–1950, Italy.

Fisch.: Friedrich Ernst Ludwig von Fischer, 1782–1854, U.S.S.R.

Fisch. & Mey.: Friedrich Ernst Ludwig von Fischer and Carl Anton von Meyer.

Flaksb.: Konstantin Andreevich Fliaksberger (Flaksberger), 1880–1942, U.S.S.R. (*Triticum*).

Flerov: Aleksandr Fedorovich Flerov, 1872–1960, U.S.S.R.

Floderus: Bjorn Gunnar Floderus, 1867–1941, Sweden (*Salix*).

Fluegge: Johan Fluegge, 1775–1816, Germany.

Focke: Wilhelm Olbers Focke, 1834–1922, Germany (*Rubus*).

Fomin: Alexander Vasilievic Fomin, 1867–1935, U.S.S.R. (ferns).

Forbes.: Francis Blackwell Forbes, 1839–1908, U.S.A. (China).

Forbes & Hemsl.: F. B. Forbes, and W. B. Hemsley, United Kingdom (enum. Chinese plants).

Forselles: Jacob Heinrich Forselles, 1785–1855, Finland.

Forsk.: Pehr Forskål, 1736–1768, Finland (Egypt, Arabia).

Forst.: Johann Reinhold Forster, 1729–1798, Germany (Pacific Isl.).

Forst. f.: Johann Georg Adam Forster, 1754–1794, Germany.

Fort.: Robert Fortune, 1812–1880, United Kingdom.

Franch.: Adrien Franchet, 1834–1900, France (flora of Asia).

Fr. & Sav.: A. Franchet and L. Savatier (flora of Japan).

Fresen.: Johann Baptist Georg Wolfgang Fresenius, 1808–1866, Germany.

Freyn: Joseph Franz Freyn, 1845–1903, Czechoslovakia.

Fries: Elias Magnus Fries, 1794–1878, Sweden.

Th. Fries: Theodor Magnus Fries, 1832–1913, Sweden (*Hieracium*).

Fritsch: Karl Fritsch, 1864–1934, Austria (Gesneriaceae).

Froederstr.: Harold Froederstroem, 1876–1944, Sweden (*Sedum*).

Froel.: Joseph Aloys Froelich, 1766–1841, Germany.

Frye & Rigg: Theodore Christian Frye, 1869–1962, and George Burton Rigg, 1872–1961, U.S.A.

Fukuyama: Noriaki Fukuyama, 福 山 伯 明 , 1912–1946, Japan (Orchidaceae).

Furtado: Caetano Xavier Dos Remedios Furtado, 1897–, Malaya.

Furumi: Masatomi Furumi, 古 海 正 福 , 1888–1930, Japan (Scrophulariaceae).

Fuss: Johann Michael Fuss, 1814–1883, Transylvania (Jugoslavia).

Gaertn.: Joseph Gaertner, 1732–1791, Germany.

Gaertn. f.: Karl Friedrich von Gaertner, 1772–1850, Germany.

Gaertn., Mey. & Scherb.: Philipp Gottfried Gaertner, 1754–1825, Bernhard Meyer, 1767–1836, and Johannes Scherbius, 1769–1813, Germany.

Gagnep.: François Gagnepain, 1866–1952, France.

Gamble: James Sykes Gamble, 1847–1925, United Kingdom (India).

Gand.: Abbé Michel Gandoger, 1850–1926, France.

Garcke: Friedrich August Garcke, 1819–1904, Germany.

Gardn.: George Gardner, 1812–1849, United Kingdom (India).

Gaudich.: Charles Gaudichaud-Beaupré, 1789–1864, France.

Gaudin: Jean François Gottlieb Philippe Gaudin, 1766–1833, Switzerland (Gramineae).

Gawl.: See Ker-Gawl.

J. Gay: Jacques Étienne Gay, 1786–1864, France.

Geel: Petrus Cornelis van Geel, 1796–1838, Belgium (horticulture).

Gelert: Otto Carl Leonor Gelert, 1862–1899, Denmark.

Georgi: Johann Gottlieb Georgi, 1729–1802, U.S.S.R.

Giesenh.: Karl Giesenhagen, 1860–1928, Germany (ferns).

Gilg: Ernst Gilg, 1867–1933, Germany.

Gilib.: Jean Emmanuel Gilibert, 1741–1814, France.

Ging.: Frédérick Charles Jean Gingins de Lassaraz, 1790–1863, Switzerland (*Viola*).

Gleditsch: Johann Gottlieb Gleditsch, 1714–1786, Germany.

Glehn: Peter von Glehn, 1837–1876, U.S.S.R. (Sakhalin).

Glueck: Christian Maximilian Hugo Glueck, 1868–1940, Germany (hydrophytes).

Gmel.: Samuel Gottlich Gmelin, 1743–1774, U.S.S.R. (Siberia).

J. F. Gmel.: Johann Friedrich Gmelin, 1748–1804, Germany.

Godr.: Dominique Alexandré Godron, 1807–1880, France.

Goepp.: Johann Heinrich Robert Goeppert, 1800–1884, Germany.

GOERING: Philip Friedlich Wilhelm Goering (Göring), 1809–1876, Germany (Collector in Japan).

GOMBÓCZ: Endré Gombócz, 1882–1945, Hungary.

GONTSCH.: Nikolai Fedorovich Gontscharov, 1900–1942, U.S.S.R.

GOODEN.: Rev. Samuel Goodenough, 1743–1827, United Kingdom.

GORD.: George Gordon, 1806–1879, Ireland (Coniferae).

GORODK.: Boris Nikolaevich Gorodkov, 1890–1953, U.S.S.R. (*Carex*).

GORSKI: Stanislaw Batys Gorski, 1802–1864, Lithuania (Gramineae).

GRAEBN.: Karl Otto Robert Peter Paul Graebner, 1871–1933, Germany.

A. GRAY: Asa Gray, 1810–1888, U.S.A.

S. F. GRAY: Samuel Frederick Gray, 1766–1828, United Kingdom.

P. S. GREEN: Peter Shaw Green, 1920—, United Kingdom (*Osmanthus*).

GREENE: Edward Lee Greene, 1842–1915, U.S.A.

GREN.: Jean Charles Marie Grenier, 1808–1875, France.

GREV.: Robert Kaye Greville, 1794–1866, United Kingdom.

GRIFF.: William Griffith, 1810–1845, United Kingdom (India).

GRIG.: Yurii Sergeevich Grigoriev (G. Grigorjev), 1905—, U.S.S.R.

GRIMM: Johann Friedrich Karl Grimm, 1737–1821, Germany (*Stellaria*).

GRISEB.: August Heinrich Rudolph Grisebach, 1814–1879, Germany.

GRONOV.: Johannes Fredericus Gronovius, 1690–1762, Netherlands.

GROSS: Hugo Gross, 1888—, Germany (*Polygonum*).

GROSSHEIM: Alexandr Alfonsovich Grossheim, 1888–1948, U.S.S.R.

GRUBOV: Valery Ivanovitch Grubov, 1917—, U.S.S.R.

GUERKE: Robert Louis August Max Guerke, 1854–1911, Germany.

GUILL.: André Guillaumin, 1885–1952, France.

GUSS.: Giovanni Gussone, 1787–1866, Italy.

HACK.: Eduard Hackel, 1850–1926, Austria (Gramineae).

HAENKE: Thaddeus Haenke, 1761–1817, Czechoslovakia (traveler).

HAGERUP: Olaf Hagerup, 1889–1961, Denmark.

HAGIWARA: Tokio Hagiwara, 萩原時雄, 1896—, Japan (Convolvulaceae).

HAGSTR.: Johan Oskar Hagström, 1860–1922, Sweden (*Potamogeton, Ruppia*).

HAINES: Henry Hazelfoot Haines, 1867–1945, United Kingdom (India).

HALLER: Albert von Haller, 1708–1777, Switzerland.

HALLER F.: Albert von Haller fil., 1758–1823, Switzerland.

HALLIER: Hans Gottfried Hallier, 1868–1932, Netherlands.

HAMAYA: Toshio Hamaya, 濱谷稔夫, 1928—, Japan (Thymelaeaceae).

HAMET: Raymond-Hamet, 1890—, France.

HAMILT.: See Buch.-Ham.

HANCE: Henry Fletcher Hance, 1827–1886, United Kingdom (China).

HANCOCK: William Hancock, 1847–1914, United Kingdom (collector in Japan and China).

HAND.-MAZZ.: Heinrich von Handel-Mazzetti, 1862–1940, Austria (China).

HARA: Hiroshi Hara, 原寛, 1911—, Japan.

HARMS: Hermann August Theodor Harms, 1870–1942, Germany.

HARPER: Roland McMillan Harper, 1878—, U.S.A.

HARTM.: Carl Johan Hartman, 1790–1849, Sweden.

HASSK.: Justus Karl Hasskarl, 1811–1894, Germany (Java).

HATUS.: Sumihiko Hatusima, 初島住彦, 1906—, Japan (dendrology).

HAUSSKN.: Heinrich Carl Haussknecht, 1838–1903, Germany (*Epilobium*).

HAW.: Adrian Hardy Haworth, 1768–1833, United Kingdom (Cactaceae, *Euphorbia*).

HAYATA: Bunzo Hayata, 早田文藏, 1874–1934, Japan (Formosa).

HAYASHI: Yasaka Hayashi, 林彌榮, 1911—, Japan (dendrology).

HAYEK: August von Hayek, 1871–1928, Austria (flora of Orient, *Anemone*).

HAYNALD: Ludwig von Haynald, 1816–1891, Hungary.

H. B. K.: Baron Friedrich Wilhelm Heinrich Alexander von Humboldt, Aimé Jacques Alexandre Bonpland, and Carl Sigismund Kunth (flora of trop. Amer.).

HEDL.: Johann Theodor Hedlund, 1861–1953, Sweden.

HEGELM.: Christof Friedrich Hegelmaier, 1833–1906, Germany (hydrophytes).

HEGI: Gustav Hegi, 1876–1932, Switzerland.

HEIM.: Anton Heimerl, 1857–1942, Austria (*Achillea*).

HEIST.: Lorenz Heister, 1683–1758, Germany.

HELLER: Amos Arthur Heller, 1867–1944, U.S.A.

HEMSL.: William Botting Hemsley, 1843–1924, United Kingdom.

HENR.: Jan Theodor Henrard, 1881—, Netherlands (Gramineae).

HENRY: Augustine Henry, 1857–1930, United Kingdom (collector in China).

L. HENRY: Louis Henry, 1853–1903, France (cultivated plants).

HERB.: Hon. William Herbert, 1778–1847, United Kingdom (Amaryllidaceae).

HERBICH: Franz Herbich, 1791–1865, Austria; Poland.

HERD.: Ferdinand Gottfried Theobald Maximilian von Herder, 1828–1896, U.S.S.R.

HERINCQ: François Herincq, 1820–1891, France.

HERRM.: Rudolf Albert Wolfgang Herrmann, 1885–, Germany.

HERT.: Wilhelm Gustav Herter, 1884–1958, Germany.

HESSE: Hermann Albrecht von Hesse, 1852–1937, Germany (nurseryman).

VAN HEURCK: Henri Ferdinand van Heurck, 1838–1909, Belgium.

HEYNE: Benjamin Heyne, —1819, Germany (India).

HEYNH.: Gustav Heynhold, 1800–, Germany.

HEYWOOD: Vernon H. Heywood, 1927–, United Kingdom.

HIERN: William Philip Hiern, 1839–1925, United Kingdom.

HIERON.: Georg Hans Emo Wolfgang Hieronymus, 1846–1921, Germany (vascular cryptogams).

HIKINO: Hiroshi Hikino, ヒキノヒロシ, 1931–, Japan.

HILL: Sir John Hill, 1716–1775, United Kingdom.

A. W. HILL: Sir Arthur William Hill, 1875–1941, United Kingdom.

HILLEBR.: Wilhelm Hillebrand, 1821–1886, Germany (Hawaii).

HIROE: Minosuke Hiroe, 廣江美之助, 1914–, Japan (Umbelliferae).

HISAUCHI: Kiyotaka Hisauchi (or Hisauti), 久內清孝, 1884–, Japan.

HITCHC.: Albert Spear Hitchcock, 1865–1935, U.S.A. (Gramineae).

E. HITCHC.: Edward Hitchcock, 1793–1864, U.S.A. (*Botrychium simplex*).

HIYAMA: Kozo Hiyama, 檜山庫三, 1905–, Japan.

HOCHREUT.: Bénédict Pierre Georges Hochreutiner, 1873–1959, Switzerland.

HOCHST.: Christian Friedrich Hochstetter, 1787–1860, Germany (Africa).

HOFFM.: Georg Franz Hoffmann, 1760–1826, Germany (Umbelliferae).

O. HOFFM.: Karl August Otto Hoffmann, 1853–1909, Germany (Compositae).

HOFFMANNS.: Count Johann Centurius von Hoffmannsegg, 1766–1849, Germany.

HOLM: Herman Theodor Holm, 1854–1932, U.S.A.

HOLMBERG: Otto Rudolf Holmberg, 1874–1930, Sweden.

HOLTTUM: Richard Eric Holttum, 1895–, United Kingdom (Malaya).

HONDA: Masaji Honda, 本田正次, 1897–, Japan.

HOOK.: Sir William Jackson Hooker, 1785–1865, United Kingdom.

HOOK. & ARN.: W. J. Hooker and G. A. Walker-Arnott.

HOOK. F.: Sir Joseph Dalton Hooker, 1817–1911, United Kingdom.

HOPPE: David Heinrich Hoppe, 1760–1846, Germany.

HORIKAWA: Tomiya Horikawa, 1920–1956, 堀川富彌, Japan (*Quercus*).

HORKEL: Johann Horkel, 1769–1847, Germany.

HORNEM.: Jens Wilken Hornemann, 1770–1841, Denmark.

HORNIBR.: Murray Hornibrook, 1874–1949, United Kingdom (Coniferae).

HORNSTEDT: Cladius Fredrik Hornstedt, 1758–1809, Sweden.

HORT.: Hortorum, of Gardens.

HOSOKAWA: Takahide Hosokawa, 細川隆英, 1909–, Japan (ecology, Formosa).

HOST: Nicolaus Thomas Host, 1761–1834, Austria.

HOTTA: Teikichi Hotta, 堀田禎吉, 1899–, Japan (*Morus*).

HOUSE: Homer Doliver House, 1878–1949, U.S.A.

HOUTT.: Martinus Houttuyn, 1720–1798, Netherlands.

HOUZ. DE LEHAIE: Jean Houzeau de Lehaie, 1820–1888, Belgium (bamboo).

HOVEY: C. M. Hovey, 1810–1887, U.S.A.

HOWELL: Thomas Jefferson Howell, 1842–1912, U.S.A.

HU: Hsien-Hsu Hu, 胡先驌, 1894–, China (dendrology).

S. Y. HU: Shiu-Ying Hu, 胡秀英, 1910–, China; U.S.A.

C. E. HUBB.: Charles Edward Hubbard, 1900–, United Kingdom (Gramineae).

F. T. HUBB.: Frederic Tracy Hubbard, 1875–1962, U.S.A. (Gramineae).

HUDS.: William Hudson, 1730–1793, United Kingdom.

HUGHES: Dorothy K. Hughes (Mrs. Wilson Popenoe), 1899–1932, U.S.A.

HULT.: Oskar Eric Gunnar Hultén, 1894–, Sweden.

HUMB.: Baron Friedrich Wilhelm Heinrich Alexander von Humboldt, 1769–1858, Germany.

HURUSAWA: Isao Hurusawa, 古澤潔夫, 1916–, Japan (Euphorbiaceae).

HUTCHINS.: John Hutchinson, 1884–, United Kingdom (classification, Ranunculaceae).

HUTH: Ernst Huth, 1854–1897, Germany (Ranunculaceae).

HYLANDER: Nils Hylander, 1904—, Sweden.

IINUMA: Yokusai Iinuma, 飯沼　慾斎, 1782–1865, Japan ("Somoku Dzusetsu").

IKENO: Seiichiro Ikeno, 池野成一郎, 1866–1943, Japan (*Plantago*).

ILJIN: Modese Mikhailovich Iljin (Iliin), 1889—, U.S.S.R. (Compositae).

IMAMURA: Shun'ichiro Imamura, 今村駿一郎, 1903—, Japan (Podostemaceae).

INMAN: Ondess Lamar Inman, 1890–1942, U.S.A.

INOKUMA: Taizo Inokuma, 猪熊泰三, 1904—, Japan (dendrology).

IRMSCH.: Edgar Irmscher, 1887—, Germany (*Begonia, Saxifraga*).

IRVING: Walter Irving, 1867—, United Kingdom.

ISHIDOYA: Tsutomu Ishidoya, 石戸谷勉, 1884–1958, Japan.

H. ITŌ: Hiroshi Itō, 伊藤　洋, 1909—, Japan (ferns).

K. ITŌ: Koji Itō, 伊藤浩司, 1934—, Japan.

T. ITŌ: Tokutarō Itō, 伊藤篤太郎, 1868–1941, Japan.

IWATA: Jiro Iwata, 岩田次郎, 1906—, Japan.

K. IWATSUKI: Kunio Iwatsuki, 岩槻邦男, 1934—, Japan.

JACK: William Jack, 1795–1822, United Kingdom.

JACKS.: Benjamin Dayton Jackson, 1846–1927, United Kingdom.

JACOB-MAKOY: Lamert Jacob-Makoy, 1790–1873, Belgium (cultivated plants).

JACQ.: Nicolaus Joseph Baron von Jacquin, 1727–1817, Austria.

JACQUEM.: Victor Jacquemont, 1801–1831, France.

JANCZ.: Eduard Janczewski, Ritter von Glinka, 1846–1918, Poland (*Ribes*).

JANSEN: Pieter Jansen, 1882–1955, Netherlands (Gramineae).

JOHNSTON: Ivan Murray Johnston, 1898–1960, U.S.A. (Boraginaceae).

JORD.: Alexis Jordan, 1814–1897, France.

JOTANI: Yukio Jōtani, 常谷幸雄, 1904—, Japan.

JUEL: Hans Oskar Juel, 1863–1931, Sweden ("Plantae Thunberg").

JUNGH.: Franz Wilhelm Junghuhn, 1809–1864, Germany (Malaysia).

JUSS.: Antoine Laurent de Jussieu, 1748–1836, France.

A. JUSS.: Adrien Henri Laurent de Jussieu, 1797–1853, France.

B. JUSS.: Bernard de Jussieu, 1699–1777, France.

JUZEP.: Sergei Vasilievich Juzepczuk (Jusepczuk), 1893–1959, U.S.S.R.

KAEMPFER: Engelbert Kaempfer, 1651–1716, Germany (traveler in Japan).

KALENIC.: Johann Kaleniczenkow (Ivan Osipavich Kaleniczenkow), 1805–1876, U.S.S.R.

KAMIKOTI: Sizuka Kamikōti, 上河内靜, 1910—, Japan (Lauraceae of Formosa).

KANEH.: Ryōzō Kanehira, 金平亮三, 1882–1947, Japan (dendrology, Formosa.).

KANITZ: Agost Kanitz, 1843–1896, Hungary.

KAREL.: Grigorii Silych Karelin, 1801–1872, U.S.S.R.

KARST.: Gustav Karl Wilhelm Hermann Karsten, 1817–1908, Germany.

KAULF.: Georg Friedrich Kaulfuss, 1786–1830, Germany.

KAWAKAMI: Takiya Kawakami, 川上瀧彌, 1871–1915, Japan.

KAWANO: Syo'ichi Kawano, 河野昭一, 1936—, Japan (flora of Hokkaido).

KEARNEY: Thomas Henry Kearney, 1874–1956, U.S.A. (Malvaceae).

KEISSLER: Karl von Keissler, 1872—, Austria.

KELLER: Robert Keller, 1854–1939, Switzerland (*Rosa, Rubus, Hypericum*).

KELSO: Leon Kelso, 1907—, U.S.A. (*Glyceria, Salix*).

H. KENG: Hsuan Keng, 耿　煊, 1923—, China; Formosa.

Y. L. KENG: Yi-Li Keng, 耿以禮, 1898—, China (Gramineae).

KER-GAWL.: John Bellenden Ker, or John Ker Bellenden, John Gawler (before 1804), or Bellenden Ker, 1765–1842, United Kingdom.

KIKUCHI: Akio Kikuchi, 菊地秋雄, 1883–1951, Japan (pomology, *Pyrus*).

M. KIKUCHI: Masao Kikuchi, 菊地政雄, 1908—, Japan.

KIMURA: Arika Kimura, 木村有香, 1900—, Japan (*Salix*).

Y. KIMURA: Yōjirō Kimura, 木村陽二郎, 1912—, Japan (*Hypericum*).

KING: Sir George King, 1840–1909, United Kingdom (India).

KIRCHN.: G. Kirchner, 1837–1885, Germany.

KITAG.: Masao Kitagawa, 北川政夫, 1909—, Japan.

KITAIB.: Paul Kitaibel, 1757–1817, Germany (flora of Hungary).

KITAM.: Sirō Kitamura, 北村四郎, 1906—, Japan (Compositae).

KITTLITZ: Friedrich Heinrich von Kittlitz, 1799–1874, Germany.

KLENZE: ——Klenze, 19th century, Germany (*Pyrola japonica*).

KLETT: Gustav Theodor Klett, 1808–1882, Germany.

KLOTZSCH: Johann Friedrich Klotzsch, 1805–1860, Germany.

KNIGHT: Josef Knight, 1781 (?)–1855, United Kingdom.

KNOLL: Fritz Knoll, 1883–, Austria (*Astilbe*).

R. KNUTH: Reinhard Gustav Paul Knuth, 1874–1957, Germany (Geraniaceae, *Dioscorea*).

KOBUSKI: Clarence Emmeren Kobuski, 1900–1963, U.S.A. (Theaceae).

C. KOCH: See K. KOCH.

K. KOCH: Karl Heinrich Emil Koch, 1809–1879, Germany (dendrology).

W. KOCH: Wilhelm Daniel Joseph Koch, 1771–1847, Germany (Umbelliferae).

KODAMA: Shinsuke Kodama, 兒玉親輔, 1884–, Japan (Pteridophyta).

KOEHNE: Bernhard Adalbert Emil Koehne, 1848–1918, Germany (dendrology).

KOELER: Georg Ludwig Koeler, 1765–1807, Germany.

KOENIG: Johann Gerhard Koenig, 1728–1789, Latvia (traveler and naturalist se. Asia).

KOERN.: Friedrich Koernicke, 1828–1908, U.S.S.R. (Eriocaulaceae, *Deutzia,* cereals).

KOIDZ.: Gen'ichi Koidzumi, 小泉源一, 1883–1953, Japan.

H. KOIDZ.: Hideo Koidzumi, 小泉秀雄, 1886–1945, Japan (*Taraxacum*).

KOMAR.: Vladimir Leontievitch Komarov, 1869–1946, U.S.S.R. (e. Asia).

KOMAT.: Shunzō Komatsu, 小松春三, 1879–1932, Japan (Ericaceae).

KOORD.: Sijfert Hendrik Koorders, 1863–1919, Netherlands (Java).

KORIBA: Kwan Koriba, 郡場　寛, 1882–1957, Japan (Podostemaceae).

KORSH.: Sergyei Ivanovitch Korshinsky, 1861–1900, U.S.S.R.

KORT: Antoine Kort, 1874–1951, Belgium (cultivated plants).

KORTH.: Pieter Willem Korthals, 1807–1892, Netherlands.

KOSO-POL.: Boris Mikhailovich Koso-Polianski, 1890–1957, U.S.S.R. (Umbelliferae).

KOSTEL.: Vincenz Franz Kosteletzky, 1801–1887, Czechoslovakia.

KOYAMA: Mitsuo Koyama, 小山光男, 1885–1935, Japan (dendrology).

T. KOYAMA: Tetsuo Koyama, 小山鐵夫, 1933–, Japan.

KRAENZL.: Fritz Wilhelm Ludwig Kraenzlin, 1874–1934, Germany (Orchidaceae).

H. KRASCHEN.: Hippolit Mikhailovich Krasheninnikov (H. Krascheninnikov), 1884–1947, U.S.S.R.

S. KRASCHEN.: Stefan Petrovich Krasheninnikov, 1713–1755, U.S.S.R. (traveler in Kamchatka, Kuriles).

KRAUSE: Kurt Krause, 1883–, Germany (Liliaceae).

V. KRECZ.: Vitalii Ivanovitch Kreczetovicz, 1901–1942, U.S.S.R. (*Carex*).

KRYLOV: Porfirii Nikitovich Krylov, 1850–1931, U.S.S.R.

KUDO: Yushūn Kudō, 工藤祐舜, 1887–1932, Japan.

KUDO & SUSAKI: Y. Kudō and Chusuke Susaki, 須崎忠輔, 1866–1933, Japan.

KUEKENTH.: Georg Kuekenthal, 1864–1956, Germany (Cyperaceae).

KUHN: Friedrich Adalbert Maximilian Kuhn, 1842–1894, Germany (ferns).

KUMAZAWA: Masao Kumazawa, 熊澤正夫, 1904–, Japan (palynology, *Ranunculus*).

KUNG: Hsien-Wu Kung, 孔憲武, 1897–, China.

KUNTH: Carl Sigismund Kunth, 1788–1850, Germany.

O. KUNTZE: Carl Ernst Otto Kuntze, 1843–1907, Germany.

KUNZE: Gustav Kunze, 1793–1851, Germany (ferns).

KURATA: Satoru Kurata, 倉田　悟, 1922–, Japan (dendrology).

KUROSAWA: Sachiko Kurosawa, 1937–, 黒澤幸子, Contemporary, Japan.

KURZ: Wilhelm Sulpiz Kurz, 1834–1878, Germany (Burma; India).

KUSAKA: Masao Kusaka, 草下正夫, 1915–, Japan (dendrology).

KUSN.: Nikolai Ivanovitch Kusnetzov, 1864–1932, U.S.S.R. (*Gentiana*).

V. KUSN.: Vladimir Alexandrovich Kusnetzov, 1877–1940, U.S.S.R. (*Beckmannia*).

L.: Carolus Linnaeus (Carl von Linné), 1707–1778, Sweden.

L.F.: Carl von Linné fil., 1741–1783, Sweden.

LABILL.: Jacques Julien Houtton de la Billardière, 1755–1834, France.

LACKSCH.: Paul Lackschewitz, 1865–1936, U.S.S.R.

LAEST.: Lars Levi Laestadius, 1800–1861, Lapland.

LAG.: Mariano Lagasca y Segura, 1776–1839, Spain.

LAHARPE: Jean de Laharpe, 1802–1863?, France (Juncaceae).

LALLAVE.: Canónigo Pâblo de La Llave, 1773–1883, Mexico.

LALLEM.: Julius Leopold Edward Avé-Lallemant, 1803–1867, Germany.

LAM: Herman Johannes Lam, 1892–, Netherlands (Verbenaceae).

LAM.: Jean Baptiste Pierre Antoine de Monet de Lamarck, 1744–1829, France.

LAMB.: Aylmer Bourke Lambert, 1761–1842, United Kingdom.

LAMOTTE: Martial Lamotte, 1820–1883, France.

LANGE: Johan Martin Christian Lange, 1818–1898, Denmark.

LANGSD.: Georg Heinrich von Langsdorff, 1774–1854, Germany.

LA PYL.: Auguste Jean Marie Bachelot de la Pylaie, 1786–1856, France.

K: LARSEN: Kai Larsen, 1926—, Denmark.
LAUTERB.: Carl Adolf Georg Lauterbach, 1864–1937, Germany.
LAVALL.: Alphonse Lavallée, 1835–1884, France (dendrology).
LAVRENKO: Eugenii Mikhailovich Lavrenko, 1900—, U.S.S.R.
LAWRANCE: Mary Lawrance, 1790–1831, United Kingdom.
LAWRENCE: George Hill Mathewson Lawrence, 1910—, U.S.A.
LAXM.: Erich Laxmann, 1737–1796, U.S.S.R.
LEBAS: E. Lebas, France?.
LECOMTE: Henri Lecomte, 1856–1934, France.
LECOQ: Henri Lecoq, 1802–1871, France.
LECOYER: C. J. Lecoyer, fl. 1875, Belgium (*Thalictrum*).
LEDEB.: Carl Friedrich von Ledebour, 1785–1851, U.S.S.R.
LEHAIE: See HOUZ. DE LEHAIE.
LEHM.: Johann Georg Christian Lehmann, 1792–1860, Germany (*Potentilla*, Boraginaceae).
LEICHTL.: Max Leichtlin, 1831–1910, Germany (horticulture).
LEJ.: Alexander Louis Simon Lejeune, 1779–1858, Belgium.
LEM.: Charles Antoine Lemaire, 1801–1871, Belgium (cultivated plants).
LEMOINE: Pierre Louis Victor Lemoine, 1823–1911, and Emile Lemoine, 1862–1942, France (cultivated plants).
LEPECH.: Ivan Lepechin, 1737–1802, U.S.S.R.
LESCHEN.: Louis Théodor Leschenault de la Tour, 1773–1826, France.
LESS.: Christian Friedrich Lessing, 1810–1880, Germany (Compositae).
LÉV.: Augustin Abel Hector Léveillé, 1863–1918, France.
LEX.: Juan Martinez de Lexarza, 1785–1824, Mexico.
LEYSS.: Friedrich Wilhelm von Leysser, 1731–1815, Germany.
L'HÉRIT.: Charles Louis L'Héritier de Brutelle, 1746–1800, France.
LIEBM.: Frederik Michael Liebmann, 1813–1856, Denmark.
LIGHTF.: Rev. John Lightfoot, 1735–1788, United Kingdom.
LILJEBL.: Samuel Liljeblad, 1761–1815, Sweden.
LIMPR.: Wolfgang Limpricht, flourished 1913–1941, Germany (*Pedicularis*, China).
H. LINDB.: Harald Lindberg, 1871—, Finland.
LINDBL.: Alexis Eduard Lindblom, 1807–1853, Sweden.
LINDEM.: Eduard von Lindemann, 1825–1900, U.S.S.R.
LINDEN: J. J. Linden, 1817–1898, Belgium.
LINDL.: John Lindley, 1799–1865, United Kingdom.
LINDM.: Carl Axel Magnus Lindman, 1856–1928, Sweden.
LINDSAY: Robert Lindsay, 1846–1913, United Kingdom.
LINGELSH.: Alexander von Lingelsheim, 1874–1937, Germany (Oleaceae).
LINK: Johann Heinrich Friedrich Link, 1767–1851, Germany.
LIOU: Tchen-Ngo Liou, 劉 慎 諤, China.
LITV.: Dmitri Ivanovitch Litvinov, 1854–1929, U.S.S.R.
LLANOS: Antonio Llanos, 1806–1881, Spain (Philippines).
LODD.: Conrad Loddiges, 1743?–1826, and George Loddiges, 1784–1846, United Kingdom.
LOES.: Ludwig Eduard Theodor Loesener, 1865–1941, Germany (Aquifoliaceae).
LOES. F.: O. Loesener, flourished about 1926, Germany (*Veratrum*).
LOISEL.: Jean Louis Augusta Loiseleur-Deslongchamps, 1774–1849, France.
LOUD.: John Claudius Loudon, 1783–1843, United Kingdom (cultivated plants).
LOUR.: Juan Loureiro, 1715–1796, Portugal (Cochin China).
LOWE: Edward Joseph Lowe, 1825–1900, United Kingdom.
LUERSS.: Christian Luerssen, 1843–1916, Germany.
LUTATI: F. Vignolo Lutati, 1878—, Italy.
MA: Yü-Ch'üan Ma, 馬 毓 泉, China.
MAACK: Richard Maack, 1825–1886, U.S.S.R.
MACBR.: James Francis Macbride, 1892—, U.S.A. (Boraginaceae, *Clintonia*).
MACKENZIE: Kenneth Kent Mackenzie, 1877–1934, U.S.A. (*Carex*).
MacNAB: William Ramsay MacNab, 1844–1889, United Kingdom (*Abies*).
MACOUN: John Macoun, 1832–1920, Canada.
MAEKAWA: Tokujirō Maekawa, 前 川 德 次 郎, 1886—, Japan (*Chrysanthemum*).
F. MAEKAWA: Fumio Maekawa, 前 川 文 夫, 1908—, Japan (*Hosta, Asarum*).
MAGNUS: Paul Wilhelm Magnus, 1844–1914, Germany.
MAKINO: Tomitarō Makino, 牧 野 富 太 郎, 1862–1957, Japan.
MALINV.: Luis Jules Ernst Malinvaud, 1836–1913, France.
MALME: Gustav Oskar Malme, 1864–1937, Sweden.

MANSF.: Rudolf Mansfeld, 1901–1960, Germany (Orchidaceae).
MARKGR.: Friedrich Markgraf, 1897—, Germany.
MARQ.: Cecil Victor Boley Marquand, 1897—, United Kingdom.
MARSILI: Giovanni M. Marsili, 1727–1795, Italy.
MARTELLI: Ugolino Martelli, 1860–1934, Italy.
MARTENS: Martin Martens, 1797–1863, Belgium.
MARTIUS: Karl Friedrich Philipp von Martius, 1794–1868, Germany (flora of Brazil).
MAR.-VICT.: Frère Marie-Victorin, 1885–1944, Canada.
MASAM.: Genkei Masamune, 正 宗 嚴 敬, 1899—, Japan (Formosa, Orchidaceae).
MAST.: Maxwell Tylden Masters, 1833–1907, United Kingdom (cultivated plants).
MATSUDA: Sadahisa Matsuda, 松 田 定 久, 1857–1921, Japan (flora of China).
MATSUM.: Jinzō Matsumura, 松 村 任 三, 1856–1928, Japan.
MATSUNO: Jūtarō Matsuno, 松 野 重 太 郎, 1868–1946, Japan.
MATTF.: Johannes Mattfeld, 1895–1951, Germany.
MAXIM.: Carl Johann Maximowicz, 1827–1891, U.S.S.R.
MAXON: William Ralph Maxon, 1877–1948, U.S.A. (ferns).
MAYR: Heinrich Mayr, 1856–1911, Germany (Coniferae).
MAZEL: Eugène Mazel, flourished 1872, France (cultivated plants).
McCLINTOCK: Elizabeth McClintock, 1912—, U.S.A. (*Hydrangea*).
McCLURE: Floyd Alonzo McClure, 1897—, U.S.A. (bamboo).
MEDIC: Friedrich Casimir Medicus, 1736–1808, Germany.
L. W. MEDIC.: Ludwig Wallrad Medicus, 1771–1850, Germany.
MEEHAN: Thomas Meehan, 1826–1901, U.S.A.
MEERB.: Nicolaas Meerburg, 1734–1814, Netherlands.
MEINSH.: Karl Friedrich Meinshausen, 1819–1899, U.S.S.R. (Cyperaceae, *Sparganium*).
MEISSN.: Carl Friedrich Meissner, 1800–1874, Switzerland.
MELCHOIR: Hans Melchoir, 1894—, Germany.
MENZ.: Archibald Menzies, 1754–1842, United Kingdom.
MERR.: Elmer Drew Merrill, 1876–1956, U.S.A. (Philippines, China, U.S.A.).
MERT.: Franz Karl Mertens, 1764–1831, Germany.
MERT. & KOCH: F. K. Mertens and W. D. J. Koch.
METCALF: Franklin Post Metcalf, 1892–1955, U.S.A. (China).
METT.: Georg Heinrich Mettenius, 1823–1866, Germany (ferns).
C. A. MEY.: Carl Anton von Meyer, 1795–1855, U.S.S.R.
E. MEY.: Ernst Heinrich Friedrich Meyer, 1791–1858, Germany.
MEYEN: Franz Julius Ferdinand Meyen, 1804–1840, Germany (China).
MEZ: Karl Christian Mez, 1866–1944, Germany (Myrsinaceae).
MICH.: Pier Antonio Micheli, 1679–1737, Italy.
M. MICHELI: Marc Micheli, 1844–1902, Switzerland (Alismataceae, Scheuchzeriaceae).
MICHX.: André Michaux, 1746–1802, France (North America).
MIDDENDORFF: Alexander Theodor von Middendorff, 1815–1894, U.S.S.R.
MIEG: Achilles Mieg, 1731–1799, Switzerland.
MIERS: John Miers, 1789–1879, United Kingdom.
MIGO: Hisao Migo, 御 江 久 夫, 1905—, Japan (China).
MIKI: Shigeru Miki, 三 木 茂, 1901—, Japan (hydrophytes).
MILDE: Carl August Julius Milde, 1824–1871, Germany (Pteridophyta).
MILL.: Philip Miller, 1691–1771, United Kingdom.
MILLSP.: Charles Frederick Millspaugh, 1854–1923, U.S.A.
MIQ.: Friedrik Anton Willem Miquel, 1811–1871, Netherlands.
MIRBEL: Charles François Brisseau de Mirbel, 1776–1854, France.
MITCHELL: John Mitchell, 1676–1768, Virginia.
MITF.: Algernon Bertram Freeman-Mitford, Lord Redesdale, 1837–1916, United Kingdom (bamboo).
MIYABE: Kingo Miyabe, 宮 部 金 吾, 1860–1951, Japan.
MIYAKE: Tsutomu Miyake, 三 宅 勉, 1880—, Japan (Sakhalin).
MIYOSHI: Manabu Miyoshi, 三 好 學, 1861–1939, Japan.
MIZUSHIMA: Masami Mizushima, 水 島 正 美, 1925—, Japan (Caryophyllaceae).
MOEHR.: Paul Heinrich Gerhard Moehring, 1710–1792, Germany.
MOELLER: Hjalmar Moeller, 1866–1941, Sweden (Podostemaceae).
MOENCH: Conrad Moench, 1744–1805, Germany.
MOHLENBROCK: Robert H., Jr., 1931—, U.S.A. (*Zornia*).
MOHNIKE: Otto Gottlieb Johan Mohnike, 1814–1887, Germany (collector in Japan).
MOLDENKE: Harold Norman Moldenke, 1909—, U.S.A. (*Verbenaceae*).

MOMIYAMA: Yasuitsi Momiyama, 籾山泰一, 1904—, Japan.

MOMOSE: Sizuo Momose, 百瀬靜男, 1906—, Japan (ferns).

MOMOTANI: Yoshihide Momotani, 桃谷好英, 1928—, Japan.

MONJUS.: Vladimir A. Monjushko, 1903–35.

MOORE: Thomas Moore, 1821–1887, United Kingdom (Pteridophyta).

S. MOORE: Spencer Le Marchant Moore, 1851–1931, United Kingdom.

MOQ.: Christian Horace Benedict Alfred Moquin-Tandon, 1804–1863, France (Chenopodiaceae, Amaranthaceae).

MOREL: F. Morel, flourished 1883–1916, France.

MORI: Kunihiko Mori, 森 邦彦, 1905—, Japan.

MORIKAWA: Kinichi Morikawa, 森川均一, 1898–1936, Japan.

MORITZI: Alexander Moritzi, 1806–1850, Switzerland.

MORONG: Rev. Thomas Morong, 1827–1894, U.S.A. (*Potamogeton*).

MORR.: Charles François Antoine Morren, 1807–1858, Belgium.

E. MORR.: Charles Jacques Edouard Morren, 1833–1886, Belgium.

MORR. & DECNE.: C. F. A. Morren and Joseph Decaisne.

MORTON: Conrad Vernon Morton, 1905—, U.S.A. (Filices).

MOUILL.: Pierre Mouillefert, 1845–1903, France (dendrology).

MUELL.-ARG.: Jean Mueller (Argoviensis, i.e., of Aargau), 1828–1896, Switzerland (Euphorbiaceae).

F. MUELL.: Baron Ferdinand Jacob Heinrich von Mueller, 1825–1896, Australia.

P. J. MUELL.: Philipp Jacob Mueller, 1832–1889, France.

MUHL.: Henry Ludwig Muhlenberg, 1756–1817, U.S.A.

MUNRO: William Munro, 1818–1880, United Kingdom.

MURATA: Gen Murata (Gen Nakai), 1927—, 村田 源 （中井源）, Japan.

MURB.: Svante Samuel Murbeck, 1859–1946, Sweden (*Gentiana*).

MUROI: Hiroshi Muroi, 室井 綽, 1914—, Japan.

MURR.: Johann Anders Murray, 1740–1791, Sweden; Germany (Coniferae).

A. MURR.: Andrew Murray, 1812–1878, United Kingdom (Coniferae).

MUTIS: José Celestino Mutis, 1732–1811, Spain (explorer in Colombia).

NAKAI: Takenoshin Nakai, 中井猛之進, 1882–1952, Japan.

G. NAKAI: Changed to Gen Murata, see MURATA.

NAKAJIMA: Sadao Nakajima, 中島定雄, Japan.

NAKANO: Harufusa Nakano, 中野治房, 1883—, Japan.

NANNF.: Johan Axel Nannfeldt, 1904—, Sweden.

NASH: George Valentine Nash, 1864–1921, U.S.A.

NATHORST: Alfred Gabriel Nathorst, 1850–1921, Sweden.

NAUD.: Charles Victor Naudin, 1815–1899, France (Cucurbitaceae, Melastomataceae).

NAVES: Andrés Naves, 1839–1910, Spain (Philippines).

NECK.: Noel Joseph de Necker, 1729–1793, France.

NEES: Christian Gottfried Daniel Nees von Esenbeck, 1776–1858, Germany.

T. NEES: Theodor Friedrich Ludwig Nees von Esenbeck, 1787–1837, Germany (medicinal plants).

NEES & EBERM.: T. F. L. Nees and Karl Heinrich Ebermaier, 1802–1870, Germany.

NEKRASSOWA: Wera Leontievna Nekrassowa, 1884—, U.S.S.R. (Saxifragaceae).

NELMES: Ernest Nelmes, 1896–1959, United Kingdom (*Carex*).

A. NELS.: Aven Nelson, 1859–1952, U.S.A.

NEMOTO: Kwanji Nemoto, 根本莞爾, 1860–1936, Japan.

NESTL.: Christian Gottfried Nestler, 1778–1832, Germany.

NEUBERT: Wilhelm Neubert, 1808–1905, Germany (Gart. Mag.).

NEUMANN: Louis Neumann, 1827–1903, France (cultivated plants).

NEVSKI: Sergei Arsenievich Nevski, 1908–1938, U.S.S.R.

NEWBOLD: Patty Thum Newbold, U.S.A.

NEWM.: Edward Newman, 1801–1876, United Kingdom.

NICHOLS.: George Nicholson, 1847–1908, United Kingdom (cultivated plants).

NIEDENZU: Franz Josef Niedenzu, 1857–1937, Germany.

NIEUWL.: Julius Aloysius Arthur Nieuwland, 1878–1936, U.S.A.

NISHIDA: Makoto Nishida, 西田 誠, 1927—, Japan (Pteridophyta).

NOISETTE: Louis Claude Noisette, 1772–1849, France.

NUTT.: Thomas Nuttall, 1786–1859, United Kingdom; U.S.A. (collector, flora of North America).

NYM.: Carl Fredrik Nyman, 1820–1893, Sweden.

ODASHIMA: Kijiro Odashima, 小田島喜次郎, Japan.

OED.: Georg Christian von Oeder, 1728–1791, Denmark.

OERST.: Anders Sandøe Oersted, 1816–1872, Denmark.

OETT.: H. von Oettingen, flourished about 1905, Latvia.

Ogata: Masasuke Ogata, 緒方正資, 1883–1944, Japan (Pteridophyta).

Ohba.: Tatsuyuki Ohba, 大場達之, 1936–, Japan.

Ohki: Kiichi Ohki, 大木麒一, 1882–, Japan.

M. Ohki: Masao Ohki, 大木正夫, 1930–, Japan (*Betula*).

Ohwi: Jisaburō Ohwi, 大井次三郎, 1905–, Japan (Cyperaceae, Gramineae).

Ōkubo: Saburō Ōkubo, 大久保三郎, 1857–1914, Japan.

Okuyama: Shunki Okuyama, 奥山春季, 1909–, Japan.

Oliv.: Daniel Oliver, 1830–1916, United Kingdom.

F. W. Oliv.: Francis Wall Oliver, 1864–1951, United Kingdom.

Onno: Max Onno, 1903–, Austria.

Ono: Motoyoshi Ono, 小野職愨, 1837–1890, Japan.

Ooststr.: S. J. van Ooststroom, 1906–, Netherlands (Convolvulaceae).

Opiz: Philipp Maximilian Opiz, 1787–1858, Czechoslovakia.

Osbeck: Pehr Osbeck, 1723–1805, Sweden (collector and traveler in China).

Ostenf.: Carl Emil Hansen Ostenfeld, 1873–1931, Denmark.

Otto: Christoph Friedrich Otto, 1783–1856, Germany.

Otto & Dietr.: C. F. Otto and A. Dietrich.

Ottol.: Kornelius Johannes Willem Ottolander, 1822–1887, Netherlands.

Oudem.: Cornelius Antoon Jan Abraham Oudemanns, 1825–1906, Netherlands.

Packer: John George Packer, 1929–, Canada (*Kengia*).

Palib.: Ivan Vladimirovitch Palibin, 1872–1949, U.S.S.R. (flora of Korea).

Pall.: Peter Simon Pallas, 1741–1811, U.S.S.R.

Palla: Eduard Palla, 1864–1922, Czechoslovakia (Cyperaceae).

Palmgr.: Alvar Palmgren, 1880–, Finland.

Pampan.: Renato Pampanini, 1875–1949, Italy (*Artemisia*).

Pantling: Robert Pantling, 1857–1910, United Kingdom.

Panzer: Georg Franz Volgang Panzer, 1755–1829, Germany.

Paoletti: Giulio Paoletti, 1865–, Italy.

Pardé: Léon Gabriel Charles Pardé, 1865–1943, France (dendrology).

Parl.: Filippo Parlatore, 1816–1877, Italy.

Pascher: Adolf Pascher, 1881–194–?, Austria (*Gagea,* Solanaceae).

Patr.: Eugene Louis Melchoir Patrin, 1742–1815, France.

Patschke: Wilhelm Patschke, 1888–, Germany.

W. Paul: Wilhelm Paul, 1822–1905, United Kingdom.

Pav.: José Antonio Pavón, 175–?–1844, Spain (flora of Peru).

Pax: Ferdinand Albin Pax, 1858–1942, Germany.

Paxt.: Sir Joseph Paxton, 1801–1865, United Kingdom.

Pears.: Henry Harold Welch Pearson, 1870–1916, United Kingdom.

Pennell: Francis Whittier Pennell, 1886–1952, U.S.A. (Scrophulariaceae).

Perkins: Janet Russell Perkins, 1853–1933, U.S.A.

Pers.: Christian Hendrick Persoon, 1755–1837, South Africa; France.

Peter-Stibal: Elfriede Peter-Stibal, 1905–, Austria.

Peterm.: Wilhelm Ludwig Petermann, 1806–1855, Germany.

Petitm.: Marcel Georges Charles Petitmengin, 1881–1908, France.

Petrov.: Sava Petrovich (Petrovič), 1839–1889, Jugoslavia.

Pfeiffer: Ludwig Georg Carl Pfeiffer, 1805–1877 (Cactaceae).

H. Pfeiff.: Hans Pfeiffer, 1890–, Germany (Cyperaceae).

Pfitz.: Ernst Hugo Heinrich Pfitzer, 1846–1906, Germany.

Phil.: Rudolph Amandus Philippi, 1808–1904, Chile.

Pierot: Jacques Pierot, 1812–1841, Netherlands (collector in Japan).

Pierre: Jean Baptiste Louis Pierre, 1833–1905, France.

Pilger: Robert Knud Friedrich Pilger, 1876–1953, Germany (Gramineae, Plataginaceae, Coniferae).

Piper: Charles Vancouver Piper, 1867–1926, U.S.A. (Leguminosae).

Planch.: Jules Émile Planchon, 1823–1888, France.

Poelln.: Karl von Poellnitz, 1896–1945, Germany.

Poir.: Jean Louis Marie Poiret, 1755–1834, France.

Poit.: Antoine Poiteau, 1766–1854, France (cultivated plants).

Pojark.: Antonina Ivanovna Pojarkova, 1897–, U.S.S.R.

Poll.: Johann Adam Pollich, 1740–1780, Germany?

Polunin: Nicholas Polunin, 1909–, United Kingdom.

Popov: Mikhail Grigorievich Popov, 1893–1955, U.S.S.R.

Porsild: Alf Erling Porsild, 1901–, Canada (Arctic flora).

Pourr.: Pierre André Pourret, 1754–1818, France.
Praeg.: Robert Lloyd Praeger, 1865–1953, Ireland (*Sedum*).
Prain: Sir David Prain, 1857–1944, United Kingdom.
Prantl: Karl Anton Eugen Prantl, 1849–1893, Germany.
Presc.: John D. Prescott, —1837, U.S.S.R. (Cyperaceae).
C. Presl: (Carel) Karel Boriweg Presl, 1794–1852, Czechoslovakia.
J. Presl: Jan Swaptopluk Presl, 1791–1849, Czechoslovakia.
Price: William Robert Price, 1886—, United Kingdom.
Printz: Karl Hendrik Oppegaard Printz, 1888—, Denmark (flora of Siberia).
Pritz.: Georg August Pritzel, 1815–1874, Germany (*Anemone*, bibliogr.).
E. Pritz.: Ernst Georg Pritzel, 1875–1946, Germany (Lycopodiaceae).
Prokhan.: Yaroslav Ivanovich Prokhanov (Prokhanoff), 1902—, U.S.S.R.
Pulle: August Adriaan Pulle, 1878–1955, Netherlands.
Pulliat: Victor Pulliat, 1827–1866, France.
Pursh: Frederick Traugott Pursh, 1774–1820, Germany; U.S.A. (flora of U.S.A.).
Quisumbing: Eduardo Quisumbing, 1895—, Philippines.
Racib.: Marian Raciborski, 1863–1917, Poland (ferns).
Radde: Gustav Ferdinand Richard Johannes Radde, 1831–1903, Germany.
Raddi: Giuseppe Raddi, 1770–1829, Italy (Brazil).
Radlk.: Ludwig Adolph Timotheus Radlkofer, 1829–1927, Germany (Sapindaceae).
Raeusch.: Ernst Adolf Raeuschel, flourished 1772–1797, Germany ("Nomenclator Botanicus").
Raf.: Constantine Samuel Rafinesque-Schmaltz, 1783–1840, U.S.A.
Ramat.: Abbé Thomas Albin Joseph d'Audibert de Ramatuelle, 1750–1794, France.
Raven: Peter Raven, Contemporary, U.S.A. (*Epilobium*).
R. Raymund: Raymund Rapaics von Ruhmwert, 1885—, Hungary.
Reching.: Karl Rechinger, 1867–1952, Austria.
Rechinger f.: Karl Heinz Rechinger, 1906—, Austria (*Rumex*).
Redout.: Pierre Joseph Redouté, 1761–1840, France (Liliaceae, flower painter).
Reeder: John Raymond Reeder, 1914—, U.S.A. (Gramineae).
Regel: Eduard August von Regel, 1815–1892, Germany; U.S.S.R.
Reg. & Til.: Eduard August von Regel and Heinrich Sylvester Theodor Tiling.
Rehd.: Alfred Rehder, 1863–1949, Germany; U.S.A. (dendrology).
Reichenb.: Heinrich Gottlieb Ludwig Reichenbach, 1793–1879, Germany (Europe, *Aconitum*).
Reichenb. f.: Heinrich Gustav Reichenbach, 1823–1889, Germany (Orchidaceae).
E. M. Reid: Eleanor Mary Reid, 1860–1953, United Kingdom.
Reinw.: Kaspar Georg Karl Reinwardt, 1773–1854, Netherlands (Malaysia).
Rendle: Alfred Barton Rendle, 1865–1938, United Kingdom.
Retz.: Anders Johan Retzius, 1742–1821, Sweden.
A. Rich.: Achille Richard, 1794–1852, France (*Hydrocotyle, Elaeagnus*).
J. Rich.: Sir John Richardson, 1787–1865, United Kingdom.
L. C. Rich.: Louis Claude Marie Richard, 1754–1821, France (collector in trop. Amer.).
Richt.: Karl Richter, 1855–1891, Austria (pl. Europe).
Ricker: Percy Leroy Ricker, 1878—, U.S.A. (*Lespedeza*).
Ridl.: Henry Nicholas Ridley, 1855–1956, United Kingdom (Malaya).
Du Rietz: Gustav Einar Du Rietz, 1895—, Sweden (*Euphrasia*, lichens, ecology).
Rigg: George Burton Rigg, 1872–1961, U.S.A.
A. & C. Riv.: Marie Auguste Rivière, 1821–1877, and Charles Marie Rivière, 1845–?, France (bamboo).
Robins.: Benjamin Lincoln Robinson, 1864–1935, U.S.A.
C. B. Robinson: Charles Budd Robinson, 1871–1913, U.S.A.; Philippines.
Rochebr.: Alphonse Trémeau de Rochebrune, 1834–1912, France.
Roem.: Johann Jacob Roemer, 1763–1819, Germany.
Rohrb.: Paul Rohrbach, 1847–1871, Germany (Caryophyllaceae).
Rolfe: Robert Allen Rolfe, 1855–1921, United Kingdom (Orchidaceae, Formosa).
Romain: Charles Romain, France.
Ronn.: Karl Ronniger, 1871–1954, Austria (*Thymus, Galium*).
Rosc.: William Roscoe, 1753–1831, United Kingdom (Zingiberaceae).
Rose: Joseph Nelson Rose, 1862–1928, U.S.A. (Cactaceae).
Rosenburgh: Cornelis Rugier Willem Karel Alderwerelt van Rosenburgh, 1863–1936, Netherlands (ferns).
Rosend.: Carl Otto Rosendahl, 1875–1956, U.S.A. (*Mitella*).
Rosenst.: Eduard Rosenstock, 1856—, (Pteridophyta).
Rosenth.: Käthe Rosenthal, flourished about 1919, Austria (*Daphniphyllum*).
Rosenvinge: Janus Lauritz Andreas Kolderup Rosenvinge, 1858–1939, Denmark.

ROSHEV.: Roman Julievich Rozhevitz (Roshevitz), 1882–1949, U.S.S.R. (Gramineae).

ROSTK.: Friedrich Wilhelm Gottlieb Rostokovius, 1770–1848, Germany.

ROTH: Albrecht Wilhelm Roth, 1737–1834, Germany.

ROTHERT: Wladislaw Adolfovich Rothert, 1863–1916, U.S.S.R. (*Sparganium*).

ROTTB.: Christen Friis Rottboell, 1727–1797, Denmark.

ROTTL.: Johann Peter Rottler, 1749–1836, Denmark.

ROUSS.: Joseph Jules Jean Jacques Rousseau, 1905–, Canada.

ROXB.: William Roxburgh, 1751–1815, United Kingdom (India).

ROYLE: John Forbes Royle, 1799–1858, United Kingdom (India).

RUDGE: Edward Rudge, 1763–1846, United Kingdom.

RUDOLPH: Johann Heinrich Rudolph, 1744–1809, U.S.S.R. (*Dicentra*).

RUEMPLER: Theodor Ruempler, 1817–1891, Germany.

RUHL.: Eugen Otto Willy Ruhland, 1878–, Germany (Eriocaulaceae).

RUIZ: Hipólito Ruiz-Lopez, 1754–1815, Spain.

RUMPH.: Georg Eberhard Rumphius, 1628–1702, Germany ("Fl. Amboina").

RUPR.: Franz Joseph Ruprecht, 1814–1870, U.S.S.R.

RYDB.: Per Axel Rydberg, 1860–1931, U.S.A. (flora of Rocky Mts.).

RYLANDS: Thomas Glazebrook Rylands, 1818–1900, United Kingdom (*Athyrium alpestre*)

SABINE: Joseph Sabine, 1770–1837, United Kingdom.

ST. HIL.: Auguste de Saint Hilaire, 1779–1853, France.

ST. JOHN: Harold St. John, 1892–, U.S.A. (Pacific Isls.).

ST. YVES: Alfred Saint-Yves, 1855–1933, France (Gramineae).

SAKAMOTO: Sadao Sakamoto, 阪本寧男, 1930–, Japan.

SALISB.: Richard Anthony Salisbury, 1761–1829, United Kingdom.

G. SAM.: Gunnar Samuelsson, 1855–1944, Sweden.

SANDER: Henry Frederick Conrad Sander, 1847–1920, United Kingdom.

SARG.: Charles Sprague Sargent, 1841–1927, U.S.A. (dendrology).

SASAKI: Syun'iti Sasaki, 佐々木舜一, 1888–1960, Japan (flora of Formosa).

SATA: Tyosyun Sata, 佐多長春, 1907–, Japan (*Ficus*).

SATAKE: Yoshisuke Satake, 佐竹義輔, 1902–, Japan (*Eriocaulon, Juncus, Boehmeria*).

SATOMI: Nobuo Satomi, 里見信生, 1922–, Japan.

SAVAT.: SAV.: Paul Amedée Ludovic Savatier, 1830–1891, France (flora of Japan).

SAVI: Gaetano Savi, 1769–1844, Italy.

SAWADA: Taketarō Sawada, 澤田武太郎, 1899–1938, Japan.

SCHANGIN: Petr Ivanovich Schangin, 1741–1816, U.S.S.R.

SCHAUER: Johan Conrad Schauer, 1813–1848, Germany (Verbenaceae).

SCHELLE: Ernst Schelle, fl. 1890–1930, Germany (Garteninspektor Univ. Tübingen).

SCHERB.: Johannes Scherbius, 1769–1813, Germany (flora of Witter.).

SCHINDL.: Anton Karl Schindler, 1879–, Germany (Leguminosae, Haloragaceae).

SCHINZ: Hans Schinz, 1858–1941, Switzerland.

SCHIPCZ.: Nikolai Valerianovich Shipchinskii (Schipczinsky), 1886–1955, U.S.S.R. (Ranunculaceae).

SCHISCHK.: Boris Konstantinovich Shishkin (Schischkin), 1886–1963, U.S.S.R. (Caryophyllaceae).

SCHK.: Christian Schkuhr, 1741–1811, Germany (*Carex*).

SCHLEICH.: Johann Christoph Schleicher, 1768–1834, Switzerland.

SCHLEID.: Matthias Jacob Schleiden, 1804–1881, Germany.

SCHLTDL.: Dietrich Franz Leonhard von Schlechtendal, 1794–1866, Germany.

SCHLTR.: Friedrich Reichardt Rudolf Schlechter, 1872–1925, Germany (Orchidaceae).

SCHMALH.: Johannes Theodor Schmalhausen, 1849–1894, U.S.S.R.

F. SCHMIDT: Friedrich Schmidt, 1832–1908, U.S.S.R. (Sakhalin, Bureja).

F. W. SCHMIDT: Franz Wilibald Schmidt, 1764–1796, Czechoslovakia.

J. SCHM.: Johann Anton Schmidt, 1823–1905, Germany.

C. K. SCHN.: Camillo Karl Schneider, 1876–1951, Germany (dendrology).

SCHNIZL.: Adelbert Schnizlein, 1814–1868, Germany.

SCHOTT: Heinrich Wilhelm Schott, 1794–1865, Austria (Araceae).

SCHOTTKY: Ernst Max Schottky, 1888–1915, Switzerland.

SCHRAD.: Heinrich Adolph Schrader, 1767–1836, Germany.

SCHRANK: Franz von Paula Schrank, 1747–1835, Germany.

SCHREB.: Johann Christian Daniel von Schreber, 1739–1810, Germany.

SCHULT.: Joseph August Schultes, 1773–1831, Germany.

SCHULT. F.: Julius Hermann Schultes, 1804–1840, Austria.

SCHULTZ-BIP.: Carl Heinrich Schultz *Bipontinus* (i.e., of Zweibrüchen), 1805–1867, Germany (Compositae).

O. E. SCHULZ: Otto Eugen Schulz, 1874–1936, Germany (Cruciferae).

K. Schum.: Karl Moritz Schumann, 1851–1904, Germany (Zingiberaceae).

Schumach.: Heinrich Christian Friedrich Schumacher, 1757–1830, Denmark.

Schur: Philipp Johann Ferdinand Schur, 1799–1878, Austria.

Schweick.: Herald Georg Wilhelm Johannes Schweickerdt, 1903—, South Africa.

Schwein.: Lewis David de Schweinitz, 1780–1834, U.S.A.

Schwerin: Graf Fritz von Schwerin, 1856–1934, Germany.

Scop.: Giovanni Antonio Scopoli, 1723–1788, Italy.

Scribn. & Smith: Frank Lamson Scribner, 1851–1938, U.S.A. (Gramineae); Jared Gage Smith, 1866—, U.S.A.

Sealy: Joseph Robert Sealy, 1907—, United Kingdom (*Camellia;* cultivated plants).

Seem.: Berthold Carl Seemann, 1825–1871, Germany (China, Araliaceae).

Seemen: Karl Otto von Seemen, 1838–1910, Germany (*Salix*).

Ser.: Nicolas Charles Seringe, 1776–1858, France (Cucurbitaceae, Saxifragaceae).

Sergiev.: Lidia Palladievna Sergievskaja, 1897—, U.S.S.R.

Serv.: Camille Servettaz, 1870–1947, France (*Elaeagnus*).

Setchell: William Albert Setchell, 1864–1943, U.S.A.

Sherff: Earl Edward Sherff, 1886—, U.S.A.

Shibata: Keita Shibata, 柴田桂太, 1877–1949, Japan (*Sasa*).

T. Shimizu: Tatemi Shimizu, 清水建美, 1932—, Japan.

Shimotomai: Naomasa Shimotomai, 下斗米直昌, 1899—, Japan (*Chrysanthemum*).

Shirai: Mitsutarō Shirai, 白井光太郎, 1863–1932, Japan.

Shiras.: Homi Shirasawa, 白澤保美, 1868–1947, Japan (dendrology).

Sibth.: John Sibthorp, 1758–1796, United Kingdom.

Sieb.: Philipp Franz von Siebold, 1796–1866, Germany; Netherlands.

Sieb. & Zucc.: P. F. von Siebold and J. G. Zuccarini.

Sieber: Franz Wilhelm Sieber, 1789–1844, Austria.

Simonk.: Lajos tól Simonkai, 1851–1910, Hungary.

Sims: John Sims, 1749–1831, United Kingdom.

Skan: Sidney Alfred Skan, 1870–1940, United Kingdom.

Skottsb.: Carl Johan Fredrik Skottsberg, 1880–1963, Sweden (flora of Pacific Isls.).

Skvort.: B. V. Skvortzov (B. W. Skvortzow), flourished 1917–1955, U.S.S.R.; China (Manchuria), Brazil.

Sledge: William A. Sledge, 1904—, United Kingdom.

Sloss.: Margaret Slosson, 1872—, U.S.A. (ferns).

Small: John Kunkel Small, 1869–1938, U.S.A.

A. C. Smith: Albert Charles Smith, 1906—, U.S.A.

Harry Smith: Harry Smith, 1889—, Sweden (China, *Saxifraga, Gentiana*).

J. Smith: John Smith, 1798–1888, United Kingdom (Pteridophyta).

J. E. Smith: Sir James Edward Smith, 1759–1828, United Kingdom.

J. J. Smith: Johannes Jacobus Smith, 1867–1947, Netherlands (Orchidaceae).

L. B. Smith: Lyman Bradford Smith, 1904—, U.S.A. (Bromeliaceae).

W. W. Smith: William Wright Smith, 1875–1956, United Kingdom (Chinese flora).

Sobol.: Gregor Federovitch Sobolewski, 1741–1807, U.S.S.R.

Soland.: Daniel Carl Solander, 1736–1782, Sweden; United Kingdom.

Solms-Laub.: Count Hermann Maximilian Carl Ludwig Friedrich Laubach, 1842–1915, Germany.

Somm.: Carlo Pietro Stefano Sommier, 1848–1922, Italy.

Sonnerat: Pierre Sonnerat, 1749–1814, France (China, India).

Soó: Károly Rezsö Soó von Bere, 1903—, Hungary (*Melampyrum,* Orchidaceae).

Sowerb.: James Sowerby, 1757–1822, and James De Carl Sowerby, 1787–1871, United Kingdom (English botany).

Spach: Édouard Spach, 1801–1879, France.

Spae: Dieudonné Spae, 1819–1879, Belgium (*Lilium*).

Spaeth: Franz Ludwig Spaeth, 1838–1913, Germany (cultivated plants).

Spenner: Fridolin Karl Leopold Spenner, 1798–1841, Germany.

Sprague: Thomas Archibald Sprague, 1877–1958, United Kingdom.

Spreng.: Kurt Polykarp Joachim Sprengel, 1766–1833, Germany.

Spring: Frédéric Antoine Spring, 1814–1872, Belgium (Pteridophyta).

Stapf: Otto Stapf, 1858–1933, Austria; United Kingdom (Gramineae).

Staunton: Sir George Leonard Staunton, 1737–1801, United Kingdom (China).

Stearn: William Thomas Stearn, 1911—, United Kingdom (cultivated plants).

Stebbins: George Ledyard Stebbins, Jr., 1906—, U.S.A.

Steinb.: Elisabeth Ivanovna Steinberg, 1884–1963, U.S.S.R. (*Aconitum*).

Steller: Georg Wilhelm Steller, 1709–1746, Germany (Siberia).

Steph.: Christian Friedrich Stephan, 1757–1814, U.S.S.R.

Sternb.: Caspar, Graf von Sternberg, 1761–1838, Austria (*Saxifraga*).

STEUD.: Ernst Gottlieb Steudel, 1783–1856, Germany.

STEV.: Christian von Steven, 1781–1863, U.S.S.R. (flora of Russia; *Pedicularis*).

STEWARD: Albert Newton Steward, 1897–1959, U.S.A.; China (Polygonaceae).

STOKER: Fred Stoker, 1878–1943, United Kingdom (cultivated plants).

STOKES: Jonathan Stokes, 1755–1831, United Kingdom.

STOUT: Arlow Burdette Stout, 1876–1957, U.S.A. (*Hemerocallis*).

SUGAYA: Sadao Sugaya, 菅谷貞男, 1917—, Japan.

SUGIMOTO: Jun'ichi Sugimoto, 杉本順一, 1901—, Japan.

SUKATSCHEV: Vladimir Nikolaevich Sukhachev (Sukatschev), 1880—, U.S.S.R.

SUMMERF.: Sörren Christian Summerfelt, 1794–1838, Norway.

SUMMERH.: Victor Samuel Summerhayes, 1897—, United Kingdom.

SURING.: Jan Valckenier Suringer, 1865–1932, Netherlands (*Rhododendron*).

SUTŌ: Tiharu Sutō, 須藤千春, 1910—, Japan (*Chrysosplenium, Carex*).

SUTTON: Rev. Charles Sutton, 1756–1846, United Kingdom.

SUZUKI: Sigeyosi Suzuki, 鈴木重良, 1894–1937, Japan (Formosa).

T. SUZUKI: Tokio Suzuki, 1911—, 鈴木時夫, Japan; Formosa (*Rhododendron, Chrysosplenium*).

SVENSON: Henry Knute Svenson, 1897—, U.S.A. (*Eleocharis*).

SW.: Olof Peter Swartz, 1760–1818, Sweden.

SWALLEN: Jason Richard Swallen, 1903—, U.S.A. (Gramineae).

SWEET: Robert Sweet, 1783–1835, United Kingdom (*Geranium,* cultivated plants).

SWINGLE: Walter Tennyson Swingle, 1871–1952, U.S.A. (*Citrus*).

SWINHOE: Robert Swinhoe, 1836–1877, United Kingdom (collector in Formosa).

SYME: John Thomas Irvine Boswell, né Syme, afterwards Boswell-Syme, 1822–1888, United Kingdom.

SZYSZ.: Ignaz Szyszylowicz, 1857–1910, Poland.

TAGAWA: Motozi Tagawa, 田川基二, 1908—, Japan (ferns).

TAGG: Harry Frank Tagg, 1874–1933, United Kingdom (*Rhododendron*).

TAKAH.: Yoshinao Takahashi, 高橋良直, flourished 1896–1915, Japan (plant pathologist).

TAKEDA: Hisayoshi Takeda, 武田久吉, 1883—, Japan.

TAKENOUCHI: Makoto Takenouchi, 竹内　亮, 1894—, Japan.

TAMURA: Michio Tamura, 田村道夫, 1927—, Japan.

TANAKA: Yoshio Tanaka, 田中芳男, 1838–1916, Japan.

T. TANAKA: Tyōzaburō (or Chōzaburō) Tanaka, 田中長三郎, 1885—, Japan (*Citrus*).

Y. TANAKA: Yuichiro Tanaka, 田中諭一郎, 1900—, Japan (*Citrus*).

TANAKA & ONO: Yoshio Tanaka and Motoyoshi Ono, Japan.

TANG: Tsin Tang (Ching T'ang), 唐　進, 1900—, China (Cyperaceae).

TARD.-BL.: Marie L. Tardieu-Blot, 1902—, France (Pteridophyta).

Y. TASHIRO: Yasusada Tashiro, 1856–1928, 田代安定, Japan.

Z. TASHIRO: Zentarō Tashiro, 1872–1947, 田代善太郎, Japan.

TATEW.: Misao Tatewaki, 館脇　操, 1899—, Japan (flora of Hokkaido).

TAUB.: Paul Hermann Wilhelm Taubert, 1862–1897, Germany (Leguminosae).

TAUSCH: Ignaz Friedrich Tausch, 1793–1848, Austria (*Hieracium*).

TEIJSM.: Johannes Elias Teijsmann (Teysmann), 1808–1882, Netherlands; Java.

TEMPLE: F. L. Temple, fl. 1887–1893, U.S.A. (nurseryman).

TENORE: Michele Tenore, 1780–1861, Italy.

TERRAC.: Achille Terraciano, 1861–1917, Italy (*Chrysosplenium*).

TEUTSCHELL: Messers. Teutschell of Colchester, United Kingdom.

TEYSM.: See TEIJSM.

THELL.: Albert Thellung, 1881–1928, Switzerland.

THOMS.: Thomas Thomson, 1817–1878, United Kingdom (India).

THORY: Claude Antoine Thory, 1759–1827, France.

THOU.: Louis Marie Aubert Du Petit Thouars, 1758–1831, France.

THUILL.: Jean Louis Thuillier, 1757–1822, France.

THUNB.: Carl Pehr Thunberg, 1743–1828, Sweden (flora of Japan).

THW.: George Henry Kendrick Thwaites, 1811–1882, United Kingdom (Ceylon).

TILING: Heinrich Sylvester Theodor Tiling, 1818–1871, U.S.S.R.

TIMM: Joachim Christian Timm, 1754–1805, Germany (*Nuphar pumilum*).

TINDALE: Mary D. Tindale, (contemporary) Australia (Pteridophyta).

TOBLER: Friedrich Tobler, 1879—, Germany (*Hedera*).

TODARO: Augustino Todaro, 1818–1892, Sicily.

TOEPFFER: Adolph Toepffer, 1853–1931, Germany.

TOLMATCHEV: Aleksandr Innokentevich Tolmachev (Tolmatchev), 1903—, U.S.S.R.

TORR.: John Torrey, 1796–1873, U.S.A.

TOYOKUNI: Hideo Toyokuni, 豐國秀夫, 1932—, Japan (*Gentiana*).

TRACY: Samuel Mills Tracy, 1847–1920, U.S.A.

TRATT.: Leopold Trattinnick, 1764–1849, Austria (*Rosa, Hosta*).

TRAUTV.: Ernst Rudolph von Trautvetter, 1809–1889, U.S.S.R.

TRÉC.: Auguste Adolphe Lucien Trécul, 1818–1896, France.

TREL.: William Trelease, 1857–1945, U.S.A. (Onagraceae, Liliaceae).

TREVIR.: Christian Ludolf Treviranus, 1779–1864, Germany.

TRIANA: José (Jérónimo) Triana, 1828–1890, Colombia (Melastomataceae).

TRIM.: Henry Trimen, 1843–1896, United Kingdom (Ceylon).

TRIN.: Carl Berhard von Trinius, 1778–1844, U.S.S.R. (Gramineae).

TRYON: Rolla Milton Tryon, 1916—, U.S.A. (Filices).

TSCHONOSKI: Tschonoski (or Chonosuke) Sukawa, 須川長之助, 1841–1925, Japan (collector for Maximowicz).

TUCKERM.: Edward Tuckerman, 1817–1886, U.S.A.

TURCZ.: Nikolai Stepanovich Turczaninov, 1796–1864, U.S.S.R.

TURRILL: William Bertram Turrill, 1890–1961, United Kingdom.

TUYAMA: Takasi Tuyama, 津山 尚, 1910—, Japan.

TZVEL.: Nikolai Nikolaievich Tzvelev (Tsvelev), 1925—, U.S.S.R. (*Dimeria*).

UCHIDA: Shigetaro Uchida, 內田繁太郎, 1885—, Japan (bamboo).

UITTIEN: Hendrik Uittien, 1898–1944, Netherlands (Cyperaceae).

ULBR.: Eberhard Ulbrich, 1879–1952, Germany (Theligonaceae, Ranunculaceae).

UNDERW.: Lucien Marcus Underwood, 1853–1907, U.S.A. (Filices).

URBAN: Ignatius (Ignatz) Urban, 1848–1931, Germany (Scrophulariaceae).

UYEKI: Homiki Uyeki, 植木秀幹, 1882—, Japan (Korea, dendrology).

VAHL: Martin Hendriksen Vahl, 1749–1804, Denmark (Cyperaceae).

VALETON: Theodoric Valeton, 1855–1929, Netherlands (Rubiaceae).

VAN HALL: Hermann Christiaan van Hall, 1801–1874, Netherlands.

VAN HOUTTE: Louis Van Houtte, 1810–1876, Belgium (cultivated plants).

VAN TIEGH.: Philippe van Tieghem, 1838–1914, France (Loranthaceae).

VAN'T.: Eugène Vaniot, —1913, France (*Carex,* Compositae).

VAN STEENIS: Cornelis Gijbert Gerrit Jan van Steenis, 1901—, Netherlands; Java ("Flora Malesiana").

VASEY: George Vasey, 1822–1893, U.S.A. (Gramineae).

VASING.: Antonina Vasilievna Vasinger-Alektorova, 1892–194-?, U.S.S.R.

VASS.: Viktor Nikolayenich Vasiliev (Vassiljev), 1890—, U.S.S.R.

VATKE: Georg Carl Wilhelm Vatke, 1849–1889, Germany.

VEITCH: John Gould Veitch, 1839–1870, United Kingdom (cultivated plants).

VELL.: José Marianno da Conceição Vellozo, 1742–1811, Brazil.

VENT.: Étienne Pierre Ventenat, 1757–1808, France.

VIDAL Y SOLAR: Sebastian Vidal y Solar, 1842–1889, Spain (Philippines).

VIEILL.: Eugéne Vieillard, 1819–1896, France.

VIERH.: Fritz Vierhapper, 1876–1902, Austria.

VILL.: Dominique Villars, 1745–1814, France.

F. VILL.: Celestine Fernandez-Villar, 1838–1907, Spain (Philippines).

VILM.: Pierre Philippe André Levéque de Vilmorin, 1746–1804, France (cultivated plants).

VIS.: Roberto de Visiani, 1800–1878, Italy.

VIV.: Domenico Viviani, 1772–1840, Italy.

VOGEL: Julius Rudolph Theodor Vogel, 1812–1843, Germany.

VOIGT: Johann Otto Voigt, 1798–1843, Germany.

VOSS: Andreas Voss, 1857–1924, Germany (cultivated plants).

VVED.: Aleksei Ivanovich Vvedenski (Vvedensky), 1898—, U.S.S.R. (Liliaceae).

WAHLENB.: Göran (Georg) Wahlenberg, 1780–1851, Sweden.

WALDST.: Count Franz Adam von Waldstein-Wartemberg, 1759–1823, Austria.

WALKER: Egbert H. Walker, 1899—, U.S.A. (Myrsinaceae, "Bibl. e. Asiat. Bot.").

WALL.: Nathaniel Wallich (Nathan Wolff), 1786–1854, Denmark (India, Nepal).

WALLR.: Carl Friedrich Wilhelm Wallroth, 1792–1857, Germany.

WALP.: Wilhelm Gerhard Walpers, 1816–1853, Germany.

WALT.: Thomas Walter, 1740–1788, U.S.A.

F. T. WANG: Fa-Tsuan Wang, 汪發纘, 1900—, China (Orchidaceae).

WANGENH.: Friedrich Adam Julius von Wangenheim, 1747–1800, Germany.

WANGER.: Walther Leonhard Wangerin, 1884–1938, Germany (Cornaceae).

WARB.: Otto Warburg, 1859–1938, Germany.

WARD: Francis Kingdon-Ward, 1885–1958, United Kingdom (collector in China).

WARREN: Fred Adelbert Warren, 1902—, U.S.A.

S. WATANABE: Sadamoto Watanabe, 渡邊定元, 1934–, Japan (*Betula*).

S. WATS.: Sereno Watson, 1858–1925, U.S.A.

WATT: George Watt, 1851–1930, United Kingdom (India, economic plants).

WAWRA: Heinrich Ritter Wawra von Fernsee, 1831–1887, Austria (collector).

WEATHERBY: Charles Alfred Weatherby, 1875–1949, U.S.A. (Filices).

WEBB: Philip Barker Webb, 1793–1854, United Kingdom.

WEBER: Georg Heinrich Weber, 1752–1828, Germany.

WEBERB.: August Weberbauer, 1871–1948, Germany; Peru.

WEDD.: Hugh Algernon Weddell, 1819–1877, France (S. America, Urticaceae).

WEHRH.: Heinrich Rudolf Wehrhahn, 1882–1940, Germany.

WEIGEL: Christian Ehrenfried Weigel, 1748–1831, Germany.

WEIHE: Carl Ernst August Weihe, 1779–1834, Germany (*Stellaria*).

WELW.: Friedrich Martin Josef Welwitsch, 1806–1872, Austria; Portugal.

WENDL.: Johann Christoph Wendland, 1755–1828, Germany (flora of Germany, *Rosa*).

H. WENDL.: Hermann A. Wendland, 1832–1903, Germany (Palmae).

WENZIG: Theodor Wenzig, 1824–1892, Germany.

WESSMAEL: Alfred Wessmael, 1832–1905, Belgium.

WETTST.: Richard Ritter von Wettstein, 1863–1931, Austria.

WEYRICH: Heinrich Weyrich, 1828–1863, U.S.S.R. (collector in Japan).

C. T. WHITE: Cyrill Tennison White, 1890–1950, Australia.

T. G. WHITE: Theodore Greely White, 1872–1901, U.S.A.

WIDDER: Felix Josef Widder, 1892–, Austria.

WIEG.: Karl McKay Wiegand, 1873–1942, U.S.A.

WIGGERS: Friedrich Heinrich Wiggers, 1746–1811, Germany.

WIGHT: Robert Wight, 1796–1872, United Kingdom (India).

W. F. WIGHT: William Franklin Wight, 1874–1954, U.S.A. (Leguminosae).

WIKSTR.: Johann Emanuel Wikstroem, 1789–1856, Sweden.

WILLD.: Karl Ludwig Willdenow, 1765–1812, Germany.

WILLIAMS: Frederic Newton Williams, 1862–1923, United Kingdom (Caryophyllaceae).

WILS.: Ernest Henry Wilson, 1876–1930, U.S.A. (dendrology, *Lilium*).

G. F. WILS.: George Fox Wilson, 1896–1951, United Kingdom (*Lilium*).

WIMM.: Christian Friedrich Heinrich Wimmer, 1803–1868, Germany.

WINKL.: Hubert Winkler, 1875–1941, Germany (Betulaceae).

C. WINKL.: Constantin Alexander Winkler, 1848–1900, U.S.S.R. (Compositae).

H. WINKL.: Hans Winkler, 1877–1945, Germany.

WIRTGEN: Philipp Wilhelm Wirtgen, 1806–1870, Germany.

DE WIT: Hendrik Cornelis Dirk de Wit, 1909–, Netherlands (Leguminosae, Gramineae).

WITH.: William Withering, 1741–1799, United Kingdom.

WITTM.: Marx Carl Ludwig Wittmack, 1839–1929, Germany.

E. WOLF: Egbert Wolf, 1860–1931, U.S.S.R. (dendrology).

T. WOLF: Franz Theodor Wolf, 1841–1924, Germany (*Potentilla*).

H. WOLFF: Karl Friedrich August Hermann Wolff, 1866–1929, Germany (Umbelliferae).

WOOD: Alphonso Wood, 1810–1881, U.S.A.

WORMSK.: Morten Wormskjöld, 1783–1845, Denmark.

WOYNAR: Heinrich Woynar, 1865–1917, Austria.

WRIGHT: Charles Henry Wright, 1864–1941, United Kingdom.

C. WRIGHT: Charles Wright, 1811–1886, U.S.A. (collector on U.S. North Pacific Exploring Expedition).

WU: Yin-Ch'an Wu, 吳韞珍, China.

WULFEN: Franz Xaver von Wulfen, 1728–1805, Hungary.

YABE: Yoshitaka Yabe, 矢部吉禎, 1876–1931, Japan (Umbelliferae, Manchuria).

YAMAMOTO: Yoshimatsu Yamamoto, 1893–1947, 山本由松, Japan (Formosa).

YAMAZAKI: Takasi Yamazaki, 山崎敬, 1921–, Japan (Scrophulariaceae).

YANAGIHARA: Masayuki Yanagihara. 柳原政之, flourished about 1941, Japan (Formosa).

YANAGITA: Yoshizo Yanagita, 柳田由藏, 1872–1945, Japan.

YATABE: Ryokichi Yatabe, 矢田部良吉, 1851–1899, Japan.

YOSHINAGA: Torama Yoshinaga (Inoue), 吉永虎馬 (井上虎馬), 1871–1946, Japan.

ZABEL: Hermann Zabel, 1832–1912, Germany (dendrology).

ZAHN: Karl Hermann Zahn, 1865–1940, Germany (*Hieracium*).

ZENKER: Jonathan Karl Zenker, 1799–1837, Germany.

ZINSERL.: Yurii Dmitrievich Tzinzerling (Y. Zinserling), 1894–1938, U.S.S.R. (*Eleocharis*).

ZOLL.: Heinrich Zollinger, 1818–1859, Switzerland (Java).

ZUCC.: Joseph Gerhard Zuccarini, 1797–1848, Germany (Japan).

INDEX OF JAPANESE PLANT NAMES

A

Abe-maki, 379
Abe-tō-hiren, 917
Abu-no-me, 796
Abu-no-me Zoku, 796
Abura-chan, 472
Abura-gaya, 202
Aburagi, 599
Aburagiku, 892
Abura-na Ka, 479
Abura-shiba, 248
Abura-susuki, 187
Abura-susuki Zoku, 187
Abura-tsutsuji, 707
Adeku, 652
Adeku Zoku, 652
Agi-nashi, 127
Agi-sumire, 641
Ai-ashi, 194
Ai-ashi Zoku, 194
Aiba-sō, 202
Aida-kugu, 197
Ai-nae, 734
Ai-nae Zoku, 734
Ainoko-hiru-mushiro, 121
Ainoko-yanagi-mo, 122
Ainu-somosomo, 163
Ainu-tachi-tsubo-sumire, 639
Ainu-wasabi, 483
Aira-tobi-kazura, 569
Aizu-hime-azami, 908
Aizu-shimotsuke, 520
Aizu-suge, 249
Aizu-tori-kabuto, 456
Ajisai, 511
Ajisai Zoku, 510
Akaba-gumi, 647
Akabana, 658
Akabana Ka, 654
Akabana-mansaku, 517
Akabana-nokogiri-sō, 888
Akabana-shimotsuke, 537
Akabana Zoku, 656
Aka-Ezo-matsu, 113
Akagashi, 377
Akagi-tsutsuji, 702
Aka-hana-warabi, 30
Akaishi-hyōtan-boku, 843
Akaishi-kōzori-na, 921
Aka-itaya, 609
Aka-jiku-hebi-noborazu, 462

Aka-kamba, 374
Aka-matsu, 115
Akame-inode, 57
Akame-gashiwa, 591
Akame-gashiwa Zoku, 591
Akame-yanagi, 364
Akami-no-budō, 620
Akami-no-inu-tsuge, 599
Akami-no-ruiyō-shoma, 456
Aka-mi-yadorigi, 397
Aka-mono, 708
Akane, 829
Akane Ka, 820
Akane-mugura, 829
Akane-suge, 253
Akane-sumire, 637
Akane Zoku, 829
Akan-kasa-suge, 253
Akan-suge, 228
Akanuma-fūro, 579
Aka-sasage, 568
Aka-shide, 371
Aka-shōma, 500
Aka-sō, 391
Aka-suge, 242
Aka-suguri, 515
Aka-todo-matsu, 112
Aka-tsume-kusa, 576
Aka-ukikusa, 107
Aka-ukikusa Zoku, 107
Aka-yashio, 702
Akaza, 415
Akaza Ka, 413
Akaza Zoku, 414
Akebi, 461
Akebi Ka, 461
Akebi Zoku, 461
Akebono-aoi, 400
Akebono-shusu-ran, 340
Akebono-sō, 741
Akebono-sumire, 636
Akebono-tsutsuji, 702
Aki-chōji, 787
Aki-gibōshi, 290
Akigiri, 778
Akigiri Zoku, 777
Aki-gumi, 647
Akihagiku, 874
Aki-karamatsu, 452
Aki-kasa-suge, 250
Akikaze-gibōshi, 290
Aki-mi-hi-shiba, 183
Aki-nire, 381

Aki-no-enokoro-gusa, 181
Aki-no-ginryō-sō, 692
Aki-no-hahakogusa, 858
Aki-no-kirin-sō, 868
Aki-no-kirin-sō Zoku, 868
Aki-no-kusa-tachibana, 747
Aki-no-nogeshi, 929
Aki-no-nogeshi Zoku, 928
Aki-no-michi-yanagi, 407
Aki-no-tamura-sō, 778
Aki-no-yahazu-azami, 915
Aki-sango, 688
Akitabuki, 878
Akita-tennan-shō, 259
Aki-tennan-shō, 259
Aki-yahazu-azami, 918
Akiyoshi-azami, 908
Akizaki-nagi-ran, 355
Akizaki-yatsushiro-ran, 338
Akkeshi-sō, 416
Akkeshi-sō Zoku, 416
Ako, 385
Aku-shiba, 712
Aku-shiba-modoki, 710
Ama, 581
Ama-cha, 511
Amachazuru, 848
Amachazuru Zoku, 848
Amadokoro, 302
Amadokoro Zoku, 301
Amagi-ama-cha, 511
Amagi-kan-aoi, 399
Amagi-ko-ajisai, 511
Amagi-tsutsuji, 701
Ama Ka, 580
Ama-kusagi, 765
Amakusa-shida, 42
Amami-goyō, 115
Amami-hitotsuba-hagi, 590
Amami-tennan-shō, 258
Ama-mo, 124
Ama-mo Ka, 124
Amamo-shishi-ran, 106
Ama-mo Zoku, 124
Amana, 299
Amana Zoku, 298
Ama-nyū, 684
Ama Zoku, 581
Amazuru, 619
Amerika-azena, 796
Amerika-kuro-suguri, 514
Amerika-nadeshiko, 432
Amerika-sendangusa, 901

Amerika-yama-bōshi, 688
Amerika-zuta, 620
Ami-shida, 77
Ami-shida Zoku, 77
Andon-mayumi, 604
Anpera-i, 214
Anpera-i Zoku, 214
Anzu, 542
Aoba-no-ki, 726
Aoba-omoto-gibōshi, 289
Aoba-suge, 241
Ao-benkei-sō, 496
Aobiyu, 418
Ao-chasen-shida, 93
Ao-chidori, 326
Ao-chidori Zoku, 326
Ao-chikara-shiba, 180
Ao-damo, 733
Ao-funabara-sō, 747
Ao-futaba-ran, 331
Aogane-shida, 94
Aogashi, 471
Ao-gayatsuri, 199
Aogeitō, 418
Aogiri, 625
Aogiri Ka, 625
Aogiri Zoku, 625
Ao-gō-so, 231
Ao-hada, 598
Ao-hakobe, 430
Ao-hie-suge, 241
Ao-horagoke, 36
Ao-horagoke Zoku, 36
Ao-hōzuki, 789
Ao-ichigo-tsunagi, 165
Ao-iga-warabi, 87
Aoigoke, 752
Aoigoke Zoku, 752
Aoi Ka, 623
Aoi-kazura, 271
Aoi-kazura Zoku, 270
Aoi-sumire, 635
Ao-jiku-mayumi, 605
Ao-jiku-su-noki, 709
Aojiku-yuzuri-ha, 588
Ao-kago-no-ki, 473
Ao-kamojigusa, 154
Ao-kamomezuru, 749
Ao-kara-mushi, 391
Ao-kazura, 612
Ao-kazura Zoku, 612
Ao-ki, 687
Ao-kinuta-sō, 832

951

Ezo-tō-uchi-sō, 539
Ezo-tsuri-suge, 245
Ezo-tsuru-kimbai, 525
Ezo-tsutsuji, 702
Ezo-uki-yagara, 203
Ezo-ukogi, 665
Ezo-usagigiku, 878
Ezo-usu-yuki-sō, 859
Ezo-waniguchi-sō, 302
Ezo-ware-mokō, 538
Ezo-wasabi, 483
Ezo-wata-suge, 205
Ezo-yama-azami, 908
Ezo-yama-hagi, 559
Ezo-yama-kōbō, 153
Ezo-yama-narashi, 362
Ezo-yama-odamaki, 454
Ezo-yama-zakura, 544
Ezo-yama-zengo, 679
Ezo-yanagi, 367
Ezo-yanagi-mo, 122
Ezo-yomogi, 897
Ezo-yomogigiku, 890
Ezo-yuzuri-ha, 588
Ezo-zentei-ka, 292

F

"Faurie"-gaya, 167
Fōrii-azami, 916
Fōrii-gaya, 167
Fōrii-gaya Zoku, 167
Fuchige-haguro-sō, 818
Fūchō-sō Ka, 479
Fude-kusa, 225
Fude-rindō, 738
Fugire-azuki-nashi, 551
Fugire-sumire, 642
Fuiri-hime-zazen-sō, 264
Fuiri-soyogo, 600
Fuiri-yuzuri-ha, 588
Fuji, 572
Fuji-aka-shōma, 500
Fuji-azami, 905
Fujibakama, 867
Fujibakama Zoku, 866
Fuji-beni-utsugi, 838
Fujigae-sō, 565
Fuji-hana-yasuri, 29
Fuji-hatazoa, 489
Fuji-ibara, 540
Fuji-kanzō, 562
Fujiki, 556
Fujiki Zoku, 556
Fuji-koke-shinobu, 34
Fuji-nadeshiko, 431
Fuji-sennin-sō, 443
Fuji-shida, 39
Fuji-shida Zoku, 38
Fuji-sumire, 638
Fuji-taigeki, 593
Fuji-tsurigane-tsutsuji, 695

Fuji-tsutsuji, 700
Fuji-utsugi, 734
Fuji-utsugi Ka, 734
Fuji-utsugi-Zoku, 734
Fuji Zoku, 572
Fuka-no-ki, 663
Fuka-no-ki Zoku, 663
Fuki, 877
Fukiya-mitsuba, 671
Fuki-yuki-no-shita, 502
Fuki Zoku, 877
Fukki-sō, 595
Fukki-sō Zoku, 595
Fukudo, 895
Fukuju-sō, 450
Fukuju-sō Zoku, 450
Fukuō-niga-na, 925
Fukuō-sō, 925
Fukuō-sō Zoku, 925
Fukuregi-shida, 81
Fukurodagaya, 175
Fukuro-mochi, 729
Fukuro-shida, 52
Fukushima-shajin, 850
Fumoto-kaguma, 38
Fumoto-shida, 38
Fumoto-shida Zoku, 38
Fumoto-sumire, 637
Funabara-sō, 747
Fū-ran, 358
Fū-ran Zoku, 358
Furasaba-sō, 801
Fūrin-tsutsuji, 707
Fūrin-ume-modoki, 599
Furi-sode-yanagi, 365
Fūro-keman, 477
Fūro-sō, 579
Fūro-sō Ka, 577
Fūro-sō Zoku, 578
Fusagaya, 144
Fusagaya Zoku, 144
Fusa-mo, 660
Fusa-mo Zoku, 660
Fusa-nakiri-suge, 249
Fusa-saji-ran, 95
Fusa-suge, 249
Fusa-sugina, 22
Fusa-tanuki-mo, 816
Fusazakura, 439
Fusazakura Ka, 438
Fusazakura Zoku, 439
Fushidaka-fūro, 579
Fushidaka-shino, 138
Fushige-chigaya, 188
Fushige-tachi-fūro, 578
Fushiguro, 434
Fushiguro-sennō, 433
Fushiguro Zoku, 433
Fushi-no-ha-awabuki, 613
Fushi-no-ki, 597
Fushizaki-sō, 900
Fushizaki-sō Zoku, 900

Futaba-aoi, 399
Futaba-mugura, 822
Futaba-mugura Zoku, 822
Futaba-ran, 332
Futaba-ran Zoku, 331
Futaba-tsure-sagi, 324
Futa-mata-ichige, 445
Futa-mata-tampopo, 926
Futa-mata-tampopo Zoku, 926
Futanami-sō, 920
Futanami-sō Zoku, 920
Futari-shizuka, 361
Futobo-naginata-kōju, 785
Futo-hiru-mushiro, 121
Futo-i, 204
Futo-kazura, 360
Futo-momo, 652
Futo-momo Ka, 652
Fuyazuta, 663
Fuyō, 624
Fuyō Zoku, 623
Fuyu-ichigo, 532
Fuyu-no-hana-warabi, 30
Fuyuzanshō, 582

G

Gaga-buta, 742
Gaga-imo, 745
Gaga-imo Ka, 744
Gaga-imo Zoku, 745
Gaju-maru, 385
Gaku-utsugi, 512
Gama, 118
Gama Ka, 118
Gama Zoku, 118
Gamazumi, 835
Gamazumi Zoku, 834
Gampi, 433, 645
Gampi Zoku, 645
Ganju-azami, 912
Gankō-ran, 596
Gankō-ran Ka, 595
Gankō-ran Zoku, 596
Gankubi-sō, 863
Gankubi-yabu-tabako, 863
Ganzeki-ran, 348
Ganzeki-ran Zoku, 347
Gasha-moku, 122
Gashangi, 765
Gassan-chidori, 325
Gejigeji-shida, 72
Gekkitsu-modoki, 584
Genge, 574
Genge Zoku, 573
Genji-sumire, 638
Genkai-iwa-renge, 498
Genkai-moegi-suge, 236
Gen-no-shōko, 579
Gettō, 318
Gibōshi-ran, 347

Gibōshi Zoku, 287
Gimbai-sō, 510
Gimbai-sō Zoku, 509
Gin-doro, 362
Gin-mokusei, 731
Gin-ran, 333
Ginrei-ka, 718
Ginryō-sō, 693
Ginryō-sō-modoki, 692
Ginryō-sō Zoku, 693
Ginsen-ka, 624
Gishi-gishi, 404
Gishi-gishi Zoku, 403
"Glehn"-suge, 245
Gokayō-ō-ren, 457
Gokidake, 138
Gokizuru, 847
Gokizuru Zoku, 847
Goma Ka, 810
Goma-ki, 835
Goma-kusa, 798
Goma-kusa Zoku, 798
Goma-na, 873
Goma-no-ha-gusa, 792
Goma-no-ha-gusa Ka, 790
Goma-no-ha-gusa Zoku, 792
Gomashio-hoshi-kusa, 267
Gomo-ju, 836
Gongen-suge, 238
Gonzui, 606
Gonzui Zoku, 606
Gosho-ichigo, 533
Goshuyu Zoku, 583
Gō-so, 230
Goto-zuri, 510
Goyō-akebi, 461
Goyō-ichigo, 536
Goyō-matsu, 115
Goyō-toga, 112
Goyō-tsutsuji, 702
Goyōzan-yōraku, 695
Gozen-tachibana, 688
Gui-matsu, 114
Gumbai-hirugao, 752
Gumbai-nazuna, 481
Gumbai-nazuna Zoku, 481
Gumbaizuru, 801
Gumi Zoku, 646
Gumi Ka, 646
Gunnai-fūro, 578
Gunnai-kimpōge, 448
Guren-suge, 245
Gyoboku, 479
Gyoboku Zoku, 479
Gyōgi-shiba, 175
Gyōgi-shiba Zoku, 175
Gyōja-azami, 909
Gyōja-ninniku, 295
Gyōja-no-mizu, 619
Gyokushin-ka, 825
Gyokushin-ka Zoku, 825
Gyō-ō-chiku, 138

INDEX OF ENGLISH NAMES

(Mostly names of families.)

MAP OF JAPAN SHOWING OLDER NAMES OF DIVISIONS OF TOKUGAWA ERA SOMETIMES CURRENTLY USED

YEZO (HOKKAIDO)

SEA OF JAPAN

HOKURIKU-DO

TOSAN-DO

TOKAI-DO

SAN'IN-DO

SAN'YO-DO

KINAI

NANKAI-DO

SAIKAI-DO

PACIFIC OCEAN